NINTH EDITION

REFRIGERATION & AIR CONDITIONING TECHNOLOGY

Eugene Silberstein • Jason Obrzut

John A. Tomczyk • William M. Johnson • William C. Whitman

CENGAGE

Australia • Brazil • Canada • Mexico • Singapore • United Kingdom • United States

CENGAGE

***Refrigeration & Air Conditioning Technology* 9E**
Eugene Silberstein, Jason Obrzut, John A. Tomczyk, William M. Johnson, William C. Whitman

SVP, Higher Education & Skills Product: Erin Joyner

Product Director: Matt Seeley

Product Manager: Vanessa Meyers

Product Assistant: Kimberly Klotz

Director, Learning Design: Rebecca von Gillern

Learning Designer: Mary Clyne, Elizabeth Berry

Marketing Manager: Scott Chrysler

Director, Content Creation: Juliet Steiner

Senior Content Manager: Jim Zayicek

Digital Delivery Lead: Elizabeth Cranston

Art Director: Erin Griffin

Cover image: iStockPhoto.com/kirstypargeter

For product information and technology assistance,
contact us at Cengage Customer & Sales Support, 1-800-354-9706 or support.cengage.com.

For permission to use material from this text or product,
submit all requests online at **www.cengage.com/permissions.**

Library of Congress Control Number: 2019915474

ISBN: 978-0-3571-2227-3

Cengage
200 Pier 4 Boulevard
Boston, MA 02210
USA

Cengage is a leading provider of customized learning solutions with employees residing in nearly 40 different countries and sales in more than 125 countries around the world. Find your local representative at **www.cengage.com.**

To learn more about Cengage platforms and services, register or access your online learning solution, or purchase materials for your course, visit **www.cengage.com.**

Notice to the Reader
Publisher does not warrant or guarantee any of the products described herein or perform any independent analysis in connection with any of the product information contained herein. Publisher does not assume, and expressly disclaims, any obligation to obtain and include information other than that provided to it by the manufacturer. The reader is expressly warned to consider and adopt all safety precautions that might be indicated by the activities described herein and to avoid all potential hazards. By following the instructions contained herein, the reader willingly assumes all risks in connection with such instructions. The publisher makes no representations or warranties of any kind, including but not limited to, the warranties of fitness for particular purpose or merchantability, nor are any such representations implied with respect to the material set forth herein, and the publisher takes no responsibility with respect to such material. The publisher shall not be liable for any special, consequential, or exemplary damages resulting, in whole or part, from the readers' use of, or reliance upon, this material.

Printed in the United States of America
Print Number: 03   Print Year: 2021

BRIEF CONTENTS

CONTENTS

SECTION 3: Automatic Controls

SECTION 5: Commercial Refrigeration

SECTION 6: Air-Conditioning (Heating and Humidification)

SECTION 9: Domestic Appliances

SECTION 10: Commercial Air-Conditioning and Chilled-Water Systems

PREFACE

Refrigeration & Air Conditioning Technology is designed and written for students in vocational-technical schools and colleges, community colleges, and apprenticeship programs. The content is in a format appropriate for students who are attending classes full-time while preparing for their first job, for students attending classes part-time while preparing for a career change, or for those working in the field who want to increase their knowledge and skills. Emphasis throughout the text is placed on the practical applications of the knowledge and skills technicians need to be productive in the refrigeration and air-conditioning industry. The contents of this book can be used as a study guide to prepare for the Environmental Protection Agency (EPA) mandatory technician certification examinations. It can be used in the HVACR field or closely related fields by students, technicians, installers, contractor employees, service personnel, and owners of businesses.

This text is also an excellent study guide for the Industry Competency Exam (ICE), the North American Technician Excellence (NATE), the HVAC Excellence, the Refrigeration Service Engineers Society (RSES), the United Association (UA) STAR certification, and the Heating, Air Conditioning, and Refrigeration Distributors International (HARDI) voluntary HVACR technician certification and home-study examinations.

The book is also written to correspond to the National Skill Standards for HVACR technicians. Previous editions of this text are often carried to the job site by technicians and used as a reference for service procedures. "Do-it-yourselfers" will find this text valuable for understanding and maintaining heating and cooling systems.

As general technology has evolved, so has the refrigeration and air-conditioning industry. A greater emphasis is placed on digital electronic controls and system efficiency.

Global warming has become a major environmental issue. When HVACR systems are working correctly and efficiently, they will greatly reduce energy consumption and greenhouse gases. Organizations like the Green Mechanical Council (GreenMech) are advocates for the HVACR industry and assist the industry in meeting with government, educational, industry, and labor interests to find solutions to the world's global-warming problem. GreenMech has created a scoring system designed to help engineers, contractors, and consumers know the "green value" of each mechanical installation. The "green value" encompasses the system's energy efficiency, pollution output, and sustainability. Realtors, building inspectors, builders, and planning and zoning officials will now have some knowledge about and guidance on how buildings and mechanical systems are performing. Green buildings and green mechanical systems are becoming increasingly popular in today's world as a way to curb global warming.

Energy audits have become an integral part of evaluating and assessing an existing building's energy performance. Higher efficiency standards for the energy performance of new buildings have been established. Higher levels of training and certification have been developed for HVACR technicians to meet the needs of more sophisticated, energy-efficient buildings and HVACR equipment.

TEXT DEVELOPMENT

This text was developed to provide the technical information necessary for a technician to be able to perform satisfactorily on the job. It is written at a level that students can easily understand. Practical application of the technology is emphasized. Terms commonly used by technicians and mechanics have been used throughout to make the text easy to read and to present the material in a practical way. Many of these key terms are also defined in the glossary. This text is updated regularly in response to market needs and emerging trends. Refrigeration and air-conditioning instructors have reviewed each unit.

Illustrations and photos have been updated and are used extensively throughout the text. Full-color treatment of photos and illustrations helps amplify the concepts presented.

No prerequisites are required for this text. It is designed to be used by beginning students, as well as by those with training and experience.

ORGANIZATION

Considerable thought and study have been devoted to the organization of this text.

The text is organized so that after completing the first four sections, students may concentrate on courses in refrigeration or air-conditioning (heating and/or cooling). If the objective is to complete a whole program, the instruction may proceed until the sequence scheduled by the school's curriculum is completed.

NEW TO THIS EDITION

The 9th edition includes more than 400 new and revised figures and images. Many of these images have been recolored to improve accessibility for learners with different forms of color-blindness.

Learning objectives have been validated and carefully mapped to content in all units. Each unit's objectives clearly align to the major topics in the unit and to the unit's review questions. This alignment is detailed for instructors in the form of a Content Map, which is now available in the instructor resource center.

Unit summaries have been updated and moved from the core reading to the instructor resource center. This placement encourages students to use the summaries as a study supplement rather than a main source of information and gives instructors full control over whether and when their students will have access to this resource.

SEVERAL SAFETY TOPICS HAVE BEEN UPDATED OR EXPANDED, INCLUDING:

- safe workplace surroundings
- working in confined spaces
- personal protection equipment (PPE)
- first aid
- ladder safety
- fire extinguisher usage
- heat-related injuries

THE FOLLOWING TECHNICAL TOPICS HAVE BEEN ADDED, UPDATED, OR EXPANDED:

New Coverage	Updated Coverage	Expanded Coverage
• System evacuation without the use of a gauge manifold • General refrigeration • Retrofitting considerations and guidelines • Combustion air • Compressor testing • Chilled beam technology	• Leak repair requirements and EPA regulations • Capacitor testing • Fan-cycling devices • Scroll compressor efficiency • Piston metering devices • Combustion and high-efficiency furnaces • Temperature difference (TD)	• Refrigerant numbering system • Azeotropic and near-azeotropic blends • Specific volume, specific gravity, and density • R-value, U-value, and calculations • Systems with combinations of metering devices • Indoor fan motor control

NEW TO THIS EDITION'S MINDTAP:

In addition to all the updates included in the printed text, the 9th edition MindTap adds the following enhancements:

- Each unit in the MindTap now includes an introductory video by author Eugene Silberstein to explain how the unit's content will support the learner's HVACR career.
- Review Questions are now auto-graded, providing immediate feedback and full explanations for learners.
- The Electricity and HVACR Troubleshooting simulations have been updated to version 4.0, with improved loading times and smoother LMS integration, enhanced user tutorials, new start screens in troubleshooting scenarios, shareable badges, and more.

HOW TO USE THIS TEXT AND SUPPLEMENTARY MATERIALS

This text may be used as a classroom text, as a learning resource for an individual student, as a reference text for technicians on the job, or as a homeowner's guide. An instructor may want to present the unit objectives, briefly discuss the topics included, and assign the unit to be read. The instructor then may want to discuss the material with students. This can be followed by students completing the review questions, which can later be reviewed in class.

The instructor resource DVD may be used to access a computerized test bank for end-of-unit review questions, teaching tips, PowerPoint® presentations, and more.

FEATURES OF THE TEXT

Objectives

Objectives are listed at the beginning of each unit. The objective statements have been stated clearly and simply to give students direction.

OBJECTIVES

After studying this unit, you should be able to

- discuss applications for high-, medium-, and low-temperature refrigeration.
- describe the term *ton of refrigeration.*
- describe the basic refrigeration cycle.
- explain the relationship between pressure and the boiling point of water or other liquids.
- describe the function of the evaporator or cooling coil.
- explain the concept of superheat as it applies to the evaporator.
- explain the purpose of the compressor.
- list three types of compressors normally used in residential and light commercial buildings.
- discuss the function of the condenser.
- explain the concept of subcooling as it applies to the condenser.
- state the purpose of the metering device.
- list four characteristics to consider when choosing a refrigerant for a system.
- list the designated cylinder colors for three commonly used refrigerants.
- discuss four commonly used refrigerants and their applications.
- describe how refrigerants can be stored or processed while refrigeration systems are being serviced.
- plot a refrigeration cycle for refrigerants (R-22 and R-134a) on a pressure/enthalpy diagram.
- plot a refrigeration cycle on a pressure/enthalpy diagram for refrigerant blends R-404A and R-410A.
- plot a refrigeration cycle on a pressure/enthalpy diagram for a refrigerant blend (R-407C) that has a noticeable temperature glide.

Safety Checklists

A Safety Checklist is presented at the beginning of each unit, when applicable, immediately following the Objectives. This checklist emphasizes the importance of safety and is included in units where "hands-on" activities are discussed.

Safety is emphasized throughout the text. In addition to the Safety Checklist at the beginning of most units, safety precautions and techniques are highlighted throughout. It would be impossible to include a safety precaution for every conceivable circumstance that may arise, but an attempt has been made to be as thorough as possible. The overall message is to work safely whether in a school shop, laboratory, or on the job and to use common sense.

SAFETYCHECKLIST

Areas in which there is the potential for refrigerant leaks should be properly ventilated.
Extra precautions should be taken to ensure that no refrigerant leaks occur near an open flame.
Refrigerants are stored in pressurized containers and should be handled with care.
Eye protection and gloves should be worn when checking system pressures or when otherwise working on air-conditioning and refrigeration systems.
Only transfer refrigerants from a system into DOT-approved containers.

Recovery/Recycling/Reclaiming/Retrofitting

Discussions relating to recovery, recycling, reclaiming, retrofitting, or other environmental issues are highlighted in blue throughout the text. In addition, one complete unit on refrigerant management is included—Unit 9, "Refrigerant and Oil Chemistry and Management—Recovery, Recycling, Reclaiming, and Retrofitting."

heat source and the heating loops does not need flow control valves. However, when the system utilizes only circulators, flow control valves are desirable. Some circulator manufacturers are now installing flow control valves right on the circulators themselves to ease the installation process, **Figure 33.50.**

Figure 33.49 A flow control valve. Courtesy Bell and Gossett

ting of 1.0 will cause the boiler to maintain a water temperature of 160°F if the outside ambient temperature is 20°F. The same control setting will cause the boiler water temperature to drop to 140°F if the outside ambient temperature rises to 40°F. *R As of September 2012, all newly installed boilers are required to have some type of reset control strategy to automatically adjust the boiler water temperature. R*

Green Awareness

As previously mentioned, global warming stemming from the uncontrolled rate of greenhouse gas emissions is a major global environmental issue. Buildings are important users of energy and materials and so are a major source of the greenhouse gases that are the by-products of energy and materials use. At the time of this writing, there are approximately 5 million commercial buildings and 125 million housing units in the United States. Surprisingly, almost every one of their mechanical systems is obsolete. Discussions relating to the green awareness movement (e.g., lowering energy costs, reducing operating and maintenance costs, increasing productivity, and decreasing the amount of pollution generated) are highlighted in green throughout the text.

good pump performance. The vacuum pump oil should be clean and clear; if it is cloudy, moisture is likely to be present. The pump should then be serviced by changing the oil. **NOTE:** *For the best performance, only approved vacuum pump oil should be used. Any other oil will have a vapor pressure that is too high and thus prevent the achievement of deep vacuums.* • *Be sure to dispose of the old oil in an environmentally appropriate manner.*

With the new oil in the pump, run the pump long enough to get up to operating temperature; it will usually be too hot to touch or hold. Drain the oil again. With fresh oil in

HVACR GOLDEN RULES

When making a service call to a business:

- Never park your truck or van in a space reserved for customers.
- Look professional and be professional.
- Before starting troubleshooting procedures, get all the information you can regarding the problem.
- Be extremely careful not to scratch tile floors or to soil carpeting with your tools or by moving equipment.
- Be sure to practice good sanitary and hygiene habits when working in a food preparation area.
- Keep your tools and equipment out of the customers' and employees' way if the equipment you are servicing is located in a normal traffic area.
- Be prepared with the correct tools and ensure that they are in good condition.
- Always clean up after you have finished. Try to provide a little extra service by cleaning filters, oiling motors, or providing some other service that will impress the customer.
- Always discuss the results of your service call with the owner or representative of the company. Try to persuade the owner to call if there are any questions as a result of the service call.

HVACR Golden Rules

Golden Rules for the refrigeration and air-conditioning technician give advice and practical hints for developing good customer relations. These "golden rules" appear in appropriate units.

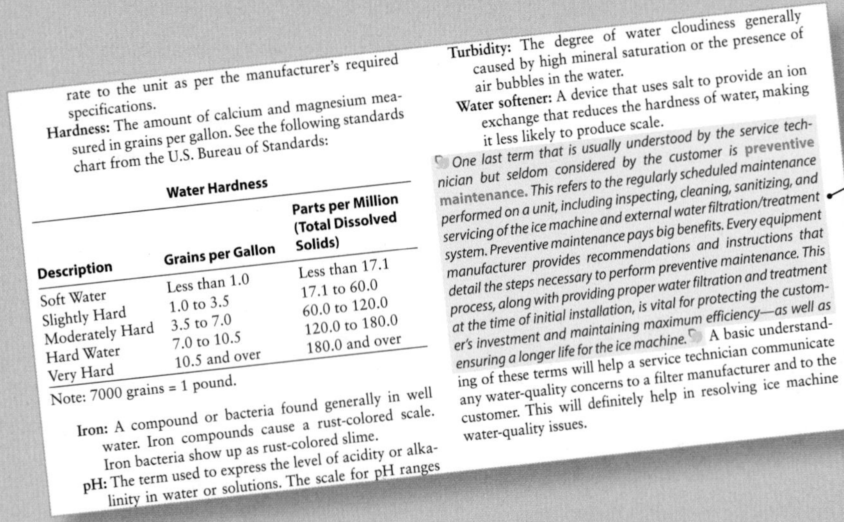

rate to the unit as per the manufacturer's required specifications.

Hardness: The amount of calcium and magnesium measured in grains per gallon. See the following standards chart from the U.S. Bureau of Standards:

Water Hardness

Description	Grains per Gallon	Parts per Million (Total Dissolved Solids)
Soft Water	Less than 1.0	Less than 17.1
Slightly Hard	1.0 to 3.5	17.1 to 60.0
Moderately Hard	3.5 to 7.0	60.0 to 120.0
Hard Water	7.0 to 10.5	120.0 to 180.0
Very Hard	10.5 and over	180.0 and over

Note: 7000 grains = 1 pound.

Iron: A compound or bacteria found generally in well water. Iron compounds cause a rust-colored scale. Iron bacteria show up as rust-colored slime.

pH: The term used to express the level of acidity or alkalinity in water or solutions. The scale for pH ranges

Turbidity: The degree of water cloudiness generally caused by high mineral saturation or the presence of air bubbles in the water.

Water softener: A device that uses salt to provide an ion exchange that reduces the hardness of water, making it less likely to produce scale.

One last term that is usually understood by the service technician but seldom considered by the customer is **preventive maintenance.** *This refers to the regularly scheduled maintenance performed on a unit, including inspecting, cleaning, sanitizing, and servicing of the ice machine and external water filtration/treatment system. Preventive maintenance pays big benefits. Every equipment manufacturer provides recommendations and instructions that detail the steps necessary to perform preventive maintenance. This process, along with providing proper water filtration and treatment at the time of initial installation, is vital for protecting the customer's investment and maintaining maximum efficiency—as well as ensuring a longer life for the ice machine.* A basic understanding of these terms will help a service technician communicate any water-quality concerns to a filter manufacturer and to the customer. This will definitely help in resolving ice machine water-quality issues.

Preventive Maintenance

Preventive Maintenance procedures are included in many units and relate specifically to the equipment presented in that unit. Technicians can provide some routine preventive maintenance service when on other types of service calls as well as when on strictly maintenance calls. The preventive maintenance procedures provide valuable information for the new or aspiring technician and homeowner, as well as for those technicians with experience.

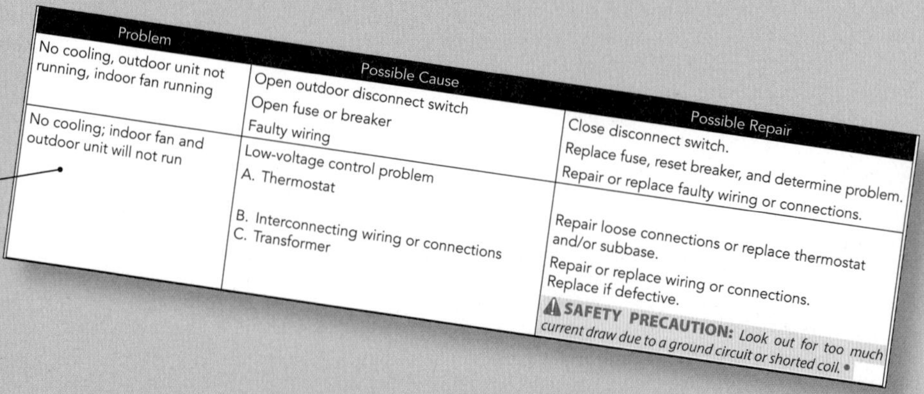

Problem	Possible Cause	Possible Repair
No cooling, outdoor unit not running, indoor fan running	Open outdoor disconnect switch	Close disconnect switch.
	Open fuse or breaker	Replace fuse, reset breaker, and determine problem.
	Faulty wiring	Repair or replace faulty wiring or connections.
No cooling; indoor fan and outdoor unit will not run	Low-voltage control problem A. Thermostat	Repair loose connections or replace thermostat and/or subbase.
	B. Interconnecting wiring or connections	Repair or replace wiring or connections.
	C. Transformer	Replace if defective. **SAFETY PRECAUTION:** *Look out for too much current draw due to a ground circuit or shorted coil.*

Diagnostic Charts

Diagnostic Charts are included at the end of many units. These charts include material on troubleshooting and diagnosis.

SERVICE CALL 1

A customer calls indicating that the boiler in the equipment room at a motel has hot water running out and down the drain all the time. Another service company has been performing service at the motel for the last few months. *The problem is that the water-regulating valve (boiler water feed) is out of adjustment. Water is seeping from the boiler's pressure relief valve,* **Figure 14.65**.

The technician arrives at the motel, parking alongside the building so as not to block the front door or the motel's

Service Technician Calls

In many units, practical examples of service technician calls are presented in a down-to-earth situational format. These are realistic service situations in which technicians may find themselves. In many instances, the solution is provided in the text, and in others the reader must decide what the best solution should be. These solutions are provided in the Instructor's Manual. The Service Technician Calls will now incorporate customer relations and technician soft skills.

REVIEW QUESTIONS

1. The control voltage that is typically used in split-type residential central air-conditioning systems is
   A. 24 V.
   B. 48 V.
   C. 96 V.
   D. 115 V.
2. What device is used to produce the control voltage that is used to power the control circuits of a split-type residential central air-conditioning system?
   A. Step-down transformer
   B. Step-up transformer
   C. Low-voltage thermostat
   D. Line-voltage thermostat
3. All of the following are true statements regarding line-voltage thermostats and control wiring except
   A. An electrician's license is often required to run line-voltage control wiring.
   B. Line-voltage thermostats and controls respond faster than low-voltage thermostats.
   C. Line-voltage thermostats and controls are not as sensitive as low-voltage thermostats.
   D. Line-voltage thermostats must be matched to the voltage and current of the circuit.
4. System lag is the result of which of the following with regards to a heating system?
   A. The residual system heat increasing the space temperature
   B. The amount of heat generated in the thermostat
7. Which of the following is likely to be found on a line-voltage thermostat?
   A. Contacts with low amperage ratings
   B. 18-gauge or 22-gauge wire connections
   C. Lightweight switches and buttons
   D. Locking or otherwise secure cover
8. If a heated space reaches a maximum temperature of 78°F and a minimum temperature of 71°F, the difference between these two temperatures is referred to as the
   A. temperature lag.
   B. system lag.
   C. system overshoot.
   D. temperature swing.
9. Why is measuring the temperature of a solid object challenging for the HVACR technician?
   A. The temperature of solid objects is not uniform throughout.
   B. The entire thermometer probe must be at the temperature of the solid.
   C. Thermometers are not designed to measure the temperature of solids.
   D. Solid objects do not transfer heat as effectively as other forms of matter.
10. What method is most likely to be used to measure the temperature of a liquid contained in an open vessel?
   A. Immersion

Review Questions

Review Questions follow the Summary in each unit and can help to measure the student's knowledge of the unit. There are a variety of question types—multiple choice, true/false, short answer, short essay, and fill-in-the-blank.

SUPPORT MATERIALS

LAB MANUAL

The Complete HVAC Lab Manual was developed out of the need for a lab manual that comprehensively covers the knowledge and skills that are required for an HVACR technician entering the industry. It was designed to support the hands-on application and practice required within the HVACR program.

INSTRUCTOR RESOURCES

- Syllabus. This is the standard course syllabus for this textbook, providing a summary outline for teaching HVACR.
- Teaching Tips. Teaching hints form a basis for presenting concepts and material. Key points and concepts can be highlighted graphically to enhance student retention.
- PowerPoint Presentation. These slides can be used to outline a lecture on the concepts and material. Key points and concepts are highlighted graphically to enhance student retention.
- Image Gallery. This database of key images (all in full color) taken from the text can be used in lecture presentations, as transparencies, for tests and quizzes, and with PowerPoint presentations.
- Test Bank. Over 1000 questions of varying levels of difficulty are provided in multiple-choice and short-answer formats for assessing student comprehension. This versatile tool allows the instructor to manipulate the data to create original tests.

VIDEO DVD SET

A seven-DVD video set addressing over 120 topics covered in the text is available. Each DVD contains four 20-minute videos. To order the seven-DVD set, reference ISBN: 978-1-111-64451-2.

MINDTAP

MindTap is well beyond an eBook, a homework solution or digital supplement, a resource center website, a course delivery platform, or a Learning Management System. MindTap is a new personal learning experience that combines all your digital assets—readings, multimedia, activities, and assessments—into a singular learning path to improve student outcomes.

INSTRUCTOR SITE

An Instructor Companion website containing supplementary material is available. This site contains an Instructor's Manual, teaching tips, syllabus, lecture outline, an image gallery of text figures, unit presentations done in PowerPoint, and testing powered by Cognero.

Cengage Learning Testing Powered by Cognero is a flexible, online system that allows you to:

- author, edit, and manage test bank content from multiple Cengage solutions
- create multiple test versions in an instant
- deliver tests from your LMS, your classroom, or wherever you want

Contact Cengage or your local sales representative to obtain an instructor account.
To access an Instructor Companion website from SSO Front Door:

1. Go to http://login.cengage.com and log in using the instructor e-mail address and password.
2. Enter author, title, or ISBN in the **Add a title to your bookshelf** search.
3. Click **Add to my bookshelf** to add instructor resources.
4. At the Product page, click the **Instructor Companion site** link.

DELMAR ONLINE TRAINING SIMULATION: HVAC 4.0

Delmar Online Training Simulation: HVAC is a 3D immersive simulation that offers a rich learning experience and mimics field performance. To address the critical area of Electricity, it offers a learning path from basic electrical concepts to real-world electrical troubleshooting. This innovative product includes dynamic interactive wiring diagrams in two modes: an open sandbox mode for exploration and experimentation, and a tutorial mode where the proper sequencing required for sound electrical practice is provided. Both modes are supported by an adaptive question engine. Learning electrical theory, and trying and testing sound electrical practice prepares the student for life-like, simulated exposure to faults with the HVACR equipment that follows. It also challenges learners to master diagnostic and troubleshooting skills across seven pieces of HVACR equipment found in the industry—Gas Furnace, Oil Furnace, Gas Boiler, Split Residential A/C, Commercial A/C, Heat Pumps, and Commercial Walk-in Freezers. Soft skills are also included within the simulation.

To create successful learning outcomes, Delmar Online Training Simulation: HVACR offers approximately 200 scenarios which allow students to troubleshoot and build diagnostic and critical thinking skills. Two modes within the simulation promote incremental learning: Training Mode and Challenge Mode. Training Mode has fixed scenarios to aid in familiarizing the user with the equipment, the problem needing attention, and the capabilities of the simulation. Challenge Mode has randomized scenarios within three levels: Beginner, Intermediate, and Advanced. Both modes require learners to diagnose a fault or faults and perform the repair successfully while materials and labor costs are tracked. An integrated digital badging system helps students track their progress and adds additional engagement and motivation. Simulation-based videos teach students key troubleshooting concepts as well as familiarize them with the simulation. The instructional design allows for full open engagement, so students do not have artificial guardrails leading them to a conclusion.

Combining sound instructional design with top-quality computer immersive technology, learners develop critical thinking skills and apply them to real-world customer service calls in a simulated, 3D, life-like setting. This performance simulation complements live training practice by reinforcing good habits, and even presenting scenarios that are impractical (dangerous, expensive, etc.) to create in labs or in a residence. Available for instant purchase on www.cengagebrain.com.

ABOUT THE AUTHORS

EUGENE SILBERSTEIN

Since entering the HVACR industry in 1980, Eugene has taken on many roles ranging from field technician and system designer to company owner, teacher, administrator, consultant, and author. Eugene is presently the Director of Technical Education and Standards at the ESCO Institute.

Eugene has 25 years of teaching experience and has taught air-conditioning and refrigeration at many institutions including high school vocational programs, proprietary post-secondary institutions, and community colleges. In December of 2015, Eugene retired from his tenured teaching position at Suffolk County Community College, in Brentwood, New York, to join the ESCO Group and relocate to Southern California.

Eugene earned his dual bachelor's degree from The City College of New York (New York, NY) and his master's of science degree from Stony Brook University, (Stony Brook, NY) where he specialized in Energy and Environmental Systems, studying renewable and sustainable energy sources such as wind, solar, geothermal, biomass, and hydropower. He earned his Certified Master HVACR Educator (CMHE) credential from HVAC Excellence and the ESCO Group in 2010. Eugene also carries ASHRAE's BEAP credential, which classifies Eugene as a Building Energy Assessment Professional. He is also an active member of many industry societies.

Eugene served as the subject matter expert on over 15 HVACR-related educational projects. His most notable work is Cengage Learning's *Refrigeration and Air Conditioning Technology title*, which is presently in its 9th Edition (2021). This book is used in over 1000 schools both in this country and abroad to help individuals learn, and master, the skills required to service, install, design, service, and troubleshoot HVACR equipment. Other book credits include *The Complete HVAC Lab Manual (2019), Refrigeration and Air Conditioning Technology, 6th and 7th Editions (2008, 2012), Residential Construction Academy: HVAC, 1st and 2nd Editions (2005, 2012), Pressure Enthalpy Without Tears (2006), Heat Pumps, 1st and 2nd Editions (2003, 2016)*, and *Psychrometrics Without Tears* (2014). In addition to his book credits, Eugene has written the production scripts for over 30 HVACR educational videos and has written articles for the industry's newspapers, magazines, and other periodicals.

Eugene was selected as one of the top three HVACR instructors in the country for the 2005/2006, 2006/2007, and 2007/2008 academic school years by the Air-Conditioning, Heating and Refrigeration Institute (AHRI) and the Air Conditioning, Heating and Refrigeration (ACHR) News.

JASON OBRZUT

In 2002, Jason Obrzut graduated from the HVAC Technical Institute in Chicago, kick-starting his impressive career in the HVACR industry. Upon graduating from school, Jason began working as a service technician and, within a short time, became an installation supervisor, and ultimately a business owner. While working during the day, he returned to the HVAC Technical Institute as a part-time, evening instructor. Teaching quickly became his passion and working in the field took the proverbial backseat as his part-time night teaching assignment morphed into a full-time teaching position. While teaching, Jason developed, wrote, and implemented the curriculum that is currently being taught at the school. His dedication to his students and the institution became evident and, after a few short years, he assumed the role as the school's Director of Education in addition to being the center's lead HVACR instructor.

In 2016, The ACHR News recognized Jason as one of the "Top 40 Under 40" in the HVACR industry. He is affiliated with numerous HVACR industry organizations including Refrigeration Service Engineers Society (RSES), the ESCO Group, and HVAC Excellence. At HVAC Excellence's 2016 National HVAC Educators and Trainers Conference (NHETC),

Jason participated in the first annual Teachers-N-Trainers (TnT) competition and took 1st place in a field of over 60 participants. Jason is one of only 125 instructors nationwide who has had the title of Certified Master HVAC/R Educator (CMHE) bestowed upon him. In addition, he possesses other industry certifications, such as Certified Subject Matter Educator (CSME) certificates from the ESCO Group. Jason has recently accepted a full-time position with the ESCO Institute, an industry partner of Cengage Learning, where he will be taking on a role as part of the content development team to help improve the quality of HVACR education and training.

JOHN TOMCZYK

John Tomczyk received his associate's degree in refrigeration, heating, and air-conditioning technology from Ferris State University in Big Rapids, Michigan; his bachelor's degree in mechanical engineering from Michigan State University in East Lansing, Michigan; and his master's degree in education from Ferris State University. Mr. Tomczyk has worked in refrigeration, heating, and air-conditioning service and technical writing consultation for both the academic and industrial fields for numerous years—enabling him to bring a wealth of experience to this text. His technical articles have been featured in many HVACR magazines and journals. Mr. Tomczyk is also the author of *Troubleshooting and Servicing Modern Air Conditioning and Refrigeration Systems.* Mr. Tomczyk has 29 years of teaching experience at the Refrigeration, Heating, and Air-Conditioning Technology program at Ferris State University and is a member of many HVACR trade organizations.

BILL JOHNSON

Now retired, Bill Johnson has taught heating, air-conditioning, and refrigeration for more than two decades in various technical colleges and factory schools. A graduate of Southern Technical Institute, a branch of Georgia Tech in Atlanta, Georgia, he has also served as Service Manager for a major manufacturer for six years. Mr. Johnson owned his own HVACR business for 10 years and has been a member of the Refrigeration Service Engineers Society and the American Society of Heating Refrigerating Air Conditioning Engineers. His authoring credits include three major textbooks on the market today as well as their ancillary materials. He also writes a monthly article—BTU Buddy—that is available online.

BILL WHITMAN

Bill Whitman (July 1, 1930–August 12, 2018)
Bill Whitman, along with co-author Bill Johnson, spent 5 years collaborating on the first edition of *Refrigeration and Air Conditioning Technology*, which published in 1986. Little did they know that the title would grow in popularity, be used in nearly 1000 career and technical education programs across the country, and would be the huge success that it is, even 30-plus years after its initial release.

Bill was a native of Bennington, Vermont and a veteran of the Korean War while in the Air Force. He graduated from Keene State College in Keene, New Hampshire with a degree in industrial education and went on to achieve a master's degree in school administration at St. Michael's College in Winooski, Vermont.

After instructing drafting classes for 3 years, Bill became the Director of Vocational Education for the Burlington Public School System, Burlington, Vermont, a position he held for 8 years. He spent 5 years as the Associate Director of Trident Technical College in Charleston, SC. Later, Bill accepted a position as head of the Department of Industry at Central Piedmont Community College, Charlotte, NC where he supervised the operation of six technical programs.

Bill was a soft spoken, easy going person that was easy to work with. His hobbies have been sailing, golf, and pool, all of which he was an expert at. He is missed by all who had the pleasure of meeting him.

ACKNOWLEDGMENTS

The authors would like to express their deepest thanks and appreciation to the following members of the Cengage team who have helped make the 9th Edition of Refrigeration and Air Conditioning Technology possible:

Vanessa Myers, Senior Product Manager, for her work with the authors and the publisher to produce a workable text that is both economical and comprehensive. Her perpetual energies and insistence on the best possible product have resulted in a quality text that is usable both in schools and in the field.

Elizabeth Berry, Learning Designer and Mary Clyne, Learning Designer, for their work with the authors and publisher to ensure the accuracy of the details in this text. Their professional, thorough, and enthusiastic handling of the text manuscript will provide the student with a well-presented, usable product.

Jim Zayicek, Senior Content Manager, for a great job of making sure all pages, artwork, and photographs are well presented and easy to follow. His professional skills and talents are greatly appreciated.

In addition, the authors would like to thank the following individuals, companies, organizations, and institutions for their valuable contributions to this text over the past 30-plus years:

Jerry Ackerman, Director of Marketing and Communications, Clearwater Systems Corporation

Rez Ambs, Manager, GeoFurnace Heating & Cooling, LLC, and CoEnergies, LLC, Traverse City, MI

Keith Avery, Regional Sales manager, Hampden Engineering, East Longmeadow, MA

Johnnie O. Bellamy, Eastfield College Continuing Education, Mesquite, TX

Ed Bottum, Sr, President, Refrigeration Research, Inc., Brighton, MI

Dave Boyd, Vice President, Appion Star Performance, Englewood, CO

Neal Broyles, Rolla Technical Institute, Rolla, MO

Barry Burkan, Apex Technical School, Queens, NY

Victor Cafarchia, El Camino College, Torrance, CA

John Campbell, Renton Technical College, Renton, WA

Tim Cary, Dwyer Instruments, Inc., Michigan City, IN

Norman Christopherson, San Jose City College, San Jose, CA

Cecil W. Clark, American Trade Institute, Dallas, TX

Hugh Cole, Gwinnett Technical Institute, Lawrenceville, GA

John Corbitt, Eastfield College Continuing Education, Mesquite, TX

George M. Cote, Erwin Technical Center, Tampa, FL

Phil Coulter, Durham College, Skills Training Center (Whitby Campus), Whitby, Ontario, Canada

John B. Craig, Sheridan Vocational Technical Center, Hollywood, FL

Mark Davis, New Castle School of Trades, Pulaski, PA

Victor DesRoches, Marketing Coordinator, LAKOS Separators and Filtration Solutions, Fresno, CA

James DeVoe, Senior Acquisitions Editor (Retired), Cengage, Clifton Park, NY

Kara DiCaterino, Content Project Manager, Cengage, Clifton Park, NY

Eugene Dickson, Indian River Community College, Fort Pierce, FL

Richard Dorssom, N.S. Hillyard AVTS, St, Joseph, MO

Paul Fauci, Slant/Fin Corporation, Greenvale, NY

Ferris State University, Big Rapid, MI

Fieldpiece Instruments Corporation, Orange, CA

John Fisher, Senior Project Manager (Retired), Cengage, Clifton Park, NY

David Foster, Managing Director, Uniweld Products, Inc., Fort Lauderdale, FL

Tina French, Marketing Manager, Fieldpiece Instruments Corporation, Orange, CA

Dr. Stanley Friedman, Phoenix, AZ

Keith Fuhrman, Del Mar College West, Corpus Christi, TX

George Gardianos, Lincoln Technical Institute, Mahwah, NJ

Mary Jo Gentry, Marketing Communications Manager, Ritchie Engineering Company—Yellow Jacket Products Division, Minneapolis, MN

Arthur Gibson, Erwin Technical Center, Tampa, FL

Brian C. Good, GI Industries, Monroe, CT

Brian Green, Spectronics Corporation, Westbury, NY

Laura Harris, WaterFurnace. Fort Wayne, IN

Rudy Hawkins, Kentucky Tech—Jefferson Campus, Louisville, KY

Jim Holstine, Manager, GeoFurnace Heating & Cooling, LLC, and CoEnergies, LLC, Traverse City, MI

Robert J. Honer, New England Institute of Technology at Palm Beach, West Palm Beach, FL

Karl Huffman, President, Hendrick Associates—Marley Cooling Towers, Grand Rapids, MI

Darren M. Jones, Meade County Area Technology Center, Brandenburg, KY

Deborah Kenny, SPX Cooling Technologies, Overland Park, KS

Tom Kiessel, Manager, CoEnergies, LLC, Traverse City, MI

Bob Kish, Belmont Technical College, St. Clairsville, OH

J. Curtis Lawson, Senior Technical Service Consultant, DuPont Refrigerants, Wilmington, DE

David Lax, Spectronics Corporation, Westbury, NY

Jason Leeds, North Central Kansas Tech, Hayes, KS

John Levey, President, Oilheat Associates, Wantagh, NY

Bill Litchy, Training Materials Manager, Scotsman Ice Systems, Vernon Hills, IL

Raul Lopez, Houston Community College, Houston, TX

Dan Mason, Danfoss Turbocore Compressors, Inc., Baltimore, MD

Dennis Matney, Ivy Tech Community College, Indianapolis, IN

Jason Mauric, Ferris State University, Big Rapids, MI

Richard McDonald, Santa Fe Community College, Gainesville, FL

Roger McDow, Senior Instructional Lab Facilitator, Central Piedmont Community College, Charlotte, NC

Marvin Maziarz, Niagara County Community College, Sanborn, NY

Arn McIntyre, Energy Center Director, Ferris State University, Big Rapids, MI

Modine Manufacturing Company, Racine, WI

Danny Moore, Director of Technical Support, Hoshizaki America, Inc., Peachtree City, GA

Joe Moravek, Lee College, Baytown, TX

Paul Morin, Technical Sales Specialist, The Energy Conservatory, Minneapolis, MN

Robert Nash, Jr., Emerson Climate Technologies, Sidney, OH

National Refrigerants, Inc., Philadelphia, PA

Philip Norris, Vatterott College, St. Louis, MO

Raymond Norris, Central Missouri State University, Warrensburg, MO

Robert Ortero, School of Cooperative Technical Education, New York, NY

Joseph R. Pacella, Associate Professor, Ferris State University, Big Rapids, MI

Joe Parsons, Vice President, Earthlinked Technologies Incorporated, Lakeland, FL

John Pastorello, CEO, Refrigeration Technologies, Anaheim, CA

Larry Penar, Refrigeration Service Engineers Society (RSES), Des Plaines, IL

John Pendleton, Central Texas College, Killeen, TX

Greg Perakes, Tennessee Technology Center at Murfreesboro, Murfreesboro, TN

Thomasena Philen, Daikin AC, Houston, TX

Lawrence D. Priest, Tidewater Community College, Virginia Beach, VA

Patrick Pung, Selfridge National Guard, Harrison Charter Township, MI

Tony Quick, Air Systems Components, Trion Division, Plano, TX

Chris Rebecki, Baran Institute, Windsor and West Haven, CT

Mitch Rens, Service Publications Manager, Manitowoc Ice, Inc., Manitowoc, WI

Brad Richmand, ACCA, Washington, DC

Terry M. Rogers, Midlands Technical College, West Columbia, SC

Jon Romanesko, Appion Corporation, Englewood, CO

John Sassen, Ranken Technical College, St. Louis, MO

Keith Satterthwaite, Lincoln Technical Institute, Union, NJ

Thomas Schafer, Macomb Community College, Warren, MI

Dennis Setchfield, Marketing Communications/e-Business/Digital Media, Sporlan Division, Parker Hannifin, Washington, MO

Dick Shaw, Air Conditioning Contractors of America (ACCA), Washington, DC

Phyllis Shaw, Marketing Communications Supervisor, Sporlan Division—Parker Hannifin Corporation, Washington, MO

Greg Skudlarek, Minneapolis Community and Technical College, Minneapolis, MN

Russell Smith, Athens Technical College, Athens, GA

Tim Snyder, Marketing Manager, The Chemours Company, Wilmington, DE

Frank Spevak, Marketing and Sales Manager, The Energy Conservatory, Minneapolis, MN

Sporlan Division, Parker Hannifin Corporation, Washington, MO

Nick Strickland, DuPont Corporation, Wilmington, DE

Michael D. Stuart, Senior Product Marketing Manager, Fluke Corporation, Everett, WA

Greg Sundheim, President, Appion Corporation, Englewood, CO

Robert Thompson, Dwyer Instruments, Inc., Michigan City, IN

Billy W. Truit, Worcester Career and Technical Center, Newark, MD

Dennis Weston, Manager, CoEnergies, LLC, Traverse City, MI

Richard Wirtz, Columbus State Community College, Columbus, OH

Larry W. Wyatt, Advanced Tech Institute, Virginia Beach, VA

Tony Young, Emerson Climate Technologies, Sidney, OH

Wayne Young, Midland College, Midland, TX

Eugene Ziegler, Sales Engineer and Training Materials Manager, Sporlan, Parker Hannifin, Washington, MO

A very special thank you goes out to the family and close friends of the authors for their continued emotional support, patience, and assistance, without which this project could not have come to fruition.

AVENUE FOR FEEDBACK

The authors would appreciate feedback from students and/or instructors. They can be reached through Cengage Learning in Clifton Park, New York, or directly via e-mail at: *Eugene_and_Jason@yahoo.com*

INTERNATIONAL CODE COUNCIL (ICC)

The International Code Council is a member-focused association. It is dedicated to developing model codes and standards used in the design, build, and compliance process to construct safe, sustainable, affordable, and resilient structures.

Most US communities and many global markets choose the I-Codes. ICC Evaluation Service (ICC-ES) is the industry leader in performing technical evaluations for code compliance, fostering safe and sustainable design and construction.

500 New Jersey Avenue, NW, 6th Floor
Washington, DC 20001-2070
Phone: 888-ICC-SAFE (888-422-7233)
Website: www.iccsafe.org

Eastern Regional Office (BIRMINGHAM)
Central Regional Office (CHICAGO)
Western Regional Office (LOS ANGELES)

PARTNERSHIP FOR AIR-CONDITIONING, HEATING, REFRIGERATION ACCREDITATION (PAHRA)

The Partnership for Air-Conditioning, Heating, Refrigeration Accreditation (PAHRA) is a partnership between HVACR educators and the HVACR industry that recognizes programs that have met or exceeded industry-validated standards.

- The purpose of PAHRA is to improve the quality of training offered at all levels of education by meeting or exceeding established industry standards in the heating, ventilation, air conditioning and refrigeration (HVACR) industry.
- The development of the evaluation standards for educational programs in HVACR is the work of all the parties interested in the HVACR Industry and the educational future of the industry's technicians.
- The PAHRA program was developed, supported, and implemented by a number of HVACR educators, administrators and vocational technical education state supervisors.

www.pahrahvacr.org

HVAC EXCELLENCE

HVAC Excellence was established in 1994 to improve the technical competency of the HVACR industry through validation of the technical education process. By creating industry-embraced standards for education, and verifying that they have been met, HVAC training programs are able to meet the challenges facing our industry by continuous improvement in the way that technicians are prepared to enter the workforce. One way that HVAC Excellence ensures that students receive high quality training is through programmatic accreditation.

Programmatic accreditation is an independent, non-governmental, third-party review of an educational program. Programmatic accreditation of HVACR educational programs readily identifies HVACR educational programs that possess the resources, and administrative support, to prepare students for entry-level positions in the HVACR industry.

Accreditation standards require a thorough examination of: mission of program, administrational responsibilities, finances and funds, student services, instruction design and program elements, physical facilities, equipment and tools, cooperative training and instructor qualifications. HVAC Excellence is the industry's oldest and largest provider of HVACR programmatic accreditation.

In addition to accrediting HAVCR training programs, HVAC Excellence and its affiliates offer entry-level technician certifications, advanced-level service technician certifications and instructor credentials. With their portable and stackable credentials, HVAC Excellence follows individuals through their entire HVACR careers.

The authors of this text have taken the initiative to ensure that the 9th Edition of Refrigeration and Air Conditioning Technology has been cross-walked to HVAC Excellence's task and competency list. This cross-walk ensures that this text can be used to help students prepare for the nationally-recognized certification examinations that they will take prior to entering the field.

P.O. Box 491
Mount Prospect, IL 60056-0491
Tel: 800 726-9696
Website: www.escogroup.org

SECTION 1

THEORY OF HEAT

INTRODUCTION

UNITS

UNIT 1	Heat, Temperature, and Pressure
UNIT 2	Matter and Energy
UNIT 3	Refrigeration and Refrigerants

INTRODUCTION

Refrigeration is a complex topic that covers a wide range of areas. Refrigeration relates to the cooling of substances to

- preserve and transport food products,
- produce ice,
- aid in the manufacturing of many commercial products, and
- aid in medical research.

In addition, refrigeration plays vital roles in many other industrial, commercial, and residential applications. Air-conditioning, a form of refrigeration, refers to space heating, cooling, dehumidifying, humidifying, air filtering, exhausting, ventilating, and improving overall indoor air quality for those in the occupied space. Very often, the term air-conditioning is associated only with cooling processes but, as you can see, it involves much more than that.

HISTORY OF REFRIGERATION AND AIR-CONDITIONING (COOLING)

Most evidence indicates that the Chinese, as early as 1000 B.C., were the first to store ice and snow in order to cool wine and other food products. Early Greeks and Romans used underground pits, which were insulated with straw and weeds, to store ice for long periods of time. The ancient people of Egypt and India cooled liquids in porous earthen jars. These jars were set out in the dry night air, and the evaporation of the liquids seeping through the porous walls provided the cooling. Some evidence indicates that ice was even produced from the vaporization of water through the walls of these jars.

In the eighteenth and nineteenth centuries, natural ice was cut from lakes and ponds in the winter in the northern United States and stored underground for use in the warmer months. Some of this ice was packed in sawdust and transported to southern states to be used for preserving food. In the early twentieth century, it was still common in the northern states for ice to be cut from ponds and then stored in open ice houses. Sawdust insulated the ice, which was then delivered to homes and businesses.

In 1823, Michael Faraday discovered that certain gases under constant pressure will condense when they cool. In 1834, Jacob Perkins, an American, developed a closed refrigeration system using liquid expansion and then compression to produce cooling. He used ether as a refrigerant, a hand-operated compressor, a water-cooled condenser, and an evaporator in a liquid cooler. He was awarded a British patent for this system. In Great Britain during the same year, L. W. Wright produced ice by the expansion of compressed air.

TIME LINE

1800

1823
Michael Faraday discovered that certain gases under constant pressure will condense when they cool.

1834
L. W. Wright produced ice by expansion of compressed air. Jacob Perkins developed a closed refrigeration system using expansion of a liquid and compression to produce cooling. David Perkins developed a closed refrigeration system using the expansion and compression of ether.

1842
John Gorrie used dripping ammonia to produce cooling.

1850

1858
Ferdinand Carré developed an ammonia mechanical refrigerator. It made blocks of ice.

1875
Raoul Pictet used sulfur dioxide as a refrigerant.

1881

Gustavus Swift was developing refrigeration railcars.

1890
Michael Cudahy improved railcar refrigeration.

In 1842, Florida physician John Gorrie placed a vessel of ammonia atop a stepladder and let the ammonia drip, which then vaporized and produced a cooling effect. This basic principle is still used in air-conditioning and refrigeration today. In 1856, Australian inventor James Harrison, an immigrant to America from Scotland, also used ammonia experimentally, but reverted to an ether compressor in equipment that had been previously constructed. In 1858, a French inventor, Ferdinand Carré, developed a mechanical refrigerator using liquid ammonia in a compression machine that produced blocks of ice. Generally, mechanical refrigeration was first designed to produce ice.

In 1875, Raoul Pictet of Switzerland first used sulfur dioxide as a refrigerant. Sulfur dioxide was not only a good refrigerant, but also served as a good lubricant for the system's compressor. This refrigerant was used frequently after 1890 and on British ships into the 1940s. Refrigeration railcars were developed by Gustavus Swift in 1881, and in 1890, Michael Cudahy had improved their design. Sulfur dioxide was also used in the Audiffren-Singrün refrigeration machine patented in 1894 by a French priest and physicist, Father Marcel Audiffren. It was originally designed to cool liquids, such as wine, for the monks.

In 1902, Willis Carrier, the "father of air-conditioning," designed a humidity control to accompany a new air-cooling system. He pioneered modern air-conditioning. In 1915, he, along with other engineers, founded Carrier Engineering, now known as the Carrier Corporation.

In 1918, the Kelvinator company, originally named the Electro Automatic Refrigeration Corporation, came into being and sold the first Kelvinator household units. The refrigerator was a remote-split type in which the condensing unit was installed in the basement and connected to an evaporator in a converted icebox in the kitchen. The Guardian Refrigerator Company developed a refrigerator they called "the Guardian." General Motors purchased Guardian in 1919 and developed the refrigerator they named Frigidaire. By 1929, refrigerator sales topped 800,000. The average price fell from $600 in 1920 to $169 in 1939. By the 1930s, refrigeration was well on its way to being used extensively in American homes and commercial establishments.

In 1923, Nizer introduced a water-cooled compressor and condensing unit for ice cream cabinets, considered to be the first commercial ice cream unit. Nizer soon merged into the Kelvinator Company. Between 1923 and 1926, units produced by Savage Arms were among the first automatically controlled commercial units. The Savage Arms compressor had no seals, no pistons, and no internal moving parts. A mercury column compressed the refrigerant gas as the entire unit rotated. The compressor was practically noiseless.

In 1928, Paul Crosley introduced an absorption-type refrigeration machine so that people could have refrigeration in rural areas where electricity was scarce. These systems, which used a mixture of ammonia and water, could lower the inside temperature to 43°F or less. Ice cubes actually could be made for a period of about 36 hours, depending on the room temperature. These machines would need periodic "recharging" by heating the system over a kerosene burner.

In 1939, the Copeland Company introduced the first successful semihermetic (Copelametic) field-serviceable

1894

The Audiffren-Singrün refrigerating machine used a hermetically sealed system.

1900

1902

Willis Carrier designed humidity control for a new air-cooling system and pioneered modern air-conditioning. Willis Carrier also originated the Carrier equation upon which the psychrometric chart and all air-conditioning is based.

1918

Kelvinator refrigeration unit introduced.

1919

FRIGIDAIRE

Guardian Refrigerator Company developed a refrigerator they called the Guardian. General Motors purchased Guardian in 1918 and the name was changed to Frigidaire.

1923

Nizer water-cooled compressor and condenser for ice cream cabinets. Nizer made the first ice cream unit for the market. Nizer soon merged into Kelvinator.

1926

Savage Arms ice cream unit introduced. It contained a unique mercury column compressor with no seals, pistons, or lubrication.

1928

Crosley Icy Ball—An absorption-type refrigerator machine provided refrigeration in rural areas without electricity.

compressor. Three engineering changes made these compressors successful:

1. Cloth-insulated motor windings were replaced with Glyptal insulation.
2. Neoprene insulation replaced porcelain enamel in the electric terminals.
3. Valves were redesigned to improve efficiency.

Many different refrigerants have been developed over the years. R-12, a chlorofluorocarbon (CFC), was developed in 1931 by Thomas Midgley of Ethyl Corporation and C. F. Kettering of General Motors. It was produced by DuPont. In 1974, two professors from the University of California, Sherwood Rowland and Mario Molina, presented the "ozone theory." Their hypothesis was that CFC refrigerants released into the atmosphere were depleting the earth's protective ozone layer. Scientists conducted high-altitude studies and concluded that CFCs were indeed linked to ozone depletion. Representatives from the United States, Canada, and more than 30 other countries met in Montreal, Canada, in September, 1987, to try to solve the problem of released refrigerants and the effect they had on ozone depletion. This meeting produced the Montreal Protocol, which by 1989 had been ratified by 100 nations. It mandated a global freeze on the production of CFCs at 1986 levels. The Protocol also froze production of hydrochlorofluorocarbon (HCFC) refrigerants at their 1986 levels, beginning in 1992. In addition, the Protocol set a schedule of taxes on CFC refrigerants. As research on ozone depletion continues today, reassessments and updates to the Montreal Protocol also continue. At the time of this writing, the most current updates are as follows:

- 1990 (November)—President George H. W. Bush signed the Clean Air Act amendments that initiated production freezes and bans on certain refrigerants.
- 1992 (July)—The EPA made it against the law to intentionally vent CFC and HCFC refrigerants into the atmosphere.
- 1993—The EPA mandated the recycling of CFC and HCFC refrigerants.
- 1994 (November)—The EPA mandated a technician certification program deadline. Current HVACR technicians had to be EPA-certified by this date.
- 1995 (November)—The EPA made it against the law to intentionally vent alternative refrigerants (HFCs and all refrigerant blends) into the atmosphere.
- 1996—The EPA made it illegal to manufacture or import CFC refrigerants.
- 1996—The EPA put into place a gradual HCFC production phaseout schedule, which will totally phase out the production of HCFC refrigerants by the year 2030.
- 1998 (June)—The EPA proposed new regulations on recovery/recycling standards, equipment leak rates, and alternative refrigerants.

TIME LINE

1929
Household refrigerator sales topped 800,000.

1931
Refrigerant R-12 was developed by Thomas Midgley and C. F. Kettering.

1939
First successful semi-hermetic (Copelametic) compressor introduced by the Copeland Company. Sales started in 1939. It was a field serviceable compressor.

1935–1965
A number of chlorofluorocarbons (CFCs) were developed during this period.

1974
Professors Rowland and Molina presented the "ozone theory" that CFCs were depleting the ozone layer.

1985

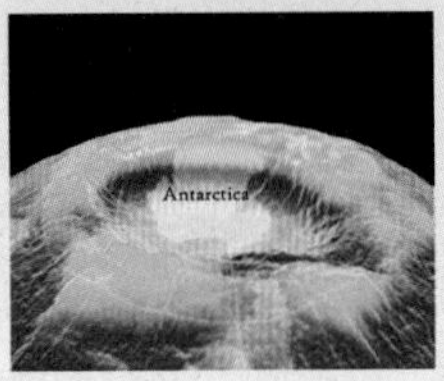

Stratospheric ozone hole discovered.

1989
Montreal Protocol ratified by 100 countries.

1990
Clean Air Act amendments. Production freezes and bans on certain refrigerants. Signed by President George H. W. Bush, November, 1990.

1992
EPA
The EPA made it illegal to intentionally vent CFC and HCFC refrigerants into the atmosphere.

1994
Mandatory EPA technician certification required to service refrigeration and A/C equipment or to buy ozone-depleting refrigerants as of November, 1994.

1995
Unlawful to intentionally vent alternative refrigerants (HFCs and all refrigerant blends) into the atmosphere as of November, 1995.

1996
Gradual HCFC production phaseout schedule with the total phaseout of HCFC refrigerants in the year 2030. Phaseout of CFC refrigerants.

- 2004—35% reduction in HCFC refrigerant production.
- 2007—HCFC reduction on production was accelerated from 65% to 75% from the baseline 1989 production year.
- 2010—HCFC-22 is banned in new equipment. No production or importing of HCFC-22 and HCFC-142b, except for use in equipment manufactured before January 1, 2010.
- 2015—90% reduction in HCFC-22 production from the baseline production year of 1989. No production or importing of any HCFC, except for use in equipment manufactured before January 1, 2010.
- 2016—EPA updates regulations on refrigerant handling and usage.
- 2018—EPA expands sales restrictions on HFC and HFO refrigerants. Daily fine for intentionally venting refrigerant increases to $44,539.
- 2019—New refrigerant leak inspection, verification test, and record-keeping requirements go into effect.
- 2020—Total ban on HCFC-22 production. No production and no importing of R-22 and R-142b.
- 2030—Total ban on all HCFC production. No production and no importing of any HCFC.

From 1997 to 2000, voluntary HVACR technician certification became a major focus of the industry. From 1998 to the present, the major players in voluntary HVACR technician certification and home-study examinations were, and continue to be, the AC&R Safety Coalition, the Air-Conditioning, Heating and Refrigeration Institute (AHRI), the Heating, Air Conditioning, and Refrigeration Distributors International (HARDI), the Carbon Monoxide Safety Association (COSA), the Green Mechanical Council, HVAC Excellence, North American Technician Excellence (NATE), the Refrigeration Service Engineers Society (RSES), and the United Association of Journeymen and Apprentices (UA).

By 2008, global warming had become a major environmental issue. A scoring system was designed to help engineers, contractors, and consumers know the "green value" of each mechanical installation. R-410A, an efficient and chlorine-free HFC-based refrigerant blend for residential and light-commercial air-conditioning applications was developed for use with the scroll compressor for greater efficiencies. Also today, every central split cooling system manufactured in the United States must have a Seasonal Energy Efficiency Ratio (SEER) rating of at least 13. This energy requirement was mandated by federal law as of January 23, 2006. The "green value" encompasses the system's energy efficiency, pollution output, and **sustainability**. Green buildings and green mechanical systems are becoming increasingly popular in today's world as a way to curb global warming.

Green awareness, global warming, energy efficiency, energy savings, sustainability, and high-performance buildings are still at the forefront of environmental and energy concerns. Energy audits have become an integral part of

1997–2000

Kyoto Protocol introduced, intended to reduce worldwide global warming gases. Global warming has become a major environmental issue.

Voluntary HVACR technician certification became a major focus.

2000

1998–2005

Puron

R-410A, an efficient and chlorine-free HFC-based refrigerant blend for residential and light-commercial air-conditioning applications, is used with the scroll compressor for greater efficiencies.

"Green awareness"—Green mechanical systems and green buildings become increasingly popular as a way to curb global warming and conserve energy. Natural refrigerants like carbon dioxide, ammonia, propane, butane, and isobutane are becoming increasingly popular because of their low GWP, zero ODP, low costs, availability, and high efficiencies.

2006

"Minimum 13 SEER" required for 1 1/2- to 5-ton unitary equipment and split/ packaged air conditioners and heat pumps.

2008–2014

Green awareness, global warming, energy efficiency, equipment efficiency, energy savings, sustainability and high performance buildings are still on the environmental and energy forefront. Energy audits have become an integral part of evaluating and assessing existing building's energy performance. Higher efficiency standards for new building's energy performance have been established. Higher levels of training and certification have been developed for HVACR technicians to meet the needs of more sophisticated, energy-efficient buildings and HVACR equipment. Carbon Dioxide (R-744) is gaining popularity in commercial refrigeration applications. This naturally occurring refrigerant does not deplete ozone and has a low global warming potential (GWP).

2015–2016

The SMART revolution is taking hold where wireless technology, applications (Apps), smart phones, notebooks, and pads, are controlling and monitoring HVACR equipment through wireless networks. Cloud computing is also becoming very popular.

2016

EPA revised regulations under its Section 608 refrigerant handling program. Visit www.epa.gov for details.

2018–2020

U.S. Department of Energy raises minimum efficiency requirements for newly installed rooftop systems in low-rise buildings by 10%

Total phaseout of R-22 and R-142b.

2030

Total phaseout scheduled for all HCFC refrigerants.

The USGBC membership, which is composed of every sector of the building industry and consists of over 9000 organizations, developed and continues to refine LEED. LEED promotes expertise in green building by offering project certification, professional accreditation, and training. LEED emphasizes state-of-the-art strategies for sustainable site development, water savings, energy efficiency, material selection, and indoor environmental quality. According to the United Nations World Commission on Environment and Development, a **sustainable design** "meets the needs of the present without compromising the ability of future generations to meet their own needs." Companies looking to utilize **green** technologies or incorporate sustainable design into their buildings and facilities, are concerned with six areas:

- Optimizing site location
- Optimizing energy use
- Protecting and conserving water
- Using environmentally preferable products
- Enhancing indoor environmental quality
- Optimizing operational and maintenance practices

There are multiple LEED rating categories that include:

- Building Design and Construction—For new construction or major renovations; includes New Construction, Core & Shell, Schools, Retail, Hospitality, Data Centers, Warehouses & Distribution Centers, and Healthcare.
- Interior Design and Construction—For complete interior fit-out projects; includes Commercial Interiors, Retail and Hospitality.
- Building Operations and Maintenance—For existing buildings that are undergoing improvement work or little to no construction; includes Existing Buildings, Schools, Retail, Hospitality, Data Centers, and Warehouses & Distribution Centers.
- Neighborhood Development—For new land development projects or redevelopment projects containing residential uses, nonresidential uses, or a mix. Projects can be at any stage of the development process, from conceptual planning to construction; includes Plan and Built Project.
- Homes—For single-family homes, low-rise multifamily (one to three stories) or mid-rise multifamily (four to six stories); includes Homes and Multifamily Lowrise and Multifamily Midrise.
- Cities and Communities—For entire cities and sub-sections of a city. Using the Arc performance platform, LEED for Cities projects can measure and manage their city's water consumption, energy use, waste, transportation, and human experience.

Visit https://new.usgbc.org/leed for more information on the LEED program.

Points are awarded in each category depending on how well the building meets the category's requirements. There are four levels of certification according to the point system:

Certified	40–49
Silver	50–59
Gold	60–79
Platinum	80–110

The purpose of LEED is to provide a third-party certification process using nationally developed and accepted minimum standards for the construction industry. It affects the design, construction, and operation phases of high-performance "green" buildings. LEED systems take into account other ways of increasing efficiencies, such as water conservation, **heat island** reduction in urban areas, incentives for use of locally manufactured materials, site preparation, and maintenance. To receive a LEED rating, the facility must be built by a team, some of whose members are LEED accredited professionals. LEED-rated projects have a higher cost than similar, non-LEED projects because the enhancements required to increase efficiencies and the certification and documentation required cost more. Many European nations have made LEED-type systems mandatory for all buildings and have instituted existing-building rating systems that monitor yearly energy consumption of all utilities in these buildings. The higher a building's energy usage or "energy utilization index" above a minimum consumption, the higher amount of penalty tax the building owner must pay. This provides an incentive for improving the building's energy footprint.

The green awareness movement isn't just a temporary "buzzword" that will fade away with time. It is one that will be rapidly gaining momentum in the coming years. If contractors want to remain competitive, they must obtain the necessary training with regard to green building and LEED certification.

An alternative to LEED certification is the Green Globes® program, which is offered by the Green Building Initiative. The Green Globes program operates on a 1000-point scale and certifications range from one to four Green Globes, with four Green Globes being their highest possible rating. Both the LEED and Green Globes programs are nationally accepted.

HISTORY OF HOME AND COMMERCIAL HEATING

Human beings' first exposure to fire was probably when lightning or another natural occurrence, such as a volcanic eruption, ignited forests or grasslands. After overcoming the fear of fire, early humans found that placing a controlled fire in a cave or other shelter could create a more comfortable living environment. Fire was often carried from one place to another. Smoke was always a problem, however, and methods needed to be developed for venting it outside. Native

Americans, for example, learned in later years to vent smoke through holes at the peak of their tepees, and some of these vents were constructed with a vane that could be adjusted to prevent downdrafts. The fireplaces common in Europe and North America were vented through chimneys.

Early stoves were found to be more efficient than fireplaces. These early stoves were constructed of a type of firebrick, ceramic materials, or iron. In the mid-eighteenth century, a jacket for the stove and a duct system were developed. The stove could then be located at the lowest place in a structure, and the heated air in the jacket around the stove would rise through a duct system and grates into the living area. This was the beginning of the development of circulating warm-air heating systems.

Boilers that heated water were also developed, and this water was circulated through pipes in duct systems. The water heated the air around the pipes, and the heated air passed into the rooms to be heated. Radiators were then developed. The heated water circulated by convection through the pipes to the radiators, and heat was passed into the room by radiation. These early systems were forerunners of modern hydronic heating systems.

Steam heat became a popular heating option at the beginning of the nineteenth century and coal was the fuel of choice for boilers. Coal was desirable because it burned hot and lasted a long time. But coal was not inexpensive and the coal dust that was ever-present resulted in health, primarily breathing, problems for many people. In the late 1920s, the oil burner was invented and was a very attractive alternative to coal. Oil was less expensive and cleaner than coal and nobody had to keep feeding coal to keep the fire burning.

Oil remained popular, and inexpensive, until the Arab oil embargo of 1973 and the Iranian Revolution in 1979. Oil prices spiked and people had to wait in lines, sometimes for hours, to get their ration of fuel for their cars. As a result, many people switched to natural gas, comprised primarily of methane. Natural gas boilers began to replace the old oil boilers, just as oil had replaced coal.

After the price shocks of the 1970s, oil prices stayed low for most of the 1980s and 1990s, with occasional moderate peaks. Oil prices then rose steadily from the period between September 11, 2001 and 2009, and continue to fluctuate today.

Today, commercial and residential heating needs are being met in a number of ways that include traditional hot water and steam, but new, more efficient technologies are becoming more attractive. These include radiant heating, radiant cooling, and geothermal heat pump systems.

CAREER OPPORTUNITIES

The HVACR industry is rapidly changing due to advancements in technology being spurred on by the need for increased energy efficiencies. The career opportunities available in HVACR for those who have acquired formal technical training coupled with field experience are unlimited. Schools that provide excellent technical training in the field are becoming easier to identify through HVACR program accreditation. As new equipment becomes more technically challenging and the existing workforce continues to age, the employment positions available will continue to outnumber applicants for the foreseeable future. This shortfall in available, competent HVACR service technicians is being addressed through the cooperative efforts of educational institutions, labor unions, employers, and manufacturers. Many organizations offer apprenticeship opportunities that can lead to high-income positions. Manufacturers are also teaming up with select educational institutions across North America to help develop the next generation of HVACR technicians.

Many newer buildings are constructed so tightly that the quality of the air must be controlled by specialized equipment. The conditions of the air must also be carefully controlled in areas that perform manufacturing processes. Heating and air-conditioning systems control the temperature, humidity, and total air quality in residential, commercial, industrial, and other types of buildings. Refrigeration systems are used to store and transport food, medicine, and other perishable items. Refrigeration and air-conditioning technicians design, sell, install, or maintain these systems. Many contractors and service companies specialize in commercial refrigeration. The installation and service technicians employed by these companies install and service refrigeration equipment in supermarkets, restaurants, hotels/motels, flower shops, and many other types of retail and wholesale commercial businesses.

Other contractors and service companies may specialize in air-conditioning. Many specialize in residential-only or commercial-only installation and service; others may install and service both residential and commercial equipment up to a specific size. Air-conditioning may include cooling, heating, humidifying, dehumidifying, ventilating, exhausting, and air cleaning. Heating equipment may rely on fossils fuels, such as natural gas, liquefied petroleum, or oil, or may be configured as electric-based or heat pump systems. The type and number of installations will vary from one part of the country to another, depending on the climate and availability of the heat source. The heating equipment may be either a furnace (which heats air) or a boiler (which heats water). The boiler heats water and pumps it to the space to be heated, where one of many types of heat exchangers transfers the heat to the air.

Technicians may specialize in installation or service of equipment, or they may be involved with both. Other technicians may design installations or work in the sales area. Sales representatives may be in the field selling equipment to contractors, businesses, or homeowners; others may work in wholesale supply stores. Still other technicians may represent manufacturers, selling equipment to wholesalers and large contractors.

Many opportunities exist for technicians to be employed in the industry or by companies owning large buildings. Technicians may be responsible for the operation of air-conditioning equipment, or they may be involved in the service

of this equipment. Opportunities also exist for employment in servicing household refrigeration and room air conditioners, which would include refrigerators, freezers, and window or through-the-wall air conditioners. Opportunities are also available for employment in a field often called transport refrigeration. This includes servicing refrigeration equipment on trucks or on large containers hauled by trucks and ships.

Most modern houses and other buildings are constructed to keep outside air from entering, except through planned ventilation. Consequently, the same air is circulated through the building many times. The quality of this air may eventually cause a health problem for people spending many hours in the building. This indoor air quality (IAQ) presents another opportunity for employment in the air-conditioning field. Technicians clean filters and ducts, take air measurements, check ventilation systems, and perform other tasks to help ensure healthy air quality. Other technicians work for manufacturers of air-conditioning equipment. These technicians may be employed to assist in equipment design, in the manufacturing process, or as equipment salespersons.

Following is a list of many career opportunities in the HVACR field:

- Field service technician
- Service manager
- Field supervisor
- Field installer
- Journeyman
- Project manager
- Job foreman
- Application engineer
- Controls technician
- Draftsperson
- Contractor
- Lab technician
- Inspector
- Facilities technician
- Instructor
- Educational administrator
- Inside/outside sales rep
- Sales manager
- New product developer
- Research engineer
- Estimator

TECHNICIAN CERTIFICATION PROGRAMS

HISTORY

Even though mandatory technician certification programs are in place today, the EPA originally did not consider them as its lead option. As a matter of fact, the EPA initially thought private incentives would ensure that technicians were properly trained in refrigerant recycling and recovery. The EPA also stated that it would play an important role through a voluntary technician certification program by recognizing those who provide and participate in voluntary technician training programs that meet certain minimum standards. The EPA also thought that a mandatory certification program would be an administrative burden. The EPA then requested public comments on a mandatory versus voluntary technician certification program. More than 18,000 comments were in favor of a mandatory program, and only 142 were in favor of a voluntary program. Most of the 18,000 in favor of the mandatory certification program were major trade organizations and technicians themselves. Manufacturers of recovery and recycling equipment, along with environmental organizations, also supported mandatory certification. They believed it would increase compliance with venting, recovery, and recycling laws and the general safe handling of refrigerants. The following were reasons given by those favoring mandatory technician certification:

- Improve refrigerant leak detection techniques
- Promote awareness of problems relating to venting, recovery, and recycling of refrigerants
- Improve productivity and cost savings through proper maintenance practices
- Ensure environmentally safe service practices
- Gain more consumer trust
- Receive more liability protection
- Ensure that equipment is properly maintained
- Educate technicians on how to effectively contain and conserve refrigerants
- Create uniform and enforceable laws
- Foster more fair competition in the regulated community

With these comments in mind, the EPA decided that mandatory technician certification would increase fairness by ensuring that all technicians were complying with today's rules. The EPA also said that a mandatory certification program would enhance the EPA's ability to enforce the rules by providing a tool to use against intentional noncompliance: the ability to revoke the technician's certification. The EPA then created a mandatory technician certification program that mandated all technicians to be certified effective November 14, 1994.

All technicians must pass an examination administered by an approved EPA testing organization in the private sector in order to purchase refrigerant and to work on equipment that contains refrigerant. This mandatory certification applies to individuals who work as installers, contractor employees, in-house service personnel, and anyone else who installs, maintains, or repairs equipment that might reasonably have the opportunity to release CFCs or HCFCs into the atmosphere. The EPA created three separate technician certification types:

- Small appliances
- High- and very high pressure appliances
- Low-pressure appliances

Persons who successfully pass a *core of questions* on stratospheric ozone protection and legislation and also pass one of the three certifications are certified in that category. If all three certifications are passed, a person will be *universally* certified. To date, the EPA is not requiring recertification. However, it will be the technicians' responsibility to keep up to date on new technologies and governmental rule changes. By creating types of certification, the EPA allowed technicians to be tested on information concerning equipment and service practices that the technicians primarily service and maintain.

Although training programs are beneficial, participation in a training program is not required currently. To create price-competitive training programs, training programs requested by technicians will be administered by the private sector. Many national educational and trade organizations—such as the AC&R Safety Coalition, the Air-Conditioning, Heating and Refrigeration Institute (AHRI), the Air Conditioning Contractors of America (ACCA), the Heating, Air Conditioning, and Refrigeration Distributors International (HARDI), the Carbon Monoxide Safety Association (COSA), the Educational Standards Corporation (ESCO), the Environmental Protection Agency (EPA), Ferris State University (FSU), the Green Mechanical Council (GreenMech), HVAC Excellence, North American Technician Excellence (NATE), the Refrigeration Service Engineers Society (RSES), and the United Association of Journeymen and Apprentices (UA)—have developed training and/or testing programs. These programs are specifically intended to help technicians comply with the July 1, 1992, refrigerant venting law. Unit 9 of this text, "Refrigerant and Oil Chemistry and Management—Recovery, Recycling, Reclaiming, and Retrofitting," gives more detailed information on the EPA's mandatory technician certification program, including details on the specific types of certification tests and specifications.

CERTIFICATION PROGRAMS

Technician certification programs can be divided into two categories:

- Mandatory technician certification programs
- Voluntary technician certification programs

Mandatory technician certification programs are covered in the preceding paragraphs and in Unit 9. Voluntary technician certification programs are becoming popular because they are industry-led and are much more comprehensive in nature when compared to mandatory certification programs. They give technicians an educational opportunity from the beginning to the end of their careers. These programs allow technicians to become recognized for their level of expertise and also allow them to achieve higher levels of competence. Their diverse nature allows almost every aspect of the industry to be covered. Voluntary certification testing is based on the courses taken for each level, with an outline and roadmap on what material will be covered on the test and where to find it. An example of a voluntary certification is the ESCO Group's R-410A Safety and Certification test, which evaluates a candidate's knowledge of the special considerations that must be taken into account when working on systems that contain this refrigerant.

WHY TECHNICIANS SHOULD BECOME CERTIFIED

As mentioned, mandatory technician certification allows the technician to purchase ozone-depleting refrigerants legally and work on equipment that contains refrigerant. Some advantages of having both mandatory and voluntary technician certifications are:

- Customers tend to ask for certified technicians because of their reliability and good workmanship.
- Equipment manufacturers develop faith in certified technicians and have a sense of well-being when they know the job has been accomplished by a certified technician.
- Higher standards are set on the job by certified technicians, giving them more respect, recognition, trust, higher pay, and a higher quality of life in the long run.
- Employers would rather hire a certified technician, because they know certified technicians care more about their reputation, customer relations, and overall professionalism.
- Certification gives the technician a status symbol for other technicians to work up to.
- Certified technicians have proven technical proficiencies with measured capabilities.

PROGRAMMATIC ACCREDITATION

Accreditation has gained popularity in the last decade for secondary and postsecondary HVACR programs. Program accreditation involves an independent, nongovernmental, nonindustry, third-party review of an educational program. At the time of this printing, the leading HVACR programmatic accreditation agency is HVAC Excellence (www.escogroup.org). The programmatic accreditation process validates that the established standards of excellence for HVACR educational programs are met or, in many cases, exceeded. These standards are designed to ensure that our future workforce receives the quality of training required to provide the skills necessary for success in the HVACR industry. The Department of Education in its Code of Federal Regulations, Title 34, sets guidelines and defines criteria for who can be an accredited body but does not do any accrediting itself.

The standards require a thorough examination of the programs:

- Mission
- Curriculum
- Plans of instruction
- Finances and funds
- Administrative responsibilities
- Equipment and tools
- Facilities
- Instructor's qualifications

Programmatic accreditation provides an HVACR program with an opportunity to recognize both its strengths and weaknesses. Some of the benefits of programmatic accreditation include the following:

- Enhanced student confidence in the quality of the program
- Assurance and confidence of employers that student graduates are properly trained
- Enhanced student placement and employer satisfaction
- Acceptability of transfer credits for students
- Self-evaluation of faculty and staff
- Enhanced articulation agreements with other HVACR programs and institutions
- Assistance in acquiring federal, state, and private funding
- Creation of self-improvement goals

NATIONAL SKILL STANDARDS

The National Skill Standards (NSS), as interpreted by the Vocational-Technical Education Consortium of States (VTECS) for Heating, Air Conditioning, and Refrigeration Technicians, were funded by the U.S. Department of Education from 1992 to 1998 as part of twenty-two projects from the National Skill Standards Board. The NSS were created by a joint effort of committees composed of heating, air-conditioning, and refrigeration industry professionals. These skill standards not only help technicians identify the skills and knowledge needed for their occupation but also assess their weaknesses and/or needs for additional training.

Skill standards are often described as workplace behaviors, technical skills, and the general body of knowledge required of technicians to be successful, productive, and competitive in today's workforce. As HVACR manufacturers increase the efficiency and sophistication of their equipment, technicians require additional, updated information as well as a sound technical skills base to maintain, install, and service this equipment. The increased number of environmental regulations concerning more energy-efficient and environmentally friendly HVACR equipment have also created a new knowledge and skills base for technicians to learn and use on their jobs. Although it is difficult to provide all users of this text with information on the vast array of issues covered in the NSS, the authors have made every effort to do so. We hope both private and public institutions as well as the industry will use our comprehensive book to provide both students and workers with the competencies needed for successful employment and advancement in the ever-changing and growing technical HVACR field.

The NSS are divided into three main areas with subdivisions as follows:

CORE KNOWLEDGE

- Communications
- Mathematics
- Science

OCCUPATIONAL-SPECIFIC SKILLS

- Core skills
- Occupational-specific skills

WORKPLACE BEHAVIORS

- Ethics
- Environment
- Communications
- Professionalism
- Problem solving

The *core skills* consist of:

- Safety and environment
- Electrical principles
- Electric motors
- Controls
- Refrigeration principles and practices
- Heating principles and practices
- Air-conditioning principles and practices
- Piping principles and practices

The *occupational-specific skills* consist of:

- Residential and light-commercial heating
- Residential and light-commercial air-conditioning
- Residential and light-commercial heat pumps
- Commercial conditioned-air systems
- Commercial refrigeration

CUSTOMER RELATIONS AND TECHNICIAN SOFT SKILLS

Customer relations are extremely important to a service business and consequently to a service technician. Without customers there will be no business and no income. The technician is a major factor in acquiring and keeping customers. This is true whether work is performed at a residence, an office, a restaurant, or a store, or whether the technician is an inside or outside salesperson for a distributor or contractor. The HVACR business and technician are dependent on the customers. All technicians should be concerned with the quality of their work because customers have the right to insist on quality. If they have had a previous unsatisfactory experience, customers may have some doubt as to whether they will get the quality service for which they are

paying. As professionals, technicians should strive to provide the best workmanship possible. Quality work will prove beneficial to the technician, to the company, and to the consumer. Customers depend on the technician for their comfort and air quality at home and at the office.

FIRST IMPRESSIONS

The impression the technician makes on the customer is very important, and the first impression is the most important. The first impression begins with the technician arriving on time. Most customers feel that their time is valuable. If the technician is going to be delayed, the customer should be called and given an explanation. An appointment should be scheduled for either later that day or another time convenient for the customer. The customer affected by a delay should be given priority in scheduling a makeup appointment. If the service call is an emergency, all efforts should be made to arrive as soon as possible.

When arriving, do not park in or block the customer's driveway unless necessary. If carrying equipment or having to make several trips to the vehicle, ask permission to park in the driveway. The customer may suggest another location. Ensure that the service vehicle is kept in a neat, clean, and orderly manner. This will help to make a good impression and provide better working conditions for the technician.

Remember the customer's name and use it frequently, preferably with Ms., Mrs., or Mr. Sir or Ma'am may also be used when appropriate. Always make eye contact with the customer and avoid talking to the ground or to a clipboard. Always wear clean clothes. If a company-issued shirt and pants are available, wear them. A name patch bearing your name and the company's name and logo also makes a good first impression. Politely ask the customer how you can help. When meeting a customer, be prepared to shake hands. In many cases, it may be appropriate to let the customer initiate the handshake. Your handshake should be firm and accompanied with a smile. A handshake that is too limp may give the impression of weakness; one that is too strong may indicate an overbearing type of person. Not all people like to shake hands. After ringing the doorbell, always maintain a distance from the door. This gives customers comfortable spacing between them and you. Make sure the customer has invited you in before making a move toward the door. Politely introduce yourself by name and then introduce the company you work for. Hand the customer your business card and then tell the customer the reason for your visit. Be friendly and always have a smile. Make sure you have a pen or pencil to write down any concerns or complicated issues the customer may have. Politely answer any questions asked by the customer. Listed below is a summary of what makes a good first impression:

- Arrive on time.
- Do not park in or block the customer's driveway unless you have permission.
- Keep your service vehicle in a neat, clean, and orderly manner.
- Remember the customer's name.
- Be prepared to shake hands.
- Always make eye contact with the customer.
- Always introduce yourself and your business.
- Tell the customer the reason for your visit.
- Hand the customer your business card.
- Always wear clean clothing.
- Wear a name tag and company logo.
- Be polite to the customer.
- Give the customer a comfortable space once at the door.
- Never enter the house until invited.

APPEARANCE

Another major factor in first impressions and maintaining good customer relations is appearance. It is almost impossible to get a second chance to make a good first impression. As a serviceman moves from one service call to another throughout the day, dirt and grime may start to accumulate on clothing. For this reason, it is important to always keep a spare pair of work pants and shirt in the service truck. Work shoes may need to be wiped clean between service calls to keep them presentable also. Appearance includes the following:

Hair—Brushed or combed, neatly trimmed. Male technicians should be clean-shaven or have a neatly trimmed beard or mustache. Female technicians with longer hair may wish to contain it in a ponytail or pinned back in some fashion.

Clothing—Neat and clean. For most uniforms, ensure that the shirttail is tucked in. A clean and neat uniform will help to make the appropriate impression. If you have an ID badge, wear it in plain sight. Make certain that your clothes are pressed to give that professional appearance.

Personal hygiene—Cleanliness is important. Hands should be washed and clean. A shower before going to bed or before going to work should be a regular habit. Your appearance and personal hygiene are major indicators of your personality and the quality of work you offer.

First impressions are so important that a service technician does not want to ruin a relationship with a customer because of poor appearance. Dirty shoes can be wiped down with a moist towel or rag if necessary. Some companies require that their service personnel put on plastic booties that cover their work boots when they enter the customer's house. Make sure the uniform you are wearing fits. A uniform that is too tight or too loose can make the service technician look uncared for. Long-sleeved shirts should be worn to protect the your arms. Long pants instead of shorts should also be worn to protect the legs and give a neat and tidy appearance. Any tattoos should be covered up by the shirt or pants. Earrings and finger rings should not be worn to work for safety reasons. Even though earrings and tattoos are in style for many younger people, they

do it. If you must report it to your supervisor, let customers know that you will do this right away and will get back to them immediately if possible. After listening carefully to the customer and resolving any complaint to the extent possible, you should be ready to start the troubleshooting process.

One of the best ways to handle an angry customer is to try to calm the customer down by taking on the responsibility yourself. Apologize and assure the customer that the problem will be fixed as soon as possible. Accept the responsibility and try to find answers to the customer's questions.

If you are on a service call with a co-worker, never argue in front of the customer with that co-worker. If you and a co-worker disagree on a topic, discuss the problem outside or in private, and then approach the customer together with a unified front. Arguing in front of a customer with a co-worker ruins the customer's confidence, trust, and credibility in both workers and their company. Even if you and your co-worker never come to an agreement, pretend that you have and tell the customer that you are looking at two possibilities for solving the problem and want to test both. No matter how the customer replies or what is said to you, never take the comments personally and cause a conflict. Always maintain your professionalism. Tell customers that they are valued and you will do your best to try and resolve the problem. Let them know that you, your co-worker, and the company value their business. You may even have to admit in some situations that the company's excellent track record in customer service may have fallen short this time.

Below is a summary of how to handle conflicts and arguments:

- Listen to customers and let them finish talking before replying.
- The customer must feel that the technician is competent.
- Never be critical of a customer, even in a joking manner.
- Be sympathetic, listen carefully, and try to determine why the customer is so upset.
- Do not take anything the customer says personally, and always avoid a conflict.
- Ask what you can do to resolve the problem.
- Try to calm the customer down by accepting the responsibility yourself, and try to find an answer to the customer's questions.
- Never argue with a co-worker in front of a customer.
- Always maintain your professionalism.
- Let customers know that they are valued and you and your company value their business.

THE SERVICE CALL

After arriving and introducing yourself, it is important to ask as many questions as needed to have a clear understanding of the problem. These questions will help to assure the customer that you are capable of solving the problem. During the service procedure you may need to talk with the customer to explain what you have found and to indicate the parts needed and possibly state the approximate costs if they may be higher than expected. If you must leave the job for any reason, tell the customer the reason and when you will return. You may need to go for parts or to another job emergency, but the customer needs to be informed. An informed customer is less likely to become angry or to complain. Keep the customer informed of all unusual circumstances. Double-check all your work. Clean the work site when finished and protect the customer's property from damage.

After the service work is completed, tell the customer what you found wrong, indicate that it has been corrected, and demonstrate when possible by turning the unit on while explaining how the problem was corrected. Customers deserve to know what they are paying for. All discussions should be in terms the customer will understand. Before leaving, give billing information to the customer. This should include a description of the work done and the costs.

Listed next is a summary of how the technician should handle a service call:

- Politely introduce yourself.
- Politely ask questions of the customer to make sure you clearly understand the problem.
- Politely explain to the customer the problems found, parts needed, and the approximate cost of fixing the problem.
- If temporarily leaving the job site, inform the customer of a return time and a reason for leaving.
- Inform the customer of any unusual circumstances.
- Double-check your work.
- Clean the work site when finished.
- Protect the customer's property from damage.
- Explain to the customer how the problem was corrected.
- Before leaving, give billing information to the customer, including a work description and the costs.

THE TECHNICIAN AS A SALESPERSON

A good technician is also a good salesperson. All options to resolve a problem should be presented in an honest and fair manner. Most customers are honest about why they are calling your company. It makes no sense for them to lie, because your time is their money. The faster you can fix the problem, the less money it will cost them in most cases. A service technician should never lie about a service problem or the price of fixing the problem. Even though customers may be disappointed with the truth, if they find out the service technician has lied about the problem, respect will be lost and satisfaction will not be gained. A customer will never be satisfied with a lie. An HVACR company will not be around very long if it is not truthful and ethical. With the truth, a customer always knows where he stands. Even though the truth may hurt at first because it is not what the customer

wants to hear, it is a temporary situation. Provide estimates and work orders in writing. A customer is buying not only service or equipment but also a solution to a problem. The customer may be offered an option not necessarily required but should not be "talked into it." The sale and installation of a new system is not always the best option for a customer. If a customer feels that he or she was talked into something that was not needed, there is a good chance that the transaction will end the relationship between the customer and the company. A company may have written recommendations for guiding the technician in presenting options. For instance if a unit is "x" years old and the repair will cost "x" amount, a recommendation to replace the unit or system may be appropriate.

In summary, what makes a technician a good salesperson is the following:

- Present all problem-solving options honestly and fairly.
- Provide estimates and work orders in writing.
- Never talk the customer into something that is not needed.
- Written company guidelines should be followed by the technician in presenting service or installation options.

UNIT 1

HEAT, TEMPERATURE, AND PRESSURE

OBJECTIVES

After studying this unit, you should be able to

- define temperature.
- make conversions between the Fahrenheit and Celsius scales.
- describe molecular motion at absolute zero.
- define the British thermal unit.
- describe heat flow between substances at different temperatures.
- explain the transfer of heat by conduction, convection, and radiation.
- describe sensible heat, latent heat, and specific heat and provide examples of each.
- state atmospheric pressure at sea level and explain why it varies at different elevations.
- describe the purpose and operation of a barometer.
- explain inches of Mercury, psig, and psia as they apply to pressure measurements.
- convert absolute pressures to gauge pressures and vice versa.
- convert pressure readings in inches of Mercury to absolute pressures and vice versa.

SAFETY CHECKLIST

- ☑ Heating, ventilation, air conditioning and refrigeration technicians are often exposed to very high and very low temperatures. Be sure to wear gloves and other appropriate pieces of personal protection equipment (PPE) to reduce the chances of getting injured.
- ☑ Many fluids that are used by the HVACR technicians are under pressure. Be sure to transport all pressure vessels vertically and be sure they are properly secured.
- ☑ Make certain that all tanks are properly capped to prevent accidental releases from the tanks.
- ☑ Make certain all test instruments are properly calibrated and fully operational to ensure accurate pressure and temperature readings.
- ☑ Make certain to properly store all test instruments when not in use to prevent damage.

1.1 HEAT, TEMPERATURE, AND PRESSURE

The terms heat, hot, and cold are often used to describe many things in our daily lives such as comfort, food, beverages, and weather. We experience heat waves and cold fronts. We drink hot tea and cold water. *Hot* and *cold* are relative terms used to describe the temperature of something with respect to some imaginary or unknown temperature. This "unknown" temperature varies from one individual to the next, so the terms hot and cold are highly subjective. Consider the situation where the temperature in a particular city has stayed in the −20°F range for two weeks straight and then, one day, the temperature reaches +30°F. The residents of the city get so happy, they start proclaiming how hot it is outside and shed their winter coats. Many may argue that 30°F is not hot but, when exposed to −20°F temperatures for so long, a 30°F temperature can appear to be a heatwave.

In the HVACR industry, the terms *hot* and *cold* are too general and, for the most part, undefined. For example, the word *cold* simply refers to the absence of heat. Those working in this industry need to be more specific when referring to heat and how heat energy is transferred within and between substances. When we need to be more specific, we often refer to temperature. Saying that it is 95 degrees outside is much more specific than simply saying that it is hot outside. Imagine trying to bake a cake with a recipe that instructs you to preheat the oven until it's hot.

In our industry, accurate temperature readings are very important because it is based on these temperature readings that we make decisions regarding system operation. Therefore, it is important to use high-quality thermometers to take these readings, **Figure 1.1**. It is important to understand that *heat* and *temperature* are not the same thing and cannot be used interchangeably. The following pages provide more details about heat and temperature.

Another term that must be understood is pressure. We use the term atmospheric pressure when we refer to weather conditions. We also refer to pressure when we talk about the air in the tires of bicycles and cars. Information regarding the pressures inside air-conditioning and refrigeration systems is very important to the HVACR technician. System operating pressures, along with various temperature readings, provide valuable information to the technician that is used to properly evaluate and troubleshoot heating and cooling equipment. System pressures are obtained by using a refrigeration gauge manifold, **Figure 1.2**.

The terms *heat*, *pressure*, and *temperature* will be a part of nearly all conversations relating to the HVACR industry, so it's very important to understand them. The sections that follow provide more insight into these concepts and how they relate to each other.

Figure 1.1 Thermocouple thermometer used to take accurate temperature readings in heating, air-conditioning, and refrigeration systems. Photo by Eugene Silberstein

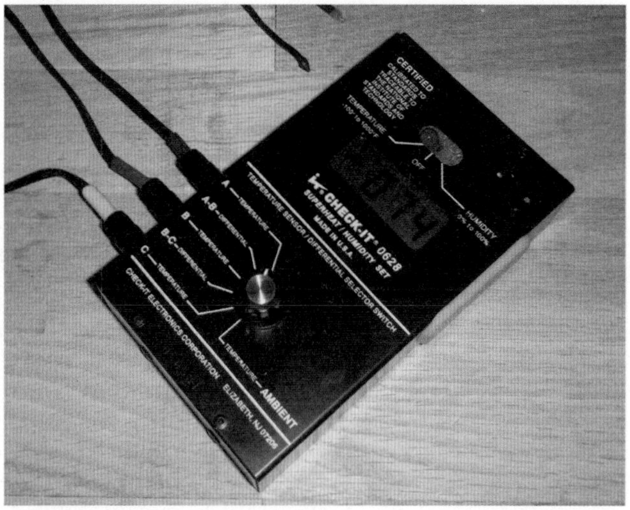

Figure 1.2 Pressure gauge used to measure operating pressures in air-conditioning and refrigeration equipment. Photo by Eugene Silberstein

1.2 TEMPERATURE

Temperature can be thought of as the level of **heat intensity**. The level of heat intensity should not be confused with the amount of heat, or heat content. As more heat is transferred to a substance, the molecular motion within the substance increases and, as a result, its temperature increases.

Most people know that the freezing point of water is 32°F and that the boiling point is 212°F, **Figure 1.3**. These points are commonly indicated on a thermometer, which is an instrument that measures temperature. However, these temperatures are only accurate when standard atmospheric conditions exist. For example, in Denver, Colorado, which is

Figure 1.3 The water in the container increases in temperature because the molecules move faster as heat is applied. When the water temperature reaches 212°F, boiling will occur.

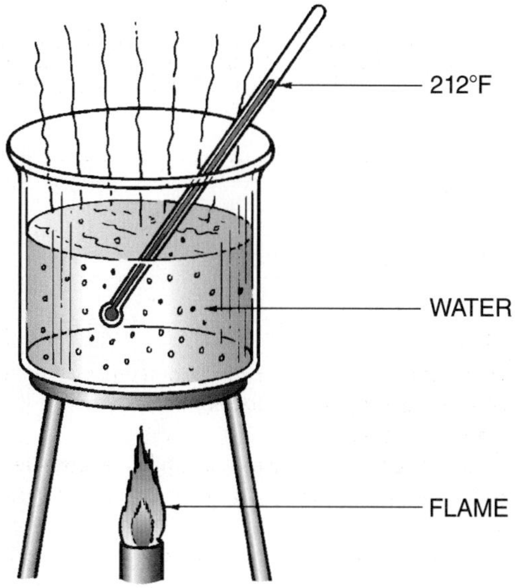

about 5600 feet above sea level, water boils at about 203°F. At the top of Mt. Everest, which is about 31,000 feet above sea level, water boils at about 167°F. Standard conditions occur at sea level with the barometer reading 29.92 in. Hg (14.696 psia). This topic is covered later in this unit as part of the discussion of pressure. It is important to understand the concept of standard conditions because these are the conditions that will be applied to actual practice later in this book.

Heat theory states that the lowest attainable temperature is −460°F. This is the temperature at which all molecular motion stops and the temperature at which there is no heat present. This temperature is referred to as absolute zero. This is a theoretical temperature because molecular motion has never been totally stopped. Scientists have actually come very close, getting within one-millionth of a degree of absolute zero in a laboratory setting.

Throughout the world, there are two commonly used temperature scales. The Fahrenheit temperature scale is part of the English measurement system used by the United States. This measurement system is also known as the I-P, or inch-pound, system. The Celsius temperature scale is part of the International System of Units (SI) or *metric system* used by most of the world. Some Fahrenheit and Celsius equivalent temperatures are provided in **Figure 1.4**. A more detailed conversion chart can be found in Appendix B of this book.

If we have a Celsius temperature that we want to convert to a Fahrenheit temperature, we can use the following formula:

$$°F = (1.8 \times °C) + 32°$$

Figure 1.4 Fahrenheit and Celsius equivalent temperatures.

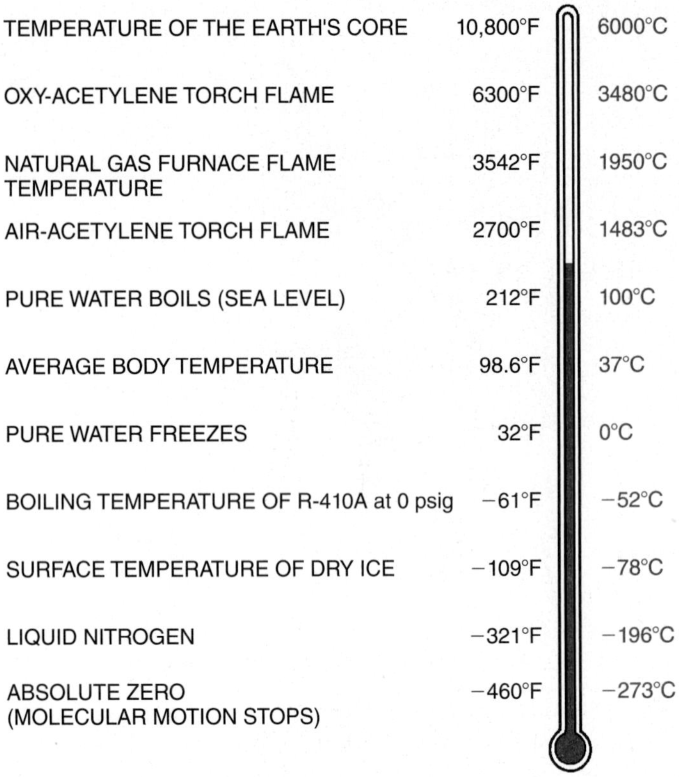

For example, if we had a Celsius temperature of 20°C, we can determine the equivalent Fahrenheit temperature by plugging the Celsius value into the formula to get:

$$°F = (1.8 \times °C) + 32°$$
$$°F = (1.8 \times 20°C) + 32°$$
$$°F = 36° + 32°$$
$$°F = 68°$$
So, 20°C = 68°F

If we have a Fahrenheit temperature that we want to convert to a Celsius temperature, we can use the following formula:

$$°C = (°F - 32°) \div 1.8$$

For example, if we had a Fahrenheit temperature of 50°F, we can determine the equivalent Celsius temperature by plugging the Fahrenheit value into the formula to get:

$$°C = (°F - 32°) \div 1.8$$
$$°C = (50°F - 32°) \div 1.8$$
$$°C = 18° \div 1.8$$
$$°C = 10°$$
So, 50°F = 10°C

Up to this point, temperature has been expressed in everyday terms. It is equally important in the HVACR industry to refer to temperature in engineering and scientific terms. Performance ratings of equipment are

Figure 1.5 (A) A Fahrenheit and Rankine thermometer. (B) A Celsius and Kelvin thermometer.

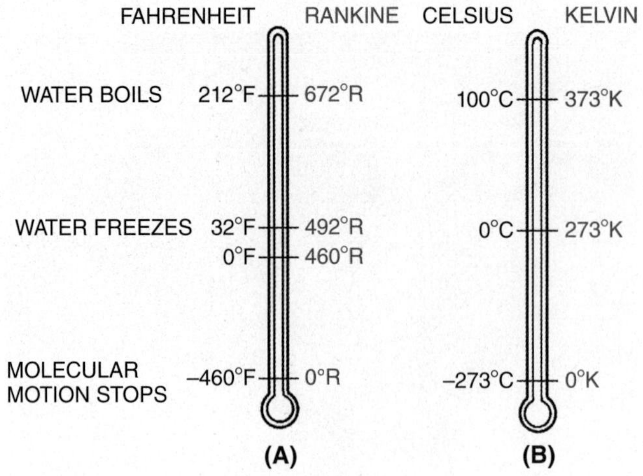

established using **absolute** temperatures. Performance ratings allow for easy comparison among equipment produced by different manufacturers. The Fahrenheit absolute scale is called the **Rankine** scale (named for W. J. M. Rankine), and the Celsius absolute scale is known as the **Kelvin** scale (named for scientist Lord Kelvin). Absolute temperature scales use zero as their lowest value and start where molecular motion starts. For instance, 0 on the Fahrenheit absolute scale is called absolute zero or 0° Rankine (0°R). Similarly, 0 on the Celsius absolute scale is called absolute zero or 0° Kelvin (0°K), **Figure 1.5**.

1.3 INTRODUCTION TO HEAT

The laws of thermodynamics can help further our understanding of heat. The first law of thermodynamics states that energy can be neither created nor destroyed, but can be converted from one form to another. This means that the "production" of heat energy is actually the conversion of other forms of energy, such as fossil fuels (gas and oil).

Temperature describes the level of heat intensity with reference to absolute zero, which was discussed earlier. The term used to describe the *quantity of heat energy* or heat content is the **British thermal unit (Btu)**, which indicates how much heat energy is contained in a substance. The rate of heat transfer can be determined by considering the time it takes to transfer a certain amount of heat energy. Air-conditioning and heating equipment is rated in Btu/h, where the "h" represents "hour." For example, an air-conditioning system operating at a rated capacity of 24,000 Btu/h has the ability to remove 24,000 Btu of heat energy from the structure every hour.

The Btu is defined as the amount of heat required to raise the temperature of 1 pound (lb) of water 1°F. For example, when 1 lb of water (about 1 pint) is heated from 70°F to 71°F,

Figure 1.6 One British thermal unit (Btu) of heat energy is required to raise the temperature of 1 lb (pound) of water 1°F.

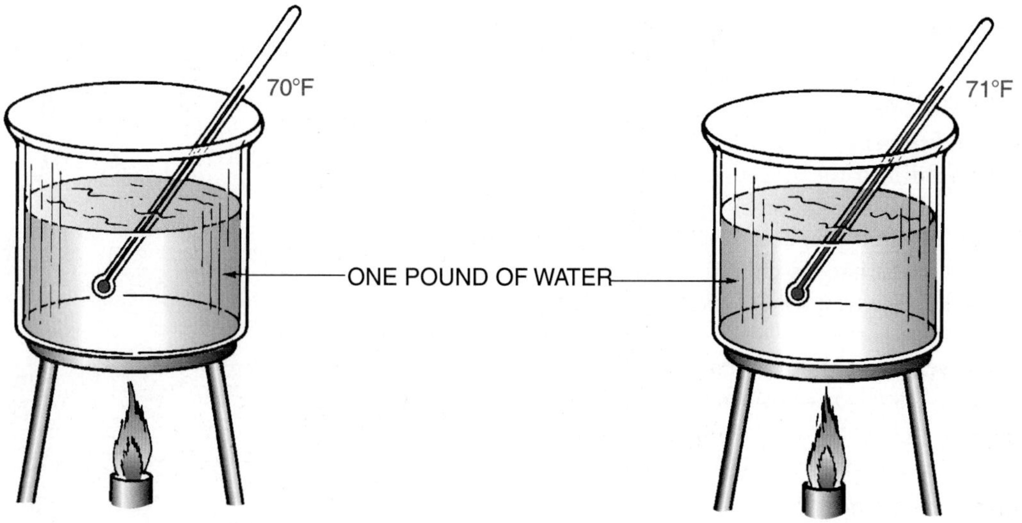

1 Btu of heat energy is absorbed into the water, **Figure 1.6.** When a temperature difference exists between two substances, a transfer of heat between the substances will occur. The next three sections in this unit address three common types of heat transfer, namely conduction, convection, and radiation. The temperature difference between two substances is the driving force behind heat transfer. The greater the temperature difference, the greater the heat transfer rate. Heat flows naturally from a warmer substance to a cooler substance. Rapidly moving molecules in the warmer substance give up some of their energy to the slower-moving molecules in the cooler substance. The warmer substance cools because the molecules have slowed. The cooler substance becomes warmer because the molecules are moving faster. So, the next time you pour warm soda into a glass filled with ice, remember that the ice isn't making the soda cold, the soda is making the ice hot!

The following example illustrates the difference between the quantity of heat and the level of heat. Consider a tank that contains 10 pounds of water at room temperature, which is about 70°F. If the water in the tank is to be heated to 170°F, 1000 Btu of heat energy must be added to the water (100 degrees × 10 pounds = 1000 Btu). Now consider a tank that contains 100,000 pounds of water at room temperature. If the water in this tank is to be heated to 170°F, 10,000,000 Btu of heat energy must be added to the water (100 degrees × 100,000 pounds = 10,000,000 Btu), **Figure 1.7**. So, even though the initial and final temperatures are the same in both situations, the water in the larger tank contains much more heat energy than the water in the smaller tank. In addition, the water in the larger tank will cool down much slower than the water in the smaller tank. This is because there is much more heat energy stored in the water in the larger tank.

Another water-related example may be helpful in differentiating between the level of heat intensity and the quantity of heat. Consider a 200-ft-deep well and a large lake with a water depth of 25 ft. Although the level of water is higher in the well than in the lake, these depth measurements cannot be used to determine the quantity of water present in either the well or the lake. The depth of water (in feet) tells us the level of water, but it in no way expresses the quantity (gallons) of water. In a similar manner, temperature relates to the level of heat, not the quantity of heat energy.

Figure 1.7 More heat energy is required to heat larger amounts of a substance.

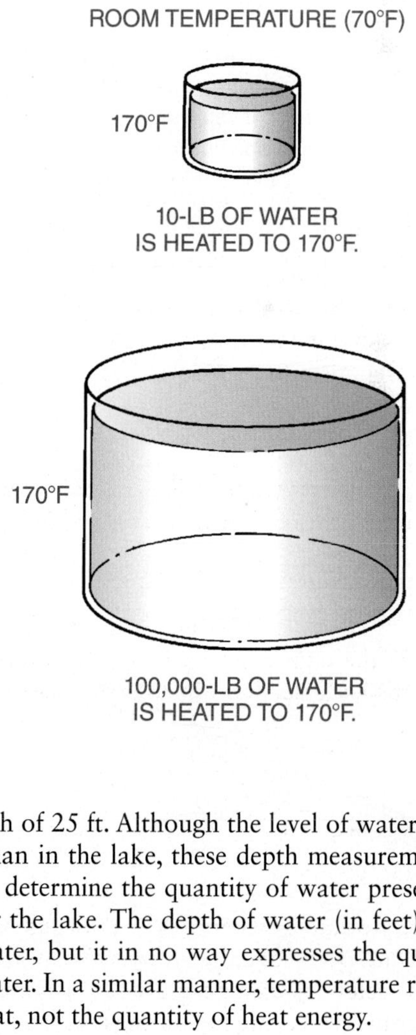

In practical terms, a piece of heating equipment is rated according to the amount of heat it will produce in a given time period. If the equipment had no such rating, it would be difficult for a buyer to choose the correct appliance. The rating of a gas or oil-fired furnace used to heat a home is permanently printed on the appliance's nameplate. Either furnace would be rated in Btu per hour, which is a *rate* at which heat energy is transferred into the structure. For now, it is sufficient to say that if a house loses, on average, 75,000 Btu of heat per hour, the selected furnace should be rated at 75,000 Btu/h. If a smaller furnace is chosen, the house will begin to get cold any time the heat loss of the house exceeds the Btu/h output of the furnace.

To describe the absence of heat, the term *cold* is often used. Because heat is expressed as a positive value in relation to no heat, cold is really an expression of comparison with no numerical value. When a person says it is cold outside, the term is being used to express a relationship to the normal expected temperature for the time of year or to the inside temperature. Cold is sometimes referred to as a "level of heat absence."

1.4 CONDUCTION

Heat transfer by conduction can be explained as the energy actually traveling from one molecule to another within a substance. As a molecule moves faster, it causes the nearby molecules to do the same. For example, if one end of a copper rod is placed in a flame, in a short time the other end will get hot as well. The heat travels up the rod from molecule to molecule, **Figure 1.8**. Heat transfer by conduction is used in many applications. For example, heat is transferred by conduction from the hot electric burner on the cookstove to the pan or pot resting on it.

Heat does not conduct through all materials at the same rate. Copper, for instance, conducts heat at a different rate

Figure 1.8 The copper rod is held in the flame only for a short time before heat is felt at the far end.

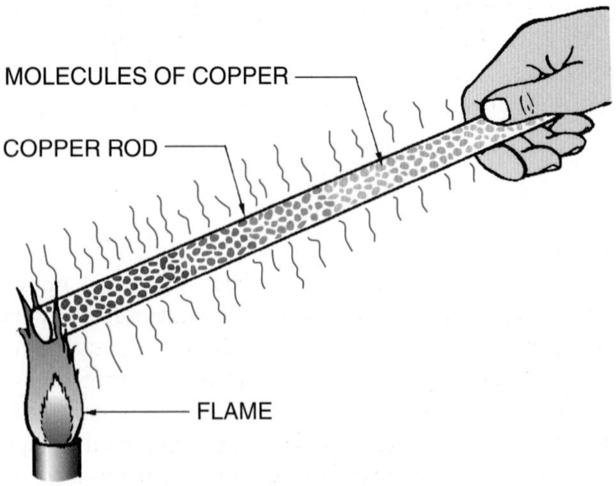

Figure 1.9 The car fender and the fence post are actually the same temperature, but the fender feels colder because the metal conducts heat away from the hand faster than the wooden fence post.

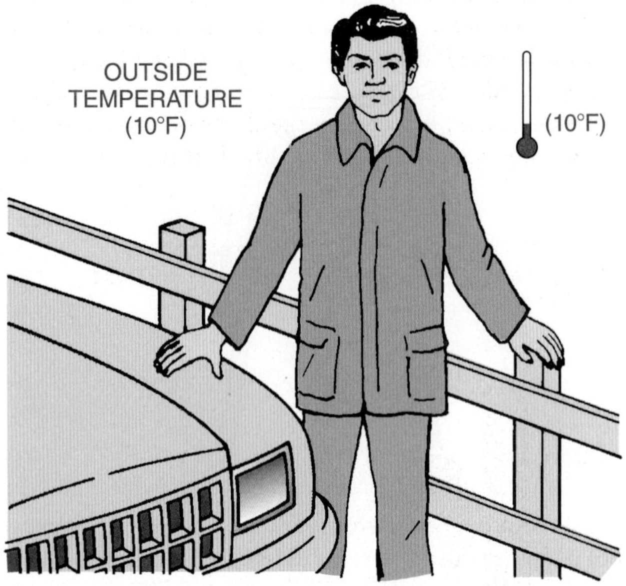

than iron. Glass is a very poor conductor of heat. Touching a wooden fence post or another piece of wood on a cold morning does not give the same sensation as a metallic surface on a car or other piece of steel. Although the steel might feel colder than the wood, in reality, the steel is not colder. Steel is a better conductor of heat than wood and, as such, causes heat to travel out of the hand at a faster rate, **Figure 1.9**. The sensation of cold is actually the feeling we experience when heat is *leaving* our body. When we get a fever, we often experience chills. This is partly due to an elevated body temperature that increases the rate of heat transfer from the body.

The different rates at which various materials conduct heat have an interesting similarity to the conduction of electricity. As a rule, substances that are poor conductors of heat are also poor conductors of electricity. For instance, copper is one of the best conductors of electricity and heat, and glass is one of the poorest conductors of both. Glass is actually used as an insulator in some electrical applications. The properties of electrical conductors will be discussed in Unit 12, "Basic Electricity and Magnetism."

1.5 CONVECTION

Heat transfer by convection relies on a fluid flow, such as air or water, to move heat from one location to another. Many large buildings have a central heating plant that heats water and pumps it throughout the building to the spaces to be heated. Notice the similarity between the words *convection* and *convey* ("to carry from one place to another"). Convection can be classified as either forced

or natural. Forced convection utilizes fans, blowers, or pumps to move the fluid, while natural convection refers to the fluid's natural tendencies to flow.

A gas-fired furnace is an example of forced convection because air from the conditioned space is drawn to the appliance by a blower. The air then passes over the furnace's heat exchanger, which transfers heat from the burning fuel to the air. The air is then forced into the ductwork and distributed to the various rooms in the structure. **Figure 1.10** shows this process in which 70°F room air is entering the furnace; 130°F air is leaving; and the blower is creating the pressure difference to force the air into the various rooms. The blower, as mentioned, provides the forced convection.

Figure 1.10 Air from the room enters the blower at 70°F. The blower forces the air across the hot heat exchanger and out into the structure at 130°F.

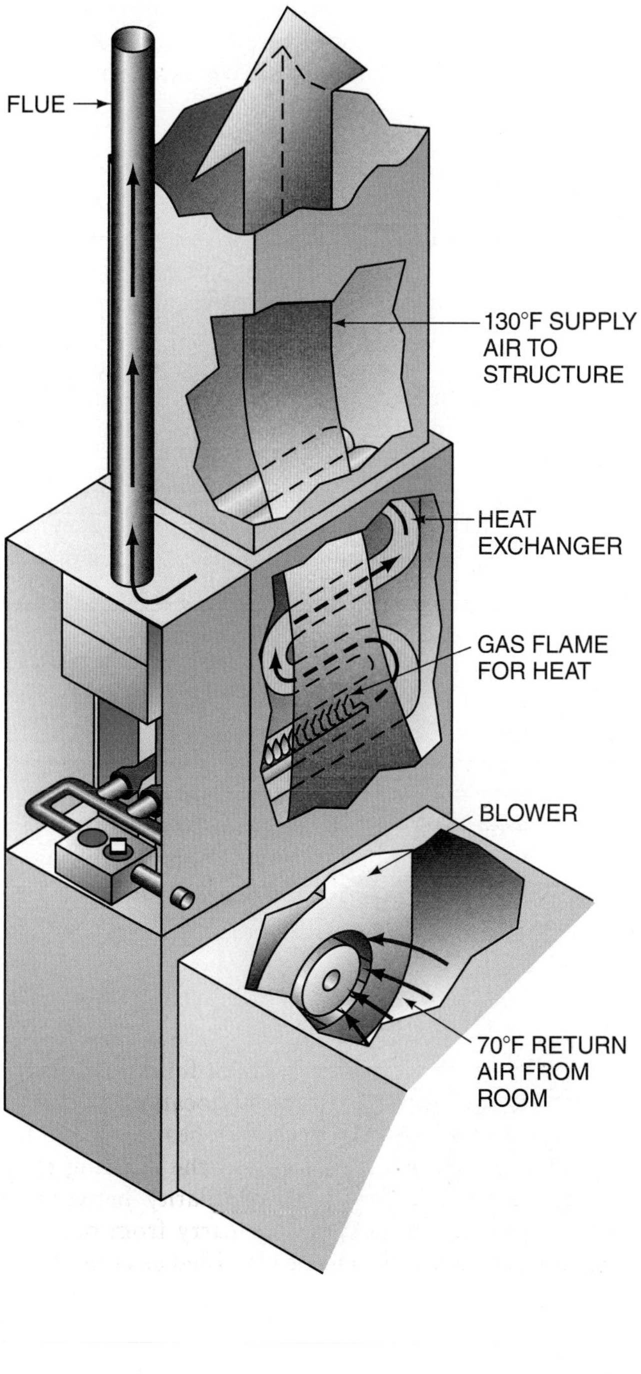

Figure 1.11 Natural convection occurs when heated air rises and cool air takes its place.

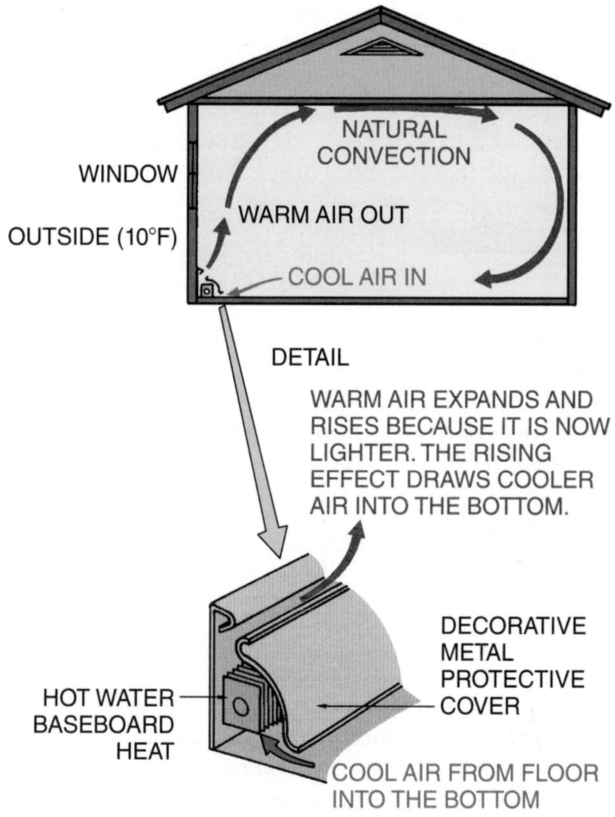

Another example of heat transfer by convection occurs when heated air rises naturally. This is referred to as natural convection. When air is heated, it expands, and the warmer air becomes lighter than the surrounding unheated air. This principle is applied in many ways in the air-conditioning industry. Sections of baseboard heaters are an example. When the air near the floor is heated, it expands and rises. This heated air is replaced by cooler air around the heater, which sets up a natural convection current in the room, **Figure 1.11**.

1.6 RADIATION

Heat transfer by radiation can best be explained by using the sun as an example of the heat source. The sun is approximately 93 million miles from earth's surface, yet we can feel its intensity. Heat transferred by radiation travels through space without heating it and is absorbed by the first solid object that it encounters. Radiation is the only type of heat transfer that can travel through a vacuum, such as space, because it is not dependent on matter as a medium of heat

Figure 1.12 The intensity of the radiant heat diminishes by the square of the distance.

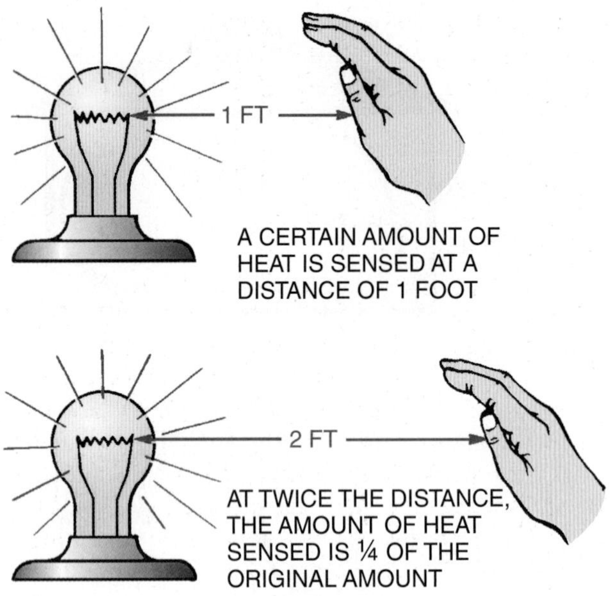

transfer. Convection and conduction require some form of matter to be the transmitting medium. The earth does not experience the total heat of the sun because heat transferred by radiation diminishes by the inverse of the square of the distance traveled. In practical terms, this means that every time the distance is doubled, the heat intensity decreases by a factor of 4. If you hold your hand close to an incandescent light bulb, for example, you feel the heat's intensity, but if you move your hand twice the distance away, you will feel only one-fourth of the heat intensity, **Figure 1.12**. Keep in mind that, because of the inverse-square-of-the-distance rule, radiant heat does not transfer the actual temperature or heat quantity value. If it did, the earth would be as hot as the sun.

Electric heaters that glow red hot are practical examples of radiant heat. The red-hot electric heater coil radiates heat into the room. It does not heat the air, but it warms the solid objects that the heat rays encounter. Any heater that glows has the same effect.

1.7 SENSIBLE HEAT

Heat level or heat intensity can readily be measured when it changes the temperature of a substance (remember the example of changing 1 lb of water from 70°F to 71°F). This change in the heat level can be measured with a thermometer. When a change of temperature occurs, we know that the level of heat or heat intensity has changed. This is called **sensible heat**.

1.8 LATENT HEAT

Not all changes in heat content result in a change of temperature. Some heat transfers result in a change of state with no temperature change. Such heat transfers are referred to as **latent** or **hidden heat** transfers. An example of a latent heat transfer is any additional heat that is added to water while it is boiling in an open container. Once water is brought to the boiling point, adding more heat only makes it boil faster; it does not raise the temperature of the water, **Figure 1.13**.

There are three other terms that are important to understand when referring to latent heat transfers: **latent heat of vaporization**, **latent heat of condensation**, and **latent heat of fusion**. Latent heat of vaporization is the amount of

Figure 1.13 Adding three times as much heat only causes the water to boil faster. The water does not increase in temperature.

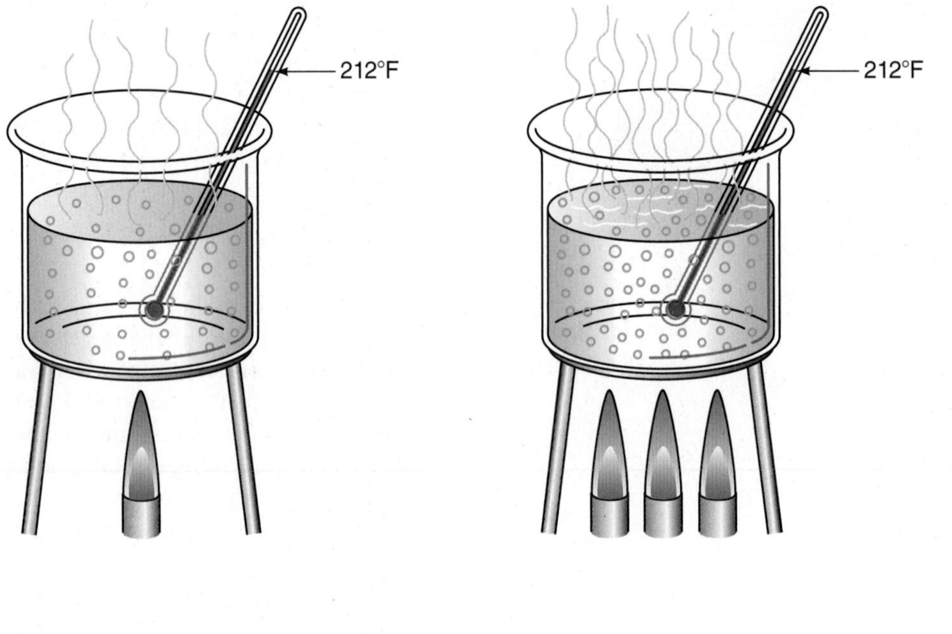

heat energy, in Btu/lb, required to change a substance into a vapor. For example, the latent heat of vaporization for water at atmospheric conditions is 970.3 Btu/lb. For ease of calculation, we often round this value to 970 Btu/lb. This means that if we have 1 lb of water at 212°F that we want to change to 1 lb of steam at 212°F, we would need to *add* 970 Btu of heat energy to the water.

Latent heat of condensation is the amount of heat energy, in Btu/lb, required to change a vapor into a liquid. This is, in effect, the opposite of latent heat of vaporization. For example, the latent heat of condensation for steam at atmospheric conditions is 970.3 Btu/lb. If we had 1 lb of steam at 212°F that we wanted to change to 1 lb of water at 212°F, we would need to *remove* 970 Btu of heat energy from the steam.

Latent heat of fusion is the amount of heat energy needed to change the state of a substance from a solid to a liquid or from a liquid to a solid. The latent heat of fusion for water is 144 Btu/lb. If we had 1 lb of ice at 32°F under atmospheric conditions, we would have to add 144 Btu of heat energy to the ice in order to melt it. If, on the other hand, we had 1 lb of water at 32°F under atmospheric conditions, we would have to *remove* 144 Btu of heat energy from the water in order to freeze it.

The following example describes the sensible heat and latent heat characteristics of 1 lb of water at standard atmospheric pressure from 0°F through the temperature range to above the boiling point. In the chart in **Figure 1.14** notice that temperature is plotted on the vertical axis, and the amount of heat energy added to the water/ice/steam throughout the example is plotted along the horizontal axis, along the bottom of the chart. Since no heat is added at the start of this example, the heat content scale starts (at point 1) at zero Btu/lb. As heat is added, the temperature will rise, except during latent- or hidden-heat processes.

The following statements should help you to understand the chart:

1. Water is in the form of ice at point 1 where the example starts. Remember that point 1 is *not* absolute zero. Point 1 represents ice at 0°F and is used as a starting point. Since no heat has been added to the ice at point 1, the heat content, as read on the heat content scale along the horizontal axis, is zero.
2. From point 1 to point 2, 16 Btu of heat energy has been added to the ice. As can be seen in the diagram, after the addition of this heat, the temperature of the ice is now 32°F. This is a sensible heat transfer because there was a measurable temperature change. Since it took 16 Btu of heat energy to raise the temperature of the ice from 0°F to 32°F, we can conclude that it takes only 0.5 Btu of heat energy to raise the temperature of 1 lb of ice 1°F.
3. When point 2 is reached, the temperature of the ice is 32°F. This is the highest temperature at which ice can exist. If more heat is added, it will be latent heat and the ice will start to melt. Because this is a latent heat transfer, the temperature of the ice will not increase. Adding 144 Btu of heat will change the 1 lb of ice to 1 lb of water at 32°F.

Figure 1.14 The heat/temperature graph for 1 lb of water at atmospheric pressure explains how water responds to heat. An increase in sensible heat causes a rise in temperature. An increase in latent heat causes a change of state, for example, from solid ice to liquid water.

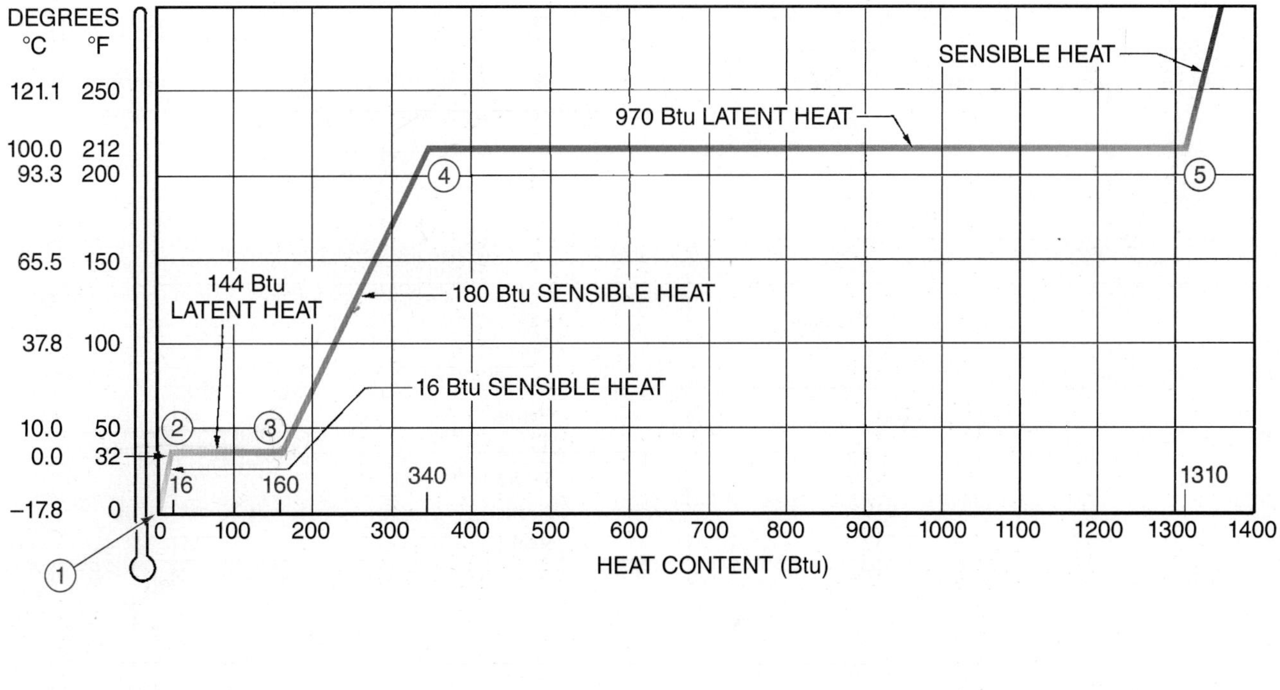

4. When point 3 is reached, the substance is now 100% water. At this point, a total of 160 Btu of heat energy has been added. 16 Btu raise the temperature of the ice from 0°F to 32°F and 144 Btu to melt the ice. Adding more heat to the water at this point will result in will cause a rise in temperature. (This is sensible heat.) Removal of any heat at point 3 results in some of the water changing back to ice. This is described as removing latent heat because there is no change in temperature.
5. Between points 3 and 4, 180 Btu of heat have been added. The water has been heated from 32°F to 212°F, which is a 180°F (212°F − 32°F = 180°F) rise. Referring back to the definition of the Btu, you will see that it takes 1 Btu of heat energy to raise the temperature of 1 lb of water 1°F. So, a 180°F temperature rise will require 180 Btu of heat energy. At point 4 a total of 340 Btu of heat energy has been added to the process (16 Btu + 144 Btu + 180 Btu).
6. Point 4 represents the 100% saturated liquid point. The water is saturated with heat to the point that the addition of any more heat will cause the water to boil (a latent heat process) and start changing to a vapor (steam). If any heat is removed from the water at this point, it will simply cool to a temperature that is lower than 212°F. Adding 970 Btu of heat energy causes the 1 lb of liquid to boil, becoming vapor.
7. Point 5 represents the 100% saturated vapor point. The water is now in the vapor state. Heat removed would be latent heat and would change some of the vapor back to a liquid. This is called *condensing the vapor.* Any heat added to the process after point 5 is sensible heat; it raises the vapor temperature above the boiling point, which is called *superheating*. Any water vapor with a temperature above the boiling point of 212°F is superheated vapor. The concept of superheat will be important in future studies. Note that in the vapor state it takes only 0.5 Btu to heat the water vapor (steam) 1°F, the same as when water was in the ice (solid) state.

Between points 1 and 5, a total of 1310 Btu of heat energy has been added to the process. A summary of the process is provided here:

Process	Latent or Sensible	# of Btu	State/Temperature Change
Point 1 to Point 2	Sensible Heat Transfer	16 Btu	Ice at 0°F to Ice at 32°F
Point 2 to Point 3	Latent Heat Transfer	144 Btu	Ice at 32°F to Water at 32°F
Point 3 to Point 4	Sensible Heat Transfer	180 Btu	Water at 32°F to Water at 212°F
Point 4 to Point 5	Latent Heat Transfer	970 Btu	Water at 212°F to Steam at 212°F

1.9 SPECIFIC HEAT

We now realize that different substances respond differently to heat. When 1 Btu of heat energy is added to 1 lb of water, it changes the temperature 1°F. This holds true only for water; other heated substances have different values. For instance, we noted that adding 0.5 Btu of heat energy to either ice or steam (water vapor) caused a 1°F rise per pound. The temperatures of ice and steam increased at twice the rate of water. Adding 1 Btu of heat energy to one pound of either ice or steam would result in a 2°F rise. This difference in heat rise is known as specific heat.

Specific heat is the amount of heat necessary to raise the temperature of 1 lb of a substance 1°F. Every substance has a different specific heat. Note that the specific heat of water is 1 Btu/lb/°F. See **Figure 1.15** for the specific heat of some other substances.

1.10 SIZING HEATING EQUIPMENT

Specific heat is significant because the amount of heat required to change the temperatures of different substances

Figure 1.15 Specific heat values for some common substances.

SUBSTANCE	SPECIFIC HEAT Btu/lb/°F	SUBSTANCE	SPECIFIC HEAT Btu/lb/°F
Air	0.24 (Average)	Ice	0.504
Aluminum	0.214	Iron	0.129
Beef, lean	0.77	Kerosene	0.50
Building Brick	0.20	Marble	0.21
Concrete	0.156	Pork, fresh	0.68
Copper	0.092	Seawater	0.94
Cucumbers	0.97	Steam	0.5
Eggs	0.76	Steel	0.116
Gold	0.0312	Water	1.00

is used to size equipment. Recall the example of the house and furnace earlier in this unit. The following example shows how this would be applied in practice. A manufacturing company may need to buy a piece of heating equipment to heat steel before it can be machined. The steel may be stored outside in the cold at 0°F and need preheating before machining. The desired metal temperature for the machining process is 70°F. How much heat must be added to the steel if the plant wants to process 1000 pounds of steel each hour?

The steel is coming into the plant at a fixed rate of 1000 lb/h, and heat has to be added at a steady rate to stay ahead of production. **Figure 1.15** gives a specific heat of 0.116 Btu/lb/°F for steel. This means that 0.116 Btu of heat energy must be added to 1 lb of steel to raise its temperature 1°F. The following formula can be used to determine the total amount of heat required to heat the steel in preparation for processing:

$$Q = W \times \text{Specific Heat} \times \Delta T$$

where Q = quantity of heat needed, W = weight of the substance, ΔT = change in temperature. Substituting in the formula, we get:

$Q = 1000 \text{ lb/h} \times 0.116 \text{ Btu/lb/°F} \times (70°\text{F} - 0°\text{F})$
$Q = 1000 \text{ lb/h} \times 0.116 \text{ Btu/lb/°F} \times 70°\text{F}$
$Q = 8120$ Btu/h required to heat the steel for machining.

The previous example has some known values and an unknown value to be found. The known information is used to find the unknown value with the help of the formula. The formula can be used when adding heat or removing heat and is often used in heat-load calculations for sizing both heating and cooling equipment.

1.11 PRESSURE

Pressure is defined as force per unit of area. This is normally expressed in pounds per square inch (psi). Simply stated, when a 1-lb weight rests on an area of 1 square inch (1 in^2), the pressure exerted downward is 1 psi. Similarly, when a 100-lb weight rests on a 1-in^2 area, 100 psi of pressure is exerted, **Figure 1.16**. If the 100-lb weight rested on a surface that was 100 in^2, the pressure exerted would be 1 psi.

When you swim under the surface of the water, there is a pressure pushing inward on your body. This pressure is the result of the weight of the water. You would feel a different sensation when flying in an unpressurized airplane cabin. Your body would be subjected to less pressure instead of more, yet you would still feel uncomfortable. It is easy to understand why the discomfort under water exists, the weight of the water pushes in. In the airplane, the situation is just the reverse. There is less pressure high in the sky than down on the ground. The pressure is greater inside your body and is pushing out.

Water weighs 62.4 pounds per cubic foot (lb/ft^3). A cubic foot (7.48 gal) of water exerts a downward pressure of 62.4 lb/ft^2 when in actual cube (12 in. × 12 in. × 12 in.) shape. How much weight is then resting on 1 in^2? The answer is simply calculated. The bottom of the cube has an area of 144 in^2 (12 in. × 12 in.) sharing the weight. Each square inch has a total pressure of 0.433 lb (62.4 ÷ 144) resting on it. Thus, the pressure at the bottom of the cube is 0.433 psi, **Figure 1.17**.

A volume of 1 ft^3 is equivalent to 1728 in^3 (12 in. × 12 in. × 12 in. = 1728 in^3), so the dimensions of the container can be changed as long as the volume remains the same. If the same 1 ft^3 of water had the shape shown in **Figure 1.18** (24 in. × 24 in. × 3 in. = 1728 in^3), the pressure at the bottom would be different. The area at the bottom would be 576 in^2, since the base is 24 in. by 24 in. (24 in. × 24 in. = 576 in^2). The resulting pressure would then be only

Figure 1.16 Both weights are resting on a 1-square-inch (1-in^2) surface. One weight exerts a pressure of 1 psi; the other a pressure of 100 psi.

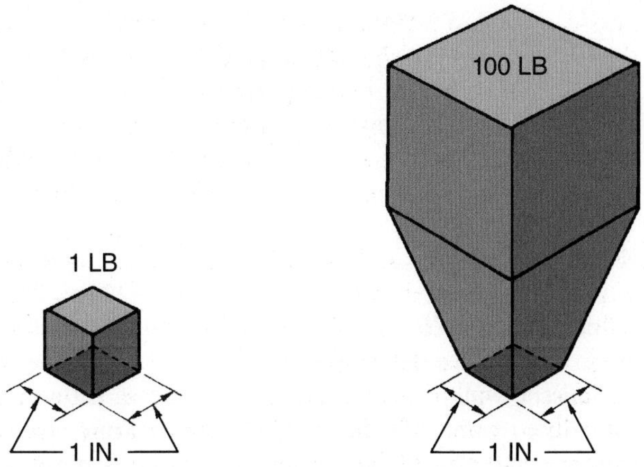

Figure 1.17 One cubic foot (1 ft^3) of water exerts a downward pressure of 62.4 lb/ft^2 on the bottom surface area of a cube.

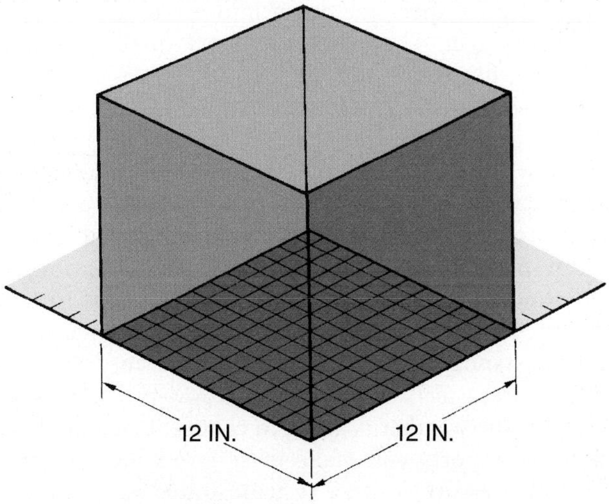

Figure 1.18 If the bottom surface is 4 ft^2, or 576 in^2, the downward pressure is 0.108 psi.

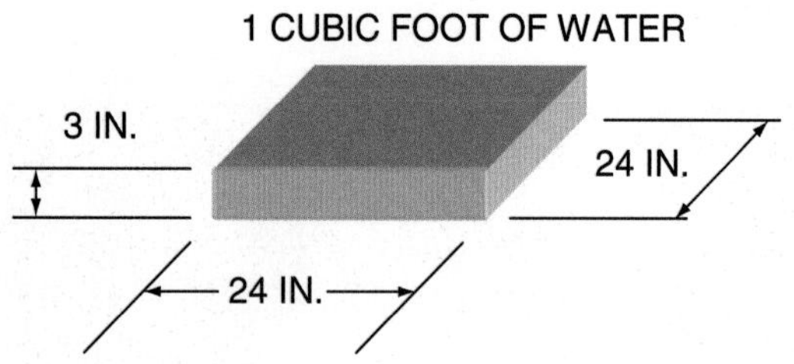

0.108 psi (62.4 ÷ 576). As the base gets larger, the pressure it exerts in the downward direction gets lower. Have you ever seen someone walking across the snow wearing snowshoes? The purpose of the snowshoes is to increase the size of the base to lower the pressure pushing down. This reduces the chance of sinking into the snow.

1.12 ATMOSPHERIC PRESSURE

The sensation of being underwater and feeling the pressure of the water is familiar to many people. The earth's atmosphere is like an ocean of air that has weight and exerts pressure. The earth's surface can be thought of as being at the bottom of this ocean of air. Different locations are at different depths. For instance, there are sea-level locations such as Miami, Florida, or mountainous locations such as Denver, Colorado. The atmospheric pressures at these two locations are different. For now, we will assume that we live at the bottom of this ocean of air. The atmosphere that we live in has weight just as water does, but not as much. Actually earth's atmosphere exerts a weight or pressure of 14.696 psi at sea level. This is known as a **standard condition**.

Atmospheric pressure can be measured with an instrument called a **barometer**, which is a glass tube about 36 in. long that is closed on one end and filled with mercury. It is then inserted open-side-down into a puddle of mercury and held upright. The mercury will try to run down into the puddle, but it will not all run out. This is because the atmosphere is pushing down on the puddle, and a vacuum is formed in the top of the tube. At sea level the mercury in the tube will fall to 29.92 in. when the surrounding atmospheric temperature is 70°F, **Figure 1.19**. This is a standard that is used for comparison in engineering and scientific work. If the barometer is located at a higher elevation, such as on a mountain, the mercury column will start to fall. It will fall about 1 in./1000 ft of altitude. When the barometer is at standard conditions and the mercury drops, the weather forecaster will talk about a low-pressure system; this means the weather is going to change. Listening closely to weather reports will make these terms more meaningful. A low-pressure weather system is often associated with cloudy, rainy, or snowy conditions, while a high-pressure system is associated with drier weather.

Figure 1.19 Mercury (Hg) barometer.

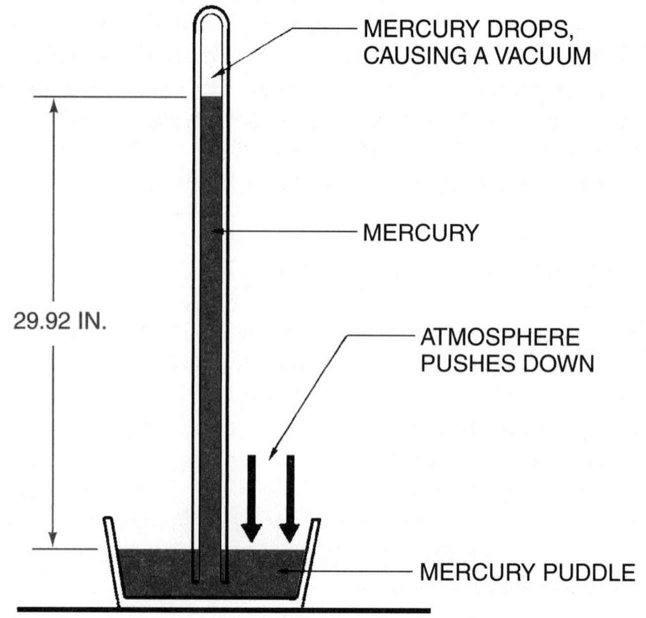

If the barometer is placed inside a closed jar and the air in the jar is removed, the mercury column will fall to the level of the puddle at the bottom, **Figure 1.20**. When the atmosphere is allowed back into the jar, the mercury again rises because a vacuum exists above the mercury column in the tube.

Figure 1.20 When the mercury barometer is placed in a closed glass jar (known as a bell jar) and the atmosphere is removed from the jar, the column of mercury drops to the level of the puddle in the dish.

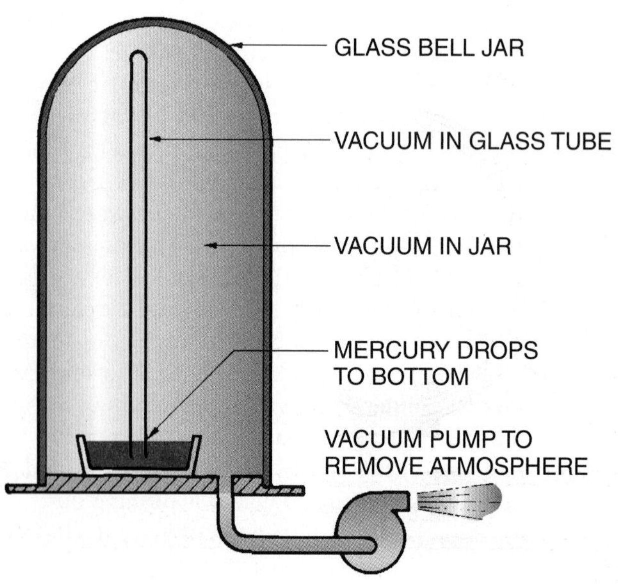

The mercury in the column has weight and counteracts the atmospheric pressure of 14.696 psi at standard conditions. A pressure of 14.696 psi, then, is equal to the weight of a column of Mercury (Hg) 29.92 in. high. The expression "inches of mercury" thus becomes an expression of pressure and can be converted to pounds per square inch. The conversion factor is 1 psi = 2.036 in. Hg (29.92 ÷ 14.696); 2.036 is often rounded off to 2 (30 in. Hg ÷ 15 psi).

1.13 PRESSURE GAUGES

Measuring pressures in a closed system requires a different method—the Bourdon tube, **Figure 1.21**. The Bourdon tube is linked to a needle and can measure pressures above and below atmospheric pressure. A common tool used in the refrigeration industry to take readings in the field or shop is a combination of a low-pressure gauge (called the low-side gauge) and a high-pressure gauge (called the high-side gauge), **Figure 1.22**. The gauge on the left reads pressures above and below atmospheric pressure and is called a compound gauge. The gauge on the right will read up to 500 psi and is called the high-pressure (high-side) gauge. **⚠SAFETY PRECAUTION:** *Working with temperatures that are above or below body temperature can cause skin and flesh damage. Proper protection, such as gloves and safety glasses,* **must be used.** *Pressures that are above or below the atmosphere's pressure can cause bodily injury. A vacuum can cause a blood blister on the skin. Pressure above atmospheric can pierce the skin or inflict damage when blowing air lifts small objects like filings.*•

These gauges read 0 psi when opened to the atmosphere. If they do not, then they should be calibrated to 0 psi. The gauges are designed to read psig (pounds-per-square-inch gauge pressure). Atmospheric pressure is used as the starting or reference point. If you want to know what the absolute pressure is, you must add the atmospheric pressure to the gauge reading. The following formula can be used: psia = psig + 14.7. This formula can be used as long as the gauge pressure is greater than or equal to 0 psig. For example, to convert a gauge reading of 50 psig to absolute pressure, you must add the atmospheric pressure of 14.696 psi to the gauge reading. Let us round off 14.696 to 15 for this example. Then 50 psig + 15 = 65 psia (pounds per square inch absolute), **Figure 1.23**. New generation gauges are of the digital variety, **Figure 1-24**. If the

Figure 1.22 The gauge on the left, called a *compound gauge*, reads pressures above and below atmospheric pressure. The gauge on the right, called the *high-pressure (high-side) gauge*, will read up to 500 psi. Photo by Eugene Silberstein

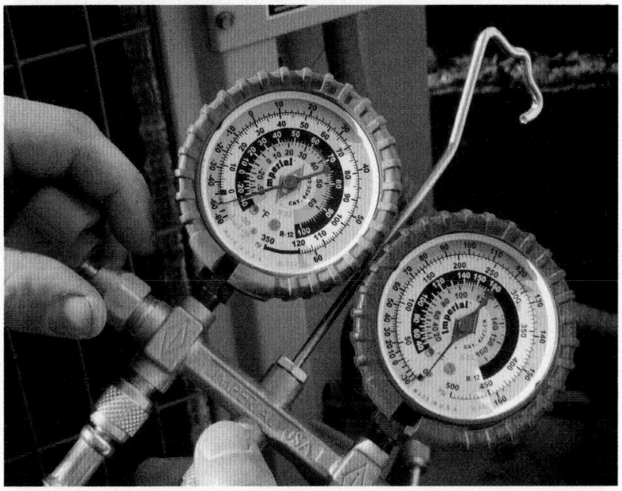

Figure 1.21 (A) Bourdon tube illustration. (B) A Bourdon tube. Photo by Eugene Silberstein

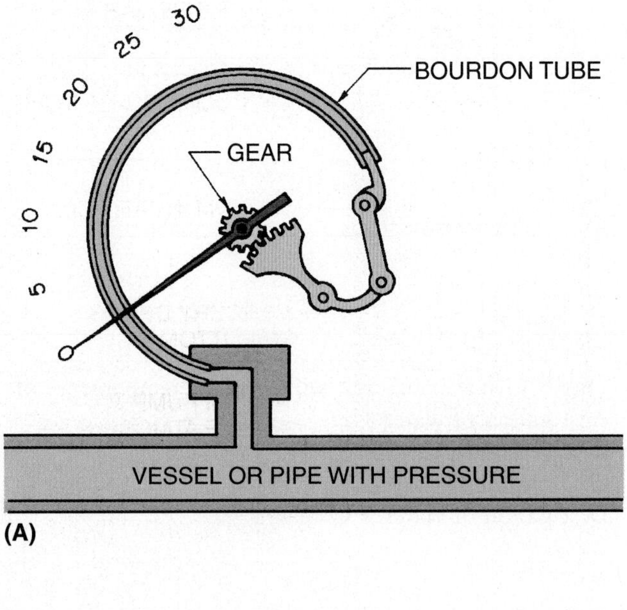

(A)

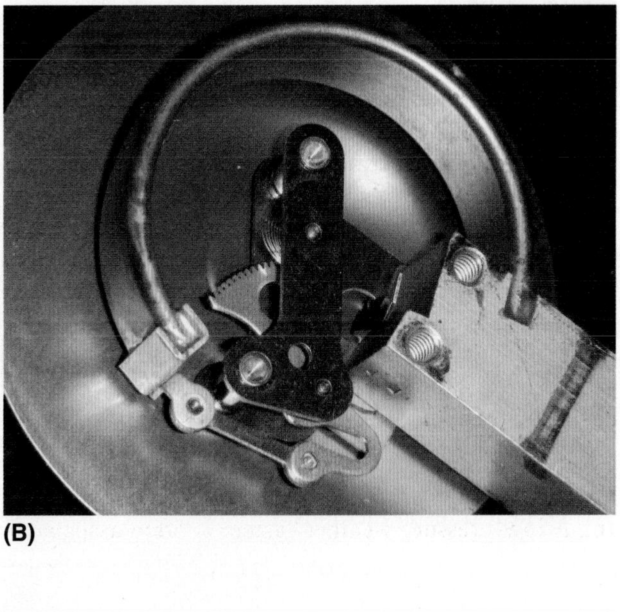

(B)

Figure 1.23 This gauge reads 50 psig. To convert this gauge reading to psia, add 15 psi (atmospheric pressure) to 50 psig to get 65 psia.

gauge pressure is below 0 psig (in a vacuum), a different formula can be used:

psia = (30 in. Hg − vacuum reading in in. Hg)/2

So, a pressure reading of 20 in. Hg can be expressed in psia by substituting this value into the formula:

psia = (30 in. Hg − 20 in. Hg)/2
psia = 10 in. Hg/2
20 in. Hg = 5 psia

Figure 1.24 Digital gauges. The compound gauge on the left reads pressures above and below atmospheric pressure. The high-pressure gauge on the right will read pressures up to 800 psi. The center gauge is a vacuum gauge, which is used for system evacuation. Gauges provided by Appion Inc. Photo by Eugene Silberstein

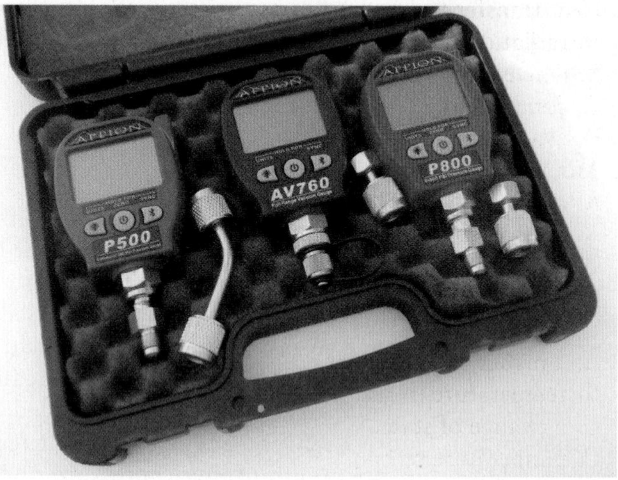

Similarly, an absolute pressure (psia) reading can be converted to its equivalent vacuum reading in psig, by using the following formula:

In. Hg = 30 in. Hg − 2(psia)

So an absolute pressure reading of 6 psia can be converted to inches of Mercury as follows:

In. Hg = 30 in. Hg − 2(6 psia)
In. Hg = 30 in. Hg − 12
12 psia = 18 in. Hg

REVIEW QUESTIONS

1. Temperature is defined as
   A. heat content.
   B. the level of heat.
   C. the British thermal unit.
   D. the absence of cold.
2. The standard conditions for water to boil at 212°F are
   A. an atmospheric pressure of 14.696 psig, or 29.92 in. Hg.
   B. an atmospheric pressure of 0 psia, or 29.99 in. Hg.
   C. an atmospheric pressure of 0 psig, or 0.92 in. Hg.
   D. an atmospheric pressure of 14.696 psia, or 29.92 in. Hg.
3. Fahrenheit, Celsius, Rankine, and Kelvin are
   A. temperature scales.
   B. pressure scales.
   C. Btu measurements.
   D. heat transfer methods.
4. Under standard conditions, water freezes at _____
   A. 0°C.
   B. 10°C.
   C. 20°C.
   D. 32°C.
5. Molecular motion stops at
   A. 212°F.
   B. −32°F.
   C. −460°F.
   D. 0°F.
6. One British thermal unit will raise the temperature of
   A. 1 gallon of water 1°F.
   B. 1 pound of water 1°F.
   C. 1 pound of water 1°K.
   D. 1 gallon of water 1°C.

7. In which direction does heat flow?
   A. From a cold substance to a cold substance
   B. From a cold substance to a warm substance
   C. From a warm substance to a warmer substance
   D. From a warm substance to a cold substance
8. Heat transfer from molecule to molecule is known as
   A. radiation.
   B. convection.
   C. conduction.
   D. evaporation.
9. An increase in sensible heat causes
   A. a higher thermometer reading.
   B. a lower thermometer reading.
   C. no change of the thermometer reading.
   D. ice to melt.
10. Latent heat causes
   A. a higher thermometer reading.
   B. temperature to rise.
   C. a change of state.
   D. temperature to fall.
11. Heat that is transferred by moving in a medium such as air or water is known as
   A. radiation. C. conduction.
   B. convection. D. evaporation.
12. Heat that is transferred through space without heating the space and that heats the first solid object that it strikes is known as
   A. radiation.
   B. convection.
   C. conduction.
   D. evaporation.
13. Specific heat is the amount of heat necessary to raise the temperature of 1 lb of which substance 1°F?
   A. Water
   B. Air
   C. Refrigerant
   D. Any substance
14. Atmospheric pressure at sea level under standard conditions is _____ inches of Mercury (Hg) or _____ pounds per square inch absolute (psia).
   A. 14.696, 29.92
   B. 29.92, 14.696
   C. 0.92, 0
   D. 29.99, 0
15. 80°F is the equivalent of
   A. 26.7°C.
   B. 35.4°C.
   C. 41.2°C.
   D. 48.6°C.
16. 22°C is the equivalent of
   A. 26.7°F.
   B. 34.5°F.
   C. 71.6°F.
   D. 81.4°F.
17. To change from psig to psia, you must
   A. add 29.92 to psig.
   B. add 14.696 to psig.
   C. subtract 0.925 from psig.
   D. subtract 3.1416 from psig.
18. 10 in. Hg is equivalent to
   A. 10 psia.
   B. 20 psia.
   C. 25 psia.
   D. 30 psia.
19. 2 psia is equivalent to
   A. 29 in. Hg.
   B. 28 in. Hg.
   C. 27 in. Hg.
   D. 26 in. Hg.
20. 125 psig is equivalent to
   A. 14.696 psia.
   B. 135 psia.
   C. 139.7 psia.
   D. 147.7 psia.
21. A barometer is used to measure
   A. atmospheric pressure.
   B. gauge pressure.
   C. absolute pressure.
   D. stratospheric pressure.
22. What size heater will be required to heat the water in a swimming pool from 70°F to 72°F in 1 hour if the appliance is 100% efficient and the pool holds 20,000 gallons of water?
   A. 166,000 Btu/h
   B. 332,000 Btu/h
   C. 86,000 Btu/h
   D. 224,000 Btu/h
23. What size heater will be required to heat the water in a swimming pool from 70°F to 80°F in 5 hours if the appliance is 100% efficient and the pool holds 20,000 gallons of water?
   A. 166,000 Btu/h
   B. 830,000 Btu/h
   C. 332,000 Btu/h
   D. 1,660,000 Btu/h
24. How much pressure is exerted downward by a 50-pound solid object that has a base area of 10 square inches?
   A. 50 psi
   B. 25 psi
   C. 10 psi
   D. 5 psi
25. If a 100-lb solid object exerts a downward force of 200 psi, what is the area of the object's base?
   A. 0.25 square inches
   B. 0.50 square inches
   C. 2 square inches
   D. 4 square inches

UNIT 2

MATTER AND ENERGY

OBJECTIVES

After studying this unit, you should be able to

- define matter.
- list the three states in which matter is commonly found.
- explain the properties of each state of matter.
- define density.
- explain the difference between mass and weight.
- discuss Boyle's law.
- state Charles' law.
- discuss the general law of perfect gas.
- discuss Dalton's law as it relates to the pressure of different gases.
- define specific gravity and specific volume.
- explain the law of conservation of energy.
- state two forms of energy important to the HVACR industry.
- explain how heat energy flows between substances at different temperatures.
- explain how magnetism can be used to convert electrical energy to mechanical energy.
- describe work and state the formula used to determine the amount of work done by performing a given task.
- define horsepower.
- convert horsepower to watts.
- convert watt-hours to British thermal units.

SAFETY CHECKLIST

- ☑ Power-consuming devices have the potential to cause injury. Be sure to de-energize all pumps, motors, and other electrical devices before working on them.
- ☑ When measuring pressures, be sure that your test instruments are fully operational and properly calibrated to avoid possible injury.

2.1 MATTER

Matter is commonly defined as a substance that occupies space and has mass. The terms mass and weight are often used interchangeably, but they are not the same. The weight of a substance is the product of the substance's mass and the force of gravity pulling down on it:

$$\text{Weight} = \text{Mass} \times \text{Gravity}$$

Matter is made up of atoms. Atoms are often described as the smallest amount of a substance and may combine to form molecules. Atoms of one substance may combine chemically with those of another to form a new substance. These new substances are referred to as **compounds**. The molecules formed cannot be broken down any further without changing the chemical nature of the compound. Matter exists in three states: **solids**, **liquids**, and **gases**. The heat content as well as the pressure exerted on matter determine its state. For instance, water is made up of molecules containing atoms of hydrogen and oxygen. Two atoms of hydrogen and one atom of oxygen are in each molecule of water, **Figure 2.1**. The chemical formula for water is H_2O.

Water in the solid state is known as ice. It exerts all of its force downward. When in the solid state, the molecules are highly attracted to each other, **Figure 2.2**. When water is heated

Figure 2.1 A water molecule is formed when two hydrogen atoms bond with one oxygen atom.

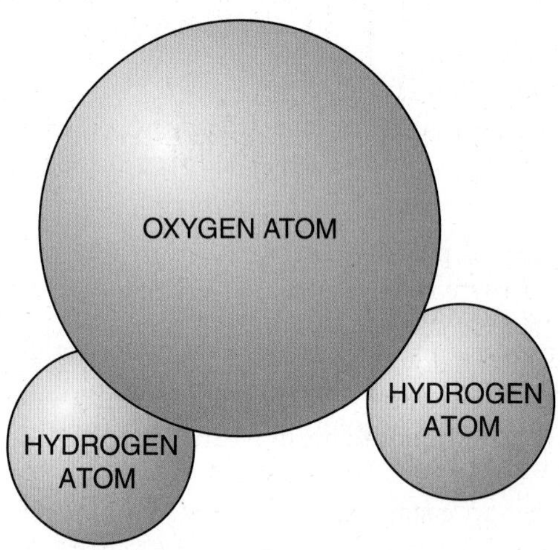

Figure 2.2 Solids exert all their pressure downward. The molecules of solid water have a great attraction for each other and hold together.

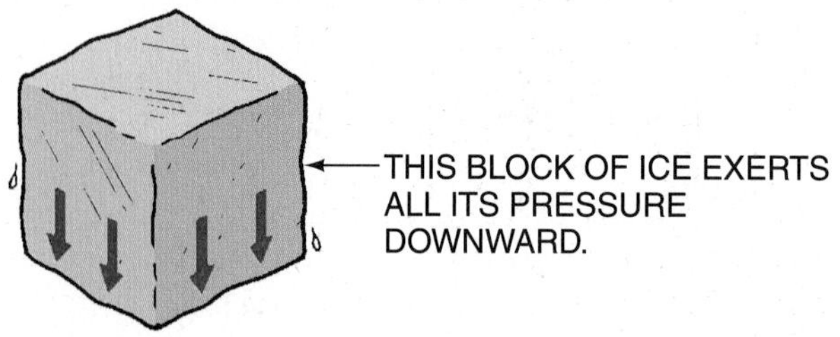

Figure 2.3 The water in the container exerts pressure outward and downward. The outward pressure is what makes water seek its own level. The pressure is proportional to the depth.

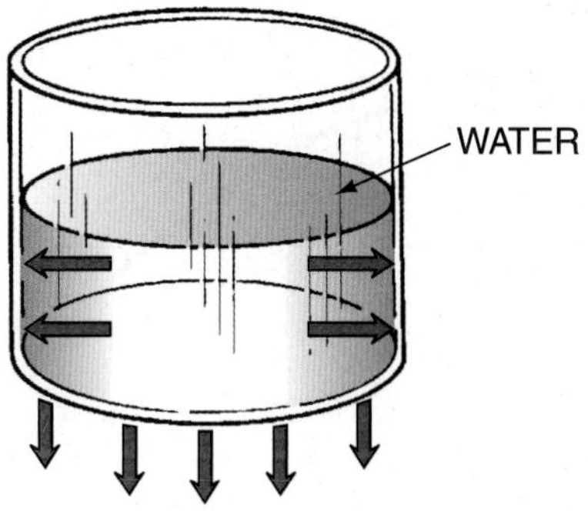

Figure 2.4 Gases push in all directions and completely fill the container.

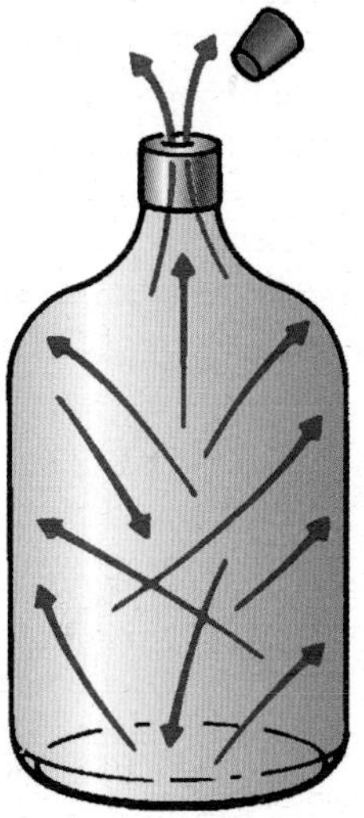

above its freezing point, it begins to change to a liquid state. The molecular activity is higher, and the water molecules have less attraction for each other. Water in the liquid state exerts a pressure outward and downward. Because water pressure is proportional to its depth, water seeks a level surface where its pressure equals that of the atmosphere above it, **Figure 2.3**. Water heated above the liquid state, 212°F at standard conditions, becomes a vapor. In the vapor state, the molecules have even less attraction for each other. The vapor exerts pressure evenly in all directions to completely fill the container, **Figure 2.4**.

In addition to the states of matter, there are other properties of matter that should be discussed to better one's understanding of how different substances react with and compare to each other.

2.2 MASS AND WEIGHT

Mass is the property of matter that responds to gravitational attraction. **Weight** is the force that matter (solid, liquid, or gas) applies to a supporting surface when it is at rest. Weight is *not*

Figure 2.5 Table of density and specific gravity.

SUBSTANCE	DENSITY LB/FT^3	SPECIFIC GRAVITY
ALUMINUM	171	2.74
BRASS (RED)	548	8.78
COPPER	556	8.91
GOLD	1208	19.36
ICE @ 32°F	57.5	0.92
TUNGSTEN	1210	19.39
WATER	62.4	1
MARBLE	162	2.596

a property of matter and is dependent on gravitational attraction. The stronger the force of gravity, the more an object will weigh. The earth has stronger gravitational attraction forces than the moon. For this reason, objects weigh more on the earth than they do on the moon. All matter has mass.

2.3 DENSITY

The **density** of a substance describes its mass-to-volume relationship. The mass contained in a particular volume is the density of that substance. It is often advantageous to compare different substances according to their density, which is expressed in pounds per cubic foot. Water, for example, has a density of 62.4 lb/ft^3. Wood floats on water because the density of wood is lower than the density of water. Iron, on the other hand, sinks in water because it is more dense than water. **Figure 2.5** lists some typical densities. Using Figure 2.5 as a guide, can you figure out why ice floats on water?

2.4 SPECIFIC GRAVITY

Specific gravity is a unitless number that relates the density of a substance with respect to the density of water. The specific gravity of a substance is calculated by dividing the density of the substance by the density of water, which is 62.4 lb/ft^3. So, for example, the specific gravity of water is 62.4 lb/ft^3 $\div$ 62.4 lb/ft^3 = 1. Notice that the units cancel because of the division. The density of red brass is 548 lb/ft^3. Its specific gravity is then 548 lb/ft^3 $\div$ 62.4 lb/ft^3 = 8.78. Figure 2.5 lists some typical specific gravities of substances. Note that a substance with a specific gravity greater than 1 will sink in water, and a substance with a specific gravity less than 1 will float on water.

2.5 SPECIFIC VOLUME

Specific volume refers to the volume that each pound of a gas occupies. The measurement requires that there must be only 1 lb of the gas. Specific volume is expressed in ft^3/lb, which differs from volume, which is expressed in ft^3. One

pound of clean, dry air has a total volume of 13.33 ft^3 at standard atmospheric conditions. Its specific volume would then be 13.33 ft^3/lb. As air is heated it becomes lighter and its specific volume increases. As air is cooled, its specific volume decreases. Hydrogen has a specific volume of 179 ft^3/lb at standard atmospheric conditions. Because more cubic feet of hydrogen are needed to make pound, it has a higher specific volume, thus it is lighter than air. Although both are gases, the hydrogen has a tendency to rise when mixed with air. Natural gas, which is a popular fuel used for heating, is also lighter than air and has a tendency to rise like hydrogen. Propane gas, another frequently used heating gas, has to be treated differently than natural gas because it is heavier than air. Propane has a tendency to fall and collect in low places and poses a danger in the event it ignites. The specific volume of propane gas is 8.62 ft^3/lb, and the specific volume of natural gas is, on average, about 22 ft^3/lb. Note: The specific volume of natural gas will vary based in the makeup of the gas. Unlike propane, which is a compound, natural gas is a mixture of different hydrocarbons. The typical range of specific volume for natural gas is between 15 ft^3/lb and 24 ft^3/lb.

Specific volume and density are inverses of one another. This means that specific volume = 1 ÷ density and that density = 1 ÷ specific volume. If one knows the specific volume of a substance, its density can be calculated and vice versa. For example, the specific volume of dry air is 13.33 ft^3/lb. Its density would then be 1 ÷ 13.33 ft^3/lb = 0.075 lb/ft^3. Notice that even the units are inverses of each other. Substances with high specific volumes are said to have low densities. Also, substances with high densities have low specific volumes. Since density and specific volumes are inverses of each other, multiplying the density of a substance by its specific volume will always yield a product of 1.

The specific volume of air is a factor in determining the fan or blower horsepower needed in air-conditioning work. As an example, moving air that has a low specific volume requires a blower motor with higher horsepower, while moving air that has a high specific volume can be accomplished with a blower motor with a lower horsepower rating. The specific volumes of various pumped gases is valuable information that enables the engineer to choose the size of the compressor or vapor pump to do a particular job. The specific volumes for vapors vary according to the pressure that is exerted on the vapor. As the pressure of the gas increases, the specific volume of the gas decreases and its density increases. As the pressure of the gas decreases, the specific volume of the gas increases and its density decreases.

The graph provided in **Figure 2.6** shows the relationship that exists among the pressure, specific volume, and density of gases. The horizontal scale across the bottom of the graph represents pressure and increases from left to right. The vertical scale on the left side of the graph represents specific volume and increases from bottom to top. The vertical scale on the right side of the graph represents density and *decreases* from bottom to top.

Figure 2.6 Graph showing the relationship among pressure, specific volume, and density.

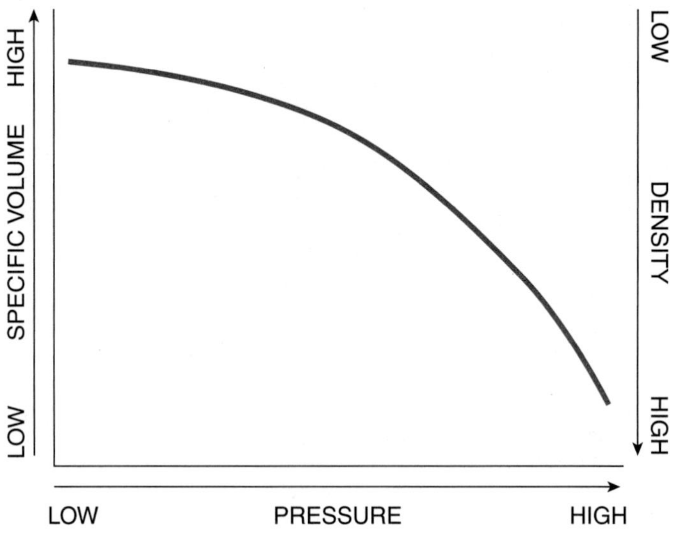

Figure 2.7 Pressure increases from point A to point B.

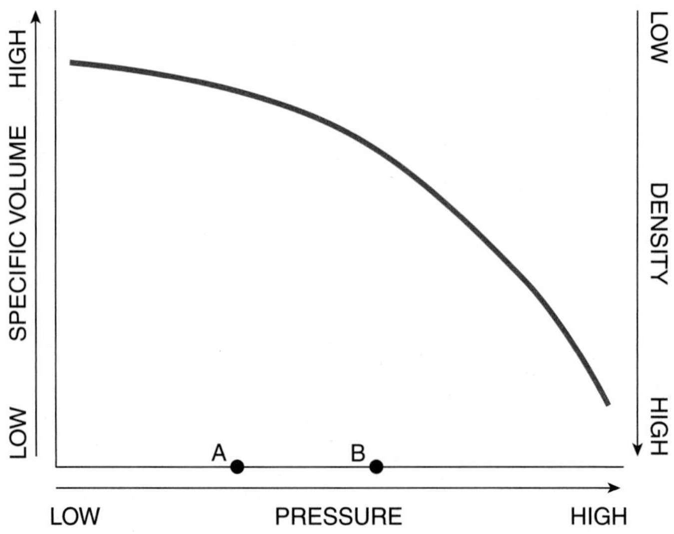

Consider a gas that experiences an increase in pressure from A to B, **Figure 2.7**. Extending these lines upward until they cross the curve identifies two additional points A′ and B′, **Figure 2.8**. Drawing horizontal lines through points A′ and B′ provides two other sets of points, namely A″ and B″, and A‴ and B‴, **Figure 2.9**. As the pressure of the gas increases from A to B, it can be seen that the specific volume decreases from A″ to B″, where B″ is lower than A″. It can also be seen that the density increases from A‴ to B‴, where B‴ is greater than A‴.

Figure 2.8 Identifying points A′ and B′ on the curve.

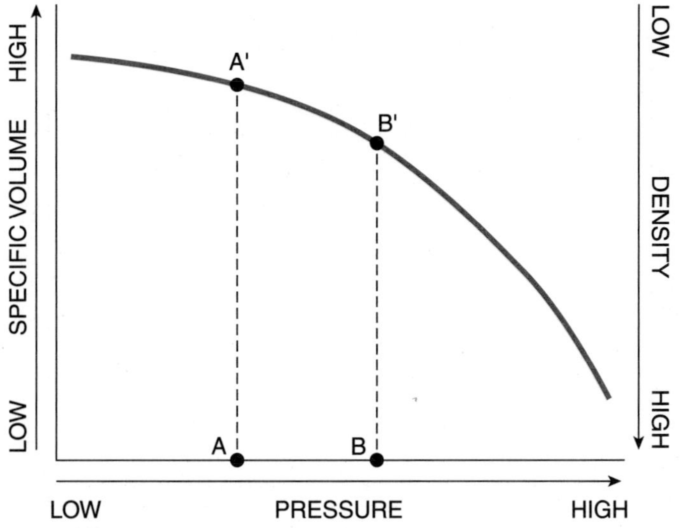

Figure 2.9 The change from A″ to B″ represents a decrease in specific volume, while the change from A‴ to B‴ represents an increase in density.

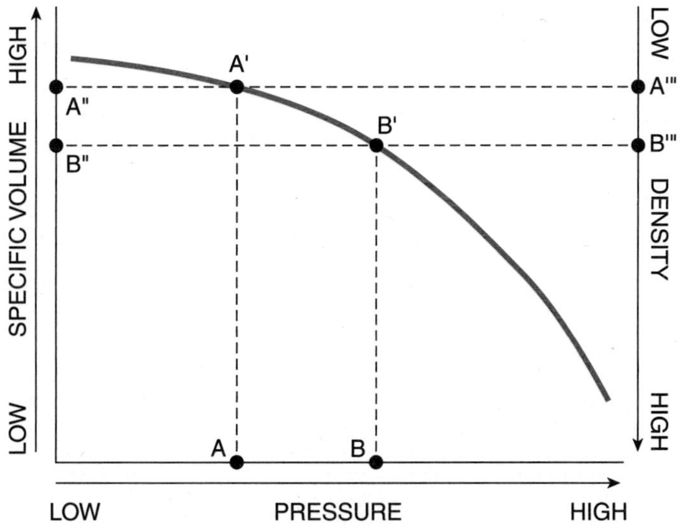

The concept of specific gravity, as discussed in Section 2.4, can also be applied to gases. Just as the density of water was used as a baseline to calculate the specific gravity of liquids and solids, the specific volume of air can be used as a baseline to determine the specific gravity of gases.

As mentioned, the specific volume of air, at standard atmospheric conditions, is 13.33 ft^3/lb, so the specific gravity of air is 1, since 13.33 ft^3/lb ÷ 13.33 ft^3/lb = 1. Hydrogen, with a specific volume of 179 ft^3/lb, will have a specific gravity of 0.074 (13.33 ft^3/lb ÷ 179 ft^3/lb). Similarly, the specific gravity of propane gas is 1.6 (13.33 ft^3/lb ÷ 8.62 ft^3/lb) and the specific gravity of natural gas, on average, is 0.6 (13.33 ft^3/lb ÷ 22 ft^3/lb). In general, gases with specific gravity values less than 1 are lighter than air, while those with specific gravity values greater than 1 are heavier than air.

2.6 GAS LAWS

It is necessary to have a working knowledge of gases and how they respond to pressure and temperature changes. Gas laws have been developed to aid in the understanding of the reaction of gases and the pressure/temperature/volume relationships in various parts of an air-conditioning or refrigeration system. Whenever using pressure or temperature in an equation like the gas laws, one has to use the absolute scales of pressure (psia) and temperature (Rankine or Kelvin), or the solutions to these equations will be meaningless. Absolute scales use zero as their starting point, because zero is where molecular motion actually begins.

BOYLE'S LAW

In the early 1600s, Robert Boyle discovered that when pressure is applied to a volume of air that is contained, the volume of air becomes smaller and the pressure greater. **Boyle's law** states that *the volume of a gas varies inversely with the absolute pressure, provided the temperature remains constant.* For example, if a cylinder with a piston at the bottom and closed at the top were filled with air and the piston moved halfway up the cylinder, the pressure of the air would double, **Figure 2.10**. That part of the law pertaining to the temperature remaining constant keeps Boyle's law from being used in practical situations. This is because when a gas is compressed some heat is transferred to the gas from the mechanical compression, and when gas is expanded heat is given up. However, this law, when combined with another, becomes practical to use.

Figure 2.10 Absolute pressure in a cylinder doubles when the volume is reduced by half.

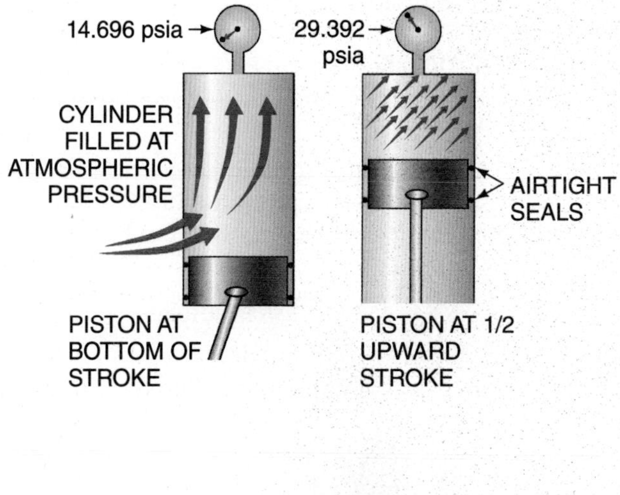

The formula for Boyle's law is as follows:

$$P_1 \times V_1 = P_2 \times V_2$$

where P_1 = original absolute pressure
V_1 = original volume
P_2 = new pressure
V_2 = new volume

For example, if the original pressure was 40 psia and the original volume was 30 in³, what would the new volume be if the pressure were increased to 50 psia? We are determining the new volume. The formula would have to be rearranged so we could find the new volume.

$$V_2 = \frac{P_1 \times V_1}{P_2}$$

$$V_2 = \frac{40 \text{ psia} \times 30 \text{ in}^3}{50 \text{ psia}}$$

$$V_2 = 24 \text{ in}^3$$

CHARLES' LAW

In the 1800s, scientist Jacques Charles made discoveries regarding the effect of temperature on gases. **Charles' law** states that *at a constant pressure, the volume of a gas varies directly as to the absolute temperature, and at a constant volume, the pressure of a gas varies directly with the absolute temperature.* Stated in a different form: When a gas is heated and if it is free to expand, it will do so, and the volume will vary directly as to the absolute temperature; if a gas is confined in a container that will not expand and it is heated, the pressure will vary directly with the absolute temperature.

This law can also be stated with formulas. Two formulas are needed because one part of the law pertains to pressure and temperature and the other part to volume and temperature.

This formula pertains to volume and temperature:

$$\frac{V_1}{T_1} = \frac{V_2}{T_2}$$

where V_1 = original volume
V_2 = new volume
T_1 = original temperature
T_2 = new temperature

If 2000 ft³ of air is passed through a gas-fired furnace and heated from 75°F room temperature to 130°F, what is the volume of the air leaving the heating unit? See **Figure 2.11**.

$V_1 = 2000 \text{ ft}^3$
$T_1 = 75°\text{F} + 460°\text{R} = 535°\text{R (absolute)}$
V_2 = unknown
$T_2 = 130°\text{F} + 460°\text{R} = 590°\text{ R (absolute)}$

Figure 2.11 Air expands when heated.

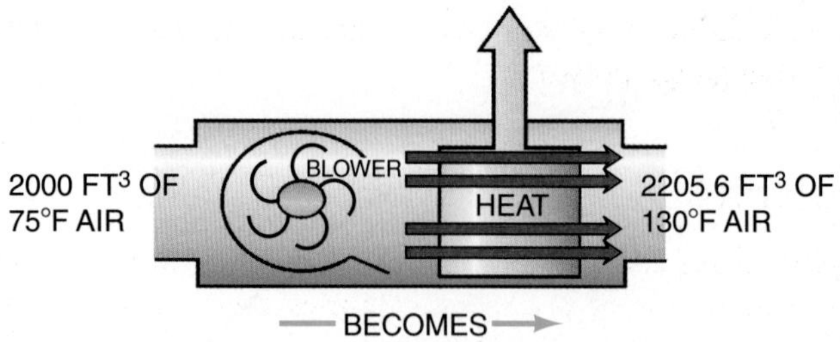

We must mathematically rearrange the formula so that the unknown is alone on one side of the equation.

$$V_2 = \frac{V_1 \times T_2}{T_1}$$

$$V_2 = \frac{2000 \text{ ft}^3 \times 590°\text{R}}{535°\text{R}}$$

$$V_2 = 2205.6 \text{ ft}^3$$

This result shows that the air expanded when heated. It is important to note that, in the previous calculation, the temperatures were converted to absolute values in order for the results to be meaningful.

The following formula pertains to pressure and temperature:

$$\frac{P_1}{T_1} = \frac{P_2}{T_2}$$

where P_1 = original pressure
T_1 = original temperature
P_2 = new pressure
T_2 = new temperature

If a large natural gas tank holding 500,000 ft³ of gas is stored at 70°F in the spring and the temperature rises to 95°F in the summer, what would the pressure be if the original pressure was 25 psig in the spring?

P_1 = 25 psig + 14.696 (atmospheric pressure) or 39.696 psia
$T_1 = 70°\text{F} + 460°\text{R}$ or 530°R (absolute)
P_2 = unknown
$T_2 = 95°\text{F} + 460°\text{R}$ or 555°R (absolute)

Again, the formula must be rearranged so that the unknown is on one side of the equation by itself.

$$P_2 = \frac{P_1 \times T_2}{T_1}$$

$$P_2 = \frac{39.696 \text{ psia} \times 555°\text{R}}{530°\text{R}}$$

$$P_2 = 41.57 \text{ psia} - 14.696 = 26.87 \text{ psig}$$

In this example, both the pressures and temperatures were converted to absolute values for calculation purposes. Once the calculation was completed, the answer was converted back to a gauge pressure.

GENERAL LAW OF PERFECT GAS

A general gas law, often called the general law of perfect gas, is a combination of Boyle's and Charles' laws. This combination law is more practical because it includes temperature, pressure, and volume.

The formula for this law can be stated as follows:

$$\frac{P_1 \times V_1}{T_1} = \frac{P_2 \times V_2}{T_2}$$

where P_1 = original pressure
V_1 = original volume
T_1 = original temperature
P_2 = new pressure
V_2 = new volume
T_2 = new temperature

For example, 20 ft^3 of gas is being stored in a container at 100°F and a pressure of 50 psig. This container is connected by pipe to one that will hold 30 ft^3, for a total volume of 50 ft^3, and the gas is allowed to equalize between the two containers. The temperature of the gas is lowered to 80°F. What is the pressure in the combined containers?

P_1 = 50 psig + 14.696 (atmospheric pressure) or 64.696 psia
V_1 = 20 ft^3
T_1 = 100°F + 460°R = 560°R
P_2 = unknown
V_2 = 50 ft^3
T_2 = 80°F + 460°R or 540°R

The formula is mathematically rearranged to solve for the unknown P_2.

$$P_2 = \frac{P_1 \times V_1 \times T_2}{T_1 \times V_2}$$

$$P_2 = \frac{64.696 \text{ psia} \times 20 \text{ ft}^3 \times 540°\text{R}}{560°\text{R} \times 50 \text{ ft}^3}$$

$$P_2 = 24.95 \text{ psia} - 14.696 \text{ (Atmospheric pressure)} = 10.26 \text{ psig}$$

DALTON'S LAW

John Dalton, a mathematics professor, made the discovery in the early 1800s that the atmosphere is made up of several different gases. He found that each gas created its own pressure and that the total pressure was the sum of these individual pressures. Officially called Dalton's Law of Partial Pressures, **Dalton's law** states that *the total pressure of a confined mixture of non-reacting (inert) gases is the sum of the pressures of each of the gases in the mixture.*

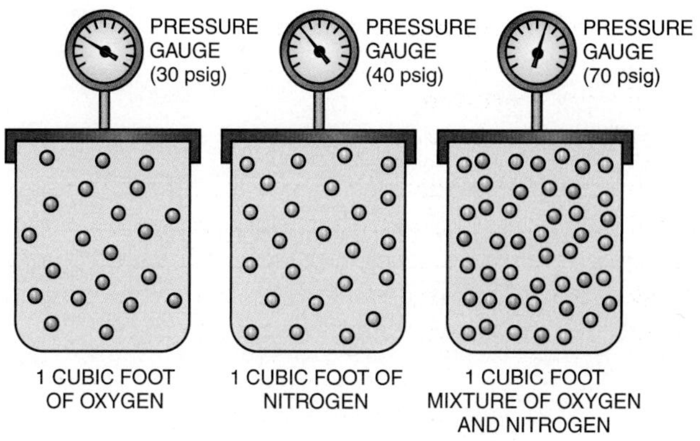

Figure 2.12 Dalton's law of partial pressures. The total pressure is the sum of the individual pressures of each gas.

Consider two containers, each with a volume of 1 ft^3, with one containing oxygen and one containing nitrogen. The pressure of the oxygen is given to be 30 psig and the pressure of the nitrogen is 40 psig. If placed together in a 1 ft^3 container, the total pressure of the mixed gases will be 70 psig (30 psig + 40 psig), **Figure 2.12.**

2.7 ENERGY

Using energy efficiently is a major goal of the HVACR industry. There are many forms of energy including electricity that drives motors, for example, and heat energy from fossil fuels such as gas, oil, and coal.

The only *new* energy we get comes from the sun, in the form of radiant heat, striking the earth. Most of the energy we use is converted from something already here (e.g., fossil fuels). The conversion from fuel to heat, for example, is accomplished by the burning, or combustion process. In a gas furnace, fuel is burned in a combustion chamber, and heat from combustion is transferred to circulated air by conduction through the walls of the furnace's heat exchanger. The heated air is then distributed throughout the heated space. This energy conversion is referred to as a direct process.

An example of indirect energy conversion is a fossil-fuel power plant. Gas may be used in the power plant to produce steam that turns a turbine to generate electricity. The electricity is then distributed by the local power company. This electrical energy can then be converted into electric heat.

Figure 2.13 Gas and oil deposits settle into depressions.

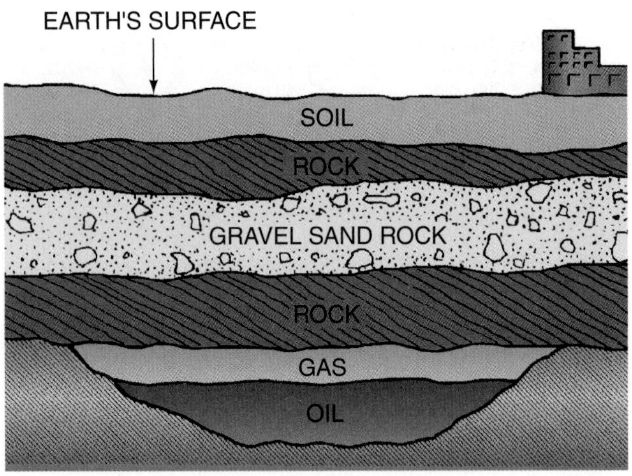

2.8 CONSERVATION OF ENERGY

The law of conservation of energy states that *energy can be neither created nor destroyed, but can be converted from one form to another.* It can thus be said that all energy can be accounted for.

Most of the energy we use is a result of the sun's support of plant growth for millions of years. Fossil fuels come from decayed vegetable and animal matter covered by earth and rock during changes in the earth's surface. This decayed matter is in various states, such as gas, oil, or coal, depending on the conditions it was subjected to in the past, **Figure 2.13**. The energy stored in fossil fuels is called chemical energy because a chemical reaction is needed to release the energy. A battery is an example of a device that changes chemical energy into electrical energy.

2.9 THERMAL ENERGY

Temperature is a measure of the degree of heat or the heat intensity, but not necessarily the amount of heat. Heat is a form of energy because of the motion of molecules. To be specific, heat is thermal energy. If two substances of different temperatures are moved close to each other, heat energy from the substance with the higher temperature will flow to the one with the lower temperature, **Figure 2.14(top)**. Because molecular motion does not stop until −460°F, energy is still available in a substance even at very low temperatures. In **Figure 2.14(bottom)** a substance at −200°F is placed next to a substance at −350°F. As we discussed earlier, heat travels from hot to cold, so heat travels from the warmer −200°F substance to the colder −350°F substance.

Most of the heat energy used in homes and industry is provided from fossil fuels, but some comes from electrical energy. As shown in **Figure 2.15**, electron flow in high-resistance wire causes the wire to become hot, thus heating the air. A moving airstream is then passed over the heated wire, allowing heat to be transferred to the air by conduction and moved to the heated space by forced convection (the blower).

Figure 2.14 (Top) If two substances of different temperatures are moved close to each other, heat from the substance with the higher temperature will flow to the one with the lower temperature. (Bottom) Heat energy is still available at these low temperatures and will transfer from the warmer substance to the colder substance.

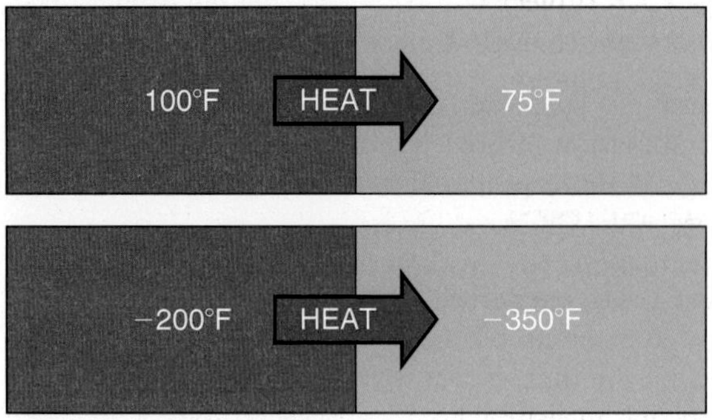

Figure 2.15 An electric heat airstream.

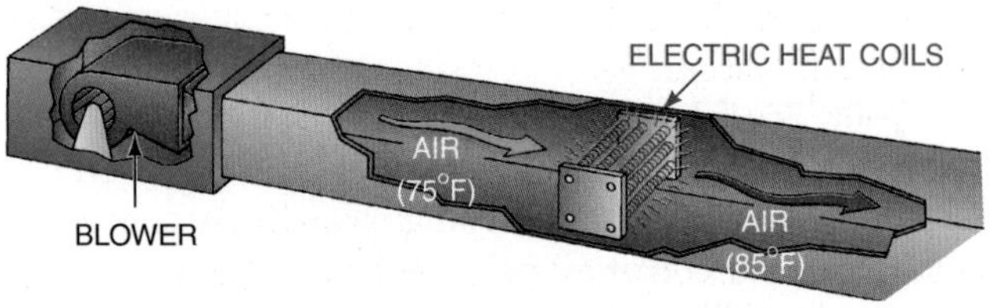

2.10 ENERGY IN MAGNETISM

Magnetism is another method of converting electron flow to usable energy. Electron flow is used to generate magnetic fields that are, in turn, used to turn motors. The motors turn the fans, pumps, and compressors that ultimately move air, water, and refrigerant. In **Figure 2.16** an electric motor turns

Figure 2.16 Electrical energy used in an electric motor is converted to work to boost water pressure to force circulation.

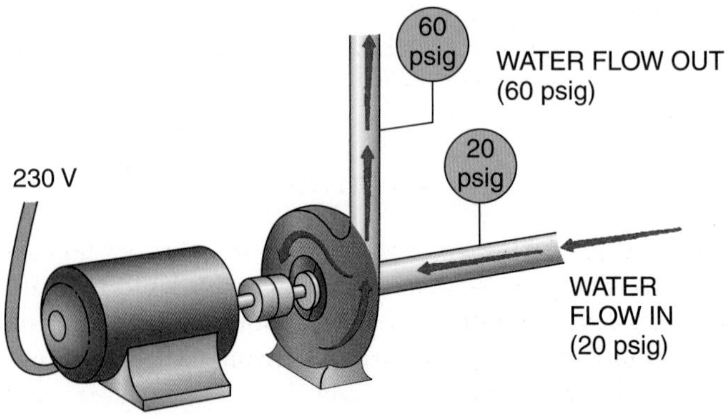

a water pump to boost the water pressure from 20 to 60 psig. This takes energy. In this case, electrical energy is converted to mechanical energy that is used to pump the water.

The preceding examples serve only as an introduction to the concepts of chemical energy, heat energy, and electrical energy. Each subject will be covered in detail later. For now, it is important to understand that all heating and cooling systems require energy.

2.11 PURCHASE OF ENERGY

In most cases, energy that is purchased is done so as either a fossil fuel or as electric power. Energy is normally purchased by the unit. Natural gas, for example, flows through a meter that measures how many cubic feet have passed during some time span, such as a month. Fuel oil is normally sold by the gallon, coal by the ton. Electrical energy is sold by the kilowatt-hour or kWh. The amount of heat each of these units contains is known, so a known amount of heat is purchased. Natural gas, for instance, has a heat content of about 1000 Btu/ft^3; the heat content of fuel oil #2 is about 139,000 Btu/gallon, and 1 kWh of electrical energy is the equivalent of 3413 Btu. There are numerous types of coal available and the heat content can range from 14 million Btu/ton up to over 26 million Btu/ton.

2.12 ENERGY USED AS WORK

Energy purchased from electrical utilities is often referred to as electric power. Power is the rate of doing work. Work can be explained as a force moving an object and is expressed in units called foot-pounds, or ft-lb. It is expressed by this formula:

$$\text{Work} = \text{Force} \times \text{Distance}$$

Force is expressed in pounds, and distance is expressed in feet. For instance, when a 150-lb man climbs a flight of stairs 100 ft high (about the height of a 10-story building), he performs work. But how much? The amount of work in this example is equivalent to the amount of work necessary to lift this man the same height. We can calculate the work by using the preceding formula.

$$\begin{aligned}\text{Work} &= 150\ \text{lb} \times 100\ \text{ft}\\ &= 15{,}000\ \text{ft-lb}\end{aligned}$$

Notice that time does not factor into this calculation. Whether this process takes 1 minute or 1 hour, the same amount of work is done. However, completing the task faster will require more power, while completing the task slower will require less power.

Figure 2.17 When a horse can lift 660 lb to a height of 50 ft in 1 min, it has done the equivalent of 33,000 ft-lb of work in 1 min, or 1 hp.

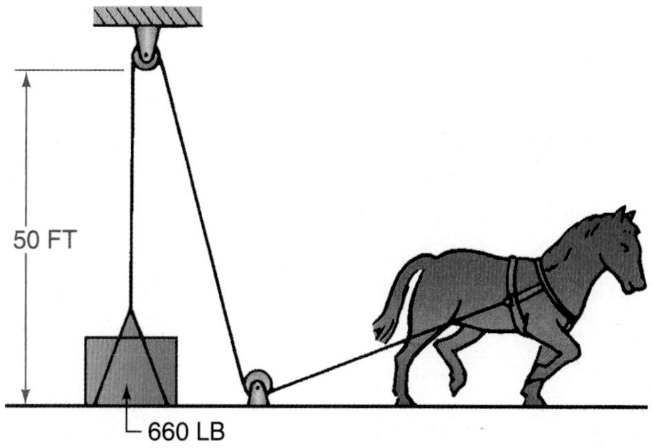

2.13 POWER

Power is the *rate* of doing work. An expression of power is horsepower (hp). Horsepower is the rate of doing work and includes a time element, and is expressed as ft-lb/min. At one time, it was determined that an average horse could lift the equivalent of 33,000 lb to a height of 1 ft in 1 min, which is the same as 33,000 ft-lb/min, or hp, **Figure 2.17**. This describes a rate of doing work because the time factor has been added. Lifting 330 lb to a height of 100 ft (330 lb × 100 ft = 33,000 ft-lb) in 1 min or lifting 660 lb to a height of 50 ft (660 lb × 50 ft = 33,000 ft-lb) in 1 min will require the same amount of power. As a point of reference, the blower motor in the average furnace can be rated at 1/2 hp.

If the man in the previous section took 1 minute to climb the stairs, the power used is the equivalent of 0.45 hp (15,000/33,000).

Our purpose in discussing these topics is to help you understand how to use power effectively and to understand how power companies determine their methods of charging money for power.

2.14 ELECTRICAL POWER—THE WATT

The unit of measurement for electrical power is the watt (W). This is the unit used by power companies. When you pay your energy bill, you are paying for the number of kilowatt hours (kWh) you used in the previous month. One kilowatt hour, or 1 kWh, is the equivalent of using 1000 watts of power for 1 hour. The relationship between watts and horsepower is 746 watts = 1 horsepower (1 hp).

Fossil-fuel energy can be compared to electrical energy, and one form of energy can be converted to the other. There must be some basis, however, for conversion to make the

comparison. The examples we use to illustrate this will not take efficiencies for the various fuels into account. Efficiencies of fossil fuels will be discussed in the corresponding Units of this text.

Some examples of conversions follow:

1. **Converting electric heat rated in kilowatts (kW) to the equivalent gas or oil heat rated in Btu/h.** Suppose that we want to know the capacity in Btu/h for a 20-kW electric heater (a kilowatt is 1000 watts).

$$1 \text{ kW} = 3413 \text{ Btu/h}$$
$$20 \text{ kW} \times 3413 \text{ Btu/kWh} = 68{,}260 \text{ Btu/h of heat energy}$$

2. **Converting Btu/h to kW.** Suppose that a gas or oil furnace has an output capacity of 100,000 Btu/h. Since 3413 Btu/h = 1 kW, we have

$$100{,}000 \text{ Btu/h} \div 3413 \text{ Btu/h/kW} = 29.3 \text{ kW}$$

3. In other words, a 29.3-kW electric heating system would be required to replace the 100,000-Btu/h furnace.

SAFETY PRECAUTION: *Any device that consumes power, such as an electric motor or gas furnace, is potentially dangerous. These devices should only be handled or adjusted by experienced people.* •

REVIEW QUESTIONS

1. Matter is a substance that occupies space and has
   A. color.
   B. texture.
   C. temperature.
   D. mass.
2. Solid, liquid, and gas are the most common
   A. states of matter.
   B. specific volumes.
   C. gravitational forces.
   D. fossil fuels.
3. In what direction does a solid exert force?
   A. Outward
   B. Upward
   C. Downward
   D. Outward, upward, and downward
4. In what direction does a liquid exert force?
   A. Outward
   B. Upward
   C. Downward
   D. Downward and outward
5. Vapor exerts pressure in what direction?
   A. Outward
   B. Upward
   C. Downward
   D. Outward, upward, and downward
6. The weight per unit volume of a substance is defined as
   A. mass.
   B. density.
   C. work.
   D. specific gravity.
7. The density of a substance compared to the density of water is known as
   A. specific gravity.
   B. specific volume.
   C. mass.
   D. weight.
8. The volume occupied by 1 pound of a fluid is known as
   A. specific gravity.
   B. specific volume.
   C. mass.
   D. weight.
9. The density of tungsten is 1210 lb/ft^3. What would be its specific volume?
   A. 0.0826 ft^3/lb
   B. 0.00826 lb/ft^3
   C. 0.000826 lb/ft^3
   D. 0.000826 ft^3/lb
10. The specific volume of red brass is 0.001865 ft^3/lb. What would be its density?
    A. 53.61 lb/ft^3
    B. 56.19 lb/ft^3
    C. 536.19 lb/ft^3
    D. 536.19 ft^3/lb
11. Aluminum has a density of 171 lb/ft^3. What would be its specific gravity?
    A. 2.74
    B. 27.4
    C. 4.72
    D. 47.2
12. Four pounds of a gas occupy 10 ft^3. What would be its density and specific gravity?
    A. 0.4 ft^3/lb
    B. 0.4 lb/ft^3
    C. 4.0 lb/ft^3
    D. 4.0 ft^3/lb
13. Whose law states that the volume of a gas varies inversely with the absolute pressure, as long as the temperature remains constant?
    A. Charles'
    B. Boyle's
    C. Newton's
    D. Dalton's

14. At a constant pressure, how does a volume of gas vary with respect to the absolute temperature?
    A. It will expand when it is heated or cooled.
    B. It will remain constant.
    C. It will expand when cooled or contract when heated.
    D. It will expand when heated or contract when cooled.
15. Whose law states that the total pressure of a confined mixture of gases is the sum of the pressures of each of the gases in the mixture?
    A. Charles'
    B. Boyle's
    C. Newton's
    D. Dalton's
16. What are the two types of energy most frequently used or considered in this industry?
    A. Solar and fossil fuels
    B. Nuclear and wind
    C. Nuclear and solar
    D. Wind and solar
17. The time rate of doing work is called
    A. weight.
    B. distance.
    C. power.
    D. force.
18. Force × distance is used to determine
    A. weight.
    B. work.
    C. power.
    D. volume.
19. If an air-conditioning compressor weighing 300 lb had to be lifted 4 ft to be mounted on a base, how much work, in ft-lb, must be done?
    A. 12,000 lb/ft^3
    B. 1200 ft^3/lb
    C. 120 foot-pounds
    D. 1200 foot-pounds
20. How many watts of electrical energy are equal to 1 hp?
    A. 1492
    B. 746
    C. 74.6
    D. 373
21. What unit of energy does the power company charge the consumer for?
    A. Kilowatt-hour
    B. Watt-hour
    C. Btu/h
    D. Therm
22. If 3000 ft^3 of air is crossing an evaporator coil and is cooled from 75°F to 55°F, what would be the volume of air, in ft^3, exiting the evaporator coil?
    A. 288.7 ft^3
    B. 28.7 ft^3
    C. 2887.8 ft^3
    D. 1392.2 ft^3
23. A gas is compressed inside a compressor's cylinder. When the piston is at its bottom dead center, the gas is initially at 10 psig, 65°F, and 10.5 in^3. After compression and when the piston is at top dead center, the gas is 180°F and occupies 1.5 in^3. What would be the new pressure of the gas in psig?
    A. 196
    B. 19.6
    C. 1.96
    D. 1960
24. How many Btu/h would be produced in a 12-kW electric heater?
    A. 4095 Btu/h
    B. 40,956 Btu/h
    C. 400,956 Btu/h
    D. 495 Btu/h
25. How many Btu of heat can be produced by 4 kWh of electricity?
    A. 136,520 Btu/h
    B. 136 Btu/h
    C. 1365 Btu/h
    D. 13,652 Btu/h
26. Why does an object weigh less on the moon than on earth?
27. Why is information regarding the specific volume of gases important to the designer of air-conditioning, heating, and refrigeration equipment?
28. Describe horsepower and list the three quantities needed to determine horsepower.
29. Explain how specific volume and density are related to each other.
30. State the law of conservation of energy and explain its importance with regards to accounting for energy.
31. Explain how heat energy flows between a substance at −125°F and a substance at −175°F.
32. Explain how magnetism can be used to help convert electrical energy to mechanical energy.

UNIT 3

REFRIGERATION AND REFRIGERANTS

OBJECTIVES

After studying this unit, you should be able to

- discuss applications for high-, medium-, and low-temperature refrigeration.
- describe the term *ton of refrigeration.*
- describe the basic refrigeration cycle.
- explain the relationship between pressure and the boiling point of water or other liquids.
- describe the function of the evaporator or cooling coil.
- explain the concept of superheat as it applies to the evaporator.
- explain the purpose of the compressor.
- list three types of compressors normally used in residential and light commercial buildings.
- discuss the function of the condenser.
- explain the concept of subcooling as it applies to the condenser.
- state the purpose of the metering device.
- list four characteristics to consider when choosing a refrigerant for a system.
- list the designated cylinder colors for three commonly used refrigerants.
- discuss four commonly used refrigerants and their applications.
- describe how refrigerants can be stored or processed while refrigeration systems are being serviced.
- plot a refrigeration cycle for refrigerants (R-22 and R-134a) on a pressure/enthalpy diagram.
- plot a refrigeration cycle on a pressure/enthalpy diagram for refrigerant blends R-404A and R-410A.
- plot a refrigeration cycle on a pressure/enthalpy diagram for a refrigerant blend (R-407C) that has a noticeable temperature glide.

SAFETYCHECKLIST

- ☑ Areas in which there is the potential for refrigerant leaks should be properly ventilated.
- ☑ Extra precautions should be taken to ensure that no refrigerant leaks occur near an open flame.
- ☑ Refrigerants are stored in pressurized containers and should be handled with care.
- ☑ Eye protection and gloves should be worn when checking system pressures or when otherwise working on air-conditioning and refrigeration systems.
- ☑ Only transfer refrigerants from a system into DOT-approved containers.

3.1 INTRODUCTION TO REFRIGERATION

This unit is an introduction to refrigeration and refrigerants. The term *refrigeration* is used here to include the cooling processes for both preserving food and comfort cooling (air-conditioning).

Preserving food is one of the most valuable applications of refrigeration. Refrigeration causes the temperature of a substance to drop, slowing the molecular motion within the substance. As the molecular motion slows down, so does the rate of food spoilage. Below the "frozen hard" point, which for most foods is approximately 0°F, food-spoiling bacteria stop growing. Most refrigeration systems fall into one of three categories. Low-temperature refrigeration systems often refer to systems that maintain products at temperatures below 0°F. Low-temperature systems are mainly intended to freeze product. Medium-temperature systems typically maintain products at temperatures between 35°F and 45°F. Medium-temperature refrigeration systems are often referred to as refrigerators and are designed primarily to keep food fresh. Refrigeration systems that operate to produce warmer temperatures are referred to as high-temperature refrigeration systems. Comfort cooling systems, commonly referred to as air-conditioning systems, are used, for example, to cool our homes and commercial spaces and are classified as high-temperature refrigeration systems. High-temperature refrigeration systems are most commonly used to keep an occupied space in the 68°F to 80°F range.

For many years, dairy products and other perishables were stored in the coldest room in the house, the basement, a well, or a spring. Early-generation refrigerators did not resemble the refrigerators we know today. These low-tech appliances were called "ice boxes," which were, quite literally boxes with ice in them. The ice melted as it absorbed heat from the food in the box, cooling the food, **Figure 3.1.**

In the early 1900s, ice was manufactured by mechanical refrigeration and sold to people with ice boxes, but still only the wealthy could afford it. Also in the early 1900s, some companies manufactured a household refrigerator. Like all new items, it took a while to become popular. Now, most houses have a refrigerator with a freezing compartment. Modern refrigerators have become state-of-the-art appliances—and some models even include automatic beverage and ice

Figure 3.1 Ice boxes were made of wood at first, then metal. The boxes were insulated with cork. If a cooling unit were placed where the ice was located, this would constitute a refrigerator.

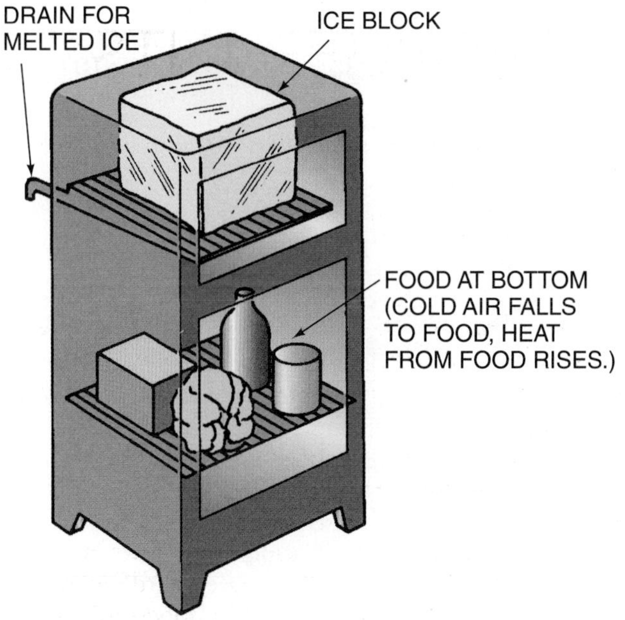

dispensers, built-in television screens, and connections to the World Wide Web.

Frozen food was just beginning to become popular about the time World War II began. Because most people did not have a freezer at this time, centralized frozen food lockers were established so that a family could have its own compartment in which to store frozen foods. Food that is frozen fresh is appealing because it stays fresh for a longer period of time. Refrigerated and frozen foods are so common now that most people take them for granted. Advances in refrigeration technologies have made it possible to enjoy perishable foods and products from all around the world year round.

3.2 REFRIGERATION

Refrigeration is the process of removing heat from a place where it is not wanted and transferring that heat to a place where it makes little or no difference. In the average household, the room temperature from summer to winter is normally between 70°F and 90°F. The temperature inside the refrigerator fresh food section, a medium-temperature application, should be about 35°F. Heat flows naturally from a warmer substance to a cooler one. Therefore, the heat contained in the air in the room is trying to flow into the cooler refrigerator. Heat flows into the refrigerator through the insulated walls of the appliance, **Figure 3.2**, and also when the refrigerator door is open, **Figure 3.3**. Since heat also enters a refrigerator when warm food is placed in the box, it is always best to allow food to cool down to room temperature before placing it in the refrigerator.

Figure 3.2 Heat transfers through the walls into the box by conduction. The walls have insulation, but this does not completely stop the heat leakage.

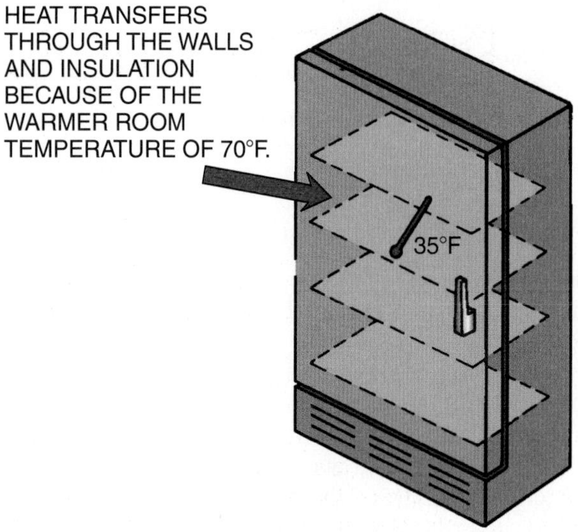

Figure 3.3 The colder air falls out of the refrigerator because it is heavier than the warmer air located outside. The cooler air is replaced with warmer air at the top. This is referred to as heat leakage.

3.3 RATING REFRIGERATION EQUIPMENT

Refrigeration equipment must have a capacity rating system so that equipment made by different manufacturers can be compared. The method for rating refrigeration equipment

capacity goes back to the days of using ice as the source for removing heat. It takes 144 Btu of heat energy to melt 1 lb of ice at 32°F. This figure is referred to as the latent heat of fusion and is the basis for determining the capacity rating of refrigeration equipment.

The term for this capacity rating is the *ton*. **One ton of refrigeration** is the amount of heat required to melt 1 ton of ice in a 24-hour period. Since it takes 144 Btu of heat to melt 1 lb of ice, it would then take 2000 times that much heat to melt 1 ton of ice (2000 lb = 1 ton):

$$144 \text{ Btu/lb} \times 2000 \text{ lb} = 288{,}000 \text{ Btu}$$

When the melting of 1 ton of ice is accomplished in a 24-hour period, the rate of heat transfer is 12,000 Btu/h (288,000 ÷ 24 = 12,000), or 200 Btu/min (12,000 ÷ 60 = 200). When heat is transferred at a rate of 12,000 Btu/h or 200 Btu/min, we refer to this as 1 ton of refrigeration. For example, an air-conditioning system that moves heat at a rate of 24,000 Btu/h is referred to as a 2-ton system, because 24,000 Btu/h divided by 12,000 Btu/h/ton = 2 tons.

3.4 THE REFRIGERATION PROCESS

The components of the refrigerator have to pump the heat up the temperature scale from the 35°F or 0°F refrigeration compartments to the 70°F room, **Figure 3.4**. Heat leaking into the refrigerator raises the air temperature inside but does not normally raise the temperature of the food an appreciable amount. If it did, the food in the box would quickly spoil. When the temperature inside the refrigerator rises to a predetermined level, the refrigeration system comes on and pumps the heat out.

The process of pumping heat out of the refrigerator could be compared to pumping water from a valley to the top of a hill. A water pump with a motor accomplishes the work. If, for instance, a gasoline engine were driving the pump, the gasoline would be burned and converted to work energy. An electric motor uses electric power as work energy, **Figure 3.5**. Refrigeration is the process of moving heat from an area of lower temperature into an area or medium of higher temperature. This takes energy that must be purchased.

The following example uses a residential window air-conditioning system to explain the basics of refrigeration. Residential air-conditioning, whether a window unit or a central system, is considered to be high-temperature refrigeration and is used for comfort cooling.

The refrigeration concepts utilized in the residential air conditioner are the same as those in the household refrigerator. The air conditioner pumps the heat from inside the house to the outside of the house, just as the household refrigerator pumps heat from the refrigerator into the kitchen. In addition, just as heat leaks into the refrigerator, heat leaks into the house and must be removed. The heat is transferred outside by the air-conditioning system. The cold air in the house is recirculated air. Room air at approximately 75°F goes into the air-conditioning unit, and air at approximately 55°F comes out. This is the same air with some of its heat removed, **Figure 3.6**.

Figure 3.4 Heat that leaks into the refrigerator from any source must be removed by the refrigerator's heat-pumping mechanism. The heat has to be pumped from the cool, 35°F interior of the refrigerator to the warmer, 70°F air in the room in which the refrigerator is located.

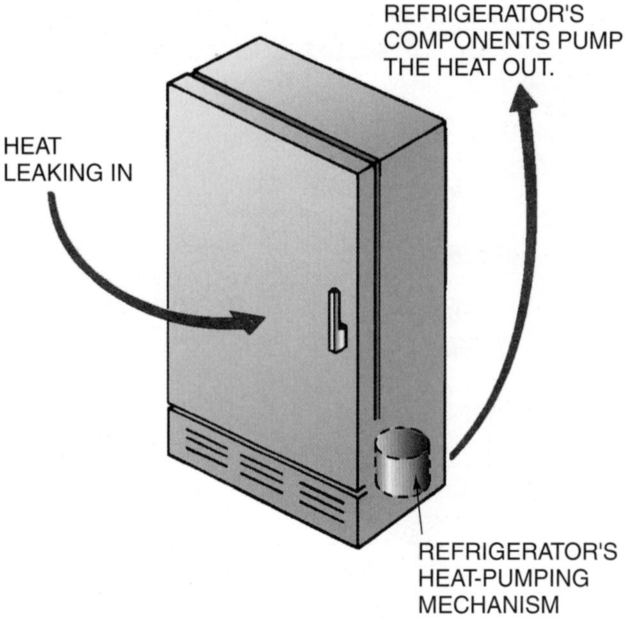

Figure 3.5 Power is required to pump water uphill. The same is true for pumping heat up the temperature scale from the interior of a 35°F refrigerated box to a 70°F room.

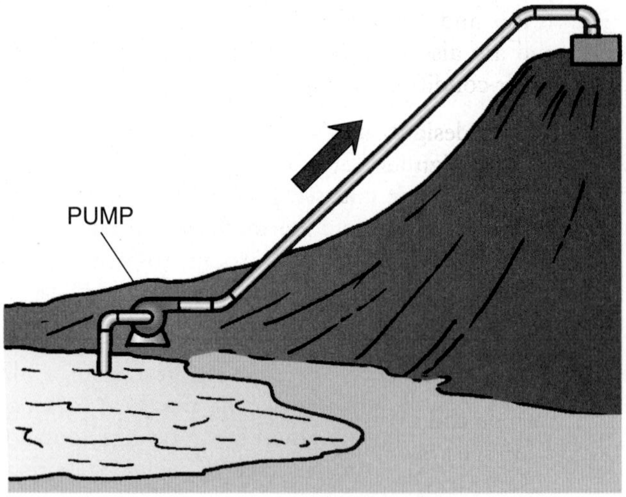

Figure 3.6 A window air-conditioning unit.

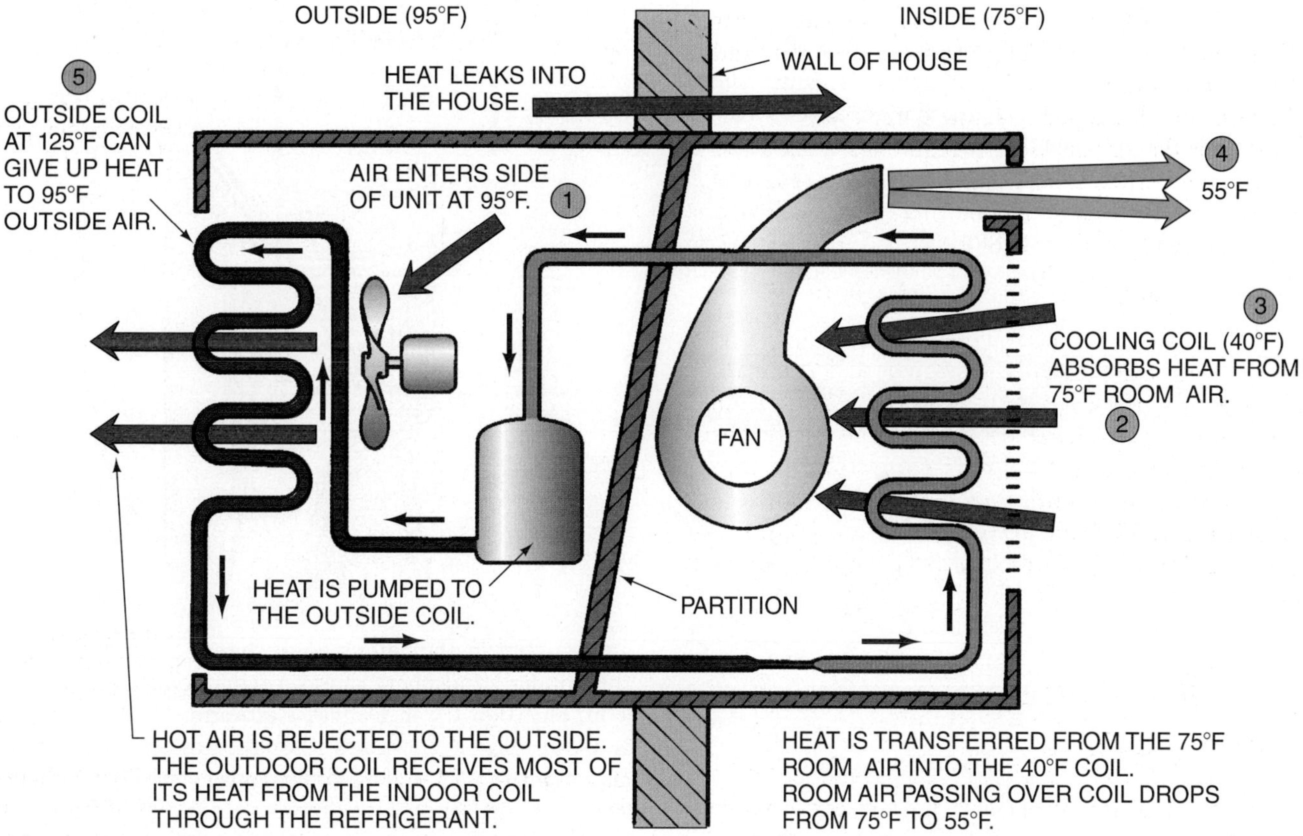

Figure 3.6 and the following statements illustrate this concept and are also guidelines to some of the design data used in the air-conditioning field.

1. The outside design temperature is 95°F.
2. The inside design temperature is 75°F.
3. The design cooling coil temperature is 40°F. This coil transfers heat from the room air into the refrigeration system. Notice that with a 75°F room temperature and a 40°F cooling coil temperature, heat will transfer from the room air into the refrigerant in the coil.
4. The heat transfer makes the air leaving the coil cooler, about 55°F. The air exits the fan at 55°F.
5. The outside coil temperature is 125°F. The coil transfers heat from the system to the outside air. Notice that when the outside air temperature is 95°F and the coil temperature is 125°F, heat will be transferred from the system to the outside air.

Careful examination of **Figure 3.6** shows that heat from the house is transferred into the inside coil and transferred to the outside air from the outside coil. The air-conditioning system is actually pumping the heat out of the house. The system capacity must be large enough to pump the heat out of the house faster than it is leaking back in so that the occupants will not become uncomfortable. The cycle, as just described, is a repeating cycle. This means that, once one full cycle has been completed, it is ready to repeat the cycle over and over again, without depleting the system refrigerant.

3.5 TEMPERATURE AND PRESSURE RELATIONSHIP

To understand the refrigeration process, we must go back to **Figure 1.15** (heat/temperature graph), where ice was changed to steam. Since water boils at 212°F at 29.92 in. Hg pressure, it is suggested that water has different boiling temperatures at different pressures. ***The boiling point of water can be changed and controlled by controlling the vapor pressure above the water.*** Understanding this concept is necessary because water is used as the heat transfer medium. The next few paragraphs are extremely important to gain a solid understanding of refrigeration.

The **temperature/pressure relationship** correlates the vapor pressure and the boiling temperature of water and is the basis for controlling a system's temperatures. So, if we are able to control the pressures in a refrigeration or air-conditioning system, we will be able to control the temperatures that the system will maintain.

Pure water boils at 212°F at sea level when the barometric pressure is at the standard value of 29.92 in. Hg. This condition exerts an atmospheric pressure of 14.696 psia (0 psig) on the water's surface. This reference point is on the last line of the table in **Figure 3.7**. Also see **Figure 3.8,** which shows the container of water boiling at 212°F while at sea level and atmospheric pressure. When this same pan of water is taken to a mountaintop, the boiling point changes, **Figure 3.9,** because the thinner atmosphere causes a reduction in pressure (about 1 in. Hg/1000 ft). In Denver, Colorado, for example, which is about 5000 ft above sea level, the atmospheric pressure is approximately 25 in. Hg. Water boils at 203.4°F at that pressure. This makes cooking some foods more difficult because they now need more time to cook. But by placing the food in a closed container that can be pressurized, such as a pressure cooker, and allowing the pressure to go up to about 15 psi above atmosphere (or 30 psia), the boiling point can be raised to 250°F, **Figure 3.10.**

Figure 3.7 The boiling point of water. The temperature at which water will boil at a specific pressure can be found on the temperature/pressure chart for water.

WATER TEMPERATURE	ABSOLUTE PRESSURE	
°F	lb/in² (psia)	in. Hg
10	0.031	0.063
20	0.050	0.103
30	0.081	0.165
32	0.089	0.180
34	0.096	0.195
36	0.104	0.212
38	0.112	0.229
40	0.122	0.248
42	0.131	0.268
44	0.142	0.289
46	0.153	0.312
48	0.165	0.336
50	0.178	0.362
60	0.256	0.522
70	0.363	0.739
80	0.507	1.032
90	0.698	1.422
100	0.950	1.933
110	1.275	2.597
120	1.693	3.448
130	2.224	4.527
140	2.890	5.881
150	3.719	7.573
160	4.742	9.656
170	5.994	12.203
180	7.512	15.295
190	9.340	19.017
200	11.526	23.468
210	14.123	28.754
212	14.696	29.921

Figure 3.8 Water boils at 212°F when the atmospheric pressure is 29.92 in. Hg.

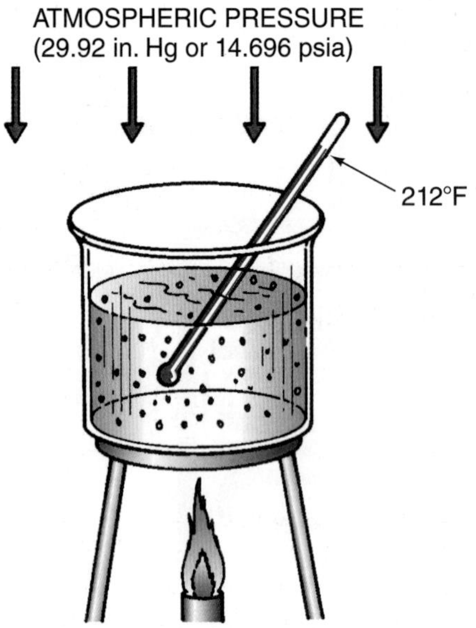

Figure 3.9 Water boils at 203°F when the atmospheric pressure is 24.92 in. Hg.

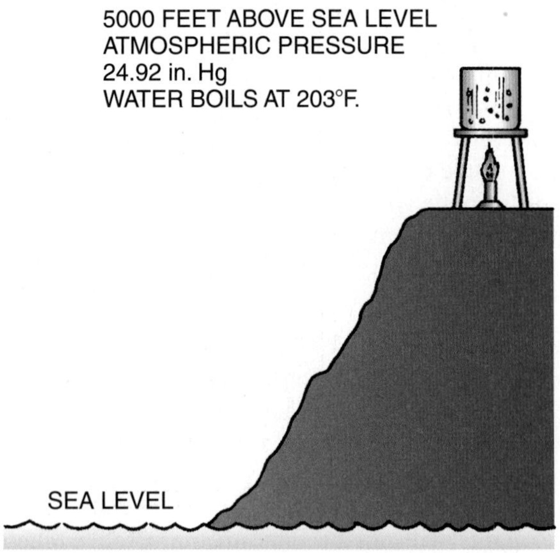

Figure 3.10 The water in a pressure cooker boils at 250°F. As heat is added, the water boils into a vapor. Since the vapor cannot escape, the vapor pressure in the pressure cooker rises to 15 psig. The water boils at a temperature that is higher than 212°F because the pressure in the vessel has risen to the level that is above atmospheric pressure.

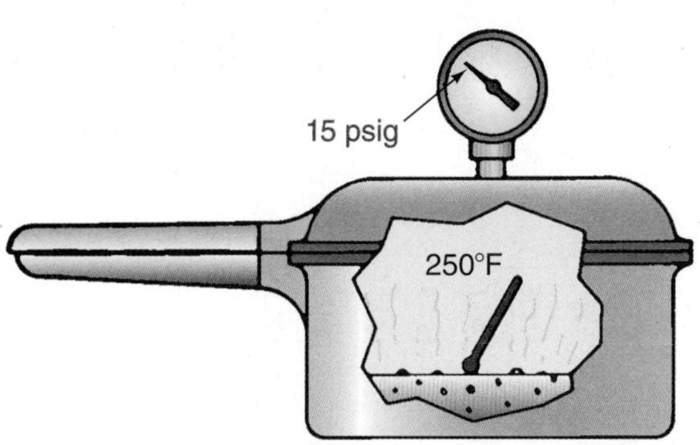

Studying the water temperature/pressure table reveals that whenever the pressure is increased, the boiling temperature increases, and that whenever the pressure is reduced, the boiling temperature is reduced. If water were boiled at a temperature low enough to absorb heat from a room, the result would be comfort cooling (air-conditioning).

Suppose we place a thermometer in a pan of pure water at room temperature (70°F), put the pan inside a bell jar with a barometer, and start the vacuum pump. When the pressure in the jar reaches the pressure that corresponds to the boiling point of water at 70°F, the water will start to boil and vaporize. This point is 0.739 in. Hg (0.363 psia). (**Figure 3.11** illustrates the container in the bell jar; the data can be found in the table in Figure 3.7.)

Figure 3.11 The pressure in the bell jar is reduced to 0.739 in. Hg. The boiling temperature of the water is reduced to 70°F because the pressure is 0.739 in. Hg (0.363 psia).

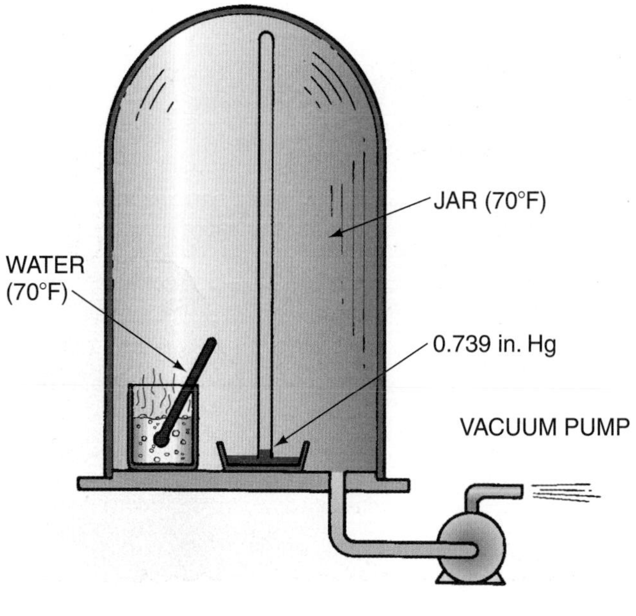

Notice, in Figure 3.7, that the temperatures are listed in the left-hand column and the pressures are found in the body of the chart, hence the name temperature/pressure chart (or table). Sometimes, however, the information is presented differently. On some charts, the pressures are located in the left-hand column and the temperatures are found in the body of the chart. These charts or tables are referred to as pressure/temperature charts. Although the information is the same, the way that it is presented is different. Make certain that you know which type of chart you are using—or inaccurate calculations and system conclusions may result.

If we were to lower the pressure in the bell jar to correspond to a temperature of 40°F, this new pressure of 0.248 in. Hg (0.122 psia) will cause the water to boil at 40°F. The water is not hot even though it is boiling. The thermometer in the pan indicates this. If we opened the jar to the atmosphere, we would find the water to be cold. Also, the pressure in the jar would rise and the boiling process would stop.

Now let us circulate this water boiling at 40°F through a cooling coil. If room air were passed over it, it would absorb heat from the room air. Because this air is giving up heat to the coil, the air leaving the coil is cold. **Figure 3.12** illustrates the cooling coil.

When water is used in this way, it is called a *refrigerant*. A refrigerant is a substance that can be changed readily to a vapor by boiling it and then changed back into a liquid by condensing it. The refrigerant must be able to make this change repeatedly without altering its characteristics. In addition, a good refrigerant is able to absorb and reject large amounts of heat as it transitions between the liquid and vapor states. Water is not normally used as a refrigerant in small applications for reasons that will be discussed later. We used it in this example because most people are familiar with its characteristics.

To explore how a real refrigeration system works, we will use refrigerant R-410A in the following examples because it is commonly used in residential air-conditioning. See **Figure 3.13** for the temperature/pressure relationship chart for several

Figure 3.12 The water is boiling at 40°F because the pressure is 0.122 psia, or 0.248 in. Hg. The room air is 75°F and gives up heat to the 40°F coil.

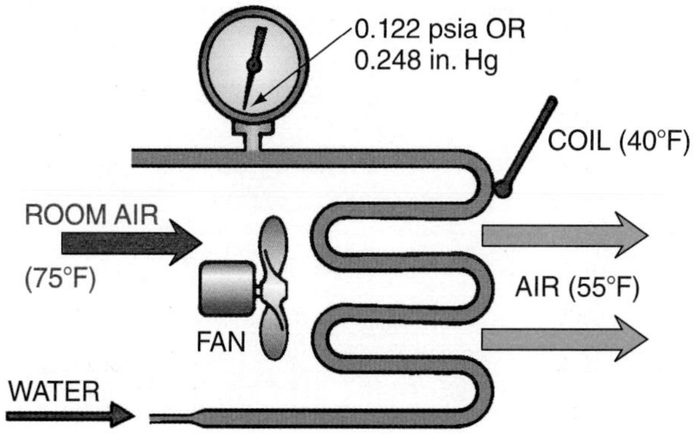

Figure 3.13 This chart shows the temperature/pressure relationship in in. Hg vacuum or psig. Pressures for R-404A, R-407C, and R-410A are average liquid and vapor pressures.

TEMPERATURE	REFRIGERANT						TEMPERATURE	REFRIGERANT						TEMPERATURE	REFRIGERANT					
°F	R-1234yf	22	134a	R-407C	404A	410A	°F	R-1234yf	22	134a	R-407C	404A	410A	°F	R-1234yf	22	134a	R-407C	404A	410A
−60		12.0		-8	6.6	0.3	12	16.1	34.7	13.2	16.1	46.2	65.3	42	40.2	71.4	37.0	40.2	89.7	122.9
−55	-8.6	9.2		-6.6	3.1	2.6	13	16.7	35.7	13.8	16.7	47.4	66.8	43	41.2	73.0	38.0	41.2	91.5	125.2
−50	-7.7	6.2		-5.3	0.8	5.0	14	17.4	36.7	14.4	17.4	48.6	68.4	44	42.3	74.5	39.0	42.3	93.3	127.6
−45	-6.7	2.7		-3.8	2.5	7.8	15	18.1	37.7	15.1	18.1	49.8	70.0	45	43.3	76.0	40.1	43.3	95.1	130.0
−40	-5.6	0.5	14.7	-2.2	4.8	9.8	16	18.7	38.7	15.7	18.7	51.0	71.6	46	44.3	77.6	41.1	44.3	97.0	132.4
−35	-4.3	2.6	12.4	-0.4	7.4	14.2	17	19.4	39.8	16.4	19.4	52.3	73.2	47	45.4	79.2	42.2	45.4	98.8	134.9
−30	-2.9	4.9	9.7	1.7	10.2	17.9	18	20.1	40.8	17.1	20.1	53.5	75.0	48	46.5	80.8	43.3	46.5	100.7	136.4
−25	-1.3	7.4	6.8	3.9	13.3	21.9	19	20.8	41.9	17.7	20.8	54.8	76.7	49	47.5	82.4	44.4	47.5	102.6	139.9
−20	0.44	10.1	3.6	6.5	16.7	26.4	20	21.5	43.0	18.4	21.5	56.1	78.4	50	48.6	84.0	45.5	48.6	104.5	142.5
−18	1.2	11.3	2.2	7.5	18.2	28.2	21	22.2	44.1	19.2	22.2	57.4	80.1	55	54.4	92.6	51.3	54.4	114.6	156.0
−16	1.9	12.5	0.7	8.7	19.6	30.2	22	23	45.3	19.9	23	58.8	81.9	60	60.5	101.6	57.3	60.5	125.2	170.0
−14	2.7	13.8	0.3	9.8	21.1	32.2	23	23.7	46.4	20.6	23.7	60.1	83.7	65	67	111.2	64.1	67	136.5	185.0
−12	3.6	15.1	1.2	11	22.7	34.3	24	24.5	47.6	21.4	24.5	61.5	85.5	70	73.9	121.4	71.2	73.9	148.5	200.8
−10	4.4	16.5	2.0	12.3	24.3	36.4	25	25.3	48.8	22.0	25.3	62.9	87.3	75	81.2	132.2	78.7	81.2	161.1	217.6
−8	5.3	17.9	2.8	13.6	26.0	38.7	26	26	49.9	22.9	26	64.3	90.2	80	89	143.6	86.8	89	174.5	235.4
−6	6.2	19.3	3.7	15	27.8	40.9	27	26.8	51.2	23.7	26.8	65.8	91.1	85	97.3	155.7	95.3	97.3	188.6	254.2
−4	7.2	20.8	4.6	16.4	30.0	42.3	28	27.6	52.4	24.5	27.6	67.2	93.0	90	1.6	168.4	104.4	1.6	203.5	274.1
−2	8.2	22.4	5.5	17.8	31.4	45.8	29	28.5	53.6	25.3	28.5	68.7	95.0	95	115.3	181.8	114.0	115.3	219.2	295.0
0	9.2	24.0	6.5	19.4	33.3	48.3	30	29.3	54.9	26.1	29.3	70.2	97.0	100	125	195.9	124.2	125	235.7	317.1
1	9.7	24.8	7.0	20.1	34.3	49.6	31	30.1	56.2	26.9	30.1	71.7	99.0	105	135.3	210.8	135.0	135.3	253.1	340.3
2	10.3	25.6	7.5	20.9	35.3	50.9	32	31	57.5	27.8	31	73.2	101.0	110	146.1	226.4	146.4	146.1	271.4	364.8
3	10.8	26.4	8.0	21.7	36.4	52.3	33	31.9	58.8	28.7	31.9	74.8	103.1	115	157.5	242.7	158.5	157.5	290.6	390.5
4	11.3	27.3	8.6	22.6	37.4	53.6	34	32.7	60.1	29.5	32.7	76.4	105.1	120	169.5	259.9	171.2	169.5	310.7	417.4
5	11.9	28.2	9.1	23.4	38.4	55.0	35	33.6	61.5	30.4	33.6	78.0	107.3	125	182	277.9	184.6	182	331.8	445.8
6	12.5	29.1	9.7	24.2	39.5	56.4	36	34.5	62.8	31.3	34.5	79.6	108.4	130	195.2	296.8	198.7	195.2	354.0	475.4
7	13.1	30.0	10.2	25.1	40.6	57.8	37	35.5	64.2	32.2	35.5	81.2	111.6	135	209	316.6	213.5	209	377.1	506.5
8	13.7	30.9	10.8	26	41.7	59.3	38	36.4	65.6	33.2	36.4	82.9	113.8	140	223.5	337.2	229.1	223.5	401.4	539.1
9	14.2	31.8	11.4	26.9	42.8	60.7	39	37.3	67.1	34.1	37.3	84.6	116.0	145	238.5	358.9	245.5	238.5	426.8	573.2
10	14.9	32.8	11.9	27.8	43.9	62.2	40	38.3	68.5	35.1	38.3	86.3	118.3	150	254.3	381.5	262.7	254.3	453.3	608.9
11	15.5	33.7	12.5	28.7	45.0	63.7	41	39.3	70.0	36.0	39.3	88.0	120.5	155	270.8	405.1	280.7	270.8	479.8	616.2

VACUUM (in. Hg) – RED FIGURES
GAUGE PRESSURE (psig) – BOLD FIGURES

refrigerants, including R-410A. This chart is similar to the one previously shown for water but uses different temperature and pressure levels. Take a moment to become familiar with the chart and that the temperatures are listed in the left column, expressed in °F, and the pressures fill the body of the chart, expressed in psig. Find 40°F in the left column, read to the right, and notice that the gauge reading is 118.3 psig for R-410A. This means that when R-410A liquid is at a vapor pressure of 118.3 psig, it will boil at a temperature of 40°F. When air is passed over the coil, it will cool just as in the previous water example.

The pressure and temperature of a refrigerant will correspond, according to its pressure/temperature relationship, when both liquid and vapor are present under two conditions:

1. When a change of state (boiling or condensing) is occurring, and
2. When the refrigerant is at equilibrium (i.e., no heat is added or removed).

When a change of state is occurring, the refrigerant is said to be **saturated**. When a refrigerant is saturated, both liquid and vapor can exist simultaneously. When they do, both the liquid and the vapor will have the same saturation temperature. **NOTE:** *The same temperature for the liquid and vapor is not true of some of the newer blended refrigerants that have a temperature glide. Temperature glide will be covered in Unit 9, "Refrigerant and Oil Chemistry and Management—Recovery, Recycling, Reclaiming, and Retrofitting."* • The saturation temperature depends on the pressure of the liquid/vapor mixture, called the **saturation pressure**. The higher the pressure, the higher the saturation temperature of the liquid and vapor mixture. The lower the pressure, the lower the saturation temperature.

Suppose that a cylinder of R-410A is allowed to sit in a room until it reaches the room temperature of 75°F. It will then be in equilibrium because no outside forces are acting on it. The cylinder and its partial liquid, partial vapor contents will now be at the room temperature of 75°F and, according to the temperature/pressure chart provided in Figure 3.13, will be at a pressure of 217.6 psig. The temperature/pressure chart is also referred to as a saturation chart because it contains saturation temperatures for different saturation pressures.

Suppose that the same cylinder of R-410A is moved into a walk-in cooler and allowed to reach the box temperature of 35°F. The refrigerant in the cylinder will then reach a new pressure of 107.3 psig because while it is cooling off to 35°F, the vapor inside the cylinder is reacting to the cooling effect by partially condensing; therefore, the pressure drops.

If we move the cylinder (now at 35°F) back into the warmer room (75°F) and allow it to warm up, the liquid inside it reacts to the warming effect by boiling slightly and creating vapor. Thus, the pressure gradually increases once again to 217.6 psig, which corresponds to 75°F.

Figure 3.14 (left) The pressure of saturated R-410A at 35°F is 107.3 psig. (center) The pressure of saturated R-410A at 75°F is 217.6 psig. (right) The pressure of saturated R-410A at 100°F is 317.1 psig.

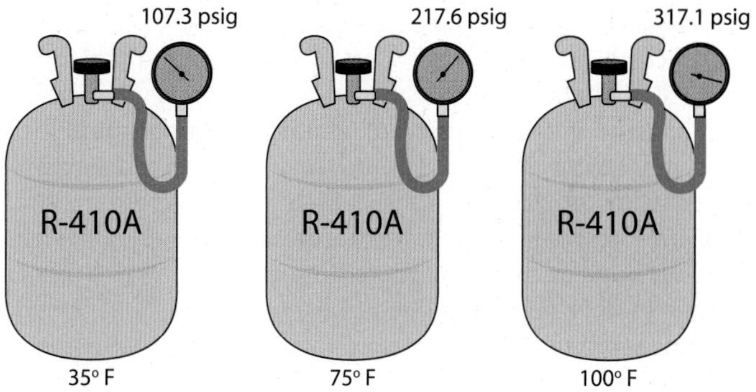

If we move the cylinder (now at 75°F) into a room at 100°F, the liquid again responds to the temperature change by slightly boiling and creating more vapor. As the liquid boils and makes vapor, the pressure steadily increases (according to the temperature/pressure chart) until it corresponds to the liquid temperature. This continues until the refrigerant in the cylinder reaches a pressure of 317.1 psig, which corresponds to a temperature of 100°F, **Figure 3.14**.

The pressure of the vapor in a vessel that pushes against the liquid is referred to as **vapor pressure**. Any time saturated vapor and liquid are together, vapor pressure is generated. Vapor pressure acts equally in all directions, and this action is what a pressure gauge reads on a refrigeration or air-conditioning system. As the temperature of the liquid/vapor mixture increases, vapor pressure increases. As the temperature of the liquid/vapor mixture decreases, vapor pressure decreases.

Further study of the temperature/pressure chart shows that when the pressure is lowered to atmospheric pressure, R-410A boils at about −60°F. This can be seen on the first line of the temperature/pressure chart, where the pressure of R-410A is 0.3 psig. So, if liquid R-410A is accidentally released from a tank, frostbite can result, as the temperature of this liquid refrigerant is extremely low.

3.6 BASIC REFRIGERATION SYSTEM COMPONENTS

The basic vapor-compression refrigeration system is made up of four main components. The four major components are the following:

1. The evaporator
2. The compressor
3. The condenser
4. The refrigerant metering device

3.7 THE EVAPORATOR

The evaporator is a heat transfer surface that absorbs heat into the system. When the refrigerant boils at a temperature that is lower than that of the substance to be cooled, it absorbs heat from the substance. The boiling temperature of 40°F was chosen in the previous air-conditioning examples because it is the design temperature normally used for air-conditioning systems. The reason is that the ideal room temperature is close to 75°F, which readily gives up heat to a 40°F coil. The 40°F temperature is also well above the freezing point of the coil, preventing it from freezing under normal system operating conditions. See **Figure 3.15** for the coil-to-air relationships.

Figure 3.15 The evaporator is operated at 40°F to be able to absorb heat from the 75°F air.

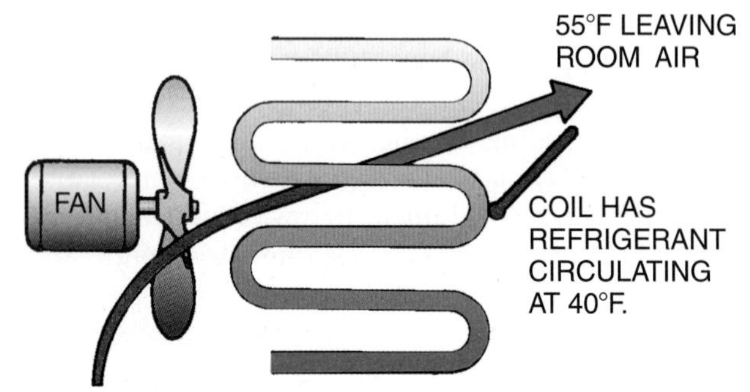

The evaporator can be thought of as a "heat sponge." Just as a dry kitchen sponge absorbs liquid from a spill because the water content of the sponge is low, the evaporator is able to absorb heat because the temperature of the coil is lower than the temperature of the medium being cooled. Continuing our analogy, the sponge will stop absorbing liquid if it becomes completely soaked. This is why we have to squeeze the sponge to remove the absorbed water. Looking back at our evaporator, if we just keep adding heat to the coil, its temperature will rise and the amount of heat it can absorb will drop. So, the heat that is absorbed into the evaporator must be removed later on to allow the system to continue to operate.

Let us see what happens as R-410A refrigerant passes through the evaporator coil, **Figure 3.16.** The refrigerant enters the coil from the bottom as a saturated mixture of about 75% liquid and 25% vapor. Evaporators are usually fed from the bottom to help ensure that no liquid leaves the top without first boiling off to a vapor. If refrigerant entered the evaporator coil from the top, liquid could quickly drop by its own weight to the bottom before it is completely boiled off to a vapor. Liquid refrigerant, if allowed to leave the evaporator, could enter, and ultimately damage, the compressor. In addition, since vapor is less dense than liquid, the vapor has a tendency to rise. Since the saturated refrigerant enters the coil from the bottom and leaves from the top, the direction of refrigerant flow is from the bottom to the top, the same as the direction of the rising vapor. If the refrigerant entered the evaporator from the top and left through the bottom, the direction of flow

Figure 3.16 The evaporator absorbs heat into the refrigeration system by boiling the refrigerant at a temperature that is lower than the temperature of the room air passing over it. The 75°F room air readily gives up heat to the 40°F evaporator coil.

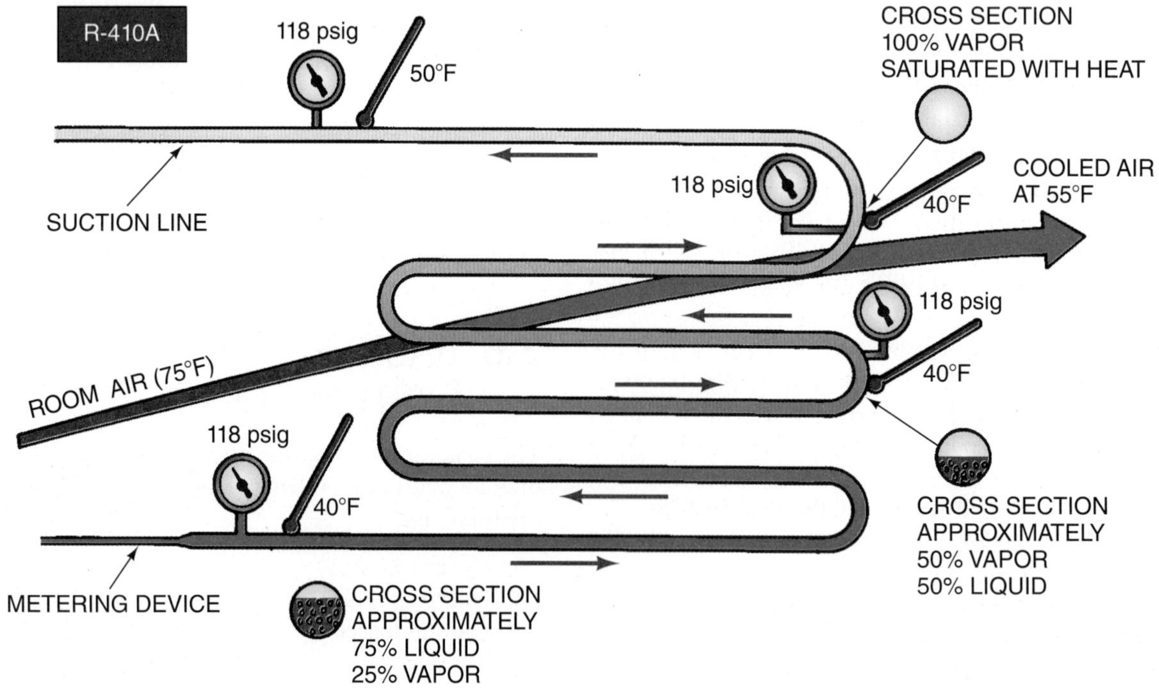

would be opposite to that of the rising vapor, which would cause the refrigerant to slow down.

The refrigerant mixture boils as it flows through the coil, with the liquid being turned to vapor because heat from the air is being added to the refrigerant. As the refrigerant continues to flow through the coil, the percentage of liquid in the mixture decreases, while the percentage of vapor increases. About halfway up the coil, the mixture becomes more vapor than liquid. The purpose of the evaporator is to boil all of the liquid into a vapor just before the end of the coil. This occurs approximately 90% of the way through the coil, leaving pure vapor. At the point where the last droplet of liquid vaporizes, we have what is called a saturated vapor. This is the point where the vapor would start to condense if heat were removed from it or become superheated if any heat were added to it. *When a vapor is superheated, it no longer corresponds to a temperature/pressure relationship. Because no liquid remains to boil off to vapor, no more vapor pressure can be generated when heat is added. The vapor will now take on sensible heat when heated, and its temperature will rise, but the pressure will remain unchanged.* Superheat is considered insurance for the compressor because it ensures that no liquid leaves the evaporator and enters the compressor. When there is superheat, there is no liquid. The amount of superheat in the evaporator is the difference between the temperature of the refrigerant at the outlet of the evaporator and the temperature at which the refrigerant boils. Referring to Figure 3.16, it can be seen that the refrigerant boils at 40°F and leaves the evaporator at a temperature of 50°F. This means that this evaporator is operating with 10° of superheat (50°F – 40°F = 10°F).

To summarize, the three main functions of the evaporator are to

1. absorb heat from the medium being cooled.
2. allow the heat to boil off the liquid refrigerant to a vapor in its tubing bundle.
3. allow the heat to superheat the refrigerant vapor in its tubing bundle.

Evaporators, **Figure 3.17**, have many design configurations. But for now just remember that they absorb the heat into the system from the substance to be cooled. Once absorbed into the system, the heat is now contained in the refrigerant and is drawn into the compressor through the suction line. The suction line simply connects the evaporator outlet to the compressor inlet and provides a path for refrigerant vapor to travel, **Figure 3.18**. The suction line typically passes through unconditioned or unrefrigerated spaces, and the surrounding air temperature is often much higher than the temperature of the refrigerant line. Heat from the surrounding air can, therefore, be readily absorbed into the refrigeration system, making the system work much harder than needed. For this reason, the suction line is insulated, **Figure 3.19**. A well-insulated suction line helps the system operate as efficiently as possible and the insulation prevents the suction line from sweating.

Figure 3.17 A typical refrigeration evaporator. Courtesy Ferris State University. Photo by John Tomczyk

Figure 3.18 The suction line connects the outlet of the evaporator to the inlet of the compressor.

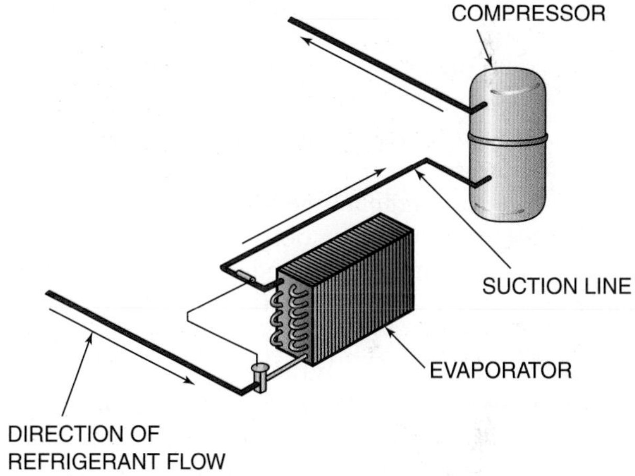

Figure 3.19 Suction lines should be well insulated to increase system efficiency and capacity and to prevent sweating. Courtesy Ferris State University. Photo by John Tomczyk

3.8 THE COMPRESSOR

The compressor is the heart of the refrigeration system. It pumps heat through the system in the form of heat-laden refrigerant. A compressor can be considered a **vapor pump**. This is because, under normal operating conditions, 100% vapor refrigerant enters the compressor and 100% vapor is discharged from the compressor. The compressor reduces the pressure on the low-pressure side of the system, which includes the evaporator, and increases the pressure on the high-pressure side of the system. This pressure difference is what causes the refrigerant to flow. All compressors in air-conditioning and refrigeration systems perform this function by compressing the vapor refrigerant, which can be accomplished in several ways with different types of compressors. The most common compressors used in residential and light commercial air-conditioning and refrigeration applications are the **reciprocating**, the **rotary**, and the **scroll**. These compressors are referred to as positive displacement compressors, since the pressure of the refrigerant is increased by physically reducing the volume of the compression space.

The reciprocating compressor uses a piston in a cylinder to compress the refrigerant, **Figure 3.20**. Valves, usually reed or flapper valves, ensure that the refrigerant flows in the correct direction, **Figure 3.21**. The valves in the reciprocating compressor open and close depending on the pressure difference across them. The suction valve will be open when the pressure in the suction line is greater than the pressure in the compression cylinder and closed when the cylinder pressure is greater than the suction line pressure. The discharge valve will be open when the cylinder pressure is greater than the pressure in the discharge line and closed when the pressure in the discharge line is greater than the pressure in the compression cylinder.

Rotary compressors are typically used for applications in the small equipment range, such as window air conditioners, household refrigerators, and some residential air-conditioning systems. These compressors are desirable for these applications because they are usually physically smaller than reciprocating compressors of the same capacity. Rotary compressors are extremely efficient and have few moving parts, **Figure 3.22**. They use a rotating drum-like piston that squeezes the refrigerant vapor out the discharge port.

The scroll compressor is the most commonly used compressor in newly manufactured residential and light commercial air-conditioning equipment. This compressor uses two nesting scrolls to facilitate the compression process, **Figure 3.23**. One scroll is stationary, while the other is not.

Figure 3.20 The crankshaft converts the circular, rotating motion of the motor to the reciprocating, or back-and-forth, motion of the piston.

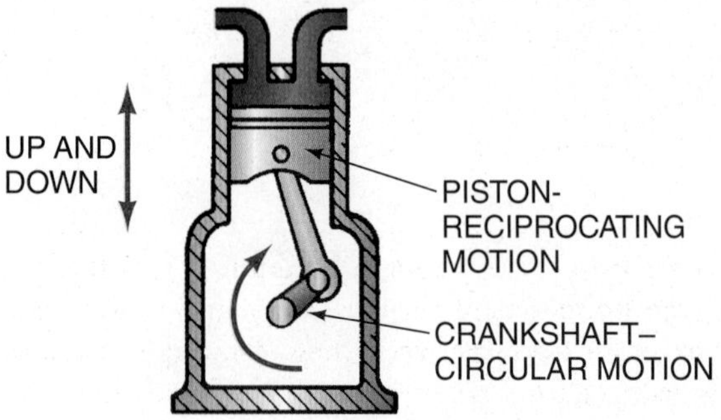

Figure 3.21 Flapper valves and compressor components.

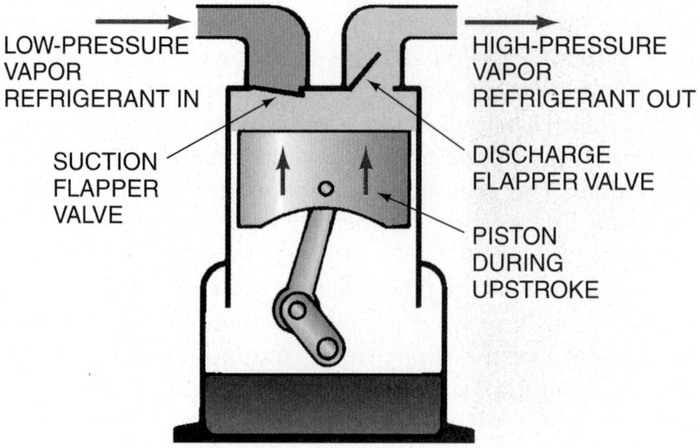

Figure 3.22 A rotary compressor with motion in one direction and no backstroke.

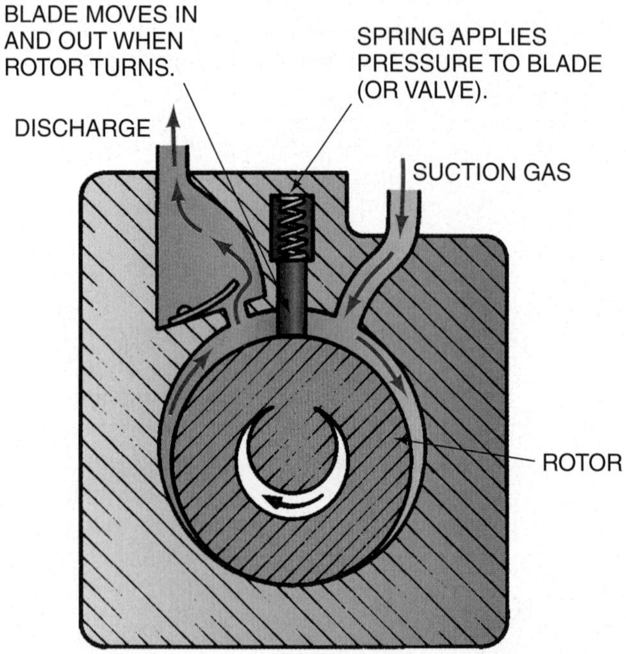

Figure 3.23 Nested scrolls. Photo by Eugene Silberstein

The movable scroll orbits inside the stationary scroll, squeezing the vapor refrigerant from the low-pressure side to the high-pressure side of the system, **Figure 3.24**. Several stages of compression are taking place within the nested scrolls at the same time, making it a very smooth running compressor with few moving parts. The scroll is sealed on the bottom and top by the rubbing action and at the tip with a tip seal. These sealing surfaces prevent refrigerant from the high-pressure side from pushing back to the low-pressure side while the compressor is running. The scrolls are capable of moving apart if unsafe pressures are present in the compressor, allowing high-pressure refrigerant to blow back through to the low-pressure side of the component. This capability makes the scroll compressor more forgiving if a little liquid refrigerant enters the compressor, reducing the possibility of compressor damage.

Large commercial systems use other types of compressors because they must move much more refrigerant vapor through the system. The centrifugal compressor is used in large air-conditioning systems. It is much like a large fan and is not positive displacement, **Figure 3.25**. The centrifugal compressor is referred to as a kinetic displacement compressor. These compressors increase the kinetic energy of the refrigerant vapor with a high-speed fan. The fan exerts centrifugal force and pushes the refrigerant toward the outer portion of the chamber, increasing its pressure. A low-pressure condition is created at the center of the chamber, allowing new suction gas to enter the compressor. This technology is also used in jet engines.

The screw compressor is another positive displacement compressor and is used in larger air-conditioning and refrigeration applications. This compressor is made up of two nesting "screws," **Figure 3.26**. A space between the screws gets smaller and smaller as the refrigerant moves from one end to the other. The vapor refrigerant enters the compressor at the point where the screw spacing is the widest. As the refrigerant flows between these rotating screws, the volume is reduced and the pressure of the vapor increases. The

Figure 3.24 An illustration of the operation of a scroll compressor mechanism.

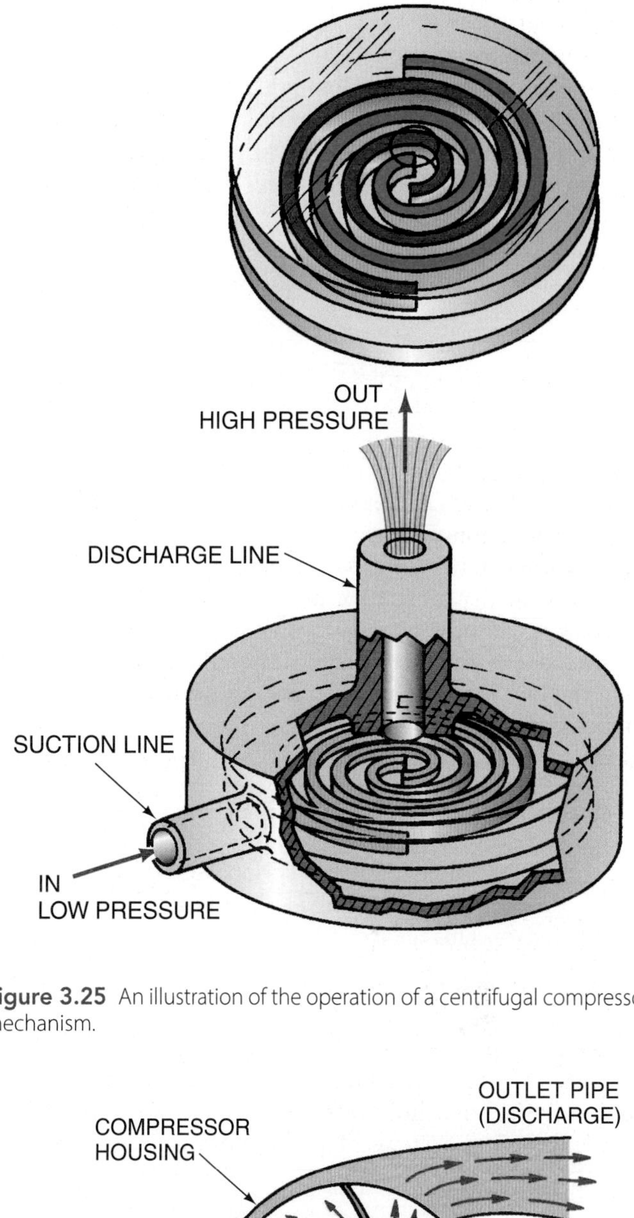

Figure 3.25 An illustration of the operation of a centrifugal compressor mechanism.

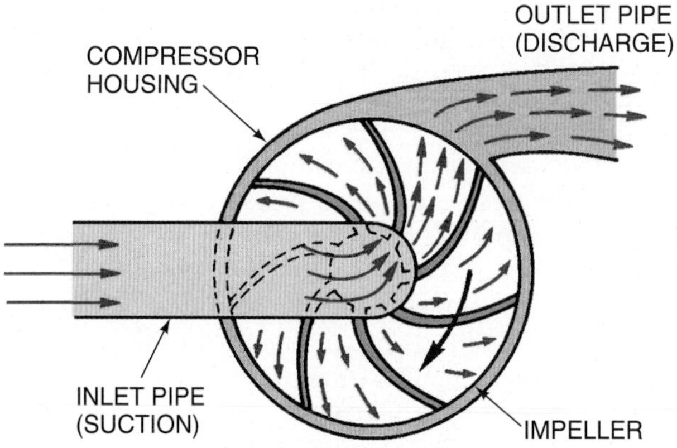

Figure 3.26 An illustration of the internal working mechanism of a screw compressor. One stage of compression as refrigerant moves through screw compressor

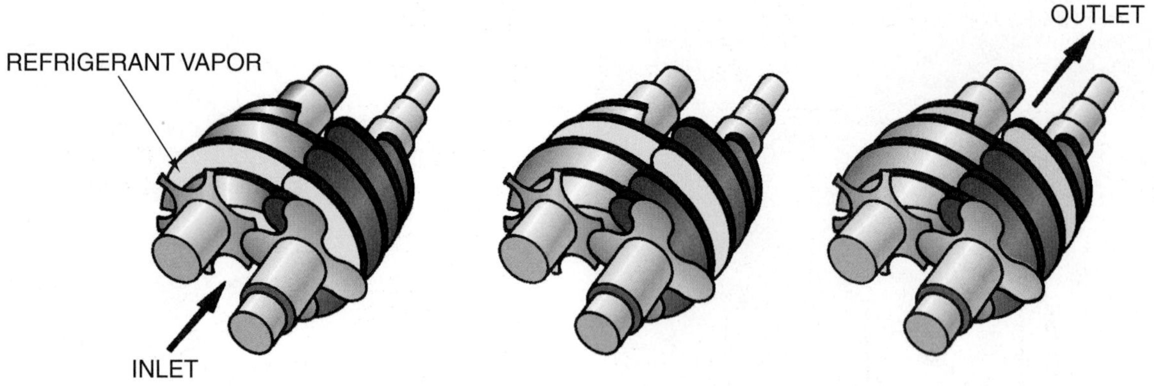

screw compressor is popular for use in low-temperature refrigeration applications.

The important thing to remember is that a compressor performs the same function no matter what type it is. For now it can be thought of as a component that increases the pressure in the system and moves the vapor refrigerant from the low-pressure side to the high-pressure side into the condenser.

3.9 THE CONDENSER

The condenser is a heat transfer surface that rejects both sensible (measurable) and latent (hidden) heat from the refrigeration system. This rejected heat is a combination of the heat absorbed by the evaporator, the heat generated and concentrated in the compressor, and any heat absorbed by the refrigerant in the suction line before it enters the compressor.

The condenser receives the hot discharge gas after it leaves the compressor through the short pipe, called the hot gas line, which connects the outlet of the compressor to the inlet of the condenser, **Figure 3.27**. This line is also referred to as the discharge line. The hot gas is forced into the top of the condenser coil by the compressor. The discharge gas from the compressor is a high-pressure, high-temperature, superheated vapor. The temperature of the hot gas from the compressor can be in the 200°F range and will change depending on the surrounding temperatures and the system application. The refrigerant at the outlet of the compressor *does not* follow a temperature/pressure relationship. This is because the refrigerant is 100% vapor and superheated. The high-side pressure reading of 446 psig in **Figure 3.28** corresponds to the 125°F condenser saturation temperature, which is the temperature at which the refrigerant will begin to condense from a vapor into a liquid. The vapor pressure of 446 psig read by the high-side gauge is actually coming from the condenser. This vapor pressure is often referred to as head pressure, high-side pressure, discharge pressure, or condensing pressure in the HVACR industry. The 200°F gas temperature coming out of the compressor is superheated by 75°F (200°F − 125°F) and, again, cannot have a temperature/pressure relationship. Therefore, the

Figure 3.27 The discharge line connects the outlet of the compressor to the inlet of the condenser. Courtesy Ferris State University. Photo by John Tomczyk

Figure 3.28 Subcooled liquid at the outlet of the condenser.

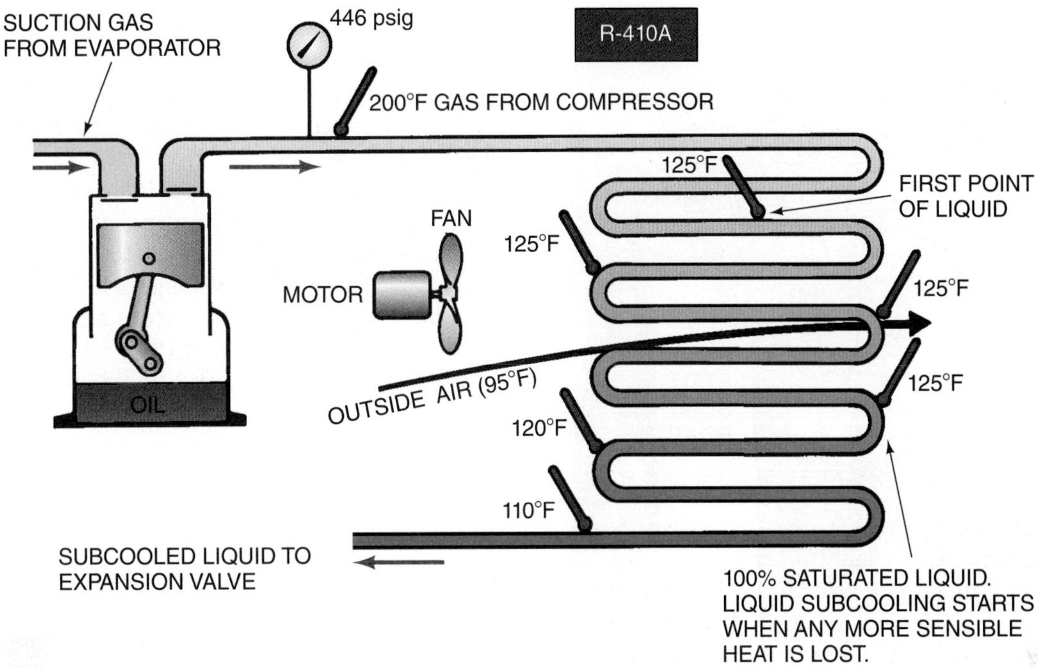

hot vapor must first cool, or desuperheat, 75° before it can begin to condense.

The gas leaving the compressor, flowing through the discharge line, and entering the condenser is so hot compared with the surrounding air that a heat exchange between the hot gas and the surrounding air begins to occur immediately. The air being passed over the condenser is 95°F, compared to the near 200°F gas entering the condenser. As the gas moves through the condenser, it begins to give up sensible heat to the surrounding air, which causes a drop in gas temperature. The gas keeps cooling off, or desuperheating, until it reaches the condensing temperature of 125°F, when the change of state begins to occur. The change of state begins slowly at first with small amounts of vapor changing to liquid and gets faster as the vapor-liquid mixture moves through the condenser. This change from vapor to liquid happens at the condensing saturation temperature and pressure of 125°F and 446 psig, respectively. This is a latent heat process, meaning that even though heat in Btu is being rejected from the refrigerant, the temperature is staying constant at 125°F. **NOTE:** *This process of constant temperature occurring during the change of state does not occur for some of the newer, blended refrigerants that have a temperature glide.* •

When the condensing refrigerant gets about 90% of the way through the condenser, the refrigerant in the pipe becomes pure saturated liquid. If any more heat is removed from the 100% saturated liquid, it will go through a sensible-heat rejection process because no more vapor is left to condense. This causes the liquid to drop below the condensing saturation temperature of 125°F. Liquid cooler than the condensing saturation temperature is called subcooled liquid, **Figure 3.28.**

The amount of subcooling in the condenser is the difference between the temperature of the refrigerant at the outlet of the condenser and the temperature at which the refrigerant condenses. Referring to Figure 3.28, it can be seen that the refrigerant condenses at 125°F and leaves the condenser at a temperature of 110°F. This means that this condenser is operating with 15° of subcooling (125°F – 110°F = 15°F).

Three important things may happen to the refrigerant in the condenser:

1. The hot gas from the compressor is desuperheated from the hot discharge temperature to the condensing temperature. Remember, the condensing temperature determines the head pressure. This is sensible heat transfer.
2. The refrigerant is condensed from a vapor to a liquid. This is latent heat transfer.
3. The liquid refrigerant temperature may then be lowered below the condensing temperature, or subcooled. The refrigerant can usually be subcooled to between 10°F and 20°F below the condensing temperature, but is system-dependent, Figure 3.28. This is a sensible heat transfer.

Many types of condensers are available, but all are designed to reject heat as they desuperheat, condense and subcool refrigerant. Very often, the condenser and the compressor are incorporated into a single piece of equipment called a *condensing unit.* **Figure 3.29** shows some typical condensing units.

Figure 3.29 (A) Typical air-cooled condensing unit found on a split-type central air-conditioning system. (B) An air-cooled hermetic condensing unit from a refrigeration system.

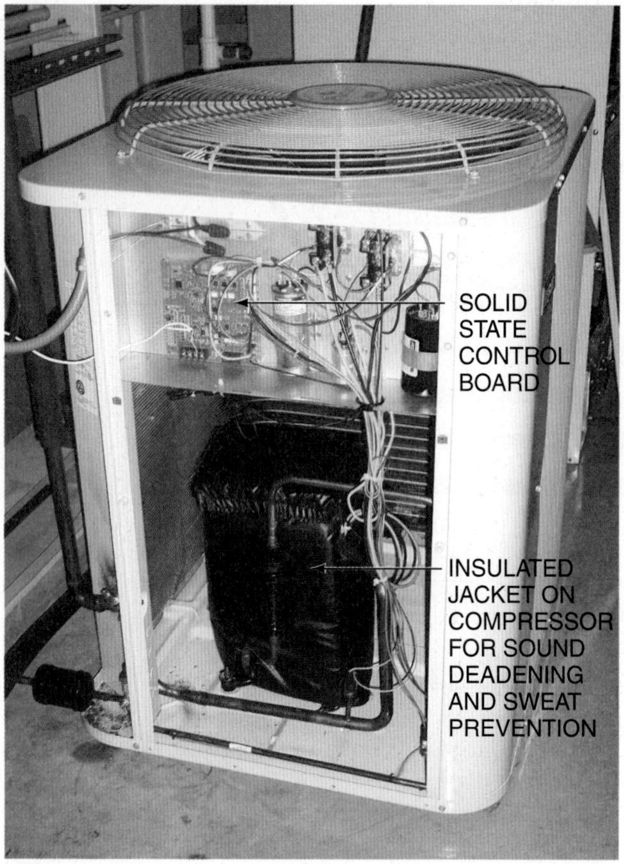

(A)

(B)

Incorporating both of these components into a single unit makes system installation an easier task. The suction line carries low-pressure, low-temperature, superheated vapor to the condensing unit, and the liquid line carries high-pressure, high-temperature subcooled liquid from the condensing unit to the metering device.

3.10 THE REFRIGERANT METERING DEVICE

From the condenser, the warm subcooled liquid flows through the liquid line toward the metering device. The temperature of the liquid refrigerant is about 110°F and may still give up some heat to the surroundings before reaching the metering device. Any heat given off to the surroundings is helpful because it will help improve system capacity and efficiency. It also brings the subcooled liquid temperature closer to the evaporator saturation temperature, which also increases system capacity.

One type of metering device is a simple fixed-size (fixed-bore) type known as an orifice, which is a small restriction of a fixed size in the liquid line, **Figure 3.30**. This device holds back the full flow of refrigerant and is one of the dividing points between the high-pressure and the low-pressure sides of the system. Only pure liquid must enter it. The pipe leading to the orifice may be the size of a pencil, and the precision-drilled hole in the orifice may be the size of a very fine sewing needle. As you can see from the figure, the liquid flow is greatly restricted here. In an R-410A system, the liquid refrigerant entering the orifice is at a pressure of 446 psig; the refrigerant leaving the orifice is a *mixture* of about 75% liquid and 25% vapor at a new pressure of 120.5 psig and a new temperature of 41°F, **Figure 3.30**. Two questions usually arise about this fact:

1. Why did approximately 25% of the liquid change to a gas?
2. How did the mixture of 100% pure liquid go from about 105°F to 41°F in such a short space?

These questions can be answered by using the analogy of a garden hose under pressure, where the water coming out feels cool. The water actually *is* cooler because some of it evaporates and turns to mist. Evaporation removes heat from the rest of the water and cools it down. When the high-pressure subcooled refrigerant passes through the orifice, it does the same thing as the water in the hose: Its pressure changes (446 psig to 120.5 psig) and some of the refrigerant flashes to a vapor (called *flash gas*). At this point, the refrigerant is a saturated mixture of liquid and vapor at 120.5 psig. Since the refrigerant is saturated, it follows the temperature/pressure relationship for that refrigerant—hence, the 41°F refrigerant temperature at the outlet of the metering device. In addition, the evaporating

Figure 3.30 A fixed-bore (orifice) metering device on an R-410A system.

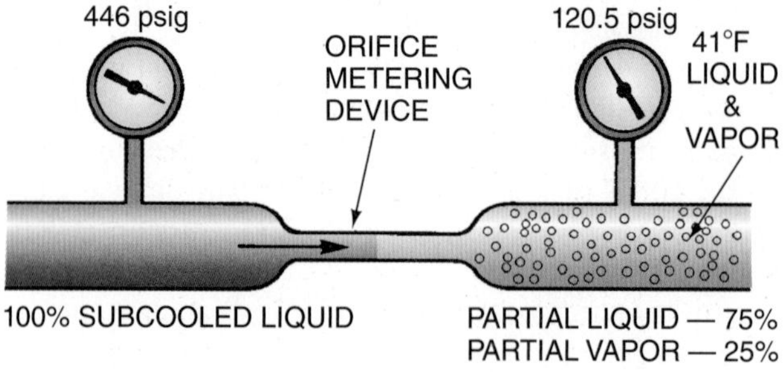

refrigerant further cools the liquid refrigerant that remains as the refrigerant flows through the metering device. This quick drop in pressure in the metering device lowers the boiling point or saturation temperature of the liquid leaving the metering device. Although flash gas helps to cool the liquid refrigerant in the evaporator, too much flash gas can result in a loss in system capacity because less liquid is now available to boil to vapor in the evaporator when cooling the refrigerated space. Because of this, flash gas should be kept to a minimum.

Several types of metering devices for many applications are available. They include the capillary tube, the automatic expansion valve, the thermostatic expansion valve, and the electronic expansion valve.

3.11 A HIGH-TEMPERATURE, MATCHED REFRIGERATION SYSTEM BY THE NUMBERS

We have described the basic components of the mechanical, vapor-compression refrigeration system according to function. These components must be properly matched for each specific application. For instance, a low-temperature compressor cannot be used in a high-temperature application because of the pumping characteristics of the compressor. Some equipment can be mixed and matched successfully by using the manufacturer's data, but only someone with considerable knowledge and experience should do so. What follows is a description of a matched, high-temperature, system correctly working at design conditions. Later we will explain malfunctions and adverse operating conditions.

A typical home air-conditioning system operating at a design temperature of 75°F inside temperature has a relative humidity (moisture content of the conditioned room air) of 50%. The air in the house gives up heat to the refrigerant. The indoor coil is also responsible for removing some of the moisture from the air to keep the humidity at an acceptable level. This is known as **dehumidifying**.

Moisture removal requires considerable energy. Approximately the same amount (970 Btu) of latent heat removal is required to condense a pound of water vapor from the air as to condense a pound of steam. All air-conditioning systems must have a method for dealing with this moisture after it has turned to a liquid. Some units drip, some drain the liquid into plumbing waste drains, some use a slinger ring on the condenser fan, and some use the liquid at the outdoor coil to help the system capacity by evaporating it at the condenser. Remember that part of the system is inside the house and part of the system is outside the house. The numbers in the following description correspond to the circled numbers in **Figure 3.31**.

Figure 3.31 A typical R-410A air-conditioning system showing temperatures and airflow. Red indicates warm/hot refrigerant; blue indicates cool/cold refrigerant.

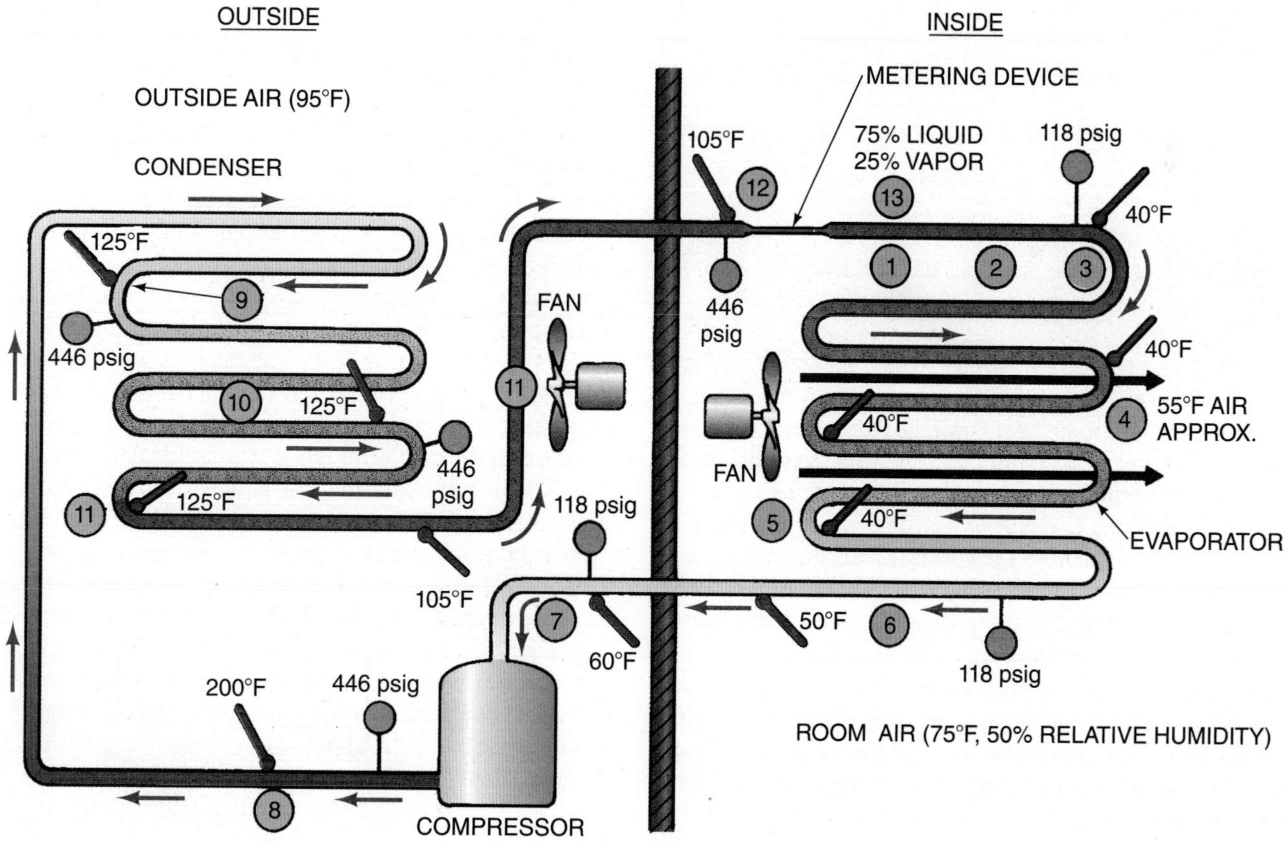

1. A mixture of 75% liquid and 25% vapor leaves the metering device and enters the evaporator.
2. The mixture is saturated R-410A at a pressure of 118 psig, which corresponds to a 40°F boiling point. It is important to remember that *the pressure is 118 psig because the evaporating refrigerant is boiling at 40°F.*
3. The mixture tumbles through the evaporator with the liquid evaporating from the 75°F heat and humidity load of the inside air as it moves along.
4. When the mixture is about halfway through the coil, the refrigerant is composed of about 50% liquid and 50% vapor and is still at the same temperature and pressure because a change of state is taking place. Remember that this is a latent heat transfer.
5. The refrigerant is now 100% vapor. In other words, it has reached the 100% *saturation* point of the vapor.

   Recall the example using the saturated water table; at various points the water was saturated with heat. That is, if any heat is removed at that point, some of the vapor changes back to a liquid; if any heat is added, the temperature of the vapor rises. This rise in temperature of the vapor makes it a *superheated* vapor. (Superheat is sensible heat.) At point 5, the saturated vapor is still at 40°F and still able to absorb more heat from the 75°F room air.

6. The pure vapor that now exists is normally superheated about 10°F above the saturation temperature. Examine the line in **Figure 3.31** at this point to see that the temperature is about 50°F. **NOTE:** *To arrive at the correct superheat reading, take the following steps:*
   A. Note the suction pressure or evaporating pressure reading from the suction, or low-side, gauge: 118 psig.
   B. Convert the suction pressure reading to suction or evaporating temperature using the temperature/pressure chart for R-410A: 40°F.
   C. Use a suitable thermometer to record the actual temperature of the suction line at the outlet of the evaporator: 50°F.
   D. Subtract the saturated suction temperature from the actual suction line temperature: 50°F − 40°F = 10°F of superheat. •

   The vapor is said to be heat laden because it contains the heat removed from the room air. The heat was absorbed into the vaporizing refrigerant that boiled off to vapor as it traveled through the evaporator. The vapor superheats another 10°F and is now 60°F as it travels down the suction line to the compressor. The superheat that is picked up only in the evaporator is referred to as *evaporator superheat*; the total superheat that is picked up in the evaporator and the suction line is referred to as *system superheat*. To calculate the system superheat, we measure the suction line temperature (from item C, above) close to the compressor's inlet instead of at the outlet of the evaporator. In this example, the evaporator superheat is 10°F (50°F − 40°F) and the system superheat is 20°F (60°F − 40°F). System superheat is often referred to as *compressor* superheat.
7. The vapor is drawn into the compressor by its pumping action, which creates a low-pressure suction. When the vapor left the evaporator, its temperature was about 50°F with 10°F of superheat above the saturated boiling temperature of 40°F. As the vapor moves along toward the compressor, it is contained in the suction line, which is usually copper and should be insulated to keep it from drawing heat into the system from the surroundings and to prevent it from sweating. However, it still picks up some heat. Because the suction line carries vapor, any heat that it picks up will quickly raise the temperature. Remember that it does not take much sensible heat to raise the temperature of a vapor. Depending on the length of the line and the quality of the insulation, the suction line temperature may be 60°F at the compressor inlet.
8. Highly superheated gas leaves the compressor through the *hot gas line* on the high-pressure side of the system. This line normally is very short because the condenser is usually close to the compressor. On a hot day the hot gas line may be close to 200°F with a pressure of 446 psig. Because the saturated temperature corresponding to 446 psig is 125°F, the hot gas line has about 75°F (200°F − 125°F) of superheat that must be removed before condensing can occur. Because the line is so hot and a vapor is present, the line will give up heat readily to the surroundings. The surrounding air temperature is 95°F.
9. The superheat has been removed and the refrigerant has cooled down to the 125°F condensing temperature. Point 9 is the point in the system where the desuperheating vapor has just cooled to a temperature of 125°F; this is referred to as a 100% saturated vapor point. If any heat is taken away or rejected, liquid will start to form. As more heat is rejected, the remaining saturated vapor will continue to condense to saturated liquid at the condensing temperature of 125°F. Now notice that the coil temperature is corresponding to the high-side pressure of 446 psig and 125°F. The high-pressure reading of 278 psig is due to the refrigerant condensing at 125°F. In fact, 446 psig is the *vapor pressure* that the vapor is exerting on the liquid while it is condensing. Remember, it is vapor pressure that the gauge reads.
10. The conditions for condensing are determined by the efficiency of the condenser. In this example we use a standard condenser, which has a condensing temperature about 30°F higher than the surrounding air used to absorb heat from the condenser. In this example, 95°F outside air is absorbing the heat, so 95°F + 30°F = 125°F condensing temperature. Some condensers will condense at 25°F above the surrounding air temperature; these are high-efficiency condensers and the high-pressure side of the system will be operating at a lower

pressure. Condensing temperatures and pressures are also dependent on the heat load given to the condenser to reject. The higher the heat load, the higher will be the condensing temperature and the corresponding pressure. It can also be concluded that as the outside temperature rises, the operating pressure on the high side of the system will also rise. For example, if the outside ambient (surrounding) temperature rises to 105°F, the condenser saturation temperature on our sample system would rise to about 135°F.

11. The refrigerant is now 100% liquid at the saturated temperature of 125°F. As the liquid continues along the coil, the air continues to cool the liquid to below the actual condensing temperature. The liquid may go as much as 20°F below the condensing temperature of 125°F before it reaches the metering device. Any liquid at a temperature below the condensing temperature of 125°F is called *subcooled* liquid. In this example, the liquid is cooled to 105°F before it reaches the metering device. The liquid now has 20°F (125°F − 105°F) of subcooling.
12. The liquid refrigerant reaches the metering device through a pipe, usually copper, from the condenser. This liquid line is often field installed and not insulated. Since the temperature of the liquid line is warmer than the temperature of the surrounding air, keeping the liquid line uninsulated allows some additional heat to be rejected by the refrigerant into the surrounding air, which helps increase the operating efficiency of the system. Since the liquid line may be long, and depending on the distance between the condenser and the metering device, the amount of additional heat being rejected may be significant. Heat given up here is leaving the system, and that is good. The refrigerant entering the metering device may be as much as 20°F cooler than the condensing temperature of 125°F, so the liquid line entering the metering device may be 105°F.
13. The refrigerant entering the metering device is 100% subcooled liquid. In the short length of the metering device's orifice (a pinhole about the size of a small sewing needle), the subcooled liquid is changed to a mixture of about 75% saturated liquid and 25% saturated vapor. The ratio of liquid to vapor leaving the metering device depends on both the system and the application. The 25% vapor, known as flash gas, is used to cool the remaining 75% of the liquid down to 40°F, the boiling temperature of the evaporator. The cooling performed by flash gas is a system loss because it cools the liquid temperature down to the 40°F evaporating temperature. The cooling is therefore wasted because it is not performed in the evaporator to lower the temperature of the inside air and remove humidity. The only way to minimize flash gas is to get the temperature of the subcooled liquid entering the metering device closer to the evaporating temperature.

The refrigerant has now completed one full cycle. It should be evident that a refrigerant does the same thing over and over, changing from a liquid to a vapor in the evaporator and back to a liquid in the condenser. The expansion device meters the flow to the evaporator, and the compressor pumps the refrigerant out of the evaporator.

The following list briefly summarizes the refrigeration cycle:

1. The evaporator absorbs heat into the system.
2. The compressor pumps the heat-laden vapor.
3. The condenser rejects heat from the system.
4. The expansion device meters the flow of refrigerant.

3.12 REFRIGERANTS

Previously, we have used water and R-410A as examples of refrigerants. Although many products have the characteristics of a refrigerant, we will cover only a few here. Unit 9, "Refrigerant and Oil Chemistry and Management—Recovery, Recycling, Reclaiming, and Retrofitting," provides more detailed information.

The following four refrigerants either can no longer be manufactured:

R-12—Used primarily in medium- and high-temperature refrigeration applications. Manufacturing and importing of this refrigerant was banned as of January 1, 1996.

R-22—Used primarily in residential, commercial, and industrial air-conditioning applications and in some commercial and industrial refrigeration. R-22 was phased out in new equipment in 2010, and total production ended in 2020.

R-500—Used primarily in older air-conditioning applications and some commercial refrigeration. Manufacturing and importing of this refrigerant was banned as of January 1, 1996.

R-502—Used primarily in low-temperature refrigeration applications. Manufacturing and importing of this refrigerant was banned as of January 1, 1996. It is an azeotropic refrigerant blend that has negligible temperature glide and behaves like a pure compound.

The following are some of the more popular refrigerants, even though some of them may be replaced in the near future with substances that have lower global warming potentials (GWPs). Note: At the time of this writing, the Environmental Protection Agency (EPA) has proposed rules for refrigerant substitutes. The rules involve a change in the listing status for certain refrigerant substitutes under EPA Section 612, the Significant New Alternatives Policy Program (SNAP). Some of the refrigerants and refrigerant blends listed below may be affected when this proposed rule takes effect. Many refrigerants and refrigerant blends that have high GWPs will no longer be allowed to be used even as retrofit refrigerants.

These refrigerants and blends will still be listed as follows because they exist in HVACR equipment from past installations and retrofits. It is important that service technicians understand what these refrigerants and blends consist of for service purposes. It is the service technician's responsibility to keep abreast with the most recent laws and rules concerning refrigerants use and service. Be sure to visit the EPA website, www.epa.gov, for the most up-to-date information regarding refrigerant-related regulations.

R-134a—Properties very similar to R-12. Used primarily in medium- and high-temperature refrigeration applications, refrigerators and freezers, and automotive air-conditioning. It is a replacement for R-12 but not a direct one because retrofitting is required. The ester-based lubricants used with R-134a are not compatible with the oils typically used in R-12 systems. R-134a does not deplete stratospheric ozone, but does contribute to global warming.

R-402A—A near-azeotropic interim HCFC-based refrigerant retrofit blend of R-125, R-290, and R-22 for low- and medium-temperature commercial refrigeration applications. Replaces R-502.

R-404A—R-404A has slightly higher working pressures than R-502. A near-azeotropic refrigerant blend with a small temperature glide. Eliminated from new supermarket refrigeration systems, remote condensing units, and some stand-alone equipment because of its high GWP. Some new lower-GWP refrigerants, such as R-448A, R-449A, R-450A, and R-513A, all have less than half the GWP of the candidates they are replacing.

R-407A—A retrofit refrigerant blend for R-22 in medium- and low-temperature commercial refrigeration applications. It is not recommended for air-conditioning applications.

R-407C—Similar properties to R-22. It is replacing R-22 in residential and commercial air-conditioning applications. It can be used as a retrofit refrigerant for R-22. A near-azeotropic refrigerant blend, R-407C, like R-134a, operates with ester-based lubricants that are not compatible with the oils typically used in R-22 systems. R-407C is a great choice when the condensing unit of a system must be replaced, as most of the refrigeration oil is contained in the compressor. R-407C can also be used in-medium temperature refrigeration systems.

R-408A—A near-azeotropic interim HCFC-based refrigerant retrofit blend of R-125, R-143a, and R-22 for low- and medium-temperature commercial refrigeration applications. Replaces R-502.

R-410A—A near-azeotropic refrigerant blend replacing R-22 in residential and commercial air-conditioning applications. It has much higher operating pressures than R-22 and special safety concerns. R-410A has a very small temperature glide and is not recommended as a retrofit refrigerant. Although R-410A does not deplete stratospheric ozone, it is classified as a super-polluter given its high GWP. **⚠SAFETY PRECAUTION:** *Never add R-410A to a system that was manufactured for use with R-22. R-22 system components are typically not manufactured to accept the higher operating pressures that are present in R-410A systems. For example, the shells of R-410A compressors are manufactured with thicker steel than are the shells of R-22 compressors.•*

R-417A—A near-azeotropic retrofit refrigerant blend for R-22 direct expansion air-conditioning systems. It is best suited to replace R-22 in smaller split air-conditioning systems. It is an efficient and cost effective, non-ozone depleting HFC-based refrigerant blend. It is compatible with mineral oil, alkylbenzene, and polyol ester lubricants. It has lower discharge temperatures and pressures than R-22.

R-438A—A near-azeotropic refrigerant blend for use as a retrofit refrigerant for R-22 with minimum system change. R-438A is a blend of five different refrigerants. It can be used for retrofitting R-22 direct expansion air-conditioning (high-temperature) applications and medium- and low-temperature refrigeration applications. It has a small percentage of hydrocarbons in its blend to help in oil return. This makes it compatible with mineral oils. R-438A is nonflammable and has an ASHRAE safety group classification of A1. Its ozone depletion potential (ODP) is zero. The evaporator and condenser temperature glide is typically in the 6°F to7°F temperature range. It is not recommended for flooded evaporators or systems with centrifugal compressors.

R-422A—A near-azeotropic retrofit refrigerant blend for R-502 and R-502 service blends in commercial refrigeration applications. Consists of R-125, R-134a, and R-600a.

R-422B—A near-azeotropic retrofit refrigerant blend for R-22 in air-conditioning.

R-422C—A near-azeotropic retrofit refrigerant blend for R-502 and R-502 service blends in commercial refrigeration applications.

R-422D—A near-azeotropic refrigerant retrofit blend for R-22 in commercial refrigeration applications.

R-427A—A near-azeotropic refrigerant retrofit blend for R-22 in air-conditioning and commercial refrigeration, but preferred for commercial refrigeration applications.

R-507—A replacement for R-502 in low- and medium-temperature refrigeration applications. R-507 is an azeotropic mixture of R-125 and R-143a. Has slightly higher pressures and capacity than R-404A.

R-507A—An azeotropic mixture of R-125 and R-143a. It has a higher efficiency than R-404A and also a

higher efficiency than low-temperature R-22. Polyol ester lubricant must be used with R-507A. Eliminated from new supermarket refrigeration systems, remote condensing units, and some stand-alone equipment because of its high GWP.

HFO-1234yf—A low GWP replacement for R-134a intended for use in mobile air-conditioning (MAC) systems in the automotive industry. It is currently being used in this application today in some MAC systems. HFO-1234yf does not deplete stratospheric ozone.

HFO-1234-ze—Intended to replace R-410A in residential and light commercial air-conditioning and heat pump applications while offering a 75% reduction in GWP. It also offers excellent energy efficiency, is cost effective, and can be used in existing equipment design with minimal changes. HFO-1234ze does not deplete stratospheric ozone. HFO-1234ze also has significant advantages over R-32, which has medium GWPs and high discharge temperatures. High discharge temperatures can affect system performance and durability in hot climates. Also, oil changes may be needed with R-32.

Hydrofluoro-olefin (HFO) refrigerants are widely recognized as the next generation of refrigerants because of their environmental friendliness, cost-effectiveness, and great energy efficiencies. Refer to Unit 9, "Refrigerant and Oil Chemistry and Management," for more information on hydrofluoro-olefin (HFO) and hydrocarbon (HC) refrigerants. 4R *As we will see later in this unit, the choice of refrigerant is becoming more important because of environmental issues. Some refrigerants, when released to the atmosphere, are causing damage to the ozone layer in the stratosphere, 7 to 30 miles above earth's surface. Refrigerants are also being blamed for global warming effects.* 4R

3.13 REFRIGERANTS MUST BE SAFE

To protect people from sickness or injury, even death, refrigerants must be safe in the event they escape from a system. For instance, using ammonia for the air-conditioning system in a public place could be a disaster even though ammonia is an efficient refrigerant. Modern refrigerants are nontoxic, and equipment is designed to use a minimum amount of refrigerant to accomplish its job. A window air conditioner, for example, normally contains less than 2 lb of refrigerant, while household refrigerators typically contain only a few ounces of refrigerant. **SAFETY PRECAUTION:** *Because refrigerants are heavier than air, proper ventilation is important. Refrigerants should not be exposed to open flames. When refrigeration equipment or cylinders are located in a room with an open gas flame, such as a pilot light on a gas water heater or furnace, the equipment must be kept leak free. If the refrigerant escapes and gets to the flame, toxic and corrosive gases can be created. These gases can burn the eyes and nose, and severely hamper the breathing of anyone in the room.•*

3.14 REFRIGERANTS MUST BE DETECTABLE

Refrigerants must be readily detectable. For detecting some large leaks, the first thing to do is to listen for the hiss of escaping refrigerant. In some cases, some leaks may be so small they cannot be heard by the human ear. However, many leaks can be found by listening. There is an ultrasonic leak-detecting device on the market that detects leaks by sound, **Figure 3.32.** This leak detector enables the technician to hear the sound of the fluid moving through the piping circuit toward the location of the leak. The pitch of the emitted sound changes as the sensor is brought closer and closer to the leak.

Soap bubbles are a practical and yet simple leak detector. Many service technicians use commercially prepared products that will blow large elastic-type bubbles at the point of the leak, **Figure 3.33.** These products are valuable when the technician knows a leak is in a certain area. Soap-bubble solution can be applied with a brush to the tubing joint to see exactly where the leak is. When refrigerant lines are below freezing temperature, a small amount of antifreeze can be added to the bubble solution. At times, a piece of equipment can be submerged in water to watch for bubbles, which is effective when applicable.

A halide leak detector, **Figure 3.34,** uses an acetylene or propane fueled flame to heat a copper disk, which is then

Figure 3.32 Ultrasonic leak detector. Photo by Bill Johnson

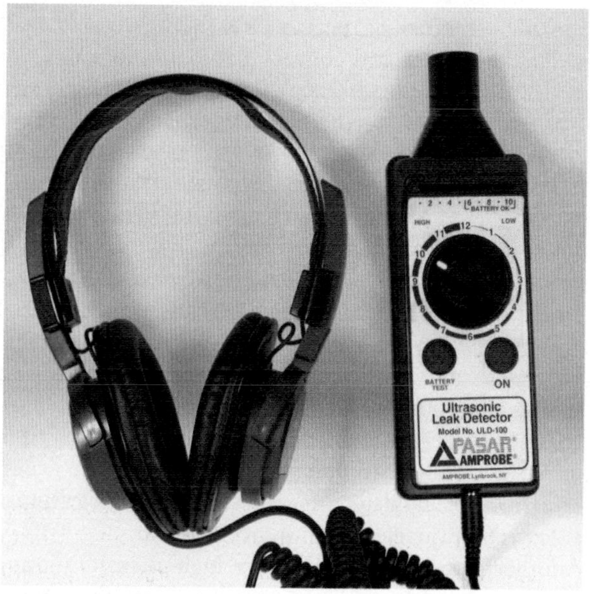

Figure 3.33 Soap-bubble solution.

Figure 3.34 Halide torch. Courtesy of Uniweld Products

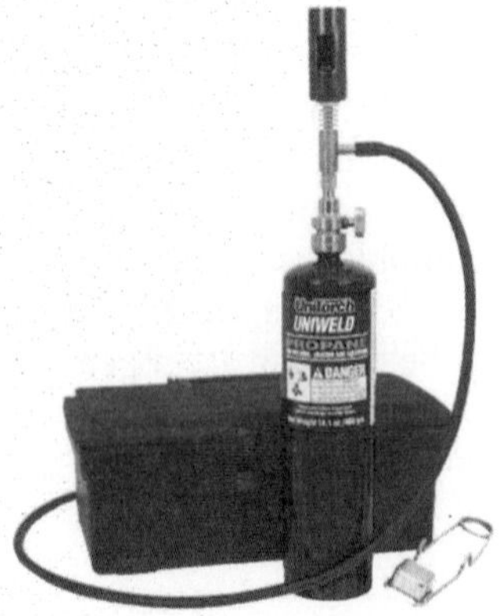

Figure 3.35 Electronic leak detector. Courtesy Ferris State University. Photo by John Tomczyk

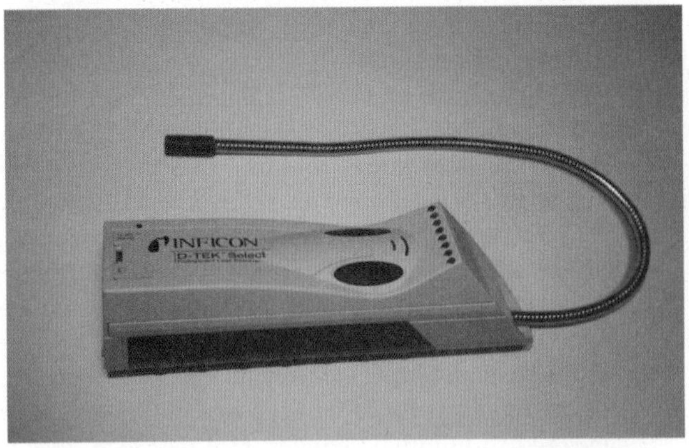

Figure 3.36 Ultraviolet leak detection equipment. Courtesy of Spectronics Corporation, Westbury, NY

used to detect chlorinated refrigerant gas leaks. It operates on the principle that when the refrigerant is exposed to an open flame in the presence of glowing copper, the flame will change color. The halide leak detector should be used in well-ventilated areas, because the resulting blue or green flame indicates that toxic gas is being produced. Although the amounts produced can be very small, the gas is an irritant—corrosive as well as toxic.

The electronic leak detector in **Figure 3.35** is battery operated, is small enough to be easily carried, and has a flexible probe. The refrigerant charge specifications for some residential air-conditioning equipment call for half-ounce accuracy. The electronic leak detectors are capable of detecting leak rates down to a quarter of an ounce per year.

Another detection system uses a high-intensity ultraviolet lamp, **Figure 3.36.** An additive introduced into the refrigerant system will show as a bright yellow-green glow under the ultraviolet lamp at the source of the leak. After repairing the leak and wiping the area clean with a general-purpose cleaner, the area can be reinspected—the additive remains in the system. Should a new leak be suspected at a later date, it will still show as a yellow-green color under the ultraviolet light. This system will detect leaks as small as a quarter of an ounce per year.

3.15 THE BOILING POINT OF THE REFRIGERANT

The boiling point of a refrigerant should be low at atmospheric pressure so that low temperatures may be obtained without going into a vacuum. For example, R-410A can be

boiled as low as −61°F before the boiling pressure goes into a vacuum, whereas R-12 can be boiled only as low as −21°F before it goes into a vacuum. Water would have to be boiled at 29.67 in. Hg vacuum just to boil at 40°F. **NOTE:** *When using the compound gauge below atmospheric pressure, the number scale is different from the one that is used above atmospheric pressure. Above atmospheric pressure, the units of pressure are psig, while below atmospheric pressure, the units of pressure (vacuum) are inches of mercury. In the vacuum region, the gauge starts at atmospheric pressure (0 psig) and continues down to a perfect vacuum (29.92 in. Hg). When possible, design engineers avoid using refrigerants that must be in a vacuum to boil at the desired temperature. When a system operates in a vacuum and a leak occurs, the atmosphere is pulled inside the system instead of the refrigerant leaking out of the system.*•

3.16 PUMPING CHARACTERISTICS

Pumping characteristics have to do with how much refrigerant vapor is pumped per amount of work accomplished. Water was disqualified as a practical refrigerant for small equipment partly for this reason. One pound of water at 40°F has a vapor, or specific, volume of 2445 ft^3 compared to about 0.5 ft^3 for R-410A. This means that the system capacity is directly related to the number of pounds of refrigerant that are circulated through the system per unit of time. This is expressed in units called pounds per minute, or lb/min. Since the vapor volume of 40°F water is so high, the compressor would have to move 2445 ft^3 of vapor to move a single pound. Thus, the compressor would have to be very large for a water system. Modern refrigerants meet all of these requirements better than any of the older types.

3.17 POPULAR REFRIGERANTS AND THEIR IMPORTANT CHARACTERISTICS

The American National Standards Institute (ANSI) and the American Society of Heating, Refrigerating and Air-Conditioning Engineers (ASHRAE) are responsible for naming refrigerants and identifying their characteristics. **Figure 3.37** is an organized list of some of the more popular refrigerants and their characteristics. For more detailed information on refrigerants, refer to Unit 9, "Refrigerant and Oil Chemistry and Management—Recovery, Recycling, Reclaiming, and Retrofitting." 4R *As mentioned earlier, environmental issues like ozone depletion and global warming have forced manufacturing phaseout dates for many refrigerants. However, these refrigerants can still be used if recovered or recycled, or if they are in an operating refrigeration or air-conditioning system. Environmental issues and phaseout dates have made many refrigerants very expensive due to the heavy taxes imposed, and so alternative (environmentally friendly) refrigerants have entered the market. It is illegal to intentionally vent any refrigerant into the atmosphere. Stiff fines of up to $44,539 and/or imprisonment can follow. Because of this, mandatory technician certification programs have educated HVACR personnel on environmental issues, alternative refrigerants, and legislation issues. Visit the EPA's website for the most up-to-date information regarding refrigerant-related regulations.* 4R

3.18 REFRIGERANT CYLINDER COLOR CODES

Each type of refrigerant is contained in a cylinder or drum that has a designated color. Following are the colors for some of the most frequently used refrigerants:

R-407B	Cream	R-114	Dark blue
R-407C	Chocolate	R-500	Yellow
R-410A	Rose	R-502	Orchid
R-11	Orange	R-717	Silver
R-12	White	R-409A	Tan
R-22	Green	R-123	Light gray-blue
R-113	Purple	R-401A	Coral red
R-134a	Light blue	R-401B	Mustard yellow
R-401C	Aqua	R-404A	Orange
R-402A	Light brown	R-406A	Light gray-green
R-402B	Green-brown	R-407A	Bright green

Some equipment manufacturers mark their compressors with color-coded tags and labels to indicate the type of refrigerant used in the system. **Figure 3.38** shows refrigerant containers for some newer refrigerants.

3.19 RECOVERY, RECYCLING, AND RECLAIMING OF REFRIGERANTS

4R *It is mandatory for technicians to recover and sometimes recycle refrigerants during installation and servicing operations to help reduce emissions of chlorofluorocarbons (CFCs), hydrochlorofluorocarbons (HCFCs), and hydrofluorocarbons (HFCs) to the atmosphere. Examples of recovery equipment are shown in* ***Figure 3.39****. Many larger systems can be fitted with receivers or dump tanks into which the refrigerant can be pumped and stored while the system is serviced. However, in smaller-capacity systems it is not often feasible to provide these components. Recovery units or other storage devices may be necessary. Most recovery and/or recycling units vary in technology and capabilities, so manufacturers' instructions must be followed carefully when using this equipment. Unit 9, "Refrigerant and Oil Chemistry and Management—Recovery, Recycling, Reclaiming, and Retrofitting," includes a detailed description of the recovery, recycling, and reclaiming of refrigerants.* 4R

Figure 3.37 A list of refrigerants and some of their characteristics. ***Note:** *Safety classifications are covered in Unit 4, "General Safety Practices."* •

ANSI/ASHRAE DESIGNATION	SAFETY* CLASSIFICATION	EMPIRICAL FORMULA	MOLECULAR FORMULA	COMPONENTS/WEIGHT PERCENTAGES CHEMICAL NAME		CYLINDER COLOR
R-11	A1	CFC	CCl_3F	Trichlorofluoromethane		Orange
R-12	A1	CFC	CCl_2F_2	Dichlorodifluoromethane		White
R-13	A1	CFC	$CClF_3$	Chlorotrifluoromethane		Light Blue
R-14	A1	PFC	CF_4	Tetrafluoromethane		Mustard
R-22	A1	HCFC	$CHClF_2$	Chlorodifluoromethane		Light Green
R-23	A1	HFC	CHF_3	Trifluoromethane		Light Gray-Blue
R-32	A2	HFC	CH_2F_2	Difluoromethane		White/Red Stripe
R-113	A1	CFC	$CCl_2F\text{-}CClF_2$	1,1, 2-Trichloro-1, 2, 2-trifluoroethane		Dark Purple (Violet)
R-114	A1	CFC	$CClF_2\text{-}CClF_2$	1, 2-Dichloro-1,1, 2, 2-tetrafluoroethane		Dark Blue (Navy)
R-115	A1	CFC	$CClF_2\text{-}CF_3$	Chloropentafluoroethane		White/Red Stripe
R-116	A1	PFC	$CF_3\text{-}CF_3$	Hexafluoroethane		Dark Gray (Battleship)
R-123	B1	HCFC	$CHCl_2\text{-}CF_3$	2, 2-Dichloro-1,1,1-trifluoroethane		Light Gray-Blue
R-124	A1	HCFC	$CHClF\text{-}CF_3$	2-Chloro-1,1,1, 2-tetrafluoroethane		Dark Green
R-125	A1	HFC	$CHF_2\text{-}CF_3$	Pentafluoroethane		Medium Brown (Tan)
R-134a	A1	HFC	$CH_2F\text{-}CF_3$	1,1,1, 2-Tetrafluoroethane		Light Sky Blue
R-143a	A2	HFC	$CH_3\text{-}CF_3$	1,1,1-Trifluoroethane		White/Red Stripe
R-152a	A2	HFC	$CH_3\text{-}CHF_2$	1,1-Difluoroethane		White/Red Stripe
R-290	A3	HC	$CH_3\text{-}CH_2\text{-}CH_3$	Propane		Red
R-500	A1	CFC	$CCl_2F_2/CH_3\text{-}CHF_2$	R-12/R-152a	73.8/26.2	Yellow
R-502	A1	CFC	$CHClF_2/CClF_2\text{-}CF_3$	R-22/R-115	48.8/51.2	Light Purple (Lavender)
R-503	A1	CFC	$CHF_3/CClF_3$	R-23/R-13	40.1/59.9	Blue-Green (Aqua)
R-507	A1	HFC	$CHF_2\text{-}CF_3/CH_3\text{-}CF_3$	R-125/R-143a	50/50	Blue-Green (Teal)
R-717	B2		NH_3	Ammonia		Silver
R-401A	A1	HCFC	$CHClF_2/CH_3\text{-}CHF_2/CHClF\text{-}CF_3$	R-22/R-152a/R-124	53/13/34	Coral Red
R-401B	A1	HCFC	$CHClF_2/CH_3\text{-}CHF_2/CHClF\text{-}CF_3$	R-22/R-152a/R-124	61/11/28	Yellow-Brown (Mustard)
R-401C	A1	HCFC	$CHClF_2/CH_3\text{-}CHF_2/CHClF\text{-}CF_3$	R-22/R-152a/R-124	33/15/52	Blue-Green (Aqua)
R-402A	A1	HCFC	$CHF_2\text{-}CF_3/CH_3\text{-}CH_2\text{-}CH_3/CHClF_2$	R-125/R-290/R-22	60/02/38	Light Brown (Sand)
R-402B	A1	HCFC	$CHF_2\text{-}CF_3/CH_3\text{-}CH_2\text{-}CH_3/CHClF_2$	R-125/R-290/R-22	38/02/60	Green-Brown (Olive)
R-403A	A1	HCFC	$CH_3\text{-}CH_2\text{-}CH_3/CHClF_2/CF_3\text{-}CF_2\text{-}CF_3$	R-290/R-22/R-218	05/75/20	Light Purple
R-404A	A1	HFC	$CHF_2\text{-}CF_3/CH_3\text{-}CF_3/CH_2F\text{-}CF_3$	R-125/R-143a/R-134a	44/52/04	Orange
R-406A	A2	HCFC	$CHClF_2/CH(CH_3)_3/CH_3\text{-}CClF_2$	R-22/R-600a/R-142b	55/04/41	Light Gray-Green
R-407A	A1	HFC	$CH_2F_2/CHF_2\text{-}CF_3/CH_2F\text{-}CF_3$	R-32/R-125/R-134a	20/40/40	Bright Green
R-407B	A1	HFC	$CH_2F_2/CHF_2\text{-}CF_3/CH_2F\text{-}CF_3$	R-32/R-125/R-134a	10/70/20	Cream
R-407C	A1	HFC	$CH_2F_2/CHF_2\text{-}CF_3/CH_2F\text{-}CF_3$	R-32/R-125/R-134a	23/25/52	Medium Brown
R-408A	A1	HCFC	$CHF_2\text{-}CF_3/CH_3\text{-}CF_3/CHClF_2$	R-125/R-143a/R-22	07/46/47	Medium Purple
R-409A	A1	HCFC	$CHClF_2/CHClF\text{-}CF_3/CH_3\text{-}CClF_2$	R-22/R-124/R-142b	60/25/15	Mustard Brown (Tan)
R-410A	A1	HFC	$CH_2F_2/CHF_2\text{-}CF_3$	R-32/R-125	50/50	Rose

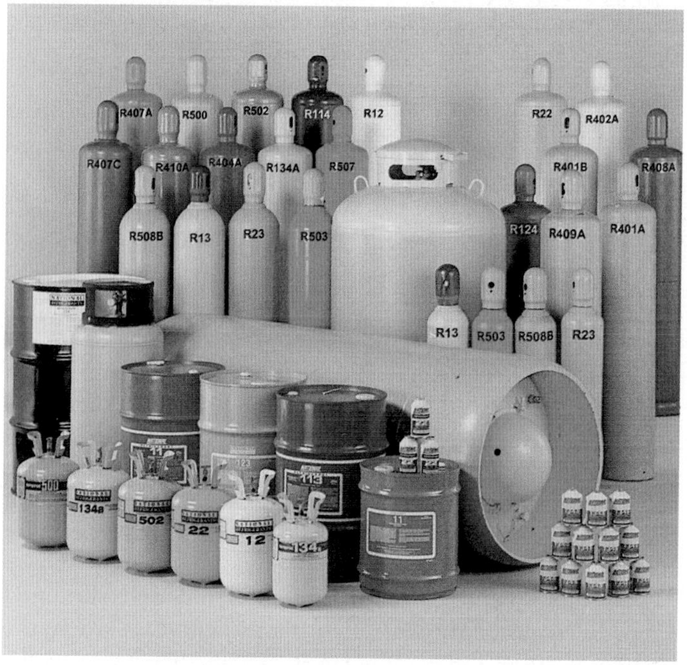

Figure 3.38 Color-coded refrigerant cylinders and drums for some of the newer refrigerants. Courtesy National Refrigerants, Inc.

Figure 3.39 Recovery units. Photo by Eugene Silberstein

3.20 PLOTTING THE REFRIGERATION CYCLE

A graphic picture of the refrigerant cycle may be plotted on a pressure/enthalpy diagram. Pressure is plotted on the left-hand side of the diagram and enthalpy on the bottom of the diagram, **Figure 3.40**. **Enthalpy** describes how much heat a substance contains with respect to an accepted reference point. Quite often, people refer to enthalpy as total heat,

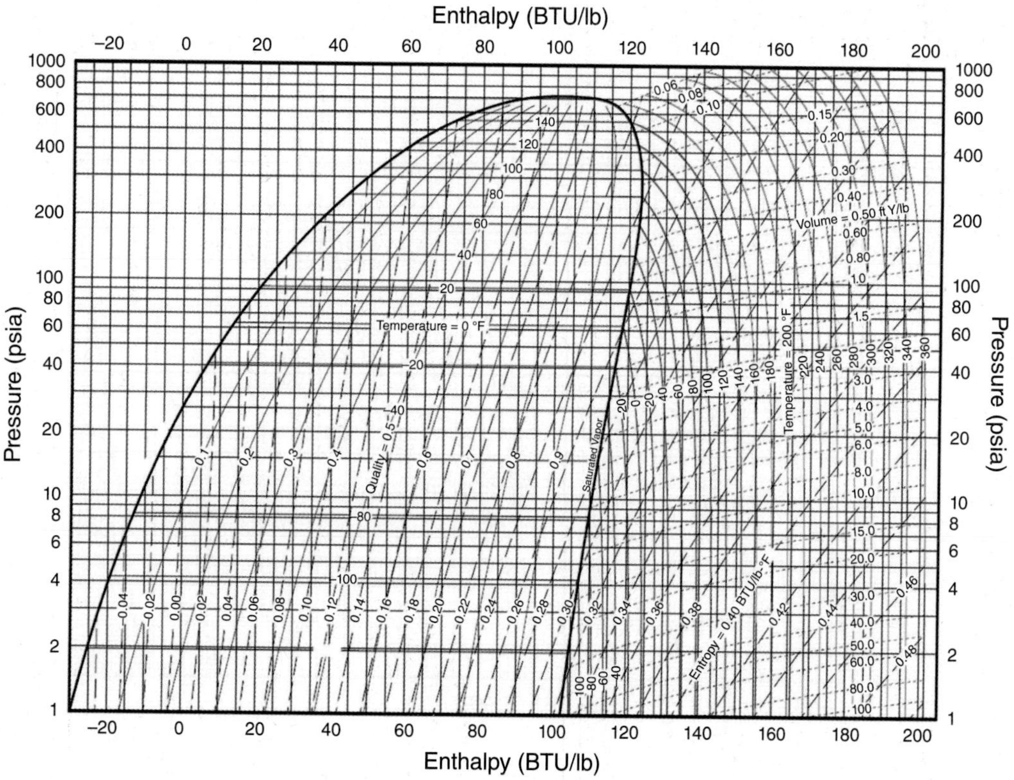

Figure 3.40 The pressure/enthalpy chart relates system operating pressures and temperatures to the heat content of the refrigerant in Btu/lb. Source: E. I. DuPont

but this is not absolutely accurate because it refers to the heat content above the selected reference point. Refer back to **Figure 1.14**, the heat/temperature graph for water. We used 0°F as the starting point of heat for water, knowing that you can really remove more heat from the water (ice) and lower the temperature below 0°F. We described the process as the amount of heat added starting at 0°F. This heat is called enthalpy. The pressure/enthalpy diagram is a similar diagram, available for all refrigerants and sometimes referred to as a p-e (pressure/enthalpy) or p-h (pressure/heat) chart. Since different refrigerants have different characteristics, properties, and temperature/pressure relationships, the pressure/enthalpy chart for each refrigerant is different. The pressure/enthalpy chart is used to plot the complete refrigeration cycle as a continuous loop.

The selected reference point for measuring the heat content in a refrigerant is −40°F. The pressure/enthalpy chart in **Figure 3.41** shows that the enthalpy, or heat content, in Btu/lb (along the bottom of the chart) has a value of 0 Btu/lb when the refrigerant is a saturated liquid at −40°F. The heat content for temperature readings below −40°F saturated liquid are indicated as being negative. As we move from left to right on the chart, the heat content per pound of refrigerant increases. As we move from right to left, the heat content per pound of refrigerant decreases.

As you inspect the pressure/enthalpy chart in **Figure 3.40**, you will notice that there is a horseshoe-shaped curve in the center of the chart. This curve is called the *saturation curve* and contains the same information that is contained on the temperature/pressure chart discussed earlier. The only difference is that the pressures on the pressure/enthalpy chart are expressed in absolute pressures, psia, instead of gauge pressures, psig.

Any time a point lies on, or under, this curve, the refrigerant is saturated and will have a corresponding temperature and pressure relationship. There are two saturation lines on the chart. The saturated liquid line makes up the left boundary of the saturation curve. If while on this line, heat is added, the refrigerant will start changing state to a vapor; if heat is removed, the liquid will be subcooled. The saturated vapor line makes up the right boundary of saturation curve. If, while on this line, heat is added, the vapor will superheat; if heat is removed, the vapor will start changing state to a liquid. The saturated liquid and vapor lines meet at the top of the saturation curve. This point is called the critical temperature or critical pressure. Above this point, the refrigerant will not condense. It is a vapor regardless of how much pressure is applied.

The area between the saturated liquid and saturated vapor lines, inside the horseshoe-shaped curve, is where the change of state occurs. Whenever a point lies between the saturation lines, the refrigerant is in the partial liquid, partial vapor state. The slanted, near-vertical lines between the saturated liquid and saturated vapor lines are the constant quality

Figure 3.41 The reference point used for measuring heat content is saturated liquid at −40°F.

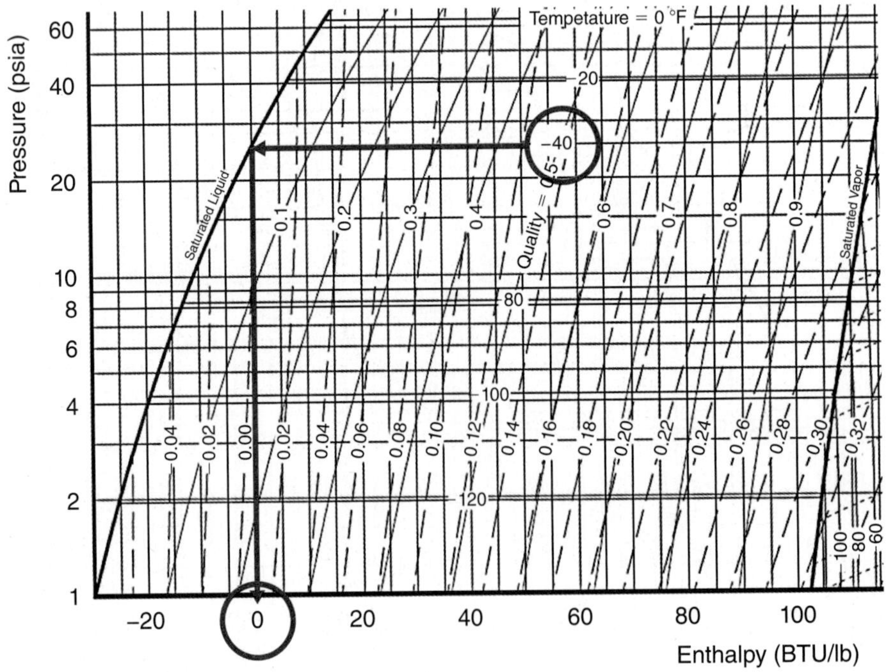

lines and describe the ratio of vapor to liquid in the mixture between the saturation points. There are nine of these lines under the saturation curve, each of which represents 10%. The saturated liquid line on the left side of the curve represents 0% vapor and 100% liquid, while the saturated vapor line on the right side of the saturation curve represents 100% vapor and 0% liquid. The nine constant quality lines that are located under the saturation curve are labeled, from left to right, 0.1 through 0.9. These figures represent the percentage of quality. *Percentage of quality* means percentage of vapor. This means that if a point falls on the 0.2 constant quality line, it would be 20% vapor and 80% liquid. If the plot is closer to the saturated liquid curve, there is more liquid than vapor. If the plot is closer to the saturated vapor curve, there is more vapor than liquid. For example, let's find a point on the chart at 60°F (horizontal line under the saturation curve) and 45 Btu/lb (along the bottom), **Figure 3.42**. This point is underneath the horseshoe-shaped curve and lies on the 0.1 constant quality line. This means that the refrigerant is 90% liquid and 10% vapor. **Figure 3.43** summarizes, in skeletal form, the important regions, points, and lines of the pressure/enthalpy diagram. For practical reasons, we will use only a few of these skeletal forms for illustrating the functions of the refrigeration cycle.

A refrigeration cycle is plotted in **Figure 3.44.** The system to be plotted is an air-conditioning system using R-410A. On this chart, five important points have been identified:

- Point A: Outlet of the condenser and the inlet of the metering device
- Point B: Outlet of the metering device and the inlet of the evaporator
- Point C: Outlet of the evaporator and the start of the suction line
- Point D: End of the suction line and inlet of the compressor
- Point E: Outlet of the compressor and start of the condenser

The system is operating at 120°F condensing temperature (417.4 psig or 432.1 psia discharge pressure) and at an evaporating temperature of 40°F (118.3 psig or 133 psia suction pressure). The cycle plotted has 20°F of subcooling, as calculated by the difference between the condenser saturation temperature of 120°F and the 100°F condenser

Figure 3.42 The quality of saturated R-410A at 60°F with a heat content of 45 Btu/lb is 10% vapor and 90% liquid.

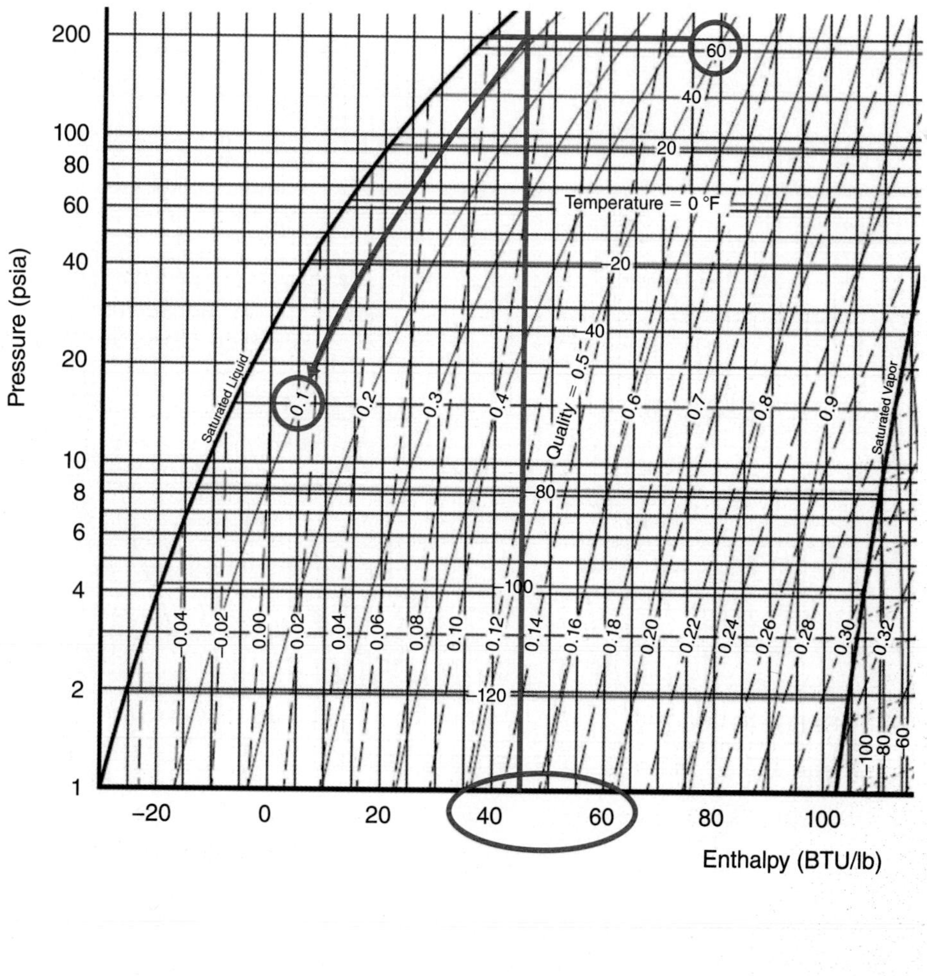

Figure 3.43 Skeletal pressure/enthalpy diagrams.

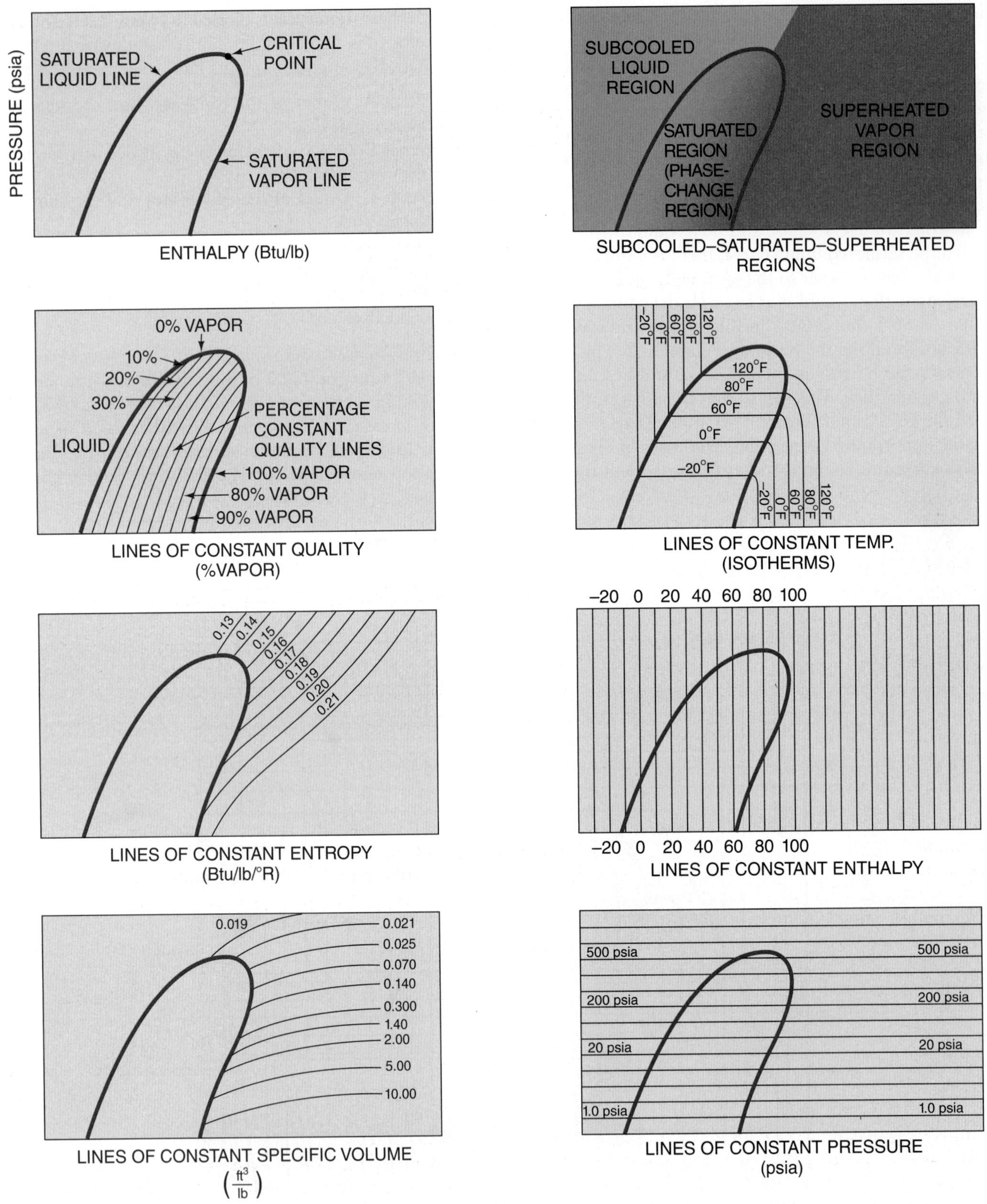

Figure 3.44 A refrigeration cycle plotted on a pressure/enthalpy chart. Courtesy E. I. DuPont

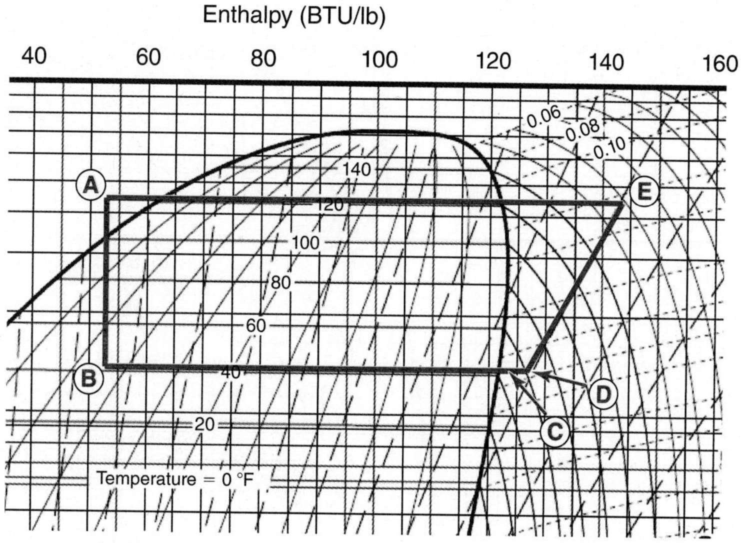

outlet, which is indicated by the point where the vertical line on the left side of the plot, crosses the saturation curve, **Figure 3.45.** There is 10°F of superheat in the evaporator with another 10°F of superheat absorbed by the refrigerant in the suction line as it returns to the compressor. The temperature of the suction gas entering the compressor is 60°F, **Figure 3.46.** The inlet of the compressor was identified as point D in **Figure 3.44.** The compressor is air-cooled, and the suction gas enters the suction valve adjacent to the cylinders. Assume that the hot gas line is very short and its heat of rejection is negligible. The following eight steps summarize the basic refrigeration cycle (refer to **Figure 3.44**):

Figure 3.45 The condenser outlet is identified by the point the expansion device line crosses the saturated liquid line on the left side of the pressure/enthalpy chart.

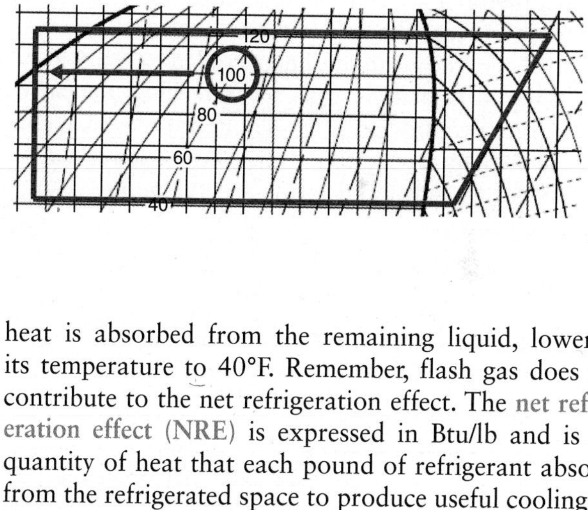

1. Refrigerant R-410A enters the expansion valve as a subcooled liquid at 432.1 psia (417.4 psig) and 100°F, point A. The heat content is 53 Btu/lb entering the expansion valve and 53 Btu/lb leaving the expansion valve. **Figure 3.44** shows the metering device represented by a vertical line at the left side of the plot. It can therefore be concluded that even though the temperature and pressure of the refrigerant drops as it flows through the expansion valve, the heat content of the refrigerant remains the same. It is important that heat content (Btu/lb) is not confused with temperature. Note that the temperature of the liquid refrigerant before the valve is 100°F, and the temperature leaving the valve is 40°F. The temperature drop can be accounted for by observing that we have 100% liquid entering the valve and about 75% liquid leaving the valve. About 25% of the liquid has changed to a vapor (flash gas). During the process of evaporating, heat is absorbed from the remaining liquid, lowering its temperature to 40°F. Remember, flash gas does not contribute to the net refrigeration effect. The **net refrigeration effect (NRE)** is expressed in Btu/lb and is the quantity of heat that each pound of refrigerant absorbs from the refrigerated space to produce useful cooling.
2. Flash gas is a result of the pressure drop across the metering device. No enthalpy is gained or lost in this process, which further explains why the expansion line from point A to point B is vertical. This expansion process is called *adiabatic expansion* because it happened at a constant enthalpy. Adiabatic processes, by definition, result in temperature and pressure changes with no change in heat content.
3. Usable refrigeration starts at point B where the refrigerant has a heat content of 53 Btu/lb. As heat is added to the refrigerant in the evaporator, the refrigerant

Figure 3.46 The compressor inlet (point D) on the pressure/enthalpy chart.

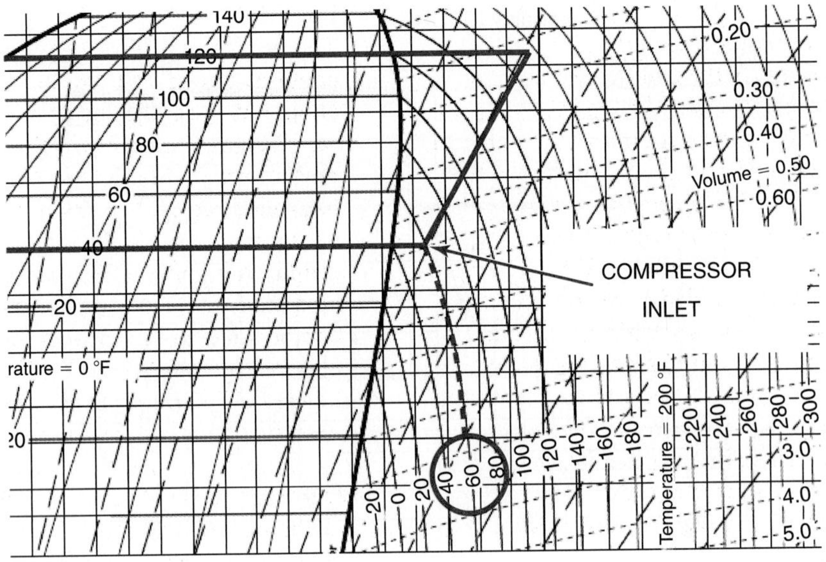

gradually changes state to a vapor. Notice that as we move toward the right from point B along a horizontal line, the pressure remains unchanged but the heat content increases. The source of this increase in enthalpy is the heat in the air that is being cooled by the evaporator. All liquid has changed to a vapor when it reaches the saturated vapor line. A small amount of sensible heat is added to the refrigerant in the form of superheat (10°F) between the saturated vapor line and point C. When the refrigerant leaves the evaporator at point C, it contains about 125 Btu/lb of heat energy. This is a net refrigeration effect of 72 Btu/lb (125 Btu/lb − 53 Btu/lb = 72 Btu/lb) of refrigerant circulated, **Figure 3.47.** The net refrigeration effect, NRE, is the heat actually extracted from the conditioned space. About 10 more Btu/lb are absorbed by the refrigerant in the suction line before reaching the compressor inlet at point D. This is not usable refrigeration because the heat does

Figure 3.47 Net refrigeration effect (NRE) and total heat of rejection (THOR) calculations.

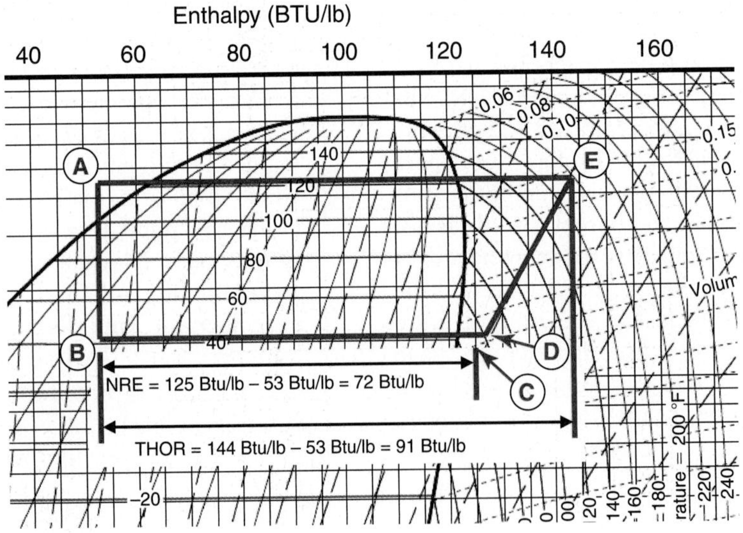

not come from the conditioned space, but is heat that must be pumped by the compressor and rejected by the condenser.

4. The refrigerant enters the compressor at point D (the end of the suction line) and leaves the compressor at point E. The heat added to the compressor during the compression process is called the heat of compression. Some of the heat of compression will be transferred to the head of the compressor and will be rejected to the surroundings. (A fully hermetic compressor with a suction-cooled motor would not plot out just like this. We have no way of knowing how much heat is added to the refrigerant by the motor, so we do not know what the temperature of the suction gas entering the compressor cylinder would be for a suction-cooled motor. Manufacturers obtain their own figures for this using internal thermometers during testing.) Notice that the line that represents the compressor is sloped up and to the right. This indicates that both the heat content and the pressure of the refrigerant are increasing.
5. The line that represents the compressor is drawn parallel to another set of lines on the chart that are referred to as lines of constant entropy, **Figure 3.43**. Entropy, in our case, represents the compression process and the relationship among the system characteristics of heat content, absolute pressure, and absolute temperature. These lines of constant entropy indicate that during the compression process, the changes in pressure and temperature are predictable. The units of constant entropy are Btu/lb/°R, where °R is an absolute temperature (as mentioned in Unit 1 "Heat, Temperature, and Pressure").
6. The refrigerant leaves the compressor at point E and contains about 144 Btu/lb. At point E, the refrigerant is now at the outlet of the compressor and traveling in the discharge line toward the condenser. This condenser must reject 91 Btu/lb (144 Btu/lb − 53 Btu/lb = 91 Btu/lb), called the *heat of rejection*. Remember that the condenser must reject all of the heat that is absorbed in the evaporator and suction line as well as the heat generated and concentrated in the compressor during the compression process. Therefore, the heat of rejection is also referred to as the *total heat of rejection*, THOR, Figure 3.47, because the condenser must reject all of the heat introduced to the system.
7. At point E we can also determine the temperature at the outlet of the compressor. We can do this by looking at the position of the point with respect to the lines of constant temperature, which are the downward-sloped, curved lines on the right-hand side of the saturation curve, **Figure 3.48**. The temperature of the discharge gas is about 180°F (see the constant temperature lines for temperature of superheated gas). When the hot gas leaves the compressor, it contains the maximum amount of heat that must be rejected by the condenser.
8. The refrigerant enters the condenser at point E as a highly superheated gas. The refrigerant condensing temperature is 120°F and the hot gas leaving the compressor

Figure 3.48 The compressor discharge temperature is determined from the constant temperature lines.

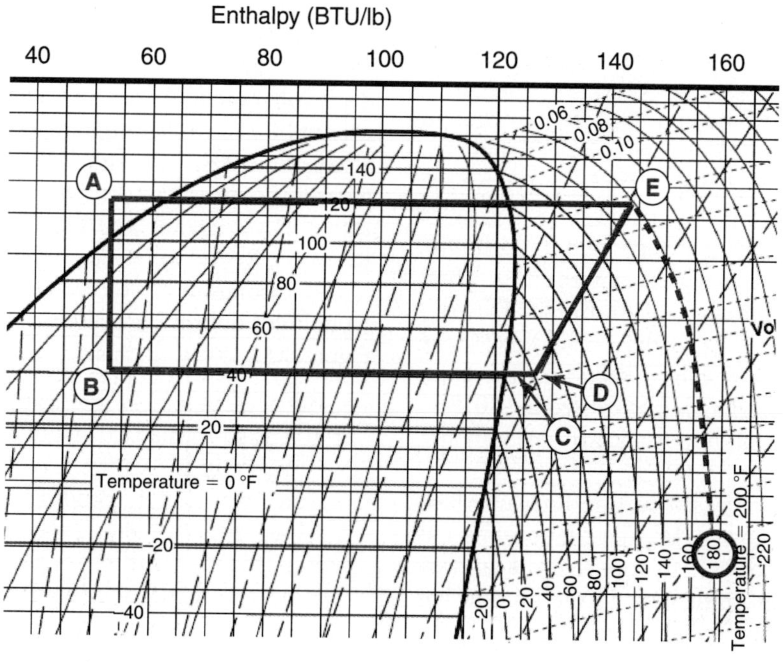

Figure 3.49 A rise in system head pressure will result in a reduction of system capacity.

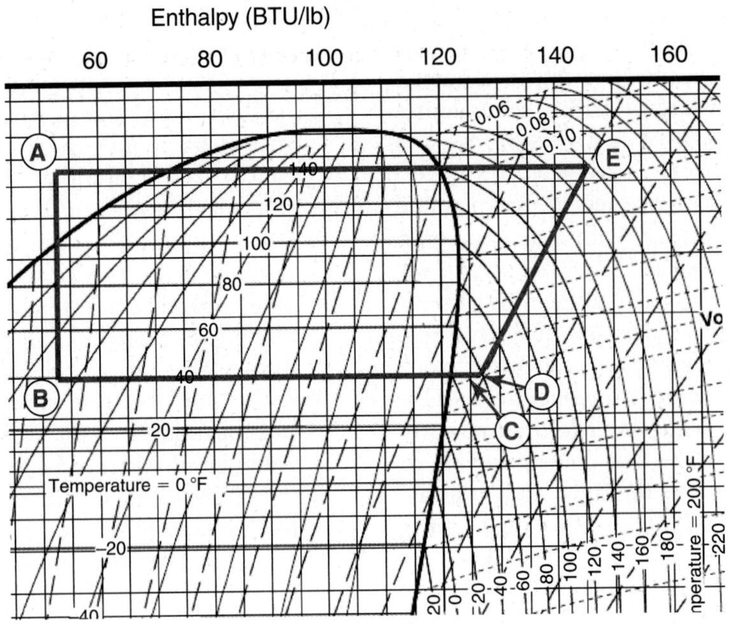

is 180°F, so it contains 60°F (180°F − 120°F = 60°F) of superheat. The condenser will first reduce the temperature of the discharge gas to the condensing temperature of 120 degrees F. (This process is referred to as desuperheating.) Then, the condenser will condense the refrigerant to a liquid at 120°F and will enter the expansion device at point A as a subcooled liquid at a temperature of 100°F.

Other conditions may be plotted on the pressure/enthalpy diagram. For example, suppose the head pressure of our R-410A system increases. Keeping all other parameters of the system constant to clarify the concept, the new system plot is provided in **Figure 3.49.** With a new, higher condensing temperature of 140°F (539.1 psig or 553.8 psia), we see the percentage of liquid leaving the expansion valve to be about 64% (36% flash gas) with a heat content of 63 Btu/lb. Using the same heat content leaving the evaporator, 125 Btu/lb, we have a net refrigeration effect of 62 Btu/lb. This is a net reduction in refrigeration effect of about 13.9% from the original example. In addition, it can be seen that the compressor discharge temperature has increased from 180°F to 200°F. This example shows the importance of keeping systems properly maintained.

Figure 3.50 shows how increased suction line superheat affects the first system in **Figure 3.44.** The suction line has not been insulated and absorbs heat. Notice that the amount of sensible heat added to the refrigerant in the suction has increased, causing an increase in the temperature of the refrigerant entering the compressor. Notice the high discharge temperature (about 200°F). This is approaching the temperature that will cause oil to break down and form acids in the system. The compressor must pump more refrigerant to accomplish the same refrigeration effect, and the condenser must reject more heat.

Before its manufacturing phase-out date in 1996, R-12 used to be a popular medium-temperature application refrigerant. Environmental issues have caused manufacturers to explore some different refrigerants to replace R-12, namely, R-134a and the newer refrigerant blends. R-134a has a zero ozone depletion potential but has oil compatibility problems and cannot be easily retrofitted into existing R-12 systems. R-134a also experiences a reduction in capacity when used in low-temperature applications. (Environmental issues will be covered in Unit 9, "Refrigerant and Oil Chemistry and Management—Recovery, Recycling, Reclaiming, and Retrofitting.")

A medium-temperature R-134a system, **Figure 3.51,** would plot out on the pressure/enthalpy chart as follows. Use **Figure 3.52** to follow this example.

1. Refrigerant enters the expansion valve at point A at 105°F, subcooled 10° from 115°F.
2. Refrigerant leaves the expansion valve at point B with a heat content of 47 Btu/lb. The quality is 33% vapor with 67% liquid. It then travels through the evaporator.
3. Vapor refrigerant leaves the evaporator at point C with a heat content of 109 Btu/lb and a net refrigeration effect of 62 Btu/lb (109 − 47 = 62).

Figure 3.50 An increase in suction line superheat results in an increase in compressor discharge temperature.

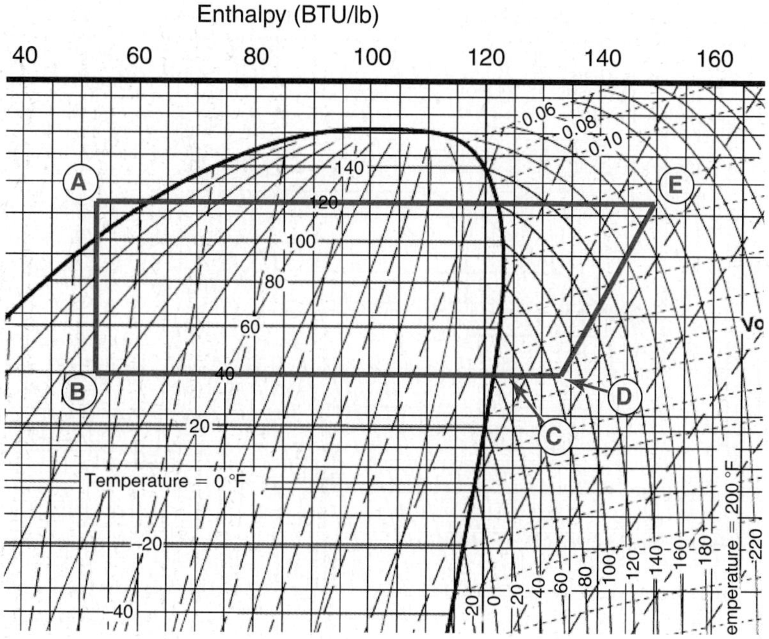

4. The refrigerant enters the compressor at point D with a superheat of 40° and a heat content of 114 Btu/lb. The vapor is compressed along the line from D to E. *Note the lower discharge temperature of 160°F.*

Figure 3.51 A medium-temperature R-134a refrigeration system showing operating temperatures and pressures.

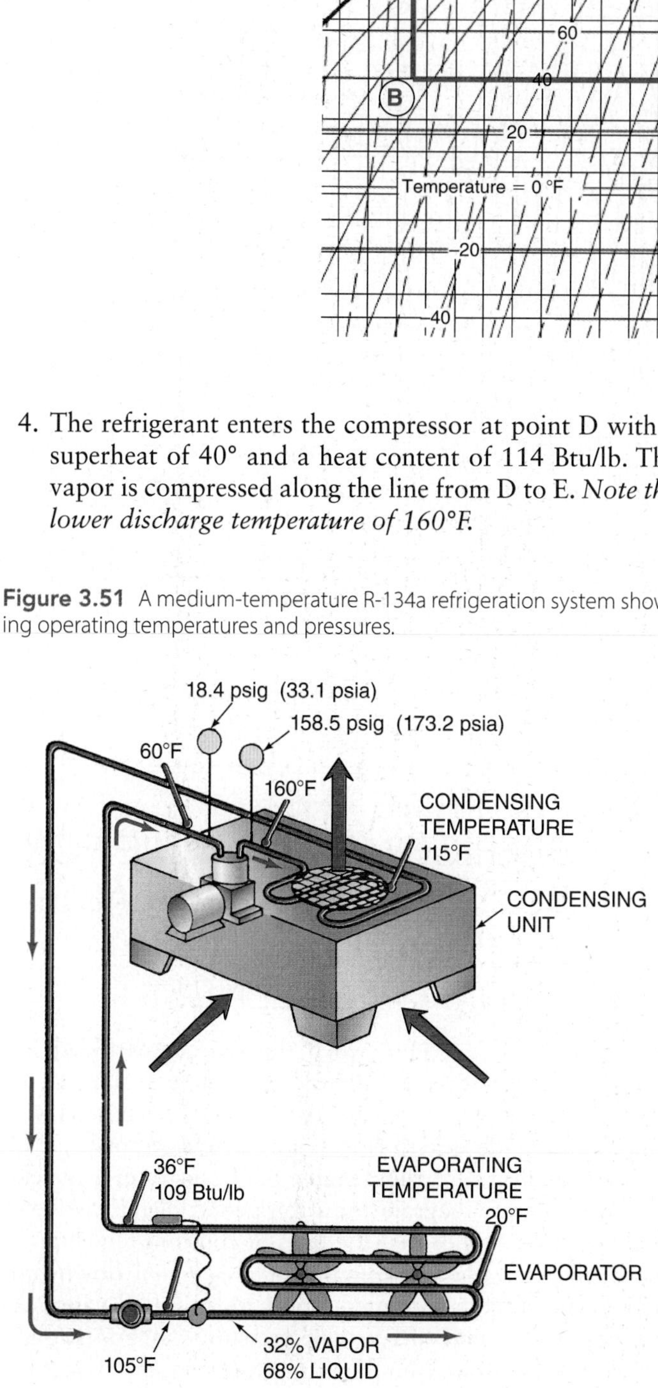

5. The refrigerant leaves the compressor at point E with a heat content of 130 Btu/lb. The HOC for this system is 16 Btu/lb.
6. The refrigerant is then desuperheated 45°F from point E to the saturated vapor line.
7. The refrigerant is now gradually condensed from the saturated vapor line to the saturated liquid line at 115°F. This is the *latent heat of condensation* being rejected.
8. The saturated liquid is now subcooled from the saturated liquid line to point A. It is subcooled 10° (115°F – 105°F) until it enters the metering device at 105°F. The process then repeats itself.

As mentioned earlier, R-502 was used primarily in medium- and low-temperature refrigeration applications. Its manufacture was banned in 1996. R-502 is an azeotropic refrigerant blend with negligible temperature glide, and it behaves like a pure compound. R-404A is its long-term replacement refrigerant and has a slightly higher working pressure than R-502. **Figure 3.53** shows a low-temperature refrigeration system plotted on a pressure/enthalpy chart incorporating R-404A as the refrigerant. The system is a commercial freezer operating with an evaporating temperature of −20°F (16.7 psig) and a 115°F (290 psig) condensing temperature.

Follow the sequence below.

1. The refrigerant enters the expansion valve at point A at 105°F, subcooled 10°F from the 115°F condensing temperature.
2. The refrigerant leaves the expansion valve at point B at −20°F (55% vapor and 45% liquid) to be totally

Figure 3.52 An R-134a pressure/enthalpy diagram. Courtesy E. I. DuPont

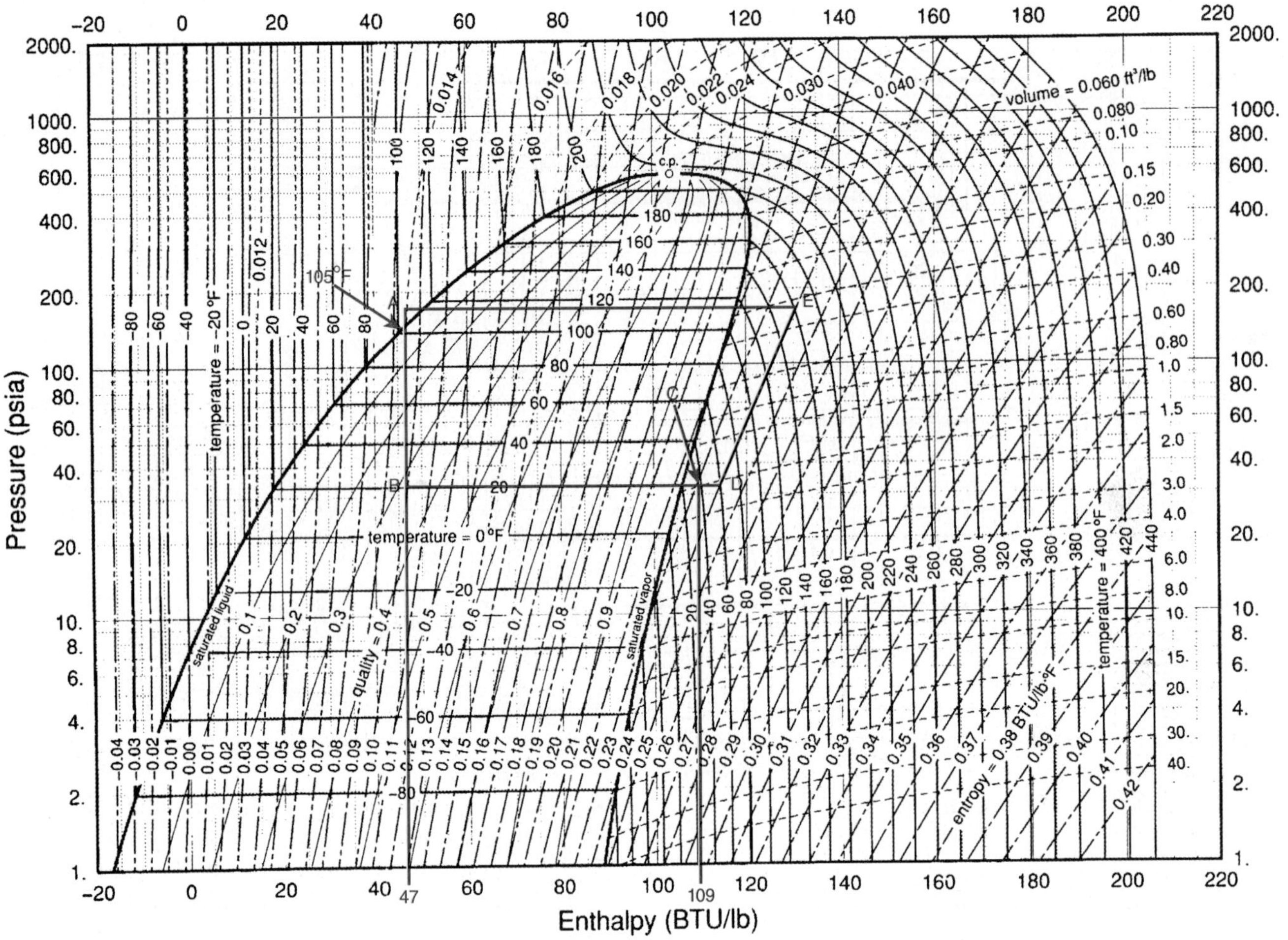

H-39917

evaporated in the evaporator. At point C, the refrigerant leaves the evaporator as 100% vapor with 10°F of evaporator superheat and at a temperature of −10°F. The difference in enthalpy between point B and point C is the net refrigeration effect.

3. The superheated vapor enters the compressor at point D with 30°F of total superheat and at a temperature of 10°F. The superheated vapor is now compressed to 170°F along the line to point E.
4. The superheated vapor leaves the compressor at point E. The refrigerant is now desuperheated from point E to the saturated vapor line.
5. The now saturated refrigerant is gradually condensed from the saturated vapor line to the saturated liquid line at 115°F. This is referred to as rejection of the latent heat of condensation.
6. The saturated liquid is subcooled 10°F (115°F − 105°F) from the saturated liquid line to point A, where it enters the metering device at 105°F. The process then repeats itself.

3.21 PLOTTING THE REFRIGERATION CYCLE FOR BLENDS WITH NOTICEABLE TEMPERATURE GLIDE (ZEOTROPIC BLENDS)

Temperature glide occurs when the refrigerant blend has many temperatures as it evaporates or condenses at a given pressure. The pressure/enthalpy diagram for refrigerants with temperature glide differs from that for refrigerants that do not have temperature glide. **Figure 3.54** illustrates a skeletal pressure/enthalpy diagram for a refrigerant blend with temperature glide. Notice that the lines of constant temperature (isotherms) are not horizontal, but are angled downward as they travel from saturated liquid to saturated vapor. As you follow the lines of constant pressure (isobars) straight across from saturated liquid to saturated vapor, more than

Figure 3.53 An R-404A low-temperature refrigeration system. Courtesy E. I. DuPont

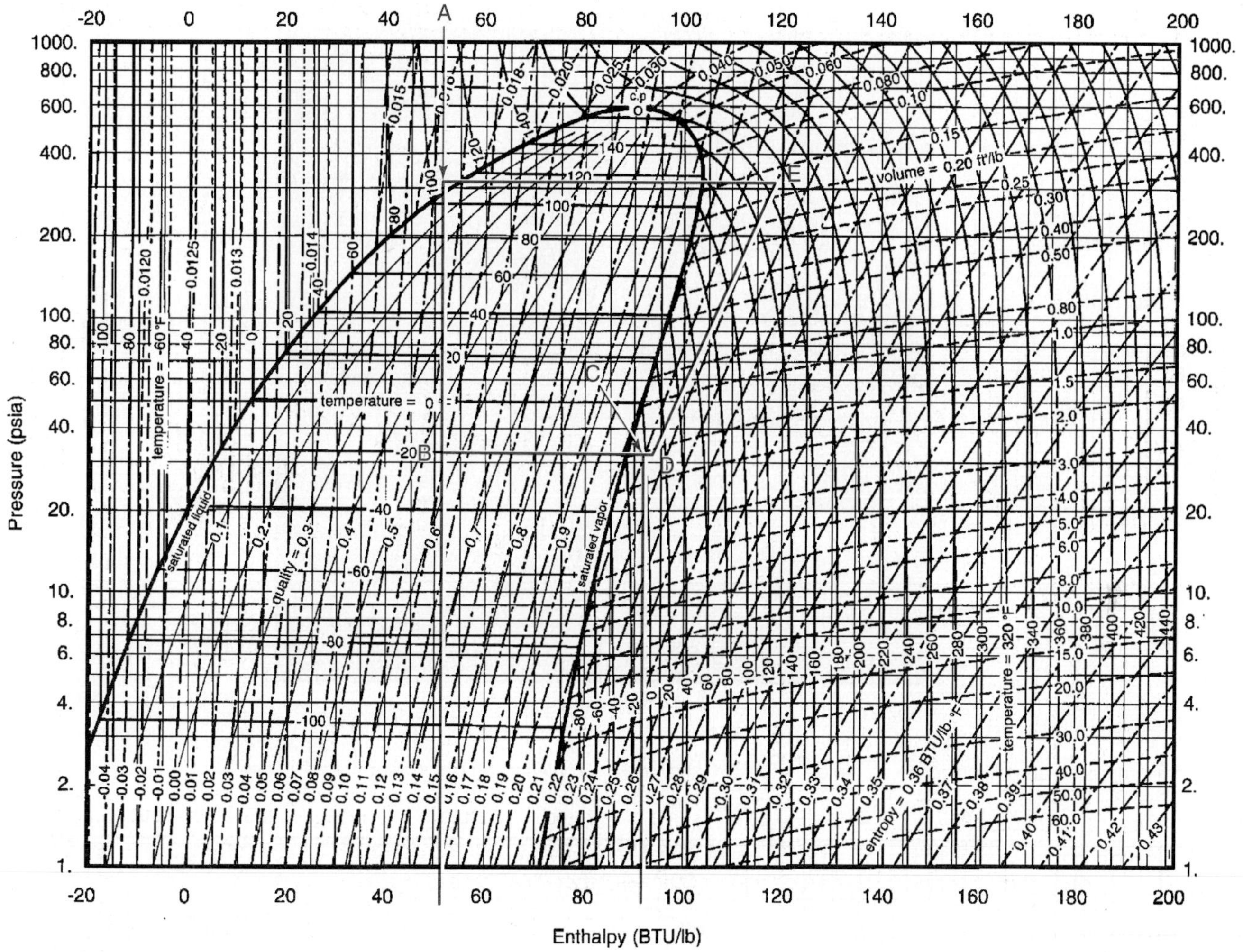

one isotherm will be intersected. Thus, for any one pressure (evaporating or condensing), there will be a range of associated temperatures—a temperature glide. For example, in **Figure 3.54**, for a constant pressure of 130 psia, the temperature glide is 10°F (70°F − 60°F). The isobar of 130 psia intersects both the 60°F and the 70°F isobars as it travels from saturated liquid to saturated vapor.

Figure 3.55 illustrates an actual pressure/enthalpy diagram for refrigerant R-407C, a zeotropic refrigerant blend with very similar properties to those of R-22 in air-conditioning equipment. It is a blend of R-32, R-125, and R-134a with a fairly large temperature glide (10°F) in the ranges of air-conditioning applications. Notice the isotherms angled downward from left to right, showing temperature glide.

The terms near-azeotropic and zeotropic refrigerant blends are used in the HVACR industry because they describe temperature glide and fractionation. Azeotropic blends have negligible franctionation and temperature glide, near-azeotropic blends have some fractionation and temperature glide, while zeotropic blends have the most fractionation and temperature glide. Refer to Unit 9, "Refrigerant and Oil Chemistry and Management—Recovery, Recycling, Reclaiming, and Retrofitting," for more in-depth information on refrigerants, refrigerant blends, temperature glide, and fractionation. Charging methods for refrigerant blends will be covered in Unit 10, "System Charging." **SAFETY PRECAUTION:** *All refrigerants that have been discussed in this text are stored in pressurized containers and should be handled with care. Consult your instructor or supervisor about the use and handling of these refrigerants. Goggles and gloves should be worn while transferring the refrigerants from the container to the system.*•

Figure 3.54 A skeletal pressure/enthalpy diagram of a refrigerant blend with a noticeable temperature glide (near-azeotropic blend). Notice the angled isotherms.

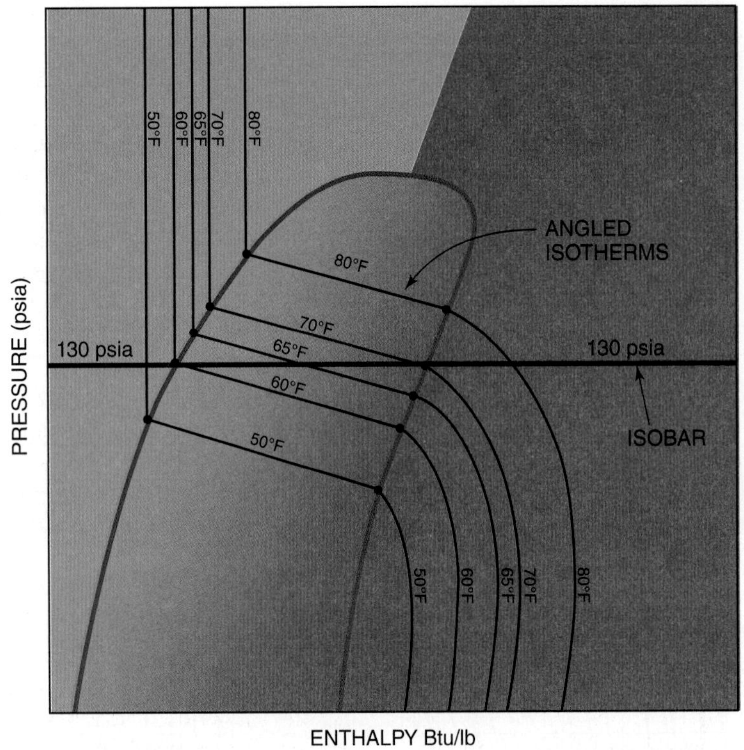

Figure 3.55 A pressure/enthalpy diagram for refrigerant R-407C showing the angled isotherms. Courtesy E. I. DuPont

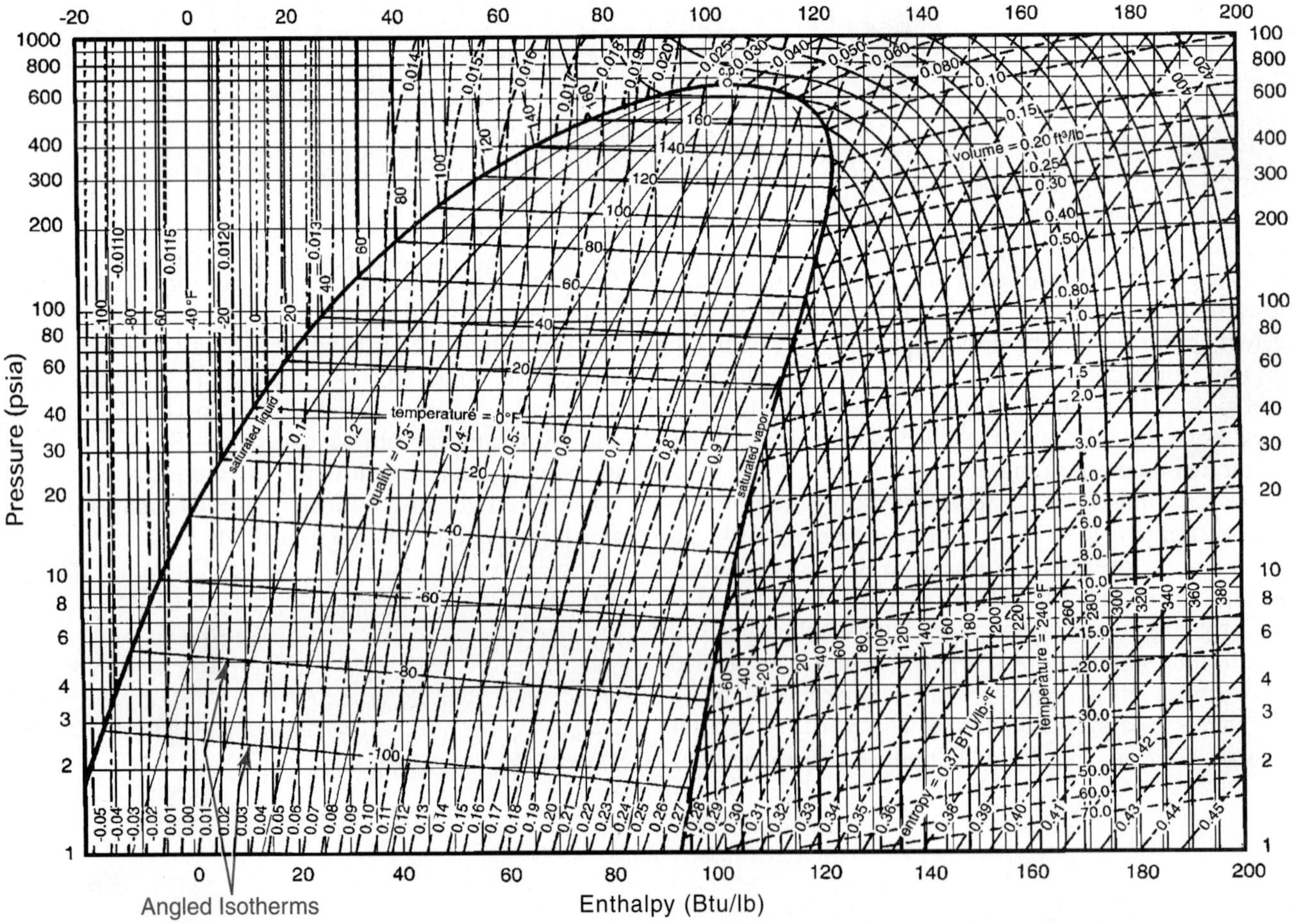

REVIEW QUESTIONS

1. A refrigeration system that maintains a space at a temperature between 35°F and 40°F would be classified as what type of system?
   A. High-temperature
   B. Medium-temperature
   C. Low-temperature
   D. Ultra-low-temperature
2. A refrigeration system that is intended to provide comfort cooling would be classified as what type of system?
   A. High-temperature
   B. Medium-temperature
   C. Low-temperature
   D. Ultra-low-temperature
3. One ton of refrigeration is the equivalent of moving heat at a rate of
   A. 1200 Btu/min
   B. 12,000 Btu/min
   C. 1200 Btu/h
   D. 12,000 Btu/h
4. The basic refrigeration cycle can best be described as a
   A. non-repeating, vapor-compression cycle.
   B. non-repeating, vapor-combustion cycle.
   C. repeating, vapor-compression cycle.
   D. repeating, vapor-combustion cycle.
5. Saturated R-410A at a pressure of 95 psig will change at a temperature of
   A. 29°F.
   B. 95°F.
   C. 118°F.
   D. 295°F.
6. The evaporating pressure for R-22 is 76 psig and the evaporator outlet temperature is 58°F. What is the evaporator superheat?
   A. 13°F
   B. 45°F
   C. 58°F
   D. 132°F
7. The state of the refrigerant at the inlet of the evaporator is approximately
   A. 50% liquid and 50% vapor.
   B. 75% liquid and 25% vapor.
   C. 100% superheated vapor.
   D. 100% subcooled liquid.
8. The state of the refrigerant at the outlet of the evaporator should be
   A. 50% liquid and 50% vapor.
   B. 75% liquid and 25% vapor.
   C. 100% superheated vapor.
   D. 100% subcooled liquid.
9. The state of the refrigerant at the outlet of the condenser should be
   A. 50% liquid and 50% vapor.
   B. 75% liquid and 25% vapor.
   C. 100% superheated vapor.
   D. 100% subcooled liquid.
10. If the evaporator outlet temperature on an R-410A system is 50°F and the evaporator superheat is 10°F, what is the evaporating pressure of the refrigerant in this system?
    A. 143 psig
    B. 118 psig
    C. 84 psig
    D. 62 psig
11. If the evaporating pressure was 76 psig for R-22 and the compressor inlet temperature was 65°F, what would be the total superheat entering the compressor?
    A. 11°F
    B. 21°F
    C. 10°F
    D. 20°F
12. When an air-conditioning system operates in a vacuum, which of the following is true?
    A. Air will leak out of the system
    B. Refrigerant will leak out of the system
    C. Air will be pulled into the system
    D. Refrigerant will be pulled into the system
13. The condensing pressure is 417.4 psig for R-410A and the condenser outlet temperature is 108°F. How much subcooling is there in the condenser?
    A. 7°F
    B. 12°F
    C. 21°F
    D. 42°F
14. The expansion device is responsible for
    A. controlling the operation of the compressor.
    B. controlling the subcooling.
    C. metering refrigerant flow into the evaporator.
    D. storing refrigerant for future use.
15. A properly operating compressor should have ___________ at its inlet and ___________ at its outlet.
    A. 100% vapor, 100% vapor
    B. 100% liquid, 100% liquid
    C. 100% vapor, 100% liquid
    D. 100% liquid, 100% vapor
16. Of the following compressors, which is a residential compressor that is forgiving when it comes to having liquid entering it?
    A. The reciprocating compressor
    B. The scroll compressor
    C. The rotary compressor
    D. The screw compressor

17. Which of the following statements is true with regards to reciprocating and rotary compressors?
    A. Rotary and reciprocating compressors with similar capacities are similarly sized.
    B. Both rotary and reciprocating compressors are equipped with suction and discharge valves.
    C. Both rotary and reciprocating compressors can be opened for service.
    D. For compressors with similar capacities, rotary compressors are smaller than reciprocating compressors.
18. The color-coding for an R-22 refrigerant cylinder is
    A. white.
    B. green.
    C. orange.
    D. brown.
19. The color-coding for an R-404A refrigerant cylinder is
    A. white.
    B. green.
    C. orange.
    D. brown.
20. Of the following refrigerants, which has the lowest global warming potential (GWP)?
    A. R-410A
    B. R-134a
    C. R-407C
    D. R-1234yf
21. The three processes commonly used to describe refrigerant handling are
    A. recovering, recycling, and reclaiming.
    B. recovering, removing, and replacing.
    C. removing, recycling, and replacing.
    D. removing, recycling, and reclaiming.
22. The term that is used to quantify the amount of heat absorbed in the evaporator is
    A. total heat of rejection.
    B. net refrigeration effect.
    C. heat of compressor.
    D. adiabatic expansion.
23. The three regions on a pressure/enthalpy chart are
    A. high-pressure, low-pressure, critical pressure.
    B. superheated vapor, superheated liquid, depressurized liquid.
    C. superheated vapor, subcooled liquid, saturated.
    D. high-pressure liquid, low-pressure gas, low-pressure liquid.
24. The thumbprint-shaped region on the pressure/enthalpy chart is called the
    A. saturation curve.
    B. saturation cycle.
    C. superheated curve.
    D. superheated region.
25. One of the challenges of plotting an R-407C system on a pressure/enthalpy chart is that this refrigerant
    A. is a single-compound refrigerant with one pressure for each temperature.
    B. has a significant temperature glide.
    C. is a binary blend of R-32 and R-125, making it unstable.
    D. is easily mistaken for R-134a and R-410A.

SECTION 2

SAFETY, TOOLS AND EQUIPMENT, AND SHOP PRACTICES

UNIT 4

GENERAL SAFETY PRACTICES

OBJECTIVES

After studying this unit, you should be able to

- explain the importance of being aware of your surroundings at all times.
- explain the importance of personal protective equipment (PPE) and provide six examples of such equipment.
- describe and demonstrate proper procedures for working with and around pressurized vessels and systems.
- describe the dangers associated with working on and around electricity.
- explain what precautions a technician should take when working with or around electricity.
- explain the importance of knowing proper first aid procedures.
- describe and demonstrate the proper use and setup of two commonly used types of ladders.
- explain the dangers associated with exposure to extreme temperatures and what can be done to reduce the potential for injury resulting from these dangers.
- describe three common classifications used to describe fire extinguishers.
- explain the acronym PASS as it refers to fire extinguisher use.
- explain the dangers associated with working on or around rotating machinery.
- describe and demonstrate proper procedures for lifting and/or moving heavy objects.
- explain the importance of proper ventilation as it applies to remaining safe on the job.
- define the term "confined spaces" and the dangers associated with them.
- explain what precautions a technician should take when working with or around chemicals.

The heating, air-conditioning, and refrigeration technician is often subjected to many dangerous conditions and situations. Exposure to chemicals, pressurized liquids and gases, electricity, rotating machinery, and excessive heat and cold all have the potential to cause severe personal injury or death. In addition, other physical injuries can result from improper lifting techniques, falling objects, tripping, inadequate ventilation, or improper use of tools and equipment. Each and every job must be completed in a manner that ensures the safety of both the workers on the job and the general population.

This unit describes some general safety practices and procedures with which all technicians should be familiar. Whether in a school laboratory, a shop, a commercial service shop, or a manufacturing plant, technicians must always be familiar with the location of emergency exits, first aid supplies, automatic external defibrillator (AED) equipment, and eyewash stations. When you are on the job, be aware of how you would get out of a building or particular location should an emergency occur. Always use common sense and be prepared to act in the event an emergency arises.

All tools of the trade should be used for their intended purposes and their limitations should never be exceeded. The technician should read all manufacturer's literature to ensure the tool is used properly. Improper use of tools can cause property damage and personal injury.

Safety goggles are one of the most overlooked pieces of personal protection equipment (PPE). Flying debris can cause an eye injury before a person can react, so it is good practice to wear goggles anytime the possibility of injury exists. Even when good work practices are followed, accidents do happen. It is an excellent suggestion for every technician to take a class on how to perform cardiopulmonary resuscitation (CPR) and how to properly use an automatic external defibrillator (AED). Classes are offered by many institutions including schools, fire departments, police stations, and the American Red Cross.

Technicians should always act professionally on the job. Countless mistakes have been made while playing around or otherwise not paying attention to the task at hand. Carelessness is a major cause for job site accidents and injuries.

4.1 WORKPLACE SURROUNDINGS AND PERSONAL PROTECTIVE EQUIPMENT (PPE)

Remaining safe on the job should be the number one concern of any individual working in the HVACR industry. Although it is not always possible to avoid dangerous conditions while working in the field, there are some things that can be done to help reduce the possibility of sustaining an injury while working. One thing that should always be done by field personnel when arriving on a jobsite is to evaluate the workplace conditions and identify any potentially hazardous conditions or situations. From loose handrails to low overhangs to trip hazards, knowing what dangerous conditions are present in your immediate surroundings will help you avoid them as you proceed to complete the task at hand.

Another thing that can be done to help prevent workplace injuries is to dress appropriately and wear personal protective equipment (PPE) as required. Personal protection equipment is a global term used to describe the various pieces of clothing and equipment that individuals wear to protect their bodies in the event an accident occurs. Depending on the jobs being performed, the recommended and/or required pieces of PPE will vary. Typical personal protective items include, but are definitely not limited to:

- Safety glasses or goggles
- Ear plugs or muffs
- Safety gloves
- Work boots
- Breathing masks or respirators
- Hard hats
- Long-sleeved shirts
- Face shields
- Reflective clothing
- Heat-resistant creams
- Sunblock

Throughout this Unit, many different types of PPE will be discussed. It is everyone's responsibility to work in a safe and professional manner to help ensure that accidents, and their resulting injuries, do not occur.

4.2 PRESSURIZED VESSELS, PIPING, AND SYSTEMS

Pressurized vessels, piping, and systems are encountered by refrigeration and air-conditioning technicians on a daily basis. For example, a cylinder of R-410A refrigerant in the back of a service truck with the sun shining on it may very well be at a temperature of 110°F on a summer day. The temperature/pressure chart indicates that the pressure inside the cylinder at 110° is 368 psig. This reading means that the cylinder has a pressure of 368 lb for each square inch of surface area. A large cylinder may have a total area of 1500 in^2. This gives a total inside pressure (pushing outward) of

$$1500 \text{ in}^2 \times 368 \text{ psi} = 552{,}000 \text{ lb}$$

This is equal to

$$552{,}000 \text{ lb} \div 2000 \text{ lb/ton} = 276 \text{ tons}$$

As long as the cylinder is protected and not damaged, the pressure is well contained and will be safe. Cylinders should not be permitted to fall over and should never be dropped. Make certain that all tanks are properly secured at all times. **SAFETY PRECAUTION:** *Move the cylinder only while a protective cap is on it, if it is designed for one,* **Figure 4.1.** • Cylinders too large to be carried should only be moved when chained to an approved cart, **Figure 4.2**. If a refrigerant cylinder ever falls off the moving cart, the protective cap will protect the valve from breaking off and becoming a projectile. If the cylinder falls accidentally, the protective cap will also prevent the vaporization of all the liquid in the cylinder, which can displace the air we breathe and cause death or personal injury. The air is displaced because the refrigerant vapor is denser than the air.

Figure 4.1 A refrigerant cylinder with a protective cap. Photo by Bill Johnson

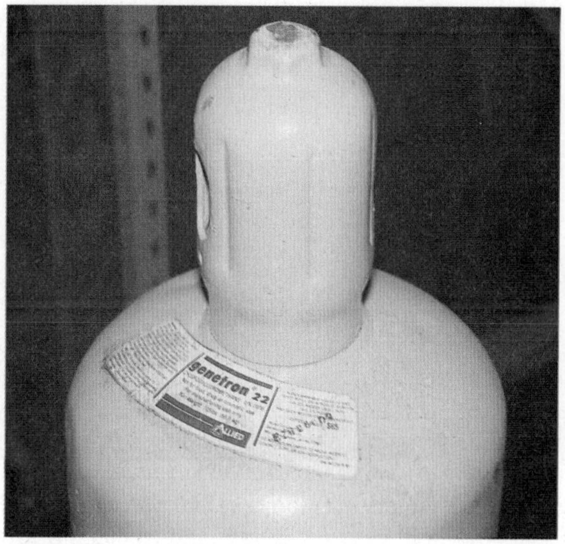

Figure 4.2 Pressurized cylinders should be chained to and moved safely on an approved cart.

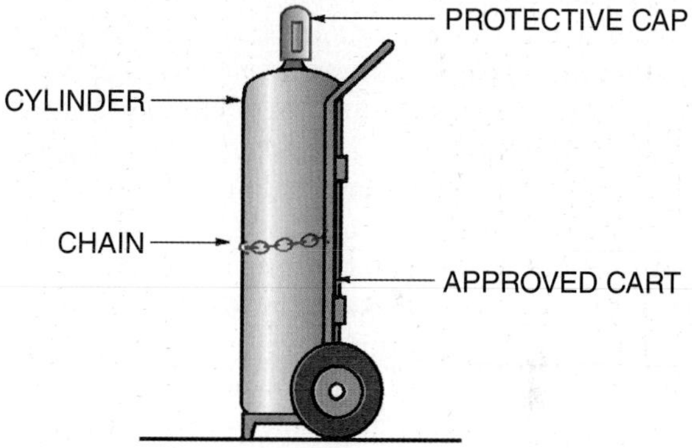

The pressure in a cylinder can be thought of as a potential danger. It will not become dangerous unless it is allowed to escape in an uncontrolled manner. The cylinder has a relief valve at the top in the vapor space, **Figure 4.3**. If the pressure builds up to the relief valve setting, the valve will start relieving vapor. As the vapor pressure is relieved, the liquid in the cylinder will begin to vaporize and absorb heat from the surrounding liquid, producing a cooling effect. This will reduce the pressure in the cylinder. Relief valves are set to open at pressure levels that are above the worst typical operating conditions, which is often higher than 400 psig.

The refrigerant cylinder has a fusible plug made of a material with a low melting temperature. The plug will melt and

Figure 4.3 Cylinder pressure relief valve. Courtesy Ferris State University. Photo by John Tomczyk

blow out if the cylinder gets too hot. This prevents the cylinder from bursting and injuring people and damaging property around it. **⚠ SAFETY PRECAUTION:** *Refrigerant cylinders should be stored and transported in the upright position to keep the pressure relief valve in contact with the vapor space, not the liquid inside the cylinder. Some technicians may apply heat to refrigerant cylinders while charging a system to keep the pressure from dropping in the cylinder. This is an extremely dangerous practice. It is recommended for the above purpose that the cylinder of refrigerant be set in a container of warm water with a temperature no higher than 90°F.•*

You will be taking pressure readings of refrigeration and air-conditioning systems. Liquid R-410A boils at −61°F when it is released to the atmosphere. If you are careless and get this refrigerant on your skin or in your eyes, it will quickly cause frostbite. Keep your skin and eyes away from any liquid refrigerant. When you attach gauges and take refrigerant pressure readings or transfer refrigerant into or out of a system, wear gloves **Figure 4.4**, and protective eye wear, **Figure 4.5**.

Figure 4.4 Safety gloves to protect the hands from excessive refrigerant temperatures. Photo by Eugene Silberstein

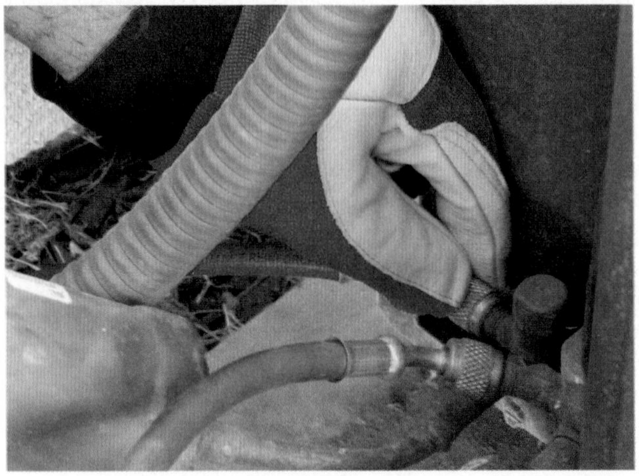

Figure 4.5 Safety glasses. Photo by Eugene Silberstein

⚠ SAFETY PRECAUTION: *If you are working in an indoor space and a refrigerant leak causes large amounts of refrigerant to be released, the best thing to do is vacate the area immediately. Only return to the area once it has been completely ventilated and it is deemed safe to re-enter the area.•*

Released refrigerant, and any particles that may become airborne with the blast of vapor, can harm the eyes. Wear approved goggles, which are vented to keep them from fogging over with condensation and which help keep the operator cool.

⚠ SAFETY PRECAUTION: *All cylinders into which refrigerant is transferred should be approved by the Department of Transportation as recovery cylinders.* **Figure 4.6** *shows cylinders with color codes that indicate they are approved as recovery cylinders—gray bodies with yellow tops. It is against the law to transport refrigerant in cylinders that are not approved for this purpose.•*

In addition to the pressure potential inside a refrigerant cylinder, there is tremendous pressure potential inside nitrogen and oxygen cylinders. They are shipped at pressures in the range of 2500 psig and must not be moved unless the protective cap is in place. They should be chained to and moved on carts designed for the purpose, as shown previously in **Figure 4.2.** Dropping a cylinder without the protective cap may break the valve off the cylinder, and the pressure inside can propel the cylinder into the air, **Figure 4.7.**

Nitrogen must have its pressure regulated before it can be used because the pressure in the cylinder is too great to be connected to a system. Nitrogen pressure is regulated with a nitrogen pressure regulator, **Figure 4.8.** If a person allowed nitrogen under cylinder pressure to enter a refrigeration system, some weak point in the system could burst, potentially causing severe personal injury. **Figure 4.8** shows a nitrogen tank pressure of approximately 2200 psi being regulated down to about 130 psi. Notice the downstream safety pressure relief valve that protects the system and the operator.

Figure 4.6 Color-coded refrigerant recovery cylinders. The bodies of the cylinders are gray; the tops are yellow. Courtesy White Industries

Figure 4.7 When a cylinder valve is broken off, the cylinder becomes a projectile until pressure is exhausted.

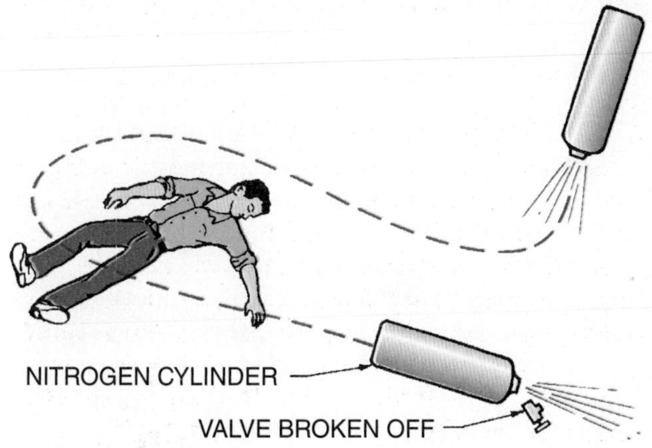

Oxygen also must be regulated because of high pressure. In addition, all oxygen lines must be kept absolutely oil free. Oil residue in an oxygen regulator connection may cause an explosion, which may blow the regulator apart in your hand. **SAFETY PRECAUTION:** *Oxygen is under a great deal of pressure but should never be used as a pressure source to pressurize lines or systems. If there is oil or oil residue in the system, adding oxygen to pressurize them can lead to an explosion.* •

Figure 4.8 A nitrogen tank with pressure regulator and safety downstream pressure relief valve. Courtesy Ferris State University. Photo by John Tomczyk

Oxygen is often used with acetylene. Acetylene cylinders are not under the same high pressure as nitrogen and oxygen but must be treated with the same respect because acetylene is highly explosive. A pressure regulator must be used, **Figure 4.9**. Always use an approved cart, chain the cylinder to the cart,

Figure 4.9 Pressure regulators on oxygen and acetylene tanks. Courtesy of Ferris State University. Photo by John Tomczyk

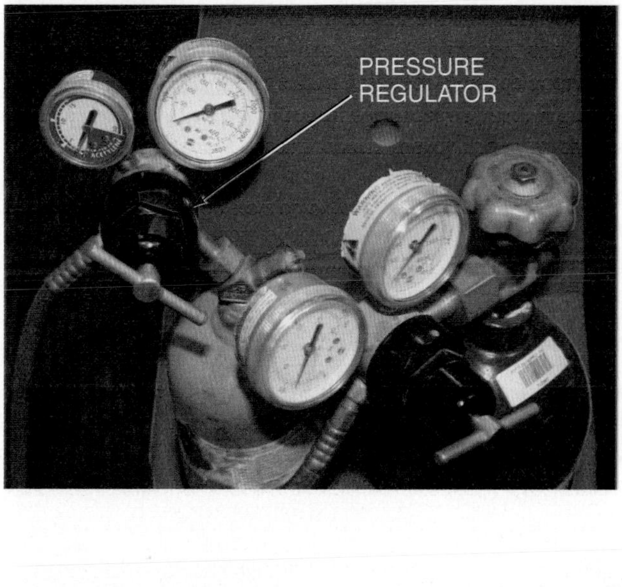

Figure 4.10 (A) Pressurized cylinders must be chained when stored. The minimum safe distance between stored fuel gas cylinders and any flammable materials is 20 ft or a wall 5 ft high. (B) Refrigerant cylinders chained to a wall for safety. Courtesy of Ferris State University. Photo by John Tomczyk

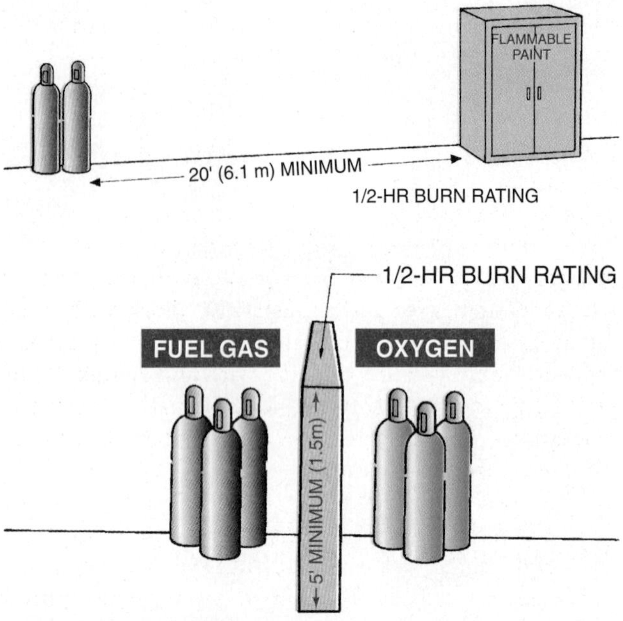

(A)

(B)

and secure the protective cap. All stored cylinders must be supported so they will not fall over, and they must be separated as required by code, **Figures 4.10(A)** and **(B)**.

4.3 ELECTRICAL SAFETY

Electrical shocks and burns are ever-present hazards. It is impossible to troubleshoot all circuits with the power off, so you must learn safe methods for troubleshooting "live" circuits. As long as electricity is contained in the conductors and the device is where it is supposed to function, there is nothing to fear. When uncontrolled electrical flow occurs, if you were to touch two live wires, for example, you are very likely to get injured.

⚠ SAFETY PRECAUTION: *Electrical power should always be shut off at the distribution panel and locked out in an approved manner when installing equipment. The Occupational Safety and Health Administration (OSHA) provides specific details for ensuring safe working conditions, including electrical lockout and tagout procedures. Whenever possible, the power should be shut off and locked out when servicing the equipment. Electrical panels have a place for a padlock for the purpose of lockout,* **Figure 4.11.** *To avoid someone turning the power on, you should keep the panel locked when you are out of sight and keep the only key with you at all times. Emergency STOP controls can be installed for added safety,* **Figure 4.12.** *Do not ever think that you are good enough or smart enough to work with live electrical power when it is not necessary.*•

However, at certain times, tests have to be made with the circuit power on. Extreme care should be taken when making these tests. Ensure that your hands touch only the meter probes and that your arms and the rest of your body stay clear of all electrical terminals and connections. Always work on electric circuits as if they are energized, even if you know that they are not.

Figure 4.11 Electrical disconnect "locks out" with padlock. Courtesy of Ferris State University. Photo by John Tomczyk

Figure 4.12 Emergency stop disconnect for electrical safety. Courtesy of Ferris State University. Photo by John Tomczyk

Working on electric circuits with the same degree of care all the time will help ensure that you remain safe on the job at all times.

SAFETY PRECAUTION: *Do not stand in a wet or damp area when working on equipment. Use only the proper pieces of test instrumentation and make certain that they are in good working condition. Wear heavy shoes with an insulating sole and heel. Intelligent and competent technicians take all precautions.•*

ELECTRICAL SHOCK

An electrical shock occurs when you become part of an electric circuit and current flows though your body. Current flowing through your body can cause damage to your heart—stop it from pumping—resulting in death if it is not restarted quickly. It is a good idea to take a comprehensive first aid course that includes instruction on various lifesaving techniques. **SAFETY PRECAUTION:** *To prevent electrical shock, avoid becoming a conductor between two live wires or a live wire and ground.•* In order for electric current to flow, there must be a complete path for the current to take. The amount of current flow through a circuit is directly related to the amount of voltage being supplied to the circuit. The higher the voltage that you are working with, the greater the potential for injury. In this industry, we work with circuits that may have voltages of 460 V or up to 560 V. Always know the power supply you are working with and act accordingly. Do not let your body create a conductive path for current to take. **Figure 4.13** illustrates a situation where a technician has become part of a circuit.

SAFETY PRECAUTION: *Technicians should use only properly grounded power tools connected to properly grounded power supplies and circuits.•* Caution should be used when using portable electric tools. Although not common, some portable electric tools are constructed with metal frames. These should all have a grounding wire in the power cord to protect the

Figure 4.13 This figure shows ways that the technician can become part of the electrical circuit and receive an electrical shock.

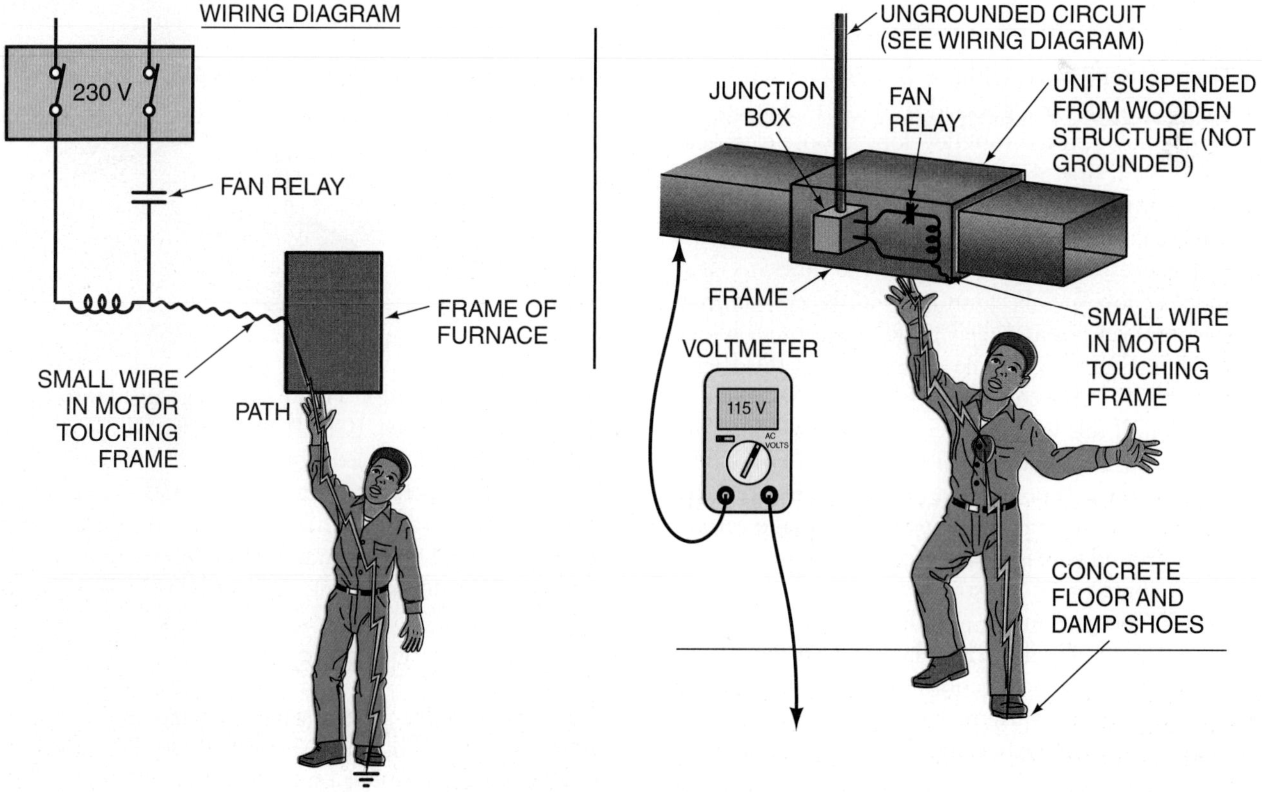

Figure 4.14 (A) An electrical circuit to ground from the metal frame of a drill. (B) Metal frame of a drill properly grounded.

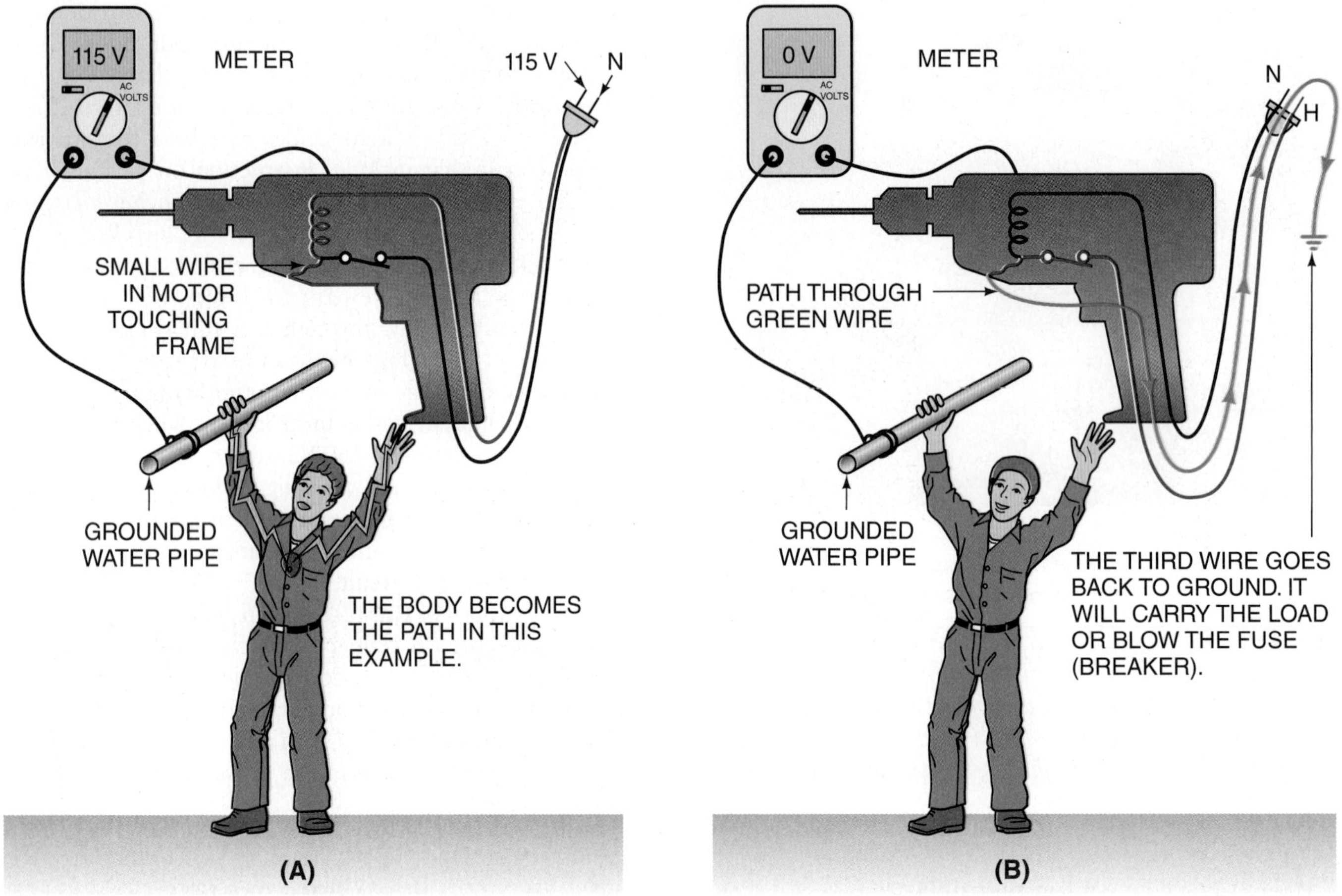

operator. The tool will work without this ground wire, but it is not safe and it is never a good idea to use tools that are not properly grounded. If the tool develops a loose electrical connection and the frame of the tool becomes electrified, the ground circuit, rather than your body, will carry the current, and a fuse or breaker will interrupt the circuit, **Figure 4.14**.

In some instances, technicians may use the three- to two-prong adapters at job sites because the wall receptacle may have only two connections and their portable electric tool has a three-wire plug, **Figure 4.15**. The adapter has a third wire that must be securely connected to a ground for the circuit to provide protection. If this wire is fastened under the wall plate screw and the screw terminates in an ungrounded box nailed to a wooden wall, there is no ground protection, **Figure 4.16**. Before using any plug-in tool, it is important to make certain that the third or ground wire is properly connected to a ground.

Advances in battery and instrument technologies have led to a dramatic increase in the popularity of plastic-cased, battery-operated tools, **Figure 4.17**. Plastic-cased tools have the motor and electrical connections insulated within the tool. Battery-operated tools use rechargeable batteries, so it is a good idea to have multiple charged backup batteries on hand, **Figure 4.18**.

Figure 4.15 A three- to two-prong adapter.

The extension cord in **Figure 4.19** is plugged into a ground fault circuit interrupter (GFCI) receptacle. GFCIs are recommended for use with corded, portable electric tools. Designed to help protect the operator from shock, GFCIs detect very small electrical leaks to ground. When such an electrical current leak is detected, the GFCI will open the circuit, preventing further current flow and greatly reducing the possibility of personal injury.

ELECTRICAL BURNS

Do not wear metallic jewelry while working on live electric circuits because they can cause shock and possible burns. Never use a screwdriver or other tool in an electrical panel when the

Figure 4.16 The wire from the adapter is intended to be fastened under the screw in the duplex wall plate. However, this will provide no protection if the outlet box is not grounded.

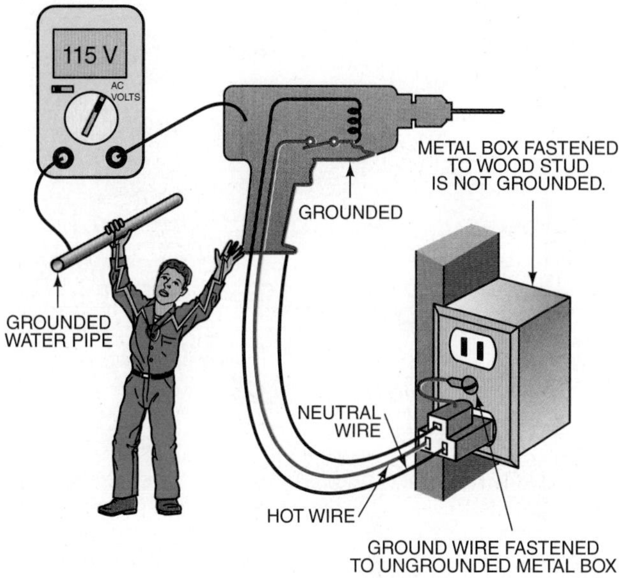

Figure 4.17 Battery-powered tools. Photo by Eugene Silberstein

power is on. Electrical burns can come from an electrical arc, or if the screwdriver slips while you were working in a panel or on an electric circuit. When such an accident occurs, large amounts of electric current flows, as the circuit created has very low resistance. Consider the following examples. When a circuit has a resistance of 10 Ω and is operated on 120 V, it would, using Ohm's law, have a current flow of

$$I = \frac{E}{R} = \frac{120\,\text{V}}{10\,\Omega} = 12\,\text{A}$$

In this calculation, *I* represents the current, *E* represents the voltage, and *R* represents the resistance. If this example is calculated again with less resistance, the current will be greater because the

Figure 4.18 Batteries and a charger. Photo by Eugene Silberstein

Figure 4.19 An extension cord with a ground fault circuit interrupter receptacle. Photo by Eugene Silberstein

voltage is divided by a smaller number. If the resistance is lowered to 1 Ω, the current flow is then

$$I = \frac{E}{R} = \frac{120\,\text{V}}{1\,\Omega} = 120\,\text{A}$$

If the resistance is further reduced to 0.1 Ω, the current flow will be 1200 A. By this time the circuit breaker will likely have already tripped, but you may have already received serious burns or an electrical shock. Current flow of 0.015 ampere or less through the body can prove fatal.

4.4 FIRST AID

As the name implies, first aid is the aid that is given first, often within moments of an individual becoming injured. This initial assistance is often performed by non-medical professionals in an attempt to stabilize the injured person and prevent an injury from getting worse while waiting for medical professionals to arrive at the scene. The first aid that is administered, or not administered, within the first few minutes of sustaining and injury can, quite literally, be the difference between life and death. Common first aid practices include, but are definitely not limited to, cleaning cuts or scratches, removing splinters, lancing blisters, applying cream to a burn, providing fluids, and applying bandages or other wound coverings. The main goals of first aid can be summarized as helping to:

- preserve life
- prevent injuries from getting worse
- relieve pain
- promote recovery
- protect unconscious individuals from further danger

When tending to an injured person that might be in severe danger, it is important to make certain that someone calls 9-1-1 for professional assistance. Quite often, bystanders assume that someone else has already notified the first responders, and do not make the call. This can delay the response time and can put the injured individual in unnecessary, additional danger. It is far better for the first responders to receive multiple calls for help than none at all.

In order to perform first aid procedures, it is strongly recommended that a well-equipped first aid kit, **Figure 4.20**, be available on the job site. A well-equipped first aid kit should contain:

- Adhesive bandages—various sizes
- Sterile gauze pads—various sizes
- Triangular bandages
- Disposable latex or rubber gloves
- Eye wash
- Eye bath
- Antiseptic pads
- Antiseptic cream
- Burn cream
- Aspirin
- Painkillers such as ibuprofen
- Scissors
- Tweezers
- Oral thermometer
- Adhesive cloth bandage tape
- Ace bandages
- Breathing barrier
- First aid instruction booklet

To ensure that the first aid kit is complete and ready for use when needed, it is recommended that the first aid kit be checked regularly so that any items that have been removed can be replaced. Please visit the OSHA and the American Red Cross websites, www.osha.gov and www.redcross.org, for more detailed information on first aid regulations, procedures, training, and supplies.

Figure 4.20 Contents of a typical first aid kit. Courtesy of Ferris State University. Photo by John Tomczyk

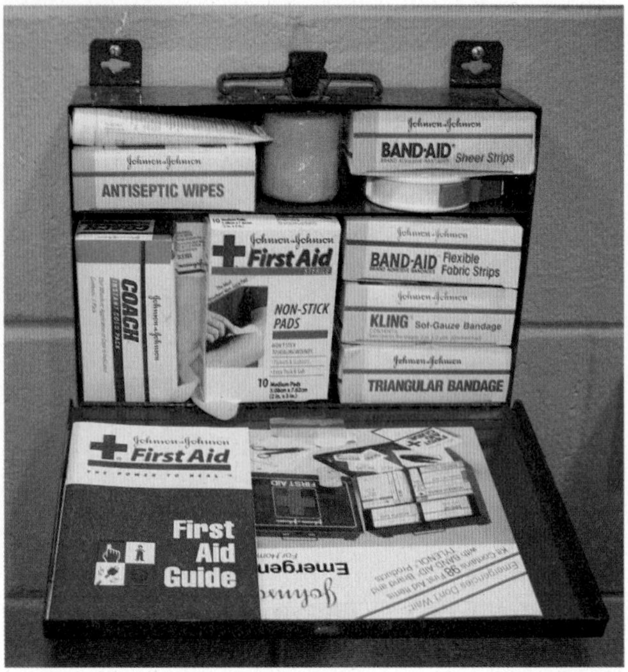

Not only is it a good idea to know how to properly administer traditional first aid to those in need, lives can be saved by learning how to perform cardiopulmonary resuscitation, CPR, and how to use an automatic external defibrillator (AED).

CPR is a lifesaving technique that is used to keep oxygenated blood flowing to the brain, and involves a combination of chest compressions and rescue breaths. Together, these actions can help prevent brain damage or death. CPR is performed on individuals who have stopped breathing or whose heart has stopped pumping. Individuals should be trained to properly perform CPR and should be retrained every few years.

Automatic external defibrillators, AEDs, **Figure 4.21**, are devices that can be used to help save the lives of individuals who go into cardiac arrest, which is the sudden loss of heart function. These devices deliver electric shocks, as required, to help restore heart function. AEDs are becoming very common in public places and, although it is a good idea to receive training on how to use the device, untrained individuals can use them in case of an emergency. Using the device is not complicated and detailed instructions are provided on the device itself. In addition, the AED uses voice prompts to guide users through the process of administering the shocks as needed.

Figure 4.21 An automatic external defibrillator (AED). Photo by Eugene Silberstein

Figure 4.22 An extension ladder, left, and an A-frame ladder, right. Photo by Jason Obrzut

4.5 LADDER SAFETY

HVACR service technicians and installers will have to use ladders to access equipment, filters, rooftops, and system components that are located out of arm's reach. When ladders are needed on the job, it is good field practice to use ladders that are made of a nonconductive material such as wood or fiberglass. Ladders made of wood are not nearly as common now as they once were. Ladders made of aluminum, although much lighter than those made of wood or fiberglass, should be avoided as they conduct electricity and can cause severe personal injury.

Two common types of ladder are the "A-frame" ladder and the extension ladder, **Figure 4.22**. The A-frame ladder is freestanding ladder, while the extension ladder is designed to be leaned up against the side of a building or other solid landing surface such as a roof hatch.

Before using any ladder, it is important to thoroughly inspect it for any physical damage. If any damage is found, the ladder should not be used. In addition, the ladder must be free of any oil or grease and the rubber feet must be in good shape. A-frame ladders are to be used only when fully opened and should never be used by leaning the closed ladder up against a wall or other object.

SAFETY PRECAUTION: *A technician may raise a ladder into a power line or place it against a live electrical hazard without realizing it. Chances should not be taken. When the technician is standing on a nonconducting ladder, it will provide protection from electrical shock to ground. However, it will not provide protection if the user makes contact between two or more electrical conductors.* •

When an extension ladder is used for access to an upper landing area such as a roof, it should extend at least 3 ft above the landing surface. Also, the horizontal distance from building to the foot of the ladder should be approximately one-fourth the working length of the ladder, **Figure 4.23**. Consider the example where an extension ladder must be used to access the roof of a 20-ft building. The foot of the ladder should be slightly more than 5 ft from the building. In this example, the actual distance can be calculated using the Pythagorean Theorem. It is left as a student exercise to determine that the actual distance from the foot of the ladder to the base of the building is 5.25 ft. Ensure that ladders are placed on a stable, level surface, not on slippery surfaces, and that they have slip-resistant feet. Ladders should be used only for their designed purpose and should not be loaded with more weight than they were designed to hold. Use other methods for raising loads to the roof when the weight is excessive. Ladders should be maintained free of oil, grease, and any other slipping hazards.

Working from heights on rooftops or from ladders is potentially very dangerous. Rooftops have different pitches or slants. When the angle becomes steep, the technician should use a

Figure 4.23 The side rails of a portable ladder should extend 3 ft above the upper landing surface. The angle of the ladder should be such that the horizontal space at the bottom is about one-fourth the distance of the working length of the ladder.

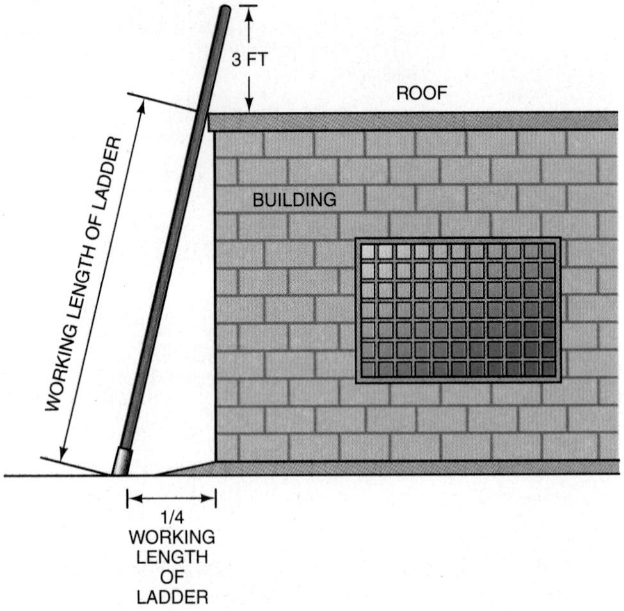

safety belt and be tied off in such a manner that a fall would be stopped. Technicians should always work in pairs when doing rooftop work or when working on a high ladder. In many cases, the ladder should be secured at the top, particularly if it is to be used for several climbs. **SAFETY PRECAUTION:** *Never use the cross-bracing on the far side of stepladders for climbing. The far side is only safe for climbing if it is provided with steps.•*

When climbing ladders, you should always be making three points of contact with the ladder. These three contact points can be two feet and one hand, or two hands and one foot. Therefore, when carrying tools and equipment up a ladder, care must be taken to use the proper techniques, methods, or tools. **Figure 4.24** shows a service technician transporting a refrigerant cylinder with a carrying strap. The carrying strap enables technicians to carry cylinders up a ladder, while still maintaining three points of contact with the ladder. If three points of contact cannot be maintained, tools and equipment should be lifted to roof by using ropes or other acceptable methods.

Figure 4.24 Carrying strap for carrying refrigerant cylinders up a ladder safely. Courtesy SC Products, Inc. Wilmington, DE

4.6 HEAT

The use of **heat** requires special care. Torches used for soldering, brazing, welding, among other things, have a high concentration of heat. Combustible materials may be in the area where torches are used. When soldering or using concentrated heat, you should have a fire extinguisher close by, and you should know exactly where it is and how to use it, **Figure 4.25**. Learn to use a fire extinguisher *before* a fire occurs. A fire extinguisher should always be included as a part of the service tools and equipment on a service truck.

Although there are many different types of fire extinguishers available, they all work in the same way. Fire extinguisher use can best be described by the acronym PASS.

- P: Pull the pin from the top of the fire extinguisher. Without removing the pin, the handle cannot be squeezed.
- A: Aim the nozzle of the fire extinguisher at the BASE of the fire. The nozzle should be about 6 to 8 feet from the fire.
- S: Squeeze the handle on the fire extinguisher.
- S: Sweep the nozzle from side to side until the fire has been extinguished.

Figure 4.25 A typical fire extinguisher. Photo by Bill Johnson

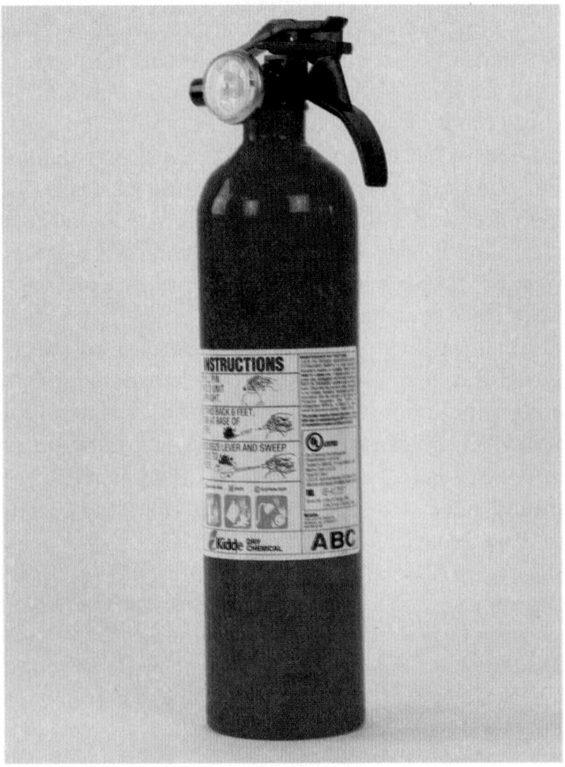

Figure 4.26 Fire extinguisher details showing the types of fires it should be used on.

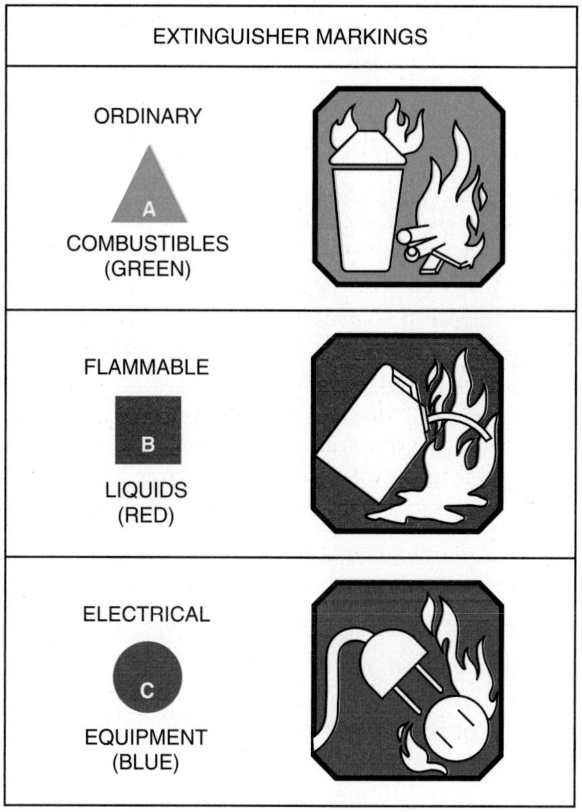

Many fire extinguishers are designed for specific types of fires. Type A is for trash, wood, and paper fires. Type B is for burning liquids and Type C is for electrical equipment fires. However, some fire extinguishers are rated for all three types of fires, **Figure 4.26**. Usually, these extinguishers are the dry-chemical type. Make sure you know what type of fire extinguisher you are using on a fire, or the fire can spread and serious personal injury and property damage can occur.

SAFETY PRECAUTION: *When a solder connection must be made next to combustible materials or a finished surface, use a shield of noncombustible material for insulation,* **Figure 4.27.** *A fire-resistant spray may also be used to decrease the flammability of wood if a torch must be used nearby. The spray retardant should be used with an appropriate shield.* • A shield is also often necessary when soldering within an equipment cabinet, for example, when an ice-maker compressor is changed or when a drier must be soldered in line. A shield should be used to protect any wires that may be close by.

SAFETY PRECAUTION: *Never solder tubing lines that are sealed. Service valves or Schrader ports should be open before soldering is attempted. When heat from a torch is added to a sealed tubing line, pressure will build up inside the tubing and may result in serious personal injury.* •

Figure 4.27 A shield used when brazing or soldering.

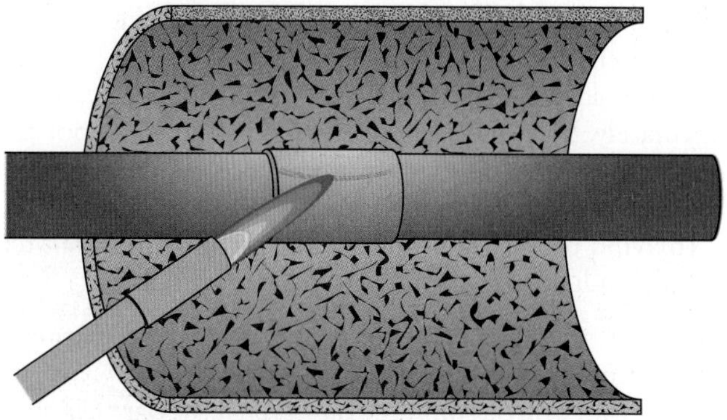

SAFETY PRECAUTION: *Hot refrigerant lines, hot heat exchangers, and hot motors can burn your skin and leave a permanent scar. Care should be used while handling them.* •

SAFETY PRECAUTION: *Working in very hot weather or in hot attics can be very hard on the body. Technicians should be aware of how their body and the bodies of those around them are reacting to the working conditions. Watch for signs of overheating, such as someone's face turning very red or someone who has stopped sweating. Get them out of the heat and cool them off. Overheating can be life-threatening; call for emergency help.* •

In warmer climates, work in attics should be performed, whenever possible, in the early morning hours, before the heat of the day creates unsafe working conditions.

When exposed to excessive heat, the body reacts by exhibiting some tell-tale signs that something is wrong. Under no circumstances should these signs be ignored. If ignored, severe personal injury can result. Two common ailments related to excessive heat exposure are heat exhaustion and heatstroke. The symptoms for heat exhaustion and heat stroke are provided in **Figure 4.28**.

If it is suspected that an individual is suffering from heat exhaustion, move the victim to a cooler place and encourage the individual to drink cool water. Applying cool

Figure 4.28 Symptoms of heat exhaustion and heat stroke.

HEAT EXHAUSTION	HEATSTROKE
Feeling faint or dizzy	Throbbing headache
Excessive sweating	No sweating
Skin is cool, pale and clammy	Skin is hot, red and dry
Nauseous feeling, vomiting	
Rapid, weak pulse	Rapid, strong pulse
Muscle cramps and weakness may occur	Confusion, slurred speech, loss of consciousness
Core body temperature: Normal to 104°F	Core body temperature: Above 104°F

compresses and taking a cold shower will help reduce the symptoms. If left untreated, heat exhaustion can become worse, possibly resulting in heatstroke, which can be life-threatening.

If heat stroke is suspected, it is important to call 9-1-1 immediately. If possible, move the individual to a cool place and, using a cool, damp rag, apply cool compresses to the skin to help further cool the victim.

To avoid becoming a victim of heat-related injuries, technicians should:

- Attempt to stay indoors as much as possible during the hottest part of the day
- Drink water throughout the day, even if they are not thirsty
- Wear loose-fitting, light-colored clothing whenever possible
- Avoid diuretics, or drinks that dehydrate, such as coffee and other caffeinated products
- Wear "breathable" fabrics, such as cotton

4.7 COLD

Cold can be as harmful as heat. Liquid refrigerant can freeze your skin or eyes instantaneously. But long exposure to cold is also harmful. Working in cold weather can cause frostbite. Wear proper clothing and waterproof boots, which also help protect against electrical shock. When cold, wet, and uncomfortable, technicians do not always think clearly and might make decisions that can put them in danger. Make it a point to stay warm. **SAFETY PRECAUTION:** *Be aware of the wind-chill when working outside. The wind can quickly take the heat out of your body, and frostbite can occur. Waterproof boots not only protect your feet from water and cold but help to protect you from electrical shock when you are working in wet weather. However, do not depend on these boots for protection against electrical hazards.*

Low-temperature freezers are just as cold in the middle of the summer as they are in the winter. Cold-weather gear must be used when working inside these freezers.• For example, an expansion valve may need changing, and you may be in the freezer for more than an hour. It is a shock to your body when you step out from a zero-degree freezer to the warm space outside, which may be as hot as 100°F. If you are servicing any low-temperature applications, carry a coat and gloves with you and be sure to wear them as needed.

4.8 MECHANICAL EQUIPMENT

Rotating equipment can damage property and cause severe personal injury. Motors that drive fans, compressors, and pumps are among the most dangerous because they have so much power. **SAFETY PRECAUTION:** *If a shirt sleeve or coat were caught in a motor drive pulley or coupling, severe injury could occur. Loose clothing should never be worn around rotating machinery,* **Figure 4.29.** *Even a small electric drill can wind a necktie up before the drill can be shut off.•*

Figure 4.29 Never wear a necktie or loose clothing when using or working around rotating equipment.

SAFETY PRECAUTION: *When starting an open motor, stand well to the side of the motor drive mechanism.•* If the coupling or belt were to fly off the drive, it would fly outward in the direction of rotation of the motor. All set screws or holding mechanisms must be tight before a motor is started, even if the motor is not connected to a load. Keep all wrenches away from a coupling or pulley. A wrench or nut thrown from a coupling can be a lethal projectile.

SAFETY PRECAUTION: *When a large motor, such as a fan motor, is coasting to a stop, do not try to stop it. If you try to stop the motor and fan by gripping the belts, the momentum of the fan and motor may pull your hand into the pulley and under the belt,* **Figure 4.30.•**

SAFETY PRECAUTION: *Never wear jewelry while working on a job that requires much movement. A ring may be caught on a nail head, or a bracelet may be caught on the tailgate of a truck as you jump down.•*

Figure 4.30 Never attempt to stop a motor or other mechanism by gripping the belt.

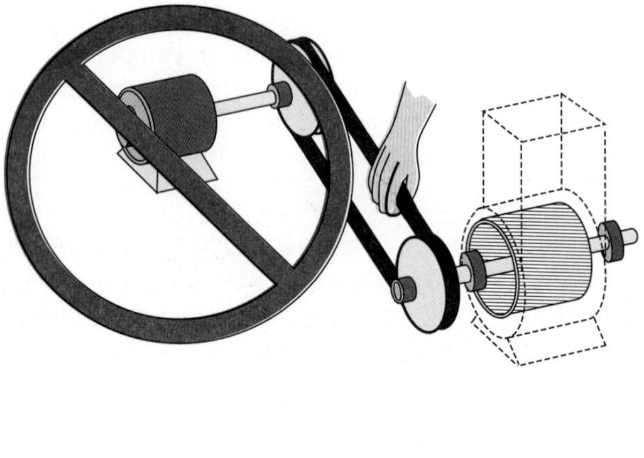

Figure 4.31 Use a face shield when grinding. Photo by Eugene Silberstein

Figure 4.32 Keep the tool rest on a grinder adjusted properly.

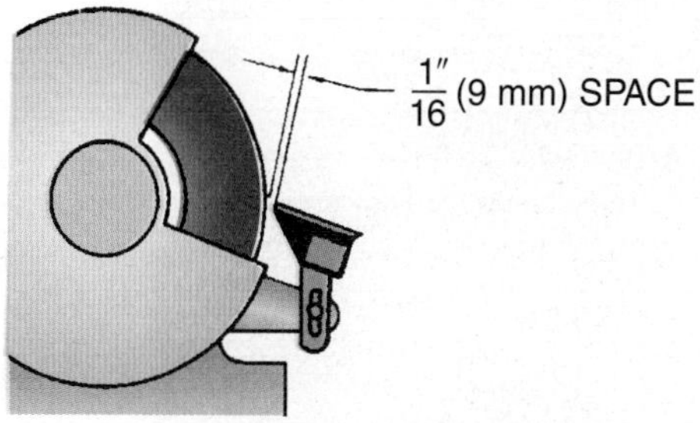

SAFETY PRECAUTION: *When using a grinder to sharpen tools, remove burrs, or for other reasons, always use a face shield,* **Figure 4.31.**• Most grinding stones are made for grinding ferrous metals, such as cast iron, steel, and stainless steel. However, other stones are made for nonferrous metals, such as aluminum, copper, or brass. Use the correct grinding stone for the metal you are grinding. The tool rest should be adjusted to approximately 1/16 in. from the grinding stone, **Figure 4.32**. As the stone wears down, keep the tool rest adjusted to this setting. A grinding stone must not be used on a grinder that turns faster than the stone's rated maximum revolutions per minute (rpm), as it may explode.

4.9 MOVING HEAVY OBJECTS

From time to time heavy objects must be moved. Think out the best and safest method for moving these objects. Never rely solely on muscle power. Special tools can help you move large and heavy equipment safely. Do not try to lift heavy equipment by yourself. Get help from another person and use tools and equipment designed for that purpose.

Figure 4.33 (A) Use your legs, not your back, to lift objects. Keep your back straight. (B) Use a back brace belt. Courtesy Wagner Products Corp.

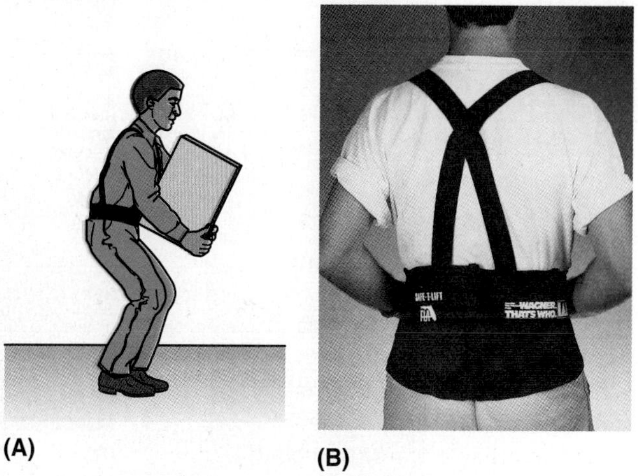

SAFETY PRECAUTION: *When you must lift, use your legs, not your back, and wear an approved back brace belt,* **Figure 4.33.**• Some available tools are a hand truck, a lift gate on a pickup truck, and a portable dolly, **Figure 4.34**.

When moving large equipment across a carpeted or tiled floor or across a gravel-coated roof, first lay down some plywood. Keep the plywood in front of the equipment as it is moved along. When equipment has a flat bottom, such as a package air conditioner, short lengths of pipe may be used to move the equipment across a solid floor, **Figure 4.35**.

4.10 REFRIGERANTS IN YOUR BREATHING SPACE

Refrigerant vapors and many other gases are heavier than air and can displace the oxygen in a closed space. Because of this, proper ventilation must be used at all times to avoid being overcome by lack of oxygen. Should you be in a confined space and the concentration of refrigerant become too great, you may not notice it until it is too late. Your symptoms would be a dizzy feeling, and your lips may become numb. If you should feel this way, move quickly to a place with fresh air.

SAFETY PRECAUTION: *Proper ventilation should be set up in advance of starting a job. Fans may be used to push or pull fresh air into a confined space where work must be performed. Cross-ventilation can help prevent a buildup of fumes.*•

SAFETY PRECAUTION *Confined spaces can be dangerous and the utmost care should be taken when working in them. A confined space is an area that has limited access and is, by Occupational Safety and Health Administration (OSHA) standards, not suitable for human inhabitants. Confined spaces have the potential to pose many hazards including injury due to the inhalation of refrigerants*

Figure 4.34 (A) A hand truck. (B) The lift gate on a pickup truck. (C) A dolly. Photos by Bill Johnson

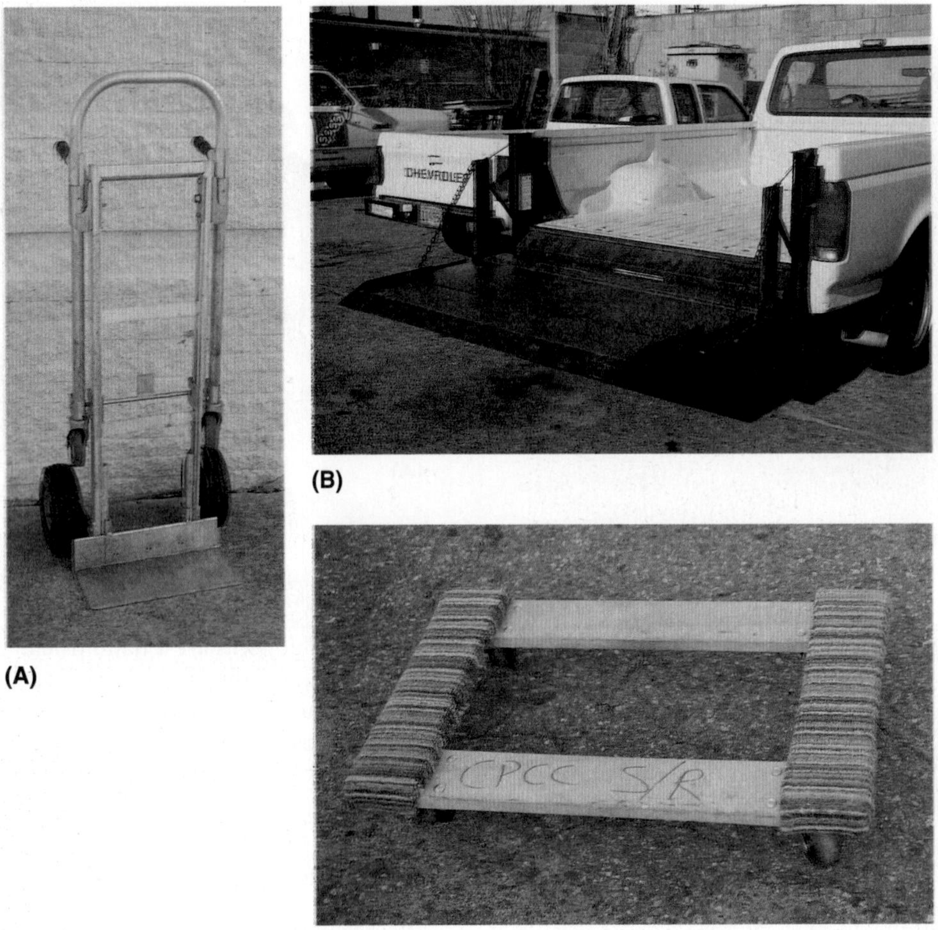

(A)

(B)

(C)

Figure 4.35 Moving equipment using short lengths of pipe as rollers.

or other substances. Other potential hazards include fires, falls, entrapment, and flooding. Always be aware of your surroundings and know how to react in the event an accident occurs. When working in tight spaces, make certain that you are not alone and that others know you are there in the event something goes wrong. •

Some installations require special leak detectors with alarms. These detectors sound an alarm well in advance of a harmful buildup. **Figure 4.36** shows a refrigerant-specific leak detector for HCFC-123 (R-123) that will sniff out certain parts per million (ppm) of refrigerant. If the programmed ppm level is reached, the detector will sound an alarm and color-coded lights will flash. When the leak-detector alarm sounds, take proper precautions, which may require special breathing gear. Turn on the ventilation system. **Figure 4.37** shows an evacuation plan for the second floor of a building in case of a refrigerant leak, fire, or any other emergency.

Refrigerant vapors that have been heated during soldering or because they have passed through an open flame (e.g., when there is a leak in the presence of a fire) are dangerous. They have a strong odor, are toxic, and may cause harm. **SAFETY PRECAUTION:** *If you are soldering in a confined place, keep your head below and to the side of the rising fumes and make sure that you have plenty of ventilation,* **Figure 4.38.** *This helps ensure that you do not inhale the fumes and are also out of the way in case molten flux or solder drips from the fitting.* •

Figure 4.36 Refrigerant-specific leak detector for R-123 with alarm lights. Courtesy of Ferris State University. Photo by John Tomczyk

Figure 4.37 An evacuation plan on building's interior wall for safety. Courtesy of Ferris State University. Photo by John Tomczyk

The American National Standards Institute (ANSI) and the American Society of Heating, Refrigerating and Air-Conditioning Engineers (ASHRAE) has developed Standard 34-2010 for Designation and Safety Classifications of Refrigerants. **Figure 4.39** is a matrix outlining Standard 34-2010, and the following explains its letter and number classifications.

Class A—For refrigerants where toxicity levels have not been identified at concentrations less than or equal to 400 ppm (parts per million).

Class B—For refrigerants where evidence of toxicity exists at concentrations below 400 ppm.

Class 1—For refrigerants that do not show flame propagation when tested in 65°F air at 14.7 psia.

Class 2—For refrigerants with a lower flammability limit of more than 0.00625 lb/ft^3 when tested at 70°F and 14.7 psia. They also have a heat of combustion of less than 8174 Btu/lb.

Figure 4.38 Keep your face below the heated area and ensure that the area is well ventilated.

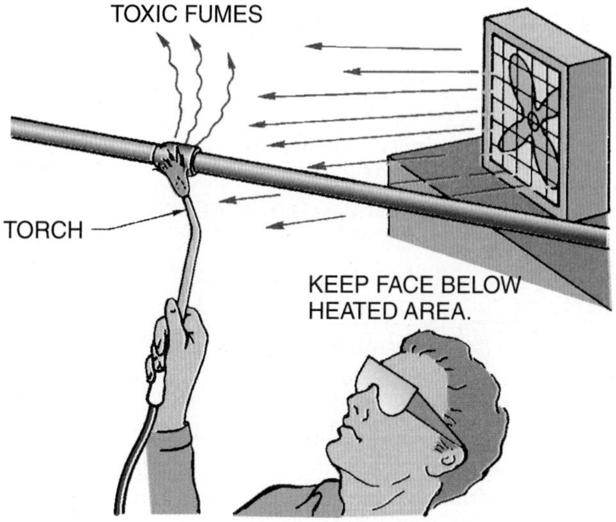

Figure 4.39 ASHRAE's Standard 34-2010 for Designation and Safety Classifications of Refrigerants.

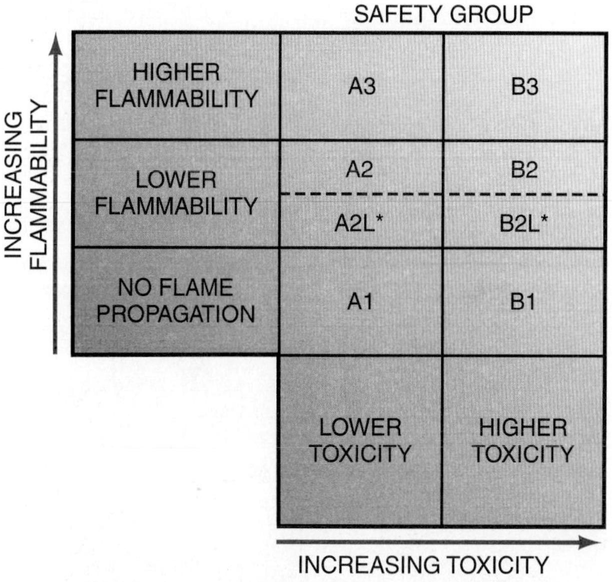

* A2L and B2L are lower flammability refrigerants with a maximum burning velocity of ≤10 cm/s (3.9 in./s).

Class 3—For refrigerants with high flammability. Their lower flammability limit is less than or equal to 0.00625 lb/ft^3 when tested at 70°F and 14.7 psia. They have a heat of combustion greater than or equal to 8174 Btu/lb.

The two subclasses (A2L and B2L) apply to lower flammability refrigerants. The capital letter "L" refers to "Lower" flammability.

4.11 USING CHEMICALS

HVACR service technicians come in contact with many different chemicals on a daily basis while performing their job. Chemicals are used to help perform a number of different tasks such as cleaning dirty condenser and evaporator coils. The chemicals that are used are typically mild, except for some harsher products that are used for water treatment. **SAFETY PRECAUTION:** *These chemicals should be handled according to the manufacturer's directions. Do not get careless. If you spill chemicals on your skin or splash them in your eyes, follow the manufacturer's first aid instructions and go to a doctor. It is a good idea to read the entire label before starting a job. It is hard to read the first aid treatment for eyes after your eyes have been damaged.* •

Following a compressor burnout, the refrigerant and oil in a compressor can be harmful, as acids can form when the compressor fails. The contaminated refrigerant and oil may be hazardous to your skin, eyes, and lungs. **SAFETY PRECAUTION:** *Keep your distance if a refrigerant line is subject to rupture or any amount of refrigerant is allowed to escape.* •

Emergency eye and body wash stations should be readily available in order to deal with accidents involving any harmful chemicals or refrigerants, **Figures 4.40** and **4.41**. Emergency call boxes can also be located throughout the building for added safety, **Figure 4.42.**

Figure 4.40 Eye wash station. Courtesy of Ferris State University. Photo by John Tomczyk

Figure 4.41 Body wash station. Courtesy of Ferris State University. Photo by John Tomczyk

Figure 4.42 Emergency call station. Courtesy of Ferris State University. Photo by John Tomczyk

REVIEW QUESTIONS

1. To help remain safe on the job, it is always recommended that field personnel
   A. test fire extinguisher operation regularly.
   B. test fire alarm operation before starting work.
   C. be aware of all potential hazards around them.
   D. work alone to help prevent careless accidents.
2. Which of the following represents three pieces of personal protection equipment (PPE)?
   A. Ear plugs, ladders, safety gloves
   B. Hard hats, ear plugs, work boots
   C. Back brace, safety glasses, plastic-banded watches
   D. Short-sleeved shirts, hard hats, work boots
3. Which of the following can happen to a full nitrogen cylinder if the top is broken off?
   A. Nitrogen will slowly leak out
   B. There will be a loud bang
   C. The tank can become airborne
   D. The nitrogen vapor will freeze
4. Which of the following tasks could potentially subject a service technician to frostbite?
   A. Connecting or removing a service hose from a suction-line service port
   B. Connecting or removing a service hose from a liquid-line service port
   C. Connecting or removing a service hose from a discharge line service port
   D. Connecting or removing a service hose from the inlet port on a vacuum pump
5. Which of the following statements is true regarding the use of nitrogen?
   A. The nitrogen tank must be outfitted with a pressure regulator.
   B. Nitrogen must only be added to systems that are in a deep vacuum.
   C. Nitrogen must only be added to fully charged HVACR systems.
   D. Dry nitrogen is a volatile gas that follows a pressure/temperature relationship.
6. What can happen when oil is mixed with oxygen under pressure?
   A. An explosion can occur
   B. The ability of the oil to lubricate will increase
   C. The mixture will solidify
   D. Methane gas will form
7. Which organ is most likely to be affected by electric shock?
   A. Liver
   B. Kidney
   C. Lungs
   D. Heart
8. When electric current flows through the body, which of the following injuries will commonly result?
   A. Hearing loss and burns
   B. Shocks and burns
   C. Death and hearing loss
   D. Burns and death
9. Which of the following factors will most directly determine the severity of an electric shock?
   A. Current flow through the victim
   B. The height of the victim
   C. The weight of the victim
   D. Voltage that the victim is subjected to
10. Which of the following statements is true regarding working on electrical circuits?
    A. Ensure that the circuit power is on whenever working on electric circuits.
    B. Metallic jewelry reduces the possibility of receiving an electric shock.
    C. Be sure to use proper lock-out and tag-out procedures on all equipment.
    D. Using aluminum ladders will protect individuals from electricity-related injuries.
11. What can be done to prevent damaging the surroundings while soldering or brazing?
    A. Use a lower-temperature torch
    B. Use a heat shield
    C. Apply extra flux to the joints
    D. Properly clean all fittings
12. The ground wire on most systems is commonly color-coded
    A. white.
    B. red.
    C. black.
    D. green.
13. A Class A fire extinguisher is to be used on which type of fire?
    A. Electrical
    B. Grease
    C. Paper
    D. Chemical
14. Which of the following refrigerant safety groups has the highest flammability and the highest toxicity?
    A. A1
    B. B1
    C. A3
    D. B3
15. When lifting a heavy object by yourself, make certain that you
    A. lift with your back.
    B. avoid wearing a back brace.
    C. lift with your legs.
    D. avoid wearing gloves.

16. Which of the following is the safest acceptable method for moving a piece of heavy equipment across a smooth, flat roof?
    A. Dragging it
    B. Using pipe rollers
    C. Pulling it
    D. Carrying it by hand
17. What should be done before using chemicals to clean a dirty evaporator coil?
    A. Read the label on the chemical container
    B. Spray a small amount of the chemical on your skin
    C. Turn off all exhaust and ventilation fans
    D. Smell the chemical to see if it burns
18. If a compressor motor experiences a burnout, which of the following can be produced?
    A. Oil
    B. Water
    C. Grease
    D. Acid
19. How far above the landing surface should an extension ladder extend?
    A. 3 feet
    B. 6 feet
    C. 1/4 of the ladder's working length
    D. 1/2 of the ladder's working length
20. If an extension ladder is used to access the roof of a 10-foot high building, the base of the ladder should be approximately how many feet from the building?
    A. 2.5 feet
    B. 5 feet
    C. 7.5 feet
    D. 10 feet
21. With respect to fire extinguishers, the acronym PASS stands for
    A. Push, Aim, Squeeze, Sweep
    B. Push, Adjust, Sweep, Squeeze
    C. Pull, Aim, Squeeze, Sweep
    D. Pull, Aim, Sweep, Squeeze
22. When working around rotating machinery, which of the following is true?
    A. Be sure to remove all protective shields while the equipment is operating.
    B. Allow all rotating equipment to come to a complete stop before servicing.
    C. It is safe to stop rotating equipment with your hand if the power is off.
    D. Avoid wearing tight-fitting clothing when working on rotating equipment.
23. All of the following are true regarding first aid except
    A. first aid is intended to help relieve pain.
    B. first aid is intended to help prevent injuries from getting worse.
    C. first aid is intended to help promote recovery.
    D. first aid is intended to be administered only by trained professionals.
24. A confined space can be defined as an area or place that
    A. is not intended for human occupancy.
    B. has an interior volume that is less than 1000 cubic feet.
    C. is intended to accommodate no more than three people.
    D. is enclosed on all sides with no access openings.
25. To prevent injury from leaking refrigerants, chemical vapors, or torch fumes, it is good field practice to
    A. only work outdoors.
    B. ensure adequate ventilation.
    C. only use chemicals with low toxicities.
    D. wear a dust mask whenever possible.

UNIT 5

TOOLS, INSTRUMENTATION, AND EQUIPMENT

OBJECTIVES

After studying this unit, you should be able to

- list three benefits of using the correct tool for the job.
- explain how to determine if a screwdriver is properly sized for the screw to be turned.
- explain how nut drivers differ from traditional straight slot and Phillips head screwdrivers.
- explain the function of a refrigeration service wrench.
- explain the differences between an open-end wrench and a box-end wrench.
- explain the purpose of the universal joint in a socket wrench set.
- explain the benefit of using a ball-tipped Allen wrench.
- list four types of hammers commonly used in the HVACR industry.
- list three types of general purpose pliers.
- list four hand tools commonly used in conjunction with electrical wiring.
- list three common types of hand saws used by HVACR field personnel.
- explain the purpose of the pilot bit on a hole saw.
- explain the purpose and uses of a chalk line.
- explain the purpose of a fish tape.
- explain the benefits of battery-powered tools.
- explain the differences between a reciprocating saw and a jigsaw.
- describe the function of a digital multimeter (DMM).
- describe the function of a clamp-on ammeter.
- list four types of refrigerant leak detectors.
- explain the basic operation of an electronic leak detector.
- list eight tools that are commonly used when working with soft-drawn tubing.
- explain the differences between flaring and swaging.
- explain why bending refrigerant tubing is preferred over using fittings.
- explain how the length of a pipe wrench handle affects its ability to tighten and loosen piping connections.
- describe the three common types of tin snips.
- explain the purpose of crimpers and notchers as they apply to sheet metal work.
- describe air-acetylene and oxyacetylene torch kits.
- compare the function of a two-valve gauge manifold with that of a four-valve gauge manifold.
- describe the function of a micron gauge.
- describe the function of a refrigerant charging scale.
- explain the purpose of a vacuum pump.
- explain the purpose of a refrigerant recovery unit.
- describe the instrument used to measure the speed of an air stream in a duct system.
- describe the units that are used to measure the pressures in a duct system.
- describe the function of a psychrometer.
- describe the purpose of a "wet" kit as it applies to combustion.
- explain the importance of performing a smoke test on an oil-fired heating appliance.
- describe three popular pieces of test instrumentation that are used by an energy auditor to evaluate duct systems and the integrity of a structure's envelope.

SAFETY CHECKLIST

- ☑ Keep tools free from grease and oil to prevent slipping, which can cause severe personal injury.
- ☑ Use sharp blades, bits, saws, and other cutting instruments. Sharp blades require less force to use and, as a result, are much safer. Dull blades, bits, and saws should not be used.
- ☑ Many tool-related accidents and injuries result from distractions. Concentrate on the task at hand to help reduce the chance of becoming injured.
- ☑ Tools are designed to perform specific tasks. Do not use tools for tasks or jobs that they were not intended to perform. Improper tool use can result in personal injury.
- ☑ Inspect the cords on all power tools before using them. Do not use corded power tools if the cord is damaged.
- ☑ Be sure to fully understand the operation of a tool before using it. Always read all literature provided with the tool prior to use.

5.1 INTRODUCTION TO TOOLS, INSTRUMENTATION, AND EQUIPMENT

Air-conditioning, heating, and refrigeration technicians must know how to properly use hand tools and specialized equipment relating to this field. Technicians must be completely aware of how a particular tool operates as well as its limitations. Technicians should read all tool-use instructions prior to using a tool and must not operate or use the tool if its intended use does not align with the task to be performed. Using the

correct tool is more efficient, saves time, and is safer. Accidents often can be prevented by using the correct tool. This unit contains a brief description of some of the general and specialized tools and equipment that HVACR technicians use in the field. Some of the tools and pieces of equipment introduced here are described in more detail throughout the book, primarily in the units where the tools are used.

Generally speaking, when purchasing tools, it is a good idea to buy the best tools that you can afford. Good hand tools are usually more expensive, but they typically perform better and will usually last longer. For example, an inexpensive screwdriver will often have an undersized handle and will be hard to grip. Also, the bit on the end will not be made of hard, tempered steel and will deform easily. The better choice would be a screwdriver with a comfortable handle and a very hard tip, which is under great stress when the technician is working with a very tight screw. Good hand tools are usually guaranteed for replacement if they are broken during regular use.

5.2 HAND TOOLS

SCREWDRIVERS AND NUT DRIVERS

By far, the most common of all hand tools is the screwdriver, **Figure 5.1**. Screwdrivers are manufactured as a shaft with a handle grip on one end and a uniquely shaped tip on the other. The tip on the end of the screwdriver is inserted into the slot or groove in the screw that needs to be turned. The most common types of screwdriver in the United States are the Phillips head, Figure 5.1(A) and the straight-slot, Figure 5.1(B). To gain access to tight-fitting spaces, offset screwdrivers are readily available, Figure 5.1(C). Other screwdriver tip configurations are shown in Figure 5.1(D). To reduce the number of tools that must be carried by a technician, multitip screwdriver tools, such as the 6-in-1 tool shown in Figure 5.1(E) are very popular. Some multitip tools have as many as twelve different interchangeable tips. When selecting a screwdriver, make certain that the screwdriver tip fits snugly into the screw slot, **Figure 5.2**.

Figure 5.2 The screwdriver tip must fit snugly into the slot. (A) This tip is far too small. (B) This tip is too thick to fit into the slot. (C) This tip is too wide. (D) This tip fits perfectly into the slot.

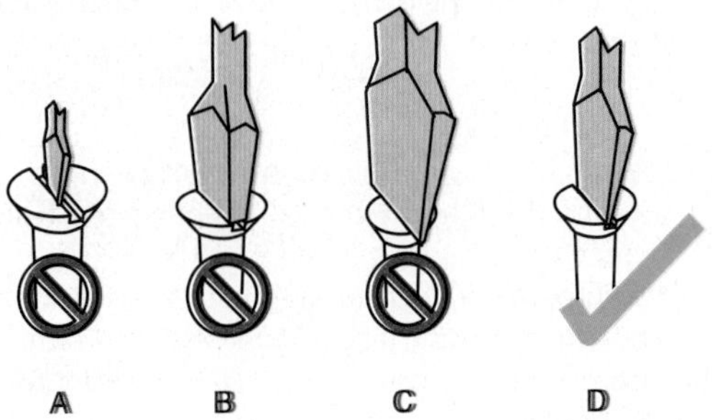

Figure 5.1 Screwdrivers. (A) Phillips tip. Photo by Bill Johnson (B) Straight or slot blade. Photo by Bill Johnson (C) Offset. Photo by Bill Johnson (D) Standard screwdriver bit types. Courtesy of Klein Tools. (E) 9-in-1 Multi Tool. Courtesy of Hilmor.

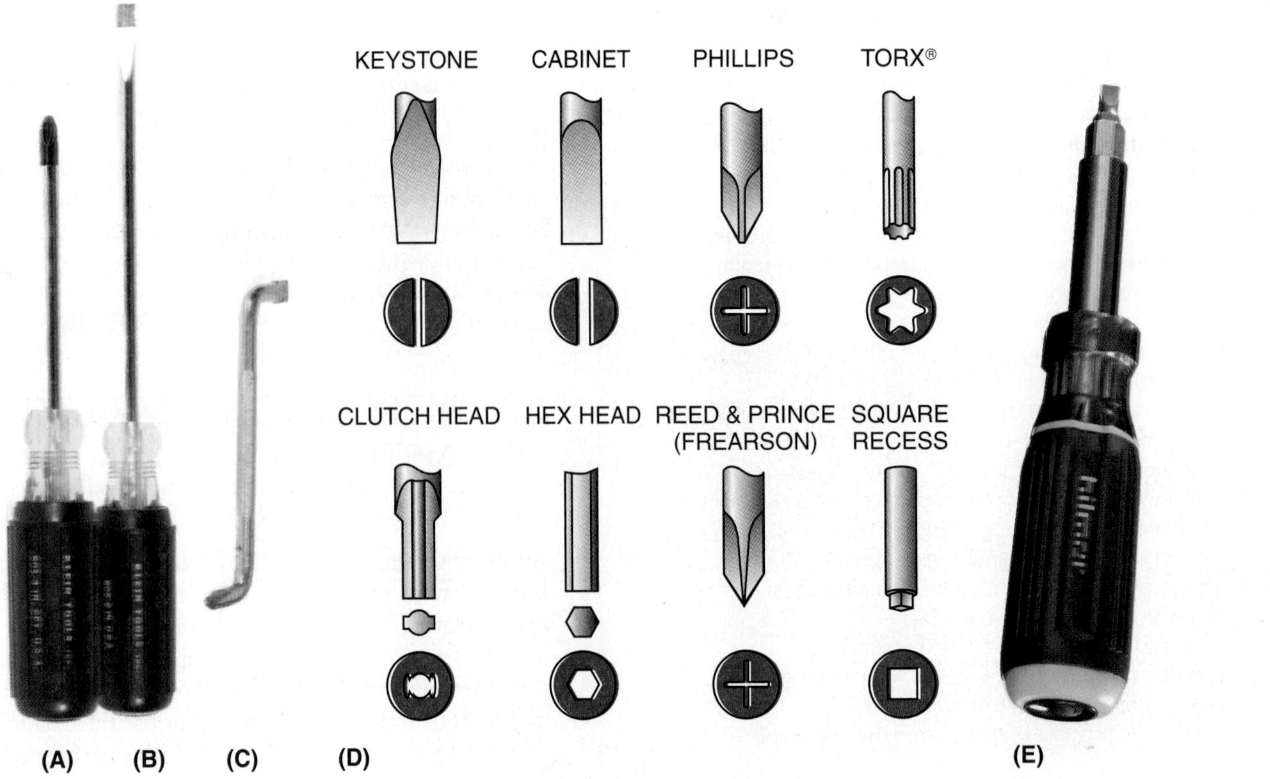

Figure 5.3 Thermostat screwdriver. Photo by Eugene Silberstein

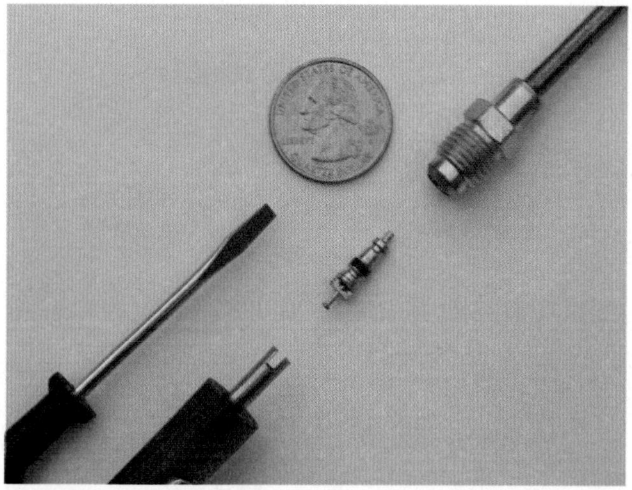

In the HVACR industry, control wires must be connected to control devices such as thermostats, so very small thermostats, referred to as thermostat screwdrivers, **Figure 5.3**, should be a part of every service technician's tool box. These small screwdrivers often have a special tool called a Schrader tool incorporated into the grip. The Schrader tool is used to loosen and tighten Schrader pins, which are part of self-sealing valves that are similar to those found in car and bicycle tires.

⚠ SAFETY PRECAUTION: *Screwdrivers are intended to tighten and loosen screws. They are not to be used as substitutes for chisels, pry bars, or any other tool. Personal injury can result if tools are improperly used.* •

Nut drivers, **Figure 5.4**, have socket heads and are used primarily to turn hex-head screws on the cabinets of HVACR equipment, **Figure 5.5**. They are available in a number of configurations including solid shaft, hollow shaft, extra-long and stubby shafts. The hollow-shaft variety allows the screw to extend into the shaft when it extends past the nut, **Figure 5.6**. Stubby shafts are used to get into tight spaces. The most commonly used nut driver sizes are 1/4 in. and 5/16 in.

Figure 5.4 Nut drivers. Photo by Eugene Silberstein

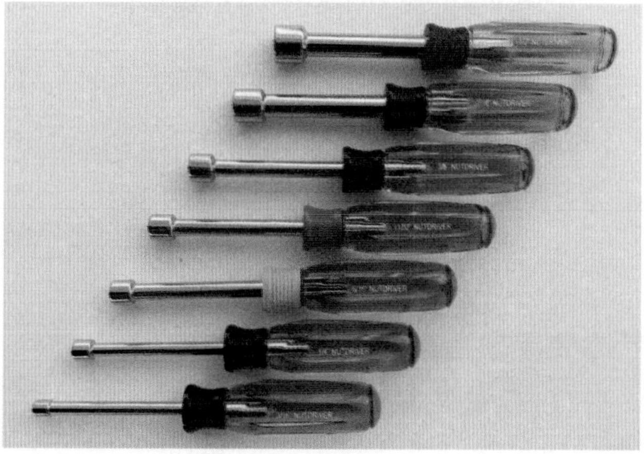

Figure 5.5 Nut drivers are designed to turn hex-headed screws and bolts.

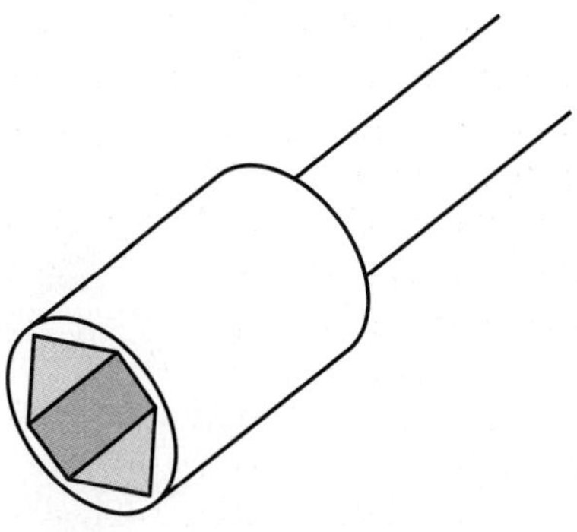

Figure 5.6 Nut drivers with hollow shafts are used when the screw or bolt extends past the nut.

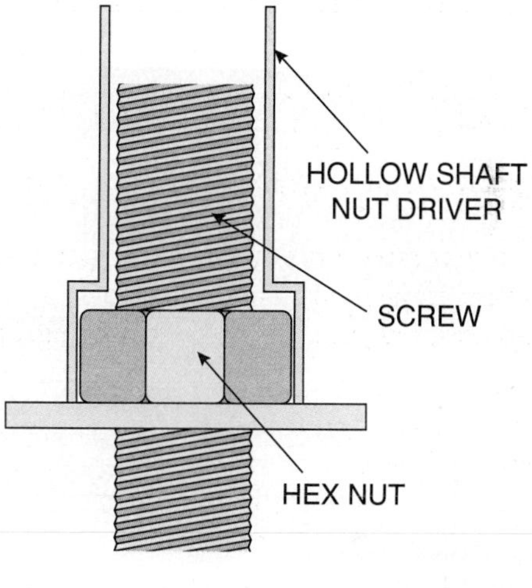

WRENCHES AND SOCKETS

ADJUSTABLE WRENCH AND REFRIGERATION SERVICE WRENCH. Two very common wrenches that can be found in the HVACR service technician's toolbox are the adjustable wrench and the refrigeration service wrench, **Figure 5.7**. The adjustable wrench is a desirable tool because it is a single tool that can be used to turn many different size hex-head bolts. When using the adjustable wrench, it is important that the tool is positioned so the user pulls on the wrench while exerting force on the stationary portion of the tool's jaw, **Figure 5.8**. The refrigeration service wrench is a ratcheting tool that is used to turn square stems on valves and other system components, **Figure 5.9**. The ratchet feature on this wrench allows the turning direction to be changed by pushing the lever-like button near the end of the wrench. Two openings on each end of the wrench allow it to be used on four sizes of valve stems or fittings. Another tool that is similar to the refrigeration service wrench is the ratcheting hex wrench, **Figure 5.10**. This wrench works in exactly the same manner as the refrigeration service wrench, but is used on hex-head bolts.

Figure 5.7 An adjustable wrench (left) and a refrigeration service wrench (right). Photo by Eugene Silberstein

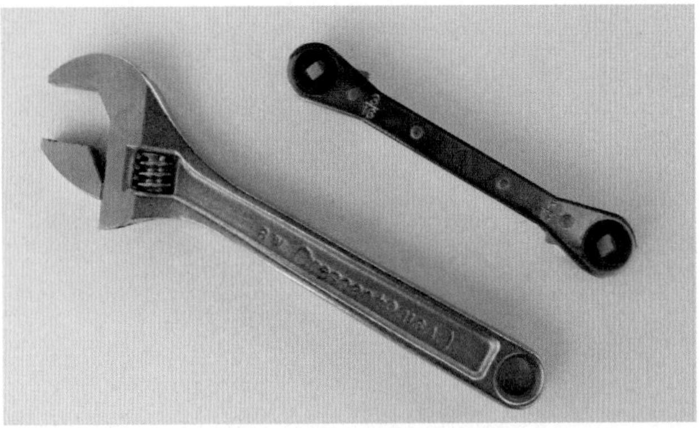

Figure 5.8 Proper use of an adjustable wrench. Photo by Eugene Silberstein

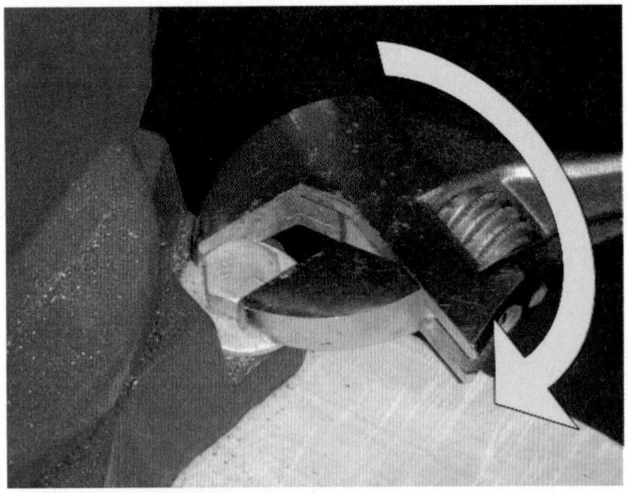

Figure 5.9 The refrigeration service wrench is used to turn square stems. Source: Ritchie Engineering Company, Inc. Yellow Jacket Products Division.

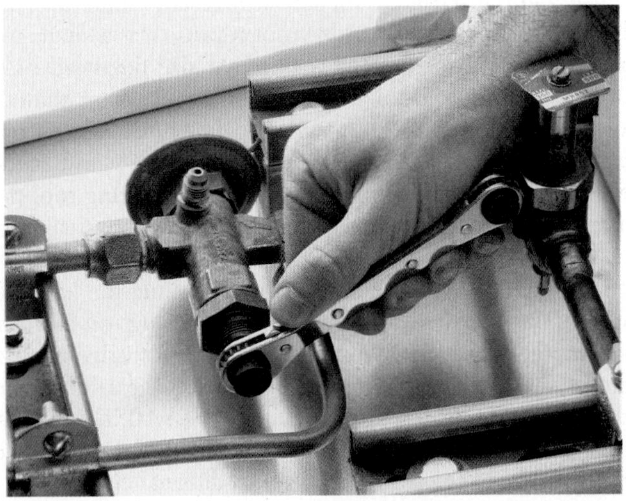

Figure 5.10 The ratcheting hex wrench is used to turn bolts with hex-heads. Source: Ritchie Engineering Company, Inc. Yellow Jacket Products Division.

OPEN-END, BOX-END, AND COMBINATION WRENCHES. Other common wrenches are the open-end, box-end, and combination wrench, **Figure 5.11**. The open-end wrench is desirable because it can be slipped onto a nut from the side but, because it has an open end, can slip. The open-end wrench is not intended to loosen extremely tight nuts or bolts or to provide the final tightening of a fastener. The box-end wrench is a general-duty wrench but provides more torque than the open-end wrench and is less likely to slip. The combination wrench has an open-end wrench on one end and a box-end wrench on the other.

FLARE NUT WRENCHES. The flare nut wrench, **Figure 5.12**, is used like a box-end wrench. The opening at the end of the

Figure 5.11 Open-end wrench (left), box wrench (center), combination wrench (right). Photo by Eugene Silberstein

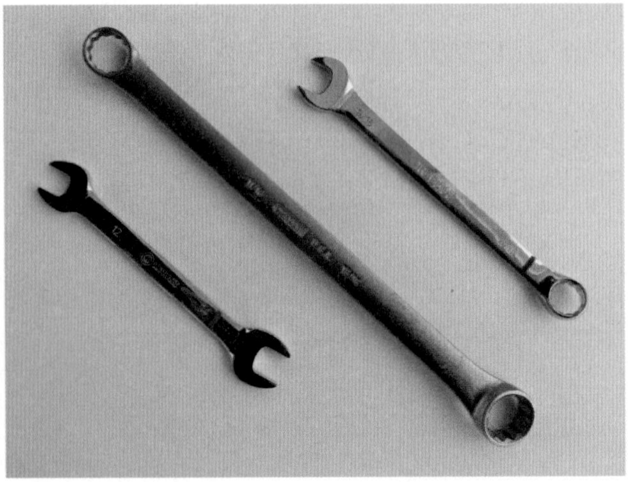

Figure 5.12 Flare nut wrenches. Photo by Bill Johnson

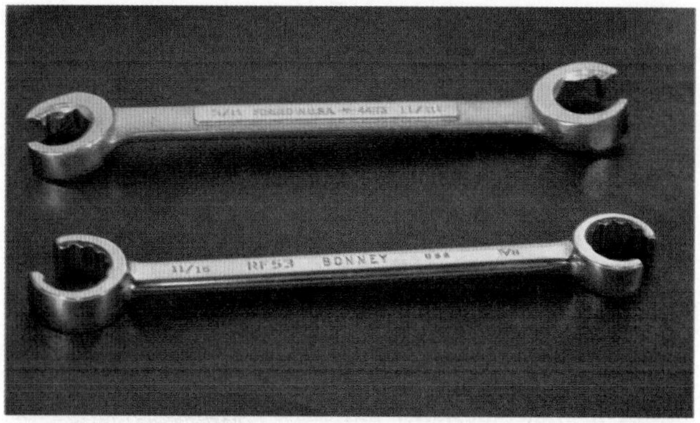

Figure 5.13 Socket set.

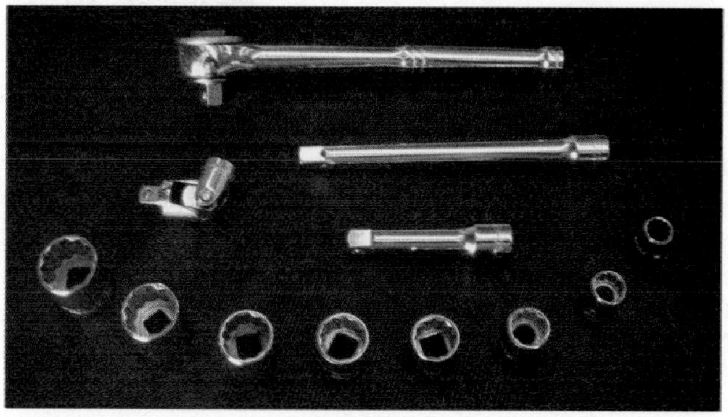

wrench allows it to be slipped over the tubing. The wrench can then be placed over the flare nut to tighten or loosen it. Adjustable wrenches should not be used on flare nuts because they may round off the corners of fittings made of soft brass.

SOCKET WRENCHES. Socket wrenches are used to turn hex-head bolts and are used in conjunction with a ratcheting handle. A socket wrench set, such as the one shown in **Figure 5.13**, is comprised of a set of socket wrenches, a reversible, ratcheting handle, at least one extension, and a universal joint socket. The universal joint socket allows the tool to be used at an angle, which is beneficial when access is limited or difficult.

ALLEN WRENCHES. Allen wrenches, also referred to as Allen keys, **Figure 5.14**, are hex-shaped tools that are used to turn screws and bolts that have heads fashioned with hex-shaped sockets. To use the Allen key, the tool is inserted into the head of the screw or bolt and then rotated to either tighten or loosen it. Allen wrenches are commonly sold as a multikey fold-out set, Figure 5.14(left), but can be purchased as a set of separate keys. Traditional Allen keys must be inserted straight into the screw, but ball-tipped Allen keys, Figure 5.14(right), can be used at an angle. This is especially helpful when the screw to be turned is not readily accessible. Allen wrenches are also available as individual T-handled tools, **Figure 5.15**.

Figure 5.14 Multikey, fold-out Allen key set (left). Ball-tipped Allen keys (right) . Photo by Eugene Silberstein

Figure 5.15 T-handle Allen wrenches. Photo by Eugene Silberstein

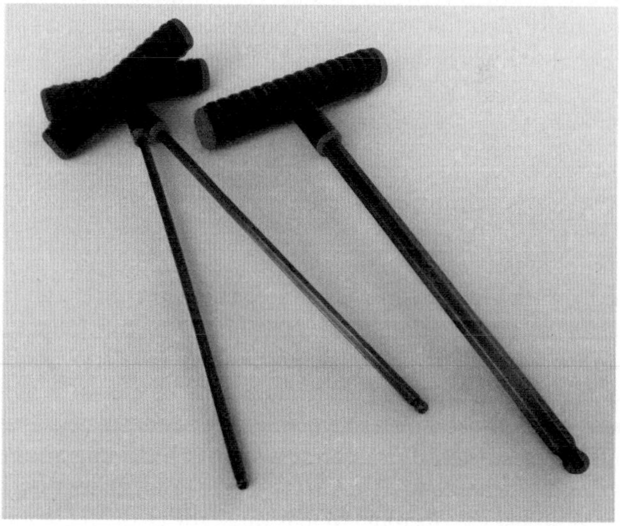

HAMMERS

There are many different types of hammers. Some of the most commonly used types are the claw hammer, the sheet metal hammer, the ball peen hammer, and the soft-faced hammer, **Figure 5.16**. Each type is intended to perform specific tasks. The claw hammer is the most common type of hammer. The head of the hammer is used to drive nails and the claw portion of the tool is used to pry and remove nails from wood. The sheet metal, or tin-knockers hammer, is typically used by individuals who fabricate and/or install ductwork. This hammer, often called a setting hammer, has a square, flat face that is used for flattening seams without causing damage to the metal. The sheet metal hammer also has a beveled, tapered peen that is used to assemble duct sections. The ball peen hammer, often referred to as a mechanic's hammer, has a head that resembles the head of a traditional claw hammer

Figure 5.16 Hammers. From left to right: Claw-hammer, sheet metal hammer, ball peen hammer, soft-faced hammer. Photo by Eugene Silberstein

and a rounded peen that resembles a ball. This rounded peen is typically used to shape metal. Soft-faced hammers or mallets are used when the surface being hammered must not be damaged. These hammers are often used when working with wood or other softer materials.

GENERAL PURPOSE PLIERS

As with other tools, there are many different types of pliers. Each type is intended to perform specific tasks, so it is best field practice to use the appropriate pair of pliers for the job. Three of the most common types of pliers used by the HVACR service technician are slip-joint pliers, pump pliers, and locking pliers, **Figure 5.17**. General purpose slip-joint pliers are used for a number of tasks including holding, turning, and bending. They have a slip joint, allowing the user to adjust the opening dimension of the jaws. These pliers are not intended to perform the final tightening of fasteners. Slip-joint and pump pliers have a tongue-and-groove mechanism that, as with the general purpose pliers, allows the user to adjust the jaws based on the job being performed. The longer handle provides more torque than general slip-joint pliers. These pliers are popular with plumbers and other tradespeople who work with larger diameter pipes. Locking pliers are the most versatile of the three types of pliers just discussed. Locking pliers can function as a pair of pliers, a wrench, or even a vise.

ELECTRICAL PLIERS

Pliers and wiring tools commonly used by electricians are available in many designs. The most common pliers used in conjunction with electrical wiring are the side-cutting pliers, the needle-nose pliers, and the linesman pliers, **Figure 5.18**. Side-cutting pliers are typically used to cut wire or other thin metal object close to a surface. Needle-nose pliers are used to gain access to tight spaces. Needle-nose pliers are popular for HVACR control wiring. Linesman pliers are heavier than needle-nose pliers and can be used to perform a number of functions. They can cut and strip wire as well as perform bending and twisting operations.

WIRING TOOLS

The HVACR service technician needs to work with electrical circuits and perform a number of tasks with wire. Common wiring tools include manual wire strippers, automatic wire strippers, and crimping tools, **Figure 5.19**. To use the manual wire stripper, the wire to be stripped is inserted into the proper slot based on the gauge of the wire. The tool is closed onto the wire and then pulled off to remove the insulation from the wire. Manual stripping tools can also cut wire. Automatic wire strippers are also popular. To use this tool, simply insert the wire into the tool while the tool's handle is not squeezed. The length of the strip is determined by the amount of wire extending

Figure 5.17 Slip-joint pliers (left), pump pliers (center), locking pliers (right). Photo by Eugene Silberstein

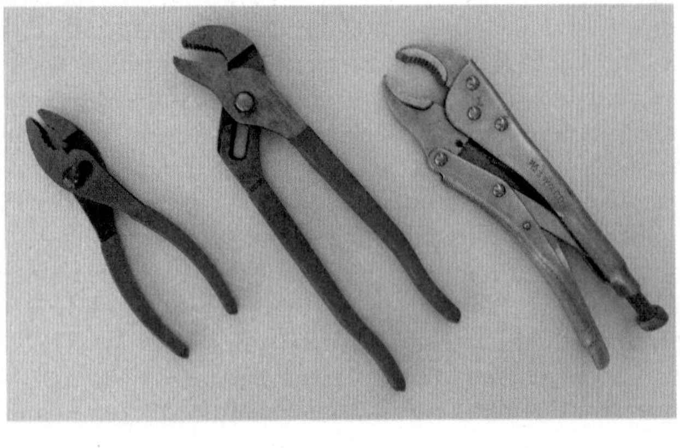

Figure 5.18 Side-cutting pliers (left), needle-nose pliers (center), linesman pliers (right). Photo by Eugene Silberstein

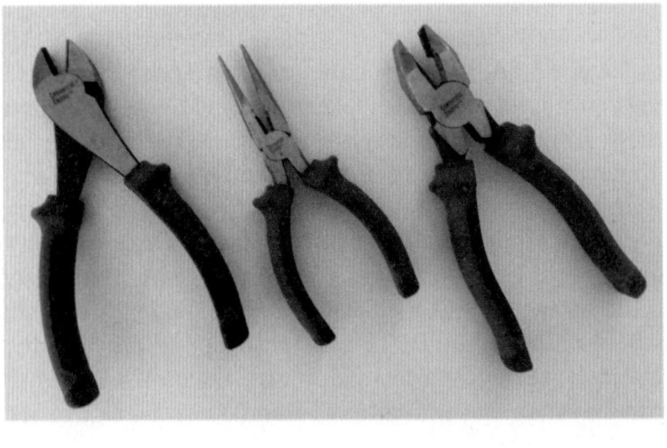

Figure 5.19 Manual wire strippers (left), automatic wire stripper (center), crimping tool (right). Photo by Eugene Silberstein

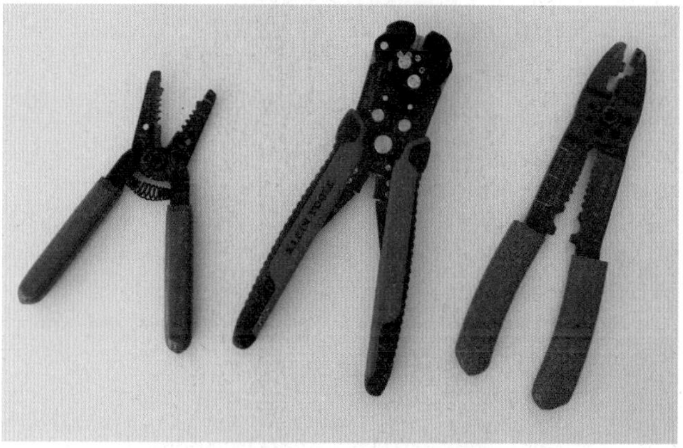

beyond the cutters. Hold the wire in one hand and squeeze the handles with the other. When the handles are released, the stripped wire can be removed. The crimping tool is a combination tool that can be used for crimping solderless connectors, stripping wire, cutting wire, and cutting small bolts.

SAWS

Saws, **Figure 5.20**, are selected based on a number of factors including the material being cut and the accessibility of the item being cut. Materials that are typically cut by HVACR field personnel are wood, drywall, and metal. Common saw types are wood saws, keyhole saws, and hacksaws. The keyhole saw, shown second from the left in Figure 5.20, is typically used to cut sheetrock. The pointed tip on the saw allows the user to simply push the saw into the sheetrock, preventing any additional damage from occurring. There are also variations on saws that make it easier to get into tighter spaces. For example, the saw in the center of Figure 5.20 is a folding

Figure 5.20 Saws. From left to right: Wood saw, keyhole saw, folding jab saw, hacksaw, compact hacksaw. Photo by Eugene Silberstein

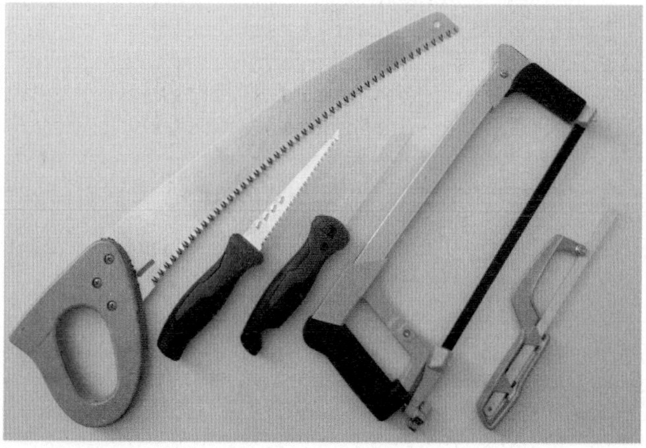

Figure 5.21 Hole saws. Photo by Eugene Silberstein

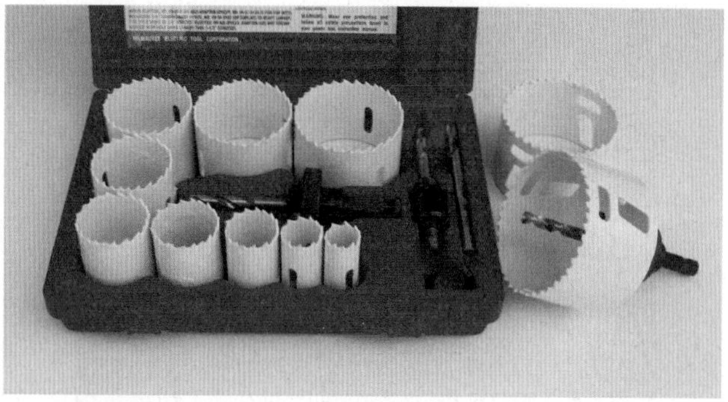

jab saw that accommodates the blade from a reciprocating saw. This allows the user to cut through material by hand with a heavy duty blade otherwise intended for use on power tools. The rightmost saw provides the user with the ability to use a hacksaw blade in spaces where a traditional hacksaw might not fit.

Although not a handsaw, hole saws, **Figure 5.21**, are included in this section. Hole saws are available in a wide range of sizes and are used in conjunction with power drills. The hole saw is mounted on an arbor, which is secured to the drill. A pilot drill bit is mounted in the arbor. This pilot bit allows the user to ensure hole location, while preventing the hole saw from drifting during use.

UTILITY HAND TOOLS

FLASHLIGHTS AND DROPLIGHTS. HVACR field personnel often work in dark locations, so providing light is extremely important to ensure that a safe environment is maintained. Battery-powered lights, **Figure 5.22**, or corded droplights,

Figure 5.22 Battery-powered light. Photo by Eugene Silberstein

Figure 5.23 Droplights. Courtesy of Klein Tools.

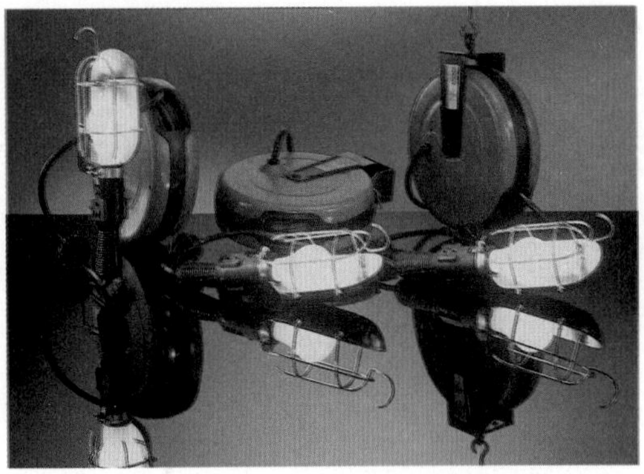

Figure 5.23, can be used. Always make certain that, if the work location is wet, only use corded droplights when they are plugged into outlets that are protected with ground fault circuit interrupters (GFCIs). To ensure continuous operation, make certain that the batteries on battery-powered lights are fully charged.

TAPE MEASURES. Typical tape measures, **Figure 5.24**(left), range from 15 feet to 25 feet in length. In the event larger distances need to be measured, longer, fabric, or plastic tapes can be used, such as the 150-foot tape measure shown in Figure 5.24(right). Digital laser distance meters can also be used, **Figure 5.25**.

UTILITY KNIVES. HVACR technicians should always have a utility knife, **Figure 5.26**, on hand. Be sure to replace the blade frequently to make certain that the knife is sharp, and be sure to retract the blade inside the handle whenever the knife is not in use.

SQUARES, LEVELS, AND CHALK LINES. When installing HVACR systems, it is important that pipes and other equipment be either level or pitched as required by the manufacturer's literature. Tools that help ensure that cuts are made square and that pipes are level or plumb are the square, the

Figure 5.24 25-foot metal tape measure (left) 150-foot plastic tape measure (right). Photo by Eugene Silberstein

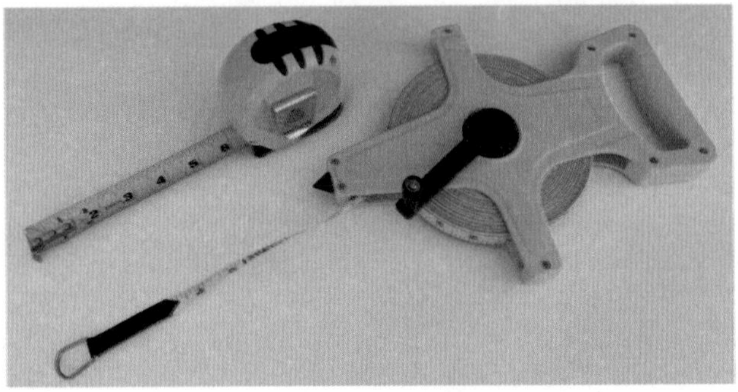

Figure 5.25 Digital distance laser meter. Courtesy of Fluke Corporation

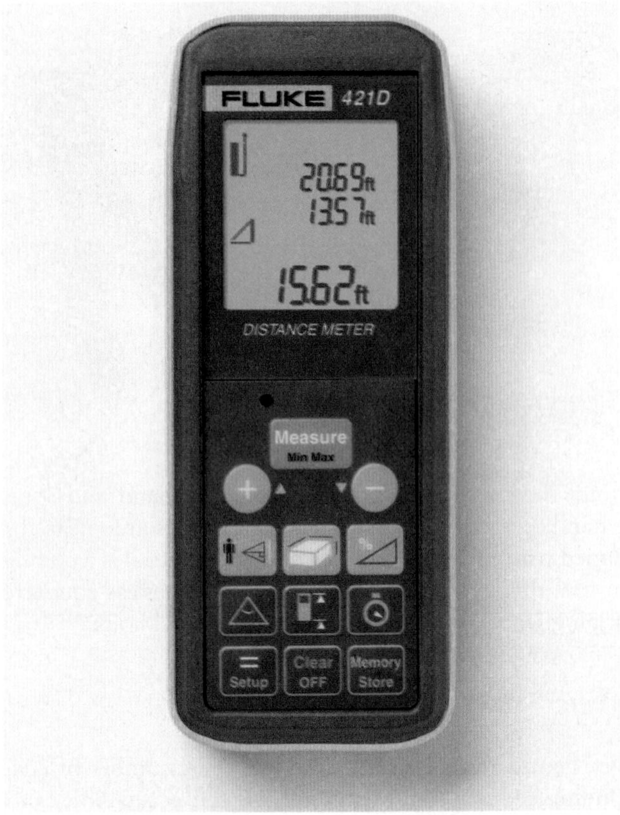

level, and the chalk line, **Figure 5.27** and **Figure 5.28**. The chalk line is comprised of a container that is filled with colored chalk dust. In the container is a roll of string that is covered with the chalk dust. In use, the string is pulled from the container and held taut up against the surface to which

Figure 5.26 Utility knife. Photo by Eugene Silberstein

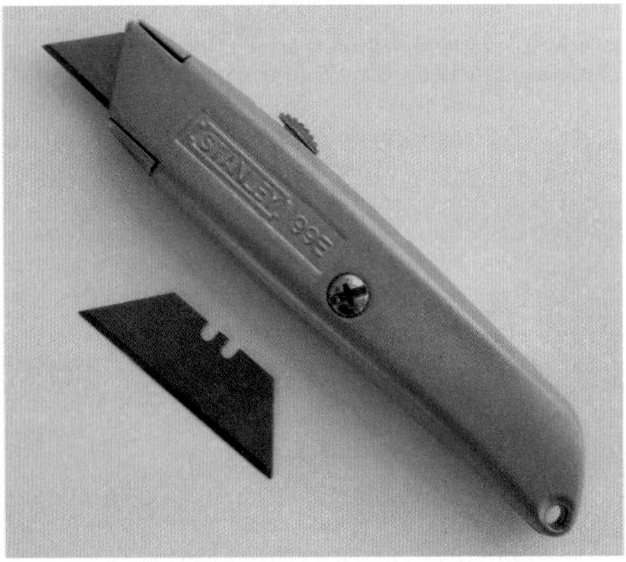

Figure 5.27 Square (left), level (right). Photo by Eugene Silberstein

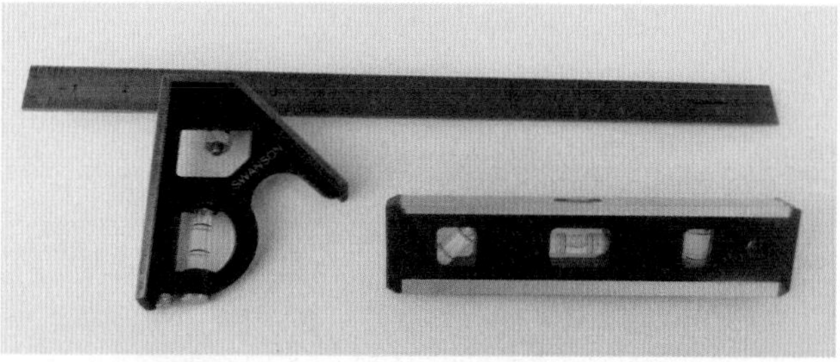

Figure 5.28 Chalk line. Photo by Eugene Silberstein

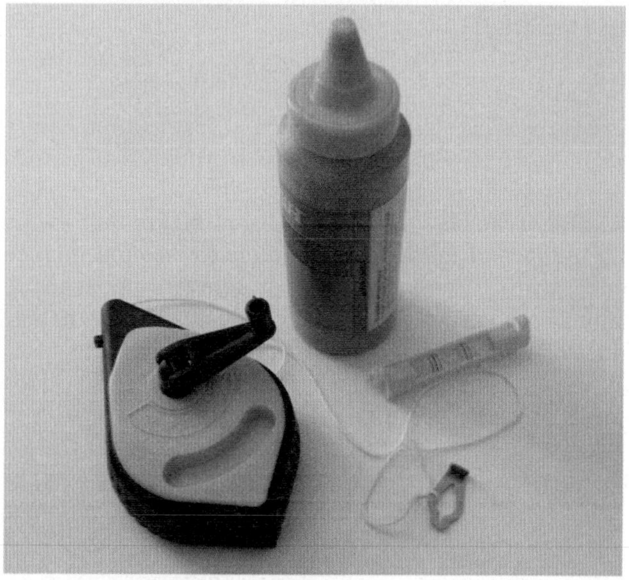

the straight line is to be drawn. While held taut, the string is "snapped" against the surface and a chalk line is left behind. This line is then used as a guide. The chalk dust container on many chalk line kits can be used as a plumb bob, which is used to help create a vertical reference line.

CLAMPS. Clamps, **Figure 5.29**, are tools used to temporarily hold objects together for a variety of purposes. The clamps shown in Figure 5.29 are referred to as C-clamps due to their shape, but other clamp types include trigger clamps, parallel clamps, spring clamps, and corner/angle clamps. Clamps are available in a number of different sizes and should be selected based on the job being performed.

FILES. Files, **Figure 5.30**, are used to remove small amounts of material, usually from wood or metal. There are a number of different configurations of files including flat, round, and half-round. When a larger amount of material is to be removed, a rasp is used. A rasp closely resembles a file, but they have teeth that are coarser than those on files.

Figure 5.29 C-clamps. Photo by Eugene Silberstein

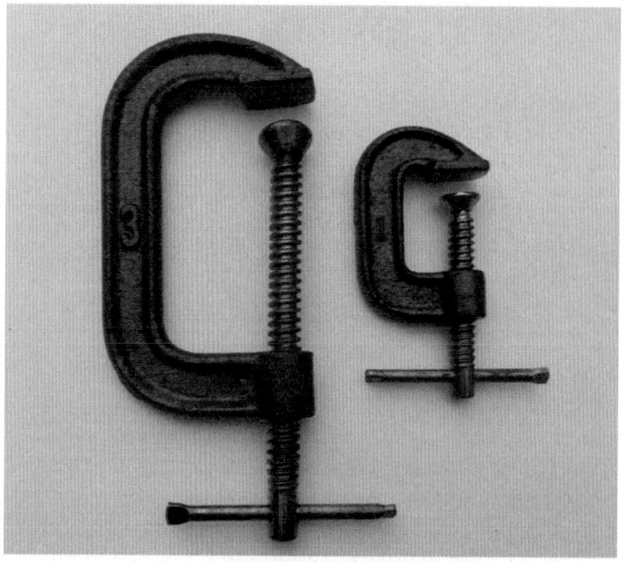

Figure 5.30 Files.

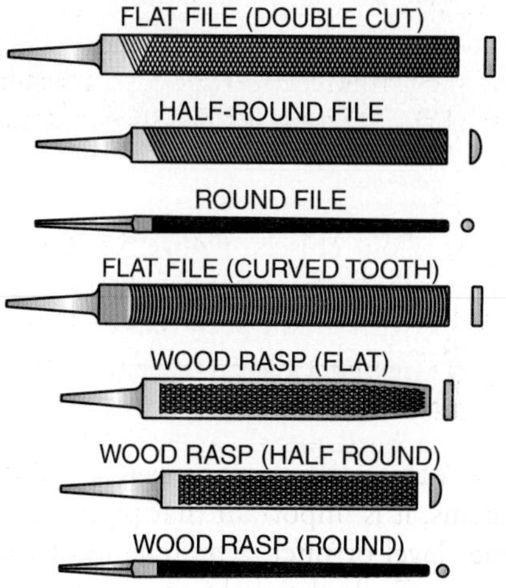

FISH TAPE. A fish tape, **Figure 5.31**, is typically used to snake electrical wires through passages such as walls and electrical piping or conduit. The fish tape is passed through the wall first and, once it has reached the end, one end of the wire is connected to the end of the fish tape. The fish tape is then carefully pulled out, pulling the wire behind it.

CAULKING GUN. The caulking gun, **Figure 5.32**, is used to extract sealants or adhesives from tubes that are inserted into the gun. By squeezing the handle on the caulking gun, the bottom of the sealant or adhesive cartridge is pushed up, causing the material to be pushed from the tube. A wide variety of sealants and adhesives are available in tube or cartridge form that fit into the caulking gun.

Figure 5.31 Electrical snake. Courtesy of Ideal Industries, Inc.

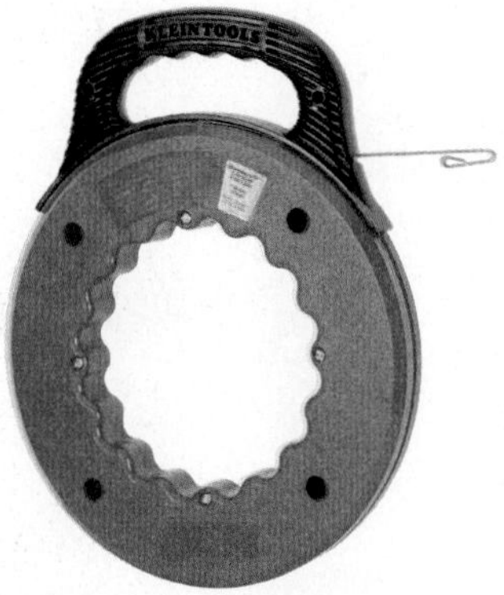

Figure 5.32 Caulking gun. Photo by Eugene Silberstein

COLD CHISELS. Cold chisels, **Figure 5.33**, are used to cut through cold metals, such as screws, bolts, and nails. Cold chisels are struck with a ball peen hammer to cut away the desired metal. Carpenter's claw hammers should not be used in conjunction with cold chisels as their faces are not tempered for striking hard metals and can chip. Always use a chisel that is slightly wider than the metal being cut.

STAPLERS. Staplers, **Figure 5.34**, and stapling tackers are used to fasten insulation and other soft materials to wood; some types may be used to secure low-voltage wiring.

Figure 5.33 Cold chisels. Photo by Eugene Silberstein

Figure 5.34 Stapler. Photo by Eugene Silberstein

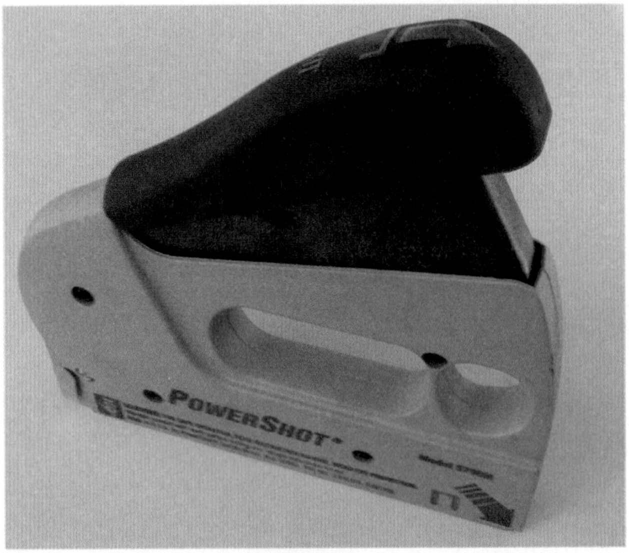

SOLDERING IRONS. A soldering iron, **Figure 5.35**, is used for a number of jobs including joining wires together, connecting wires to circuit boards or other components, and "tinning" the end of a stranded conductor so that it can be connected under a screw terminal, **Figure 5.36.**

Figure 5.35 Soldering iron. Photo by Eugene Silberstein.

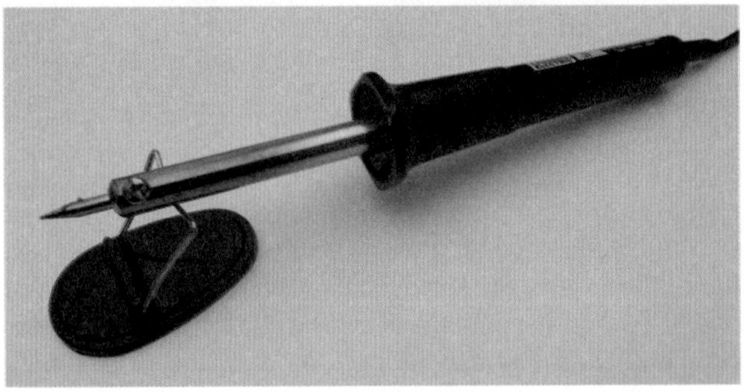

Figure 5.36 A tinned wire being secured under a screw terminal. Photo by Eugene Silberstein

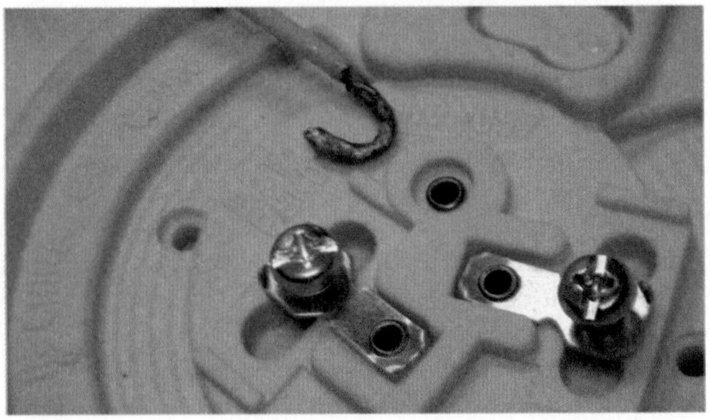

5.3 POWER TOOLS

ELECTRIC DRILLS

Portable electric drills are used extensively by HVACR technicians. They are available as power-corded (115 V) tools, **Figure 5.37**, or as cordless, battery-powered, tools, **Figure 5.38**. If the drill has a cord and is not double-insulated, it must have a three-prong plug and be used only with a grounded receptacle. Drills that have variable speed and reversing options are recommended.

Many technicians prefer the battery-operated type for its convenience and safety. Line voltage power supplies are not required to operate battery-powered tools provided that the batteries are kept charged. Electric drills are used for a number of purposes including the creation of holes. It is a good idea to have a complete set of drills bits, **Figure 5.39**, on hand to ensure that the proper size hole is drilled.

RECIPROCATING SAWS

Reciprocating saws create a back-and-forth action to cut through a variety of materials and are available as corded, **Figure 5.40**, or battery-powered, **Figure 5.41**, tools. There are

Figure 5.37 Corded drill. Courtesy of Milwaukee Electric Tool Corporation

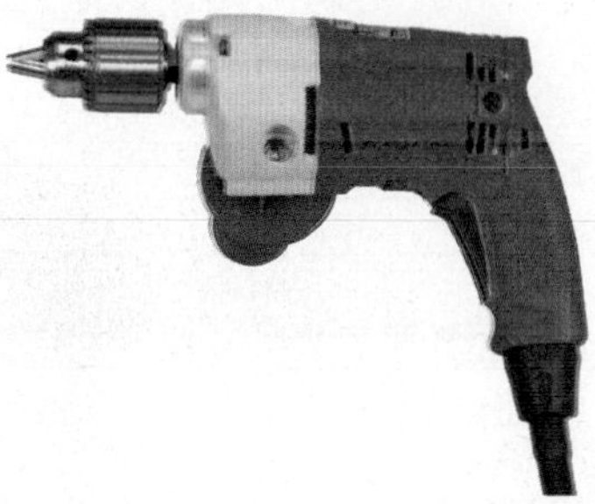

Figure 5.38 Battery-powered drills. Photo by Eugene Silberstein

Figure 5.39 Drill bits. Photo by Eugene Silberstein

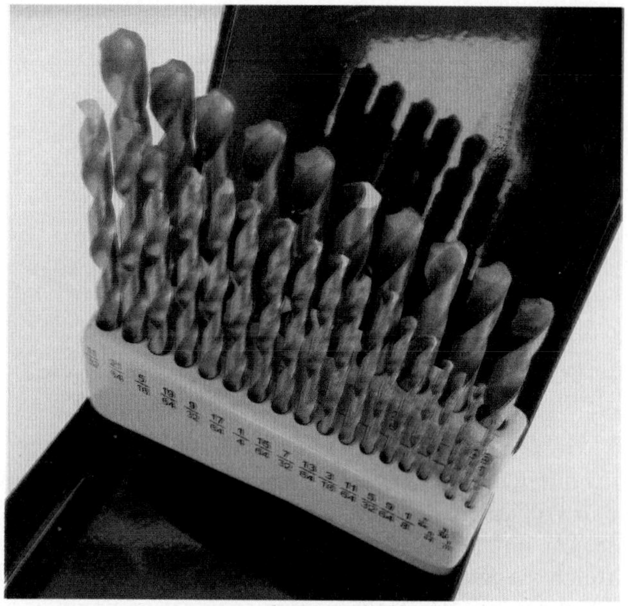

Figure 5.40 Corded reciprocating saw. Photo by Eugene Silberstein

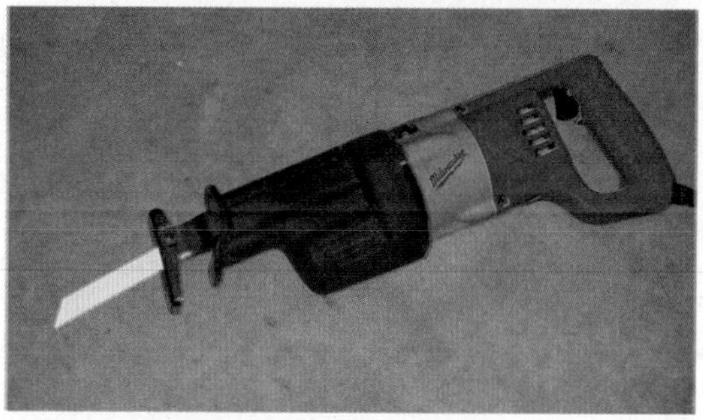

Figure 5.41 Cordless, battery-powered reciprocating saw. Photo by Eugene Silberstein

Figure 5.42 Battery-powered circular saw. Photo by Eugene Silberstein

many different types of blades available for use on these saws and are typically classified as metal, wood, and general purpose. The number of teeth on the blade determines how fine or rough the cut will be. The more teeth per inch, the finer the finished cut will be. Reciprocating saws are commonly used for demolition work.

CIRCULAR SAWS

Circular saws, **Figure 5.42**, are typically used to make straight cuts in wood sheets, such as plywood. This type of saw uses a round blade. As with other blades, the number of teeth on the blade determines how fine or rough the cut will be. The more teeth per inch, the finer the finished cut will be. As opposed to the reciprocating saw, that uses a back-and-forth motion, the circular saw uses a rotational motion.

JIGSAWS

Jigsaws are available as both corded, **Figure 5.43**, and battery-powered, **Figure 5.44**, tools.

Figure 5.43 Corded jigsaw. Photo by Bill Johnson

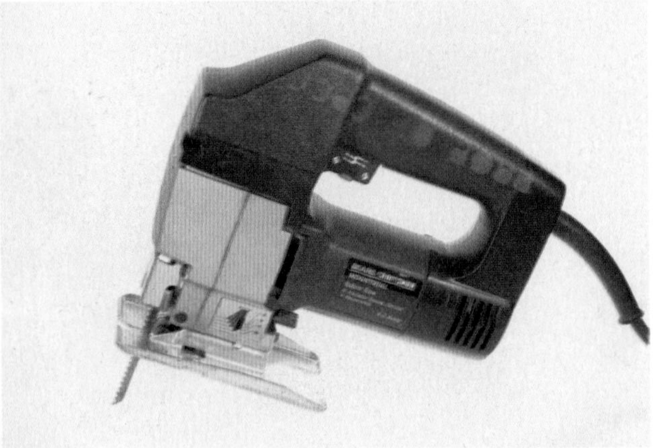

Figure 5.44 Battery-powered jigsaw. Photo by Eugene Silberstein

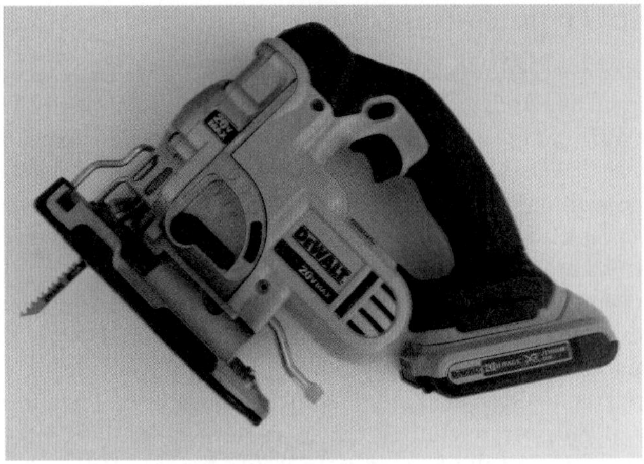

Figure 5.45 Pilot hole for jigsaw blade.

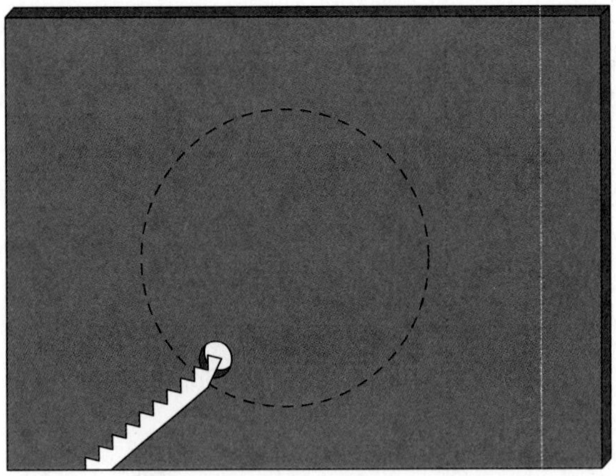

Like the reciprocating saw, the jigsaw relies on a back-and-forth action to perform the cutting. Unlike the reciprocating saw, the blades on a jigsaw are much thinner, making the tool extremely useful when making rounded cuts. When an internal cut needs to be made, a hole can first be drilled in the material to be cut, and the jigsaw blade can then be inserted into the hole, **Figure 5.45**.

5.4 ELECTRICAL TEST INSTRUMENTS

DIGITAL MULTIMETERS (DMMs)

The digital multimeter (DMM), **Figure 5.46**, often referred to as a volt-ohm-milliammeter (VOM), is a piece of electrical instrumentation that measures voltage (volts), resistance (ohms), and current (milliamperes). There are many types of DMMs available and many have the ability to measure

Figure 5.46 Digital multimeter. Courtesy of Fieldpiece Instruments.

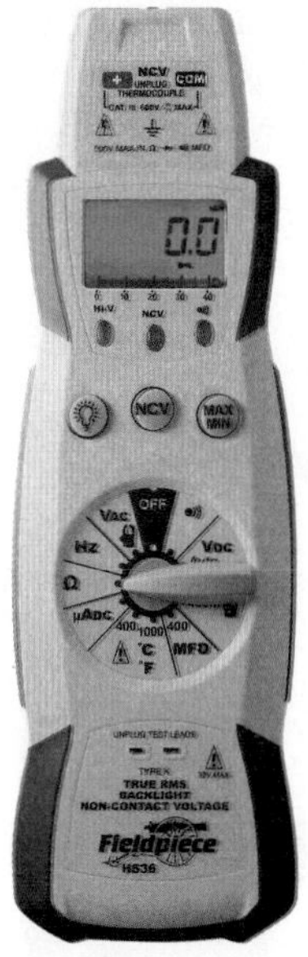

Figure 5.47 Meter test leads. Courtesy of Fieldpiece Instruments.

capacitance, continuity, and temperature, in addition to measuring voltage, resistance, and current. Some meters are equipped with a feature referred to as an NCV tester, which stands for non-contact voltage. This feature allows the instrument user to detect voltage without actually having the instrument come in physical contact with the circuit being tested. To measure voltage and resistance, meters rely on test leads, **Figure 5.47**. These test leads can be configured with handheld probes or with alligator clips for ease of use.

Figure 5.48 Voltage tester. Courtesy of Fluke Corporation.

Sometimes, a field technician only needs to know whether or not a particular circuit is energized. When such is the case, a voltage tester, such as the one shown in **Figure 5.48**, can be used.

CLAMP-ON AMMETER

The clamp-on ammeter, **Figure 5.49**, is a versatile instrument that has operable jaws that are clamped around a single conductor in an electric circuit to measure the current flow through the conductor. The reason that this particular instrument is so versatile is that it can also measure voltage and resistance. A

Figure 5.49 Clamp-on ammeter. Courtesy of Fieldpiece Instruments.

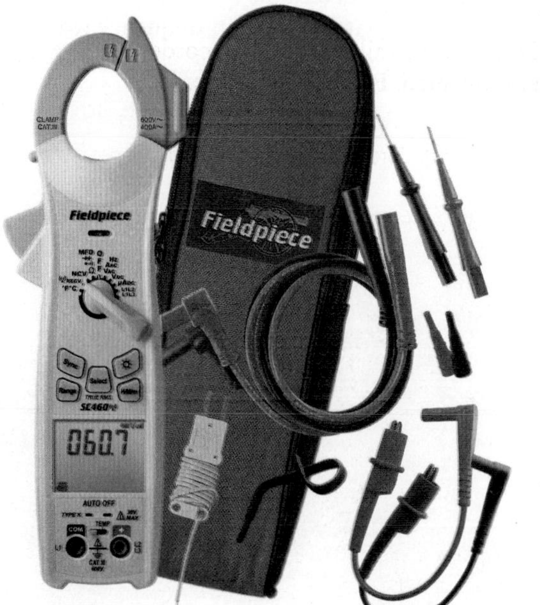

Figure 5.50 Meter being used to measure amperage. Courtesy of Fluke Corporation.

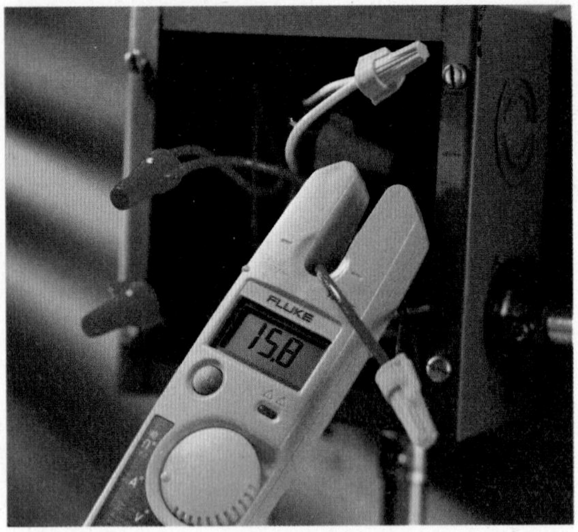

variation on the clamp-on ammeter is an instrument that can measure circuit current by simply positioning the meters non-operational jaws around the conductor, **Figure 5.50.**

5.5 REFRIGERANT LEAK DETECTORS

ELECTRONIC LEAK DETECTORS

Electronic leak detectors, **Figure 5.51**, contain an element sensitive to a particular refrigerant class. The instrument typically has an internal pump that pulls vapor into the leak detector and through a sensor. The leak detector evaluates the vapor being pulled in to determine if refrigerant is present. An audio beeping signal that increases in frequency when refrigerant is sensed in the vapor sample. Always make certain that the leak detector being used is intended to be used with the refrigerant that is to be detected. Varieties of electronic leak detectors include the heated diode refrigerant leak detector, **Figure 5.52**, the infrared refrigerant leak detector, **Figure 5.53**, and the carbon dioxide leak detector, **Figure 5.54**.

Figure 5.51 Electronic leak detector. Source: Ritchie Engineering Company, Inc. Yellow Jacket Products Division.

Figure 5.52 Heated diode refrigerant leak detector. Courtesy of Fieldpiece Instruments.

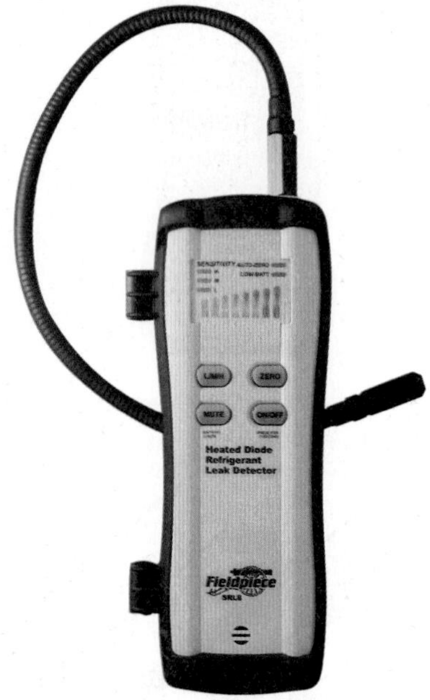

Figure 5.53 Infrared refrigerant leak detector. Courtesy of Fieldpiece Instruments.

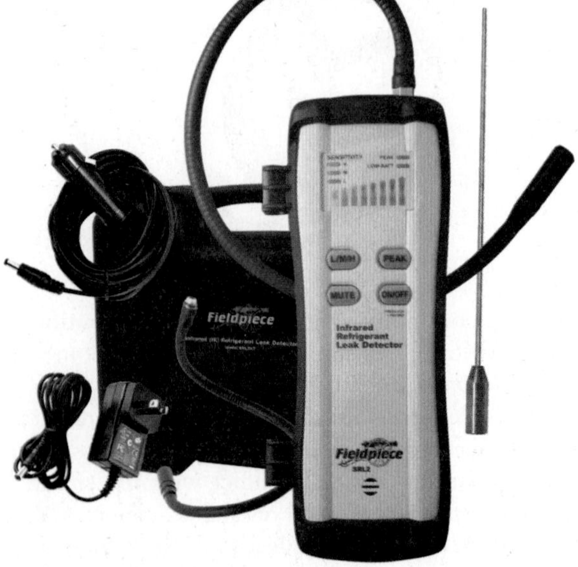

Figure 5.54 Carbon dioxide leak detector. Courtesy of Fieldpiece Instruments.

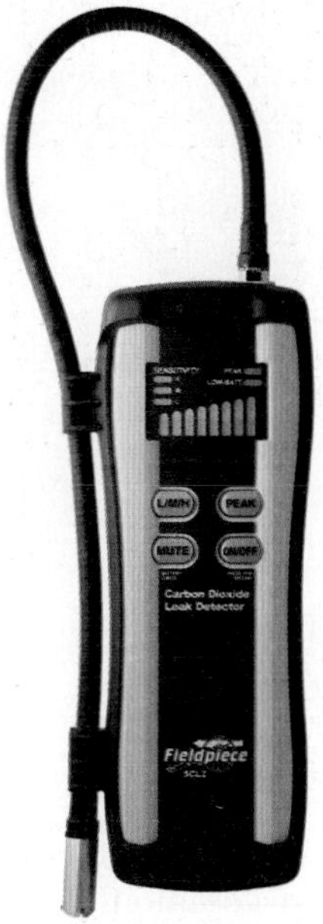

ULTRAVIOLET LEAK DETECTORS

The ultraviolet leak detection system, **Figure 5.55**, introduces an additive into the refrigerant system. The additive shows up as a bright yellow-green glow on the refrigerant piping at the leak site when viewed under ultraviolet light. The additive can remain in the system to test for a new suspected leak at a later date. The ultraviolet lights that are used to inspect the refrigerant piping are compact, handheld lights that resemble small flashlights.

Figure 5.55 Ultraviolet leak detecting system. Courtesy of Spectronics.

ULTRASONIC LEAK DETECTOR

Ultrasonic refrigerant leak detectors use the sound from the escaping refrigerant to detect a leak, **Figure 5.56**. Ultrasonic leak detectors can be used on systems that are pressurized with dry nitrogen, as the detector is not detecting the presence of particular chemicals, just the sound of moving gases within a pipe.

HALIDE LEAK DETECTORS

A halide leak detector, **Figure 5.57**, uses an acetylene- or propane-fueled flame to heat a copper disk, which is then used to

Figure 5.56 (A) Ultrasonic leak detector. Courtesy of Robinair Division, SPX Corporation. (B) Ultrasonic leak detector. Courtesy of Amprobe Instruments.

(A)

(B)

Figure 5.57 Halide leak detector. Courtesy of Uniweld Products.

detect chlorinated refrigerant gas leaks. When the detector is ignited, the flame heats a copper disk and a green flame is produced. Air for combustion is drawn through the attached hose, the end of which is passed over or near fittings or other areas where a leak may be suspected. If there is a leak, the refrigerant will be drawn into the hose and contact the copper disk. This breaks down the halogen refrigerants into other compounds and changes the color of the flame from green to blue.

5.6 TUBING TOOLS

TUBING CUTTERS

Tube cutters, **Figure 5.58**, are available in different sizes and styles. The tubing cutter at the left of Figure 5.58 is a compact

Figure 5.58 Tubing cutters. Compact cutter (left), larger cutters (center and right). Photo by Eugene Silberstein

Figure 5.59 Plastic tubing shear. Photo by Bill Johnson

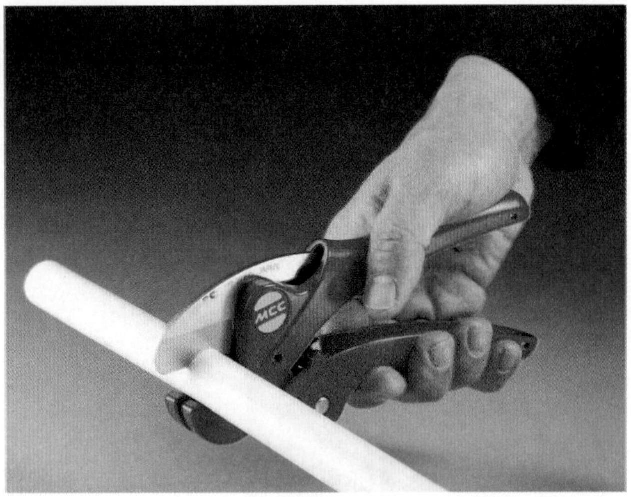

cutter that allows the user to effectively cut tubing in tight spaces. Tubing cutters are commonly used to cut copper and aluminum tubing but can be used to cut rigid copper pipe as well. Some models have a flare cut-off groove that reduces tube loss when removing a cracked or damaged flare. Many models also include a retractable reamer to remove inside burrs, and a filing surface to remove outside burrs.

Another cutter that is used to cut plastic tubing and non-wire reinforced plastic or synthetic hose is the plastic tubing shear, **Figure 5.59**. This cutter has a ratcheting handle to make the cutting process easier.

REAMERS

Reamers, **Figure 5.60**, are used to remove burrs from sections of cut copper tubing. The inner-outer reamer shown in Figure 5.60, with its three cutters, has the ability to ream both the inside and outside of a tubing section. Other reamers, **Figure 5.61**, are compact, handy tools. **Figure 5.62** shows a technician using a reamer. Some tubing cutters are equipped with reamers.

Figure 5.60 Inner-outer reamers. Photo by Bill Johnson

Figure 5.61 Reamers. Source: Ritchie Engineering Company, Inc. Yellow Jacket Products Division.

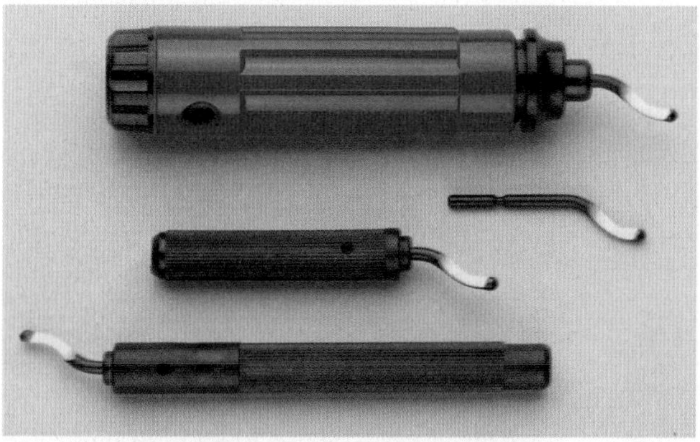

Figure 5.62 Technician using a reamer. Source: Ritchie Engineering Company, Inc. Yellow Jacket Products Division.

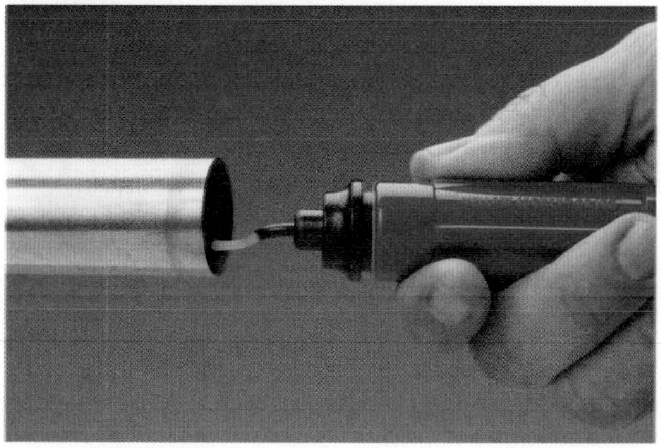

Figure 5.63 Inspection mirrors. Source: Ritchie Engineering Company, Inc. Yellow Jacket Products Division.

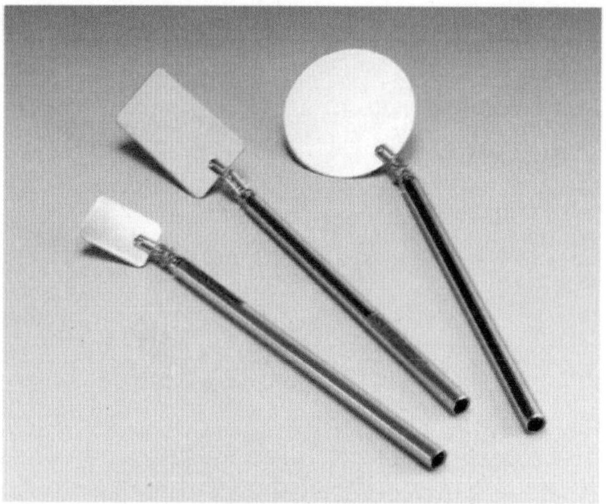

Figure 5.64 Tubing brushes. Photo by Eugene Silberstein.

INSPECTION MIRRORS

Inspection mirrors, **Figure 5.63**, are usually available in rectangular or round shapes, with stainless steel or glass reflective surfaces that are usually mounted on ball joints that swivel 360 degrees. They come with fixed or telescoping handles, some more than 30 in. long. The mirrors are used to inspect areas or parts in components that are behind or underneath other parts. Inspection mirrors are commonly used to inspect newly soldered or brazed joints to locate improperly soldered joints.

TUBING BRUSHES

Tubing brushes, **Figure 5.64**, are used to clean the outside of tubing and the inside of fittings prior to soldering. The brushes are sized to fit the tubing to be cleaned.

FLARING TOOLS

The flaring tool, **Figure 5.65**, has a flaring bar or block to hold the tubing, a slip-on yoke, and a feed screw with flaring cone and handle. Several sizes of tubing can be flared with this tool. A completed flare resembles a bell-shaped bevel on the end of the tubing. Tubing is flared so that it can be used in conjunction with flare nuts. Refer to Unit 7 for detailed information on the use of the flaring block.

SWAGING TOOLS

Swaging tools, **Figure 5.66**, are used to expand the end of a section of soft-drawn tubing so that it can accept another section of same size tubing, in preparation for soldering or brazing. Refer to Unit 7 for detailed information on the use of swaging tools.

Figure 5.65 Flaring block. Source: Ritchie Engineering Company, Inc. Yellow Jacket Products Division.

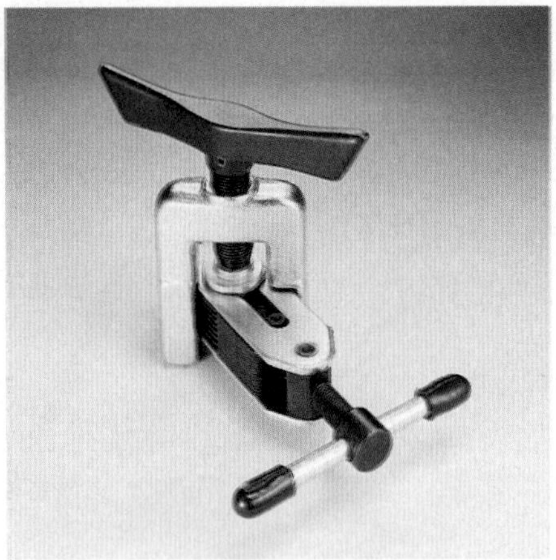

Figure 5.66 Swaging tools. Source: Ritchie Engineering Company, Inc. Yellow Jacket Products Division.

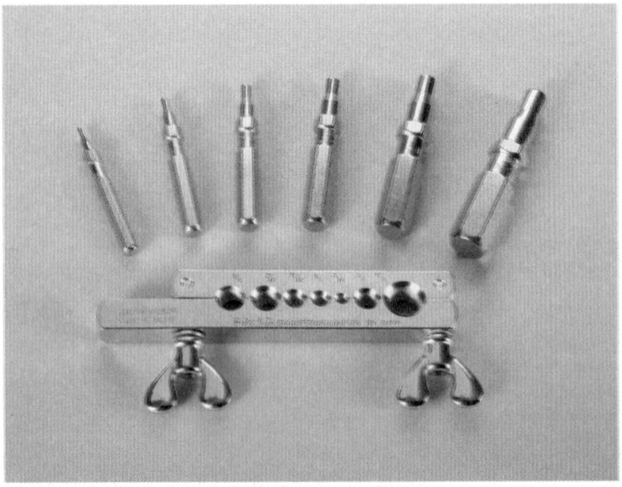

Figure 5.67 Spring bender. Source: Ritchie Engineering Company, Inc. Yellow Jacket Products Division.

Figure 5.68 Lever bender. Courtesy of Hilmor.

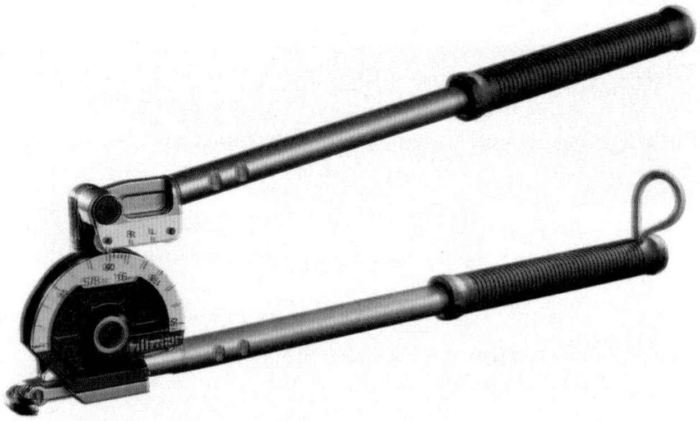

Figure 5.69 Ball-type, pinch-off tool. Photo by Bill Johnson

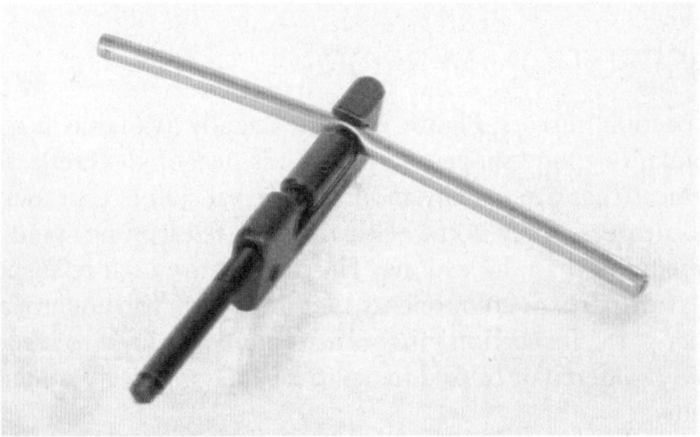

TUBING BENDERS

Tubing benders are used to bend, without kinking, sections of soft-drawn tubing. Bending soft-drawn tubing is desirable especially when installing refrigerant lines, as doing so greatly reduces the number of soldered and/or brazed joints, thus reducing the potential for system leaks. Two commonly used types of tubing benders are the spring bender, **Figure 5.67**, and the lever bender, **Figure 5.68**. Refer to Unit 7 for detailed information on the use of tubing benders.

PINCH-OFF TOOLS

A tubing pinch-off tool is used to pinch shut an access or process tube, such as the service stub on a hermetic compressor. The stub is pinched shut in preparation for being soldered. The ball-type, pinch-off tool, **Figure 5.69**, creates a rounded pinch-off, which is very sturdy. **Figure 5.70** shows two other types of pinch-off tools that create straight-line pinch-offs.

Figure 5.70 Straight-line pinch-off tools. Photo by Bill Johnson

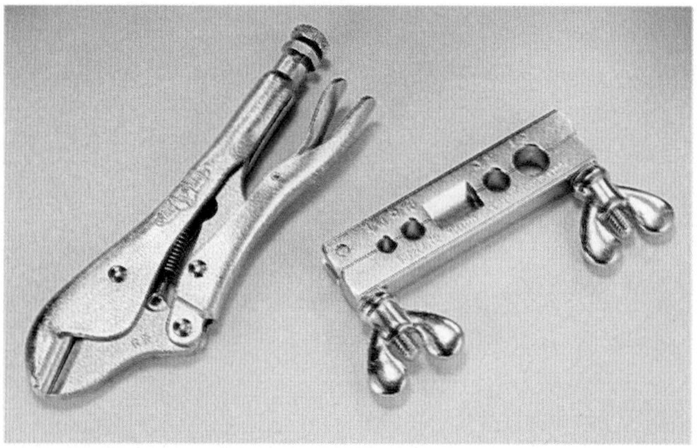

5.7 PIPING TOOLS

PIPE WRENCHES

Pipe wrenches, **Figure 5.71**, are used to turn sections of rigid pipe and related piping system components. These wrenches come in a number of different sizes, based on the size of pipe being turned. The long handles on these wrenches provide the additional torque needed to effectively tighten or loosen piping connections.

PIPE CUTTERS

A larger version of the tubing cutter, the pipe cutter, **Figure 5.72**, is used to cut sections of rigid piping made from materials such as steel.

PIPE REAMERS

Pipe reamers are used to remove burrs from sections of rigid piping. **Figure 5.73** shows a technician using a reamer on a newly cut section of black pipe.

Figure 5.71 Pipe wrench. Courtesy of Ridge Tool Company.

Figure 5.72 Pipe cutter. Courtesy of Ridge Tool Company.

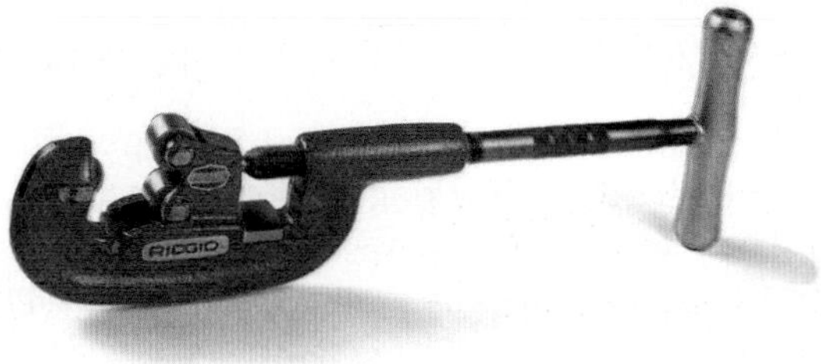

Figure 5.73 Reamer used to remove burrs from the interior of pipe sections. Photo by Eugene Silberstein

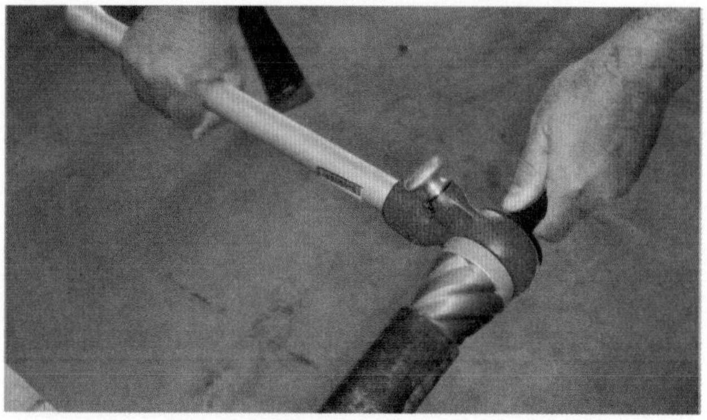

Figure 5.74 Pipe vise. Courtesy of Ridge Tool Company.

PIPE VISES

Pipe vises, **Figure 5.74**, are used to secure sections of pipe so that operations such as cutting, reaming, threading, and joining pipe sections to other pipe sections or components can be performed.

PIPE THREADERS

Pipe threaders are used to create external threads on a section of rigid pipe. **Figure 5.75** shows a manual pipe threader, which is a ratcheting tool used to create the external threads. An automatic pipe threader, **Figure 5.76**, performs the threading process

Figure 5.75 Pipe threader. Courtesy of Ridge Tool Company.

Figure 5.76 Automatic pipe threader. Courtesy of Ridge Tool Company.

automatically and is often equipped with other commonly used piping tools such as a pipe cutter and reamer.

5.8 SHEET METAL TOOLS

TIN SNIPS

Tin snips, **Figure 5.77**, are intended to cut sheet metal. Snips are configured for left-turning cuts, right-turning cuts, and straight cuts. Red tin snips are intended for straight and left-turning cuts, while green tin snips are intended for straight and right-turning cuts. Yellow snips are typically used for straight cutting, but are referred to as general purpose tin snips.

SCRATCH AWL

Although used for other tasks as well, the scratch awl, **Figure 5.78**, is commonly used when working with sheet metal. This tool is used to scratch lines in the sheet metal to mark the point where cuts, bends, or notches are to be made.

Figure 5.77 Tin snips. Red-handled snips for left-turning cuts (left), yellow-handled snips for straight cuts (center), green-handled snips for right-turning cuts (right). Photo by Eugene Silberstein

Figure 5.78 Scratch awl. Photo by Eugene Silberstein

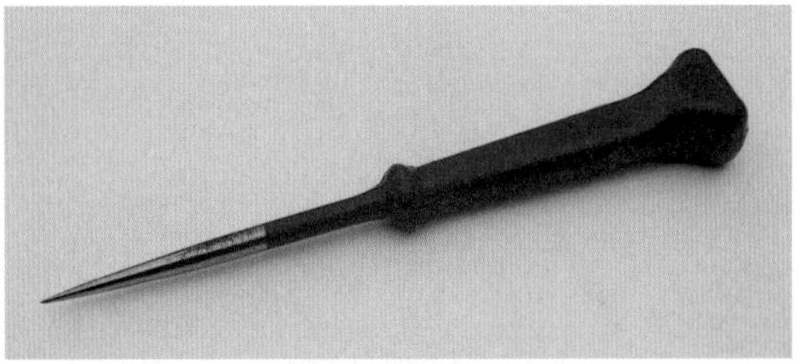

Figure 5.79 Sheet metal crimpers (left), square notchers (center), triangular notchers (right). Photo by Eugene Silberstein

CRIMPERS AND NOTCHERS

Crimpers and notchers, **Figure 5.79**, are common tools used by sheet metal workers. Crimpers are primarily used to create a ripple in the end of round stove pipe duct sections. This ripple reduces the diameter of the end of the stove pipe so that two sections of same-size stove pipe can be connected together, **Figure 5.80**. Notchers are used to create uniform notches, either square or triangular in shape, in pieces of sheet metal. These notches are created for a number of reasons including creating clinch fittings and angled tabs for slip- and drive-connected duct sections, **Figure 5.81**.

HAND SEAMERS AND FOLDING TOOLS

Hand seamers and folding tools, **Figure 5.82**, are utility folding tools used by those working with sheet metal and duct systems. These tools allow the user to create sharp bends in sheet metal, such as when flanges need to be bent in plenums to mount them onto air handlers.

Figure 5.80 Sheet metal crimpers used to reduce the diameter of a section of round duct.

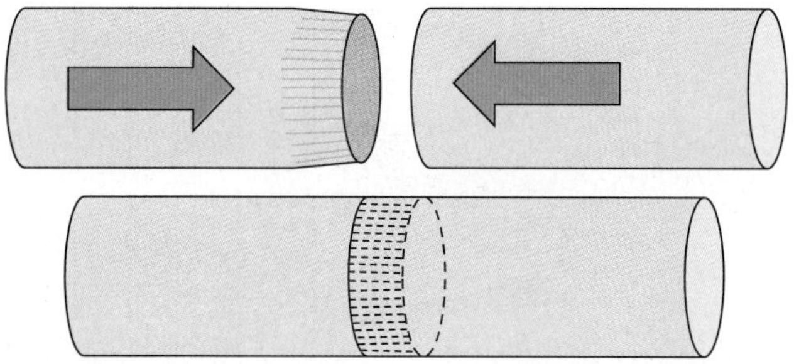

Figure 5.81 Illustration of square notches (left), and triangular notches (right).

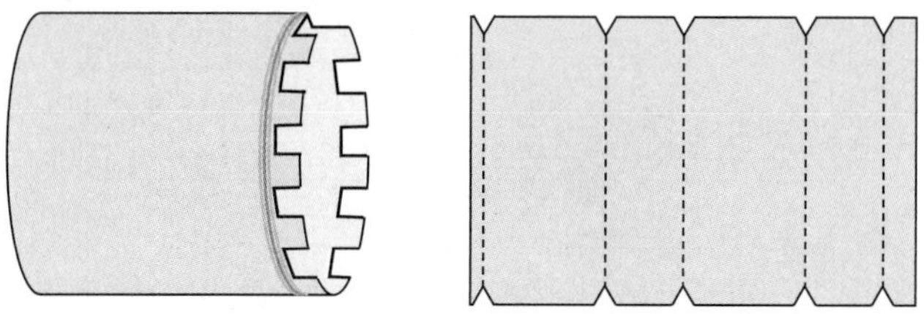

Figure 5.82 Hand seamers (left), locking hand seamers (center), sheet metal folding tool (right). Photo by Eugene Silberstein

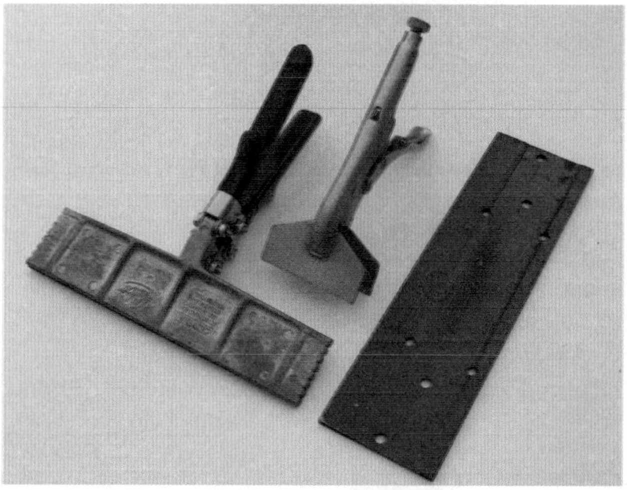

5.9 HVACR INSTALLATION AND SERVICE TOOLS AND EQUIPMENT

TORCHES

Soldering and brazing are two tasks that are commonly performed by the HVACR service technician. There are many options that are available to the technician when it comes to the source of heat. The most compact and portable option is a propane gas torch with a disposable propane gas tank, **Figure 5.83**. This is an easy torch to use, as the flame adjusts easily and can be used for many soldering operations. The propane fuel comes in small 14-ounce tanks. Another option for these torches is what is known as MAPP™ gas, which is a mixture of hydrocarbon fuels. The propane torch flame is at a temperature of approximately 3600°F, while the MAPP™ gas flame is approximately 3730°F. The air-acetylene torch kit, **Figure 5.84**, provides sufficient heat for soldering and brazing, as a

Figure 5.83 Propane torch with a disposable gas tank. Photo by Eugene Silberstein

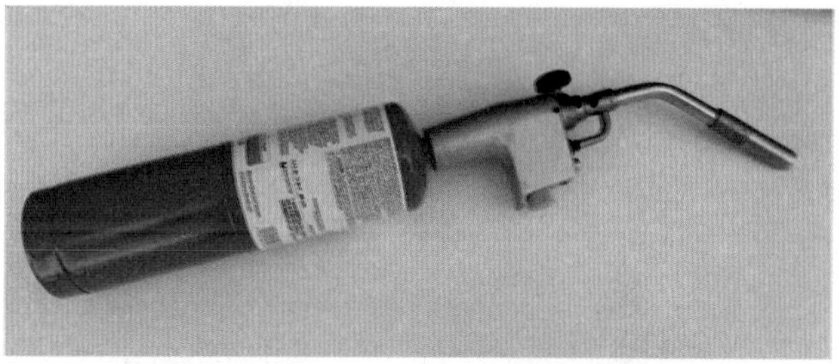

Figure 5.84 Air-acetylene torch kit. Photo by Eugene Silberstein

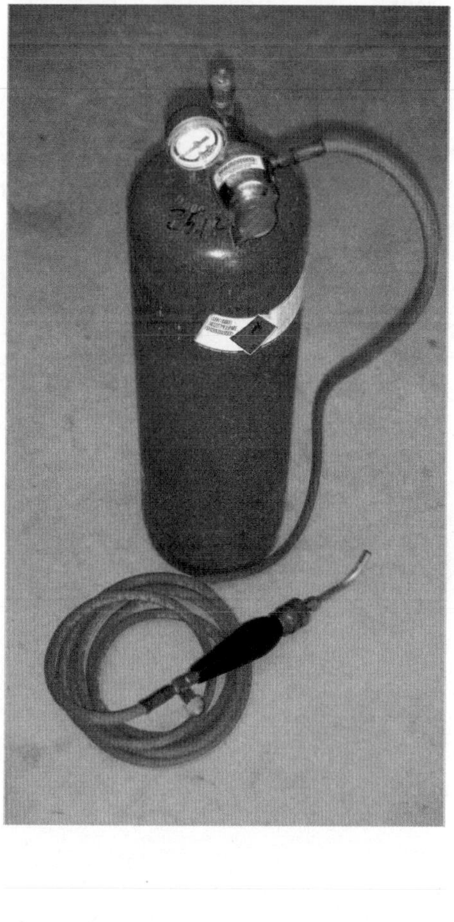

flame temperature of about 4200°F is produced. It consists of a torch, which can be fitted with several sizes of tips, and a regulator, hoses, and an acetylene tank. Oxyacetylene torch kits, **Figure 5.85**, are used in certain soldering, brazing, and welding applications. The units consist of a torch, regulators, **Figure 5.86**, hoses, oxygen tanks, and acetylene gas tanks. When the oxygen and acetylene gases are mixed in the proper proportion, a very hot flame, approximately 5600°F, is produced.

When brazing refrigerant lines, it is best field practice to purge the refrigerant lines with nitrogen to prevent oxidation in the lines. The desired nitrogen pressure is between 1 and 2 psig, which is a pressure that is too low to measure on a gauge or regulator. For this reason, a nitrogen flow indicator, **Figure 5.87**, is useful to provide visual confirmation that the correct amount of nitrogen is flowing through the piping arrangement.

Figure 5.85 Oxyacetylene torch kit. Courtesy of Uniweld Products, Inc.

Figure 5.86 Oxygen and acetylene regulators. Photo by Eugene Silberstein

Figure 5.87 Nitrogen flow indicator. Courtesy of Uniweld Products, Inc.

GAUGES AND MANIFOLDS

One of the most important of all pieces of refrigeration and air-conditioning service equipment is the gauge manifold. Gauge manifolds can be analog or digital instruments. A

Figure 5.88 Four-valve gauge manifold. Source: Ritchie Engineering Company, Inc. Yellow Jacket Products Division.

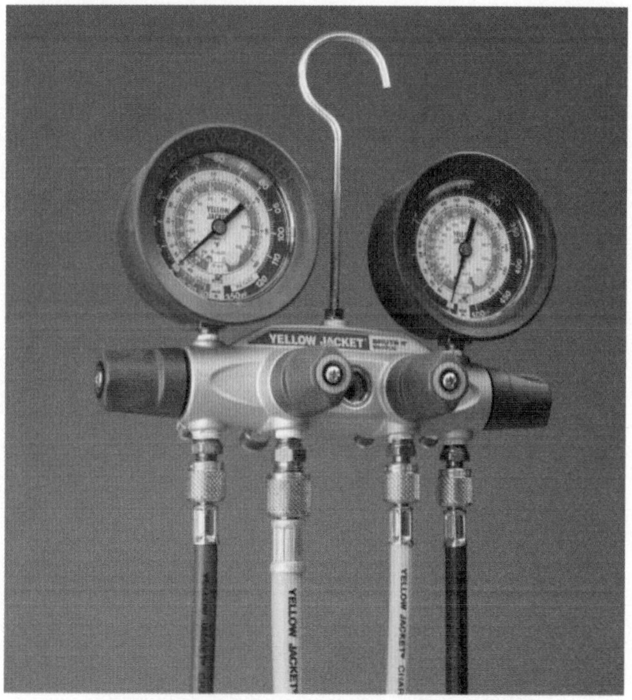

typical gauge manifold, **Figure 5.88**, includes the compound gauge (low-pressure and vacuum), the high-pressure gauge, and the manifold, valves, and hoses. The gauge manifold may be configured with two valves and three hoses or with four valves and four hoses. The two-valve manifold has a high-side hose port, a low-side hose port, and one center port. The four-valve manifold has a high-side hose port, a low-side hose port, and two center ports. Figure 5.88 shows a four-valve manifold, which has separate valves for the vacuum, low-pressure, high-pressure, and refrigerant cylinder connections. The vacuum hose on four-valve manifolds is 3/8 in. in diameter, while the other hoses are 1/4 in. in diameter. This larger hose offers less restriction to flow for faster evacuation. Digital manifolds, such as the one shown in **Figure 5.89**, in addition to being able to measure pressures, has the ability to measure temperatures and calculate evaporator superheat and condenser subcooling.

Some electronic manifolds are also available that do not have physical gauges on them at all. The manifold in **Figure 5.90** utilizes a smart device application and can measure system pressures and temperatures, calculate superheat and subcooling, and measure vacuum level in microns. It may also be used as a digital thermometer. Whenever hoses are connected to a system, they must be equipped with low-loss fittings, **Figure 5.91**. These fittings reduce the amount of refrigerant that is lost during the installation or removing of gauges from a system.

Figure 5.89 Digital gauge manifold. Courtesy of Fieldpiece Instruments.

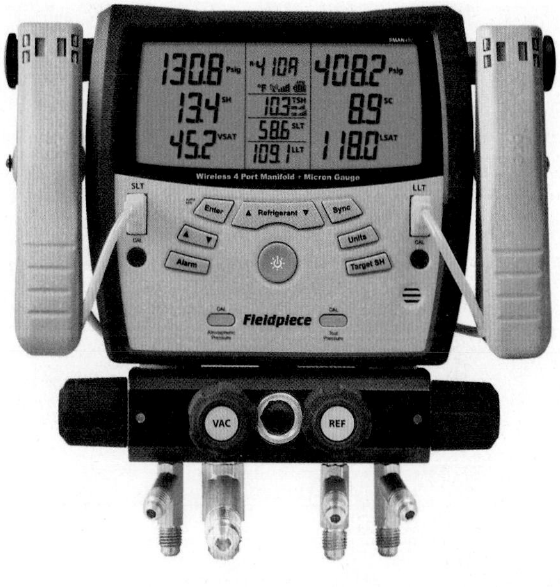

Figure 5.90 Electronic, Gaugeless manifold. Courtesy of Northpark Innovations.

Some air-conditioning and refrigeration service gauges do not have a manifold. **Figure 5.92** shows a set of digital, wireless, Bluetooth capable, stub gauges that can be connected directly to the service ports to obtain system pressure readings.

Another gauge that is commonly used by the HVACR service technician is the vacuum, or micron gauge. The micron gauge is used to measure the vacuum, in microns, created during the evacuation process. This micron gauge shown in **Figure 5.93** is a digital, wireless, stand-alone stub gauge that

Figure 5.91 Low-loss fittings.

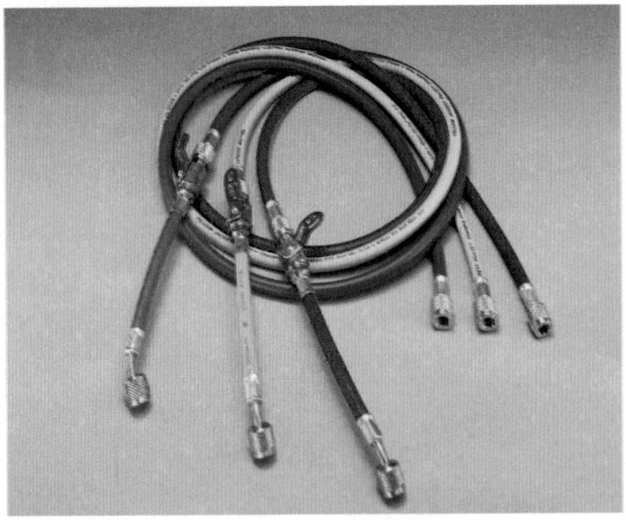

Figure 5.92 Digital, wireless, Bluetooth capable high and low side stub pressure gauges. Courtesy of Appion Inc.

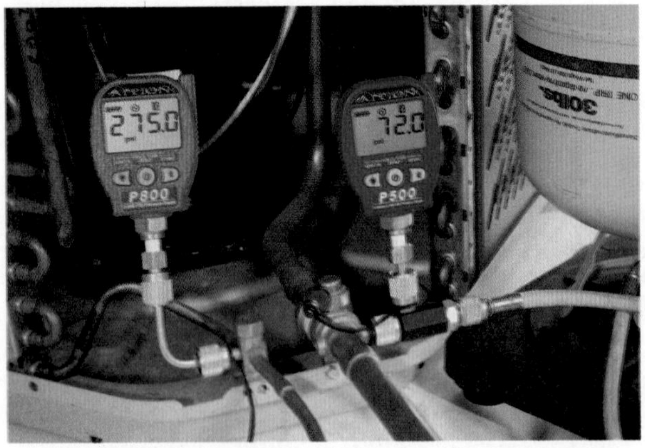

has Bluetooth capabilities for remote monitoring. The micron gauge gets connected to the system in a similar manner as the stub gauges discussed earlier, **Figure 5.94**.

SAFETY PRECAUTION: *Always make certain that the gauges, hoses, and other instrumentation being used is rated for the system's expected operating pressures. Serious personal injury could result if system pressures exceed the pressure ratings on connected devices.* •

REFRIGERANT-SIDE SYSTEM PERFORMANCE INSTRUMENTATION

Other specialty instruments are available that assist the HVACR service technician in evaluating and troubleshooting equipment. Some of these devices aid in the calculation of evaporator superheat and condenser subcooling,

Figure 5.93 Digital, wireless, stand-alone micron gauge with Bluetooth capabilities. Courtesy of Appion Inc.

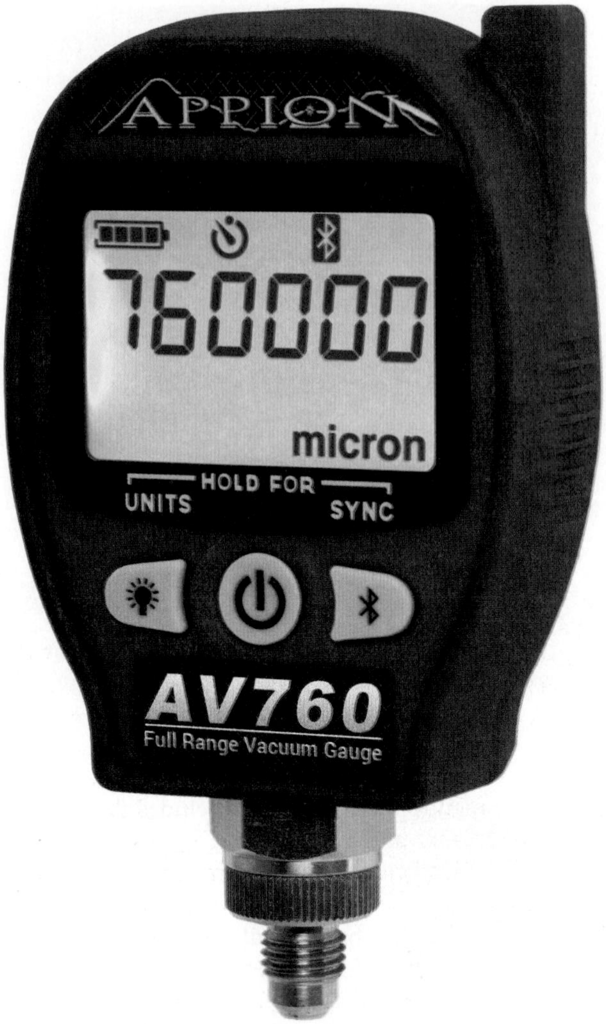

Figure 5.94 Micron gauge installed on a system. Courtesy of Appion Inc.

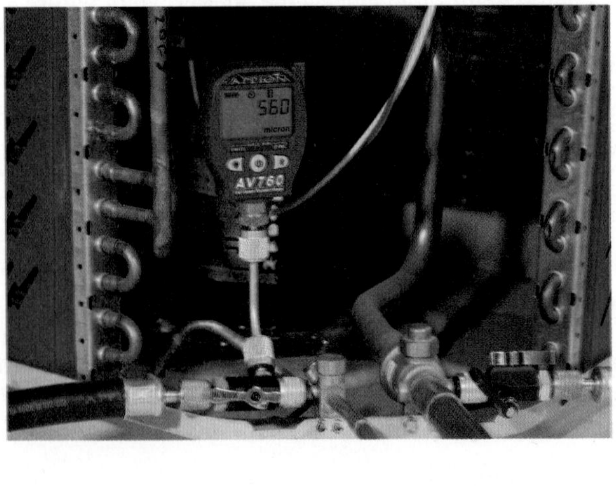

Figure 5.95 Instrument for calculating evaporator superheat and condenser subcooling. Courtesy of Fieldpiece Instruments.

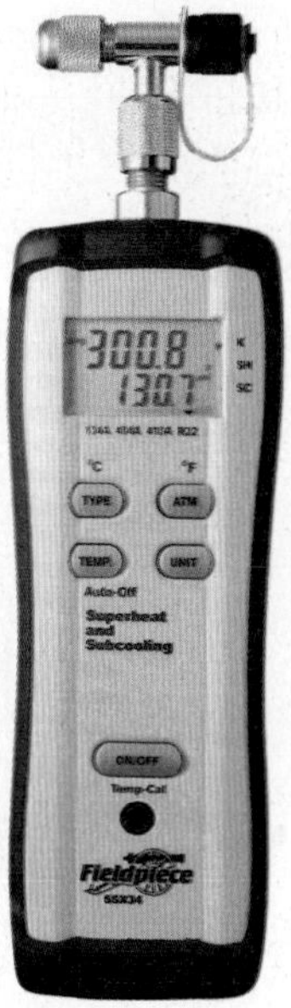

Figure 5.96 Application-based, complete system troubleshooting kit. Courtesy of Fieldpiece Instruments.

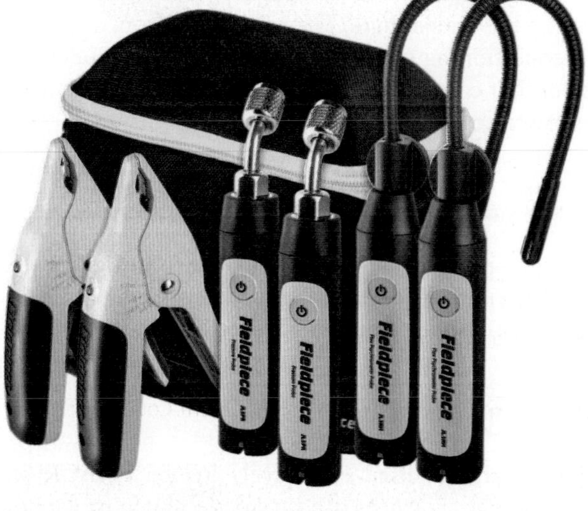

Figure 5.97 Wired, digital charging scale with alarm. Courtesy of Fieldpiece Instruments.

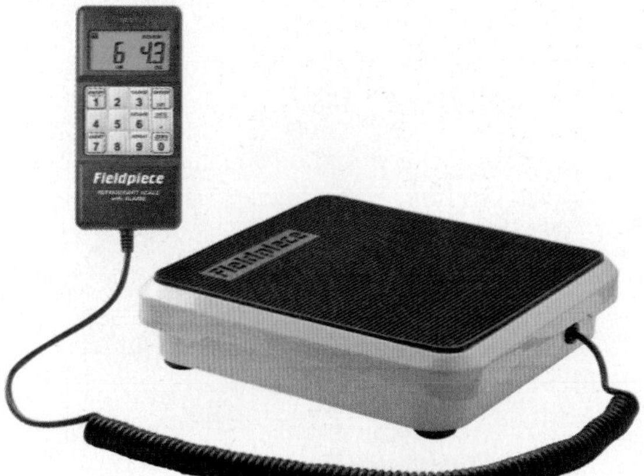

Figure 5.98 Wireless, digital charging scale. Courtesy of Fieldpiece Instruments.

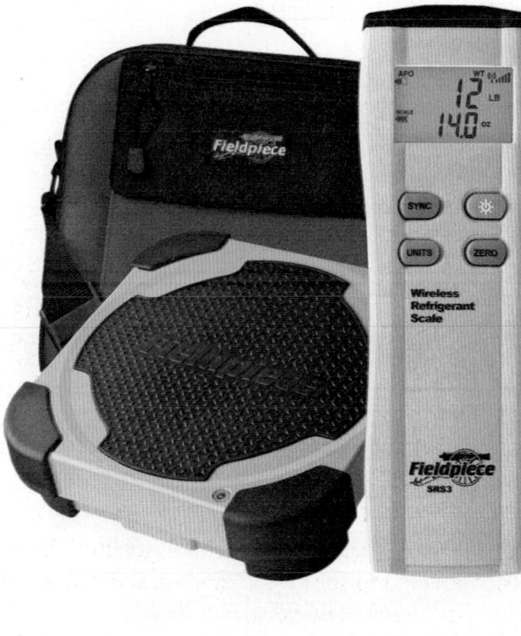

Figure 5.95. Other instruments are application-based and allow the technician to obtain pressure, temperature, and air property measurements to effectively evaluate system performance, **Figure 5.96**.

CHARGING SCALES

Digital, programmable refrigerant charging scales allow a technician to accurately introduce refrigerant to HVACR systems by weight. This can usually be done manually or automatically, and the amount of refrigerant to be charged into the system can be programmed. **Figure 5.97** is an example of a programmable charging scale that is equipped with an alarm and **Figure 5.98** shows a wireless refrigerant scale. When charging systems with blended refrigerants, the refrigerant must leave the tank as a liquid. To prevent liquid refrigerant from potentially damaging the system compressor, a liquid restrictor, **Figure 5.99**, can be used.

VACUUM PUMPS

Vacuum pumps, **Figure 5.100**, are designed to evacuate air-conditioning and refrigeration systems by removing air and

Figure 5.99 Liquid restrictor. Photo by Bill Johnson

Figure 5.100 8 cfm vacuum pump. Courtesy of Appion.

other noncondensable gases, as well as moisture, from the system. Evacuating the system is necessary because the air and noncondensable gases take up space, contain moisture, and cause excessive pressures. Vacuum pumps may be either one- or two-stage and are rated in cubic feet per minute or cfm. The two-stage pump is the one most often recommended for use. It will create a lower vacuum to improve the system vacuum in situations where moisture is an issue.

Figure 5.101 Recovery unit. Courtesy of Appion.

RECOVERY UNITS

The recovery unit, **Figure 5.101**, is used to remove refrigerant from HVACR systems. It is illegal to intentionally vent refrigerant into the atmosphere. The refrigerant from a refrigerator or air-conditioning system is pumped into a cylinder or container until it can be either charged back into the system or is surrendered to a refrigerant reclaiming facility.

FIN STRAIGHTENERS

Fin straighteners, **Figure 5.102**, are used to straighten the fins on condenser and evaporator coils. The spacing between coil fins changes, so the right straightener needs to be used.

THERMOMETERS

Thermometers are used extensively by the HVACR service technician. These devices can be simple, single probe devices, **Figure 5.103**, or can be multiprobe instruments with the ability to internally calculate temperature differentials, **Figure 5.104**. Infrared thermometers, **Figure 5.105**, are also popular. Be sure to use thermometers according to manufacturer's instructions to ensure the required accuracy is achieved.

WHEEL PULLER

Wheel pullers, **Figure 5.106**, are used for removing propeller type fan blades, squirrel cage type blower wheels, and pulleys and sheaves.

Figure 5.102 (A) Color-coded fin straighteners for different fin spaces. The white and silver fin tool can be used on coils with different fin spacing. (B) Fin tool with 6 different straighteners. Photo by Bill Johnson

(A)

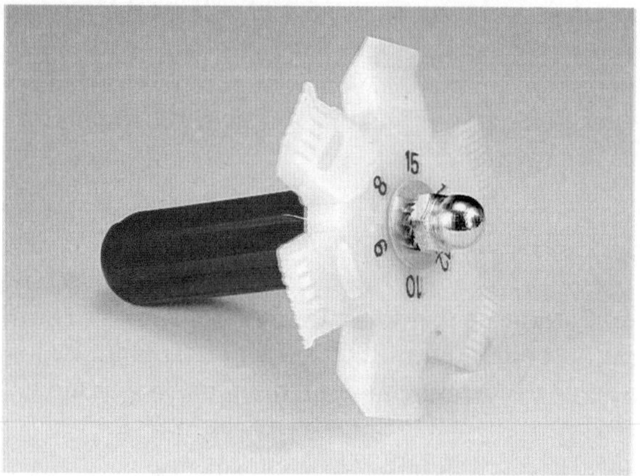

(B)

Figure 5.103 Digital, single-probe thermometer. Courtesy of Fieldpiece Instruments.

Figure 5.104 Multiprobe, differential thermometer. Courtesy of Fieldpiece Instruments.

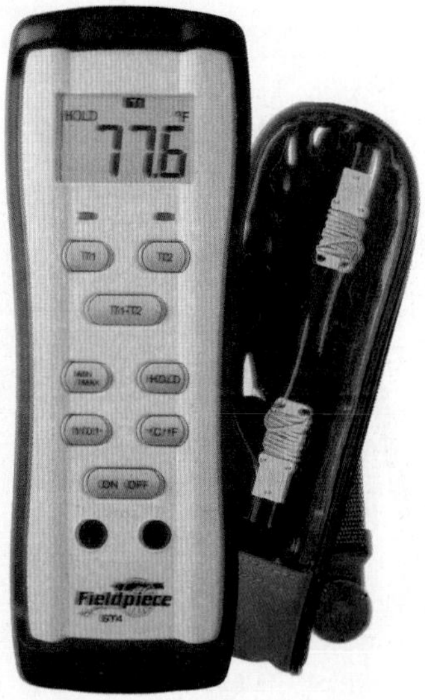

Figure 5.105 Infrared thermometer. Source: Ritchie Engineering Company, Inc. Yellow Jacket Products Division.

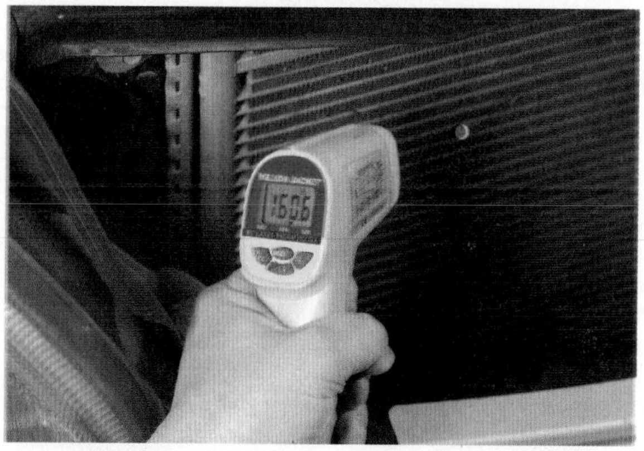

Figure 5.106 Fan blade and wheel puller. Source: Ritchie Engineering Company, Inc. Yellow Jacket Products Division.

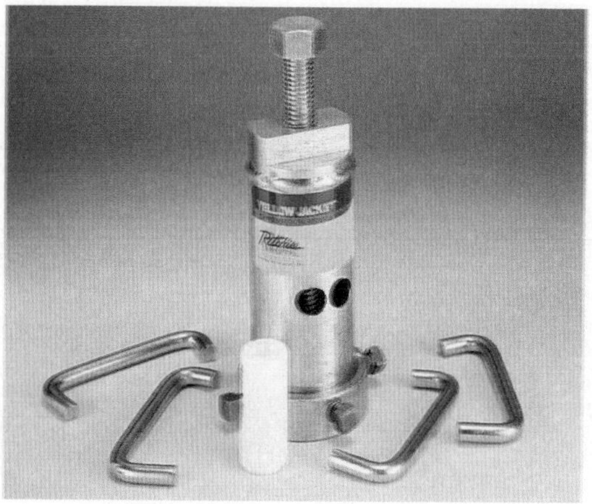

Figure 5.107 Tank heater. Source: Ritchie Engineering Company, Inc. Yellow Jacket Products Division.

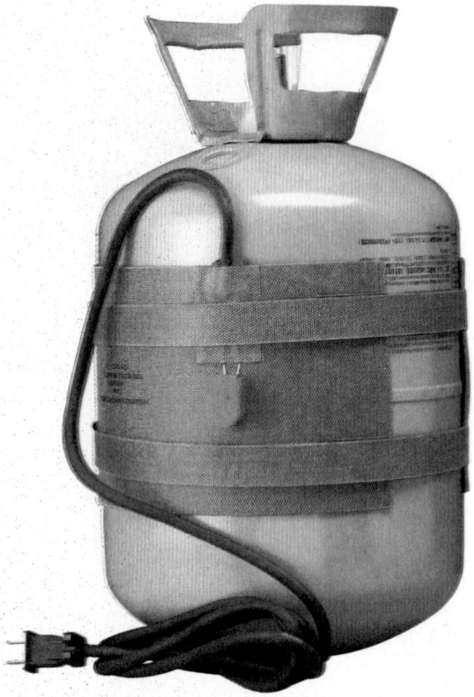

TANK HEATER

The tank heater, **Figure 5.107**, is used to add heat and pressure to refrigerant tanks for faster charging especially on cold days.

CABLE TIE TOOL

The cable tie tool, **Figure 5.108**, is used to install nylon strap clamps around a flexible duct. This tool automatically cuts the clamping strap off flush when a preset tension is reached.

Figure 5.108 Tool used to install nylon strap clamps around a flexible duct. Photo by Bill Johnson

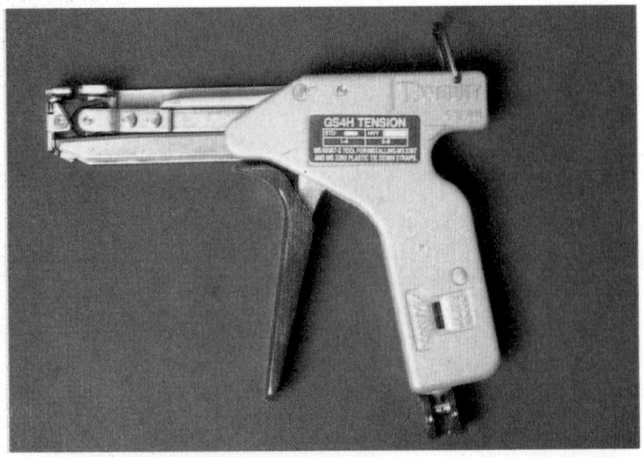

Figure 5.109 Schrader core remover and replacer. Source: Ritchie Engineering Company, Inc. Yellow Jacket Products Division.

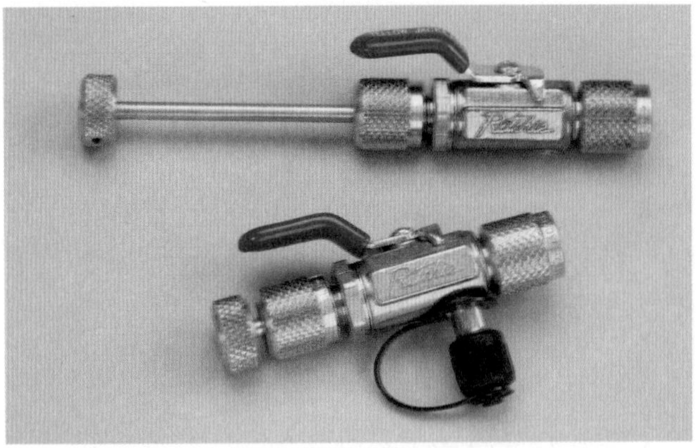

CORE REMOVAL TOOL

The core removal tool, **Figure 5.109**, allows the service technician to remove and replace Schrader cores without removing the refrigerant charge. Schrader core removal allows for less restriction when evacuating a system allowing for a faster and deeper vacuum.

REFRACTOMETER

The refractometer, **Figure 5.110**, is used for retrofitting a mineral oil system with POE lubricant. This tool can accurately predict remaining mineral oil in the system down to as low as 1%.

OIL PUMPS

Oil pumps, **Figure 5.111**, are used to manually transfer oil into and out of HVACR system compressors.

Figure 5.110 Handheld refractometer. Source: Ritchie Engineering Company, Inc. Yellow Jacket Products Division.

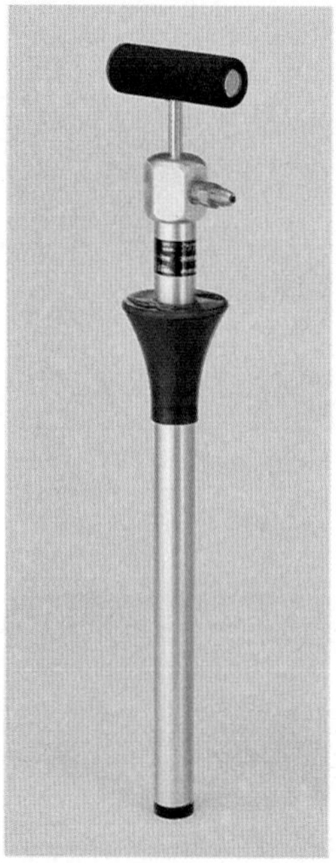

Figure 5.111 Oil pump. Source: Ritchie Engineering Company, Inc. Yellow Jacket Products Division.

5.10 AIRSIDE INSTRUMENTATION

VELOMETERS AND ANEMOMETERS

Measuring the velocity of air flowing through a duct system is an important task that the HVACR service technician performs. Tools commonly used to measure air speed are the air analyzer, **Figure 5.112**, and the microanemometer, **Figure 5.113**.

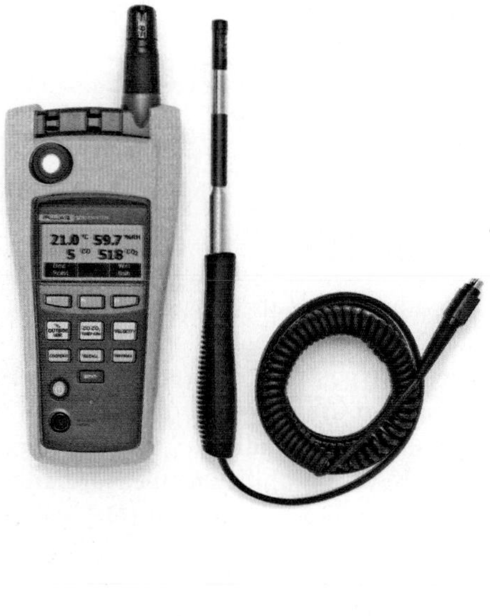

Figure 5.112 Air analyzer. Courtesy of Fluke Corporation.

Figure 5.113 Micro-anemometer. Courtesy of Fluke Corporation.

MANOMETERS

Manometers are used to measure very low pressures and express these pressure readings in inches of water column, where 27.7 IWC is equal to 1 psi of pressure. Manometers are available as single-port devices, **Figure 5.114**, or

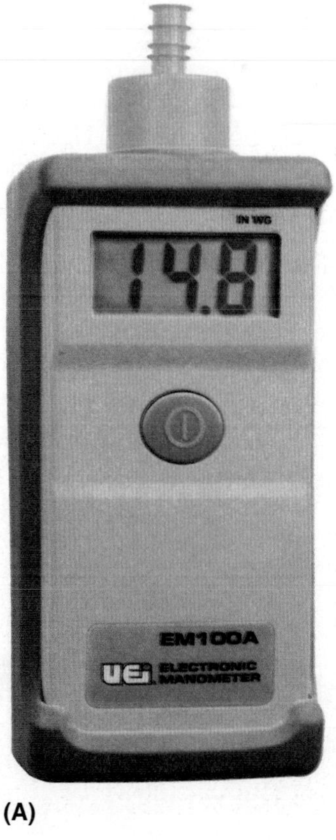

(A)

Figure 5.114 Manometer. Courtesy of UEi.

Figure 5.115 Dual-port manometer. Courtesy of Fieldpiece Instruments.

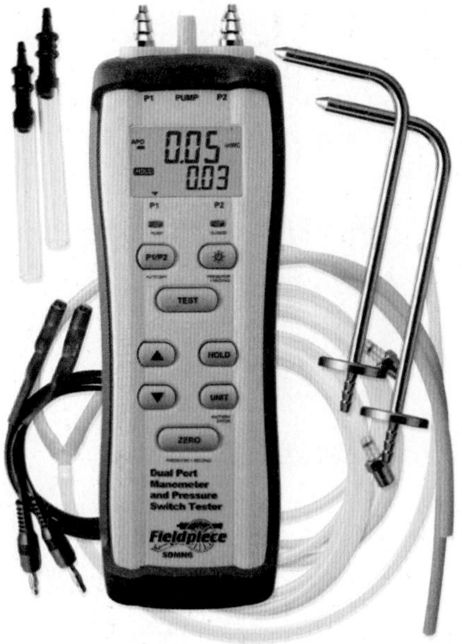

dual-port devices, **Figure 5.115**. Manometers are commonly used to measure the pressures (total, static, and velocity) in duct systems as well as fuel pressures in gas-fired heating appliances.

PSYCHROMETERS

Psychrometers are used to measure the wet-bulb and dry-bulb temperatures to determine the relative humidity of an airstream. A sling psychrometer, **Figure 5.116**, uses two thermometers, a dry-bulb and a wet-bulb thermometer. The thermometers are whirled together in the air to obtain the wet-bulb and dry-bulb temperatures. From these temperatures, the relative humidity can be determined.

Digital psychrometers, **Figure 5.117**, are also available. These devices are similar to the sling psychrometer, but these devices do not have to be spun in an air stream and often provide more information about the air sample, including, in some instances, enthalpy.

Figure 5.116 Sling psychrometer. Photo by Bill Johnson.

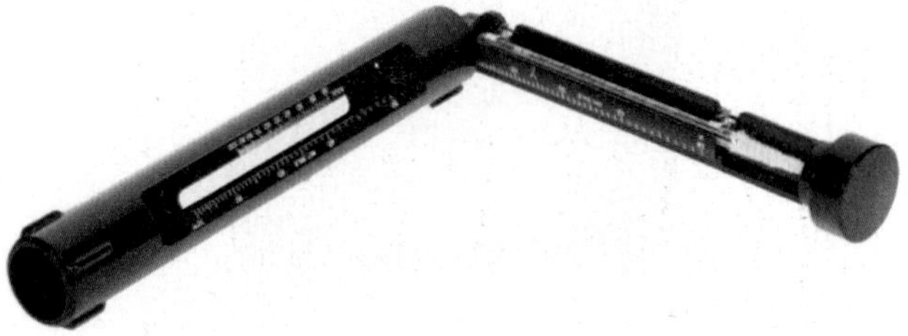

Figure 5.117 Digital psychrometer. Photo by Eugene Silberstein.

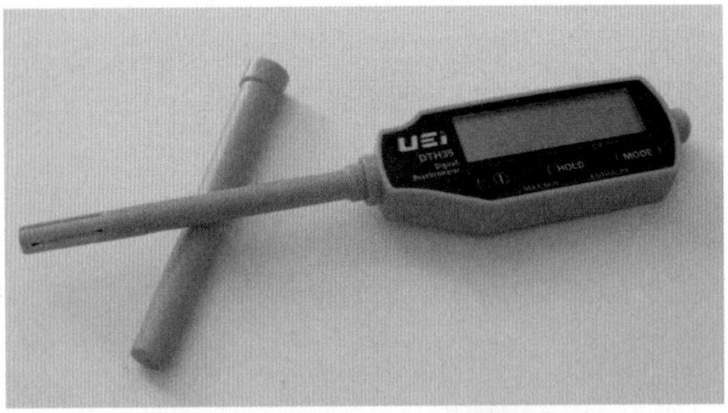

Figure 5.118 Flow hood. Source: Alnor Instrument Company.

FLOW HOODS

Flow hoods, **Figure 5.118**, are used to measure airflow, in cfm, through supply and return registers and grilles.

5.11 COMBUSTION TESTING EQUIPMENT

CARBON DIOXIDE AND OXYGEN TEST INSTRUMENTS

Carbon dioxide (CO_2) and oxygen (O_2) test instruments are used to analyze flue gases to determine the combustion efficiency of gas- or oil-fired furnaces, **Figure 5.119**. The test kit shown in Figure 5.119 is referred to as a "wet" kit. Digital combustion analyzers, such as the one shown in **Figure 5.120**, have the ability to measure and display concentrations in flue gases of oxygen (O_2), carbon monoxide

Figure 5.119 Carbon dioxide and oxygen indicators. Source: Bacharach, Inc. Pittsburgh, PA, USA

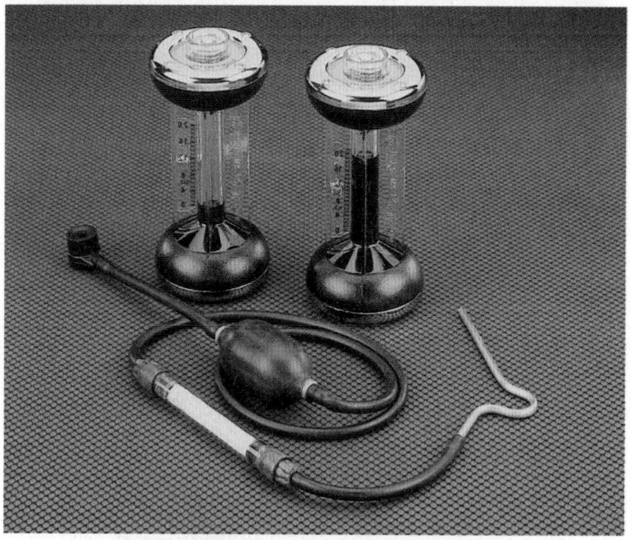

Figure 5.120 Digital combustion analyzer. Source: Bacharach, Inc. Pittsburgh, PA, USA

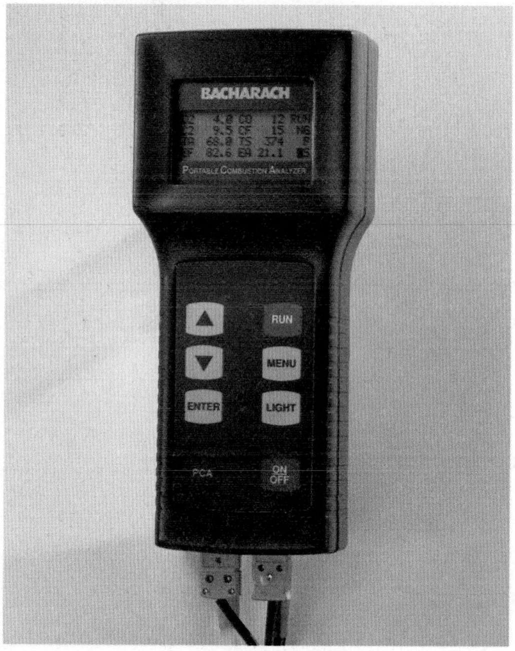

(CO), oxides of nitrogen (NOx), and sulfur dioxide (SO_2). They can also measure and display stack gas temperature, while computing and displaying combustion efficiency, stack loss, and excess air. These digital devices are very popular, but they do need to be factory calibrated each year. Some analyzers have the ability to print the results of the combustion test for office records as well as for the customer, **Figure 5.121**.

CARBON MONOXIDE (CO) DETECTORS

A CO detector is used to determine the percentage of CO present, **Figure 5.122**.

Figure 5.121 Combustion analyzer printer. Source: Bacharach, Inc. Pittsburgh, PA, USA

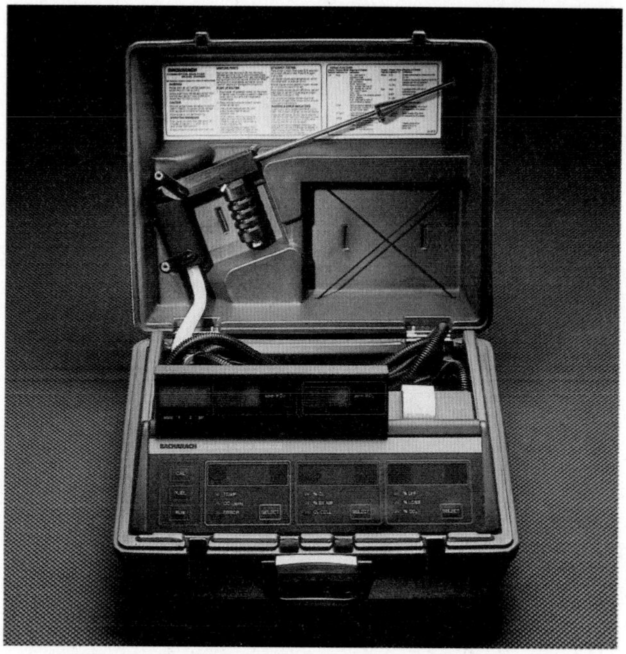

Figure 5.122 Carbon monoxide (CO) detector. A technician checking a furnace for carbon monoxide spillage with a digital carbon monoxide test instrument. Courtesy of Fluke Corporation.

Figure 5.123 Combustible gas leak detector. Courtesy of UEi.

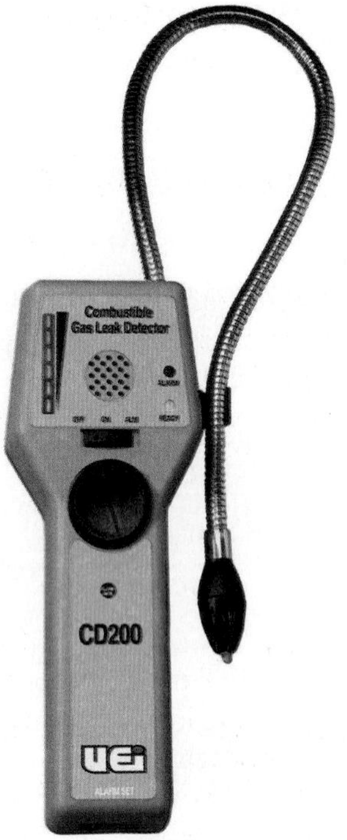

Figure 5.124 Draft gauge. Source: Bacharach, Inc. Pittsburgh, PA, USA.

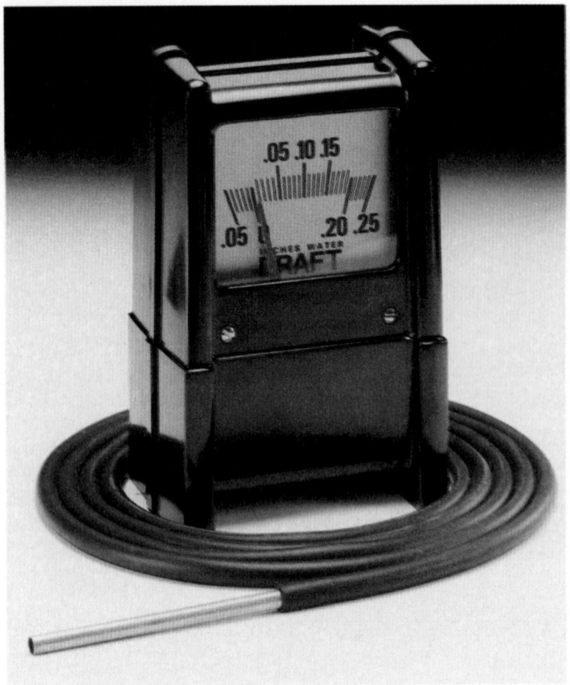

COMBUSTIBLE GAS LEAK DETECTOR

The combustible gas leak detector, **Figure 5.123**, has an audible alarm and provides visual leak detection by means of light-emitting diode (LED) indicators. A red LED light on the tip illuminates the search area.

DRAFT GAUGES

A draft gauge, **Figure 5.124**, is used to check the pressure of the flue gas in gas and oil furnaces to ensure that the flue gases are moving up the flue at a satisfactory speed. The flue gas is normally at a slightly negative pressure.

SMOKE TESTERS

The smoke tester, **Figure 5.125**, is used to determine the amount of soot that is contained in the flue gases of an oil-fired heating appliance. Using the tool, a predetermined amount of flue gases is drawn through a filter paper, which is then compared to a color scale to determine the degree of sooting that is taking place. It is strongly recommended that a smoke test be performed before a combustion test is conducted to protect the instrumentation from damage caused by excessive smoke. Refer to Unit 32 for more information on the procedure to properly perform a smoke test.

Figure 5.125 Smoke tester. Photo by Bill Johnson

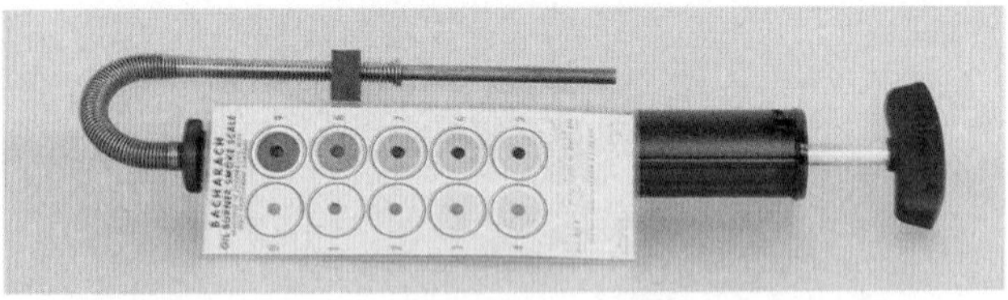

5.12 ENERGY AUDITING INSTRUMENTATION

Although many tools and instruments have been discussed throughout this unit, the following three pieces of test equipment are very popular when evaluating the integrity of the air distribution (duct) system and the building envelope.

THERMAL IMAGING CAMERAS

Thermal imaging cameras, **Figure 5.126**, are used to capture heat signatures as visible light. This allows the energy auditor to identify areas where heat is either escaping from a heated structure, or leaking into a cooled structure. These

Figure 5.126 Infrared thermal imaging camera. Courtesy of Fluke Corporation.

instruments are extremely useful because they have the ability to detect changes in temperature through walls and other solid objects. Refer to Unit 39 for more information on how thermal imaging cameras are used.

DUCT PRESSURIZATION TEST EQUIPMENT

Duct pressurization equipment, **Figure 5.127**, is used to pressurize, or de-pressurize, air distribution systems to determine how tight the duct system is. Refer to Unit 39 for more information on how duct pressurization equipment is used.

BLOWER DOOR TEST EQUIPMENT

Blower door test equipment, **Figure 5.128**, is used by an energy auditor to determine how tight or how leaky a structure's envelope is. Refer to Unit 39 for more information on how blower door test equipment is used.

Figure 5.127 A technician performing a duct pressurization test on a residential furnace. Courtesy of The Energy Conservatory.

Figure 5.128 A technician performing a blower door test on a residential dwelling. Courtesy of The Energy Conservatory.

REVIEW QUESTIONS

1. Screwdrivers can be safely used to
   A. tighten screws and chisel wood.
   B. tighten and loosen screws.
   C. loosen screws and pry/lift heavy objects.
   D. chisel metal and wood.
2. A screwdriver is properly sized when the blade
   A. fits snugly and completely into the screw's slot.
   B. is half as wide as the screw's slot.
   C. is twice as wide as the screw's slot.
   D. is thicker than the screw's slot.
3. Nut drivers are used to turn screws and bolts with
   A. hex-shaped slots in their heads.
   B. round-shaped heads.
   C. square-shaped heads.
   D. hex-shaped heads.
4. The refrigeration service wrench is used to turn
   A. round valve stems.
   B. hex-shaped valve stems.
   C. square valve stems.
   D. triangular valve stems.
5. Which of the following type of wrench is most prone to slipping?
   A. Open-end wrench
   B. Pipe wrench
   C. Allen wrench
   D. Box-end wrench
6. Which of the following tools is most likely to be used correctly for the final tightening of a bolt?
   A. Allen wrench
   B. Refrigeration service wrench
   C. Open-end wrench
   D. Box-end wrench
7. The universal joint is typically part of a(n)
   A. socket wrench set.
   B. Allen key set.
   C. hole saw set.
   D. nut driver set.
8. The universal joint
   A. extends the length of a wrench's handle.
   B. allows the technician to use socket wrenches at an angle.
   C. can be used in place of a nut driver.
   D. is used in conjunction with an adjustable wrench.
9. Why do some Allen wrenches have rounded or ball tips?
   A. To increase the torque of the wrench
   B. To prevent the ends of the wrench from shattering
   C. To allow them to turn round screws
   D. To allow them to be used at an angle
10. Which of the following hammers has a rounded peen for shaping metal?
    A. Sheet metal hammer
    B. Carpenter's claw hammer
    C. Soft-faced mallet
    D. Ball peen hammer
11. Which of the following hammers is less likely to damage delicate surfaces such as wood?
    A. Sheet metal hammer
    B. Carpenter's claw hammer
    C. Soft-faced mallet
    D. Ball peen hammer
12. The most common type of pliers used by a plumber is the
    A. general purpose slip-joint pliers.
    B. pump pliers.
    C. locking pliers.
    D. linesman pliers.
13. The type of pliers that can be used as a vice is the
    A. general purpose slip-joint pliers.
    B. pump pliers.
    C. locking pliers.
    D. linesman pliers.
14. The tool specifically designed to remove small amounts of insulation from a conductor is the
    A. utility knife.
    B. side cutter.
    C. locking pliers.
    D. wire stripper.
15. The crimping tool can be used to perform all of the following tasks except
    A. securing solderless connectors to stranded wire.
    B. cutting small bolts and screws.
    C. cutting wire.
    D. securing solderless connectors to solid wire.
16. The type of hand saw that would most likely be used to cut a small hole in a drywall wall or ceiling is the
    A. keyhole saw.
    B. reciprocating saw.
    C. hacksaw.
    D. jigsaw.
17. The hand saw that is used primarily for cutting metal is the
    E. keyhole saw.
    F. reciprocating saw.
    G. hacksaw.
    H. Jigsaw.
18. What part of the hole saw ensures that the hole is positioned exactly where it needs to be?
    A. The circular blade
    B. The drill chuck
    C. The pilot bit
    D. The arbor
19. Which of the following tools is used to help ensure level, straight, and plumb installations?
    A. The tape measure
    B. The plumb bob
    C. The chalk line
    D. The carpenter's square

20. A fish tape is used to
    A. measure the distance between two points.
    B. ensure level piping installations.
    C. push wires through a wall or conduit.
    D. pull wires through a wall or conduit.
21. One of the benefits of battery-powered tools is that they
    A. can be used for longer periods of time than corded tools.
    B. can be used in areas where live receptacles are not available.
    C. must be charged periodically for continual use.
    D. can be simultaneously powered by line voltage circuits and by batteries.
22. Which of the following saws is most likely used for demolition work?
    A. Reciprocating saw
    B. Jigsaw
    C. Hacksaw
    D. Hole saw
23. Which of the following saws uses a back-and-forth motion and is best suited to cut freeform shapes with tight turns?
    A. Jigsaw
    B. Circular saw
    C. Reciprocating saw
    D. Hacksaw
24. All of the following are electrical characteristics that are measured by a simple digital multimeter (DMM) except
    A. A/C voltage.
    B. power.
    C. D/C voltage.
    D. resistance.
25. When referring to a digital multimeter (DMM), the acronym NCV stands for
    A. normally closed value.
    B. normally closed voltage.
    C. non-contact value.
    D. non-contact voltage.
26. Which feature of an electric meter allows the technician to measure circuit amperage without accessing the circuit's bare conductors?
    A. Clamp-on jaws
    B. Alligator clips
    C. Meter's test probes
    D. Meter's auto-range feature
27. The type of refrigerant leak detector that utilizes an open flame is the ___________ leak detector.
    A. halide
    B. electronic
    C. ultrasonic
    D. ultraviolet
28. The type of refrigerant leak detector that uses an internal pump to pull in vapor samples is the ________ leak detector.
    A. halide
    B. electronic
    C. ultrasonic
    D. ultraviolet
29. The tubing reamer is used to
    A. expand the end of a section of soft-drawn tubing.
    B. remove burrs from a section of soft-drawn tubing.
    C. cut soft-drawn tubing to the desired length.
    D. bend copper tubing without kinking it.
30. Swaging is the process of
    A. expanding the end of a section of soft-drawn tubing.
    B. flaring the end of a section of soft-drawn tubing.
    C. bending a section of soft-drawn tubing.
    D. soldering a section of soft-drawn tubing.
31. Flaring is the process of
    A. expanding the end of a section of soft-drawn tubing.
    B. creating a bell-shaped bevel on the end of a section of soft-drawn tubing.
    C. bending a section of soft-drawn tubing.
    D. soldering a section of soft-drawn tubing.
32. In order to minimize the potential for leaks in a refrigerant piping arrangement, which of the following processes is desirable?
    A. Brazing
    B. Soldering
    C. Flaring
    D. Bending
33. The long handles on a pipe wrench are intended to
    A. allow the user to access tight/difficult spaces.
    B. increase the torque of the wrench.
    C. decrease the torque of the wrench.
    D. allow the wrench to be used as a pry bar.
34. Red-handled tin snips are typically used to make
    A. right- and left-turning sheet metal cuts.
    B. straight and left-turning sheet metal cuts.
    C. straight and right-turning sheet metal cuts.
    D. only straight sheet metal cuts.
35. The sheet metal tool that is commonly used to reduce the diameter of a section of round stove pipe is the
    A. crimper.
    B. notcher.
    C. tin snip.
    D. hand seamer.
36. Which of the following torches uses both an oxygen and an acetylene regulator?
    A. Air-acetylene torch
    B. Propane torch
    C. Oxyacetylene torch
    D. MAPP gas torch
37. The air-acetylene torch flame is about
    A. 3600°F
    B. 3730°F
    C. 4200°F
    D. 5600°F
38. Which of the following gauge manifold hoses is typically larger than the others?
    A. High-side hose
    B. Low-side hose
    C. Evacuation hose
    D. Refrigerant charging hose

39. The digital vacuum gauge measures a vacuum in which units?
    A. Microns
    B. Inches of water column (IWC)
    C. Psig
    D. Psia
40. Which of the following tools monitors the amount of refrigerant added to or removed from an HVACR system?
    A. Refrigerant recovery unit
    B. Refrigerant charging scale
    C. Vacuum gauge
    D. Micron gauge
41. What two functions does a vacuum pump perform?
    A. Charging and recovering
    B. Pushing and pulling
    C. Pressurizing and de-pressurizing
    D. Dehydrating and degassing
42. Refrigerant is mechanically removed from an HVACR system using a
    A. refrigerant recovery unit.
    B. vacuum pump.
    C. refrigerant charging scale.
    D. refrigerant gauge manifold.
43. A velometer is used to measure the
    A. speed of refrigerant in an HVACR system.
    B. volume of refrigerant in an HVACR system.
    C. speed of an air stream in a duct system.
    D. volume of airflow in a duct system.
44. Duct system pressures are measured in
    A. psig.
    B. psia.
    C. inches of water column (IWC).
    D. microns.
45. The instrument that measures wet-bulb and dry-bulb temperatures is the
    A. anemometer.
    B. velometer.
    C. psychrometer.
    D. refractometer.
46. A "wet" combustion test kit measures which two substances?
    A. Soot and carbon monoxide
    B. Carbon dioxide and oxygen
    C. Carbon monoxide and oxygen
    D. Carbon monoxide and carbon dioxide
47. A smoke test is used to determine the amount of
    A. soot in flue gases.
    B. carbon dioxide in flue gases.
    C. carbon monoxide in flue gases.
    D. soot in a heat exchanger.
48. To perform a smoke test,
    A. room air is pulled through a filter paper.
    B. room air is pulled through a filter drier.
    C. flue gas is pulled through a filter paper.
    D. flue gas is pulled through a filter drier.
49. The piece of test instrumentation that converts heat energy into visible light is the
    A. thermal imaging camera.
    B. psychrometer.
    C. refractometer.
    D. anemometer.
50. A blower door test is used to determine
    A. the amount of airflow through a duct system.
    B. the heat content in an air stream.
    C. how leaky an air distribution system is.
    D. how leaky a building envelope is.

UNIT 6

FASTENERS

OBJECTIVES

After studying this unit, you should be able to

- identify common fasteners used with wood.
- explain how staples are used as fasteners in the HVACR industry.
- identify common fasteners used with sheet metal.
- identify and describe typical screw dimensions.
- identify typical machine screw heads.
- identify common fasteners used on hollow walls.
- identify and describe fasteners used in masonry applications.
- explain how threaded rod can be used to support HVACR equipment.
- describe hanging and securing devices for piping, tubing, and ductwork.
- describe various pins, clamps, and rings that are commonly used as fasteners.
- describe solderless terminals and screw-on wire connectors.

SAFETY CHECKLIST

- ☑ When staples are used to fasten a wire in place, do not hammer them too tightly as they may damage the wire.
- ☑ When using fasteners, make sure that all materials are strong enough for the purpose for which they are being used.
- ☑ Wear goggles whenever drilling holes and when cleaning out holes.
- ☑ Do not use powder-actuated systems without the proper training.
- ☑ When installing fasteners, be sure to follow all installation instructions carefully.

As an air-conditioning (heating and cooling) and refrigeration technician, you need to know about different types of fasteners and various fastening systems. Knowing about different fasteners, including their intended use as well as their limitations, will help make sure that you use the right fastener or system to securely install and mount all equipment, system components, and materials.

6.1 NAILS

Probably the most common fastener used in wood is the nail. Nails are available in many styles and sizes. Common nails are large, flat-headed wire nails with a specific diameter for each length, **Figure 6.1(A)**. A finishing nail, **Figure 6.1(B)**, has a very small head so that it can be driven below the surface of the wood; it is used where a good finish is desired.

The term *penny*, which originated in England, is used to size nails. Ten-penny nails, for example, cost 10 pence (an English coin) for 100 nails; four-penny nails cost four pence per 100 nails. "Penny" is represented by the letter "d," which is the symbol for English pence. For instance, 8d describes the shape and size of the common nail. These nails can be purchased plain or coated, usually with zinc or resin. The coating protects against corrosion and helps to prevent the nail from working its way out of the wood.

A roofing nail is used to fasten shingles and other roofing materials to the roof of a building. It has a large head to keep the shingles from tearing and pulling off. A roofing nail can sometimes be used on metal strapping material to hang ductwork or tubing, **Figure 6.2(A)**. Masonry nails are made of hardened steel and can be driven into masonry, **Figure 6.2(B)**.

Figure 6.1 (A) Common nails. (B) A finishing nail.

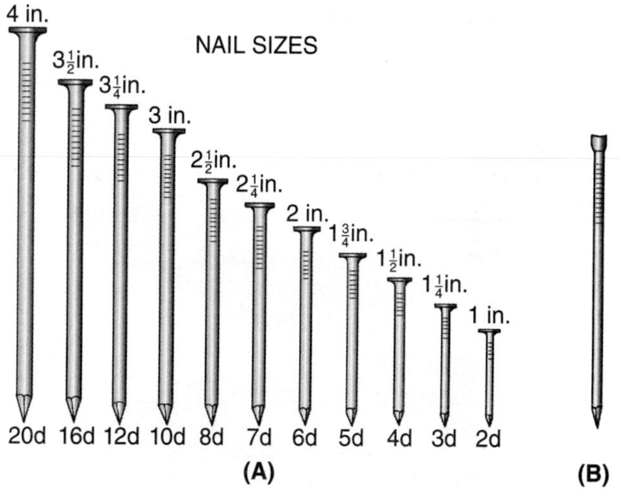

Figure 6.2 (A) A roofing nail. (B) A masonry nail.

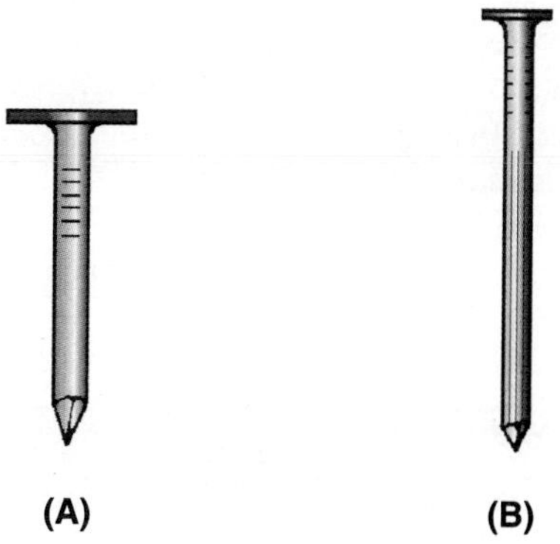

6.2 STAPLES AND RIVETS

STAPLES

A staple, **Figure 6.3**, is a fastener made somewhat like a nail but shaped like a U with a point on each end. Staples are available in different sizes for different uses, one of which is to fasten wire in place. The staples are simply hammered in over the wire. **SAFETY PRECAUTION:** *Never hammer staples too tightly because they can damage the wire.* Other types of staples may be fastened in place with a stapling tacker, **Figure 6.4**. By depressing the handle of a tacker, a staple can be driven through paper, fabric, or insulation into the wood. Outward clinch tackers will anchor staples inside soft materials. The staple legs spread outward to form a strong, tight clinch, **Figure 6.5**. They can be used to fasten insulation around heating or cooling pipes and ducts and to install ductboard. Other models or tackers are used to staple low-voltage wiring to wood.

Figure 6.3 Staples used to fasten wire to wood.

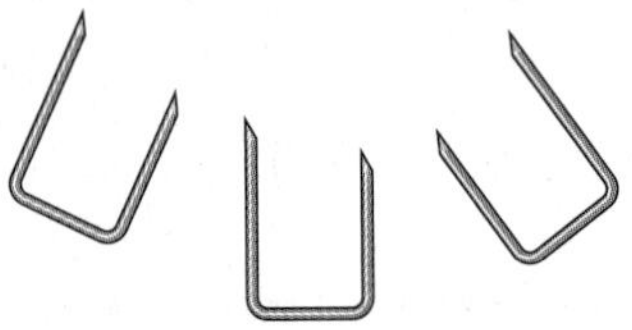

Figure 6.4 A stapling tacker. Photo by Bill Johnson

Figure 6.5 A staple that is clinched outward is often used to fasten soft materials together.

RIVETS

Pin rivets or blind rivets are used to join two pieces of sheet metal. They are actually hollow rivets assembled on a pin, often called a mandrel. **Figure 6.6** illustrates a pin rivet assembly, how it is used to join sheet metal, and a riveting tool or gun. The rivets are inserted and set from only one side of the metal. Thus, they are particularly useful when there is no access to the back side of the metal being fastened.

To use a pin rivet, you first drill a hole in the metal the size of the rivet and insert the pin rivet in the hole with the pointed end facing out. Then you place the nozzle of the riveting tool over the pin and squeeze the handles. Jaws grab the pin and pull the head of the pin into the rivet, expanding it and forming a head on the other side. Continue squeezing the handles until the head on the reverse side forms a tight joint. When this happens, the pin breaks off and is ejected from the riveting tool. **Figure 6.7** shows what a finished rivet looks like. The rivet can only be removed by drilling it out with a drill bit that is the same size as the rivet. When the rivet is drilled out, the back side of the rivet will fall on the inside of

Figure 6.6 (A) A pin rivet assembly. (B) Using pin rivets to fasten sheet metal. (C) A riveting gun. (A) and (B) Courtesy of Duro Dyne Corp. (C) Photo by Eugene Silberstein

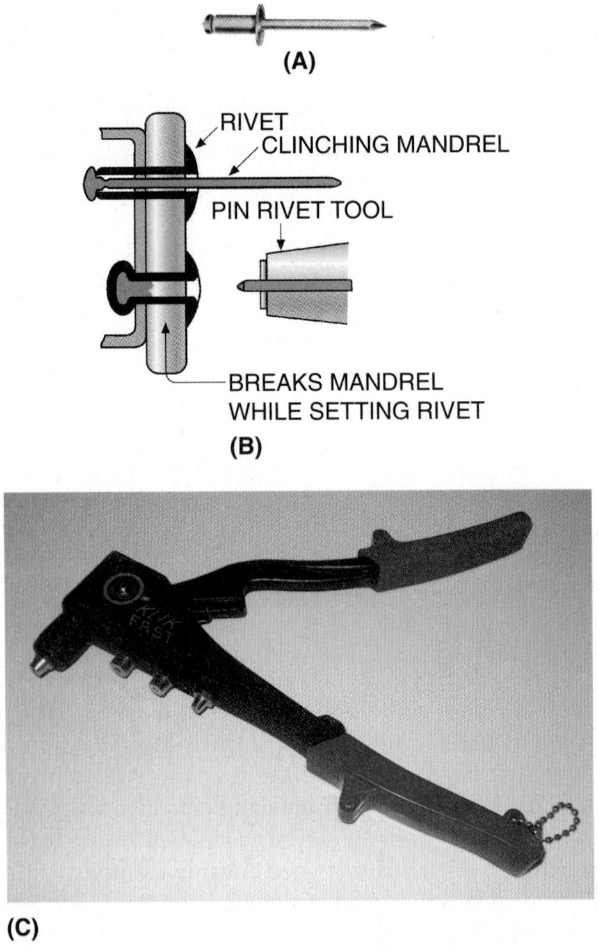

Figure 6.7 A completed rivet. Photo by Eugene Silberstein

the sheet metal pieces that are fastened together. **⚠ SAFETY PRECAUTION:** *If you are working on an electrical panel, be sure the power is off.* • Another rivet of the same size can be used to refasten the panels.

6.3 SCREWS

WOOD SCREWS

Wood screws are used to fasten many types of materials to wood. They generally have a flat, round, or oval head, **Figure 6.8**. The portion of flat- and oval-head screws beneath the head is at an angle. Holes for these screws should be countersunk so that the angular portion is recessed. Wood screws can also have a straight, slotted head or a Phillips recessed head.

Wood screw sizes are specified by length (in inches) and shank diameter (a number from 0 to 24). The larger the number, the larger the diameter or gauge of the shank. The number that represents the shank diameter can be converted to inches by using the following formula:

$$\text{Shank Diameter in Inches} = (\text{Shank Number} \times 0.013) + 0.06$$

For example, a number 8 screw has a larger shank diameter (0.1643) than a number 6 screw (0.1383). An 8-1″ wood screw and a 6-1″ wood screw are both 1 in. long, but the number 8 screw is wider. The numbering system described here is used for smaller screws up to 1/4 in. in diameter; then the system changes to fractions of an inch. Larger sizes of wood screws (screws with a tapered shaft) are called lag screws. They often come with a hexagonal or a square head and are tightened with a wrench, not a hand driver. These are typically used to fasten timber or large pieces of wood together, **Figure 6.9**.

Figure 6.8 (A) A flat-head wood screw. (B) A round-head wood screw. (C) An oval-head wood screw.

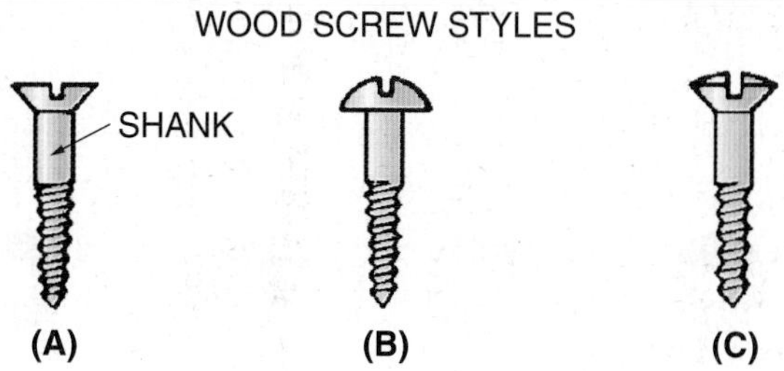

TAPPING SCREWS

Tapping or sheet metal screws are used extensively by service technicians. These screws may have a straight slot, a Phillips head, or a straight slot and hex head, **Figure 6.10**. To use a tapping screw, you drill a hole into the sheet metal the approximate size of the root diameter of the thread. Then you turn the screw into the hole with a conventional screwdriver or electric drill with a screwdriver bit.

Figure 6.11 shows a tapping screw, often called a self-drilling screw, that can be turned into sheet metal with an electric drill using a chuck similar to the one shown. This screw has a special point used to start the hole. It can be used with light to medium gauges of sheet metal. Note that self-tapping and self-drilling screws are not the same. Self-tapping screws require a hole to be drilled in the sheet metal prior to installing the screw; the self-drilling screw makes its own hole. Notice the difference in the tips of the screws in **Figure 6.10** and **Figure 6.11**.

Figure 6.9 Lag screw or bolt. Photo by Eugene Silberstein

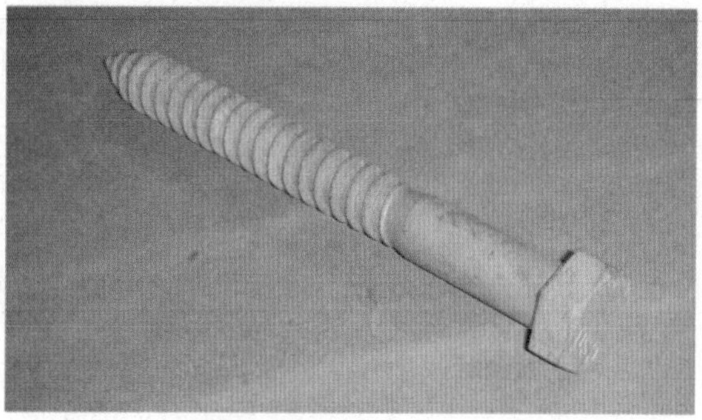

Figure 6.10 A tapping screw.

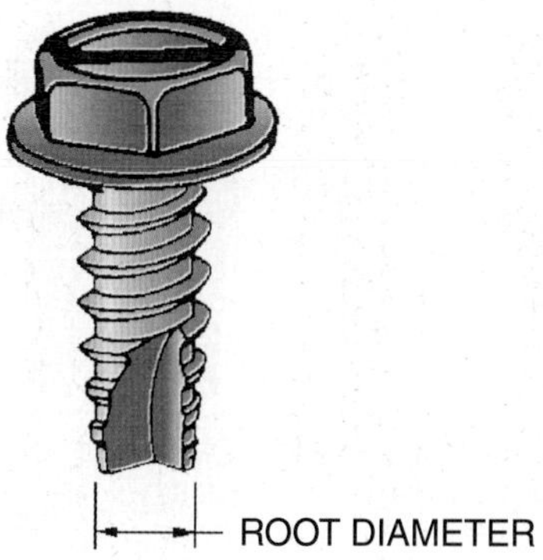

Figure 6.11 Self-drilling screws and a drill chuck. Photo by Eugene Silberstein

Tapping screws are identified or labeled in a specific way. The following is a typical identifying label:

6-20 × 1/2 Type AB, Slotted hex
6 = Screw Size Number
20 = Number of threads per inch
1/2 = Length of the screw
Type AB = Type of point
Slotted hex = Type of head

MACHINE SCREWS

There are many types of machine screws. They normally are identified by their head styles. Some of these are indicated in **Figure 6.12**. A thread dimension or label indicates the following information:

5/16–18 UNC -2
5/16 = Outside thread diameter
18 = Number of threads per inch
UNC = Unified thread series (Unified National Coarse)
2 = Class of fit (the amount of play between the internal and external threads)

Figure 6.12 Machine screws.

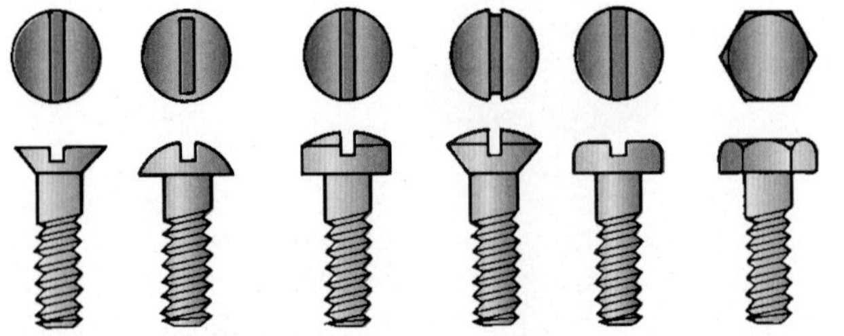

SET SCREWS

Set screws have points, as illustrated in **Figure 6.13**, and have square heads, hex heads, or no heads. The headless type may be slotted for a screwdriver, or it may have a hexagonal or fluted socket. These screws are used to keep a pulley from turning on a shaft and for other, similar applications. The set screw that is used to hold a pulley tight on a shaft is made of hardened steel; the pulley shaft is made of soft steel. When the set screw is tightened onto the shaft, it indents the shaft with its tip.

FASTENER HEADS

Figure 6.14 is an illustration of several fastener heads and their driver tips.

Figure 6.13 Styles of set screws.

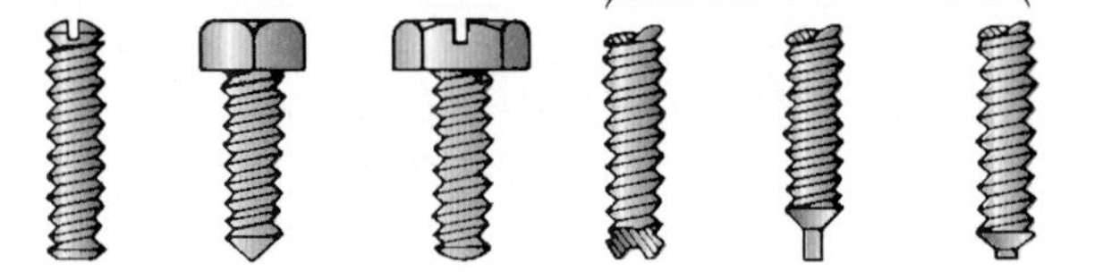

Figure 6.14 Several fastener heads and their driver tips. Courtesy of Klein Tools

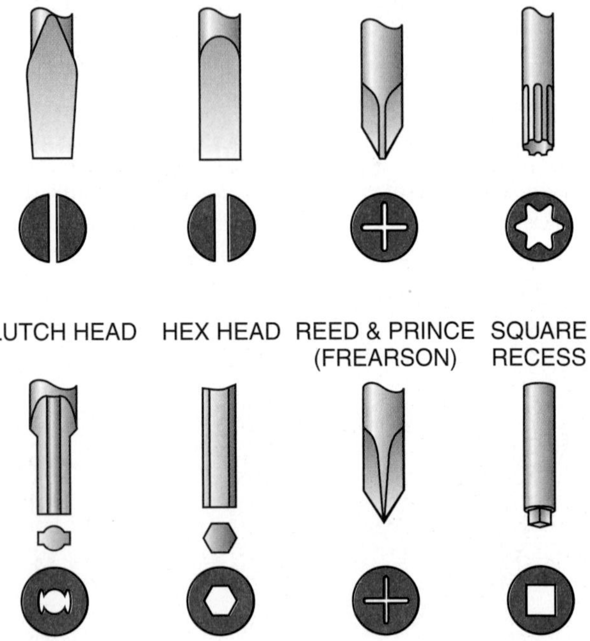

6.4 HOLLOW WALL FASTENERS

WALL ANCHORS

Hollow wall anchors, **Figure 6.15**, can be used in plaster, wallboard, gypsum board, and similar materials. Once the anchor has been set, the screw may be removed as often as necessary without affecting the anchor.

TOGGLE BOLTS

Toggle bolts provide secure anchoring in hollow tiles, building block, plaster over lath, and gypsum board, **Figure 6.16**. You drill a hole in the wall large enough for the toggle in its folded position to go through, making certain that the object being secured has been positioned on the bolt before screwing the toggle on the bolt, **Figure 6.17**. Push the toggle through the hole and use a screwdriver to turn the bolt head, **Figure 6.18**. You must maintain tension on the toggle or it will not tighten. **Figure 6.19** shows the completed toggle bolt installation.

Figure 6.15 A hollow wall anchor. Courtesy of Rawlplug Company, Inc.

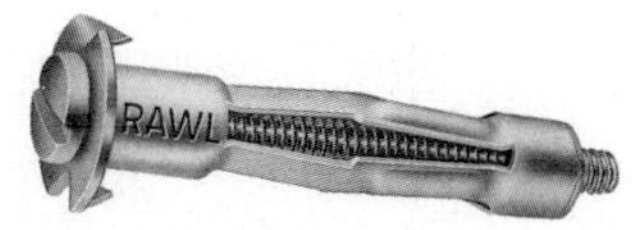

Figure 6.16 A toggle bolt. Courtesy of Rawlplug Company, Inc.

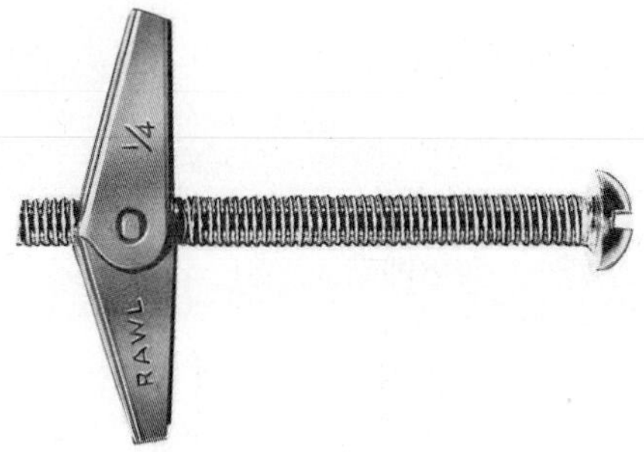

Figure 6.17 Make certain that the item being secured is placed on the bolt before the toggle is screwed onto the bolt. Photo by Eugene Silberstein

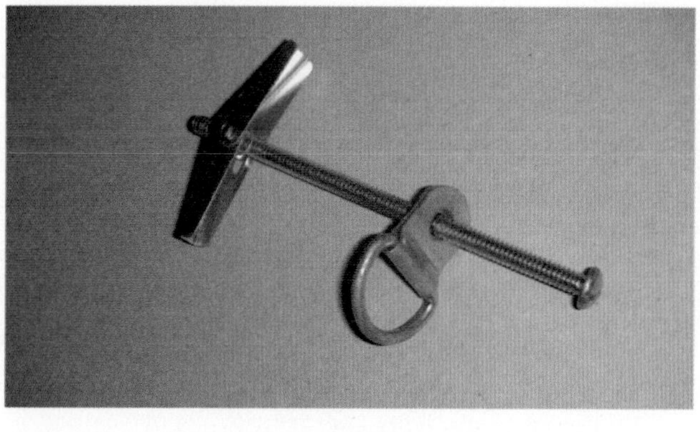

Figure 6.18 Inserting the folded toggle into the hole in the gypsum board. Photo by Eugene Silberstein

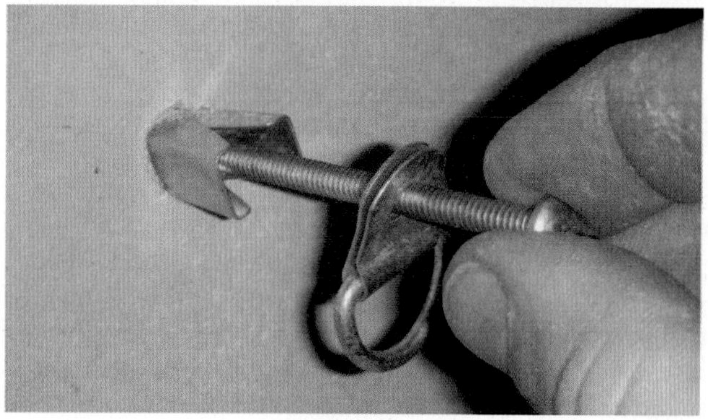

Figure 6.19 (A) Front view of completed toggle bolt installation. (B) Back view of completed toggle bolt installation. Photos by Eugene Silberstein

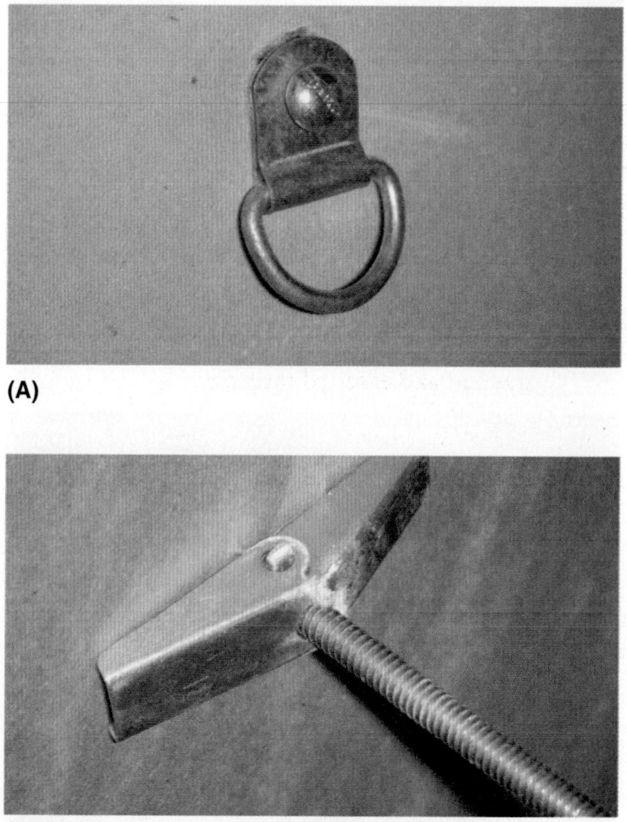

(A)

(B)

PLASTIC TOGGLE DRYWALL ANCHORS

One variation on the toggle bolt is the plastic toggle, **Figure 6.20**. Once a hole of the proper size has been made in the gypsum board, the plastic toggle is folded in half and inserted into the hole, **Figure 6.21**. A special pin that comes with the anchor is inserted in the anchor to ensure that it opens inside the wall, **Figure 6.22**. Once the anchor is in place, the screw is fed through the item being secured and tightened into the anchor, **Figure 6.23**. Unlike the traditional toggle bolt, constant tension on the anchor is not required for tightening.

SELF-DRILLING DRYWALL ANCHORS

Another option for drywall applications is the self-drilling drywall anchor, **Figure 6.24**. This type of anchor does not require a predrilled hole. You simply press the tip of the anchor into the gypsum board and turn it clockwise until it is flush with the surface, **Figure 6.25**. Once the anchor is in place, the screw is fed through the item being secured and tightened into the anchor.

Figure 6.20 Plastic toggle. Photo by Eugene Silberstein

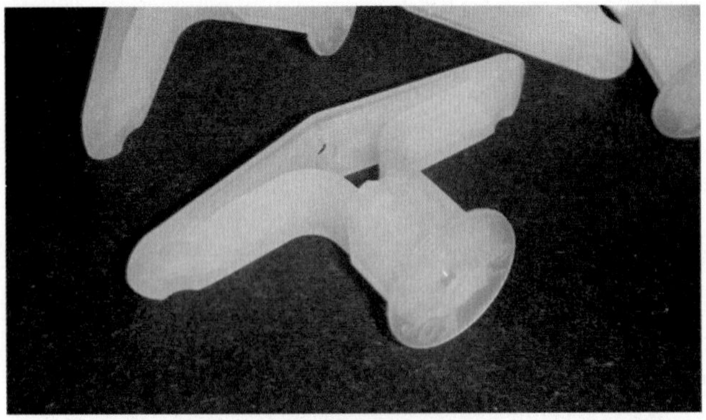

Figure 6.21 The folded plastic toggle inserted into the hole in the gypsum board. Photo by Eugene Silberstein

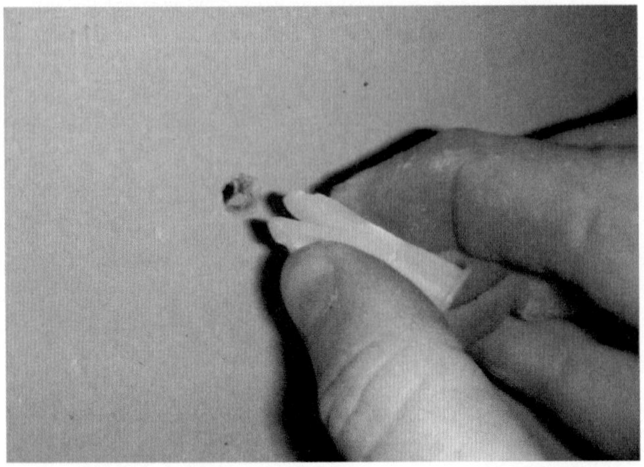

Figure 6.22 (A) Using a pin to open the toggle. (B) Back view of opened plastic toggle. Photos by Eugene Silberstein

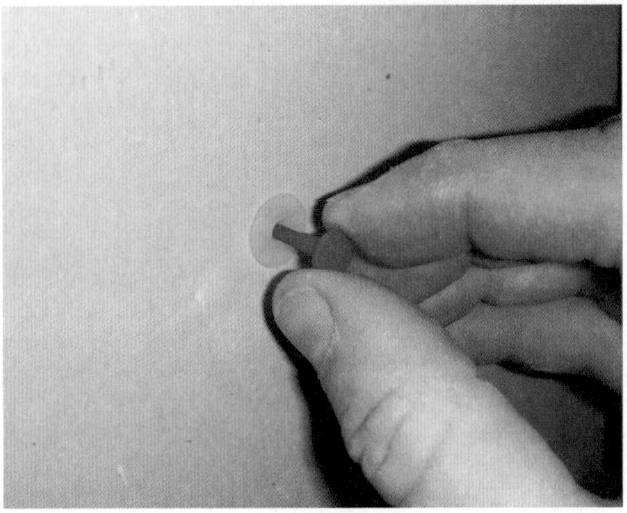

(A)

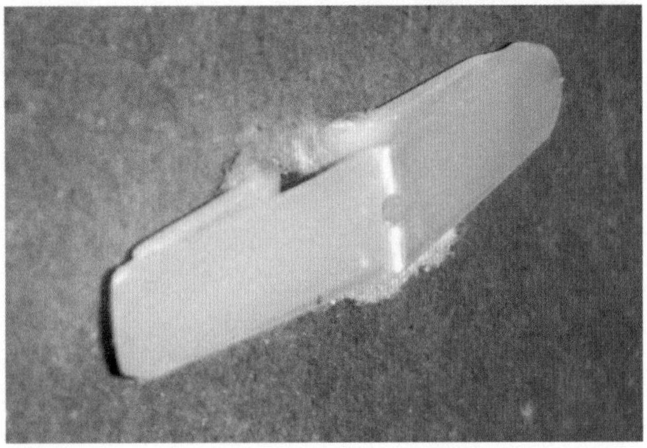

(B)

Figure 6.23 Securing object to the wall. Photo by Eugene Silberstein

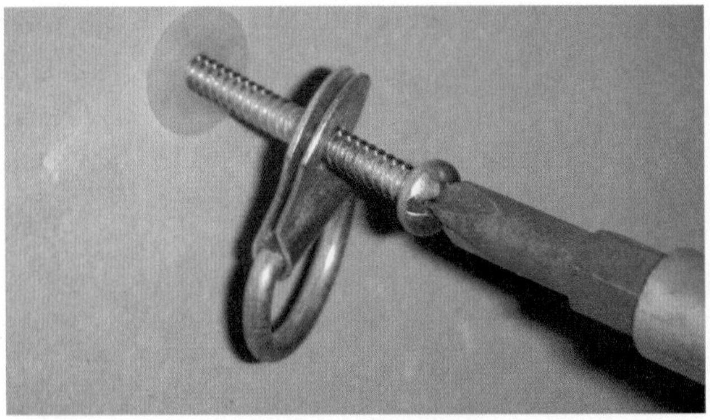

Figure 6.24 Self-drilling drywall anchors. Photo by Eugene Silberstein

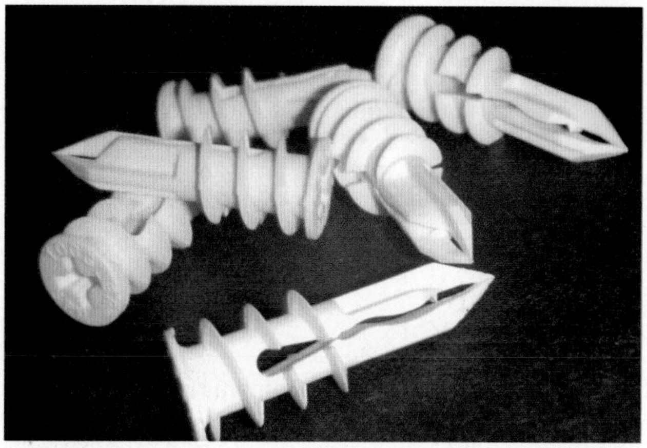

Figure 6.25 Installing the self-drilling drywall anchor. Photo by Eugene Silberstein

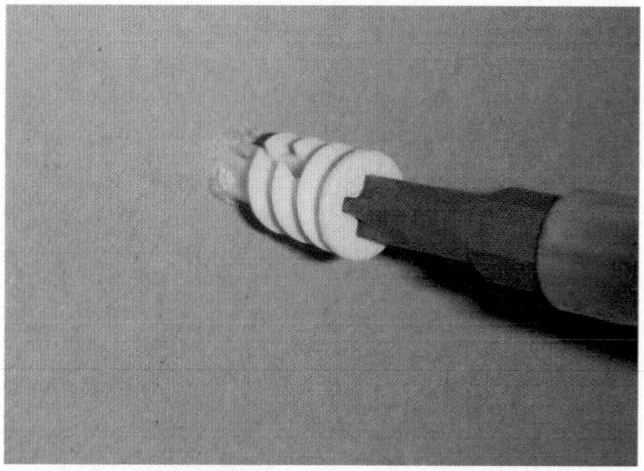

6.5 CONCRETE FASTENERS

SCREW FASTENERS FOR CONCRETE

A concrete screw fastener may be used in poured concrete, concrete block, or brick, **Figure 6.26**. A hole must first be drilled in the base material. It is important to use the drill diameter recommended by the manufacturer for each screw size. Drill the hole slightly deeper than the length of the screw that will extend into the base material. Blow out the dust and other material from the hole to the extent possible. **⚠ SAFETY PRECAUTION:** *Wear safety goggles during this drilling and cleaning process.•* The screw may then be installed

Figure 6.26 A concrete screw fastener.

with an electric screwdriver or drill using a screwdriver bit or drill chuck, depending on the type of head on the screw.

ANCHOR SHIELDS

Anchor shields with bolts or screws are used to fasten objects to masonry or, in some instances, hollow walls. **Figure 6.27** illustrates a multipurpose steel anchor bolt used in masonry material. You drill a hole in the masonry the size of the sleeve, tap the sleeve and bolt into the hole with a hammer, and turn the bolt head. This expands the sleeve and secures the bolt in the masonry, **Figure 6.28**. A variety of head styles are available.

LAG SHIELD ANCHORS

A two-part shield, generally used with lag screws, has tapered internal threads for part of its length, **Figure 6.29**. To install them, drill a hole of the diameter recommended by the manufacturer for the size of shield. The hole should be drilled as deep as the length of the shield plus approximately

Figure 6.27 An anchor shield with a screw. Courtesy of Rawlplug Company, Inc.

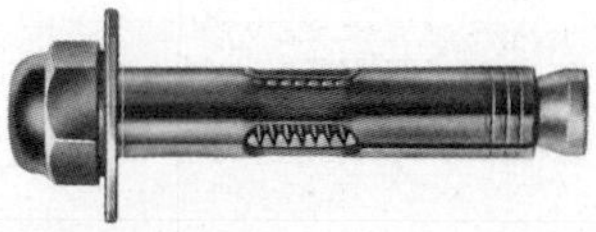

Figure 6.28 Image showing the anchor shield before and after expansion. Photo by Eugene Silberstein

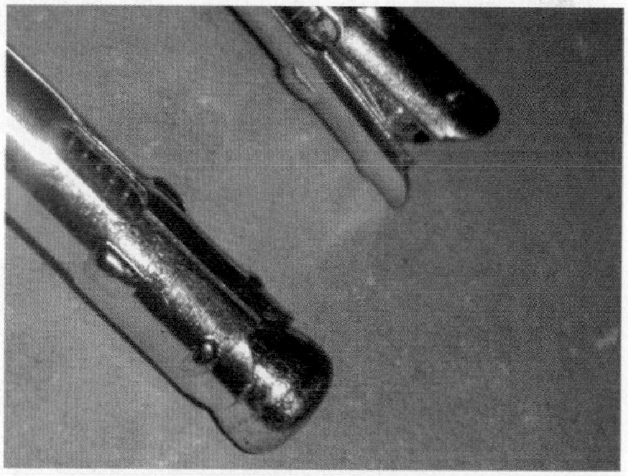

Figure 6.29 A lag shield anchor.

1/2 in. Clean dust or loose concrete material out of the hole. **SAFETY PRECAUTION:** *Wear goggles when drilling and cleaning.* Tap the shield in with a hammer until it is flush with the surface. Position the material to be fastened with the hole over the shield and turn in the lag screw.

POWDER-ACTUATED FASTENER SYSTEMS

These systems provide another method of fastening materials to concrete, **Figure 6.30**. **SAFETY PRECAUTION:** *This type of system is very dangerous. Operators must have specific training and in many instances must be licensed. We only mention it here to make you aware that there is such a system.* A powder load, **Figure 6.31**, is used in a powder-actuated tool (PAT). When actuated, it forces a pin, threaded stud, or other fastener into the concrete. The powder loads are color-coded based on the amount of power needed for a particular job. **SAFETY PRECAUTION:** *Do not try this without the training and the license to operate this equipment.*

Many other types of fasteners are used to fasten material to concrete. Those described here are but a few examples.

Figure 6.30 A powder-actuated fastener system. Photo by Eugene Silberstein

Figure 6.31 Color-coded powder loads. Photo by Eugene Silberstein

6.6 THREADED ROD

THREADED ROD AND ANGLE STEEL

Threaded rod and angle steel can be used to make custom hangers for pipes or components such as an air handler, **Figure 6.32**. Threaded rod can be attached to steel I-beams with a beam clamp, **Figure 6.33**. **SAFETY PRECAUTION:** *Be sure that all materials are strong enough to adequately support the equipment.*

THREADED ROD AND STEEL CHANNEL

Threaded rod can also be used in combination with steel channel to fabricate custom hanging arrangements, **Figure 6.34**.

Figure 6.32 Using threaded rod and angle steel to fabricate a hanger.

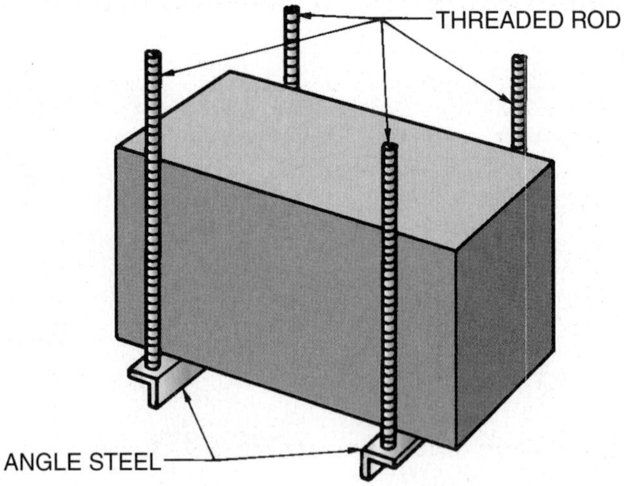

Figure 6.33 Beam clamp used to secure threaded rod. Photo by Eugene Silberstein

Common components in this type of hanger include the steel channel, threaded rod, saddle plates, square washers, round washers, and nuts, **Figure 6.35**.

Figure 6.34 Threaded rod and steel channel used to make custom hangers. Photo by Eugene Silberstein

Figure 6.35 (A) Steel channel, threaded rod, saddle plate, square washer, round washer, and nuts. (B) Exploded view of a typical assembly. Photos by Eugene Silberstein

(A)

(B)

Figure 6.36 Steel channel and pipe clamp. Photo by Eugene Silberstein

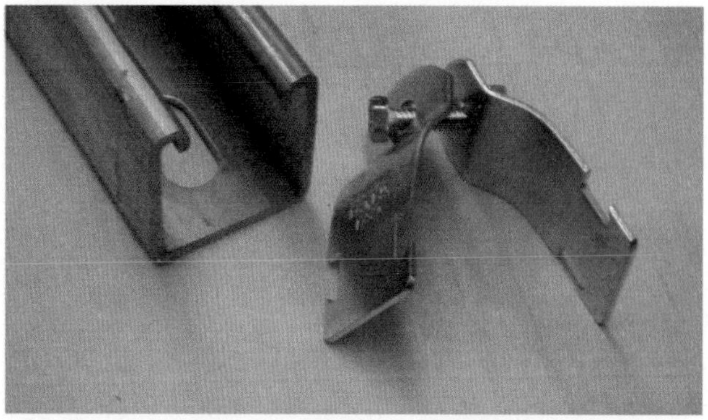

Figure 6.37 Steel channel and pipe clamp used to hold piping material. Photo by Eugene Silberstein

STEEL CHANNEL AND PIPE CLAMPS

Steel channel can also be used in conjunction with pipe clamps, **Figure 6.36**, on a number of different piping materials used in both refrigeration and electrical applications, **Figure 6.37**.

6.7 PIPING AND DUCTWORK FASTENERS

PIPE HOOK

A wire pipe hook is a hardened steel wire bent into a U with a point at an angle on both ends. The pipe or tubing rests in the bottom of the U, and the pointed ends are driven into wooden joists or other wood supports, **Figure 6.38**.

PIPE STRAP

A pipe strap is used for fastening pipe and tubing to joists, ceilings, or walls, **Figure 6.39(A)**. The arc on the strap should be approximately the same as the outside diameter of the

Figure 6.38 (A) A wire pipe hook. (B) Pipe hook installation. (B) Photo by Eugene Silberstein

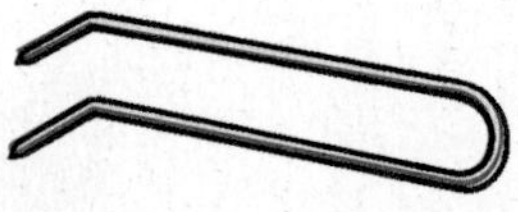

(A)

(B)

Figure 6.39 (A) A pipe strap. (B) Perforated strapping material.

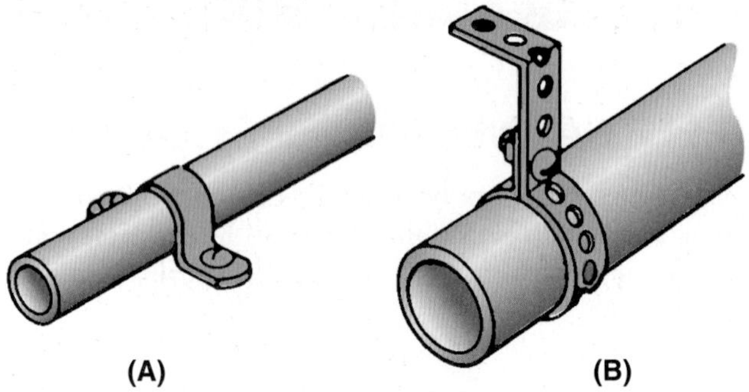

(A) (B)

pipe or tubing. These straps are normally fastened with round-head screws. Notice that the underside of the head of a round-head screw is flat and provides good contact with the strap.

PERFORATED STRAP

A perforated strap may also be used to support pipe and tubing. Round-head stove bolts with nuts fasten the straps to themselves, and round-head wood screws fasten the straps to wood supports, **Figure 6.39(B)**. These straps are so named because of the many holes that are provided in the material. These holes allow the strap to be used on pipes of many different sizes.

Figure 6.40 A system for fastening round flexible duct to a sheet metal collar. Courtesy of Panduit Corporation

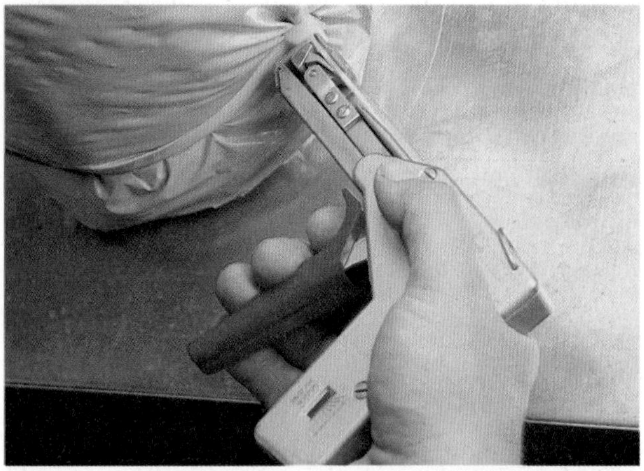

NYLON STRAP

To fasten round flexible duct to a sheet metal collar, apply sealer to the inner liner of the duct and clamp it with a nylon strap. A special tool is manufactured to install this strap, which applies the correct tension and cuts the strap off. The insulation and vapor barrier are positioned over the collar, which can then be secured with another nylon strap. Duct tape is applied over this strap and duct end to further seal the system. **Figure 6.40** illustrates this procedure.

6.8 PINS, CLAMPS, AND RINGS

COTTER PINS AND HAIRPINS

Cotter pins, **Figure 6.41**, are commonly used to secure objects on shafts. The cotter pin is inserted through the hole in the shaft and the ends of the pin are spread to keep it in place, **Figure 6.42**. Another type of clip that is similar to the cotter pin is the hairpin, **Figure 6.43**. This type of pin is also inserted through a hole in a shaft, but the configuration of the pin holds it in place, **Figure 6.44**.

Figure 6.41 Cotters pin. Photo by Eugene Silberstein

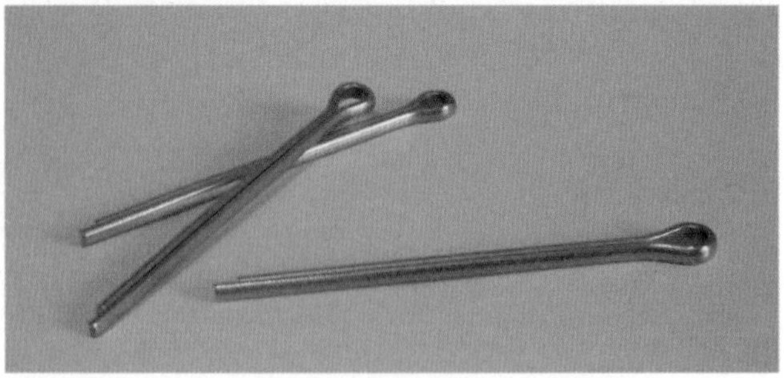

Figure 6.42 Ends of cotter pin are separated to hold the pin in place. Photo by Eugene Silberstein

Figure 6.43 A hairpin. Photo by Eugene Silberstein

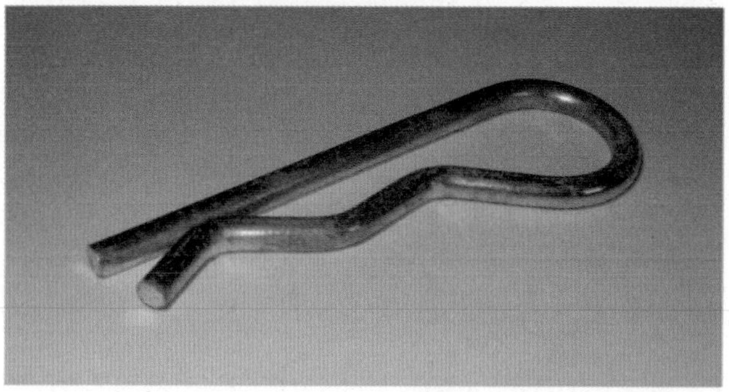

Figure 6.44 Hairpin installed on a shaft. Photo by Eugene Silberstein

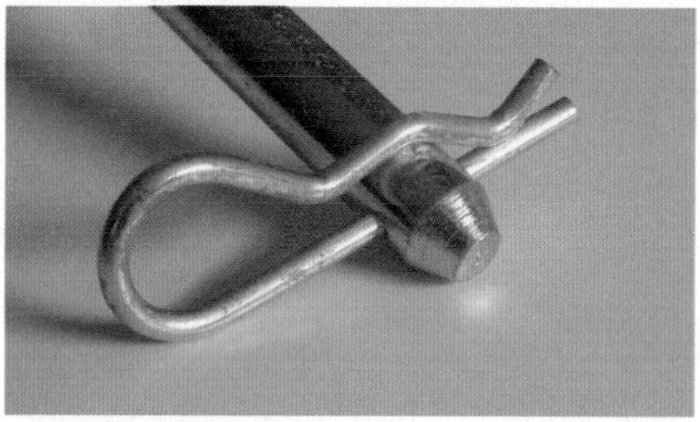

RETAINING RINGS

Retaining rings, also referred to as snap-rings, are fasteners that hold a component either onto a shaft or inside some type of housing. Rings that are used to hold a wheel on a

Figure 6.45 External retaining rings. Photo by Eugene Silberstein

Figure 6.46 Internal retaining rings. Photo by Eugene Silberstein

Figure 6.47 (A) Groove in the end of the shaft. (B) Retaining ring in the groove.

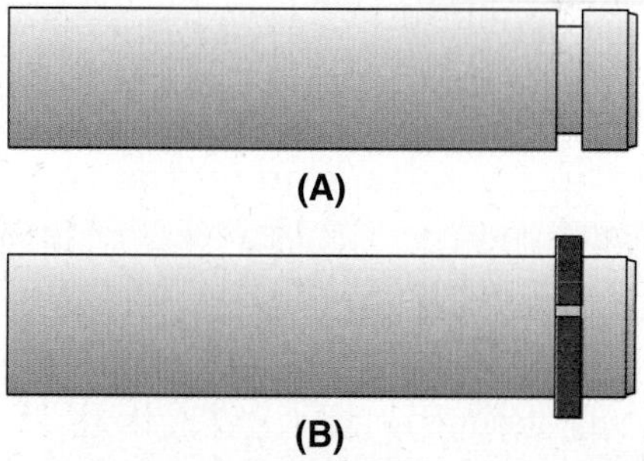

shaft, for example, are referred to as external retaining rings, **Figure 6.45**, while those that are used to keep a component inside a shaft or other housing are called internal retaining rings, **Figure 6.46**. Retaining rings are intended to be positioned inside grooves in a shaft, **Figure 6.47**. When properly installed, the portion of the ring that extends past the shaft creates a shoulder that holds the wheel or component in place. Special pliers are used to remove and install retaining rings, **Figure 6.48**. The pair of pliers in **Figure 6.48** has interchangeable heads that allow the user to get into tight areas. In addition, the pliers has two sets of pins on which a head can be

Figure 6.48 Retaining ring pliers. Photo by Eugene Silberstein

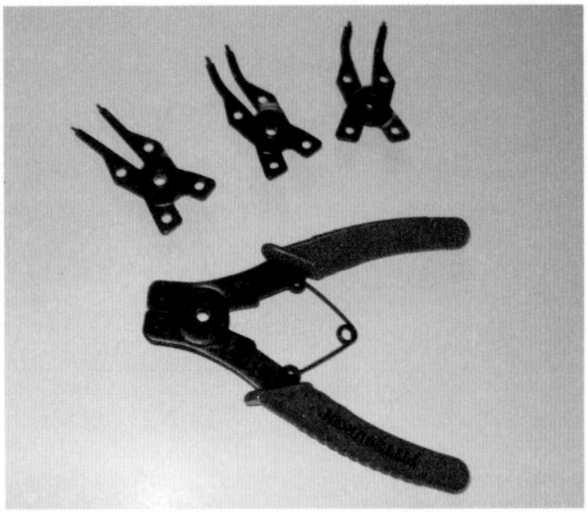

Figure 6.49 Two sets of mounting pins on the pliers. Photo by Eugene Silberstein

Figure 6.50 Squeezing the pliers opens the jaws. Photo by Eugene Silberstein

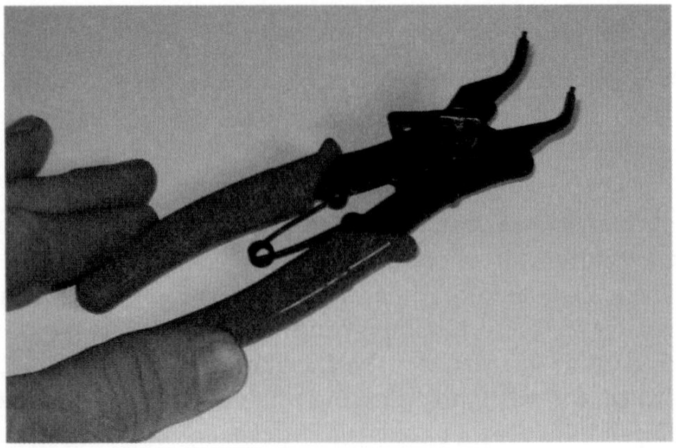

mounted, **Figure 6.49**. The head is mounted to the upper pins when the tool is to be used on externally mounted retaining rings. Squeezing the pliers will cause the jaws to open, allowing the tool to be used to open the ring, **Figure 6.50**.

Figure 6.51 Jaws close when the pliers are released. Photo by Eugene Silberstein

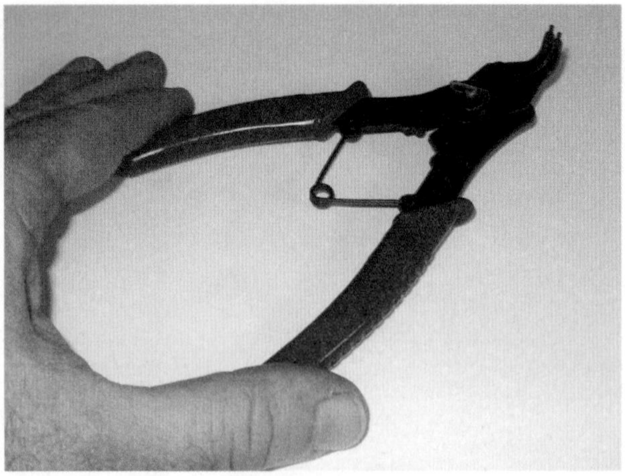

Figure 6.52 Squeezing the pliers closes the jaws. Photo by Eugene Silberstein

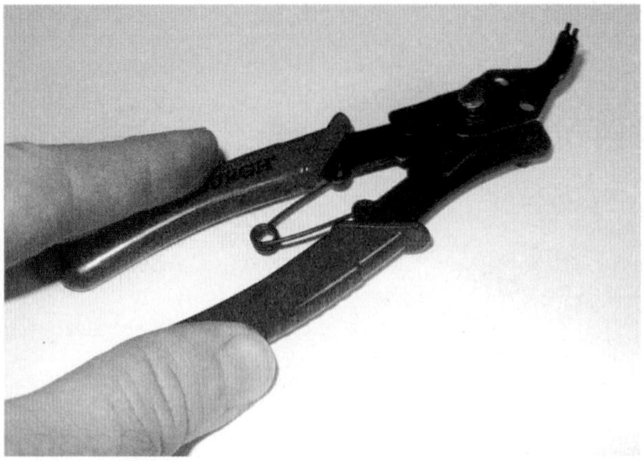

When released, the jaws close, **Figure 6.51**. The head is mounted to the lower pins when the tool is to be used on internally mounted retaining rings. Squeezing the pliers will cause the jaws to close, allowing the tool to be used to pull the ring closed, **Figure 6.52**. When released, the jaws open, **Figure 6.53**. Another type of retaining clip is the C-clip, which is commonly used to hold a wheel on a shaft, **Figure 6.54**.

HOSE CLAMPS

Hose clamps are used to attach one end of a hose to a fitting, component, or section of pipe. Two types of hose clamps that are commonly used in the HVACR industry are the screw-type and the spring-type. The screw-type hose clamp, **Figure 6.55**, is a galvanized or stainless steel band into which a series of notches have been cut. One end of the band has a fitting that receives the loose end of the strap. This fitting has a stationary screw connected to

Figure 6.53 Jaws open when the pliers are released. Photo by Eugene Silberstein

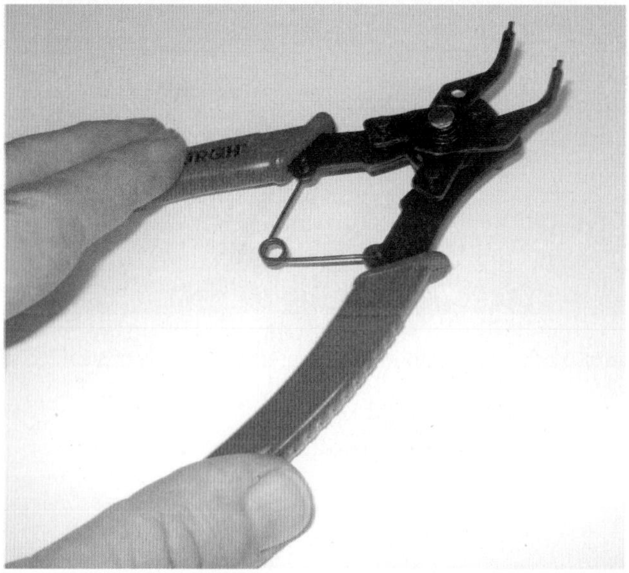

Figure 6.54 C-clip holding a wheel on a shaft. Photo by Eugene Silberstein

Figure 6.55 Screw-type hose clamp. Photo by Eugene Silberstein

Figure 6.56 Spring-type hose clamp. Photo by Eugene Silberstein

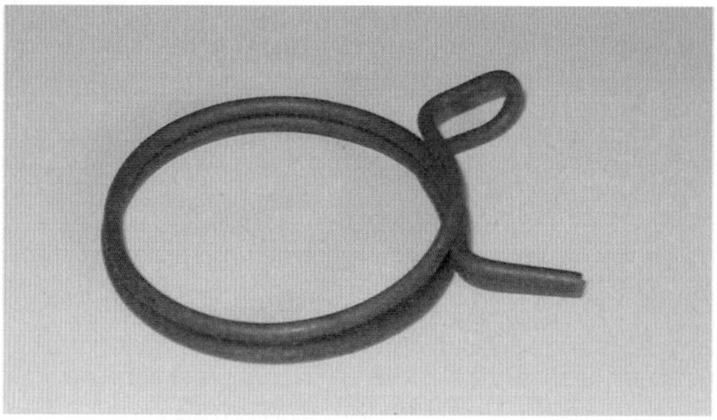

it. When the screw is turned, it pulls the loose end of the strap through the fitting, and tightens the clamp. For convenience, the head of the screw is often a hex-head screw, so a nut driver can be used to quickly and easily tighten or loosen the hose clamp.

The spring-type hose clamp, **Figure 6.56**, is made from a strip of spring-steel or from spring-steel wire. The clamp has two tabs on it and, when the tabs are squeezed together, the clamp opens and the diameter of the clamp is increased. At this point, the clamp is placed over the tubing to be secured. When the tabs are released, the clamp tightens and secures the tubing in place. Pliers are usually used to squeeze the tabs on the clamp together.

6.9 ELECTRICAL FASTENERS

SOLDERLESS TERMINALS

Solderless terminals are used to fasten stranded wire to various terminals or to connect two lengths of stranded wire together. Various styles of these terminals are shown in **Figure 6.57**. **Figure 6.58** shows a typical terminal before the wire has been inserted. To connect wire to a terminal, strip a short end of the insulation. A stripping and cutting tool similar to the one in **Figure 6.59** may be used. Insert wire into the terminal as indicated in **Figure 6.60**. A crimping tool, **Figure 6.61**, is then used to crimp the solderless connector onto the wire.

CONNECTING STRANDED WIRE UNDER SCREW TERMINALS

When connecting stranded wire under screw terminals it is recommended that the solderless connectors just described be used, as they provide a tight connection that will help ensure proper circuit operation. Stranded wire should not be directly connected under a screw terminal because the strands of

Figure 6.57 Typical solderless connectors. Courtesy of Klein Tools

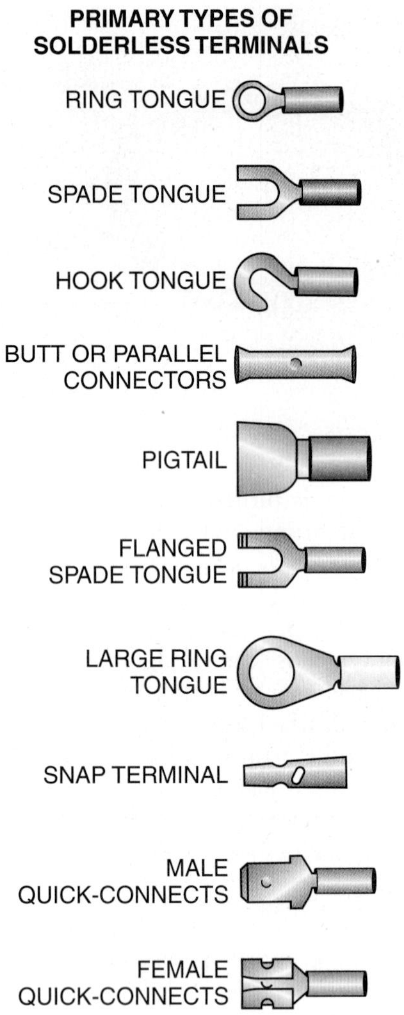

wire can separate, resulting in a poor electrical connection, **Figure 6.62**. If stranded wire is to be connected under a screw terminal, the end of the wire should be tinned, which is the process of soldering the strands of a wire together so they will not separate, **Figure 6.63**. The tinned wire can then be secured under a screw terminal, **Figure 6.64**.

Figure 6.58 Cutaway view of a typical solderless connector. Courtesy of Klein Tools

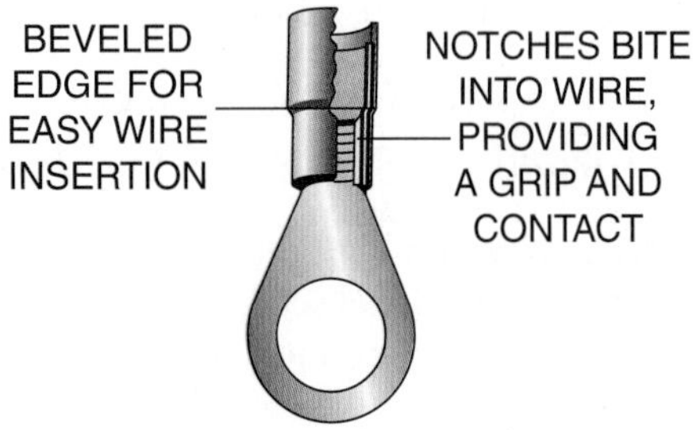

Figure 6.59 A wire stripping and cutting tool. Photo by Bill Johnson

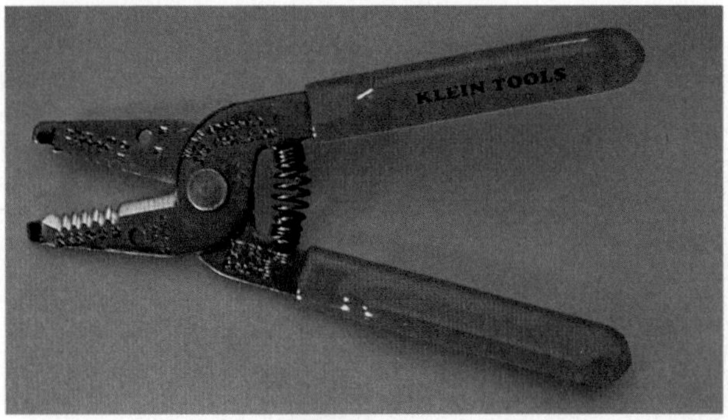

Figure 6.60 A connector with wire inserted. Courtesy of Klein Tools

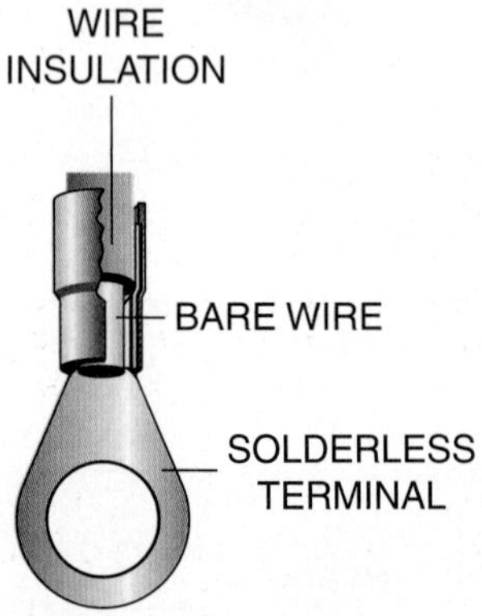

Figure 6.61 A crimping tool. Photo by Bill Johnson

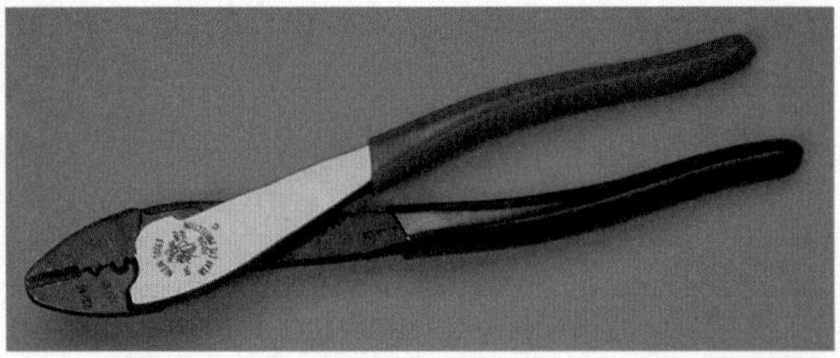

Figure 6.62 Stranded wire should not be directly connected under a screw terminal. Photo by Eugene Silberstein

Figure 6.63 The end of a stranded wire being tinned. Photo by Eugene Silberstein

Figure 6.64 The tinned wire being secured under a screw terminal. Photo by Eugene Silberstein

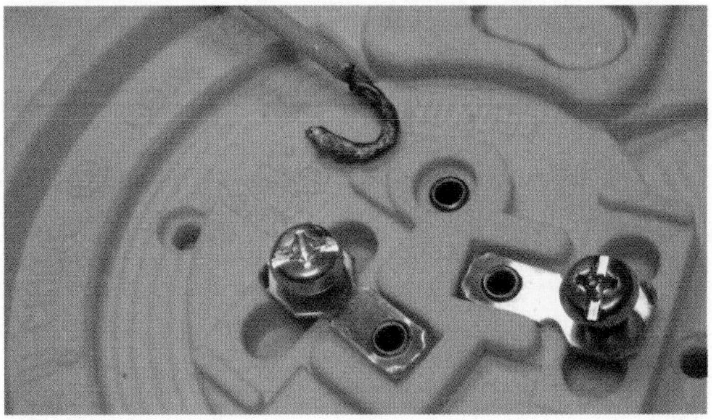

Figure 6.65 Screw-on wire connectors. Photo by Eugene Silberstein

Figure 6.66 Installing screw-on wire connectors.

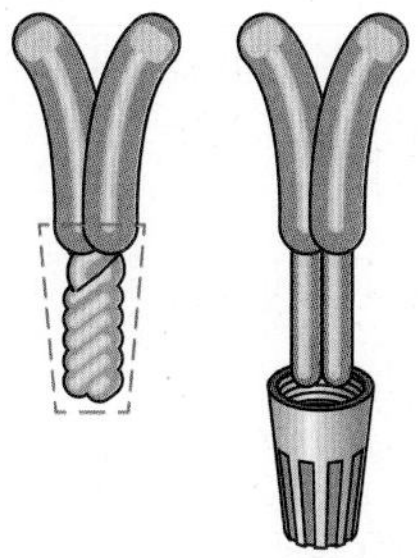

SCREW-ON WIRE CONNECTORS. Screw-on wire connectors, commonly referred to as wire nuts, are used to connect two or more wires, **Figure 6.65**. They come in various sizes for different sizes and numbers of wires. Be sure to follow the manufacturer's instructions when installing these connectors. Following is a summary of one or more manufacturers' instructions:

1. Strip each wire to the depth of the connector.
2. Insert the wires into the connector and twist it on, **Figure 6.66**.
3. The screw action will twist the wires together, pressing them into the threads of the connector and forming a strong bond.

SAFETY PRECAUTION: *Ensure that the connector completely covers the bare wire. Refrigeration and air-conditioning technicians normally will use these connectors on control voltage or other low-voltage circuits unless licensed for higher-voltage applications. It is important to use a connector of the proper size.*•

REVIEW QUESTIONS

1. All of the following are common types of nails except
   A. common.
   B. finishing.
   C. carpentry.
   D. roofing.
2. Which term is used to indicate nail size?
   A. Penny
   B. Nickel
   C. Dime
   D. Quarter
3. The letter used to indicate nail size is
   A. a.
   B. b.
   C. c.
   D. d.

4. An outward clinch tacker is used to
   A. bend wire on a pipe hook.
   B. install a pin rivet.
   C. drive roofing nails.
   D. spread the ends of a staple.
5. The fastener commonly used to join two pieces of flat sheet metal is called a
   A. pin rivet.
   B. toggle bolt.
   C. anchor shield.
   D. cotter pin.
6. Hollow rivets are assembled in a pin that is referred to as
   A. a nozzle.
   B. a mandrel.
   C. an anchor.
   D. a clinch.
7. A tapping screw that is identified as an 8-20 × 1 Type AB, slotted hex is how long?
   A. 8/20 in. long
   B. 1 in. long
   C. 8 in. long
   D. 20 in. long
8. A machine screw that is identified as a 1/4–18 UNC-2 is how long?
   A. 1/4 in.
   B. 18 in.
   C. 2 in.
   D. Cannot be determined from the information provided
9. Which of the following sets of driver tips are the most similar?
   A. Clutch head and square recess
   B. Keystone and hex head
   C. Cabinet and keystone
   D. Cabinet and Phillips
10. Toggle bolts are best suited for securing objects to
    A. hollow walls.
    B. solid concrete slabs.
    C. steel "I" beams.
    D. brick walls.
11. When using a toggle bolt, it is important to make certain that the
    A. toggle is pushed completely through the hole before threading the bolt onto it.
    B. object being secured is placed on the bolt before threading the toggle onto it.
    C. length of the bolt is longer than the thickness of the wall.
    D. diameter of the hole created is smaller than the width of the folded toggle.
12. Powder-actuated fastener systems are most commonly used to secure objects to
    A. hollow walls.
    B. steel "I" beams.
    C. concrete slabs and walls.
    D. gypsum board walls.
13. A lag shield anchor is best described as
    A. a two-part shield with tapered internal threads.
    B. a two-part shield with tapered external threads.
    C. a three-part shield with tapered internal threads.
    D. a three-part shield with tapered external threads.
14. Threaded rod can be attached to steel I-beams with
    A. hose clamps.
    B. beam clamps.
    C. toggle bolts.
    D. lag shields.
15. The pipe-hanging device that is hammered or driven into a wood support is the
    A. pipe clamp.
    B. pipe strap.
    C. perforated strap.
    D. pipe hook.
16. Perforated strapping material
    A. has limited applications because it can only be used on one size of pipe.
    B. is a two-hole strap that is sized according to the outside diameter of the pipe.
    C. is used in conjunction with nuts and bolts to secure the piping material.
    D. is a two-hole strap that is sized according to the inside diameter of the pipe.
17. Retaining rings are typically classified as either
    A. round or oblong.
    B. solid or split.
    C. rigid or flexible.
    D. internal or external.
18. Which of the following is true regarding cotter pins?
    A. They are secured in external grooves in a shaft.
    B. They are inserted into a shaft hole and bent around to secure.
    C. They are inserted into a shaft hole and do not need to be bent.
    D. They are secured in internal grooves in a shaft or channel.
19. The ring tongue, hook tongue, and snap terminal are all examples of
    A. screw-on solderless terminals.
    B. screw-on soldered terminals.
    C. crimped-on solderless terminals.
    D. crimped-on soldered terminals.
20. Which of the following is true regarding the use of screw terminals?
    A. Only stranded wire should be secured under screw terminals.
    B. Only solid wire should be secured under screw terminals.
    C. Either solid or stranded wire can be secured under screw terminals.
    D. Neither solid nor stranded wire should be secured under screw terminals.

UNIT 7

TUBING AND PIPING

OBJECTIVES

After studying this unit, you should be able to

- explain the purpose and importance of tubing and piping in the HVACR industry.
- list the different types of tubing used in heating, air-conditioning, and refrigeration applications.
- explain why some piping and tubing runs are insulated.
- describe a refrigerant line set.
- describe two common ways of cutting copper tubing.
- list procedures used for bending tubing.
- list five common heat sources used to solder and braze tubing.
- discuss procedures used for soldering tubing.
- discuss procedures used for brazing tubing.
- describe two methods for making flared joints.
- state procedures for making swaged joints.
- explain how a compression fitting is made.
- describe procedures for preparing and threading steel pipe ends.
- list four types of plastic pipe and describe uses for each.
- describe alternative, mechanical methods for joining pipe sections.

SAFETY CHECKLIST

- ✔ Use care while cutting, threading, and reaming tubing and piping; the burrs that are created can cut your skin.
- ✔ Be careful while using a hacksaw; the blade is sharp.
- ✔ Do not allow your bare skin near the flame of a torch or near material that has just been soldered, brazed, or welded; the 5000°F and higher temperatures can cause severe burns.
- ✔ Always wear eye protection while performing any task where particles may be in the air.
- ✔ Use extra caution with the oxyacetylene heat source.
- ✔ Do not allow any oil around the fittings or allow it to enter the hoses or regulators.
- ✔ Be aware of flammable materials in the area where you are soldering or brazing.
- ✔ Do not breathe excessive amounts of the adhesive used for plastic fittings.
- ✔ Only use chemicals that give off dangerous fumes in well-ventilated areas.
- ✔ Follow all manufacturers' instructions when using soldering and brazing materials.

7.1 PURPOSE OF TUBING AND PIPING

The correct size, layout, and installation of tubing, piping, and fittings helps to keep a refrigeration or air-conditioning system operating properly and efficiently and prevents refrigerant loss.

The piping system provides the path for refrigerant flow between system components. It also provides the way for oil to drain back to the compressor. Tubing, piping, and fittings are used in numerous applications, such as fuel lines for oil and gas burners and water lines for hot water heating. The tubing, piping, and fittings must be of the correct material and the proper size; the system must be laid out properly and installed correctly.

SAFETY PRECAUTION: *Careless handling of the tubing and poor soldering or brazing techniques may cause serious damage to system components. You must prevent contaminants, such as moisture and dirt, from entering air-conditioning and refrigeration systems.*•

7.2 TYPES AND SIZES OF TUBING

Copper tubing is generally used for plumbing, heating, and refrigerant piping. Steel and wrought iron pipe are used for gas piping and frequently for steam heating. Plastic pipe is often used for waste drains, condensate drains, water supplies, water-source heat pumps, and venting high-efficiency gas furnaces.

Copper tubing is available as soft- or hard-drawn copper. Soft copper tubing, **Figure 7.1**, may be bent or used with elbows, tees, and other fittings. Hard-drawn tubing is not intended to be bent and is used only with fittings to obtain the necessary configurations. This tubing is available in four standard weights: K, L, M, and DWV. K-type copper is heavy duty with a thick tubing wall. L-type copper is considered medium weight and is the type used most frequently. M is a lightweight copper and is not commonly used in the industry. DWV copper—drain, waste, and vent—is also a lightweight copper not typically used in the HVACR industry so far as refrigerant-carrying applications are concerned. Copper tubing used for refrigeration or air-conditioning is called ACR (air-conditioning and refrigeration) tubing and is sized by its outside diameter (OD). Therefore, if a specification calls for

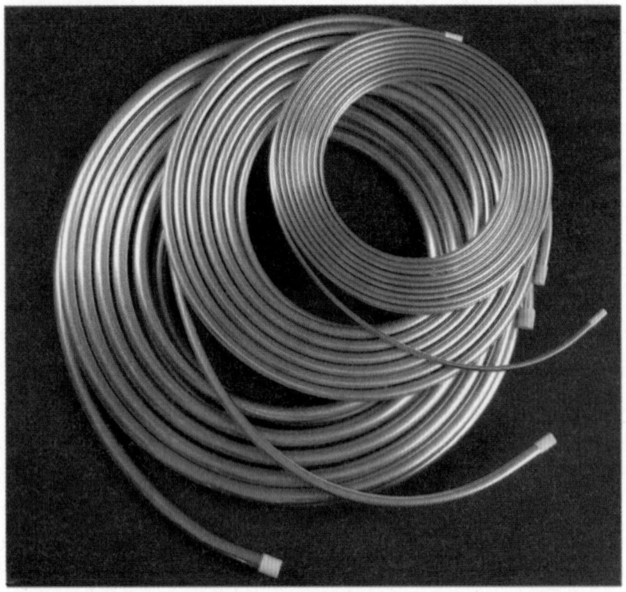

Figure 7.1 Soft-drawn copper tubing typically comes in 50-ft rolls. Note that the ends of the rolls are capped to keep the interior of the tubing clean.

1/2-in. ACR tubing, the outside diameter of the tubing would measure 1/2 in., **Figure 7.2**. Tubing used for plumbing and heating applications, called nominal-sized tubing, is sized by its inside diameter (ID). The OD for most applications in the plumbing and heating industry will be 1/8 in. larger than the size indicated. A 1/2-in. nominal-sized tubing would have an OD of 5/8 in., **Figure 7.2**. Copper tubing normally is available in diameters from 3/16 in. to greater than 6 in.

Soft copper tubing is normally available in 25-ft or 50-ft rolls and in diameters from 3/16 in. to 7/8 in. It can be special ordered in 100-ft lengths. ACR tubing is capped on each end to keep the inside clean and dry and it often has a charge of nitrogen to keep it free of contaminants. Proper practice should be used to remove tubing from the coil. Place it on a flat surface and unroll it, **Figure 7.3**. Cut only what you need and recap the ends, **Figure 7.1**. Hard-drawn copper tubing is available in lengths up to 20 ft and in larger diameters than soft copper tubing. Be as careful with hard-drawn copper as with soft copper, and recap the ends when the tubing is not in use.

Figure 7.2 Copper tubing (nominal) used for plumbing and heating is sized by its inside diameter. ACR tubing is sized by its outside diameter.

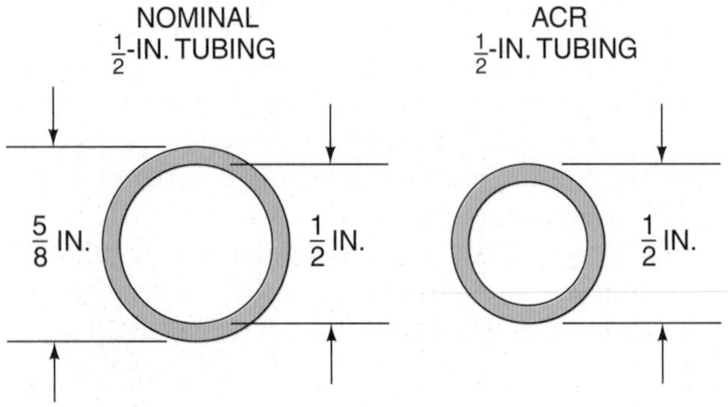

Figure 7.3 A roll of soft-drawn copper tubing. Place on a flat surface and unroll. Photo by Eugene Silberstein

Do not bend or straighten the tubing more than necessary because it will harden. This is called **work hardening**. Work-hardened tubing can be softened by heating and allowing it to cool slowly. This is called **annealing**. When annealing, do not use a high concentrated heat in one area, but use a flared flame over 1 ft at a time. Heat to a cherry red and allow to cool slowly.

7.3 TUBING INSULATION

ACR tubing is often insulated on the low-pressure side of an air-conditioning or refrigeration system between the evaporator and the compressor to keep the refrigerant from absorbing excess heat, **Figure 7.4**. Insulation also prevents condensation from forming on the lines. The closed-cell

Figure 7.4 ACR tubing with insulation. Photo by Eugene Silberstein

Figure 7.5 When joining two ends of tubing insulation, use an adhesive intended for this purpose. Photo by Eugene Silberstein

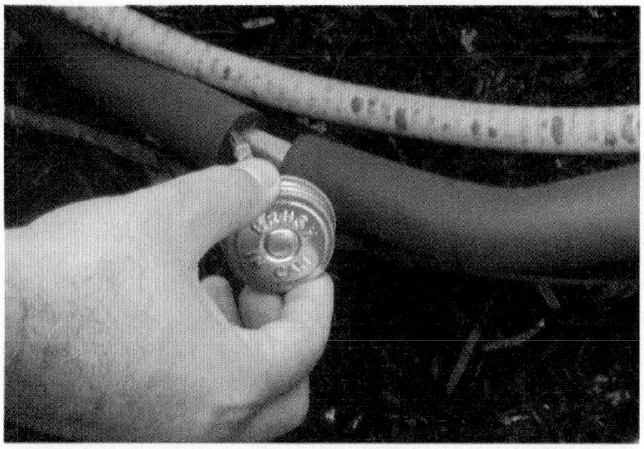

structure of this insulation eliminates the need for a vapor barrier. The insulation may be purchased separately from the tubing, or it may be factory installed. If you install the insulation, it is easier, where practical, to apply it to the tubing before assembling the line. The inside surface of the insulation is usually powdered to allow easy slippage even around most bends. You can buy adhesive to join the ends of the insulation together, **Figure 7.5**.

It is not always possible or practical, however, to install the insulation on the tubing section before installing the line. When installing insulation on a section of tubing that has already been installed, you will have to slit the insulation. To help ensure a neat-looking job, here are some suggestions:

- Slit the insulation along the markings on the insulation, **Figure 7.6**.
- Slide the insulation over the tubing section, **Figure 7.7**.
- Once the insulation has been placed on the tubing, use adhesive to seal the insulation, **Figure 7.8**.
- Do not use tape, plastic wire/cable ties, or other methods to hold the insulation together.
- Rotate the insulation so that the insulation markings and the seam are not visible, **Figure 7.9**.

Figure 7.6 Pipe insulation being slit with a sharp razor knife. Photo by Eugene Silberstein

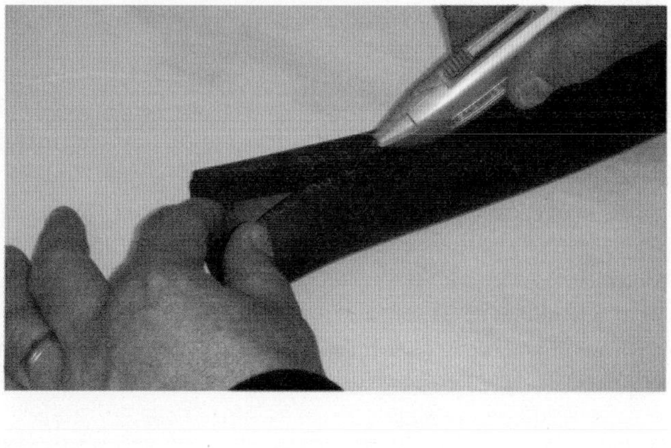

Figure 7.7 Pipe insulation being positioned over tubing section. Photo by Eugene Silberstein

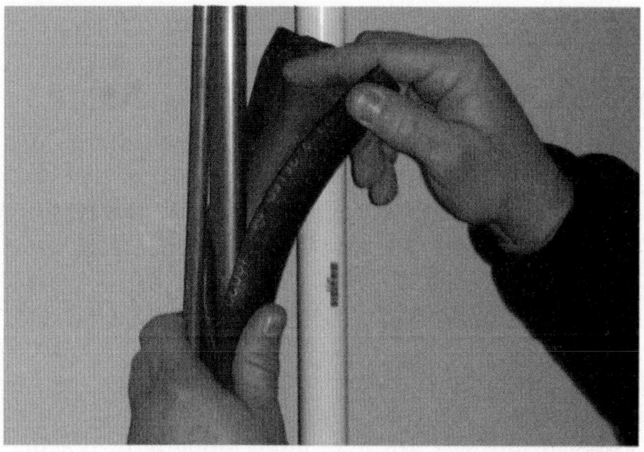

Figure 7.8 Adhesive being applied to both surfaces of the cut insulation. Photo by Eugene Silberstein

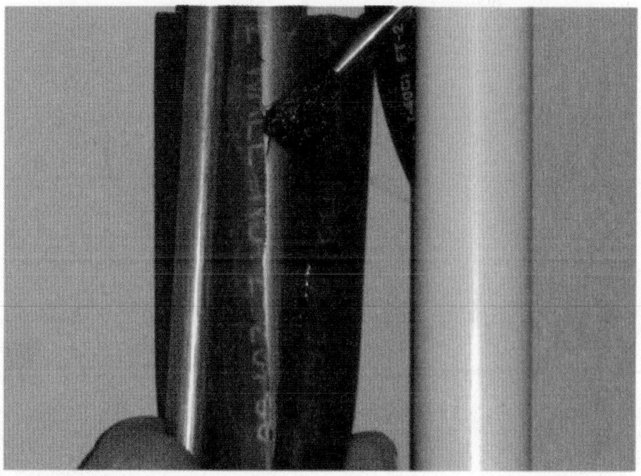

Figure 7.9 Insulation being rotated to conceal the seam. Photo by Eugene Silberstein

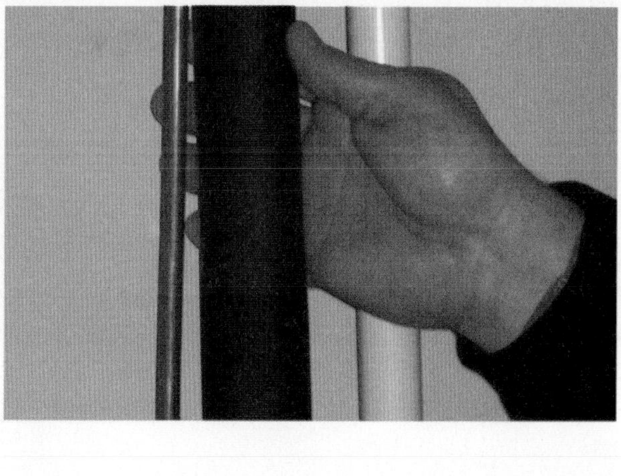

Do not stretch tubing insulation because that will reduce the wall thickness of the insulation, which will reduce the effectiveness of the insulation, and the adhesive may fail to hold.

7.4 LINE SETS

Tubing can be purchased in bundles called line sets, which often consist of a liquid line and an insulated suction line. One type of line set comes precharged with refrigerant and each of the lines is sealed at both ends. Precharged line sets have quick-connect fittings installed on the ends for quicker, neater, and easier installation, **Figure 7.10**. The fittings on the lines must be properly aligned with the mating fittings on the equipment before tightening to ensure that refrigerant is not released into the atmosphere. Since precharged line sets are purchased along with precharged system components, the system does not have to be evacuated prior to putting it into operation. In addition, the refrigerant charge should be perfect, since the refrigerant is added at the factory according to the particular requirements of the system. As of January 1, 2010, it became illegal to manufacture or import precharged line sets that contain R-22 or R-142b. This law covers equipment components such as line sets, condensing units, and evaporator coils. Precharged line sets, for example, can be used to service existing equipment as long as the line set was manufactured before January 1, 2010.

Figure 7.10 (A) A typical line set. (B) Quick-connect fittings. (A) Photo by Bill Johnson (B) Courtesy Aeroquip Corporation

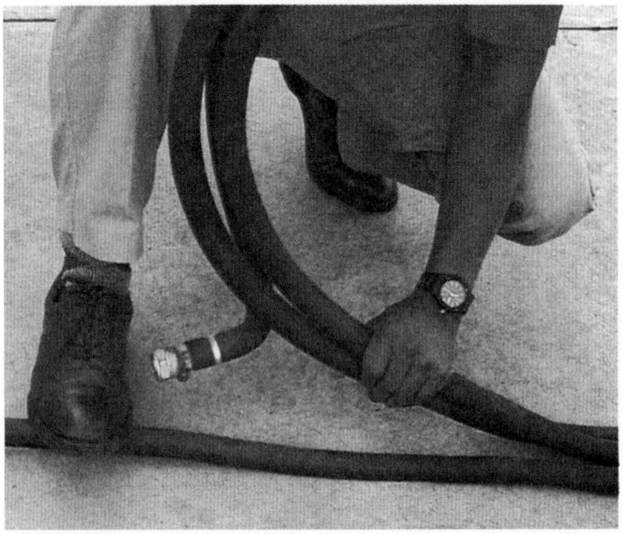

(A)

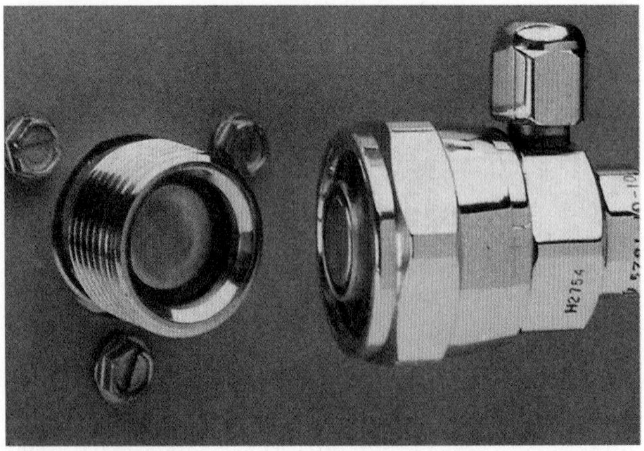

(B)

Another type of line set is nitrogen-charged, just like standard ACR tubing. The suction line in this line set, however, is insulated. This type of line set may be purchased with various fitting options, but if the fittings need to be changed, they can be cut off and replaced. Since these line sets are not precharged with refrigerant, the nitrogen must be released and the system must be properly evacuated prior to putting the system into operation.

7.5 CUTTING TUBING

Tubing is normally cut with a tube cutter or a hacksaw. The tube cutter is most often used with soft tubing and smaller-diameter hard-drawn tubing. A hacksaw may be used with larger-diameter hard-drawn tubing. To cut the tubing with a tube cutter, follow these steps, as shown in **Figure 7.11**.

A. Place the tubing in the cutter and align the cutting wheel with the cutting mark on the tube. Tighten the adjusting screw until a moderate pressure is applied to the tubing.
B. When properly tightened, the cutter should be held in place on the tubing and there should be no cutter wheel indentations in the copper tubing. Overtightening the cutter on the tubing will result in oblong or out-of-round edges that can cause refrigerant leaks.
C. Revolve the cutter around the tubing, keeping a moderate pressure on the tubing by gradually turning the adjusting screw, roughly 1/8 of a turn per rotation around the tubing.
D. Continue until the tubing is cut. *Do not apply excessive pressure because it may break the cutter wheel and/or constrict the opening in the tubing.*

When the cut is finished, the excess material (called a burr) pushed into the pipe by the cutter wheel must be removed, **Figure 7.12**. Burrs can restrict the fluid or vapor passing through the pipe. When removing burrs from the end of a cut pipe section, make every attempt to prevent the burrs from falling into the pipe. Burrs that remain in the system can restrict refrigerant flow and affect system operation and performance.

Cuts made with a hacksaw should be at a 90° angle to the tubing. After cutting, ream the tubing and file the end.

Figure 7.11 (A) Position the cutter around the tubing. (B) Tighten the cutter just enough to hold the cutter in place if released. (C) Rotate the cutter around the tubing unit the cutter is loose. (D) Slightly tighten the cutter on the tubing and continue to rotate the cutter around the tubing. Photos by Eugene Silberstein

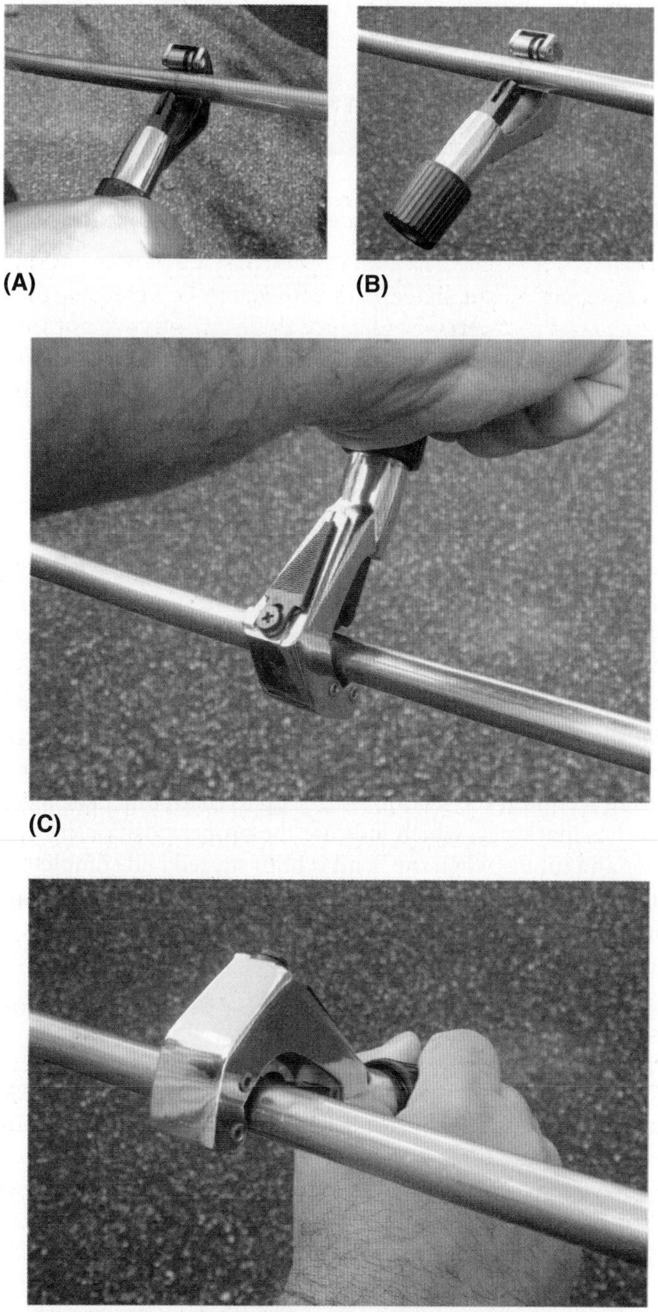

(A) (B) (C) (D)

Remove all chips and filings, making sure that no debris or metal particles get into the tubing. Since using a hacksaw under normal conditions produces a large amount of copper filings and burrs, this method is not preferable when using a tubing cutter is an option. Given certain conditions, however, it may be the only choice available.

Figure 7.12 Removing burrs from inside the tubing.

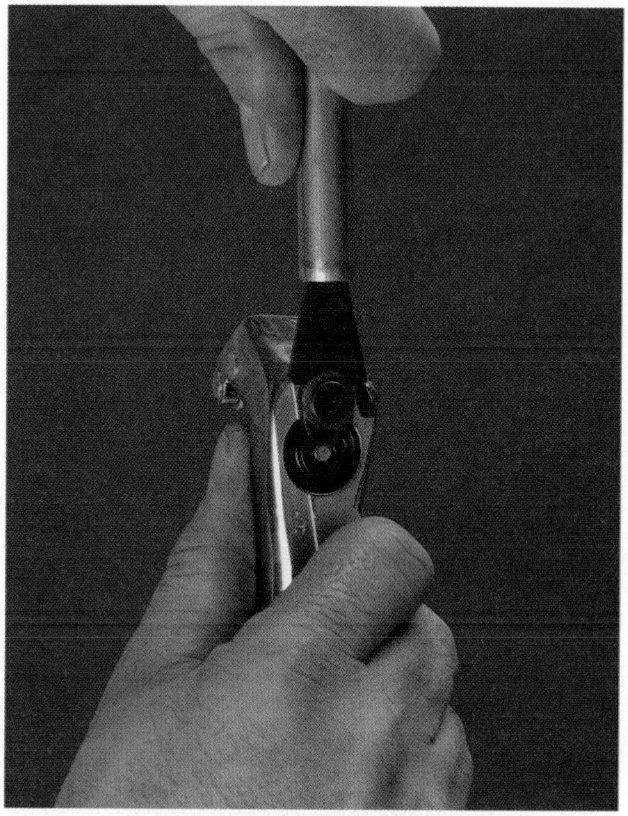

7.6 BENDING TUBING

To ensure refrigerant stays in the system, the number of solder joints and mechanical fittings should be kept to a minimum. Each solder joint, fitting, and mechanical connection represents a potential leak. One way to reduce the number of solder joints in a piping system is to bend the tubing material into the desired configuration. Typically, only soft tubing should be bent. Use as large a radius bend as possible, **Figure 7.13(A)**. All areas of the tubing must remain round; *do not allow it to flatten or kink,* **Figure 7.13(B)**. Carefully bend the tubing, gradually working around the radius.

Tube bending springs may be used to help make the bend, **Figure 7.14**, and can be placed either inside or outside the tube. They are available in different sizes for different diameters of tubing, **Figure 7.15**. To remove the spring after the bend, you might have to twist it. If you use a spring on the OD, bend the tube before flaring so that the spring may be removed.

Lever-type tube benders, **Figure 7.16**, which are available in different sizes, are used to bend soft copper, aluminum, and thin-walled steel tubing. There are many different types of benders available; we describe briefly here the operation of the lever-type bender.

Figure 7.13 (A) Small-diameter tubing can be bent by hand. (B) Larger tubing can easily kink when bent by hand.

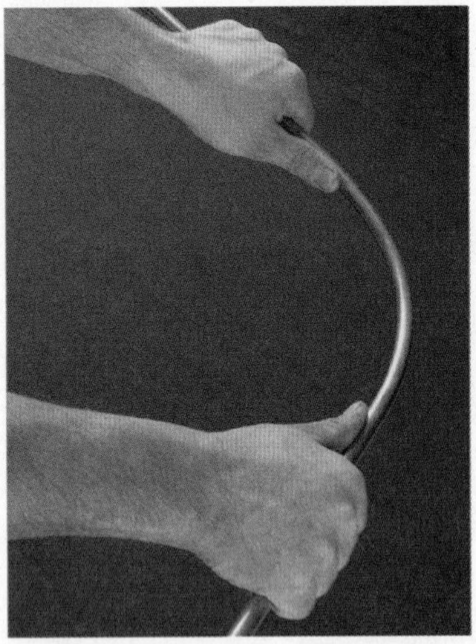

(A)

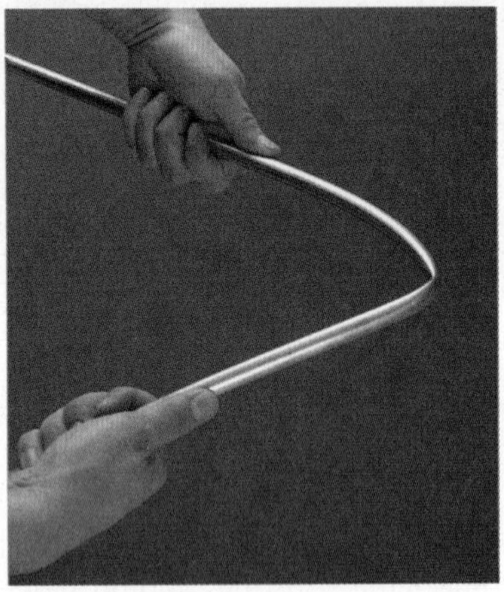

(B)

Figure 7.14 Tube bending springs are used to help prevent kinking the tubing while bending. Be sure to use the proper size. Photo by Bill Johnson

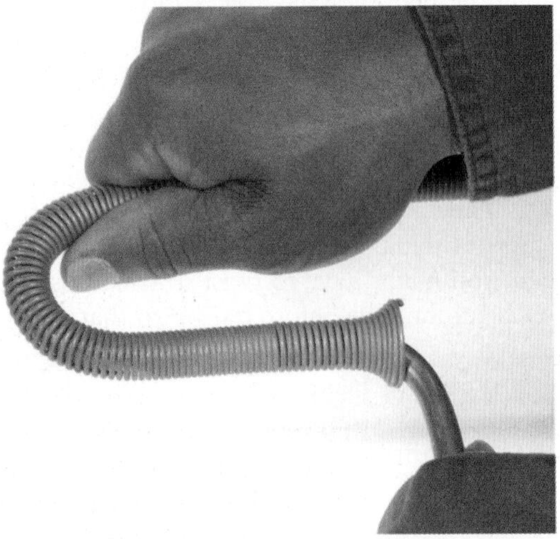

Figure 7.15 Different-sized bending springs.

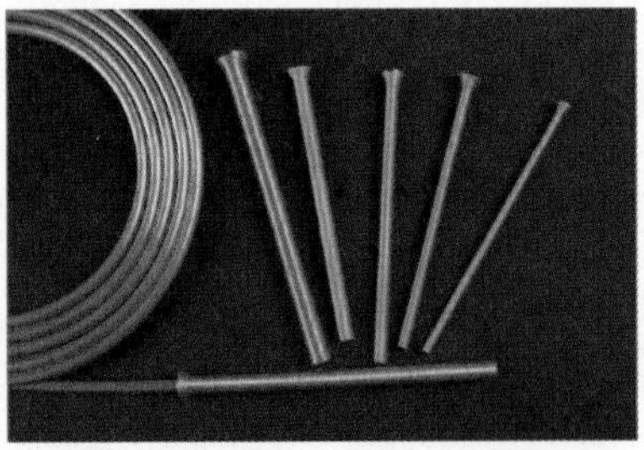

Figure 7.16 Lever-type tube bender. Photo by Eugene Silberstein

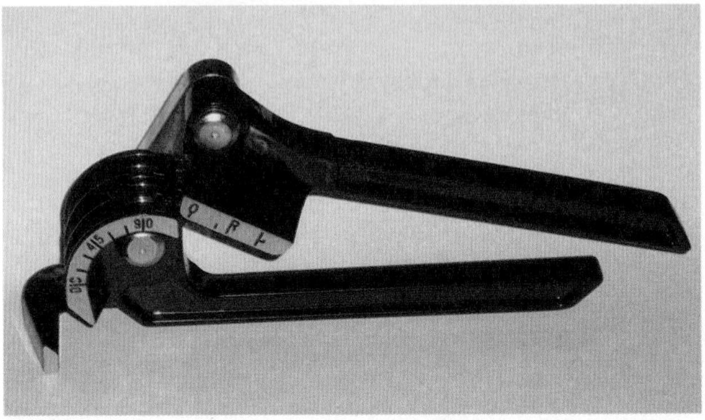

The lever-type bender is made up of two parts, the body and the lever, **Figure 7.17**. The body has markings that represent the amount of bend, in degrees, that will be put into a tubing section if the section is bent up to that point. The lever also has markings, which indicate the proper positions of the lever and tubing when the bend is both started and completed. The tubing position in the bender depends on whether the piping is extending from the left or the right side of the bender.

Figure 7.17 Markings on the lever and body of the lever-type tube bender. Photo by Eugene Silberstein

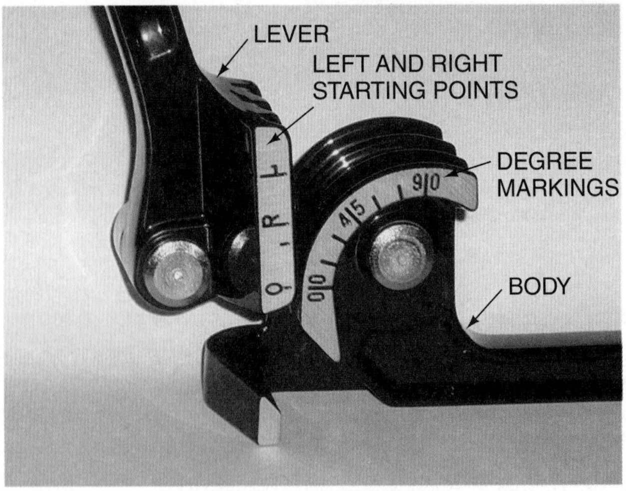

Figure 7.18 Mark tubing at the desired point. Photo by Eugene Silberstein

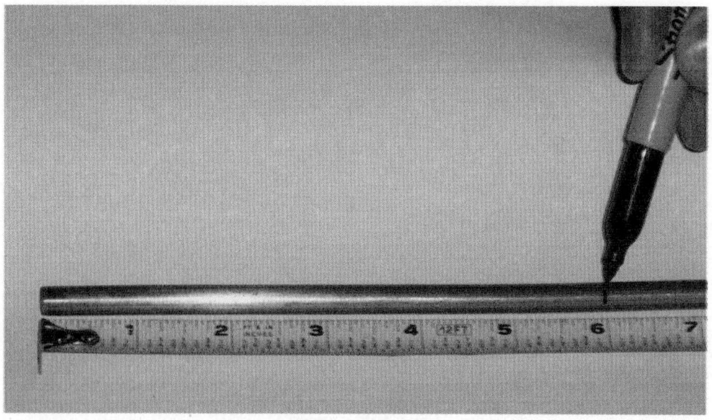

Figure 7.19 Proper tubing and lever positions at the beginning of the bending process. Photo by Eugene Silberstein

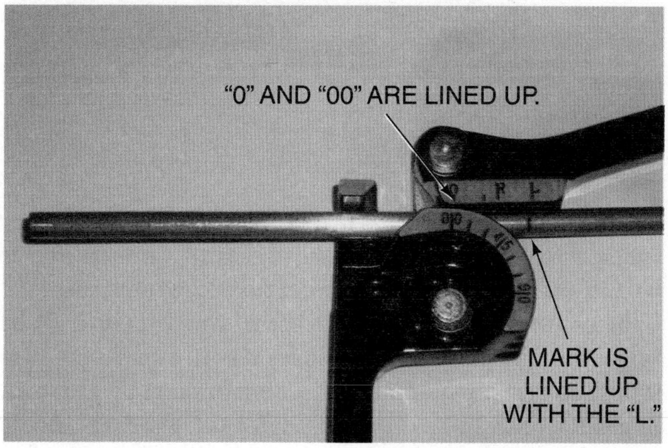

Let's say you wanted to put a 90-degree bend in a piece of soft-drawn copper tubing so that the length from the unbent end of the tubing to the center of the bent end is 6 in. In this case, you would first measure and mark the tubing at the 6-in. point, **Figure 7.18**. Once marked, you insert the tubing into the bender and line up the mark with the "L," which indicates that the measured portion of tubing is extending from the left side of the bender, **Figure 7.19**. You also want to make certain that the "0" on the lever lines up with the "00" on the body of the bender. This ensures that the lever is in the proper position prior to bending. While holding the handle on the body of the bender, pull on the lever until the "0" on the lever is lined up with the "90" on the body. This ensures that the bend will be 90 degrees, **Figure 7.20**.

If, however, you wanted to make a 45-degree bend in the tubing instead of a 90-degree bend, the lever would be positioned so that the "0" on the lever is lined up with the "45" on the body, **Figure 7.21**. Be sure to practice on smaller, scrap sections of tubing to gain confidence in creating tubing bends at the desired positions. It takes a little time to get it just right, but the results will be well worth the effort.

Figure 7.20 Completing a 90-degree bend. Photo by Eugene Silberstein

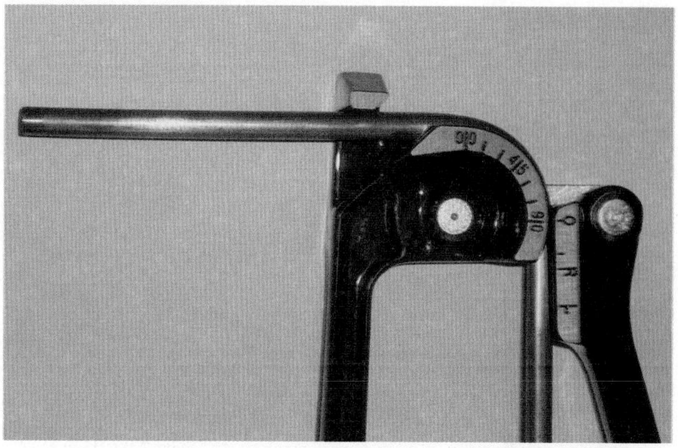

Figure 7.21 A 45-degree bend. Photo by Eugene Silberstein

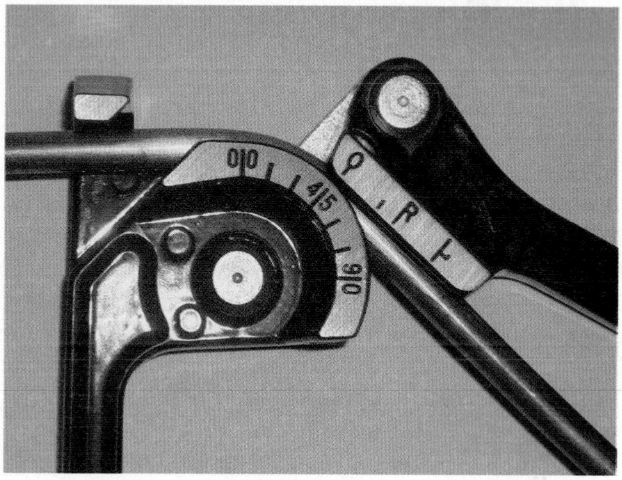

7.7 SOLDERING AND BRAZING PROCESSES

Soldering is a process that joins piping and tubing to fittings. It is used primarily in plumbing and heating systems utilizing copper and brass piping and fittings. According to the American Welding Society, soldering, often called soft soldering, utilizes filler materials that melt at temperatures under 842°F, usually in the 361°F to 500°F range, **Figure 7.22**.

When soldering or brazing, a gap is left between the two surfaces that are to be fastened together. For example, when two lengths of copper tubing are to be fastened, a fitting or coupling may be placed between them into which each pipe slides. This requires a clearance between the tubing and the coupling or fitting. A filler metal must be used to fill this space. As it fills the gap, the filler metal actually adheres to the tubing and fitting to be joined. Only the filler metal melts.

Figure 7.22 Temperature ranges for soldering and brazing.

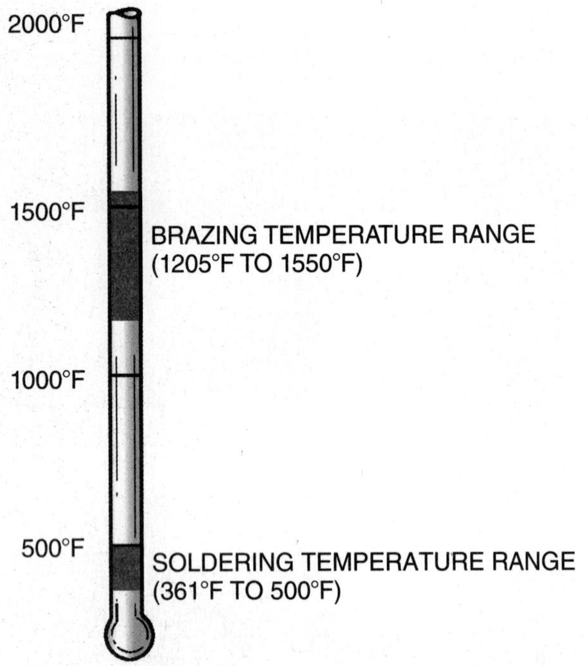

Two common filler materials used for soldering are 50/50 and 95/5. The 50/50 tin–lead solder (50% tin–50% lead) is a suitable filler metal for moderate pressures and temperatures. A 50/50 tin–lead solder cannot be used for water-supply or potable water lines because of its lead content. For higher-pressure potable water-supply lines, or where greater joint strength is required, 95/5 tin–antimony solder (95% tin–5% antimony) can be used.

Brazing is similar to soldering but requires higher temperatures. Often called silver brazing or hard soldering, it is used to join tubing and piping in air-conditioning and refrigeration systems. Do not confuse this with welding brazing. Brazing utilizes filler materials that melt at temperatures above 842°F. The differences in temperature are necessary due to the different combinations of alloys used in the filler metals. Brazing filler metals suitable for joining copper tubing are alloys containing 15% to 60% silver (BAg) or copper alloys containing phosphorous (BCuP). Brazing filler metals are sometimes referred to as hard solders or silver solders.

In soldering and brazing, the base metal (the piping, tubing, and/or fitting) is heated to the melting point of the filler material. *The piping and tubing must not melt.* When two close-fitting, clean, smooth metals are heated to the point where the filler metal melts, the molten metal is drawn into the close-fitting space by **capillary attraction**. Capillary attraction occurs when a liquid rises within a small tube or travels between two closely spaced surfaces. This is due to the interaction of forces between the liquid and the surfaces. If the soldering is properly done, the molten solder will be absorbed into the pores of the base metal, adhere to all surfaces, and form a bond, **Figure 7.24**. On the left are two pieces of glass spaced closely together. When the pieces of glass are inserted in water, capillary attraction draws the water into the space between them. The water molecules

Figure 7.23 Two examples of capillary attraction.

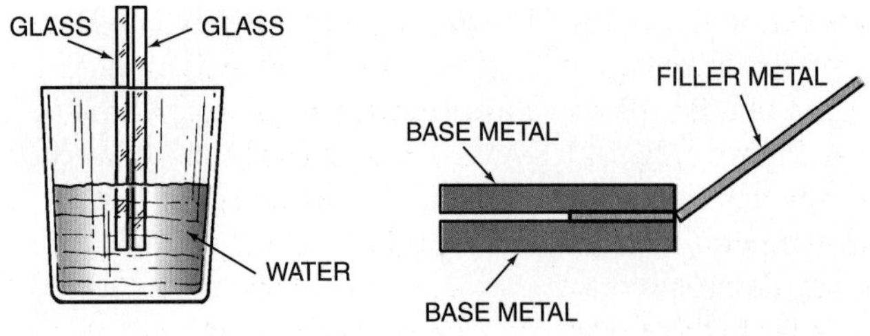

Figure 7.24 The molten solder in a soldered joint will be absorbed into the surface pores of the base metal.

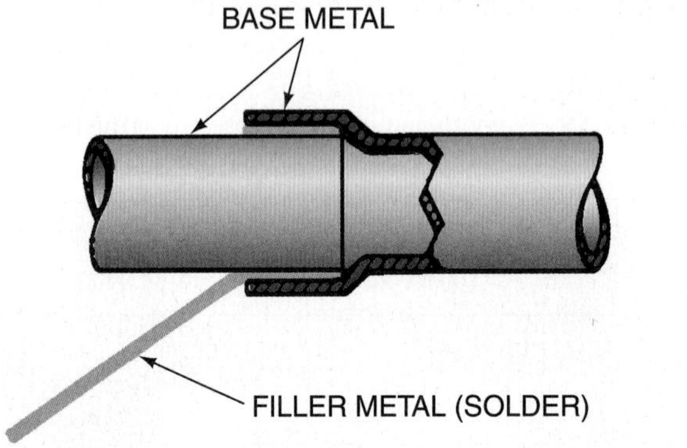

have a greater attraction for the glass than they do for each other. On the right is an illustration showing molten filler material being drawn into the space between the two pieces of base metal. The molecules in the filler material have a greater attraction for the base metal than they do for each other. These molecules work their way along the joint, first "wetting" the base metal and then filling the joint.

7.8 HEAT SOURCES FOR SOLDERING AND BRAZING

Propane, butane, MAPP™ gas, air-acetylene, and oxyacetylene torches are the most common sources of heat for soldering or brazing. A **propane** or **butane** torch can be ignited easily and adjusted to the type and size of joint being soldered, **Figure 7.25**. A propane torch can produce flame temperatures in the range of 3600°F, while butane torches produce flames at

Figure 7.25 A propane torch with a typical soldering tip. Photo by Eugene Silberstein

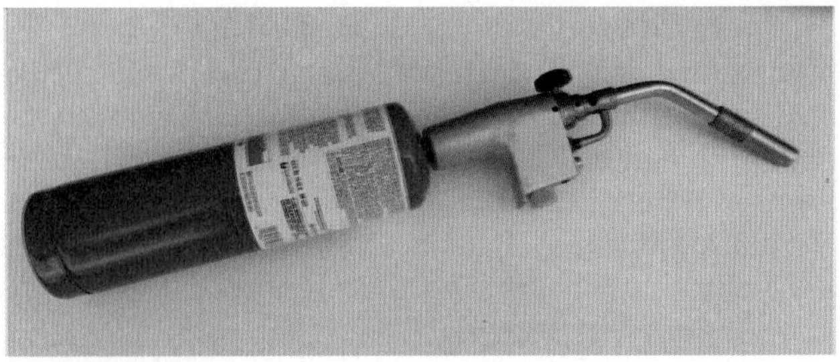

about 2600°F. MAPP™ gas is a mixture of fuels and produces flame temperatures of about 3730°F. MAPP™ gas is covered in more detail in Section 7.12. An **air-acetylene** unit is often used by air-conditioning and refrigeration technicians. It usually consists of a B tank of acetylene gas, a regulator, a hose, and a torch, **Figure 7.26.** Various sizes of standard tips are available. The smaller tips are used for small-diameter tubing and for high-temperature applications. A high-velocity tip may be used to provide more concentration of heat. **Figure 7.27** illustrates a popular high-velocity tip. Air-acetylene torch produces flame temperatures of about 4200°F. Oxyacetylene torches produce the hottest flames, often as high as 5600°F.

Follow these procedures when setting up, igniting, and using an air-acetylene unit, **Figure 7.28.**

Figure 7.26 A typical air-acetylene torch setup. Photo by Eugene Silberstein

Figure 7.27 A high-velocity torch tip. Courtesy Thermadyne Industries, Inc.

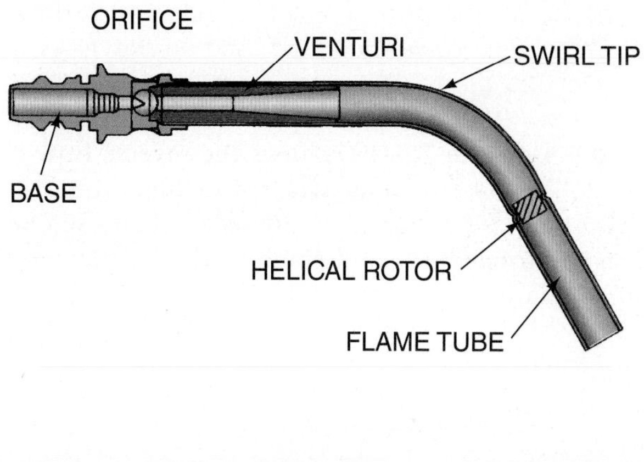

Figure 7.28 (A) Clean tank connection and briefly open tank valve. (B) Connect regulator and hose assembly to tank. (C) With torch handle valve closed, open tank valve one-half turn. (D) Spray bubble solution. (E) Check for leaks.(F) Light the torch. (G) Adjust the flame. . Photos (A) to (F) by Eugene Silberstein. Photo (G) by Bill Johnson

(A)

(B)

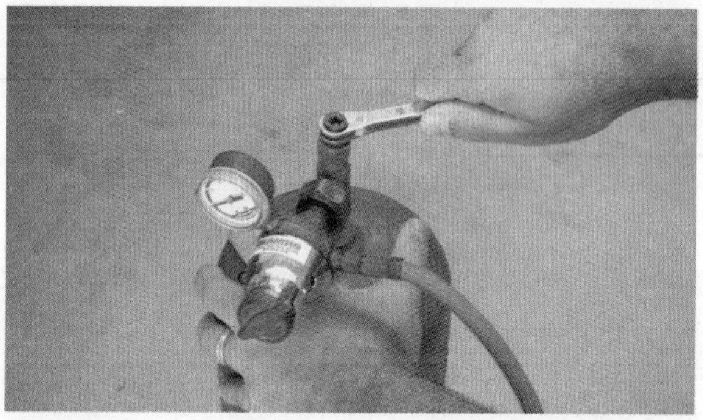

(C)

(D)

Figure 7.28 *(Continued)*

(E)

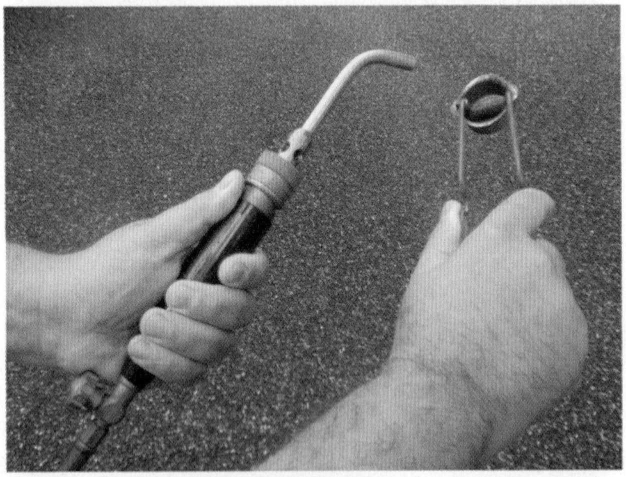

(F)

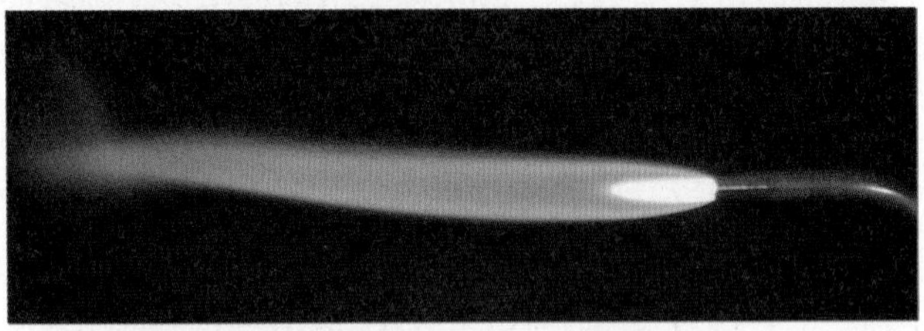

(G)

A. Before connecting the regulator to the tank, quickly open and close the tank valve slightly to blow out any dirt that may be lodged at the valve. **SAFETY PRECAUTION:** *Stand back from the tank and wear safety goggles when blowing dirt from the tank valve.•*

B. Connect the regulator with hose and torch to the tank. Be sure that all connections are tight.

C. Open the tank valve one-half turn. **SAFETY PRECAUTION:** *There is no need to open the tank valve more than one-half turn—this will give adequate flow. Leave the wrench on the tank valve stem. The valve can then be quickly closed in the event of an emergency.•* Adjust the regulator valve to about midrange.

D. Spray a soap bubble solution on the hose, regulator, and torch handle.

E. Look for bubbles on the hose, regulator, and torch handle that could indicate a leak.

F. Open the needle valve on the torch slightly, and ignite the gas with the striker or spark lighter. **SAFETY PRECAUTION:** *Do not use matches or cigarette lighters.•*

G. Adjust the flame using the needle valve at the handle until the flame is correct. The ideal flame on the air-acetylene torch is a carburizing flame, which has a well-defined inner cone and provides excess acetylene. The excess acetylene helps remove oxidation from the surface of the copper pipes and fittings. **SAFETY PRECAUTION:** *After each use, shut off the valve on the tank and bleed off the acetylene in the hose by opening the valve on the torch handle. Bleeding the acetylene from the hoses relieves the pressure when the hoses are not in use.•*

Some technicians prefer oxyacetylene torches, particularly when brazing large-diameter tubing or in other applications requiring higher temperatures. The addition of pure oxygen produces a much hotter flame. **SAFETY PRECAUTION:** *This equipment can be extremely dangerous when not used properly. You must thoroughly understand what the proper procedures are for using oxyacetylene equipment before attempting to use it. When you first use the equipment, do so only under the close supervision of a qualified person.•*

The following is a brief description of oxyacetylene welding and brazing equipment. As mentioned, oxyacetylene brazing and welding processes use a high-temperature flame, which is produced by mixing oxygen with acetylene gas. The equipment includes oxygen and acetylene cylinders, oxygen and acetylene pressure regulators, hoses, fittings, safety valves, torches, and tips, **Figure 7.29**.

The regulators each have two gauges, one to register tank pressure and the other to register pressure to the torch. The pressures indicated on these regulators are in pounds per square inch gauge (psig). **Figure 7.30** is a photo of an oxygen regulator and an acetylene regulator. These regulators must be used only for the gases and service for which they are intended. **SAFETY PRECAUTION:** *All connections must be free from dirt, dust, grease, and oil. Oxygen can produce an explosion when in contact with grease or oil. A reverse flow valve should be used on the hoses. These valves allow the gas to flow in one direction only and prevent the two gases from mixing in the hose, which could be very dangerous.•* **Figure 7.31(A)** shows the reverse flow check valve. Follow instructions when attaching these valves because some are designed to attach to the hose connection on the torch body and some to the hose connection on the regulator.

Figure 7.29 An oxyacetylene torch kit. Photo by Eugene Silberstein

Figure 7.30 Acetylene (left) and oxygen (right) regulators. Photo by Eugene Silberstein

Figure 7.31 (A) Reverse flow check valve with an oxyacetylene torch body and tip. (B) An assortment of oxyacetylene torch tips. Photos by Bill Johnson

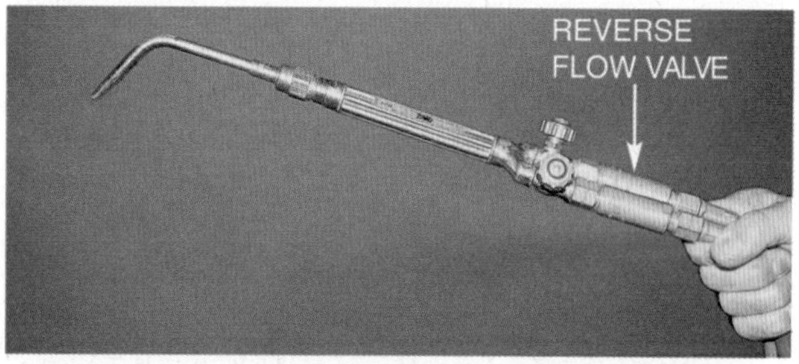

(A)

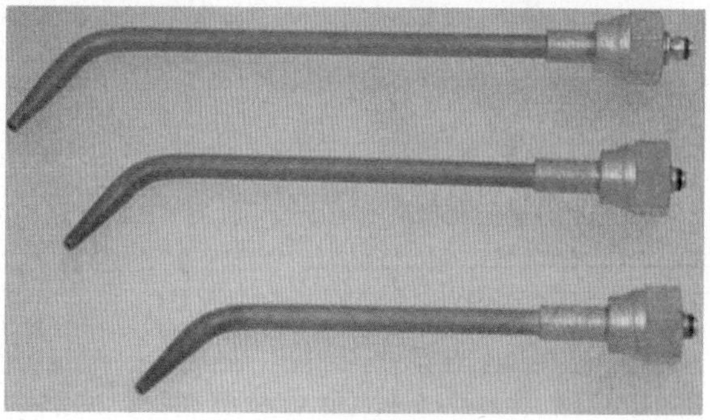

(B)

The red hose is attached to the acetylene regulator with left-handed threads, and the green hose is attached to the oxygen regulator with right-handed threads. The torch body is then attached to the hoses and the appropriate tip to the torch, **Figure 7.31(A)**. Many tip sizes and styles are available, **Figure 7.31(B)**. Some are designed to provide heat around the tubing and fittings, allowing the entire circumference of the tubing to be heated at one time, **Figure 7.32**. Another type of tip removes and replaces a four-way or reversing valve for a heat pump, **Figure 7.33**. All three of the copper fittings from this valve can be heated simultaneously, and the tubing can be pulled from the valve when the brazing alloys reach the proper temperature. You should learn through your specialized training how to use this equipment and which tips to use for particular applications.

The following procedure may be used to set up an oxyacetylene torch system for use. **⚠SAFETY PRECAUTION:** *Always turn the fuel gas (acetylene) on or off first.•* With the regulators and hoses fastened to the tank and the torch tip and the T handles on the regulators turned counterclockwise so that no pressure will go to the hoses (they may feel loose):

1. Turn the acetylene cylinder valve one-half turn so that pressure is introduced to the regulator. **⚠SAFETY PRECAUTION:** *Always stand to the side of the regulator. If the regulator were to blow out, it would likely blow toward the T handle. The wrench should remain on the stem of the*

Figure 7.32 Tips that can be used for heating the entire circumference of a fitting at the same time. Courtesy Uniweld Products, Inc.

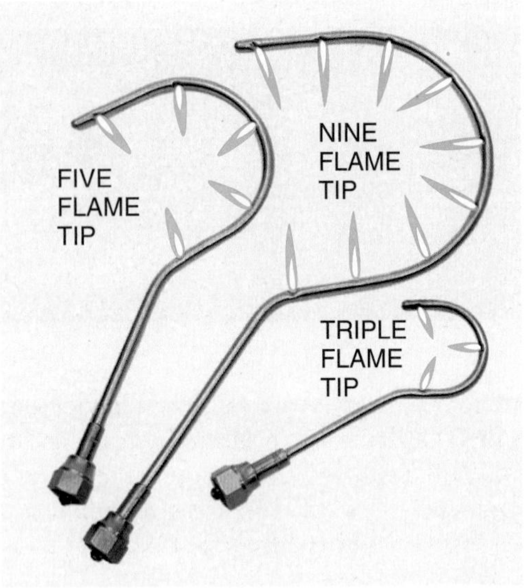

Figure 7.33 A tip that can be used for removing and replacing the four-way reversing valve on a heat pump system. Courtesy Uniweld Products, Inc.

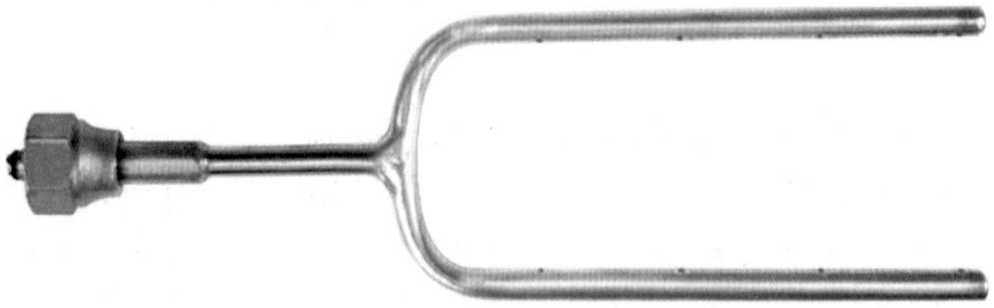

cylinder valve so the valve can be quickly turned off in case of an emergency. You should be able to see the pressure register on the acetylene cylinder gauge, **Figure 7.34**.

2. Slowly turn on the oxygen cylinder valve to the oxygen regulator. **SAFETY PRECAUTION:** *Again, stand to the side.* You should be able to see the pressure register on the oxygen cylinder pressure gauge, **Figure 7.35**.

Figure 7.34 Pressure registers on the acetylene cylinder gauge. Photo by Eugene Silberstein

Figure 7.35 Pressure registers on the oxygen cylinder gauge. Photo by Eugene Silberstein

3. Slightly open the valve on the torch handle that controls the acetylene (the red hose). With this valve slightly open, adjust the T handle on the acetylene regulator until the gauge on the red hose reads 5 psig, **Figure 7.36**. **SAFETY PRECAUTION:** *Now shut off the valve at the torch.* This part of the torch is ready.
4. Slightly open the valve on the torch handle that controls the oxygen (the green hose). With the valve slightly open, adjust the T handle on the oxygen regulator until the gauge reads 10 psig, **Figure 7.37**. **SAFETY PRECAUTION:** *Now shut off the valve at the torch.* This part of the torch is ready.

The following procedure may be used to light the torch. **SAFETY PRECAUTION:** *Before lighting, always point the torch tip away from you or any flammable substance. Make sure that you point the torch away from the cylinders and hoses. Failure to do so could result in tragedy.* While holding the torch handle in your left hand:

1. Slightly open the valve on the torch handle on the acetylene side and allow it to flow a moment to get any air out of the line.
2. Using only an approved lighter, light the fuel gas, acetylene. You will have a large yellow-orange flame that smokes, **Figure 7.38**.

Figure 7.36 This acetylene regulator gauge indicates 5 psig. Photo by Eugene Silberstein

Figure 7.37 This oxygen regulator gauge indicates 10 psig. Photo by Eugene Silberstein

Figure 7.38 An air-acetylene flame.

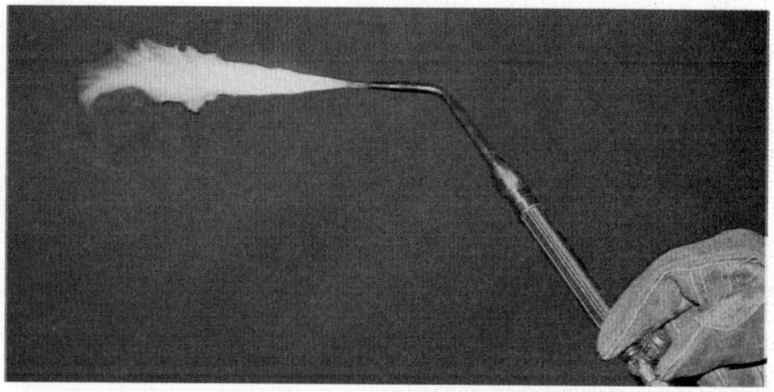

3. Slightly open the valve on the torch handle on the oxygen side. The flame will begin to clear up and turn blue, **Figure 7.39**. You will have to adjust the valves until you get the flame you want. The flame should be blue and be setting firmly on the torch tip, not blowing away from it.

To shut the system down, follow these procedures:

1. Shut off the fuel gas (acetylene) valve at the torch first.
2. Shut off the oxygen valve at the torch.
3. Turn off the cylinder valve on the fuel gas (acetylene).
4. Open the valve on the torch handle on the fuel gas side to relieve the pressure on the hose. Both regulators will lose their pressure to 0 psig.
5. Turn the T handle on the fuel gas regulator counterclockwise until it appears to be loose. This side of the system has now been relieved of pressure. Shut off the valve on the torch handle.
6. Turn off the cylinder valve on the oxygen cylinder.
7. Open the valve on the torch handle on the oxygen side to relieve the pressure on the hose. Both regulators will lose their pressure to 0 psig. This side of the system has been bled of pressure. Turn off the torch handle valve on the torch handle.
8. Turn the T handle on the oxygen regulator counterclockwise until it feels loose.

Figure 7.39 An oxyacetylene flame.

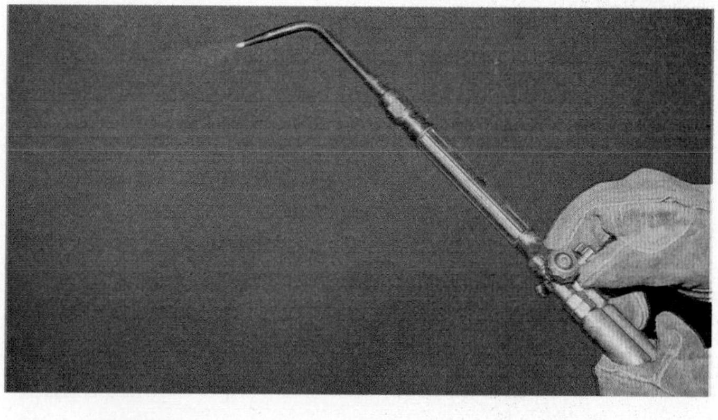

Figure 7.40 (A) A neutral flame. (B) A carbonizing flame. (C) An oxidizing flame. Photos by Bill Johnson

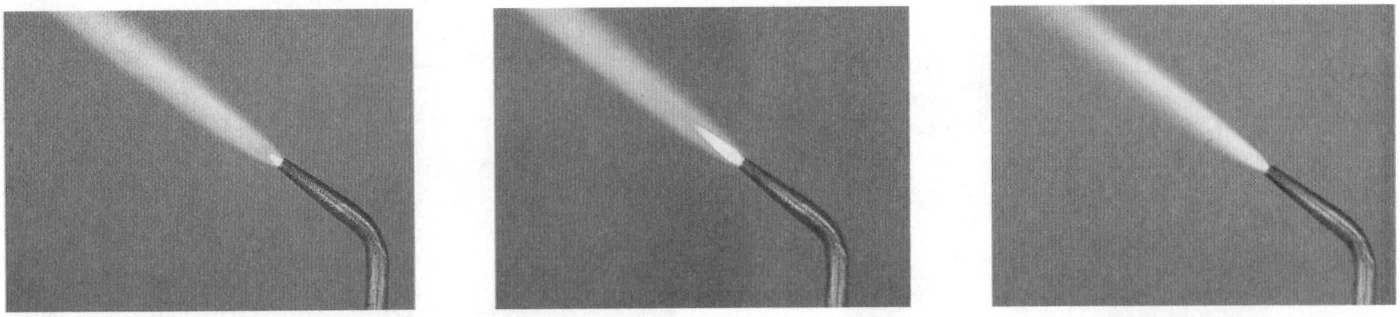

(A) **(B)** **(C)**

The system is now ready for storage and/or transport.

SAFETY PRECAUTION: *If the oxyacetylene system is to be transported, it must be moved using an approved hand cart or carrier with the tanks properly secured. If it is to be transported on the road in a truck, caps must be used on the cylinders.•*

A neutral flame should be used for most operations. See **Figure 7.40** for photos of a neutral flame, a carbonizing flame (too much acetylene), and an oxidizing flame (too much oxygen).

SAFETY PRECAUTION: *As stated previously, there are many safety precautions that should be followed when using oxyacetylene equipment. Be sure that you follow all of these precautions. You should become familiar with them during your training in how to use this equipment.•*

7.9 FLUXING

Before getting into the specific techniques for soldering and brazing, we need to discuss one important topic, namely fluxing. In terms of soldering and brazing, the main function of flux is to prevent oxidation from forming on the surfaces of the pipes being joined and in the filler materials themselves, **Figure 7.41**. Filler materials, such as solders and brazing rods, will adhere very well to the piping materials, but not the oxides that form during the soldering and brazing processes. By limiting the amount of oxides that form, and absorbing the oxides that do form when the pipes are heated, flux helps facilitate the flow of the filler material into the joint. So, to help ensure a leak-free joint, flux is used whenever needed.

SAFETY PRECAUTION: *Brazing and soldering materials, including fluxes, can produce dangerous fumes when used during soldering and brazing. Be sure to protect yourself from these fumes by wearing proper breathing masks and by working in well-ventilated areas. Always follow the manufacturers' and Material Safety Data Sheet (MSDS) instructions when soldering and brazing.•*

Figure 7.41 Flux is used to reduce the amount of oxidation on the surfaces of the pipes being soldered or brazed. Photo by Eugene Silberstein

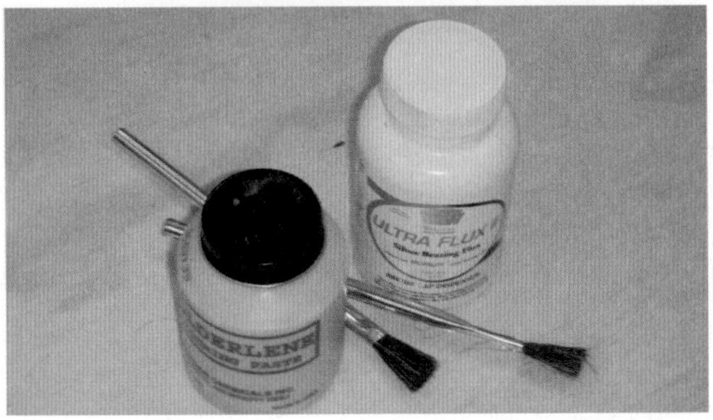

Although fluxing is important, too much of a good thing can have negative effects on the quality of the connection. Apply flux only to the male portion of the fitting. This will prevent excessive flux from entering the refrigerant lines, which is important because flux typically contains between 15% and 35% water and every effort should be made to limit the amount of moisture in an air-conditioning or refrigeration system. To ensure proper flux coverage, it is good field practice to rotate the male portion of the joint in the fitting, or to rotate the fitting around the pipe. Some brazing rods are coated with flux. Although this seems to be solely for convenience, there is another, more important reason why this type of product is desirable. As the rod is heated, the flux melts and flows before the filler material is introduced to the joint. This also helps avoid using too much flux.

Be sure to follow the guidelines for soldering and brazing carefully; applying too much heat to the joint can also have negative effects on the quality of the connection. If flux is used and the joint is overheated, the flux can become saturated with oxides and prevent the filler material from flowing properly. Note also that fluxes are corrosive and need to be removed after the joint has been completed.

The following sections in this unit provide detailed information about both soldering and brazing, and more

information regarding flux and how it affects the soldering and brazing processes.

7.10 SOLDERING TECHNIQUES

The mating diameters of tubing and fittings are designed or sized to fit together properly. For good capillary attraction, the space between the metals should be approximately 0.003 in. After the tubing has been cut to size and deburred, you must do the following for good joint soldering:

1. Clean mating parts of the joint.
2. Apply a flux to the male connection.
3. Assemble the tubing and fitting.
4. Heat the joint and apply the solder.
5. Wipe the joint clean.

CLEANING

The end of the copper tubing and the inside of the fitting must be absolutely clean. Even though the surfaces may look clean, they may contain fingerprints, dust, or oxidation. A fine sand cloth, a cleaning pad, or a special wire brush may be used for cleaning. When the piping system is to be used for a hermetic compressor, the sand cloth should be an approved nonconducting type, **Figure 7.42**.

FLUXING

Apply flux soon after the surfaces are cleaned. For soft soldering, flux may be a paste, jelly, or liquid. Apply the flux with a *clean* brush or applicator. Do *not* use a brush that has been used for any other purpose. Apply the flux only to the area to be joined and avoid getting it into the piping system. Too much flux will make it easier for the molten filler material to flow into the pipe itself, possibly restricting flow through the pipe. Be sure to apply flux sparingly and only to the male end of the joint. This will help ensure that the filler material remains in the space between the mating surfaces.

ASSEMBLY

Soon after the flux is applied, assemble and support the joint so that it is straight and will not move while being soldered.

HEATING AND APPLYING SOLDER

When soldering, heat the tubing near the fitting first for a short time. Then move the torch from the tubing to the fitting. Keep moving the torch from the tubing to the fitting. Heat the entire joint, not just one side. Keep moving the torch to spread the heat evenly and do not overheat any area. Do not point the flame into the fitting socket. Hold the torch so that the inner

Figure 7.42 Cleaning and fluxing. (A) Clean the tube with sand cloth. (B) Clean the fitting with a tubing brush. (C) Clean the fitting with sand cloth. (D) Apply flux to the male portion of the fitting. Photos by Bill Johnson

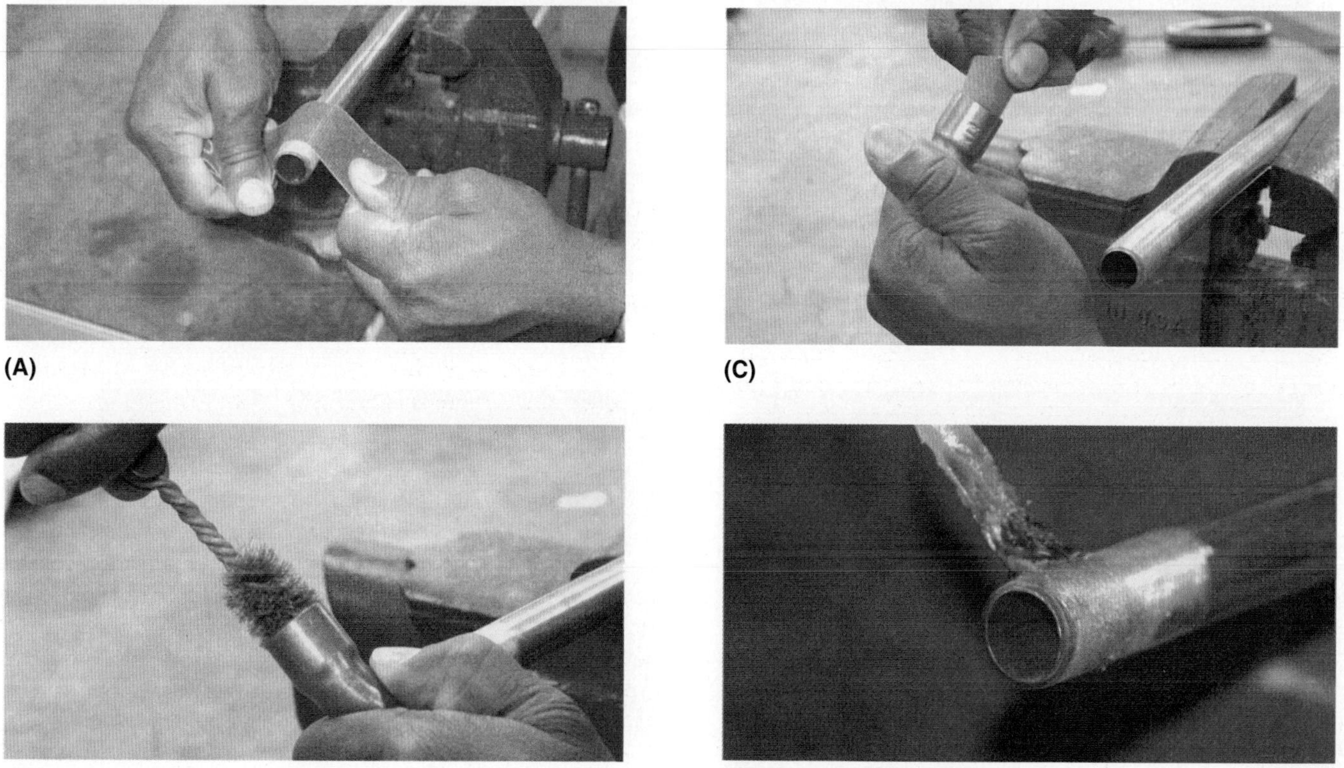

(A) **(C)**

(B) **(D)**

cone of the flame just touches the metal. After briefly heating the joint, touch the solder to the joint. If it does not readily melt, remove it and continue heating the joint. Continue to test the heat of the metal with the solder. Do *not* melt the solder with the flame; use the heat in the metal.

When the solder flows freely from the heat of the metal, feed enough solder in to fill the joint. Watching the appearance of the flux will help determine when to apply the filler material to the joint. When the flux "melts" and starts to take on the appearance of water, it is time for the filler material to be applied. Do not use excessive solder. Using more than the recommended amount of filler material will not make the joint stronger. The excess material can accumulate inside the tubing and create flow restrictions.

Different-size pipes will require different amounts of filler material. A connection that is joining two sections of 2-in. pipe will require more filler material than a joint that is made up of two 1-in. pipes. For smaller joints, the estimated amount of filler material is about twice the diameter of the joint. For example, if a soldered joint is being made to join two sections of ½-in. pipe, about 1 in., or twice the diameter of the pipe, of filler material will be used. In the 3-in. to 6-in. range, however, the amount of filler material will be about four times the diameter of the pipe. So, a 4-in. solder joint will require about 16 in. of filler material. **Figure 7.43** shows a step-by-step procedure for heating the joint and applying the solder.

For horizontal joints, apply the filler metal first to the bottom, then to the sides, and finally on the top, making sure the operations overlap, **Figure 7.44**. On vertical joints, it does not matter where the filler is first applied.

WIPING

While the joint is still hot, you may wipe it with a rag to remove excess solder. This is not necessary for producing a good bond, but it improves the appearance of the joint. Be careful not to break the joint when wiping. In addition, it is important to remember that fluxes are corrosive and need to be removed once the joint has been completed.

Figure 7.43 The proper procedures for heating a joint and applying solder. (A) Start by heating the tubing. (B) Keep the flame moving. Do not point the flame into the end of the fitting. (C) Touch the solder to the joint to check for proper heat. Do not melt the solder with the flame. (D) When the joint is hot enough, solder will flow. Photos by Bill Johnson

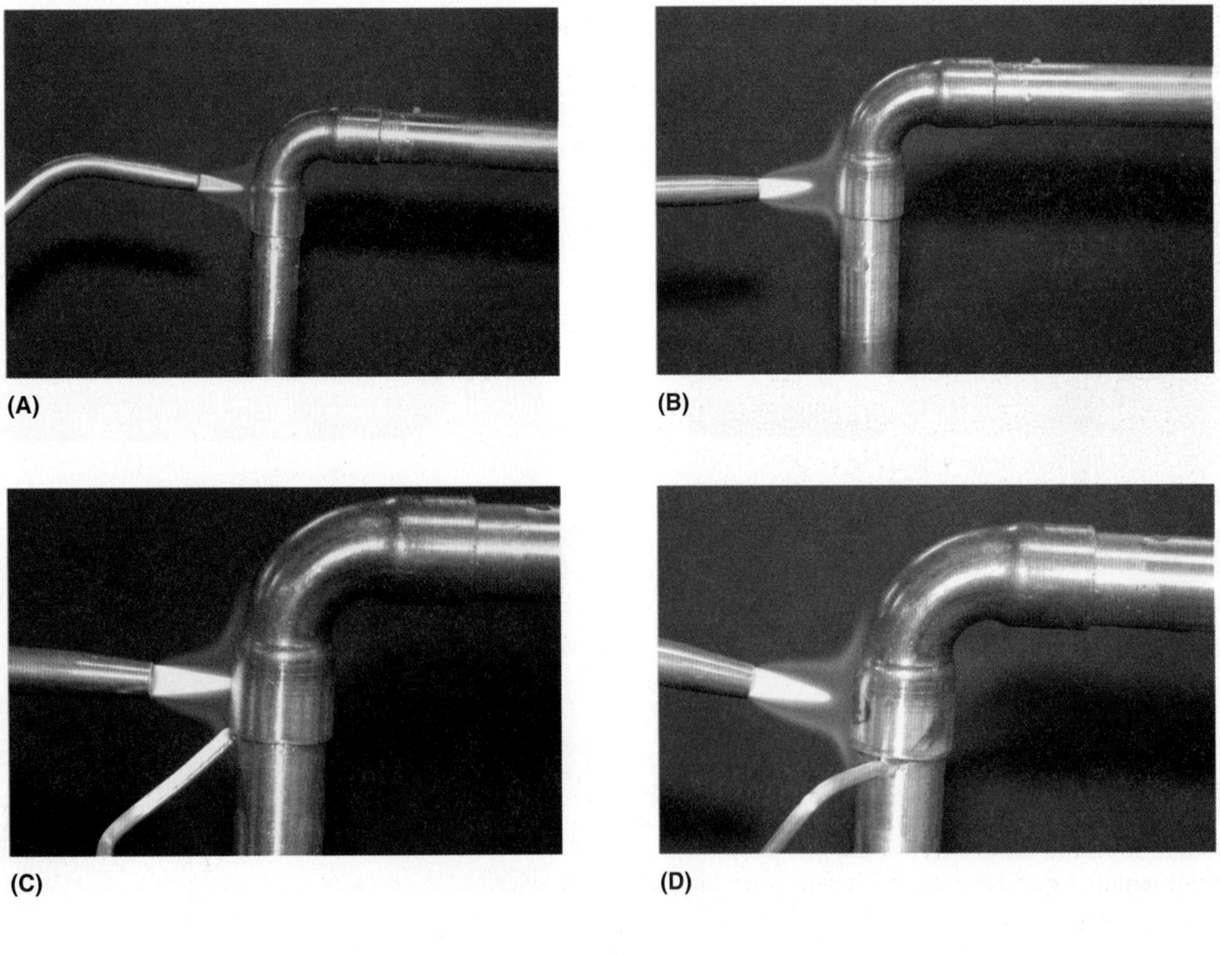

(A) (B) (C) (D)

Figure 7.44 When making horizontal soldered or brazed joints, (A) apply the filler material to the bottom, (B) then to the two sides, and (C) finally on the top, making certain that the operations overlap. Photos by Bill Johnson

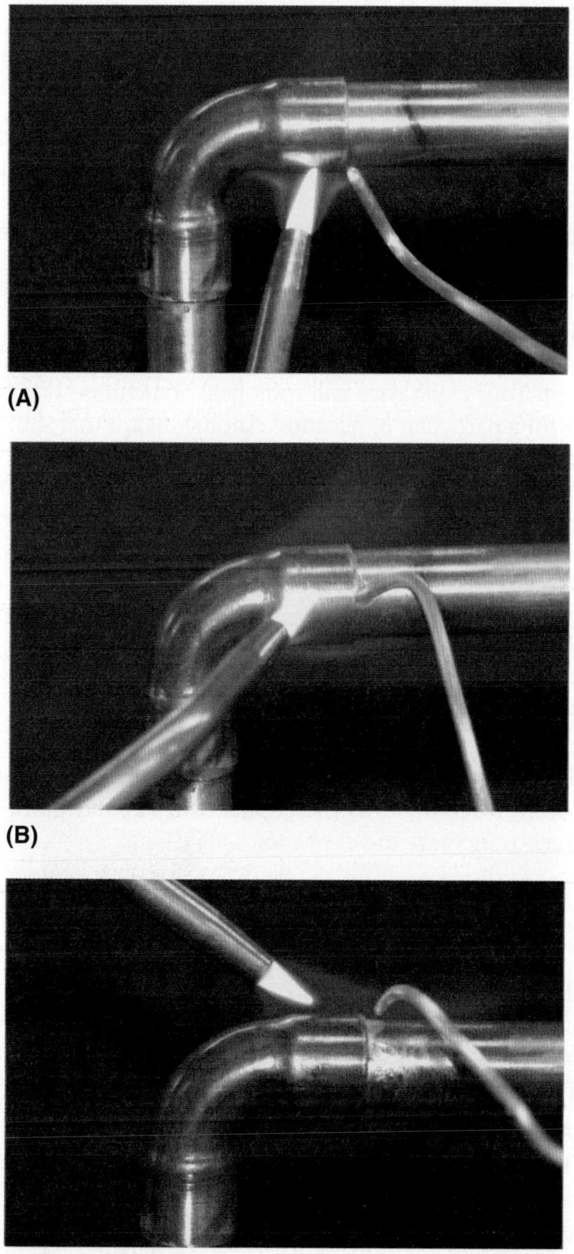

(A) (B) (C)

7.11 BRAZING TECHNIQUES

CLEANING

The cleaning procedures for brazing are similar to those for soldering. The brazing flux is applied with a brush to the cleaned area of the tube end. Avoid getting flux inside the piping system. Some brands of silver or copper–phosphorous alloys do not require extensive cleaning or flux when brazing copper to copper. This is because, at the higher temperatures required for brazing, the mating surfaces are cleaned by the torch flame. In addition, these filler materials are typically comprised of phosphorous and copper and are self-fluxing, so no additional flux is needed. Always follow the instructions from the filler-material manufacturer.

APPLYING HEAT FOR BRAZING

Before you heat the joint, it is good practice to introduce nitrogen into the system to purge the air (containing oxygen) and reduce the possibility of oxidation to a minimum. A pressure of 1 to 2 psig is enough to sufficiently purge the piping. Because 1 or 2 psig of pressure is very difficult to read on the gauge manifold, a nitrogen flow indicator, **Figure 7.45**, can be used. This device allows the technician to maintain the desired nitrogen flow rate and also, at a quick glance, ensure that nitrogen is indeed flowing. Since more and more systems are operating with R-410A, the number of air-conditioning systems that operate with ester-based lubricants is increasing. Ester-based lubricants are more likely to separate oxidation from the inside surfaces of the piping materials than are mineral or alkylbenzene lubricants, which

Figure 7.45 A nitrogen flow indicator. The floating ball indicates flow. Indictor provided by Uniweld Products, Inc. Photo by Eugene Silberstein

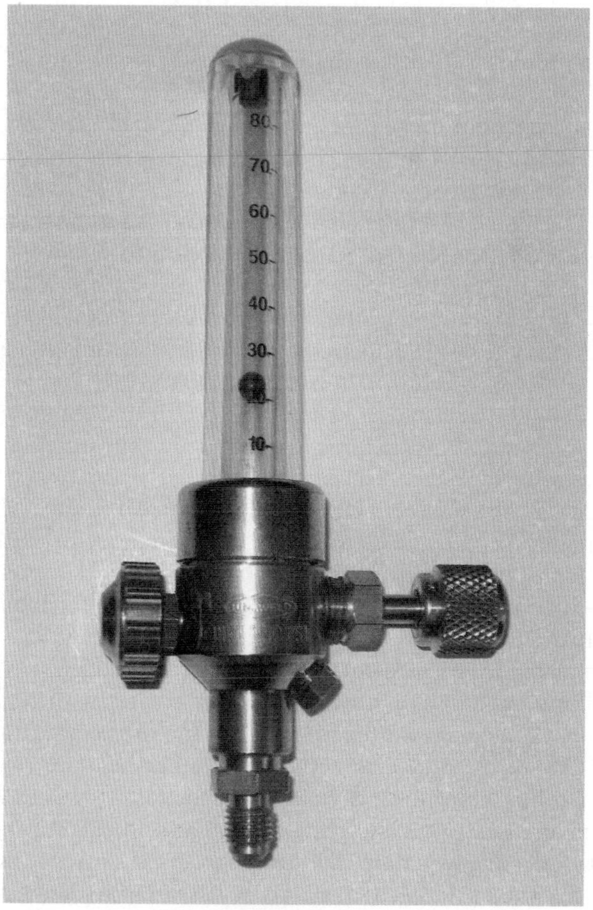

increases the potential for clogging metering devices, filter driers, and other refrigerant flow controls. By allowing a small amount of nitrogen to flow through the piping circuit while the brazing process is being performed, the amount of oxidation will be greatly reduced.

Apply heat to the parts to be joined with an air-acetylene or oxyacetylene torch. Heat the tube first, beginning about 1 in. from the edge of the fitting, sweeping the flame around the tube. It is very important to keep the flame in motion and not overheat any one area. Then switch the flame to the fitting at the base of the cup. Heat uniformly, sweeping the flame from the fitting to the tube. Apply the filler rod or wire at a point where the tube enters the socket. When the proper temperature is reached, the filler metal will flow readily by capillary attraction into the space between the tube and the fitting. As in soldering, do not heat the rod or wire itself. The temperature of the metal at the joint should be hot enough to melt the filler metal. When the joint is at the correct temperature, it will be cherry red in color. The procedures are the same as those used for soldering except for the materials used and the higher heat applied. The flux used in the brazing process will cause oxidation.

7.12 PRACTICAL SOLDERING AND BRAZING TIPS

LOW TEMPERATURE (SOLDERING)

When soldering, the surfaces between the male and female parts being joined must be clean. Clean surfaces are necessary to ensure leak-free connections. It takes much longer to prepare surfaces for soldering than it does to actually make the soldered joint. If a connection to be soldered is cleaned in advance, it may need a touch-up cleaning when it is actually soldered. Oxidation can occur, making it more difficult to solder. Some fluxes may be applied after cleaning to prevent oxidation until the tubing and fittings are ready for soldering.

Only the best solders should be used for low-temperature solder connections for refrigeration and air-conditioning. Systems were soldered successfully for many years with 95/5 solder. If the soldered connection is completed in the correct manner, 95/5 can still be used, but it should never be used on the high-pressure side of the system close to the compressor. The high temperature of the discharge line and the vibration will very likely cause it to leak. A better choice than 95/5 for low-temperature solder would be a solder with a high strength and low temperature. Low-temperature solders with a silver content offer greater strength with low melting temperatures.

One of the problems with most low-temperature solders is that the melting and flow points are too close together. This is evident when you are trying to use the solder; it flows too fast, and you have a hard time keeping it in the clearance in the joint. Some of the silver-type, low-temperature solders have a wider melt and flow point and are easier to use. They also have the advantage of being more elastic during the soldering procedure. This allows gaps between fittings to be filled more easily.

HIGH-TEMPERATURE (BRAZING)

Several choices of high-temperature brazing materials are available. Some have a high silver content (45% silver) and must always be used with a flux, and some do not have high silver content (15% silver). These may not require flux when making copper-to-copper connections; however, flux may be used. Other brazing materials have no silver content. Filler materials comprised solely of phosphorous and copper (no silver) are typically over 90% copper and are intended for connections that are not subject to movement or vibrations. So if the joint will be subjected to vibration, an alternative filler material should be selected. Your experience will soon help you choose which you prefer for a particular application. Another factor that should be considered is the expansion of the materials being joined.

When brazing copper to copper, the expansion rate of the copper is not an issue since both parts of the connection will expand at the same rate when heated. On the other hand, when brazing copper to steel, the expansion rates must be taken into account since copper will expand at a faster rate than the steel or any other ferrous material. If the joint involves a copper pipe in a steel female connection, the initial fit should be very loose, as the copper will expand more than the steel when heated and reduce the clearance. If the joint involves a steel pipe and a female copper fitting, the joint can almost be force-tight, as the copper will expand more than the steel to increase the clearance.

DIFFERENT JOINTS

The type of joint dictates which solder or brazing materials to use. All connections are not copper-to-copper. Some may be copper-to-steel, copper-to-brass, or brass-to-steel. These are called dissimilar metal connections; the following are examples.

1. **A copper suction line to a steel compressor or connection.** The logical choice is to use 45% silver content because of the strength of the connection combined with the high melting temperature. Do not use filler materials that contain phosphorous on steel joints. When used on steel pipe, phosphorous filler materials produce phosphides, which are brittle and prone to cracking. This can lead to piping system leaks.
2. **A copper suction line to a brass accessory valve.** From a strength standpoint, the best choice would be 45% silver. Another choice would be a solder with a silver content that has a high strength and a low melting temperature. The valve body will not have to be heated to the high melting temperature of the 45% silver solder.
3. **A copper liquid line to a steel filter drier.** Although 45% silver brazing material is a good choice, it requires a lot of heat. A low-temperature solder with silver content

may be the best choice. It also gives you the option of easily removing the drier for replacement at a later date. Once again, on steel connections avoid using filler materials that contain phosphorous.

4. **A large copper suction line connection using hard-drawn tubing.** A high-temperature brazing material with low silver content is the choice of many technicians, but you may not want to take the temper out of the hard-drawn tubing. A low-temperature solder with a silver content will give the proper strength, and the low melting temperature will not take the temper from the pipe.

Any time copper tubing is heated to a temperature that causes the tubing to become red-hot, heavy oxidation occurs on the inside and the outside of the pipe. This can be noted by the black scale that accumulates on the outside of a pipe when it is brazed. The pipe looks the same on the inside, where there is black scale. The oxygen that is in the air inside the pipe causes this black scale to form. Displacing the oxygen with dry nitrogen can prevent this. All technicians should sweep the system with dry nitrogen when making high-temperature connections. When a system is made of many connections, the scale can amount to a considerable quantity. It is trapped in either the oil, the filter driers, or the metering device—whichever it reaches first.

CHOICE OF HEAT FOR SOLDERING AND BRAZING

Air-acetylene or oxyacetylene torch kits may be used as the heat source for soldering and brazing. It is important, however, to choose the best torch for the job to be done. This decision should be made, in part, by the temperature of the flames produced as well as what is referred to as the target temperature. The flame temperature is just what it sounds like: it's the temperature of the torch flame. For an air-acetylene torch, the flame temperature is approximately 4200°F, while the temperature of an oxyacetylene torch flame is in the range of 5600°F. The target temperature is significantly lower than the flame temperature, as the target temperature refers to the temperature that can actually act on the object being heated. The target temperature for an air-acetylene torch is about 2700°F, while the target temperature for an oxyacetylene torch is about 4700°F. In any event, it is important that the user of the torch keep the flame at the proper distance from the target to avoid melting the materials (base metals) being joined. Copper, for example, melts at about 2000°F. In addition to the heat source being used, the correct torch tip must be used for the solder type and pipe size. **Figure 7.46** includes a photo and table of several tip sizes for different pipe sizes and solder combinations.

Figure 7.46 Different tip sizes may be used for different pipe sizes and solder combinations. Courtesy of Thermadyne Industries, Inc.

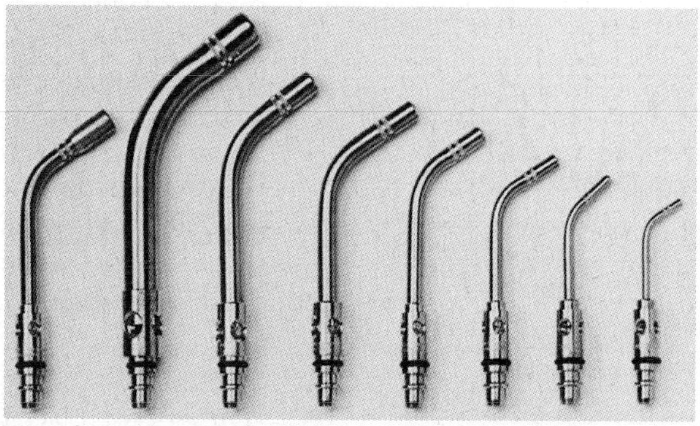

ACETYLENE TORCH TIPS

Tip No.	Tip Size in.	Tip Size mm	Gas Flow @ 14 psi ft^3/hr	Gas Flow @ 0.9 Bar m^3/hr	Copper Tubing Size Capacity Soft Solder in.	Soft Solder mm	Silver Solder in.	Silver Solder mm
A-2	3/16	4.8	2.0	0.17	1/8-1/2	3-15	1/8-1/4	3-10
A-3	1/4	6.4	3.6	0.31	1/4-1	5-25	1/8-1/2	3-12
A-5	5/16	7.9	5.7	0.48	3/4-1 1/2	20-40	1/4-3/4	10-20
A-8	3/8	9.5	8.3	0.71	1-2	25-50	1/2-2	15-30
A-11	7/16	11.1	11.0	0.94	1 1/2-3	40-75	7/8-1 5/8	20-40
A-14	1/2	12.7	14.5	1.23	2-3 1/2	50-90	1-2	30-50
A-32*	3/4	19.0	33.2	2.82	4-6	100-150	1 1/2-4	40-100
MSA-8	3/8	9.5	5.8	0.50	3/4-3	20-40	1/4-3/4	10-20

*Use with large tank only.

NOTE: For air-conditioning, add 1/8 in. for type L tubing.

Another heat source is MAPP™ gas. It is a composite gas that is similar in nature to propane and may be used with air. Composite gases are made up of multiple gases. The term MAPP comes from the chemical make-up of the gas, namely methylacetylene-propadiene propane. The flame temperature of MAPP gas in air is about 3730°F, while the temperature of a propane flame is about 3600°F. MAPP gas is supplied in containers that are much easier to transport and use than oxyacetylene torch kits, **Figure 7.47**.

SOLDERING AND BRAZING TIPS

1. Clean all surfaces to be soldered or brazed.
2. Keep filings, burrs, and flux from inside the pipe.
3. When making upright soldered joints, apply heat to the top of the fitting, causing the filler to rise.
4. When soldering or brazing fittings of different weights, such as soldering a copper line to a large brass valve body, most of the heat should be applied to the large mass of metal, the valve.

Figure 7.47 A MAPP™ gas kit. Courtesy of Thermadyne Industries, Inc.

Figure 7.48 Make a bend in the end of the solder so you will know when to stop. Photo by Bill Johnson

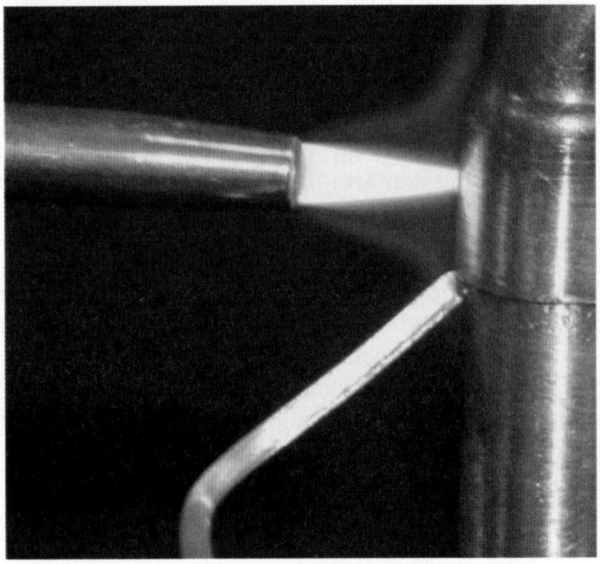

5. Do not overheat the connections. The heat may be varied by moving the torch closer to or farther from the joint. Once heat is applied to a connection, it should not be completely removed because the air moves in and oxidation occurs.
6. Do not apply excessive amounts of low-temperature solders to a connection. It is a good idea to mark the length of solder you intend to use with a bend, **Figure 7.48**. When you get to the bend, stop or you will overfill the joint, and the excess may be in the system.
7. When using flux with high-temperature brazing material, always chip the flux away when finished. **SAFETY PRECAUTION:** *Wear eye protection.*• Flux is hard and appears like glass on the brazed connection, **Figure 7.49**. This hard substance may cover a leak and be blown out later.

Figure 7.49 Flux used with high-temperature brazing materials will form a glaze that looks like glass.

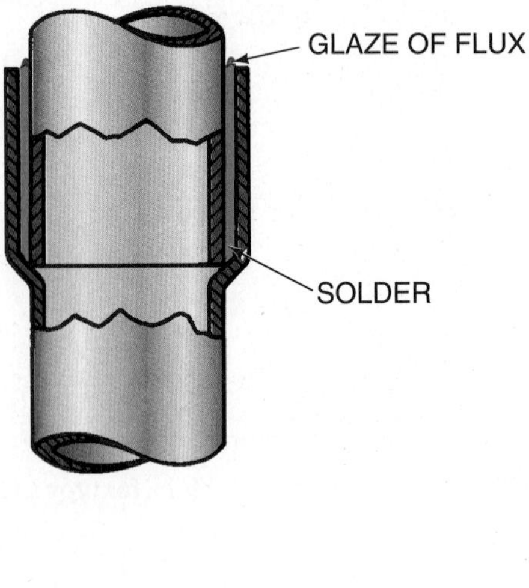

8. When using any flux that will corrode the pipe, such as some fluxes for low-temperature solders, wash the flux off the connection or corrosion will occur. If this is not done, it will soon look like a poor job.
9. Talk to the expert at a supply house for your special solder needs.
10. When using an oxyacetylene torch for brazing, be sure to use a neutral flame, not an oxidizing flame. An oxidizing flame results when too much oxygen is supplied to the flame. This causes more oxidation to form on the surfaces of the pipes being joined.

Figure 7.50 Components of a flare joint. Photo by Bill Johnson

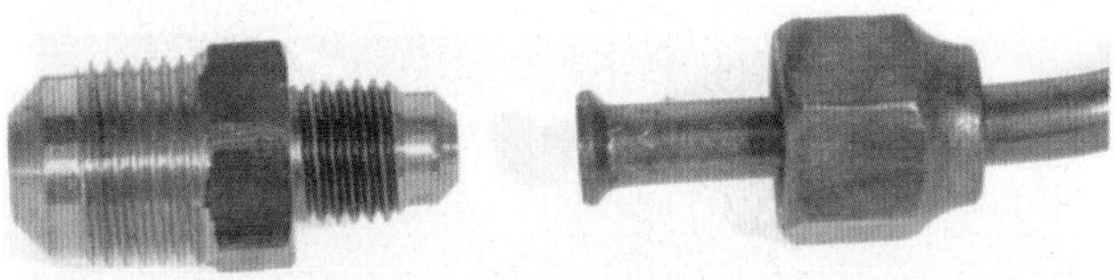

7.13 MAKING FLARE JOINTS

Another method of joining tubes and fittings is the flare joint. A flare on the end of the tubing is made against an angle on a fitting and is secured with a flare nut behind the flare on the tubing, **Figure 7.50**. The flare on the tubing can be made with a screw-type flaring tool using the following procedure:

1. Cut the tube to the correct length.
2. Ream to remove all burrs and clean all residue from the tubing.
3. Slip the flare nut or coupling nut over the tubing with the threaded end facing the end of the tubing.
4. Clamp the tube in the flaring block, **Figure 7.51(A)**. Adjust it so that the tube is slightly above the block (about one-third of the total height of the flare).
5. Place the yoke on the block with the tapered cone over the end of the tube. Many technicians use a drop or two of refrigerant oil to lubricate the inside of the flare while it is being made, **Figure 7.51(B)**.

Figure 7.51 (A) Secure the tubing in the flaring block. (B) Position the yoke on he block, centered over the tubing. (C) Tighten the yoke to form the flare. (D) Inspect the completed flare. Photos by Bill Johnson

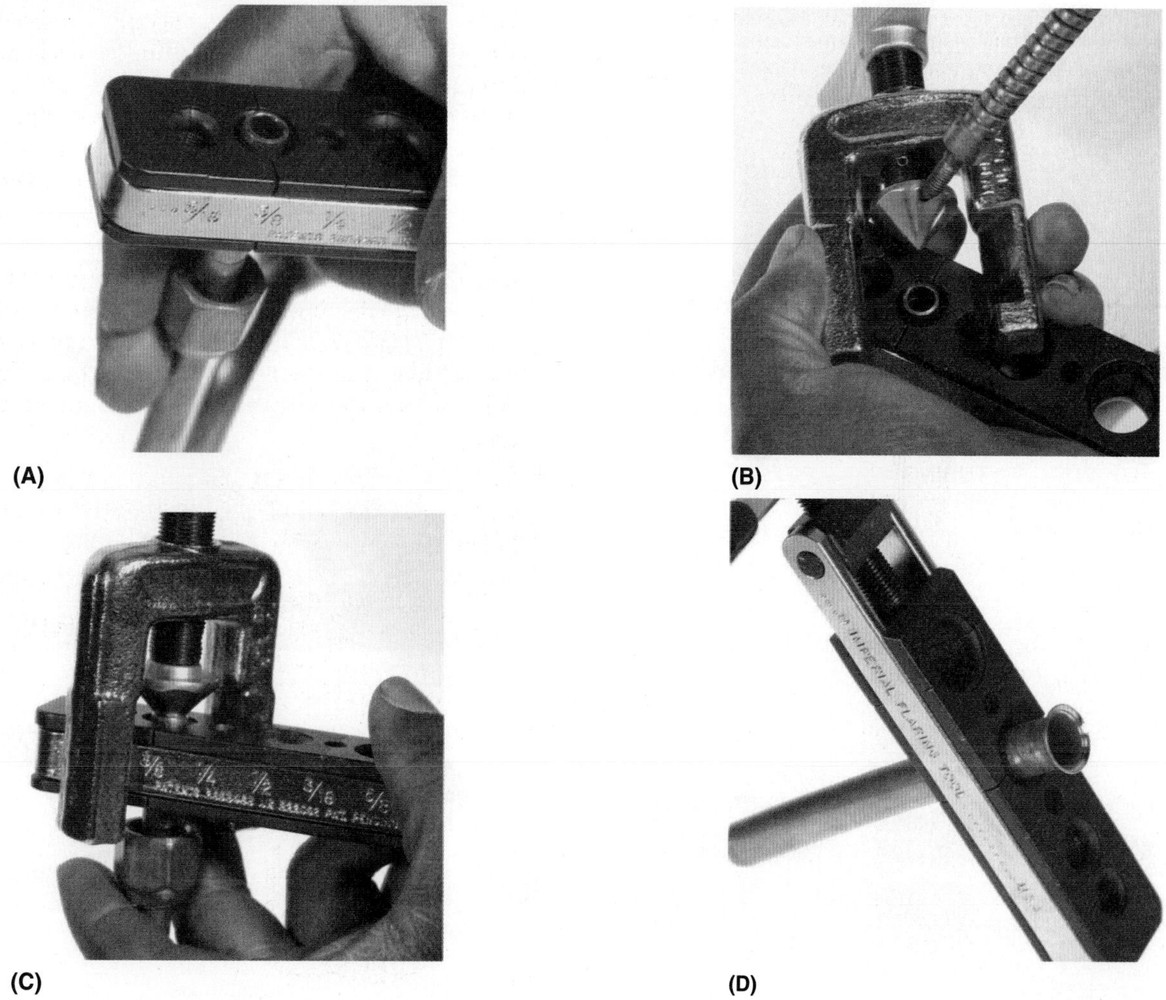

6. Turn the screw down firmly, **Figure 7.51(C)**. Tighten and loosen the screw several times during the flaring process to prevent the work from hardening. Continue until the flare is completed.
7. Remove the tubing from the block, **Figure 7.51(D)**. Inspect for defects. If you find any, cut off the flare and start over.
8. Assemble the joint.

MAKING A DOUBLE-THICKNESS FLARE

A double-thickness flare provides more strength at the flare end of the tube. This is a two-step operation. Either a punch and block or combination flaring tool is used. **Figure 7.52** illustrates the procedure for making double-thickness flares with the combination flaring tool. Many fittings are available to use with a flare joint. Each of the fittings has a 45° angle on the end that fits against the flare on the end of the tube, **Figure 7.53**.

Figure 7.52 Procedure for making a double-thickness flare. (A) Place the adapter of the combination flaring tool over the tubing in the block. Screw down to bell out the tubing. (B) Remove the adapter, place a cone over the tubing, and screw down to form the double flare.

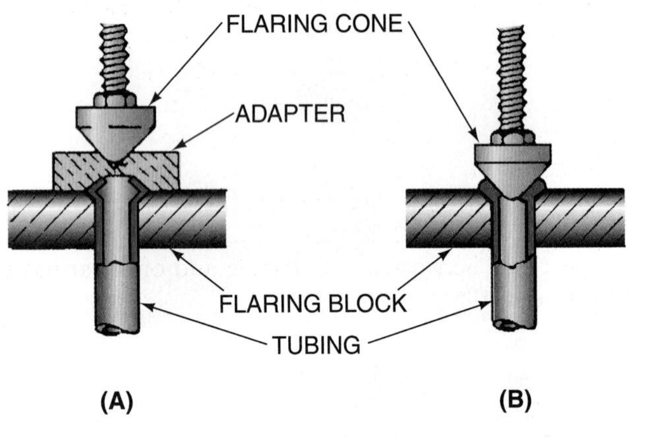

Figure 7.53 Flare fittings. Photo by Bill Johnson

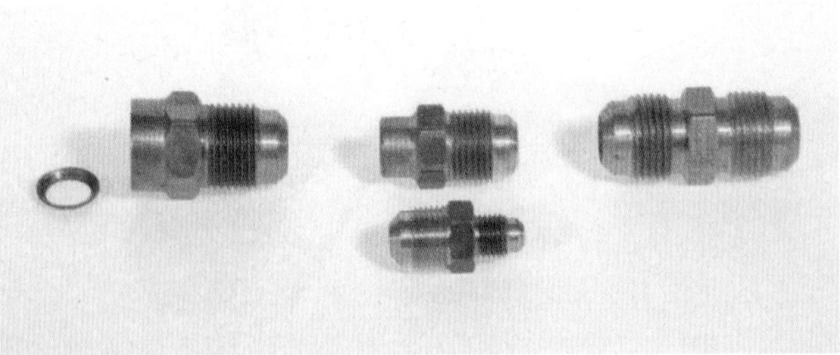

7.14 SWAGING TECHNIQUES

Swaging is not as common as flaring, but you should know how to make a swaged joint. **Swaging** is the joining of two pieces of copper tubing of the same diameter by expanding or stretching the end of one piece to fit over the other so that the joint may be soldered or brazed, **Figure 7.54**. As a general rule, the length of the joint that fits over the other is equal to the approximate OD of the tubing.

You can make a swaged joint by using a punch or a lever-type tool to expand the end of the tubing, **Figure 7.55**. Place

Figure 7.54 A swage connection.

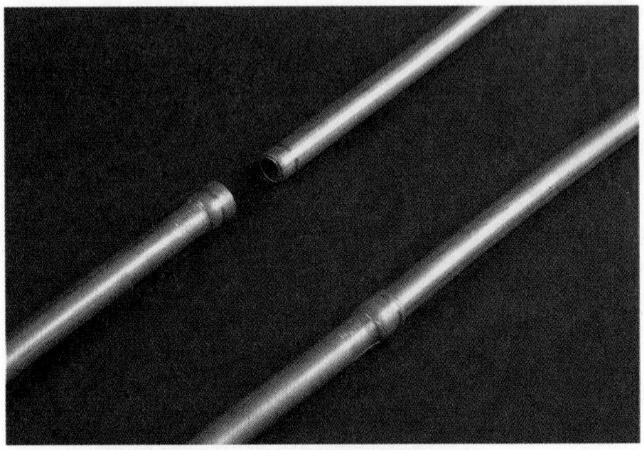

Figure 7.55 (A) A swaging punch. (B) A lever-type swaging tool. Photo (A) by Bill Johnson

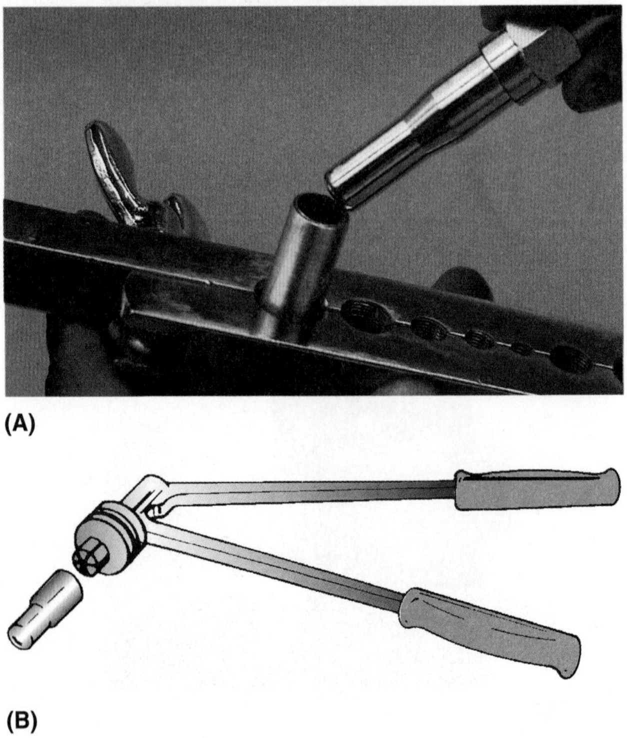

the tubing in a flare block or an anvil block that has a hole equal to the size of the OD of the tubing. The tube should extend above the block by an amount equal to the OD of the tube plus approximately 1/8 in., **Figure 7.56(A)**. Place the correct-size swaging punch in the tube and strike it with a hammer until the proper shape and length of the joint has been obtained, **Figure 7.56(B)**. Follow the same procedure with screw-type or level-type tools. A drop or two of refrigerant oil on the swaging tool will help but it must be cleaned off before soldering. Assemble the joint. The tubing should fit together easily. Always inspect the tubing after swaging to see whether there are cracks or other defects. If any are seen or suspected, cut off the swage and start over.

SAFETY PRECAUTION: *Field fabrication of tubing is not done under factory-clean conditions, so you need to be observant and careful that no foreign materials enter the tubing. When the tubing is applied to air-conditioning or refrigeration, remember that any foreign matter will cause problems. Utmost care must be taken.•*

Figure 7.56 Making a swaged joint. (A) A tube secured in a block. (B) Striking the swage punch to expand the metal. Photos by Bill Johnson

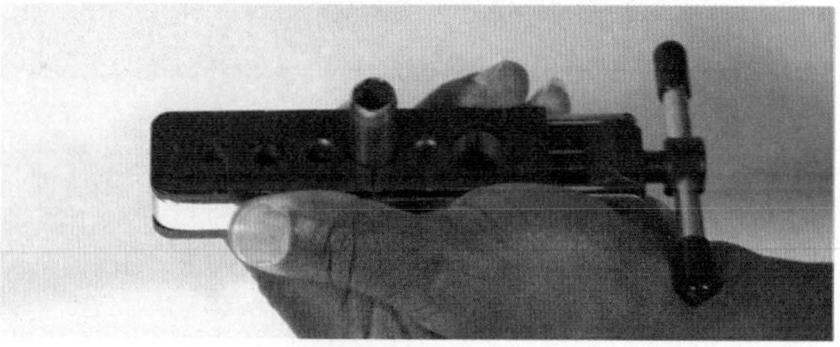

(A)

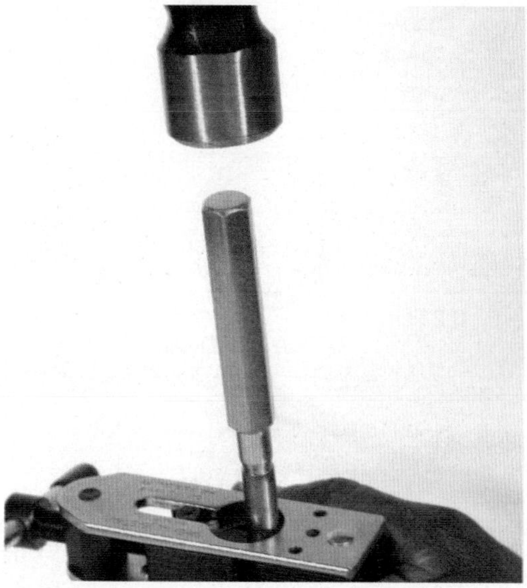

(B)

Here are some tips to help ensure that the swage is correct, neat, and straight:

- If using a swage punch, rotate the flaring block 180° every three or four strikes. This will keep the tubing on the other side of the block straight as the swage is formed.
- Do not tap the swage punch from side to side with the hammer. This will expand the swage too much and the space between the mating surfaces will be too large.
- Avoid placing the flaring block in a vise to hold the tubing while striking the swage punch.

7.15 COMPRESSION FITTINGS

Another popular method for joining sections of soft-drawn tubing materials is the compression fitting. This type of mechanical connection is primarily used on piping arrangements that carry water and are not intended for refrigerant piping circuits. The compression fitting is made up of a compression nut, a compression ring (also known as the ferrule), and the fitting itself. **Figure 7.57** shows the component parts of a compression union, which is used to join two sections of soft-drawn tubing.

To assemble a compression fitting, the compression nut is first placed over the tubing, with the female thread portion of the nut facing toward the end of the tubing section being joined, **Figure 7.58**. Then, the compression ring is placed over the tubing, **Figure 7.59**. When making the connection, it is important that the tubing be fully inserted into the fitting to ensure a proper seal. With the tubing fully inserted, the compression nut can be tightened to the fitting. When the nut is tightened, the ferrule is compressed between the nut and the fitting. This causes the ferrule to clamp onto the tubing, expand, and create a leak-free seal between the tubing, the nut, and the fitting itself. **Figure 7.60** shows the difference in appearance between a new ferrule and one that has been compressed.

Figure 7.57 Component parts of a compression fitting. Photo by Eugene Silberstein

Figure 7.58 Placing the compression nut on the tubing. Photo by Eugene Silberstein

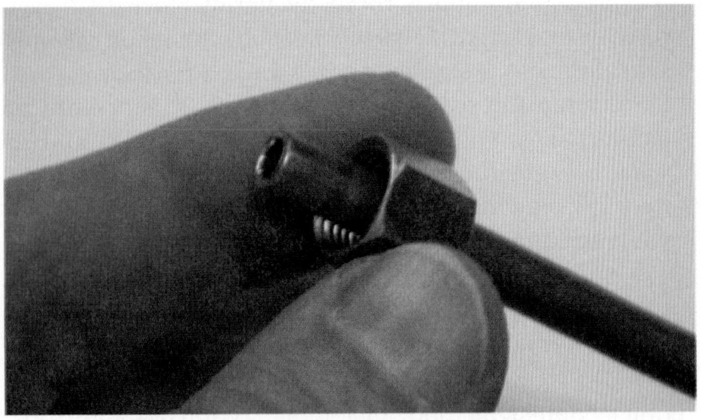

Figure 7.59 Placing the ring/ferrule on the tubing. Photo by Eugene Silberstein

Figure 7.60 A compressed ring (left) and a new, uncompressed ring (right). Photo by Eugene Silberstein

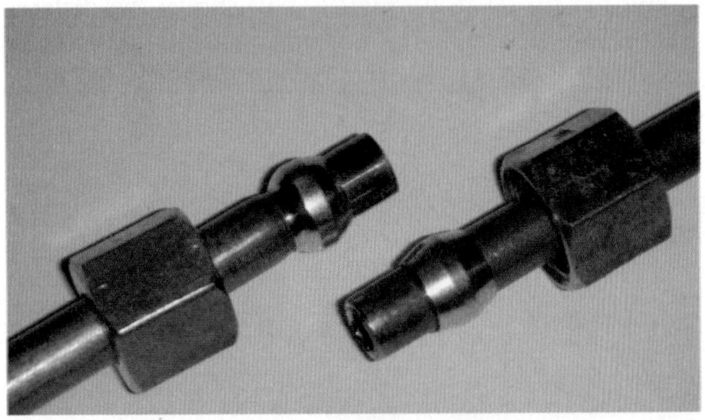

Teflon tape or other thread sealers should not be used on compression fittings because they can actually cause the joint to leak. If a compression fitting leaks, it is a good idea to replace the ferrule. Quite often, the ferrule can become damaged if the compression nut is tightened too much. This can cause the ferrule to become deformed and the connection to leak.

7.16 STEEL AND WROUGHT IRON PIPE

The terms *steel pipe*, *wrought steel*, and *wrought iron pipe* are often used interchangeably and incorrectly. When you want wrought iron pipe, specify "genuine wrought iron" to avoid confusion.

When manufactured, the steel pipe is either seam welded or produced without a seam by drawing hot steel through a forming machine. This pipe may be painted, left black, or coated with zinc (galvanized) to help resist rusting. Steel pipe is often used in plumbing, hydronic (hot water) heating, and gas heating applications. The size of the pipe is referred to as the **nominal** size. For pipe sizes 12 in. or less in diameter, the nominal size is approximately the size of the ID of the pipe. For sizes larger than 12 in. in diameter, the OD is considered the nominal size. The pipe comes in many wall thicknesses but is normally furnished in standard, extra-strong, and double-extra-strong sizes. **Figure 7.61** is a cross section showing the different wall thicknesses of a 2-in. pipe. Steel pipe is normally available in 21-ft lengths.

Figure 7.61 A cross section of standard, extra-strong, and double extra-strong steel pipe.

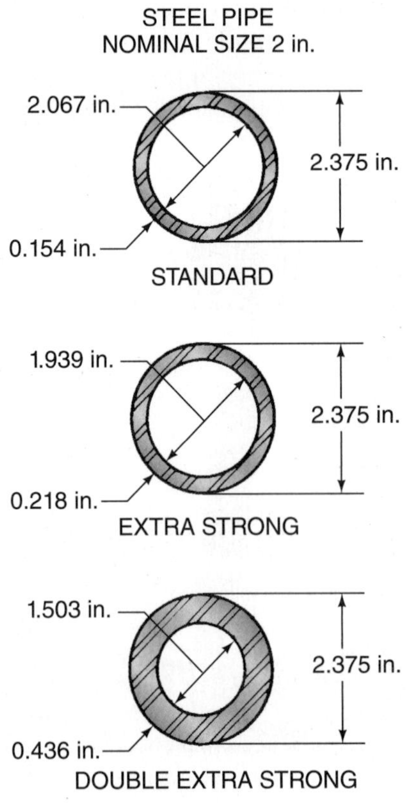

JOINING STEEL PIPE

Steel pipe is joined (with fittings) either by welding or by threading the end of the pipe and using threaded fittings. The two types of American National Standard pipe threads are tapered pipe and straight pipe. In the HVACR industry only the tapered threads are used because they produce a tight joint and help prevent the pressurized gas or liquid in the pipe from leaking. Pipe threads have been standardized. Each thread is V-shaped with an angle of 60°. The diameter of the thread has a taper of 3/4 in./ft or 1/16 in./in. There should be approximately seven perfect threads and two or three imperfect threads for each joint, **Figure 7.62**. Perfect threads must not be nicked or broken, or leaks may occur. Thread diameters refer to the approximate ID of the steel pipe. The nominal size then will be smaller than the actual diameter of the thread. **Figure 7.63** shows the number of threads per inch for some pipe sizes. A thread dimension is written as follows: first the diameter, then the number of threads per inch, then the letters *NPT*, **Figure 7.64**. You also need to be familiar with various fittings. Some common fittings are illustrated in **Figure 7.65**.

Four tools are needed to cut and thread pipe:

- A hacksaw (one with 18 to 24 teeth per inch) or pipe cutter is generally used to cut the pipe, **Figure 7.66**. A pipe cutter is best because it makes a square cut, but there must be room to swing the cutter around the pipe.
- A **reamer** removes burrs from the inside of the pipe after it has been cut. The burrs must be removed because they restrict the flow of the fluid or gas, **Figure 7.67**.

Figure 7.62 The cross section of a pipe thread.

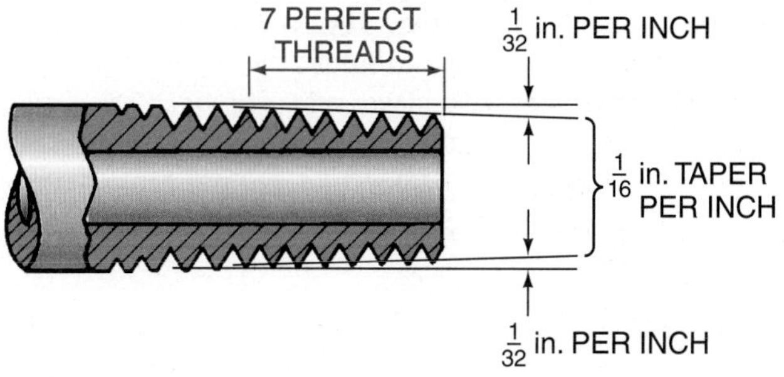

Figure 7.63 Number of threads per inch for some common pipe sizes.

PIPE SIZE (INCHES)	THREADS PER INCH
$\frac{1}{8}$	27
$\frac{1}{4}, \frac{3}{8}$	18
$\frac{1}{2}, \frac{3}{4}$	14
1 - 2	$11\frac{1}{2}$
$2\frac{1}{2}$ - 12	8

Figure 7.64 A thread specification.

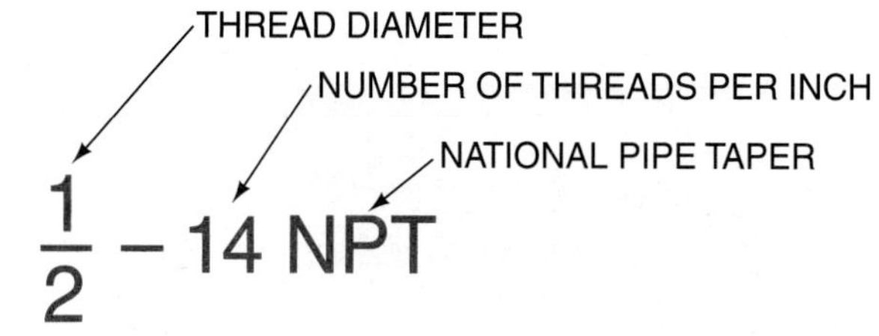

Figure 7.65 Steel pipe fittings. (A) A 90° elbow. (B) A union. (C) A coupling.

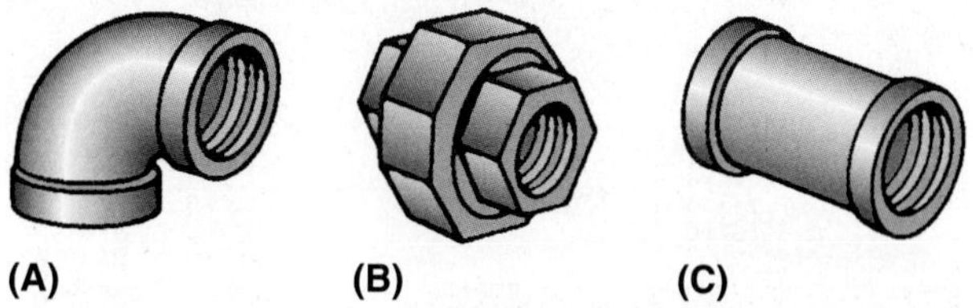

Figure 7.66 (A) A hacksaw. (B) and (C) Three- and four-wheel pipe cutters. Courtesy of Ridge Tool Company

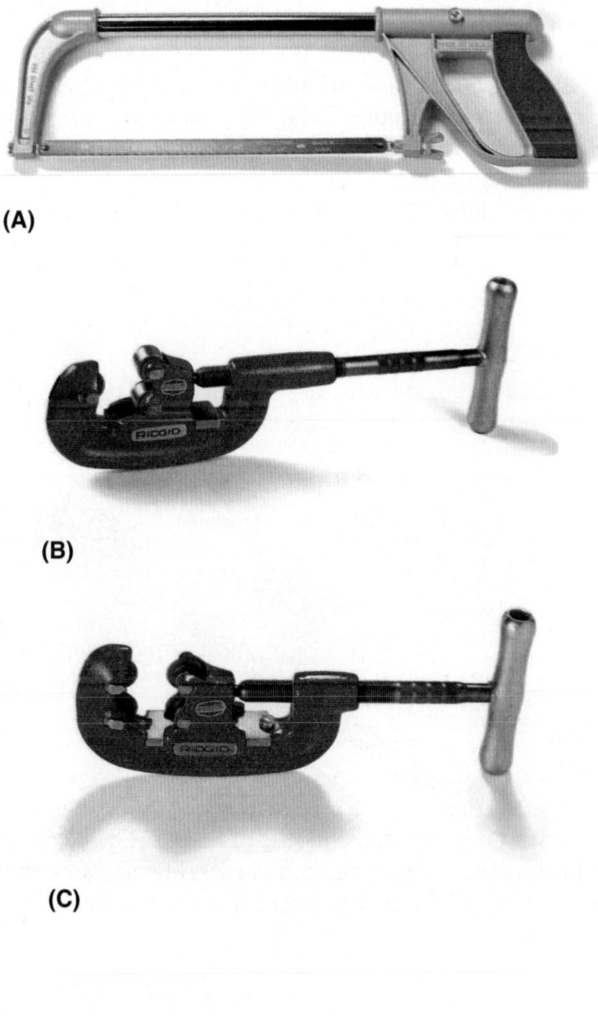

Figure 7.67 (A) A burr inside a pipe. (B) Using a reamer to remove a burr. Photo by Bill Johnson

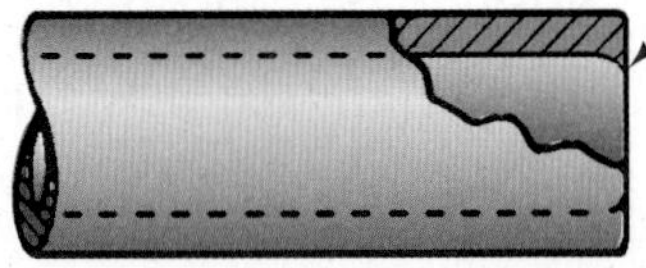

(A)

(B)

- A threader is also known as a **die**. The threader often has interchangeable heads to allow for the threading of different size pipes, **Figure 7.68**. The heads have sets of teeth that cut the threads in the end of the pipe, **Figure 7.69**.
- Holding tools such as the chain vise, yoke vise, and pipe wrench, **Figure 7.70**, are also needed.

When large quantities of pipe are cut and threaded regularly, special, automatic pipe-threading machines can be used.

Figure 7.68 A pipe threader with interchangeable heads. Photo by Eugene Silberstein

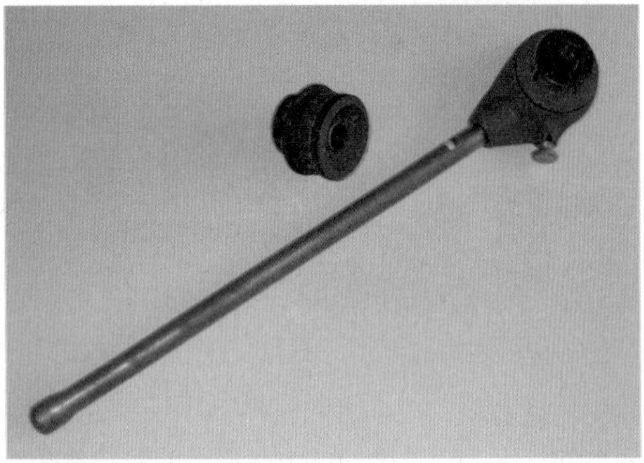

Figure 7.69 Teeth inside the threader heads that create the threads on the end of the pipe. Photo by Eugene Silberstein

CUTTING

The pipe must be cut square to be threaded properly. If there is room to revolve a pipe cutter around the pipe, you can use a one-wheel cutter, **Figure 7.71**. Otherwise, use one with more than one cutting wheel. Hold the pipe in the chain vise or yoke vise if it has not yet been installed. Place the cutting wheel directly over the place where the pipe is to be cut. Adjust the cutter with the T handle until all the rollers or cutters contact the pipe. Apply moderate pressure with the T handle and rotate the cutter around the pipe. Turn the handle about one-quarter turn for each revolution around the pipe. **SAFETY PRECAUTION:** *Do not apply too much pressure because it will cause a large burr inside the pipe and excessive wear of the cutting wheel.*•

To use a hacksaw, start the cut gently, using your thumb or a holding fixture to guide the blade, **Figure 7.72**. **SAFETY PRECAUTION:** *Keep your thumb away from the teeth.*• A hacksaw will only cut on the forward stroke. Do not apply pressure on the backstroke. Do not force the hacksaw or apply excessive pressure. Let the saw do the work.

REAMING

After the pipe is cut, put the reamer in the end of the pipe. Apply pressure against the reamer and turn clockwise. Ream only until the burr is removed, **Figure 7.67(B)**.

THREADING

To thread the pipe, place the die over the end and make sure it lines up square with the pipe, **Figure 7.73**. Apply cutting oil to the pipe and turn the die once or twice. Then reverse the die approximately one-quarter turn. This is accomplished by pulling out and reversing the direction of the arrowed knob on the threader, **Figure 7.74**.

Figure 7.70 (A) A tri-stand with a chain vise. (B) A tri-stand with a yoke vise. (C) A pipe wrench. Courtesy of Ridge Tool Company.

(A) **(B)** **(C)**

Figure 7.71 A single-wheel pipe cutter. Photo by Eugene Silberstein

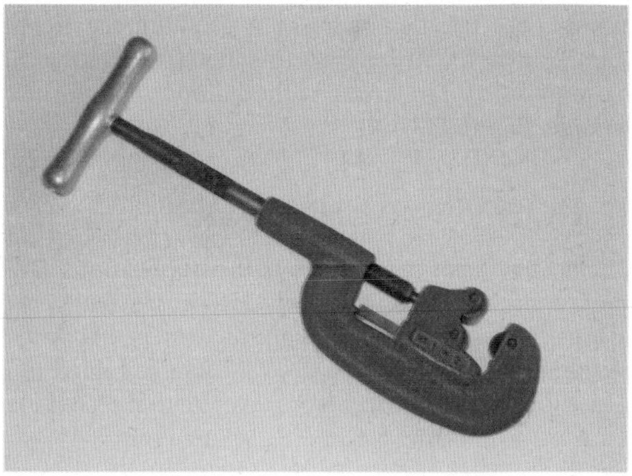

Figure 7.72 Cutting steel pipe with a hacksaw and a holding fixture. Photo by Bill Johnson

Figure 7.73 Threading pipe. Photo by Bill Johnson

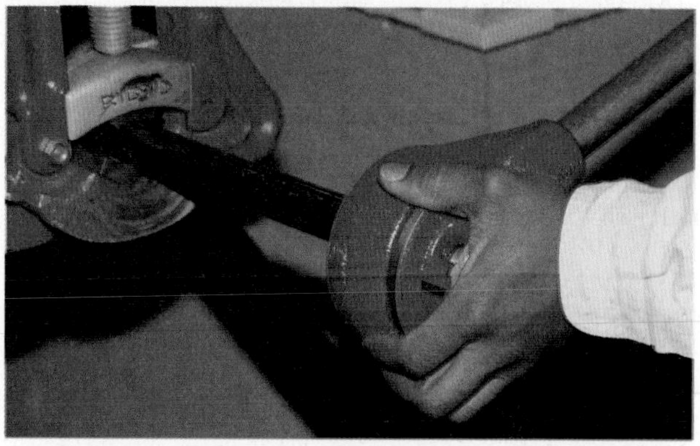

Figure 7.74 Knob used to reverse the direction of the threader. Photo by Eugene Silberstein

Rotate the die one or two more turns, and reverse again. Continue this procedure and apply cutting oil liberally until the end of the pipe is flush with the far side of the die.

7.17 INSTALLING STEEL PIPE

When installing steel pipe, hold or turn the fittings and the pipe with pipe wrenches. These wrenches have teeth set at an angle so that the fitting or pipe will be held securely when pressure is applied. Position the wrenches in opposite directions on the pipe and fitting, **Figure 7.75**.

When assembling the pipe, use the *correct* pipe thread dope on the male threads. Pipe dope is a thread-sealing compound that acts to both lubricate and seal the pipe connection. Do not apply the dope closer than two threads from the end of the pipe, **Figure 7.76**; otherwise, it might get into the piping system. All state and local codes must be followed. You should continually familiarize yourself with all applicable codes. **NOTE:** *The technician should also realize the importance of installing the pipe size specified. A designer has carefully studied the entire system and has indicated the size that will deliver the correct amount of gas or fluid. Pipe sizes other than those specified should never be substituted without permission of the designer.* •

Figure 7.75 Holding the pipe and turning the fitting with pipe wrenches. Photo by Bill Johnson

Figure 7.76 Applying pipe dope.

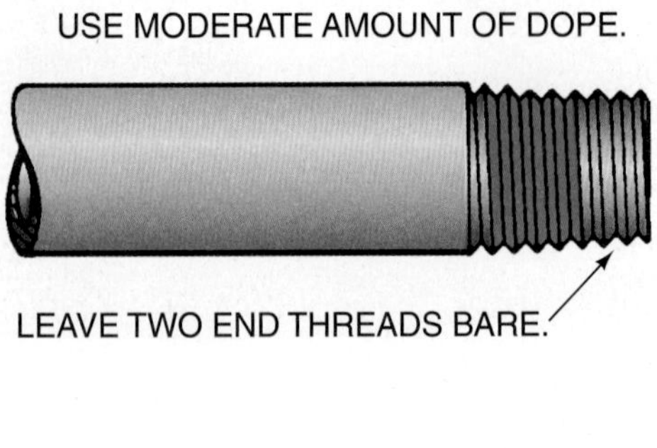

7.18 PLASTIC PIPE

Plastic pipe is used for many plumbing, venting, and condensate applications. You should be familiar with the following types.

ABS (ACRYLONITRILE BUTADIENE STYRENE)

ABS is used for water drains, waste, and venting. It can withstand heat to 180°F without pressure. Use a solvent cement to join ABS with ABS; use a transition fitting to join ABS to a metal pipe. ABS is rigid and has good impact strength at low temperatures.

PE (POLYETHYLENE)

PE is used for water, gas, and irrigation systems. It can be used for water-supply and sprinkler systems and water-source heat pumps. PE is not used with a hot water supply, although it can withstand heat under pressure. The pressure rating of PE tubing is typically lower than that of other plastic tubing types. It is flexible and has good impact strength at low temperatures. It normally is attached to fittings with two hose clamps. Place the screws of the clamps on opposite sides of the pipe, **Figure 7.77**.

PVC (POLYVINYL CHLORIDE)

PVC can be used in high-pressure applications at low temperatures. It can be used for water, gas, sewage, certain industrial processes, and in irrigation systems. It is a rigid pipe with a high impact strength. PVC can be joined to PVC fittings with a solvent cement, or it can be threaded and used with a transition fitting for joining to metal pipe.

CPVC (CHLORINATED POLYVINYL CHLORIDE)

CPVC is similar to PVC except that it can be used with temperatures up to 180°F at 100 psig. It is used for both hot and

Figure 7.77 The position of clamps on PE pipe.

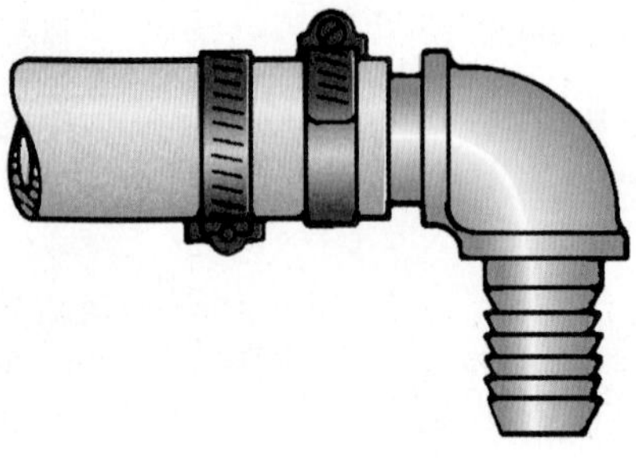

Figure 7.78 Cutting and joining PVC or CPVC pipe (A) – (D). Photos by Bill Johnson

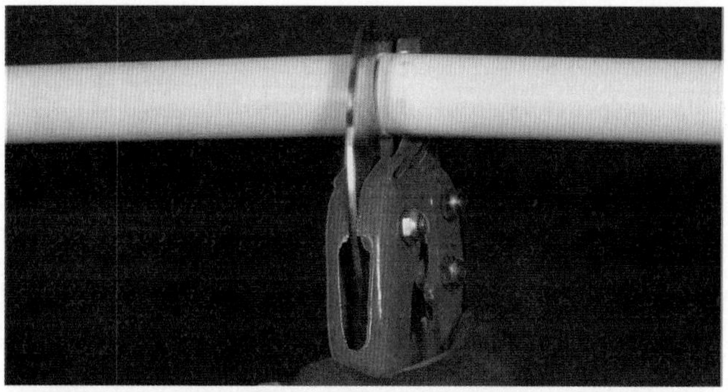

(A)

(B)

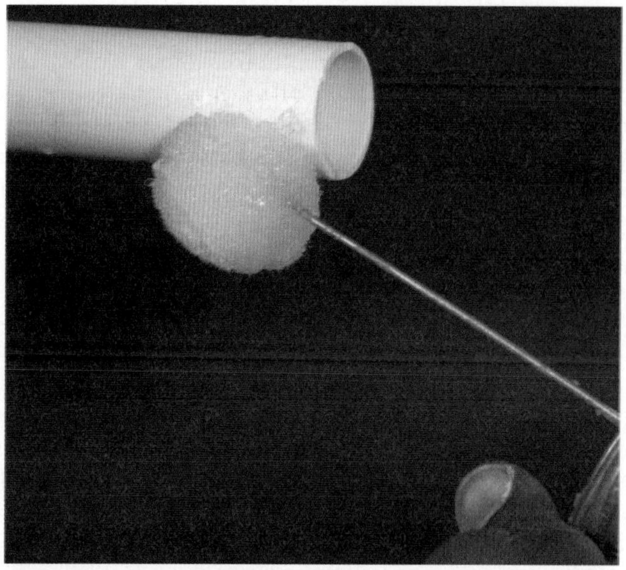

(C)

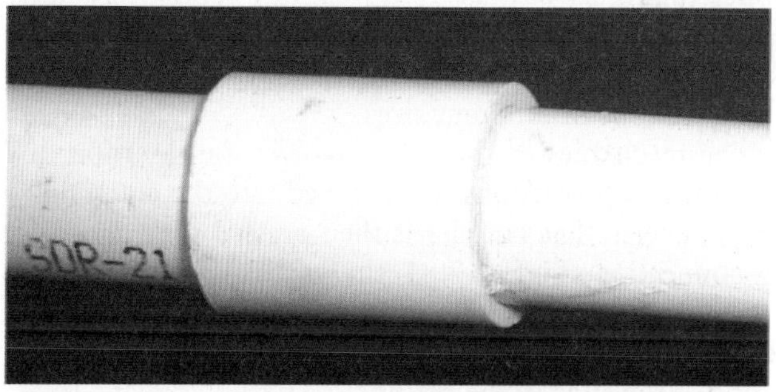

(D)

cold water supplies and is joined to fittings in the same manner as PVC. The following list and **Figure 7.78** describe how to prepare PVC or CPVC for joining.

A. Cut the end square with a plastic tubing shear, a hacksaw, or a tube cutter. The tube cutter should have a special wheel for plastic pipe.
B. Deburr the pipe inside and out with a knife or half-round file.
C. Clean the pipe end. Apply primer if required and cement to both the outside of the pipe and the inside of the fitting. (One-step primer/cement products are available for some applications. Follow instructions on primer and cement containers.) **NOTE:** *For certain PVC piping installations, typically household plumbing, some local codes require the use of purple primer,* **Figure 7.79**, *which will allow an inspector to quickly verify that the plastic pipe sections were primed before connected. Be sure to check the codes in your area to make certain you are using the correct primer.* •
D. Insert the pipe all the way into the fitting. Turn approximately one-quarter turn to spread the cement and allow it to set (dry) for about 1 min.

Schedule #80 PVC and CPVC can be threaded. A regular pipe thread die can be used. **NOTE:** *Do not use the same die for metal and plastic pipe. The die used for metal will become too dull to be used for plastic. The teeth on the plastic pipe die must be kept very sharp. Always follow the manufacturers' directions when using any plastic pipe and cement* •

Figure 7.79 A PVC connection that was primed with purple primer. Photo by Eugene Silberstein

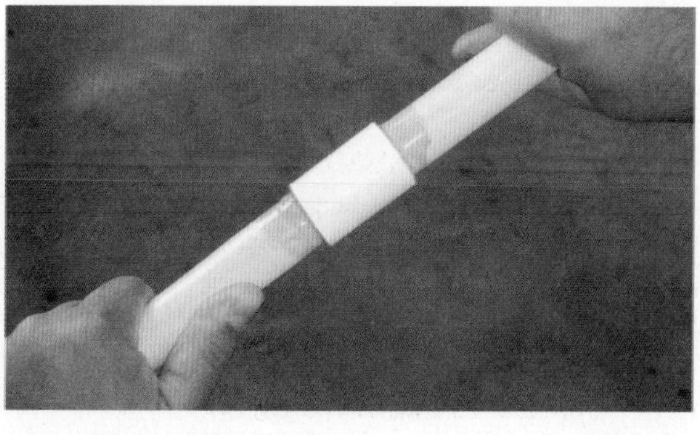

7.19 ALTERNATIVE MECHANICAL PIPING CONNECTIONS

In addition to the piping-connection methods already mentioned in this unit, a number of other alternative methods have become popular. Although each method is unique in that each is intended for specific purposes and applications, the methods are also similar in that they all require special fittings and/or equipment for making the necessary connections. Quite often, the equipment and the required fittings are far more expensive than traditional pipe fittings and connections, but the manufacturers of such product lines stress that these products result in substantial time savings that can easily translate into reliable and profitable piping systems. It is not the intention of this text to determine the reliability of these items, but to present them to the reader for the sake of comprehensiveness. Please refer to the manufacturers' literature for important information regarding applications, operating pressure ratings, and other safety-related issues.

One such mechanical piping connection is shown in **Figure 7.80**. This connector can be used with two sections of copper tubing as well as to join copper tubing to a section made of a different material, such as aluminum. There are adapters available that enable the service technician to join tubing sections of different sizes.

The mechanics of the joint involves the insertion of the ends of the tube sections to be joined into the coupling/connection device. The coupler is then pressed together to form the connection. **Figure 7.81** shows the connection before (top) and after (bottom) the fitting has been compressed. **Figure 7.82** illustrates how an installation tool is used to compress the ends of the fitting.

Other mechanical methods are intended to eliminate the use of soldering. As with the previous method, they both use special fittings to make the connection. The fitting shown in **Figure 7.83** is intended to receive sections of pipe that are simply pushed into it. The tabs located around the interior circumference of the fitting, along with the gasket material inside the fitting, provide the seal, **Figure 7.84**.

Figure 7.80 Fitting used to mechanically join two tubing sections. Photo by Eugene Silberstein

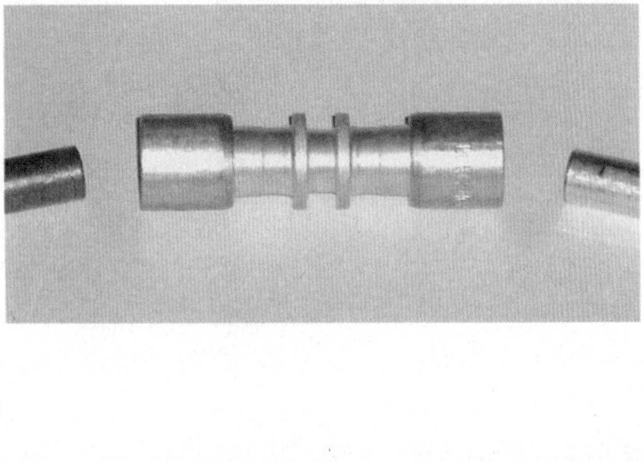

Figure 7.81 Tubing connector before (top) and after (bottom) compression. Photo by Eugene Silberstein

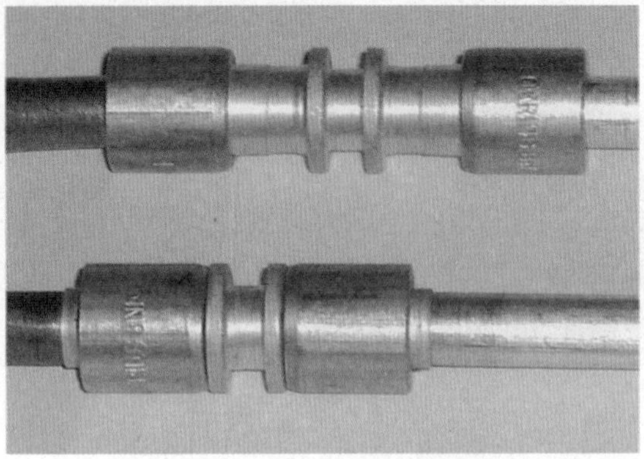

Figure 7.82 Process of compressing fitting onto two piping sections (A) – (C).

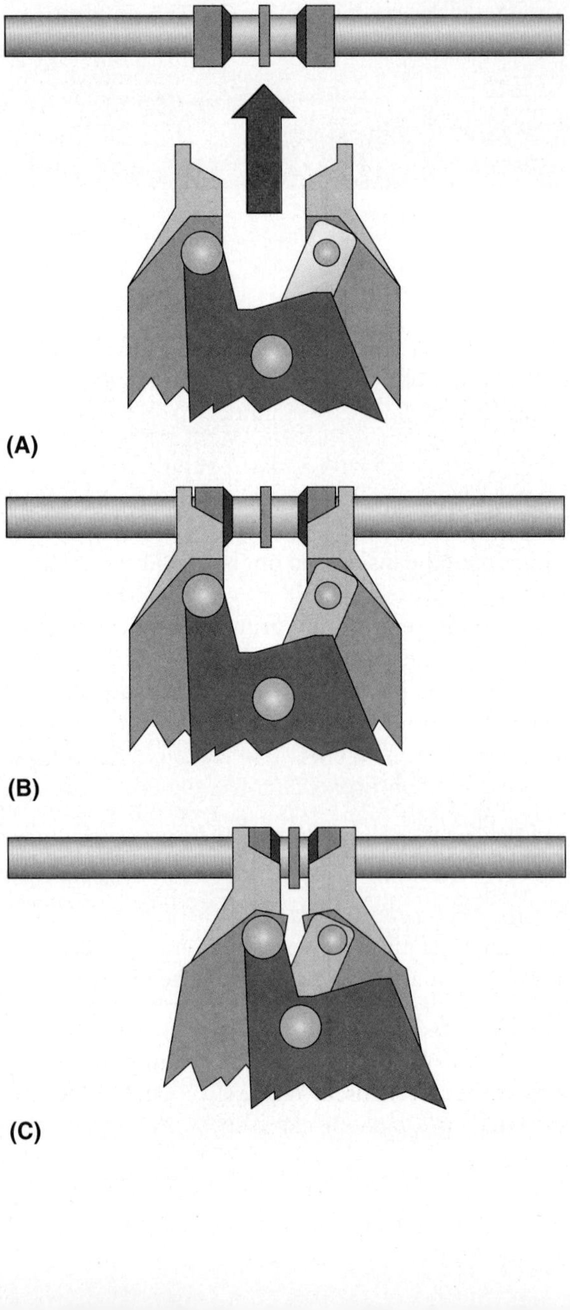

Figure 7.83 Push-on fitting with internal gasket and metal teeth for grabbing onto the pipe section. Photo by Eugene Silberstein

Figure 7.84 Pipe section pushed into the fitting. Photo by Eugene Silberstein

The connection shown in **Figure 7.85** is a press-on fitting. The fitting is placed on the pipe section and is then squeezed onto the pipe using a special tool, **Figure 7.86**. Internal gaskets in the fitting help reduce the possibility of leaks. The crimping tool used has interchangeable jaws, **Figure 7.87**, so that the same tool can be used on a number of different pipe sizes. As mentioned at the beginning of this section, please check with the individual manufacturer for information on pressure ratings and applications. Also, when installing mechanical fittings, be sure to follow the manufacturer's instructions carefully.

Figure 7.85 Cutaway view of press-type fitting installed on pipe. Courtesy of Viega.

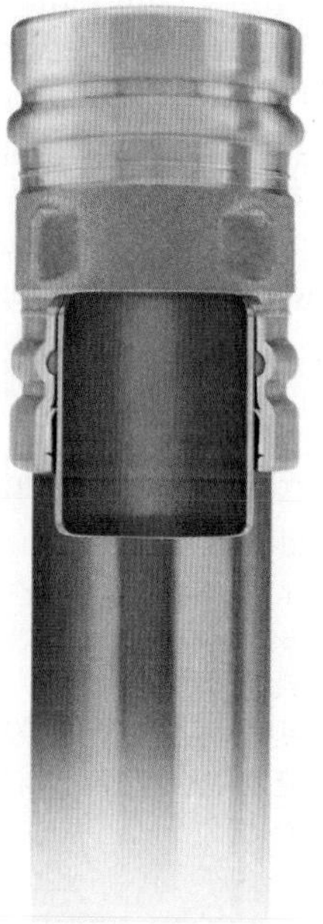

Figure 7.86 Crimping the press-type fitting to join two pipe sections. Courtesy of Ridge Tool Company.

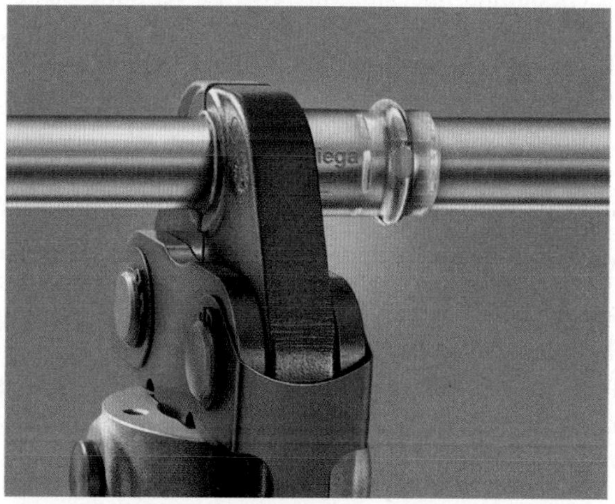

Figure 7.87 Various jaw sizes used for installing press-type fittings on pipes of different sizes. Courtesy of Ridge Tool Company.

REVIEW QUESTIONS

1. Correctly installed refrigerant piping circuits help prevent
   A. refrigerant loss and oil-related compressor failure.
   B. compressor oil return and refrigerant loss.
   C. refrigerant loss and excess refrigerant oil in the system.
   D. oil return and excess refrigerant in the system.
2. The standard weight of copper tubing used most frequently in the heating and air-conditioning industry is
   A. K.
   B. L.
   C. DWV.
   D. M.
3. The size of 1/2 in. would refer to the ________ with regard to copper tubing used in plumbing and heating.
   A. ID
   B. OD
   C. length
   D. fitting length
4. The size of 1/2 in. would refer to the ________ with regard to ACR copper tubing.
   A. ID
   B. OD
   C. length
   D. fitting length
5. The suction line on an air-conditioning or refrigeration system is insulated to
   A. encourage line sweating and prevent the absorption of additional heat.
   B. encourage line sweating and allowing additional heat absorption.
   C. prevent line sweating and prevent the absorption of additional heat.
   D. prevent line sweating while allowing the absorption of additional heat.
6. All of the following statements are true regarding line sets except
   A. they come with special quick-connect fittings.
   B. it is illegal to manufacture precharged line sets that contain R-22.
   C. they must be cut and brazed into the system.
   D. the set consists of an uninsulated liquid line and an insulated suction line.
7. The spring-type and lever-type are two kinds of tubing
   A. cutters.
   B. reamers.
   C. benders.
   D. crimpers.
8. Why is a hacksaw not the preferred tool to use to cut copper tubing?
   A. It takes longer to cut tubing with a hacksaw than with a tubing cutter.
   B. The hacksaw blade is not compatible with copper piping materials.
   C. The hacksaw cannot get into places that a tubing cutter can.
   D. The hacksaw cut produces a large amount of burrs and filings.
9. Of the following heat sources, which of them burns with the hottest flame?
   A. Handheld propane torch
   B. Handheld MAPP gas torch
   C. Air-acetylene torch
   D. Oxyacetylene torch
10. Flux is used to
    A. make the filler material melt at a lower temperature.
    B. make the filler material melt at a higher temperature.
    C. help prevent oxidation from forming.
    D. increase the rate of oxidation formation.
11. A common filler material used for brazing is composed of
    A. 50/50 tin–lead.
    B. 95/5 tin–antimony.
    C. 15% to 60% silver.
    D. cast steel.
12. The portion of a compression fitting that actually tightens against the tubing to create the leak-free seal is the
    A. union.
    B. coupling.
    C. compression nut.
    D. ring or ferrule.
13. What is the most common cause of leaking compression fittings?
    A. Cracked compression nut
    B. Overtightening the compression nut
    C. An improperly sized ring or ferrule
    D. Pushing the tubing all the way into the fitting
14. Which of the following statements is true regarding the flaring of copper tubing?
    A. The flare nut is placed on the end of the tubing after the flare has been created.
    B. The flare nut is placed on the end of the tubing before the flare has been created.
    C. The flare is created by hammering the end of the soft-drawn tubing.
    D. The flare is created by extending the tubing a distance equal to the diameter of the tubing from the flaring block.
15. The process of expanding one end of a section of tubing so that another section of same size tubing can be inserted into it is called
    A. flaring.
    B. bending.
    C. brazing.
    D. swaging.
16. When swaging the tubing should stick out the flaring block a distance that is approximately equal to
    A. the thickness of a nickel plus 1/2 in.
    B. the thickness of the tubing wall minus 1/16 in.
    C. the outside diameter of the tubing plus 1/8 in.
    D. the radius of the tubing plus 1/4 in.

17. Steel pipe is referred to as galvanized pipe when it is
    A. painted black.
    B. coated with zinc.
    C. coated with rust.
    D. left bare.
18. For black pipe sizes under 12 in., the nominal pipe size refers most closely to the
    A. thickness of the pipe wall.
    B. inside diameter of the pipe.
    C. outside diameter of the pipe.
    D. length of the pipe.
19. How many perfect threads should there typically be at the end of a threaded section of steel pipe?
    A. 5
    B. 7
    C. 9
    D. 11
20. Which two types of plastic pipe can typically withstand temperatures up to 180°F?
    A. ABS and CPVC
    B. ABS and PVC
    C. CPVC and PE
    D. PE and PVC
21. When joining sections of PVC pipe together with a union, it is always recommended to
    A. deburr the ends of the pipe and clean the pipe and fittings with primer.
    B. deburr the ends of the pipe and avoid using primer.
    C. leave remaining burrs in place and then clean the pipe and fittings with primer.
    D. leave remaining burrs in place and cement the sections together without cleaning.
22. The most important thing to keep in mind when using a press-type mechanical pipe connecting tool is to
    A. avoid inserting the pipe sections completely into the fittings.
    B. carefully read and follow all tool manufacturer's instructions.
    C. ensure that the pipe and fittings are out-of-round to ensure a leak-free connection.
    D. use a hammer to ensure that the pipe sections are secure in the fittings.

UNIT 8

LEAK DETECTION, SYSTEM EVACUATION, AND SYSTEM CLEANUP

OBJECTIVES

After studying this unit, you should be able to

- describe a standing pressure test.
- describe the six classes of leaks.
- explain the test procedures for evaporator and condenser section leaks.
- explain the test procedures for suction and liquid-line leaks.
- explain the test procedures for temperature-, pressure-, and vibration-dependent leaks.
- choose a leak detector for six types of leaks.
- describe the EPA regulations pertaining to system leak rates.
- explain the evacuation process and what it does to a refrigeration system.
- describe two reasons for performing a system evacuation.
- describe a deep vacuum.
- describe two different types of evacuation.
- describe two different types of vacuum measuring instruments.
- describe how wireless, Bluetooth technology can be used in an evacuation process.
- choose the correct vacuum pump for the evacuation process.
- list three of the proper evacuation practices and techniques.
- describe a deep-vacuum single evacuation.
- describe a triple evacuation.
- explain the process involved in cleaning a system after a hermetic motor burnout.

SAFETY CHECKLIST

- ✔ Care should be used while handling any of the products from a contaminated system because acids may be encountered.
- ✔ Do not place your hand over any opening that is under high vacuum because the vacuum may cause a blood blister on the skin.
- ✔ Wear goggles and gloves while transferring refrigerant.
- ✔ Do not allow mercury from any instrument to escape. It is a hazardous material.

8.1 LEAKS

All sealed systems leak. The leak could be 1 pound per second or as low as 1 oz every 10 years. Every pressurized system leaks because "flaws" exist at every joint fitting, seam, or weld. These flaws may be too small to detect with even the best of leak detection equipment. But given time, vibration, temperature, and environmental stress, these flaws become larger, detectable leaks, **Figure 8.1**.

It is technically incorrect to state that a unit has no leaks. All equipment has leaks to some degree. A sealed system which has operated for 20 years without ever having to be recharged is called a "tight system." The equipment still has leaks, but not large enough to read on a gauge or affect cooling performance. No pressurized machine is perfect. A leak is not some arbitrary reading on a meter. Gas escapes

Figure 8.1 (A) Electron micrograph of a "clean" silver soldered joint. Note the crack lines and other impurities of the metal. Source: Refrigeration Technologies, Anaheim, CA. (B) Further magnification shows actual metal separation. Source: Refrigeration Technologies, Anaheim, CA

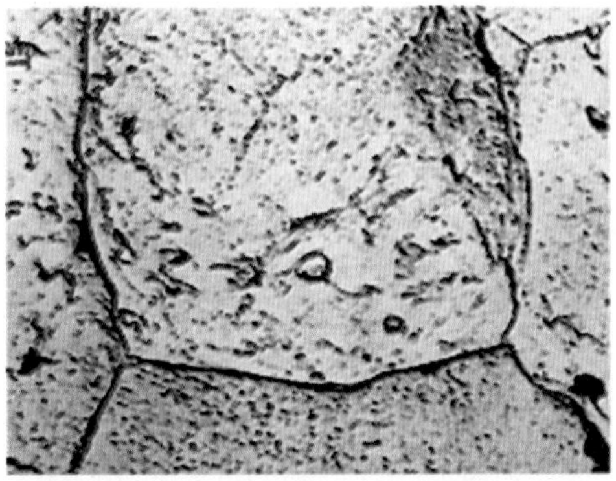

(A)

(B)

at different times and at different rates. In fact, some leaks cannot be detected at the time of the leak test. Leaks may plug, and then reopen under certain conditions. A leak is a physical path or hole, usually of irregular dimensions. The leak may be the tail end of a fracture, a speck of dirt on a gasket, or a microgroove between fittings.

EXPOSING THE LEAK SITE

Refrigerant vapor can flow under layers of paint, flux, rust, slag, and pipe insulation. Often, the refrigerant gas may show up quite a long distance from the leak site. This is why it is important to clean the leak site by removing loose paint, slag, flux, or rust. Pipe insulation, oil, and grease must also be removed from the site because they will contaminate the delicate detection tips of electronic detectors.

TYPES OF LEAKS

1. **Standing leaks.** Standing leaks are leaks that can be detected while the unit is at rest or off, including freezer evaporator coils warmed up by the defrost cycle. Standing leaks, fortunately, are the most common of all leaks.
2. **Pressure-dependent leaks.** Pressure-dependent leaks can only be detected as the system pressure increases. Nitrogen is used to pressurize systems to around 150 psig (low side) and 450 psig (high side). Never use air or pure oxygen. When necessary, a refrigerant trace gas can be introduced into the system along with the nitrogen. The trace gas allows electronic leak detectors to detect the vicinity of the leak. Refrigerant trace gas will be covered in more detail later in the unit. Pressure-dependent leak testing should be performed if no leaks are discovered by the standing leak test. A microfoam solution can also be used to locate pressure-dependent leaks, **Figures 8.2(A)** and **(B)**.

NOTE: *Mixtures of nitrogen and a trace gas of refrigerant, usually of the system's refrigerant, can be used as a leak test for gases, because in these cases the trace gas is not used as a refrigerant for cooling. However, a technician cannot avoid recovering refrigerant by adding nitrogen to a charged system. Before nitrogen is added, the system must be recovered and then evacuated to appropriate levels. Otherwise, the charge vented along with the nitrogen will be considered a refrigerant. This will constitute a violation of the prohibition on venting. The use of a chlorofluorocarbon (CFC) as a trace gas is not permitted.* •

3. **Temperature-dependent leaks.** Temperature-dependent leaks are associated with the heat of expansion. They are usually due to high-temperature ambient air or condenser blockages or occur during a defrost period.
4. **Vibration-dependent leaks.** Vibration-dependent leaks only occur during unit operation. The mechanical strain of motion, rotation, refrigerant flow, or valve actuation are all associated with vibration-dependent leaks.

Figure 8.2 (A) Microfoam solution used to detect and pinpoint leaks. Source: Refrigeration Technologies, Anaheim, CA (B) A foam "cocoon" that has formed on a receiver when the microfoam solution contacted a slow leak. Source: Refrigeration Technologies, Anaheim, CA

(A)

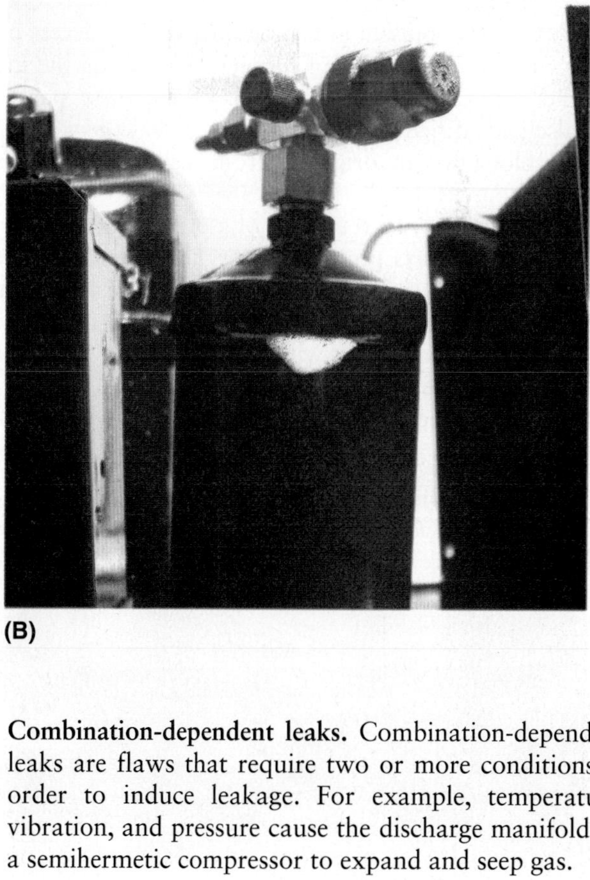

(B)

5. **Combination-dependent leaks.** Combination-dependent leaks are flaws that require two or more conditions in order to induce leakage. For example, temperature, vibration, and pressure cause the discharge manifold on a semihermetic compressor to expand and seep gas.

6. **Cumulative microleaks.** Cumulative microleaks are all the individual leaks that are too small to detect with standard tools. The total refrigerant loss over many years of operation slightly reduces the initial refrigerant charge. A system having many fittings, welds, seams, or gasket flanges will probably have a greater number of cumulative microleaks.

8.2 BASIC REFRIGERANT LEAK DETECTION

SPOTTING REFRIGERANT OIL RESIDUE FROM STANDING LEAKS

Successful leak detection depends solely on the careful observation made by the testing technician. Fortunately, all refrigeration systems internally circulate compressor oil with the refrigerant. Oil will blow off with the leaking refrigerant gas and "oil mark" the general area of leakage. Oil spots appear wet and have a fine coating of dust, **Figure 8.3**. The technician must determine whether the wetness is oil and not condensate. This can be accomplished by rubbing the area with your fingers and feeling for oil slickness. Oil spotting is the technician's first quick check but is not always reliable, for the following reasons:

1. Oil is always present at Schrader valves and access ports due to the discharging of refrigerant hoses on the manifold and gauge set, **Figure 8.4**. Often, these parts are falsely identified as the main point of leakage.
2. Oil blotches can originate from motors, pumps, and other sources.
3. Oil residue may be the result of a previous leak.
4. Oil is not always present at every leak site. It may take months, even years, of unit operation to cause enough oil blow-off to accumulate on the outer side.
5. Oil may not be present with microleaks.
6. Oil may not reach certain leak positions.
7. Oil may not be present on new start-ups.

Figure 8.3 A vibration absorber that has formed an oil spot with a fine coating of dust caused by a leak. Source: Refrigeration Technologies, Anaheim, CA

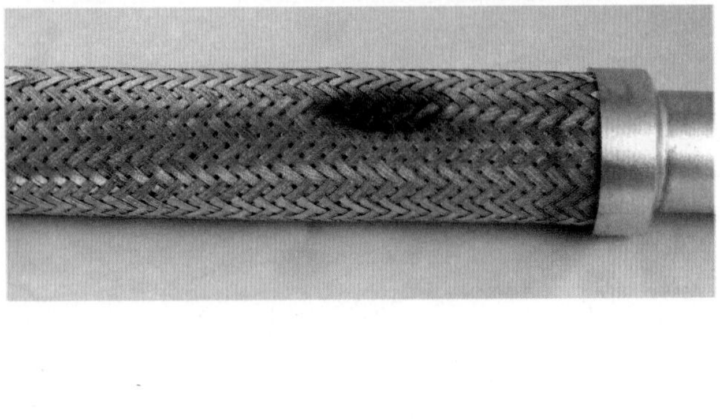

Figure 8.4 Oil on a service valve's access port. Source: Refrigeration Technologies, Anaheim, CA

TESTING FOR EVAPORATOR SECTION LEAKS

Many leaks that go undetected are in the evaporator coil. This is because most evaporator sections are contained in cabinets, buttoned-up, or framed into areas that do not allow easy access. To avoid the time-consuming labor of stripping off covers, ducting, and blower cages or the unloading of product, there is an easy electronic screening method that can be used:

1. Turn off all system power, including evaporator fan motors.
2. Equalize high- and low-side pressures in the refrigeration or air-conditioning system and defrost any frozen evaporator coils. (If the system does not have any pressure, add a refrigerant trace gas and then add nitrogen to generate a practical test pressure.) Most system low sides have a working pressure between 150 and 350 psig, but always read the name plate on the evaporator section for test pressure specifications, **Figure 8.5**.
3. Calibrate an electronic leak detector to its highest sensitivity.
4. Locate the evaporator drain outlet or downstream drain trap.
5. Position the electronic leak detector probe at the drain opening, **Figure 8.6**. Be careful that the leak detector probe does not come into contact with any water.
6. Test with an electronic leak detector for a minimum of 10 minutes or until a leak is sensed. Recalibrate the device and test again. Two consecutive "positive" tests confirm an evaporator leak. Two consecutive "negative" tests rule out an evaporator section leak.

Remember, refrigerant gas is heavier than air. Gravity will cause the gas to flow to the lowest point. If the evaporator

Figure 8.5 Design test pressure specification on an evaporator nameplate of 300 psig. Courtesy Ferris State University. Photo by John Tomczyk

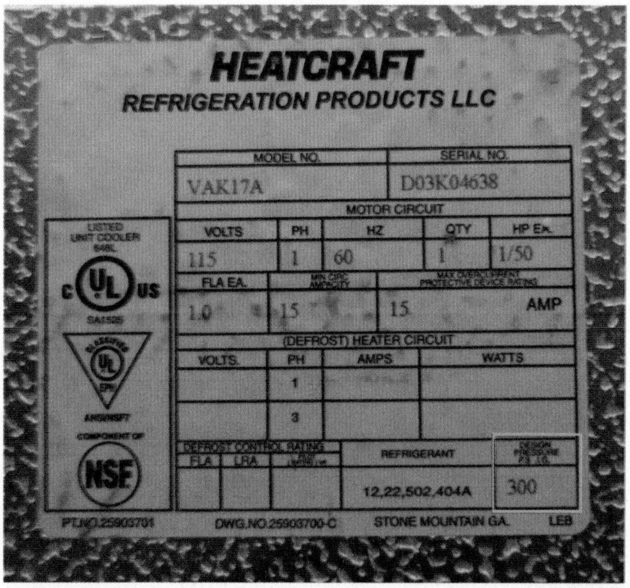

Figure 8.6 A service technician positioning an electronic leak detector at the drain opening of an evaporator drain. Source: Refrigeration Technologies, Anaheim, CA

section tests positive, the technician should expose the coil and spray coat all surfaces with a specially formulated bubble/foam promoter. Bubble/microfoam solutions are cost-effective and have been very successful in leak detection. They are superior to a soap and water solution for bubble checking because, as research has shown, a soap and water solution does not have the coagulants and wet adhesives the microfoam solutions do. In addition, household detergents often contain chlorides that will pit and corrode brass and iron. The specially formulated and patented bubble solutions form a foam "cocoon" when in contact with a leak, **Figure 8.2(B)** and **Figure 8.10**. When the solution is applied over the suspected leak area, bubbles or foam will tell the technician of the presence of a leak and its exact location. The bubbles must be allowed to stand for at least 10 to 15 minutes if small leaks are suspected. Even small leaks of less than a couple of ounces per year can often be found with these solutions. Bubble solutions can also be used when nitrogen or refrigerants are pressurizing the system. If the evaporator section tests negative for leaks, continue on to leak test the condensing unit.

NOTE: *Whenever applying a leak detection solution to any surface of a refrigeration or air-conditioning system, care must be taken to thoroughly wash the area with water. Simply wiping the area with a rag will not suffice. Depending on what the surface area is composed of, many leak detection solutions, like some soaps and water, will either oxidize, corrode, or rust the area of application. Often, brass and copper will turn green and oxidize when exposed to some soap solutions if they are not thoroughly washed with water and then dried. Also, any residue of some leak detection fluids may collect dust and appear to be a system leak.* •

TESTING FOR CONDENSING SECTION LEAKS

1. Calibrate an electronic leak detector to its highest sensitivity and place the probe at the base of the unit, usually under the compressor. The unit should be fully pressurized.
2. Cover the condensing unit with a cloth tarp or bed sheet to serve as a barrier against any outside air movement and to trap refrigerant gas, **Figure 8.7**. Do not use a plastic material because some plastics may set off some electronic leak detectors and give a false reading.
3. Monitor for leakage for 10 minutes or until a leak is sensed. Recalibrate and test again. Two consecutive positive tests confirm condensing section leakage. Two consecutive negative tests rule out a detectable leak.

Figure 8.7 A service technician has covered the condensing unit with a sheet to serve as a barrier against any outside air movement before using an electronic leak detector. Never use a plastic cover because some plastics will set off certain electronic leak detectors and give a false reading. Source: Refrigeration Technologies, Anaheim, CA

Figure 8.8 A service technician using an electronic leak detector to check for leaks on the bellows of a pressure control. Courtesy Ferris State University. Photo by John Tomczyk

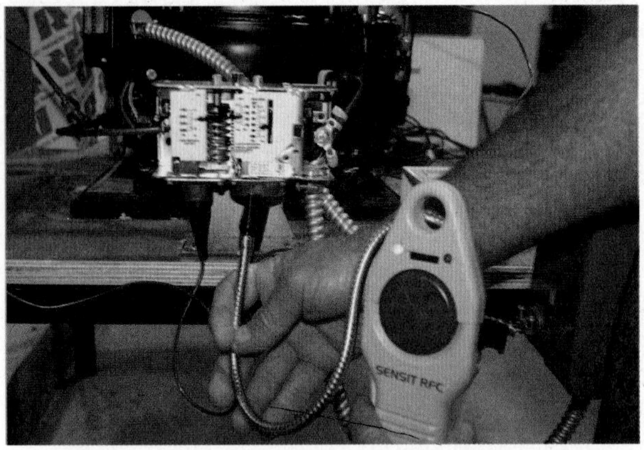

4. Use the electronic leak detector to check for leaks on the bellows of the pressure controls, **Figure 8.8**. Remove the control box cover and place the probe within the housing. Cover the control tightly with a cloth barrier and monitor for 10 minutes, as above.
5. If the results are positive, uncover the equipment and begin spray coating with a microfoam solution, **Figure 8.2(A)**. If the results are negative, continue to the suction/liquid-line leak test described next.

TESTING FOR SUCTION- AND LIQUID-LINE LEAKS

The longer the tubing runs are between the evaporator and the condensing unit, the greater the odds of defects. Do not rule out any possibilities, whether it be a typical sight glass–drier connection leak to a poor solder joint hidden under pipe insulation.

1. The suction line can be screened by calibrating an electronic leak detector to its highest sensitivity.
2. Tuck the probe underneath the pipe insulation, **Figure 8.9**. Monitor for 10-minute intervals while the system is at rest and fully pressurized to equalization. It may be necessary to insert the probe at several downstream points.
3. If a leak is sensed, strip off the insulation and apply a bubble/foam solution to all surfaces. If no leak is positively screened, test the liquid line.

Figure 8.9 The probe of an electronic leak detector tucked underneath refrigerant pipe insulation sniffing for refrigerant leaks. Source: Refrigeration Technologies, Anaheim, CA

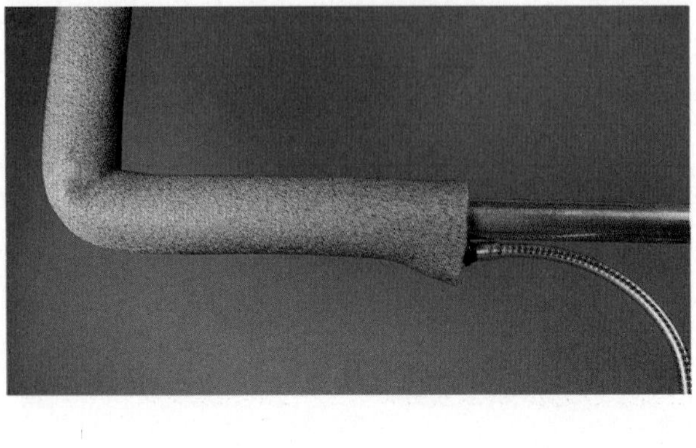

8.3 ADVANCED LEAK DETECTION

TESTING FOR PRESSURE-DEPENDENT LEAKS

1. When testing for pressure-dependent leaks, the unit data plate should be referenced to determine the maximum allowable test pressure for each side of the system. On systems where the high and low sides cannot be isolated from each other, the entire system should be pressurized with dry nitrogen to the lowest pressure listed on the data plate.

SAFETY PRECAUTION: *Never use pure oxygen or air to raise the pressure in a refrigeration system. Pure air contains about 20% oxygen. The pure oxygen and/or the oxygen in the air can combine with refrigerant oil to form an explosive mixture. Pure oxygen and the oxygen in the air will oxidize the system's oil rapidly. In a closed system, pressure from the oxidizing oil can build up rapidly and may generate pressures to the point of exploding. Some refrigerants can become explosive under pressure when mixed with air or oxygen. For example, HFC-134a (R-134a) is combustible at pressures as low as 5.5 psig (at 350°F) when mixed with air at concentrations of more than 60% air by volume. At lower temperatures, higher pressures are required to support combustion. Therefore, air should never be mixed with R-134a for leak detection.* •

2. Always bubble test by thoroughly saturating all surfaces with a microfoam solution. Allow up to 15 minutes for the microfoam to expand into a visible white "cocoon" structure, **Figure 8.2(B)**. Use an inspection mirror to view any undersides and a light source for dark areas.
3. Starting at the compressor, coat all suspected surfaces. Continue to coat all suction-line connections back to the evaporator section.
4. Spray coat all fittings starting at the discharge line at the compressor to the condenser coil. Spray coat all soldered condenser coil U-joints.
5. From the condenser, continue to spray coat all liquid-line connections, including the receiver, valves, seams, pressure taps, and any mounting hardware. Continue the liquid-line search back to the evaporator section.
6. Any control line taps to the sealed system must be spray coated the entire length of their run all the way back to the bellows device.
7. Expose the evaporator section and coat all connections, valves, and U-joints.

Notice that the first sequence of searching started with the compressor and suction line due to their large surface areas.

The next sequence began with the discharge line, went across the condenser to the liquid-line connection, and then to the evaporator section. The evaporator section is the last and least desirable component to pressure test in the field.

TESTING FOR TEMPERATURE-DEPENDENT LEAKS

All mechanical connections expand when heated. The connections on refrigeration and air-conditioning systems are usually made of soft metals, such as copper, brass, or aluminum. These metals actually warp when heated, then contract and seal when heat is removed.

1. Start the unit and raise the operating temperature by partially blocking the condenser's air intake. Warm water may also be used to pressurize the system. For example, water chillers are usually pressurized with controlled warm water by turning off the valves for the condenser and evaporator water circuits and introducing controlled warm water to the evaporator tube bundle. This causes the rate of vaporization of the refrigerant to increase, causing higher pressures in the evaporator. One must slowly control the amount of warm water introduced to avoid temperature shock to the evaporator. The rupture disc on the evaporator may open if the pressures are raised too high. Special fittings are available from chiller manufacturers to equalize pressure inside and outside of the rupture disc to prevent rupture. Please consult with the manufacturer before attempting to service or leak check any chiller.

   An electronic leak detector may be used while the system is running. However, running systems usually produce a lot of fast air currents from fans and motors that may interfere with electronic detection. It helps to cover the unit with a blanket or sheet to try to collect escaping refrigerant gases. The leaking refrigerant will be easier to pick up with an electronic detector if it can collect somewhere, instead of being dissipated by air currents.
2. Spray coat all metal connections one at a time with a microfoam solution, **Figure 8.2(A)**, and observe for leakage. Rewet any extremely hot surfaces with water to keep the fluid from evaporating too quickly.
3. When testing evaporator components, heat may be induced by placing the unit into defrost.

TESTING FOR VIBRATION-DEPENDENT LEAKS

Leaks that only occur while the unit is in operation are the rarest of all leaks. However, studies have shown that certain components and piping on refrigeration units will develop vibration leaks. They are caused by cracks that open and close from physical shaking.

An electronic leak detector or a microfoam solution can be used while the unit is running. Again, drafts have to be minimized when the unit is running. If an electronic detector is used before the microfoam, a blanket or sheet should be used to help collect escaping gases and minimize air currents. When using microfoam, start the unit and spray coat the areas in the following list with the solution. Look for foam cocoon formations or large bubbles, **Figure 8.10** and **Figure 8.11**. Large bubbles will form on larger leaks, and foam cocoons will form on small leaks.

- All compressor bolts and gasket edges
- Suction-line connection at compressor

Figure 8.10 Foam "cocoons" of a microfoam solution on a small leak on the U-bends of a coil. Some microfoam solutions will detect and pinpoint leaks as small as 0.5 oz/year. Source: Refrigeration Technologies, Anaheim, CA

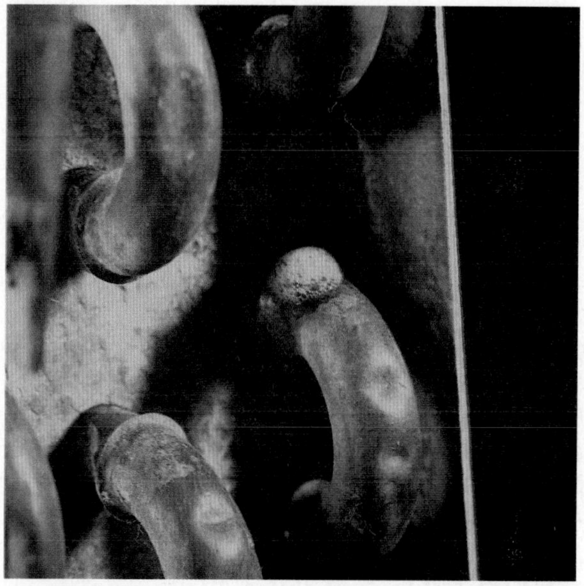

Figure 8.11 A microfoam solution pinpointing a larger leak at the outlet of a TXV. The presence of dirt, oil, or water does not affect bubble production with this microfoam solution. Source: Refrigeration Technologies, Anaheim, CA

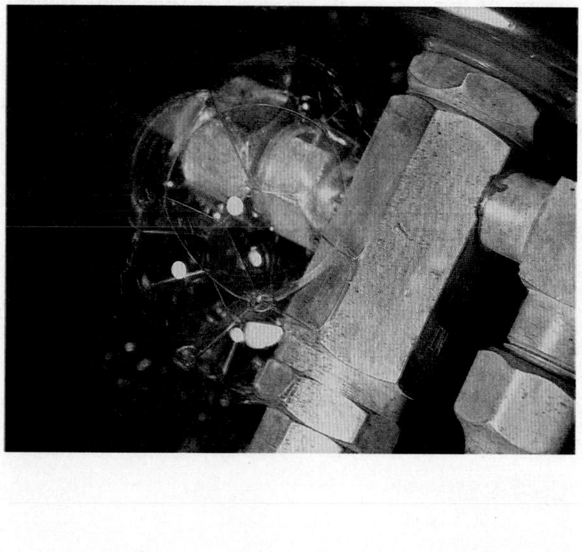

- Suction-line connection at evaporator
- Discharge-line connection at compressor
- Discharge-line connection at condenser
- Vibration eliminators
- Any joint or fitting on unsupported pipe runs
- Expansion and solenoid valves
- Capillary tube connections
- Sight glass
- All threaded or flare connections

TESTING FOR COMBINATION-DEPENDENT LEAKS

Combination-dependent leak checking involves overlapping the procedures already mentioned. At least two, and usually three, procedures should be merged into one. This type of testing requires a high order of skill and observation techniques. Each suspected component must be isolated and tested in the following manner:

1. Subject a valve or fitting to high pressure.
2. Spray coat the valve or fitting.
3. Tap the component repeatedly with a rubber mallet to induce vibration. If no leaks are present, proceed to the next step.
4. Gently add heat to the component. If no leaks are found, continue on to another component.

SAFETY PRECAUTION: *Do not use flames or halide torches to search for leaks. Open flames (halide flame torches or brazing torches) can release large quantities of acidic compounds in the presence of all refrigerants, and these compounds can be hazardous. Halide torches are not effective on HFC refrigerants because they are used to detect chlorine, which is not present in HFC refrigerants.* •

TESTING FOR CUMULATIVE MICROLEAKS

Cumulative microleaks are detected using a helium mass spectrometer. Such superfine leak testing is beyond the normal operations of the service technician. At this time, microleaks are considered an acceptable amount of leakage in the industry. *Installation and service personnel must learn good field piping and assembly practices to ensure that systems will run efficiently as long as possible. Many field-installed systems last for 50 or more years because of good practices. It all begins with technicians knowing what they are doing when they are installing systems. After the installation or service, a proper leak test must be performed.*

SAFETY PRECAUTION: *Never use handheld leak detectors to check for breathable air in enclosed working spaces. These detectors are not designed to determine if the air is safe to breathe. Use oxygen monitors to ensure that adequate oxygen is available to sustain life.* •

SAFETY PRECAUTION: *If you detect a visible change in the size or color of the flame when using brazing torches to repair equipment, stop work immediately and leave the area. Ventilate the work area well and stop any refrigerant leaks before resuming work. These flame effects may be an indication of very high refrigerant concentrations, and continuing to work without adequate ventilation may result in injury or death. Overexposure to high concentrations of refrigerant vapor can cause asphyxiation and cardiac arrest. Please read all safety information before handling any refrigerant.* •

8.4 STANDING PRESSURE TEST

The standing pressure test is probably the most frequently used method of leak detection. The technician can start with a standing pressure test for a newly assembled system to see if it will hold pressure. At the same time, the system can be checked with a sensitive leak detector. The technician should assemble the system first and then give it a visual test to make sure that all connections look as they should.

- The solder connections should have no gaps.
- All flanges and threaded connectors should be tight.
- All control valves must be installed in the correct direction and correctly set.
- All valve covers must be on.

There are several leak-checking procedures. The first test is to charge a trace amount of refrigerant into the system. Most technicians start at 0 psig and introduce a refrigerant trace gas into the system until the pressure reads about 10 psig. The trace gas should be the same refrigerant that is used in the system to avoid any cross-contamination of residual refrigerant once the leak-checking process is complete. Using the system's refrigerant type as a trace gas also avoids any refrigerant and oil incompatibility. The refrigerant trace gas should enter the system from the middle hose of the gauge manifold, and be evenly distributed in both the low and high sides of the system. This practice is acceptable to the Environmental Protection Agency (EPA) because there must be some refrigerant in the system to detect a leak electronically. The refrigerant trace gas is being used only to detect leaks, not as a refrigerant.

When the trace refrigerant has entered the system, dry nitrogen is then introduced to the level of a practical test pressure. The low-pressure side of most systems has a working pressure of 150 psig. Some of the new refrigerants have a higher working pressure; therefore, the system components have a higher working pressure, **Figure 8.5**. The technician should check each system for the proper working pressure before pressurizing it. The components on the low-pressure side of the system are the evaporator, suction line, and compressor. The only parts of the compressor that are considered to be on the high-pressure side are the compressor head and the discharge piping, most of which is internal to the compressor. The compressor is considered a low-pressure device because the compressor shell is on the low-pressure side of the system. The high-pressure side of the system consists of

Figure 8.12 Isolating the compressor to pressurize the condenser and evaporator.

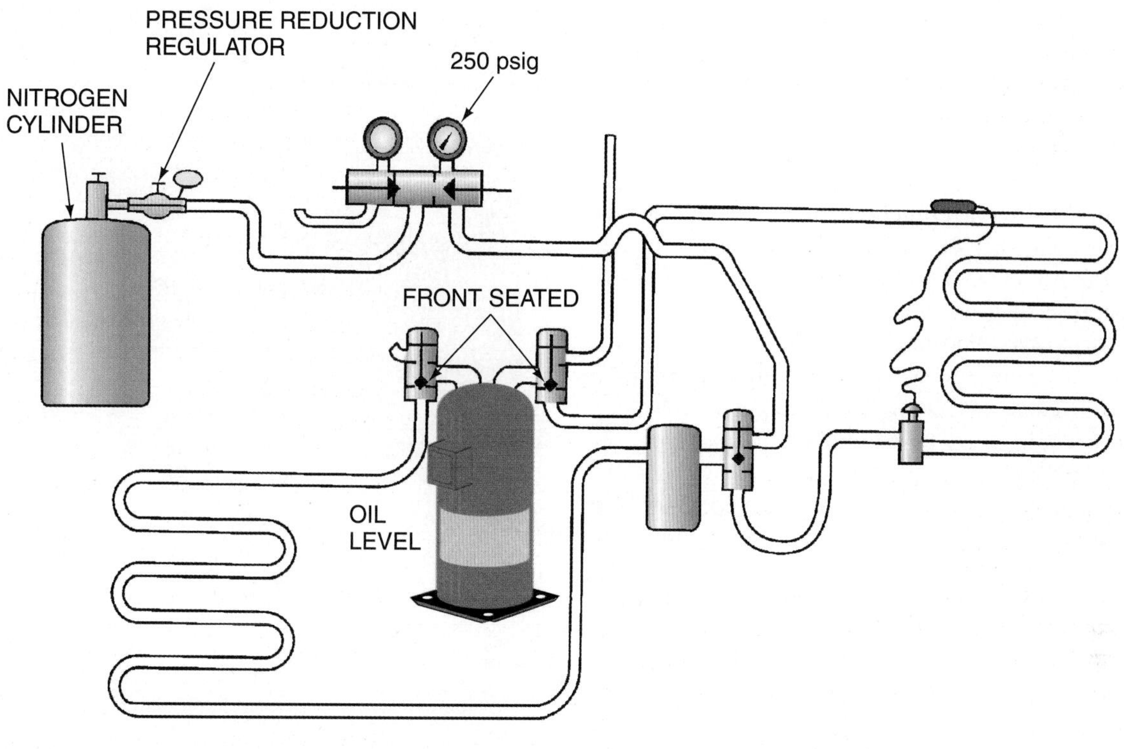

the discharge line to the condenser, the condenser, and the liquid line. The test pressure for the high-pressure side for most refrigerants is at least 350 psig when the compressor shell is not pressurized. Systems using R-410A may have working pressures up to 450 psig. If the compressor has service valves, it can be isolated, **Figure 8.12.**

The pressure used to pressurize a complete system must not exceed the lowest test pressure for the system. Once pressurized, the system should be allowed to stand for about 10 minutes to allow pressures in the system to equalize. At this time, the technician should listen to each connection for an obvious leak. Once the pressures have equalized, the high-pressure gauge should be marked after tapping it to settle the needle down.

The standing pressure test should be left on the equipment for as long as practical. A 1½-ton package unit that has a small refrigeration circuit may need to stand for 1 hour. A larger system, such as a 10-ton split system, should be allowed to stand overnight, or for 24 hours, if possible. The longer the system stands under pressure without a drop in pressure, the more confidence the technician may have that the system is leak-free. If there is a drop in pressure, a very close leak test should be conducted unless there is a significant drop in the ambient (surrounding) temperature. With a refrigerant trace gas in the system, an electronic leak detector can now be used to locate the vicinity of the leak. Once the vicinity of the leak is found with the electronic leak detector, a bubble/microfoam solution can pinpoint its exact location.

LEAK DETECTION METHODS

The most basic but effective method of detecting leaks is to just listen for the rush of nitrogen from the connections. Often, leaks can be heard but not pinpointed. When a leak is large enough to be heard, you can often find it by wetting the back of your hand and moving it around where you suspect the leak to be. The leak will spray vapor on your wet hand and evaporation will cool your hand and help locate the leak.

Halide leak detectors, like the one shown in **Figure 8.13,** are capable of detecting a leak as small as about 1 oz per year.

Figure 8.13 A halide torch for leak detection. Photo by Bill Johnson

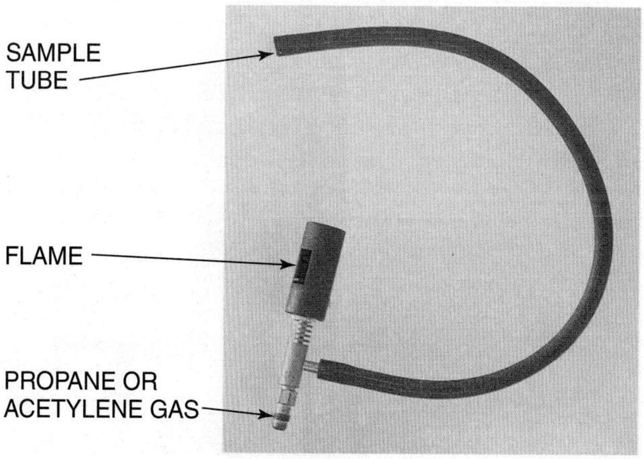

They are powered by a propane or acetylene gas flame which creates a draft that draws a sample of whatever is in the vicinity of the sample tube. The halide detector, which has been around for a long time, is considered primitive by many technicians. A sample of refrigerant, for example, passes over a glowing copper element and causes the flame to change color. This change in flame color can best be viewed where there is very little light. This detector is not very effective in bright sunlight and can only be used with refrigerants that have chlorine as part of their formula, such as CFCs and HCFCs. The halide leak detector will not detect hydrofluorocarbon (HFC)-based refrigerants because they do not contain chlorine. It cannot be used in an explosive atmosphere, such as around gasoline or other flammables, because of the open flame. **⚠ SAFETY PRECAUTION:** *Research has shown that some refrigerants and refrigerant blends are flammable when mixed with air. Never use a halide leak detector with these refrigerants. Never use a halide leak detector with hydrocarbon (HC) and hydrofluoro-olefin (HFO) refrigerants.*• Also, open flames can release large quantities of acidic compounds in the presence of all refrigerants, and these compounds can be hazardous. It is for these reasons that the use of the halide torch for leak checking is not strongly recommended, especially when ultrasensitive electronic leak detectors and microfoam bubble solutions are available.

Ultrasonic leak detectors have earphones, a microphone, and an amplification system to listen for the high-pitched sound of a leak. These detectors would be very good in a place where there is a lot of refrigerant in the air, **Figure 8.14**.

Ultraviolet (UV) fluorescent leak detectors work by means of a special fluorescent dye that is charged into the system. The dye mixes with the oil that circulates with the refrigerant and glows under black light when it escapes through the leak, **Figure 8.15**.

Figure 8.14 An ultrasonic leak detector that listens for leaks. Photo by Bill Johnson

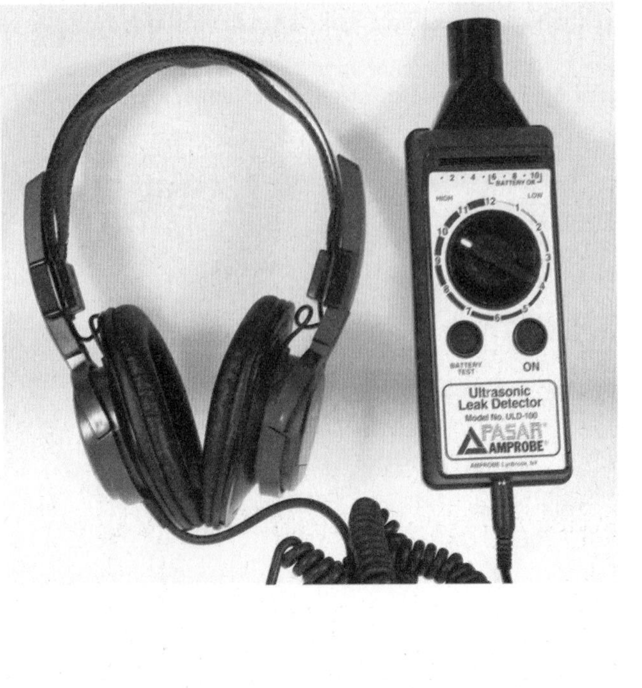

Figure 8.15 (A) An ultraviolet leak detector kit that uses black light and a special oil-specific dye. (B) A service technician injecting the oil-specific dye into the refrigeration or air-conditioning system through the manifold gauge set. (C) A service technician using a black light to locate a refrigerant leak. All photos Courtesy Ritchie Engineering Company, Inc., Yellow Jacket Products Division.

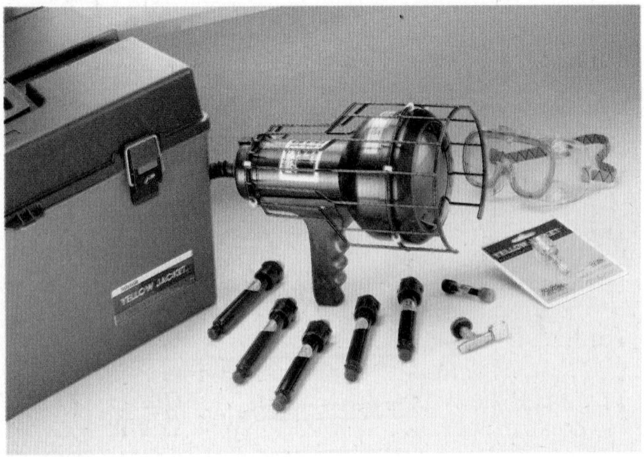

(A)

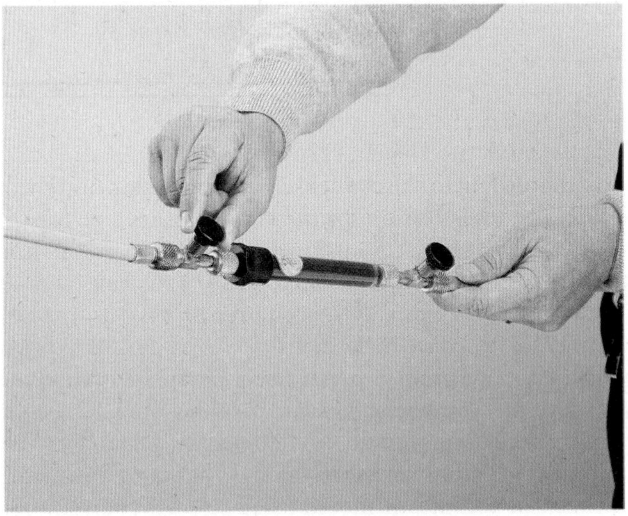

(B)

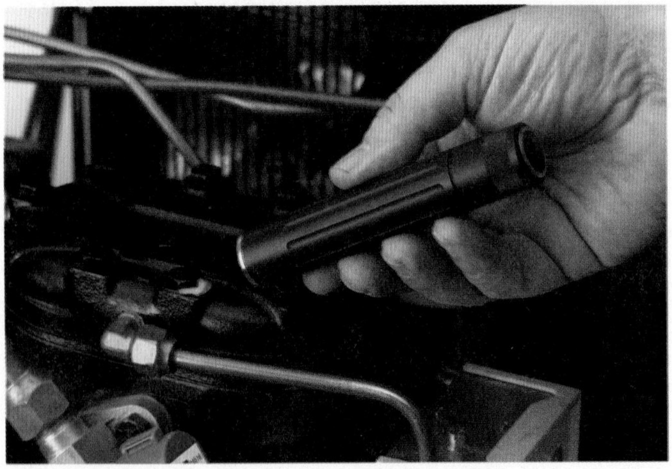

(C)

Developed for the auto industry to find leaks in automatic transmissions, this technology is very effective, particularly in older systems that may be oil-soaked from prior leaks.

Dye-saturated oil is a leak detection system that causes a red dye residue to seep out wherever there is a leak and escaping oil. The dye is normally charged into the system with a special factory-charged refrigerant cylinder. Before performing this test, you should consult with the compressor manufacturer to make sure the dye is approved, particularly while the system is under warranty. If the compressor is beyond its warranty, it belongs to the customer. Most dyes are lubricant-specific, meaning a different type of dye has to be used with mineral oil, alkylbenzene, and ester-type lubricants. Notice the different oil-specific fluorescent dye tubes in the foreground of **Figure 8.15(A)**. **Figure 8.15(B)** shows a technician injecting an oil-specific fluorescent dye from a gauge line into a refrigeration system for leak detection.

Electronic leak detectors, the most accurate and probably the most often used detectors, will detect leaks as low as about 1/4 oz per year, **Figure 8.16**. They are great for certain applications. When there is a lot of refrigerant in the air, electronic detectors may have a hard time locating where to start unless they have a feature that allows them to correct for background refrigerant. The sensitivity of these instruments can be their greatest asset or their greatest liability, depending on whether the technician takes the time to understand them. Most of these detectors have very small air pumps that pull a sample of air, or air and refrigerant, over a sensor. These sensors are sensitive to moisture. They must not come in contact with water, such as in drain lines or on a wet coil.

The leak detector in **Figure 8.16(E)** features a state-of-the-art infrared sensor with a 10-year life expectancy. It also features three sensitivity levels, down to 0.1 oz/year. This leak detector has a replaceable water-blocking filter that keeps water and other contaminants out. It will not give false signals when in contact with water or oil and will detect CFC, HFC, and HCFC refrigerants along with HCFC-based refrigerant blends. The detector also features an 8-hour lithium ion battery and visual alert capabilities for noisy environments and will automatically recalibrate for working in refrigerant-contaminated areas.

8.5 LEAK DETECTION TIPS

Both a split system that is new and has not been operating and a system that has been operating can be checked for leaks. The field-installed portion of a new split system should be leak checked (as described previously) using nitrogen and a standing leak check. Leak checking a system that has been operating may require additional steps to be taken. This system may or may not have refrigerant in it. If it is cooling at all, at least some liquid refrigerant is in the system, and

Figure 8.16 (A) A modern electronic leak detector. (B) A service technician using a modern electronic leak detector on a flare fitting of a semihermetic compressor. (C) A service technician leak-detecting a fully hermetic compressor housed within a rooftop air-conditioning unit's enclosure. Photos (A, B, C) Courtesy Ritchie Engineering Company, Inc., Yellow Jacket Products Division.

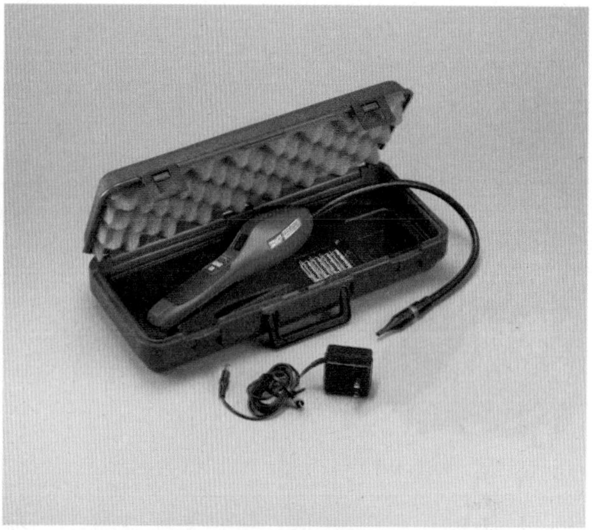

(A)

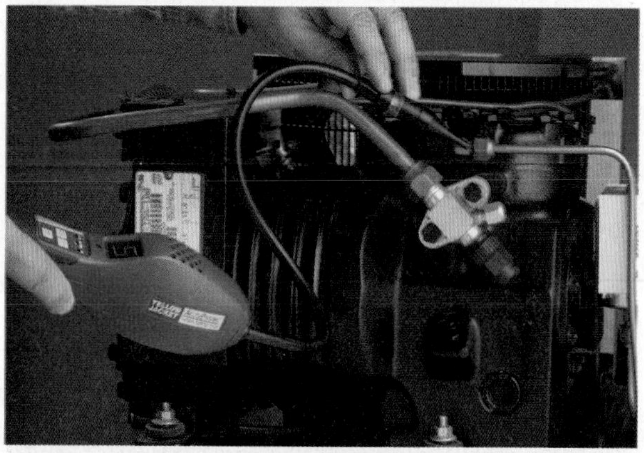

(B)

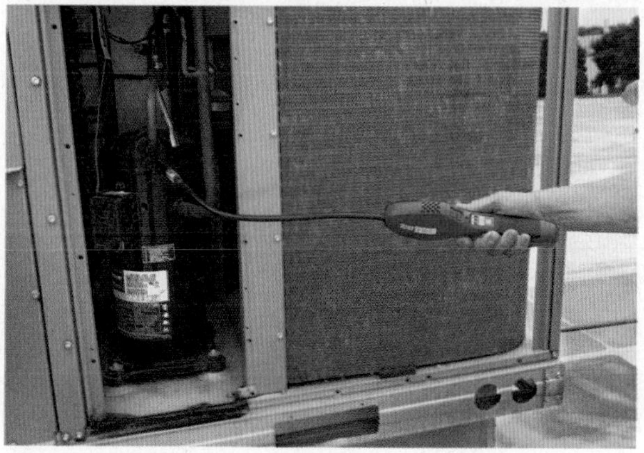

(C)

Figure 8.16 *(Continued)* (D) Two modern electronic leak detectors for CFC, HCFC, and HFC refrigerants. Courtesy Ferris State University. Photo by John Tomczyk (E) A modern, state-of-the-art leak detector with infrared sensing technology to detect CFC-, HFC-, HCFC-based refrigerants and HCFC-based refrigerant blends. Courtesy Fieldpiece Instruments

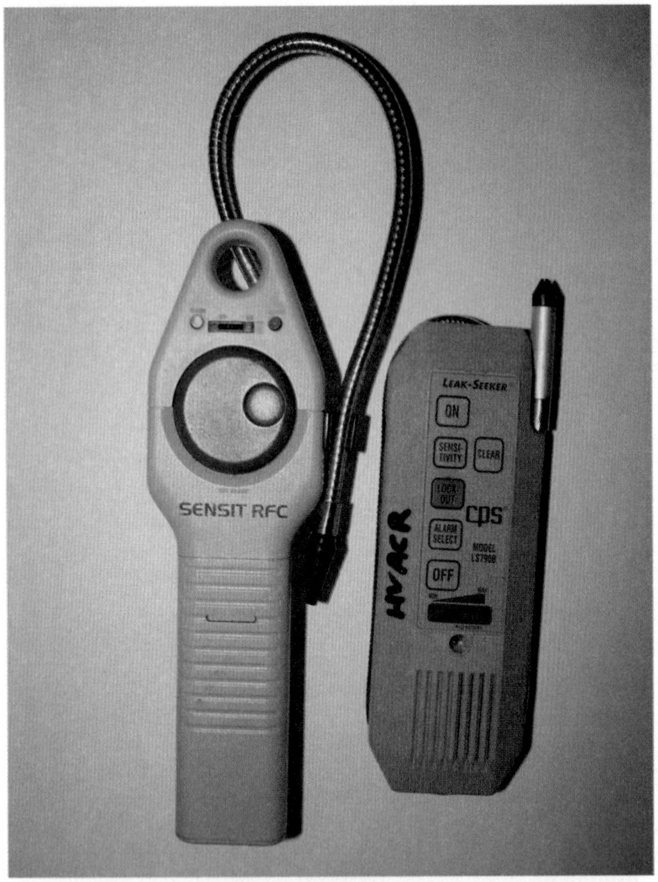

(D)

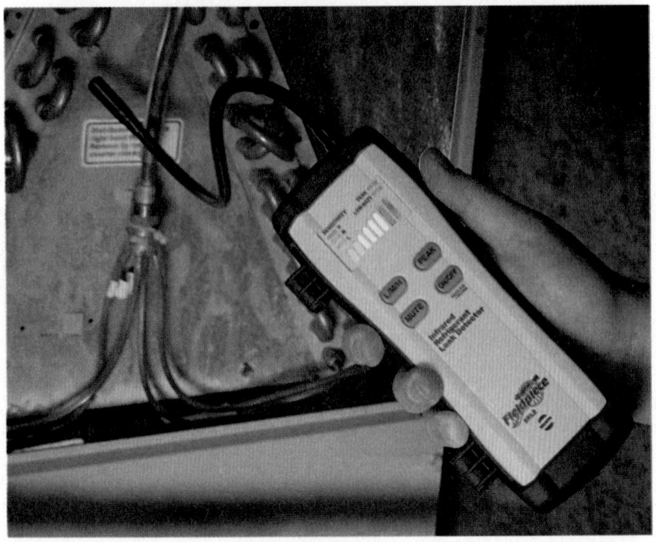

(E)

nitrogen should not be added unless the system can meet certain vacuum-level requirements established by the Environmental Protection Agency (EPA).

To leak check a system that has refrigerant in it, the technician should shut down the compressor and let the evaporator fan run. This will boost the low-pressure side of the system up to the temperature of the air-conditioned space. Low-temperature systems may not produce pressure that is high enough for an adequate leak test, but for an air-conditioning application where the indoor pressure corresponds to the indoor temperature, the pressure may be high enough for a leak test. In either situation, it is recommended that the refrigerant be properly recovered and the system pressured with nitrogen for a high-pressure leak test. This can be very time-consuming but will produce the best results.

When investigating a persistent leak in a split system, always do a visual check to look for fresh oil or dust spots where dust collects on oil. Leaks are often caused by heat, vibration, or temperature. For example, the discharge line of the compressor is an area of high vibration and possibly great temperature changes, and the gas line on a heat pump is at high pressure and temperature in the winter and low pressure and temperature in the summer. Look at any threaded connections, such as quick-connect fittings, where expansion and contraction due to temperature difference may occur.

The system parts may be isolated by disconnecting the liquid line and the suction line at the outdoor unit and pressurizing the entire low-pressure side of the system, which would include the greater part of the field-installed system as well as the evaporator, **Figure 8.17**. This will isolate the leak to the indoor or the outdoor unit. The indoor portion of the system can now be pressurized for a standing pressure and leak detection test. Never pressurize the low side of the system to a pressure greater than the low side test pressure specification. This test pressure specification is

Figure 8.17 The low-pressure side of the system is isolated and pressurized.

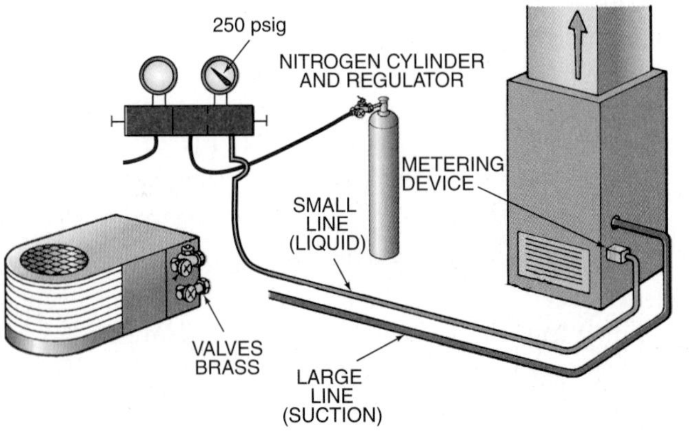

usually stamped on the evaporator section's nameplate, **Figure 8.5**. The outdoor unit may be draped with a cloth tarp or bed sheet and pressurized. The leak detector probe may be inserted under the edge of the drape after a set period of time. If a leak is detected, then the search is on for the exact location.

It is good practice to leak check any gauge connection ports before gauges are installed. First, if the system is cooling, there is some refrigerant in it. If the gauge ports are leaking you know where part of the refrigerant went. Many technicians will install gauges, pressurize the system, and not be able to find a leak, only to discover later that the gauge ports were the problem in the first place. **SAFETY PRECAUTION:** *When servicing Schrader valve ports, use only an approved cap, the one that came with the valve. If you screw a brass cap down too tight, you may create a real problem by crushing the delicate seat in the valve so far down that the valve core cannot ever be removed.* •

As mentioned, many technicians use a special soap or bubble solution/microfoamer to pinpoint leaks. The special additive in the soap keeps it from evaporating and makes it much more elastic so that larger bubbles develop. Some bubble solutions have a microfoam additive that forms a small cocoon of foam (resembling shaving cream) at the leak, **Figure 8.2(B)**. Remember, a very small leak of 1/2 oz a year will take time to blow a bubble. Bubbles do not work well on hot surfaces, such as the compressor discharge line. **NOTE:** *A thorough cleanup must be performed after using soap for leak detecting. The soap residue will turn the copper pipe green and it will start to corrode. Painted surfaces will start to peel if not cleaned. Carefully wiping surfaces with a wet rag or rinsing them with a hose will usually remove the residual soap.* •

Refrigerant is heavier than air and has a tendency to fall away from the leak source. It is important to start at the highest point and work downward. The technician can check an evaporator coil by stopping the evaporator fan, letting the system stand for a few minutes, and then placing the detector probe in the vicinity of the condensate drain line connection and letting it remain there for about 10 minutes (making sure the probe does not contact water in the drain line). If the probe sounds the refrigerant leak alarm, the test should be repeated. If the alarm sounds again, there is a leak in the evaporator section. Removing panels and going over the connections one at a time may find the leak.

Technicians should not rush when trying to locate a leak. Leak detection with any device is a slow process. It takes time for the instrument to respond, so the probe must be moved very slowly—about 1 in. every 2 sec is a good rate of speed. The probe tip must be directly over the leak to detect a very small leak, **Figure 8.18**. If the technician moved too fast, the probe may pass over the leak or indicate a leak that is several inches back.

Figure 8.18 (A) The electronic leak detector is not sensing the small pinhole leak because it is spraying past the detector's sensor. (B) The sensor will detect a refrigerant leak.

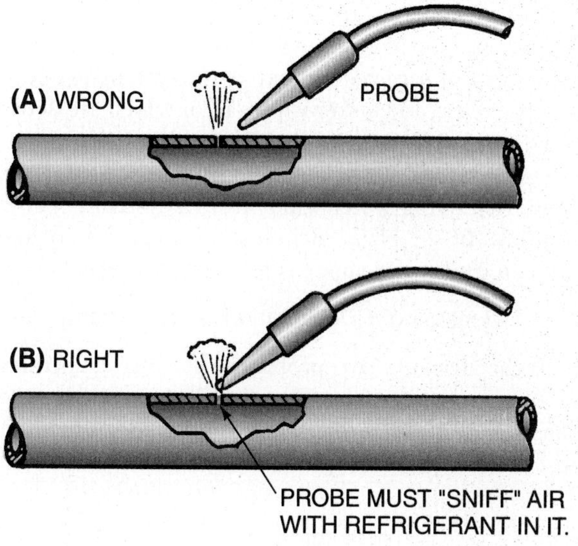

8.6 REPAIRING LEAKS

Repairing a very small leak in a small piece of equipment is usually not as economical as it is to add refrigerant. Therefore, the technician must know when to add refrigerant and when to repair a leak.

It is generally accepted as a best field practice to repair system leaks whenever possible. These leaks will have a negative impact on the system's capacity, energy consumption, and efficiency. This is true whether the system has a 2-pound charge or a 200-pound charge. There are requirements established by the EPA regarding when leaks must be repaired. These requirements may depend on the type of system, the refrigerant used, and/or the total system charge. Since regulations change periodically, be sure to visit the EPA's website for the most up-to-date information regarding leak repair requirements.

The technician must put together a system that does not leak on initial installation. Most residential systems hold less than 10 lb of refrigerant and do not require leak repair. Many homeowners think that air-conditioning equipment actually uses or consumes refrigerant because the technician just "tops off the charge" each year without really looking for the leak, but this is not good practice. Often, the leak is in the Schrader valve connection, and with just a little effort it could be repaired. Environmentally conscious service technicians should always do everything possible to prevent HVACR equipment from losing refrigerant. Global warming and ozone depletion are major environmental issues, and leak-free HVACR systems will help keep the amount of released refrigerant to a minimum.

8.7 SYSTEM EVACUATION

THE PURPOSE OF EVACUATION

Refrigeration systems are designed to operate with only refrigerant and oil circulating inside them. When systems are assembled or serviced, air enters the system. Air contains oxygen, nitrogen, and water vapor, all of which are detrimental to the system. Removing air and/or other noncondensable gases from a system with a vacuum pump is called degassing a system. Removing water vapor from a system is known as dehydration. In the HVACR industry, the process of removing both air and water vapor is referred to as evacuation.

Degassing + Dehydration = Evacuation

Gases in the air cause two problems. The nitrogen contained in air is a noncondensable gas. It will not condense in the condenser and move through like the liquid; instead, it will occupy condenser space that would normally be used for condensing refrigerants thus causing a rise in head pressure and a resulting increase in discharge temperatures and compression ratios, which produce undesirable inefficiencies. The other gases cause chemical reactions that produce acids within the system, which cause deterioration of the system's parts, electroplating of the running gear, and the breakdown of motor insulation. **Figure 8.19(A)** is an illustration of a condenser with noncondensable vapors inside.

The acids produced by noncondensable gases can be very mild to quite strong. Electroplating that occurs in a refrigeration system, like electroplating that occurs in other systems, requires electrical current, acid, and dissimilar metals. A refrigeration system has copper, brass, cast iron, and steel, with aluminum in some systems. Electroplating in refrigeration systems seems to only plate from copper to steel—copper from the pipes and steel on the crankshaft and bearing surfaces, **Figure 8.19(B)**. The bearing surfaces have close tolerances, so when a small amount of copper is plated onto the steel, the bearings become tight and bind. The equally serious

Figure 8.19 (A) A condenser containing noncondensable gases. (B) Copper plating of a crankshaft. Courtesy Appion Inc.

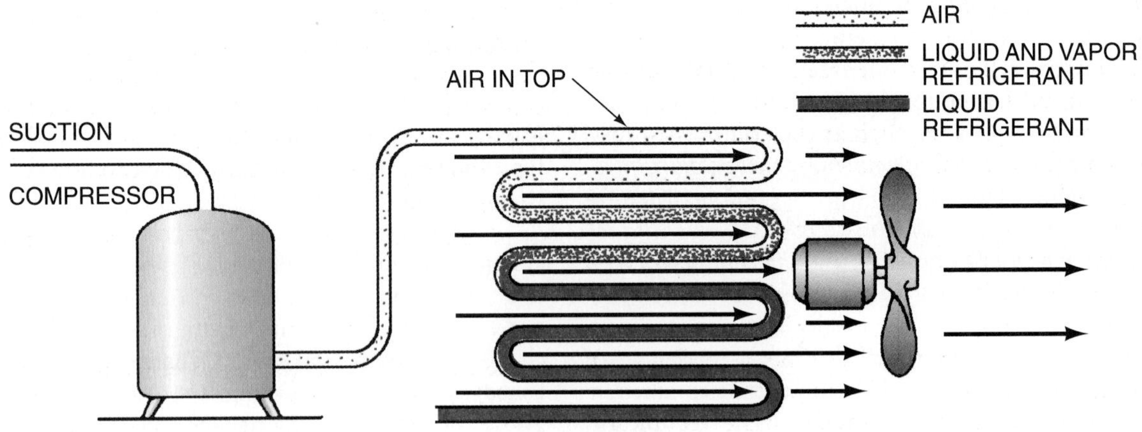

(A)

(B)

problem of the breakdown of electric motor insulation will cause electrical short circuits and either ground or short the motor from phase to phase.

When moisture (water vapor), heat, and refrigerant are all present in a system, acids will start to form after a short period of time. Refrigerants, such as R-12, R-502, R-22, R-134a, and the newer refrigerant blends—R-404A, R-407C, R-422D, R-410A, R-507, R-422A, R-438A, to name a few—contain either chlorine or fluorine and will *hydrolyze* (a chemical reaction) with water, forming hydrochloric or hydrofluoric acids and more water. After acids form, motor windings may deteriorate, and metal corrosion and sludge can occur. Because noncondensables in a system cause system head pressures and discharge temperatures to rise, oxygen from the air in a system will react with the refrigeration oil to form organic solids. This reaction of oil and oxygen usually occurs at the discharge valve because this is the hottest place in the entire system. Sludge is a tightly bound mixture of water, acid, and oil.

Moisture + Acid + Oil = Sludge

Acids may remain in a system for years without causing obvious problems. Then, the motor may burn out or copper deposited on the crankshaft from the copper in the system will cause it to become slightly oversized and bind, scoring the rubbing surfaces and wearing them down prematurely. Sludge and corrosion cause expansion devices, filter driers, and strainers to plug and malfunction.

To avoid corrosion and sludge problems, moisture must be kept out through good service practices and effective preventive maintenance. The only sure way to rid the cooling system of moisture is through effective evacuation procedures using a high-capacity vacuum pump. Once sludge has formed, standard cleanup procedures involving oversized driers specifically designed for sludge removal must be used. Vacuum pumps are not designed to remove solids such as sludge, and deep vacuum procedures will not take the place of liquid-line or suction-line driers for the same reason. Sludge and solids can be removed only by correct filtration.

Noncondensable gases must be removed from the system if it is to have a normal life expectancy. Many systems operate for years with small amounts of gas-produced products inside them, but they will not last or give the reliability the customer pays for. After the system is leak checked, noncondensable gases are removed by vacuum pumps. The pressure inside the system must be reduced to an almost perfect vacuum for this to be accomplished.

EVACUATION THEORY

To *pull a vacuum* means to lower the pressure in a system below the pressure of the atmosphere. The atmosphere exerts a pressure of 14.696 psia (29.92 in. Hg) at sea level. Vacuum is commonly expressed in millimeters of mercury (mm Hg); it can also be expressed in inches of mercury and microns, a unit of length. There are 25,400 microns in 1 in. The atmosphere will support a column of mercury 760 mm (29.92 in.) high. To pull a perfect vacuum in a refrigeration system, for example, the pressure inside the system would have to be reduced to 0 psia (29.92 in. Hg vacuum). Removing all of the atmosphere, that is, pulling a perfect vacuum, has never been achieved.

A compound gauge is often used to measure vacuum level. The gauge scale starts at 0 in. Hg vacuum and reduces to 30 in. Hg vacuum. The phrase "in. Hg vacuum" applies to the compound gauge; "in. Hg" without "vacuum" applies to a manometer or barometer. Gauges used for refrigeration systems have scales graduated in in. Hg vacuum.

We can use the bell jar illustrated in **Figure 8.20** to describe a typical system evacuation. Like the bell jar, a refrigeration system contains a volume of gas. The only difference is that a refrigeration system is composed of many small chambers connected by piping. These chambers include the cylinders of the compressor, which may be partially sealed off from the system by a reed valve. Removing the atmosphere from the bell jar is often called "pulling a vacuum" or "degassing" the bell jar. Similarly, removing noncondensables from refrigeration equipment is often called degassing the system. Evacuation, as mentioned earlier, is the process of taking both noncondensables (degassing) and water vapor (dehydration) out of the HVACR system. As the atmosphere is pulled out of the bell jar, the barometer inside the jar changes, **Figure 8.20**. The standing column of mercury begins to drop. When the column drops down to 1 mm, only a small amount of the atmosphere is still in the jar (1/760 of the original volume).

Figure 8.20 The mercury barometer in this bell jar illustrates how the atmosphere will support a column of mercury. As the atmosphere is removed from the jar, the column of mercury will begin to fall. If all of the atmosphere could be removed, the mercury would be at the bottom of the column.

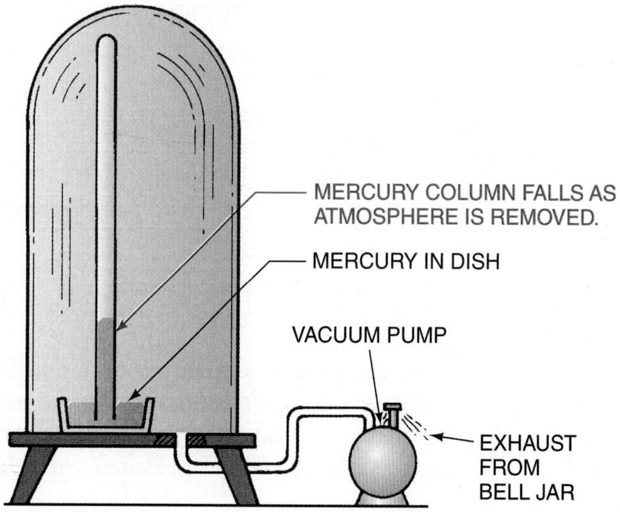

Figure 8.21 Pressure and temperature relationships for water below atmospheric pressure.

ATMOSPHERIC PRESSURES, ABSOLUTE VALUES				COMPOUND GAUGE READING in. Hg VACUUM	SATURATION POINTS OF H_2O (BOILING—CONDENSING) °F
psia	in. Hg	mm Hg	microns		
14.696	29.921	759.999	759,999	00.000	212.00
14.000	28.504	724.007	724,007	1.418	209.56
13.000	26.468	672.292	672,292	3.454	205.88
12.000	24.432	620.577	620,577	5.490	201.96
11.000	22.396	568.862	568,862	7.526	197.75
10.000	20.360	517.147	517,147	9.617	193.21
9.000	18.324	465.432	465,432	11.598	188.28
8.000	16.288	413.718	413,718	13.634	182.86
7.000	14.252	362.003	362,003	15.670	176.85
6.000	12.216	310.289	310,289	17.706	170.06
5.000	10.180	258.573	258,573	19.742	162.24
4.000	8.144	206.859	206,859	21.778	152.97
3.000	6.108	155.144	155,144	23.813	141.48
2.000	4.072	103.430	103,430	25.849	126.08
1.000	2.036	51.715	51,715	27.885	101.74
0.900	1.832	46.543	46,543	28.089	98.24
0.800	1.629	41.371	41,371	28.292	94.38
0.700	1.425	36.200	36,200	28.496	90.08
0.600	1.222	31.029	31,029	28.699	85.21
0.500	1.180	25.857	25,857	28.903	79.58
0.400	0.814	20.686	20,686	29.107	72.86
0.300	0.611	15.514	15,514	29.310	64.47
0.200	0.407	10.343	10,343	29.514	53.14
0.100	0.204	5.171	5,171	29.717	35.00
0.000	0.000	0.000	0.000	29.921	—

NOTE: psia × 2.035 = in. Hg   psia × 51.715 = mm Hg   psia × 51,715 = microns

Figure 8.21 shows comparative scales for the saturation points of water. There will be more in this unit on how to use a vacuum to both degas and dehydrate a system. For now, realize that the compound pressure gauge on the refrigeration manifold is only an indicator of the vacuum present. More accurate methods of measuring vacuum must be used.

When pulling a vacuum on a refrigeration system, the technician should attach the vacuum pump to the high- and the low-pressure sides of the system. This will prevent air from being trapped in one side of the system, such as when only one line is used. All large systems have gauge ports on both the high and low sides of the system.

MEASURING THE VACUUM

When the pressure in the bell jar is reduced to 1 mm Hg, the mercury column is hard to see, so another pressure measurement called the micron is used (1000 microns = 1 mm, 1 in. = 25,400 microns). Microns are measured with electronic instruments.

Accurately measuring and proving a vacuum in the low micron range can be accomplished with an electronic instrument. **Figure 8.22** shows typical electronic vacuum gauges. Several companies manufacture these gauges; a reliable supply house should be able to name the most popular. Micron gauge displays can be electronic analog, digital, or light-emitting diode (LED). Today's modern solid-state technology favors either digital or LED. Some digital micron gauges today come with Bluetooth technology capabilities.

Figure 8.25 shows a digital, wireless, dual-sensor, standalone micron vacuum gauge. **Figure 8.23** shows this same gauge connected to a system using a large diameter vacuum hose for less resistance to flow, allowing deep vacuum levels to be reached faster. Only certified, large diameter, vacuum rated hoses should be connected from the HVACR system directly to the vacuum pump to prevent leaks and lower resistance to flow. This method avoids manifold gauge and hose attachments, which minimizes potentials for leaks. It is not necessary to have a gauge manifold and hoses involved in the evacuation process. This digital vacuum gauge can measure from

Figure 8.22 (A, B) Modern handheld, digital electronic micron gauges. Courtesy Ferris State University. Photos by John Tomczyk

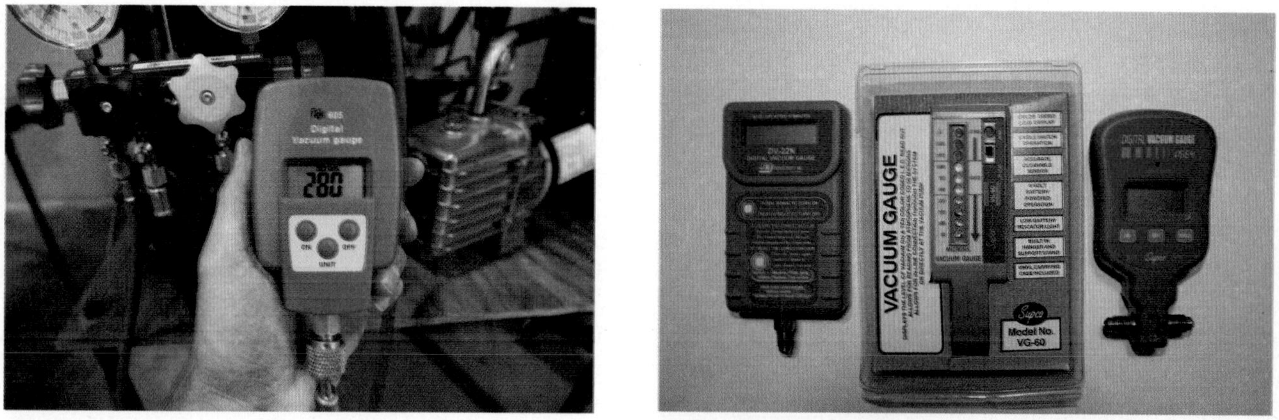

(A) **(B)**

atmosphere pressure (760,000 microns) down to 1 micron with a +/−5% reading accuracy and a 1 micron resolution.

Bluetooth technology is a wireless standard that exchanges data wirelessly over short distances (100 meters) using radio wave transmissions. It is built into many products like computers, mobile phones, and medical devices. Bluetooth technology was created as an open standard to allow connectivity and collaboration between different products and industries. This technology is bringing everyday devices into the digital world, allowing them to connect to one another

Figure 8.23 A digital, Bluetooth capable, dual-sensor, standalone micron vacuum gauge hooked up to an HVACR system. Large evacuation hoses should be used to reach deeper vacuum levels.

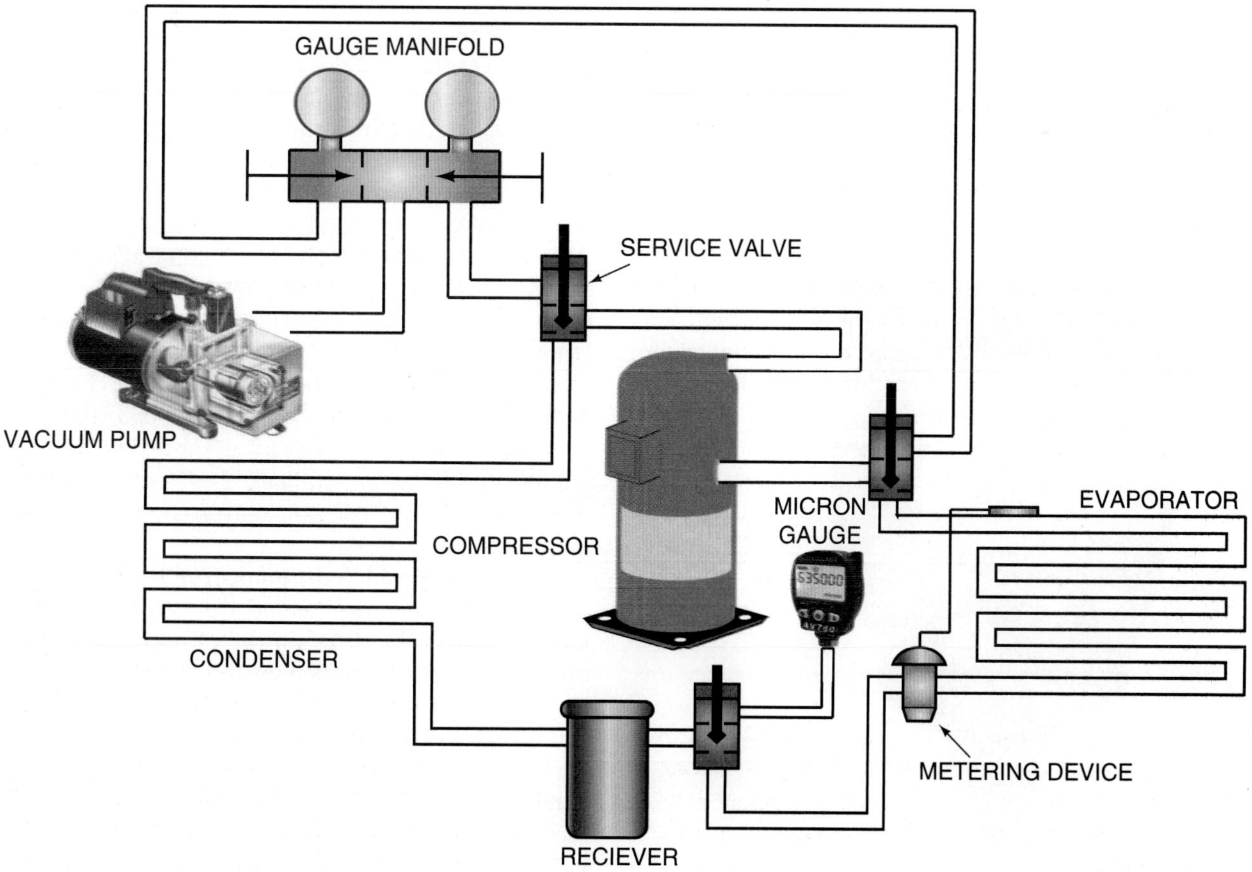

Figure 8.24 An electronic vacuum gauge sensor is located on the system as far as possible from the vacuum pump. Notice that it can be disconnected at the "tee" fitting to prevent pressure from entering the sensor. Courtesy Appion Inc.

without wires. Today, just about all top-selling tablets, personal computers, headsets, computer mice, keyboards, and smart-phones support Bluetooth technology.

The electronic vacuum gauge measures the pressure inside the system. It has a separate sensor that attaches to the system and that is connected to the gauge. Some modern, wireless, digital micron vacuum gauges have Bluetooth capabilities and need no wires, **Figure 8.25**. The sensor portion of the instrument must not be exposed to system pressure, so it is advisable to have a valve between it and the system. The sensor may also be installed using gauge lines; it is disconnected from the system before refrigerant is added, **Figure 8.24**. *The sensor should always be located in an upright position so that any oil in the system will not drain into the sensor.*

Figure 8.25 The electronic vacuum sensor must be mounted upright to prevent oil from entering the sensor. Courtesy Appion Inc.

All micron gauges should be located on the HVACR system as far as possible from the vacuum pump. The Bluetooth gauge shown in **Figures 8.24** and **8.25** offers extended functionality including wireless remote vacuum measurements, a rate of change display, data-logging for trend analysis, custom alarms, and a pressure temperature library. When the vacuum gauge reaches the vacuum desired, usually about 500 to 300 microns, the vacuum pump should then be valved off and the reading marked. The instrument reading may rise for a very short time, about 1 min, then it should stabilize. When the reading stabilizes, the true system reading can be recorded. If it continues to rise, either an atmospheric leak or moisture is present and is boiling to create pressure. A rising pressure could also mean refrigerant vapor is still seeping from the refrigeration system. Refrigerant could be dissolved in the compressor's crankcase oil. Often, heating the crankcase with a heat lamp or gently tapping it with a rubber hammer will cause enough heat and/or vibration to release the dissolved refrigerant.

One of the advantages of the electronic vacuum gauge is its rapid response to even very small pressure rises. The smaller the system, the faster the rise. A very large system will take time to reach new pressure levels. These pressure differences may be seen instantly on the electronic instrument. *Be sure you allow for the first rise in pressure mentioned above before deciding a leak is present.* Again, it is the measurement from the meter that determines when an evacuation is complete.

The vacuum pump used to evacuate systems can often become contaminated with whatever is evacuated from a system. All contaminants seem to settle in the oil sump of the vacuum pump, which can contain acids and moisture. The oil in the vacuum pump must be changed regularly for good pump performance. The vacuum pump oil should be clean and clear; if it is cloudy, moisture is likely to be present. The pump should then be serviced by changing the oil.

***NOTE:** For the best performance, only approved vacuum pump oil should be used. Any other oil will have a vapor pressure that is too high and thus prevent the achievement of deep vacuums.* • *Be sure to dispose of the old oil in an environmentally appropriate manner.*

With the new oil in the pump, run the pump long enough to get up to operating temperature; it will usually be too hot to touch or hold. Drain the oil again. With fresh oil in the pump, repeat the vacuum test. If it is still not satisfactory, change the oil again and observe what the oil looks like. If it is discolored, add new oil and run a new test. The pump may need several changes to remove the contaminants. Fortunately, most vacuum pumps use only small quantities of oil, so there is not much expense or much oil to dispose of.

Vacuum pump oil has a very low vapor pressure and it is difficult to keep it from being contaminated while in storage. If you store your vacuum pump in the truck, it is open to the atmosphere and a certain amount of contamination will occur as moisture from the atmosphere enters the pump. In the daytime when the sun is out, the air temperature rises and the pump gets up to daytime temperatures. At night when the air cools down, the pump cools, and this moist air migrates into the pump. If the pump is allowed to sweat due to weather changes, even more moisture will enter the pump. Oil stored in a can that is not properly sealed goes through the same process. All of this means that to keep a vacuum pump in top condition for very deep vacuums, it should not be charged with fresh oil until it is time to use it. This is often not practical as you move from job to job, so frequent oil changes are required. Always cap the top of the vacuum pump oil container and vacuum pump as soon as possible to prevent atmospheric moisture from migrating to the oil.

Clean oil and unrestricted flow from the HVACR system to the vacuum pump maximize evacuation speed. Clean and pure vacuum pump oil has a very low vapor pressure. A vacuum pump can only pull as deep as the vapor pressure of its sealing oil. The vacuum pump oil is a liquid seal between the vanes and the cylinder walls of all two-stage rotary vacuum pumps. When moisture enters the vacuum pump sealing oil, it reduces the oil's ability to maintain a low micron seal. As the oil becomes saturated with moisture, its vapor pressure rises. This in turn raises the vacuum level that can be achieved by the vacuum pump. The oil must be changed as its gets cloudy or dirty to remove the initial contamination, or the rate of evacuation will slow and a deep vacuum will never be reached. A micron gauge is the only true indicator as to when the oil needs to be changed. The vacuum pump will pull at a slower rate and the micron gauge will not reach deep vacuums with contaminated oil.

A vacuum pump needs the maximum flow to pump at its full capacity. Removing Schrader valve cores and core depressors in service hoses and using as many ½ in. or larger hoses as your system can handle will provide more evacuation channels at less resistance to flow. **Figure 8.24** shows a valve core remover tool with a digital, Bluetooth micron gauge connected. Schrader cores can cover 90% of the internal diameter of the access port. Valve core removal tools allow the service technician to remove valve cores for less resistance to flow during an evacuation process. The core can then be replaced after the evacuation is complete without disturbing the HVACR system.

A 1/4 in. service hose will only allow a vacuum pump to pull at 0.8 CFM, regardless of the vacuum pump's CFM rating. Never use 1/4 in. hoses for HVACR system evacuations. Using 1/2 in. hoses greatly reduces the resistance to flow from the HVACR system through the vacuum pump. Always ensure the connecting equipment from the HVACR system to the vacuum pump is vacuum rated to the capability of your pump. Rotary vane vacuum pumps' capabilities are 20 to 100 microns. Avoid using manifold gauge sets for pulling vacuum because they increase the chances for leaks. Also, many manifold gauge sets and hoses offer a lot of restriction to flow from the HVACR system to the vacuum pump. It is not necessary to have a manifold gauge set involved in the evacuation process. **Figure 8.24** shows evacuation accesses from both the high and low side of the HVACR system with large diameter hoses coming from the system. These large diameter hoses are connected to the HVACR system through a valve core removal tool. Both large diameter evacuation hoses are vacuum rated and connect to the vacuum pump. Also shown is a digital, wireless, Bluetooth micron vacuum gauge connected to the HVACR system, not to a gauge manifold or vacuum pump.

Some modern vacuum pumps have special features that allow a technician to monitor oil quality and to change wet or dirty oil on the fly in 5 seconds without breaking the vacuum, **Figure 8.26(D)**. A clear debris monitoring tube prevents contamination from entering and damaging the vacuum pump. The debris tube is also a positive pressure relief to eliminate accidental oil showers. The pump also has an oil "anti-suckback" feature that will not let vacuum pump oil get sucked back into the refrigeration system in case of a power failure. Also, massive aluminum heat sinks keep the oil cool and reduce viscosity breakdown. The vacuum pump in **Figure 8.26(E)** also has a remote exhaust with a garden hose fitting, an easy-to-clean debris filter screen, and adjustable multiport hose input fittings for ¼-, ⅜-, and ½-in. hose connections for pulling single or multiple vacuums. Used in conjunction with short, large, nonrestrictive vacuum hoses and a large manifold set, both industrial systems and residential systems can be evacuated in one-quarter of the time it takes with a traditional vacuum system, **Figure 8.48(C)**.

RECOVERING REFRIGERANT

4R *If there is or has been refrigerant in a system, the technician must remove the refrigerant before evacuating the system to remove contaminants or noncondensable gases. This must be done with EPA-approved recovery equipment. The amount of vacuum to be achieved when removing the refrigerant depends on the size of the system, the type of refrigerant, and whether or not the recovery equipment was manufactured before or after November 15, 1993. The recovery of refrigerant is discussed in more detail in Unit 9, "Refrigerant and Oil Chemistry and Management—Recovery, Recycling, Reclaiming, and Retrofitting."* 4R

THE VACUUM PUMP

A vacuum pump capable of reducing the atmosphere in the system down to a very low vacuum is necessary. The vacuum pumps usually used in the refrigeration field are manufactured with rotary compressors, and the pumps that produce

Figure 8.26 (A) A modern two-stage rotary vacuum pump. Courtesy Robinair SPX Corporation. (B) A modern vacuum pump. Courtesy R Courtesy Ritchie Engineering Company, Inc., Yellow Jacket Products Division. (C) Modern vacuum pumps of various sizes. Courtesy Ritchie Engineering Company, Inc., Yellow Jacket Products Division. (D) and (E) Modern state-of-the-art vacuum pump that features an on-the-fly 5-second oil change cartridge, a clear debris monitoring tube, an oil anti-suckback feature, remote exhaust using a garden hose fitting, adjustable multiport hose input fittings, oil quality monitoring, and massive heat sinks to keep the oil cool and prevent viscosity breakdown. Photos (D, E) Courtesy Appion Inc.

(C)

(A)

(D)

(B)

(E)

the lowest vacuums are two-stage rotary vacuum pumps, **Figure 8.26**. These pumps are capable of reducing the pressure in a leak-free vessel down to a very low micron range. It is not practical to pull a vacuum below 20 microns in a field-installed system because the refrigerant oil in the system may boil slightly and create a vapor. The usual vacuum required by manufacturers is approximately 500 microns, although some may require a vacuum as low as 250 microns.

Two-stage vacuum pumps are made up of two single-stage vacuum pumps in series. They are almost always of the rotary style because rotary vacuum pumps do not require any head clearance like piston-type pumps. This allows them to have a much higher volumetric efficiency. Two-stage vacuum pumps can pull much lower vacuums because their second stage experiences a much lower intake pressure, which is the result of the exhaust of the first-stage vacuum pump being exhausted into the intake of the second-stage vacuum pump instead of into the atmospheric pressure. This gives the first-stage pump less back pressure and, therefore, a higher efficiency. The second-stage vacuum pump begins pulling at a lower pressure, so it can pull to a lower vacuum. Two-stage vacuum pumps have the best track record because they consistently pull lower vacuums and are much more efficient when removing moisture.

When moisture is in a system, a very low vacuum will cause it to boil to a vapor. The vapor will be removed by the vacuum pump and exhausted to the atmosphere. Small amounts of moisture can be removed this way, but it is not practical to remove large amounts with a vacuum pump because boiling water produces a large amount of vapor. For example, 1 lb of water (about a pint) in a system will turn into 867 ft^3 of vapor if boiled at 70°F.

Figure 8.27 This vacuum pump has pulled a vacuum on a large system.

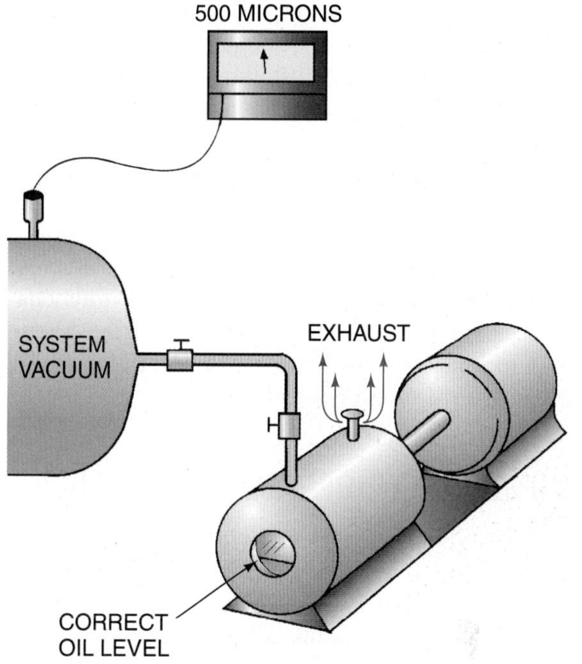

DEEP VACUUM

The deep vacuum method involves reducing the pressure in the system to about 250 to 500 microns. When the vacuum reaches the desired level, the vacuum pump is valved off and the system is allowed to stand for some time to see whether the pressure rises. If the pressure rises and stops at some point, a material such as water is boiling in the system. It could also mean that dissolved refrigerant is still coming out of the oil in the compressor crankcase. If this occurs, continue evacuating. If the pressure continues to rise, there is a leak and the atmosphere is seeping into the system. In this case, the system should be pressured and leak checked again.

When pressure is reduced to 250 to 500 microns and remains constant, no noncondensable gas or moisture is left in the system. Reducing the system pressure to 250 microns is a slow process, because when the vacuum pump pulls the system pressure to under about 5 mm (5000 microns), the pumping process slows down. The technician should have other work planned and let the vacuum pump run. Most technicians plan to start the vacuum pump as early as possible and finish other work while the pump does its work. Some technicians let the vacuum pump run all night because of the time involved in reaching a deep vacuum. However, if the vacuum pump's oil is saturated with moisture, even letting it run all night will not guarantee a deep vacuum.

When the vacuum pump pulls a vacuum, the system becomes a large volume of low pressure; the vacuum pump is between this volume and the atmosphere, **Figure 8.27**. If the pump shuts off during the night due to a power failure, oil may be sucked into the refrigeration system because of the strong pressure difference, **Figure 8.28**. When the power is restored and the vacuum pump starts back up, it will be without adequate lubrication and could be damaged. Having

Figure 8.28 This system under vacuum has pulled the oil out of the vacuum pump.

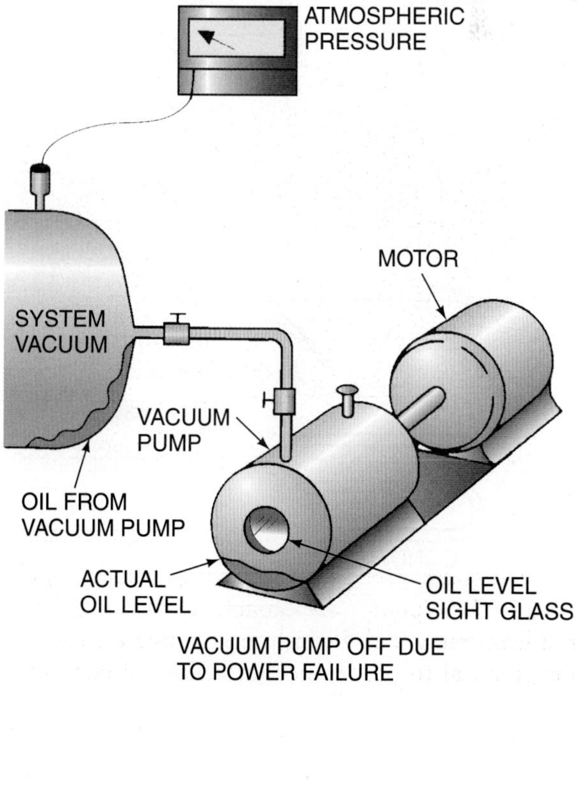

vacuum pump oil in the refrigeration or air-conditioning system can also damage the system because vacuum pump oil is designed for pumps only. Mixing vacuum pump oil with the cooling system's oil will do serious damage to the cooling system. This can be prevented by installing a large solenoid valve in the vacuum line entering the vacuum pump and wiring the solenoid valve coil in parallel with the vacuum pump motor. The solenoid valve should have a large port so that it doesn't restrict the flow. If the power fails, or if someone disconnects the vacuum pump (a good possibility at a construction site), the vacuum will not be lost and the pump will not lose its lubrication. Most vacuum pumps manufactured today have a built-in check valve to prevent this from occurring. However, the service technician should check with the pump manufacturer to verify if the pump has a check valve.

MULTIPLE OR TRIPLE EVACUATION

Multiple or triple evacuation is used by many technicians to evacuate a system to a deep vacuum level. It is normally accomplished by evacuating a system to a low vacuum, about 1500 microns, and then bleeding a small amount of dry nitrogen into the system. The system is then evacuated until the vacuum is again reduced to 1500 microns. This procedure is repeated three times, with the last vacuum level reaching a deep vacuum of 250 to 500 microns, which is held for 10 to 15 minutes. The following is a detailed description of a multiple or triple evacuation, the latter being performed three times. **Figure 8.29** is a diagram of the system connections.

1. Attach an electronic micron gauge to the system. The best place is as far from the vacuum port as possible. For example, on a refrigeration system, the pump may be attached to the suction and discharge service valves and the micron gauge to the liquid receiver valve port. Then start the vacuum pump.
2. Let the vacuum pump run until the micron gauge reaches 1500 microns, **Figure 8.30**.
3. Allow a small amount of dry nitrogen to enter the system until the vacuum is about 20 in. Hg. **Figure 8.31** shows the manifold gauge reading. This small amount of dry nitrogen will fill the system and absorb and mix with other vapors.
4. Open the vacuum pump valve and start the vapor moving from the system again. Let the vacuum pump run until the vacuum is again reduced to 1500 microns. Then repeat step 3.
5. When the nitrogen has been added to the system the second time, open the vacuum pump valve and again remove the vapor. Operate the vacuum pump until the vacuum on the electronic micron gauge reads from 250 to 500 microns.

Figure 8.29 This system is ready for multiple evacuations. Notice the valve arrangements and where the micron gauge is installed.

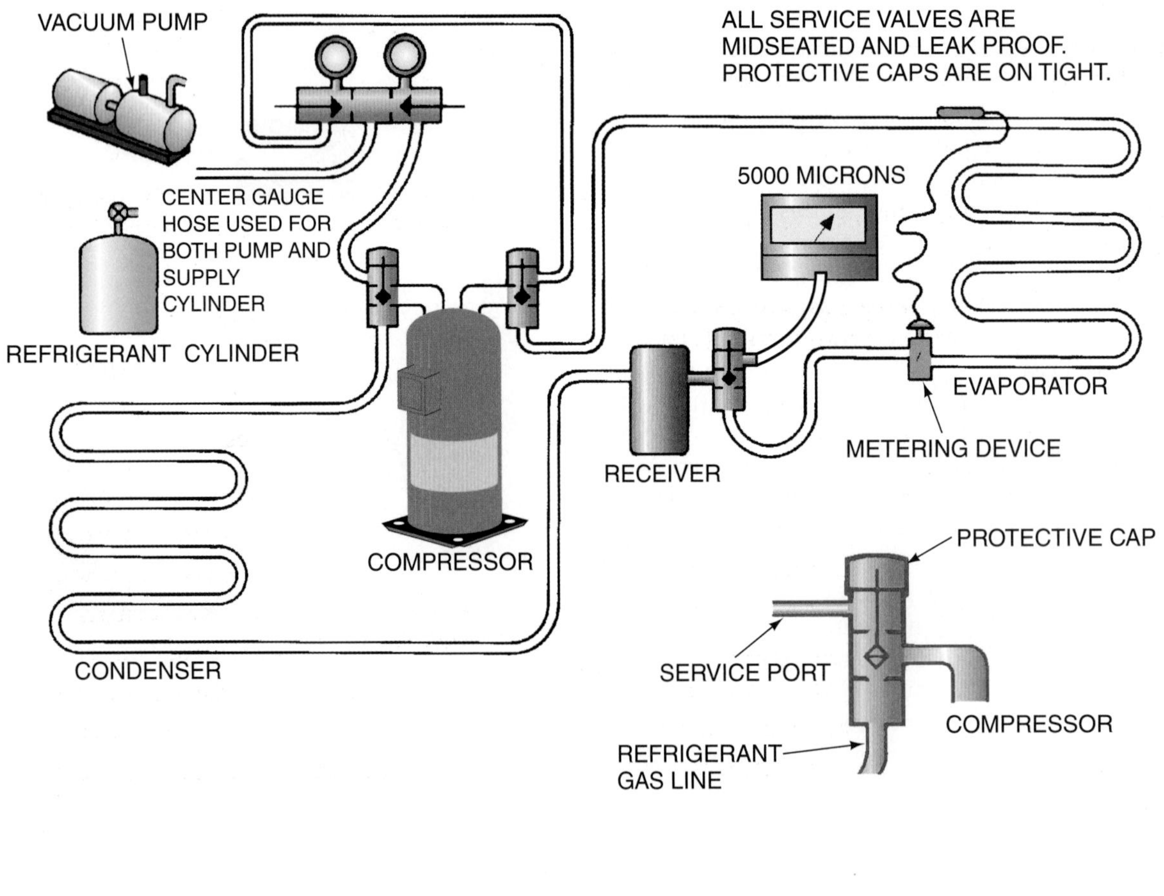

Figure 8.30 A digital micron gauge. Courtesy Ferris State University. Photo by John Tomczyk

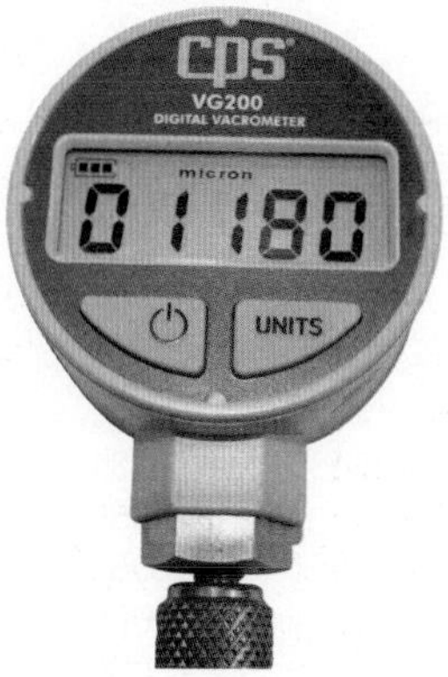

Figure 8.31 Manifold gauge reading 20 in. Hg vacuum. Courtesy Ferris State University. Photo by John Tomczyk

Once the micron gauge reads the desired vacuum (250 to 500 microns) for 10 to 15 minutes, isolate the micron gauge and charge the system with the proper refrigerant. Adding liquid refrigerant to the high side of the system is the safest and quickest method of charging a system while under a vacuum. This will prevent any liquid from entering the low side and getting to the compressor before vaporizing. **NOTE:** *It is of utmost importance that the third vacuum pulled be a deep vacuum of 250 to 500 microns and held for 10 to 15 minutes.* •

LEAK DETECTION

We mentioned that if a leak is present in a system, the vacuum gauge will start to rise if the system is still under vacuum, indicating a pressure rise. **NOTE:** *Many technicians use this as an indicator that a leak is still in the system, but this is not a recommended leak test procedure. It allows air to enter the system, and the technician cannot determine from the vacuum where the leak is. Also, when a vacuum is used for leak checking, it is only proving that the system will not leak under a pressure difference of 14.696 psi.* •

Checking for a leak using a vacuum is making use of a reverse pressure (the atmosphere trying to get into the system) of only 14.696 psi. The system, when fully loaded, typically has an operating pressure of 350 psig + 14.696 psi = 364.696 psia for an R-22 air-cooled condenser on a very hot day, **Figure 8.32**.

Using a vacuum for leak checking also may hide a leak. For example, the vacuum will tend to pull flux buildup into a pin-sized hole in a solder connection and may even

Figure 8.32 These two systems are being compared under different pressure situations. One is evacuated and the atmosphere, under atmospheric pressure, is trying to get into the system. The other has 350 psig + 14.696 psia. The one under the most pressure is under the most stress. Using a vacuum as a leak test is not a proper leak test of the system.

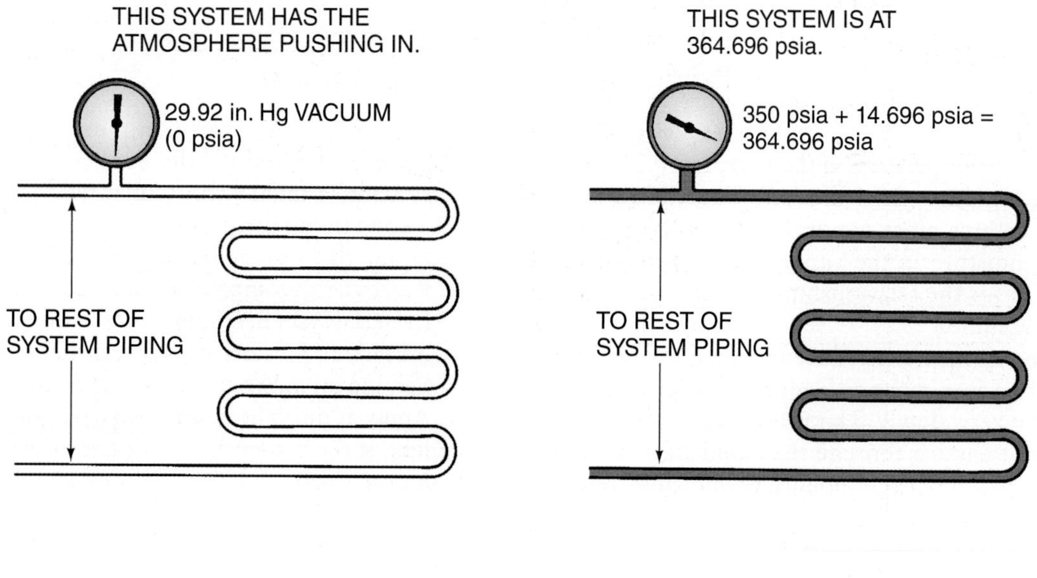

Figure 8.33 This system was leak checked under a vacuum.

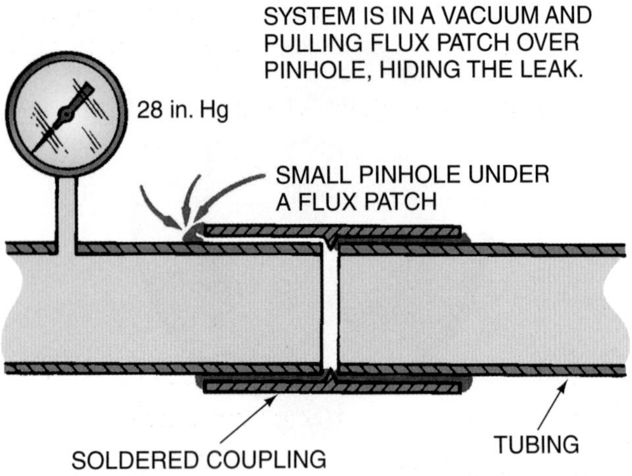

Figure 8.34 This partial temperature/pressure relationship table for water shows the specific volume of water vapor that must be removed to remove a pound of water from a system.

TEMPERATURE		SPECIFIC VOLUME OF WATER VAPOR	ABSOLUTE PRESSURE		
°C	°F	ft³/lb	lb/in.²	kPa	in. Hg
−12.2	10	9054	0.031	0.214	0.063
−6.7	20	5657	0.050	0.345	0.103
−1.1	30	3606	0.081	0.558	0.165
0.0	32	3302	0.089	0.613	0.180
1.1	34	3059	0.096	0.661	0.195
2.2	36	2837	0.104	0.717	0.212
3.3	38	2632	0.112	0.772	0.229
4.4	40	2444	0.122	0.841	0.248
5.6	42	2270	0.131	0.903	0.268
6.7	44	2111	0.142	0.978	0.289
7.8	46	1964	0.153	1.054	0.312
8.9	48	1828	0.165	1.137	0.336
10.0	50	1702	0.178	1.266	0.362
15.6	60	1206	0.256	1.764	0.522
21.1	70	867	0.363	2.501	0.739
26.7	80	633	0.507	3.493	1.032
32.2	90	468	0.698	4809	1.422
37.8	100	350	0.950	6.546	1.933
43.3	110	265	1.275	8.785	2.597
48.9	120	203	1.693	11.665	3.448
54.4	130	157	2.224	15.323	4.527
60.0	140	123	2.890	19.912	5.881
65.6	150	97	3.719	25.624	7.573
71.1	160	77	4.742	32.672	9.656
76.7	170	62	5.994	41.299	12.20
82.2	180	50	7.512	51.758	3
87.8	190	41	9.340	64.353	15.29
93.3	200	34	11.526	79.414	5
98.9	210	38	14.12	97.307	19.017
100.0	212	27	3	101.255	23.46

hide it to the point where a deep vacuum can be achieved, **Figure 8.33**. Then when pressure is applied to the system, the flux will blow out of the pinhole and a leak will appear.

REMOVING MOISTURE

A vacuum pump can be used to remove moisture (dehydrate) from a system. When the moisture is in the vapor state, it is easy to remove. When it is in the liquid state, removing it is much more difficult. The example given earlier in this unit showed that 867 ft³ of vapor at 70°F must be pumped to remove 1 lb of water. This explanation is not complete, because as the vacuum pump begins to remove the moisture, the water will boil and the temperature of the trapped water will drop. For example, if the water temperature drops to 50°F, 1 lb of water will then boil to 1702 ft³ of vapor that must be removed. This is a pressure level in the system of 0.362 in. Hg or 9.2 mm Hg (0.362 × 25.4 mm/in.). The vacuum level is just reaching the low ranges. As the vacuum pump pulls lower, the water will boil more (if the vacuum pump has the capacity to pump this much vapor), and the temperature will decrease to 36°F. The water will now create a vapor volume of 2837 ft³. This a vapor pressure in the system of 0.212 in. Hg, or 5.4 mm Hg (0.212 × 25.4 mm/in.). Thus, lowering the pressure level creates more vapor. It takes a large vacuum pump to pull moisture in the vapor state out of a system. (See **Figure 8.34** for the relationships between temperature, pressure, and volume.)

If the system pressure is reduced further, the water will turn to ice and be even more difficult to remove. The ice will turn to a vapor very slowly. The process of a solid turning to vapor without passing through the liquid state is referred to as sublimation. If large amounts of moisture must be removed from a system with a vacuum pump, the following procedure will help:

1. Start with a large vacuum pump. Using a large vacuum pump can speed up the evacuation process. For larger systems, a higher cfm (cubic feet per minute) pump or second pump can be used.
2. Drain the system in as many low places as possible. Remove the compressor and pour out the water and oil from the system. **Do not add the oil back until the system is ready to be started, after evacuation. If you add it earlier, the oil may become wet and hard to evacuate.**
3. Apply as much heat as possible without damaging the system. If the system is in a heated room, the room may be heated to 90°F without fear of damaging the room and its

furnishings or the system, **Figure 8.35**. The vacuum gauge should be connected at the system as far away from the vacuum pump as possible. This assures an accurate measurement of the vacuum in the system. If the vacuum gauge is connected close to the vacuum pump, it will give you the vacuum reading of the vacuum pump, not the system. If part of the system is outside, use a heat lamp, **Figure 8.36**. The entire system, including the interconnecting piping,

Figure 8.35 When a system has moisture in it and is being evacuated, heat may be applied to the system. This will cause the water to turn to vapor, and the vacuum pump will remove it.

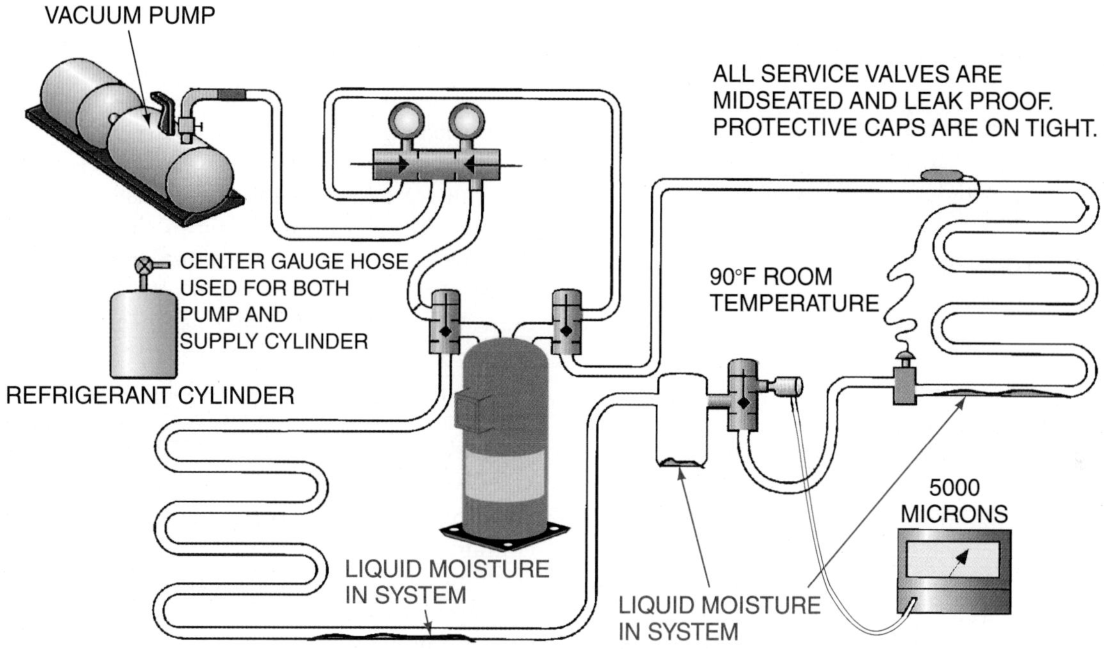

Figure 8.36 When heat is supplied to a large system with components inside and outside, the entire system must be heated. If not, the moisture will condense where the system is cool.

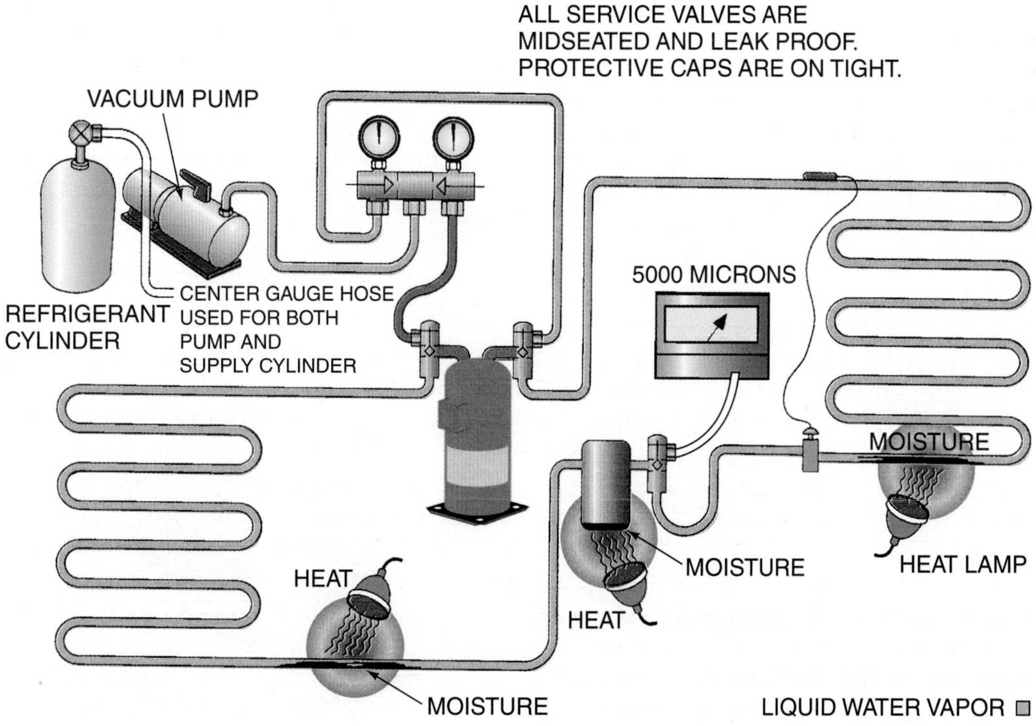

Figure 8.37 This vacuum pump has a feature called gas ballast, which allows a small amount of relatively dry atmospheric air to enter between the first and second stage. This prevents some moisture from condensing in the pump and contaminating the oil. Courtesy of Robinair Division, SPX Corporation

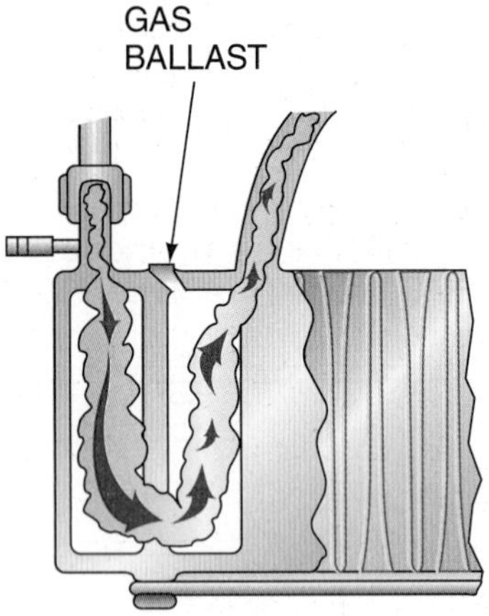

must be heated to a warm temperature or the water will boil to a vapor where the heat is applied and condense where the system is cool. For example, if water is in the evaporator inside the structure and heat is applied to the evaporator, the water will boil to a vapor. If it is cool outside, the water vapor may condense outside in the condenser piping. The water is only being moved around.

4. Start the vacuum pump and observe its oil level. At the beginning of the evacuation, water vapor is quickly removed from the system. Because of this, if a system is heavily laden with moisture, the pump's oil can become quickly contaminated with moisture that has condensed in its crankcase. The oil then loses its lubricating and sealing qualities, and the vacuum pump will be unable to pull a deep vacuum. Some vacuum pumps have a feature called *gas ballast* that introduces some relatively dry atmospheric air between the first and second stages of the two-stage pump, **Figure 8.37**. This prevents some of the moisture from condensing in the crankcase. Regardless of the vacuum pump, the quality of the oil and the pump oil level should be monitored. **NOTE:** *The water will displace the oil and push the oil out of the pump. Soon, water may be the only lubricant in the vacuum pump crankcase, and damage may occur.* •

8.8 GENERAL EVACUATION PROCEDURES

Some general rules apply to deep vacuum and deep multiple evacuation procedures. If the system is large enough or if you must evacuate the moisture from several systems, you can construct a cold trap to use in the field. The cold trap is a refrigerated volume in the vacuum line between the wet system and the vacuum pump. When the water vapor passes through the cold trap, the moisture freezes to the walls of the trap, which is normally refrigerated with dry ice (CO_2), a commercially available product. The trap is heated, pressurized, and drained periodically to remove the moisture, **Figure 8.38**. **NOTE:** *The cold-trap container must be able to withstand atmospheric pressure, 14.696 psi, pushing in, when in a deep vacuum. A light-duty container can collapse.* • The cold trap can save a vacuum pump from moisture contamination.

Noncondensable gases and moisture may be trapped in a compressor and are as difficult to release as a vapor that can be pumped out of a system. A compressor has small chambers, such as cylinders, that may contain air or moisture. Only flapper valves sit on top of these chambers, but there is nothing to force the air or water to move out of the cylinder while it is under a vacuum. At times it is advisable to start up a semihermetic or open-type compressor after a vacuum has been tried, which is easy to do when using the triple evacuation method. After the first vacuum has been reached, nitrogen can be charged into the system until it reaches atmospheric pressure. The compressor can then be started for a few seconds. All chambers should

Figure 8.38 A cold trap.

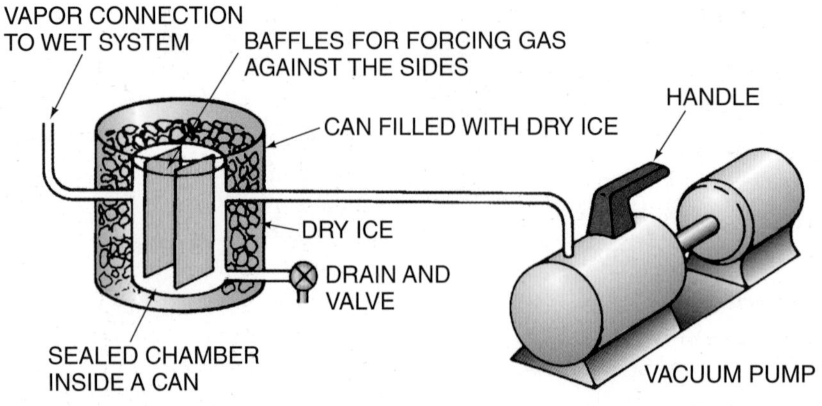

Figure 8.39 Vapor trapped in the cylinder of the compressor.

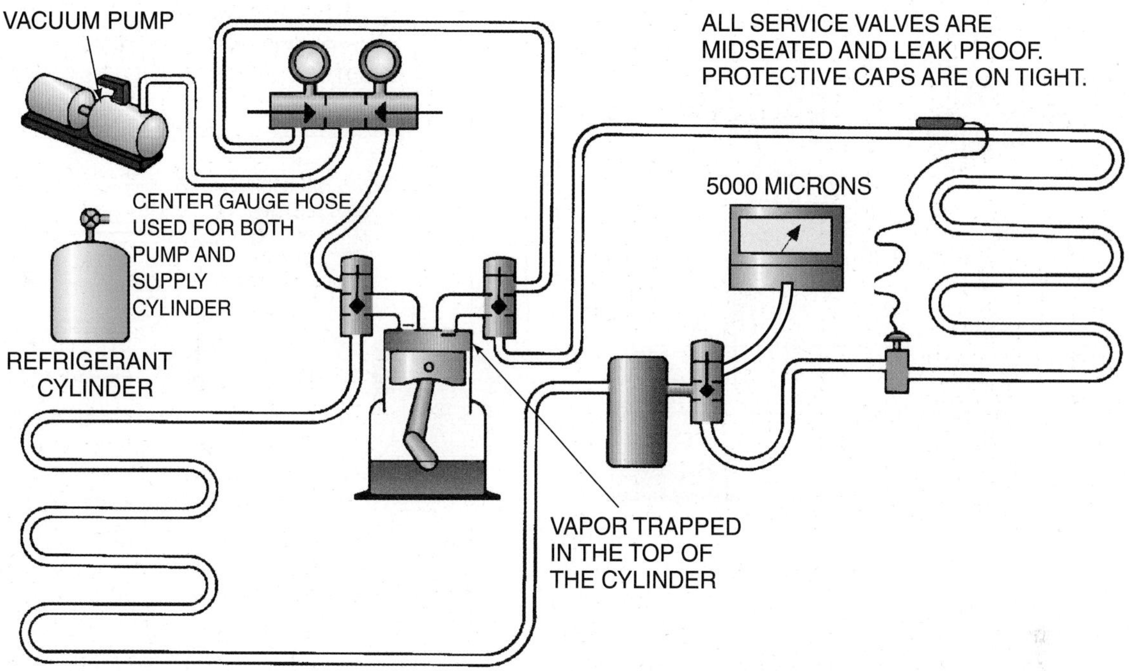

be flushed at this time. **NOTE:** *Do not start a hermetic compressor while it is in a deep vacuum. Motor damage may occur.* • The fusite terminals or studs that protrude into the shell of a hermetic compressor that are connected to power must be insulated from one another with refrigerant vapor and/or oil vapor. Both refrigerant and oil are good dielectrics, meaning they are good insulators and will not conduct electricity. When a vacuum is pulled on a hermitic compressor, these dielectrics will disappear and no longer insulate the three fusite terminals inside the compressor. A short circuit will occur and damage the motor windings and fusite terminals of the compressor motor. **Figure 8.39** shows an example of vapor trapped in the cylinder of a compressor.

Water can be trapped in a compressor under the oil. Oil will float on the water because it is less dense. Because the oil has surface tension, the moisture may stay under it even under a deep vacuum. But the oil surface tension can be broken with vibration, such as the vibration that occurs when striking the compressor housing with a soft-face hammer. Any kind of movement that causes the oil's surface to shake will work, **Figure 8.40**. Applying heat to the compressor crankcase will also release the water, **Figure 8.41**.

The technician who evacuates many systems must use time-saving procedures. For example, a typical gauge manifold may not be the best choice because it has very small valve ports that slow the evacuation process, **Figure 8.42**. However, some gauge manifolds are manufactured with large valve ports and a special large hose for the vacuum pump connection, **Figure 8.43** and **Figure 8.48 (B)** and **(C)**. The gauge manifold in **Figure 8.44** has four valves and four hoses. The two extra valves are used to control the refrigerant and the vacuum pump lines. When using this manifold, the vacuum pump need not be disconnected and the hose line switched to the refrigerant cylinder to charge refrigerant into the system. When the time comes to stop the evacuation and charge refrigerant into the system, one valve is closed and the other opened, **Figure 8.44**. This is a much easier and cleaner method of changing from the vacuum line to the refrigerant line. With a three-hose manifold, air is drawn into the hose when a gauge line is disconnected from the vacuum pump. This air must be purged from the hose at the top, near the manifold. It is impossible to get all of the air out of the manifold because some will be trapped and pushed into the system.

Most gauge manifolds have valve stem depressors in the ends of the gauge hoses. The depressors are used for servicing systems with Schrader access valves. These valves are much like the valve and stem on an automobile tire. The depressors restrict the evacuation process and slow down vacuum speed considerably. Many technicians erroneously use oversized vacuum pumps and undersized connectors because they do not realize that the vacuum can be pulled much faster with large connectors. The valve depressors can be removed from the ends of the gauge hoses, and adapters can be used when valve depression is needed. **Figure 8.45** is a small valve that can be used on the end of a gauge hose; it will even give the technician a choice as to when to depress the valve stem.

Figure 8.40 A compressor with water under the oil in the crankcase.

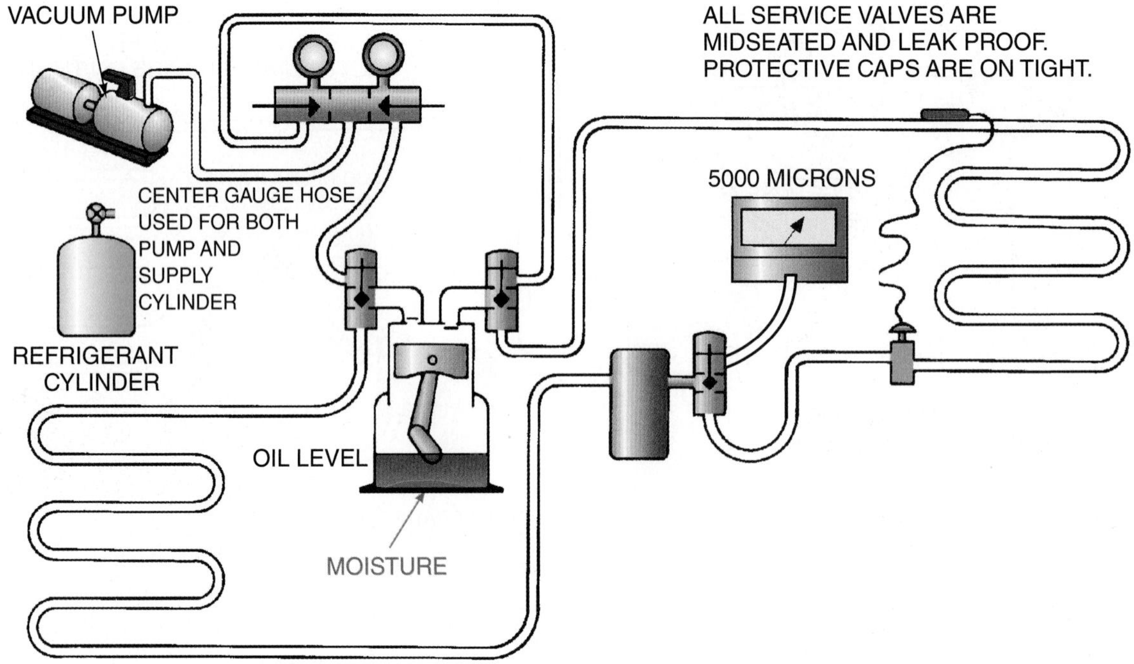

Figure 8.41 Heat is applied to the compressor to boil water under the oil.

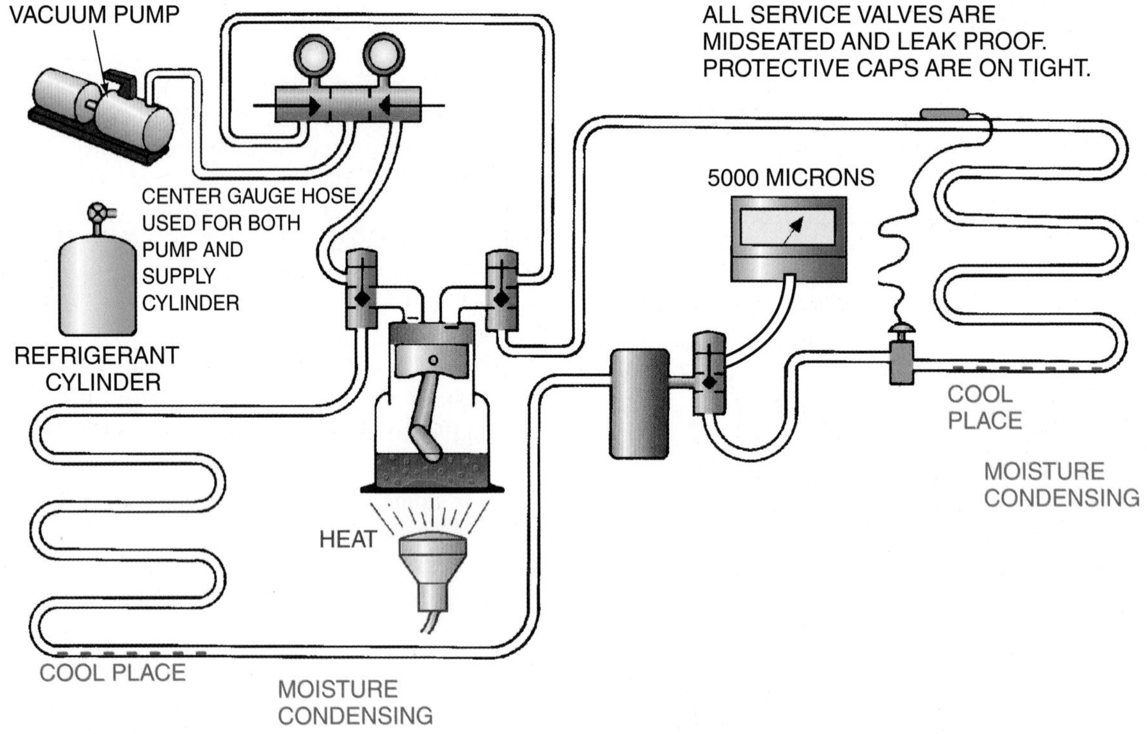

SYSTEMS WITH SCHRADER VALVES

A system having Schrader valves for gauge ports will take much longer to evacuate than a system with service valves. The reason is that the valve stems and the depressors act as restrictions. An alternative is to remove the valve stems during evacuation and replace them when evacuation is finished. It will take a great deal of time to evacuate water or moisture from a system if there are Schrader valve stems in the service ports. These valve stems are designed to be

Figure 8.42 A gauge manifold with small ports.

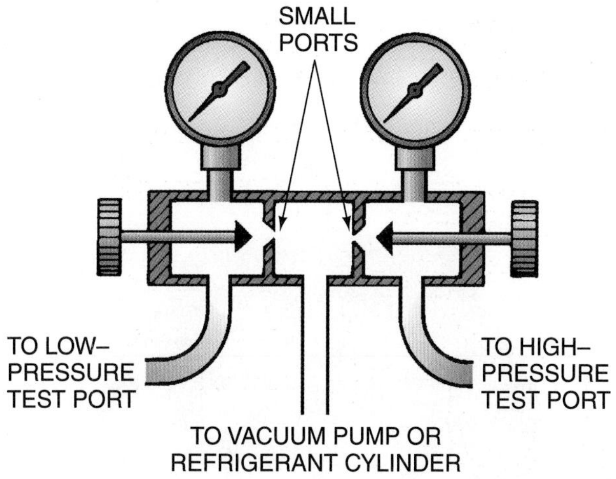

Figure 8.43 A gauge manifold with large ports.

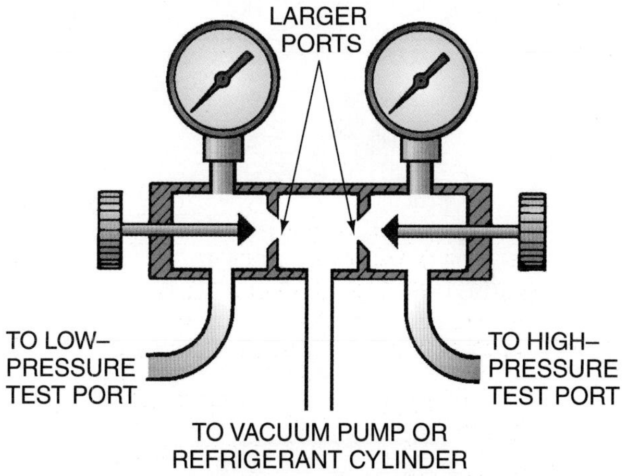

Figure 8.44 A manifold with four valves and four gauge hoses. Photo by Jason Obrzut

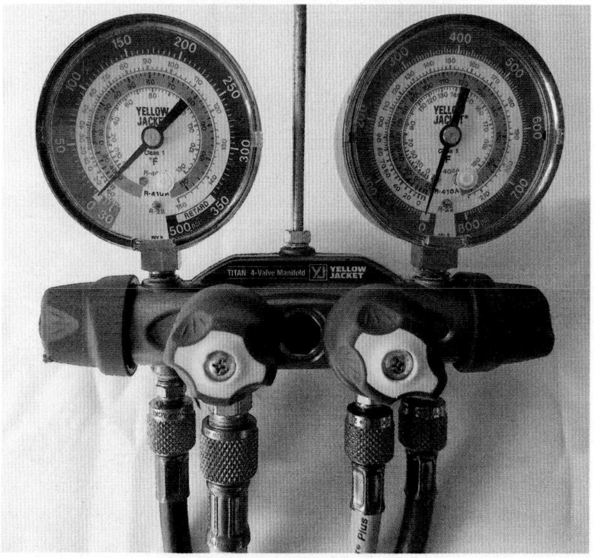

Figure 8.45 This small valve can also be used for controlled gauge readings. To read a pressure in a Schrader port, the adapter valve may be used by turning the valve handle. Photo by Jason Obrzut

removed and replaced, so they should be removed for evacuation, **Figure 8.46**. A special tool, called a core removal tool, can be used to replace Schrader valve stems under pressure, or it can be used as a control valve during evacuation. The tool has a valve arrangement that allows evacuation of a system through it with the stem backed out of the Schrader valve, **Figure 5.109** in Unit 5, "Tools, Instrumentation, and Equipment." The stem is replaced when the evacuation is completed.

Schrader valves are shipped with a special cap, which is used to cover the valve when it is not in use. The cover has a soft gasket and should only be to cover Schrader valves.

Figure 8.46 A Schrader valve assembly. Source: J/B Industries.

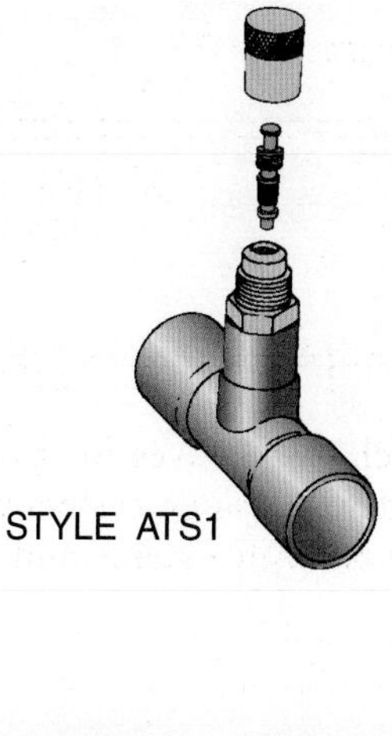

Figure 8.47 This fitting has a sensitive sealing shoulder on top. The sealing cap that comes with it has a soft neoprene gasket. If a brass flare cap is used instead and tightened down, the top of the fitting will become distorted and the valve stem cannot be removed.

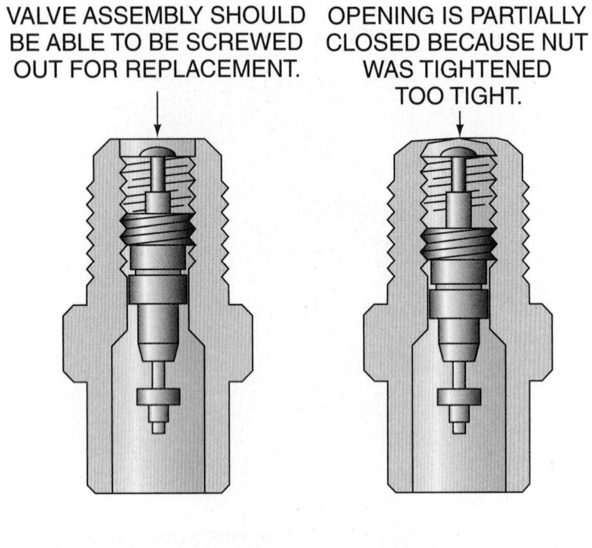

If a standard brass flare cap is used and overtightened, the Schrader valve top will be distorted and valve stem service will be difficult, if it can be done at all, **Figure 8.47.**

GAUGE MANIFOLD HOSES

The standard gauge manifold has flexible hoses with connectors on the ends. These hoses sometimes develop pinhole leaks, usually around the connectors, which may leak while under a vacuum but not when the hose has pressure inside it, because the hose swells when pressurized. If there is trouble finding a leak while pulling a vacuum, substitute soft copper tubing for the gauge lines, **Figure 8.48A.** Some manufacturers market gauge manifolds with large

Figure 8.48 (A) A gauge manifold with copper gauge lines. Photo by Jason Obrzut

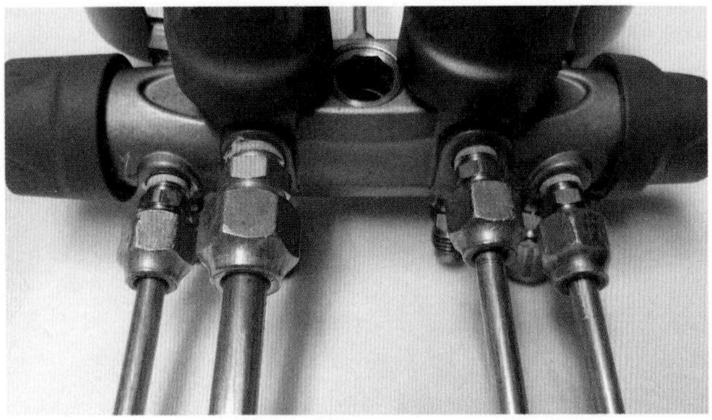

(A)

Figure 8.48 (*Continued*) (B) and (C) Gauge manifolds with large, short hoses to minimize pressure drop and to increase the speed of the evacuation process by increasing flow through the vacuum pump. Photos (B, C) Courtesy Appion Inc.

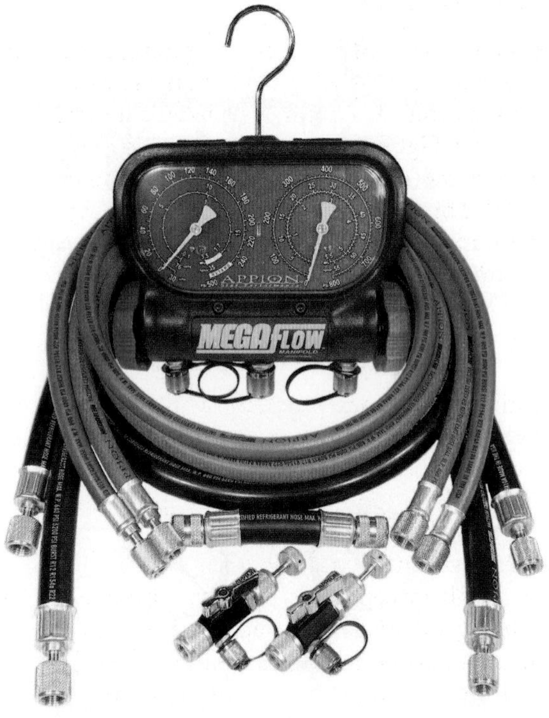

(B)

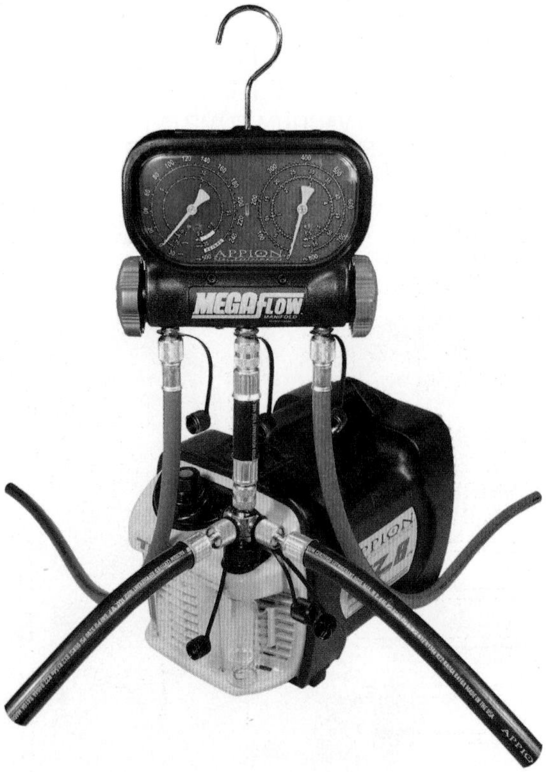

(C)

diameter, short-length hoses to minimize pressure drop and to speed the evacuation process by increasing flow to the vacuum pump, **Figures 8.48(B)** and **(C)**.

SYSTEM VALVES

For a system with many valves and piping runs, including multiple evaporator systems, verify that the valves are open before evacuation. A system may have a normally closed solenoid valve, **Figure 8.49**, which may trap air in the liquid line between the expansion valve and the solenoid valve, **Figure 8.50**. This valve must be opened for complete evacuation. A thorough inspection should be done to ensure that all valves on the system are in the open position.

USING DRY NITROGEN

Maintaining good work practices while assembling or installing a system can make system evacuation an easier task. *When piping is field-installed, sweeping dry nitrogen through the refrigerant lines can keep the atmosphere pushed out and clean the pipe. It is relatively inexpensive to use a dry nitrogen setup, and it saves time and money.*

Also, slowly sweeping dry nitrogen through pipes while they are being soldered or brazed will prevent oxidation on the inside and outside of the brazed joints. The 20% oxygen content of the air causes the oxidation on pipes when brazing if dry nitrogen is not used. When a system has been open to the atmosphere, it needs to be evacuated. The task can be shortened by sweeping the system with dry nitrogen before evacuation. **Figure 8.51** shows how this is done.

Figure 8.49 A solenoid valve. Photo by Jason Obrzut

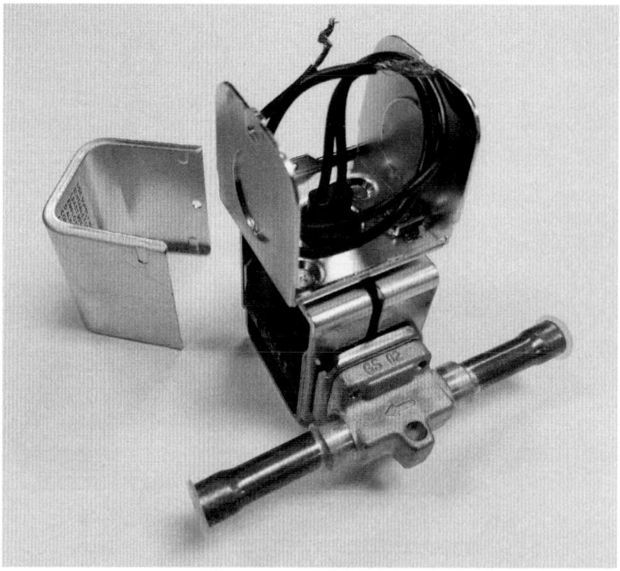

EVACUATION WITHOUT A MANIFOLD

Through advancements in tools and technology, a quicker, more efficient method of system evacuation is gaining popularity. This method involves evacuating a system without

Figure 8.50 A closed solenoid valve trapping air in the system liquid line.

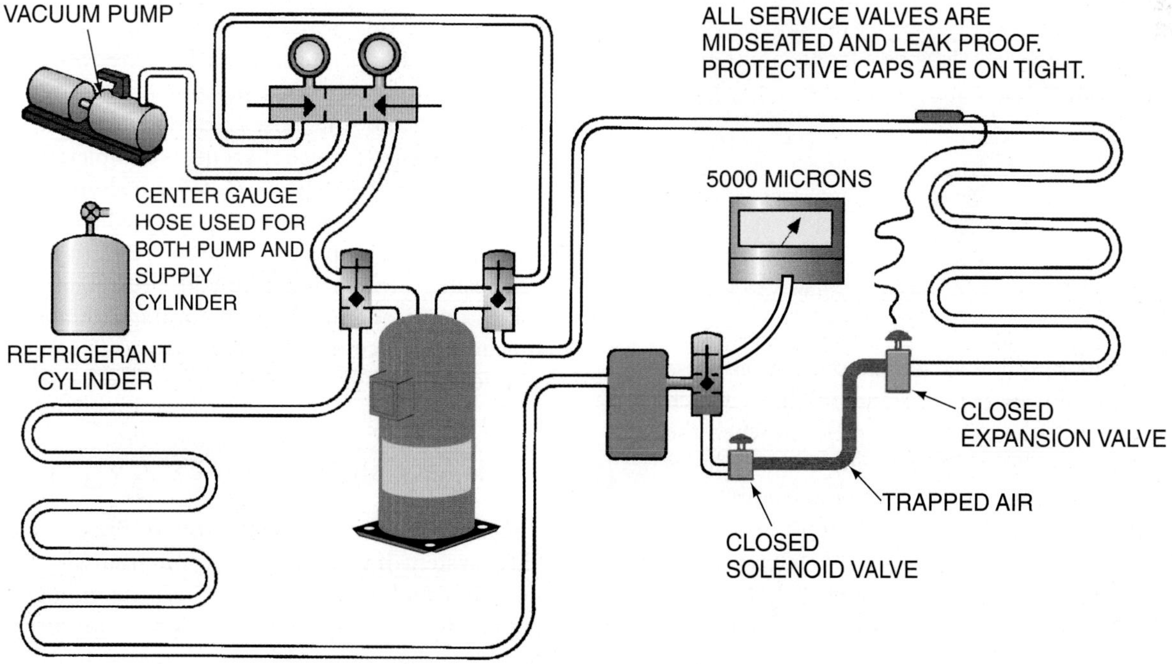

Figure 8.51 The technician is using dry nitrogen to sweep this system before evacuation.

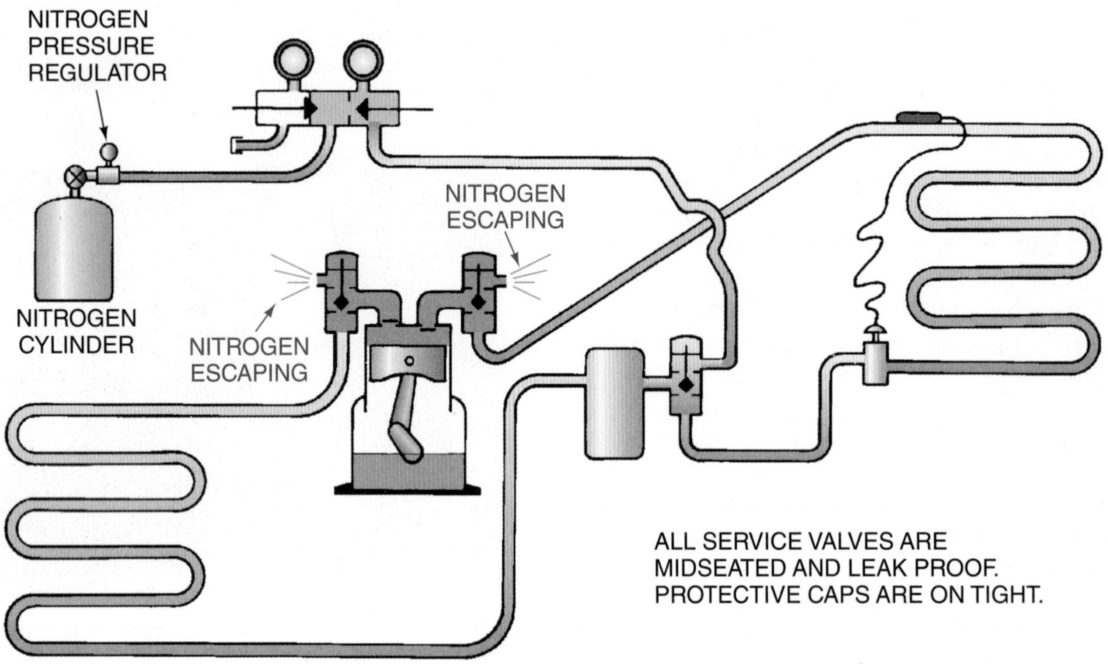

Figure 8.52 Evacuation without a gauge manifold. Photo by Jason Obrzut

the use of a gauge manifold. Evacuation without the use of a gauge manifold reduces the resistance to flow, which can increase the speed of the evacuation process, when the proper equipment is used. Proper equipment usage includes the use of tools that can remove Schrader pins from valves, which add resistance to flow, **Figure 8.52**. Larger hoses also help increase the speed of system evacuation. These larger hoses, typically ½ in. in diameter, are connected from the core removal tools to the tee on the vacuum pump. An electronic micron gauge can be connected to the side port of one, or both, core removers to monitor the evacuation process. This setup creates a less restrictive passage to the vacuum pump, shortening evacuation time and producing deeper vacuum levels. The core removal tools are equipped with ball valves that are used to isolate the system from the pump to perform a time-to-rise test. The time-to-rise test evaluates the rise in system pressure with respect to the time it takes for the system pressure to rise. Upon completion of the evacuation, refrigerant can be added directly to the system through the side port of the core removal tool, or by first connecting a gauge manifold to the core tools. Whichever method is ultimately chosen, remember to purge any air from the manifold and hoses prior to charging and to replace the Schrader pins once charging is completed. Micron gauges should be removed from the system shortly after the charging process has been started, ideally when the system pressure is just above atmospheric pressure (2 to 3 psig), to prevent exposing the instrument to high pressure and potential damage. Do not disconnect a micron gauge from a system that is under a vacuum as this may cause air to be pulled into the system.

8.9 CLEANING A DIRTY SYSTEM

The technician should be aware of how to use a vacuum to clean a dirty system of several types of contaminants. We have discussed water and air, but they are not the only contaminants that can form in a system. The hermetic motor inside a sealed system is the source of heat in a motor burnout. The

heat can raise the refrigerant and oil to temperatures that will break them down to acids, soot (carbon), and sludge that cannot be removed with a vacuum pump. The high temperature arc of a motor burnout inside a compressor also causes a portion of the refrigerant and oil mixture to break down into carbonaceous sludge, corrosive acids, and water. These contaminants must be removed from the system, especially from the compressor crankcase.

Heat + Refrigerant + Moisture = Acids

Moisture + Acid + Oil = Sludge

System contaminants can be in the solid, liquid, or vapor state. Most contaminants will be in the compressor's oil, **Figure 8.53(B)**. It is because of these contaminants that a heavy-duty suction-line filter or filter drier is recommended for every new field installation. These filters will catch contaminants before they get to other parts of the system. A suction-line filter drier will also capture any acids generated during a compressor burnout before they cause any additional system damage. **SAFETY PRECAUTION:** *Whenever working with a burned-out system, extreme caution must be used when handling and/or operating the system. Adequate ventilation, butyl gloves, and safety glasses are required for protection against acids. The oil from a burnout can cause serious skin irritation and possible burns. In some cases, the fumes can be toxic.* •

Let us use the example of a bad motor burnout to demonstrate how most manufacturers would expect you to clean a system. Suppose a 5-ton air-conditioning system with a fully hermetic compressor were to have a severe motor burn while running. When a motor burn occurs while the compressor is running, some of the soot, sludge, and acid from the hot oil moves into the refrigeration or air-conditioning system, **Figure 8.53 (A)**. The following steps should be taken to clean the system after a hermetic motor burnout:

1. Front seat (close) the inoperative compressor's service valves to isolate the compressor from the rest of the system. Recover the refrigerant from the inoperative compressor and remove it. Install the replacement compressor.
2. Take a sample of oil from the replacement compressor. Seal it in a glass bottle for comparison purposes after the cleaning procedure is complete.
3. With the compressor isolated (both service valves front seated) from the rest of the system, evacuate the compressor only. After evacuation, open the compressor service valves and close the liquid-line valve. This will minimize the amount of refrigerant handled during pump-down. Pump down the system. Some contaminants will return to the compressor during the pump-down, but the compressor will not be harmed by the short period of operation required for pump-down. The contaminants will be removed by the filter driers.
4. Inspect all system components such as expansion valves, solenoid valves, check valves, reversing valves, and

Figure 8.53 (A) Oil being drained from a fully hermetic compressor which has had a motor burn. Courtesy Ferris State University. Photo by John Tomczyk (B) An acid test kit used for checking the acidity of compressor oil after a suspected motor burn. Courtesy Ferris State University. Photo by John Tomczyk (C) A combination sight glass and moisture indicator for installation in the liquid line after the filter drier. Courtesy Sporlan Valve Division, Parker Hannifin Corporation

(A)

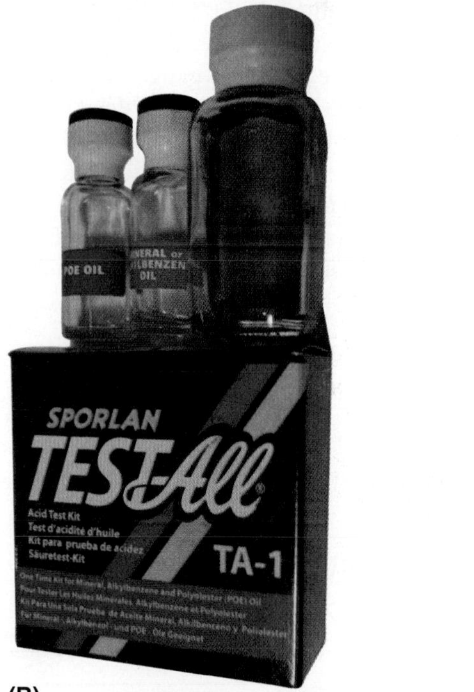

(B)

(C)

contactors. Clean and replace if necessary. Remove or replace any filter driers previously installed in the system, and clean and/or replace any filters and strainers. Install a good-quality moisture indicator if the system does not have one, **Figure 8.53(C)**.

In systems without service valves, the refrigerant must be recovered. Recovery should take place through a filter drier to protect the recovery machine from acids and contamination. The inoperative compressor should then be replaced with an operative compressor. The entire system should then be evacuated. Charge the recovered refrigerant back into the system through a filter drier. Add additional refrigerant as needed. The previous and following procedures should then be performed.

1. Install the recommended size of filter drier in the suction line. Install an oversize filter drier in the liquid line. Both liquid and suction lines will require a special temporary filter drier manufactured especially for system burnout cleaning procedures.
2. Start the compressor and run the system. As contaminants in the system are filtered out, the pressure drop across the filter driers will increase. Observe the pressure differential across the filters for a minimum of 4 hours. Use the same gauge on both sides of the drier to minimize error in gauges (driers come with pressure taps on both ends), **Figure 8.54**. For the suction line filter drier, if the pressure drop exceeds the maximum limit shown in the table in **Figure 8.55** for a temporary installation, replace the filter drier and restart the system. Note that the table in **Figure 8.55** shows the pressure drop for both permanent and temporary filter driers. During the cleanup procedure, technicians will encounter oversize filter driers.
3. Allow the system to run for 48 hours. Isolate the new compressor and take a sample of its oil. Check the odor (smell cautiously) and compare the color of the oil with the sample previously taken. If an acid test kit is available, test for acid content, **Figure 8.53(B)**. If the oil is discolored, acidic, has an acrid odor or if the moisture indicator indicates a high moisture content in the system, change the filter driers, **Figure 8.53(C)**. The compressor oil may also be changed if desired. Allow the system to run for another 48 hours, and recheck as before. Repeat until the oil remains clean and odor-free and the color approaches that of the original sample taken.
4. Replace the liquid line filter drier with one of normal size and for normal applications. Remove the suction-line filter drier, and replace it with a normal-size, permanent suction-line filter or filter drier.
5. After the cleaning procedure is completed, recheck in approximately 2 weeks to make sure the system conditions and operation are completely satisfactory. (Note: This system cleanup procedure has been reprinted with permission from Emerson Climate Technologies.)

This system cleaning procedure has been used in countless installations over the years, and when it has been properly followed, there have been no instances of a second failure.

When recovering refrigerant from a system, certain evacuation requirements must be met. Unit 9, "Refrigerant and Oil Chemistry and Management—Recovery, Recycling,

Figure 8.54 Checking the pressure drop across a suction-line filter drier.

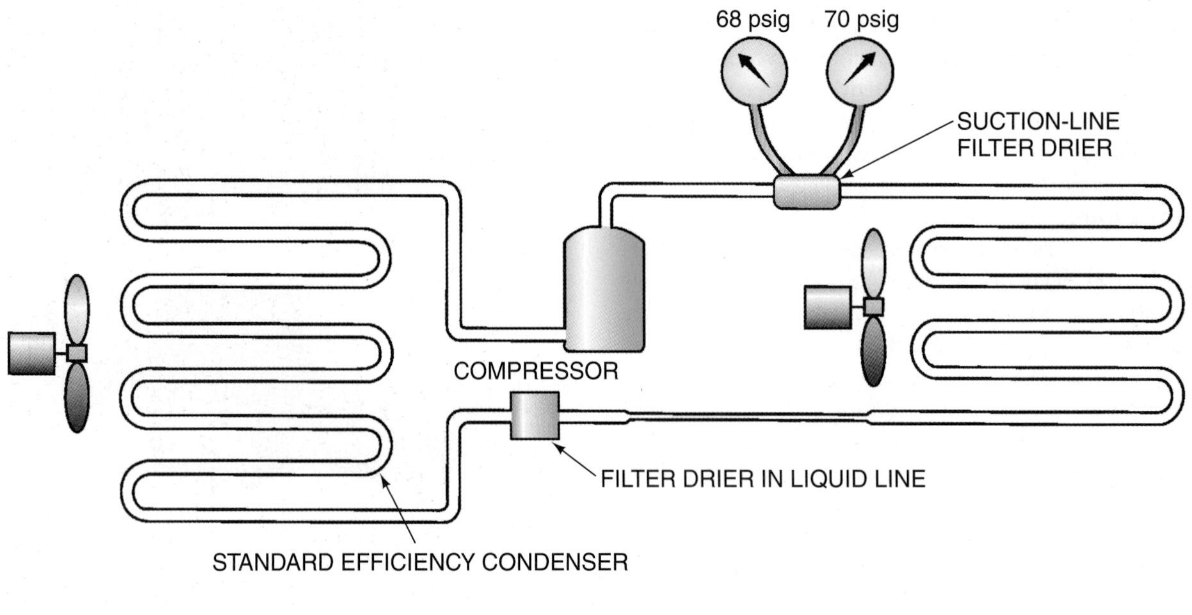

Figure 8.55 Maximum recommended pressure drop for suction-line filter driers, psi (bar). Suction-line filter driers are used when performing a cleanup procedure on a dirty system that usually has experienced a motor burn. Courtesy Emerson Climate Technologies

SYSTEM	PERMANENT INSTALLATION		TEMPORARY INSTALLATION	
	REFRIGERANT			
	R-22, R-404A, R-407C, R-410A, R-502, R-507	R-12, R-134a	R-22, R-404A, R-407C, R-410A, R-502, R-507	R-12, R-134a
Air-Conditioning	3 (0.21)	2 (0.14)	8 (0.55)	6 (0.41)
Commercial	2 (0.14)	1.5 (0.10)	4 (0.28)	3 (0.21)
Low Temperature	1 (0.07)	0.5 (0.03)	2 (0.14)	1 (0.07)

Reclaiming, and Retrofitting," discusses these requirements for air-conditioning, refrigeration, and recovery/recycling equipment.

Filter-drier manufacturers have done a good job of developing filter media that will remove acid, moisture, and carbon sludge. They claim that nothing can be created inside a system by oil and refrigerant that the filter driers will not remove. It is of utmost importance that all foreign materials be removed from a system during the initial field installation. Without good suction- and liquid-line filters or filter driers, metal filings, shavings, dirt, flux, solder, metal chips, sand from sandpaper, bits of steel wool and wire from cleaning brushes could get into the compressor. Many of these contaminants are so small they will pass through a fine mesh screen that is found on some compressor suction inlets. In fact, many compressors returned to the factory for replacement could have been saved if all contaminants had been removed by filter driers during the initial field installation. Returned compressors are at times reported as motor failures, but in reality the return is due to damaged bearings, valve reeds, or connecting rods caused by contaminants, which in turn caused motor damage as a secondary effect. There are many chemical flushing kits for system cleanup on the market today. However, always check to make sure they have been approved and tested by the compressor manufacturer before using.

REVIEW QUESTIONS

1. The low-pressure side of the system must not be pressurized to more than ________ when testing for leaks.
   A. 75 psig
   B. 200 psig
   C. the low-side design pressure
   D. atmospheric pressure
2. Standing, pressure-dependent, temperature-dependent, vibration-dependent, combination-dependent, and micro are all types of
   A. charging methods.
   B. leak methods.
   C. evacuation methods.
   D. electronic micron gauges.
3. The ____________ is both the greatest asset and the greatest liability of an electronic leak detector.
   A. range
   B. battery life
   C. cost
   D. sensitivity
4. A deep vacuum can most accurately be measured in
   A. inches of mercury.
   B. inches of water column.
   C. pounds per square inch.
   D. microns.
5. Which of the tools used during the evacuation process can be wireless and use Bluetooth technology?
   A. Micron gauge
   B. Vacuum pump
   C. Core removal tool
   D. Evacuation hoses
6. A __________ can remove any type of foreign matter that enters a refrigeration system.
   A. vacuum pump
   B. filter drier
   C. recovery machine
   D. compressor
7. Moisture, acid buildup, and copper plating can all be caused by which of the following?
   A. Refrigerant and oil mixing
   B. A clogged filter dryer
   C. An undersized evaporator
   D. Air in the system
8. The only two substances that should be circulating in a refrigeration system are
   A. liquid and vapor.
   B. refrigerant and nitrogen.
   C. refrigerant and oil.
   D. nitrogen and oil.

9. In order to remove moisture from a refrigeration system using a vacuum pump, it must be in a ____________state.
   A. solid
   B. heated
   C. liquid
   D. vapor
10. When checking for a leak in the evaporator coil, the leak detector can be placed
    A. in the airstream.
    B. in the drain outlet.
    C. in the return duct.
    D. next to the evaporator fan.
11. Two types of evacuation methods used by technicians are
    A. active recovery and passive recovery.
    B. system-dependent and self-contained.
    C. deep vacuum and triple evacuation.
    D. nitrogen sweep and standing pressure test.
12. An evacuation process that is performed a number of times with nitrogen being used to break the vacuum, and ends with a deep vacuum is known as
    A. triple evacuation.
    B. refrigerant recovery.
    C. system flush.
    D. nitrogen sweep.
13. A system that is equipped with a ____________ may require additional steps to ensure that the entire system is evacuated.
    A. TXV
    B. relief valve
    C. solenoid valve
    D. receiver
14. The most accurate type of gauge for measuring a deep vacuum is a
    A. compound gauge.
    B. digital manometer.
    C. mercury manometer.
    D. digital micron gauge.
15. Proper evacuation without the use of a gauge manifold requires which of the following?
    A. Recovery machine.
    B. Nitrogen and a pressure regulator.
    C. Core removal tool.
    D. Electronic leak detector.
16. Which of the following conditions requires a system cleanup, an oil test, and oversized filters to be added to the system?
    A. Noncondensables in the system
    B. A vibration-dependent leak in the evaporator
    C. A temperature-dependent leak in the compressor discharge line
    D. A compressor motor burnout
17. A large refrigeration system that contains a significant amount of moisture should be evacuated with
    A. a high CFM vacuum pump.
    B. a low CFM vacuum pump.
    C. the system compressor running.
    D. a recovery machine then a nitrogen sweep.
18. Proper maintenance for a vacuum pump would include which of the following?
    A. Replacing all of the caps and plugs with new ones
    B. Changing the oil as often as necessary
    C. Replacing the oil sight glass
    D. Installing a new motor on the pump
19. When a deep vacuum is reached (500 microns), the system is valved off and observed for 10 to 15 minutes. A rise in pressure could indicate which of the following?
    A. It is normal and does not indicate anything.
    B. There may be a leak or the presence of air and/or moisture.
    C. The system charge was not properly recovered prior to evacuation.
    D. There may be acid or sludge in the system.
20. One of the advantages of a microfoam leak detection solution over standard bubble solutions is that it
    A. will not pit or corrode the components.
    B. contains chlorides.
    C. can detect leaks as small as 0.1 oz per year.
    D. will react with refrigerant oil and become luminescent.
21. An oily dust spot on a condenser or evaporator coil could indicate
    A. a defective fan motor.
    B. noncondensables in the system.
    C. the presence of acid.
    D. a refrigerant leak.
22. When leak checking the suction line with an electronic leak detector, the probe should be placed
    A. near the receiver.
    B. under the pipe insulation.
    C. in the condenser air stream.
    D. in the evaporator air stream.
23. Leaks that occur only when the unit is in operation are called
    A. combination-dependent leaks.
    B. temperature-dependent leaks.
    C. cumulative microleaks.
    D. vibration-dependent leaks.
24. Leaks that require two or more conditions in order to induce leakage are known as
    A. cumulative microleaks.
    B. combination-dependent leaks.
    C. temperature-dependent leaks.
    D. standing pressure leaks.

UNIT 9

REFRIGERANT AND OIL CHEMISTRY AND MANAGEMENT—Recovery, Recycling, Reclaiming, and Retrofitting

OBJECTIVES

After studying this unit you should be able to

- explain why refrigerant releases to the atmosphere are an environmental concern.
- describe ozone depletion and global warming.
- discuss how chlorine depletes the earth's ozone layer.
- differentiate between CFCs, HCFCs, HFCs, HFOs, and HCs.
- discuss four popular refrigerants and their applications.
- explain how refrigerants are numbered.
- discuss the terms "temperature glide" and "fractionation" as they apply to refrigerant blends.
- discuss refrigerant oils and their applications.
- discuss EPA regulations as they relate to refrigerants.
- define the terms *recover, recycle,* and *reclaim*.
- describe the three certification types that are available under Section 608 of the EPA Clean Air Act.
- describe methods of recovering refrigerants, including active and passive methods.
- discuss oil-less recovery machines.
- identify a DOT-approved recovery cylinder.
- describe a quick-connect or quick-coupler as they apply to refrigerant conservation.
- discuss the general considerations and guidelines for retrofitting an air-conditioning or refrigeration system.

SAFETY CHECKLIST

- ✔ Wear gloves and goggles any time you transfer refrigerant from one container to another.
- ✔ Transfer refrigerants only into DOT-approved containers.
- ✔ When retrofitting refrigerants and oils, always follow retrofit guidelines.
- ✔ Never use R-22 or any other refrigerant's service equipment on R-410A equipment. Service equipment must be rated to handle the higher operating pressures of R-410A.
- ✔ Use room sensors and alarms to detect R-123 refrigerant leaks, per ASHRAE Standard 15.
- ✔ Ensure that all refrigeration systems have some sort of pressure relief device, which should be vented to the outdoors to prevent the buildup of excessive system pressures.
- ✔ Be aware of a high-pressure reading in a recovery tank; it indicates air or other noncondensables in the tank.
- ✔ Never overheat the system's tubing while the system is pressurized. This can cause a blowout and serious injury. When certain refrigerants are exposed to high temperatures, a flame, or a glowing hot metal, they can decompose into hydrochloric and/or hydrofluoric acids and phosgene gas.
- ✔ Never energize the compressor when under a deep vacuum. This can cause a short in the motor windings at the fusite terminals or at a weak spot in the windings, which may damage the compressor.
- ✔ Always check the nameplate of the system into which dry nitrogen will enter. Never exceed the design low-side factory test pressure on the nameplate. Never pressurize a system with oxygen or compressed air because dangerous pressures can be generated from a reaction with the oil as it oxidizes.
- ✔ When using dry nitrogen, always use a pressure relief valve and pressure regulator on the nitrogen tank.
- ✔ Be sure to visit the Environmental Protection Agency's (EPA) website for the most up-to-date refrigerant-related regulations.

9.1 REFRIGERANTS AND THE ENVIRONMENT

As the earth's population grows and demands increase for comfort and newer technologies, more and more chemicals are produced and used in various combinations. Many of these chemicals reach the earth's atmospheric layers and produce different types of pollution. Some of the chemicals are the refrigerants that preserve our food and cool the spaces in which we live and work. Some of these refrigerants were developed in the 1930s; many of the refrigerants manufactured a few decades ago are now banned because of their negative effect on the natural environment. The trend has been toward a gradual phasing out of environmentally harmful refrigerants and a search for more environmentally friendly substitutes. While contained within a system, refrigerants are stable, are not pollutants, and produce no harmful effects on the earth's atmospheric layers. Over the years, however, quantities have leaked from systems and have been allowed to escape by the willful purging or venting of systems during routine service procedures. It is illegal to knowingly release certain refrigerants to the atmosphere, as established by EPA regulations.

9.2 OZONE DEPLETION

Ozone is a gas that is found in both the stratosphere and troposphere, **Figure 9.1**. It is a form of oxygen gas that is made up of three oxygen atoms (O_3), **Figure 9.2(A)**, instead of the two atoms in the standard oxygen molecule that we breathe (O_2), **Figure 9.2(B)**. Stratospheric ozone, which is found in a shell 7 to 30 miles above the earth, is considered good ozone because it prevents harmful ultraviolet radiation (UV-B) from reaching the earth. Tropospheric ozone, which is found within 7 miles of the earth's surface, is undesirable ozone because it is a pollutant.

Figure 9.1 Atmospheric regions.

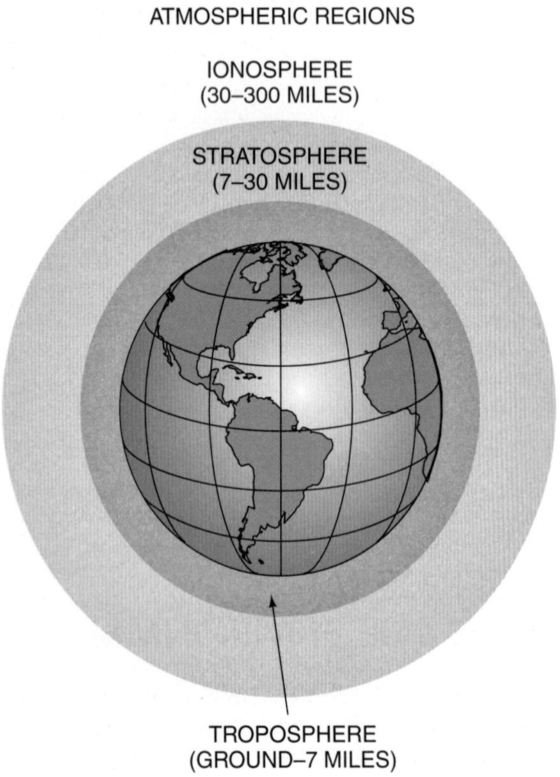

Figure 9.2 (A) An ozone molecule (O_3) and (B) a standard oxygen molecule (O_2).

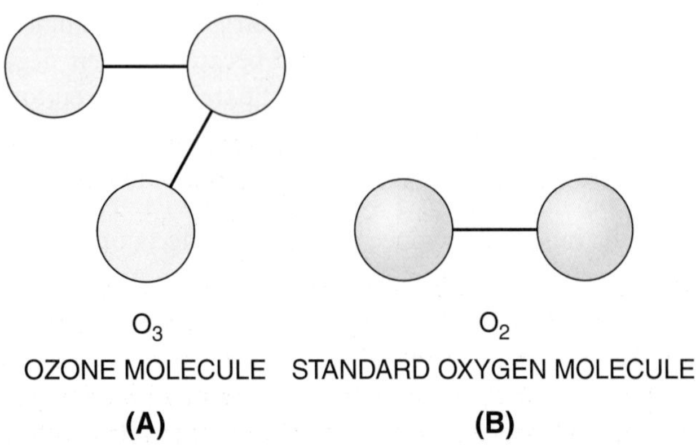

Stratospheric ozone shelters the earth from UV-B radiation, a protection essential for human health. Stratospheric ozone accounts for more than 90% of the earth's total ozone and is rapidly being depleted by man-made chemicals containing chlorine, including CFC and HCFC refrigerants.

The troposphere, or lower atmosphere that extends upward from ground level up to about 7 miles, holds only about 10% of the total ozone and contains 90% of our atmosphere, which is well mixed by weather patterns. Sunlight acting on chemicals in the troposphere creates bad ozone. This ozone has a bluish color when seen from earth and may irritate mucous membranes when inhaled. A popular term for tropospheric ozone pollution is *smog*.

Once stratospheric ozone is formed, much of it is easily destroyed by the sun's ultraviolet radiation, which is very powerful and has a frequency almost as high as that of X-rays. The sun's radiation breaks down the ozone (O_3) molecule into a standard oxygen (O_2) molecule and a free oxygen (O) atom, **Figure 9.3**. At the same time, more ozone is produced through plant photosynthesis and the bonding of standard oxygen (O_2) molecules with free oxygen (O) atoms. Thus, ozone is constantly being made and destroyed in the stratosphere, a balance that has been going on for millions of years. However, in the last several decades man-made chlorine-containing chemicals have disturbed this delicate balance.

Once in the stratosphere, ultraviolet radiation will separate a chlorine atom (Cl) from a CFC or HCFC molecule, which then goes on to destroy many ozone molecules. One chlorine atom (Cl) can destroy up to 100,000 ozone molecules. The depleted ozone molecules in the stratosphere let more UV-B radiation reach the earth, causing an increase in skin cancers and cataracts in humans and animals, a weakening of the human immune system, and a decrease in plant and marine life. Under the United Nations Environment Programme (UNEP) Montreal Protocol, an index called the ozone depletion potential (ODP) has been used for regulatory purposes. The higher the ODP, the more damaging the chemical is to the ozone layer in the stratosphere.

Figure 9.3 The breaking up of an ozone molecule by ultraviolet radiation.

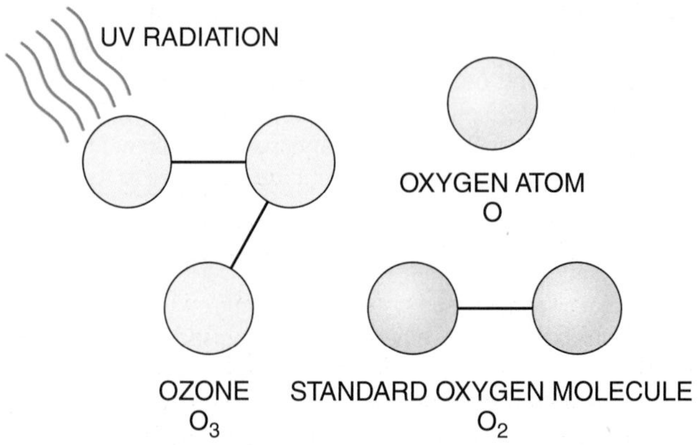

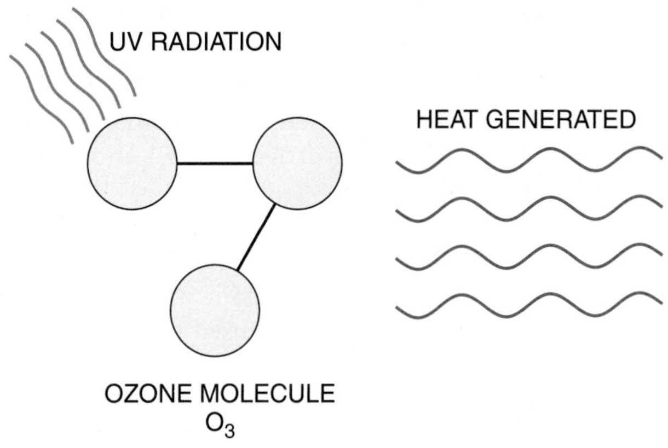

Figure 9.4 Heat generated from the breakup of an ozone molecule.

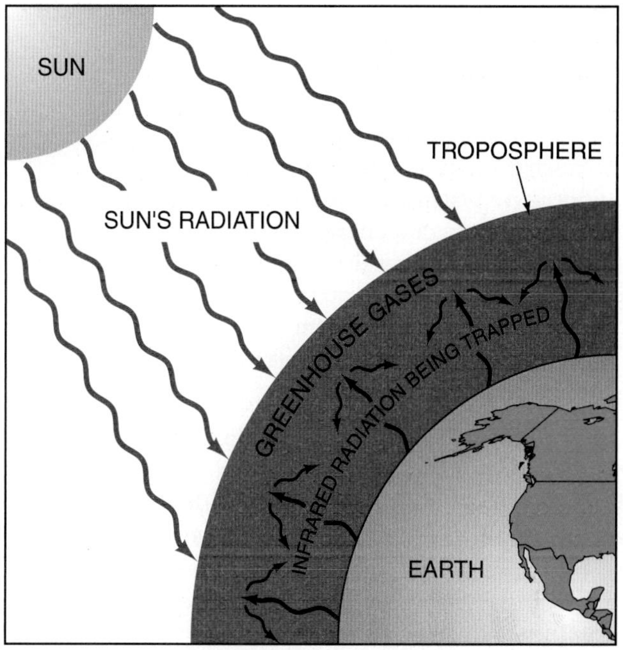

Figure 9.5 Greenhouse gases in the troposphere absorbing, refracting, and reflecting infrared radiation given off by the earth. This causes the greenhouse effect, or global warming.

When stratospheric ozone intercepts ultraviolet light, heat is generated, **Figure 9.4.** This heat is the force behind stratospheric winds, which affect weather patterns on earth. By changing the amount of ozone, or even its distribution in the stratosphere, the temperature of the stratosphere can be affected. This can seriously affect weather here on earth.

9.3 GLOBAL WARMING

Most of the sun's energy reaches the earth as visible light. After passing through the atmosphere, part of this energy is absorbed by the earth's surface and, in the process, is converted into heat energy. The earth, now warmed by the sun, radiates heat energy back into the atmosphere toward space. Naturally occurring gases and tropospheric pollutants like CFCs, HCFCs, and HFCs—as well as carbon dioxide, water vapor, and many other chemicals—absorb, reflect, and/or refract the earth's infrared radiation and prevent it from escaping the lower atmosphere, **Figure 9.5.** These gases, often referred to as *greenhouse gases*, slow the earth's heat loss, making its surface warmer than it would be if the heat energy had passed unobstructed through the atmosphere into space. The warmer earth's surface, in turn, radiates more heat until a balance is established between incoming and outgoing energy. This warming process, caused by the atmosphere's absorption of the heat energy radiated from the earth's surface, is called the *greenhouse effect*, or global warming.

The direct effects of global warming are created by chemicals that are released directly into the atmosphere. These direct emissions are measured by an index referred to as global warming potential (GWP). Refrigerants leaking from a refrigeration or air-conditioning system contribute to global warming. Their direct effects on global warming are measured by comparing them to the global warming effect of carbon dioxide (CO_2), which has a baseline GWP of 1.

Carbon dioxide is the number-one contributor to global warming. Human needs have resulted in an increase in the amount of carbon dioxide in the atmosphere by 25% in excess of its usual concentration. Most of this increase is due to the combustion of fossil fuels, as carbon dioxide is a by-product of the combustion of fossil fuels that is required by the modern world's need for electricity. In fact, of the electricity generated is used to power refrigeration and air-conditioning equipment. *More efficient HVACR equipment will require less electrical energy. However, even refrigeration or air-conditioning systems with relatively small refrigerant charges that never leak may have a great impact on global warming if the equipment is improperly charged. The equipment would be very inefficient under these conditions, and would contribute to global warming more than the leaking of refrigerant. This is because the longer run times would require more electricity, resulting in more carbon dioxide produced from the combustion of more fossil fuels. In the United States, fossil-fuel combustion is used for much of the generation of electricity. This is an example of an indirect effect of global warming.*

The indirect effects of global warming are what concern scientists. Most of the newer refrigerants introduced to the HVACR industry are more energy-efficient than the ones they are replacing; however, some are not as efficient. Even though some of the newer refrigerants do not contribute to ozone depletion, they still can contribute to the direct and indirect effects of global warming. Refrigerant HFC-134a is an example. It has a 0 ozone depletion index but contributes to global warming directly via refrigerant leaks and indirectly through being used in an inefficient system with a longer run time caused by refrigerant leaks. Many other refrigeration or

air-conditioning system problems may create higher current draws, higher power usage, and/or longer run times and thus contribute to the indirect effects of global warming. Some of system problems include:

- Dirty condensers and/or evaporators
- Frosted evaporator coils
- Incorrect or mixed refrigerants
- Inefficient compressors
- Excessively long or improperly sized line sets
- Inoperative system fans and blowers
- Incorrect refrigerant charge
- Dirty or clogged filter driers and/or metering devices
- Improper TXV settings
- Air or moisture in the system
- Incorrect refrigeration oil levels
- Unmatched system components

The total equivalent warming impact (TEWI) takes into consideration both the direct and indirect global warming effects of refrigerants. Refrigerants with the lowest global warming and ozone depletion potentials will have the lowest TEWI. Using HFC and HCFC refrigerants in the place of CFCs will also reduce the TEWI. All of the newer refrigerant alternatives that have been introduced have a much lower TEWI.

9.4 REFRIGERANTS

Most refrigerants are made from two molecules, methane and ethane, **Figure 9.6**. These two molecules contain hydrogen (H) and carbon (C) and are referred to as pure hydrocarbons (HCs). Pure hydrocarbons were at one time considered good refrigerants, but because of their flammability they were not used after the 1930s to any large degree. However, hydrocarbons are making a comeback in Europe, where they are used in domestic refrigerators, and are also making a comeback in the United States as a small percentage of the mixture in refrigerant blends for non-flammable applications. An example of a refrigerant that is not methane- or ethane-based is ammonia. Ammonia contains only nitrogen and hydrogen (NH_3) and is not an ozone-depleting substance.

Figure 9.6 Most refrigerants originate from (A) methane and (B) ethane molecules. Both molecules are hydrocarbons.

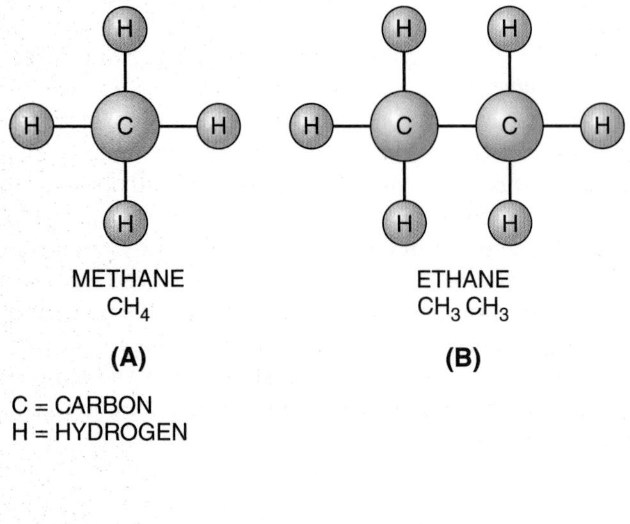

Any time some of the hydrogen atoms are removed from either the methane or ethane molecule and replaced with halogens such as chlorine or fluorine, the new molecule is said to be halogenated. Abbreviations are used to describe refrigerants chemically and to make it simpler for technicians to differentiate between them. The following are some common abbreviations:

- Chlorofluorocarbons (CFCs)
- Hydrochlorofluorocarbons (HCFCs)
- Hydrofluorocarbons (HFCs)
- Hydrocarbons (HCs)
- Hydrofluoro-olefins (HFOs)

⚠ SAFETY PRECAUTION: *Most refrigerants are heavier than air. Because of this, they can displace air and cause suffocation. Always vacate and ventilate the area if large refrigerant leaks occur. In fact, some sort of self-contained breathing apparatus should be available in the event of a large refrigerant leak. Inhalation of large amounts of refrigerant can cause cardiac arrest and eventually death.*•

9.5 CFC REFRIGERANTS

CFCs contain chlorine, fluorine, and carbon and are considered the most damaging to stratospheric ozone because their molecules are not destroyed before they reach the stratosphere. Because of their stable chemical structure, CFC molecules have a very long life when exposed to the atmosphere. They can be blown up into the stratosphere by atmospheric winds, where they react with ozone molecules and cause destruction. CFCs also contribute to global warming. It became illegal to intentionally vent CFC refrigerants into the atmosphere on July 1, 1992. Following is a chart showing the chemical name and formula for some CFC refrigerants.

Chlorofluorocarbons (CFCs)

Refrigerant No.	Chemical Name	Chemical Formula
R-11	Trichlorofluoromethane	CCl_3F
R-12	Dichlorodifluoromethane	CCl_2F_2
R-113	Trichlorotrifluoroethane	CCl_2FCClF_2
R-114	Dichlorotetrafluoroethane	$CClF_2CClF_2$
R-115	Chloropentafluoroethane	$CClF_2CF_3$

All of the refrigerants in the CFC group were phased out at the end of 1995. One important refrigerant is R-12, because it is commonly used for residential and light commercial refrigeration and for centrifugal chillers in some commercial buildings. The other refrigerants are used in other applications. For example, R-11 is used for many centrifugal chillers in office buildings and also as an industrial solvent to clean parts. R-113 is used in smaller commercial chillers in office buildings and also as a cleaning solvent. R-114 has been used in some household refrigerators in the past and in centrifugal chillers for marine applications. R-115 is part of the blend of refrigerants that make up R-502 used for low-temperature refrigeration systems. Because R-502 is a refrigerant blend that contains a CFC, it was also included in the January 1, 1996 CFC phaseout.

9.6 HCFC REFRIGERANTS

The second group of refrigerants in common use is the HCFC group. HCFCs do have some global-warming potential, but it is much lower than most CFCs. HCFCs contain hydrogen, chlorine, fluorine, and carbon. The amount of chlorine is small, but they also have hydrogen, which makes them less stable in the atmosphere. Although these refrigerants have much less potential for ozone depletion because they tend to break down in the atmosphere, releasing the chlorine before it reaches and reacts with the ozone in the stratosphere, the HCFC group is scheduled for a total phaseout by the year 2030. As with CFCs, it became illegal to intentionally vent HCFC refrigerants into the atmosphere on July 1, 1992. HCFC-22 (R-22) has an earlier phaseout date of 2010 for new equipment and a total production phaseout of 2020 under the Montreal Protocol. Total phaseout means no production and no importing of the refrigerants. At the time of this writing, all HCFCs will be totally banned by the year 2030.

Even though HCFCs have lower ODPs than chlorofluorocarbons (CFCs), it wasn't low enough, so the United States began to phase out HCFCs. In fact, the Montreal Protocol was modified in 2007 and the HCFC phaseout schedule was accelerated. Below is a history and timeline of the schedule:

1987 Montreal Protocol sets reduction rates for HCFCs and other ozone-depleting substances.

1992 Illegal to intentionally vent HCFCs into the atmosphere.

1994 Mandatory technician certification for servicing A/C or refrigeration equipment.

1996 Capping of HCFC levels under the Montreal Protocol goes into effect.

2004 HCFC refrigerant production reduced by 35%.

2007 Reduction of HCFC production was accelerated from 65% to 75% from the baseline production year of 1989.

2010 HCFC-22 is banned in new equipment. No production or importing of HCFC-22 and HCFC-142b, except for use in equipment manufactured before January 1, 2010.

2015 90% cap on HCFC-22 production from the baseline production year of 1989. No production or importing of any HCFC, except for use as refrigerants in equipment manufactured before January 1, 2010.

2020 Total ban on HCFC-22 production. No production and no importing of R-22 and R-142b.

2030 Total ban on all HCFC production. No production and no importing of any HCFC.

These accelerated rulings affect the developed countries, such as the United States, Canada, and countries in Europe, but the new agreement also affects developing countries, such as China, India, and Mexico, which will now phase out production of HCFCs by 2030 instead of 2040. These new rulings mean that all new refrigeration and air-conditioning equipment produced in the United States will be required to use a 0-ODP refrigerant. Among this group of soon-to-be-phased-out refrigerants are:

Hydrochlorofluorocarbons (HCFCs)

Refrigerant No.	Chemical Name	Chemical Formula
R-22	Chlorodifluoromethane	$CHClF_2$
R-123	Dichlorotrifluoroethane	$CHCl_2CF_3$
R-124	Chlorotetrafluoroethane	$CHClFCF_3$
R-142b	Chlorodifluoroethane	CH_3CClF_2

9.7 HFC REFRIGERANTS

The third group of refrigerants is the HFC group, which contains hydrogen, fluorine, and carbon atoms. HFC molecules contain no chlorine atoms and will not deplete the earth's protective ozone layer, but they do have the potential to contribute to global warming. HFCs are the long-term replacements for many CFC and HCFC refrigerants. On November 15, 1995, it became unlawful to intentionally vent HFC refrigerants into the atmosphere. The original plan was to replace R-12 with R-134a, but this plan is complicated by the fact that R-134a is not compatible with any oil left in an R-12 system. It is also not compatible with some of the materials used to construct R-12 systems. However, an R-12 system can be retrofitted to operate with R-134a, but this is a task that must be performed by an experienced technician because it requires a complete oil change. The residual oil left in the system pipes must be removed, and all system gaskets must be checked for compatibility. To be sure the compressor materials are compatible, they must be checked by contacting the manufacturer.

Following is a list of refrigerants in the HFC group:

Hydrofluorocarbons (HFCs)

Refrigerant No.	Chemical Name	Chemical Formula
R-125	Pentafluoroethane	CHF_2CF_3
R-134a	Tetrafluoroethane	CH_2FCF_3
R-23	Trifluoromethane	CHF_3
R-32	Difluoromethane	CH_2F_2
R-125	Pentafluoroethane	CHF_2CF_3
R-143a	Trifluoroethane	CH_3CF_3
R-152a	Difluoroethane	CH_3CHF_2
R-410A	Blend	$CH_2F_2/CHF_2\text{-}CF_3$

9.8 HYDROFLUORO-OLEFIN (HFO) REFRIGERANTS

Hydrofluoro-olefin (HFO) refrigerants are referred to as fourth-generation refrigerants for the twenty-first century. HFO refrigerants are actually unsaturated hydrofluorocarbon (HFC) refrigerants and are widely recognized as the next generation of refrigerants because of their environmental friendliness, cost effectiveness, and great energy efficiencies. HFOs are distinguished from HFCs by being derivatives of olefins rather than alkanes (paraffins). Olefins have carbon atoms linked by a double bond where alkanes have single bonds between carbons atoms. Hydrofluorocarbon (HFC) refrigerants like R-134a, R-125, R-143a, R-152a and HFC-based refrigerant blends like R-507, R-407A, R-407B, R-407C, and R-410A are all composed of hydrogen, fluorine, and carbon connected by single bonds between the atoms. Hydrofluoro-olefin (HFO) refrigerants are also composed of hydrogen, fluorine, and carbon atoms, but contain at least one double bond between the carbon atoms, **Figure 9.7**.

Two popular HFO refrigerants are HFO-1234-yf and HFO-1234ze. Both HFO-1234yf and HFO-1234ze have a zero ODP with extremely low global warming potentials (GWPs). Refer to the following chart for ODP and GWP values. These refrigerants have much shorter lifecycles in the atmosphere, which results in the low GWP values. These two HFO refrigerants have low toxicity but are slightly flammable. Safety ratings for refrigerants have been established and are presented in table form later in this Unit in **Figure 9.14**. The safety rating for refrigerants such as HFO-1234yf and HFO-1234ze, which are slightly flammable and have low toxicities, is A2L.

HFO-1234yf is a low GWP replacement for R-134a intended for use in mobile air-conditioning (MAC) systems in the automotive industry. HFO-1234ze is intended to replace R-410A in residential and light commercial air-conditioning and heat pump applications while offering a 75% reduction in GWP. It can be used in both air-cooled and water-cooled chillers in supermarkets and commercial buildings. Other applications include vending machines, refrigerators, beverage dispensers, air dryers, and carbon dioxide cascade systems in commercial refrigeration. HFO-1234ze also offers excellent energy efficiency, is cost effective, and can be used in existing equipment design with minimal changes. HFO-1234ze also has significant advantages over R-32 which has a medium GWP, an A2 refrigerant safety rating, and high discharge temperatures. High discharge temperatures can affect system performance and durability in hot climates. Also, oil changes may be needed with R-32.

HFOs are miscible in polyol ester (POE) oils. The miscibility of HFOs with POE lubricants is comparable to that of R-134a. HFOs are not soluble in mineral oil or alkylbenzene lubricants.

Because HFOs are slightly flammable, extensive studies have been conducted to determine whether or not these refrigerants are safe for use in automobiles, centrifugal chillers, and other stationary applications. A key consideration for widespread acceptance of HFO refrigerants is the development of standards and codes for safe use of these mildly flammable A2L refrigerants.

Figure 9.7 HFC refrigerant molecules have only single atomic bonds (left), while HFO refrigerant molecules have at least one double atomic bond (right).

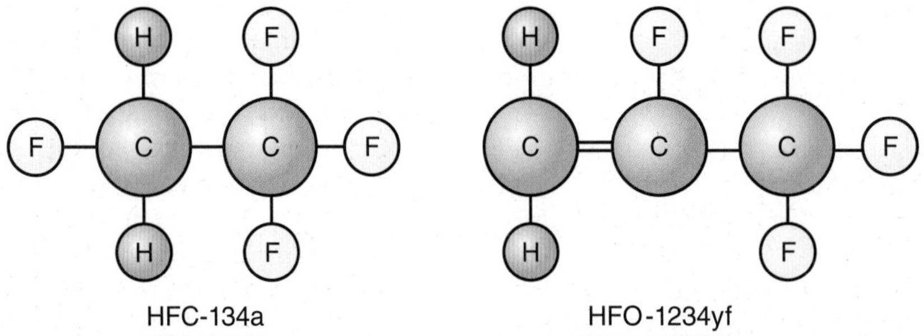

HFC-134a (CH_2FCF_3)

HFO-1234yf ($CH_2{=}CFCF_3$)

9.9 HYDROCARBON (HC) REFRIGERANTS

The fourth group of refrigerants is the hydrocarbon (HC) group. HC refrigerants are classified by ASHRAE in the A3 safety group. Refrigerants in the A3 safety group are highly flammable. Refer to **Figure 9.14** for safety classification of refrigerants. Pure HC refrigerants have no chlorine or fluorine in their molecules as they contain only hydrogen and carbon. Since HCs do not contain any chlorine, they have a zero ODP. As an example, two hydrocarbon molecules, methane and ethane were illustrated in **Figure 9.6**. Most CFC, HCFC, and HFC refrigerants originate from these two base molecules. HC refrigerants do contribute to global warming, but their global warming potential (GWP) is very low when compared to those of CFC, HCFC, and HFC refrigerants. Listed below are some GWP values for comparison purposes.

Refrigerant	Type	ODP	GWP	Level
R-32	HFC	zero	675	Medium
R-134a	HFC	zero	1430	Medium
R-404A	HFC	zero	3922	High
R-410A	HFC	zero	2088	Medium
R-290	HC	zero	3.0	Low
R-600a	HC	zero	3.0	Low
R-1234yf	HFO	zero	4.0	Low
R-1234ze	HFO	zero	6.0	Low

Note: GWP values are from the IPCC 4th Assessment Report, 2007, (100 year) carbon dioxide = 1

HC refrigerants have better energy efficiencies and reduced energy usage when compared to currently used HFC refrigerants. Hydrocarbon refrigerants are created by nature, not a chemical company, and are often referred to as natural refrigerants. The most efficient and environmentally safe refrigerants that exist in the world today are all natural refrigerants. These refrigerants include hydrocarbons, ammonia, air, water, and carbon dioxide. Hydrocarbon refrigerants are used as stand-alone refrigerants in Europe for many applications. Listed below are some of these applications:

- Commercial refrigerators and bottle coolers
- Split air-conditioning units
- Domestic refrigerators and freezers

In the United States, only some HC refrigerants have been approved by the EPA under its Significant New Alternative Policy (SNAP) program. The SNAP program has approved HC refrigerant use in the United States only in new equipment, in limited conditions, and with limited refrigerant charge amounts because of their high flammability. For many years, small percentages of HC refrigerants have been used in many HCFC- and HFC-based refrigerant blends in the United States to assist in oil return to the compressor. However, these refrigerant blends containing a small amount of HC refrigerants are not flammable. This is because the HC refrigerant constitutes only a small percentage (less than 3%) of the total refrigerant blend. Some hydrocarbon (HC) refrigerants are:

- Propane
- Butane
- Isobutane
- Isopentane
- Methane
- Ethane
- R-441A (HC blend of ether, propane, isobutene, and n-butane)

⚠ **SAFETY PRECAUTION:** *Many of these chemicals may be used for multiple purposes. However, when used as a refrigerant, a refrigerant-grade purity level is required. Never use anything but refrigerant-grade HCs when using them as a refrigerant in a system. For example, propane is used both as a fuel and as a refrigerant. However, different purity levels are required for each application. Never use fuel-grade (LPG) propane in a refrigerant system. The quality of fuel grade propane's composition varies substantially and can have high levels of moisture and unsaturated hydrocarbons. Also, fuel-grade propane contains a stenching agent, methyl mercaptan, which smells like rotten eggs to make it easily detectable by humans. Methyl mercaptan will cause unwanted by-products and corrosion in an HC refrigeration system, leading to premature system failure.*•

Some refrigerant blends that incorporate HC refrigerants in small percentages for oil return purposes include:

- R-402A
- R-402B
- R-403A
- R-406A
- R-438A

FLAMMABLE REFRIGERANT APPLICATIONS AND CHARGE AMOUNTS

At time of this writing, only R-600a (isobutane), R-290 (propane), R-170 (ethane), HFC-32 (difluoromethane), and R-441A (HC blend) have been approved by the EPA's SNAP program for use in the United States. HFC-32 is a flammable refrigerant but is not an HC refrigerant. The EPA also exempted all of the substances listed below, except HFC-32 from the Clean Air Act venting prohibition. Uses for these flammable refrigerants include:

- R-290 (propane) can be used in household refrigerators, freezers, combination refrigerators and freezers, vending machines, and room air-conditioning units as long as the refrigerant charge does not exceed 5.3 ounces (150 grams).

- R-600a (isobutane) can be used in retail food refrigeration (stand-alone commercial refrigerators and freezers) and vending machines as long as the refrigerant charge size does not exceed 2.0 ounces (57 grams).
- Ethane (HC-170) can be used in very low-temperature refrigeration and non-mechanical heat transfer.
- R-441A can be used in retail food refrigeration (stand-alone commercial refrigerators and freezers), vending machines, and room air-conditioning units as long as the refrigerant charge size does not exceed 2.0 ounces (57 grams).
- HFC-32 (difluoromethane) in room air-conditioning units.

Notice that the EPA's SNAP program only allows some HC and flammable refrigerant use in the United States in limited applications and for new equipment only. Retrofitting equipment containing HC refrigerants and other flammable refrigerants is not allowed in the United States. Only use authorized refrigerants listed on the compressor and system's nameplate in each specific piece of equipment. Even replacement parts for equipment containing a flammable refrigerant must be compatible for that specific refrigerant. Local regulations governing flammable refrigerant usage as a refrigerant may also apply in certain municipalities. It is the responsibility of the service technician to stay abreast of local, national, and global regulations that govern flammable refrigerants. It is recommended that service technicians working on flammable refrigerants receive some sort of training and/or certification on safety and handling of these refrigerants. Education is the answer to working smartly and safely when working around or with these flammable refrigerants. Contact a local trade organization for information on safety training and certification when working with flammable refrigerants.

NAMEPLATES AND LABELS

One of the first steps in working safely with flammable HC refrigerants is to identify the refrigerant being used in the piece of equipment that is being serviced. As mentioned, use only the HC refrigerant on the system and/or compressor's nameplate or serial label. The serial label or nameplate on the compressor will indicate what HC refrigerant can be used in that system and compressor. Often, both warning labels and flammable refrigerant labels will be displayed on the compressor. Specific labeling that identifies the refrigerant used is a requirement for equipment incorporating a flammable refrigerant, **Figure 9.8**.

Other sections to be labeled warning the service technician or equipment operator include:

- Evaporator sections
- The compressor or condensing unit compartment
- The exterior of the refrigerator
- Exposed refrigerant tubing

The labeling is required in order to be SNAP approved. Below are examples of such labeling:

DANGER—Risk of fire or explosion. Flammable refrigerant used. Do not use mechanical devices to defrost evaporator. Do not puncture refrigerant tubing.

DANGER—Risk of fire or explosion. Flammable refrigerant used. To be repaired only by trained service personnel. Do not puncture refrigerant tubing.

CAUTION—Risk of fire or explosion. Flammable refrigerant used. Consult the repair manual or owner's guide before attempting to service this product. All safety precautions must be followed.

CAUTION—Risk of fire or explosion. Flammable refrigerant used. Dispose of properly in accordance with federal or local regulations.

CAUTION—Flammable refrigerant used. Risk of fire or explosion due to puncture of refrigerant tubing. Follow handling instructions carefully.

CAUTION—Sudden release of refrigerant in the proper concentration with air can cause flash fire, explosion, or a sustained fire.

SERVICE ACCESS AND TUBE PAINTING

All refrigerant piping through which the refrigerator or freezer is serviced must be colored red. This red must be Pantone Matching System (PMS) #185. The red color must be present on the refrigerant process tubing and service ports or wherever a service technician may puncture the tubing for service purposes. This red piping must extend at least 1 in. or 2.54 centimeters from the compressor and at least 1 in. on each side of the process tubes or service ports. If the color has been worn or has been removed, it must be repainted on by the service technician.

SAFETY PRECAUTION: *Data from both the cabinet and compressor or condensing unit should match when identifying the type of refrigerant. If it doesn't, an unauthorized retrofit may have been done. Remember, retrofitting a system with an HC or other flammable refrigerant is not allowed in the United States. An immediate stoppage of work should follow if an HC or other flammable refrigerant system is suspected to be retrofitted and appropriate authorities should be notified.•*

COMPRESSOR TERMINAL VENTING

Figure 17.35 in Unit 17, "Types of Electric Motors," illustrates a hermetic compressor motor's terminal pins extending from the shell of the compressor. These pins often represent run, start, and common in single-phase compressors and Line 1, Line 2, and Line 3 in three-phase compressors. These terminal pins are connected to the compressor motor's electrical windings. The windings are in an environment of refrigerant and oil under pressure within the shell

Figure 9.8 Specific labeling which identifies a refrigerant used is a requirement for HVACR equipment incorporating a flammable refrigerant. Photo courtesy of Tecumseh Products Company.

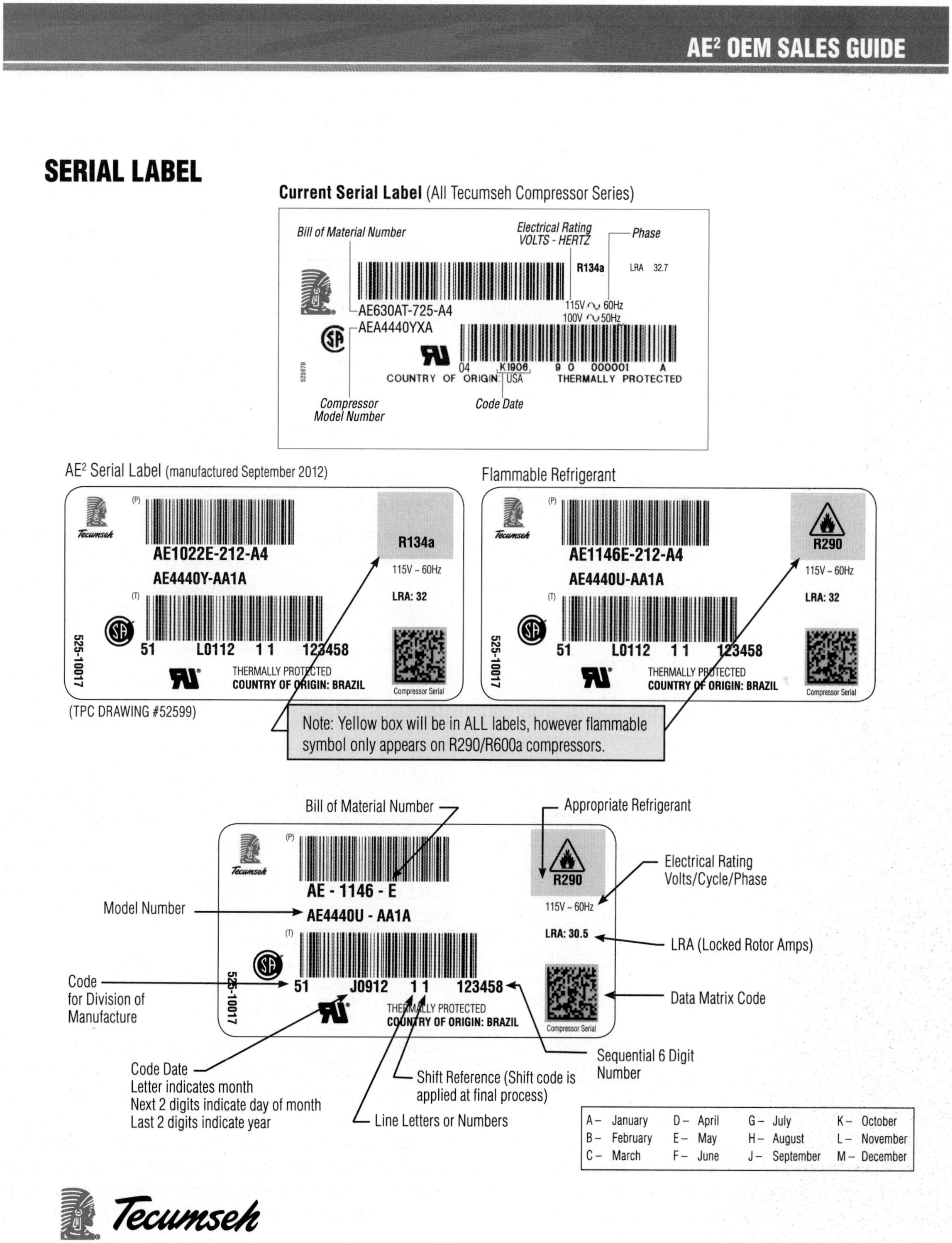

of the compressor. Often, an electrical short can occur from the motor windings to ground or from one electrical line to another causing terminal venting. Terminal venting is when one or more terminal pins blows out of its ceramic fusite terminal, allowing a pressurized refrigerant and oil spray to escape into the atmosphere. This refrigerant and oil spray can become ignited by the electrical sparking from electrical shorts or from other ignition sources nearby causing flash fires, severe injury, and even death. These terminal pins can also become damaged from misuse or from a physical jolt. Following the safety precautions below can help prevent terminal venting:

- Make sure all power supplies are disconnected before removing the compressor's protective cover.
- Make sure all power legs are open.
- If you hear any electrical-type noises coming from the compressor, disconnect power to the unit and leave the area immediately.
- Make sure the compressor's terminal covers are in place before energizing the compressor.
- Make sure the compressor and entire chassis system are properly grounded.
- Always check for a ground fault interrupt.
- Always check for blown fuses or tripped circuit breakers, which may indicate improper grounding.

Figure 9.9 A list of some known CFC, HCFC, HFC, HC, HFO, and natural refrigerants with their safety classifications and boiling points at atmospheric pressure for comparison purposes.

REFRIGERANT	BOILING POINT AT ATMOSPHERIC PRESSURE	ASHRAE 34 SAFETY CLASSIFICATION
R-11	74.7°F (23.7°C)	A1
R-12	−21.6°F (−29.8°C)	A1
R-22	−41.5°F (−40.8°C)	A1
R-123	82.1°F (27.8°C)	B1
R-1234yf	−21.2°F (−29.6°C)	A2L
R-1234ze	−2.1°F (−18.9°C)	A2L
R-134a	−14.9°F (−26.1°C)	A1
R-290 (propane)	−43.8°F (−42.1°C)	A3
R-401A	−16.9°F (−27.2°C)	A1
R-404A	−49.8°F (−45.4°C)	A1
R-407C	−33.9°F (−36.6°C)	A1
R-410A	−60.5°F (−51.4°C)	A1
R-422D	−37.0°F (−38.3°C)	A1
R-441A	−4.6°F (−20.3°C)	A3
R-507A	−52.1°F (−46.7°C)	A1
R-600a (isobutane)	10.9°F (−11.7°C)	A3
R-717 (ammonia)	−28.0°F (−33.3°C)	B2L
R-718 (water)	212°F (100°C)	A1
R-744 (carbon dioxide)	N/A	A1
R-1150	−155°F (−104°C)	A3

BOILING POINTS AND SAFETY CLASSIFICATIONS

Hydrocarbon-based refrigerants, such as HCFCs and HFCs, have low boiling points that are desirable for refrigeration and/or air-conditioning applications. However, because of these low boiling points, safety precautions must be taken. **⚠ SAFETY PRECAUTION:** *Exposing the skin to liquid refrigerants can cause severe frostbite. Exposure to the eyes can cause blindness.* **Figure 9.9** *lists some known CFC, HCFC, HFC, HFO, HC, and natural refrigerants with their safety classifications and boiling points at atmospheric pressure for comparison purposes.•*

HC REFRIGERANT HANDLING AND WORK SAFETY. Because HC refrigerants are flammable, certain safety precautions must be followed. All service technicians should be familiar with the safety precautions for flammable refrigerants. Safety Data Sheets (SDS), previously known as Material Safety Data Sheets (MSDS), are required to be available to all employees where HC and other flammable refrigerants are being used. SDS are also required to be available to local fire departments.

The lower exposure limit (LEL) of a gas is the lowest concentration of the gas in air capable of producing a flash or fire in the presence of an ignition source. Propane (R-290) has an LEL of 2.1% concentration in air. Isobutane (R-600a) has an LEL 1.8% concentration in air. Concentrations below the LEL are too "lean" to burn. The upper exposure limit (UEL) of a gas is the highest concentration of a gas in air capable of producing a flash or fire in the presence of an ignition source. Propane has a UEL of 9.5% concentration in air. Isobutane has a UEL of 8.5% concentration in air. Concentrations above the UEL are too "rich" to burn.

Flame or sparks from any HVACR equipment or tools could cause a fire or explosion when using HC refrigerants. Always check with the local authorities having jurisdiction for other safety precautions. Listed below are some important handling and safety practices for HC refrigerants and related accessories:

- Sparks from a vacuum pump switch, pressure switch, contactors, electronic leak detector, halide torch, light switch, defrost timers, or thermostats can cause explosions when exposed to the proper concentrations of HC gas and air.
- Never disconnect power or unplug anything or begin a service procedure before checking for flammable refrigerants in the area with a combustible gas leak detector or monitor, **Figure 5.123**.

Figure 9.10 An enclosed relay and overload. Photo courtesy of Tecumseh Products Company.

- Service technicians must: ***Monitor—Ventilate—Eliminate.*** Monitor the area for combustible gasses. Ventilate the area to get rid of any combustible gasses. Eliminate any sources of ignition.
- Always use non-ignitable type leak detectors for HC gases such as ultrasonic detectors, liquid solutions, or ultraviolet additives.
- Electronic leak detectors for flammable gases are manufactured which are non-ignitable.
- Make sure all relays and overloads are sealed. **Figure 9.10** illustrates an enclosed relay and overload. Many manufacturers are using positive temperature coefficient resistors (PTCR) because they are spark free, **Figure 17.41.**
- If a component has to be replaced, it is best to replace the part with an original equipment manufacturer (OEM) part.
- Service technicians should use spark-proof tools. Many brass tools are spark-proof.
- Flammable gas leak detectors and/or exposure limit monitors should be used before entering a confined space where HC refrigerants are used. Try to eliminate all ignition sources and ventilate the area. Do not turn the monitor off until you leave the service area.
- Never use a torch to remove refrigerant components. Always use tubing cutters.
- Always use approved components when replacing parts.
- Never smoke at or near the confined space where HC refrigerants exist.
- Recovery equipment must be rated for HC refrigerants and only equipment designed for flammable refrigerants should be used.
- All original and replacement parts must be listed by the manufacturer and must be spark free when in use.

TRANSPORTATION OF HC REFRIGERANT CYLINDERS

- The vehicle must be labeled to let the general public know it is carrying a flammable gas.
- The contents of the vehicle must be inventoried.
- HC cylinders must be transported in the upright position.
- Some municipalities and/or states require venting space to the exterior and that cylinders be stored in explosion-proof cabinets.
- All required signage and placarding must be used.
- HC and other flammable refrigerants must be transported in approved containers.
- Always check with the local department of transportation for more detailed requirements.

HYDROCARBON CYLINDER STORAGE

- Store cylinders away from any ignition source.
- A combustion gas monitor must be installed.
- Access must be limited to authorized personnel.
- HC cylinders must be stored away from all air intakes.
- Cylinders must be stored at ground level.
- Cylinders must be stored in a secured and locked cage.

9.10 REFRIGERANT NUMBERING

An easier way to refer to refrigerants is by number instead of by their complex chemical names. The DuPont Company has created an easy system for naming refrigerants, depending upon their chemical composition. This DuPont numbering system works with refrigerants that have either 0 or 1 as their left-most digit, which is the third number from the right. The three steps to using this numbering system are as follows:

1. The first digit from the right is the number of fluorine atoms.
2. The second digit from the right is one more than the number of hydrogen atoms.
3. The third digit from the right is one less than the number of carbon atoms.
   - For example, if this number is zero, there is one carbon atom and the refrigerant is methane-based.

Example 1

HCFC-123 $CHCl_2CF_3$ Dichlorotrifluoroethane
It can be seen that the third number from the right is a 1, indicating that there are two carbon atoms in an R-123 molecule. From **Figure 9.6(B)**, it can be determined that the base molecule for R-123 is ethane, as the ethane molecule, like R-123, has two carbon atoms. From the diagram in **Figure 9.6(B)**, it can also be seen that there are six atoms that can bond to these two carbon atoms. The right number is a 3, indicating that there are three fluorine atoms in an R-123 molecule. Since the second number from the right is a 2, we can determine that there is only one hydrogen atom in an R-123 molecule. At this point, there are three fluorine atoms and one

hydrogen atom bonded to the two carbon atoms, leaving two "empty" spaces. These empty spaces will be filled by chlorine atoms. The number of chlorine atoms in the molecule can also be found by subtracting the sum of fluorine and hydrogen atoms from the total number of atoms that can be connected to the carbon atom(s). In this example the sum of fluorine and hydrogen atoms is 4, and there are 6 total spaces for atoms to bond to the carbon atoms. Since 6 − 4 = 2, there are 2 chlorine atoms in an R-123 molecule. Since we have determined that there are hydrogen, chlorine, fluorine, and carbon atoms present in an R-123 molecule, we can conclude that R-123 is a hydrochlorofluorocarbon (HCFC) refrigerant.

Example 2

CFC-12 CCl_2F_2 Dichlorodifluoromethane

For R-12, it can be seen that the third number from the right is a 0, indicating that there is only one carbon atoms in an R-12 molecule. From **Figure 9.6(A)**, it can be determined that the base molecule for R-12 is methane, as the methane molecule, like R-12, has only one carbon atom. From the diagram in **Figure 9.6(A)**, it can also be seen that there are four atoms that can bond to the single carbon atom. The right number is a 2, indicating that there are two fluorine atoms in an R-12 molecule. Since the second number from the right is a 1, we can determine that there are no hydrogen atoms in an R-12 molecule. At this point, there are two fluorine atoms and no hydrogen atoms bonded to the two carbon atoms, leaving two "empty" spaces. These empty spaces will be filled by chlorine atoms. Since we have determined that there are chlorine, fluorine, and carbon atoms present in an R-12 molecule, we can conclude that R-12 is a chlorofluorocarbon (CFC) refrigerant.

The small letters "a, b, c, d, . . ." at the end of the numbering system represent how symmetrical the molecular arrangement is. An example is R-134a and R-134. These are completely different refrigerants with completely different properties. Both have the same number of atoms of the same elements but differ in how they are arranged. These molecules are said to be isomers of one another. In **Figure 9.11**, notice that the R-134 molecule has perfect symmetry. As isomers of this molecule start to get less and less symmetrical, the small letters a, b, c, d, . . . are assigned to their ends. Refer to the next section on the numbering of blended refrigerants.

9.11 REFRIGERANT BLENDS

Ozone depletion, global-warming, and other environmental issues have stimulated much research on **refrigerant blends**. Refrigerant blends have been used successfully in both comfort cooling and refrigeration. Blends can have as many as five refrigerants mixed together to provide properties and efficiencies similar to the refrigerants they will replace. HCFC-based blends are short-term replacements for many CFC and HCFC refrigerants, whereas HFC-based blends are considered to be long-term replacements.

Azeotropes or azeotropic blends are made up of two or more refrigerants. When these component refrigerants are mixed together, they behave like single-compound refrigerants, such as R-12 and R-22, when phase changing between the liquid and vapor states. This means that there is only one boiling point and/or one condensing point for each given system pressure. An illustration of the behavior of an azeotropic blend is provided in **Figure 9.12**. When the refrigerant is 100% vapor, such as when refrigerant is being discharged from a compressor, the refrigerant is superheated at a high discharge temperature and pressure (point 1). As the refrigerant flows toward the condenser, it desuperheats, but it is still at the same high side pressure (point 2). In a similar manner, when refrigerant is 100% liquid, the addition or removal of heat will result in a change of temperature with no change in pressure (points 3 and 4). When the refrigerant is saturated, there is a singular pressure/temperature relationship. So, at any given pressure, there is one and only one temperature. Similarly, at any given temperature, there is one and only one pressure (points 5 and 6). Azeotropic blends are not new to the HVACR industry. Both R-500 and R-502 are azeotropic blends that have been used in the industry for years. R-500 is made up of 73.8% R-12 and 26.2% R-152a (by weight).

Figure 9.11 R-134 and R-134a are isomers.

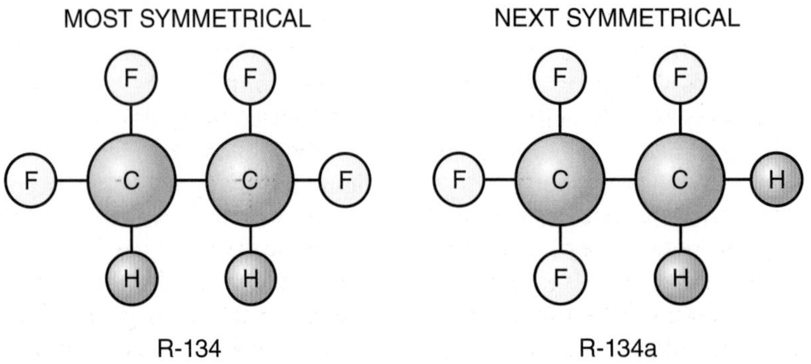

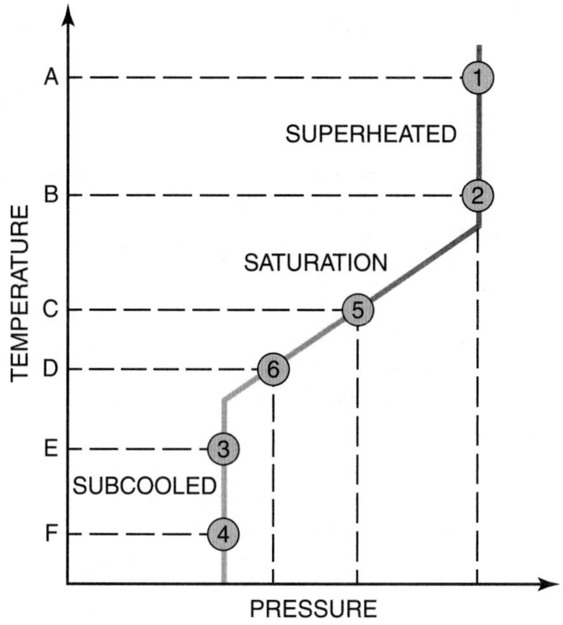

Figure 9.12 Graphic representation of the behavior of an azeotropic refrigerant blend.

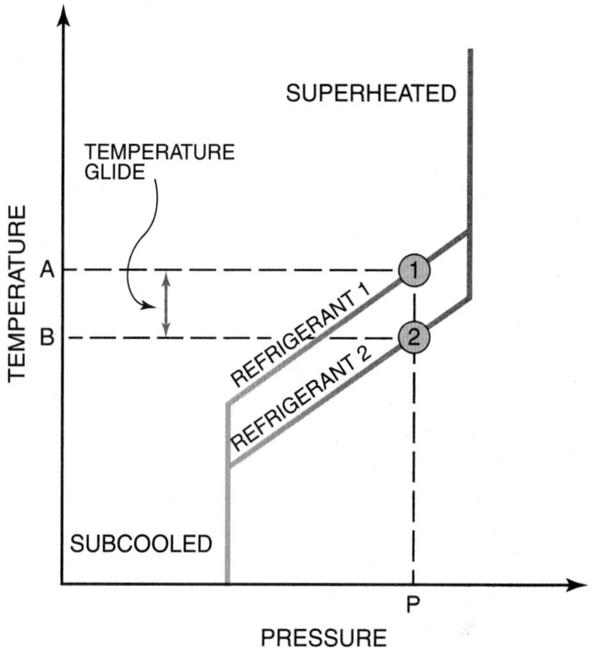

Figure 9.13 Graphic representation of the behavior of a near-azeotropic refrigerant blend.

R-500 is used in some cooling systems. R-502 is made up of 48.8% R-22 and 51.2% R-115 and has been an excellent low-temperature refrigerant since the 1960s. Both R-500 and R-502 were phased out on January 1, 1996.

However, many refrigerant blends are not azeotropic but are near-azeotropic mixtures. Such blends have a tendency to separate into individual refrigerants. These blends act differently because molecules of the different component refrigerants are present in any one sample of liquid or vapor. This is why there is a difference between the temperature/pressure relationships of the near-azeotropic blends versus single-compound refrigerants like CFC-12, HCFC-22, HFC-134a, and azeotropic blends.

Near-azeotropic blends experience what is known as temperature glide. Temperature glide occurs when the blend has many temperatures as it evaporates and condenses at a given pressure. An illustration of the behavior of a near-azeotropic blend is provided in **Figure 9.13**. Think of this illustration as two single-compound refrigerant graphs, like the one shown in **Figure 9.12**, superimposed on top of each other. When superheated or subcooled, the refrigerants behave in a similar manner. However, during the change of state, the component refrigerants behave differently, as each changes state at a different temperature at the same pressure. For example, at pressure P, refrigerant 1 changes state at temperature A (point 1), while refrigerant 2 changes state at temperature B (point 2). There is, therefore, a range of saturation temperatures for a given pressure. This range is the temperature glide. So, for the conditions in **Figure 9.13**, the temperature glide can be calculated as temperature A – temperature B. Zeotropic refrigerant blends also experience temperature glides, but are much greater than those associated with near-azeotropic refrigerant blends.

Near-azeotropic refrigerant blends may also experience **fractionation**. Fractionation occurs when one or more of the refrigerants in the blend condense or evaporate at different rates than the other refrigerants. This different rate is caused by the slightly varied vapor pressures of each refrigerant in the blend. Fractionation will be discussed in detail in Unit 10, "System Charging." Zeotropes, or zeotropic blends, also have temperature glide and fractionation but to a larger extent than near-azeotropic blends. The terms *zeotropic blends* and *near-azeotropic blends* are used in the industry because they both undergo temperature glide and fractionation. Charging methods for refrigerant blends will be covered in Unit 10, "System Charging."

"R" NUMBERS FOR THE BLENDS

- The 400-series blends are the near-azeotropic blends. If a blend has an R-410 designation, it is a near-azeotropic blend. The 10 indicates that it is the tenth blend to be commercially produced. All 400-series blends have at least some temperature glide and can fractionate.
- The 500-series blends represent the azeotropic blends, for example, R-502. The 2 means that it is the second one produced on the market.
- Blends can also be named by the percentage of each refrigerant that makes up the blend. The refrigerant with the lowest boiling point at atmospheric pressure will be listed

first. For example, a blend of 20% (R-12) and 80% (R-22) would be represented as R-22/12 (80/20). Because R-22 has the lowest boiling point, it is listed first.

- Blend names can also have capital letters at the end. The capital letters at the end of R-401A, R-401B, and R-401C mean that the same three refrigerants make up these near-azeotropic blends, but their individual percentages differ.

9.12 POPULAR REFRIGERANTS AND THEIR COMPATIBLE OILS

The following are some of the more popular refrigerants and refrigerant blends used in the HVACR industry. **NOTE:** *At the time of this writing, the Environmental Protection Agency (EPA) has proposed rules for refrigerant substitutes. The rules involve a change in the listing status for certain refrigerant substitutes under the Significant New Alternatives Policy Program (SNAP). Some of the refrigerants and refrigerant blends listed as follows may be affected when this proposed rule becomes mandatory. Many refrigerants and refrigerant blends that have high global warming potentials (GWPs) will no longer be allowed even as retrofit refrigerants. These refrigerants and blends will still be listed because they exist in HVACR equipment from past installations and retrofits. It is important that service technicians understand what these refrigerants and blends consist of for service purposes. It is the service technician's responsibility to keep abreast of the most recent laws and rules concerning refrigerant use and service. For the latest up-to-date information regarding refrigerant-related regulations, be sure to visit the EPA's website.•*

HFC-134a (R-134a), with temperature/pressure and capacity relationships similar to those of R-12, has been the alternative refrigerant of choice to replace **CFC-12** in many medium- and high-temperature stationary refrigeration and air-conditioning and automotive air-conditioning applications. However, other refrigerants like hydrofluoro-olefins (HFOs) are being used as alternatives to R-134a because of their lower GWP. R-134a has a safety classification of A1, **Figure 9.14**, and is a pure compound, not a refrigerant blend. The R-134a molecule is smaller than that of R-12 and, because of this, will tend to leak more readily than R-12. Consequently, it is important that proper leak detection techniques be employed. R-134a also experiences small capacity losses when used as a low-temperature refrigerant. R-134a contains no chlorine in its molecule and thus has a 0 ODP; its GWP is 1430. Polarity differences between commonly used organic mineral oils and HFC refrigerants make R-134a insoluble and thus incompatible with mineral oils used in many refrigeration and air-conditioning applications today. R-134a systems must employ synthetic polyol ester (POE) or polyalkylene glycol (PAG) lubricants. R-134a has also replaced R-12 and R-500 in many centrifugal chiller applications. Ester-based lubricants are also required in these applications because of R-134a's incompatibility with organic mineral oils. In many

Figure 9.14 ANSI/ASHRAE Standard 34-2013 Designation and Safety Classification of Refrigerants.

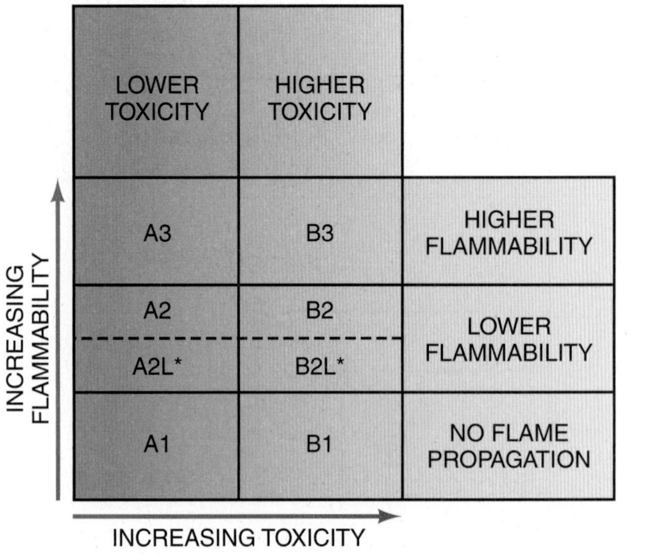

* A2L and B2L are lower flammability refrigerants with a maximum burning velocity of < 10 cm/s (3.9 in/s).

centrifugal chiller applications, efficiencies have improved but not without some reductions in capacity. **SAFETY PRECAUTION:** *R-134a is not a direct drop-in replacement for R-12. Retrofit guidelines must be followed carefully when converting a system from R-12 to R-134a.•* An example of a retrofit guideline from a major manufacturer is included toward the end of this unit.

HCFC-22 (R-22) is a pure compound containing only one molecule, and therefore not a refrigerant blend. Its safety classification is A1, meaning it has very low toxicity and no flame propagation properties. The molecule contains hydrogen, carbon, chlorine, and fluorine. Its chemical name is chlorodifluoromethane, meaning its molecule has one chlorine atom, two fluorine atoms, and its base molecule is methane. HCFC-22 is classified as a hydrochlorofluorocarbon (HCFC). January 1, 2010 was the phaseout date that prevented R-22 from being used in new equipment, with a total production phaseout of 2020. For many years, it has been the primary refrigerant used in commercial and residential air-conditioning systems. It has also been used in large, industrial, centrifugal-chiller air-conditioning applications. Industrial cooling equipment also uses R-22. Before its phaseout for new equipment in 2010, R-22 was replacing R-502 in low-temperature commercial supermarket refrigeration. However, because of its high discharge temperatures in low-temperature commercial refrigeration applications, these systems required compound compression systems with interstage cooling or liquid injection for compressor heat protection.

HFC-407A (R-407A) is a near-azeotropic refrigerant blend of R-32, R-125, and R-134a. It is an R-22 retrofit refrigerant blend for medium- and low-temperature commercial refrigeration systems. It is not recommended for air-conditioning applications. HFC-407A has capacities similar to those of R-22. The blend has a temperature glide of 9°F to 10°F, which gives it fractionation potential. Its safety classification is A1. R-407A has a bit higher head pressure than R-22 but has lower compressor discharge temperatures. These lower discharge temperatures reduce the need for compressor heat protection, such as liquid injection. This blend also works well with heat reclaim systems. R-407A works with existing capillary tubes or TXVs; however, TXV adjustments may be necessary. POE lubricant must be used with R-407A, but it will operate with partial POE lubricant when retrofitting.

HFC-407C (R-407C) is a near-azeotropic refrigerant blend of R-32, R-125, and R-134a. R-407C is a long-term replacement refrigerant for R-22 in residential and commercial air-conditioning and medium- and high-temperature commercial refrigeration applications. Its safety classification is A1. It is being used by manufacturers in new equipment and can also be used as a retrofit refrigerant blend replacing R-22. R-407C works well with R-22 expansion devices, which include TXV, capillary tube, and orifice tube. Some TXV adjustments may be necessary. Always follow retrofit guidelines when retrofitting from R-22 to R-407C. POE lubricant must be used with R-407C. However, when retrofitting an air-conditioning system containing R-22 to R-407C, research has shown that only one oil change is necessary. After a single oil change, the remaining system's mineral oil mixed well with the added POE lubricant, and the combination circulated well with HFC refrigerants. This simplifies the retrofit process for HFC refrigerants. Critical elastomeric seals must be changed for this retrofit.

HFC-410A (R-410A) is a near-azeotropic refrigerant blend of R-32 and R-125. R-410A is a higher-efficiency, long-term replacement refrigerant for R-22 in new residential and light commercial air-conditioning applications. Its safety classification is A1. Retrofitting an R-22 system to an R-410A system must not be done because R-410A operates at much higher pressures (40% to 70% higher) in air-conditioning applications than R-22. If an air-conditioning system is operating at a 45°F evaporating temperature and a 110°F condensing temperature, the corresponding pressures would be as follows:

R-410A .... 130 psig evaporating pressure
365 psig condensing pressure
R-22 .... 76 psig evaporating pressure
226 psig condensing pressure

Figure 9.15 illustrates pressure/temperature comparisons of R-410A with R-22 in tabular and graphical format.

Figure 9.15 Pressure/temperature comparison of R-410A with R-22 in (A) tabular and (B) graphical format.

TEMP. (°F)	PRESSURE (psig)	
	R-410A	R-22
0	48.3	24.0
20	78.4	43.0
40	118.2	68.5
60	169.9	101.6
80	235.4	143.6
100	317.1	195.9
120	417.4	259.9
140	539.1	337.2

(A)

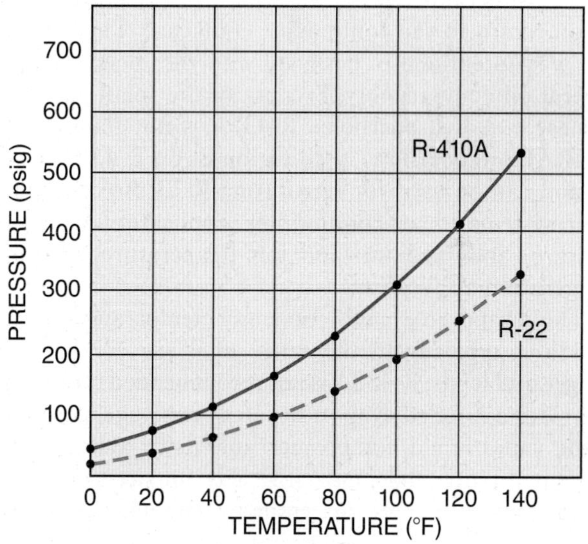

(B)

Gauge manifold sets, hoses, recovery equipment, and recovery storage cylinders all have to be designed to handle the pressures of R-410A. Different gauge hose connections are used for the two refrigerants, helping to prevent the technician from making a mistake. In fact, the gauge and manifold set for R-410A is required to range up to 800 psig on the high side and 250 psig on the low side with a 550-psig low-side retard. All R-410A hoses must have a service rating of 800 psig. Even filter driers for R-410A must be rated for and compatible to the higher pressures. **SAFETY PRECAUTION:** *Never use any other refrigerant service equipment on R-410A equipment. Service equipment must be rated to handle the higher operating pressures of R-410A. Air-conditioning equipment has been redesigned and safer service tools have been introduced to handle the higher pressures of R-410A. As with all refrigerants, safety glasses and gloves must always be worn when handling R-410A.•* Because R-410A is a near-azeotropic refrigerant blend, it has a very small evaporator temperature glide

(0.3°F) and has negligible fractionation potential. R-410A systems use POE lubricant in their crankcases.

HFC-437A (R-437A) is a near-azeotropic retrofit refrigerant blend for CFC-12 and CFC-12 service blends, like R-401A and R-409A, in domestic and commercial refrigeration applications.

HFC-438A (R-438A) is an R-22 retrofit refrigerant blend for direct-expansion residential and commercial air-conditioning and medium- and low-temperature commercial refrigeration systems. It consists of R-32/R-125/R-134a/R-600/R-601a. R-600 is the hydrocarbon (HC) butane and R-601a is the hydrocarbon isopentane. These two hydrocarbons constitute a very small percentage of the total blend (1.7% and 0.6%, respectively), but they help thin the mineral oil so that it has a lower viscosity, enhancing oil return to the compressor crankcase. This characteristic was intentionally designed into the blend to achieve mineral oil compatibility. Because of the small percentages of these two hydrocarbons, R-438A is not flammable. Its ASHRAE safety group classification is A1. R-438A is versatile and can be used for retrofitting R-22 direct-expansion (DX) systems in air-conditioning applications (high temperature) and medium- and low-temperature commercial refrigeration applications.

The evaporator and condenser temperature glide of R-438a is typically in the range of 6° to 7°F. Like most refrigerant blends, R-438A is not recommended for use in systems with a flooded evaporator or a centrifugal compressor, as the vapor/liquid composition difference associated with the temperature glide may inversely impact performance. R-438A has pressure and enthalpy characteristics similar to those of R-22 in low- and medium-temperature commercial refrigeration systems and air-conditioning applications. It is compatible with mineral oil, alkylbenzene, and polyol ester lubricants through all of the temperature ranges mentioned above, and oil changes may not be necessary when retrofitting.

Oil changes are system-dependent. If the system has a complicated piping arrangement, a liquid receiver, or a suction accumulator acting as a low-pressure receiver and oil-return problems are a concern, replacement of all, or part (10% to 25%) of the compressor's oil charge with POE lubricant is recommended. For most systems, R-438A may have 5% to 10% lower capacities with a similar energy efficiency ratio (EER) when compared to R-22. Because most R-22 systems in service today have excess compressor capacity, this small percentage in capacity loss will not be significant. The compressor simply runs a bit longer each cycle. Existing TXVs can be used with minimal adjustment to the superheat setting. R-438A can be used as a retrofit refrigerant blend in air-conditioning and refrigeration applications with minimal system changes. When retrofitting R-22 systems to R-438A, or any HFC refrigerant, it is recommended that critical elastomeric seals, such as Schrader valve core seals, liquid-level receiver gaskets, solenoid valves, ball valves, and flange seals be changed. **Critical seals** are defined as those components that are difficult to isolate and/or service while operating or that require removal of the entire refrigerant charge. Always consult with the compressor manufacturer or follow its specific retrofit guidelines and procedures before performing a retrofit on any R-22 system or any other refrigeration system.

HFC-417A (R-417A) is a near-azeotropic refrigerant blend of R-134a, R-125, and R-600. R-600 is the hydrocarbon butane. It is an efficient and cost effective, non-ozone depleting HFC-based refrigerant blend. Its safety classification is A1. R-417A has been used primarily in new and existing R-22 systems utilizing direct-expansion metering devices. R-417A has also been used as an R-22 direct drop-in replacement refrigerant. Its applications range from large-tonnage industrial refrigeration to commercial packaged air-conditioning units. It has been best suited to replace R-22 in smaller split air-conditioning systems. It is a non-ozone-depleting, HFC-based refrigerant blend with a low global-warming impact. It has a moderate temperature glide. R-417A is compatible with all standard refrigeration oils which include mineral oil, alkylbenzene, and polyol ester lubricants. R-417A operating pressures are slightly lower than R-22 operating pressures. It also has lower discharge temperatures than R-22.

HFC-417C (R-417C) is a near-azeotropic retrofit refrigerant blends for CFC-12 and CFC-12 service blends, like R-401A and R-409A, in domestic and commercial refrigeration applications.

HFC-404A (R-404A) is a near-azeotropic refrigerant blend of R-125, R-143a, and R-134a. Its safety classification is A1. R-404A had been used primarily in the medium- and low-temperature commercial refrigeration markets. R-404A used to be the long-term substitute refrigerant for R-502, but it is not a direct drop-in replacement for R-502. It was possible to retrofit an R-502 system to an R-404A system, but retrofit guidelines had to be followed carefully. R-404A has slightly higher working pressures than R-502. R-404A, as with most HFC-based refrigerants, uses POE lubricants in the crankcase. R-404A has a small evaporator temperature glide (1.5°F) over its refrigeration operating range and has a small fractionation potential over these same temperature ranges.

HFC-422A (R-422A) is a near-azeotropic refrigerant blend consisting of HFC-125, HFC-134a, and R-600a. It has been replacing R-502 and R-502 service refrigerant blends like R-402A and R-408A in commercial refrigerant applications, and R-22 in certain low-temperature applications. It has similar performance to R-404A. It is compatible with traditional lubricants. However, because of its high GWP, it may soon be an unacceptable refrigerant and be replaced with lower GWP refrigerants.

HFC-407F (R-407F) is a near-azeotropic refrigerant blend of R-32, R-125, and R-134a. Its safety classification is A1. It is an alternative to R-22 and R-404A in low- and medium-temperature commercial refrigeration applications. There is a slight decrease in capacity and efficiency compared to R-22, however, R-407F can be used in essentially the same equipment to perform the same job. New systems built with R-407F must have POE lubricants. Retrofitted R-22 systems would need their residual oil flushed with POE lubricant.

HFC-422B (R-422B) is a near-azeotropic refrigerant blend of R-125, R-134a, and HC-600a. Its safety classification is A1. R-600a is the hydrocarbon isobutane used for oil-return purposes. R-422B has been an R-22 retrofit blend for air-conditioning. Its retrofit guidelines will call for possible POE lubricant additions or replacement of mineral oil and possible component changes.

HFC-422C (R-422C) is a near-azeotropic refrigerant blend of R-124, R-134a, and HC-600a with different percentages than R-422A and R-422B. Its safety classification is A1. R-422C has been a retrofit refrigerant blend for R-22, R-502, R-402A, R-402B, R-408A, R-404A, and R-507. It has very similar properties to R-404A in commercial refrigeration applications. Its retrofit guidelines will call for possible POE lubricant addition or replacement of mineral oil. Also, certain component changes will be necessary. The hydrocarbon component (HC-600a) in the blend will help promote oil return in systems containing mineral oil or alkylbenzene lubricant.

HFC-422D (R-422D) is a near-azeotropic refrigerant blend of R-125, R-134a, and R-600a with different percentages than both R-422B and R-422C. Its safety classification is A1. It has been a R-22 retrofit blend for both air-conditioning and commercial refrigeration. It has a moderate temperature glide and its typical lubricant is POE and mineral oil. Its retrofit from R-22 will call for possible POE lubricant additions or replacement of mineral oil and possible component changes.

HFC-427A (R-427A) is a near-azeotropic refrigerant blend of R-32, R-125, R-143a, and R-134a. Its safety classification is A1. It has been a retrofit refrigerant blend for R-22 in both air-conditioning and commercial refrigeration applications, but has been preferred for commercial refrigeration applications. Its retrofit guidelines will call for partial POE replacement of mineral oil. No component changes will be necessary.

HFC-507 (R-507) is an azeotropic refrigerant blend of R-125 and R-143a. R-507 has been used primarily in medium- and low-temperature commercial refrigeration applications. Its safety classification is A1. R-507 has been a replacement refrigerant for R-502, but it is not a direct drop-in replacement. R-507 has slightly higher pressures than both R-502 and R-404A and has slightly higher capacities compared to R-404A. Because R-507 is an azeotropic blend of refrigerants, it has negligible temperature glide and fractionation potential. R-507 requires POE lubricant in the compressor crankcase.

HFC-507A (R-507A) is an azeotropic mixture of R-125 and R-143a. It has a higher efficiency than R-404A and also a higher efficiency than low-temperature R-22. Polyol ester lubricant must be used with R-507A. It is a very close match to R-502.

The following are some of the refrigerants still in use in today's refrigeration and air-conditioning systems. However, because of current or near-future bans, these refrigerants are, or soon will be, no longer manufactured. Even though it is unlawful to produce these refrigerants, they can still be used when recovered or reclaimed from an existing system, as long as certain Environmental Protection Agency (EPA) guidelines are followed. More detailed information on recovering and reclaiming refrigerants is provided later in this unit.

CFC-11 (R-11) is a low-pressure refrigerant used in centrifugal chillers. Its safety classification is A1. R-11 has been in the HVACR industry since 1932. Centrifugal chillers employing R-11 as the refrigerant have low operating costs and are one of the most inexpensive ways to supply large amounts of chilled water to large commercial and industrial buildings for air-conditioning purposes. In 1996, environmental concerns about ozone depletion and global warming forced the production of R-11 to be banned. R-11 centrifugal chiller systems operate in a vacuum on their low sides. Because of this, any leaks will introduce air into the system. Efficient purge systems have been designed to rid these systems of unwanted air. The interim replacement refrigerant for R-11 is R-123. However, a retrofit procedure is required when retrofitting an R-11 system to an R-123 system. **SAFETY PRECAUTION:** *Always consult with the original equipment manufacturer for retrofit guidelines when retrofitting any system.•* Newer chiller systems may come with R-123, R-134a, or ammonia. Lithium bromide and water absorption systems may also be used in newer chiller applications.

CFC-12 (R-12) has been widely used in the HVACR industry since it was invented in 1931. Its safety classification is A1. R-12 has been used extensively in small- and large-scale refrigeration and air-conditioning system applications. Systems ranging in size from small hermetic systems to large centrifugal positive-pressure chiller applications have incorporated R-12. Its main applications have been automotive air-conditioning; high-, medium-, and low-temperature refrigeration; and other commercial and industrial refrigeration and air-conditioning applications. In 1996, environmental concerns about ozone depletion and global warming forced the production of R-12 to be banned. R-134a has been the most popular refrigerant to replace R-12 in medium- and high-temperature stationary refrigeration and air-conditioning systems. R-134a has been the refrigerant replacing R-12 in automotive air-conditioning systems. However, hydrofluoro-olefin (HFO) refrigerants may be the future for automotive air-conditioning because of their low

GWP. R-134a is not a direct drop-in replacement for R-12 in any application. Many interim refrigerant blends that are listed here have also entered the market to replace R-12, with minimal retrofit. Always refer to the manufacturer's retrofit guidelines and procedures when retrofitting any system.

CFC-502 (R-502) is an azeotropic refrigerant blend of R-22 and R-115. Its safety classification is A1. R-502 was first manufactured in 1961. In 1996, environmental concerns about ozone depletion and global warming forced a production ban on R-502 also. R-22 has been used as an interim replacement for R-502, but it usually requires a compound compression system with intercoolers for low-temperature refrigeration applications. As mentioned earlier, R-22 is considered an interim replacement refrigerant because it has a current phaseout date for new equipment of 2010 and a total production phaseout of 2020. R-502 has been a popular low-temperature refrigerant for low-temperature refrigeration applications. One of the main advantages of R-502 over R-22 in low-temperature refrigeration is its ability to operate in positive pressure even with a −40°F evaporating temperature. Also, because of its relatively higher pressures, R-502 operates with lower compression ratios than R-12 in low-temperature refrigeration applications. Lower compression ratios lead to improved capacity as compared with R-12 and R-22 systems. R-502 also has much lower discharge temperatures as compared to R-22. Because of these advantages, R-502 can perform low-temperature refrigeration with single-stage compression and relatively inexpensive compressors.

HCFC-123 (R-123) is an interim replacement refrigerant for R-11 in low-pressure centrifugal chiller applications. The American Society of Heating, Refrigerating and Air-Conditioning Engineers (ASHRAE) has given R-123 a safety classification rating of B1, meaning there is evidence of toxicity at concentrations below 400 ppm, **Figure 9.14**. This means levels of exposure are reduced for persons exposed to R-123 in the course of normal daily operations in machine rooms and service conditions. Special precautions for R-123 and machine room requirements are covered in ASHRAE Standard 15, "Safety Code for Mechanical Refrigeration." **SAFETY PRECAUTION:** *ASHRAE Standard 15 requires the use of room sensors and alarms to detect R-123 refrigerant leaks. The machine room shall activate an alarm and mechanical ventilation equipment before the concentration of R-123 exceeds certain limits.•* When retrofitting an R-11 system to an R-123 system, retrofit guidelines must be followed. R-123 systems may require new seals, gaskets, and other system components to prevent leaks and get maximum performance and capacity.

HCFC-401A (R-401A) is a near-azeotropic refrigerant blend of R-22, R-152a, and R-124. Its safety classification is A1. R-401A has been intended for use as a retrofit refrigerant for R-12 or R-500 systems with only minor modifications. It is especially designed for systems with evaporator temperatures ranging from 10°F to 20°F, because the pressure/temperature curve closely matches that of R-12 in these ranges. There is a drop in capacity at lower temperatures as compared with R-12 systems. An 8°F temperature glide in the evaporator is typical when R-401A is being used. There is an increase in discharge pressure and a slight increase in discharge temperature as compared with R-12. An oil retrofit to an alkylbenzene lubricant is recommended when retrofitting from R-12 to R-401A.

HCFC-402A (R-402A) is a near-azeotropic refrigerant blend of R-22, R-125, and R-290 (propane). The hydrocarbon propane is added to enhance oil solubility and circulation with the refrigerant for oil-return purposes. Its safety classification is A1. R-402A has been intended for use as an R-502 retrofit refrigerant with minimal retrofitting. R-402A will have 25- to 40-psi increases in discharge pressure over R-502 but will not significantly increase the discharge temperature. An oil retrofit to an alkylbenzene lubricant is recommended when retrofitting from R-502 to R-402A.

HCFC-402B (R-402B) is a refrigerant blend of R-22, R-125, and R-290 that contains more R-22 and less R-125 than R-402A does. Its safety classification is A1. R-402B has been intended for use as an R-502 ice-machine retrofit refrigerant because of its lower discharge pressures and higher discharge temperatures as compared with R-502. The higher discharge temperatures make for quicker and more efficient hot-gas defrosts in ice-machine applications. An oil retrofit to an alkylbenzene lubricant is recommended when retrofitting from R-502 to R-402B.

HCFC-409A (R-409A) is a refrigerant blend of R-22, R-142b, and R-124. Its safety classification is A1. R-409A was designed for retrofitting R-12 systems having an evaporator temperature between 10°F and 20°F. R-409A systems usually experience a 13°F temperature glide in the evaporator. R-409A has been used in retrofitting R-12 and R-500 direct-expansion refrigeration and air-conditioning systems. R-409A operates with higher discharge pressures and temperatures than R-12 and has a capacity similar to that of R-12 down to −30°F evaporator temperatures. R-409A mixes well with mineral oil lubricants down to 0°F, so oil retrofitting may not be necessary above this temperature. However, an alkylbenzene lubricant should be used in place of the mineral oil lubricant in evaporator temperatures below 0°F.

SAFETY PRECAUTION: *Always consult with the original equipment manufacturer for retrofit guidelines when retrofitting any system.•*

SAFETY PRECAUTION: *All refrigeration systems are required to have some sort of pressure relief device, usually a pressure relief valve, which should be vented to the outdoors to prevent the buildup of excessive system pressures. Pressure relief valves should always be installed in parallel with one another, never in series,* **Figure 9.16.**•

SAFETY PRECAUTION: *Any refrigerant can be hazardous if used improperly. Hazards include liquid or vapor under pressure and frostbite from the escaping liquid.•*

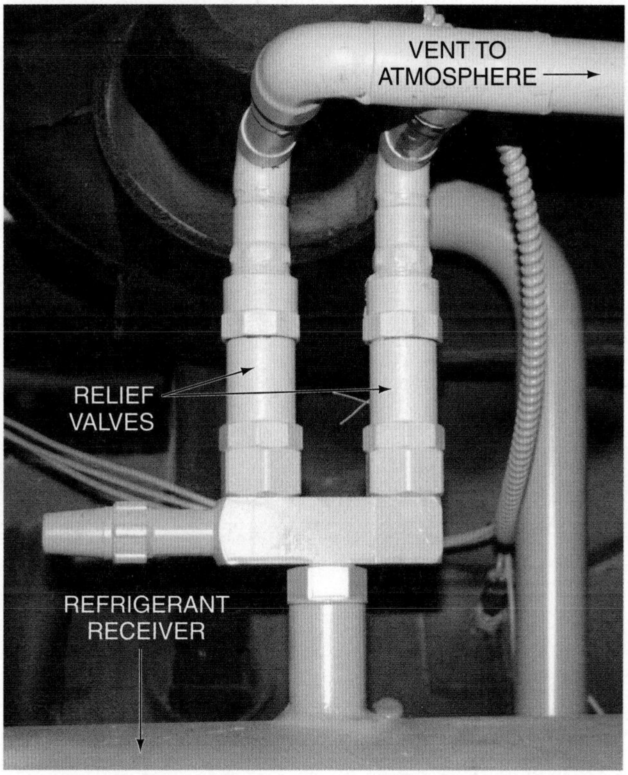

Figure 9.16 Dual pressure relief valves installed in parallel. Courtesy Ferris State University. Photo by John Tomczyk

9.13 REFRIGERANT OILS AND THEIR APPLICATIONS

Today, almost as many new oils are on the market as there are refrigerants. The compressor manufacturers specify what oil to use in any application. If you are unsure about which oil to use for a particular application, contact the manufacturer or use its specification data and performance curve literature.

OILS FALL INTO THREE BASIC GROUPS:

- Animal
- Vegetable
- Mineral

Neither animal nor vegetable oils are good refrigeration or air-conditioning lubricants. However, mineral oils are proven lubricants for these applications. The three groups of mineral oils are:

- Paraffinic
- Naphthenic
- Aromatic

Of the three mineral oil groupings, naphthenic oils are used most often with refrigeration and air-conditioning applications.

Synthetic oils for refrigeration applications have been used successfully for many years. Three popular synthetic oils are:

- Alkylbenzenes
- Glycols
- Esters

HCFC-based blends work best with alkylbenzene lubricants. Today, alkylbenzenes are used quite often in refrigeration applications. The HFC-based blends perform best with an ester lubricant, and manufacturers are using a synthetic ester-based lubricant for many HFC-based refrigerant blends. Extensive oil flushing is required when retrofitting a CFC/mineral oil system to an HFC/ester oil system. Mineral oils do not mix completely with the HCFC-based refrigerant blends, which are soluble with mineral oil up to a 20% concentration. Mineral oil systems retrofitted with HCFC-based refrigerant blends will not require a lot of oil flushing. However, some HFC-based refrigerant blends may only require one oil change from mineral oil to a POE lubricant when retrofitting from an HCFC refrigerant to a HFC-based blend. This is often referred to as a partial-POE oil retrofit, because there is still some residual mineral oil left in the system when the POE lubricant is added. For example, when retrofitting from an HCFC-22 mineral oil system to an HFC-based blend like HFC-407A or HFC-407C, research has shown that after a single oil change, the remaining mineral oil mixed well with the added POE lubricant and the combination circulated well with the HFC-based refrigerant blend. Always follow manufacturer's guidelines when performing any retrofit procedure.

SAFETY PRECAUTION: *Always follow the refrigerant manufacturer's retrofit guidelines when performing any refrigerant/oil retrofit procedure. Retrofit guidelines vary from refrigerant to refrigerant and from manufacturer to manufacturer.*•

As mentioned earlier, a popular glycol-based lubricant is polyalkylene glycol (PAG). PAGs are used in automotive applications. PAG lubricants attract and hold moisture readily and will not tolerate chlorine. When retrofitting a mineral oil system to a PAG system, a labor-intensive retrofit is involved.

Ester-based oils are also synthetic oils. They are wax-free oils and are used with HFC refrigerants like R-134a and HFC-based refrigerant blends. R-134a is a non-ozone-depleting refrigerant with properties very similar to R-12 at medium- and high-temperature applications. Notice the increase in capacity at the higher-temperature applications and the decrease in capacity at lower temperatures. R-134a is not compatible with current refrigerant mineral oils. Systems will have to come from the factory as virgin systems with ester oil, or the technician will have to go through a step-by-step flushing process in an effort to eliminate the residual mineral oil. The system cannot contain more than 1% to 5% mineral oil after retrofitting. **Figure 9.17** shows a refractometer that a technician can use in the field to make sure the system has no more than 1% to 5% residual mineral oil left

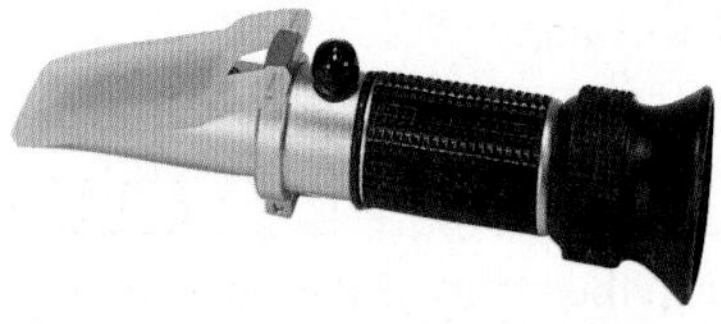

Figure 9.17 A refractometer used in the field for checking the percentage of residual oil left in the system. Source: Nu-Calgon Wholesaler, Inc.

in it. **SAFETY PRECAUTION:** *Even though the EPA has ruled that refrigeration oils are not hazardous waste unless they are contaminated, disposing of used oil in a careless manner is against the law. Check with your local waste-disposal codes when disposing of refrigeration oil.•*

Polyol esters, ester-based lubricants, have become very popular recently. POEs do, however, absorb atmospheric moisture more rapidly than mineral oils, **Figure 9.18**. POEs not only absorb moisture more readily but also hold it more tightly. Refrigeration-system evacuation procedures may require longer times because of this. Containers of POE oil, therefore, should not be exposed to the atmosphere any longer than possible to avoid moisture contamination. Many manufacturers are using metal oil containers for POE oils instead of plastic ones to prevent moisture penetration through the walls of the oil container. **Figure 9.19** lists the appropriate oils for many popular refrigerants. However, always consult with the compressor or condensing unit manufacturer when unsure of what oil to use.

Polyvinylether (PVE) is a non-POE synthetic lubricant for HFC commercial HVACR equipment. PVE is a polymer-based lubricant consisting of chains of monomers. This lubricant first started to be used by original equipment manufacturers (OEMS) in the United States in 2010. Since then, more and more manufacturers are using PVE lubricant extensively in the United States. Many Chinese, Japanese, and Korean air-conditioning manufacturers are using PVE lubricant in their compressors. PVE is very popular in Asia for carbon dioxide

Figure 9.18 The moisture-absorbing capability of mineral oil versus polyol ester lubricant.

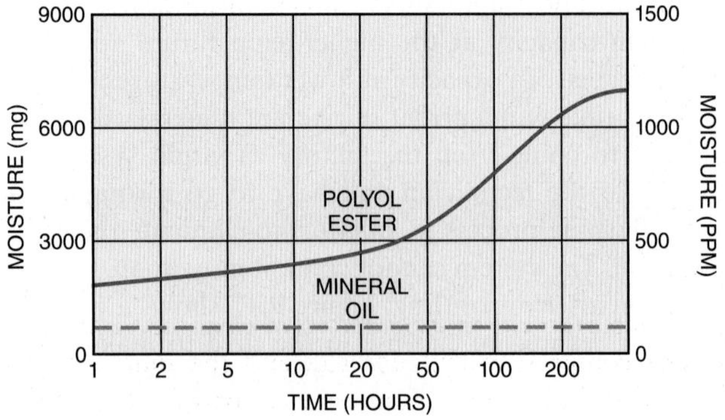

Figure 9.19 A list of refrigerants with their appropriate lubricants. POE = polyol ester, AB = alkylbenzene, MO = mineral oil, and PAG = polyalkylene glycol (automotive applications).

CFC REFRIGERANTS	APPROPRIATE LUBRICANT		
	MINERAL OIL	ALKYLBENZENE	POLYOL ESTER
R-11	+		
R-12	+	+	
R-13	+	+	
R-113	+	+	
R-114	+	+	
R-115	+	+	
R-500	+	+	
R-502	+	+	
R-503	+	+	

HCFC REFRIGERANTS	APPROPRIATE LUBRICANT		
	MINERAL OIL	ALKYLBENZENE	POLYOL ESTER
R-22	+	+	
R-123	+	+	
R-124		+	
R-401A	+	+	
R-401B			
R-401C			
R-402A		+	
R-402B	+	+	
R-403A			
R-403B			
R-405A			
R-406A	+		
R-408A		+	
R-409A	+	+	

HFC REFRIGERANTS	APPROPRIATE LUBRICANT		
	MINERAL OIL	ALKYLBENZENE	POLYOL ESTER
R-23			+
R-32			+
R-125			+
R-134a			+
R-143a			+
R-152a			+
R-404A			+
R-407A			+
R-407B			+
R-407C			+
R-410A			+
R-410B			+
R-507			+

+ Good Suitability Applications with Limitations

NOTE: Always consult the compressor manufacturer for the appropriate lubricant.

refrigeration systems and R-410A air-conditioning applications. Because of its increased usage, the commercial availability of PVE lubricant is expanding. Some main advantages of using PVE in an HVACR system are:

- No hydrolysis (reaction with water)
- Compatible with all HFC refrigerants
- Resists capillary tube blockages due to no hydrolysis
- Excellent anti-wear properties
- Excellent lubrication properties
- Excellent electrical resistivity
- Excellent chemical stability
- Excellent miscibility with HCF refrigerants
- Adaptive to many HVACR applications

Some of the larger compressor and chemical manufacturers have written retrofit guidelines for changing a system over to a newer, alternative refrigerant and its appropriate oil. Always consult with the compressor manufacturer before retrofitting any system. Later in this unit a section named "Refrigerant Retrofitting" will cover retrofit guidelines for retrofitting an existing R-22 system to an R-407C system.

9.14 REGULATIONS

The United States, Canada, and more than 30 other countries met in Montreal, Canada, in September 1987 to try to solve the problems caused by released refrigerants. This conference, known as the Montreal Protocol, agreed to reduce by 1999 the production of refrigerants believed to be harmful ozone-destroying chemicals by 50%. The agreement would take effect when ratified by at least 11 countries representing 60% of the global production of these chemicals. Additional meetings have been held since 1987 and further reduction of these refrigerants has been agreed on. The use of refrigerants as we have known them has changed to some extent and will change dramatically in the near future.

The United States Clean Air Act Amendments of 1990 regulate the use and disposal of CFCs and HCFCs. The Environmental Protection Agency (EPA) is charged with implementing the provisions of this legislation. The act affects the technician servicing refrigeration and cooling equipment, and according to the act, technicians may not "knowingly vent or otherwise release or dispose of any substance used as a refrigerant in such an appliance in a manner which permits such substance to enter the environment." This prohibition became effective July 1, 1992. Severe fines and penalties can be imposed, including prison terms. For the first offense, the act empowers the EPA to obtain an injunction against the offending party that prohibits the discharge of refrigerant to the atmosphere. The second level is a daily fine that, at the time of this printing, is $44,539 per violation, and a prison term not exceeding 5 years. Also, rewards of up to $10,000 are offered to persons furnishing information that leads to the conviction of a person in violation. New legislation made it illegal to manufacture CFC refrigerants after January 1, 1996. For the most up-to-date information regarding refrigerant-related regulations, visit the EPA's website.

9.15 RECOVER, RECYCLE, OR RECLAIM

You should become familiar with these three terms: *recover*, *recycle*, and *reclaim*—terms the industry and its technicians must understand. *Technicians must contain all refrigerants except those used to purge lines and tools of the trade, such as the gauge manifold or any device used to capture and save the refrigerant. The Clean Air Act Amendments of 1990 charged the EPA with issuing regulations to reduce the use and emissions of refrigerants to the lowest achievable level. The law states that the technician shall not "knowingly vent or otherwise release or dispose of any substance used as a refrigerant in such an appliance in a manner which permits such substance to enter the environment.* De minimis releases associated with good-faith attempts to recapture and recycle or safely dispose of any such substance shall not be subject to prohibition set forth in the preceding sentence." This prohibition became effective July 1, 1992. *De minimis* releases means the minimum amount possible under the circumstances.

Some situations requiring the removal of refrigerant from a component or system are the following:

1. A compressor motor is burned.
2. A system is being removed to be replaced. It cannot be disposed of at a salvage yard with refrigerant in the system.
3. A repair must be made to a system, and the refrigerant cannot be pumped using the system compressor and captured in the condenser or receiver.

When refrigerant must be removed, the technician must study the system and determine the best procedure for removal. For example, the system may have several different valve configurations, the compressor may or may not be operable, or sometimes the system compressor may be used to assist in refrigerant recovery. We will consider only the refrigerants and systems covered in this text. The following situations where refrigerant must be removed from a system may be encountered by the technician in the field:

1. The compressor will run and the system has no service valves or access ports, for example, a unit like a domestic refrigerator, freezer, or window unit being discarded.
2. The compressor will *not* run and the system has no access ports, for example, a burnout on a domestic refrigerator, freezer, or window unit.
3. The compressor will run and the unit has service ports (Schrader ports) only, as in a central air conditioner.
4. The compressor will *not* run and the unit has service ports (Schrader ports) only, as in a central air conditioner.

5. The compressor will run and the system has some service valves, for example, a residential air-conditioning system with liquid- and suction-line isolation valves in the line set.
6. The compressor will *not* run and the system has some service valves, for example, a residential air-conditioning system with liquid- and suction-line isolation valves in the line set.
7. The compressor will run and the system has a complete set of service valves, for example, a refrigeration system.
8. The compressor will not run and the system has a complete set of service valves, for example, a refrigeration system.
9. A heat pump with any combination of service valves.

The technician trying to recover the refrigerant in these cases requires a certain amount of knowledge and experience. The condition of the refrigerant in the system must also be considered. It may be contaminated with air, another refrigerant, nitrogen, acid, water, or motor burn particulates. Cross-contamination to other systems must never be allowed because it can damage the system into which the contaminated refrigerant is transferred. This damage may occur slowly and may not be known for long periods of time. Warranties of compressors and systems may be voided if contamination can be proven. Food loss from inoperative refrigeration systems and money loss in places of business when the air-conditioning system is off can be a result.

The words *recover, recycle,* and *reclaim* may be used to mean the same thing by many technicians, but they refer to three entirely different procedures.

RECOVER REFRIGERANT

"To remove refrigerant in any condition from a system and store it in an external container without necessarily testing or processing it in any way" is the EPA's definition of recovering refrigerant. The refrigerant may be contaminated with air, another refrigerant, nitrogen, acid, water, or motor-burn particulates. *In fact, an Industry Recycling Guideline (IRG-2) states that the only way recovered refrigerant can change hands from one owner to another is when it is brought up (cleaned or known to be cleaned) to the Air Conditioning, Heating and Refrigeration Institute (AHRI) Standard 700 purity levels by an independent EPA-certified testing agency. AHRI 700 purity standards are most often met only when the refrigerant has been reclaimed by a chemical-processing facility and has been restored to like-new conditions. If contaminated refrigerant is to be put back into an appliance owned by the same entity, this refrigerant can be recycled by the contractor using the appropriate pieces of equipment designed to perform such purification processes. The recycled refrigerant must not exceed the maximum contaminant levels listed in* **Figure 9.20.** The purity levels in **Figure 9.20** can be reached by field recycling and testing, followed by documentation. Another option is to send the refrigerant to a certified refrigerant reclaimer. The following are four options for handling recovered refrigerant:

Figure 9.20 Maximum contaminant levels of recycled refrigerant in the same owner's equipment.

CONTAMINANT	CONTAMINANT LEVEL
Acid content (by weight)	1.0 ppm
Moisture (by weight)	20 ppm
Noncondensable gas (by volume)	2.0%*
High-boiling residue (by volume)	0.02%**
Chlorides by silver nitrate test	no turbidity
Particulates	visually clean
Other refrigerants	2.0%

*Not applicable for low-pressure systems

**1.0% for low-pressure systems

- Charge the recovered refrigerant back into the same owner's equipment without recycling it.
- Recycle the recovered refrigerant to meet AHRI 740 standards and charge it back into the same system or back into another system owned by the same person or company.
- Send the recovered refrigerant to a certified reclaimer. At the reclaiming facility, the refrigerant will likely be either destroyed via incineration or reprocessed to meet AHRI 700 purity standards.

If the technician is removing the refrigerant from an operable system, it may be charged back into the system. For example, assume a system has no service valves and a leak occurs. The remaining refrigerant may be recovered and reused in this system only. It may not be sold to another customer without meeting the AHRI 700 specification standard. Analysis of the refrigerant to the AHRI 700 standard may be performed only by a qualified chemical laboratory. This test is expensive and is usually performed only on larger volumes of refrigerant. Some owners of multiple equipment may reuse refrigerant in another unit they own if they are willing to take the chance with their own equipment. It is up to the technician to use an approved, clean, refrigerant cylinder for recovery to avoid cross-contamination.

RECYCLE REFRIGERANT

"To clean the refrigerant by oil separation and single or multiple passes through devices, such as replaceable core filter

driers, which reduce moisture, acidity and particulate matter" is the EPA's definition of recycling refrigerant. This term usually applies to procedures implemented at the job site or at a local service shop. If the refrigerant is suspected to be dirty with certain contaminants, it may be recycled, which means to filter and clean. Recycled refrigerants must meet AHRI 740 purity standards. Filter driers may be used only to remove acid, particles, and moisture from a refrigerant. In some cases, the refrigerant will need to pass through the driers several times before complete cleanup is accomplished. Purging air from the system may be done only when the refrigerant is contained in a separate container or a recycling unit, **Figure 9.21**.

A recycling unit removes the refrigerant from the system, cleans it, and returns it to the system. Some units have an air purge that allows any air introduced into the system to be purged before transferring the refrigerant back into the system.

RECLAIM REFRIGERANT

"To process refrigerant to new product specifications by means which may include distillation" is the EPA's definition of reclaiming refrigerant. "It will require chemical analysis of the refrigerant to determine that appropriate product specifications are met. This term usually implies the use of processes or procedures available only at a reprocessing or manufacturing facility." The refrigerant is recovered at the job site, stored in approved cylinders, and shipped to the reprocessing site, where it is chemically analyzed and declared able to be reprocessed (reclaimed). It is then reprocessed and shipped as reclaimed refrigerant because it meets the AHRI 700 standard. At this time, there is no way of testing to the AHRI 700 standard in the field. When reclaimed refrigerant is sold, it is not subject to an excise tax because the tax applies only to newly manufactured refrigerant. This tax saving is an incentive to save and reclaim old refrigerant.

Figure 9.21 Refrigerant recovery and recycling machine.

9.16 METHODS OF RECOVERY

Several methods of refrigerant recovery are used in the systems mentioned in this text. All of these refrigerants are regarded as high-pressure refrigerants; they are also called low-boiling-point refrigerants. Some refrigerants, such as R-11 and R-123 are classified as low-pressure refrigerants and have high boiling points. R-11, for example, boils at 74.9°F at atmospheric pressure, while R-123 boils at 82.2°F at atmospheric pressure. It is very hard to remove these refrigerants from a system because of their high boiling points, and a very low vacuum must be pulled to boil it all from the system. Refrigerant 134a has a boiling point of −15.7°F; this makes it much easier to remove from a system. The boiling points of some refrigerants at atmospheric pressure are as follows:

Refrigerant	Boiling Point (°F)
R-11	+74.9
R-12	−21.62
R-22	−41.36
R-123	+82.2
R-125	−55.3
R-134a	−15.7
R-500	−28.3
R-502	−49.8
R-410A	−61
R-404A	−51
R-402A	−54
R-402B	−51
R-407A	−49.9
R-407C	−43.6
R-417A	−43.2
R-422B	−40.5
R-422C	−50.7
R-422D	−45.8
R-427A	−44.8
R-438A	−44.2

The following chart shows that most of the common refrigerants discussed in this text have high vapor pressures and will be relatively easy to remove from systems. The vapor pressures at a room temperature of 70°F would be as follows:

Refrigerant	Vapor Pressure at 70°F
R-12	70 psig
R-22	121 psig
R-125	158 psig
R-134a	71 psig
R-500	85 psig
R-502	137 psig
R-410A	201 psig
R-404A	148 psig
R-402A	158 psig
R-402B	147 psig

The boiling point is important because the EPA has established rules on how low the pressure in a system must be reduced to meet refrigerant recovery guidelines.

PASSIVE RECOVERY

The service technician's use of the internal pressure of the system or the system's compressor to aid in the refrigerant recovery process is called passive recovery or system-dependent recovery. If the compressor is inoperative and passive recovery is to be performed, the refrigerant must be recovered from both the low and the high side of the system to meet the recovery efficiency requirements. Service technicians should gently tap the side of the compressor's shell with a soft wood, rubber, or leather mallet. This action will agitate the crankcase oil and cause any refrigerant dissolved in it to vaporize, thus increasing the efficiency of the recovery process. Because the compressor is not working, the service technician can also use a vacuum pump as long as it does not discharge into a pressurized container but only into the atmosphere. If the compressor is working, the technician can recover refrigerant only from the high side of the system. All manufacturers of appliances containing Class I (CFC) or Class II (HCFC) refrigerants must include a service aperture or process stub on the system, and nonpressurized containers must be used to capture the refrigerant. Passive recovery is used often on small appliances that contain 5 lb or less of refrigerant in their systems. The recovery techniques for small appliances are covered in this unit under the subsection "Recovering Refrigerant from Small Appliances."

ACTIVE RECOVERY

Active recovery with a certified, self-contained refrigerant recovery unit is the method most often used by technicians. The compressor of the refrigeration or air-conditioning system is not used as part of the active recovery process. To speed the process, always recover at the highest possible ambient temperature to induce a higher pressure in the system being recovered. **SAFETY PRECAUTION:** *A high-pressure reading in a recovery tank is an indication of air or other noncondensables in the tank, which could result in dangerous pressures. After the tank has cooled to the ambient temperature, determining the temperature/pressure relationship to the ambient temperature will indicate what the recovery tank pressure should be. Before starting a recovery, it is always important to know what type of refrigerant is in the system.•* Unintentional mixing of refrigerants can also cause high pressures in a recovery tank, so always make sure that this does not occur. EPA regulations do not allow for refrigerants to be mixed in a recovery cylinder or tank. The following are some suggestions for how to avoid mixing refrigerants in a recovery tank:

- Use pressure/temperature charts for refrigerant in the recovery cylinder.
- Make sure the recovery unit has a self-clearing, or purge, feature.
- Dedicate a lubricant-specific recovery unit to specific refrigerants.
- Dedicate refrigerant cylinders to a specific refrigerant and mark the cylinder appropriately.
- Use commercially available refrigerant identifiers and air-detection devices on the recovery tank.
- Keep appropriate records of refrigerant in inventory.

Effective August 12, 1993, all companies have had to certify to the EPA that they have the recovery equipment capable of performing adequate recovery of the refrigerants for the systems they service. At the time of this printing, it is no longer required for contracting companies to provide verification to the EPA that they have purchased refrigerant recovery equipment. The required minimum vacuum levels that must be reached during the refrigerant recovery process are provided in **Figure 9.22**. This chart provides the most up-to-date information for recovery equipment manufactured both before and after November 15, 1993. Recovery equipment manufactured after November 15, 1993 must be approved by an AHRI-approved, third-party laboratory to meet AHRI Standard 740. Since regulations change periodically, be sure to visit the EPA's website for the most up-to-date information regarding refrigerant handling-related issues. Referring to Figure 9.22, the EPA defines the appliance classifications as follows:

Low-Pressure Appliances: Appliances that operate with refrigerants that have liquid saturation pressures below 30 psig at a liquid saturation temperature of 104°F. Examples of low-pressure refrigerants include HCFC-123, HFC-245fa, and HFO-1234zd.

Medium-Pressure Appliances: Appliances that operate with refrigerants that have liquid saturation pressures between 30 psig and 155 psig at a liquid saturation temperature of 104°F. Examples of medium-pressure refrigerants include HCFC-409A, HFC-134a, and HFO-1234yf.

Figure 9.22 Required levels of evacuation for air-conditioning, refrigeration, and recovery/recycling equipment.

REQUIRED LEVELS OF EVACUATION FOR AIR-CONDITIONING, REFRIGERATION, AND RECOVERY/RECYCLING EQUIPMENT (EXCEPT FOR SMALL APPLIANCES, MVACs, AND MVAC-LIKE EQUIPMENT) INCHES OF Hg VACUUM		
Type of Air-Conditioning or Refrigeration Equipment	**Using Recovery or Recycling Equipment Manufactured before November 15, 1993**	**Using Recovery or Recycling Equipment Manufactured on or after November 15, 1993**
Very high-pressure appliances	0	0
High-pressure appliances, or isolated component of such appliances with a refrigerant charge of less than 200 pounds	0	0
High-pressure appliances, or isolated component of such appliances with a refrigerant charge of 200 pounds or more	4	10
Medium-pressure appliances, or isolated component of such appliances with a refrigerant charge of less than 200 pounds	4	10
Medium-pressure appliances, or isolated component of such appliances with a refrigerant charge of 200 pounds or more	4	15
Low-pressure appliances	25 mm Hg absolute	25 mm Hg absolute

NOTE: MVAC = Motor Vehicle Air Conditioning

High-Pressure Appliances: Appliances that operate with refrigerants that have liquid saturation pressures between 155 psig and 340 psig at a liquid saturation temperature of 104°F. Examples of high-pressure refrigerants include HCFC-22, HFC-410A, and HFC-407C.

Very High-Pressure Appliances: Appliances that operate with refrigerants that have liquid saturation pressures above 340 psig at a liquid saturation temperature of 104°F. R-744 (carbon dioxide) is an example of a very high-pressure refrigerant.

The exception to the vacuum level requirements outlined in **Figure 9.22** would be a system that could not be reduced to these pressure levels; for example, when the system has a large leak, there is no need to pull the system full of air.

Since November 14, 1994, technicians have required certification to purchase refrigerant. The EPA defines a technician as: "Any person who performs maintenance, service, or repair that could reasonably be expected to release Class I (CFC) or Class II (HCFC) substances into the atmosphere, including but not limited to installers, contractor employees, in-house service personnel, and, in some cases, owners." These restrictions were extended to HFC refrigerants on November 15, 1995. There are four different certifications for the different types of service work. A person must take an EPA-approved proctored test and pass both the core section and either the Type I, Type II, or Type III certification sections to become certified. Certification types are discussed in the following paragraphs. A person who passes the core and certification Types I, II, and III is automatically Universally Certified. A company's training official, industry chapters, or local wholesale supply houses can supply training information, testing dates, and test sites. Once a technician is certified, it is the certified technician's responsibility to comply with any changes in the law. Technicians can obtain the most up-to-date information regarding refrigerant-related regulations by visiting the EPA's website.

TYPE I CERTIFICATION. Small Appliance—*Manufactured, charged, and hermetically sealed with 5 lb or less of refrigerant.* Includes refrigerators, freezers, room air

conditioners, *package terminal heat pumps,* dehumidifiers, under-the-counter ice makers, vending machines, and drinking-water coolers.

TYPE II CERTIFICATION. Medium, High, and Very High-Pressure Appliance—Uses refrigerants with pressures above 30 psig at a liquid saturation temperature of 104°F.

TYPE III CERTIFICATION. Low-Pressure Appliance—Uses refrigerants with pressures below 30 psig at a liquid saturation temperature of 104°F.

UNIVERSAL CERTIFICATION. Certified in Types I, II, and III. The assumption is that certification should ensure that the technician knows how to handle the refrigerant in a safe manner without exhausting it into the atmosphere.

Refrigerant may be removed from the system as either a vapor or a liquid or in a partial liquid and vapor state. It must be remembered that lubricating oil is also circulating in the system with the refrigerant. If the refrigerant is removed in the vapor state, the oil is more likely to stay in the system. This is desirable from two standpoints:

1. If the oil is contaminated, it may have to be handled as a hazardous waste. Much more consideration is necessary, and a certified hazardous-waste technician must be available.
2. If the oil remains in the system, it will not have to be measured and then replaced in the system. Time may be saved in either case, and time is money. The technician should pay close attention to the management of the oil from any system.

The cylinder must be clean and in a deep vacuum before the active recovery or recycle process is begun. One suggestion is that the cylinder be evacuated to at least 1000 microns (1 mm Hg) before recovery is started. With this level of vacuum, refrigerant from a previous job will be removed. All lines to the cylinder must be purged of air before connections are made.

SAFETY PRECAUTION: *Refrigerant must be transferred only to Department of Transportation (DOT)–approved refrigerant cylinders. The approved cylinders are recognizable by their color and valve arrangement: They are yellow on top with gray bodies and have a special valve that allows liquid or vapor to be added to or removed from the cylinder,* **Figure 9.23.**•

Figure 9.23 DOT-approved (Department of Transportation) refrigerant recovery cylinders. Source: National Refrigerants, Inc.

9.17 MECHANICAL RECOVERY SYSTEMS

Among the many mechanical recovery systems available, some recover refrigerant only, while some are designed to recover and recycle. AHRI certifies the equipment specifications, and the machines are expected to meet AHRI 740 specifications. Some of the equipment is designed to be used in the shop where refrigerant is brought in for recycling. For recovery and recycle equipment to be useful in the field, however, it must be portable, able to both recover and recycle, and capable of being easily moved to rooftops, where many systems are located. This presents a design challenge. Equipment must be small enough to be hauled in the technician's truck, and units for use on rooftops should not weigh more than approximately 50 lb, because anything heavier may require two technicians, rather than one, to haul it up a ladder.

Many units have wheels so they may be rolled to the top floor, **Figure 9.24**, while some recovery units can be easily carried by the technician, **Figure 9.25**. Manufacturers are continuing their attempts to provide a unit that is small as well as practical, yet will still perform the task of removing the refrigerant from the system properly.

The *fastest* method of removing refrigerant from a system is to take it out in the liquid state. This is because liquid refrigerant is more dense than vapor. If the system is large enough, it may have a liquid receiver where most of the charge is collected. Many systems, however, are small, and if the compressor is not operable it cannot pump the refrigerant. Although liquid recovery is fast, it can result in the loss of some of the system oil, which is often pulled from the system along with the liquid refrigerant. The *slowest* method of removing refrigerant is to remove it in the vapor state. A recovery/recycle unit will remove the vapor faster if the hoses and valve ports are not restricted and if a greater pressure difference can be created. The warmer the system is, the warmer and denser the vapor is, and the compressor in the recovery/recycle unit will

Figure 9.24 A wheeled recovery/recycle unit. Photo courtesy of Robinair Division, SPX Corporation

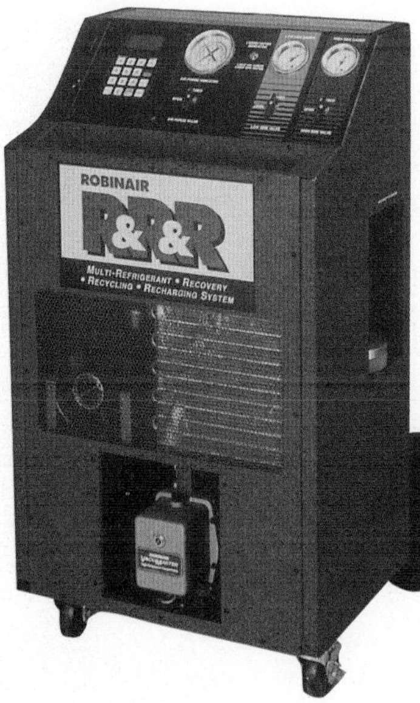

Figure 9.25 A portable refrigerant recovery unit. Courtesy Ritchie Engineering Company, Inc., Yellow Jacket Products Division.

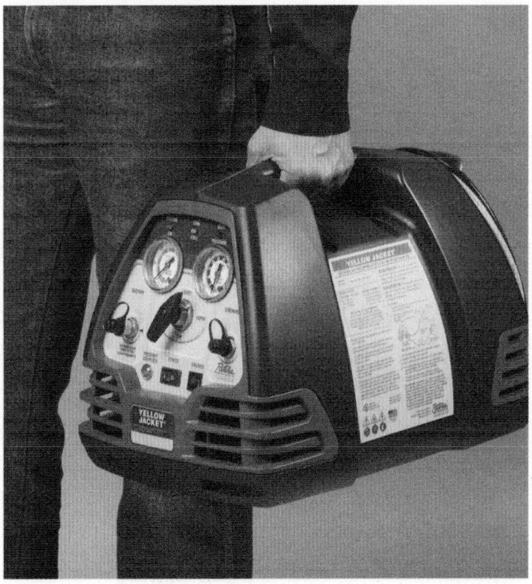

be able to pump more pounds in a minute. As the vapor pressure in the system is reduced, the vapor becomes less dense, and the recovery unit capacity is reduced. Although vapor recovery is a slower process than liquid recovery, it is often preferred as there is very little to no refrigerant oil loss from the system when only vapor is recovered. More time will be required to complete the refrigerant recovery process as the system pressure drops. For example, when removing R-134a from a system, if saturated vapor is removed at 70 psig, only 0.566 ft^3 of refrigerant vapor must be removed to remove 1 lb of refrigerant. When the pressure in the system is reduced to 20 psig, 1.353 ft^3 of vapor must be removed to remove 1 lb of refrigerant. When the pressure is reduced to 0 psig, there is still refrigerant in the system, but 3.052 ft^3 of refrigerant vapor must be removed to remove 1 lb of refrigerant. The unit recovery rate will slow down as the refrigerant pressure drops because the compressor pumping rate is a constant.

Recovery/recycling unit manufacturers provide the typical pumping rate for their equipment in the specifications. The rate at which the equipment is capable of removing refrigerant is generally 2 to 6 lb/min, which will be more constant and faster while refrigerant is being removed in the liquid state. More factors must be considered when removing vapor. When the system pressure is high, the unit will be able to remove vapor relatively fast. When the system pressure is lowered, the rate of vapor removal will slow down. With a small compressor, the rate of removal will become very slow when the system vapor pressure reaches about 20 psig.

The compressor in the recovery/recycle unit is a vapor pump. Liquid refrigerant cannot be allowed to enter the compressor because it will cause damage. Different methods may be used to prevent this from occurring. Some manufacturers use a push-pull method of removing refrigerant by connecting the liquid-line fitting on the unit to the liquid-line fitting on the cylinder. A connection is then made from the unit discharge back to the system. When the unit is started, vapor is pulled out of the recovery cylinder from the vapor port and condensed by the recovery unit. A very small amount of liquid is pushed into the system, where it flashes to a vapor to build pressure and push liquid into the receiving cylinder, **Figure 9.26**. A sight glass is used to monitor the liquid, and

Figure 9.26 Push-pull method of removing liquid refrigerant from a system.

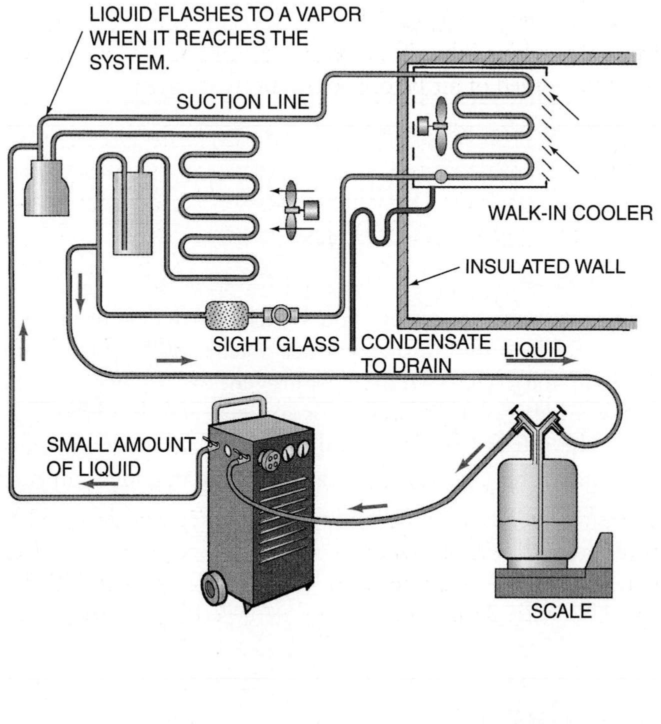

Figure 9.27 The hoses are reconnected to remove the system's vapor.

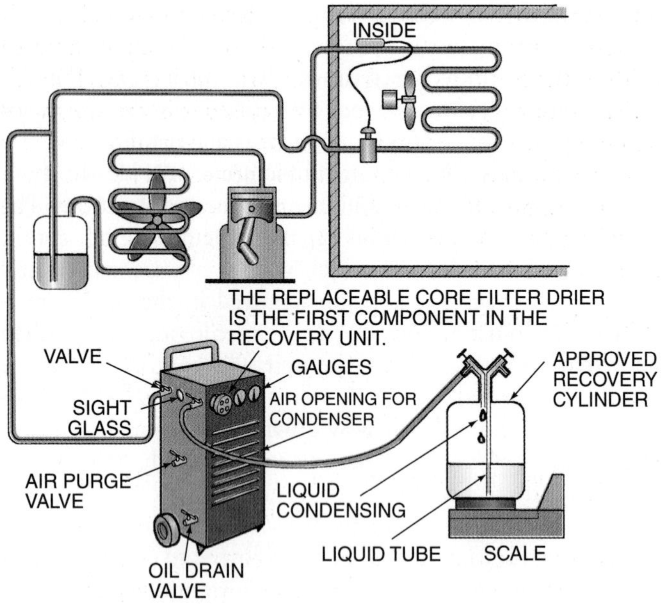

the recovery cylinder is weighed to prevent overfilling. When it is determined that no more liquid may be removed, the suction line from the recovery unit is reconnected to the vapor portion of the system, and the unit discharge is fastened to the refrigerant cylinder, **Figure 9.27**. The unit is started, and the remaining vapor is removed from the system and condensed in the cylinder. When using any recovery unit, be sure to read the manufacturer's literature and operating instructions carefully. Some recovery units can handle liquid refrigerant recovery directly, but others cannot. Never exceed the operational limits of the recovery unit to ensure continued successful operation and reliability. Before performing any refrigerant recovery procedure, an acid check of the system's oil should be performed if there is any question as to its quality, such as after a compressor motor burnout.

RECOVERY CYLINDER REQUIREMENTS AND CALCULATIONS

The DOT-approved cylinder used for some of the recovery/recycle systems has a float and switch that will shut the unit off when the liquid in the cylinder reaches 80% of the cylinder's capacity. This switch helps prevent overfilling the cylinder with refrigerant. It also helps compensate for oil that may be in the refrigerant. When oil and refrigerant are added to a cylinder, the amount of weight that the cylinder can hold changes because oil is much lighter than refrigerant. If a cylinder contains several pounds of oil due to poor practices, it can easily be overfilled with refrigerant if the cylinder is filled only by weight. The float switch prevents this from happening, **Figure 9.28**. **NOTE:** *The recovery cylinder must be clean, empty, and in a vacuum of a minimum of 1000 microns before recovery is started.* • Other methods besides a float device are being used to make sure recovery cylinders are not overfilled. These include electronic thermistors built into the recovery cylinder's wall and simply weighing the recovery cylinder before and after the refrigerant recovery process. Always make sure the recovery tank is free of oil when using the weighing method.

Figure 9.28 (A) Recovery cylinder with a float switch to prevent overfilling. (B) A float switch for a recovery tank. Courtesy Ferris State University. Photo by John Tomczyk

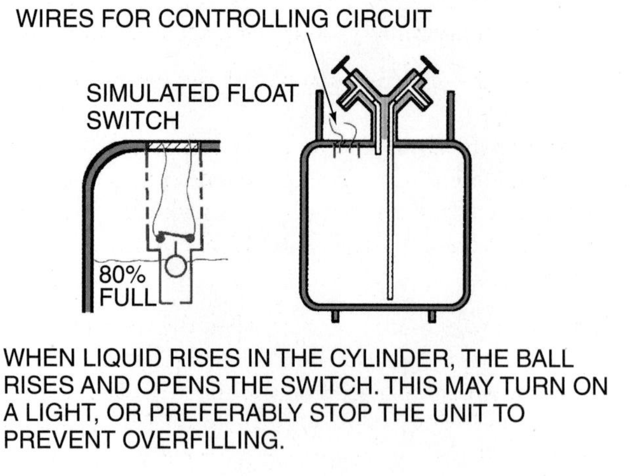

(A)

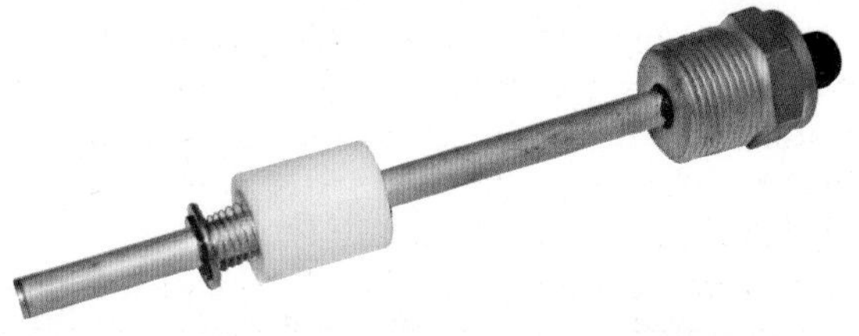

(B)

As mentioned earlier, overfilling a recovery cylinder with liquid can create a dangerous situation. Hydrostatic pressures can quickly build inside the tank when the temperature rises if the tank is overfilled with liquid. The recovery cylinder must have at least a 20% vapor head, which acts as a cushion for expansion when the tank's temperature changes. If there is no vapor head inside the recovery tank and the tank is full of liquid from overfilling, any slight rise in the temperature of the tank will cause an extremely large pressure increase inside. Because of this, a service technician must know and understand the calculations involved in filling a recovery cylinder to no more than 80% of its internal volume.

As mentioned earlier, weighing the cylinder before and after the recovery process is a proven way to safely fill a recovery cylinder. This can be accomplished by placing the recovery cylinder on an electronic scale and monitoring the

weight of the cylinder during the recovery process. When the cylinder reaches a weight equal to 80% of its total capacity, the recovery process must be stopped. However, the service technician must know what 80% of the cylinder's weight carrying capacity (maximum cylinder weight) is. To find out:

1. Calculate the internal volume of the recovery cylinder being used.
2. Determine the liquid density of the refrigerant being used.
3. Determine the tare weight of the recovery cylinder.
4. Maximum Cylinder Weight = (Cylinder Volume × Liquid Refrigerant Density @ 130°F × 0.80) + (Cylinder Tare Weight).

The 130°F temperature is used when the recovered refrigerant is to be transported. This ensures that the recovery cylinder and its contents will be safe to transport, as long as the vessel is leak free, even when exposed to high ambient temperatures. If the recovered refrigerant is not going to be transported, but simply removed from a system and then reintroduced after a system repair has been completed, the 130°F value can be changed to 77°F according to AHRI Guideline K-2015. To be on the safe side, the 130°F value can be used in all cases, as less refrigerant will be introduced to the recovery cylinder. To calculate the internal volume of the recovery cylinder, divide the water capacity of the recovery cylinder by the density of water, which is 62.5 lb/ft^3. The water capacity of the recovery cylinder is the weight of the recovery cylinder if it were full of water. The water capacity (W.C.) of the cylinder is stamped on the outside of the cylinder and its units are in pounds. **Figure 9.29** shows a recovery cylinder with a W.C. of 47.6 lb. So the internal volume of the recovery cylinder is:

Internal Cylinder Volume = (47.6 lb) ÷ (62.4 lb/ft^3) = 0.76 ft^3

Determine the liquid density of the refrigerant to be recovered at a saturation temperature of 130°F. This data is published by refrigerant manufacturers, usually in the form of tables. Most of the time, the saturated liquid density of a refrigerant can be extracted from a thermodynamic table. If the refrigerant is R-404A, from a thermodynamic table of R-404A, the saturated liquid density at 130°F is 53.18 lb/ft^3.

The tare weight of the recovery cylinder is the weight of the empty recovery cylinder. The tare weight of a recovery cylinder, stamped on the outside of the recovery cylinder, is rated in pounds. **Figure 9.29** shows a recovery cylinder with a tare weight (T.W.) of 28 lb. The maximum cylinder weight can now be calculated.

Maximum Cylinder Weight = (Cylinder Volume × Liquid Refrigerant Density @ 130°F × 0.80) + (Cylinder Tare Weight)

Maximum Cylinder Weight = (0.76 ft^3 × 53.18 lb/ft^3 × 0.8) + (28 lb) = 60.33 lb

SAFETY PRECAUTION: *The service technician should stop the refrigerant recovery process once the recovery cylinder weight has exceeded 60.33 lb. Otherwise, hydrostatic pressures will build rapidly and cause a very dangerous situation. Moreover, some of the system's oil will also be removed by the recovery machine and will enter the recovery cylinder. Before using the calculation above, always make sure the recovery cylinder is free of oil.*•

Figure 9.29 Refrigerant recovery tank specifications stamped on the protective head of the tank, showing tare weight and water capacity. Courtesy Ferris State University. Photo by John Tomczyk

RECOVERY UNIT SELECTION QUESTIONS AND CONSIDERATIONS

Manufacturers of recovery equipment try to incorporate features in their units that make them lightweight, dependable, and efficient. Always follow the manufacturers' instructions; they know the capabilities of their equipment. Technicians should ask certain questions and choose carefully when purchasing a recovery unit:

Recovery Units

1. Will the unit run under the conditions faced in the majority of the jobs to be done? For example, will it operate properly on the rooftop in hot or cold weather?
2. Will it recover both liquid and vapor if necessary?
3. What is the pumping rate in the liquid and vapor state for the refrigerants to be recovered?
4. Are the drier cores standard, and can they be purchased locally?
5. Is the unit portable enough to conveniently perform service jobs?
6. Is the recovery unit oil-less? If oil-less, it can be used on systems with different lubricants such as mineral oils, alkylbenzenes, esters, and glycols. Oil changes and oil contamination are no longer a problem with an oil-less recovery unit.

Figure 9.30 Three portable, oil-less refrigerant recovery units. Courtesy Ferris State University. Photo by John Tomczyk

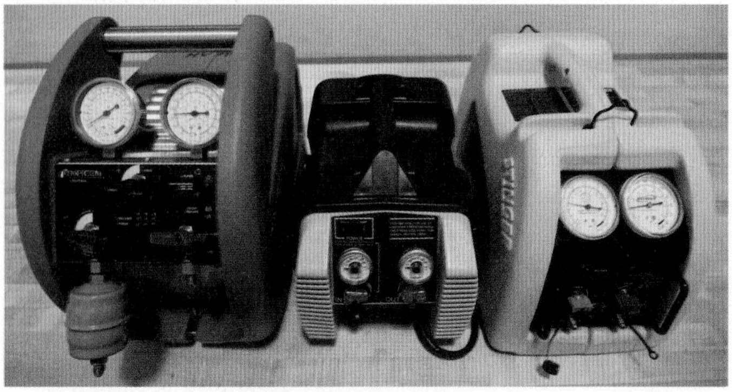

Oil-less recovery units operate on the principle that there is enough residual oil vapor coming back to the recovery unit's compressor from the system under recovery to provide proper lubrication of the recovery compressor's moving parts. The recovery unit's compressor crankcase carries no reservoir of oil and is referred to as "oil-less." This makes the recovery system more versatile and prevents cross-contamination of oil within the refrigeration systems on which it is used. These units can be used on most types of refrigerant and oil systems, and they have very few moving parts. **Figure 9.30** shows three oil-less recovery units used in the HVACR industry today. **Figure 9.31** shows the internal structure of an oil-less recovery unit's compressor and related parts. Most modern refrigerant recovery units manufactured today are oil-less.

Figure 9.31 An oil-less refrigerant recovery unit showing motor, crankcase, piston, cylinder, and cylinder head. Courtesy Ferris State University. Photo by John Tomczyk

Recovery/Recycle Units

1. Will the unit run under the conditions faced in the majority of the jobs to be done?
2. Will it recycle enough refrigerant for the types of jobs encountered?
3. What is the recycle rate, and does it correspond with the rate needed for the size of the systems to be serviced?
4. Are the drier cores standard, and can they be purchased locally?
5. Is the unit portable enough for the jobs and transportation available?

9.18 RECOVERING REFRIGERANT FROM SMALL APPLIANCES

A small appliance is defined as a system that is manufactured, charged, and hermetically sealed with 5 lb or less of refrigerant. This includes refrigerators, freezers, room air conditioners, packaged terminal heat pumps, dehumidifiers, under-the-counter ice makers, vending machines, and drinking-water coolers. Recovering refrigerant from small appliances may be easier than from larger systems because not as much refrigerant is involved. Recovery in small appliances can be either active or passive. As mentioned earlier, active recovery makes use of a recovery machine that has its own built-in compressor. Passive recovery uses the refrigeration system's compressor or internal vapor pressure for the recovery process. However, in passive recovery, the refrigerant must be recovered in a nonpressurized container, **Figure 9.32.** *No matter what kind of recovery device is used, an environmentally conscious service technician should do a daily leak check with a refrigerant leak detector on the recovery device.* To speed up the recovery process, a technician can:

- Put the recovery cylinder in cold water to lower its vapor pressure.

Figure 9.32 Removing refrigerant to a special plastic bag.

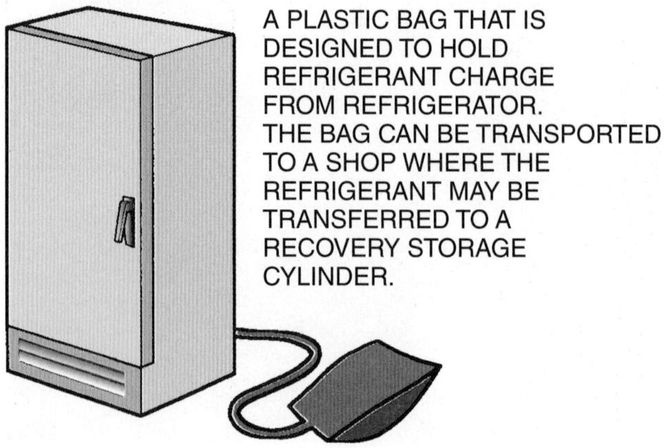

- Heat the refrigeration system to cause the vapor pressure inside the system to increase. Heating the compressor's crankcase will surely release a lot of refrigerant dissolved in the oil. Heat lamps, electric blankets, and heat guns work best. Use of the system's defrost heater will also suffice. Never use an open flame to heat the system.
- Make sure the ambient temperature around the system being recovered is as high as possible. This will ensure high vapor pressures inside the system.
- Softly tap the compressor's crankcase with a soft wood, rubber, or leather mallet to agitate the crankcase oil and release refrigerant.

LOW-LOSS FITTINGS

Any recovery equipment manufactured after November 15, 1993, must have low-loss fittings. These fittings are intended to minimize the loss of refrigerant to the atmosphere when gauge hoses are connected to, or removed from, system service valves. **Figure 9.33** shows an example; there are many different types of low-loss fittings on the market today. The fittings can be either closed off manually or automatically when a service technician takes them off a pressurized refrigeration system or recovery device.

SCHRADER VALVES

Some small appliances are equipped with Schrader valves, **Figure 9.34**. 4R *To prevent refrigerant leaks, Schrader valves must be periodically examined for defects like bends, cracks, or loose cores and caps.* 4R If leaks do occur in the Schrader valve, replace the valve core and make sure the cap is on finger tight. Schrader valve caps help prevent leaks and also help prevent the valve core from accumulating dirt and dust. Accidental depression of the valve core can also be prevented by placing a cap on the valve.

Figure 9.33 Quick-couples or quick-connects to help reduce refrigerant loss. Courtesy Ritchie Engineering Company, Inc., Yellow Jacket Products Division.

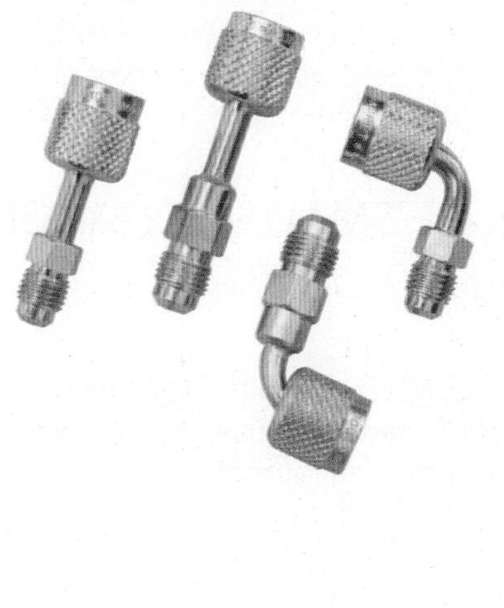

Figure 9.34 A Schrader valve with core and cap. Courtesy Ferris State University. Photo by John Tomczyk

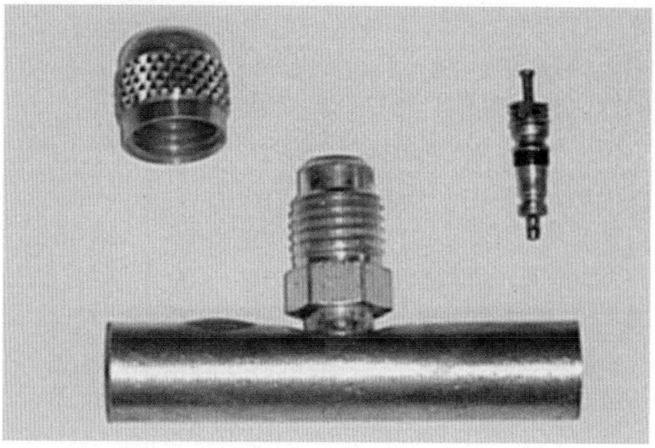

PIERCING VALVES

Many service technicians access small appliances by using a piercing valve, **Figure 9.35**. The sharp pin of the piercing valve is driven through the system's tubing, and when the needle is retracted, access is gained to the system. Piercing valves come as both clamp-on types and solder-on types. The clamp-on piercing valves are used where temporary system access is needed, such as a recovery and charging process. All clamp-on valves should be removed when service is finished because their soft rubber gaskets can rot over time and cause leaks. It is a good idea to leak check these valves before the system's tubing is pierced. Piercing valves are not recommended for use on steel tubing.

Solder-on piercing valves are used for permanent access to a system. Once the valve is soldered onto the system's tubing, leak check the assembly before piercing the tubing. This can be done by pressurizing the assembly.

Figure 9.35 (A) A gasket-type piercing valve. (B) A solder-on piercing valve. Photos courtesy Ferris State University. Photos by John Tomczyk

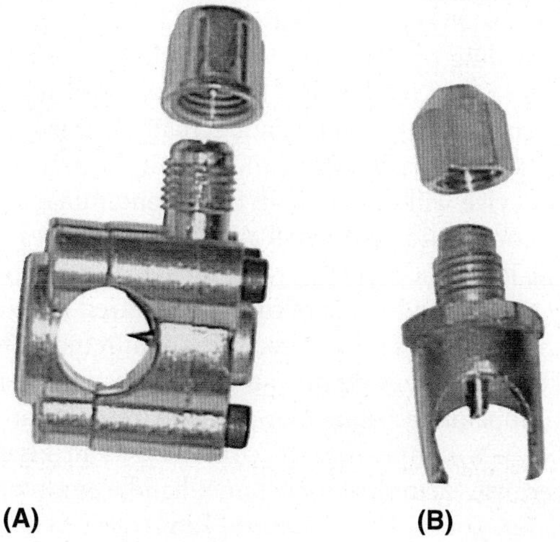

Figure 9.36 Small-appliance recovery requirements. Source: U.S. EPA

RECOVERY EFFICIENCY REQUIREMENTS FOR SMALL APPLIANCES*		
Recovery Efficiencies Required	Recovered Percentages	Inches of Mercury Vacuum
For active and passive equipment manufactured after November 15, 1993, for service or disposal of small appliances with an operative compressor on the small appliance.	90%	4*
For active and passive equipment manufactured after November 15, 1993, for service or disposal with an inoperative compressor on the small appliance.	80%	4*
For grandfathered active and passive equipment manufactured before November 15, 1993, for service or disposal with or without an operating compressor on the small appliance.	80%	4*

*AHRI 740 Standards
*NOTE: Small appliances are products that are fully manufactured, charged, and hermetically sealed in a factory and that have 5 lb or less of refrigerant.

SAFETY PRECAUTION: *Never heat the system's tubing to a cherry red color. This can cause a blowout and serious injury. When certain refrigerants are exposed to high temperatures, a flame, or a glowing-hot metal, they can decompose into hydrochloric and/or hydrofluoric acids and phosgene gas.•* The recovery process on a system with 0 psig pressure should not be continued. A pressure reading of 0 psig could mean that there is a leak in the system or that the line has not been successfully pierced. If the system has leaked and is not actually at 0 psig, noncondensables like air with water vapor are in the system and the recovery device will only take in these contaminants.

If the compressor of a small appliance is inoperative, both high- and low-side access is recommended to speed up the recovery process and to make sure the required vacuum levels are reached. **Figure 9.36** lists recovery requirements for small appliances. A vacuum pump can assist in a passive recovery with an inoperative compressor, but the refrigerant must be recovered in a nonpressurized container, as in **Figure 9.32.** This is because vacuum pumps cannot handle pumping against anything but atmospheric pressure. Too large a vacuum pump will freeze any moisture in the system due to a speedy drop in system pressure. The technician should then introduce dry nitrogen to increase pressure and avoid moisture freezing.

SAFETY PRECAUTION: *Never energize the compressor when under a deep vacuum. This can result in a short in the motor windings at the fusite terminals or at a weak spot in the windings and cause compressor damage.•* Compressor damage can happen because under a vacuum there is no oil vapor or refrigerant vapor in the compressor's crankcase to insulate the fusite terminals from one another. Both oil and refrigerant are excellent dielectrics, meaning they are good electrical insulators. For accuracy in pulling a vacuum with a vacuum pump, the system's vacuum gauge should be located as close to the system tubing and as far away from the vacuum pump as possible. To determine how long it will take to pull a deep vacuum (500 microns), the vacuum pump's capacity and the diameter of the service hose coming from the vacuum pump are two important factors. Make sure the vacuum pump is turned off or valved off when checking the final vacuum.

Never recover from a small appliance in the following situations:

- Systems having sulfur dioxide, methyl chloride, or methyl formate
- 0 psig pressure in the system to be recovered
- Refrigerants like water, hydrogen, or ammonia, which are found in recreational vehicles and absorption systems
- A strong odor of acid, as in a compressor burnout
- Nitrogen mixed with R-22 as a trace gas for electronic leak detection, or nitrogen used as a holding charge

Even though it is not necessary to repair a leak on a small appliance, every service technician should find and repair leaks because of environmental concerns, which include ozone depletion and global warming. Also, a service technician cannot avoid recovering refrigerant by adding nitrogen to a system that already has a refrigerant charge in it. The system must first be evacuated to the levels in **Figure 9.36** *before nitrogen can be added.*

SAFETY PRECAUTION: *Always check the nameplate of the system that dry nitrogen will enter. Never exceed the design low-side factory test pressure on the nameplate,* **Figure 9.37**. *When using dry nitrogen, always use a pressure relief valve and pressure regulator on the nitrogen tank because of its high pressures (over 2000 psi). Pressure relief valves must never be installed in series. Never pressurize a system with oxygen or compressed air because dangerous pressures can be generated from a reaction with the oil as it oxidizes.•*

DISPOSAL OF SMALL APPLIANCES

The following must be done before disposing of a small appliance:

- Recover the refrigerant from the appliance.
- Secure a signed statement that the refrigerant has been recovered, with the date of recovery.

Figure 9.37 A condensing unit nameplate showing factory test. Photo by John Tomczyk

Refrigerant bags are available for recovery of refrigerant from small appliances. These plastic bags will hold the charge of several refrigerators, **Figure 9.32**. The technician must have enough room in the service truck to haul the bag. When it is full, it may be taken to the shop and the refrigerant transferred into the reclaim cylinder for later pickup by or shipment to the reclaim company.

9.19 RECLAIMING REFRIGERANT

Because refrigerant must meet the AHRI 700 standard before it is resold or reused in another customer's equipment, all reclaimed refrigerant must be sent to a reclaim company to be checked and stored (provided it is in good enough condition to be used). Many service companies will not recycle refrigerant in the field because of the complications and the possibility of cross-contamination of products. Since the refrigerant cannot be vented, reclaiming it is another option. Refrigerant that meets the AHRI 700 specification can be purchased from reclaim companies, without the added tax. The reclaim companies will accept only large quantities of refrigerant (cylinders of 100 lb and larger), so the service shop will need a cylinder of the correct size for each of the refrigerants that will be shipped. Only DOT-approved cylinders with the proper labels can be used. Some companies buy recovered refrigerant and accumulate it for reclaiming.

The refrigerant in the reclaim cylinders must be of one type. The refrigerants cannot be mixed, or the reclaim company cannot reclaim it. As mentioned, mixed refrigerants cannot be separated. When the reclaim company receives refrigerant that cannot be reclaimed, it must be destroyed by incinerating it in such a manner that the fluorine in the refrigerant is captured. This is an expensive process, so technicians must be careful not to mix the refrigerants.

9.20 REFRIGERANT RETROFITTING

Most new air-conditioning systems today will be manufactured with R-410A refrigerant, and many air-conditioning systems will be retrofitted with a refrigerant blend. However, there is no one refrigerant blend that is a "perfect" drop-in replacement for retrofitting R-22 systems. Each blend has advantages and disadvantages, which have to be balanced against each other to choose the best overall blend for the specific application. There are many retrofit guidelines and procedures for a variety of refrigerants and lubricants. It would be impossible to cover them all in detail here. Major chemical companies and compressor manufacturers can provide these guidelines, and they can also be accessed on the Internet. The service technician can print them out to develop a retrofit library.

At the time of this writing, U.S. regulations state that there can be no production and no importing of HCFC-22 (R-22), except for use in equipment manufactured before January 1, 2010, and no production or importing of "new" equipment that uses R-22. Because of these rulings, many existing R-22 refrigeration and air-conditioning systems will be retrofitted with replacement refrigerants due to price and availability issues. What follows are some general considerations for retrofitting an HCFC-22 system with an HFC-based refrigerant. We will use R-22 in this example because of the recent regulations and its gradual decline in use and production.

HCFC-22 (R-22) RETROFITTING

As mentioned, most new air-conditioning systems will be manufactured with HFC-410A (R-410A). However, technicians and equipment owners often ask what the best retrofit refrigerant blend for existing R-22 systems is. There is no one quick-fix solution or answer to this question. A few key factors will contribute to the decision as to what replacement refrigerant should be used. These factors include:

- Evaporator temperatures (air-conditioning vs. refrigeration)
- Application of the equipment
- Size of the equipment

AIR-CONDITIONING VS. REFRIGERATION EVAPORATOR TEMPERATURES. Refrigerant retrofit blends will behave differently when exposed to different evaporator temperatures. Some refrigerant blends will closely match R-22 properties for air-conditioning evaporating temperatures (35°F to 5°F), while others will be better suited for the lower evaporating temperatures of medium- and low-temperature refrigeration applications (20°F to −20°F).

EQUIPMENT CAPACITY. Equipment capacity and run time also influences what refrigerant blend to incorporate in a system. One blend may have a lower capacity than another. Often, air-conditioning systems have an "off" cycle. The

systems are designed for the hottest days of the year, but on milder days the system will cycle off. By having longer run or "on" times, these systems can tolerate a lower-capacity blend. This also holds true for refrigeration storage facilities. However, refrigerants in systems that run all of the time are sized to load match, or exactly match, the capacity of the load. The solution here is to use a refrigerant blend with a higher capacity so the system can run at peak performance.

EQUIPMENT SIZE. How easy a system is to retrofit is a major factor in selecting a refrigerant retrofit blend. Retrofitting larger refrigeration or air-conditioning systems may require lubricant changes, component replacement, adjustment of controls, and maybe even seal and gasket changes. This often requires planned shut-downs and will generally be time-consuming and expensive. Choosing a retrofit blend based on performance and property matches that minimize component changes will be more important than "ease of retrofit" issues such as changing oil. On the other hand, smaller equipment may not provide direct access to the oil and may not have parts that can be individually changed. In this case, retrofit blends that minimize the work to be done during retrofits may be desirable even though performance may suffer. Often, depending on the equipment's age and condition, simply replacing the entire unit may be the more economical solution with smaller equipment.

ADDITION OF POLYOL ESTER (POE) LUBRICANT OR PARTIAL POE LUBRICANT REPLACEMENT. The retrofit refrigerant blends just listed are all HFC-based blends. Some contain hydrocarbons (HC) like R-600 (butane) and R-600a (isobutane). The small percentages of hydrocarbons help with oil return to the compressor. The hydrocarbons will soak into the mineral oil and thin it so it can be more easily returned to the compressor's crankcase on the low side of the system. However, for larger refrigeration and air-conditioning systems with receivers, oil that escapes the compressor will eventually form a layer on top of the liquid refrigerant in the receiver. These hydrocarbons will not mix with the liquid oil layer in the receiver on the high side. The solution is the addition of some POE lubricant or partial POE replacement of the mineral oil to help dissolve the oil back into the liquid refrigerant in the receiver and keep it in circulation through the refrigeration or air-conditioning system. Eventually, this lubricant will make it back to the compressor's crankcase.

In conclusion, no refrigerant blend will be a perfect drop-in replacement for retrofitting R-22 systems. Each refrigerant blend has advantages and disadvantages. These advantages and disadvantages have to be balanced to choose the best overall choice for your specific application.

NOTE: *Always consult with the refrigerant manufacturer or the compressor manufacturer and follow their specific retrofit guidelines before performing a retrofit on any R-22 system or any other refrigeration and/or air-conditioning system.* •

GENERAL RETROFITTING CONSIDERATIONS

When the retrofitting of an air-conditioning or refrigeration system is being considered, there are a number of factors that need to be carefully evaluated in order to make the best possible decision. The decision process will involve whether or not to actually perform the retrofit and, if it is determined that retrofitting is the best option, which refrigerant is the best option for the changeover. Some of the important factors that must be considered will be discussed now.

Before retrofitting, the compatibility of the plastics and elastomers in the system should be considered. Industry studies have shown that there is no one plastic or elastomer material that will be compatible with all replacement refrigerants. Components to be checked include shaft seals, O-rings, and gaskets. Since different gasket and O-ring materials will swell at different rates when exposed to different substances and conditions, refrigerant leaks may result when the system refrigerant is changed.

Lubricant selection is also a major concern. Different refrigeration oils behave differently under different conditions and can affect oil return to the compressor and an oil's lubricity, which is an oil's ability or capacity to reduce friction. In addition, there might be issues with the compatibility of the various materials in the system, such as with the gaskets and O-rings, as just described. In addition, in many cases, when switching over from one type of oil to another, care must be taken to ensure that any residual oil is removed from the system. When changing from a mineral oil to a polyol ester lubricant, for example, the mineral oil must be removed to the extent that no more than 5% of the original oil charge, on average, remains in the system. Always follow the manufacturer's guidelines as they can vary greatly from one refrigerant replacement product to another.

When evaluating a retrofit, system performance-related issues must not be overlooked. There will very likely be changes in operating pressures and temperatures, as well as capacity and efficiency changes. Even small percentage changes to system capacity and efficiency can have a significant effect on a system's ability to effectively and efficiently condition a space.

When retrofitting, be aware that most refrigerant replacements are not done on a pound-for-pound basis. Simply stated, this means that the amount of retrofitted refrigerant that is added to a system will most likely not equate to the amount of refrigerant that was removed from the system. This can pose a problem on split-type systems where the total, optimal, refrigerant charge is not known. In the case of packaged equipment, the unit's factory charge is indicated on the data plate and can be used to calculate the charge for the new refrigerant.

Retrofitting can be a challenge for relatively new equipment. For new equipment, warranties may be impacted if the

system is retrofitted. Be sure to check with the compressor and/or equipment manufacturer to determine if the warranty will be affected if the system is retrofitted.

GENERAL RETROFITTING GUIDELINES

Once the determination has been made to retrofit a system, here is a generic procedure that can be followed in most instances. Of course, be sure to refer to, and follow, the manufacturer's guidelines when working with their specific products.

1. Before starting the system retrofit, identify any existing system conditions that might adversely affect the operation of the system. These conditions include, but are not limited to dirty air filters, dirty evaporator or condenser coils, failed fan motors, airflow restrictions, damaged or missing suction-line insulation, improperly operating expansion devices, and damaged or other failed system components.
2. Resolve any conditions identified in item #1 before proceeding. Failure to do so can result in a system that is not operating to its full potential.
3. Obtain baseline data for the present system. This information includes, at a minimum, operating pressures and temperatures, evaporator superheat, condenser subcooling, ambient conditions, indoor air conditions, temperature splits, and system capacity and efficiency calculations.
4. Remove existing compressor oil, if required, to the level required by the manufacturer. Keep track of how much refrigeration oil has been removed from the system. Oil removal is often accomplished by replacing oil, running the system, and then changing oil again until remaining oil concentrations are below the required levels. This may require three or more oil changes to reach the required levels and concentrations.
5. Recover existing system refrigerant, making certain to record the amount of refrigerant removed from the system. Be sure to reach the required vacuum levels during the refrigerant recovery process as determined and outlined by the EPA.
6. Replace system filter driers.
7. Pressurize the system with an inert gas such as dry nitrogen and leak check the system.
8. Evacuate the system, pulling the equipment into a vacuum of at least 500 microns.
9. Wait. Make certain that the system has reached, and holds, the required vacuum level.
10. Charge system according to the manufacturer's guidelines and instructions.
11. Adjust thermostatic expansion valve settings, if required.
12. Adjust any other refrigerant flow controls, such as head pressure controls, evaporator and crankcase pressure regulators (EPRs and CPRs), if applicable, as needed.
13. Record system operating parameters.
14. Label the system so that future service personnel will be aware that the system is operating with a new/different refrigerant. On this label, be sure to include the new total refrigerant charge in the system.

REVIEW QUESTIONS

1. The willful purging or venting of certain refrigerants to the atmosphere is
   A. essential to the preservation of the stratospheric ozone layer.
   B. illegal as set forth by EPA regulations.
   C. encouraged by the EPA as it helps to protect tropospheric ozone.
   D. desirable as this process helps improve food preservation efforts.
2. What is the difference between an oxygen molecule and an ozone molecule?
   A. An oxygen molecule has three oxygen atoms, while an ozone molecule has two.
   B. An oxygen molecule has one oxygen atom, while an ozone molecule has two.
   C. An oxygen molecule has two oxygen atoms, while an ozone molecule has three.
   D. An oxygen molecule has two oxygen atoms, while an ozone molecule has one.
3. Tropospheric ozone is
   A. as important to protect as stratospheric ozone.
   B. found in the shell that extends between 7 and 30 miles above the earth's surface.
   C. close to the earth and is often referred to as bad ozone or smog.
   D. the ozone that protects us from harmful ultraviolet radiation.
4. It is estimated that a single chlorine atom can destroy up to ______ ozone molecules?
   A. 1000
   B. 10,000
   C. 100,000
   D. 1,000,000
5. Which of the following sets of refrigerant classifications has a zero ODP?
   A. CFCs and HFCs
   B. HCFCs and HFCs
   C. CFCs and HFOs
   D. HFOs and HFCs

6. An increase in the natural greenhouse effect that leads to the heating of the earth is called
   A. global warming.
   B. planetary effect.
   C. planetary warming.
   D. global effect.
7. What index takes into account the combined effects of global warming potential and ozone depletion potential?
   A. Total equivalent warming impact (TEWI)
   B. Total greenhouse effect impact (TGEI)
   C. Total global warming number (TGWN)
   D. Total equivalent greenhouse number (TEGN)
8. CFCs are more harmful to the stratospheric ozone layer than HFCs because CFCs contain
   A. fluorine.
   B. chlorine.
   C. hydrogen.
   D. carbon.
9. HFO refrigerants are classified as having
   A. low GWPs and low ODPs.
   B. low GWPs and high ODPs.
   C. high GWPs and low ODPs.
   D. high GWPs and high ODPs.
10. The total phaseout for all HCFC refrigerants is scheduled for the year
   A. 2025.
   B. 2030.
   C. 2035.
   D. 2040.
11. Which of the following classes of refrigerants is potentially dangerous because of its high flammability?
   A. HFO
   B. HCFC
   C. CFC
   D. HC
12. The safest overall rating with respect to both toxicity and flammability is
   A. A1.
   B. B1.
   C. A3.
   D. B3.
13. Of the following refrigerants, which is best suited as a retrofit refrigerant for an R-22 residential air-conditioning system?
   A. R-134a
   B. R-410A
   C. R-407C
   D. R-123
14. Refrigerants R-11, R-12, and R-22 all have what in common?
   A. They are all methane-based refrigerants.
   B. They are all ethane-based refrigerants.
   C. They are all CFC refrigerants.
   D. They are all HCFC refrigerants.
15. If a refrigerant experiences a temperature glide, which of the following can be correctly concluded?
   A. The refrigerant is a single-compound substance.
   B. The refrigerant is a blended refrigerant.
   C. The refrigerant is classified as an HCFC.
   D. The refrigerant is classified as a CFC.
16. The term "fractionation" refers to the
   A. separation of a refrigerant blend into its component parts.
   B. ability of refrigeration oil to lubricate effectively.
   C. range of saturation temperatures at a given pressure.
   D. effects of released refrigerants on part of the atmosphere.
17. The most common type of refrigeration oil used in conjunction with HFC refrigerants is
   A. animal.
   B. vegetable.
   C. alkylbenzene.
   D. esters.
18. Which of the following represents a complete list of the elements contained in CFC refrigerants?
   A. Hydrogen, chlorine, fluorine, and carbon
   B. Hydrogen, chlorine, and carbon
   C. Chlorine, fluorine, and carbon
   D. Hydrogen and carbon
19. Which of the following represents a complete list of the elements contained in HFC refrigerants?
   A. Hydrogen, chlorine, fluorine, and carbon
   B. Hydrogen, fluorine, and carbon
   C. Chlorine, fluorine, and carbon
   D. Hydrogen and carbon
20. Which of the following represents a complete list of the elements contained in HFO refrigerants?
   A. Hydrogen, chlorine, fluorine, and carbon
   B. Hydrogen, fluorine, and carbon
   C. Chlorine, fluorine, and carbon
   D. Hydrogen and carbon
21. The term used to describe the process of removing refrigerant from a system and storing it without testing or processing it is
   A. restoring.
   B. reclaiming.
   C. recycling.
   D. recovering.
22. The term used to describe the process of removing refrigerant from a system and filtering and/or cleaning it to meet AHRI 740 standards is
   A. restoring.
   B. reclaiming.
   C. recycling.
   D. recovering.

23. The term used to describe the process of removing refrigerant from a system and surrendering it for reprocessing to meet AHRI 700 standards is
    A. restoring.
    B. reclaiming.
    C. recycling.
    D. recovering.
24. The term used to describe a recovery process that uses a self-contained recovery unit is
    A. active.
    B. passive.
    C. system-dependent.
    D. temperature-dependent.
25. Which type of EPA Section 608 certification is required to work on small appliances?
    A. Type I
    B. Type II
    C. Type III
    D. Type IV
26. What term is used to describe the certification that is awarded when a technician passes all of the parts of an EPA certification exam?
    A. Global
    B. Stellar
    C. Universal
    D. Galactical
27. Low-loss fittings are intended to ensure that
    A. no refrigerant is trapped in the hoses when removed from a system.
    B. no nitrogen is trapped in the hoses when removed from a system.
    C. refrigerant losses are minimized when removing hoses from system.
    D. refrigerant losses are maximized when removing hoses from system.
28. The standard that a refrigerant must meet after it has been reclaimed is
    A. AHRI 7000.
    B. AHRI 700.
    C. AHRI 7400.
    D. AHRI 740.
29. The standard that a refrigerant must meet after it has been recycled is
    A. AHRI 7000.
    B. AHRI 700.
    C. AHRI 7400.
    D. AHRI 740.
30. A DOT-approved refrigerant recovery cylinder is color-coded
    A. with yellow and gray stripes.
    B. gray with a yellow top.
    C. yellow with a gray top.
    D. yellow with gray triangles.
31. How are oil-less compressors on recovery units lubricated?
    A. They still use oil, just less.
    B. The compressor shaft floats so no lubrication is needed.
    C. They rely on residual oil in the recovered refrigerant.
    D. The compressor is magnetic, so no oil is required.
32. All of the following tasks are part of a refrigerant retrofitting process except
    A. label the system with information about the new refrigerant and system charge.
    B. adjust all refrigerant flow devices such as TXVs, CPRs, and EPRs, as needed.
    C. obtain a complete set of baseline data for the system prior to retrofitting.
    D. ensure that the evaporator fan motor is off during system charging.
33. The Montreal Protocol is best described as a
    A. set of laws that put an end to CFC and HCFC production.
    B. treaty or agreement among nations to address ozone depletion.
    C. law put in place by the EPA to override the Clean Air Act.
    D. treaty between the EPA and the U.S. government to stop venting.

UNIT 10

SYSTEM CHARGING

OBJECTIVES

After studying this unit, you should be able to

- describe how refrigerant is charged into systems in its vapor and liquid states.
- state the advantage of using electronic scales for weighing refrigerant into a system.
- describe the terms "temperature glide" and "fractionation" as they apply to blended refrigerants.
- charge air-conditioning systems with fixed orifice, capillary tube, or piston-type metering devices using charging charts and curves.
- use the subcooling method of charging to charge air-conditioning and heat pump systems that incorporate a thermostatic expansion valve (TXV) as the metering device.
- use temperature/pressure charts to help properly charge air-conditioning and refrigeration systems.
- differentiate between "bubble point" and "dew point" values on a temperature/pressure chart and properly use these values to calculate superheat and subcooling values on systems that operate with blended refrigerants.

SAFETY CHECKLIST

- ✓ Never use a torch to apply heat to a refrigerant cylinder. If needed, place the refrigerant tank in a tub of warm water no hotter than 90°F.
- ✓ Do not charge liquid refrigerant into the suction line of a system unless it is vaporized before it reaches the compressor.
- ✓ Use a charging device that allows liquid refrigerant to flash into a vapor before entering the compressor. Liquid refrigerant can cause severe damage to the compressor.
- ✓ When charging blended refrigerants into a system, the refrigerant must leave the tank as a liquid to avoid fractionation.
- ✓ Wear safety glasses and gloves when working with refrigerant. Refrigerant, when it comes in contact with the skin, can cause frostbite.

10.1 CHARGING A REFRIGERATION SYSTEM

System charging refers to the adding of refrigerant to an air-conditioning or refrigeration system. *The correct charge must be added so that a refrigeration system can operate as effectively and efficiently as it was designed to. This is not always an easy task to accomplish.* Some refrigerants, namely single-compound refrigerants, may be added to the system in either the vapor or liquid states, while blended refrigerants must be charged in the liquid state. A scale should be used to determine the amount of refrigerant that is added to a system. The system's charge is evaluated by using superheat and subcooling calculations or manufacturer-supplied charging tables and charts. When using system charging charts and tables to evaluate the system charge, a number of factors should be taken into account. These factors include, but are not limited to, high- and low-side saturation pressures and temperatures, ambient temperatures, evaporator superheat, condenser subcooling, compressor amperage draw, return air wet-bulb temperature, and temperature differentials across heat exchange surfaces.

This unit describes how to add the refrigerant correctly and safely in the vapor and the liquid state. The correct charge for a particular system will be discussed in the unit where that system is covered. Heat pump charging, for example, is covered in Unit 43, "Air Source Heat Pumps."

10.2 VAPOR REFRIGERANT CHARGING

The vapor refrigerant charging of a system is accomplished by allowing vapor to flow out of the refrigerant cylinder and into the refrigeration system. When the pressure in the system is lower than the pressure in the refrigerant tank—for example, when the system has just been evacuated—refrigerant can be added to both the high- and low-pressure sides of the system. This is because the pressure on both sides of the system is lower than the pressure in the refrigerant tank. Remember that high pressure flows to low pressure.

When the system is operating, however, refrigerant must be added to the low-pressure side of the system. This is because the compressor raises the pressure of the refrigerant on the high side of the system and lowers the pressure on the low side. The high-side pressure of an operating refrigeration system is typically higher than the pressure of the refrigerant in the tank. For example, an R-134a system may have a head pressure of 185 psig on a 95°F day (this is determined by taking the outside temperature of 95°F and adding 30°F, which gives a condensing temperature of 125°F, which corresponds to a pressure of 185 psig for R-134a). The cylinder is exposed to the same ambient temperature of 95°F but the refrigerant, assuming that it is saturated, is at a pressure of only 114 psig (from the temperature/pressure chart), **Figure 10.1**. It should be noted that, if valve B in **Figure 10.1** is opened, refrigerant will flow toward

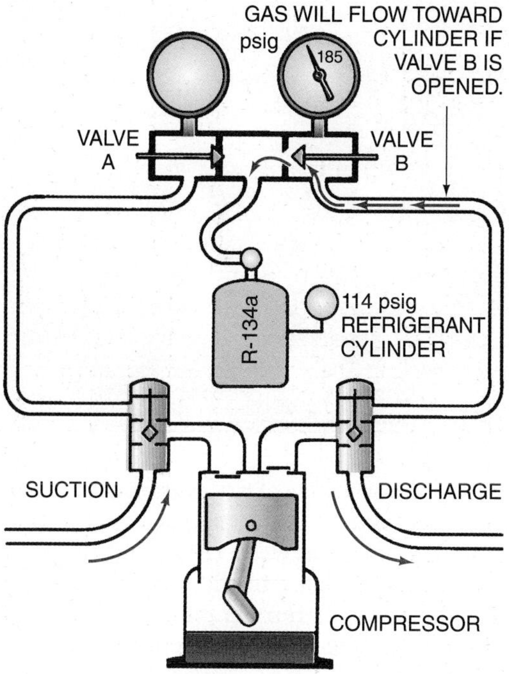

Figure 10.1 This refrigerant cylinder has a pressure of 114 psig. The high side of the system has a pressure of 185 psig. Pressure in the system will prevent the refrigerant in the cylinder from flowing into the system.

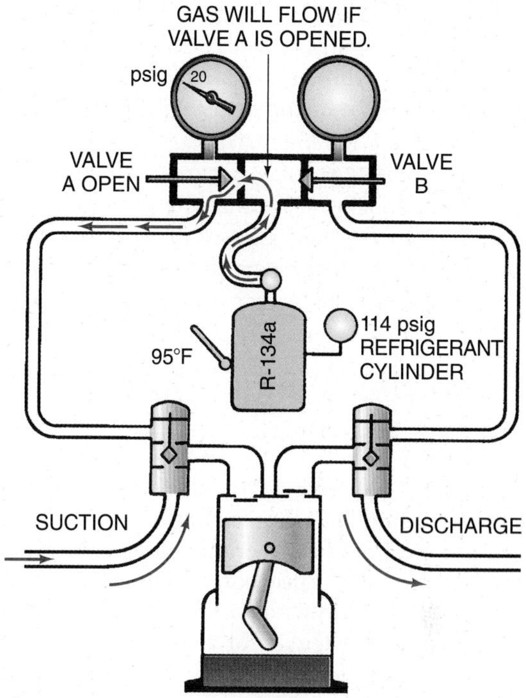

Figure 10.2 The temperature of the cylinder is 95°F. The pressure inside the cylinder is 114 psig. The low-side pressure is 20 psig.

the refrigerant tank. If the tank is a "one-trip" tank that is used only for new, "virgin" refrigerant, the refrigerant will not enter the tank. This is because one-trip tanks are manufactured with check valves that only allow refrigerant to leave the tank, not enter it. If the tank is a reusable, recovery-type, tank, there are no check valves and system refrigerant can enter the tank as long as the system pressure is higher than the pressure in the tank.

The low-side pressure in an operating system is much lower than the cylinder pressure if the cylinder is warm. For example, on a 95°F day, the cylinder of saturated R-134a will have a pressure of 114 psig, but the evaporator pressure may be only 20 psig. Refrigerant will easily move into the system from the cylinder, **Figure 10.2**. In the winter, however, the cylinder may have been in the back of a truck all night, and its pressure may be lower than the low side of the system. In this case, the cylinder will have to be warmed to get refrigerant to move from the cylinder to the system. It is a good idea to have a cylinder of refrigerant stored in the equipment room of large installations. It will always be there in case you have no refrigerant in the truck, and the cylinder will be at room temperature even in cold weather.

When vapor refrigerant is pulled out of a refrigerant cylinder, the vapor pressure is reduced and the liquid boils to replace the vapor that is leaving. This reduction in vapor pressure causes a lower saturation temperature for both the liquid and the vapor in the cylinder because of the associated temperature/pressure relationship. As more and more vapor is released from the cylinder, the liquid in the bottom of the cylinder continues to boil and its temperature decreases. If enough refrigerant is released, the cylinder pressure will decrease to the low-side pressure of the system. Heat will have to be added to the liquid refrigerant to keep the pressure up. **⚠ SAFETY PRECAUTION:** *Never use concentrated heat from a torch. Gentle heat, such as the heat from a tub of warm water, is safer. The water temperature should not exceed 90°F. Electric heating blankets manufactured for this purpose are also safe to use.•* This will maintain a cylinder pressure of 105 psig for R-134a if the refrigerant temperature is the same as the temperature of the water in the tub, **Figure 10.3**. A water temperature of 90°F is desired because it is the approximate temperature of the human hand. If the water feels warm, it is too hot.

Figure 10.3 This refrigerant cylinder is in warm water to keep the pressure up.

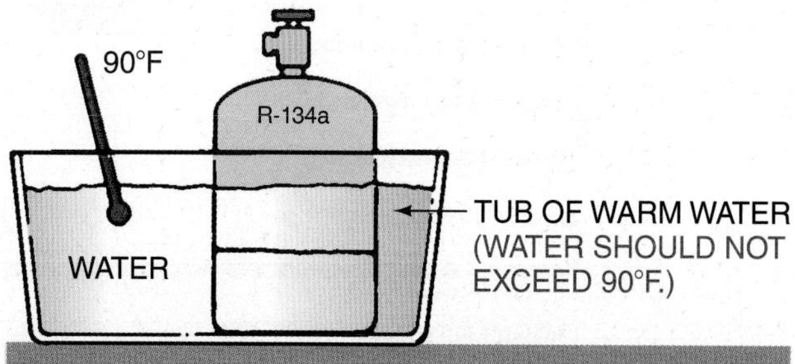

The larger the volume of the liquid refrigerant in the bottom of the cylinder, the longer the cylinder will maintain the pressure. When large amounts of refrigerant must be charged into a system, use the largest cylinder available. For example, do not use a 30-lb cylinder to charge 25 lb of refrigerant into a system if a 125-lb cylinder is available.

If you are planning on charging vapor into a system that is operating with a blended refrigerant, care must be taken to ensure that the blend does not fractionate, or separate into its component parts. Refer to Section 10.8 in this unit as well as Unit 9, "Refrigerant and Oil Chemistry and Management—Recovery, Recycling, Reclaiming, and Retrofitting," for important information regarding blended refrigerants and the charging procedures that must be used when working with them.

10.3 LIQUID REFRIGERANT CHARGING

Liquid refrigerant charging of a system is normally accomplished in the liquid line. For example, when a system is out of refrigerant, liquid refrigerant can be charged into the liquid line service valve, also commonly referred to as the king valve, in the liquid line or on the receiver, **Figure 10.4**. If the system is in a vacuum or has been evacuated, you can allow liquid refrigerant to enter the system until flow has nearly stopped. Upon entering the system, the liquid refrigerant will move toward the evaporator and condenser coils. When the system is started, the refrigerant will be about equally divided between them, and there is no danger of liquid flooding back to the compressor, **Figure 10.5**. When large amounts of refrigerant are needed, adding the liquid through the king valve is preferable to other methods because it saves time.

Figure 10.4 Liquid-line service valve located on the refrigerant receiver. Photo by Eugene Silberstein

Figure 10.5 This system is being charged while it has no refrigerant in it. The liquid refrigerant moves toward the evaporator and the condenser. No liquid refrigerant will enter the compressor.

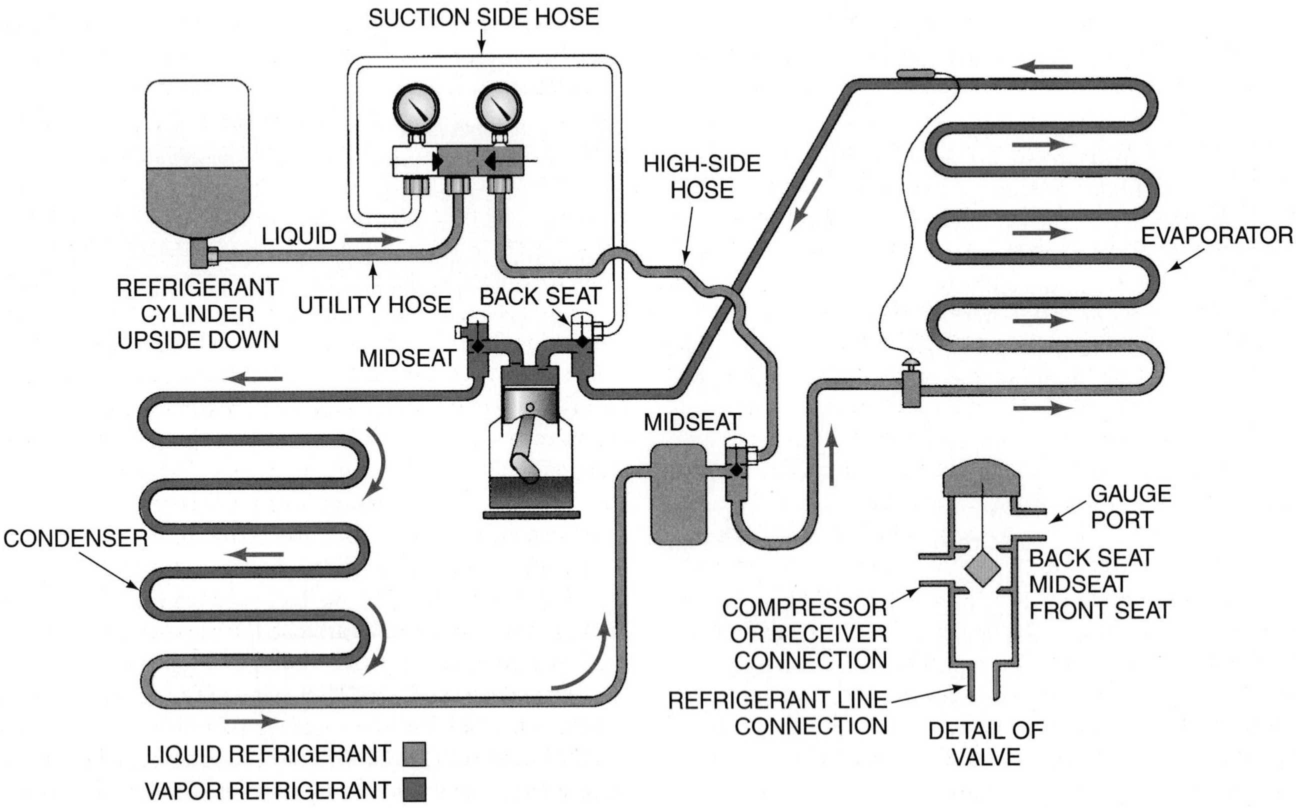

Figure 10.6 Cross-sectional view of a service valve. (A) Back-seated position. (B) Cracked-off-the-back-seat position. (C) Midseated position. (D) Front-seated position.

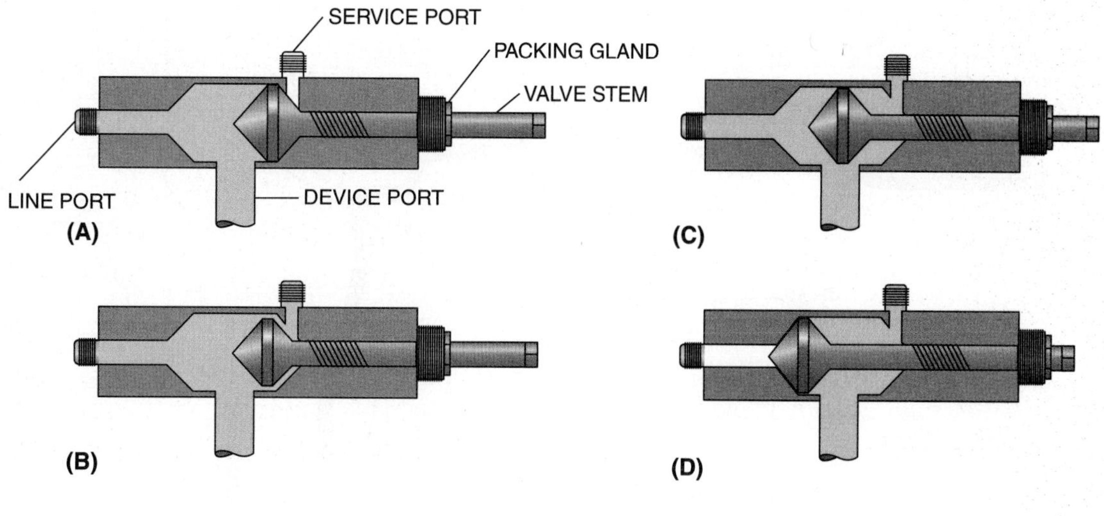

The king valve is used to help perform a number of service-related tasks that include taking pressure readings on the high side of the system and trapping all of the system refrigerant in the receiver. In the past, king valves were used primarily as safety valves on ammonia and sulfur dioxide systems. They were located outside the mechanical equipment rooms. By positioning the stem on the king valve so that the liquid line was sealed off, refrigerant could not leave the receiver, causing the remaining system refrigerant to become stored in the receiver. This process is referred to as "pumping down" and, in the case of sulfur dioxide systems, could be done from outside the equipment room so that the technician would not have to enter a toxic environment in the event of a refrigerant leak. Service valves that have their stems turned all the way clockwise into the valve to seal off the liquid line are referred to as *front seated*. The king valve has four positions: back seated, cracked-off-the-back-seat, midseated, and front seated, **Figure 10.6**.

The back-seated position, **Figure 10.6(A)**, is the normal operating position of the valve. Refrigerant flows freely from the receiver to the liquid line, while the service port is closed off. This is the port where the high-side gauge hose is connected. When the valve is cracked-off-the-back-seat, the service port is open to the system and the technician can take operating pressure readings from the high side of the system, **Figure 10.6(B)**. While in the cracked-off-the-back-seat position, refrigerant still flows freely from the receiver to the liquid line. The midseated position, **Figure 10.6(C)**, is used for performing standing pressure tests and for system evacuation. The front-seated position, **Figure 10.6(D)**, is used for system pumpdown. In this position, the refrigerant flows to the receiver but cannot leave it. This traps all of the system refrigerant in the receiver, enabling the technician to service the low-pressure side of the system.

When a system has a king valve, it may be front seated while the system is operating and the low-side pressure of the system will drop. This happens because the rest of the liquid line, expansion valve, evaporator, suction line, and compressor are now being starved of refrigerant. Liquid from the cylinder may be charged into the system at this time through an extra charging port. The liquid from the cylinder is actually feeding the expansion device. *Be careful not to overcharge the system,* **Figure 10.7.** The low-pressure control may have to be bypassed during charging to keep it from shutting the system off. Be sure to remove the electrical bypass when charging is completed, **Figure 10.8.** **NOTE:** *Precautions must be taken when charging liquid into the suction line of a system. Liquid must be vaporized once it leaves the charging cylinder to prevent any liquid from entering the compressor, as damage can result if the compressor attempts to compress liquids, which are not compressible. When working with blended refrigerants, note that the refrigerant must leave the cylinder as a liquid in order to prevent fractionation.•*

On split-type air-conditioning systems, there are service valves located in the liquid and suction lines at the points where these lines are connected to the condensing unit. These valves are configured so that when the valves are front-seated (the position they are in when the condensing unit is shipped) the service port is open to the line port. This allows the technician to pressurize the field-installed refrigerant lines for leak-testing purposes and evacuate the system without disturbing the unit's factory charge of refrigerant that is contained in the condensing unit.

Some commercially available charging devices allow the refrigerant cylinder to be attached to the suction line to charge an operating system with liquid refrigerant. These orifice-metering devices, **Figure 10.9**, act as restrictions between the gauge manifold and the system's suction line. These devices meter liquid refrigerant into the suction line, where it flashes into a vapor. The same thing may be accomplished using the

Figure 10.7 This system is being charged by front seating the king valve and allowing liquid refrigerant to enter the liquid line.

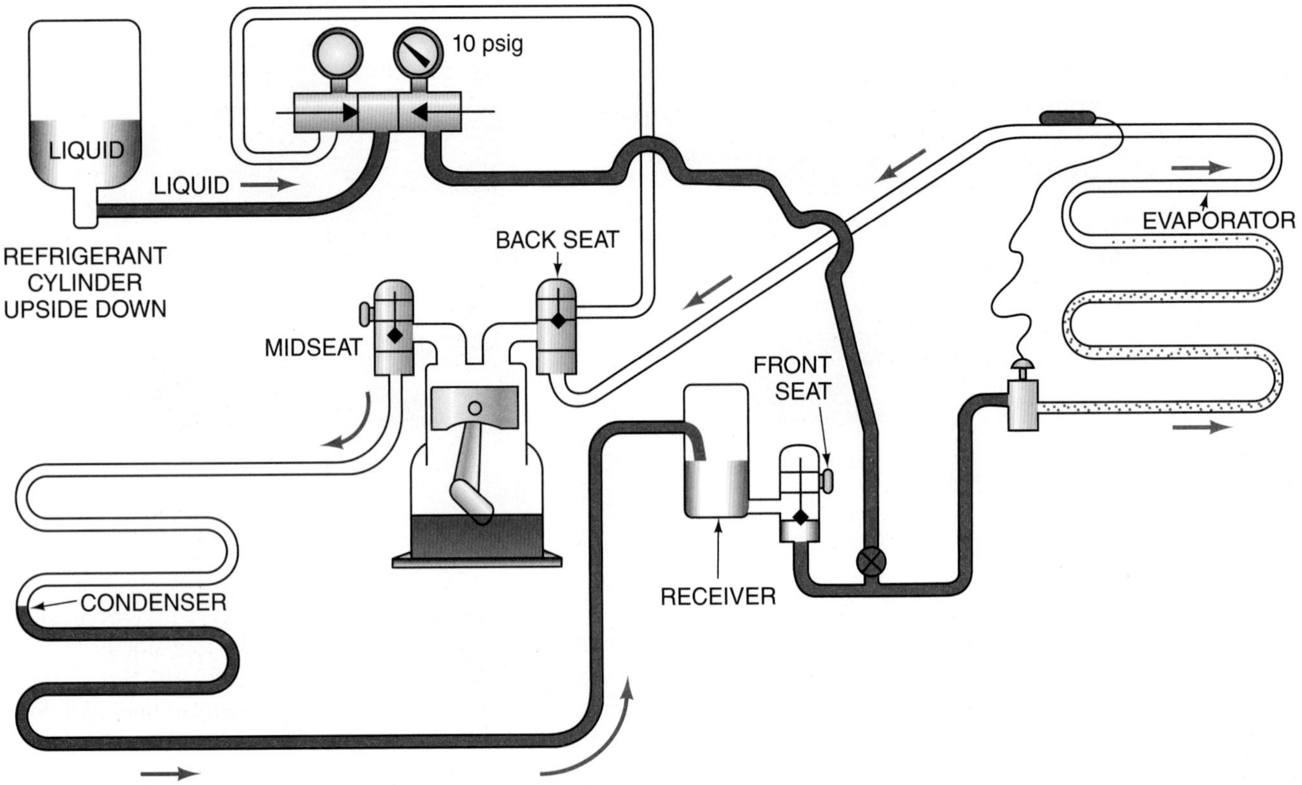

Figure 10.8 Bypassing the low-pressure control for charging purposes. Courtesy Ferris State University. Photo by John Tomczyk

Figure 10.9 A charging device in the gauge line between the liquid refrigerant in the cylinder and the suction line of the system. Device provided by Uniweld Products. Photo by Eugene Silberstein

gauge manifold valve, **Figure 10.10**. The pressure in the low-pressure hose on the gauge manifold is maintained at not more than 10 psig higher than the system suction pressure. This will meter the liquid refrigerant into the suction line, where it will vaporize before reaching the compressor. During the process of throttling the liquid refrigerant through the gauge manifold, the portion of the gauge manifold where the center hose is connected should be warm, while the portion of the manifold where the low-side hose is connected should be cold. The temperature difference that exists between the center and low-side connections on the gauge manifold is due to the fact that the vaporizing liquid will absorb heat from the manifold. **NOTE:** *The gauge manifold valve should be used only on compressors where the suction gas passes over the motor windings. This will boil any small amounts of liquid refrigerant that may reach the compressor. If the lower portion of the*

Figure 10.10 A manifold gauge set used to accomplish the same purpose as that in Figure 10.9.

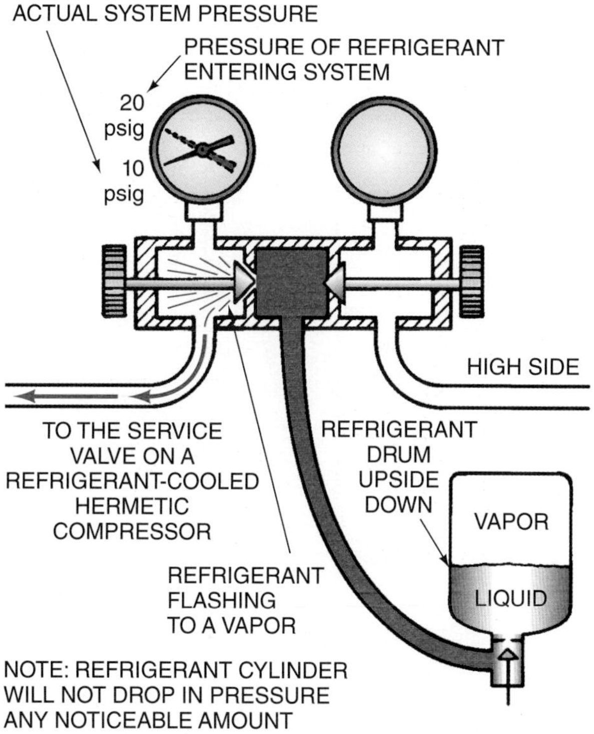

compressor casing becomes cold, stop adding liquid. This process should be performed only under the supervision of an experienced person. Throttling liquid refrigerant from a charging cylinder, however, must be done when charging certain refrigerant blends to avoid fractionation, regardless of whether the compressor is refrigerant cooled or air cooled.•

When a known amount of refrigerant must be charged into a system, it should be weighed into the system. The recommended charge for package systems such as air conditioners and refrigerated cases will be printed on the nameplate. Attempting to weigh refrigerant into the system when it contains a partial charge will lead to inaccurate charging and improper system operation. The remaining charge must first be recovered and then the system must be properly evacuated before the nameplate charge can be added to the system.

10.4 WEIGHING REFRIGERANT

Weighing refrigerant may be accomplished using various scales. Bathroom and other inaccurate scales should *not* be used. The most popular type of scale used today is the electronic scale, **Figure 10.11**. These scales can be adjusted to zero with a full cylinder, **Figure 10.12**, so as refrigerant is added to the system the scales read a positive value. For example, if 28 oz of refrigerant is needed in a system, put the refrigerant cylinder on the scale and set the scale at 0. As the refrigerant leaves the cylinder, the scale counts upward. When 28 oz is reached, the refrigerant flow can be stopped. Electronic scales are available with different features; some have automatic charging capability, **Figure 10.13**, which enables the technician to program the amount of refrigerant to be charged into the system. A solid-state microprocessor automatically stops the charging process once the desired amount of refrigerant

Figure 10.11 Digital scale used to measure the amount of refrigerant added to or removed from the system. Photo by Eugene Silberstein

Figure 10.12 The scale's display can be zeroed out to reduce the number of manual calculations. Photo by Eugene Silberstein

Figure 10.13 Automatic charging scales can be programmed to charge a precise quantity of refrigerant into the system. Photo by Eugene Silberstein

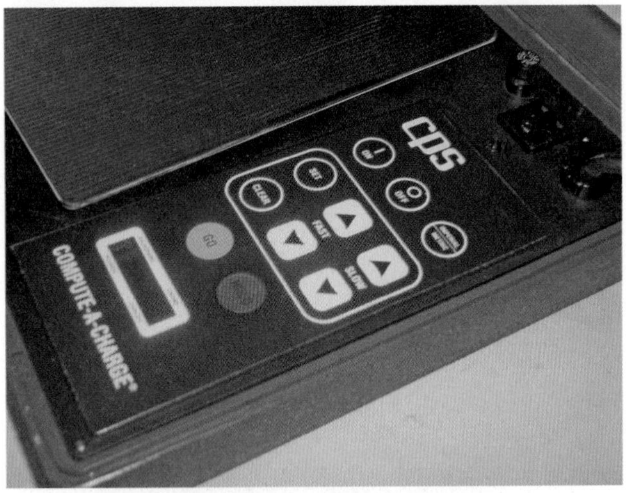

Figure 10.14 Digital refrigerant charging scale that is controlled by a wireless remote. Scale provided by Fieldpiece Instruments, Inc. Photo by Eugene Silberstein

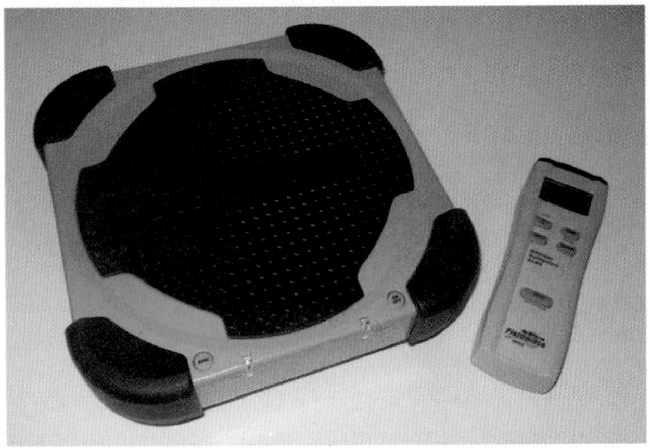

has been introduced. Some instruments can sound an audible alarm once the desired amount of refrigerant has been introduced to the system. Other electronic scales, such as the one shown in **Figure 10.14**, have unique features that include control by wireless remote.

10.5 GRADUATED CHARGING CYLINDERS

Graduated cylinders, **Figure 10.15**, are often used to add refrigerant to smaller air-conditioning and refrigeration systems, such as domestic refrigerators and window or through-the-wall air conditioners. These appliances have relatively small refrigerant charges ranging from a few ounces to about 2 pounds, and the actual refrigerant charge is provided on the appliance's nameplate. Charging cylinders are clear so that you can observe the liquid level inside the cylinder and have a pressure gauge located at the top from which the saturation pressure and temperature of the refrigerant is read. The saturation temperature gives the temperature of the surrounding ambient air, which is needed to properly prepare the cylinder for use. The volume of liquid refrigerant in the cylinder changes as the surrounding air temperature changes, so the cylinder must be calibrated before each use. The outer casing, or shroud, on the charging cylinder is rotated until the pressure indicated on the shroud corresponds to the pressure of the refrigerant in the charging cylinder. This calibration compensates for any changes in refrigerant volume due to temperature changes. The amount of liquid refrigerant in the cylinder, in pounds and ounces, can then be read from the graduated scale on the sides of the cylinder.

Figure 10.15 A graduated charging cylinder.

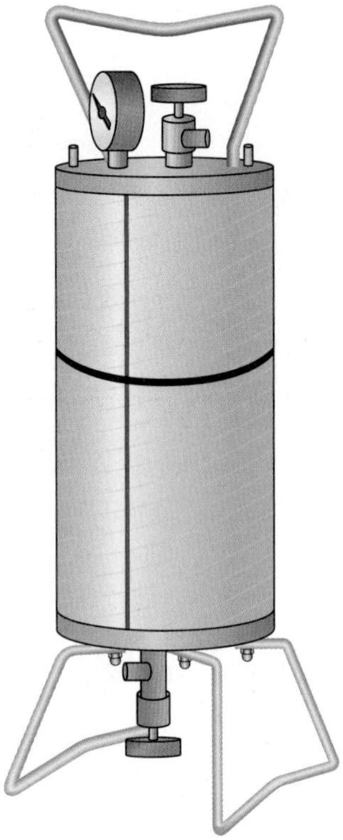

Once calibrated, the amount of liquid refrigerant in the cylinder is recorded. The amount of refrigerant that must be added to the system is then subtracted from the amount in the cylinder. The result is the amount of refrigerant that must remain in the cylinder after the charging process has been completed. This level is marked with a sliding ring, **Figure 10.16(A)**. The final liquid level inside the cylinder must then be calculated.

Suppose a properly calibrated graduated cylinder contains 4 lb 4 oz of R-134a and is to be used to charge 28 oz

Figure 10.16 (A) The ring is set to the refrigerant level desired once the charging process has been completed. (B) When the refrigerant level is even with the position of the ring, the system has been charged.

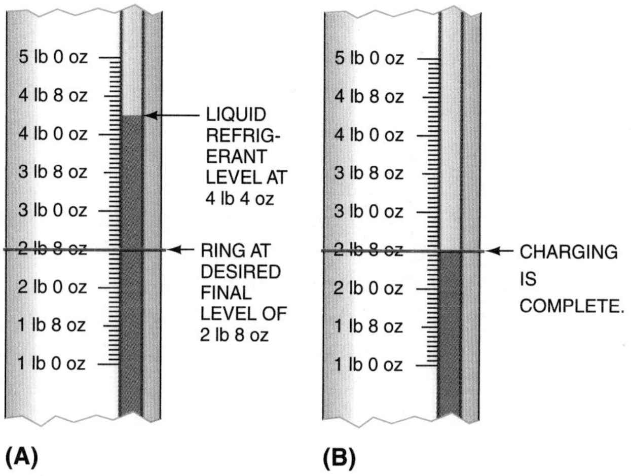

of refrigerant into a unit. The system charge of 28 oz is subtracted from the 4 lb 4 oz as follows:

$$\text{Cylinder contains } 4 \text{ lb } 4 \text{ oz} = 68 \text{ oz}$$

Then

$$68 \text{ oz} - 28 \text{ oz} = 40 \text{ oz}$$

is the final cylinder level:

$$40 \text{ oz} \div 16 \text{ oz/lb} = 2.5 \text{ lb} = 2 \text{ lb } 8 \text{ oz}$$

The sliding ring is then positioned at the 2 lb 8 oz mark on the cylinder. When the charging process is completed, the level of the liquid refrigerant in the graduated charging cylinder will be level with the sliding ring, **Figure 10.16(B)**. Some graduated cylinders have heaters in the bottom to keep the refrigerant temperature from dropping when vapor is pulled from the cylinder or when it is being used in a cold surrounding temperature.

When selecting a graduated cylinder for charging purposes, be sure to select one that is large enough for the systems with which you will be working. It is difficult to use a cylinder twice for one accurate charge. When charging systems with more than one type of refrigerant, you need a charging cylinder for each type.

10.6 CHARGING CHARTS

Some manufacturers supply charts or curves, called *charging charts* or charging curves, **Figure 10.17**, to assist the service technician in correctly charging air-conditioning and heat pump units. The style and type of charging chart or curve will likely vary with the manufacturer. Often, the type of charging chart supplied depends on what type of metering device the equipment incorporates. In the HVACR industry,

Figure 10.17 A charging curve used for charging a split-type air-conditioning system that incorporates a capillary tube metering device.

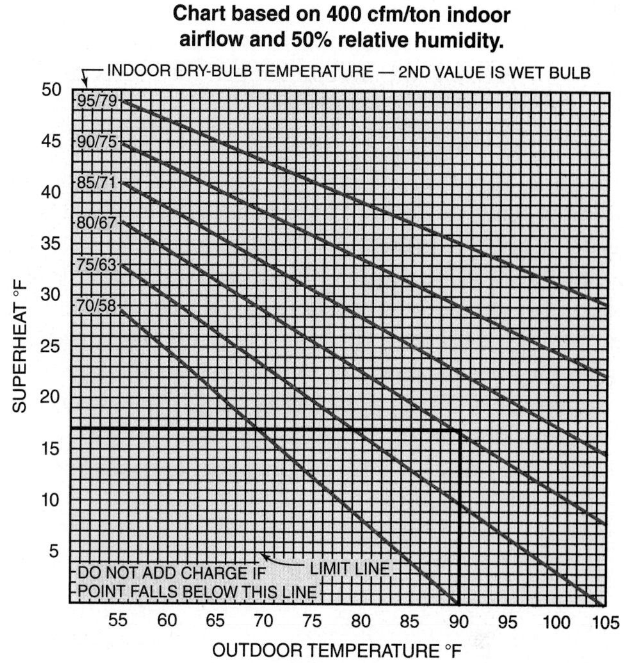

charging charts or curves that utilize superheat measurements, such as the one shown in **Figure 10.17**, are typically used in conjunction with fixed-bore metering devices—capillary tubes, fixed orifices, and pistons (short-tube devices). Charging curves that rely on subcooling measurements are typically used on systems that are equipped with thermostatic expansion valves. The same basic underlying principles hold for just about all charging curves or charts. The safest and most accurate way to charge any unit is to follow the manufacturer's charging instructions on the unit or in a manual supplied with the unit.

The charging curve in **Figure 10.17** is based on 400 cfm/ton and a return air relative humidity of 50%. To use this specific charging curve to charge a split air-conditioning system that incorporates a capillary tube for a metering device, follow these steps:

Step 1 Measure the indoor dry-bulb temperature (DBT). This is the return air at the air handler. **NOTE:** *Use wet-bulb temperature (WBT) if the percentage of relative humidity is above 70% or below 20%.* •

Step 2 Measure the outdoor DBT at the outdoor unit. This is the temperature of the air entering the condenser coil.

Step 3 Measure the suction pressure at the service valve on the outdoor condensing unit and convert to a temperature using a temperature/pressure chart.

Step 4 Measure the suction-line inlet temperature on the suction line near the service valve on the outdoor condensing unit.

Step 5 Calculate the amount of superheat by subtracting the temperature obtained in step 3 from the temperature obtained in step 4.

Step 6 Find the intersection where the outdoor temperature and the indoor temperature meet and read the degrees of superheat.

The superheat that is being calculated here is *not* the evaporator superheat. Since the suction-line temperature reading is being taken from a point near the compressor, the sensible heat being picked up in the suction line is included in the calculation. This superheat calculation is often referred to as the system superheat. If the superheat of the system is more than 5°F above what the charging chart reads, add refrigerant into the low side of the operating system until the superheat is within 5°F of the chart. If the superheat of the system is more than 5°F below what the chart reads, recover some refrigerant from the system until the superheat is within 5°F of the chart.

NOTE: *Always let the system run for at least 15 min after adding refrigerant or recovering refrigerant before recalculating system superheat. To avoid fractionation in the charging cylinder when using a refrigerant blend, liquid refrigerant has to be taken from the charging cylinder and then throttled into the low side of the system as a vapor while the system is running. For a review of refrigerant blends with temperature glide and fractionation potential, refer to Unit 9, "Refrigeration and Oil Chemistry and Management—Recovery, Recycling, Reclaiming, and Retrofitting."* •

Example: R-22 capillary tube or fixed orifice system

Indoor DBT = 80°F
Outdoor DBT = 90°F
Suction pressure at service valve of condensing unit = 60 psig or 34°F using a P/T chart
Suction-line temperature at service valve = 54°F
System superheat is (54°F − 34°F) = 20°F

The intersection of the 90°F outdoor temperature with the 80°F indoor temperature indicates that there should be about 17°F of superheat. The system has 20°F of superheat, which is within 5°F of the superheat chart, so the system is properly charged. The technician would not have to add any refrigerant to the system.

The theory behind these charging curves is simple. Let us take the charging curve in **Figure 10.17** as an example. As you move to the right on the bottom axis, the outdoor temperature rises. Notice that for a constant indoor DBT or WBT (lines that slant downward from left to right), the operating system superheat decreases as the outdoor ambient temperature increases. The reason for this is that there is now more head pressure on the high side of the system pushing the subcooled liquid out of the condenser's bottom through the liquid line and the capillary tube metering device. This will force more refrigerant into the evaporator, causing the superheat to drop. This explains why some air-conditioning systems incorporating capillary tubes or fixed orifice metering devices will flood or slug liquid at the compressor on a day with hot outdoor ambient temperatures when the devices are overcharged. The superheat curve will prevent this from occurring if it is followed properly.

Referring to **Figure 10.17**, if we assume a constant indoor DBT of 75°F across the evaporator coil and increase the outdoor DBT from 70°F to 105°F, we can see that the system superheat will fall from 23°F to 0°F. The hotter outdoor ambient temperature results in higher head pressures, pushing more liquid through the capillary tube and into the evaporator. So it is normal for the system to run with 23°F of compressor superheat when the outdoor ambient temperature is 70°F. Do not add any refrigerant to this system, because if the outdoor ambient temperature climbed to 95°F later in the day, the system's compressor would slug liquid or flood.

NOTE: *Regarding the manufacturer's charging curve in* ***Figure 10.17****, if the relative humidity is above 70% or below 20%, use WBTs instead of DBTs across the evaporator coil to compensate for the varying latent (moisture) loads.* •

If the outdoor temperature stays constant and the indoor DBT or indoor WBT increases, the operating superheat will increase. This loading of the indoor coil with either sensible or latent heat or both will cause a more rapid vaporization of refrigerant in the evaporator. This will cause high compressor superheat, a normal occurrence. Many technicians will incorrectly decide to add refrigerant in this case, unintentionally overcharging the system. It is completely normal for a capillary or fixed orifice metering device system to run with high superheat during periods of high evaporator load.

Again referring to **Figure 10.17**, if the outdoor ambient temperature stays constant at 95°F, and the indoor DBT rises from 75°F to 95°F, the operating superheat will rise from 6°F to 33°F. At a 95°F indoor air DBT and a 95°F outdoor air DBT, the superheat should normally be 33°F, according to the chart. These conditions make for an inefficient system and an inactive evaporator, which is the greatest disadvantage of a fixed orifice metering device. However, this is the only way to prevent slugging and flooding of refrigerant with the varying indoor and outdoor loads that air-conditioning systems often experience.

Although charging capillary tube, fixed orifice, and piston (short-tube) systems in air-conditioning applications follow much the same underlying idea, it is strongly recommended that the technician consult with the equipment manufacturer to determine the exact methods for charging, following their charging curves and tables. Some manufacturers use different curves and tables for different models of their equipment. Some manufacturers have eliminated the need for a WBT because of custom-made charging curves that represent their laboratory tests on the equipment.

Figure 10.18 A charging chart from the side of a 5-ton, unitary air-conditioning system. Courtesy Ferris State University. Photo by John Tomczyk

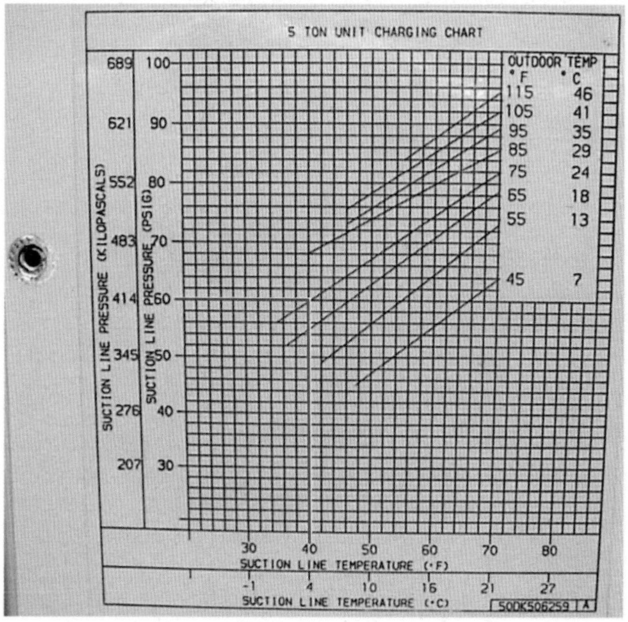

Figure 10.18 is a charging chart for a 5-ton, unitary air-conditioning system using a capillary tube as a metering device. This charging chart was affixed to the side of the unit by the manufacturer to help the service technician service and charge the unit. There are only three variables on the chart: the outdoor temperature, suction-line temperature, and suction pressure. The service technician must first establish what the outdoor ambient temperature is and then read the suction-pressure gauge at the condensing unit's service valve. With these two conditions known and the charging chart handy, the technician can determine what the appropriate suction-line temperature entering the condensing unit at the service valve should be. If the suction-line temperature is high, add some refrigerant; if the suction-line temperature is low, recover some refrigerant.

Example: R-22 system

Outdoor temperature = 75°F
Suction pressure = 60 psig
Suction-line temperature at the condensing unit's service valve should be 40°F (see **Figure 10.18**).

If the suction-line temperature measured at the service valve is 60°F, the service technician would have to add refrigerant. If the temperature measured is 32°F, the service technician would have to recover some refrigerant from the unit. Always give the unit an ample amount of time (15 to 20 min) to reach equilibrium when adding or removing refrigerant and before using the charging chart again. It is much easier to add refrigerant to reach the correct charge than to remove refrigerant to reach the correct charge level.

Although the goal of charging charts, curves, and tables is the same, their appearance may differ. Some manufacturers provide two charts that, when used together, allow the technician to properly charge the system. The charts in **Figure 10.19** and **Figure 10.20** are for an R-410A system. What follows is a set of sample guidelines only; be sure to check with the manufacturer of the equipment for specific charging procedures to make certain you get it right the first time!

1. Start up the system and allow it to operate for at least 10 min.
2. Obtain the suction pressure at the inlet of the condensing unit.
3. Obtain the suction-line temperature at the condensing unit using a high-quality thermistor-type thermometer.
4. Measure the outside air temperature using a high-quality thermometer.
5. Measure the wet-bulb temperature of the air entering the evaporator coil with a sling psychrometer or similar instrument.
6. Using the table in **Figure 10.19**, find the intersection point between the outdoor air temperature and the wet-bulb temperature of the air entering the evaporator coil.

Figure 10.19 This table provides superheat information based on outdoor air temperature and the wet-bulb temperature of the air entering the evaporator.

OUTDOOR TEMP	EVAPORATOR ENTERING AIR (WET BULB °F)								
°F	60	62	64	66	68	70	72	74	76
75	12	15	18	21	24	28	31	34	37
80	8	12	15	18	21	25	28	31	35
85	--	8	11	15	19	22	26	30	33
90	--	5	9	13	16	20	24	27	31
95	--	--	6	10	14	18	22	25	29
100	--	--	--	8	12	15	20	23	27
105	--	--	--	5	9	13	17	22	26

Figure 10.20 This table determines the desired suction-line temperature based on superheat and suction pressure.

	SUCTION PRESSURE AT LOW SIDE SERVICE PORT (psig)								
SUPERHEAT TEMP °F	107.8	112.2	116.8	121.2	126.0	130.8	138.8	140.8	145.8
6	41	43	45	47	49	51	53	55	57
8	43	45	47	49	51	53	55	57	59
10	45	47	49	51	53	55	57	59	61
12	47	49	51	53	55	57	59	61	63
14	49	51	53	55	57	59	61	63	65
16	51	53	55	57	59	61	63	65	67
18	53	55	57	59	61	63	65	67	69
20	55	57	59	61	63	65	67	69	71
22	57	59	61	63	65	67	69	71	73
24	59	61	63	65	67	69	71	73	75
26	61	63	65	67	69	71	73	75	77
28	63	65	67	69	71	73	75	77	79
30	65	67	69	71	73	75	77	79	81
32	67	69	71	73	75	77	79	81	83
34	69	71	73	75	77	79	81	83	85

7. Record the superheat value from the chart at the intersection point.
8. Using the table in **Figure 10.20**, locate the superheat value recorded in step 7 in the left-hand column of the table.
9. Locate the column that corresponds to the system's suction pressure found in step 2.
10. Identify the required suction-line temperature by identifying the intersection point of the row and column located in steps 8 and 9.
11. Compare this value to the suction-line temperature obtained in step 3.
12. If the actual suction-line temperature is higher than the suction-line temperature in the chart, add refrigerant to the system until the charted temperature is reached.
13. If the actual suction-line temperature is lower than the suction-line temperature in the chart, recover refrigerant from the system until the charted temperature is reached.

Example: R-410A system

Outdoor air temperature: 90°F
Evaporator entering air: 70°FWB
Suction pressure at suction service port: 126 psig
Suction-line temperature at suction service port: 75°F

Using the table in **Figure 10.19**, locate the superheat value in the body of the table by using the 90°F outside temperature line and the 70°FWB column. The obtained value is 20°F. Using the table in **Figure 10.20**, locate the required suction-line temperature by using the 20°F superheat row and the 126 psig column. The required suction-line temperature is 63°F. Because the actual temperature reading is 75°F, refrigerant must be added to the system.

10.7 SUBCOOLING CHARGING METHOD FOR TXV SYSTEMS

The previous charging examples were for fixed restrictor-type metering devices such as capillary tubes, fixed orifices, or piston (short-tube) devices. The main advantage of fixed orifice metering devices is their inexpensive cost and simplicity. However, some air-conditioning units, especially high-efficiency units, come with thermostatic expansion valves (TXVs) as their metering devices. When systems have a TXV, the device will maintain a constant amount of evaporator superheat and keep the evaporator active under both high- and low-heat loadings. TXV systems also maintain a constant superheat in the evaporator under changing outdoor ambient temperatures. For a review of TXVs, refer to Unit 24, "Expansion Devices."

The use of TXVs as metering devices in air-conditioning applications requires a refrigerant charging method known as the "subcooling method." This method involves measuring the amount of subcooling exiting the outdoor condensing unit, which will include condenser subcooling and a small part of the liquid line that the outdoor condensing unit employs. The subcooling method for charging a cooling system that employs a TXV follows these steps:

Step 1 Operate the system for at least 15 min before taking any pressure or temperature measurements. If the outdoor temperature is below 65°F, make sure the condensing temperature is at least 105°F. This would be a minimum pressure of 340 psig for R-410A and 210 psig for R-22. This can be done by covering up part of the outdoor condenser's coil and restricting some of its airflow.

Step 2 Measure the high-side (condensing) pressure at the outdoor condensing unit's service valve using a pressure gauge. Convert the high-side pressure to a saturation temperature using a temperature/pressure chart.

Step 3 Measure the liquid-line temperature at the outdoor condensing unit's service valve where the high-side pressure reading was taken.

Step 4 Refer to the unit's data plate or technical manual to find the required subcooling value for the unit. In the absence of any data plate or manual, use a subcooling value of approximately 15°F.

Step 5 Subtract the subcooling value (read from the data plate or technical manual) from the saturation temperature (corresponding to the high-side pressure in step 2) to determine the desired liquid-line temperature at the outdoor condensing unit.

Step 6 If the measured liquid-line temperature is higher than the value in step 5, add refrigerant to the unit. If the liquid-line temperature is lower than the value in step 5, recover refrigerant from the unit. (Only a 3°F to 4°F variance from the temperature desired is necessary when using this method.) Adding or removing refrigerant from the unit should be done in small increments, allowing at least 15 min before the next reading of pressures and temperatures. This allows the system to reach a new equilibrium.

Refer to **Figure 10.21**, which shows a high-efficiency, split air-conditioning system incorporating R-410A as the refrigerant. This system has a TXV as the metering device. The technician lets the air-conditioning system run for 15 min and takes a pressure reading (390 psig) at the liquid service valve on the outdoor condensing unit. By using a temperature/pressure chart for R-410A, the service technician converts this pressure to a saturation temperature (115°F) and then measures the liquid-line temperature (110°F) at the same place the pressure reading was taken at the service valve of the outdoor condensing unit. Referring to the data plate specifications on the unit, the technician finds out that this unit should have 15°F of subcooling when charged properly. The technician then

Figure 10.21 The subcooling charging method for a high-efficiency, split R-410A air-conditioning system with a TXV metering device.

DATA PLATE

SPECIFICATIONS
R-410A HIGH EFFICIENCY FLA: 10 AMPS LRA: 55 AMPS VOLTS 208/230 SINGLE PHASE SUBCOOLING 15°F

R-410A TEMPERATURE/PRESSURE CHART

°F	PRESSURE (psig)
95	295 psig
100	317 psig
105	340 psig
110	365 psig
115°	390 psig
120	417 psig
125	446 psig
130	475 psig
135	506 psig
140	539 psig

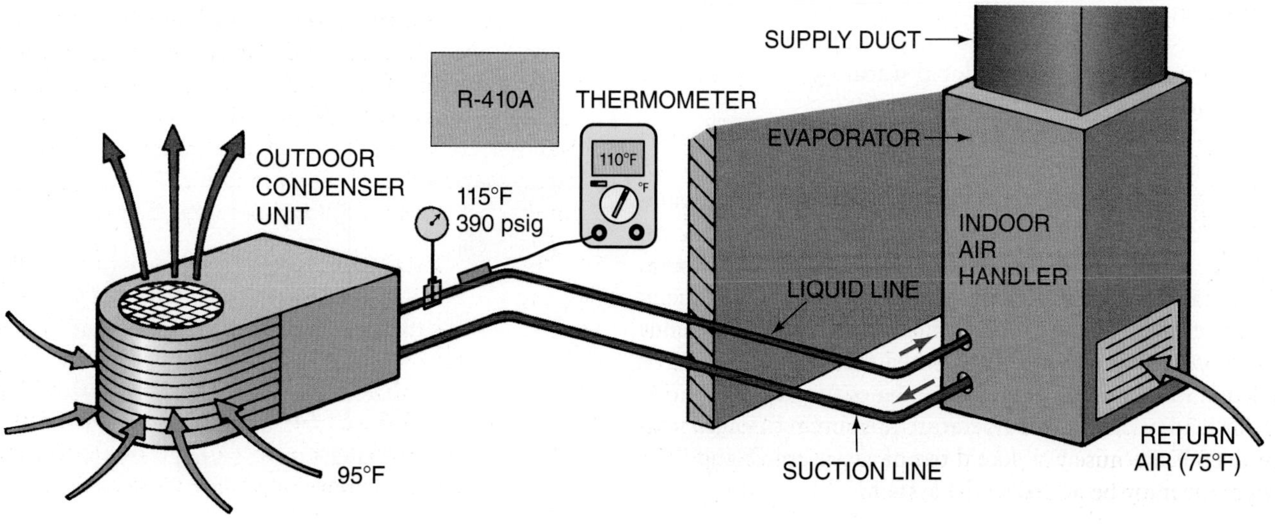

subtracts this 15°F value from the saturation temperature of 115°F to get the desired liquid-line temperature (100°F) at the outdoor condensing unit. See the following equation:

Saturation Temperature	−	Data Plate Subcooling Value	=	Desired Liquid-Line Temperature
115°F	−	15°F	=	100°F

The technician then compares the measured liquid-line temperature of 110°F with the desired liquid-line temperature of 100°F. Since the measured temperature is higher than the desired temperature, the technician adds refrigerant to the unit while it is running. Because R-410A is a near-azeotropic refrigerant blend, refrigerant must leave the charging cylinder as a liquid to prevent fractionation. It must then be throttled (vaporized) before it reaches the compressor. The technician must wait at least 15 min for this added refrigerant charge to stabilize in the system. The pressure- and temperature-measuring process must then be repeated until the measured temperature and desired temperature at the liquid line of the outdoor condensing unit are within 3°F to 4°F of one another.

10.8 CHARGING NEAR-AZEOTROPIC (ZEOTROPIC) REFRIGERANT BLENDS

Refrigerant blends and their characteristics were covered in Unit 9, "Refrigerant and Oil Chemistry and Management—Recovery, Recycling, Reclaiming, and Retrofitting." A more in-depth look at refrigerant blends and their characteristics is needed when charging them into a refrigeration or air-conditioning system.

As mentioned in Unit 9, refrigerant blends are nothing more than two or more refrigerants blended to create another refrigerant. This new refrigerant will have properties completely different from those of the original refrigerants. Two azeotropic blends that were used for many years in the HVACR industry are CFC-502 and CFC-500, which consist of azeotropic mixtures of HCFC-22/CFC-115 and CFC-12/HFC-152a, respectively. However, environmental concerns such as ozone depletion and global warming caused both of these blends to be phased out in 1996.

Azeotropic blends are mixtures of two or more liquids that when mixed together behave like the commonly known refrigerants R-12, R-22, and R-134a. Each system pressure will have only one boiling point and/or one condensing point. For a given evaporator or condensing saturation pressure, there is only one corresponding temperature. However, many refrigerant blends in use today, called *near-azeotropic* blends or *zeotropes,* act very differently when they evaporate and/or condense. Each sample of liquid combines two or three molecules instead of one. The temperature/pressure relationships of these blends are not at all like those of azeotropic blends and other refrigerants, such as CFC-12, HCFC-22, or HFC-134a.

Zeotropic blends have a temperature glide when they boil and condense. Temperature glide occurs when the refrigerants that make up a blend evaporate and condense at different temperatures at a given pressure—actually, the liquid and vapor temperatures for each refrigerant will be different for one given pressure. **Figure 10.22(A)** shows a zeotropic blend (R-401A) as it is phase changing in a condenser tube. Notice that for a given pressure of 85.3 psig, the liquid and vapor temperatures are different. An examination of **Figure 10.22(B)** clearly shows that the liquid phase and vapor phase of this blend have two distinct temperatures for one given pressure. In fact, at a condensing pressure of 85.3 psig, the liquid temperature (L) is 67.79°F and the vapor temperature (V) is 75.08°F. Its condensing temperature glide range is then from 75.08°F to 67.79°F. Its condensing temperature glide would be 7.29°F (75.08°F − 67.79°F). This means that the condensing process in the condenser is taking place through a range of temperatures. The condensing temperature is an average of the condensing temperature glide range, or (75.08°F + 67.79°F) ÷ 2 = 71.43°F. Temperature glide happens only when near-azeotropic (zeotropic) blends change phase.

One temperature no longer corresponds to one pressure as with refrigerants such as R-12, R-22, or R-134a.

Figure 10.22 (A) and (B) The temperature/pressure relationship for R-401A, a near-azeotropic refrigerant blend. R-401A is made up of R-22, R-152a, and R-124 at percentages of 53%, 13%, and 34%, respectively.

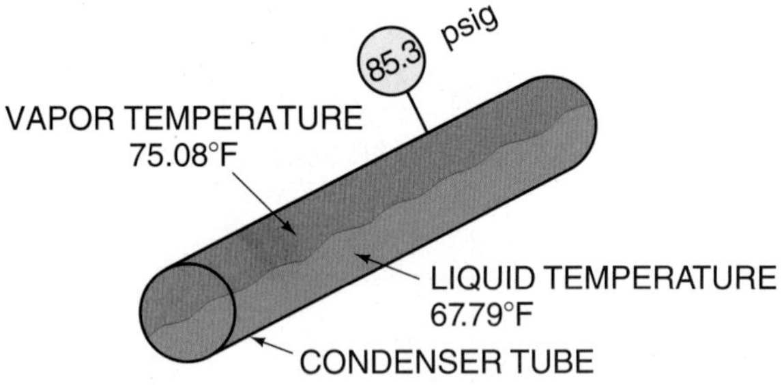

(A)

R-401A

	°F	PRESSURE PER SQUARE INCH (psig)
V	75.08	85.3
L	67.79	85.3
V	76.30	87.3
L	69.04	87.3
V	77.51	89.3
L	70.27	89.3
V	78.69	91.3
L	71.48	91.3
V	79.86	93.3
L	72.67	93.3
V	81.02	95.3
L	73.85	95.3

V = VAPOR
L = LIQUID

(B)

Temperature glides may range from 0.2°F to 14°F in refrigeration and air-conditioning applications, depending on the specific blend. Usually, the higher-temperature applications like air-conditioning will have a greater temperature glide. Most blends have low temperature glides with little effect on the refrigeration system. However, because of temperature glide, calculating superheat and subcooling is done differently. (These methods will be covered in later units.)

Examples of some near-azeotropic (zeotropic) refrigerant blends are R-401A, R-401B, R-402A, R-402B, R-404A, R-407C, and R-410A. Even though R-410A is listed as a near-azeotropic refrigerant blend, it has a very low temperature glide of 0.3°F, which can be ignored in air-conditioning applications. Refrigerants like R-12, R-22, R-134a, or azeotropic refrigerant blends like R-500 and R-502, however, boil and condense at a constant temperature for each pressure. Notice in **Figure 10.23**, R-134a has both a liquid and vapor temperature of 79.0°F at 85.3 psig as it changes phase in the condenser tube. This is because R-134a has no temperature glide.

New temperature/pressure charts make it easier for the service technician to check a system for the right amount of refrigerant charge when dealing with near-azeotropic refrigerant blends that have a temperature glide. For example, the pressure/temperature chart in **Figure 10.24** instructs the service technician who is checking for a superheat value on the

Figure 10.23 (A) and (B) The temperature/pressure relationship of R-134a.

R-134a

	°F	PRESSURE PER SQUARE INCH (psig)
V	79.0	85.3
L	79.0	85.3
V	80.5	87.3
L	80.5	87.3
V	81.7	89.3
L	81.7	89.3
V	82.6	91.3
L	82.6	91.3
V	83.8	93.3
L	83.8	93.3
V	85.0	95.3
L	85.0	95.3

V = VAPOR
L = LIQUID

(A)

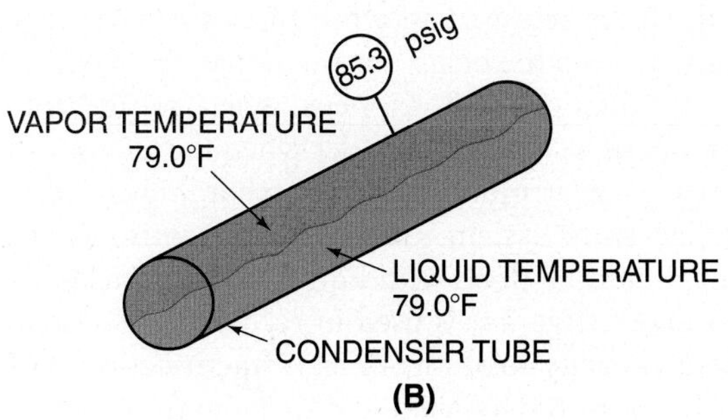

(B)

Figure 10.24 A pressure/temperature chart for refrigerant blends that have temperature glides. Courtesy Sporlan Valve Company

PRESSURE-TEMPERATURE CHART

Temperature, °F — REFRIGERANT – (Sporlan Code)

PSIG	Pink: MP39 or 401A (X) Bubble Point	Pink: MP39 or 401A (X) Dew Point	Sand: HP80 or 402A (L) Bubble Point	Sand: HP80 or 402A (L) Dew Point	Orange: HP62 or 404A (S) Bubble Point	Orange: HP62 or 404A (S) Dew Point	Green: KLEA 60 or 407A Bubble Point	Green: KLEA 60 or 407A Dew Point	Lt. Brown: 9000 or KLEA 66 407C (N) Bubble Point	Lt. Brown: 9000 or KLEA 66 407C (N) Dew Point	Brown: FX-56 or 409A Bubble Point	Brown: FX-56 or 409A Dew Point
5*	-23		-59		-57		-45		-40		-22	
4*	-22		-58		-56		-43		-39		-20	
3*	-20		-56		-54		-42		-37		-19	
2*	-19		-55		-53		-41		-36		-17	
1*	-17		-54		-52		-39		-35		-16	
0	-16		-53		-51		-38		-34		-15	
1	-13		-50		-48		-36		-31		-12	
2	-11		-48		-46		-33		-29		-9	
3	-9		-45		-43		-31		-27		-7	
4	-6		-43		-41		-29		-24		-5	
5	-4		-41		-39		-27		-22		-2	
6	-2		-39		-37		-25		-20		0	
7	0		-37		-35		-23		-18		2	
8	2		-36		-33		-21		-17		4	
9	4		-34		-32		-20		-15		6	
10	6		-32		-30		-18		-13		8	
11	8		-30		-28		-16		-12		9	
12	9		-29		-27		-15		-10		11	
13	11		-27		-25		-13		-8		13	
14	13		-26		-23		-12		-7		14	
15	14		-24		-22		-10		-5		16	
16	16		-23		-20		-9		-4		17	
17	17		-21		-19		-8		-3		19	
18	19		-20		-18		-6		-1		20	
19	20		-19		-16		-5		0		22	
20	21		-17		-15		-4		1		23	
21	23		-16		-14		-2		3		25	
22	24		-15		-12		-1		4		26	
23	25		-14		-11		0		5		27	
24	27		-12		-10		1		6		29	
25	28		-11		-9		2		8		30	
26	29		-10		-8		4		9		31	
27	30		-9		-6		5		10		32	
28	32		-8		-5		6		11		34	
29	33		-7		-4		7		12		35	
30	34		-6		-3		8		13		36	
31	35		-5		-2		9		14		37	
32	36		-4		-1		10		15		38	
33	37		-2		0		11		16		39	
34	38		-1		1		12		17		40	
35	39		0		2		13		18		41	
36	40	30	0		3		14		19		43	
37	42	31	1		4		15		20		44	
38	43	32	2		5		16		21		45	30
39	44	33	3		6		17		22		46	31
40	45	34	4		7		18		23		47	32
42	46	36	6		8		19		25		48	34
44	48	38	8		10		21		26		50	36
46	50	40	10		12		23		28			38
48		42	11		14		24		30			39
50		44	13		16		26		31			41
52		45	14		17		28		33			43
54		47	16		19		29		34			45
56		49	18		20		31		36			46
58		50	19		22		32		37			48
60		52	20		23		33		39			50
62		53	22		25		35		40			51
64		55	23		26		36		42	30		53
66		56	25		27		38		43	32		54
68		58	26		29		39		44	33		56
70		59	27		30	29	40	30	46	34		57
72		61	29		32	31	41	31	47	36		58
74		62	30		33	32	43	32	48	37		60
76		64	31		34	33	44	34	49	38		61
78		65	32	30	35	34	45	35		39		63
80		66	34	31	37	36	46	36		41		64
85		69	37	34	40	39	49	39		44		67
90		73	40	37	42	42		42		46		70
95		76	42	40	45	44		45		49		73
100		78	45	43	48	47		47		52		76
105		81	48	45		50		50		54		79
110		84	50	48		52		53		57		82
115		87		50		55		55		59		84
120		89		53		57		57		62		87
125		92		55		59		60		64		89
130		94		57		62		62		66		92
135		96		60		64		64		69		94
140		99		62		66		66		71		96
145		101		64		68		68		73		99
150		103		66		70		70		75		101
155		105		68		72		72		77		103
160		108		70		74		74		79		105
165		110		72		76		76		81		107
170		112		74		78		78		82		109
175		114		75		80		80		84		111
180		116		77		82		81		86		113
185		117		79		83		83		88		115
190		119		81		85		85		90		117
195		121		82		87		87		91		119
200		123		84		88		88		93		121
205		125		86		90		90		95		123
210		127		87		92		91		96		124
220		130		91		95		94		99		128
230		133		94		98		97		102		131
240		136		97		101		100		105		134
250		140		99		104		103		108		137
260		143		102		107		106		111		141
275		147		106		111		110		115		145
290		151		110		115		114		119		149
305		155		114		118		117		123		153
320		159		118		122		121		126		157
335		163		121		126		124		130		161
350		167		125		129		128		133		165
365		170		128		132		131		137		169

*Inches mercury below one atmosphere

system to use dew point values. Dew point is the temperature at which saturated vapor first starts to condense. For a subcooling value, the chart instructs the technician to use the bubble point values only. Bubble point is the temperature at which the saturated liquid starts to boil off its first bubble of vapor.

Example

A service technician takes the following reading on a low-temperature refrigeration system incorporating R-404A as the refrigerant.

Head pressure = 250 psig
Suction pressure = 33 psig
Condenser outlet temperature = 90°F
Evaporator outlet temperature = 10°F

The technician wants to determine the condenser subcooling and the evaporator superheat. According to the pressure/temperature chart in **Figure 10.24**, the corresponding condensing temperature for a 250-psig head pressure is 104°F. Notice that the chart will let the technician use only the bubble point temperature for the pressure/temperature relationship. This way there is no confusing a liquid or vapor temperature from the chart when subcooling or superheat readings are taken. The condenser subcooling amount would then be:

Condensing Temperature	−	Condenser Outlet Temperature	=	Condenser Subcooling
104°F	−	90°F	=	14°F

To find the evaporator superheat, the technician would again refer to the chart in **Figure 10.24** and find that the corresponding evaporating temperature for a 33-psig suction pressure is 0°F. Notice again that the chart will let the technician use only the dew point temperature for the pressure/temperature relationship. Again, this ensures that there is no way of confusing a liquid with a vapor temperature from the chart when subcooling or superheat readings are taken. The evaporator superheat amount would then be:

Evaporator Outlet Temperature	−	Evaporating Temperature	=	Evaporator Superheat
10°F	−	0°F	=	10°F

The pressure/temperature chart in **Figure 10.25** is less confusing than the chart in **Figure 10.22(B)**, which lists both a liquid temperature and a vapor temperature for the same pressure.

Another property of zeotropic refrigerant blends is fractionation. As mentioned, because zeotropic blends are made up of different refrigerants and mix together as several molecules, they act differently from refrigerants like R-12, R-22, R-134a, and azeotropic blends when changing phase in the evaporator and condenser. Fractionation occurs when some of the blend condenses or evaporates before the rest of the blend does. Phase changes actually will happen at different rates

Figure 10.25 A cylinder of a near-azeotropic refrigerant blend showing a liquid dip tube extending to the bottom of the cylinder while the cylinder is in the upright position. This allows the charging of liquid only while the cylinder is in the upright position. Courtesy Worthington Cylinders

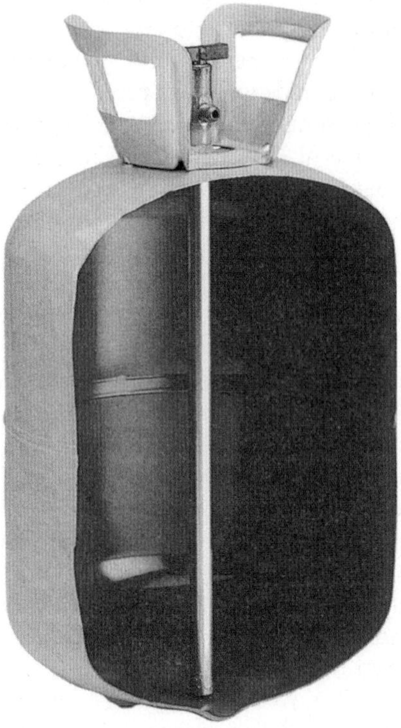

within the same blend. The many different refrigerants in the blend all have vapor pressures that are slightly different, which causes them to condense and evaporate at different rates.

If a zeotropic blend leaks (vaporizes), the refrigerant in the blend with the lowest boiling point (highest vapor pressure) will leak faster than the other refrigerants in the blend. The refrigerant in the blend with the next lowest boiling point (next higher vapor pressure) will also leak out, but at a slower rate than the refrigerant with the lowest boiling point. Many blends are made up of three refrigerants (ternary blends), resulting in three different leakage rates all at once. This is a problem because once the leak is found and repaired, the refrigerant blend that is left in the system after the leak will not have the same percentages of refrigerants that it had when initially charged. Fractionation happens only when the leak occurs as a vapor leak; it will not occur when the system is leaking pure liquid. This is because vapor pressures are involved, and liquid and vapor need to coexist for fractionation to occur.

Fractionation also happens when refrigerant vapor is taken out of a charging cylinder when the technician is charging with a zeotropic blend. The charging or recharging of a refrigeration system using a near-azeotropic or zeotropic blend should be done with liquid refrigerant. As long as virgin liquid refrigerant is used in recharging a system that has leaked, capacity losses from leak fractionation will be small and harmless. Remove liquid only from the charging cylinder

to ensure that the proper blend of percentages or composition enters the refrigeration system. If vapor is removed from the charging cylinder, fractionation will occur. However, once the right amount of liquid is removed from the charging cylinder to another holding cylinder, the refrigerant can be charged as vapor as long as all of the refrigerant vapor is put in the system. This will ensure that the proper percentages of the blend enter the system.

Some refrigerant cylinders containing zeotropic refrigerant blends have dip tubes that run to the bottom of the cylinder to allow liquid to be removed from the cylinder when upright, **Figure 10.25**. This should ensure that liquid will always be removed from the cylinder when a system is charged. The cylinder has to be tipped upside down to remove vapor. Of course, the technician cannot look inside the refrigerant tank to determine whether or not there is a dip tube in the cylinder. Refer to the side of the refrigerant tank for information regarding the correct positions of the tank for liquid or vapor removal. Arrows on the tank will indicate how the tank must be positioned for use.

When adding liquid refrigerant to the suction line of a system that is running, the liquid, once out of the charging cylinder, has to be restricted and vaporized into the system to avoid any damage to the compressor. Some sort of a restricting valve must be used to make sure that the liquid refrigerant is vaporized before it enters the compressor. **Figure 10.9** and **Figure 10.10** show examples. Keep an ammeter on the compressor's power lead and monitor the running current of the compressor when charging this way, **Figure 10.26**. These

Figure 10.26 Taking a current reading at the compressor. Photo by Eugene Silberstein

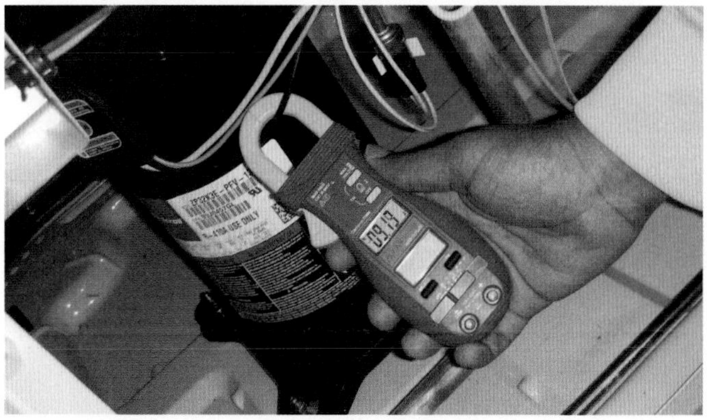

precautions avoid overloading the compressor and creating nuisance overload trips caused from refrigerant vapors that are too dense when they enter the compressor. If the current increases too much over the nameplate rating current, discontinue charging until the current stabilizes at a lower reading. The gauge manifold could also be used to restrict the liquid further, which would introduce the refrigerant into the system more slowly. This would result in less-dense vapors entering the compressor, which would create a lower current draw. The compressor's three-way suction service valve can also be cracked just off its back seat, which will create a throttling effect and help vaporize any liquid refrigerant trying to enter the compressor.

REVIEW QUESTIONS

1. When adding refrigerant to a newly evacuated system, which of the following is the best method?
   A. Add liquid refrigerant to the system through the suction-line service valve
   B. Add liquid refrigerant to the system through the liquid-line service valve
   C. Add vapor refrigerant to the system through the suction-line service valve
   D. Add vapor refrigerant to the system through the liquid-line service valve
2. When charging a system with vapor refrigerant, what can be done to increase the pressure in the refrigerant tank?
   A. Turn the tank upside down
   B. Heat the tank with an oxyacetylene torch
   C. Place the tank in a bath of warm water
   D. Open both the high-side and low-side valves on the gauge manifold
3. As a system is charged with vapor refrigerant, the refrigerant tank will likely
   A. experience a pressure drop and a temperature rise.
   B. Experience a pressure rise and a temperature rise.
   C. experience a pressure drop and a temperature drop.
   D. experience a pressure rise and a temperature drop.
4. A system's total refrigerant charge can be found on
   A. the nameplate of a package-type air-conditioning system.
   B. the nameplate of the outdoor unit of a split-type air-conditioning system.
   C. the nameplate of the indoor unit of a split-type air-conditioning system.
   D. the nameplate of the outdoor unit of a split-type heat pump system.

5. Automatic charging capability, auto-stop features, alarms, and wireless remotes are features that are found on some
   A. digital gauge manifolds.
   B. refrigerant charging scales.
   C. analog gauge manifolds.
   D. analog charging scales.
6. Before a full graduated charging cylinder is used to charge a system, it should be
   A. evacuated to at least 500 microns.
   B. turned upside down for liquid charging.
   C. calibrated based on the vacuum level.
   D. calibrated based on the cylinder pressure.
7. Which of the following terms would be an example of a refrigerant with no temperature glide?
   A. A single-compound refrigerant
   B. A binary refrigerant blend
   C. A ternary refrigerant blend
   D. A zeotrope
8. The term "blend" refers to a refrigerant that
   A. has a boiling temperature below 0°F at 0 psig.
   B. has a boiling temperature above 0°F at 0 psig.
   C. is made up of two or more different refrigerants.
   D. has a single saturation temperature at a given pressure.
9. If the components of a binary refrigerant blend have saturation temperatures of 35°F and 42°F at a pressure of 120 psig, what is the approximate temperature glide associated with this refrigerant?
   A. 7°F
   B. 35°F
   C. 42°F
   D. 77°F
10. When a portion of a refrigerant blend evaporates or condenses before the rest of the same blend, what is occurring?
    A. Fractionation
    B. Sublimation
    C. Pressurization
    D. Condensation
11. Which of the following statements is true regarding azeotropes and zeotropes?
    A. They both have no temperature glide.
    B. They are both refrigerant blends.
    C. Neither one experiences fractionation.
    D. They can both be charged as a vapor.
12. When refrigerant leaves a storage tank as a liquid and enters the system as a vapor, what process is taking place between the system and the gauge manifold?
    A. Fractioning
    B. Restricting
    C. Gliding
    D. Condensing
13. When charging an air-conditioning system that is operating with a thermostatic expansion valve, which of the following is used as the primary guideline to ensure that the system is properly charged?
    A. Evaporator superheat
    B. System superheat
    C. Condenser subcooling
    D. Compressor amperage
14. A service technician is charging an R-410A air-conditioning system that has a TXV metering device. The high-side saturation pressure is 365 psig and the unit's technical manual specifies 10°F of liquid subcooling. What would the desired liquid-line temperature be at the outlet of the condensing unit?
    A. 10°F
    B. 100°F
    C. 110°F
    D. 120°F
15. A technician is working on an R-404A refrigeration system and determines that the head pressure is 205 psig and the condenser outlet temperature is 80°F. The technician correctly concludes that the amount of condenser subcooling is
    A. 10°F.
    B. 20°F.
    C. 80°F.
    D. 90°F.
16. A technician is working on an R-404A refrigeration system and determines that the suction pressure is 30 psig and the evaporator outlet temperature is 10°F. The technician correctly concludes that the amount of evaporator superheat is
    A. −3°F
    B. 3°F
    C. 10°F
    D. 13°F

UNIT 11

CALIBRATING INSTRUMENTS

OBJECTIVES

After studying this unit, you should be able to

- list five types of test instrumentation used by technicians in the HVACR industry.
- explain the importance of calibrating test instrumentation.
- calibrate a thermometer at the low- and high-temperature ranges.
- check an ohmmeter for accuracy.
- describe the comparison test for an ammeter and a voltmeter.
- describe procedures for checking pressure instruments above and below atmospheric pressure.
- explain how a reference gas can be used to ensure proper operation of an electronic leak detector.
- check flue-gas analysis instruments.

11.1 CALIBRATION

The service technician cannot always see or hear what is occurring within a machine or piece of equipment. Instruments such as multimeters, pressure gauges, and thermometers help the technician discover what the internal conditions are. Therefore, these instruments must be reliable. Although they should be calibrated by the manufacturer, test instruments do not always remain in calibration. They should be checked before use and on a periodic basis, as continued use, exposure to moisture, vibration, and other factors can affect the accuracy of the instruments. For example, instruments may be transported in a truck over rough roads to the job site and stay in the truck through extremes of hot and cold weather; the instrument compartment may sweat (due to condensation of humidity from the air). All of these things cause stress to the instruments. Technicians are very dependent on tools and equipment and should always take proper care of them. There is no substitute for good common sense, careful use, and proper storage of tools and equipment. **⚠ SAFETY PRECAUTION:** *Some instruments are used to check for voltage to protect from electrical shock. They must function correctly for safety's sake.•*

Some instruments can be readily calibrated; some must be returned to the manufacturer for calibration; and some cannot be calibrated at all. **Calibration** means changing the instrument's output or reading to correspond to a standard or correct reading. For example, if a speedometer shows 55 mph for an automobile actually traveling at 60 mph, the speedometer is out of calibration. If the speedometer's settings, circuitry, or sensors can be adjusted so that it records the correct speed, it can be calibrated. Some instruments designed for field use will stay calibrated longer than others. Typically, electronic instruments with digital readout features stay in calibration longer than analog (needle) types and are more appropriate for field use.

This unit describes the instruments most commonly used for HVACR system troubleshooting. These instruments measure temperature, pressure, voltage, amperage, and resistance; check for refrigerant leaks; and make flue-gas analyses. To check and calibrate instruments, you must have reference points to which the instrument's readings are compared. We recommend whenever buying an instrument that you check some readings against known values. If the instrument is not within the manufacturer's standards, it should be returned. (Save the box the instrument came in as well as the directions and warranty to save time and energy.)

11.2 TEMPERATURE-MEASURING INSTRUMENTS

Temperature-measuring instruments measure the temperature of vapors, liquids, and solids. Commonly, the temperature of air, water, flue gases, and refrigerant in copper lines are measured. Regardless of the medium to be measured, the methods for checking instrument accuracy are similar.

Refrigeration technicians should have thermometers that are accurate over a wide range to measure the temperature of refrigerant lines as well as the temperatures being maintained inside the coolers. Higher temperatures are encountered when measuring ambient temperatures as well as refrigerant line temperatures on the high-pressure side of the system. Heating and air-conditioning technicians must typically measure vapor temperatures from 0°F to 150°F and liquid temperatures as high as 220°F for normal service, which can require a wide range of instruments. For temperatures above 250°F—for example, flue-gas analysis in gas- and oil-burning equipment—special thermometers are used. The stack thermometer, **Figure 11.1**, for example, is included as part of a flue-gas analysis kit.

Figure 11.1 The thermometer is included in the flue-gas analysis kit. Note the high temperature range of the thermometer. Photo by Eugene Silberstein

A popular instrument that is widely used in the HVACR industry to obtain accurate temperature readings is the electronic thermocouple thermometer, **Figure 11.2**. Electronic thermocouple thermometers are easy to use, economical, and extremely accurate. Digital, pocket thermometers, **Figure 11.3**, are very convenient to carry with you and can be used effectively for taking quick field readings.

Figure 11.2 A digital-type electronic thermometer. Photo by Eugene Silberstein

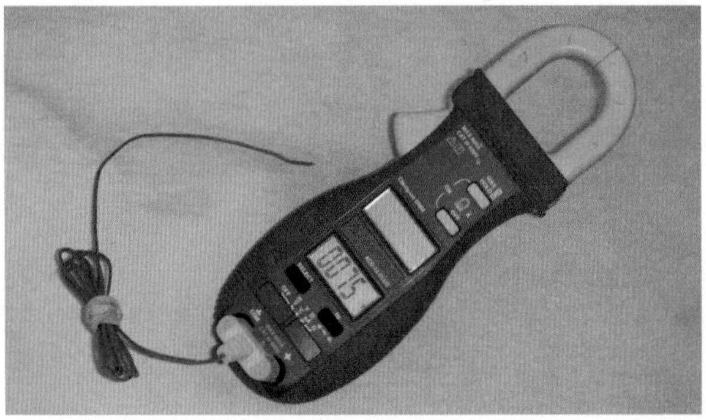

Figure 11.3 A pocket digital stem thermometer. Photo by Eugene Silberstein

Figure 11.4 Three reference points that a service technician may use.

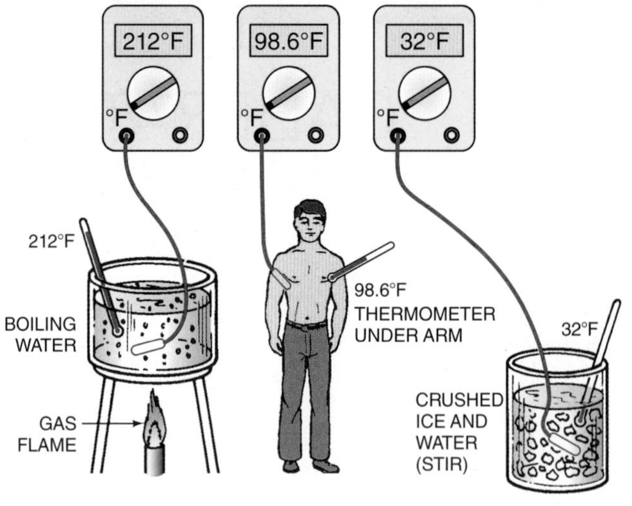

Checking the accuracy of thermometers is a straightforward and direct process, as known reference temperatures are readily available. Three reference points for checking temperature-measuring instruments are 32°F (ice and water), 98.6°F (body temperature), and 212°F (boiling point of water at standard atmospheric conditions), **Figure 11.4**. The reference points used should be close to the temperature range in which the technician is working. When using any of these as a reference for checking the accuracy of a temperature-measuring device, remember that a thermometer indicates the temperature of the sensing element. The sensing element must, therefore, make good contact with the object or fluid that is being checked. Many inexperienced technicians merely set a thermometer lead on a copper line and read the temperature, but the thermometer's temperature-sensing element has more contact with the surrounding air than with the copper line, **Figure 11.5**. For an accurate reading, the thermometer must be in contact with the medium being measured long enough for the sensor to become the same temperature as the

Figure 11.5 Simply holding the temperature-sensing element to a pipe will not yield accurate temperature measurements. Photo by Eugene Silberstein

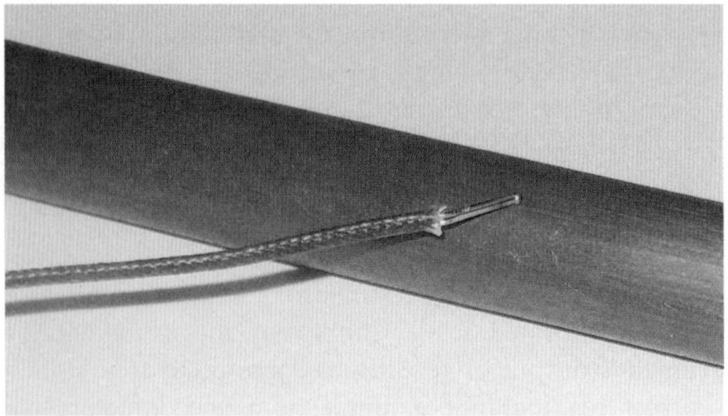

Figure 11.6 Wrapping the temperature-sensing element will insulate it from the surrounding air, giving a more accurate temperature reading. Photo by Eugene Silberstein

medium. When reading the temperature of a refrigerant line, it is also beneficial to wrap the sensor and the line with foam insulation tape. This will help ensure that the probe is sensing the temperature of the line and not the surrounding air, **Figure 11.6.** When measuring the temperature of a liquid, it is desirable to completely submerge the probe in the liquid.

One method of checking temperature instruments is to submerge the sensing element into a known temperature condition (such as ice and water while the change of state is occurring) and allow the sensing element to reach the known temperature. The following is a method for checking an electronic thermometer with four plug-in leads that can be moved from socket to socket.

1. Fasten the four leads together as shown in **Figures 11.7** and **11.8.** Something solid can be attached to them so that they can be stirred in ice and water.
2. For a low-temperature check, crush about a quart of ice, preferably made from purified or distilled water. If purified or distilled water is not available, bottled water is a good alternative. If tap water must be used, make sure the water is as clean as possible, as any foreign substances in the water will affect its freezing point. The ice must be crushed very fine (wrap it in a towel and pound it with a hammer), or there may be warm spots in the mixture.
3. Pour enough water, pure if possible, over the ice to almost cover it. **Do not cover the ice completely with water** or it will float and may be warmer on the bottom of the mixture. The ice must reach the bottom of the vessel.
4. Stir the temperature leads in the mixture of ice and water, where the change of state is taking place, for at least 5 min. The leads must have enough time to reach the temperature of the mixture.
5. If the temperature readings of each lead vary, note which leads are in error and by how much. The leads should be numbered and the temperature differences marked on the instrument case or the leads themselves.
6. For a high-temperature check, put a pan of water on a stove-top heating unit and bring the water to a boil. Make sure the thermometers you are checking are capable of registering the temperature of boiling water. If they are, immerse the sensors in the boiling water. **NOTE:** *Do not let them touch the bottom of the pan or they may conduct heat directly from the pan bottom to the lead,* **Figure 11.9.•** Stir the thermometers for at least 5 min and check the readings. It is not critical that the thermometers be accurate to a perfect 212°F because at these temperature levels, a degree or two one way or the other does not make a big difference. If any lead reads more than 4°F from 212°F, it is defective. Remember that water boils at 212°F at sea level under standard conditions. Any altitude above sea level will make a slight difference. If the instrument is to be used more than 1000 ft above sea level, we highly recommend a laboratory glass thermometer as a standard instead of the boiling water temperature.

Figure 11.7 The four leads to the temperature tester are fastened at the ends so they can all be submerged in water at the same time.

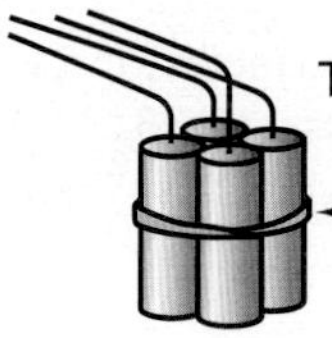

Figure 11.8 A pencil is attached to the group to give the leads some rigidity so that they can be stirred in the ice and water. Note that the ice must reach the bottom of the container.

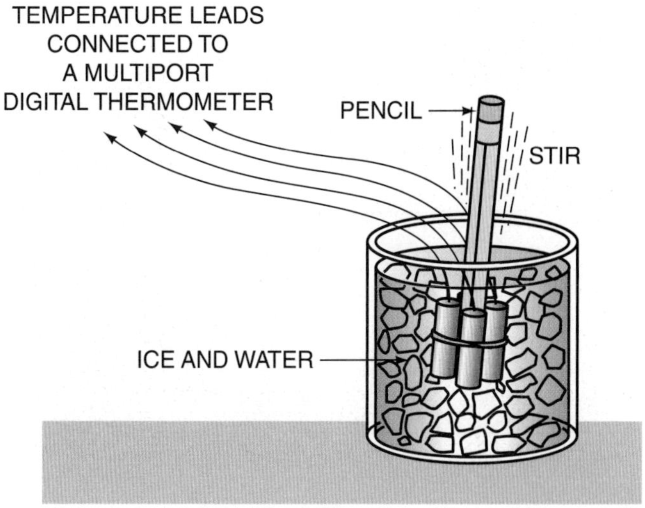

Accuracy is more important in the lower temperature ranges where small temperature differences can translate into larger percentage errors. A 1°F error does not sound like much until one lead is off +1°F and another is off −1°F and you try

Figure 11.9 A high-temperature test for the accuracy of four temperature leads.

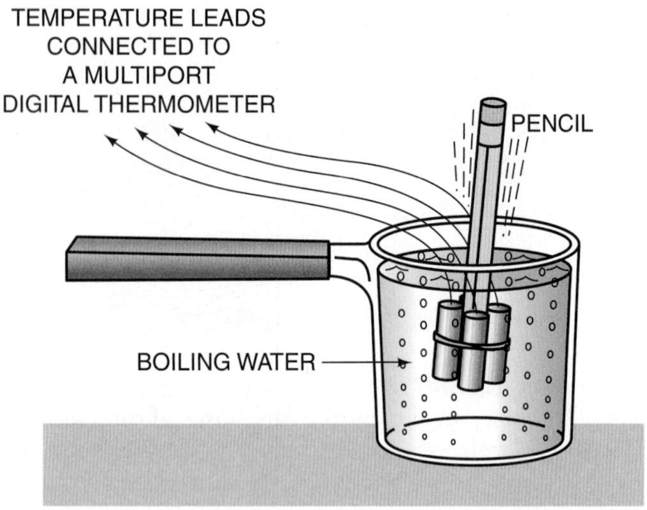

to measure accurately a temperature drop across a water heat exchanger that operates with a 10°F difference across it.

Consider the situation where a thermometer has two probes, A and B. Probe A is always off by −1°F and probe B is always off by +1°F. The thermometer is being used to measure the temperature difference across a heat exchanger and the actual temperatures of the lines are 45°F and 55°F, **Figure 11.10**. If probe A is attached to the 45°F line and probe B is attached to the 55°F line, the temperature registered by probe A will be 44°F, while the temperature registered by probe B will be 56°F. The calculated temperature difference across the heat exchanger will be 12°F, **Figure 11.11**. However, if probe A is attached to the 55°F line and probe B is attached to the 45°F line, the temperature registered by probe A will be 54°F, while the temperature registered by probe B will be 46°F. The calculated temperature difference across the heat exchanger will now be 8°F, **Figure 11.12**.

Figure 11.10 Two ends of a heat exchanger. One end is at a temperature of 45°F and the other end is at a temperature of 55°F. The temperature difference across the heat exchanger is 10°F.

Figure 11.11 Although the actual temperature difference across the heat exchanger is 10°F, these readings indicate a temperature difference of 12°F.

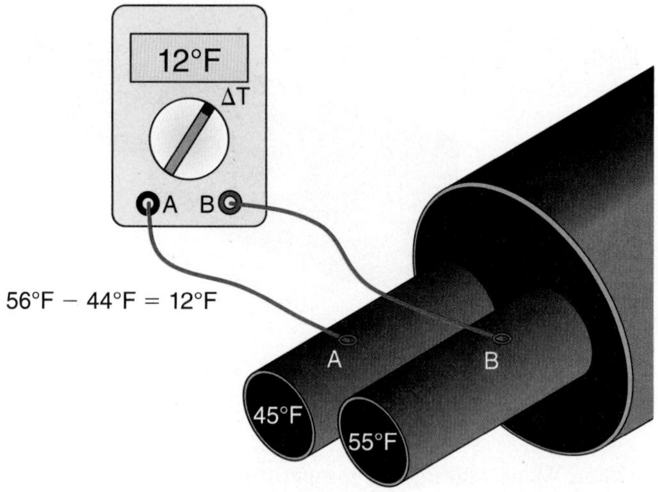

Figure 11.12 Although the actual temperature difference across the heat exchanger is 10°F, these readings indicate a temperature difference of 8°F.

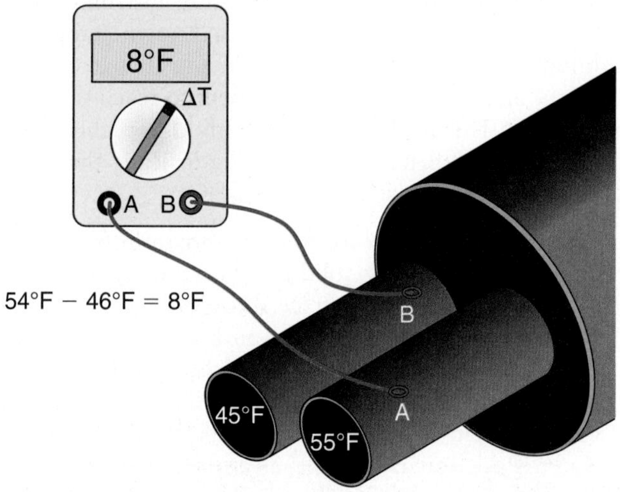

Glass thermometers often cannot be calibrated because the graduations are etched on the stem. However, graduations printed on the back of the instrument may be adjustable. A laboratory-grade glass thermometer is certified as to its accuracy and may be used as a standard for calibrating field instruments. It is a good investment for use in calibrating electronic thermometers, **Figure 11.13**. Many dial-type thermometers can be easily adjusted. These instruments may be tested for accuracy and calibrated, if possible, as we have described.

Figure 11.13 A laboratory-grade glass thermometer. Photo by Bill Johnson

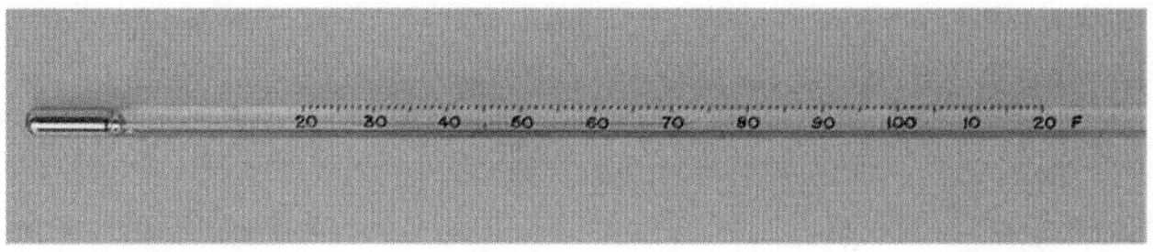

Figure 11.14 Using the human body as a standard.

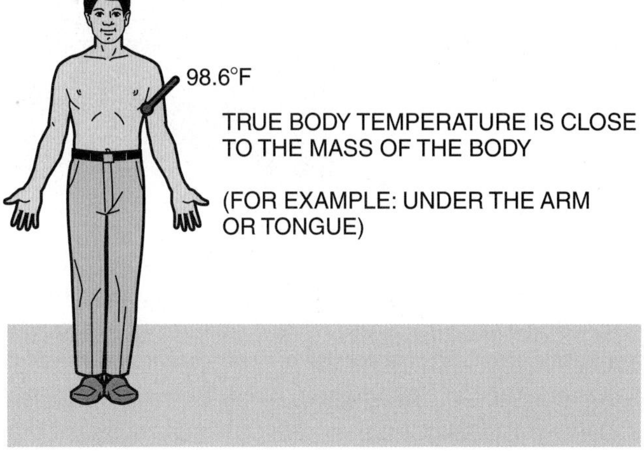

Body temperature may also be used as a standard when applicable. Remember that the outer extremities, such as the hands, are not at body temperature. The body is, on average, 98.6°F in the main blood flow, next to the trunk of the body, **Figure 11.14**.

11.3 PRESSURE TEST INSTRUMENTS

Pressure test instruments register pressures above and below atmospheric pressure. The gauge manifold and its construction were discussed in Unit 1, "Heat, Temperature, and Pressure." The technician must be able to rely on these gauges and have some reference points with which to check them periodically. This is particularly important when there is reason to doubt gauge accuracy. The gauge manifold is used frequently and is subject to considerable abuse, **Figure 11.15**.

The pressure/temperature relationship of gauge readings taken from a cylinder of refrigerant can be compared in the following manner. Open the gauge manifold to the atmosphere, and check both gauges to see that they read 0 psig. *It is impossible to determine a correct gauge reading if the gauge is not set at 0 at the start of the test.* Connect the gauge manifold to a cylinder of fresh new refrigerant that has been in a room at a fixed temperature for a long time. Purge the gauge manifold and hoses of air. Note that using an old refrigerant cylinder or a recovery tank may lead to errors due to refrigerant impurities or contaminants. If the cylinder pressure is not correct due to refrigerant contaminants, it *cannot* be used to check the gauges. The cylinder pressure is the standard and must be reliable. New virgin/disposable refrigerant cylinders have check valves that allow the refrigerant to leave the cylinder—but nothing to enter it. It is not likely that one of these cylinders will contain any contaminants.

Figure 11.15 Refrigeration gauge manifold. Photo by Eugene Silberstein

If the cylinder temperature is known, the refrigerant should have a certain pressure. For example, if the refrigerant is R-134a and the room is 75°F, the cylinder and refrigerant should be at 75°F if they have been left in the room for a long enough time. When the gauges are connected to the cylinder and purged of any air, the gauge reading should compare to 75°F and read 78.7 psig. (Refer to **Figure 3.13**, which shows the temperature/pressure chart for R-134a.) The cylinder can be connected to the center hose of the gauge manifold so that both gauges (the low and high sides) may be checked at the same time. The same test can be performed with R-22; the reading will be higher: at 75°F it should be 132 psig. Performing the test with both refrigerants checks the gauges at two different pressures.

Checking the low-side gauge in a vacuum is not as easy as checking the gauges above atmosphere because there is no readily available known vacuum. One method is to open the gauge to atmosphere and make sure that it reads 0 psig. Then connect the gauge to a two-stage vacuum pump and start the pump. When the pump has reached its lowest vacuum, the gauge should read close to 30 in. Hg (29.92 in. Hg vacuum), **Figure 11.16**. If the gauge is correct at atmospheric pressure and at the bottom end of the scale, you can assume that it is correct in the middle of the scale. If more accurate vacuum readings are needed for monitoring a system that runs in a vacuum, you should buy a larger, more accurate vacuum gauge, **Figure 11.17**.

A digital manometer and an electronic micron gauge may be checked together in the following manner. This field test is

Figure 11.16 When evacuated, the gauge should read 30 in. Hg vacuum. If this is the case, all points between 0 psig and 30 in. Hg should be correct.

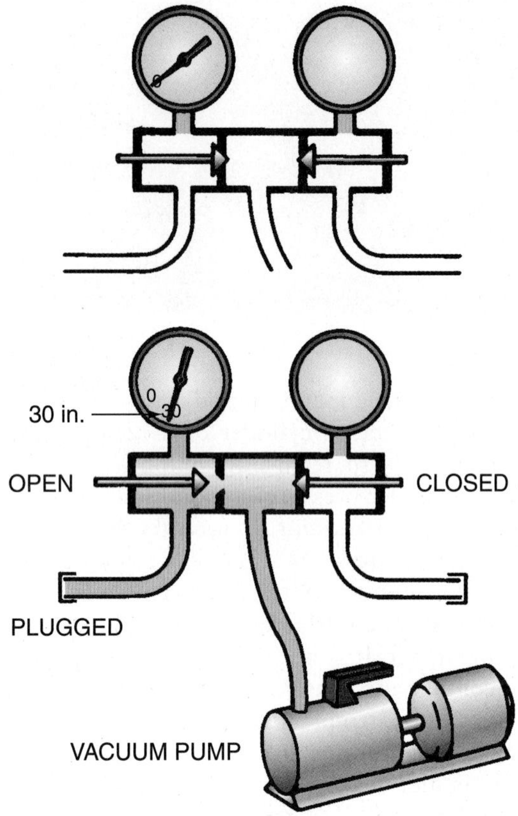

Figure 11.17 A large gauge is more accurate for monitoring systems that operate in a vacuum because the needle moves farther from 0 psig to 30 in. Hg.

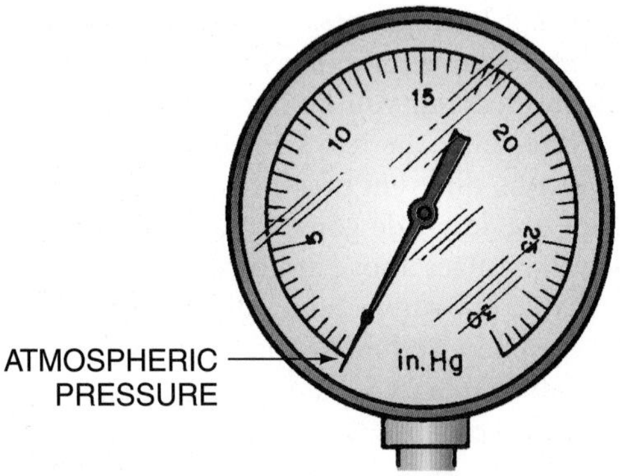

not 100% accurate, but it is sufficient to tell the technician whether the instruments are within a working tolerance or not.

1. Prepare a two-stage vacuum pump for the lowest vacuum that it will pull. Change the oil to improve the pumping capacity. Connect a gauge manifold to the vacuum pump with the digital manometer and the micron gauge as shown in **Figure 11.18**. The low-pressure gauge, the micron gauge, and the digital manometer may be compared at the same time. As the vacuum pump starts to operate, the readings on the micron gauge and the digital manometer can be compared to each other. The approximate relationship between microns and inches of water column, IWC, is as follows:

   1000 microns = 0.54 IWC
   2000 microns = 1.07 IWC
   3000 microns = 1.6 IWC
   4000 microns = 2.14 IWC
   5000 microns = 2.7 IWC

   If the readings on the two instruments agree with each other, this simply means that either both instruments are calibrated or both instruments are not.
2. When the vacuum pump has evacuated the manifold and gauges, observe the readings. The micron gauge should read between 0 and 1000 microns.

Figure 11.18 The setup for checking an electronic manometer and an electronic micron gauge.

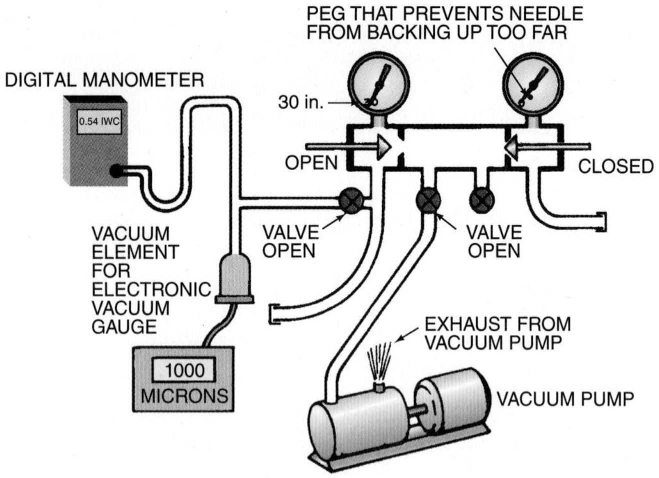

11.4 ELECTRICAL TEST INSTRUMENTS

Electrical test instruments are not as easy to calibrate; however, they may be checked for accuracy. The technician must know that the ohm scale, the volt scale, and the ammeter scale are correct. The milliamp scale on the meter is seldom used and must be checked by the manufacturer or compared with another meter.

There are many grades of electrical test instruments. When electrical testing procedures must be very accurate, it pays to buy a good-quality instrument. If exact accuracy is not required, a less expensive instrument may be satisfactory. Electrical test instruments should be checked periodically

Figure 11.19 Checking an ohmmeter at various ranges with resistances of known values.

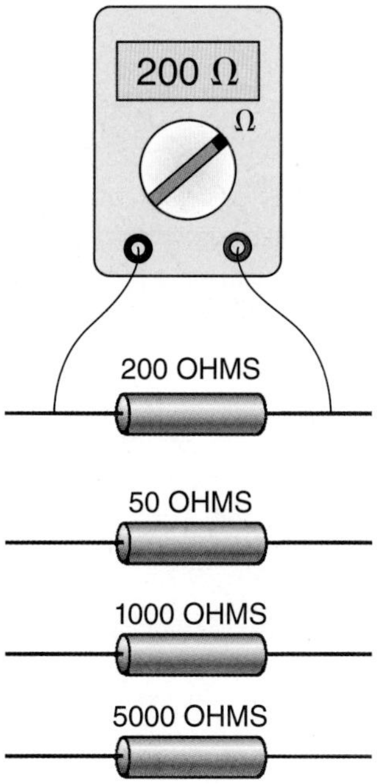

(at least once a year). One way of doing this is to compare the instrument reading against known values.

The ohmmeter feature of a digital multimeter (DMM) can be checked by using several high-quality resistors of a known resistance. The actual resistance of a particular resistor will be within a percentage range, either above or below the rating of the component. Less expensive resistors, like those with a silver band, have a 10% tolerance, which means that a 100-ohm resistor will have an actual resistance between 90 and 110 ohms. More expensive resistors can have tolerances as low as 0.1% of the resistor's rating. A tolerance of 0.1% means that a 100-ohm resistor will have an actual resistance between 99.9 and 100.1 ohms. A resistor with a 2% or 5% tolerance is usually acceptable for the needs of the HVACR technician. Different values of resistors will allow you to test the ohmmeter at different ranges, **Figure 11.19**. **NOTE:** *If the meter is out of calibration, check the batteries and change them if necessary. If the instrument will not read zero with fresh batteries, check the leads. The leads must be connected perfectly, particularly with digital meters. You may want to use the alligator clips on the end to ensure a perfect connection. Leads can become frayed internally. Good meter leads cost a little more, but they have many strands of very small wire for flexibility and will last longer and be more reliable. Be sure to start each test with a zero adjustment check of the ohmmeter.*•

The volt scale is not as easy to check as the ohm scale. A friend at the local power company or technical school may allow you to compare your meter to a high-quality bench meter. This is recommended at least once a year or whenever you suspect your meter readings are incorrect, **Figure 11.20**.

Figure 11.20 A DMM being compared with a quality bench meter in the voltage mode.

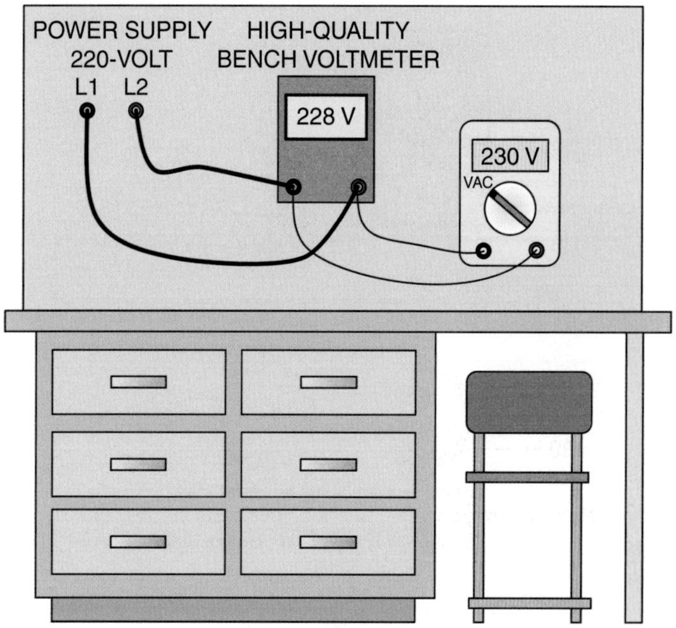

Figure 11.21 A clamp-on ammeter being compared with a quality bench ammeter.

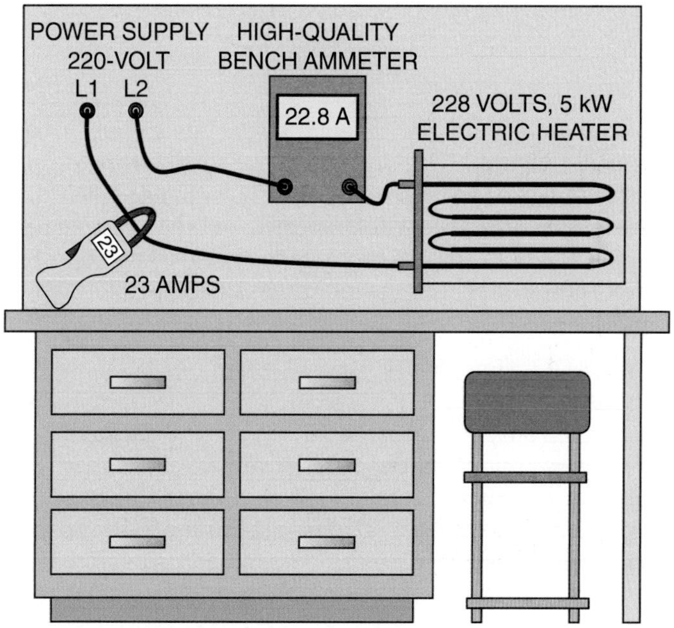

The clamp-on ammeter is used most frequently for amperage checks. This instrument clamps around one conductor in an electrical circuit. Like the voltmeter, it can be compared with a high-quality bench meter, **Figure 11.21**. Some amount of checking can be done by using Ohm's law and comparing the amperage reading with a heater circuit of known

Figure 11.22 Using an electric heater to check the calibration of an ammeter.

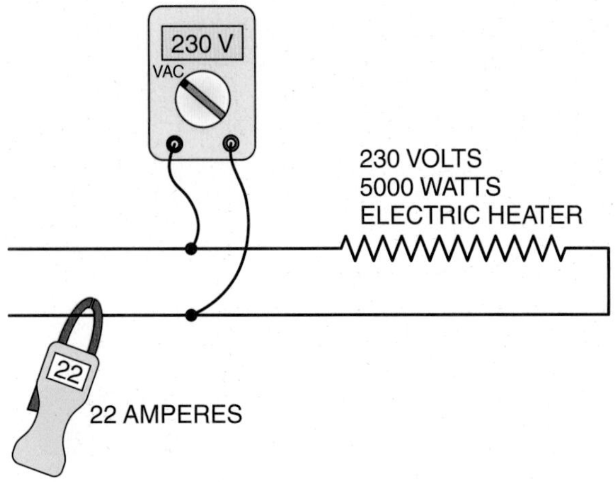

resistance. For example, Ohm's law states that current (I) is equal to voltage (E) divided by resistance (R or Ω for ohms):

$$I = \frac{E}{R}$$

If a heater has a resistance of 10 Ω and an applied line voltage of 228 V, the amperage (current) on this circuit should be

$$I = \frac{228\ \text{V}}{10\ \Omega} = 22.8\ \text{A}$$

Remember to read the voltage at the same time as the amperage, **Figure 11.22.** You will notice small errors because the resistance of the electric heaters will change when they get hot. The resistance will be greater, and the exact ampere reading will not compare precisely to the calculated one. But the purpose is to check the ammeter. An acceptable value is typically +/−10%.

NOTE: *It is important to note that Ohm's law, as written, will hold for DC-powered circuits that are used to operate resistive-type loads. When there is an AC power supply, a power factor must be used to correct the calculated values. For example, it is not uncommon for a 115-volt circuit to produce one-half of the power that was calculated. Refer to Unit 12, "Basic Electricity and Magnetism," for more on this topic.*•

11.5 ELECTRONIC REFRIGERANT LEAK DETECTION DEVICES

Electronic leak detectors can detect leak rates as low as 1/4 oz per year. The detector samples air; if the air contains refrigerant, the detector either sounds an alarm or lights the probe end. These devices are typically battery-powered. Some units may have a pump to pull the sample across the sensing element, and some have the sensing element located in the head of the probe. Some electronic leak detectors can be adjusted to compensate for background refrigerant. Some equipment rooms may have many small leaks, and a small amount of refrigerant may be in the air all the time. Unless the detector has a feature that accounts for this background refrigerant, it will constantly indicate a leak.

Figure 11.23 (A) This electronic leak detector is not sensing refrigerant escaping from the small pinhole in the tubing. This is because the refrigerant is spraying past the detector's sensor. (B) In this position, the sensor will detect the leaking refrigerant.

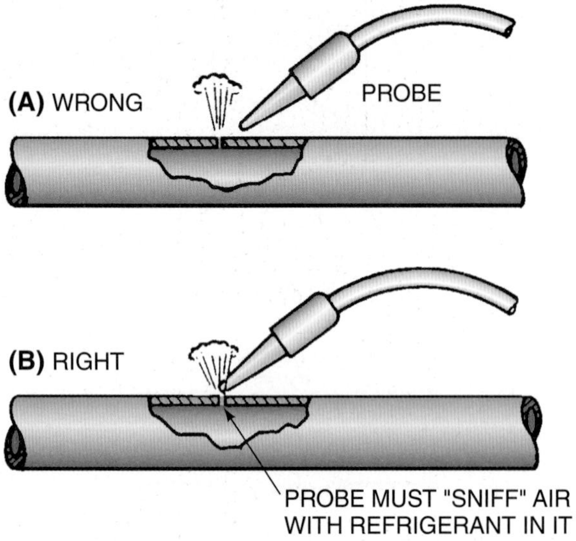

No matter what style of leak detector you use, you must be confident that the detector will actually detect a leak. Remember that the detector only indicates what it samples. If the detector is in the middle of a refrigerant cloud and the sensor senses air, it will not sound an alarm or light up. For example, the probe of a leak detector can pass by a pinhole leak in a pipe. The sensor is sensing air next to the leak, not the leak itself, **Figure 11.23.**

Some manufacturers furnish a reference leak canister, which is used to confirm the correct operation of the leak detector. **NOTE:** *Never spray pure refrigerant into the sensing element. Damage will occur.*• If you do not have a reference leak canister, a gauge line under pressure from a refrigerant cylinder can be loosened slightly, and the refrigerant can be fanned to the electronic leak detector sensing element. Doing it this way mixes air with the refrigerant to dilute it.

11.6 FLUE-GAS ANALYSIS INSTRUMENTS

Flue-gas analysis instruments analyze products of combustion from fossil-fuel–burning equipment, such as oil and gas furnaces and boilers, to determine the combustion efficiency of the appliance. These instruments are normally sold in kit

form with a carrying case, and can be of either the "electronic," **Figure 11.24**, or "wet kit" variety, **Figure 11.25**. **SAFETY PRECAUTION:** *The red fluid, potassium hydroxide, in the "wet kit" must not be allowed to contact any tools or other instruments. It is acidic and can cause severe burns! The chemicals are intended to stay within the container or the instrument to which they belong. A valve at the top of the instrument is a potential leak source and should be checked periodically. It is best to store and transport the kit in the upright position so that if the valve does develop a leak, the chemical will not run out of the instrument.* •

Flue-gas instruments are expensive, precision devices that deserve special care and attention. The draft gauge, on the left side of **Figure 11.25**, for example, is highly sensitive. The kit should only be hauled around in a truck when you intend to use it. The potassium hydroxide in the wet kit should be changed according to the manufacturer's guidelines, which is roughly after 200 tests. A calibration check of wet kits is not necessary because they are direct-reading instruments and cannot be calibrated. The only adjustment to be made is the zero adjustment on the sliding scale, **Figure 11.26**, which is adjusted at the beginning of the flue-gas test. The thermometer in the kit, **Figure 11.27**, is used to measure very high temperatures, up to 1000°F. There is no easily available reference point for it except the boiling point of water, 212°F, which may be used as the reference even though it is near the bottom of the scale, **Figure 11.28**. Electronic combustion analyzers are easier to use than wet kits, but should be calibrated annually. In order to have the instrument calibrated, it needs to be sent to the manufacturer. Turnaround time is usually 2 or 3 weeks, so it is a good idea to have backup instruments at the ready.

Figure 11.24 An electronic combustion efficiency analyzer. Courtesy Oilheat Associates, Inc. Photo by John Levey

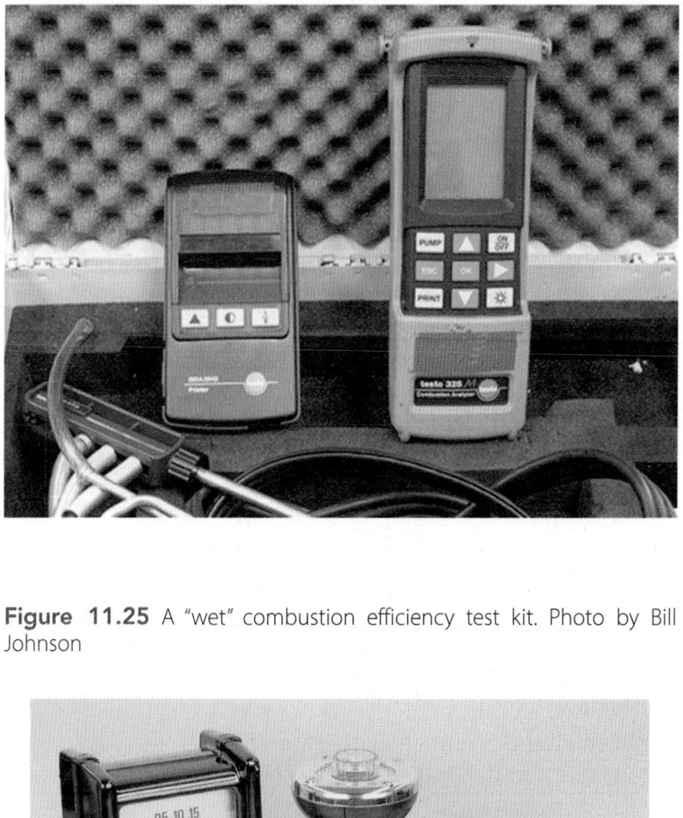

Figure 11.25 A "wet" combustion efficiency test kit. Photo by Bill Johnson

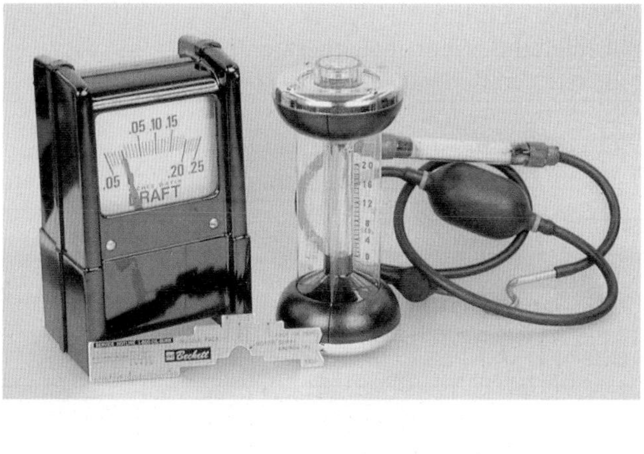

Figure 11.26 The zero adjustment on the sliding scale of the flue-gas analysis kit. Photo by Bill Johnson

Figure 11.27 High-temperature thermometer. Photo by Bill Johnson

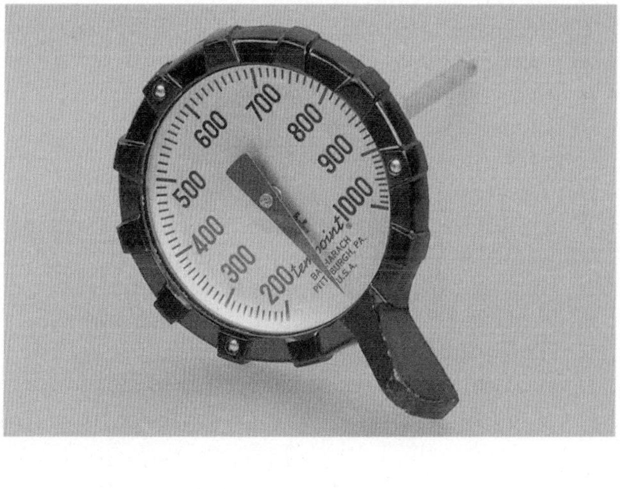

Figure 11.28 A flue-gas thermometer being checked in boiling water.

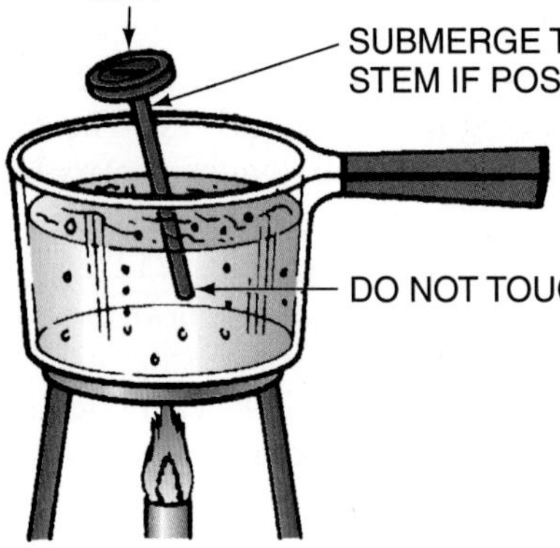

11.7 GENERAL MAINTENANCE

Any instrument with a digital readout will have batteries, and these must be maintained. *Buy the best batteries available; inexpensive batteries may cause problems. A good battery will not leak acid on the instrument components if it is left unattended and goes dead.*

The instruments you use extend your senses; the instruments, therefore, must be maintained so that they can be believed. Airplane pilots sometimes have a malady called vertigo. They become dizzy and lose their orientation and relationship with the horizon. Suppose the pilot were in a storm and being tossed around, even upside down at times. The pilot can be upside down and have a sensation that the plane is right side up and climbing. The plane may be diving toward the earth while the pilot thinks it is climbing. Instruments must be believable, and the technician must have reference points to maintain faith in the instruments.

REVIEW QUESTIONS

1. The term that is used to describe the changing of an instrument's output to provide correct and accurate readings is
   A. aligning.
   B. calibrating.
   C. fluctuating.
   D. replacing.
2. All of the following are good reference points for checking the accuracy of a thermometer except
   A. boiling water.
   B. body temperature.
   C. room temperature water.
   D. ice-water.
3. All of the following instruments can be calibrated except a(n)
   A. laboratory-grade glass thermometer.
   B. electronic combustion analyzer.
   C. analog pressure gauge.
   D. dial-type stem thermometer.
4. Thermometers are intended to be used in conjunction with all of the following except
   A. solids.
   B. liquids.
   C. gases.
   D. pressures.
5. Which of the following statement is true regarding laboratory-grade glass thermometers?
   A. They are easily calibrated by the end user.
   B. They can be used to calibrate field instruments.
   C. Their accuracy cannot be certified.
   D. They cannot be used to calibrate field instruments.
6. Thermometer accuracy is more important in which temperature range?
   A. Lower temperature ranges
   B. Medium temperature ranges
   C. High temperature ranges
   D. Very high temperature ranges
7. To test temperature probes at a low temperature reference point,
   A. crushed ice should be used.
   B. a crushed ice and water mixture should be used.
   C. a crushed ice, salt, and water mixture should be used.
   D. a crushed ice, sugar, and water mixture should be used.
8. Pressure-testing instrumentation, in general, can measure pressures
   A. both above and below atmospheric pressure.
   B. only below atmospheric pressure.
   C. only above atmospheric pressure.
   D. neither above nor below atmospheric pressure.
9. A tank of virgin refrigerant at a known temperature can be used to determine the accuracy of a
   A. digital thermocouple thermometer.
   B. refrigeration gauge.
   C. refrigerant leak detector.
   D. combustion analyzer.
10. When exposed to atmospheric pressure, a refrigeration gauge should read
    A. 14.7 psig.
    B. 0 psia.
    C. 0 psig.
    D. −14.7 psia.

11. When using a resistor with a +/−5% tolerance, a 200 Ω resistor will have an acceptable range of
    A. 195 Ω to 205 Ω.
    B. 190 Ω to 210 Ω.
    C. 180 Ω to 220 Ω.
    D. 175 Ω to 225 Ω.
12. The accuracy of the voltmeter function of a DMM should be tested by
    A. checking the voltage at a 120-volt receptacle.
    B. comparing the meter's display to that of another field instrument.
    C. comparing the meter's display to that of a high-quality bench meter.
    D. measuring the voltage on the line side of a power distribution panel.
13. To measure amperage, a clamp-on ammeter is
    A. connected in series with the circuit components.
    B. clamped around both power lines in a circuit.
    C. connected in parallel with the circuit load.
    D. clamped around one conductor in a circuit.
14. High-quality refrigerant leak detectors can detect leaks as small as
    A. 4 ounces per year.
    B. 2 ounces per year.
    C. 1 ounce per year.
    D. 0.25 ounces per year.
15. It is recommended that digital combustion analyzers be factory calibrated
    A. once a year.
    B. before every use.
    C. once a month.
    D. once every 6 months.

SECTION 3

AUTOMATIC CONTROLS

UNITS

UNIT 12

BASIC ELECTRICITY AND MAGNETISM

OBJECTIVES

After studying this unit, you should be able to

- describe the structure of an atom.
- explain the law of charges.
- explain the difference between conductors and insulators.
- describe how magnetism is used to produce electricity.
- state the differences between alternating current and direct current.
- list three common units of measurement for electricity.
- describe the components of a basic electric circuit.
- explain the differences between series and parallel circuits.
- state three forms of Ohm's law.
- state the formula for determining electrical power.
- describe a solenoid.
- explain inductance.
- describe the construction of a transformer and how voltage is induced in a secondary circuit.
- describe how a capacitor works.
- describe a sine wave.
- state the reasons for using proper wire sizes.
- describe the physical characteristics and the function of the diode and the transistor.
- describe the three types of opposition to current flow.
- describe procedures for making electrical measurements.
- describe fuses, circuit breakers, and ground fault circuit interrupters as they apply to circuit protection.

SAFETY CHECKLIST

- ✔ Before using any piece of electrical test equipment, become completely familiar with the instrument's features and carefully read the manufacturer's literature.
- ✔ Avoid becoming part of an electric circuit.
- ✔ Use non-conducting ladders, such as fiberglass or wood.
- ✔ Wear rubber-soled work boots to insulate yourself from wet floors and help prevent serious injury from electric shock.
- ✔ When servicing electric circuits, deenergize the power supply whenever possible.
- ✔ Be sure to follow lockout/tagout procedures to ensure your safety on the job.
- ✔ Be aware that some pieces of HVACR equipment have more than one power source. Be sure they are all deenergized before working on the equipment.
- ✔ Make certain that your electrical test instruments are in good working order.
- ✔ Make certain all system capacitors are properly discharged before working on equipment.
- ✔ Be sure to inspect the cords on all power tools. Do not use power tools that have damaged cords.
- ✔ Never cut the ground prong off a power cord under any circumstances.
- ✔ Never jump out safety devices.
- ✔ Make certain that the conductors used in a circuit are sized properly and protected by fuses and circuit breakers that are rated for the conductor size.
- ✔ Make certain that ground fault circuit interrupters (GFCIs) are used in wet, or potentially wet, locations.

12.1 ATOMIC STRUCTURE

To understand the theory of how electric current flows, you must understand something about matter and what it is made of. Matter is made up of atoms, which are the smallest amounts of a substance that can exist. The three main components of atoms are protons, neutrons, and electrons. Protons and neutrons are located at the center (or nucleus) of the atom. Protons have a positive charge. Neutrons have no charge and have little or no effect as far as electrical characteristics are concerned. Electrons have a negative charge and orbit around the nucleus, in a manner very similar to the way the planets orbit around the sun. Electrons move around the nucleus in different shells, which are at different distances from the nucleus of the atom. Although electrons in the same orbit are the same distance from the nucleus, they do not follow the same orbital paths, **Figure 12.1**. The number of electrons in an atom is equal to the number of protons in the atom.

Figure 12.1 The orbital paths of electrons.

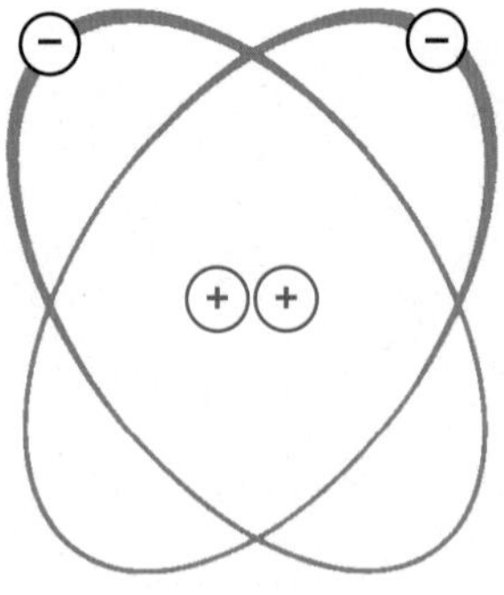

Figure 12.2 A hydrogen atom with one electron and one proton.

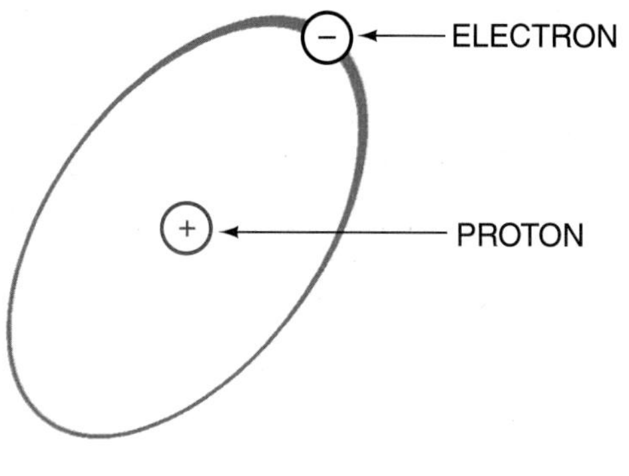

The hydrogen atom is simple to illustrate because it has only one proton and one electron, **Figure 12.2.** Not all atoms are as simple as the hydrogen atom, as they contain many electrons, protons, and neutrons. Copper, for example, which is commonly used to construct electric circuits, has 29 protons and 29 electrons, **Figure 12.3.** Some of the electron orbits in a copper atom are farther away from the nucleus than others, and each orbit has the ability to hold a different number of electrons. The formula that is used to determine how many electrons can occupy each shell is:

$$\text{Number of Electrons} = 2s^2$$

where "s" is the shell number. So, in the case of copper, the first shell can hold 2 electrons (2×1^2), the second shell can hold 8 electrons (2×2^2), and the third shell can hold 18 electrons (2×3^2). The fourth shell can hold up to 32 electrons, 2×4^2, but there is only 1 electron left over, so the fourth shell is occupied by the one remaining electron.

Figure 12.3 A copper atom with 29 protons and 29 electrons.

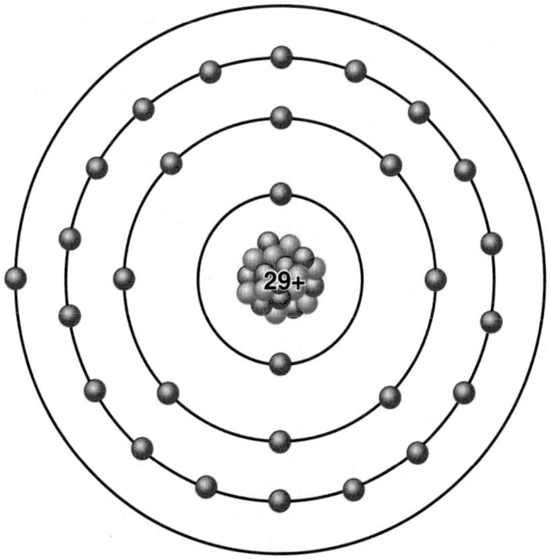

It is this single electron in the outermost orbit or shell that makes copper a good conductor of electricity. We will now see why this is so.

The goal of an atom is to become as stable as possible. Being stable refers to having either a completely full or a completely empty outermost shell. Since the copper atom has only one electron in its outer shell, the atom can become stable if it is able to give up this outermost, or *valence*, electron. The term valence electrons refers to the electrons in the outermost, or valence, shell. For this reason, the copper atom, and other metallic elements and substances as well, are eager to give up their "loose" electrons. Since electric current is described as the flow of electrons, metallic substances facilitate current flow by not holding onto their valence electrons tightly and by allowing them to move freely from atom to atom.

12.2 THE LAW OF CHARGES

When sufficient energy or force is applied to an atom, the outer electrons become free and can move from atom to atom. When electrons leave an atom, the atom will contain more protons than electrons. As a result, since protons are positively charged, the atom will have a net positive charge, **Figure 12.4(A).** The atom the electron joins will contain more electrons than protons, so it will have a net negative charge, **Figure 12.4(B).**

The "law of charges" states that like charges repel each other and unlike or opposite charges attract each other. So, two protons (or electrons) will repel each other, while a proton and an electron will be attracted to each other. An electron in an atom with a surplus of electrons (negative charge) will be attracted to an atom with a shortage of electrons (positive charge). An electron entering an orbit with a surplus of electrons will tend to repel an electron already there and cause it to become a free electron.

Figure 12.4 (A) This atom has two protons and one electron. It has a shortage of electrons and a positive charge. (B) This atom has two protons and three electrons. It has an excess of electrons and a negative charge.

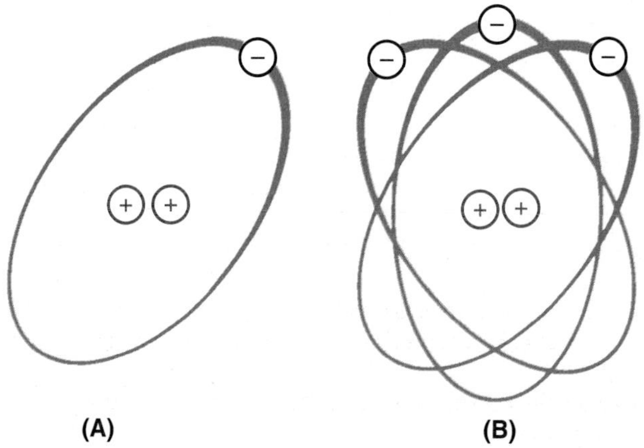

12.3 CONDUCTORS

Good conductors of electricity are those with few electrons in the outer orbit, or valence shell. Copper, silver, and gold are good conductors, and each has one electron in the outer orbit. Other metallic substances, such as mercury and aluminum, are also good conductors, even though they have two and three electrons, respectively, in their outermost electron shells. Information regarding the amount of current that can safely flow through a conductor is provided by the American Wire Gauge, AWG. Thicker wires or conductors can carry more current safely than thinner wires or conductors. Good conductors of electricity are typically also good conductors of heat.

12.4 INSULATORS

Insulators, unlike conductors, do not allow the free flow of electrons, and are often described as the opposite of conductors. Insulators are substances that are made up of atoms with several electrons in their outermost shell, making them poor conductors. The valence electrons are difficult to free, making it very difficult for the electrons to move from one atom to another. Glass, air, rubber, and plastic are examples of good insulators. Generally speaking, metallic substances make much better conductors and non-metallic substances make better insulators.

12.5 ELECTRICITY PRODUCED FROM MAGNETISM

Electricity can be produced in many ways, for example, from chemicals, pressure, light, heat, and magnetism. The electricity that air-conditioning and heating technicians are more involved with is produced by magnetism.

Magnets are common objects with many uses. They have poles, usually designated as the north (N) pole and the south (S) pole, and they also have fields of force, or magnetic lines of force. **Figure 12.5** shows the lines of the field of force around a permanent bar magnet. This field causes the like poles of two magnets to repel each other and the unlike poles to attract each other.

Figure 12.5 A permanent magnet with lines of force.

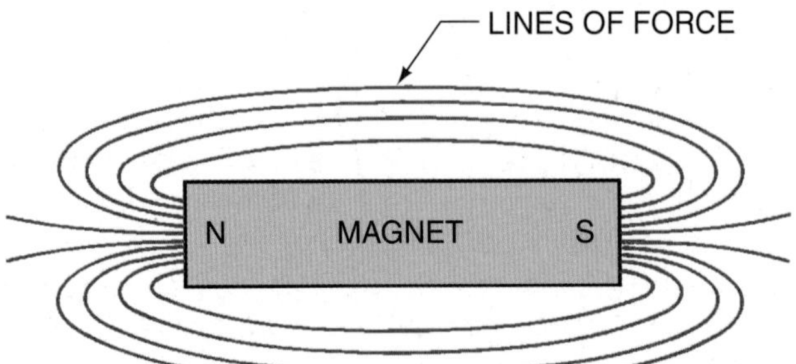

Figure 12.6 The movement of wire up and down cuts the lines of force, generating voltage, which in turn facilitates current flow.

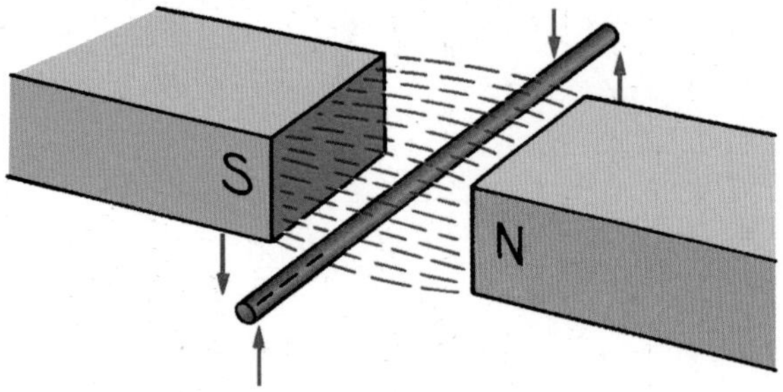

If a conductor, such as a copper wire, crosses the lines of force, the outer electrons in the atoms in the wire are freed and begin to move from atom to atom. This movement of electrons results in an excess of electrons on one end of the conductor and a deficiency of electrons on the other. This difference in charge is what we will later describe as voltage, potential difference, or electrical potential. It is this electrical potential that causes electric current to flow. It does not matter if the wire moves or if the magnetic field moves. It is only necessary that the conductor crosses through, or cuts, the lines of force, **Figure 12.6**. This movement of electrons in one direction produces current. Current is, in its simplest terms, an impulse transferred from one electron to the next. For example, if you pushed a golf ball into a tube already filled with golf balls, one would be ejected instantly from the other end, **Figure 12.7**. The energy of electromagnetic waves travels in a similar manner at speeds up to 300,000 km/s, or 186,000 miles/s. The electrons themselves do not travel through a wire at speeds even close to this. The electrons flowing through a battery-powered lamp, for example, move at a speed of about 3 in. per hour!

An electrical generator has a large magnetic field and many turns of wire crossing the lines of force. The wires are in the form of coils that are rotated within the magnetic field. A large magnetic field will produce more current than a smaller magnetic field, and many turns of wire passing through a magnetic field will produce more current than a few turns of wire. The magnetic force field for generators is usually produced by electromagnets. Electromagnets have characteristics similar to those of permanent magnets and are discussed later in this unit. **Figure 12.8** shows a simple generator. Large industrial generators and power plants can use the force of water flow, wind, or steam to rotate the coils of wire within the magnetic field.

Figure 12.7 A tube filled with golf balls; when one ball is pushed in, one ball is pushed out.

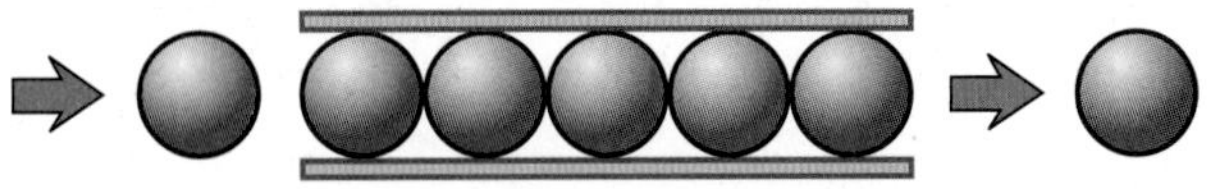

Figure 12.8 A simple generator.

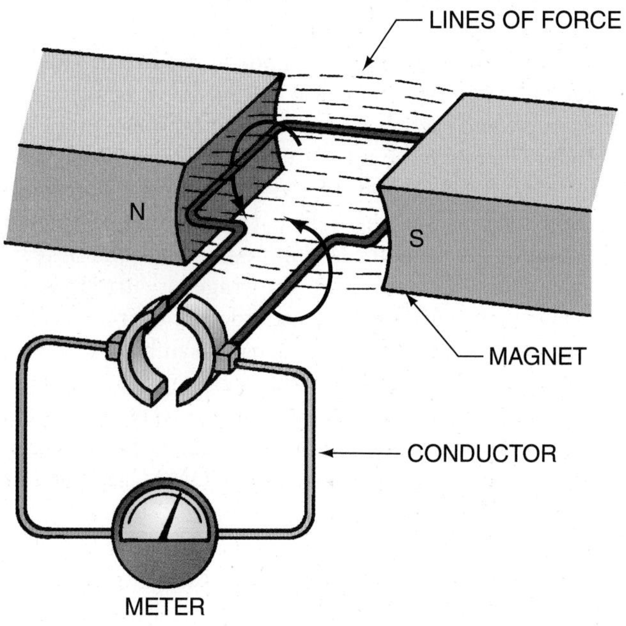

12.6 CURRENT

DIRECT CURRENT

Direct current (DC) is so-named because it travels in one direction. Since electrons have a negative charge and travel to atoms with a positive charge, DC is considered to flow from negative to positive. DC power supplies provide a constant voltage to the circuit being powered, **Figure 12.9**. DC voltage is generated primarily by chemical reactions and is commonly known as the battery-power source used in wireless and portable devices. Direct current can also be created from alternating current (AC) power sources by utilizing solid-state or electronic circuits referred to as rectifiers. Rectifiers are discussed in more detail later on in this unit.

Figure 12.9 Representation of DC power.

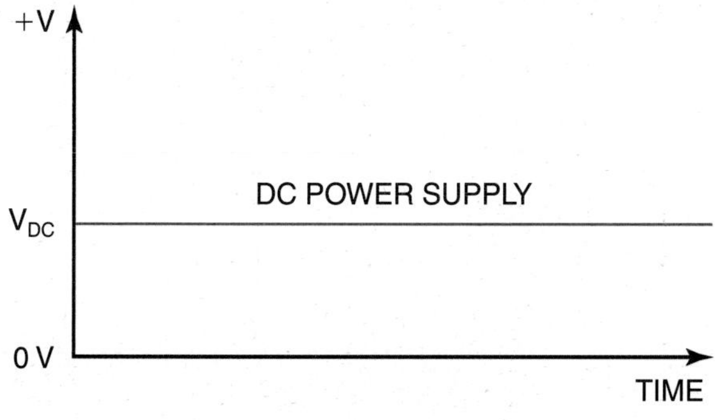

Figure 12.10 Representation of AC power.

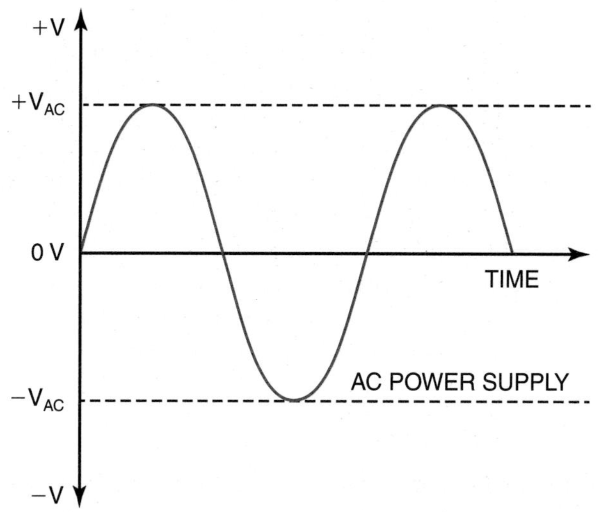

ALTERNATING CURRENT

Alternating current (AC) continually and rapidly reverses its polarity. The charge at the power source (generator) is continually changing direction; thus, the current continually reverses itself. As a result, the voltage, discussed in the next section, continuously varies between a positive AC value and a negative AC value, **Figure 12.10**. Most electrical energy generated for public use is AC. It is much more economical to transmit electrical energy long distances in the form of AC, as the transmission losses are very low compared to those associated with DC. The voltage of this type of current can be readily changed, so it has many more uses. DC still has many applications, but it is usually obtained by changing AC to DC or by producing the DC locally where it is to be used.

12.7 UNITS OF ELECTRICAL MEASUREMENT

Electromotive force (emf), or voltage (V), is used to indicate the potential difference between two charges. When an electron surplus builds up on one side of a circuit and a shortage of electrons exists on the other side, a difference in potential, or emf, is created. This potential difference is referred to as voltage. The unit used to measure this force is the volt. The ampere is the unit used to measure the quantity of electrons moving past a given point in a specific period of time (electron flow rate). When 6.24×10^{18} electrons pass a given point per second, we say that 1 ampere of electric current is flowing. One coulomb is defined as 6.24×10^{18} electrons. So, we can say that 1 ampere is equal to a current flow of 1 coulomb of electrons per second. *Amperage* and *ampere(s)* are sometimes termed “amp(s)” for short. The abbreviation for ampere is A.

All materials oppose or resist the flow of an electrical current to some extent. In good conductors this opposition or resistance is very low. In poor conductors the resistance is high. The unit used to measure resistance is the ohm. The symbol for ohm is the Greek letter Omega, Ω. A resistance of 1 ohm is present when a force of 1 volt causes a current of 1 ampere to flow in an electric circuit. As the resistance of an electric circuit increases, the current flow through the circuit drops. Similarly, as the resistance of a circuit drops, the current flow through the circuit will increase.

Volt = Electrical force or pressure (V)
Ampere = Quantity of electron flow rate (A)
Ohm = Resistance to electron flow (Ω)

12.8 THE BASIC ELECTRIC CIRCUIT

An electric circuit must have a power source, a load or device to consume the power, as well as a complete conductive path to and from the power source. There is also generally a means for starting and stopping the current flow through the circuit. A simple electric circuit is shown in **Figure 12.11**. This diagram shows an electrical generator providing power to the circuit, a complete conductive path, a light bulb for the load, and a switch for opening and closing the circuit.

The generator produces the electrical potential, or voltage, by passing many turns of wire through a magnetic field. The wire provides the path for the electricity to flow to the bulb and complete the circuit. The electrical energy is converted to heat and light energy by the bulb. The switch is used to open and close the circuit. When the switch is open, no current will flow. When it is closed, the bulb element will produce heat and light because current is flowing through it.

Figure 12.11 An electric circuit.

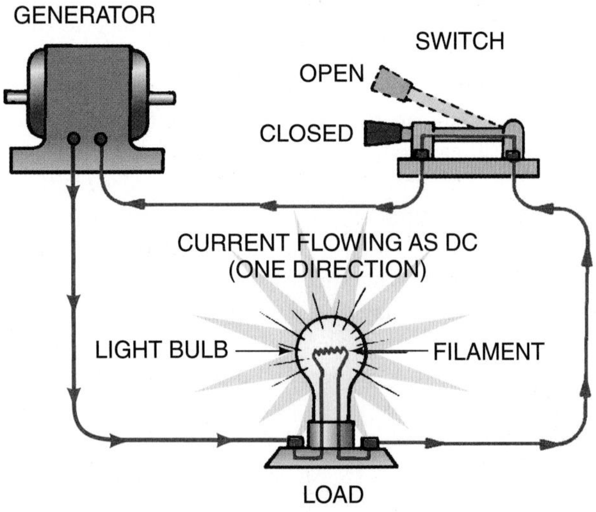

12.9 MAKING ELECTRICAL MEASUREMENTS

Electrical measurements can be made in the circuit illustrated in **Figure 12.11** to determine the voltage (emf) being supplied to the circuit and the amount of current (amperes) flowing in the circuit. To make the measurements, **Figure 12.12**, a digital multimeter (DMM) set to measure voltage, is connected across the terminals of the bulb without interrupting the circuit. A clamp-on ammeter is clamped around one of the circuit conductors to measure the amperage, or current draw, of the circuit. **Figure 12.13** illustrates the same circuit, without the DMM and clamp-on ammeter, using symbols. Commonly used electrical symbols can be found in **Figure 12.14**. Digital multimeters (DMMs) can be used to measure the potential difference between two points in an electric circuit when set to read voltage. The test leads on the DMM are connected

Figure 12.12 Voltage is measured across the load (in parallel). Amperage is measured with a clamp-on ammeter.

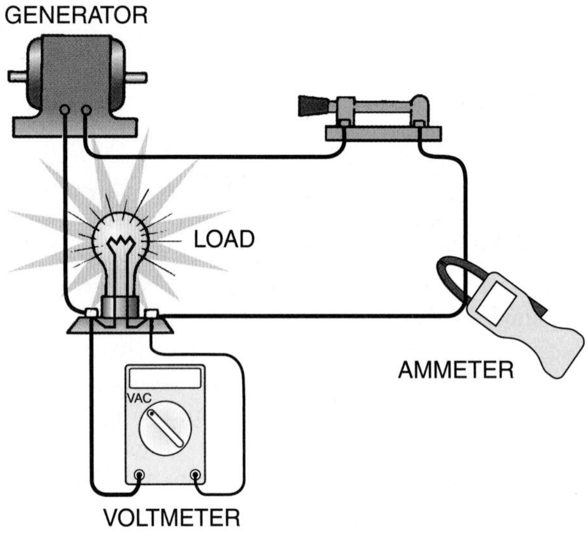

Figure 12.13 The same circuit as in Figure 12.12, illustrated with symbols, without the test instruments.

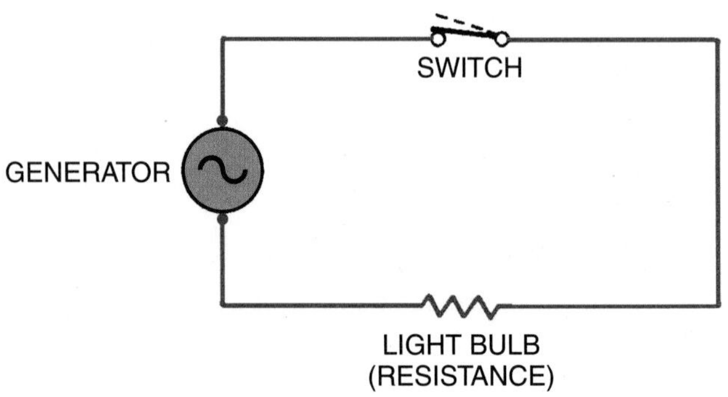

Figure 12.14 Commonly used electrical symbols.

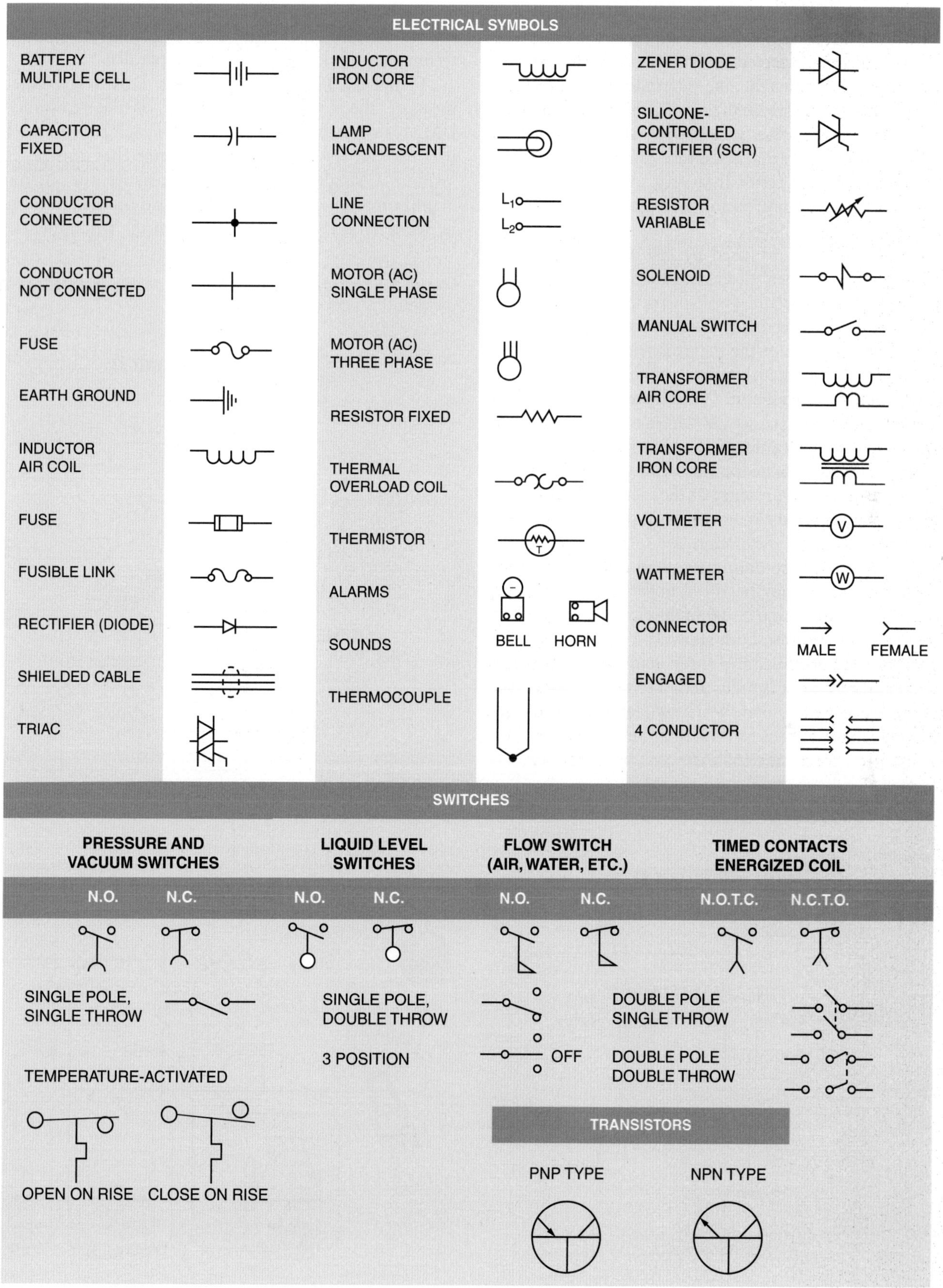

to the two points between which the voltage reading is to be taken. For example, the DMM in Figure 12.12 is connected to the two light bulb terminals, so the voltage reading that is obtained will be the potential difference across the bulb. The DMM is, therefore, connected in parallel with the portion of the circuit being evaluated. The internal resistance of the DMM when set to measure voltage is very high. As discussed in Section 12.7, current flow will be low when resistance is high, so the very high internal resistance of the DMM (when set to read voltage) will result in a very small current flow through the instrument, preventing the meter from affecting normal circuit operation.

Clamp-on ammeters are used by HVACR field personnel to measure the intensity of current flow in a circuit or a particular branch in a circuit. These instruments are used by clamping the jaws of the meter around one of the circuit's conductors. The strength of the magnetic field sensed by the clamp-on ammeter will translate into the amount of current flowing through the conductor. When bench-testing equipment, it is not uncommon to see ammeters connected in series with the equipment being tested. The internal resistance of ammeters intended for connection in series with circuit loads is very low. This is so the presence of the instrument does not affect the resistance of the circuit which, in turn, would affect the current flow through the circuit.

Often, a circuit will contain more than one resistance or load. These resistances may be wired in series or in parallel, depending on the application or use of the circuit. **Figure 12.15** shows three loads in series, both shown pictorially and with symbols. **Figure 12.16** illustrates three loads wired in parallel. In circuits where devices are wired in series, all of the current passes through each load. When two or more loads are wired in parallel, the current is divided among the loads. This is explained in more detail later in this unit. Power-passing devices, referred to as switches, are wired in series with the devices or loads they are controlling. Most resistances or loads (power-consuming devices) that air-conditioning and heating technicians work with are wired in parallel.

Figure 12.17 illustrates how a DMM is connected to measure the voltage across each of the resistances. The DMM is connected in parallel with each resistance. The figure also shows a clamp-on ammeter that can be clamped around a single conductor to measure the current. The clamp-on ammeter, **Figure 12.18**, is discussed in more detail in Section 12.19, "Electrical Measuring Instruments."

Figure 12.15 Multiple resistances (small heating elements) in series.

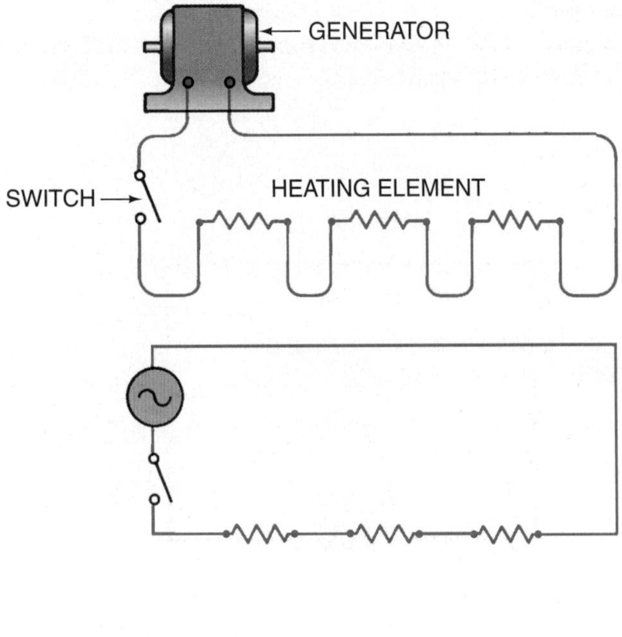

Figure 12.16 (A) Multiple resistances in parallel. (B) Three resistances in parallel using symbols.

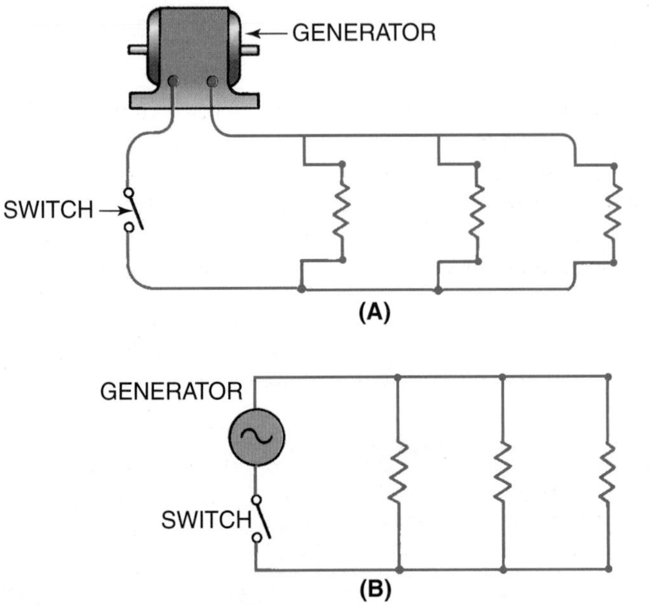

Figure 12.17 Voltage readings are taken across the resistances in the circuit. Amperage readings are taken by using a clamp-on ammeter.

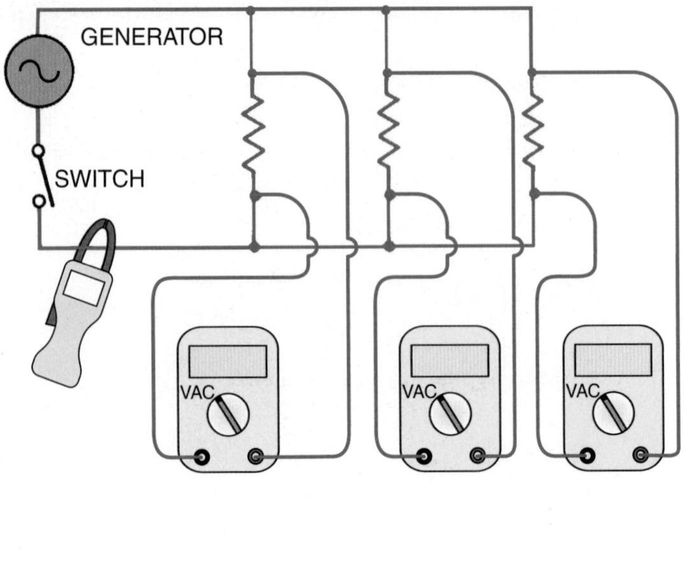

Figure 12.18 A clamp-on ammeter. Photo by Eugene Silberstein

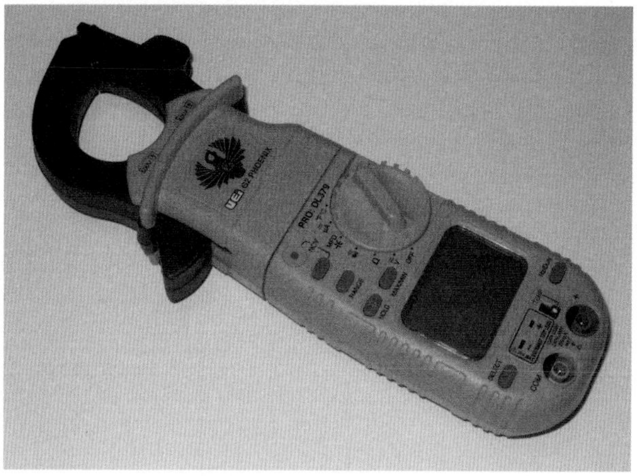

12.10 OHM'S LAW

In the early 1800s, the German scientist George S. Ohm did considerable experimentation with electrical circuits. He determined that a relationship exists among the voltage, current, resistance, and power in an electrical circuit. This relationship is called Ohm's law, which is described below. Letters are used to represent the different electrical factors.

E or V = Voltage (emf)
I = Amperage (current)
R = Resistance (load)

The "E" stands for electromotive force, the "V" stands for voltage, the "I" stands for intensity of current, and the "R" stands for resistance. Ohm's law appears in three common forms. The voltage equals the amperage times the resistance:

$$E = I \times R$$

The amperage equals the voltage divided by the resistance:

$$I = \frac{E}{R}$$

The resistance equals the voltage divided by the amperage:

$$R = \frac{E}{I}$$

Figure 12.19 shows a convenient way to remember these formulas. This figure shows the "E" for voltage at the top, while the "I" for current and the "R" for resistance are side-by-side underneath. To determine the formula for an unknown value, simply cover the unknown value. What remains will be the formula for the unknown value, **Figure 12.20**. Some examples of how Ohm's law can be used will now be provided.

Figure 12.19 Graphic representation of Ohm's law.

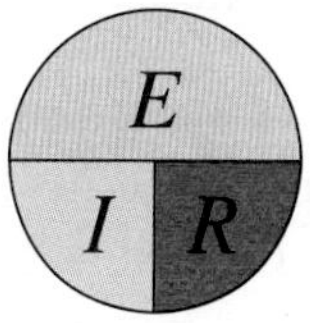

Figure 12.20 To determine the formula for the unknown quantity, cover the letter that represents the unknown.

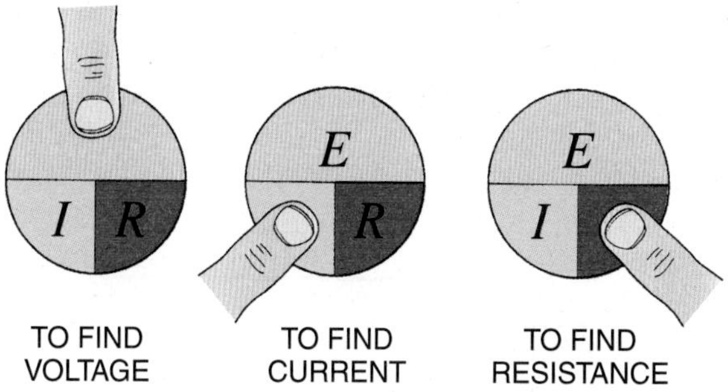

Example 1:
In **Figure 12.21**, the resistance of the heating element can be determined as follows:

$$R = \frac{E}{I} = \frac{120\text{ V}}{3\text{ A}} = 40\ \Omega$$

Figure 12.21 The resistance of the heating element can be determined by using the voltage and amperage indicated.

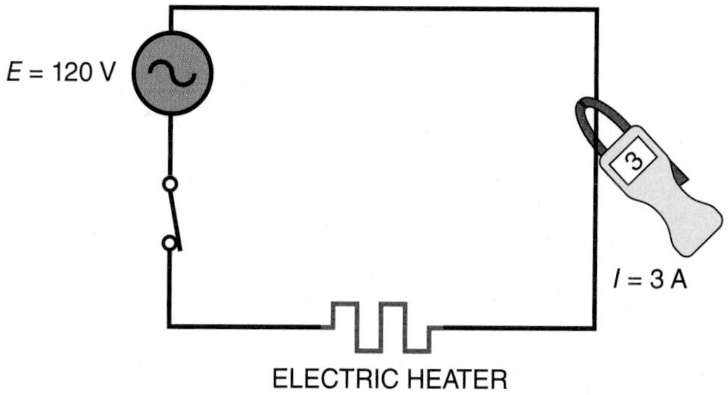

Example 2:
In **Figure 12.22**, the voltage supplied to the circuit can be calculated as follows:

$$E = I \times R = 2\text{ A} \times 60\ \Omega = 120\text{ V}$$

Figure 12.22 The voltage across the resistance can be calculated using the information in this diagram.

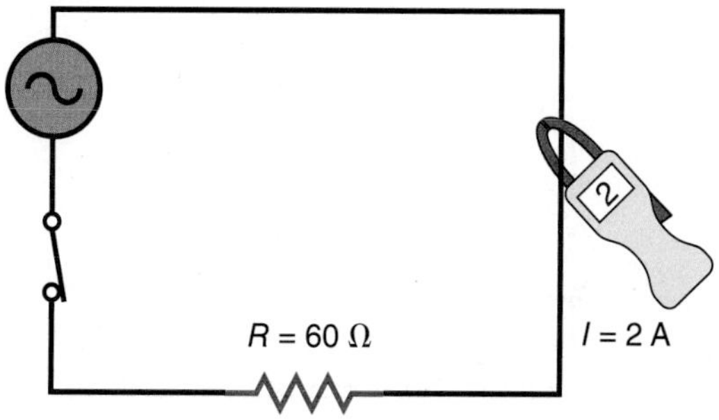

Example 3:
Figure 12.23 indicates the voltage to be 120 V and the resistance to be 20 Ω. The formula for determining the current flow is

$$I = \frac{E}{R} = \frac{120\ \text{V}}{20\ \Omega} = 6\ \text{A}$$

Figure 12.23 Voltage is 120 V and the resistance is 20 Ω. The amperage of the circuit can be determined using this information.

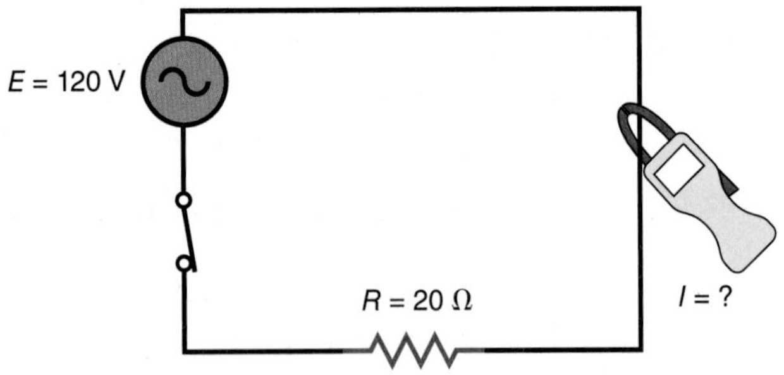

In series circuits with more than one resistance, simply add the resistances together to find the total resistance of the circuit. In **Figure 12.24**, there is only one path for the current to follow (a series circuit), so the resistance is 40 Ω (20 Ω + 10 Ω + 10 Ω). The amperage in this circuit will be

$$I = \frac{E}{R} = \frac{120\ \text{V}}{40\ \Omega} = 3\ \text{A}$$

The resistances, in ohms, of the individual loads can be determined by disconnecting the resistance to be measured from the circuit and reading the resistance with a DMM set to read resistance as illustrated in **Figure 12.25**. **NOTE:** *Resistance should only be measured on circuits and components that are deenergized. Also, be sure that the component being evaluated is disconnected from the circuit to avoid inaccurate or false readings.* •

SAFETY PRECAUTION: *Do not use any electrical measuring instruments without specific instructions from a qualified person. Follow the manufacturer's instructions. The use of electrical measuring instruments is discussed in Section 12.19.* •

Figure 12.24 The current in a series circuit has only one possible path to follow.

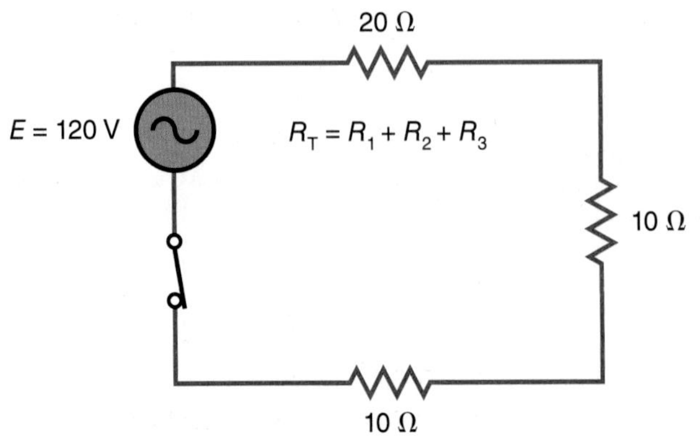

Figure 12.25 To determine the resistance of a circuit component, turn off the power and disconnect the component from the circuit. Then, check the component with a DMM set to measure resistance.

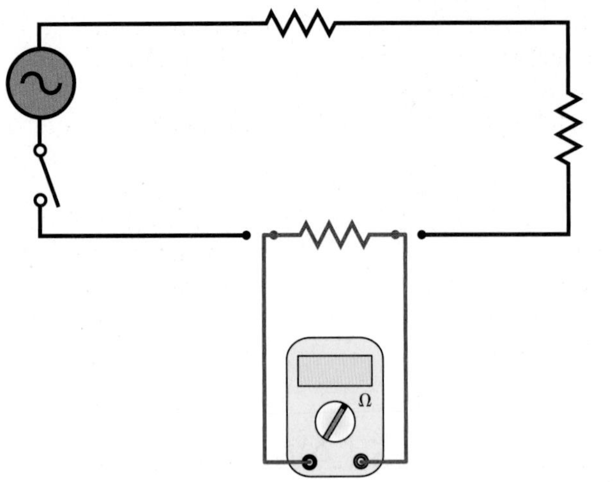

12.11 CHARACTERISTICS OF SERIES CIRCUITS

In series circuits,

- the voltage is divided across the different resistances. The mathematical formula is as follows:

$$E_{total} = E_1 + E_2 + E_3 + \ldots$$

This means that the total voltage being supplied to the circuit will be equal to the sum of the voltages that are measured across each device in the circuit.

- the total current flows through each resistance or load in the circuit. The mathematical formula is as follows:

$$I_{total} = I_1 = I_2 = I_3 = \ldots$$

Simply stated, this means that the current flowing in a series circuit will be the same no matter where in the circuit the amperage is being measured. The current flowing through the first circuit device is the same as the current flowing through the second circuit device, and so on.

- the resistances are added together to obtain the total resistance. The formula for calculating total resistance in a series circuit is as follows:

$$R_{total} = R_1 + R_2 + R_3 + \ldots$$

12.12 CHARACTERISTICS OF PARALLEL CIRCUITS

In parallel circuits,

- the total voltage is applied across each circuit branch. The mathematical formula is as follows:

$$E_{total} = E_1 = E_2 = E_3 = \ldots$$

This means that the total voltage being supplied to the circuit will be the same voltage that is supplied to branch 1, which will be the same as the voltage being supplied to branch 2, and so on.

- the current is divided between the different circuit branches according to their individual resistances, and the total current is equal to the sum of the currents in each branch. The mathematical formula is as follows:

$$I_{total} = I_1 + I_2 + I_3 + \ldots$$

- the total equivalent resistance of a parallel circuit will always be less than the value of the lowest resistance. As more and more resistive paths or branches are added to the parallel circuit, the total resistance of the circuit gets lower and lower.

Calculating the total resistance in a parallel circuit requires a different procedure than simply adding them together, as in a series circuit. A parallel circuit allows current flow along two or more paths at the same time. This type of circuit applies equal voltage to all loads. The general formula used to determine total resistance in parallel circuits is as follows:

For two resistances

$$R_{total} = \frac{R_1 \times R_2}{R_1 + R_2}$$

For more than two resistances

$$\frac{1}{R_{total}} = \frac{1}{R_1} + \frac{1}{R_2} + \frac{1}{R_3}$$

The total resistance of the circuit in **Figure 12.26** is determined as follows:

$$\frac{1}{R_{total}} = \frac{1}{R_1} + \frac{1}{R_2} + \frac{1}{R_3}$$

$$\frac{1}{R_{total}} = \frac{1}{10\ \Omega} + \frac{1}{20\ \Omega} + \frac{1}{30\ \Omega}$$

$$\frac{1}{R_{total}} = \frac{6}{60} + \frac{3}{60} + \frac{2}{60}$$

$$\frac{1}{R_{total}} = \frac{11}{60}$$

$$R_{total} = \frac{60}{11}$$

$$R_{total} = 5.45\ \Omega$$

To determine the total current draw, use Ohm's law:

$$I = \frac{E}{R} = \frac{120\ \text{V}}{5.45\ \Omega} = 22\ \text{A}$$

We can confirm this value for the total current in the circuit by determining the current flow through the individual loads

Figure 12.26 The total resistance of this circuit can be determined using the information in the diagram.

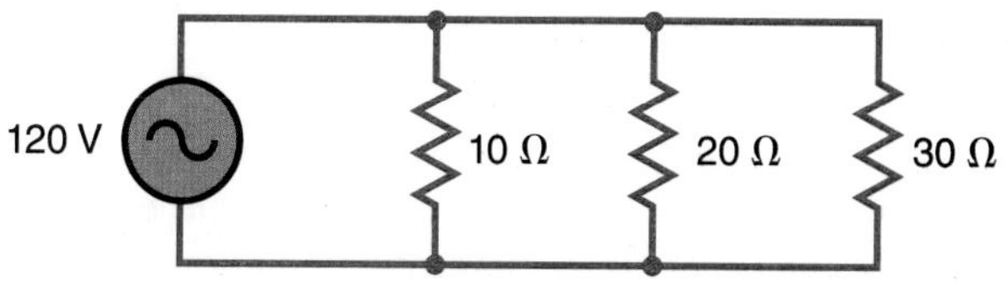

or resistances and then adding them up. The current flow through the 10-Ω resistance is determined here:

$$I_{10} = E \div R$$

$$I_{10} = 120\ \text{V} \div 10\ \Omega = 12\ \text{A}$$

The current flow through the 20-Ω resistance is determined here:

$$I_{20} = E \div R$$

$$I_{20} = 120\ \text{V} \div 20\ \Omega = 6\ \text{A}$$

The current flow through the 30-Ω resistance is determined here:

$$I_{30} = E \div R$$

$$I_{30} = 120\ \text{V} \div 30\ \Omega = 4\ \text{A}$$

By adding these three individual branch currents together, we get

$$I_{total} = I_1 + I_2 + I_3$$

$$I_{total} = 12\ \text{A} + 6\ \text{A} + 4\ \text{A}$$

$$I_{total} = 22\ \text{A}$$

12.13 ELECTRICAL POWER

Electrical power (*P*) is measured in watts. A watt (W) is the amount of power used when 1 ampere flows with a potential difference of 1 volt. Therefore, power can be determined by multiplying the voltage and the circuit amperage together.

$$\text{Watts} = \text{Volts} \times \text{Amperes}$$

or

$$P = E \times I$$

The consumer of electrical power pays the electrical utility company according to the number of kilowatts (kW) used for a certain time span, usually billed as kilowatt hours (kWh). A kilowatt is equal to 1000 W. To determine the power being consumed, divide the number of watts by 1000:

$$P\ (\text{in kW}) = \frac{E \times I}{1000}$$

In the circuit shown in **Figure 12.26**, the power consumed can be calculated as follows:

$$P = E \times I$$

$$P = 120\ \text{V} \times 22\ \text{A} = 2640\ \text{W}$$

$$\text{kW} = P \div 1000$$

$$\text{kW} = 2640\ \text{W} \div 1000 = 2.64\ \text{kW}$$

If the circuit in **Figure 12.26** is in operation for one hour, 2.64 kWh of energy have been consumed.

12.14 MAGNETISM

Magnetism was briefly discussed earlier in the unit to explain how electrical generators are able to produce electricity. Magnets are classified as either permanent or temporary. Permanent magnets are used in only a few applications that air-conditioning and refrigeration technicians would work with, such as electronically commutated motors (ECM), but electromagnets, a type of temporary magnet, are used in many electrical components of air-conditioning and refrigeration equipment.

A magnetic field is generated around a wire whenever an electrical current flows through it, **Figure 12.27**. If the wire or conductor is formed into a loop, the strength of the magnetic field will be increased, **Figure 12.28**, and if the wire is wound into a coil, an even stronger magnetic field will be created, **Figure 12.29**. This coil of wire carrying an electrical current is called a solenoid. This solenoid or electromagnet will attract or pull an iron bar into the coil, **Figure 12.30**. If an iron bar is inserted permanently in the coil, the strength of the magnetic field will be increased even more.

The magnetic field can be used to perform a number of tasks including generating electricity and causing electric motors to operate. The magnetic attraction can also open and close electrical contacts and valves, as in the case of many controls and switching devices, such as solenoids, relays, and contactors, **Figures 12.31(A)–(C)**. **Figure 12.32** is a cutaway view of a solenoid.

Figure 12.27 This cross section of a wire shows a magnetic field around the conductor.

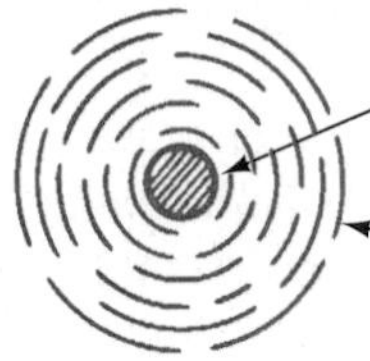

Figure 12.28 The magnetic field around a loop of wire. This is a stronger field than that around a straight wire.

Figure 12.29 There is a stronger magnetic field surrounding wire formed into a coil.

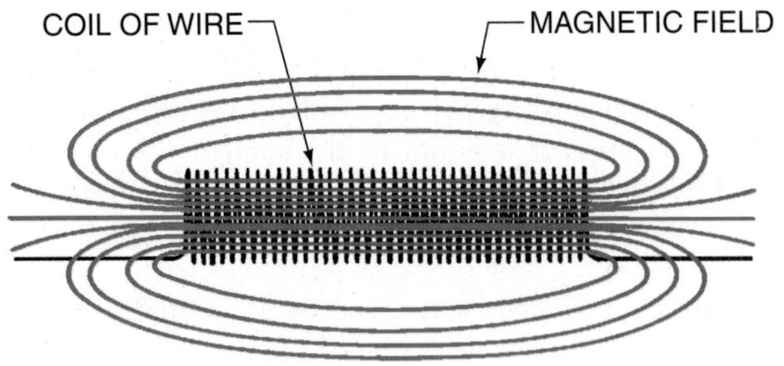

Figure 12.30 When current flows through a coil, the iron bar will be pulled into it.

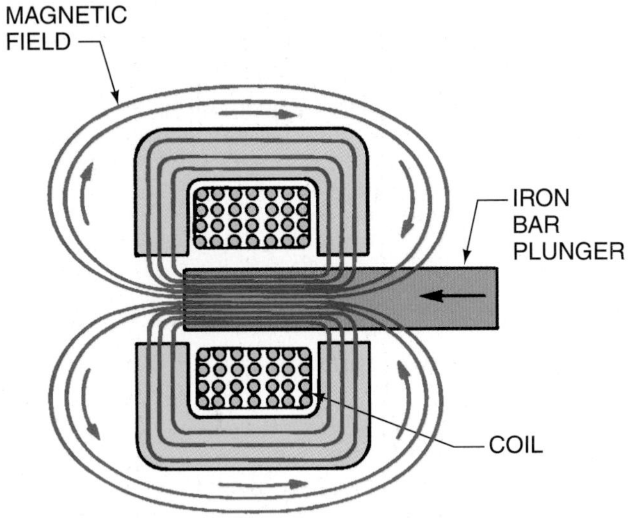

Figure 12.31 (A) A solenoid. Photo by Eugene Silberstein

(A)

Figure 12.31 (*Continued*) (B) A relay. Photo by Bill Johnson (C) Contactors. Courtesy Honeywell, Inc.

(B)

(C)

Figure 12.32 Cutaway view of a solenoid. Courtesy Parker Hannifin Corporation

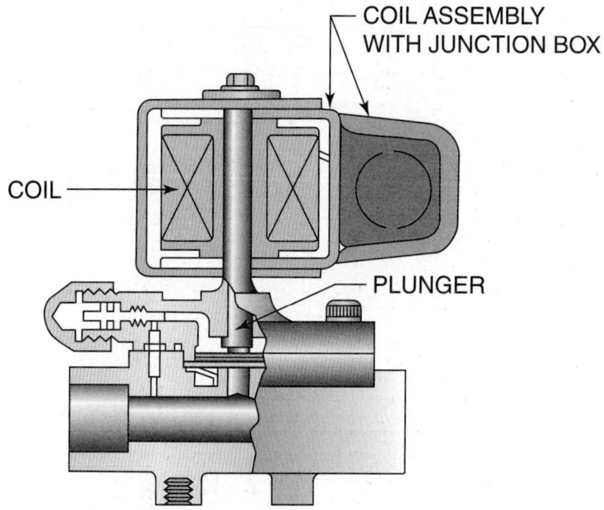

12.15 INDUCTANCE

As mentioned previously, when current flows through a conductor, a magnetic field is produced around the conductor. In an AC circuit, the current is continually changing direction. This causes the magnetic field to continually build up and collapse. When these lines of force build up and collapse, they cut through the wire or conductor and produce an emf, or voltage that opposes the existing voltage in the conductor.

Figure 12.33 A straight conductor with a magnetic field.

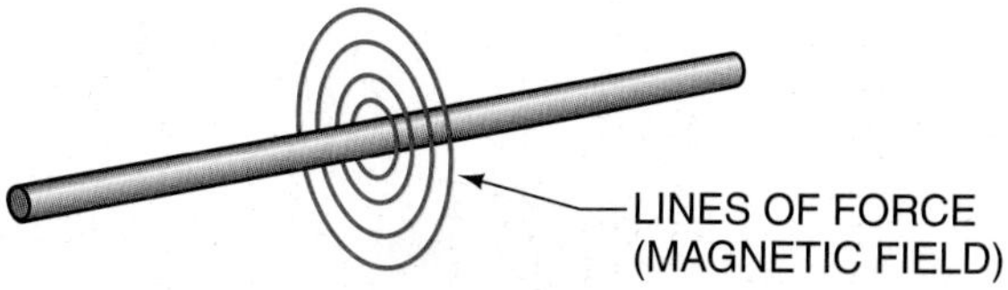

Figure 12.34 A conductor formed into a coil with lines of force.

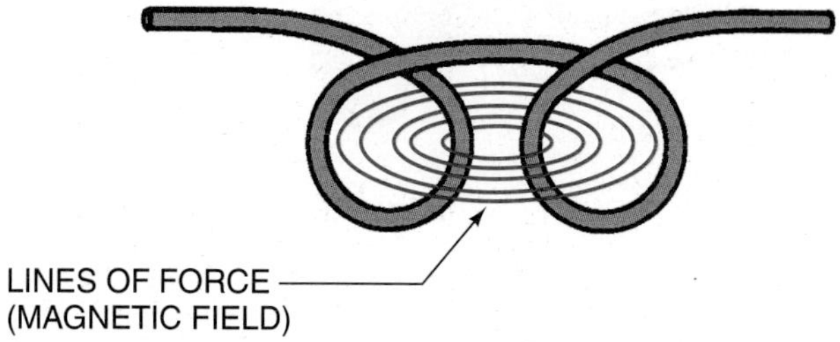

Figure 12.35 Electrical symbols for a coil.

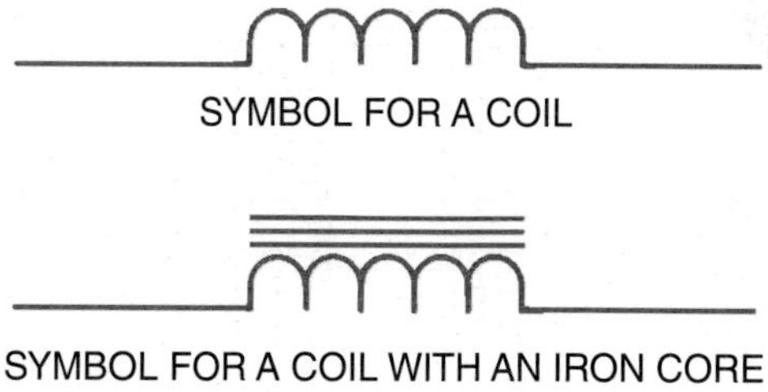

In a straight conductor, the induced voltage is very small and is usually not considered, **Figure 12.33.** However, if a conductor is wound into a coil, the lines of force overlap and reinforce each other, **Figure 12.34,** which develops an emf or voltage that is strong enough to provide opposition to the existing voltage. This opposition is called inductive reactance and is a type of resistance in an AC circuit. Coils and transformers are examples of components that produce inductive reactance. **Figure 12.35** shows symbols for the electrical coil. Electric motors have coils of wire in them and are, therefore, referred to as inductive loads. Inductive loads primarily generate a magnetic field, unlike resistive loads such as incandescent light bulbs and electric heaters that generate primarily light and/or heat.

12.16 TRANSFORMERS

Transformers, **Figure 12.36,** are electrical devices that produce voltage in a second circuit through electromagnetic induction. When AC power is applied to the primary winding of the transformer, a voltage will be produced across the

Figure 12.36 A transformer. Photo by Bill Johnson

Figure 12.37 Voltage applied across terminals produces a magnetic field around an iron or steel core.

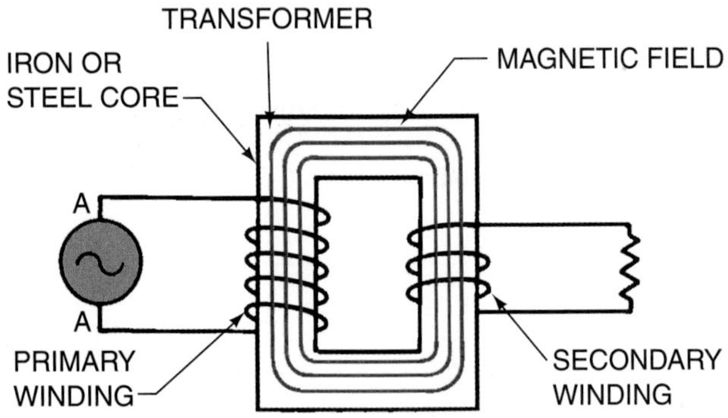

Figure 12.38 Electrical symbols for transformers.

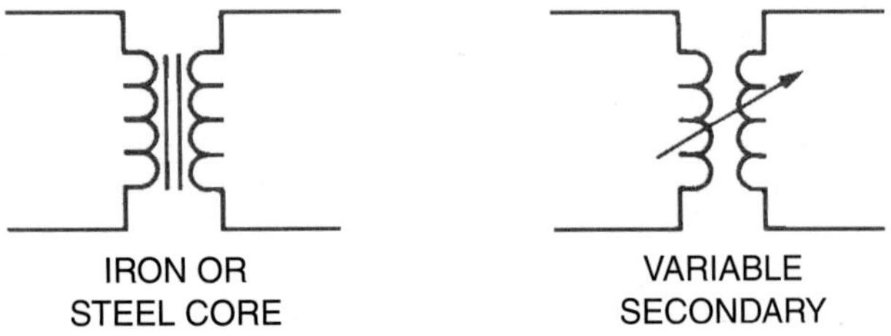

transformer's secondary winding, **Figure 12.37**. AC current supplied to the primary winding creates a magnetic field that continually builds up and collapses as the current reverses. This causes the magnetic field around the core in the secondary winding to cut across the conductor wound around it, inducing voltage in the secondary winding. If this induced voltage is then connected to a separate, otherwise complete, circuit, current will flow in that circuit. Two commonly used symbols for transformers are shown in **Figure 12.38**.

Transformers have a primary winding, a core usually made of thin plates of steel laminated together, and a secondary winding. There are step-up and step-down transformers. A step-down transformer contains more turns of wire in the primary winding than in the secondary winding. The voltage at the secondary is directly proportional to the number of turns of wire in the secondary (as compared to the number of turns in the primary) winding. For example, **Figure 12.39** shows a transformer with 1000 turns in the primary winding and 500 turns in the secondary winding. When a voltage of 120 V is applied to the primary, the voltage induced in the secondary is 60 V. (Actually, the secondary voltage will be slightly lower than 60 V due to some loss into the air of the magnetic field and because of resistance in the wire.)

A step-up transformer has more windings in the secondary than in the primary. This causes a larger voltage to be induced in the secondary. In **Figure 12.40**, with 1000 turns in the primary, 2000 in the secondary, and an applied voltage of 120 V, the voltage induced in the secondary is roughly

Figure 12.39 A step-down transformer.

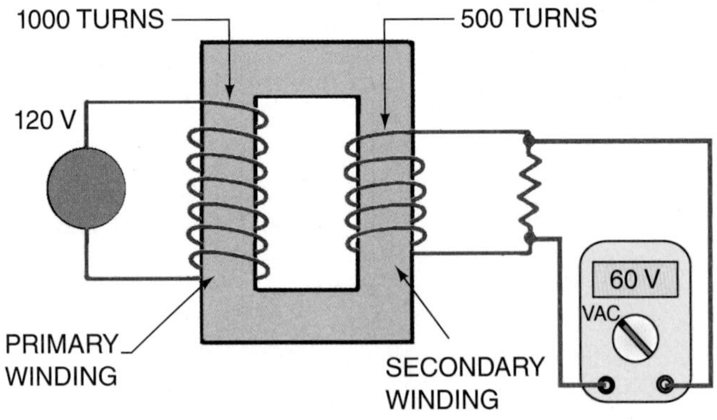

Figure 12.40 A step-up transformer.

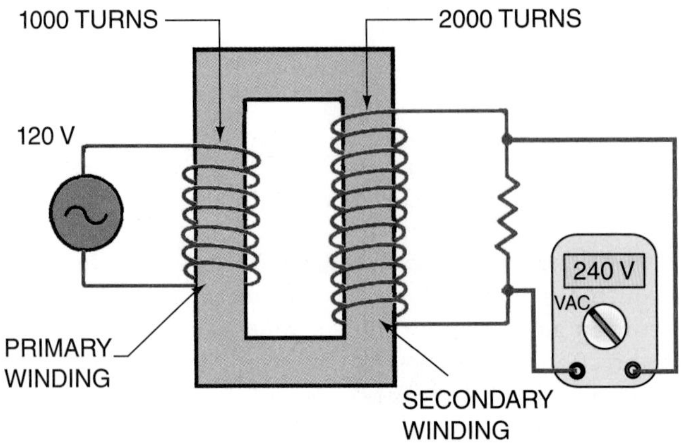

double, or approximately 240 V. The same power (watts) is available at the secondary as at the primary (except for a slight loss). If the voltage produced at the secondary winding is reduced to one-half that at the primary, the current capacity nearly doubles. Step-up transformers are commonly used to generate sparks to ignite air/fuel mixtures in heating applications.

Step-up transformers are also used at generating stations to increase voltage to more efficiently deliver electrical energy over long distances to substations or other distribution centers. At the substation, the voltage is reduced by a step-down transformer for ultimate distribution to the end users of the power. At a residence, the voltage may be reduced to 240 V or 120 V. Additional step-down transformers may be used with air-conditioning and heating equipment to produce the 24 V commonly used in thermostats and other control devices.

12.17 CAPACITANCE

A capacitor is a device in an electric circuit that stores electrical energy for later use. Two commonly used types of capacitors are the start capacitor and the run capacitor, **Figure 12.41**. A simple capacitor is composed of two plates with insulating material between them, **Figure 12.42**. This insulating material, referred to as a dielectric, can be air, paper, or other nonconductive material. The capacitor can store a charge of electrons on one plate. The amount of charge that can be stored can be fixed or variable. Electrical symbols for capacitors are shown in **Figure 12.43**. Since the plates in a capacitor are separated by a dielectric, current cannot readily flow through a capacitor when a circuit is initially energized. Instead, electrons accumulate on one of the plates, increasing the potential difference, or voltage, across the capacitor. Once the voltage across the capacitor reaches a level that is high enough to overcome the resistance of the dielectric, current will flow through the capacitor. When in the active electric circuit, capacitors will continuously charge and discharge.

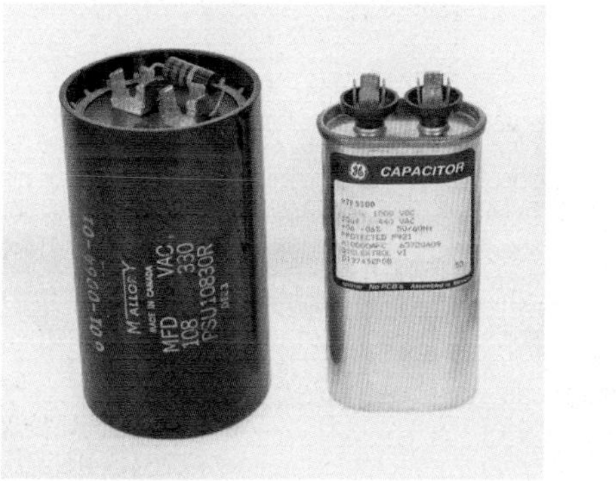

Figure 12.41 A start capacitor, left, and a run capacitor, right. Photo by Bill Johnson

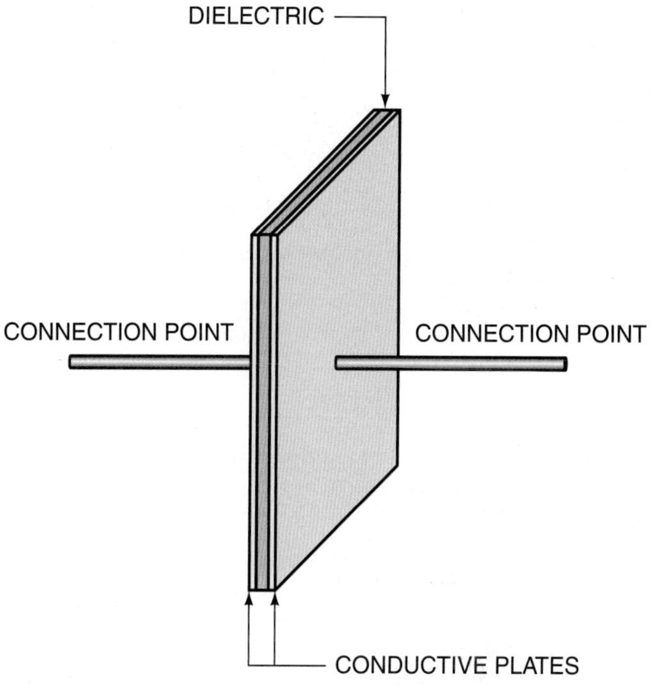

Figure 12.42 Cross-sectional illustration of a capacitor.

Figure 12.43 Electrical symbols for capacitors.

A capacitor has capacitance, which is the amount of charge that can be stored. The capacitance is determined by the following physical characteristics of the capacitor:

1. Distance between the plates, in meters
2. Surface area of the plates, in square meters
3. Dielectric material between the plates

Capacitors are rated in farads. However, farads represent such a large amount of capacitance that the unit microfarad is normally used, which is one-millionth (0.000001) of a farad. The symbol for micro is the Greek letter µ (mu) and the symbol for farad is the capital letter F, so a microfarad is represented as µF. Capacitors can range from 2 or 3 microfarads up to several hundred microfarads.

The formula for capacitance is given by

$$C = \varepsilon_0 KA/D$$

where ε_0 is a constant that is equal to 8.84×10^{-12} (0.00000000000884), K = the dielectric constant, A is the area of the capacitor's plates in square meters (m^2), and D is the distance between the plates in meters. Common dielectrics are air, paper, and glass and the dielectric constants for these materials are as follows:

Dielectric Constant for Air = 1.0

Dielectric Constant for Paper = 3.5

Dielectric Constant for Glass = 3

If we had a capacitor that had a paper dielectric, and 0.33 square meters of plates that are separated by 0.000002 meters, the capacitance can be calculated as:

$$C = \varepsilon_0 KA/D$$

$$C = (8.84 \times 10^{-12})(3.5)(0.333\ m^2)/0.000002\ m$$

$$C = 5 \times 10^{-6}\ \text{Farads, or 5 microfarads}$$

Although the calculation shown above looks complicated, the great news is that HVACR service technicians do not make these calculations. This work is done when the capacitor is designed. Service technicians can, however, check a particular capacitor to be sure that it is working properly. Many pieces of electrical test equipment have a feature that allows the technician to check capacitors, **Figure 12.44**. By setting the DMM to the microfarad, μFd, setting and connecting the DMM's test leads across the capacitor, the technician can read the capacitor's capacitance on the DMM's display. Before using any instrument to check a capacitor, be sure that the capacitor has been removed from the circuit and properly discharged to prevent personal injury and/or instrument damage.

A capacitor opposes current flow in an AC circuit similar to a resistor or to inductive reactance. This opposition or type of resistance is called **capacitive reactance**. The capacitive reactance depends on the frequency of the power supply and the microfarad rating of the capacitor.

Figure 12.44 This DMM is set to the appropriate scale to measure capacitance. The MFD stands for microfarads and the electrical symbol for a capacitor can also be seen. Photo by Eugene Silberstein.

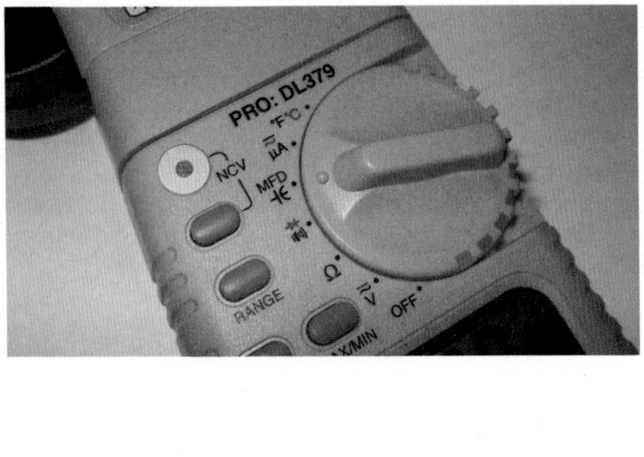

12.18 IMPEDANCE

We have learned that there are three types of opposition to current flow in an AC circuit. There is pure resistance, inductive reactance, and capacitive reactance. The total effect of these three is called *impedance*. The voltage and current in a circuit that has only resistance are in phase with each other. The voltage leads the current across an inductor and lags behind the current across a capacitor. **Figures 12.62** and **12.63** in Section 12.20, "Sine Waves," illustrate these conditions. Inductive reactance and capacitive reactance can cancel each other. Impedance is a combination of the opposition to current flow produced by these characteristics in a circuit.

12.19 ELECTRICAL MEASURING INSTRUMENTS

A digital multimeter, commonly referred to simply as a DMM, **Figure 12.45**, is an instrument that measures voltage, current, resistance, and, on some models, temperature. It can be used for making AC or DC measurements in several ranges. It is the instrument used most often by HVACR technicians. Some DMMs have the ability to measure AC amperage, such as the instrument shown in **Figure 12.18**. DMMs have different features depending on the manufacturer.

Most DMMs have a function switch and a range switch. The function switch allows the user to set the DMM to the electrical characteristic (voltage, current, or resistance) that is being measured. The range switch allows the user to enter what the expected reading should be, **Figure 12.46**. For example, when measuring the potential of a 2-V battery, it would be a good idea to set the DMM to the 6 VDC range, **Figure 12.47**. If, however, the technician was measuring a much higher DC voltage, he might choose to set the meter to the 600 VDC range, **Figure 12.48**. If uncertain about the magnitude of the measurement, it is always best to start at the highest range; it can always be lowered later on. When taking voltage readings or when making resistance measurements, the test leads must be plugged into the DMM, **Figure 12.49**.

Figure 12.45 A digital multimeter (DMM). Photo by Eugene Silberstein

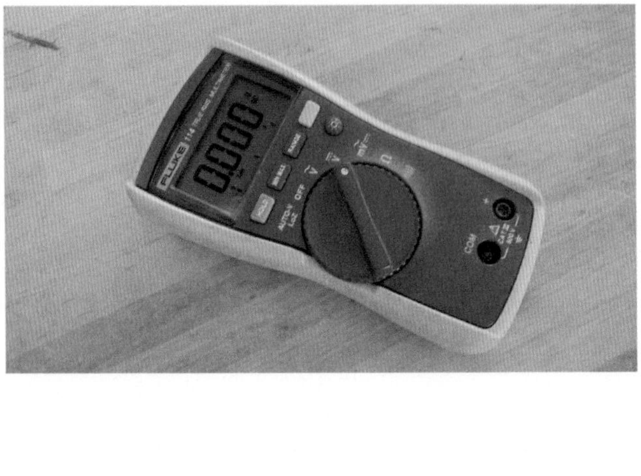

Figure 12.46 Range button on a DMM. Photo by Eugene Silberstein

Figure 12.47 6 VDC range on the DMM. Photo by Eugene Silberstein

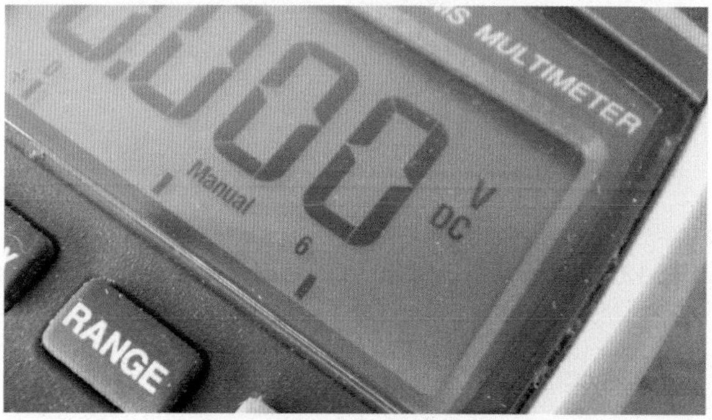

Figure 12.48 600 VDC range on the DMM. Photo by Eugene Silberstein

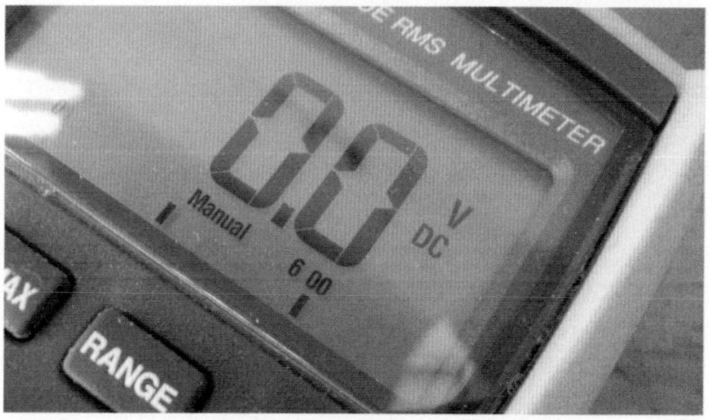

To check the voltage in the AC circuit in **Figure 12.50**, follow these steps:

1. Turn off the power.
2. Set the function switch to read AC Voltage.
3. Push the range button (if the DMM has one) until the desired range appears on the DMM's display.
4. Plug the black test lead into the common (−) jack and the red test lead into the positive (+) jack.
5. Connect the test leads across the load as shown in **Figure 12.50.**

Figure 12.49 Plugging the test leads into the DMM. Photo by Eugene Silberstein

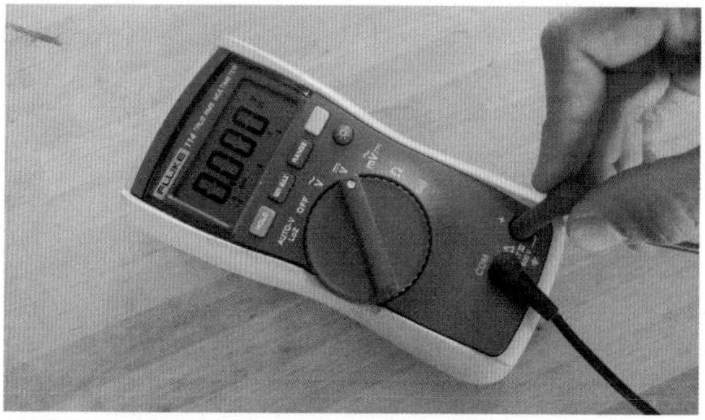

Figure 12.50 Connecting test leads across a load to check voltage.

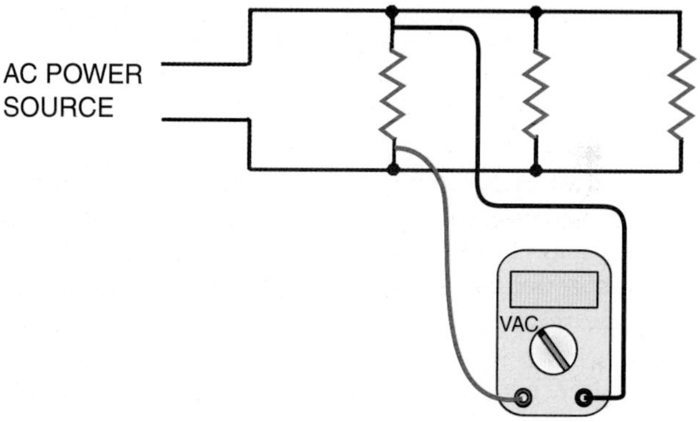

6. Turn on the power.
7. Read the voltage on the DMM's display.

To determine the resistance of a load, disconnect the load from the circuit. Make sure all power is off while doing this. If the load is not entirely disconnected from the circuit, the voltage from the battery of the DMM may damage the instrument's internal solid-state components. Then do the following:

1. Set the DMM's function switch to read resistance.
2. Plug the test leads into the DMM.
3. Connect the test leads to the device being tested.
4. Observe the reading on the DMM's display.

The clamp-on ammeter has a clamping feature that can be placed around a single wire in a circuit, and the current flowing through the wire can be read as amperage on the instrument's display, **Figure 12.51.**

SAFETY PRECAUTION: *Do not perform any of the preceding or following tests in this unit without approval from an instructor or supervisor. These instructions are simply a general introduction to DMMs. Be sure to read the operator's manual for the particular piece of test instrumentation you are using.* •

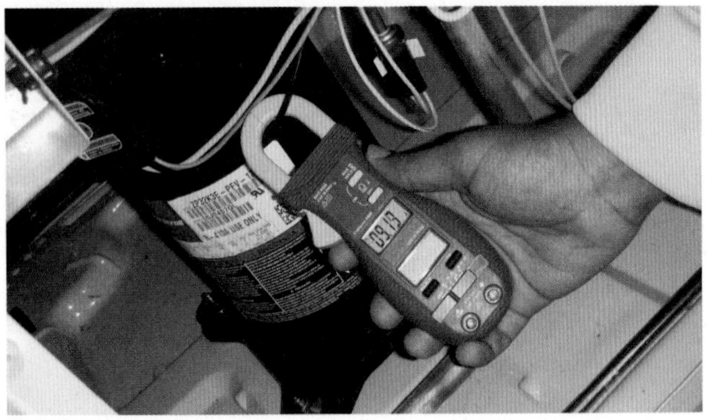

Figure 12.51 Measuring amperage by clamping the jaws of the clamp-on ammeter around the conductor. Photo by Eugene Silberstein

Figure 12.52 Units of voltage and amperage.

VOLTAGE
1 volt = 1000 millivolts (mV)
1 volt = 1,000,000 microvolts (μV)

AMPERAGE
1 ampere = 1000 milliamperes (mA)
1 ampere = 1,000,000 microamperes (μA)
Note that the symbol for micro or millionths is μ.

It is often necessary to determine voltage or amperage readings to a fraction of a volt or ampere, **Figure 12.52.** Many styles and types of DMMs are available for making these types of electrical measurements.

Electrical troubleshooting is performed one step at a time. **Figure 12.53** is a partial wiring diagram of a furnace's blower circuit. The blower motor does not operate because of an open motor winding. The technician could make the following voltage checks to determine where the failure is.

1. This is a 115 VAC circuit. The technician would set the DMM selector switch to the appropriate AC voltage scale. One of the DMM's test leads (black) is connected to a neutral terminal, N, at the power source, while the other test lead (red) is connected to the line-side power source terminal, L1. DMM "A" reads 115 VAC indicating that voltage is being supplied to the circuit, **Figure 12.54.**

Figure 12.53 Partial wiring diagram of furnace's blower circuit.

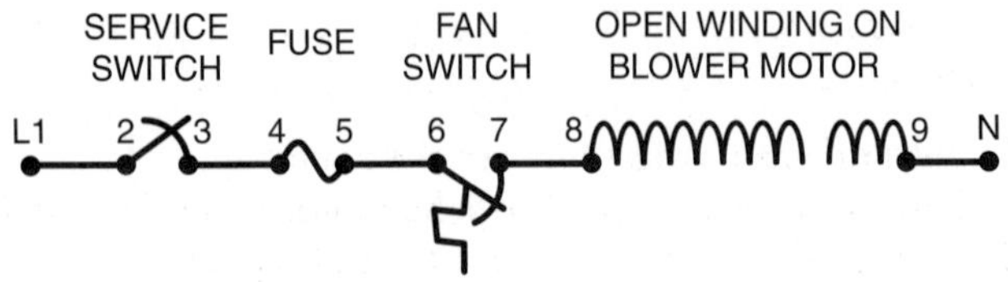

Figure 12.54 Measuring the voltage between L1 and neutral indicates that power is being supplied to the circuit.

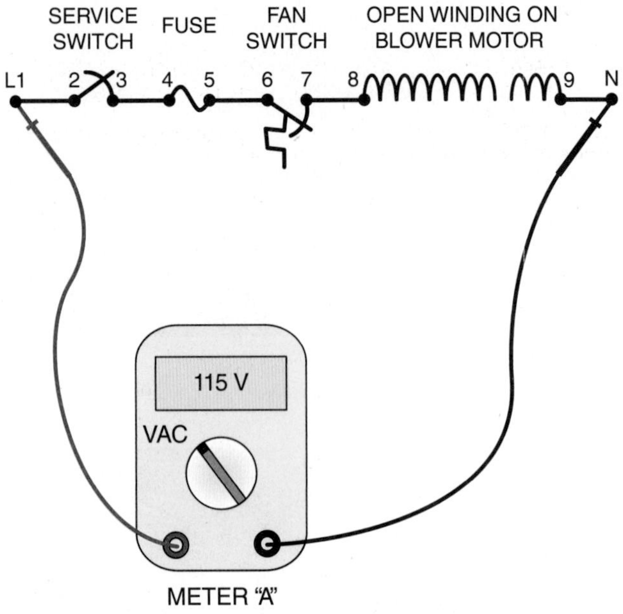

Figure 12.55 A reading of 115 VAC indicates that the service switch is in the closed position and power is passing from L1 to terminal 3.

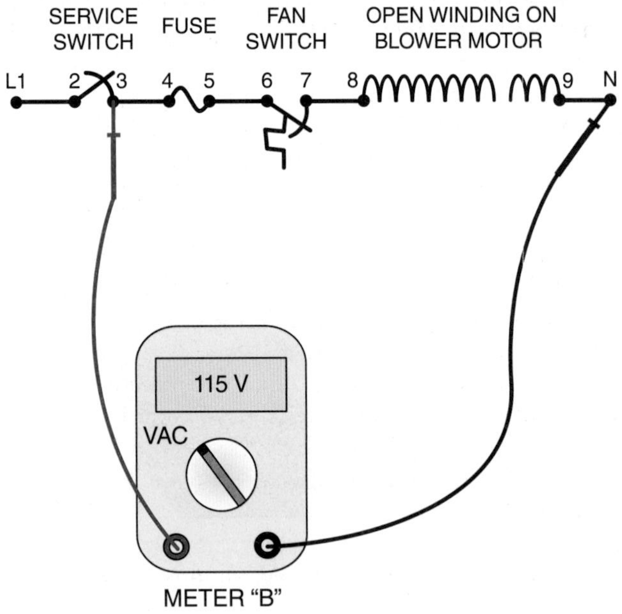

2. The positive (red) lead is then moved to the load side of the service switch at point 3, DMM "B" in **Figure 12.55.** DMM "B" reads 115 VAC indicating that power is passing through the service switch. If the voltage reading was 0 VAC, this would indicate that the service switch is open or the wiring between L1 and the switch is defective.

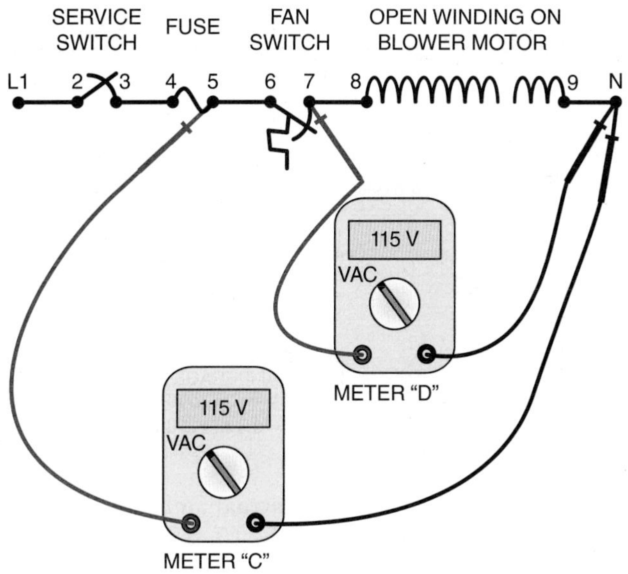

Figure 12.56 A reading of 115 VAC on DMM "C" indicates that power is passing through the fuse and a similar reading on DMM "D" indicates that power is also being passes though the fan switch. There is potential (voltage) between terminals 7 and N.

3. The positive (red) lead is then moved to the load side of the fuse at point 5. Service switch at point 3, DMM "C" in **Figure 12.56.** DMM "C" reads 115 VAC indicating that power is passing through the fuse. If the voltage reading was 0 VAC, this would indicate that the fuse is bad or the wiring between points 3 and 4 is defective.
4. The positive (red) lead is then moved to the load side of the fan switch at point 7, DMM "D" in **Figure 12.56.** DMM "D" reads 115 VAC indicating that power is passing through the fan switch. If the voltage reading was 0 VAC, this would indicate that the fan switch is open or the wiring between points 5 and 6 is defective.
5. The positive (red) lead is then moved to the line side of the motor at point 8, DMM "E" in **Figure 12.57.** This DMM reads 115 VAC indicating that power is being supplied to the motor.
6. To ensure there is power through the wire on the neutral side of the motor, the neutral probe is moved to the neutral side of the motor at point 9 leaving the positive probe on the line side of the motor. The reading on DMM "F" should be 115 V, indicating that 115 VAC is being supplied to the motor, **Figure 12.58.**

All of the checks above read 115 VAC, but the motor does not run. It would be appropriate to conclude that the motor is defective.

The motor winding can be checked with a DMM set to read resistance. The motor winding must have a measurable resistance for it to function properly. To check this resistance, the technician may do the following, **Figure 12.59:**

1. Turn off the power source to the circuit.
2. Disconnect the terminals on the motor from the circuit.
3. Set the DMM selector switch to read resistance (ohms).

Figure 12.57 A reading of 115 VAC on DMM "E" indicates that there is power being supplied to the motor at point 8.

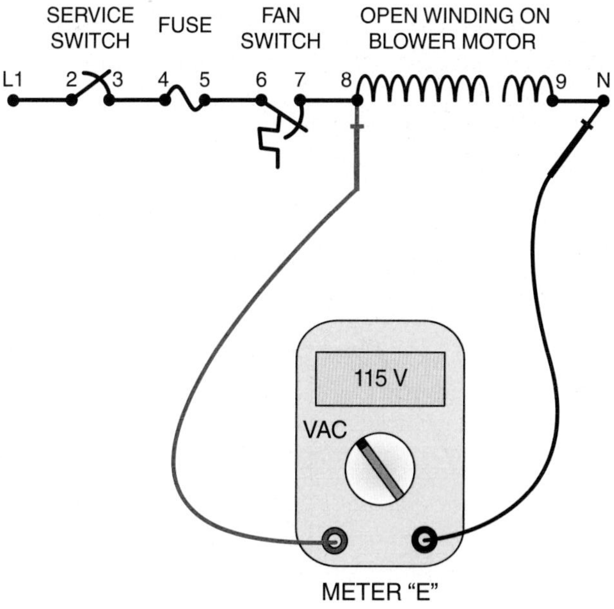

Figure 12.58 Connecting DMM "F" between terminals 8 and 9 and obtaining a reading of 115 VAC indicates that the wiring between terminal 9 and N is good and that the motor winding is being powered.

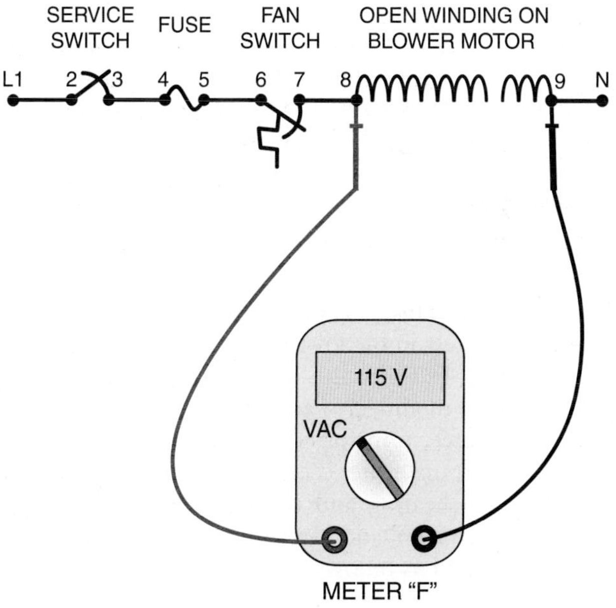

Figure 12.59 The resistance of the motor winding is checked by shutting off the power to the motor and removing the winding from the electric circuit to be sure that the resistance readings obtained are accurate.

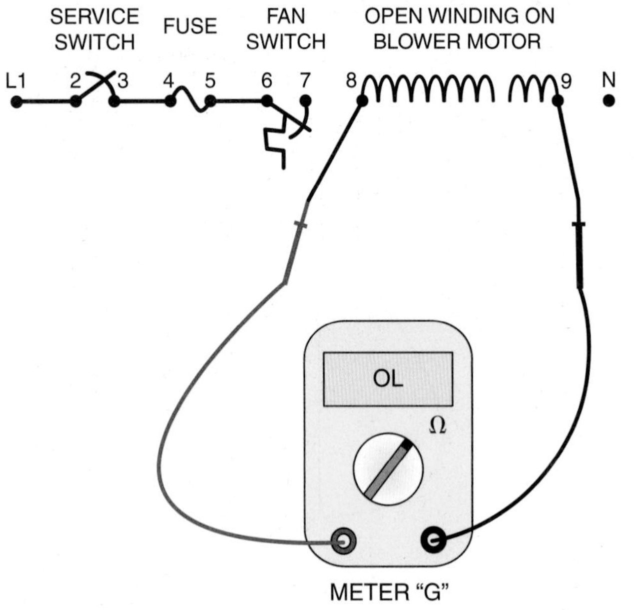

4. Touch the DMM's probes together to verify that the test leads are good and that the reading on the instrument's display indicates a resistance value very close to zero ohms.
5. Touch one DMM probe to one motor terminal and the other to the other terminal. DMM "G" reads infinity. On DMMs, infinite resistance is often displayed as "OL" which stands for "open line" or "open lead." This is the same reading you would get by holding the DMM's probes apart in the air. There is no circuit through the windings, indicating that they are open.

Although not very common in today's HVACR industry, analog clamp-on ammeters are still found in the toolboxes of some field technicians. Analog, or non-digital, ammeters do not read accurately in the lower amperage ranges, such as in the 1-ampere-or-below range. However, an analog clamp-on ammeter can be modified to accurately measure low amperages. For instance, transformers are rated in volt-amperes (VA). A 40-VA transformer is often used in the control circuit of combination heating and cooling systems. These transformers produce 24 V and can carry a maximum of 1.67 A. This is determined as follows:

$$\text{Output in amperes} = \frac{\text{VA rating}}{\text{Voltage}}$$

$$I = 40\ \text{VA}/24\ \text{V}$$

$$= 1.67\ \text{A}$$

Figure 12.60 illustrates how an analog clamp-on ammeter can be used with 10 wraps of wire to multiply the amperage

Figure 12.60 An illustration showing the use of a 10-wrap multiplier with an ammeter to obtain accurate, low-amperage readings.

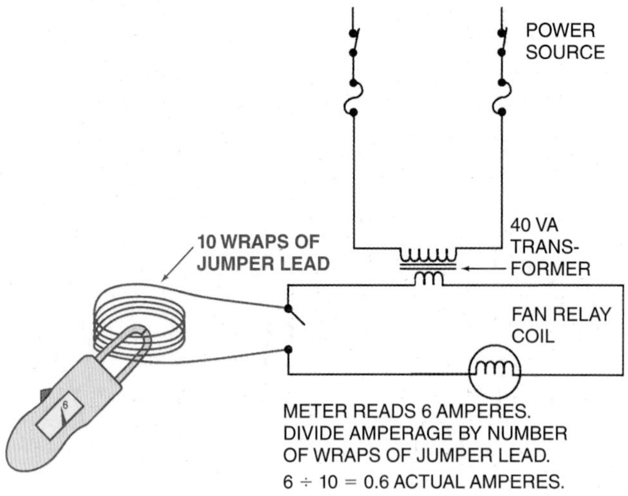

reading by 10. To determine the actual amperage, divide the amperage indicated on the meter by 10. If a digital clamp-on ammeter is used, there is no need to create the 10 wraps, as these meters are able to read low amperages accurately. Digital multimeters have the ability to measure low amperages, so the 10-wrap is not necessary.

12.20 SINE WAVES

As mentioned earlier, alternating current continually reverses direction. An oscilloscope is an instrument that measures the amount of voltage over a period of time and can display this voltage on a screen as a waveform. There are many types of waveforms, but we will discuss the one refrigeration and air-conditioning technicians encounter the most, the sine wave, **Figure 12.61**.

Figure 12.61 A sine wave displayed on an oscilloscope.

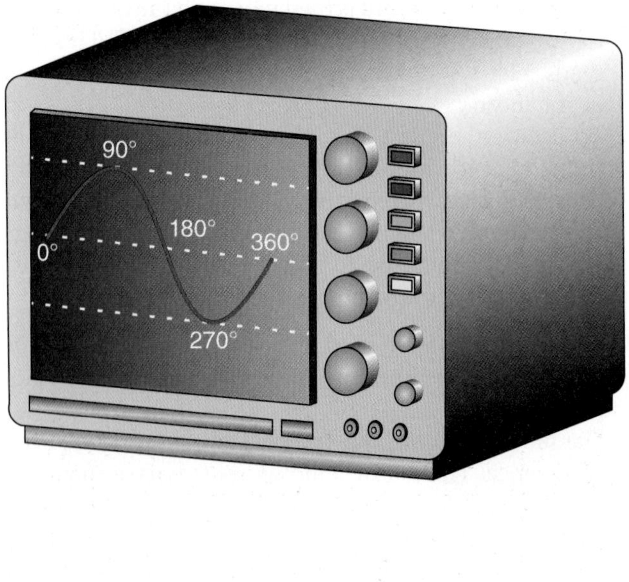

The sine wave displays the voltage of one cycle through 360°. In the United States and Canada, the standard voltage in most locations is produced with a frequency of 60 cycles per second. Frequency is measured in hertz. Therefore, the standard frequency in this country and Canada is 60 Hz. **Figure 12.61** shows a complete sine wave as it might be observed on an oscilloscope's display screen. At the 90° point, the voltage reaches its peak (positive); at 180° it is back to 0; at 270° it reaches its negative peak; and at 360° it is back to 0. If the frequency is 60 Hz, this cycle would be repeated 60 times every second. The sine wave is a representation of an alternating current cycle.

Figure 12.62 shows the peak and peak-to-peak values. As the sine wave indicates, the voltage is at its peak value only briefly during the cycle. Therefore, the peaks of the peak-to-peak values are not the effective voltage values. The effective voltage is the RMS voltage, **Figure 12.63**. The RMS stands for root-mean-square value. This is the alternating current value measured by most electrical test instruments. The RMS voltage is 0.707 × the peak voltage. If the peak voltage were 170 V, the effective voltage measured by a DMM would be 120 V (170 V × 0.707 = 120.19 V).

Figure 12.62 Peak and peak-to-peak AC voltage values.

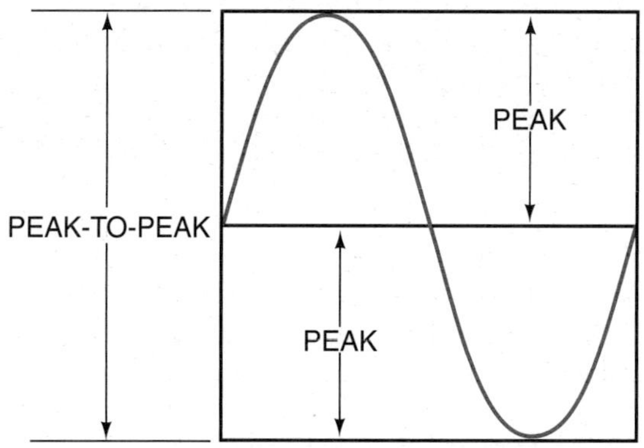

Figure 12.63 The root-mean-square (RMS) or effective voltage value.

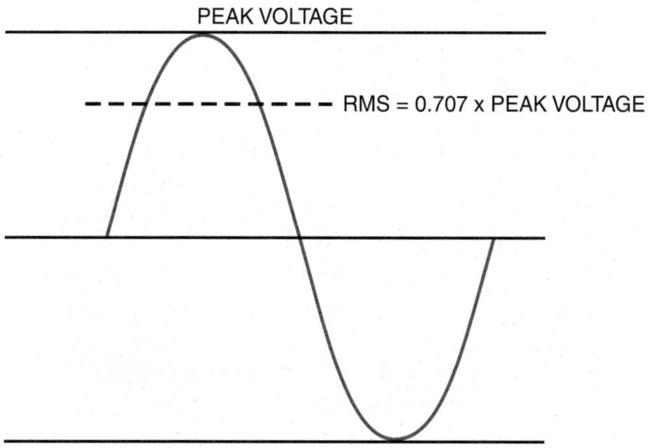

Figure 12.64 This sine wave represents both the voltage and current in phase with each other in a pure resistive circuit.

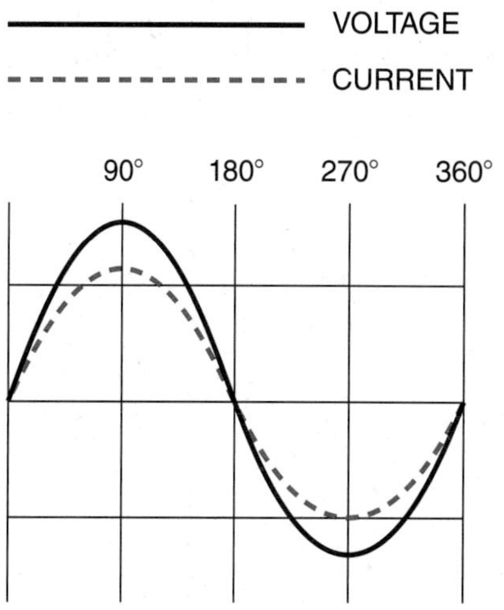

Figure 12.65 This sine wave represents the current lagging the voltage (out of phase) by 90 degrees in an inductive circuit.

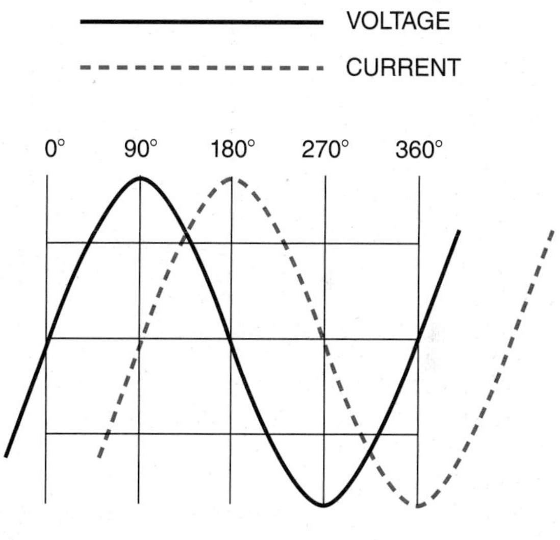

Sine waves can illustrate a cycle of an AC electrical circuit that contains only a pure resistance, for example, a circuit with electrical heaters. In a pure resistive circuit such as this, the voltage and current will be in phase, **Figure 12.64**; notice that the voltage and current reach their negative and positive peaks at the same time.

Sine waves can also illustrate a cycle of an AC electrical circuit that contains a fan relay coil, which will produce an inductive reactance. The sine wave will show the current lagging the voltage in this circuit, **Figure 12.65**.

Figure 12.66 shows a sine wave illustrating an AC cycle of a purely capacitive circuit. In this case the current leads the voltage.

Figure 12.66 This sine wave represents a capacitive AC cycle with the voltage lagging the current by 90 degrees. The voltage and current are out of phase.

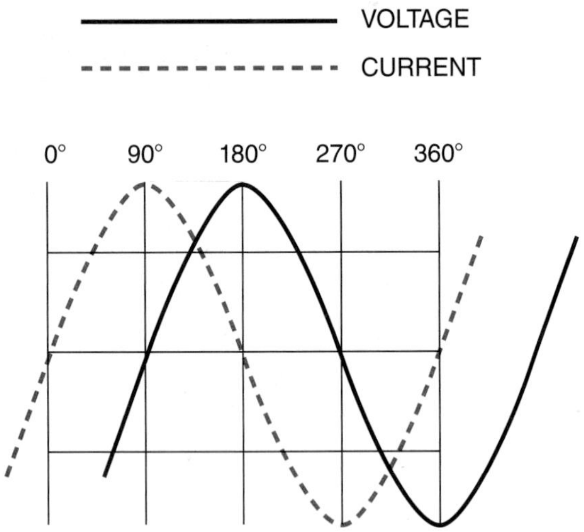

12.21 WIRE SIZES

All conductors have at least some resistance. The resistance depends on the conductor material, the cross-sectional area of the conductor, its temperature, and its length. A conductor with low resistance allows more current to flow through it than a conductor with high resistance.

The proper wire (conductor) size must always be used. The size of a wire is determined by its diameter. A large-diameter wire has lower resistance and has more current-carrying capacity than a small-diameter wire. **SAFETY PRECAUTION:** *If a wire is too small for the current passing through, it will overheat and possibly burn the insulation and cause a fire.* Standard copper wire sizes are identified by American Standard Wire Gauge (AWG) numbers and measured in circular mils. (A circular mil is the area of a circle 1/1000 in. in diameter.) Temperature is also considered because resistance increases as temperature increases. Increasing wire-size numbers indicate *smaller* wire diameters and greater resistance. For example, number 12 wire is smaller than number 10 wire and has less current-carrying capacity.

The technician should not select the size of a conductor nor install the wire unless licensed to do so. The technician should, however, be able to recognize an undersized conductor and bring it to the attention of a qualified person. As mentioned previously, an undersized wire may cause voltage to drop, breakers or fuses to trip, and conductors to overheat.

The conductors are sized by their amperage-carrying capacity. This is called ampacity. **Figure 12.67** contains a small part of a chart from the *National Electrical Code*® (*NEC*®) and a partial footnote for one type of conductor. As shown, this chart and the footnote should not be used to determine a wire size; it is included here only to familiarize you with the way in which it is presented in the *NEC*. The footnote actually reduces the amount of amperes for the number 12 and number 14 wire listed in the *NEC* table in **Figure 12.67**. The reason is that number 12 and number 14 wire are used in residential houses where circuits are often overloaded unintentionally by the homeowner. The footnote exception simply adds more protection. The following discussion is an example of a procedure that might be used for calculating the wire size for the outdoor unit of a heat pump, **Figure 12.68**.

Figure 12.67 This section of the *National Electrical Code*® shows an example of how a wire is sized for only one type of conductor. It is not the complete and official position, but a representative example. Courtesy *National Electrical Code*®

WIRE SIZE	AMPACITY
AWG or kcmil	60°C (140°F)
	Types TW, UF
COPPER	
18	—
16	—
14*	20
12*	25
10*	30
8	40
6	55
4	70
3	85
2	95
1	110

*Small Conductors. Unless specifically permitted in (e) through (g), the overcurrent protection shall not exceed 15 amperes for No. 14, 20 amperes for No. 12, and 30 amperes for No. 10 copper.

The outdoor unit is a 3½-ton unit (the 42A in the model number means 42,000 Btu/h or 3½ tons). The electrical data show that the unit compressor draws 19.2 FLA (full-load amperage) and the fan motor draws 1.4 FLA, for a total of 20.6 FLA. The specifications round this up to an ampacity of 25. In other words, the conductor must be sized for 25 amperes. According to the *NEC*, the wire size would be 10. The specifications for the unit indicate that a maximum fuse or breaker size would be 40 A. This difference allows for the compressor's locked-rotor amperage (LRA) draw at start-up. If the circuit were wired and protected to accommodate only 10 A of current, the circuit breaker would trip every time the compressor started. Some manufacturers state the ampacity and the recommended wire size in the directions or print this information on the unit's nameplate.

Figure 12.68 This specification sheet for one unit shows the electrical data that are used to size the wire to the unit. Courtesy Trane

General Data **4TWX4042A1000A**

OUTDOOR UNIT ①②	4TWX4042A1000A
SOUND RATING (DECIBELS) ②⑨	77/75
POWER CONNS. — V/PH/HZ ③	208/230/1/60
MIN. BRCH. CIR. AMPACITY	25
BR. CIR. PROT. RTG. MAX. (AMPS)	40
BR. CIR. PROT. RTG. MIN. (AMPS)	40
COMPRESSOR	CLIMATUFF® -SCROLL
NO. USED - NO. SPEEDS	1 - 1
VOLTS/PH/HZ	208/230/1/60
R.L. AMPS ⑦ - L.R. AMPS	19.2 - 104
FACTORY INSTALLED	
START COMPONENTS ⑧	NO
INSULATION/SOUND BLANKET	YES
COMPRESSOR HEAT	YES
OUTDOOR FAN — TYPE	PROPELLER
DIA. (IN.) - NO. USED	27.6 - 1
TYPE DRIVE - NO. SPEEDS	DIRECT - 2
CFM @ 0.0 IN. W.G. ④	4200
NO. MOTORS - HP	1 - 1/6
MOTOR SPEED R.P.M.	825
VOLTS/PH/HZ	200/230
F.L. AMPS	1.4
OUTDOOR COIL — TYPE	SPINE FIN™
ROWS - F.P.I.	1 - 24
FACE AREA (SQ. FT.)	27.81
TUBE SIZE (IN.)	5/16
REFRIGERANT CONTROL	EXPANSION VALVE
REFRIGERANT	
LBS. — R-410A (O.D. UNIT) ⑤	7/04 -LB/OZ
FACTORY SUPPLIED	YES
LINE SIZE - IN. O.D. GAS ⑥	3/4
LINE SIZE - IN. O.D. LIQ. ⑥	3/8
FCCV	
RESTRICTOR ORIFICE SIZE	0.071
DIMENSIONS	H X W X D
OUTDOOR UNIT - CRATED (IN.)	53.4 X 35.1 X 38.7
UNCRATED	SEE OUTLINE DWG.
WEIGHT	
SHIPPING (LBS.)	315
NET (LBS.)	267

① **CERTIFIED IN ACCORDANCE WITH THE AIR-SOURCE UNITARY HEAT PUMP EQUIPMENT CERTIFICATION PROGRAM WHICH IS BASED ON A.R.I. STANDARD 210/240.**
② **RATED IN ACCORDANCE WITH A.R.I. STANDARD 270.**
③ CALCULATED IN ACCORDANCE WITH NATIONAL ELECTRIC CODE. ONLY USE HACR CIRCUIT BREAKERS OR FUSES.
④ STANDARD AIR - DRY COIL - OUTDOOR
⑤ THIS VALUE APPROXIMATE. FOR MORE PRECISE VALUE SEE UNIT NAMEPLATE AND SERVICE INSTRUCTION.
⑥ MAX. LINEAR LENGTH: 80 FT WITH RECIPROCATING COMPRESSOR - 60 FT WITH SCROLL. MAX. LIFT - SUCTION 60 FT; MAX LIFT - LIQUID 60 FT. FOR GREATER LENGTH REFER TO REFRIGERANT PIPING SOFTWARE PUB. NO. 32-3312-01.
⑦ THE VALUE SHOWN FOR COMPRESSOR RLA ON THE UNIT NAMEPLATE AND ON THIS SPECIFICATION SHEET IS USED TO COMPUTE MINIMUM BRANCH CIRCUIT AMPACITY AND MAXIMUM FUSE SIZE. THE VALUE SHOWN IS THE BRANCH CIRCUIT SELECTION CURRENT.
⑧ NO MEANS NO START COMPONENTS
YES MEANS QUICK START KIT COMPONENTS
PTC MEANS POSITIVE TEMPERATURE COEFFICIENT STARTER.
⑨ RATED IN ACCORDANCE WITH ARI STANDARD 270/SECTION 5.3.6.

SPLIT SYSTEM

c UL US

OUTDOOR UNIT WITH HEAT PUMP COILS

12.22 CIRCUIT PROTECTION DEVICES

⚠ **SAFETY PRECAUTION:** *Electric circuits must be protected from current overloads. If too much current flows through the circuit, the wires and components will overheat, resulting in damage and possible fire. Circuits are normally protected with fuses or circuit breakers.*•

FUSES

A fuse is a simple device used to protect circuits from overloading and overheating. Most fuses contain a strip of metal that has a higher resistance than the conductors in the circuit. This strip also has a relatively low melting point. Because of its higher resistance, it will heat up faster than the conductor. When the current exceeds the rating on the fuse, the strip melts and opens the circuit. Fuses are one-time devices and must be replaced when the strip or element melts. The cause for the circuit overload must be identified and corrected before replacing the fuse.

PLUG FUSES. Plug fuses have either an Edison base or a type S base, **Figure 12.69(A)**. Edison-base fuses are used in older installations and may be used for replacement only. Type S fuses can be used only in a type S fuse holder specifically designed for the fuse; otherwise, an adapter must be used, **Figure 12.69(B)**. Each adapter is designed for a specific amperage rating, and these fuses cannot be interchanged. The amperage rating determines the size of the adapter. Plug fuses are rated up to 125 V and 30 A.

Figure 12.69 (A) A type S base plug fuse. (B) A type S fuse adapter. Reprinted with permission by Bussman Division, McGraw-Edison Company

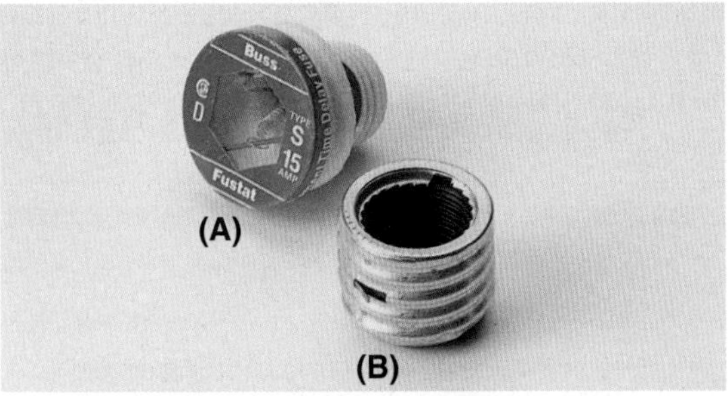

Figure 12.70 A dual-element plug fuse. Courtesy Cooper Bussman Division

DUAL-ELEMENT PLUG FUSES. Many circuits have electric motors as the load or part of the load. Motors draw more current when starting and can cause a plain (single-element) fuse to burn out or open the circuit. Dual-element fuses are frequently used in this situation, **Figure 12.70**. One element in the fuse will melt when there is a large overload, such as a short circuit. The other element will melt and open the circuit when there is a smaller current overload lasting more than a few seconds. This allows for the higher starting current of an electric motor.

CARTRIDGE FUSES. The ferrule cartridge fuse is used for 230-V to 600-V service up to 60 A, **Figure 12.71(A)**. Knife-blade cartridge fuses can be used, **Figure 12.71(B)**, from 60 A to 600 A. A cartridge fuse is sized according to its amperage rating to prevent the use of a fuse with an inadequate rating. Many cartridge fuses have an arc-quenching material around the element to prevent damage from arcing in severe short-circuit situations, **Figure 12.72**.

CIRCUIT BREAKERS

A circuit breaker can function as a switch as well as a means for opening a circuit when a current overload occurs. Most modern installations in houses and many commercial and industrial installations use circuit breakers rather than fuses for circuit protection. Circuit breakers protect the circuit in two ways. One is a bimetal strip that heats up with a current overload and trips the breaker, opening the circuit. The other

Figure 12.71 (A) A ferrule-type cartridge fuse. (B) A knife-blade cartridge fuse. Courtesy Cooper Bussman Division

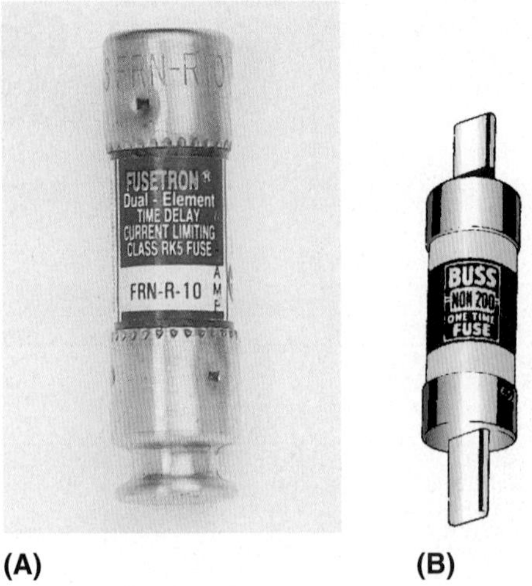

Figure 12.72 A knife-blade cartridge fuse with arc-quenching material.

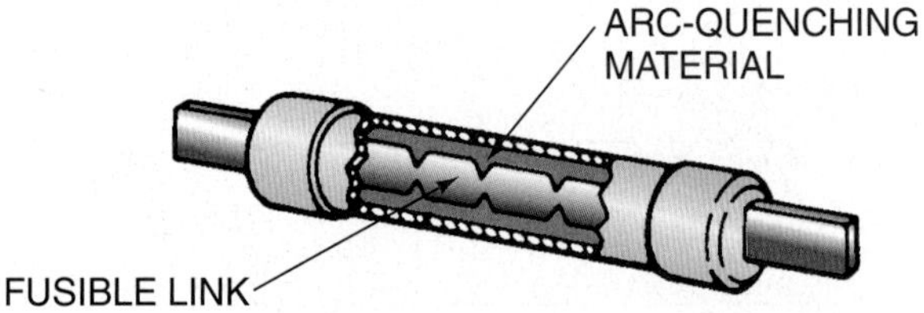

is a magnetic coil that causes the breaker to trip and open the circuit when there is a short circuit or other excessive current overload over a short period of time. Circuit breakers that break a single power line are referred as single-pole breakers, while two-pole breakers control two lines, as in the case of a 220-volt power supply. Three-phase circuits are controlled by three-pole circuit breakers, **Figure 12.73**. Two-pole and three-pole circuit breakers are configured with a common-trip mechanism that causes all controlled power lines in the circuit to open in the event a single power line experiences an overcurrent condition. This common-trip feature, **Figure 12.74**, is often a small metal rod that passes through the switches on the circuit breaker, but can be configured differently, depending on the manufacturer's specific design.

GROUND FAULT CIRCUIT INTERRUPTERS (GFCIs)

⚠ SAFETY PRECAUTION: *Ground fault circuit interrupters, known as GFCIs or simply GFIs, help protect individuals against shock in addition to providing current overload protection. The GFCI,* **Figure 12.75**, *detects even a very small current leak to a ground, which may not be detected by a conventional circuit breaker. Under certain conditions, this leak may cause severe personal injury. The leak will cause the GFCI to open the circuit.•*

Figure 12.73 Circuit breakers. Single-pole circuit breaker, (left). Two-pole circuit breaker, (center). Three-pole circuit breaker, (right). Photo by Jason Obrzut

Figure 12.74 A two-pole circuit breaker with a common trip. Photo by Jason Obrzut

Figure 12.75 A ground fault circuit interrupter. Photo by Jason Obrzut

A GFCI senses the difference between the current flowing into a circuit and the current flowing back from the circuit. If these two current values are the same, the GFCI will continue to allow current to flow. If, however, the GFCI senses an imbalance between the two current values being compared, the GFCI will trip. Current differences as small as 20 milliamps, or 0.02 A, will cause the GFCI to interrupt current flow.

Circuit protection is essential to avoid overloading the conductors in the circuit. If one of the circuit power-consuming devices were to cause an overload due to a short circuit, the circuit protector would stop the current flow before the conductor became overloaded and hot. Remember, a circuit consists of a power supply, the conductor, and the power-consuming device. The conductor must be sized large enough that it does not operate beyond its rated temperature, typically 140°F (60°C) while in an ambient of 86°F (30°C). For example, a circuit may be designed to carry a load of 20 A. As long as the circuit is drawing less than this amperage, overheating is not a potential hazard. Proper understanding of circuit protection is a lengthy process. More details can be obtained from the *NEC* and from further study of electricity.

12.23 SEMICONDUCTORS

Technological advances in the fields of semiconductors and solid-state component design have brought about major changes in the electrical devices and controls found in HVACR systems. Semiconductor-based components are generally small and lightweight and can be mounted in circuit boards, **Figure 12.76**. In this section we describe some of the individual solid-state devices and some of their uses. HVACR technicians do not replace solid-state components on circuit boards, but they should have some knowledge of these components and should be able to determine when one or more

Figure 12.76 A circuit board with semiconductors. Photo by Bill Johnson

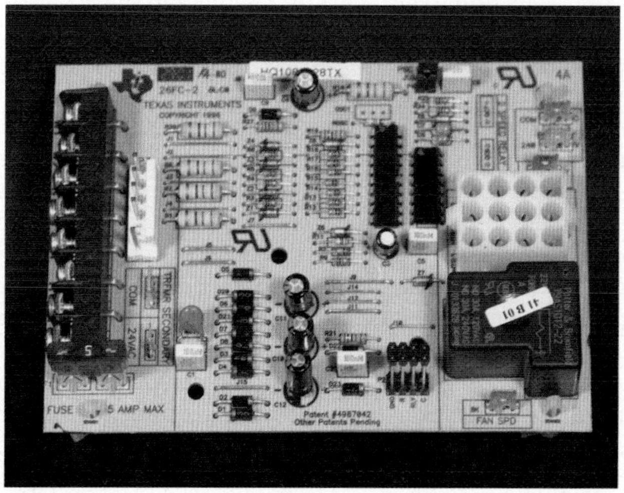

Figure 12.77 A representation of a silicon crystal. Each silicon atom has 4 valence electrons that are shared with the surrounding atoms. The surrounding atoms also share their electrons to nearby silicon atoms. Notice that the center silicon atoms have 8 electrons, indicated by red circles, in the electron orbit.

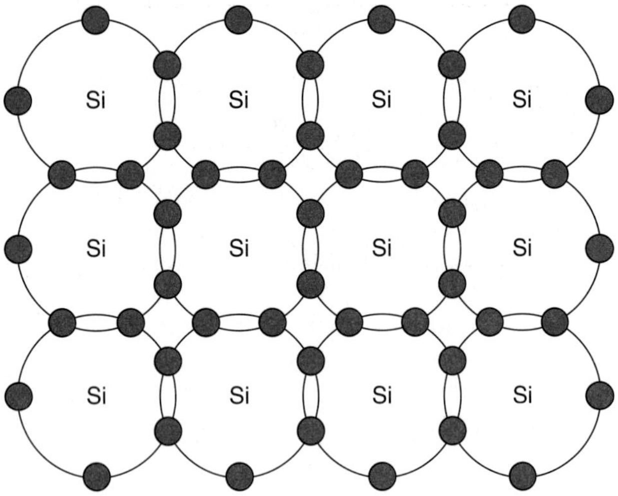

of the circuits in which they are used are defective. In most cases, when a component is defective, the entire board will need to be replaced. Most manufacturers have incorporated troubleshooting algorithms in their control modules and circuitry, making it easier for the technician to determine whether or not the circuit board is functioning correctly.

Common semiconductor materials used to create the components found on circuit boards are silicon and germanium. In their pure form, semiconductors, as their name implies, do not conduct electricity very well. This is because pure silicon and germanium atoms each have four valence electrons and bond together to form crystals that hold onto their electrons tightly, **Figure 12.77**. Each silicon atom shares its four valence electrons with its neighboring atoms, while the neighboring atoms do the same. This allows each atom to have, in effect, a full outermost electron shell. To be of value, however, semiconductors must conduct electricity in some controlled manner. To accomplish this, impurities are added to the crystal-like structures of the silicon or germanium. This process is called *doping*. One type of doping involves adding small amounts of boron or gallium to the silicon. Since boron and gallium atoms each have three valence electrons, the addition of this impurity creates a hole in the material where an electron should be, **Figure 12.78**. Because the hole replaces an electron (which has a negative charge), the material has fewer electrons, or a net result of a positive charge. This is called *P-type* material. A different type of doping involves adding small amounts of phosphorus or arsenic to the silicon. Since phosphorous and arsenic atoms each have five valence electrons, the addition of this impurity creates an excess of electrons, giving the semiconductor a negative charge, **Figure 12.79**. This creates what is referred to as N-type material. When a voltage is applied to a P-type material, electrons fill the holes and move from one hole to the next, still moving from negative to positive. This makes it appear that the holes are moving in the opposite direction (from positive to negative) as the electrons move from hole to hole. N-type material has

Figure 12.78 Silicon doped with boron, which has three valence electrons, indicated by the blue circles. Notice the hole that is present, creating P-type material.

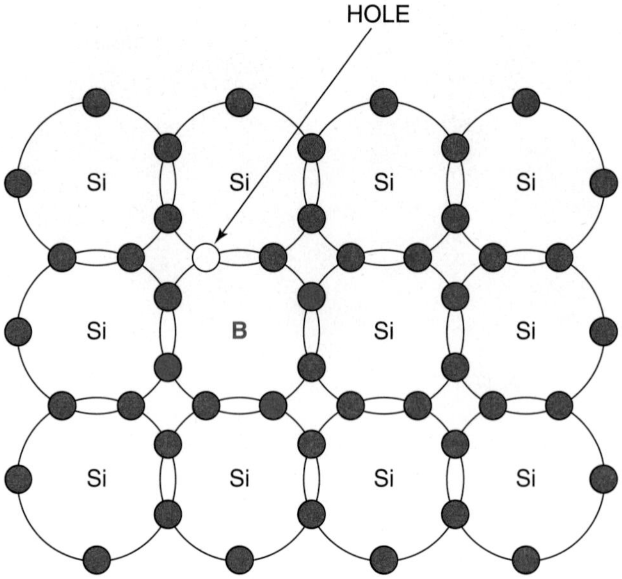

Figure 12.79 Silicon doped with phosphorous, which has five valence electrons, indicated by the green circles. Notice the extra electron that is present, creating N-type material.

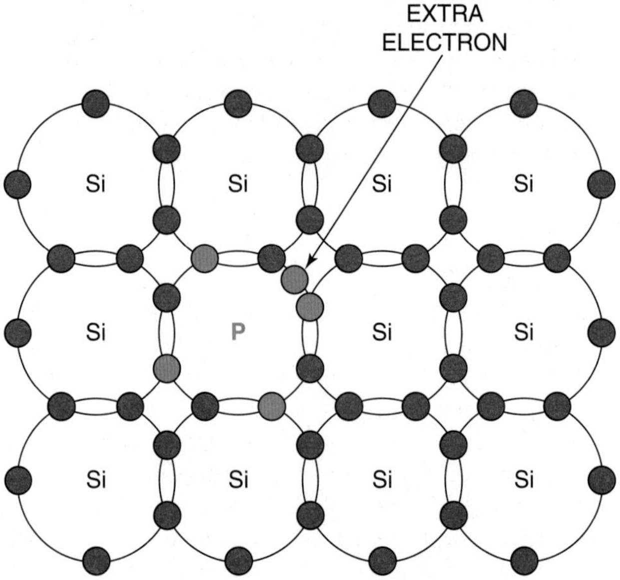

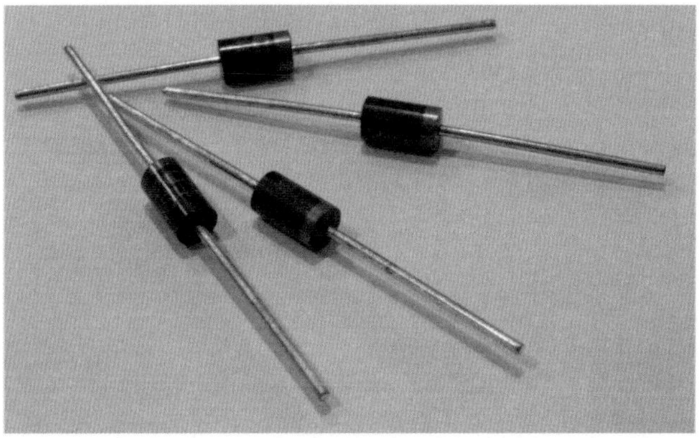

Figure 12.80 Diodes. Photo by Eugene Silberstein

Figure 12.81 A pictorial diagram of a diode.

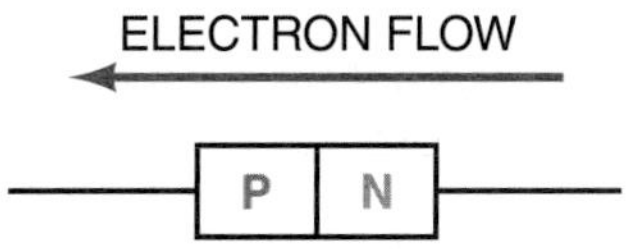

an excess of electrons that move from negative to positive when a voltage is applied.

Solid-state components are made from various combinations of N-type and P-type substances. The manner in which the materials are joined together, the thickness of the materials, and other factors determine the type of solid-state component and its electronic characteristics.

DIODES

Diodes, **Figure 12.80**, are simple solid-state devices. They consist of P- and N-type material connected together. When this combination of P- and N-type material is connected to a power source one way, it will allow current to flow and is said to have forward bias. When reversed, it is said to have reverse bias and no current will flow. **Figure 12.81** is a drawing of a simple diode. One of the connections on the diode is called the cathode and the other the anode, **Figure 12.82**. The cathode is the N-type material side of the diode, while the anode is the P-type material side of the diode. When a diode acts as a closed switch and conducts electricity, the device is said to be forward-biased. When a diode acts as an open switch and does not conduct electricity, the device is said to be reverse-biased. Therefore, if a diode is to be connected to a battery so as to have forward bias (current flow), the negative terminal on the battery should be connected to the cathode, **Figure 12.83**. Connecting the negative terminal of the battery to the anode will produce reverse bias (no current flow), **Figure 12.84**.

Figure 12.82 (A) The schematic symbol for a diode. (B) Identifying markings on the diode.

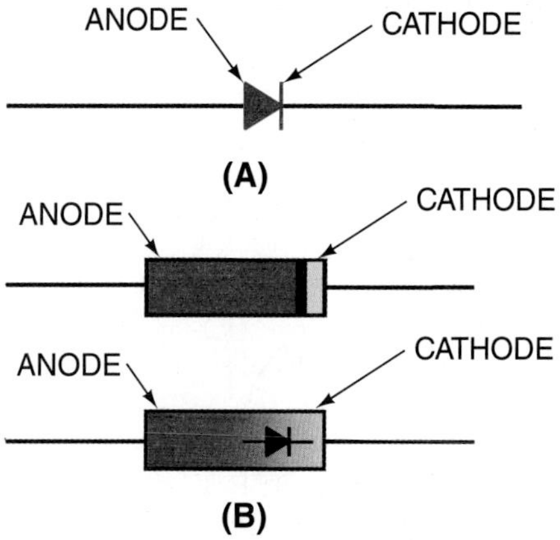

Figure 12.83 A simple diagram with a diode indicating forward bias.

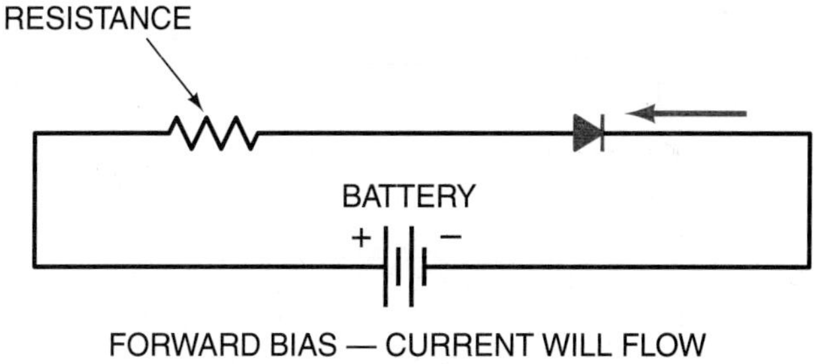

Figure 12.84 A circuit with a diode indicating reverse bias.

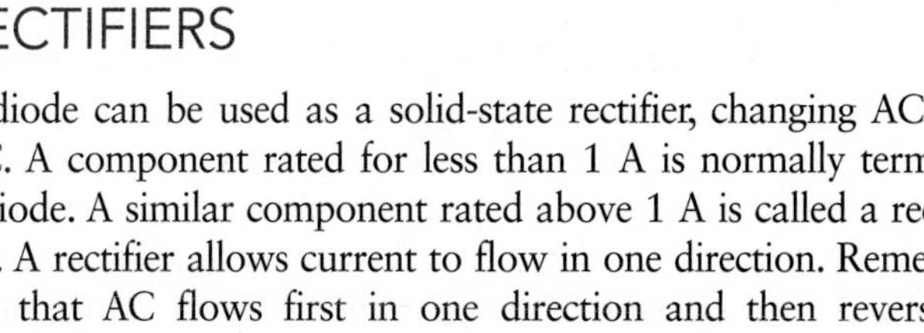

RECTIFIERS

A diode can be used as a solid-state rectifier, changing AC to DC. A component rated for less than 1 A is normally termed a diode. A similar component rated above 1 A is called a rectifier. A rectifier allows current to flow in one direction. Remember that AC flows first in one direction and then reverses, **Figure 12.85(A)**. The rectifier allows the AC to flow in one direction but blocks it from reversing, **Figure 12.85(B)**. Therefore, the output of a rectifier circuit is in one direction, or direct current. This is called a half-wave rectifier because it allows only that part of the AC moving in one direction to pass through. **Figure 12.86** illustrates a rectifier circuit. Waveform "A" in

Figure 12.85 (A) A full-wave AC waveform. (B) A half-wave DC waveform.

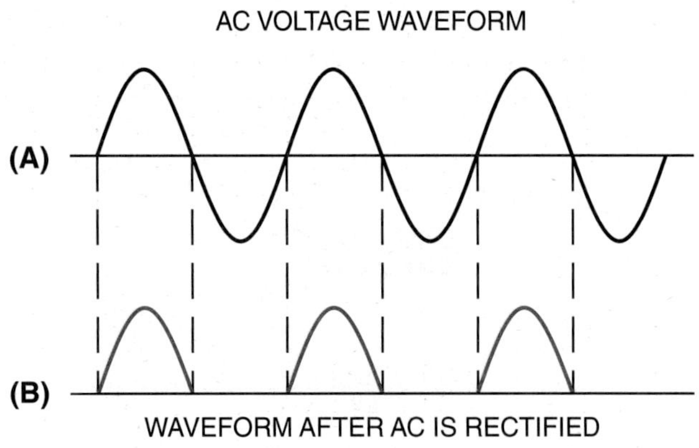

Figure 12.86 A diode rectifier circuit.

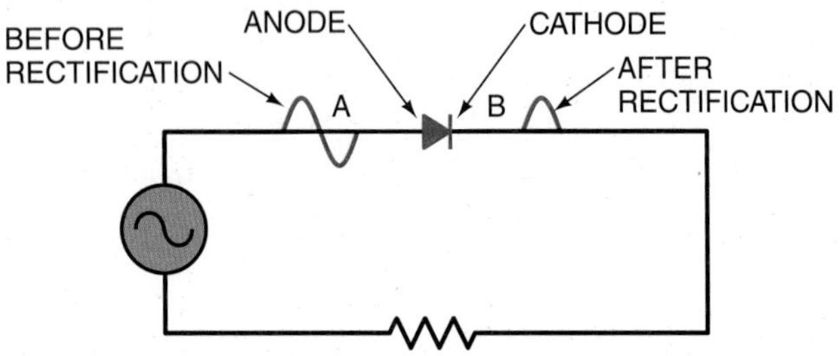

Figure 12.87 A full-wave bridge rectifier.

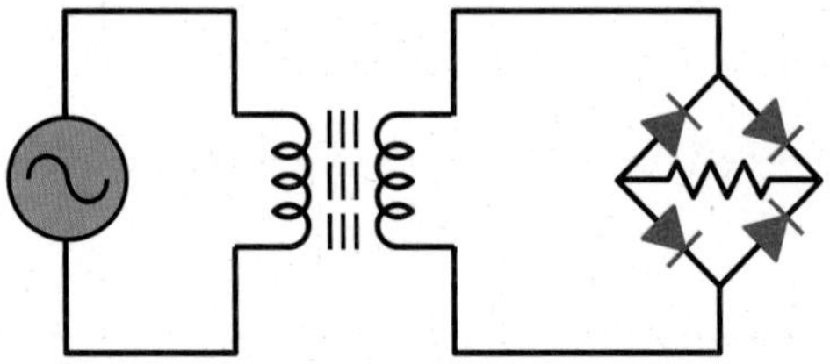

Figure 12.88 (A) A full-wave AC waveform. (B) A full-wave DC waveform.

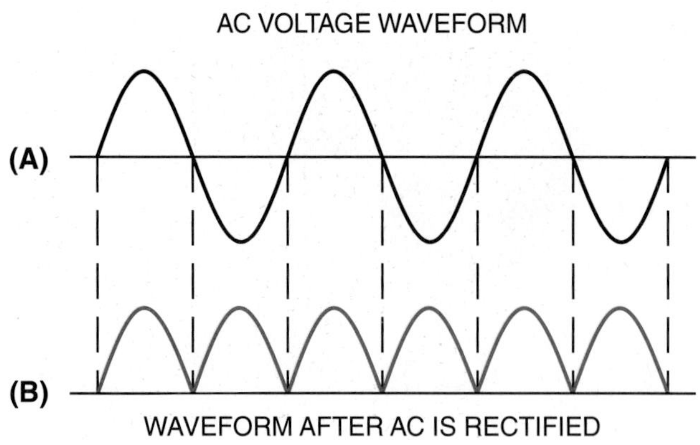

Figure 12.89 (A) Pictorial and (B) schematic drawings of a silicon-controlled rectifier.

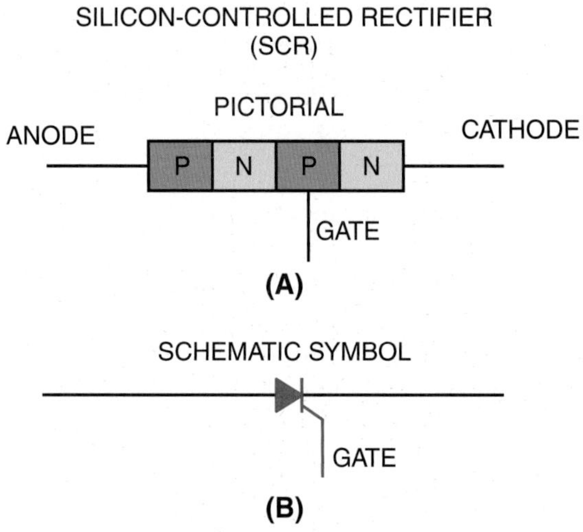

Figure 12.86 shows the AC before it is rectified and Waveform "B" in **Figure 12.86** shows the AC after it is rectified. Full-wave rectification can be achieved by using a more complicated circuit such as the one in **Figure 12.87**. This is a full-wave bridge rectifier, which is commonly used in the industry. Unlike the half-wave rectifier, which acts as an open switch and does not allow current to flow for half of the cycle, the full-wave rectifier allows current to flow for the entire cycle. Instead of "turning off" during the negative portion of the cycle, the full-wave rectifier takes the negative portion of the cycle and flips it to the positive side, **Figure 12.88**.

SILICON-CONTROLLED RECTIFIER. Silicon-controlled rectifiers (SCRs) consist of four sections of semiconductor material bonded together. These form a PNPN junction, **Figure 12.89(A)**. **Figure 12.89(B)** illustrates the schematic symbol for an SCR. Notice that the schematic is similar to that for the diode except for the gate. The gate is the control for the SCR. The SCR is used to control devices that may use large amounts of power, for example, to control the speed of motors or to control the brightness of lights.

TRANSISTORS

Transistors are made of three pieces of semiconductor materials sandwiched together. Transistors are either of the NPN or PNP type. An NPN transistor is made up of two sections of N-type material and one section of P-type material. A PNP transistor is made up of two sections of P-type material and one section of N-type material. **Figure 12.90** shows diagrams and **Figure 12.91** shows the

Figure 12.90 Drawings of NPN and PNP transistors.

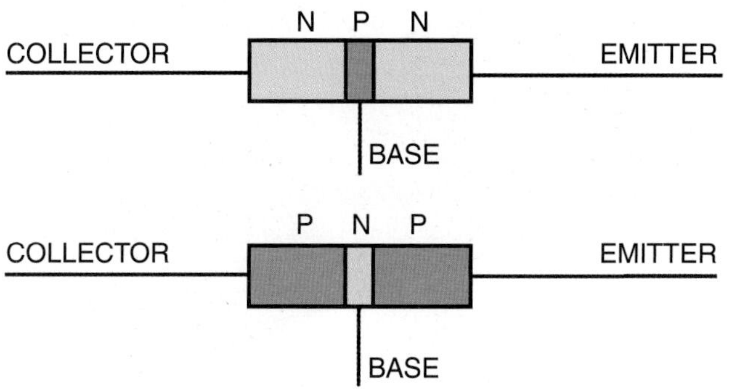

Figure 12.91 Schematic drawings of NPN and PNP transistors.

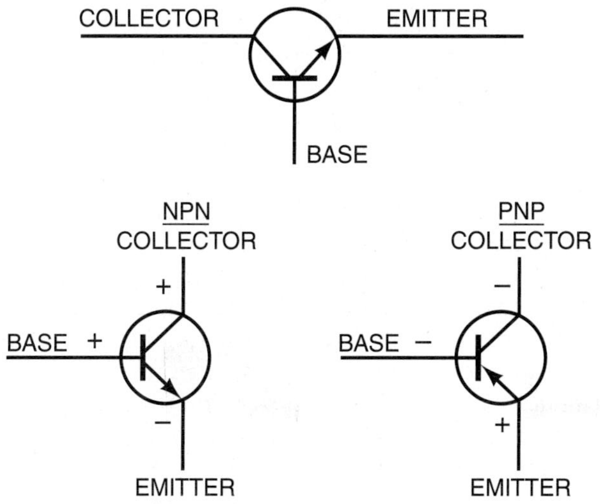

schematic symbols for the two types. As the symbols show, each transistor has a base, a collector, and an emitter. In the NPN type, the collector and the base are connected to the positive; the emitter is connected to the negative. The PNP transistor has a negative base and collector connection and a positive emitter. The base must be connected to the same polarity as the collector to provide forward bias.

The transistor may be used as a switch or as a device to amplify an electrical signal. One application of a transistor in an air-conditioning control circuit would be to amplify a small signal to provide enough current to operate a switch or relay. Current will flow through the base-emitter and through the collector-emitter. The base-emitter current is the control, and the collector-emitter current produces the action. A very small current passing through the base emitter may allow a much larger current to pass through the collector-emitter junction. A small increase in the base-emitter junction can allow a much larger increase in the current flow through the collector-emitter.

THERMISTORS

A thermistor is a type of resistor that is sensitive to temperature, **Figure 12.92**. The resistance of a thermistor changes with a change in temperature. A positive temperature coefficient (PTC) thermistor causes the resistance of the thermistor to increase when the temperature increases. A negative temperature coefficient (NTC) thermistor causes the resistance to decrease with an increase in temperature. **Figure 12.93** illustrates the schematic symbol for a thermistor.

One thermistor application is to provide motor overload protection. The thermistor is embedded in the windings of a motor. When the winding temperature exceeds a predetermined amount, the thermistor resistance changes. This change in resistance is detected by an electronic circuit that causes the motor circuit to open. Another application is providing start assistance in a PSC (permanent split-capacitor) electric motor. This thermistor, a positive temperature coefficient device, allows full voltage to reach the start windings during start-up of the motor. The thermistor heats during start-up and creates resistance, turning off power to the start winding at the appropriate time. This does not give the motor the starting torque that a start capacitor does, but it is advantageous in some applications because of its simple construction and lack of moving parts.

Figure 12.92 A thermistor. Photo by Eugene Silberstein

Figure 12.93 The schematic symbol for a thermistor.

Figure 12.94 Schematic symbols for a diac.

Figure 12.95 A simple diac circuit.

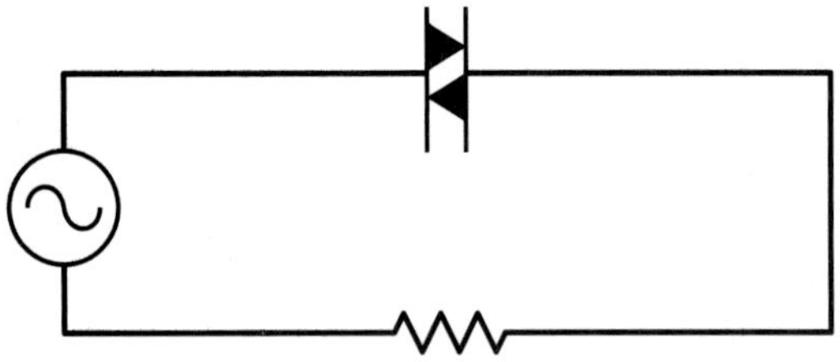

Figure 12.96 The schematic symbol for a triac.

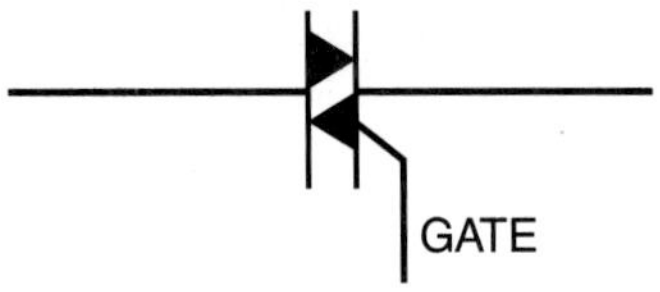

DIACS

The diac is a two-directional electronic device that can operate in an AC circuit; its output is AC. It is a voltage-sensitive switch that operates in both halves of the AC waveform. When a voltage is applied, it will not conduct (act as an open switch) until the voltage reaches a predetermined level. Assuming this predetermined level to be 24 V, when the voltage in the circuit reaches 24 V, the diac will begin to conduct or fire. Once it fires it will continue to conduct even at a lower voltage. Diacs are designed to have a higher cut-in voltage and lower cut-out voltage. If the cut-in voltage is 24 V, and assuming that the cut-out voltage is 12 V, in this case the diac will continue to operate until the voltage drops below 12 V—at which time it will cut off. **Figure 12.94** shows two schematic symbols for a diac. **Figure 12.95** illustrates a diac in a simple AC circuit. Diacs are often used as switching or control devices for triacs.

TRIACS

A triac is a switching device that will conduct on both halves of the AC waveform; its output is AC. **Figure 12.96** shows the schematic symbol for a triac. Notice that it is similar to the diac but has a gate lead. A pulse supplied to the gate lead can cause the triac to fire or to conduct. Triacs were developed to provide better AC switching. As mentioned previously, diacs often provide the pulse to the gate of the triac. A common use of a triac is as a motor speed control for AC motors.

Figure 12.97 A heat sink.

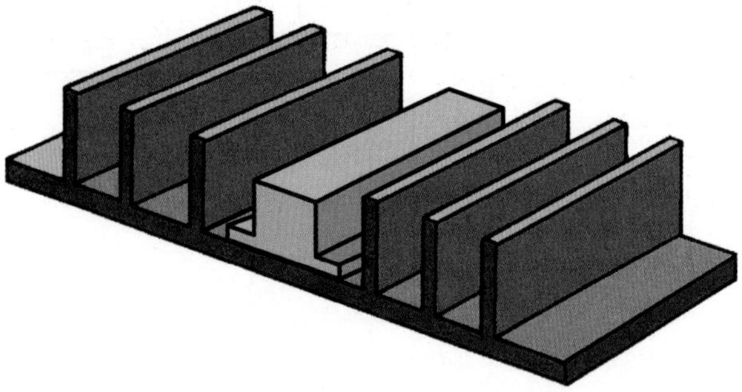

HEAT SINKS

Some solid-state devices may appear a little different from others of the same type. This is due to the differing voltages and currents they are rated to carry and the purposes for which they are designed and used. Some produce much more heat than others and will operate only at the rating specified if kept within a certain temperature range. Heat that could change the operation of the device or destroy it must be dissipated. This is done by adhering the solid-state component to an object called a *heat sink* that has a much greater surface area, **Figure 12.97**. Heat will travel from the device to the object with the large surface area, allowing the excess heat to be dissipated into the surrounding air.

PRACTICAL SEMICONDUCTOR APPLICATIONS FOR TECHNICIANS

This has been a short introduction to semiconductors. We have provided enough information so that you will have some idea of the purpose of these control devices and encourage you to pursue the study of these components because solid-state electronics are being used extensively to control the systems you will be working with. Each manufacturer has a control sequence procedure that you must use to successfully check their controls. Make it a practice to attend seminars and factory schools in your area to increase your knowledge of all segments in this ever-changing field. Unit 15, "Troubleshooting Basic Controls," contains many exercises that involve reading electrical schematic drawings.

REVIEW QUESTIONS

1. The ________ is that part of an atom that moves from one atom to another.
   A. electron C. positron
   B. proton D. neutron
2. The nucleus of an atom contains which particles?
   A. Electrons and neutrons
   B. Positrons and protons
   C. Electrons and positrons
   D. Neutrons and protons
3. If an atom has a net positive charge, which of the following is correct?
   A. The atom has lost protons.
   B. The atom has lost electrons.
   C. The atom has gained electrons.
   D. The atom has gained protons.
4. The law of charges states that
   A. like charges attract each other and unlike charges repel each other.
   B. like charges repel each other and unlike charges attract each other.
   C. like charges repel each other and unlike charges repel each other too.
   D. like charges attract each other and unlike charges attract each other too.
5. All of the following statements are true regarding good conductors of electricity except
   A. good conductors of electricity are also good conductors of heat.
   B. good conductors of electricity pass protons from atom to atom.
   C. good conductors of electricity have atoms with few electrons in the outermost shell.
   D. good conductors of electricity are typically metallic elements.
6. Which of the following is a correct formula for the total resistance in a series circuit?
   A. $R_{total} = R_1 \times R_2 \times R_3 \times \dots$
   B. $R_{total} = R_1 + R_2 + R_3 + \dots$
   C. $\mathrm{R}_{total} = R_1 - R_2 - R_3 - \dots$
   D. $R_{total} = R_1 \div R_2 \div R_3 \div \dots$
7. If three resistances of values 3 Ω, 4 Ω, and 12 Ω are wired in parallel with each other, what is the total resistance of the circuit?
   A. 1.5 Ω C. 12 Ω
   B. 3 Ω D. 19 Ω
8. If three resistances of values 3 Ω, 4 Ω, and 12 Ω are wired in series with each other, what is the total resistance of the circuit?
   A. 1.5 Ω C. 12 Ω
   B. 3 Ω D. 19 Ω
9. Which of the following represents Ohm's law?
   A. $E = I \div R$
   B. $E = R \div I$
   C. $\mathrm{R} = E \div I$
   D. $I = E \times R$
10. Which of the following represents an accurate formula for power?
    A. $P = I \div R$
    B. $P = R \times I$
    C. $P = E \div I$
    D. $P = E \times I$
11. All of the following are required in an electric circuit except a
    A. power source.
    B. switch.
    C. complete conductive path.
    D. load.
12. All of the following are true regarding DC power except
    A. DC power can be created from AC power.
    B. a common source of DC power is batteries.
    C. DC power continually reverses its polarity.
    D. the term DC stands for direct current.
13. A residential AC power supply is often represented by a
    A. horizontal line.
    B. sine wave.
    C. triangular-shaped wave.
    D. set of two parallel lines.
14. If there were a current flowing of 5 A in a 120-V circuit, what would the resistance be?
    A. 25 Ω C. 600 Ω
    B. 24 Ω D. 624 Ω
15. If the resistance in a 120-V circuit was 40 Ω, what would the current be in amperes?
    A. 4800 A C. 4 A
    B. 48 A D. 3 A
16. Electrical power is measured in
    A. amperes.
    B. volts.
    C. watts.
    D. ohms.
17. The unit of measurement for the charge a capacitor can store is the
    A. volt.
    B. microfarad.
    C. ohm.
    D. joule.
18. Two common types of capacitors are the
    A. run and stop.
    B. start and run.
    C. short and start.
    D. stop and short.

19. The most common method used to measure the amperage in a circuit is to
    A. deenergize the circuit and use a clamp-on ammeter to clamp around one conductor in the circuit.
    B. deenergize the circuit and use a clamp-on ammeter to clamp around both conductors in the circuit.
    C. energize the circuit and use a clamp-on ammeter to clamp around one conductor in the circuit.
    D. energize the circuit and use a clamp-on ammeter to clamp around both conductors in the circuit.
20. Which of the following statements can best be associated with inductance?
    A. A magnetic field is generated when current flows through a conductor.
    B. A magnetic field is generated when voltage flows through a conductor.
    C. A voltage is generated when current flows through a conductor.
    D. A voltage is generated when power flows through a conductor.
21. A solenoid can be described as
    A. the electrical contacts on a switch.
    B. a coil of wire.
    C. an energy-storing device.
    D. a power-passing device.
22. Voltage can be produced when a conductor passes through a(n)
    A. vacuum.
    B. magnetic field.
    C. dielectric.
    D. insulator.
23. The opposition to current flow in a circuit is commonly referred to as
    A. capacitance.
    B. inductance.
    C. resistance.
    D. reactance.
24. Impedance is the total effect of what three electrical characteristics?
    A. Resistance, inductive reactance, and capacitive reactance
    B. Resistance, voltage, and current
    C. Power, inductive reactance, and capacitive reactance
    D. Capacitance, reactance, and inductance
25. If a step-down transformer has windings with 1000 and 500 turns, what is the induced voltage at the secondary winding if 100 volts is being applied to the primary winding?
    A. 200 volts
    B. 100 volts
    C. 50 volts
    D. 25 volts
26. If a step-up transformer has windings with 1000 and 500 turns, what is the induced voltage at the secondary winding if 100 volts is being applied to the primary winding?
    A. 200 volts
    B. 100 volts
    C. 50 volts
    D. 25 volts
27. Referring to Figure 12.67 in the text, what size (gauge) wire should be used if the expected amperage draw of a circuit is 60 amperes?
    A. 4
    B. 6
    C. 8
    D. 10
28. Undersized wires can cause
    A. voltage surges, tripped circuit breakers, and overheating.
    B. voltage surges, decreased resistance, and overheating.
    C. voltage drops, tripped circuit breakers, and overheating.
    D. voltage drops, decreased resistance, and overheating.
29. A diode is a solid-state circuit component that
    A. allows current to flow when reverse-biased.
    B. prevents current flow when forward-biased.
    C. prevents current flow in both directions.
    D. allows current to flow in only one direction.
30. An NPN transistor is made up of
    A. one section of P-type material and one section of N-type material.
    B. one section of P-type material and two sections of N-type material.
    C. two sections of P-type material and two sections of N-type material.
    D. two sections of P-type material and one section of N-type material.
31. What is the main difference between a fuse and a circuit breaker?
    A. Fuses are designed to handle larger loads than circuit breakers.
    B. Fuses must be replaced, while circuit breakers can be reset.
    C. Circuit breakers must be replaced, while fuses can be reset.
    D. Fuses are designed to protect multi-pole circuits.
32. Which of the following statements is true regarding circuit protection devices?
    A. GFCIs are intended for use in wet or potentially wet locations.
    B. Fuses must be rated at an amperage that is lower than the expected current draw of a circuit.
    C. Circuit breakers must be rated at an amperage that is lower than the expected current draw of a circuit.
    D. Fuses and circuit breakers are intended to protect the device being controlled.

UNIT 13

INTRODUCTION TO AUTOMATIC CONTROLS

OBJECTIVES

After studying this unit, you should be able to

- explain the difference between an operational control and a safety control.
- describe the function of a thermostat.
- describe the function of a bimetal device.
- make general comparisons between different bimetal applications.
- describe the rod and tube.
- describe fluid-filled controls.
- describe partial liquid/partial vapor-filled controls.
- distinguish among the bellows, diaphragm, and Bourdon tube.
- explain how thermocouples and thermopiles function.
- describe a thermistor.
- describe three types of pressure transducers.

SAFETYCHECKLIST

✓ Pressure switches are often connected directly to the refrigerant circuit, so removing them could result in the complete loss of refrigerant from the system. Be aware of how these devices are connected and whether or not they can be removed from the system safely without releasing the refrigerant charge.

✓ Some older controls contain mercury, which is toxic. Be sure to follow all local and federal laws and guidelines when discarding devices that contain mercury.

✓ Be sure to understand how a particular control operates before attempting to service or repair it.

✓ Before working on any control system, make certain that the circuit is deenergized whenever possible.

13.1 TYPES OF AUTOMATIC CONTROLS

The HVACR field requires many types of automatic controls to start, stop, and adjust various components in response to the needs of the system. Some items that need to be controlled are the flow of electricity through circuits and the flow of various fluids (liquids and gases) through the piping, duct, and distribution systems. For example, modulating controls are used to vary the speed of a blower motor to increase or decrease the amount of airflow through a duct system.

Controls can be classified as electrical, mechanical, electromechanical, electronic or digital, and pneumatic. Pneumatic controls are discussed in Unit 16. Electrical controls are electrically operated and are typically used to control electrical devices. Mechanical controls are often operated by pressure or temperature and used to control fluid flow. Electromechanical controls are driven by pressure or temperature to provide electrical functions or they are driven by electricity to control fluid flow. Electronic controls use electronic circuits and devices to perform the same functions that electrical and electromechanical controls perform. Electronic, or digital, controls are becoming much more common, even on the smallest of systems.

The automatic control of a system is intended to maintain stable or constant conditions without excessive human interference, while also protecting people and equipment. The system must regulate itself within the design boundaries that have been established. If the equipment is allowed to operate outside of its design boundaries, the system components may become damaged, human comfort may be sacrificed and, in extreme cases, personal injury can result. The main purpose of the HVACR industry is to control space or product conditions by controlling important factors such as temperature, humidity, air quality, air movement, and air cleanliness.

Controls can be classified as either operational devices or safety devices. Operational devices and controls function to maintain the normal operating conditions of a system. An example of an operational control is a cooling thermostat. The thermostat will open and close its contacts during normal system operation to maintain the conditioned space at the desired temperature. When the space temperature rises above the preset temperature, the thermostat will close to energize the cooling system. When the space temperature drops below the preset temperature, the thermostat will open, causing the cooling system to cycle off. Safety devices, on the other hand, are intended to protect the system in the event unsafe conditions arise. Safety devices are typically normally closed and will remain closed as long as the system is operating properly. Safety devices are not intended to open and close during normal system operation.

13.2 DEVICES THAT RESPOND TO THERMAL CHANGE

Automatic controls in the HVACR industry usually provide some method of controlling temperature. Temperature control is used to maintain space or product temperature and to protect equipment from damaging itself. When used to control temperature, the control is called a **thermostat**; when used to protect equipment, it is known as a safety device. A good example of both applications is a household refrigerator. The refrigerator maintains the space temperature in the fresh food section at about 35°F. When food is placed in the box and stored for a long time, it becomes the same temperature as the space. If the space temperature is allowed to go much below 35°F, the food begins to freeze. Foods such as eggs, tomatoes, and lettuce are not edible after being frozen.

The thermostat, **Figure 13.1**, automatically stops and starts the refrigeration cycle to maintain the box at the desired temperature. Without an automatic control such as the thermostat, the appliance owner would have to manually turn the unit on and off in an effort to maintain the refrigerator at the desired temperature.

The compressor in the refrigerator has a protective device that keeps it from overloading and damaging itself, **Figure 13.2**. This device is called an overload. This automatic overload control is designed to open its contacts in the event that an overload or power problem might damage the compressor. One such example would be when the power goes off and comes right back on while the refrigerator is running. The overload control will open its contacts and prevent the compressor from operating until it is ready to go back to work again. Overloading often occurs when the refrigerator's low-torque, fractional-horsepower compressor motor tries to start against unequal system pressures. When this happens, the motor will draw locked rotor amperage (LRA). The overload control would then open the electrical circuit and protect the compressor. After the compressor goes through a cool-down period, the control resets and closes the circuit when, hopefully, system pressures are nearing equalization and the motor can start safely.

Figure 13.1 Refrigerator thermostat. Photo by Bill Johnson

Figure 13.2 Compressor overload device. Courtesy Ferris State University. Photo by John Tomczyk

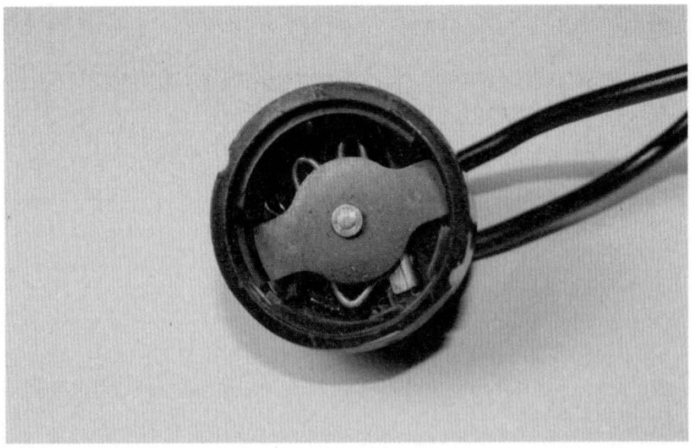

The system overload control may be electrical, thermal, or a combination of both. An electrical overload happens when the electric circuit is drawing more current than it is designed for, due to an electrical problem such as a short circuit, improper electrical connections, or a faulty electrical component. A thermal system overload may occur if a specific electrical or mechanical component is not cooled properly, resulting in overheating. System overloads may also be due to mechanical reasons, causing too much current that in turn produces excess heat. For example, restricted airflow through a condensing coil of a central air conditioner would produce high head pressure, requiring the compressor motor to work harder. **Figure 13.3** shows one combination control, a small heater located under

Figure 13.3 Overload heater. Courtesy Ferris State University. Photo by John Tomczyk

the bimetallic element of a compressor overload control. When either an electrical or thermal overload occurs, the heater warps the bimetallic element and opens the circuit.

Some common automatic control devices are

- Temperature controls in household refrigerator fresh and frozen food compartments
- Temperature controls in residential and office cooling and heating systems
- Water heater temperature controls
- Electric oven temperature controls
- Garbage disposal overload controls
- Fuses and circuit breakers that control current flow in home electric circuits

Automatic controls in the air-conditioning and refrigeration industry monitor temperatures. Some controls that respond to temperature changes are used to monitor electrical overloads by sensing temperature changes in the wiring circuits. This response is usually a change in dimension or electrical characteristic of the control-sensing element.

Figure 13.4 A basic bimetal strip made of two unlike metals such as brass and steel fastened back to back.

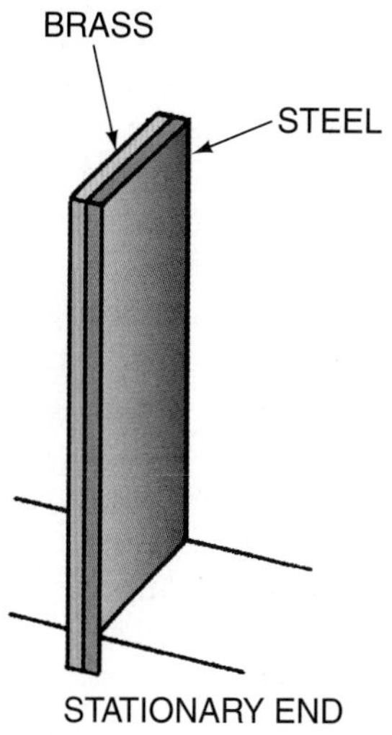

13.3 THE BIMETAL DEVICE

The **bimetal** device is probably the most common device used to detect thermal change. In its simplest form, it consists of two unlike metal strips—often brass and steel—attached back to back. Each of these metals has a different rate of expansion and contraction when heated or cooled, **Figure 13.4**. When the device is heated, the brass expands faster than the steel, which causes the device to warp. The motion of the bimetal strip is then used to perform some useful task. Bimetal controls can be attached to an electrical component or valve to stop, start, or modulate electrical current or fluid flow, **Figure 13.5**. Bimetal controls are limited in application by the amount of warp that can be accomplished with a temperature change. For example, when the device is heated, the bimetal strip bends a certain amount per degree of temperature change, **Figure 13.6**.

To make bimetal devices practical over a wider temperature range, the strip is lengthened. A longer strip is normally coiled into a circle, shaped like a hairpin, wound into a helix,

Figure 13.5 A basic bimetal strip used as a heating thermostat.

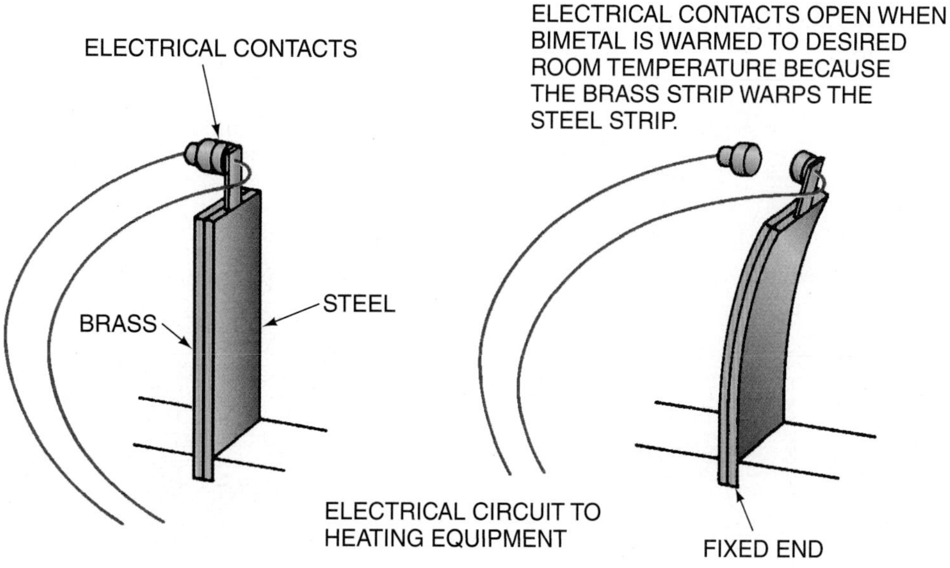

Figure 13.6 This bimetal is straight at 70°F. The brass side contracts more than the steel side when cooled, causing a bend to the left. On a temperature rise, the brass side expands faster than the steel side, causing a bend to the right. The degree of bending is predictable per degree of temperature change. The longer the strip, the greater the bend.

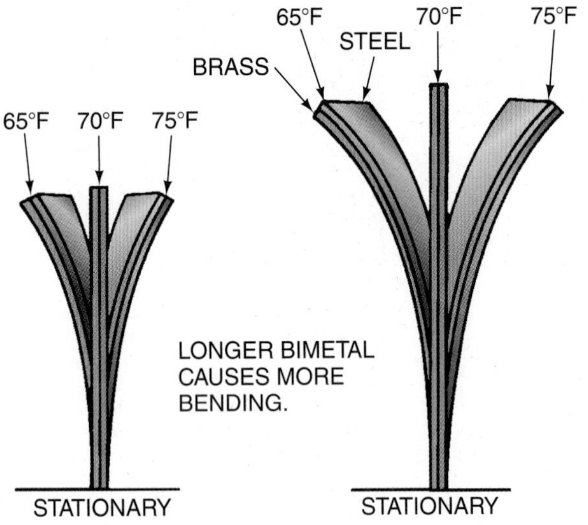

Figure 13.7 Adding length to the bimetal. (A) Coiled. (B) Wound into helix. (C) Hairpin shape. (D) Worm shape.

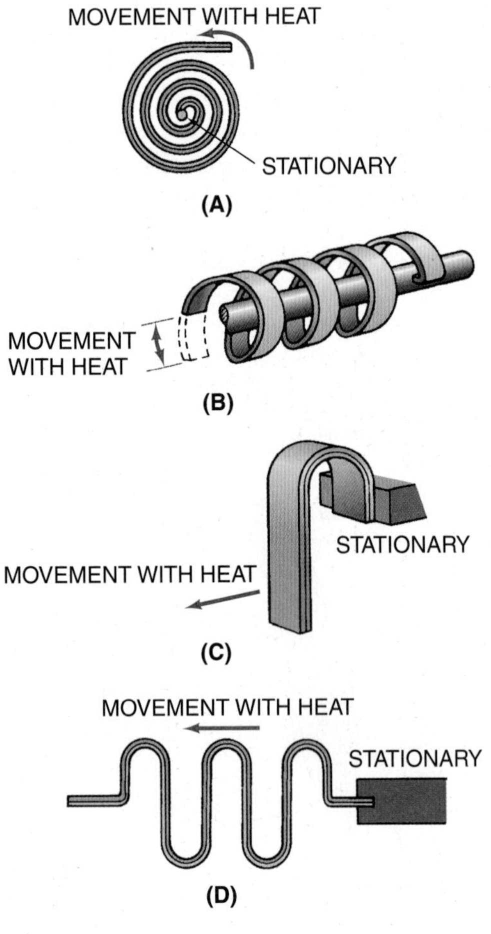

Figure 13.8 Movement of the bimetal strip due to changes in temperature opens or closes electrical contacts.

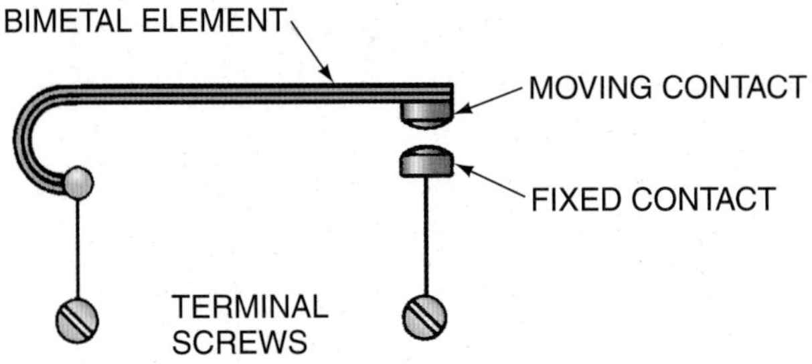

or formed into a worm shape, **Figure 13.7**. The movable end of the coil or helix can be attached to a pointer to indicate temperature, a switch to stop or start current flow, or a valve to modulate fluid flow. One basic control application is shown in **Figure 13.8**.

ROD AND TUBE

The **rod and tube** is another type of control that uses two dissimilar metals that react differently to changes in temperature. The rod and tube could more accurately be called the "rod *in* tube." It has an outer tube of metal with a high expansion rate and an internal rod of metal with a low expansion rate, **Figure 13.9**. Used for years in residential gas water heaters, the tube, which is inserted into the tank, provides very accurate sensing of water temperature. As the tank water temperature changes, the tube pushes the rod and opens or closes the gas valve to start or stop the heating of the water in the tank, **Figure 13.10**.

Heat motor valves are another application of the rod and tube device. A small heater is wrapped around the tube portion of the mechanism. When voltage is applied, the tube heats up and expands, which pulls the rod and opens the valve for water flow. When the heat is removed, the tube shrinks and closes the valve, **Figure 13.11**.

Another common application for the rod and tube type of control is in a duct-mounted limit control, **Figure 13-12**. This type of device will "lock out" or prevent an air-conditioning

Figure 13.9 The rod and tube.

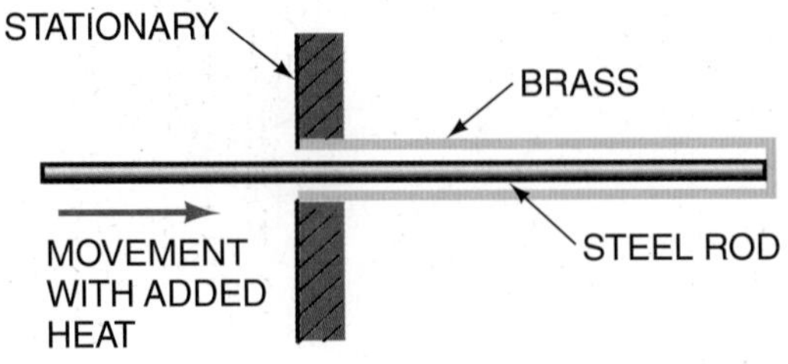

Figure 13.10 The rod-and-tube type of control consists of two unlike metals, with the faster-expanding metal enclosed in a tube, and is normally inserted in a fluid environment such as a hot water tank.

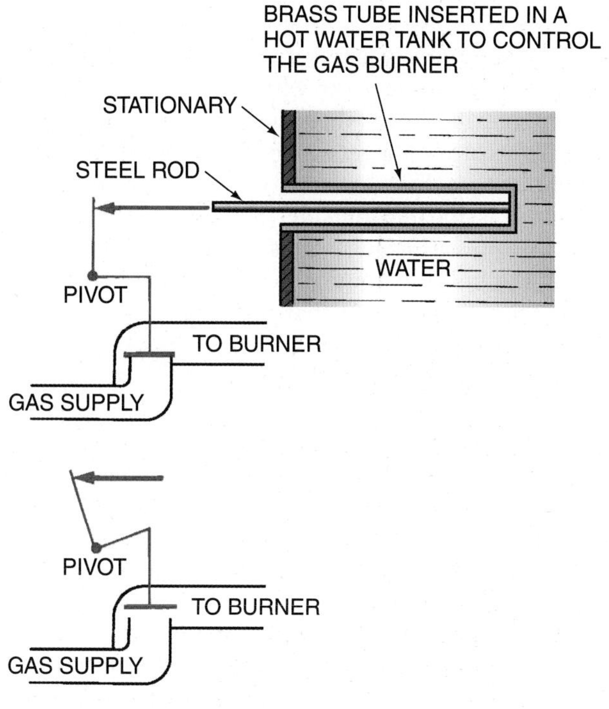

Figure 13.11 (A) When there is no voltage signal, the tube is not expanded and the valve is closed. (B) When heated, the tube expands, pulling the seat off the valve.

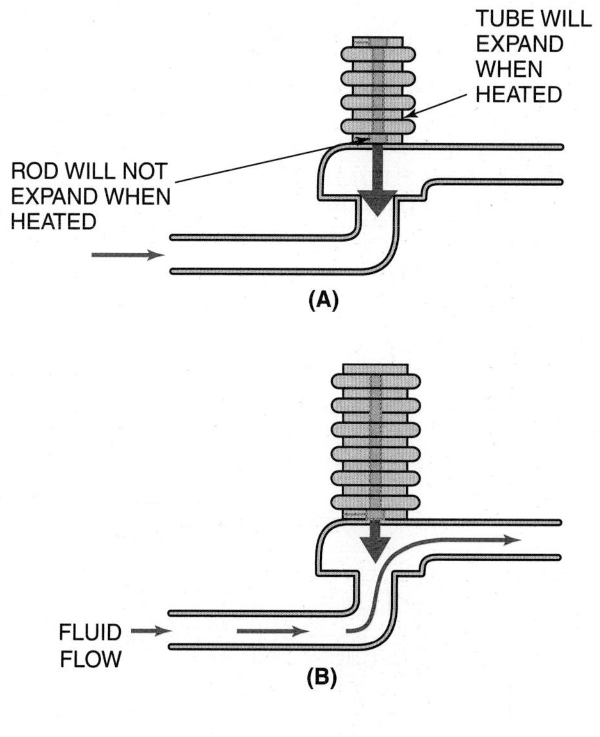

Figure 13.12 A rod and tube device used to sense return air temperature and deenergize the blower in the event of a fire.

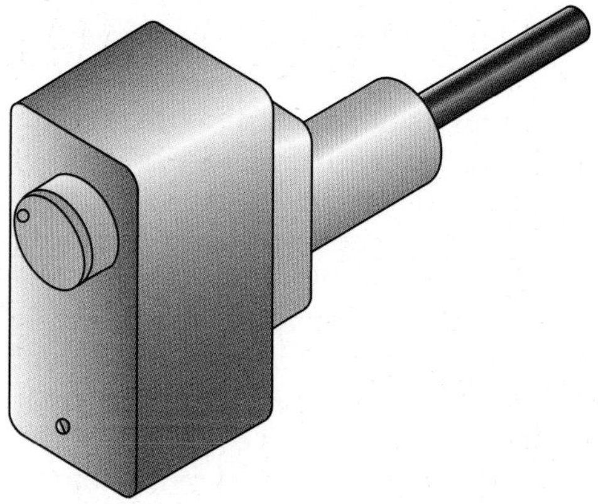

system's blower from operating in the event the return air temperature gets too high, such as when there is a fire. This device is mounted to the return duct of an air-conditioning system, and the sensing element penetrates into the duct through drilled holes. If the sensing element gets too hot, the rod pushes a set of electrical contacts open to deenergize the blower motor. When used in this manner, the device will need to be manually reset once the problem has been identified and resolved.

SNAP-DISC

The snap-disc is another type of bimetal device used to sense temperature changes in some applications. This control is treated separately from the bimetal because its snap characteristic gives it a quick open-and-close feature. Some sort of snap-action feature should be incorporated into all controls that stop and start electrical loads, **Figure 13.13**.

Figure 13.13 When heated, the disc snaps to a different position. (A) Open circuit. (B) Closed circuit.

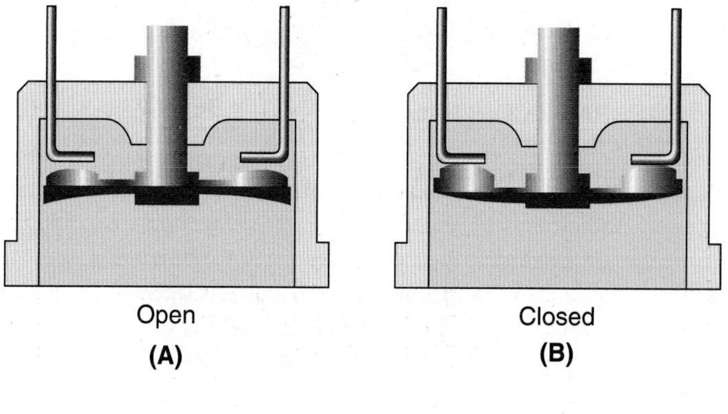

13.4 CONTROL BY FLUID EXPANSION

Fluid expansion is another way to sense temperature change. A thermometer, for example, can be described as a column of temperature-sensitive liquid rising and falling within a hollow stem in response to a sensed temperature or level of heat intensity. As the temperature of the liquid changes, the volume of the liquid will change in response. The level of the liquid in the hollow stem of the thermometer is determined by the temperature of the liquid in the thermometer's bulb. The liquid in the thermometer must rise and fall because it has no place else to go. When heated, the liquid in the thermometer expands, and so it has to rise up in the tube; when it cools, it naturally falls down the tube. This same idea can be used to transmit a signal to a control device, indicating that a temperature change is occurring. The lines or tubes that transmit these signals are referred to as transmission lines.

The liquid rising up the transmission line or tube has to act on some device to generate usable motion. One such device is the diaphragm, which is a thin, flexible metal (typically steel) disc with a large area. Pressure changes underneath the disc cause it to flex up and down, **Figure 13.14**. When a bulb is filled with liquid and connected to a diaphragm by piping, the temperature of the bulb can be transmitted to the diaphragm by the expanding liquid.

More accurate control at the actual bulb location is achieved by using a bulb partially filled with a liquid that will boil and make a vapor, which is then transmitted to the diaphragm at the control point, **Figure 13.15**. The liquid is much more sensitive to temperature change than the vapor that is used to transmit the pressure. An R-134a walk-in refrigerated box is an example of how this works. (**Figure 13.16** presents a progressive explanation of this example.) The box temperature is maintained by shutting the refrigeration system off when its temperature reaches 35°F and starting it up when the box temperature reaches 45°F. A control with a remote bulb is used to regulate the box temperature. The bulb is located inside the box, and the control is located outside the box so that it can be adjusted. A pressure gauge is installed (for illustration purposes) in the bulb to monitor the pressures as the temperature changes. At the point that the unit needs to be cycled off, the bulb temperature is 35°F, which corresponds to a pressure of 30 psig for R-134a. (Refer to **Figure 3.13** for the temperature/pressure chart for R-134a.) A control mechanism that operates on a diaphragm can open an electrical circuit and stop the unit at this point. When the cooler temperature rises to 45°F, the pressure inside the control is 40 psig and it is time to restart the unit. The same mechanism that stops the refrigeration system can be used to close the electrical circuit and restart it.

A diaphragm travels a limited amount but has a great deal of power during its movement. The travel of a liquid-filled control is limited, however, by the degree to which the liquid in the bulb expands in the temperature range within which it is working. When more travel is needed, another device, called a bellows, can be used. The bellows is much like an accordion—it has a lot of internal volume

Figure 13.14 The diaphragm is a thin, flexible, movable membrane that converts pressure changes to movement. This movement can stop, start, or modulate controls.

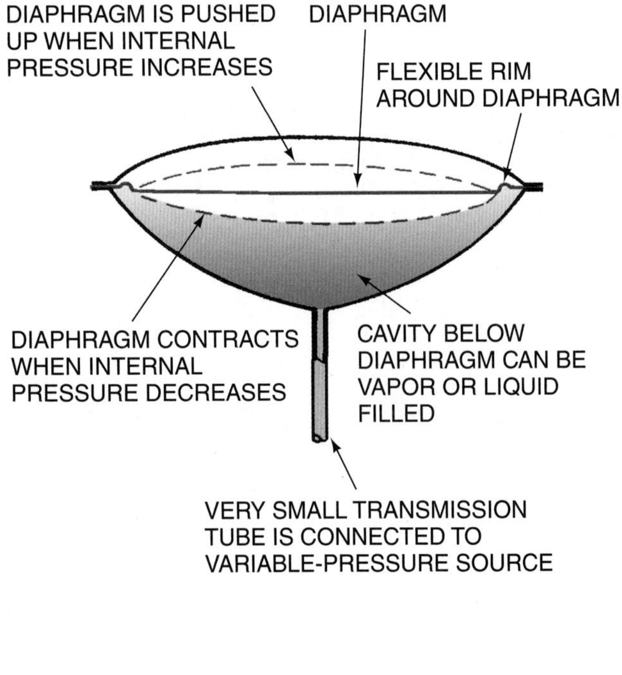

Figure 13.15 A large bulb partially filled with a volatile liquid, one that boils and creates vapor pressure when heated. An increase in vapor pressure forces the diaphragm to move outward. When cooled, the vapor condenses and the diaphragm moves inward.

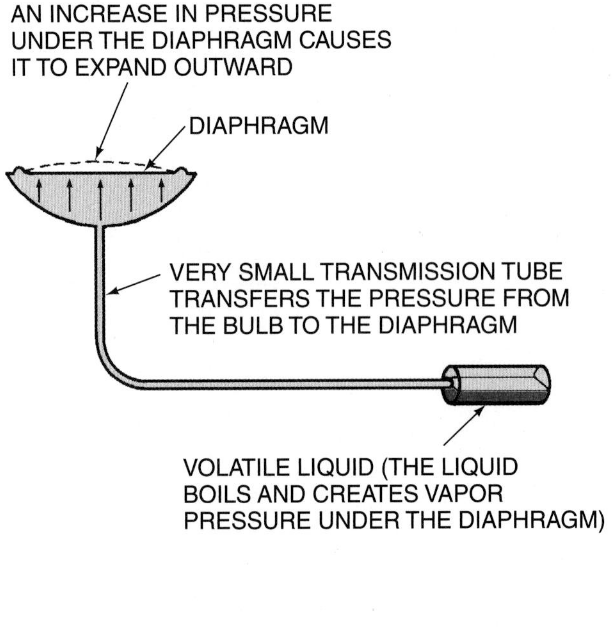

Figure 13.16 (A) and (B) A remote bulb transmits pressure to the diaphragm based on the temperature in the cooler.

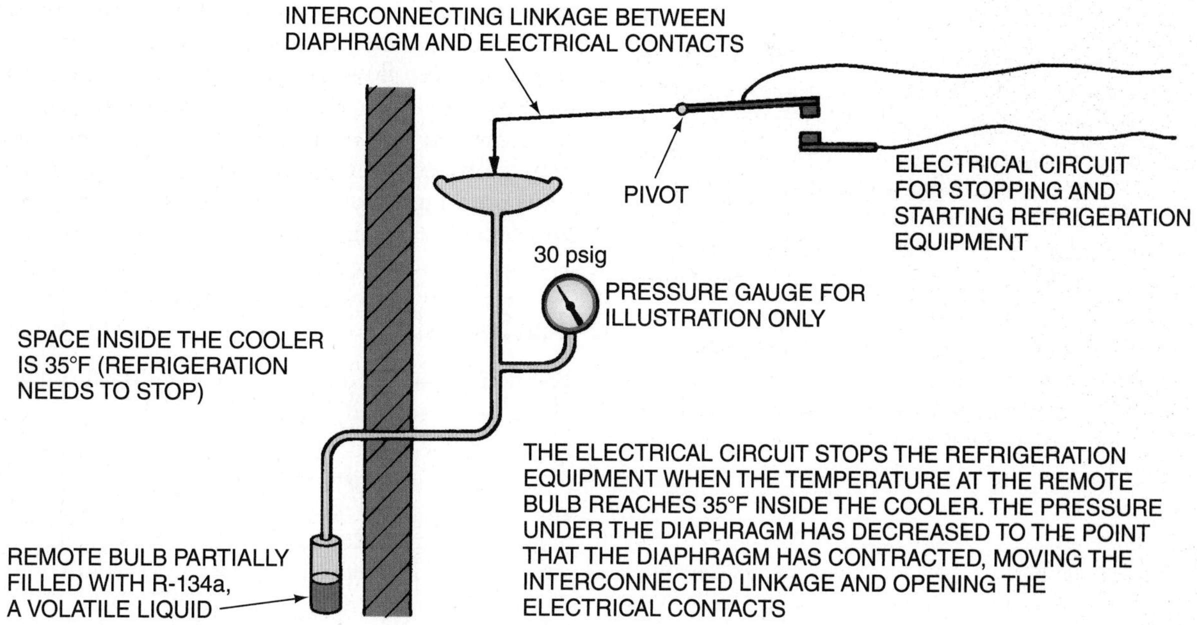

(A)

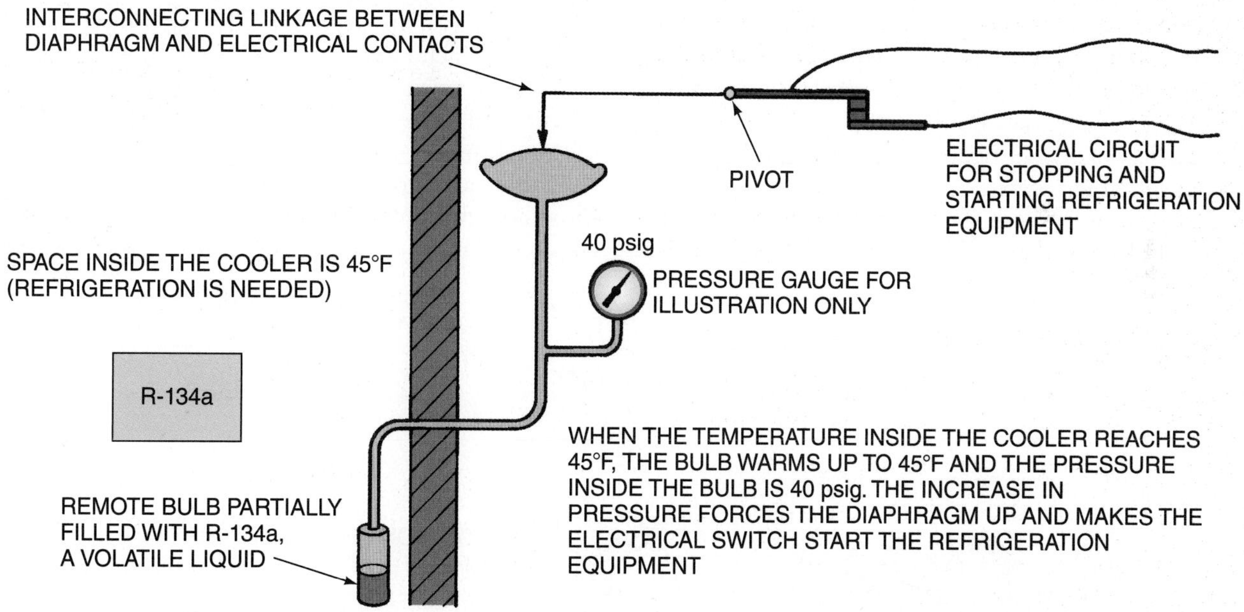

(B)

Figure 13.17 The bellows is employed where more movement per degree is desirable. This control would normally have a partially filled bulb with vapor pushing up in the bellows section.

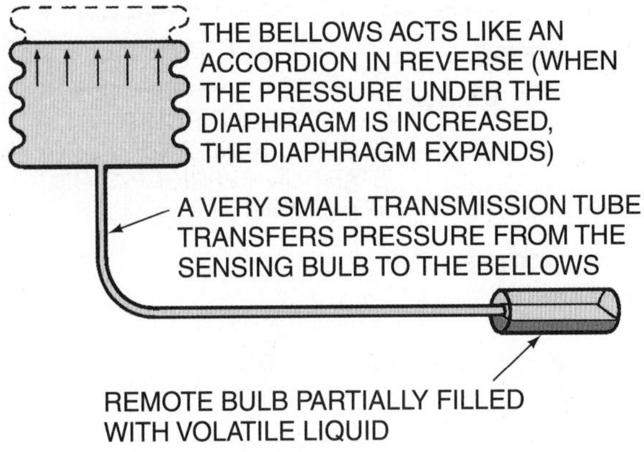

and a lot of travel, **Figure 13.17**. Normally, the bellows is used with a vapor inside instead of a liquid. The Bourdon tube is used in the same manner as the diaphragm and the bellows to monitor fluid expansion, in this case to indicate temperature. The Bourdon tube straightens out with an increase in sensed vapor pressure. This movement moves a needle on a calibrated dial to indicate temperature, **Figure 13.18**.

Controls employing partially filled bulbs are widely used in the industry because they are reliable, simple, and economical. This type of control has been around since the industry began, and it has many configurations. **Figure 13.19** shows some remote bulb thermostats.

The descriptions of these mechanisms have been simplified and are not practical because each electrical switch must also have a snap action built into it. As explained earlier, when an electric circuit is closed or opened, but especially when opened, an electrical arc occurs. This arc is a small version of an electric welding arc. Electrons are flowing and will try to keep flowing when the switch is opened, creating the arc. The faster the opening, the longer the switch will last. There are many types of snap-action mechanisms. Some switches use an over-center device to accelerate the opening. Some use a puddle of mercury to create the quick action, and some may use a magnet.

13.5 THE THERMOCOUPLE

The **thermocouple** does not use expansion to control thermal change; instead, it uses electrical principles. The thermocouple consists of two unlike metals joined together at one end (usually wire made of unlike metals such as iron and constantan), **Figure 13.20**. When heated on the fastened end, an electrical current flow is started due to the temperature difference between the two ends of the device, **Figure 13.21**. Thermocouples can be made of many different unlike metal combinations, and each one has a different characteristic. Each thermocouple has a *hot* junction and a *cold* junction. The hot junction, as the name implies, reaches a higher temperature than the cold junction. This difference in temperature is what starts the current flowing. Heat will cause an electrical current to flow in one direction in one metal and in the opposite direction in the other. When these metals are connected, they form an electrical circuit, and current will flow when heat is applied to the hot junction.

Tables and graphs for various types of thermocouples show how much current flow can be expected under different conditions for hot and cold junctions. The current flow

Figure 13.18 This remote bulb is partially filled with liquid. When heated, the expanded vapor is transmitted to a Bourdon tube that straightens out with an increase in vapor pressure. A decrease in pressure causes the Bourdon tube to curl inward.

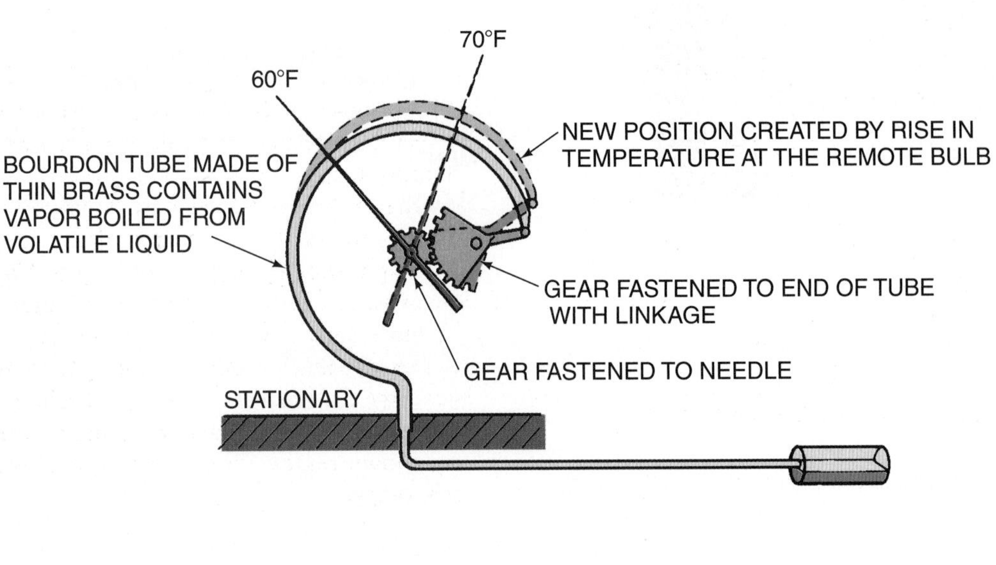

Figure 13.19 (A)–(B) Remote bulb refrigeration temperature controls. Courtesy Ferris State University. Photos by John Tomczyk

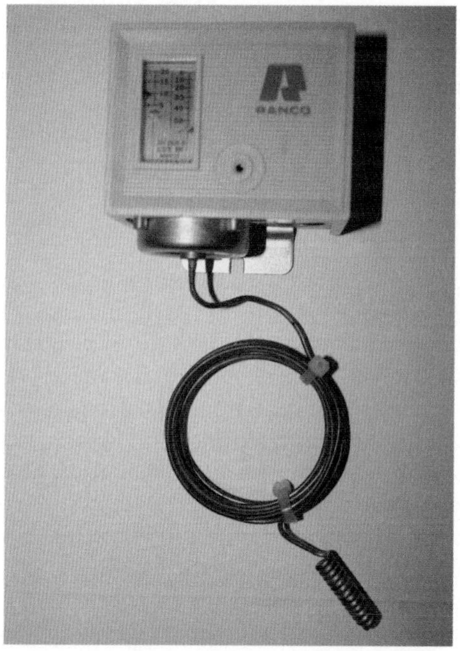

(A)

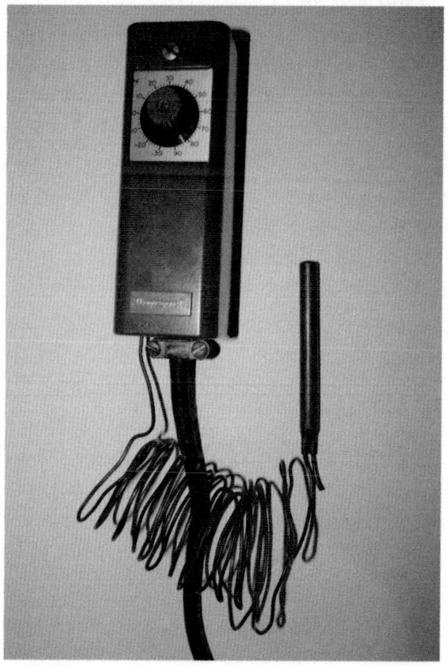

(B)

Figure 13.20 A thermocouple used to detect a pilot light in a gas-burning appliance. Courtesy Robertshaw Controls Company

Figure 13.21 The thermocouple on the top is made of two wires of different metals welded together at one end. It is used to indicate temperature. The thermocouple on the bottom is a rigid device and is used to detect a pilot light.

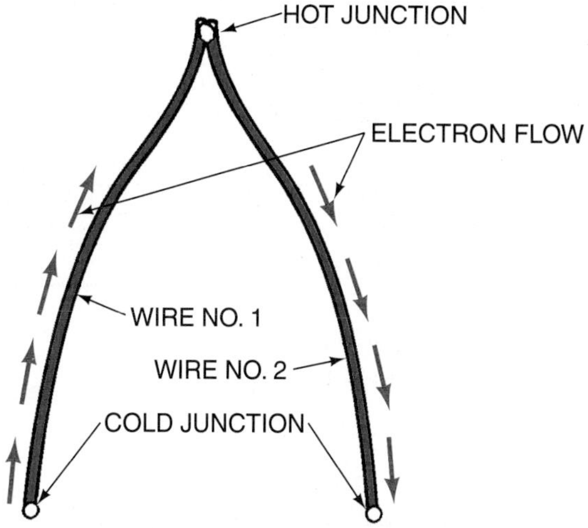

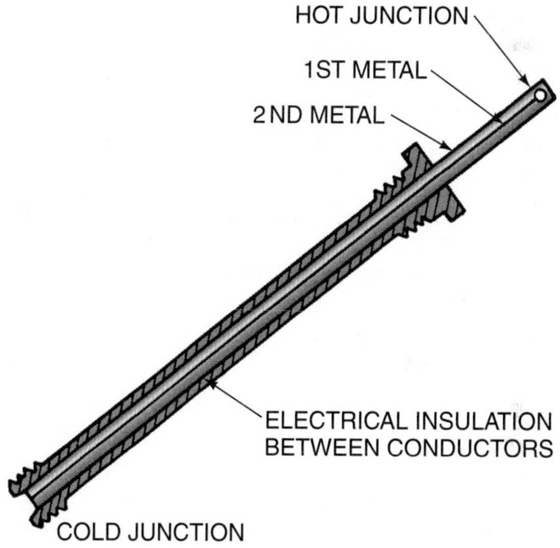

can be monitored by an electronic circuit and used for many temperature-related applications, such as a thermometer, a thermostat to stop or start a process, or a safety control. The thermocouple has been used extensively for years in gas furnaces for safety purposes to detect the pilot light flame. This application is found on older pieces of equipment, since it works best with standing pilot systems, which are, for the most part, not being produced anymore. The changeover to intermittent pilots (pilot lights that are extinguished each time the burner goes out and relit each time the room thermostat calls for heat) and other newer types of ignition is the main reason for the phaseout of older standing pilot technology.

Pilot-light thermocouples have an output voltage of about 20 to 30 millivolts (DC), all that is needed to control a safety circuit to prove there is flame, **Figure 13.22**. Multiple thermocouples can be wired in series with each other to provide a higher output voltage and, when configured like this, are called thermopiles, **Figure 13.23**. The thermopile is used as the only power source on some gas-burning equipment. This type of equipment has no need for power other than the control circuit, so the power supply is very small (about 500 millivolts). The thermopile has also been used in remote areas to operate radios using the sun or heat from a small fire.

Figure 13.22 A thermocouple and control circuit used to detect a gas flame. When the flame is lit, the thermocouple generates voltage. This voltage energizes an electromagnet that holds the gas valve open. When the flame is out, the thermocouple stops generating electricity and the valve closes.

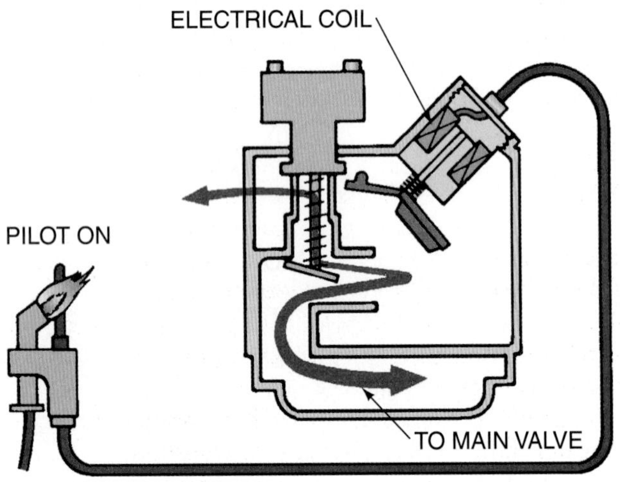

Figure 13.23 A thermopile consists of a series of thermocouples in one housing. Photo by Bill Johnson

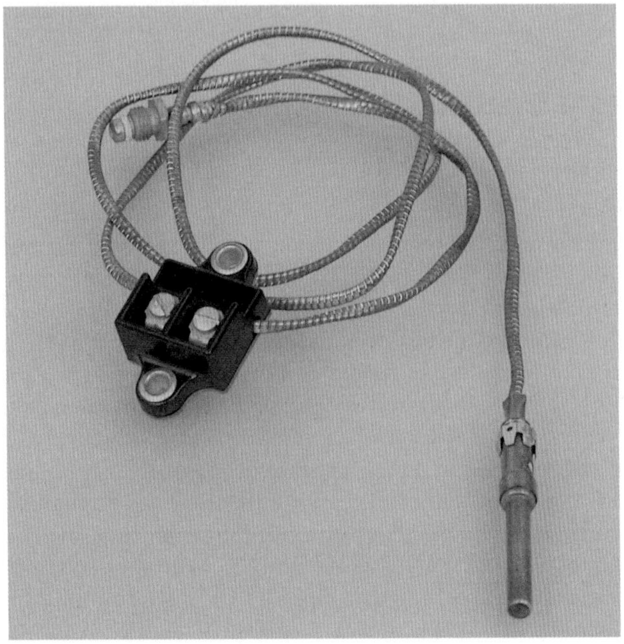

13.6 ELECTRONIC TEMPERATURE-SENSING DEVICES

Electronic instrumentation and control devices have become increasingly popular in the HVACR industry. The single component that has changed the face of electronic temperature-sensing devices is the thermistor. The thermistor varies its resistance to current flow based on its temperature. Thermistors are incorporated into solid-state circuit boards and respond to small temperature changes. The changes in current flow in the device are monitored by special electronic circuits that can stop, start, and modulate machines or provide a temperature readout, **Figure 13.24**.

Thermistors are usually made of cobalt oxide, nickel, or manganese and from a few other materials. These materials are mixed and milled in very precise proportions and then hardened for durability. A thermistor can have either a positive or a negative coefficient of resistance. If the thermistor increases its resistance as the temperature increases, it has a positive temperature coefficient (PTC) of resistance. If it decreases its resistance as the temperature increases, it has a negative temperature coefficient (NTC) of resistance. Also, some thermistors can double their resistance with as little as 1 degree of temperature change.

Thermistors are used in solid-state motor protectors, motor-starting relays, room thermostats, electronic expansion valves, duct sensors, micron gauges for vacuum pumps, heat pump controls, and in many other control schemes. Thermistors can be located on refrigerant or water lines, **Figure 13.25**, to monitor the temperature of the fluid flowing through the piping arrangement. **Figure 13.26** shows how a PTC thermistor assists in starting a permanent split-capacitor (PSC) motor. At the instant of start, the PTC thermistor has a very low resistance, which allows a large inrush of current to help increase the starting torque of the motor. However, after a fraction

Figure 13.24 A thermometer probe using a thermistor to measure temperature. Photo by Bill Johnson

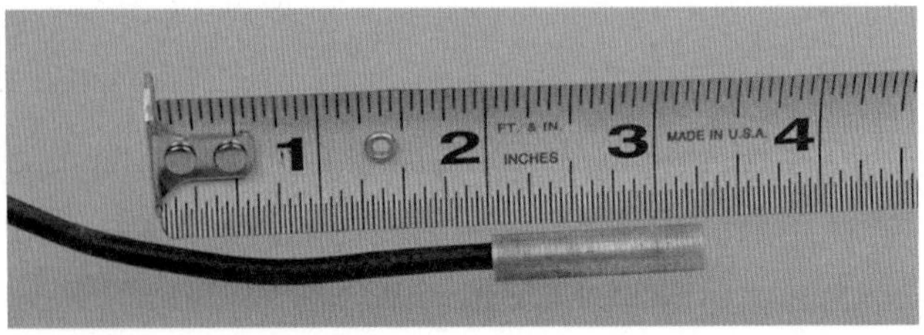

Figure 13.25 A thermistor application. Photo by Bill Johnson

Figure 13.26 A PTC thermistor wired in parallel with a run capacitor. This increases motor starting torque on this permanent split-capacitor (PSC) motor.

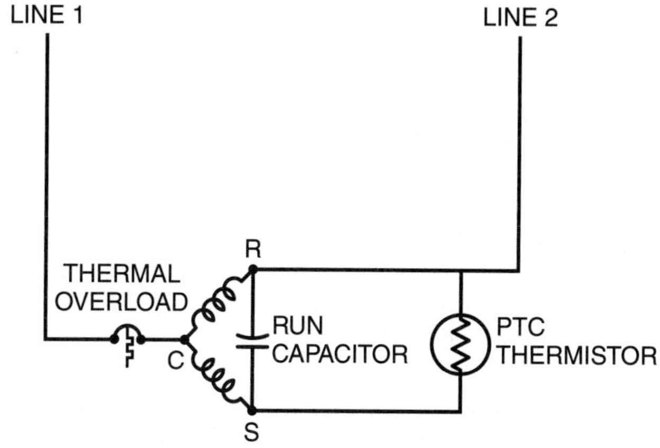

Figure 13.27 A bridge circuit showing thermistors used for micron readings when a vacuum is being measured.

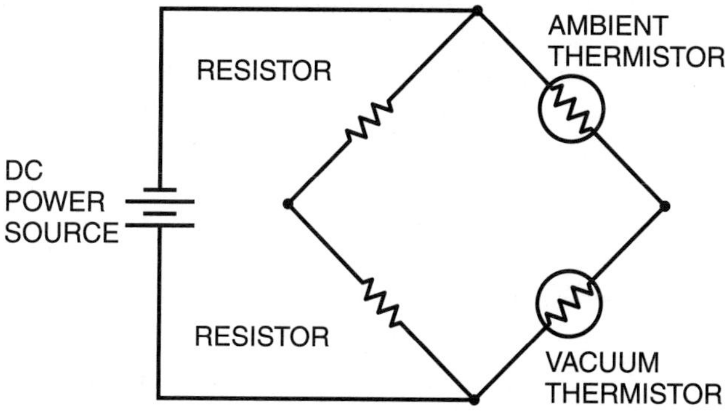

of a second, the thermistor's resistance increases because its temperature increases from the high starting current. Now the thermistor is like an open circuit, and the motor is running like a normal PSC motor.

Thermistors can also be used in a bridge circuit when measuring the vacuum in a refrigeration system, **Figure 13.27**. One thermistor is placed in the vacuum, while the other is placed in the outside ambient. As the vacuum gets deeper and deeper, the heat cannot be as easily transferred away from the thermistor placed in the vacuum because fewer molecules are surrounding it. This difference in resistance of the two thermistors can now be sensed by the bridge circuit and calibrated into microns for vacuum measurement.

13.7 ELECTRONIC PRESSURE-SENSING DEVICES

Digital gauges and other pieces of electronic instrumentation are fast replacing many of the analog pressure-sensing devices that have been in use for decades in the HVACR industry. The component that makes these digital pressure-sensing instruments possible is the pressure transducer, **Figure 13.28**.

Pressure transducers play an important role in the HVACR industry as they are able to covert a sensed pressure into a corresponding electrical output. The output from the transducer is typically a value of resistance, capacitance, or inductance. As the sensed pressure changes, the output value changes. This change in output is then processed, as the transducer is part of a larger, more complicated electronic circuit, to provide the appropriate circuit response. For example, digital gauges that have become very popular in our industry rely on pressure transducers to convert the sensed pressures in an air-conditioning or refrigeration system to the pressure readings that are displayed on the instrumentation.

Although the different types of transducers change different electrical properties, they all utilize some type of diaphragm or bellows that the sensed pressure pushes against. This diaphragm pushes an armature that facilitates the change in the configuration of the transducer. In the case of the potentiometric-type transducer, which is one that changes its resistance with changes in pressure, the armature pushes against a conducting wiper that increases or decreases the resistance of the device, **Figure 13.29**. One end of the wiper is stationary, while the other end is free to slide back and forth along the surface of the resistor. As the sensed pressure increases, the wiper is pushed further in, and the resistance of the transducer changes, **Figure 13.30**. As can be seen in **Figure 13.30**, as the sensed pressure increases, the resistance between terminals

Figure 13.28 A pressure transducer.

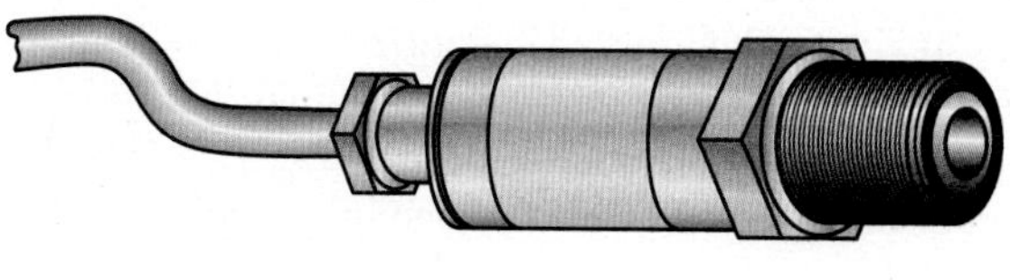

Figure 13.29 Potentiometric-type transducer.

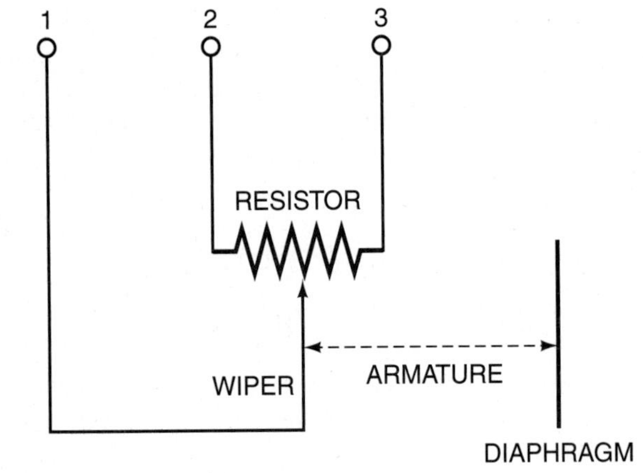

Figure 13.30 As the sensed pressure increases, the wiper is pushed further in.

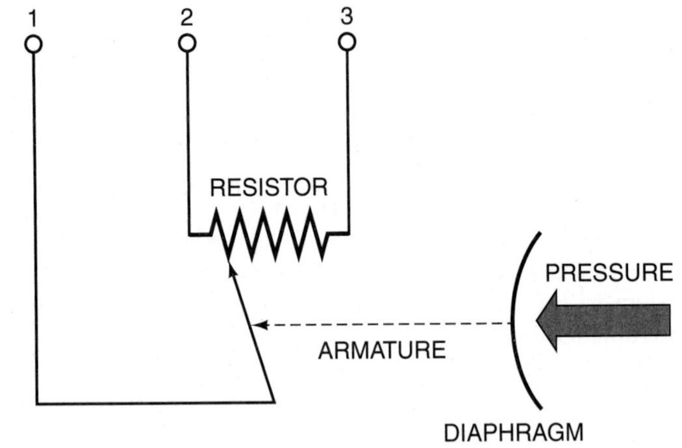

1 and 3 will increase, while the resistance between terminals 1 and 2 will decrease. The resistance between terminals 2 and 3 will always be the same and will be equal to the sum of the resistance between terminals 1 and 3 and the resistance between terminals 1 and 2.

In the case of the capacitance-type transducer, which is one that changes its capacitance with changes in pressure, the armature pushes against one of the conducting plates on the capacitor to change the capacitor's value, **Figure 13.31**. In this figure, the capacitor plates are connected to terminals 1 and 2. As the sensed pressure increases, the movable plate is pushed upward and the gap between the conducting plates decreases, reducing the capacitance, **Figure 13.32**.

In the inductive-type transducer, the sensed pressure pushes against a bellows that controls the linear motion of an iron core within a magnetic field, changing its position. As the position of the core changes, the voltage that is induced at the outlet of the transducer also changes, **Figure 13.33**.

Figure 13.31 Capacitance-type pressure transducer.

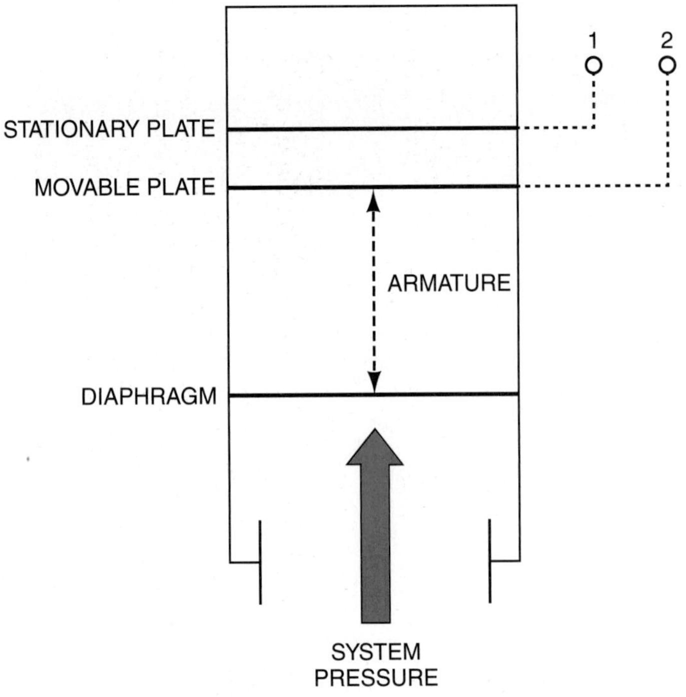

Figure 13.32 As the sensed pressure increases, the gap between the capacitor's plates decreases.

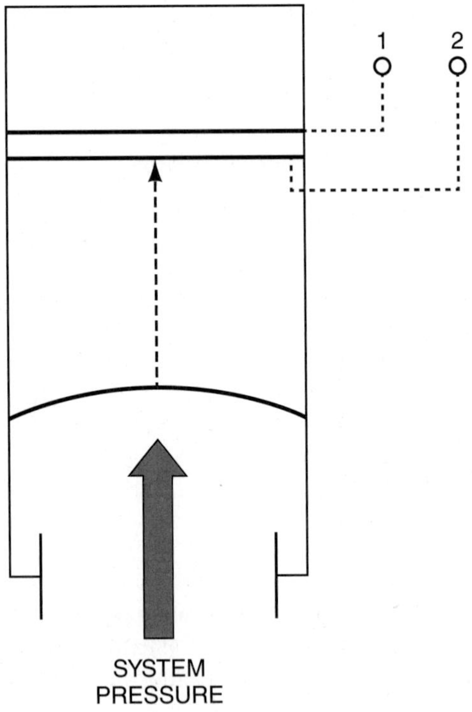

Figure 13.33 (A) When the sensed pressure is low, the iron core is not fully extended into the coil. (B) As the sensed pressure increases, the iron core extends further into the coil.

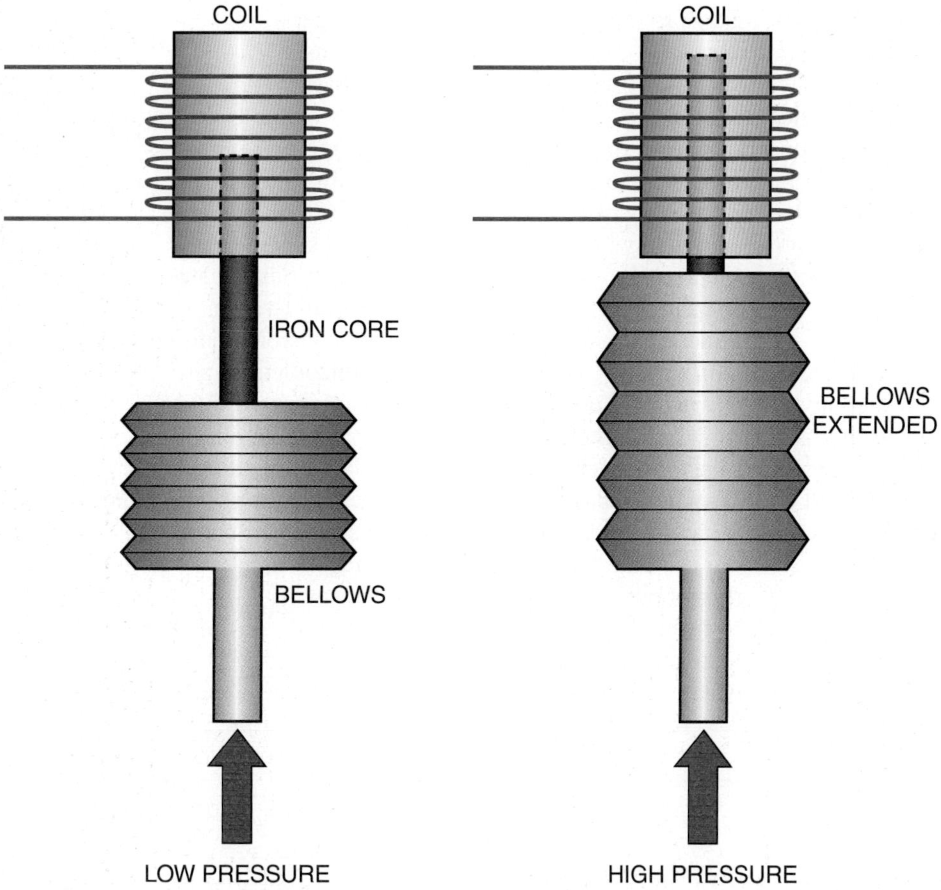

REVIEW QUESTIONS

1. A control device that opens and closes during the normal operation of a system is referred to as a(n)
   A. thermostat.
   B. pressure control.
   C. operational device.
   D. safety device.
2. A control device that typically remains closed during normal system operation and opens when dangerous conditions are present is referred to as a(n)
   A. thermostat.
   B. pressure control.
   C. operational device.
   D. safety device.
3. A thermostat is a(n)
   A. automatic control device that responds to changes in temperature.
   B. automatic control device that responds to changes in heat content.
   C. manual control device that responds to changes in temperature.
   D. manual control device that responds to changes in heat content.
4. What is the device that consists of two dissimilar metal strips that have different rates of expansion and contraction?
   A. Thermistor
   B. Thermocouple
   C. Diaphragm
   D. Bimetal
5. All of the following devices rely on the principle that different metals expand and contract at different rates except the
   A. rod and tube.
   B. snap disc.
   C. thermocouple.
   D. heat motor valve.
6. Two metals that are commonly used to create bimetal strips are
   A. copper and aluminum.
   B. steel and brass.
   C. steel and aluminum.
   D. copper and brass.

7. In a "rod and tube"-type control device,
   A. the rod pushes a mechanical linkage to open or close a valve or set of electrical contacts.
   B. the tube pushes a mechanical linkage to open or close a valve or set of electrical contacts.
   C. the rod, when heated, pushes against the tube to open or close a valve or set of electrical contacts.
   D. the mechanical linkage pushes against the tube to open or close a valve or set of electrical contacts.
8. Which of the following is a device consisting of two dissimilar metal wires fastened at one end that, when heated on the fastened end, produces a DC current?
   A. Thermocouple C. Diaphragm
   B. Thermistor D. Bimetal
9. A thermopile is best described as multiple
   A. thermistors connected in parallel with each other.
   B. thermistors connected in series with each other.
   C. thermocouples connected in parallel with each other.
   D. thermocouples connected in series with each other.
10. The component part of a control that operates on the fluid expansion concept is the
    A. thermocouple. C. thermopile.
    B. diaphragm. D. transducer.
11. The component part of a fluid-expansion control that serves to connect the location where temperature is sensed to the point of control is called the
    A. capillary tube.
    B. cold junction.
    C. transmission line.
    D. thermocouple.
12. The component part of a fluid-expansion control that resembles an accordion is the
    A. thermocouple.
    B. bellows.
    C. capillary tube.
    D. hot junction.
13. The Bourdon tube reacts to changes in
    A. pressure.
    B. temperature.
    C. voltage.
    D. current.
14. The thermistor changes its resistance in response to changes in
    A. pressure.
    B. temperature.
    C. voltage.
    D. current.
15. Which of the following best describes the makeup of a diaphragm?
    A. Thin flexible steel
    B. Thin rigid steel
    C. Thick rigid steel
    D. Thick flexible steel
16. The liquid-vapor mix in a partial liquid/partial vapor-filled control is referred to as a
    A. subcooled liquid.
    B. superheated fluid.
    C. volatile liquid.
    D. flammable fluid.
17. When referring to electronic temperature-sensing devices the "N" in the acronym NTC stands for
    A. negative.
    B. normal.
    C. neutral.
    D. nucleus.
18. Pressure transducers can take a pressure input and create a varying output by changing all the following except
    A. resistance.
    B. power.
    C. capacitance.
    D. inductance.
19. Which of the following is created when a set of electrical contacts opens or closes?
    A. An electric arc or spark
    B. A complete electric circuit
    C. An overcurrent condition
    D. A power surge
20. A combination overload device responds to changes in
    A. current and pressure.
    B. current and temperature.
    C. temperature and pressure.
    D. resistance and pressure.

UNIT 14

AUTOMATIC CONTROL COMPONENTS AND APPLICATIONS

OBJECTIVES

After studying this unit, you should be able to

- discuss space temperature control.
- describe the mercury control bulb.
- describe system lag, overshoot, and temperature swing.
- describe the difference between low- and high-voltage controls.
- list four benefits of using low-voltage controls.
- name two ways motors are protected from high temperature.
- describe two methods that can be used to measure the temperature of fluids.
- describe the characteristics of diaphragm-type and bellows-type controls.
- describe two types of relief valves commonly found in HVACR systems.
- explain the operation of a pressure regulator.
- describe a high-pressure control.
- describe a low-pressure control.
- discuss the range and differential of a control.
- describe pressure transducers.
- explain the operation of a differential pressure control.
- describe two controls that can effectively be used to sense air and gas pressure.
- describe the functions of mechanical and electromechanical controls.
- explain why it is important to routinely inspect mechanical and electromechanical controls.

SAFETY CHECKLIST

- ☑ Line-voltage thermostats are connected to electrical boxes. If an electrical arc due to overload or short circuit occurs, it is enclosed in the conduit or box. Do not confuse these thermostats with low-voltage devices.
- ☑ When a hot compressor is encountered, shut off the compressor with the space temperature thermostat "off" switch and return the next day. This gives the compressor time to cool and keeps the crankcase heat on.
- ☑ If a compressor must be cooled quickly, turn the power off and cover electric circuits with plastic. Then, use water to cool the compressor. Do not stand in this water when you make electrical checks.
- ☑ When used as safety devices, pressure controls should be installed in such a manner that they are not subject to being isolated or valved off.
- ☑ The high-pressure control provides safety for equipment and surroundings. It should not be tampered with.
- ☑ A gas pressure switch is a safety control and should never be bypassed except for troubleshooting and then only by experienced service technicians.
- ☑ A pressure relief valve is a safety device that is set at the factory; this setting should not be changed or tampered with.
- ☑ Exercise extreme caution when working with control devices that contain mercury. Mercury is toxic! Be sure to follow all local and federal laws and guidelines when disposing these devices.

This unit describes how various controls look and operate. Recognizing a control, understanding its function and what component it influences are vitally important. Doing this will help eliminate confusion when reading a wiring diagram or troubleshooting a system. It will be easier to see a control on a circuit diagram and then recognize it on the equipment once the technician becomes familiar with its description, function, and appearance.

14.1 TEMPERATURE CONTROLS

Temperature is controlled in many ways and for many reasons. For instance, space temperature is controlled for comfort, while the motor-winding temperature in a compressor motor is controlled to prevent heat-related damage to the motor. The motor could overheat and damage could occur from the very power that operates it. Both types of controls require some device that will sense temperature, with some type of sensing element, and respond accordingly. The space temperature control is an operating control or operational device. The motor temperature control serves as a safety device.

The space temperature control has two different actions, depending on whether the heating or cooling mode is needed. In winter, the control must *break a circuit* to stop the heat when the space temperature rises to a particular set point. In summer, the conditions are reversed: The control must *make a circuit* and start the cooling when the temperature rises to a certain point. The heating thermostat opens on a rise in temperature, and the cooling thermostat closes on a rise in temperature. **NOTE:** *In both cases the control is functioning on a rise. This terminology is important because it is commonly used in the industry.* • The motor temperature cut-out performs the

same circuit action as the heating thermostat. It opens the circuit on a rise in motor temperature to stop the motor. The heating thermostat and a motor-winding thermostat may operate the same way under the same conditions, but they do not physically resemble each other, **Figure 14.1**.

Another difference between the two thermostats is the medium they sense. The motor-winding thermostat must be in close contact with the motor winding, so it is fastened to the winding itself. The space temperature thermostat is mounted on a wall and responds to random air currents passing over it. Another important design concept is the current-carrying characteristics of the various controls. In the space temperature application, the stopping or starting of a heating system, such as a gas or oil furnace, involves stopping and starting low-voltage (24-V) components and line-voltage (115-V or 230-V) components. The gas or oil furnace normally has a low-voltage gas valve or relay and a line-voltage blower motor.

There is no firm rule for using one voltage or another in any specific application. However, the stopping and starting of a 3-hp compressor requires a larger switching mechanism than the one used to control a simple gas valve. A 3-hp compressor could operate with a running amperage of 18 A and a starting amperage of 90 A, whereas a simple gas valve might draw only 0.5 A. If a bimetal were large enough to carry the current for a 3-hp compressor, the control would be so large that it would be slow

Figure 14.1 Both thermostats open on a rise in temperature, but they serve two different purposes. (A) The motor-winding thermostat measures the temperature of iron and copper while in close contact with the motor windings. (B) The space thermostat measures air temperature from random air currents. Photo by Eugene Silberstein

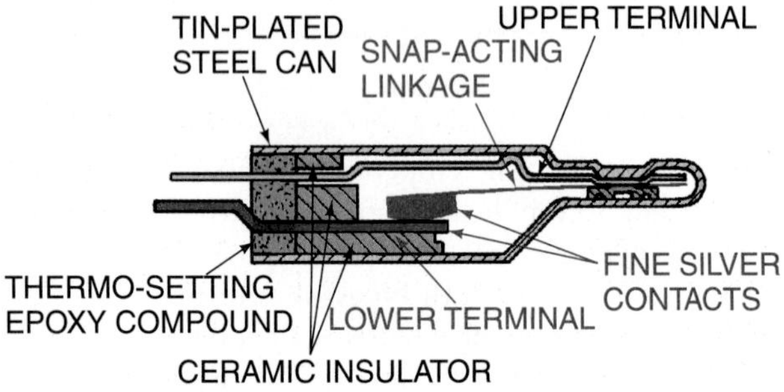

(A)

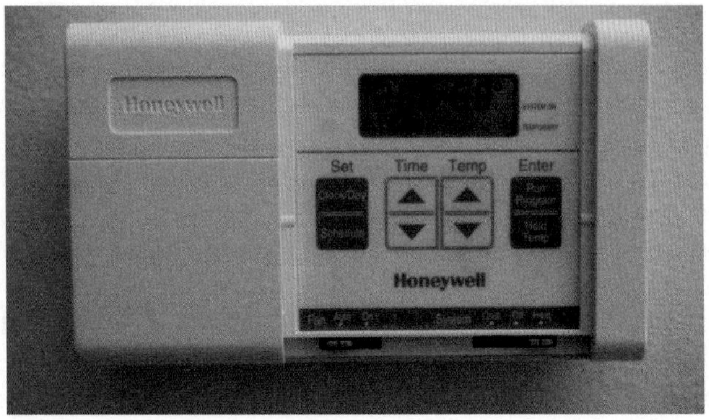

(B)

Figure 14.2 A typical transformer used in residential and light commercial buildings to change 115 V to 24 V (control voltage). Photo by Eugene Silberstein

to respond to temperature changes. This is one reason for using low-voltage controls to stop and start high-voltage components.

Residential systems usually have low-voltage control circuits. There are four reasons for this:

1. Economy
2. Safety
3. More precise control of relatively still air temperature
4. In many states, there is no need for an electrician's license to install and service low-voltage wiring

The low-voltage thermostat receives its voltage from the residential power supply that is reduced to 24 V with a small step-down transformer usually furnished with the equipment, **Figure 14.2**. On an air-conditioning system, for example, the control transformer is typically located within the air handler, wired in parallel with the indoor blower motor, **Figure 14.3**.

Figure 14.3 The typical transformer circuit used in residences to change 115 V to 24 V (control voltage).

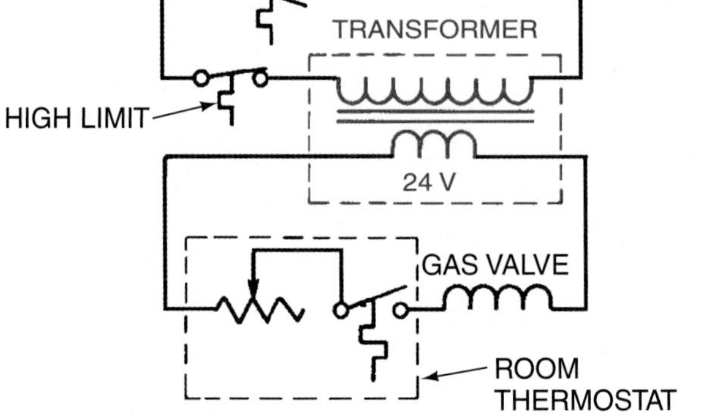

14.2 LOW-VOLTAGE SPACE TEMPERATURE CONTROLS

The low-voltage space temperature control (thermostat) normally regulates other controls and does not carry much current—seldom more than 2 A. For a 40-VA transformer, the maximum current is 1.7 Amps (40 VA/24V = 1.7 Amps). The thermostat consists of the following components:

- Electrical contacts
- Heat Anticipator
- Cold Anticipator
- Thermostat cover
- Thermostat assembly
- Subbase

ELECTRICAL CONTACTS

In *older* low-voltage thermostats, the mercury bulb is probably the most common component used to make and break electric circuits. The mercury bulb is inside the thermostat, **Figure 14.4.** It consists of a glass bulb filled with an inert gas (a gas that will not support oxidation) and a small puddle of mercury free to move from one end to the other, **Figure 14.5.** **NOTE:** *New thermostats do not contain mercury. When replacing older thermostats that contain mercury, be sure to handle them carefully to avoid breaking the glass bulb.•*

Figure 14.4 A wall thermostat with the cover off and the mercury bulb exposed.

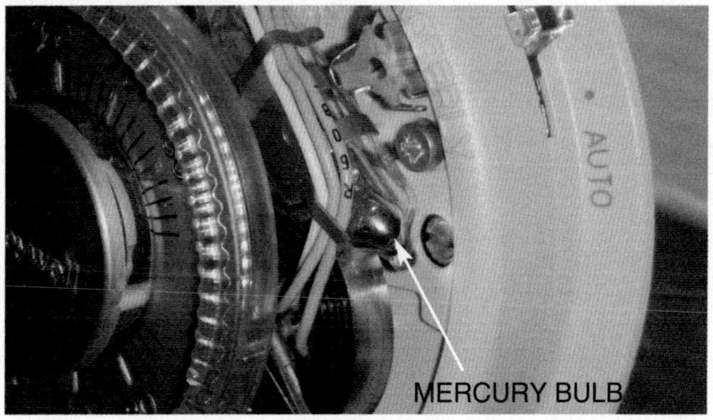

Figure 14.5 A detail of the mercury bulb. Note the very fine wire that connects the bulb to the circuit.

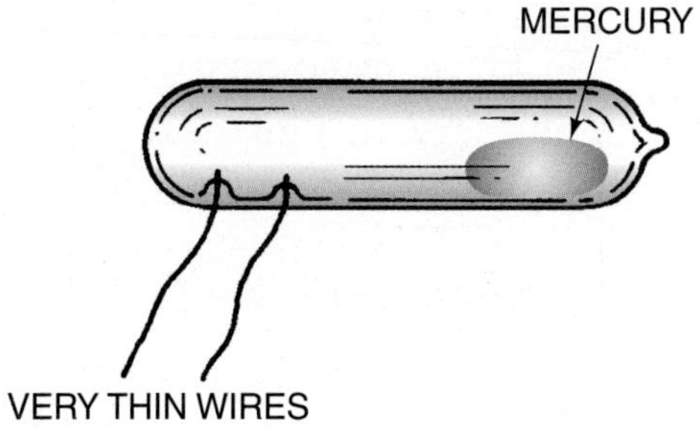

Figure 14.6 New, smart thermostats utilize digital circuitry and control strategies. Photo by Eugene Silberstein

The mercury bulb is fastened to the movable end of a bimetal strip, so it is free to rotate with the movement of the bimetal. The wires that connect the mercury bulb to the electrical circuit are very thin, so they will have little effect on the movement of the bulb. The mercury cannot be in both ends of the bulb at the same time, so when the bimetal rolls the mercury bulb to a new position, the mercury rapidly makes or breaks the electric current flow. This is called *snap* or *detent action.* Devices that contain mercury bulbs must be mounted level to ensure proper operation and temperature control. *Mercury is dangerous to the environment! When disposing of devices that contain mercury, be sure to comply with all local and federal laws! There are organizations that specialize in the collecting and recycling of thermostats that contain mercury. An Internet search in your specific area will provide options for safe and legal mercury disposal.*

New thermostats, **Figure 14.6,** use electronic circuitry instead of mercury bulbs to control circuit operation. These thermostats do not utilize bimetal strips to sense changes in temperature and are not position-sensitive like the devices that contain mercury. Even though level installation is not required for accurate system control, these thermostats should still be mounted level for aesthetic reasons.

HEAT ANTICIPATOR

The **heat anticipator** is a small resistor, which is usually adjustable, located close to the bimetal sensing element. This resistor generates a small amount of heat and is used to shut off the heating equipment prematurely, **Figure 14.7.** Consider that when an oil or a gas furnace has heated a home to just the right cutoff point, the furnace itself, weighing several hundred pounds, could still be hot. When the combustion is stopped, the furnace still radiates a great amount of heat. The heat left in the furnace is enough to drive the house temperature past the comfort point, as the furnace still runs at the end of the heating cycle.

Figure 14.7 The heat anticipator as it appears in an actual thermostat. Photo by Eugene Silberstein

The heat anticipator is wired in series with the thermostat's heating contacts and shuts the furnace off early to dissipate heat before the space temperature rises to an uncomfortable level, **Figure 14.8.** The temperature rise of the room above the set point of the thermostat is called **system overshoot**. The heat anticipator gives off a small amount of heat and causes the bimetal to become warmer than the room temperature. If there was no heat anticipator, the thermostat would open its contacts when its bimetal reaches the setpoint of, for example, 73°F. However, the furnace blower will still move the residual heat from the heat exchanger into the room for a few minutes before it shuts off, resulting in overshoot. If the thermostat has a differential of 3°F, the thermostat might be set at 73°F, but the furnace might cycle off at 76°F. Now, taking the system overshoot into account, the room might get as warm as 78°F. The heat anticipator helps reduce this condition. When the system cycles on, a similar situation arises.

Figure 14.8 The heat anticipator is usually a wire-wound, slide-bar-type variable resistor.

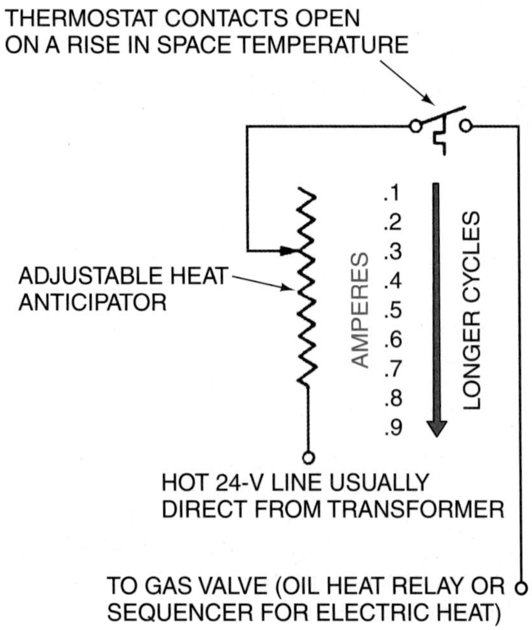

Let us say that the thermostat is set for the furnace to come on at 73°F. However, the temperature of the room may fall below 73°F because the heat source in the furnace must first warm the heat exchanger before the blower comes on and delivers heat to the room. Otherwise, objectionable colder air would be delivered to the room. The temperature drop of the heated room below the set point of the thermostat is referred to as **system lag**. System lag may cause the room to drop in temperature to 71°F before heated air is delivered to the room.

As you can see, even though the thermostat was set to close at 73°F and open at 76°F, the room temperature could fall as low as 71°F and rise as high as 78°F. This causes a large room **temperature swing** of 7°F (78°F − 71°F) from system lag and system overshoot. **Figure 14.9,** which has been exaggerated for illustrative purposes, illustrates system lag and system overshoot.

It is obvious from the figure that there needs to be some means to reduce system overshoot and cause the thermostat to open its contacts earlier, allowing the residual heat in the furnace to be delivered to the heated space. This is where the heat anticipator comes into play. The heat anticipator actually fools the bimetal of the thermostat into opening its contacts sooner by placing a small heat load on it. This causes the temperature sensed by the thermostat to be slightly higher than the room temperature when the thermostat contacts are closed. The thermostat will thus open sooner, and system overshoot will be reduced.

In order to properly set the heat anticipator (on non-digital thermostats that have one), the technician must determine what the current flow is through the heating control

Figure 14.9 Unanticipated thermostat cycles showing system lag and system overshoot causing a large temperature swing.

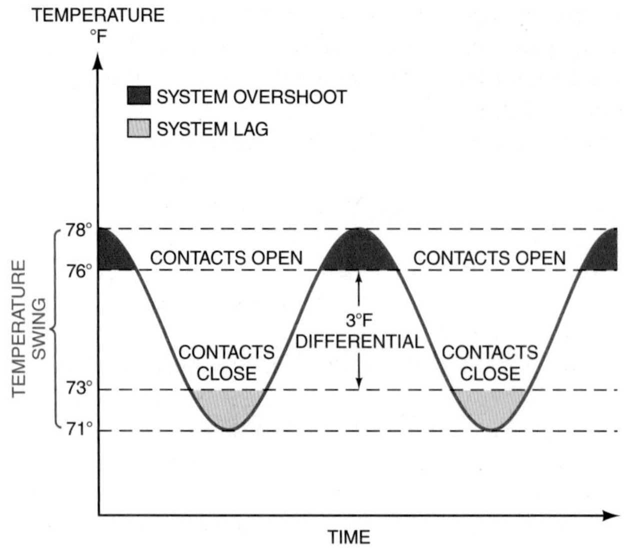

Figure 14.10 Using a digital clamp-on ammeter to measure control circuit amperage. Photo by Eugene Silberstein

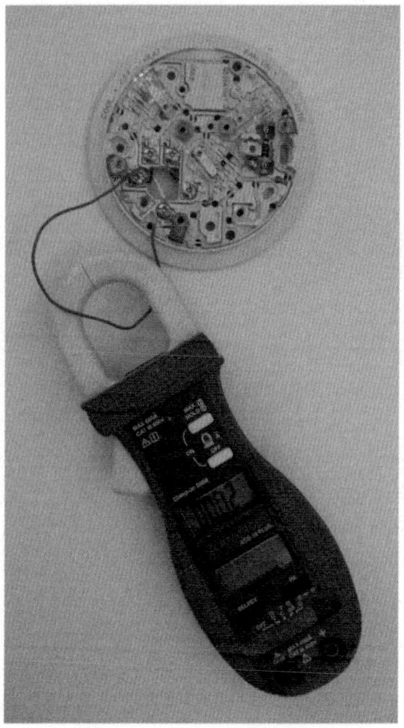

circuit when the furnace is operating in a *steady-state condition.* Furnace steady-state conditions occur when the furnace is fired and operating. Using a digital clamp-on ammeter to measure the current between the R and W terminals of the thermostat's subbase, **Figure 14.10,** is the most accurate way to measure the current through the heat anticipator. When the steady-state current is determined, the technician should adjust the heat anticipator to that current. If the heat anticipator is ever set at a higher amperage setting than what it is supposed to be, the smaller resistance in the anticipator would dissipate less heat to the thermostat's bimetal and greater system overshoot would occur. Heat anticipators do not eliminate system overshoot and temperature swings; they simply make them smaller and less objectionable in terms of human comfort. Directions that come with a thermostat showing how to set the anticipator should be followed.

Electronic or microelectronic thermostats have different anticipation circuits. Usually, these thermostats have a built-in factory microprocessor with software that will determine the cycle rate. A factory-installed thermistor wired into an electronic circuit can also determine the cycle rate. In either case, there is no need for the technician to set the anticipator on these types of thermostats.

Given the technological advances in HVACR controls, along with the reduced cost of these hi-tech devices, it is wise to replace older, non-programmable, analog thermostats with newer models. Digital, programmable thermostats have the potential to save the end-user large amounts of energy and money. Some new "smart" thermostats, such as the one shown in **Figure 14.6,** are able to learn the behavior and comfort requirements of the occupants and adjust the system accordingly to meet these needs automatically.

COLD ANTICIPATOR

A cooling system should be started a few minutes early to allow the air-conditioning system to get up to capacity when needed. If the air-conditioning system is not started until it is needed, it would take 5 to 15 min before it would reach full capacity, which is enough time to cause a noticeable temperature rise in the conditioned space. The cold anticipator causes the system to start early, to compensate for this capacity lag. Normally, the cold anticipator is a high-resistance device that is not adjustable in the field, **Figure 14.11.** It is wired in parallel with the thermostat's cooling contacts, **Figure 14.12.** When the cooling contacts are closed, the current bypasses the cool anticipator. When the cooling contacts are open, a small amount of current flows through the high resistance cool anticipator. This generates a small amount of heat that acts on the thermostat's bimetal strip. This will fool the bimetal and keep it at a temperature slightly above room temperature. This causes the thermostat contacts to close a few minutes early, reducing the cooling system's lag.

Figure 14.11 The cold anticipator is usually a fixed resistor similar to resistors found in electronic circuitry. Photo by Eugene Silberstein

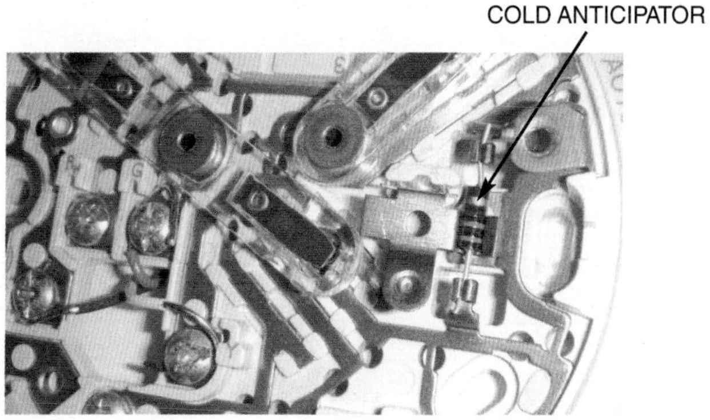

Figure 14.12 The cold anticipator is wired in parallel with the cooling contacts of the thermostat. This allows the current to flow through it during the off cycle.

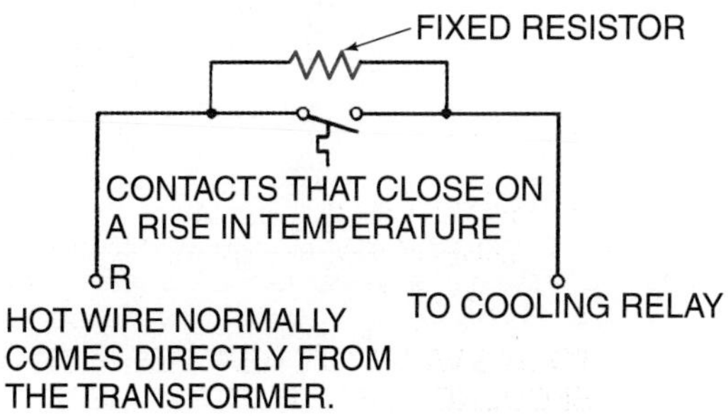

Because the cold anticipator has a high resistance, it will have a significant voltage drop across it and prevent the cooling relay coil from being energized when the thermostat contacts are open. When the thermostat's contacts close, the cold anticipator is bypassed which allows the cooling relay to be energized and cooling to begin.

THERMOSTAT COVER

The thermostat cover is intended to be decorative and protective. A thermometer is usually mounted on it to indicate the surrounding (ambient) temperature. The thermometer is functionally separate from any of the controls and would serve the same purpose if it were hung on the wall next to the thermostat. Thermostats come in many shapes, **Figure 14.13.**

THERMOSTAT ASSEMBLY

The *thermostat assembly* houses the thermostat components already mentioned and is normally mounted on a subbase fastened to the wall. This assembly could be called the brain of the system. In addition to mercury bulbs, sensors, circuitry, and anticipators, the thermostat assembly includes the levers, buttons, and slide switches that are used to adjust the temperature and, in many cases, program the device.

Figure 14.13 Decorative thermostat covers. Courtesy Honeywell

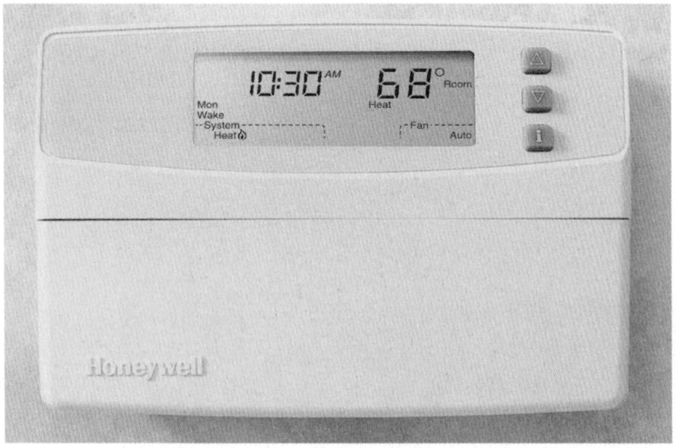

(A)

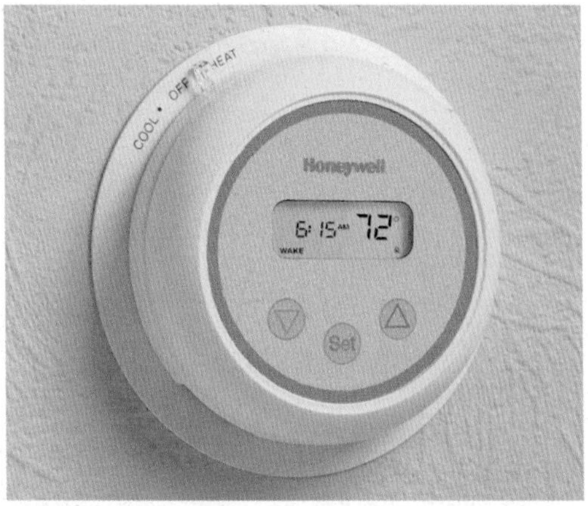

(B)

Figure 14.14 The subbase normally mounts on the wall, and the wiring is fastened inside the subbase on terminals. When the thermostat is fastened to the subbase, electrical connections are made between the two. Courtesy Ferris State University. Photo by John Tomczyk

THE SUBBASE

The **subbase**, which is usually separate from the thermostat, contains the selector switching levers, such as the FAN ON-AUTO switch or the HEAT-OFF-COOL switch, **Figure 14.14.** The subbase is important because the thermostat is mounted to it. The subbase is first mounted on the wall, then the interconnecting wiring is attached to the subbase. When the thermostat is attached to the subbase, the electrical connections are made between the two components. Often, cold anticipators are located on the subbase of the thermostat.

14.3 LINE-VOLTAGE SPACE TEMPERATURE CONTROLS

Sometimes it is desirable to use line-voltage thermostats to stop and start equipment. Some self-contained pieces of equipment do not need remote thermostats. To add a remote thermostat would be an extra expense and would only increase the potential for trouble. The window air conditioner is a good example of this. It is self-contained and needs no remote thermostat. If it had a remote thermostat, the window unit would no longer be a plug-in appliance, since it would require

Figure 14.15 The window air conditioner has a line-voltage thermostat that starts and stops the compressor. On some models with energy-saving modes, the thermostat cycles both the compressor and the fan motor. Photo by Eugene Silberstein

the installation of the thermostat. Note that the remote-bulb-type thermostat is normally used with the bulb located in the return airstream, **Figure 14.15.** The fan usually runs all the time and keeps a steady stream of room return air passing over the bulb, which provides more sensitivity.

The line-voltage thermostat is used in many types of installations. The household refrigerator, reach-in coolers, and free-standing package air-conditioning equipment are just a few examples. All of these have something in common—the thermostat is a heavy-duty type and may not be as sensitive as the low-voltage type. When replacing these thermostats, an exact replacement or one recommended by the equipment manufacturer should be used.

The line-voltage thermostat must be matched to the voltage and current that the circuit is expected to use. For example, a reach-in cooler for a convenience store has a compressor and fan motor that need to be controlled. The combined running-current draw for both components is 16.2 A at 115 V (15.1 A for the compressor and 1.1 A for the fan motor). The locked-rotor amperage for the compressor is 72 A. (Locked-rotor amperage is the inrush current that the circuit must carry until the motor starts turning. The inrush is not considered for the fan because it is so small.) This is a 3/4-hp compressor motor. A reliable, long-lasting thermostat chosen to operate this equipment should have a control rated at 20 A running current and 80 A locked-rotor current, **Figure 14.16.**

Usually, thermostats are not designed to handle more than 25 A because they would physically be too large. This limits the size of the compressor that can be started directly with a line-voltage thermostat to about 1.5 hp on 115 V or 3 hp on 230 V. Remember, the same motor would draw exactly half of the current when the voltage is doubled. When larger current-carrying capacities are needed, a motor starter is normally used with a line-voltage thermostat or a low-voltage thermostat. Line-voltage thermostats usually consist of the switching mechanism, the sensing element, the cover, and the subbase.

Figure 14.16 The line-voltage thermostat used to stop and start a compressor that draws up to 20 A full-load current and has an inrush current of up to 80 A. Courtesy Ferris State University. Photo by John Tomczyk

THE SWITCHING MECHANISM

Some line-voltage thermostats use mercury as the contact surface, but most switch contacts are constructed from a silver-coated base metal. The silver helps conduct current at the contact point. The silver contact point takes the real load in the circuit and is the component in the control that wears first, **Figure 14.17.**

THE SENSING ELEMENT

The sensing element is normally a bimetal, a bellows, or a liquid-filled remote bulb, **Figure 14.18,** located where it can sense the space temperature. If sensitivity is important, a low-velocity air stream should cross the element. The levers or knobs used to adjust the thermostat are attached to the main thermostat.

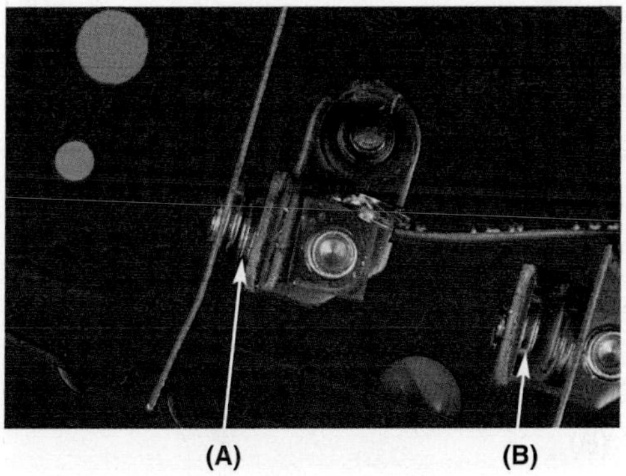

Figure 14.17 Silver-coated line-voltage contacts. (A) Closed. (B) Open. Photo by Bill Johnson

Figure 14.18 Bimetal, bellows, and remote-bulb sensing elements used in line-voltage thermostats. The levers or knobs used to adjust these controls are arranged to apply more or less pressure to the sensing element to vary the temperature range.

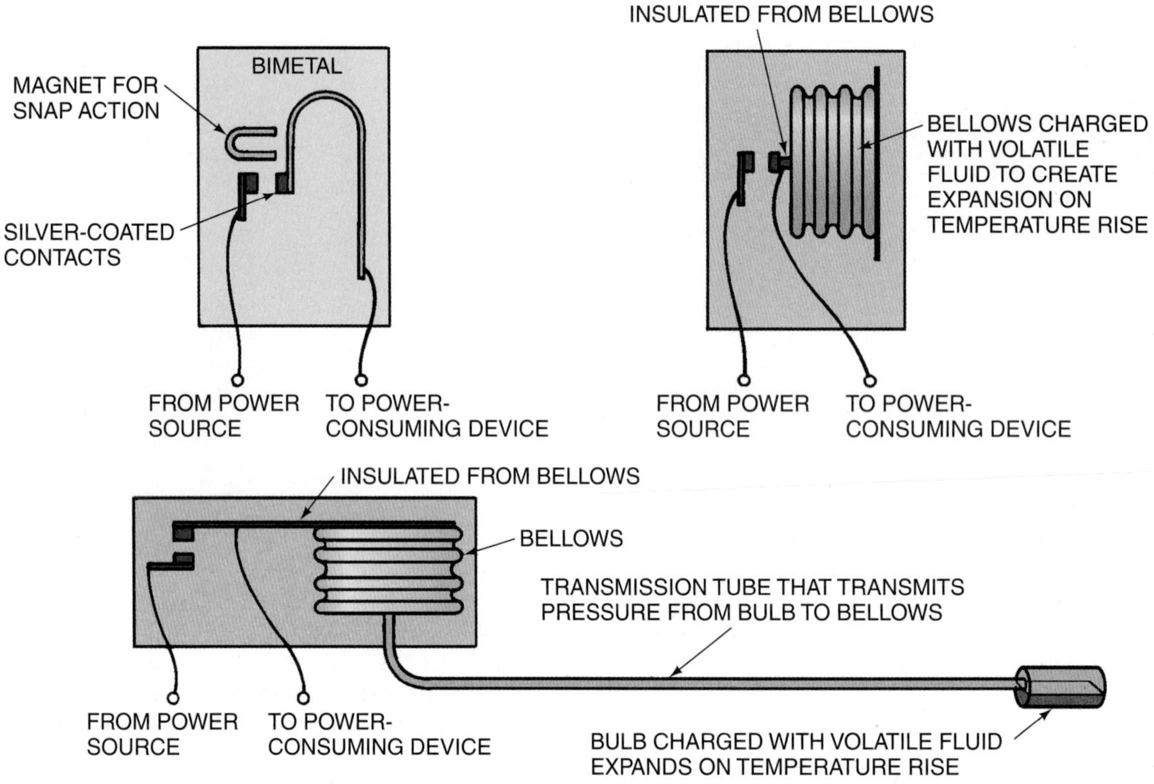

THE COVER

Because of the line-voltage nature of the controls, the cover is usually attached with some sort of fastener to discourage easy entrance, **Figure 14.19.** If the control is applied to room space temperature, such as in a building, a thermometer may be mounted on the cover to read the room temperature. If the control is used to control a box space temperature, such as a reach-in cooler, the cover might be just a plain protective cover.

Figure 14.19 A line-voltage, wall-mounted thermostat. Photo by Eugene Silberstein

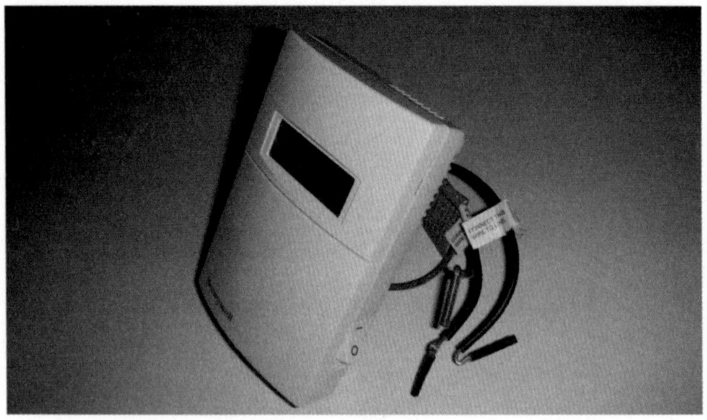

SUBBASE

When the thermostat is used for room space temperature control, there must be some way to mount it to the wall. A subbase that fits on an electrical outlet box is usually used. The wire leading into a line-voltage control is normally a high-voltage wire routed between points in a conduit. The conduit is connected to the box that the thermostat is mounted on. **SAFETY PRECAUTION:** *If an electrical arc due to overload or short circuit occurs, it is enclosed in the conduit box. This reduces fire hazard,* **Figure 14.20.**•

Figure 14.20 The box for a line-voltage, wall-mounted thermostat.

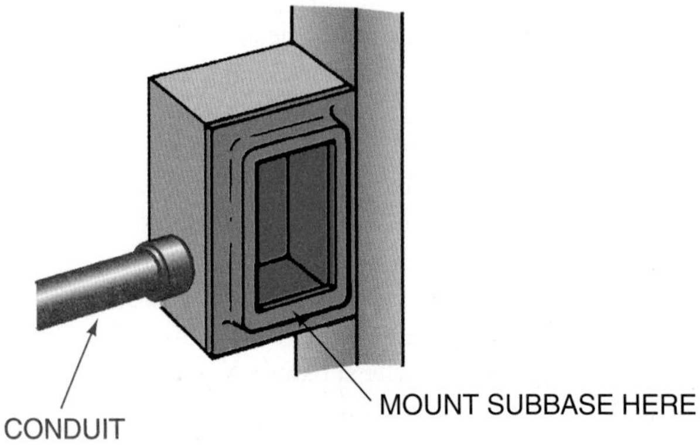

As mentioned, as far as function is concerned, a room thermostat for the purpose of controlling a space-heating system is the same as a motor temperature thermostat. They both open or interrupt the electrical circuit on a rise in temperature. However, the difference in the appearance of the two controls is considerable. One control is designed to sense the temperature of slow-moving air, and the other is designed to sense the temperature of the motor winding. The winding has much more mass, and the control must be in contact to get the desired response.

14.4 SENSING THE TEMPERATURE OF SOLIDS

A key point to remember is that any sensing device indicates or reacts to the temperature of the sensing element. A thermometer indicates the temperature of the bulb at the end of the thermometer, not the temperature of the substance in which it is submerged. If it stays in the substance long enough to reach the temperature of the substance, an accurate reading will be achieved, **Figure 14.21.**

To accurately determine the temperature of solids, the sensing element must reach the temperature of the object as soon as practical. It can be difficult to get the rounded bulb of a thermometer close enough to a flat piece of metal so that it accurately measures the temperature of the metal. Only a fraction of the bulb will be in contact with the flat metal at any time, **Figure 14.22.** This leaves most of the area of the bulb exposed to the surrounding (ambient) air. Placing insulation over the thermometer to hold it tightly on the plate and to shield the bulb from the ambient air helps give a more accurate reading. A gum-type substance can be used to hold the thermometer's bulb against the surface and insulate it from the ambient air, **Figure 14.23.** A more permanent installation may require a well into which the thermometer is inserted, **Figure 14.24.**

Most sensing devices will have similar difficulties in reaching the temperature of the substance to be sensed. Some are designed to fit the target surface. The external motor temperature-sensing element is a good example, as it is manufactured flat to fit close to the motor housing,

Figure 14.21 The mercury thermometer rises or falls depending on the expansion or contraction of the mercury in the bulb. The bulb is the sensing device.

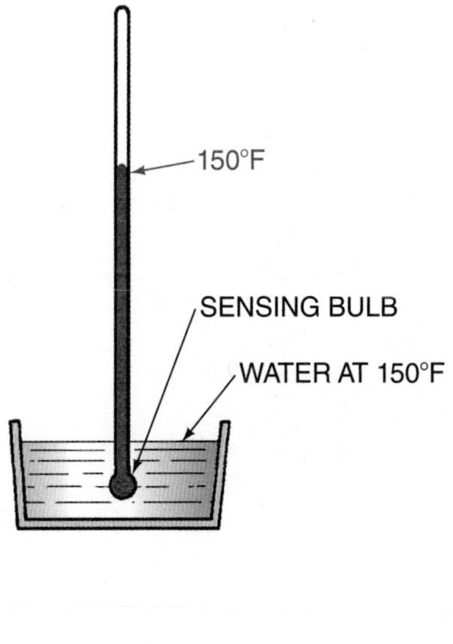

Figure 14.22 Only a fraction of the mercury-bulb thermometer can touch the surface of the flat iron plate at one time. This leaves most of the bulb exposed to the surrounding temperature.

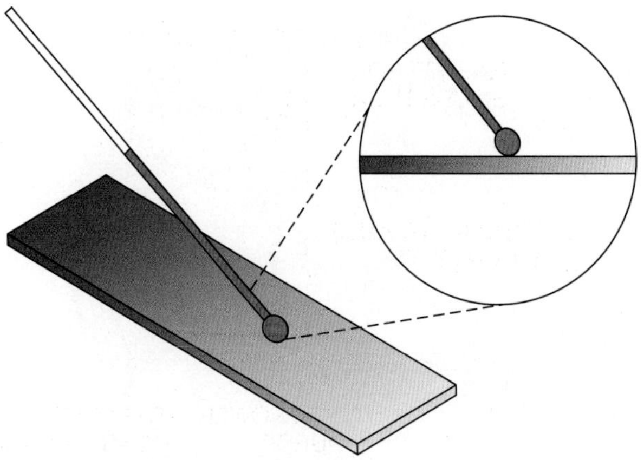

Figure 14.23 An example of temperature sensing often used for field readings. The insulating gum holds the bulb against the metal and insulates it from the ambient air.

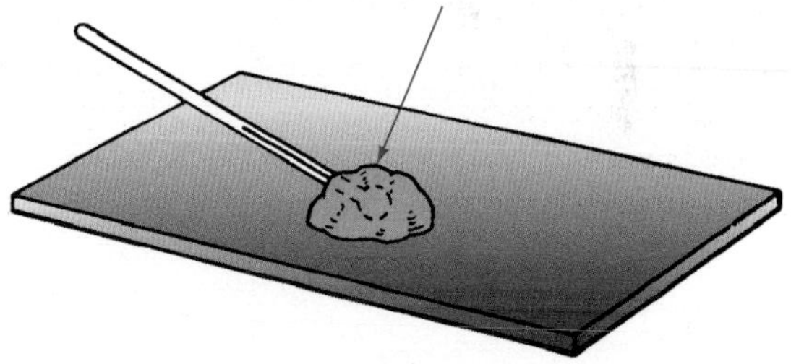

Figure 14.24 An example showing how a well for a thermometer is designed. The well is fastened to the metal plate so that heat will conduct both into and out of the bulb.

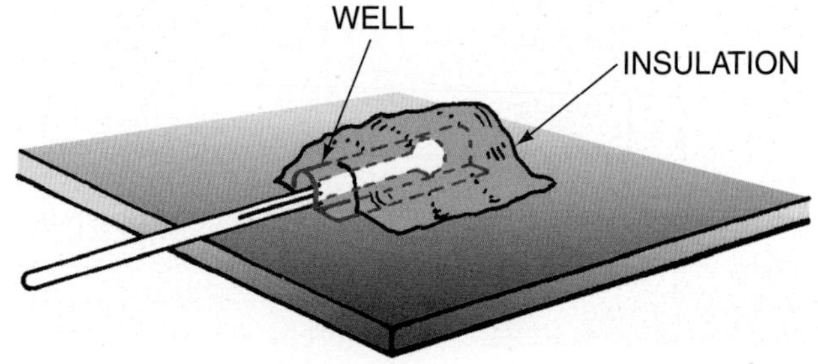

Figure 14.25 A motor temperature-sensing thermostat. Courtesy Ferris State University. Photo by John Tomczyk

Figure 14.25. This control is normally mounted inside the terminal box of the motor or compressor to shield it from the ambient temperature.

The protection of electric motors is important in the HVACR industry because motors are the prime movers of refrigerant, air, and water. Motors, especially the compressor motor, are the most expensive components in the system. They are normally made of steel and copper and need to be protected from heat and from an overload that will cause heat. All motors build up heat as a normal function of work. The electrical energy that passes to the motor is intended to be converted to magnetism and work, but some of it is converted to heat. All motors have some means of detecting and dissipating this heat under normal operating conditions. If the heat becomes excessive, it must be detected and dealt with. Otherwise, the motor will overheat and be damaged.

Motor high-temperature protection is usually accomplished with some variation of the bimetal device or the thermistor. The bimetal device can either be mounted on the outside of the motor, usually in the terminal box, or be embedded in the windings themselves, **Figure 14.26.** If the overload device is located outside of the compressor shell, it will be easier to evaluate the device. If the overload is open and the device is located within the compressor shell, there will be an infinite resistance reading between the S and C terminals as well as between the R and C terminals, **Figure 14.27(A).** There will be a measurable resistance between the S and R terminals. If the overload is open and the device is located outside the compressor shell, there will be an infinite resistance reading across the overload device, but there will be measurable resistances between the S and C, R and C, and S and R terminals, **Figure 14.27(B).** The details regarding how these devices actually protect motors are covered later on in the text. The thermistor-type of motor protection is normally embedded in the windings. This close contact with the windings gives fast, accurate response, but it also means that the wires must be brought to the outside of the compressor, which involves extra terminals in the compressor terminal box. The thermistor-type of temperature-monitoring device uses an electronic monitoring circuit to check the temperature of the motor windings. Thermistors are positioned in close contact with the windings. When the temperature reaches a predetermined temperature, the monitor interrupts the circuit to the contactor coil and stops the compressor, **Figure 14.28.**

Because of its size and weight, a motor can take a long time to cool after overheating. If the motor is an open type, a fan or moving airstream can be used to cool it more

Figure 14.26 (A) Permanent split-capacitor motor with internal protection. (B) Permanent split-capacitor motor with external protection.

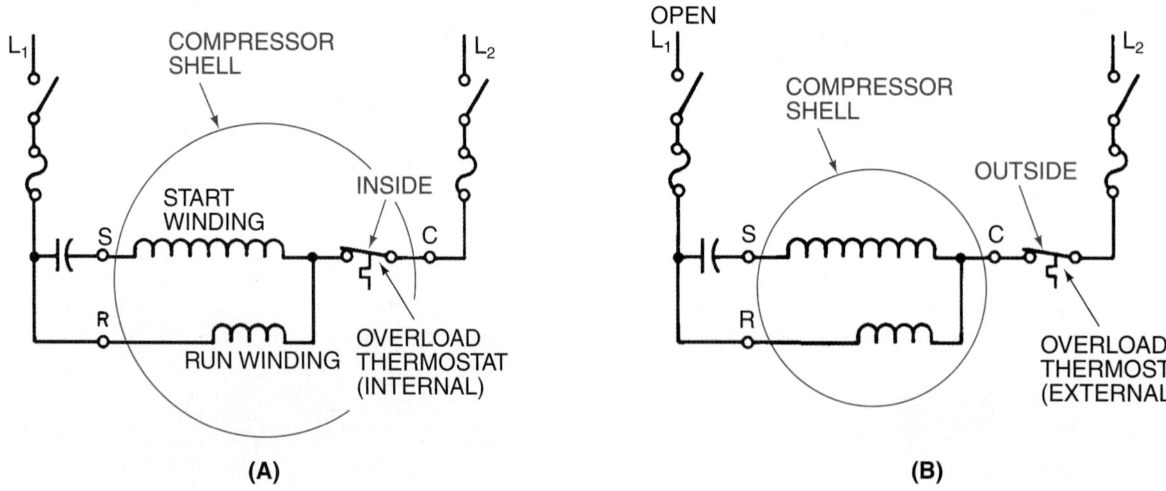

Figure 14.27 (A) Internal motor protection device in the open position. (B) External motor protection device in the open position.

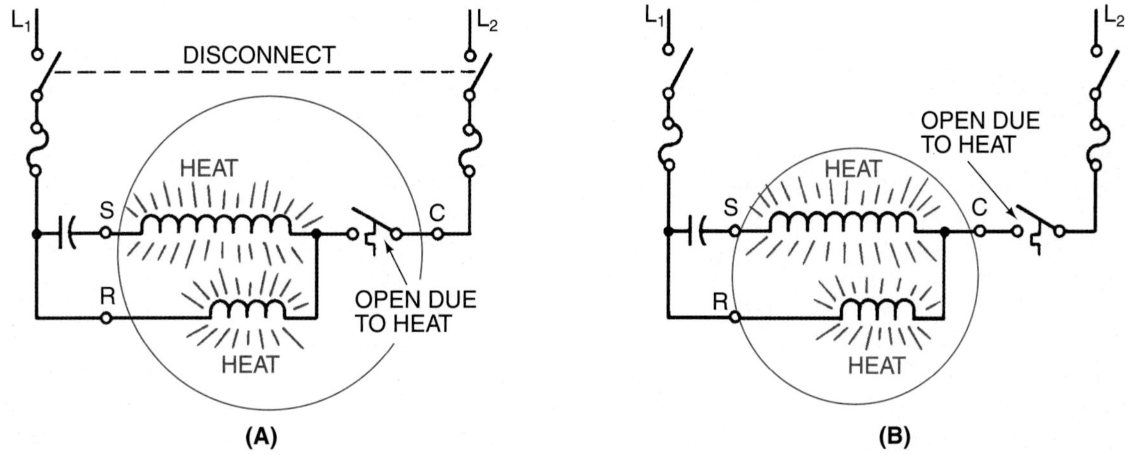

Figure 14.28 The thermistor type of temperature-monitoring device uses an electronic monitoring circuit to check the temperature at the thermistor. When the temperature reaches a predetermined high, the monitor interrupts the circuit to the contactor coil and stops the compressor.

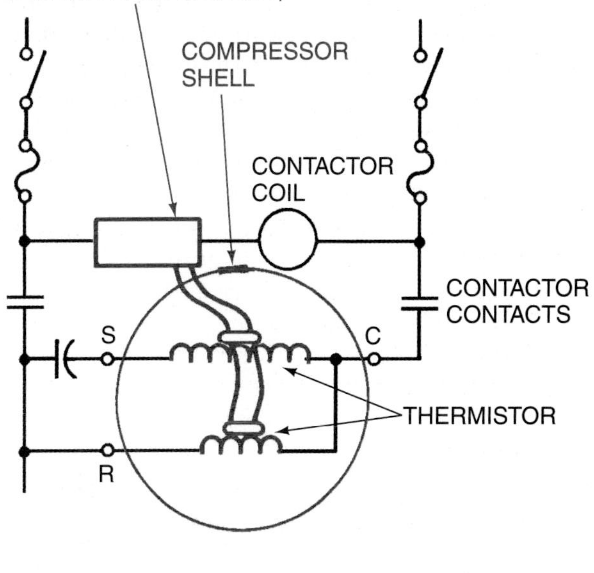

Figure 14.29 The compressor and motor are suspended in vapor space in the compressor shell. The vapor conducts heat slowly.

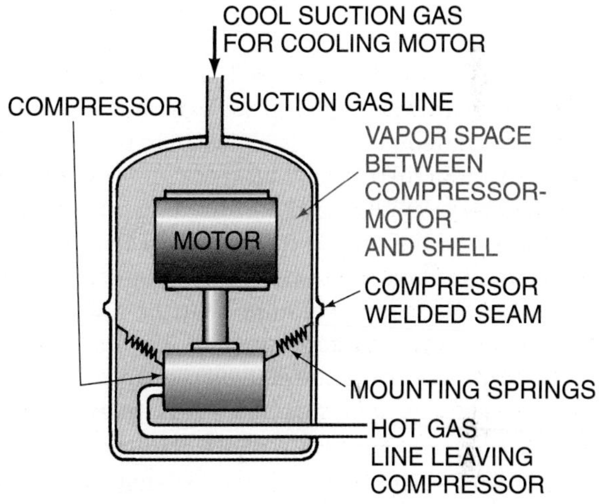

quickly. If the motor is inside a compressor shell, it may be suspended from springs inside the shell, which means that the actual motor and the compressor may be hot and hard to cool even though the shell does not feel hot. There is a vapor space between the outside of the shell and the actual heat source, **Figure 14.29.** **⚠ SAFETY PRECAUTION:** *The unit must have time to cool. If you are in a hurry, set up a fan, or even cool the compressor with water, but do not allow water to get into the electric circuits. DO NOT try to cool a cast-iron compressor with cold water. The shock may cause a casting failure. Turn the power off and cover electric circuits with plastic before using water to cool a hot compressor. Be careful when restarting. Use a clamp-on ammeter to check for an overcurrent condition and gauges to determine the charge level of the system.*• Most hermetic compressors are cooled by suction gas. If there is a refrigerant undercharge, the system superheat will be high and the temperature of the suction gas reaching the compressor will be high as well. Compressor motor overheating is very likely to occur.

⚠ SAFETY PRECAUTION: *The best procedure when a hot compressor is encountered is to shut off the compressor with the space thermostat and return the next day. This gives the compressor enough time to cool and keeps the crankcase heat on. Many service technicians have diagnosed an open winding in a compressor that was only hot, and they later discovered that the winding was open because of internal thermal protection.*•

14.5 MEASURING THE TEMPERATURE OF FLUIDS

The term *fluid* applies to both the liquid and vapor states of matter. Liquids are heavy and change temperature very slowly. For example, it takes 1 Btu of heat energy to change the temperature of 1 pound of water 1°F, but it only take 0.24 Btu of heat energy to change the temperature of 1 pound of air 1°F. The sensing element must be able to reach the temperature of the medium to be measured as soon as practical. Because liquids are contained in vessels (or pipes), the measurement can be made either by contact with the vessel or by some kind of immersion. When a temperature reading is taken from the outside of the vessel by contact, care must be taken that the ambient temperature does not affect the reading.

A good example of readings being affected by ambient temperature is a thermostatic expansion valve's sensing bulb that is used to control the operation of the valve. The sensing bulb must sense refrigerant gas temperature accurately to keep liquid from entering the suction line. Often, the technician will strap the sensing bulb to the suction line in the correct location but fasten it incorrectly. The technician may forget to insulate the bulb from the ambient temperature when it needs to be. In this case, the bulb will sense the ambient temperature and average it in with the line temperature. Since the ambient temperature is higher than the temperature of the refrigerant in the suction line at the outlet of the evaporator, the pressure in the bulb will be higher than desired. This will push the thermostatic expansion valve open, overfeeding the evaporator. Make sure when mounting a sensing bulb to a line for a contact reading that the bulb is making the very best possible contact, **Figure 14.30**. Brass or copper mounting straps, not plastic, should be used to secure sensing bulbs to the suction line or anywhere else freezing temperatures occur. Brass and copper are much stronger than plastic and can withstand the freezing and thawing cycles, which will prevent breakage from expansion when moisture gets in between the suction line and the sensing bulb and freezes.

When a temperature reading is needed from a larger pipe in a permanent installation, different arrangements are made.

Figure 14.30 The correct way to get good contact with a sensing bulb and the suction line. The bulb is mounted on a straight portion of the line with a strap that holds it securely against the line. Photo by Jason Obrzut

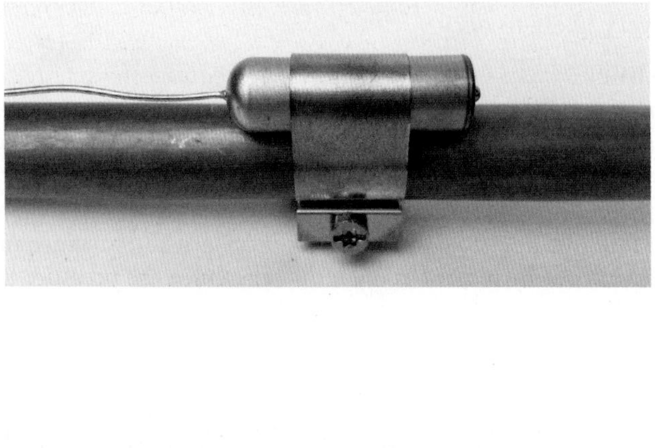

Figure 14.31 (A) A thermometer that is designed be mounted in a well. (B) A well that can be connected to a piece of equipment or installed in a piping circuit. (C) A thermometer installed in a well on a piece of equipment.

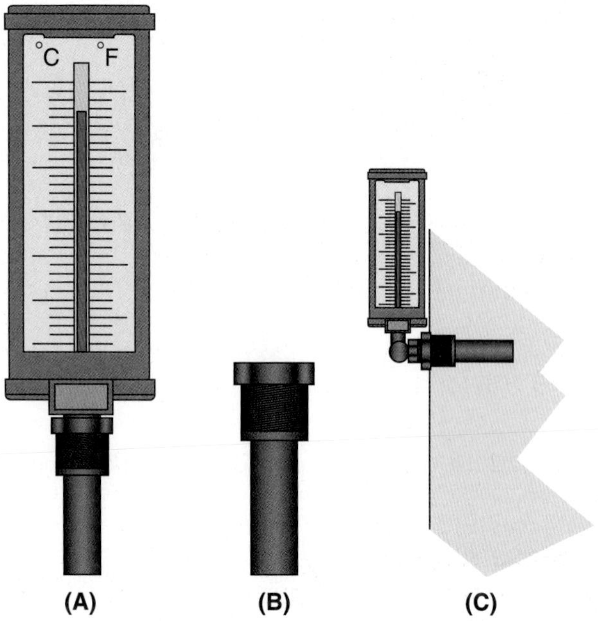

A well can be installed into the pipe so that a thermometer or controller-sensing bulb can be inserted into it, **Figure 14.31**. The well must be matched to the sensing bulb to get a good contact fit, or the reading will not be accurate. Sometimes, for troubleshooting purposes, it is desirable to remove the thermometer from the well and insert an electronic thermometer. The inside diameter of the well is much larger than the probe, but it can be packed so that the probe is held firm against the wall of the well to obtain an accurate reading, **Figure 14.32**.

Figure 14.32 The well is packed to get the probe against the wall of the well. A method for obtaining a leaving-water temperature reading from a small boiler.

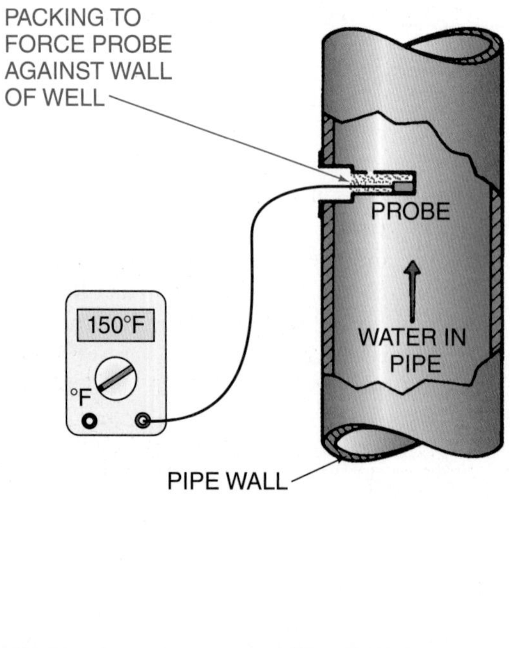

Figure 14.33 A snap-disc temperature sensor that can be used to sense air temperature in ductwork and furnaces. Photo by Bill Johnson

Sensing temperature in fast-moving airstreams such as ductwork and furnace heat exchangers is usually done by inserting the sensing element into the actual airstream. A bimetal device such as the flat-type snap-disc or the helix coil is usually used. **Figure 14.33** shows a snap-disc. Refer to **Figure 13.7(B)** for the configuration of a helix coil.

14.6 PRESSURE-SENSING DEVICES

Pressure-sensing devices are normally used when measuring or controlling the pressure of refrigerants, air, gas, and water. They are sometimes strictly pressure controls, or they can be used to operate electrical switching devices. The terms **pressure control** or **pressure switch** are often used interchangeably in the field. The actual application of the component should indicate if the device is being used to control a fluid or to operate an electrical switching device.

Mechanical controls contain a bellows, a diaphragm, or a Bourdon tube to create movement when the pressure being sensed by the device changes. Pressure switches can be used to stop and start electrical loads, such as motors, **Figure 14.34**, but they can also be attached to switches or valves, **Figure 14.35**. When used as a switch, the bellows, Bourdon tube, or diaphragm is attached to the linkage that operates the electrical contacts. The electrical contacts are the components that actually open and close the electrical circuit.

Figure 14.34 The electrical circuit of a refrigeration compressor with a high-pressure control. This control has a normally closed circuit that opens on a rise in pressure.

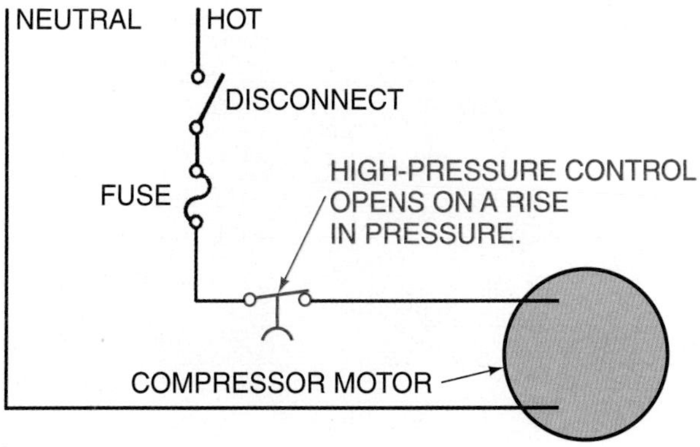

Figure 14.35 The moving part of most pressure-type controls.

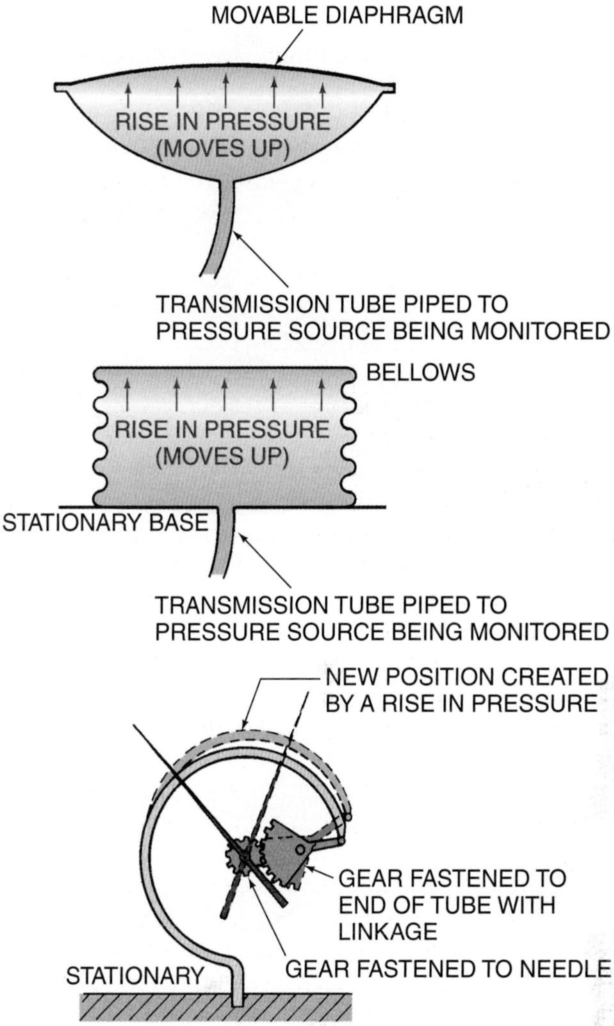

The electrical contacts either open or close with snap action on a rise in pressure, **Figure 14.36**. The pressure setting at which the control interrupts the electrical circuit is known as the **cut-out**. The point or pressure setting at which the electrical circuit is made is known as the **cut-in**. The difference between the two settings is known as the **differential**.

Pressure controls can either open or close on a rise in pressure and are classified as "open on pressure rise" or "close on pressure rise" devices. **Figure 14.37** shows the electrical symbols for some commonly used pressure switches. The pressure control can also sense a pressure differential and be designed to open or close a set of electrical contacts, depending on the sensed pressure difference. Refer to Section 14.10 for an example of the oil pressure safety switch, which is a pressure differential switch.

Pressure switches can function as operational devices or safety controls. A pressure relief valve, **Figure 14.38**, for example, is a safety control and will open if the pressure in a hot water system reaches unsafe levels. Once the pressure has

Figure 14.36 Snap-action device. Photo by Bill Johnson

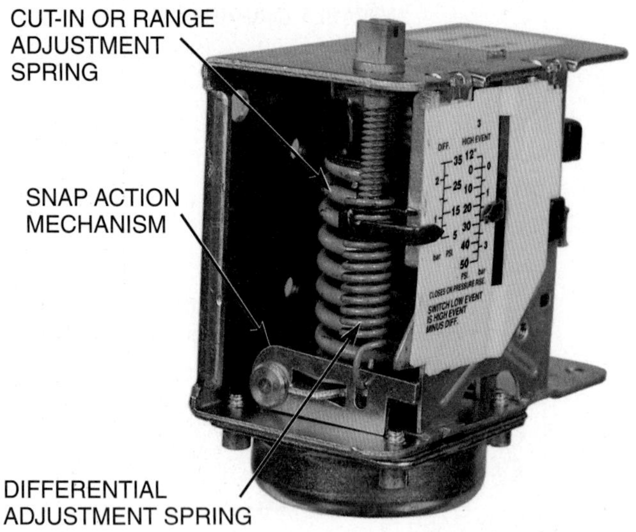

Figure 14.37 (A) Normally open pressure switch that closes on a rise in pressure. (B) Normally closed pressure switch that closes on a rise (opens on a drop) in pressure. (C) Normally open pressure switch that opens on a rise (closes on a drop) in pressure. (D) Normally closed pressure switch that opens on a rise in pressure.

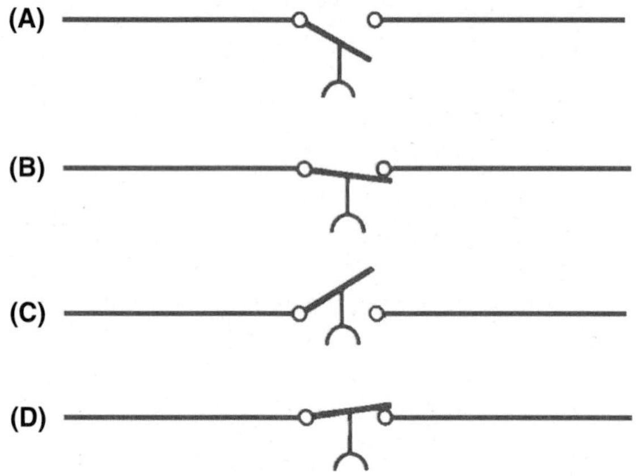

dropped to a safe pressure, the valve will close automatically. A water-regulating valve, **Figure 14.39**, is an operational device. This type of valve is commonly found on water-cooled condensers and will open and close in order to maintain the high side pressure of an air-conditioning or refrigeration system within an acceptable range. This type of valve has a spring pressure that pushes against the high side pressure of the air-conditioning system. If the high side pressure is too low, the spring pressure pushes the valve closed and the high side pressure will increase. If the high side pressure is too high, the high side pressure pushes the valve open and more water flows through the condenser to lower the high side pressure. When used to control fluid flow, the pressure controls are normally attached directly to the valve.

The pressure control can operate either at low pressures (even below atmospheric pressure) or high pressures, depending on the design of the control mechanism. Pressure controls are easily recognized by the small tubes or transmission lines connected to them. These lines are connected to the system at the point where the pressure is being sensed.

The high-pressure and low-pressure controls in refrigeration and air-conditioning equipment are the two most widely used pressure controls in this industry. Some controls are automatic reset, **Figure 14.40**, and some must be reset manually, **Figure 14.41**. Some pressure switches

Figure 14.38 The relief valve on a boiler acts as a safety control. Photo by Bill Johnson

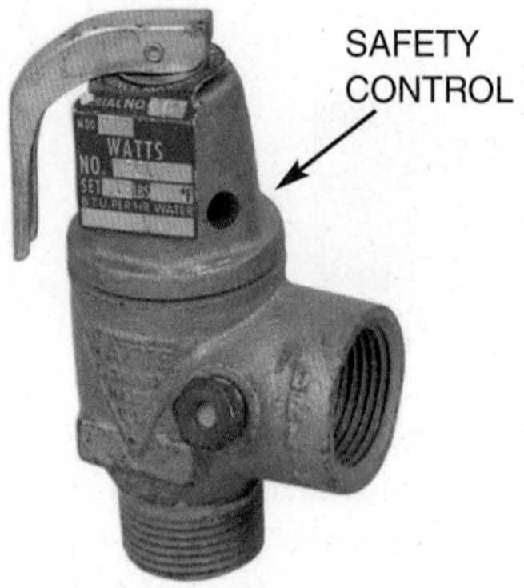

Figure 14.39 A water-regulating valve on a water-cooled condenser acts as an operational device. Photo by Eugene Silberstein

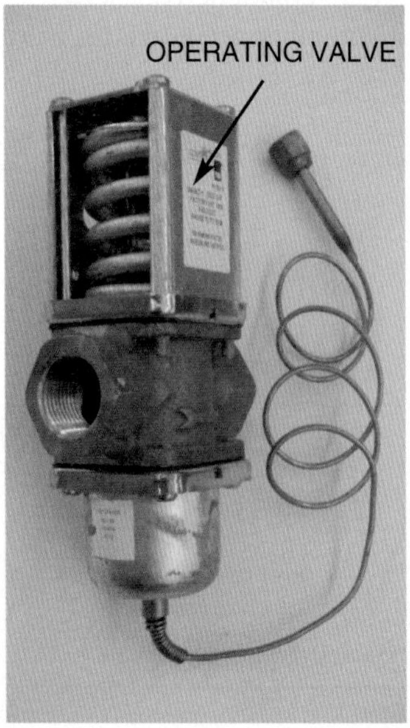

Figure 14.40 An automatic-reset high-pressure switch commonly found on central air-conditioning systems. Photo by Eugene Silberstein

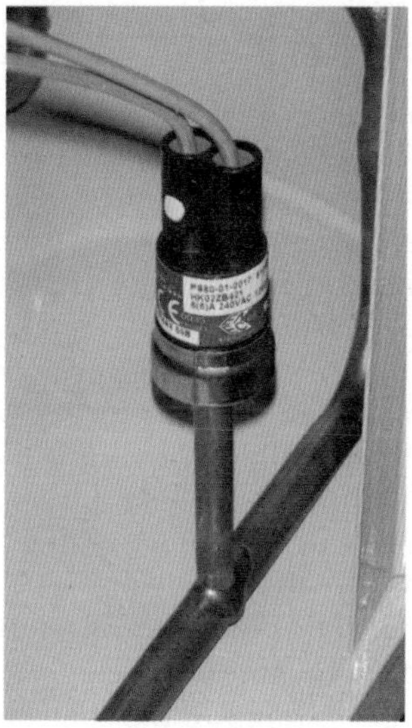

Figure 14.42 A commonly used high-pressure control. Photo by Bill Johnson

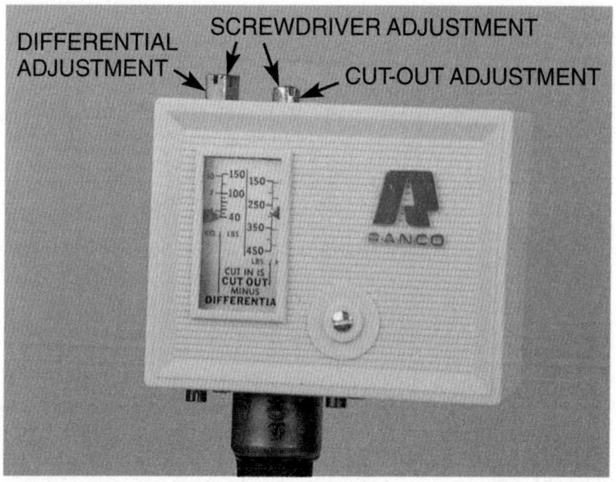

Figure 14.43 The control has two bellows acting on one set of contacts. Either control can stop the compressor. Photo by Bill Johnson

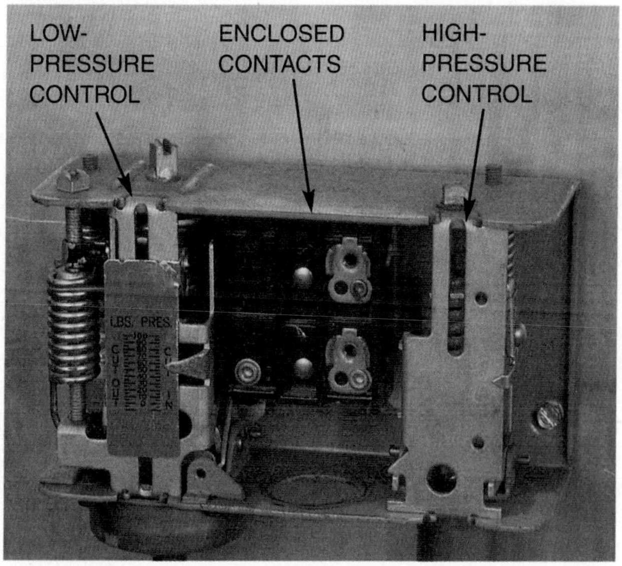

are adjustable, **Figure 14.42**, and some are not. The pressure switch in **Figure 14.40** is not adjustable. In some pressure controls, the high-pressure and the low-pressure controls are built into one housing. These are called dual-pressure controls, **Figure 14.43**. Pressure controls are usually located near the compressor on air-conditioning and refrigeration equipment. **⚠ SAFETY PRECAUTION:** *When used as safety controls, pressure controls should be installed in such a manner that they are not subject to being valved off, or otherwise isolated from the system.•*

In the HVACR field, both pressure and temperature controls incorporate cut-out, cut-in, and differential adjustments. As mentioned earlier, the cut-out of a control interrupts or opens the electric circuit. The cut-in closes the electric circuit, and the differential is the difference between the cut-in and cut-out points. This logic can be represented in the following formula:

$$\text{Cut-in} - \text{Cut-out} = \text{Differential}$$

Another way to write the equation is:

$$\text{Cut-in} - \text{Differential} = \text{Cut-out}$$

If a pressure control has a cut-in of 20 psi and a cut-out of 5 psi, the differential would be 15 psi (20 psi – 5 psi). **Figure 14.44** shows this in graphical form. Note that this control language can be used for both pressure and temperature controls. Some control manufacturers refer to the cut-out as the "low event" and the cut-in as the "high event."

Figure 14.41 The control on the left is a manual reset control. Notice the push lever on the left-hand control. Photo by Bill Johnson

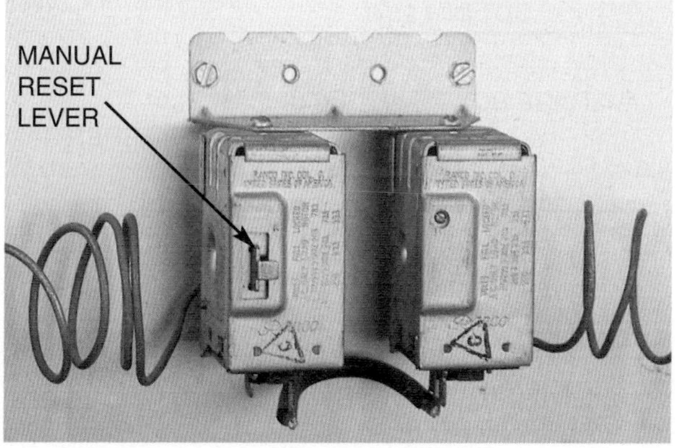

Figure 14.44 Cut-in, cut-out, and differential points shown in a graphical format for (A) a pressure controller and (B) a temperature controller.

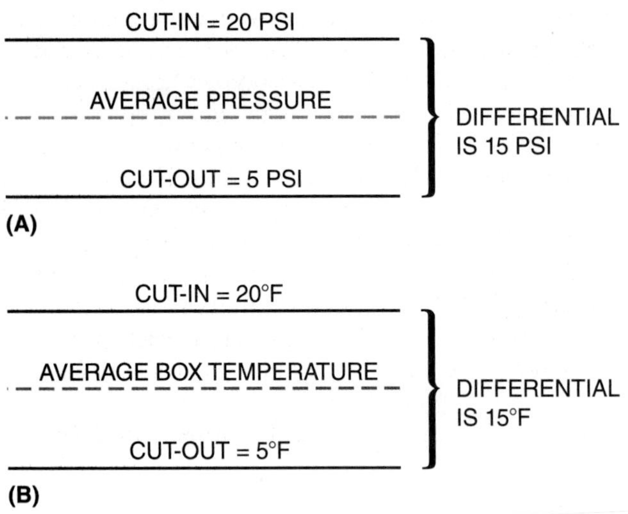

As you can see, the differential controls the pressure or temperature difference between the cut-in and cut-out settings. The differential adjustment is usually not easily accessible to the owner or operator when making a temperature or pressure change with the control. It is usually protected or locked up by some sort of mechanism that takes time and knowledge to access. In most cases, only a service technician can make an adjustment to the control's differential. If the differential is set too low, the compressor motor or other power-consuming devices it is controlling may short cycle. Short cycling can overheat motor windings and prematurely wear and pit electrical contacts. On the other hand, if the differential is set too high, the motor may never shut off and may consume large amounts of power. Some differential controls can be nonadjustable and built into the control as a fixed differential. Most high-pressure controls and inexpensive box thermostats are nonadjustable.

Changing the cut-in setting without changing the differential will automatically change the cut-out setting, **Figure 14.45**. The **range** of the control is increased and so is the average box temperature. In the figure, the beginning range is from 5°F to 20°F. By changing the cut-in adjustment of the control to cut in at 25°F, the cut-out point is automatically moved to 10°F because the differential stays constant at 15°F. Notice that both the cut-in and cut-out temperatures are higher, but the difference between the two temperatures (differential) does not change. This gives the control a new range, from 10°F to 25°F. It also increases the average box temperature. The range provides for the correct minimum and maximum pressures or temperatures when it is controlled automatically by a controller. There must be a range of temperatures because it is very difficult to control the box temperature of a refrigerated case at one exact temperature.

The cut-in adjustment of a control is often referred to as the range adjustment because changing the cut-in causes both the cut-out and cut-in (range of the control) to change, as in **Figure 14.45**. If the differential of the control is changed, the range of the control will also be changed because the differential is the distance between the two settings. The range (cut-in) adjustment spring always acts directly on the control bellows that operates the temperature- or pressure-control electrical switch. The differential adjustment or spring is usually set to the side of the bellows and adds resistance to the mechanism that the bellows of the control is acting on. This action makes it either easier or harder for the electrical contact points to open or close, **Figure 14.36**. Applications and more detailed examples of cut-in, cut-out, and differential settings will be covered in Unit 25, "Special Refrigeration System Components."

There are many electronic pressure controls on the market, such as the one shown in **Figure 14.46**. They often have small, on-board microprocessors that are hardwired to an electronic pressure transducer. These versatile electronic

Figure 14.45 A change in the cut-in adjustment by 5°F in (B) will automatically change the cut-out setting by 5°F. This changes the range. Notice that the differential is the same in both examples (A) and (B).

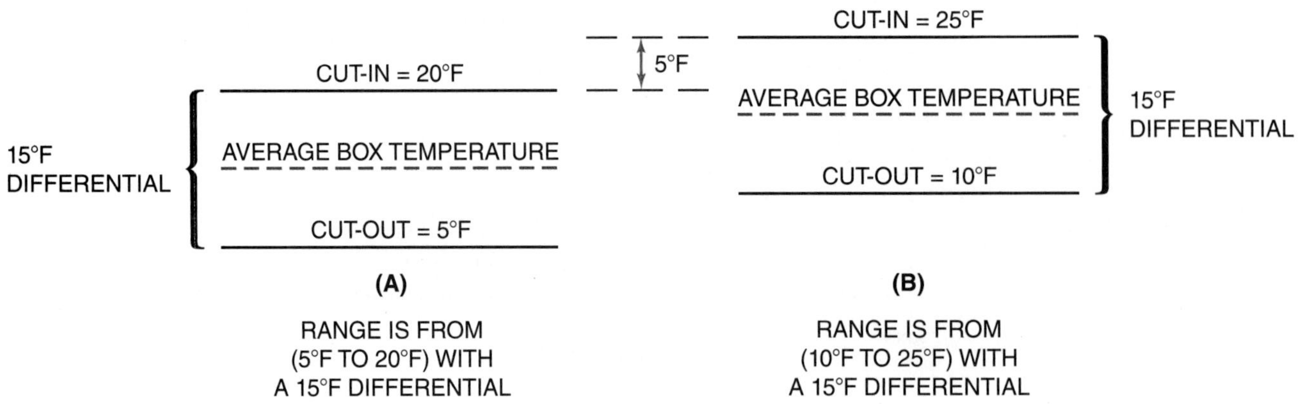

Figure 14.46 An electronic pressure control with a connecting pressure transducer. Courtesy Ferris State University. Photo by John Tomczyk

controls have easy-to-read digital liquid crystal displays. Some of their features include the following:

- Selectable pressure ranges
- Anti-short–cycling devices
- Remote pressure transducer (easily mounted)
- Wiring harness (no capillary tubes)
- Lockable touch pads (prevents tampering)
- Ability to open contacts on an increase or decrease in pressure

14.7 PRESSURE TRANSDUCERS

Pressure transducers are devices that measure pressure and then convert the sensed pressure into an electronic signal that is then processed by a microprocessor. Two modern pressure transducers are shown in **Figure 14.47**. Most pressure transducers today are capacitive-type pressure transducers. They are usually constructed of a durable, compact housing containing two closely spaced parallel plates, **Figure 14.48**, The parallel plates are made of conductive material and are electrically isolated from each other with an insulator, usually air or a vacuum, referred to as a dielectric. One of the plates is stationary; the other acts like a diaphragm that slightly flexes when pressure is applied to its surface. A slight mechanical flexing of the movable plate caused by the applied pressure will change the gap between the plates.

Figure 14.47 Two pressure transducers. Courtesy Ferris State University. Photo by John Tomczyk

Figure 14.48 The cross section of a capacitive-type pressure transducer showing a movable diaphragm, which creates a variable capacitance in a circuit.

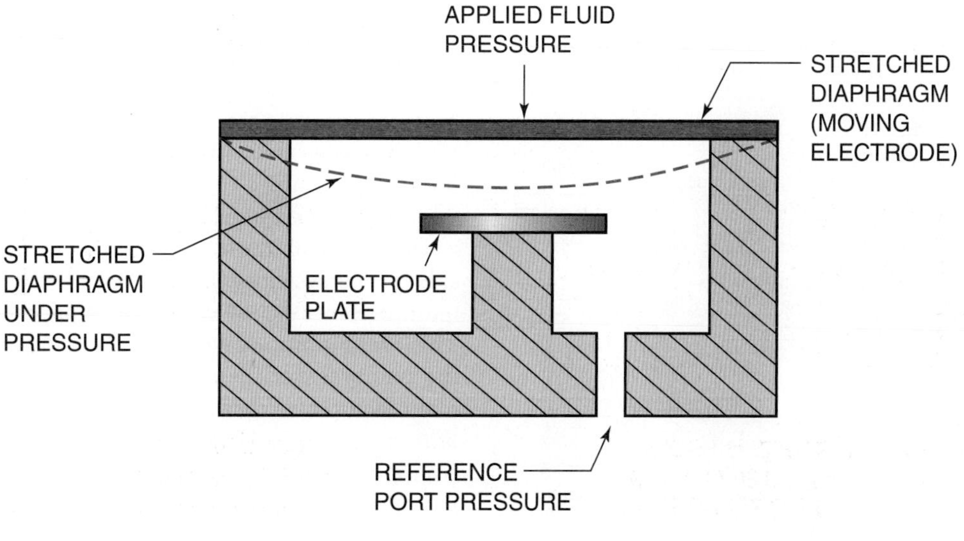

Figure 14.49 A working pressure transducer on a parallel compressor rack's suction line. Courtesy Ferris State University. Photo by John Tomczyk

This motion causes the capacitance to change as more or less pressure is applied to the flexible plate or diaphragm. The change in capacitance is sensed by solid-state, linear comparative circuitry that is usually linked to a microprocessor. The solid-state circuit analyzes the signal and then amplifies its proportional output signal. Capacitive-type pressure transducers can sense pressures in the ranges of 0.1 in. of water column, IWC, (0.0036 psig) to 10,000 psig. Other types of pressure transducers react to changes in pressure by changing the device's resistance or inductance.

Figure 14.49 shows a working pressure transducer connected to the common suction line of a parallel compressor system. The transducer will sense the common suction pressure in the suction line and convert this signal to a voltage signal for a microprocessor to analyze. This signal is used to cycle different-size compressors on and off to more accurately match the heat loads on the refrigerated cases in the store. There is also solid-state circuitry in the head of the transducer, as shown in **Figure 14.50**. The transducer usually has three wires connected to it. Two wires provide the input voltage, which is usually a DC voltage, and the other wire is the output signal, which can also be a DC voltage. The transducer senses pressure and converts the signal to an output voltage signal that is processed by the microprocessor board, which operates relays that either start or stop power-consuming devices.

Figure 14.50 Solid-state circuitry in the head of the pressure transducer. Courtesy Ferris State University. Photo by John Tomczyk

14.8 HIGH-PRESSURE CONTROLS

The high-pressure control on an air-conditioning system stops the compressor if the pressure on the high side becomes excessive. A high-pressure condition can occur, for example, when the condenser fan motor on an air-cooled air-conditioning system fails. In a wiring diagram this control appears as a normally closed control that opens on a rise in pressure, **Figure 14.37(D)**. The manufacturer may set the upper limit of operation for a particular piece of equipment and furnish a high-pressure cut-out control to ensure that the equipment does not operate above these limits. High-pressure controls on HVACR equipment usually have a built-in differential that often ranges from 50 psig to 100 psig. For example, if the control opens at a pressure of 400 psig, it will usually reset itself (close contacts) at a pressure between 300 and 350 psig.

⚠ SAFETY PRECAUTION: *The high-pressure control is one method of ensuring safety for the equipment and the surroundings. Some compressors are strong enough to burst a pipe or a container. On semihermetic compressors, pressures can often increase to the point of blowing out a head gasket or a valve plate gasket. The results can be a large refrigerant loss and spoiled product. The overload device in the compressor offers some protection, but it is really secondary because it does not directly respond to pressures. The motor overload device may also be a little slow to respond to pressure.•*

14.9 LOW-PRESSURE CONTROLS

The low-pressure control is commonly used in the air-conditioning field as a low-charge protection device, **Figure 14.51**. In commercial refrigeration applications, low-pressure controls are also used to cycle compressors on and off in response to evaporator pressures that indirectly control box temperatures. They can also be used in ice makers to initiate the harvest cycle. When equipment loses some of the refrigerant from the system, the pressure on the low-pressure side of the system will fall. Manufacturers may set a minimum pressure below which the equipment will not operate. This is the point at which the low-pressure control cuts off the compressor. These settings vary from manufacturer to manufacturer, so their recommendations should be followed. For example, R-410A air-conditioning systems are often equipped with low-pressure controls that cycle the compressor off if the low-side pressure falls below 50 psig.

Figure 14.51 The low-pressure control. Photo by Bill Johnson

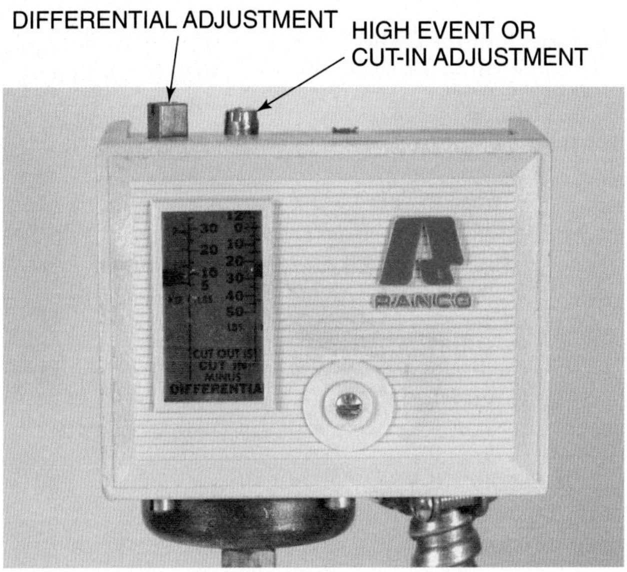

Special care must be taken when installing a low-pressure control on systems that are equipped with capillary tube metering devices, or other fixed-bore-type metering devices that do not have the ability to adjust refrigerant flow. These fixed-bore devices allow system pressures to equalize during the off cycle. If the low-pressure control is not properly selected or set, the device can cause the compressor to short cycle. To prevent short-cycling from occurring, some systems are equipped with a time-delay circuit that will not allow the compressor to restart for a predetermined time period.

It is undesirable to operate a system without an adequate charge for two good reasons:

1. The compressor motors used most commonly in the industry, particularly in air-conditioning, are cooled by the refrigerant. Without this cooling action, the motor will build up heat when the charge is low. The motor temperature cut-out is used to detect this condition and often takes the place of the low-pressure cut-out by sensing motor temperature.
2. If the refrigerant escapes from a leaking system, the system may very well operate in a vacuum. When a vessel is in a vacuum, the atmospheric pressure is greater than the vessel pressure. This causes the atmosphere to be sucked into the system, where it is sometimes hard to detect. The reference point is atmospheric pressure. If air in the system is not removed, it will be pumped through the evaporator and the compressor. Because air is a noncondensable gas, it will not condense as refrigerant will and will take up valuable volume in the condenser and cause unwanted high head pressures. These higher head pressures cause high compression ratios and low efficiencies. The compressor's discharge temperature will elevate to a point where oil may break down. Also, the combination of oxygen and moisture in the air, excess heat, and the refrigerant will create an oil sludge and acid formations.

14.10 OIL PRESSURE SAFETY CONTROLS

The oil pressure safety control is used to ensure that the compressor has sufficient oil pressure when operating, **Figure 14.52.** If the oil pressure is too low, the control will open its contacts and deenergize the compressor. Used on larger compressors, this control has a different sensing arrangement than the high- and low-pressure controls, which are single-diaphragm or single-bellows controls used to compare atmospheric pressure with the pressures inside the system.

The oil pressure safety control is a pressure differential control. It measures a difference in pressure to determine if positive oil pressure is present. The compressor crankcase pressure (where the oil pump suction inlet is located) is the same as the compressor suction pressure, **Figure 14.53.** The

Figure 14.52 An oil pressure safety control. Courtesy Ferris State University. Photo by John Tomczyk

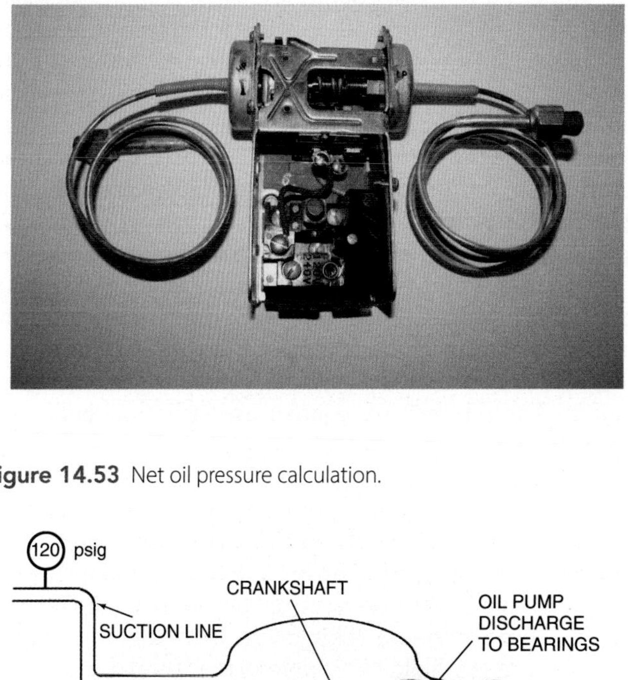

Figure 14.53 Net oil pressure calculation.

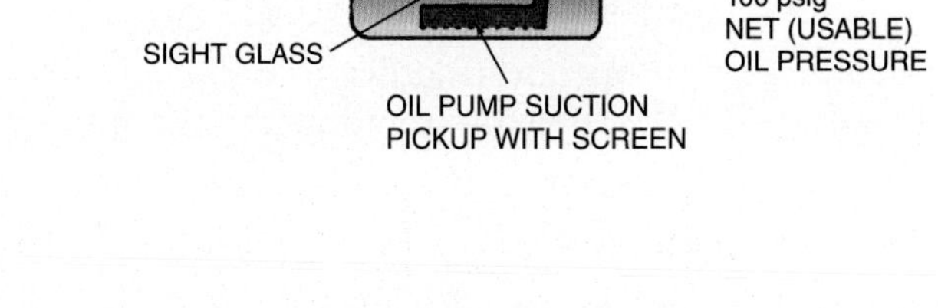

Figure 14.54 The oil pressure control is used to protect the compressor by ensuring adequate lubrication. The oil pump that lubricates the compressor is driven by the compressor crankshaft. Therefore, a time delay is necessary to allow the compressor to start up and build oil pressure.

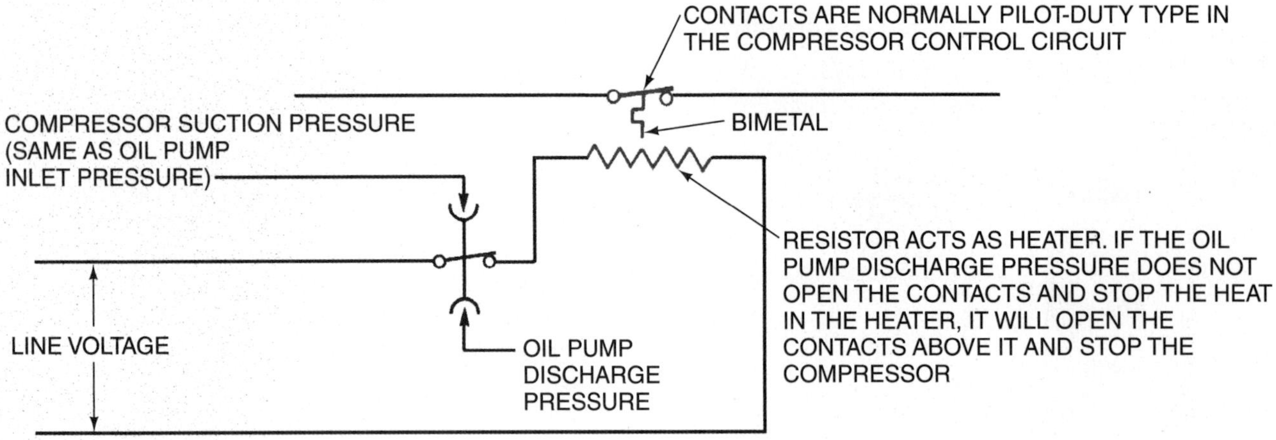

suction pressure will vary from the off or standing reading to the actual running reading. For example, a system using R-410A as the refrigerant may have the pressure readings similar to 210 psig while standing, 120 psig while operating, and 40 psig during a low-charge situation. A plain low-pressure cut-out control would not function at all of these levels, so a control had to be devised that would sensibly monitor pressures under all of these conditions. Most compressors need at least 30 psig of actual oil pressure for proper lubrication. This means that whatever the suction pressure is, the oil pump discharge pressure has to be at least 30 psig above the oil pump inlet pressure, because the oil pump inlet pressure is the same as the suction pressure. For example, if the suction pressure is 120 psig, the oil pump outlet pressure must be at least 150 psig. This difference in the suction pressure and the oil pump outlet pressure is called the **net oil pressure**.

In the basic low-pressure control, there is pressure under a diaphragm or bellows and atmospheric pressure on the other side. The atmospheric pressure is considered a constant because it does not vary by more than a small amount. The oil pressure control uses a double bellows—one bellows opposing the other to detect the net or actual oil pressure. These bellows are positioned opposite to each other, either physically or by linkage. The pump inlet pressure is under one bellows, and the pump outlet pressure is under the other bellows. The bellows with the higher pressure is the pump outlet bellows, and it overrides the bellows with the lower pressure. This override in effect is` the net pressure, and the bellows is attached to a linkage that can stop the compressor when the net pressure drops for a predetermined time.

When the compressor is off, the net oil pressure is zero and, under these conditions, the compressor should not be allowed to start. This poses a problem since we need the compressor to operate in order to build up oil pressure and we need oil pressure to allow the compressor to operate. A time delay is therefore built into the control to allow the compressor to get started, build up the necessary oil pressure, and prevent unnecessary cut-outs when the oil pressure varies for only a moment. The time delay is normally about 90 sec and is accomplished by means of a heater circuit and a bimetal device or electronic circuitry. **NOTE:** *The manufacturer's instructions should be consulted when working with any compressor that has an oil safety control,* **Figure 14.54.**•

Oil safety controls can also incorporate a pressure transducer for sensing the combination of oil pump discharge pressure and suction (crankcase) pressure, **Figure 14.55.** The difference between the two pressures (net oil pressure)

Figure 14.55 An electronic oil safety controller pressure transducer that senses net oil pressure. Courtesy Ferris State University. Photo by John Tomczyk

Figure 14.56 The pressure transducer connected by wires to the electronic oil safety controller. Courtesy Ferris State University. Photo by John Tomczyk

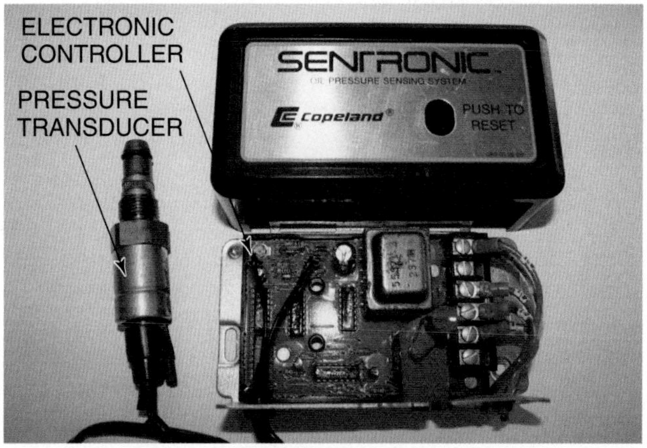

Figure 14.57 An air pressure differential switch used to detect ice on a heat pump outdoor coil. Photo by Bill Johnson

Figure 14.58 A sail switch detects air movement. Photo by Jason Obrzut

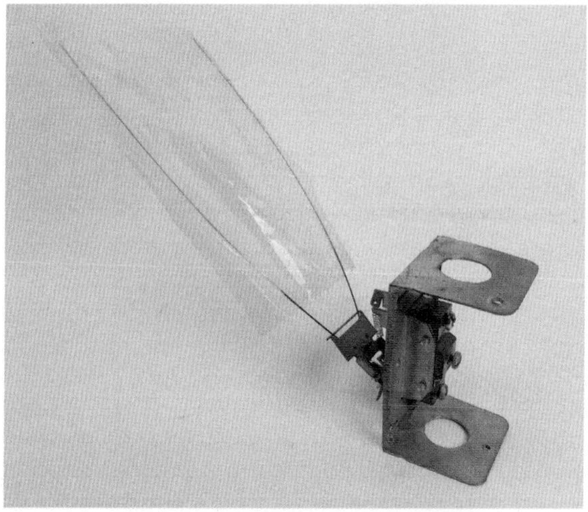

is measured by the transducer, which has two separate sensing ports. As shown in **Figure 14.56,** the pressure transducer is mounted directly in the oil pump and is connected to an electronic controller by wires. The pressure transducer transforms a pressure signal to an electrical signal for the electronic controller to process. One advantage of an electronic oil safety controller over a mechanical bellows-type controller is that it eliminates capillary tubes; thus, there is less chance of refrigerant leaks. Also, the electronic clock and circuitry are much more accurate and reliable. A more detailed explanation of both the mechanical bellows and the electronic oil safety controllers, with accompanying circuitry, will be covered in Unit 25, "Special Refrigeration System Components."

14.11 AIR PRESSURE CONTROLS

Air pressure switch sensors need a large diaphragm to sense the very low pressures in air systems. Mounted on the outside of the diaphragm is a small switch (microswitch) capable only of stopping and starting control circuits. Air pressure controls (switches) are used in several applications.

1. Ice buildup on the outdoor coil of a heat pump will cause an air pressure drop through the coil. When a predetermined pressure drop is detected, there is sufficient ice buildup and a defrost cycle is needed. When used in this manner, the air pressure control is an operating control, **Figure 14.57.** Not all manufacturers use air pressure as an indicator for defrost.
2. When electric heat is used in a remote duct as terminal heat, a sail switch is sometimes installed to ensure that the blower is passing air through the duct before heat is allowed to be generated. Thus, this control is a safety switch and should be treated as such, **Figure 14.58.** The switch has a very lightweight and sensitive sail. When air is passing through the duct, the sail is pushed to close a set of electrical contacts, allowing the heat source to be activated. When the blower stops, the sail returns to its original position, ensuring that the heat source is turned off. Some sail switches have both normally closed and normally open sets of contacts that can be used depending on the desired application.
3. Air switches or sail switches are also used to prove that the combustion blower motor is operating before energizing the main gas valve for ignition.

14.12 GAS PRESSURE SWITCHES

Gas pressure switches are similar to air pressure switches but usually smaller. They detect the presence of gas pressure in gas-burning equipment before burners are allowed to ignite. **⚠ SAFETY PRECAUTION:** *This is a safety control and should never be bypassed except for troubleshooting and then only by experienced service technicians.•*

14.13 SWITCHLESS CONTROL DEVICES

The pressure relief valve can be considered a pressure-sensitive device. It is used to detect excess pressure in any system that contains fluids (water or refrigerant). Boilers, water heaters, hot water systems, refrigerant systems, and gas systems all use pressure relief valves, **Figure 14.59.** This device can be either pressure- and temperature-sensitive (called a p & t valve) or just pressure-sensitive. The p & t valves are normally used in water heaters. The type considered here is the pressure-sensitive type.

Figure 14.59 Pressure relief valves. (A) Boiler relief valve. Photo by Eugene Silberstein (B) Refrigerant relief valve. Courtesy Superior Valve Company

(A)

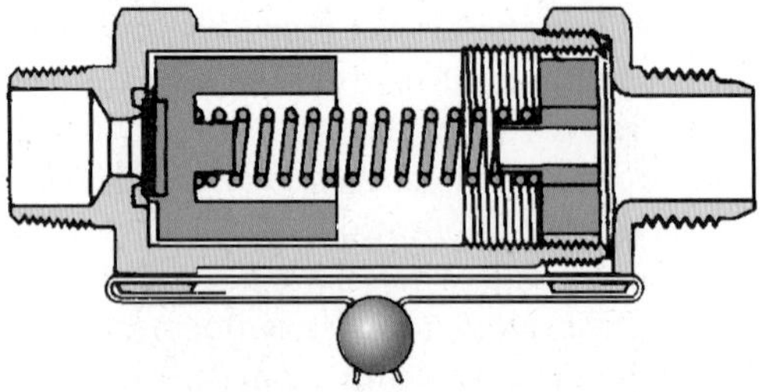

(B)

The pressure relief valve can usually be recognized by its location. On boilers it is often placed at the highest point on the boiler. **⚠ SAFETY PRECAUTION:** *This valve is a safety device that must be treated with the utmost respect. Some have levers on top that can be raised to check the valve for flow. Most have a visible spring that pushes the seat downward toward the system's pressure. A visual inspection reveals that the spring holds the seat against the valve body. The valve body is connected to the system, so, in effect, the spring is holding the valve seat down against the system pressure. When the system pressure becomes greater than the spring tension, the valve opens and relieves the pressure and then reseats.•* **⚠ SAFETY PRECAUTION:** *The pressure setting on a relief valve is set at the factory and should never be tampered with. Normally, the valve setting is sealed or marked on the valve in some manner so that the correct replacement valve can be selected.•* Most relief valves automatically reset. After they relieve, they seat back to the original position. A seeping or leaking relief valve should be replaced. **⚠ SAFETY PRECAUTION:** *Never adjust or plug the valve. Doing this can be extremely dangerous.•*

14.14 WATER PRESSURE REGULATORS

Two types of water pressure regulators are commonly used: the pressure regulating valve for system pressure and the pressure regulating valve for head pressure control in water-cooled refrigeration systems. Air-conditioning and refrigeration systems operate at lower pressures and temperatures when water-cooled, but the water circuit requires a special kind of maintenance. At one time water-cooled equipment was in widespread use. Because it is easy to maintain, air-cooled equipment has now become dominant. Therefore, although water pressure valves are still present in air-conditioning and refrigeration equipment, they are not as common as they once were.

Water pressure regulating valves are used in two basic applications.

1. They reduce supply water pressure to the operating pressure in a hot water heating system. This type of valve has an adjustment screw on top that increases the tension on the spring regulating the pressure, **Figure 14.60.** A great many boilers in home or business hydronic systems use circulating hot water at about 15 psig. The supply water may have a working pressure of 75 psig or more, which must be reduced to the system working pressure. If the supply pressure were to be allowed into the system, the boiler pressure relief valve would open. The water-regulating valve is installed in the supply water makeup line that adds water to the system when some leaks out. Most water-regulating valves have a manual valve

Figure 14.60 Adjustable water-regulating valve for a boiler. Photo by Eugene Silberstein

arrangement that allows the service technician to remove the valve from the system and service it without stopping the system. Most systems also have a manual feedline to allow the system to be filled by bypassing the water-regulating valve. **SAFETY PRECAUTION:** *Care must be taken not to overfill the system.*•

2. The water-regulating valve controls water flow to the condenser on water-cooled equipment to regulate head pressure. This valve takes a pressure reading from the high-pressure side of the system and uses it to control the water flow in order to establish a predetermined head pressure, **Figure 14.39.** For example, if an ice maker were to be installed in a restaurant where the noise of the fan would be objectionable, a water-cooled ice maker could be used. This might be a "wastewater" type of system in which the cooling water is allowed to go down the drain. In the winter the water may be very cool, and in the summer the water may get warm. In other words, the ice maker may have a winter need and a summer need. The water-regulating valve would modulate the water in both cases to maintain the required head pressure. An added benefit when the system is off for any reason is that the head pressure is reduced and the water flow is stopped until needed again. All of this is accomplished without an electrical connection.

The water-regulating valve is controlled by two pressures: the high-side pressure of the refrigeration system and the spring pressure on the water-regulating valve itself. The high-side pressure pushes to open the valve and the valve's spring pressure pushes to close the valve. If the high-side pressure is too high, the valve opens to push more water through the condenser. If the head pressure is too low, the spring pressure on the valve acts to close the valve so less water is pushed through the condenser.

14.15 GAS PRESSURE REGULATORS

The gas pressure regulator is used in all gas systems to reduce the gas transmission pressure to usable pressure at the burner. The gas pressure at the street in a natural gas system could be 5 psig, and the burners could be manufactured to burn gas at 3.5 in. WC (this is a pressure-measuring system that indicates how high the gas pressure can push a column of water). **SAFETY PRECAUTION:** *Street pressure must be reduced to the burner design. This is done with a pressure-reducing valve that acts like the water pressure regulating valve in the boiler system,* **Figure 14.61.**• A device on top of the valve can be used by qualified personnel to adjust the gas pressure. A propane tank can have as much as 150 psig in pressure that needs to be reduced to the burner design pressure of 11 in. WC. A pressure regulator is normally located on the tank, **Figure 14.62.**

Figure 14.61 Gas pressure regulators. (A) Gas pressure regulator at the meter. (B) Gas pressure regulator at the appliance. Photo by Bill Johnson

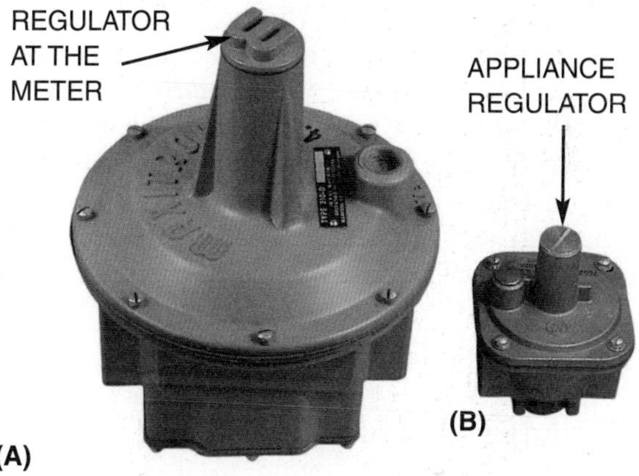

Figure 14.62 The gas pressure regulator on a bottled-gas system. This regulator is on the tank. Photo by Eugene Silberstein

14.16 MECHANICAL AND ELECTROMECHANICAL CONTROLS

A water pressure regulating valve is an example of a mechanical control. The valve maintains a preset water pressure in a boiler circuit without electrical contacts or connections, **Figure 14.60.** It acts independently from any of the other controls, yet the system depends on it as part of the team of controls that make the system function trouble-free. The technician must be able to recognize and understand the function of each mechanical control. This may be more difficult than it is for electrical controls because control diagrams do not always describe mechanical controls as well as they do electrical controls.

Electromechanical controls convert a mechanical movement into some type of electrical activity. A high-pressure switch is an example of an electromechanical control. The switch contacts are the electrical part, and the bellows or diaphragm is the mechanical part. The mechanical action of the bellows is transferred to the switch to stop a motor when high pressures occur, **Figure 14.63.** Electromechanical controls normally are shown on the electrical diagram with a symbol adjacent to the electrical contact describing what the control does, **Figure 14.64.** The symbols used are typically standardized in the industry, but some manufacturers use some variations of the symbols shown in this text.

14.17 MAINTENANCE OF MECHANICAL CONTROLS

Mechanical controls are used to control the flow of fluids such as water, refrigerant, liquids and vapors, or natural gas. The fluids flowing to the controls must be contained within the system, but the diaphragms, bellows, and gaskets that are typically part of the controls are subject to leakage over time.

Figure 14.63 This high-pressure control is considered an electromechanical control. Photo by Eugene Silberstein

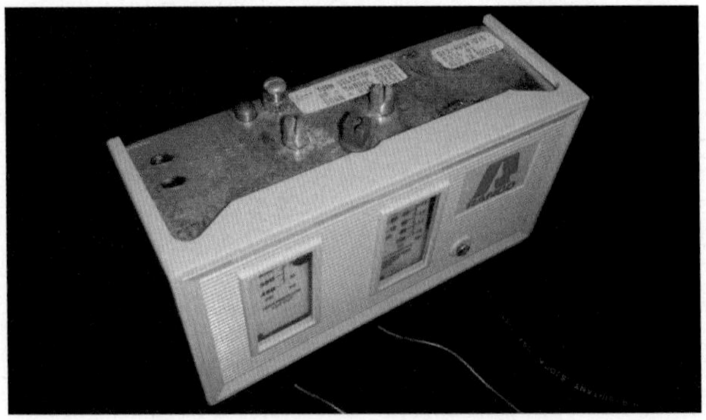

Figure 14.64 Table of electromechanical control symbols.

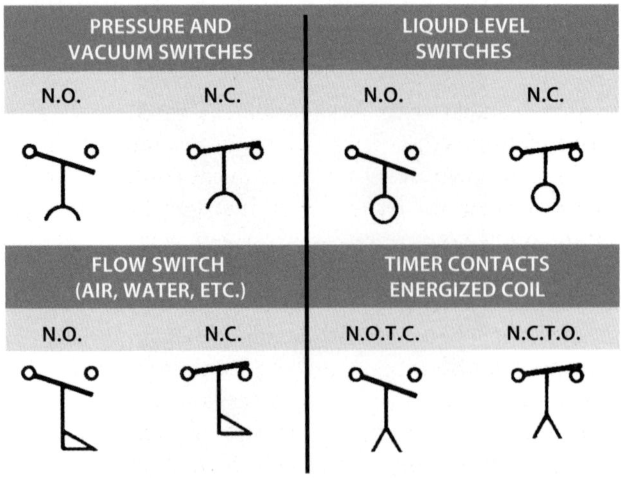

Water is one of the most difficult substances to contain and is likely to leak through even the smallest of openings. All water-regulating valves should be inspected for leaks by looking for wet spots or rust streaks. Water circulating in a system will also leave mineral deposits in the piping and valve seats or mechanisms. For example, a water-regulating valve for a hot water system, **Figure 14.39,** is subject to problems that may cause water leaks or control set-point drift. Some valves are made from brass to prevent corrosion, while some are made of cast iron. The cast iron valves are less expensive but are more likely to rust. Rust formation may bind the valve's moving parts. Mineral deposits or rust inside a valve can prevent the valve from opening and closing, which might result in overfeeding or underfeeding the system. Water valves also have a flexible diaphragm that moves every time the valve functions. The diaphragm is subject to deterioration from flexing and age. If the diaphragm leaks, water will escape to the outside and drip on the floor.

Because pressure relief valves on boilers are a safety control, they are constructed of material that will not corrode. These valves must move freely to ensure that they will not stick shut and lose their function. If a relief valve sticks in the closed position and excessive pressure is present in the boiler, an explosion can result. Many technicians from time to time pull the lever on the top of safety relief valves and let them relieve a small amount of steam or hot water. This ensures that the valve port is free from deposits that may prevent it from functioning. On some occasions, the valve may not seat properly when the lever is released and the valve may seep. Pulling the lever again will usually clear this up by blowing out any trash in the valve seat. If a valve cannot be stopped from leaking, it must be replaced. The fact that pulling the lever may start a leak is the reason many technicians do not "test" relief valves.

14.18 MAINTENANCE OF ELECTROMECHANICAL CONTROLS

Electromechanical controls have both a mechanical and an electrical action. When water is involved, many of the same procedures used to maintain mechanical controls should be followed. Inspect for leaks. Electromechanical pressure controls are often connected to the system with transmission lines that resemble the capillary tube metering device. This small-diameter tube, usually made of copper, is the cause of many leaks in refrigeration systems because of misapplication or poor installation. Consider the example of a low- or high-pressure control. These devices are located near the compressor, but the transmission lines or control tubes must be routed between the control and the compressor to properly sense system pressures. The control must be mounted securely to the frame and the control tube routed in such a manner that it does not touch the frame or any other component. If the control tubes are installed improperly, leaks can result. The vibration of the compressor will vibrate the control tube and rub a hole in the tube or one of the refrigerant lines if they are not kept separate.

The electrical section of the control will usually have a set of contacts or a mercury switch to stop and start some component in the system. The electrical contacts are often enclosed and cannot be viewed, so visual inspection is not possible. When a problem occurs, look for frayed wires or burned wire insulation adjacent to the control. The mercury in mercury switches may be viewed through the clear glass enclosure. If the mercury becomes dark, the tube is allowing oxygen to enter the space where the mercury makes and breaks the electrical contact and oxidation is occurring. In this case, the switch should be replaced because the switch may conduct across the oxidation and so be closed at all times. This could prevent a boiler from shutting off, allowing overheating to occur.

14.19 SERVICE TECHNICIAN CALLS

SERVICE CALL 1

A customer calls indicating that the boiler in the equipment room at a motel has hot water running out and down the drain all the time. Another service company has been performing service at the motel for the last few months. *The problem is that the water-regulating valve (boiler water feed) is out of adjustment. Water is seeping from the boiler's pressure relief valve,* **Figure 14.65**.

The technician arrives at the motel, parking alongside the building so as not to block the front door or the motel's registration parking areas. When the property manager comes into the office to greet the technician, the technician introduces himself and hands him his company's business card. After asking some preliminary questions about the system, the technician and the manager go to the boiler room. Once there, the tech notices that, as the customer said, water is seeping out of the relief valve pipe that terminates in the drain. This is heated water, causing a very inefficient situation. The needle on the boiler gauge is in the red; the pressure is 30 psig. No wonder the valve is relieving the pressure; it is too high. The boiler is hot, but the technician cannot tell if it is too hot. The burner is not operating, so the technician decides that this may be a pressure problem instead of a temperature problem.

The technician looks for and finds the water pressure regulating valve, which is in the supply water line entering the boiler. Actually, this valve keeps the supply water pressure from reaching the boiler and feeds water into the system when there is a loss of water due to a leak. The supply water pressure could easily be 75 psig or more. The boiler working pressure is 30 psig with a relief valve setting of 30 psig. If the supply water were allowed directly into the boiler, it would push the pressure past the design pressure of the system, which must be less than 30 psig. If the pressure were above this, it would cause the pressure relief valve to relieve the pressure and dump water from the system until the pressure is lowered.

The technician shuts off the water valve leading to the boiler makeup system and bleeds some water from the system until the pressure at the boiler is down to 15 psig. He then adjusts the water regulating valve to a lower pressure and opens the supply water valve to allow water to enter the system. The technician can hear water entering through the water regulating valve. In a few minutes, the water stops entering. The pressure in the system is now adjusted to 18 psig and all is well.

In this example, the problem is that a mechanical automatic control is causing another control to show that the

Figure 14.65 This illustration indicates how one control can cause another to look defective. The water pressure regulating valve adjustment had been changed by mistake, allowing more than operating pressure into the system. The water pressure regulating valve is designed to regulate the water pressure down to 25 psig. The valve is necessary to allow water to keep the system full automatically should a water loss develop due to a leak.

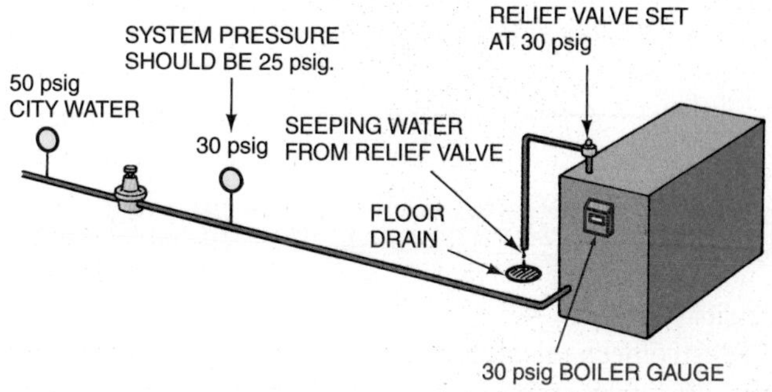

system has problems. If left unexamined, the problem might seem to be that the relief valve is defective because water is seeping out of it. This is not the case; the relief valve is doing its job. The water regulating valve is at fault.

The technician shows the motel manager what the problem was, how it was corrected, and completes the paperwork before leaving for another call. The technician thanks the manager for choosing to let his company service the equipment and asks that his company be called for future service and repairs.

SERVICE CALL 2

A customer in a service station calls to say the hot water boiler in the equipment room is blowing water out from time to time. A technician is quickly dispatched to the job because of the potential hazard. *The problem is the gas valve will not close. The boiler is overheating. The relief valve is relieving water periodically.* The season is mild, so the boiler has a small load on it.

Because the relief valve is relieving intermittently, a periodic pressure buildup in the boiler might be to blame. If the problem were the same as in Service Call 1, where a constant, bleeding relief occurred, the water regulating valve or the relief valve might be suspect. This is not the case.

The technician speaks with the lead mechanic at the service station since the manager is out to lunch. Once at the boiler, the technician reads the gauge; it is in the red and about to relieve again. The technician then looks at the flame in the burner section, notices that it is not a full flame but small, and decides that the gas valve may be stuck open, allowing fuel to enter the burner all the time. The in-line gas hand valve is shut off to put out the flame so the technician can investigate the problem.

A check with a voltmeter shows no voltage at the gas valve terminals, so the thermostat is trying to shut the gas off. The valve must be stuck open. Before replacing the gas valve, the technician calls the office, reports his findings, and gets approval to replace the part. This was very important, since the service station manager is not available.

The technician obtains a new gas valve and installs it. The boiler is started up and allowed to run. It is cold after being off for some time for the gas valve change, so it takes a while for it to get up to temperature. The set point of the thermostat is 190°F and the gas valve shuts off when this temperature is reached.

These service calls are examples of how a trained technician might approach service problems. Technicians develop their own approaches to problems. The point is that no diagram was consulted. A technician either must have prior knowledge of the system or must be able to look at the system and make accurate deductions based on a knowledge of basic principles.

Each control has distinguishing features that indicate its purpose. There are hundreds of controls and dozens of manufacturers (some of whom are out of business). If a technician has no information as to what a control does, the manufacturer should be consulted. If this is impossible, speaking with someone who has experience with the equipment would likely be helpful. In general, when examined from the proper perspective, the design parameters of the control can provide clues. For example, what maximum temperature or pressure is practical? What minimum temperature or pressure is practical? Ask these questions when running across a strange control.

Solutions for the following service calls are not provided here. Please discuss the solutions with your instructor.

SERVICE CALL 3

A residential customer reports that a heating system was heating until early morning, but for no apparent reason it stopped. The furnace, a gas-fired appliance located in the basement, seems to be hot, but no heat is coming out of the air outlets.

The technician arrives at the house, making certain to park on the street, not in the driveway. When the customer answers the doorbell, the technician introduces himself and the company he represents and tells the homeowner why he is there. The technician asks the homeowner for details regarding her concerns. From the description he believes the problem may be with the blower. The technician first checks to see if there is power at the furnace and notices that although the furnace is hot to the touch, no heat is coming out of the duct. Further investigation shows that the blower is not running.

The technician knows there is power to the furnace because it is hot and the burner ignites from time to time. The problem must be in the blower circuit. A voltage check shows that there is 115 VAC between points A and J, as well as 115 VAC between points K and L. The blower motor is not running. See the diagram in **Figure 14.66** for direction.

What is the problem and the recommended solution?

SERVICE CALL 4

A customer calls to say that the central air-conditioning system in their small office building will not cool. When the service technician arrives he turns the thermostat to COOL and sets the thermostat to a temperature that is lower than the present temperature in the space. The indoor blower motor starts. This indicates that control voltage is present and will send power on as requested for cooling. Only recirculated air is coming out of the ducts.

The technician checks the outside unit, notices that it is not running, and removes the control panel cover to see what controls will actually prevent the unit from starting, which will give some indication of what the possible problem may be. Use the diagram in **Figure 14.67**.

Figure 14.66 This diagram can be used for Service Call 3.

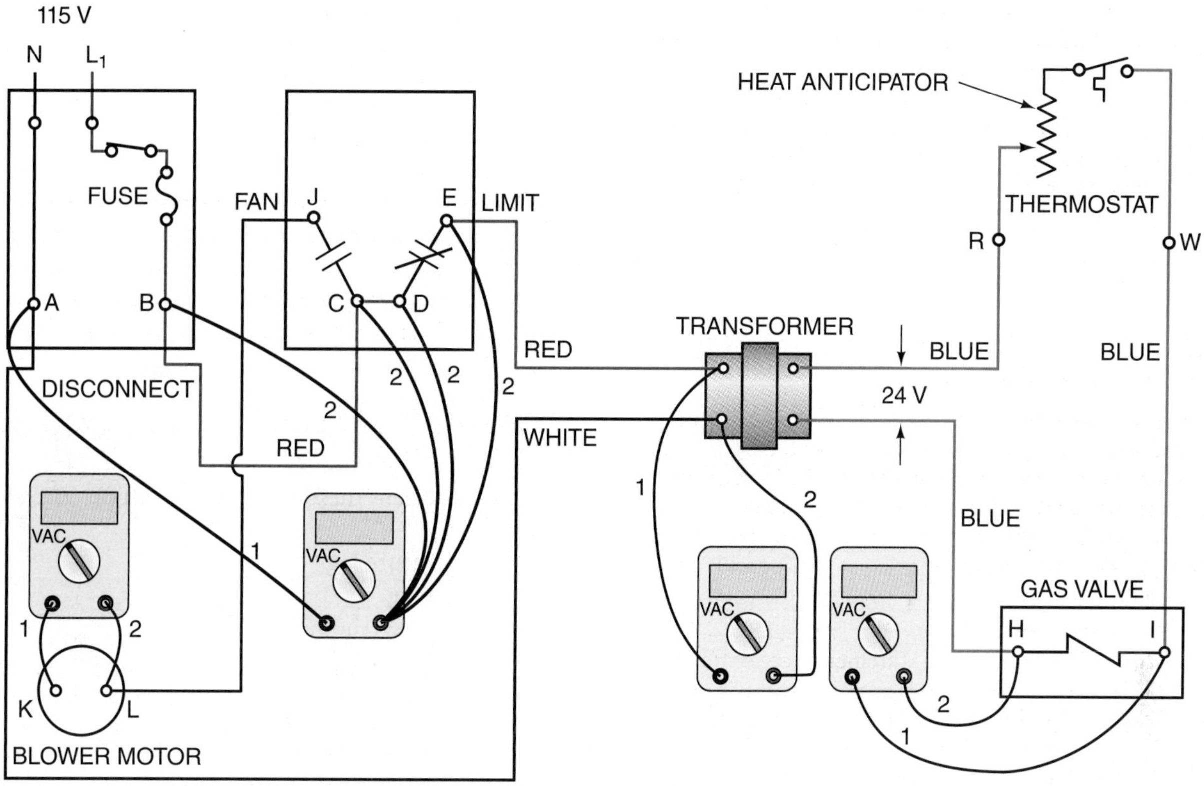

Figure 14.67 This diagram can be used for Service Call 4.

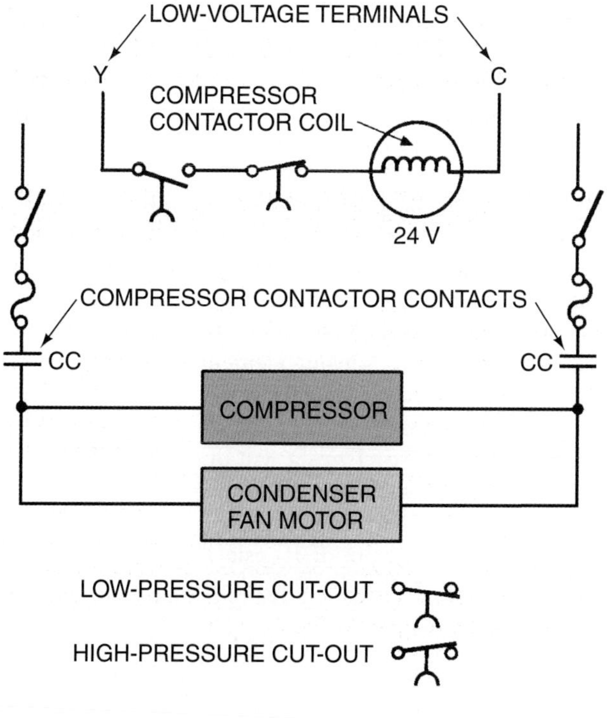

The technician installs pressure gauges to obtain the system pressures. New air-conditioning systems use R-410A. If the system were standing at 80°F, the system should have 235 psig of pressure, according to the temperature/ pressure chart for R-410A at 80°F. **NOTE:** *Care should be used in the standing pressure comparison. Part of the unit is inside, at 75°F air, and part of the unit is outside. Some of the liquid refrigerant will go to the coolest place if the system valves will allow it to.•* The gauge reading may correspond to a temperature between the indoor and outdoor temperature but not less than the cooler temperature. Some systems use fixed-bore metering devices that allow the system pressures to equalize and the refrigerant to migrate to the coolest place in the system.

In this example the pressure in the system is 105 psig, the refrigerant is R-410A, and the temperature is about 80°F. A typical pressure control setup may call for the control to interrupt the compressor when the system pressure gets down to 42 psig and to make the control start the compressor when the pressure rises back to a temperature corresponding to 70°F.

What is the problem and the recommended solution?

The foregoing typical service problems show how a service technician might arrive at the correct diagnosis, using the information at hand. Always save and study the manufacturer's literature. There is no substitute for having access to the manufacturer's specifications.

REVIEW QUESTIONS

1. The control voltage that is typically used in split-type residential central air-conditioning systems is
   A. 24 V.
   B. 48 V.
   C. 96 V.
   D. 115 V.
2. What device is used to produce the control voltage that is used to power the control circuits of a split-type residential central air-conditioning system?
   A. Step-down transformer
   B. Step-up transformer
   C. Low-voltage thermostat
   D. Line-voltage thermostat
3. All of the following are true statements regarding line-voltage thermostats and control wiring except
   A. An electrician's license is often required to run line-voltage control wiring.
   B. Line-voltage thermostats and controls respond faster than low-voltage thermostats.
   C. Line-voltage thermostats and controls are not as sensitive as low-voltage thermostats.
   D. Line-voltage thermostats must be matched to the voltage and current of the circuit.
4. System lag is the result of which of the following with regards to a heating system?
   A. The residual system heat increasing the space temperature
   B. The amount of heat generated in the thermostat
   C. The time it takes for a heating system to start producing heat
   D. The amount of current flowing through the thermostat
5. System overshoot is the result of which of the following with regards to a heating system?
   A. The residual system heat increasing the space temperature
   B. The amount of heat generated in the thermostat
   C. The time it takes for a heating system to start producing heat
   D. The amount of current flowing through the thermostat
6. The low-voltage control circuit for a typical residential air-conditioning system typically draws a maximum of how much amperage?
   A. 0.5 A
   B. 1.0 A
   C. 1.7 A
   D. 2.5 A
7. Which of the following is likely to be found on a line-voltage thermostat?
   A. Contacts with low amperage ratings
   B. 18-gauge or 22-gauge wire connections
   C. Lightweight switches and buttons
   D. Locking or otherwise secure cover
8. If a heated space reaches a maximum temperature of 78°F and a minimum temperature of 71°F, the difference between these two temperatures is referred to as the
   A. temperature lag.
   B. system lag.
   C. system overshoot.
   D. temperature swing.
9. Why is measuring the temperature of a solid object challenging for the HVACR technician?
   A. The temperature of solid objects is not uniform throughout.
   B. The entire thermometer probe must be at the temperature of the solid.
   C. Thermometers are not designed to measure the temperature of solids.
   D. Solid objects do not transfer heat as effectively as other forms of matter.
10. What method is most likely to be used to measure the temperature of a liquid contained in an open vessel?
    A. Immersion
    B. Adhesion
    C. Absorption
    D. Repulsion
11. To measure the temperature of a liquid flowing in a pipe without accessing the actual liquid in the pipe, which of the following methods can be used effectively?
    A. Insulating gum or a well
    B. Strap-on sensing element or insulation tape
    C. Well or strap-on sensing element
    D. Insulation tape or insulation
12. A pressure switch that has a cut-in pressure of 30 psig and a differential of 10 psig will operate with
    A. an average pressure of 20 psig.
    B. a cut-out pressure of 20 psig.
    C. an average pressure of 10 psig.
    D. a cut-out pressure of 10 psig.

13. Which of the following best describes a high-pressure control that is used as a safety device on an air-conditioning system?
    A. A normally open switch that closes on a rise in system pressure
    B. A normally open switch that opens on a rise in system pressure
    C. A normally closed switch that closes on a rise in system pressure
    D. A normally closed switch that opens on a rise in system pressure
14. The control device that converts a sensed pressure into an electrical signal is the
    A. pressure transducer.
    B. pressure converter.
    C. low-pressure switch.
    D. high-pressure switch.
15. Which of the following best describes a pressure control that is used as a low-charge protection device on an air-conditioning system?
    A. A normally open switch that closes on a drop in system pressure
    B. A normally open switch that opens on a drop in system pressure
    C. A normally closed switch that closes on a drop in system pressure
    D. A normally closed switch that opens on a drop in system pressure
16. Consider an R-410A air-conditioning system that is operating with a suction pressure of 118 psig and an oil pump discharge pressure of 185 psig. What is the net oil pressure?
    A. 67 psig
    B. 118 psig
    C. 185 psig
    D. 303 psig
17. A sail switch typically accomplishes which of the following tasks?
    A. Prevents an air-conditioning compressor from operating if the air filter is dirty
    B. Prevents electric strip heaters from operating if airflow in the duct is insufficient
    C. Allows a system compressor to operate if the indoor fan motor is inoperative
    D. Allows electric strip heaters to operate if the indoor fan motor is inoperative
18. What toxic substance can be found in older-generation control devices?
    A. Tin
    B. Antimony
    C. Mercury
    D. Asbestos
19. The diaphragm on a pressure-sensing device is best described as a
    A. large, thin sheet of steel.
    B. large, thick sheet of steel.
    C. small, thin sheet of steel.
    D. small, thick sheet of steel.
20. If a temperature control has a cut-in temperature of 38°F and a differential of 10°F, the range of the control is
    A. 10°F.
    B. 28°F.
    C. 38°F.
    D. 56°F.
21. When used to control water flow through a water-cooled condenser, which of the following statements is correct regarding a water-regulating valve?
    A. The spring pushes to open the valve, while the high-side pressure pushes to close the valve.
    B. The spring pushes to open the valve, while the low-side pressure pushes to close the valve.
    C. The spring pushes to close the valve, while the high-side pressure pushes to close the valve.
    D. The spring pushes to close the valve, while the high-side pressure pushes to open the valve.
22. Explain why "exercising" a pressure relief valve on a boiler can lead to problems for the HVACR service technician.
23. Explain how heating and cooling thermostats react to changes in sensed temperature.
24. Explain why a gas pressure switch should never be jumped out.
25. Explain the purpose of a gas pressure regulator.
26. Explain the difference between a mechanical control and an electromechanical control.
27. Explain why the pressure setting on a boiler's relief valve should not be adjusted.
28. Explain what can be done to properly inspect an electromechanical control.

UNIT 15

TROUBLESHOOTING BASIC CONTROLS

OBJECTIVES

After studying this unit, you should be able to

- describe the operation of power- and non-power-consuming devices.
- describe how the functions of a digital multimeter are used to troubleshoot electrical circuits.
- explain how to troubleshoot a thermostat.
- identify some typical problems in an electrical circuit.
- identify the components in a heat-cool electrical circuit.
- follow the sequence of electrical events in a heat-cool electrical circuit.
- differentiate between a pictorial and a ladder-type electrical wiring diagram.

SAFETYCHECKLIST

- ✔ When troubleshooting electrical malfunctions, disconnect electrical power unless power is necessary to make appropriate checks. Lock and tag the panel where the disconnect is made, and keep the only available key on your person.
- ✔ When checking continuity or resistance, turn off power and disconnect both leads from the component being checked.
- ✔ Ensure that meter probes come in contact with only the intended terminals or other contacts.
- ✔ Make certain that all electrical connections are tight.
- ✔ Make certain that all unused wires are properly capped to prevent them from coming into contact with other conductors or with the metal casing of the equipment.

15.1 INTRODUCTION TO TROUBLESHOOTING

Each control must be evaluated as to its function (major or minor) in the system. Recognizing the control and its purpose requires that you understand its use in the system. Studying the purpose of a control before you take action will save a great deal of troubleshooting time. Look for pressure lines going to pressure controls and temperature-activated elements on temperature controls. Determine if the control stops and starts a motor, opens and closes a valve, or provides some other function. As mentioned previously, controls are either electrical, mechanical, or a combination of electrical and mechanical. (For now, electronic controls are considered the same as electrical controls.)

Electrical devices can be considered as power-consuming or non-power-consuming so far as their function in the circuit is concerned. Power-consuming devices use power and have either a magnetic coil or a resistance circuit. They are more commonly referred to as *loads*. Non-power-consuming devices pass power on to power-consuming devices. Non-power-consuming, or power-passing, devices are commonly referred to as *switches*. They are used to simply pass power to the loads that are ultimately being controlled.

Both power-consuming and power-passing devices are wired in series with the power supply. Multiple loads are often wired in parallel with each other, and the switches that are used to control each load are connected in series with that load. If multiple switches are to be used to control the operation of a single load, the switches are wired in series with each other and in series with the load being controlled.

The concepts of the load and switch can be easily clarified using a simple light bulb circuit with a switch. The light bulb in the circuit is the load and actually consumes the power, whereas the switch passes the power to the light bulb. The object is to wire the light bulb to both sides (hot and neutral) of the power supply, **Figure 15.1**. This will then ensure that a complete circuit will be established. The switch is a power-passing device and is wired in series with the light bulb. For any power-consuming device to consume power, it must have a potential voltage across it. A potential voltage is the voltage indicated on a digital multimeter between two power legs (such as line 1 to line 2) of a power supply. This can be thought of as electrical pressure in a home; for example, the light bulb has to be wired from the hot leg (the wire that has a fuse or breaker in it) to the neutral leg (the grounded wire that actually is wired to the earth ground).

The following statements may help you understand electron flow in an electrical circuit. We have taken some liberties in this unit in explaining electrical circuits, but this has been done to aid your understanding of the concepts involved.

1. It does not make any difference which way the current flows in an alternating current (AC) circuit when the object is to get the electrons (current) to a power-consuming device and complete the circuit.
2. The electrons may pass through many power-passing devices before getting to the power-consuming device(s).

Figure 15.1 A power-consuming device (light bulb) and a non-power-consuming device (switch). The switch passes power to the light bulb.

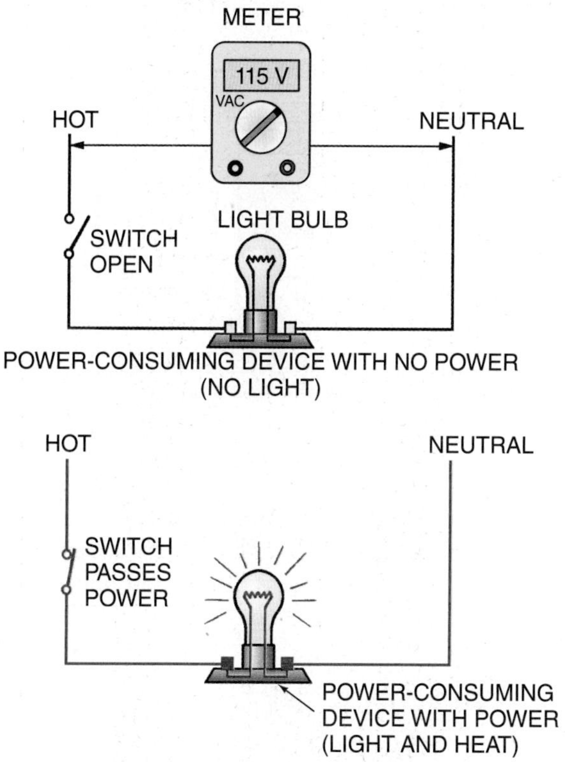

3. Devices that do not consume power pass power.
4. Devices that pass power can be either safety devices or operating devices.

Suppose the light bulb mentioned above was to be used to add heat to a well pump to keep it from freezing in the winter. A thermostat is installed to prevent the bulb from burning all the time. A fuse must also be installed in the line to protect the circuit, and a switch must be wired to allow the circuit to be serviced, **Figure 15.2**. Note that the power now must pass through three power-passing devices to reach the light bulb. These three switches are connected in series with each other. The fuse is a safety control, and the thermostat turns the light bulb on and off. The switch is not a control but a service convenience device. Notice that all switches are on one side of the circuit. Let us suppose that the light bulb does not light up when it is supposed to. Which component is interrupting the power to the bulb: the switch, the fuse, or the thermostat? Or is the light bulb itself at fault? In this example, the thermostat (switch) is open, **Figure 15.3**.

Figure 15.2 When closed, the switch passes power to the fuse, then to the thermostat, and on to the light bulb.

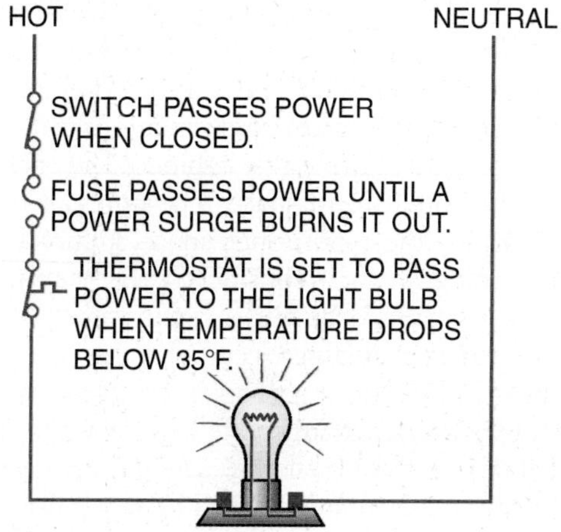

Figure 15.3 The troubleshooting procedure is to establish the main power supply of 115 V from the hot line to the neutral line. When this power supply is verified, the lead on the hot side is moved down the circuit toward the light bulb. The voltage is established at all points; when point 6 is reached, no voltage is present. The thermostat contacts are open.

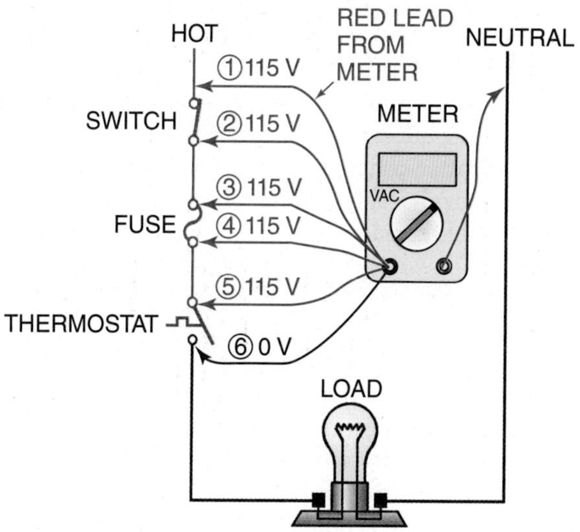

15.2 TROUBLESHOOTING A SIMPLE CIRCUIT

A digital multimeter may be used to troubleshoot the circuit in the previous paragraph. Turn the meter selector switch to the AC voltage function; most digital multimeters have an auto range feature. In this example, the supply voltage should be 115 V. Follow the diagram in **Figure 15.3** as you read the following:

1. Place the red lead of the multimeter on the hot line and the black lead on the neutral. The meter will read 115 V.
2. Place the red lead, the lead being used to find and detect power in the hot line, on the load side of the switch. (The "load" side of the switch is the side of the switch that the load is connected to. The other side of the switch, where the line is connected, is the "line" side of the switch.) The black lead should remain in contact with the neutral line. The meter will read 115 V.

3. Place the red lead on the line side of the fuse. The meter will read 115 V.
4. Place the red lead on the load side of the fuse. The meter will read 115 V.
5. Place the red lead on the line side of the thermostat. The meter will read 115 V.
6. Place the red lead on the load side of the thermostat. The meter will read 0 V. There is no power available to energize the bulb. The thermostat contacts are open. Now ask the question, "Is the room cold enough to cause the thermostat contacts to make?" If the room temperature is below 35°F, the circuit should be closed; the contacts should be made.

NOTE: *The red lead was the only one moved. It is important to note that if the meter had read 115 V at the light bulb connection when the red lead was moved to this point, then further tests should have been made.* •

Another step is necessary to reach the final conclusion. Let us suppose that the thermostat is good and 115 V is indicated at point 6.

7. The red meter lead can now be moved to the terminal on the light bulb, **Figure 15.4**. Suppose it reads 115 V. Now, move the black lead to the light bulb terminal on the right. If there is no voltage, the neutral wire is open between the source and the bulb.

If there is voltage at the light bulb and it will not burn, the bulb is defective, **Figure 15.5**. *When there is a power supply and a path to flow through, current will flow.* The light bulb filament is measurable resistance and should be the path for the current to flow through.

The power supply used in the preceding example was 115 V with neutral as one side of the circuit. Neutral and earth ground are usually the same thing. With 115-V circuits, the technician need have only a ground source, such as a water pipe, for one side of the meter. When troubleshooting 208-V or higher circuits, one meter lead must be placed on the other side of the line feeding the control circuit. This lead may be labeled line 1 and line 2, line 2 and line 3, or line 1 and line 3.

Figure 15.4 The thermostat contacts are closed, and 115 volts is the measurement when one lead is on the neutral and one lead is on the light bulb terminal (see position 7). When the black lead is moved to the light bulb at position 2, there is no voltage reading. The neutral line is open.

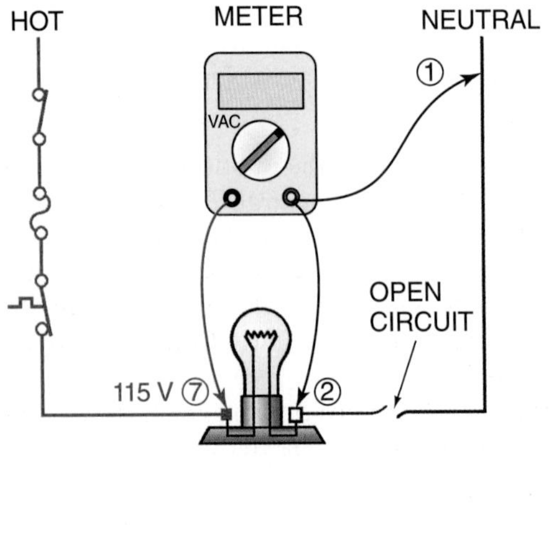

Figure 15.5 Power is available at the light bulb. The hot line on the left side and the neutral on the right side complete the circuit. The light bulb filament is burned out, so there is no current path through the bulb and no current flows through the circuit.

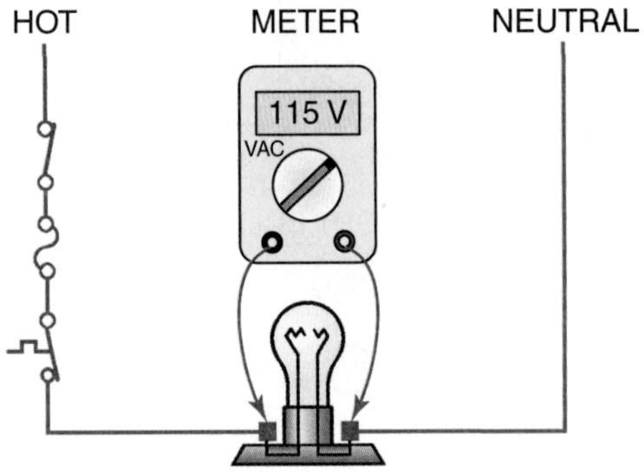

15.3 TROUBLESHOOTING A COMPLEX CIRCUIT

The following example is a progressive circuit that would be typical of a combination heating-cooling unit. Because each manufacturer may design its circuits differently, this circuit is not standard. However, although they may vary, they will be similar in nature.

The unit in this example is a package air conditioner with 1½ tons of cooling capacity and 5 kW of electric heat. It resembles a window air conditioner because all of the components are in the same cabinet. The unit can be installed through a wall or on a roof, and the supply and return air can be ducted to the conditioned space. The reason for using this unit as an example is that it comes in many sizes, from 1½ tons of cooling (18,000 Btu/h) to very large systems. The unit is popular with shopping centers because several roof units give a store good zone control. If one unit is not functioning, other units could help to hold the heating/cooling conditions steady. Also, all of the control components, except the room thermostat, are within the unit's cabinet. The thermostat is mounted in the conditioned space. The unit can be serviced without disturbing the conditioned space, **Figure 15.6**.

The first thing we will consider is the thermostat. The thermostat is not standard, but a version used for illustration purposes. It is a useable circuit that has been designed to illustrate troubleshooting. This thermostat is equipped with a selector switch for either HEAT or COOL. When the selector switch is in the HEAT position and heat is called for, the **electric heat relay** is energized. The fan must run in the

Figure 15.6 A small package air conditioner. Based on Climate Control

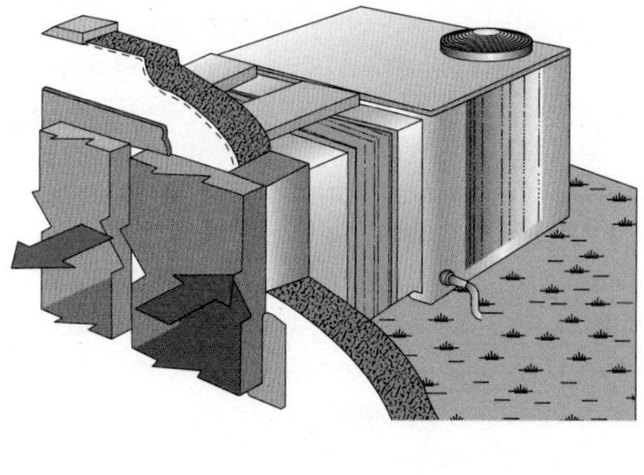

Figure 15.8 This is the same diagram as shown in Figure 15.7, but a heat anticipator has been added in series with the thermostat contacts. The selector switch is set for heat, and the thermostat contacts are closed, calling for heat.

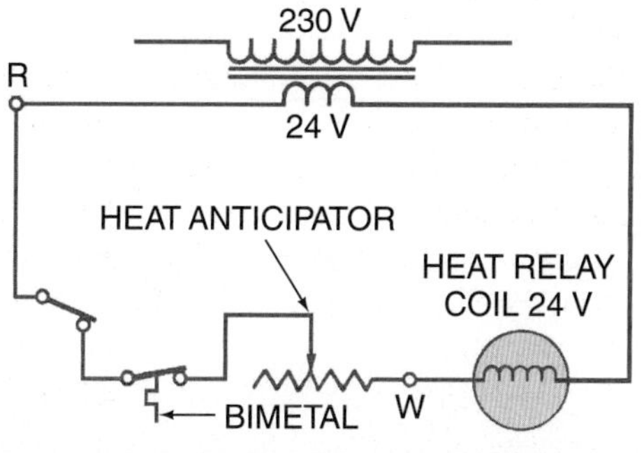

NOTE: THE HEAT ANTICIPATOR IS A VARIABLE RESISTANCE THAT IS DESIGNED TO CREATE A SMALL AMOUNT OF HEAT NEAR THE BIMETAL. THIS FOOLS THE THERMOSTAT INTO CUTTING OFF EARLY. THE FAN WILL CONTINUE TO RUN AND DISSIPATE THE HEAT REMAINING IN THE HEATERS.

heating cycle and will be started and run in the high-voltage circuit. (This is discussed later in this unit.)

Figure 15.7 begins the control explanation and illustrates the simplest of thermostats, having only a set of heating contacts and a selector switch. Follow the power from the R terminal through the selector switch to the thermostat heat contacts and on to the W terminal. The W terminal designation is not universal but it is common. When the heat contacts are closed, we can say that the thermostat is calling for heat. The contacts pass power to the heat relay coil. Power for the other side of the heat relay coil comes directly from the control transformer. When this coil is energized (24 V), the heat should come on. The coil is going to close a set of contacts in the high-voltage circuit to pass power to the electric heat element.

It should be noted that the circuit in **Figure 15.7** has two switches and one load. The two switches are the manual heat switch and the other is the set of heat contacts. The manual heat switch is opened and closed by the operator of the equipment, hence the term *manual*. The other switch, the set of heat contacts, is an automatic switch that opens and closes its contacts depending on the temperature sensed by the contacts. In order for the heating system to be energized, both the manual switch and the automatic temperature-sensing switch must be in the closed position. If the occupied space is very cold but the manual heat switch is off, the heating system will not operate. Similarly, if the manual heat switch is on and the space is warm, the system will not operate. As with any circuit that has multiple switches connected in series with each other, all of the switches must be in the closed position in order to pass power to the load being controlled.

Figure 15.7 A simple thermostat with a HEAT OFF/HEAT ON position. The selector switch is closed, calling for heating. When the heat relay is energized, the heating system starts.

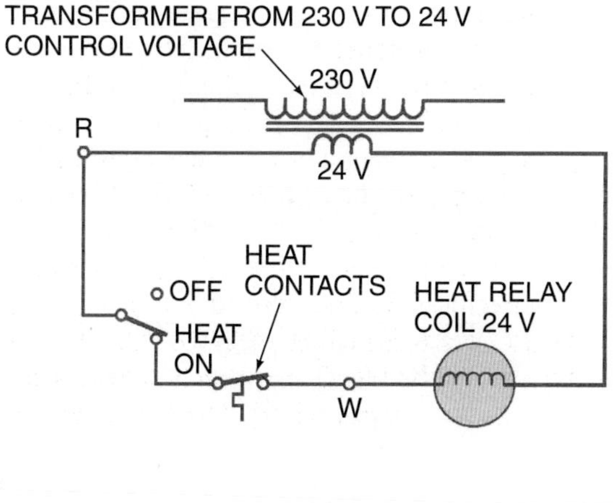

In **Figure 15.8** the heat anticipator is added to the circuit. The heat anticipator is in series with the heat relay coil and has current passing through it any time the heat relay coil is energized. The current passing through the anticipator creates a small amount of heat in the vicinity of the thermostat bimetal sensing element, which causes the bimetal to break the thermostat contacts early to dissipate the heat in the heat exchanger. This is done because the fan will blow hot air into the heated space for a short time after the heat relay is deenergized while cooling off the heat exchanger. If the contacts to the heater did not open until the space temperature was actually up to the cutoff point of the room thermostat, the extra heat would overheat the space; this is referred to as a system overshoot. The fan that moves the heat is stopped and started in the high-voltage circuit (discussed later in this unit).

Often the heat anticipator is a variable resistor and must be adjusted to the actual system. The current must be matched to the current used by the heat relay coil. All thermostat manufacturers explain this in the installation instructions for specific thermostats. Heat anticipators are covered in detail in Unit 14, "Automatic Control Components and Applications."

Figure 15.9 illustrates how a fan-starting circuit can be added to a thermostat. Notice the addition of the G terminal,

Figure 15.9 The addition of the fan relay to the thermostat allows the owner to switch the fan on for continuous operation. The fan must be operating during the heating cycle. (Automatic fan operation will be explained later in this unit.)

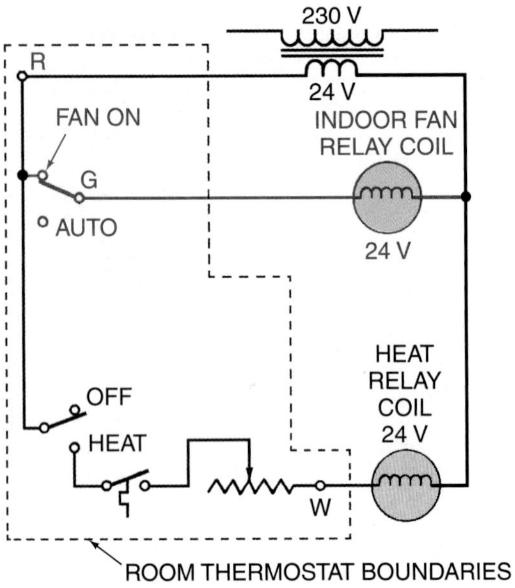

which is used to start the indoor fan. Although not universal, the G designation is common. Follow the power from the R terminal through the fan selector switch to the G terminal and on to the indoor fan relay coil. Power is supplied to the other side of the coil directly from the control transformer. When this coil has power (24 V), it closes a set of contacts in the high-voltage circuit and starts the indoor fan.

Figure 15.10 completes the circuitry in the thermostat by adding a Y terminal for cooling. Again, this is not the only letter used for cooling but it also is common. Follow the power from the R terminal down to the cool side of the selector switch through to the contacts and on to the Y terminal. When these contacts are closed, power will pass through them and go two ways:

- One path will be through the fan AUTO switch to the G terminal and on to start the indoor fan. The indoor fan has to run in the cooling cycle.
- The other path is straight to the cool relay. When the cool relay is energized (24 V), it closes a set of contacts in the high-voltage circuit to start the cooling cycle.

When the fan is switched to ON, the indoor fan will run all the time, even when the system switch is in the OFF position. This thermostat is equipped with a FAN ON or FAN AUTO position, so the fan can be switched on manually for air circulation when desired. When the fan switch is in the AUTO position, the fan will start on a call for cooling. When

Figure 15.10 (A) Cooling has been added to **Figure 15.9**. (B) The cooling circuit has a cooling anticipator in parallel with the cooling contacts. Current flows through the cooling anticipator during the OFF cycle. **NOTE:** *This is one of the few times that there are two loads connected in series with each other.* •

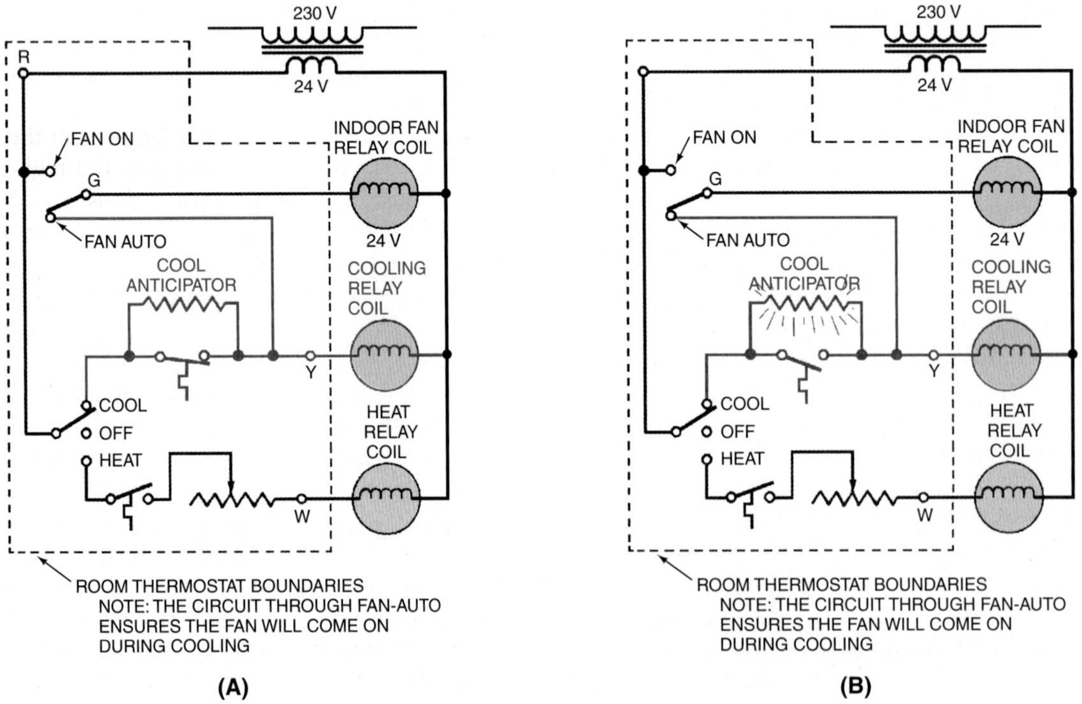

the selector switch is in the cooling mode, the cooling system will start on a call for cooling. The indoor fan will start through the AUTO mode on the fan selector switch. The outdoor fan must run also, so it is wired in parallel with the compressor.

NOTE: *The cooling anticipator is in parallel with the cooling contacts. This means that current will flow through this anticipator when the thermostat is satisfied (or when the thermostat's circuit is open). The cool anticipator is normally fixed and nonadjustable. It causes the cooling cycle to start early, which enables the system to be up to capacity at the time the room warms to above the thermostat set point.* •

15.4 TROUBLESHOOTING THE THERMOSTAT

During equipment malfunction, the thermostat is often a misunderstood and frequently suspected component. But it does a straightforward job of monitoring temperature and distributing the power leg of the 24-V circuit to the correct components to control the temperature. Service technicians should remind themselves as they approach the job that "one power leg enters the thermostat and it distributes power where called." Every technician needs a technique for checking the thermostat for circuit problems.

It should be restated that the thermostat is nothing more than a set of switches. There is a manual selector switch that selects the mode of operation, which can be OFF, COOL, or HEAT, and the automatic temperature-sensing component. When the conditions are correct, the thermostat will pass power on through the device to the system components to be controlled. When the system is calling for heat, power will be passed to the heat relay. When the system is calling for cooling, power will be passed to the cooling relay and the indoor fan relay. If power enters the thermostat and does not pass through the device, either the thermostat is not calling for the desired mode of operation or there is a problem with the control.

One way to troubleshoot a thermostat on a heating and cooling system is to first turn the fan selector switch to the FAN ON position to see if the fan starts. If the fan does not start, the problem may be with the control voltage. Control voltage must be present for the thermostat to operate. Assume for a moment that the thermostat will cause the fan to come on when turned to FAN ON but will not operate the heat or cooling cycles. The next step may be to take the thermostat off the subbase and jump the circuit out manually with an *insulated jumper*. Jump from R to G, and the fan should start. Jump from R to W, and the heat should come on. Jump from R to Y, and the cooling should come on. If the circuit can be made to operate without the thermostat, the thermostat is defective, **Figure 15.11**. Jumping these circuits out one at a time is not a worry because only one leg of power comes

Figure 15.11 Terminal designation in a heat-cool thermostat subbase. The letters R, G, Y, and W are common designations. A jumper wire placed between R and G should start the indoor fan motor. A jumper wire placed between R and Y should start the cooling cycle. A jumper wire placed between R and W should start the heating cycle.

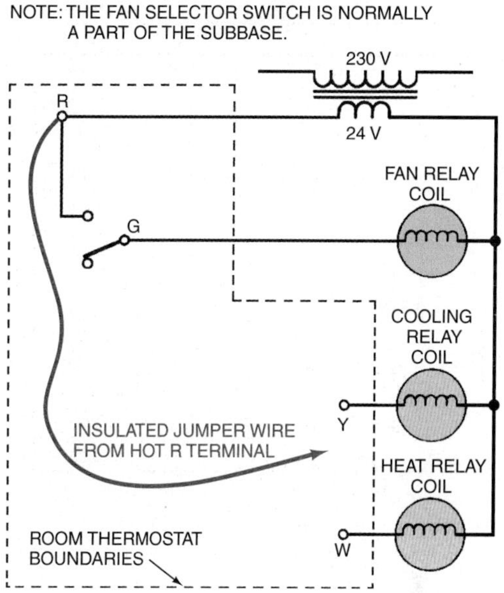

to the thermostat. Some more sophisticated thermostats may have a clock or indicator lights and may have the common side of the circuit wired to the thermostat, which means caution must be used. If all circuits were jumped out at one time in these thermostats, the heating and cooling would run at the same time. Of course, this should not be allowed to continue.

SAFETY PRECAUTION: *Jumping all circuits at once should be done only under the supervision of the instructor. Do not restart the air conditioner until 5 min have passed unless a hard-start kit is employed on the compressor. This allows the system pressures to equalize.* •

We have just covered what happens in the basic thermostat. The next step is to move into the high-voltage circuit and see how the thermostat's actions actually control the fan, cooling, and heating. We will again use the progressive circuit.

The first thing to remember is that the high voltage (230 V) is the input to the transformer. Without high voltage there is no low voltage. When the service disconnect is closed, the potential voltage between line 1 and line 2 is the power supply to the primary of the transformer. The primary input then induces power to the secondary winding of the transformer.

Figure 15.12 illustrates the high-voltage operation of the fan circuit. In the high-voltage circuit, the power-consuming device is the fan motor. The fan relay contact is in the line 1 circuit. It should be evident that the fan relay contacts have to be closed to pass power to the fan motor. These contacts close when the fan relay coil is energized

Figure 15.12 When the fan switch is turned to the ON position, a circuit is completed to the fan relay coil. When this coil is energized, it closes the fan relay contacts, passing power to the indoor fan motor.

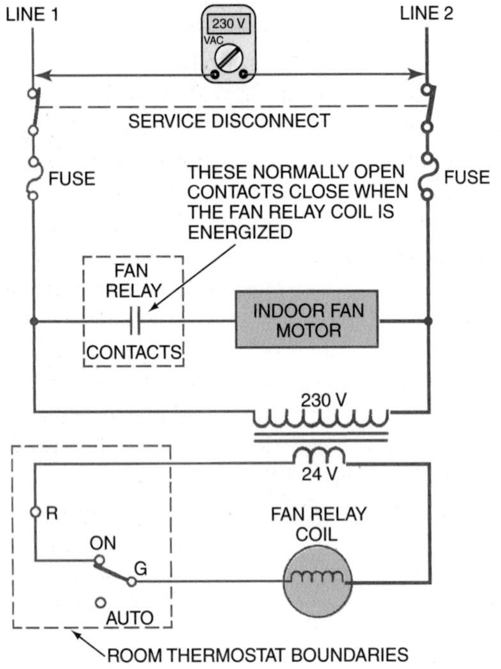

Figure 15.13 When the selector switch is in the HEAT position and the thermostat contacts close, the heat relay coil is energized. This action closes two sets of contacts in the heat relay. One set starts the fan because it is wired in parallel with the fan relay contacts. The other set passes power to the auto reset limit switch (set at 165°F). Power reaches the other side of the heater element through the fusible link. When power reaches both sides of the heating element, the heating system is functioning.

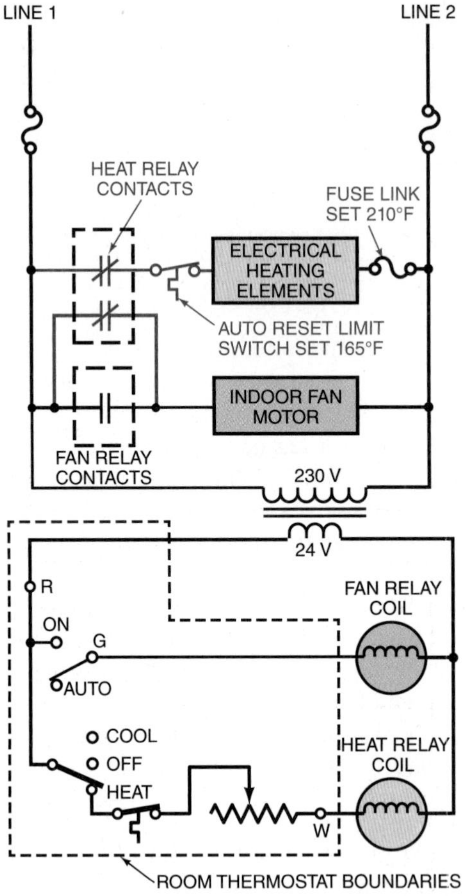

in the low-voltage control circuit. Although not drawn together, the fan relay coil in the low-voltage circuit causes the fan relay contacts in the line-voltage circuit to open and close. The magnetic field that is generated when current flows through the fan relay coil is the force that causes the relay contacts to close.

Figure 15.13 illustrates the addition of the electric heat element and the electric heat relay. The electric heat element and the fan motor are the power-consuming devices in the high-voltage circuit. When the HEAT-COOL selector switch is moved to the HEAT position in the low-voltage circuit, power is passed to the thermostat contacts. When there is a call for heat, these contacts close and pass power through the heat anticipator to the heat relay coil. This closes the two sets of contacts in the high-voltage circuit. One set starts the fan, and the other set passes power to the limit switch and then through the limit switch to the line 1 side of the heater. This limit switch (165°F) is the primary overheat safety control. The circuit is completed through the fuse link in line 2. The fuse link is the secondary overheat safety control (210°F).

In **Figure 15.14** cooling, or air-conditioning, has been added to the circuit. For this illustration the heating circuit was removed to reduce clutter and confusion. Three components have to operate in the cooling mode: the indoor fan, the compressor, and the outdoor fan. The compressor and the outdoor fan are wired in parallel and, for practical purposes, can be thought of as one component.

The line 2 side of the circuit goes directly to the three power-consuming components. Power for the line 1 side of the circuit comes through two different relays. The cooling relay contacts start the compressor and outdoor fan motor; the indoor fan relay starts the indoor fan motor. In both cases, the relays pass power to the components on a call from the thermostat.

When the thermostat selector switch is set to the COOL position, power is passed to the contacts in the thermostat. When these contacts are closed, power passes to the cooling relay coil. When the cooling relay coil is energized, the cooling relay contacts in the high-voltage circuit are closed. This passes power to the compressor and outdoor fan motor.

The indoor fan must run whenever there is a call for cooling. When power passes through the contacts in the thermostat, a circuit is made through the AUTO side of the selector switch to energize the indoor fan relay coil and start the

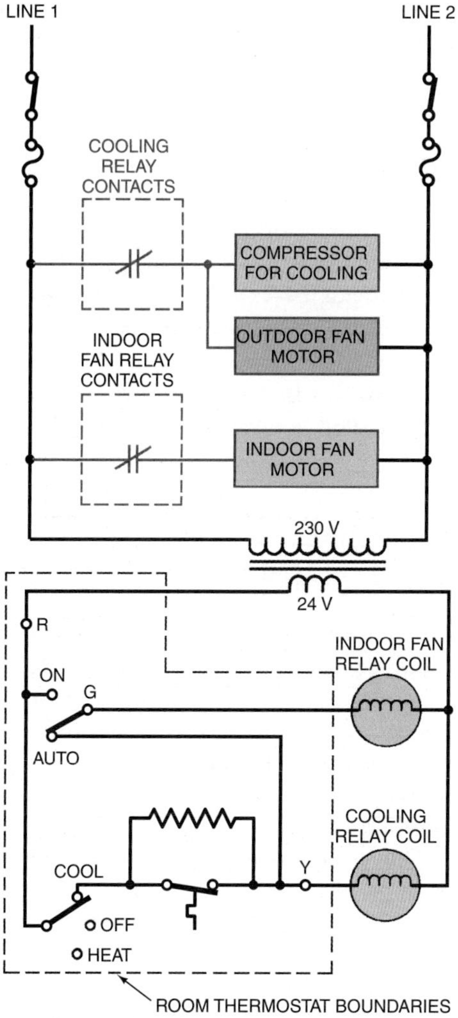

Figure 15.14 Addition of cooling to the system. When the thermostat selector switch is set to COOL and the thermal contacts close, the cooling relay coil is energized, closing the high-voltage contacts. These high-voltage contacts feed power to the compressor and condenser fan motor. Notice that the circuit in the thermostat starts the indoor fan motor through the AUTO position of the fan selector switch.

indoor fan. Note that if the fan selector switch is in the FAN ON position, the fan would run all the time and would still be on for cooling.

15.5 TROUBLESHOOTING AMPERAGE IN THE LOW-VOLTAGE CIRCUIT

Transformers are rated in volt-amperes, commonly called VA. This rating can be used to determine whether the transformer is underrated or drawing too much current. For example, it is quite common to use a 40-VA transformer for the low-voltage power source on combination cooling and heating equipment. This tells the technician that at 24 V, the maximum amperage the transformer can be expected to carry is 1.66 A. This is determined as follows:

$$\frac{40\,\text{VA}}{24\,\text{V}} = 1.66\,\text{A}$$

Older, analog clamp-on ammeters will not readily measure such a low current with any degree of accuracy, so adjustments must be made to determine an accurate current reading. Using a jumper wire coiled 10 times (called a 10-wrap amperage multiplier), place the ammeter's jaws around the 10-wrap loop and place the jumper in series with the circuit, **Figure 15.15. The reading on the ammeter will have to be divided by 10.** For instance, if the cooling circuit coil amperage reads 7 A, it is really carrying only 0.7 A. Newer digital multimeters equipped with an amp clamp can read small amperages accurately and do not need a 10-wrap coil of wire around them. Refer to Unit 5, "Tools, Instrumentation, and Equipment," for an illustration of a modern digital multimeter equipped with an amp clamp.

Some older ammeters have attachments for readings in ohms and volts. The volt attachments can be helpful for voltage readings, but the ohm attachment should be checked to

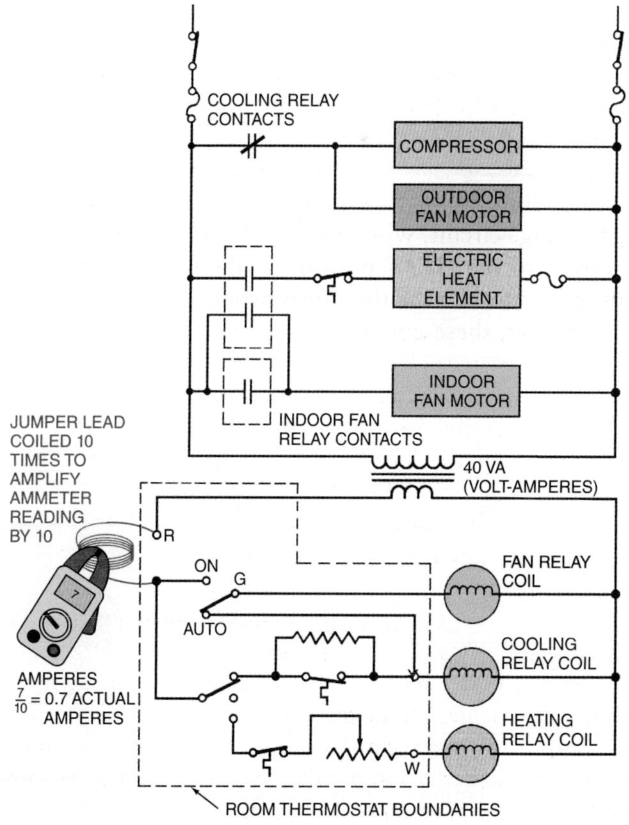

Figure 15.15 A clamp-on ammeter is used to measure the current draw in the 24-V control circuit.

make sure that it will read in the range of ohms needed. Some of the ohmmeter attachments will not read very high resistance. For instance, they may show an open circuit on a high-voltage coil that has a considerable amount of resistance.

15.6 TROUBLESHOOTING VOLTAGE IN THE LOW-VOLTAGE CIRCUIT

A digital multimeter can have many functions including volts, amps, milliamps, resistance, microfarads, and temperature. The most common applications are checking voltage and resistance. The volt function can be used to check for the presence of voltage, and the ohms function can be used to check for resistance. A high-quality meter will give accurate readings for each measurement.

Remember that any power-consuming device in the circuit must have the correct voltage applied to it in order to operate. The digital multimeter can be applied to any power-consuming device for a voltage check by placing one probe on one side of the device and the other probe on the other side. When power is present at both points, the component should function. If not, a check of resistance through the device is the next step, using the multimeter. Again, keep in mind that when voltage is present and a path is provided, current will flow. If the flowing current does not cause the load to do its job, it is defective and will have to be changed.

The multimeter cannot be used to much advantage at the actual thermostat because the thermostat is a closed device. When the thermostat is removed from the subbase, the method discussed earlier of jumping the thermostat terminals is usually more effective. With the thermostat removed from the subbase and the subbase terminals exposed, the digital multimeter can be used in the following manner. Turn the multimeter selector switch to the AC voltage function. Attach the meter lead to the hot leg feeding the thermostat, **Figure 15.16**. This terminal is sometimes labeled R or V; there is no standard letter or number. When the other lead is placed on the other terminals one at a time, the circuits will be verified.

Figure 15.16 The multimeter can be used at the thermostat location with the thermostat removed from the subbase.

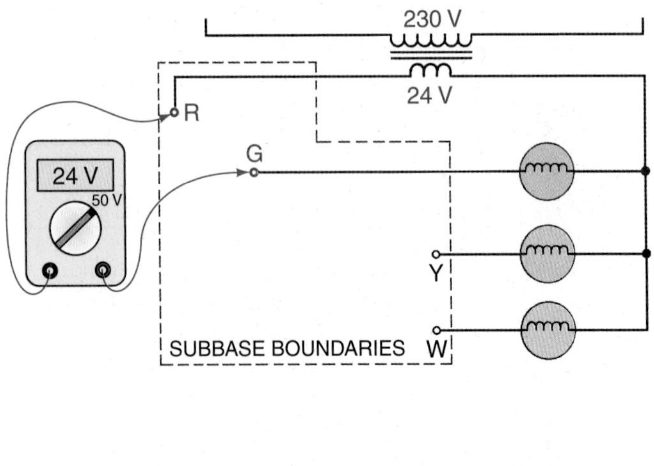

For example, with one lead on the R terminal and one on the G terminal, the terminal normally used for the fan circuit, there will be a voltage reading of 24 V. The meter is reading one side of the line straight from the transformer and the other side of the line through the coil on the fan relay. In this case, the fan relay coil is not energized and simply acts as a coil of wire. It will now let the other line of the transformer feed through to the G terminal of the thermostat's subbase. This phenomenon is sometimes referred to as voltage feedback. The multimeter now will be reading each side of the transformer, or 24 V. When the meter probe is moved to the circuit assigned to cooling (this terminal could be lettered Y), it reads 24 V. When the meter probe is moved to the circuit assigned to heating (sometimes the W terminal), 24 V again appears on the meter. The fact that the voltage reads through the coil in the respective circuits is evidence that a complete circuit is present.

15.7 TROUBLESHOOTING SWITCHES AND LOADS

Service technicians often encounter electrical problems when troubleshooting equipment. These problems are often nothing more than electrical switches that are either opened or closed. However, when power-consuming devices (electrical loads) are in series with these switches, matters can become complicated. Most of the time electrical switches are in series with one another and are relatively simple to troubleshoot electrically. The complications arise when these electrical switches are in parallel with one another.

Figure 15.17 shows a pressure switch in series with a motor (electrical load, or power-consuming device). In this case, the electrical load is a permanent split-capacitor (PSC) motor. The potential difference (voltage) between line 1 and line 2 is 230 V. This means that if the two leads of a multimeter were placed between line 1 and line 2, it would read a voltage of 230 V. If a technician measured the voltage across the open switch, the multimeter would also read a voltage of 230 V. This happens because line 1 ends at the left side of the open switch and line 2 simply bleeds through the run winding of the motor, through the closed overload, and ends at the right side of the switch. Since the motor is not running or consuming power, the windings are nothing but conductive wire for line 2 to bleed through. If a technician were to measure the voltage across the run winding (between R and C) of the PSC motor in **Figure 15.17**, the voltage would be 0 V because the motor winding is simply passing line 2 when it is not running. Line 2 would be at both the R and C terminals of the motor, and the potential difference or voltage difference between line 2 and line 2 is 0 V. However, if the technician

Figure 15.17 Voltage readings for a switch in series with a motor.

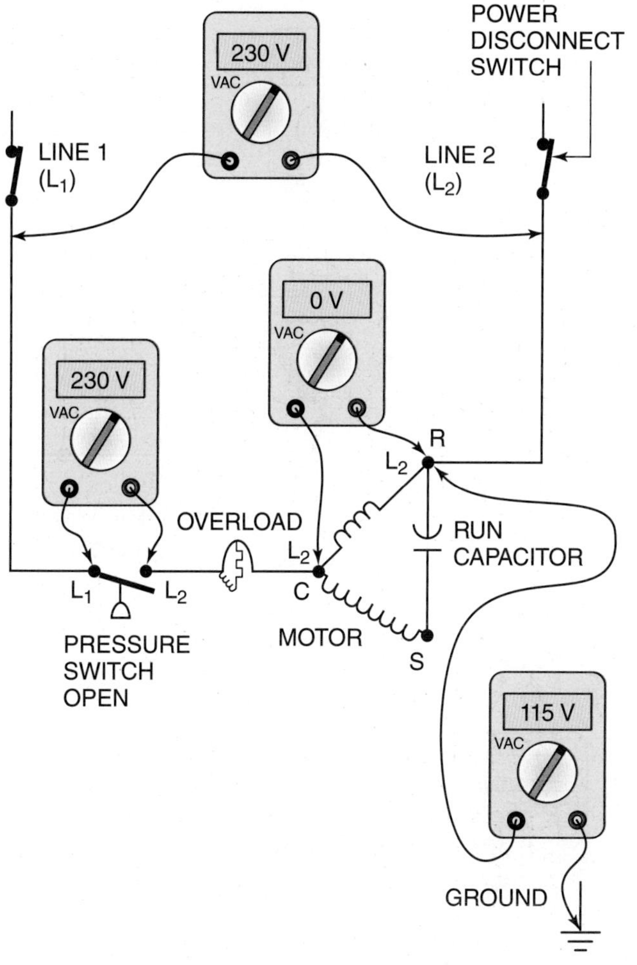

Figure 15.18 Voltage readings for a closed switch in series with an operating motor.

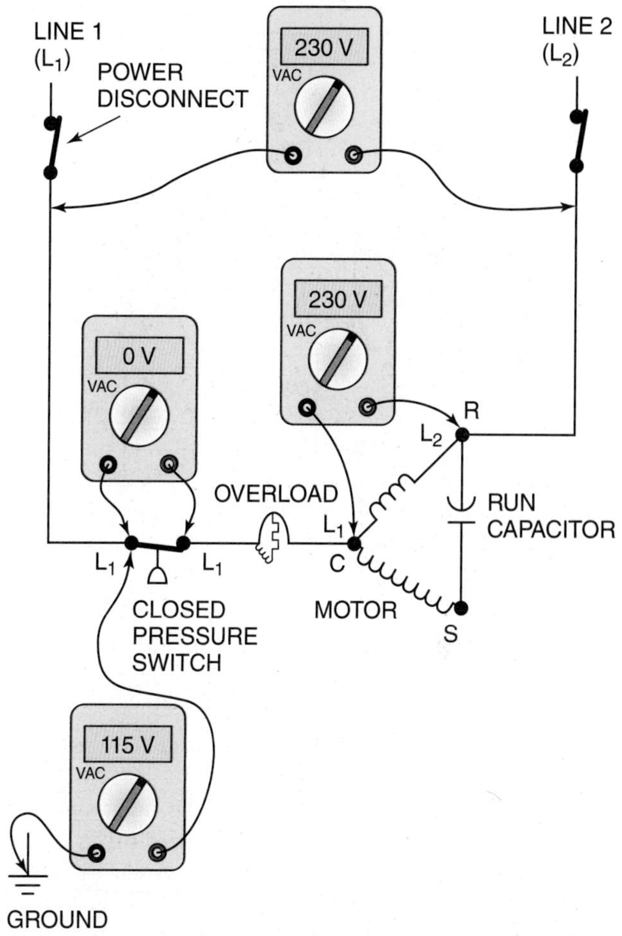

references either the R or C terminal to ground, the voltage would be 115 V because line 2 relative to ground is 115 V.

In **Figure 15.18** the pressure switch is closed and the motor is running. The motor is now consuming power and would read 230 V across its run winding (terminals R and C) while it is running. A voltage measurement across the closed switch will read 0 V. This is because line 1 ends at the C terminal of the motor when it is running. The switch experiences line 1 at both its terminals, and the potential difference or voltage difference between line 1 and line 1 is 0 V. If a technician measures from one terminal of the switch to ground, the voltage reading will be 115 V. In **Figure 15.17** the multimeter across the open switch reads 230 V, whereas in **Figure 15.18** the multimeter across the closed switch reads 0 V. It is incorrect to conclude that open switches always read some voltage and closed switches always read zero voltage. The parallel switches in **Figure 15.19** will clarify this concept.

When the switches are in parallel, the technician faces a greater challenge. **Figure 15.19** shows two switches that are in parallel but at the same time in series with a motor (power-consuming device). The top switch is closed, but the bottom switch is open. A multimeter across terminals A and B of the top switch will read 0 V because it is measuring a potential difference between line 1 and line 1. Since the motor is running, it has 230 V applied across the run winding (C and R terminals) of the PSC motor. However, a multimeter across terminals C and D of the bottom switch will also read 0 V. This happens because line 1 actually extends to the common terminal of the motor when it is running. This would make terminals C and D of the bottom switch both line 1, and the voltage difference between line 1 and line 1 is 0 V. Actually, points A, B, C, and D are all line 1. This example is a scenario where both an open switch and a closed switch read 0 V. *When doing electrical troubleshooting with a digital multimeter, technicians must ask themselves where line 1 and line 2 are, not whether the switch is open or closed.* Measuring across the same line with a multimeter, whether it be line 1 or line 2, will always give 0 V. Measuring from line 1 to line 2

Figure 15.19 Voltage readings for two switches wired in parallel with each other. These switches are both wired in series with the operating motor.

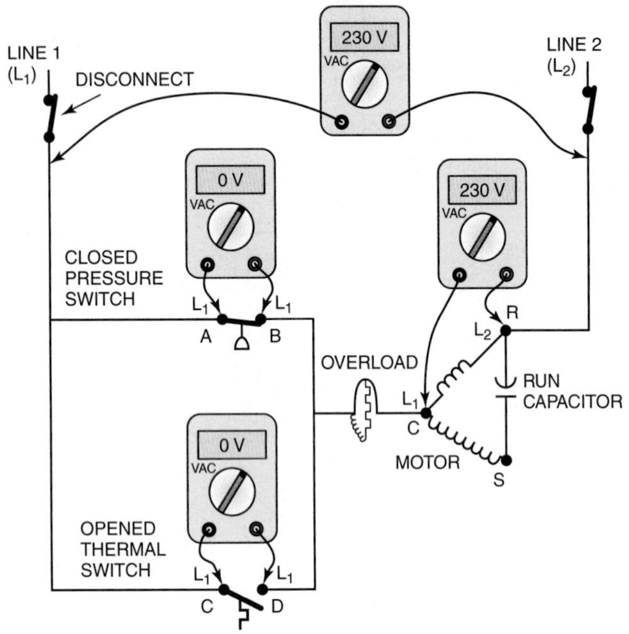

Figure 15.20 A meter reading 230 V across the run winding of a motor even when the run winding is open.

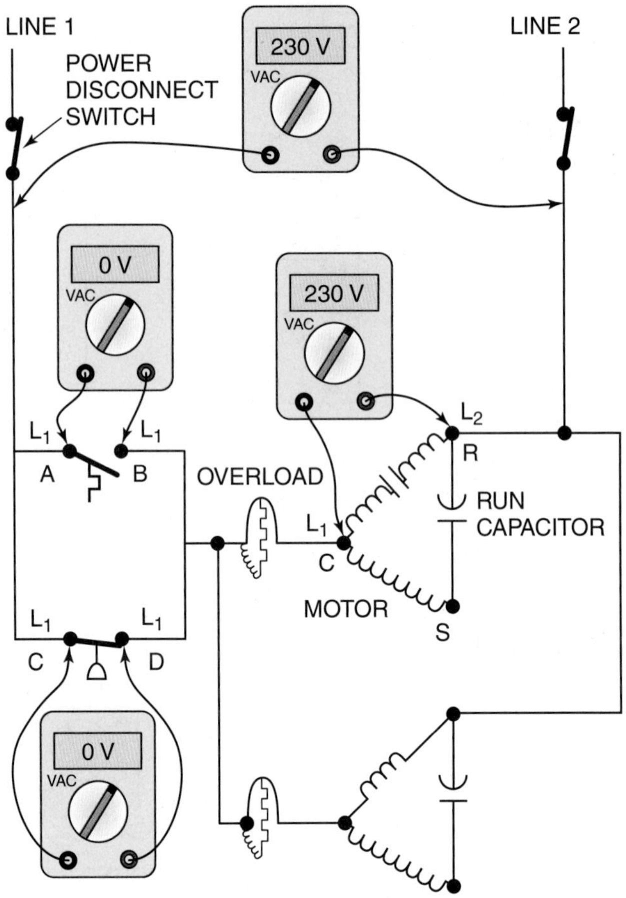

will always give the total circuit voltage, which in these examples was 230 V.

There will be times when a technician must switch the meter to ohms and shut the electrical power off in order to solve the problem. In **Figure 15.20**, what would be the voltage between R and C if the run winding between R and C opened, causing the motor to stall and draw locked-rotor amperage (LRA)? This is, of course, before the overload has opened. Notice that a multimeter placed across the R and C terminals of the motor (the opened winding) will again read 230 V. No matter if the motor is running properly or if it has an opened run winding, the voltage will still read 230 V across R and C. So, how does the service technician determine whether the run winding is opened or not? The answer is, testing for resistance. The service technician must shut the electrical power down and disconnect one wire from the R or C terminal of the motor, **Figure 15.21.** Disconnecting a wire will prevent electrical feedback from the meter's internal voltage source (battery) through another parallel electrical circuit. The technician must then place the multimeter across the R and C terminals of the motor. The measurement will read O.L. (open line), which indicates infinite ohms (∞), if the winding is open. This is the only way the service technician can tell whether the winding is opened or not.

Notice in **Figure 15.21** that the only potential current path exists between terminal C and terminal R. Even the run capacitor has been disconnected. In the event that the run capacitor had a short in it, the multimeter would read the resistance in the path through the run capacitor and the start winding, instead of the infinite resistance of the open path between terminals C and R. **Figure 15.22** shows a feedback circuit from the meter's battery if a wire is not disconnected from the motor terminals. In this case, the multimeter reading would be 3 ohms. This resistance comes from the resistance of the run winding in the feedback circuit of the PSC motor in parallel with it. This would fool the technician into thinking the winding was still good and not opened.

Figure 15.23 is a 230-V, single-phase electrical schematic of a typical commercial refrigeration system. The diagram includes a timer assembly with a defrost termination solenoid (DTS), evaporator fans, defrost heaters, a temperature-activated defrost termination/fan delay (DTFD) switch, a low-pressure control (LPC), a high-pressure control (HPC), a compressor contactor assembly, and a compressor/potential relay assembly. Although this diagram may look a little complicated, it is actually multiple series circuits connected to a single power supply. We will now examine each of these branch circuits individually. The system is drawn

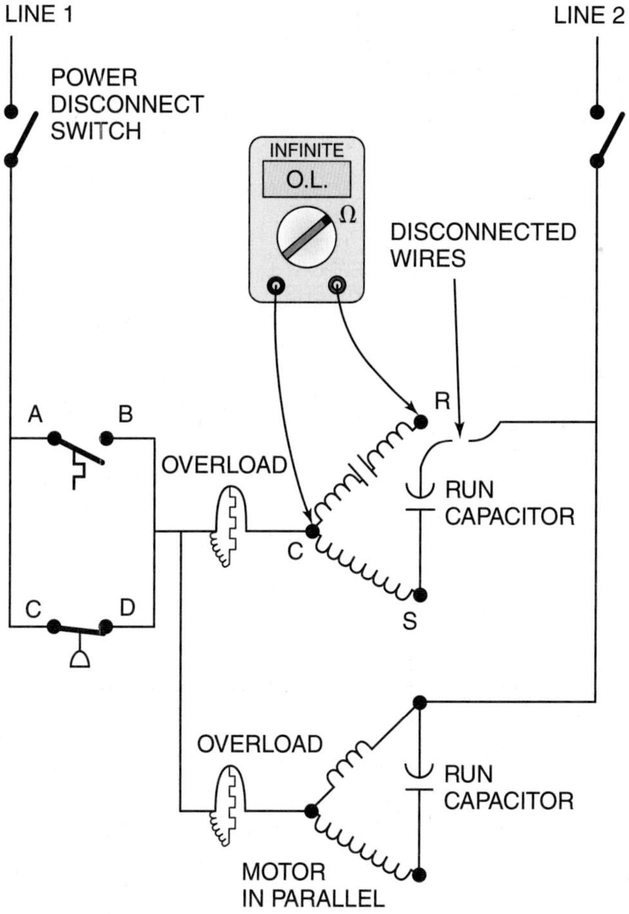

Figure 15.21 A meter reading of infinity (O.L.) across an open motor run winding. Notice that the wires have been disconnected from the R terminal.

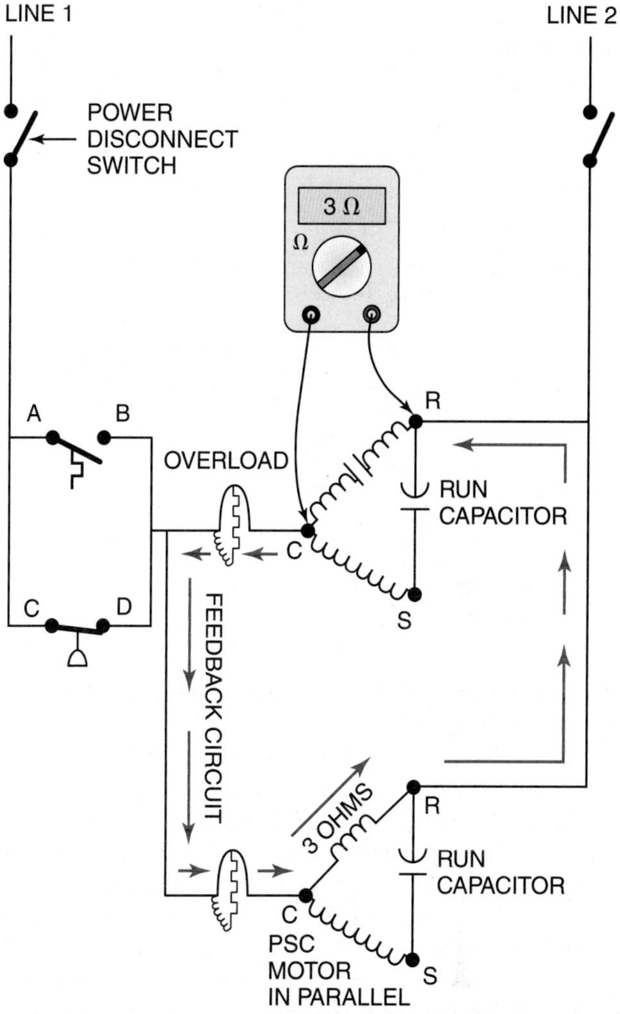

Figure 15.22 A feedback circuit from the meter's battery when a wire is not disconnected from the motor being tested. The multimeter will now measure the 3 ohms of the run winding of the PSC motor that is wired in parallel with the motor being tested.

in the refrigeration mode and shows where line 1 (L_1) is in relation to line 2 (L_2) for ease of understanding the measured voltages.

The first load we will look at in this circuit is the timer motor. It can be seen from **Figure 15.24** that the timer motor is connected directly to the 230-V power supply. As long as the power to the system is on, there should be a reading of 230 V across the timer motor. If the power is on and 230 V is not being supplied to the timer motor, there is a problem with the wiring that is connecting the power supply to the timer motor. If 230 V is being supplied to the timer motor—and the motor is not operating—there is a problem with the timer motor.

The next circuit we will examine is the defrost heater circuit, **Figure 15.25**. By starting at L_1 and tracing a path through the defrost heaters and back to L_2, we can see that there are a number of switches in series with the heaters. The first switch is the set of contacts that is located between contact points 1 and 3 on the timer, which is the component that is outlined with the dashed line. When the system is operating in the refrigeration mode, the heater contacts are in the open position, as we do not want them to be energized when the system is providing refrigeration. When the system goes into defrost, these contacts will close and the heaters will become energized as long as the second switch in the circuit and all interconnecting wiring is correct. The second switch in the circuit is the defrost limit switch. This is a safety device and will remain closed as long as the temperature it senses is below the safe operational limit. If the temperature rises too high, the limit switch will open and the heaters will be turned off. A reading of 230 V across the heaters indicates that the heaters are being powered. An amperage reading must then be taken to ensure that the heaters are indeed energized. If the heaters are being powered but the amperage draw of the heaters is 0 A, there is a problem with the heaters. A positive amperage reading in the heater circuit indicates that the heaters are energized.

Figure 15.23 A 230-V electrical schematic of a typical refrigeration circuit showing relative positions of line 1 (L_1) and line 2 (L_2).

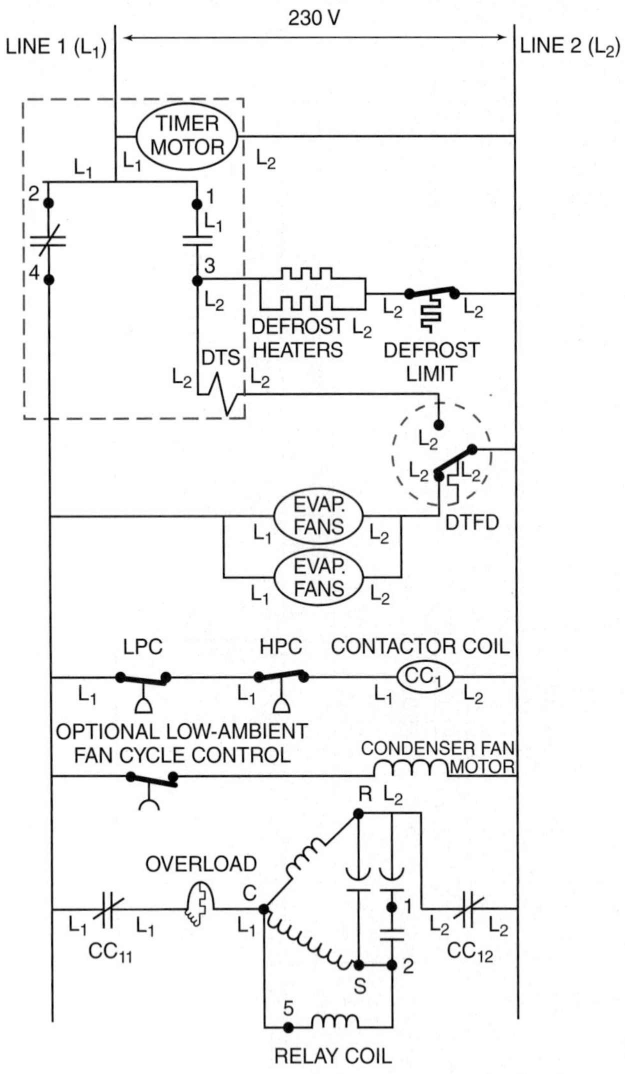

Figure 15.24 The timer motor circuit.

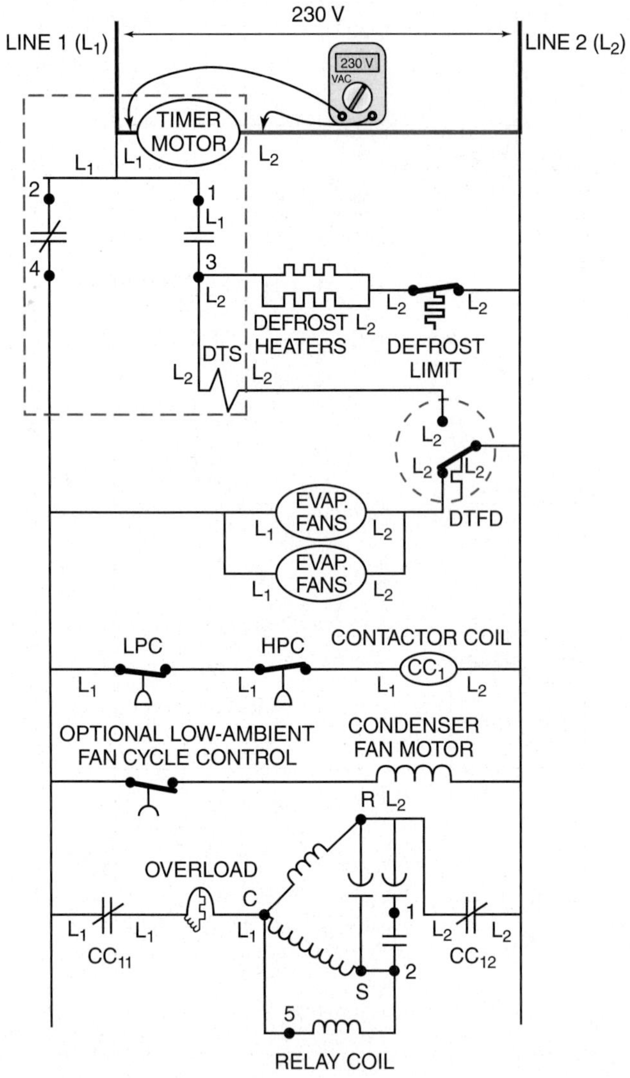

The next circuit controls the operation of the defrost termination solenoid (DTS), **Figure 15.26**. Just as with the defrost heaters, the defrost solenoid is controlled by a set of time-controlled contacts as well as a set of temperature-controlled contacts. It just so happens that the set of time-controlled contacts is the same set used to control the defrost heaters: the contacts between terminals 1 and 3 on the defrost timer. The temperature-controlled switch has a set of normally open and a set of normally closed contacts. The normally open set of contacts are sometimes referred to as the "HOT" contacts, and the normally closed set of contacts are often referred to as the "COLD" contacts. The temperature-controlled switch performs two functions:

- It takes the system out of defrost if all of the evaporator frost has been removed.
- It controls the evaporator fan motors so they can only operate when the evaporator is cold enough to provide refrigeration. This helps prevent an overload on the compressor's motor after defrost from high suction pressure due to a very warm evaporator.

In order for the defrost termination solenoid to be energized, the system must be in defrost and the temperature sensed by the defrost termination thermostat must be warm enough. When the DTS is energized, the closed contacts on the timer will open and the open contacts will close, putting the system back into the refrigeration mode.

The defrost termination solenoid is also often referred to as a "Z-coil." Taking amperage readings in the Z-coil circuit is difficult, as the coil is never energized for more than a fraction of a second. Here's why (let's say the system is in defrost):

- The set of contacts between terminals 1 and 3 is closed.
- There is frost on the evaporator coil, so the defrost termination/fan delay switch (DTFD) is in the position shown in **Figure 15.26**.

Figure 15.25 The defrost heater circuit.

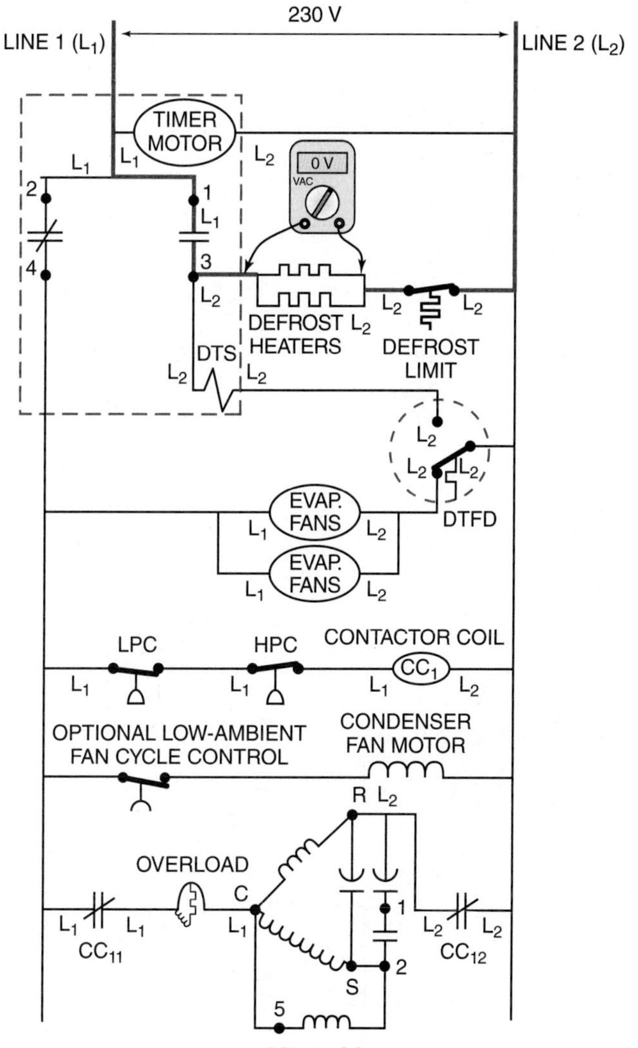

Figure 15.26 The defrost termination solenoid circuit.

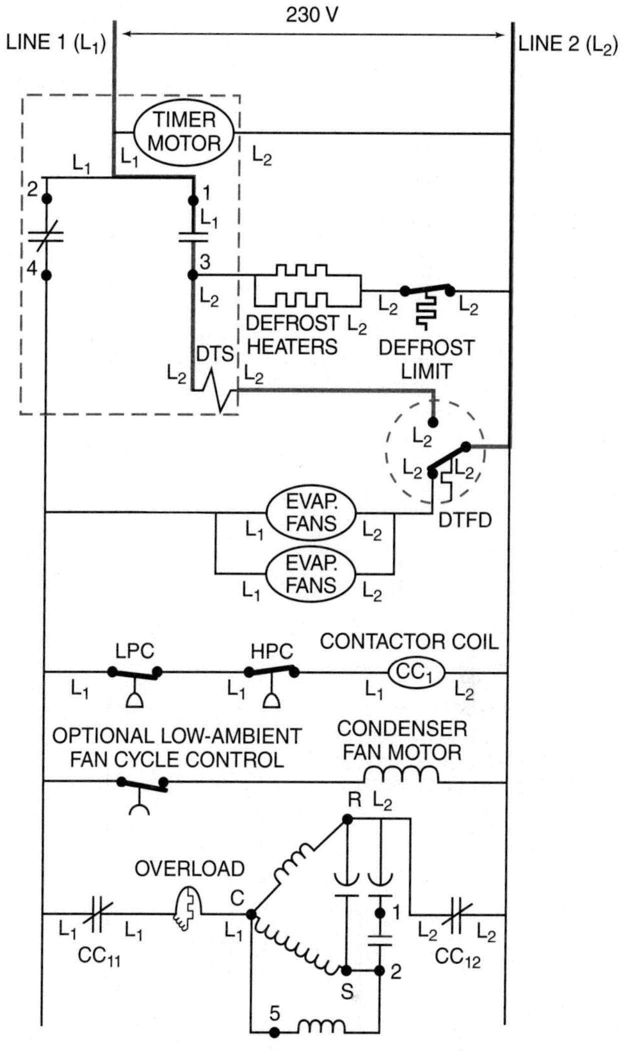

- So the Z-coil circuit is not energized.
- When the frost is melted, the DTFD changes its position and energizes the DTS.
- The instant the DTS is energized, it opens the closed contacts on the defrost timer (1 and 3) and closes the open contact (2 and 4).
- This opens the very contacts that are energizing the DTS in the first place, in this example, the contacts between 1 and 3.

For this reason, it is possible to take voltage readings, but amperage readings will not prove to be very useful for troubleshooting purposes.

We will now take a look at the evaporator fan circuit, **Figure 15.27**. There are two control devices that cycle the evaporator fans on and off. One is a set of contacts on the defrost timer that is located between terminals 2 and 4 on the timer. These contacts are in the closed position when the system is in the refrigeration mode. The other control device is the defrost termination/fan delay switch. When the evaporator is cold enough for refrigeration, the DTFD will close the contacts in series with the evaporator fan motors. If there is no voltage being supplied to the evaporator fan motors, the technician should check for voltage across the timer contacts 2 and 4 as well as for voltage across the DTFD. A reading of line voltage across contacts 2 and 4 indicates that the system is not in the refrigeration mode, while a line-voltage reading across the DTFD is an indication that the evaporator coil temperature is too high for refrigeration.

The next part of the wiring diagram is the circuit that energizes and deenergizes the compressor contactor coil, **Figure 15.28**. In this circuit, the contactor coil is the load and there are three switches that are wired in series with this coil. The three power-passing devices are

- the set of defrost timer contacts between terminals 2 and 4;
- the low-pressure control; and
- the high-pressure control.

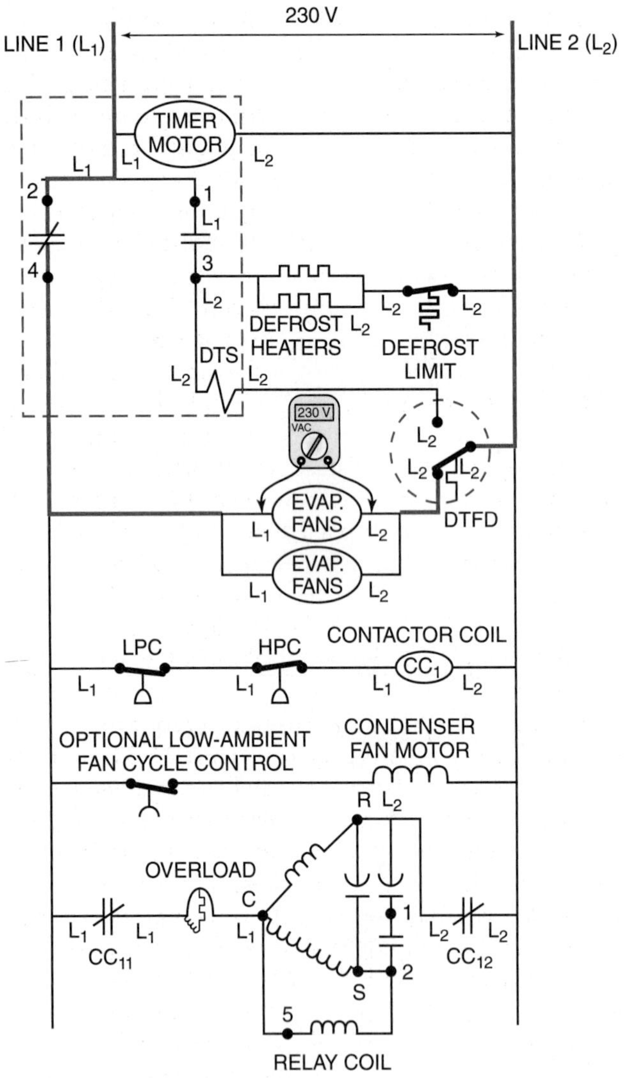

Figure 15.27 The evaporator fan motor circuit.

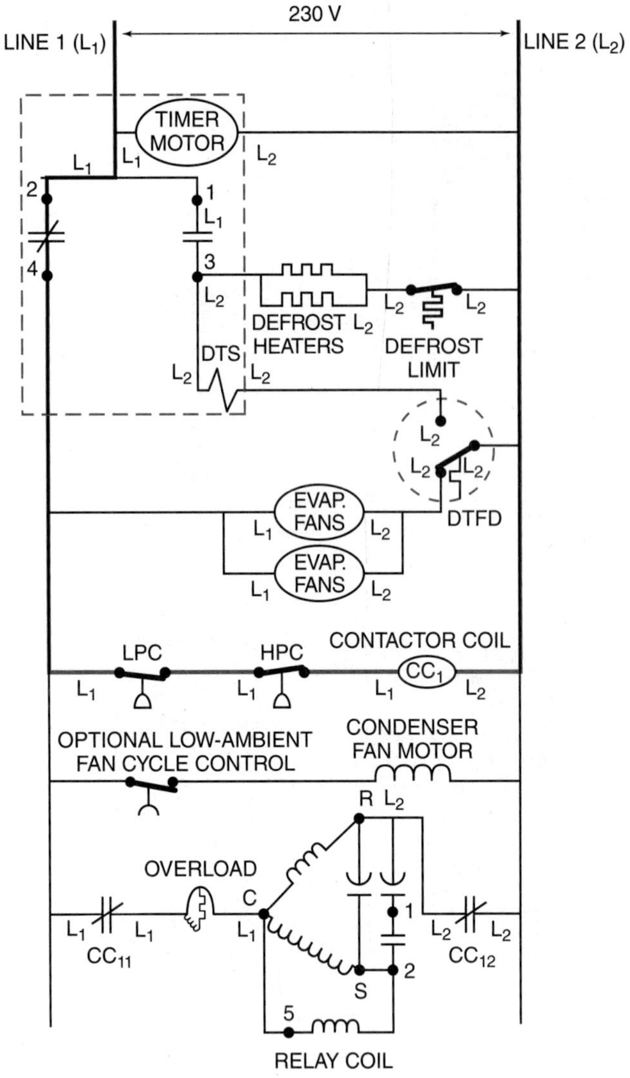

Figure 15.28 The contactor coil circuit.

So, in order for the contactor coil to be energized, the system's defrost timer must be calling for refrigeration and the system pressures must be within the safe or desired operating range. If the system pressure is too low, the low-pressure control will open and deenergize the coil. If the high-side pressure is too high, the high-pressure switch will open and deenergize the coil, as well. If there is line voltage being supplied to the contactor coil, the voltage readings across each of the three switches in the circuit will be 0 V. A reading of 0 V across the contactor coil most likely indicates that one of the three switches is in the open position. For example, a 230-V reading across contacts 2 and 4 on the defrost timer indicates that the contacts are open and that the defrost timer is not calling for refrigeration.

The last circuit in this diagram is the compressor circuit, **Figure 15.29**. Once again, the load is wired in series with a number of switches. In this case, there are four sets of contacts that must be closed in order for the compressor to operate. The four switches are

- The set of defrost timer contacts between terminals 2 and 4
- The contactor contacts CC_{11}
- The contactor contacts CC_{12}
- The compressor overload

We already know that in order for contacts 2 and 4 to be closed, the system has to be calling for refrigeration. The two sets of contactor contacts, CC_{11} and CC_{12}, are controlled by the contactor coil in the circuit branch we just discussed, **Figure 15.28**. The magnetic field generated by the contactor coil will cause these contacts to close. The last switch is the overload, which is a safety device. This device will open if there is excessive current flow in that circuit branch or if the compressor overheats. A voltage reading of 230 V across the overload indicates that there is an overcurrent, overload, or

Figure 15.29 The compressor circuit.

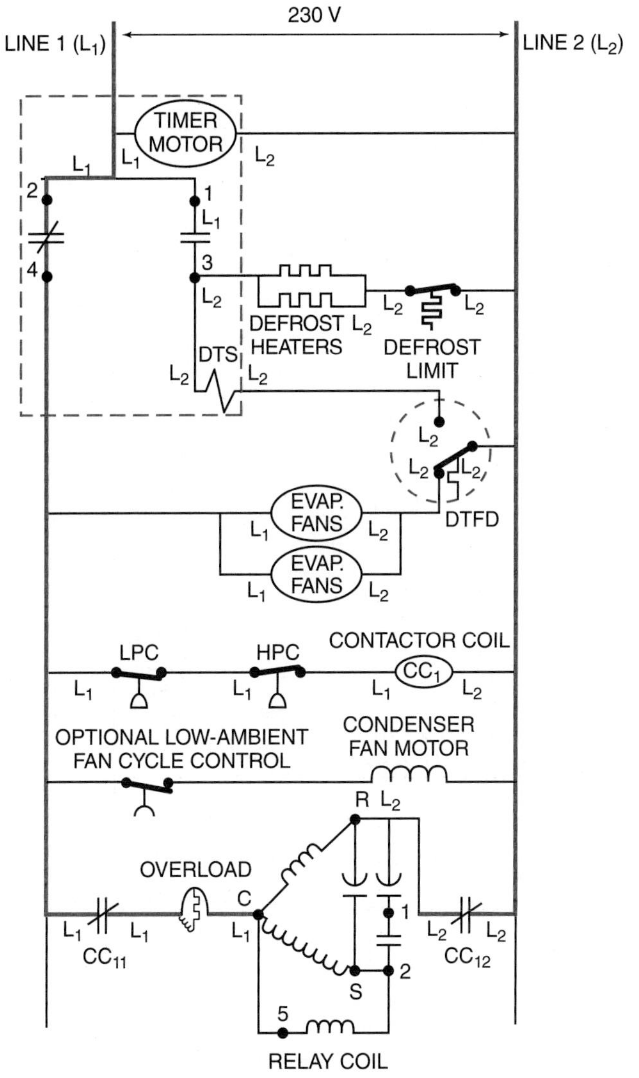

overheating condition. The cause for this condition must be identified and corrected to put the system back into operation.

15.8 PICTORIAL AND LADDER DIAGRAMS

The previous examples were all about troubleshooting the basic circuit, which is recommended to start with. The control circuits that use circuit boards work much the same way. The same rules apply but with more emphasis on manufacturers' directions. The service technician must have a good mental picture of the circuit or a good diagram to work from. Two distinct types of diagrams are furnished with equipment: the pictorial wiring diagram and the ladder wiring diagram. Equipment may come with one or both diagrams.

The pictorial diagram is used to locate the different components in the circuit, **Figure 15.30**. **Figure 15.30(A)** is a pictorial diagram of an integrated furnace control (IFC) board for a two-stage gas furnace. It contains connections for an electronic air cleaner (EAC), a humidifier (HUM), line 1 and neutral, fan connections for heating and cooling, secondary connections for the 24-V transformer, a 12-pin connector plug, a low-voltage thermostat terminal board, flame and power status lights, and an ignition transformer. Also built into the board through the 12-pin connector is a flame sensor, limit, rollout, overtemperature switches, high- and low-pressure controls, and a heat-assisted limit control. A blower door switch is connected to the integrated furnace control through a 2-pin connector. The induced draft motor is connected to the board through a 3-pin connector. This diagram is organized just as the technician would see it with the panel door open. For example, if the flame and power status lights are in the upper left-hand corner of the diagram, they will be in the upper left-hand corner of the control board when the door is opened. This is useful when you do not know what a particular component looks like. Study the pictorial diagram in **Figure 15.30(A)** until you find a control and then locate it in the actual integrated furnace control board in **Figure 15.30(B)**. Notice that the pictorial diagram also gives wire colors to further verify the components.

The ladder diagram is the easier diagram to use for following the logic behind the circuit, **Figure 15.31**. Normally, the diagram need only be studied briefly to determine the circuit functions. Notice that the power-consuming devices are to the right of the diagram. Switches and controls are usually to the left in a ladder diagram. Most of the time, a technician must know what the sequence of events is for the IFC in order to follow the logic when the furnace is called on to ignite. A technical manual from the manufacturer will sometimes outline the sequence of events, which will assist the technician in troubleshooting. The manuals often come with the furnace, or the homeowner may have one from the day of purchase. If the technician does not have such a manual, an Internet search or phone call to the furnace manufacturer may be needed before going any further in troubleshooting. Often, one of the diagrams on the inside of the front cover of the furnace will assist the service technician in figuring out the sequence of events for the furnace.

Pictorial and ladder diagrams are examples of the way most manufacturers illustrate the wiring in their equipment, but each has its own way of illustrating points of interest. The only standard the industry seems to have established is the symbols used to mark the various components, but even these tend to vary from one manufacturer to another. Anyone studying electrical circuits could benefit by first using a colored pencil for each circuit. A skilled person can divide every diagram into circuits. The colored pencil will allow an unskilled person to make a start in dividing the circuits into segments.

The following service calls are examples of actual service situations. The problem and solution are stated at the beginning of the first three service calls so that you will have a better

Figure 15.30 (A) A pictorial diagram showing the relative positions of the components as they actually appear. Courtesy Rheem Manufacturing

(A)

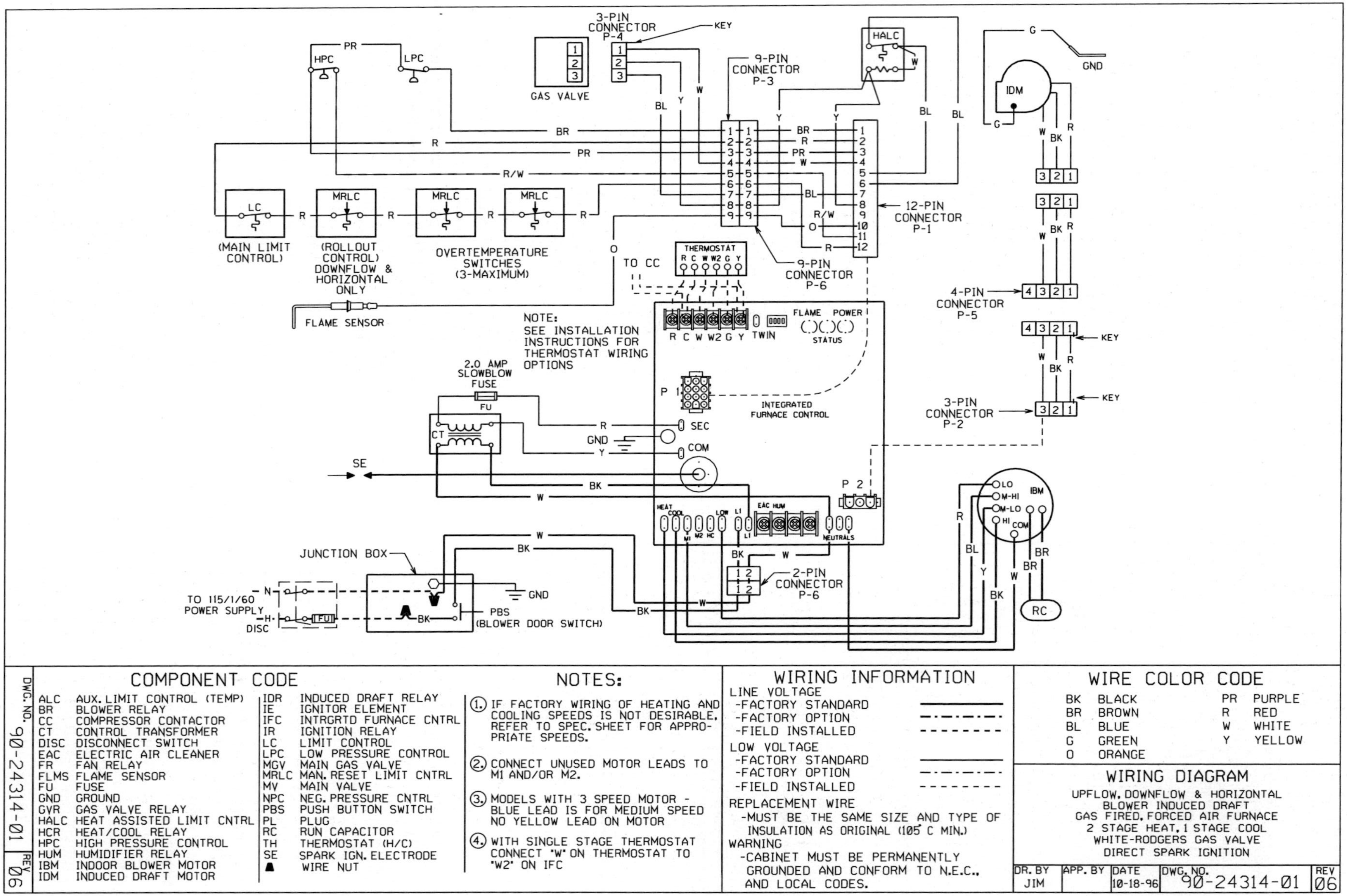

Figure 15.30 (*Continued*) (B) An actual circuit board. Courtesy Ferris State University. Photo by John Tomczyk

Figure 15.31 Ladder diagram of the circuit shown in **Figure 15.30(A)**. Notice the arrangements of the components. The right side has no switches. The power goes to the integrated furnace controller (IFC) and then to the power-consuming devices.

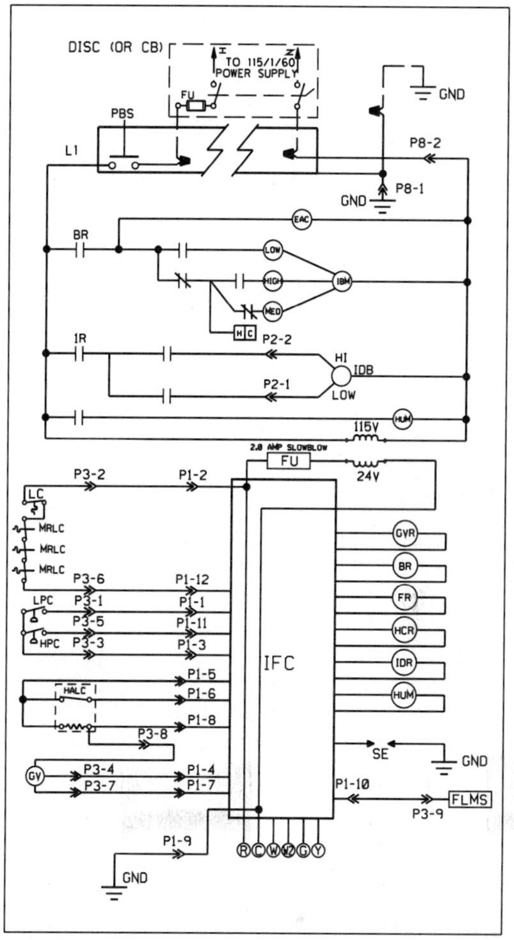

understanding of the troubleshooting procedures. The last four service calls describe the troubleshooting procedures up to a point. Solutions for these service calls are not provided here. Please use your critical thinking abilities or class discussion to determine the solutions and discuss further with your instructor.

15.9 SERVICE TECHNICIAN CALLS

SERVICE CALL 1

A customer with a package air conditioner calls and tells the dispatcher that the unit is not cooling. *The problem is that the control transformer has an open circuit in the primary,* **Figure 15.32.**

Before approaching the house, the technician makes sure there is no dirt or grease on himself from the previous job. He quickly wipes his shoes clean with a rag kept in the service truck for this purpose. After ringing the doorbell, the technician maintains a distance from the door that is comfortable for the customer, politely introduces himself and his company, and maintains eye contact with the customer. The technician allows the customer to finish talking before replying. After asking permission to enter the living room where the thermostat is located, the technician puts on shoe covers to protect the customer's floors.

The technician turns the indoor fan switch on the indoor thermostat to the FAN ON position. The fan will not start, which possibly indicates there is no control voltage. There must be high voltage before there is low voltage. The technician then goes to the outdoor unit and, with the multimeter selector switch set to measure AC voltage, checks for high voltage. There is power.

The technician removes the cover from the control compartment (usually this compartment can be located easily because the low-voltage wires enter close by). A check for high voltage at the primary of the transformer shows that there is power. A check for power at the secondary of the low-voltage transformer indicates no power. The transformer must be defective.

To prove this, the main power supply is turned off. One lead of the low-voltage transformer is removed and the ohms function of the multimeter is used to check the transformer windings for continuity. *Here the technician finds the primary circuit is open.* A new transformer must be installed.

Figure 15.32 The control transformer has an open circuit.

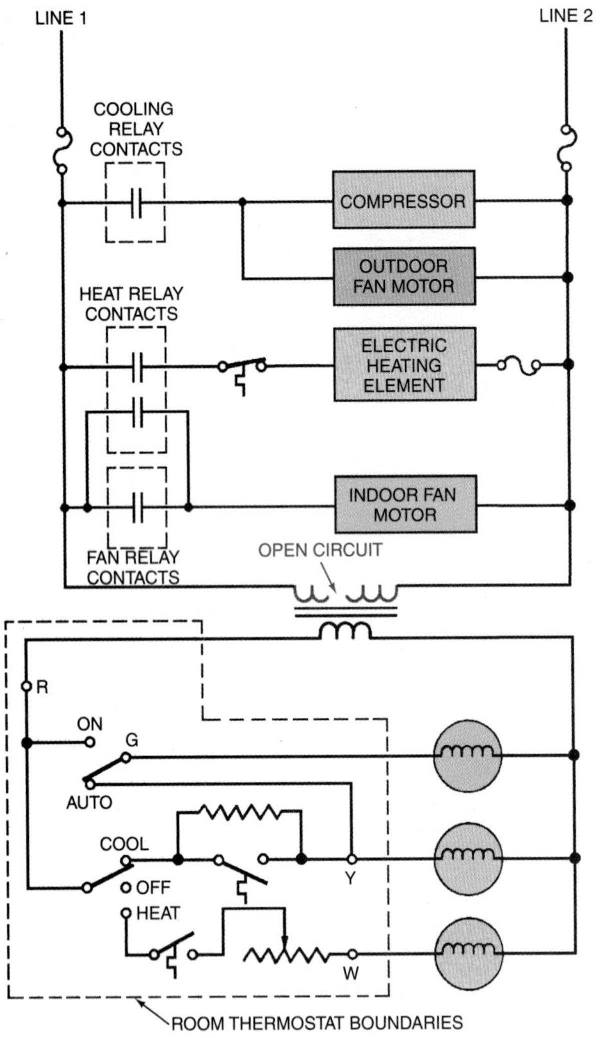

Figure 15.33 The cooling relay has an open circuit and will not close the cooling relay contacts. Notice that the indoor fan motor remains operational.

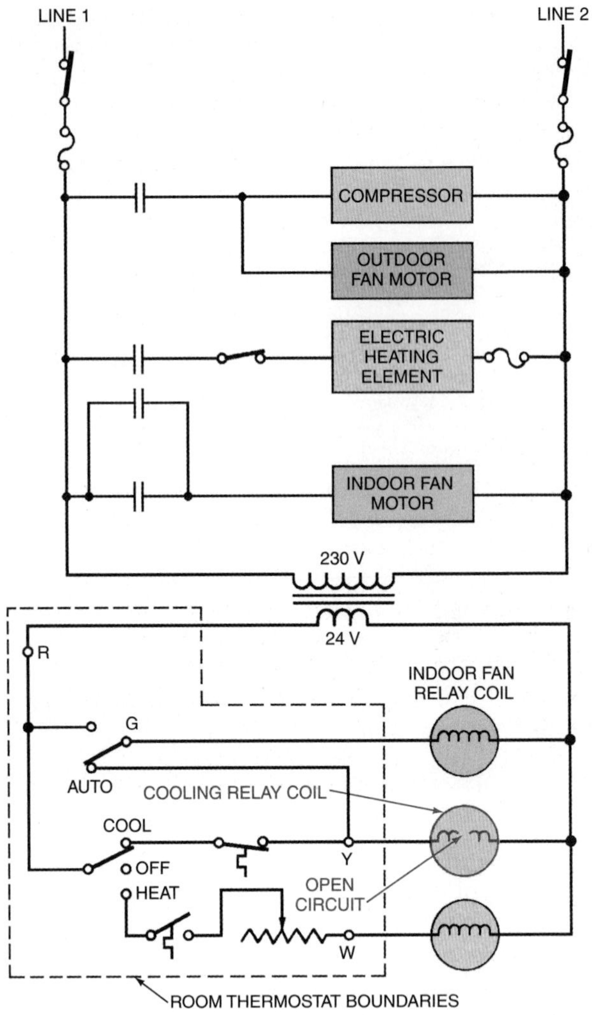

The technician installs the new transformer and turns the power on. The system starts and runs correctly. After double-checking his work, the technician cleans the work site and explains to the customer how the problem was found. The technician now gives the customer billing information with a work description and costs. The technician then lets the customer know that he is a valued customer and leaves after shaking hands.

SERVICE CALL 2

A customer calls with a "no cooling" complaint. *The problem is that the cooling relay coil is open,* **Figure 15.33**.

As the technician arrives at the customer's house, he makes sure he is not blocking the driveway with the service van; reviews the customer's name and complaint; makes sure his uniform is neat, clean, and tucked in; wipes down his shoes with a clean rag; then rings the doorbell and maintains a comfortable distance while introducing himself. The service technician politely listens to the customer, and once the problem is understood, covers his shoes with shoe covers to protect the customer's floors. The technician now heads over to the thermostat and tries the fan circuit. The fan switch is moved to FAN ON and the indoor fan starts. Then the thermostat is switched to the COOL position and the fan runs. This means that the power is passing through the thermostat, so the problem is probably not there. The thermostat is left in position to call for cooling while the technician goes to the outdoor unit. Since the primary for the low-voltage transformer comes from the high voltage, the power supply is established.

A look at the diagram shows the technician that the only requirement for the cooling relay to close its contacts is that its coil be energized. A check for 24 V at the cooling relay coil shows 24 V. The coil must be defective. To be sure, the

put back in place. The system is started with the room thermostat and it operates normally. Once the work is double-checked, the technician cleans the job site. He explains the problem and solution to the customer in everyday language and gives a written work description and billing information to the customer and explains them politely. The service technician then thanks the customer and leaves the house.

Solutions for the following service calls are not provided here. Please discuss the solutions with your instructor.

SERVICE CALL 4

A customer calls and indicates that the air conditioner quit running after the cable TV repairman had been under the house installing TV cable.

When the technician enters the house it is obvious that the air conditioner is not running. The technician can hear the fan running, but it is hot inside. A check outside shows that the condensing unit is not running. The technician then proceeds through the following checklist. Use the diagram in **Figure 15.35** to follow the checkout procedure.

1. Low voltage is checked at the condenser; there is none.
2. The technician then goes to the room thermostat, removes it from the subbase, and jumps from R to Y; the condensing unit does not start.
3. Leaving the jumper in place, the technician then goes to the condenser, checks voltage at the contactor, and finds no voltage.

What is the problem and the recommended solution?

Figure 15.35 Use this diagram for the discussion of Service Call 4.

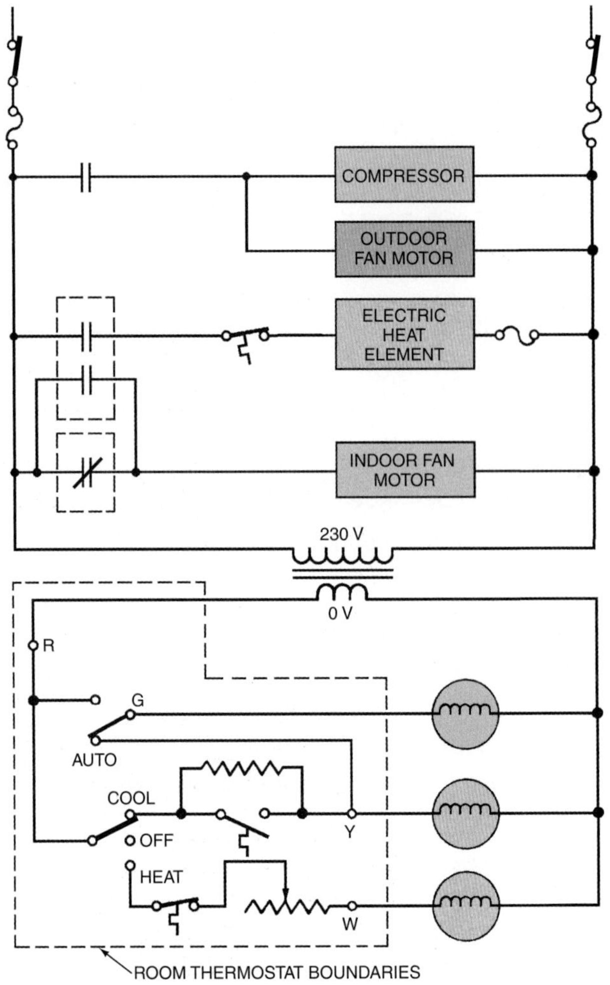

SERVICE CALL 5

A shoe store manager calls to report that the heat is off. This store has an electric heat package unit on the roof.

The service technician politely introduces himself and his company to the customer. While maintaining eye contact with the customer, the technician asks for the customer's account of the problem. The service technician should always make customers feel like they are helping to find the solution to the problem. Once all of the customer's questions are answered in everyday language the customer can understand, the technician puts on protective shoe covers and starts to troubleshoot. The technician uses the following checkout procedures to determine the problem. You can follow the procedure using the diagram in **Figure 15.36**.

1. The store manager says the system worked well until about an hour ago when the store seemed to be getting cold.

Figure 15.36 Use this diagram for the discussion of Service Call 5

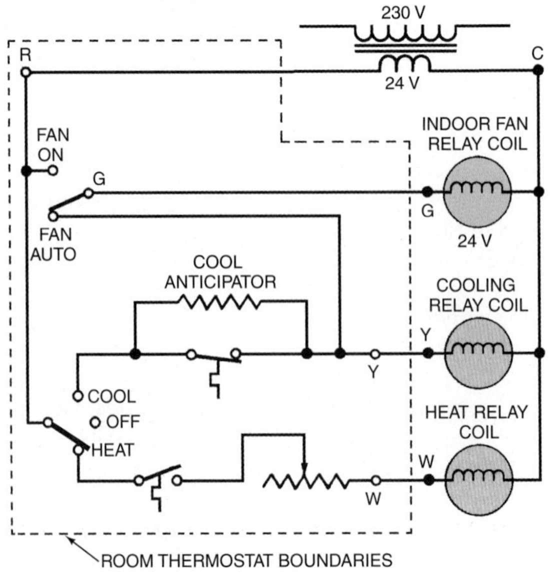

Figure 15.37 Use this diagram for the discussion of Service Call 6.

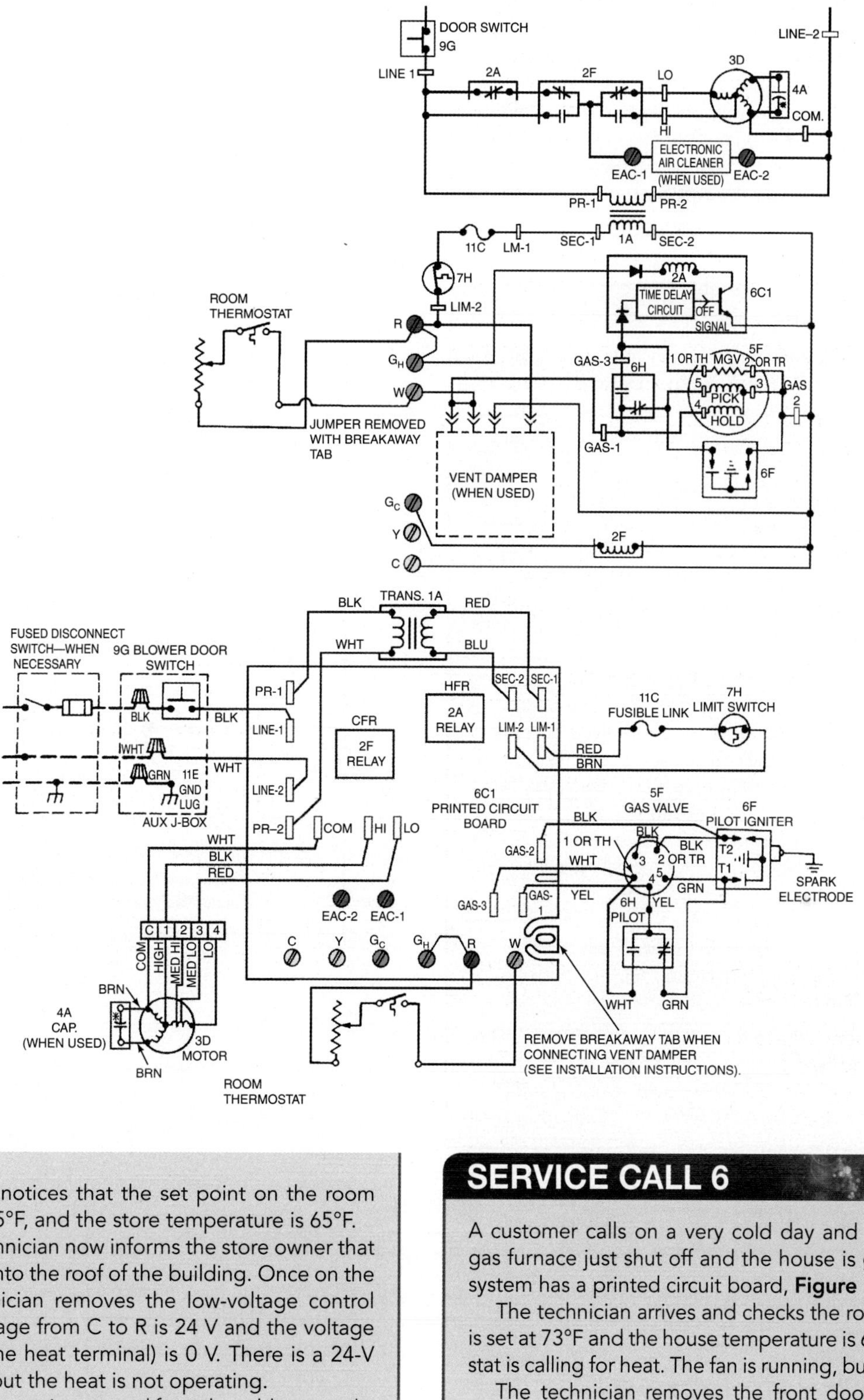

2. The technician notices that the set point on the room thermostat is 75°F, and the store temperature is 65°F.
3. The service technician now informs the store owner that he has to get onto the roof of the building. Once on the roof, the technician removes the low-voltage control panel. The voltage from C to R is 24 V and the voltage from C to W (the heat terminal) is 0 V. There is a 24-V power supply, but the heat is not operating.
4. The room thermostat is removed from the subbase, and a jumper is applied from R to W. The heating system starts.

What is the problem and the recommended solution?

SERVICE CALL 6

A customer calls on a very cold day and explains that the gas furnace just shut off and the house is getting cold. The system has a printed circuit board, **Figure 15.37**.

The technician arrives and checks the room thermostat. It is set at 73°F and the house temperature is 65°F. The thermostat is calling for heat. The fan is running, but there is no heat.

The technician removes the front door to the furnace, which stops the furnace fan because of the door switch (see 9G at the top of the diagram). The technician tapes down the door switch so he can troubleshoot the circuit.

The technician places one meter lead on SEC-2 (the common side of the circuit) and the other on SEC-1 and finds that there is 24 V. The meter lead on SEC-1 is then moved to the R terminal, and 24 V is found there. The lead is then moved to W, where the signal from the room thermostat passes power to the circuit board, and there is power.

The technician then removes the lead from W and places it on the GAS-1 terminal, where the technician finds 24 V.

What is the problem and the recommended solution?

SERVICE CALL 7

A service technician is called to a small market where the customer is complaining of products spoiling in a dairy case.

After introducing himself and his company and handing the store owner a business card, the service technician carefully listens to the owner's account of the problem. The technician then politely asks the customer some pertinent questions in clear, concise, everyday language. Even though the questions are somewhat technical, the technician tries to not make the customer feel inept about any technical topics. Once the store owner shows where the dairy cases are located, the technician starts working. The technician takes the box temperature of the dairy case with a thermometer and finds it to be 47°F. The case is supposed to be holding product at 35°F to 39°F. While measuring the temperature of the dairy case, the technician notices that there is no airflow on one side of the case. A quick inspection of the evaporator section and evaporator fans indicates that one of the three PSC evaporator fan motors is not operating.

The three fans are wired in parallel and operate off 230 V, **Figure 15.38**. The run winding of the inoperative PSC fan motor has burned open. The technician suspects that the inoperative fan motor probably has an open winding. The service technician then opens the power disconnect switch and places a multimeter across the run winding of the inoperative fan motor. The meter reads 2 ohms.

Did the service technician take the right steps when troubleshooting the open winding in the inoperative evaporator fan motor with the ohms function of the digital multimeter? Explain your answer.

Figure 15.38 The fan circuit for a dairy case showing three PSC motors in parallel with one another.

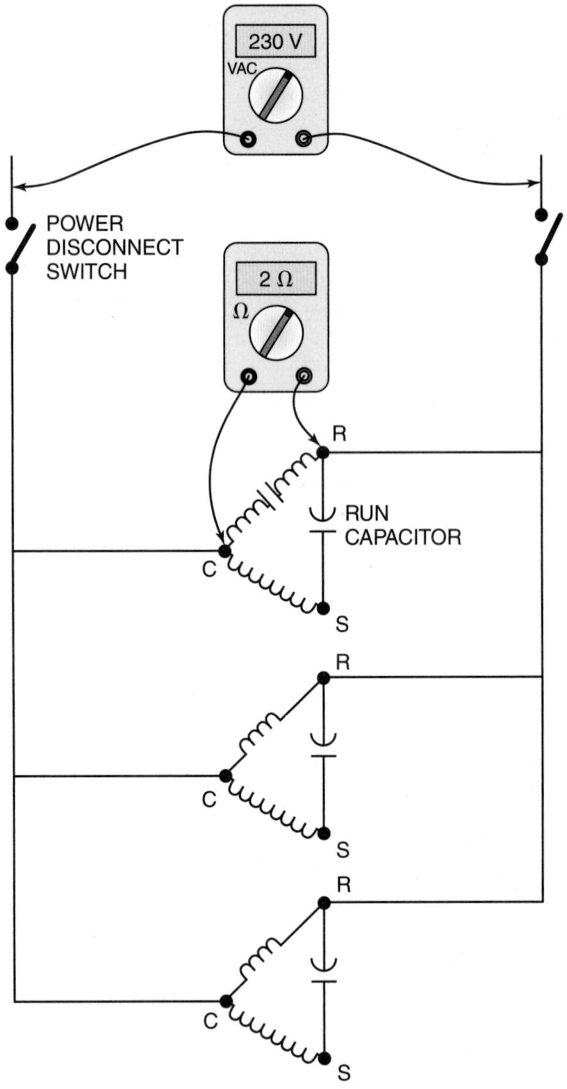

REVIEW QUESTIONS

1. Electrical, mechanical, and electromechanical are all types of __________.
   A. wiring diagrams
   B. test instruments
   C. automatic controls
   D. voltage feedback

2. Devices that pass power can be __________ devices.
   A. pressurized or non-pressurized
   B. power-passing or power-consuming
   C. switch or load
   D. safety or operating

3. The thermostat Y terminal energizes the __________ circuit.
   A. cooling
   B. heating
   C. fan
   D. ventilation
4. The component in the system that is energized when the G terminal is energized is the _______.
   A. compressor
   B. fan motor
   C. gas valve
   D. electric heat strips
5. The _____ terminal must be energized in the thermostat for the heat to start.
   A. Y
   B. W
   C. G
   D. H
6. The 24-V power circuit in a low-voltage thermostat is connected to the _____terminal.
   A. G
   B. Y
   C. W
   D. R
7. A heat anticipator is wired ________ the thermostat contacts.
   A. in parallel with
   B. directly opposite of
   C. in series with
   D. far away from
8. The cooling anticipator is wired __________.
   A. in parallel with the cooling contacts
   B. next to the liquid line
   C. in series with the cooling contacts
   D. inside the indoor unit
9. Which anticipator is adjustable?
   A. Cooling anticipator
   B. Both heat and cooling anticipators
   C. Heat anticipator
   D. Neither anticipator
10. During the ______ cycle, the indoor fan must always run for the entire cycle.
    A. final
    B. initial
    C. heating
    D. cooling
11. When the control transformer is not working, the indoor fan _________.
    A. will only run with the fan switch in the ON position
    B. will not run at all
    C. will only run during the cooling cycle
    D. will only run during the heating cycle
12. Which of the following should be turned off while working on the compressor circuit?
    A. The indoor unit fan
    B. The unit disconnect
    C. The thermostat
    D. The main power supply to the building
13. The ________diagram is used to locate different components on the diagram in the unit.
    A. pictorial
    B. ladder
    C. thermostat
    D. condenser
14. To trace the circuits in the unit, the _______ diagram is used.
    A. thermostatic
    B. system
    C. pictorial
    D. ladder
15. Electrical devices can be considered as either _______ ______devices.
    A. power-passing or power-consuming
    B. pressurized or non-pressurized
    C. digital or electronic
    D. mechanical or digital
16. Describe how amperage is measured in a live circuit.
17. Explain why an ohmmeter, and not a voltmeter, should be used to assist in finding an opened run winding on a motor.
18. Explain why a service technician must always know where line 1 and line 2 are in relation to one another when voltage troubleshooting a schematic wiring diagram.
19. Explain how a digital multimeter can be used to troubleshoot a switch.
20. Explain the functions of a digital multimeter as they relate to troubleshooting an electric circuit.

UNIT 16

ADVANCED AUTOMATIC CONTROLS—Direct Digital Controls (DDCs) and Pneumatics

OBJECTIVES

After studying this unit, you should be able to

- discuss advanced control terminology.
- describe control applications.
- describe electronic control circuits.
- describe pneumatic control circuits.
- discuss a control loop.
- understand direct digital control systems.
- discuss sensitivity or gain in controls.
- describe residential electronic controls.

16.1 CONTROL APPLICATIONS

In the beginning of the HVACR industry, there was only manual control of all the components in a unit. For example, coal furnaces required a person called a "stoker" to put the coal in at the correct time to heat a structure. When the furnace was overfired, it was too hot. The windows had to be opened to release some of the heat. If the furnace was underfired, it took time to get the temperature up to a comfortable level. The first big improvement was the automatic stoker, which was a conveyor belt operated by a timer. Every so many minutes, it started up and ran for a prescribed length of time, dumping coal into the furnace. This method had no connection to the actual temperature of the heated space, but it eliminated a person from the process. As time passed, the process was improved by using a thermostat in the conditioned space. Then it was further improved with an outside thermostat that anticipated heating needs. All of these control methods would be considered primitive and unacceptable by today's standards. Today, we expect the temperature to be well within the bounds of comfort, summer and winter. This can be accomplished with modern systems and a small number of technicians, even in large buildings.

If you walked through a large modern building with the best control systems and noticed warm and cool spots, you would be correct in thinking that the temperature level should be much more even. Many buildings have adequate systems and control arrangements but do not have a staff of people who understand how to operate the controls. The control system must be thoroughly understood to be set up to operate correctly. Even after the system is correctly set up, the controls may drift out of calibration from time to time and have to be readjusted. Buildings are designed to be divided into zones and into different styles and sizes of office space. Walls are easily removed and new ones installed. Many times this is done with no regard to the heating and cooling system. It is not unusual to find the return air inlet in one office, the supply air outlet in another office, and the thermostat in yet another office. Sometimes construction workers put the thermostat in the ceiling while they are building walls and forget to remount it. All of this leads to unbalanced systems.

The accomplished technician should be able to look at the building blueprints and interpret the intentions of the control designer. Many control systems do not control as well as expected, but most will be satisfactory if set up correctly.

16.2 TYPES OF CONTROL SYSTEMS

The very first control systems only stopped and started the components to be controlled. These controls were discussed in Unit 14, "Automatic Control Components and Applications." Control technology has been expanded not only to switch components on and off but also to modulate them. An example of modulation is the throttle on an automobile. Can you imagine a car that had only on and off as controls? You would be either accelerating at full speed or decelerating. The accelerator is properly used to slowly move up to the required speed and to slow down by applying less pressure to the gas pedal. Modern control systems can accomplish these same functions with automatic controls.

Control systems can be broken down into three basic components. These components make up a **control loop**. All control systems can be subdivided into control loops, no matter how many controls there are:

- The sensor measures the change in conditions.
- The controller controls an output signal to the controlled device. The sensor and controller are often in the same device.

Figure 16.1 Typical pneumatic thermostat. Photo by Bill Johnson.

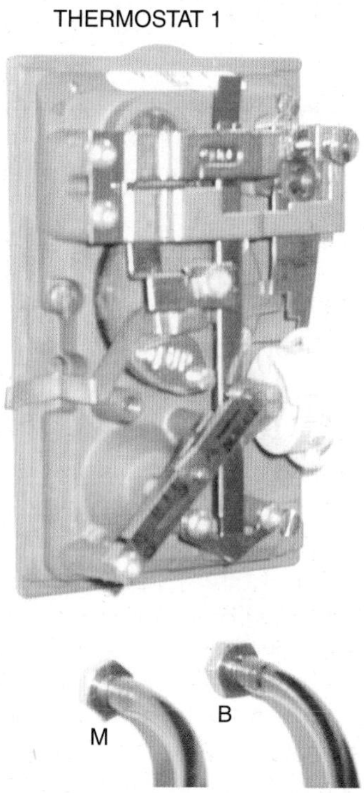

- The controlled device reacts to stop, start, or modulate the flow of water, air, or refrigerant to the system or the conditioned space.

The sensor is the device that responds to a change in conditions. In this field, the sensors respond to temperature, velocity of air, humidity, water level, and pressure. **Figure 16.1** shows the inside of a pneumatic room thermostat. The sensor sends a signal to the controller, which in turn sends the correct signal to the device to be controlled. The controller may be built into the sensor. Common connection points on a pneumatic control are "M," which stands for "Main Air" and "B" which stands for "Branch Air."

The controls that control the temperature in a home will serve as a simple example of a basic control loop. The wall thermostat may be located in the hall close to the return air where it senses the space temperature. When the temperature rises above the set point, for example to 75°F, the bimetal in the thermostat will move and cause the mercury contacts to close. This is the action of the sensor coupled to the controller. When the mercury contacts close, a 24-V signal goes to the compressor contactor (the controlled device), which passes power to the compressor and starts to cool the house. It will continue to run until the mercury contacts open the circuit to the compressor contactor.

The technician must understand that all systems can be divided into "control loops" and that they can be identified. A large building will have a control print that shows the basic control loops as individual loops, **Figure 16.2**. The legend that shows abbreviated words and symbols is at the beginning of the mechanical print section. You may drive past a 200-room motel and think you could never understand what

Figure 16.2 A building is divided into control loops for each area. One area may be larger than another, but the control loop may be the same.

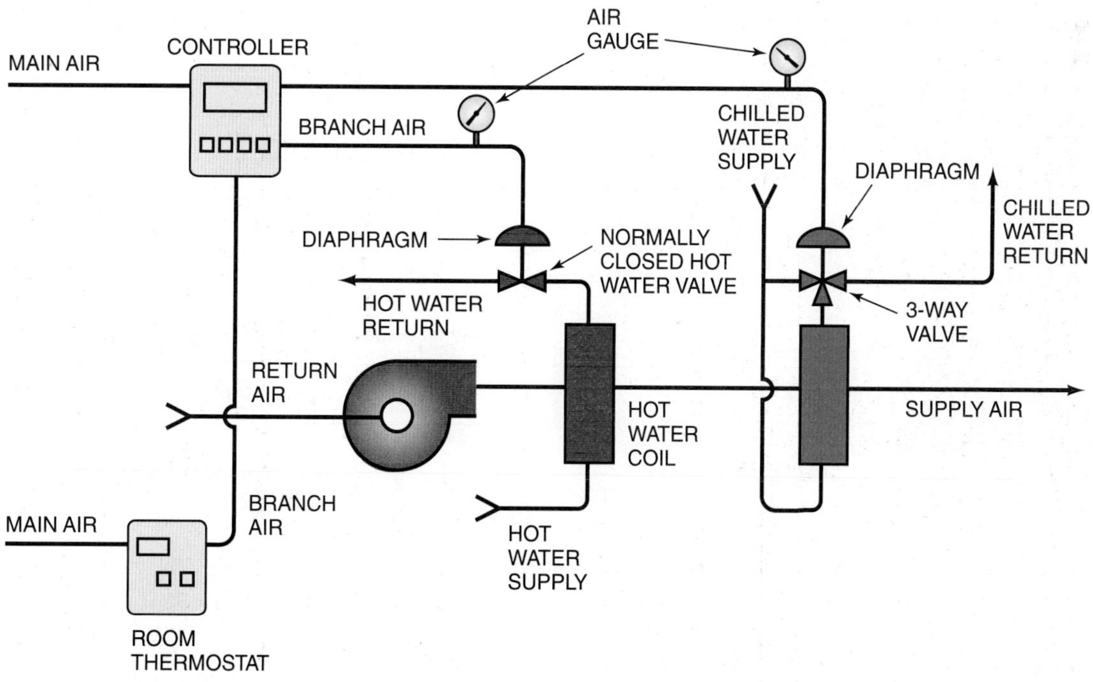

goes on inside to operate the air-conditioning and heating systems. When you look further, you will find that it is actually one plan that has been implemented 200 times. Each room is practically identical as far as the utilities are concerned. They each have the same sink, toilet, lights, and air-conditioning system. When you learn one of them, you have learned all of them because it is a control loop for a portion of the system. The control loop in an office building may be much the same. When you learn one area, you have a basic understanding of the next one.

The basic types of controls being installed today to control large systems are pneumatic and digital electronic controls. These systems can also be divided into control loops for the purpose of design, installation, and troubleshooting. Pneumatic control systems were used before electronic ones. Pneumatic controls are mechanical in nature and use air to transmit signals from the sensor controller to the controlled device. Electronic controls are quickly taking the place of pneumatic controls and are the controls of the future. However, many installations already set up and operating with pneumatic controls will be used for many years. Electronic control systems can be integrated into the pneumatic systems to employ the best features of both control systems.

16.3 PNEUMATIC CONTROLS

In the past, many buildings and industrial installations were supplied with pneumatic controls. A good mechanic can easily master these simple controls. The control signal is air, which is run in small pipes, typically 3/8 in. or 1/4 in. This eliminates the need for an electrical license and expertise in installing control wiring. Some advantages of pneumatic controls are as follows:

- They are simple to install and understand. A person with a mechanical background can soon adapt to them.
- They are safe and explosion-proof around combustible materials.
- They can be easily adapted to provide either on/off or modulating control functions.
- They are versatile and can perform many control sequences.
- They are very reliable.
- The installation cost may be lower than the cost of running long electrical lines and conduit.

The starting point for understanding pneumatic controls is to understand the air system that supplies the power to operate the controls. Most systems use a dedicated air compressor and tank with a rather elaborate filtering and drying system to supply the air. Pneumatic controls must have a permanent air supply that is **clean** and **dry**. The actual control portion of the system will have many components with small air passages called orifices, which are so small that they cannot tolerate a dirt particle or a drop of water. Because some of the controls have a constant bleed rate with air escaping all the time, the systems use air continuously. **Figure 16.3** shows a drawing of a typical dedicated single-compressor system for the air supply. Some systems have two compressors of the same capacity so that there is always a backup compressor. Either compressor can supply the total air requirements for

Figure 16.3 An air compressor system for a pneumatic control system. Source: Honeywell

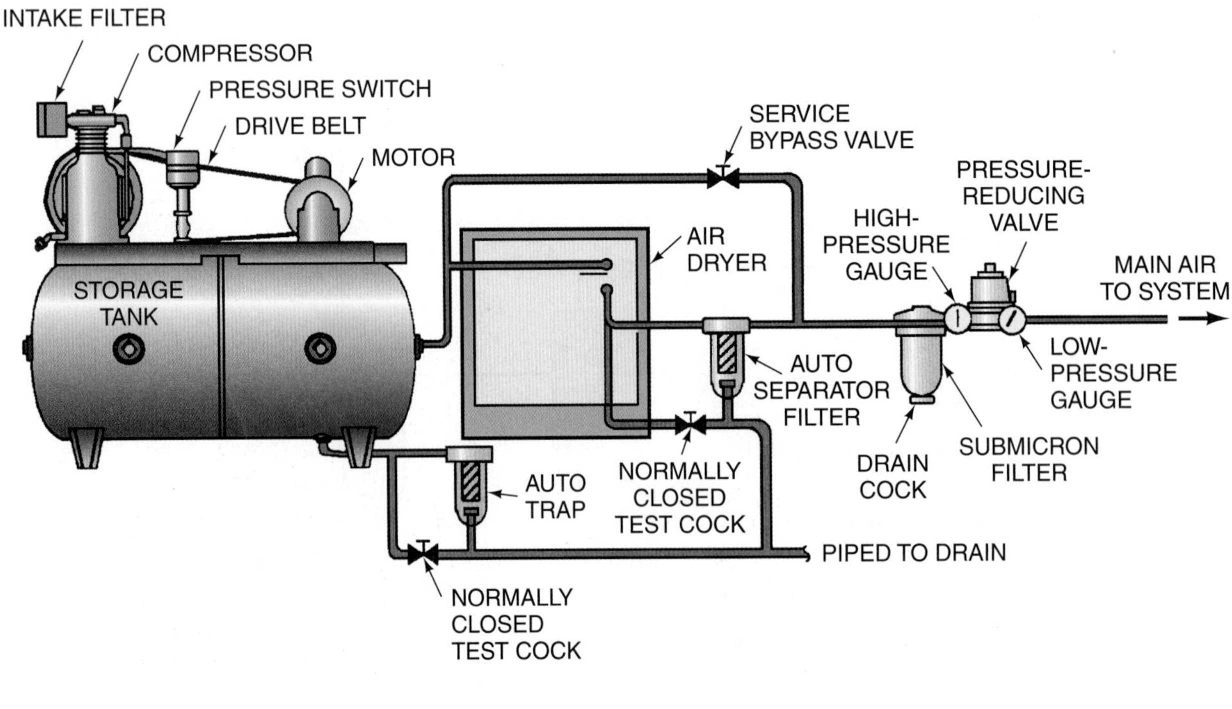

the entire system by running only about half the time. The second compressor operates when there is a great demand. Most system compressors provide 100 to 125 psig of air pressure to the storage tank and are controlled with a pressure switch. When the air pressure drops to about 75 psig, the compressor starts; when the pressure rises to about 100 or 125 psig, the compressor stops.

16.4 CLEANING AND DRYING CONTROL AIR

The cleaning of the air starts with the prefilter to the compressor. It is much like the air filter on an automobile. It may be a dry-type filter that filters all entering air. This must be kept clean, or pressure will drop, causing excess running times.

The air entering the compressor usually has moisture, or humidity, entrained in it. Most parts of the country have some humidity, and many states have high humidity, particularly in the summer. When air with high humidity is compressed from atmospheric pressure to 100 psig, the outlet air from the compressor into the storage tank may be 100% saturated with moisture. Often, the air passes through a small coil over which air from the blades on the compressor flywheel passes. This coil will condense some of the moisture to a liquid, where it will quickly drop to the bottom of the storage tank. The storage tank normally will have a float-operated automatic bleed valve that will drain water from the tank, **Figure 16.4**.

Figure 16.4 As water accumulates in the float chamber, the float rises and drains the water to the floor drain.

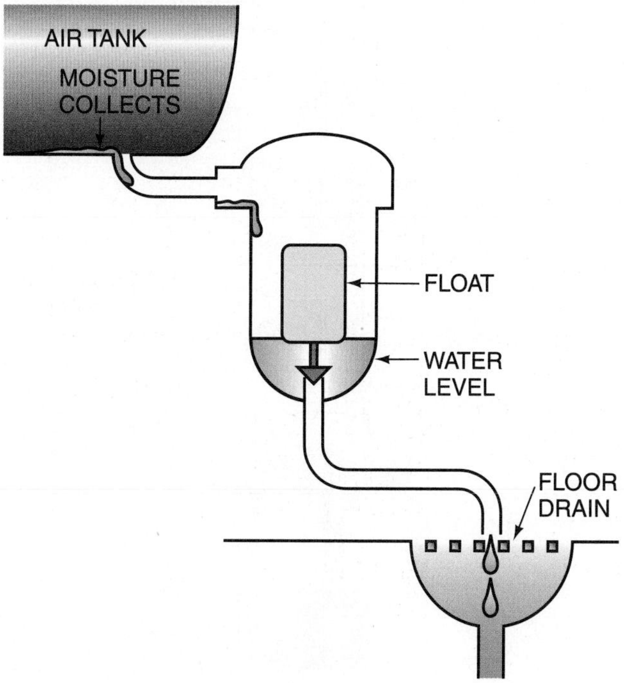

The remaining air in the storage tank normally will still have too much moisture to be used for control air, so it must be further dried. Usually one of four types of drier is used:

- Local water supply and coil
- Chilled water coil
- Desiccant driers
- Refrigerated air drier

The local water supply may be used if it is cold enough to remove the correct amount of moisture from the air. This is not likely in most applications. Chilled water from the refrigeration system may be used; however, if the chiller is not operating, the air may not be dried enough, and problems may occur. Desiccant driers use some of the same products that refrigerant driers use—silica gel or molecular sieve as an adsorbent, which removes moisture much like table salt attracts moisture. The desiccant may be regenerated by means of heat, so there must be two driers—one in operation and one being regenerated at all times. Regeneration may be accomplished with heated dry air. Using some portion of the dried air from the system that is in operation may be part of the process.

Refrigerated air driers are probably most common, **Figure 16.5**. These use a small refrigerated system that passes the compressed air through a refrigerated heat exchanger, which is operated below the dew point temperature of the compressed air. This condenses much of the moisture from the air, where it is automatically drained from the system with a float-operated drain device. When the water level rises in the float chamber, a valve is opened and the water goes down the drain.

When the air is dried, it may pass through another filter that is so fine that oil vapor or liquid cannot pass through it. This is a high-density filter to prevent any leftover particles from entering the control system. The system has virtually a zero tolerance for contaminants.

Figure 16.5 A small refrigerated air drier.

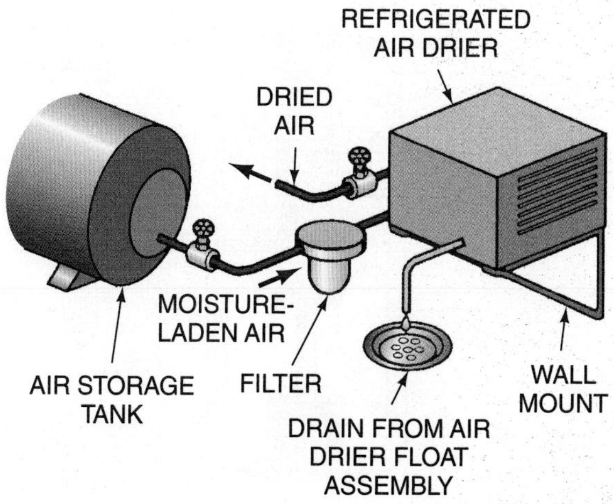

The control systems require an air supply that is usually no more than 20 psig. A pressure-reducing regulator is located in the line after the air-drying device. This pressure reducer or regulator reduces the air pressure from about 100 psig to 20 to 30 psig, **Figure 16.3**. A relief valve is located either in the pressure reducer or in the downstream side of the system to ensure that the control air pressure does not exceed about 30 psig. The reason for this is that the controls themselves are not rated for high pressure. High pressures will destroy the components.

16.5 CONTROL COMPONENTS

The 20-psig control air is used to position various components in the system. We will use a valve actuator assembly as an example. Remember that we will be working with an air pressure of 20 psig, that is, 20 pounds per square inch. Main air pressure is used in this example. This is important because we will use the 20 psig to open a valve with a large diaphragm that has a diameter of 4 in. This diaphragm has an area of 12.56 in^2 (area $= \pi \times r^2$, or $3.14 \times 2 \times 2 = 12.56$ in^2). When 20 psig of pressure is applied to the top of the diaphragm, there are 251.2 lb of force (12.56 $in^2 \times$ 20 psi = 251.2 pounds of force), **Figure 16.6**. The diaphragm is held in the upward location with a strong spring, and the air above the diaphragm overcomes the spring. Therefore, when the air pressure is released, the diaphragm will be repositioned to the upward location.

Figure 16.6 This diaphragm assembly can exert a pressure of up to 250 lb because 20 psi of air pressure is exerted on a 4-in. diaphragm.

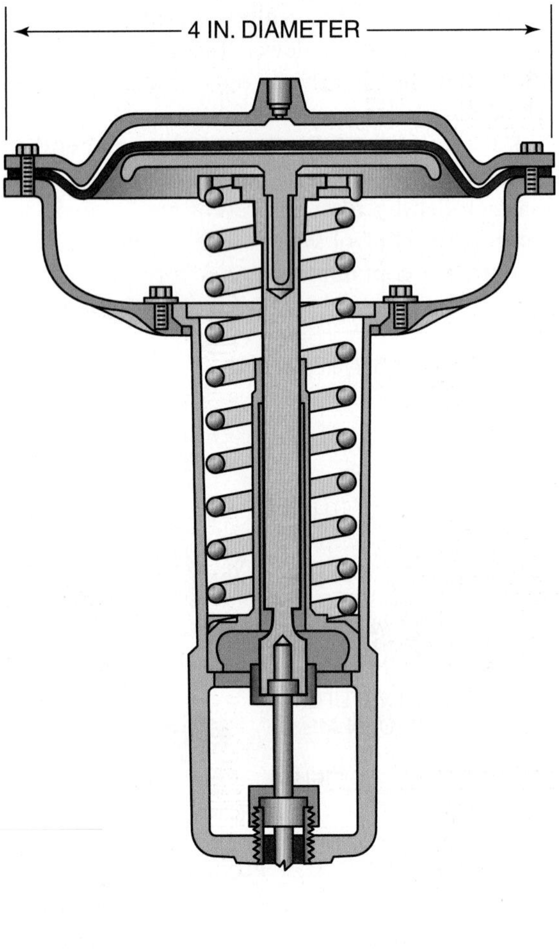

The diaphragm and its plunger can be fastened to any device that we want to move with the air pressure, such as a steam or water valve or a set of air dampers, **Figure 16.7**. We can move a rather large load with 251.2 lb of energy. The air pressure, which is the force, can be piped to the various components using small copper or plastic tubing. The 20-psig main air lines are typically run using 3/8-in. lines, and the branch control pressures are run with 1/4-in. lines. If you have ever watched a service station operator push an air valve and lift a 4000-lb car with the air lift, it is the same principle, only with about 100 to 125 psig of air pressure.

The pneumatic control components—the sensors such as thermostats, humidistats, pressure controls, relays, and switches—perform the same functions that other controls do. The components to be controlled are valves, dampers, pumps, compressors, and fans. These components work to control the conditioned space. There are as many combinations of

Figure 16.7 This diaphragm and plunger can be fastened to any number of devices that need to be positioned to control conditions.

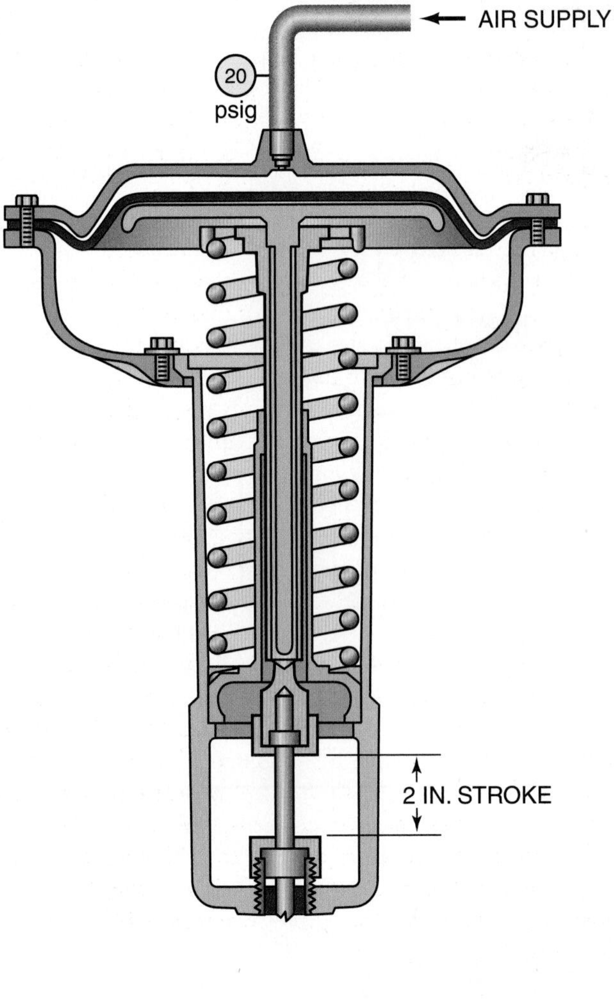

Figure 16.8 This control loop is used to control room temperature.

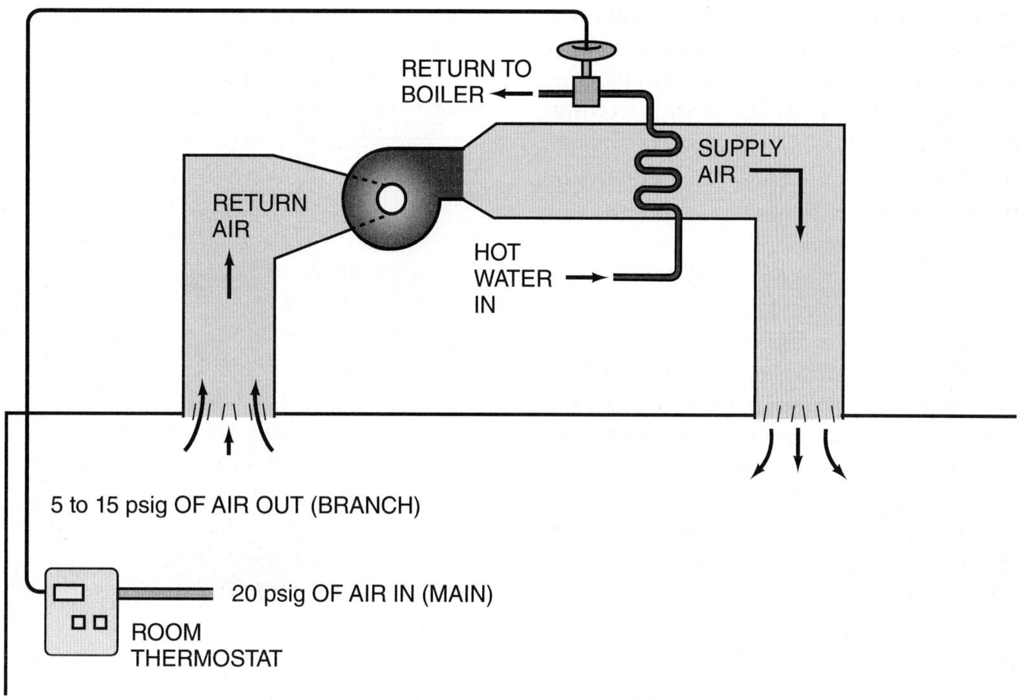

controls in various systems as there are systems. We will outline some basic processes.

The system in **Figure 16.8** can control the room temperature in many increments, depending on the sensitivity of the thermostat. This is a basic control loop: When the room conditions change and cause a change in the room thermostat, the outlet air changes and that changes the room thermostat. This is called a closed loop because all changes happen within the loop. This control setup is found in many office buildings and motels. When you change a thermostat and then hear air movement, it is a pneumatic control system that uses a normally closed diaphragm valve and air pressure to overcome the spring to allow hot water to pass to the coil. Many other applications use a normally open valve that is held closed by air pressure. When the room conditions call for heat, it allows air pressure to bleed off the diaphragm and open the valve. This is called reverse acting. It has the advantage that if the systems fail for any reason, the building will overheat rather than have no heat. This can be considered a safety feature in cold climates because it prevents building freezing conditions over a holiday.

The following is an example of how a room thermostat controls chilled water flow by modulating it based on the room's needs. The thermostat will provide a branch line pressure, called control air, of between 0 and 15 psig. The control air positions the valve for the correct chilled water flow through the coil to maintain even temperatures in the conditioned space. With electric controls, only a solenoid valve may be used to divert water around the chilled water coil or through the coil, **Figure 16.9**. Pneumatic controls can be used to position the valve for the correct water flow by modulating

Figure 16.9 The electrical solenoid can only provide on/off control. This causes a swing in room temperature.

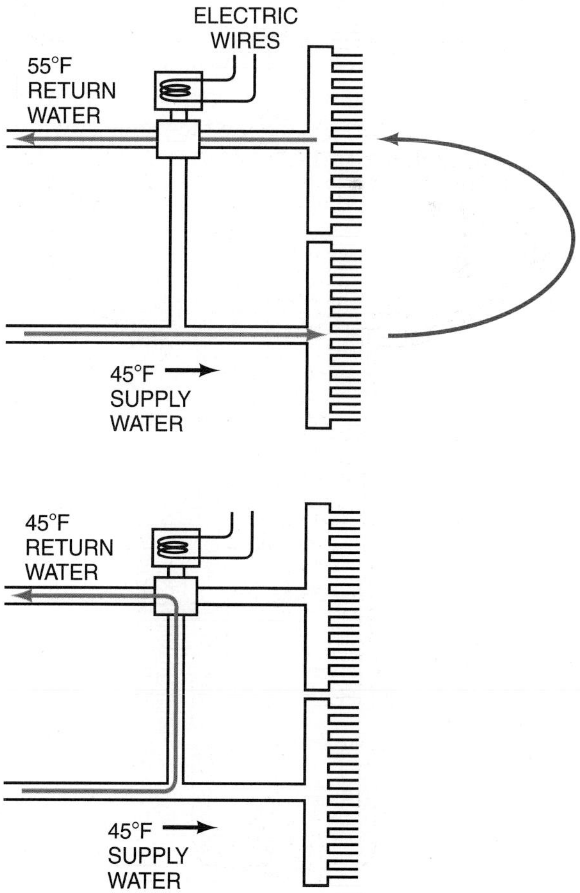

the air pressure to the valve that controls the water. The valve is the controlled device; the pneumatic device is the controller (controlled by the sensor). This high-air-volume valve may be located some distance from the room thermostat. A device called a pilot positioner is used to control the large volume of air. The pilot positioner receives the signal from the thermostat and regulates the air pressure to the air cylinder, which modulates the water valve anywhere from fully closed to fully open. The pilot positioner uses main air pressure, 20 psig, to move an air cylinder piston. The piston is held in the retracted position by a coil spring within the air cylinder body, and the shaft can only be moved with air pressure on the opposite side of the diaphragm, **Figure 16.10**.

The pilot positioner has a main air supply (the largest line) that is governed by the internal diaphragm, which enables it to handle the large volume of air required to fill the air cylinder chamber, **Figure 16.11**. The thermostat now uses a small volume of air, which can be piped a greater distance from the pilot positioner. **Figure 16.12** shows the pilot positioner and the air cylinder it operates. For example, suppose that the room thermostat calls for maximum cooling. It will provide 15 psig of air to the pilot positioner. The pilot positioner will send 20 psig to the air cylinder, causing a full stroke of the air motor and full water flow will pass through the coil, providing maximum cooling. When the room temperature gets close to the 72°F set point—for example, 75°F—the air pressure from the output of the thermostat, the branch line pressure into the pilot positioner, drops to 7 1/2 psig, **Figure 16.13**. The pilot positioner would, in turn, govern the main air to the valve. The spring would move the valve to halfway, reducing the water flow through the coil to 50%, **Figure 16.14**.

Figure 16.10 The air-operated cylinder is commonly used to control valves and dampers.

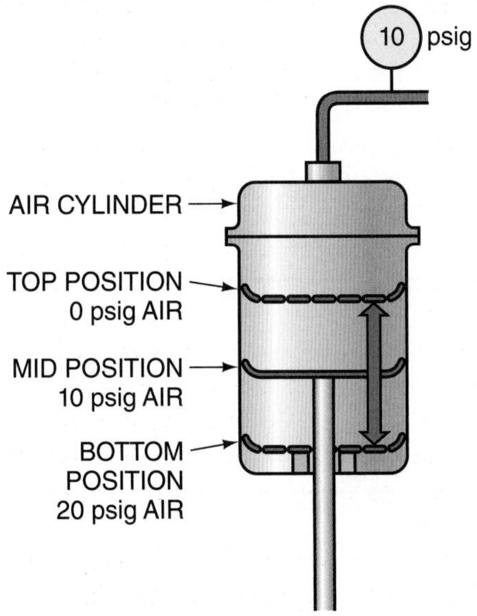

Figure 16.11 The pilot positioner in this system uses the thermostat, which operates at a lower air pressure than that in the main line, to regulate the flow of higher pressure air from the main line to the air cylinder.

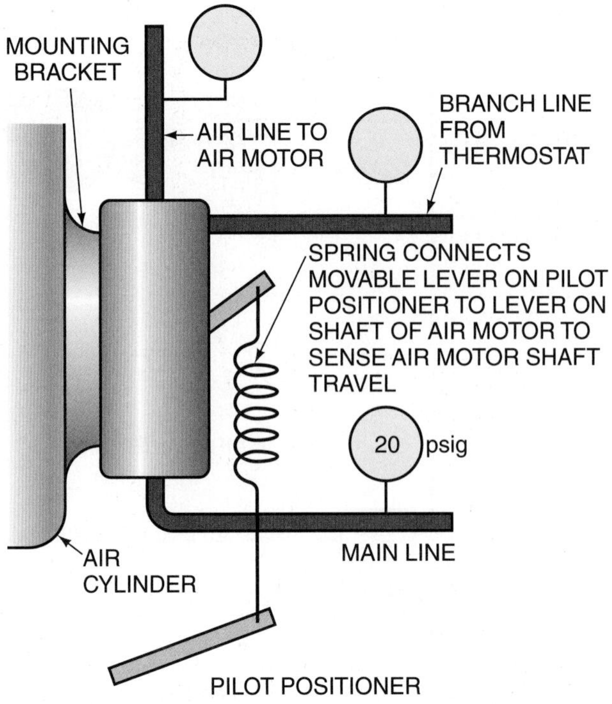

Figure 16.12 The room thermostat is supplying 15 psig of air to the pilot positioner, causing full stroke of the air motor. This will result in full water flow through the coil, providing maximum cooling. The room temperature is 78°F and the thermostat is set for 72°F.

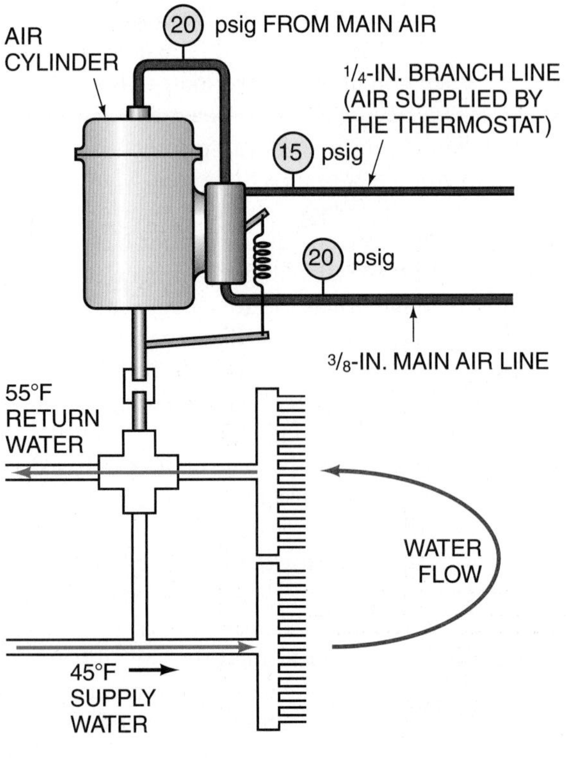

Figure 16.13 Air cylinder controlled with room thermostat and pilot positioner. The room thermostat sends control air to the pilot positioner. The pilot positioner regulates the main air to the air cylinder.

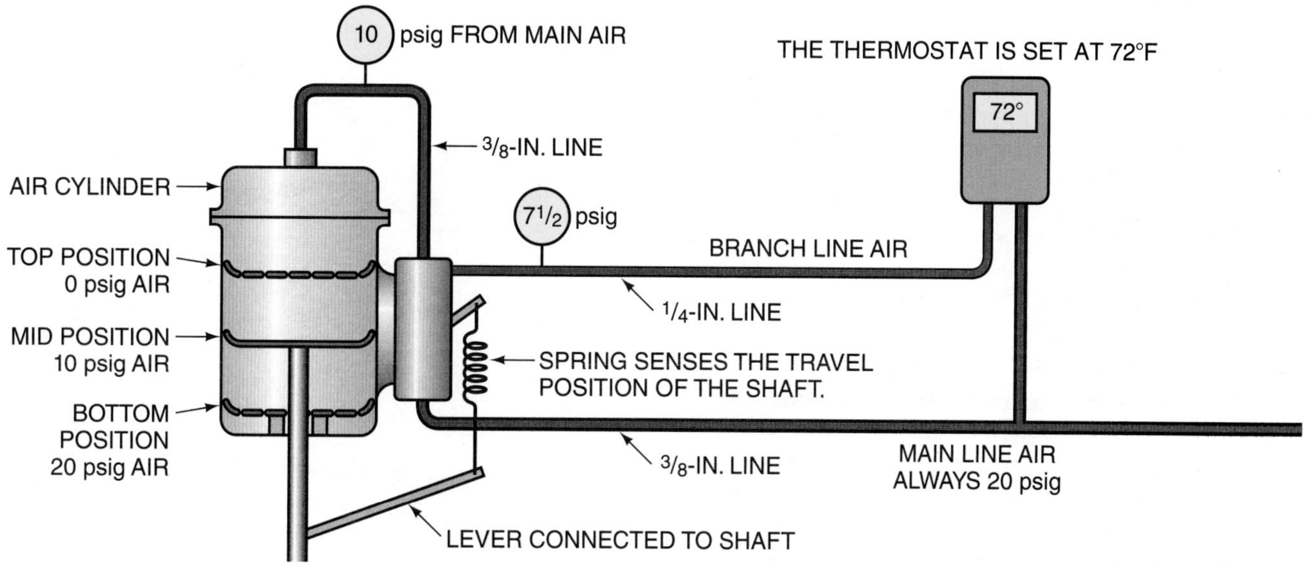

Figure 16.14 The room thermostat is set to control the room temperature at 72°F. The room is at 75°F so the thermostat is sending 7 1/2 psig of air pressure to the pilot positioner. This positions the valve to allow 50% water flow through the coil.

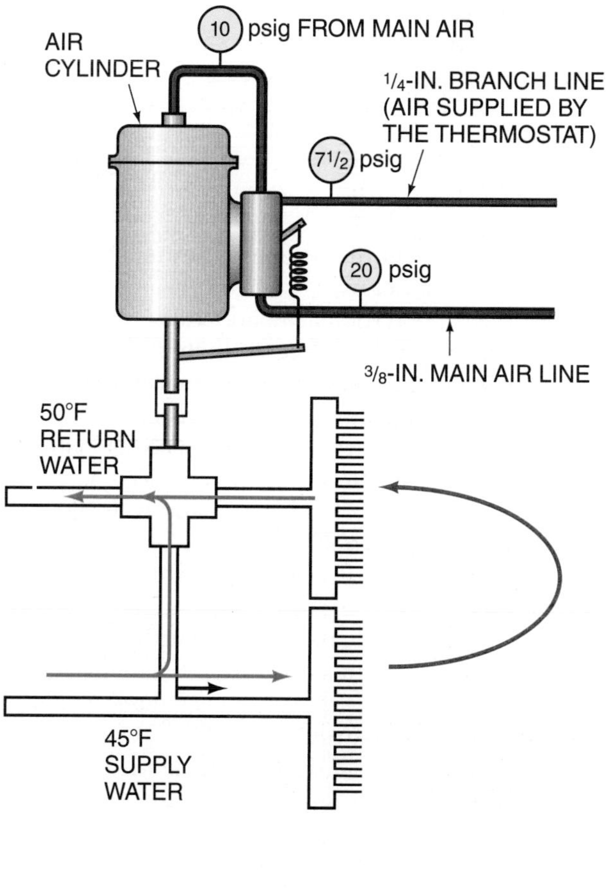

When the room thermostat is satisfied at about 72°F, the valve will be positioned to completely bypass water around the coil and back into the return water to the chiller, **Figure 16.15**. This allows the thermostat to adapt the water flow to the room's needs, rather than an on/off flow. This is called **modulating flow**. Much like the accelerator on an automobile, it modulates the flow of fuel.

Notice in the preceding explanation that at 75°F we had 50% flow and a branch line pressure of 7 1/2 psig, and at 72°F we had no flow. The pressure should be 0 psig. When the temperature reaches 78°F, we should have 15 psig branch line pressure and full flow. We have a temperature difference of 78°F − 72°F = 6°F and an air pressure change of 15 psig. The **sensitivity**, or **gain** as it is sometimes called, is 2.5 psig per 1°F (15 psig/6°F = 2.5 psig per 1°F temperature change). When there is 1°F temperature change, the thermostat will change the branch line pressure 2.5 psig, a variable called for in the design of the application. If sensitivity is too great, the control may have a tendency to hunt, back and forth, as if it cannot find equilibrium or a steady state. If sensitivity is too little, there will be a noticeable difference in the room temperature that will not be much better than on/off control. The job application dictates what the sensitivity should be; the previous example is typical for cooling applications. The three-way valve is used in the cooling application to maintain a constant water flow through the chiller. If a building has 20 chilled water coils similar to the one in the example and they all bypass water, there is still water flow through the chiller, and the chiller will shut off due to low load.

The tools of the pneumatic technician are a pressure gauge that can be plugged into the various control points,

Figure 16.15 The room temperature is now 72°F. The room thermostat is satisfied and has an output of 0 psig. The valve is now bypassing all water around the coil. If the room temperature drops much more, heating will be required. A hot water coil may then be allowed to provide heat in a similar manner.

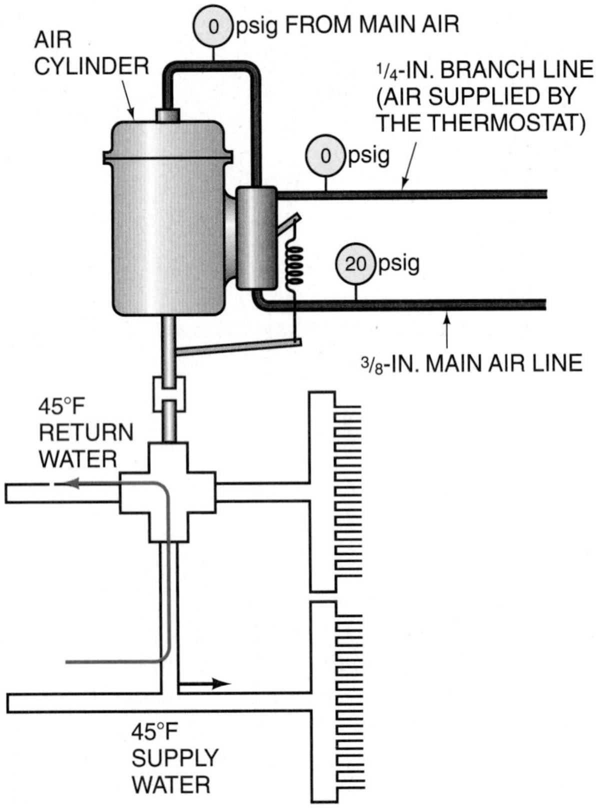

an Allen wrench to make adjustments, and a good pocket thermometer, **Figure 16.16**. The pressure gauge has a quick-connect fitting that will fit the particular brand of pneumatic control, available from the manufacturer. Many technicians carry fittings for each brand.

The technician may be called on to check a room thermostat for calibration. Leaving the pocket thermometer in the air long enough to adjust to the air temperature and detect the temperature in the vicinity (ambient air temperature) of the thermostat, the technician then records the air pressure at the thermostat. Suppose the air temperature is 75°F, the thermostat is set for 72°F, and the air pressure is 3 psig. The thermostat indicates that cooling should be occurring, but the air pressure does not call for cooling. The thermostat has drifted out of calibration. The thermostat can be adjusted to call for 7 1/2 psig under this condition and cooling will start.

Pneumatic controls may be used to switch electrical circuits, or electrical circuits may be used to switch pneumatic control circuits on or off. The pneumatic device that is used to switch electrical circuits is called a pneumatic-electric (PE) device. **Figure 16.17** shows a PE switch that is much like a

Figure 16.16 (A) Pneumatic calibration test kit. Photo by Bill Johnson (B) A digital pocket thermometer. Courtesy Fieldpiece Instruments

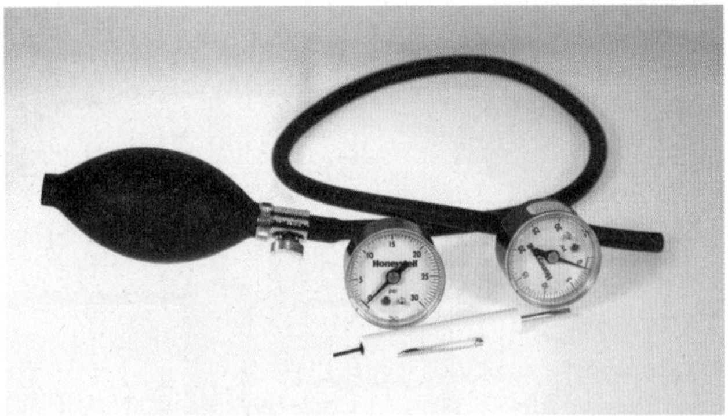

(A)

(B)

pressure control except that it operates at much lower pressures. The electric device used to switch pneumatic devices on and off is called an electric-pneumatic (EP) device.

Many different types of pneumatic controls and arrangements are possible, but they all operate using the principles described. Pneumatic controls can be recognized by the air pressure line going to them. When they change conditions or positions, you will often hear the air bleeding from the control. Technicians who have pneumatic controls in their maintenance program should attend any pneumatic control classes available, try to watch a seasoned technician, and ask a lot of questions. The control sequence provided by the building blueprint is invaluable for showing how the building controls are supposed to function. Control manufacturers are very cooperative with regard to supplying duplicate diagrams and material on their controls. Many times, when the building blueprint has been lost, the original control company or architect will have a control blueprint on file.

Figure 16.17 This switch changes an air pressure difference to an electrical signal. It can be thought of as a pressure switch, only operating at low pressures. Courtesy Ferris State University. Photo by John Tomczyk

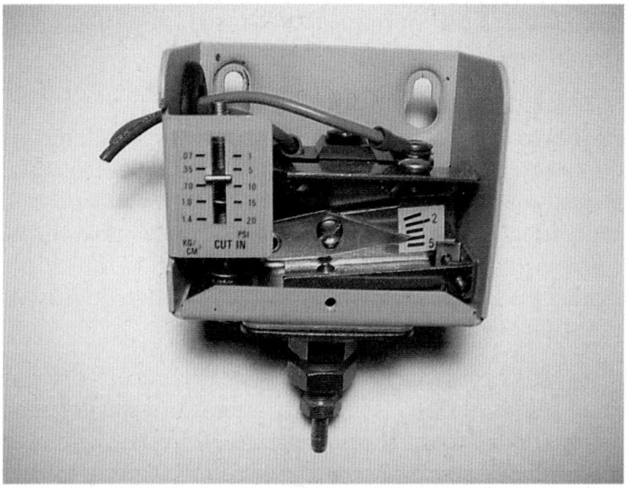

16.6 DIRECT DIGITAL CONTROLS (DDCs)

Electronic controls have taken over the majority of today's control systems. These controls, commonly called direct digital controls (DDCs), have the advantage of being able to operate on very low voltages. In some states the control wiring may be installed without an electrical license. DDCs can also be modulated like pneumatic controls. They are much more compact than pneumatic controls; however, they do the same sort of duty. They may need to open, close, or modulate a water valve, steam valve, or an air damper in an airstream. They can also modulate motor speeds, as will be seen in Unit 17, "Types of Electric Motors."

DDC is a control process in which a microprocessor, acting as a digital controller, constantly updates a database of internal information. Through sensors, the microprocessor updates this information by watching and monitoring information from an external controlled environment or conditioned space. The microprocessor then continually produces corrections and outputs this corrected information back to the controlled environment. The controlled environment could be controlled by an array of controlled output devices, such as dampers, fans, variable-frequency drives (VFDs), cooling coil valves, heating coil valves, relays, and motors—to name a few. The sensors can sense many things like pressure, temperature, light, humidity, velocity, or flow rate. They can also do a status check to see if something is energized or running or not running. A conventional control system (pneumatic, electric, or electromechanical) works very similarly to a DDC system except that digital (binary) information can be processed much faster and is much more accurate at all times and under all conditions. Binary information consists of nothing but zeros and ones (0 or 1) and often represents one of two elements in opposition to each other, such as on/off, high/low, or yes/no.

The DDC system actually changes or conditions a signal in three steps, **Figure 16.18**. The following is an explanation of the three steps:

- The input information from sensors is converted from analog (modulating) to a digital (on/off) binary format. Signal converters, usually housed internally inside the controller, convert the information from either analog to digital (A/D) or digital to analog (D/A) format, **Figure 16.18**. This allows the microprocessor or digital controller to read the incoming data.
- Control commands (algorithms) stored in the digital controller's central processing unit (CPU) are performed by the microprocessor with this newly converted digital data.
- The correct output response of the microprocessor or digital controller is reconverted from digital data to analog data for analog output devices. Digital data is the output

Figure 16.18 Three-step procedure for conditioning a signal on a DDC control system.

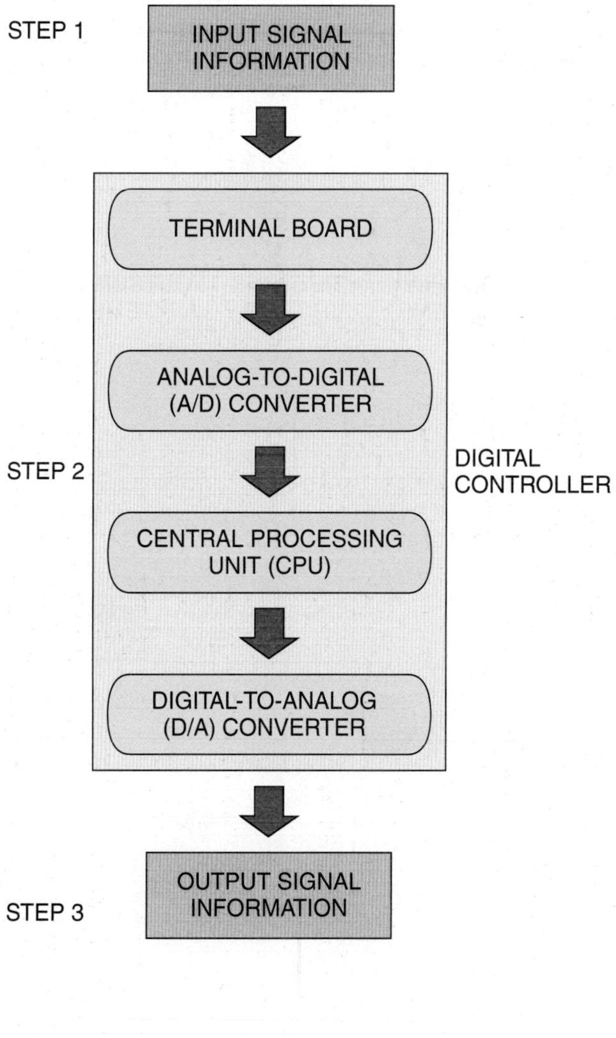

of the digital controller when the controlled output device is digital. This allows the appropriate controlled output devices to recognize the controller's output signals and respond accordingly.

Figure 16.19 shows this process in graphical form. The digital controller (microprocessor) will address each **control point** in the system through an input/output (I/O) card. The signal then passes through an analog/digital converter, which gives the signal a binary value. This allows the digital controller to understand the input signal and will allow the internal instructions (**algorithms**) stored in its memory to act upon the information. The controller's output response is converted from digital to analog format through a digital-to-analog converter. Notice that the signal is now back into its original analog format. The signal can now be sent to the controlled devices, however, control output devices today can receive either analog or digital signals. So digital controllers can be programmed to send the output signal that is appropriate to the device under control.

The microprocessor or digital controller must remember its tasks, what information is available for its use, what this information does, and the results of what it has done. The digital controller must also remember a number of rules and mathematical functions for controlling the system. **Memory** is the part of the digital controller where this information is stored. An algorithm (program of instructions) resides in the memory of the digital controller and serves as the instructions or step-by-step operations sequence for how the digital controller should act (take action) on the data received from the inputs. The controlled output devices can now implement this action and make changes to the controlled environment. The following lists some of the major advantages a DDC control system has over a traditional control system:

- Speed and accuracy
- Ability to control many control sequences simultaneously
- Ability to manage many control loops simultaneously
- Repeatable control of system set points
- Operating functions of the controller adjustable through software
- Ability to automatically adapt to changing conditions in the controlled environment

DDC controls are being used worldwide to reduce energy costs. Current trends in DDC technology are toward smaller, modular, stand-alone DDC panels, **Figure 16.20**. Today, with microprocessor-based controls located right on the equipment being controlled, component failure does not put at risk an entire building's HVACR system. The systems are said to be intelligent, self-compensating, self-maintaining, and self-diagnosing.

Basically, a control system, no matter how complex, consists of the following:

- Sensors
- Controller
- Controlled devices

Figure 16.19 Digital controller showing inputs and final control devices.

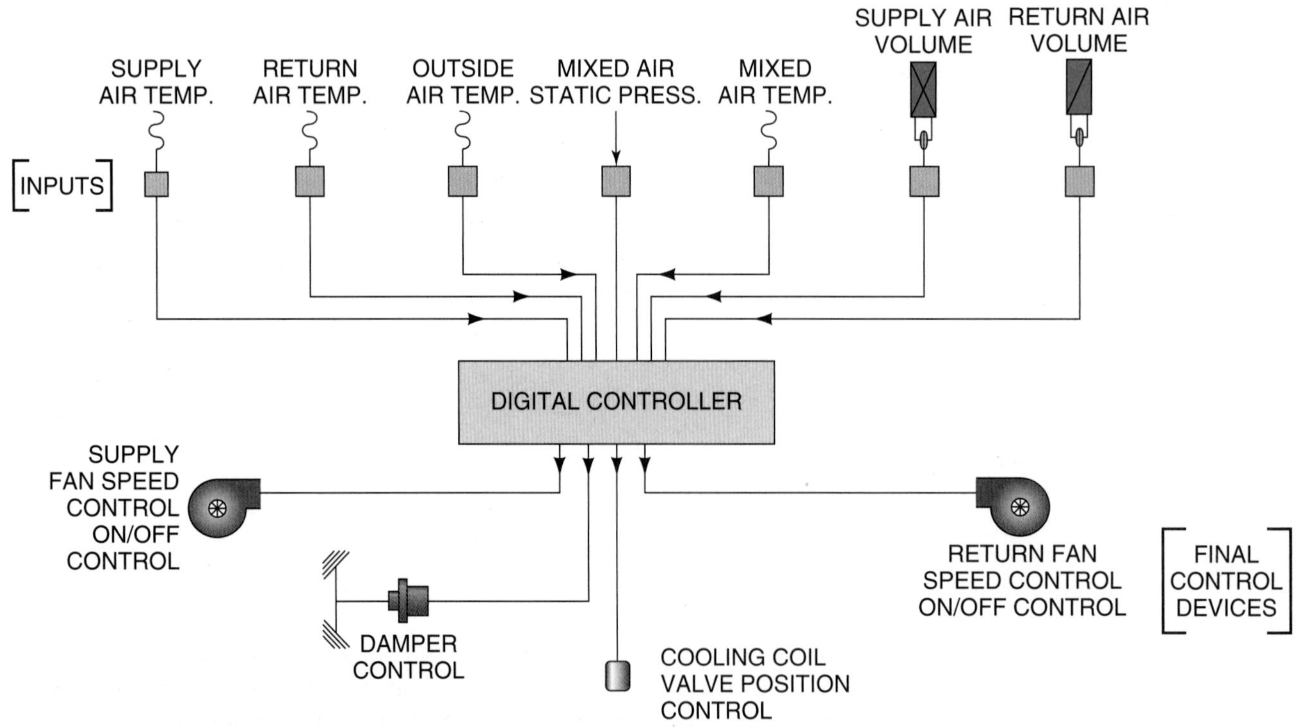

Figure 16.20 Small, modular, stand-alone DDC controller. DI = digital inputs, DO = digital outputs; AI = analog inputs, AO = analog outputs. Courtesy Ferris State University. Photo by John Tomczyk

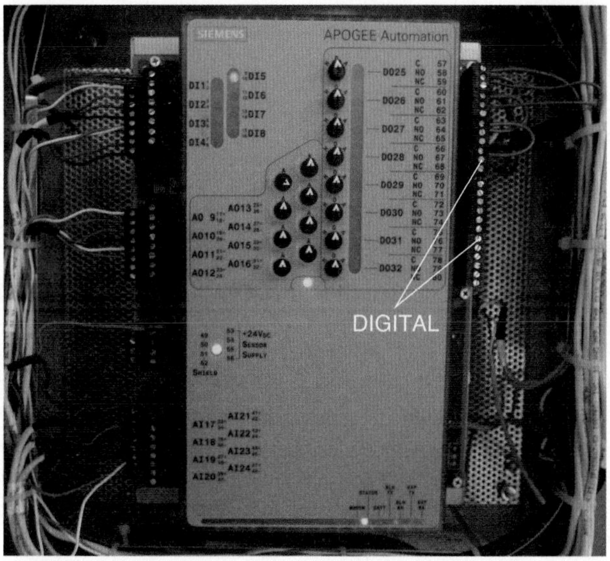

Sensors are devices that can measure some type of environmental parameter or a controlled variable and convert this parameter to a value that the controller can understand. Sensors can measure parameters such as the following:

- Temperature
- Pressure
- Humidity
- Light
- Position
- Moisture
- Velocity
- Enthalpy
- Conductivity

Since a sensor measures the current values of controlled variables, it is the most critical element of a control system, **Figure 16.21**. For accurate control, a change in a controlled variable must be

- sensed or measured;
- transmitted to a controller for interpretation; and
- acted on to elicit the correct response from the controlled output device.

Sensors can be divided into two basic categories. They are

- active sensors.
- passive sensors.

Passive sensors send information back to the controller in terms of resistance (ohms). A typical temperature sensor is a 1000-Ω nickel sensor. This sensor converts temperature to resistance. The resistance is measured by the controller and is translated back to temperature. Passive sensors are typically

Figure 16.21 Sensors for a DDC controller. (A) Thermistor (analog). (B) and (C) Pressure transducers (analog). (D) Current sensor (digital). (E) Humidity sensor (analog). Courtesy Ferris State University. Photo by John Tomczyk

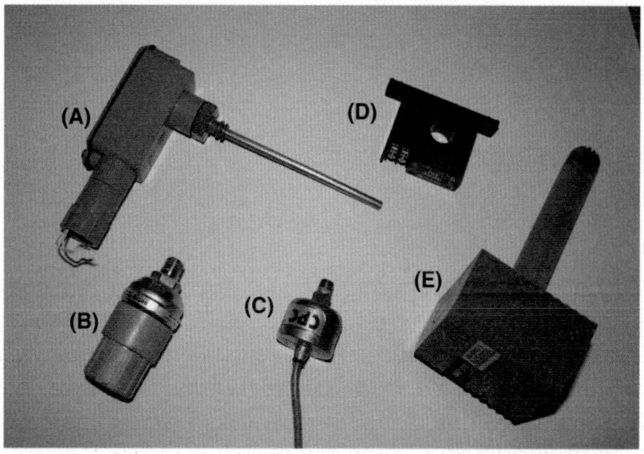

less expensive than other types of sensors. One disadvantage is that they are limited in how far they can be located from the controller because of the resistance of the wire connecting the sensor to the controller. Active sensors send information back to the controller in terms of milliamps (mA) or volts. The sensor converts the input signal at the sensor to mA or volts before being sent to the controller, which then measures mA or volts. These sensors are not affected by the length of wire to the controller. Typical output signals sent back to the controller from the active sensors are 0 to 10 VDC or 4 to 20 mA.

Sensors have both binary (digital) and analog inputs to the controller. Digital inputs report back to the controller as one of two opposing conditions. Examples of digital inputs are the following:

- Yes/no
- Open/closed
- On/off
- Start/stop

In the HVACR field, digital or binary inputs can be a set of contacts that are either open or closed or a motor that is either on or off. They can be used as indicators of status, proof, safety, and limits.

Analog inputs have a range of values over any given time period, **Figure 16.22**. Examples of analog inputs are the following:

- Temperature
- Pressure
- Humidity
- Flow
- Light

In the HVACR field, analog inputs can measure room temperature, discharge air temperature, return air temperature, mixed air temperature, outdoor air temperature, chilled

Figure 16.22 Analog, or modulating, input from a temperature sensor.

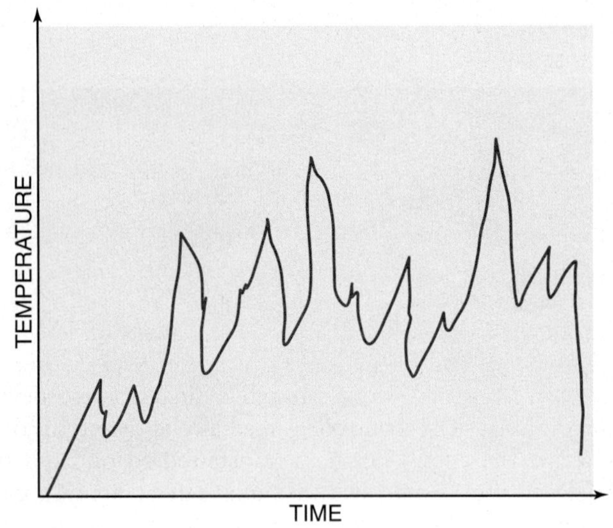

water temperature, hot water temperature, static pressure, chilled water flow, outdoor air humidity, return air humidity, or light levels—to name a few. Probably the most recognized analog inputs to a controller are thermistors and resistance temperature devices (RTDs). Thermistor sensors measure temperature using silicone instead of a metal. However, the thermistor does not establish a linear relationship between resistance and temperature, **Figure 16.23(A)**. So the digital controller has to be able to handle this type of sensor. RTDs, on the other hand, have a more linear and predictable output resistance whenever their temperature changes. The controller has to analyze a much more linear temperature/resistance-to-voltage relationship, **Figure 16.23(B)**.

Controlled output devices can be either analog or digital. When analog, the final control device has a range of operation such as "full open to full close" or "full speed to stop." Analog output devices are typically operated with either a 0- to 10-VDC or a 4- to 20-mA signal range. The controller has to be able to handle one of these two outputs or both. Typical analog outputs in the HVACR field are valves, dampers, and VFDs (variable-frequency drives) that control motor speed.

Digital or binary output devices have two states of operation—such as on/off, open/closed, start/stop, and yes/no, **Figures 16.24(A)** and **(B)**. Binary outputs are simply an electronic switch (triac) that is operated by the controller. Examples of binary outputs in the HVACR fields are fans (start/stop), pumps (start/stop), direct-expansion (DX) cooling (on/off), and gas heat (on/off)—to name a few.

Any time there is a group of control devices interconnected in such a way that they control a single process at a certain desired condition, a control loop is said to exist. An example of this is a group of sensors, controllers, valve actuators, dampers, VFDs, and signal converters. Processes

Figure 16.23 (A) A thermistor used as a temperature sensor for the controller's input. Courtesy Ferris State University. Photo by John Tomczyk (B) This temperature/voltage graph shows a simplified relationship of temperature to voltage.

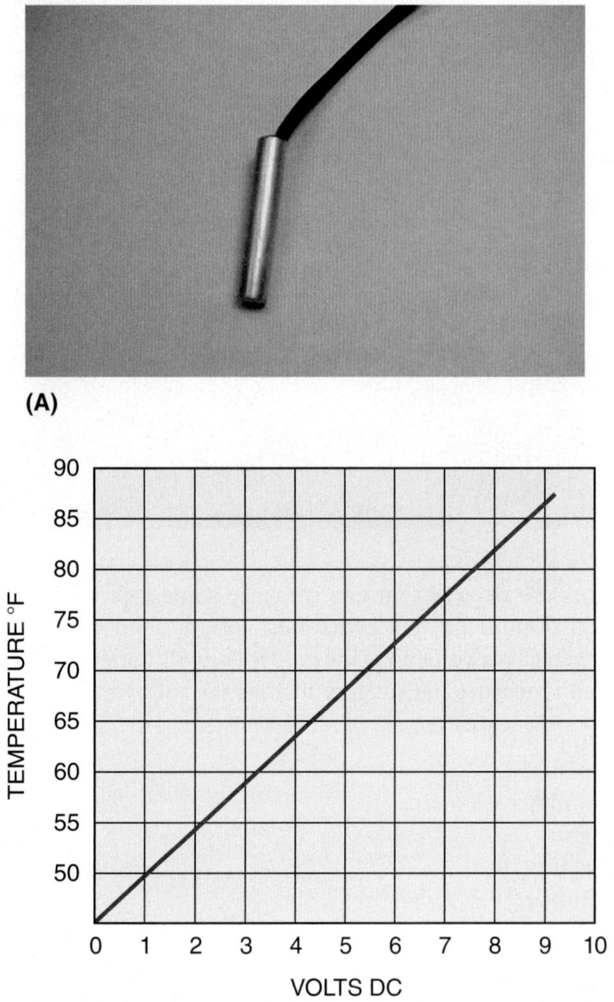

(B)

controlled by a control loop in the HVACR fields are usually comfort-related. Certain control loops can maintain temperature, pressure, lighting, flow rates, or humidity at a certain set point. Others simply make sure a process or control point never exceeds a predetermined system design limit. The difference between the two loops will be discussed later in this section.

The information that any control loop handles starts with the sensors. The sensors pass the information on to the controller, which alters the information and sends it to the final control device. Control loops can only maintain one pressure, temperature, humidity, or lighting value. So it would take four control loops to maintain a space requiring specific pressure, temperature, humidity, and lighting values. Individual control loops can be combined with other control loops

Figure 16.24 (A) Binary, or digital, input—often referred to as a two-position control. (B) A binary (digital) final control device. An open/closed two-position hot water valve feeding a coil in ductwork. Courtesy Ferris State University. Photo by John Tomczyk

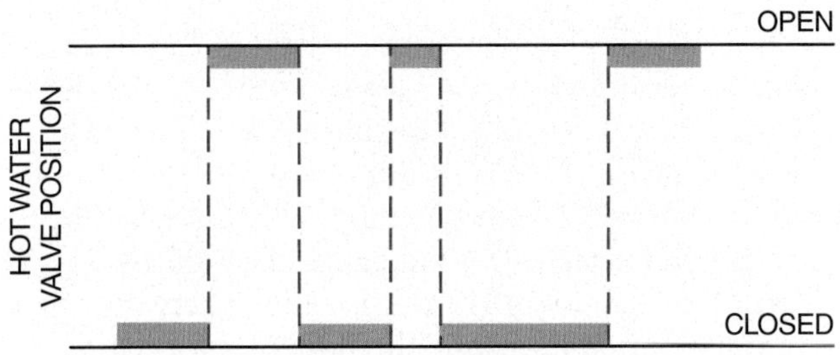

(A)

(B)

to form a control system. There are two basic configurations for control loops. They are the following:

- Open loop
- Closed loop

Open- and closed-loop configurations do not describe how a loop's control devices are connected, but how the loop is connected to the HVACR process. It is actually the sensor's location that determines whether the loop is open or closed.

In an open-loop control configuration, the sensor is upstream of the location where the control agent causes a change in the controlled medium, **Figure 16.25**. The control agent is the fluid that transfers energy or mass to the controlled medium. It is the control agent (such as hot or cold water) that will flow through a final control device (water valves). The controlled medium is what absorbs or releases the energy or mass transferred to or from the HVACR process. Controlled mediums can be air in a duct or water in a pipe. In **Figure 16.25**, the control agent is the hot water in the coil, the controlled medium is the air in the duct, the control point is the outdoor air temperature, and the final control device is the hot water valve. **Figures 16.26(A)–(C)** show three analog (modulating) final control devices. The flow of information in an open-loop configuration travels in only one direction. Because the sensor is located upstream of the control agent, it never senses the results of the action of the final control device (hot water valve). Open-loop control systems cannot maintain a balance between the energy transferred to or from the system and the process's heating or cooling loads. The sensor does

Figure 16.25 An open-loop control configuration.

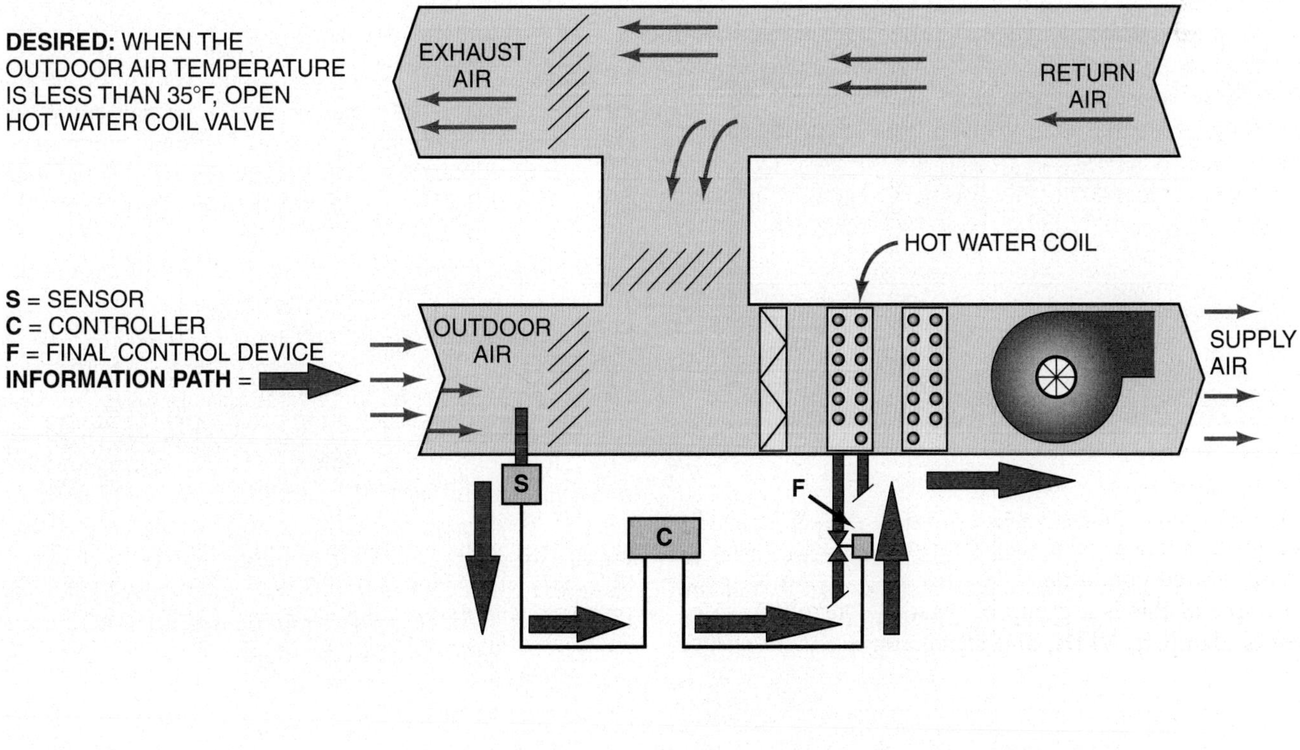

Figure 16.26 (A) Mixing valve acting as an analog final control device. (B) Return air damper acting as an analog final control device. (C) Two variable-frequency drives (VFDs) acting as analog final control devices for supply-and-return air fans. Courtesy Ferris State University. Photos by John Tomczyk

(A)

(B)

(C)

not monitor a control point like supply air temperature, but can monitor a variable like the outdoor air temperature in **Figure 16.25**. The sensor is simply making sure that if the outdoor air temperature is ever less than 40 degrees, the hot water valve will open. This opening or closing of the hot water valve would be considered a binary, or two-position, response; it is simply preventing the controlled variable (outdoor air temperature) from exceeding a predetermined design limit.

A **closed-loop control configuration** can maintain the controlled variable, like the air in the duct in **Figure 16.27**, at its set point because the sensor is now located in the air being heated or cooled (controlled medium). The sensor is located downstream of the water coils or point where energy or mass is being introduced. The sensor can now sense what the final control device (water valve) did and make adjustments. This is called a **feedback loop**. The feedback loop actually "closes" the information loop, causing the flow of information to be in a circle. This is where the term *closed loop* originated. This type of control loop can now maintain a controlled medium at a set point—whether the control is a modulating (analog) or digital (binary) type. With the modulating closed-loop control in **Figure 16.27**, there can now be a balance between the HVACR process load and the energy transferred from the controlled agent. Refer to Section 24.22, "Stepper Motor Expansion Valves," and Section 24.23, "Algorithms and PID Controllers," in Unit 24, "Expansion Devices," for an example of a closed-loop control configuration in which you can apply the principles you have just learned about DDC. These sections discuss DDC applied to an expansion valve for a commercial refrigeration system used in today's supermarkets.

DDC control systems in large buildings or supermarkets can be responsible for the following:

- Total building energy management
- HVACR functions
- Building lighting
- Fire protection
- Temperature setback
- Refrigeration

DDCs use a typical personal computer or laptop to assimilate and distribute information to other, small computers in the system, such as control boards. The small computers that are part of the DDC system for each section of the building are tied to the main personal computer. There may be multiple sensors and controllers for each portion of the building feeding information to the individual control computers. An advantage is that the technician can control the system from wherever there is an Internet connection. Security codes prevent others from accessing the system. A technician who gets a call about a building temperature being out of control can get on the Internet or a wireless cable modem and view all the building conditions, what is running and what is not running, and what the conditions are in all parts of the system. The corrective action can often be entered from a remote location. This gives building maintenance personnel some extra freedom and eliminates some overtime calls for the owners. Reliable sensors and their control wiring are the greatest challenge in this type of system.

Figure 16.27 A closed-loop control configuration.

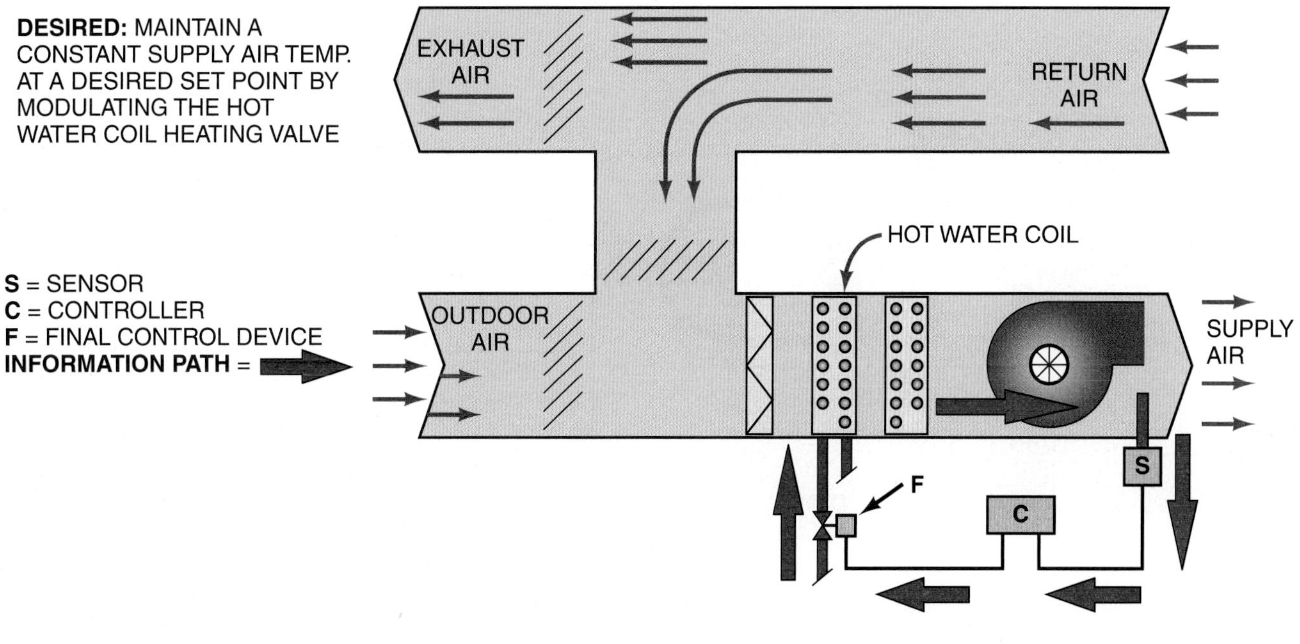

*Building management may want to perform what is known as **load shedding** or **duty cycling** to lower electrical power costs. Electrical load shedding uses management methods to reduce the power consumption.* For example, suppose the air conditioner is turned off at exactly 5:00 PM rather than 5:30 PM. The people exiting the building may notice that the space temperature is rising, but it does not matter because they are leaving. Duty cycling is shutting down nonessential equipment during high peak loads; for example, water coolers, decorative water fountains, ice makers, water heaters, and other nonessential devices can be shut down temporarily to reduce the electrical load. Remember, every Btu that is allowed into the building must be removed by the air-conditioning system in the summer. It pays to shut nonessential devices off when they are not needed.

Power companies use a system called **demand metering** to charge for electricity in commercial buildings. Demand metering bills the owner at the power rate for the month based on the highest usage for some time span, usually 15 or 30 min. Power companies try to collect in a way that charges for the demand on their systems. To understand how demand billing works, suppose a building's air-conditioning system had not been operated for a 30-day period, and on the next day, it becomes very hot toward the last part of the day. If the system operated at maximum capacity for the last 30 min of the last day of the month, the building owners would be billed for the entire month at that rate, even the previous 30 days. A sharp operator or a computer-analyzed control and power system would discover this and not allow the system to operate at full capacity or would shed some nonessential loads. The power bill for the month would then be greatly reduced. In fact, a well-designed system can monitor a building's power consumption by the minute for the entire year.

A system can be designed to stop at a certain time at night and start early in the morning. This allows the equipment to bring the building conditions to temperature at a lower power consumption as opposed to running at full load for a short period of time. The pull-down time in the summer and the heat-up time in the winter can also be lengthened to reduce energy consumption. The computer system can keep a record of the pull-down time for the building and allow for extra time or less time as needed during future events. It keeps a history of how the system responds to specific condition changes and works to make it more efficient. Weather patterns can be tracked to predict fast temperature changes that may be needed and phase them in at lower energy cost. Different solar zone requirements can also be predicted and adjusted. The energy requirements are different for the east side of a building than for the west side at different times of day because of the changing position of the sun. In actuality, there are so many variations that can be accomplished with DDC and a good computer system that they cannot all be mentioned here.

Figure 16.28 is a simple example of a closed-loop control system that controls a chilled water coil to a conditioned space using DDC. The thermostat in the return air from the conditioned space is checking the air temperature. Suppose we want to maintain the temperature at 75°F. The digital signal voltage range is from 0- to 10-VDC and we want to have a sensitivity (often called *gain* in electronics) of 1.7 V per degree of temperature change. The voltage should read

Figure 16.28 The analog signal from the return air sensor is changed to a digital signal and analyzed by the computer to allow more or less chilled water to flow through the chilled water coil to maintain space temperature.

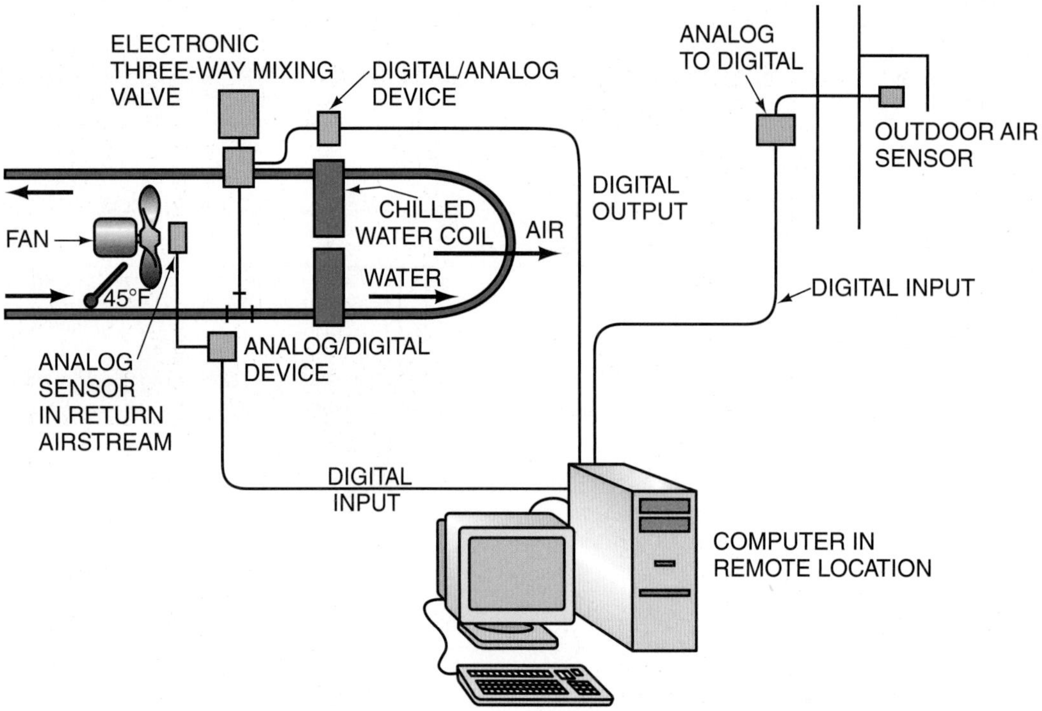

5 V at 75°F. The three-way mixing valve should be bypassing about 50% of the water, and 50% should be passing through the coil to cool the air. This should result in about a 5°F rise in the water temperature if it has a 10°F rise at full load. The coil is operating at about 50% capacity. If the return air temperature goes up to 76°F, the voltage will rise to 6.7 V and the three-way mixing valve will begin to allow more chilled water through the coil, bypassing less water. When the load begins to reduce, the sensor will alert the computer to tell the valve to bypass more water. The technician can use the DC voltage feature of the voltmeter to monitor these voltages and make adjustments. After checking the voltage at various points to determine how the controller is supposed to react, the technician can make changes accordingly. The reason for using a three-way mixing valve for this system is that the central water chiller must have a constant flow through it. It is not like the boiler and hot water coil used in the pneumatic example. If too many chilled water coils were to shut down using straight-through valves instead of bypassing the water, the chiller would be in danger of approaching freezing conditions that could result in damage. It is customary to use mixing valves for chilled water circuits.

There is a lot of versatility in these control arrangements, particularly for large buildings, **Figure 16.29**. These control advantages are beginning to be incorporated into residential systems because the commercial systems have been proven to work. Electronic control components, like computers, are becoming more affordable for the average person.

Many old buildings are still outfitted with pneumatic controls, and it is not cost-effective to replace them. This is what makes the DDC system so versatile. DDC systems can be meshed in with the older systems for less cost than a complete building update. Both control systems are reliable and work well together when the system is designed correctly. Again, it is best to have the designer's prints to know what is really expected of a system.

16.7 RESIDENTIAL ELECTRONIC CONTROLS

Technicians have been dealing with electronic controls in residential systems for many years, often without knowing it. The residential oil burner control board is an example. Some technicians seem to be afraid of the word *electronics*, but it is not necessary to have the knowledge of a radio or TV technician to work with the electronic controls in the HVACR industry. We do not check out the individual components like the TV technician; we work with complete circuit boards. We use manufacturers' diagnostic systems or treat the circuit board as a switch in many cases. Many systems use error messages at the thermostat or blinking light-emitting diodes (LEDs) as codes for problems. Manufacturers' literature has become vital for troubleshooting. Attending manufacturers'

Figure 16.29 This pictorial shows the layout of a typical system. Notice that there is a start-up phase for when a building is first built. Then there is a growth phase where new equipment may be added as a building expands. The "Integrate" part of the figure shows that different types of systems may be operated with the system. The "Access" part shows that the system can be accessed away from the site. The "Manage" portion will allow off-property management of the system by an outside company. Courtesy Siemens Building Technologies, Inc.

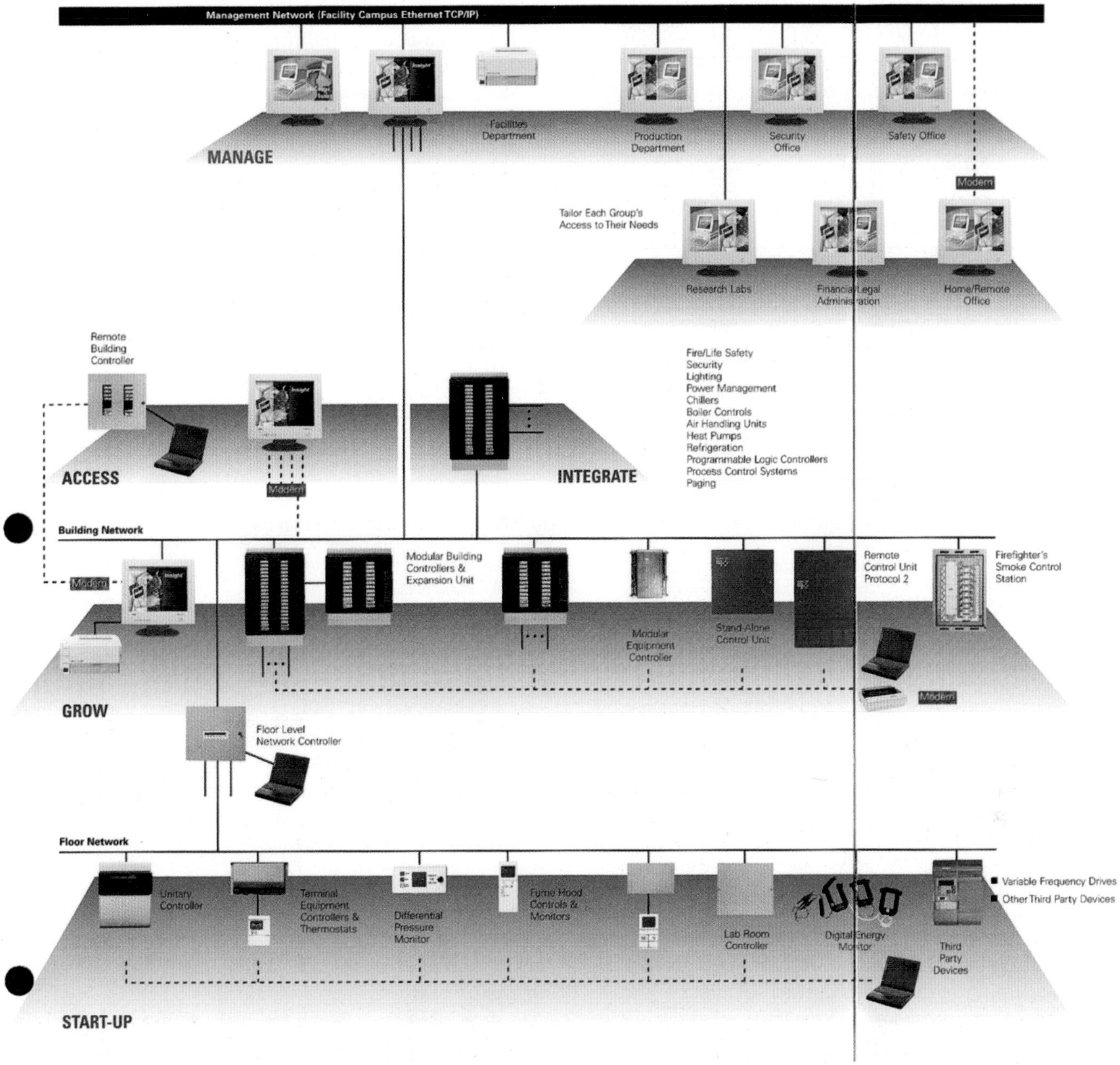

service classes will help you understand their goals for the equipment and controls.

Electronics are used in virtually every type of equipment today. The electronic thermostat often uses thermistors to control temperature. Recall the discussion about thermistors from Unit 12, "Basic Electricity and Magnetism." A thermistor changes resistance with a change in temperature. Thermistors are very small and respond quickly to a temperature change, and they can send a different signal for a different temperature. **Figure 16.30** shows a thermistor temperature and resistance graph. The thermistor can be tested with a quality ohmmeter. Just record the temperature of the actual element with an accurate thermometer and check out the plot on a graph for that thermistor. The change in temperature of the thermistor can be used in electronic circuits to monitor temperatures throughout the system and send

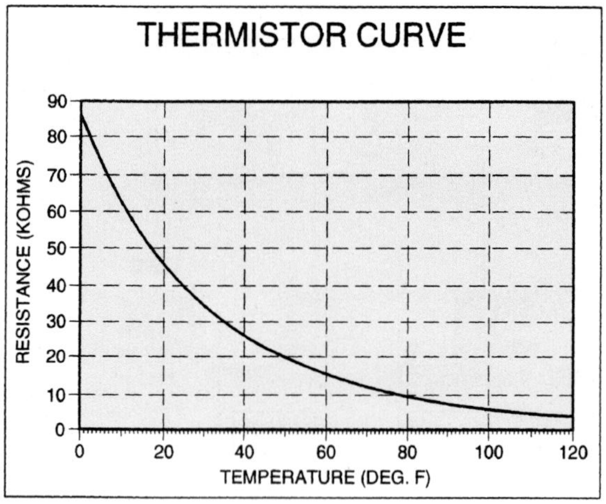

Figure 16.30 This is a temperature-resistance graph for a thermistor. The technician can use this graph, a thermometer, and an ohmmeter to test a thermistor for accuracy.

information to the circuit board that contains a computer chip, a small computer. It can, in turn, respond and cause a change in conditions. Space temperature is not the only temperature that is important in modern systems. Other temperatures that may be important are the following:

- Discharge-line temperature
- Motor-winding temperature
- Return air temperature
- Suction-line temperature
- Liquid-line temperature
- Flue-gas temperature
- Outdoor temperature
- Supply air temperature
- Indoor temperature

Figure 16.31 shows an electronic thermostat that performs many functions other than just turning the equipment on and off. This thermostat can be used with a two-stage

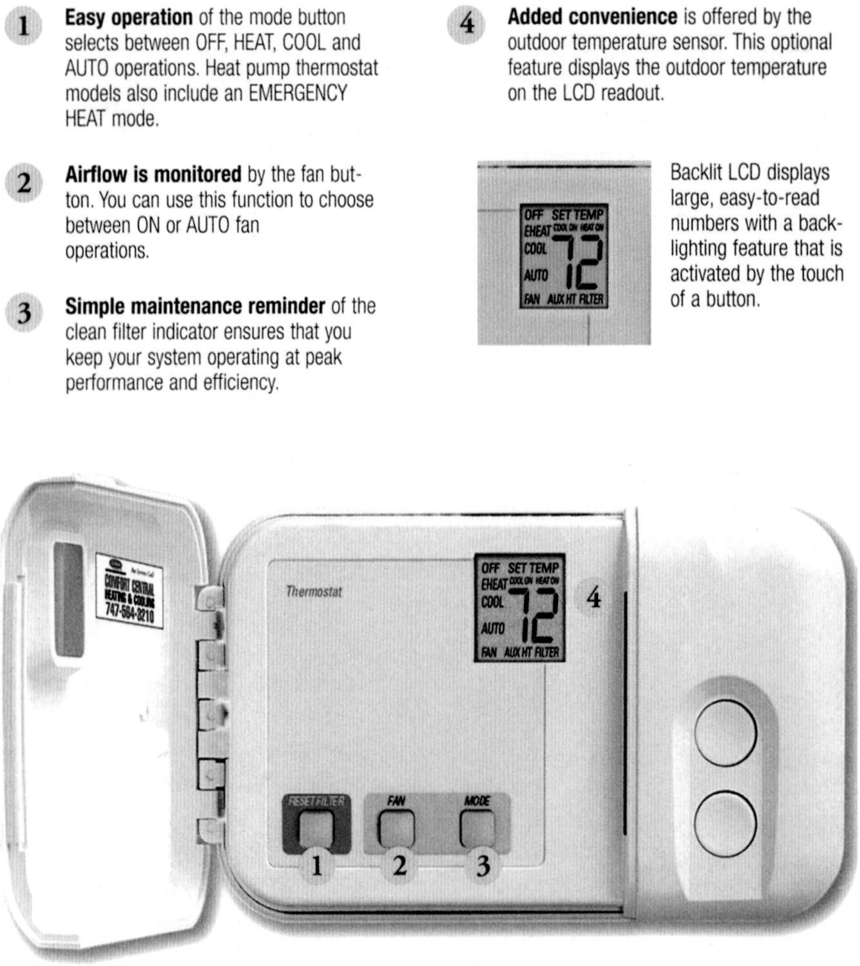

Figure 16.31 The electronic thermostat can perform many functions that could not be accomplished with earlier thermostats. Source: Carrier Corporation

cooling or heating system and is fully programmable for different events 7 days per week. *It is suitable for a working family that wants to change the space temperature at night and during the day while at work and then have the space temperature brought back to the comfortable set point while at home. It can be programmed to stay at the comfortable set point for weekends and to go to setback functions for vacations. This is typical of many versatile thermostats on the market.*

Some of the functions of this thermostat are

- 3-minute compressor off time to prevent compressor short cycling
- 15-minute cycle timer to prevent the heating system from short cycling
- 15-minute stage timer to prevent the second-stage heat from starting up too soon
- 3-minute minimum on timer for staged heat
- Heat/cool set points for desired temperature
- Auto changeover from heat to cool and back
- Emergency heat mode for heat pumps
- Power-on check
- Low-voltage check
- Error codes for thermostat operation, line-voltage drops, and outdoor temperature sensor problems
- Outdoor temperature smart recovery that tells the system when to restart the equipment after night or day setback.

The thermostat has some small switches in the back that are called dual in-line pair (DIP) switches. DIP switches generally have two positions, on and off. These allow the technician to set the thermostat up for the particular piece of equipment it is controlling. Using DIP switches and following the wiring diagram for the particular installation will give the customer the desired type of control over the equipment. **Figure 16.32** is the wiring diagram and DIP switch setup for a two-speed heat pump operated with a variable-speed fan coil. The two-speed heat pump (two speeds are achieved with a combination two- and four-pole compressor motor) allows the system to operate at 50% compressor capacity in cooling and heating, and the reduced fan speed is used to keep the air temperature up in winter and control humidity removal in the summer. This adds a lot of capabilities to the system, and it is all accomplished with electronics. Other versions of this thermostat have an actual humidity control that varies the fan speed to remove more or less moisture, which is a great advantage because the equipment is shipped all over the country, to desert low-humidity regions and to the southern coastal regions where humidity is very high.

The outdoor unit for a heat pump also has some electronics that the technician will be interested in. The circuit board in the outdoor unit receives a signal from the thermostat asking it to either heat or cool, at either high or low speed. When all is well, there is nothing to worry about. Suppose the technician arrives at the job because of a service complaint of no heat. What should be done first?

Figure 16.32 This diagram shows the field control wiring diagram and the DIP switches that must be set up for the various configurations that the thermostat can accomplish. Based on Carrier Corporation

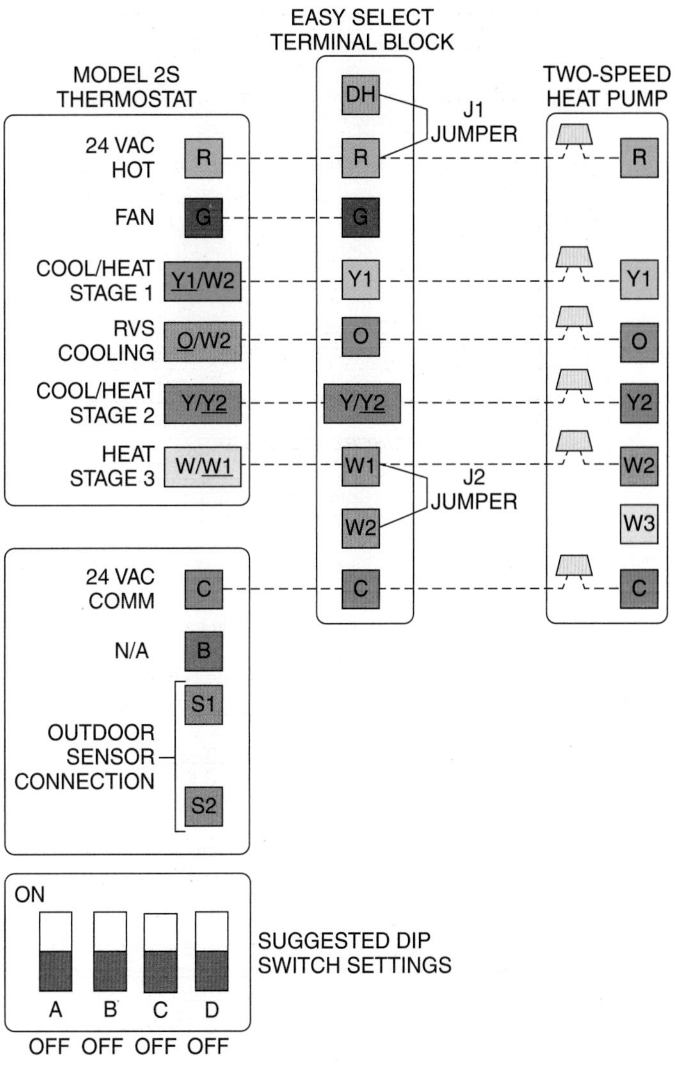

- Check the thermostat to see if the indoor fan will run. If not, check the power supply.
- Make sure there is power to the indoor unit, which may be under the house or in the attic.
- When power is established, make sure the indoor fan will run; then verify power to the outdoor unit. Make sure the thermostat is calling for it to operate.

When all signs seem to call for operation at the outdoor unit and it will not run, remove the cover and look at what the manufacturer recommends. **Figure 16.33** is a table called an

Figure 16.33 This chart shows the control function light codes for this particular system. Based on Carrier Corporation

NOTE: A Signal (code) is not sent through the L lead to thermostat unless a failure has occurred.

LED Control Function Light Code		
CODE	**DEFINITION**	*
Constant flash No pause	No demand Stand by	9
1 flash w/pause	Low-speed operation	8
2 flashes w/pause	High-speed operation	7
3 flashes w/pause	Outdoor ambient thermistor failure	6
4 flashes w/pause	Outdoor coil thermistor failure	5
3 flashes pause 4 flashes	Thermistor out of range+	4
5 flashes w/pause	Pressure switch trip (LM1/LM2)	3
6 flashes w/pause	Compressor PTC's out of limit	2
Constant light No pause No flash	Board failure	1

* Function light signal order of importance in case of multiple signal request; 1 is most important.

+ Check both thermistors to determine which is faulty.

LED control function light code that shows the status lights for this particular manufacturer's equipment. This chart is typical of what the technician can expect. During the service call, the technician can look at the status lights and see what is supposed to be happening. If the equipment is supposed to be running and is not running, more investigation is in order.

The manufacturer will supply a troubleshooting chart to take the technician to the next step. Manufacturers are working to design more efficient equipment with unique features. Installers and service technicians must keep up-to-date. Most contractors try to work with one or two brands of equipment so that their installers and service technicians can be efficient with those systems.

REVIEW QUESTIONS

1. A ________ measures the change in the conditions being monitored.
   A. signal converter
   B. control point
   C. controlled device
   D. sensor
2. Pneumatic controls are operated by ________.
   A. water
   B. electricity
   C. electronics
   D. air
3. The internal instructions stored in the memory of a digital controller are called ________.
   A. a data base
   B. signal converters
   C. algorithms
   D. binary inputs
4. A sensor, a controller, and a controlled device are used to form a ________.
   A. controlled output device
   B. control loop
   C. control point
   D. signal converter
5. Pneumatic air must ________.
   A. be clean and dry
   B. have very high pressure
   C. encompass a large volume
   D. be heated
6. The air supplied to a diaphragm is 20 psig, and the diaphragm has a diameter of 6 in. How much pressure can this diaphragm apply to a plunger?
   A. 56.52 pounds
   B. 62.56 pounds
   C. 625.5 pounds
   D. 565.2 pounds
7. When a large volume of air is needed at a diaphragm, the thermostat may feed branch line pressure to the ________.
   A. air cylinder
   B. air compressor
   C. pilot positioner
   D. electric solenoid
8. Which are the typical tools that a pneumatic control technician would use?
   A. Gauge manifold and nitrogen cylinder
   B. Flue-gas analyzer and thermometer
   C. Flaring tools, adjustable wrench, and screwdriver
   D. Pressure gauge, Allen wrench, and pocket thermometer

9. Electronic controls operate on__________.
   A. time or temperature
   B. high or low air pressure
   C. low voltages or milliamps
   D. air pressure differential
10. An analog control signal __________.
    A. has an infinite number of control steps
    B. is very limited in the number of control steps
    C. makes a hissing noise when the position is changed
    D. is not used in this industry
11. Another term used to describe sensitivity in electronic controls is _________.
    A. temperature
    B. gain
    C. rpm
    D. pressure
12. One of the advantages of DDCs is __________.
    A. total building control management
    B. that they are simple
    C. that no air compressor is needed
    D. that they operate on high voltage
13. Many residential digital controls use _________ as codes for problems.
    A. binary inputs
    B. analog outputs
    C. passive sensors
    D. blinking LEDs
14. One of the major components of a residential electronic control circuit that measures temperature is a __________.
    A. thermometer
    B. thermistor
    C. capacitor
    D. filter
15. After a control system is correctly set up, the controls __________.
    A. may drift out of calibration and have to be adjusted
    B. should be replaced on a regular basis
    C. must be adjusted daily to ensure proper operation
    D. should be locked out to prevent the settings from being adjusted
16. Describe a final control device and give three examples of these devices used in a DDC system in today's HVACR industry.
17. Describe what a sensor does in a DDC system used in today's HVACR industry, and list five examples of sensors.
18. What is the difference between an active and a passive sensor used in DDC control systems?
19. A control system consists of what three basic components?
20. List six advantages that DDC control systems have over traditional control systems.
21. Describe the difference between the analog and digital signals in a DDC control system.
22. Describe the difference between an open-loop and a closed-loop control configuration in a DDC system.

SECTION 4

ELECTRIC MOTORS

UNIT 17

TYPES OF ELECTRIC MOTORS

OBJECTIVES

After studying this unit, you should be able to

- describe the components of an electric motor and how they function.
- discuss synchronous speed and motor slip.
- describe motor start and run windings and how they affect starting torque.
- identify the most common electrical power supplies used in the United States.
- describe the characteristics and operation of open split-phase motors and centrifugal switches.
- describe the operation of capacitor-start and capacitor-start, capacitor-run motors.
- describe a multispeed, permanent split-capacitor (PSC) motor and indicate how the different speeds are obtained.
- describe the characteristics and operation of a shaded-pole motor.
- describe various types of motor applications.
- explain the operation and applications of a three-phase motor.
- describe hermetically sealed compressor motors and their motor terminal connections.
- explain the function of a potential relay and how to troubleshoot it.
- describe the operation and characteristics of a current relay.
- discuss positive temperature coefficient resistors (PTCRs), how they are used and how to troubleshoot them.
- describe the use of variable-speed motors, inverters, variable frequency drives (VFDs), and electronically commutated motors (ECMs).
- describe two basic types of DC converters and how they function.
- describe how various electric motors are cooled.

17.1 USES OF ELECTRIC MOTORS

Electric motors are used to turn the prime movers of air, water, and refrigerant, which are the fans, pumps, and compressors, **Figure 17.1**. Several types of motors, each with its particular use, are available. For example, some applications need motors that will start under heavy loads and develop their rated work horsepower under continuous running conditions. Some motors run for years in dirty operating conditions, and others operate in a refrigerant atmosphere. These are a few of the typical applications of motors in the HVACR industry. The technician must understand which motor is suitable for each job in order to troubleshoot effectively and, if necessary, to replace a motor with the proper type. But the basic operating principles of an electric motor must first be understood. Although there are many types of electric motors, most operate on similar principles.

Figure 17.1 (A) Fans move air. (B) Pumps move water. Photos by Jason Obrzut

(A)

(B)

17.2 PARTS OF AN ELECTRIC MOTOR

Electric motors have a **stator** with windings, a **rotor**, **bearings**, **end bells**, **housing**, and some means to hold these parts in the proper position, **Figure 17.2** and **Figure 17.3**. The stator is a winding that, when energized, will generate a magnetic field—because there will be current flowing through it. The rotor is the rotating portion of the motor and is made of iron

Figure 17.2 A cutaway of an electric motor. Courtesy of Hampden Engineering

Figure 17.3 Individual electric motor parts.

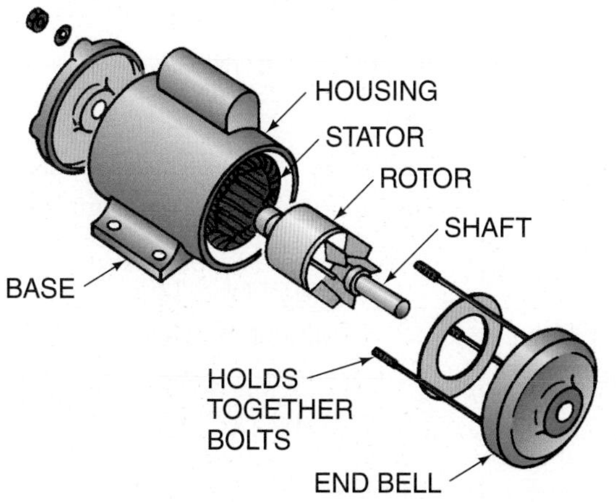

or copper bars bound on the ends with aluminum. The rotor is not wired to the power source like the stator is. The motor shaft is connected to the rotor; therefore, when the rotor turns, so does the motor shaft. The motor's bearings allow for low-friction rotation of the rotor. Bearings help reduce the friction and the heat generated by the motor's moving and rubbing parts and surfaces. The motor housing, end bells, and base hold the motor components in place and provide a means for securely mounting the motor itself.

17.3 ELECTRIC MOTORS AND MAGNETISM

Electricity and magnetism are used to create the rotation in an electric motor to drive the fans, pumps, and compressors. Magnets have two different electrical poles, north and south. Unlike poles of a magnet attract each other, and like poles repel each other. If a stationary horseshoe magnet were placed

Figure 17.4 Poles (north and south) on a rotating magnet will line up with the opposite poles on a stationary magnet.

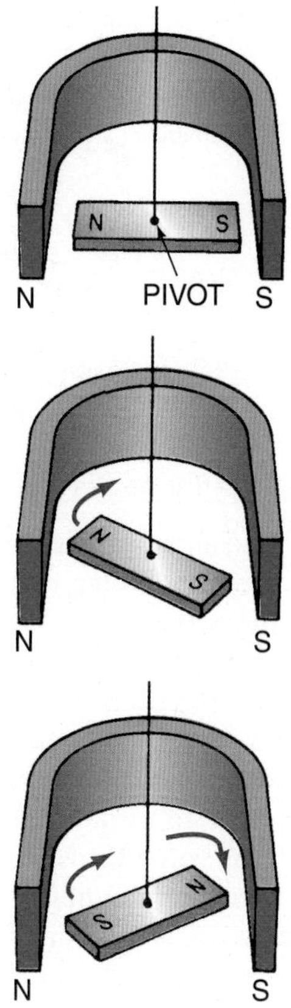

with its two poles (north and south) at either end of a free-turning magnet as in **Figure 17.4**, one pole of the free-rotating magnet would line up with the opposite pole of the horseshoe magnet. If the horseshoe magnet were an electromagnet and the wires on the battery were reversed, the poles of this magnet would reverse, and the poles on the free magnet would be repelled, causing it to rotate until the unlike poles again were lined up. This is the basic principle of electric motor operation. In this example, the horseshoe magnet acts as the stator, and the free-rotating magnet acts as the rotor.

In a two-pole split-phase motor, the stator has two poles with insulated wire windings, called the **run windings**. When an electrical current is applied, these poles become an electromagnet with the polarity changing constantly. In normal 60-cycle operation, the polarity changes 120 times per second.

The rotor may be constructed of bars, **Figure 17.5**, a type called a **squirrel cage rotor**. The rotor is positioned between the run windings. When an alternating current (AC) is applied to these windings, a magnetic field is produced in the windings and is also induced in the rotor. The bars in the rotor actually form a coil. This is similar to the field induced in a transformer secondary by the magnetic field in the transformer primary. The field induced in the rotor has a polarity

Figure 17.5 A simple sketch of a squirrel cage rotor.

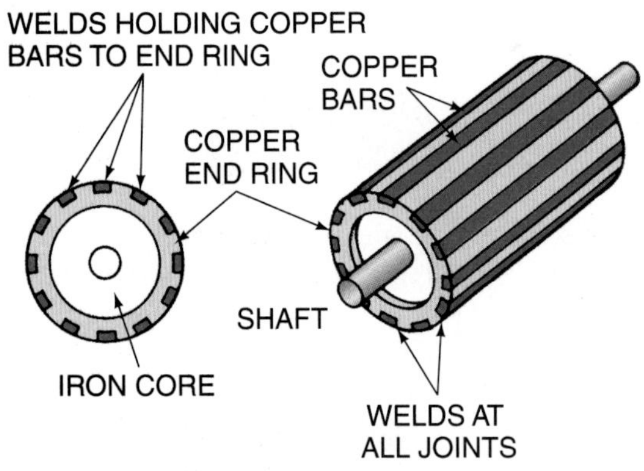

Figure 17.6 The direction of current flow changes twice during each cycle.

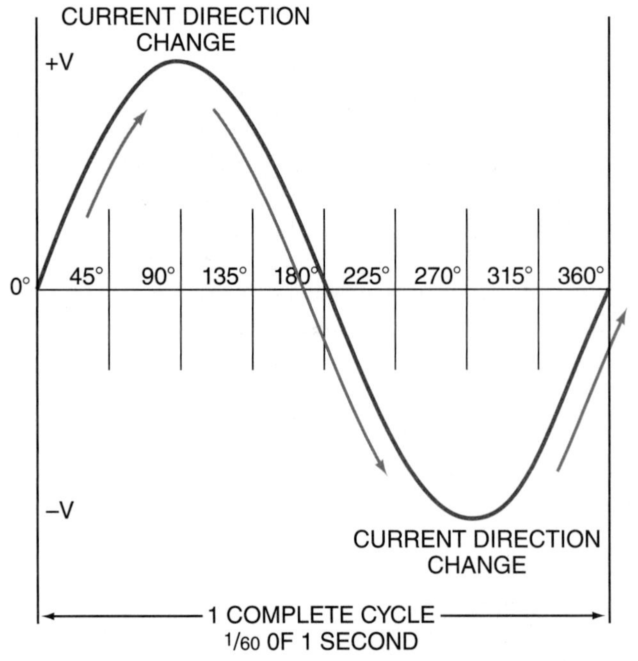

opposite to that in the run windings. The opposite poles of the run winding are wound in different directions. If one pole is wound clockwise, the opposite pole would be wound counterclockwise. This would set up opposite polarities for the attraction and repulsion forces that cause rotation.

The attracting and repelling action between the poles of the run windings and the rotor sets up a rotating magnetic field and causes the rotor to turn. Since this is AC reversing 60 times per second, the rotor turns, in effect "chasing" the changing polarity in the run windings. The motor will continue to rotate as long as power is supplied. The motor starting method determines the direction of motor rotation. The motor will run equally well in either direction.

17.4 DETERMINING A MOTOR'S SPEED

In the United States, AC power is supplied at a frequency of 60 hertz (Hz, or cycles per second). This means that in a 1-sec time interval, 60 complete sine waves are generated. From this, we can conclude that one cycle is completed in 1/60 (one-sixtieth) of a second. During this very brief instant, the direction of current flow changes twice, **Figure 17.6**. This means that every second, the current changes direction 120 times.

Referring back to our horseshoe magnet example, **Figure 17.4**, the rotating bar magnet will come to rest when the north pole of the bar magnet is aligned with the south pole of the horseshoe magnet and when the south pole of the bar magnet is aligned with the north pole of the horseshoe magnet. So, in order for the bar magnet to turn 360 degrees, or one full rotation, the horseshoe magnet must change the polarity of its poles twice. We are discussing AC power; therefore, during each sine wave, the current changes direction twice each cycle. So, the rotor of a motor with two poles, as in our bar magnet example, will turn 60 times every second, or 3600 times a minute. The more poles a motor has, the slower the motor will turn.

The following formula can be used to determine the synchronous speed (without load) of motors in rpm, or revolutions per minute:

$$S\text{ (rpm)} = \frac{\text{Frequency} \times 120}{\text{Number of poles}}$$

NOTE: *The magnetic field builds and collapses twice each cycle (each time it changes direction); the 120 is a time conversion that changes seconds into minutes. It is also a conversion from the motor's pole pairs (poles/2) to the actual number of poles.* •

$$\text{Speed of Two-Pole, Split-Phase Motors} = \frac{60 \times 120}{2} = 3600\text{ rpm}$$

$$\text{Speed of Four-Pole, Split-Phase Motors} = \frac{60 \times 120}{4} = 1800\text{ rpm}$$

The speeds indicated above are the calculated speeds and not the actual speeds at which the motor will turn. Internal motor component friction and the mechanisms or drives to which the motor is connected add a certain amount of resistance to rotation. This resistance, referred to as the motor slip, is often expressed as a percentage. Motor slip represents the difference between the motor's calculated and actual speeds. Our 3600 rpm motor actually turns at a speed of about 3450 rpm. We can calculate slip as follows:

Motor Slip (%)

$$= \frac{\text{Calculated Motor speed} - \text{Actual Motor Speed}}{\text{Calculated Motor Speed}} \times 100$$

In this example we have the following:

$$\text{Motor Slip (\%)} = \frac{3600 - 3450}{3600} \times 100$$
$$= \frac{150}{3600} \times 100$$
$$= 4.17\%$$

It is left as an exercise for the reader to verify that the motor slip for an 1800-rpm motor that is turning at 1725 rpm is also 4.17%. Excessive motor slip can be an indication of improper motor lubrication, defective motor bearings, poorly adjusted belts (too tight), or an improperly selected motor for the desired application (overloaded).

17.5 START WINDINGS

The preceding explanation does not include how the motor rotation is started. This is accomplished by using a separate motor winding called the start winding. The start winding is wound next to the run winding but is a few electrical degrees out of phase with the run winding. This design is much like the pedals on a bicycle—if both pedals were at the same angle, it would be very hard to get started. The start winding is in the circuit only for as long as it takes the rotor to get close to its rated speed, and it then is electrically disconnected, **Figure 17.7**. The start winding has more turns than the run winding and is wound with a smaller diameter wire. Opposite poles of the start winding are wound in different directions. If one pole is wound clockwise, then the opposite pole would be wound counterclockwise. This sets up opposite polarities for the attraction and repulsion forces that cause rotation and produces a larger magnetic field and greater resistance, which helps the rotor to start turning and determines the direction in which it will turn. Because this winding is located between the run winding, it changes the phase angle between the voltage and the current in the windings.

We have just described a two-pole split-phase induction motor, which is rated to run at 3600 revolutions per minute (rpm). It actually turns at a slightly slower speed when running under full load. When the motor approaches its normal speed, the start winding is often removed from the circuit, leaving only the run windings energized. Many split-phase motors have four poles and run at 1800 rpm.

Figure 17.7 The placement of the start and run windings inside the stator.

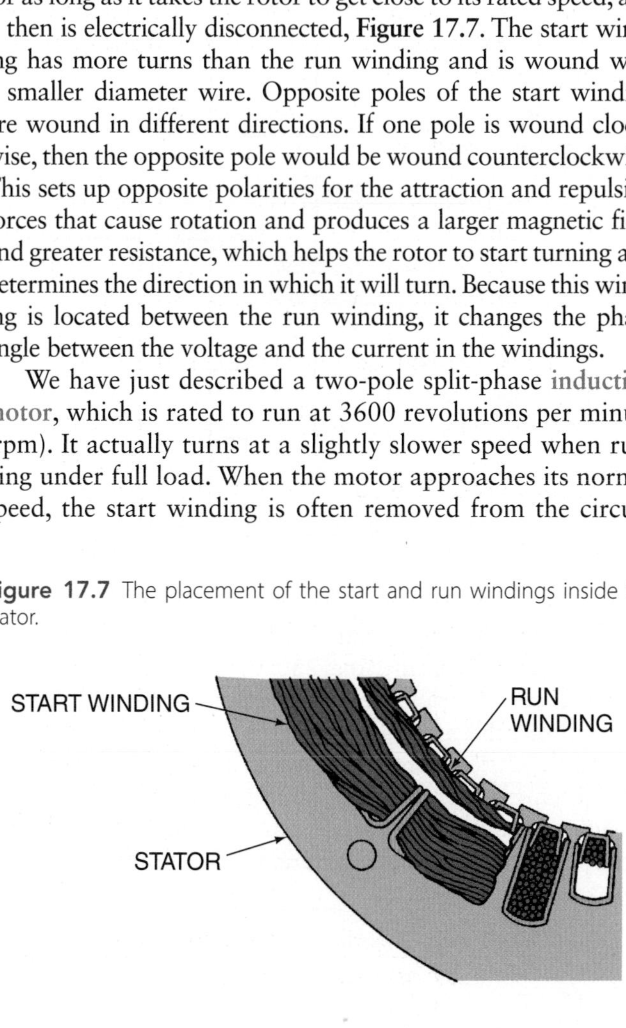

17.6 STARTING AND RUNNING CHARACTERISTICS

Two major considerations in electric motor applications are the starting and running characteristics. A motor that runs a refrigeration compressor must have a high starting torque—it must be able to start under heavy starting loads. Torque is the twisting power of the motor shaft. The starting torque is the power to turn the shaft from the stopped position. The motor must also have enough torque to operate under the load of the application. Some motors must have a great deal of starting torque to turn the motor from the stopped position but do not need a great deal of torque to maintain speed. For example, systems equipped with capillary tubes can typically use compressors with low starting torque, as the system pressures can equalize during the off cycle. Some systems with thermostatic expansion valves (TXV) require greater starting torque, as there may be a pressure differential across the compressor even when the compressor is not operating.

Motors have two different current ratings, full-load or rated-load amperes (FLA or RLA) and locked-rotor amperes (LRA). Locked-rotor amperage is often referred to as inrush current. When a motor is still, it takes a great deal of torque to get it turning, particularly if it has a load on start-up. Typically, the LRA for a motor is about five times the FLA/RLA. For example, suppose a compressor operates on 25 A when at full load. This same compressor will pull about 125 A for just a moment on start-up. As the motor gets up to speed, the amperage will reduce to FLA, 25 A. Since this inrush of current is of short duration, it does not figure in the wire sizing for the motor. A refrigeration compressor may have a head pressure of 155 psig and a suction pressure of 5 psig and be required to start in systems where the pressures do not equalize. The pressure difference of 150 psig is the same as saying that the compressor has a starting resistance of 150 psi of piston area. If this compressor has a 1-in.-diameter piston, the area of the piston is 0.78 in^2 ($A = \pi r^2$, or $3.14 \times 0.5 \times 0.5$). This area multiplied by the pressure difference of 150 psi is the starting resistance for the motor (117 lb), comparable to a 117-lb weight resting on top of the piston when it tries to start, **Figure 17.8**.

A motor does not need as much starting torque to start a small fan. The motor must simply overcome the friction needed to start the fan moving. There is no pressure difference because the pressures equalize when the fan is not running, **Figure 17.9**. A high starting load, called starting torque, puts more load on the compressor compared to the light load of the fan. These are two different motor applications and require two different types of motors. These applications will be discussed individually in this unit when we discuss the different types of motors.

Figure 17.8 A compressor with a high-side pressure of 155 psig and a low-side pressure of 5 psig.

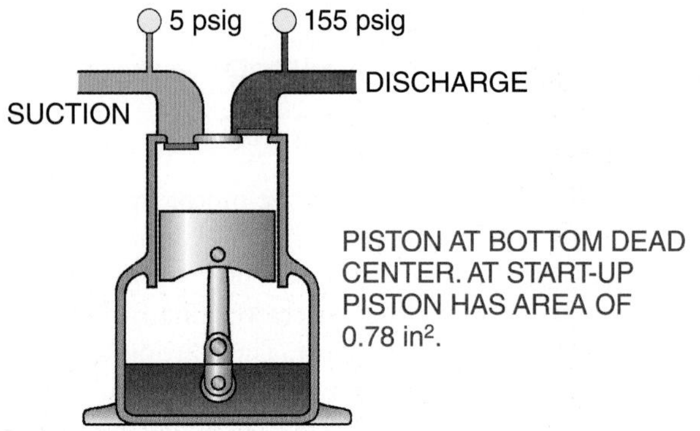

Figure 17.9 This fan has no pressure difference to overcome while starting. When the fan stops, the air pressure equalizes.

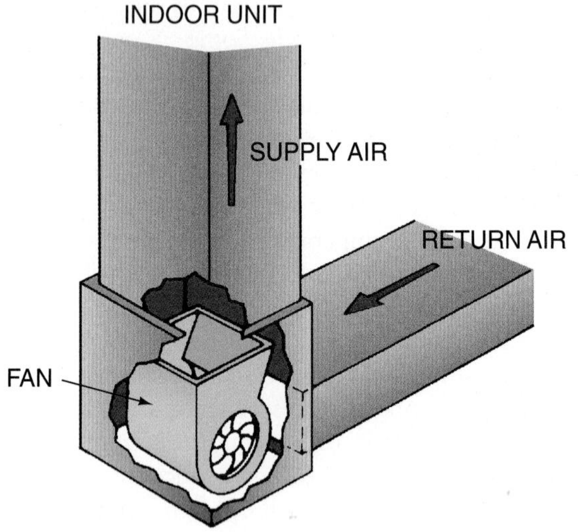

17.7 ELECTRICAL POWER SUPPLIES

The power company furnishes power to customers and determines the type of power supply. Power is moved across the country by means of high-voltage transmission lines to transformers that reduce the power to the voltage needed at the consumer level. Residences are typically furnished with a single-phase power supply (ϕ is the symbol often used for phase). The power pole to the house may look like the one in **Figure 17.10**. Several homes may be supplied by the same power-reducing transformer, which has three wires that pass through the meter base and on to the house electrical panel, where the power is distributed through the circuit protectors (circuit breakers or fuses) to the various circuits in the house and then to the power-consuming equipment, **Figure 17.11**. The power panel in the house would look like the illustration

Figure 17.10 A single-phase power supply.

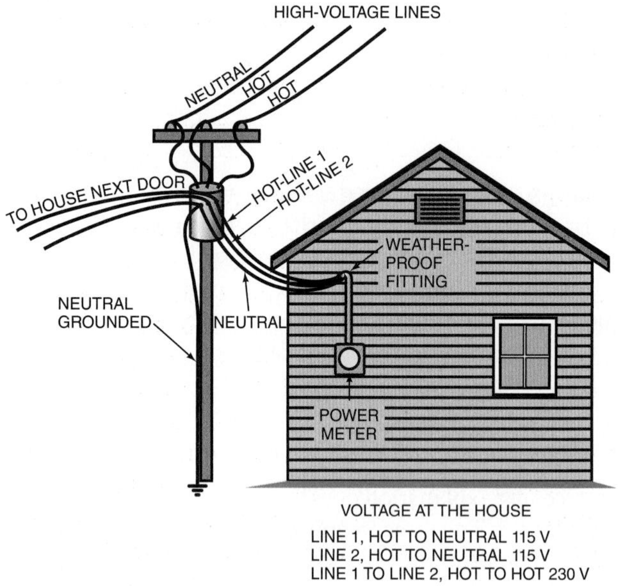

Figure 17.11 A wiring diagram of a main circuit breaker panel for a typical residence.

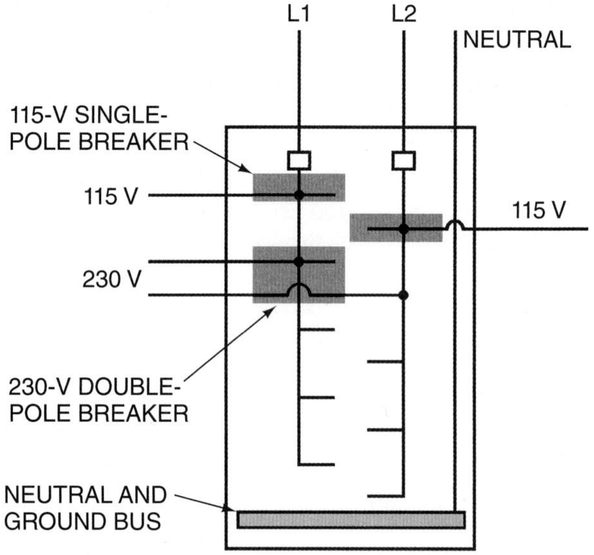

in **Figure 17.11**. Notice that it will furnish both 115-V and 230-V service. Typically, the 230-V service is used for the electric dryer, electric range, electric oven, many electric heat devices, and the air-conditioning. All other appliances are typically operated from the 115-V circuits.

Equipment in commercial buildings and factories is large enough to require three-phase power. A three-phase power supply may furnish several different voltage options:

A. 115-V single-phase for common appliances
B. 230-V single-phase for heavy-duty appliances

C. 230-V three-phase for large loads such as electric heat or motors
D. 460-V three-phase for large loads such as electric heat or motors
E. 277-V single-phase for lighting circuits (may be obtained between 460 V and neutral with certain systems)
F. 560-V three-phase for special industrial systems

The wiring for a typical three-phase system from the pole, which is furnished by the power company, may look like the illustration in **Figure 17.12**. The building could very well have made use of a 230-V three-phase power supply. One of the reasons for using 460-V three-phase is that it reduces the wire size to all components in the building, which reduces the materials and labor cost of the installation. The same load using 230 V would use twice the current flow, so the wire would have to be sized up to twice the current-carrying capacity. Many of the components in the building will need to operate from 115 V and maybe even 230 V, so a step-down transformer is used for equipment like small fans, computers, and office equipment. The step-down transformer enables the building to take advantage of the benefits of 460 V, 277 V, 230 V, and 115 V in the same system, **Figure 17.13**. **SAFETY PRECAUTION:** *A technician must be very careful with any live power circuits. 460-V circuits are particularly dangerous, and the technician must take extra care while servicing them.* •

Figure 17.12 A 460-V, three-phase power supply to a commercial building.

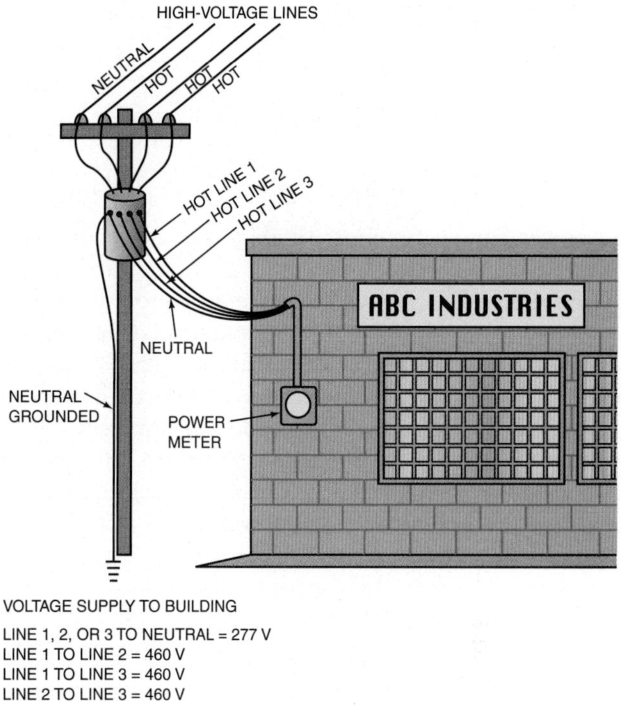

Figure 17.13 When a building has a 460-V power supply, it will have a step-down transformer to provide 115-V circuits to power office machines and small appliances.

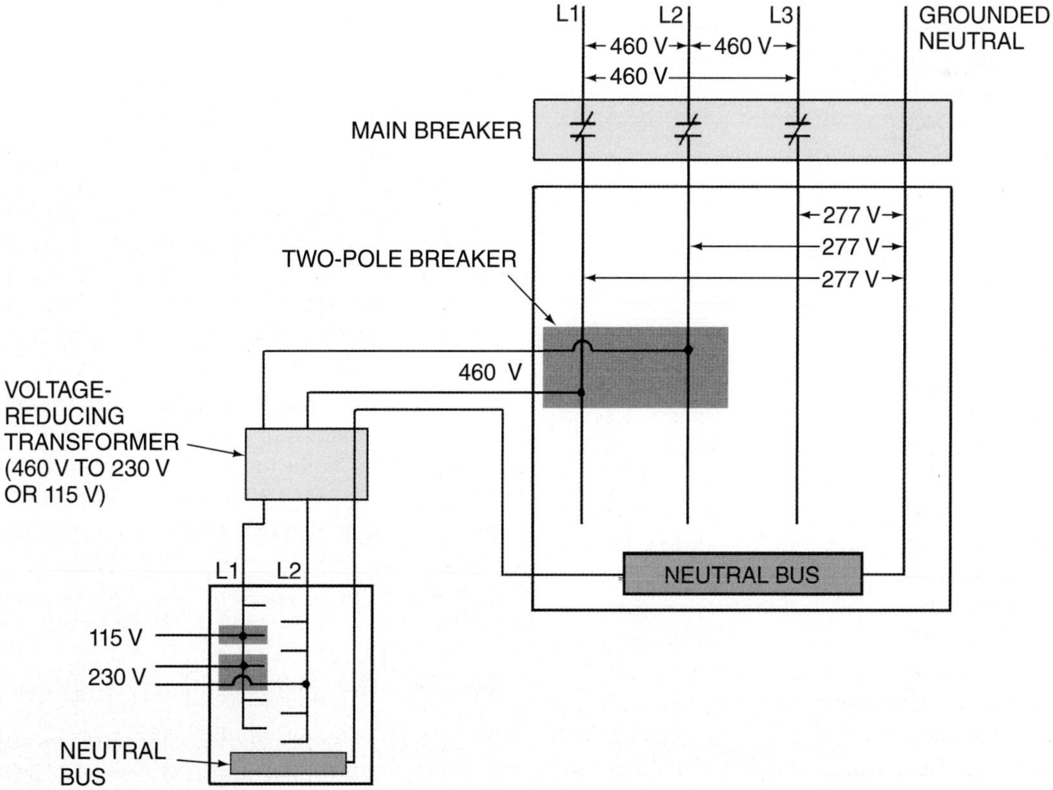

Following is a description of some motors currently used in the heating, air-conditioning, and refrigeration industry. We emphasize the electrical characteristics, not the working conditions. Some older motors are still in operation, but they are not discussed in this text.

17.8 SINGLE-PHASE OPEN MOTORS

The power supply for most single-phase induction motors is either 115 V or 208 V to 230 V. A home furnace uses a power supply of 115 V, whereas the air conditioner outside uses a power supply of 230 V. A commercial building may use either 208 V, 230 V, or 460 V, depending on the power company. Some single-phase motors can operate in dual voltage. They have two run windings and one start winding. The two run windings have the same resistance, and the start winding has a high resistance. The motor will operate with the two run windings in parallel in the low-voltage mode. When the high-voltage mode is required, the technician changes the numbered motor leads according to the manufacturer's instructions. This wires the run windings in series with each other and delivers an effective voltage of 115 V to each winding. The motor windings are actually only 115 V because they operate only on 115 V, no matter which mode they are in. The technician can change the wiring connections at the motor terminal box, **Figure 17.14**. Some commercial and industrial installations may use a 460-V power supply for large motors and reduce that to a lower voltage to operate smaller motors. The smaller motors may be single-phase and must operate from the same power supply, **Figure 17.13**.

Figure 17.14 The wiring diagram of a dual-voltage motor. This motor is designed to operate at either 115 V or 230 V, depending on how the motor is wired in the field. (A) A 230-V wiring diagram. (B) A 115-V wiring diagram.

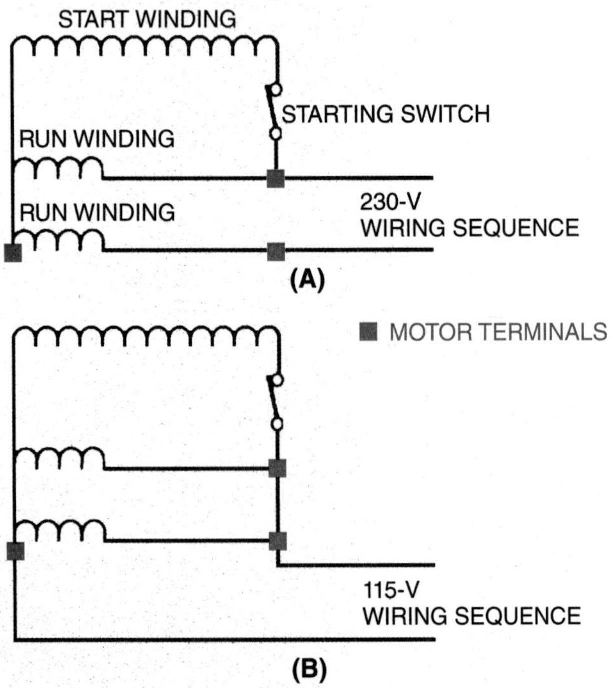

Figure 17.15 The direction of rotation of a single-phase motor can be reversed by changing the connections in the motor terminal box. This will change the direction of current flow through the start winding.

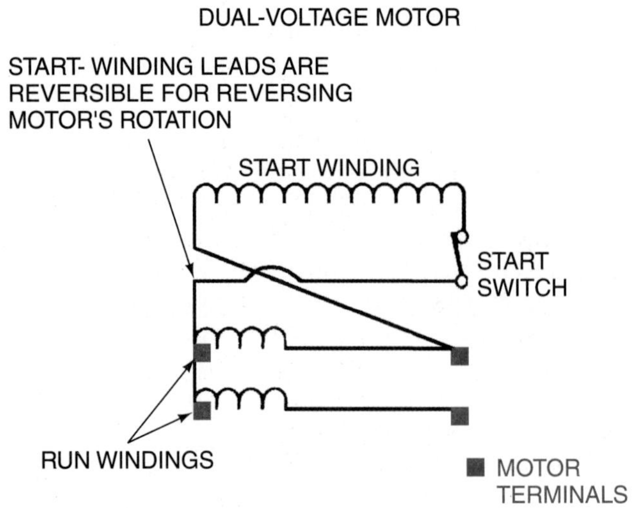

A single-phase motor can rotate clockwise or counterclockwise. Some are reversible from the motor terminal box, **Figure 17.15**. By reversing the start winding leads, the direction of current flow through the start winding is reversed, changing the charges of the poles. This will cause the motor to rotate in the opposite direction.

17.9 SPLIT-PHASE MOTORS

Split-phase motors are induction motors that have two distinctly different windings, **Figure 17.16**. They have a medium amount of starting torque and a good operating efficiency and are normally used for operating fans in the fractional horsepower range. The normal operating ranges are 1800 rpm and 3600 rpm. An 1800-rpm motor will operate at 1725 rpm to 1750 rpm under a load. If the motor is loaded to the point where the speed falls below 1725 rpm, the current draw will climb above the rated amperage. Motors rated at 3600 rpm will normally slip in speed to about 3450 rpm to 3500 rpm, **Figure 17.17**. Some of these motors are designed to operate at either speed, 1750 rpm or 3450 rpm. The speed of the motor is determined by the number of motor poles and by the method of wiring the motor poles. The technician can change the speed of a two-speed motor at the motor terminal box.

17.10 THE CENTRIFUGAL SWITCH

All split-phase motors have a start and run winding. The start windings must be disconnected from the circuit within a very short period of time or they will overheat. Several methods

Figure 17.16 Diagram of the split-phase motor start and run windings.

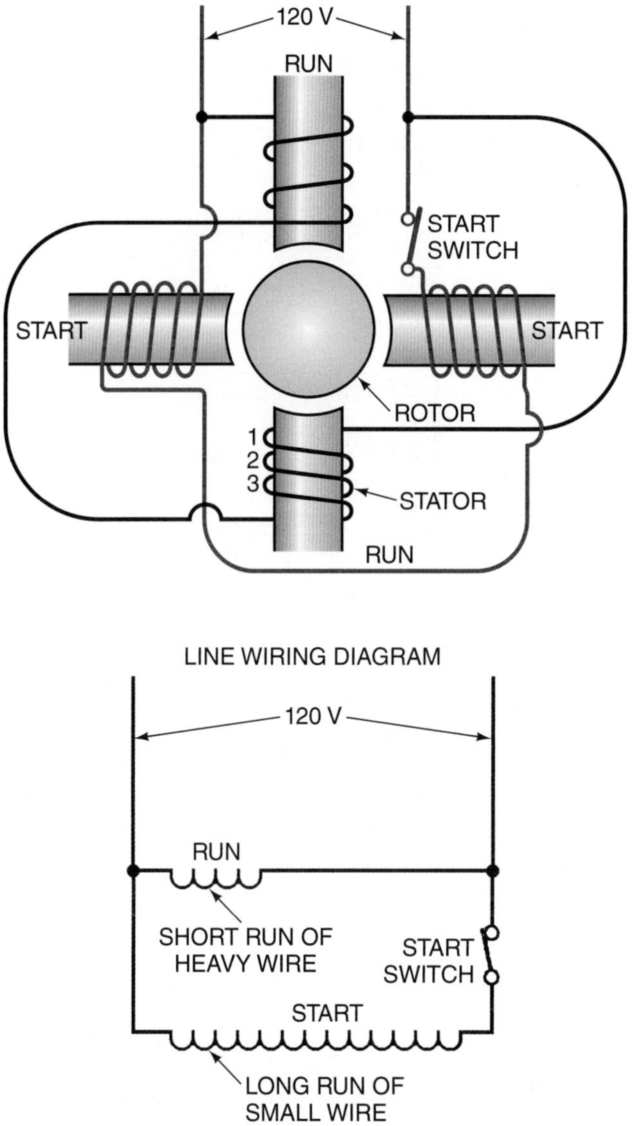

Figure 17.17 Motor speed changes depend on the load on the split-phase motor.

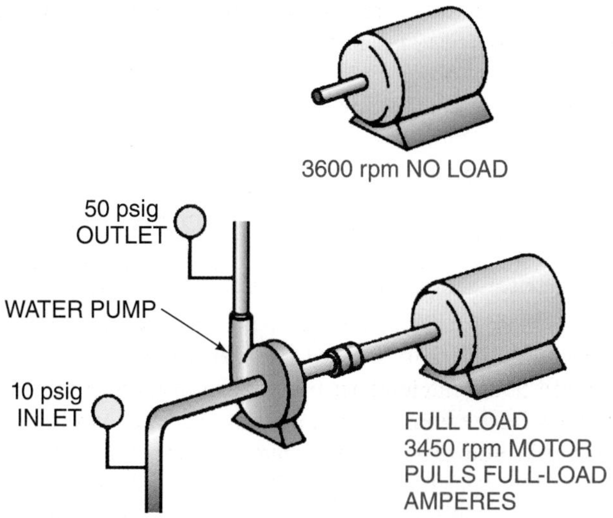

are used to disconnect the start windings. The centrifugal switch is the most common method in open motors; however, electronic start switches are sometimes used.

The **centrifugal switch** is used to disconnect the start winding from the circuit when the motor reaches approximately 75% of the rated speed. Motors described here are those that run in the atmosphere. (Hermetic motors that run in the refrigerant environment are discussed later in this unit.) When a motor is started in the air, the arc from the centrifugal switch will not harm the atmosphere. (It will harm the refrigerant, so there must be no arc in a refrigerant atmosphere.)

A centrifugal switch is a mechanical device attached to the end of the shaft with weights that will sling outward when the motor reaches approximately 75% speed. For example, if the motor has a rated speed of 1725 rpm, the centrifugal weights will change position at 1294 rpm (1725×0.75) and open a switch to remove the start winding from the circuit. This switch is under a fairly large current load, so a spark will occur. If the switch fails to open its contacts and remove the start winding, the motor will draw too much current, and the overload device will cause it to stop. When the motor is deenergized, it will slow down and the centrifugal switch will close its contacts in preparation for the next motor starting attempt. The more the switch is used, the more its contacts will burn from the arc. If this type of motor is started many times, the first thing that will likely fail is the centrifugal switch. This switch makes an audible sound when the motor starts and stops, **Figure 17.18.**

Figure 17.18 The centrifugal switch is located at the end of the motor.

17.11 THE ELECTRONIC RELAY

The electronic relay is used with some motors to open the start windings after the motor has started. This solid-state device is designed to open the start winding circuit when the design speed has been obtained. Other devices are also used to perform this function; they are described in the discussion of hermetic motors.

17.12 CAPACITOR-START MOTORS

The capacitor-start motor is basically the same motor as the split-phase induction motor, **Figure 17.19**. It has two distinctly different windings for starting and running. However, a start capacitor is wired in series with the start windings to give the motor more starting torque. **Figure 17.20** shows voltage and current cycles in an induction motor. In an inductive circuit the current *lags* the voltage. In a capacitive circuit, the current leads the voltage. The amount by which the current leads or lags the voltage is the *phase angle*. A capacitor is chosen to create a phase angle that is the most efficient for starting the motor, **Figure 17.21**. The capacitor is not designed to be used while the motor is running, and it must be switched out of the circuit soon after the motor starts. This is done at the same time the windings are taken out of the circuit, and with the same switch.

Once the start winding and the start capacitor have been removed from the circuit, the motor will continue to operate with only the run winding energized. The rotating motor shaft and rotor will create, or induce, the needed imbalance in the magnetic field to keep the motor turning. For this reason, capacitor-start motors are often referred to as CSIR motors, which stands for capacitor-start-induction-run. A simplified wiring diagram for a CSIR motor is shown in **Figure 17.22**.

Figure 17.19 A capacitor-start motor. Courtesy W. W. Grainger, Inc.

Figure 17.20 The current lags the voltage in an inductive circuit.

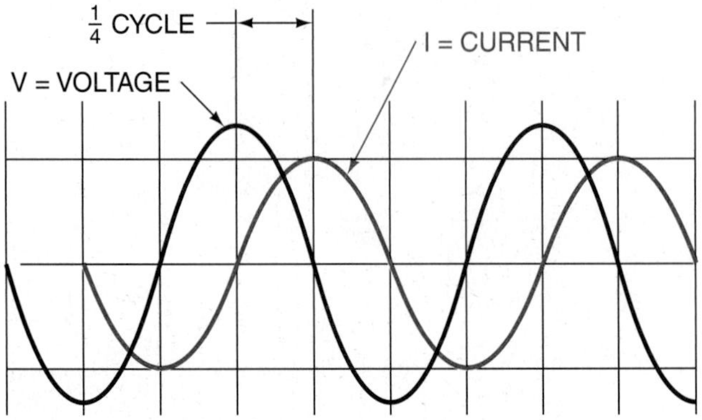

Figure 17.21 A start capacitor.

Figure 17.22 Wiring diagram of a capacitor-start motor.

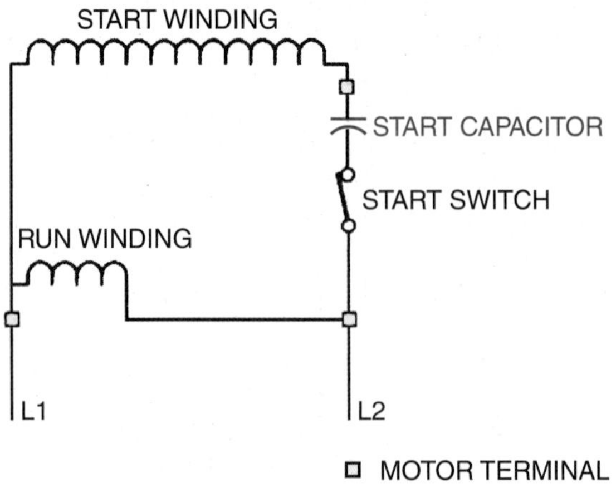

17.13 CAPACITOR-START, CAPACITOR-RUN MOTORS

Capacitor-start, capacitor-run motors are much the same as the split-phase induction motor. A run capacitor is wired into the circuit to provide the most efficient phase angle between the current and voltage when the motor is running and is in the circuit at any time the motor is running. Both the run and start capacitors are wired in series with the start winding but are in parallel to one another, **Figure 17.23**. The microfarad rating of the run capacitor is much lower than that of the start capacitor for any given motor application. The capacitances of two capacitors in parallel add their values the same as resistors do in series. If the run capacitor has a capacitance of 10 microfarads and the start capacitor has a capacitance of 110 microfarads, their total capacitance would add to be 120 microfarads. During start-up, this added capacitance in series with the start winding causes a greater phase angle between the run and start windings, which gives the motor more starting torque.

When the start switch opens, the start capacitor is taken out of the circuit. It is used for nothing but adding starting torque. However, the run capacitor and start winding stay in the circuit. The run capacitor stays in series with the start winding for extra running torque. The capacitor, being in series with the start winding during the running mode, also limits the current through the start winding so the winding will not get hot. If a run capacitor fails because of an open circuit within the capacitor, the motor may start, but the running amperage will be about 10% too high and the motor will get hot if operated at full load. The motor is actually a permanent split-capacitor (PSC) motor when running (PSC motors will be discussed in the next section). These motors are specially designed to have the start winding energized whenever the motor is energized, so be sure to use and wire the motor according to the manufacturer's recommendations and guidelines.

The capacitor-start, capacitor-run motor is one of the most efficient motors used in refrigeration and air-conditioning equipment. It is normally used with belt-drive fans and compressors. Belt-driven motor and blower assemblies offer a great deal of resistance to the motor when it initially starts because the motor must turn the pulley, which is connected to a belt, which is then connected to a pulley and blower assembly. The added starting torque is needed to help overcome this resistance. Once the motor starts, however, keeping it turning does not require the same amount of torque.

Figure 17.23 Wiring diagram of a capacitor-start, capacitor-run motor. The start capacitor is in the circuit only during motor start-up, whereas the run capacitor is in the circuit whenever the motor is energized.

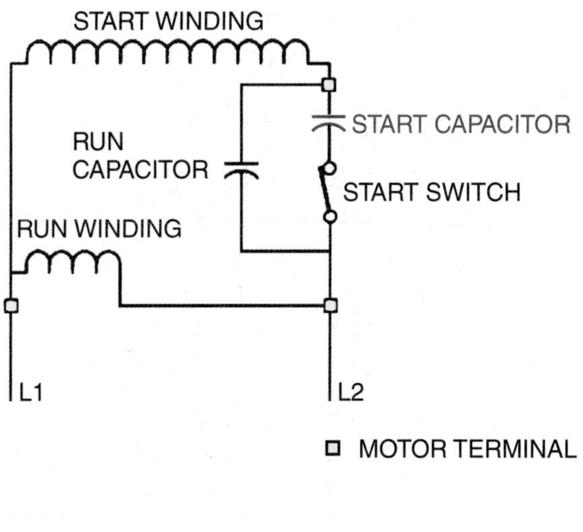

17.14 PERMANENT SPLIT-CAPACITOR (PSC) MOTORS

The PSC motor has windings very similar to those of the split-phase motor, **Figure 17.24**, but it does not have a start capacitor. Instead it uses a run capacitor wired into the circuit similar to the run capacitor in the capacitor-start, capacitor-run motor. The run capacitor is connected to the L2 sides of the start and run windings. The PSC motor does not utilize a start switch as both the start and run windings, along with the run capacitor, are in the active electric circuit whenever the motor is energized. This is the simplest split-phase motor. It is very efficient and has no moving parts to start the motor; however, the starting torque is very low, so it can be used only in low-starting-torque applications, **Figure 17.25**.

A multispeed PSC motor can be identified by the many wires at the motor electrical connections, **Figure 17.26** and **Figure 17.27**. As the electrical resistance of the motor winding decreases, the speed of the motor increases. When more resistance is wired into the circuit, the motor speed decreases. Most manufacturers use this motor in the fan section of air-conditioning and heating systems. Motor speed can be changed by switching the wires. Earlier systems used a capacitor-start, capacitor-run motor and a belt drive, and

Figure 17.24 A permanent split-capacitor (PSC) motor. Source: Universal Electric Company

Figure 17.25 This open-type PSC motor can be used to turn a fan. Source: Universal Electric Company

Figure 17.26 A multispeed PSC motor.

Figure 17.27 This diagram shows how the windings of a three-speed PSC motor are configured. As the winding resistance increases, the motor speed decreases.

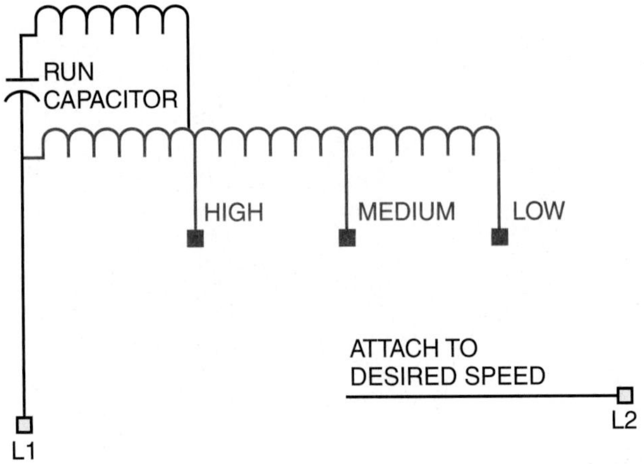

air volumes were adjusted by varying the drive pulley diameter. Many PSC motors are manufactured in 2-, 4-, 6-, and 8-pole designs. The speed in rpm of these motors depends on the frequency of the AC source and the number of poles wired in the motor. The higher the frequency, the faster the motor. The more poles there are, the slower the rpm of the motor. For 60-Hz operation, the rated or synchronous speeds would be 3600, 1800, 1200, and 900 rpm, respectively. However, due to motor slip, the actual speeds would be 3450, 1725, 1075, and 825 rpm.

The PSC motor may be used to obtain slow fan speeds during the winter heating season for higher supply air temperatures for gas, oil, and electric furnaces. The fan speed can be increased by switching to a different resistance in the winding using a relay. This will provide more airflow in summer to satisfy cooling requirements, **Figure 17.28**. The PSC fan motor also has another advantage over the split-phase motor in fan applications: It has a very soft start-up. When a split-phase motor is used for a belt-drive application, the motor, belt, and fan must get up to speed very quickly, which often creates a start-up noise. The PSC

Figure 17.28 Diagram showing how a multispeed PSC motor is wired to operate at low speed (lower air volume) in the winter and high speed (higher air volume) in the summer.

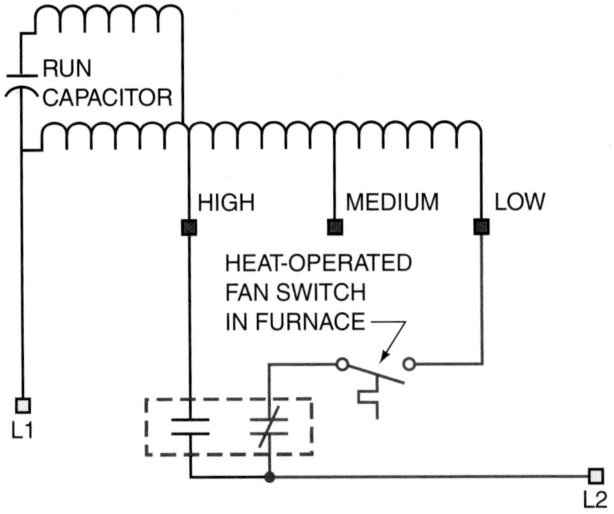

motor starts up very slowly, gradually getting up to speed, which is very desirable if the fan is close to the return air inlet to the duct system.

17.15 SHADED-POLE MOTORS

The shaded-pole motor has small starting windings or shading coils at the corner of each pole that help the motor start by providing an induced current and a rotating field, **Figure 17.29**. The location of the shading coil on the pole face determines the direction of rotation. For years the shaded-pole motor has been used in air-cooled condensers and evaporators to turn the fans, **Figures 17.30(A)** and **(B)**. On most shaded-pole motors, rotation can be reversed by disassembling the motor and turning the stator over, **Figure 17.30(C)**. This moves the shaded coil to the side opposite to the one the pole faces and reverses rotation. The motor has very little starting torque and is not as efficient as the PSC motor, so it is used only for light-duty applications. From the standpoint of initial cost, it is an economical motor and is normally manufactured in the fractional horsepower range.

17.16 THREE-PHASE MOTORS

Three-phase motors, which range from about 1 hp into the thousands of horsepower, have some characteristics that make them popular for many applications. The motors are very efficient and require no start assist for high-torque applications—they have no starting windings or capacitors. They are used mainly on commercial equipment, and the building power supply must have three-phase power available. (Three-phase power is seldom found in a home.)

Figure 17.29 Wiring diagram of a shaded-pole motor.

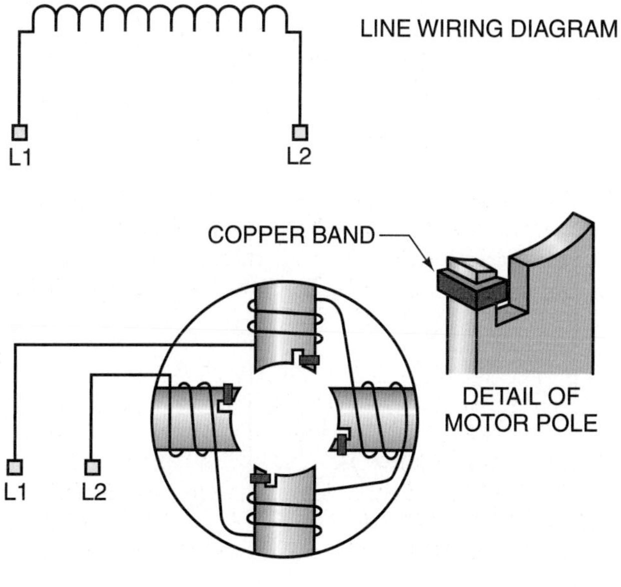

Figure 17.30 (A) The shaded-pole motor. (B) A small shaded-pole motor showing the copper-banded shaded poles. Courtesy Ferris State University. Photo by John Tomczyk (C) A shaded-pole motor with the rotor removed in preparation for reversing rotation. Courtesy Ferris State University. Photo by John Tomczyk

(A)

(B)

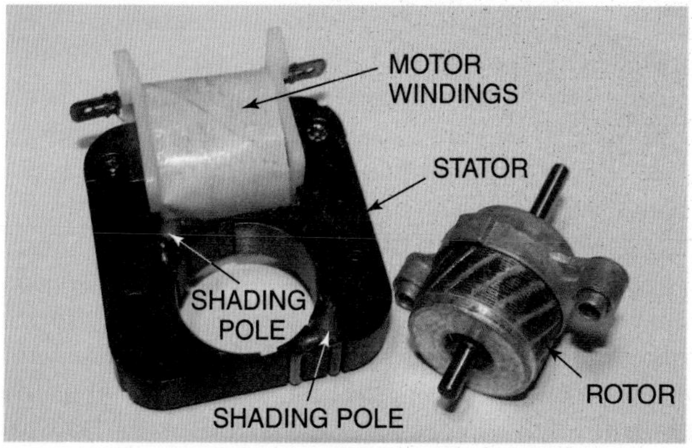

(C)

Figure 17.31 Wiring diagram of a three-phase power supply.

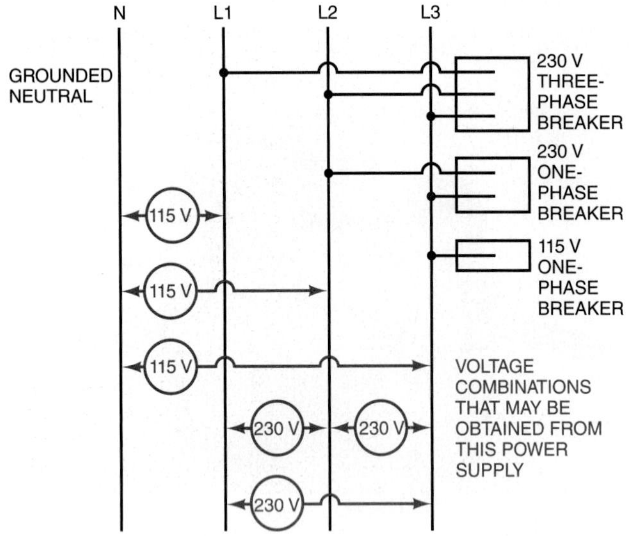

Figure 17.32 (A) Diagram of a three-phase power supply. (B) Diagram of a typical single-speed, three-phase motor.

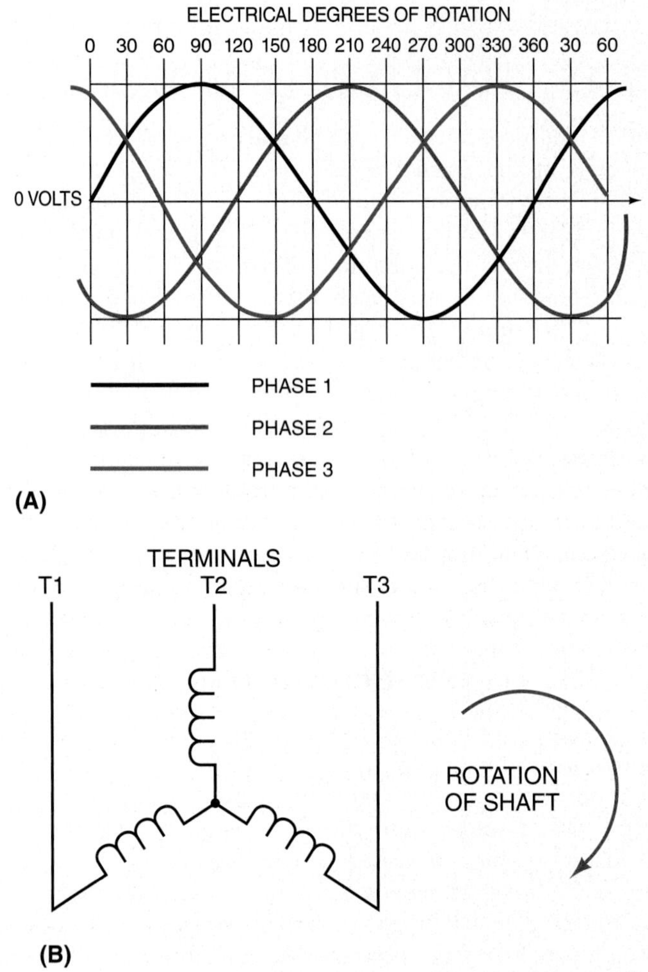

The motors can be thought of as having three single-phase power supplies, **Figure 17.31**, with each of the phases having either two or four poles. A 3600-rpm motor will have three sets, each with two poles (total of six), and an 1800-rpm motor will have three sets, each with four poles (total of 12). Each phase changes the direction of current flow at different times but always in the same order. A three-phase motor has high starting torque because of the three phases of current under which it operates. At any given part of the rotation of the motor, one of the windings is in position for high torque. This makes starting large fans and compressors very easy, **Figure 17.32.** The three-phase motor rpm slips to about 1750 rpm and 3450 rpm when under full load. The motor is not normally available with dual speed; it is either an 1800-rpm or a 3600-rpm motor.

Three-phase power supply consists of three single-phase power supplies that are generated 120 electrical degrees out of phase with each other. As can be seen in **Figure 17.32(A)**, phase number 1, indicated by the black sine wave, starts at "0" electrical degrees of rotation and has a potential of 0 volts. At 90 electrical degrees, phase number 1 has reached its maximum potential, and at 360 electrical degrees it has completed a full cycle. It can be seen that phases 2 and 3 start at 120 and 240 electrical degrees of rotation, respectively. All three of these phases are 120 degrees out of phase with each other. It should be noted that 120 degrees after the start of phase number 3 (which is at 360 electrical degrees), phase number 1 is ready to start another cycle.

The rotation of a three-phase motor may be changed by switching any two motor leads, **Figure 17.33.** The rotation must be carefully observed when it drives three-phase fans. If a fan rotates in the wrong direction, it will move only about half as much air. If this occurs, reverse the motor leads, and the fan will turn in the correct direction.

Figure 17.33 The direction of rotation of a three-phase motor can be reversed by switching any two power leads.

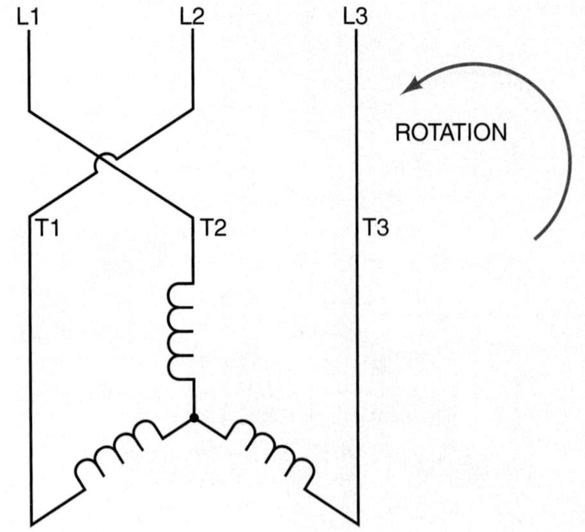

Motors have other characteristics that must be considered when selecting them for a particular application, for example, the motor mounting. Is the motor solidly mounted to a base, or is there a flexible mount to minimize noise? The sound level is another factor. Will the motor be used where ball bearings would make too much noise? If so, sleeve bearings should be used. Still another factor is the operating temperature of the motor surroundings. A condenser fan motor that pulls the air over the condenser coil and then over the motor requires a motor that will operate in a warmer atmosphere. The best advice is to replace any motor with exactly the same type whenever possible.

17.17 HERMETICALLY SEALED MOTORS

A hermetically sealed motor is one that is completely enclosed inside of a metal shell, such as a compressor. The shell is welded closed, which makes servicing the motor inside impossible. The motors that are most commonly used in a hermetic setting are the split-phase, capacitor-start, the capacitor-start, capacitor-run, and the permanent split-capacitor (PSC). A hermetic motor is designed to operate in a refrigerant vapor atmosphere. Materials used in the construction of a hermetic compressor motor must be compatible with the refrigerant and oil circulating in the system. The coatings on the windings, ties for the motor windings, and the papers used as wedges must be of the correct material. The motor is assembled in a dry, clean atmosphere. It is undesirable for liquid refrigerant to enter the shell—for example, from an overcharge of refrigerant. Single-phase hermetic compressors are usually manufactured up to 5 hp, **Figure 17.34**. If more capacity is needed, multiple systems or larger three-phase units are used. Hermetic motors are started in much the same way as the other motors we have described. The start windings may be removed from the circuit when the motor nears normal operation, but not in the same way as for an open motor because the windings are in a refrigerant atmosphere. Open single-phase motors are operated in air, and a spark is allowed when the start winding is disconnected. This cannot be allowed in a hermetic motor because the spark will cause the refrigerant to deteriorate. Special devices determine when the compressor motor is drawing the desired current or running at the correct speed to disconnect the start winding from outside the compressor shell. Because the hermetic motor is enclosed in refrigerant, the motor leads must pass through the compressor shell to the outside. A terminal box on the outside houses the three motor terminals, **Figure 17.35**, one for the run winding, one for the start winding, and one for the line common to the run and start windings. See **Figure 17.36** for a wiring diagram of a three-terminal

Figure 17.34 A typical motor for a hermetic compressor. Courtesy of Hampden Engineering

Figure 17.35 The motor terminal box on the outside of the compressor shell.

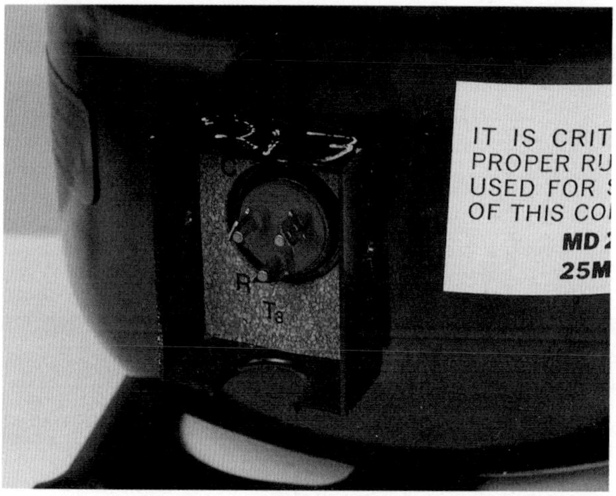

Figure 17.36 A wiring diagram of the terminal and internal wiring connections on a single-phase compressor.

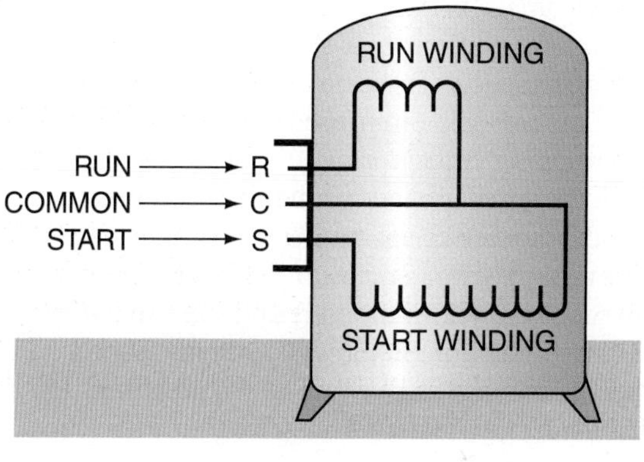

Figure 17.37 These motor terminals use neoprene O rings as insulators between the terminals and the compressor housing. Source: Trane Company

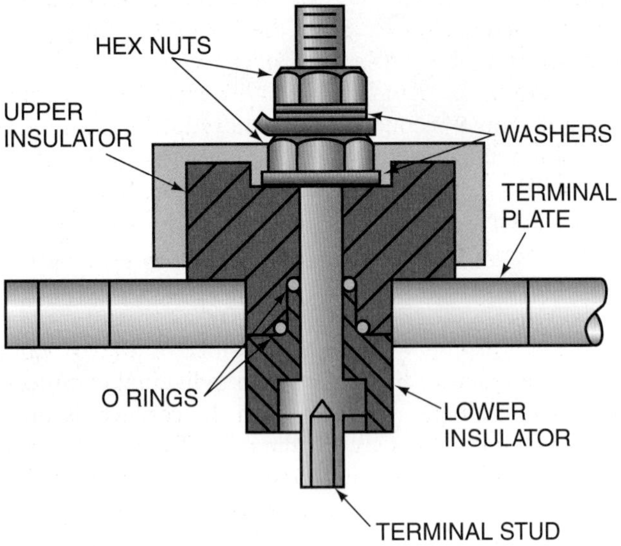

compressor. It can be seen in this figure that the start winding has more resistance than the run winding. The motor leads are insulated from the steel compressor shell. Neoprene was the most popular insulating material for years. However, if the motor terminal becomes too hot, due to a loose connection, the neoprene may eventually become brittle and possibly leak, **Figure 17.37**. Today, many compressors use a ceramic material to insulate the motor leads.

17.18 THE POTENTIAL RELAY

Potential relays are used with single-phase capacitor-start, capacitor-run motors that need relatively high starting torque, **Figure 17.38(A)**. Their main function is to assist in starting the motor. These relays are sometimes referred to as voltage relays or potential magnetic relays (PMRs). Potential relays consist of a high-resistance coil and a set of normally closed contacts, **Figure 17.38(B)**. The coil is wired between terminals 2 and 5, and the normally closed contacts are wired between terminals 1 and 2, **Figure 17.38(C)**. Other terminal designations on the relay, sometimes referred to as accommodation terminals, serve only as wire connections. These dummy terminals are often used for wiring such devices as condenser fans or capacitors.

When power is applied through the compressor's contactor contacts, both the run and start windings are energized and the motor's rotor starts to turn. The start winding is wired in series with both the start and run capacitors. The run and start capacitors are in parallel with one another, but both are in series with the start winding, **Figure 17.38(C)**. Parallel capacitors add like resistors in series, so their capacitances add, giving the motor high starting torque. The large metal mass of the motor's rotor turning at high speeds with motor windings in close proximity has a voltage-generating effect, more across the start winding because it is wound with a longer and thinner wire than the run winding. This generated voltage is often referred to as induced voltage, or back electromotive force (BEMF). This BEMF, measured with a voltmeter across the start winding, opposes line voltage. It is higher than line voltage and can range over 400-V AC in

Figure 17.38 (A) A potential relay. Courtesy Ferris State University. Photo by John Tomczyk (B) Internals of a potential relay showing the coil and contacts. Courtesy Ferris State University. Photo by John Tomczyk

(A)

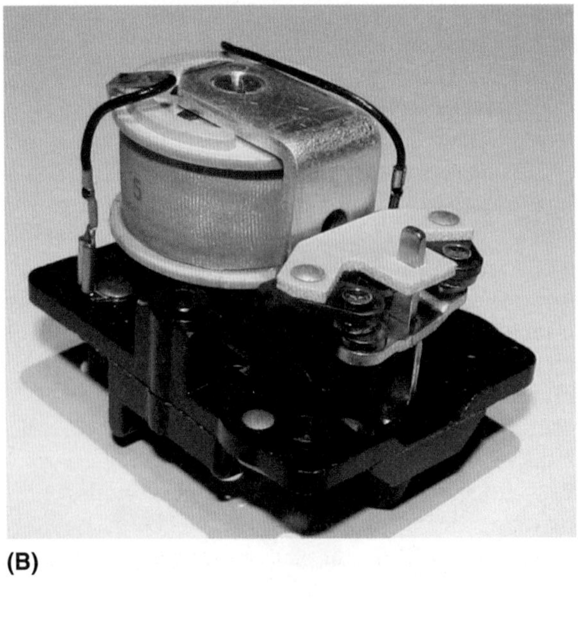

(B)

Figure 17.38 *(Continued)*(C) Diagram illustrating the higher induced voltage that is measured across the start winding in a typical motor.

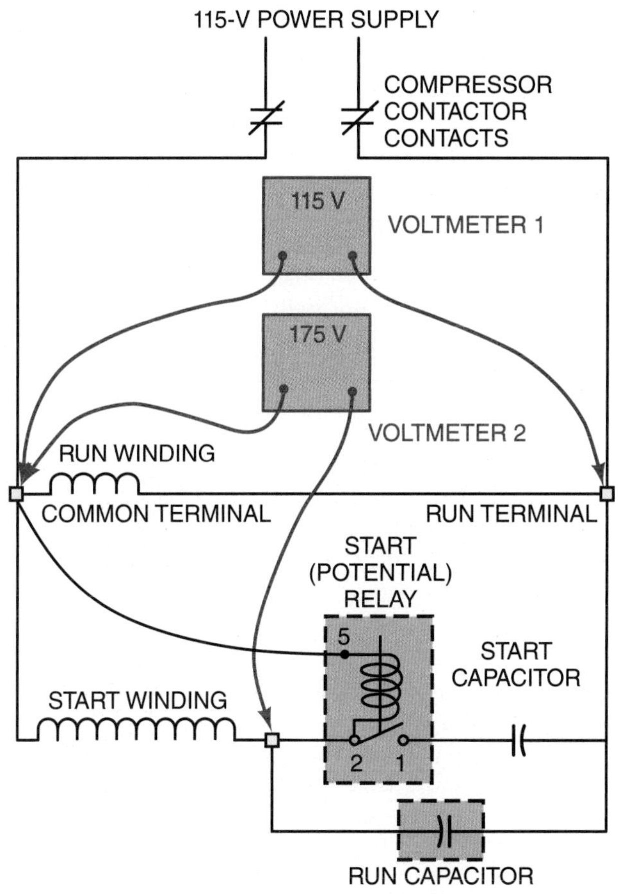

a 230-V circuit, **Figure 17.38(C)**. All motors have different magnitudes of BEMF.

The BEMF generated across the start winding causes a small current to flow in the start winding and potential relay coil because they are in parallel with one another. When the BEMF has built up to a high enough value—referred to as "pick-up voltage"—the contacts between terminals 1 and 2 will be picked up or "opened." This will take the start capacitor out of the start winding circuit. The start winding is still in the BEMF circuit keeping the relay coil energized while the motor is running at full speed. Pickup voltage usually occurs at about 3/4 motor speed. The run capacitor causes a small phase shift between the run and start winding for better running torque. It also limits the current through the start winding so overheating does not occur during the running cycle.

When the compressor contactor contacts open, line voltage is taken away from the motor. The motor's rotor now decreases in speed; thus the BEMF generated across the start winding decreases in magnitude. The relay coil now sees a decreased BEMF and can no longer generate enough magnetism in its iron core to keep the contacts between terminals 1 and 2 open. The contacts will return to their originally closed position by spring power as the motor coasts to a stop. The voltage at which the contacts return to their original closed position is called the "drop-out voltage." One benefit of using the PMR as a starting relay is that the start winding can be brought back into the active electric circuit if the motor should encounter a temporary load increase. If the motor slows down and the normally closed contacts between terminals 1 and 2 close, this will bring the start capacitor back into the circuit and provide additional torque to get the motor back up to speed. Once the induced voltage rises (indicating that the motor is operating at a satisfactory speed), the contacts will open to remove the start capacitor from the circuit.

Because these motors generate different BEMF values, potential relays must be sized to each individual compressor. The pick-up voltage needed to open the contacts is one of the considerations for selecting the relay. The actual current-carrying capacity of the relay contacts, the drop-out voltage, and the continuous coil voltage are other important considerations when selecting the relay. Consult a service manual, supply house, or the compressor manufacturer to select the correct potential relay for a specific compressor.

POTENTIAL RELAY SPECIFICATIONS

The pick-up voltage for a specific potential relay will be listed as a minimum and a maximum, **Figure 17.38(A)**. For proper operation, the pick-up voltage must stay within this range. (The pickup voltage is the BEMF voltage generated across the start winding by the motor's rotor when it is up to about 3/4 speed.) If the pick-up voltage generated by BEMF is under the minimum, the contacts between terminals 1 and 2 will never open. The start capacitor will then stay in the circuit, which will cause high amp draws and open the motor's protective devices. If the pick-up voltage generated by BEMF is above the maximum, the relay coil stands a good chance of overheating and opening the circuit. Again, the contacts between terminals 1 and 2 would stay closed, causing high amp draws if the relay coil opens the circuit. Potential relays also have a continuous coil voltage rating. This is the maximum EMF that the relay's coil can tolerate "continuously" without overheating and opening the circuit.

As mentioned, potential relays also have a drop-out voltage rating. The drop-out voltage is the BEMF voltage that must be generated across the relay coil to "hold" the contacts open once they have been picked up (opened). Note that it takes more BEMF (pick-up voltage) to pick up and open the contacts than it does to hold them open (drop-out voltage). Once the cycling control opens the circuit, the rotor will decrease in speed, thus generating less BEMF across the start winding and relay coil. As the BEMF drops below the drop-out voltage, the contacts between terminals 1 and 2 will return to their normally closed position and be ready for the next starting cycle.

Because of these three voltage rating specifications, potential relays must be sized to each individual compressor. Consult a service manual, the compressor manufacturer, or a supply house for information on selecting the correct potential relay. Replacement relays from different manufacturers can be cross-referenced in convenient tables. Whenever possible, the model number on the old relay should be used when ordering a new one. Adjustable and universal potential relays are also available that can serve as replacements for many potential relays presently being manufactured.

17.19 TROUBLESHOOTING

Please refer to **Figure 17.38(C)** for troubleshooting the potential relay. A simple ohmmeter is all that is needed to troubleshoot. After taking all connecting wires off the relay, measure the resistance across the 1 and 2 terminals. The resistance should read close to zero, since they are normally closed contacts. If the ohmmeter reads infinity, the contacts are stuck open and the relay should be discarded and replaced. Open contacts will prevent the start capacitor and start winding circuit from being energized, which will lock the rotor and cause locked-rotor amps (LRA), thus opening the compressor's protection device. In time, short cycling on the compressor's compression device can overheat and open a winding. Often the relay contacts could be stuck or arced in the closed position. In this case, the start capacitor would never be taken out of the circuit, and high amp draws would open motor protective devices. If the contacts are stuck in the closed position, the relay will have to be checked in the running mode, since the contacts between 1 and 2 are normally closed when not in operation.

Once the motor is up and running, use a multimeter to measure the voltage across terminals 1 and 2 of the relay. A voltage reading of zero volts would prove that the contacts are not opening. Also, high amp draws from the start capacitor still being in series with the start winding circuit is a telltale sign that the contacts have not opened.

After disconnecting all wires from the relay, test the resistance of the coil located between terminals 2 and 5. It is not uncommon to have the resistance read in the many thousands of ohms. If the ohmmeter reads infinity, the relay coil is open. The relay should be discarded and a new one installed. An open relay coil will prevent the contacts between 1 and 2 from opening due to no magnetism in the iron core. High amp draws will again be experienced. When replacing a potential relay, always use the model number on the old relay when ordering a new one, or cross-reference it to a different manufacturer's relay. This will ensure that all relay specifications are met.

17.20 THE CURRENT RELAY

Current relays are used on single-phase, fractional horsepower motors requiring low starting torque. Their main function is to assist in starting the motor. Current relays, also referred to as current magnetic relays, or CMRs, are often seen when capillary tubes or fixed orifices are used as metering devices, because systems employing capillary tubes or fixed-orifice metering devices equalize pressure during their off cycles. This causes a lower starting torque condition than in systems that do not equalize pressure during their off cycles, as with the conventional thermostatic-expansion or automatic-expansion metering devices. Some typical applications for current relays include domestic refrigeration compressors, drinking fountains, small window air conditioners, small ice makers, and small unitary supermarket display cases.

Current starting relays consist of a low-resistance coil (1 Ω or less) and a set of normally open contacts. The coil is short in length and large in diameter. In fact, the current relay may always be identified by the size of the wire in the relay coil. This wire is large because it must carry the full-load current of the motor. The coil is wired between terminals L and M, and the contacts usually are wired between terminals L and S, **Figure 17.39**. The standard designation is L for line, S for start winding, and M for main winding.

When power is applied, both the run (main) winding and relay coil are at LRA. This happens because the relay coil is in series with the run winding. For this reason, the relay coil is designed to be of very small resistance; otherwise, it would drop too much voltage and interfere with the run winding's power consumption needs. The start winding will not experience the LRA because the contacts between L and S are normally open.

Figure 17.39 Wiring diagram of a current magnetic relay. The "L" indicates line voltage, the "M" refers to the main, or run, winding, and the "S" refers to the start winding. The coil is connected between the L and M terminals; the relay contacts are connected between the L and S terminals.

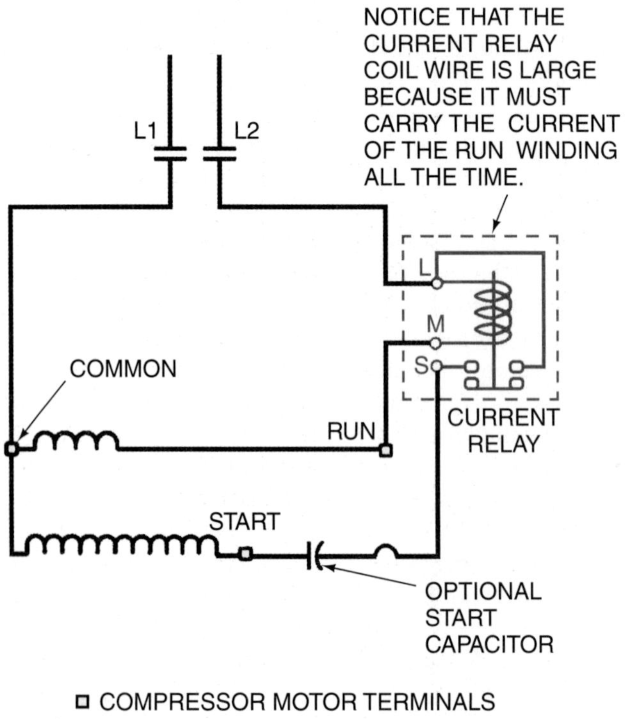

Figure 17.40 The current magnetic relay is easily identified by the large wire used to wind the holding coil. Photo by Jason Obrzut

Once the relay coil experiences LRA, a strong electromagnetic field is generated around it. This will make an electromagnet out of the iron core that the coil surrounds. This magnetic force will close the contacts between L and S and energize the start winding. The motor's rotor then starts to turn, **Figure 17.40**.

Once the start winding circuit is closed, the motor will quickly accelerate in speed. After the motor starts, the current draw of the run winding will decrease because of a BEMF generated in its winding. It is this reduced current draw that will decrease the magnetism in the iron core of the relay. Now spring pressure or gravity forces the contacts between L and S back to their normally open position because the magnetic field generated by the lower FLA/RLA is not strong enough to keep the relay contacts closed. This will take the start winding out of the electrical circuit and allow the compressor to operate with only the run winding energized. When power is taken away from the circuit, the motor will coast to a stop and the contacts will already be open and waiting for the next run cycle.

These starting methods are used on many compressors with split-phase motors that need high starting torque. If a system has a capillary tube metering device or a fixed-bore orifice metering device, the pressures will equalize during the off cycle and a high starting torque compressor may not be necessary.

17.21 POSITIVE TEMPERATURE COEFFICIENT RESISTOR (PTCR)

Positive temperature coefficient resistors (PTCRs) are compressor starting devices. They are variable resistors that vary their resistance when the surrounding temperature changes. Because they have a positive temperature coefficient, as their temperature increases so does their resistance. Also, as their temperature decreases, so does their resistance. PTCRs have a low resistance over a wide temperature range. However, when they reach a certain higher temperature, their resistance greatly increases. This very high resistance can act as an open circuit and stop all current flow through a circuit. When the heat source on the PTCR is removed, its resistance will return to its initial room temperature resistance. PTCRs are made from a very pure semiconducting ceramic material, **Figure 17.41(B)**.

This type of variable resistor is often used in the HVACR industry in place of current and potential relays. In fact, the ice-making industry is using PTCRs in place of potential relays in commercial ice machines. PTCRs are also used in air-conditioning and heat pump systems having either an orifice metering device or TXV with a bleed port to equalize pressure during the off cycle. PTCRs make wiring much easier and get rid of a lot of nonessential, high-maintenance moving parts like relay contacts, coils, and springs.

The PTCR is wired in parallel with the run capacitor and in series with the start winding in a permanent split capacitor (PSC) motor, **Figure 17.41(A)**. PTCRs can supply the extra starting torque needed with a single-phase, PSC compressor motor, usually on systems that can somewhat equalize system pressures before starting. When the cycling control brings power to the motor, both the run and start winding experience a heavy inrush of current. The run (main) winding sees heavy current flow because its low resistance winding is wired across Line 1 and Line 2 directly. The start (auxiliary) winding also sees heavy current flow because its low resistance winding also sees Line 1 and Line 2 directly for a split instant. By increasing the current flow in the start winding, additional starting torque is provided by the compressor motor because the PTCR is at room temperature and its resistance is very low. It actually shunts out the run capacitor because it acts like a straight wire with very low resistance when it is at room temperature. Since both the run and start winding see high current that is a bit out of phase from their different inductance values, the motor will have torque to start. The high current flow through the PTCR will now increase its temperature and it will produce a very high resistance in less than 1 sec. This will act as an open circuit in the PTCR circuit. The run capacitor will no longer be shunted out and the motor will now run as a permanent split capacitor (PSC) motor for the remainder of its running cycle. The PTCR will remain hot and at a high resistance value as long as voltage remains on the circuit.

To assist in PSC motor starting, some ice machine manufacturers that use PTCRs as starting relays incorporate a hot gas solenoid (harvest) valve to assist in compressor motor start-up. Energizing a hot gas solenoid valve located between the compressor's discharge line and the entrance to the evaporator prior to starting will somewhat equalize system pressures and lead to much easier starting. Many times manufacturers will allow the hot gas solenoid valve to remain energized for up to 45 sec to ensure system equalization before

Figure 17.41 (A) A PTCR in a motor starting circuit. (B) Two positive temperature coefficient resistors (PTCRs). Courtesy Ferris State University. Photo by John Tomczyk

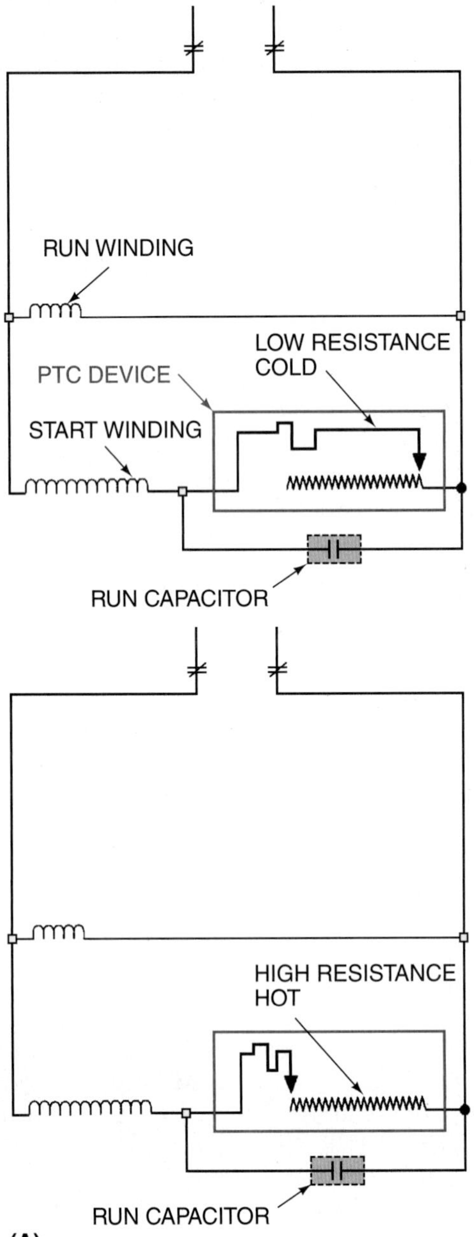

(A)

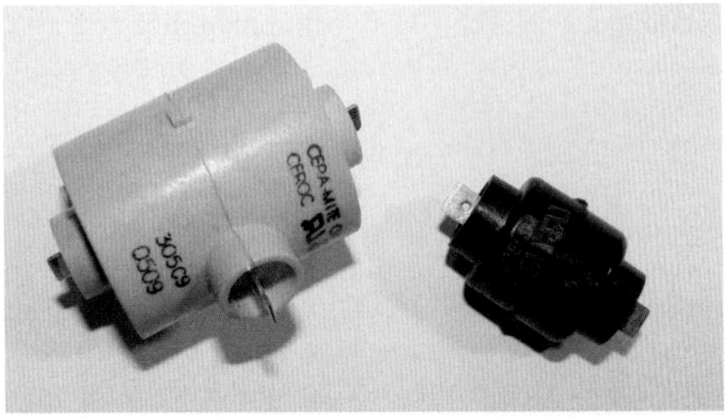

(B)

compressor starting. To ensure the compressor is started and will not stall, often the hot gas solenoid valve is energized for 5 to 10 sec after the compressor has started.

It is important that the PTCR be cooled down to near ambient temperature before attempting to restart the compressor. If the cool-down process is not completed, the high starting torque of the PSC compressor motor may not last long enough. Often, a good PTCR may be too hot to operate properly at start-up because the compressor motor has short cycled or the compressor overload has opened. A low voltage situation at the compressor on start-up may also cause the compressor to not have enough starting torque and it will stall.

17.22 TROUBLESHOOTING THE PTCR

In normal operation, a PTCR may reach over 200°F while the compressor is running. To check for a bad PTCR, visually inspect it for physical signs of damage. Next, wait for at least 10 to 15 min for the PTCR to cool down to room temperature with the unit off. Isolate the PTCR from the unit and use an ohmmeter to measure its resistance. The manufacturer of the refrigeration unit should provide resistance values for room-temperature PTCRs. If the resistance value falls outside the acceptable range specified by the manufacturer, the PTCR should be replaced.

17.23 TWO-SPEED COMPRESSOR MOTORS

Two-speed compressor motors are used by some manufacturers to control the capacity of small compressors. For example, a residence or small office building may have a 5-ton air-conditioning load at the peak of the season and a 2 1/2-ton load as a minimum. Capacity control is desirable in this application, which may be accomplished with a two-speed compressor. Two-speed operation is obtained by wiring the compressor motor to operate as a two-pole motor or a four-pole motor. The automatic changeover is accomplished through the space temperature thermostat and the proper compressor contactor for the proper speed. For all practical purposes, these motors can be considered as two motors in one compressor housing. One motor turns at 1800 rpm, the other at 3600 rpm. The compressor uses either motor, based on capacity needs. The compressor has more than three motor terminals to operate its two motors.

17.24 SPECIAL APPLICATION MOTORS

Although some special application single-phase motors may have more than three motor terminals, they may not be two-speed motors. Some manufacturers design an auxiliary winding in the compressor to give the motor more efficiency. These motors are normally in the 5-hp and smaller range. Other

Figure 17.42 This compressor has extra winding thermostat terminals. In the event that one winding thermostat goes bad, the others can still be used.

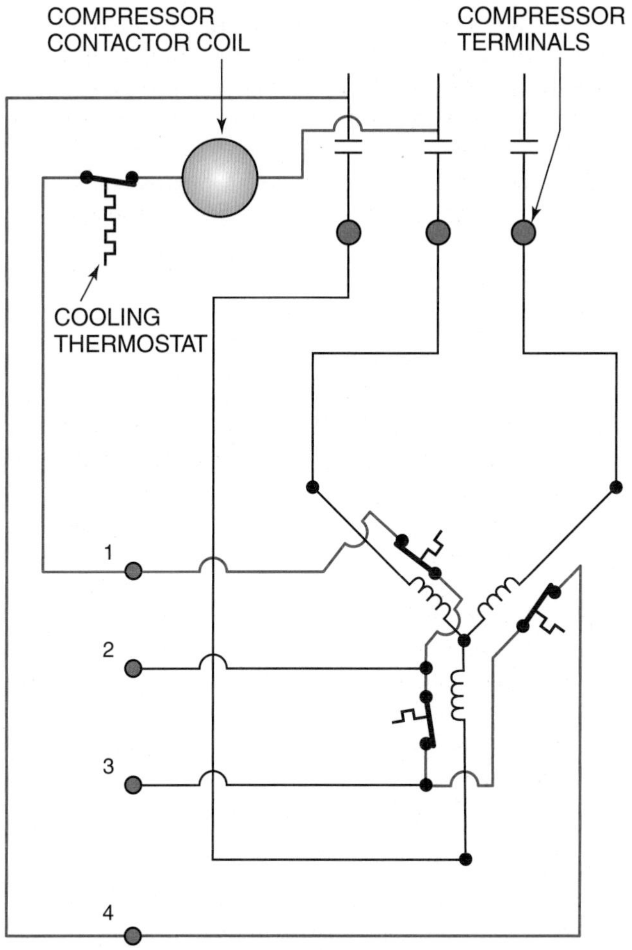

special motors may have a winding thermostat wired through the shell with extra terminals. A large three-phase compressor motor may have a winding thermostat for each winding. This calls for three thermostats within the compressor housing. Using a combination of four small terminals, they can be wired in series so that any one of them will shut the compressor off. If one of them fails, the extra terminals would still allow the others to be used for some motor protection, **Figure 17.42**.

17.25 THREE-PHASE COMPRESSOR MOTORS

Large commercial and industrial installations will have three-phase power for air-conditioning and refrigeration equipment. Three-phase compressor motors normally have three motor terminals, but the resistance across each winding is the same, **Figure 17.43**. As explained earlier, three-phase motors have a high starting torque and, consequently, should experience no starting problems. As mentioned previously, three-phase motors do not require any start assist components.

For many years, welded hermetic compressors were limited to 7 1/2 tons but are now being manufactured in sizes up to about 50 tons. The larger welded hermetic compressors are traded back to the manufacturer when they fail and are then remanufactured. They must be cut open for service, **Figure 17.44**.

Figure 17.43 A three-phase compressor motor with three leads for the three windings. Unlike a single-phase motor, the resistance of all three windings is the same.

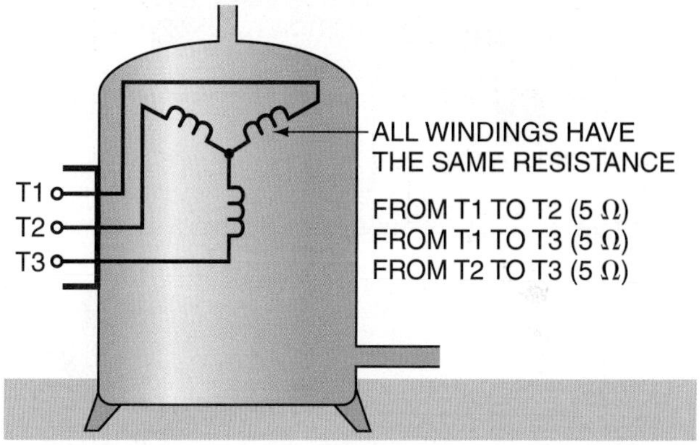

Figure 17.44 A large welded hermetic compressor. Photo by Jason Obrzut

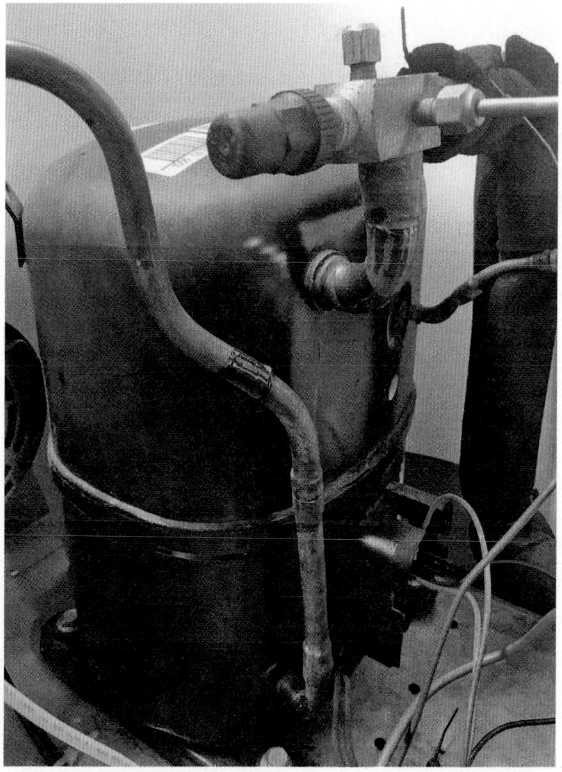

Serviceable hermetic compressors, referred to as semi-hermetic, of the reciprocating type are manufactured in sizes up to about 125 tons, **Figure 17.45**. These compressors may have dual-voltage motors for 208-V to 230-V or 460-V operation, **Figure 17.46**. When the motor fails, the compressors are normally rebuilt or remanufactured, so an overhaul is considered upon motor failure. The compressors may be rebuilt in the field or traded in for remanufactured ones. Companies that can rebuild the compressor to the proper specifications are located in most large metropolitan areas. (The difference between rebuilding and remanufacturing is that one is done by an independent rebuilder and the other by the original manufacturer or an authorized rebuilder.)

Figure 17.45 Serviceable compressor. Courtesy of Hampden Engineering

17.26 VARIABLE-SPEED MOTORS

The desire to provide more efficient fans, pumps, and compressors has led the industry to explore the development and use of variable-speed motors. Most motors do not need to operate at full speed and load except during the peak temperature of the season

Figure 17.46 A dual-voltage compressor motor wiring diagram. This motor may be used for either 208 V/230 V or 460 V.

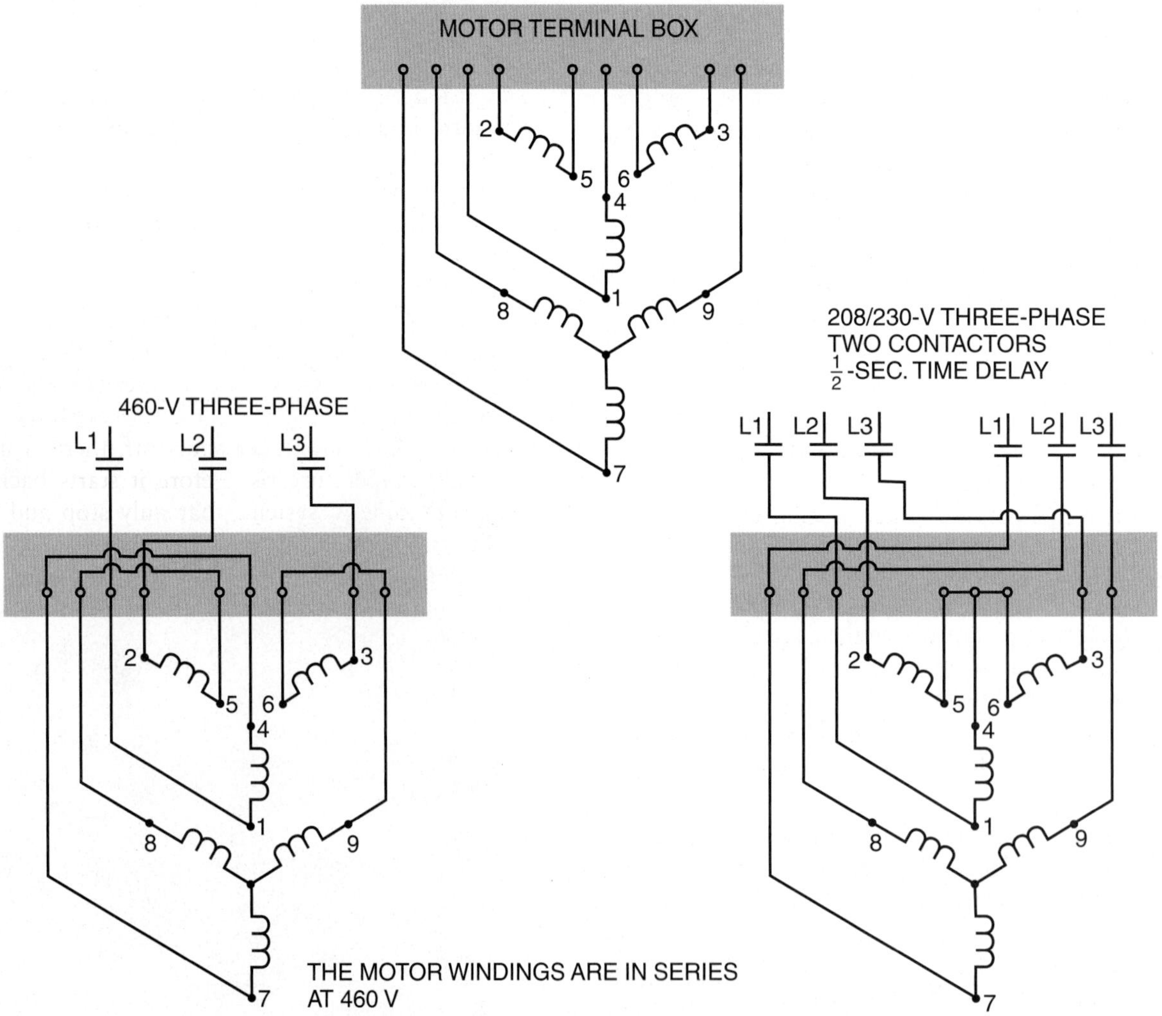

and could easily satisfy the heating or air-conditioning load at other times by operating at a slower speed. When the motor speed is reduced, the power to operate the motor is reduced proportionately. For example, if a home or building needs only 50% of the capacity of the air-conditioning unit to satisfy the required space temperature, it would be advantageous to reduce the capacity of the unit rather than stop and restart it. When power consumption can be reduced in this manner, the unit becomes more efficient. As mentioned earlier, when motors are initially started, they draw locked-rotor amperage, which can be five to seven times greater than the full-load or rated-load amperage of the motor. By limiting the number of times the motor starts, the rate of power consumption is reduced. In addition, having (for example) a compressor motor operate at a lower speed for a longer period of time will result in more even cooling of an occupied space.

The frequency (cycles per second) of the power supply and the number of poles determine the speed of a conventional motor. Some new motors can operate at different speeds by using electronic circuits. There are several methods of varying the frequency of the power supply, depending on the type of motor. Depending on the needs of the system, compressor motors and the fan motors may be controlled by any number of speed combinations.

Unit 12, "Basic Electricity and Magnetism," covered some of the fundamentals of AC electricity. All of the electricity furnished in the United States is AC and the current frequency is 60 cycles per second, or 60 hertz (Hz). It is much more efficient to distribute AC than direct current (DC), moreover, AC can be converted to DC by means of rectifiers. Many applications for motors in the industry require variable-speed motors, which traditionally have been DC. These motors are typically much more complicated to work with because they have both a field (stator) and an armature (rotor) winding and use brushes to carry power to the rotor. Brushes are carbon connectors that rub on the armature and create an arc. The brushes on the armature cannot be used inside a refrigerant atmosphere; therefore, DC motors with brushes cannot be used for hermetic compressors.

In the past, variable-speed systems using DC have used open-drive DC motors, but the hermetic compressor is the most common variable-speed application. In addition, variable-speed motors can be used to drive many types of equipment, such as pumps and fans and fractional horsepower equipment. The two types of motors found in equipment today are the squirrel cage induction motor and the ECM (electronically commutated) DC motor. Instead of brushes rubbing on an armature, the motor is electronically commutated, **Figure 17.47**. This motor can run a fan as well as fractional horsepower equipment. Electronic components now make variable-speed AC motor operation possible. The electronic components can switch power on and off in microseconds without creating an arc; therefore, open-type contacts are not necessary for switching purposes. The no-arc components have virtually no wear, have a long life cycle, and are reliable. **Figure 17.48** shows a transistor that is essentially an electronic switch with no contacts to arc.

Figure 17.47 An electronically commutated motor (ECM). Courtesy General Electric

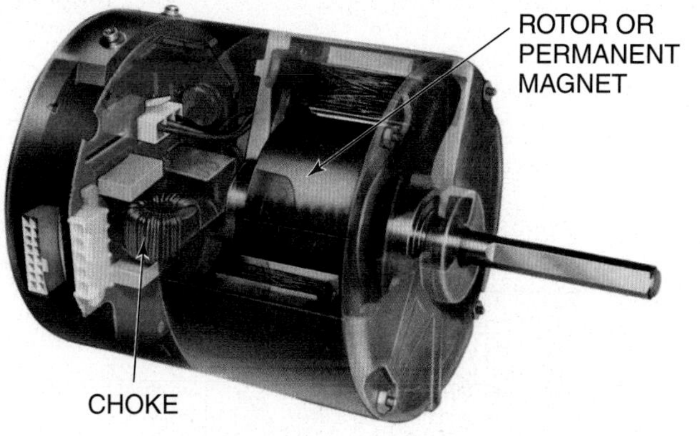

The air-conditioning load on a commercial or industrial building varies during the season and during each day. The central air-conditioning system in a house or other building would have many of the same operating characteristics. Let us use a house as an example. Starting at noon, the outside temperature may be 95°F and the system may be required to run at full load all the time to remove heat as fast as it is entering the house. As the house cools off in the evening, the unit may start to cycle off and then back on, based on the space temperature. Remember, every time the motor stops and restarts, there is wear at the contactor contacts and a burden is put on the motor in the form of starting up. Motor inrush current stresses the bearings and windings. Most motor bearing wear occurs in the first few seconds of start-up because the bearings are not lubricated until the motor is turning. It would be best not to ever turn the motor off and instead just keep it running at a reduced capacity.

When an air conditioner shuts off, there is normally a measurable temperature rise before it starts back up. This is very noticeable in systems that only stop and start. The humidity rises during this period. In the winter, the typical

Figure 17.48 This power transistor is basically an electronic switch that does not arc when it changes position and has the ability to turn on and off very quickly.

Figure 17.49 This typical ON/OFF furnace thermostat has a 3°F temperature swing during its cycle. This is enough temperature swing to be noticeable to the occupants of the space.

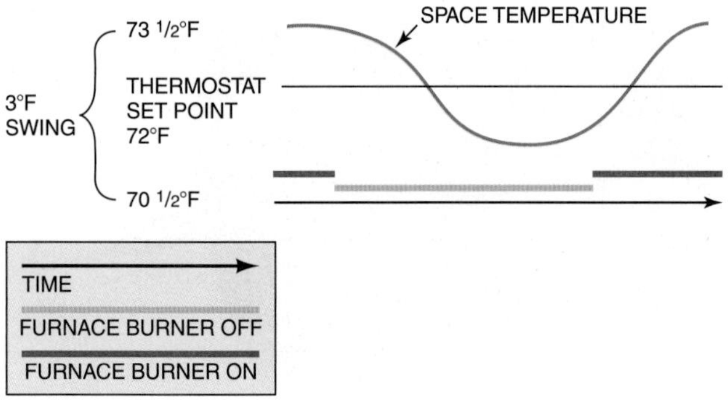

gas or oil furnace does the same thing. It starts up and runs until the thermostat is satisfied and then it shuts off. There is a measurable temperature rise before it shuts off and a measurable temperature drop before it starts back up. The actual space temperature at the thermostat may look like the graph in **Figure 17.49.** When variable-speed motor controls are used along with a variable firing rate for a furnace, the temperature graph may look more like the one in **Figure 17.50.** The same temperature curve profile would be true for the cooling season—a flatter profile with fewer temperature and humidity variations.

Variable-speed motors can smooth out air conditioner operation by running for longer periods of time. In any building you may notice when the system thermostat is satisfied and the unit shuts off. Suppose we could just keep the unit running at a reduced capacity that matched the building load. If we could gently ramp the motor speed down as the load reduces and then ramp it up as the load increases, the temperature and the humidity would be more constant in the summer. This can be accomplished with modern electronics and variable-speed motor drives.

For example, a light bulb turns on and off 120 times per second when used with a typical household 60-cycle current. **Figure 17.51** shows a sine wave of the current that is furnished to the home. The voltage goes from 0 V to 120 V

Figure 17.50 This furnace is controlled by a variable-speed fan motor and firing rate. It has only a 1°F temperature swing during its cycle. This will not be noticeable.

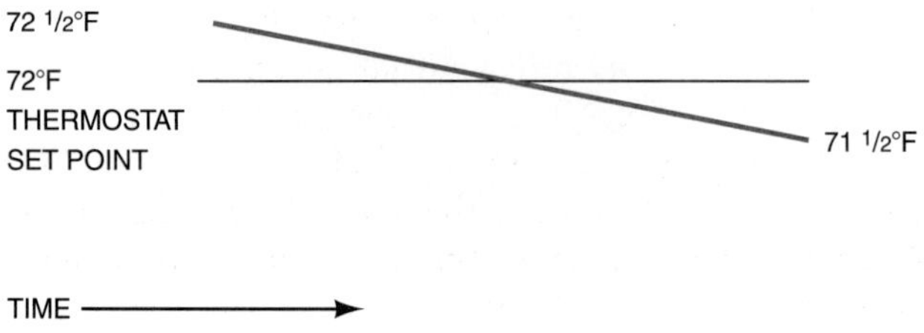

Figure 17.51 This sine wave is generated in one-sixtieth (1/60) of a second.

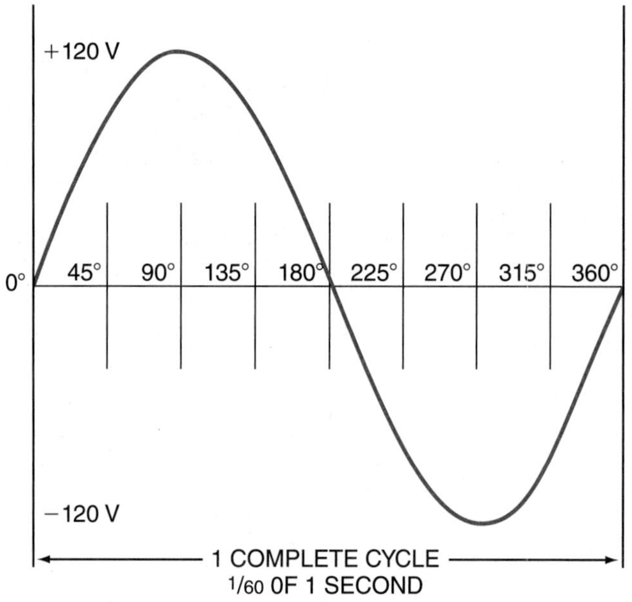

to 0 V and back to 120 V 60 times per second; thus it is a 60-cycle current. Current to the light bulb element is interrupted 120 times in that second. The light from the bulb comes from the glowing element. Because it does not cool off between the cycles, you do not notice the interruptions in the current. Electronic devices can detect any part of the sine wave and switch off the current at any place in time.

If a switch could be installed to make even more interruptions in the current to the light bulb, it would produce a lower light output. **Figure 17.52** shows a simple rectifier in the form of a diode installed in the line that will interrupt the current flow to the light bulb on half of the sine wave cycle. The light bulb will now glow at half of its light output. This diode does not draw any current or have any moving parts but interrupts half of the current flow to the light bulb. The light bulb not only uses half power, it also gives off half as much light. The diode is like a check valve in a water circuit; it allows current flow in only one direction. The term *alternating current* tells you that current flows in two directions, one direction in one part of the cycle and the other direction in the other part of the cycle. When the current tries to reverse for the other half of the cycle, the diode stops it from flowing. This saves power and gives the owner some control over the light output.

AC electric motors can be controlled in a similar manner with electronic circuits. AC motor speed is directly proportional to the cycles per second (hertz, Hz). If the cycles per second are varied, the motor speed will vary. The voltage must also be varied in proportion to the cycles per second for the motor to remain efficient at all speeds. Once the voltage is converted to DC and filtered, it then goes through

Figure 17.52 This light dimmer uses a diode to cut the voltage in half at low light.

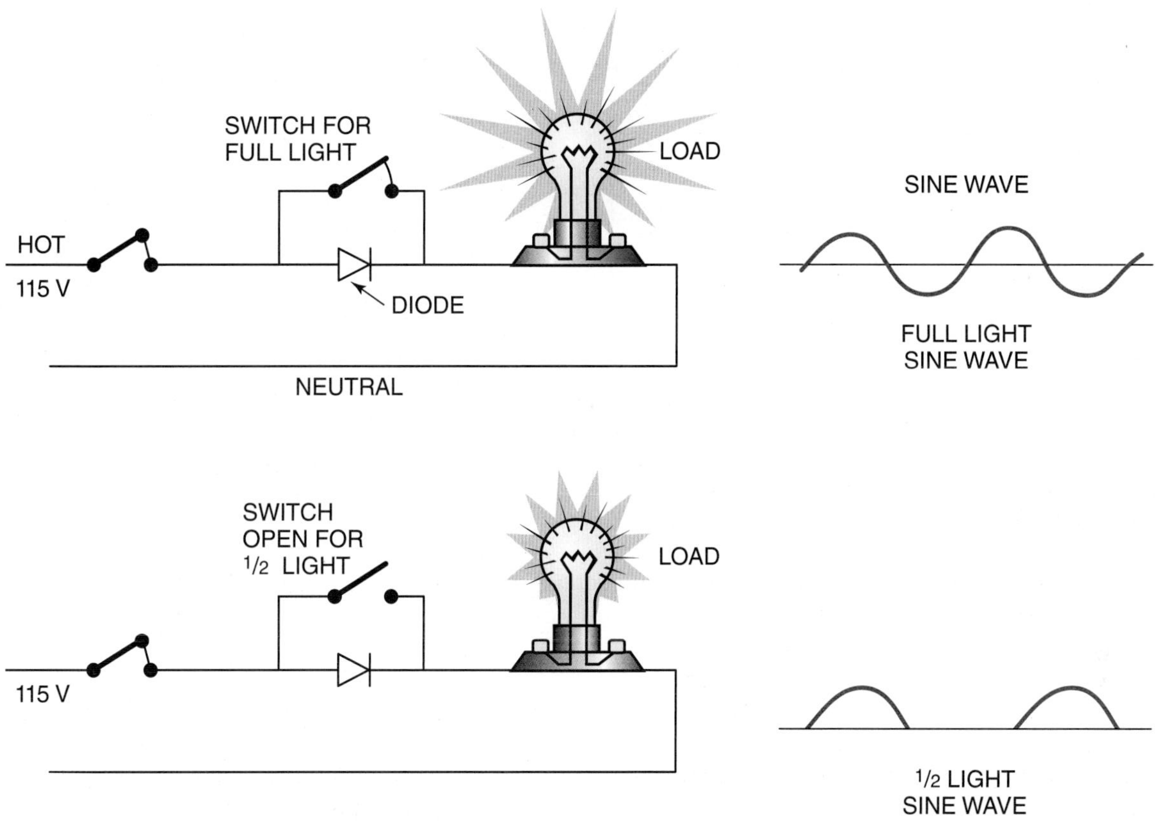

an inverter to change it back to controllable AC. (Actually, this is still pulsating DC.) The reason for all of this is to change the frequency (cycles per second) and voltage at the same time. As the frequency is reduced, the voltage must also be reduced at the same rate. If only the voltage is reduced, the motor will overheat. For example, if you have a 3600-rpm, 230-V, 60-cycle-per-second motor and you want to reduce the speed to 1800 rpm, the voltage and the cycles must be reduced in proportion. By reducing the voltage to 115 V and the frequency to 30 cycles per second, the motor will have the correct power to operate at 1800 rpm. It will not lose its torque characteristics. The motor will also operate on about half the power requirement of the full-load operation.

AC supplied from the power company is very hard to regulate at the motor, so it must be altered to make the process easier and more stable. This involves changing the incoming AC voltage to DC with a device called a converter or rectifier, which is much like a battery charger that converts AC to 14-V DC to charge an automobile battery. This DC voltage is actually pulsating DC voltage, so it is then filtered using capacitors to create a more pure DC voltage. A simplified diagram of a circuit used to accomplish the voltage conversion is shown in **Figure 17.53**. There are many components in an actual system, but the technician should think about the system one component at a time for the purposes of understanding and troubleshooting it.

17.27 DC CONVERTERS (RECTIFIERS)

There are two basic types of converters, the *phase-controlled rectifier* and the *diode bridge rectifier.* The phase-controlled rectifier receives the AC supplied by the power company and converts it to variable voltage DC using silicon-controlled rectifiers (SCRs) and transistors that can be turned off and back on in microseconds. **Figure 17.54** shows the waveform of AC furnished by the power company that enters the phase-controlled rectifier and the DC current that leaves the device. Notice the connection for turning these diodes or transistors on and off. The DC voltage leaving this rectifier is varied within the rectifier to coincide with the motor speed. The frequency of the power will be adjusted to the required motor speed in the inverter, which is between the converter and the motor. Remember, both the voltage and the frequency must be changed to achieve an efficient motor speed adjustment.

Figure 17.53 A simple diagram of a variable-speed motor drive.

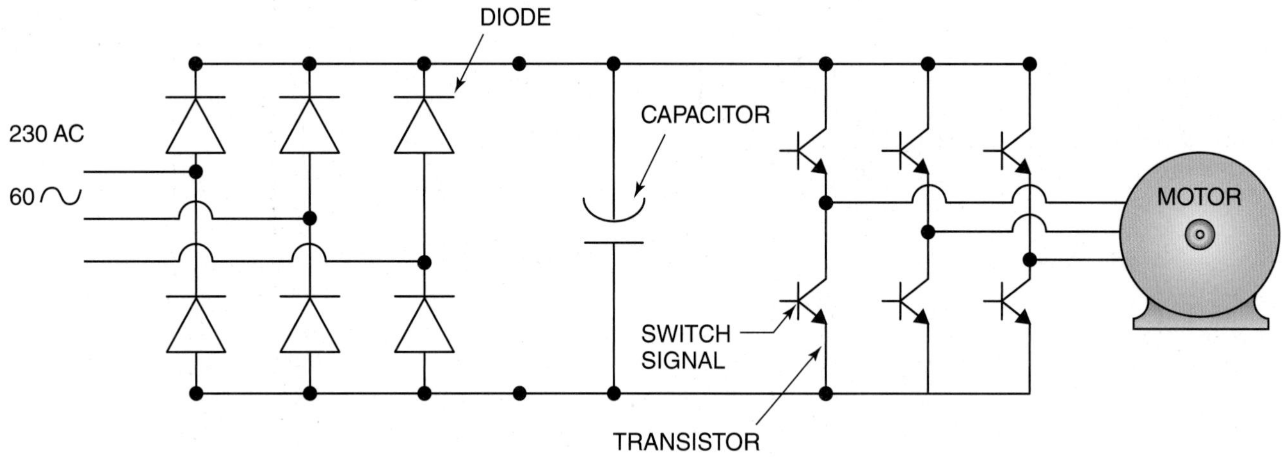

Figure 17.54 The full-wave rectifier moves the entire sine wave to the positive side of the graph.

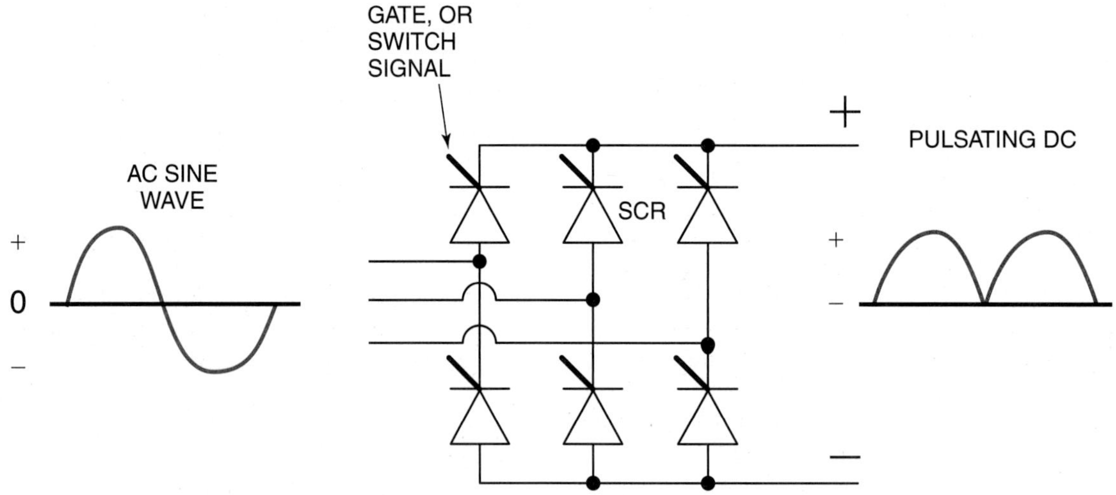

The other component in the system is a capacitor bank used to smooth out the DC voltage. The rectifier turns AC into a pulsating DC voltage that appears as though all of the AC voltage is on one side of the sine curve. When it leaves the capacitor bank, the voltage looks more like pure DC voltage, **Figure 17.55.** This type of capacitor bank is used with any rectifier to create a better DC profile.

The diode bridge rectifier is a little different in that it does not regulate the DC voltage. The diodes used in this rectifier are not controllable. The voltage and the frequency will be adjusted at the inverter of the system. Constant pure DC voltage is produced after it has been filtered through the capacitor bank. The diode bridge rectifier has no connection for switching the diodes on and off.

Figure 17.55 Pulsating DC enters the capacitor bank and straight-line DC leaves.

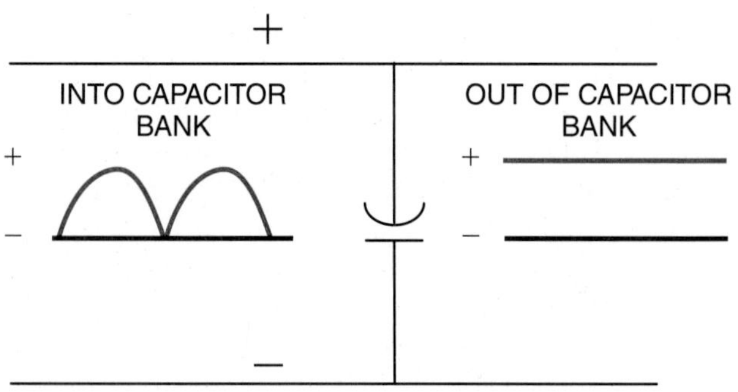

17.28 INVERTERS AND VARIABLE FREQUENCY DRIVES (VFDs)

Inverters produce the correct frequency for the motor speed desired. The speed of a conventional motor is controlled by the number of poles, and the frequency is a constant 60 Hz. In variable-speed motors, inverters can control motor speeds down to a fraction of the rated speed at 60 Hz and up to about 120% of the rated speed by adjusting the hertz to above the 60-Hz standard.

There are different types of inverters. A common one is a *six-step* inverter, and there are two variations. One controls voltage and the other controls current. The six-step inverter has six switching components, two for each phase of a three-phase motor. This inverter receives regulated voltage from the converter, such as the phase-controlled inverter, and the frequency is regulated in the inverter. The voltage-controlled six-step inverter has a large capacitor source at the output of the DC bus that maintains the output voltage, **Figure 17.56**. Notice that the controllers are transistors that can be switched on and off. The current-controlled six-step inverter also receives controlled voltage at the input. It uses a large coil, often called a choke, in the DC output bus, **Figure 17.57**. This helps stabilize the current flow in the system.

The **pulse-width modulator** (PWM) inverter receives a fixed DC voltage from the converter and then pulses the voltage to the motor. At low speeds, the pulses are short; at high speeds, the pulses are longer. The PWM pulses are sine-coded so that they are narrower at the part of the cycle close to the ends. This makes the pulsating signal look more like a sine wave to the motor. **Figure 17.58** shows the signal the motor receives. This motor speed can be controlled very closely.

Four variable frequency drives (VFDs) for motors are shown in **Figure 17.59**. The world's first mass-produced VFD was introduced in 1968. Computerization, the reduced cost of

Figure 17.56 When the control system (computer) sends a signal to the base connection on the transistor, the transistor turns on and allows current to flow through the device. When the signal is dropped, the transistor turns off.

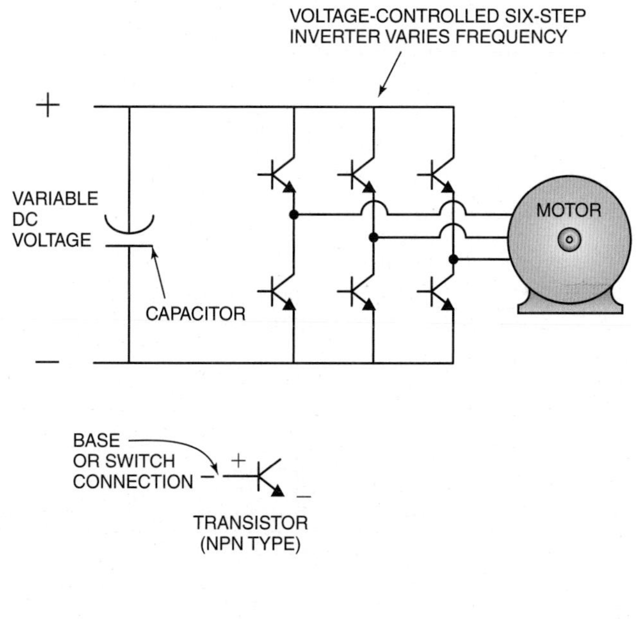

Figure 17.57 Current-controlled six-step inverter with choke coil controls frequency. The choke coil stabilizes current flow.

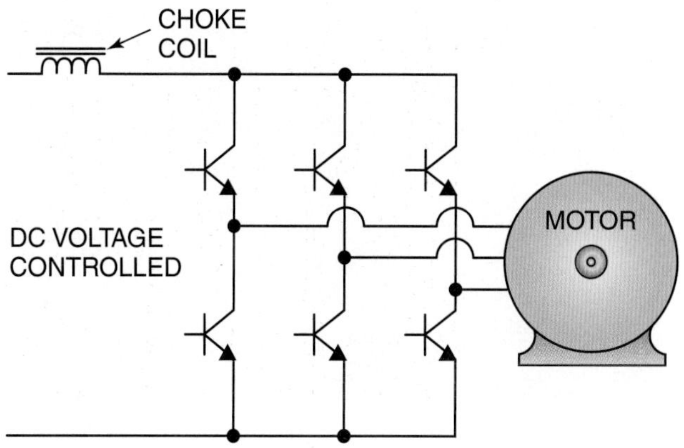

Figure 17.58 Sine-coded pulse-width modulation.

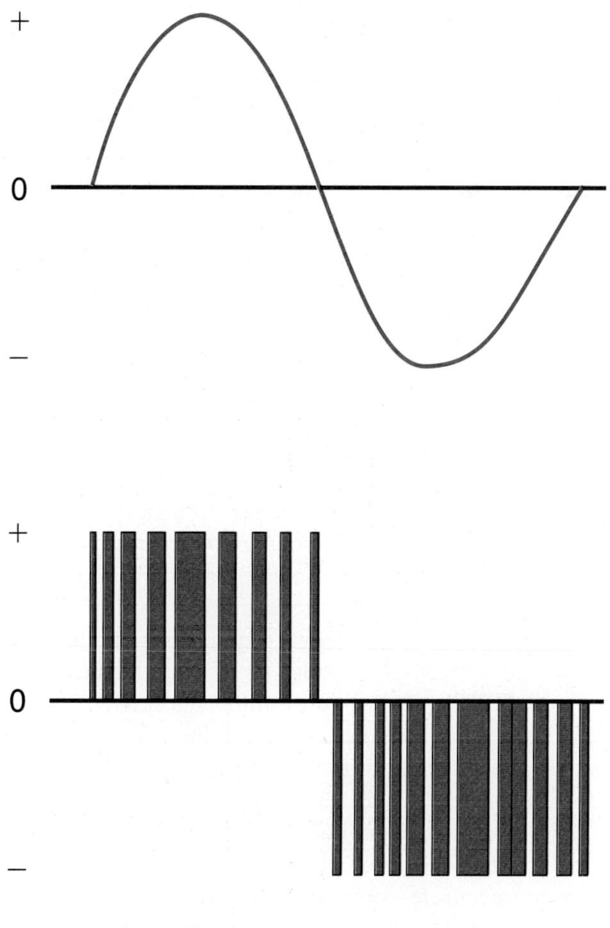

Figure 17.59 Four **variable frequency drives (VFDs) for motors. Courtesy Ferris State University. Photo by John Tomczyk

electronics, and the interfacing with direct digital control (DDC) systems, along with their energy savings, have made VFDs even more popular today. In fact, VFDs for evaporator and condenser fans and compressor motors are often standard equipment in many newer, high-efficiency rooftop air-conditioning systems that incorporate greener, more efficient, HFC-based refrigerants like R-410A. The payback period for retrofitting HVACR systems with VFDs ranges from 1 to 2 years, depending on the type, size, and application of the system. This is a very short time and yields fast cost savings, considering that most commercial HVAC systems have life cycles of 15 to 20 years. It is estimated that motors incorporating VFDs, and linked to the building's DDC system, are 65% to 75% more efficient than motors operating at a constant speed at line voltage.

It is usually only on the hottest and coldest days of the year that HVACR systems must operate at 100% capacity with full fan, compressor, and/or centrifugal pump speeds. Most of the time, HVACR system motors can operate much more energy efficiently at reduced capacities and speeds. HVACR systems with variable-speed fan motors (VFDs) have the ability to deliver variable air volume (VAV) flows, which allows the airflows to exactly match the system's heating and cooling demands and provides an opportunity to save electrical energy and money. VFDs also contribute to the overall comfort level within the building by regulating air and water flows according to the instantaneous heating or cooling load. This allows closer control of system set points for temperatures and humidity, resulting in higher human comfort levels. Noise levels are also reduced when VFDs are used on HVACR system motors. A VFD can also eliminate motor line starting shock (in-rush current) by soft starting the motor and gradually ramping it up to the speed required for the heating or cooling load at that time, instead of starting the motor at full speed and drawing locked-rotor amperage (LRA). Soft starting can reduce maintenance costs and downtimes. VFDs also eliminate short cycling of motors, which will result in longer life for the motors and driven equipment.

Major compressor manufacturers are also incorporating VFDs to control the speed of compressor motors, which provides more accurate control of the capacity of the system. HVACR systems that incorporate VFD technology on a combination of motors, including evaporator and condenser fans, compressors, and centrifugal pumps for chilled water systems, optimize cost savings and energy efficiency. Other opportunities to save costs within HVACR systems include the following motors:

- Cooling tower fans
- Cooling tower water pumps
- Make-up air fans
- Exhaust fans
- Air handler fans
- Booster fans
- Centrifugal hot water pumps
- Centrifugal cool water pumps

The following formula can be used to determine the no-load (synchronous speed) of an AC motor in revolutions per minute (rpm):

$$\text{rpm} = \frac{(\text{Hz}) \times (60\ \text{sec/min})}{(\#\ \text{of Pole Pairs})}$$

where:

- Hz = the frequency of the voltage in cycles/second
- rpm = revolutions/minute, or motor speed

Notice that units of revolutions/minute are on one side of the equation, but (Hz) is in units of cycles/seconds. Seconds are converted to minutes by multiplying by 60 seconds/minute. After this time conversion from minutes to seconds, it is apparent that rpm is governed by the frequency (Hz) and the number of poles. As one can see from the equation, the more poles the motor has, the slower it will turn. Also, as the motor's frequency decreases, the rpm, or speed, of the motor will decrease. Conversely, as frequency increases, the motor's speed will increase. Conventional motor speeds are controlled by the number of poles, and the frequency is constant at 60 Hz. It is much easier to change the frequency (Hz) of the voltage coming into the motor with electronics than it is to increase or decrease the number of poles in the motor. This is where VFDs come into play.

The equation can be rewritten in the simpler form that follows:

$$\text{rpm} = \frac{(\text{Hz})\ (60\ \text{sec/min})}{(\#\ \text{of Poles/2})}$$

By multiplying the numerator and denominator (top and bottom) of the equation by 2, the equation now becomes

$$\text{rpm} = \frac{(\text{Hz})\ (120\ \text{sec/min})}{(\#\ \text{of Poles})}$$

Rewriting the equation again without the units, it becomes

$$\text{rpm} = \frac{(\text{Hz})120}{(\#\ \text{of Poles})}$$

This is the form of the equation used in most books. However, do not get the 120 confused with voltage. It does not represent 120 volts!

OPERATION OF THE VFD

A VFD has three separate electronic sections. Their functions and accompanying electronics components are:

- Rectification (diodes) or converter section
- Filtering (capacitors and inductors) or DC bus section
- Switching (transistors) or inverter section

RECTIFICATION. **Figure 17.60** shows a circuit diagram of a three-phase VFD. Notice that all three lines of the three-phase power go through diodes in the form of a bridge rectifier. Diodes let current pass in only one direction. The bridge circuit of diodes rectifies (changes) the three-phase AC voltage to pulsating DC current. The diodes reconstruct the negative half of the waveform into the positive half. So the DC bus section sees a fixed DC voltage. The output of this section is actually 12 half-wave pulses, which are electronically 60 degrees apart.

FILTERING. The filtering section of the VFD makes the pulsating DC smoother or filters out its imperfections. This section can have more than two capacitors wired in parallel with one another, but in series with the bridge rectifier and an inductor. The capacitance of capacitors adds when they are in parallel. The charging and discharging of the capacitors is synchronous with the three-phase input voltage. This makes a pure DC signal from the half-wave signal of the bridge rectifier. The capacitors in parallel filter the voltage wave and the inductor filter the current wave. Both the inductor and capacitors work together to filter out any AC component of the DC waveform. The smoother the DC waveform, the cleaner the output waveform from the drive.

SWITCHING. The switching or transistor section of the VFD produces an AC voltage at just the right frequency for motor speed control. This section, also called the **inverter** section, converts the DC back to AC. Inverters consist of an array of transistors that can be switched on and off, with two transistors for each output phase that act as switches for current flow. When the control system sends a signal to the base connection of the transistor, the transistor turns on and allows current to flow through it. When the signal is dropped, the transistor turns off and no current will flow. The base or controlling part of the transistor is controlled by a preprogrammed microprocessor that fires the transistors in six steps at appropriate times. Each set of transistors is connected to a positive and negative side of the filtered DC line, **Figure 17.60**.

Almost all modern inverters use transistors to switch the DC bus on and off at specific intervals, a process also

Figure 17.60 Circuit diagram of a three-phase variable-frequency drive.

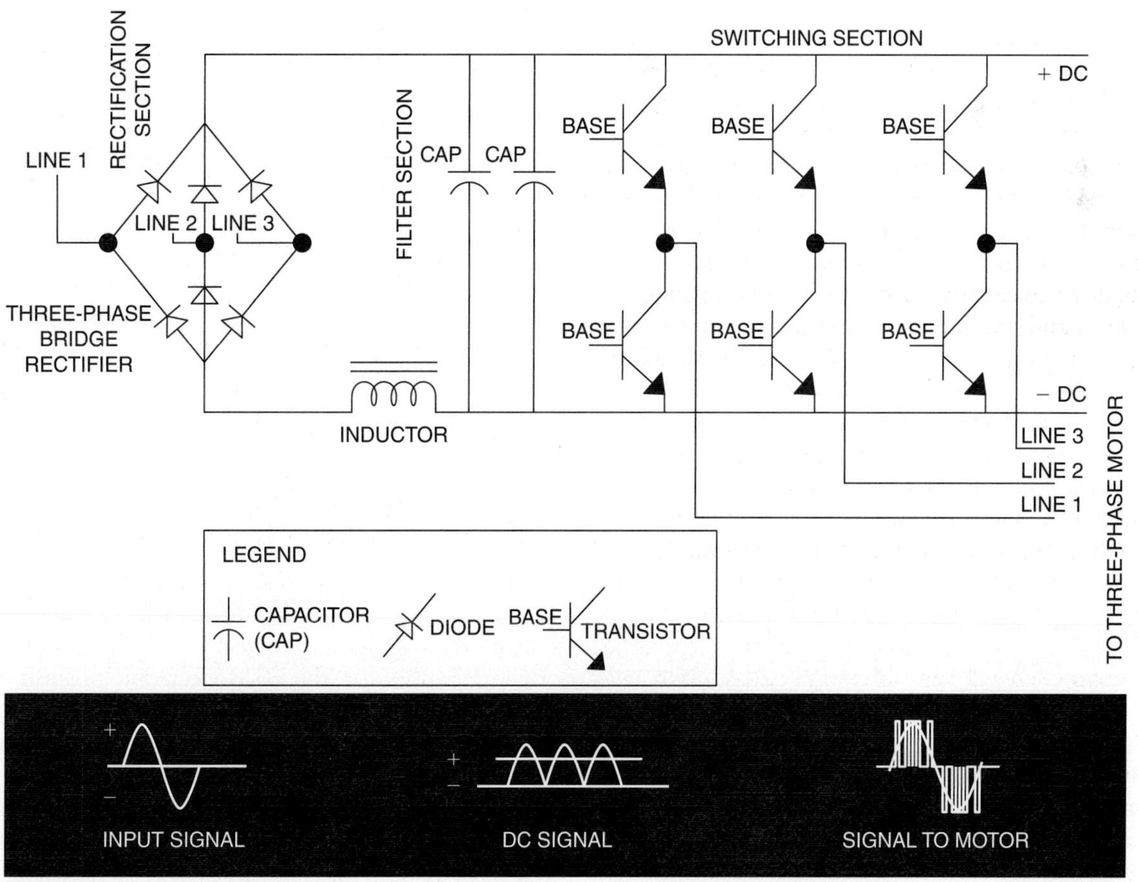

Figure 17.61 The microprocessor or controller signals the power device to "turn on" the waveform's positive half or negative **half wave pulses.

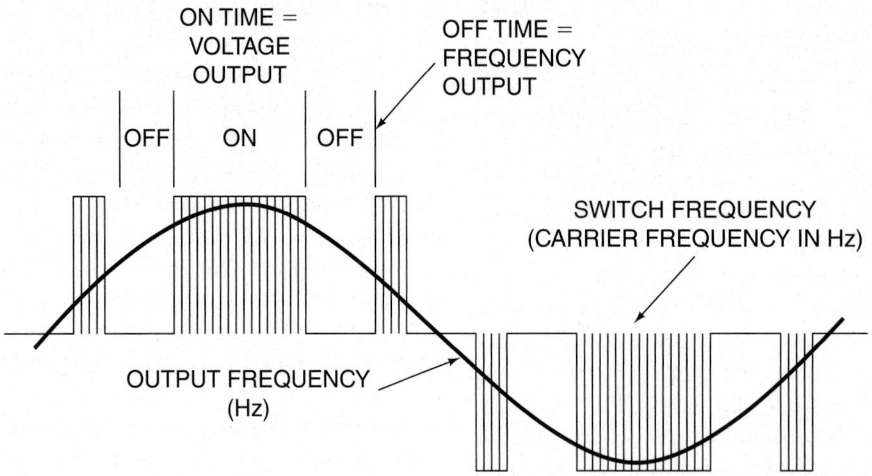

known as pulse-width modulation (PWM). Thus, inverters produce the correct frequency of voltage and current to the motor for the desired speed. This is the origin of the term *variable frequency drive* (VFD). In other words, the inverter creates a variable AC voltage and frequency output. Inverters can actually control motor speed down to a fraction of their rated speed and up to about 120% of their 60-hertz rated speed. Always consult with the specific motor manufacturer before using a VFD controller on any motor.

MOTOR SPEED. The motor's speed is controlled by supplying the stator (stationary) coils with small voltage pulses. At low speeds, the voltage pulses are short; at high speed, the pulses are longer. The PWM pulses are sine-coded, meaning that they are narrower at the part of the cycle close to the ends. The output wave is not an exact replica of the AC input sine waveform, but the pulsating signal looks like a sine wave to the motor, **Figure 17.58**. The VFD's microprocessor or controller signals the power device to "turn on" the waveform's positive half or negative half. The longer the power device remains on, the higher the output voltage will be. The less time the power device is on, the lower the output, **Figure 17.61**. The speed at which power devices switch on and off is referred to as the switch frequency or carrier frequency. As the switch frequency increases, the resolution and smoothness of the output waveform increases. However, as the switch frequency increases, the heat in the power device also increases.

17.29 ELECTRONICALLY COMMUTATED MOTORS (ECMs)

The ECM is used for applications involving open-drive fans that are 1 hp and smaller, **Figure 17.47**. The ECM is reliable and produces energy savings that are worth the difference in cost over time. The technology electronically commutates a DC motor, eliminating the need for brushes, as mentioned earlier. The DC motor has always been more efficient than the AC motor, but it required DC power to magnetize the armature with the brushes. Now, the armature can be commutated (magnetized) with permanent magnets that are attached to the armature with reliable high-tech adhesives.

The ECM motor is calibrated at the factory to suit the piece of equipment. Speed, torque, airflow, and external static pressure relationships are programmed into the motor for a particular application, which helps the air handler supply the correct airflow under different field conditions. For example, restricted system air filters or closed supply registers reduce airflow. With the ECM, the motor simply speeds up to accommodate the reduced airflow and stabilizes at the correct level. This is made possible by the relationships of airflow to torque, speed, and external static pressure already set up by the manufacturer. The ECM can also be applied to any system that can furnish the correct 24-V signal to the motor for load demand. The technology has been incorporated in variable-speed motors for gas furnaces, air-conditioning systems, oil burner systems, and heat pumps. It is also used in the refrigeration industry for variable loads in refrigerated cases, **Figure 17.62**.

The ECM is actually a two-piece motor: the controls and the motor section. If the technician suspects that the motor is defective, the controls can be removed and the motor may be checked using an ohmmeter in the conventional manner. If the motor turns out to be electrically sound, a test module (furnished by the manufacturer) can be fastened to the motor and operated manually at variable speeds, **Figure 17.63**. When the motor will operate correctly with the test module and not operate without it, the ECM electronic portion of the motor can be checked for correct input and tight connections. If the motor cannot be made to operate with the correct signal to the ECM control portion, the ECM controls can be changed and the same motor used. **NOTE:** *Each motor has its own ECM control package, which is not interchangeable with any other ECM control package from the same manufacturer. Remember,*

Figure 17.62 This very small motor may be used for a refrigerated case. Courtesy General Electric

Figure 17.63 This analyzer is connected to the motor and control board. The technician can diagnose whether the problem lies within the motor or the board itself. Courtesy Ferris State University. Photo by John Tomczyk

the control package is programmed for the specific piece of equipment. If it is to be replaced, an exact replacement must be used. •

All variable-speed motors must have very close fault control. They all tend to give off a lot of electronic noise, peaks, and valleys to the input supply AC voltage because they are constantly stopping and starting the input current flow to the converter. If a large current flow is stopped in the middle of the entering sine wave cycle, the power furnished by the power company will have a tendency to fluctuate, which will create a spike in voltage. The same thing will happen if a large current load is started, like the inrush current on start-up of a motor. The fluctuations must be filtered to prevent causing problems with computers and other electronic equipment in the vicinity. The manufacturer of the equipment has taken special precautions to avoid this and all of the manufacturer's directions should be followed.

One of the advantages of variable-speed motors is that they can be started at a reduced speed, which will reduce inrush current. As discussed earlier, the inrush current is about five times the running current. When a large motor is started in an office building, the inrush current can cause the voltage in the entire building to drop to a lower value, causing lights to blink and computers to cycle off if the drop is extreme. There are methods of softening the start-up of large motors. Variable-speed motors have what is called a soft start-up, at a reduced speed. Then the motor speed is ramped up to full load and full current.

The compressor motor is the largest motor in the system. When the load is reduced, the indoor fan and the condenser fan speeds must be reduced in proportion to the load on them. By reducing the indoor fan speed, the humidity in the conditioned space can be controlled. If the indoor fan speed is not reduced when the compressor speed is reduced, the suction pressure will rise and proper humidity removal will not be accomplished. By reducing the indoor fan speed, the suction pressure will be reduced and humidity removal will continue. The same is true of the condenser fan motor. If the compressor load is reduced, the load on the condenser is reduced. If the condenser fan is still operating at full speed, the head pressure will become too low and operating problems may occur.

As electronic components become more cost-effective, they will be used in smaller equipment, which will create very efficient systems. Each manufacturer has established a troubleshooting procedure, which should be followed. Always check the incoming power supply to any device to make sure it is what it should be before you decide the electronics are the problem. It is possible to have one fuse blown on a three-phase circuit. A quick check with a voltmeter will verify whether the power supply is good.

Some benefits of variable-speed technology are

- Power savings
- Load reduction based on demand
- Soft starting of the motor (no LRA)
- Better space temperature and humidity control
- Solid-state motor starters without open contacts
- Oversized units for future expansion that can run at part load until the expansion
- Load and capacity matching.

17.30 COOLING ELECTRIC MOTORS

All motors must be cooled because part of the electrical energy input to the motor is given off as heat. Most open motors are air-cooled; hermetic motors may be cooled by air or refrigerant gas, **Figure 17.64**. Small- and medium-sized motors are water-cooled. An air-cooled motor has fins on the surface to help give the surface more area for dissipating heat. These motors must be located in a moving airstream. To cool properly, refrigerant gas-cooled motors must have an adequate refrigerant charge.

(A)

(B)

Figure 17.64 All motors must be cooled or they will overheat. (A) This compressor is cooled by the refrigerant gas that passes over the motor windings. (B) This compressor is cooled by air from the fan. The air-cooled motor has fins on the compressor to help dissipate heat. Courtesy Copeland Corporation

REVIEW QUESTIONS

1. Two common operating voltages for residential equipment are ______________.
   A. 115 V and 24 V
   B. 115 V and 230 V
   C. 24 V and 230 V
   D. 230 V and 277 V
2. When an open motor gets up to speed, the __________ switch takes the start winding out of the circuit.
   A. centrifugal
   B. inverter
   C. rectifier
   D. transistor
3. In a single-phase motor, the motor windings that have the highest resistance are the _________ windings.
   A. rotor
   B. shaft
   C. run
   D. start
4. Which of the following devices may be wired into the starting circuit of a motor to improve the starting torque?
   A. Another winding
   B. Start capacitor
   C. Thermostat
   D. Centrifugal switch
5. Which of the following devices may be wired into the running circuit of a motor to improve the running efficiency?
   A. Start capacitor
   B. Run capacitor
   C. Both a start and run capacitor
   D. Current relay
6. The voltage generated by an electric motor's rotor turning within a motor is known as __________.
   A. drop-out voltage
   B. pick-up voltage
   C. back electromotive force (BEMF)
   D. continuous coil voltage
7. A(n) ____________ motor is manufactured from special materials that are compatible with refrigerants and oils used in a refrigeration system.
   A. hermetic compressor
   B. open split-phase
   C. single-phase open
   D. centrifugal pump
8. Two types of motors that are commonly used for hermetic compressor motors are ____________.
   A. open split-phase and hermetic split-phase
   B. single-phase open and shaded pole
   C. split-phase and permanent split capacitor (PSC)
   D. shaded pole and split-phase

9. Power enters a hermetic compressor shell and passes to the motor inside using ____________.
A. insulated motor terminals
B. separate motor windings
C. DC converters
D. start and/or run capacitors
10. Two types of relays that can be used when starting hermetic compressors are ______________ relays.
A. induction and current
B. centrifugal and potential
C. induction and centrifugal
D. current and potential
11. A PTCR device is used ________________________.
A. as a compressor starting device
B. as a motor protection device
C. as a pressure protection device
D. to increase motor running efficiency
12. Three-phase motors have ________________.
A. low starting torque
B. start capacitors wired into the starting circuit
C. high starting torque
D. run capacitors wired into the running circuit
13. Large motors with a horsepower greater than 5 are typically __________ motors.
A. single-phase
B. three-phase
C. split-phase
D. low-voltage
14. Two-speed compressor motors are used by some manufacturers to _________.
A. run at varying voltages
B. control capacity
C. run on single-phase and three-phase power supplies
D. increase starting torque
15. The rotating portion of a motor that is made of iron or copper bars bound on both ends with aluminum is called the _____________.
A. stator
B. end bell
C. rotor
D. winding
16. A motor that has two sets of windings, start and run, and has a start capacitor wired in series with its start windings is a _______________ motor.
A. shaded-pole
B. three-phase
C. permanent split-capacitor
D. capacitor-start
17. A motor that has both the start and run windings, along with a run capacitor, energized whenever the motor is energized is a ______________ motor.
A. permanent split-capacitor
B. capacitor-start
C. shaded-pole
D. variable-speed
18. The difference between a motor's calculated speed and its actual speed is known as ___________.
A. starting torque
B. motor slip
C. full-load amperes
D. induction rate
19. Capacitors are used after the converter to ____________________.
A. smooth out the DC current
B. suppress the noise
C. change the voltage frequency
D. reduce the voltage
20. The converter ________________________.
A. changes the frequency of the power
B. changes AC to DC
C. is a switch
D. is used only with DC motors
21. The inverter ____________________.
A. varies the frequency
B. adjusts the DC voltage
C. smooths out the AC power
D. contains a capacitor bank
22. A hermetic motor that is cooled by suction gas must have a(n) _____________ to be cooled properly.
A. liquid receiver
B. adequate refrigerant charge
C. refrigerant overcharge
D. accumulator
23. An electronically commutated motor (ECM) is made up of two components, the ___________________.
A. motor and the shaft
B. stator and the rotor
C. motor mount and the motor
D. controls and the motor
24. On most ____________ motors, the rotation can be reversed by disassembling the motor and turning the stator over.
A. permanent split-capacitor (PSC)
B. three-phase
C. shaded-pole
D. capacitor-start
25. A(n) ____________ is used commonly in open split-phase motors to disconnect the start winding from the electrical circuit when the motor reaches approximately 75% of its rated speed.
A. squirrel cage rotor
B. winding thermostat
C. end bell
D. centrifugal switch

UNIT 18

APPLICATION OF MOTORS

OBJECTIVES

After studying this unit, you should be able to

- explain service factor amperage (SFA).
- explain full-load amperage (FLA) and rated-load amperage (RLA).
- explain compressor amperage performance data.
- identify the proper power supply for a motor.
- describe the application of three-phase versus single-phase motors.
- describe how motors are selected for a specific set of working conditions.
- explain how the noise level in a motor can be isolated from the conditioned space.
- describe four types of motor mounts.
- identify two types of motor drive mechanisms.

SAFETYCHECKLIST

☑ Ensure that electric motor ground straps are connected properly when appropriate so that the motor will be grounded.

☑ Never touch a motor drive belt when it is moving. Be sure that your fingers do not get between the belt and the pulley.

18.1 MOTOR APPLICATIONS

Because electric motors perform so many different functions, choosing the proper motor is necessary for safe and effective performance. Usually, the manufacturer or design engineer for a particular job chooses the motor for each piece of equipment. However, as a technician you often will need to substitute a motor when an exact replacement is not available, so you should understand the reasons for choosing a particular motor for a job. For example, when a fan motor burns out in the air-conditioning condensing unit, the correct motor must be obtained or another failure may occur. **NOTE:** *In an air-cooled condenser, the air is normally pulled through the hot condenser coil and passed over the fan motor. This hot air is used to cool the motor. You must be aware of this, or you may install the wrong motor.* • The motor must be able to withstand the operating temperatures of the condenser air, which may be as high as 130°F, **Figure 18.1**.

Open motors are discussed in this unit because they are the only ones from which a technician has a choice of selection. Some design differences that influence the application are as follows:

Figure 18.1 This motor is operating in the hot airstream after the air has been pulled through the condenser.

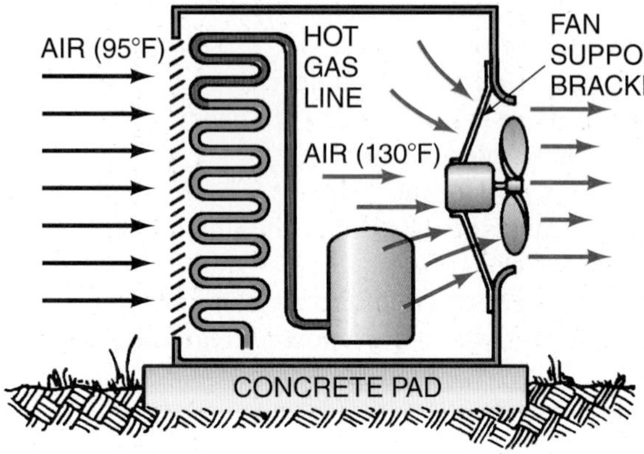

- Power supply
- Work requirements
- Motor insulation type or class
- Bearing types
- Mounting characteristics
- Cooling requirements

18.2 THE POWER SUPPLY

The power supply must provide the correct voltage and sufficient current. For example, the power supply in a small shop building may be capable of operating a 5-hp air compressor. But suppose air-conditioning is desired. If the air-conditioning contractor prices the job expecting the electrician to use the existing power supply, the electrical service for the whole building may have to be changed. The motor equipment nameplate and the manufacturer's catalogs provide the information needed for the additional service, but someone must put the whole project together. The installing air-conditioning contractor may have that responsibility. See **Figure 18.2** for a typical motor nameplate and **Figure 18.3** for a part of a page from a manufacturer's catalog that includes the electrical data. **Figure 18.4** is an example of a typical electrical panel rating.

Figure 18.2 A motor nameplate. Photo by Bill Johnson

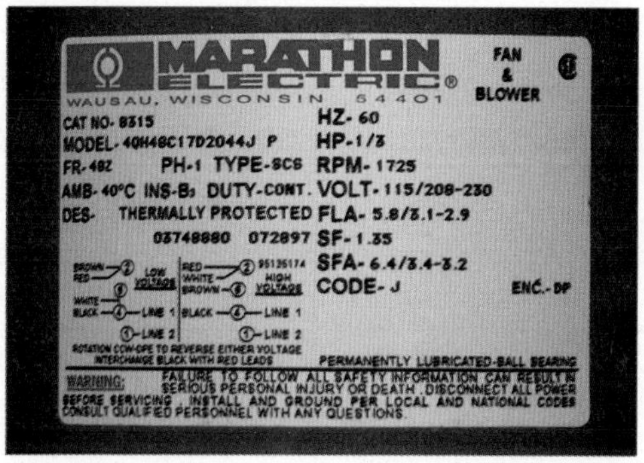

Figure 18.3 Part of a page from a manufacturer's catalog. Source: Century Electric, Inc.

TYPE CS • NEMA SERVICE FACTOR • 60 HERTZ • CUSHION BASE					
HP	RPM	Bearings	Overload Protector	Full Load .(12) Amps	Frame
$\frac{1}{6}$	1800	Sleeve	Auto	4.4	K48
$\frac{1}{4}$	1800	Sleeve	Auto	2.7	K48
		Ball	Auto	2.7	K48
$\frac{1}{3}$	3600	Sleeve	Auto	3.1	L48
	1800	Sleeve	Auto	2.9	L48
		Sleeve	Auto	2.6	J56
		Ball	No	2.9	L48
		Ball	No	2.6	J56
		Ball	Auto	2.6	J56
	1200	Sleeve	No	3.0	K56
$\frac{1}{2}$	3600	Sleeve	Auto	4.0	L48
	1800	Sleeve	No	3.6	J56
		Sleeve	Auto	3.6	J56
		Ball	No	3.6	J56
		Ball	Auto	3.6	J56
$\frac{3}{4}$	3600	Sleeve	Auto	4.6	J56
	1800	Sleeve	No	5.2	K56
		Sleeve	Auto	5.2	K56
		Ball	No	5.2	K56
		Ball	Auto	5.2	K56
1	3600	Sleeve	Auto	6.0	K56
		Ball	No	6.0	K56
	1800	Sleeve	Auto	6.5	L56
		Ball	No	6.5	L56
		Ball	Auto	6.5	L56
$1\frac{1}{2}$	3600	Ball	Auto	8.0	L56
	1800	Ball	Auto	7.5	M56
2	3600	Ball	No	9.5	L56

The power supply data contain the following:

1. The voltage (115 V, 208 V, 230 V, 460 V)
2. The current capacity in amperes

Figure 18.4 A nameplate on an electrical panel used for a power supply. Photo by Jason Obrzut

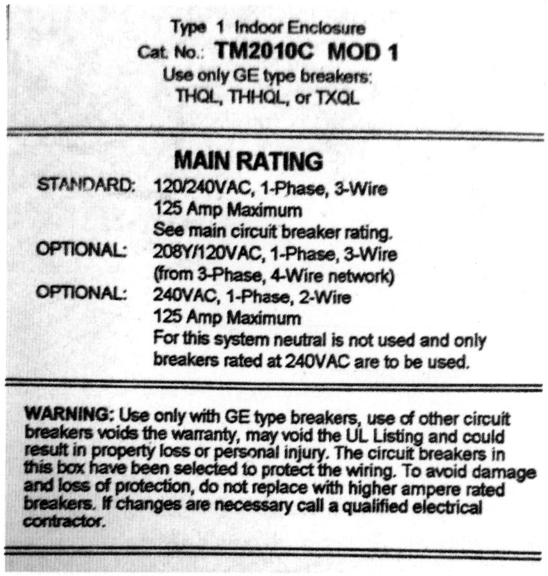

3. The frequency in hertz or cycles per second (60 cps in the United States and 50 cps in many foreign countries)
4. The phase (single or three phase)

Motors and electrical equipment must fit within the system's total electrical capacity, or failures may occur. Inadequate voltage and current are most often encountered. The technician can check them, but it is usually preferable to have a licensed electrician make the final calculations.

VOLTAGE

The voltage of an installation is important because every motor operates within a specified voltage range, usually within ±10%. **Figure 18.5** gives the upper and lower limits of common voltages. If the voltage is too low, the motor will draw a high current. For example, if a motor is designed to operate on 230 V but the supply voltage is really 200 V, the

Figure 18.5 This table shows the maximum and minimum operating voltages of typical motors.

208 VOLT-RATED MOTOR	+10%	228.8 VOLTS
	−10%	187.2 VOLTS
230 VOLT-RATED MOTOR	+10%	253 VOLTS
	−10%	207 VOLTS
208–230 VOLT-RATED MOTOR	+10%	253 VOLTS
	−10%	187.2 VOLTS

Figure 18.6 A motor operating under a low-voltage condition.

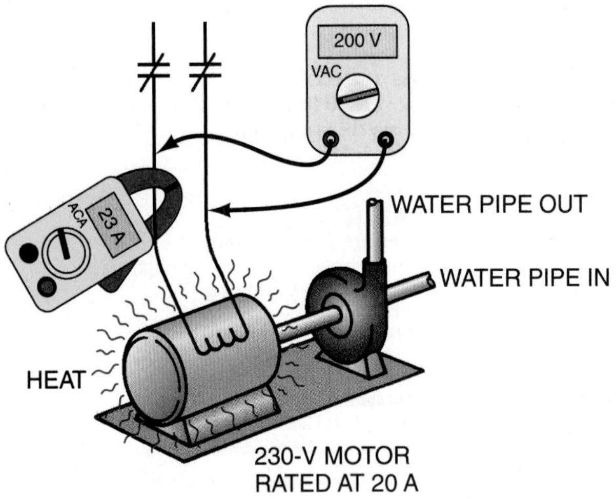

motor's current draw will go up. The motor is trying to do its job, but it lacks the power and it will overheat, **Figure 18.6**.

If the applied voltage is too high, the motor may develop local hot spots within its windings, but it will not always experience high amperage. The high voltage will actually give the motor more power than it can use. A 1-hp motor with a voltage rating of 230 V that is operating at 260 V is running above its 10% maximum. This motor may be able to develop 1 1/4 hp at this higher-than-rated voltage, but the windings are not designed to operate at that level. The motor can overheat and eventually burn out if it continually runs overloaded. This can happen *without* drawing excessive current, **Figure 18.7**.

CURRENT CAPACITY

There are two current ratings for a motor. The full-load amperage (FLA) is the current the motor draws while operating at a full-load condition at the rated voltage. When the full-load torque and horsepower of a motor is reached, the amperage the motor is drawing is referred to as the full-load amperage (FLA). FLA is determined under laboratory conditions and is printed on the motor's nameplate. For example, a 1-hp motor will draw approximately 16 A at 115 V or 8 A at 230 V in a single-phase circuit. **Figure 18.8** shows approximate amperages for some typical motors.

Figure 18.7 A motor rated at 230 V and operating at 260 V.

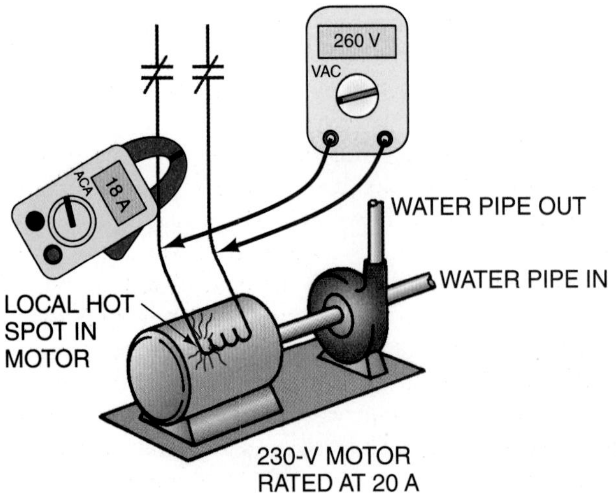

Figure 18.8 This chart shows approximate full-load amperage values. Source: BDP Company

APPROXIMATE FULL-LOAD AMPERAGE VALUES FOR ALTERNATING CURRENT MOTORS					
Motor	Single Phase		3-Phase-Squirrel Cage Induction		
HP	120V	230V	230V	460V	575V
$\frac{1}{6}$	4.4	2.2			
$\frac{1}{4}$	5.8	2.9			
$\frac{1}{3}$	7.2	3.6			
$\frac{1}{2}$	9.8	4.9	2	1.0	0.8
$\frac{3}{4}$	13.8	6.9	2.8	1.4	1.1
1	16	8	3.6	1.8	1.4
$1\frac{1}{2}$	20	10	5.2	2.6	2.1
2	24	12	6.8	3.4	2.7
3	34	17	9.6	4.8	3.9
5	56	28	15.2	7.6	6.1
$7\frac{1}{2}$			22	11.0	9.0
10			28	14.0	11.0

Note: Does not include shaded pole.

Another amperage rating that may be given for a motor is the locked-rotor amperage (LRA). The LRA is the amperage the motor draws at start-up before the rotor has started to turn. The two current ratings are available for every motor and are stamped, for an open motor, on the motor nameplate. By taking amperage readings at start-up and while the motor is operating under load and comparing these readings with those on the motor nameplate, the technician can determine whether the motor is operating within its design parameters. Both ratings may not be printed on the nameplate of some motors. Normally, the LRA is about five times the FLA. For example, a motor that has a FLA of 5 A will normally have a LRA of about 25 A. If the LRA is given on the nameplate and the FLA is not, divide the LRA by 5 to get an approximate FLA. For example, if a compressor nameplate shows a LRA of 80 A, the approximate FLA is $80/5 = 16$ A.

Rated-load amps (RLA) is a mathematical calculation used to get Underwriters Laboratories (UL) approval for a compressor motor. It should not be confused with full-load amps (FLA). Full-load amps (FLA) have not been used by compressor manufacturers since 1972, when UL changed the rating to rated-load amps (RLA).

Refrigeration and air-conditioning service technicians deal with compressor motors whose loads and amp draws vary and are dependent on:

- Surrounding ambient temperature
- Condensing and evaporating temperatures
- Compressor return gas temperatures
- Liquid temperatures entering the metering device
- Voltage at the compressor

In these cases, the service technician must obtain compressor performance data from the compressor manufacturer. Many times, this data can be found on the Internet at the manufacturer's website. **Figure 18.9(A)** illustrates compressor data in table form. Notice in **Figure 18.9(A)** that the amperage drawn by the compressor depends on both the condensing and evaporating temperatures. This type of data also comes in curve form, often referred to as a rating chart or performance data for a particular compressor. Compressor manufacturers publish a rating or performance chart for each compressor they manufacture. Let's use **Figure 18.9(A)** to help solve a compressor field problem.

Problem: Let's assume the R-22 compressor in **Figure 18.9(A)** is operating at 230 V and has an evaporator temperature of −15°F and a condensing temperature of 100°F. Determine the correct current (amp) draw for this compressor using the table.

Solution: Using the current (amperage) table in **Figure 18.9(A)**, the intersection of the 100°F condensing temperature with a −15°F evaporating temperature results in an amp draw of 29.4 A. This data tells us that this compressor operating under these pressures should draw close to 29.4 A at 230 V. The allowable tolerance is ±10%. If the compressor was operating at 460 V, the amperage values in the table would have to be multiplied by 1/2. Remember, the compressor must be operating at the proper voltage for the table to be meaningful.

If the service technician now actually measures the compressor amps with an ammeter at the compressor and finds the amps to be very close to 29.4 A, he has a pretty good idea that the compressor is up to specifications. If, however, the technician measures the current draw of the compressor and finds that it is pulling only 13 A under these same conditions, there is a problem somewhere. The following is a list of reasons why a compressor may draw low amperage:

- Bad suction and/or discharge valve
- Starved evaporator
- Worn rings
- Leaking head gasket or valve plate gasket
- Restricted liquid line or filter drier
- Undercharged system
- Weak motor
- Metering device starving evaporator

The service technician must now systematically check each of these causes to find the one that is the reason for the low amp draw. Amperage tables like the one used in this example are invaluable tools in systematically troubleshooting and diagnosing compressor problems.

Notice in **Figure 18.9(A)** that as the condensing temperature decreases, so do the amps, assuming a constant evaporating temperature. Also, as the evaporating temperature decreases, so do the amps, assuming a constant condensing pressure. Decreased condensing temperatures mean lower condensing pressures and a lower compression ratio. The compressor's piston now operates with less pressure against it as it compresses the suction gases. This, in turn, places less stress on the motor and draws less amperage. As the evaporating temperature decreases, so do the evaporating pressures. These lower pressures entering the compressor's cylinders mean the vapors filling the cylinders are less dense, and the compressor pumps less mass flow rate of refrigerant vapor. This causes a decreased amp draw.

NOTE: *Operating conditions other than just evaporating and condensing temperatures (pressures) are used in these tables. The conditions in the upper left-hand corner of* **Figure 18.9(A)** *must be met for the values to be 100% accurate. The conditions are:*

- *65°F return gas temperature*
- *0°F liquid subcooling*
- *95°F ambient air over the condenser*
- *60-Hz operation*

If these conditions are not met, the table will not be 100% accurate. However, for purposes of systematic field troubleshooting, these inaccuracies are not large enough to have any major effect on a good service technician's judgment. •

The point is, when troubleshooting a refrigeration or air-conditioning compressor's motor, the RLA or FLA value should not be used to determine if the compressor is good or bad or if it is drawing the correct amps. Compressor performance data like those in **Figure 18.9(A)** must be used for a specific compressor. The value of RLA or FLA to a refrigeration or air-conditioning service technician is simply to size wires, fuses, overloads, and contactors for the compressor. The nameplate RLA does not change as the compressor motor load changes with temperatures and pressures. And the RLA should not be used by the service technician to determine the proper refrigerant charge or to determine if the compressor is running properly. In summary, always refer to performance or rating data from the manufacturer for that compressor. More detailed coverage of compressor performance data in table and curve form will be provided in Unit 23, "Compressors."

Every motor has a service factor, which may be listed in the manufacturer's literature. This service factor is actually reserve horsepower. A service factor of 1.15 means that the motor can periodically operate at 15% over the nameplate horsepower before it is out of its design parameters. A motor operating with a variable load and above normal conditions for short periods of time should have a larger service factor. If at a particular installation the voltage varies, a motor with a high service factor may be chosen. The service factor is standardized by the National Electrical Manufacturer's Association (NEMA).

Figure 18.9 (A) Compressor data in table form showing the compressor's amperage draw at different system conditions.

CONDITIONS
65°F Return Gas Temperature
0°F Liquid Subcooling
95°F Ambient Air Over Condenser
60 Hz Operation

ABC COMPRESSOR COMPANY
MODEL #ABCDWXYZ567
LOW TEMPERATURE
208/230/460 VOLT (3 PHASE)
HCFC-22 (R-22)

	EVAPORATOR CAPACITY (BTU/HR)					EVAPORATING TEMPERATURE			
Cond. Temp. Degrees F (Degrees C)	−40 (−40)	−35 (−37.2)	−30 (−34.4)	−25 (−31.7)	−20 (−28.9)	−15 (−26.1)	−10 (−23.3)	−5 (−20.6)	0 (−17.8)
70 (21.1)	32800	39200	46200	53800	62100	71100	80800	91400	103000
80 (36.7)	29100	35400	42300	49800	58000	66800	76500	87000	98300
90 (32.2)	25400	31600	38400	45700	53700	62500	71900	82200	93400
100 (37.8)	21500	27600	34200	41400	49200	57700	67000	77100	88100
110 (43.3)	17400	23200	29600	36600	44200	52500	61500	71400	82100
120 (48.9)	12800	18400	24500	31200	38600	46600	55400	64900	75400
130 (54.4)	7490	12800	18700	25100	32100	39900	48300	57600	67700

	POWER (WATTS)					EVAPORATING TEMPERATURE			
Cond. Temp. Degrees F (Degrees C)	−40 (−40)	−35 (−37.2)	−30 (−34.4)	−25 (−31.7)	−20 (−28.9)	−15 (−36.3)	−10 (−23.3)	−5 (−20.6)	0 (−17.8)
70 (21.1)	5520	6400	6420	6800	7140	7480	7800	8140	8510
80 (26.7)	5630	6200	6710	7170	7590	7990	8390	8790	9200
90 (32.2)	5720	6400	7010	7570	8090	8570	9050	9520	10000
100 (37.8)	5730	6530	7260	7930	8550	9140	9720	10300	10900
110 (43.3)	5570	6510	7380	8170	8920	9630	10300	11000	11700
120 (48.9)	5160	6260	7280	8220	9100	9940	10800	11500	12300
130 (54.4)	4440	5710	6880	7990	9020	10000	11000	11900	12800

	CURRENT (AMPS)				@ 230 VOLTS (MULTIPLY AMPS BY 0.5 FOR 460 VOLTS)				
Cond. Temp. Degrees F (Degrees C)	−40 (−40)	−35 (−37.2)	−30 (−34.4)	−25 (−31.7)	−20 (−28.9)	−15 (−36.3)	−10 (−23.3)	−5 (−20.6)	0 (−17.8)
70 (21.1)	22.0	22.8	23.8	24.4	25.2	25.8	26.8	27.4	28.0
80 (26.7)	22.2	23.2	24.2	25.2	26.0	27.0	27.8	28.8	29.6
90 (32.2)	22.4	23.6	24.8	26.0	27.2	28.2	29.2	30.4	31.6
100 (37.8)	22.4	24.0	25.4	26.8	28.2	29.4	30.8	32.2	33.4
110 (43.3)	21.2	24.0	25.6	27.4	29.0	30.6	32.2	33.8	35.4
120 (48.9)	21.4	23.4	25.6	27.4	29.4	31.2	33.2	35.0	37.0
130(54.4)	19.9	22.4	24.8	27.0	29.2	31.6	33.8	36.	38.4

	EVAPORATOR MASS FLOW (LBS/HR)					EVAPORATING TEMPERATURE			
Cond. Temp. Degrees F (Degrees C)	−40 (−40)	−35 (−37.2)	−30 (−34.4)	−25 (−31.7)	−20 (−28.9)	−15 (−36.3)	−10 (−23.3)	−5 (−20.6)	0 (−17.8)
70 (21.1)	382	458	540	629	727	834	950	1080	1210
80 (26.7)	351	428	512	604	704	813	932	1060	1200
90 (32.2)	318	397	482	576	677	789	910	1040	1190
100 (37.8)	281	361	447	542	646	759	883	1020	1170
110 (43.3)	237	317	404	500	605	720	846	984	1130
120 (48.9)	182	262	350	447	553	669	797	937	1090
130 (54.4)	112	192	280	377	484	601	731	873	1030

(A)

Figure 18.9(B) Service factors for different horsepower motors at different synchronous speeds.

Horsepower (hp)	SERVICE FACTOR (SF) Synchronous Speed (RPM)			
	3600	800	1200	900
1/6, 1/4, 1/3	1.35	1.35	1.35	1.35
1/6	1.25	1.25	1.25	1.25
3/4	1.25	1.25	1.15	1.15
1.0	1.25	1.15	1.15	1.15
1.5 and up	1.15	1.15	1.15	1.15

(B)

The chart in **Figure 18.9(B)** shows different service factors for drip-proof motors at synchronous speed (rpm). These service factors are actually "periodic" overload capacities at which the motor can operate without damage, overheating, or overloading. The service factor is not intended to be a general increase in the motor's horsepower rating. A motor that operates "continuously" at a service factor greater than 1 will have a reduced life expectancy compared to when operating at its rated nameplate horsepower. Also, the bearing and insulation life will be reduced if the motor operates at its service factor load. Never rely on a motor to perform continuous duty if it is running on its service factor loading. The appropriate use of the service factor rating is to handle short-term or occasional overloads.

Any motor that has a service factor greater than 1 must have the service factor indicated on its nameplate. In some instances, the current the motor is drawing at its service factor loading is also indicated on the nameplate. This is usually expressed as **service factor amperes (SFA)**.

FREQUENCY

The frequency in cycles per second (cps) is the frequency of the electrical current that the power company supplies. The technician has no control over this. Most motors are 60 cps in the United States but could be 50 cps in another country. Most 60-cps motors will run on 50 cps, but they will develop only five-sixths of their rated speed (50/60). If you believe that the supply voltage is not 60 cps, contact the local power company. When motors are operated with local generators as the power supply, the generator's speed will determine the frequency. A cps meter is normally mounted on the generator and can be checked to determine the frequency.

PHASE

The number of phases of power to be supplied by the power company is determined by the type of equipment to be used at a particular installation. For equipment using three-phase motors, three-phase power must be supplied. Normally, single-phase power is supplied to residences and three-phase power is supplied to commercial and industrial installations when needed. Single-phase motors will operate on two phases of three-phase power, **Figure 18.10**. Three-phase motors will not operate on single-phase power, **Figure 18.11**.

The power company furnishes the power to the respective services to meet the specifications of the particular job. For example, if the job or system calls for a large electrical load, such as a manufacturing plant or high-rise building, it will consult with the engineers about what they recommend.

Figure 18.10 This wiring diagram shows how single-phase motors are wired into a three-phase circuit.

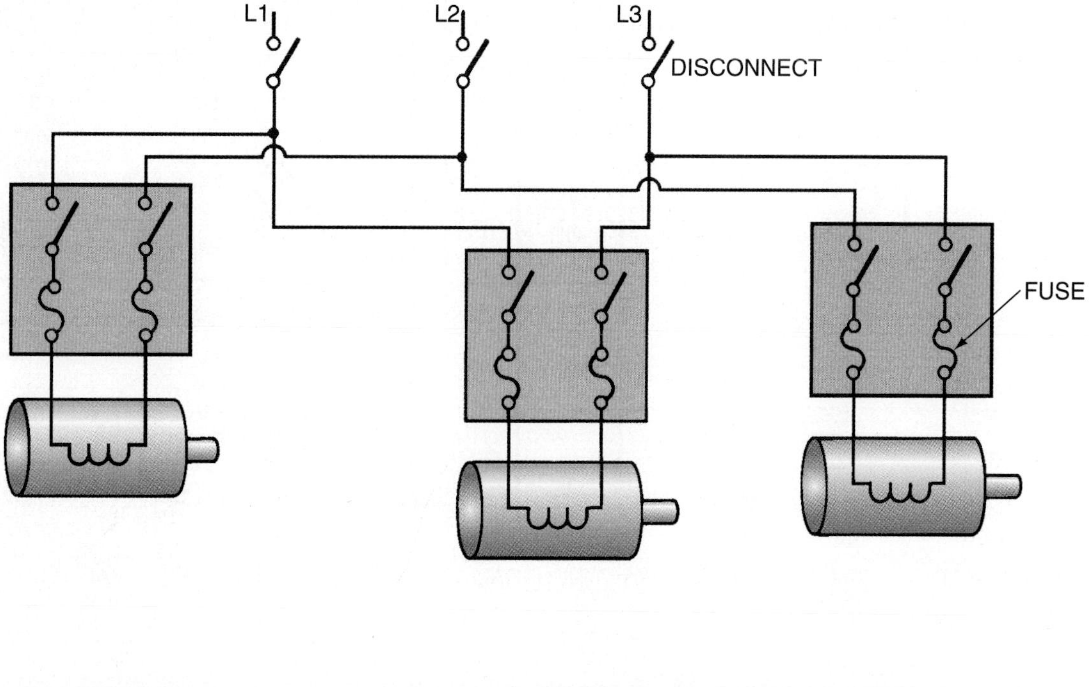

Figure 18.11 A three-phase motor with three power leads cannot be connected to a single-phase power supply.

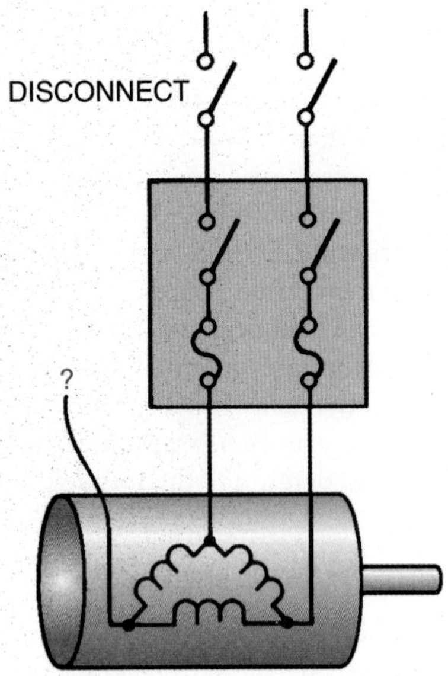

Power companies rate the voltage furnished according to the nominal voltage furnished, which is typically 480 V, 240 V, 208 V, and 120 V. Motors and appliances are typically rated at 460 V, 230 V, 208 V, and 115 V. The technician must work with the rated motor and appliance voltages. Usually, the power supply will be close to the power company rating and still be within the motor's rating.

The technician needs to know the characteristics of the voltage supplied. The supplied voltage depends on the power company's transformer and the way it is connected. The two transformer types commonly used for commercial and industrial applications are the wye, or star, transformer and the delta transformer, **Figure 18.12**.

The wye transformer is configured in a Y-shaped connection with three windings going to a center tap. The center tap is considered the neutral or grounded neutral leg. The consumer typically will have a power supply of 208 V at the secondary terminals from phase to phase and 115 V (120 V) from any phase to the grounded neutral.

Another option for the wye transformer is to furnish 460 V (480 V) phase to phase and 277 V phase to the

Figure 18.12 Two different transformers for commercial and industrial applications.

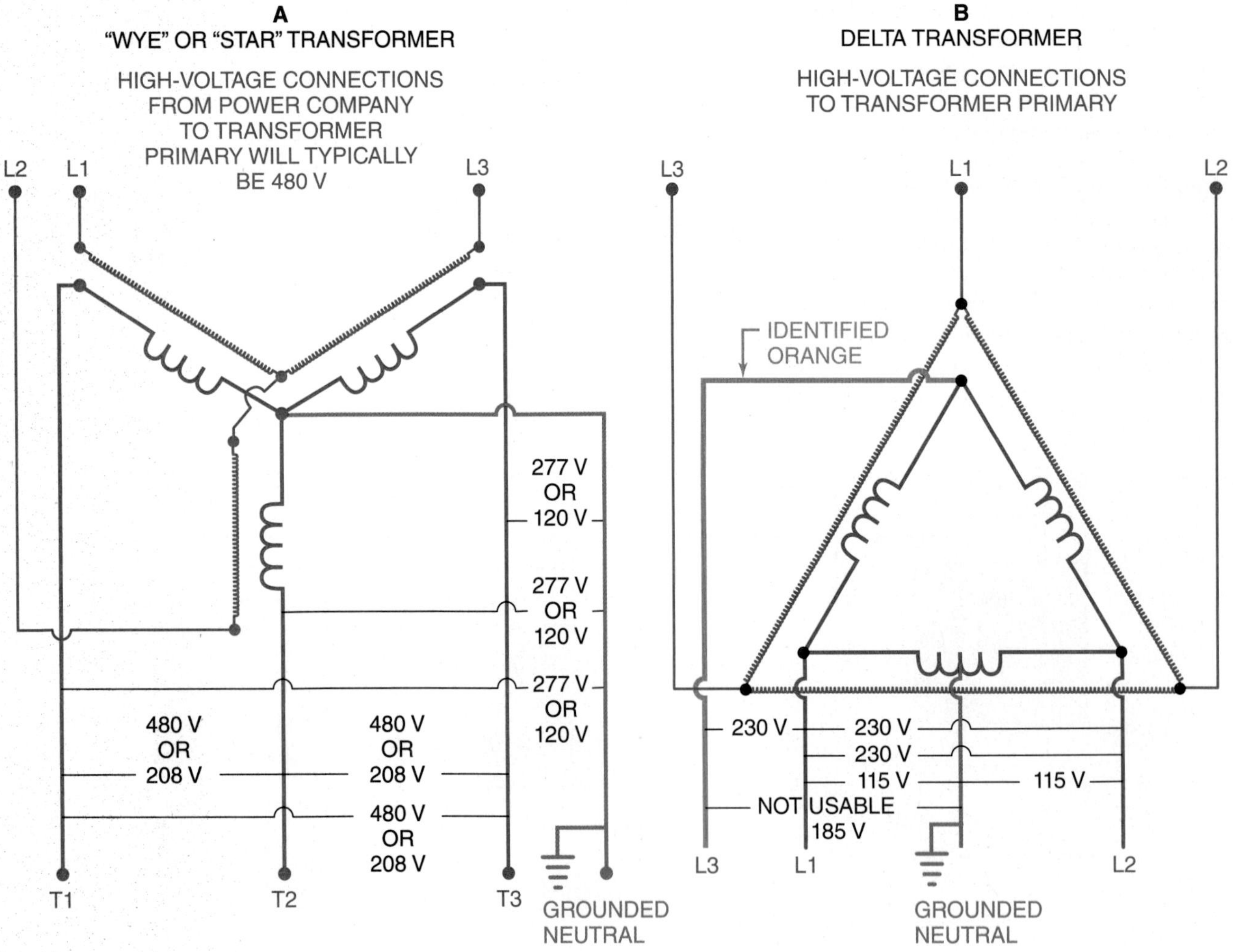

grounded neutral. The 277 V to the neutral is often used for the building's lighting service. Smaller wire sizes can be used in the lighting circuits. When 460-V (480 V) and 277-V grounded neutral are used, there must be a further reduction of voltage for any 115-V (120-V) circuits in the building. This is accomplished with step-down transformers at locations near the point of consumption. The wye transformer connection service is popular because it is easier to balance the load of each phase-to-phase transformer connection. This is necessary to keep the load on the power supply balanced.

The delta transformer service has been popular in some localities in the past. This service furnishes 230 V and can have a "center-tapped" leg to furnish 115 V. Notice that this service furnishes 230 V phase to phase, 115 V from phase 1 to neutral, and 115 V from phase 2 to neutral. It is hard to balance the loads on this system because of the two phases that have a relationship to neutral. **NOTE:** *The relationship from the third phase to neutral is not usable. This is often called a "wild leg" or a "high leg" and is identified with orange along its length to prevent confusion. This leg often has 185 V or more to neutral or ground. If a 115-V appliance is connected to this leg, it will be burned out in short order.* •

The technician is advised to carry and use a quality voltmeter and to always check the power supply voltages to know exactly what is available at the job site.

18.3 ELECTRIC-MOTOR WORKING CONDITIONS

The motor's working conditions determine which motor is the most economical for a particular installation. An open motor with a centrifugal starting switch (single-phase) for an air-conditioning fan may not be used in a room containing explosive gases. When the motor's centrifugal switch opens to interrupt the power to the start winding, the gas may ignite. Instead, an explosion-proof motor enclosed in a housing must be used, **Figure 18.13**. Local codes should be checked and adhered to. Because the explosion-proof motor is too expensive to be installed in a standard office building, another type of motor should be chosen. A motor operated in a very dirty area may need to be enclosed, which would give no ventilation for the motor windings. This motor must have some method of dissipating the heat from the windings, **Figure 18.14**. A drip-proof motor should be used where water may fall on the motor; it is designed to shed water, **Figure 18.15**.

Figure 18.13 An explosion-proof motor. Photo by Jason Obrzut

Figure 18.14 A totally enclosed motor. Photo by Bill Johnson

Figure 18.15 A drip-proof motor. Photo by Bill Johnson

18.4 INSULATION TYPE OR CLASS

The insulation type or class of a motor describes how hot the motor can safely operate in a particular ambient temperature condition. The earlier example of the motor used in an air-conditioning condensing unit is typical. This motor must be designed to operate in a high ambient temperature. Motors are classified by the maximum allowable operating temperatures of the motor windings, **Figure 18.16.**

For many years, motors were rated by allowable temperature rise of the motor above the ambient temperature. Many motors still in service are rated this way. A typical motor has a temperature rise of 40°C. If the maximum ambient temperature is 40°C (104°F), the motor should have a winding temperature of 40°C + 40°C = 80°C (176°F). When troubleshooting these motors, the technician may have to convert from Celsius to Fahrenheit or may need a conversion table if the temperature rating is given only in Celsius. If the temperature of the motor winding can be determined, the technician can tell if a motor is running too hot for the conditions. For example, a motor in a 70°F room is allowed a 40°C rise on the motor. The maximum winding temperature is 142°F (70°F = 21°C; 21°C + 40°C rise = 61°C, which is 142°F), **Figure 18.17.**

Figure 18.16 Temperature classifications for typical motors.

Class A	221°F	(105°C)
Class B	266°F	(130°C)
Class F	311°F	(155°C)
Class H	356°F	(180°C)

Figure 18.17 This motor is being operated in an ambient temperature that is equal to the maximum allowable for its insulation with a 40°C rise.

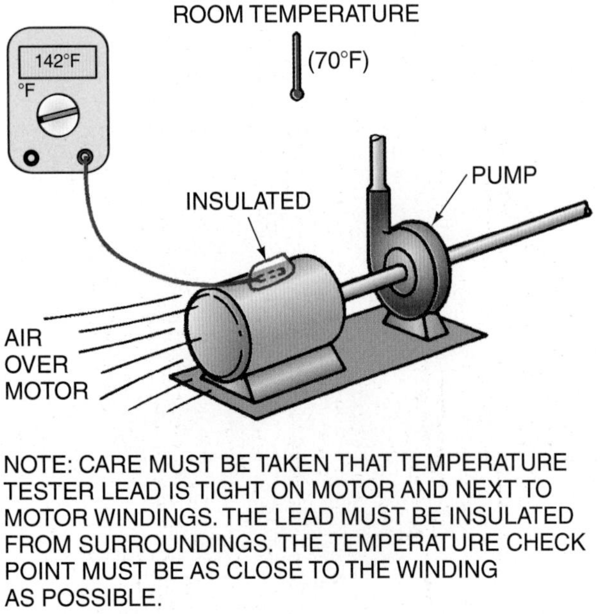

18.5 TYPES OF BEARINGS

Load characteristics and noise level determine the type of bearings that should be selected for the motor. Two common types are the *sleeve* bearing and the *ball* bearing, **Figure 18.18.** Large motors use a type of ball bearing called a *roller bearing,* which has cylindrically shaped rollers instead of balls.

The sleeve bearing is used where the load is light and the noise must be low (e.g., a fan motor on a residential furnace). A ball-bearing motor would probably make excessive noise in the conditioned space. Metal ductwork is an excellent sound carrier, and any noise in the system is carried throughout the entire system. For this reason, sleeve bearings are normally used in smaller applications, such as in residential and light commercial air-conditioning systems. They are quiet and dependable, but they cannot stand great pressures

Figure 18.18 (A) Sleeve bearings with an oil port. (B) Ball bearings. Photos by Jason Obrzut

(A)

(B)

Figure 18.19 This motor is working against two conditions: the normal motor load and the fact that the fan is trying to fly downward while pushing air upward.

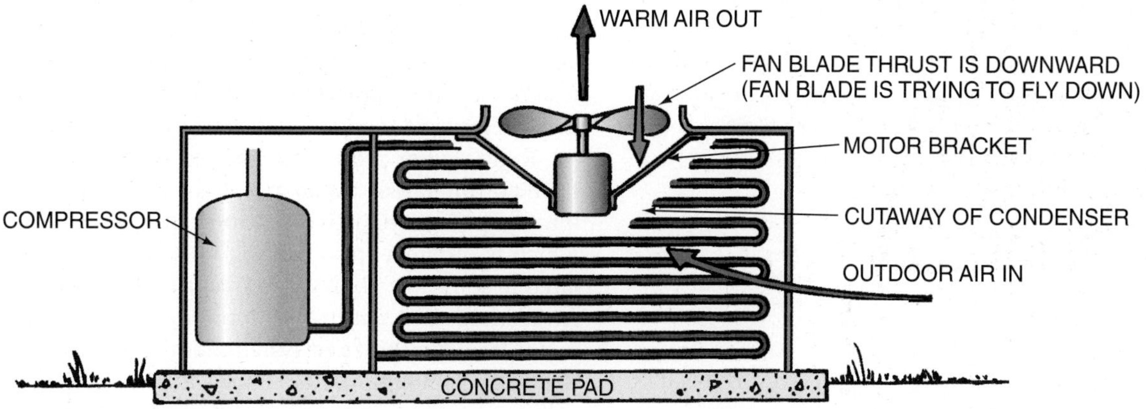

(e.g., if fan belts are too tight). Motors that use sleeve bearings can have either vertical or horizontal shafts. The typical air-cooled condenser has a vertical motor shaft and pushes the air out the top of the unit. This results in a downward thrust on the motor bearings, **Figure 18.19**. A furnace fan has a horizontal motor shaft, **Figure 18.20**. These two types may not look very different, but they are. The vertical condenser fan is trying to fly downward into the unit to push air out the top. This puts a real load on the end of the bearing (called the **thrust surface**), **Figure 18.21**.

Sleeve bearings are made from material that is softer than the motor shaft. The bearing must have lubrication—an oil film between the shaft and the bearing surface. The shaft actually floats on this oil film and should never touch the bearing surface. The oil film is supplied by the lubrication system. Two types of lubrication systems for sleeve bearings are the *oil port* and the *permanently lubricated* bearing.

The oil port bearing has an oil reservoir that is filled from the outside by means of an access port. This bearing must be lubricated at regular intervals with the correct type of oil, which is usually 20-weight nondetergent motor oil or an oil specially formulated for electric motors. If the oil is too thin, it will allow the shaft to run against the bearing surface. If the oil is too thick, it will not run into the clearance between the shaft and the bearing surface. The correct interval for lubricating a sleeve bearing depends on the design and use of the motor. The manufacturer's instructions will indicate the recommended interval. Some motors have large reservoirs and do not need lubricating for years. This is good if there is limited access to the motor.

The permanently lubricated sleeve bearing is constructed with a large reservoir and a wick to gradually feed oil to the bearing. This bearing truly does not need lubrication until the oil deteriorates, which will happen if the motor has been running hot for many hours. Shaded-pole motors operating in the heat and weather have these bearing systems, and many have operated without failure for years.

Figure 18.20 A furnace fan motor mounted in a horizontal position.

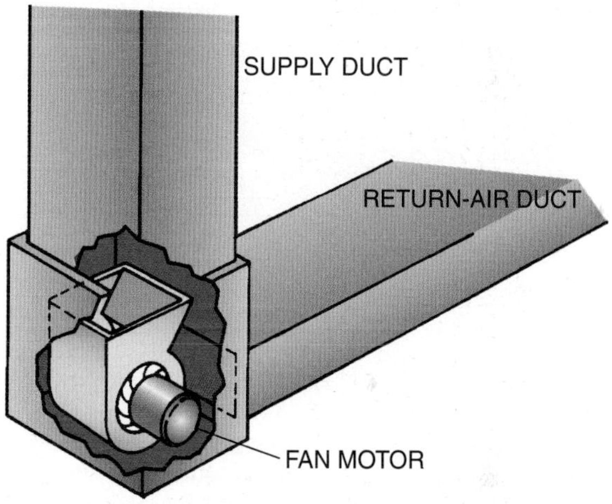

Figure 18.21 The thrust surface on a fan motor bearing.

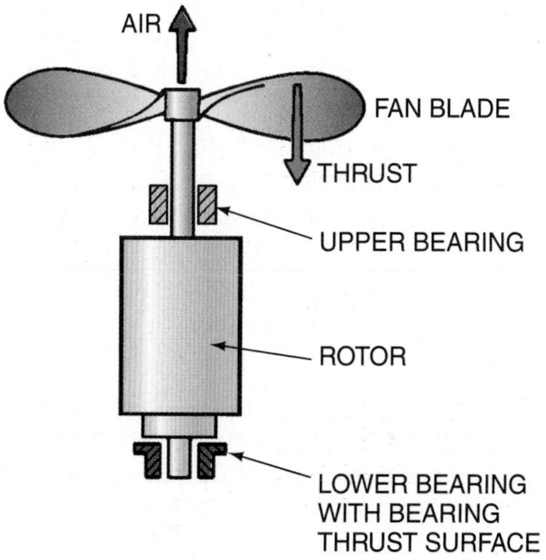

Figure 18.22 A fitting through which the bearing is greased.

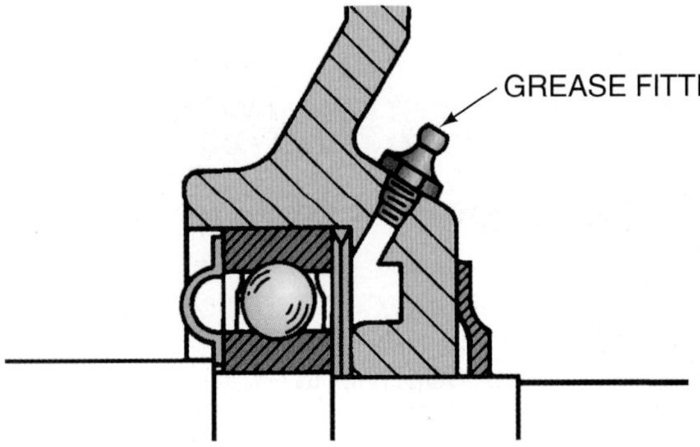

Figure 18.23 A motor bearing using grease for lubrication. Notice the relief screw.

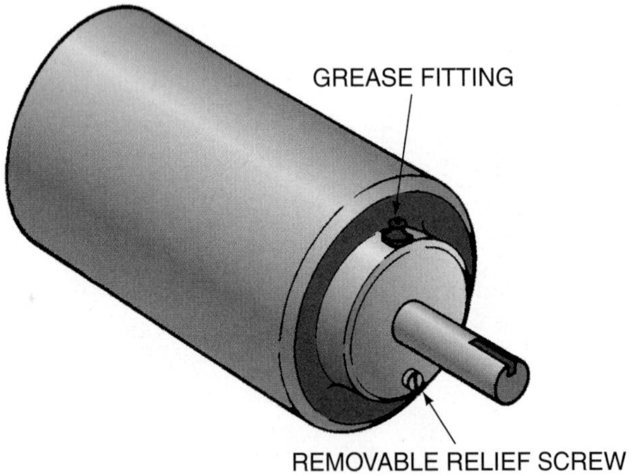

Ball-bearing motors are not as quiet as sleeve-bearing motors and are used in locations where their noise levels will not be a problem. Large fan motors and pump motors are normally located far enough from the conditioned space that the bearing noises will not be noticed. These bearings are made of very hard material and usually lubricated with grease rather than oil. Motors with ball bearings generally have permanently lubricated bearings or grease fittings. Permanently lubricated ball bearings are similar to permanently lubricated sleeve bearings, except that they have reservoirs of grease sealed in the bearing. They are designed to last for years with the lubrication furnished by the manufacturer, unless the conditions in which they operate are worse than those for which the motor was designed.

Bearings needing lubrication have grease fittings so a grease gun can force grease into the bearing. This is often done by hand, **Figure 18.22**. Only an approved grease should be used. In **Figure 18.23**, the slotted screw at the bottom of the bearing housing is a *relief screw*. When grease is pumped into the bearing, this screw must be removed or the pressure of the grease may push the grease seal out, and grease will leak down the motor shaft.

18.6 MOTOR MOUNTING CHARACTERISTICS

The mounting characteristics of a motor determine how it will be secured during its operation. When mounting a motor, the technician must consider noise level. The two primary mountings are the rigid mount and the resilient, or rubber, mount.

Rigid-mount motors are bolted, metal to metal, to the frame of the fan or pump and will transmit any motor noise into the piping or ductwork. Motor hum is an electrical noise, which is different from bearing noise, and must also be isolated in some installations. **Resilient-mount motors** use different methods of isolating the motor and bearing noise from the metal framework of the system. Notice the ground strap on the resilient-mount motor, **Figure 18.24**. This motor is electrically and mechanically isolated from the metal frame. If the motor were to become grounded (circuit from the hot line to the frame of the motor), the motor frame would be electrically hot without the ground strap. **SAFETY PRECAUTION:** *When replacing a motor, always connect the ground strap properly or the motor could become dangerous,* **Figure 18.25.** •

Figure 18.24 The grounding strap to carry current from the frame if the motor has a grounded winding. Photo by Bill Johnson

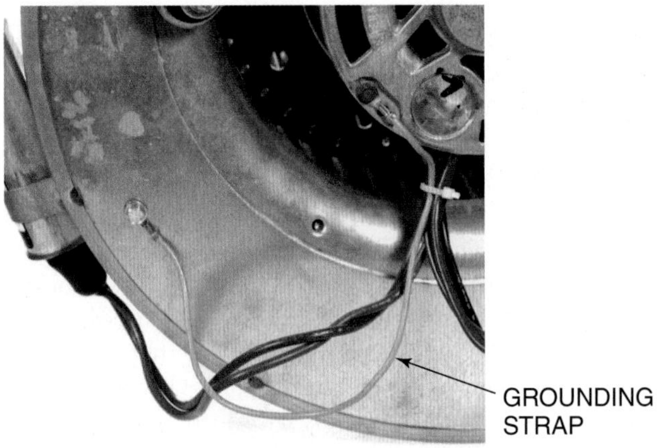

Figure 18.25 This illustration shows how the grounding strap works.

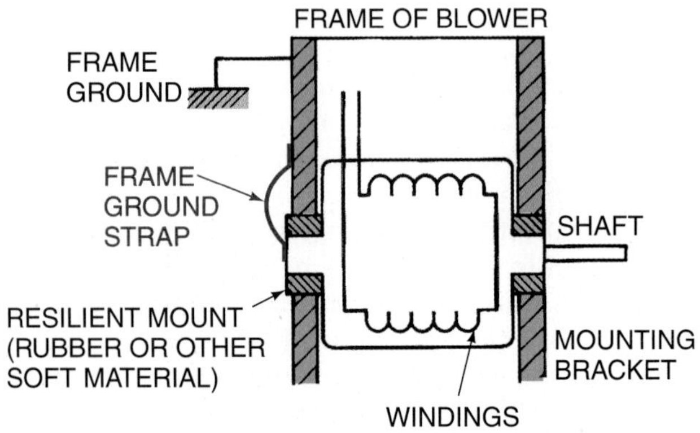

Figure 18.26 Dimensions of typical motor frames.

MOTOR DIMENSIONS FOR NEMA FRAMES

Standardized motor dimensions as established by the National Electrical Manufacturers Association (NEMA) are tabulated below and apply to all base-mounted motors listed herein which carry a NEMA frame designation.

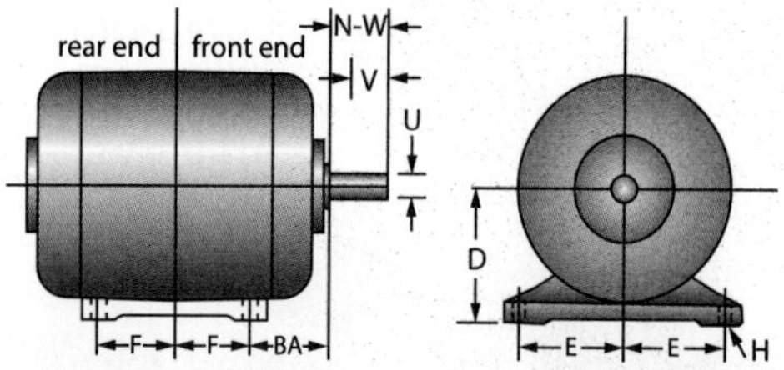

NEMA FRAME	D(*)	2E	2F	BA	H	N-W	U	V(§) Min.	Key Wide	Key Thick	Key Long	NEMA FRAME
42	$2\frac{5}{8}$	$3\frac{1}{2}$	$1\frac{11}{16}$	$2\frac{1}{16}$	9/32 slot	$1\frac{1}{8}$	3/8	—	—	21/64 flat	—	42
48	3	$4\frac{1}{4}$	$2\frac{3}{4}$	$2\frac{1}{2}$	11/32 slot	$1\frac{1}{2}$	1/2	—	—	29/64 flat	—	48
56 56H	$3\frac{1}{2}$	$4\frac{7}{8}$	3 3&5(‡)	$2\frac{3}{4}$	11/32 slot	$1\frac{7}{8}$(†)	5/8(†)	—	3/16(†)	3/16(†)	$1\frac{3}{8}$(†)	56 56H
56HZ	$3\frac{1}{2}$	**	**	**	**	$2\frac{1}{4}$	7/8	2	3/16	3/16	$1\frac{3}{8}$	56HZ

The four basic mounting styles are cradle mount, rigid-base mount, end mount (with tabs or studs or flange mount), and belly-band mount, all of which fit standard dimensions established by NEMA and which are distinguished from each other by *frame numbers*. **Figure 18.26** shows some typical examples.

CRADLE-MOUNT MOTORS

Cradle-mount motors are used in either direct-drive or belt-drive applications. A cradle fits the motor end housing at each end, and the end housing is held down with a bracket, **Figure 18.27**. The cradle is fastened to the equipment or pump base with machine screws, **Figure 18.28**. Cradle-mount motors are available only in the small horsepower range. A handy service feature is that the motor can be removed easily.

RIGID-BASE-MOUNT MOTORS

Rigid-base-mount motors are similar to cradle-mount motors except that the base is fastened to the motor body, **Figure 18.29**. It is the belt, if one is used, which drives the

Figure 18.27 A cradle-mount motor. Photo by Jason Obrzut

Figure 18.28 A cradle fastened to the base of a pump. Photo by Jason Obrzut

Figure 18.29 A rigid-mount motor. Photo by Bill Johnson

Figure 18.30 A direct-drive motor with flexible coupling.

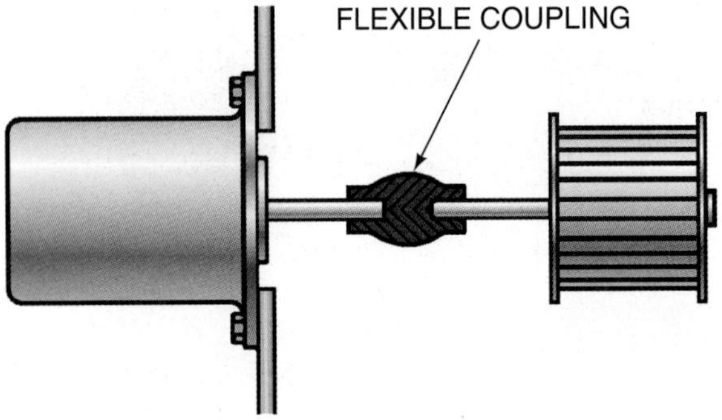

prime mover that provides the sound isolation for this motor. The belt is flexible and dampens motor noise. This motor is often used as a direct drive to turn a compressor or pump. A flexible coupling is used between the motor and prime mover in a direct-drive installation, **Figure 18.30**.

END-MOUNT MOTORS

End-mount motors are very small motors mounted to the prime mover with tabs or studs fastened to the motor housing, **Figure 18.31**. Flange-mounted motors have a flange as a part of the motor housing, for example, an oil burner motor, **Figure 18.32**.

BELLY-BAND-MOUNT MOTORS

Motors with belly-band mounts have a strap that wraps around the motor to secure it with brackets mounted to the strap. These motors are often used in air-conditioning air handlers. Several universal types of motor kits are belly-band-mounted and will fit many different applications. These motors are all direct drive, **Figure 18.33**.

Figure 18.31 A motor that is end-mounted with tabs and studs. Photo by Bill Johnson

Figure 18.32 The motor for an oil burner with a flange on the end to hold the motor to the equipment that is being turned. Photo by Bill Johnson

Figure 18.33 A belly-band-mount motor. Photo by Jason Obrzut

18.7 MOTOR DRIVES

Motor drives are devices or systems that connect a motor, the driving device, to the driven load, a fan for instance. Motors drive their loads by belts or direct drives through couplings, or the driven component may be mounted on the motor shaft. Gear drives, a form of direct drive, will not be covered in this text because they are used mainly in large industrial applications.

The drive mechanism is intended to transfer the motor's rotating power or energy to the driven device. For example, a compressor motor is designed to transfer the motor's power to the compressor, which compresses the refrigerant vapor and pumps refrigerant from the low side to the high side of the system. Efficiency, speed of the driven device, and noise level are some factors involved in this transfer. It takes energy to turn the belts and pulleys on a belt-drive system in addition to the compressor load. Therefore, a direct-drive motor may be better suited for this application. **Figure 18.34** shows an example of both direct and belt drives.

Belt-driven systems have been used for years to drive both fans and compressors. A feature of the belt-drive system is its versatility: both pulley sizes and the speed of the driven device

Figure 18.34 Motor drive mechanisms. (A) Belt drive. Photo by Eugene Silberstein (B) Direct drive. Photo by Jason Obrzut

(A)

(B)

Figure 18.35 Belt-driven compressors using different size drive pulleys.

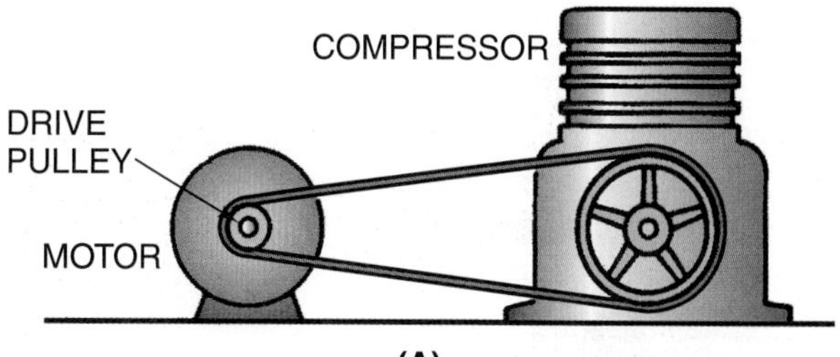

(A)

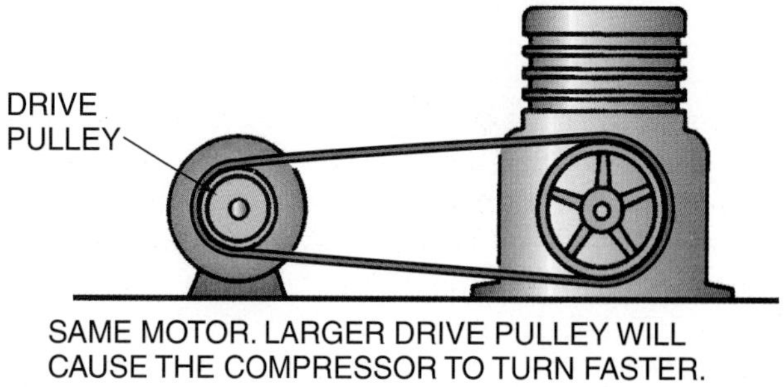

(B)

may be changed, **Figure 18.35**. This can be a great advantage if the capacity of a compressor or a fan speed needs to be altered. However, the changes must be made according to the capacity of the drive motor.

There is a definite relationship between the speed of the motor and the speed of a blower, which involves the diameters of the drive and of the driven pulleys. The drive pulley is the pulley connected to the motor shaft; the driven pulley is the pulley connected to the blower shaft. The relationship is as follows:

$$\text{Motor Speed} \times \text{Drive Pulley Diameter} = \text{Blower Speed} \times \text{Driven Pulley Diameter}$$

This formula assumes that the belts are positioned at the outermost edge of the pulleys, **Figure 18.36**. If the belts are resting in the pulley grooves with the outermost edge of the belt well within the pulley itself, use this formula: diameter of the pulley minus twice the distance between the top of the belt and the edge of the pulley, **Figure 18.37**.

Consider the pulley arrangement in **Figure 18.38**. The motor is turning at an actual speed of 3000 rpm and has a 4-in. pulley connected to its shaft. The blower shaft has a 6-in. pulley attached to it. How fast is the blower turning? From the formula, we get:

$$\text{Motor Speed} \times \text{Drive Pulley Diameter} = \text{Blower Speed} \times \text{Driven Pulley Diameter}$$

$$3000 \text{ rpm} \times 4 \text{ in.} = \text{Blower Speed} \times 6 \text{ in.}$$

$$\text{Blower Speed} = [(3000 \text{ rpm})(4 \text{ in.})] \div 6 \text{ in.}$$

$$\text{Blower Speed} = 2000 \text{ rpm}$$

Figure 18.36 A belt riding at the outer edge of the pulley.

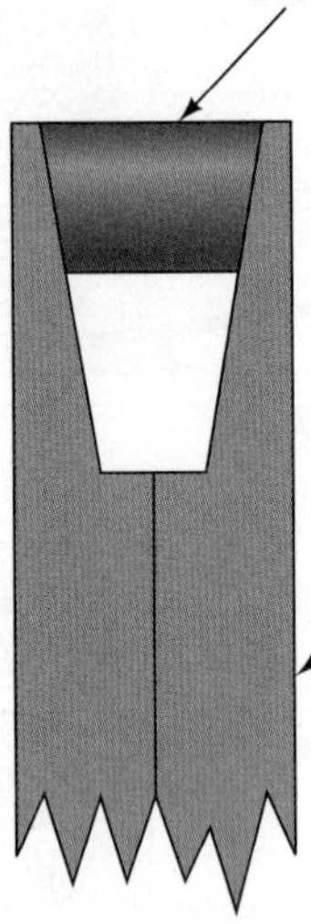

Figure 18.37 The belt is riding inside the pulley, so adjustments need to be made to the calculations.

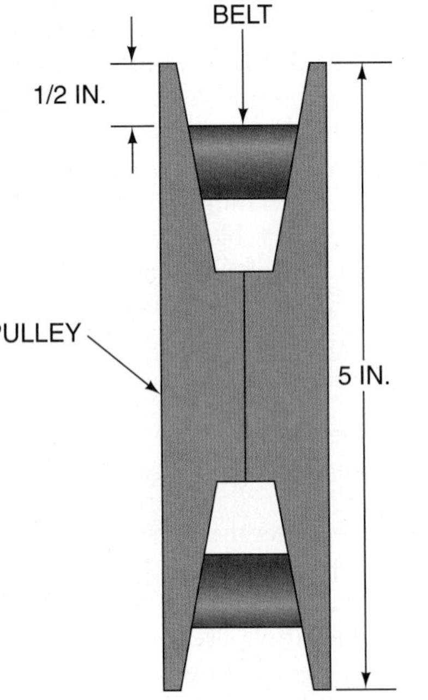

$$5 \text{ IN.} - \frac{1}{2} \text{ IN.} - \frac{1}{2} \text{ IN.} = 4 \text{ IN.}$$

Figure 18.38 A blower-speed example.

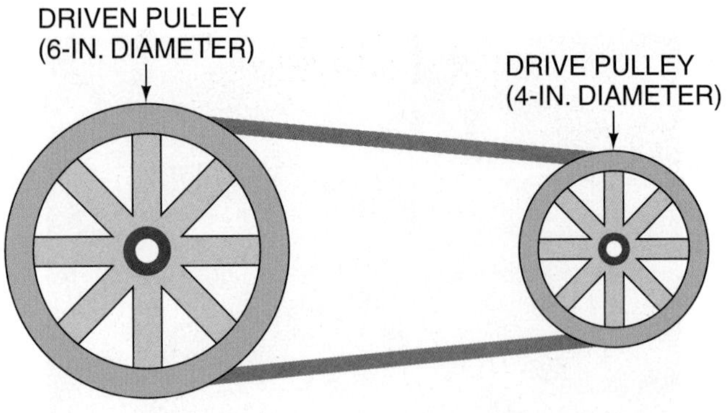

Figure 18.39 The correct method for installing belts over a pulley. The adjustment is loosened to the point that the belt may be passed over the pulleys.

Figure 18.40 A- and B-width belts.

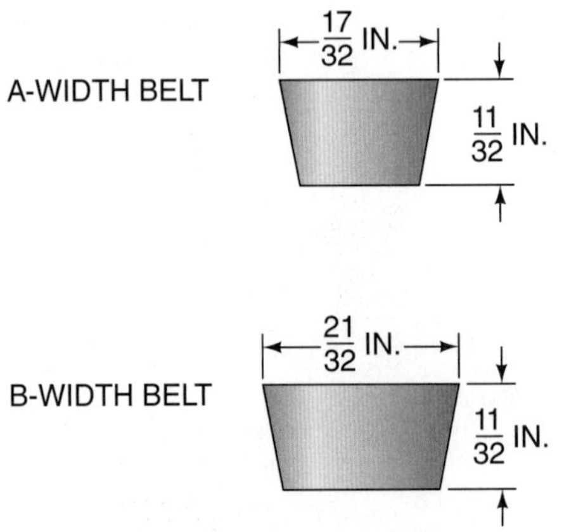

As long as three of the four values are known, the fourth can be computed using this very useful formula.

Belts are manufactured in different types and sizes. Some are made with different fibers to prevent stretching. **SAFETY PRECAUTION:** *Handle belts carefully during installation. A belt designed for minimum stretch must not be installed by forcing it over the side of the pulley because it may not stretch enough. Fibers will break and the belt will be weakened,* **Figure 18.39**. *Do not get your fingers between the belt and the pulley. Never touch the belt when it is moving.*•

Belt widths are denoted "A" and "B." An A-width belt must not be used with a B-width pulley nor vice versa, **Figure 18.40**. Belts can also have different grips, **Figure 18.41**. When a drive has more than one belt, the belts must be matched. Two belts marked with the same length are not necessarily matched.

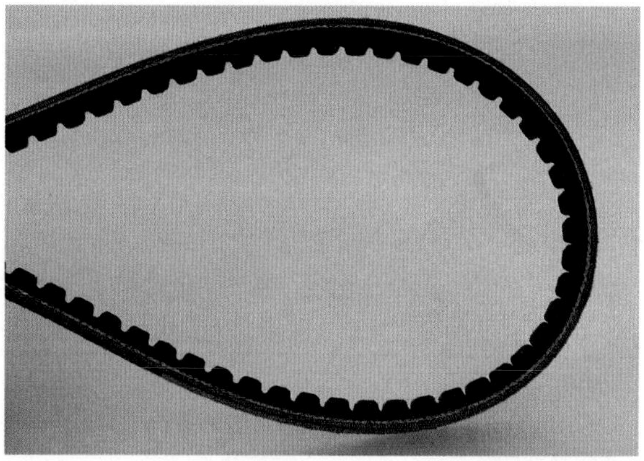

Figure 18.41 This belt with grooves has a tractor-type grip. Photo by Bill Johnson

They may not be exactly the same length. A *matched* set of belts means the belts are *exactly* the same length. A set of 42-in. belts *marked* as a matched set means each belt is exactly 42 in. If the belts are not marked as a matched set, one may be 42 1/2 in. and the other may be 41 3/4 in. Thus the belts will not pull evenly—one belt will take most of the load and will wear out first.

Belts and pulleys wear like any moving or sliding surface. When a pulley begins to wear, the surface roughens and wears out the belts. Normal pulley wear is caused by use or running time. Belt slippage will cause premature wear. Pulleys must be inspected occasionally, **Figure 18.42**.

Belts must have the correct tension, or they will cause the motor to operate in an overloaded condition. A belt tension gauge should be used to correctly adjust belts to the proper tension. This gauge is used in conjunction with a chart that gives the correct tension for different types of belts of various lengths, **Figure 18.43**.

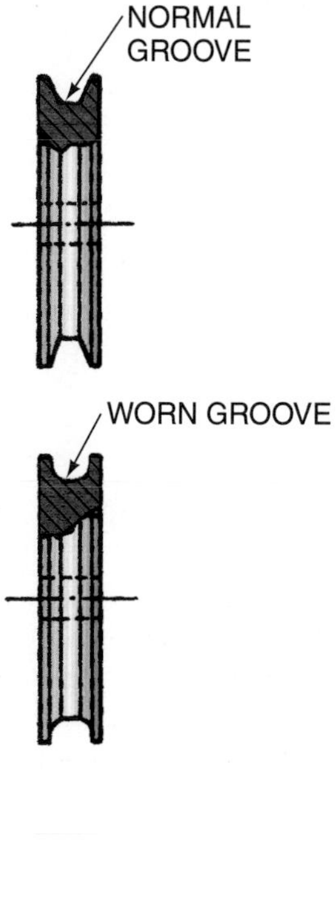

Figure 18.42 A normal and a worn pulley.

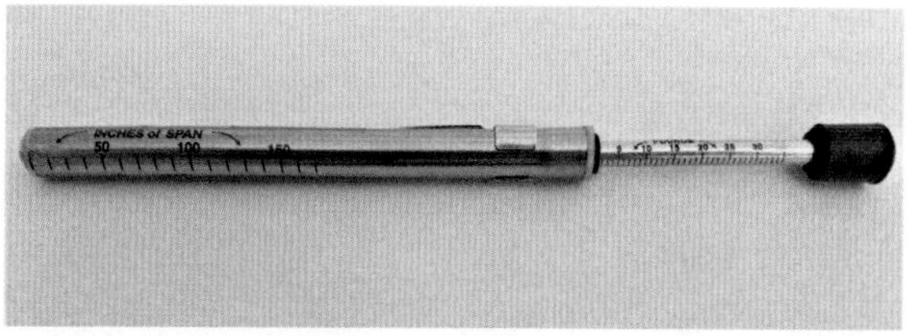

Figure 18.43 Belt tension gauge. Photo by Jason Obrzut

Direct-drive motors are normally used with fans, pumps, and compressors. Small fans and hermetic compressors are direct drive, but the motor shaft is actually an extended shaft with the fan or compressor on the end. The technician can do nothing to alter these. When used in an open-drive application, some sort of coupling must be installed between the motor and the driven device. Some couplings are connected with springs to absorb small amounts of vibration from the motor or pump.

A more complicated coupling is used between the motor and a larger pump or a compressor, **Figure 18.44**. This coupling and shaft must be in very close alignment or vibration

(A)

Figure 18.44 (A) A more complicated coupling used to connect larger motors to large compressors and pumps. Source: Lovejoy, Inc.

Figure 18.44 *(Continued)* (B) Flexible coupling for connecting compressors to motors. Courtesy Ferris State University; Photo by John Tomczyk

(B)

will occur. The motor shaft must be parallel to the compressor or pump shaft. Alignment is a very precise operation and is done by experienced technicians using a dial indicator. If two shafts are aligned to within tolerance while the motor and driven mechanism are at room temperature, the alignment must be checked again after the system is run long enough to get it up to operating temperature. The motor may not expand and move the same distance as the driven mechanism, and the alignment may need to be adjusted to the warm value, **Figure 18.45**.

Figure 18.45 The alignment of the two shafts must be very close for the system to operate correctly.

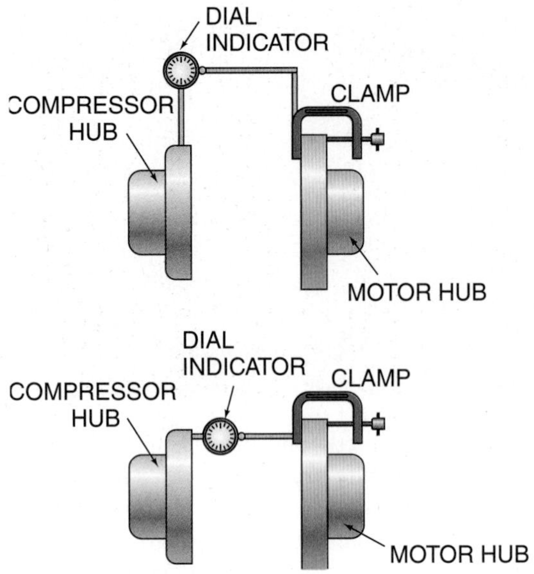

When an old motor must be replaced, try to use an exact replacement. An exact replacement may be found at a motor supply house or it may have to be obtained from the original equipment manufacturer. If the motor is not normally in stock, much time can be saved by taking the old motor to the distributor or supply house so that the motor's specifications can be matched.

REVIEW QUESTIONS

1. Power supply, work requirements, motor insulation class, bearing types, and mounting characteristics are all factors that can influence ________.
   A. motor selection
   B. motor rotation
   C. belt selection
   D. pulley diameter
2. Power supply data should include which of the following?
   A. Bearing types, mounting characteristics, and drive mechanism
   B. Voltage, service factor, shaft rotation, and capacitance
   C. Service factor, service factor amperage, insulation class, and capacitance
   D. Voltage, current capacity, frequency, and phase
3. The voltage range for an electric motor is usually within + or − ________%.
   A. 1
   B. 5
   C. 10
   D. 20
4. If an electric motor is rated for 208 V, the maximum and minimum allowable voltages would be ________.
   A. 208 and 240
   B. 187.2 and 228.8
   C. 197.7 and 215.3
   D. 187.2 and 238.8
5. Using evaporator pressure, condensing pressure, and compressor current, which of the following can be used to verify proper compressor operation?
   A. Amperage performance chart
   B. Superheat calculator
   C. Belt tension tool
   D. Power supply data
6. In the United States, the frequency in cycles per second (cps) at which an AC electric motor operates under normal conditions is _______.
   A. 50
   B. 60
   C. 120
   D. 208

7. The motors most often used in a residential application are __________ motors.
   A. 460 V
   B. 277 V
   C. three-phase
   D. single-phase
8. The two most common types of bearings used on small motors are __________ bearings.
   A. resilient and rigid
   B. cradle and sleeve
   C. sleeve and ball
   D. ball and rigid
9. An open single-phase motor with a centrifugal starting switch may not be used in __________.
   A. an air-conditioning condenser fan motor
   B. an air-conditioning indoor air handler
   C. an area where explosive gases exist
   D. a ventilating fan motor
10. A term used to describe a motor's reserve horsepower is ________.
    A. thrust surface
    B. service factor
    C. full-load amps
    D. locked-rotor amps
11. The motor parameter that is used to determine how hot the motor can be safely operated is __________.
    A. insulation class or type
    B. service factor (SF)
    C. revolutions per minute (rpm)
    D. service factor amperage (SFA)
12. A ________ bearing surface must be used in a vertical shaft motor application.
    A. sleeve
    B. ball
    C. greased
    D. thrust
13. Two primary electric-motor mount characteristics are __________ mounts.
    A. sleeve and ball
    B. oil port and permanently lubricated
    C. rigid and resilient
    D. belt and direct drive
14. A(n) __________ is used to verify that the belt on a belt-driven motor is installed properly.
    A. ammeter
    B. digital multimeter
    C. flexible coupling
    D. belt tensioning tool
15. Two types of motor drives are ________ drives.
    A. rigid and resilient
    B. belt and direct
    C. sleeve and ball
    D. oil port and permanently lubricated
16. A motor that has a service factor of 1.10 means that it can operate _____% of its rated horsepower for a limited period of time.
    A. 10
    B. 0.10
    C. 1.0
    D. 110
17. The current that a motor is drawing at its service factor loading is referred to as _________.
    A. full-load amperage (FLA)
    B. locked-rotor amperage (LRA)
    C. service factor amperage (SFA)
    D. rated-load amperage (RLA)
18. The two transformer configurations commonly used for commercial and industrial applications are ________.
    A. 50 and 60 cycles per second (cps)
    B. single-phase and three-phase
    C. wye and delta
    D. primary and secondary
19. An instrument often used to check the alignment of two shafts is the
    A. micrometer.
    B. dial indicator.
    C. manometer.
    D. sling psychrometer.
20. Which type of motor mount requires a ground strap to carry current from the motor frame should the motor become grounded?
    A. Rigid-mount
    B. Cradle-mount
    C. End-mount
    D. Resilient-mount
21. A belt-driven blower assembly is equipped with a motor that turns a 5-in. drive pulley at a speed of 3450 rpm. The blower is turning at a speed of 1725 rpm. Determine the size of the driven pulley.
22. Briefly describe what is meant by the term service factor as it applies to a motor.
23. Briefly explain what is meant by the term full-load amperage (FLA) as it applies to an electric motor.
24. Briefly explain what is meant by the term rated-load amperage (RLA) as it applies to compressor motors.
25. Briefly describe how a compressor amperage performance chart is used.

UNIT 19

MOTOR CONTROLS

OBJECTIVES

After studying this unit, you should be able to

- describe the differences between a relay, a contactor, and a starter.
- state how the locked-rotor current of a motor affects the choice of a motor starter.
- list the basic components of a contactor and starter.
- describe the purpose and function of motor protection devices.
- compare two types of external motor overload protection.
- describe conditions that must be considered when resetting safety devices to restart electric motors.

SAFETY CHECKLIST

- ✓ When a motor is stopped for safety reasons, do not restart it immediately. If possible, determine the reason for the overload condition before restarting.
- ✓ Conductor wiring should not be allowed to pass too much current or it will overheat and cause conductor failure or fire.
- ✓ When replacing or servicing line voltage components, ensure that the power is off, that the panel is locked and tagged, and that you have the only key.

19.1 INTRODUCTION TO MOTOR CONTROL DEVICES

This unit concerns those automatic components used to close and open the power supply circuit to the motor. These devices are called relays, contactors, and starters.

For example, a compressor in a residential central air conditioner is controlled in the following manner. With a temperature rise in the space to be conditioned, the thermostat contacts close and pass low voltage to energize the coil on the compressor contactor. This closes the contacts in the compressor contactor, allowing the applied or line voltage to pass to the compressor motor windings. Relays, contactors, and starters are the different names given to these motor controls, even though they perform the same function (but with some simple differences). Relays are designed to control loads that draw less than 20 A of electric current; contactors are designed to control loads that draw more than 20 A of current. Motor starters are, simply put, three-pole contactors with motor overload protection built into the device. More details will be presented as we discuss each of these system components.

Figure 19.1 is a diagram of a typical motor circuit. In this circuit there is a contactor as well as a potential magnetic relay (PMR) that is acting as a motor-starting relay. The contactor passes power to the motor on an external call for motor operation, and the starting relay helps the motor start on a call for operation. Notice the contactor is a major part of the motor circuit even though it is external to the motor windings. Power is fed to the motor windings and starting relay from the contactor.

The size of the motor and the application usually determine the type of switching device used (relay, contactor, or starter). For example, a small manual switch can start and stop a handheld hair drier, and a person operates the switch. A large 100-hp motor that drives an air-conditioning compressor must start, run, and stop unattended. It will also consume much more current than the hair drier. The components that start and stop large motors must be more elaborate than those that start and stop small motors.

Figure 19.1 Typical motor wiring diagram utilizing a contactor and a potential magnetic relay (PMR) as a motor-starting relay.

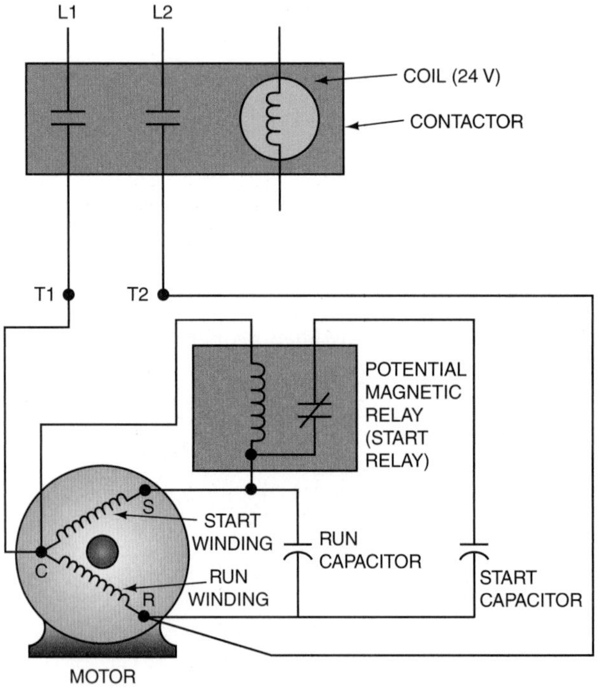

19.2 FULL-LOAD AND LOCKED-ROTOR AMPERAGE

Electric motors have two current (amperage) ratings: the full-load amperage (FLA), and the locked-rotor amperage (LRA). As the load or torque on a motor increases, the amperage draw required to power the motor will also increase. When the full-load torque and horsepower on the motor is reached, the corresponding amperage the motor is drawing is referred to as the full-load amperage (FLA). This value is determined in a test laboratory and recorded on the nameplate of the motor. It is used to select the correct wire size, overload, and motor starter devices to serve and protect the motor. The LRA is the current drawn by the motor just as it begins to start but before the rotor begins to rotate. The LRA of a motor is typically five times higher than the FLA of the motor and can be as high as seven times the FLA. Both currents must be considered when choosing the component (relay, contactor, starter) that passes the line voltage to the motor. Refer to Unit 18, "Application of Motors," Section 18.2 "The Power Supply" for a more detailed explanation of RLA, FLA, and LRA.

19.3 THE RELAY

The relay has a magnetic coil that opens or closes one or more sets of contacts, **Figure 19.2**. It is considered a throwaway device because parts are not available for rebuilding it when it no longer functions properly.

Relays are designed for light-duty applications. *Pilot-duty* relays can switch (on and off) larger contactors or starters. These relays are not designed to directly start motors. Relays designed for starting motors are not really suitable as switching relays because they have more resistance in the contacts. The pilot-duty relay contacts are often made of a fine silver alloy and designed for low-level current switching. Use on a larger load would melt the contacts. Heavier-duty motor switching relays are often made of silver cadmium oxide with a higher surface resistance and are physically larger than pilot-duty relays.

If a relay starts the indoor fan in the cooling mode of a central air-conditioning system, it must be able to withstand the inrush current of the fan motor on start-up, **Figure 19.3**. (A motor normally has a starting current of five times the running current, but only for a very short time.) Relays are often rated in horsepower: If a relay is rated for a 3-hp motor, it will be able to stand the inrush or locked-rotor current of a 3-hp motor as well as the run current.

A relay may have more than one type of contact configuration. It could have two sets of contacts that close when the magnetic coil is energized, **Figure 19.4**, or it may have two sets of contacts that close and one set that opens when the coil is energized, **Figure 19.5**. A relay with a single set

Figure 19.2 General-purpose relay typically used for starting a motor that has a relatively low amperage draw. This relay has a holding coil, one normally open set of contacts, and one normally closed set of contacts. Both sets of contacts change position when the holding coil is energized. Photo by Jason Obrzut

Figure 19.3 A throwaway fan relay that may be used to start an evaporator fan motor on a central air-conditioning system. Photo by Jason Obrzut

Figure 19.4 A double-pole–single-throw relay with two sets of contacts that close when the coil is energized.

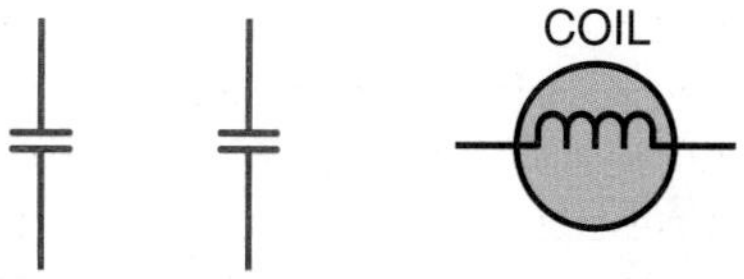

Figure 19.5 A relay with two sets of contacts that close and one set of contacts that opens when the coil is energized. It has two normally open (NO) sets and one normally closed (NC) set.

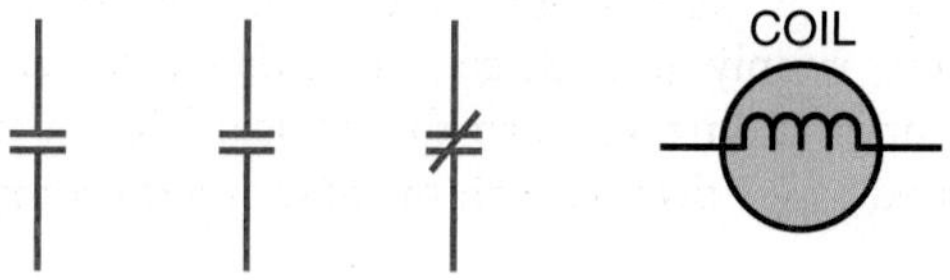

Figure 19.6 Some of the more common configurations of contacts found on relays.

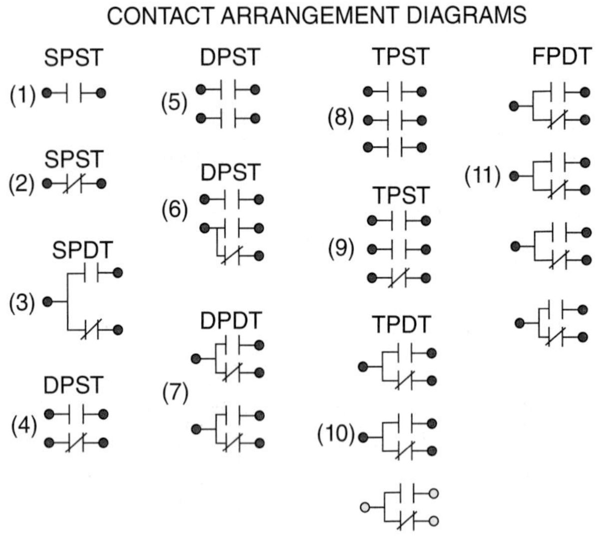

of contacts that close when the coil is energized is called a *single-pole–single-throw, normally open relay* (spst, NO). A relay with two contacts that close and one that opens is called a *triple-pole–single-throw,* with two *normally open* contacts and one *normally closed contact* (tpst). **Figure 19.6** shows some different relay contact arrangements.

Most relays have wiring diagrams printed on the side of the device. These diagrams identify the coil terminals as well as the normally open and normally closed contacts. **Figure 19.7** shows the wiring diagram of the relay shown in **Figure 19.2.** By referring to the diagram and identifying these marked terminals on the relay itself, replacing one relay with another in the event of component failure is not a complicated process. Many relays use terminals 1 and 3 for the coil, terminals 2 and 4 for a normally open set of contacts, and terminals 2 and 5 for a normally closed set of contacts.

Figure 19.7 Wiring diagram of the relay shown in **Figure 19.2.** The coil is connected between terminals 1 and 3. Terminals 2 and 4 are connected to the NO contacts and terminals 4 and 5 are connected to the NC contacts.

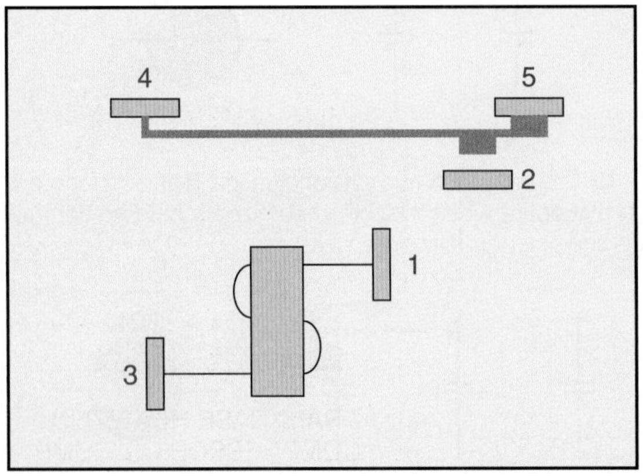

19.4 THE CONTACTOR

The **contactor** is a larger version of a relay, **Figure 19.8.** It can be rebuilt if it fails, **Figure 19.9.** A contactor has movable and stationary contacts. The holding coil can be designed for various operating voltages: 24-V, 115-V, 208-V to 230-V, or 460-V operation.

Figure 19.8 (A) A common double pole contactor. Photo by Bill Johnson (B) Contactors for larger motors. There are circuit breakers and current sensing coils before each contactor. The current sensors monitor current flow through the motors for feedback to a microprocessor providing running status, overload protection, and remote monitoring. Source: Meijers Corporation. Photo by John Tomczyk

(A)

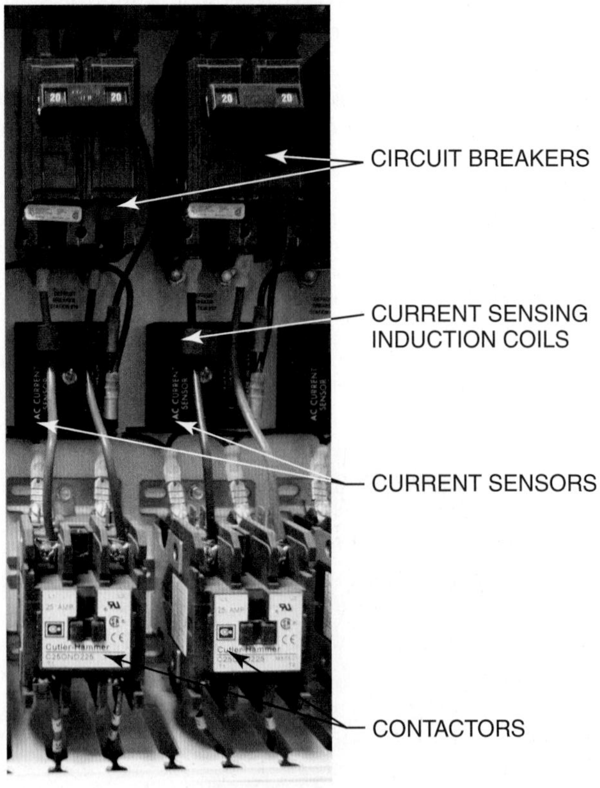

(B)

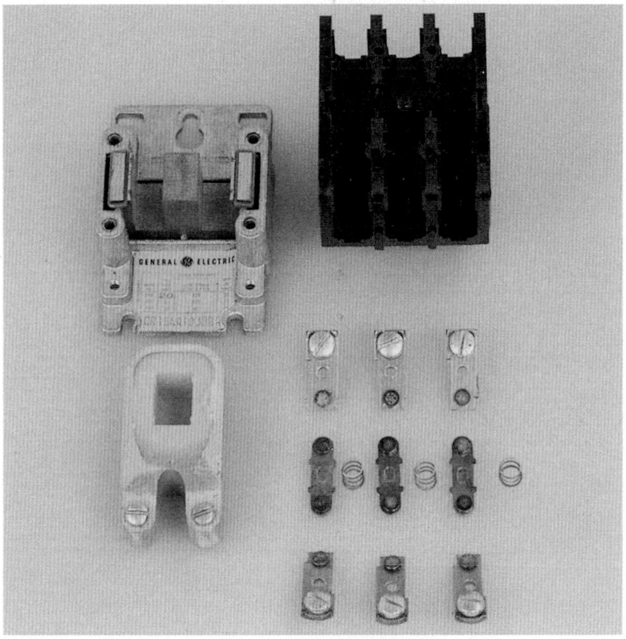

Figure 19.9 The parts of a contactor that can be replaced: the contacts (both moving and stationary), the holding coil, and the contact holding springs. Photo by Bill Johnson

Contactors can be very small or very large, but all have similar parts. A contactor may have many configurations of contacts, from the most basic, which has just one contact, to more elaborate types with several. The single-contact contactor is used on many residential air-conditioning units. **Figure 19.8(B)** shows a larger compressor motor contactor that incorporates a current induction coil. This coil monitors current flow through the motor for feedback to a microprocessor for overload protection and remote monitoring of the compressor motor. It interrupts power to one leg of the compressor, which is all that is necessary to stop a single-phase compressor. Sometimes a single-contact, single-pole, contactor is needed to provide crankcase heat to the compressor. A trickle charge of electricity passes through line 2 to the motor windings and crankcase heater when the contacts are open, during the off cycle, **Figure 19.10**. If you use a contactor with two contacts for a replacement, the compressor would not have crankcase heat. Once again, an exact replacement is the best choice. **Figure 19.11** shows another method of supplying crankcase (oil sump) heat only during the off cycle.

Some contactors have as many as five or six sets of contacts. Generally, for large motors there will be three heavy-duty sets to start and stop the motor; the rest (*auxiliary contacts*) can be used to switch auxiliary circuits.

Auxiliary contacts are rated at lower amperages than the main contacts. The auxiliary contacts are typically used to open and close control circuits and to energize and deenergize other, smaller system loads that have low current requirements. Auxiliary contacts can be normally open or normally closed and are often mounted to the side of the contactor, where its

Figure 19.10 This contactor (a single-pole contactor) has only one set of contacts. Only one line of power needs to be broken to stop a motor like the one in the diagram. This is a method commonly used to supply crankcase heat to the compressor during the off cycle. When the contacts close, current will bypass the heater.

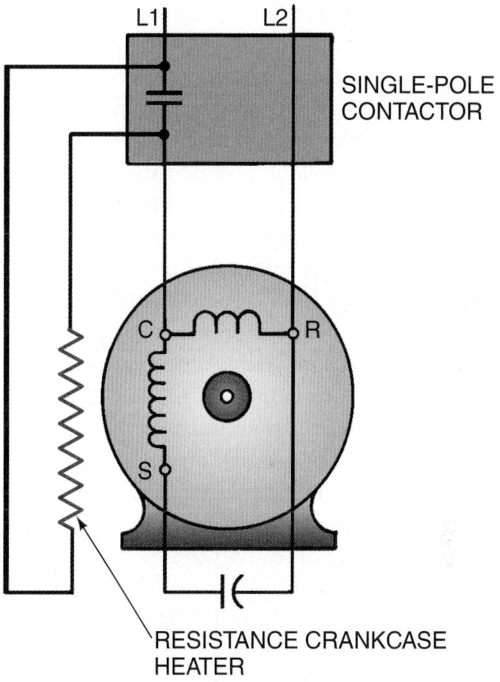

Figure 19.11 A contactor with auxiliary contacts for switching other circuits.

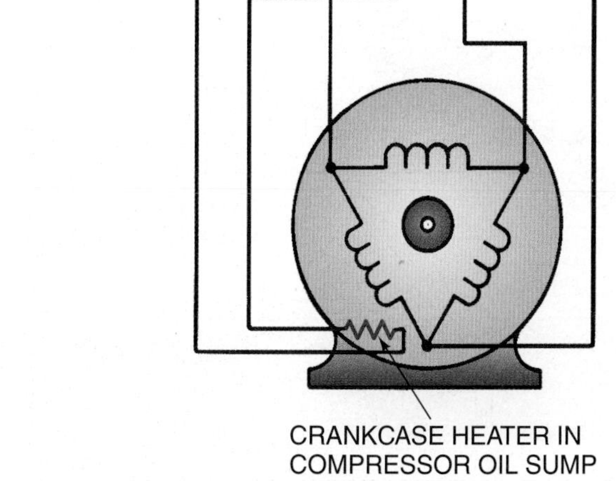

moving armature can change the position of the auxiliary contacts at the same moment the main contacts change position.

19.5 MOTOR STARTERS

The *starter,* or motor starter as it is sometimes called, is similar to the contactor. In some cases, a contactor may be converted to a motor starter. The motor starter differs from a contactor because it has motor overload protection built into its framework, **Figures 19.12** and **19.13**. A motor starter may be rebuilt and is available in a wide range of sizes.

Motor protection is discussed more fully later in this unit, but you should know that overload protection protects a particular motor. **Figure 19.14** shows a melting alloy-type overload heater. The fuse or circuit breaker cannot be wholly relied on for motor protection because it protects the entire circuit, not the circuit's individual components. In some cases, the motor-starter protection can indicate a motor problem better than can the motor-winding thermostat protection.

The contact surfaces become dirtier and more pitted with each motor starting sequence, **Figure 19.15**. Some technicians believe that these contacts may be buffed or cleaned with a file or sandpaper. **NOTE:** *Contacts may be cleaned, but this should be done as a temporary measure only. Filing or sanding exposes base metal under the silver plating and speeds deterioration. Replacing the contacts is the only recommended repair.* • If the device is a relay, the complete relay must be replaced. If the device is a contactor or a starter, the contacts may be

Figure 19.12 A motor starter equipped with motor overload heaters that are interchangeable and can be sized to fit motors of different horsepower. The overload contact can be manually reset. The manual reset feature prevents excessive wear and tear on the motor by preventing short cycling. Courtesy Ferris State University. Photo by John Tomczyk

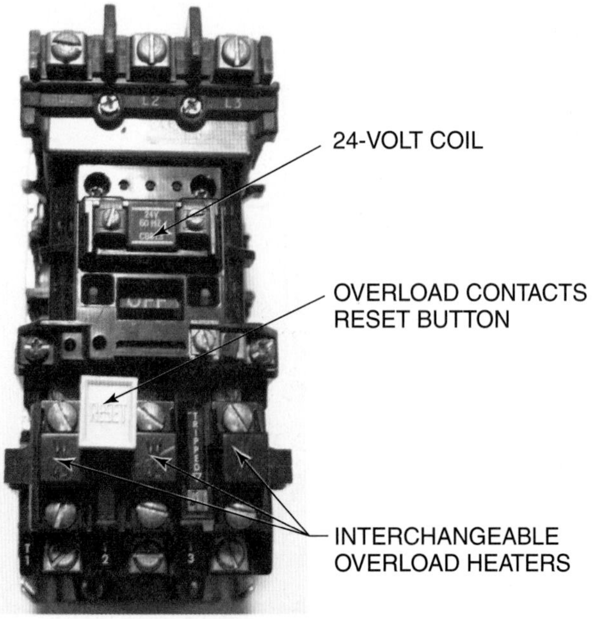

Figure 19.13 Components that may be changed on the motor starter: contacts (both moving and stationary), the springs, the coil, and the overload protection devices (heaters and switches). Photo by Bill Johnson

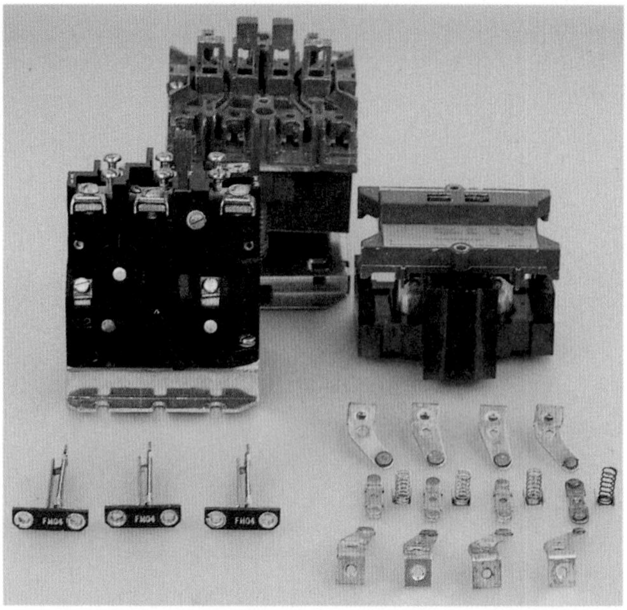

Figure 19.14 A melting alloy-type overload heater. Photo by Jason Obrzut

Figure 19.15 Clean contacts contrasted with a set of dirty, pitted contacts. Source: Square D Company

replaced with new ones, **Figure 19.15**. Both movable and stationary contacts and springs hold tension on the contacts.

You can use a digital multimeter to check resistance across a set of contacts under full load. When the meter leads are placed on each side of a set of contacts and measure a voltage, there is resistance in the contacts. When the contacts are new and have no resistance, the meter should read no voltage. An old set of contacts will have some slight voltage drop and will produce heat, due to resistance in the contact's surface, where the voltage drop occurs. **Figure 19.16** shows an example of how contacts may be checked with a multimeter.

Figure 19.16 Checking contacts with a digital multimeter shows resistance in the contacts caused by dirty surfaces. Make certain that there is no current flowing through the contacts when checking resistance.

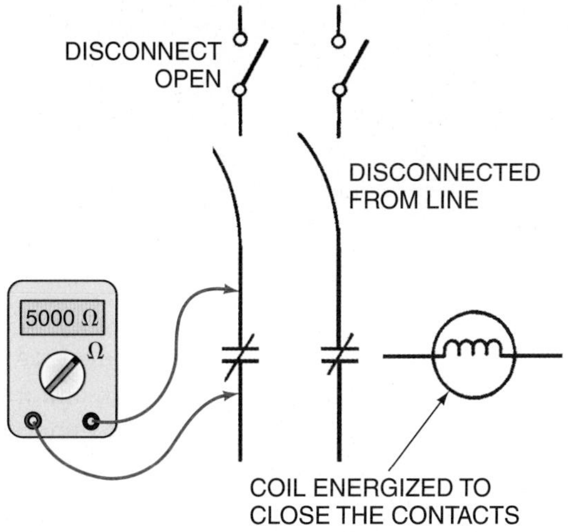

Each time a motor starts, the contacts will be exposed to the inrush current (the same as LRA). The contacts are under a tremendous load for this moment. When the contacts open, the breaking of the electrical circuit causes an arc. The contacts must make and break as fast as possible. A magnet pulls the contacts together with a mechanism to take up any slack. For example, three large movable contacts in a row may all need to be held equally tight against the stationary contacts. The springs behind the movable contacts keep this pressure even, **Figure 19.17**. If there is a resistance in the contact surface, the contacts may get hot enough to take the tension out of the springs, which will make the contact-to-contact pressure even less and the heat greater. The tension in these springs must be maintained.

19.6 MOTOR PROTECTION

The electric motors used in air-conditioning, heating, and refrigeration equipment are the most expensive single components in the system. These motors consume large amounts of electrical energy, and the motor windings undergo considerable stress. Therefore, they deserve the best protection possible within the economic boundaries of a well-designed system. The more expensive the motor, the more expensive and elaborate the protection should be.

Fuses are used as circuit (not motor) protectors. **SAFETY PRECAUTION:** *The conductor wiring in the circuit must not be allowed to pass too much current or it will overheat and cause conductor failures or fires.* • A motor may be operating in an overloaded condition that would not cause the conductor to be overloaded; hence, the fuse will not open the circuit. Let us

Figure 19.17 When energized, springs will hold the three movable contacts tightly against the stationary contacts.

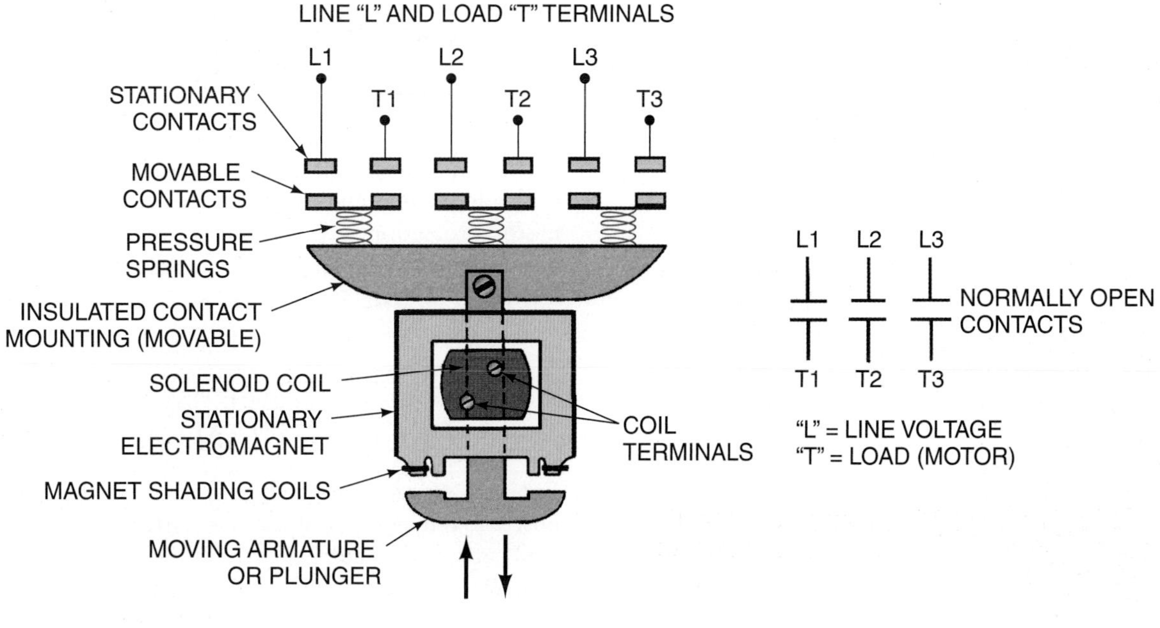

use a central air-conditioning system as an example. There are two motors in the condensing unit: the compressor (the largest) and the condenser fan motor (the smallest). In a typical unit, the compressor may have a current draw of 23 A and the fan motor may use 3 A. The fuse protects the total circuit. The circuit will be fused to protect the circuit wiring. If one motor is overloaded, the fuse may not open the circuit, **Figure 19.18**. Each motor needs to be protected within its own operating range. The condenser fan motor and the compressor will have an inherent (internal) motor protector that protects the fan motor at 3 A and the compressor at 23 A.

Motors can operate without harm for short periods under a slight overcurrent condition. The overload protection is designed to disconnect the motor at some current draw value that is slightly more than the FLA value so that the motor can be operated at its full-load design capacity. Time is involved in this value in such a manner that the higher the current value above FLA, the more quickly the overload should react. In other words, the amount of the overload and the time are both figured into the design of the particular overload device.

Overload current protection is applied to motors in different ways. Overload protection, for example, is not needed for some small motors that will not cause circuit overheating or will not damage themselves. Some small motors do not have overload protection because they will not consume enough power to damage the motor unless shorted from winding to winding or from the winding to the frame (to ground).

Figure 19.18 Two motors in the same circuit are served by the same conductors and have different overload requirements. The main circuit fuses protect the entire circuit and not the individual circuit components.

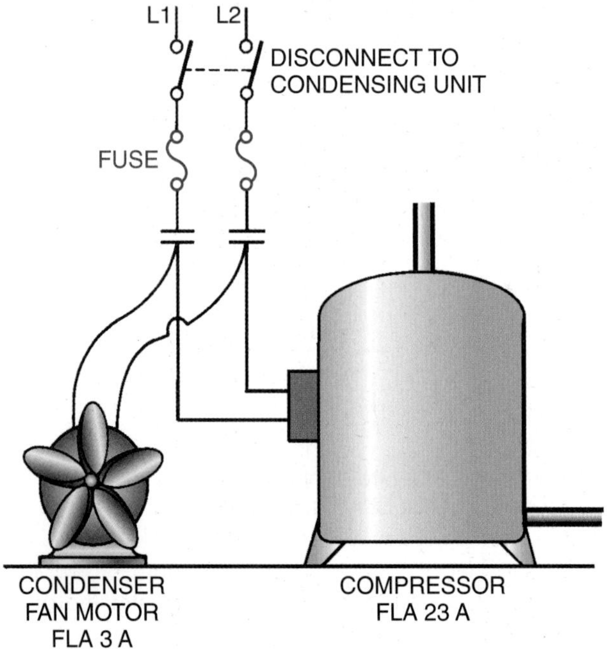

Figure 19.19 A small condenser fan motor that does not pull enough amperage at locked-rotor amperage (LRA) to create enough heat to be a problem. This is known as impedance motor protection. Courtesy Ferris State University. Photo by John Tomczyk

See **Figure 19.19** for an example of a small condenser fan motor that does not draw enough amperage at the LRA condition to overheat. It is not an expensive motor, and it does not have overload protection; it is described as "impedance protected." It may only generate 10 W of power, the same as a 10-W light bulb. If the motor fails because of a burnout, the current draw will be interrupted by the circuit protector.

Overload protection is divided into inherent (internal) protection and external protection.

19.7 INHERENT MOTOR PROTECTION

Inherent (internal) overload protection is provided by internal thermal overloads either embedded in, or located near, the motor's windings, **Figure 19.20(B)**. External overload motor protection is provided by thermally activated snap-discs (bimetals) located on the compressor's exterior shell, **Figures 19.20(A)**, **(C)**, and **(D)**.

19.8 EXTERNAL MOTOR PROTECTION

External protection is often applied to the device passing power to the motor contactor or starter. These devices are normally actuated by current overload and break the circuit to the contactor coil. The contactor stops the motor. When started

Figure 19.20 (A) External bimetal overload protectors. Courtesy Ferris State University. Photo by John Tomczyk (B) A three-phase internal overload protector usually located in the motor's shell near the windings. Courtesy Ferris State University. Photo by John Tomczyk (C and D) Internal overload protector with an external manual reset. Courtesy Ferris State University. Photo by John Tomczyk

(A)

(B)

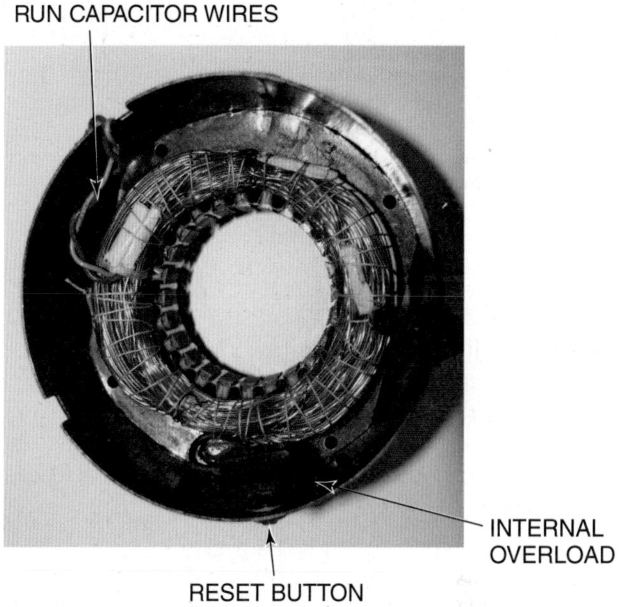

(C)

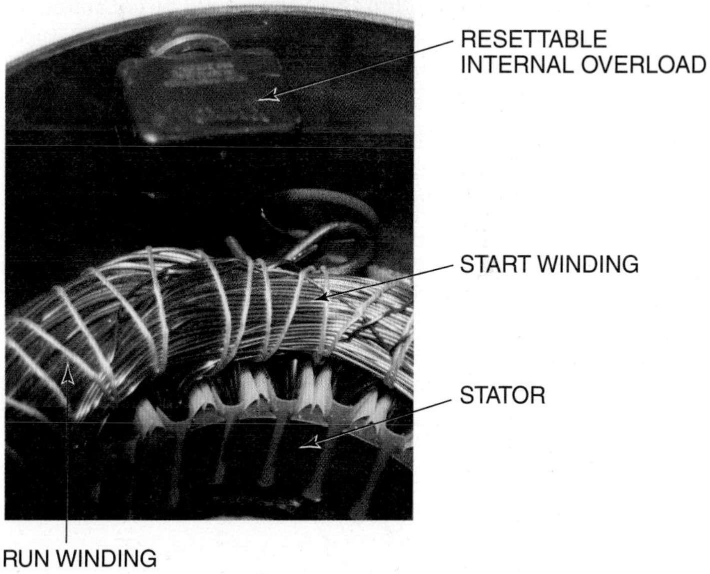

(D)

with a relay, the motor is normally small and has only internal protection, **Figure 19.21**. Contactors are used to start larger motors, and either inherent protection or external protection is used. Large motors (above 5 hp in air-conditioning, heating, and refrigeration systems) use starters and overload protection built into either the starter or the contactor's circuit.

The value (trip point) and type of the overload protection are normally chosen by the system design engineer or by the manufacturer. The technician checks the overload devices when there is a problem, such as random shutdowns because of an overload tripping. The technician must be able to understand the designer's intent with regard to the motor's operation and the operation of the overload device because they are closely related in a working system.

Many motors have a *service factor*; this is the reserve capacity of the motor and is expressed as a percentage above the full-load amperage (FLA) of the motor. The motor can operate above the FLA and within the service factor without harm. Typical service factors are 1.15 to 1.40. A service factor of 1.15 indicates that the motor can operate at an amperage that is 15% higher than the nameplate operating current, whereas a motor with a service factor of 1.40 can operate at amperages that are 40% higher than nameplate. The smaller the motor, the larger the service factor. For example, a motor with a FLA of 10 A and a service factor of 1.25 can operate at 12.5 A (10 A × 1.25 = 12.5 A) without damaging the motor. The overload protection for a particular motor takes the service factor into account.

Figure 19.21 A fan motor that is normally started with a relay. Motor protection is internal. Photo by Jason Obrzut

19.9 NATIONAL ELECTRICAL CODE® STANDARDS

The *National Electrical Code®* (NEC) sets the standard for all electrical installations, including motor overload protection. The code book published by the National Electrical Manufacturer's Association (NEMA) should be consulted for any overload problems or misunderstandings that may occur regarding correct selection of the overload device.

The purpose of the overload protection device is to disconnect the motor from the circuit when an overload condition occurs. Detecting the overload condition and opening the circuit to the motor can be done by an overload device mounted on the motor starter or by a separate overload relay applied to a system with a contactor. **Figure 19.22** shows an example of a thermal overload relay.

Figure 19.22 An overload using a resistor-type heater that heats a low-temperature solder.

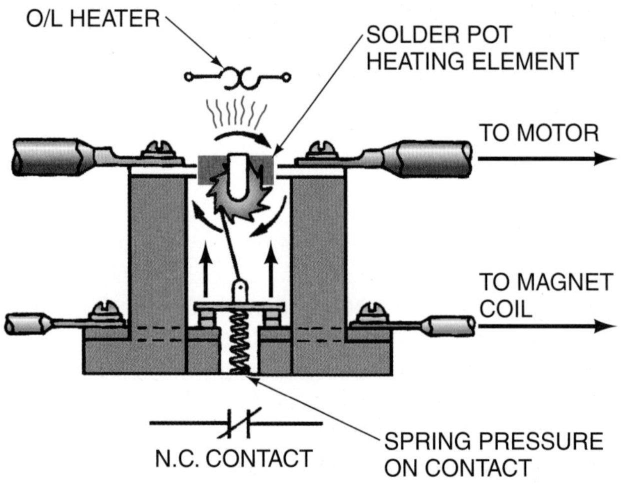

19.10 TEMPERATURE-SENSING DEVICES

Various sensing devices are used to detect overcurrent situations. The most popular ones are those sensitive to temperature changes.

The bimetal element is an example of such sensing devices. The line current of the motor passes through a heater (which can be changed to suit a particular motor amperage) that heats a bimetal strip. When the current is excessive, the heater warps the bimetal, opening a set of contacts that interrupt power to the contactor's coil circuit. All bimetal overload devices are designed with snap action to avoid excessive arcing. These thermal-type overloads are sensitive to any temperature and conditions around them, such as high ambient temperature and loose connections. **Figure 19.23** shows an example of a thermal overload with a loose connection.

A low-melting solder, called a **solder pot**, may be used in place of the bimetal. Heat caused by an overcurrent condition will melt the solder. The overload heater is sized for the particular amperage draw of the motor it is protecting. The overload control circuit will interrupt the power to the motor contactor coil and stop the motor in case of overload. The solder melts, and the overload mechanism turns because it is spring-loaded. It can be reset when it cools, **Figure 19.22**.

Both of these overload protection devices are sensitive to temperature. The temperature of the heater causes them to function. Heat from any source, even if it has nothing to do with motor overload, makes the protection devices more sensitive. For example, if the overload device is located in a metal control panel in the sun, the heat from the sun may affect the performance of the overload protection device. A loose connection on one of the overload device leads

Figure 19.23 The overload is tripping because of a loose connection.

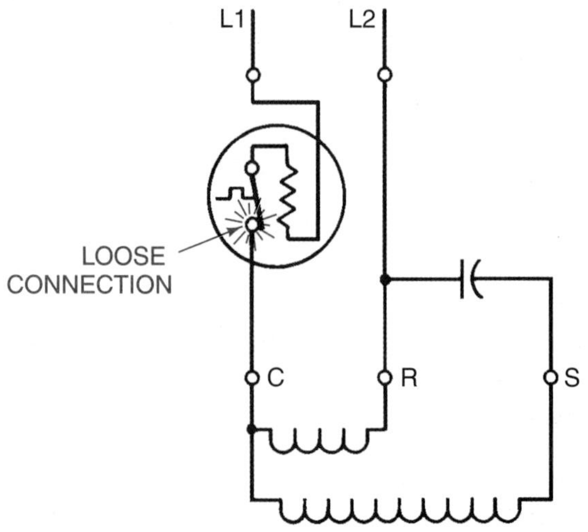

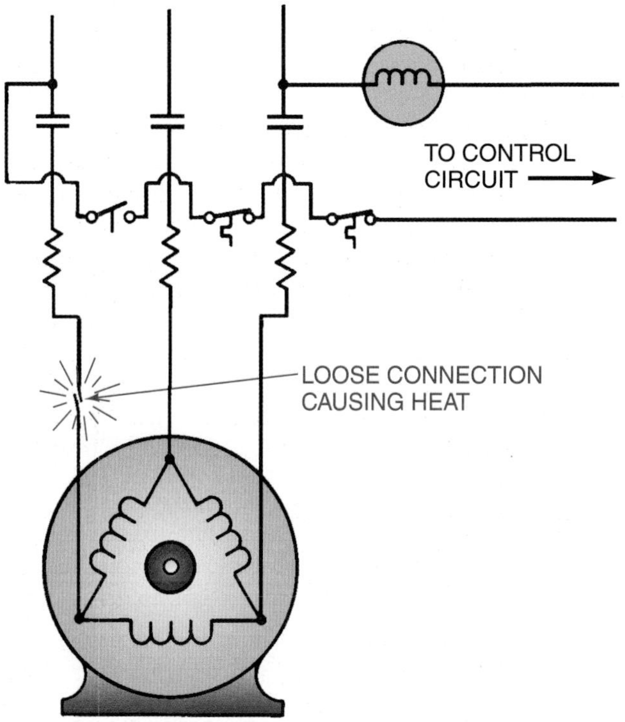

Figure 19.24 Local heat, such as from a loose connection, will influence the thermal-type overload.

will cause local heat and may cause it to open the circuit to the motor even though there is actually no overload, **Figure 19.24**.

19.11 MAGNETIC OVERLOAD DEVICES

Magnetic overload devices are separate components and are not attached to the motor starter. This type of component is very accurate and not affected by ambient temperature, **Figure 19.25**, so a magnetic overload device can be located in, say, a hot cabinet on the roof and will not be affected by the temperature. It will shut the motor off at an accurate ampere rating regardless of the temperature.

Figure 19.25 A magnetic overload device. Photo by Bill Johnson

19.12 RESTARTING THE MOTOR

SAFETY PRECAUTION: *When a motor has been stopped for safety reasons, such as an overload, do not restart it immediately. Look around for the possible problem before a restart.* • When a motor stops because it is overloaded, the overload condition at the instant the motor is stopped is reduced to 0 A. This does *not* mean that the motor should be restarted immediately. The cause of the overload may still exist, and the motor could be too hot and may need to cool.

There are various ways of restarting a motor after an overload condition has occurred. Some manufacturers design their control circuits with a manual reset to keep the motor from restarting, and some use a time delay to keep the motor off for a predetermined time. Others use a relay that will keep the motor off until the thermostat is reset. The units that have a manual reset at the overload device may require someone to go to the roof to reset the overload if that is where the unit is located. See **Figure 19.12** for an example of a manual reset. When the reset is in the thermostat circuit, the protection devices may be reset from the room thermostat. This is convenient, but several controls may be reset at the same time, and the technician may not know which control is being reset. When the manual reset button is pushed and a restart occurs, there is no doubt which control has been reset to start the motor. Time-delay reset devices keep the unit from short cycling but may reset themselves even though a problem condition still exists.

REVIEW QUESTIONS

1. The recommended repair for a defective relay is to ________.
   A. replace it
   B. send it to a repair shop
   C. repair it yourself
   D. bypass it

2. When rebuilding a contactor or starter, which of the following components can be changed?
   A. The contacts and springs only
   B. The coil and contacts only
   C. The contacts and coil only
   D. The contacts, coil, and springs

3. The two types of relays are ______.
   A. thermal and magnetic
   B. pilot duty and solder pot
   C. pilot duty and motor starting
   D. motor starting and solder pot
4. The two amperages that influence the choice for replacing a motor starter are ______.
   A. rated-load amperage (RLA) and locked-rotor amperage (LRA)
   B. rated-load amperage (RLA) and reverse-rotation amperage (RRA)
   C. locked-rotor amperage (LRA) and start-up amperage (SA)
   D. service factor amperage (SFA) and reverse-rotation amperage (RRA)
5. What is the difference between a contactor and a starter?
   A. A relay has overload protection built into, a starter does not.
   B. A starter has overload protection built into it, a relay does not.
   C. A relay can be repaired, a starter cannot.
   D. A relay is used to start large motors, a starter is used to start smaller motors.
6. Which of the following is considered when the overload protection for a motor is designed?
   A. The rotation of the motor shaft
   B. The size of the motor capacitor
   C. The motor mount that is used
   D. The motor's service factor
7. The contact surfaces of relays, contactors, and starters are made of ________.
   A. magnesium
   B. steel
   C. silver
   D. iron
8. An overload can be opened by ___________.
   A. heat or amperage
   B. pressure or heat
   C. amperage or pressure
   D. timer or heat
9. Which is not a typical operating voltage used for relays, contactors, and motor starters?
   A. 12 V
   B. 24 V
   C. 115 V
   D. 230 V
10. Why should contactor contacts not be filed or sanded?
    A. The shavings or filings will cause increased electrical resistance.
    B. The silver coating will be removed and cause more resistance in the contact surface.
    C. The contactor would have to be disassembled and it is not recommended to do so.
    D. It may cause the contactor coil to burn.
11. Why is it not a good idea to use circuit protection devices to protect large motors from overload conditions?
12. Describe how an external overload can shut down a motor.
13. Describe the difference between inherent and external overload protection.
14. What is the purpose of thermal overload protection at the motor?
15. What must be done before a motor overload can be reset?

UNIT 20

TROUBLESHOOTING ELECTRIC MOTORS

OBJECTIVES

After studying this unit, you should be able to

- describe different types of electric motor problems.
- describe the electrical instrumentation commonly used to evaluate motors.
- Identify the symptoms of a motor with an open winding.
- Identify the symptoms of a motor with a shorted winding.
- Identify the symptoms of a grounded motor.
- explain how loose electrical connection can affect the operation of a motor.
- explain the differences between a start and run capacitor.
- explain how to properly test a capacitor.
- explain how to evaluate a belt-driven motor assembly with a belt that is too tight or too loose.
- describe the symptoms of a belt-driven motor assembly with improperly aligned pulleys.
- explain the difference between troubleshooting a hermetic motor problem and an open motor problem.

SAFETY CHECKLIST

- ✓ If it is suspected that a motor has electrical problems, pull the motor disconnect to prevent further damage or an unsafe condition, lock and tag the panel, and keep the only key on your person.
- ✓ Before checking a motor capacitor, short from one terminal to the other with a 20,000-Ω, 5-W resistor to discharge the capacitor. This practice is recommended even if the capacitor has its own bleed resistor. Use insulated pliers.
- ✓ When wiring a motor-run capacitor be sure to connect the lead that feeds power to the capacitor to the terminal identified for this purpose.
- ✓ Turn the power off before trying to turn the open drive compressor over using a wrench.
- ✓ Never try to stop a spinning motor shaft with your hand, allow it to come to a complete stop on its own.

20.1 MOTOR TROUBLESHOOTING

Electric motor problems are either mechanical or electrical. Mechanical problems may appear to be electrical. For example, a bearing dragging in a small, permanent split-capacitor (PSC) fan motor may not make any noise. The motor may not start, and it appears to be an electrical problem. The technician must know how to diagnose the problem correctly. This is particularly true with open motors because if the driven component is stuck, a motor may be changed unnecessarily. If the stuck component is a hermetic compressor, the whole compressor must be changed; if it is a serviceable hermetic compressor, the motor can be replaced or the compressor running gear can be rebuilt.

20.2 MECHANICAL MOTOR PROBLEMS

Mechanical motor problems normally occur in the bearings or the shaft where the drive is attached. The bearings may be tight or worn due to lack of lubrication. Grit can easily get into the bearings of some open motors and cause them to wear. Overtightened belts can also put excessive stress on the motor's bearings, causing them to wear and fail prematurely.

Problems with large motors are not usually repaired by heating, air-conditioning, and refrigeration technicians. They are handled by technicians trained in rebuilding motors and rotating equipment. A motor vibration may require you to seek help from a qualified balancing technician. Explore every possibility to ensure that the vibration is not caused by a field problem, such as a fan loaded with dirt or liquid flooding into a compressor.

Motor bearing failure with roller and ball bearings can often be identified by the bearing noise. When sleeve bearings fail, they normally lock up (will not turn) or sag to the point that the motor is out of its magnetic center. At this point the motor will not start. When motor bearings fail, they can be replaced. If the motor is small, the motor is normally replaced because it would cost more to change the bearings than to purchase and install a new motor. The labor involved in obtaining bearings and disassembling the motor can take too much time to be profitable. This is particularly true for fractional horsepower fan motors. These small motors almost always have sleeve bearings pressed into the end bells of the motor, and special tools may be needed to remove and install new bearings, **Figure 20.1**.

When deciding whether to repair or replace a motor, a number of factors should be considered. These factors include the availability of a new motor versus the availability of the parts needed to repair the motor, the cost of repairing the motor, the cost of the new motor, the labor involved in repairing as opposed to replacing the motor, and the warranty/guaranty on the new and repaired motors. If the cost of repairing the motor is close to the cost of replacing the

Figure 20.1 A special tool for removing bearings.

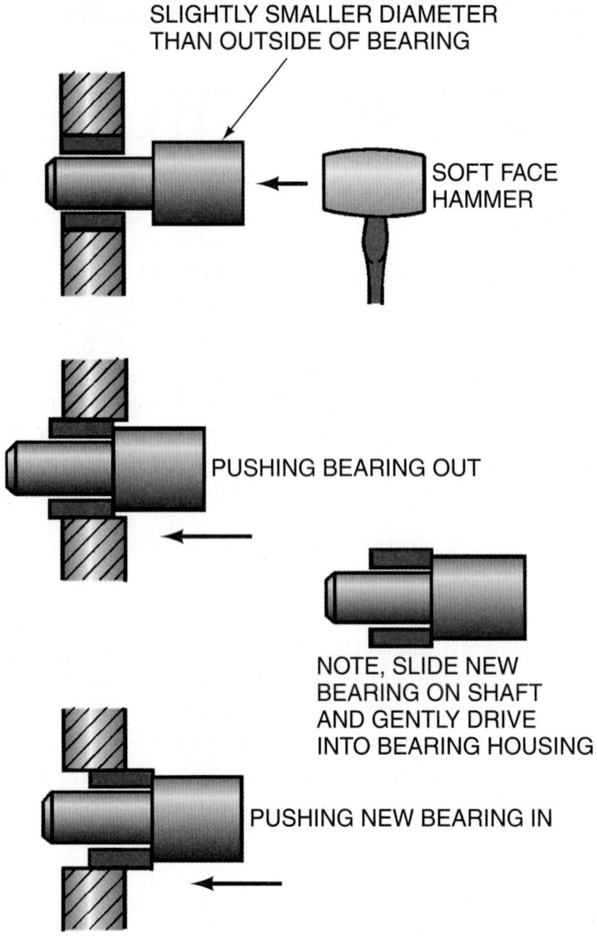

motor, for example, it would definitely pay to replace the motor if the new component carries a one-year warranty and the repaired motor carries only a 90-day guaranty.

20.3 REMOVING DRIVE ASSEMBLIES

To remove the motor, you need to remove the pulley, coupling, or fan wheel from the motor shaft. The fit between the shaft and whatever assembly it is fastened to may be very tight. **SAFETY PRECAUTION:** *Removing the assembly from the motor shaft must be done with care.* The assembly may have been running on this shaft for years, and there may be rust between the shaft and the assembly. You must remove the assembly without damaging it. Special pulley pullers, **Figure 20.2**, will help, but other tools or procedures may be required.

Most assemblies are held to the motor shaft with setscrews threaded through the assembly and tightened against the shaft. A flat spot is usually provided on the shaft for the seating of the setscrew and to keep it from damaging the shaft surface, **Figure 20.3**. The setscrew is made of very hard steel, much harder than the motor shaft. Larger motors with more torque normally have a matching keyway machined into the shaft and assembly. This keyway with a key provides a better bond between the motor and assembly, **Figure 20.4**. A setscrew is then often tightened down on the top of the key to secure the assembly to the motor shaft.

Many technicians make the mistake of trying to drive a motor shaft out of the assembly fastened to the shaft. In doing so, they blunt or distort the end of the shaft. When it is distorted, the motor shaft will never go through the assembly without damaging it, **Figure 20.5**. The shaft is made from mild steel and can be damaged easily. If the shaft must be driven, you may need to use as the driving tool a similar shaft with a slightly smaller diameter, **Figure 20.6**. Before attempting to

Figure 20.2 A pulley puller.

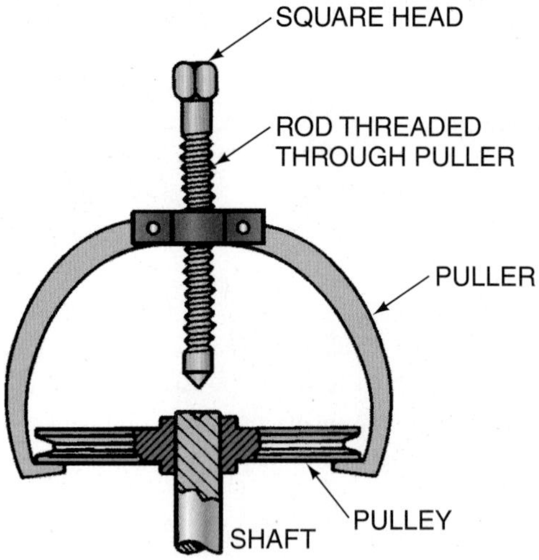

Figure 20.3 A flat spot on the motor shaft where the pulley setscrew is tightened. Photo by Jason Obrzut

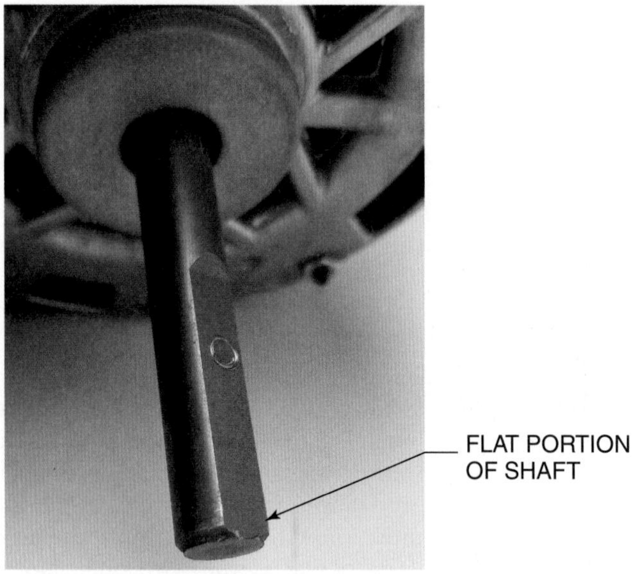

Figure 20.4 A pulley with a groove cut in it that matches a groove in the shaft. A key is placed in these grooves and the setscrew is often tightened down on top of the key.

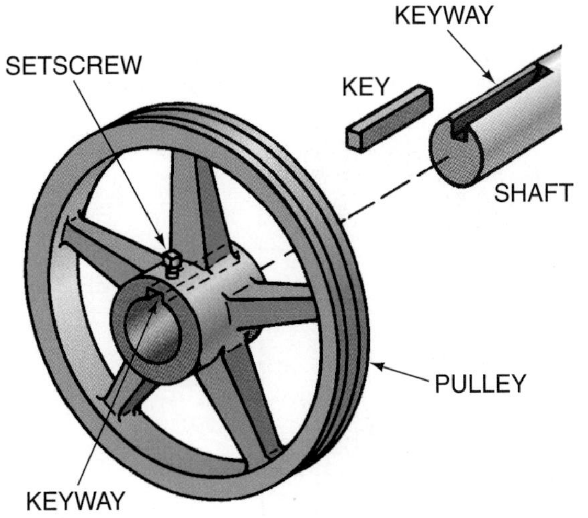

Figure 20.5 A motor shaft damaged from trying to drive it through the pulley. Photo by Jason Obrzut

Figure 20.6 A shaft driven through the pulley with another shaft as the contact surface. The shaft that is used as the contact surface has a smaller diameter than the original shaft.

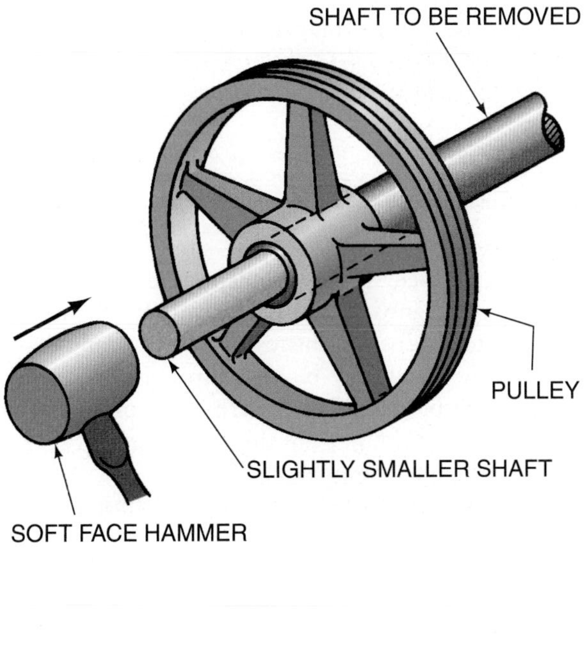

remove a belt from a motor shaft, it is a good idea to clean, sand, and remove all debris from the shaft. In addition, spraying the shaft with a rust-dissolving solution or penetrating oil will help make the removal process much easier.

20.4 BELT TENSION

Many motors fail because of overtightened belts and incorrect alignment. The technician should be aware of the specifications for the motor belt tension on belt-drive systems. A belt tension gauge will ensure properly adjusted belts when the gauge manufacturer's directions are followed. Belts that are too tight strain the bearings so that they wear out prematurely, **Figure 20.7**. **Figure 20.8** shows one type of belt tension gauge.

The belt tension gauge uses two O-rings. When using the gauge, the center-to-center measurement must be taken between the two shafts, **Figure 20.9**. The larger O-ring is positioned on the gauge so that it is lined up with the

Figure 20.7 A belt that is too tight.

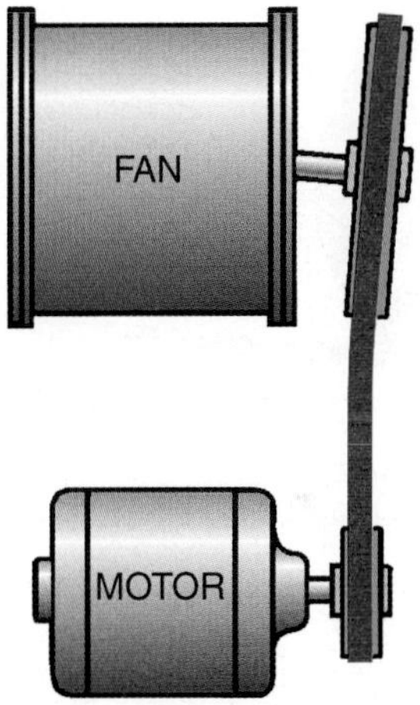

Figure 20.8 A belt tension gauge.

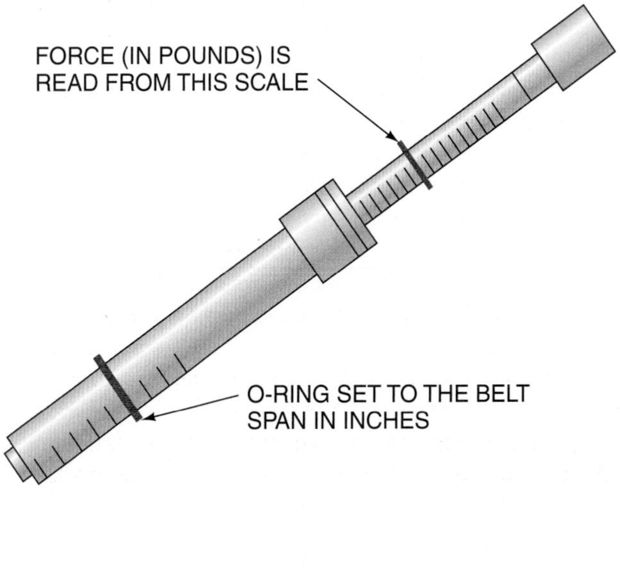

Figure 20.9 The measurement between the centers of the drive and driven shafts.

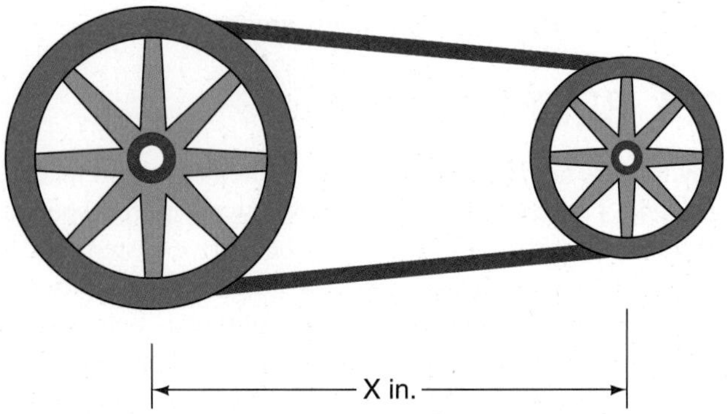

Figure 20.10 Using the belt tension gauge.

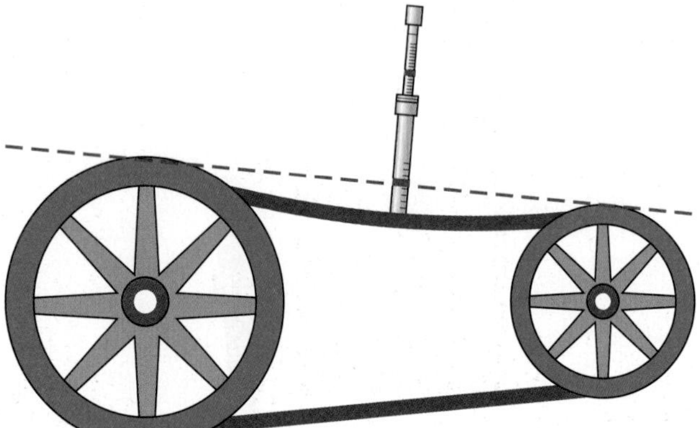

corresponding distance, in inches, between the two shafts. A straightedge is then rested across the two pulleys. The gauge is then positioned at the center of the belt span between the two pulleys and pushed toward the other end of the pulley until the large O-ring is even with the straightedge, **Figure 20.10**. The smaller O-ring will register a force, in pounds, on the top portion of the gauge. By comparing this value with the chart that accompanies the gauge, the technician can determine if the belt is too tight or too loose. A low force reading indicates that the belt is too loose, whereas a force reading that is too high indicates that the belt is too tight.

Belts that are too loose can cause problems—as can those that are too tight. Loose belts have a tendency to slip, and slipping belts are a cause of reduced airflow through the air distribution system. Since motors often rely on the air moving over them to cool the windings, a reduction in airflow can cause motors to overheat. As a result, the internal thermal overload on a motor may open and give the technician reason to believe that there is an electrical problem with the motor.

20.5 PULLEY ALIGNMENT

Pulley alignment is very important. If the drive pulley and driven pulley are not in line, a strain is imposed on the drive mechanisms of the shafts. The pulleys may be aligned with

Figure 20.11 Pulleys must be in proper alignment or excessive belt and bearing wear will occur. The pulleys can be aligned using a straightedge.

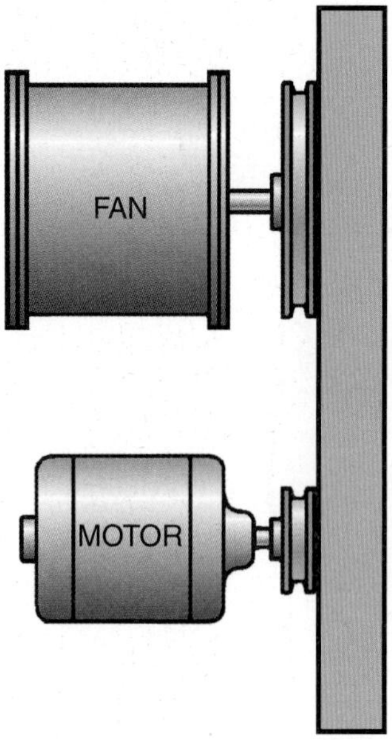

the help of a straightedge, **Figure 20.11**. On small motors, a certain amount of adjustment tolerance is built into the motor base, and this may be enough to allow the motor to be out of alignment if the belt becomes loose. Aligning shafts on some pieces of equipment may not be easy to do, but it must be done or the motor or drive mechanism will not last.

When mechanical problems occur with a motor, the motor is normally either replaced or taken to a motor repair shop. Bearings can sometimes be replaced in the field by a competent technician, but it is generally better to leave this type of repair to motor experts. When the problem is the pulley or drive mechanism, the air-conditioning, heating, and refrigeration technician may be responsible for the repair. **NOTE:** *Proper tools must be used for motor repair, or shaft and motor damage may occur.* •

20.6 ELECTRICAL PROBLEMS

Electrical motor problems are the same for hermetic and open motors. Open motor problems are a little easier to understand or diagnose because they can often be seen. When an open motor burns up, this can often be easily diagnosed because the winding may be seen through the end bells. It may also smell burned. With a hermetic motor, instruments must be used because they are the only means of diagnosing problems inside the compressor, which are not visible. There are three common electrical motor problems: (1) an open winding, (2) a short circuit from winding to winding, and (3) a short circuit from the winding to ground.

20.7 OPEN WINDINGS

Open windings in a motor can be detected with the ohms function on a digital multimeter. On every motor there should be a known, measurable resistance from terminal to terminal for it to run when power is applied to the windings. Single-phase motors must have the applied system voltage at the run winding to run and at the start winding during starting, **Figure 20.12. Figure 20.13** is an illustration of a motor with an open start winding.

A single-phase motor that is in good working order may have the following resistance readings between the motor terminals:

Common to start: 3 Ω
Common to run: 2 Ω
Run to start: 5 Ω

This situation is shown in **Figure 20.14.**

Figure 20.12 A wiring diagram with run and start windings.

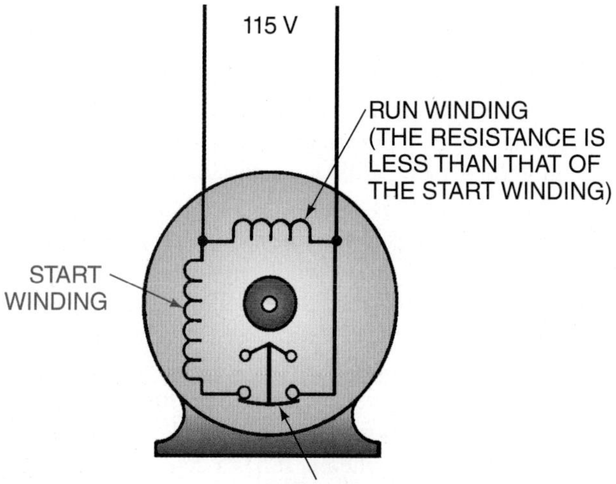

Figure 20.13 A motor with an open winding.

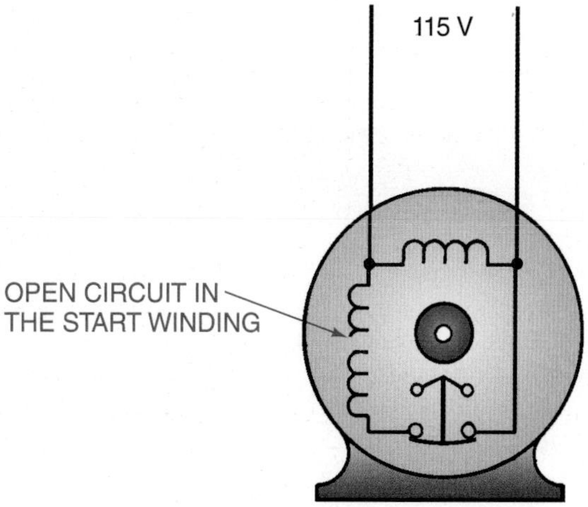

Figure 20.14 Resistance readings between the terminal pairs on a hermetic motor.

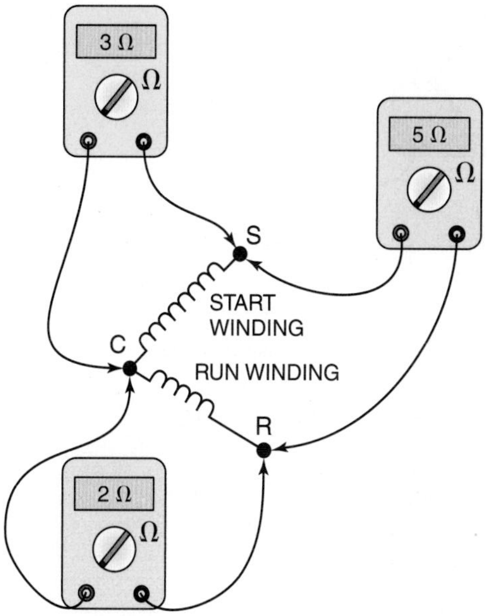

The resistance reading from the common terminal to the run terminal provides the technician only with the resistance reading of the run winding. The resistance reading from the common terminal to the start terminal provides the technician only with the resistance reading of the start winding. The resistance between the start and run terminals gives the sum of the resistances of both the start and run windings. If the start winding in this example were to open, the resistance readings for the same motor would be as follows:

Common to start: O.L. (open line or infinite resistance)
Common to run: 2 Ω
Run to start: O.L. (open line or infinite resistance)

The open winding would affect the resistance reading of the start winding as well as the resistance reading between the run and start terminals, **Figure 20.15.**

20.8 SHORTED MOTOR WINDINGS

Short circuits in windings occur when the conductors in the winding touch each other where the insulation is worn or in some way defective. This creates a short path through which the electrical energy flows. This path has a lower resistance and increases the current flow in the winding. Although motor windings appear to be made from bare copper wire, they are coated with an insulator to keep the copper wires from touching each other. The measurable resistance mentioned in the previous paragraph is known for all motors. Some resistances for motor windings have been published. The best way to check a motor for electrical soundness is to know what the measurable resistance should be for a particular winding and verify it with a high-quality

Figure 20.15 Resistance readings between terminal pairs on a hermetic motor with an open start winding.

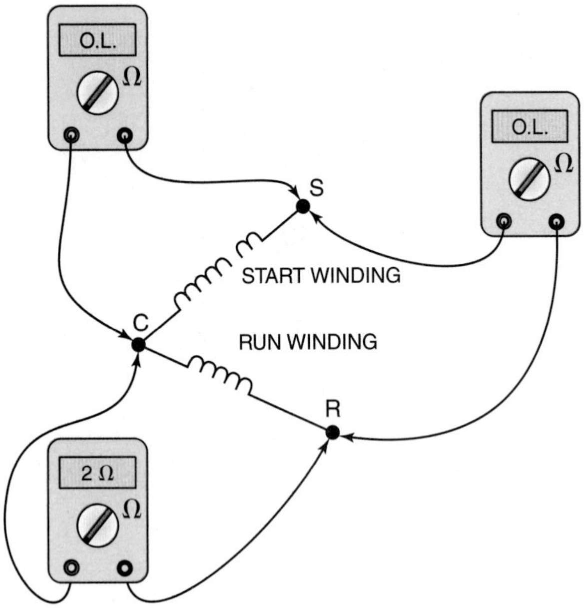

digital multimeter, **Figure 20.16**. This measurable resistance will be less than the rated value when a motor has short-circuit problems. The decrease in resistance causes the current to rise, which causes motor overload devices to open the circuit and possibly even trips the circuit overload protection. If the resistance does not read within these tolerances, there is a problem with the winding. A table is helpful when you troubleshoot a hermetic compressor. Tables may not be easy to obtain for open motors, and the windings are not as easy to check because the individual windings do not all come out to terminals as they do on a hermetic compressor.

If the decrease in resistance in the windings is in the start winding, the motor may not start. The reason for this is that, in order to start, the motor relies on an imbalance in the magnetic field. If the start winding is partially shorted, the resistance in that winding will be closer to the resistance of the run winding. This will reduce, or possibly even eliminate, the magnetic field imbalance that is needed to start the motor, **Figure 20.17**. If the decrease in resistance is in the run winding, the motor may start and draw too much current while running. If the resistance of the run winding is lowered, the imbalance in magnetic fields between the start and the run winding will be increased, and the starting torque of the motor will actually increase, **Figure 20.18**, but the motor may very well overheat. If the motor winding resistance cannot be determined, then it is hard to know whether a motor is overloaded or whether it has a defective winding when only a few of its coils are shorted.

Figure 20.17 Resistance readings between terminal pairs on a hermetic motor with a partially shorted start winding.

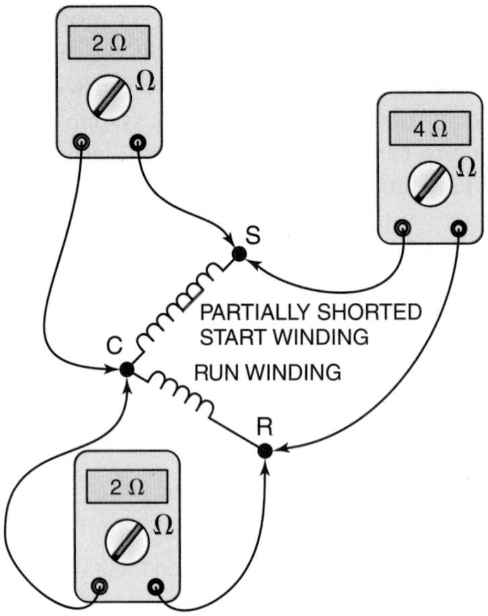

Figure 20.16 Resistances for some typical hermetic compressors. Courtesy Copeland Corporation

Compressor Model	Voltage	MOTOR AMPS Full Winding Rated Load	Full Winding Locked Rotor	1/2 Winding Rated Load	1/2 Winding Locked Rotor	FUSE SIZE Recommended Max Fusetron	Std.	Winding Resistance in Ohms
9RA - 0500 - CFB	230/1/60	27.5	125.0			FRN–40	50	Start 1.5 Run 0.40
9RB TFC	208–230/3/60	22.0	115.0			FRN–25	40	0.51-0.61
9RJ TFD	460/3/60	12.1	53.0			FRS–15	15	2.22-2.78
9TK TFE	575/3/60	7.8	42.0			FRS–10	15	3.40-3.96
MRA FSR	200–240/3/50	17.0	90.0	8.5	58.0	FRN–25	35	0.58-0.69
MRB FSM	380–420/3/50	9.5	50.0	4.8	32.5	FRS–15	20	1.80-2.15
MRF								

Figure 20.18 Resistance readings between terminal pairs on a hermetic motor with a partially shorted run winding.

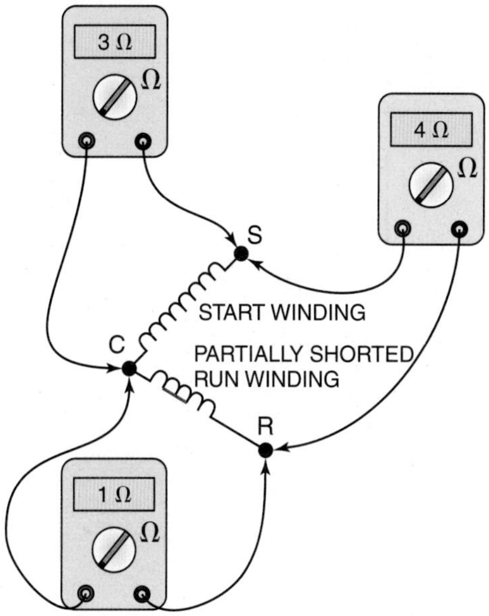

Figure 20.19 The coupling was disconnected between this motor and the pump because it was suspected that the motor or pump was locked up and would not turn.

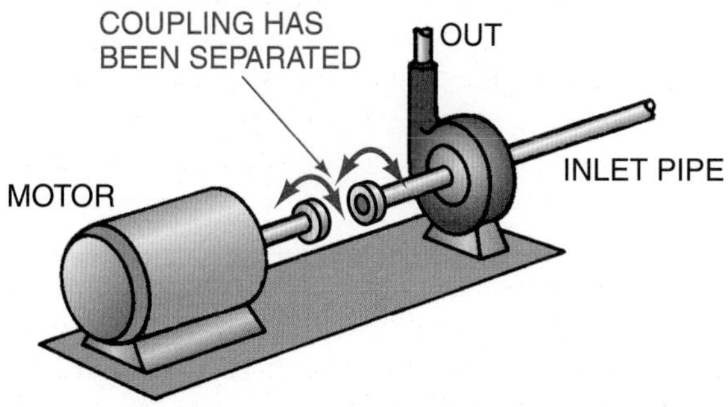

When stretched out, the windings of a motor may be many feet long. This insulated winding wire is wound into coils. The shorting of a winding wire may make it a shorter winding; thus a higher amperage reading and a localized "hot spot" will occur where the windings are shorted together. For example, suppose the total winding length is 50 ft and the factory ohm reading is supposed to be 30 Ω. When it is shorted to 45 ft where two turns are touching due to insulation failure, the ohm reading may be reduced to 27 Ω. There would be a hot spot where they touch and the amperage would go up slightly.

When the motor is an open motor, the load can be removed. For example, the belts can be removed, or the coupling can be taken apart, and the motor can be started without the load, **Figure 20.19**. If the motor starts and runs correctly without the load, the load may be too great.

Figure 20.20 A wiring diagram of a three-phase motor. The resistance is the same across all three windings.

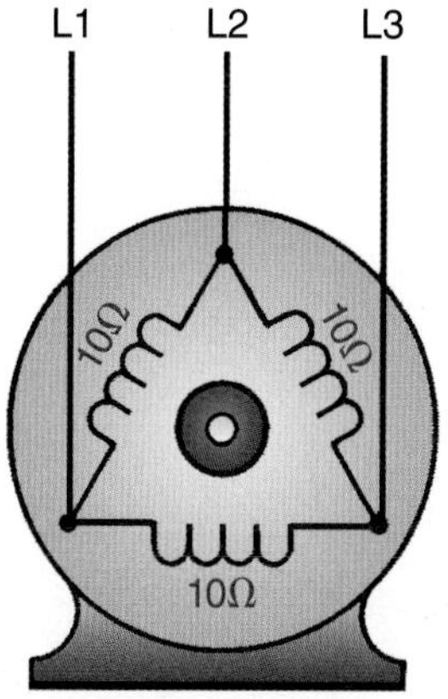

Three-phase motors must have the same resistance for each winding. (There are three identical windings.) Otherwise, there is a problem. A meter check will quickly reveal an incorrect winding resistance, **Figure 20.20**.

20.9 SHORT CIRCUIT TO GROUND (FRAME)

A short circuit from winding to ground or to the frame of the motor may be detected with a good ohmmeter. No circuit should be detectable from the winding to ground. The copper suction line on a compressor is a good source for checking to the ground. **NOTE:** *"Ground" or "frame" are interchangeable terms because the frame should be grounded to the earth ground through the building electrical system,* **Figure 20.21.** •

To check a motor for a ground, use a good digital multimeter set to the ohms function. Special instruments for finding very high resistances to ground are used for larger, more sophisticated motors. Most technicians use digital multimeters. High quality instruments and megohmmeters can detect a ground in the 10,000,000-Ω and higher range. The megohmmeter has an internal high-voltage, direct-current (DC) supply to help create conditions for detecting the ground, **Figure 20.22(A)**.

Megohmmeters (meggers) are electrical meters used to check the resistance and condition of the motor windings and the condition of the refrigeration and oil environment around motor windings. A megger is nothing but a giant ohmmeter that creates a very large DC voltage (usually 500 VDC) from its internal battery. The meter will read out in megohms (millions of ohms). Any motor winding or electrical coil can be checked with a megger. A megohmmeter's main function is to detect weak motor winding insulation and moisture accumulation and acid formations from the motor windings to ground before they can cause more

damage to the winding insulation. With HVACR hermetic and semihermetic compressor motors, as contaminants in the refrigerant and oil mixture increase, the electrical resistance from the motor windings to ground will decrease. Regular preventive maintenance checks can be made with a megohmmeter, which can signal early motor winding breakdown from a contaminated system when accurate records are kept.

Figure 20.21 This building electrical diagram shows the relationship of the earth ground system to the system's piping.

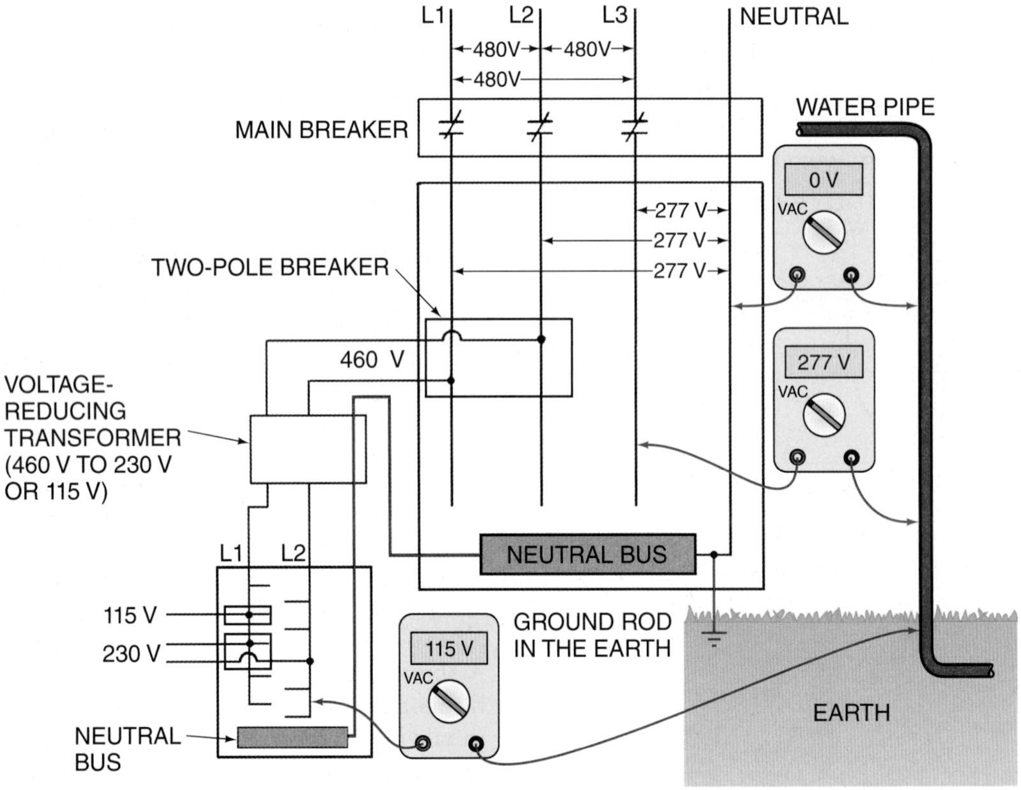

Figure 20.22 (A) Megohmmeters can measure resistances in the millions of ohms. This instrument is often referred to as megger. Photo by Jason Obrzut (B) Megohmmeter readings with varying degrees of deterioration and motor winding breakdown.

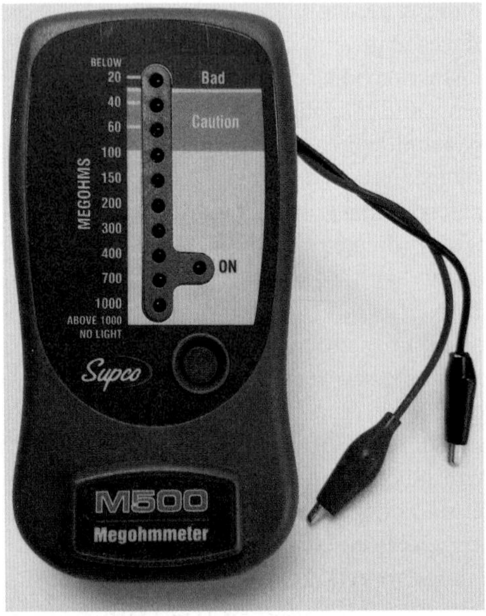

(A)

Resistance Reading (megohms)	Condition Indicated	Required Preventive Maintenance	Percent of Winding in Field
Over 100	Excellent	None	30%
100–50	Some moisture present	Change filter drier	35%
50–20	Severe moisture and/or contamination	Several filter drier changes; change oil if acid is present	20%
20–0	Severe contamination	Check entire system and make corrections. Consider an oversized filter drier, refrigerant and oil change, and reevacuation. System burnout and clean-up procedures required.	15%

(B)

One probe of the megger is connected to one of the motor winding terminals, and the other probe to the shell of the compressor (ground). **NOTE:** *Make sure metal is exposed at the shell of the compressor where the probe is attached so that the compressor's shell paint is not acting as an insulator to ground.* • When a button is pushed and held on the megger, it will apply a high DC voltage between its probes and measure all electrical paths to ground. It is important to disconnect all wires from the compressor motor terminals when megging a compressor motor. Also, read the instructions that come with the meter to determine how long to energize the megger when checking winding or coils. If possible, it is a good idea to run the motor for at least one hour, disconnect power, disconnect all electrical leads, and then quickly connect the megger to the motor. This will give a more meaningful comparison between readings for the same compressor on different days because of the approximately similar winding temperatures.

Good motor winding readings should have a resistance value of a minimum of 100 megohms relative to ground. In fact, good motor winding resistance should be between 100 megohms and infinity. **Figure 20.22(B)** lists megohm readings with varying degrees of contamination and motor winding breakdown. Because of the very high resistance of the motor winding insulation, a digital multimeter cannot be used in place of the megger. A multimeter does not generate enough voltage from its internal battery to detect high resistance problems like deteriorated winding insulation, moisture, or other system contamination.

Following are some other important tips service technicians should know about the use of a megohmmeter:

- Never use a megger if the motor windings are under a vacuum.
- Meggers can be used for electrical devices other than electric motors. Always consult with the meter manufacturer or user's manual for detailed instructions on megging other electrical devices such as coils.
- Meggers are often used in preventive maintenance programs, especially before a contractor signs a preventive maintenance contract, to determine the condition of the electrical devices.
- Any megger with a higher voltage output than 500 VDC should be used only by an experienced technician. A high voltage for too long a time may further weaken or fail motor windings and the winding insulation could be damaged by the testing procedure.

A typical multimeter will detect a ground of about 1,000,000 Ω. The general rule is that if a measurable resistance is present with one meter lead touching the motor terminal and the other touching a ground (such as a copper suction line), the motor should be started with care. **SAFETY PRECAUTION:** *If a low resistance is measured, do not start the motor,* **Figure 20.23.**• When a meter reads a very slight resistance to ground, the windings may be dirty and damp if it is an open motor. Clean the motor, and the ground will probably be eliminated. Some open motors operating in a dirty, damp atmosphere may indicate a slight circuit to ground in damp weather. Air-cooled condenser fan motors are an example. When the motor is started and allowed to run long enough to get warm and dry, the ground circuit may disappear.

Hermetic compressors may occasionally have a slight ground due to the oil and liquid refrigerant in the motor splashing on the windings. The oil may have dirt suspended in it and show a slight ground. Liquid refrigerant makes this condition worse. If the digital multimeter shows a slight ground but the motor starts, run the motor for a little while and check again. If the ground persists, the motor is probably going to fail soon if the system is not cleaned. A suction-line filter drier may help remove particles that are circulating in the system and causing the slight ground.

The main functions of a multimeter used for troubleshooting electric motors are amperage and resistance. If the resistance

Figure 20.23 A multimeter detecting a circuit to ground in a compressor winding.

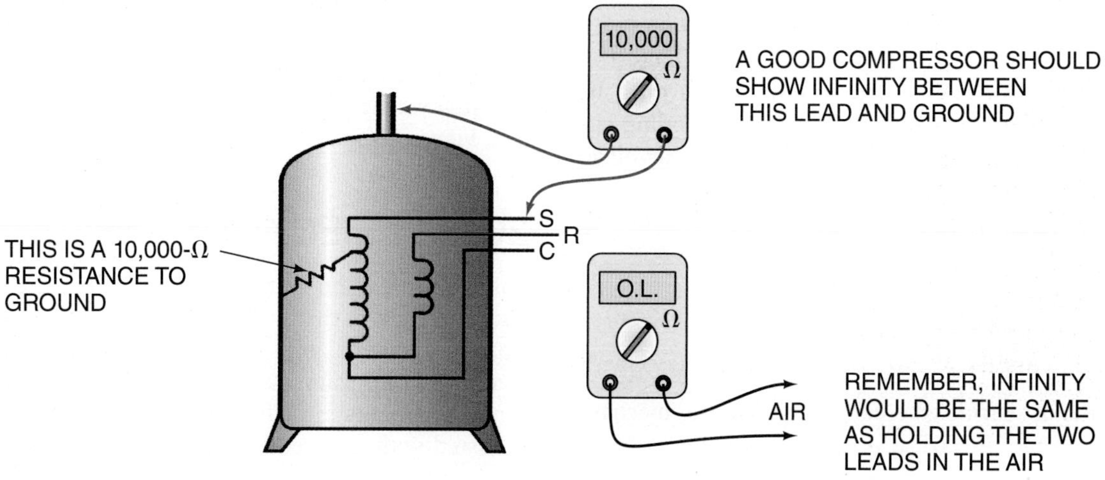

is correct, the motor is electrically sound. Other problems may be found using the multimeter's amp-clamp.

20.10 SINGLE-PHASE MOTOR STARTING PROBLEMS

Single-phase open electric motors are fairly easy to troubleshoot for starting problems because they can be seen and heard. It is very frustrating when turning the power on to any motor if it fails to start. First, make sure that you really have full power to the motor leads. This entails fastening meter leads to the motor terminals and reading the meter when the power is applied. Low voltage can occur at any place in the electrical system, such as at loose connections, loose fuse holders, undersized wire, or low voltage from the main electrical panel. The only way to be sure is to measure it at the motor terminals. Even at start-up when the motor is pulling LRAs, the voltage should be ±10% of the rated voltage. For example, if the motor is a combination 208/230-V fan motor, the minimum voltage should be 187 V (208 × 0.90 = 187.2) and the maximum should be 253 V (230 × 1.10 = 253). The motor's voltage must be within these limits during starting or running.

To start, the start winding and the run winding of open motors must receive power at the same time. The start winding is then disconnected from the circuit as the motor reaches about three-fourths of its rated speed. On some motors, such as capacitor-start, capacitor-run (CSCR) motors, the start winding remains in the circuit through the run capacitor. The run capacitor thus limits the current through the start winding to prevent it from getting too hot and burning up. The run capacitor is a low-microfarad capacitor. Disconnecting the start winding can be just as much a part of troubleshooting as operating the motor itself. When the open motor with a centrifugal switch is started and starts to turn, you should hear the audible disconnect of the start winding by the switch. When the motor is stopped, the switch makes an audible sound when the contacts close. If you hear the centrifugal switch function and the motor still will not start, you must apply meter leads to the start winding in the end bell (start to common terminals) of the motor to make sure full voltage is reaching the start winding during start-up. If the motor still will not start, perform the same test on the run windings (run to common terminals). The electronic start switch must be checked with a multimeter applied to the start winding circuit in the end bell of the motor. This is a little more involved than just listening to the centrifugal switch.

The following are symptoms of electric-motor starting problems:

- The motor hums and then shuts off.
- The motor runs for short periods and shuts off.
- The motor will not try to start at all.

The technician must decide whether there is a motor mechanical problem, a motor electrical problem, a circuit problem,

Figure 20.24 An example of how a pump or compressor coupling can be disconnected, and the component's shaft turned by hand to check for a hard-to-turn shaft or a stuck component. **⚠ SAFETY PRECAUTION:** *Turn off the power to the motor and lock the panel out before servicing the motor.•*

Figure 20.25 The end bell of this motor has been removed to examine the start switch and the windings. Photo by Bill Johnson

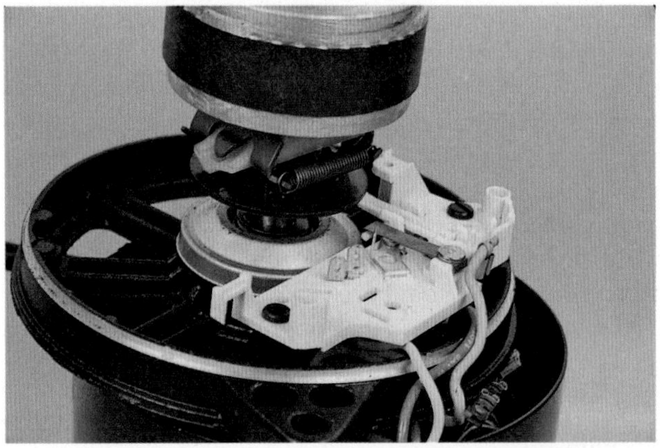

or a load problem. **⚠ SAFETY PRECAUTION:** *If the motor is an open motor, turn the power off. Lock and tag the panel. Try to rotate the motor by hand. If it is a fan or a pump, it should be easy to turn,* **Figure 20.24.** *If it is a compressor, the shaft may be hard to turn. Use a wrench to grip the coupling when trying to turn a compressor. Be sure the power is off.•*

If the motor and the load turn freely, examine the motor windings and components. If the motor is humming and not starting, the starting switch may need replacing, or the windings may be burned. If the motor is open, you may be able to visually check them; if you cannot, remove the motor end bell, **Figure 20.25.**

20.11 CHECKING CAPACITORS

Capacitors can be tested with a digital multimeter while under load, with the motor running, or they can be removed from the circuit and tested without a load. Whichever method a technician uses, safety should always be a top priority. Capacitors must be discharged prior to being handled or tested with a meter. Some start capacitors have a resistor

across the terminals to bleed off the charge during the off cycle, **Figure 20.26**. A capacitor's stored charge can damage a test instrument or cause severe injury to a technician. A digital multimeter with a microfarad function or a capacitor analyzer may be used, **Figures 20.27(A)** and **(B)**. **⚠ SAFETY PRECAUTION:** *A capacitor should be discharged by shorting one terminal to the other with a 20,000-Ω, 5-W resistor.* • On a dual capacitor, both the fan (FAN) and compressor (HERM) terminals should be shorted to the common (C) terminal, **Figure 20.28(A)**. It is recommended to short across the terminals with the previously mentioned resistor for the sake of safety. If the resistor across the start capacitor is defective, it may not bleed the charge from the capacitor. If a capacitor is to be tested under load, care should be taken as the voltage present can be significantly higher than the source voltage due to the back EMF of the motor.

Figure 20.26 A start capacitor with a bleed resistor across the terminals to bleed off the charge during the off cycle. **⚠ SAFETY PRECAUTION:** *Make certain that capacitors are completely discharged before you come into contact with the terminals.* • Photo by Jason Obrzut

When testing a run capacitor under load, use the following formula: (Start Winding Amperage × 2652) ÷ Voltage. With the system running, the amperage reading should be taken between the capacitor and the start winding terminal, **Figure 20.28(B)**. The voltage measurement should be done at the capacitor terminals, **Figure 20.28(C)**. For example, a 208-V hermetic compressor motor with a 45-microfarad run capacitor is drawing 5.2 A on its start winding. The voltage measured at the run capacitor terminals is 306.5. Multiplying the amperage, 5.2, by 2652 gives us 13,790.4. Dividing this number by the voltage, 306.5, gives a measured capacitance of 44.9. This is within the manufacturer's tolerance which

Figure 20.27 (A) A digital multimeter with a capacitance function. (B) Digital capacitor tester. Photos by Jason Obrzut

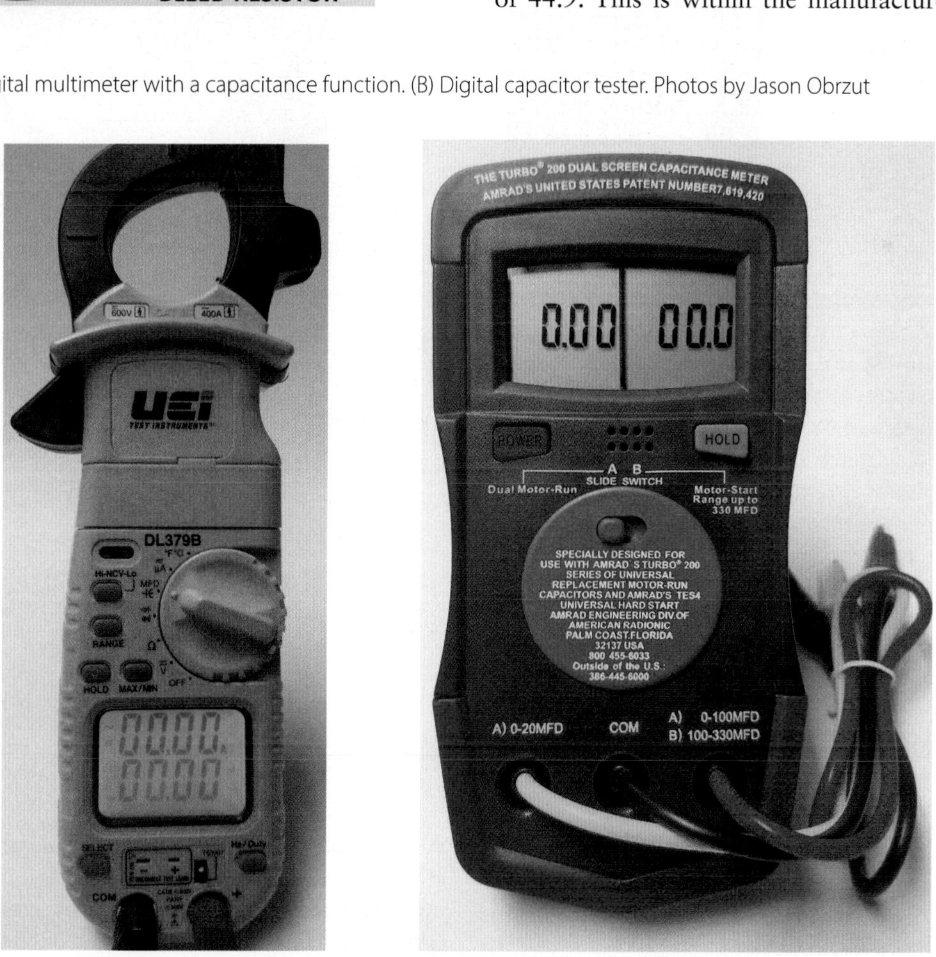

(A) (B)

Figure 20.28 (A) A 20,000-Ω, 5-W resistor is used to discharge a dual run capacitor. (B) A multimeter is used to test the amperage of the start winding. (C) A voltage check is done at the capacitor terminals. (D) A digital multimeter can be used to test a capacitor. (E) A capacitor analyzer used to test both sides of a dual run capacitor at the same time. (F) The bleed resistor is removed from a start capacitor prior to being tested. Photos by Jason Obrzut.

is typically ±6 to 10%. A capacitor found to be outside the recommended range must be replaced.

When testing a capacitor without a load, the system power should be turned off and locked out at the disconnect. The capacitor is removed from the circuit and discharged using a 20,000-Ω, 5-W resistor and insulated pliers. The meter is connected to the capacitor terminals, one lead to each terminal, **Figure 20.28(D)**. Dual capacitors, which are essentially two capacitors in the same housing, must be checked on both sides, **Figure 20.28(E)**. The bleed resistor that is installed on some start capacitors must be removed before the capacitance is tested, **Figure 20.28(F)**. The resistor may affect the meter reading and the true value may not be displayed. The measured capacitance must be within the manufacturer's specified tolerance or the capacitor must be replaced.

20.12 CAPACITOR IDENTIFICATION

A run capacitor is contained in a metal can that is oil filled. If this capacitor gets hot due to overcurrent, it will often swell, **Figure 20.29**. The capacitor should then be changed. **⚠ SAFETY PRECAUTION:** *Run capacitors have an identified terminal to which the lead that feeds power to the capacitor should*

be connected. When the capacitor is wired in this manner, a fuse will blow if the capacitor is shorted to the container. If the capacitor is not wired in this manner and a short occurs, current can flow through the motor winding to ground during the off cycle and overheat the motor, **Figure 20.30.**•

There are devices on the market that can act as universal replacement motor run capacitors, **Figure 20.29(C)**. The one shown in **Figure 20.29(C)** has six built-in capacitance values in one housing with a common center tap. It can replace over 200 single- and dual-value capacitors with a rating of 370 VAC or 440 VAC by using the five supplied jumper wires. Larger kits offering more variations are also available. This simplifies the service technician's task of carrying a diverse supply of individual capacitors.

The start capacitor is a dry type and may be contained in a shell of paper or plastic. Paper containers are no longer used, but you may find one in an older motor. If this capacitor has been exposed to overcurrent, it may have a bulge in the vent at the top of the container, **Figure 20.31**. The capacitor should then be changed.

Figure 20.29 (A) A swollen (bad) capacitor next to a good capacitor. (B) Various run capacitors. (C) Universal replacement motor run capacitors that can replace up to 200 single- and dual-value capacitors. Photos by Jason Obrzut

(A)

(B)

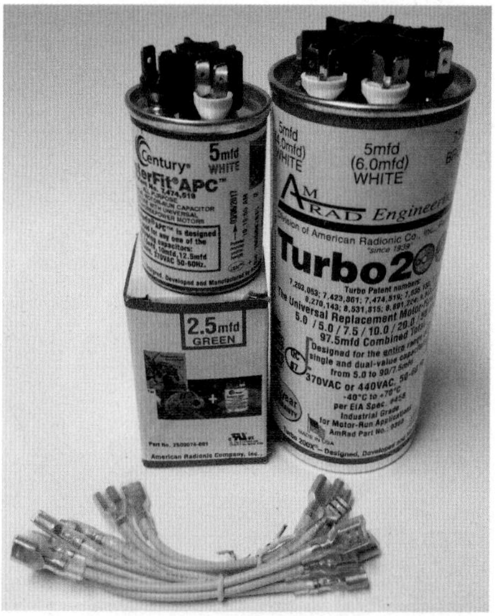

(C)

Figure 20.30 The wiring to a run capacitor for proper fuse protection of the circuit.

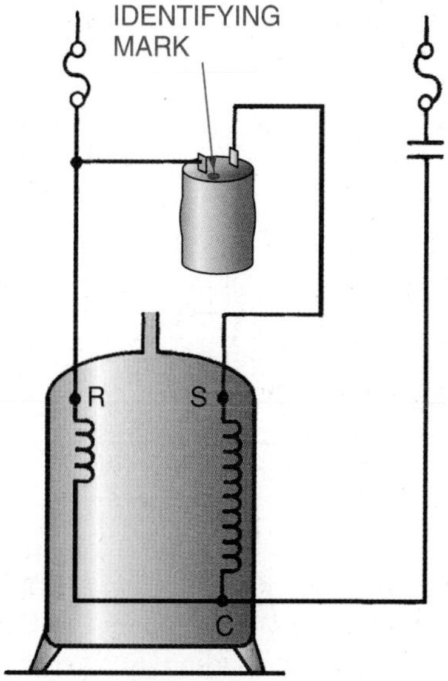

Figure 20.31 When the start capacitor has overheated, the small rubber diaphragm in the top of the component will bulge through the vent. The vent is shown here. Photo by Jason Obrzut

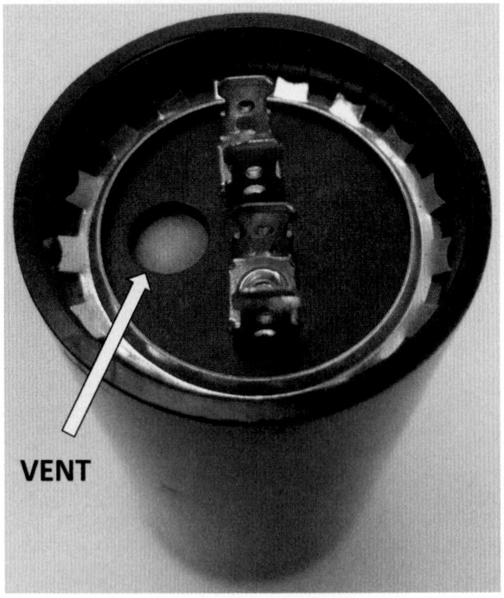

20.13 WIRING AND CONNECTORS

The wiring and connectors that carry the power to a motor must be in good condition. When a connection becomes loose, oxidation of the copper wire occurs. The oxidation acts as an electrical resistance and causes the connection to heat up more, which in turn causes more oxidation. This condition will only get worse. Loose connections will result in low voltages at the motor and in overcurrent conditions. Loose connections will appear the same as a set of dirty or burnt contacts and may be located with a multimeter, **Figure 20.32**. If a connection is loose enough to create an overcurrent condition, it can often be located by a temperature rise at the connection.

Figure 20.32 When multimeter leads are applied to both sides of a set of dirty or burnt contacts, it will read a voltage drop of 10 V or more.

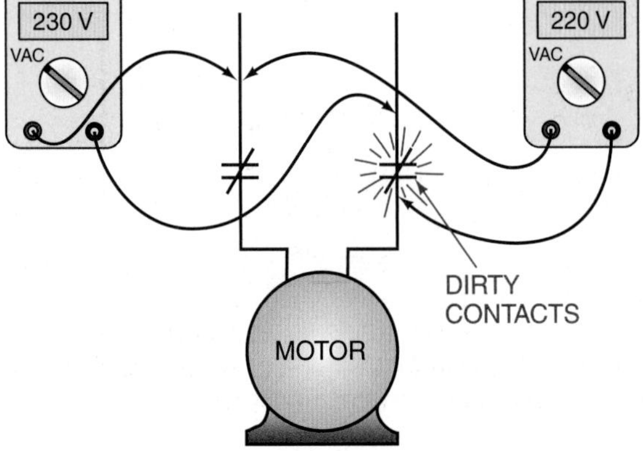

20.14 TROUBLESHOOTING HERMETIC MOTORS

Diagnosing hermetic compressor motor problems differs from diagnosing open motor problems because the motor is enclosed in a shell and cannot be seen, and the motor sound level may be dampened by the compressor shell. Motor noises that are obvious in an open motor may become hard to hear in a hermetic compressor.

The motor inside a hermetic compressor can be checked electrically only from the outside. It will have the same problems as an open motor, that is, open circuit, short circuit, or grounded circuit. The technician must give the motor a complete electrical checkout from the outside of the shell; if it cannot be made to operate, the compressor must be changed. A motor checkout includes the starting and running components for a single-phase compressor, such as the run and start capacitors and the start relay.

As mentioned earlier, compressors operate in remote locations and are not normally attended. A compressor can throw a rod, tearing it up, and the motor may keep running until damaged. The compressor parts damage the motor, which may then burn out. The compressor damage is not detected because the damage cannot be seen. When a compressor is changed because of a bad motor, you should suspect mechanical damage as well.

20.15 SERVICE TECHNICIAN CALLS

SERVICE CALL 1

The manager of a retail store calls. *They have no air-conditioning and the piping going into the unit is frozen solid. The air handler is in the stockroom. The evaporator fan motor winding is open, and the outdoor unit has continued to run without air passing over the evaporator, freezing it solid.*

Upon arriving at the store, the service technician takes a minute to make sure his uniform is clean and tucked in and checks his face for dirt and grease smudges and his hair for neatness in the van's mirror. The technician politely introduces himself and his company to the store owner while handing the customer a business card. The technician writes the customer's name down so he can remember it later, especially when the job is complete and he has to bill the customer. After shaking hands, the technician asks the store owner for his account of the problem. After patiently listening, the technician politely asks the store owner some pertinent questions without being too technical. The technician remembers to never interrupt when the customer is answering his questions or simply talking to him. On arriving at the air handler in the stockroom and noticing that the fan is not running, the technician turns the cooling thermostat to OFF to stop the condensing unit, then turns the fan switch to ON.

SAFETY PRECAUTION: *The technician then turns off the power, locks and tags the panel, and removes the fan compartment cover.•* The fan motor body is cool to the touch, indicating the motor has been off for some time. The technician then turns on the power and checks power at the motor terminal block; the motor has power. The power is turned off and locked out. The technician checks the motor winding with an ohmmeter; the motor windings are open. The technician now provides the store owner with written work orders and an estimate.

The fan motor is changed. The unit cannot be started until the coil is allowed to thaw. The technician instructs the store manager to allow the fan to run until air is felt at the outlet in the men's room. This ensures the airflow has started. Then the technician waits 15 min and sets the thermostat to COOL and the fan switch to AUTO. The technician cleans up the work site, double-checks his work, politely tells the store owner of the service problem, and then informs the store owner of the detailed work description and gives him the billing information with costs. The technician informs the store owner that he will call him later in the day for feedback on the air-conditioning call. A call back later in the day indicates that the system is now cooling correctly.

SERVICE CALL 2

A residential customer calls. *There is no heat. The fan motor on the gas furnace has a defective run capacitor. The furnace burner is cycling on and off with the high-limit control. The fan will run for a few minutes each cycle before shutting off from overtemperature. The defective capacitor causes the fan motor current to run about 15% too high.*

After making sure the service van is not blocking the customer's driveway, the technician makes sure his uniform, shoes, and overall appearance looks good. He then rings the doorbell, remembering to keep a comfortable distance between himself, the door, and the customer. After he politely introduces himself and his company and handing the customer a business card, the homeowner invites the technician into the home. The technician patiently listens as the customer describes the problem, and, once the customer is done talking, he politely asks a few questions, being careful to talk clear and concisely and not too technically. The technician must use everyday language that the customer can understand to make him feel like he is helping with the solution to the problem. The technician asks permission to go under the house where the horizontal furnace is located, to which the customer agrees, and the technician leaves for under the house.

The furnace is hot. The technician hears the burner ignite and watches it for a few minutes. The burner shuts off, and about that time the fan comes on. The technician removes the fan compartment door and sees the fan turning very slowly. **SAFETY PRECAUTION:** *The power is shut off. The panel is locked and tagged.•* The motor is very hot. It seems to turn freely, so the bearings must be normal. The technician removes the run capacitor and uses a multimeter for a capacitor check, **Figure 20.27**. The capacitor is open, as the resistance reading across its terminal is infinite. The capacitor is replaced. After shutting the fan compartment door, power is restored. It is evident from the sound of the motor that it is turning faster than before the capacitor was changed. The service technician now double-checks his work, turns off the power, oils the motor, replaces the air filter, and cleans the work site. Remembering the customer's name, the technician politely provides a written and verbal explanation of the problem solved, the work completed, billing information, and costs. The technician and customer shake hands and the technician leaves the house for his service van.

SERVICE CALL 3

A retail store customer calls. *There is no air-conditioning. The compressor, located outside, starts up, then shuts off. The condenser fan motor is not running. The dispatcher tells the customer to shut the unit off until the technician arrives. The PSC fan motor for the condenser has bad bearings. The motor feels free to turn the shaft, but if power is applied to the motor when it is turning, the motor will stop.*

The technician arrives and talks to the store manager. The technician goes to the outdoor unit and disconnects the power so that the system can be controlled from outside. The room thermostat is set to call for cooling.

The technician goes back outside and turns the disconnect on and observes. The compressor starts, but the fan motor does not. **SAFETY PRECAUTION:** *The power is turned off. The panel is locked and tagged. The compressor leads are disconnected at the load side of the contactor.•* The technician turns the disconnect switch on again and, using the amp-clamp on the multimeter, checks the current going to the fan motor. It is drawing current and trying to turn. The technician does not know at this point if the motor or the capacitor is defective. The power is shut off. The technician spins the motor and turns the power on while the motor is turning. The motor acts like it has brakes and stops. This indicates bad bearings or an internal electrical motor problem.

The motor is changed, and the new motor performs normally even with the old capacitor. The compressor is reconnected and power is resumed. The system now cools normally. The technician changes the air filter before leaving.

The following service calls do not describe the solution. Please discuss the solutions with your instructor.

SERVICE CALL 4

An insurance office customer calls. *There is no air-conditioning. This system has an electric furnace with the air handler (fan section) mounted above the suspended ceiling in the stockroom. The condensing unit is on the roof. The low-voltage power supply is at the air handler.*

After politely introducing himself, the service technician puts on shoe covers to protect the customer's floors. After carefully listening to the customer's account of the problem and answering any questions, the technician goes to the room thermostat. It is set on cooling, and the indoor fan is running. Before going out of the building and onto the roof, the service technician informs the customer why he is leaving. The technician goes to the roof and discovers the breaker is tripped. Before resetting the breaker, the technician decides to find out why the breaker tripped. The cover to the electrical panel to the unit is removed. A voltage check shows that all the voltage to the breaker is off. The contactor's 24-V coil is the only thing energized because its power comes from downstairs at the air handler.

The technician checks the motors for a ground circuit by placing one lead on the load side of the contactor and the other on the ground with the meter set on ohms. The meter shows a short circuit to ground (no resistance to ground). See **Figure 20.23** for an example of a motor ground check. From this test, the technician does not know if the problem is the compressor or the fan motor. The wires are disconnected from the line side of the contactor to isolate the two motors. The compressor shows normal, no reading. The fan motor shows 0 Ω resistance to ground.

What is the problem and the recommended solution?

SERVICE CALL 5

A commercial customer calls. *There is no air-conditioning in the upstairs office. This is a three-story building with an air handler on each floor and a chiller in the basement.*

The service technician arrives and politely introduces himself to the customer. After handing the customer a business card and describing the service he can provide, the technician asks some pertinent questions using everyday language the customer can understand. The technician then carefully listens and makes the customer feel like part of the solution to the service call problem. Once the technician has permission to enter the building, he goes to the fan room where the complaint is. There is no need to go to the chiller or the other floors because there would be complaints if the chiller was not providing cooling to those locations.

The chilled water coil piping feels cold; the chiller is definitely running and furnishing cold water to the coil. The fan motor is not running. ⚠ **SAFETY PRECAUTION:** *Because the motor may have electrical problems, the technician proceeds with caution by opening, locking, and tagging the electrical disconnect.*• The technician pushes the reset button on the fan motor starter and hears the ratchet mechanism reset. The motor will not try to start while the technician is pushing the reset button because the disconnect switch is open. The unit must have been pulling too much current for the overload to trip.

Using a digital multimeter, the technician checks for a ground by touching one lead to a ground terminal and the other to one of the motor leads on the load side of the starter. The meter is set on ohms and will detect a fairly high resistance to ground. See **Figure 20.23** for an example of a motor ground check. ⚠ **SAFETY PRECAUTION:** *Any measurable resistance displayed when the meter's leads are connected between ground and the motor lead would indicate a ground—so caution is necessary. If the meter reads little to no resistance, the circuit should not be energized until the ground is cleared up or physical damage may occur.*• The motor is not grounded. The resistance between each winding is the same, so the motor appears to be normal. The technician then turns the motor over by hand to see whether the bearings are too tight; the motor turns normally.

The technician shuts the fan compartment door and then fastens the amp-clamp on the multimeter to one of the motor leads at the load side of the starter. The electrical disconnect is then closed to start the motor. When it is closed, the motor tries, but will not start and pulls a high amperage. The motor seems to be normal from an electrical standpoint and turns freely, so the power supply is now suspected.

⚠ **SAFETY PRECAUTION:** *The technician quickly pulls the disconnect to the off position and gets a digital multimeter.*• The continuity through each fuse is checked. The fuse in L2 is open. The fuse is replaced and the motor is started again. The motor starts and runs normally with normal amperage on all three phases. The question is, why did the fuse blow?

What is the problem and the recommended solution?

SERVICE CALL 6

A customer who owns a truck stop calls. *There is no cooling in the main dining room. The customer does some of the maintenance, so the technician can expect some problem related to this maintenance program.*

The technician arrives and can hear the condensing unit on the roof running; it shuts off about the time the technician gets out of the truck. The customer had just serviced the system by changing the filters, oiling the motors, and checking the belts. Since then, the fan motor has been shutting off. It can be reset by pressing the reset button at the fan contactor, but it will not run long.

Once arriving on the job on time and making sure the service van is not blocking any restaurant customer parking, the service technician makes sure his face, hair, and uniform are looking neat. He now enters the restaurant, explains to a waitress why he is there, and asks to talk to the owner. Once out of the way of the restaurant customers, the technician politely introduces himself to the owner while handing him a business card, always making eye contact. The fan is not running, and the condensing unit is shut down from low pressure. The technician fastens the amp-clamp on the multimeter on one of the motor leads and restarts the motor. The motor is pulling too much current on all three phases, and the motor seems to be making too much noise.

What is the problem and the recommended solution?

REVIEW QUESTIONS

1. If a sleeve bearing fails in a small, fractional horsepower motor, ______________.
   A. the motor should be replaced
   B. the bearings must be replaced
   C. a larger capacitor should be used to prevent future failures
   D. the motor rotation should be reversed, and the motor restarted
2. An open winding in an electric motor means that ________________.
   A. the winding is making contact with the motor frame
   B. one winding is making contact with another winding
   C. a wire in one winding is broken
   D. the centrifugal switch to the start winding is open
3. There is a short circuit in an electric motor winding when ___________.
   A. there is an adverse weather condition
   B. a winding conductor is worn bare and touches another part of the conductor that is also bare
   C. there is a defect in the centrifugal start switch
   D. the motor will not start
4. An open winding or a shorted winding can be determined with which of the following functions on a digital multimeter?
   A. Volts
   B. Amperage
   C. Ohms
   D. Microfarads
5. A decrease in resistance will cause the amperage to __________.
   A. decrease
   B. increase
   C. decrease or increase
   D. remain the same
6. A ________ may be used to detect resistances in the millions of ohms.
   A. megohmmeter
   B. digital multimeter
   C. capacitor analyzer
   D. psychrometer
7. What size of resistor is recommended for shorting across an electrical motor capacitor before checking with an ohmmeter?
   A. 20,000 ohms, 5 watts
   B. 2000 ohms, 5 watts
   C. 5 ohms, 20,000 watts
   D. 20,000 ohms, 5000 watts
8. To determine the capacitance of a capacitor, a ________ should be used.
   A. psychrometer
   B. megger
   C. capacitor analyzer
   D. capacitor generator
9. What are some physical differences between a run and a start capacitor?
   A. Run capacitors are housed in paper or plastic, start capacitors in a metal can.
   B. Run capacitors are dry-type, start capacitors are oil-filled.
   C. Run capacitors typically have 5 terminals, start capacitors have 4.
   D. Run capacitors are oil-filled, start capacitors are dry-type.
10. In a three-phase motor in good operating condition, the resistance ___________.
    A. should be the same across all three windings
    B. should vary from a lower resistance in the first winding to a higher resistance in the third winding
    C. across each winding is not important
    D. should add up to the applied voltage
11. What may happen to an electric motor if the drive belt is too tight?
    A. The motor will operate more efficiently.
    B. The bearings will wear out prematurely.
    C. The internal overload will malfunction.
    D. The mechanism being driven will break apart.
12. How are the electrical components checked on a hermetic compressor?
    A. The compressor should be cut open, and the windings tested.
    B. The compressor should be pulled into a vacuum and tested with a megger.
    C. Hermetic compressors should just be replaced if a failure is suspected.
    D. The electrical terminals on the outside of the compressor can be tested.
13. A loose connection may result in ________ voltage and ________ current at the motor.
    A. higher, lower
    B. lower, higher
    C. no measurable change in, higher
    D. higher, no measurable change in
14. The current can be checked with the ________ function on a digital multimeter.
    A. amperage
    B. ohms
    C. voltage
    D. millivolt
15. Electric motor roller and ball bearing failure often can be determined by the ________.
    A. voltage drop
    B. noise
    C. resistance of the run winding
    D. resistance of the start winding

16. Which of the following electric motor problems can cause the motor housing to become energized?
    A. Shorted windings
    B. Open windings
    C. Grounded windings
    D. Closed windings

17. On a belt-driven motor assembly, if the drive pulley and the driven pulley are not aligned, ____________.
    A. the motor will deliver more airflow than it is rated for
    B. start and run capacitors are required
    C. the motor windings will become grounded
    D. strain will be placed on the drive mechanisms

18. Explain how a motor tensioning tool is used.

SECTION 5

COMMERCIAL REFRIGERATION

UNITS

UNIT 21

EVAPORATORS AND THE REFRIGERATION SYSTEM

OBJECTIVES

After studying this unit, you should be able to

- describe the refrigeration process.
- define high-, medium-, and low-temperature applications.
- discuss the functions of an evaporator in a refrigeration circuit.
- explain heat exchange characteristics of liquids and vapors.
- describe a refrigerant's boiling point and its relation to evaporator superheat.
- describe the operating characteristics of a dry-type evaporator.
- describe the operating characteristics of a flooded evaporator.
- identify four types of evaporator designs.
- discuss evaporator performance and operating conditions.
- explain how an evaporator removes latent heat or moisture.
- describe pressure drop in multiple- and single-circuit evaporators.
- list three evaporator defrost methods.
- explain the functions of an evaporator efficiency controller.

SAFETYCHECKLIST

✔ Wear goggles and gloves when attaching or removing gauges to transfer refrigerant or to check pressures.
✔ Wear warm clothing when working in a walk-in cooler or freezer.

21.1 REFRIGERATION

Refrigeration is the process of removing heat from a place where it is not wanted and transferring that heat to a place where it makes little or no difference. Commercial refrigeration is similar to the refrigeration that occurs in your household refrigerator. The food that you keep in the refrigerator is stored at a temperature lower than the room temperature. Typically, the fresh-food compartment temperature is about 35°F. Heat from the room (typically 75°F) moves through the walls of the refrigerator to the cooler temperature in the refrigerator. Heat travels normally and naturally from a warm to a cool medium.

If the heat that is transferred into the refrigerator remains in the refrigerator, it will warm the food products and spoilage will occur. This heat may be removed from the refrigerator by mechanical means using the refrigerator's refrigeration equipment, which requires energy, or work. **Figure 21.1** shows how heat is removed with the compression cycle. Because it is 35°F in the box and 75°F in the room, the mechanical energy in the compression cycle actually pumps the heat to a warmer environment from the box to the room.

Figure 21.1 Heat normally flows from a warm place to a colder place. When it is desirable for heat to be removed from a colder to a warmer place, the heat must be moved by force. The compressor in the refrigeration system is the pump that forces the heat up the temperature scale. These compressors are normally driven by electric motors.

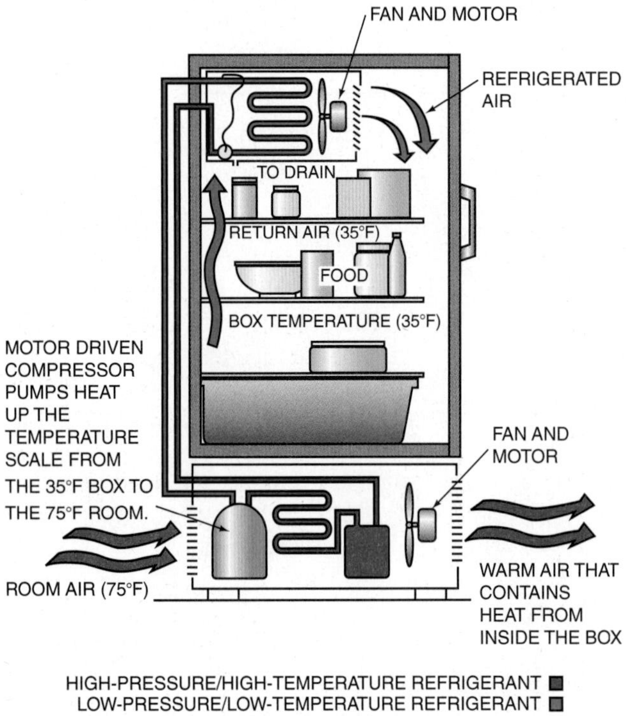

The heat in the refrigerator is transferred into a cold refrigerant coil and pumped by the system compressor to the condenser, where it is released into the room. This is much like using a sponge to move water from one place to another. When a dry sponge is allowed to absorb water in a puddle and you take the wet sponge to a container and squeeze it, you exert energy, much like a compressor in the refrigeration system, **Figure 21.2**.

Another example of refrigeration is a central air-conditioning system in a residence. It absorbs heat from the

Figure 21.2 A sponge absorbs water. The water can then be carried in the sponge to another place. When the sponge is squeezed, the water is rejected to another place. The squeezing of the sponge may be considered the energy that it takes to pump the water.

home by passing indoor air at about 75°F over a coil that is cooled to about 40°F. Heat will transfer from the room air to the coil cooling the air. This cooled air may be mixed with the room air, lowering its temperature, **Figure 21.3**. This process is called air-conditioning, but it is also refrigeration at a higher temperature than that of the household refrigerator, so it is frequently called high-temperature refrigeration.

Commercial refrigeration is used in commercial business locations. The grocery store, convenience store, fast-food restaurant, flower shop, and food processing plant are only a few of the applications. Some of the commercial systems are plug-in appliances, such as a small, reach-in ice storage bin at the local convenience store. The system is entirely located in the same housing. Some systems consist of individual boxes with single remote condensing units, and some are complex systems with several compressors in a rack serving several reach-in display cases, as in a supermarket. Most commercial refrigeration equipment is installed and serviced by a specialized group of technicians who work only with commercial refrigeration and the food-service business.

Figure 21.3 An air-conditioning example of refrigeration.

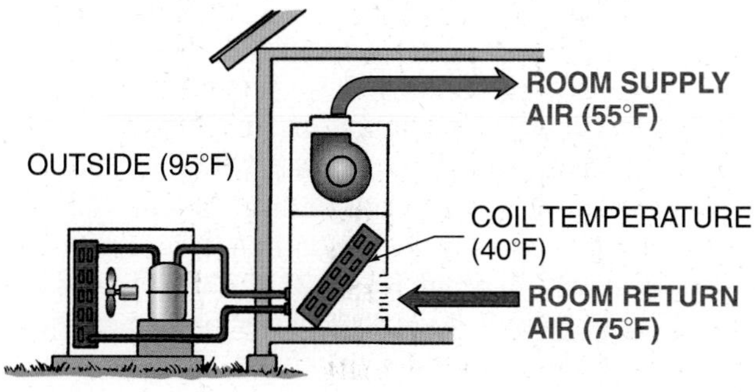

21.2 TEMPERATURE RANGES OF REFRIGERATION

The temperature ranges for commercial refrigeration may refer to the temperature of the refrigerated box or the boiling temperature of the refrigerant in the evaporator coil. The following temperatures illustrate some of the guidelines used in the industry when discussing box temperatures.

HIGH-TEMPERATURE APPLICATIONS

High-temperature refrigeration applications will normally involve box temperatures of 47°F to 60°F. Storing such products as flowers and candy may require these temperatures.

MEDIUM-TEMPERATURE APPLICATIONS

The household refrigerator fresh-food compartment is a good example of medium-temperature refrigeration, which typically ranges from 28°F to 40°F. Many different products are stored at the medium-temperature range. For most products, the medium-temperature refrigeration range is above freezing; few products are stored below 32°F. Items such as eggs, lettuce, and tomatoes lose their appeal if they freeze in a refrigerator.

LOW-TEMPERATURE APPLICATIONS

Low-temperature refrigeration produces temperatures below the freezing point of water, 32°F. One of the higher low-temperature applications is the making of ice.

Low-temperature food storage applications generally start at 0°F and go as low as −20°F. At this temperature ice cream would be frozen hard. Frozen meats, vegetables, and dairy products are only a few of the foods preserved by freezing. Some foods may be kept for long periods of time and are appetizing when thawed for cooking, provided they are frozen correctly and kept frozen.

21.3 THE EVAPORATOR

The evaporator in a refrigeration system is responsible for absorbing heat into the system from whatever medium is to be cooled. This heat-absorbing process is accomplished by maintaining the evaporator coil at a lower temperature than the medium to be cooled. For example, if a walk-in cooler is to be maintained at 35°F to preserve food products, the coil in the cooler must be maintained at a lower temperature than the 35°F air that will be passing over it. **Figure 21.4** shows the refrigerant in the evaporator boiling at 20°F, which is 15°F lower than the entering air. The evaporator operating at these low temperatures removes latent and sensible heat from the cooler. Operating at 20°F, as in the preceding example, the evaporator will collect moisture from the air in the cooler, latent heat. The removal of sensible heat reduces the food temperature.

Figure 21.4 The relationship of the coil's boiling temperature to the air passing over the coil while it is operating in the design range.

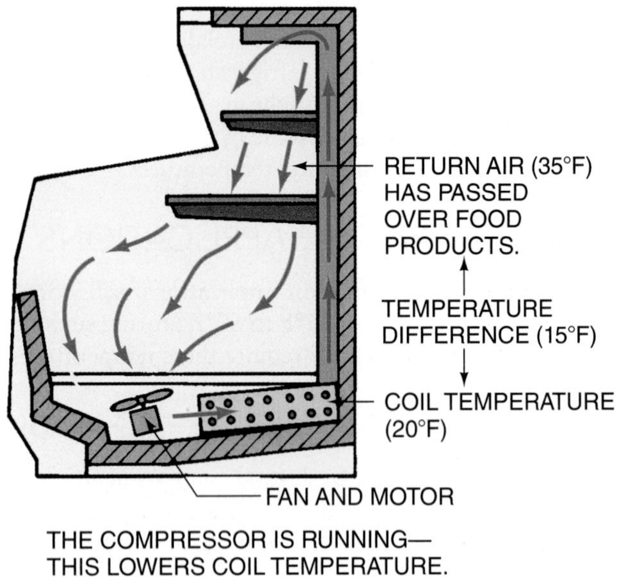

21.4 BOILING AND CONDENSING

Two important factors in understanding refrigeration are the refrigerant (1) boiling temperature and (2) condensing temperature. The boiling temperature and its relationship to the system involve the evaporator. The condensing temperature involves the condenser and will be discussed in the next unit. These temperatures can be followed by using the temperature/pressure chart in conjunction with a set of refrigeration pressure gauges, **Figures 3.13** and **21.5**.

Figure 21.5 Many pressure gauges have temperature/pressure relationships printed on the gauge. Photo by Jason Obrzut

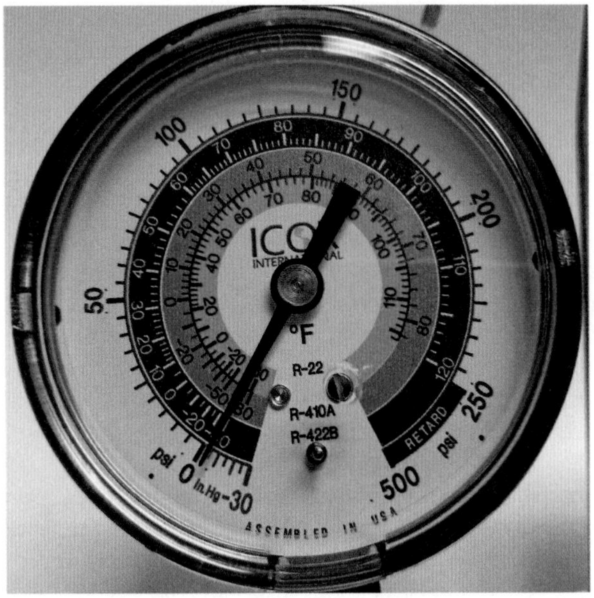

21.5 THE EVAPORATOR AND BOILING TEMPERATURE

The boiling temperature of the liquid refrigerant determines the coil operating temperature. In an air-conditioning system a 40°F evaporator coil with 75°F air passing over it produces conditions used for air-conditioning or high-temperature refrigeration. Boiling is normally associated with high temperatures and water. Unit 3, "Refrigeration and Refrigerants," discussed the fact that water boils at 212°F at atmospheric pressure. It also discussed the fact that water boils at other temperatures, depending on the pressure. When the pressure is reduced, water will boil at 40°F. This is still boiling—changing a liquid to a vapor. In a refrigeration system, the refrigerant may boil at 20°F by absorbing heat from the 35°F food.

The service technician must be able to determine what operating pressures and temperatures are correct for the various systems being serviced under different load conditions. Much of this knowledge comes from experience. When taking readings from thermometers and gauges, the readings must be evaluated. There can be as many different readings as there are changing conditions. Guidelines can help the technician know the pressure and temperature ranges at which the equipment should operate. There are relationships between the entering air temperature and the evaporator for each system. These relationships are similar from installation to installation.

21.6 REMOVING MOISTURE

Dehumidifying the air means to remove the moisture, and this is frequently desirable in refrigeration systems. Moisture removal is similar from one refrigeration system to another. Knowing what the coil-to-air relationship is can help the technician know what conditions to look for. The load on the coil would rise or fall accordingly as the return-air temperature rises or falls. Warmer return air in the box will also have more moisture content, which imposes further load on the coil. If the cooler is warm due to food added to it, the coil would have more heat to remove because it has more load on it. It would be much like boiling water in an open pan on the stove. The water boils at one rate with the burner on medium and at an increased rate with the burner on high. The boiling pressure stays the same in the boiling water in a pan because the pan is open to the atmosphere. When this same boiling process occurs in an enclosed coil, the pressures will rise when the boiling occurs at a faster rate. This causes the operating pressure of the whole system to rise, **Figure 21.6**.

When the evaporator removes heat from air and lowers the temperature of the air, sensible heat is removed. When moisture is removed from the air, latent heat is removed. The moisture is piped to a drain, **Figure 21.7**. Latent heat is called hidden heat because it does not register on a thermometer, but it is heat, like sensible heat, and it must be removed, which takes energy.

Figure 21.6 The coil-to-air temperature relationship under increased coil load.

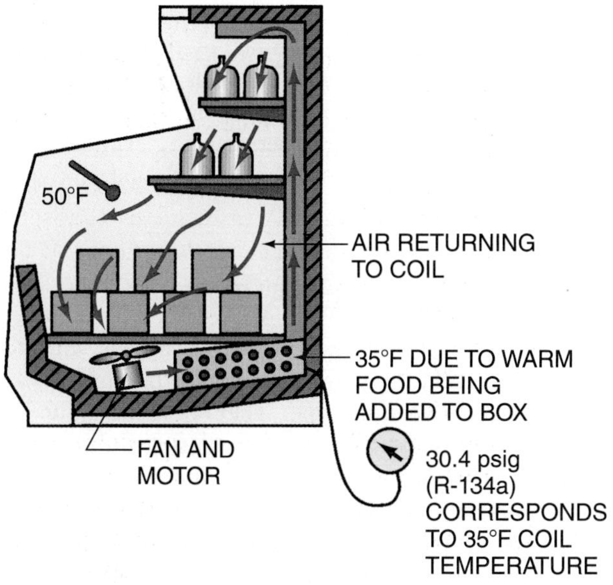

Figure 21.7 The cooling coil condenses moisture from the air.

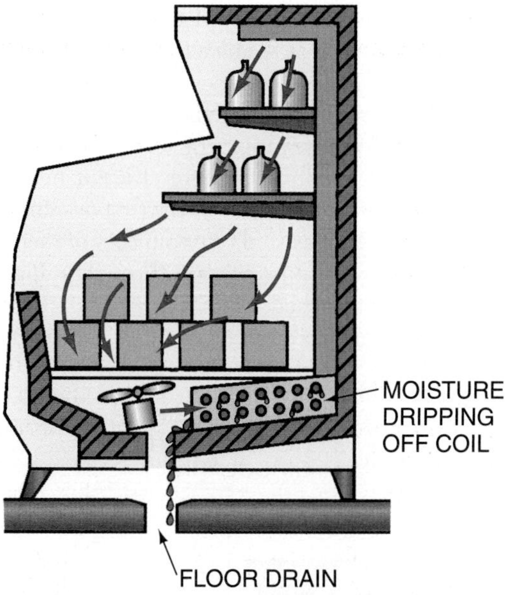

The difference in temperature between the air being cooled and the evaporator is directly related to the amount of moisture (humidity) that will be removed. The greater the difference in temperature, the more moisture is removed. A small difference in temperature will result in less moisture being removed. In some produce refrigeration applications, the amount of moisture being removed must be monitored and controlled. If too much moisture is removed, the produce will dehydrate and become wilted. Dehydrated produce looks less appealing and weighs less. Since a lot of produce items are sold by the pound, dehydration also reduces profits for the seller. The evaporator will also accumulate an excessive amount of frost which may require a manual defrost. In a comfort-cooling application, a high temperature difference (about 35°F) is beneficial. This ensures adequate humidity removal from the conditioned space.

21.7 HEAT EXCHANGE CHARACTERISTICS OF THE EVAPORATOR

The following are conditions that govern the rate of heat exchange:

1. The evaporator *material* through which the heat has to be exchanged. Evaporators may be manufactured from copper, steel, brass, stainless steel, or aluminum. Corrosion is one factor that determines what material is used. For instance, when acidic materials need to be cooled, copper or aluminum coils would be eaten away. Stainless steel may be used instead, but stainless steel does not conduct heat as well as copper. Some evaporators are even coated with a plastic-like substance to protect the metal underneath from rust or oxidation. This application is often seen in restaurants with smaller commercial, medium-temperature coolers that store salad preparation materials, which are acidic because they often have a vinegar base for added flavor and for extended shelf life.
2. The *medium* to which the heat is transferred. Giving heat up from air to refrigerant is an example, **Figure 21.8**. The best heat exchange occurs between two liquids, such as water to liquid refrigerant. This is because liquids are more dense than vapors and usually have a higher specific heat. However, this is not always practical because heat frequently has to be exchanged between air and vapor refrigerant. The vapor-to-vapor exchange is slower than the liquid-to-liquid exchange, **Figure 21.9**.
3. The film factor. This is a relationship between the medium giving up heat and the heat exchange surface. The film factor relates to the velocity of the medium

Figure 21.8 The heat exchange relationship between air and refrigerant.

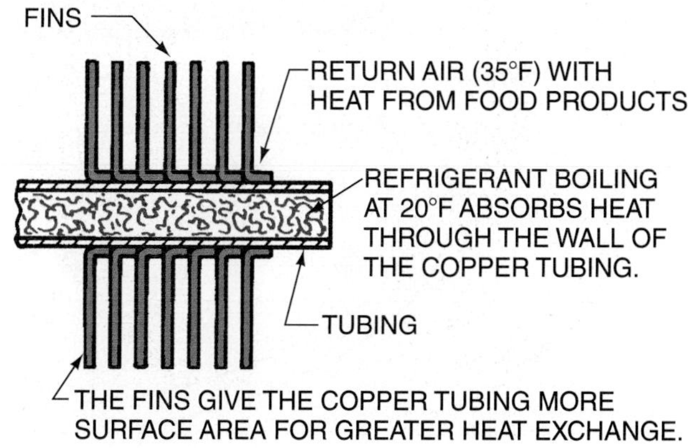

Figure 21.9 The heat exchange relationship between a liquid in a heat exchanger and the refrigerant inside the coil.

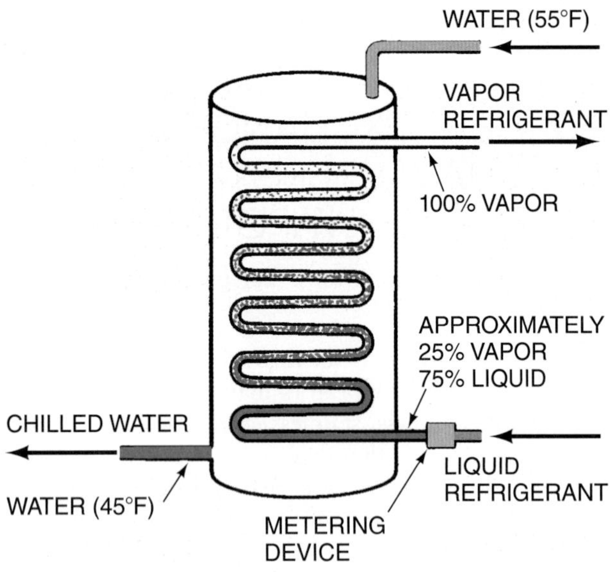

Figure 21.10 One of the deterring factors in a normal heat exchange. The film factor is the film of air or liquid next to the tube in the heat exchanger.

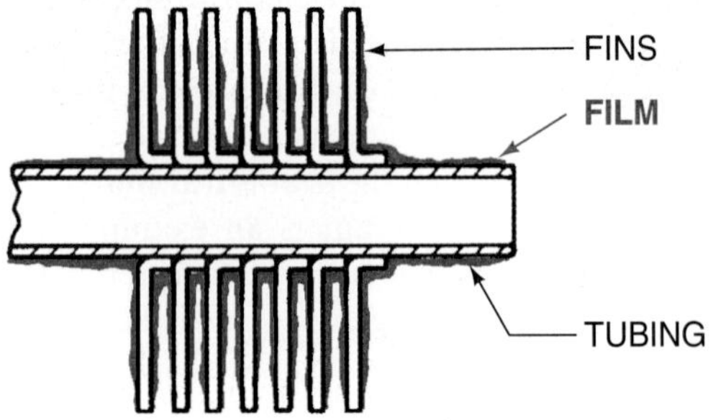

Figure 21.11 A bare-pipe evaporator.

passing over the exchange surface. When the velocity is too slow, the film between the medium and the surface becomes an insulator and slows the heat exchange. The velocity keeps the film to a minimum, **Figure 21.10**. The correct velocity is determined by the manufacturer.

4. The *temperature difference* between the two mediums in which the heat exchange is taking place. The greater the temperature difference between the evaporator coil and the medium giving up the heat, the faster the heat exchange will occur.

21.8 TYPES OF EVAPORATORS

Numerous types of evaporators are available, and each has its purpose. The first evaporators for cooling air were of the natural-convection type. They were actually bare-pipe evaporators with refrigerant circulating through them, **Figure 21.11**. This evaporator was used in early walk-in coolers and was mounted high in the ceiling. It relied on the air being cooled, falling to the floor, and setting up a natural air current. The evaporator had to be quite large for the particular application because the velocity of the air passing over the coil was so slow. Natural-convection evaporators are still occasionally used today. The use of a blower to force or induce air over the coil improved the efficiency of the heat exchange. This meant that smaller evaporators could be used to do the same job. Design trends in the industry have always been toward smaller, more efficient equipment, **Figure 21.12**.

The expansion of the evaporator surface to a surface larger than the pipe itself produces a more efficient heat exchange. The *stamped evaporator* was one of the first designs to create a larger pipe surface. It consisted of two pieces of metal stamped with the impression of a pipe passage through it, **Figure 21.13**.

A pipe with fins attached, called a *finned-tube evaporator*, is today used more than any other type of heat exchanger between

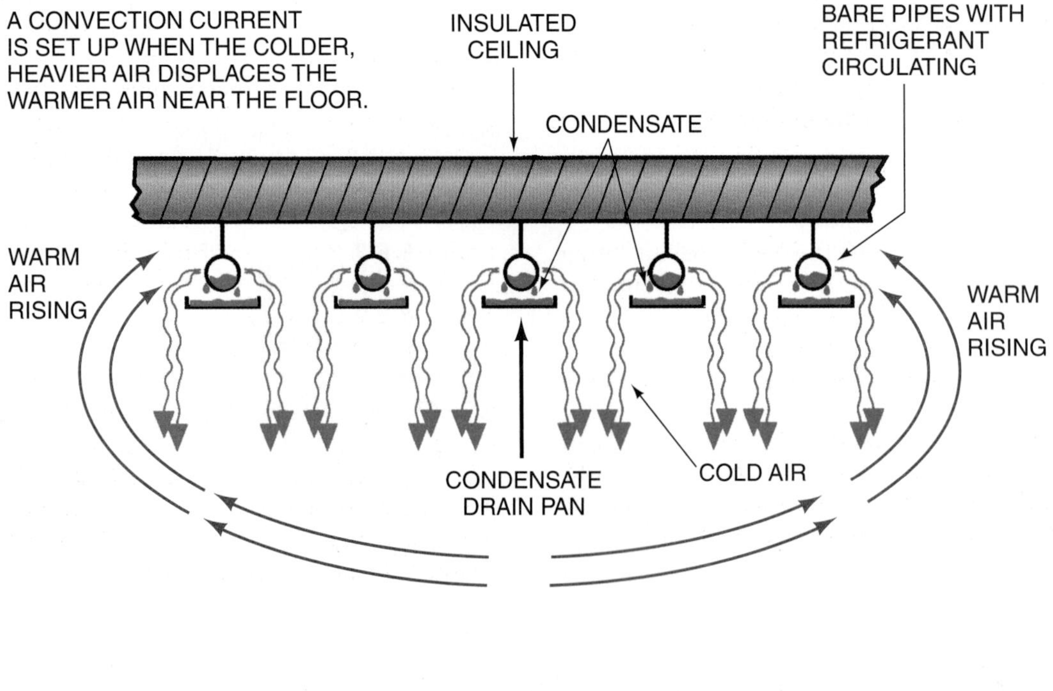

Figure 21.12 A forced-draft evaporator. Photo by Jason Obrzut.

Figure 21.13 A stamped evaporator. Courtesy Sporlan Valve Company

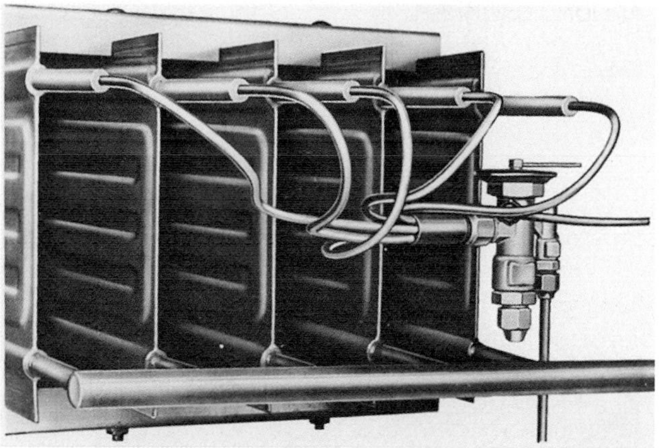

Figure 21.14 (A) A finned-tube evaporator for a medium-temperature commercial refrigeration case. (B) Finned-tube evaporator for a low-temperature, commercial, well-type freezer. Notice the two different fin spacings in the freezer's evaporator. Courtesy Ferris State University. Photos by John Tomczyk

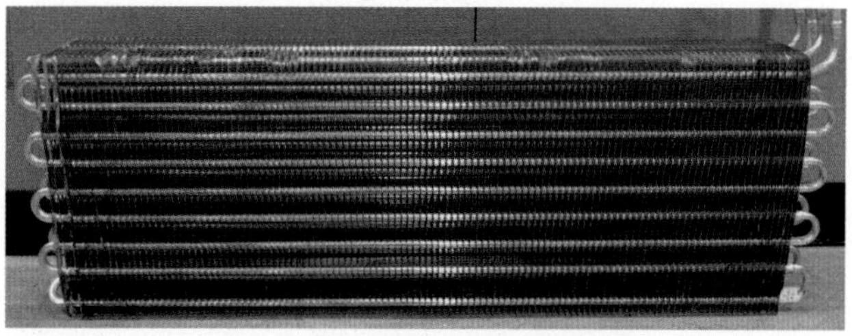

(A)

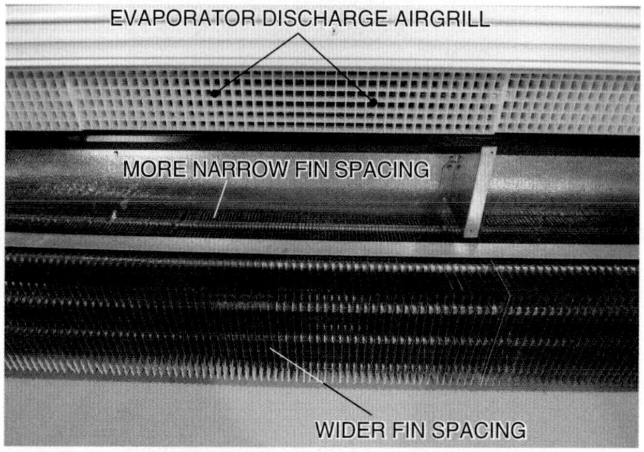

(B)

air and refrigerant. This heat exchanger is efficient because the fins are in good contact with the pipe carrying the refrigerant. **Figure 21.14(A)** shows an example of a finned-tube evaporator. **Figure 21.14(B)** shows a finned-tube evaporator used in a low-temperature, commercial, well-type freezer. The leading fins come in contact with the air first and must have wider spacing between the fins. It is these leading fins that are more prone to frost accumulation. The wider fin spacing can accumulate more frost before becoming iced over and impeding the airflow. The fins toward the back of the evaporator then encounter drier air with less frost and so are more closely spaced.

Multiple circuits improve evaporator performance and efficiency by reducing pressure drop inside the evaporator. Even though the pipes inside the evaporator might be polished smooth, they still offer resistance to the flow of both liquid and vapor refrigerants. The shorter the evaporator is, the less resistance there is to this flow. The "U" bends at the ends of the evaporator also offer a great deal of resistance to the flow of refrigerant. As evaporators become longer, they have more and more pressure drop associated with refrigerant flowing through them. The manufacturer will design the evaporator so that the tubing bundles are in parallel to one another, **Figure 21.15**.

The evaporator for cooling liquids or making ice operates under the same principles as one for cooling air but is designed differently. It may be strapped on the side of a cylinder with liquid inside, submerged inside the liquid container, or be a double-pipe system with the refrigerant inside one pipe and the liquid to be cooled circulated inside an outer pipe, **Figure 21.16**.

Most equipment manufacturers are looking for more efficient evaporator and condenser designs, more efficient compressors and fan motors, and more sophisticated control systems in order to meet new energy-efficiency requirements.

One such evaporator design incorporates an aluminum parallel-flow, flat-plate-and-fin configuration with small parallel channels inside the flat plate. The plates are flattened, streamlined tubes each one of which is split into smaller, parallel ports, **Figure 21.17**. *Refrigerant will phase change or evaporate from a liquid to a vapor inside the channels in the plate, while strategically shaped fins (extended surfaces) will enhance heat transfer from the air into the evaporator. The plates and fins are bonded or soldered to increase heat transfer and to eliminate any contact resistance (air gaps) that will reduce heat transfer.* Headers at the inlet and outlet of the heat exchanger are also bonded to the plates through soldering.

Figure 21.15 (A) A multicircuit evaporator fed with a distributor to reduce pressure drop. Courtesy Sporlan Valve Company. (B) and (C) A two-pass, multicircuit, finned-tube evaporator. There are actually two evaporators in parallel to reduce pressure drop. Courtesy Ferris State University. Photo by John Tomczyk

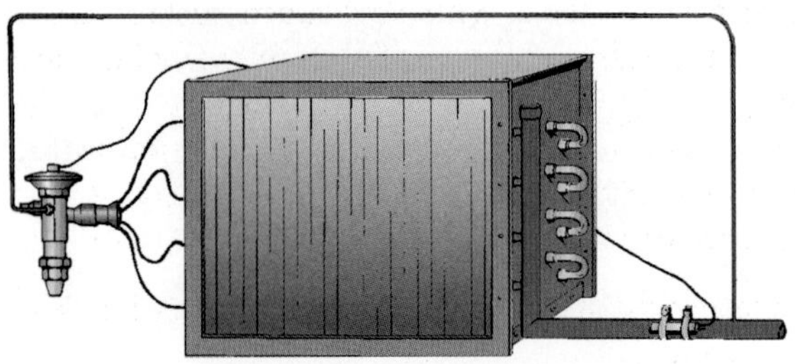

(A)

(B)

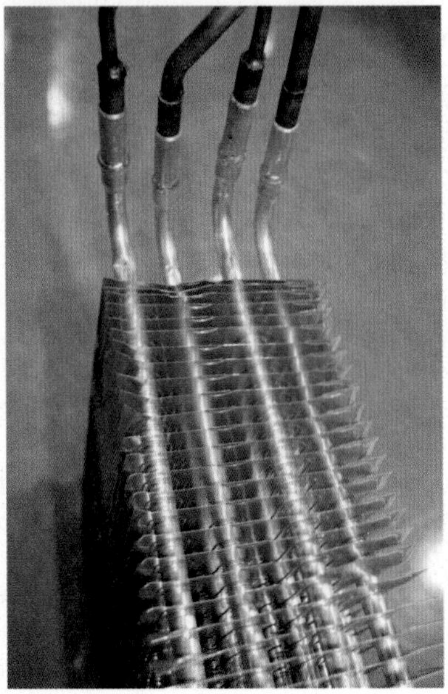

(C)

Figure 21.16 Liquid heat exchangers. (A) A drum-type evaporator. (B) A plate-type evaporator in a tank. (C) A pipe-in-pipe evaporator. Source: Noranda Metal Industries, Inc.

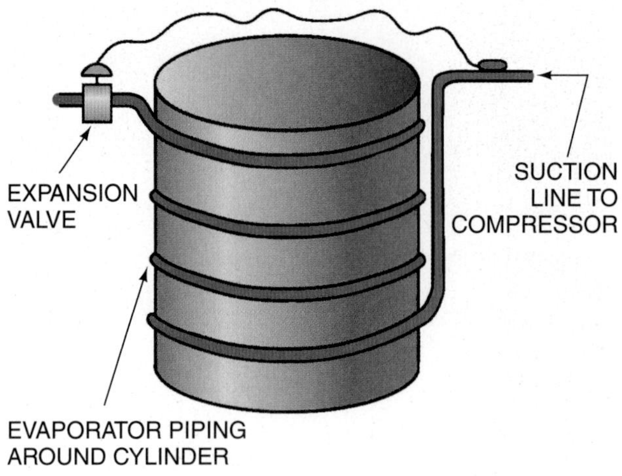

(A)

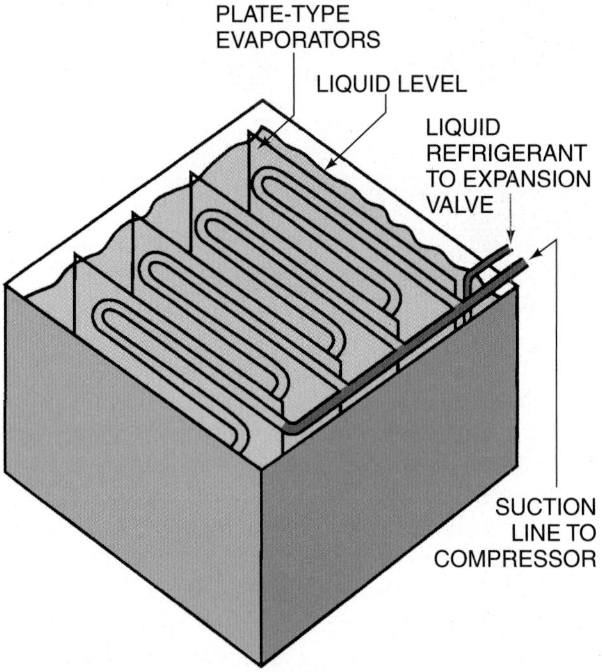

(B)

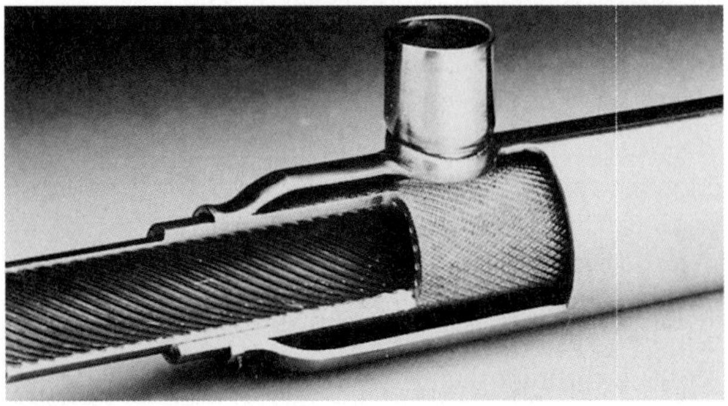

(C)

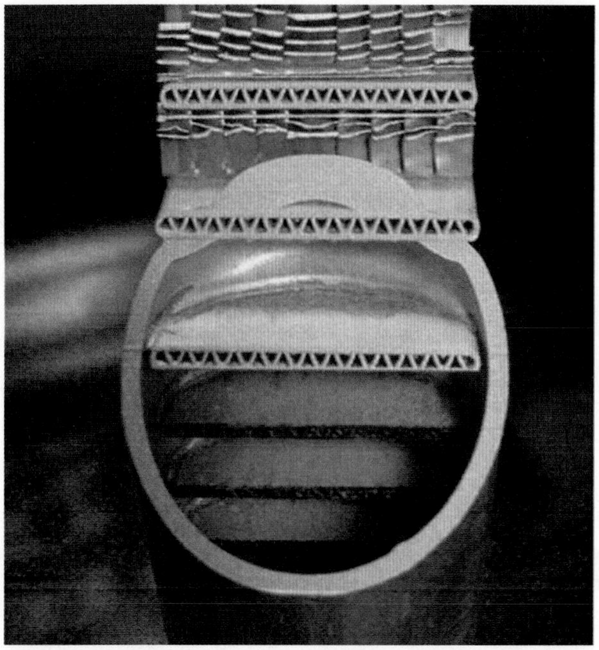

(A)

(B)

Figure 21.17 (A) and (B) An aluminum parallel-flow, flat-plate-and-fin heat exchanger. Source: Modine Manufacturing Co., Racine, WI.

Heat is transferred from the air to the evaporating refrigerant in three steps, as follows:

1. Air side—between the fins and the air to be cooled
2. Heat conduction—between the fins and the tubes
3. Refrigerant side—between the tubes and the evaporating refrigerant

The air side of the heat exchange can be enhanced through fin geometry. Louvres, lances, and rippled edges all increase heat transfer, **Figure 21.18**. *The conduction between the fins and the tubes is enhanced through the application of a metallic bond (soldering) that eliminates any air gaps.* The refrigerant side of the heat transfer deals with how much surface area of the inside of the tubes will come into contact with the phase-changing refrigerant. This internal surface area is often referred to as a *wetted perimeter*. As the internal surface area of the tubes increases, the heat transfer increases. Internal surface area can be increased by

- increasing the number of parallel channels inside the flat plates, and/or
- increasing the number of flat plates (decreasing the spacing between them).

The capacity (tonnage) of the heat exchanger can vary with its height and length. The plates can be oriented vertically

Figure 21.18 (A) and (B) Fin geometry for increasing heat transfer. Source: Modine Manufacturing Co., Racine, WI.

(A)

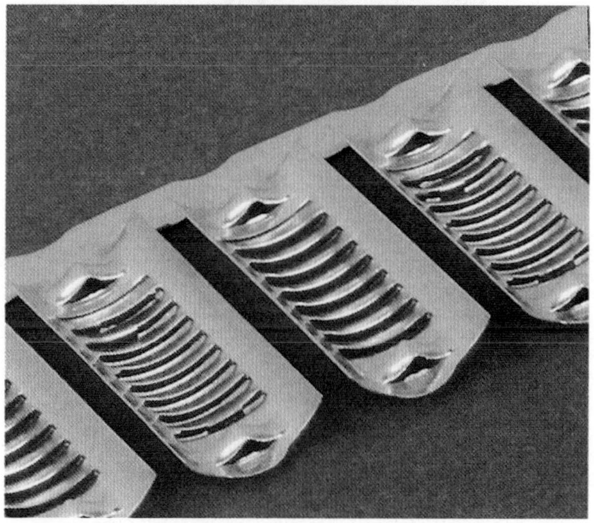

(B)

for an evaporator application or horizontally for condenser applications. The vertical orientation of the flat plates allows condensate removal to occur naturally, alleviating any water drainage issues from the evaporator, **Figure 21.19**. This technology is being used with condensers as well as evaporators, **Figure 21.20**. Applications in the HVACR field include residential air-conditioning, rooftop air-conditioning, chillers, geothermal heat pumps, electronic cooling, packaged terminal air conditioners (PTACs), ice machines, beverage dispensers, refrigerated display cases, and food-service refrigeration. Some of the benefits of this parallel-flow, plate-and-fin heat exchanger technology are as follows:

- Reduces static pressure through the coil—which means fewer fan watts and less horsepower

Figure 21.19 (A) and (B) Condensate removal for vertical and horizontal header configuration. Source: Modine Manufacturing Co., Racine, WI.

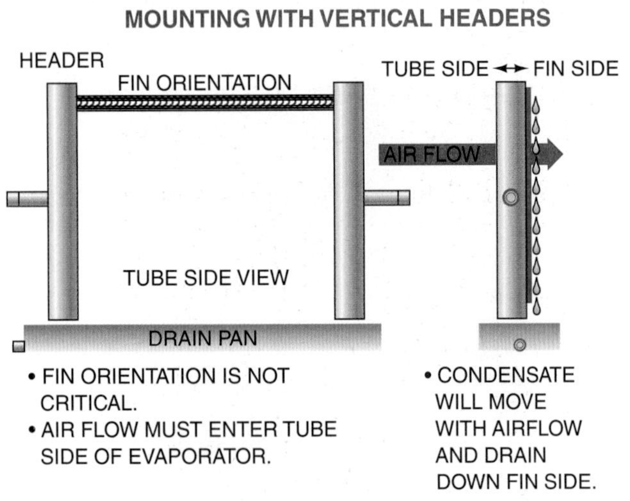

(A)

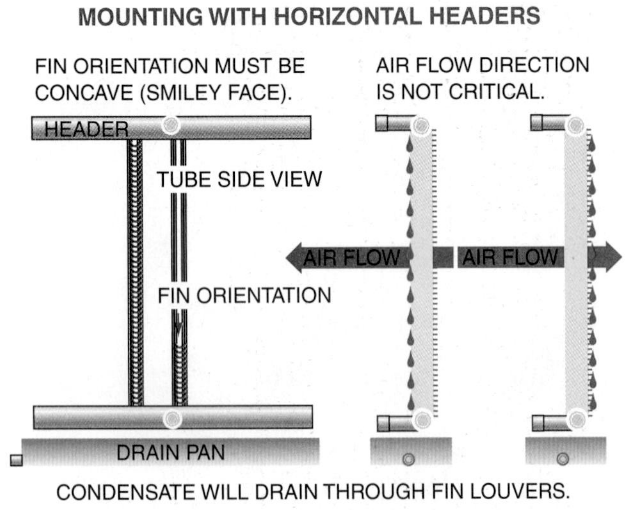

(B)

Figure 21.20 (A) Condenser. (B) Evaporator. Source: Modine Manufacturing Co., Racine, WI.

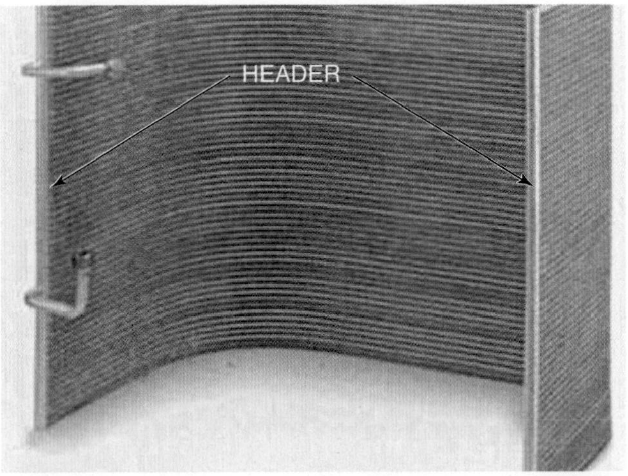

(A)

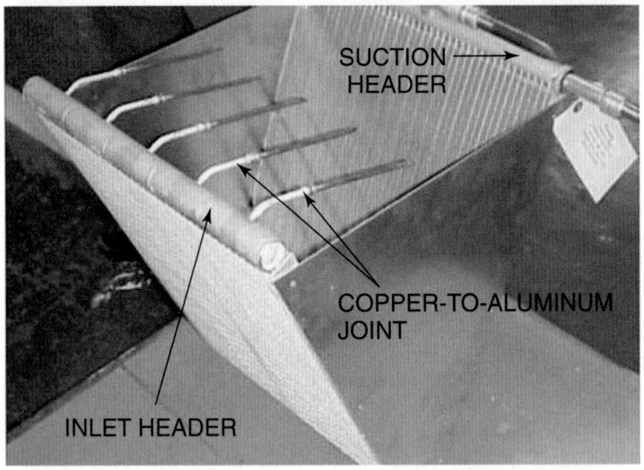

(B)

- Reduces coil depth for the evaporator and condenser, which leads to easier cleaning and less air-side static pressure
- Reduced internal volume reduces refrigerant charge
- Reduced face area of condenser and evaporator
- Smaller footprint for the condensing unit
- A 30% reduction in coil weight and size
- A reduction in packing costs, sizes, and weights
- The all-aluminum coil, header, and fins enhance corrosion resistance
- Lower system costs
- Higher system efficiencies than a round, copper-tube heat exchanger with aluminum fins, **Figure 21.15(C)**
- Lower operating costs
- Quieter operations

Field repair of leaking heat exchangers can be accomplished by

- recovering the refrigerant,
- cleaning the leaking area with a solution,

- brushing the area with a wire brush,
- using a utility knife blade to remove any fins that may be in the local area,
- pulling a vacuum with a vacuum pump,
- applying a two-part epoxy that will be sucked into the flat plate where the leak exists,
- applying heat with an electric blow drier until the epoxy is cured,
- evacuating to a 500-micron vacuum,
- charging with the appropriate refrigerant.

Field cleaning heat exchangers can be accomplished using the same methods as for a standard round copper-tube heat exchanger with aluminum fins:

- Elevate the temperature of the mixed cleaner to 120°F.
- Use a power washer with a broad spray pattern.
- Use nonacidic cleaners (pH < 10.5).
- When the heat exchanger is clean, rinse the coil with clean water.

21.9 EVAPORATOR EVALUATION

Knowing the design considerations helps in *evaporator evaluation*. When the service technician arrives at the job, it may be necessary to evaluate whether a particular evaporator is performing properly. This can be considered one of the starting points in organized troubleshooting. The evaporator absorbs heat, the compressor pumps it, and the condenser rejects it. The following example pertains to a medium-temperature walk-in box. However, the procedure would be about the same for any typical application.

Evaporator Specifications

1. Copper-pipe coil
2. Aluminum fins attached to the copper-pipe coil
3. Forced draft with a prop-type fan
4. One continual refrigerant circuit (single-circuit)
5. R-134a
6. Evaporator to maintain space temperature at 35°F
7. Evaporator clean and in good working condition

First we describe how the evaporator functions when it is working correctly.

As seen in **Figure 21.21**, warm liquid passes through the small opening in the expansion device into the low pressure (18.4 psig) of the evaporator side. Due to the pressure drop across the expansion device, approximately 25% of the liquid entering the expansion device at the evaporator is changed to a vapor and cools the remainder of the liquid to the evaporator's boiling temperature (20°F). So, entering the evaporator is a partial liquid-partial vapor, or saturated, mixture at 20°F and 18.4 psig; it is approximately 75% liquid and 25% vapor.

As the saturated mixture moves through the evaporator, more of the liquid changes to a vapor. This is called *boiling* and is a result of heat absorbed into the coil from whatever medium the evaporator is cooling. Finally, near the end of

Figure 21.21 When the 110°F liquid passes through the expansion valve orifice, some of the liquid flashes to a vapor and cools the remaining liquid to the evaporator temperature of 20°F.

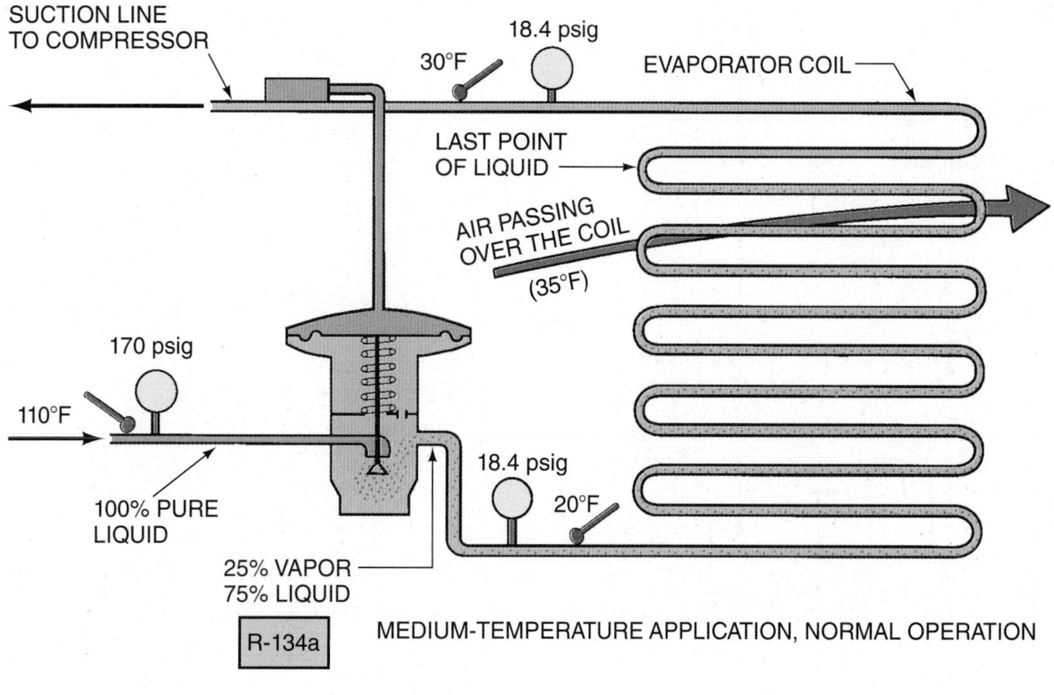

the evaporator, the liquid is all boiled away to a vapor. At this point the refrigerant is known as saturated vapor. This means that the refrigerant vapor is saturated with heat. If any more heat is added to it, it will rise in temperature. If any heat is taken away from it, it will start changing back to a liquid. This vapor is saturated with heat, but it is still at the evaporating temperature corresponding to the boiling point, 20°F. *This is a most important point in the function of an evaporator because all of the liquid must be boiled away as close to the end of the coil as possible. This is necessary to (1) keep the coil efficiency up and (2) ensure that liquid refrigerant does not leave the evaporator and move into the compressor.* For efficient operation, the evaporator must be as full of liquid as possible without allowing liquid to reach the outlet of the coil, because the best heat exchange is between the liquid refrigerant and the air passing over the coil.

The pressure/enthalpy chart in **Figure 21.22** shows graphically what happens inside the evaporator of the preceding walk-in cooler example. The refrigerant enters the evaporator at point B_1 (after leaving the expansion valve). The liquid pressure is 18.4 psig and contains 48.7 Btu/lb of heat at this point. Approximately 25% of the liquid flashed to a vapor when passing through the expansion valve. As the liquid proceeds through the evaporator, moving left to right on the chart, it is changing to a vapor. All of it has changed to a vapor at point B_2, but the vapor temperature is still 20°F and capable of absorbing heat, in the form of superheat. The vapor temperature starts to rise while it is still in the evaporator until the temperature is 30°F (containing 10°F of superheat). The vapor leaves the evaporator at point C with a heat content of 108.1 Btu/lb. The usable refrigeration in the evaporator is from points B_1 to C, where the refrigerant absorbed 59.4 Btu/lb (108.1 Btu/lb − 48.7 Btu/lb). Refer to Unit 3, "Refrigeration and Refrigerants," for more information on pressure/enthalpy charts. You only need to know how many Btu/h capacity is needed to determine the amount of refrigerant that needs to be circulated. For example, if the evaporator needs to have a capacity of 35,000 Btu/h, it must have 589.2 lb of refrigerant circulate through it per hour (35,000 Btu/h ÷ 59.4 Btu/lb = 589.2 lb/h). This sounds like a lot of refrigerant, but it is only 9.82 lb/min (589.2 lb/h ÷ 60 min/h = 9.82 lb/min). The size of the compressor and the system operating conditions determine how much refrigerant can be pumped.

Figure 21.22 The net refrigeration effect in the evaporator. Source: E. I. Dupont

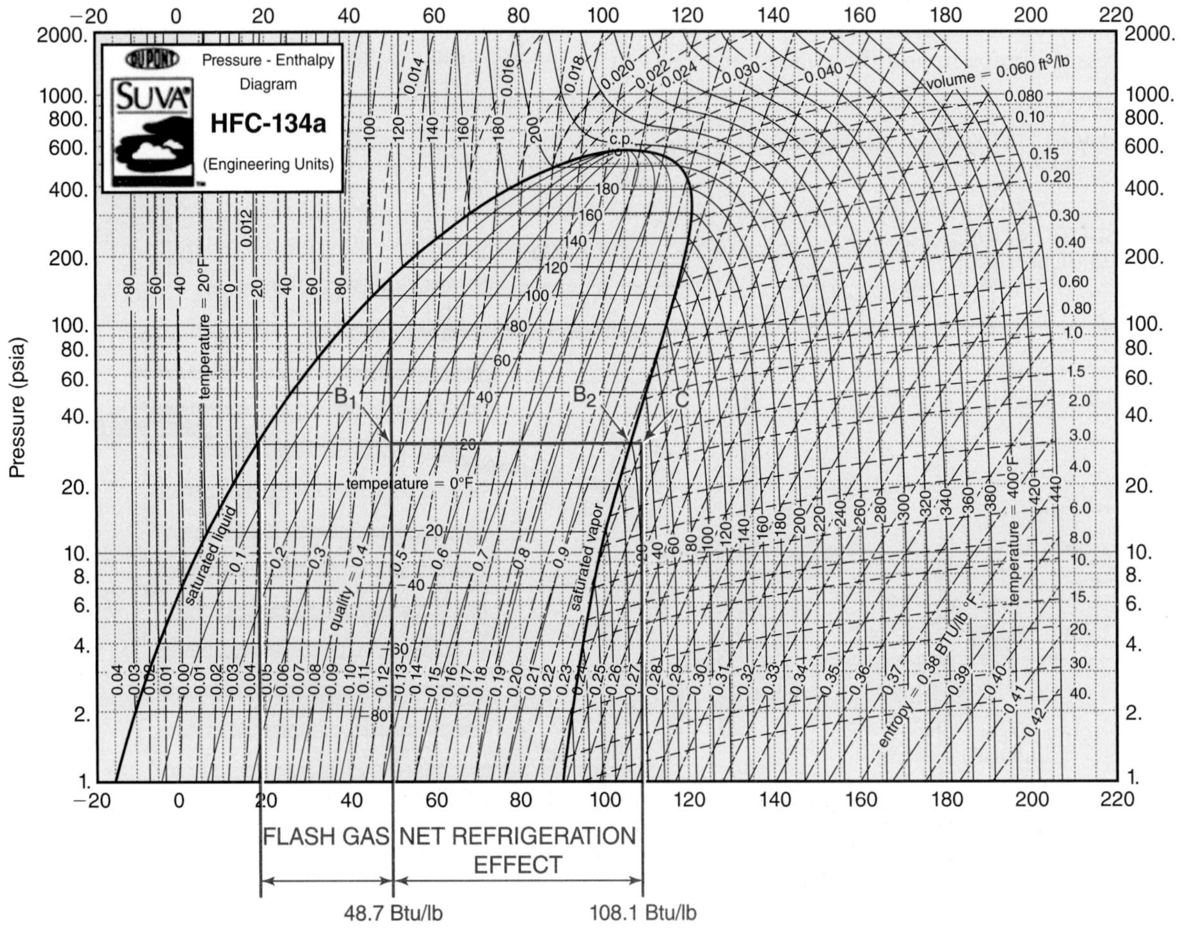

21.10 LATENT HEAT IN THE EVAPORATOR

The latent heat absorbed during the change of state is much more concentrated than the sensible heat that is added to the vapor leaving the coil. Refer to the example in Unit 1, "Heat, Temperature, and Pressure," Section 1.3, "Introduction to Heat," that showed how it takes 1 Btu to change the temperature of 1 lb of 68°F water to 69°F water. Section 1.8 also showed that it takes 970 Btu to change 1 lb of 212°F water to 212°F steam. The change of state is where the greatest amount of heat is absorbed into the system. The preceding example showed that 59.4 Btu of heat were absorbed for every 1 lb of refrigerant that was circulated (59.4 Btu/lb). This happened at a boiling temperature of 20°F, without a change in pressure.

21.11 THE FLOODED EVAPORATOR

To get the maximum efficiency from the evaporator heat exchange, some evaporators are operated full of liquid, or flooded, and are equipped with a device to keep the liquid refrigerant from passing to the compressor. These flooded evaporators are specially made and normally use a float-type metering device to keep the liquid level as high as possible in the evaporator. This text will not go into detail about this system because it is not a device often encountered. The manufacturer's literature should be consulted for any special application.

When an evaporator is flooded, it would operate much like water boiling in a pot with a compressor taking the vapor off the top of the liquid. There would always be a liquid level. If the evaporator is not flooded, that is, when the refrigerant starts out as a partial liquid and boils away to a vapor in the heat exchange pipes, it is known as a *dry-type,* or direct-expansion, evaporator.

21.12 DRY-TYPE EVAPORATOR PERFORMANCE

To check the performance of a dry-type evaporator, the service technician would first make sure that the refrigerant coil is operating with enough liquid inside the coil. To determine this, the technician must calculate the evaporator superheat. This is generally done by comparing the boiling temperature of the refrigerant inside the coil with the line temperature leaving the coil. The difference in temperatures is usually 8°F to 12°F. For example, in the coil pictured in **Figure 21.23**, the superheat in the coil was arrived at by converting the coil pressure (suction pressure) to a saturation temperature. In this example, the pressure is 18.4 psig, which corresponds to a 20°F saturation temperature. The suction pressure reading is important to the technician because the boiling temperature must be known to arrive at the superheat reading for the coil. In the following example, the evaporator superheat reading is 10°F (30°F − 20°F).

Figure 21.23 The evaporator operating under normal load.

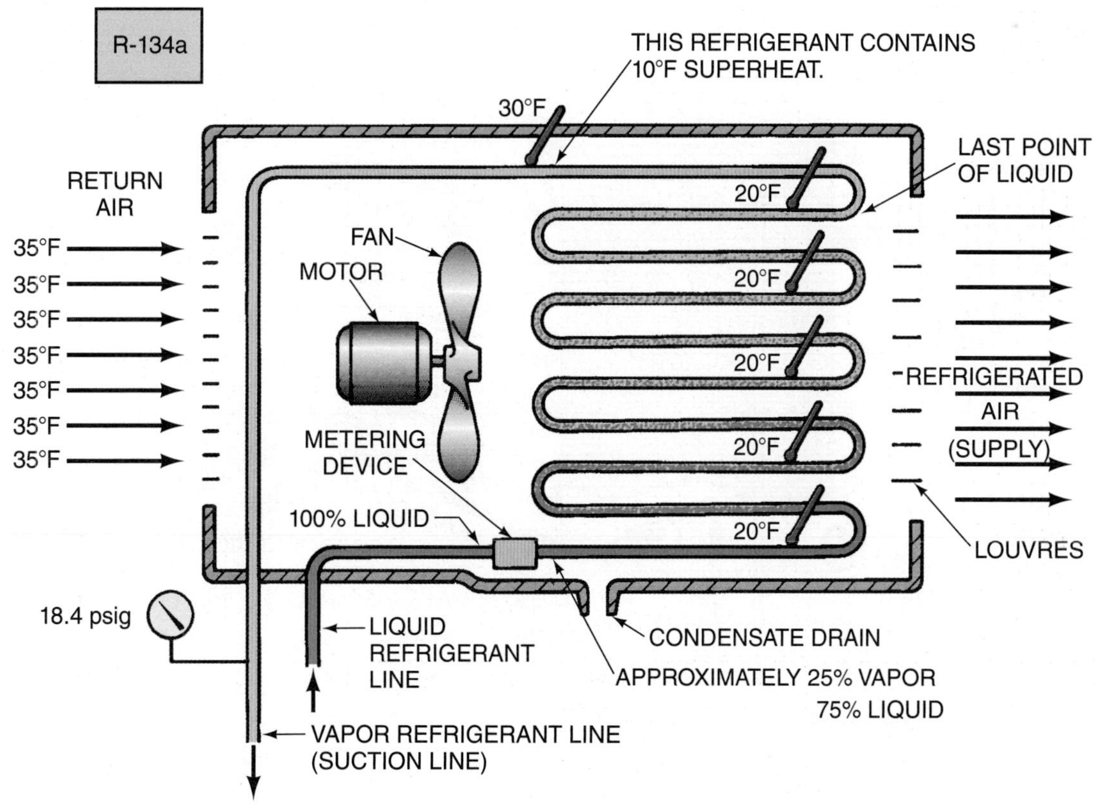

21.13 EVAPORATOR SUPERHEAT

The difference in temperature between the boiling refrigerant temperature and the evaporator outlet temperature is known as evaporator superheat. Superheat is the sensible heat added to the vapor refrigerant after the change of state has occurred. Superheat is the best method of checking to see when a refrigerant coil has a proper level of refrigerant. When a metering device is not feeding enough refrigerant to the coil, the coil is said to be a starved coil, and the superheat is greater, **Figure 21.24**. It can be seen from the example that all of the refrigeration takes place at the beginning of the coil. The suction pressure is very low, below freezing, but only a portion of the coil is being used effectively. This coil would freeze solid and no air would pass through it. The freeze line would creep upward until the whole coil was a block of ice, and the refrigeration would do no good. The refrigerated box temperature would rise because ice is a good insulator.

21.14 HOT PULLDOWN (EXCESSIVELY LOADED EVAPORATOR)

When the refrigerated space has been allowed to warm up considerably, the system must go through a hot pulldown. On a hot pulldown the evaporator and metering device are not expected to act exactly as they would in a typical design condition. For instance, if a walk-in cooler supposed to maintain 35°F were allowed to warm up to 60°F and had some food or beverages inside, it would take an extended time to pull the air and product temperature down. The coil may be boiling the refrigerant so fast that the superheat may not come down to 8°F to 12°F until the box has cooled down closer to the design temperature.

A superheat reading on a hot pulldown should be interpreted with caution, **Figure 21.25**. The reading will be correct only when the coil is at or near design conditions. However, many modern thermostatic expansion valves (TXVs) have wide temperature control ranges. Some can control evaporator superheat from +20°F to −20°F, and they are advertised to do this effectively even when under heavy or light heat loadings of the evaporator. TXVs should control superheat under most normal conditions. However, when a system is under a hot pulldown, the technician should let the system get past this heavy load period and reach a somewhat stabilized condition before trying to calculate an evaporator superheat reading. Hot pulldowns are not considered normal conditions, and the technician must be patient when calculating evaporator superheat. It takes time for a TXV to fill out the evaporator with refrigerant even when it is wide open during a hot pulldown.

When a dry-type coil is fed too much refrigerant, not all the refrigerant changes to a vapor. This coil is thought of as a *flooded* coil—flooded with liquid refrigerant, **Figure 21.26. Do not confuse this with a coil flooded by design.** This is a symptom that can cause real trouble because unless the liquid in the suction line boils to a vapor before it reaches

Figure 21.24 A starved evaporator coil showing 38°F (40°F − 2°F) of evaporator superheat.

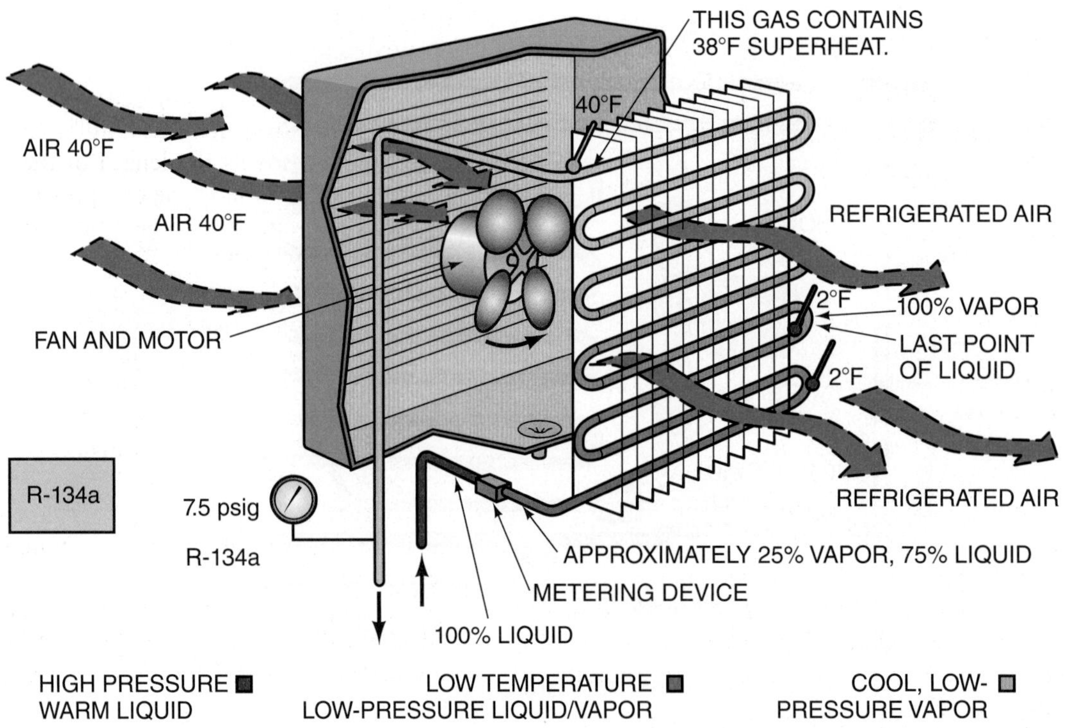

Figure 21.25 Hot pulldown with a coil. This is a medium-temperature evaporator that should be operating at 18.4 psig (R-134a, 20°F). The return air is 55°F instead of 35°F. This causes the pressure in the coil to rise. The warm box boils the refrigerant at a faster rate. The thermostatic expansion valve is not able to feed the evaporator quickly enough to keep the superheat at 10°F. The evaporator has 15°F of superheat.

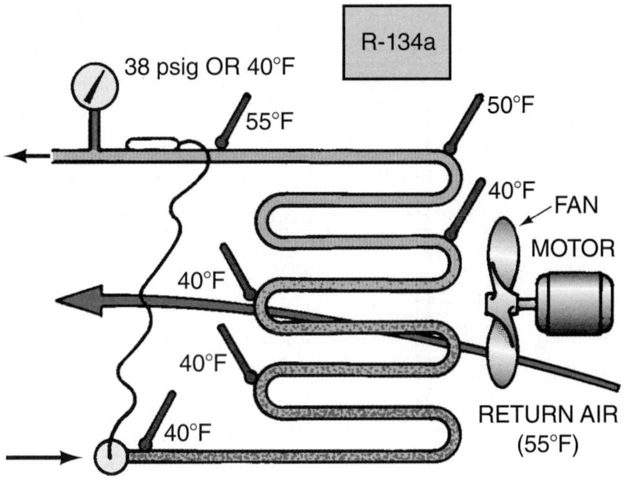

Figure 21.26 The evaporator is flooding because the thermostatic expansion device is not controlling refrigerant flow properly.

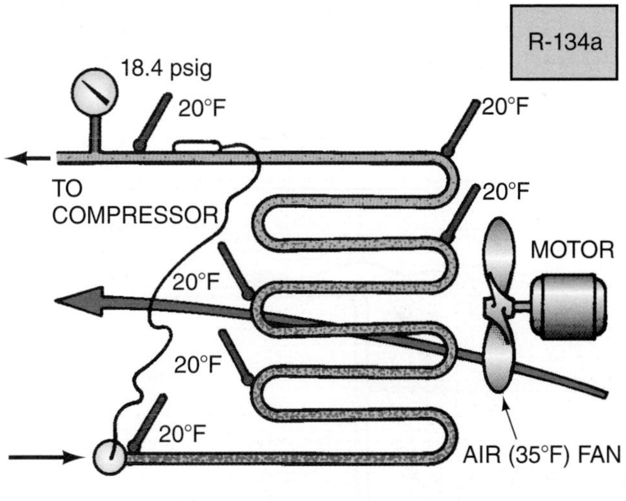

the compressor, compressor damage may occur. **Remember, the evaporator is supposed to boil all of the liquid to a vapor. Therefore, a thermostatic expansion device that is not operating correctly can cause compressor failure.**

21.15 PRESSURE DROP IN EVAPORATORS

Multicircuit evaporators are used when the coil would become too long for a single circuit, **Figure 21.27**. The same evaluating procedures hold true for a multicircuit evaporator as for a single-circuit evaporator.

Figure 21.27 A multicircuit evaporator. Courtesy Sporlan Valve Company

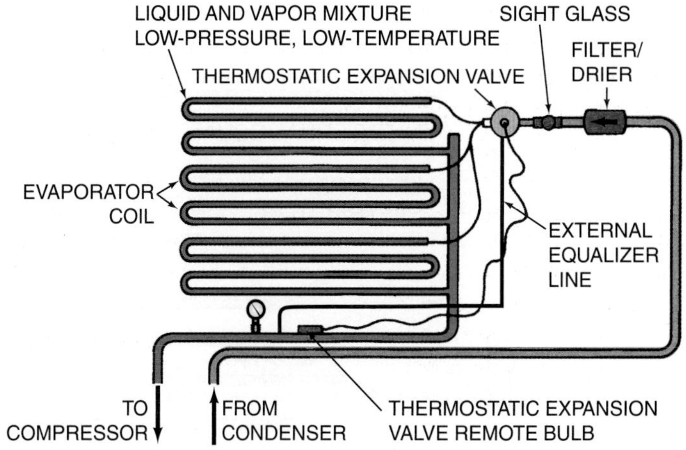

A dry-type evaporator has to be as full as possible with refrigerant to be efficient. Each circuit should be feeding the same amount of refrigerant. If this needs to be checked, the service technician can check the common pressure tap for the boiling pressure, which can be converted to temperature. Then the temperature will have to be checked at the outlet of each circuit to see whether any circuit is overfeeding or starving, **Figures 21.28** and **21.29**.

Some reasons for uneven feeding of a multicircuit evaporator are the following:

- Blocked distribution system
- Dirty coil
- Uneven air distribution
- Coil circuits of different lengths

In larger commercial and industrial-type evaporators, an associated pressure drop usually is caused by friction as the refrigerant travels the length of the evaporator and down a long suction line to the compressor. This causes the pressure at the compressor to be a bit lower than the pressure at the evaporator outlet. With larger evaporators and longer suction lines, it is important to measure the refrigerant's pressure at the evaporator outlet—not at the compressor service valves—when measuring evaporator superheat, **Figures 21.27** and **21.28**. It is best to measure the refrigerant pressure at the same location that the evaporator outlet temperature is taken when measuring evaporator superheat. This will give the service technician a more accurate evaporator superheat reading and, therefore, better evaporator efficiencies. Schrader taps are often provided at the outlet of larger evaporators for this reason. Line taps also can be used to gain access to evaporator outlet pressure. This method also will protect the compressor from flooding or slugging problems caused by inaccurate evaporator superheat readings.

Figure 21.28 The appearance of a multicircuit evaporator on the inside when it is feeding correctly. It resembles several evaporators piped in parallel.

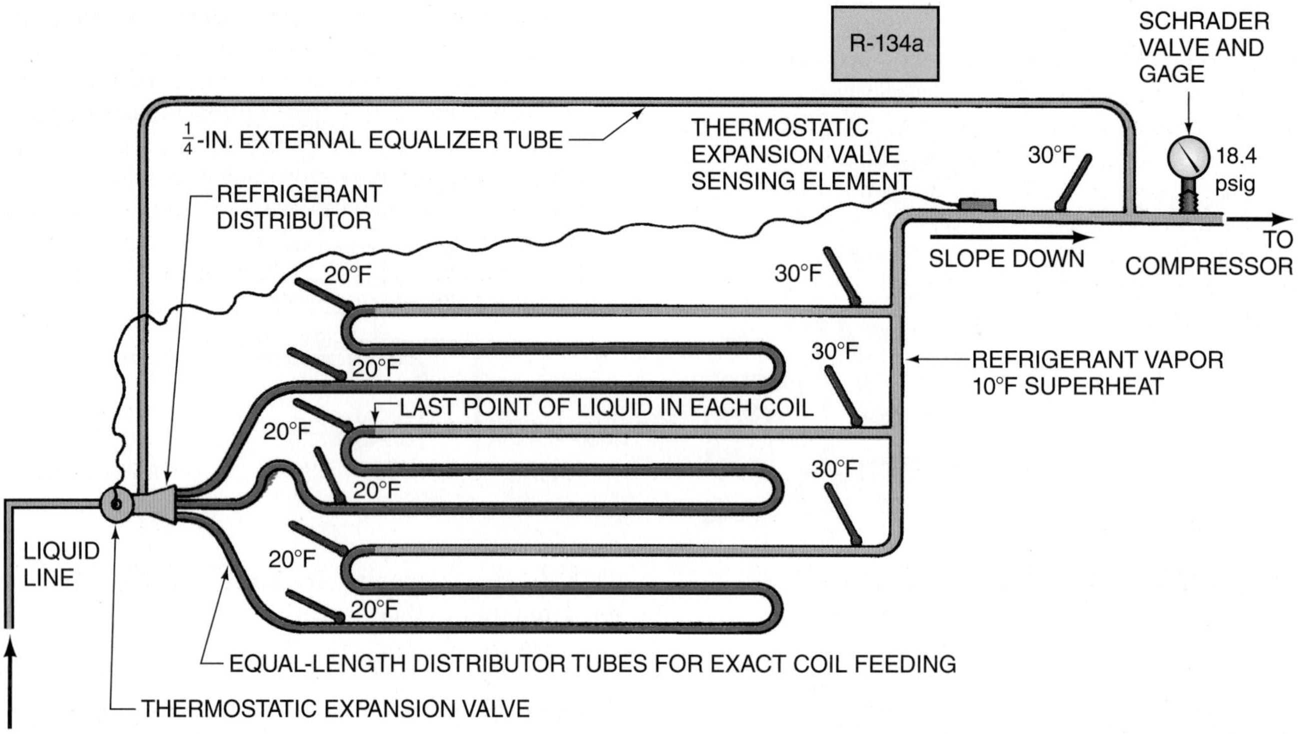

Figure 21.29 The appearance of a multicircuit evaporator on the inside when it is not feeding evenly.

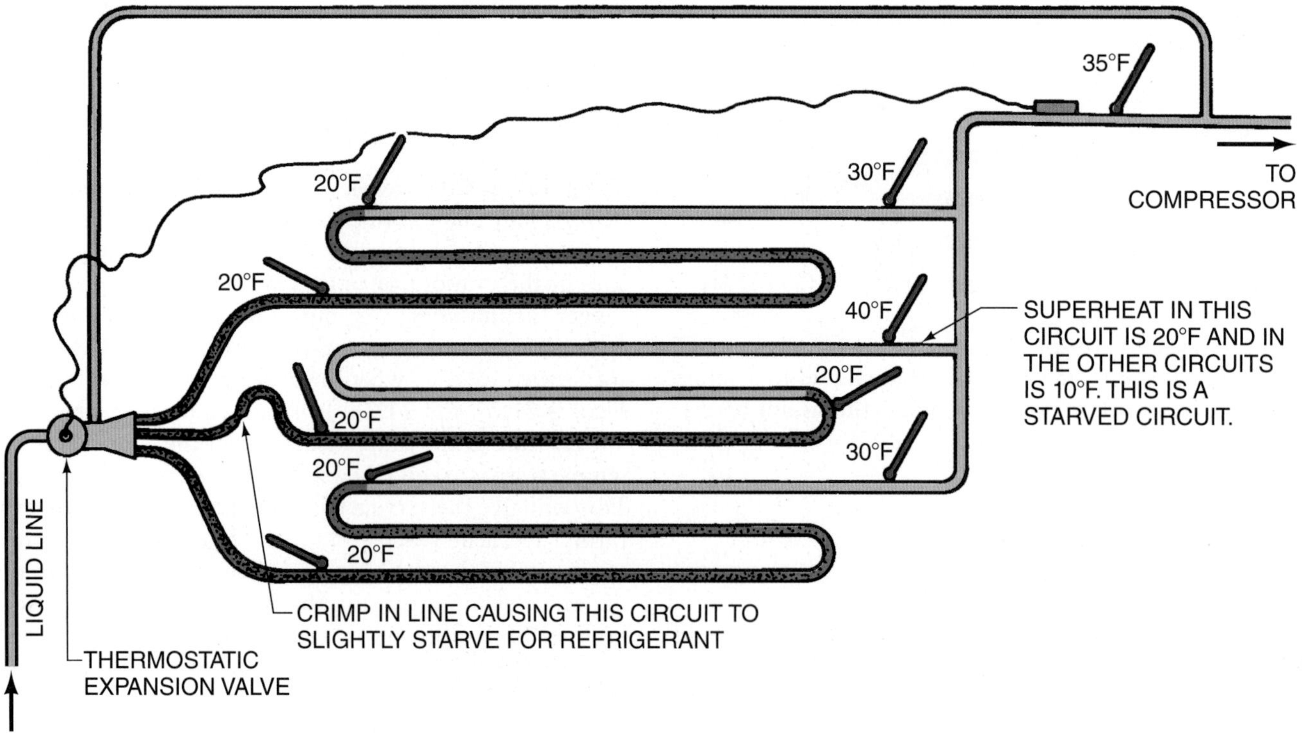

21.16 LIQUID COOLING EVAPORATORS (CHILLERS)

A different type of evaporator is required for liquid cooling. It functions much like the one for cooling air and is normally a dry-type expansion evaporator in smaller systems, **Figure 21.30(A)**. Evaporators for larger-tonnage chillers are usually the flooded type. They have saturated liquid/vapor refrigerant in the shell and the water to be chilled flows in the tube bundles, **Figure 21.30(B)**. They use a low side float to meter the refrigerant into the shell of the evaporator to maintain the proper refrigerant level.

Liquid cooling evaporators have more than one refrigerant circuit to prevent pressure drop. These evaporators sometimes have to be checked to see whether they are absorbing heat as they should. Using refrigeration gauges and some accurate method for checking the temperature of the suction line are very important. These evaporators have a normal superheat range similar to air-type evaporators (8°F to 12°F). When the superheat is within this range and all circuits in a multicircuit evaporator are performing alike, the evaporator is doing its job on the refrigerant side. However, this does not mean that it will cool properly. The liquid side of the evaporator must be clean so that the liquid will come in proper contact with the evaporator.

Figure 21.30(A) A direct-expansion evaporator used for cooling liquids.

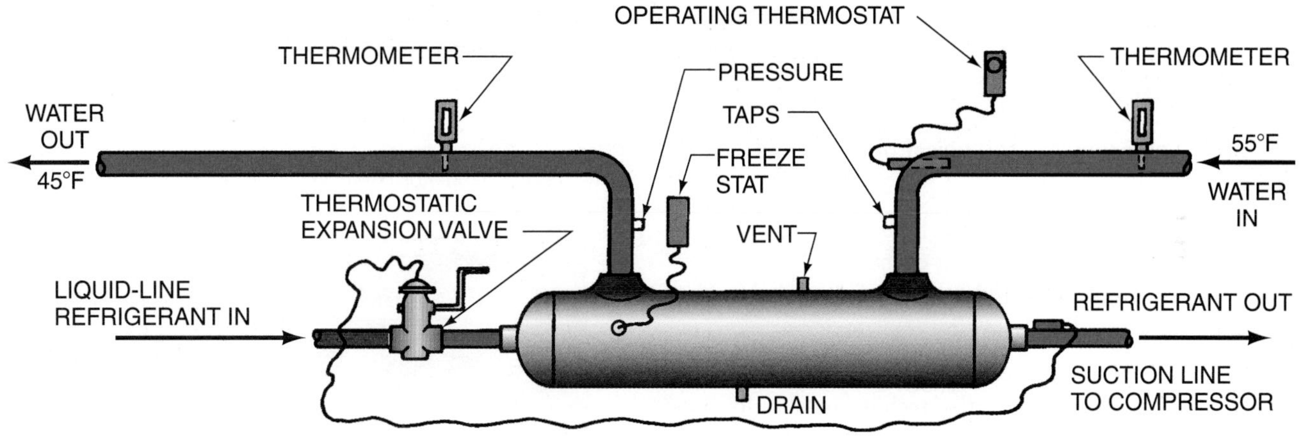

Figure 21.30(B) A 100-ton (R-134a) flooded-type evaporator used to chill water for cooling a large building. Courtesy Ferris State University. Photo by John Tomczyk

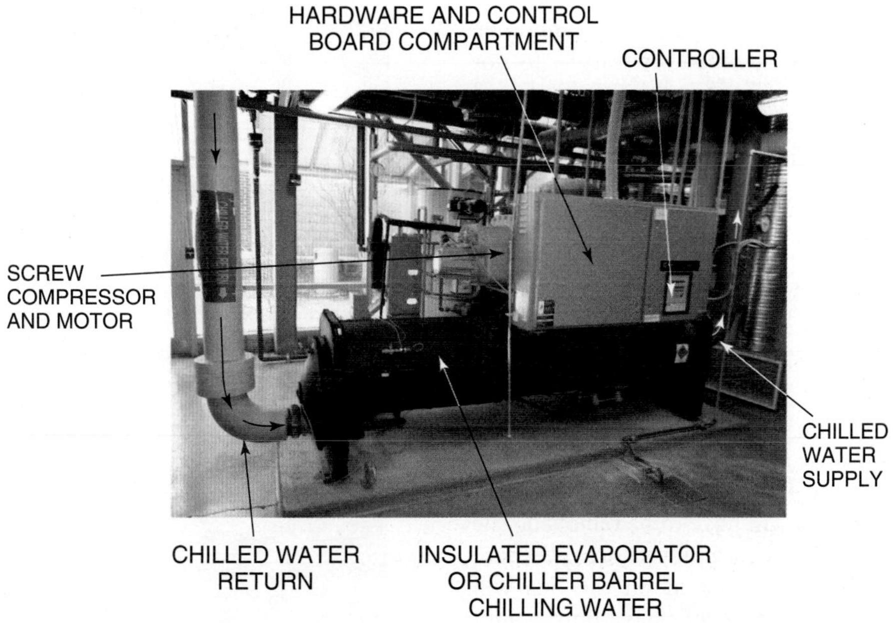

Figure 21.31 A hot pulldown on a liquid evaporator giving up its heat to refrigerant. This evaporator normally has 55°F water in and 45°F water out. The hot pulldown with 75°F water instead of 55°F water boils the refrigerant at a faster rate. The expansion valve may not be able to feed the evaporator quickly enough to maintain 10°F superheat. No conclusions should be made until the system approaches design conditions.

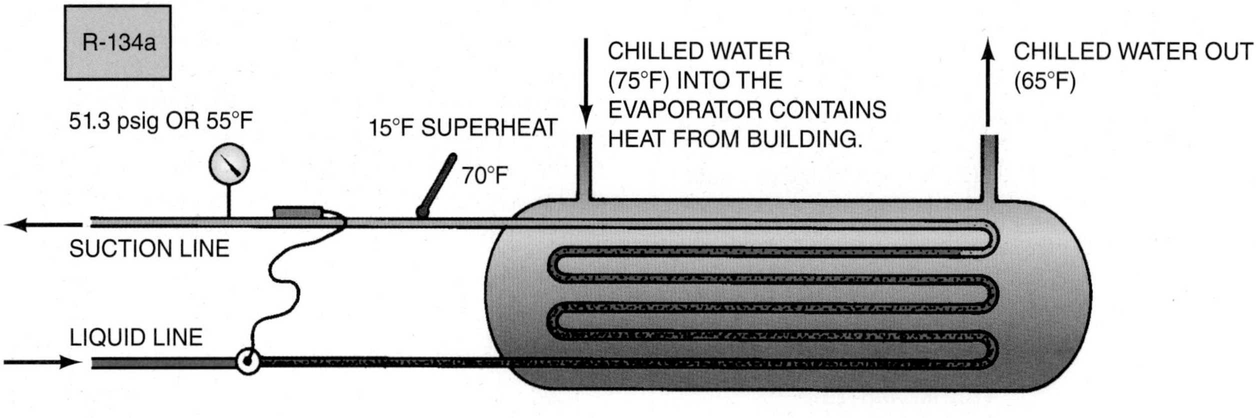

The following problems are typical on the liquid side of the evaporator:

1. Mineral deposits built up on the liquid side, acting as an insulator and causing a poor heat exchange
2. Poor circulation of the liquid to be cooled where a circulating pump is involved

When the superheat is correct and the coil is feeding correctly in a multicircuit system, the technician should consider the temperature of the liquid. The superheat may not be within the prescribed limits if the liquid to be cooled is not close to the design temperature. On a hot pulldown of a liquid product, the heat exchange can be such that the coil appears to be starved for refrigerant because it is so loaded up that it is boiling the refrigerant faster than normal. The technician must be patient because a pulldown cannot be rushed, **Figure 21.31**. Air-to-refrigerant evaporators do not have quite the pronounced difference in pulldown that liquid heat exchange evaporators do because of the excellent heat exchange properties of the liquid to the refrigerant.

21.17 EVAPORATORS FOR LOW-TEMPERATURE APPLICATIONS

Low-temperature evaporators used for cooling space or product to below freezing are designed differently because they require the coil to operate below freezing.

In an airflow application, the condensate that accumulates on the coil will freeze and will have to be removed. The design of the fin spacing must be carefully chosen, because a very small amount of ice accumulated on the fins will restrict the airflow. Low-temperature coils have fin spacings that are wider than medium-temperature coils, **Figure 21.32**.

Figure 21.32 Fin spacing. (A) Low-temperature evaporator. (B) Medium-temperature evaporator. Photos by Jason Obrzut

(A)

(B)

Other than the airflow blockage due to ice buildup, these low-temperature evaporators perform much the same as medium-temperature evaporators. They are normally dry-type evaporators and have one or more fans to circulate the air across the coil. The defrosting of the coil has to be done by raising the coil temperature above freezing to melt the ice. Then the condensate water has to be drained off and kept from freezing. Defrost is sometimes accomplished with heat from outside the system. Electric heat can be added to the evaporator to melt the ice, but this heat adds to the load of the system and needs to be pumped out after defrost.

21.18 DEFROST METHODS

Defrost of low-temperature evaporators can be accomplished with heat from inside the system using the hot gas from the discharge line of the compressor or by using electric resistance heaters. Hot gas defrost is accomplished by routing a hot gas line from the compressor discharge line to the outlet of the expansion valve and installing a solenoid valve to control the flow. When defrost is needed, hot gas is released inside the evaporator, which will quickly melt any ice, **Figure 21.33**.

When the hot gas enters the evaporator, it is likely that liquid refrigerant will be pushed out of the evaporator and into the suction line toward the compressor. In fact, when hot gas enters an evaporator and starts to cool as it melts ice or frost, it will soon lose all of its superheat and turn to liquid or condense. This liquid will go to the accumulator and fall to its bottom. Dense saturated vapors will be drawn into the compressor's suction stroke. The compressor will see an increased load from these dense vapors and may draw a higher amp than during the normal running cycle. If the system does not have an accumulator, this condensed liquid may flood the compressor's crankcase and foaming of the oil in the crankcase may occur, **Figure 21.34**. This can lower the oil level in the crankcase and cause scoring of bearing surfaces in the compressor. This condition is often referred to as **bearing washout**.

Flooding the compressor's crankcase with liquid refrigerant can also cause the foaming refrigerant and oil mixture to pressurize the crankcase, which causes the mixture of liquid and vapor refrigerant and oil foam to be forced through any crevice available, including the compressor's piston rings. The mixture is often pumped into the high side of the system by the compressor. The compressor's discharge temperature will

Figure 21.34 (A) A compressor's clear oil sight glass. (B) A sight glass from a flooding crankcase foams with liquid refrigerant. Photos, courtesy of Tecumseh Products Company.

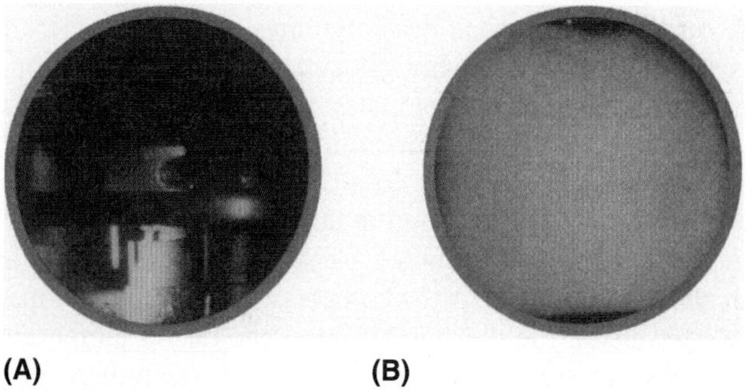

Figure 21.33 Using hot gas to defrost an evaporator.

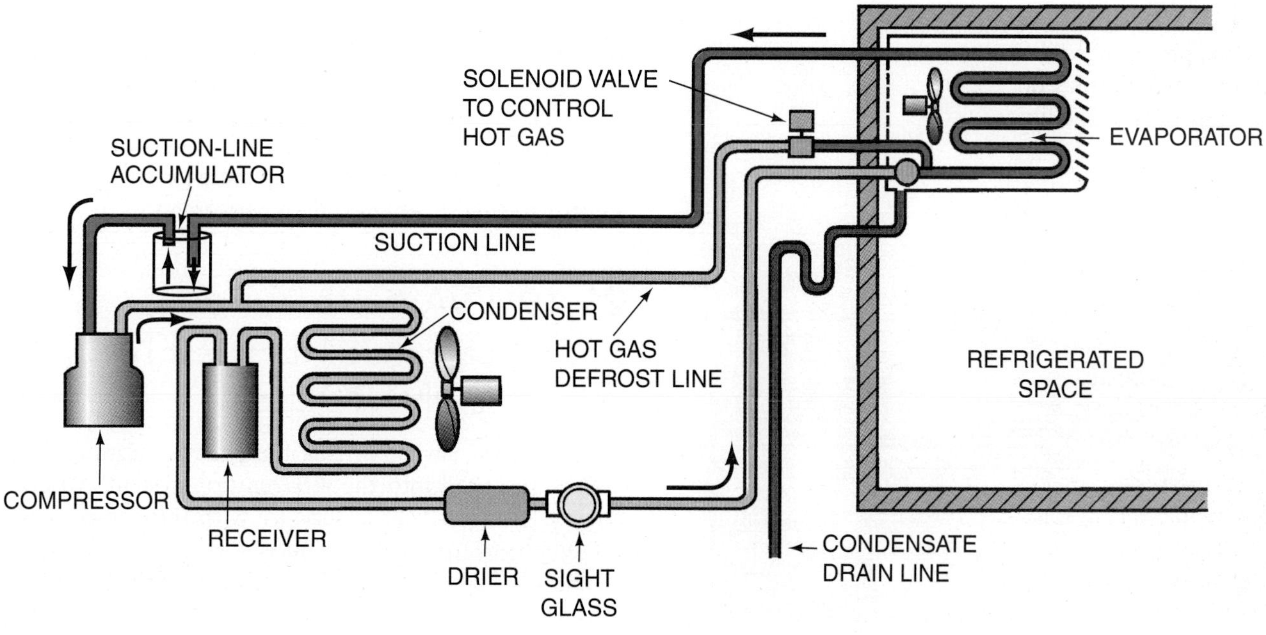

decrease from the wet compression of the rich mixture of refrigerant and oil foam. As soon as this mixture is compressed, it will vaporize and absorb heat away from the cylinder walls. This is what causes a lower-than-normal discharge temperature on the compressor's discharge line. Some manufacturers of systems incorporating hot gas defrost place a thermistor on the discharge line of the compressor to sense this cooler-than-normal discharge temperature while in the defrost mode. The thermistor relays a message to a control circuit to deenergize the hot gas solenoid; closing the hot gas solenoid temporarily prevents any more hot gas from entering the evaporator and thus turning to liquid. As the compressor's discharge temperature rises again, the thermistor senses the rise in temperature and relays a message to the control circuit to continue with hot gas defrost by energizing the hot gas solenoid again.

This sequence of events may continue until defrost is complete. The sequence simply protects the compressor from bearing washout and wet compression while in defrost. It may also protect the compressor during the refrigeration cycle by shutting down the compressor if the discharge temperature gets too low, indicating wet compression. To prevent this liquid from entering the compressor, often a suction line accumulator will be added to the suction piping, **Figure 21.33**.

The hot gas defrost system is economical because power does not have to be purchased for defrost using external heat, such as electric heaters that will heat the evaporator. The heat is already in the system.

Electric defrost is accomplished using electric heating elements located at the evaporator. The compressor is stopped and the heaters are energized on a call for defrost and allowed to operate until the frost is melted from the coil, **Figure 21.35**. These heaters are often embedded in the actual evaporator fins and cannot be removed if they burn out. Frequently, in the event that the heaters do burn out, hot gas defrost can be added to the system and the electric heat defrost procedures discontinued.

Figure 21.35 A heater used for electric defrost of low-temperature evaporators. Photo by Jason Obrzut

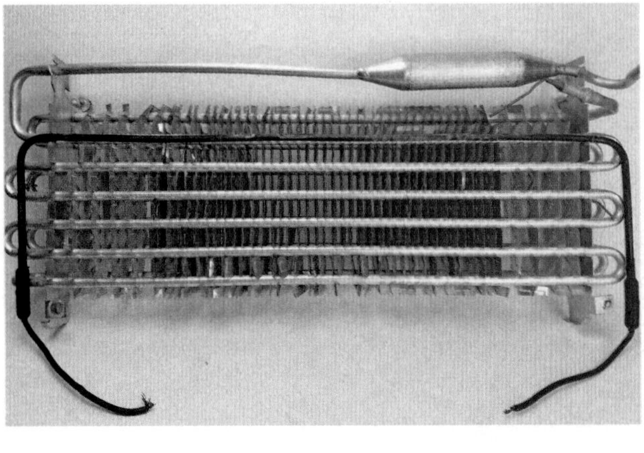

When either system is used for defrost, the evaporator fan is often turned off during defrost; if it is not, two things will happen:

1. The heat from defrost will be transferred directly to the conditioned space.
2. The cold, conditioned air will slow down the defrost process.

However, some manufacturers design their open frozen-food cases so that the evaporator fans are left on during defrost. This allows the supply-air and return-air ducts to be defrosted along with the coil. The warm defrost air is discharged from the supply duct and rises out of the case, having little effect on the product temperature. Fans are always shut off on closed, glass-door cases because of the problem of fogging on glass or mirrored surfaces.

Evaporators in some ice-making processes have similar defrost methods. They must have some method of applying heat to the evaporator to free the ice. Sometimes the heat is electric or hot gas. When the evaporator is being used to make ice, the makeup water for the ice maker is sometimes used for defrost. Evaporator defrosting is covered in more detail in Section 25.7, "Defrost Controls," of Unit 25, "Special Refrigeration System Components."

Medium-temperature refrigeration systems operate with evaporator temperatures below 32°F and box temperatures above 32°F. Because the box temperature is above freezing, these systems are defrosted using random, or off-cycle defrost. These evaporators will develop frost due to the fact that they operate below freezing. Defrost occurs when the thermostat is satisfied, and the refrigeration system cycles off. The evaporator fans run continuously, circulating the ambient air over the evaporator. With the average box temperatures above freezing, the circulating air is warm enough to defrost the evaporator before the refrigeration system cycles back on. Under normal operating conditions, the off-cycle defrost will keep the evaporator from icing up. However, if the evaporator does accumulate large amounts of ice due to conditions such as an evaporator fan failure, dirty coil, or an undercharge condition, it will have to be defrosted manually with a heat gun or warm water. Failure to remove excessive amounts of ice from the evaporator coils on medium-temperature systems will result in a noticeable reduction in system performance.

21.19 EVAPORATOR AND DEFROST EFFICIENCY CONTROLLER

As mentioned earlier in this unit, the evaporator is where heat is absorbed into the refrigeration system. An evaporator is basically a heat exchanger with refrigerant tubes and extended surfaces. The extended surfaces are often referred to as fins. The evaporator's function is to efficiently transfer heat from the surrounding area to the phase-changing

Figure 21.36 Evaporator efficiency controller. Courtesy KE2 Therm Solutions, Inc.

refrigerant flowing through the evaporator. The refrigerant will be changing its phase from a liquid to a vapor as it travels through the evaporator absorbing heat. As the evaporator's heat load changes, an expansion type metering device at the entrance of the evaporator will control the flow of refrigerant through the evaporator. The metering device will also ensure that all of the liquid in the evaporator has fully vaporized prior to exiting the evaporator.

THE EVAPORATOR AND ENERGY SAVINGS

The evaporator is a major component in the refrigeration system, and an efficient evaporator plays a key role in saving energy. Many low-temperature commercial and industrial evaporator assemblies consist of:

- The evaporator coil
- Fans
- An expansion device
- Defrost heaters
- A liquid line solenoid valve

By carefully timing, managing, and controlling each one of these components with an evaporator efficiency controller, **Figure 21.36**, the evaporator is able to provide the maximum output with the minimum amount of energy input.

FROST

Although frost is unavoidable on an evaporator, it is the most common cause of inefficiency in evaporator coils. When the evaporator's coil temperature drops below the dew point temperature, moisture will begin to collect on the evaporator's cool surface. The dew point temperature is the temperature where moisture starts to form. If the temperature of the evaporator coil continues to drop below the freezing point of water, the moisture on the evaporator's coil will begin to solidify, forming a thin layer of ice.

Depending on humidity, evaporator temperature, and air flow, the subsequent layers of frost that accumulate on the evaporator may be more crystalline or snow-like. These layers of frost will hold more entrained air and will be very porous. This type of frost, often referred to as radiant frost or hoar frost, will insulate the evaporator and severely reduce its heat-absorbing ability. **Figure 21.37** illustrates how the performance of an evaporator is enhanced initially, but declines over time as hoar frost accumulates on the coil, insulating it. As the hoar frost accumulates, the evaporator will see less heat from the refrigerated space. This will cause the evaporator or suction pressure to drop and the evaporator coil will become colder. A drop in suction pressure will result in a higher compression ratio. Higher compression ratios cause

Figure 21.37 Evaporator performance as the evaporator coil builds frost. Courtesy KE2 Therm Solutions, Inc.

lower compressor volumetric efficiencies, less dense refrigerant return gases entering the compressor, reduced mass flow rates of refrigerant, and loss of system capacity. A result of this insulating effect is the temperature of the evaporator coil must be reduced to maintain the same desired refrigerated space temperature. Now, the temperature difference (TD) between the evaporator coil and the refrigerated space will be greater.

PUMP-DOWN CYCLE

The control system is based around a refrigerant pump-down cycle. Once the refrigerated space temperature is satisfied, the controller deenergizes (closes) the liquid line solenoid valve and turns the evaporator fans off. The compressor continues to run until any liquid refrigerant in the evaporator and suction line is fully vaporized and pumped to the high side of the refrigeration system. Automatic pump-down systems are covered in detail in Unit 25, "Special Refrigeration System Components." During this refrigerant pump down, the temperature of the evaporator coil will continue to drop because of the low pressure created in the coil from the pump-down process, and from the refrigeration effect of the vaporizing refrigerant remaining in the evaporator coil during pump down. This residual refrigeration effect, or flywheel effect, will drop the coil temperature below the refrigerated space temperature. The evaporator's controller is able to capture this energy and give it back to the refrigeration system by controlling the evaporator fans. The controller will now cycle the evaporator fans back on without starting the compressor. The refrigerated space temperature is now maintained by only the fans blowing air over a very cold evaporator coil. This greatly reduces the amount of energy used to maintain the refrigerated space temperature. Only when this stored "flywheel effect" energy is depleted out of the evaporator coil, will the compressor be cycled back on. By cycling the fans, the evaporator controller extends the time between compressor run times, thus saving energy.

FROST REMOVAL BY SUBLIMATION

During pump down, the evaporator fans are blowing air through a very cold evaporator coil with frost accumulated on its surfaces. The frost will change its phase to vapor through the process of sublimation. Sublimation is a process where a solid changes directly to a vapor, skipping the liquid phase. This sublimation process reduces the frost on the coil naturally. As the frost changes to vapor, it also returns valuable moisture to the refrigerated space. Higher humidity levels maintained in the refrigerated space reduce product shrinkage. Most defrost systems run the melted frost and residual energy down a drain and out of the refrigerated case. Also, many traditional defrost systems use three or four defrosts per day, each lasting perhaps for 45 to 60 min. When melting frost comes in contact with a 300°F defrost heater, it can flash into steam creating a fogging effect throughout the refrigerated case. This fog will re-condense on other cold surfaces in the refrigerated room, causing unwanted ice. **Figures 21.38** and **21.39** show a walk-in refrigerated space before and after an evaporator efficiency controller with an advanced defrost control algorithm was installed.

Figure 21.38 Walk-in refrigerated space before installation of an evaporator efficiency controller. Courtesy KE2 Therm Solutions, Inc.

Figure 21.39 Walk-in refrigerated space after the installation of an evaporator efficiency controller. Courtesy KE2 Therm Solutions, Inc.

EVAPORATOR EFFICIENCY CONTROLLER AND EVAPORATOR TEMPERATURE DIFFERENCE (TD)

The evaporator efficiency controller uses a proprietary defrost control algorithm that eliminates the dependency on "time" for defrost initiation. Instead, the controller monitors evaporator coil efficiency to initiate defrost and also minimizes the effects of defrost on space temperature. Maintaining a lower TD between the evaporator coil and the refrigerated case throughout the run time will reduce the rate of frost formation on the coil. This will lead to extended time between defrosts and save energy. Below are some important functions of the evaporator efficiency controller:

- Monitors evaporator efficiency and performance
- Creates an initial evaporator profile from a series of measurements
- Identifies the temperature relationship between the evaporator coil and refrigerated space temperature through two sensor locations, **Figure 21.40**
- Initiates a defrost once the evaporator coil efficiency is less than 90%
- Monitors coil temperature once in defrost. Once a predetermined temperature is reached, the heaters are turned off allowing the heat in the heater elements to be transferred to the evaporator coil. Once the heat is dissipated, the heaters are again energized to continue defrost. This saves energy, quickens defrost time, and eliminates fogging.
- Contain Ethernet network devices supporting industry standard protocols

Figure 21.40 Sensor locations to identify the temperature relationship between the evaporator coil and the refrigerated space. Courtesy KE2 Therm Solutions, Inc.

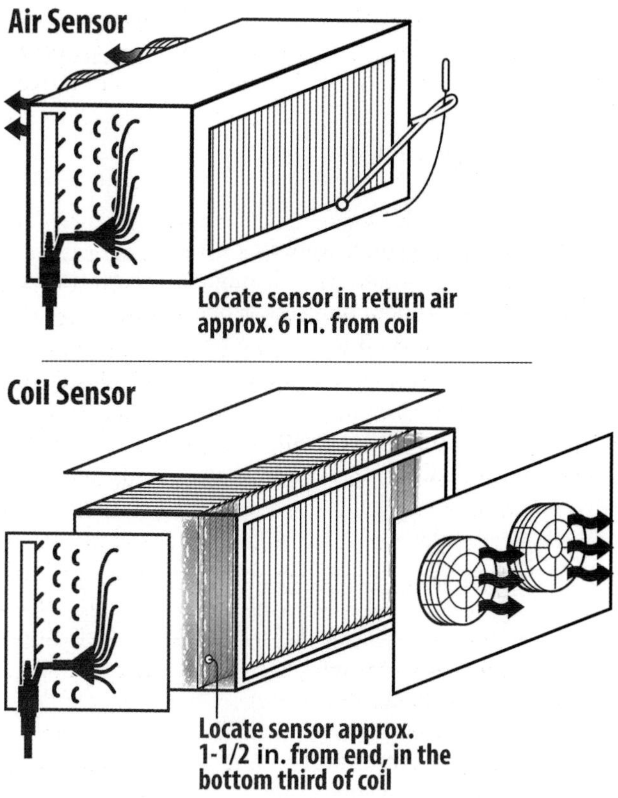

In summary, when checking an evaporator remember that its job is to absorb heat into the refrigeration system.

REVIEW QUESTIONS

1. What is the function of the evaporator in the refrigeration system?
   A. To add heat to the space or product being cooled.
   B. To add humidity to the space being refrigerated.
   C. It condenses superheated refrigerant vapor to a liquid.
   D. It absorbs heat into the refrigeration system.
2. Refrigerant in the evaporator ___________.
   A. changes from vapor to liquid
   B. changes from liquid to vapor
   C. stays in the vapor state
   D. stays in the liquid state
3. The sensible heat that is added to a saturated vapor after all of the liquid has boiled away is referred to as ________.
   A. superheat
   B. latent heat
   C. subcooling
   D. specific heat
4. What determines the pressure on the low-pressure side of the system?
   A. The condensing temperature of the refrigerant in the condenser.
   B. The length of the liquid line feeding the TXV.
   C. The boiling temperature of the refrigerant in the evaporator.
   D. The relative humidity percentage in the conditioned space.
5. A refrigeration system's evaporator typically runs at about ________ of superheat.
   A. 15°F to 20°F
   B. 12°F to 15°F
   C. 8°F to 12°F
   D. 3°F to 8°F

6. What does a high evaporator superheat indicate?
   A. A flooded evaporator
   B. Proper system operation
   C. Low evaporator heat load
   D. A starved evaporator
7. A low evaporator superheat may indicate a(n) __________.
   A. undercharge
   B. system restriction
   C. overcharge
   D. high heat load
8. Why is a multicircuit evaporator used?
   A. To reduce pressure drop through larger evaporators.
   B. To ensure proper compressor operation.
   C. To prevent ice buildup on the fins.
   D. They are easily defrosted in low-temperature applications.
9. Flooded evaporators typically use a(n) _______ type of expansion device.
   A. automatic
   B. float
   C. TXV
   D. fixed-bore
10. An evaporator that does not operate flooded is thought of as what type of evaporator?
    A. A dry-type evaporator
    B. A starved evaporator
    C. A float-type evaporator
    D. A low-pressure evaporator
11. When an evaporator experiences a heat-load increase, the suction pressure __________.
    A. remains constant
    B. decreases
    C. varies up and down
    D. increases
12. What is commonly used to defrost the ice from a low-temperature evaporator?
    A. Random or off-cycle defrost.
    B. Hot gas or electric resistance heat defrost.
    C. Off-cycle or manual defrost.
    D. Low-temperature evaporators do not require a defrost.
13. A medium-temperature refrigeration box typically operates within what temperature range?
    A. 28°F to 40°F
    B. 40°F to 60°F
    C. 0°F to 220°F
    D. 0°F to 50°F
14. The best heat exchange occurs between __________.
    A. a vapor and a liquid
    B. two liquids
    C. two solids
    D. a liquid and a solid
15. The greater the difference in temperature between the evaporator and the air being cooled, ____________.
    A. the less moisture will be removed from the air
    B. the less moisture will be added to the air
    C. the more moisture will be added to the air
    D. the more moisture will be removed from the air
16. A finned-tube evaporator equipped with fans is called a(n) _______ evaporator.
    A. bare-tube
    B. stamped
    C. induced- or forced-draft
    D. liquid-cooling
17. Which of the following is not a function of an evaporator efficiency control?
    A. Monitors evaporator efficiency and performance.
    B. Initiates a defrost when coil efficiency drops below 90%.
    C. Contains an Ethernet network device that supports industry standard protocols.
    D. Regulates condensate removal by activating a condensate pump.
18. A medium-temperature evaporator uses a(n)________ defrost method.
    A. hot gas
    B. random or off-cycle
    C. electric resistance heat
    D. liquid-injection
19. The difference in the amount of heat (in Btu/lb) that the refrigerant contains at the point where it enters the evaporator and the point at which it leaves the evaporator is known as ________.
    A. mass flow rate
    B. total heat of rejection
    C. the net refrigeration effect
    D. the heat of compression
20. Low-temperature refrigeration systems typically operate at what temperature range?
    A. −20°F to 0°F
    B. −50°F to −20°F
    C. 0°F to 35°F
    D. 28°F to 40°F
21. List seven advantages that an aluminum parallel-flow, plate-and-fin evaporator has over a standard round, copper-tube plate-and-fin evaporator.

UNIT 22

CONDENSERS

OBJECTIVES

After studying this unit, you should be able to

- explain the function of the condenser in a refrigeration system.
- describe the differences between the operating characteristics of water-cooled and air-cooled condensers.
- describe the heat exchange that takes place in a condenser.
- explain the difference between a tube-within-a-tube coil-type condenser and a tube-within-a-tube serviceable condenser.
- describe the difference between a shell-and-coil condenser and a shell-and-tube condenser.
- describe a wastewater system.
- describe a recirculated water system.
- describe the functions of a cooling tower.
- explain the relationship between the condensing refrigerant and the condensing medium for water-cooled systems.
- compare an air-cooled, high-efficiency condenser with a standard efficiency air-cooled condenser.
- describe the operation of head pressure control valves, fan cycling controls, VFD condenser fan motors, dampers, condenser flooding, and condenser splitting.
- explain the operation of a system that uses a floating head pressure.
- describe a heat reclaim system and its uses.

SAFETY CHECKLIST

- ✔ Wear goggles and gloves when attaching or removing gauges to transfer refrigerant or to check pressures.
- ✔ Wear warm clothing when working in a walk-in cooler or freezer. A technician does not think properly when chilled.
- ✔ Keep hands well away from moving fans. Do not try to stop a fan blade as it slows after the power has been turned off.
- ✔ Do not touch the hot gas line.

22.1 THE CONDENSER

The condenser is a heat exchange device similar to the evaporator; it rejects from the system the heat absorbed by the evaporator. In the first passes of the condenser the heat is rejected from a hot, superheated vapor. The middle of the condenser rejects latent heat from saturated vapor and liquid, which is in the process of changing phase to a 100% saturated liquid. The last passes of the condenser reject heat from subcooled liquid. This further subcools the liquid to below its condensing temperature. In fact, the condenser performs three main functions on the refrigerant flowing through it. They are in order of occurrence:

1. Desuperheating the vapors
2. Condensing vapors to liquid
3. Subcooling the liquid

The greatest amount of heat is absorbed in the system when the refrigerant is at the point of changing state (liquid to a vapor). The same thing happens, in reverse, in the condenser. The point where the change of state (vapor to a liquid) occurs is where the greatest amount of heat is rejected.

The condenser operates at higher pressures and temperatures than the evaporator and is often located outside. The same principles apply to heat exchange in the condenser as in the evaporator. The materials a condenser is made of and the medium used to transfer heat make a difference in the efficiency of the heat exchanger.

In addition to the heat absorbed in the evaporator, the condenser must also reject heat picked up in the suction line, the compressor, and the heat of compression. For this reason, condensers are typically sized at 15,000 Btu/ton instead of the typical 12,000 Btu/ton that is used to size other systems components.

22.2 WATER-COOLED CONDENSERS

The first commercial refrigeration condensers were water-cooled. Compared with modern water-cooled devices, these condensers were crude, **Figure 22.1.** Water-cooled condensers can operate at much lower condensing temperatures. Because water has a higher specific heat and density than air, water-cooled condensers are more efficient than air-cooled condensers. Water-cooled equipment comes in several styles. Three of the most common are the tube within a tube (double tube), shell and coil, and shell and tube.

Figure 22.1 An early water-cooled condensing unit. Source: Tecumseh Products Company

Figure 22.2 Two types of tube-within-a-tube condensers. (A) A tube-within-a-tube coiled condenser. Photo by Jason Obrzut. (B) A flanged condenser. The flanged condenser can be cleaned by removing the flanges. Removal of the flanges opens only the water circuit, not the refrigerant circuit. Photo by Bill Johnson

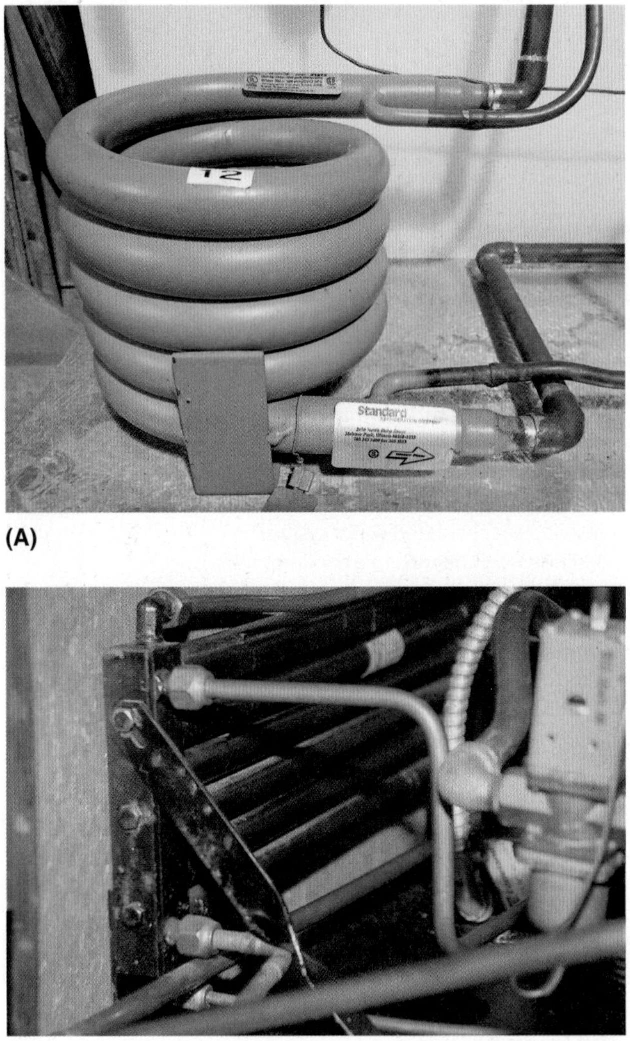

(A)

(B)

22.3 TUBE-WITHIN-A-TUBE CONDENSERS

The tube-within-a-tube condenser comes in two styles: the coil type and the cleanable type with flanged ends, **Figure 22.2.** The HVACR industry refers to these condensers as tube within a tube, double tube, and co-axial.

The tube within a tube that is fabricated into a coil is manufactured by slipping one pipe inside another and sealing the ends in such a manner that the outer tube becomes one container and the inner tube becomes another container, **Figure 22.3.** The two pipes are then formed into a coil to save space. The heat exchange occurs between the fluid inside the outer pipe and the fluid inside the inner pipe, **Figure 22.4.** *Some manufacturers will put extended surfaces or fins around the inner water tube to enhance heat transfer between the water and refrigerant. The fins add extra surface area to the inner tube, which enhances heat transfer. The fins also cause the refrigerant in the outer tube to have a more turbulent flow, which also enhances heat transfer,* **Figure 22.3(C).** Because water flows in the inner tube and hot refrigerant flows in the outer tube, the hot refrigerant also can reject some of its heat to the surrounding air. However, most of the heat goes from the refrigerant to the water because of the higher density, flow rate, and specific heat of the water.

The refrigerant in the outer tube and the water in the inner tube flow in opposite directions. This is referred to as a counter-flow arrangement. Counter-flowing the refrigerant with the water keeps a more evenly distributed, or consistent, temperature difference between the two fluids throughout the length of the heat exchange, as compared to fluids flowing in the same direction. At the beginning of the condenser heat exchanger, the superheated compressor discharge gas, which is the warmest refrigerant, experiences the warmest water. At the end of the condenser heat exchange, the coolest water experiences the coolest refrigerant, or subcooled, liquid. Because of their small internal volume, tube-in-tube condensers require a separate receiver.

22.4 MINERAL DEPOSITS

Even the cleanest water will leave mineral deposits and scale in a tube due to the heat in the vicinity of the discharge gas, which causes any minerals in the water to deposit onto the tube surface. This is a slow process, but it will happen in time to any water-cooled condenser. The mineral deposits act as an insulator between the tube and the water and must be kept to a minimum, **Figure 22.3(D).**

Figure 22.3 (A) A tube-within-a-tube condenser constructed by sliding one tube through another tube. The tubes are sealed in such a manner that the inside tube is separated from the outside tube. Source: Noranda Metal Industries, Inc. (B) A cutaway view of a tube-within-a tube condenser. Courtesy Ferris State University. Photo by John Tomczyk. (C) A cutaway view of a tube-within-a tube condenser showing extended surfaces or fins around the inner water tube. Courtesy Ferris State University. Photo by John Tomczyk. (D) Scale deposits on the inner water tube of a tube-within-a tube condenser. Courtesy Ferris State University. Photo by John Tomcyzk

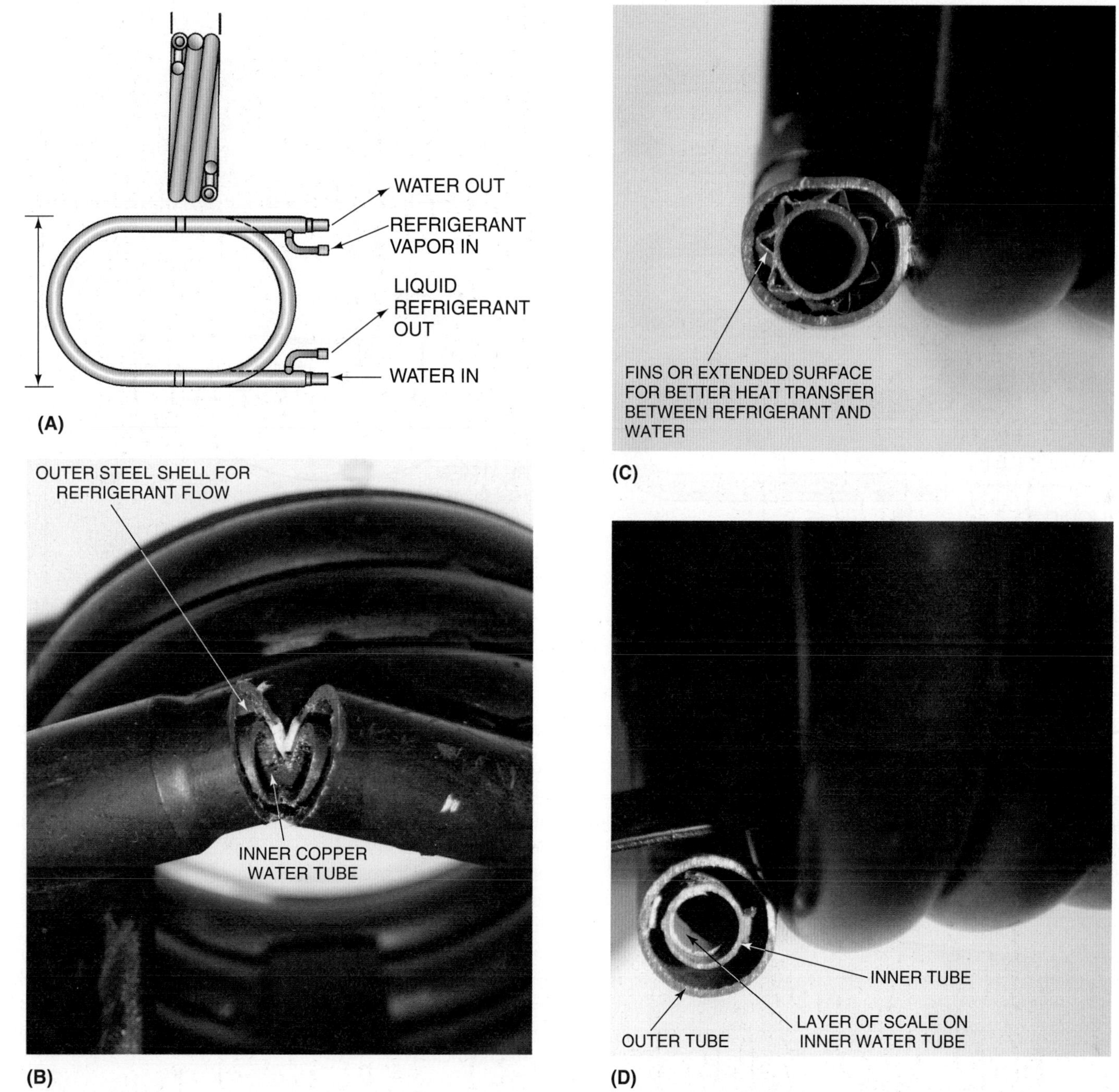

The water can be treated to help prevent this buildup of mineral scale. The treatment is normally added at the tower or injected into the water by chemical feed pumps. **Figure 22.5** shows treatment being added to a tower. **Figure 22.6** is an example of treatment being pumped into the water piping.

In some mild cases of scale buildup, putting more water into circulation will improve the heat exchange. Variable water-flow controls, introduced later in this unit, step up the water flow through the condenser on an increase in head pressure caused by a poor heat exchange due to mineral deposits. High head pressure in a dirty water-cooled

Figure 22.4 Fluid flow through the condenser. The refrigerant is flowing in one direction, water in the other. This is referred to as a counter-flow arrangement.

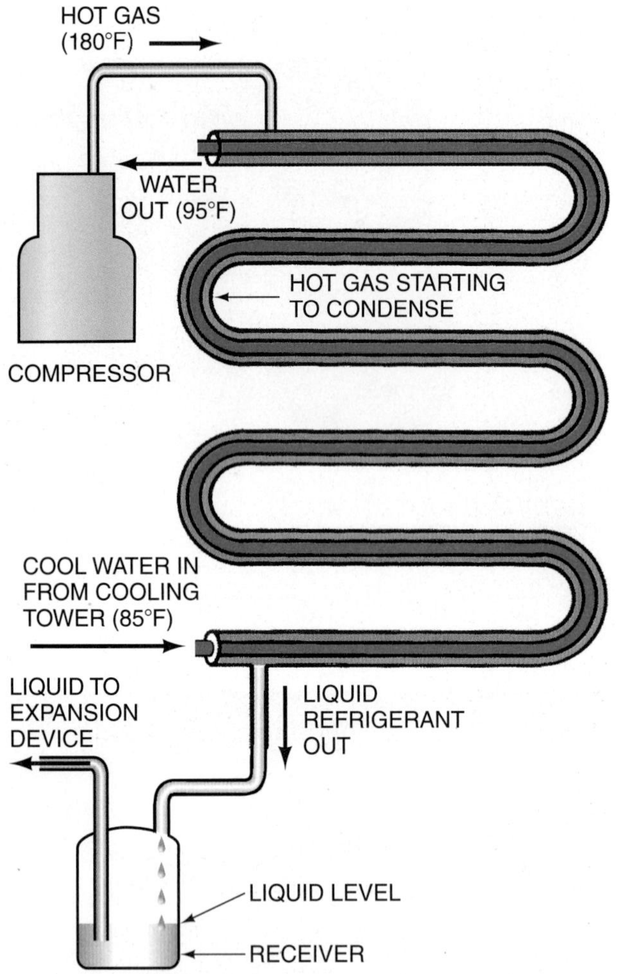

Figure 22.5 A method of adding water treatment to a cooling tower. Treatment is being metered at a rate that will last about a month so the operator will not have to be in the tower constantly. The tower water bleeds continuously to the drain to avoid an overconcentration of minerals in the water. Source: Calgon Corporation

Figure 22.6 A system for automatically feeding treatment chemicals to the water. It includes a monitor that determines when chemicals need to be added. This system is normally used on larger installations because of its cost-effectiveness.

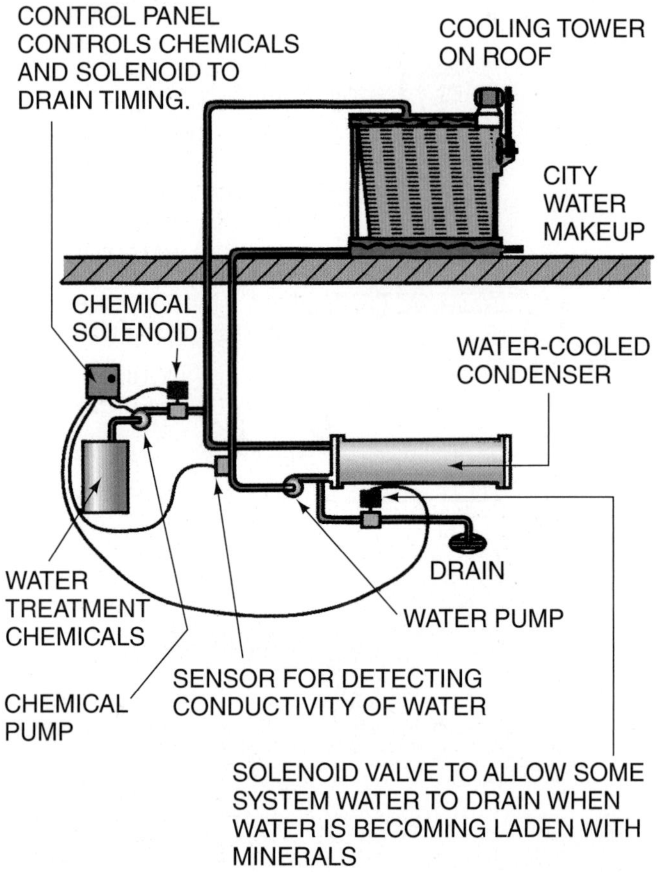

condenser requires more energy to run. If the water is wasted, instead of being cooled and used again, the water bill would go up before the operator notices there is a condenser problem.

The tube-within-a-tube condenser that is made into a coil cannot be cleaned mechanically with brushes. These condensers must be cleaned with chemicals that are designed not to harm the metal, which is normally copper or steel and sometimes stainless steel or copper and nickel. Getting professional help from a company that specializes in water treatment is recommended when a condenser must be cleaned with chemicals.

22.5 CLEANABLE TUBE-WITHIN-A-TUBE CONDENSERS

The tube-within-a-tube condenser that is fabricated with flanges on the end can be mechanically cleaned. The flanges can be removed, and the tubes can be examined and brushed with an approved brush, **Figure 22.7**. Consult the manufacturer for the correct brush; fiber or nylon is usually preferable.

Figure 22.7 (A) This condenser is flanged for service. When the flanges are removed, the refrigerant circuit is not disturbed. Photo by Bill Johnson (B) A coiled tube-within-a-tube condenser is cleaned using a circulator pump and chemicals. Courtesy Ferris State University. Photo by John Tomczyk. (C) Parallel bypass circuit with manual shutoff valves for condenser cleaning. Courtesy Ferris State University. Photo by John Tomczyk

(A)

(B)

(C)

Figure 22.8 (A) Brushes being pushed through the water side of the condenser. Use only approved brushes. Source: Tools Corporation. (B) Cutaway view of cleaning brush.

(A)

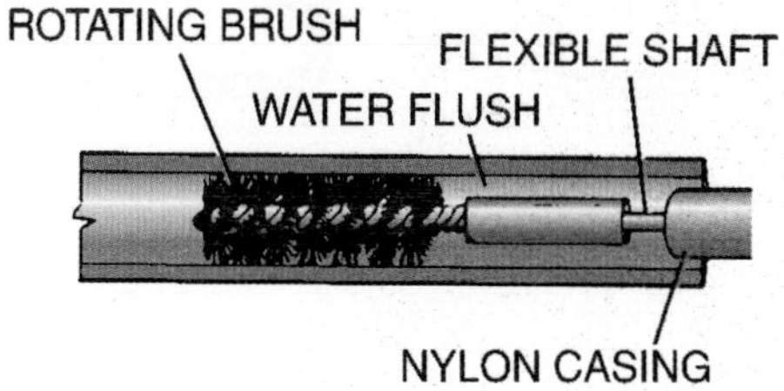

(B)

The flanges and gaskets on this type of condenser are in the water circuit; the refrigerant flows around the water tubes in a circuit that is not opened when the water tubes are cleaned. **Figure 22.8** shows how straight water tubes are cleaned. Cleanable water-cooled condensers with straight tubes are more expensive, but they are more easily serviced. **Figures 22.7(B)** and **(C)** show how a coiled tube-within-a-tube condenser is cleaned using a circulator pump and chemicals to descale the condenser's water tube circuit. A circulator pump circulates a chemical bath from a holding tank through the water tubes of the condenser and then back to the holding tank. The condenser's water circuit has a special bypass circuit that parallels the normal flow of water and contains manual shutoff valves for use in cleaning. Always consult with the manufacturer of the condenser or water treatment company to make sure the circulated chemicals are safe to use with the materials used to manufacture the water-cooled condenser.

22.6 SHELL-AND-COIL CONDENSERS

The shell-and-coil condenser is similar to the tube-within-a-tube coil condenser. A coil of tubing (usually copper) is packed into a shell (usually steel) and then closed and welded. Normally, the refrigerant gas is discharged into the top of the steel shell and condenses as it comes in contact with the water coil. Subcooled liquid refrigerant is then picked up from the bottom with a standpipe or bottom drain. Water is circulated in the copper coil located within the shell. The shell of the condenser serves as a receiver storage tank for the extra refrigerant in the system, so the refrigeration system does not require a separate receiver. This condenser is not mechanically cleanable because the coil is not straight, **Figure 22.9**; it must be cleaned chemically.

Figure 22.9 The shell-and-coil condenser. The hot refrigerant gas is piped into the shell, and the water is contained inside the tubes.

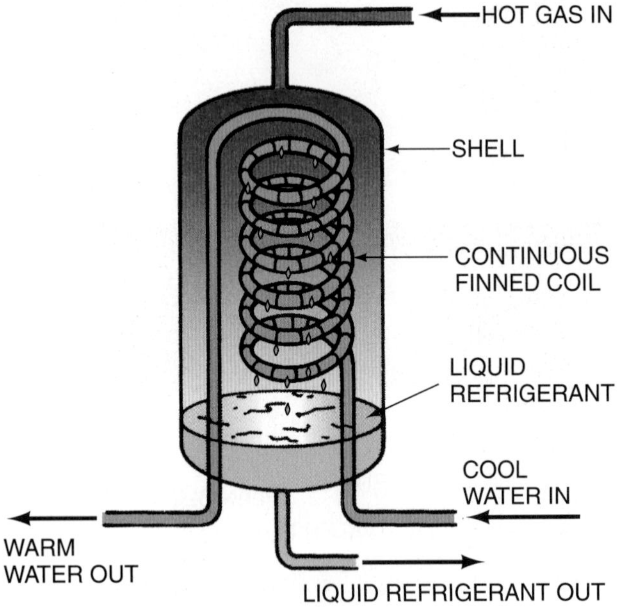

22.7 SHELL-AND-TUBE CONDENSERS

Shell-and-tube condensers are the most expensive condenser and are normally used in larger installations, but they can be cleaned mechanically with brushes. Straight copper tubes are fastened into an end sheet within the steel shell. The number of water passes will vary with the end plate design. The shell acts as a receiver storage tank for extra refrigerant. The refrigerant is discharged into the shell, and water is circulated through the tubes. The ends of the shell are like end caps (known as water boxes) and have water circulating in them, **Figure 22.10**. The end caps can be removed so the tubes can be inspected and brushed out if needed.

The water-cooled condenser is used to remove heat from the refrigerant; after removal, it remains in the water, which gets warmer. Two things can be done at this point: (1) waste the water down a drain or (2) pump the water to a remote place, remove the heat, and reuse the water.

22.8 WASTEWATER SYSTEMS

Wastewater systems are just what the name implies. The water is used once and then disposed of—wasted down the drain, **Figure 22.11**. This is acceptable if the water is free

Figure 22.10 (A) Water can be circulated back and forth through the condenser by using the end caps to give the water the proper direction. This is a four-pass shell-and-tube condenser. The number of water passes will vary with the end plate design. (B) Disassembled shell-and-tube condenser showing the water tubes. Courtesy Ferris State University. Photo by John Tomczyk

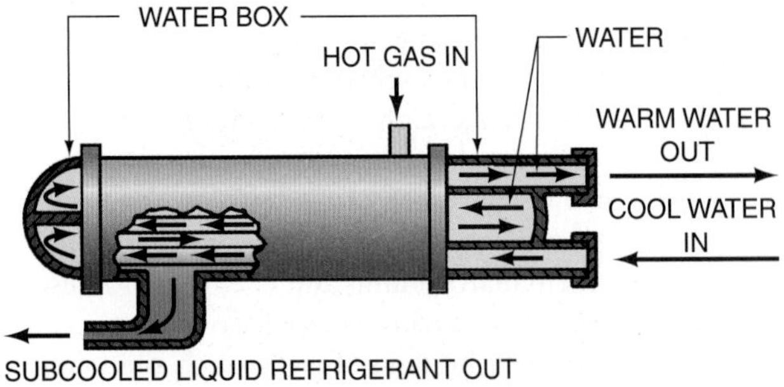

(A)

(B)

Figure 22.10 (*Continued*) (C) Large-tonnage, two-pass shell-and-tube condenser showing cooling tower water in and out. Courtesy Ferris State University. Photo by John Tomczyk. (D) Shell-and-tube condenser showing end plate, hot gas entering the top, and king valve with subcooled liquid refrigerant exiting the bottom. Courtesy Ferris State University. Photo by John Tomczyk. (E) Large shell-and-tube condenser showing the entire steel shell and the liquid line on the bottom. Courtesy Ferris State University. Photo by John Tomczyk. (F) Small-tonnage shell-and-tube condenser with end cap removed. Courtesy Ferris State University. Photo by John Tomczyk. (G) Small-tonnage shell-and-tube condenser with water tubes being brushed cleaned with a rotating nylon brush and a water/acid flush. Courtesy Ferris State University. Photo by John Tomczyk. (H) Inside view of an end cap of a two-pass shell-and-tube condenser. Courtesy Ferris State University. Photo by John Tomczyk

(C)

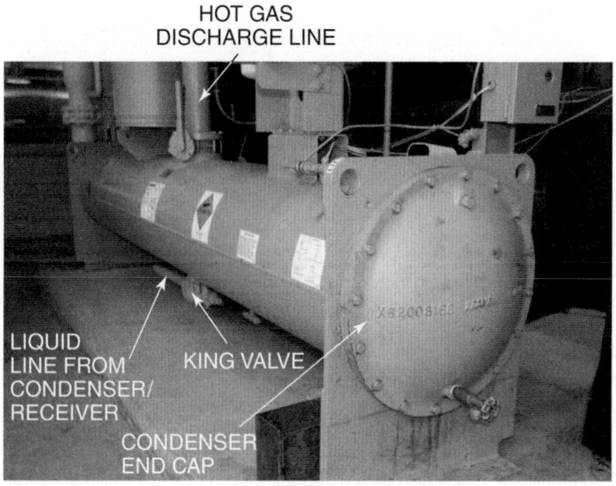

(D)

(E)

(F)

(G)

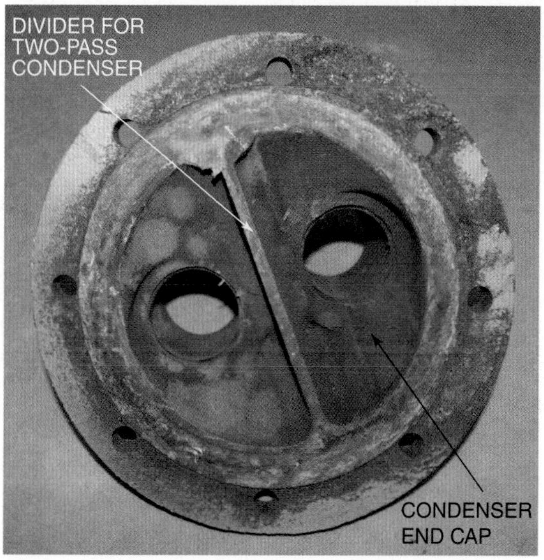

(H)

Figure 22.11 Wastewater systems are used when water is plentiful and at a low cost, such as from a well or lake.

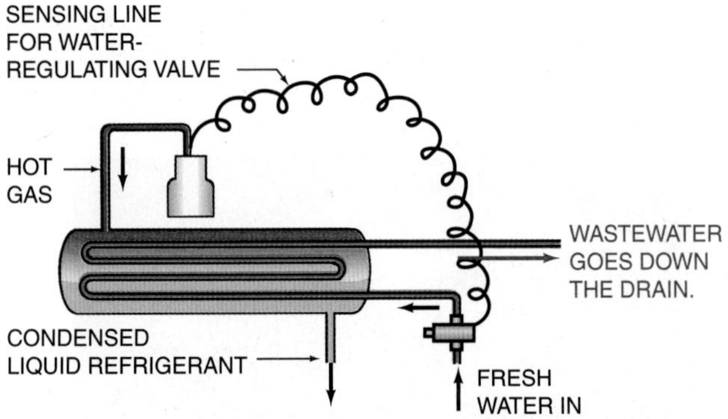

or if only a small amount is used. Where large amounts of water are involved, it is probably more economical to save the water, cool it in an outside water tower, and reuse it.

The water supplied to systems that use the water only once and waste it has a broad temperature range. For instance, the water in summer may be 75°F out of the city mains and as low as 40°F in the winter, **Figure 22.12**. Water piped through a building with long pipe runs may be warm in the beginning from standing in the pipes even though the water temperature in the main may be quite low. This change in water temperature has an effect on the refrigerant head pressure (condensing temperature). A typical wastewater system with 75°F entering water uses 1.5 gal/min

Figure 22.12 A wastewater condenser system under two sets of conditions. (A) Summer with warm water entering the system. (B) Winter with cooler water entering the system.

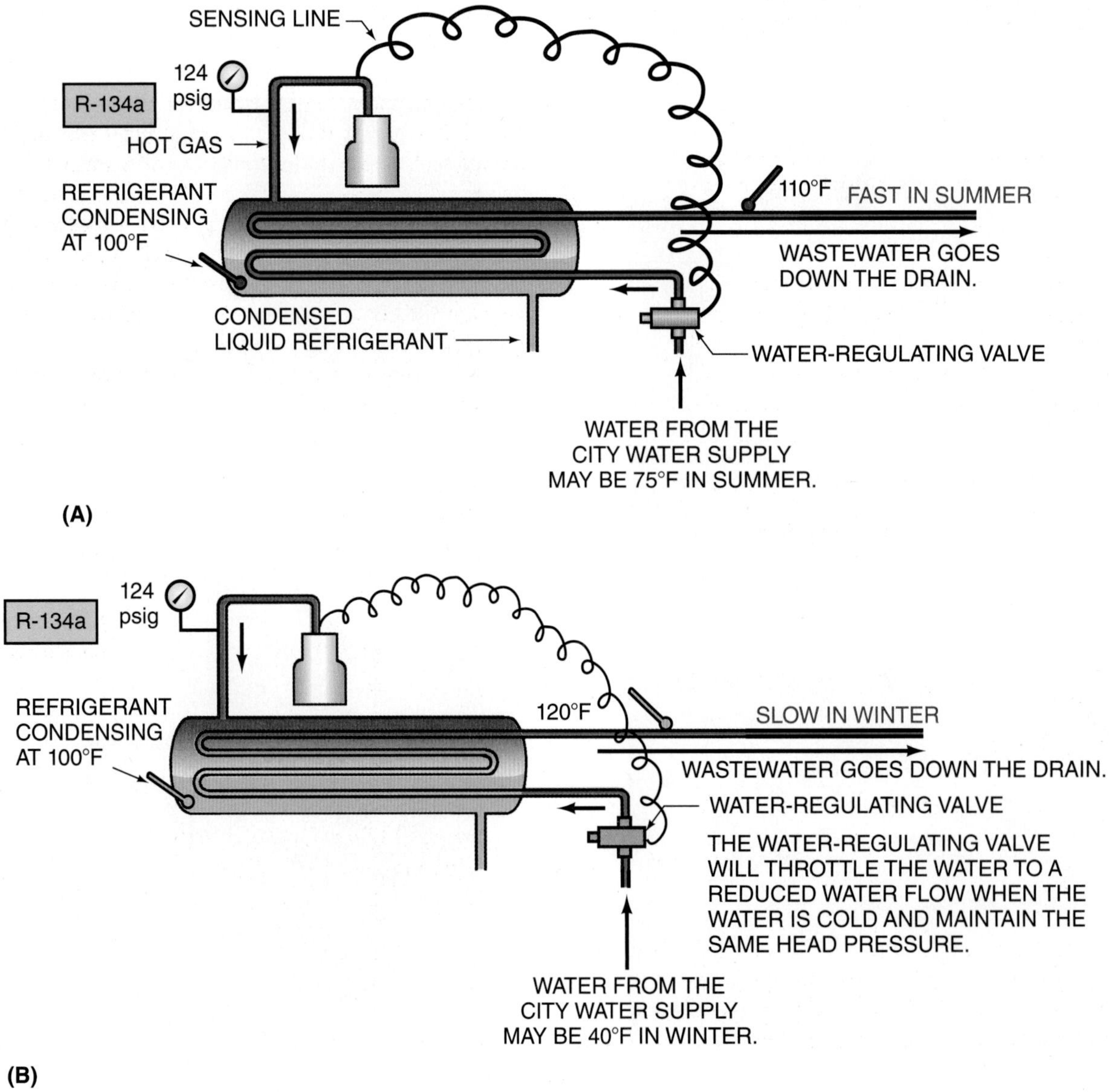

per ton of refrigeration. In the winter, when the water temperature is lower, the water flow required will be lower.

For a designer's purpose, the amount of water needed to flow through a water-cooled condenser in gallons per minute per ton of refrigeration depends on the temperature difference between the water in and out of the condenser. If a cooling tower is being used, a 10-degree temperature difference is often used for design purposes. However, if city water is being used through the condenser and the water is then discharged down a drain, a 20-degree temperature difference is often used for design purposes. The formula below governs how many gallons of water per minute per ton of refrigeration (GPM/ton) is required for a water-cooled condenser. A standard cooling tower can only cool water to within 7 to 8 degrees of the surrounding wet-bulb temperature (WBT). For this reason, water from the cooling tower to the water-cooled condenser is usually warmer than if it were from a city water source. This is why there is a designed 10-degree temperature difference for a cooling tower compared to a 20-degree temperature difference using a city water or well water source.

Formula:

$$\text{GPM/Ton} = \frac{15{,}000\ \text{Btuh/Ton}}{(\text{Specific heat of water})(\text{Temperature difference}) \times (8.33\ \text{lbs/gallon})(60\ \text{min/h})}$$

Where:

- 15,000 Btuh/ton is a standard condenser ton.
- Specific heat of water is 1 Btu/lb per degree Fahrenheit temperature difference.
- 8.33 lbs/gallon is the density of water.
- Temperature difference is the temperature difference between water in and out of the water-cooled condenser.

Using the above formula for a cooling tower:

$$\text{GPM/Ton} = \frac{15{,}000\ \text{Btuh/Ton}}{(1\ \text{Btu/lb/F})(10\ \text{degrees F})(8.33\ \text{lbs/gallon})(60\ \text{min/h})}$$

$$\text{GPM/Ton} = 3.0$$

Using the above formula for a city or well water wastewater system:

$$\text{GPM/Ton} = \frac{15{,}000\ \text{Btuh/Ton}}{(1\ \text{Btu/lb/F})(20\ \text{degrees F})(8.33\ \text{lbs/gallon})(60\ \text{min/h})}$$

$$\text{GPM/Ton} = 1.5$$

Example:
How much water per day will a water-cooled condenser use if it is using city or well water and the system's capacity is 5 tons? The system is a wastewater system where the discharge water from the condenser goes down a drain. It has an average run time of 16 hours per day.

Answer:
(1.5 GPM/Ton)(5 ton)(16 hrs/day)(60 min/h) = 7200 gallons of water per day. If the condenser was using cooling tower water with a 10-degree temperature difference and a flow rate of 3 GPM/Ton, the water usage would 14,400 gallons per day. However, this water would be re-circulated water to and from the cooling tower and water-cooled condenser.

22.9 REFRIGERANT-TO-WATER TEMPERATURE RELATIONSHIP FOR WASTEWATER SYSTEMS

In a wastewater system, the water flow can be varied to suit the need by means of a water-regulating valve. This valve has a pressure tap that fastens the bellows in the control to the high-pressure side of the system. When the head pressure goes up, the valve opens and allows more water to flow through the condenser to keep the head pressure correct, **Figure 22.13(A)**. Constant head pressure is maintained by letting more or less water into the water-cooled condenser as the head pressure changes. **Figures 22.13(A)** and **(B)** show a direction arrow stamped on the body of the valve to let the service technician know that the valve must be installed so that the direction of the valve is in the direction of the water flow. **Figure 22.13(B)** shows how an increase in head pressure puts more pressure on the flexible bellows to push up on the moveable rod and open the valve to let more water through the condenser. This in turn keeps the head pressure constant at the spring's set point.

The spring pressure in the valve can be adjusted by turning the tension adjustment located on top of the valve with

Figure 22.13 (A) This water-regulating valve controls water flow during different demand periods. Photo by Jason Obrzut

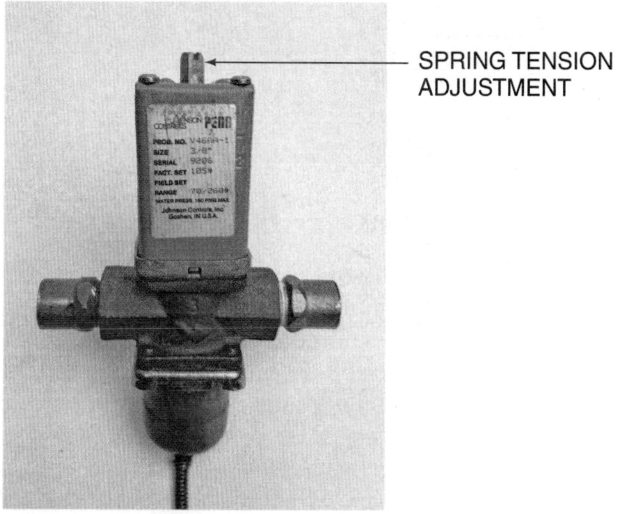

(A)

Figure 22.13 *(Continued)* (B) Internal components of a water-regulating valve. Courtesy Ferris State University. Photo by John Tomczyk (C) Top of a water-regulating valve showing spring, spring adjustment, threaded stud, and retainer plate. Courtesy Ferris State University. Photo by John Tomczyk. (D) Specification plates for a water-regulating valve. Courtesy Ferris State University. Photo by John Tomczyk. (E) Flushing a water-regulating valve. Courtesy Ferris State University. Photo by John Tomczyk

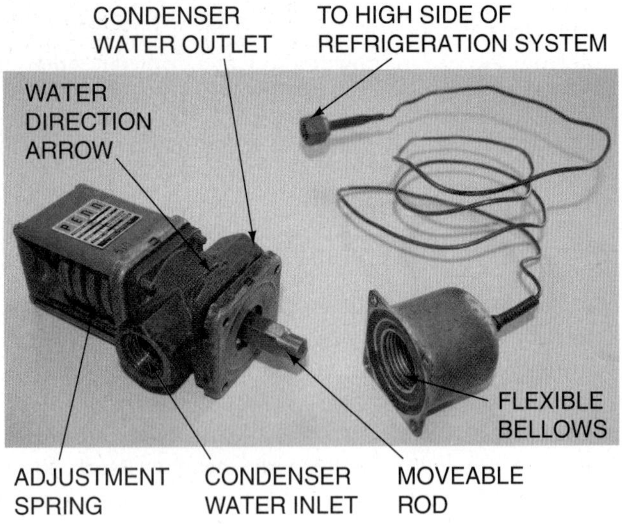

(B)

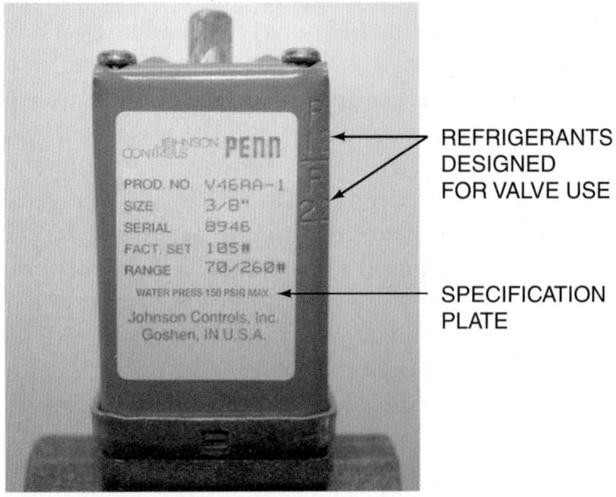

(D)

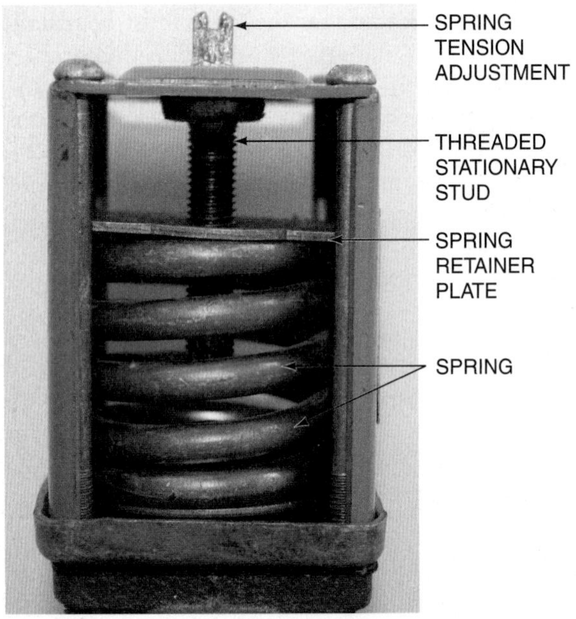

(C)

(E)

a refrigeration service wrench, standard slot screwdriver, or adjustable crescent wrench. A spring retainer plate with female threads rides up and down on a threaded stationary stud when the spring tension is adjusted, **Figure 22.13(C)**. An arrow on the top of the valve indicates that a counterclockwise turn will increase the head pressure setting by letting in less water. A pressure gauge connected to the high side of the refrigeration system while the spring tension of the water-regulating valve is being changed will show the head pressure change almost instantaneously. **Figure 22.13(D)** shows the valve specification plate and the refrigerants the valve is designed to be used with. A maximum water pressure and refrigerant pressure range is stamped on the specification plate for safety and operational reasons. Notice that the valve in the illustration can handle a water pressure of 150 psig maximum and that its adjustment, or refrigerant pressure, range is 70 to 260 psig. **Figure 22.13(E)** shows a water-regulating valve operating with a tube-within-a-tube water-cooled condenser. The water-regulating valve can be flushed by prying up on the bottom of the spring with two standard slot screwdrivers. This will rid the valve body of any loose scale deposits, dirt, or sand.

As mentioned, water inlet temperatures in wastewater systems often vary with the season. **Figure 22.12** illustrates a water-regulating valve maintaining a constant head pressure of 124 psig in an R-134a refrigeration system. When the water temperature entering the condenser is 75°F, the flow

of water through the condenser is much greater and faster because the water-regulating valve is opened more to maintain a constant condensing temperature of 100°F (124 psig) for the R-134a system. Notice that the water exiting the condenser is 110°F, which is 10°F hotter than the condensing temperature of 100°F. This happens because the water exiting the condenser comes in contact with hot, superheated discharge gases from the compressor before it exits down the drain. In the winter months, when the entering water temperature is 40°F, the water regulating valve will reduce the water flow rate to maintain the same 100°F (124 psig) condensing temperature for the system. Because of the reduced flow rate, the water stays in contact longer with the condenser's internal tubing. Notice that this causes the water exiting temperature to be 120°F.

Because of the variable flow and variable temperature of water flowing through a wastewater system during different seasons of the year, there is no refrigerant-to-water temperature relationship or rule of thumb for the technician to go by. The service technician must simply rely on the water-regulating valve to throttle more or less water through the condenser at different temperatures to maintain a constant head pressure of 124 psig.

22.10 RECIRCULATING WATER SYSTEMS

When a system is large enough to make saving water a concern, then a system that recirculates the water is considered. A recirculating water system will circulate approximately 3 gal of water per min per ton of refrigeration. Just as in the wastewater system, recirculating systems use the condenser to reject heat to water. The water is then pumped to an area away from the condenser where the heat is removed, **Figure 22.14**. The temperature relationship between the water and the refrigerant is known: The refrigerant will normally condense at a temperature about 10°F higher than the leaving-water temperature, **Figure 22.15**. The typical design temperature for the entering water is 85°F, and most systems have a 10°F rise across the condenser. Therefore, you could look for 85°F entering and 95°F leaving water temperature when the system is under full load. Since the R-134a refrigerant condenses at a temperature about 10°F higher than the leaving-water temperature, the approximate condensing temperature would be 105°F (95°F + 10°F). The head pressure for R-134a corresponding to this 105°F condensing temperature is 135 psig.

Figure 22.14 This water-cooled condenser absorbs heat from the refrigerant and pumps the water to a cooling tower at a remote location. The condenser is located close to the compressor, and the tower is on the roof outside the structure.

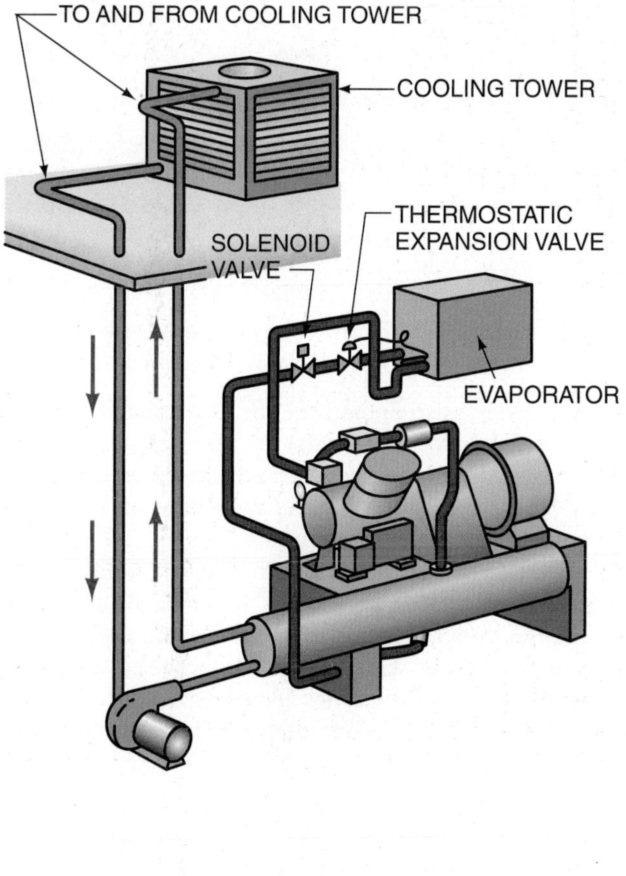

Figure 22.15 The relationship of the condensing refrigerant to the leaving-water temperature.

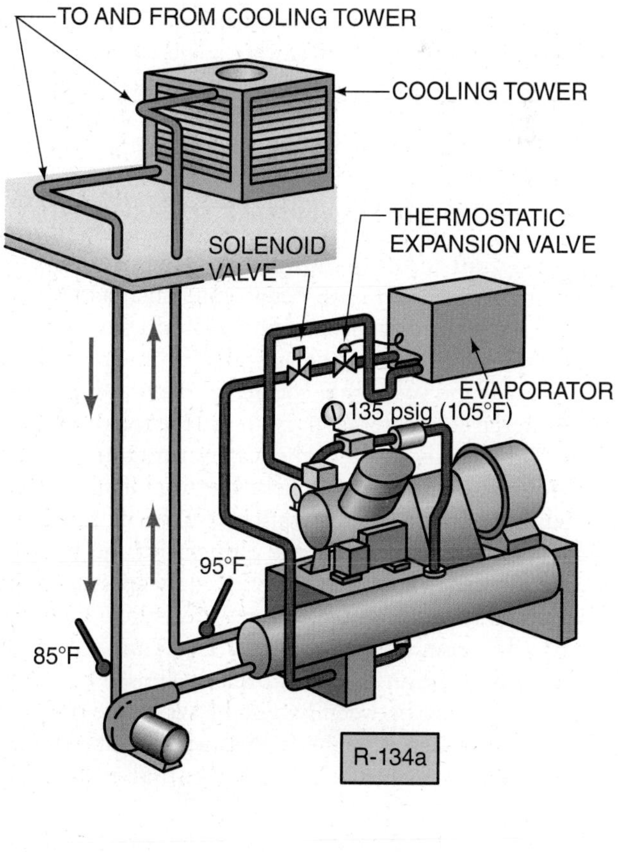

Figure 22.16 The relationship of a forced-draft cooling tower to the ambient air. Cooling tower performance depends on the wet-bulb temperature of the air. This relates to the humidity and the ability of the air to absorb moisture.

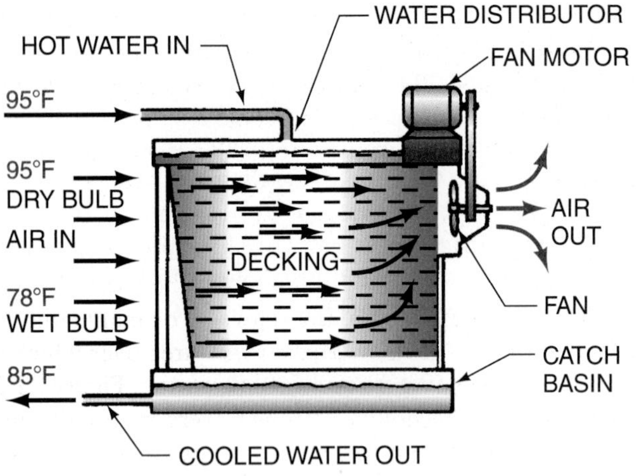

22.11 COOLING TOWERS

The cooling tower is a device that passes outside air over water to remove the system heat. Its capacity is limited by the amount of evaporation that occurs at any given time. The evaporation rate is linked to the wet-bulb temperature of the outside air (humidity). Usually, a cooling tower can cool the water that returns to the condenser to within 7°F of the wet-bulb temperature of the outside air, **Figure 22.16.** With a 78°F wet-bulb temperature air entering the cooling tower, the water in the tower can be cooled to 85°F. Towers come in sizes that can handle about 2 tons of refrigeration and up and can be either (1) natural draft, (2) forced draft, or (3) evaporative.

22.12 NATURAL-DRAFT TOWERS

The natural-draft tower is customarily made of some material that the weather will not deteriorate, such as redwood, fiberglass, or galvanized sheet metal. It does not have a blower to move air through it but relies on the natural prevailing breezes and so must be located in the prevailing wind. Water from the condenser is sprayed into the top of the tower through spray heads; some of the water evaporates as it falls to the bottom of the tower where the rest of the water is collected in a basin. The evaporation removes heat from the water remaining and adds to the capacity of the tower. To make up for the evaporated water, a makeup system with a float assembly is connected to the water supply and automatically adds fresh water, **Figures 22.17(A)** and **(B)**.

The tower location must be carefully chosen. If a tower is placed in a corner between two buildings where the breeze cannot blow through it, higher-than-normal water temperatures will occur, which will cause higher-than-normal head pressures, **Figure 22.18.** Towers also have two weather-related limitations that must be considered:

1. The tower must operate in the winter, and the water can freeze in some climates if freeze protection is not provided. Heat can be added to the water in the basin of the tower; antifreeze will also prevent a freeze-up, **Figure 22.19.**
2. The water can get cold enough to cause a drop in head pressure. A water-regulating valve can be installed to prevent this. Also, it is possible that winds can be strong enough to blow water out of the tower. Natural-draft towers on top of buildings sometimes look like they are made of slats. These slats help keep the water from blowing out of the tower, **Figure 22.20.**

Figure 22.17 (A) A makeup water system in a cooling tower. Because the cooling tower performance depends partly on evaporation of water from the tower, the makeup system is necessary. (B) Makeup water with a ball-and-water valve assembly. Plastic honeycomb fill material is shown in the background. Courtesy Ferris State University. Photo by John Tomczyk

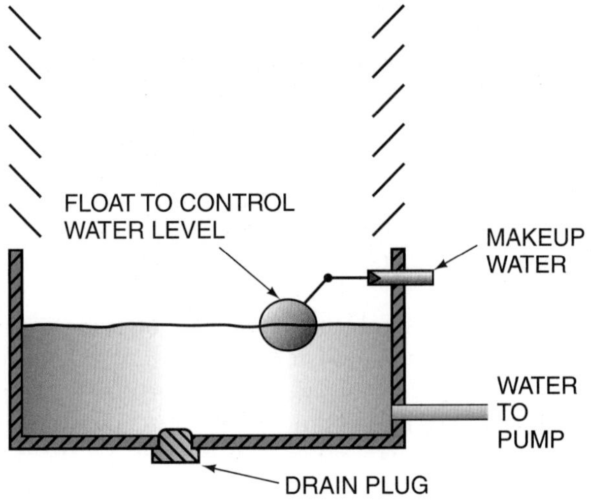

(A)

(B)

Figure 22.18 A natural-draft cooling tower. (It must be located in the path of prevailing winds.)

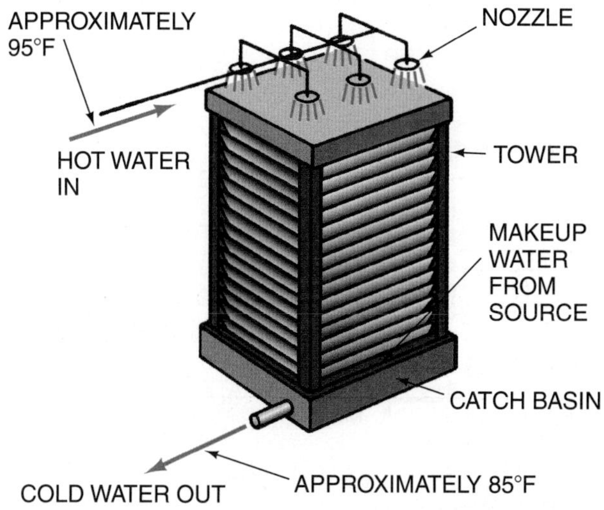

Figure 22.19 Applying heat to keep the water in the basin from freezing in winter. The heat can be controlled thermostatically so it can be turned off when not needed.

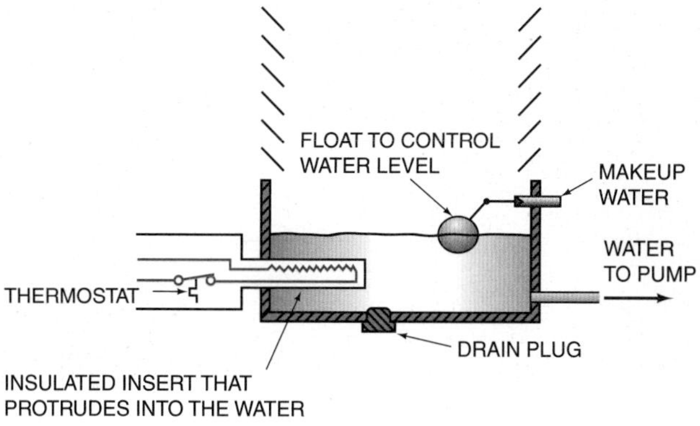

Figure 22.20 Slats on the sides of a natural-draft cooling tower keep the water inside the tower when the wind is blowing.

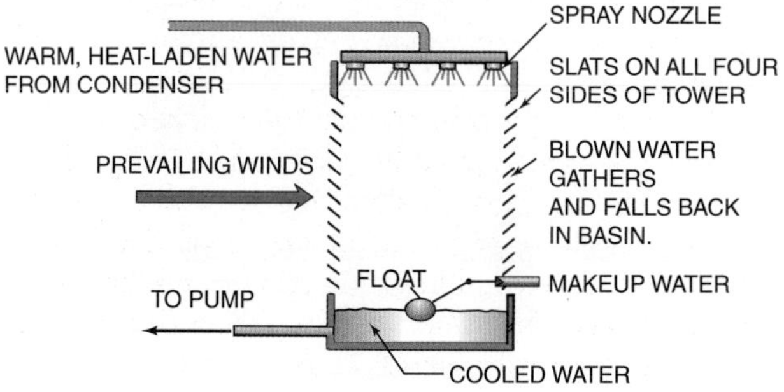

22.13 FORCED- OR INDUCED-DRAFT TOWERS

Forced-draft or induced-draft cooling towers differ from natural-draft towers because they have a fan to move air over a wetted surface, **Figure 22.21**. They are typically designed so that the warm water from the condenser is pumped into a flat basin at the top of the tower. Calibrated holes are drilled into the basin to allow a certain amount of water to pass downward through the fill material, **Figure 22.22(A)**. The holes in the basin often become plugged with scale from minerals in the water, **Figure 22.22(B)**, which can cause an erratic water flow through the fill material and make the tower less efficient. The fill material is usually redwood or manmade fiber and is used to enlarge the wetted surface area over which the fan blows air to evaporate and cool the water, **Figure 22.23**. Modern fill materials used in cooling towers are usually made of plastic formed into a honeycomb pattern for higher efficiencies and ease of cleaning, **Figure 22.17(B)**. As the water is

Figure 22.21 (A) A forced-draft tower. Courtesy of Baltimore Aircoil Company, Inc. (B) An induced-draft tower. Courtesy of Baltimore Aircoil Company, Inc.

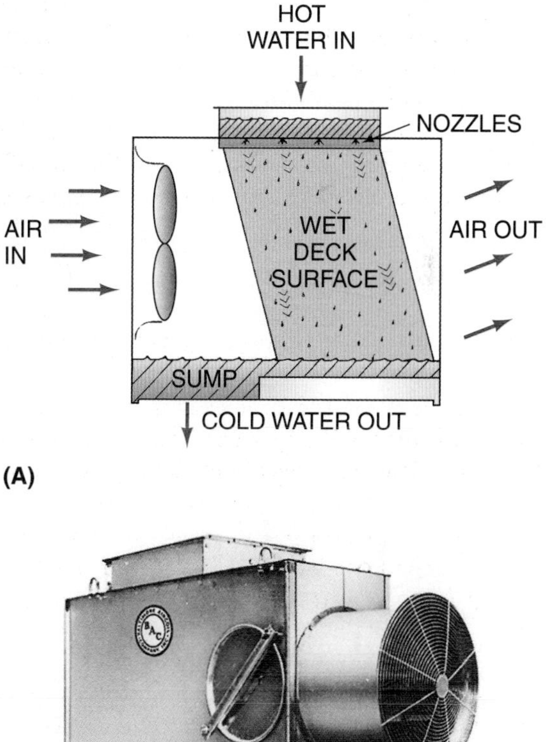

(A)

(B)

Figure 22.22 (A) Calibrated holes at the top of an induced-draft tower distribute the water over the fill material below. (B) Calibrated holes at the top of a cooling tower showing significant rust. Photo by Jason Obrzut

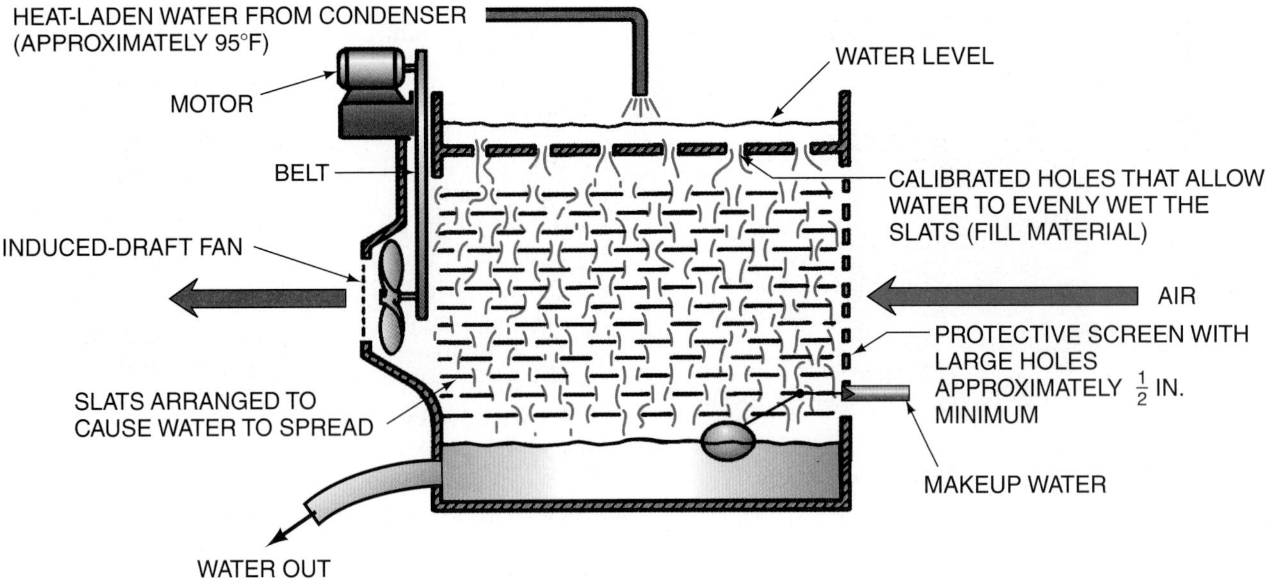

(A)

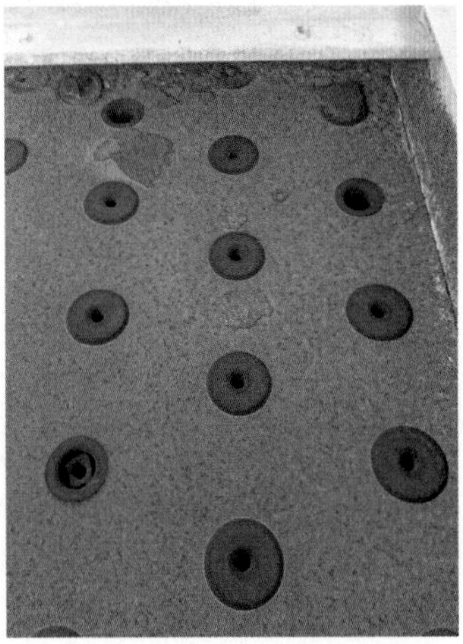

(B)

Figure 22.23 (A) The water trickles down through the fill material. Source: Marley Cooling Tower Company. (B) Plastic honeycomb fill material in a modern cooling tower. Courtesy Ferris State University. Photo by Jason Obrzut

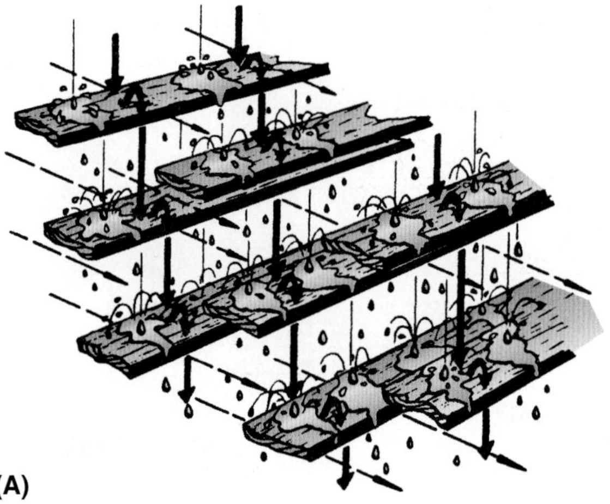

(A)

(B)

Figure 22.24 An induced-draft tower located inside a building with air ducted to the outside.

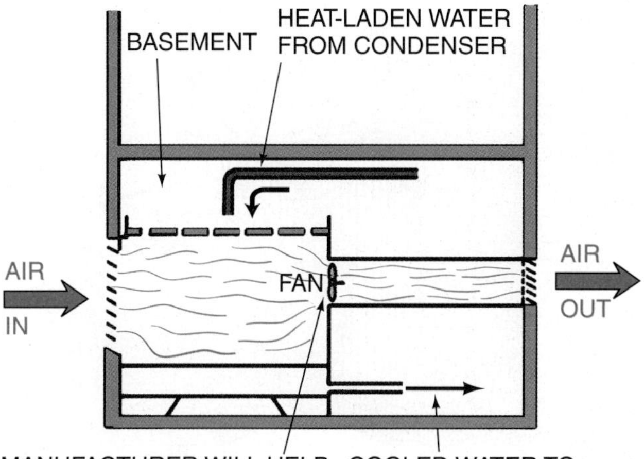

evaporated, it is replaced by a water makeup system using a float, similar to the system used with the natural-draft tower.

Forced- or induced-draft towers can be located almost anywhere because the fan can move the air. They can even be located inside buildings, where the air is brought in and out through ducts, **Figure 22.24**. The tower is fairly enclosed to the prevailing winds, so a water-regulating valve is typically not necessary. The fan can be cycled off and on to control the water temperature and thus control the head pressure. The mass of the water in the tower provides a long cycle between the time the fan starts and stops. Forced-draft towers are small compared to natural-draft towers. They are versatile because of the forced movement of air. **NOTE:** *An induced-draft tower is similar to a forced-draft tower, but the air is pulled, not pushed, across the wetted surface.* •

22.14 EVAPORATIVE CONDENSERS

Although often confused with cooling towers, **Figure 22.25**, **evaporative condensers** are a different type altogether from other condensers because they are actually located inside a cooling tower. In the cooling towers just discussed,

Figure 22.25 Water re-circulates in the evaporative condenser. The condenser tubes are in the tower, rather than in a condenser shell located in a building.

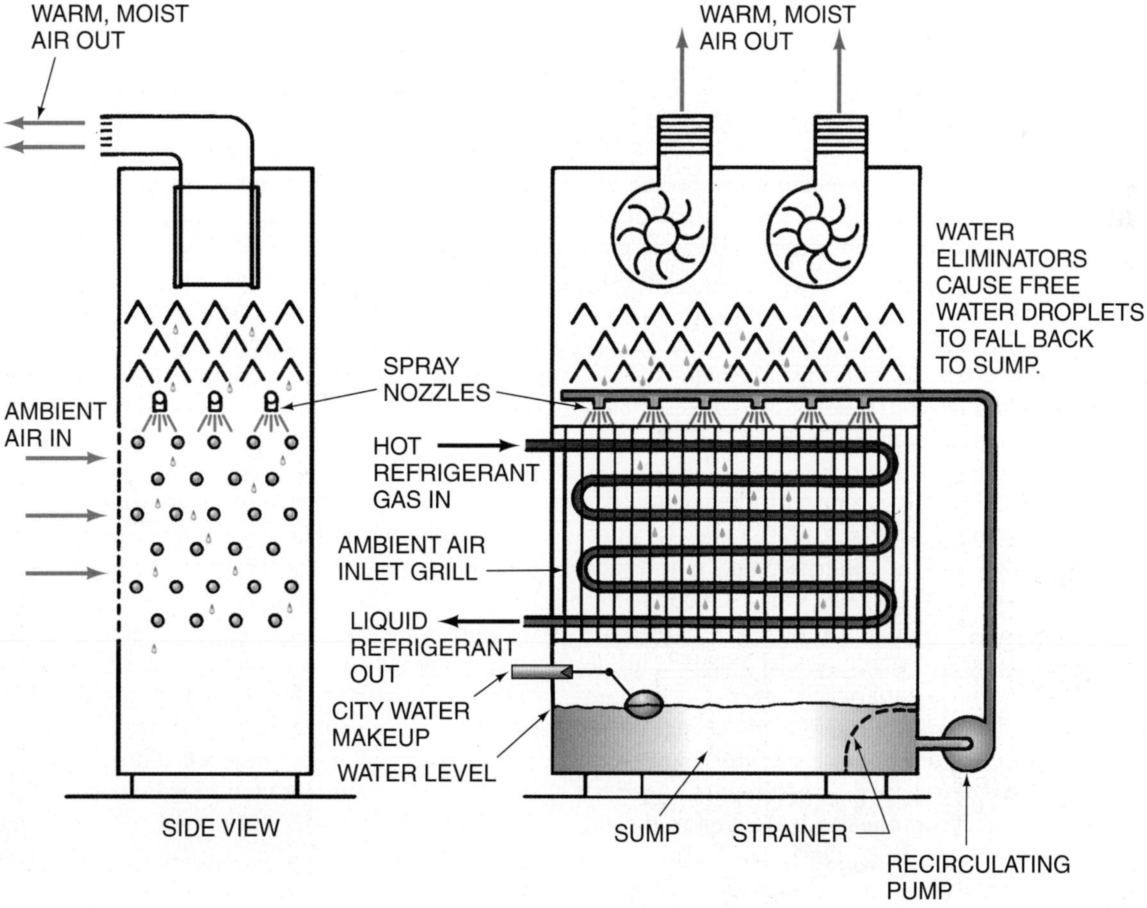

the condenser containing the refrigerant was remote from the tower and the water was piped through the condenser to the tower. The evaporative condenser uses the same water over and over by means of a pump located at the tower. As the water is evaporated, it is replaced by a makeup system using a float, as with the other towers. In cold climates, freeze protection must be provided in winter. In an evaporative condenser, both air and water are used to desuperheat, condense, and subcool the refrigerant. The water is sprayed over the condenser and absorbs heat from the warmer refrigerant. The air is pulled in from the side, passes over the condenser, and absorbs heat from the condensing refrigerant. The process is based on the principle of evaporative cooling, because as the water spray evaporates, much more heat can be absorbed from the refrigerant.

As water is evaporated from any cooling tower system, minerals will become more concentrated in the remaining water. If these minerals are allowed to become overconcentrated, they will begin to deposit on the condenser surface and cause head pressure problems. To prevent this from happening, water must be allowed to escape the system on a continuous basis. Water that is allowed to escape down the drain, called blowdown, is made up with fresh water from the float system. It is not unusual for people who do not know the purpose of blowdown to shut off the drain line for what appears to be clean water being wasted down the drain. Problems will result. Technicians must establish a regular procedure for checking the rate of blowdown. Cooling towers and blowdown are discussed in more detail in Unit 48, "Cooling Towers and Pumps."

22.15 AIR-COOLED CONDENSERS

Air-cooled condensers use air as the medium into which the heat is rejected. This can be advantageous where it is difficult to use water. The first air-cooled condensers were bare pipe over which air from the compressor flywheel was blown. At that time the compressors were open drive. To improve the efficiency of the condenser and to make it smaller, the surface area was then extended with fins, normally steel as was the condenser, **Figure 22.26**. These condensers resembled radiators and were sometimes referred to as radiators.

Figure 22.26 Fins designed to give the coil more surface area.

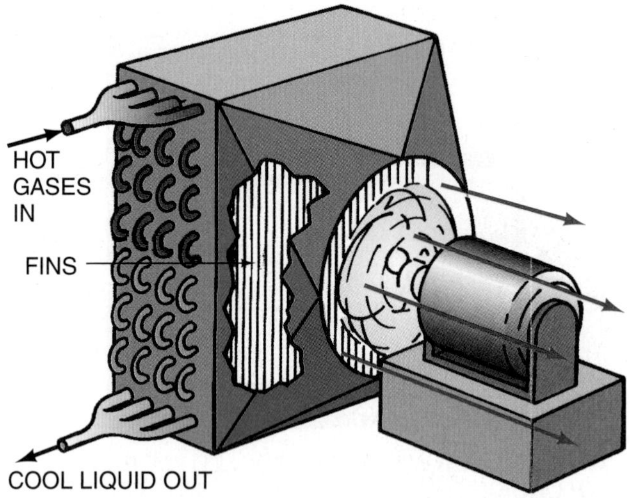

Steel air-cooled condensers are still used in refrigeration installations. **Figure 22.27(A)** is a photo of a condenser for a large refrigeration system. **Figures 22.27(B)** and **(C)** illustrate some newer technology in condenser and evaporator coil design to improve system efficiency. One such condenser design incorporates an aluminum, parallel-flow, flat-plate-and-fin configuration that has small parallel channels inside the flat plate. These plates are flattened, streamlined tubes split into smaller, parallel ports. Refrigerant will change phase or condense from a vapor to a liquid inside the channels in the plate, while strategically shaped fins (extended surfaces) will enhance heat transfer from the condenser coil to the air. The plates and fins are bonded or soldered to increase heat transfer and to eliminate any contact resistance (air gaps) that will reduce heat transfer. Headers at the inlet and outlet of the heat exchanger are also bonded to the plates through soldering. (Refer to Unit 21, "Evaporators and the Refrigeration System," for a more detailed explanation about using the parallel-flow, flat-plate-and-fin condenser and evaporator configuration to increase the efficiency of a refrigeration or air-conditioning system.)

Air-cooled condensers come in a variety of styles. In some, the air blows horizontally through, subject to prevailing winds, **Figure 22.28**. In other air-cooled condensers, the airflow pattern is vertical. They take air into the bottom and discharge it out the top. The prevailing winds do not affect these condensers to any extent, **Figure 22.29**. Another style of air-cooled condenser takes the air in at the sides and discharges it out the top. This condenser can also be affected by prevailing winds, **Figure 22.30**.

Smaller refrigeration systems are often located in a conditioned space, such as a restaurant or store. Their air-cooled condensers normally have widely spaced steel fins on a steel coil, which allows more time before the coil will stop up with dust and other airborne material. The hot gas normally enters the air-cooled condenser at the top, so the beginning tubes of the condenser receive the hot gas, which is highly superheated, straight from the compressor. (Remember that superheat is heat that is added to the refrigerant after the change of state in the evaporator.) When the superheated refrigerant from the evaporator reaches the compressor and is compressed, more superheat is added to the gas. Part of the energy applied to the compressor transfers into the refrigerant in the form of heat energy. This additional heat added by the compressor causes the refrigerant leaving the compressor to be heavily laden with heat. On a hot day (95°F) the hot gas leaving the compressor could easily reach 250°F. **⚠SAFETY PRECAUTION:** *DO NOT TOUCH the hot gas discharge line from the compressor. It will burn your fingers.* • The condenser must remove heat from the refrigerant until it reaches the condensing temperature before any condensation can occur.

Figure 22.27 (A) This larger refrigeration air-cooled condenser resembles an air-conditioning condenser. Source: Heatcraft, Inc., Refrigeration Products Division. (B) An aluminum parallel-flow, flat-plate-and-fin heat exchanger. Source: Modine Manufacturing Co., Racine, WI. (C) An aluminum parallel-flow, flat-plate-and-fin condenser. Source: Modine Manufacturing Co., Racine, WI.

(A)

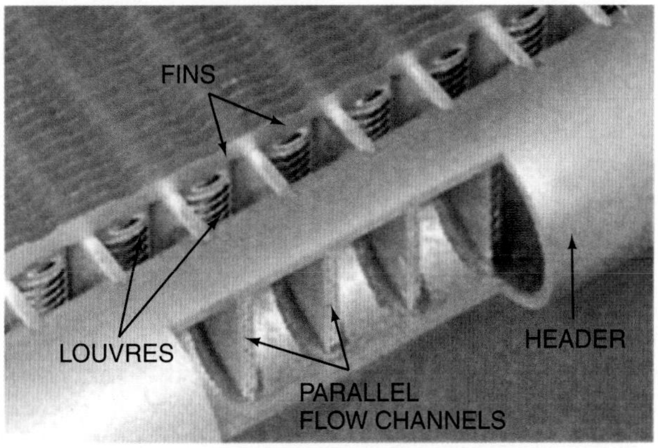

(B)

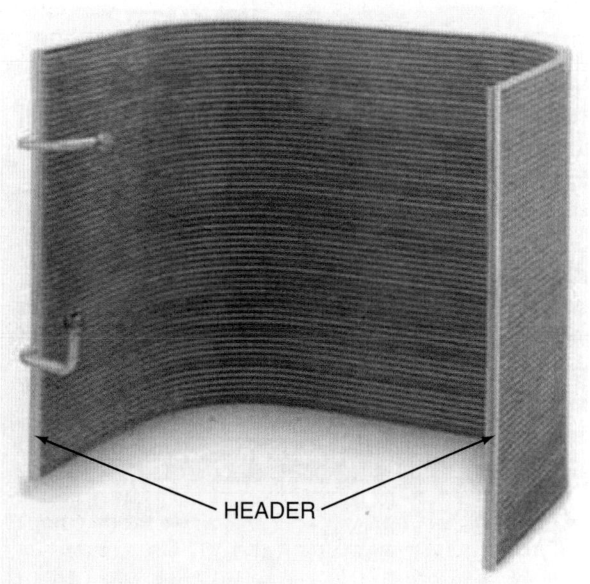

(C)

Figure 22.28 A horizontal air-cooled condenser subject to prevailing winds blowing through it. Courtesy Copeland Corporation

Figure 22.29 A condenser with a vertical airflow pattern. Air enters the bottom and is blown out the top. It is unaffected by prevailing winds. Source: Photo by Jason Obrzut

Figure 22.30 This condenser takes the air in at the sides and discharges it out the top. The prevailing winds could affect it. Photo by Jason Obrzut

Air-cooled condensers have a relationship to the temperature of the air passing over them (much like the evaporator in the previous unit). For instance, the refrigerant inside the coil will normally condense at a 30°F higher temperature than the air passing over it (also known as the ambient air). This is true for most standard-efficiency condensers that have been in service long enough to have a typical dirt deposit on the fins and tubing. Efficiency can be improved by adding condenser surface area. With an outside air temperature of 95°F, the condensing temperature will be about 125°F. With the refrigerant R-134a condensing at 125°F, the head pressure or high-pressure gauge should read 184 psig. (See the temperature/pressure chart for R-134a, **Figure 3.13.**) This is important because it helps the service technician establish what the head pressure should be.

See **Figure 22.31** for an illustration of the following description of an air-cooled R-134a condenser located outside a store. The condenser is responsible for rejecting the heat absorbed inside a medium-temperature walk-in cooler. This cooler has reach-in doors typical of those in a convenience store. In this box the beverages are on shelves and store personnel can stock the shelves from the walk-in portion behind the shelves. The cooler is maintained at 35°F. The outside air temperature is 95°F. The refrigeration system must absorb heat at 35°F and reject the same heat to the outside, where the condensing medium is 95°F.

1. The hot gas is entering the condenser at 200°F. The condensing temperature is going to be 30°F warmer than the outside (ambient) air.
2. The outside air temperature is 95°F. The condensing temperature is 95°F + 30°F = 125°F. The refrigerant must be cooled to 125°F before any actual condensing occurs. Thus the condenser has to lower the hot gas temperature 75°F (200°F − 125°F) in the first part of the coil. This is called *desuperheating*. It is the first job of the condenser.
3. Partway through the coil the superheat is removed and the refrigerant reaches the actual condensing temperature of 125°F; liquid begins to form in the coil. Liquid will continue to form at the condensing temperature of 125°F until all of the vapor has turned to 100% liquid. This is called the latent heat of condensation.
4. The liquid can now reject sensible heat and *subcool* to below the 125°F condensing temperature. Any liquid below the 125°F condensing temperature is referred to as a subcooled liquid. When the subcooled liquid refrigerant gets to the end of the coil, the condenser tubes will be full of liquid that is pumped into the receiver.
5. The liquid in the bottom of the condenser that is pumped into the receiver may cool to 110°F. This would give the liquid 15°F (125°F − 110°F) of subcooling.

In commercial and industrial applications in which the condensers are large, a pressure drop will occur as the refrigerant flows through the condenser. This pressure drop is caused by friction. When measuring condenser subcooling in these cases, the refrigerant pressure has to be read close to the condenser outlet, not at the compressor service valve, **Figure 22.32.** The condenser outlet pressure can be read at the receiver's king valve or at the receiver charging valve. In fact, the refrigerant

Figure 22.31 The following points are illustrated: (1) The hot gas flowing into the first part of the condenser is highly superheated. The condensing temperature is 125°F, and the discharge gas must be cooled from 200°F to 125°F before any condensing will occur. (2) The best heat exchange is between the liquid and the air on the outside of the coil. More heat is removed during the change of state than during the desuperheating of the vapor. (3) When all of the liquid is condensed and pure liquid is in the coil, the liquid can be subcooled to below the condensing temperature.

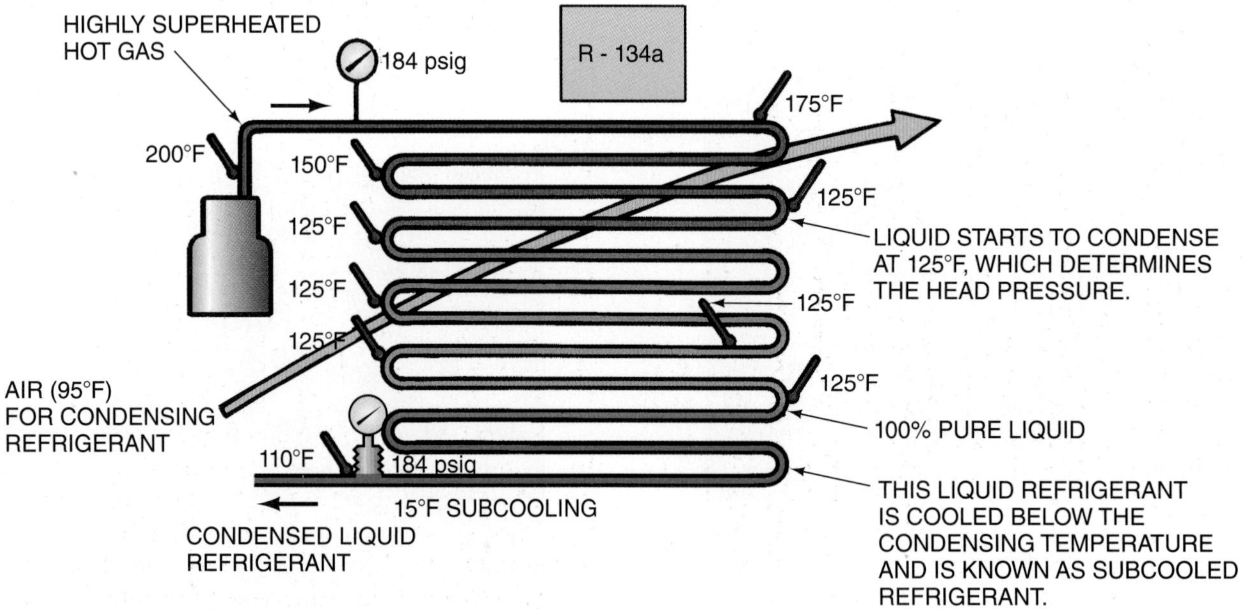

Figure 22.32 The air-cooled condenser has a 13-psig pressure drop from where the hot gas enters to where the subcooled liquid leaves the condenser. This is due to friction loss in the piping and is typical of larger condensers. The subcooling must be measured close to the same pressure location as where the temperature pressure is measured.

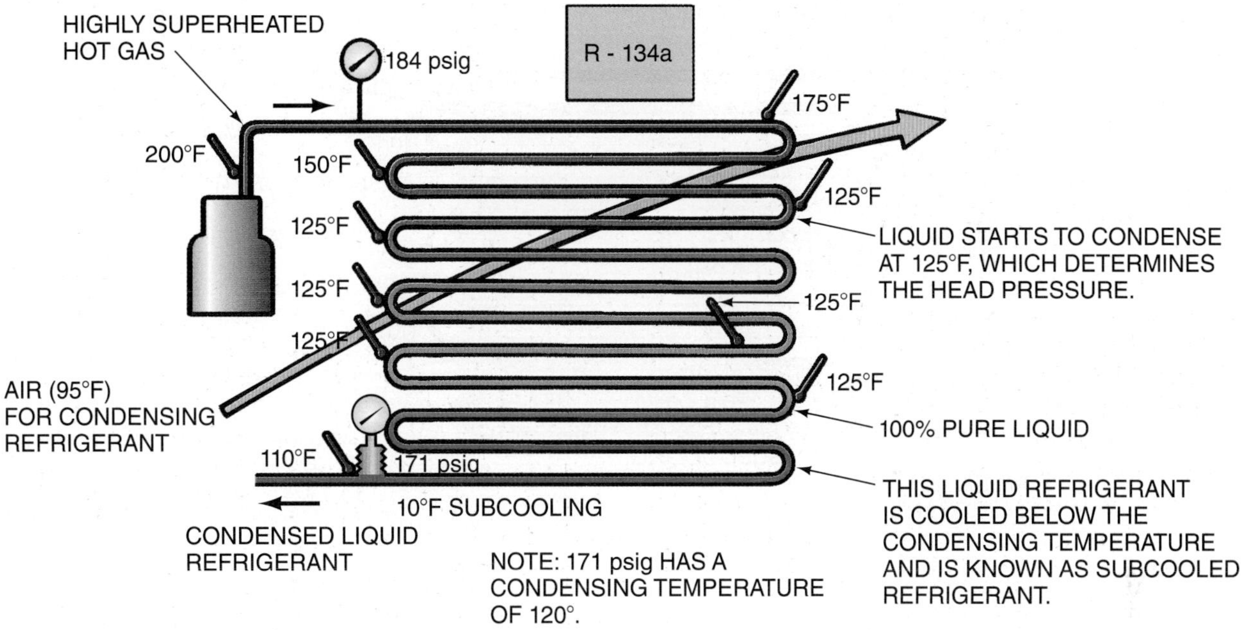

pressure should be taken very near the same place the condenser outlet temperature is taken, or condenser subcooling calculations will be inaccurate. If the condenser in **Figure 22.32** was large, and the pressure was measured at the condenser outlet, the condenser subcooling calculation would be 120°F − 110°F = 10°F of condenser subcooling. The lower pressure of 171 psig at the condenser's outlet gives a lower condensing temperature of 120°F when the condenser pressure drop of 13 psig (184 psig − 171 psig) is taken into consideration.

All air-cooled condensers do not have the same 30°F relationship with the ambient air. It is good practice to determine and record the temperature relationship between the air and refrigerant at start-up time. Then if the relationship were to change, you would suspect trouble, such as a dirty condenser or an overcharge of refrigerant that has been added without your knowledge.

22.16 HIGH-EFFICIENCY CONDENSERS

Condensing temperatures can be reduced by increasing the condenser surface area. The larger the condenser surface area, the closer the condensing temperature is to the ambient temperature. A perfect condenser would condense the refrigerant at the same temperature as the ambient, but would be so large that it would not be practical. In the previous examples, we used a temperature difference of 30°F between the ambient and the condensing temperatures. **Figure 22.33** shows an example of a condenser condensing at 110°F, 15°F above the ambient temperature. The head pressure for R-134a condensing at 110°F would be 146.4 psig. (R-404A would be 271.4 psig.) *The reduction of head pressure provides for a more efficient system—less power is consumed for the same amount of usable refrigeration.*

Some condensers, particularly in extra-low-temperature systems, can reduce condensing temperatures to within 10°F of the ambient.

Compressors are affected by a decrease in head pressure and will be discussed in the next unit.

22.17 THE CONDENSER AND LOW AMBIENT CONDITIONS

The previous example described how an air-cooled condenser operates on a hot day. An example of a condenser operating under different conditions might be in a supermarket with a small package-display case located inside the store. Because this is a package-display case, the condenser is also located inside the store, **Figure 22.34**. The inlet for air to the case is in the store itself and the air may be quite cool if the air conditioner is operating—70°F or even cooler at times. This reduces the operating pressure on the high-pressure side of the system. When the condenser relationship temperature rule is applied, we see that the new condensing temperature would be 70°F + 30°F = 100°F. The head pressure would be 124 psig. This may be enough to affect the performance of the expansion device (covered in more detail later). The low-side pressure for a medium-temperature fixture using

Figure 22.33 The operating conditions for a high-efficiency condenser.

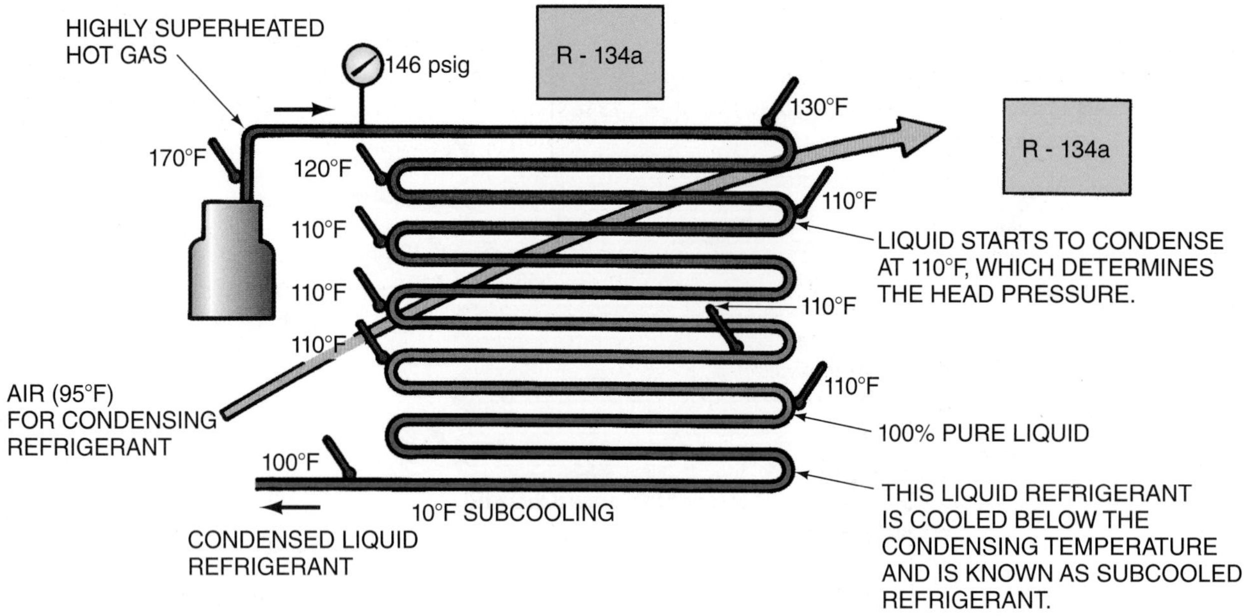

Figure 22.34 The package-display case located inside a store with the compressor and the condenser located inside the cabinet. This is a plug-in, self-contained device; no piping is required.

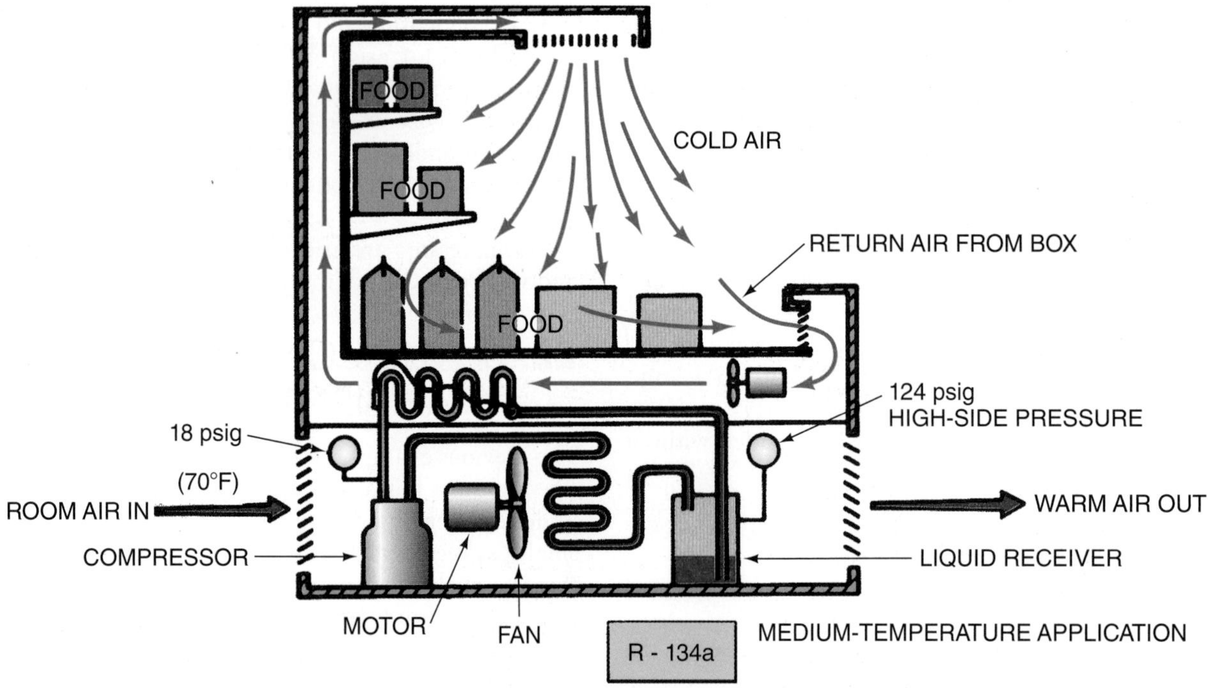

R-134a will be in the neighborhood of 18 psig at the lowest point. When the relationship of the low-side pressure of 18 psig is subtracted from the high-side pressure of 124 psig for a condenser with 70°F air passing over it, the difference is 106 psig. Should this fixture be exposed to temperatures lower than 70°F, a starved evaporator coil may be the result.

If the medium-temperature box were moved outside the store, the fixture may quit working to capacity during the winter. If the air temperature over the coil drops to 50°F outside, the head pressure is going to drop to 50°F + 30°F = 80°F condensing temperature, which corresponds to 87 psig, **Figure 22.35**. The pressure difference of 87 psig − 18 psig = 69 psig may

not be sufficient to feed enough liquid refrigerant through the expansion device to properly feed the coil, **Figure 22.35**. In the figure, notice the pressure and temperature difficulties encountered. The head pressure must be regulated.

Another example of a condenser operating outside design parameters would be an ice-holding box like those found at service stations and convenience stores. These fixtures hold ice made at another location. They must keep the ice hard in all types of weather and often operate at about 0°F to 20°F inside the box. It may be 30°F outside the box where the small air-cooled condenser is rejecting the heat. The condenser would be operating at about 30°F + 30°F = 60°F, or at a head pressure of 57 psig, **Figure 22.36**. The evaporator should be operating at about −15°F, or at a suction pressure

Figure 22.35 The fixture in **Figure 22.34** is relocated outside the store. Now the condenser is subject to winter conditions. Its performance will fall off if some type of head pressure control is not installed, because the head pressure is so low it cannot push enough liquid refrigerant through the expansion valve. The evaporator is starved. The unit has a reduced capacity that would be evident if a load of warm food were placed in the box. It might not pull the food temperature down, and the coil would ice up because there would be no off-cycle defrost.

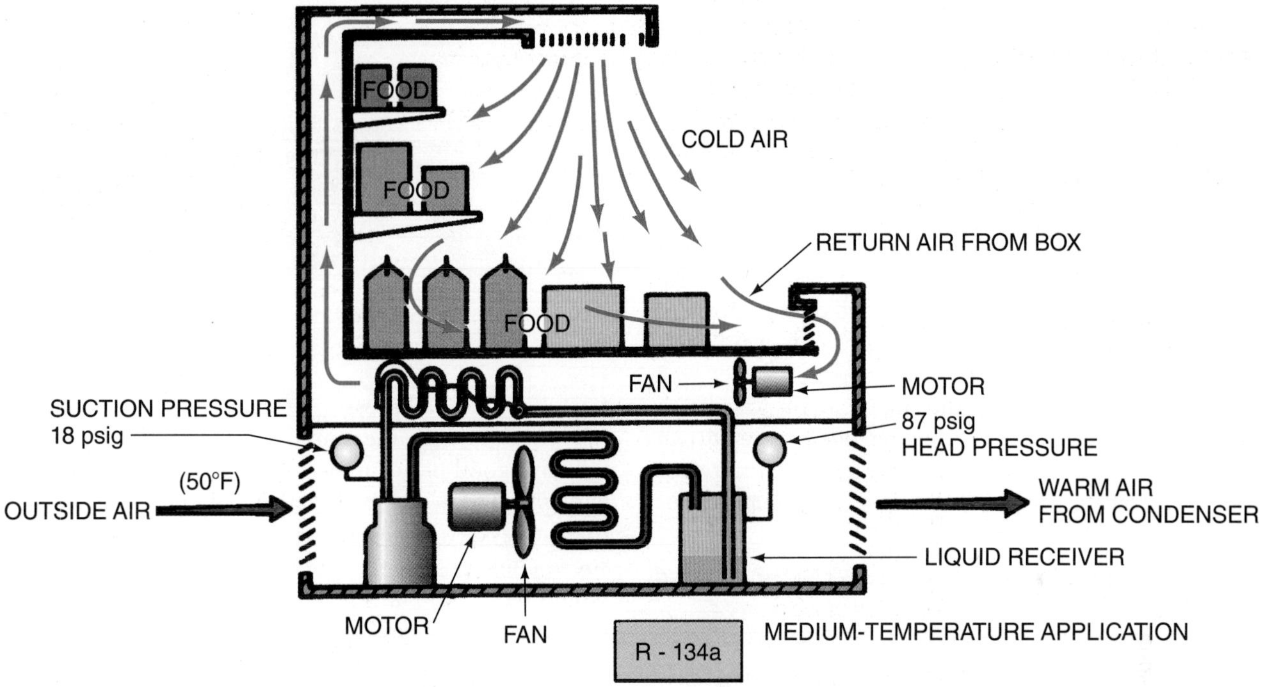

Figure 22.36 An illustration of how the condenser acts in a low ambient condition. Notice that this condenser has much more liquid refrigerant in it and the head pressure is very low. It may be so low that it may starve the expansion device. The condenser relies on head pressure to force liquid refrigerant through its metering device.

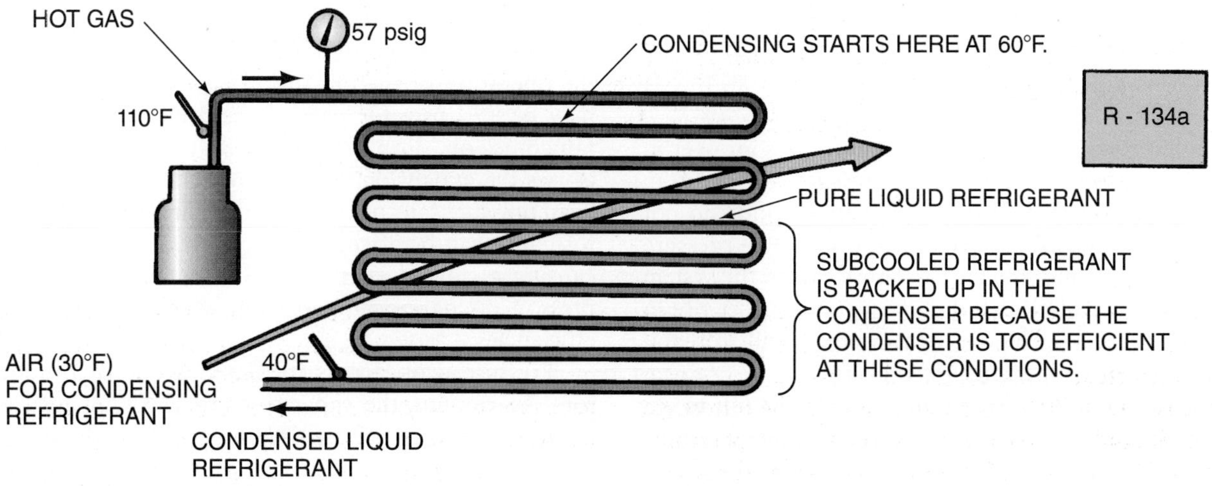

of 0 psig to maintain 0°F. This gives a pressure difference of 57 psig − 0 psig = 57 psig, which will starve the evaporator. This unit has to run to keep the ice frozen, so something has to be done to get the head pressure up.

One thing that helps prevent problems in some equipment operating in low ambient conditions is that the load is reduced. The ice-holding box, for example, does not have to run as much at 30°F weather to keep the ice frozen at 0°F. It may perform poorly, but you will not know it unless it must run for a long period of time for some reason, such as if a load of ice that is barely frozen is put into it. The unit may have run a long cycle to pull the load temperature down, and the evaporator may freeze solid.

22.18 HEAD PRESSURE CONTROLS

As the outdoor ambient temperature falls, and the temperature difference between the condensing refrigerant and the ambient temperature becomes greater, the condenser capacity will increase. In low ambient conditions, the system's condenser becomes very efficient and can be considered oversized. An oversized condenser means lower head pressures, lower compression ratios, and reduced consumption of electricity. However, there are limits to how low head pressures can be allowed to fall in order to make sure metering devices are being properly fed, evaporators remain active with proper superheats, mass flow rates or refrigerant are correct for proper oil returns to the compressor, and compressor discharge temperatures are not excessive from compressor return gas temperatures being too hot.

Practical methods to maintain the correct workable head pressure automatically and not cause equipment wear are:

- *fan-cycling controls,*
- *variable-frequency drive (VFD) condenser fan motor controls,*
- *dampers,*
- *condenser flooding, and*
- *condenser splitting.*

Each control method has different characteristics and features. Manufacturers may recommend various head pressure controls for their own equipment.

FAN-CYCLING DEVICES

The air-cooled condenser uses a small fan to pass air over it. When this fan is cycled off, the head pressure will go up in any kind of weather, provided the prevailing winds do not take over and do the fan's job. To cycle the fan off, a pressure control can be piped into the high-pressure side of the system that will close a set of electrical contacts on a preset rise in pressure, **Figure 22.37**. The electrical contacts will stop and start the fan motor with changes in pressure.

A common setup for R-134a may call for the fan to cut off when the head pressure falls to 135 psig and to restart the fan when the head pressure reaches 190 psig. This setting will not interfere with the summer operation of the system and will give good performance in the winter. The settings are far enough apart to keep the fan from short cycling any more than necessary. When the thermostat calls for the compressor to come on and the ambient air is cold, the head pressure would be so low that the system would never get up to good running capacity because the condenser fan would be moving more cold air over the condenser. The condenser fan-cycling control keeps the condenser fan off until the head pressure is within the correct operating range.

Figure 22.37 A condenser fan-cycling device. This control performs the same action (make on a rise in pressure) as the low-pressure control except that it operates at a much higher range. Photo by Jason Obrzut

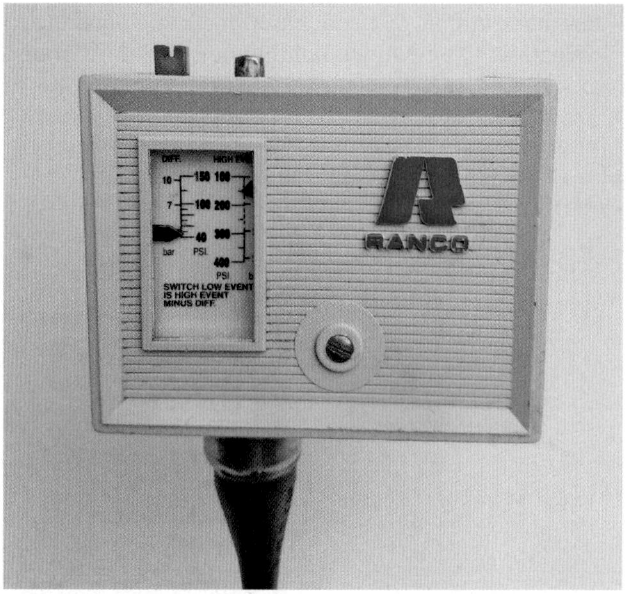

The fan-cycling device is one method of maintaining a correct operating head pressure range. It can be added to the system without much expense, and the piping normally does not have to be altered. One problem with the device is that it has a tendency to cause the head pressure to swing up and down as the fan is stopped and started. This can also affect the operation of the expansion device because the pressure may be 190 psig for part of the cycle and then, when the fan comes on, drop rapidly to 135 psig. The pressure swing causes the expansion device to operate erratically. When the head pressure falls below certain minimums, the system's performance can be affected in three ways: underfeeding TXVs causing a starved evaporator, oil logging, and higher compressor discharge temperatures which reduce compressor efficiencies.

Pressure drop across the TXV is one of the main factors determining the capacity of the TXV. A head pressure reduced below a certain minimum will reduce the liquid

pressure feeding the TXV which will cause the TXV to close and underfeed the evaporator. The evaporator will become less active and starved of refrigerant, causing a low net refrigeration effect. When an evaporator is starved by low head pressures feeding the TXV, high superheats are experienced in the evaporator and coming back to the compressor. These higher-than-normal temperatures coming back to the compressor decrease the density of vapors being pumped. This in turn results in a low mass flow rate of pumped refrigerant. A decreased mass flow rate of refrigerant will make it difficult for oil to return to the compressor. Because oil and refrigerant do not completely mix, a minimum refrigerant velocity in the suction line is required for proper oil return. It is refrigerant velocity that brings oil back to the compressor from the suction line and suction-line risers. The higher temperature of refrigerant coming into the compressor will also cause a higher compressor discharge temperature. In fact, for every 1-degree rise in suction gas temperature returning to the compressor, there will be a 1-degree rise in the compressor's discharge temperature.

There are other means of controlling the head pressure at a steady state on air-cooled condensers. When a condenser has more than one fan, one fan can be put in the lead and the other fans can be cycled off by temperature or pressure, **Figure 22.38**. The lead fan can then be cycled off by pressure, like a single fan. Cycling multiple fans can help prevent pressure swings from being as close together as they are in the single-fan application. For example, when three fans are used, the first can cycle off at approximately 70°F, and the second can cycle off at approximately 60°F. The remaining fan may be controlled by the pressure or by a temperature sensor located on the liquid line. If a condenser had eight fans, each set of two fans could be staged off by a pressure control. Below is a typical fan-cycle control scheme with set points using pressure controls.

Condenser Set Points

Fan #	Cut-in Pressure (psig)	Cut-out Pressure (psig)
1 and 2	180	160
2 and 3	190	170
3 and 4	200	180
4 and 5	210	190

With this pressure control strategy cycling four sets of fans, the condensing pressure would never fall below 160 psig, unless the condenser experienced extremely cold winds blowing through it.

A disadvantage of this type of head pressure control is that as more fans cycle off, the head pressure (saturated condensing temperature) and receiver pressure will decrease. Any saturated liquid refrigerant in the receiver will then try to reduce its temperature by flashing into vapor to comply with the new lower condensing pressure and temperature. As some of the liquid in the receiver flashes, it absorbs heat from the surrounding liquid as it tries to reduce its temperature to the new saturation temperature in the condenser. Depending on how much liquid is in the receiver, how much subcooled liquid is in the liquid line, and how quickly the condensing pressure is falling from the cycling fans, the receiver and liquid line may not be able at all times and conditions to deliver vapor-free liquid to the TXV. There may be premature liquid flashing (liquid-line flash gas) in the liquid line all the way to the TXV. This will greatly reduce the capacity of the TXV, starve evaporators, and cause erratic TXV operation with very poor evaporator superheat control.

Figure 22.38 A multiple-fan condenser. Source: Heatcraft, Inc., Refrigeration Products Division

VARIABLE-FREQUENCY DRIVE (VFD) CONDENSER FAN MOTORS

As mentioned, one problem with cycling condenser fan motors is their tendency to cause the head pressure to swing up and down as the fan motor is started and stopped. This can cause erratic and inconsistent operation of the TXV. However, the variable-speed condenser fan motors used on larger condensing units can vary their speed gradually according to head pressure or outside ambient temperature changes. Many of these fan motors are equipped with variable-frequency drives (VFDs) controlled by electronic inverters. VFDs operate on the principle that when the frequency, or hertz (Hz), of the voltage coming into the fan motor changes, the speed of the fan motor will also change. As the voltage frequency increases, the motor speed increases; as the voltage frequency decreases, the motor speed decreases. The following formula governs the speed (rpm) of a motor. Notice that the numerator is the frequency or hertz of the incoming voltage in cycles/sec,

multiplied by a factor for converting from minutes to seconds. The denominator is the number of motor pole pairs.

$$\text{rpm} = \frac{(\text{Frequency})(60\ \text{sec/min})}{(\text{Number of Motor Poles} \div 2)}$$

This simplifies to

$$\text{rpm} = \frac{(\text{Frequency})(120)}{(\text{Number of Motor Poles})}$$

Since we cannot change the number of motor poles, the frequency is the easiest variable to alter for a change in the rpm of the motor.

Through special electronic circuitry called an inverter, the voltage frequency being fed to a condenser fan motor can be altered with changing head pressure by linking the refrigeration system's high-side pressure transducer to the inverter. Another control scheme is to measure changing outdoor ambient temperatures with a temperature sensor near the condenser's air inlet. These signals can then be linked to an electronic inverter. **Figure 22.39** shows a VFD for changing the frequency of the voltage signal to a motor. The frequency change can be done in small increments, thus changing the speed of the condenser fan motors in small increments. This type of control will keep a more consistent head pressure as the outside ambient temperature changes. It will also prevent the head pressure from swinging up and down erratically, which occurs when the fan motors are cycled on and off. VFDs are covered in more detail in Unit 17, "Types of Electric Motors."

AIR SHUTTERS OR DAMPERS

Shutters may be located either at the inlet to the condenser or at the outlet. The air shutter has a pressure-operated piston that pushes a shaft to open the shutters when the head pressure rises to a predetermined pressure, **Figure 22.40**. This pressure-operated piston extends to open the shutters when the pressure rises, **Figure 22.41**. Shutters open and close slowly and provide even head pressure control. With a single fan, the shutter is installed over the inlet or over the outlet to the fan. When there are multiple fans, the shutter-covered fan is operated all the time and the other fans can be cycled off by pressure or temperature. This arrangement can give good head pressure control down to low temperatures.

Figure 22.39 A variable-frequency drive (VFD) used in changing the voltage frequency to a motor. This will change the motor's speed in very small increments. Courtesy Ferris State University. Photo by John Tomczyk

Figure 22.40 A condenser with an air shutter. With one fan it is the only control. With multiple fans the other fans can be cycled by temperature with the shutter controlling the final fan. Source: Trane Company

Figure 22.41 A piston-type shutter operator. The high-pressure discharge gas is on one side of the bellows, and the atmosphere is on the other side. When the head pressure rises to a predetermined point, the shutters begin to open. In the summer, the shutters will remain wide open during the running cycle. Courtesy Robertshaw Controls Company

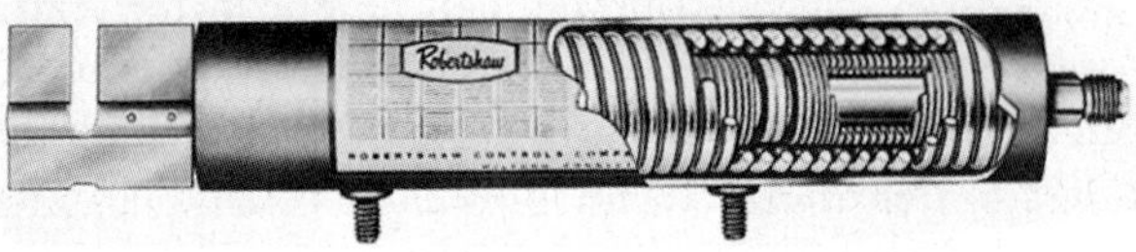

CONDENSER FLOODING

Flooding the condenser with liquid refrigerant causes the head pressure to rise just as though the condenser was covered with a piece of cardboard or plastic tarp. Flooding is accomplished by having enough refrigerant in the system to flood the condenser in both mild and cold weather to keep the head pressure constant. This will also assure a consistent liquid pressure for feeding the metering device in low ambient conditions. Flooding requires a large refrigerant charge and an oversized liquid receiver to store the extra charge in warmer weather when the condenser doesn't need to be flooded. In addition to the charge, there must be a valve arrangement that allows the refrigerant liquid to fill the condenser during

both mild and cold weather. The flooding method is designed to maintain the correct head pressure in the coldest weather during start-up and while operating.

One big problem designers face when designing air-cooled refrigeration and air-conditioning equipment is the changing temperatures of the changing seasons. A properly designed condenser will operate inefficiently during extremely high ambient conditions. The inefficiencies come from the higher head pressures and thus higher compression ratios associated with high ambient conditions. However, it is the low ambient conditions that cause more serious problems. Moreover, it is a fact that most units will be required to operate at temperatures below their design dry-bulb temperature during the fall, winter, and spring seasons. It is because of these changing ambient conditions that head pressure control valves were designed. Without these valves, head pressures would fluctuate as widely as ambients do during the changing seasons. Both running-cycle and off-cycle problems can occur if head pressure controls are not used.

Low head pressures from low ambients can cause insufficient refrigerant flow rates through metering devices, which in turn can starve evaporators. Low suction pressures, iced coils, short cycling, and inefficient cooling can also result from low head pressures. Also, any system using hot-gas defrost or hot-gas bypass for compressor capacity control must have a minimum head pressure for proper defrosting and capacity control. With outdoor forced-air condensers, low ambient conditions also cause refrigerant to migrate to the condenser, compressor, and suction line during the off cycle.

Hard starting may also occur in a low ambient condition, causing low head pressures. Often, refrigeration systems are not able to start because most of the refrigerant has migrated to the colder condenser. The evaporator has a hard time building vapor pressure and may never get to the cut-in pressure of its low-pressure control. Even if the compressor does eventually start by reaching the cut-in pressure, it may soon short cycle. This is caused by low refrigerant flow through the metering device at the lower head pressure in the low ambient condition.

The solution to the low head pressure problem is to install a pressure-actuated "holdback" valve at the outlet of the condenser, **Figures 22.42(A)** and **(B)**. The valve throttles shut when the condenser pressure reaches a preset minimum for low ambient conditions. This allows liquid refrigerant in the condenser to be held back and actually flood portions of the condenser. The partial flooding will cause a portion of the condenser to become inactive and cause it to have a smaller internal volume. At this point, desuperheating and condensation must take place. Condensing pressures will rise, thus giving sufficient liquid-line pressures and pressure differences across the thermostatic expansion valve (TXV) for normal system operation in the colder ambient.

The valve shown in **Figure 22.42(A)**, referred to as an ORI (open on rise of inlet pressure) valve, is an inlet-pressure regulating valve that responds to changes in inlet pressure (condensing pressure) only. A decrease in condensing pressure causes less pressure to act on the bottom of the seat disk. This action throttles the valve more in the closed position and starts to back up liquid refrigerant in the bottom of the condenser. Soon the head pressure will start to rise as a result of the condenser's reduced internal volume. Any increase in inlet (condensing pressure) above the valve setting will tend to open the valve.

The condensing pressure is opposed by the force of an adjustable spring that is on top of the valve seat disc. The setting can be changed simply by either increasing or decreasing the tension of the spring with a screwdriver or an Allen wrench. Increasing the spring pressure by turning the fitting clockwise will increase the minimum opening pressure of the valve. The outlet pressure of the valve is then cancelled out and has no bearing on valve movement because the outlet pressure is exerted on top of the bellows and on top of the seat disc simultaneously. Since the effective area of the bellows is equal to the area of the top of the seat disc, the pressures cancel one another and do not affect the valve movement, **Figure 22.43**. Only changes in condensing pressure can throttle the valve either open or closed.

An ORI valve is usually used in conjunction with an ORD (open on rise of differential pressure) valve, **Figures 22.44(A)** and **(B)**. The ORD valve is located between the discharge line and the receiver inlet, **Figure 22.42(B)**. It responds to changes in pressure differences across the valve. The ORD valve is thus dependent on the ORI valve for its operation. The ORD valve will bypass hot compressor discharge gas from the compressor to the receiver inlet when it senses a preset determined pressure difference across the valve. As the ORI valve senses a drop in condenser pressure and starts to throttle shut, a pressure difference is created across the ORD valve. The pressure difference is created by the reduced flow to the receiver due to the throttling action of the ORI valve. One outlet of the ORD valve senses receiver pressure, and the inlet side senses discharge pressure from the compressor. When the ORI valve starts to throttle shut, the receiver is still supplying refrigerant to the TXV and receiver pressure will eventually drop. If the receiver pressure drops too low, its ability to keep feeding the liquid line and the TXV will diminish. Something has to keep the receiver pressure up while the ORI valve is throttling liquid from the condenser. This is when the function of the ORD valve comes into play.

When the ORD valve senses a factory preset pressure difference of 20 psi between the receiver and compressor discharge, it will start to open and bypass hot compressor discharge gas to the receiver's inlet. If the pressure difference across the ORD valve ever reaches 30 psi, the valve will be fully open. The hot gas from the compressor, which flows through the ORD valve, serves to heat up any cold liquid refrigerant being throttled through the ORI valve at the receiver's inlet. This hot gas entering the receiver will also

Figure 22.42 (A) Head pressure control valve designed for low ambient conditions. Courtesy Parker Sporlan. (B) Open on rise of inlet pressure (ORI) head pressure control valve shown at the outlet of the condenser. Courtesy Parker Sporlan.

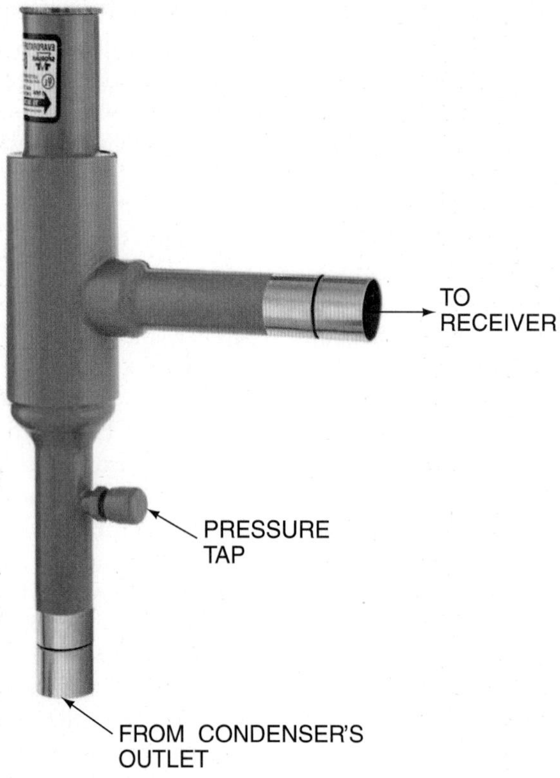

(A)

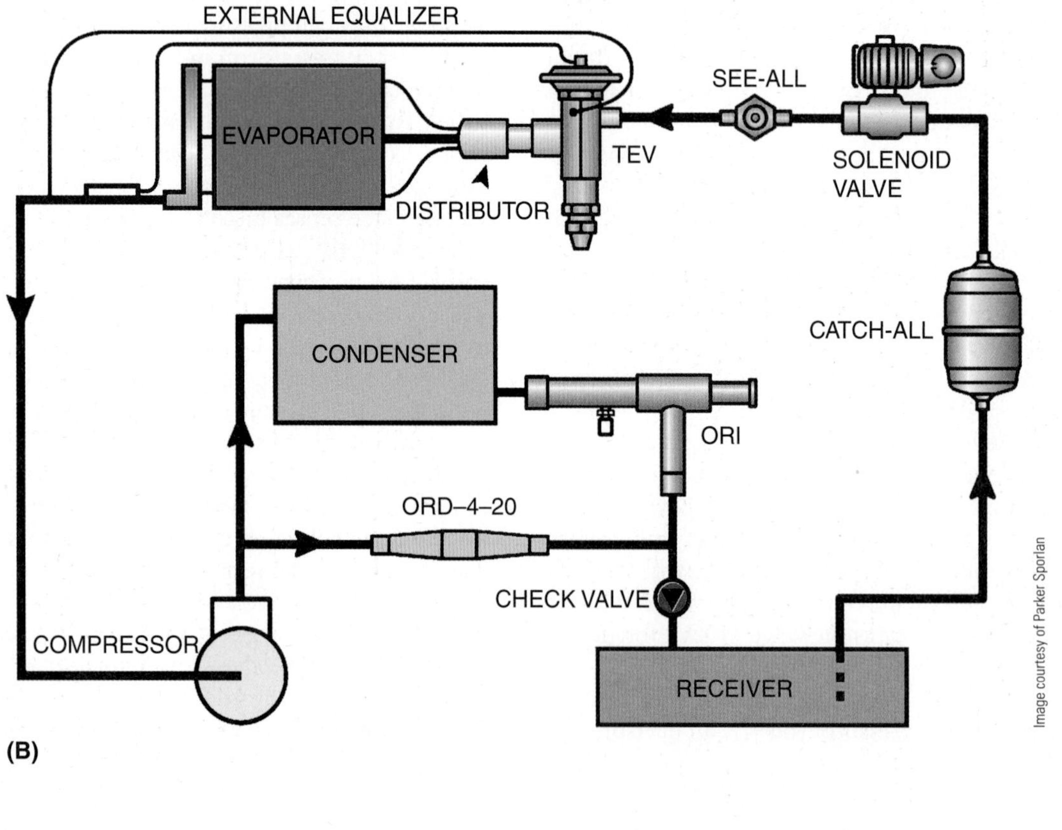

(B)

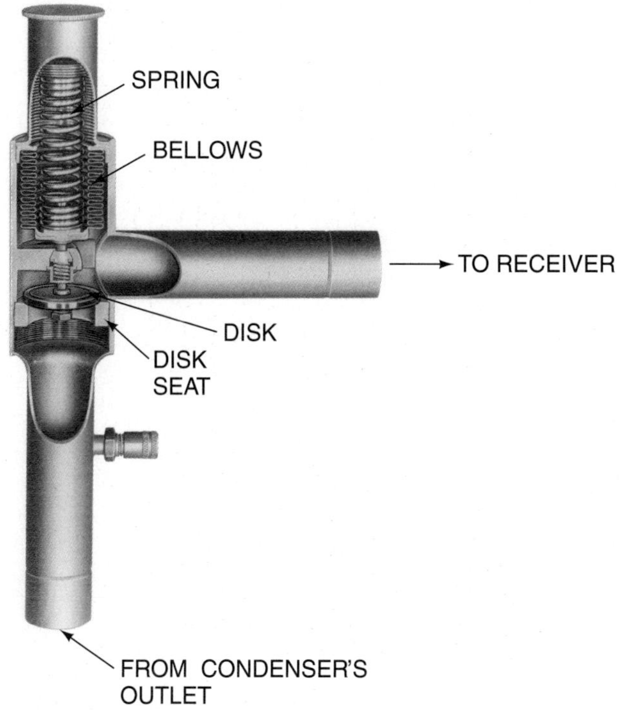

Figure 22.43 Cutaway view of an ORI head pressure control valve. Courtesy Sporlan Division, Parker Hannifin Corp.

increase the pressure of the receiver and allow it to deliver liquid to the liquid line and TXV when the ORI valve is throttling shut on the outlet of the condenser.

Both the ORI and ORD valves will automatically work in conjunction with one another to maintain proper receiver pressure, regardless of the outside ambient conditions. The ORD valve also acts as a check valve to prevent reverse flow from the receiver to the compressor's discharge line during the off cycle. A combination ORI/ORD nonadjustable valve arrangement can also be used—which simplifies the piping arrangement, **Figures 22.45(A)** and **(B)**. These valves also limit the flow of refrigerant from the condenser to the receiver and at the same time regulate the flow of the compressor's hot gas to the receiver. The combination valve has a round dome at the top that is pressure charged. The pressurized dome will be explained in the paragraphs to come.

Figures 22.46(A) and **(B)** show another style of head pressure control valve for low ambient conditions that is often referred to as a low-ambient control (LAC) valve. It has a round, pressurized dome at the top. The charge in the dome is independent of the refrigerant charge in the actual refrigeration system. The valve is located in the condensing unit that is in the outside ambient temperature and so the charge expands and contracts in volume and acts on an internal diaphragm along with changes in the outside ambient. The expansion and contraction of the pressure charge moves the diaphragm, which in turn moves a piston in the valve and modulates the valve to either a more open or closed position. When the temperature of the condenser is above 70°F, the refrigerant flow from the compressor is directed by the mixing valve through the condenser and into the receiver. If the outside ambient temperature drops below 70°F, the condensing pressure will fall and the pressure of the liquid coming from the condenser will also fall to a point below that of the bellows in the dome of the valve. This causes a piston to move in the valve and partially restrict the flow of refrigerant leaving the condenser. The condenser will now partially flood with refrigerant to maintain a certain condensing pressure. At the same time, discharge gas will bypass the condenser and flow directly to

Figure 22.44 (A) Open on rise of differential pressure (ORD) head pressure control valve. Courtesy Sporlan Division, Parker Hannifin Corp. (B) Cutaway view of an ORD head pressure control valve. Courtesy Sporlan Division, Parker Hannifin Corp.

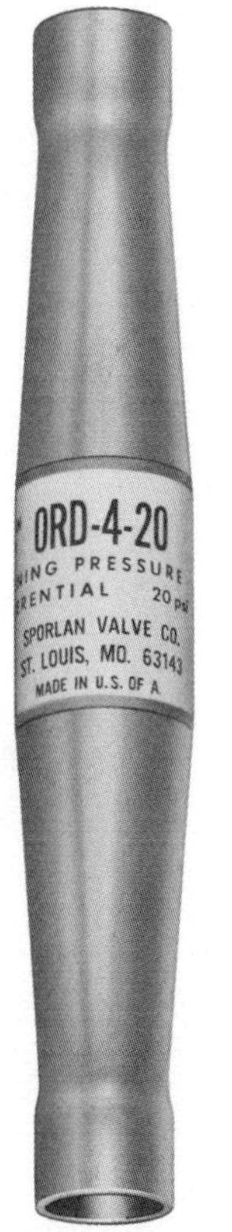

(A)

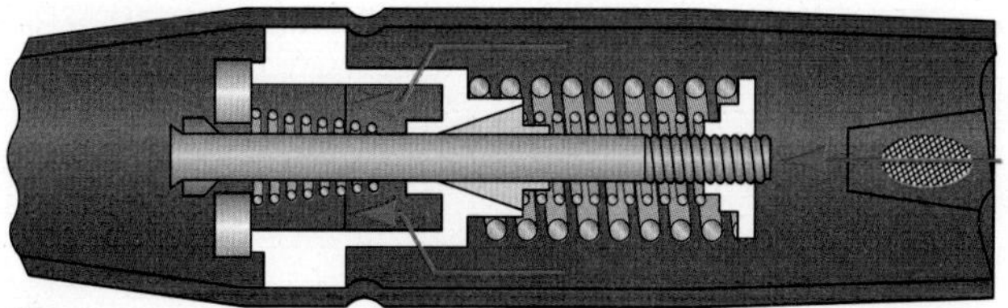

(B)

Figure 22.45 (A) Combination ORI/ORD head pressure control valve. Courtesy Sporlan Division, Parker Hannifin Corp. (B) Internal view of a combination ORI/ORD head pressure control valve. Courtesy Sporlan Division, Parker Hannifin Corp.

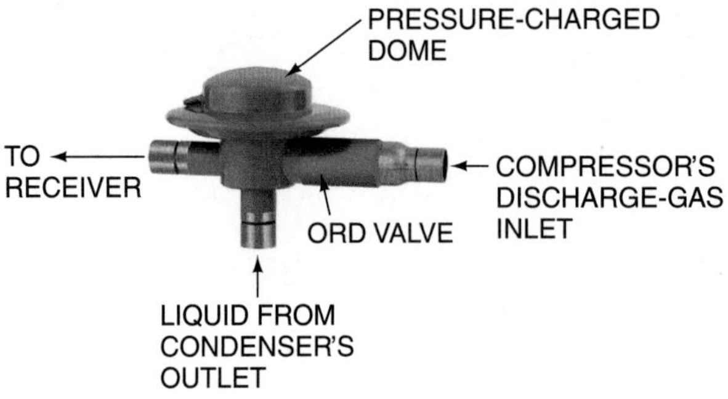

(A)

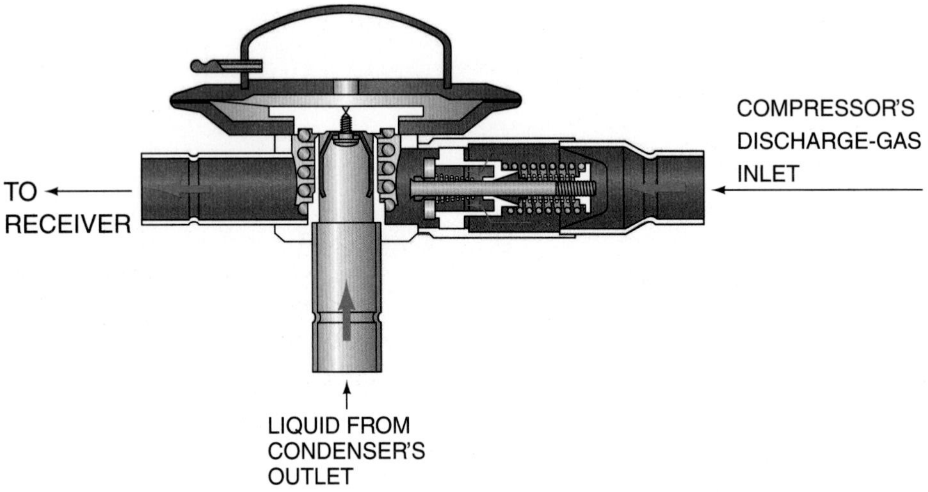

(B)

Figure 22.46 (A) Low-ambient control (LAC) valve with a pressurized dome. Courtesy Sporlan Division, Parker Hannifin Corp. (B) Internal view of a low-ambient control (LAC) valve. Courtesy Sporlan Division, Parker Hannifin Corp.

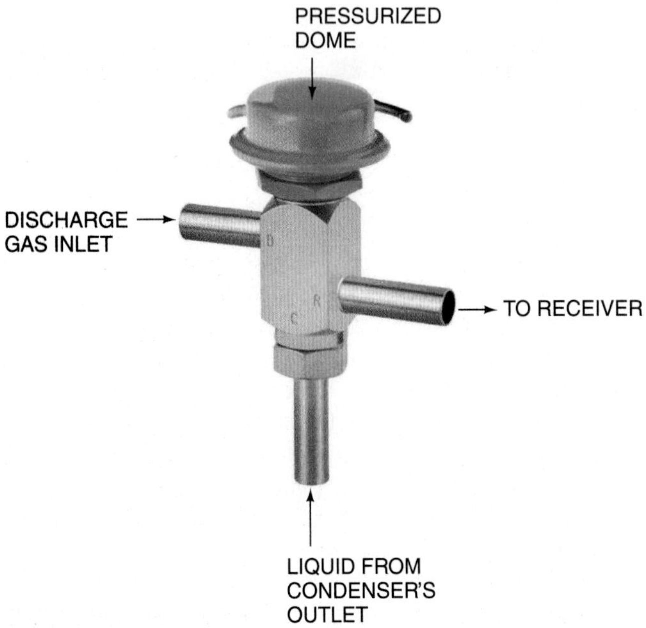

(A)

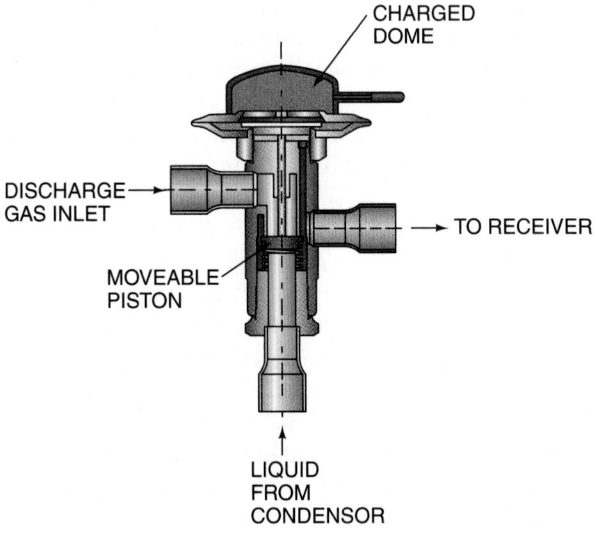

(B)

Figure 22.47 System diagram showing location of the low-ambient control (LAC) head pressure control valve. Courtesy Sporlan Division, Parker Hannifin Corp.

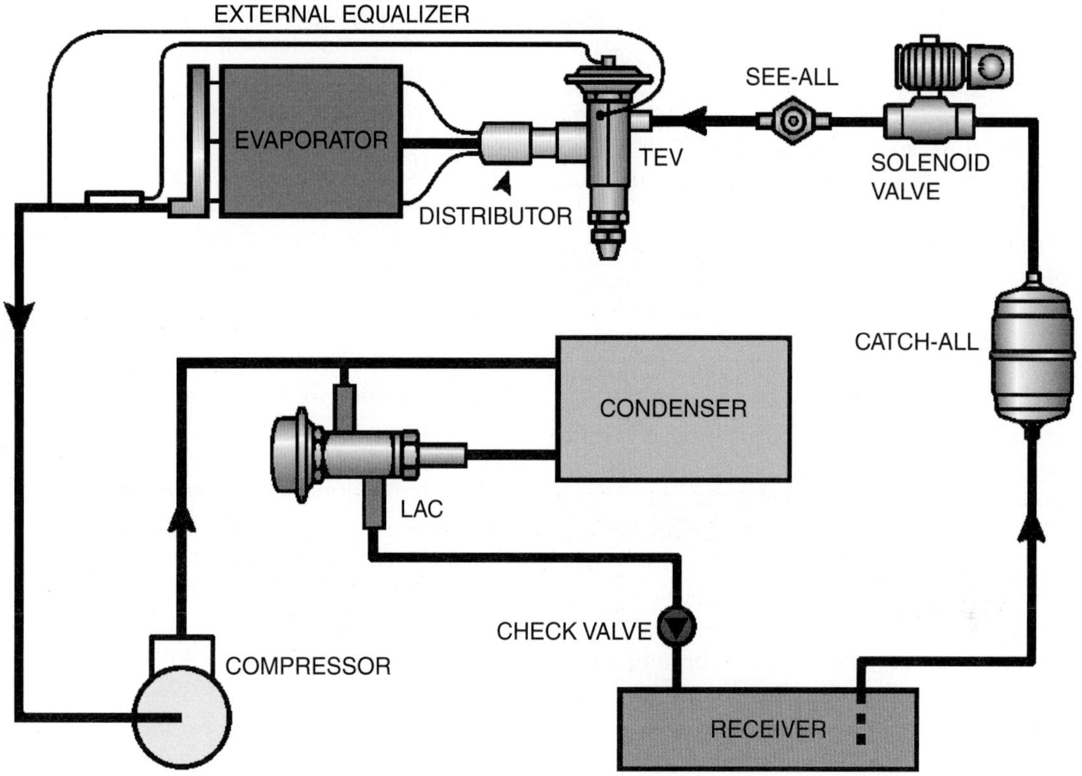

the receiver, **Figure 22.47**. The hot, superheated discharge gas going to the receiver now mixes with liquid refrigerant coming from the condenser. This keeps the receiver pressure up and helps keep metering devices fed with refrigerant.

In low ambient conditions, these mixing valves will maintain a head pressure of approximately 190 psi for R-502 systems, a head pressure of either 240 psi or 190 psi for R-404A systems (depending on the model of the valve), and a head pressure of approximately 180 psi for R-22 systems. The dome charges can also be custom-ordered from the manufacturer to meet specific system requirements, which allows LAC valves to be used on *floating head pressure* systems. Floating head pressure is covered later in this section.

Receivers on systems having head pressure control valves must be large enough to hold the normal operating charge plus the additional charge that is necessary to totally flood the condenser for wintertime operation. In fact, the receivers should be sized so that they are about 80% full when they contain the entire system charge. This allows for a 20% vapor head for safety when pumping the system down. Any refrigeration receiver should be able to hold all of the system's refrigerant charge and still have a 20% vapor head for safety. **Figure 22.48** shows a refrigeration system with a head pressure control mixing valve and an oversized receiver. Most manufacturers publish recommendations in the form of charts and tables of system charges that show the service technician how much refrigerant to add to systems having head pressure control valves designed to flood the condenser

Figure 22.48 Low-ambient control (LAC) head pressure control valve shown piped to condenser and receiver. Courtesy Ferris State University. Photo by John Tomczyk

in colder weather. Nowadays, these tables and charts can be accessed on the Internet with a personal computer or mobile device. If the receiver of a refrigeration or air-conditioning system hasn't been oversized to accommodate the extra refrigerant needed for flooding the condenser during low ambient conditions, the technician will have to remove refrigerant every spring to prevent high head pressures at a design ambient—only to add it back again in the fall when it will be required for flooding the condenser for head pressure control, a technique often referred to as a winter/summer charge procedure. However, as one can see, oversizing the receiver can save the technician the added labor.

As mentioned earlier, one of the advantages of condenser flooding is to keep consistent liquid pressure for feeding the metering device in low ambient conditions. Manufacturers do supply technical information on how much extra refrigerant is needed for flooding a condenser for a certain low ambient condition. However, in extreme low ambient conditions, it may be necessary to flood 80% to 90% of the condenser. On larger systems, this could require several hundred pounds of refrigerant. This is the main disadvantage of flooding a condenser. With the rising price of refrigerant and the environmental concerns of global warming and ozone depletion, condenser flooding can become quite expensive and environmentally unsound if not managed and serviced properly.

CONDENSER SPLITTING

One way to reduce the amount of extra refrigerant charge needed for condenser flooding is to split the condenser into two separate and identical condenser circuits, **Figure 22.49**. This method is referred to as **condenser splitting**. The splitting is done by installing a pilot-operated, three-way solenoid valve in the discharge line from the compressors, **Figure 22.50**. The splitting is done in such a way that only one-half of the condenser is used for winter operation and both halves are used for summer operation. The top half of the condenser is referred to as the "Summer/Winter" condenser, and the bottom half of the condenser is referred to as the "Summer" condenser, **Figure 22.49**. The three-way solenoid valve controlling the splitting of the condensers can be energized and deenergized by a controller sensing outside ambient, an outdoor thermostat, or a high-side pressure control. *During summertime operations, the added surface area and volume of both condensers are needed to maintain a reasonable head pressure at higher ambient conditions.* When the pilot-operated, three-way solenoid valve is deenergized, the main piston inside the valve is in position to let refrigerant flow from the compressor's discharge line to the three-way valve's inlet port and then equally to the valve's two outer ports. In other words, the refrigerant will flow to both of the condenser halves equally.

Figure 22.49 Parallel compressor supermarket refrigeration system showing split condensers. Condenser halves are labeled "Winter/Summer" and "Summer." Courtesy Sporlan Division, Parker Hannifin Corp.

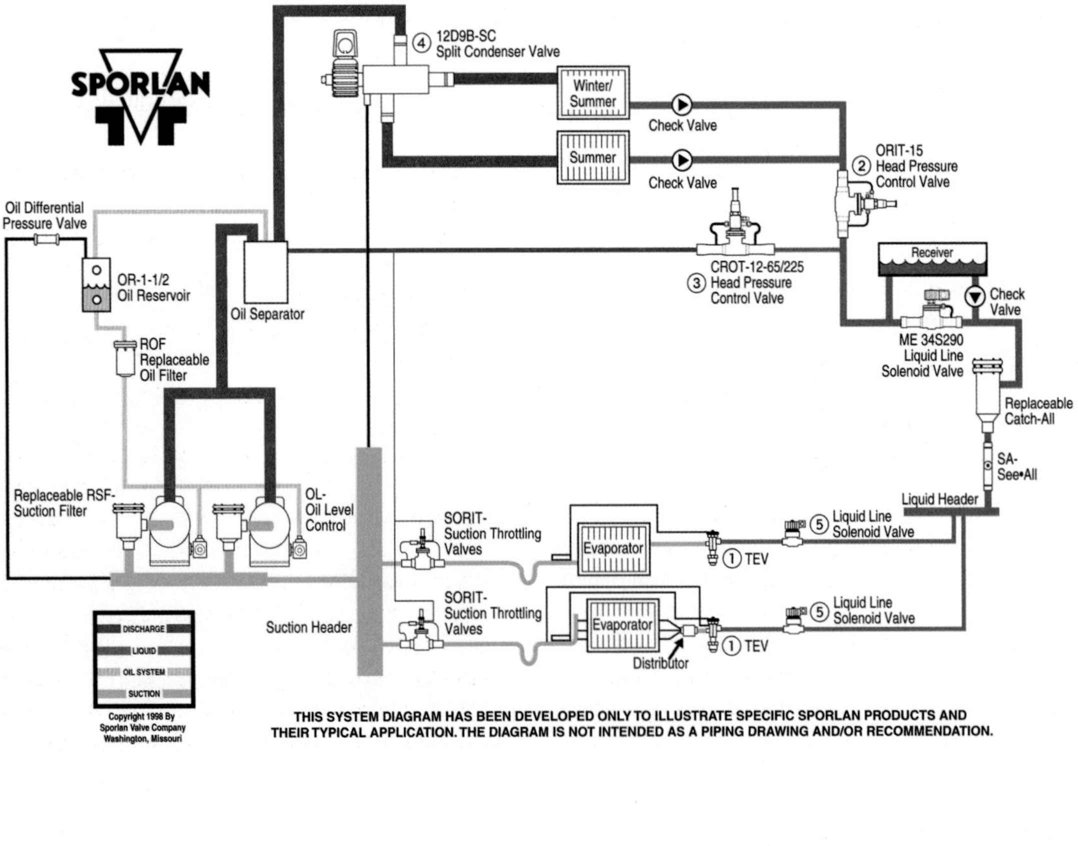

Figure 22.50 Pilot-operated, three-way solenoid condenser splitting valve. Courtesy Sporlan Division, Parker Hannifin Corp.

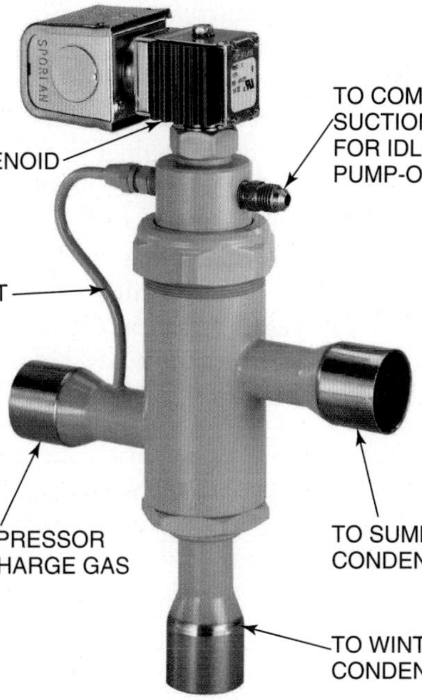

In low ambient conditions, the summer portion of the condenser can be taken out of the active refrigeration system by the three-way valve. When the coil of the pilot-operated, three-way solenoid valve in **Figure 22.50** is energized, the sliding piston inside the valve will move and close off the flow of refrigerant to the port on the bottom of the valve that feeds the summer condenser. This action will render the summer condenser inactive or idle, and the minimum head pressure can be maintained by flooding the Summer/Winter half of the condenser with conventional refrigerant-side head pressure control valves, as explained in the condenser flooding section of this unit.

In fact, during winter operations, the system's head pressure is best maintained with a combination of condenser splitting, refrigerant-side head pressure controls, and air-side controls like fan cycling or fan variable-speed devices. This combination of refrigerant-side and air-side controls will minimize the refrigerant charge even more while splitting the condenser. These combinations will also maintain the correct head pressure for better system efficiencies.

The refrigerant that is trapped in the idle summer condenser during low ambient conditions will flow back into the active system through a bleed hole in the piston of the three-way valve. The trapped refrigerant will flow through the piston's bleed hole, into the valve's pilot assembly, and back to the suction header through a small copper line that feeds all parallel compressors, **Figure 22.49**. Another scheme to rid the idle summer condenser of its refrigerant is to install a dedicated pump-out solenoid valve that will open when energized and vent the trapped refrigerant to the common suction header through a capillary tube restriction. Both the bleed hole in the piston and the capillary tube ensure that the refrigerant experiences a restriction and is mostly vaporized before reaching the common suction header, which is under low-side (common-suction) pressure. A check valve is located at the summer condenser's outlet to prevent any refrigerant from entering it while it is idle and under a low pressure condition. While not needed for back-flow prevention, a check valve is also located at the outlet of the Summer/Winter condenser simply to make the pressure drops equal in both halves of the condenser when both are being used simultaneously in summertime operations.

Troubleshooting a low-ambient head pressure control valve with a pressurized dome can be easy if the proper steps are taken. If the outdoor ambient is below 70°F and the head pressure is low, feel the line between the valve and the receiver, **Figure 22.47**. If this line is cold, the valve is not allowing the compressor's discharge gas into the receiver. The valve is defective and should be replaced. This line should be warm at an ambient below 70°F. If the head pressure is low and the ambient is above 70°F—and the line between the valve and the receiver is hot—the valve is also defective and should be replaced. The line should be a little warmer than the ambient because of the subcooled liquid coming from the condenser's outlet to the receiver at the warmer outside ambient. Discharge gas flow to the receiver should be shut off by the valve. However, the system could also be low on charge—causing a low condensing pressure that fools the valve. So, the defective valve should only be replaced after verifying that the system charge is correct.

To replace a defective low-ambient control valve with a pressurized dome charge, follow these steps:

1. Recover refrigerant.
2. Cut the process tube on the valve's charged dome to remove the pressure charge.
3. With a torch, heat the valve until the solder liquifies, then remove the valve.
4. Wrap the valve's body with a heat sink to prevent damage and silver-solder a new valve in place.
5. Install a new filter drier.
6. Leak check the system and evacuate to a deep vacuum (500 microns).
7. Charge the system and leak check again.

Systems that use condenser flooding for head pressure control have large refrigerant charges, as mentioned earlier. The total charge for the system can be determined by following the manufacturer's calculations. When a service technician is working on a unit that has a partial charge, it is very hard to charge the system without some guidelines. When the system is small—for example, under 10 tons—the technician would be advised to recover the charge that is in the system and measure in the charge that is recommended by the manufacturer. Then the system should perform in all temperature conditions.

Large systems will normally have two sight glasses on the liquid receiver, one at the bottom and one at the 80%-full level. The proper procedure for charging one of these systems is to turn off the system until the load builds up and the compressor or compressors can be run at full load. Then block the air over the condenser until the head pressure is that of a 95°F day. This will assure you that there is no excess liquid refrigerant in the condenser. Then charge the system until there is a liquid level in the sight glass at the 80% level of the liquid receiver. If the weather is warm outside, you may have to uncover the condenser if the head pressure begins to rise too high. Now you know that there is enough refrigerant to flood the condenser in cold weather and that there is not an excess of refrigerant in the condenser. The cover can then be removed from the condenser, and the system should get into balance on its own.

22.19 USING THE CONDENSER SUPERHEAT

Air-cooled condensers have high discharge line temperatures even in winter. This characteristic can be an advantage because the heat can be captured and redistributed as heat for the structure in winter or to heat water. The refrigeration system is rejecting heat from the refrigerated box, heat that leaked into the box from the structure. This heat has to be rejected to a place that is unobjectionable. In summer, the heat should be rejected to the outside of the structure or possibly into a domestic hot-water system, **Figure 22.51**. Water temperatures of 140°F are obtainable for use and can lead to big energy savings.

Figure 22.51 A heat exchanger to capture the heat from the highly superheated discharge line and use it to heat domestic water. This line can easily be 210°F and can furnish 140°F water in a supply limited by the size of the refrigeration system. The larger the system, the more heat is available. Source: Noranda Metal Industries, Inc.

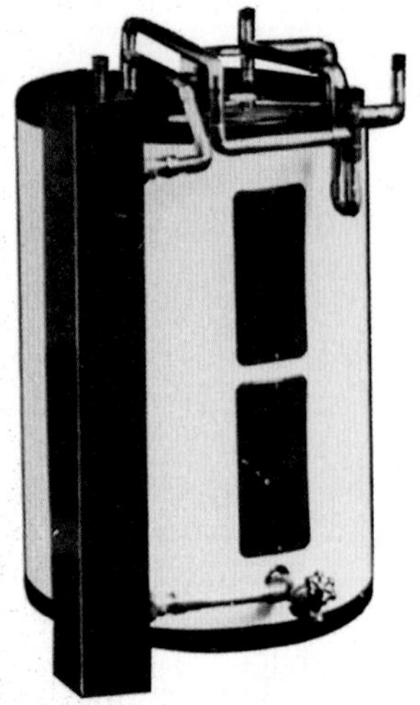

22.20 HEAT RECLAIM

In winter, structures need heat. If heat could be rejected to the inside, heating costs could be reduced. Any heat recovered from the system is heat that does not have to be purchased, **Figure 22.52.** Heat recovery can be accomplished easily with air-cooled equipment. The discharge gas can be passed to a rooftop condenser or to a coil mounted in the ductwork that supplies heat to a structure. The condensing temperature of air-cooled equipment is high enough to be used as heat, and the quantity of heat is sizable enough to be important. Some businesses in moderate climates will be able to extract enough heat from the refrigeration system to supply the full amount of heat required. Heat reclaim is covered in detail in Unit 26, "Applications of Refrigeration Systems."

22.21 FLOATING HEAD PRESSURES

Historically, high condensing pressures and temperatures in a refrigeration system were artificially maintained with head pressure controls so that the system would function properly at low ambient temperatures. These higher pressures were considered mandatory in order for the metering device to feed the evaporator properly. However, as condensing pressures increase, the refrigeration system draws more electrical power and becomes more inefficient. At today's power costs, these inefficiencies are becoming more and more unacceptable.

In the past, designers of air-conditioning and refrigeration systems assigned an outdoor design condition to a system. The outdoor design condition typically was a temperature that would not be reached more than 2% of the time during the life of the equipment. The condenser was then selected based on this seldom-reached condition. Today, designers of refrigeration and air-conditioning systems often try to attain the lowest possible condensing pressures and temperatures and design condensers accordingly. An industry term for attaining the lowest possible condensing pressure is **floating head pressure**. Floating head pressure systems simply let the condensing pressures follow the ambient temperature as the seasons swing from summer to winter and back again. This means that in the fall, winter, and spring, the condensing pressures will be at their lowest. It is not uncommon for condensing temperatures to be 30°F to 40°F in the winter months.

In fact, a majority of the outdoor temperatures in the United States are below 70°F more than they are above. The compressor capacity increases about 6% for every 10°F drop in condensing temperatures. Working with metering device suppliers, researchers discovered that **thermostatic expansion valves (TXVs)** would work with much less pressure drop across them than expected in the past, as long as pure liquid was supplied to them. Some newer metering

Figure 22.52 A system that can supply heat to a structure.

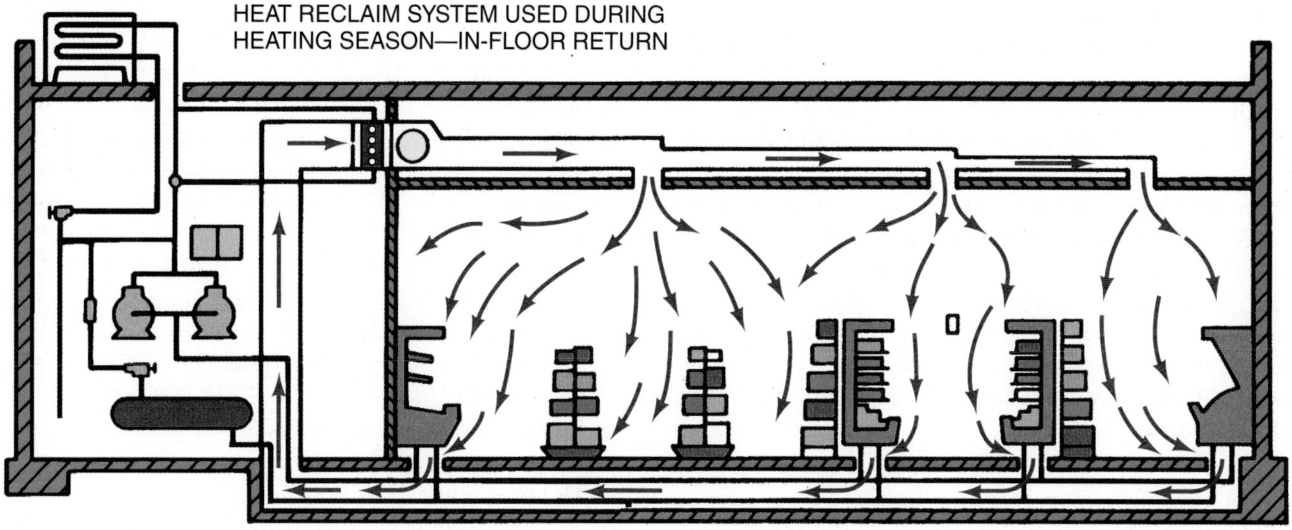

devices can operate with as little as a 30-psi pressure drop. *With this new knowledge, designers are allowing condensing pressures to float downward with the ambient temperature in the spring, fall, and winter months. This gives the system higher efficiencies with less power consumption.* Low ambient controls can still be used on these systems when extreme cold temperatures are experienced. Often, their settings will have to be modified for the specific application and extreme cold ambient conditions.

When head pressures are floated in the fall, winter, and spring, condensing pressures will be lowered. This causes less heat to be available for the heat reclaim coils to heat buildings. However, it is much more efficient to heat a building with a fossil-fuel furnace than by artificially elevating head pressures with head pressure controls on a refrigeration or air-conditioning system and causing inefficiencies. Because of this, fewer heat reclaim systems are being installed.

REVIEW QUESTIONS

1. Why are some water-cooled condensers cleaned with brushes and others with chemicals?
   A. Some water-cooled condensers will accumulate scale while others do not.
   B. Some condenser tubes are not straight and will not allow brushes to pass through them.
   C. All water-cooled condensers are hermetically sealed and cannot be cleaned with brushes.
   D. Chemicals are only used on outdoor condensers, indoor condensers are cleaned with brushes.
2. The three materials of which air-cooled condensers are normally made are _________.
   A. brass, aluminum, and cast iron
   B. B. brass, copper, and steel
   C. copper, aluminum, and steel
   D. copper, cast iron, and brass
3. One of the differences between air-cooled and water-cooled condensers is the _________.
   A. type of refrigerant used
   B. refrigerant condensing temperature
   C. type of compressor used
   D. compressor discharge line temperature
4. When is the most heat removed from the refrigerant in the condensing process?
   A. During the change in state from a vapor to a liquid
   B. When the refrigerant vapor enters the receiver
   C. When the liquid refrigerant leaves the condenser
   D. During the change in state from a liquid to a vapor
5. After heat is absorbed into the water in a water-cooled condenser, the water can be _________.
   A. filtered and used for drinking water
   B. treated with chemicals and piped to a chiller
   C. filtered and then treated with chemicals before being reused domestically
   D. wasted down the drain or sent to a cooling tower

6. A water-cooling tower capacity is governed by what aspect of the ambient air?
   A. Dry-bulb temperature
   B. Wet-bulb temperature
   C. Enthalpy of the air
   D. Altitude
7. When a standard-efficiency air-cooled condenser is used, the condensing refrigerant will normally be _______ above the ambient air temperature.
   A. 15°F C. 45°F
   B. 30°F D. 60°F
8. Fan cycling, air shutters or dampers, condenser flooding, and variable-frequency drive (VFD) condenser fan motors are all methods for controlling __________.
   A. suction pressure
   B. evaporator capacity
   C. electrical efficiency
   D. head pressure
9. An outdoor air-cooled condenser that has a vertical coil can be affected by the ___________.
   A. prevailing winds
   B. municipal water temperature
   C. indoor ambient temperature
   D. refrigerant oil type
10. Air-cooled condensers are ____________.
    A. much more efficient than water-cooled condensers
    B. less efficient than water-cooled condensers
    C. being phased out due to their inefficiencies
    D. the most efficient condensers on the market
11. The type of condenser that has straight copper tubes that are fastened into an end sheet within a steel shell that is typically used in large installations is the ________ condenser.
    A. air-cooled
    B. shell-and-coil
    C. shell-and-tube
    D. tube-within-a-tube
12. On a system that uses a floating head pressure, ______________.
    A. the condenser must be located outdoors
    B. the amount of heat discharged from the condenser is extremely high
    C. less system heat is available for heat reclaim coils
    D. more system heat is available for heat reclaim coils
13. When the hot discharge gas enters the compressor, it is first _________.
    A. vaporized C. subcooled
    B. condensed D. desuperheated
14. A 2-ton wastewater system that operates approximately 16 hours per day would use about _______ gallons of water per day.
    A. 2880 C. 7880
    B. 5760 D. 11,520
15. A 10-ton water-cooled system that recirculates water through a cooling tower averages 16 hours of run time per day. The amount of water that would be circulated in a day would be about ________ gallons.
    A. 14,400 C. 36,400
    B. 28,800 D. 57,600
16. In a water-cooled system that recirculates water through a cooling tower, the refrigerant in the condenser typically condenses at ______ above the leaving water temperature.
    A. 10°F C. 25°F
    B. 20°F D. 30°F
17. A system that uses heat from the compressor discharge line to provide heated air or water to a conditioned space is called a(n) _________ system.
    A. heat reclaim
    B. energy recovery ventilator
    C. water recirculation
    D. latent boost
18. A typical water-cooling tower can cool the water to within ______ of the ambient air wet-bulb temperature.
    A. 3°F to 4°F C. 7°F to 8°F
    B. 5°F to 6°F D. 9°F to 10°F
19. Due to the seasonal temperature variations of municipal water, a wastewater system ___________.
    A. has a defined refrigerant-to-water temperature relationship
    B. has a constant water flow rate to maintain an adequate head pressure
    C. should have an auxiliary air-cooled condenser to stabilize system pressures
    D. has no definite refrigerant-to-water temperature relationship
20. The component that varies the flow of water through a water-cooled condenser to maintain a stable head pressure is called a(n) _____________.
    A. ORD head pressure control
    B. water-regulating valve
    C. ORI head pressure control valve
    D. low-ambient control valve
21. Briefly describe what is meant by floating head pressure and tell why it is used in refrigeration and air-conditioning systems.
22. Explain how to flush impurities from a water-regulating valve that is on a water-cooled condenser.
23. Explain the basic operation of a combination ORI and ORD low-ambient head pressure control valve.
24. Name two problems a refrigeration system may have if the condensing unit is exposed to a low ambient condition when not equipped with a low-ambient head pressure control valve.
25. Briefly explain what is meant by condenser splitting and list its main advantage.

UNIT 23

COMPRESSORS

OBJECTIVES

After studying this unit, you should be able to

- explain the function of the compressor in a refrigeration system.
- discuss compression ratio.
- describe five different methods of compression.
- explain the difference between a hermetic compressor and a semihermetic compressor.
- discuss the differences between a belt-drive and a direct-drive compressor.
- describe the various working parts of reciprocating and rotary compressors.
- explain how compressor valves and clearance volume affect the efficiency of a reciprocating compressor.
- explain the characteristics and operation of a scroll compressor.
- discuss the conditions that can lead to liquid slugging a compressor.
- describe system maintenance steps that can improve a compressor's operating efficiency.

SAFETY CHECKLIST

- ☑ Wear goggles and gloves when attaching or removing gauges to transfer refrigerant or to check pressures.
- ☑ Wear warm clothing when working in a walk-in cooler or freezer.
- ☑ Do not touch the compressor discharge line with your bare hands.
- ☑ Be careful not to get your hands or clothing caught in moving parts such as pulleys, belts, or fan blades.
- ☑ Wear an approved back brace belt when lifting and use your legs, keeping your back straight and upright.
- ☑ Observe all electrical safety precautions. Be careful at all times and use common sense.

23.1 THE FUNCTION OF THE COMPRESSOR

The compressor is considered the heart of the refrigeration system. The term that best describes a compressor is vapor pump, because it actually increases the suction pressure level to the discharge pressure level. For example, in a low-temperature R-404A refrigerant system, the suction pressure may be 21 psig and the discharge pressure may be 236 psig. The compressor increases the pressure 215 psig (236 − 21 = 215), **Figure 23.1.** Another system may have a different increase in pressure. A medium-temperature system may have a suction pressure of 56 psig with a discharge pressure of 236 psig. This system has an increase of 180 psig (236 − 56 = 180), **Figure 23.2.**

Compression ratio is the technical expression for pressure difference; it is the high-side absolute pressure divided by the low-side absolute pressure. Absolute pressures rather than gauge pressures are used when figuring compression ratios in order to keep the calculated compression ratio from becoming a negative number. Absolute pressures keep compression ratios positive and meaningful. For example, when a compressor is operating with R-404A, a head pressure of 331.8 psig (125°F), and a suction pressure of 19.6 psig (−16°F), the compression ratio would be

$$\text{Compression Ratio} = \frac{\text{Absolute Discharge}}{\text{Absolute Suction}}$$

$$CR = \frac{331.8\text{ psig} + 14.7\text{ atmosphere}}{19.6\text{ psig} + 14.7\text{ atmosphere}}$$

$$CR = \frac{346.5}{34.3}$$

$$CR = 10.1 \text{ to } 1 \text{ or } 10.1{:}1$$

Figure 23.1 The pressure difference between the suction and discharge side of the compressor on an R-404A, low-temperature system.

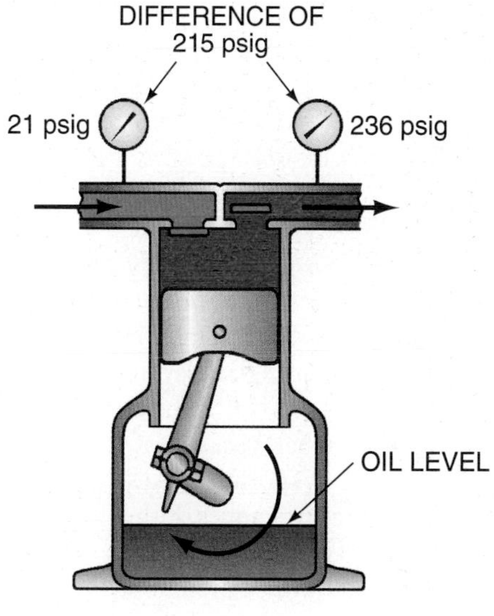

Figure 23.2 The pressure difference between the suction and discharge side of the compressor on an R-404A, medium-temperature system. The compressor only has to raise the pressure 180 psig.

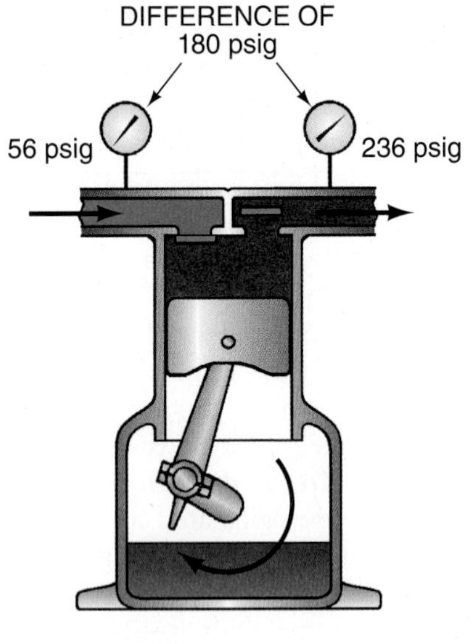

A compression ratio of 10.1:1 would indicate to a service technician that the absolute or true discharge pressure is 10.1 times as great as the absolute suction pressure.

Assume now that this same refrigeration system is operating with R-134a as the refrigerant. With the same condensing temperature of 125°F (184.6 psig) and an evaporating temperature of −16°F (0.7 in. Hg vacuum), the compression ratio would be

$$CR = \frac{184.6 \text{ psig} + 14.7 \text{ atmosphere}}{(29.92 \text{ in. Hg} - 0.7 \text{ in. Hg}) \div 2.036}$$

$$CR = \frac{199.3}{14.35}$$

$$CR = 13.89 \text{ to } 1 \text{ or } 13.89{:}1$$

NOTE: *Because the suction (evaporating) pressure of the R-134a system is in a vacuum of 0.7 in. Hg, a different calculation has to be applied to convert the vacuum reading to absolute pressure in psia. First, the gauge's vacuum measurement of 0.7 in. Hg has to be converted to in. Hg absolute. This is done by subtracting the vacuum gauge reading from the standard atmospheric pressure of 29.92 in. Hg. The result is then divided by 2.036 to convert in. Hg to psi. There are 2.036 in. Hg in every 1 psi.* •

Notice that the compression ratio for the R-134a system is higher than that for the R-404A system even though both systems have the same condensing and evaporating temperatures of 125°F and −16°F, respectively. The compression ratio for the R-134a system is higher because of the higher condensing pressure and lower evaporating pressure associated with R-134a at these temperature ranges. Either an increase in head pressure or a decrease in suction pressure will cause higher compression ratios. Any time a system has high compression ratios the compressor's discharge temperature will also be elevated. This is because of the higher heat of compression during the compression stroke associated with higher compression ratios. When compression ratios are high, more energy is needed to raise the pressure of the suction gases to the discharge pressure. This means more heat is generated as heat of compression during the compression stroke.

Compression ratios are used to compare pumping conditions for a compressor. When compression ratios become too high—above approximately 12:1 for a hermetic reciprocating compressor—the refrigerant gas temperature leaving the compressor rises to the point that oil for lubrication may become overheated. Overheated oil may turn to carbon and create acid in the system. Compression ratios can be reduced by two-stage compression. One compressor discharges into the suction side of the second compressor. In **Figure 23.3**, the first-stage compressor has a compression ratio of 3.2:1 (114.7 psia ÷ 35.7 psia), and the second-stage compressor has a compression ratio of 1.6:1 (183.7 psia ÷ 114.7 psia). Both of these compression ratios are acceptable and will result in good compressor efficiencies and low compressor discharge temperatures. However, if only one compressor were used, the compression ratio would be 5.14:1. Even though a compression ratio of 5.14:1 is acceptable, the compressor would be less efficient and have higher discharge temperatures than if two compressors were used. The previous example is for illustrative purposes only. Two-stage or compound compression is not generally used until the compression ratio has exceeded 10:1. This is often experienced with low-temperature freezing applications in commercial and industrial storage facilities.

Cool refrigerant passes through the suction valve of the compressor to fill the cylinders. This cool vapor contains the heat absorbed in the evaporator. The compressor pumps this heat-laden vapor to the condenser, where the heat can be rejected from the system. The vapor leaving the compressor can be very warm. With a discharge pressure of 169 psig, the discharge line at the compressor could easily be 200°F or

Figure 23.3 Two-stage compression. Notice that the second stage of the compressor is smaller than the first stage.

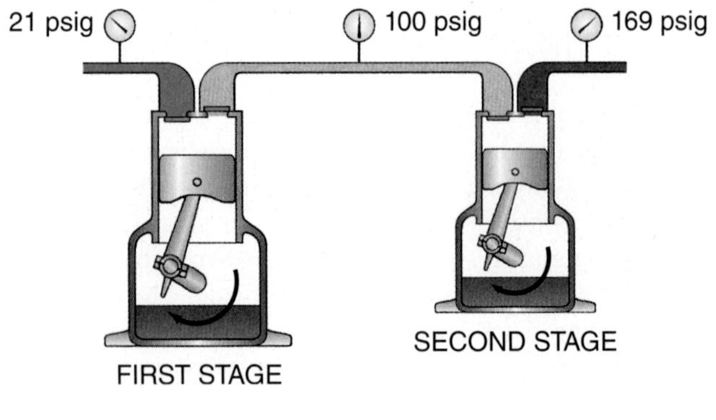

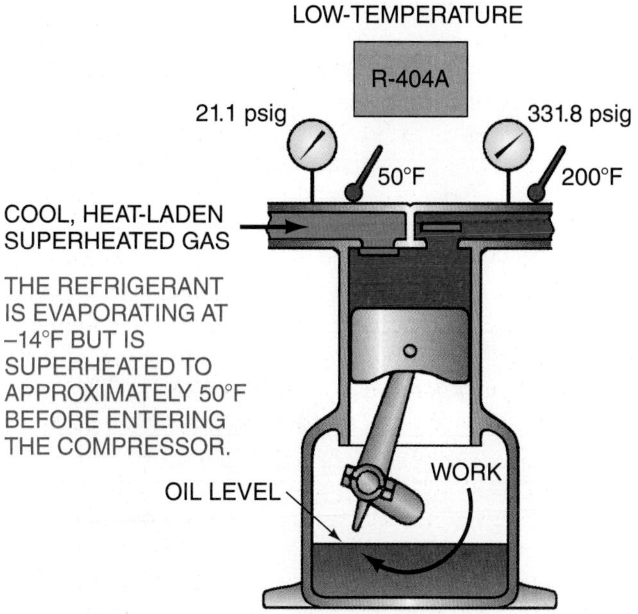

Figure 23.4 The refrigerant entering the compressor is called "heat-laden" because it contains the heat that was picked up in the evaporator from the boiling process. The gas is cool but full of heat that was absorbed at a low level of pressure and temperature. When this gas is compressed in the compressor, the heat concentrates. In addition to the heat absorbed in the evaporator, the act of compression also converts some energy to heat.

higher. **⚠ SAFETY PRECAUTION:** *Do not touch the compressor discharge line—you can burn your fingers.* • The vapor is compressed with the heat from the suction gas concentrated in the gas leaving the compressor, **Figure 23.4**.

23.2 TYPES OF COMPRESSORS

Five major types of compressors are used in the refrigeration and air-conditioning industry. These are the reciprocating, screw, rotary, scroll, and centrifugal. The reciprocating compressor, **Figure 23.5**, is used most frequently in small- and medium-sized commercial refrigeration systems and will be described in detail in this unit. The screw compressor, **Figure 23.6**, is used in large commercial and industrial systems and will be described only briefly here because it is dealt with later in this text. The rotary, **Figure 23.7**, and the scroll, **Figure 23.8**, along with the reciprocating compressor, are used in residential and light commercial air-conditioning. Centrifugal compressors, **Figure 23.9**, are used extensively for air-conditioning in large buildings and are described in Section 48.10, "Low Pressure Chillers."

Figure 23.5 A reciprocating compressor. Courtesy Copeland Corporation

Figure 23.6 A screw compressor. Source: Frick Company

Figure 23.7 A rotary compressor. Based on Motors and Armatures, Inc.

Figure 23.8 A scroll compressor. Courtesy Hampden Engineering

Figure 23.9 A centrifugal compressor. Source: York International Corp

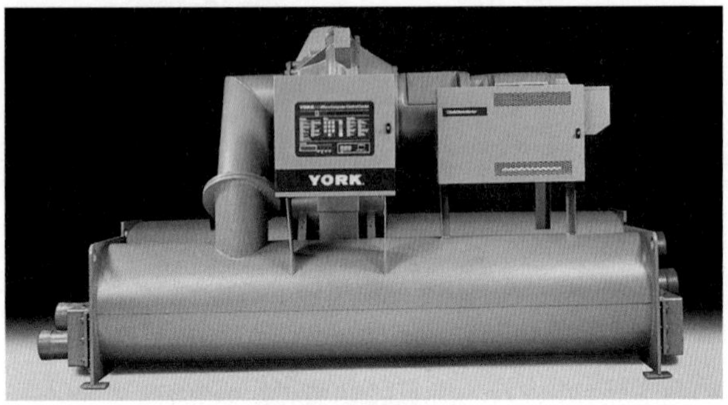

Figure 23.10 (A) An open-drive compressor. Source: Trane Company 2000. (B) A serviceable hermetic compressor. Courtesy Hampden Engineering

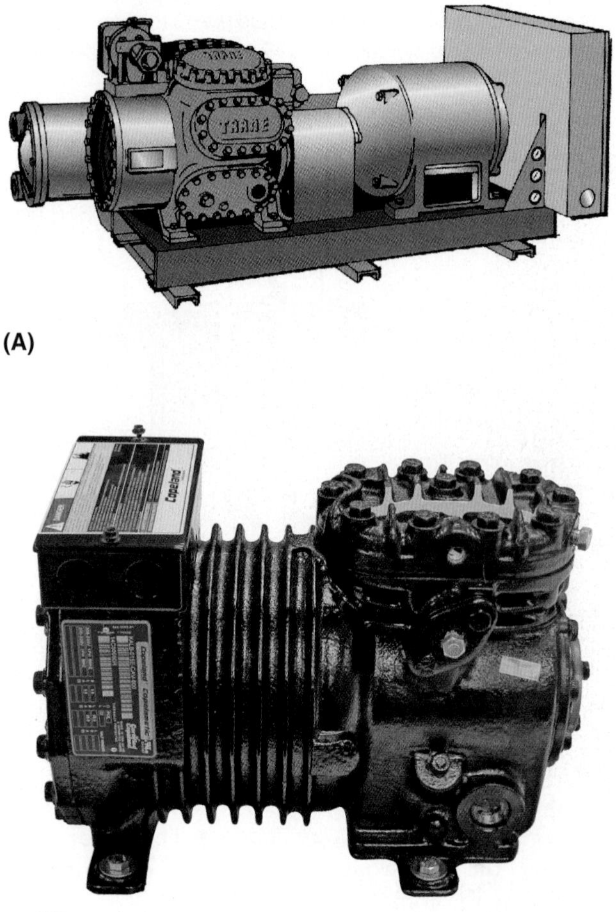

(A)

(B)

THE RECIPROCATING COMPRESSOR

Reciprocating compressors are categorized by their housing and by their drive mechanisms. The two housings are open and hermetic, **Figure 23.10**. "Hermetic" refers to the type of housing that contains the compressor and is either fully welded, **Figure 23.11**, or serviceable or semihermetic, **Figure 23.10(B)**. The drive mechanisms may be either inside the shell or outside the shell. A hermetic compressor has a direct-drive mechanism. The same shaft drives the compressor and motor inside the compressor shell.

FULLY WELDED HERMETIC COMPRESSORS. When a welded hermetic compressor is manufactured, the motor and compressor are contained inside a single shell that is welded closed, **Figure 23.11**. This unit is sometimes called the *tin can* compressor because it cannot be serviced without cutting open the shell. Characteristics of the fully welded hermetic compressor are as follows:

1. The only way to access the inside of the shell is by cutting the shell open.
2. Hermetic compressors can only be opened by the very few companies that specialize in this type of work. Otherwise, unless the manufacturer wants the compressor back for examination, it is a throwaway compressor.
3. The motor shaft and the compressor crankshaft are one shaft.
4. The hermetic compressor is usually considered to be a low-side device because the suction gas is vented to the whole inside of the shell, which includes the crankcase. The discharge (high-pressure) line is normally piped to the outside of the shell, so the shell only has to be rated at the low-side working-pressure value.
5. Generally, these compressors are cooled with suction gas.

Figure 23.11 The welded hermetic compressor is typical of smaller compressors, from fractional horsepower to about 50 hp. In most welded hermetic compressors, the suction line is usually piped directly into the shell and is open to the crankcase. The discharge line is normally piped from the compressor inside the shell to the outside of the shell. The compressor shell is typically thought of as a low-side component. Source: Bristol Compressors, Inc.

6. Hermetic compressors usually have a pressure lubrication system, but smaller hermetics are splash lubricated.
7. The combination motor and crankshaft are customarily in a vertical position with a bearing at the bottom of the shaft next to the oil pump. The second bearing is located about halfway up the shaft between the compressor and the motor.
8. The pistons and rods work outward from the crankshaft, so they are working at a 90° angle in relation to the crankshaft, **Figure 23.12**.

SERVICEABLE HERMETIC COMPRESSORS. The motor and compressor of a serviceable hermetic (semihermetic) compressor are contained inside a single shell that is bolted together. This unit can be serviced by removing the bolts and opening the shell at the appropriate place, **Figure 23.13**. Following are the characteristics of the serviceable hermetic compressor:

1. The unit is bolted together at locations that facilitate service and repair. Gaskets separate the mating parts connected by the bolts.
2. The housing is normally cast iron and may also have a steel housing fastened to it. These compressors are usually heavier than the fully welded type.

Figure 23.12 The internal workings of a welded hermetic compressor. Photo Courtesy of Tecumseh Products Company.

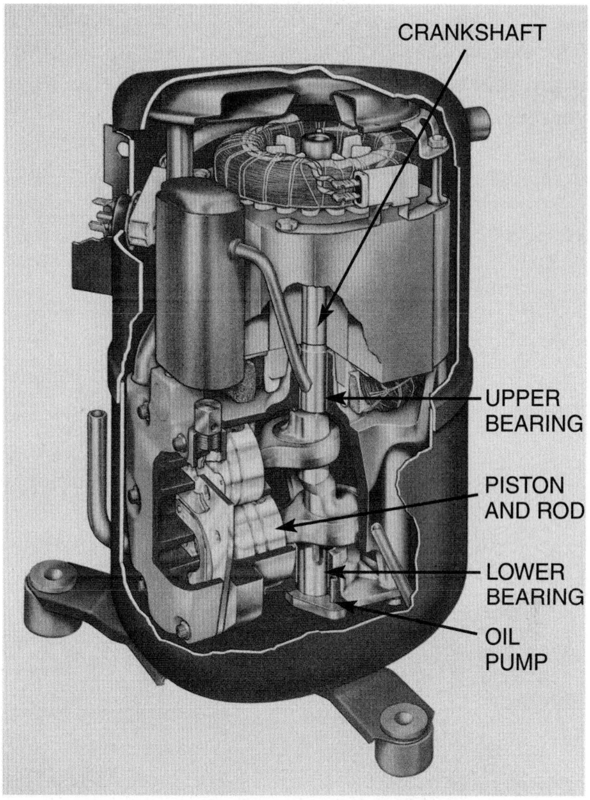

Figure 23.13 A serviceable hermetic (semihermetic) compressor designed in such a manner that it can be serviced in the field. Courtesy Copeland Corporation

3. The motor and crankshaft combination are similar to the motor and crankshaft in the fully welded compressor except that the crankshaft is usually horizontal.
4. Smaller compressors of this type generally use a splash lubrication system; larger compressors use pressure lubrication.

5. Serviceable compressors are often air-cooled and can be identified by the fins in the casting or extra sheet metal on the outside of the housing that give the shell more surface area.
6. The piston heads are normally at the top or near the top of the compressor and work up and down from the center of the crankshaft, **Figure 23.14**.

Figure 23.14 The working parts of a serviceable hermetic compressor. The crankshaft is in the horizontal position, and the rods and pistons move in and out from the center of the shaft. The oil pump is on the end of the shaft and draws the oil from the crankcase at the bottom of the compressor. Courtesy Copeland Corporation

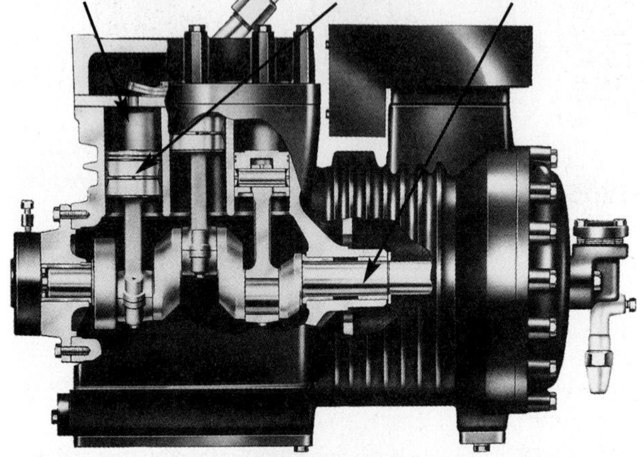

OPEN-DRIVE COMPRESSORS. Open-drive compressors are manufactured in two styles: belt drive and direct drive, **Figure 23.15**. Any compressor with the drive on the outside of the casing must have a shaft seal to keep the refrigerant from escaping to the atmosphere. This seal arrangement has not changed much in many years. Open-drive compressors are bolted together and may be disassembled for servicing the internal parts.

BELT-DRIVE COMPRESSORS. The first type of compressor, the belt-drive compressor, is still used to some extent. In the belt-drive unit, the motor and its shaft are beside the compressor and parallel to the shaft of the compressor. Notice that since the compressor and motor shaft are parallel, their sideways pull tightens the belts. This puts strain on both shafts and requires the manufacturer to compensate for the strain in the shaft bearings. **Figure 23.16** indicates correct and incorrect alignments of belt-drive compressors and motors.

DIRECT-DRIVE COMPRESSORS. The direct-drive compressor differs from the belt-drive in that the compressor shaft is placed end to end with the motor shaft. The coupling between the shafts has a small amount of flexibility, so the two shafts have to be in very good alignment to run correctly, **Figure 23.17**.

THE ROTARY SCREW COMPRESSOR

The rotary screw compressor is another mechanical method of compressing refrigerant gas and is used in larger installations. Instead of a piston and cylinder, this compressor uses two

Figure 23.15 (A) A belt-drive compressor. Source: Carrier Corporation. (B) A direct-drive compressor. The belt-drive compressor may have different speeds, depending on the pulley sizes. The direct-drive compressor turns at the speed of the motor because it is attached directly to the motor. Common speeds are 1750 rpm and 3450 rpm. The coupling between the motor and the compressor is slightly flexible. Source: Carrier Corporation

(A)

(B)

Figure 23.16 Correct and incorrect alignments of a belt-drive compressor with its motor. The belts used in a multiple-belt installation are matched at the factory for the correct length.

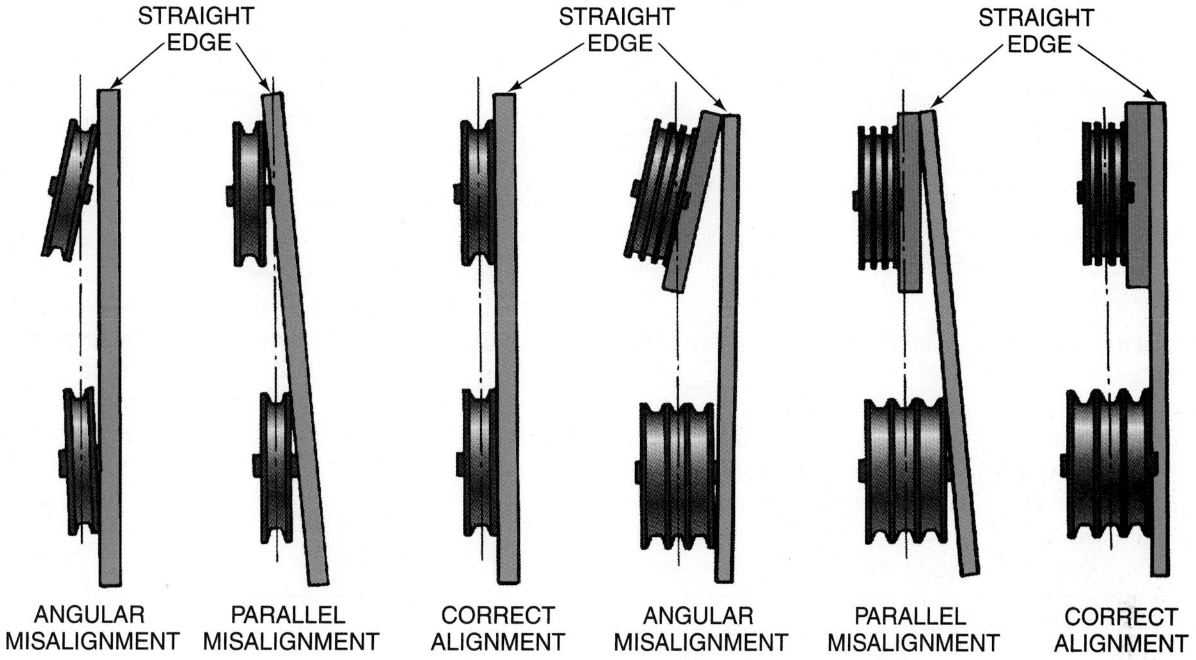

Figure 23.17 (A) and (C) Alignment of a direct-drive compressor and motor. The alignment must be within the compressor manufacturer's specifications. Source: Trane Company. (B) These compressors turn so fast that if the alignment is not correct, a seal or bearing will soon fail. When the correct alignment is attained, the compressor and motor are normally fastened permanently to the base on which they are mounted. In larger installations, the motor and compressor can be rebuilt in place. Source: Trane Company

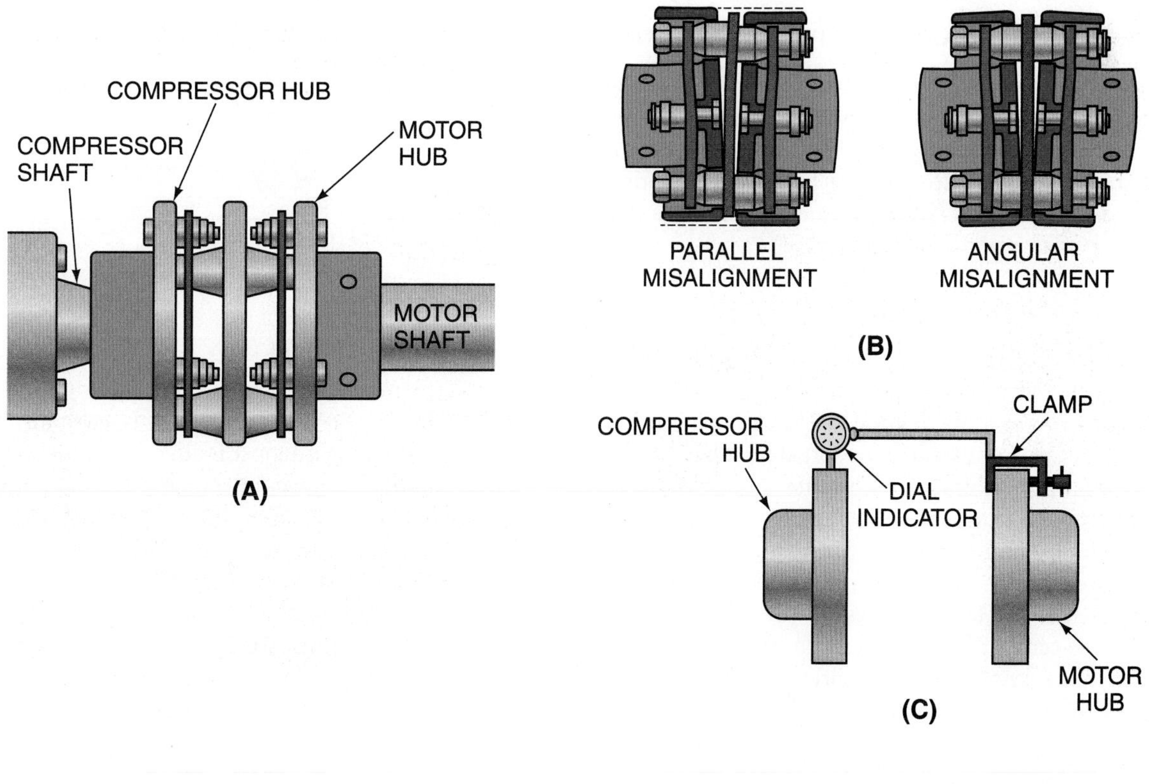

Figure 23.18 Tapered, machined, screw-type gears in a screw compressor.

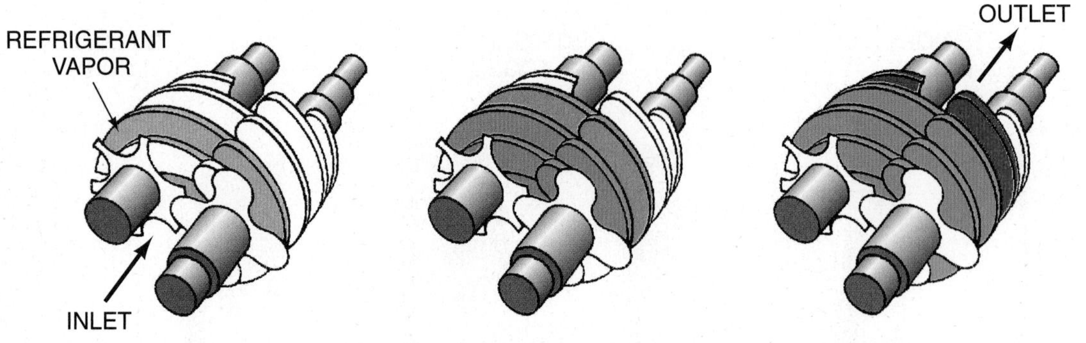

tapered, machined, screw-type matching gears that squeeze the refrigerant vapor from the inlet to the outlet, **Figure 23.18.** The rotary screw compressor has an open motor instead of the hermetic design. A shaft seal traps the refrigerant in the compressor housing where the rotating shaft leaves the housing. A flexible coupling is used to connect the motor shaft to the compressor shaft to prevent minor misalignment from causing seal or bearing damage.

Screw-compressors are used in large systems (see **Figure 23.6**). The refrigerant may be any of the common refrigerants. The operating pressure on the low and high sides of the system are the same as for a reciprocating system in a similar application.

Figure 23.19 The relationship of the pistons, rods, and crankshaft.

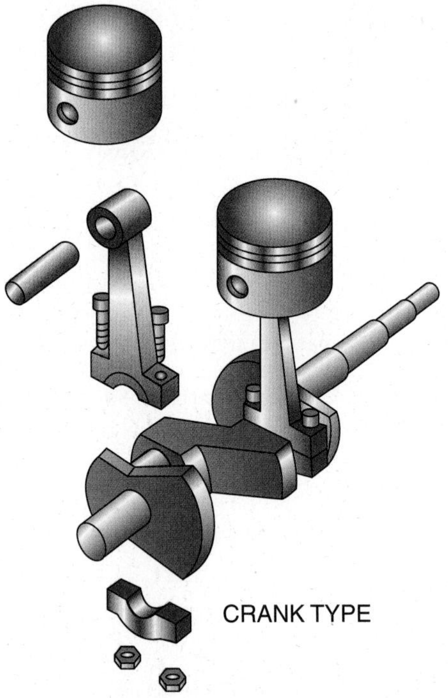

23.3 RECIPROCATING COMPRESSOR COMPONENTS

Reciprocating compressors house a vast array of moving parts. Most of these component parts have been precision machined with very tight tolerances for proper compressor efficiency and operation. Following is a list and description of the major component parts of a reciprocating compressor.

THE CRANKSHAFT

The *crankshaft* of a reciprocating compressor transmits the circular motion of the rods into a back-and-forth (reciprocating) motion for the pistons, **Figure 23.19.** Crankshafts are normally manufactured of cast iron or soft steel. They can be cast (molten metal poured into a mold) into a general shape and machined into an exact size and shape for the application. In this case, the shaft is made of cast iron, **Figure 23.20.** The machining process is critical because the throw (the off-center part where the rod fastens) does not turn in a circle (in relationship to the center of the shaft) when placed in a lathe. The machinist must know how to work with this type of setup. The off-center shafts normally have two main-bearing surfaces in addition to the off-center rod-bearing surfaces: One is on the motor end of the shaft, and one is on the other end. The bearing on the motor end is normally the largest because it carries the greatest load.

Some shafts are straight and have a cam-type arrangement called an eccentric. This allows the shaft to be manufactured straight and from steel. The shaft may not be any more durable than the cast-iron shaft, but it is easier to machine. The eccentrics can be machined off-center to the shaft to accomplish the reciprocating action, **Figure 23.21.** Notice that the rod for the eccentric shaft is different because the end of the rod has to fit over the large eccentric on the crankshaft.

Figure 23.20 Crankshafts are cast into a general shape and machined to the correct shape in a machine shop. This crankshaft is in an open-drive compressor but is also typical of hermetic compressors. Based on Trane Company 2000.

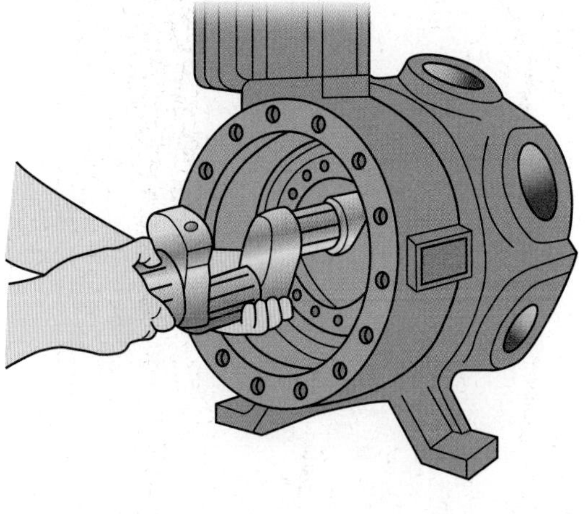

Figure 23.21 This crankshaft obtains the off-center action with a straight shaft and an eccentric. The eccentric is much like a cam lobe. The rods on this shaft have large bottom throws and slide off the end of the shaft. This means that to remove the rods, the shaft must be taken out of the compressor.

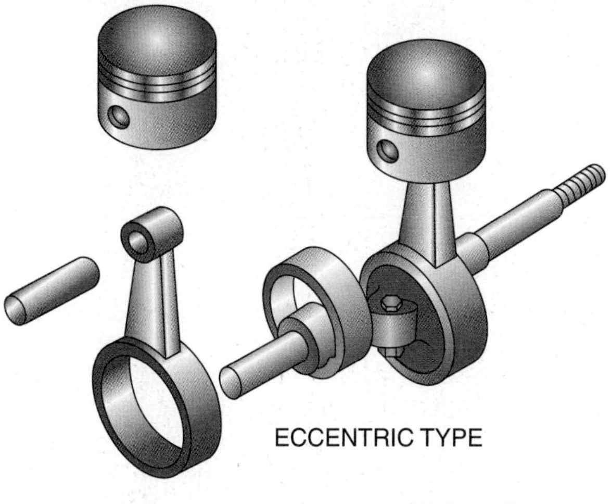

Figure 23.22 A splash lubrication system that splashes oil up onto the parts.

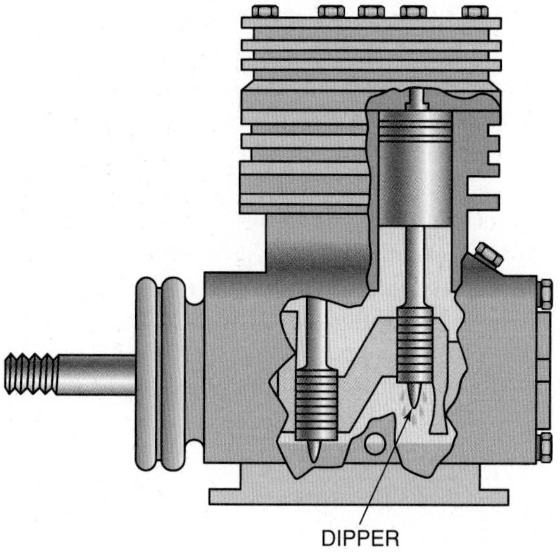

Figure 23.23 A crankshaft drilled so that an oil pump can force the oil up the shaft to the rods, then up the rods to the wrist pins. Magnetic elements are sometimes placed along the passage to capture iron filings. Source: Trane Company 2000.

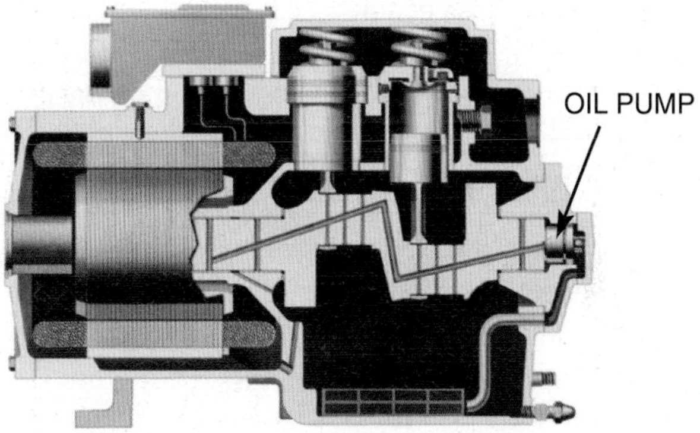

All of these shafts must be lubricated. Smaller compressors using the splash system may have a catch basin to catch oil and cause it to flow down the center of the shaft, **Figure 23.22.** When the compressor runs, the oil is then slung to the outside of the crankshaft surface. To be properly lubricated, compressors with splash-type lubrication systems must usually rotate in only one direction. This causes the oil to move to the other parts, such as the rods on both ends. Some shafts are drilled and lubricated by a pressure lubrication system. These compressors have an oil pump mounted on the end of the crankshaft that turns with the shaft, **Figure 23.23.**

NOTE: *There is no pressure lubrication when the compressor first starts. The compressor must be running up to speed before the lubrication system is fully effective.* •

Because some compressors have vertical shafts and some have horizontal, manufacturers have been challenged to provide for proper lubrication where needed. You may have to consult the compressor manufacturer for questions about a specific application.

CONNECTING RODS

Connecting rods connect the crankshaft to the piston. These rods are normally made in two styles: one fits the crankshaft with off-center throws and the other fits the eccentric crankshaft, **Figures 23.19** and **23.21.** Rods can be made of several different metals, such as iron, brass, and aluminum. The design of the rod is important because it takes a lot of the

load in the compressor. If the crankshaft is connected directly to the motor and the motor is running at 3450 rpm, the piston at the top of the rod is changing direction 6900 times per minute. The rod is the connection between this piston and the crankshaft and is the link between the changes of direction.

Rods with large holes in the shaft end are for eccentric shafts. They cannot be taken off with the shaft in place. The shaft has to be removed to take the piston out of the cylinder. Rods with smaller holes are for the off-center shafts; these are split and have rod bolts. These rods can be separated at the crankshaft, and the rod and piston can be removed with the crankshaft in place. They are small on the piston end and fasten to the piston differently than rods made for eccentric shafts. They normally have a connector called a *wrist pin* that slips through the piston and the upper end of the rod. This almost always has a *snap ring* to keep the wrist pin from sliding against the cylinder wall.

THE PISTON

The piston is the part of the cylinder assembly exposed to the high-pressure gas during compression. During the upstroke, pistons have high-pressure gas on top and suction or low-pressure gas on the bottom. They have to slide up and down in the cylinder in order to pump and must have some method of preventing the high-pressure gas from slipping by to the crankcase. Piston rings like those used in automobile engines are used on the larger pistons and are of two types: compression and oil. Smaller compressors use the oil on the cylinder walls as a seal. A cross-sectional view of these rings can be seen in **Figure 23.24**.

REFRIGERANT CYLINDER VALVES

The valves in the top of the compressor determine the direction the gas entering the compressor will flow. (See the cutaway view of a compressor cylinder in **Figure 23.25**.)

Figure 23.24 The piston rings for a refrigeration compressor resemble the rings used on automobile pistons. Based on Trane Company 2000.

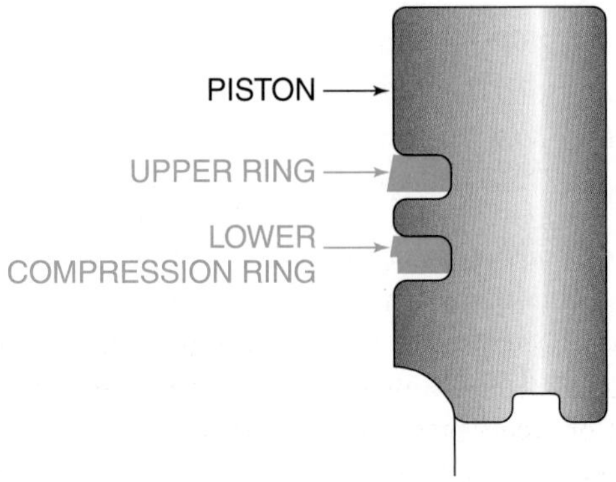

Figure 23.25 This cutaway view of a compressor shows a typical cylinder. Photo Courtesy of Tecumseh Products Company.

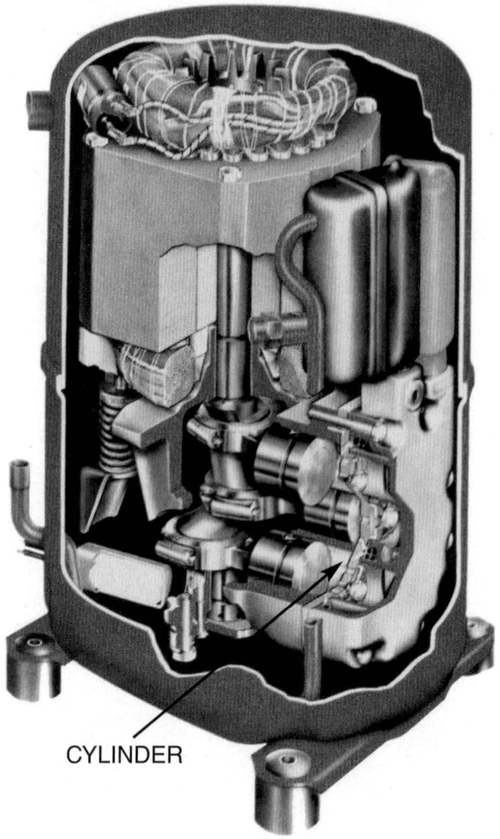

The valves are made of very hard steel. The majority of the valves on the market are either *ring valves* or flapper (reed) valves. They serve both the suction and the discharge ports of the compressor. The ring valve is circular and has springs under it. If ring valves are used for the suction and the discharge, the larger valve will be the suction valve, **Figure 23.26**. Flapper valves have been made in many different shapes. Each manufacturer has their own design, **Figure 23.27**.

THE VALVE PLATE

The valve plate holds the suction and discharge flapper valves. It is located between the head of the compressor and the top of the cylinder wall, **Figure 23.28**. Many different methods have been used to hold the valves in place without taking up any more space than necessary. The bottom of the plate actually protrudes into the cylinder. Any volume of gas that cannot be pumped out of the cylinder because of the valve design will reexpand on the downstroke of the piston. This makes the compressor less efficient.

Figure 23.26 (A) Ring valves. (B)–(D) They normally have a set of small springs to close them. Source: Trane Company 2000.

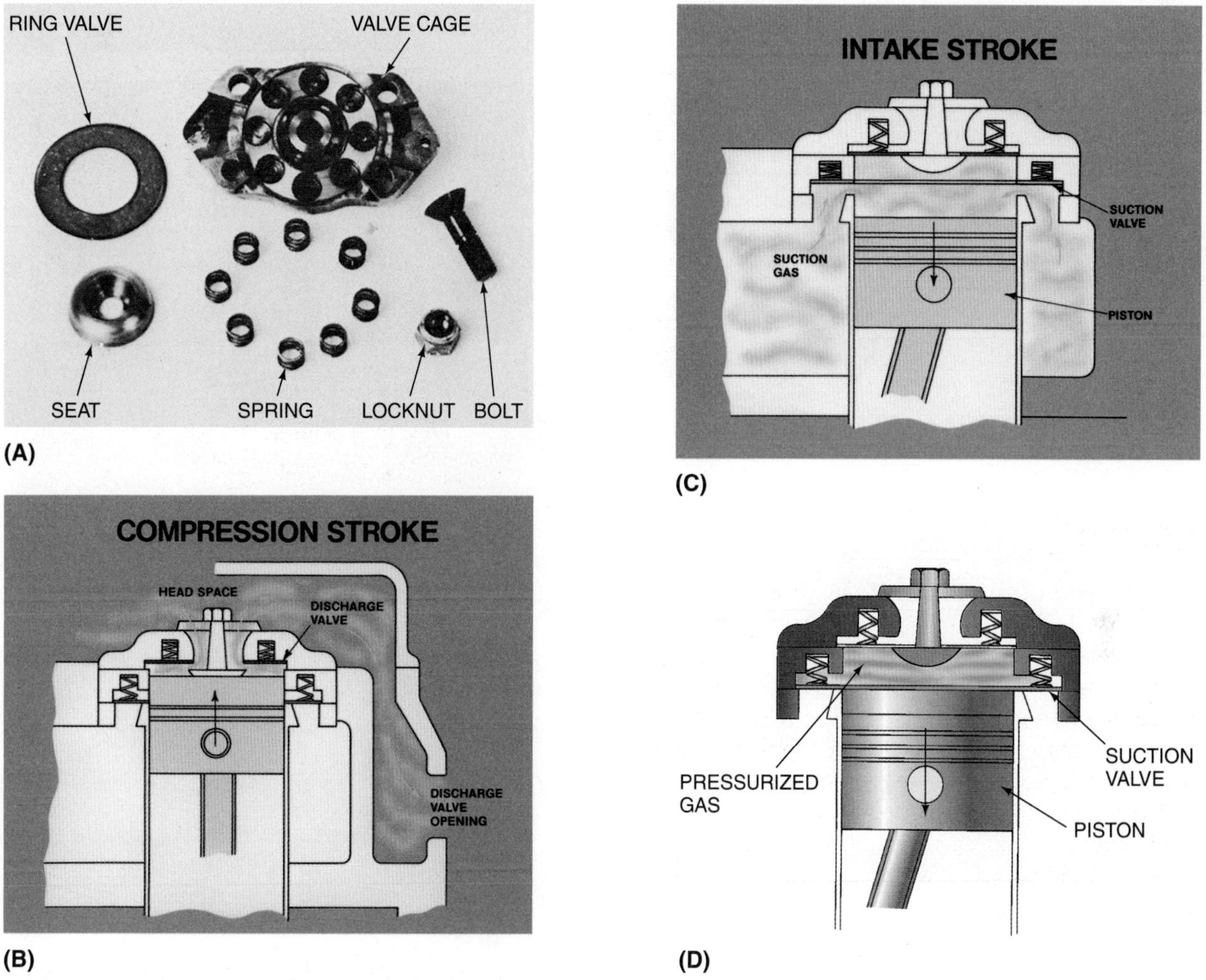

Figure 23.27 Reed or flapper valves held down on one end. This provides enough spring action to close the valve when reverse flow occurs.

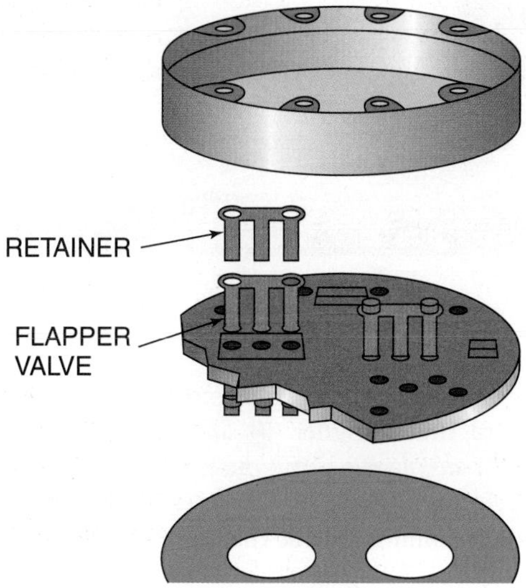

There are other versions of the crankshaft and valve arrangements than those listed here, but they are not used widely enough in the refrigeration industry to justify coverage here. If you need more information, contact the manufacturer.

THE HEAD OF THE COMPRESSOR

The component that holds the top of the cylinder and its assembly together is the *head*. Sitting on top of the cylinder, it contains the high-pressure gas from the cylinder until it moves into the discharge line. The head often contains the suction chamber, which is separated from the discharge chamber by a partition and gaskets. Heads have many different design configurations, but they all need to accomplish two things: hold the pressure in and hold the valve plate on the cylinder. In some welded hermetic compressors, the heads are made of steel; they are made of cast iron in serviceable hermetic compressors. These cast iron heads may be in the

Figure 23.28 (A) Valve plates typical of those used to hold the valves. They can be replaced or rebuilt if not badly damaged. There is a gasket on both sides. Source: Carrier Corporation. (B) Valve plate with discharge reed valves under discharge valve backers. Courtesy Ferris State University. Photo by John Tomczyk. (C) Suction reed valves. Courtesy Ferris State University. Photo by John Tomczyk. (D) Valve plate with broken discharge reed valves and damaged pistons from liquid slugging. Courtesy Ferris State University. Photo by John Tomczyk

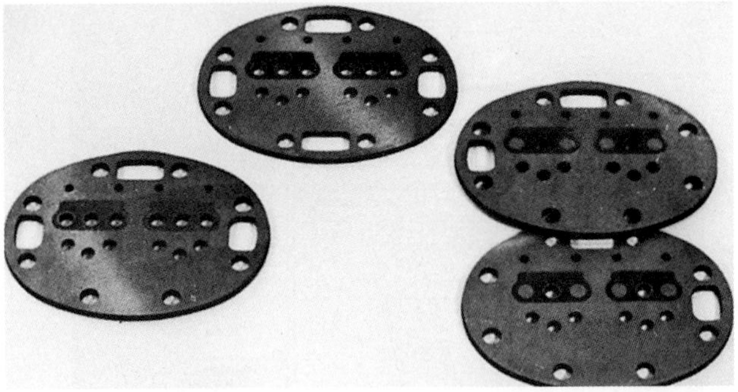

(A)

(B)

(C)

(D)

moving airstream and have fins to help dissipate the heat from the top of the cylinder, **Figure 23.29**.

MUFFLERS

Mufflers are used in many fully hermetic compressors to muffle compressor pulsation noise. Audible suction and discharge pulsations can be transmitted into the piping if they are not muffled. Mufflers must be designed to have a low pressure drop and still muffle the discharge pulsations, **Figure 23.30**.

THE COMPRESSOR HOUSING

The *housing* holds the compressor and sometimes the motor. It is made of stamped steel for the welded hermetic compressor and of cast iron for serviceable hermetic compressors.

The welded hermetic compressor is designed so that the compressor shell is under low-side pressure; it will often have a working pressure of 150 psig. Many compressors manufactured today must use the newer refrigerants that operate under much higher pressure; therefore, compressor shell working pressures are much higher. **⚠ SAFETY PRECAUTION:** *The technician should always check the working pressures of all system parts before applying high pressures for leak checking.* • The compressor is mounted inside the shell, and the discharge line is normally piped to the outside of the shell. This means the

Figure 23.29 Typical compressor heads. (A) A suction-cooled compressor. Courtesy Copeland Corporation. (B) Air-cooled compressors that have air-cooled motors must be located in a moving airstream or overheating will occur.

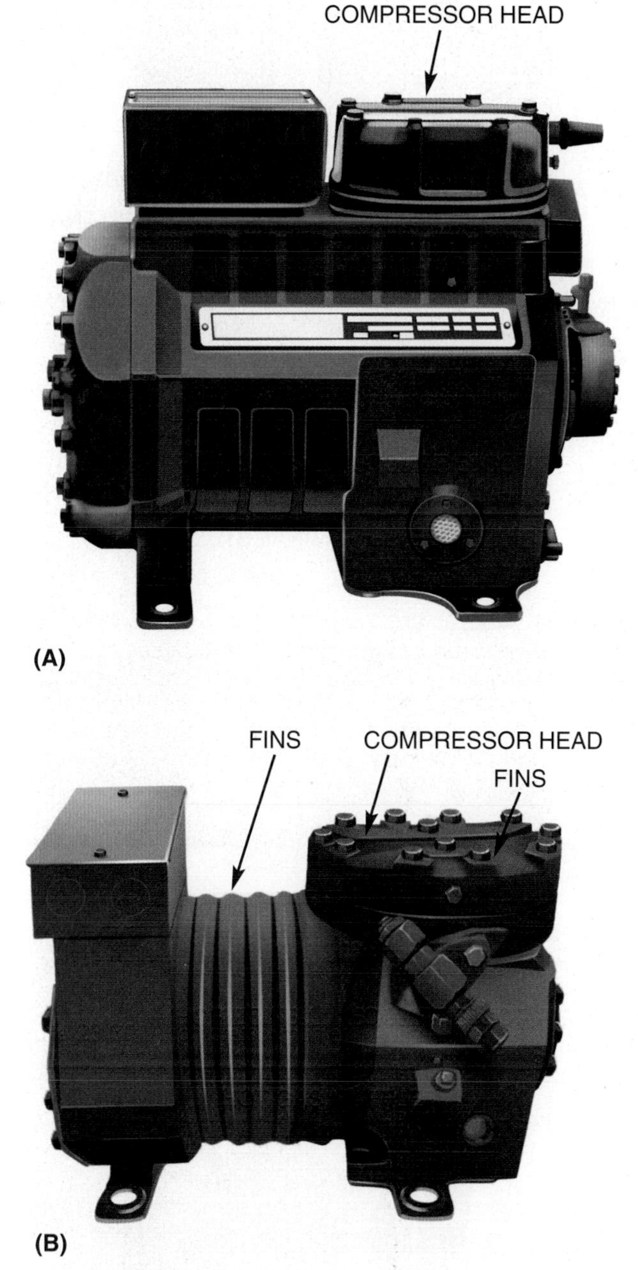

Figure 23.30 A compressor muffler.

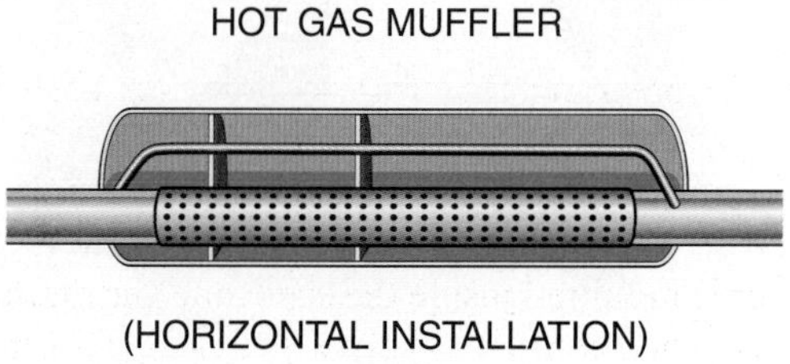

Figure 23.31 A compressor motor pressed into its steel shell. It requires experience to remove the motor. The compressor has springs under the mounting feet to help eliminate vibration. This compressor is shipped with a bolt tightened down through the springs. This bolt must be loosened by the installing contractor.

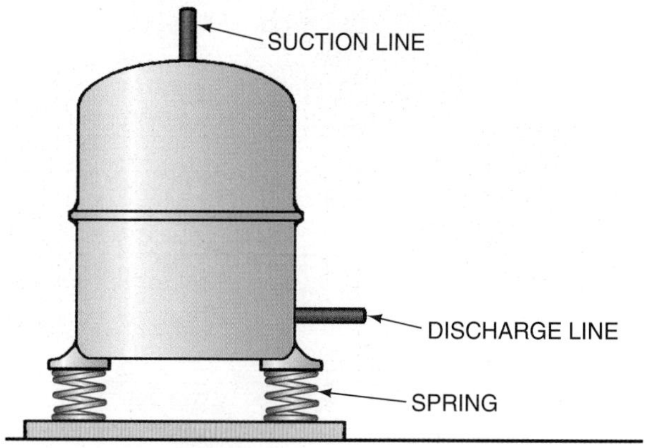

shell does not need to have a test pressure as high as the high-side pressure. A cutaway view of a hermetic compressor inside a welded shell and the method used to weld the shell together are shown in **Figure 23.25**.

The two methods used to mount the compressor inside the shell are *rigid mounting* and *spring mounting*. Rigid-mounted compressors were used for many years. The compressor shell was mounted on external springs that had to be bolted tightly for shipment. The springs were supposed to be loosened when the compressor was installed, **Figure 23.31**, but occasionally they were not, and the compressor vibrated because, without the springs, it was mounted rigidly to the condenser casing. External springs can also rust, especially where there is a lot of salt in the air.

In internal spring-mounting, compressors are actually suspended from springs inside the shell. There are ways of keeping these compressors from moving too much during shipment. Sometimes a compressor will come loose from one or two of the internal springs. When this happens, it will run and pump normally just like it is supposed to but will make a noise on start-up or shutdown or both. If the compressor comes off the springs and they are internal, there is nothing that can be done to make a repair in the field, **Figure 23.32.**

COMPRESSOR MOTORS IN A REFRIGERANT ATMOSPHERE

MOTORS IN A REFRIGERANT ATMOSPHERE. Special consideration must be given to compressor motors operating inside a refrigerant atmosphere. Motors for hermetic compressors differ from standard electric motors. The materials used to make them are not the same as those that would be used for a fan or pump motor to be run in air. Hermetic motors must

Figure 23.32 A compressor mounted inside a welded hermetic shell on springs. The springs have guides that allow them to move only a certain amount during shipment. Photo Courtesy of Tecumseh Products Company

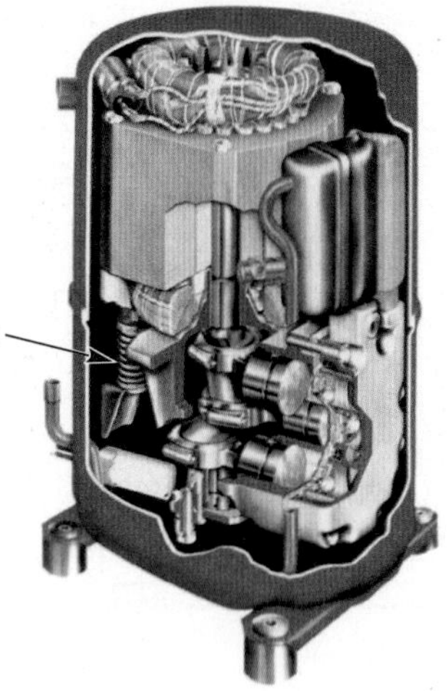

be manufactured of materials compatible with the system refrigerants. For instance, rubber cannot be used because the refrigerant would dissolve it. The motors are assembled in a very clean atmosphere and kept dry. When a hermetic motor malfunctions, it cannot be repaired in the field.

MOTOR ELECTRICAL TERMINALS. There must be some conductor to carry the power from the external power supply to the internal motor, and power must be carried through the compressor housing in a way that prevents the refrigerant from leaking. The connection also has to insulate the electric current from the compressor shell. On smaller compressors, terminals are sometimes fused glass with a terminal stud through the middle. When large terminals are required, the terminals are sometimes placed in a fiber block that is sealed with an O-ring, **Figure 23.33**. A fused-glass terminal block is hard to repair. It will stand more heat, but there is a limit to how much. The O-ring and fiber type of terminal board can tolerate less heat, but when they are damaged, they can be replaced with new parts. Refrigerant loss can result, however, before the problem is discovered. **NOTE:** *Care must be taken with electrical terminals to prevent overheating (due to loose electrical connections). Should the terminal overheat, a leak could occur.* •

INTERNAL MOTOR PROTECTION DEVICES. Internal overload protection devices in hermetic motors protect the motor from overheating. These devices are embedded in or near the windings and are wired in two different ways. One style breaks the line circuit inside the compressor. Because it is internal and carries the line current, this style is limited to smaller compressors. It has to be enclosed to prevent the electrical arc from affecting the refrigerant, **Figure 23.34(A)**. If a

Figure 23.33 Motor terminals. The power to operate the compressor is carried through the compressor shell but must be insulated from the shell. This is a fiber block used as the insulator. O-rings keep the refrigerant in. Based on Trane Company.

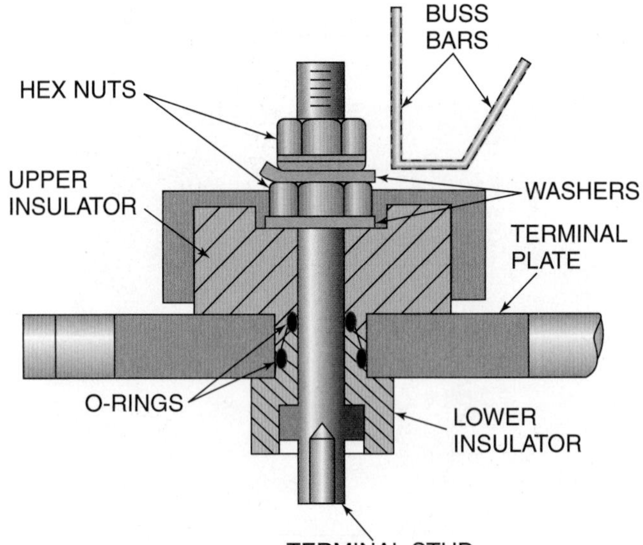

Figure 23.34 (A) An internal compressor overload protection device that breaks the line circuit. Because this set of electrical contacts is inside the refrigerant atmosphere, they are enclosed inside a hermetic container of their own. If the electrical arc were allowed inside the refrigerant atmosphere, the refrigerant would deteriorate in the vicinity of the arc. Based on Tecumseh Products Company.

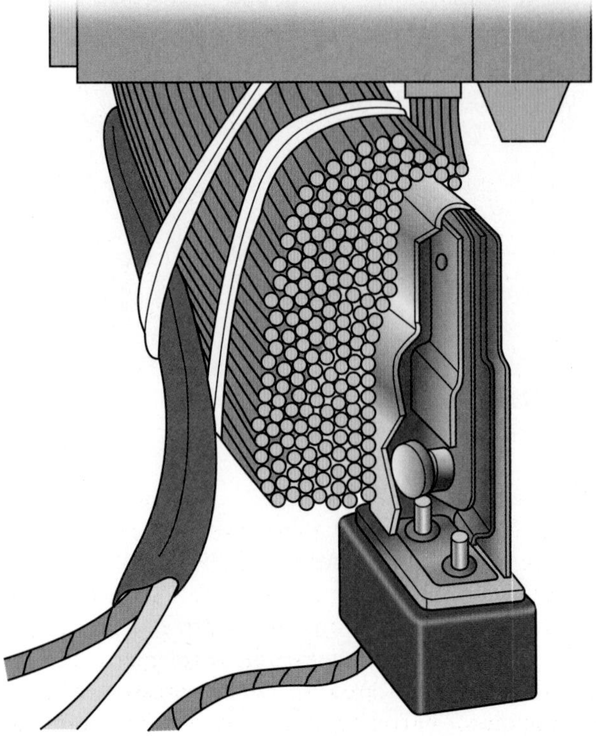

(A)

Figure 23.34 (*Continued*) (B) A three-phase wye-wound motor with internal thermal overload protection. All three internal overload contacts will open at the same time to prevent single phasing of the three-phase motor. (C) A three-phase internal overload protector. Courtesy Ferris State University. Photo by John Tomczyk (D) A three-phase wye-wound motor with internal thermistor protection linked to a solid-state electronic motor protection module. The motor sensors are thermistors. Courtesy of Tecumseh Products Company

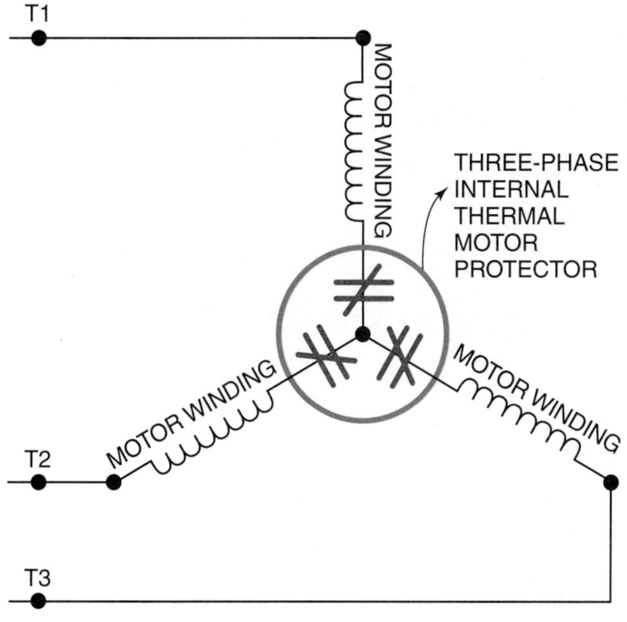

(B)

(C)

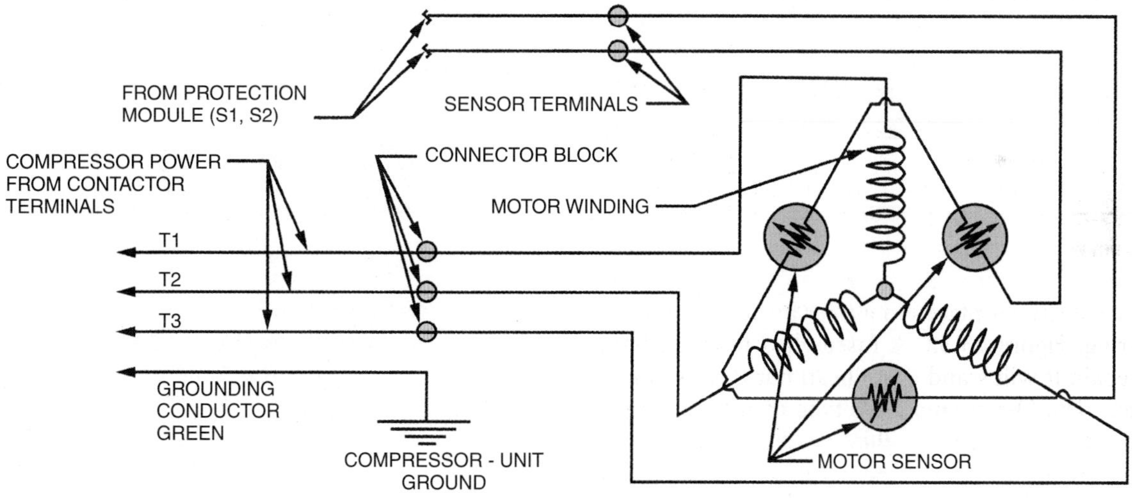

(D)

contact in this line remains open, the compressor cannot be restarted. The compressor would have to be replaced.

Figure 23.34(B) illustrates a three-phase internal thermal overload. All three of the overload contacts will open at the same time if a motor overheating problem occurs. This prevents single phasing of the three-phase motor. The overload device is located in the compressor's shell, usually in the motor barrel compartment over the motor windings. **Figure 23.34(C)** is a photo of a three-phase overload protector that has been taken out of a working three-phase motor.

Figure 23.34(D) shows a three-phase motor with internal thermistor protection. Thermistors are resistors that vary their resistance according to their temperature. If a motor overheating problem occurs, the thermistors will sense the heat, change resistance, and relay this change in resistance to a solid-state motor protection module. The thermistor circuit is usually a 24-V control circuit. The motor protection

Figure 23.34 (*Continued*) (E) The wiring diagram of a typical electronic three-phase motor protection package. Courtesy of Tecumseh Products Company.

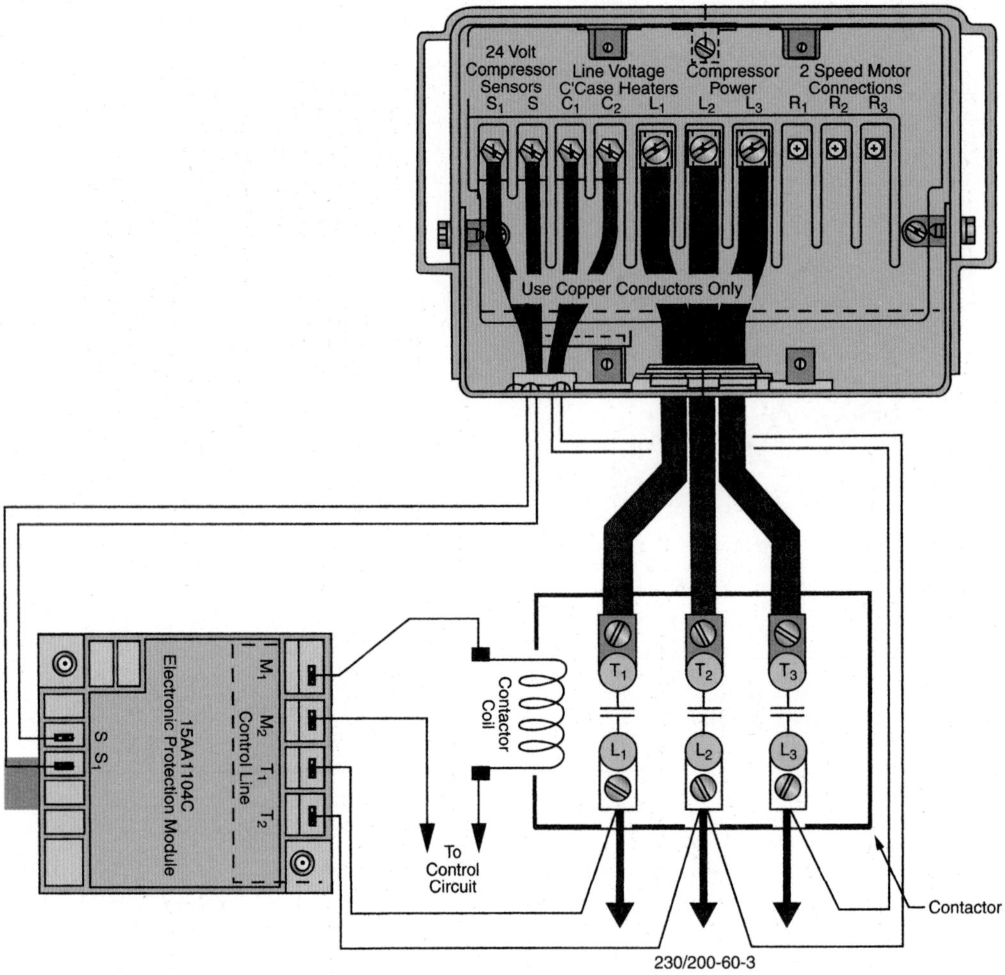

(E)

module controls the coil of the contactor or motor starter and can turn the motor off when necessary for a cool-down period. **Figure 23.34(E)** is a wiring diagram of a typical electronic motor protection package showing the motor contactor and electronic protection module.

Another type of motor overload protection device breaks the control circuit. This is wired from the outside of the compressor to the control circuit. If the pilot-duty type wired to the outside of the compressor were to remain open, an external overload device could be substituted.

THE SERVICEABLE HERMETIC COMPRESSOR. The serviceable hermetic compressor motor is rigidly mounted to the shell, and the compressor must be externally mounted on springs or other flexible mounts to prevent vibration. The compressor usually has a cast-iron shell and is considered a low-side device, **Figure 23.29.** Because of the piping arrangement in the head, the discharge gas is contained either under the head or is expelled out the discharge line. The serviceable hermetic design is used exclusively for larger compressor sizes because it can be rebuilt. The compressor components are much the same as the components in the welded hermetic type.

OPEN-DRIVE COMPRESSOR. Open-drive compressors are manufactured with the motor external to the compressor shell. The shaft protrudes through the casing to the outside, where either a pulley or a coupling is attached. The motor is either mounted end to end with the compressor shaft or alongside the compressor and uses belts to turn the compressor. This compressor is normally heavy duty in nature and must be mounted firmly to a foundation.

THE SHAFT SEAL

The pressure inside a compressor crankcase can be either a vacuum (below atmosphere) or a positive pressure. If the unit is an extra-low-temperature unit using R-134a as the

refrigerant, the crankcase pressure could easily be a vacuum, so if the shaft seal were to leak, the atmosphere would be pulled into the crankcase. When the compressor is sitting still, a high positive pressure could be exerted on it. For example, when R-507A or R-404A is used and the system is off for extended periods (the whole system may get up to 100°F in a hot climate), the crankcase pressure may go over 200 psig. The crankcase shaft seal must be able to hold refrigerant inside the compressor under all of these conditions and while the shaft is turning at high speed, **Figure 23.35**.

Figure 23.35 The shaft seal is responsible for keeping the refrigerant inside the crankcase and allowing the shaft to turn at high speed. This seal must be installed correctly. When the seal is installed on a belt-drive compressor, the belt tension is important. When it is installed on a direct-drive compressor, the shaft alignment is important.

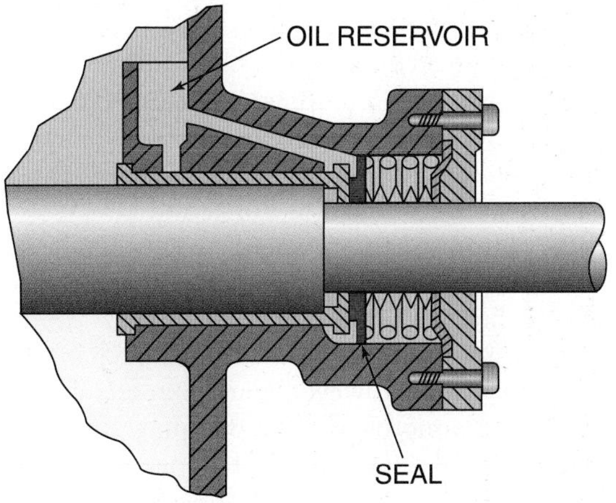

The shaft seal has a rubbing surface designed to keep the refrigerant and the atmosphere separated. This surface is usually made of a carbon or ceramic material that rubs against a steel surface. If assembled correctly, these two materials can rub together for years and not wear out. Correct assembly normally involves aligning the shafts correctly on a system where the bearings are in good working order. The belts must also have the correct tension if the unit is belt-driven. If the unit is direct-drive, the shafts have to be aligned according to the manufacturer's instructions.

23.4 BELT-DRIVE MECHANISM CHARACTERISTICS

A belt-drive compressor has the motor mounted at the side of the compressor and a pulley mounted on the motor as well as on the compressor. The pulley on the motor is called the *drive pulley,* and the pulley on the compressor is called the *driven pulley.* The drive pulley is sometimes adjustable, which allows the compressor speed to be adjusted. The drive pulley can also be changed to a different size to vary the compressor speed. This can be advantageous when a compressor is too large for a job (too much capacity) and needs to be slowed down to compensate. The compressor can also be sped up for more capacity if the motor has enough horsepower in reserve (if the motor is not already running at maximum horsepower). If a change in pulley size is needed, consult the compressor manufacturer to be sure that the design limits of the compressor are not exceeded. Different pulleys are shown in **Figure 23.36.**

The formula for determining the size of the drive pulley is determined using information gathered from the

Figure 23.36 Different pulleys used on belt-drive compressors. They are single-groove and multiple-groove pulleys with different belt widths. Some pulleys are adjustable in order to change the compressor speed.

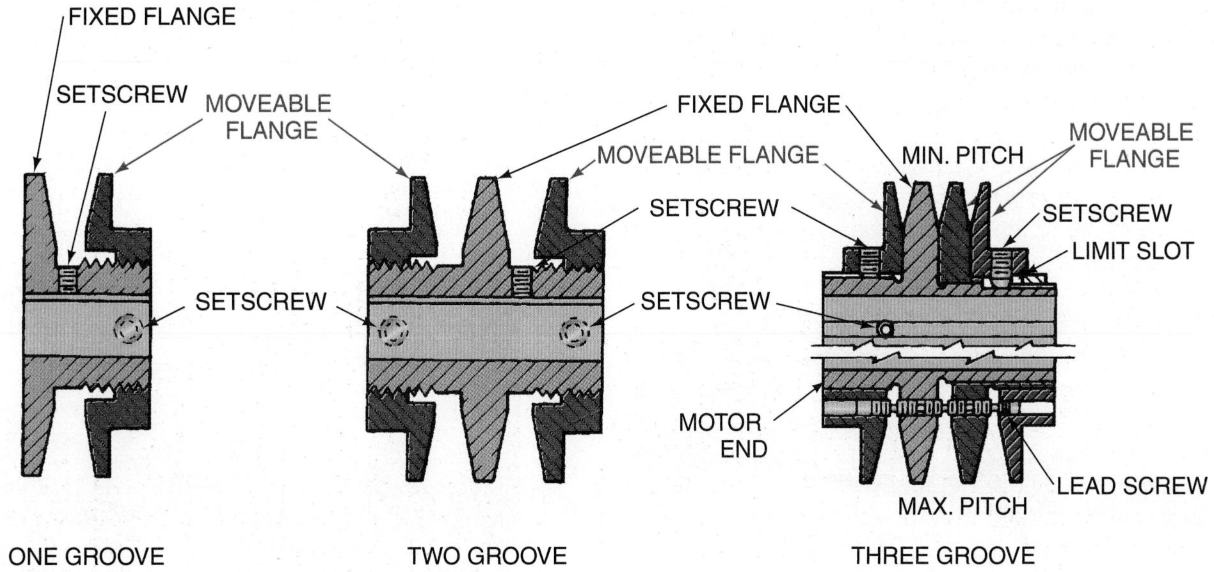

existing system. The motor speed is normally fixed, 1725 or 3450 rpm, and the pulley (flywheel) on the compressor is also normally a fixed size. The only variable is the drive pulley. A typical problem may be a motor with an rpm of 1725 and a drive pulley of 4 in. used to drive a compressor at 575 rpm with a pulley size of 12 in. It is desirable to reduce the compressor speed to 500 rpm to reduce the capacity. What would the new drive pulley size be? The formula is:

$$\text{Drive Pulley Size} \times \text{Drive Pulley rpm} =$$
$$\text{Driven Pulley Size} \times \text{Driven Pulley rpm}$$

or

$$\text{Pulley 1} \times \text{rpm 1} = \text{Pulley 2} \times \text{rpm 2}$$

Solve the problem for pulley 1, restated as:

$$\text{Pulley 1} = \frac{\text{Pulley 2} \times \text{rpm 2}}{\text{rpm 1}}$$

$$\text{Pulley 1} = \frac{12 \text{ in.} \times 500 \text{ rpm}}{1725 \text{ rpm}}$$

$$\text{Pulley 1} = 3.48 \text{ or } 3\ 1/2 \text{ in.}$$

Most compressors are not designed to operate over a certain rpm, information that is available from the compressor manufacturer. A pulley supplier can help choose the correct pulley size. Belt sizes also must be calculated. Choosing the correct belts and pulleys is important and should be done by an experienced person.

When adjusting or repairing belt-drive compressors, the compressor and motor shafts must be in correct alignment and the proper tension must be applied to the belts. The motor base and the compressor base must also be tightened rigidly so that no dimensions will vary during operation. Several belt combinations may have to be considered if the compressor drive mechanism has multiple belts, **Figure 23.37**. Even if the compressor is small and has a single belt, there are different types of belts to consider. The width of the belt, the grip type, and the material all have to be taken into account. Multiple belts have to be bought as matched sets. For example, if a particular compressor and motor drive has four V belts, which are of B width and are 88 in. long, you should order four 88-in. belts of B width that are factory matched for the exact length. These are called *matched belts* and must be used on multiple-belt installations.

Figure 23.37 A multiple-belt compressor. Photo Courtesy of Tecumseh Products Company.

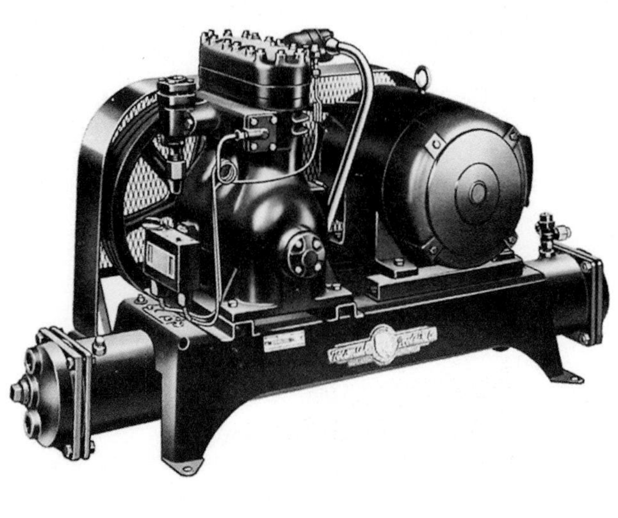

23.5 DIRECT-DRIVE COMPRESSOR CHARACTERISTICS

The direct-drive compressor is limited to the motor speed that the drive motor is turning. In this type of installation the motor and compressor shafts are end to end. These shafts have a slightly flexible coupling between them but must be in very close alignment, or the bearings and seal will fail prematurely, **Figure 23.38.** This compressor and motor combination is mounted on a common rigid base.

Both the motor and compressor are customarily manufactured so that they can be rebuilt in place. Thus, once the shafts are aligned, the shell of the motor and the shell of the compressor can be fastened down and always remain in place. If the motor or compressor has to be rebuilt, the internal parts may be removed for rebuilding, then the shafts will automatically line back up when reassembled.

Reciprocating compressor design has not changed significantly over the years. However, manufacturers continuously try to improve the motor and the pumping efficiencies. The valve arrangements can make a difference in pumping efficiency and are being studied for improvement.

23.6 RECIPROCATING COMPRESSOR EFFICIENCY

A compressor's efficiency is determined by its design. Efficiency starts with the filling of the cylinder. The following sequence of events takes place inside a reciprocating compressor during the pumping action. Our example for the

Figure 23.38 A flexible coupling. An extensive procedure is used to obtain the correct shaft alignment. This must be done, or bearings and seal will fail prematurely. Source: Lovejoy, Inc.

Figure 23.39 An illustration of what happens inside the reciprocating compressor while it is pumping. When the piston starts down, a low pressure is formed under the suction reed valve. When this pressure becomes less than the suction pressure and the valve spring tension, the cylinder will begin to fill. Gas will rush into the cylinder through the suction reed valve.

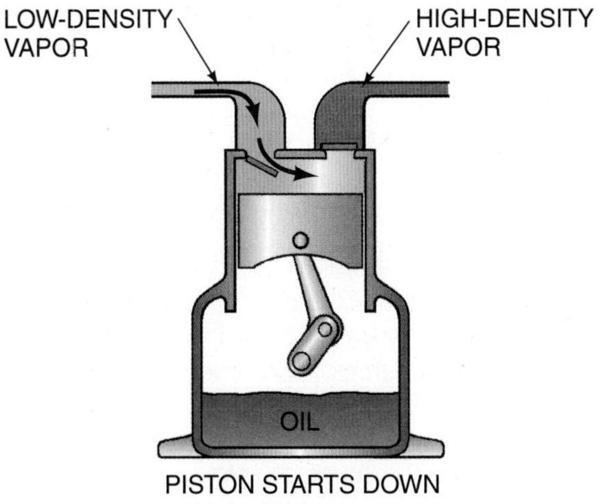

pumping sequence is a medium-temperature application with R-134a refrigerant, a suction pressure of 18 psig, and a discharge pressure of 185 psig.

1. **Piston at the top of the stroke and starting down.** When the piston has moved down far enough to create less pressure in the cylinder than is in the suction line, the intake flapper valve will open and the cylinder will start to fill with gas, **Figure 23.39.**
2. **Piston continues to the bottom of the stroke.** At this point the cylinder is nearly as full as it is going to get. There is a very slight time lag at the bottom of the stroke as the crankshaft carries the rod around the bottom of the stroke, **Figure 23.40.**
3. **Piston is starting up.** The rod throw is past bottom dead center, and the piston starts up. When the cylinder is as full as it is going to get, the suction flapper valve closes.

Figure 23.40 When the piston gets near the bottom of the stroke, the cylinder is nearly full. There is a short time lag as the crankshaft circles through bottom dead center, during which a small amount of gas can still flow into the cylinder.

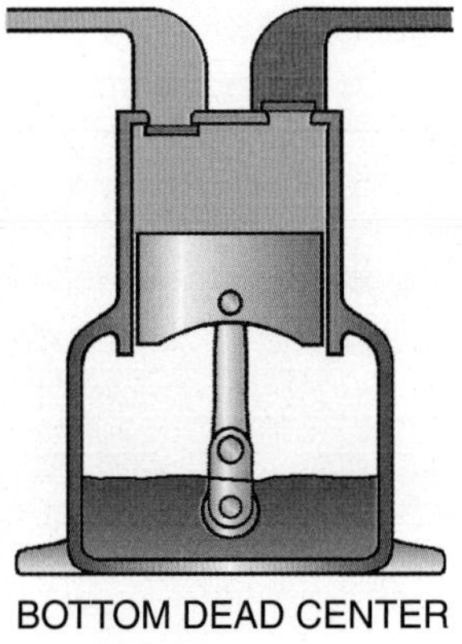

Figure 23.41 When the piston starts back up and gets just off the bottom of the cylinder, the suction valve will have closed, and pressure will begin to build in the cylinder. When the piston gets close to the top of the cylinder, the pressure will start to approach the pressure in the discharge line. When the pressure inside the cylinder is greater than the pressure on the top side of the discharge reed valve, the valve will open, and the discharge gas will empty out into the high side of the system.

4. **The piston proceeds to the top of the stroke.** When the piston reaches a point that is nearly at the top, the pressure in the cylinder becomes greater than the pressure in the discharge line. If the discharge pressure is 185 psig, the pressure inside the cylinder may have to reach 195 psig to overcome the discharge valve's weight and spring tension, **Figure 23.41.**
5. **The piston is at exactly top dead center.** The piston is as close to the top of the head as it can go. There has to be a certain amount of clearance in the valve assemblies and between the piston and the head, or they would touch. This clearance is known as **clearance volume**. The piston is going to push as much gas out of the cylinder as time and clearance volume will allow. There will be a small amount of gas left in the clearance volume, **Figure 23.42.** This gas will be at the discharge pressure mentioned earlier. When the piston starts back down, the gas will reexpand and the

Figure 23.42 A reciprocating compressor cylinder cannot completely empty because of the clearance volume at the top of the cylinder. Manufacturers try to keep this clearance volume to a minimum but cannot completely do away with it.

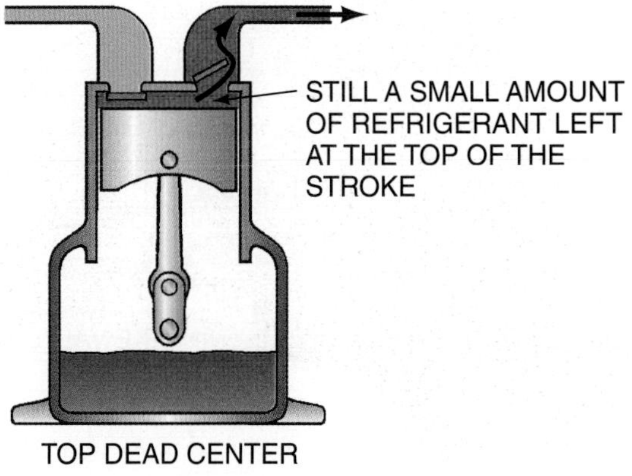

cylinder will not start to fill until the cylinder pressure is lower than the suction pressure of 18 psig. This reexpanded refrigerant is part of the reason that the compressor is not 100% efficient. Valve design and the short period of time the cylinder has to fill at the bottom of the stroke are other reasons the compressor is not 100% efficient.

23.7 DISCUS VALVE DESIGN

The design of the discus valve allows a closer tolerance inside the compressor cylinder at top dead center. The closer tolerance makes the compressor more efficient because it reduces the clearance volume. The discus valve also has a larger bore and allows more gas through the port within a short period of time. **Figure 23.43** shows a discus compressor, a conventional valve design, a discus valve design, and a cutaway view of a discus compressor. Discus compressors are refrigeration compressors used in supermarkets, walk-in coolers and freezers, and industrial applications. They typically range in horsepower from 5 to 60 hp.

Discus compressors are available with capacity modulation that ranges from 10% to 100%. This provides the ability to match compressor capacity to the load desired for the refrigeration equipment. Capacity modulation reduces the suction pressure and temperature variation of the refrigerated

Figure 23.43 (A) A discus compressor. Courtesy Copeland Corporation. (B) A discus valve design compared to a conventional valve design. Courtesy Copeland Corporation. (C) The cutaway view of a discus compressor. Courtesy Copeland Corporation. (D) Discus suction valves. Courtesy Ferris State University. Photo by John Tomczyk

(A)

(C)

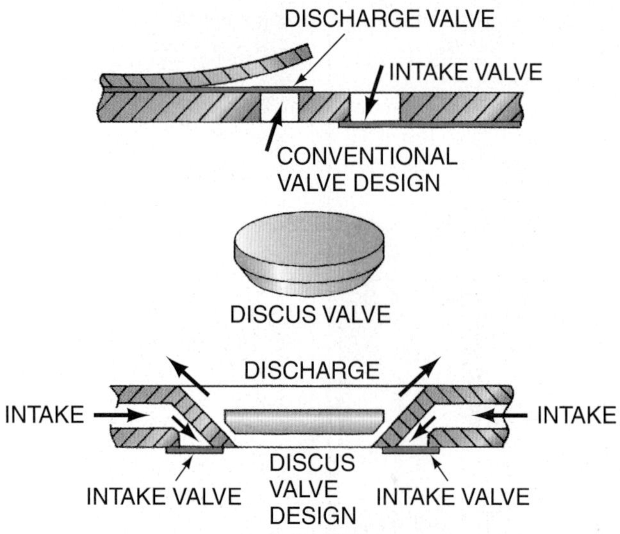

(B)

(D)

Figure 23.43 (*Continued*) (E) Discus valve plate showing discharge discus valve. Courtesy Ferris State University. Photo by John Tomczyk. (F) Discus valve plate and discus discharge valve cage. Courtesy Ferris State University. Photo by John Tomczyk

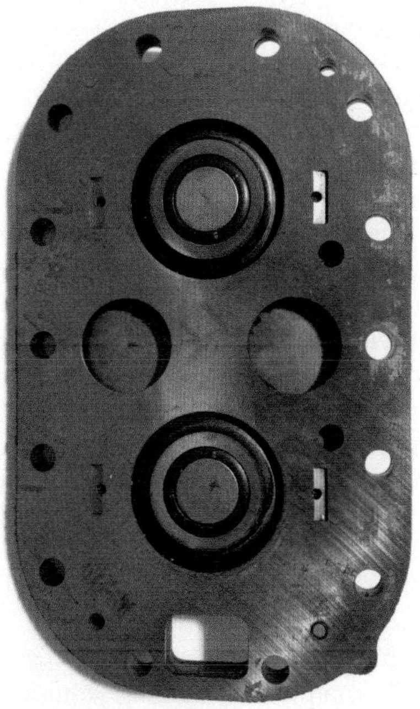

(E)

(F)

space and allows a decrease in the compressor's cycling rate. A reduced cycling rate increases the compressor's reliability. Modulation is accomplished with a blocked suction technology by feeding a variable voltage to a solenoid for its open and closed intervals.

23.8 NEW TECHNOLOGY IN COMPRESSORS

Discus compressor technology also offers onboard diagnostics. This technology makes possible real-time monitoring to let the service technician know the status of what is happening inside the compressor before any major problems develop. Today, compressor monitoring operations can be centralized off-site, which allows service technicians to systematically troubleshoot compressor problems before arriving at the site. It also improves troubleshooting accuracy and speed of service. Centralized monitoring technology can gather data, transmit operating information, and visually display compressor status and alarm codes on the front control box. It also can record and retain a history of the compressor's operating information and past alarms and dispatch service technicians automatically if an alarm problem arises. **Figure 23.44** shows a modern discus compressor that incorporates onboard diagnostics. This new technology has the following advantages:

- Monitors the compressor's discharge temperature
- Provides contactor protection
- Enables remote diagnostics
- Integrates the compressor's system electronics—including the high- and low-pressure controls, the cooling and temperature control, oil pressure monitoring, motor protection devices, and input/output (I/O) boards

Figure 23.44 An Intelligent Store™ discus compressor with integrated system electronics and onboard diagnostics. Courtesy Emerson Climate Technologies

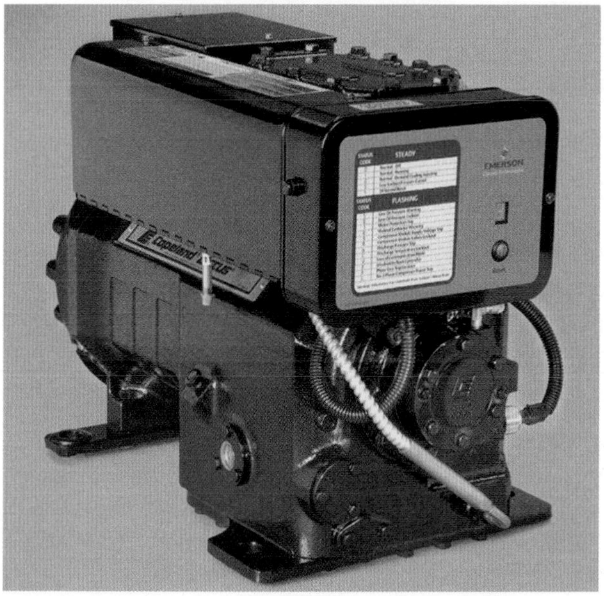

- Reduces the number of brazed joints on the compressor that can develop leaks
- Guarantees consistent field installation due to less wiring and fewer components

THE SCROLL COMPRESSOR

The concept of compressing a gas by turning one scroll against another around a common axis isn't new. Scroll compressor technology has been around for 100 years, but it did not become commercially available and cost-effective until the mid 1980s. Today, it is becoming even more fine-tuned and is available in many more applications. A major hurdle in making the scroll compressor a viable product has been the difficulty in achieving a balance between the requirement for high-volume precision manufacturing and the need for consistent high performance and efficiency, low sound levels, and great reliability. **Figure 23.45** shows a fifth-generation scroll compressor with its very few moving parts.

It is the scroll compressor's ability to reach higher Seasonal Energy Efficiency Ratio (SEER) levels that has helped the HVACR industry navigate both new and evolving government-mandated regulations. The U.S. Department of Energy (DOE) has established "Regional Standards" for HVACR equipment. These standards set minimum HVACR system efficiency requirements based on geographic location and climate zone. SEER is calculated, in part, based on the total amount of cooling (in Btu) the system will provide over an entire season, divided by the total number of watt-hours it will consume. Higher SEER ratings reflect a more efficient cooling system. Most equipment manufacturers are investing heavily to design and create more efficient evaporators, condensers, compressors, fan motors, and more sophisticated control systems in order to meet and, in many cases, exceed new energy-efficiency requirements. Refer to Unit 21, "Evaporators and the Refrigeration System," and Unit 22, "Condensers," for more detailed information on evaporator and condenser designs.

Figure 23.45 Fifth-generation Copeland scroll compressor showing very few moving parts. Courtesy Emerson Climate Technologies

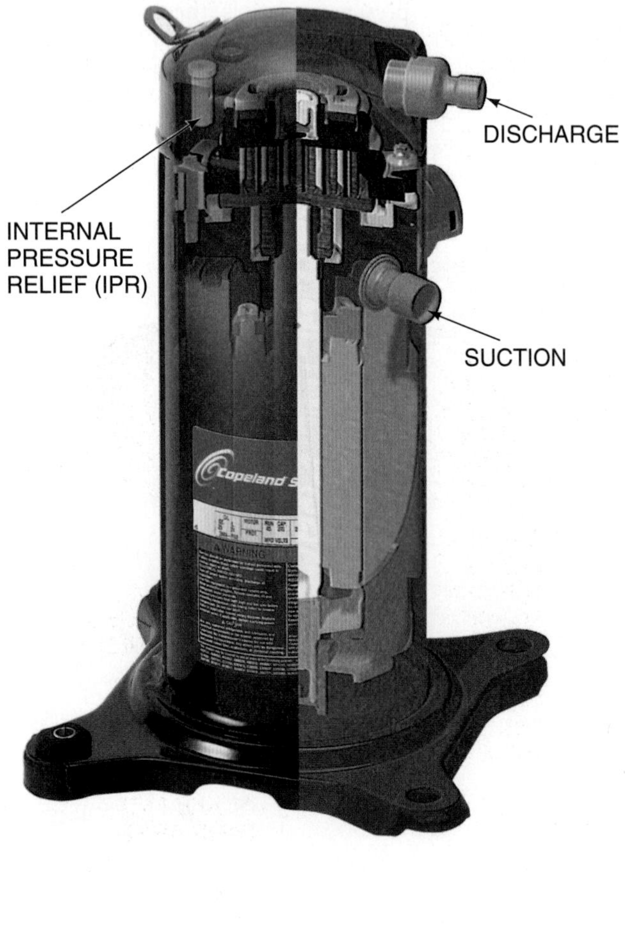

Because of the scroll compressor's fewer moving parts, it operates much more quietly than other compressor designs and technologies. Less noise and higher efficiencies have become two key selling points for installing the scroll compressor in residential air-conditioning units. With advancements in technology, manufacturers began using scroll compressors in commercial cooling equipment. A major development for commercial scroll compressors has been the advent of modulating technology. Modulating technology is able to maintain ideal comfort levels in buildings such as churches, where rooms can be empty for hours and then suddenly fill to capacity. Modulation technology operates in several ways, which will be covered later in the unit. Some high-end systems now have communication systems onboard that enable components to talk to each other, **Figure 23.44.**

Refrigeration has also become a major application for the scroll compressor due to its reliability and durability. It is used in walk-in coolers, reach-ins, and distributed-refrigeration systems. Beyond coolers and freezers, the scroll compressor has found its way to soft-serve ice cream machines, frozen carbonated beverage machines, and other types of machines. Scroll compressors are also being utilized in cryogenic equipment, magnetic resonance imaging (MRI) machines, and even some dental equipment.

SCROLL COMPRESSOR OPERATION

The operation of a scroll compressor is relatively simple. Two spiral-shaped scrolls fit inside one another, **Figure 23.46.** These two mating parts are often referred to as *involute spirals*. One of the spiral-shaped parts stays stationary while the other orbits around the stationary member. The orbiting motion is created by the offset operation of the centers of the journal bearings and the motor, **Figure 23.47.**

The motion is a true orbital motion, not a rotational motion. The orbiting motion causes continuous crescent-shaped gas pockets to form, **Figure 23.48,** then draws gas into the outer pocket and seals it as the orbiting continues.

Figure 23.46 Two spiral-shaped scrolls fitting inside one another. Courtesy Emerson Climate Technologies

Figure 23.47 A scroll compressor's drive train showing center bearing and motor shaft offset. Courtesy Ferris State University. Photo by John Tomczyk

The continuous orbiting motion reduces the crescent-shaped gas pocket to a smaller and smaller volume as it nears the center of the scroll form. When the gas pocket is fully compressed, it is discharged out of a port of the nonorbiting (fixed) scroll member, **Figure 23.49.** Several crescent-shaped gas pockets are compressed at the same time, which provides for a smooth and continuous compression cycle. Thus, the scroll compressor conducts its intake, compression, and discharge phases simultaneously.

The scroll compressor always takes a fixed volume of gas at suction pressure and then decreases the same gas volume, which increases its pressure. In fact, during the discharge phase, the scroll compressor compresses the discharge gas to a zero volume, eliminating any carryover of trapped discharge gas into a clearance volume that is characteristic of piston-type compressors. Because of this, the scroll compressor is often referred to as a "fixed compression ratio" compressor. Piston-type compressors are often referred to as "variable compression ratio" compressors. Reexpansion of discharge gas contributes to low volumetric efficiencies in piston-type compressors.

SCROLL COMPRESSOR ADVANTAGES

The ways in which the two scroll members interact and operate provide several advantages. When liquid refrigerant, oil, or small solid particles enter between the two scrolls, the mating scroll parts can actually move apart in a sideways direction, referred to as "radial" movement. Radial movement eliminates high-stress situations and allows for just the right amount of contact force between mating scroll surfaces. This action allows the compressor to handle some liquid. When the compressor experiences a liquid slug, the scroll's mating

Figure 23.48 Continuous crescent-shaped gas pockets being formed in the scroll. Courtesy Emerson Climate Technologies

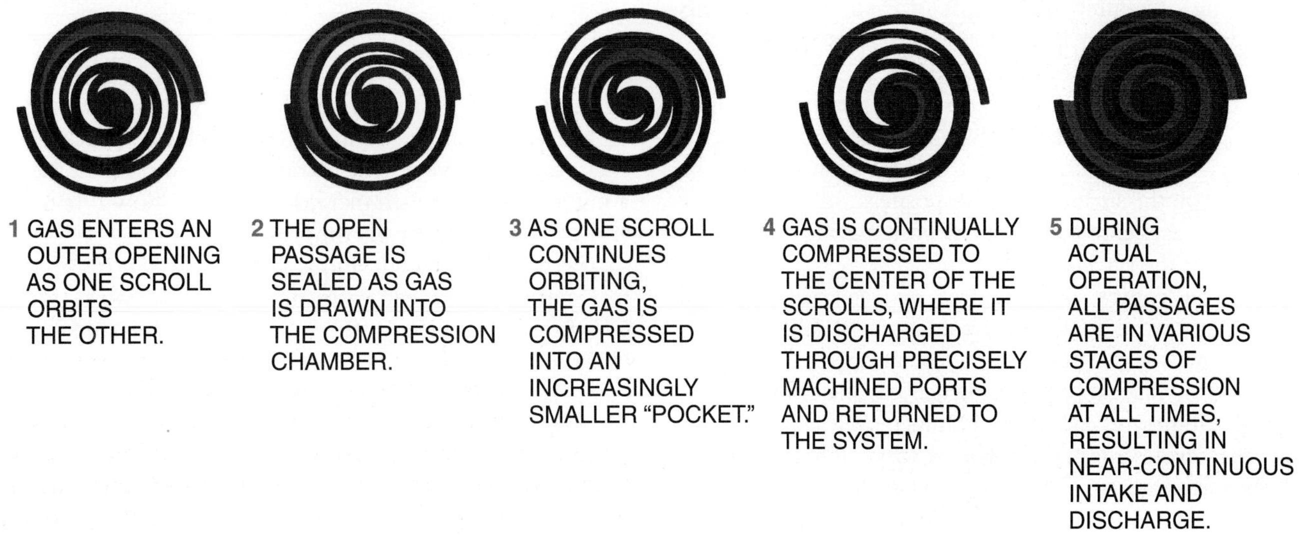

Figure 23.49 (A) Non-orbiting or "fixed" scroll member showing discharge port. Courtesy Ferris State University. Photo by John Tomczyk. (B) Top side of non-orbiting or "fixed" scroll member showing discharge valve. Courtesy Ferris State University. Photo by John Tomczyk

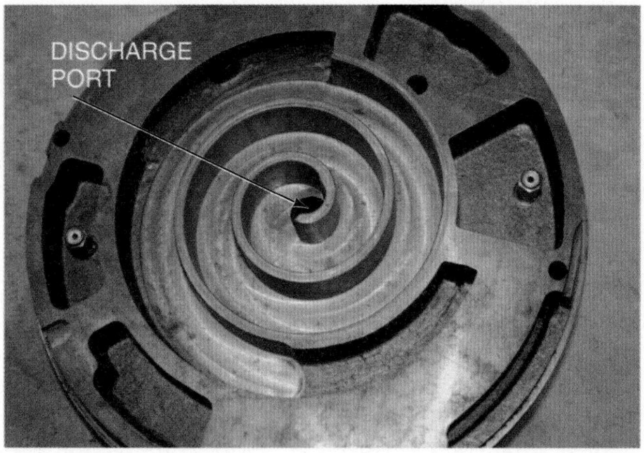

(A)

(B)

parts will separate slightly and allow the pressurized gas to vent to suction pressure. This allows the liquid slug to be swept from the mating scroll surfaces to the suction pressure and be vaporized. A gurgling noise may be heard during this process. The compressor may even stop pumping briefly and then restart as the excess liquid is purged from the scrolls. Although scroll compressors handle liquid better than other compressors, they can still require additional accessories like crankcase heaters and suction-line accumulators for added protection.

As the orbiting scroll orbits, centrifugal forces on the sides of the mating scrolls, along with some lubricating oil, form a seal that prevents gas-pocket leakage. This is often referred to as "flank sealing," a major contributor to the scroll's high efficiency. A small amount of lubricating oil is usually entrained in the suction gases and, along with the centrifugal forces, provides the flank sealing. Tight up-and-down mating or sealing of the tips of each scroll prevents any compressed gas-pocket leakage and adds to efficiency. Up-and-down sealing is often referred to as "axial" sealing. Some scroll manufacturers use tip seals for axial sealing, **Figure 23.50.** Scroll-tip seals serve the same purpose as the piston rings in a reciprocating piston-type compressor. The tip seals ride on the surface of the opposite scroll and provide a seal so that gases cannot escape between mating scroll parts and the tips of the scrolls.

The scroll compressor requires no reed valves, so it does not have valve losses that contribute to inefficiencies as piston-type compressors do. As mentioned, there is no reexpansion of discharge gases, which can be trapped in a clearance volume and cause low volumetric efficiencies. This is why the scroll compressor has a very high capacity in high-compression-ratio applications. In addition, a considerable distance separates the scroll compressor's suction and discharge port locations, which greatly reduces the transfer of heat between the suction and discharge gases. Because suction gases have less heat transferred into them, they have a higher density, which increases the mass flow rate of refrigerant through the scroll compressor.

Figure 23.50 Scroll member showing a tip seal for axial sealing. Courtesy Ferris State University. Photo by John Tomczyk

Because of its continuous compression process and the fact that it has no reed valves to create valve noise, the scroll compressor has very soft gas-pulsation noises and very little vibration as compared with piston-type compressors. Scroll compressors also have a check valve in the discharge chamber that prevents high-side gas from flowing into the low side when the compressor stops. This valve opens by pressure difference when the compressor restarts. Finally, the simplicity of the scroll compressor's design requires only the stationary

and the orbiting scroll for compressing gas. Piston-type compressors require about 15 parts to do the same task.

In summary, the main reasons scroll compressors are gaining popularity over piston-type compressors in energy efficiency, reliability, and quieter operation are as follows:

- There are no volumetric losses through gas reexpansion as with piston-type compressors.
- The scroll compressor requires no reed valves, so it does not have valve losses that contribute to inefficiencies as piston-type compressors do.
- Separation of suction and discharge gases reduces heat-transfer losses.
- Centrifugal forces within the mating scrolls maintain nearly continuous compression and constant, leak-free contact.
- Radial movement eliminates high-stress situations and allows for just the right amount of contact force between mating scroll surfaces. This action allows the compressor to handle some liquid.
- Scroll compressors maintain a continuous compression process and have no reed valves to create valve noise. This creates very soft gas-pulsation noises and very little vibration as compared with piston-type compressors.

TWO-STEP CAPACITY CONTROL FOR SCROLL COMPRESSORS

A cooling system that can modulate capacity within the compressor and a variable-speed blower motor for the conditioned air will deliver much tighter temperature control and a reduced humidity level within the conditioned space than will a standard cooling system. The overall comfort level for occupants will be much higher because the compressor will run longer at part load to reduce humidity levels and maintain very precise temperature levels.

Advancements in scroll compressor design allow the compressor to have a two-step capacity of either 67% or 100%. A modern, two-step scroll compressor, shown in **Figure 23.51**, has an internal unloading mechanism in the compressor that opens a bypass port or vent at the end of the first compression pocket. This internal unloading mechanism is a direct-current (DC) solenoid controlled by the second stage of a conditioned space thermostat in either the heating or cooling modes. The DC solenoid, which is controlled by a rectified, external, 24-V alternating-current (AC) signal initiated by the conditioned space thermostat, moves a slider that covers and uncovers the bypass ports or vents, **Figures 23.52(A)** and **(B)**. The compressed gas is then vented into the beginning of a suction pocket within the scroll. When the bypass or vent port is opened by deenergizing the DC solenoid, the effective displacement of the scroll is reduced to 67%. When the DC solenoid is energized by the second stage of the conditioned space thermostat, the bypass ports are blocked or closed and the scroll is at 100% capacity. Again, this opening and closing of the bypass ports or vents is controlled by an internal, electrically operated DC solenoid. The unloading and loading of the two-step scroll compressor is done while the compressor's motor is running—and without cycling the motor on and off. The compressor motor is a single-speed, high-efficiency motor that will continue to run while the scroll alternates between the two capacity steps. Whether in the high-capacity (100%) or low-capacity (67%) mode, the two-step scroll operates like a standard scroll compressor.

Figure 23.51 A modern, two-step modulating scroll compressor with either 100% or 67% capacity. Courtesy Emerson Climate Technologies

As mentioned earlier, the internal DC solenoid in the compressor that operates the internal unloading mechanism is energized by the second stage of a conditioned space thermostat. It is expected that the majority of run hours will be at low capacity (unloaded at 67%). It is in this mode that the solenoid is deenergized. This allows the two-stage thermostat to control capacity through the second stage of the thermostat in both cooling and heating, **Figure 23.52(C)**. An extra external electrical connection is made with a molded plug assembly that contains a full-wave rectifier to supply direct current to the solenoid unloaded coil, **Figure 23.51.** The rectifier is actually located in the external power plug on the compressor.

Figure 23.52 (A) Internal unloading mechanism showing modulating ring and solenoid coil assembly. Courtesy Emerson Climate Technologies. (B) Bypass ports shown open and closed. Courtesy Emerson Climate Technologies. (C) Electrical diagram of internal unloading and compressor motor circuit. Courtesy Emerson Climate Technologies

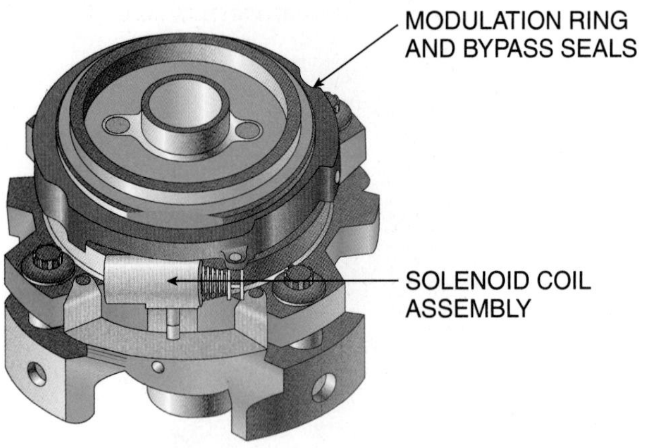

(A)

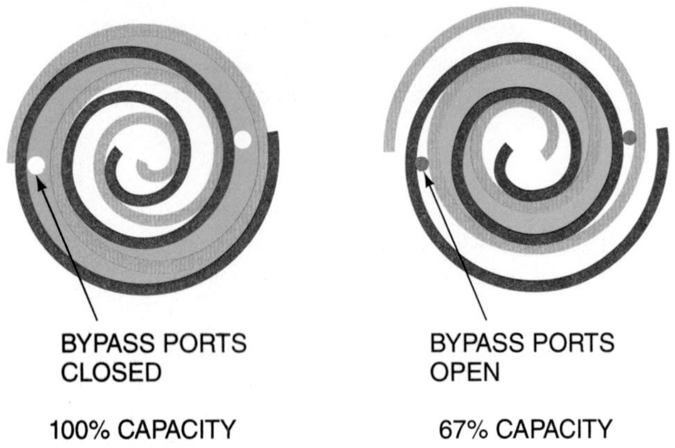

(B)

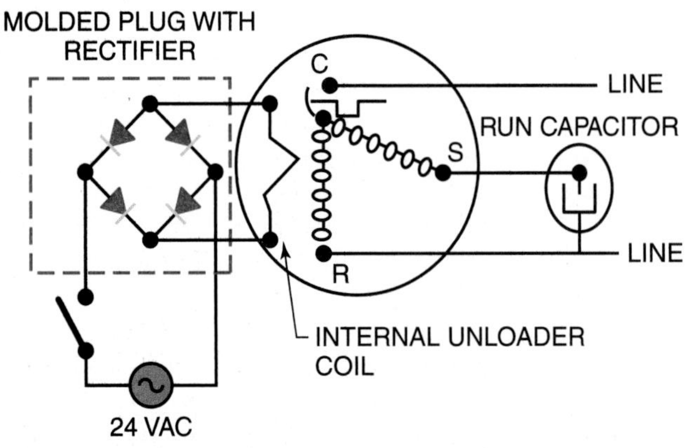

(C)

DIGITAL CAPACITY CONTROL FOR SCROLL COMPRESSORS

As mentioned, compressor capacity control is desirable for optimum system performance when loads vary over a wide range. Compressor capacity control through modulation can reduce power consumption, produce better dehumidification control, and reduce compressor cycling along with starting currents in smaller compressors. This is especially important when trying to cool an area that has different cooling needs throughout the day. With a digital capacity scroll doing the cooling, constant temperatures of ±0.5°F can be accomplished in every room, at any time of the day, whether the room is mostly empty or standing-room-only.

If the mating scroll members of a scroll compressor are separated axially, no refrigerant gas will be compressed and there will be only 10% power usage. If axial separation of the mating scrolls can be controlled by varying the amount of time they are separated, capacity control can be achieved at percentages between 10% and 100%. The separation of the mating scrolls is achieved by bypassing a controlled amount of discharge gas to the suction side of the compressor through a solenoid valve, **Figure 23.53.** When the pressure in the modulating chamber is lowered by energizing the solenoid valve, discharge gas will be metered through a bleed hole and the scrolls will separate axially, **Figure 23.54.** No flow of refrigerant gas will take place when the mating scrolls are separated, which will unload the scroll compressor. The scroll compressor is shown in both the loaded and unloaded positions in **Figure 23.55.**

A modulating capacity is achieved by either energizing or deenergizing the solenoid valve. An energized solenoid will unload the compressor by axially separating the scrolls, making the compressor's capacity zero. When the solenoid valve is deenergized, the compressor's capacity is 100%. Solenoid cycle times of between 10 and 30 seconds should be used to minimize solenoid-valve cycling and to make the system more responsive. One complete "cycle time" is a combination of solenoid-valve energized (unloaded) time plus deenergized (loaded) time. It is suggested that the solenoid never be deenergized less than 10% of the cycle time; this will ensure that there is enough refrigerant gas flow for motor cooling, since the digital scroll is a refrigerant-cooled compressor.

Example: If you have a 30-second cycle time and the solenoid is deenergized for 20 seconds and then energized for 10 seconds, the resulting capacity will be

$$20 \div 30 = 66.6\%$$

Any normal control parameter (including the surrounding air temperature, humidity, or suction pressure) can unload the compressor. A compressor discharge-line thermistor is required with a cut-out temperature of 280°F. The solenoid valve must have 15 watts of power at the appropriate voltage.

Figure 23.53 (A) Digital scroll compressor showing a disengaging solenoid valve for bypassing discharge gas to the suction side of the compressor. Courtesy Emerson Climate Technologies. (B) Digital scroll with internal piping and enclosed solenoid. Courtesy Emerson Climate Technologies

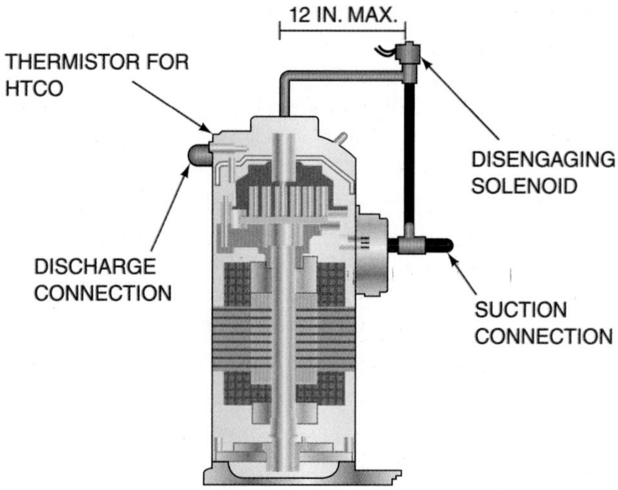

(A)

(B)

Figure 23.54 An internal view of the head of a digital scroll compressor showing modulation chamber, bleed hole, lift piston assembly, spring, and solenoid valve. Courtesy Emerson Climate Technologies

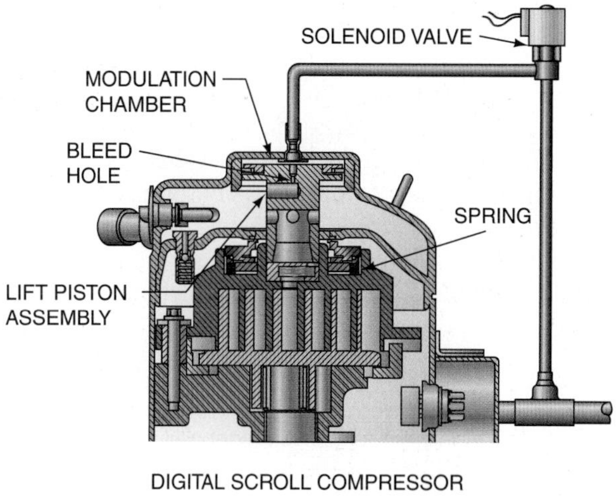

Figure 23.55 (A) Digital scroll compressor shown in the "loaded" position. The solenoid is deenergized. Courtesy Emerson Climate Technologies. (B) Digital scroll compressor shown in the "unloaded" position. Courtesy Emerson Climate Technologies

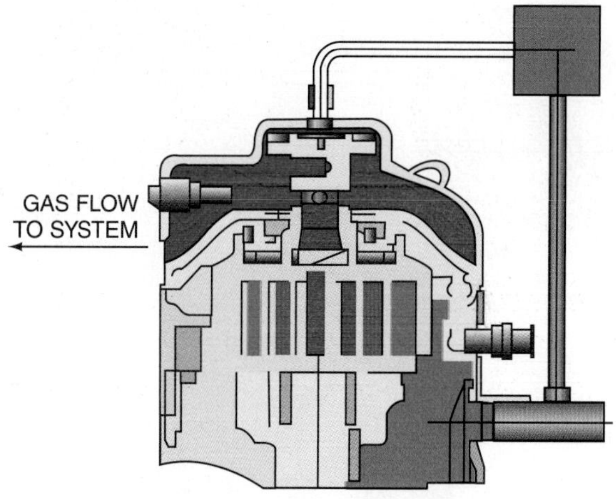

(A)

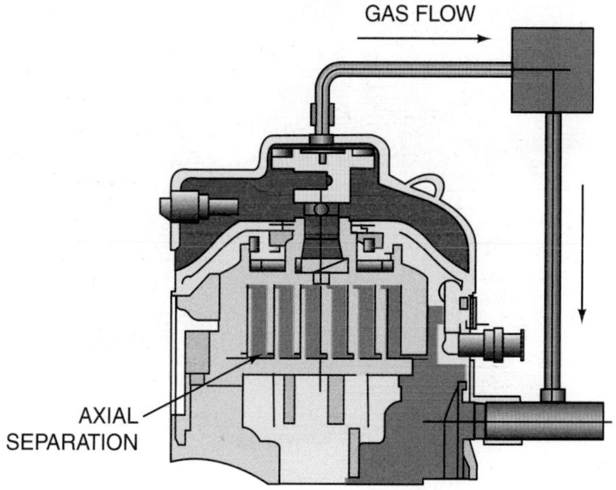

(B)

A controller that is used with the digital scroll compressor offers many protective features, such as phase control, short-cycling control, amperage and voltage imbalance controls, and high-amperage monitoring, to name a few. The controller can also supply a variable voltage to the unloading solenoid valve for open/closed time intervals, **Figure 23.56(A)**.

Figure 23.56(B) illustrates an electronic compressor protection module that uses the compressor as a sensor. The module monitors and interprets system and electrical information within the compressor, shutting it down before damage occurs. The module alerts service technicians to the most serious faults first, so they can accurately and rapidly correct problems and prevent compressor failures that could suddenly generate product spoilage. The electronic module also offers a safeguard against insufficient oil pressure and high motor temperatures, and is also capable of high compressor discharge temperature protection. User-friendly diagnostics optimize the service technician's time, saves money, and ultimately frees operators to focus on other priorities. Fault history and LED indicators allow service technicians to diagnose the past and present state of the system for rapid, accurate troubleshooting and increased uptime. When the module senses that the compressor is running in an unsustainable condition, it responds by shutting down the compressor to avoid damage. Based on fault severity, the compressor will either reset itself when the condition is cleared, or require a service technician to troubleshoot the fault before restarting. Its communication capability also makes it possible to retrieve information and take action remotely.

Figure 23.56 (A) A controller used with a digital scroll compressor. Courtesy Emerson Climate Technologies (B) A second-generation, modern, electronic compressor protection controller. Courtesy Emerson Climate Technologies

(A)

(B)

In summary, the digital scroll compressor can offer the following advantages:

- Ability to hold a precise temperature and humidity level
- Full-load and part-load efficiencies
- Ability to be 30% more efficient than traditional compressor modulation methods
- Less compressor cycling for longer compressor life
- Delivery of maximum comfort, efficiency, and reliability in one compressor

Many refrigeration and air-conditioning compressors today can employ electronic variable frequency drives (VFDs) to vary the frequency feeding their electric motors. VFDs work by converting the motor's input alternating current (AC) to direct current (DC). From the direct current, the VFD will generate a simulated AC signal at varying frequencies. Motor speed is directly proportional to the electrical frequency used by the motor. So, by varying the frequency to the compressor's motor, the compressor's motor speed, thus compressor capacity, can be controlled. This is a very effective way to accurately match compressor capacity to a load requirement. Refer to Unit 17, "Types of Electric Motors," for more in-depth coverage on variable frequency drives. Matching a compressor's capacity to the evaporator heat loading offers many advantages:

- Better efficiency at part load
- Less cycling on and off, which contributes to longer compressor life
- Less fluctuation in evaporator temperatures and pressures
- Gradual speed increases at start-up, meaning less liquid refrigerant and oil return to the compressor

Many modern VFDs can generate frequencies from 3.0 to over 350 hertz, which is outside the operating ranges of modern refrigeration and air-conditioning compressors. This is why compressor manufacturers will state minimum and maximum allowable frequencies and frequency ranges for their compressor motors when using a VFD. Many compressor operational characteristics can be affected if a compressor

motor does not operate within a certain frequency range. These may include:

- Compressor lubrication and cooling at low speeds
- Proper oil pump operation
- Compressor overheating
- Increased electrical motor losses
- Increased vibration
- Reed valve performance

NOTE: *Always consult with the specific compressor manufacturer before using a VFD controller on any compressor.* •

SCROLL COMPRESSOR PROTECTION

Many modern scroll compressors incorporate a variety of internal safety controls that can activate the internal line-break motor protection, **Figure 23.57.** Some safety features found in an air-conditioning scroll with less than 7 tons of capacity may include the following:

- Temperature-operated disc (TOD)—A bimetallic disc that senses compressor discharge temperature and opens at 270°F.
- Internal pressure relief (IPR)—Opens at an approximate 400 ± 50 psi differential between high- and low-side pressures for R-22, and at a 500 to 625 psi differential for R-410A.
- Floating seal—Separates the high side from the low side. Also prevents the compressor from drawing into a deep vacuum and damaging the fusite electrical terminal.
- Internal motor protection—Senses both internal temperatures and amperages.

Figure 23.57 Variety of internal safety controls for a modern scroll compressor. Courtesy Emerson Climate Technologies

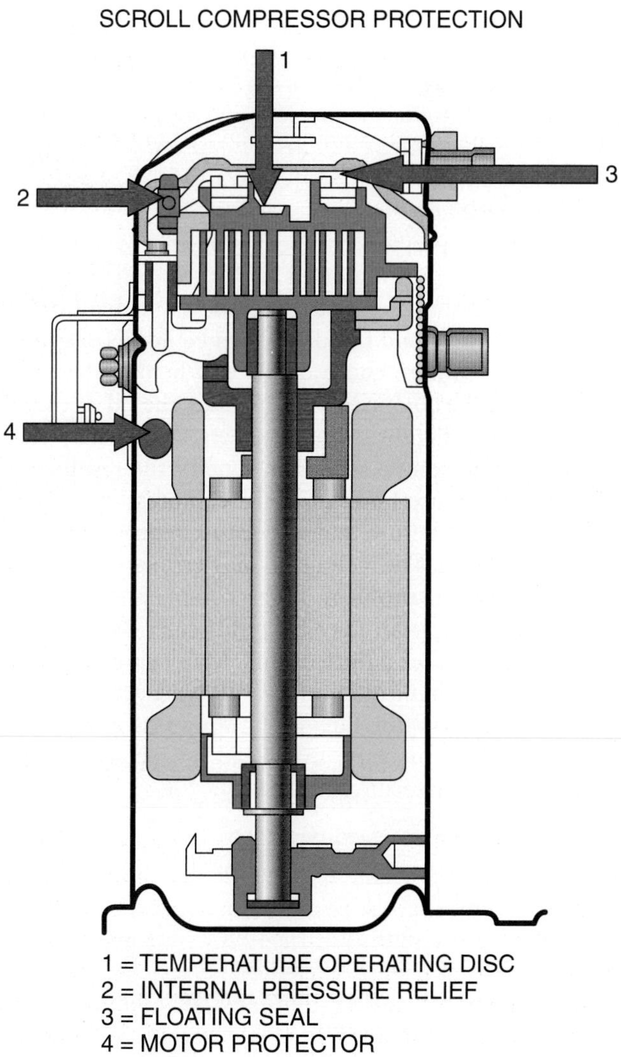

Another innovative scroll compressor protection device is used to ensure better reliability, diagnostic accuracy, speed of service, and a reduction in the number of callbacks when the technician is systematically troubleshooting the compressor. The diagnostic controller is installed in the electrical box of a commercial condensing unit or residential unit or inside a rooftop unit. It is completely self-contained and has no external sensors, **Figure 23.58.** In fact, it actually uses the compressor as a sensor because the compressor's electrical lines

Figure 23.58 Diagnostic controller with no external sensors. Courtesy Emerson Climate Technologies

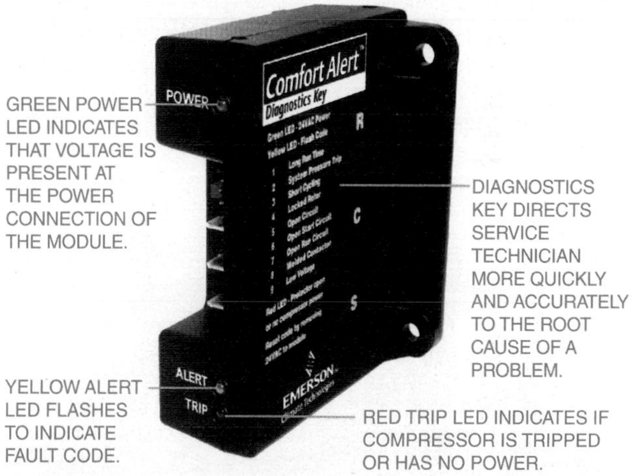

Commercial Comfort Alert Diagnostics Codes

ALERT CODE	SYSTEM CONDITION	ALERT INDICATOR BLINKS	LOCKOUT
Code 2	System pressure trip	2 times	Yes
Code 3	Short-cycling	3 times	Yes
Code 4	Locked rotor	4 times	Yes
Code 5	Open circuit	5 times	No
Code 6	Missing phase	6 times	Yes
Code 7	Reverse phase	7 times	Yes
Code 8	Welded contactor	8 times	No
Code 9	Low voltage	9 times	No

run through the device, which acts as a current transformer. It monitors vital information from the scroll compressor that will help pinpoint the root cause of cooling system problems, such as electrical problems, compressor defects, and other general system faults. Phase dropouts, miswiring, and short cycling can also be detected with this device. A flashing light-emitting diode (LED) will quickly communicate the alert code and direct the service technician to the problem.

CENTRIFUGAL COMPRESSOR

A revolutionary oil-free centrifugal compressor has been designed for the HVACR industry. This centrifugal compressor has magnetic bearings, variable speeds, digital electronic technology, and achieves very high efficiencies, **Figure 23.59.**

The rotor shaft and impellers are the only moving parts in this compressor. The shaft acts as the rotor for the permanent-magnet synchronous motor. The rotor shaft has magnetically levitated bearings for frictionless operation, **Figures 23.60(A)** and **(B)**. The magnetic bearings and bearing sensors are composed of both permanent magnets and electromagnets. They precisely control the frictionless compressor shaft rotation on a levitated magnetic cushion, **Figure 23.60(B)**. Bearing sensors, located at each magnetic bearing, feed back rotor orbit information in real time to the bearing control system, **Figure 23.61.**

The permanent-magnet synchronous motor is powered by a pulse-width-modulated (PWM) voltage supply. High-speed variable frequency drive (VFD) operation allows for high efficiency, compactness, and soft-start capability. Soft starting reduces high in-rush currents at start-up. The variable frequency drive has an inverter that converts a DC voltage into an adjustable three-phase AC voltage. Signals from the motor/bearing controller determine the inverter output frequency, torque, and voltage phase—all of which regulate the motor speed. The compressor motor can reduce speed as the condensing temperatures or heat loads are reduced. This allows the variable-speed compressor to optimize energy performance from 100% to 20% of rated capacity.

The centrifugal compressor has onboard digital control electronics that oversee its operation. The electronic package also provides for Web-enabled monitoring access and external control. This provides valuable performance and reliability information. It also reduces costs because it replaces the power and control panel functions seen on traditional chillers and rooftop packages.

Figure 23.59 Oil-free centrifugal compressor with variable frequency drive (VFD), onboard digital electronic controls, and magnetically levitated bearings. Source: Danfoss Torbocor Compressors, Inc.

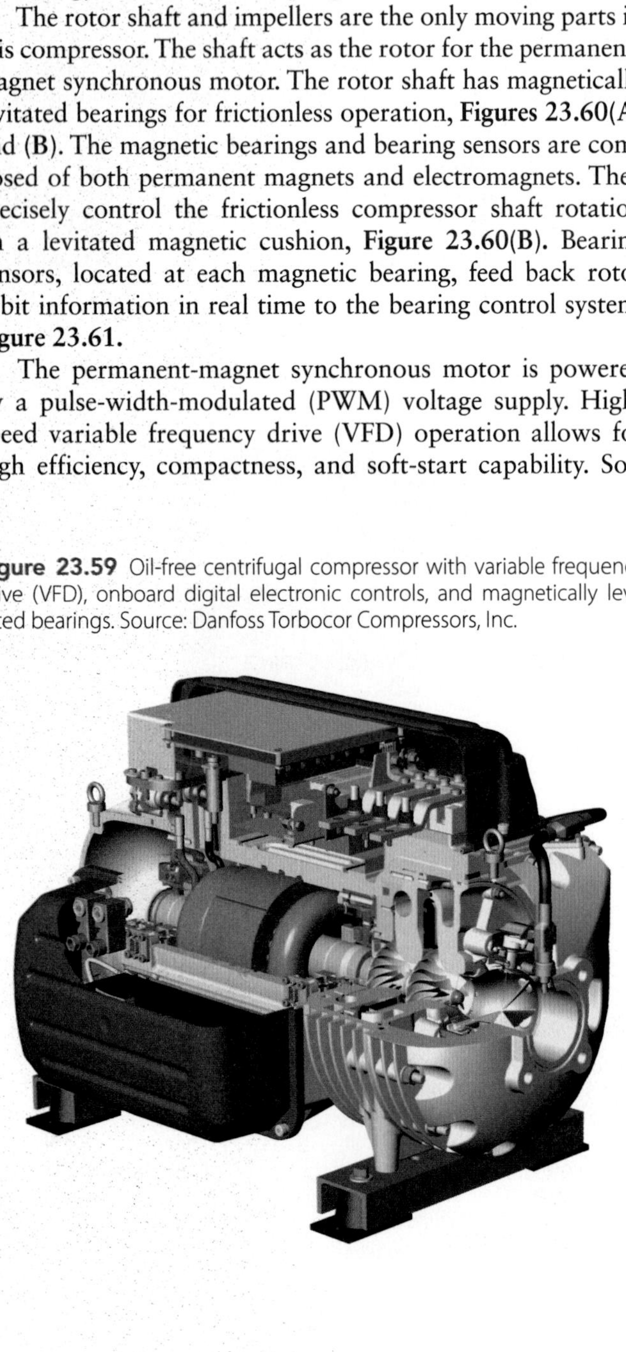

23.9 LIQUID IN THE COMPRESSOR CYLINDER

The piston-type reciprocating compressor is known as a positive displacement device. This means that when the cylinder starts on the upstroke, it is going to empty itself or stall. **SAFETY PRECAUTION:** *If the cylinder is filled with liquid refrigerant that does not compress (liquids are not compressible), something is going to break. Piston breakage, valve breakage, and rod breakage can all occur if a large amount of liquid reaches the cylinder,* **Figure 23.62.** •

Liquid in compressors can be a problem from more than one standpoint. Large amounts of liquid, called a *slug* of liquid or **liquid slugging**, usually cause immediate damage. Small amounts of liquid floodback can be just as detrimental but take more time to cause damage. When small amounts of liquid enter the compressor, they dilute the oil. If the compressor has no oil lubrication protection, this may not be noticed until the compressor fails. One of the results is that marginal oil pressure may cause the compressor to throw a rod, a mechanical failure that may cause motor damage and burn out the motor. The technician may erroneously diagnose this as an electrical problem. If the compressor is a welded hermetic, the technician may have difficulty diagnosing the problem until another failure occurs. Many electrical failures are caused by mechanical failures. Following are some causes that may lead to liquid slugging a compressor with liquid refrigerant, which in turn may lead to electrical failures because of a motor seizure:

- Metering device overfeeding
- Low air circulation over evaporator
- Low heat load
- Evaporator fan motor not working
- Dirty or iced-up evaporator coil
- Defrost heaters or defrost timer not working
- Dirty or blocked evaporator filters

Figure 23.60 (A) Rotor shaft, centrifugal impellers, and magnetically levitated bearings. Source: Danfoss Torbocor Compressors, Inc. (B) Rotor shaft shown cushioned on magnetically levitated bearings. Based on Danfoss Torbocor Compressors, Inc.

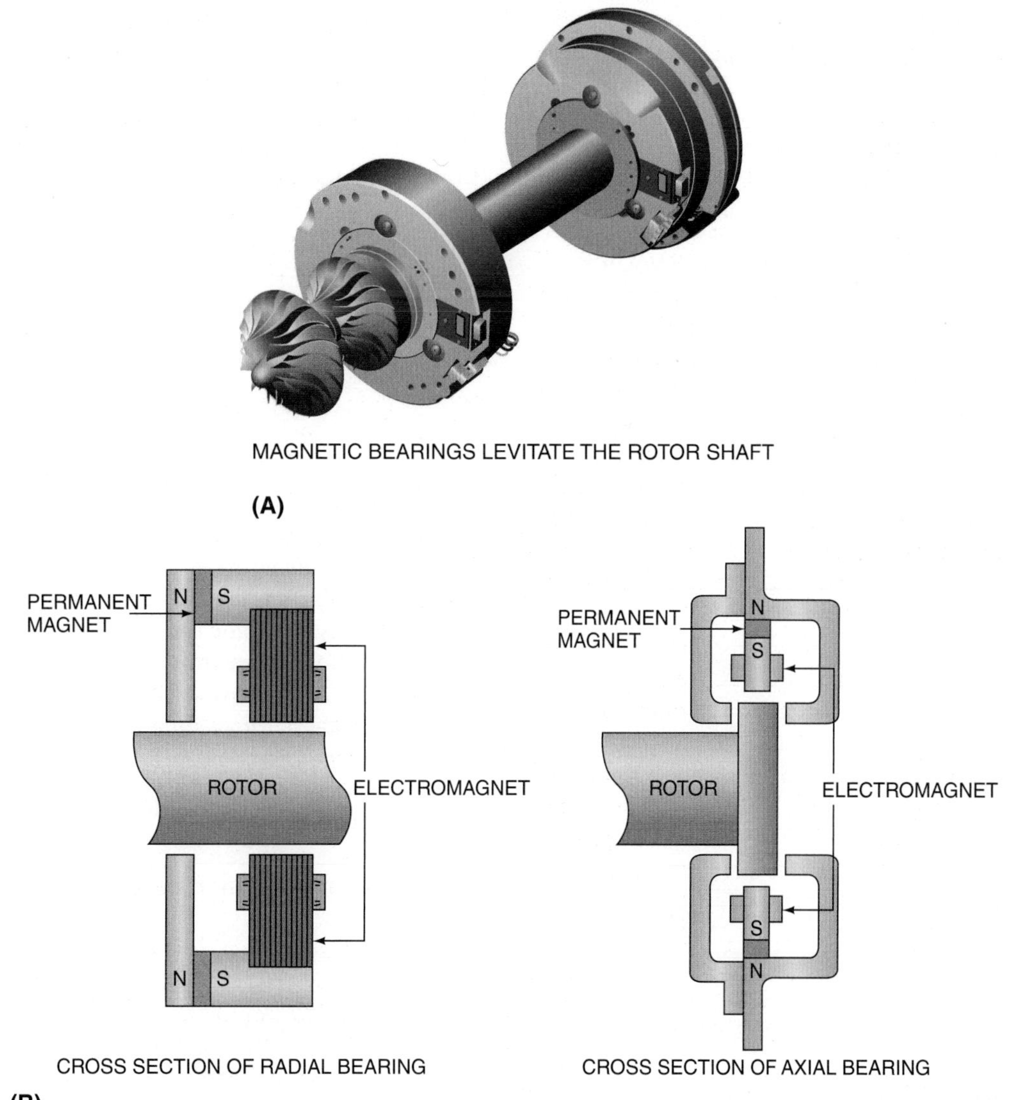

Figure 23.61 Bearing control system and sensors. Based on Danfoss Torbocor Compressors, Inc.

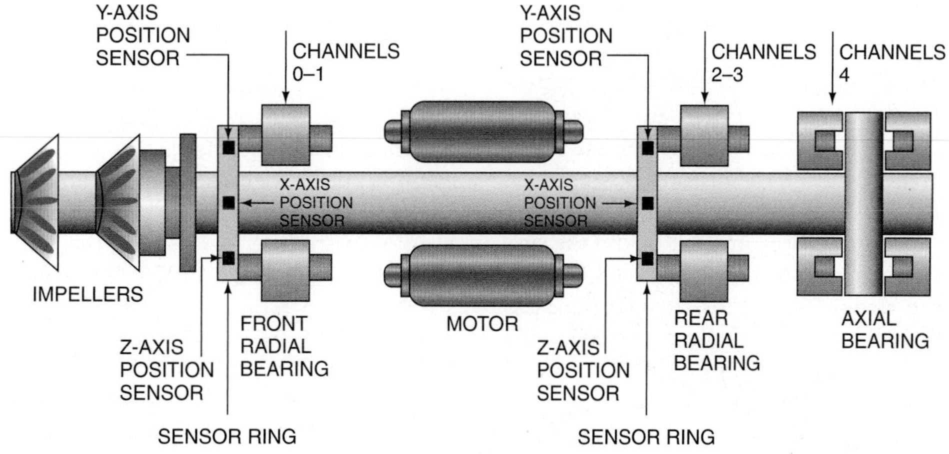

Figure 23.62 A cylinder trying to compress liquid.

NOTE: *Never take a compressor failure for granted. Give the system a thorough check on start-up and actively look for a problem.* •

23.10 SYSTEM MAINTENANCE AND COMPRESSOR EFFICIENCY

The compressor's overall efficiency can be improved by maintaining the correct working conditions. This involves keeping the suction pressure as high as practical and the head pressure as low as practical within the design parameters.

A dirty evaporator will cause the suction pressure to drop. When the suction pressure goes below normal, the vapor that the compressor is pumping becomes less dense and gets thin (sometimes called rarified vapor). The compressor performance declines. Low suction pressures also cause the high-pressure gases caught in the clearance volume of the compressor to expand more during the piston's downstroke. This gives the compressor a lower volumetric efficiency. These gases have to expand to a pressure just below the suction pressure before the suction valve will open. Because more of the downward stroke is used for reexpansion, less of the stroke can be used for suction. Suction ends when the piston reaches bottom dead center.

In the past, not much attention was paid to dirty evaporators. Technicians are now beginning to realize that low evaporator pressures cause high compression ratios. For example, if an ice cream storage walk-in cooler is designed to operate at −5°F room temperature, it would result in a coil temperature of about −20°F (using a coil-to-return-air temperature of 15°F). If the refrigerant is R-134a, the suction pressure would be 3.6 psig, slightly above atmospheric pressure. Assume the condenser will operate with a 25°F temperature difference between the outside air and the condensing temperature. If the outside air is 95°F, the condensing temperature would be 120°F (95°F + 25°F). The condensing pressure would then be 171.2 psig. The following calculation would show the compression ratio to be 10:1, **Figure 23.63(A)**.

Compression Ratio = Absolute Discharge Pressure
÷ Absolute Suction Pressure

CR = [171.2 psig + 14.7 atmosphere]
÷ [3.6 psig + 14.7 atmosphere]

CR = 10.15:1

If the coil were old and dirty, **Figure 23.63(B)**, and the coil temperature reduced to −30°F, assuming the head pressure remained the same, a new refrigerant boiling temperature of −30°F would produce a suction pressure of 9.7 in. Hg vacuum. This is below atmospheric pressure. It is an absolute pressure of 9.93 psia. (This was arrived at by converting inches of mercury vacuum to inches of mercury absolute, then converting this to psia.)

29.92 in. Hg absolute − 9.7 in. Hg vacuum
= 20.22 in. Hg absolute

20.22 in. Hg absolute ÷ 2.036 in. Hg per psi = 9.93 psia

The new compression ratio is 18.72:1

CR = [171.2 + 14.7 atmosphere] ÷ 9.93 psia

CR = 18.72:1

This compression ratio is too high. Most manufacturers recommend that the compression ratio not be more than 12:1.

The same situation could occur if the store manager were to turn the refrigerated box thermostat down to −15°F. The coil temperature would go down to −30°F and the compression ratio would be 18.72:1. This is too high. Many low-temperature coolers are operated at temperatures far below what is necessary. This condition often causes compressor problems.

Dirty condensers also cause compression ratios to rise, but they do not rise as fast as with a dirty evaporator. For example, if the ice cream storage box above were operated at 3.6 psig and the condenser became dirty to the point that the head pressure rose to 190 psig, the compression ratio would be 11.18:1, **Figure 23.63(C)**. This is not good but not as bad as the 18.72:1 ratio in the previous example.

CR = [190 psig + 14.7 atmosphere]
÷ [3.6 psig + 14.7 atmosphere]

CR = 11.18:1

If both situations were to occur at the same time, as in **Figure 23.63(D)**, the compressor would have a compression ratio of 20.61:1. The compressor would probably not last very long under these conditions.

CR = [190 psig + 14.7 atmosphere] ÷ 9.93 psia

CR = 20.61:1

The technician has a responsibility to maintain equipment to obtain the greatest efficiency and to otherwise protect the equipment with a maintenance program. Clean coils are part of this program. The owner usually does not know the difference. A dirty condenser makes the head pressure rise. This causes the amount of refrigerant in the clearance volume (at the top of the compressor cylinder) to be greater than the design conditions allow and the compressor efficiency to drop. If there is a dirty condenser (high head pressure) and a dirty evaporator (low suction pressure), the compressor will run longer to keep the refrigerated space at the design temperature. The overall efficiency drops. A customer with a lot of equipment may not be aware of this if only part of the equipment is not efficient. When the efficiency of a compressor drops, the owner is paying more money for less refrigeration. A good maintenance program is economical in the long term.

Figure 23.63 (A) A system operating with a normal compression ratio. Notice that the condenser is large enough to keep the head pressure low. (B) This system has a dirty evaporator.

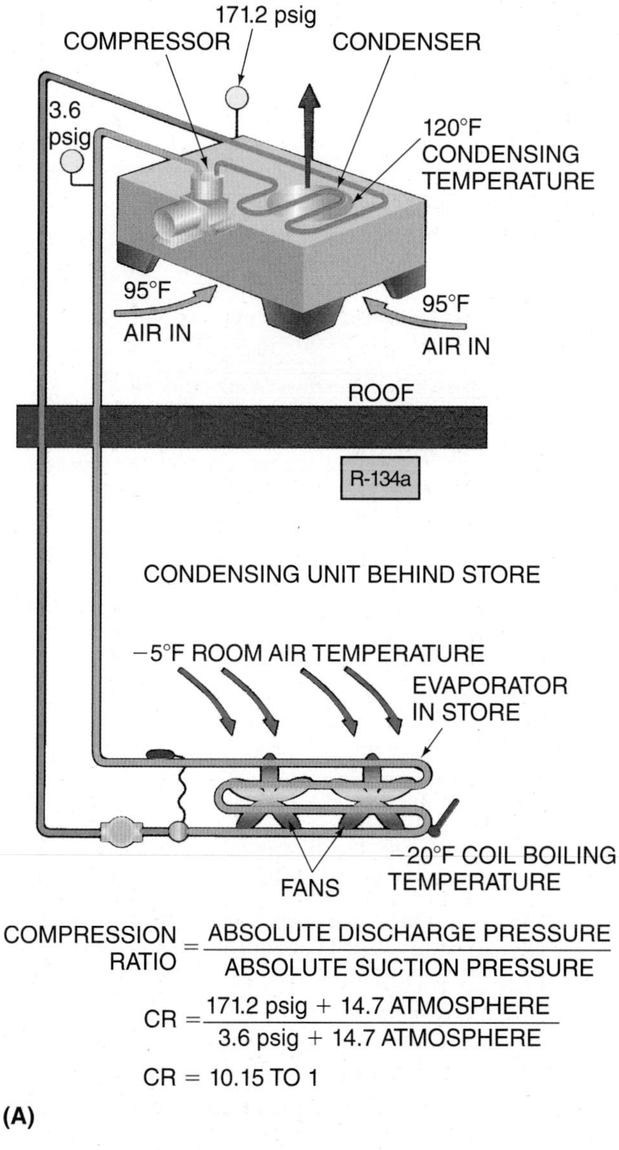

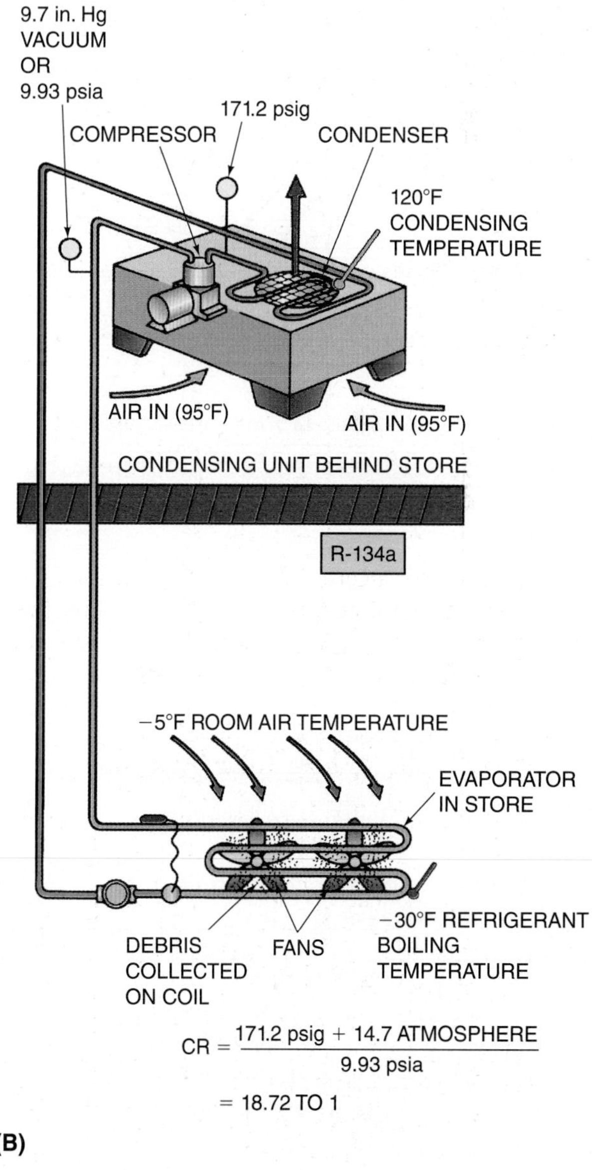

Figure 23.63 *(Continued)* (C) A system with high head pressure because the condenser is dirty. (D) This system has a dirty condenser and a dirty evaporator.

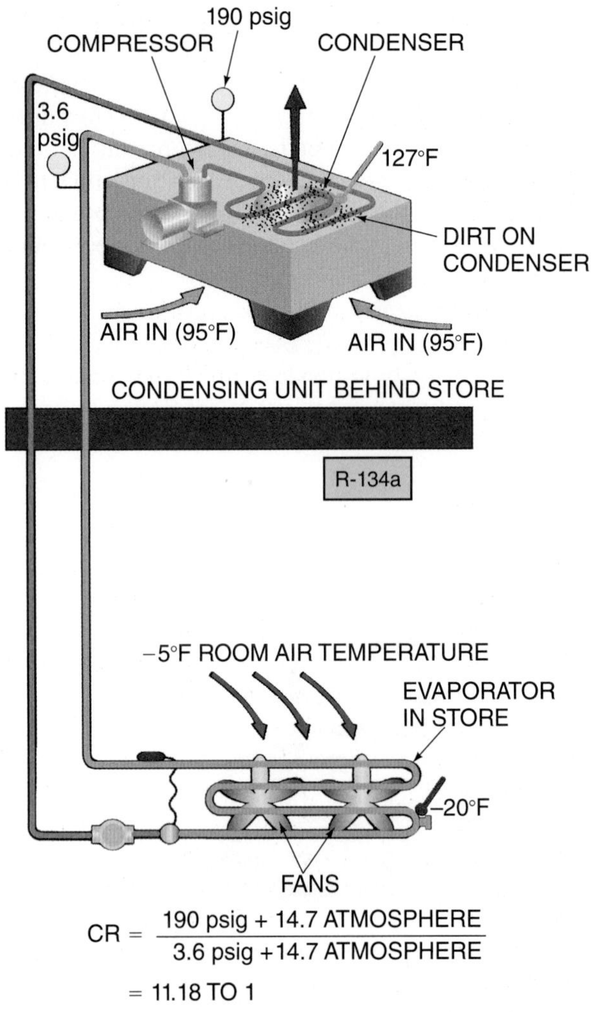

(C)

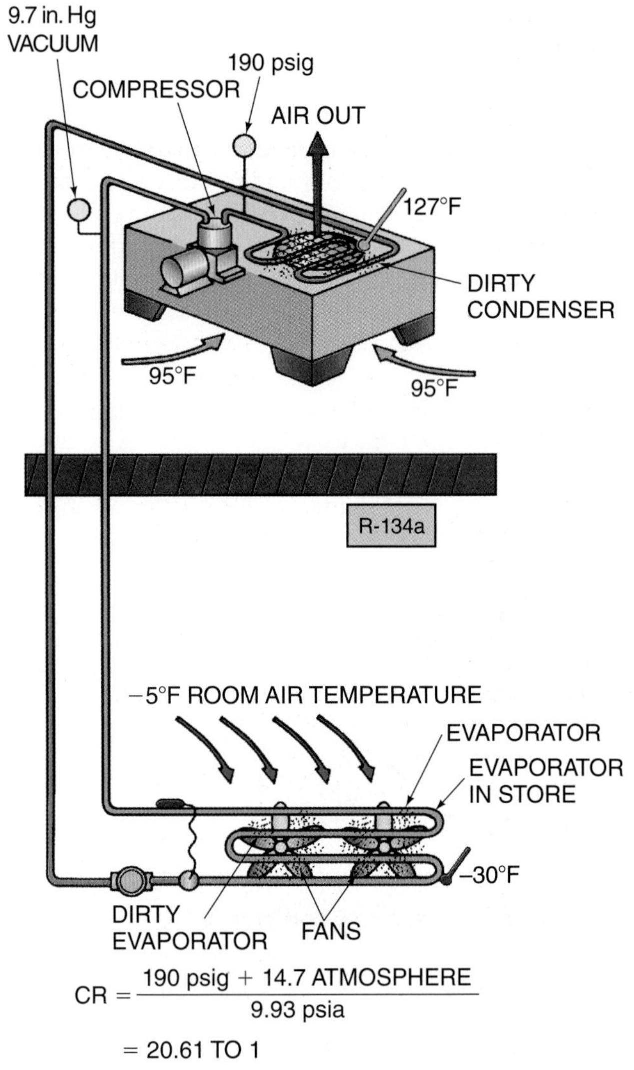

(D)

REVIEW QUESTIONS

1. Which of the following best describes the function of a compressor?
   A. The compressor restricts the flow of refrigerant that enters the evaporator.
   B. The compressor pumps liquid refrigerant into the condenser to transfer heat.
   C. The compressor is used to throttle a thermostatic expansion valve (TXV) open or closed.
   D. The compressor increases the pressure in the system from the low-side pressure to the high-side pressure.
2. Five types of compressors are ____________.
   A. rotating, analog, scroll, screw, and binary
   B. reciprocating, rotary, scroll, screw, and centrifugal
   C. bolted, screw, scroll, binary, and centrifugal
   D. belt-drive, direct-drive, rotating, binary, and scroll
3. Which of the following is true of HVACR compressors?
   A. A compressor cannot compress a liquid.
   B. A compressor can compress a liquid.
   C. A compressor requires liquid refrigerant for lubrication.
   D. A compressor takes in vapor and discharges liquid.

4. What would be considered a higher-than-normal temperature for a discharge gas line on a reciprocating compressor?
   A. 190°F
   B. 200°F
   C. 250°F
   D. 175°F
5. Which compressor type uses pistons to compress the gas?
   A. Scroll
   B. Reciprocating
   C. Rotary
   D. Screw
6. Which compressor type uses tapered, machined gear components to trap the gas for compression?
   A. Scroll
   B. Reciprocating
   C. Rotary
   D. Screw
7. Which style of compressor uses belts to turn the compressor?
   A. Open drive
   B. Closed drive
   C. Chassis drive
   D. Hermetic drive
8. What would be done to increase the compressor speed on a belt-driven compressor?
   A. Decrease drive pulley size
   B. Increase drive pulley size
   C. Increase driven pulley size
   D. Both A and C
9. Which of the following design factors can affect the efficiency of a reciprocating compressor?
   A. The two tapered, machined, screw-type gears.
   B. The connecting rod and the muffler.
   C. The piston and the internal springs.
   D. The valve plate design and clearance volume.
10. A medium-temperature R-134a system that has a suction pressure of 22 psig and a discharge pressure of 185 psig would have a compression ratio of _________.
    A. 8.40 to 1
    B. 5.44 to 1
    C. 10.1 to 1
    D. 13.89 to 1
11. A direct-drive compressor typically runs at _______ rpm.
    A. 1750 or 3450
    B. 3450 or 4500
    C. 17,500 or 34,500
    D. 1800 or 4800
12. A piston, connecting rod, crankshaft, valves, valve plate, head, shaft seal, internal motor overload device, pilot-duty motor overload device, and coupling are parts that may be found in a __________ compressor.
    A. scroll
    B. reciprocating
    C. variable capacity
    D. screw
13. What effect does a slight amount of liquid refrigerant have on a compressor over a long period of time?
    A. It causes excessive compressor cooling which may freeze the crankshaft.
    B. It adds to the compressor lubrication.
    C. It dilutes the oil and causes wear on any moving parts.
    D. It causes a rise in pressure which may open the internal pressure relief valve.
14. What lubricates the refrigeration compressor?
    A. Refrigerant
    B. Moisture
    C. Desiccants
    D. Oil
15. What can the service technician do to keep the refrigeration system operating at peak efficiency?
    A. Recover the refrigerant charge and replace it annually.
    B. Replace the compressor oil every 3 years or 3000 hours of operation.
    C. Replace the compressor crankshaft and connecting rods regularly.
    D. Clean the evaporator and condenser coils to keep pressures within design parameters.
16. A _________ compressor is usually considered to be a low-side device because the suction gas is vented to the inside of the shell, including the crankcase.
    A. screw
    B. hermetic
    C. open-drive
    D. belt-drive
17. State the main advantage that a discus compressor's valve plate design has over a conventional piston-type compressor.
18. Briefly describe how the scroll compressor compresses refrigerant gas.
19. List four main advantages a scroll compressor has over a conventional piston-type compressor.
20. What are two main reasons why a compressor should have modulating capacity control?
21. Briefly describe how the digital capacity control for a scroll compressor works.
22. List the three main advantages of having digital capacity control on a scroll compressor.

UNIT 24

EXPANSION DEVICES

OBJECTIVES

After studying this unit, you should be able to

- discuss the operating characteristics of a thermostatic expansion valve.
- describe the components of a thermostatic expansion valve.
- discuss the four types of thermostatic expansion valve bulb charges.
- explain the difference between an internally and externally equalized thermostatic expansion valve.
- describe the operation of pressure-limiting and balanced-port expansion valves.
- discuss the selection, installation, and service of expansion valves.
- describe how stepper motors, electronic expansion valves, and their controllers operate.
- explain the operating characteristics of an automatic expansion valve.
- discuss the operating characteristics of systems equipped with a capillary tube or piston metering device.
- discuss the system charge requirements for equipment with a fixed-bore metering device.

SAFETY CHECKLIST

✔ Wear warm clothing when working in a walk-in cooler or freezer.

24.1 EXPANSION DEVICES

The **expansion device**, often called the metering device, is the fourth component necessary for the compression refrigeration cycle to function. The expansion device is not as visible as the evaporator, the condenser, or the compressor. Generally, it is concealed inside the evaporator cabinet and not obvious to the casual observer. The device can be either a valve or a fixed-bore. **Figure 24.1** illustrates an expansion valve installed inside an evaporator cabinet.

The expansion device is one component that divides the high side of the system from the low side (the compressor is the other). **Figure 24.2** shows its location. It is responsible for metering the correct amount of refrigerant to the evaporator. The evaporator performs best when it is as full of liquid refrigerant as possible, with none left by the time it reaches the suction line. Liquid refrigerant that enters the suction line may reach the compressor because only a small amount of heat is added in the suction line. Later in this section we will discuss a suction-line heat exchanger for special applications that is used to boil away liquid that may be in the suction line. Usually, liquid in the suction line is a problem.

Figure 24.1 Metering devices installed on a refrigerated case. The valve is not out in the open and is not as visible as the compressor, condenser, or evaporator. Courtesy Ferris State University. Photo by john Tomczyk

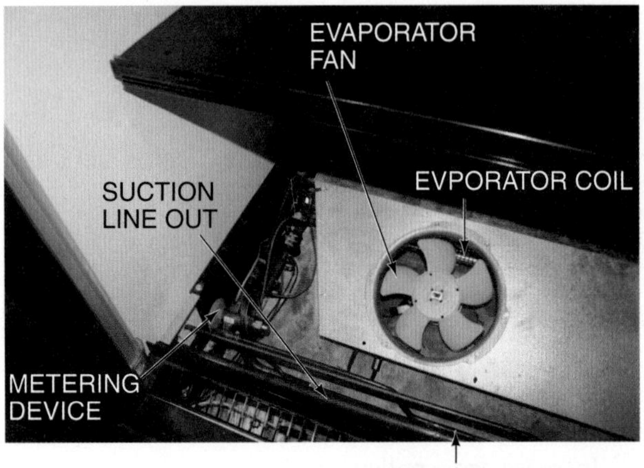

The expansion device is normally installed in the liquid line between the condenser and the evaporator. On a hot day, the liquid line may be warm to the touch and thus can be followed quite easily to the expansion device, where there is a pressure drop and an accompanying temperature drop. For example, the liquid line entering the expansion device may be 110°F on a hot day. In a low-temperature cooler using R-134a, the low-side pressure on the evaporator side may be 3 psig at a temperature of −8°F. This dramatic temperature drop can be easily detected when found. The device may be warm on one side and frosted on the other, **Figure 24.3**. Because some expansion devices are valves and some are fixed-bore devices, the temperature change can occur in a very short distance—less than an inch on a valve or a more gradual change on some fixed-bore devices.

24.2 THERMOSTATIC EXPANSION VALVE

Expansion devices can be one of the following different types: (A) thermostatic expansion valve, (B) automatic expansion valve, and (C) fixed bore, such as the capillary

Figure 24.2 The complete refrigeration cycle with the four basic components: compressor, condenser, evaporator, and expansion device.

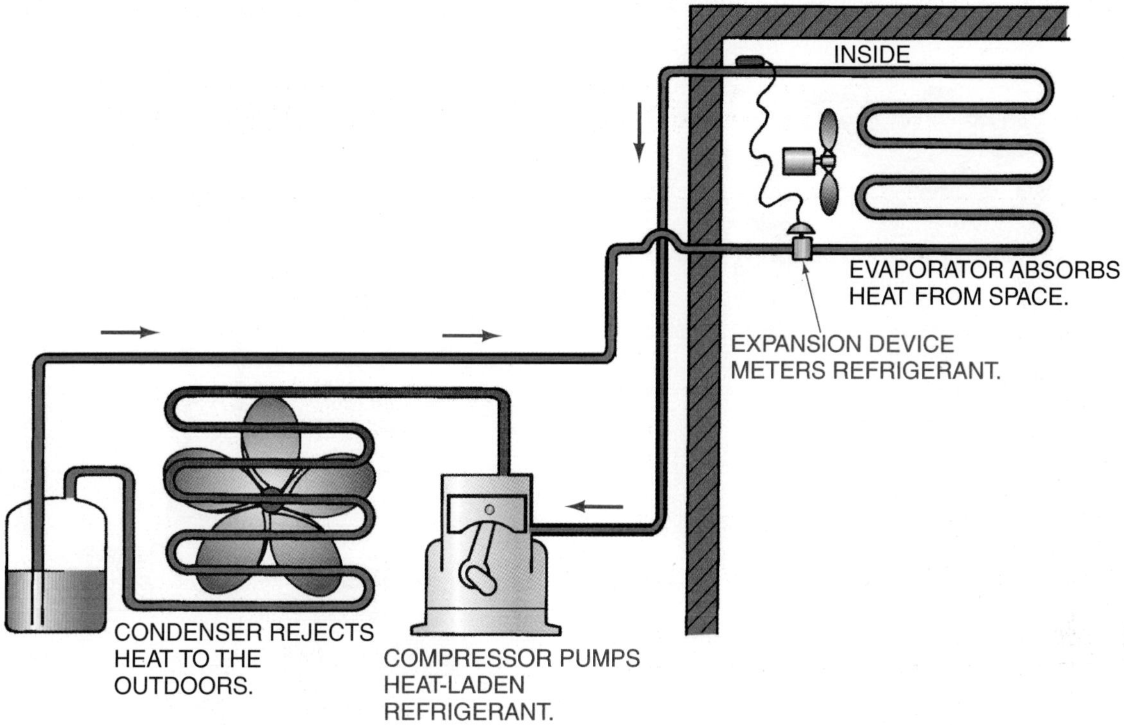

Figure 24.3 The expansion device has a dramatic temperature change from one side to the other. Based on Parker Hannifin Corp.

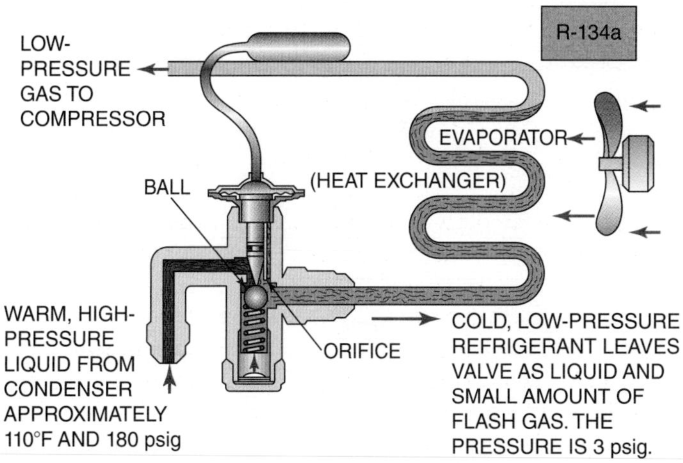

tube or piston, **Figure 24.4**. The *thermostatic expansion valve* (TXV) meters the refrigerant to the evaporator by using a thermal sensing element to monitor the superheat. This valve opens or closes in response to a thermal element. The TXV maintains a constant superheat in the evaporator. Remember, when there is superheat, there is no liquid refrigerant. Excess superheat is *not* desirable, but a small amount is necessary with this valve to ensure that no liquid refrigerant leaves the evaporator. *The TXV can be adjusted to maintain a low superheat to ensure that the majority of the evaporator surface is being used. This will give the refrigeration system a higher net refrigeration effect and higher capacities and efficiencies.*

The service technician should always check evaporator superheat to make sure the evaporator is being fully utilized. When adjusting the TXV, care should be taken to do so slowly. The superheat spring adjustment should be turned in the counterclockwise direction about one-half to one full turn at a time. The system should then be allowed to operate for approximately 15 min before making additional adjustments. Evaporator superheat measurements should be taken after each adjustment to help ensure that liquid refrigerant is not leaving the evaporator. Recommended approximate evaporator superheat settings are as follows:

- Comfort cooling (40°F) 10–16°
- High-temperature evaporators (30°F) 10–12°
- Medium-temperature evaporators (0°–30°F) 5–10°
- Low-temperature evaporators (below 0°F) 2–5°

The lower the evaporator temperature, the more necessary it is to keep the evaporator active with liquid refrigerant and to reduce superheat. At the lower evaporator temperatures, system efficiencies and capacities are at their lowest and there is a greater need for an active evaporator.

Figure 24.4 Three metering devices. (A) Thermostatic expansion valve. Courtesy Sporlan Division, Parker Hannifin Corp. (B) Automatic expansion valve. Photo by Bill Johnson (C) Capillary tube with liquid-line drier. Photo by Jason Obrzut

(A)

(B)

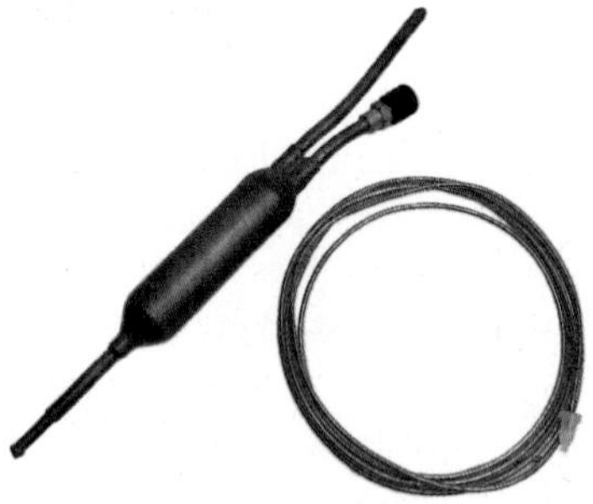

(C)

24.3 TXV COMPONENTS

The TXV consists of the valve body, diaphragm, needle and seat, spring, adjustment and packing gland, and the sensing bulb and transmission tube, **Figure 24.5**.

Figure 24.5 An exploded view of the TXV. All parts are visible in the order in which they go together in the valve. Courtesy Sporlan Division, Parker Hannifin Corp.

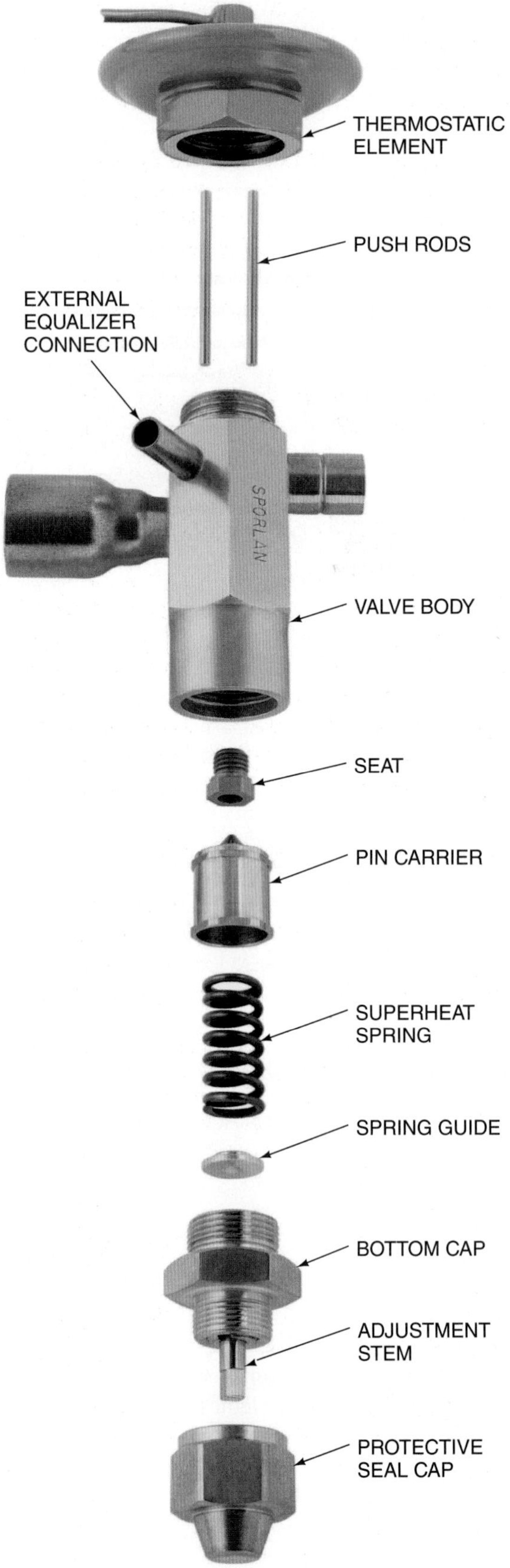

24.4 THE VALVE BODY

On common refrigerant systems the *valve body* is an accurately machined piece of solid brass or stainless steel that holds the rest of the components and fastens the valve to the refrigerant piping circuit, **Figure 24.6**. Notice that the valves have different configurations. Some of them are single-body pieces and cannot be disassembled, and some are made so that they can be taken apart.

TXVs may be fastened to a system by one of three methods: flare, solder, or flange. Future service should be considered when installing a valve in a refrigeration system, so a flare connection or a flange-type valve should be used, **Figure 24.7**. If the connection is to be soldered, a valve that can be disassembled and rebuilt in place is desirable, **Figure 24.8**. The valve often has an inlet screen with a very fine mesh to strain out any small particles that may obstruct the needle and seat, **Figure 24.9**.

Some valves have a third connection called an external equalizer on the side of the valve close to the diaphragm, **Figure 24.10**. This connection is normally a 1/4-in. flare or 1/4-in. solder. As will be discussed later, the diaphragm of the expansion valve registers the evaporator pressure. A pressure connection at the end of the evaporator supplies the evaporator pressure to the diaphragm. If an evaporator has a very long circuit, a pressure drop may occur; an external equalizer is used to correct this. Some evaporators also have several circuits and the method of distributing the refrigerant will cause a pressure drop between the expansion valve outlet and the evaporator inlet. This type of installation also must have an external equalizer for the expansion valve to provide the proper flow of refrigerant to the evaporator, **Figure 24.11**.

24.5 THE DIAPHRAGM

The diaphragm is located inside the valve body and moves the needle in and out of the seat in response to system load changes. The diaphragm is made of thin metal and is under the round, dome-like top of the valve, **Figure 24.12**.

Figure 24.6 TXVs have remote sensing elements. (A) Some of these valves are one-piece valves that are thrown away when defective. Photo by Jason Obrzut. (B) Some of the remote sensing elements can be replaced. This feature is particularly good when the valve is soldered into the system. Courtesy Sporlan Division, Parker Hannifin Corp.

(A)

(B)

Figure 24.7 (A) The flare and (B) the flange-type valve. Either can be removed from the system and replaced easily if it is installed where it can be reached with wrenches. Images, courtesy of Parker Sporlan.

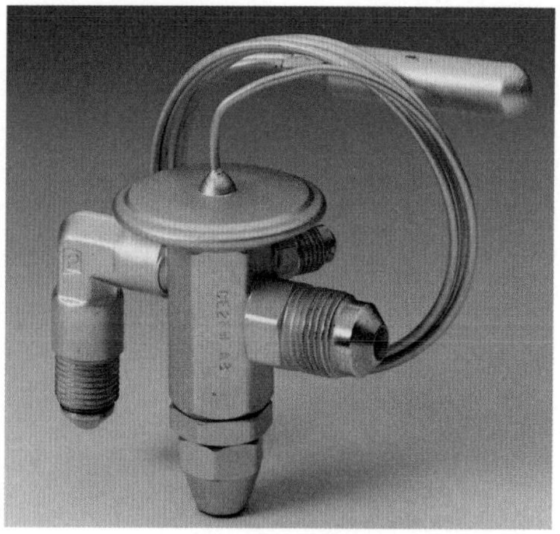

(A)

(B)

Figure 24.8 A solder-type valve that can be disassembled and rebuilt without taking it out of the system. This valve can be quite serviceable in some situations. Courtesy Sporlan Division, Parker Hannifin Corp.

Figure 24.9 Most valves have some sort of inlet screen to strain any small particles out of the liquid refrigerant before they reach the very small opening in the expansion valve. Courtesy Ferris State University. Photo by John Tomczyk

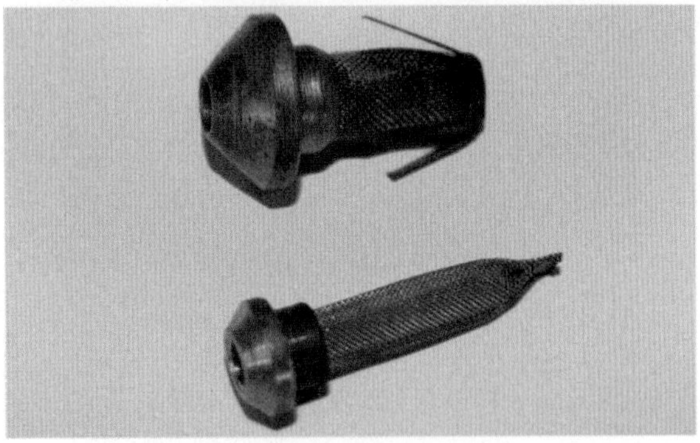

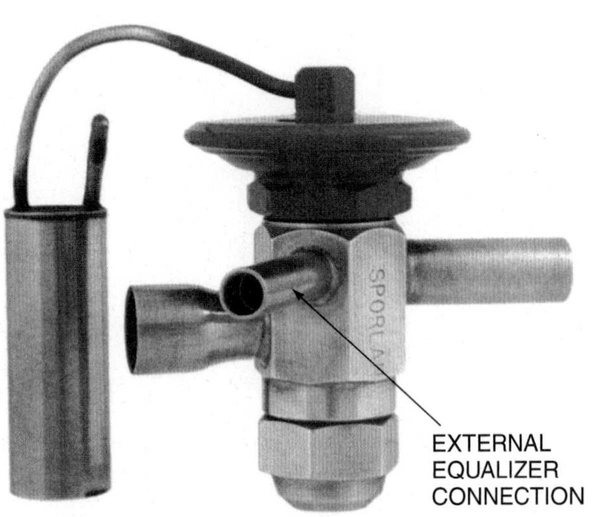

Figure 24.10 The third connection on this expansion valve is called the external equalizer. Courtesy Sporlan Division, Parker Hannifin Corp.

Figure 24.11 An evaporator with circuits and an external equalizer line to account for pressure drop. When an evaporator becomes so large that the length would create pressure drop, the evaporator is divided into circuits, and each circuit must have the correct amount of refrigerant.

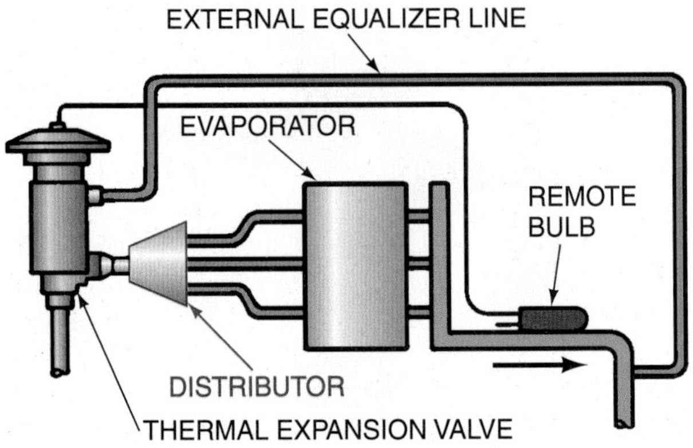

Figure 24.12 The diaphragm in the expansion valve is a thin membrane that has a certain amount of flexibility. It is normally made of a hard metal such as stainless steel. Based on Parker Hannifin Corp.

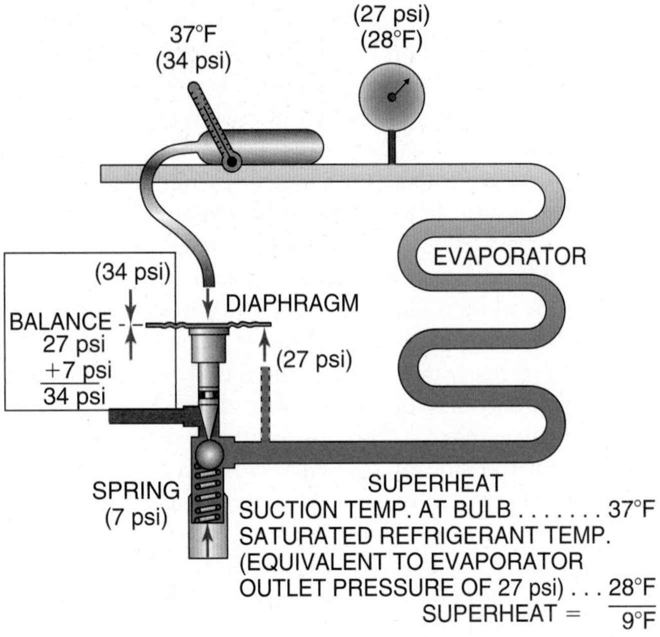

24.6 NEEDLE AND SEAT

The *needle* and seat control the flow of refrigerant through the valve. They are used in a metering device to maintain close control of the refrigerant, **Figure 24.13**, and are normally made of some type of very hard metal, such as stainless steel, to prevent the passing refrigerant from eroding the seat. Some valve manufacturers make needle and seat mechanisms that can be changed for different capacities or to correct a problem.

The size of the needle and seat determines how much liquid refrigerant will pass through the valve under a specific pressure drop. For example, when the pressure is 170 psig on one side of the valve and 2 psig on the other side, a certain measured and predictable amount of liquid refrigerant will pass through the valve. If this same valve were used when the pressure is 100 psig and 3 psig, the valve would not be able to pass as much refrigerant. So the conditions under which the valve will operate must be considered when selecting the valve. The manufacturer's manual for a specific valve is the best place to get the proper information to make these decisions.

The pressure difference from one side of the valve to the other is not necessarily the discharge versus the suction pressure. Pressure drop in the condenser and interconnecting piping may be enough to cause a problem if it is not taken into consideration, **Figure 24.14**. Notice in the figure that the actual discharge and suction pressure are not the same as the pressure drop across the expansion valve. The pressure drop across the distributor would be 25 psig (62 psig−37 psig). The pressure drop across the TXV, not considering the distributor, would be 58 psig (120 psig−62 psig).

Figure 24.13 Needle and seat devices used in expansion valves. Source: (A) Singer Controls Division. (B) Sporlan Valve Company. (C) Sporlan Valve Company

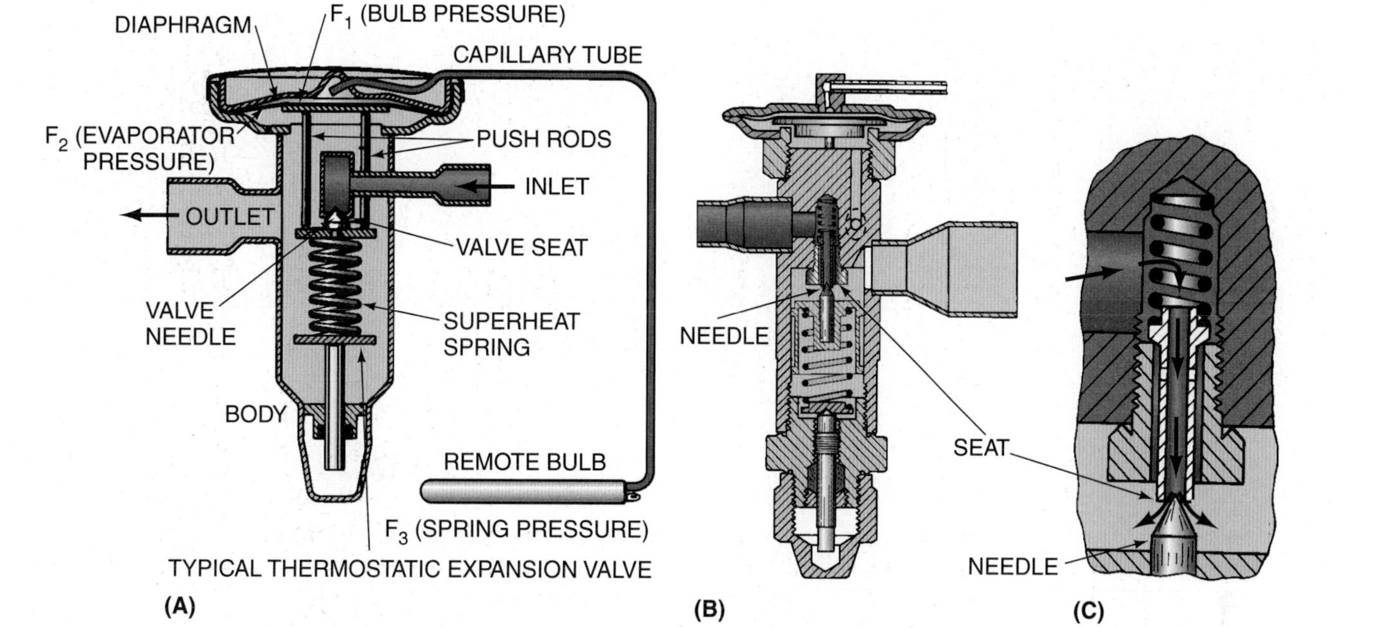

Figure 24.14 Real pressures as they would appear in a system. The pressure drop across the expansion valve is not necessarily the head pressure versus the suction pressure. The pressure drop through the refrigerant distributor is one of the big pressure drops that has to be considered. The refrigerant distributor is between the expansion valve and the evaporator coil. Courtesy Sporlan Valve Company

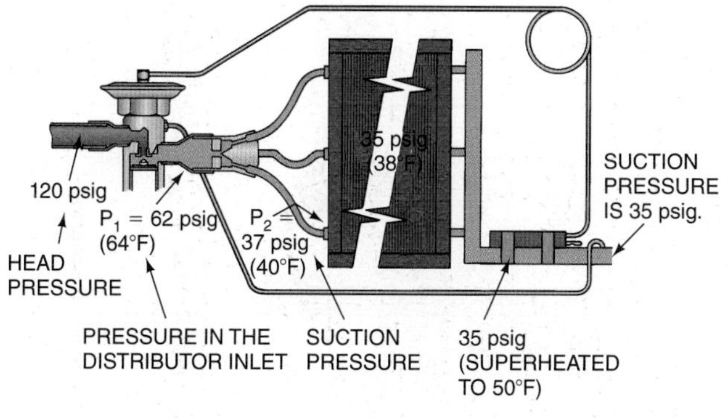

The capacity of thermostatic expansion valves is rated in tons of refrigeration under a particular pressure drop condition. To determine the appropriate size of the valve, the capacity of the system and the working conditions must be known. For example, using the manufacturer's catalog data for sizing a TXV in the table in **Figure 24.15,** we see that a 1-ton R-134a TXV is needed in a medium-temperature cooler with a 20°F evaporator (18.4 psig) and a 1-ton (12,000 Btu/h) capacity. This system is going to operate inside a store, which is expected to stay at 75°F year-round. The head pressure is expected to remain constant at 124 psig (100°F). Assuming negligible pressure drops in the associated piping and accessories in the system, the TXV would be operating with a pressure drop across the valve of 105.6 psig (124 psig − 18.4 psig). Since the 100-psig column in the table is closest to the 105.6-psig calculated pressure drop, that is the one we will use. A 1-ton nominal capacity valve has a capacity of 1.34 tons under these conditions.

If the same cooler is moved outside where the temperature is warmer and the condensing pressure rises to 145 psig, the new pressure drop across the TXV would be 126.6 psig (145 psig − 18.4 psig). We will use the 120-psig pressure drop column in the table because it is closest to our calculated 126.6-psig pressure drop across the valve. A 1-ton nominal capacity valve has a capacity of 1.47 tons under this new condition of 120-psig pressure drop across the valve. The increase in capacity of the 1-ton nominal TXV from 1.34 to 1.47 is due to the increase in head pressure, which causes an increase in the pressure drop across the valve.

These examples of TXV capacity ratings assumed that the subcooled liquid temperature coming into the TXV was 100°F. The bottom of the table in **Figure 24.15** includes liquid correction factors for liquid temperatures other than 100°F. Notice that the correction factor for 100°F liquid is 1.00. Correction factors greater than 1.00 will increase the valve's capacity, and correction factors less than 1.00 will decrease the valve's capacity. In our first example, the nominal 1-ton TXV at a 20°F evaporating temperature and a 100-psig pressure drop across the valve had a capacity of 1.34 tons. This is assuming 100°F liquid temperatures entering the valve. However, if the liquid temperature entering the valve were 80°F, the valve would have a new capacity of 1.34 tons × 1.14 = 1.52 tons. A correction factor of 1.14 from the table was used for the 80°F liquid temperature entering the valve.

Figure 24.15 The manufacturer's table shows the capacity of valves at different pressure drops. Notice that the same valve has different capacities at different pressure drops. The higher the pressure drop, the more capacity a valve has. (This is a partial table and not intended for use in the design of a system.) Courtesy Sporlan Valve Company

TEV CAPACITIES in TONS for R-134a (HFC)

JCP60 & JC THERMOSTATIC CHARGES

Condition 1 (20°, 100 psi) Condition 2 (20°, 120 psi)

VALVE TYPES	NOMINAL CAPACITY tons	EVAPORATOR TEMPERATURE (°F) / PRESSURE DROP ACROSS VALVE (PSI): 40°						20°						0°					
		40	60	80	100	120	140	60	80	100	120	140	160	60	80	100	120	140	160
NI-F-G-EG	1/8	0.12	0.15	0.17	0.19	0.21	0.23	0.13	0.15	0.17	0.18	0.20	0.21	0.12	0.13	0.15	0.16	0.18	0.19
NI-F-G-EG	1/4	0.26	0.31	0.36	0.40	0.44	0.48	0.30	0.35	0.39	0.42	0.46	0.49	0.28	0.32	0.36	0.39	0.43	0.45
NI-F-G-EG	1/2	0.49	0.60	0.70	0.78	0.85	0.92	0.52	0.60	0.67	0.73	0.79	0.85	0.47	0.54	0.60	0.66	0.71	0.76
NI-F-G-EG	1	0.98	1.21	1.39	1.56	1.70	1.84	1.04	1.20	1.34	1.47	1.58	1.69	0.93	1.07	1.20	1.32	1.42	1.52
F-G-EG	1-1/2	1.57	1.93	2.23	2.49	2.73	2.95	1.66	1.91	2.14	2.34	2.53	2.71	1.49	1.72	1.92	2.11	2.28	2.43
F(Ext)-G & EG(Ext)-S	2	1.97	2.41	2.78	3.11	3.41	3.68	2.07	2.39	2.67	2.93	3.17	3.38	1.86	2.15	2.40	2.63	2.84	3.04
C-S	2-1/2	2.46	3.01	3.48	3.89	4.26	4.60	2.59	2.99	3.34	3.66	3.96	4.23	2.33	2.69	3.00	3.29	3.56	3.80
C-S	3	2.95	3.62	4.18	4.67	5.11	5.52	3.11	3.59	4.01	4.40	4.75	5.08	2.79	3.22	3.61	3.95	4.27	4.56
C&S(Ext)	5	4.92	6.03	6.96	7.78	8.52	9.21	4.32	4.98	5.57	6.10	6.59	7.05	3.53	4.08	4.56	4.99	5.39	5.77
S(Ext)	6	5.91	7.23	8.35	9.34	10.2	11.0	5.18	5.98	6.69	7.33	7.91	8.46	4.24	4.89	5.47	5.99	6.47	6.92
H	1-1/2	1.57	1.93	2.23	2.49	2.73	2.95	1.66	1.91	2.14	2.34	2.53	2.71	1.49	1.72	1.92	2.11	2.28	2.43
H	3	2.95	3.62	4.18	4.67	5.11	5.52	3.11	3.59	4.01	4.40	4.75	5.08	2.57	2.97	3.32	3.63	3.92	4.19
H	4	3.94	4.82	5.57	6.22	6.82	7.37	4.14	4.79	5.35	5.86	6.33	6.77	3.42	3.95	4.42	4.84	5.23	5.59
H	5	4.92	6.03	6.96	7.78	8.52	9.21	5.18	5.98	6.69	7.33	7.91	8.46	4.28	4.94	5.53	6.05	6.54	6.99
P-H	8	7.38	9.04	10.4	11.7	12.8	13.8	7.77	8.97	10.0	11.0	11.9	12.7	6.42	7.41	8.29	9.08	9.81	10.5
P-H	12	11.5	14.1	16.3	18.2	19.9	21.5	12.1	14.0	15.6	17.1	18.5	19.8	10.0	11.6	12.9	14.2	15.3	16.4
M	13	12.8	15.7	18.1	20.2	22.2	23.9	13.5	15.6	17.4	19.0	20.6	22.0	10.7	12.4	13.8	15.1	16.4	17.5
M	15	15.3	18.7	21.6	24.1	26.4	28.5	16.1	18.5	20.7	22.7	24.5	26.2	12.8	14.7	16.5	18.1	19.5	20.9
M	20	19.7	24.1	27.8	31.1	34.1	36.8	20.7	23.9	26.7	29.3	31.7	33.8	16.5	19.0	21.3	23.3	25.2	26.9
M	25	24.6	30.1	34.8	38.9	42.6	46.0	25.9	29.9	33.4	36.6	39.6	42.3	20.6	23.8	26.6	29.1	31.5	33.6

Liquid Temperature (°F)	0°	10°	20°	30°	40°	50°	60°	70°	80°	90°	100°	110°	120°	130°	140°
Correction Factor	1.70	1.63	1.56	1.49	1.42	1.36	1.29	1.21	1.14	1.07	1.00	0.93	0.85	0.78	0.71

These factors include corrections for liquid refrigerant density and net refrigerating effect and are based on an average evaporator temperature of 0°F. However they may be used for any evaporator temperature from 40°F to 0°F since the variation in the actual factors across this range is insignificant.

EXAMPLE: 2 ton valve at 20°F evaporator, 80 psi pressure drop, 90°F liquid temperature = 2.39 × 1.07 = 2.56 tons.

The colder the liquid temperature entering the TXV, the more capacity in tons the TXV will have. *As the liquid temperature gets colder, the less flashing (loss effect) there will be at the entrance of the evaporator in cooling the liquid down to the evaporating temperature. The more liquid subcooling there is, the lower the loss effects at the evaporator entrance. This will increase the net refrigeration effect and the capacity of the system.* Because increased subcooling increases the refrigeration effect, many systems put the suction lines in contact with the liquid line to increase the subcooling effect. This also helps reduce the possibility of liquid refrigerant entering the compressor by adding heat to the refrigerant in the suction line.

24.7 THE SPRING

The *spring* is one of the three forces that act on the diaphragm. It raises the diaphragm and closes the valve by pushing the needle into the seat. When a valve is adjusted, more or less pressure is applied to the spring to change the tension for different superheat settings. The spring is often referred to as the superheat spring. The spring tension is usually factory set for a predetermined superheat of 8°F to 12°F, **Figure 24.16**.

Figure 24.16 The spring used in the TXV. Based on Singer Controls Division.

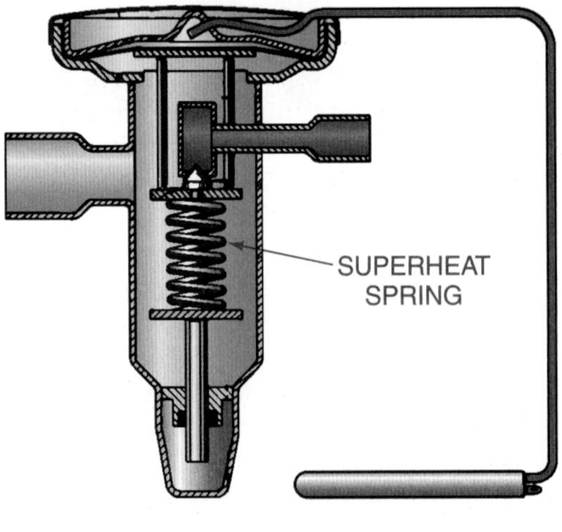

The adjustment part of the valve can have either a screw slot or a square-headed shaft. Both are normally covered by a cap to prevent water, ice, or other foreign matter from collecting on the stem. The cap also serves as backup leak prevention. Most adjustment stems on expansion valves have a

(A)

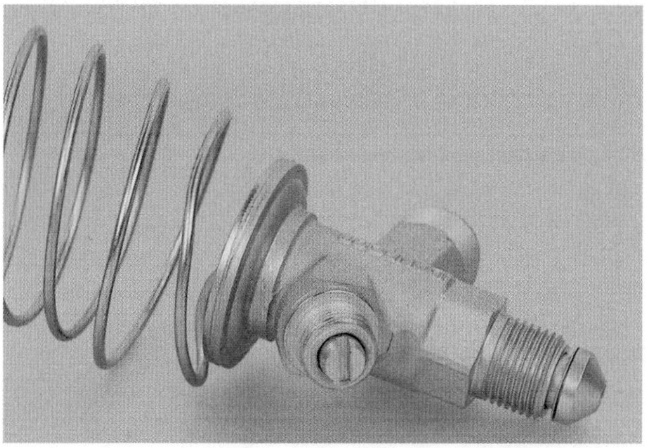

(B)

Figure 24.17 Adjustment stems on expansion valves. Some of them are adjusted with (A) a valve wrench and some with (B) a screwdriver. Photos by Bill Johnson

packing gland that can be tightened to prevent refrigerant from leaking. The cap covers the gland as well, **Figure 24.17**. One complete turn of the stem usually changes the superheat reading significantly, depending on the manufacturer. It is good field practice not to adjust the spring unless you are a seasoned technician. If the valve is adjusted improperly, compressor damage can occur.

24.8 THE SENSING BULB AND TRANSMISSION TUBE

The *sensing bulb* and *transmission line* are extensions of the valve diaphragm. The bulb detects the temperature at the end of the evaporator on the suction line and transmits this temperature, converted to pressure, to the top of the diaphragm. The bulb contains a fluid, such as refrigerant, that follows the relationships in a temperature/pressure chart just like any refrigerant. When the suction line temperature goes up, the temperature change occurs inside the bulb as well. As the temperature of the TXV's sensing bulb increases, the valve will gradually open, letting more refrigerant into the evaporator. This increases both the evaporating pressure and the compressor's suction pressure, while attempting to reduce the amount of superheat in the evaporator. When the pressure changes, the transmission line (which is nothing more than a small-diameter hollow tube) allows the pressure between the bulb and diaphragm to equalize back and forth, **Figure 24.18**.

Figure 24.18 An illustration of the diaphragm, the bulb, and the transmission tube.

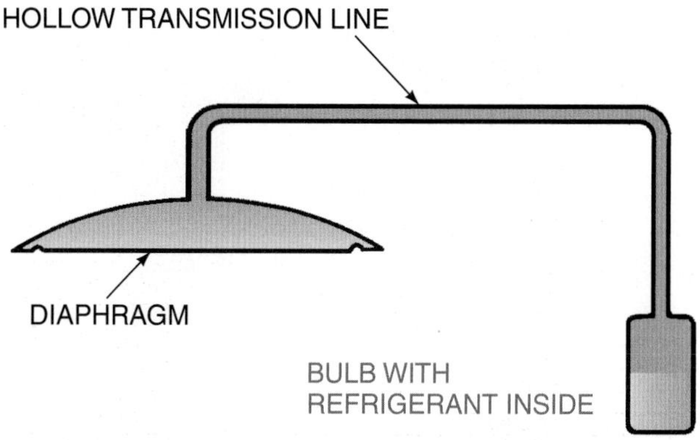

The seat in the TXV is stationary in the valve body, and the needle is moved by the diaphragm. One side of the diaphragm gets its pressure from the bulb, and the other side gets its pressure from the evaporator and the spring. The diaphragm moves up and down in response to three different pressures, which act at the proper time to open, close, or modulate the valve needle between open and closed. The three pressures are the *bulb pressure*, the *evaporator pressure*, and the *spring pressure*. They all work as a team to position the valve needle correctly to meet the load conditions at any particular time. **Figure 24.19** is a series of illustrations that show how the TXV functions under different load conditions.

24.9 TYPES OF BULB CHARGE

The fluid inside the expansion valve bulb is known as the *charge* for the valve. Four types of charge can be used in the TXV:

- Liquid charge
- Cross liquid charge
- Vapor charge
- Cross vapor charge

24.10 THE LIQUID CHARGE BULB

The **liquid charge bulb** is charged with a fluid characteristic of the refrigerant in the system. The diaphragm and bulb are not actually full of liquid, but they have enough so as to always have some liquid inside them. The liquid will not entirely boil away.

Figure 24.19 All components work together to hold the needle and seat in the correct position to maintain stable operation with the correct superheat. (A) Valve in equilibrium. (B) Valve opening. (C) Valve closing.

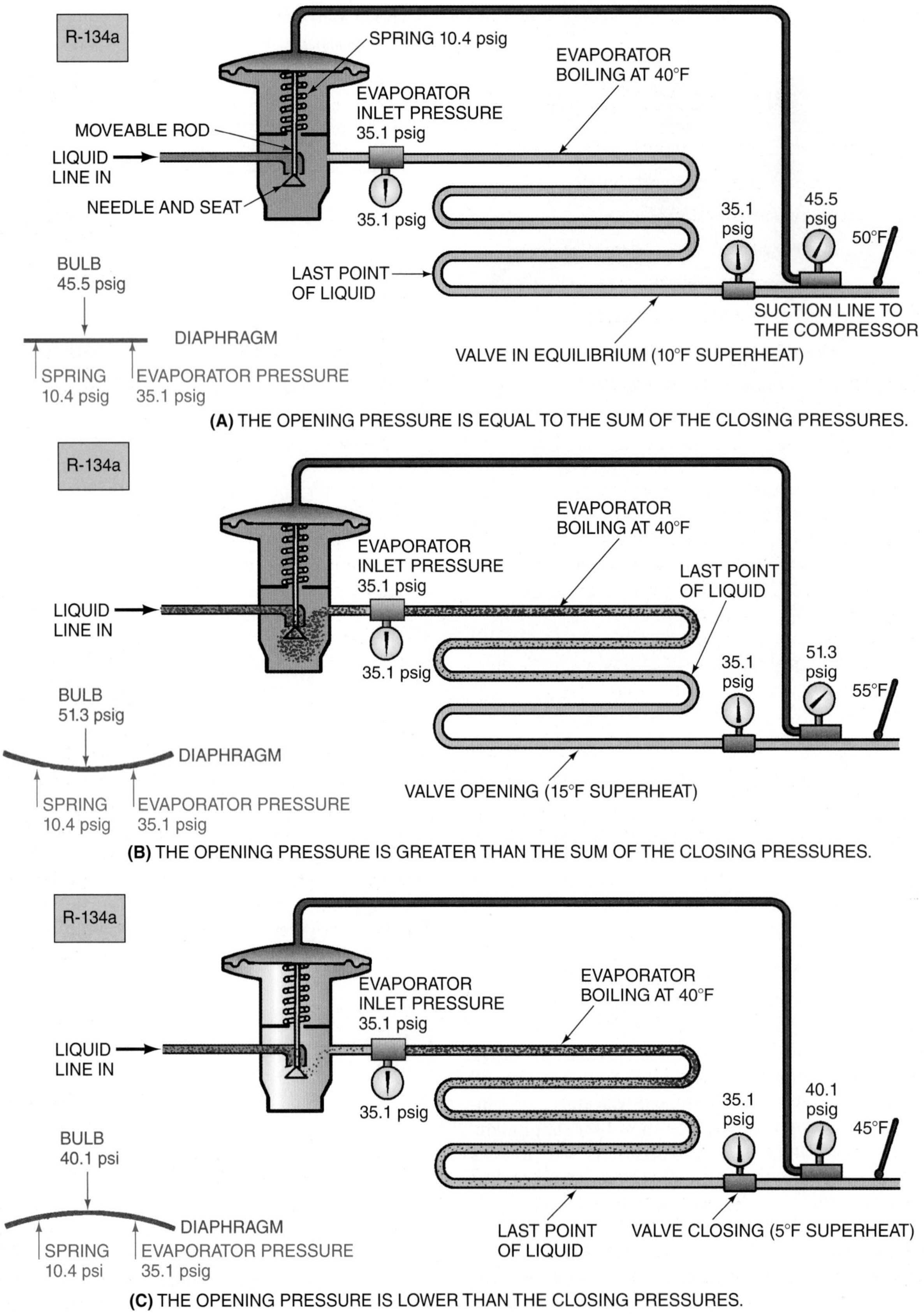

Figure 24.20 The valve is open because the evaporator suction line is warm after the defrost cycle. This allows the liquid in the evaporator to spill over into the suction line before the expansion valve can gain control. The bulb can become quite hot in an extended defrost cycle, and the remaining liquid can move on to the suction line, causing possible compressor damage.

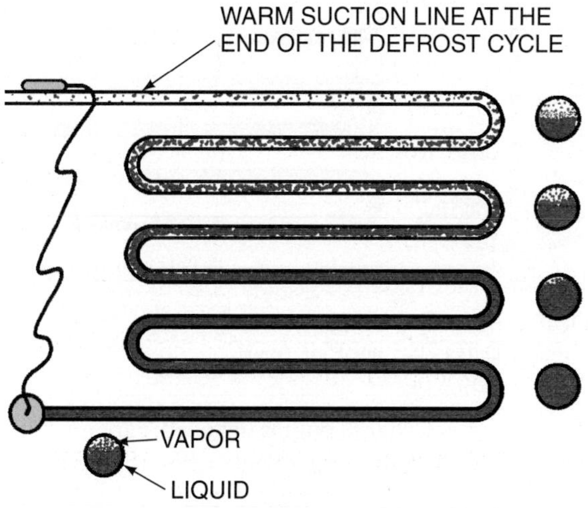

The temperature/pressure relationship in the bulb is almost a straight line on a graph. When the temperature goes up a degree, the pressure goes up a specific amount that can be followed on a temperature/pressure chart. When high temperatures are encountered, high pressures will exist. During defrost, the expansion valve bulb may reach high temperatures, which can cause two things to happen. The pressures inside the bulb can cause excessive pressures over the diaphragm, and the valve will open wide when the bulb gets warm. This will overfeed the evaporator and can cause liquid to flood the compressor when defrost is terminated, **Figure 24.20**. This can lead the service technician or manufacturer to change to the cross liquid charge bulb. The liquid charge bulb is usually limited to applications that have narrow operating ranges or to special functions like desuperheating on the high side of the system.

24.11 THE CROSS LIQUID CHARGE BULB

The cross liquid charge bulb is filled with a fluid that is different from the system fluid. It does not follow the system's temperature/pressure relationship but has a flatter curve and will close the valve faster on a rise in evaporator pressure. The cross-charged liquid bulb on a TXV will not open the valve as fast when the remote bulb gets real warm. Cross-charged liquid bulbs are good for low-temperature applications. They will usually give high evaporator superheats under high evaporator-heat load conditions. Normal evaporator superheats will be experienced within the operating range of the valve. The valve closes during the off cycle when the compressor shuts off and the evaporator pressure rises. This helps prevent liquid refrigerant from flooding over into the compressor at startup, **Figure 24.21**.

Figure 24.21 This cross charge valve has a flatter temperature/pressure curve. The pressure rise after defrost or at the end of the cycle will have a tendency to close the valve and prevent liquid floodback problems. Courtesy Sporlan Valve Company

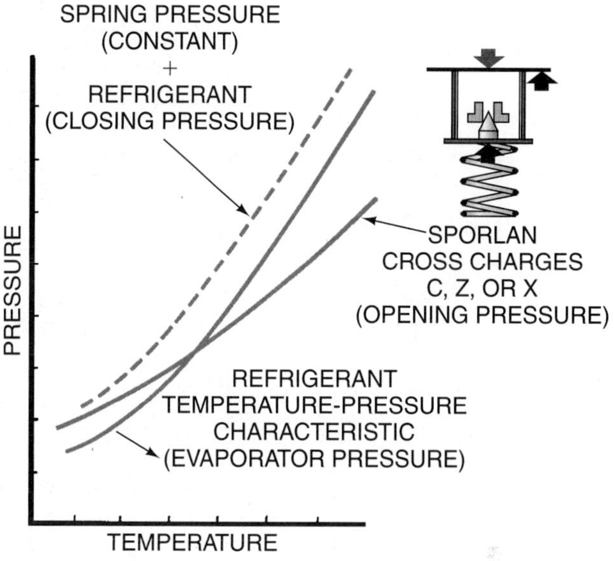

24.12 THE VAPOR (GAS) CHARGE BULB

The vapor charge bulb, *or gas-charged bulb*, has only a small amount of liquid refrigerant in the bulb, **Figure 24.22**. It is sometimes called a *critical charge bulb*. When the bulb temperature rises, more and more of the liquid will boil to a vapor until there is no more liquid. When this point is reached, an increase in temperature will no longer bring an increase in pressure. The pressure curve will be flat, **Figure 24.23(A)**. This acts to limit the maximum operating pressure the valve will allow the evaporator to experience, and so these valves are also often referred to as maximum operating pressure (MOP) valves. Notice that the TXV in **Figure 24.23(B)** has a MOP of 15. This means that when the evaporator pressure reaches 15 psig, all of the liquid in the remote bulb will be vaporized and no more opening pressure will be exerted on the diaphragm of the valve. The evaporator pressure will never go higher than 15 psig. Limiting the evaporator pressure will protect the compressor

Figure 24.22 This gas charge bulb is actually a liquid charge bulb with a critical charge. When the bulb reaches a predetermined temperature, the liquid is boiled away, and the pressure will not rise any further.

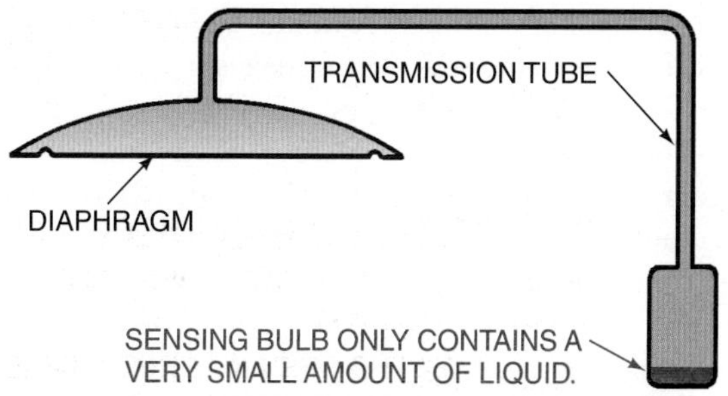

Figure 24.23 (A) This graph shows pressures inside the valve when the sensing bulb is in a hot area. Courtesy Sporlan Valve Company. (B) TXV showing a maximum operating pressure (MOP) of 15 psi for compressor protection. Courtesy Ferris State University. Photo by John Tomczyk

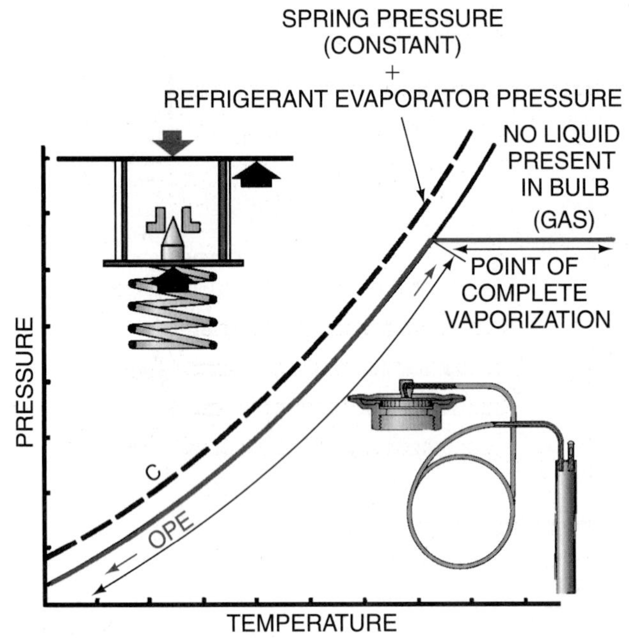

(A)

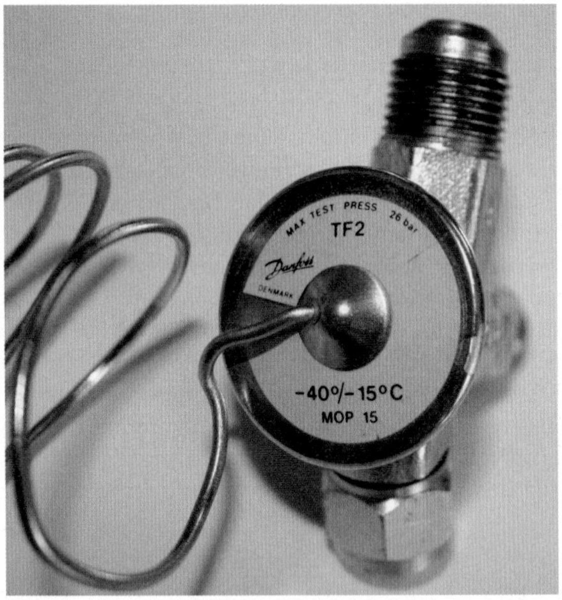

(B)

from an overloading condition resulting from excessively high pressures. Pressure entering the compressor that is too high causes the density of suction gases entering the compressor to increase. This in turn causes an increase in the mass flow rate of refrigerant through the compressor. High mass flow rates of refrigerant will overload the compressor by causing high compressor-motor amperage draws, which can overheat compressor motors and open motor overload protection devices, which will take the compressor out of service temporarily.

Figure 24.24 The liquid can migrate to the valve body when the body becomes cooler than the bulb. When this happens, the valve body can be heated with warm cloth, and the liquid will move back to the bulb.

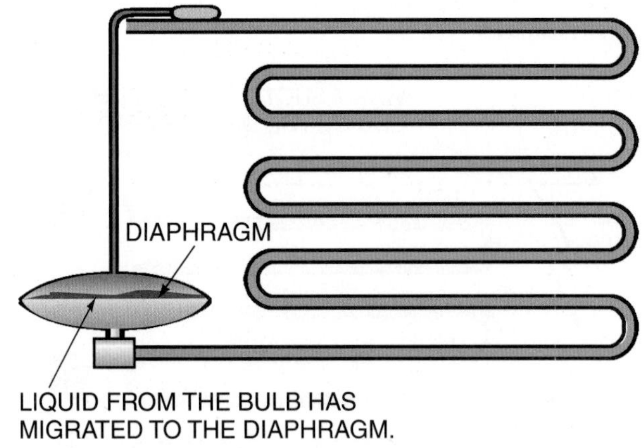

When the vapor charge bulb is installed, care must be taken that the valve body does not get colder than the bulb; if it does, the vapor in the bulb will condense to liquid above the diaphragm. When this happens, the control at the end of the evaporator will migrate to the valve diaphragm area, **Figure 24.24**. Since the valve is operated by the temperature of the liquid at the diaphragm, control will be lost. Small heaters have been installed at the valve body to keep this from happening.

24.13 THE CROSS VAPOR CHARGE BULB

The cross vapor charge bulb is similar to the vapor charge bulb, but is charged with a fluid different from the refrigerant in the system. This creates a different pressure and temperature relationship under different conditions. These special valves are applied to special systems. Manufacturers or suppliers should be consulted when questions are encountered.

24.14 EXAMPLE OF A TXV FUNCTIONING WITH AN INTERNAL EQUALIZER

When all of these components are assembled, the expansion valve will function as follows. **NOTE:** *This example is based on a liquid-filled bulb.* •

1. **Normal load conditions.** The valve is operating in equilibrium (no change, stable), **Figure 24.25**. The evaporator is operating in a medium-temperature application and just before the cutoff point. The suction pressure is 18.4 psig, and the refrigerant, R-134a, is boiling (evaporating) in the evaporator at 20°F. The expansion valve is maintaining 10°F of superheat, so the suction-line temperature is 30°F at the bulb location. The bulb has been on the line long

Figure 24.25 A TXV under a normal load condition. The valve is said to be in equilibrium. The needle is stationary.

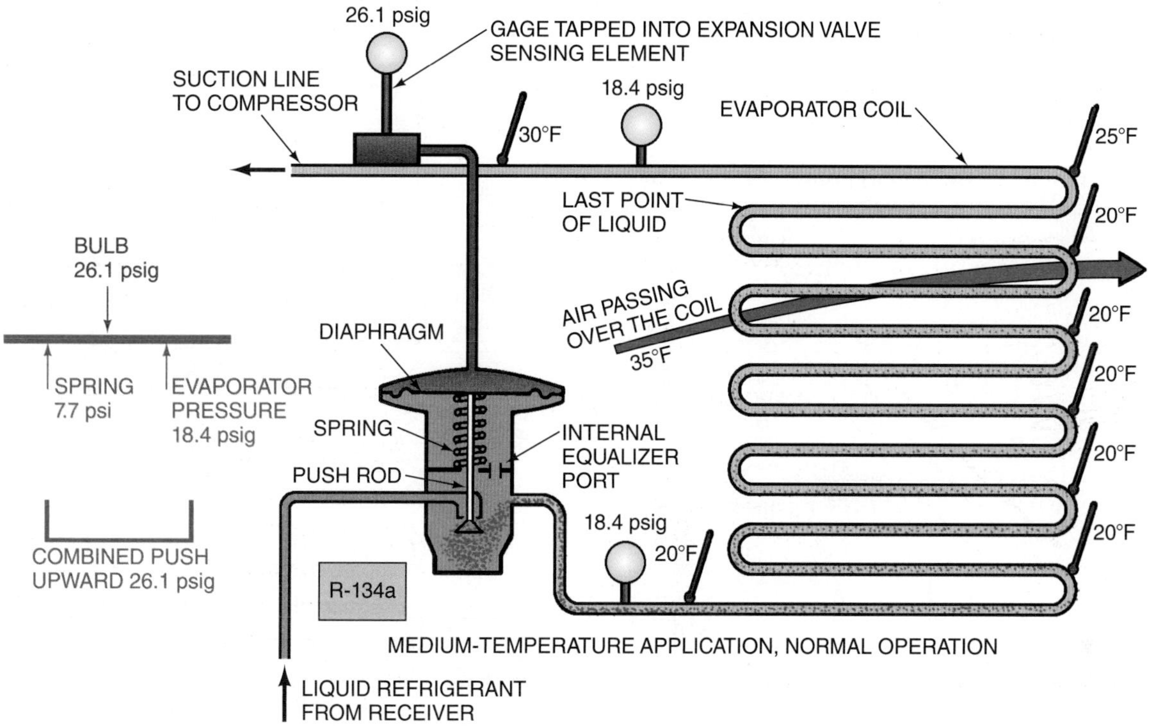

enough so that it is the same temperature as the line, 30°F. For now, suppose that the pressure of the liquid in the bulb is 26.1 psig, corresponding to 30°F. The spring is exerting a pressure equal to the difference in pressure to hold the needle at the correct position in the seat to maintain this condition. The spring pressure in this example is 7.7 psi.

2. **Load changes with food added to the cooler.** See **Figure 24.26(A)**. When food that is warmer than the inside of the cooler is added to the cooler, the load on the evaporator changes. The warmer food warms the air inside the cooler, and the load is added to the refrigeration coil by the air. This warmer air passing over the coil causes the liquid refrigerant inside the coil to boil faster. The suction pressure will also rise. The net effect of this condition will be that the last point of liquid in the coil will be farther from the end of the coil than it is when the coil is under normal working conditions. The coil will start to starve for liquid refrigerant. The TXV will start to feed more refrigerant to compensate for this shortage. This happens because the increased superheat in the evaporator causes the temperature of the remote bulb to rise. Since the pressure in the remote bulb is the only opening pressure in the TXV, the valve will open to feed more refrigerant to the evaporator. The rate of vaporization of the refrigerant in the evaporator will increase and cause the evaporator pressure to increase. In fact, the pressure on the whole low side of the system will increase all the way down to the compressor. When this condition of increased load has gone on for an extended time, the TXV will stabilize and reach a new point of equilibrium in the feeding of refrigerant, where no adjustment occurs.
3. **Load changes with food removed from the cooler.** See **Figure 24.26(B)**. When a large portion of the food is removed from the cooler, the load will decrease on the evaporator coil. There will no longer be enough load to boil the amount of refrigerant that the expansion valve is feeding into the coil, so the expansion valve will start overfeeding the coil. The coil is beginning to flood with liquid refrigerant. The TXV needs to throttle the refrigerant flow. When the TXV has operated at this condition for a period of time, it will stabilize and reach another point of equilibrium at or near 10°F of evaporator superheat. When the condition exists for a long enough time, the thermostat will stop the compressor because the air in the cooler is reduced to the thermostat cut-out point.

24.15 TXV FUNCTIONING WITH EXTERNAL EQUALIZERS

Evaporators are often designed and applied with pressure drop from the inlet to the outlet. This pressure drop may be created by a distributor located after the expansion valve or a long piping circuit. An external equalizer is required whenever the pressure drop in the evaporator exceeds:

- 3 psig (air-conditioning applications).
- 2 psig (medium-temperature refrigeration applications).
- 1 psig (low-temperature applications).

Figure 24.26 (A) When food is added to the cooler, the load on the coil goes up. The extra Btus added to the air in the cooler are transferred into the coil, which causes a temperature increase in the suction line. This causes the valve to open and feed more refrigerant. If this condition is prolonged, the valve will reach a new equilibrium point with the needle remaining steady, not moving. (B) When there is a load decrease, such as when some of the food is removed, the valve will reach a new point of equilibrium at a low-load condition. The thermostat will cut the unit off if this condition goes on long enough. The equilibrium point in (A) and (B) is the valve superheat set point of 10°F superheat. The valve should always return to this 10°F set point with prolonged steady-state operation.

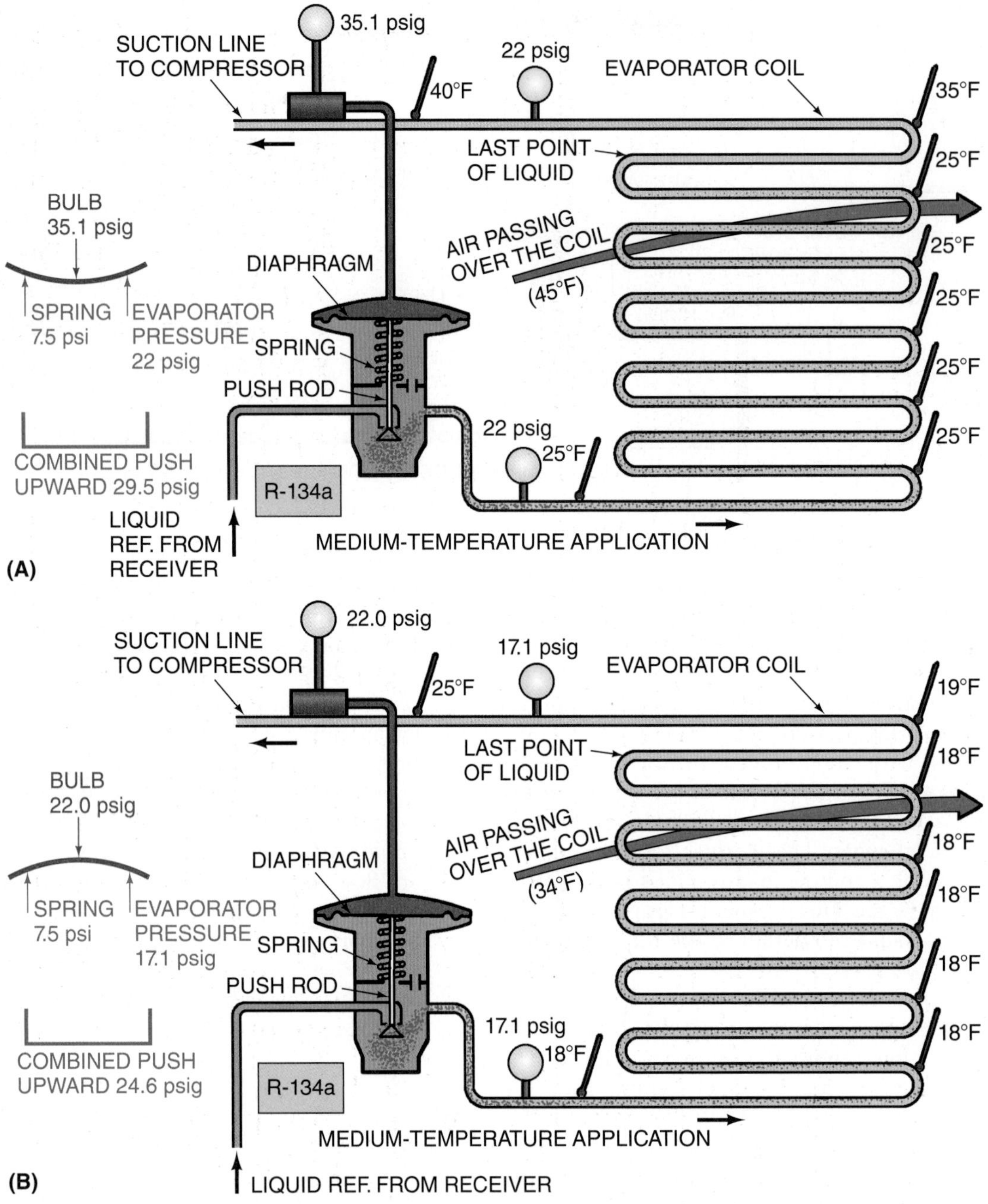

The external equalizer on a TXV is used to compensate for, not get rid of, pressure drops from the inlet to the outlet of the evaporator. As mentioned, the pressure drops could be caused by a refrigerant distributor located after the TXV or simply by long evaporator piping circuits with a lot of U bends. **An external-equalized TXV must always be installed when a refrigerant distributor is used because of the distributor's large pressure drop characteristics.** Excess pressure drop in a coil with a TXV will cause the valve to starve the coil of refrigerant, **Figure 24.27**. When evaporator outlet pressure is sensed, less pressure is on the bottom of the TXV diaphragm than if evaporator inlet

Figure 24.27 (A) This evaporator has 7-psig pressure drop, which starves the evaporator. (B) When a TXV with an external equalizer is added, the coil has the correct amount of refrigerant for best coil efficiency.

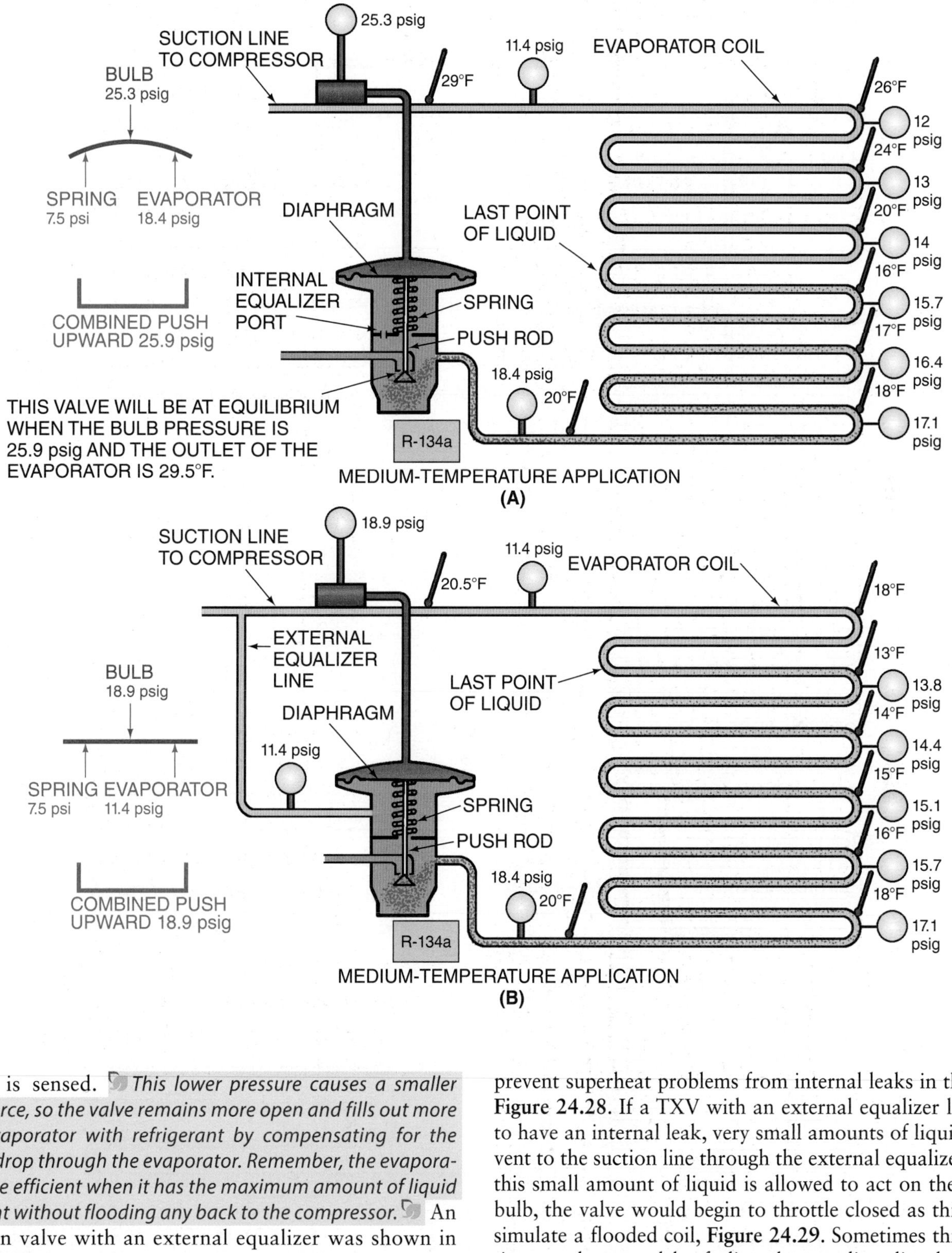

pressure is sensed. *This lower pressure causes a smaller closing force, so the valve remains more open and fills out more of the evaporator with refrigerant by compensating for the pressure drop through the evaporator. Remember, the evaporator is more efficient when it has the maximum amount of liquid refrigerant without flooding any back to the compressor.* An expansion valve with an external equalizer was shown in **Figure 24.10.**

The external equalizer line should always be piped to the suction line after the expansion valve sensing bulb to prevent superheat problems from internal leaks in the valve, **Figure 24.28.** If a TXV with an external equalizer line were to have an internal leak, very small amounts of liquid would vent to the suction line through the external equalizer line. If this small amount of liquid is allowed to act on the sensing bulb, the valve would begin to throttle closed as this would simulate a flooded coil, **Figure 24.29.** Sometimes this condition can be sensed by feeling the equalizer line leading to the suction line. It should not become cold as it leaves the expansion valve.

Figure 24.28 This illustration shows the correct method for connecting an external equalizer line.

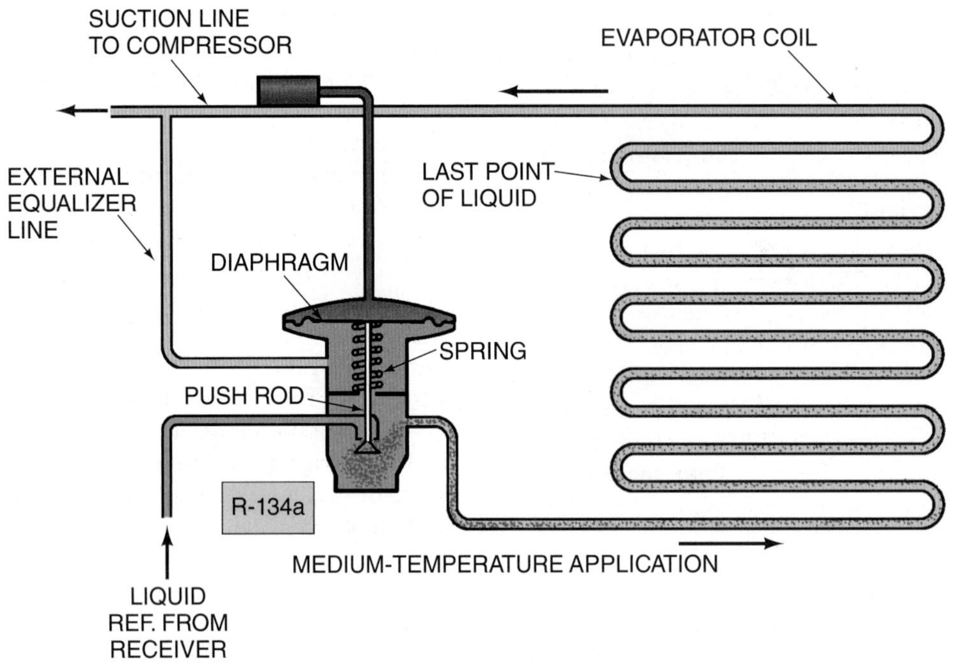

Figure 24.29 When the equalizer line is connected before the sensing bulb, a small leak in the internal parts of the valve will simulate a flooded evaporator.

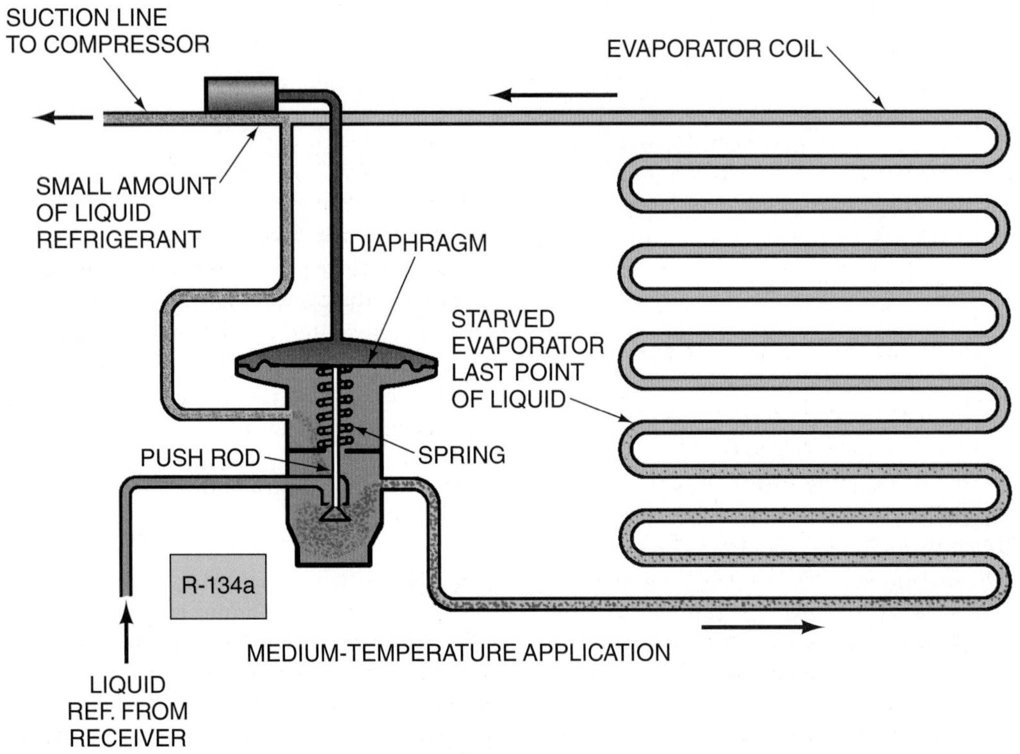

24.16 TXV RESPONSE TO LOAD CHANGES

The TXV responds to a change in load in the following manner. When the load is increased—for example, when a load of food warmer than the cooler (refrigerated box) is placed inside the cooler—the TXV opens, allowing more refrigerant into the coil. The evaporator needs more refrigerant at this time because the increased load is evaporating the refrigerant in the evaporator faster. The suction pressure increases, **Figure 24.26(A)**. When there is a load decrease—for example, when some of the food is removed from the cooler—the liquid in the evaporator evaporates more slowly and the suction pressure decreases. At this time the TXV will throttle back by slightly closing the needle and seat to maintain the correct superheat. **The TXV responds to a load increase by metering more refrigerant into the coil. This causes an increase in suction pressure from the refrigerant boiling at a faster rate.**

24.17 TXV SELECTION

The TXV must be carefully chosen for a particular application. **Each TXV is designed for a particular refrigerant or group of refrigerants.** Many newer TXVs entering the market are manufactured to handle some of the newer, alternative refrigerants and refrigerant blends and are designed to be used with more than one refrigerant. This helps in the selection process and makes it a lot easier for service technicians when retrofitting a system.

The capacity of the system is very important. If the system calls for a 1/2-ton expansion valve and a 1-ton valve is used, the valve will not control correctly because the needle and seat are too large. If a 1/4-ton valve were used, it would not pass enough liquid refrigerant to keep the coil full, and a starving coil condition would occur.

24.18 BALANCED-PORT TXV

In response to need, manufacturers have developed TXVs that will feed refrigerant at the same rate when ambient temperatures are low. When the head pressure is low in mild weather, these valves do not reduce refrigerant flows, so the evaporator then has the correct amount of refrigerant and is able to operate at design conditions under lower outdoor temperatures. The flow of liquid on the face of the needle valve on larger TXVs tends to open the valve, which can cause erratic superheat control. With the balanced-port TXV, the liquid pressure is actually cancelled out because it acts on equal areas but in opposite directions. **Figure 24.30(A)** illustrates the liquid force on the face of the needle valve. For a larger TXV, this can be a significant opening force because of the larger surface area associated with the needle valve faces. This liquid or opening force also becomes significant with a large pressure drop across the valve from either high head pressures, low suction pressures, or a combination of both. **Figure 24.30(B)** illustrates how the liquid force on a balanced-port TXV is cancelled out. If this force is not cancelled out, the TXV would operate erratically, and the valve could not hold its original superheat setting. This can cause compressor damage if certain system parameters change, such as head pressure or suction pressure.

As compared to a conventional TXV, balanced-port valves have a large orifice or needle valve area. The needle valve also operates very close to its seat, which provides a very stable control of evaporator superheat with a minimum movement of the valve. This allows a large orifice or needle valve assembly to handle both large and small loads. This type of design can help refrigeration equipment pull down faster. The larger orifice or needle valve assembly can also handle some liquid-line bubbles or flash gas if they happen. This is where the term *balanced port* originates. Balanced-port TXVs should be used if any of the following conditions exist:

- Large varying head pressures
- Larger varying pressure drops across the TXV
- Widely varying evaporator loads
- Very low liquid-line temperatures

You cannot tell a balanced-port expansion valve from a regular expansion valve by its appearance. It will have a different model number, and you can look it up in a manufacturer's catalog to determine the type of valve it is.

24.19 THE PRESSURE-LIMITING TXV

The pressure-limiting TXV has, in addition to the diaphragm, a bellows that will only allow the evaporator pressure to rise to a predetermined level. Above this pressure, the valve will throttle the flow of liquid. This valve is desirable on low-temperature applications because it keeps the suction pressure to the compressor down during a hot pulldown that could overload the compressor. For example, when a low-temperature cooler is started with the inside of the box hot, the compressor will operate under an overloaded condition until the box cools down. The pressure-limiting TXV will prevent this from happening, **Figure 24.31**. The vapor charge bulb TXV also has pressure-limiting qualities, **Figure 24.23(B)**.

24.20 SERVICING THE TXV

Care should be taken when selecting any TXV that the valve will be serviceable and perform correctly. Several things should be considered: (1) type of fastener (flare, solder, or flange), (2) location of the valve for service and performance,

Figure 24.30 (A) The liquid force on the face of the needle valve, which can be a significant opening force for the TXV. Source: Copeland Corporation and Emerson Flow Controls. (B) Equal liquid forces, but opposite in direction, can cancel one another out in a balanced-port TXV. Source: Copeland Corporation and Emerson Flow Controls

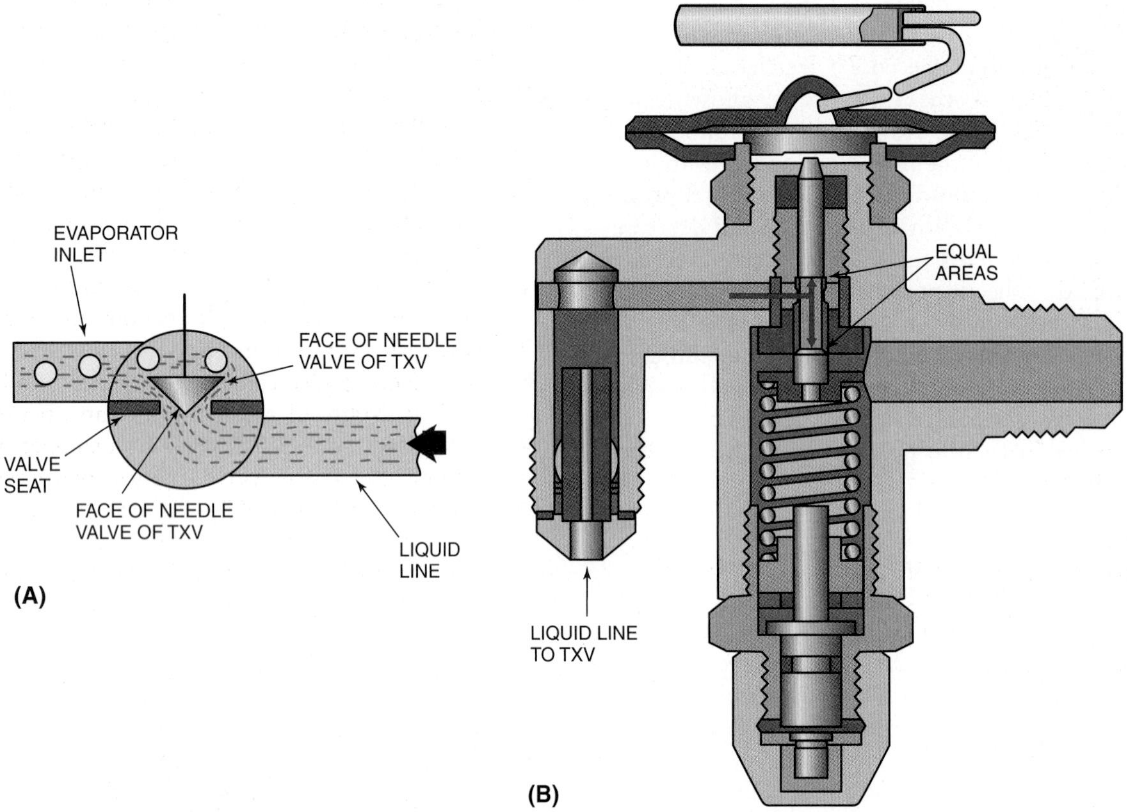

Figure 24.31 A pressure-limiting expansion valve. When the pressure is high in the evaporator, such as during a hot pulldown, this valve will override the thermostatic element with a pressure element and throttle the refrigerant. Courtesy Sporlan Valve Company

and (3) location of the expansion valve sensing bulb. The valve has moving parts that are subject to wear. When a valve must be replaced, an exact replacement is usually best. When this is not possible, a supplier can furnish you with the information needed to select another valve.

Many technicians (and owners) will adjust a TXV. This may cause problems because pressures and temperatures in the coil will be changed. If the valve sensing element is mounted properly and in the correct location where it can sense the superheat in the suction line, the valve should work correctly as shipped from the factory. The valves are very reliable and normally do not require any adjustment if sized properly. If there are signs that the valve has been adjusted, look for other problems, because the valve probably did not need adjustment to begin with. Many manufacturers are producing nonadjustable valves to prevent people from adjusting the valve unnecessarily.

24.21 INSTALLING THE SENSING ELEMENT

Particular care should be taken when installing the expansion valve sensing element. Each manufacturer has a recommended method for installation, but all are similar. The valve sensing bulb has to be mounted at the end of the evaporator on the suction line. The recommended location for the thermal bulb on smaller suction lines is near the top of the line. The best location for larger suction lines is near the bottom of the line on a horizontal run where the bulb can be mounted flat and not be raised by a fitting, **Figure 24.32**. The bulb should not be located at the bottom of the line because returning oil will act as an insulator to the sensing element. The objective of the sensing element is to sense the temperature of the refrigerant gas in the suction line. To do this, the line should be very clean and the

Figure 24.32 The best positions for mounting the expansion valve sensing bulb. Based on ALCO Controls Division, Emerson Electric Company.

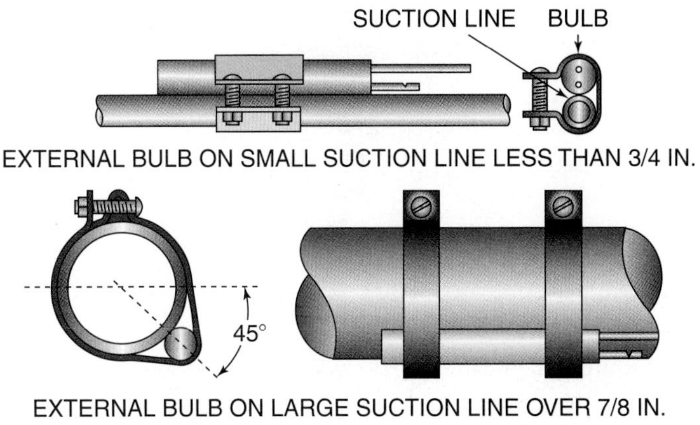

bulb should be fastened to this line very securely. Normally, the manufacturer suggests that the bulb be insulated from the ambient temperature if the ambient is much warmer than the suction-line temperature, because the bulb will be influenced by the ambient temperature as well as by the line temperature.

24.22 STEPPER MOTOR EXPANSION VALVES

The *stepper motor valve* uses a small motor to control the valve's port, which in turn controls evaporator superheat. A solid-state controller instructs the stepper motor to rotate a fraction of a revolution, or step, for each signal sent by the controller, **Figure 24.33**. The controller has a temperature sensor, or thermistor, as one of its inputs, **Figure 24.34**. Thermistors, solid-state devices that will change their electrical resistance in response to a change in temperature, are used because they are accurate, readily available, and reasonably priced. Once the electrical signal is taken away, the motor stops; it does not rotate continuously. The motor can return the valve to any previous position at any time because the controller can remember the number of steps it has taken. Many of these motors can operate at a rate of 200 steps per second. Most modern stepper motors used in electronic expansion valve applications have a bipolar motor design. Multiple electromagnets, as many as 100, are placed around a permanent magnet acting as the motor's rotor, **Figure 24.35**. They can produce step increments as small as 3.6° in rotation (this figure comes from a total of

Figure 24.33 A solid-state stepper motor controller. Courtesy Sporlan Valve Company

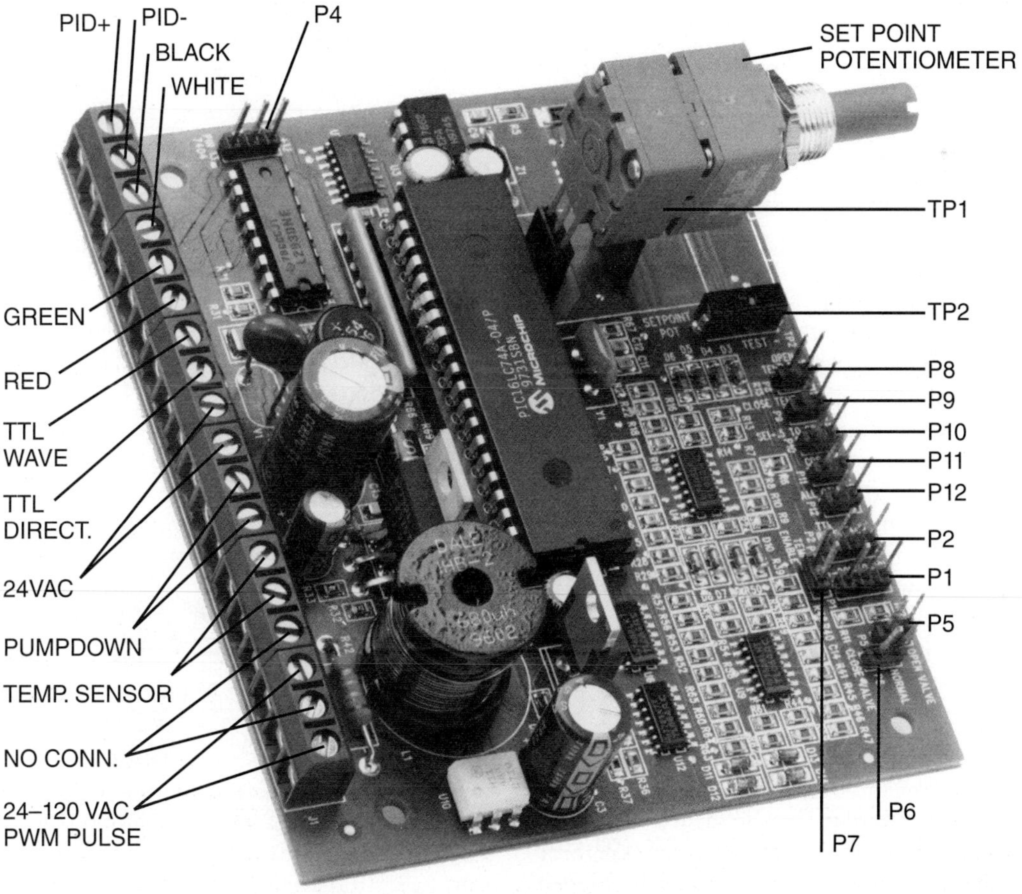

Figure 24.34 A thermistor used as a temperature sensor for the controller input. Courtesy Ferris State University. Photo by John Tomczyk

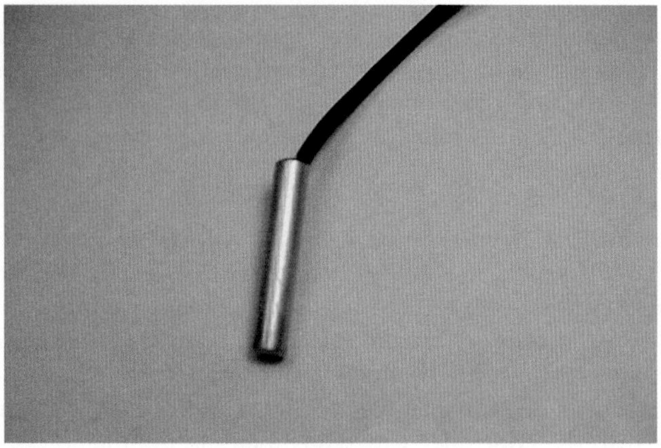

Figure 24.35 Electromagnets placed around a permanent magnet rotor. Courtesy Sporlan Valve Company

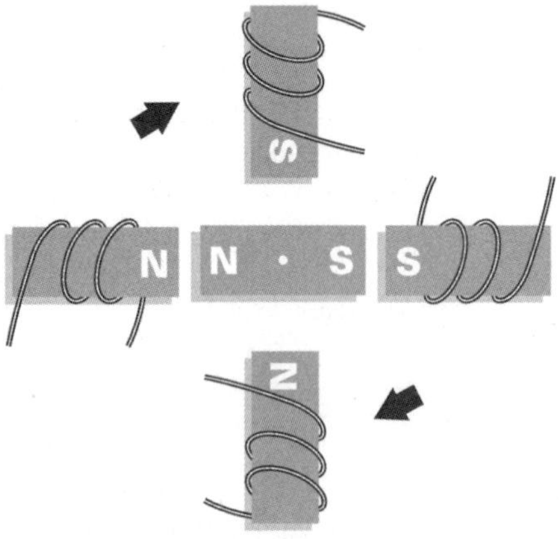

360° divided by 100 magnets). The pin or valve port piston of the electronic stepper motor expansion valve can move in a linear motion of 0.0000783 in./step.

The motor is powered by signals that change in polarity. This creates a push/pull action within the motor's magnetic fields and increases the torque and efficiency of the motor and so permits a very small motor. The torque, or rotational force, is then translated into linear motion by a digital linear actuator, **Figure 24.36**. The motor and valve thus make up a true modulating device comparable to the dimmer switch on a typical household light, which has thousands of steps. The stepper motor is not a pulsing digital motor or a common light switch that has only two steps, on or off.

The controller consists of many transistors for each switching function. Transistors are solid-state switches that are made from a solid chip of silicon and have no moving parts. They act as relays by using small electrical signals to turn a larger electrical signal on and off. The small electrical signal enters the base of the transistor, which allows flow from emitter to collector. A small microcomputer in the controller has a programmed

Figure 24.36 A digital linear actuator. Courtesy Sporlan Valve Company

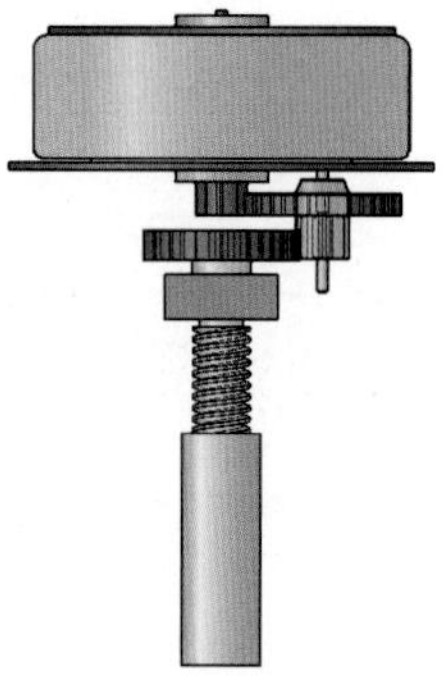

algorithm, which can control or sequence the electrical signal given to the base of each transistor, **Figure 24.37**. An algorithm is simply a program or set of instructions stored in the microprocessor of the controller. The transistors are usually turned on and off in pairs, which step the electronic valve in either the open or the closed direction in small increments to control evaporator superheat.

The controls of electronic expansion valves (EEVs) use algorithms, and a set of electronic instructions, along with a *proportional, integral, and derivative (PID) controller.* (PID controllers will be explained in the next section.) The valve's hardware includes the stepper motor, the controller, transistors, and the microprocessor. As mentioned earlier, algorithms, or sets of instructions, often referred to as software, must be programmed in the microprocessor.

Feedback loops are used to let the controller know that the process of controlling evaporator superheat needs to be changed or modified. This means that when the controller opens the electronic expansion valve (EEV) too much, which in turn overcools the thermistor sensor on the evaporator outlet, the sensor will feed back this information so the controller can start to close the valve, **Figure 24.38**. The sensor in the feedback loop is usually a thermistor that is located on the evaporator outlet. Often, two thermistors, located on the evaporator's inlet and outlet, are used to calculate evaporator superheat. The difference in temperature would be the calculated evaporator

Figure 24.37 The multiple transistors that make up the stepper motor. Courtesy Sporlan Valve Company

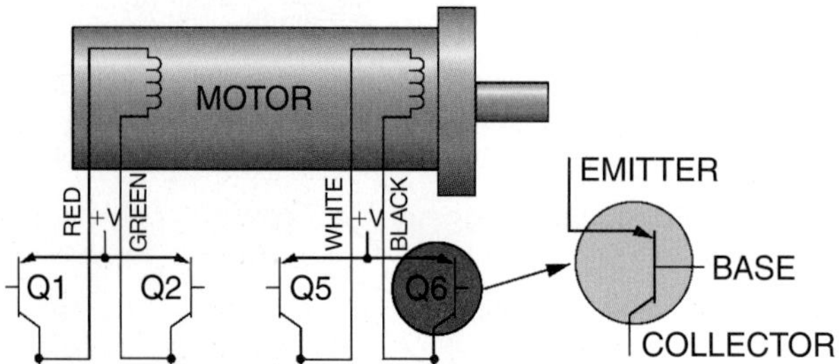

Figure 24.38 (A) The electronic expansion valve (EEV) feedback loop. Courtesy Sporlan Valve Company. (B) Different types and styles of electronic expansion valves (EEVs). Courtesy Sporlan Division, Parker Hannifin Corp.

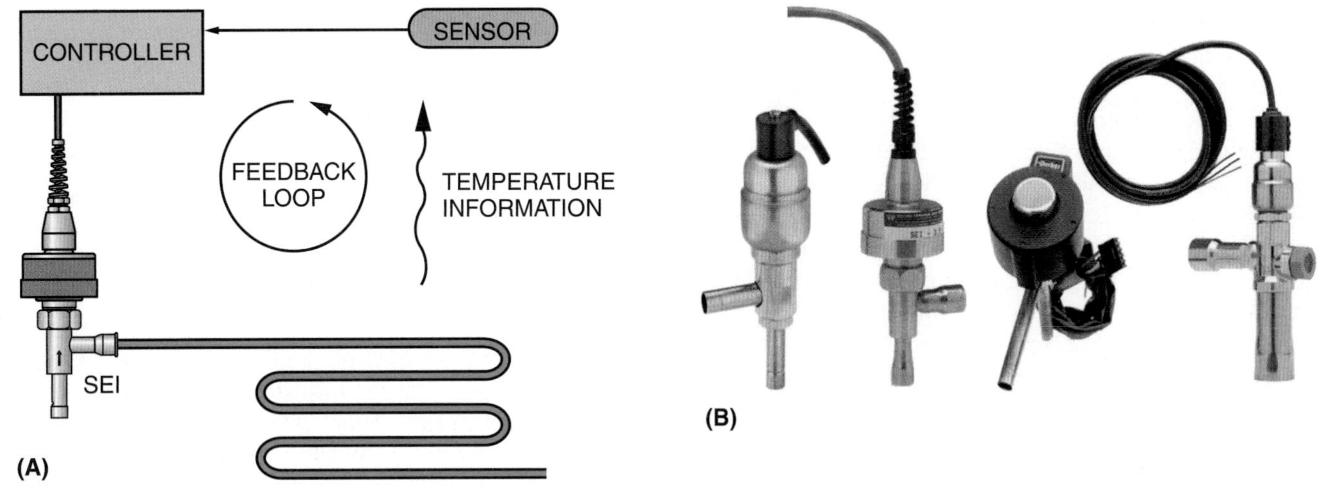

superheat. With this method, there is no need for temperature/pressure tables or for a pressure transducer to sense pressure. This greatly simplifies the superheat calculation process.

24.23 ALGORITHMS AND PID CONTROLLERS

As mentioned earlier, the EEV controls use many different algorithms or sets of instructions along with a PID controller. The following paragraphs will explain the PID controller. Although proportional, integral, and derivative controls are covered separately here, in reality they work together to modify the controller's output signal to the EEV for less error.

Proportional controllers refer to a modulating controller mode. Proportional controllers generate an immediate change in the position of the expansion valve in response to changes in the evaporator superheat of the system. They generate an analog (modulating) output signal instead of a binary (on/off) output signal. The output signal of a proportional controller is always present, but its magnitude will vary. Proportional modes can only be used in closed-loop, feedback applications like that of the EEV. In proportional mode, the actual evaporator superheat will "approach" the evaporator superheat set point of the controller but may never reach it in all applications. The difference between the superheat set point of the controller and the actual superheat is called offset or error, **Figure 24.39**. Proportional-mode controllers change or modify the controller's output signal to the expansion valve in proportion to the size of the change in the error. It cannot sense, and has no control over, any timespan that the superheat error has been in existence. For example, if the thermistor on the evaporator's outlet is very warm, meaning high superheat to the controller, a large movement of the expansion valve in the opening direction would follow. In this example, the controller saw a large error between the evaporator superheat set point and the actual evaporator superheat and compensated for it with a large valve movement.

Figure 24.39 This temperature/time graph shows set point and offset, or "error." Courtesy Sporlan Valve Company

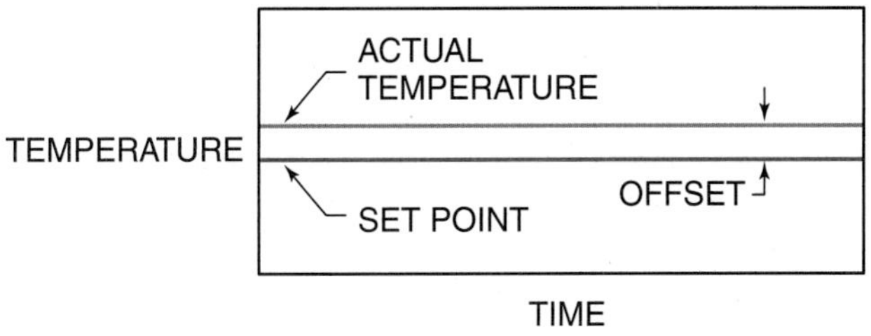

A typical programmed algorithm for a proportional controller follows:

If evaporator superheat is 20°F, then open the valve 175 steps.

If evaporator superheat is 15°F, then open the valve 125 steps.

If evaporator superheat is 10°F, then open the valve 0 steps.

If evaporator superheat is 6°F, then close the valve 75 steps.

If evaporator superheat is 0°F, then close the valve 1000 steps.

This algorithm will change the output (motor steps) in direct relation to the input (superheat). However, if the algorithm is used, the valve will slowly reach the controller's superheat set point, especially if the stepper motor has over 5000 steps. This is why a more sophisticated algorithm including integral and derivative functions must be used.

If the offset or error was a constant value, as shown in **Figure 24.39**, this constant difference could simply be preprogrammed into the controller. However, offset will change over time as refrigeration system parameters change with fluctuating heat loads. This is why some means of predicting the offset must be used to lessen its value and let the controller reach its evaporator superheat set point value. This is where the integral controller comes into play.

Integral controller modes can be added to a proportional controller to modify the controller's output signal, which will help force the offset value error to zero. The integral mode will change the controller's output signal in proportion to the size of the error or offset and the length of time it exists. The integral mode will calculate the amount of error that exists over a specific time interval. The integral controller actually calculates the area under the curve of the temperature/time graph, **Figure 24.40**. In mathematical terms, finding an area under a curve is often called integration. Integration, or the integral mode, of a controller will sense the average deviation of the actual evaporator superheat from the controller's superheat set point. Following is an example of a typical integral algorithm:

If the average superheat for 60 seconds is 6°F high, then open the valve 50 steps.

Thus, the offset or error in superheat is constantly being changed as heat load and refrigeration system parameters change. The changing amount of offset is calculated by the control algorithm and is then added to the set point. Another name for integral control is reset control. The calculation of the time-weighted offset (error), or the area under the curve of the time/temperature graph, occurs over a period of time referred to as the reset time. The controller will constantly try to reset the measured evaporator superheat to the superheat set point of the controller.

Derivative (differential) controller modes can also be added to a proportional controller, which will ensure even closer evaporator superheat control. The terms *derivative* and *differential* are used interchangeably. Derivative algorithms estimate the slope, or rate of change, of the temperature/time curve, **Figure 24.41**. If the rate of change (slope) is great, then the controller will make larger changes. If the slope is small, the controller will make small changes. Derivative control actually calculates the instantaneous rate of change of the error with respect to time. It can tell how fast the process error is changing. The controller

Figure 24.40 The temperature/time graph of an integral controller showing set point, offset, and reset. Courtesy Sporlan Valve Company

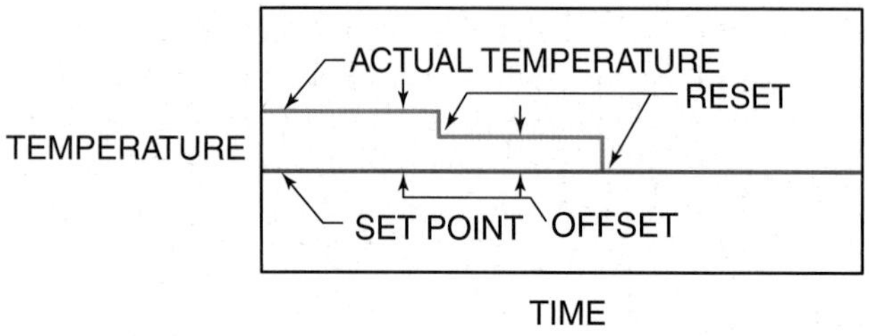

Figure 24.41 The temperature/time graph of a derivative, or differential, controller showing slopes or rates of change. Courtesy Sporlan Valve Company

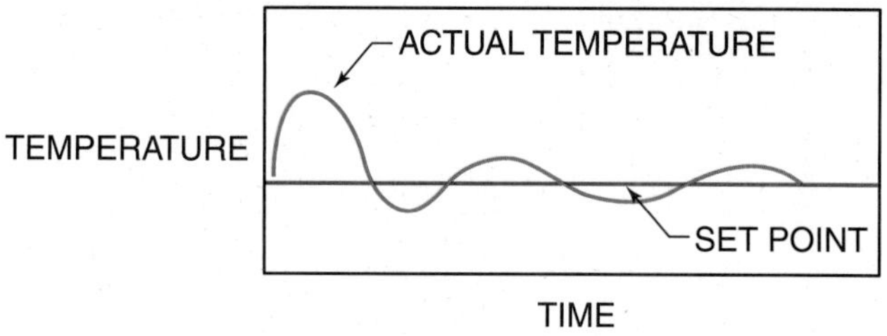

can then adjust the magnitude of its output signal to the EEV in proportion to the rate of change of the error in evaporator superheat of the system. This type of control reacts quickly to changes and will reduce any effect that changes in evaporator loads or system parameters have on the evaporator superheat set point. A typical derivative algorithm is shown here:

If superheat has dropped 0.2°F in 10 seconds, then close the valve 15 steps.

If superheat has dropped 2.0°F in 2 seconds, then close the valve 50 steps.

24.24 THE AUTOMATIC EXPANSION VALVE

The automatic expansion valve (AXV) is an expansion device that meters the refrigerant to the evaporator by using a pressure-sensing device. This device is also a valve that changes in inside dimension in response to its sensing element. **The AXV maintains a constant pressure in the evaporator. Notice that superheat is not mentioned.** This device has a needle and seat like the TXV that is fastened to a diaphragm, **Figure 24.42**. One side of the diaphragm is common to the

Figure 24.42 The automatic expansion valve uses the diaphragm as the sensing element and maintains a constant pressure in the evaporator but does not control superheat.

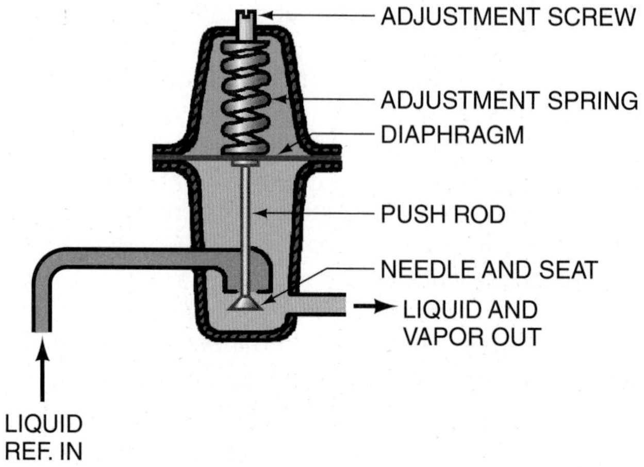

Figure 24.43 An automatic expansion valve. This valve resembles the TXV, but it does not have the bulb for sensing temperature at the suction-line end of the evaporator. Image Courtesy of Parker Sporlan

evaporator and the other side to the atmosphere. When the evaporator pressure drops for any reason, the valve begins to open and feed more refrigerant into the evaporator.

The AXV is built much the same as the TXV except that it does not have a sensing bulb. The body is normally made of machined brass. The adjustment for this valve is normally at the top of the valve. There may be a cap to remove, or there may be a cap to turn. This adjustment changes the spring tension that supports the atmosphere in pushing down on the diaphragm. When the tension is increased, the valve will feed more refrigerant and increase the suction pressure, **Figure 24.43**.

24.25 AUTOMATIC EXPANSION VALVE RESPONSE TO LOAD CHANGES

The AXV responds differently than the TXV to load changes. It actually acts in reverse. When a load is added to the coil, the suction pressure starts to rise. The AXV will start to throttle the refrigerant by closing enough to maintain the suction pressure at the set point. This has the effect of starving the coil slightly. A large increase in load will cause more starving. When the load is decreased and the suction pressure starts to fall, the AXV will start to open and feed more refrigerant into the coil. If the load reduces too much, liquid could actually leave the evaporator and proceed into the suction line, **Figure 24.44**.

Because the AXV can maintain constant pressure, it cannot be used with a low-pressure motor control. When the AXV is used with a thermostat, the thermostat should be adjusted

Figure 24.44 An automatic expansion valve under varying load conditions. This valve responds in reverse to a load change. (A) Normal operation. (B) When the load goes up, the valve closes down and starts to starve the coil slightly to keep the evaporator pressure from rising. (C) When the load goes down, the valve opens up to keep the evaporator pressure up. This valve is best applied where the load is relatively constant. When applied to a water-type evaporator, the valve's freeze protection can be a big advantage.

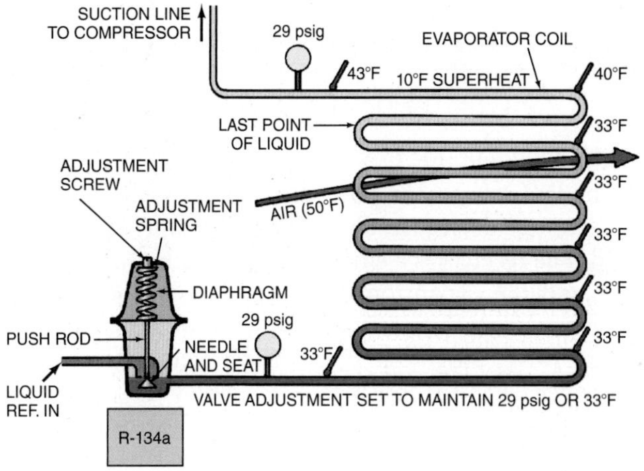

(A)

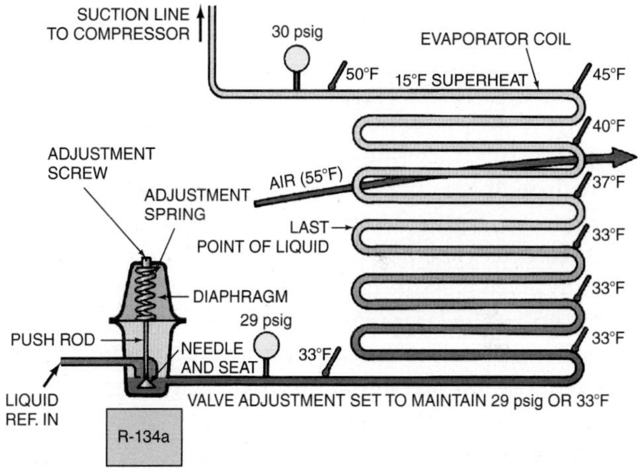

(B)

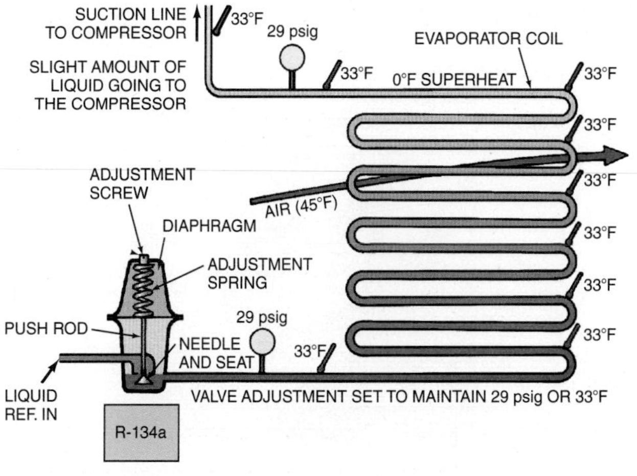

(C)

to shut the compressor off before flooding of the entire evaporator occurs. The AXV should be adjusted to maintain an evaporator pressure that corresponds to the lowest evaporator temperature desired through the entire running cycle of the compressor. AXVs are usually found on smaller equipment with somewhat constant evaporator loads, including ice cream freezers and makers and drinking fountains.

24.26 SPECIAL CONSIDERATIONS FOR THE TXV AND AXV

Both the TXV and the AXV are expansion devices that allow more or less refrigerant flow, depending on the load. Both need a storage device (receiver) for refrigerant when it is not needed. The receiver is a small tank located between the condenser and the expansion device. Normally, the condenser is close to the receiver. It has a *king valve* that functions as a service valve. This valve stops the refrigerant from leaving the receiver when the low side of the system is serviced. The receiver can serve both as a storage tank for different load conditions and as a tank into which the refrigerant can be pumped when servicing the system, **Figure 24.45**.

24.27 THE CAPILLARY TUBE METERING DEVICE

The capillary tube metering device controls refrigerant flow by pressure drop. It is a copper tube with a very small, calibrated, inside diameter, **Figure 24.46(A)**. The diameter and the length of the tube determine how much liquid will pass through the tube at any given pressure drop, **Figure 24.46(B)**. It is much like a garden hose: The larger the hose, the more gallons of water will pass at 100 psig pressure at the hose inlet, **Figure 24.46(C)**. The water hose has pressure drop all along its length, so a longer hose will have less pressure at the outlet even though the pressure at the inlet is 100 psig. This length-to-pressure drop relationship is used by manufacturers to arrive at the pressure drop that will allow the correct amount of refrigerant to pass through the capillary tube to correctly fill the evaporator. The capillary tube can be quite long on some installations and may be wound in a coil to store the extra tubing length.

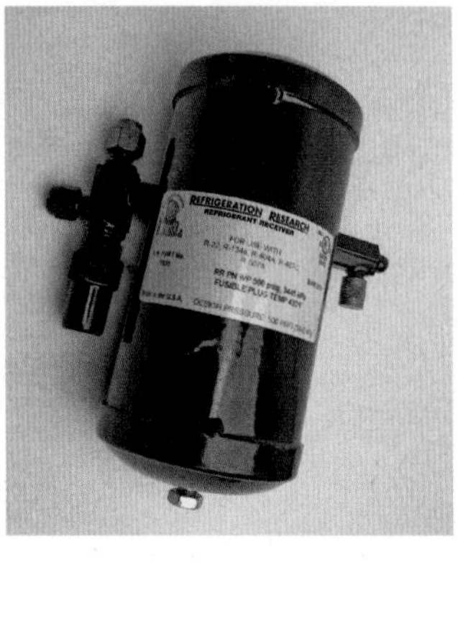

Figure 24.45 A refrigerant receiver. When the load increases and more refrigerant is needed, it moves from the receiver into the system. Photo by Jason Obrzut

(A)

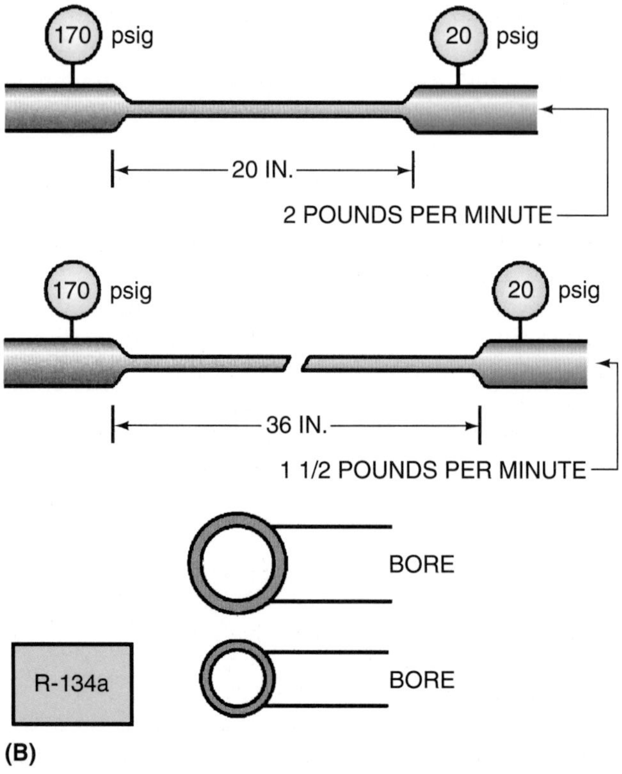

(B)

Figure 24.46 (A) A roll of capillary tubing. Photo by Jason Obrzut (B) The length of the capillary tube as well as the bore determines the flow rate of the refrigerant, which is rated in pounds per minute.

Figure 24.46 (*Continued*) (C) The longer the hose, the less water pressure at the end. A capillary tube functions the same way.

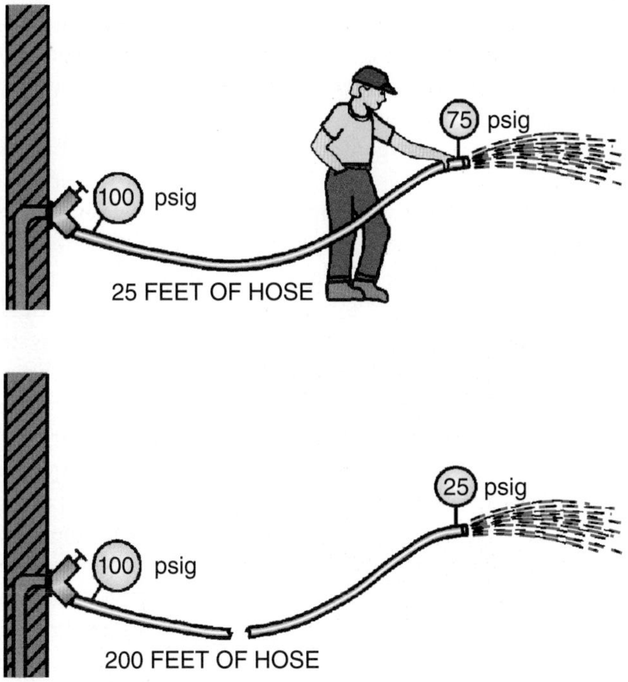

(C)

Figure 24.47 (A) Typical operating conditions for a medium-temperature application. (B) A strainer-drier protects the capillary tube from circulating particles.

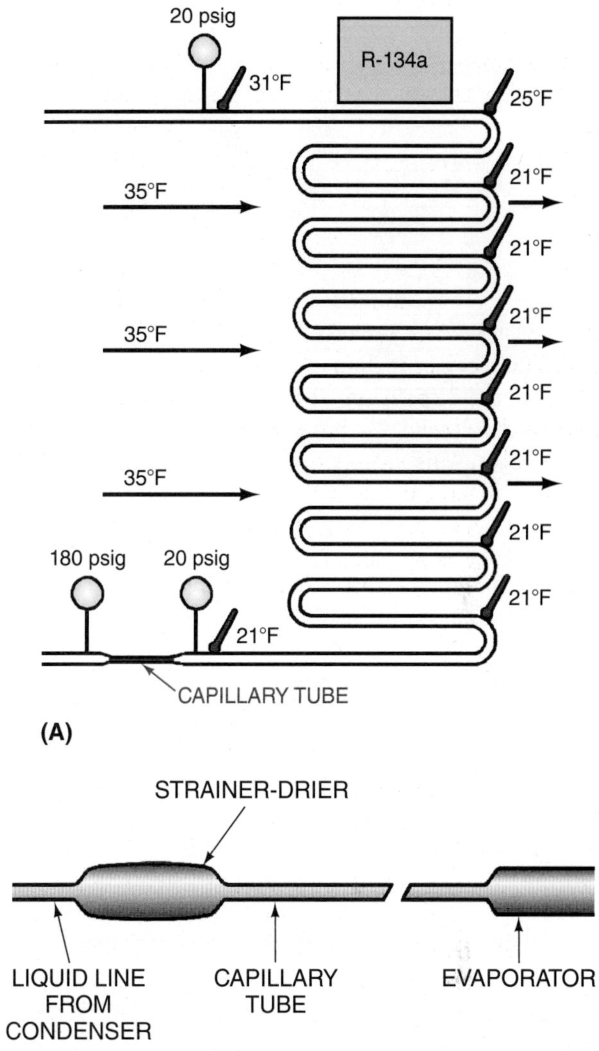

The capillary tube does not control superheat or pressure. It is a fixed-bore device with no moving parts. Because this device cannot adjust to load change, it is usually used where the load is relatively constant with no large fluctuations. It is an inexpensive device for the control of refrigerant and is often used in small equipment. This device does not have a valve and does not stop the liquid from moving to the low side of the system during the off cycle, so pressures will equalize during the off cycle. This reduces the motor-starting torque requirements for the compressor. **Figure 24.47(A)** illustrates a capillary tube in place at the inlet of the evaporator.

Because the capillary tube has no moving parts, it will not wear out. About the only problems it may have would come from small particles that may block or partially block the tube. The bore is so small that a small piece of flux, carbon, or solder would cause a problem if it were to reach the tube inlet. Manufacturers always place a strainer or strainer-drier just before the capillary tube to prevent this from happening, **Figure 24.47(B)**. Capillary tube metering devices can also become oil logged, which can cause a partial or complete blockage of the tube. Too much oil, or the wrong viscosity of oil, in the refrigeration system can cause oil logging. Another cause of oil logging is liquid refrigerant entering the crankcase of the compressor and then flashing. This is referred to as flooding the crankcase when it occurs during the run cycle. It is called migration when it occurs during the off cycle. The flashing of refrigerant in the warm compressor's crankcase causes the oil in the crankcase to foam and be sucked into the suction valve of the compressor during the run cycle. The oil is then circulated until it eventually reaches the capillary tube. Oil-logged capillary tubes cause sluggish operations and systems usually run with high superheats as a result, causing inefficiencies and damage to compressors from overheating due to lack of refrigerant in circulation for cooling. Often, long running cycles will return the oil to the compressor's crankcase. If oil in the capillary tube is suspected, force the system to run for several hours. This may resolve the problem of an oil-logged capillary tube. There are hydraulic capillary tube cleaners on the market, but they should be used as a last resort, **Figure 24.48**.

Figure 24.48 Hydraulic capillary-tube cleaning kit. Courtesy Ferris State University. Photo by John Tomczyk

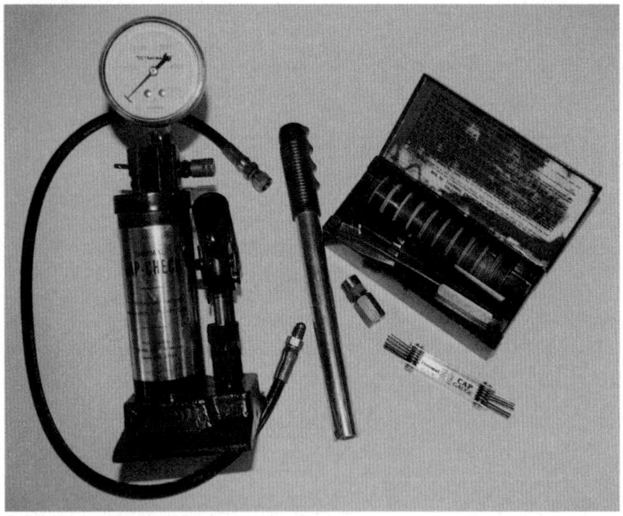

24.28 THE PISTON METERING DEVICE

The fixed orifice, or piston, metering device functions similar to the capillary tube metering device, **Figure 24.49**. It has a precisely sized hole bore through it which controls the flow of refrigerant to the evaporator. The smaller the bore, the less refrigerant flow through it. Pistons are available in various sizes and can be easily changed to a different size, which will change the capacity. The piston housing can be opened, and the piston cleaned if it becomes obstructed. Pistons are typically used in residential air-conditioning and heat pump systems.

24.29 OPERATING CHARGE FOR THE FIXED-BORE SYSTEM

The fixed-bore system requires only a small amount of refrigerant because it does not modulate (feed more or less) refrigerant according to the load. These systems are known

Figure 24.49 A piston type metering device most commonly found in residential cooling systems. Photo by Jason Obrzut

as critically charged systems. When the refrigerant charge is analyzed and when the unit is operating at the design conditions, a specific amount of refrigerant is in the evaporator and a specific amount of refrigerant is in the condenser. This is the amount of refrigerant required for proper refrigeration. Any other refrigerant that is in the system is in the pipes for circulating purposes only.

The amount of refrigerant in the system is critical in capillary tube systems. It is easy for technicians to overcharge if they are not careful and familiar with the system. In most capillary tube systems, the charge is printed on the nameplate of the equipment. The manufacturer always recommends measuring the refrigerant into these systems either by scales or by liquid-charging cylinders. A properly charged system would maintain about 10°F superheat at the end of the coil once the system is pulled down to the desired box temperature, **Figure 24.47(A)**. A characteristic of a capillary tube metering device system is to run high evaporator superheat under high evaporator heat loads. However, once the system is pulled down near the desired refrigerated box temperature, an evaporator superheat of about 8°F to 12°F will be reached if the system is critically charged properly.

In many capillary tube systems, the capillary tube is fastened to the suction line between the condenser and the evaporator to exchange heat between the capillary tube and the suction line. To troubleshoot a possible capillary tube problem, the correct superheat reading must be taken. The capillary tube should hold the refrigerant to about 10°F superheat at the end of the evaporator once the box temperature is pulled down and the heat load is at a minimum. *To obtain the correct superheat reading it must be taken before the capillary tube/suction line heat exchanger,* **Figure 24.50**.

The capillary tube is very slow to respond to load changes or charge modifications. For example, if a technician were to add a small amount of refrigerant to the capillary tube system, it would take at least 15 min for the charge to adjust. The reason for this is that the refrigerant moves from one side of the system to the other through the small bore, and this takes time. When refrigerant is added to the low side of the system, it moves into the compressor and is pumped into the condenser. It now must move to the evaporator before the charge is in balance. A technician that is in a hurry can end up overcharging this type of system. Manufacturers do not recommend adding refrigerant to top off the charge; they recommend starting over from a deep vacuum and measuring a complete charge into the system.

Under high evaporator loads, capillary tube systems will often operate with high evaporator superheats of 40°F or 50°F. This is because the refrigerant in the evaporator will vaporize rapidly, and the last point of liquid will be toward the middle of the evaporator. This makes the system very inefficient. Many technicians have the urge to add more refrigerant to the system at these high loads because of the high superheat readings. Adding refrigerant at this

Figure 24.50 Checking the superheat before the suction–liquid heat exchange.

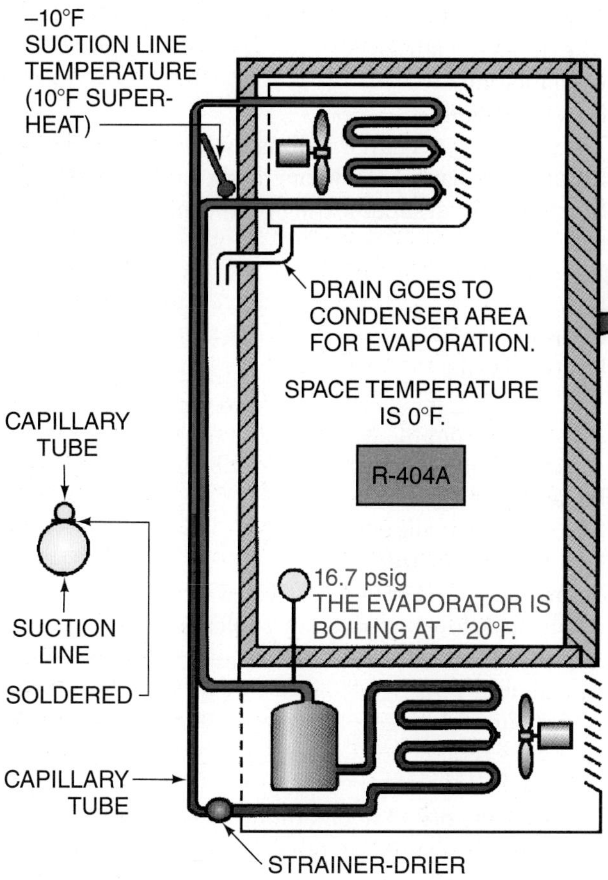

point will, in time, only overcharge the system. Before adding refrigerant, check the superheat at the normal evaporator load to see whether it is normal. It should typically range from 8°F to 12°F at the normal evaporator load. If unsure, recover the refrigerant, evacuate the system, and add the specified nameplate critical charge of refrigerant. Under low evaporator heat loads, the refrigerant in the evaporator will not vaporize as fast. Low evaporator loads are usually experienced just before the desired box temperature is reached. The last point of liquid will now be toward the end of the evaporator. The system will still have a normal superheat of about 8°F to 12°F because of its critical charge characteristics.

The capillary tube metering device is used primarily on small fractional-horsepower refrigeration systems that have somewhat constant evaporator heat loads. These systems are hermetically sealed; have no bolted, gasketed connections; and are leak-free systems. They are factory assembled in a very clean environment and should run trouble-free for many years.

Fixed orifice or piston systems should be charged using a superheat calculator or the charging charts provided by the manufacturer. Refer to Unit 10, "System Charging," for more information on using charging charts or curves. Systems with a piston metering device typically experience high evaporator superheats under high loads. The refrigerant in the evaporator will vaporize faster, moving the last point of liquid further back in the evaporator. When the system is close to its design temperature, the typical superheat range is about 10°F to 16°F.

REVIEW QUESTIONS

1. Which is *not* a pressure or force acting on a TXV diaphragm?
   A. Head pressure
   B. Evaporator pressure
   C. Spring pressure
   D. Bulb pressure
2. The needle and seat of a TXV are normally made of ________.
   A. copper
   B. aluminum
   C. stainless steel
   D. brass
3. The TXV tries to maintain a constant ________ in the evaporator.
   A. pressure
   B. superheat
   C. subcooling
   D. frost line
4. Liquid, vapor, cross liquid, and cross vapor are all types of ________.
   A. TXV bulb charges
   B. AXV bulb charges
   C. pistons
   D. capillary tube bores
5. The bulb of the TXV is mounted on the ___________.
   A. compressor inlet
   B. compressor outlet
   C. evaporator inlet
   D. evaporator outlet
6. When should an externally equalized TXV be used?
   A. When the system is a low- or medium-temperature application.
   B. When there is excessive pressure drop through the evaporator.
   C. When the TXV is used in a low ambient condition on a single-circuit evaporator.
   D. When the system is equipped with a semihermetic compressor.
7. An AXV tries to maintain a constant ________ in the evaporator.
   A. pressure
   B. superheat
   C. subcooling
   D. frost line

8. The TXV responds to an increased heat load by _________.
   A. decreasing refrigerant flow
   B. keeping refrigerant flow constant
   C. increasing refrigerant flow
   D. not reacting to changing heat loads
9. The three factors that determine the capacity of a TXV are: evaporator temperature, liquid temperature entering the valve, and ________.
   A. number of evaporator circuits
   B. evaporator pressure
   C. length of the sensing bulb transmission line
   D. pressure drop across the valve
10. How does the automatic expansion valve respond to a load increase?
   A. It opens, decreasing the refrigerant flow.
   B. It closes, decreasing the refrigerant flow.
   C. It opens, increasing the refrigerant flow.
   D. It closes, increasing the refrigerant flow.
11. On a _________ TXV, the liquid pressure is cancelled out because it acts on equal areas but in opposite directions.
   A. balanced-port
   B. pressure-limiting
   C. cross vapor
   D. cross liquid
12. If a system with a capillary tube metering device is properly charged, when it nears the desired box temperature it will have a superheat of approximately ______.
   A. 2°F to 5°F
   B. 15°F to 20°F
   C. 10°F to 15°F
   D. 8°F to 12°F
13. A capillary tube metering device controls refrigerant flow by________.
   A. using superheat
   B. pressure drop
   C. using a stepper motor
   D. temperature drop
14. On a PID controller, the difference between the superheat set point of the controller and the actual superheat is called ______.
   A. offset
   B. proportion
   C. integral
   D. derivative
15. An EEV uses a _______ to control the valve's port, which in turn controls evaporator superheat.
   A. sensing bulb and transmission line
   B. capillary tube
   C. spring
   D. stepper motor
16. The _______ is located inside the valve body and moves the needle in and out of the seat in response to system load changes.
   A. transmission line
   B. external equalizer
   C. diaphragm
   D. sensing bulb
17. The _________ charge bulb is usually limited to applications that have narrow operating ranges or to special functions like desuperheating on the high side of the system.
   A. vapor
   B. liquid
   C. cross vapor
   D. cross liquid
18. The ________ on a TXV is used to compensate for pressure drops from the inlet to the outlet of the evaporator.
   A. sensing bulb and transmission line
   B. needle and seat
   C. external equalizer
   D. internal equalizer
19. To prevent an overloaded compressor during a hot pulldown on a low-temperature cooler, a ________ expansion valve can be used.
   A. balanced-port
   B. pressure-limiting
   C. cross vapor
   D. cross liquid
20. The two factors that determine how much liquid will pass through a capillary tube at any given pressure drop are the ________.
   A. length and diameter
   B. type of refrigerant and oil
   C. evaporator and suction-line size
   D. spring and bulb pressure
21. Explain why the entering liquid pressure has no effect on the opening force of a balanced-port TXV.
22. Explain how a PID controller works.
23. Explain the following terms as they apply to electronic expansion valves:
   Stepper motor
   Offset
   Algorithm
   Microprocessor
   Feedback loop
24. Explain what is meant by a maximum operating pressure (MOP) TXV.
25. Explain why systems equipped with a TXV or AXV require a receiver.

UNIT 25

SPECIAL REFRIGERATION SYSTEM COMPONENTS

OBJECTIVES

After studying this unit, you should be able to

- distinguish between operational and safety controls.
- explain how multiple evaporators at different pressures can be controlled by a single compressor.
- describe the function of a partition wall in a compressor.
- describe four methods that can be used to control head pressure during periods of low ambient temperatures.
- describe two common functions of a low pressure control.
- explain the operation of an oil pressure safety control.
- describe the automatic pumpdown cycle.
- describe the evaporator conditions that would require a defrost cycle.
- list and describe five methods of evaporator defrost.
- explain the concept of temperature-terminated defrost.
- explain the importance of filter driers.
- explain how suction-line accumulators and liquid-line refrigerant receivers function in an air-conditioning or refrigeration system.
- explain the two benefits of a suction/liquid-line heat exchanger.
- describe the importance of suction and discharge service valves to the HVACR service technician.

SAFETY CHECKLIST

- ✔ Wear goggles and gloves when attaching or removing gauges to transfer refrigerant or to check pressures.
- ✔ Wear warm clothing when working in a walk-in cooler or freezer. A technician does not think properly when cold.
- ✔ Be careful not to get your hands or clothing caught in moving parts such as pulleys, belts, or fan blades.
- ✔ Wear an approved back brace or belt when lifting and use your legs, keeping your back straight and upright.
- ✔ Observe all electrical safety precautions. Be careful at all times and use common sense.
- ✔ Never work near live electrical circuits unless absolutely necessary. Turn the power off at the nearest disconnect, lock and tag the panel, and keep the only key in your possession.

25.1 INTRODUCTION TO SPECIAL REFRIGERATION SYSTEM COMPONENTS

The basic vapor-compression refrigeration cycle, and the four main components that make it up, has been discussed throughout this text. These four components, namely the compressor, the condenser, the evaporator, and the expansion device allow the cycle to continually repeat itself as long as it is operating, without depleting or otherwise affecting the system refrigerant. However, relying solely on these four system components does not ensure that the system will operate as effectively or efficiently as it can. To enhance system operation, reliability, and performance, additional devices and components can be added to the refrigeration system. These additional devices and components are commonly referred to as special refrigeration system components and can be categorized in a number of different ways. The two broadest categories for special HVACR system components are controls and accessories.

Controls can be classified as either operational or safety devices. Operational controls open and close during normal system operation, while safety controls are intended to stop system operation in the event unsafe conditions are present. Control devices can be mechanical, electrical, electromechanical, or electronic devices. Mechanical controls are generally used to stop, start, or modulate fluid flow and can be operated by pressure, temperature, or other means. These controls can usually be identified by their location, because they are almost always found in the piping. Accessories are intended to enhance or improve overall system performance. Because there are so many different classifications of HVACR special system components and components themselves, some devices may fall in more than one category. It is important to realize, therefore, that because one particular system component is discussed in one section of this unit, it is not necessarily excluded from other sections.

25.2 EVAPORATOR CONTROLS

EVAPORATOR PRESSURE REGULATOR

The evaporator pressure-regulating (EPR) valve, **Figure 25.1**, is a mechanical control that keeps the refrigerant pressure in the evaporator from going below a predetermined point. Two pressures, the evaporator outlet pressure, which is the valve's inlet pressure, and the valve's spring pressure, control the positioning of the valve. The evaporator pressure and spring pressure counteract one another. The spring pressure is a closing force, and the evaporator pressure is an opening force. The outlet pressure is effectively cancelled out and does not affect the valve operation because the area of the bellows is equal to the area of the inlet port. The outlet pressure acts on these equal areas, but in opposite directions, thus they cancel one another, **Figure 25.2**. EPR valves are also referred to as open on rise of inlet pressure (ORI) valves, because they will throttle open on a rise in their inlet (evaporator) pressure. When the evaporator is under a high load, the EPR valve will be fully opened—in effect, act as if it were not there. The spring pressure can be adjusted with a screwdriver or Allen wrench to allow the setting of a minimum evaporator pressure. Remember, these valves simply establish a minimum evaporator pressure; they do not hold a constant pressure or limit maximum pressure in the evaporator.

Figure 25.1 The EPR valve modulates the flow of vapor refrigerant leaving the evaporator. It limits the pressure in the evaporator and keeps it from dropping below the set point. Courtesy Sporlan Division, Parker Hannifin Corporation

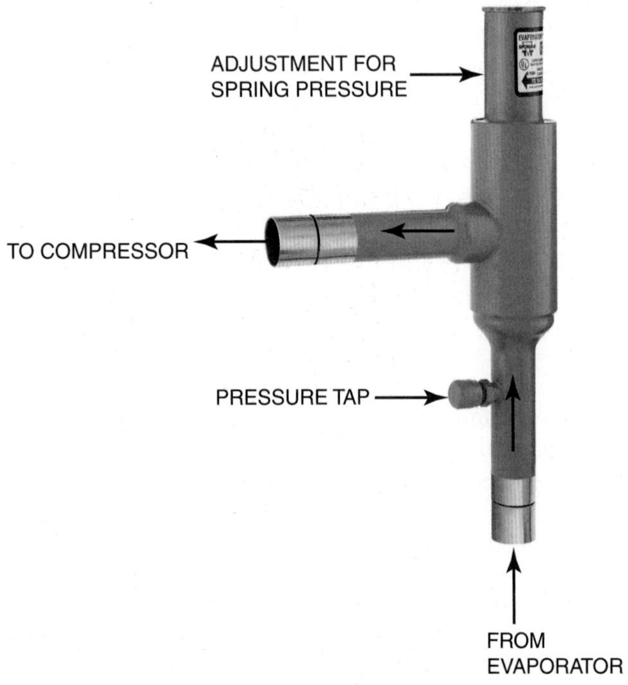

Figure 25.2 The cutaway view of a direct-acting evaporator pressure-regulating (EPR) valve. Courtesy Sporlan Division, Parker Hannifin Corporation

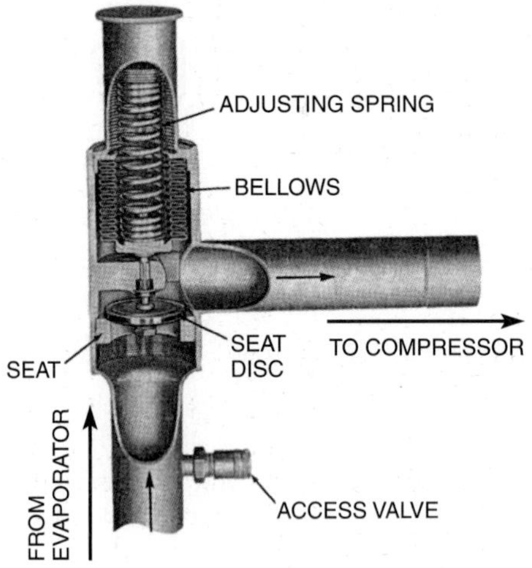

The EPR valve is installed in the suction line at the outlet of the evaporator. The bellows in the EPR valve senses evaporator pressure and throttles (modulates) the suction gas to the compressor. This will then allow the evaporator pressure to go as low as the pressure setting on the valve. When the EPR valve is used with the thermostatic expansion valve (TXV), the system can maintain a constant superheat and keep the pressure from going too low. If the pressure in the evaporator is lower than the EPR valve's set point, the EPR valve will close, causing refrigerant to back up into the evaporator coil until the minimum pressure level in the evaporator has been reached. As the evaporator pressure rises above the minimum pressure setting, the EPR valve will throttle open to allow excess pressure to flow from the evaporator coil into the suction line and on to the compressor.

The EPR valve can be used in a system that cools water to keep the evaporator from going below a predetermined point, which could be freezing. For example, if there is a load on the system at start-up, the EPR valve will be wide open, and the TXV will throttle the flow of liquid refrigerant into the evaporator. When the evaporator pressure reaches the point at which the EPR valve is set, the valve will begin to throttle to slow the flow of refrigerant. If the setting is just above freezing, the valve will throttle just enough to keep the evaporator from freezing until the thermostat responds and shuts the system off, **Figure 25.3**. The EPR valve is often called a "holdback valve" in the industry because it holds the refrigerant back in the evaporator to prevent the suction pressure in the evaporator from going below a set point.

Figure 25.3 The TXV keeps the proper level of refrigerant in the evaporator while the EPR valve keeps the evaporator from getting too cold. The EPR valve may be used on a water-type evaporator to keep it from operating below freezing. The thermostat should cut off the compressor soon after the EPR valve starts to throttle.

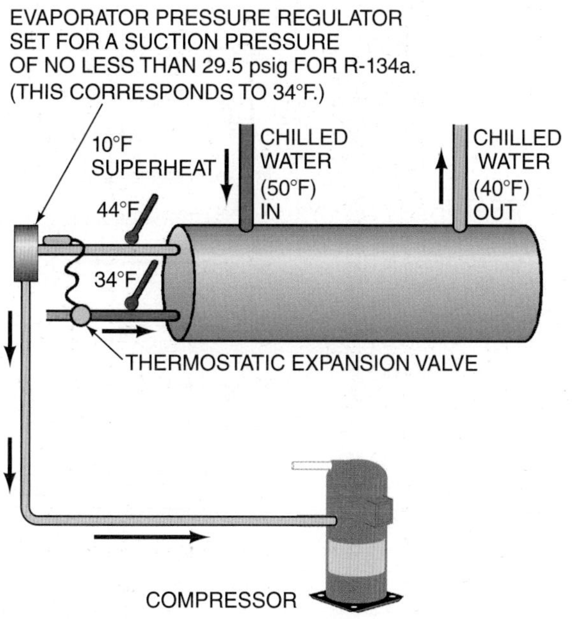

The EPR valve may operate for an extended period of time while in the throttled-down position in order to satisfy the cooling load and keep the pressure in the evaporator from falling below a minimum set point. Many times, the evaporator pressure will be close to the EPR's set point pressure, but some refrigeration is still needed due to a small heat load on the evaporator. The flow of refrigerant will be minimal, and the TXV may not be able to keep a constant evaporator superheat because of the reduced mass flow rate of refrigerant. High evaporator superheat readings are often encountered at these throttled-down conditions. However, once the heat load on the evaporator increases and the EPR valve opens more to increase the mass flow rate of refrigerant, the TXV will regain control of evaporator superheat.

Larger Btu capacities, as in supermarket installations, call for larger EPR valves or pilot-operated valves, **Figure 25.4**. Some pilot-operated EPR valves are equipped with additional features such as a suction stop and an automatic close feature, **Figure 25.5**. Some normally open pilot EPR valves operate by high-pressure gas. These valves are held open by a spring and are forced closed by high-side gas pressure, **Figure 25.6**. Three pressures control the modulation of the main valve piston: the high-pressure source, the evaporator pressure, and a spring pressure. A pilot port that is controlled by a pilot regulator valve indirectly controls the operation of the main regulating valve. These valves include a solenoid stop feature when hot gas defrost is used because the hot gas is injected into the suction line between the EPR valve and the coils being defrosted. A solenoid valve in the EPR valve will be deenergized, shutting off the suction inlet pressure. This allows the internal pilot spring to force the EPR valve

Figure 25.4 Pilot-operated EPR valve. Courtesy Sporlan Division, Parker Hannifin Corporation

Figure 25.5 Pilot-operated EPR valve with suction stop and automatic close feature. Courtesy Sporlan Division, Parker Hannifin Corporation

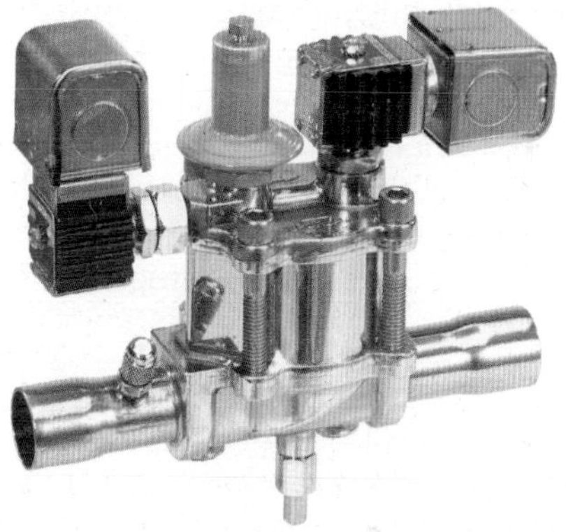

Figure 25.6 An internal cutaway view of a pilot-operated EPR valve showing springs, high-pressure source connection, pilot regulator, main piston, and solenoid stop feature. Courtesy Sporlan Division, Parker Hannifin Corporation

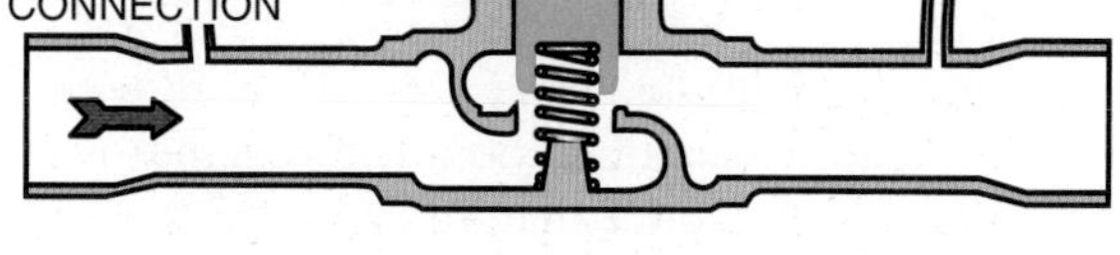

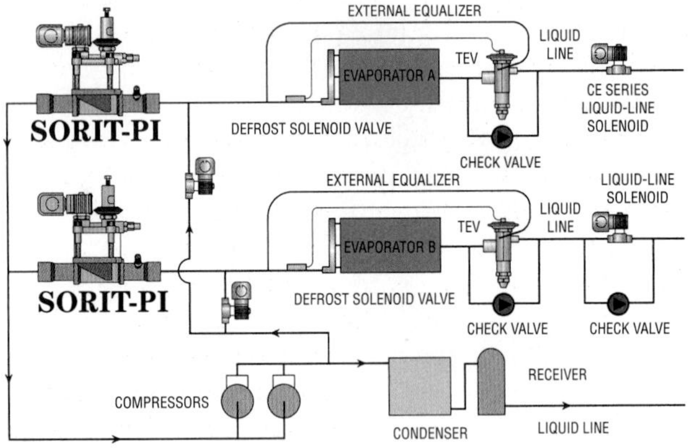

Figure 25.7 A multiple-evaporator circuit showing defrost solenoid valves, EPR valves, and defrost circuit. Courtesy Sporlan Division, Parker Hannifin Corporation

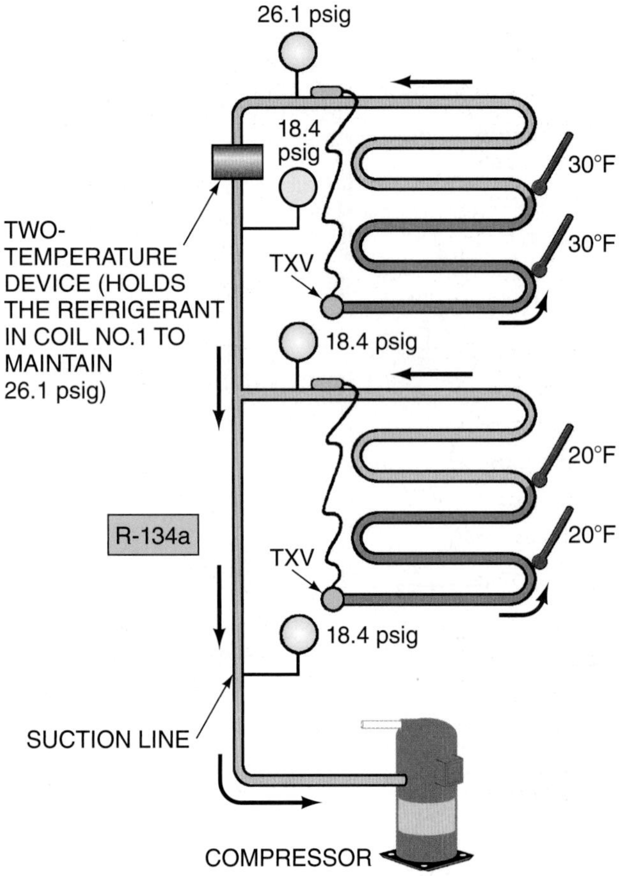

Figure 25.8 Two-temperature operation. Two evaporators are operating with one compressor. One evaporator operates at 20°F (18.4 psig), the other at 30°F (26.1 psig).

closed. At the same time, a hot gas defrost solenoid valve will open; the evaporator fans will stop; and the hot gas will flow in reverse through the evaporators, around the expansion valves, through a check valve, and then around a liquid-line solenoid valve. The condensed hot gas, which is now a cool liquid, will then feed into a liquid header to supply other evaporators with subcooled liquid refrigerant, **Figure 25.7**.

MULTIPLE EVAPORATOR SYSTEMS

Some refrigeration systems operate with a single compressor that is used in conjunction with multiple evaporators. These evaporators, depending on the system, might be located in separate refrigerated boxes and might operate at different temperatures and pressures. When such is the case, EPR valves are used to allow each evaporator to operate at its own, unique pressure and temperature. Consider the example when one R-134a compressor is used with two evaporators, one operating at a pressure of 18.4 psig (20°F for R-134a) and the other operating at a pressure of 26.1 psig (30°F for R-134a). To allow these two evaporators, that share a suction line, to operate at different pressures, an EPR valve is needed at the outlet of the higher-temperature evaporator, **Figure 25.8**. Please note that **Figure 25.8** is showing only the low pressure side of the refrigeration system. The true suction pressure for the system will be the lower value of 18.4 psig, but the higher-temperature evaporator will operate at a pressure of 26.1 psig. Several evaporators with different pressure requirements can be piped together using this method, **Figure 25.9**. Notice that, in **Figure 25.9**, EPR valves are installed at the outlet of all evaporators, with the exception of the lowest temperature evaporator which, in this case, is operating at a pressure of 15 psig. Please note that **Figure 25.9** is showing only the low pressure side of the refrigeration system.

It is desirable to know the actual pressure in the evaporator because that is not the same as the true suction pressure. Normally, a gauge port, known as a Schrader valve port, is permanently installed in the EPR valve body, which allows the service technician to take gauge readings on the evaporator side of the valve, **Figure 25.1**. The common suction-line pressure can be obtained at the compressor service valve.

ELECTRIC EVAPORATOR PRESSURE-REGULATING (EEPR) VALVE

As a result of regulatory changes in the U.S. Food and Drug Administration's (FDA) Food Code concerning product temperature in supermarkets, more sophisticated and accurate ways to control the discharge air temperatures in the evaporator section of display cases were put in place. Electric evaporator pressure-regulating (EEPR) valves, ***Figure 25.10****, designed for close temperature control, move in very small increments in response to the sensed temperature of the evaporator's discharge air.*

Figure 25.9 Evaporators piped together and one compressor used to maintain a suction pressure for the lowest-pressure evaporator. The evaporators that have pressures higher than the true suction pressure of the lowest-pressure evaporator have EPR valves set to their individual needs.

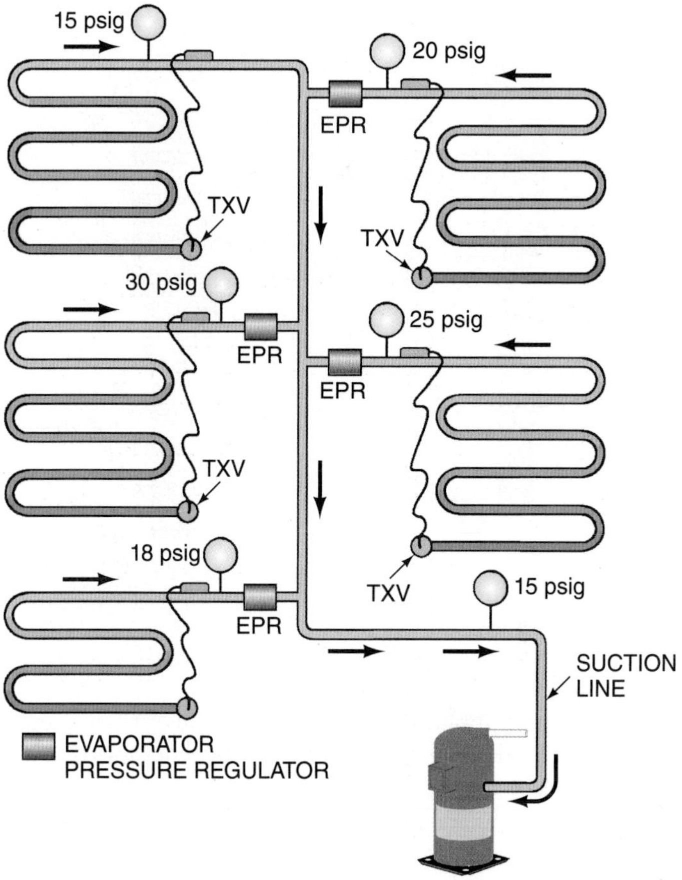

Figure 25.10 An electric evaporator pressure-regulating (EEPR) valve. Courtesy Sporlan Division, Parker Hannifin Corporation

As with traditional EPR valves, EEPR valves are located at the evaporator outlet for single evaporator applications. In multiple evaporator applications, they are located on the suction line before the common suction header as in parallel compressor systems, **Figure 25.11**. EEPR valves and a refrigerated case discharge air sensor are wired to a microprocessor board for control purposes.

Although EEPR valves are referred to as evaporator pressure regulators, they do not serve the same purpose as the standard EPR valve mentioned earlier. A standard mechanical EPR valve prevents pressure, and thus temperature, from falling below a minimum in the evaporator. EEPR valves, on the other hand, control the discharge air temperature in the refrigerated case. They regulate the evaporator pressure in order to provide the temperature required at the discharge air sensor. EEPR valves can deliver a quick temperature pulldown since they are sensing discharge air instead of evaporator pressure. They will remain fully open until the temperature of the discharge air in the case drops to the valve's set point. At this point, the valve will start to throttle closed to maintain the desired discharge air temperature.

A bipolar stepper motor operates the EEPR valve. The stepper motor operates by moving the motor in discrete steps instead of in a continuous rotation. The motor can step in either direction because it has two different windings. A stepper motor may have as many as 100 electromagnets around the rotor, resulting in an angular rotation that can be as small

Figure 25.11 EEPR valves in suction lines on a parallel compressor system. Courtesy RSES Journal

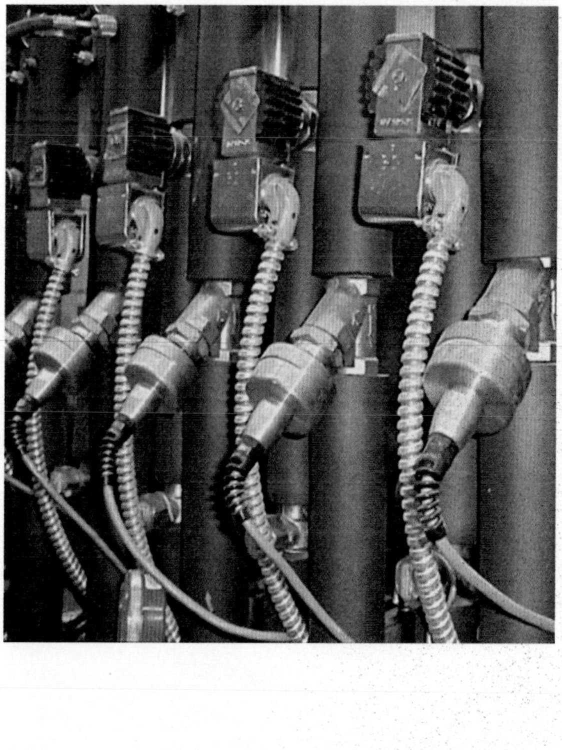

Figure 25.12 A stepper motor showing how the lead screw is driven through gear reduction for providing additional power to the valve.

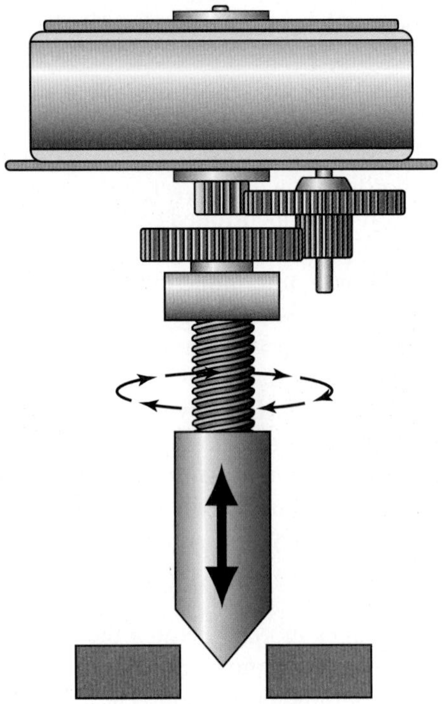

Figure 25.13 Three CPR valves that prevent the compressor from running in an overloaded condition during a hot pulldown. Courtesy Sporlan Division, Parker Hannifin Corporation

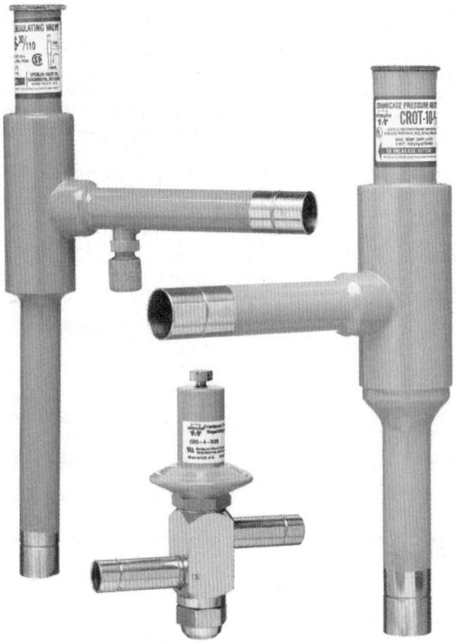

as 3.6° per step. Since a full circle is made up of 360°, an angular rotation of 3.6° represents 1/100th of a rotation. The motor drives a lead screw that converts the angular motion into straight-line, or linear, motion, **Figure 25.12**, with a resolution that can be as fine as 0.0000783 in. per step. An algorithm programmed into a controller or microprocessor controls the direction and position of the stepper motor. Refer to Unit 24, "Expansion Devices," for a more detailed explanation of bipolar stepper motor operation and control.

25.3 COMPRESSOR CONTROLS AND OIL-RELATED COMPONENTS

CRANKCASE PRESSURE REGULATOR

The crankcase pressure-regulating (CPR) valve, **Figure 25.13**, looks very much like the EPR valve, but it performs a different function. The CPR valve is used to limit the pressure that returns to the compressor in the suction line. The CPR valve, like the EPR valve, is installed in the suction line, but it is usually located close to the compressor rather than at the evaporator outlet. The CPR valve sensing bellows are on the true compressor suction side of the valve and would normally have a gauge port on the evaporator side of the valve, as shown in **Figure 25.13.**

A common application for the CPR valve is to keep a low-temperature compressor from overloading during a hot pulldown. A hot pulldown can occur:

- when the compressor has been off for a long enough time and the foodstuff has risen in temperature,
- on start-ups with a warm box temperature, or
- after a defrost period.

In any case, the temperature in the refrigerated box or the evaporator influences the suction pressure. When the temperature is high, the suction pressure is high. When the suction pressure goes up, the density of the suction gas goes up. The compressor is a constant-volume pump and does not know when the gas it is pumping is dense enough to create a motor overload. When a compressor is started and the refrigerated cooler is warmer than the design range of the compressor, an overload can occur. The refrigeration compressor motor can be operated at about 10% over its rated capacity without harm during this pulldown. For example, when a motor has a full-load amperage rating of 20 A, it could run at 22 A for an extended pulldown and no motor harm would result. If the temperature in the cooler were to be 75°F or 80°F, the motor would be overloaded to the point that its overcurrent protection would shut it off. But this protection device automatically resets, so the compressor would try to restart immediately, causing it to short cycle. This type of operation could continue until the manual reset motor control, the fuse, or the breaker stopped the compressor, or until the system

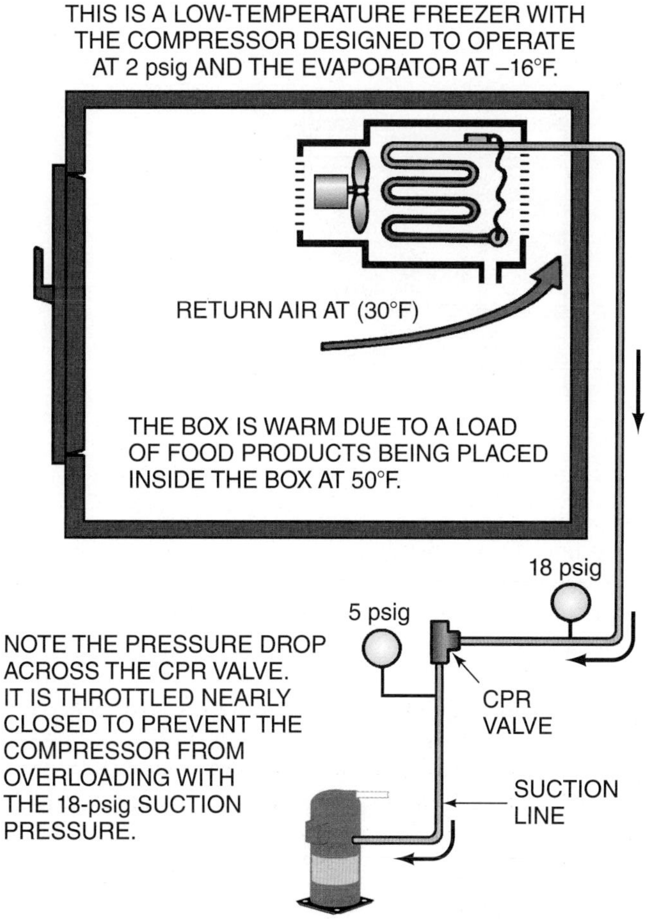

Figure 25.14 The CPR valve throttles the vapor refrigerant entering the compressor.

slowly pulls down during the brief running times. The CPR valve throttles the suction gas entering the compressor to keep the compressor current at no more than the rated value, **Figure 25.14**.

The CPR valve is often referred to as a close on rise of outlet pressure (CRO) valve. This is because it closes when its outlet (crankcase) pressure rises to a predetermined pressure. The CPR valve is controlled by crankcase (outlet) pressure and its spring pressure. The crankcase pressure is a closing force, and the spring pressure is an opening force. The inlet pressure of the CPR valve is actually canceled out because of the equal areas of the bellows and the seat disc. The inlet pressure exerts an equal pressure on both the bellows and seat disc, but these pressures are in opposite directions and cancel one another out, **Figure 25.15**. The spring pressure can be adjusted with a screwdriver or an Allen wrench to set the maximum desired crankcase pressure.

During periods of high load, the CPR valve will be partially closed, holding some refrigerant back in the suction line. When such is the case, there will be a pressure drop across the CPR valve. When the load on the system is low and the low-side pressure is below the set point on the CPR valve, the valve will be fully open and the pressure at the valve's outlet will be equal to the pressure at its inlet. This, of course, will only happen if the valve is properly adjusted.

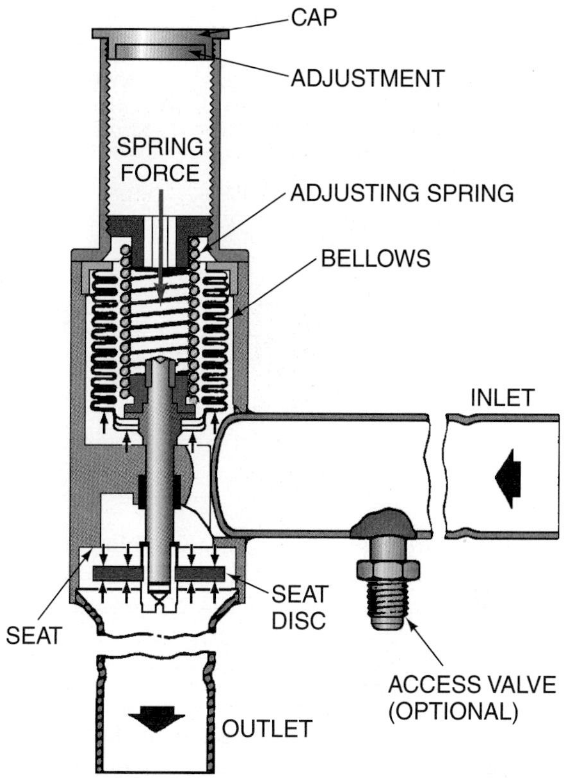

Figure 25.15 The cutaway view of a crankcase pressure-regulating (CPR) valve. Courtesy Sporlan Valve Company

ADJUSTING THE CPR VALVE

The setting of the CPR valve should be accomplished on a hot pulldown or at least when the compressor has enough load that it is trying to run overloaded. This can be done by shutting off the unit until the box or cooler is warm enough to create a load on the evaporator. This loads up the compressor enough to cause it to run at a high current. It is important to use an ammeter while adjusting the CPR valve to ensure that the amperage draw of the compressor does not exceed its full-load amperage rating. Adjusting the CPR valve is accomplished by changing the spring pressure on the valve. Increasing the spring pressure will increase the maximum suction pressure that can return to the compressor, while decreasing the spring pressure decreases the maximum suction pressure to the compressor.

For example, if a compressor is supposed to draw 20 amps at fully loaded conditions and it is presently drawing 23 amps, reduce the CPR valve's spring pressure until the compressor amperage drops to 20 A. Once this point is reached, the valve is properly set. Under no circumstances will this compressor

Figure 25.16 This is a float-action valve that will return liquid refrigerant as well as oil. It must be kept warm to keep refrigerant from condensing in it during the off cycle. Courtesy AC & R Components, Inc.

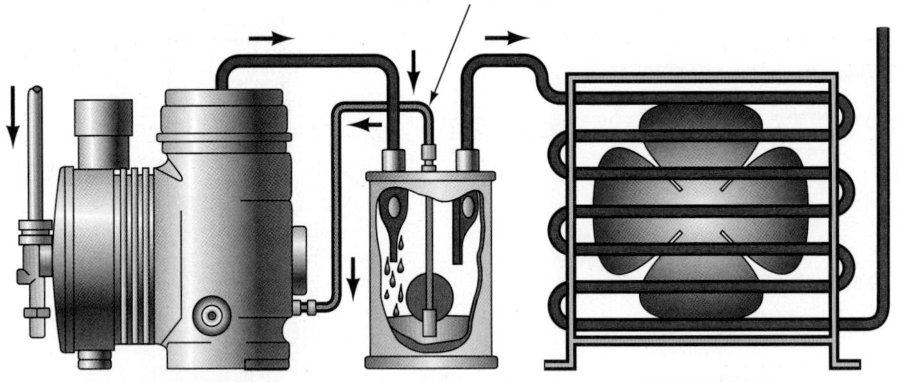

draw more than 20 amps. It is important to note that it is not possible to properly adjust a CPR valve without running, and monitoring the amperage draw of, the compressor.

OIL SEPARATORS

The oil separator, *Figure 25.16, is installed in the discharge line to separate the oil from the refrigerant and return the oil to the compressor crankcase.* All reciprocating and rotary compressors allow a small amount of oil to pass through into the discharge line. Once the oil leaves the compressor, it would have to go through the entire system to get back to the compressor crankcase by way of the system piping and coils. The oil separator has a reservoir and float assembly that allows oil to return to the crankcase shortly after it leaves the compressor.

Oil separators work on the principle of low refrigerant and oil velocity. As oil-laden discharge gas enters the oil separator, its velocity is immediately slowed. This slowing of velocity happens because of an abrupt change of direction, impingement screens, internal baffling, or a large opening or transitioning of the piping at the entrance of the oil separator. Oil is usually mixed with the hot discharge gas in the form of a fog. This refrigerant/oil fog runs into the internal baffling or screens, which cause the refrigerant and oil fog to change direction and rapidly slow down. Very fine particles of oil collide with one another and form larger, heavier particles. Fine mesh impingement screens separate the oil and refrigerant even more, causing large oil droplets to form and drop to the bottom of the separator. The oil collects in a sump or reservoir at the bottom of the separator. A magnet is often connected to the bottom of the oil sump to collect any metallic particles.

When the oil level becomes high enough to raise a float, an oil return needle valve is opened, and the oil is returned to the compressor crankcase or oil reservoir through a small return line. This return line should be just above room temperature most of the time. If it is always hot, the float could be stuck open, allowing hot gas to enter the crankcase, which would also cause excessively high crankcase pressures and could overheat the compressor, resulting in serious damage. Because the oil separator is in the high side of the system and the crankcase is in the low side, a pressure difference naturally exists. This pressure difference drives the oil to travel from the separator to the crankcase when the float is in the right position. Only a small amount of oil is needed to actuate the float mechanism. This ensures that only a small amount of oil is ever absent from the compressor crankcase at any given time.

Another type of oil separator is the helical oil separator, **Figure 25.17**. Helical oil separators are about 98% efficient in separating oil. When the oil and refrigerant fog enters the oil separator, it hits the leading edge of a helical flighting. The fog mixture is centrifugally forced along the spiral path of the helix. This causes the heavier oil particles to be forced to the perimeter of the separator's shell. This is where impingement with a screen layer occurs. The screen layer acts as an oil stripping and draining medium. The liquid oil then flows through a baffle and into an oil collection area at the bottom of the separator, and the oil-free refrigerant gas exits the separator just below the flighting. A float-activated oil return valve allows the captured oil to return to the compressor's crankcase or oil reservoir.

The oil separator should be kept warm to keep liquid refrigerant from condensing in it during the off cycle. The float mechanism cannot distinguish between oil and liquid refrigerant. If liquid refrigerant were in the separator, it would return the refrigerant to the compressor crankcase, which would dilute the oil and cause marginal lubrication. If liquid refrigerant is returning to the compressor crankcase, the return line between the separator and the crankcase will

Figure 25.17 A helical oil separator. Courtesy AC & R Components, Inc.

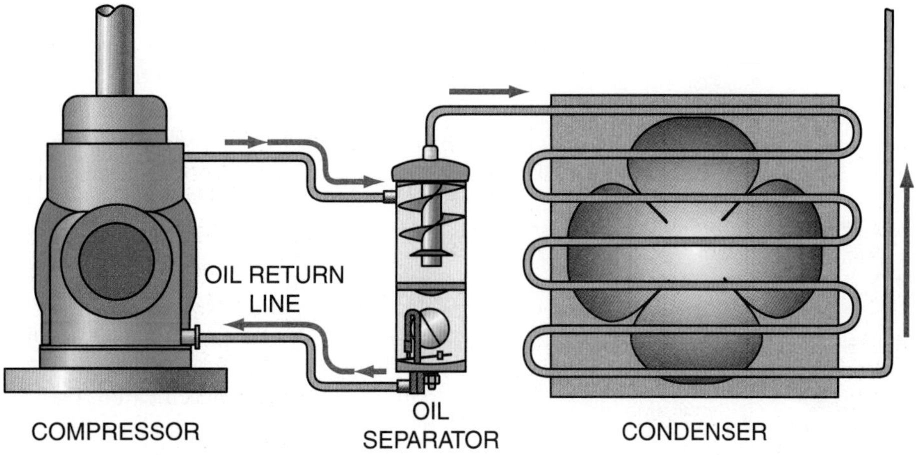

be cold from refrigerant vaporizing due to the pressure drop. A damaged, leaking, or collapsed float in the oil separator will cause the orifice to stay closed and excessive amounts of oil will accumulate in the oil separator. This situation could rob the system of its valuable lubricant and cause compressor damage.

VIBRATION ELIMINATORS

Compressors produce enough vibration while running that it is often necessary to protect the tubing at the suction and discharge lines. Vibration can be eliminated on small compressors successfully with vibration loops. Large tubing cannot be routed in loops, so special vibration eliminators, **Figure 25.18**, are often installed. These are constructed with a bellows-type lining and a flexible outer protective support, **Figure 25.19**. These devices must be installed correctly, or other problems, such as extra vibration, may occur. The typical recommendation is that they be installed close to the compressor and parallel to the compressor's crankshaft, **Figure 25.20**. If mounted crossways or perpendicular to the crankshaft, excess vibration may occur. Follow the manufacturer's directions.

Figure 25.18 Suction and discharge line vibration eliminators on a semihermetic compressor. Courtesy Ferris State University. Photo by John Tomczyk

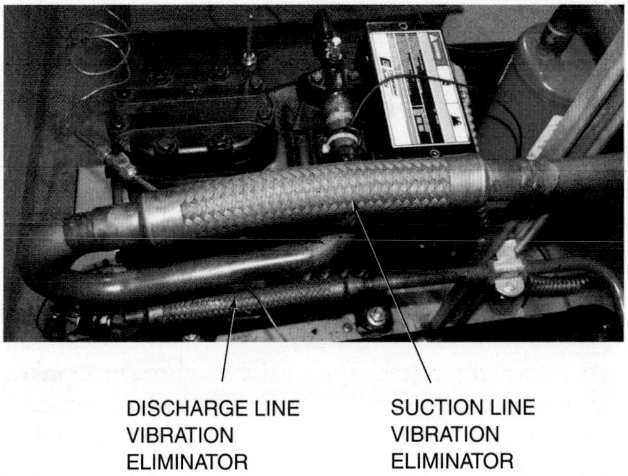

Figure 25.19 Cutaway view of a vibration eliminator.

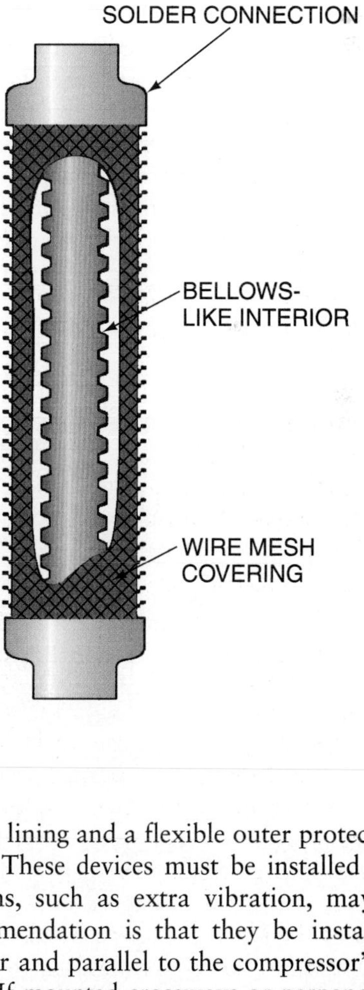

Figure 25.20 Vibration eliminators installed parallel to the crankshaft.

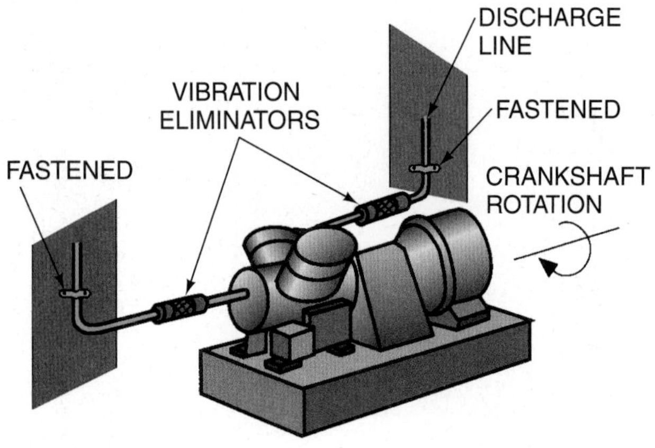

CRANKCASE HEAT

The crankcase heater, **Figure 25.21**, is designed to help protect the compressor from the effects of refrigerant migration. Migration occurs when refrigerant liquid or vapor returns to the compressor's crankcase and mixes with the oil during the off-cycle. If the compressor is started when liquid refrigerant is present in the crankcase, compressor damage can result. On start-up, the crankcase pressure reduces as soon as the compressor starts to turn. The refrigerant will boil and the oil will foam. The oil and refrigerant, some of which may be in the liquid state, will be pumped out of the compressor. Valve and bearing damage may occur and the compressor may be operated with a limited oil charge until the oil returns to the crankcase from the evaporator. The function of the crankcase heater, therefore, is to keep the oil in the compressor's crankcase at a temperature higher than the coldest part of the system. This helps ensure that any liquid refrigerant entering the crankcase will be vaporized and driven back into the suction line. It is recommended that crankcase heaters be energized for at least 24 hours before allowing the compressor to operate.

Figure 25.21 Strap-on crankcase heater. Photo by Jason Obrzut

The crankcase heater is a resistance heater that is strapped on, inserted, or connected in some way to the compressor's crankcase. Crankcase heat can be applied in several ways. Sometimes it is applied by heaters that are inserted into a well located on the shell of the compressor's crankcase where the compressor's oil can be heated, **Figure 25.22**, and sometimes by external heaters at the base of the compressor adjacent to the oil, **Figure 25.23**. When the heater is external, the heating element must be in good contact with the compressor housing or it will overheat the element and not transfer heat into the compressor. It is not uncommon to see the strap-on type heater as shown in **Figure 25.21**. It is important that any external crankcase heaters be installed correctly and tightly to ensure that adequate heat is applied to the crankcase.

Since compressor oil is warm when the compressor is operating, crankcase heaters are often wired in a manner that

Figure 25.22 Heater for inserting into a well inside the crankcase. Courtesy Ferris State University. Photo by John Tomczyk

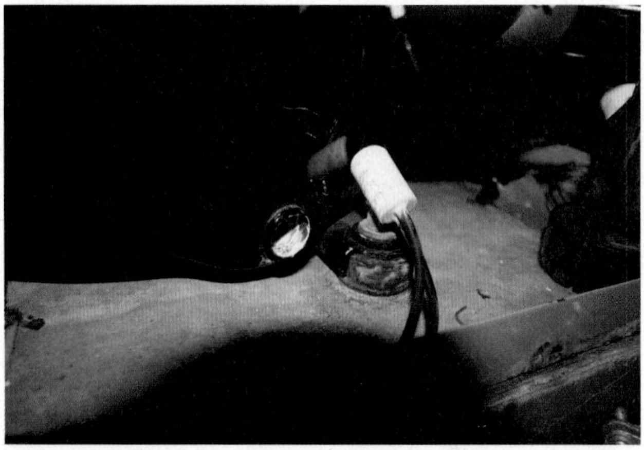

Figure 25.23 External loop heater for placement under a compressor's crankcase. Courtesy Ferris State University. Photo by John Tomczyk

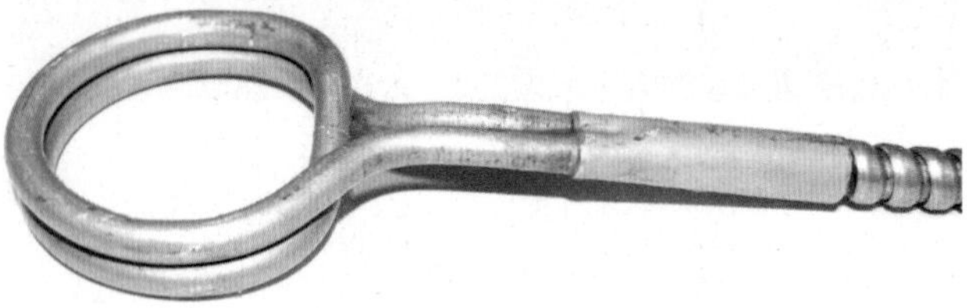

Figure 25.24 A contactor with auxiliary contacts to control a crankcase heater.

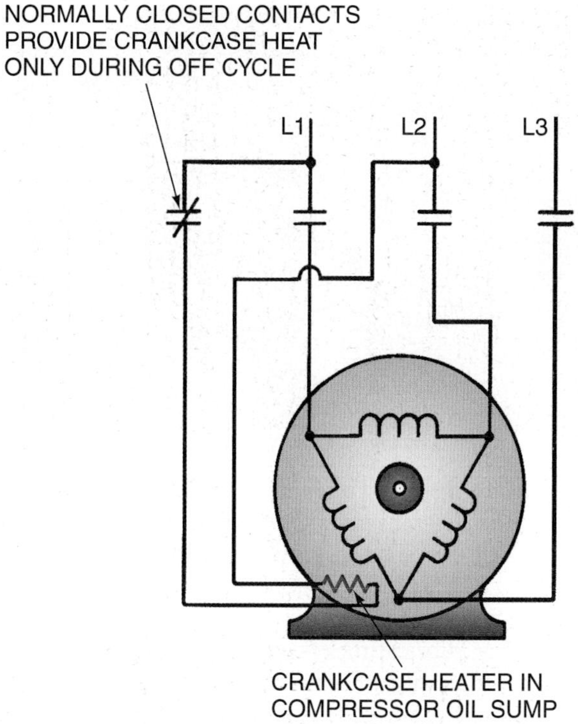

energizes them only when the compressor is not operating. Manufacturers typically use relays to shut the crankcase heat off during the running cycle, as heat added during the running cycle reduces system efficiency. This can also be accomplished with a set of normally closed auxiliary contactor contacts that energize the crankcase heater when the compressor is off. The heater contacts open when the main contactor contacts close to energize the compressor, **Figure 25.24**.

Crankcase heat is common in air-conditioning systems, but is only beneficial when the heater is energized during the off cycle. In some instances, the homeowner shuts off the disconnect at the outdoor condensing unit for the winter, leaving the compressor without crankcase heat. If the homeowner then starts the unit without allowing some time for the heat to boil the refrigerant out of the oil, damage will likely occur to the compressor. Compressors used on refrigeration typically operate year-round and are not subject to seasonal shutdown. However, if they are shut down for a period of time, the preceding situation will occur.

Crankcase heaters, as was seen, do help in combating refrigerant migration. They do not, however, prevent slugging at start-ups or liquid floodback to compressors once the compressor is running. Crankcase heaters, when used in conjunction with an automatic pumpdown cycle and/or suction-line accumulator, will help remedy many of these start-up problems.

The technician should be well aware of crankcase heat. If a compressor has been off for a long time, the housing should be warm to the touch if it has crankcase heat. If the manufacturer has installed crankcase heat, do not start the compressor when it is cold, or damage may occur. In order to avoid carbonization of the oil from excessive heat, the wattage inputs of the crankcase heater must be limited. In very cold ambient or when exposed to cold winds, the crankcase heater may have a limited effect and refrigerant migration to the compressor's crankcase may still occur.

OIL PUMPS

Smaller refrigeration compressors, typically under 3 horsepower, usually rely on a splash-type lubrication system and do not require a physical oil pump. These smaller compressors often have an oil scoop that resembles a paddle that flings oil throughout the compressor's crankcase causing an oil fog that lubricates the bearings, wrist pins, pistons, and rings. However, larger compressors in the refrigeration and air-conditioning fields have mechanical, forced lubrication systems. These compressors are usually over 3 horsepower. They contain an oil pump located at the end of the compressor's crankshaft, **Figure 25.25**. The oil pump is actually keyed into the compressor's crankshaft which supplies power to turn the oil pump, **Figure 25.26**. Oil pumps can be of the gear or eccentric type. **Figure 25.27** shows a gear type oil pump. The star gear in this type of pump actually squeezes the oil to a higher pressure. The added pressure that the star gear puts into the oil can be referred to as the net oil pressure. **Figure 25.28** shows the star gear disassembled from the oil pump and out of its seat within the oil pump.

One has to remember that the oil pump picks up the oil from the compressor's crankcase (under crankcase pressure), adds pressure to the oil, and then delivers the oil into drilled

Figure 25.25 Oil pump located at end of the compressor and turned by the compressor's crankshaft. Courtesy Ferris State University. Photo by John Tomczyk

Figure 25.26 Oil pump is actually keyed into a slot into the compressor's crankshaft. Courtesy Ferris State University. Photo by John Tomczyk

Figure 25.27 Gear type oil pump. Courtesy Ferris State University. Photo by John Tomczyk

Figure 25.28 Oil pump with star gear disassembled. Courtesy Ferris State University. Photo by John Tomczyk

Figure 25.29 Eccentric type oil pump. Courtesy Ferris State University. Photo by John Tomczyk

holes in the crankshaft. The oil is then delivered to the compressor's bearings and connecting rods through these drilled holes or oil galleys. The oil now drops to the crankcase to be picked up again by the oil pump. Notice that the oil was first picked up by the oil pump from the compressor's crankcase, and after lubricating essential compressor parts, ends up back in the crankcase. **Figure 25.29** shows an eccentric type oil pump which uses an eccentric type of mechanism to squeeze the oil to a smaller and smaller volume, which adds pressure to the oil. Again, this squeezing action is referred to as net oil pressure.

COMPRESSOR OIL CHECK VALVE AND PARTITION WALL

Many compressors have a partition wall which separates the crankcase from the motor barrel. On refrigerant gas-cooled compressors, the motor barrel is where the motor windings and stator reside and where oil entrained suction gases enter the compressor's end bell through the suction service valve. Because the motor barrel has a large diameter compared to the suction line, oil-entrained suction gases will lose velocity and slow down significantly as they enter the motor barrel. This loss of velocity will cause the oil in the suction gases to drop out and collect at the bottom of the motor barrel. The partition wall separating the motor barrel from the compressor's crankcase has an oil check valve which will only let oil pass from the compressor's motor barrel to its crankcase, **Figure 25.30**. When the compressor is running and no system problems exist, there is usually a 1 to 2 psi pressure drop between the motor barrel and the crankcase. This pressure difference allows the oil check valve to open, which allows any collected oil in the bottom of the motor barrel to return to the compressor's crankcase. The returned oil can now be picked up by the oil pump and delivered throughout the compressor's moving parts to provide lubrication.

Figure 25.30 Compressor's crankcase showing partition wall, oil pickup screen, and oil check valve. Courtesy Ferris State University. Photo by John Tomczyk

However, during start-up periods or when some system problems occur, the compressor's crankcase could experience a slightly higher pressure than the motor barrel. When this occurs, the oil check valve will be closed, preventing any oil return to the crankcase for use by the oil pump. An accumulation of oil will now occur in the motor barrel which will prevent the compressor's crankcase from receiving any new oil from the motor barrel. If the system problem is not remedied soon after start-up, the oil check valve will remain closed and the crankcase will soon be starved of oil. This will result in a loss of lubrication and a tripping of the oil safety control. Some system problems that may cause this are:

- Liquid refrigerant floodback during the on cycle from the motor barrel to the crankcase.
- Compressor start-ups with liquid refrigerant in the crankcase.
- Refrigerant migration to the crankcase from the suction line and motor barrel during the off cycle causing start-up pressurization of the crankcase. Refrigerant loves oil and will migrate to it because of the oil's lower vapor pressure.
- No automatic pumpdown system to prevent refrigerant migration.
- Compressor located outside in a cold ambient, causing refrigerant migration during the off cycle from suction line through the motor barrel to the crankcase. This causes oil foaming and crankcase pressurization at start-ups.

25.4 LOW-AMBIENT CONTROLS

Since refrigeration systems typically operate year-round, low-ambient controls are an important part of a system's control strategy. When the condenser is located outside, the head pressure will go down in the winter to the point that the expansion valve will not have enough pressure drop across it to feed refrigerant correctly. When this happens, some method must be used to keep the head pressure up to an acceptable level. The most common methods are as follows:

- Fan cycling using a pressure control
- Fan speed control
- Air volume control using shutters and fan cycling
- Condenser flooding and condenser splitting

This section provides an overview of commonly used low-ambient controls. For more detailed coverage, please refer to Section 22.18, "Head Pressure Controls" in Unit 22, "Condensers."

FAN-CYCLING HEAD PRESSURE CONTROLS

Fan cycling is a simple and reliable method to control and maintain a refrigeration system's head pressure. When there are multiple condenser fans, some may be controlled by temperature, while the last fan can be controlled by head pressure. In this configuration, the control used is a pressure control that closes on a rise in pressure to start the fan and opens on a fall in pressure to stop the fan when the head pressure falls, **Figure 25.31**. When fan cycling is used, the fan will cycle on and off in the winter, but will run constantly in hot weather. On multiple fan condensing units, one or more ambient temperature controls may accomplish condenser fan cycling. The controls are usually thermal disks located somewhere on the condensing unit's frame, **Figure 25.32**. As the outside ambient temperature drops, the fans begin to turn off. As the outside ambient temperature rises, the fans begin to turn on. Frequently, on smaller condensing units, one fan will run continuously and the other fan will be cycled on and off with an ambient temperature controller.

Using fan controls can be hard on a motor because of all the cycling. The fan motor needs to be a type that does not have a high starting current. In addition, the control must

Figure 25.31 A pressure control with the same action as a low-pressure control, but at a higher pressure range than the low-pressure control.

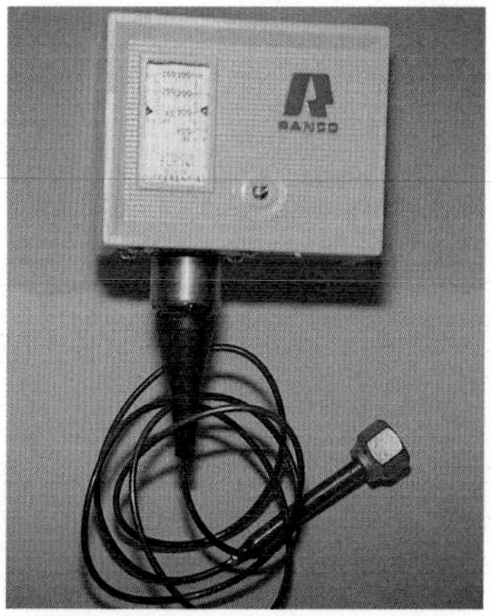

Figure 25.32 A thermal disk-type fan speed control. Courtesy Ferris State University. Photo by John Tomczyk

be rated to handle the current draw of the motor. To verify this, the motor's specifications should be carefully compared to the control's limitations and capabilities. Fan cycling can vary the pressure to the expansion device a great deal. The technician must choose whether to have the control set points close together for the best expansion device performance or far apart to keep the fan from short cycling. One application for fan-cycling devices is the use of multiple fans, where one fan operates continuously to minimize pressure fluctuation.

Vertical condensers may be affected by the prevailing winds. If these winds are directed into the coil, they can provide airflow similar to that of a fan. Sometimes a shield is installed to prevent this, **Figure 25.33**. When an air-cooled condenser is installed either on the ground or on the roof, the direction of airflow across the condenser's face should be in the prevailing wind direction. This way, when the condenser's fan is operating, it will not be bucking the prevailing winds and decreasing airflow through the condenser.

Figure 25.33 When fan-cycling controls are used, care must be taken if the condenser is vertical. The prevailing winds can cause the head pressure to be too low even when the fan is off. A shield can be used to prevent the wind from affecting the condenser.

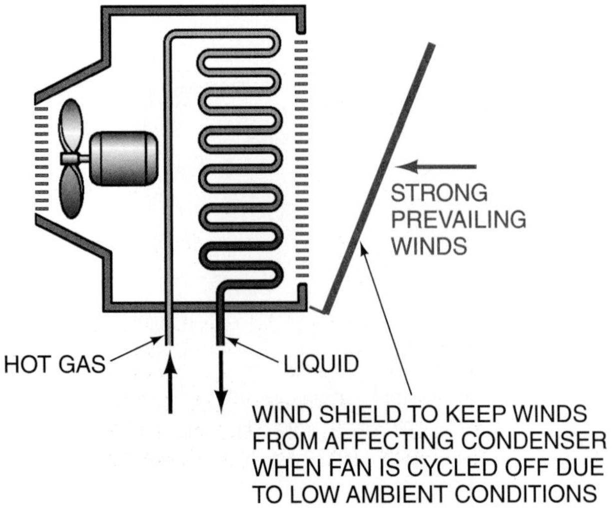

FAN SPEED CONTROL FOR CONTROLLING HEAD PRESSURE

Devices that control fan speed have been used successfully in many installations. These devices can be used with multiple fans, where the first fans are cycled on and off by temperature and the speed of the last fan is controlled by head pressure or condensing temperature. The device that controls the last fan is normally a transducer that converts a pressure or temperature signal to an electrical signal that feeds a motor speed controller. As the temperature drops on a cool day, the motor speed is reduced. As the temperature rises, the motor speed increases. At some predetermined point the additional fans are started. On a hot day all fans will be running, with the variable-speed fan running at maximum speed. Some fan speed controls use a temperature sensor to monitor the condenser's temperature. Newer generation systems use variable frequency drives (VFDs) to control condenser fan motor speed, **Figure 25.34**. A condenser fan motor is much more energy efficient at reduced speeds. Fan motors with VFDs have the ability to deliver variable air volume (VAV) flows. This allows the condenser fan to provide the correct amount of airflow through the condenser coil that is needed under a certain ambient condition. VFDs can control the speed of a condenser fan motor by controlling the frequency of the power signal supplied to the motor. The lower the frequency, the slower the motor, and vice versa. For a more detailed explanation of VFD-controlled motors, refer to Section 17.28, "Inverters and Variable Frequency Drives (VFDs)" in Unit 17, "Types of Electric Motors."

Figure 25.34 Four VFDs used to control the speed of a motor by varying the frequency delivered to the motor. Courtesy Ferris State University. Photo by John Tomczyk

AIR VOLUME CONTROL FOR CONTROLLING HEAD PRESSURE

Air volume control using shutters can be accomplished with a piston-driven damper controlled by the high-pressure refrigerant, **Figure 25.35**. As the system's high-side pressure rises, the dampers are pushed open to allow more air to pass through the condenser coil. As the system's high-side pressure drops, the dampers are pushed closed, reducing the amount of air passing through the condenser coil.

When there are multiple fans, the first fans can be cycled off using temperature, and the shutter can be used on the last fan, **Figure 26.36**. This system results in a steady head pressure, as with fan speed control. The expansion valve inlet pressure does not fluctuate as it does when one fan is cycled on and off. The shutter can be located either on the fan inlet or on the fan outlet. Be careful that the fan motor is not overloaded if the damper is positioned on the discharge of the fan.

CONDENSER FLOODING AND CONDENSER SPLITTING FOR CONTROLLING HEAD PRESSURE

Condenser flooding devices, **Figure 25.37**, are used in both mild and cold weather to cause the refrigerant to move from an oversized receiver into the condenser. This three-port valve controls the flow of liquid refrigerant from the condenser while

Figure 25.35 Installation using dampers to vary the air volume instead of fan speed. This is accomplished with a refrigerant-operated damper that operates the shutters. This application modulates the airflow and gives steady head pressure control.

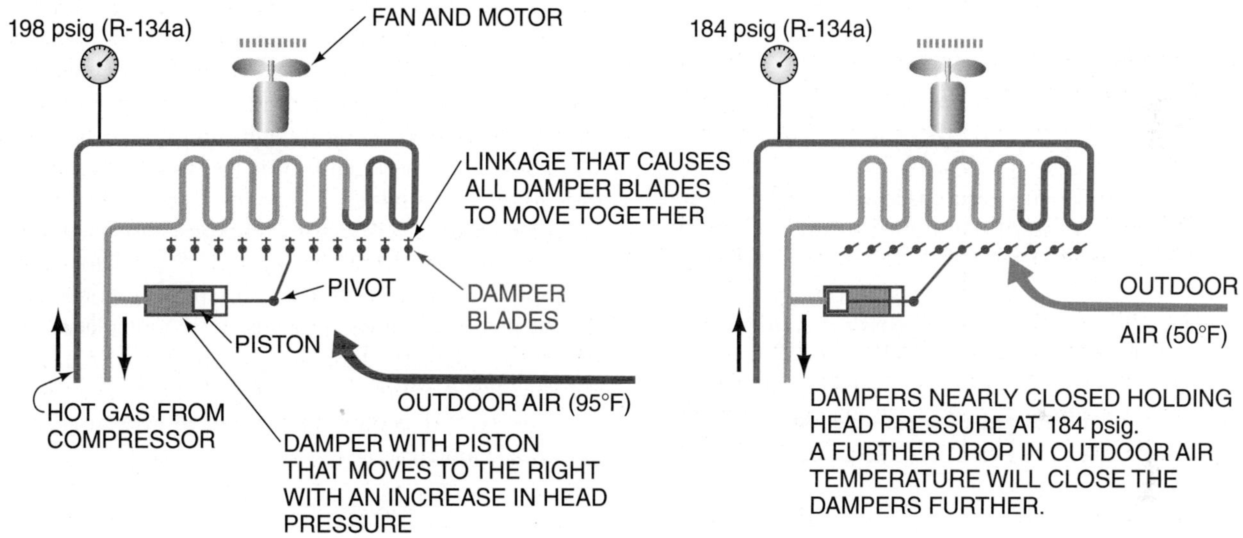

Figure 25.36 If more than one fan is used, the first fans can be cycled off based on ambient temperature. Care should be taken not to overload the condenser fan when the shutters are placed at the fan outlet.

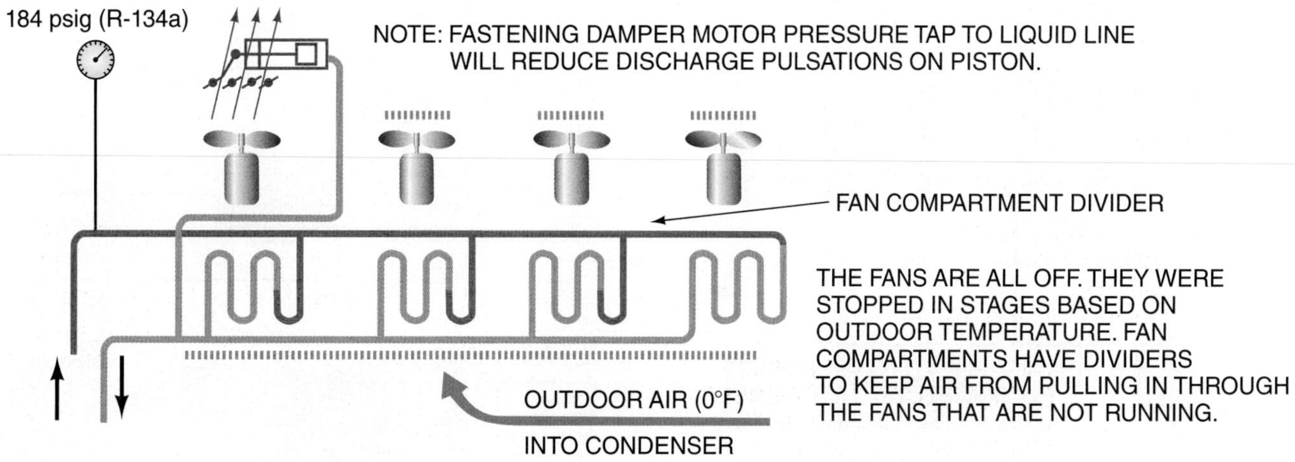

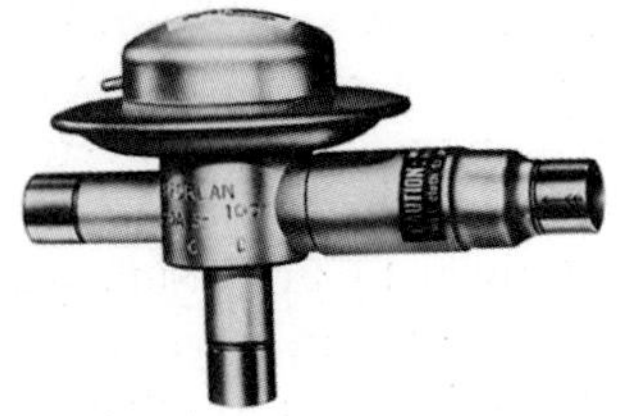

Figure 25.37 A condenser flooding valve. Courtesy Sporlan Valve Company

regulating the flow of hot compressor discharge gas, around the condenser, directly to the receiver, **Figure 25.38**. The sensed pressure at the outlet, receiver side, of the valve controls the position of the valve. As the high-side pressure increases, during periods of warm ambient temperature, the valve throttles open to allow more refrigerant to flow from the condenser outlet to the receiver and reduces the flow of discharge gas from the compressor directly to the receiver. When the ambient temperature is very high, no hot discharge gas bypasses the condenser. The OROA notation (open on rise of outlet) on the valve indicates that it opens the connection between the condenser and the receiver on a rise of outlet pressure.

When the ambient temperature is low, the high-side pressure drops and the valve port connected to the outlet of the condenser starts to throttle closed. At the same time, the connection between the hot discharge gas port and the receiver throttles open, allowing hot gas to bypass the condenser coil. This causes the high-side pressure to rise. During periods of low-ambient temperature, refrigerant backs up into the condenser. This excess refrigerant in the condenser acts like an overcharge and causes the head pressure to be much higher than it normally would be on a mild or cold day. On warmer days, the excess refrigerant is directed back to the receiver. This method of head pressure control gives very steady pressure control while the system is running.

The system requires a large amount of refrigerant because, in addition to the normal operating charge, there must be enough refrigerant to flood the condenser in the winter. A large receiver is used to store the refrigerant in the summer, when valves will divert the excess liquid into the receiver. Condenser flooding has one added benefit in the winter. Because the condenser is nearly full of refrigerant, the liquid refrigerant that is furnished to the expansion devices is well below the condensing temperature. Remember, subcooling will help improve the efficiency of the system. The liquid may be subcooled to well below freezing for systems operating in a cold climate.

One way to reduce the amount of extra refrigerant charge needed for condenser flooding is to split the condenser into two separate and identical condenser circuits, **Figure 25.39**. This is referred to as condenser splitting. The splitting of the condensers is done with the addition of a pilot-operated, three-way solenoid valve installed in the discharge line from the compressor. The splitting of the identical condensers is done in such a way that only one half of the condenser is used for winter operation, and both halves are used for summer operation. The top half of the condenser is referred to as the "Summer/Winter" condenser, and the bottom half of the condenser is referred to as the "Summer" condenser. The three-way solenoid valve controlling the splitting of the condensers can be energized and deenergized by a controller sensing outside ambient, an outdoor thermostat, or a high-side pressure control. For a more detailed explanation of head pressure controls, refer to Section 22.18, "Head Pressure Controls" in Unit 22, "Condensers."

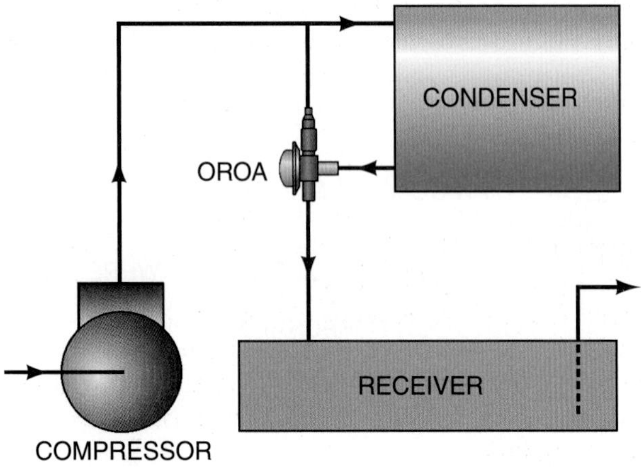

Figure 25.38 The gas going to the receiver keeps the pressure up for correct expansion valve operation. The gas going to the condenser is changed to a liquid and subcooled in the condenser. It then returns to the receiver to subcool the remaining liquid. Courtesy Sporlan Valve Company

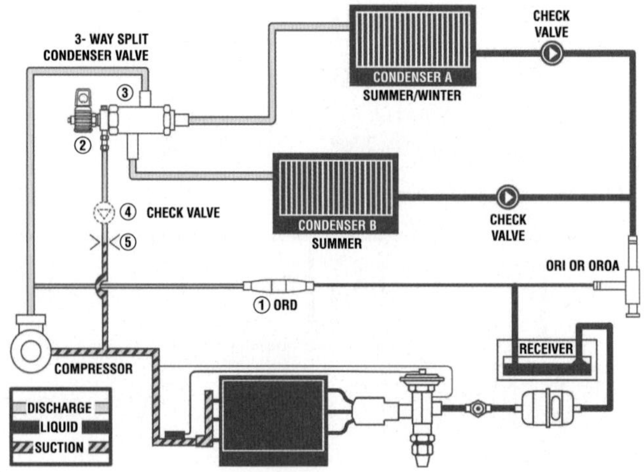

Figure 25.39 Split condensers with a three-way, solenoid-operated split condenser valve incorporated into a refrigeration system. Courtesy Sporlan Division, Parker Hannifin Corporation

25.5 PRESSURE SWITCHES AND CONTROLS

Pressure switches are used to stop and start electrical current flow to refrigeration components. They play a very important part in the function of the equipment. The typical pressure switch can be

- a low-pressure switch—closes on a rise in pressure,
- a high-pressure switch—opens on a rise in pressure,
- a low-ambient control—closes on a rise in pressure, or
- an oil safety switch—has a time delay; opens on a rise in pressure.

LOW-PRESSURE SWITCH

The low-pressure switch or control, **Figure 25.40**, is used in the following two major refrigeration applications:

- Low-charge protection
- Control of space temperature

The low-pressure control can be used as a low-charge protection by setting the control to cut out at a value that is below the typical evaporator operating pressure. For example, in a medium temperature cooler the air temperature may be expected to be no lower than 34°F. When this cooler uses R-134a for the refrigerant, the lowest pressure at which the evaporator would be expected to operate would be around 18 psig because the coil would normally operate at about 15°F colder than the air temperature in the coil (34°F − 15°F = 19°F). The refrigerant in the coil should not operate below this temperature of 19°F, which converts to 18 psig. The low-pressure cut-out should be set below the expected operating condition of 18 psig and above the atmospheric pressure of 0 psig. A setting that would cut off the compressor at 5 psig would keep the low-pressure side of the system from going into a vacuum in the event of refrigerant loss. This setting would be well below the typical operating condition. The control would normally be automatically reset. When a low-charge condition exists, the control would cut the compressor off and on and maintain some refrigeration. A store owner may call with a complaint that the unit is cutting off and on but not lowering the cooler temperature properly.

Figure 25.40 The low-pressure control. The control contacts close on a rise in pressure and open on a fall in pressure. Photo by Bill Johnson

The cut-in setting for this application should be a pressure that is below the pressure corresponding to the highest temperature the cooler is expected to experience in a typical cycle and just lower than the pressure corresponding to the thermostat cut-in point. For example, the thermostat may be set at 45°F, which would mean that the pressure in the evaporator may rise as high as 40 psig. The low-pressure control should be set to cut in at about 25 psig. The previous example of a low-pressure control set to cut out at 5 psig and to cut in at 40 psig means that the control differential is 35 psig. Remember that the compressor is not running. The wide differential helps to prevent the compressor from coming back on too soon or from short cycling.

LOW-PRESSURE CONTROL APPLIED AS A THERMOSTAT

The low-pressure control setup described in the previous example is for low-charge protection only, to keep the system from going into a vacuum. The same control can be set to operate the compressor to maintain the space temperature in the cooler and to serve as a low-charge protection as well.

Using the same temperatures as in the preceding example as the operating conditions, 34°F and 45°F, the low-pressure control can be set to cut out when the low pressure reaches 18 psig. This corresponds to a coil temperature of 19°F. When the air in the cooler reaches 34°F, the coil temperature should be 19°F, with a corresponding pressure of 18 psig. This system has a room-to-coil temperature difference of 15°F (34°F room temperature − 19°F coil temperature = 15°F temperature difference). When the compressor cuts off, the air in the cooler is going to raise the coil temperature to 34°F and a corresponding pressure of 30 psig. As the cooler temperature goes up, it will raise the temperature of the refrigerant. When the temperature of the air increases to 45°F, the coil temperature should be 45°F with a corresponding pressure of 40 psig. This could be the cut-in point of the low-pressure control. The settings would be cut out at 18 psig and cut in at 40 psig. This is a differential of 22 psig and would maintain a cooler temperature of 34°F to 45°F. You could say that you are using the refrigerant in the evaporator coil as the thermostat sensing fluid.

Figure 25.41 This table is to be used as a guide for setting low-pressure controls for the different applications. Courtesy Tecumseh Products Company

APPROXIMATE PRESSURE CONTROL SETTINGS
Pressure — Pounds Per Square Inch Gauge

APPLICATION	TEMP RANGE (°F)	EVAP TD (°F)	REFRIGERANT 22		134a		404A		507	
			Out	In	Out	In	Out	In	Out	In
Beverage Cooler Floral Cooler Produce Cooler	35 to 38	15	41	66	17	33	53	82	56	86
Smoked Meat Cooler Meat Reach Thru Service Deli Seafood	32 to 35	15	38	62	15	30	49	77	52	81
Multi-Deck Fresh Meat	26 to 29	15	32	54	11	25	42	68	45	72
Frozen Glass Door Frozen Walk-In	-10 to 0	10	9	24	—	—	15	33	16	35
Frozen Ice Cream Frozen Food - Open Type	-30 to -20	10	0	10	—	—	4	16	4	18

Pressure control settings assume a suction line pressure loss equivalent to 2°F

One of the advantages of this type of control arrangement is that no interconnecting wires must be installed between the inside of the cooler and the condensing unit. If a thermostat is used to control the air temperature in the cooler, a pair of wires must be run between the condensing unit and the inside of the cooler. In some installations, the condensing unit is a considerable distance from the cooler, which makes running wires impractical. With the temperature being controlled at the condensing unit, an owner is less likely to adjust the control and potentially cause problems.

As many low-pressure control settings exist as there are applications. Different situations call for different settings. **Figure 25.41** is a chart of settings recommended by one company.

Low-pressure controls are rated by their pressure range and the current draw of the contacts. A low-pressure control that is suitable for R-134a may not be suitable for use with R-404A because of their different pressure ranges. For the same application in the previous paragraph using R-404A, the cut-out would be 55 psig and the cut-in would be 95 psig. The correct control for the pressure range must be chosen. Some controls are single-pole–double-throw, which can make or break on a rise. Such controls can serve as one component serving two different functions. The contact rating for a pressure control has to do with the size of electrical load the control can carry. If the pressure control is expected to start a small compressor, the inrush current should also be considered. Normally a pressure control used for refrigeration is rated so that it can directly start up to a 3-hp single-phase compressor. If the compressor is any larger, or three phase, a contactor or motor starter is normally used. The pressure control then can control the contactor or motor starter coil, **Figure 25.42**.

Figure 25.42 When a compressor that is larger than the electrical rating of the pressure control contacts is encountered, the compressor has to be started with a contactor. The wiring diagram shows how this is accomplished.

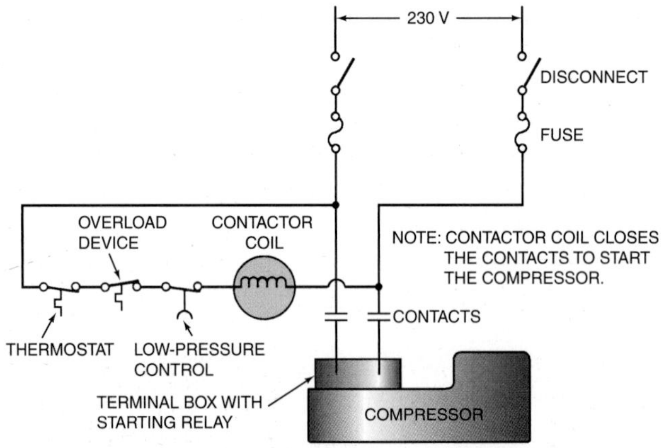

HIGH-PRESSURE CONTROL

The high-pressure control, **Figure 25.43**, is usually not as complicated as the low-pressure control. It is used to keep the compressor from operating with a high head pressure. This control is necessary on water-cooled equipment because an interruption of water is more likely than an interruption of air. This control opens on a rise in pressure and should be set above the typical high pressure that the machine would normally encounter. The high-pressure control may be either automatic or manual reset. With air-cooled condensers, the condenser saturation temperature should be no more than 30°F higher than the temperature

Figure 25.43 A high-pressure switch to keep the compressor from running when high system pressures occur.

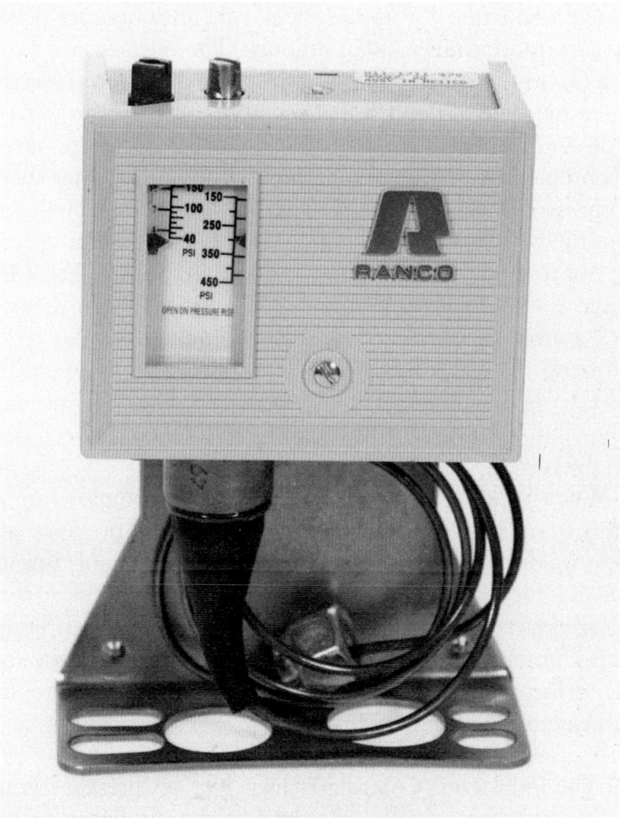

of the ambient air. If the ambient air is 95°F, the condenser would be operating at about 185 psig if the system used R-134a (95°F + 30°F = 125°F condensing temperature, which corresponds to 185 psig). The high-pressure control should be set well above 185 psig for the R-134a system. If the control were set to cut out at 250 psig, there should be no interference with normal operation, and it still would give good protection. Most condensers operate at a 30°F temperature difference as shipped from the factory and will continue to operate under this condition when properly maintained. Many condensers for low-temperature applications may operate at temperature differences as low as 15°F to maintain low compression ratios.

The control cut-in point must be above the pressure corresponding to the ambient temperature of 95°F. If the compressor cuts off and the outdoor fan continues to run, the temperature inside the condenser will quickly reach the ambient temperature. For example, if the ambient is 95°F, the pressure will quickly fall to 114 psig for an R-134a system. If the high-pressure control were set to cut in at 135 psig, the compressor could come back on with a safe differential of 115 psig (cut-out 250 – cut-in 135 = 115 psig differential).

Some high-pressure controls have a fixed differential of 50 psig. This means they will cycle the compressor back on at 50 psig lower than the cut-out setting. For example, if a system cycles off on the high-pressure control at 250 psig, it will come back on at 200 psig (250 psig – 50 psig).

Some manufacturers specify a manual reset high-pressure control. When this control cuts out, someone must press the reset button to start the compressor. This calls attention to the fact that a problem exists. The manual reset control provides better equipment protection, but the automatic reset control may save the food by allowing the compressor to run at short intervals. An observant owner or operator should notice the short cycle of the automatic control if the compressor is near the workspace.

LOW-AMBIENT FAN CONTROL

The low-ambient fan control, **Figure 25.44**, has the same switch action as the low-pressure control but operates at a higher pressure range. This control stops and starts the condenser fan in response to head pressure. This control must be coordinated with the high-pressure control to keep them from working against each other. The high-pressure control stops the compressor when the head pressure gets too high, and the low-ambient control starts the fan when the pressure gets to a predetermined point before the high-pressure control stops the compressor.

When a low-ambient control is used, the high-pressure cut-out should be set to a pressure that is higher than the cut-in point of the low-ambient control. For example, if the low-ambient control is set to maintain the head pressure between 125 and 175 psig, a high-pressure control setting of 250 psig should not interfere with the low-ambient control setting.

Figure 25.44 The low-ambient control used to open the contacts on a drop in head pressure to stop the condenser fan.

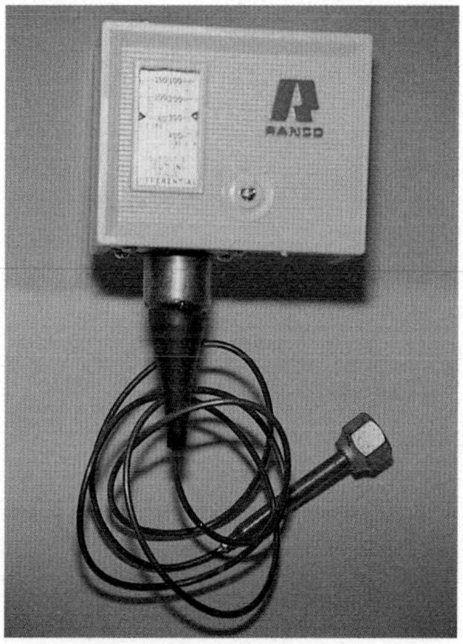

This can easily be verified by installing a gauge and stopping the condenser fan to make sure that the high-pressure control is cutting out where it is supposed to.

With the gauge on the high side, the fan action can also be observed to see that the fan is cycling off and on as it should. The fan will operate all of the time on high-ambient days, and the fan control will not stop the fan. The low-ambient pressure control can be identified by the terminology on the control's action. It is described as "close on a rise" in pressure. This is the same terminology used on a low-pressure control. The difference is that the low-ambient fan control has a higher operating pressure range.

OIL PRESSURE SAFETY CONTROL

Many larger refrigeration and air-conditioning compressors have forced oiling systems. These compressors are usually over 3 hp. They contain an oil pump located at the end of the compressor crankshaft, **Figure 25.27**. The crankshaft is connected to the oil pump and supplies power, which turns the oil pump, which can be either of the gear or eccentric type. **Figure 25.27** shows a gear type oil pump and **Figure 25.29** illustrates an eccentric type oil pump. The oil pump forces oil through holes drilled in the crankshaft and delivers it to bearings and connecting rods. The oil then drops to the crankcase to be picked up again by the oil pump. Smaller compressors usually have a splash-type oiling system with a scoop that scoops and flings the oil throughout the crankcase, causing an oil fog as the crankshaft turns.

The oil pressure safety switch, **Figure 25.45**, *is used to ensure that the compressor has oil pressure when operating.*

The oil pressure control is used on larger compressors and has a different sensing arrangement than do the high- and low-pressure controls. The high- and low-pressure controls are single-diaphragm or single-bellows controls because they are comparing atmospheric pressure with the pressures inside the system. Atmospheric pressure can be considered a constant for any particular locality because it does not vary more than a small amount. The oil pressure safety control, on the other hand, is a pressure differential control. It measures the difference between two pressures to determine if a net positive oil pressure is present. One pressure is the compressor crankcase pressure, which is the same as the compressor suction pressure. The suction pressure will vary from the off or standing reading to the actual running reading, not to mention the reading that would occur when a low charge is experienced. For example, when a system is using R-22 as the refrigerant, the pressures may be similar to the following: 125 psig while standing, 70 psig while operating, and 20 psig during a low-charge situation. The other pressure is the oil pump's outlet pressure. The difference between these two pressure is called the net oil pressure.

When dealing with compressors that employ an oil pump, many service technicians confuse net oil pressure with oil pump discharge pressure. However, it is of utmost importance that technicians understand the difference between these two pressures when servicing compressors with oil pumps. The oil pump's rotating gear or eccentrics add a certain pressure to the oil pumped through the crankshaft. This pressure is considered net oil pressure.

Figure 25.45 Oil safety controls have a bellows on each side. They are opposed in their forces and measure the net oil pressure. Photo by Bill Johnson

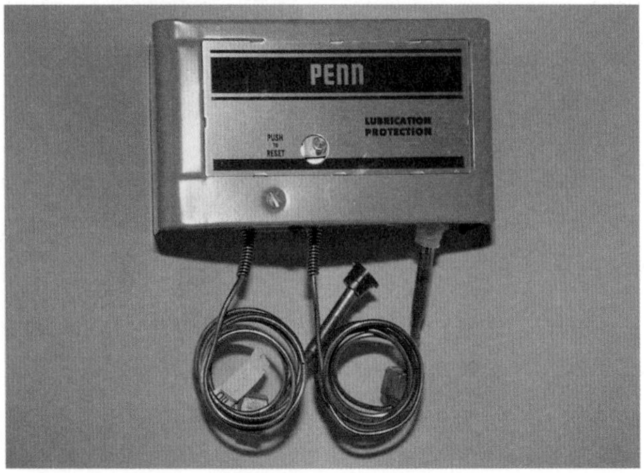

NET OIL PRESSURE CALCULATION. Net oil pressure is not the pressure that can be measured at the discharge or outlet of the oil pump. The oil pump picks up oil at suction or crankcase pressure from the compressor's crankcase through a screen or filter. The oil pump discharge port's pressure includes both crankcase pressure and the oil pump gear pressure it adds to the oil. This is why net oil pressure cannot be measured directly with a gauge. A gauge at the oil pump's discharge port would register a combination of crankcase pressure and oil pump gear pressure. The technician must subtract the crankcase pressure from the discharge port pressure to get the net oil pressure. The following is an equation for net oil pressure:

$$\text{Oil Pump Discharge Pressure} = \text{Crankcase Pressure} + \text{Net Oil Pressure}$$

Example: The oil pump discharge pressure is 80 psig. The crankcase pressure is 20 psig. What would be the net oil pressure?

Solution: Simply subtract the crankcase pressure from the oil pump discharge pressure to get net oil pressure.

$$80 \text{ psi} - 20 \text{ psi} = 60 \text{ psid net oil pressure}$$

This means the oil pump is actually putting 60 psid of pressure into the oil when delivering it into the crankshaft's drilled passages. Notice that the net oil pressure is expressed in pounds per square inch differential (psid) because it is a difference between two pressures.

Sometimes the compressor's crankcase may be operating in a vacuum. In this case, the crankcase pressure is negative. Remember that every 2 in. of mercury vacuum is equivalent to approximately 1 psi.

Example: What is the net oil pressure if the oil pump discharge pressure is 35 psig and the crankcase pressure is 6 in. of vacuum (−3 psi)?
Solution: Again, we must subtract the crankcase pressure from the oil pump discharge pressure to get the net oil pressure.

$$35 \text{ psi} - (-3 \text{ psi}) = 38 \text{ psid net oil pressure}$$

This means that the oil pump is delivering 38 psid of net oil pressure through the crankshaft and bearings. The terms suction pressure and crankcase pressure are often used interchangeably in the HVACR industry. However, on some larger refrigerant-cooled compressors there is a partition wall between the crankcase and motor barrel, **Figure 25.30**. During start-up periods or when system problems occur, there could be a slight difference in pressure between crankcase pressure in the crankcase and suction pressure in the motor barrel. Because of this, it is important to use the crankcase pressure, not the suction pressure, when connecting an oil safety controller to these types of compressors.

In the basic low-pressure control, the pressure is under the diaphragm or bellows and the atmospheric pressure is on the other side. The oil pressure control uses a double bellows—one bellows opposing the other—to detect the net or actual oil pressure, **Figure 25.46**. The pump inlet pressure is under one bellows, and the pump outlet pressure is under the other bellows. These bellows are opposite each other either physically or by linkage. The bellows with the most pressure is the oil pump outlet, and it overrides the bellows with the least amount of pressure. This override reads out in net pressure and is attached to a linkage that can stop the compressor when the net pressure drops for a predetermined time.

Figure 25.46 Internal view of an oil pressure safety control. Photo by Bill Johnson

OIL PRESSURE SWITCH TIME DELAY. Because the control needs a pressure differential to allow power to pass to the compressor, it must have some means of allowing the compressor to initially start. There is no pressure differential until the compressor starts to turn because the oil pump is attached to the compressor crankshaft. Therefore, a time delay is built into the control to allow the compressor to get started and to prevent nuisance cut-outs when oil pressure may vary for only a moment. This time delay is normally about 90 to 120 sec. It is accomplished with a heater circuit and a bimetal device or electronic circuitry, **Figure 25.47**. The delay also gives the crankcase time to clear any unwanted refrigerant during periods when refrigerant migration or flooding has occurred. Furthermore, it avoids shutdowns during short fluctuations in net oil pressure on start-up.

Net oil pressures vary from compressor to compressor. They usually range from 20 to 40 psi. Most oil pressure safety controllers will shut the compressor down if the net oil pressure falls below 10 psi. The following variables affect the net oil pressure:

- Compressor size
- Oil viscosity
- Oil temperature
- Bearing clearance
- Percent of refrigerant in the oil

Larger compressors need more net oil pressure because they have more surface areas to lubricate, and the oil pumps must pump and carry the oil greater distances. As the oil gets hotter and its viscosity drops, the net oil pressure will also usually drop. As a compressor wears, its tolerances become greater and it becomes easier for the oil to escape through these clearances.

Figure 25.47 The oil pressure control is used for lubrication protection for the compressor. The oil pump that lubricates the compressor is driven by the compressor crankshaft.

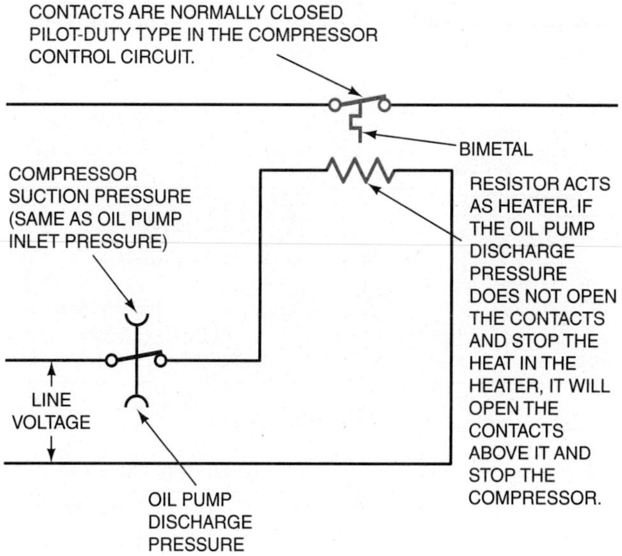

TRANSDUCER-TYPE OIL SAFETY CONTROLLERS. Transducer-type oil safety controllers use a pressure transducer, which senses a combination of oil pump discharge pressure and crankcase pressure. The pressure transducer has two separate ports to sense both crankcase pressure and oil pump discharge pressure. The difference between these two pressures, the net oil pressure, is mechanically determined by the pressure transducer. The transducer is connected to an electronic controller by wires, **Figure 25.48**. The pressure transducer is shown connected to the compressor in **Figure 25.49**. It then transforms a pressure signal to an electrical signal for the electronic

Figure 25.48 The pressure transducer wired to an electronic controller. Courtesy Ferris State University. Photo by John Tomczyk

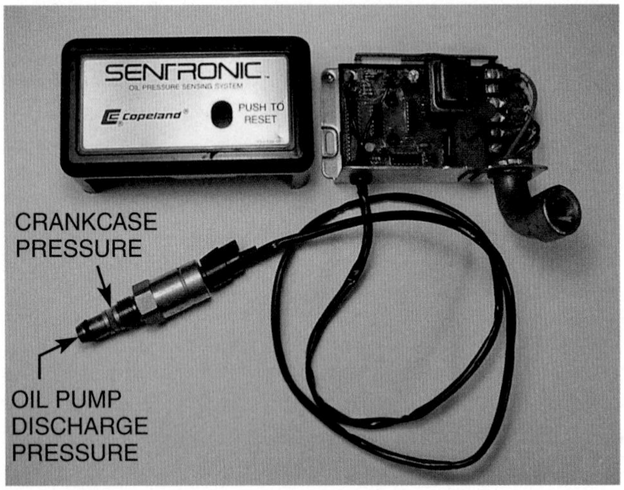

Figure 25.49 The pressure transducer connected to the compressor. Courtesy Ferris State University. Photo by John Tomczyk

Figure 25.50 The schematic diagram of an oil safety control circuit and a three-phase motor with internal overload.

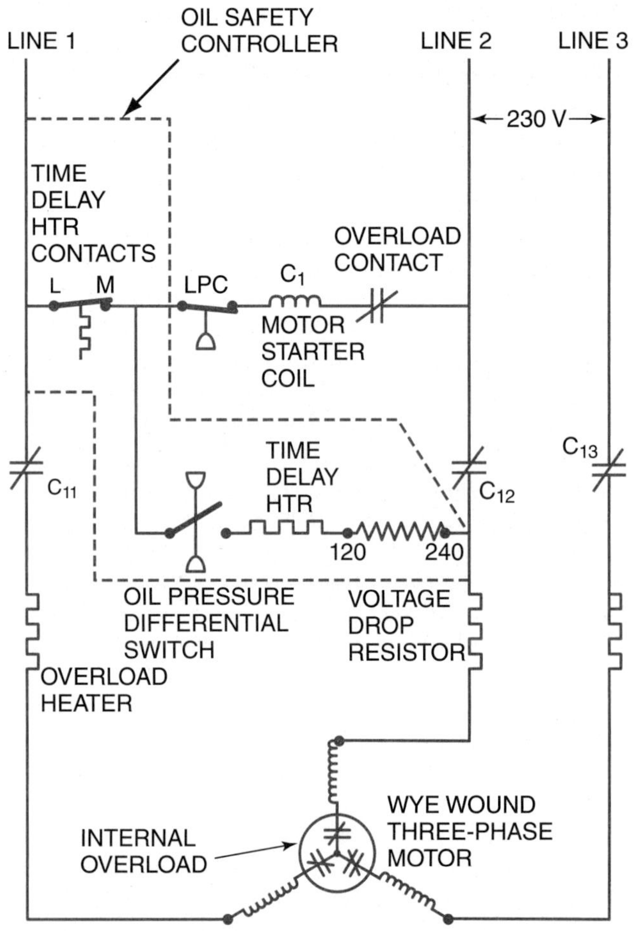

controller to process. The advantage of an electronic oil safety controller is that it eliminates the need for the transmission lines that are found on traditional oil pressure safety switches. This reduces the chance of refrigerant leaks. Also, the electronic clock and circuitry are much more accurate and reliable. Remember, when the compressor is off, the net oil pressure is 0 psi and the differential pressure switch contacts are closed. The heater in the oil safety controller will not be energized during the off cycle because it is wired to the load side of the motor starter contacts, **Figure 25.50**. When the motor starter contacts are opened, L2 is taken out of the heater circuit. At start-up, when the motor starter contacts close and the compressor starts, the differential pressure switch contacts will stay closed and the heater will be energized until at least 9 psid of net oil pressure is developed on most controllers. As discussed previously, this time delay will prevent nuisance trips to the controller at compressor start-ups or when system problems like floodback and migration occur.

INTERNAL OVERLOADS If a motor is equipped with both an internal inherent motor protector and an oil safety controller, the oil safety controller may trip on some systems due

to a motor overheating or overloading problem. When the internal overload opens, the motor is shut off but the motor starter coil remains energized with contacts closed. The 2-min timing circuit would be activated due to the lack of net oil pressure. This will trip the oil safety controller in a matter of 2 min because the heater will still be energized through the load side of the motor starter, **Figure 25.50.** Remember, there is no net oil pressure when the compressor is shut off by its internal overload protector. In time, the compressor will cool off and the internal overload will reset (close). But the compressor will not restart until pushing the reset button manually resets the oil safety controller. This condition can be prevented by the use of a current sensing relay.

The current sensing relay, **Figure 25.51**, allows the compressor to cycle on the internal overload while not affecting the operation of the oil safety controller. The current sensing relay is wired to the load side of the contactor or motor starter. It acts like an inductive-type ammeter. One leg of the load side of the compressor motor is passed through the relay inductive coil. The inductive coil controls a normally open set of contacts in series with the time delay heater, **Figure 25.52.**

Figure 25.51 A current sensing relay. Photo by John Tomczyk

Figure 25.52 An oil safety circuit showing a current sensing relay with motor and internal overload.

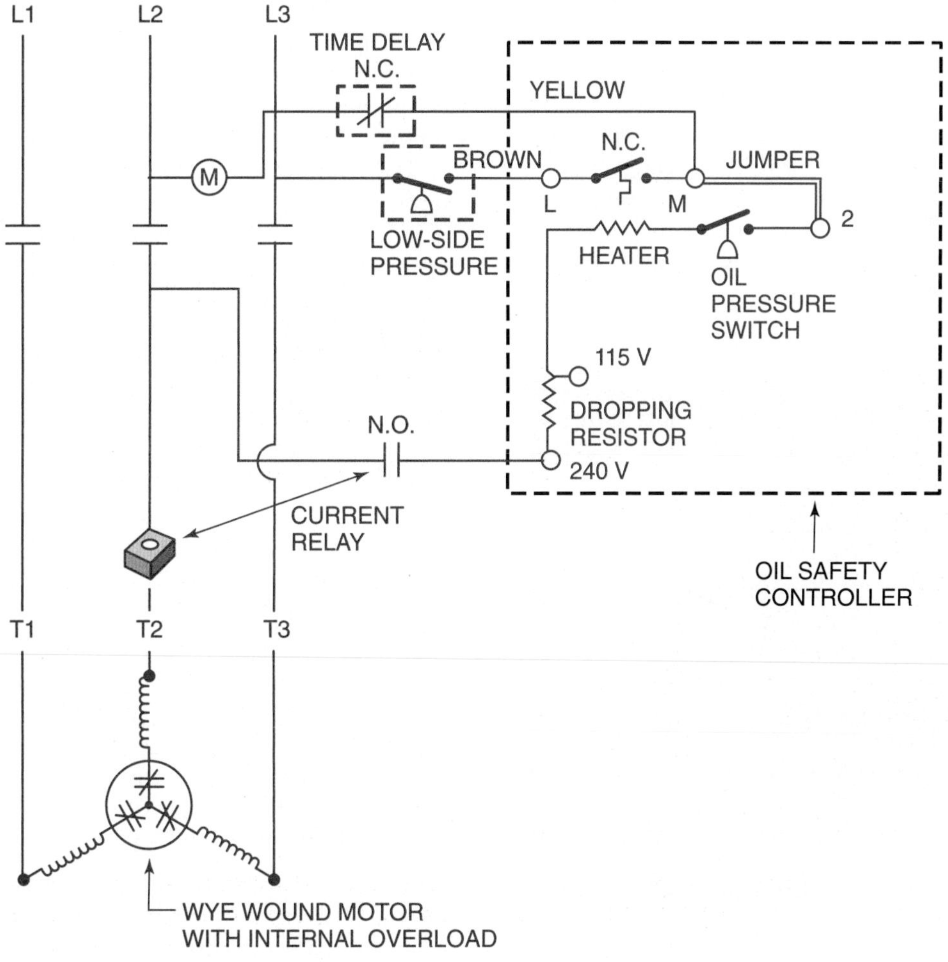

One leg of the compressor motor is shown passing through the inductive coil in **Figure 25.51**. The contacts are closed when normal motor current is sensed. When the current drops quite low, or reduces to zero because of the motor being off, the contacts of the current relay will open and take power away from the time delay heater. This prevents a tripping of the oil safety controller in case the internal overload opens (compressor off) but the motor starter's coil is still energized. The voltage dropping resistor in **Figure 25.52** makes the control more versatile for use on 230-V or 115-V systems.

In these situations, it is often necessary to incorporate another field-wired relay when dealing with an electronic oil safety controller. The reason for this is that the electronic oil safety module requires power to the module for proper reset. The added relay simply provides this power to the module during an oil safety trip. However, always consult with the compressor and control manufacturer before applying or wiring any of these devices or when working with any compressor that has an oil safety control.

25.6 SOLENOID VALVES AND THE AUTOMATIC PUMPDOWN CYCLE

THE SOLENOID VALVE

The solenoid valve, **Figure 25. 53**, is the component most frequently used to control fluid flow. The valve has a magnetic coil that, when energized, will lift a plunger into the coil, **Figure 25.54**. This valve can be either normally open (NO) or normally closed (NC). The NC valve is closed until energized; then it opens. The NO valve is normally open until energized; then it closes. The plunger is attached to the valve so that the plunger action moves the valve. Solenoid valves must be sized correctly for their particular application. For the refrigerant liquid line, they are sized according to refrigerant tonnage and a pressure drop through the valve that is acceptable. If a valve needs to be replaced, the manufacturer's data will help you choose the correct valve.

Figure 25.53 Electrically operated solenoid valves. The valve is moved by the plunger attached to the seat. The plunger moves into the magnetic coil when the coil is energized to either open or close the valve. Courtesy Sporlan Division, Parker Hannifin Corporation

Figure 25.54 Cutaway view of an electrically operated solenoid valve. Courtesy Sporlan Division, Parker Hannifin Corporation

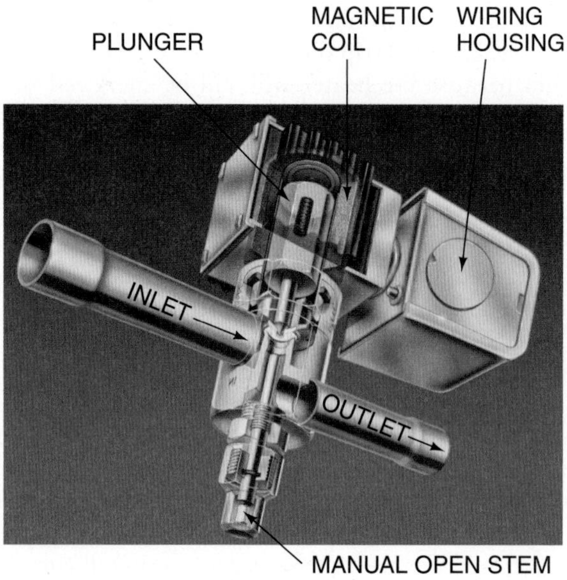

Solenoid valves can be used to control either liquid or vapor flow. They are snap-acting valves and open and close very fast when electrical energy is applied to the coil. The snap action can cause liquid hammer when installed in a liquid line, however, so care must be taken when locating the valve. Liquid hammer occurs when the fast-moving liquid is shut off by the solenoid valve, causing the liquid to stop abruptly. The solenoid valve must be fastened securely to the refrigerant line by flare, flange, or solder connections. Most valves have to be serviced at some time. The valves that are soldered in the line can be serviced easily if they can be disassembled. The manufacturer's instructions as to the location and placement of solenoid valves should be followed. In order to operate properly, the solenoid valve must be oriented in the correct direction to the fluid flow or the valve may not close tightly, **Figure 25.55**. The valve has an arrow that must point in the direction of flow. When positioned correctly, the pressure of the fluid in the valve helps to close the valve. If the high pressure is under the valve seat, the plunger may have a tendency to lift the valve off the seat. Most solenoid valves have a heavy plunger that is lifted to open the valve. When the plunger is not energized, its weight holds the valve on its sealing seat. If this type of valve is installed on its side or upside down, the valve will remain in the energized position when it should be closed, **Figure 25.56**.

Figure 25.55 The fluid flow in a solenoid valve must be in the correct direction, or the valve will not close tight. If the high-pressure fluid is under the seat, it will have a tendency to raise the valve off the seat. When installed correctly, the fluid helps to hold the valve on the seat.

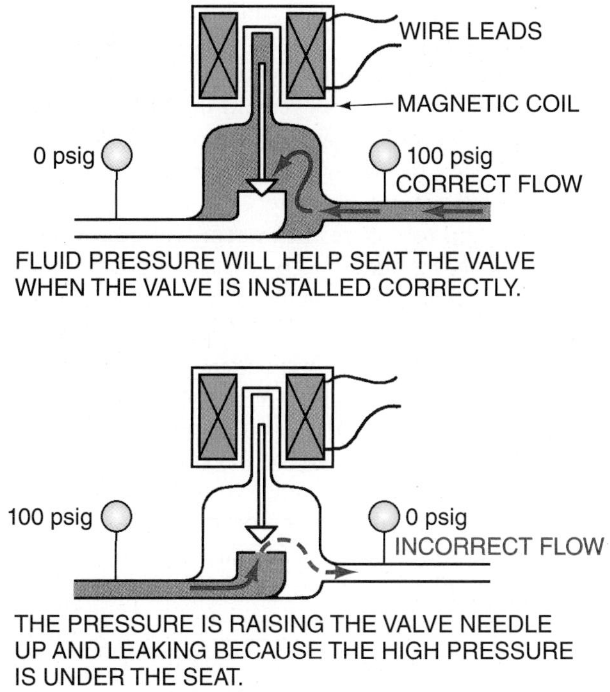

Figure 25.56 A solenoid valve installed in the correct position. Most valves use gravity to seat, and a magnetic force raises the valve off its seat. If the valve is installed on its side or upside down, the valve will not seat when deenergized.

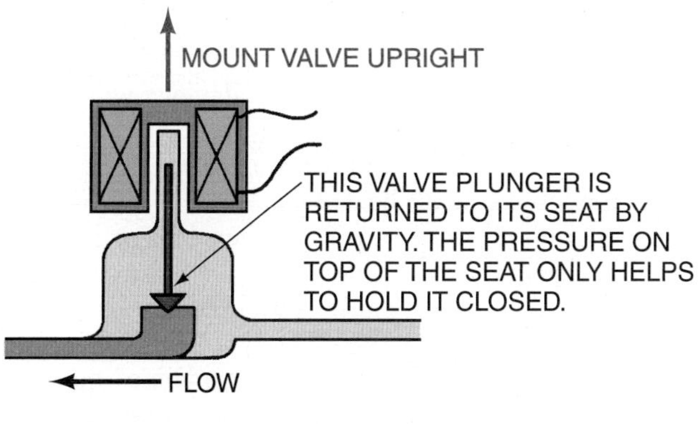

Figure 25.57 A pilot-operated solenoid valve. The magnetic coil controls a small line that directs the high-pressure refrigerant to one end of a sliding piston. The magnetic coil has to do only a small amount of work, and the difference in pressure does the rest. Based on Alco Controls Division, Emerson Electric Company.

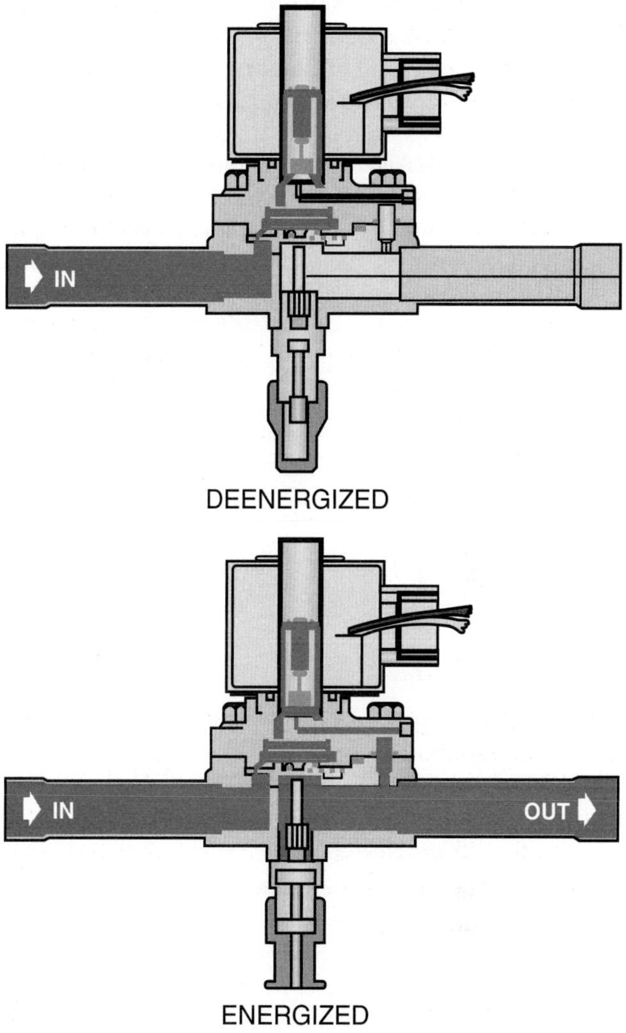

A special solenoid valve known as pilot-operated valve is available for controlling fluid flow in larger applications. This valve uses a very small valve seat to divert high-pressure gas that causes a larger valve to change position, thus using the difference in pressure to cause a large movement. In pilot-operated valves, the solenoid's magnetic coil only has to lift a small seat. An electrical coil large enough to cause the switching action of a large valve would be very big and draw too much current for practical use, **Figure 25.57**. Pilot action reduces the size of the electrical coil and the overall size of the valve. Pilot-acting valves are used when large vapor or liquid lines must be switched and can have more than one inlet and one outlet. Some are known as four-way valves; others as three-way valves.

AUTOMATIC PUMPDOWN SYSTEMS

The **automatic pumpdown system** consists of a normally closed liquid-line solenoid valve installed in the liquid line of a refrigeration system. The normally closed solenoid valve is controlled by a thermostat that is located somewhere in the refrigerated space. When the desired box temperature is reached in the refrigerated space, the thermostat will open and deenergize the liquid-line solenoid valve, causing the valve to close.

This will stop the flow of refrigerant into the metering device and evaporator. The compressor will continue to run, pumping any refrigerant downstream of the closed solenoid valve to the high side, causing the low-side pressure to drop. **Figure 25.58** is an electrical schematic diagram of an automatic pumpdown system using a liquid-line solenoid valve, thermostat, and low-pressure control. **Figure 25.59** is a system diagram showing the electrical hookup of an automatic pumpdown system.

Refrigerant pumped by the compressor from the low side of the system will be stored in the condenser and receiver on the high side of the refrigeration system. The compressor will shut off by the action of a low-pressure control set to open at about 5 to 10 psig, which ensures that no refrigerant migrates to the compressor during the off cycle. On a call for cooling, the thermostat closes and energizes the liquid-line solenoid valve. This action sends liquid refrigerant to the expansion valve and into the empty evaporator, thus increasing

Figure 25.58 An electrical schematic diagram of an automatic pumpdown system.

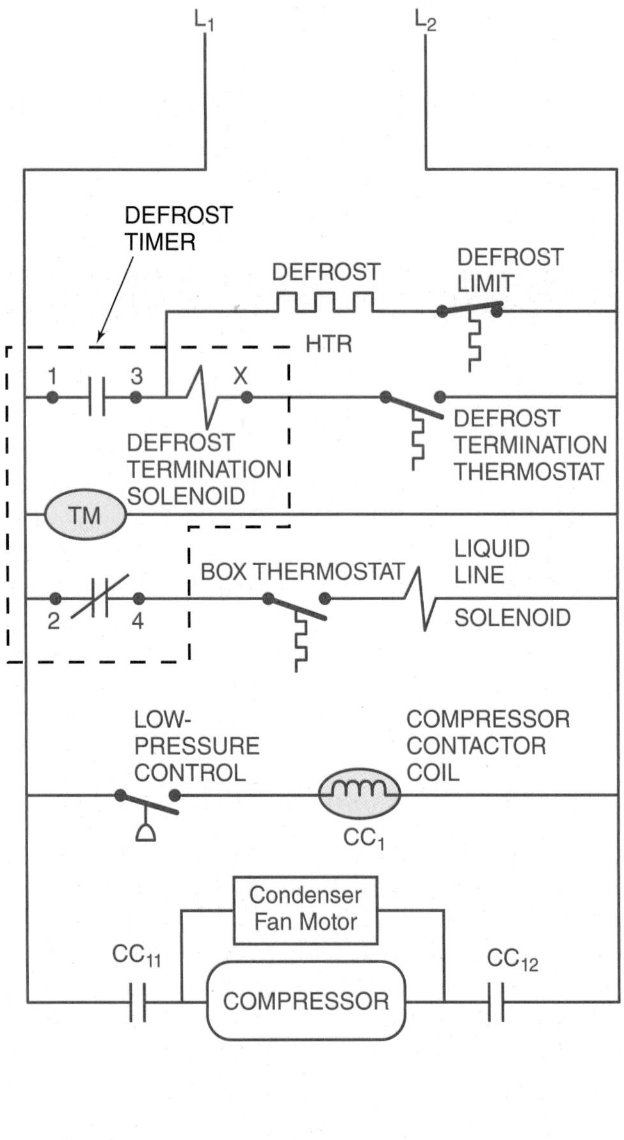

evaporator and suction pressure. Once the cut-in pressure of the low-pressure control is reached, the compressor will easily start and resume a normal refrigeration cycle. Automatic pumpdown systems do not need a larger receiver since the receiver is designed to hold the entire refrigerant charge and still have a 10% to 20% vapor head for safety.

The cut-in pressure of the low-pressure control should be set high enough to ensure that the system will not short cycle if residual pressure does remain in the low side of the system once it is pumped down. Compressor short cycling can be devastating to motor windings and starting controls because of overheating. However, the cut-in pressure has to be low enough to ensure the system will cut in once the liquid-line solenoid is energized by the thermostat to start the next on cycle. These pressures are dependent on refrigerant type and desired box temperatures. A pressure setting guide like the one in **Figure 25.41** should be sufficient to determine the correct pressures for an automatic system. Compressor short cycling during the off cycle can also be caused by a leaky liquid-line solenoid valve or the compressor discharge valve leaking high pressure back into the compressor during the pumpdown phase. An important thing to remember when leak testing a system with automatic pumpdown is that the solenoid circuit must be energized to prevent trapping of refrigerant in the condenser and receiver.

It is a good practice to install automatic pumpdown systems on all refrigeration systems employing large amounts of refrigerant and oil. Pumpdown systems are not typically used on systems with small amounts of oil or refrigerant because there is not enough refrigerant to migrate to the oil and cause any damage if migration does occur during the off cycle. For example, a domestic refrigerator or freezer may have a refrigerant charge of only 8 to 16 oz and an oil charge from 12 to 20 oz. There is often more oil than refrigerant in these systems. Even if the entire refrigerant charge migrates to the crankcase during an off cycle, no real damage will be done because the refrigerant flashes off during the next on cycle. So, only if the refrigerant-to-oil ratio is large should automatic pumpdown systems be employed.

Figure 25.60 illustrates a wiring scheme for an automatic pumpdown system that will not short cycle the compressor during a pumpdown phase caused by a leaking solenoid valve or compressor discharge valve. This type of system is often referred to as a non-short-cycling or "onetime" automatic pumpdown system. Only the thermostat calling again for cooling will take the refrigeration system out of pumpdown. The diagram shows the system in a pumpdown with the compressor off. On a call for cooling, the thermostat closes and energizes the liquid-line solenoid valve. This allows the low side of the system to experience refrigerant pressure, which allows the low-pressure switch to close, **Figure 25.61**. The latching relay coil (LRC) is energized through the normally closed (NC) contacts A and B. Once the LRC is energized, it closes contacts B and C and contacts C and D, while opening

Figure 25.59 The electrical and mechanical hookup of an automatic pumpdown system. Based on ESCO Press.

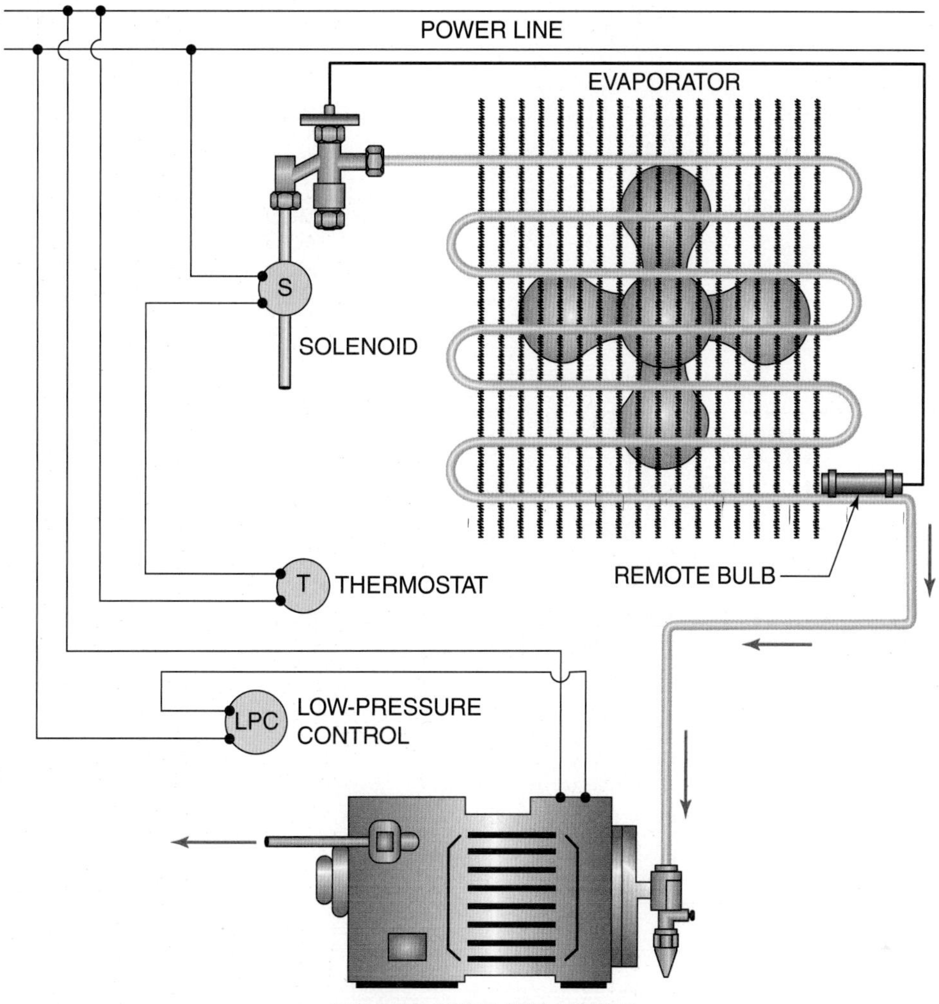

contacts A and B. With the LRC energized, the compressor contactor coil (CC) is energized through contacts C and D and the low-pressure switch. This starts the compressor and then the system is in a normal cooling mode, **Figure 25.62**.

When the thermostat opens, the liquid-line solenoid valve is deenergized, **Figure 25.63**. This physically closes off the liquid line and initiates an automatic pumpdown. The low-pressure switch now opens from the low pressure, deenergizing the LRC coil, which opens contacts C and D and B and C. The opening of the low-pressure switch also deenergizes the compressor contactor coil (CC), which shuts off the compressor. The system is pumped down and the compressor remains off until there is another call for cooling from the thermostat. However, if the low-pressure switch closes because of a leaking component while the thermostat is still open and not calling for cooling, the compressor contactor coil (CC) will not be energized because of the normally open (NO) contacts C and D of the latching relay. This prevents the compressor from short cycling while in the pumpdown mode. The LRC will also not be energized because of the NO contacts B and C of the latching relay. In summary, when components leak, these latching relay contacts prevent the compressor from short cycling during pumpdown, which causes low-side pressures to increase and low-pressure switches to close prematurely.

The following are three main reasons automatic pumpdown systems are employed on refrigeration systems:

- *To rid the evaporator, suction line, and crankcase of refrigerant before the off cycle and defrost cycle so that migration of refrigerant to the compressor and/or compressor crankcase cannot occur*
- *To prevent surges of liquid refrigerant from entering the suction port of the compressor (slugging) during start-ups*
- *To rid the crankcase of refrigerant to prevent oil from foaming during start-ups and robbing lubrication from the compressor's mechanical parts*

Figure 25.60 A non-short-cycling, or "one-time," automatic pumpdown system using a latching or lockout relay. Based on ESCO Press.

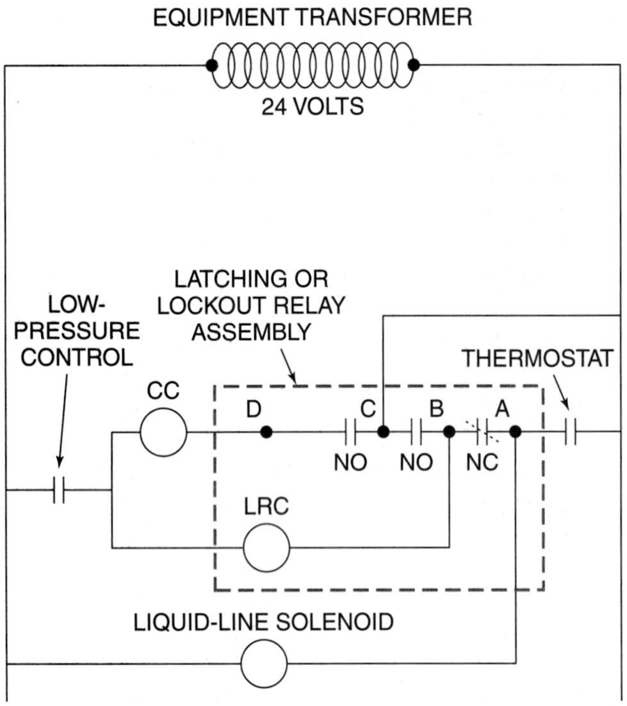

Figure 25.61 The thermostat provides power to the liquid-line solenoid, and refrigerant begins to flow. When the low-side pressure reaches the cut-in point on the low-pressure control, the batching relay closes, starting the compressor.

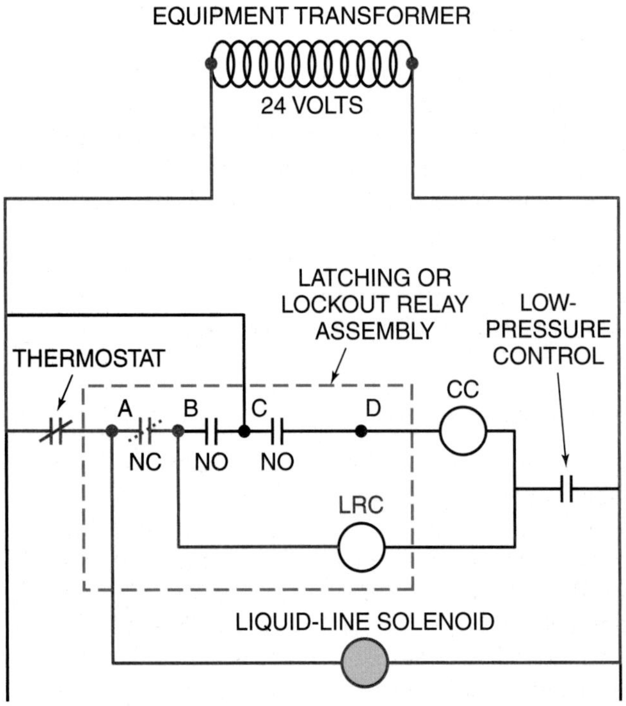

Figure 25.62 The compressor is now running.

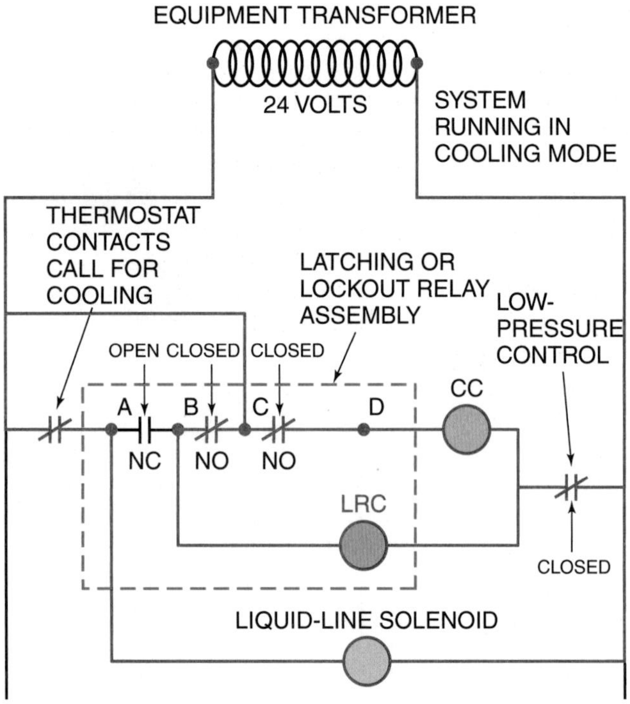

Figure 25.63 The system will run until the low-pressure control contacts open.

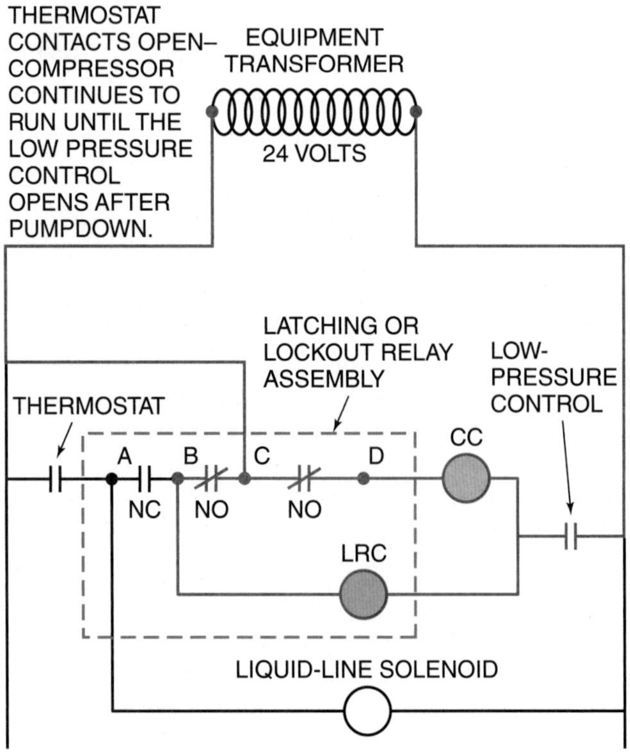

25.7 DEFROST CONTROLS

THE DEFROST CYCLE

Any evaporator that operates below 32°F will freeze and will need some sort of method of defrosting. This also holds true when products must be stored above freezing temperatures, because the evaporators that cool them are often below freezing. For example, dairy products are often kept refrigerated at 35°F to 36°F, while their evaporator surface temperatures will range between 25°F and 30°F. When the surface of the evaporator falls below freezing, it will begin to build a coating of ice or frost. Frost can occur on the surface of the evaporator in a number of forms. The layer of frost closest to the evaporator coil will be a very hard form of ice, similar to what you would see in an ice cube. The other layers of frost that form on this hard surface of ice covering the coil often resemble snow or flaked ice and will hold a lot of entrained air. This layer is often referred to as "hoar frost." The amount of hoar frost depends on:

- Humidity levels
- Airflow rates
- Evaporator temperatures

As humidity levels and airflow rates increase, so does the amount of hoar frost. As evaporator temperatures decrease, the hoar frost will also increase. Although ice will conduct heat four times as well as water, the air entrained in the porous hoar frost will insulate the evaporator and severely reduce its heat-absorbing abilities. The suction pressure will now drop and the evaporator will become even colder. The colder evaporator and lower suction pressures cause the refrigeration system to have a higher compression ratio. Higher compression ratios cause lower volumetric efficiencies within the compressor. Lower suction pressures also cause the return gas coming back to the compressor to be less dense, reducing the mass flow rates of refrigerant. If the evaporator is severely insulated in frost to the point that hardly any heat load is reaching the coil, the TXV metering device may lose control and flood the compressor.

Evaporator fin spacing is very important in handling hoar frost. The colder the operating temperature of the evaporator, the more space there needs to be between the fins. This provides space for frost to accumulate and at the same time not slow down or block the evaporator's airflow. Some typical fin spacing for different evaporators are as follows:

- Low-temperature evaporators: 4 to 6 fins per inch
- Medium- and high-temperature evaporators: 6 to 10 fins per inch
- Air-conditioning evaporators: 12 to 15 fins per inch

However, an evaporator that has fewer fins per inch will need to be larger. This is because these evaporators have less surface area for heat transfer to take place. So, if frost can be minimized, fins can be more closely spaced and a smaller, more economical evaporator can be used. Defrosting the evaporator and eliminating frost from its surface will accomplish this task.

Laboratory testing has shown that a thin layer of frost will actually increase heat transfer in an evaporator. The thin layer of frost will somewhat reduce the area for airflow, causing a slight increase in the air's velocity which will increase heat transfer to the evaporator. Also, a thin layer of frost will increase the surface area of the evaporator tubes, which will also enhance heat transfer. These same tests also show that the best time to defrost is when the evaporator has lost about 10% of its efficiency. It is for these reasons that service technicians must follow the manufacturer's recommendation for defrost period intervals and defrost period durations to get maximum efficiency out of an evaporator surface and a defrost period. The defrost cycle in refrigeration is divided into medium-temperature and low-temperature ranges; the components that serve these defrost cycles are different.

MEDIUM-TEMPERATURE REFRIGERATION

The medium-temperature refrigeration coil normally operates below freezing and rises above freezing during the off cycle. A typical temperature range would be from 34°F to 45°F inside the cooler. The coil temperature is normally 10°F to 15°F cooler than the space temperature in the refrigerated box. This means that the coil temperatures would normally operate as low as 19°F (34°F − 15°F = 19°F). The air temperature inside the box will always rise above the freezing point during the off cycle and can be used for the defrost. This is called off-cycle defrost and it can be either random or planned.

RANDOM OR OFF-CYCLE DEFROST

Random or off-cycle defrost is the most common type of defrost. Random defrost will occur when the refrigeration system has enough reserve capacity to cool more than the load requirement. When the system has reserve capacity, it will be shut down from time to time by the thermostat, and the air in the cooler can defrost the ice from the coil. When the compressor is off, the evaporator fans will continue to run and circulate the air in the cooler. The air that is circulated must be warmer than 32°F to defrost the coil. The main advantage of off-cycle defrost is its simplicity and economy of installation. When the refrigeration system does not have enough capacity or the refrigerated box has a constant load, there may not be enough off time to accomplish defrost. This is when the defrost cycle must be planned.

PLANNED DEFROST

Planned defrost is accomplished by forcing the compressor to shut down for short periods of time so that the air in the cooler can defrost the ice from the coil. A timer that can be programmed stops the compressor during times that the refrigerated box is under the least amount of load. For example, a restaurant unit may defrost at 2 AM and 2 PM to avoid the rush hours, **Figure 25.64**. Again, always consult with the refrigerated case manufacturer for defrost period intervals and durations.

Figure 25.64 A timer to program off-cycle defrost. This is a 24-hour timer that can have several defrost times programmed for convenience. Photo by Bill Johnson

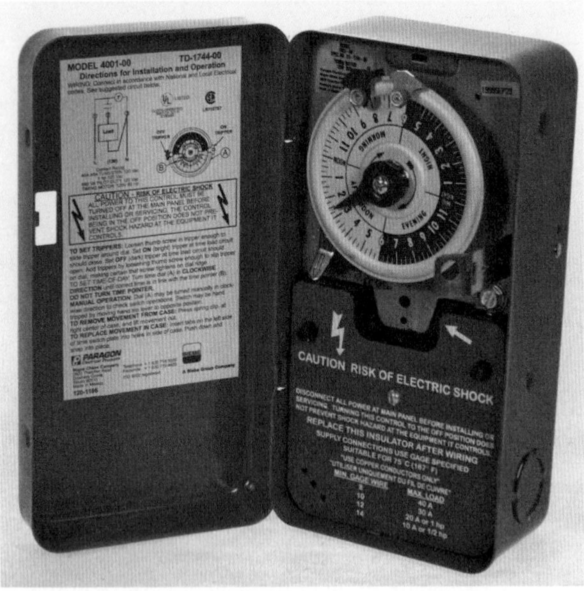

LOW-TEMPERATURE EVAPORATOR DEFROST

Low-temperature evaporators all operate below freezing and must have planned defrost. Because the air inside the refrigerated box is well below freezing, heat must be added to the evaporator for defrost, which is accomplished using either internal heat or external heat.

INTERNAL HEAT DEFROST (HOT GAS AND COOL GAS DEFROST). Hot and cool gas defrost is the least common type of defrost when compared to random or off-cycle defrost and electric defrost. In cool gas defrost, the gas used for defrosting the evaporator comes from the top of the receiver. Even though this gas has a cooler temperature, it often has more heat content, or enthalpy, because it is a more dense, saturated gas. The internal heat method of defrost normally uses the hot gas from the compressor. The hot gas can be introduced into the evaporator from the compressor discharge line to the inlet of the evaporator and allowed to flow until the evaporator is defrosted, **Figure 25.65**. *Because a portion of the energy used for hot gas defrost is available in the system, this method is attractive from an energy-saving standpoint.* Injecting hot gas into the evaporator is rather simple if the evaporator is a single-circuit type because inserting a T in the expansion valve outlet is all that is necessary. With a multicircuit evaporator, the hot gas must be injected between the expansion valve and the refrigerant distributor. This gives an equal distribution of hot gas to defrost all of the coils equally. Often, the hot gas is 200°F or above, and as it rejects its heat to the frosted evaporator coil from the inside of the coil, it will cool enough to condense to a liquid. The liquid is often caught by the accumulator in the suction line before it enters the compressor, **Figure 25.65**. Once the frost on the evaporator is melted to water, the water must have time to drip and drain off the evaporator. It is important to keep the fans off during this "**drip time**" and for some time after the defrost to prevent moisture from being blown off the coil and on to the product. In space temperature applications, where forced air is used to cool the product, the defrost cycle is normally started with a timer. The cycle can be terminated by time or temperature. The amount of time it

Figure 25.65 Hot gas defrost.

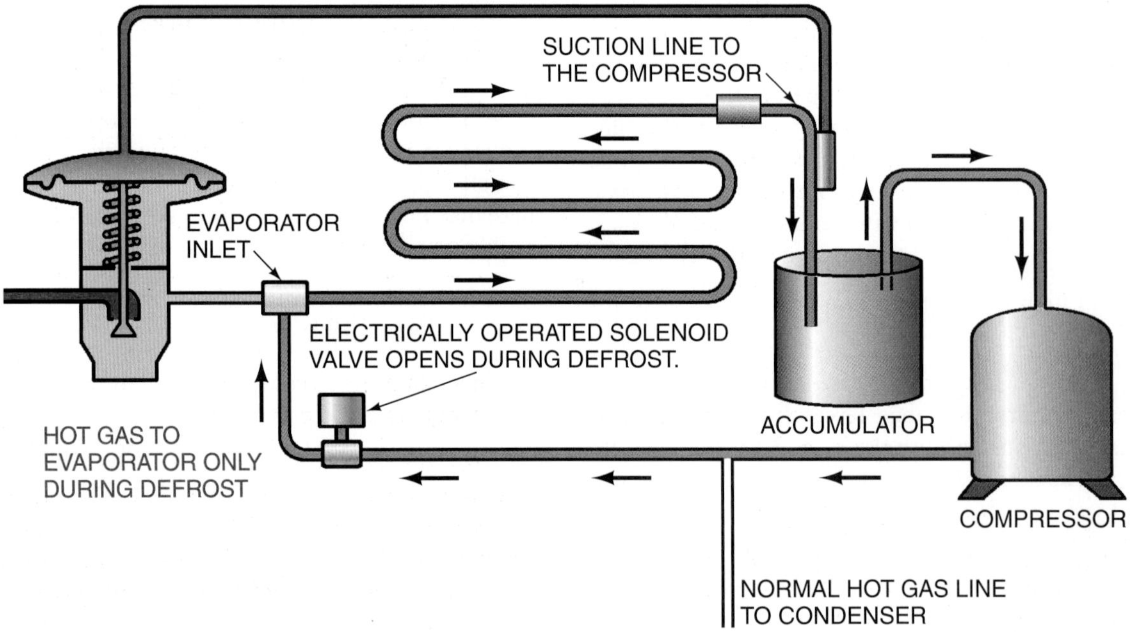

takes to defrost the coil must be known before time alone can be used efficiently to terminate defrost. Because this can vary from one situation to another, if the timer is set for too long a time the unit would run in defrost when it is not desirable, causing energy loss and possible product spoilage. Defrost can also be started with time and terminated by temperature.

A temperature-sensing device is used to determine that the coil is above freezing. The hot gas entering the evaporator is stopped, and the system goes back to normal operation. **Figure 25.66** is a wiring diagram that shows how a time clock and a defrost termination switch work together to terminate the defrost cycle.

During the hot gas defrost cycle, several things must happen at the same time. The timer is used to coordinate the following functions:

1. The hot gas solenoid must open.
2. The evaporator fans must stop, or cold air will keep defrost from occurring.
3. The compressor must continue to run.
4. A maximum defrost time must be determined and programmed into the timer in the event the defrost termination switch fails to terminate defrost.
5. Drain pan heaters may be energized.

EXTERNAL HEAT DEFROST. The external heat method of defrost is often accomplished by electric heating elements that are factory mounted next to the evaporator coil. Electric defrost is the second most common type after random or off-cycle defrost and hot gas or cool gas defrost. It is also a planned defrost that is controlled by a timer. The external heat method is usually not as efficient as the internal heat method because energy must be purchased for it. Even though it is one of the simpler methods to install, electric defrost can be very expensive in terms of energy usage. The electric heaters often use 1 kW per foot of evaporator length, and there can be 3 to 5 defrost periods per day, of 45 to 60 min in duration. The heaters often reach temperatures in excess of 300°F and can damage components near the evaporator if not carefully controlled. Often, melted frost that touches the element will flash to steam and recondense on any cold surface, as evidenced by layers of ice on the ceilings of refrigerated rooms. This may require extending the defrost periods and run into unwanted costs. Thus, electric defrost must be closely controlled to make it safe, efficient, and effective. However, if long runs are required for the hot gas lines, electric defrost may be the more efficient method. The long runs can cause refrigerant to condense, slowing the defrost process, and liquid refrigerant could reach the compressor, causing slugging and compressor damage.

When electric defrost is used, it is critical that the cycle be terminated at the earliest possible time. The timer controls the following events:

1. Stopping the evaporator fan in most cases
2. Stopping the compressor (with a possible pumpdown cycle to pump the refrigerant out of the evaporator to the condenser and receiver)

Figure 25.66 A wiring diagram of a timer with a mechanism that can stop defrost with an electrical signal from a temperature-sensing element.

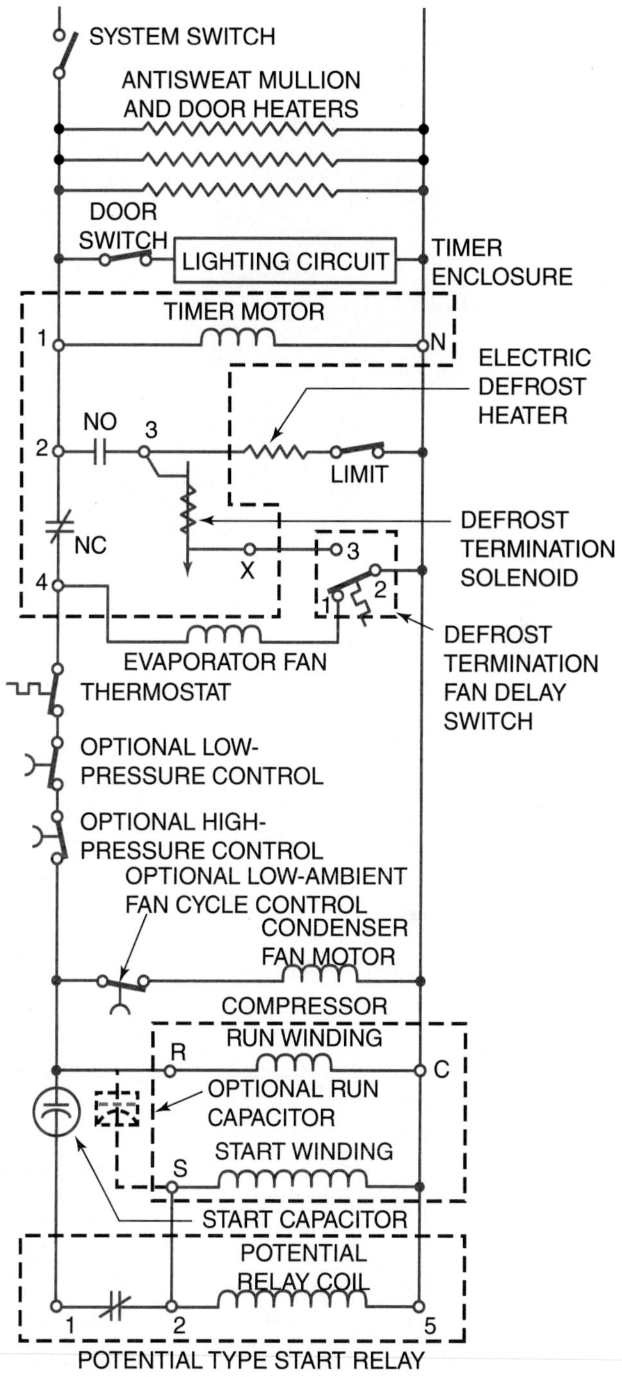

3. Energizing the electric heaters
4. Energizing drain pan heaters

A temperature sensor may be used to terminate defrost when the coil is above freezing. A maximum defrost time should be programmed into the timer in the event the defrost termination switch, **Figure 25.67**, fails to terminate defrost. The programmed defrost duration time is often referred to as the **fail-safe** time.

Figure 25.67 A defrost termination and fan delay (DTFD) switch. The switch is a single-pole–double-throw device. It has a hot contact (made from the common to the terminal that is energized on a rise in temperature) and a cold contact (made from the common to the terminal that is energized when the coil is cold). Courtesy Ferris State University. Photo by John Tomczyk

Figure 25.68 (A) A defrost termination/fan delay control. Based on Ranco Controls Division. (B) A defrost termination/fan delay control installed on a forced air evaporator. Based on Ranco Controls Division.

(A)

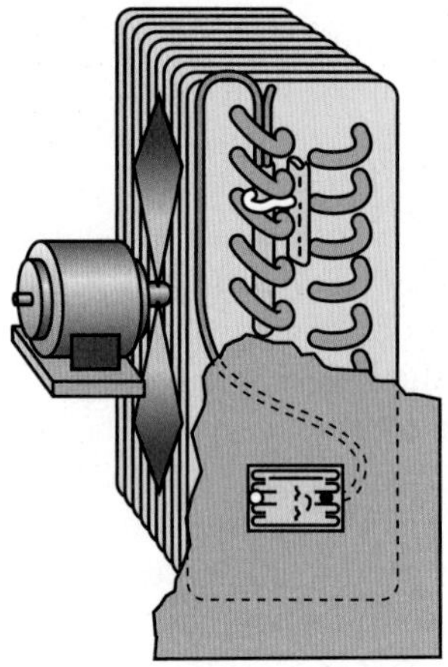

(B)

DEFROST TERMINATION AND FAN DELAY CONTROL

The defrost termination/fan delay control is a temperature-activated, single-pole–double-throw switch controlled by a remote sensing bulb. The control shown in **Figure 25.68(A)** also happens to be adjustable. The control installed on an evaporator is shown in **Figure 25.68(B)** and is wired into the refrigeration circuit as shown in **Figure 25.66**. The control's remote sensing bulb is located high on the evaporator where the frost is likely to clear last. The function of this temperature-activated switch is to terminate defrost when the evaporator coil has been defrosted and also to delay the evaporator fans from coming on immediately after defrost.

Defrost time clocks can be programmed to control the length of defrost periods in increments of minutes. For example, a defrost time clock on a freezer could be programmed to defrost every 6 hours (four times daily) for periods of 40 min. However, there will be times throughout the year when the coil does not need the entire 40 min of defrost heat, for example, when low freezer usage keeps door openings to a minimum or when the humidity is low and not much frost accumulates on the coil. This is where the defrost termination part of the control comes into play. The following paragraphs explain the defrost termination function.

Let us assume that the system does not have a defrost termination/fan delay control. Once the normally open (NO) contacts of the defrost time control have closed and the unit is in defrost, the defrost heaters will emit heat and frost will melt off the evaporator coil, **Figure 25.66**. If it takes only 10 min for all of the frost to leave the evaporator coil, there will still be 30 min left (40 min − 10 min) in the programmed defrost period. If the system has an optional heater safety switch (sometimes referred to as a defrost limit control) wired in series with the defrost heater, as shown in **Figure 25.66**, the limit switch will open and take the defrost heaters out of the active circuit. However, the system's defrost timer will still have 30 min left in the defrost period. The system will simply sit idle, and because there will be no refrigeration for 30 min the product load will suffer an increase in temperature. After 30 min, the defrost timer switches over to refrigeration mode and the fans will start immediately. They will blow the moist, residual defrost heat through the refrigerated space and the evaporator coil while the system is in refrigeration mode. This puts

the system, and thus the compressor, under an extremely high load from high suction pressures. The compressor will experience high suction pressures and dense vapors coming to its cylinders, which will cause high amperage draws and may overload the compressor to a point where its internal or external overload may open. *To prevent a long defrost period and compressor overload after defrost, a defrost termination/fan delay switch can be installed on the system, as shown in* **Figure 25.66.** If it takes only 10 min for the ice to leave the coil, the remote bulb of the defrost termination control will sense the defrost heat and close the contacts between 2 and 3 on the control. This energizes a defrost termination solenoid (release solenoid) in the time clock, which will mechanically put the system back into refrigeration mode. The solenoid action and levers will mechanically close the NC contacts and open the NO contacts of the defrost time control, **Figure 25.66**. This action prevents the system from sitting idle for 30 min in defrost with the heaters off. It terminates the defrost mode and puts the refrigeration mode back into service. The fan delay part of the sequence will be explained in the following paragraph.

Now that the system is in refrigeration mode, the evaporator fans will be delayed until the contacts between 2 and 1 of the defrost termination/fan control close. This usually happens at about 120°F to 130°F, a setting that is adjustable on most controls, and is sensed and controlled by the control's remote bulb. This allows the evaporator coil to prechill itself and get rid of some of the defrost heat still in the coil. *Delaying the fans prevents the suction pressure from getting too high after defrost and overloading the compressor when an automatic pumpdown cycle is not being used. It also prevents warm, moist air and droplets of moisture from accumulating on the evaporator coil. The end of the fan delay period is often referred to as a "drip time." It allows the evaporator coil to drip and drain its melted frost from the coil and keep it from being blown on the product in the refrigerated space.*

25.8 REFRIGERATION ACCESSORIES

RECEIVERS

The receiver is located in the liquid line and is used to store the liquid refrigerant after it leaves the condenser. The receiver should be lower than the condenser so that the refrigerant has an incentive to flow into it naturally, but this is not always possible. The receiver is a tank-like device that can be either upright or horizontal, depending on the installation, **Figure 25.69**.

Receivers can be quite large on systems that need to store large amounts of refrigerant. The receiver inlet and outlet connection can be at almost any location on the outside of the tank body. On the inside of the receiver, however, the

Figure 25.69 A refrigerant receiver. Photo by Jason Obrzut

refrigerant must enter the receiver at the top in some manner. The refrigerant that is leaving the receiver must be taken from the bottom to ensure that it is 100% liquid. This is accomplished with a dip tube, **Figure 25.70**, if the receiver outlet is at the top.

Figure 25.70 Diagram showing the dip tube in a refrigerant receiver.

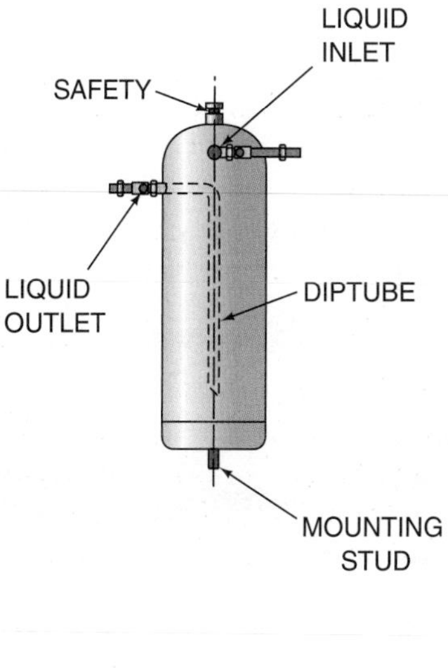

THE KING VALVE ON THE RECEIVER

The king valve is located in the liquid line between the receiver and the expansion valve. It is often fastened to the outlet of the receiver tank, **Figure 25.71**. The king valve is important in service work because, when it is front-seated, no refrigerant can leave the receiver. If the compressor is operated with this valve front-seated, all the refrigerant will be pumped into the condenser and receiver. Most of it will flow into the receiver. The receiver should be sized to hold approximately 20% more than the system's full charge of refrigerant. This leaves a vapor head for safety reasons during system pumpdowns. The other valves in the system can then be closed, and the low-pressure side of the system can be opened for service. When service is complete and the low side of the system is ready for operation, the king valve can be opened, and the system can be put back into operation. The king valve has a pressure service port to which a gauge manifold can be fastened. When the valve stem is turned away from the back seat, this port will give a pressure reading in the liquid line between the expansion valve and the compressor's high-side gauge port. When the valve is front-seated and the system is being pumped down, this gauge port is on the low-pressure side of the system common to the liquid line (this becomes the low side because the king valve is trapping the high-pressure refrigerant in the receiver). If there is any pressure in the liquid line, removing the gauge line may be difficult until the king valve is back-seated after the repair is completed, **Figure 25.71**. Another component that can be found in many systems is a solenoid valve. This was covered in the section "The Solenoid Valve," earlier in this unit. It is a valve that stops and starts the liquid flow.

LIQUID-LINE FILTER DRIERS

The refrigerant liquid-line filter drier, **Figure 25.72**, can be found at any point on the liquid line after the king valve.

Figure 25.71 A king valve piped in the circuit. The back seat is open to the gauge port when the valve is front-seated. This valve has to be back-seated to remove the gauge if refrigerant is in the system.

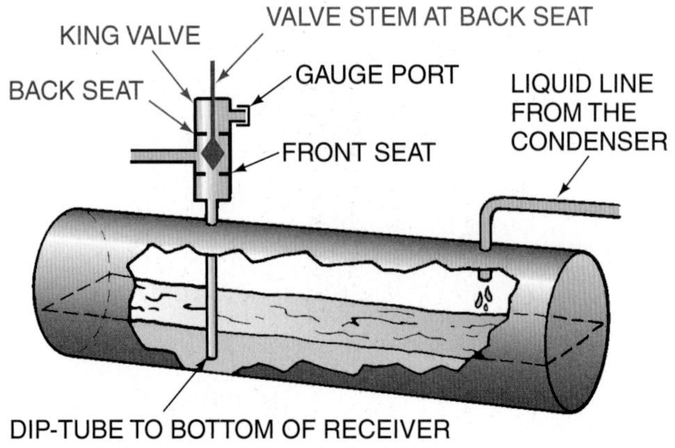

Figure 25.72 A filter drier. This device removes particles and moisture from a liquid line. The moisture removal is accomplished with a desiccant in bead or block form inside the drier shell. Courtesy Sporlan Division, Parker Hannifin Corporation

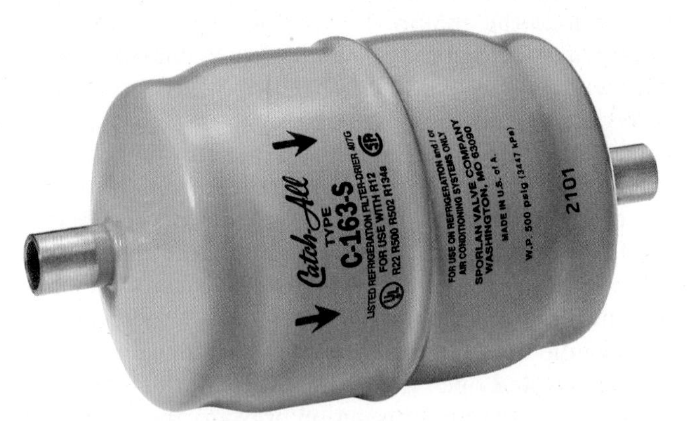

The filter drier is a device that removes foreign matter from the refrigerant. This foreign matter can be dirt, flux from soldering, solder beads, filings, moisture, parts, and acid caused by moisture. Some sources of moisture in refrigeration systems include poor service practices, air trapped from improper evacuation techniques, leaky systems, and improper handling of polyol ester (POE) lubricants. The device has a fine screen at the outlet to catch any small particles that may be moving in the system.

Filter-drying operations are accomplished with a variety of materials that are packed inside the drier, **Figure 25.73**.

Figure 25.73 Cutaway view of a filter drier showing the desiccant core and prefilters. Courtesy Sporlan Division, Parker Hannifin Corporation

Some manufacturers use chemical beads and some use porous blocks made from a drying agent. The most common agents found in filter driers are activated alumina, molecular sieve, or silica gel. Silica gel, a noncrystalline material with molecules formed by bundles of polymerized silica, forms a weaker bond with water, which is why silica gels are not widely used in modern filter driers.

Molecular sieve desiccants are sodium alumina-silicates in cubic crystalline form. The crystal structure is honeycombed with cavities that are uniform in size that can selectively absorb molecules based on their charge (polarity) and size. These properties allow water to be absorbed while still permitting the larger molecules like refrigerants and their lubricants to pass by. The surface of the desiccant is coated with cations, which are positively charged and act like magnets to attract polarized molecules such as water. Molecular sieves retain the highest amount of water and keep the concentration of water in the refrigerant low. The molecular sieve filter is recommended for use as liquid-line filter driers because of these good water-removal capabilities. The strong bond between water and the molecular sieve thus keeps the occurrence of system corrosion, acid formation, and **freezeups very** low.

Inorganic acids like hydrochloric and hydrofluoric acids are formed when high system temperatures break down refrigerants and water. These acids can attack metal surfaces in the system and destroy the crystalline structure of the molecular sieve. In the presence of the wrong desiccant and water, organic acids can then form from the broken down refrigerant lubricant. Organic acids, a sludgelike material, can plug up metering devices. Activated alumina is more effective than the molecular sieve for removing organic acid molecules. Activated alumina is formed from aluminum oxide. It is not a highly crystalline structure like a molecular sieve; its pores vary in size and do not absorb molecules based on pore size. Because of this, it can absorb much larger molecules like refrigerants and lubricants along with water. Activated alumina is even more effective in removing organic acids when it is used in the suction line of the refrigerant system. This is why some manufacturers recommend using a suction-line filter drier made of activated alumina for acid cleanups.

Filtration, another main function of a filter drier, can also be done with a screen. A screen will filter or catch particles that cannot fit through its wire mesh. As particles and contaminants start to cover the mesh of the screen, they act as a fine filter that removes even smaller particles. The accumulation of particles will soon cause enough pressure drop that the refrigerant will flash into a vapor. This will often cause a local cold spot on the filter drier and it may even sweat.

Filters may also include the following:

- Bonded desiccant cores
- Fiberglass pads
- Rigid fiberglass with a resin coating

Bonded desiccant cores have smaller openings than fiberglass pads. As refrigerant passes through, the channels trap the particles. As the channels fill, a pressure drop occurs and the refrigerant will vaporize or flash. Fiberglass pads are less compressed than rigid fiberglass with resins. As refrigerant and particles flow through the pads, their speed decreases and they deposit themselves in the openings of the fiberglass. As the pads fill with more particles, the filtration becomes finer, soon becoming similar to a rigid fiberglass filter. A fiberglass pad can hold up to four times as many particles and contaminants as the bonded desiccant core drier can and still retain the same filtration capacity.

The filter drier comes in two styles: permanent and replaceable core. Both types of driers can be used in the suction line as well as the liquid line, and this application will be discussed in more detail later. Filter driers in the smaller sizes up to 5/8 in. can be fastened to the liquid line by a flare connection. Solder connections can be used for a full range of sizes, from 1/4 to 1 5/8 in. The larger, soldered filter driers have replaceable cores. Replaceable-core filter driers as well as plain filters can be easily serviced, **Figure 25.74**. The king valve is front-seated to pump the refrigerant into the condenser and receiver to replace the drier cores. An evacuation to at least 500 microns with a vacuum pump is then applied to the pumped-down part of the system before putting the new filter or core back into service.

Figure 25.74 This suction filter has a removable core. Its casing or shell is left in the system and only the core is replaced, once the system is pumped down, by removing the bolts and end cap. Courtesy Sporlan Division, Parker Hannifin Corporation

Figure 25.75 Refrigerant sight glasses. Photo by Jason Obrzut

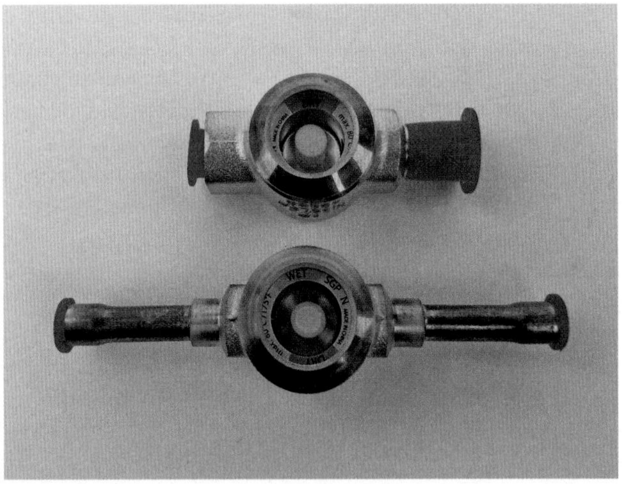

Figure 25.76 A multicircuit refrigerant distributor used to feed equal amounts of liquid refrigerant to the different refrigerant circuits. This is a precision-machined device. Courtesy Sporlan Division, Parker Hannifin Corporation

Figure 25.77 Examples of metering devices with distributors. Courtesy Sporlan Division, Parker Hannifin Corporation

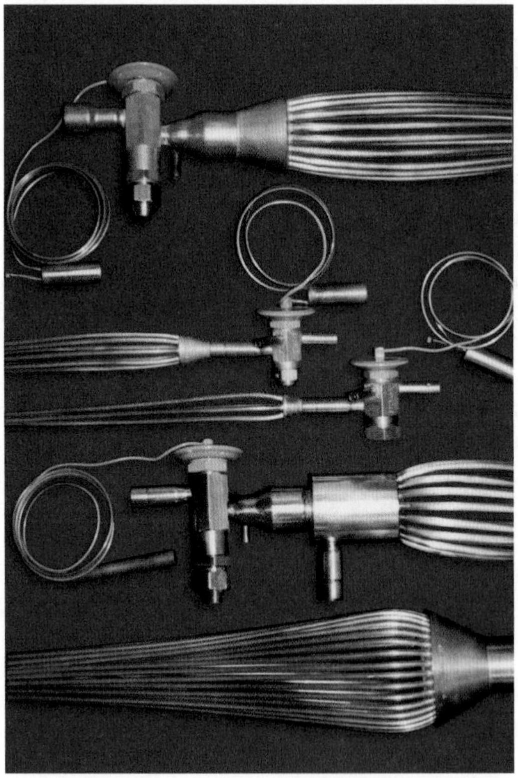

Filter driers can be installed anywhere in the liquid line. The closer it is to the metering device, the better it cleans the refrigerant before it enters the tiny orifice in the metering device. The closer it is to the king valve, the easier it is to service. Service technicians and engineers choose their own placement, based on experience and judgment.

REFRIGERANT SIGHT GLASSES

The refrigerant sight glass, **Figure 25.75**, is normally located anywhere liquid flow exists and anywhere it can serve a purpose. *When it is installed just prior to the expansion device, the technician can be assured that a solid column of liquid is reaching the device. When it is installed at the condensing unit, it can help with troubleshooting.* More than one sight glass, one at each place, is sometimes a good investment. Sight glasses come in two basic styles: plain and with a moisture indictor. The plain sight glass is used to observe the refrigerant as it moves along the line. The sight glass with a moisture indicator can tell the technician what the moisture content is in the system. It has a small element in it that changes color when moisture is present.

LIQUID REFRIGERANT DISTRIBUTORS

The refrigerant distributor, **Figure 25.76**, is used on multicircuit evaporators. It is fastened to the outlet of the expansion valve and distributes the refrigerant to each individual evaporator circuit, **Figure 25.77**. This is a precision-machined device that ensures that the refrigerant is divided equally to each circuit. This is not a simple task because the refrigerant is not all liquid or all vapor. It is a mixture of liquid and vapor. A mixture has a tendency to stratify and feed more liquid to the bottom and the vapor to the top. Choosing the correct size of distributor for a system can be a complex decision because the distributor must have the exactly right pressure drop to perform correctly. The expansion valve and the distributor must be considered together because of the pressure reduction through the distributor. For a new installation,

Figure 25.78 The same type of distributor as in Figure 25.76 except that it has a side inlet for allowing hot gas to enter the evaporator evenly during defrost. Courtesy Sporlan Valve Company

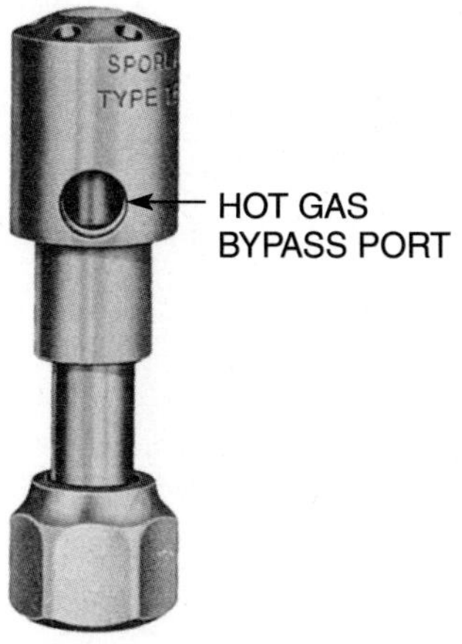

the manufacturer determines the sizing of the distributor. *In the future, as chlorofluorocarbon refrigerants are changed to the new refrigerants, the refrigerant distributor will need to be considered. Without a close relationship with manufacturers, distributors may be mismatched to expansion valves.* Some distributors have a side inlet for the injection of hot gas for hot gas defrost, **figure 25.78.**

HEAT EXCHANGERS

An additional heat exchanger is sometimes placed in or at the suction line leaving the evaporator. The heat exchange takes place between warmer refrigerant in the liquid line and the cooler refrigerant in the suction line. This heat exchanger, which can be as simple as securing the two lines together, **Figure 25.79**, performs two important functions. *First, it helps to increase the capacity of the evaporator by subcooling the liquid refrigerant entering it, which can allow a smaller evaporator to be used in the refrigerated box. Second, it helps prevent liquid refrigerant from moving out of the evaporator to the suction line and into the compressor.*

Figure 25.79 Capillary tube fastened to the suction line; it accomplishes the same thing as the heat exchanger.

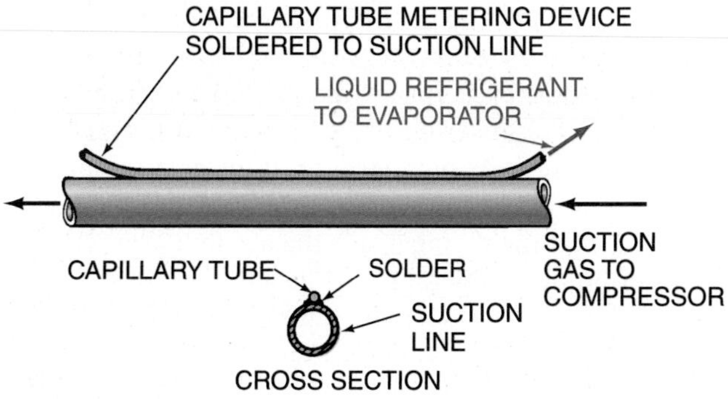

Figure 25.80 Heat exchangers. Plain liquid-to-suction type.

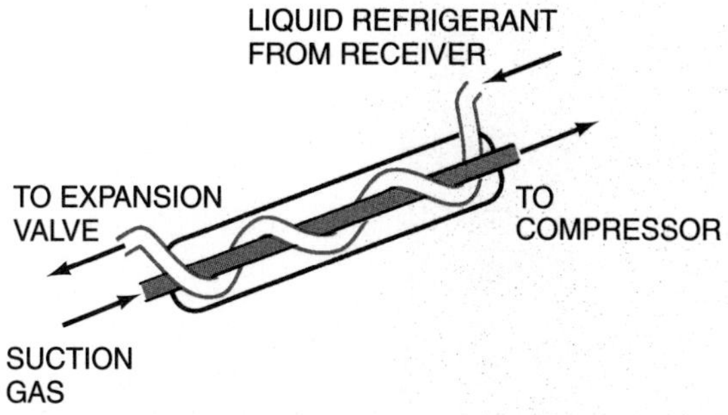

By cooling the liquid refrigerant entering the expansion device, the net refrigeration effect in the evaporator will increase. This is not necessarily a net system efficiency improvement because the heat is given up to the suction gas and is still in the system. The heating of the suction gas expands the gas, giving it a lower density before it enters the cylinders of the compressor. This can reduce the refrigerant flow rate pumped by the compressor. There is the benefit, however, of ensuring that liquid refrigerant does not enter the compressor.

Most heat exchangers are simple and straightforward. The heat exchanger has no electrical circuit or wires. On larger systems, the heat exchanger can be easily recognized by its four piping connections, two from the suction line and two from the liquid line, **Figure 25.80**.

SUCTION-LINE ACCUMULATORS

The suction-line accumulator, **Figure 25.81**, is located in the suction line to prevent liquid refrigerant from passing into the compressor. As discussed earlier, liquid refrigerant floodback causes serious problems for a compressor. Floodback dilutes the compressor's oil with liquid refrigerant and causes foaming in the crankcase, resulting in bearing washout. Under certain conditions, usually during defrost, liquid refrigerant may leave the evaporator. Since the suction line is insulated, the refrigerant in this line will not absorb much heat and will not be able to evaporate much refrigerant. *The suction-line accumulator collects this liquid and gives it a place to vaporize before continuing on to the compressor.*

As mentioned, liquid entering a compressor during the running cycle is called liquid floodback. The following are some causes of floodback:

- Wrong TXV setting (low or no compressor superheat)
- Overcharge
- Evaporator fan not operating
- Low load on evaporator
- End of cycle (lowest load)

Figure 25.81 A suction-line accumulator. Courtesy AC & R Components, Inc.

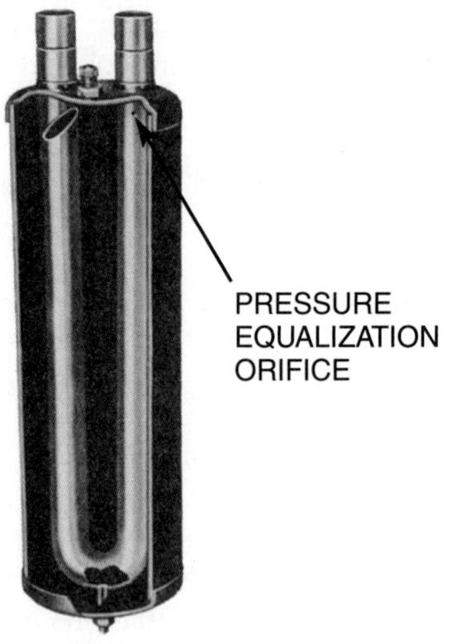

Figure 25.82 A suction-line accumulator with heat exchanger.

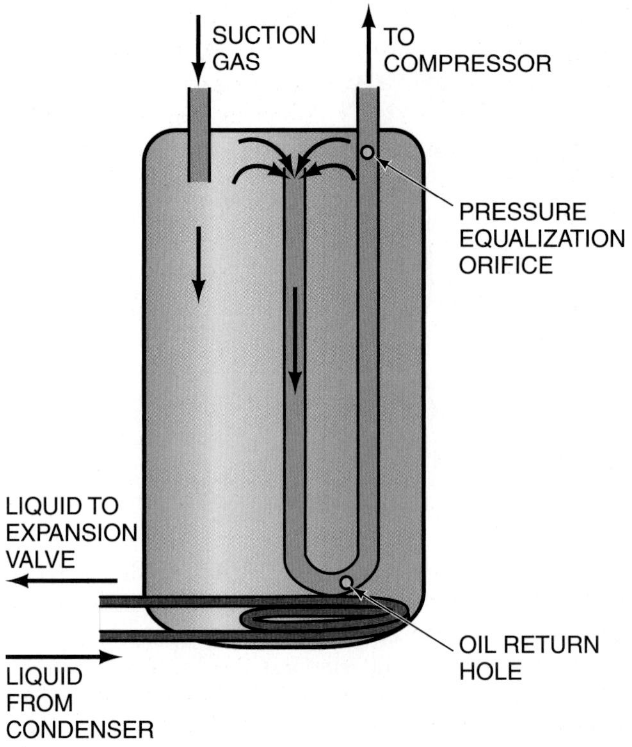

- Defrost clock or heater not operating, causing an iced coil
- Dirty or blocked evaporator coil
- Capillary tube overfeeding
- Capillary tube system overcharged
- Loose TXV bulb on evaporator outlet
- Oversized expansion valve
- Flooding after hot gas termination
- Heat pump changeover
- Defrost termination

When liquid is present in the bottom of the accumulator, most accumulators are designed to meter both the liquid refrigerant and the oil back to the compressor at an acceptable rate that will not damage compressor parts or cause oil foaming in the crankcase. This is done with a small metering orifice at the bottom of the outlet tube while the compressor is running, **Figure 25.82.** However, even though suction-line accumulators assist in refrigerant flooding and migration problems, if flooding is severe the accumulator may also flood and cause compressor damage. The only 100% safe accumulator is one that can hold the entire system charge. Always consult with the manufacturer when sizing a suction accumulator for specific catalog numbers and tonnage ratings.

High compressor amperage draws may occur when liquid refrigerant is returning to the compressor from the bottom of an accumulator. When such is the case, the compressor is now being exposed to dense, saturated or near-saturated gas coming into it. With little or no superheat being present at the compressor inlet, the mass flow rate of the compressor will increase, which can overload a compressor to the point of overheating.

Many indoor compressor installations have problems with suction accumulators sweating and dripping on floors. The only way around this problem is for the accumulator to be insulated completely and vapor sealed to prevent condensate from forming under the insulation. Because they are made of steel, accumulators can rust if exposed to moisture for any long period of time. The rusting usually happens at the seams or at the piping stubs after about 6 to 8 years of use.

Although manufacturers do supply accumulators with rust preventive paints, during the welding processes these paints can be burned off, leaving exposed metal. If an accumulator is resoldered because of pinhole leaks, usually near the seams or piping stubs, the residual soldering flux should be washed away and the area cleaned with emery cloth or steel wool. The surfaces should then be covered with a silicone rubber or roofing tar for a waterproof seal. The suction lines connected to the accumulator should also be insulated to prevent sweat from forming on them and dripping onto the accumulator.

During an off cycle, some liquid refrigerant may move into the accumulator by gravity from the evaporator and suction line, especially when the system does not have an automatic pumpdown cycle for its off cycles. This refrigerant will flow through the metering orifice for liquid refrigerant and oil return and seek its own level both inside and outside the U-tube formation of the suction piping of the accumulator. This means that the piping inside the accumulator will have a column of liquid refrigerant in it during the off cycle simply because of liquid inside the accumulator seeking its own level. When the compressor starts up, this column of liquid refrigerant would be sucked out of the accumulator and into

the compressor if it were not for a pressure equalization orifice at the outlet of the accumulator, **Figures 25.81 and 25.82.** This orifice will equalize pressure on both sides of the liquid column when the compressor is on or off. In other words, the liquid column will have accumulator pressure on both sides of it because of the orifice. This prevents the column of liquid from being sucked out of the U-tube of the accumulator and possibly damaging the compressor during the running cycle. The liquid column will momentarily hang in the U-tube and rapidly vaporize on a compressor start-up. This is a time where saturated gas could enter the compressor. Once the column of liquid in the accumulator is cleared or vaporized, the compressor will again see superheated gas.

Since compressors are extremely susceptible to liquid coming back to their crankcases or valves, accumulators add an extra safety precaution. Suction accumulators can assist in flooding and migration conditions. However, if flooding or migration problems are severe, suction-line accumulators have been known to flood and fail to prevent compressor damage. This is why the only 100% safe accumulator is one that can hold 100% of the entire system's refrigerant charge.

Also, refrigerant vapor can migrate through the accumulator during the off cycle and still get to the compressor and condense. So, the accumulator is not the cure-all for refrigerant migration problems.

Accumulators should be selected with three basic considerations in mind. Consult with the manufacturer for specific catalog numbers and tonnage ratings.

1. The accumulator should have adequate liquid holding capacity, not less than 50% of the entire system charge.
2. The accumulator should not add excessive pressure drop to the system.
3. The selection of an accumulator should not be based on the line size of the suction line. Many times an accumulator will not have the same line size as the suction line.

The accumulator is usually located close to the compressor. It should not be insulated so that any liquid that is in it may have a chance to vaporize. The suction-line accumulator may have a small coil in the bottom where liquid from the liquid line may be routed to boil off the refrigerant in the bottom of the accumulator, **Figure 25.82.** Otherwise, part of the refrigerant charge for the system will stay in the accumulator until the heat from the vapor passing through boils it to a vapor. Actually, not much heat will exchange unless liquid is in the accumulator, so there is not much effect on the system while only vapor is in the accumulator. The liquid-line coil in the accumulator also will help subcool the liquid refrigerant before it reaches the metering device. This will increase the net refrigeration effect in the evaporator. Suction-line accumulators are typically made of steel. They also sweat, which causes rust. Older accumulators should be observed from time to time for refrigerant leaks. Cleaning and painting the solder connection will help to prevent rust.

SUCTION-LINE FILTER DRIERS

The suction-line filter drier, **Figure 25.83,** is similar to the liquid-line filter drier. It is rated for suction-line use. A filter drier is placed in the suction line to protect the compressor—and it is good insurance in any installation. It is essential to install a suction-line filter drier after any failure that contaminates a system. A motor burnout in a compressor shell usually moves acid and contamination into the whole system. When a new compressor is installed, a suction-line filter drier can be installed to clean the refrigerant and oil before they reach the compressor. Driers that have the capability to remove high levels of acid are typically used after motor burnout problems.

Most suction-line filter driers have some means for determining the pressure drop across the cores. This is important because even a small pressure drop can cause the suction pressure to decrease to the point that the compression ratio is increased, which will reduce the efficiency of the compressor. The drier in **Figure 25.83** shows the pressure fittings on a typical suction-line drier. From a sizing standpoint, the suction line is the most critical refrigerant line. Any accessories—such as evaporator pressure regulators, suction filters, suction accumulators, or crankcase pressure regulators—must also be sized correctly when installed in the suction line. The right refrigerant velocity must be maintained in the suction line, whether it is a level line or a suction-line riser, to make sure there is proper oil return. Refrigerant velocity is what sweeps oil through the suction line and back to the compressor. If the suction line is oversized, refrigerant velocity will be low and oil return will be hindered. If the suction line or its accessories are undersized, excessive pressure drop will reduce suction pressure, which will cause higher compression ratios and lower volumetric efficiencies.

Figure 25.83 A suction-line filter drier with access ports for checking pressure drop. Courtesy Sporlan Division, Parker Hannifin Corporation

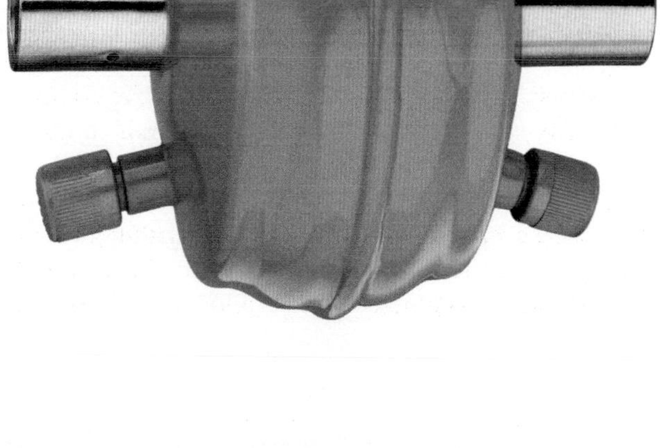

25.9 SYSTEM VALVES AND SERVICE ACCESS DEVICES

REFRIGERANT CHECK VALVES

A check valve, **Figure 25.84**, is a device that will allow fluid to flow in only one direction. It is used to prevent refrigerant from backing up in a line by means of several types of mechanisms. Check valves have arrows imprinted on them to indicate the direction of flow through the valve.

Two common types of check valves are the ball check and magnetic check. When fluid tries to flow in the direction opposite to that as indicated on the valve, the valve will close, preventing flow, **Figure 25.85**. When fluid flows in the same direction as that indicated on the valve, the valve will open and allow flow, **Figure 25.86**. Both of these check valves, even when open, cause some pressure drop, depending on the flow rate. The technician must consult the manufacturer's catalog for flow rate and application.

Figure 25.84 A check valve. Photo by Bill Johnson

SUCTION SERVICE VALVES

Refrigeration equipment usually has service valves. This is not always the case on air-conditioning equipment. The suction service valve is normally attached to the compressor on one side. The refrigerant piping from the evaporator fastens to the other side of the valve. The piping can be either flanged, flared, or soldered where fastened to the service valve. The valve can never be totally closed because of the valve's seat design. The service valve is either back-seated, front-seated, or midseated. A position of "just cracked" off the back seat is often used for reading system pressures and charging. An example of these functions is shown in **Figure 25.87**. The suction service valve is used

1. as a gauge port,
2. to throttle the gas flow to the compressor,

Figure 25.85 Check valves in the closed position.

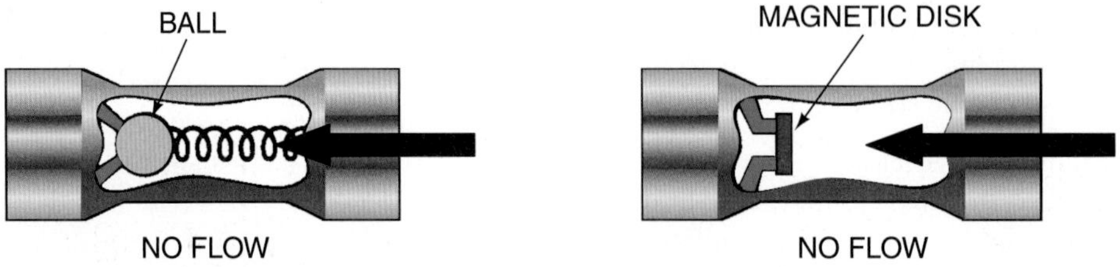

Figure 25.86 Check valves in the open position.

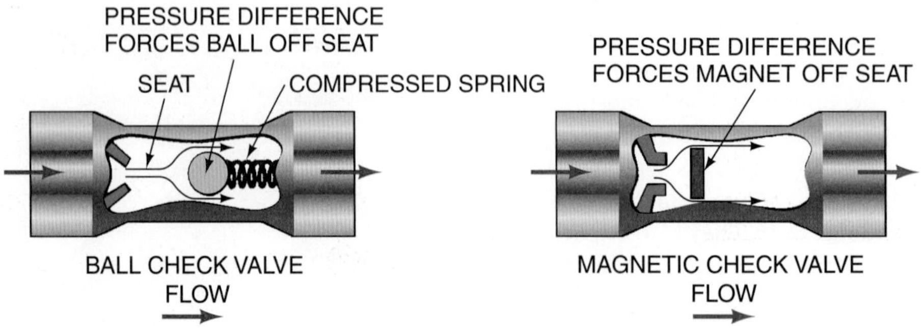

Figure 25.87 Three positions for the suction service valve (A)–(C).

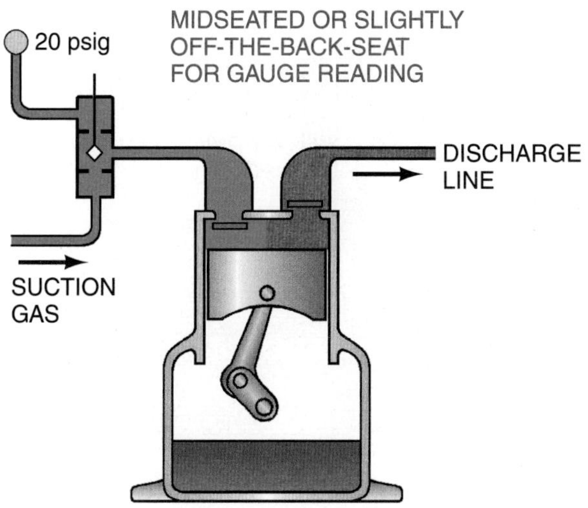

(A)

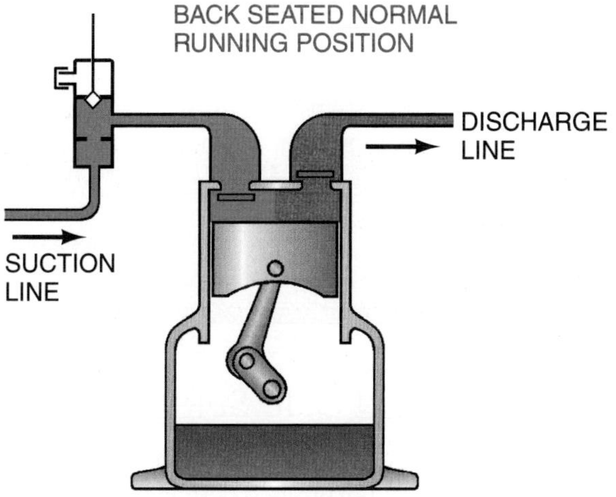

(B)

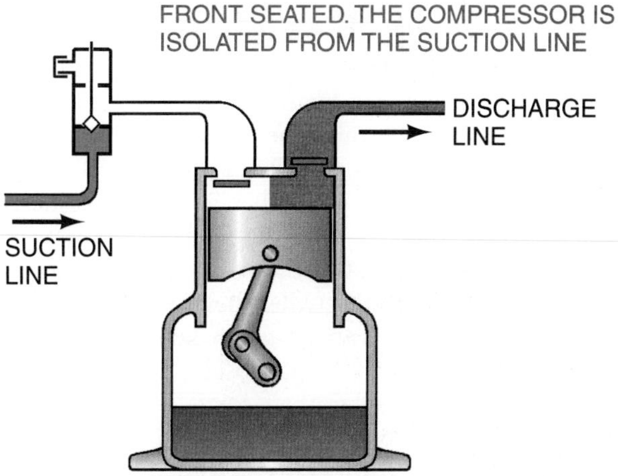

(C)

3. to valve off the compressor from the evaporator for service, and
4. to charge the system with refrigerant.

The service valve consists of the valve cap, valve stem, packing gland, inlet, outlet, and valve body, **Figure 25.88**. The valve cap is used as a backup to the packing gland to prevent refrigerant from leaking out around the stem. It is normally made of steel and should be kept dry on the inside. An oil coating on the inside will help prevent rust. Valve covers should always be in place except during service operations. The valve stem has a square head to accommodate the valve wrench. This stem is normally steel and should be kept rust

Figure 25.88 (A) The suction and discharge valves are made the same except that the suction valve is often larger. Based on Henry Valve Company. (B) Components of a service valve. Based on Henry Valve Company.

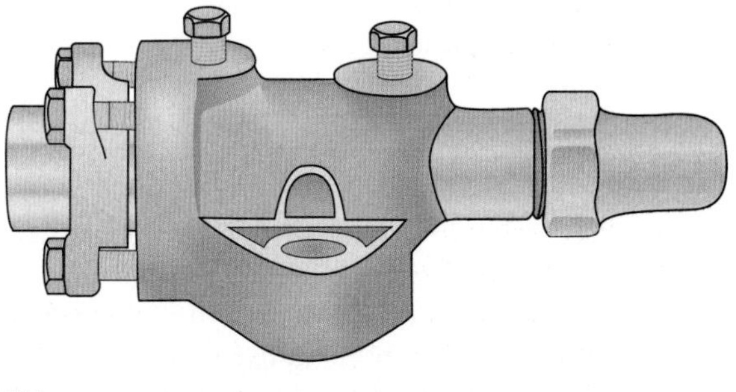

(A)

PCE NO.	DESCRIPTION	QUAN.
1	BODY	1
2	STEM	1
3	SEAT DISC	1
4	RETAINER RING	1
5	DISC SPRING	1
6	GASKET	1
7	CAPSCREW	4
8	PACKING WASHER	1
9	PACKING	1
10	PACKING GLAND	1
11	CAP	1
12	CAP GASKET	1
13	FLANGE	1
14	ADAPTER	1

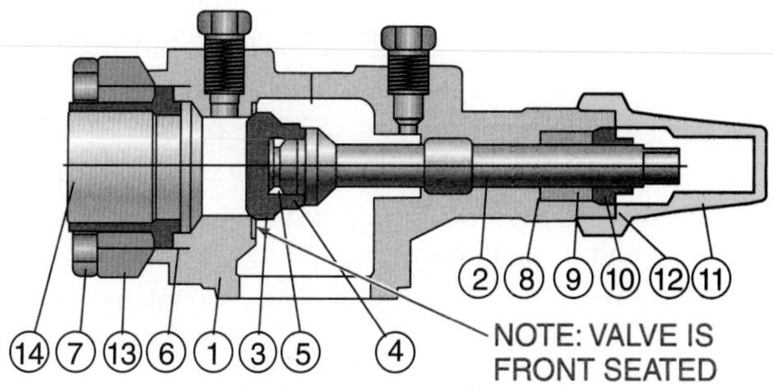

(B)

Figure 25.89 A service valve showing packing material. This can be replaced if it leaks. Based on Henry Valve Company.

PCE NO.	DESCRIPTION	QUAN.
1	BODY	1
2	STEM	1
3	SEAT DISC ASS'Y	1
4	DISC SPRING	1
5	DISC PIN	4
6	RETAINER RING	1
7	PACKING WASHER	1
8	PACKING	2*
9	PACKING GLAND	1
10	CAP	1
11	CAP GASKET	1
12	FLANGE	1
13	ADAPTER	1
14	GASKET	1
15	CAPSCREW	4
16	PIPE PLUG	2

free. If rust does build up on the stem, a light sanding and a coat of oil will help. The valve turns in and out through the valve packing gland. Rust on the stem will destroy the packing in the gland. The packing gland can be either permanent or adjustable. If the gland is not adjustable and a leak starts, only the valve cap will stop. If the gland is adjustable, the packing can be replaced. Normally, packing is graphite rope, **Figure 25.89**. Any time the valve stem is turned, the gland should be first loosened if it is adjustable. This will keep the valve stem from wearing the packing when the stem is turned. Because the suction valve sweats, it is not uncommon for rust to cause pits in the valve stem, but this occurs only when poor service practices have been followed. A pitted valve stem should be carefully sanded smooth, then coated with refrigerant oil before turning. If the stem is not smoothed, the packing will be damaged. The technician should always dry the valve stem and the inside of the protective cap and apply a light coat of oil before placing the cap on the valve stem.

DISCHARGE SERVICE VALVES

The discharge service valve is the same as the suction service valve except that it is located in the discharge line. This valve can be used as a gauge port and to valve off the compressor for service. Compressor service valves are used for many service functions. One of the most important of these functions is the changing out of a compressor. When both service valves are front-seated, the refrigerant may be recovered from between the valves where it is trapped inside the compressor, **Figure 25.90**. When the refrigerant is recovered, the isolated compressor can be removed. When a new one is installed, the only part of the system that must be evacuated is the new compressor. A compressor can be changed in this fashion with no loss of refrigerant. **SAFETY PRECAUTION:** *The compressor cannot be operated with this valve front-seated. An exception is a closed-loop capacity check that takes place under experienced supervision. Extremely high pressures will result if the test equipment is not properly connected to the valve.* •

REFRIGERATION LINE SERVICE VALVES

Refrigeration line service valves are hand-operated specialty valves used for service purposes. These valves can be used in any line that may have to be valved off for any reason. They come in two types: the diaphragm valve and the ball valve.

DIAPHRAGM VALVES. The diaphragm valve has the same internal flow pattern that a "globe" valve does. The fluid has to rise up and over a seat, **Figure 25.91**. There is a measurable pressure drop through this type of valve. It can be tightened by hand enough to hold back high pressures. The valve can be installed in a system by either a flare or soldered connection. When soldered, care should be taken that it is not overheated because most of these valves have seats made of materials that would melt when being soldered into a line. The diaphragm valve can be taken apart and the internal parts removed for soldering purposes.

BALL VALVES The ball valve is a straight-through valve with little pressure drop. This valve can also be soldered into the line, but temperature has to be considered. All manufacturers furnish directions that show how to install this

Figure 25.90 This compressor has been isolated by using the service valves. The refrigerant can be recovered; then the compressor can be replaced.

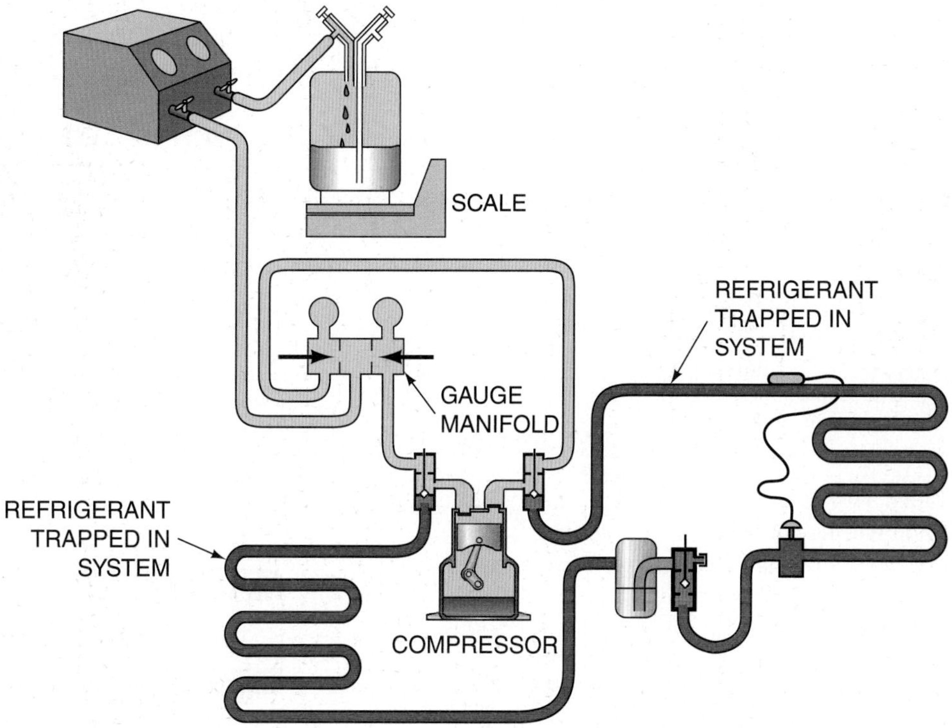

Figure 25.91 The diaphragm hand valve used when servicing a system. Care should be taken that the valve is not overheated when being soldered. Based on Henry Valve Company.

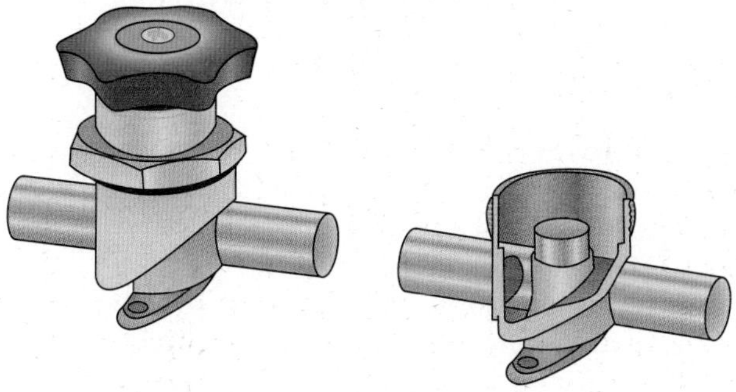

Figure 25.92 The ball valve. This valve is open straight through and creates very little pressure drop or resistance to the refrigerant flow. Photo by Jason Obrzut

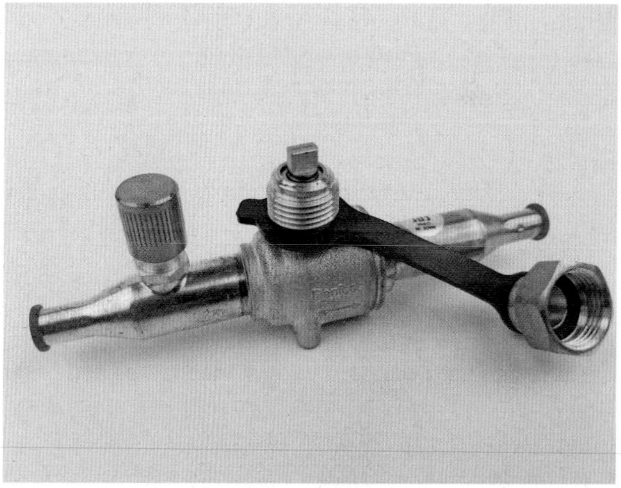

valve, **Figure 25.92.** One of the advantages of the ball valve is that it can be opened or closed easily because it takes only a 90° turn to do either. Thus it is known as a "quick open or close" valve.

PRESSURE ACCESS PORTS

Pressure access ports provide a method of taking pressure readings at places that do not have service ports. Several types can be used effectively. Some can be attached to a line while the unit is operating. This can be helpful when a pressure reading is needed in a hurry and the system needs to keep on running. One of these types has a gasket and is bolted on the line. When the valve is bolted in place, a pointed plunger is forced through the pipe to make a very small hole in the line, just large enough to take pressure readings and transfer small amounts of refrigerant, if needed, **Figure 25.93.** The other type of valve that can be installed while the system is running requires the valve to be soldered onto the line with

Figure 25.93 Pressure tap devices used to obtain pressure readings when there are no gauge ports. Courtesy Ferris State University. Photo by John Tomczyk

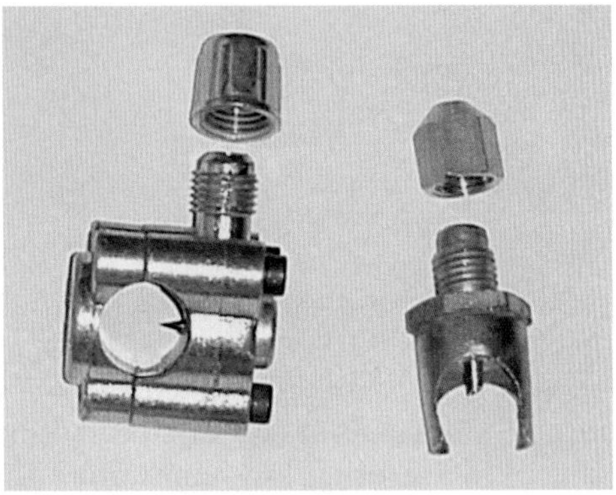

a low-temperature solder. This can only be done on a vapor line because a liquid line will not heat up as long as there is liquid in it. Manufacturers claim that this type of soldering application does no damage to the refrigerant in the line. This valve may be more free of leaks because it is soldered. After the valve is soldered onto the line, a puncture is made in a manner similar to the valve previously described, **Figure 25.94.**

Line-piercing valves should be installed according to the manufacturer's directions or a leak is likely to occur. It is good practice not to locate a line tap valve on the hot gas line if a gasket is involved because, in time, the heat will deteriorate the gasket and a leak will occur. If a high-side pressure reading is required, use the liquid line; it is much cooler and less likely to leak. A line tap valve with a gasket should not be left on the line as a permanent gauge port. For other valves used as ports a hole must be drilled in the refrigerant line. All copper filings should be removed from the line to prevent system damage. A copper tee also can be installed in the line to house the valve. This can be done without drilling. The valve stem is inserted into the line and soldered when there is no pressure in the line, **Figure 25.95.**

Figure 25.94 This valve can be soldered on a vapor line while the system has pressure in it. Low-temperature solder is used. It cannot be soldered to a line with liquid in it because the liquid will not allow the line to get hot enough. Courtesy J/B Industries

Figure 25.95 This valve port must have a hole drilled in the line and can be installed only with the system at atmospheric pressure. Courtesy J/B Industries

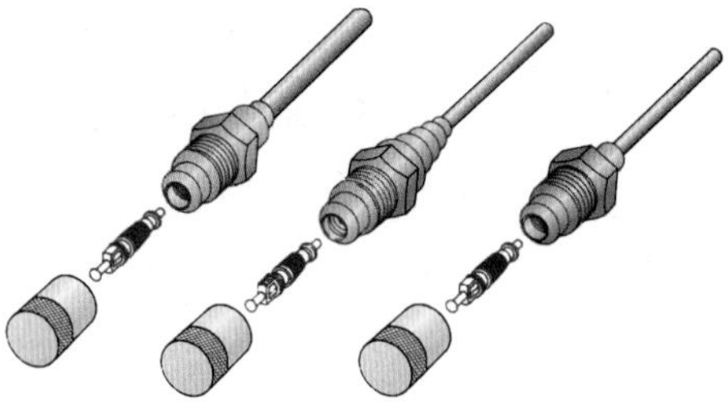

Figure 25.96 The Schrader valve is similar to the valve on a car tire. It has threads to accept the service technician's gauge line threads, (A)–(B). Courtesy J/B Industries

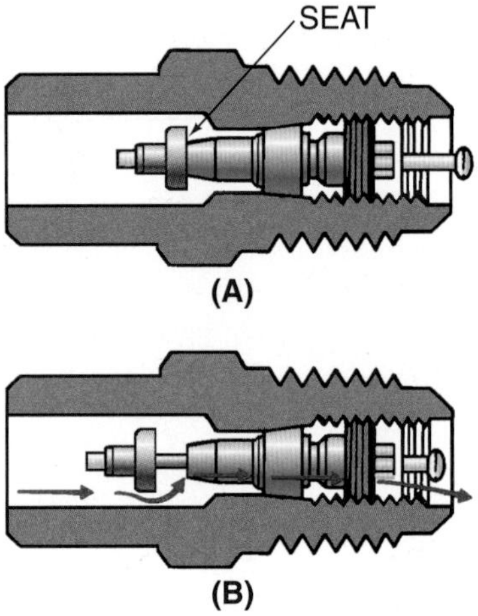

Attaching a gauge hose to these pressure access ports can be done in two ways. Some of the valves have handles that shut off the valve to the atmosphere. The others normally use a Schrader connection, which is like a tire valve on a car or bicycle except that it has threads that accept the 1/4-in. gauge hose connector from a gauge manifold, **Figure 25.96.**

RELIEF VALVES

Relief valves are designed to release refrigerant from a system when predetermined high pressures exist. Refrigerant relief valves come in two different types: the spring-loaded type that will reseat and the one-time type that does not close back. Spring-loaded relief valves are mainly used in today's applications because of the increased cost of refrigerants and ozone depletion scares.

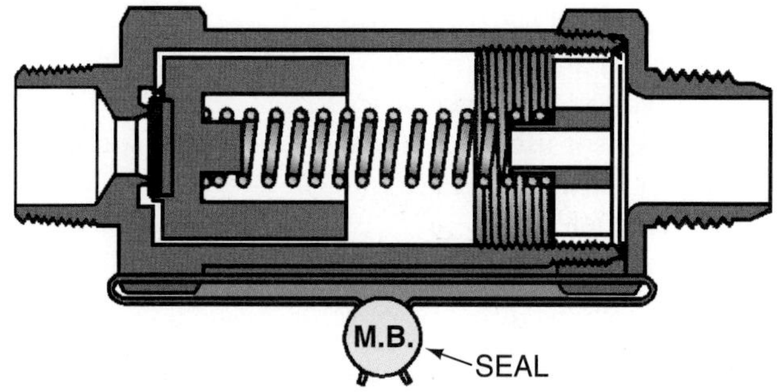

Figure 25.97 The spring-loaded relief valve used to protect the system from very high pressures. Courtesy Superior Valve Company

The spring-loaded relief valve is normally brass with a neoprene seat. This valve is piped so that it is in the vapor space of the condenser or receiver. The relief valve must be in the vapor space, not in the liquid space, for it to relieve pressure. The object is to let vapor off in order to vaporize some of the remaining liquid and lower its pressure, **Figure 25.97**. The top of the valve normally is threaded so that the refrigerant can be piped to the outside of the building. **SAFETY PRECAUTION:** *When relief valves are used to protect vessels in a fire, it is desirable for the refrigerant to be removed from the area. Refrigerant gives off a noxious gas when it is burned. Some systems pipe the relief valve into a large, evacuated tank. The refrigerant will not be vented if the relief valve opens in this case.* •

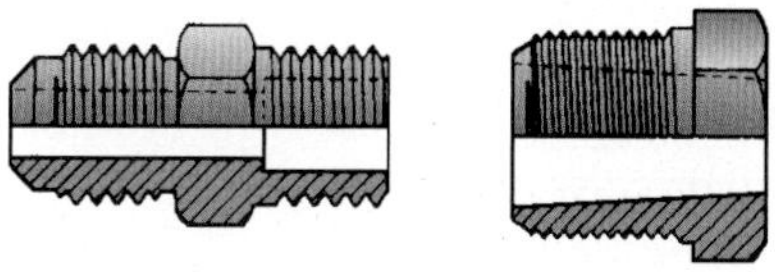

Figure 25.98 The fusible relief valve may be a low-temperature solder patch that covers a hole drilled in a fitting or pipe. Care should be taken when soldering in the vicinity of this relief device, or it will melt. Courtesy Mueller Brass Company

Temperature causes the one-time relief valve to relieve. These valves, often called fusible plugs, are designed in one of the following ways: with a fitting filled with a low melting temperature solder, a patch of copper soldered on a drilled hole in the copper line using low-temperature solder, or a fitting that has been drilled out at the end with a spot of low-temperature solder over the hole, **Figure 25.98**. Sometimes the melting temperature will be very low, about 220°F. When soldering around the compressor, make sure that the solder in the fusible plug is not melted away. This type of device will normally never relieve unless there is a fire. It can often be found on the suction side of the system close to the compressor to protect the system and the public. This will keep the compressor shell from experiencing high pressures and temperatures and rupturing during a fire. Remember, the compressor shell can have a working pressure as low as 150 psig. The plug blows at approximately 220°F.

REVIEW QUESTIONS

1. Which of the following is not a major component of the compression refrigeration cycle?
   A. Compressor
   B. Condenser
   C. Evaporator
   D. Suction-line drier
2. To throttle or regulate the flow of fluid as opposed to just starting and stopping it is called ________.
   A. pulsing C. rationing
   B. modulating D. mixing
3. The purpose of the crankcase pressure-regulating valve is to
   A. maintain constant pressure in the crankcase.
   B. maintain constant pressure in the evaporator.
   C. prevent a compressor from going into a vacuum.
   D. prevent a compressor overload condition.
4. The _______ valve can move in very small increments or steps to either open or close the suction line in response to the discharge air temperature of a refrigerated case.
   A. electronic evaporator pressure-regulating
   B. thermostatic expansion
   C. electronic expansion
   D. crankcase pressure-regulating
5. An automatic pumpdown system may be used to
   A. maintain a constant evaporator pressure during low-load conditions.
   B. raise system head pressure during low-ambient conditions.
   C. prevent surges of liquid refrigerant from entering the compressor (slugging) during start-up.
   D. prevent liquid refrigerant from becoming trapped in the receiver.
6. On a "one-time" or non-short-cycling pumpdown system
   A. a leaking solenoid valve or compressor discharge valve will cause the system to turn back on.
   B. only the thermostat calling again for cooling will take the refrigeration system out of pumpdown.
   C. the refrigerant is stored in a liquid receiver and must be manually released back into the system.
   D. once the system has been pumped down, it will lock out, requiring service.
7. Two types of relief valves are ________.
   A. check and spring-loaded
   B. manual and ball
   C. spring-loaded and one-time
   D. one-time and multiuse

8. To prevent the head pressure from dropping to the point that the expansion valve will not feed correctly
   A. a pumpdown system is used.
   B. an evaporator pressure regulator is used.
   C. low-ambient controls are used.
   D. a high-pressure control is used.
9. Which of the following is not a method of low-ambient control on air-cooled equipment?
   A. Modulating TXV
   B. Fan-cycling controls
   C. Condenser flooding
   D. Fan speed control
10. Which of the following is used as a protection against operation with a low charge or as a temperature control for some refrigeration evaporators?
    A. Oil safety switch
    B. Thermostat
    C. High-pressure switch
    D. Low-pressure switch
11. Which of the following is used to protect the compressor and high-side system components from operating with excessive pressure?
    A. Oil safety switch
    B. Low-ambient controls
    C. High-pressure switch
    D. Low-pressure switch
12. Which of the following is a pressure differential control which protects the compressor from operating without sufficient lubrication?
    A. Oil safety switch
    B. Compressor thermistor
    C. High-pressure switch
    D. Low-pressure switch
13. A ________ allows the compressor to cycle on internal overload without affecting the operation of the oil safety control.
    A. oil pressure transducer
    B. current sensing relay
    C. low-pressure switch
    D. lockout relay
14. A type of defrost used by medium-temperature systems that uses the evaporator fans to circulate air in cooler over the coil when the compressor is deenergized is called
    A. hot gas defrost.
    B. electric heat defrost.
    C. radiant defrost.
    D. off-cycle defrost.
15. Two methods for accomplishing defrost on low-temperature refrigeration systems are
    A. radiant and electric defrost.
    B. hot gas and electric defrost.
    C. planned and random defrost.
    D. hot gas and off-cycle defrost.
16. Two types of off-cycle defrost used on medium-temperature refrigeration systems are
    A. radiant and electric defrost.
    B. hot gas and electric defrost.
    C. planned and random defrost.
    D. hot gas and off-cycle defrost.
17. A component used to remove moisture and foreign matter from refrigerant is a(n)
    A. filter drier.
    B. receiver.
    C. sight glass.
    D. accumulator.
18. The component, located on the suction line, that is used to prevent liquid refrigerant from entering the compressor is called a(n)
    A. filter drier.
    B. receiver.
    C. refrigerant distributor.
    D. accumulator.
19. The component, located on the liquid line, that is used to store liquid refrigerant as it leaves the condenser is called a(n)
    A. filter drier.
    B. receiver.
    C. refrigerant distributor.
    D. accumulator.
20. The function of the evaporator pressure regulator is to maintain a
    A. constant evaporator pressure.
    B. minimum pressure in the evaporator.
    C. constant evaporator temperature.
    D. maximum pressure in the evaporator.
21. The component used to prevent compressor oil from having to circulate through the entire system before returning to the compressor crankcase is called a(n)
    A. oil safety control.
    B. accumulator.
    C. crankcase pressure regulator.
    D. oil separator.
22. A(n) ________ is often placed in or on the suction line leaving the evaporator and improves the evaporator capacity by increasing the liquid subcooling.
    A. heat exchanger
    B. evaporator pressure regulator
    C. refrigerant distributor
    D. oil separator
23. The component located on the outlet of the receiver that can be used to trap the refrigerant into the receiver in order to perform system service is called a(n)
    A. Schrader valve.
    B. refrigerant distributor.
    C. king valve.
    D. electronic evaporator pressure regulator.

24. Condenser flooding maintains head pressure by
    A. filling the receiver with liquid refrigerant, raising head pressure.
    B. splitting the condenser into two equally sized condensers, referred to as summer and summer/winter.
    C. storing liquid refrigerant in the evaporator during the off-cycle.
    D. backing up liquid refrigerant into the condenser, acting like an overcharge.
25. A component that has a magnetic coil, can be normally open or normally closed, and is used to control fluid flow is a
    A. solenoid valve.
    B. current sensing relay.
    C. refrigerant check valve.
    D. lockout relay.
26. Describe the function of a defrost termination/fan delay (DTFD) switch.
27. What is the function of the pressure equalization orifice at the outlet of an accumulator?
28. Explain the function of the various refrigeration system service valves.
29. Why is there a time delay in an oil safety control?
30. Describe the function of an oil separator.

UNIT 26

APPLICATIONS OF REFRIGERATION SYSTEMS

OBJECTIVES

After studying this unit, you should be able to

- describe three common types of refrigeration display equipment.
- describe a parallel refrigeration system.
- describe a secondary-fluid refrigeration system.
- discuss the two common types of carbon dioxide refrigeration systems.
- describe a pressurized liquid system.
- describe a distributed refrigeration system.
- discuss the operation of multiple-evaporator systems.
- describe package versus remote-condensing applications.
- discuss mullion heaters and their applications.
- explain how hot gas can be used to defrost a multi-evaporator refrigeration system.
- discuss the applications for reach-in and walk-in refrigeration systems.
- explain basic refrigerated air-dry unit operation.

SAFETY CHECKLIST

- ✔ Wear goggles and gloves when attaching or removing gauges and when using gauges to transfer refrigerant or check pressures.
- ✔ Be careful not to get your hands or clothing caught in moving parts such as pulleys, belts, or fan blades.
- ✔ Observe all electrical safety precautions. Be careful at all times and use common sense.
- ✔ Follow manufacturers' instructions when using any cleaning or other types of chemicals. Many of these chemicals are hazardous, so make sure that they are kept away from ice, drinking water, and other edible products.

26.1 APPLICATION DECISIONS

When refrigeration equipment is needed, a decision must be made as to the specific equipment to install. The factors that enter into the decision process are, first, purchase cost of the equipment, the conditions to be maintained, the operating cost of the equipment, the location of the equipment, and the long-term intent of the installation. When purchase cost alone is considered, the equipment chosen may not perform acceptably. For example, if the wrong equipment is chosen for storing meat, the humidity may be too low in the cooler and dehydration may occur—too much moisture may be taken out of the meat, causing a weight loss. This weight loss is actually part of the condensate that goes down the drain during defrost.

This unit provides information regarding some of the different options to consider when equipment is purchased or installed. The service technician should know these options in order to help make service decisions.

26.2 REACH-IN REFRIGERATION

Retail stores use reach-in refrigeration units for merchandising their products. Customers can go from one section of the store to another and choose items to purchase. These reach-in display cases are available in high-, medium-, and low-temperature ranges. Each has open-display and closed-display cases from which to choose. The open and closed styles may be of the chest type, upright type with display shelves, or upright type with doors. The boxes can be placed end-to-end for a continuous display. Thus, the frozen food is kept together and fresh foods (medium-temperature applications) are grouped, **Figures 26.1(A), 26.1(B), and 26.1(C).**

Figure 26.1 (A) Display cases are used to safely store food displayed for sale. Some display cases are open to allow the customer to reach in without opening a door; some have doors. Closed cases are the most energy efficient. Courtesy Hill PHOENIX Corporation.

(A)

Figure 26.1 (*Continued*) (B) Medium-temperature open display cases displaying an array of fresh produce. Courtesy Meijer Corporation, Big Rapids, MI. Photo by John Tomczyk. (C) A medium-temperature, glass-door, closed deli case. Courtesy Meijer Corporation, Big Rapids, MI. Photo by John Tomczyk.

(B)

(C)

All of these combinations of reach-in displays add to the customer's convenience and enhance the sales appeal of the products. The two broad categories of open and closed displays are purely for merchandising the product. A closed case is more efficient than an open case, but the open case is more appealing to the customer because the food is more visible and easier to reach. Open-display cases can maintain temperature conditions because the refrigerated air is heavy and settles to the bottom. Even upright display cases can store low-temperature products. This is accomplished by designing and maintaining air patterns to keep the refrigerated air from leaving the case, **Figure 26.2**. When reach-in refrigerated cases are used as storage, such as in restaurants, they do not display products, although they have the same components, and doors are used to keep the room air out. Restaurant equipment is usually located in a hot, humid kitchen, whereas supermarket equipment is located in a temperature- and humidity-controlled atmosphere.

Figure 26.2 Open display cases maintain even, low-temperature refrigeration because cold air can be controlled. It falls naturally. Air patterns are formed to create air curtains for shelves that are up high. These air curtains must not be disturbed by the air-conditioning system's air discharge. All open cases have explicit directions as to how to load them, with load lines conspicuously marked. Courtesy Tyler Refrigeration Corporation.

SELF-CONTAINED REACH-IN FIXTURES

Reach-in refrigeration fixtures can be either self-contained, with the condensing unit built inside the box, or they may have a remote condensing unit. Deciding which type of fixture is best to purchase depends on several factors.

Self-contained equipment rejects heat at the case. The compressor and condenser are located at the fixture, and the condenser rejects its heat back into the store, **Figure 26.3**. This can be acceptable in the winter however, in summer it is desirable in warmer climates to reject the heat to the outside. Self-contained equipment can also be moved around to new locations without much difficulty. It normally plugs into either a 120-V or 230-V electrical outlet. Only the outlet may have to be moved. **Figure 26.4** shows examples of wiring diagrams for medium- and low-temperature fixtures. When service problems arise, only one fixture is affected, and the food can be moved to another cooler. However, self-contained equipment is located in the conditioned space, and the condenser is subject to any

Figure 26.3 This self-contained refrigerated box has its condensing unit built in. Refrigeration tubing does not have to be run. This box rejects the heat it absorbs back into the store. It is moveable because it needs only an electrical outlet and, sometimes, a drain, depending on the box. Courtesy Hill Refrigeration.

airborne particles in that space. Keeping several condensers clean at a large installation may be difficult. In a restaurant kitchen, for example, large amounts of grease will deposit on the condenser fins and coil, which will then collect dust. The combination of grease and dust is not easy to clean in the kitchen area without contaminating other areas. The versatility of self-contained equipment is good for a kitchen, but remote condensing units may be easier to maintain. The warm condensers are also a natural place for pests to locate.

All refrigeration equipment must have some means for disposing of the condensate that is gathered on the evaporator. The condensate must be either piped to a drain or evaporated. Self-contained equipment can use the heat that the compressor gives off to evaporate the condensate, provided the condensate can be drained to the area of the condenser.

Figure 26.4 Wiring diagrams for self-contained reach-in boxes. (A) Low-temperature with hot gas defrost. (B) Low-temperature with electric defrost.

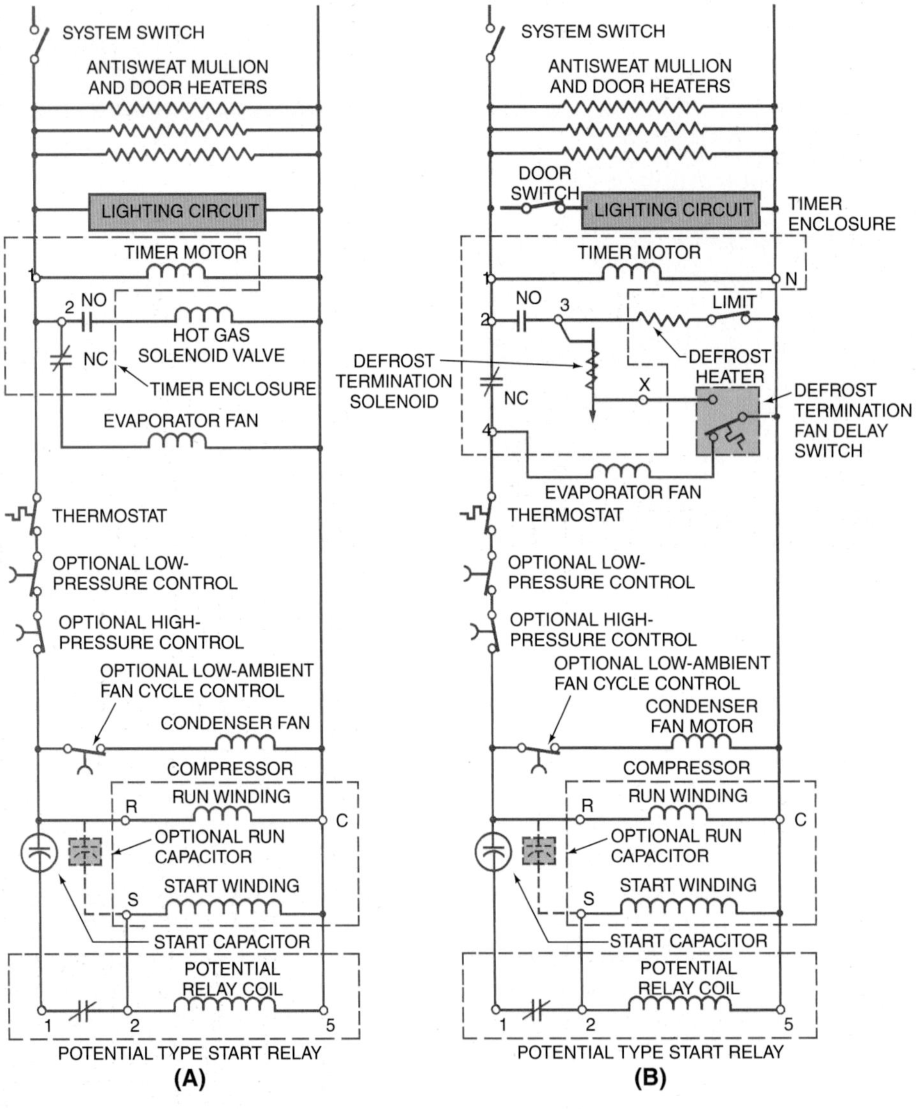

Figure 26.4 (*Continued*) (C) Medium-temperature with planned off-cycle defrost. (D) Medium- or high-temperature with random defrost.

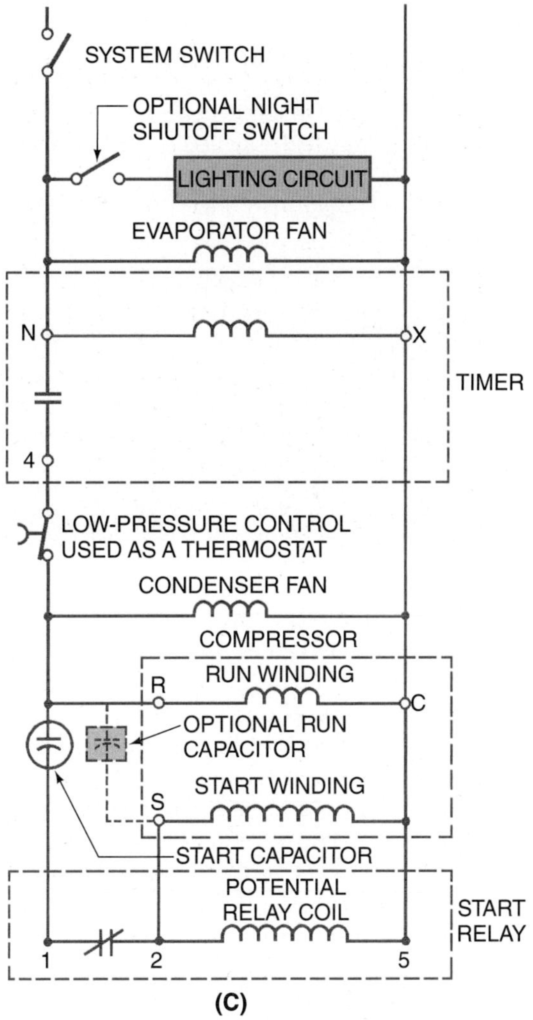

This also helps to desuperheat the hot gas leaving the compressor, which will lower discharge line temperatures. To do this, the condenser must be lower than the evaporator. This is not easy to arrange when the condenser is on top of the unit, **Figure 26.5**.

INDIVIDUAL CONDENSING UNITS

When individual condensing units are used and trouble is encountered with the compressor, or when a refrigerant leak occurs, only one system is affected. The condensing unit can be located outside, with proper weather protection, or in a common equipment room, where all equipment can be observed at a glance for routine service. The temperature of the air in the room can be controlled with dampers to maintain head pressure control. During the colder months, heat rejected by the condensers in the equipment room can, in some instances, be used to offset the cost of heating the room.

Figure 26.5 The evaporator can be high enough on a self-contained cooler so that the condensate will drain into a pan and be evaporated by heat from the hot gas line. No drain is needed in this application.

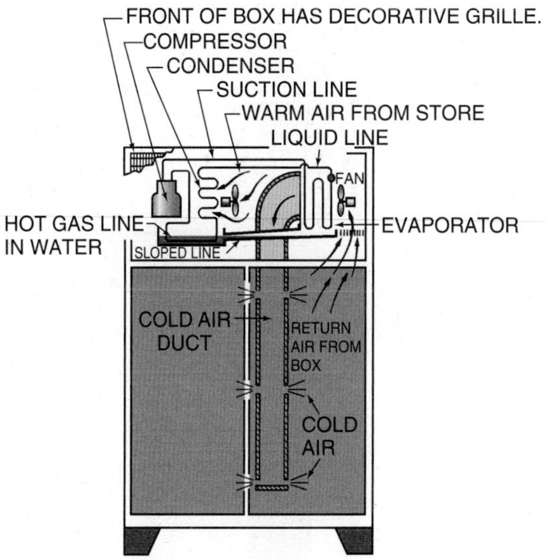

26.3 SINGLE-COMPRESSOR APPLICATIONS AND MULTIPLE EVAPORATORS

Using one compressor for several fixtures has its advantages:

1. The compressor motors are more efficient because they are larger.
2. More heat from the equipment can be captured for use in heating the space or for hot water, because larger compressors give off more heat.

The three-way heat reclaim valves in most heat reclaim systems are activated by a pilot-operated solenoid, **Figure 26.6**. Pilot operation means that the valve uses the refrigeration system's pressures to move a sliding mechanism when the solenoid is energized. Because the heat reclaim valve experiences very hot compressor discharge gases, it must be mounted properly and securely to counteract any expansion and contraction forces from heating and cooling effects, **Figure 26.7**. Its operation is simple. The first stage of a two-stage thermostat in the store calls for heat. *The three-way heat reclaim valve is energized and enables hot gas from the discharge header to flow to an auxiliary condenser or heat reclaim coil located in the store's ductwork,* ***Figure 26.8.*** If this reclaimed heat is not enough to satisfy the first call for heat, the second stage of the thermostat will call. This will initiate the primary heating equipment, which could be a fossil-fuel furnace, a heat pump, or electric heat. The heat reclaim coil, cooling coil, air filters, and primary heating source are located in the ductwork, usually in series with one another, **Figure 26.9**. Heat reclaim coils can be used in parallel or in series with the normal condensers.

Figure 26.6 This three-way heat reclaim valve is pilot operated and solenoid activated. Courtesy Sporlan Division, Parker Hannifin Corporation.

Figure 26.7 A heat reclaim valve shown mounted securely to handle any expansion and contraction forces. Source: Meijer Corporation. Photo by John Tomczyk.

Figure 26.10 shows a heat reclaim system both in parallel and in series. In a parallel heat reclaim system, the reclaim condenser must be sized to handle 100% of the refrigeration system's rejected heat when in the heat reclaim mode. The normal condenser is inactive during the heat reclaim mode and will not hold any refrigerant. It is pumped out through a normally closed solenoid valve that is energized during the heat reclaim mode. The refrigerant that is pumped out of the normal or inactive condenser also passes through a restrictor before entering the suction line. This restrictor vaporizes any liquid refrigerant coming out of the inactive condenser before it enters the suction line and also controls the pumpout rate of the inactive condenser. By pumping out the inactive condenser in the heat reclaim mode, a proper system refrigerant charge is maintained and refrigerant is also prevented from sitting idle and condensing in the inactive condenser. If the liquid were left in the inactive coil to condense, once the system switched out of the heat reclaim mode the mixing of hot gas with subcooled or saturated liquid could cause liquid hammer, which can create tremendous forces, causing refrigerant lines and valves to rupture. Refrigerant must also be pumped out of the heat reclaim coil when the system is not calling for heat reclaim. Some heat reclaim valves have bleed ports built into the valve for the sole purpose of removing refrigerant from the heat reclaim coil when it is not being used, **Figure 26.11**. A check valve is usually installed in the heat reclaim pumpout or bleed line to prevent migration of refrigerant to the coldest location when the heat reclaim coil is exposed to temperatures that are lower than the saturated suction temperatures of the refrigeration system itself, **Figure 26.10**.

Figure 26.8 (A) The heat from a common discharge line can be captured and reused. This heat can be discharged from a condenser on the roof or from a coil mounted in the ductwork. (B) This is a hot water heat reclaim device. It uses the heat from the hot gas line to heat water.

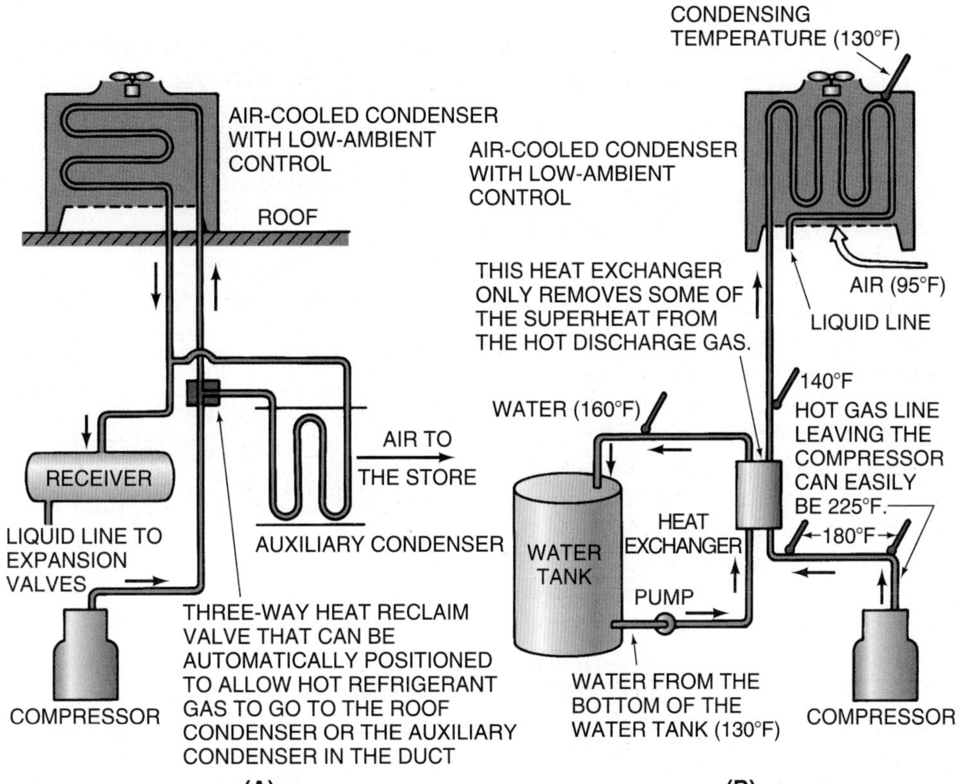

Figure 26.9 A supermarket HVACR system showing filter, cooling coil, heat reclaim coil, fan, and primary heating source in series with one another within the ductwork.

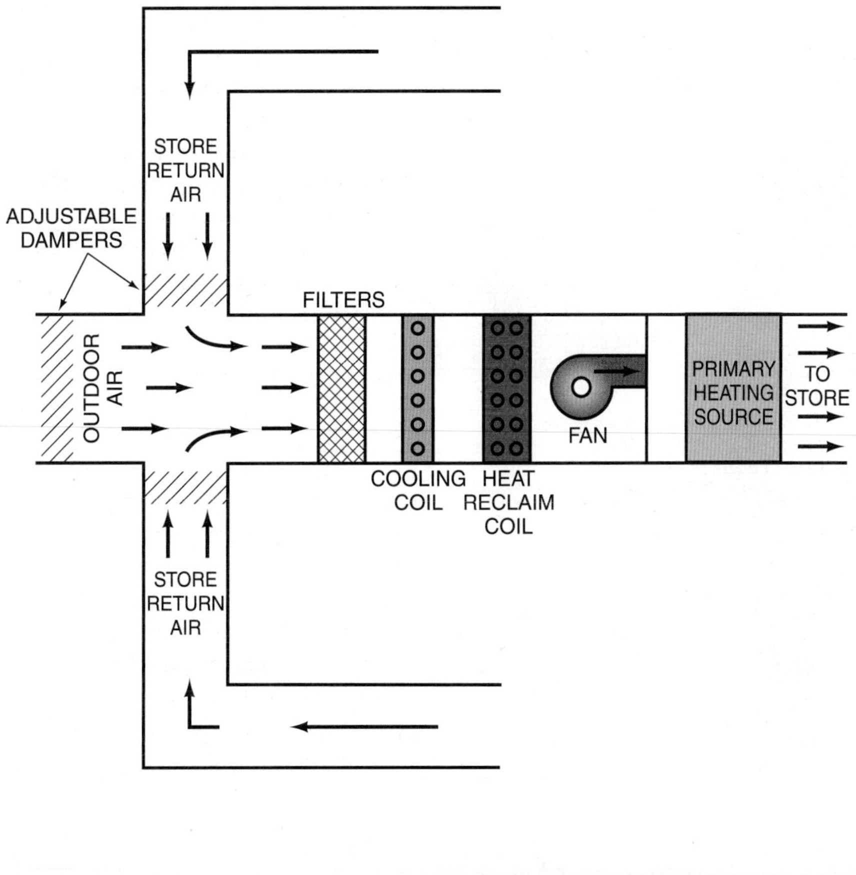

Figure 26.10 (A) Series and (B) parallel heat reclaim systems. (A) and (B) Courtesy Sporlan Valve Company.

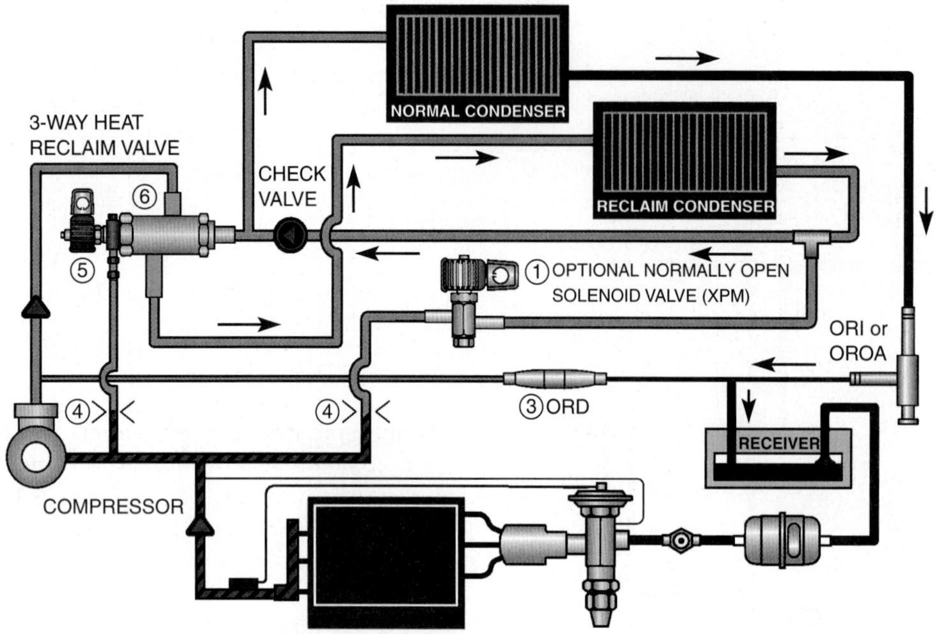

(A)

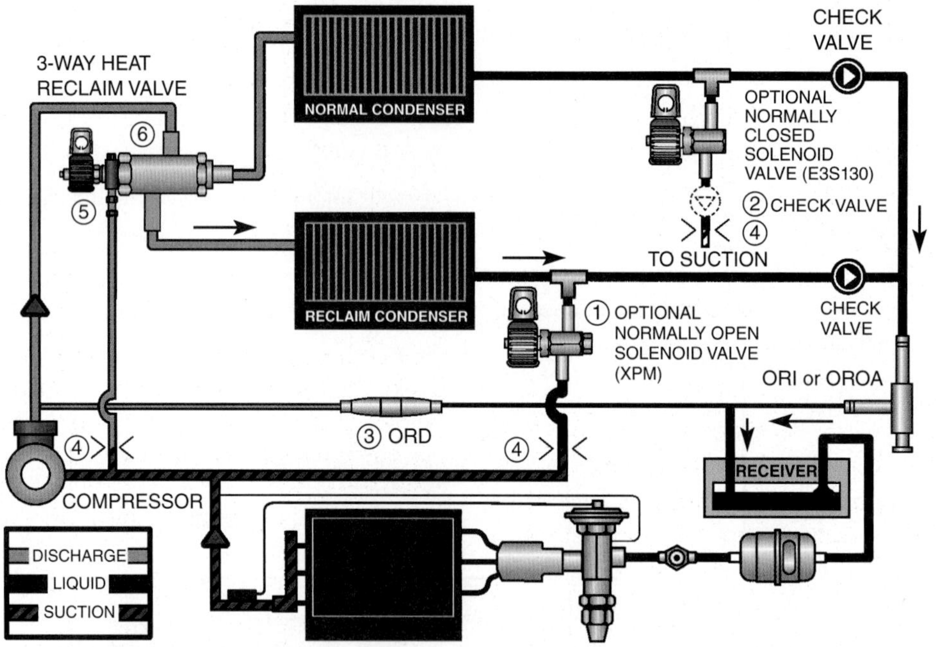

① Use optional solenoid valve and piping if pump-out is required and "C" model Heat Reclaim Valve is used; see Note 4. It is optional to omit this solenoid valve and piping on systems using "B" model Heat Reclaim Valve.

② This check valve is required if lowest operating ambient temperature is lower than evaporator temperature.

③ Not used with OROA (B, C, or D).

④ Restrictor, Part #2449-004, may be required to control pump-out rate on inactive condenser.

⑤ Pilot suction line must be open to common suction whether or not Heat Reclaim Coil is installed at time of installation and regardless of Heat Reclaim Valve model/type.

⑥ Proper support of heat reclaim valve is essential. Concentrated stresses resulting from thermal expansion or compressor vibrations can cause fatigue failure of tubing, elbows, and valve fittings. Fatigue failures can also result from vapor-propelled liquid slugging and condensation-induced shock. The use of piping brackets close to each of the 3-way valve fittings is recommended.

(B)

Figure 26.11 A three-way heat reclaim valve showing the bleed port for pumping down the heat reclaim coil when it is not in use. Courtesy Sporlan Valve Company.

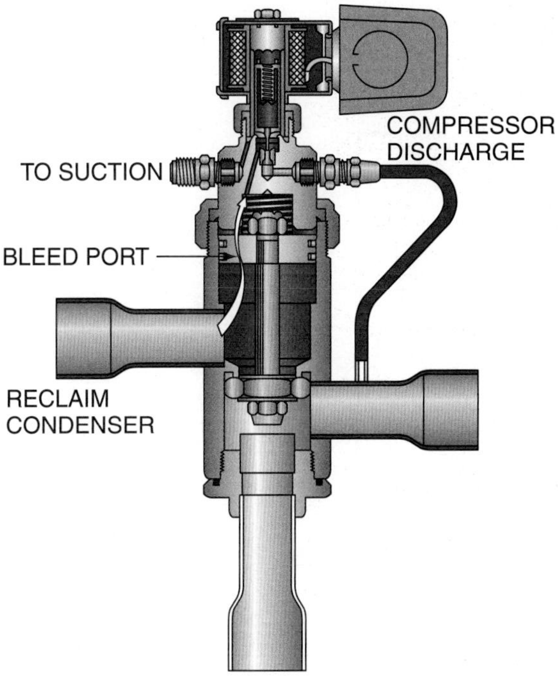

In a series heat reclaim system, the normal condenser and the heat reclaim condenser are used simultaneously when in the heat reclaim mode. The hot, superheated gas from the compressors enters the heat reclaim condenser first. However, in series with the heat reclaim condenser is the normal condenser, and it also may receive some hot gas. Most of the condensing of the liquid takes place in the normal condenser downstream from the heat reclaim coil. Because of this series arrangement, the heat reclaim coil may not be sized to handle 100% of the heat rejected from the refrigeration system.

In any heat reclaim system, the selection of a parallel or series system and the design of the piping arrangement will depend on the control scheme and the size of the heat reclaim coil being used for the particular application. These installations are designed in two basic ways. One way is to use one compressor for several refrigeration cases. A 30-ton compressor may serve 10 or more cases. This particular application may have cycling problems at times when the load varies unless the compressor has a cylinder unloading design, a method to vary the capacity of a multicylinder compressor by stopping various cylinders from pumping at predetermined pressures. The following are some disadvantages of using one compressor for several cases:

1. The refrigeration load cannot be closely matched even when compressor cylinder unloaders are used.
2. Starting and stopping larger compressors will draw a larger locked-rotor amperage (LRA) and is hard on starting components and compressor windings.
3. Short cycling larger compressors may increase both electrical demand and consumption charges.
4. Even the slightest increase in refrigeration load may call upon a large compressor to cycle on where a smaller compressor would suffice.
5. If the compressor malfunctions, many boxes will be inoperable, as opposed to only one box not working if its individual compressor malfunctions.

Because of the disadvantages of single-compressor systems, parallel compressor systems have become popular.

26.4 PARALLEL COMPRESSOR SYSTEMS

Parallel compressor systems apply two or more compressors to a common suction header, a common discharge header, and a common receiver, **Figures 26.12(A)–(C)**. These systems are often referred to as rack systems because they are mounted on a steel rack, **Figure 26.13**. The compressors can be of the reciprocating, scroll, screw, or rotary type.

Advantages of parallel compressor systems are

- load matching,
- diversification,
- flexibility,
- higher efficiencies,
- lower operating costs, and
- less compressor cycling.

Some disadvantages of parallel compressor systems are

- leaks will affect the entire compressor rack,
- compressor burnouts that contaminate compressor oil and will affect the entire compressor rack oiling system.

The primary advantage of parallel compressor systems is their ability to match the refrigeration load nearly exactly, often referred to as ***load matching.*** As the refrigeration load varies, compressors of varied capacities are turned on or off in response to the ever-changing load. The refrigeration load is sensed by a pressure transducer mounted on the common suction line or manifold, **Figure 26.14(A)**. The pressure transducer is wired to a microprocessor and an input/output control board on the compressor rack. The common suction pressure on the parallel system varies as case thermostats in the store call for cooling or as *evaporator pressure-regulating (EPR)* valves on the individual suction lines, which come into the parallel rack, throttle open and closed, according to the load of each individual line of refrigerated cases, **Figure 26.14(B)**. The pressure transducer senses this pressure increase or decrease and sends a voltage message to a computer or microprocessor-based control board located on the parallel compressor rack, **Figures 26.15(A)** and **(B)**. The pressure transducer changes the pressure signal to a voltage signal so that the microprocessor can process it. The pressure transducer receives its power from the same input/output board that powers the microprocessor, **Figure 26.15(B)**.

Figure 26.12 (A) A parallel refrigeration system. Source: AC & R Components, Inc. (B) A supermarket application of a parallel compressor refrigeration system showing the oil management system, heat reclaim system, head pressure control system, liquid subcooling system using an electronic expansion valve (EEV), various suction line controls, and hot gas defrost system. Courtesy Sporlan Division, Parker Hannifin Corporation

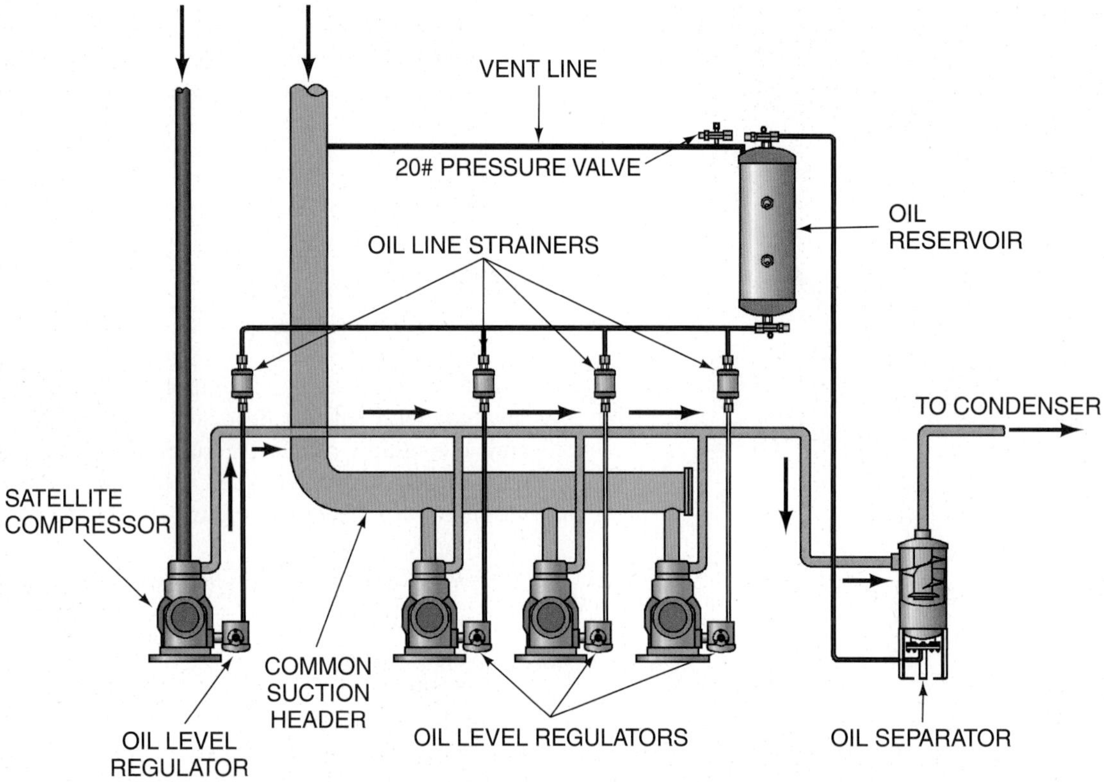

(A)

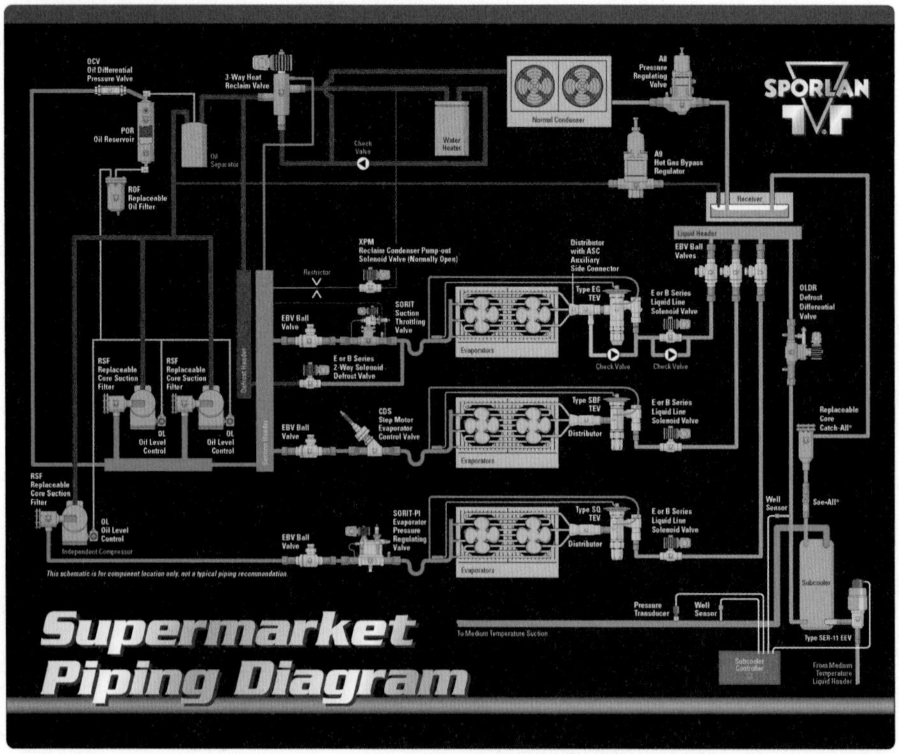

(B)

Figure 26.12 (*Continued*) (C) A supermarket application of a parallel compressor refrigeration system showing an auxiliary refrigeration system for mechanical subcooling of liquid and a diverting valve and receiver bypass, which converts the flow-through receiver to a surge-type receiver during low-ambient conditions. Courtesy Sporlan Division, Parker Hannifin Corporation.

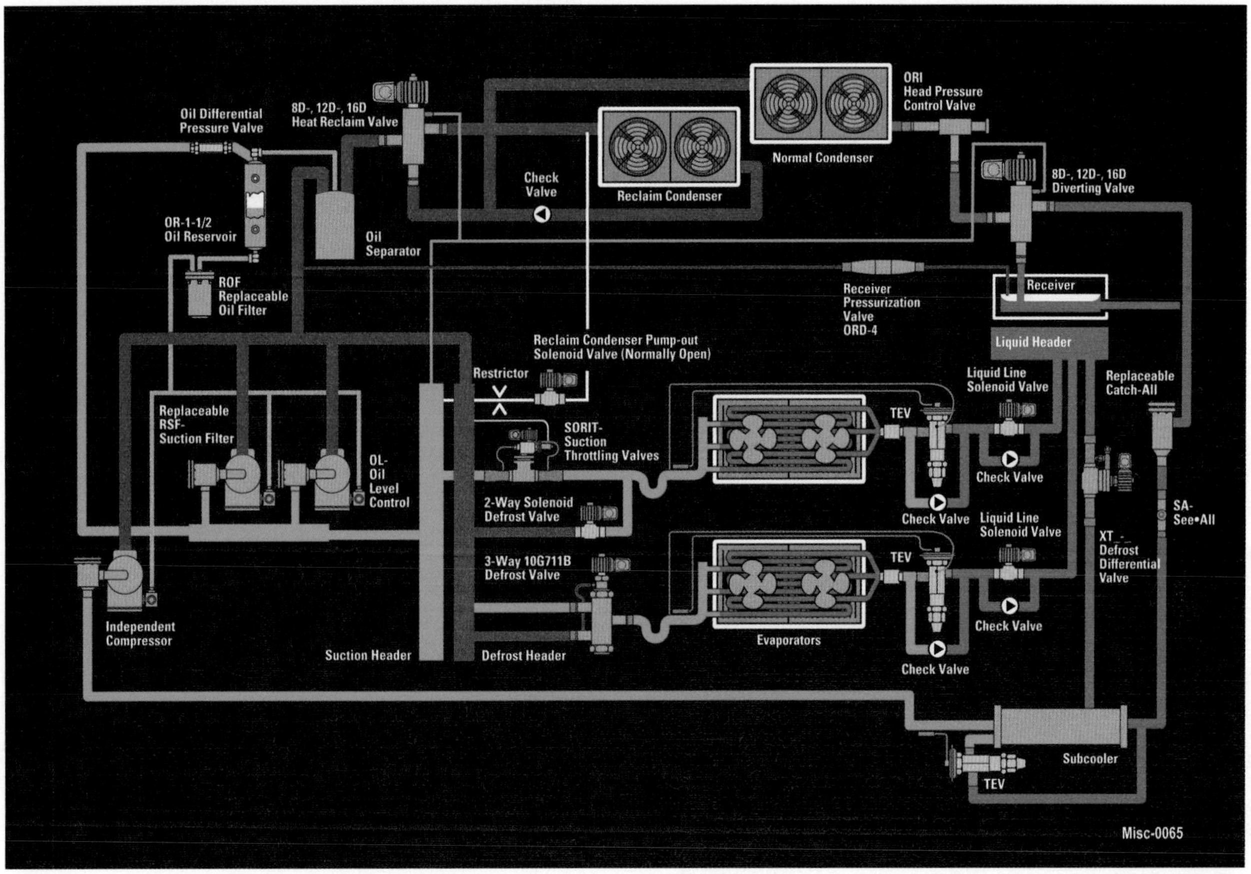

(C)

Figure 26.13 A parallel compressor system with load-matching capabilities. The microprocessor-based electronic controller is shown as part of the package. Courtesy Meijer Corporation, Big Rapids, MI. Photo by John Tomczyk.

Refer to Section 14.7, "Pressure Transducers," in Unit 14, "Automatic Control Components and Applications," for more detailed information on pressure transducers.

As the refrigeration load changes, so does the common suction pressure in the suction manifold, and, thus, the voltage coming from the pressure transducer to the microprocessor. Capacity steps that have been programmed into the microprocessor now control the cycling of the parallel compressor rack. Capacity steps can be changed with programming plugs or computer software and can be monitored on-site or remotely by a personal computer. For example, if the parallel compressor rack had four compressors of unequal sizes, increasing from compressor #1 through compressor #4, 10 capacity steps would be programmed into the microprocessor, **Figure 26.16**. As the pressure transducer on the common suction line sensed the common suction pressure increasing from the increased refrigeration loads, it would send a higher voltage signal to the microprocessor. This would trigger the capacity steps to progress from step 1. Soon the capacity of the compressor rack would

Figure 26.14 (A) A pressure transducer mounted on a common suction header. This transducer transforms a pressure signal to an electrical signal for the controller to process. This cycles the compressors to their next capacity step. (B) Evaporator pressure-regulating (EPR) valves on suction lines of a parallel compressor rack. As the EPR valves open and close according to the load on the cases, the pressure transducer on the common suction line will sense the change in pressure. Courtesy Village Market. Photo by John Tomczyk. (C) Modern step-motor-type, electronic evaporator pressure-regulator (EEPR) valves on the common suction line of a parallel compressor system. They react to a thermistor in the refrigerated case's evaporator discharge air. Courtesy RSES Journal.

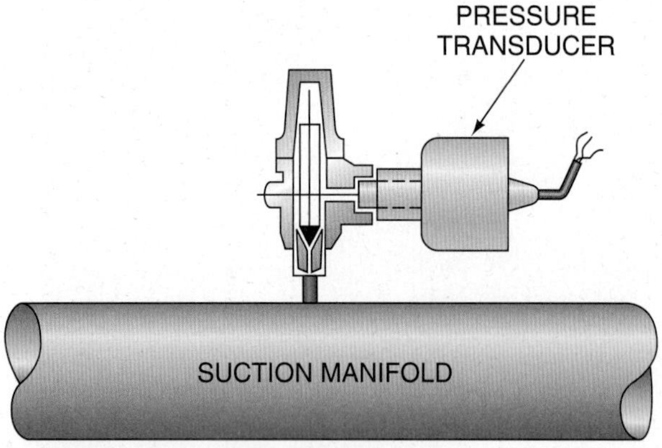

(A)

(B)

(C)

match the refrigeration load. This could happen at capacity step 6, where only compressors #2 and #4 were operating. At this time, the common suction pressure would stop rising and the parallel compressor rack would not increase or decrease in capacity steps. The same thing would happen if the load decreased, but the pressure transducer would sense a decrease in pressure. It would now relay a decreased voltage to the microprocessor. This would cause the parallel rack's capacity steps to decrease until the load was matched with the right compressor capacity. This could happen at capacity step 3 with just compressor #3 running. **Figure 26.17** shows line-voltage hardware, which includes circuit breakers and motor starters with current-type relays, used in cycling the parallel compressors.

Parallel compressor racks usually come with compressors of unequal size and are referred to as uneven parallel systems. They are used for widely varying loads. Compressors of equal size, referred to as even parallel systems, are for refrigeration loads that are relatively constant or have known increases or decreases in capacity, as in an industrial process. Compressor cycling logic for these systems may provide for even run time among all compressors, resulting in even wear on all compressors through the system's life. The parallel compressor system may have a split suction line. This enables the same parallel compressor rack to handle two different suction pressures by using two different pressure transducers. A check valve will separate the two suction lines having different saturated suction temperatures. The split suction option enables one parallel compressor rack to handle both low- and medium-temperature applications, thus making it a more diverse refrigeration system.

OIL DISTRIBUTION SYSTEM

A sophisticated oil system was designed to maintain the correct amount of oil in each compressor of the parallel system. This oil system usually consists of four major components:

- Oil separator
- Oil reservoir

Figure 26.15 (A) The computer-based control for a parallel refrigeration system. Source: Meijer Corporation, Big Rapids, MI. Photo by John Tomczyk. (B) An internal view of the computer-based controller showing smaller, microprocessor-based controllers. Source: Meijer Corporation, Big Rapids, MI. Photo by John Tomczyk.

(A)

(B)

Figure 26.16 Capacity steps programmed into the microprocessor-based controller for load-matching capabilities.

	CAPACITY STEPS										
	0	1	2	3	4	5	6	7	8	9	10
COMPRESSOR #1 (10 HP)	OFF	ON	OFF	OFF	OFF	ON	OFF	OFF	ON	OFF	ON
COMPRESSOR #2 (12 HP)	OFF	OFF	ON	OFF	OFF	OFF	ON	OFF	OFF	ON	ON
COMPRESSOR #3 (15 HP)	OFF	OFF	OFF	ON	OFF	OFF	OFF	ON	ON	ON	ON
COMPRESSOR #4 (20 HP)	OFF	OFF	OFF	OFF	ON	ON	ON	ON	ON	ON	ON
	OPERATING CAPACITY IN HORSEPOWER										
	DECREASING CAPACITY ← → INCREASING CAPACITY										
OPERATING CAPACITY IN HORSEPOWER	0	10	12	15	20	30	32	35	45	47	57

Figure 26.17 Line-voltage hardware, which includes circuit breakers and motor starters with current relays used for cycling the parallel compressors. Source: Meijer Corporation, Big Rapids, MI. Photo by John Tomczyk.

- Pressure differential valve
- Oil level regulators

A large oil separator is located in the common discharge line to trap the pumped oil, ***Figure 26.12.*** Oil is not returned directly from the oil separator to the compressors, as in conventional systems, but is pumped by differential pressure to an oil reservoir. Often, the oil separator and the oil reservoir are combined into one vessel, **Figure 26.18**. From the reservoir, the oil passes to an oil-level regulator float valve, which is mounted on each compressor at its sight glass location, **Figure 26.19**. The oil-level regulator float valve senses crankcase oil levels and opens and closes as necessary to ensure an adequate oil level in each compressor crankcase. A pressure differential valve located on the oil reservoir usually maintains a 5- to 20-psi pressure differential between reservoir pressure and common suction pressure, depending on what pressure differential valve is used, and is vented to the suction line, **Figure 26.12**. This valve ensures that the oil reservoir is pressurized each time an oil regulator opens. The factory pressure setting is system-dependent, which allows the oil-level regulator orifice to be calibrated to a certain pressure difference over suction pressure. This prevents excessive inlet pressures at the oil-level regulator, which may overfill the compressor crankcase with oil before the mechanical float valves can react. The pressure differential valve is shown in **Figure 26.20(B)**, and an oil reservoir is shown in **Figure 26.20(C)**.

Oil-level-regulating float valves can be fixed or adjustable and can be purely mechanical, **Figure 26.20(A)**, or electromechanical. Electromechanical valves can use a liquid-level

Figure 26.18 A combination oil separator and oil reservoir. Photo by John Tomczyk.

Figure 26.20 (A) Adjustable and fixed oil-level regulators. Source: AC & R Components, Inc (B) A differential pressure check valve. (C) An oil reservoir and a cutaway. Images Courtesy of Parker Sporlan.

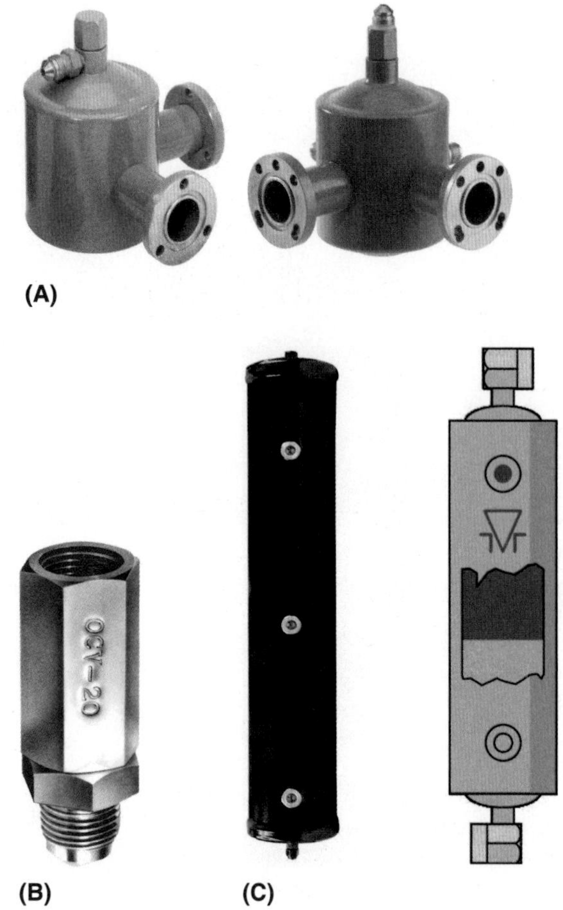

switch, a solenoid valve, and a magnetic reed float switch to detect oil levels, **Figure 26.21(A)**. These systems also offer alarm and system shutdown control if oil levels get dangerously high or low. Some electromechanical liquid-level switches detect oil level by sensing light refracting through a glass prism, **Figure 26.21(B)**. Their advantage is that they can be mounted on a system without disturbing the system internally.

Figure 26.19 A compressor float-operated oil-level regulator. Source: AC & R Components, Inc.

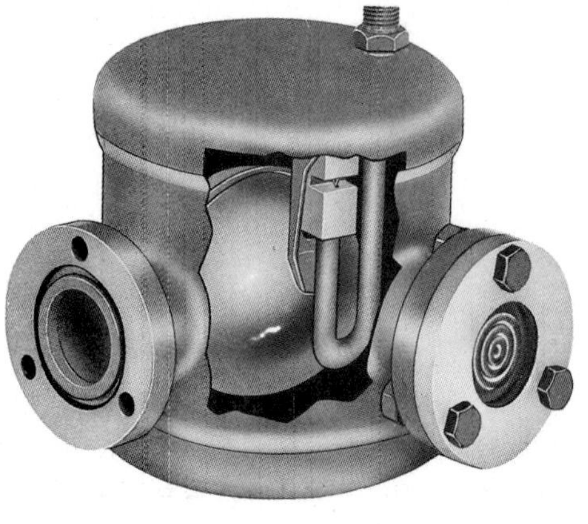

SATELLITE OR BOOSTER COMPRESSORS

Often, parallel compressor systems employ a **satellite** *or* **booster compressor**, **Figure 26.12**. *Satellite compressors can be large or small and are often dedicated to the coldest evaporators. This prevents the entire parallel rack from operating at the lower suction pressures inherent to lower-temperature evaporators. This improves the efficiency of the entire parallel compressor rack by letting it run at a lower compression ratio, resulting in higher efficiency.* Satellite compressors can be valved into the parallel rack. This enables the rack of compressors to assist the satellite compressor in pulling down. A check valve will

Figure 26.21 (A) An electromechanical oil-level regulator. Source: AC & R Components, Inc. (B) A compressor oil-level detector, which uses light refracting through a glass prism to detect crankcase oil levels. Source: AC & R Components, Inc.

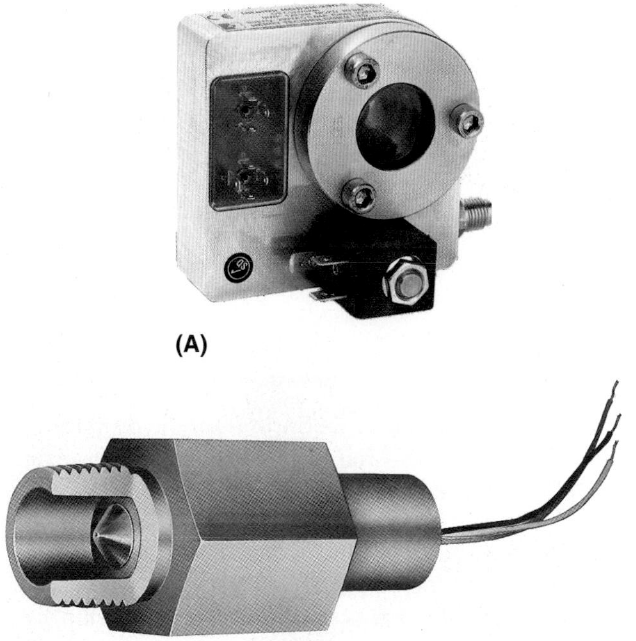

(A)

(B)

Figure 26.22 Inverters or variable-speed drives that change the alternating current's frequency, thus changing the satellite compressor's speed and capacity for better load matching. Source: Meijer Corporation, Big Rapids, MI. Photo by John Tomczyk.

then separate the parallel rack from the satellite compressor, allowing the satellite compressor to pull down further. *Satellite compressors often have variable-speed drives called inverters controlling their motor speeds, thus compressor capacity, to save energy,* **Figure 26.22**. *The refrigeration load can then be matched even more closely by varying the frequency of an electronically altered sine wave, which affects the speed of alternating current motors.* As the load decreases, the motor slows down, giving the compressor less capacity. As the load increases, the motor increases in speed, giving the compressor more capacity. Refer to Unit 17, "Types of Electric Motors," for more specific information on inverters. Cylinder unloading is another method of controlling the capacity of satellite compressors and the parallel compressors on the rack.

PRESERVING LIQUID SUBCOOLING

Subcooling can be accomplished in a variety of ways in commercial parallel refrigeration systems. Listed here are the three main methods:

- Liquid/suction-line heat exchangers
- Auxiliary refrigeration systems
- Natural condenser subcooling

Liquid/suction-line heat exchangers bond the liquid line to the suction line. This allows for the transfer of heat from the higher-temperature liquid line to the suction line. Many different types and arrays of commercial liquid/suction heat exchangers can be purchased to fit system needs. The heat exchangers provide for more subcooled liquid, which will increase the net refrigeration effect (NRE) in the evaporator by decreasing the amount of flash gas at the evaporator's inlet. However, the heat exchange between the liquid line and the suction line is not without cost to system capacity. As the suction line experiences more heat from the liquid line, the refrigerant vapors within the suction line will expand and become less dense. This will decrease the mass flow rate of refrigerant through the compressor and have some negative effects on system capacity. The type of subcooling has to be designed and managed carefully for the system to have a net gain in capacity. For more detailed information on liquid/suction heat exchangers, refer to Section 25.8, "Refrigeration Accessories," in Unit 25, "Special Refrigeration System Components."

Auxiliary refrigeration systems use some of the liquid from the liquid-line header to feed an auxiliary expansion valve and evaporator. This auxiliary expansion valve and evaporator's heat load is the remaining liquid flowing to the liquid header. Often, but not always, this mechanical subcooler is a plate-and-frame or shell-and-tube heat exchanger, and the liquid feeding the expansion valve comes from a different-temperature application parallel refrigeration system in the same equipment room. The suction line coming off the heat exchanger cooling the liquid refrigerant often returns to a different-temperature application parallel refrigeration system's suction header, **Figure 26.12(B)**. In **Figure 26.12(C)**, the liquid feeding the expansion valve for the mechanical subcooler branches off the same refrigeration system's liquid line. Also, the suction line coming off the shell-and-tube

subcooler's heat exchanger returns to an independent compressor's suction line. This type of subcooling is actually using some of its own, or another refrigeration system's capacity and is dedicated to subcool liquid for use in a liquid header feeding multiple expansion devices and evaporators. It does not, however, come completely free and without additional system expenses and costs.

Natural condenser subcooling is free subcooling when the condenser is intentionally designed with extra volume for a liquid subcooling loop at its bottom. The added subcooling loop does not back up liquid in the normal condenser or increase head pressure. The reason is that the additional volume for this extra liquid is intentionally designed into the condenser. Significant increases in system capacity can be realized with this free liquid subcooling in the cooler ambient temperatures, especially when used in conjunction with high-side head pressure controls. However, this free liquid subcooling is only available when the outdoor ambient conditions are cool enough. During summer operation, subcooling is lower because of the warmer air flowing across the condenser.

One way to preserve liquid subcooling coming from a condenser and keep it cool before it reaches the liquid header is to avoid allowing the liquid to enter and flow through a conventional flow-through receiver. Flow-through receivers can store the cool, subcooled liquid refrigerant for some time before releasing it to the liquid line and liquid header. This can cause a gain in temperature within the liquid and subcooling can be lost. Also, when high-side head pressure controls are used in low-ambient conditions, hot gas bypassed from the compressor's discharge line to the receiver actually helps warm the subcooled liquid coming from the condenser to a higher saturated condition. This allows the receiver to have a higher pressure when feeding metering devices and evaporators. At the same time, the condenser's outlet is being restricted by a head pressure control valve, causing flooding of the condenser. The action of the head pressure control valve will maintain a decent head pressure in low-ambient conditions, while keeping receiver pressure high enough to satisfactorily feed metering devices, **Figure 26.12(C)**. Refer to the Section 22.18, "Head Pressure Controls," in Unit 22, "Condensers," for more detailed information on head pressure control devices.

Figure 26.12(C) shows a piping diagram for a parallel refrigeration system that allows the receiver to be converted from a "flow-through receiver" to a "surge-type receiver." This type of arrangement preserves the liquid subcooling achieved in the condenser during the cooler ambient months when free subcooling is available. A solenoid-operated diverting valve is opened and the subcooled liquid coming from the condenser's bottom is closed off to the receiver. However, it is allowed to bypass the receiver and flow directly to a mechanical subcooler and then to the liquid header to feed expansion valves and evaporators. This method prevents subcooled refrigerant in the receiver from being warmed and returning

Figure 26.23 The Enviroguard Refrigerant Control System with an SPR valve. Courtesy Tyler Refrigeration Corporation

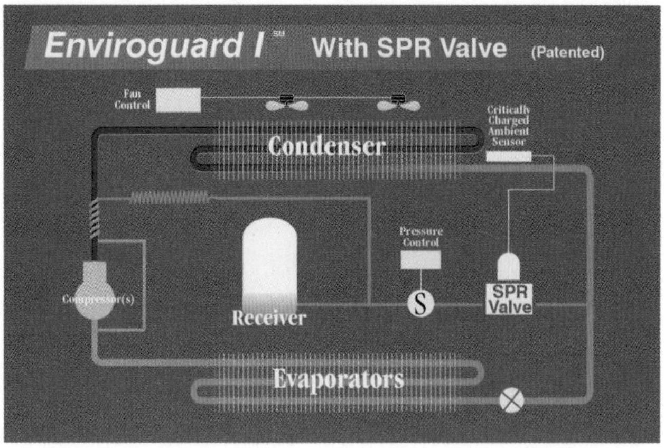

to a saturated condition. A thermostat on the drop leg of the outdoor condenser is what controls the receiver-diverting valve in low-ambient conditions. All liquid lines must be insulated to prevent any unnecessary loss of liquid subcooling, especially when the lines pass through a conditioned space.

Another refrigeration system scheme that uses parallel compressors for capacity control and load matching also uses a **system pressure-regulating (SPR) valve** *located between the condenser and the receiver,* ***Figure 26.23.*** This system has a receiver that is in parallel with the active refrigeration system, instead of a series receiver as in a conventional refrigeration system. The SPR valve has a remote, liquid-filled bulb located in the condenser's inlet air, sensing outdoor ambient temperatures. The sensor is connected to the SPR valve by a small copper tube. A more recent design uses sensors for determining outdoor ambient temperatures, liquid-line temperatures, and other temperatures strategic to the operation of the system. Inputs to an electronic controller operate condenser fans, a vapor relief valve out of the receiver, a liquid bleed line to the receiver, and liquid pressure to the evaporator, **Figure 26.24.**

Figure 26.24 The Enviroguard II Refrigerant Control System with an electronic controller. Courtesy Tyler Refrigeration Corporation.

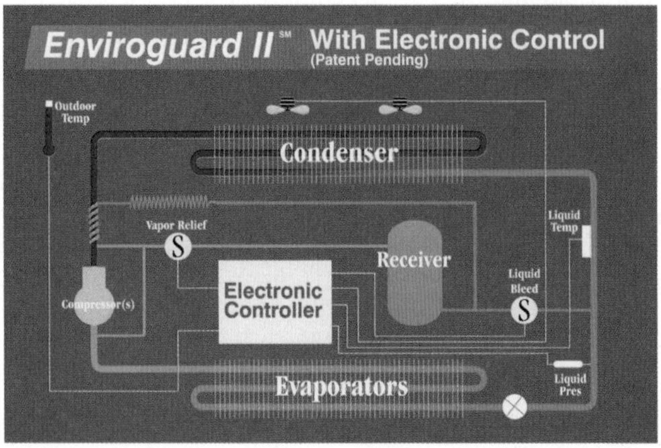

A complex computer program in the electronic controller keeps the system operating at peak performance. *The SPR valve controls the amount of liquid refrigerant used in the active refrigeration system. The active system's refrigerant charge is reduced by taking the receiver out of the active refrigeration circuit and allowing liquid to return directly to the liquid manifold, which feeds the branch refrigeration circuits. With the receiver out of the active refrigeration system, a minimum receiver charge is no longer required as with a conventional system.* The receiver is now simply used as a storage vessel to hold the condenser charge between summer and winter operations. On hot days, the SPR valve will bypass refrigerant to the receiver. This lowers the liquid levels in the condenser and gives it more internal volume for the hotter weather conditions. On colder days, the SPR valve throttles down, which bypasses less refrigerant to the receiver and keeps more in the condenser. This partially floods the condenser and keeps condensing pressures from dropping too low. The receiver has a bleed circuit that is opened to the suction manifold. This circuit bleeds refrigerant from the receiver back to the system for use, which enables the refrigerant working charge in the system to seek a level of equilibrium relative to the ambient temperature. This means that in some conditions, the receiver has no refrigerant and is at the same pressure as the suction line.

If the condenser should become damaged, fouled internally or externally, or lose a fan motor, an elevated condensing pressure will occur. This elevated pressure will cause the SPR valve to bypass refrigerant into the receiver because a design temperature difference between the ambient and the condensing temperature has been exceeded. *This will prevent high compression ratios and high discharge temperatures with lower volumetric efficiencies. These occurrences often go unnoticed for long periods of time and stress compressor systems.* Eventually, branch circuit evaporator temperatures will rise, because refrigerant is being bypassed out of the working system. The results will be a starved refrigeration system. Rising evaporator and product temperatures are noticed sooner, and the problem can be remedied immediately. *Refrigerant leaks will also be noticed sooner, which means less refrigerant is lost to the atmosphere.*

26.5 SECONDARY-FLUID REFRIGERATION SYSTEMS

With ozone depletion and global warming legislation phasing out some refrigerants, manufacturers are researching, and some are using, a primary (HFC or HCFC) refrigerant to cool a **secondary fluid**. The secondary fluid in this case is also considered a refrigerant. The secondary fluid or refrigerant is usually a low-temperature antifreeze solution, which stays in the liquid state and does not vaporize when it is circulated through the system by centrifugal pumps.

Secondary fluids include **propylene glycol** and **hydrofluoroethers (HFEs)**. Propylene glycol is the most popular secondary fluid used today, but among the newest secondary fluids are HFEs. *They have zero ozone-depletion potential and small global-warming potential, and they are nontoxic.* HFEs are, however, one of the most expensive secondary fluids. Along with other popular secondary fluids, HFEs can be used for all temperature ranges found in supermarket refrigeration, from dairy to frozen-food cases. This secondary fluid, which is the more abundant compound, is circulated instead of the primary (HCFC or HFC) refrigerant by centrifugal pumps to the display cases in supermarkets. A heat exchange then occurs between the primary and secondary fluids.

Figure 26.25(A) illustrates the difference between a secondary-fluid refrigeration system and a direct-expansion refrigeration system. **Figure 26.25(B)** is a photo of chillers and circulating pumps for a secondary-fluid refrigeration system. **Figure 26.25(C)** shows evaporator coils piped for chilled glycol as a secondary fluid. **Figure 26.25(D)** is a photo of a deli case with chilled glycol pans. This system operates in much the same way as a direct-expansion chiller but uses much less HCFC or HFC to accomplish the same task. These systems eliminate the long liquid and suction lines that are usually installed throughout the store, which reduces the overall system refrigerant charge and thousands of joints, which are potential leak points.

Defrost is accomplished by heating the same HFE fluid used in the secondary loop by a controlled process. It is done through a third piping loop with solenoid-controlled valving. The heat for defrost comes from the heat of compression in the discharge gases from the compressor. This heat is then transferred to the same secondary fluid in a third piping loop used for defrost purposes at the case coils.

The advantages of chilled secondary-fluid systems are:

- *large refrigerant reductions,*
- *greatly reduced leak potential,*
- *simpler installations,*
- *less critical piping,*
- *reduced brazing and nitrogen use,*
- *minimized vacuum requirements to the equipment room,*
- *few or no oil return concerns,*
- *fewer expansion valve superheat adjustments,*
- *less compressor (total) superheat in the primary refrigerant loop because of shorter suction lines, which adds to the thermodynamic efficiency of the system,*
- *simpler and shorter defrosts,*
- *no requirement for defrost limit thermostats,*
- *faster defrost periods and quicker case pulldown times after defrost, which result in lower energy requirements and improved product temperatures,*
- *better and more stable product temperatures from higher mass flow rates in the secondary fluid,*

Figure 26.25 (A) A secondary-fluid refrigeration system. Courtesy Second Nature© System by Hill Phoenix. Adaptation by Refrigeration Training Service. (B) A secondary-fluid refrigeration system showing chillers and circulating pumps. Photo by Dick Wirz. (C) Evaporator coils piped for chilled glycol. Photo by Dick Wirz.

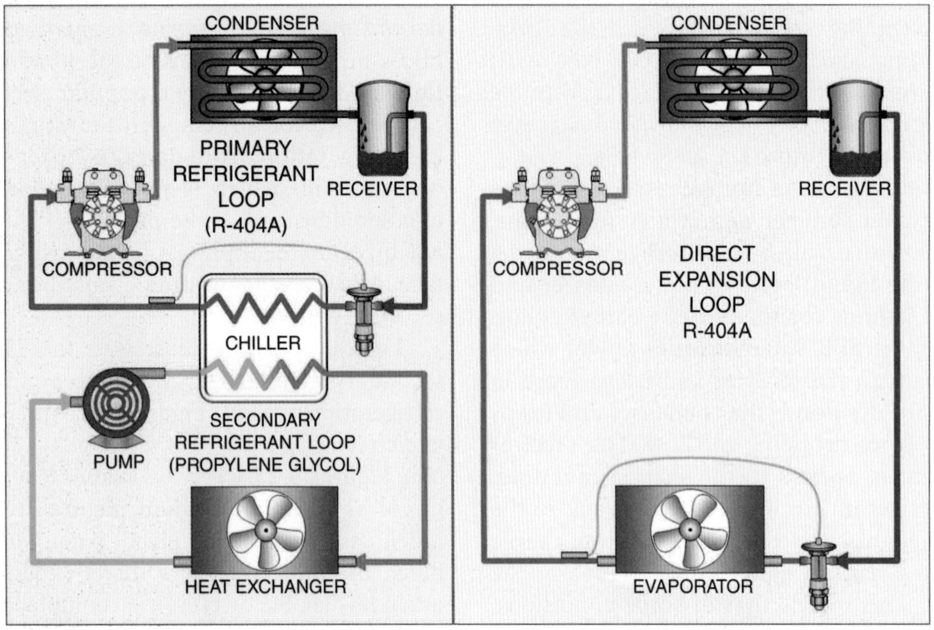

(A)

(B)

(C)

- *more efficient coil surface use, resulting in better and more stable product temperatures,*
- *simpler control strategies,*
- *higher refrigerant coil operating temperatures in secondary loop,*
- *lower maintenance,*
- *lower total equivalent warming impact (TEWI) than conventional refrigeration systems,*
- *the use of PVC piping or K-type copper tubing in the secondary loop for long line runs instead of ACR-type copper tubing used with direct-expansion refrigerants,*
- *need to locate only the low-pressure secondary loop in the store.*

A few system-dependent disadvantages are:

- *higher cost of original equipment,*
- *requiring the use of industrial-grade solenoid valves for case control in the secondary loop,*
- *higher electrical consumption from the extra heat exchange energy penalty—and the added cost of secondary-fluid pumping, which, however, is often offset by the higher refrigerant coil operating temperatures,*

Figure 26.25 (*Continued*) (D) Deli cases with chilled glycol pans. Photo by Dick Wirz.

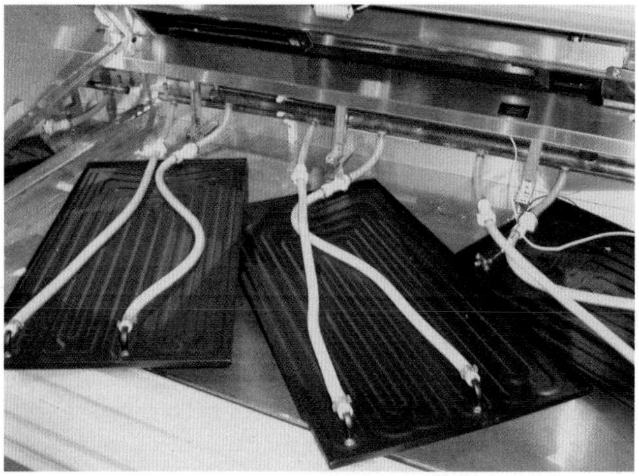

(D)

- *thicker and completely sealed pipe insulation needed in the secondary loop to prevent temperature and system loss, and*
- *higher cost of using the low-viscosity secondary fluid that is required for low-temperature applications.*

26.6 CARBON DIOXIDE (R-744) REFRIGERATION SYSTEMS

Carbon dioxide (CO_2) is not a new refrigerant. Carbon dioxide (R-744) was used in the early days of the refrigeration industry. It was gradually phased out when chlorofluorocarbons (CFCs), often referred to as "synthetic refrigerants," came into existence in the 1920s. Synthetic refrigerants are man-made. Another reason for the gradual phaseout of CO_2 was its very high operating pressures. Finding refrigeration system components that could handle these higher system operating pressures was very difficult back then. However, modern technology has developed refrigeration components that can handle the very high operating pressures of CO_2 in certain applications when required.

Since the 1920s and 1930s, synthetic refrigerants like CFCs and HCFCs have dominated the refrigeration and air-conditioning industry. Synthetic HFC and HFO refrigerants were introduced later on. However, with the ever-growing environmental concerns about ozone depletion and global warming, natural refrigerants like CO_2 (R-744), ammonia (R-717), propane (HC-290), butane (HC-600), and isobutane (HC-600a), to name a few, are becoming more popular as refrigerants. These refrigerants occur in nature and can be used effectively as refrigerants. Driven by economics and regulations, some hydrocarbon (HC) refrigerants like propane and butane have become very popular in domestic applications like refrigerators and freezers and some smaller commercial applications in Europe, China, Japan, and India. The Unites States is following close behind in these technologies. The natural refrigerants have a zero ozone depletion potential (ODP), with a very small global warming potential (GWP), if any. The following chart compares the GWPs of two natural refrigerants with those of two synthetic HFC refrigerants.

Global Warming and Ozone Depletion Potentials

Refrigerant	GWP	ODP
(R-744) Carbon Dioxide (CO_2)	1	0
(R-717) Ammonia	0	0
HFC-134a	1300	0
HFC-404A	3260	0

CO_2 occurs naturally in the atmosphere at around 350 ppm and its long-term influence on the environment has been well researched. It is not self-identifying because it does not have a distinctive odor or color. Facilities using CO_2 must be equipped with sensors that trigger an alarm when concentrations exceed 5000 ppm. CO_2 is heavier than air so the detecting sensors and ventilators must be located in low areas. CO_2 is a by-product of combustion. Since there are many power plants in the United States, Europe, and throughout the world, it makes sense to capture it from the power industry's combustion processes and use it for refrigeration. CO_2 is very readily available, if not in surplus in the atmosphere, and is also very inexpensive.

Other than its high operating pressures, however, another disadvantage to using CO_2 as a refrigerant is its low critical point of 88°F/1067 psia. The critical temperature of a gas is the highest temperature the gas can have and still be condensable by the application of pressure. Above the critical temperature, a gas cannot be saturated regardless of the amount of pressure applied. **Figure 26.26(A)** shows the location of the critical point on the pressure/enthalpy diagram. For CO_2, no liquid refrigerant can be produced above its critical temperature of 88°F. At the critical temperature, the density of the liquid and the vapor states is equal, the distinction between the two phases disappears, and a new phase exists, called the supercritical phase. The supercritical phase exists in the region above the critical point. In this region, the pressure of CO_2 is not solely dependent on its temperature. So, on the condenser side, CO_2 used in refrigeration should be kept well below the critical point (10–20 degrees below) so that condensation of the refrigerant can take place.

CO_2 can exist as all three phases (vapor, liquid, and solid) simultaneously in equilibrium at its triple point, −69.9°F/75 psia. So, on the evaporator side, refrigeration applications must be kept well above the triple point to avoid plugging the refrigeration lines with solid CO_2 (dry ice). The refrigeration cycles that have been covered so far in this

Figure 26.26 (A) Illustration of a subcritical CO_2 refrigeration application in a pressure/enthalpy diagram. This diagram shows the locations of the critical point and triple point. Notice that the cycle takes place below the critical point and above the triple point for CO_2. (B) Illustration of a transcritical CO_2 refrigeration application on a pressure/enthalpy diagram. Notice that evaporation takes place below the critical point and above the triple point for CO_2. However, condensation will not take place because the CO_2 gas has been compressed into the supercritical region. Both its temperature and pressure have been elevated above the critical point.

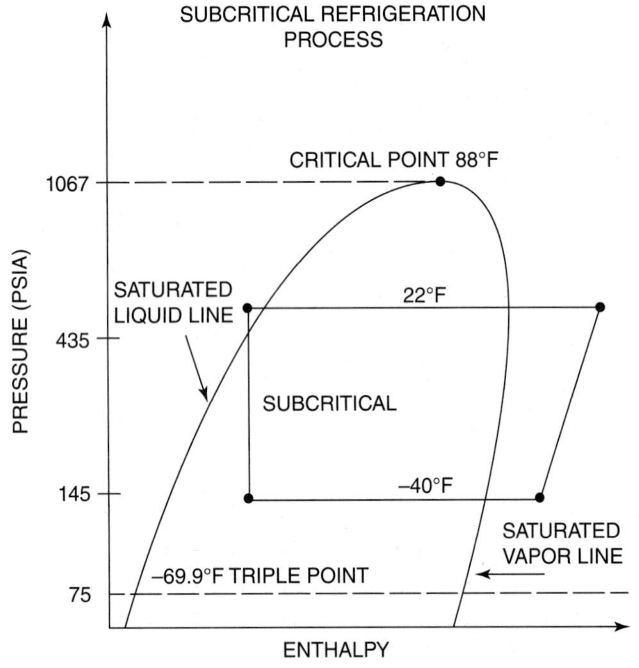

(A)

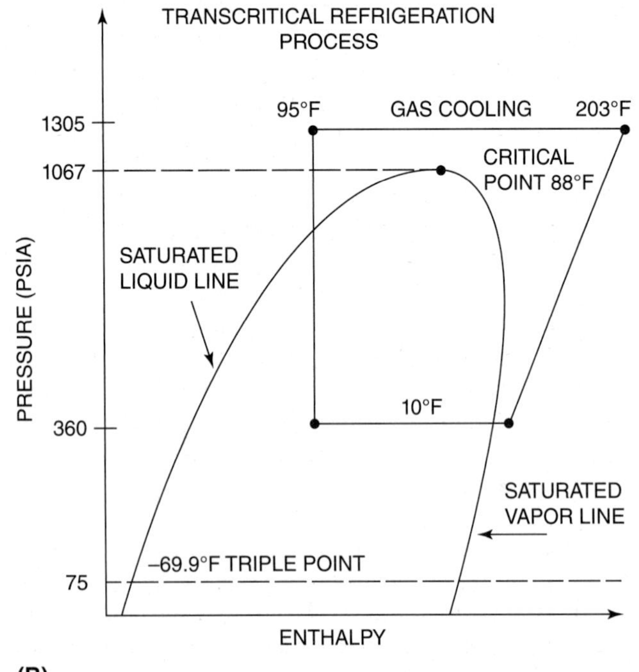

(B)

book all take place above the triple point and below the critical point of the refrigerants used.

Refrigeration systems using CO_2 as a refrigerant can be divided into two main processes or cycles. These two processes or cycles are

1. the subcritical cycle and
2. the transcritical (supercritical) cycle.

SUBCRITICAL CO_2 CYCLE

As mentioned, the subcritical cycle refers to CO_2 refrigeration cycles that take place above the refrigerant's triple point and below the refrigerant's critical point. **Figure 26.26(A)** illustrates a subcritical CO_2 refrigeration application on a pressure/enthalpy diagram. The subcritical refrigeration cycle for CO_2 is the same as that for any subcritical refrigeration cycle. There are four processes, including

1. compression,
2. condensation with subcooling,
3. expansion, and
4. evaporation with superheat.

Notice in **Figure 26.26(A)** that the critical point is located at the dome of the pressure/enthalpy diagram. Also notice that both the latent heat of condensation and the latent heat of vaporization occur below the critical point and above the triple point for CO_2.

TRANSCRITICAL CO_2 CYCLE

On the other hand, **Figure 26.26(B)** illustrates a transcritical CO_2 refrigeration application on a pressure/enthalpy diagram. *Transcritical* simply means that the thermodynamic refrigeration processes span above the subcritical region and into the supercritical region of the pressure/enthalpy diagram. The four processes of the transcritical refrigeration cycle include

1. compression,
2. gas cooling,
3. expansion, and
4. evaporation.

Notice in **Figure 26.26(B)** that evaporation takes place below the critical point and above the triple point for CO_2. However, condensation will not take place because the CO_2 gas has been compressed into the supercritical region. Its temperature and pressure have been elevated above its critical point. **Figure 26.26(B)** shows that superheated CO_2 gas above 88°F cannot be condensed, so it is cooled from 203°F to 95°F in a gas cooler and remains in the supercritical region. This heat of rejection takes place in the supercritical region above the critical point for CO_2 of 88°F/1067 psia. For the transcritical cycle, the heat of rejection is therefore called gas

cooling, and the heat exchanger used is referred to as a gas cooler, not a condenser.

Very high pressures can be generated in this region. High-side pressures of 1500 psi can be encountered, with pressures as high as 4000 psi in certain instances. Because of these extremely high pressures, system components have to be designed to handle them. Compressor shells, heat exchangers, pressure relief valves, rings, valves, and microchannel heat exchangers will all be exposed to these higher system pressures. It is because of these extremely high system pressures that designers have favored systems in the subcritical regions. For a review of the applications of pressure/enthalpy diagrams, please refer to Section 3.20, "Plotting the Refrigeration Cycle," in Unit 3, "Refrigeration and Refrigerants."

Following is a comparison of system saturation pressures, when they can be reached, for low- and medium-temperature refrigeration systems employing CO_2, HFC-404A, and HFC-134a.

Saturated Temperatures/Pressures for Low-Temperature Refrigeration

Refrigerant	Saturated Suction Temperature (SST)	Saturated Condensing Temperature (SCT)	Corresponding Saturated Pressures (psig)
CO_2	−25°F	85°F	181/1017
HFC-404A	−25°F	85°F	13/187

Saturated Temperatures/Pressures for Medium-Temperature Refrigeration

Refrigerant	Saturated Suction Temperature (SST)	Saturated Condensing Temperature (SCT)	Corresponding Saturated Pressures (psig)
CO_2	20°F	85°F	407/1017
HFC-134a	20°F	85°F	18/95

SUBCRITICAL CO_2 REFRIGERATION SYSTEMS

Because CO_2 is environmentally friendly, is readily available, has very low toxicity, is nonflammable, is very inexpensive, and has a high vapor density, it has become more popular in supermarket refrigeration systems. Its safety group classification is A1, meaning CO_2 has very low toxicity and no flame propagation. One popular CO_2 system for supermarket refrigeration is an indirect system that uses CO_2 as a secondary refrigerant. Another popular system uses CO_2 as one of the cascade refrigerants in a cascade refrigeration system. The following paragraphs and diagrams explain these two systems.

Figure 26.27 An indirect refrigeration system where CO_2 is a secondary refrigerant used for freezing applications, with HFC-404A used as the primary refrigerant. Courtesy Refrigeration Training Services and Hill Phoenix. Adaptation by Refrigeration Training Services.

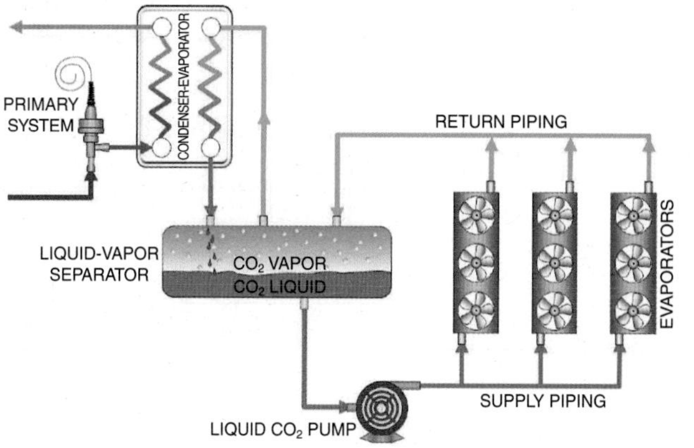

INDIRECT SYSTEMS. The main reasons for using an indirect CO_2 system are its simplicity and the possibility of using other refrigerant system components in the construction of the system. **Figure 26.27** illustrates an indirect refrigeration system that uses CO_2 as a secondary refrigerant for freezing applications and HFC-404A as the primary refrigerant. Chilled liquid CO_2 is pumped by a centrifugal pump through the evaporators and absorbs heat in the refrigerated cases. As the CO_2 absorbs heat in the evaporators, only a portion of its liquid evaporates. The liquid and vapor mixture of CO_2 exits the evaporators and enter a liquid-vapor separator. Once in the separator, CO_2 vapor from the separator's top is pulled into the condenser/evaporator heat exchanger. This is where a primary refrigerant, like HFC-404A, is evaporated and used to cool and condense the CO_2 vapor back to a liquid. The chilled liquid CO_2 falls to the bottom of the separator and is again pumped by a centrifugal pump through the evaporators to absorb heat from the refrigerated cases.

Because CO_2 vapor has the major advantage of a high vapor density, it has a low-volume flow rate, which gives it an important edge compared to brine-based systems. Also, as the CO_2 exits the evaporator, it is about 50% liquid and 50% vapor by design. This will always assure a wet evaporator inner surface, which enhances heat transfer even under changing evaporator heat loads. The centrifugal pump is designed to pump a certain circulation ratio, making the system flexible in the face of changing evaporator heat loads while still maintaining a wet evaporator.

CASCADE SYSTEMS. **Figure 26.28** illustrates a supermarket cascade refrigeration system where CO_2 is being used as the low-stage (primary), low-temperature refrigerant in the lower portion of the system. Liquid CO_2 actually flows through expansion valves and is then evaporated in the

Figure 26.28 A supermarket cascade refrigeration system where CO_2 is being used as the primary (low-stage) low-temperature refrigerant in the lower portion of a cascade system. HFC-404A is used as a secondary (high-stage) refrigerant. Courtesy Refrigeration Training Services and Hill Phoenix. Adaptation by Refrigeration Training Services.

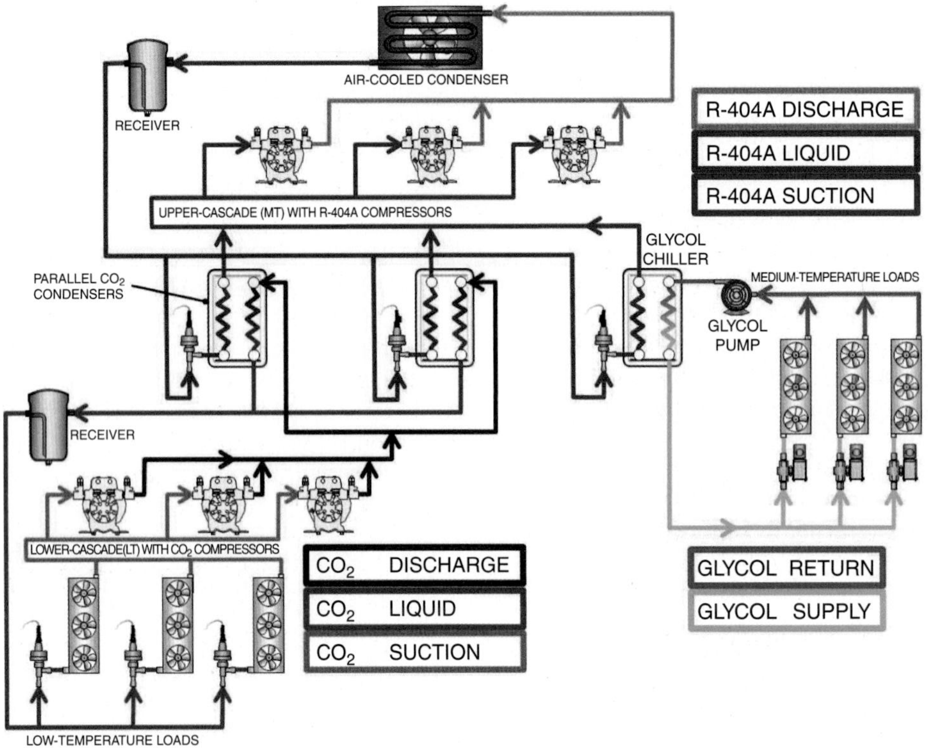

low-temperature evaporators. CO_2 compressors are used in the lower portion of the cascade system to send hot, superheated vapor to a condenser/evaporator heat exchanger.

HFC-404A is being used as a high-stage (secondary) refrigerant; propane and ammonia have also been used successfully as high-stage refrigerants. The HFC-404A refrigerant system has two cooling functions. One function is to act as a high-stage (secondary) refrigerant in the upper portion of the cascade system to cool the low-stage refrigerant's (CO_2) condensers. Its second function is to chill a glycol solution in the glycol chiller heat exchanger. This glycol solution is then circulated by a centrifugal pump through the evaporators of the medium-temperature case. There is no phase change in the glycol solution. Subcritical cascade CO_2 systems keep the condensing pressure low, so common, off-the-shelf components can be used. **Figure 26.29** shows two subcritical, cascade parallel compressor CO_2 systems for use in supermarket refrigeration. Compressors for subcritical refrigeration applications can be of many different types: reciprocating, scroll, screw, or rotary.

However, the refrigeration cycle would have to run continuously unless some other method of pressure control is used to prevent high pressures from occurring during the off periods. When the CO_2 refrigerant is not being condensed, as when the system is turned off or there is a power failure, it starts to expand as it changes back to a vapor. System pressures would rise (1017 psi at 88°F) to the point where system components or personnel could be endangered or harmed, so relief valves start to open to prevent the failure or bursting of the weakest part of the system. To prevent the system pressures from exceeding the system's design pressures, expansion tanks are often employed on systems with a large enough volume. Larger industrial/commercial CO_2 systems employ a small, sometimes self-powered, refrigeration system that will keep the liquid CO_2 temperatures at levels where the saturation pressures are less than system design pressures. If pressures do elevate in the system and relief valves do blow off CO_2, the valves must be big enough and positioned so as to not let liquid CO_2 pass through or else the triple point might be reached in the blow-off process. Once the triple point is reached (−69.9°F/75 psia), solid CO_2 (dry ice) will form and could plug the relief valves if they are not sized and positioned carefully.

TRANSCRITICAL (SUPERCRITICAL) CO_2 REFRIGERATION SYSTEMS

Figure 26.30 illustrates a transcritical CO_2 refrigeration system. The pressure of the superheated CO_2 gas is increased by the compressor. Through the desuperheating of the vapors, the heat of rejection of the superheated gas is in the

Figure 26.29 (A) A subcritical, cascade parallel compressor CO_2 system. Courtesy Hill PHOENIX Corporation. (B) A subcritical, cascade parallel compressor CO_2 system installation in a supermarket. Courtesy Hill PHOENIX Corporation.

(A)

(B)

supercritical region. No condensation takes place because the CO_2 gas is above its critical point of 88°F/1067 psia. In fact, the temperature of the CO_2 gas will continually decline in the heat rejection process, but its pressure will remain the same, **Figure 26.26(B)**.

The CO_2 gas is then cooled in the gas cooler and enhancing heat exchanger and then expands by passing through a variable or fixed expansion device where it achieves a lower pressure. The expansion process puts the CO_2 into the subcritical region and over the 100% saturated liquid line, allowing it to enter the evaporator as a mixture of liquid and vapor. The CO_2 mixture now absorbs heat and boils back to a 100% saturated vapor which then becomes superheated. The CO_2 gas is slightly superheated before it enters the compressor, **Figure 26.26(B)**. Because there is no saturation temperature or pressure on the high side of transcritical cycle, the pressure at the compressor's discharge depends on how much CO_2 charge is in the system. Medium-temperature transcritical refrigeration applications usually employ single-stage reciprocating compressors. Two-stage reciprocating compressors are mostly used for low-temperature applications and some medium-temperature applications.

The expansion device at the inlet of the evaporator does not control evaporator superheat as in a conventional subcritical refrigeration system. Its function is to inject refrigerant into the evaporator and control the gas cooler pressure and temperature. The superheat in the evaporator and in the suction line returning to the compressor is determined by the system charge and system balance at the time.

All system components have to be designed and selected to handle the very high pressures of transcritical CO_2 systems. Typical medium-temperature transcritical CO_2 systems may have suction pressures of 350 to 500 psig and discharge pressures in the 1250 to 1500 psig ranges. Because transcritical systems are designed for these higher pressures and the system is critically gas charged, there is no need to compensate for off-cycle gas expansion or provide for triple-point protection as in subcritical systems.

26.7 PRESSURIZED LIQUID SYSTEMS

This technology uses a small centrifugal pump to pressurize the liquid entering the liquid line, **Figure 26.31(B)**. (A normal system is shown in **Figure 26.31(A)** for comparison purposes.) The pressurized amount is equivalent to the pressure loss between the condenser outlet and the TXV inlet on receiverless systems or between the receiver and the TXV inlet on TXV/receiver systems. By increasing the pressure of liquid refrigerant, the saturation temperature is raised but the actual temperature of the liquid remains the same. The liquid becomes subcooled and will not flash if exposed to pressure drops in the liquid line on its way to the metering device. The centrifugal pump can be sized to fit any system design.

Subcooling liquid refrigerant by this method gives the condenser more internal volume for condensing refrigerant. The condenser no longer has to use its bottom passes as a subcooling loop. More condensing volume lowers the condensing temperatures and pressures, which lowers compression ratios and boosts efficiency. In fact, because the

going green

THE GREENCHILL PARTNERSHIP©

At the time of this writing, most refrigerants used in supermarket refrigeration systems have a potentially large environmental impact in terms of ozone depletion and/or global warming when they leak from these large systems. Because of the environmental impacts, the Environmental Protection Agency (EPA), supermarkets, and other supermarket industry participants have formed a partnership to help alleviate the problem. This partnership is "GreenChill©," whose members include

- store architects and designers,
- service technician certification organizations,
- trade organizations,
- component manufacturers,
- service contractors and technicians,
- environmental regulators,
- food retailers,
- chemical manufacturers,
- data system providers, and
- advanced refrigeration technology manufacturers.

The GreenChill© partnership's main mission is to reduce refrigerant emissions from supermarkets and decrease the industry's impact on the ozone layer and climate change by

- transitioning to environmentally friendly refrigerants,
- lowering refrigerant charge sizes and eliminating refrigerant leaks, and
- adopting "green" refrigeration technologies, strategies, and practices.

GreenChill's© partners measure emissions and set annual goals along with facilitating reduction of store refrigerant emissions. The partners also promote the use of advanced refrigeration technologies, strategies, and practices and award certifications to stores for achieving emissions reduction targets. The three award levels are Platinum, Gold, and Silver. Some objective criteria for measuring achievement are:

1. Refrigerant charge
2. Refrigerant leak rate
3. Type of refrigerant used
4. Installation leak tightness (new stores only)

For more information on GreenChill©, go to www.epa.gov/greenchill.

Figure 26.30 Illustration of a transcritical CO_2 refrigeration system. Based on Robert Nash Jr., Senior Engineer, Emerson Climate Technologies.

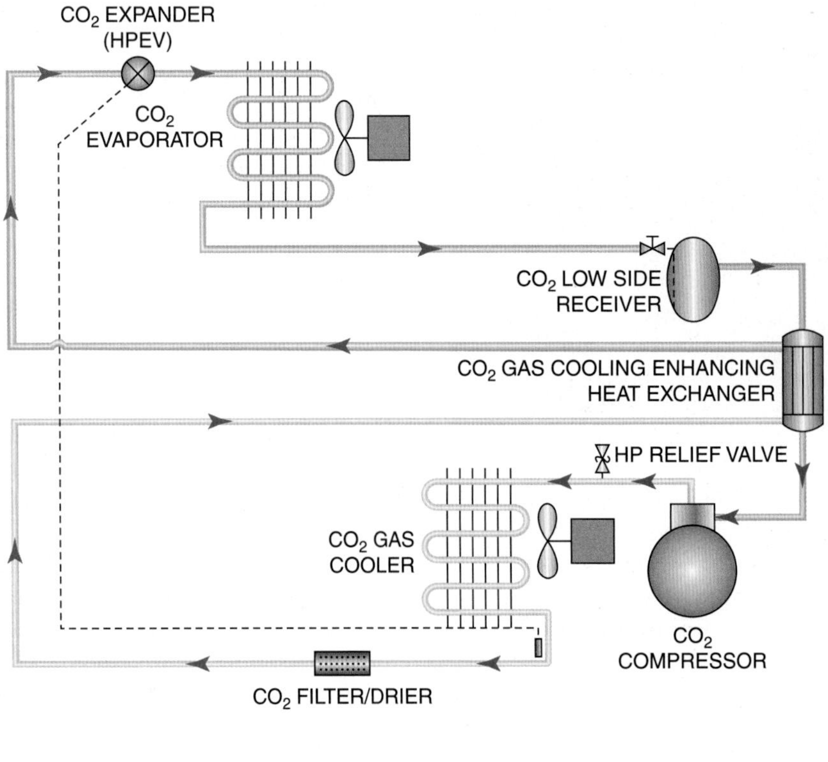

Figure 26.31 (A) A normal refrigeration system. (B) A pressurized liquid system. (C) A pressurized liquid system with liquid injection. Based on DTE Energy Technologies.

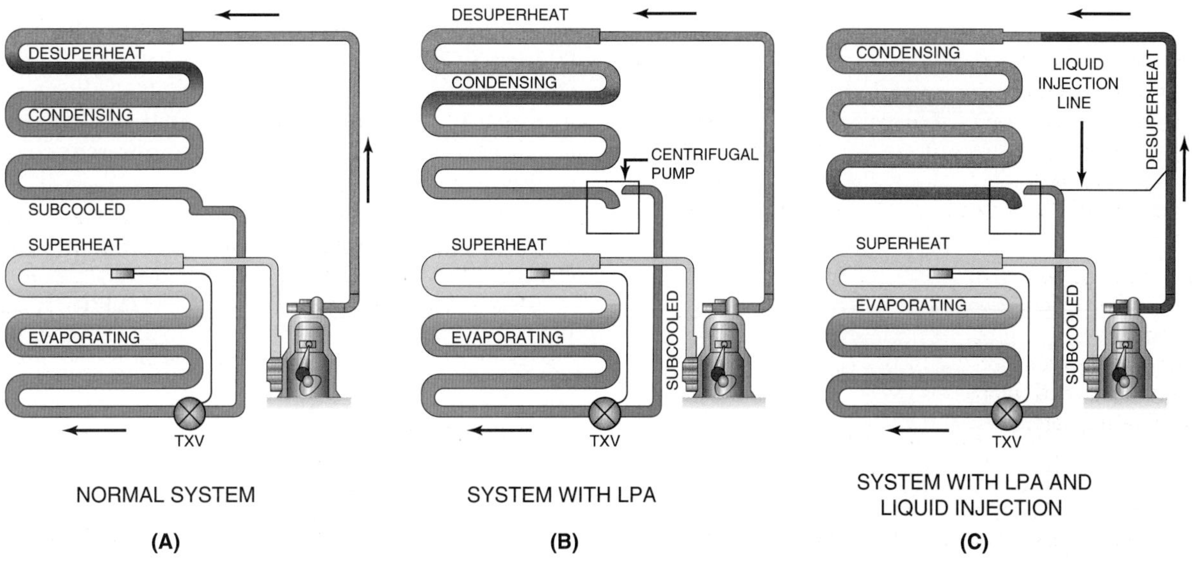

centrifugal pump is doing the liquid subcooling, head pressures are allowed to *float* with the outside ambient temperature. "Floating" the head pressure means attaining the lowest possible head pressure by taking advantage of cool ambient temperatures that are used as the condensing medium. Very low condensing pressures can be attained with these systems. TXVs available today can operate with as little as a 30-psi pressure drop across them and still successfully feed the evaporator if supplied with 100% liquid.

Listed below are four advantages of a pressurized liquid system:

- *They eliminate liquid line flashing by overcoming line pressure losses.*
- *They reduce energy costs because pumping liquid refrigerant is more than 40 times more efficient than using head pressure from the compressor to do the same work.*
- *They increase the evaporator capacity along with the net refrigeration effect.*
- *They lower compression ratios and reduce stress on compressors.*

In conjunction with the pressurized liquid system, liquid refrigerant can be injected into the compressor's discharge line or the condenser inlet. This liquid flashes to a vapor while cooling the superheated discharge gas closer to its condensing temperature. The reduced surface area required for desuperheating makes for a more efficient condenser, which increases the overall performance of the system. The injected liquid comes from the centrifugal pump, which delivers pressurized liquid to the metering device, **Figure 26.31(C)**. Although some efficiency gains are seen at low-ambient temperatures, the greatest gains are realized at higher ambient temperatures.

26.8 DISTRIBUTED REFRIGERATION SYSTEMS

Attractive, multiple-compressor, distributed supermarket refrigeration modules that are located very close to the display cases they refrigerate are becoming increasingly popular, **Figures 26.32** and **26.33**. These modules can be cleverly hidden in the sales display area, mounted on top of display cases, or simply placed behind a wall close to the display cases. Microprocessor-based electronic controllers on each module manage the compressor cycling (load matching) and defrost schedules either on-site or remotely by personal computer. These modules have small, plate-type condensers that are cooled by a closed-loop fluid cooler. Heat reclaim can be performed with these systems for both

Figure 26.32 Unitary distributed refrigeration systems located close to the display cases they are refrigerating. Source: Hussman Corporation.

Figure 26.33 An attractive and compact commercial refrigeration module utilizing four hermetic scroll compressors. Source: Hussman Corporation.

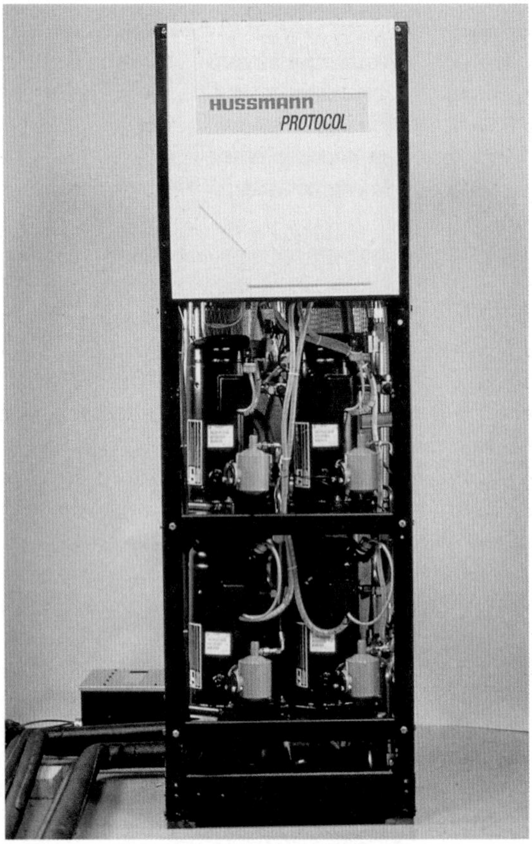

store and hot water heating. Modules come in either vertical or horizontal designs and take up very little floor area. A remote mechanical room is not needed.

Other advantages of these modular stand-alone refrigeration systems are that they

- *reduce refrigerant charge by 75% to 80%,*
- *use 75% less piping,*
- *require 75% fewer brazed joints,*
- *reduce the likelihood of refrigerant leaks,*
- *decrease or eliminate the need for EPR valves,*
- *lower installation time, costs, and complexity, and*
- *provide load matching with backup capacity protection, and*
- *provide energy savings when compared with conventional back-room systems.*

26.9 EVAPORATOR TEMPERATURE CONTROL

When multiple evaporators are used, they may not all have the same temperature rating. For example, the coldest evaporator may require a suction pressure of 28 psig. EPR valves can be located in each of the higher-temperature evaporators. Check valves are typically used at the outlet of the low-pressure evaporators to prevent a rise in pressure during the off cycle. When the load varies, the compressors can be cycled on and off to maintain a suction pressure of 20 psig. The compressors can be cycled using low-pressure control devices or a pressure transducer located in the common suction line. The pressure transducer transforms a pressure signal to a voltage signal and then relays the signal to a microprocessor-based compressor controller for processing.

26.10 INTERCONNECTING PIPING IN MULTIPLE-EVAPORATOR INSTALLATIONS

When the fixtures are located in the store and the compressors are in a common equipment room, the liquid line must be piped to the fixture and the suction line must be piped back to the compressor area. The suction line should be insulated to prevent it from picking up heat on the way back to the compressor. This can be accomplished by preplanning and by providing a pipe chase in the floor. Preferably, the pipe chase would be accessible in case of a leak, **Figure 26.34**. In some installations, refrigerant lines are routed through individual plastic pipes in the floor, **Figure 26.35(A)** and **(B)**. If a leak occurs, the individual lines may be replaced by pulling the old ones and replacing them with new lines. In modern stores, this may be a more popular method than the trench method because it gives the designer more versatility for locating equipment. It is hard to plan a large trench to every fixture, and long refrigerant lines can cause oil return problems. Individual runs can reduce the line

Figure 26.34 A pipe chase is an effective method of running piping from the fixture to the equipment room. With a chase like this, the piping can be serviced if needed. Courtesy Tyler Refrigeration Corporation.

Figure 26.35 (A) An access pit for refrigeration piping that is run under the floor. Source: Meijer Corporation, Big Rapids, MI. Photo by John Tomczyk. (B) Refrigeration piping and electrical wires within the access pit under the floor. Source: Meijer Corporation, Big Rapids, MI. Photo by John Tomczyk. (C) The ventilation system for under-floor refrigeration piping. Source: Meijer Corporation, Big Rapids, MI. Photo by John Tomczyk.

(A)

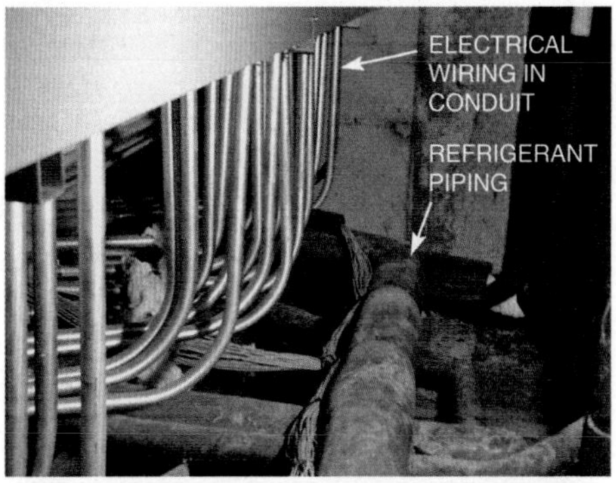

(B)

VENTILATOR PIPING

(C)

length. Design factors must be carefully considered to ensure correct oil return. Under-floor refrigerant pipes are ventilated to prevent mold and mildew. **Figure 26.35(C)** shows a venting system that directs the store's ceiling air to the back of the cases, through the under-floor refrigeration piping system, and through the access pit in the mechanical room.

26.11 FIXTURE TEMPERATURE CONTROL

Remote medium-temperature applications can be controlled without installing interconnecting wiring between the fixture and the equipment room. A power supply for the fixture must be located at the fixture, and planned off-cycle defrost can be accomplished at the equipment room through a time clock and a liquid-line solenoid valve. The clock and solenoid valve can be located at the case, but the equipment room provides a more central location for all controls. The clock can deactivate the solenoid, closing the valve for a predetermined time. The refrigerant will be pumped out of the individual fixture, the evaporator fan will continue to run, and the air in the fixture will defrost the coil. When the proper amount of time has passed, the solenoid-activated valve will open, and the coil will begin to operate normally. The compressor will continue to run during the defrost of the individual cases. The defrost times can be staggered by offsetting them. This is sometimes accomplished with a master time clock with many circuits. Use of electronic timing devices enables many different events to be controlled at the same time.

Low-temperature installations must have a more elaborate method of defrost because heat must be furnished to the coil in the fixture. This can be accomplished at the fixture with a time clock and heating elements when the power supply for the fixture is in the vicinity. It could also be accomplished with hot gas, but a hot gas line must then be run from the compressor to the evaporator, which would mean a third line to be run to each case that must also be insulated to keep the gas hot until it reaches the evaporator. As in the medium-temperature application, the defrost for each case can be staggered by different time clock settings, **Figures 26.36(A)** and **(B)**. This is the most efficient way to defrost because the heat for defrost comes from the other cases; however, electric defrost typically has fewer problems and is easier to troubleshoot. These methods of defrost are typical, but by no means the only, methods. Different manufacturers devise their own methods of defrost to suit their equipment.

26.12 THE EVAPORATOR AND MERCHANDISING

The evaporator is the device that absorbs heat into the refrigeration system. Because the evaporator is located on the outside of the system, it is visible to the store customer. At best, an evaporator is bulky, but it can be built in several ways. Manufacturers and their engineering staff must do a certain amount of planning to provide attractive fixtures that are also functional, **Figure 26.37**. Customer appeal must be considered when choosing equipment, which is why the

Figure 26.36 (A) Hot gas defrost for multiple cases. The condensed liquid from the evaporator in defrost is returned to the liquid line through the bypass and check valve around the expansion valve. The liquid then feeds the evaporators in the refrigeration mode. Three-way defrost diverting valves control the refrigerant flow in the defrost and refrigeration modes. (B) All three evaporators shown in the refrigeration mode.

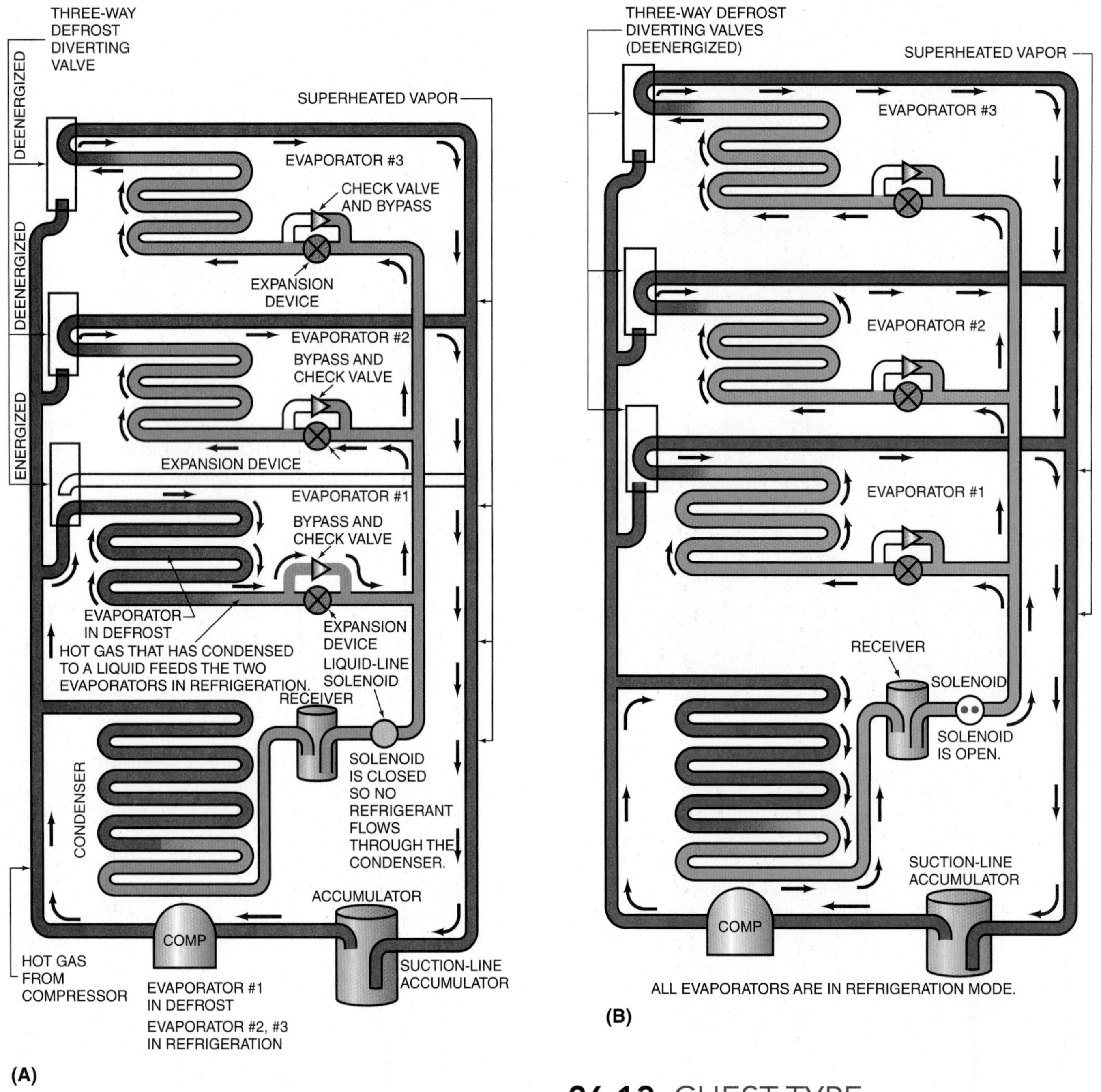

service technician may not understand the way in which some equipment is installed; customer convenience or merchandising may have played a major role in the decision. Each supermarket chain is trying to stay ahead of the competition in the marketplace, and customer convenience is part of that effort. Display fixtures are available in the following types: (1) chest (open, open with refrigerated shelves, closed) and (2) upright (open with shelves, with doors).

26.13 CHEST-TYPE REFRIGERATION FIXTURES

Chest-type reach-in equipment can be designed with an open top or lids and the fixture may have its own lighting so that the product can be seen better. Vegetables, for example, can easily be displayed in an open chest. The vegetables can be stocked, rotated, and kept damp with an automatic sprinkler system, **Figure 26.38**. Meat that is stored in the open is normally

Figure 26.37 A dairy case showing the location of the evaporator. Courtesy Hill PHOENIX Corporation.

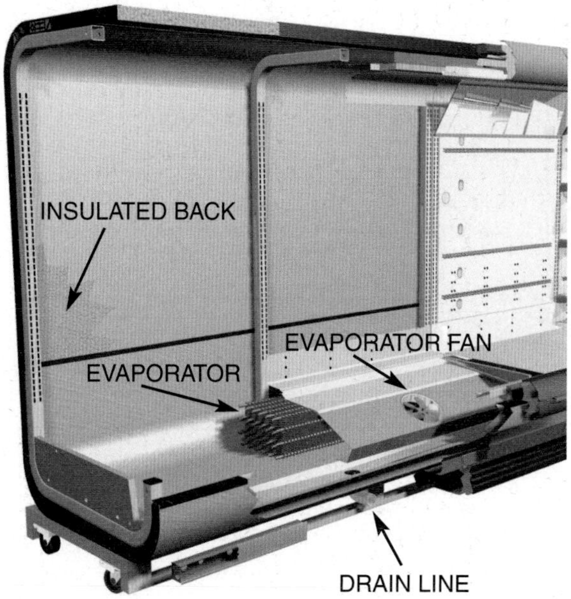

Figure 26.38 An open display case used to store fresh vegetables. The case has a drain, so the vegetables can be moistened by hand or with an automatic sprinkler. Source: KES Science and Technology, Inc.

packaged in clear plastic. At night, the product can be covered with plastic lids or film. In this type of fixture the evaporator may be at the bottom of the box. Fans blow the cold air through grills to give good air circulation. On open systems with high shelves, sinking cold air creates a curtain effect, which helps the case maintain its temperature. Service for the coil components and fan is usually performed through removable panels under the product storage or on the front side of the fixture. This type of appliance is normally placed against a wall, **Figure 26.39**.

The chest-type fixture can have a built-in condensing unit or be designed so that the condenser is located remotely, such as in an equipment room. If the condensing unit is furnished with the cabinet, it is typically located at the bottom of unit and near the front so that it may be serviced by removing the

Figure 26.39 The cutaway side view of a chest-type display case. The fans can be serviced from inside the case. The case is normally located with its back to a wall or another fixture. Based on Hill Phoenix.

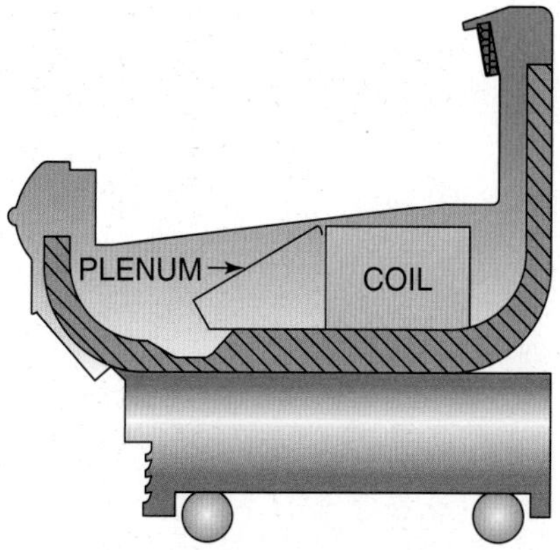

front panel. When the condensing unit is furnished with a fixture, it is called self-contained.

In some chest-type fixtures, refrigerated shelves are located at the top. These shelves must have correct airflow around them or they must have plate-type evaporators to maintain the correct food conditions. When evaporators are at the top, they are normally piped in series with the evaporators at the bottom, **Figure 26.40**.

The horizontal closed chest is a low-temperature fixture that normally stores ice cream or frozen foods. The lids may be lifted off, raised, or slid from side to side. These boxes

Figure 26.40 Multiple evaporators in a single case because refrigeration is needed in more than one place. Courtesy Tyler Refrigeration Corporation.

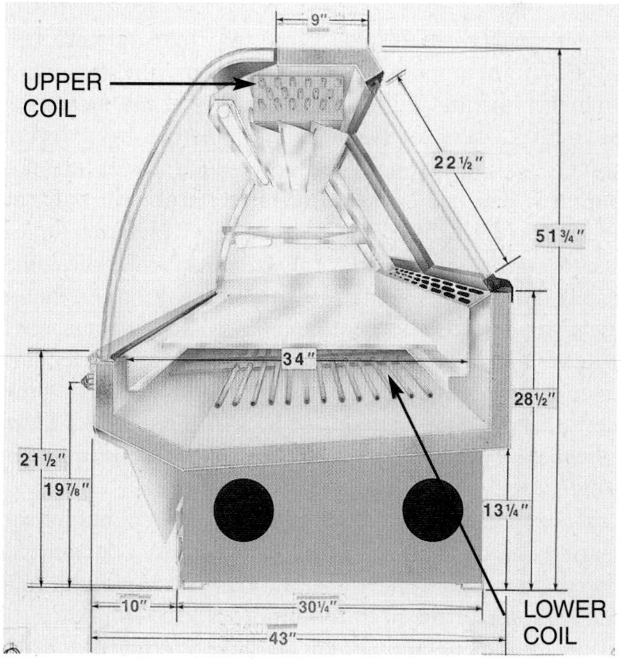

Figure 26.41 A display case with a walk-in cooler as its back wall. The display case can be supplied from inside the cooler. Courtesy Hill PHOENIX Corporation.

are not as popular now as in the past because the lids are a barrier to the customer. The upright closed display has doors through which the customer can look to see the product. This type of cooler may have either a self-contained or remote condensing unit. It can also be piped into a system with one compressor. This display is sometimes used as one side of the wall for a walk-in cooler. The display shelves can then be loaded from inside the walk-in cooler, **Figure 26.41**. These fixtures are found in many convenience stores.

26.14 CONTROLLING SWEATING ON FIXTURE CABINETS

Cabinet surfaces of display cases that may operate below the dew point temperature of the room (the temperature at which moisture will form) must have some means (usually small heaters) of keeping this moisture from forming. Cabinet heating normally occurs around doors on closed equipment. The colder the refrigerated fixture, the more need there is for the protection. The heaters, which are typically resistance-type wire run just under the surface of the cabinet, are called *mullion* heaters. *Mullion* means "division between panels." In refrigeration equipment, it is the panel between the doors that becomes cold, but any surface that may collect moisture from sweating needs a heater. In some equipment, there can be a large network of heaters, **Figure 26.42**. *Some of them can be thermostatically controlled or controlled by a humidistat based on the humidity in the store.*

More than ever before, air-conditioning (cooling) systems are used for controlling humidity in stores. You may have noticed how cold some supermarkets are. Air-conditioning removes some of the heat and moisture load from the refrigeration equipment, and at a higher level of efficiency. Because the air-conditioning system has a much lower compression ratio than any of the refrigeration systems, it will remove more moisture. Store managers realize that with low humidity in the store, they do not have to use as much mullion heat, and coils operating below freezing do not need as much defrost time.

Figure 26.42 A wiring diagram for mullion heaters on a closed display case. Several circuits can be applied to the heaters. This unit has two power supplies, 120 V and 208 V or 230 V.

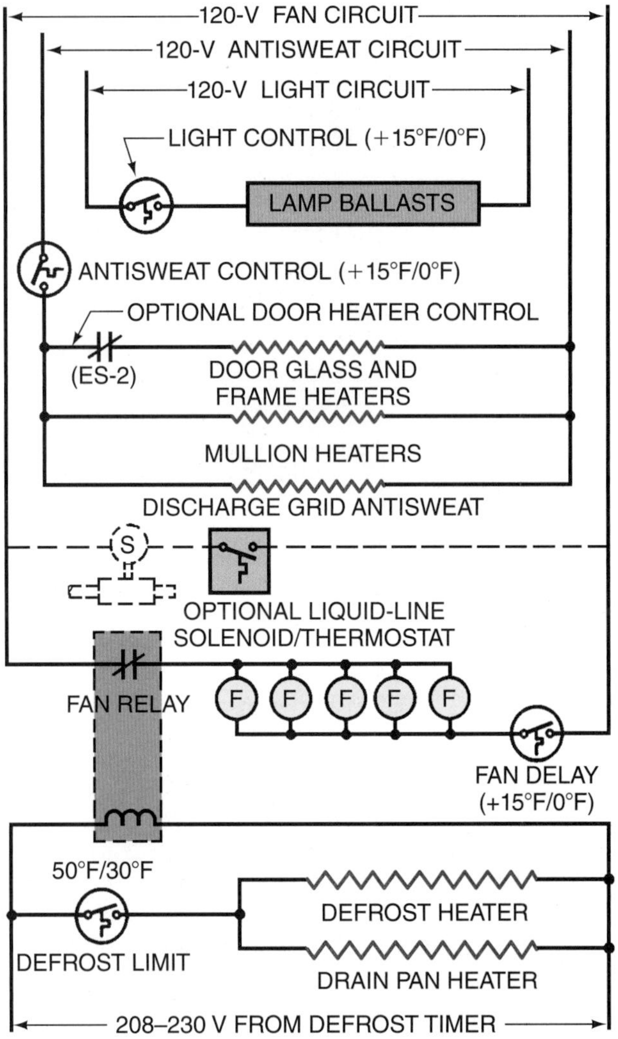

26.15 MAINTAINING STORE AMBIENT CONDITIONS

Humidity is taken out of a store in warm weather by the air-conditioning system and the display cases. The more humidity removed by the air-conditioning system, the less the refrigeration fixtures have to remove. This means less defrost time. Some stores maintain a positive air pressure with makeup air that is conditioned through the air-conditioning system. This

Figure 26.43 Planned infiltration known as ventilation. Courtesy Tyler Refrigeration Corporation.

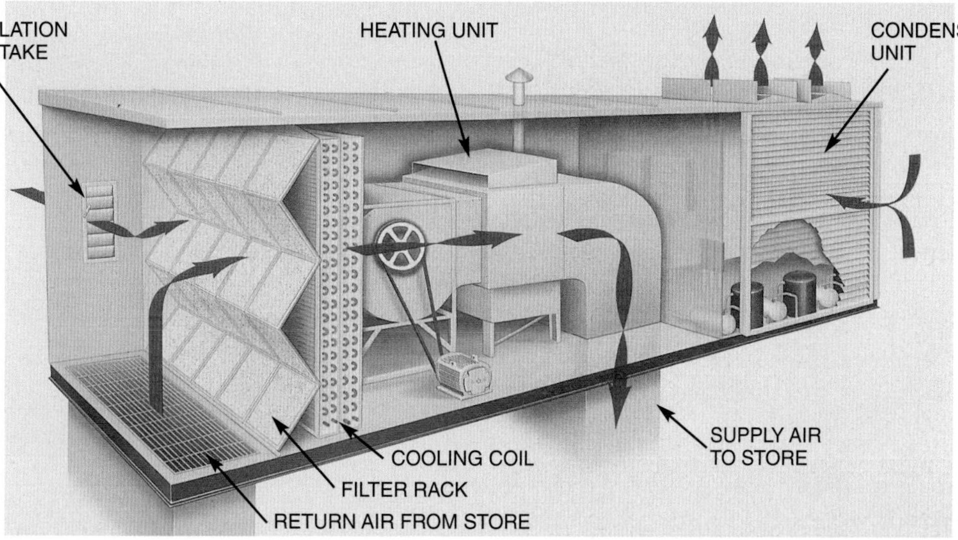

Figure 26.44 These doors are rugged and reliable. They are constructed of double-pane glass sealed on the edges to keep them from sweating between the glass. Based on Hill Phoenix.

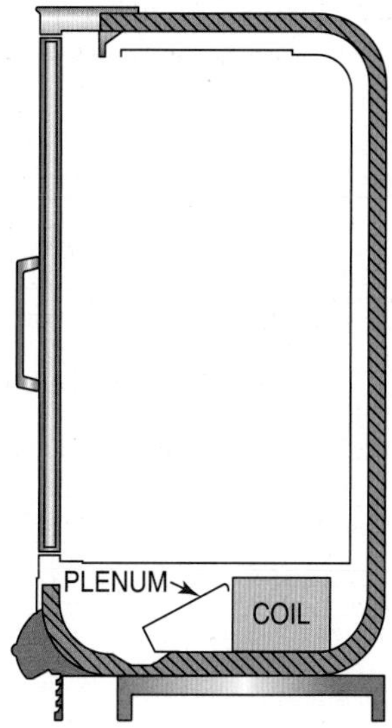

is different from taking in outside air randomly when the doors are opened by customers. It is a more carefully planned approach to the infiltration of the outside air, **Figure 26.43.** You will notice this system when the front doors are opened and a slight volume of air blows in your face.

In addition, the doors on display cases are usually constructed of double-pane glass and sealed around the edges to keep moisture from entering between the panes. These doors must be rugged because any fixture that the public uses is subject to abuse, **Figure 26.44.** The cabinets must be made of a strong material—stainless steel, aluminum, porcelain, or vinyl—that is easy to clean. The most expensive cabinets are made of stainless steel, and they are the longest lasting.

26.16 WALK-IN REFRIGERATION

Walk-in refrigeration equipment is either permanent or of the knock-down type. The permanently erected refrigerated boxes cannot be moved. Very large installations are permanent. When installing a walk-in fixture, the type of floor that it is going to be set on must be taken into consideration. Coolers can be installed on existing concrete floors as long as they are level. Due to the low temperature of freezers, they should have an insulated floor installed on top of the concrete.

KNOCK-DOWN WALK-IN COOLERS

Knock-down walk-in coolers are constructed of interlocking panels 1-in. to 4-in. thick, depending on the temperature needed inside the cooler. The construction is of a sandwich type, with metal on each side and foam insulation between. The metal in the panels may be galvanized sheet metal or aluminum, and the panels are strong enough that no internal support steel is needed for small coolers. They are shipped disassembled on flats and can be assembled at the job site. This type of cooler can be moved from one location to another and can be reassembled, **Figure 26.45.**

Walk-in coolers come in a variety of sizes and applications. One wall of the cooler can be a display, with the shelves filled from inside the cooler. The coolers are normally waterproof and can be installed outdoors. The outside finish may be aluminum or galvanized sheet metal. When the panels are pulled together with their locking mechanism, they become a prefabricated structure, **Figure 26.46.**

Figure 26.45 A knock-down walk-in cooler that can be assembled on the job. It can be moved at a later time if needed. The panels are foam with metal on each side, creating a structure that needs no internal braces. This prefabricated box is weatherproof and can be located inside or outside. Courtesy Bally Case and Cooler, Inc.

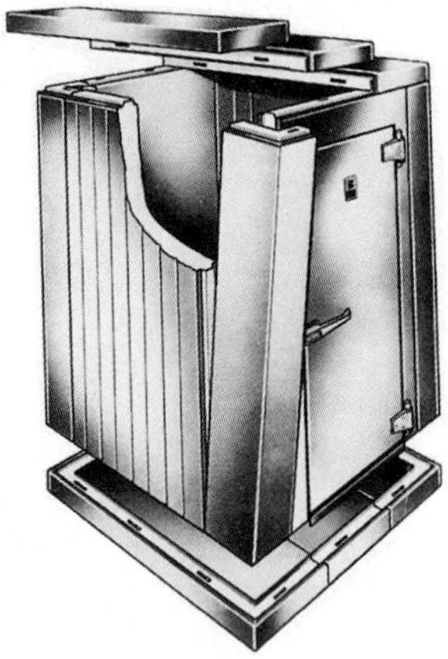

Figure 26.46 A locking mechanism for prefabricated walk-in cooler panels. They can be unlocked when the cooler needs to be moved. Courtesy Bally Case and Cooler, Inc.

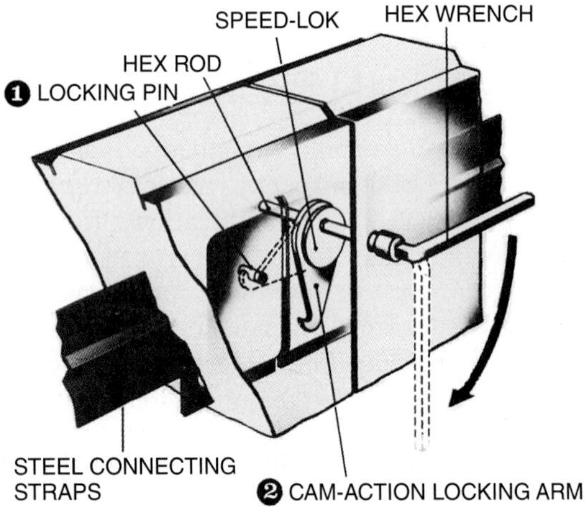

⚠ SAFETY PRECAUTION: *Walk-in cooler doors are very durable and must have a safety latch on the inside to allow anyone trapped on the inside to get out,* **Figure 26.47.** •

EVAPORATORS IN A WALK-IN COOLER

Refrigerating a walk-in cooler is much like cooling any large space. The methods that are used today take advantage of evaporator fans to improve air circulation and make evaporators compact. The following is a list of the types of systems used to refrigerate walk-in coolers:

1. Evaporators piped to condensers using field-assembled pipe
2. Evaporators with precharged piping
3. Package units, wall-hung or top-mounted, with condensing units built in

Figure 26.47 Walk-in cooler doors are rugged. They have a safety latch and can be opened from the inside. Courtesy Bally Case and Cooler, Inc.

Evaporators should be mounted in such a manner that the air currents blowing out of them do not blow all of the air out the door when the door is opened. They can be located on a side wall or in a corner. In most cases, the cooler is setup so that the fans do not blow directly on the product being stored. Also, the product being refrigerated should not be stored in a manner that will block the air from circulating through the cooler properly. These evaporators are normally in aluminum cabinets in which the expansion device and electrical connections are accessed at the end panel or through the bottom of the cabinet, **Figure 26.48**. Some installations have a fan switch that shuts the fan off when the door is opened to prevent cold air from being pushed out of the cooler. This works well if the door is not propped open for long periods of time. When the door is open, the compressor is operating without the fans; this can cause liquid flooding to the compressor. Some coolers deenergize the liquid-line solenoid valve and pump the refrigerant from the evaporator to the condenser and receiver while the fans are off. This works also but is an added step in the control sequence.

Figure 26.48 A fan coil evaporator used in walk-in coolers and freezers. Photo by Jason Obrzut.

Figure 26.49 This wall-hung "saddle unit" actually hangs on the wall of a cooler. The weight is distributed on both sides. This is a package unit that only has to be connected to the power supply to be operable. Courtesy Bally Case and Cooler, Inc.

PACKAGE REFRIGERATION FOR WALK-IN COOLERS

Wall-hung or ceiling-mount units are package units, **Figure 26.49**. They can be installed by personnel who do not understand refrigeration, somewhat like installing a window air conditioner. However, proper installation techniques must be followed. Thus, package units are very popular for some applications because costs can be minimized. Care must be used as to where the condenser air is discharged outside. It must have room to discharge without recirculating to the air inlet, or head pressure problems will occur. These units are factory assembled and require no field evacuation or charging. They come in high-, medium-, and low-temperature ranges. Only the electrical connections need to be made in the field.

26.17 CONDENSATE REMOVAL

The bottom of the evaporator cabinet contains the drain pan for the condensate that must be piped to the outside of the cooler. When the inside is below freezing, heat has to be provided to keep the condensate in the line from freezing, **Figure 26.50**. Normally, this heat is provided by an electrical resistance heater that can be installed in the field. The line is piped to a drain and must have a trap to prevent the atmosphere from being pulled into the cooler. The drain lines also should be sloped downward about ¼ in. per foot for proper drainage. The line and trap must be heated if the line is run through below-freezing surroundings. These drain line heaters sometimes have their own thermostats to keep them from using energy during warm weather. Thermostatically controlled heaters are used when the drain line is run outside. The heaters can cycle off when the outside temperature is warm. When the drain line is located inside the box, the heater is energized all of the time. Heat tape is usually used in this case.

Figure 26.50 Drain pan heaters to keep pans from freezing. Source: Larkin Coils, Inc.

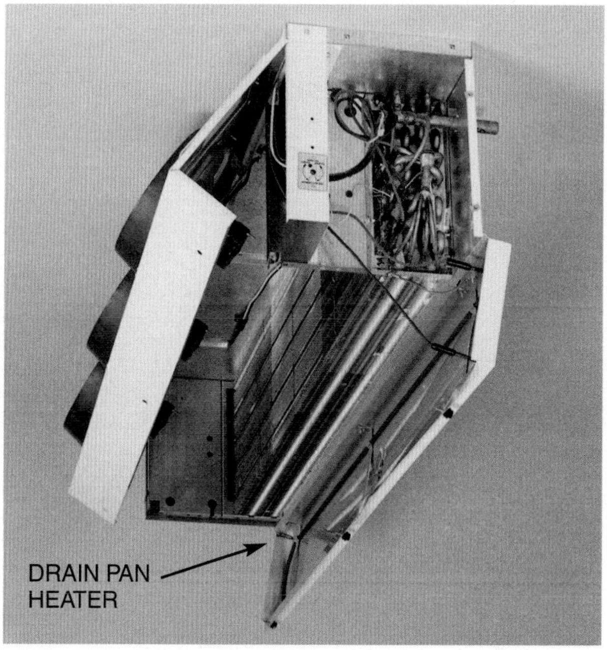

26.18 REFRIGERATION PIPING

There are two methods of installing refrigeration piping for walk-in coolers. One method is to pipe the cooler in the conventional manner. The other is to use precharged tubing, called a *line set*. In the conventional method, the installing contractor usually furnishes the copper pipe and fittings as part of the agreement. This piping can have straight runs and factory elbows for the corners. When the piping is completed, the installing contractor has to leak check, evacuate, and charge the system. This requires the service of an experienced technician.

Precharged tubing is furnished by the equipment manufacturer. It is sealed on both ends and has quick-connect fittings. This piping has no fittings for corners and must be handled with care when bends are made. The condenser and the evaporator come with their operating charges already installed, and the tubing has the correct operating charge for the length chosen for the job. This has some advantages because the installation crew does not have to balance the operating charge in the system. The system is factory sealed with field connections, **Figure 26.51**; no soldering or flare connections have to be made. The system can be installed by someone with limited experience.

If, because of miscalculation, the piping is too long for the installation, a new line set can be obtained from the supplier or the existing line set can be cut to fit. If the existing line set is to be altered, refrigerant can be recovered, the line cut, soldered, leak checked, evacuated, and charged

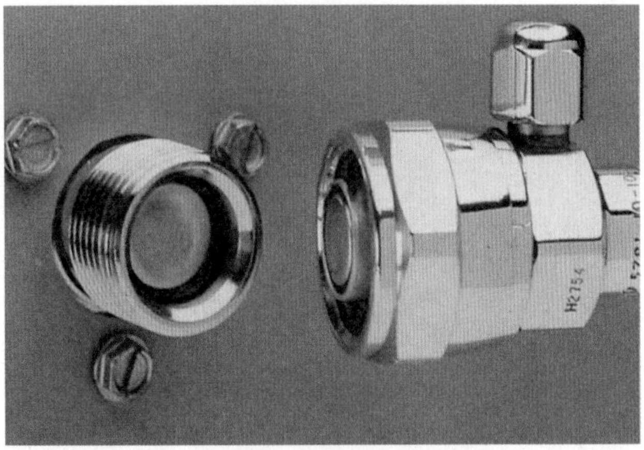

Figure 26.51 Quick-connect fittings providing the customer with a system that is factory sealed and charged. Courtesy Aeroquip Corporation

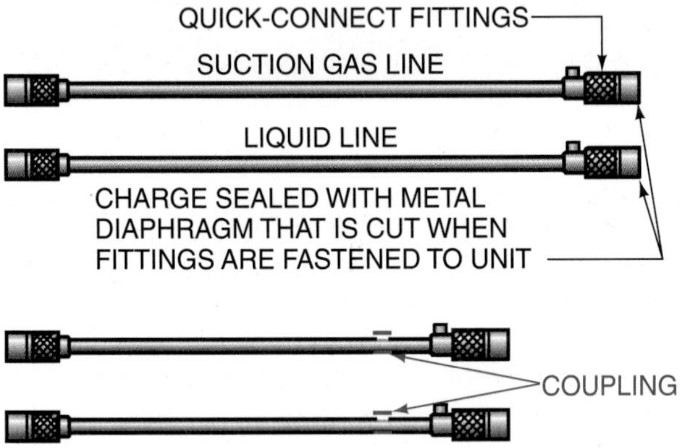

Figure 26.52 Altering quick-connect lines.

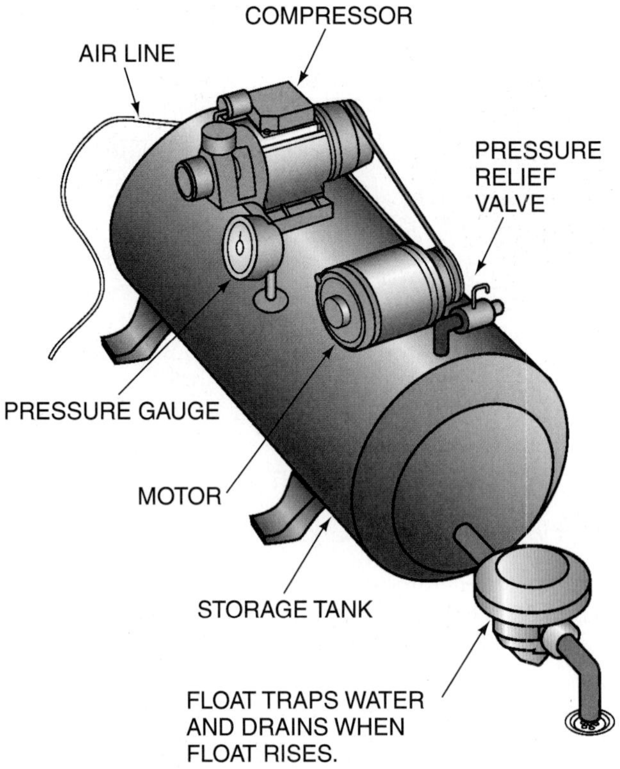

Figure 26.53 An air compressor showing a moisture-trapping system.

to specifications. If the line set is too long, the extra tubing should be coiled and placed in a horizontal position to ensure good oil return. **Figure 26.52** is an illustration of how the line set can be altered.

NOTE: *The suction line is charged with vapor. The liquid line has a liquid charge. It contains about as much liquid as may be pulled into it under a deep vacuum. Recover the charge in each tube and cut to the desired length. Fasten together with couplings.* •

SAFETY PRECAUTION: *Leak check, evacuate, and charge (1) the suction line with vapor, and (2) the liquid line with liquid. It is now a short line set with the correct charge. See Unit 8, "Leak Detection, System Evacuation, and System Cleanup," and Unit 10, "System Charging."* •

26.19 REFRIGERATED AIR DRIERS

Many applications use compressed atmospheric air; this must be dehydrated. Compressed air leaves a compressor saturated with moisture. Some of this moisture condenses in the air storage tank and is exhausted through a float, but the air is still very close to saturated as it leaves the storage tank. The air used for special manufacturing processes or air used for controls, called "pneumatic control," must be dry. Air may be dehydrated using refrigeration.

Refrigerated air driers are basically refrigeration systems located in the air supply, after the storage tank. The air may be cooled in a heat exchanger and then moved to the storage tank, where much of the water will separate from the air. The water may be drained using a float that rises and provides for drainage at a predetermined level, **Figure 26.53**. The air then passes through another heat exchanger where its temperature is reduced to below the dew point temperature (the point where water will condense from the air). The heat exchanger consists of an evaporator that normally operates at a temperature just above freezing, **Figure 26.54**.

The system typically runs 100% of the time because it does not know when there will be a demand for air flowing through the system. When there is no airflow, there is virtually no load on the refrigeration system. The only time there is a load on the system is when air is passing through that needs to be dried. Because the refrigeration system is running, in case there is a demand, a false load must be placed on the evaporator when there is no load. This

Figure 26.54 A refrigerated air drier installation.

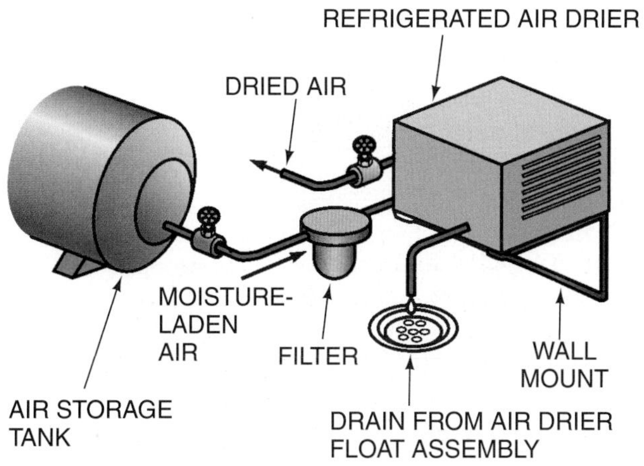

Figure 26.55 The flow schematic for a refrigerated air drier. Courtesy Van Air Systems, Inc.

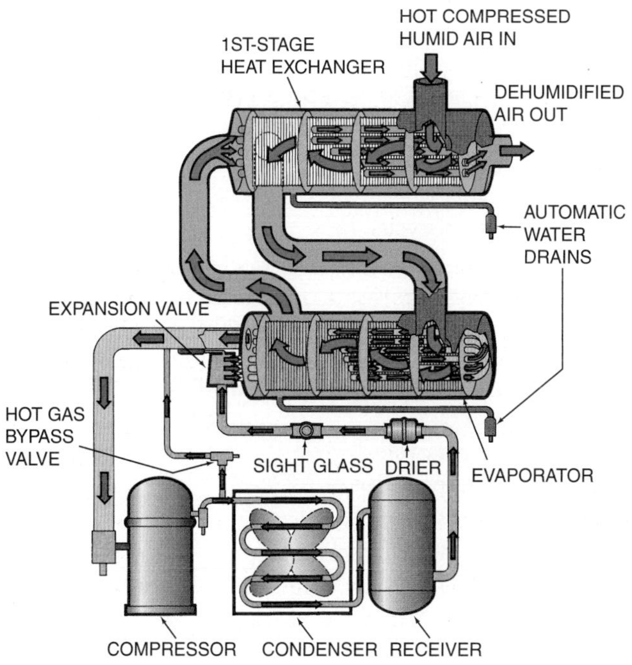

is usually accomplished with a special refrigeration valve called a hot gas bypass valve, **Figure 26.55**. This valve is situated between the hot gas line and the inlet to the compressor or the evaporator. It monitors the pressure in the evaporator and allows hot gas to enter the suction line at any time there is no or reduced load on the evaporator by monitoring the suction pressure with a pressure-sensitive valve. The hot gas bypass valve is called a pressure regulator because it does not allow the pressure in the evaporator to drop below the set point. The difference between this pressure regulator and others discussed in this text is that it is located between the high- and low-pressure sides of the system. The set point is usually just above the freezing point of water. For example, for R-22, the set point of the hot gas bypass valve may be 61.5 psig (35°F). The valve will not allow the evaporator pressure to go below 35°F, so the evaporator will not freeze during any no- or reduced-load situations.

Figure 26.56 Refrigerated air drier with accompanying air compressor, controller, and pressurized air storage tank. Courtesy Ferris State University. Photo by John Tomczyk.

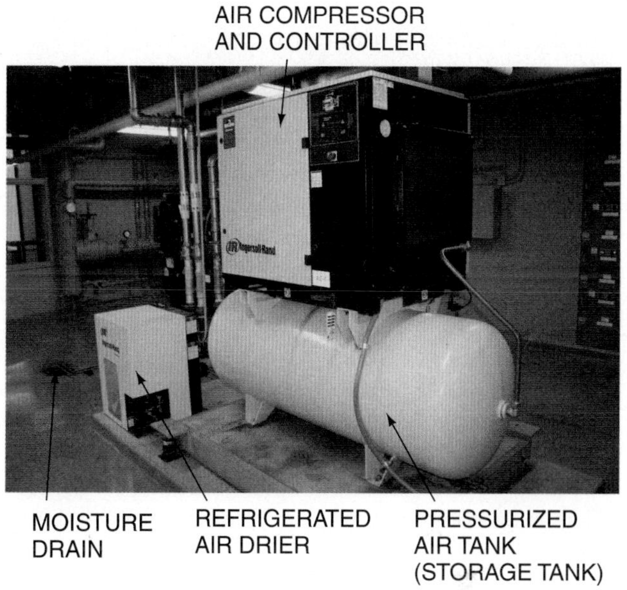

Refrigerated air driers are normally self-contained refrigeration systems that may be air- or water-cooled, **Figure 26.56**. The smaller driers can have a hermetic compressor; semihermetic units are used for larger driers. The air line is piped to and through the evaporator of the air drier and then moves on to the system where it is to be used. The water is collected and exhausted to a convenient drain. The evaporator may be one of several types: a pipe within a pipe or a honeycomb type are common.

The service technician should understand the manufacturer's intent for each application. At times, the technician must improvise when replacing components that are not available, which may happen as equipment becomes outdated.

REVIEW QUESTIONS

1. The two broad categories of display cases are
   A. open- and closed-type.
   B. high- and low-pressure systems.
   C. vertical and horizontal cases.
   D. indoor and outdoor fixtures.
2. How are conditions maintained in open display cases when there are high shelves?
   A. Evaporating water from misting systems removes heat.
   B. A continuous run cycle handles the constant heat load.
   C. Sinking cold air creates a curtain effect.
   D. Warm air rises out of the cooler, leaving only cool air.
3. A mullion heater is used to
   A. prevent ice formation on the TXV or capillary tubing.
   B. defrost the evaporator of medium-temperature fixtures.
   C. prevent the formation of moisture on refrigerated fixtures.
   D. keep the receiver warm in low-ambient conditions.
4. The most common temperature ranges for refrigeration systems are
   A. low- and ambient-temperature.
   B. high-, medium-, and low-temperature.
   C. high- and low-temperature.
   D. ambient- and variable-temperature.
5. The two common methods for rejecting heat from refrigerated cases are
   A. inside the store (recirculated-type) or outside the store (ambient-dependent).
   B. inside the store (ambient-dependent) or outside the store (geothermal).
   C. inside the store (split-system) or outside the store (package-system).
   D. inside the store (package-system) or outside the store (split-system).
6. When the compressors are located in the equipment room, the piping is routed
   A. in floor chases or in plastic pipes under the floor.
   B. in suspended ceilings.
   C. between the walls.
   D. in back of the walls.
7. When one large compressor is used, what desirable feature should it have?
   A. Discharge muffler
   B. Suction muffler
   C. Cylinder unloaders or variable-speed drive
   D. Hot gas bypass
8. Parallel compressor systems that apply two or more compressors to a common suction header, a common discharge header, and a common receiver are called
   A. loaded systems.
   B. rack systems.
   C. booster systems.
   D. split systems.
9. The __________ of a gas is the highest temperature that the gas can have and still be condensable by the application of pressure.
   A. critical point
   B. triple point
   C. dew point
   D. flash point
10. CO_2 (R-744), ammonia (R-717), propane (HC-290), butane (HC-600), and isobutane (HC-600a) are all examples of
    A. HFCs.
    B. HFOs.
    C. secondary fluids.
    D. natural refrigerants.
11. The term used to describe the point at which CO_2 exists as all three phases (vapor, liquid, and solid) simultaneously in equilibrium is
    A. critical point.
    B. triple point.
    C. transcritical point.
    D. subcritical point.
12. A CO_2 refrigeration cycle that takes place above the refrigerant's triple point and below the refrigerant's critical point is called a
    A. subcritical cycle.
    B. transcritical cycle.
    C. natural cycle.
    D. supercritical cycle.
13. A system that operates at the lowest possible head pressure by taking advantage of cool ambient temperatures that are used as the condensing medium, is said to operate with a
    A. transcritical cycle.
    B. fixed head pressure.
    C. floating head pressure.
    D. subcritical cycle.
14. In low-temperature walk-in systems, the condensate drain line
    A. must terminate inside of the box.
    B. must be heated to prevent freezing.
    C. must be piped above the evaporator for proper drainage.
    D. should be pitched toward the evaporator.
15. A type of cooler that is made up of interlocking panels that are assembled at the jobsite, which can be disassembled to be relocated, is called a(n)
    A. package cooler.
    B. reach-in fixture.
    C. upright chest fixture.
    D. knock-down cooler.

16. Refrigerated air driers typically use a(n) _______ to place a "false load" on the refrigeration circuit when there is no heat load.
   A. evaporator pressure regulator
   B. condenser flooding device
   C. hot gas bypass valve
   D. condenser air recycler
17. On a system with multiple evaporators that are kept at different temperatures, what component is used on the higher temperature evaporators to prevent their pressures from dropping below the desired setting?
   A. Evaporator pressure regulator
   B. Hot gas bypass valve
   C. Pressure transducer
   D. Crankcase pressure regulator
18. A secondary fluid or refrigerant is usually
   A. a methane-based refrigerant.
   B. a type of distilled water solution.
   C. a low-pressure HFC refrigerant.
   D. a low-temperature antifreeze solution.
19. Exactly matching the refrigeration load with the capacity of refrigeration compressors is referred to as
   A. capacity control.
   B. load shedding.
   C. load paralleling.
   D. load matching.
20. Compressors on parallel compressor racks, which can be large or small and are usually dedicated to the coldest cases, are called
   A. satellite or booster compressors.
   B. remote compressors.
   C. off-site or off-location compressors.
   D. auxiliary compressors.
21. Describe a parallel compressor system and list five advantages of using one.
22. What are three disadvantages of parallel compressor systems?
23. What is the difference between an even and an uneven parallel compressor system?
24. List some advantages of using modular stand-alone refrigeration systems.
25. Briefly explain what is meant by a surge-type receiver.
26. Briefly explain what is meant by a subcritical refrigeration cycle and list its four processes.
27. Briefly explain what is meant by a transcritical refrigeration cycle and list its four processes.
28. Explain the function and advantages of a secondary-fluid system.
29. Explain the main idea behind pressurized liquid systems and list their advantages.
30. Explain how the oil separator, oil reservoir, pressure differential valve, and oil-level regulators work together to maintain the correct oil level in parallel compressors.
31. Explain the difference between a synthetic refrigerant and a natural refrigerant.
32. Explain the function of a system pressure-regulating valve (SPR).
33. Describe the most common defrost methods used in low- and medium-temperature refrigeration systems.
34. Explain how mullion heaters prevent refrigeration fixtures from sweating.
35. Discuss applications that require the use of a refrigerated air-dry unit.

UNIT 27

COMMERCIAL ICE MACHINES

OBJECTIVES

After studying this unit, you should be able to

- describe the basic refrigeration cycle for ice flake machines.
- discuss basic troubleshooting for ice flake machines.
- state the purpose of the water fill system in an ice flake machine.
- explain the purpose of a flush cycle in an ice flake machine.
- state the purpose of a bin control in an ice flake machine.
- read and interpret ice production and performance charts for ice machines.
- describe how crescent-shaped ice is made.
- describe how cell-type ice cubes are made.
- explain the sequence of operation of an ice machine.
- describe the purpose of a harvest cycle in an ice machine.
- state the purpose of microprocessor controls in ice machines.
- explain what is meant by input/output troubleshooting for microprocessors.
- discuss the importance of water and ice quality in ice making.
- discuss the difference between cleaning and sanitizing an ice machine.
- define water filtration and treatment.

SAFETY CHECKLIST

- ✔ Follow manufacturers' instructions when using cleaning and sanitizing chemicals.
- ✔ Wear protective clothing and eyewear when handling cleaning and sanitizing chemicals.
- ✔ Be careful not to get your hands or clothes caught in any moving ice maker parts.
- ✔ Ice machines are located in wet environments, so extra precautions must be taken to protect against electrical shock and slipping.
- ✔ Observe all electrical safety precautions.
- ✔ Be careful at all times and use common sense when working on or around ice-making equipment.

27.1 PACKAGED-TYPE ICE-MAKING EQUIPMENT

Ice making is accomplished in a temperature range that is somewhat different from conventional low-, medium-, or high-temperature refrigeration systems. The ice is made with evaporator temperatures of about 10°F with the ice at 32°F. Most refrigeration applications use tube and fin evaporators and a defrost cycle to clear the ice buildup from the evaporator. Ice making is accomplished by accumulating the ice on some type of evaporator surface and then catching and saving it. The process of storing this ice is called the harvest cycle.

Large, commercial, block ice makers use cans and freeze the ice in the cans. In this unit, we will discuss only small ice-making machines, such as those found in commercial kitchens, motels, and hotels. These ice makers are usually of the package type. The power is supplied to these appliances by a power cord, or in some cases the machines are hard-wired directly to the electrical power distribution panel. Some ice makers may be of the split-type with the condensing unit located remotely.

Package ice machines store their own ice at 32°F in a bin below the ice maker. This bin is refrigerated by the melting ice in the bin (which explains the storage temperature of 32°F), so there is some melting. On hotter days, more melting will occur. A drain must be provided to properly direct the melted ice and any water that may overflow during the ice-making process. Do not confuse an ice-making machine with an ice-holding machine such as those seen on the outside of a convenience store or service station. Ice-holding machines hold ice, usually in bags, at a temperature well below freezing. This ice is made and bagged at one location and then stored and dispensed at the retail location.

Most package ice makers are air-cooled and must be located where the correct airflow across the condenser can be maintained. Some package ice makers are water-cooled wastewater systems where the water used to cool the condenser is used once and then passed down the drain.

The installation of a package machine is relatively simple. The following four steps outline the basic installation of these systems:

1. Set and level the machine.
2. Supply power, plug, or hard wire.
3. Provide a water supply.
4. Provide a drain for the ice bin.

Package ice machines make two types of ice, flake ice and several forms of cube (solid) ice.

27.2 MAKING FLAKE ICE

Flake ice is normally made in a vertical refrigerated cylinder surrounded by the evaporator on the outside. Flake ice is very thin pieces of ice often seen in restaurants and vending

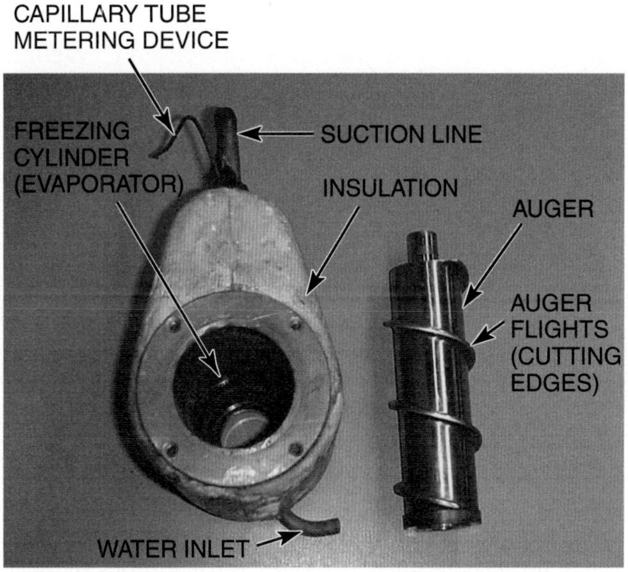

Figure 27.1 An evaporator, freezing cylinder, and auger of a flake ice machine. Courtesy Ferris State University. Photo by John Tomczyk

machine cups. Flake ice melts fast because it is very light and the surface area is large. Solid forms of ice last longer because there is more weight with less surface area. Recall from Unit 1 that it takes 144 Btu to make or melt one pound of ice. It takes less ice, by weight, to fill a cup with flake ice. Solid ice costs more, and some of it is likely to be thrown out.

Flake ice is formed by maintaining a predetermined water level inside a cylinder. An auger is constantly turning inside the cylinder, scraping the formed ice off the evaporator surface, **Figure 27.1**. The auger forces the ice upward and out a chute on the side where it moves by gravity to the ice storage bin. The auger's cutting surfaces, called *flights,* cut or shave the ice off the interior of the flooded evaporator cylinder. In some modern flake ice machines, the flights on the auger add pressure to the ice toward the top of the auger. This pressurized ice can then be cut or extruded through small openings near the top of the auger to make many shapes and forms of ice. This type of ice-making machine has a constant harvest cycle that continues as long as the compressor and auger are operating.

A motor and gear assembly is used to turn the auger. The motor may be located on the top or the bottom of the auger. **Figure 27.2** shows a gear motor machine located at the bottom. The auger is in a vertical position and must have a shaft seal where the auger leaves the gearbox. If the gear motor is on top of the auger, the shaft seal prevents gear oil from seeping into the ice, **Figure 27.3**. When the gearbox is on the bottom of the auger, the shaft seal prevents water from entering the gearbox, **Figure 27.4**.

The level of the water in the evaporator is determined by a float chamber located at the top of the evaporator. This float level is critical. If the level is too low, the suction pressure will drop, making the evaporator colder. A colder evaporator will

Figure 27.2 A flake ice maker, its evaporator, and its auger. The auger turns and shaves the ice off the evaporator. The tolerances are very close. Based on Scotsman Ice Systems.

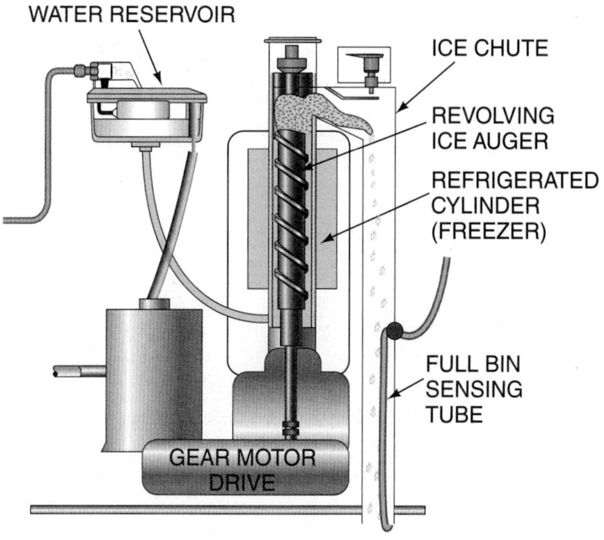

Figure 27.3 The auger is on top of the ice maker. The shaft seal prevents oil from seeping into the ice.

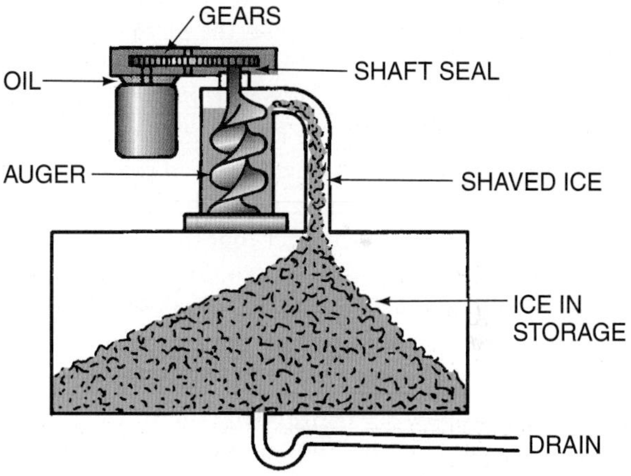

Figure 27.4 The gearbox is on the bottom of the auger. A seal leak would allow water to enter the gearbox.

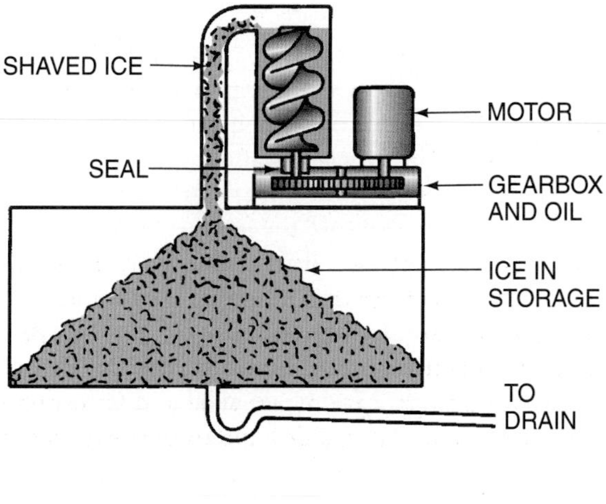

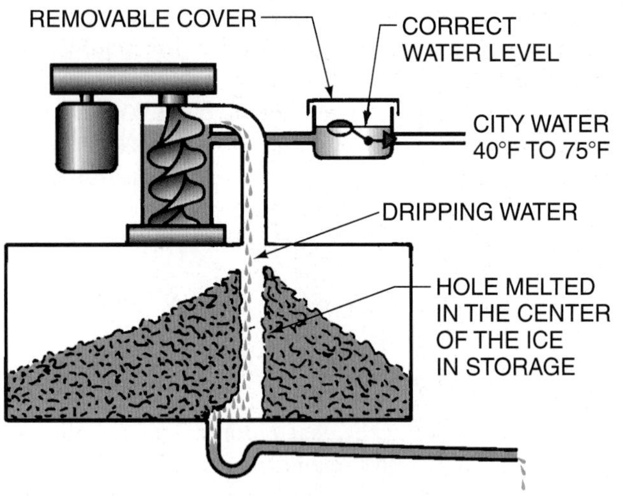

Figure 27.5 When the float level is too high, water will run over and melt ice that has been made. The fresh water must be refrigerated to freezing after start-up.

create hard ice, which the auger must cut. This will put an increased load on the motor. The ice may be too hard, making the ice maker less efficient. In extreme cases, the auger may freeze to the evaporator. In this case, the power of the motor and gears may turn the evaporator out of alignment. If the water level is too high in the evaporator, the auger is doing more work because there is more ice to cut. The motor driving the auger may have a manual reset overload that will shut the motor off and stop the compressor. When the motor reset button must be reset often, suspect that the water level is the problem. Check the seat in the float chamber. A float that will not properly control the water level may allow water to pour over the top of the chamber during the off cycle. This would impose an extra load on the refrigeration capacity because it must refrigerate this extra water to the freezing point to make ice. Water overflowing the evaporator will drip into the ice bin and melt ice that is already made, causing a loss of ice capacity, **Figure 27.5**.

THE REFRIGERATION CYCLE

The refrigeration cycle for ice flakers is simple and straightforward. A TXV, AXV, or capillary tube can be used as the metering device. However, most modern ice flakers incorporate TXVs for efficiency and capacity reasons. If a TXV is used, low evaporator superheat readings of 3°F to 4°F are common to make sure the ice sheet is full. Often, the evaporator is just a tube coiled around a cylinder. Sometimes, multiple expansion valves are used to create parallel circuits for better refrigerant distribution throughout the evaporator. The tube and cylinder are then both covered with thick, rigid insulation, **Figure 27.1**. Once insulated, the evaporator tubes are not visible. The rest of the refrigeration system is no different from any other refrigeration system containing a suction line, compressor, water- or air-cooled condenser, liquid line, metering device, and evaporator. An EPR valve is sometimes used to establish a minimum pressure in the evaporator. An AXV is used when a constant evaporator pressure is to be maintained.

Flake ice machines are usually critically charged systems. Weighing in the correct charge allows the ice machine to operate at its greatest potential under all conditions. Charging by the frost line, sight glass, or pressure method will likely lead to system inefficiencies, especially when a critical charge is called for by the manufacturer. Symptoms of an undercharged system, usually from a refrigerant leak, include an overheated compressor motor or reduced ice production.

Often, flake ice machines can be retrofitted to a parallel refrigeration system used in larger supermarkets. In this situation, the compressor and condenser from the parallel refrigeration system are used. Pressurized liquid from the liquid header of the parallel system is routed into the ice flaker. The pressurized liquid then flows to the metering device of the packaged ice flaker. The refrigeration effect takes place in the flooded evaporator freezing cylinder, and ice is cut off the freezing cylinder by the auger. An EPR valve is used on the suction line in this application to prevent too low an evaporator pressure from occurring. Liquid-line solenoid valves are also used to start and stop the liquid flow into the expansion valve. The suction gas exiting the evaporator is routed back to the suction header of the parallel compressor system. Parallel refrigeration systems are covered in detail in Unit 26, "Applications of Refrigeration Systems."

BASIC TROUBLESHOOTING

When water is frozen in the evaporator freezing cylinder, minerals from the water are left behind and often become attached to the interior cylinder walls and other surface areas of the ice flaker. Using an approved ice machine cleaner will help dissolve these minerals. Always follow the manufacturer's instructions when using an ice machine cleaner or damage to the ice machine can result. When minerals do build up on the interior of the evaporator, they add resistance to the flow of ice that is being cracked, cut, or broken away by the auger and then pushed out of the cylinder. This dirty condition causes the evaporator to run at a lower suction pressure and thus a colder evaporator temperature, which results in the ice being very hard. Often, noises like two pieces of rigid Styrofoam being rubbed together, a crunching sound, or a loud squeal will come from the freezing cylinder when minerals are allowed to build up. A regular clicking sound usually indicates a mechanical problem, which could be in the drivetrain and is probably a chipped gear. A high-pitched noise probably indicates a worn auger drive motor. It is a good idea to start troubleshooting by simply observing and listening to the ice flaker. Ask the owner if there have been any strange or unusual

noises coming from the ice machine. If the answer is yes, a cleaning may be needed. The auger and freezing cylinder may have to be removed and inspected, **Figures 27.6(A)** and **(B)**. An ice flaker's evaporator cylinder is made of either stainless steel or a more porous material such as brass or copper. If made of copper, the cylinder must be plated as required by the National Sanitation Foundation (NSF). Nickel is typically used as the plating material. Stainless steel is much more durable and resists corrosion better than copper or brass. However, copper facilitates more efficient heat transfer. Manufacturers who use stainless steel for the ice-making surfaces usually incorporate design features that enable the heat transfer rate of the evaporator surface to closely match that of copper.

Figure 27.6 (A) A service technician inspecting the freezing cylinder in an ice flaker. Source: Hoshizaki America, Inc. (B) A service technician removing the auger from the freezing cylinder of an ice flaker. Source: Hoshizaki America, Inc.

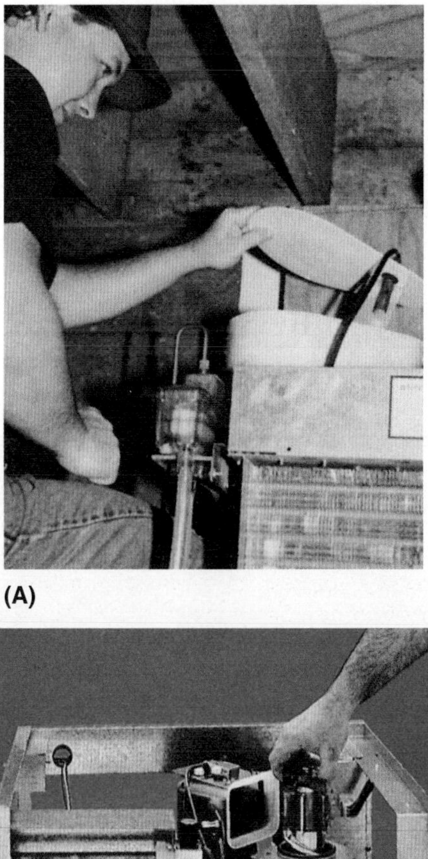

(A)

(B)

The auger and evaporator surface must be very clean and free from any score marks or scars. If a bearing failure has occurred, and the auger has touched the evaporator or freezing cylinder's surface, usually either the evaporator or the auger, or sometimes both, must be replaced. Always follow the manufacturer's instructions for disassembly. Drain the freezing cylinder and allow it to dry before inspecting it. A wet piece of equipment may look clean, but once it has had time to dry, mineral buildup will be obvious. Mineral deposits also cause stress to the auger motor, auger bearings, and the gear reducer. It is these components that cause most of the service calls for ice flakers, not the refrigeration system.

Most ice flakers have a gear reduction drive that reduces the auger drive motor's speed from about 1725 rpm to 9 to 16 rpm. These drives are generally maintenance-free. However, if water has entered the gear case, it should be disassembled and inspected. Often, there is a coupling between the auger and the output shaft of the gear reducer. This coupling may need to be lubricated on a regular basis. All ice flakers have bearings that allow the auger to rotate. The bearings can be either a sleeve bearing or a ball bearing. Bearings are a high-wear item and annual inspection is recommended. Consult the manufacturer's maintenance manual to become familiar with the ice machine before starting this procedure. **Figure 27.7** shows a self-contained ice flake machine with one side cover removed. Notice the water reservoir, the gear drive, the insulated evaporator, and the ice chute that delivers ice to the storage bin. **Figure 27.8** is an exploded view of most of the mechanical parts that make up an ice flaker.

The gear motor assembly is more prone to failure than any other part in the ice flaker because of the pressures and stresses acting on it. To make the servicing and repairing of these appliances easier, some manufacturers decided to use

Figure 27.7 A self-contained ice flaker. Courtesy Scotsman Ice Systems

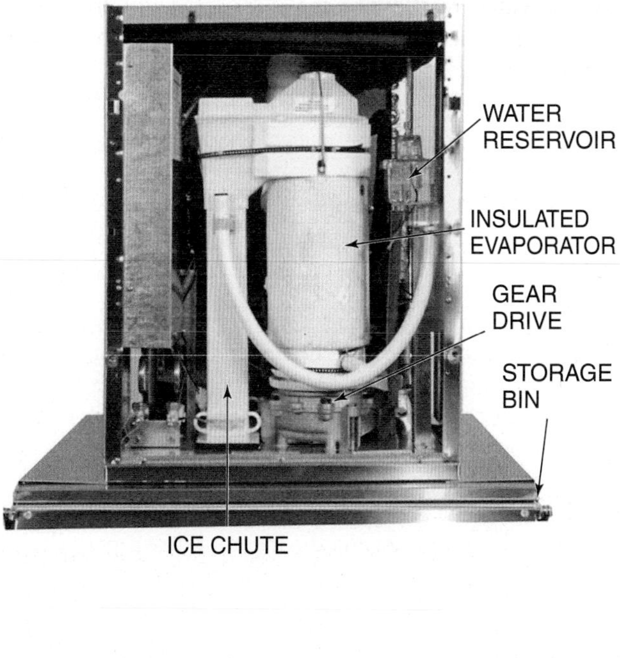

Figure 27.8 An exploded view of most of the mechanical parts of an ice flaker. Source: Hoshizaki America, Inc.

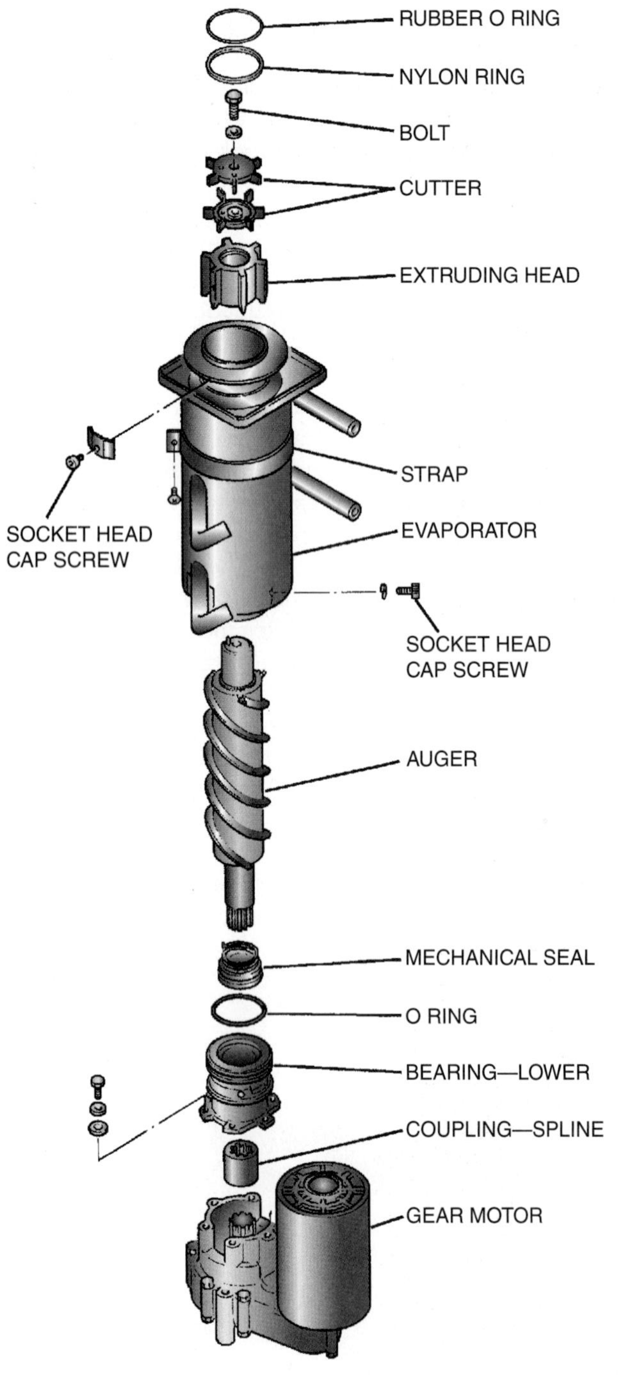

open-type gear assemblies. This means the gear assembly is open to the atmosphere. Usually, there is a hole in the housing of the gear assembly for hot grease to exit during long run times with extreme stress on the drive motor, which causes high amperage draw and excessive heat. The drawback to the open-type gear drive assembly is that moisture can enter the hole in the drive housing during cool-down periods. This deteriorates the lubricating effect of the grease in the drive housing. Excessive gear wear and shorter gear life will be the outcome. Some manufacturers use gears made of a fibrous material designed to strip or break when the motor and gearbox are put under heavy loads.

It is important that the auger does not wobble within the freezing cylinder or evaporator. Sleeve-type alignment bearings made from a slippery poly material with a graphite base keep the auger from wobbling on some machines. Others use grease-lubricated steel roller bearings. Grease can leak out, causing the bearings to overheat. This usually causes auger wobble. Auger wobble causes the cutting edges of the auger to scrape the freezing cylinder's wall. Excessive wear on all surfaces, and metal shavings in the ice, which can pose health problems, will result.

Manufacturers of ice flakers install thermal overload controls in the motor windings or a manual-reset, current-type overload control. In fact, most modern ice machines have solid-state control boards with status indicators that act as self-diagnostic tools. There are usually light-emitting diodes (LEDs) or indicator lights to inform the service technician whether the ice flaker is operating normally, **Figure 27.9**. The lights may indicate the status of the auger drive motor, auger relay, compressor relay, low- or high-pressure control, reservoir level, bin level, starting mode, condenser temperature, discharge line temperature, and more. The type of control board and its control scheme will vary from model to model. When the

Figure 27.9 A solid-state control board with light-emitting diode (LED) status indicator lights used for self-diagnostics. Courtesy Scotsman Ice Systems

drive motor is stressed from mineral buildup on the evaporator freezing cylinder or lack of lubrication, the extra motor torque causes higher amperage draws, which create excessive heat. The overload devices will shut the unit down, and a manual reset will often be needed. An annual cleaning with an approved ice machine cleaner along with a bearing inspection will keep the ice flake machine running longer and quieter. Remember, the buildup of mineral scale on the freezing cylinder's surface, caused by poor water quality or lack of a preventive maintenance cleaning schedule, leads to most service problems.

THE WATER FILL SYSTEM

One method of controlling the amount of water coming into an ice flaker's water reservoir is a dual float switch and a water control valve, **Figure 27.10(A)**. Another manufacturer may use

Figure 27.10 (A) The water system diagram of an ice flake machine. Source: Hoshizaki America, Inc. (B) A conductivity probe in the water reservoir. Courtesy Scotsman Ice Systems

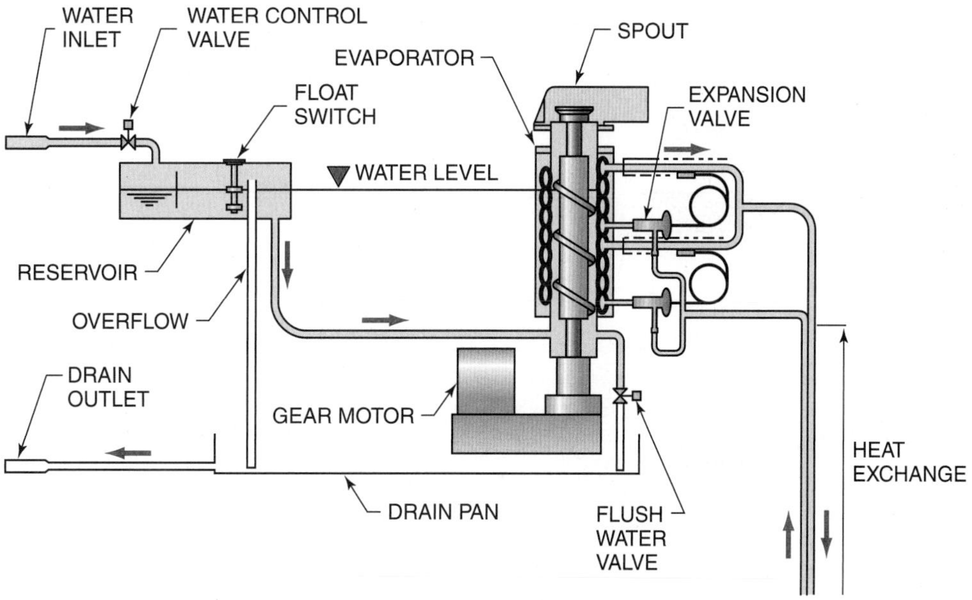

(A)

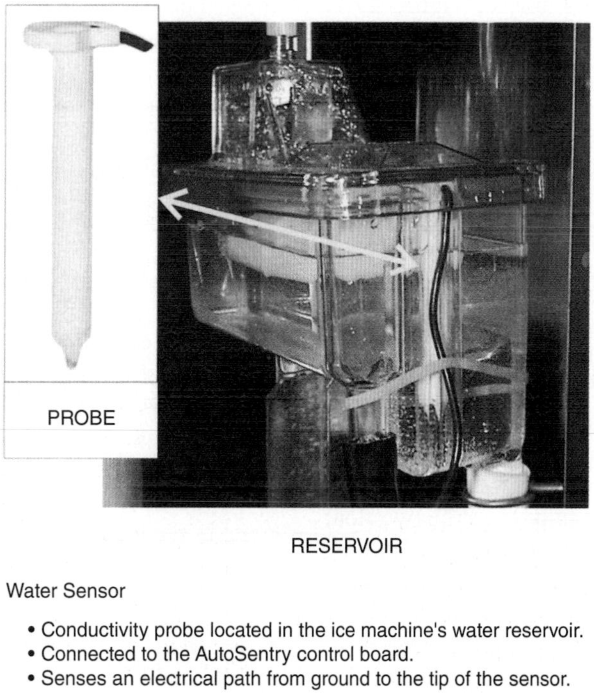

(B)

Figure 27.11 A dual float switch that controls reservoir water level and low-water shutdown timing. Source: Hoshizaki America, Inc.

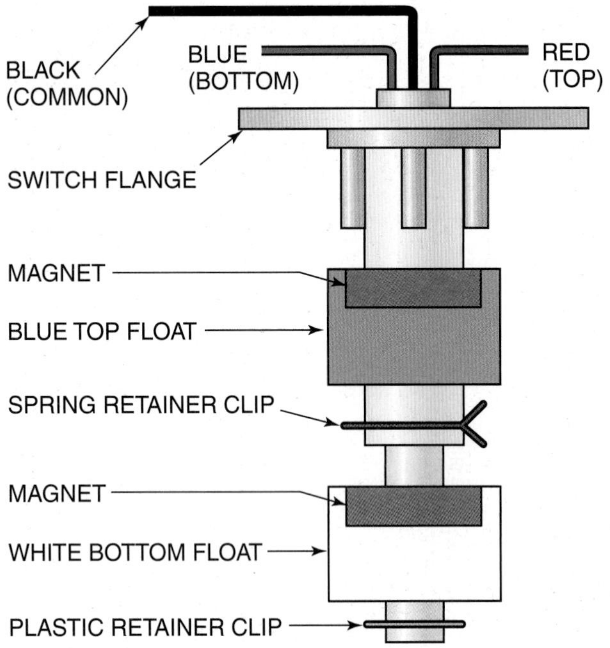

a conductivity probe or water sensor located in the ice machine's water reservoir, **Figure 27.10(B)**. This probe senses an electrical path from ground to the tip of the sensor. It is connected to an electronic control board. The sensor eliminates low- or no-water conditions and cannot be affected by adverse water conditions.

With a dual float switch, the water inlet control valve is solenoid-operated and energized through the switch, **Figure 27.11**. The solid-state control board, **Figure 27.9**, monitors both of these components. The reservoir feeds water by gravity to the evaporator freezing cylinder. The dual float switch maintains the proper level of water in the reservoir and also causes the ice machine to shutdown in case of low water level. The dual float switch is usually made of two reed switches and two floats. The reed switches are opened and closed by magnets positioned inside the floats. As ice is made, the reservoir's water level drops, and the top float opens a set of latching relay contacts. Now the bottom float has control of the water control relay and the water inlet control solenoid. As the reservoir's water level continues to drop, another set of contacts opens by the action of the bottom float switch. This action deenergizes the water control relay, which also closes a circuit to the water inlet valve solenoid. At the same time, the control board energizes a low-water shutdown timer. The reservoir is now filled, and the two float switches swap jobs. The bottom switch becomes a latching relay and the top switch reenergizes the water control relay. As the top float rises, the low-water timer is shut off and the water inlet control solenoid valve is deenergized. This type of arrangement keeps a proper amount of water in the reservoir, prevents short cycling of the water inlet solenoid valve, and provides for low-water safety shutdown. **Figure 27.12** illustrates a wiring diagram of a modern ice flake machine and shows the control water valve, water control relay, control timer, and water inlet solenoid valves.

Water levels can also be sensed and controlled by a probe located in the water reservoir. The probe senses water level and relays the message to an electric water valve. The probe should be removed from the reservoir and cleaned at least twice a year.

FLUSH CYCLES

Many manufacturers build in an automatic water flush cycle for ice flakers. When water freezes, minerals are left behind. As mentioned earlier, these minerals can build up in the evaporator freezing cylinder and cause maintenance problems and poor ice quality. A way to get rid of this unwanted buildup of minerals is to flush out the water system on a periodic basis. One manufacturer's control scheme shuts down the refrigeration cycle for 20 min every 12 hours to allow the water system to drain. When to initiate these flushing periods may be adjusted. The control board energizes a solenoid-operated water flush valve, which opens a drain on the bottom of the evaporator freezing cylinder, **Figure 27.10(A)**. This action drains the reservoir, freezing cylinder, and connecting piping of all impurities. The reservoir is then refilled with clean, fresh water. A mechanical or solid-state timer can initiate the flush cycle, which energizes a gravity-flow flush solenoid valve. Manual flushing can also be performed for service purposes. The wiring diagram in **Figure 27.12** shows the flush timer, flush switch, and flush water solenoid valve. A service technician must understand the sequence of operation when servicing ice flakers. It is important to remember the flush timer when diagnosing a problem if the unit will not start. Manufacturers often provide a cam wheel, which can be turned manually to advance the flush timer. This allows the unit to start if it had been in a periodic flush mode.

Some models of ice flakers drain the entire water system each time the unit cycles off. This off-cycle draining can be done without the use of a timer motor. However, when run time is heavy, flushing is not initiated as often, and trace amounts of minerals may build up in the water system.

BIN CONTROLS

Once the ice-holding bin is full of ice, some sort of automatic control must terminate the operation of the ice flake machine. If this is not done, flaked ice may pile up in the bin and ice chute and overflow. The ice may spill into the ice flake machine's mechanical parts and eventually spill onto the floor. Water from the melting ice may also cause damage to mechanical parts in the ice machine itself or to nearby flooring.

There are many ways to control the overflow condition. The following are some of the methods used for bin control:

- Mechanical flapper-type device controlling a microswitch
- Infrared electric eye
- Sonar
- Thermostats
- Liquid-filled bulbs

Figure 27.12 The wiring diagram of an ice flake machine. Source: Hoshizaki America, Inc.

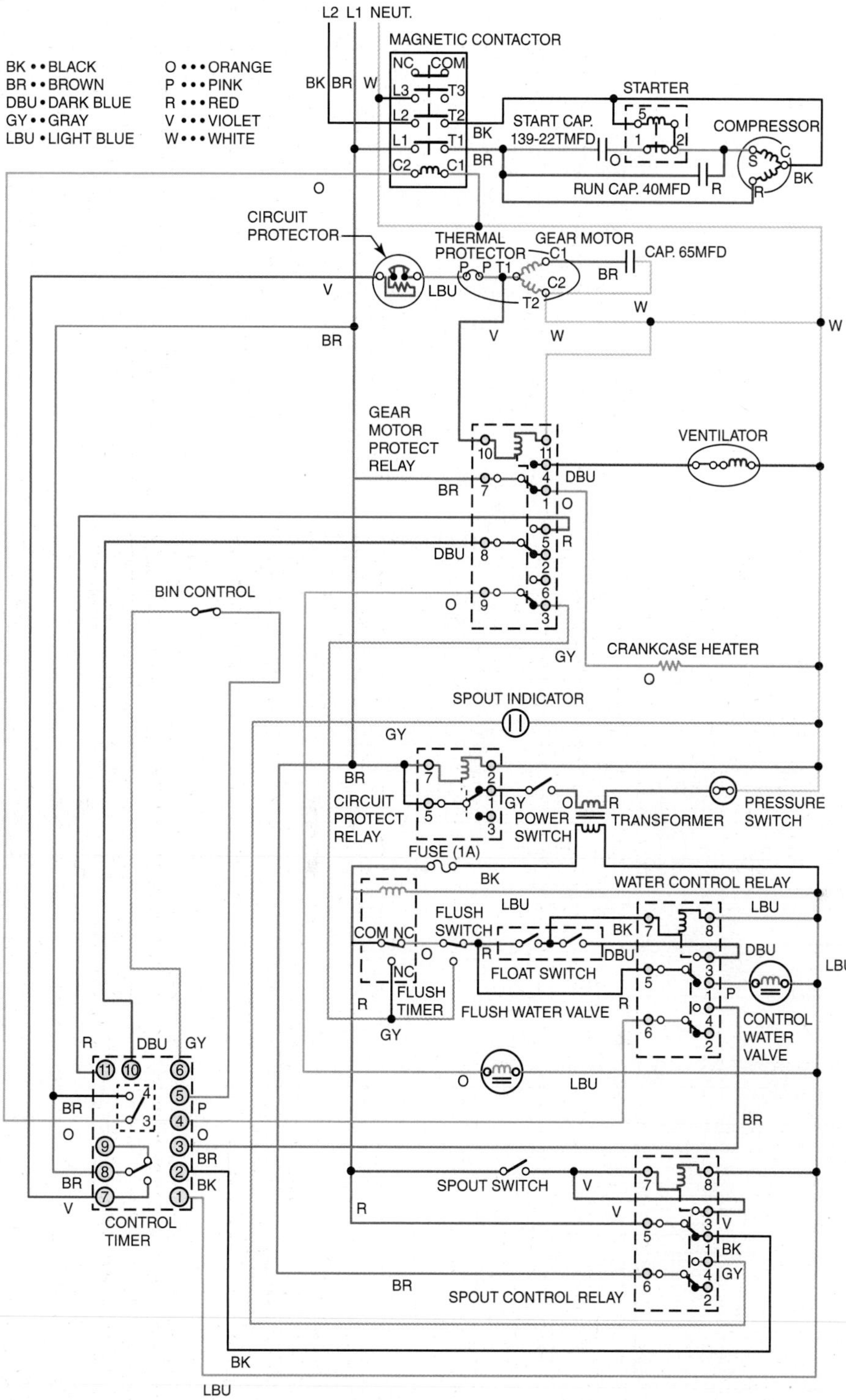

Mechanical flapper-type devices are usually located in the spout. They simply sense backed-up ice by pressure and activate a microswitch. The microswitch shuts down the operation of the ice flake machine. When in the normal position, they are an electronic control with a known amount of resistance. When the paddle is pushed out of its normal position by ice, the resistance value in the electronic control changes and the unit shuts down after a short, predetermined time delay period.

Infrared electric eyes can also be used to sense the buildup of ice. This control allows the ice machine to shut down before the ice chokes the spout opening. Usually, two infrared sensors are mounted on the outside of the ice chute's base, and they are powered and controlled by a solid-state controller. The sensors must be kept clean so that they can see one another. During normal operation, ice passes between two ice-level sensors and momentarily interrupts the infrared beam,

keeping the machine running. However, if ice builds up and blocks the path between two infrared sensors for 6 sec or longer, the machine shuts down. If the path between the infrared sensors remains clear for more than 10 sec, the ice machine will restart. In most cases, this type of bin control must be removed at least twice a year and cleaned with a soft cloth.

Sonar controls use sonic or sound waves and vibrations, which are reflected back to the control from the ice. One disadvantage of infrared and sonic controls is condensate forming on the lenses, making these controls less accurate.

Thermostats often have liquid-filled sensing bulbs placed somewhere in the ice chute or simply at the top of the ice bin. As the ice touches the sensing bulb of the thermostat, the ice machine terminates its freezing cycle, usually within 10 sec. Most bin thermostats are adjustable and can be checked by placing a handful of ice on the remote bulb and timing the shutdown.

ICE PRODUCTION AND PERFORMANCE DATA

When flake ice machines operate at the proper refrigerant charge, ice production depends on two main factors:

- Inlet water temperature
- Temperature (ambient) surrounding the ice machine

Notice in the top portion of **Figure 27.13** that the ice production in pounds per 24 hours decreases as the ambient temperature around the ice machine gets warmer. The second column shows four ambient temperatures ranging from 70°F to 100°F. Assuming a constant 50°F entering water temperature, at a 70°F ambient, the ice production is 2010 lb per day. However, if the ambient rises to 100°F, ice production falls to 1570 lb per day. Italicized values are in kilograms of ice per day.

Figure 27.13 Ice production and other system performance data for an ice flake machine. Source: Hoshizaki America, Inc.

	Ambient Temp. (°F)	Inlet (Entering) Water Temp. (°F)					
		50		70		90	
Approximate Ice Production per 24 hr. lbs./day (kg/day)	70	2010	*(912)*	1950	*(845)*	1895	*(860)*
	80	1845	*(837)*	1795	*(814)*	1750	*(794)*
	90	1700	*(771)*	1695	*(769)*	1610	*(730)*
	100	1570	*(712)*	1525	*(692)*	1410	*(640)*
Approximate Electric Consumption watts	70	2850	—	2850	—	2855	—
	80	2855	—	2860	—	2860	—
	90	2865	—	2865	—	2875	—
	100	2890	—	2890	—	2910	—
Approximate Water Consumption per 24 hr. gal./day (l/day)	70	241	*(912)*	234	*(845)*	228	*(860)*
	80	222	*(837)*	216	*(814)*	210	*(794)*
	90	204	*(771)*	203	*(769)*	194	*(730)*
	100	188	*(712)*	183	*(692)*	169	*(640)*
Evaporator Outlet Temp. °F (°C)	70	14	*(−10)*	14	*(−10)*	14	*(−10)*
	80	14	*(−10)*	14	*(−10)*	14	*(−10)*
	90	14	*(−10)*	14	*(−10)*	16	*(−9)*
	100	16	*(−9)*	16	*(−9)*	16	*(−9)*
Head Pressure psig (kg/sq.cmG)	70	219	*(15.4)*	219	*(15.4)*	219	*(15.4)*
	80	230	*(16.2)*	230	*(16.2)*	230	*(16.2)*
	90	241	*(16.9)*	241	*(16.9)*	241	*(16.9)*
	100	271	*(19.0)*	271	*(19.0)*	271	*(19.0)*
Suction Pressure psig (kg/sq.cmG)	70	25	*(1.8)*	25	*(1.8)*	25	*(1.8)*
	80	26	*(1.8)*	26	*(1.8)*	25	*(1.8)*
	90	27	*(1.9)*	27	*(1.9)*	27	*(1.9)*
	100	29	*(2.0)*	29	*(2.0)*	29	*(2.0)*
Condenser Volume		214 in^3					
Heat of Rejection from Condenser		16890 Btu/h (AT 90°F /WT 70°F)					
Heat of Rejection from Compressor		2860 Btu/h (AT 90°F /WT 70°F)					

Note: The data without *marks should be used for reference.

The reduction in ice production at warmer ambient temperatures happens because of a higher rate of heat transfer into the system. For air-cooled machines, a hotter ambient causes a higher head pressure for the refrigeration system. This causes an increase in compression ratios and a loss of efficiency, and thus less ice production. Also notice in **Figure 27.13** that, as the water temperature coming into the ice machine increases from 50°F to 90°F, the ice production drops from 2010 to 1895 lb per day. This is assuming a constant 70°F ambient air temperature. This happens because it now takes more energy and refrigeration effect to cool the warmer water down to freezing temperatures. The figure also includes electricity consumption, water consumption, evaporator outlet temperature, head pressure, and suction pressure, as they depend on the surrounding ambient air temperature and inlet water temperature. Notice that head pressure and suction pressure both increase as the ambient temperature increases.

ADJUSTING ICE FLAKE SIZE

Some ice flakers can be adjusted to alter the size or shape of the flake ice to meet the needs of the customer. Flake ice is used in packing produce, fish, and poultry and to fill a salad bar or display case because ice flakes pack tightly and keep the products cool and fresh. One method of adjustment is to rotate an ice cutter at the top of the auger. Since the pressure applied to the ice comes from the last couple of inches of the auger flights on some machines, aligning a small, medium, or large cutter opening with the pressure points on the flights of the auger will change ice size. Small, medium, or large flakes can come from one ice machine. Another method is to change the cutting heads located on the upper side of the auger, **Figure 27.14**. One cutter head will produce fine flakes and another will produce coarse flakes.

The extrusion head can also be changed on some ice flakers, which allows one ice machine to produce both cubelet and flake ice. The size of the cubelets or nuggets can be increased or decreased. Cubelet or nugget ice is about the size of a small marble. The flake ice is extruded through small openings in the extrusion head, squeezing out additional water, which makes the ice more dense and compact. Cubelet or nugget ice lasts much longer than flake ice when submerged in a drink. The health care industry uses cubelets or nuggets because they are much easier to chew than larger cubes, and sick and elderly people are less likely to choke when chewing them.

Figure 27.14 Fine and coarse ice-cutting heads for a flake ice machine. Source: Hoshizaki America, Inc.

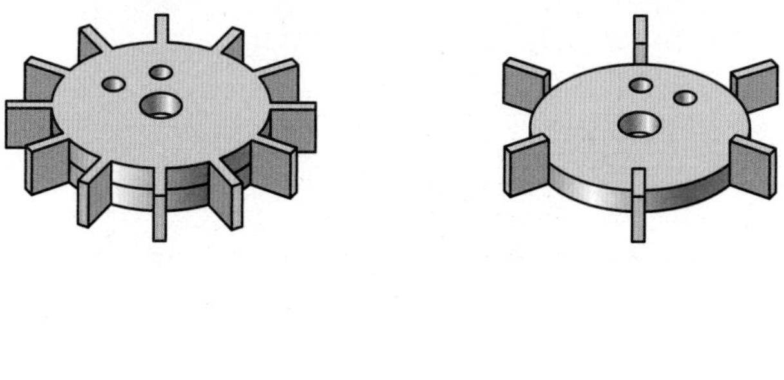

27.3 MAKING CUBE ICE

Several methods are used to make several types of cube ice in package ice makers. Cube ice should be clear, not cloudy. Clear ice is usually preferred for cooling beverages. Ice with a high mineral content or that is aerated (has minute air bubbles inside) will appear cloudy and is not as desirable to some people because of its appearance. Some of the design features of the cube ice maker result in ice that is very clear. Clear versus cloudy ice will be covered in Section 27.5, "Water and Ice Quality" later in this unit.

CUBE ICE

CRESCENT-SHAPED ICE CUBES. Making a crescent-shaped ice cube involves running water over an evaporator plate in which the copper evaporator tubing is soldered onto the plates at regular intervals. **Figure 27.15** shows two stainless-steel evaporator plates and connected tubing for making crescent-shaped ice cubes. Stainless steel is often used on evaporators because it is durable, very sanitary, and resists corrosion. It is also much less porous than copper, brass, or nickel. The copper serpentine tubes soldered to the evaporator are oval to increase heat transfer; the shape provides more surface area in contact with the stainless plates.

During the freeze cycle, water from the reservoir is circulated to the outside freezing surface; no other water is allowed to enter. Once the water reaches 32°F, ice starts to form on the plates. The coldest point on the evaporator is where the copper serpentine coil contacts the evaporator plate. This is where ice starts to form first and continues to grow outward in a crescent shape. As the ice grows, the reservoir water level

Figure 27.15 An evaporator that produces crescent-shaped ice cubes. Source: Hoshizaki America, Inc.

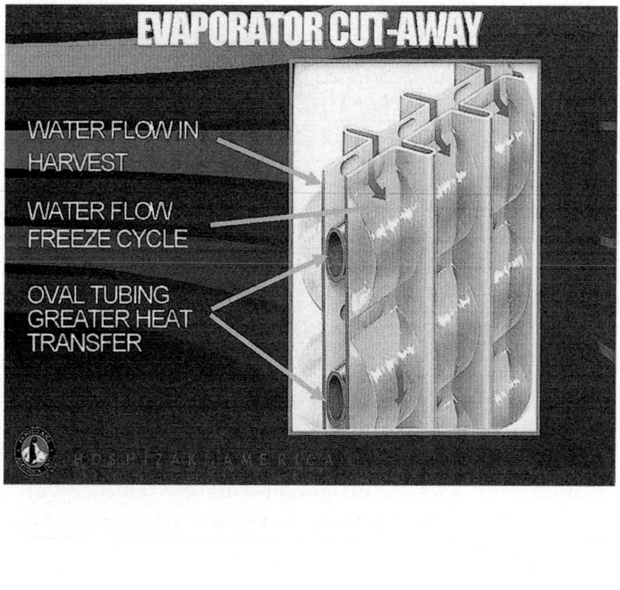

Figure 27.16 The vertical evaporator of a cell-type ice machine. Courtesy Ferris State University. Photo by John Tomczyk

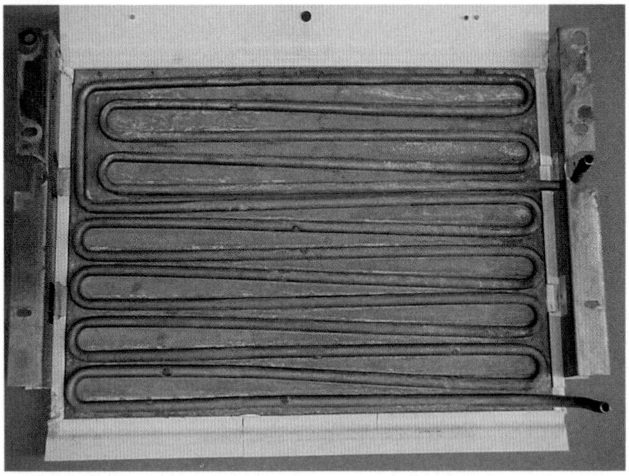

Figure 27.17 A serpentine coil soldered to the back of a cell-type ice cube evaporator. Courtesy Ferris State University. Photo by John Tomczyk

falls because some of the water is freezing into ice. Once the crescent cubes are formed, a float switch in the water sump will open contacts and the harvest cycle begins.

Hot, superheated gas from the discharge of the compressor is circulated through the serpentine coil during defrost. As the heat is transferred to the stainless-steel evaporator, the cubes begin to melt away from the evaporator plate. There will be a film of water between the cube and the plate, which will cause a capillary action and attract the cubes to the plate as they slide down the plate by gravity. As the cubes slide down the plate, they will contact dimples, which break the capillary attraction. This will release the cubes from the plate, and they will fall into the ice bin. Water is allowed to flow during the defrost cycle, but it flows on the opposite evaporator plate. This absorbs heat from the copper coils where hot gas is being circulated and transfers this heat to the freezing surface. This is referred to as water-assisted defrost.

CELL-TYPE ICE CUBES. Cell-type ice cube machines usually have vertical evaporator plates with water flowing over the individual cells, **Figure 27.16**. A plastic water curtain covers the evaporator to prevent splashing, and a serpentine evaporator coil is soldered to the back of the evaporator plate, **Figure 27.17**. The water is distributed over the cells by a water pump through a water distributor tube, shown at the top of **Figure 27.16**, and the tube has evenly spaced holes for even water distribution. The tube is usually removable for cleaning and repair purposes, **Figure 27.18**, and has an adjustable restrictor to let the service technician adjust the water flow rate, **Figure 27.19**. If the water flow is too slow, minerals in the water freeze, causing cloudy ice. Ideally, only water should be allowed to freeze on the evaporator plate. This requires the right water velocity over the plates, which can be adjusted with the restrictor on the water tube. Ice will gradually form on each individual cell until it bridges with the neighboring cells. Ice bridging forms one large ice slab the size of the evaporator itself. Once the ice slab has the proper

Figure 27.18 A water distribution tube that is removable for cleaning and repairs. Based on Manitowoc Ice, Inc.

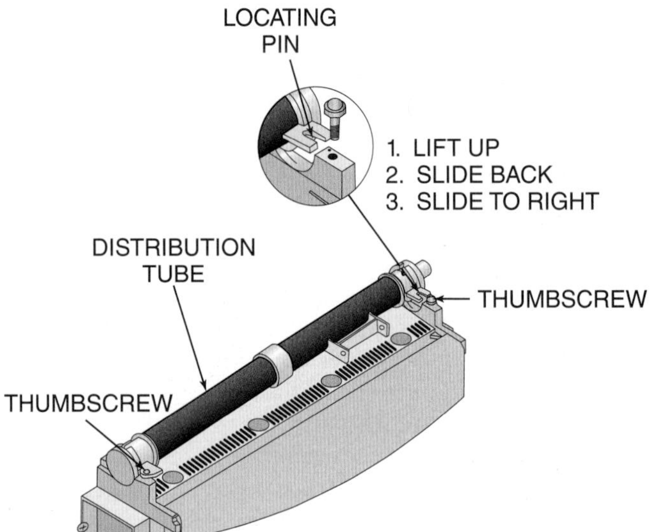

Figure 27.19 A restrictor in the water line from the water pump allows for varying the water flow rate. Courtesy Ferris State University. Photo by John Tomczyk

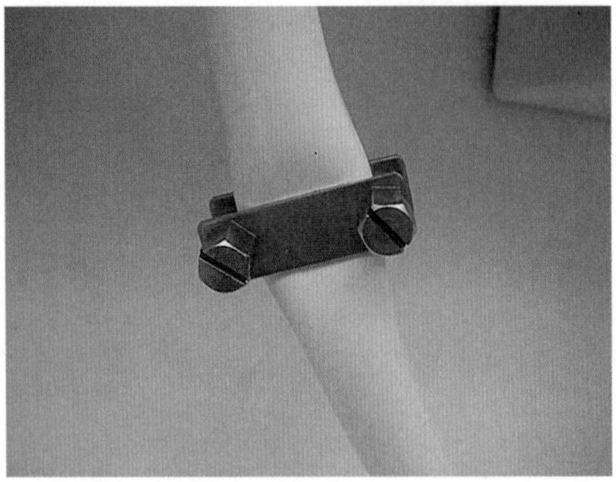

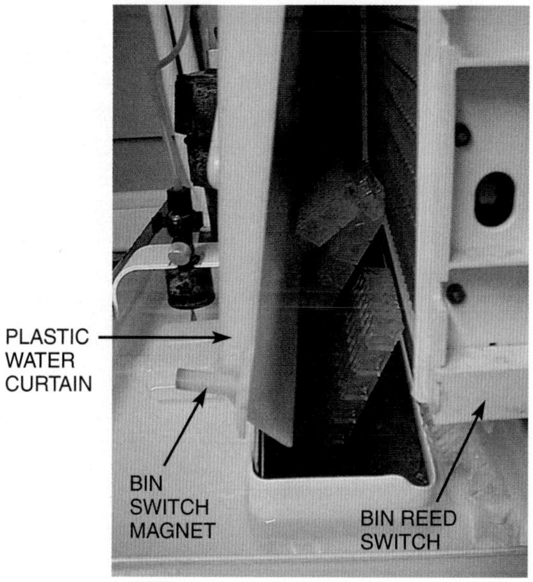

Figure 27.20 An ice slab partially breaking up as it collides with other ice in the storage bin. Courtesy Ferris State University. Photo by John Tomczyk

bridge thickness, a harvest or defrost cycle is initiated and the ice slab falls into the ice storage bin. Ice harvests are covered in detail later in this unit.

The proper bridge thickness is important for the individual ice cubes to be easily separated when scooped out of the ice storage bin. When a server taps the ice slab in the storage bin with an ice scoop, it should easily break up into individual cell cubes. Most of the time, when the ice slab drops off the evaporator during a harvest period, it partially breaks up as it collides with other ice in the storage bin, **Figure 27.20.** However, as the ice storage bin starts to fill, the distance the ice slab has to fall is reduced. If the bridge thickness is too great, the ice slab will not easily break up into individual cubes. A bridge thickness of 1/8 in. is not uncommon. The ice bridge thickness is adjustable through an ice thickness probe's adjustment screw, **Figures 27.21(A)** and **(B). Figure 27.21(C)** and **(D)** show a different style of ice thickness probe. **Figure 27.16** also shows the ice thickness probe and its location near the cell-type evaporator.

The centrifugal water pump, which gets its water from a water reservoir, is generally a fractional horsepower, alternating current motor, **Figure 27.22(A).** The water pump may

Figure 27.21 (A) An ice thickness probe showing a direction for changing ice bridge thickness. Courtesy Ferris State University. Photo by John Tomczyk. (B) An ice thickness probe and ice slab showing a 1/8-in.-thick ice bridge. Based on Manitowoc Ice, Inc. (C) and (D) Different style of ice thickness probe. Courtesy Ferris State University. Photo by John Tomczyk

(A)

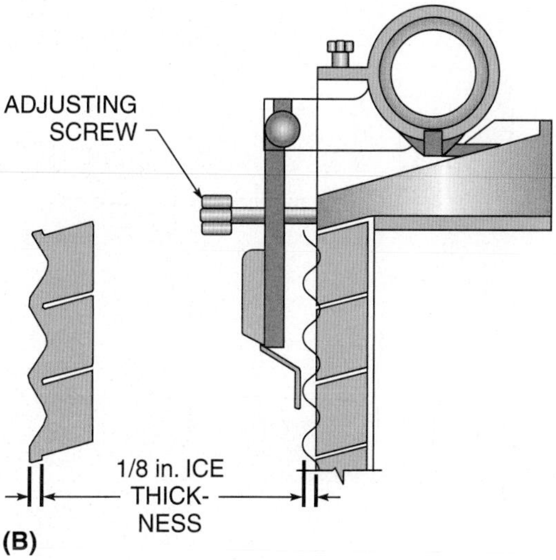

(B)

(C)

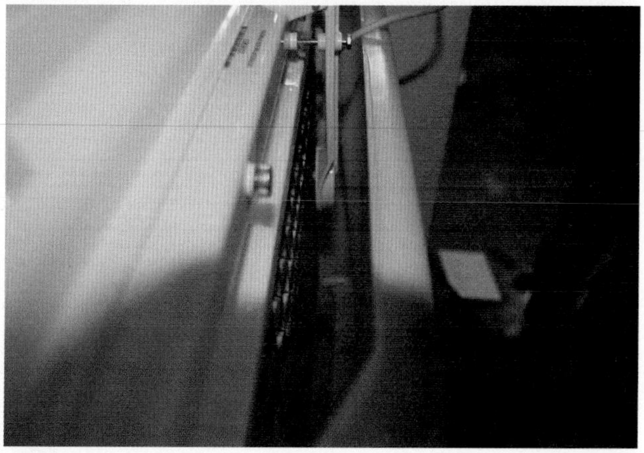

(D)

Figure 27.22 (A) A fractional horsepower, alternating current, centrifugal water pump. Courtesy Ferris State University. Photo by John Tomczyk. (B) A water level sensor to control cube size and initiate harvest through the use of infrared technology. Courtesy Scotsman Ice Systems

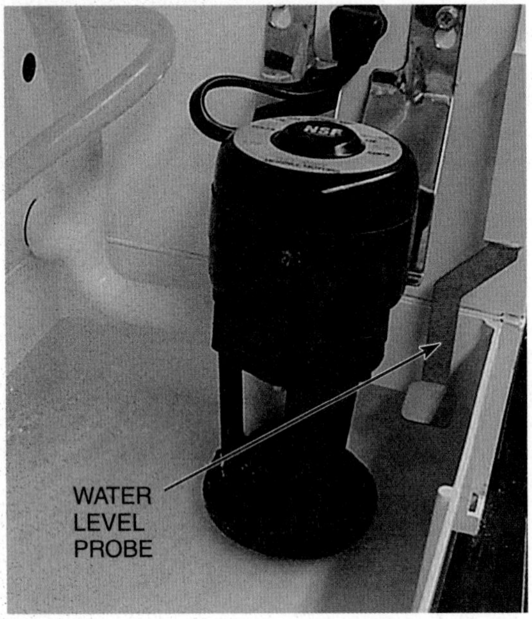

(A)

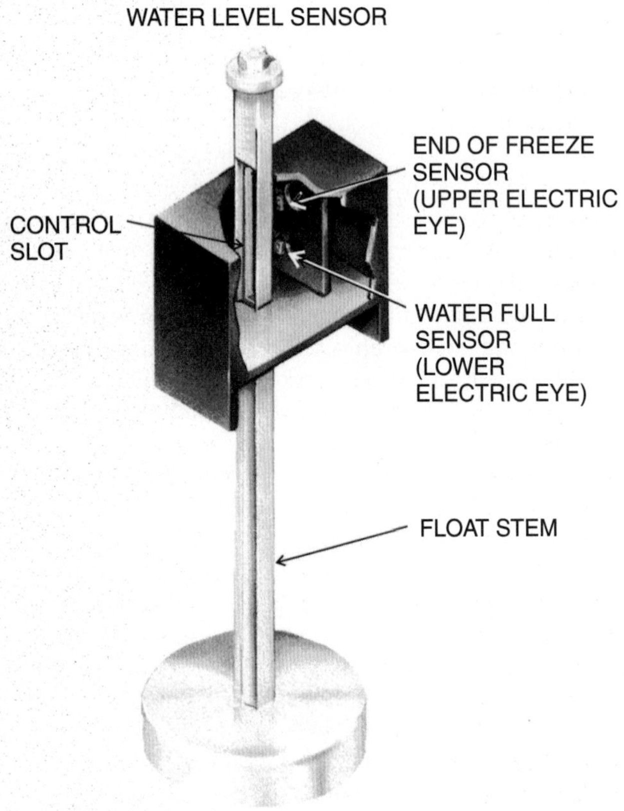

(B)

deliver water to more than one evaporator, **Figure 27.23.** The water reservoir's water level can be controlled by a float mechanism, **Figure 27.24.** The float controls the amount of incoming fresh water to the water reservoir. The float control is adjustable and also controls the level of water in the reservoir. As the water freezes on the evaporator's cells, the water reservoir level starts to drop. Instantly, the float or water level probe senses this drop in water level and slightly opens a valve to let more fresh water into the reservoir. This action keeps just the right level of water in the reservoir for the water pump to use. Too low a water level in the water reservoir allows the water pump to suck a mixture of water and air, which has a detrimental effect on ice quality. The water that does not freeze on the evaporator's cells will trickle into a water trough and eventually make its way into the water reservoir by gravity.

A water level probe, **Figure 27.22(A)**, can also control water levels in the reservoir. It is usually nonadjustable and controls an electric, solenoid-operated water fill valve. One manufacturer uses a water level sensor that provides an electronic controller with two important pieces of information, **Figure 27.22(B)**. As the machine freezes ice, the water level sensor's top photoelectric eye receives infrared light through an open control slot in the float system. When the water falls

Figure 27.23 A water pump serving two evaporators. Based on Manitowoc Ice, Inc.

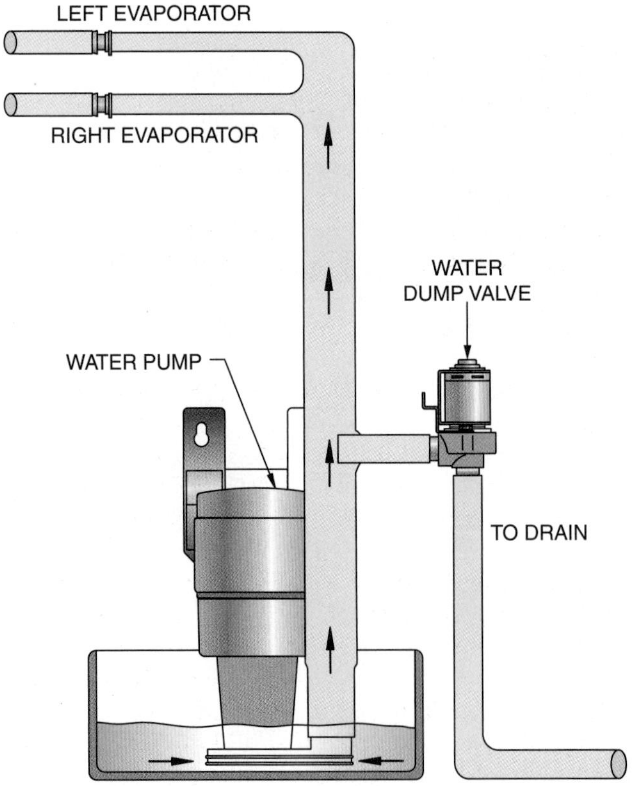

Figure 27.24 A water reservoir, float mechanism, incoming water line, and water pump. Courtesy Ferris State University. Photo by John Tomczyk

far enough, the float stem's slot is too low to allow the light to pass, and the stem breaks the top electric eyebeam. This signals the controller to initiate a harvest cycle. Also, when water flows into the reservoir, the float lifts the float stem. When the bottom electric eyebeam is blocked by the solid plastic body of the float stem, the controller knows that the reservoir is full. Water will continue to flow for a predetermined amount of time to overfill and rinse the reservoir.

SEQUENCE OF OPERATION (FREEZE AND HARVEST CYCLE)

Figure 27.25(A) is a refrigeration system diagram of a cell-type ice maker that has two evaporators. **Figures 27.25(B)–(D)** show the water and refrigeration system, freeze cycle, and harvest cycle on one manufacturer's ice maker. **Figure 27.26** is a refrigeration system diagram of a single-evaporator system in the freeze cycle. Every manufacturer seems to have its own control schemes and sequence of operations, so it would be impossible to cover all control schemes in this unit. A typical control scheme with its sequence of operation for a cell-type ice maker is described in the following paragraphs. *The refrigeration compressor comes on for about 30 sec before the water pump turns on. This allows the evaporator to be prechilled. Prechilling makes for a very cold evaporator when it finally experiences water flow and helps prevent slush from forming in the water sump. Slush may interfere with the water flow to the water pump and may also cause an ice dam.* The water pump comes on after about a 30-sec prechill. Water is forced through the water distribution tube and evenly distributed onto the cell-type evaporator where it begins to freeze. Either a float mechanism or an electric, solenoid-operated water fill valve keeps the water in the reservoir at the proper level. Once the right thickness of ice has formed on the evaporator, water flowing over the ice contacts the ice-thickness probe. *Water has to contact the ice-thickness probe for at least 7 sec before an ice harvest is initiated. This keeps the harvest cycle from prematurely initiating due to splashing. Often, ice machines cannot begin harvest until their freeze cycles have been underway for a certain amount of time. A period of 6 to 7 min is not uncommon. This time period is referred to as a freeze-lock time. Freeze lock allows enough ice to form on the evaporator before a harvest is initiated and prevents an evaporator cookout or overheating from a premature harvest. Overheating can cause damage to the metallic plating on some evaporator surfaces.*

The diagram in **Figure 27.27** shows a system during the harvest cycle. Once the ice-thickness probe contacts water for more than 7 sec, the hot gas solenoid valve located on the compressor's discharge line is energized, **Figure 27.28(A)**. The compressor continues to pump superheated gas through the evaporator. This hot gas warms the evaporator, causing the sheet of cubes to slide into the storage bin. One manufacturer uses a harvest bypass valve that is open for just a few seconds each harvest period, **Figure 27.28(B)**. The purpose of this valve is to add the correct amount of refrigerant to the harvest circuit to promote a good release of ice. *At the same time, the water pump continues to run and the electric, solenoid-operated water dump valve,* **Figure 27.29**, *is energized for a specified amount of time, usually 45 sec.* **Figure 27.30(A)** *shows sump water being routed down a drain during the harvest cycle when the dump valve is energized. This action purges the water from the sump and gets rid of any residual minerals from the previous freeze cycle, which is very important for clear, hard ice.* The water fill valve will energize for the last 15 sec of the 45-sec water sump purge time. This time may be adjustable on an electronic microprocessor-based unit. If a float mechanism is used, the float valve will automatically fill the water sump to the proper level.

As the ice sheet falls off the evaporator, a water curtain swings out and opens a bin switch. This opening and closing of the bin switch ends the harvest cycle and initiates the next freeze cycle. The bin switch is a reed switch that is operated by a magnet. The magnet is attached to the water curtain, and the reed switch is attached to the evaporator mounting bracket, **Figure 27.20**. The bin switch has three functions: to terminate the harvest cycle, to return the ice machine to the freeze cycle, or to cause an automatic shutoff.

If the ice storage bin is full at the end of a harvest cycle, the sheet of ice cubes will not let the water curtain close, **Figure 27.31**. If after a certain time period, usually 7 sec, the bin switch is still open, the ice machine will shut off automatically. It will stay off until the ice in the bin is removed and the water curtain swings back to close the bin switch. A minimum off period of 3 min is generally needed before the ice machine can automatically restart.

On an initial start-up, or after an automatic shutoff, the water pump and the water dump valve are energized

Figure 27.25 (A) A refrigeration system diagram of a cell-type ice maker with two evaporators. Based on Manitowoc Ice, Inc. (B) The water and refrigeration system for a cell-type ice maker. Based on Manitowoc Ice, Inc.

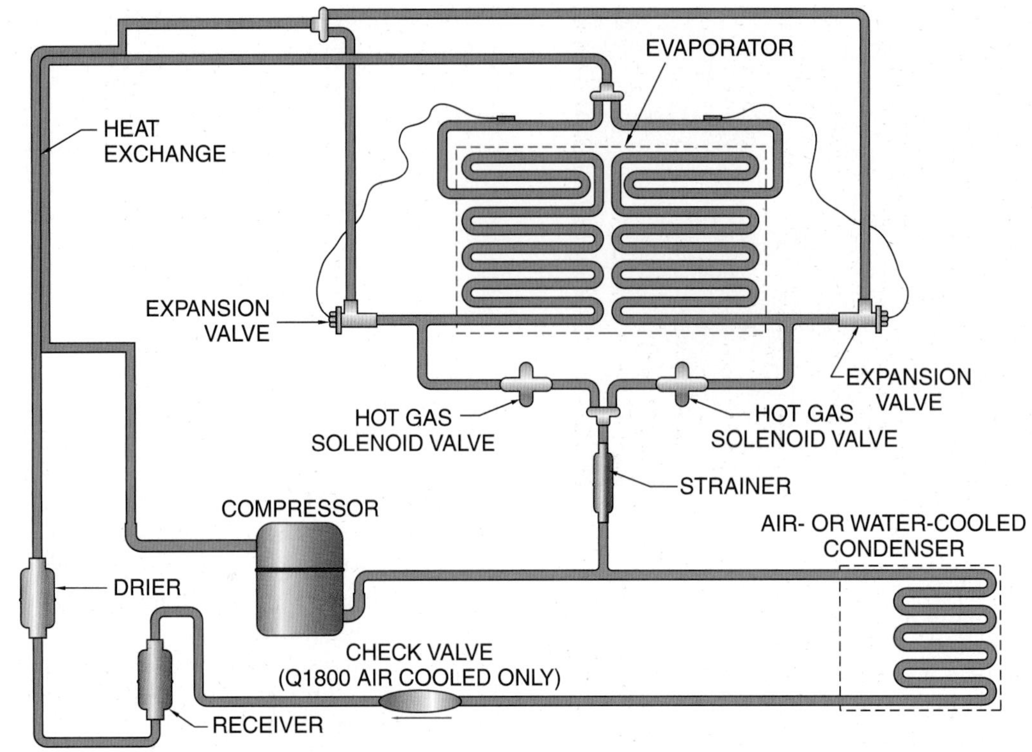

(A)

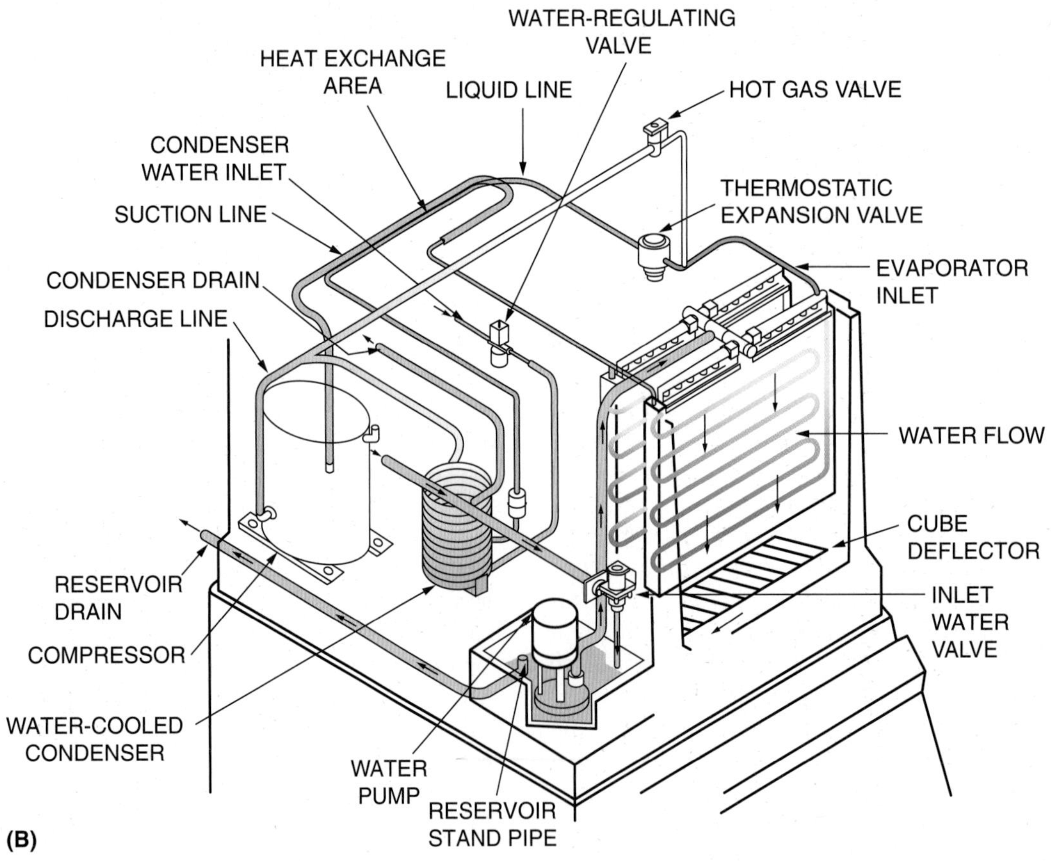

(B)

Figure 27.25 (*Continued*) (C) An ice machine during the freeze cycle. Courtesy Scotsman Ice Systems

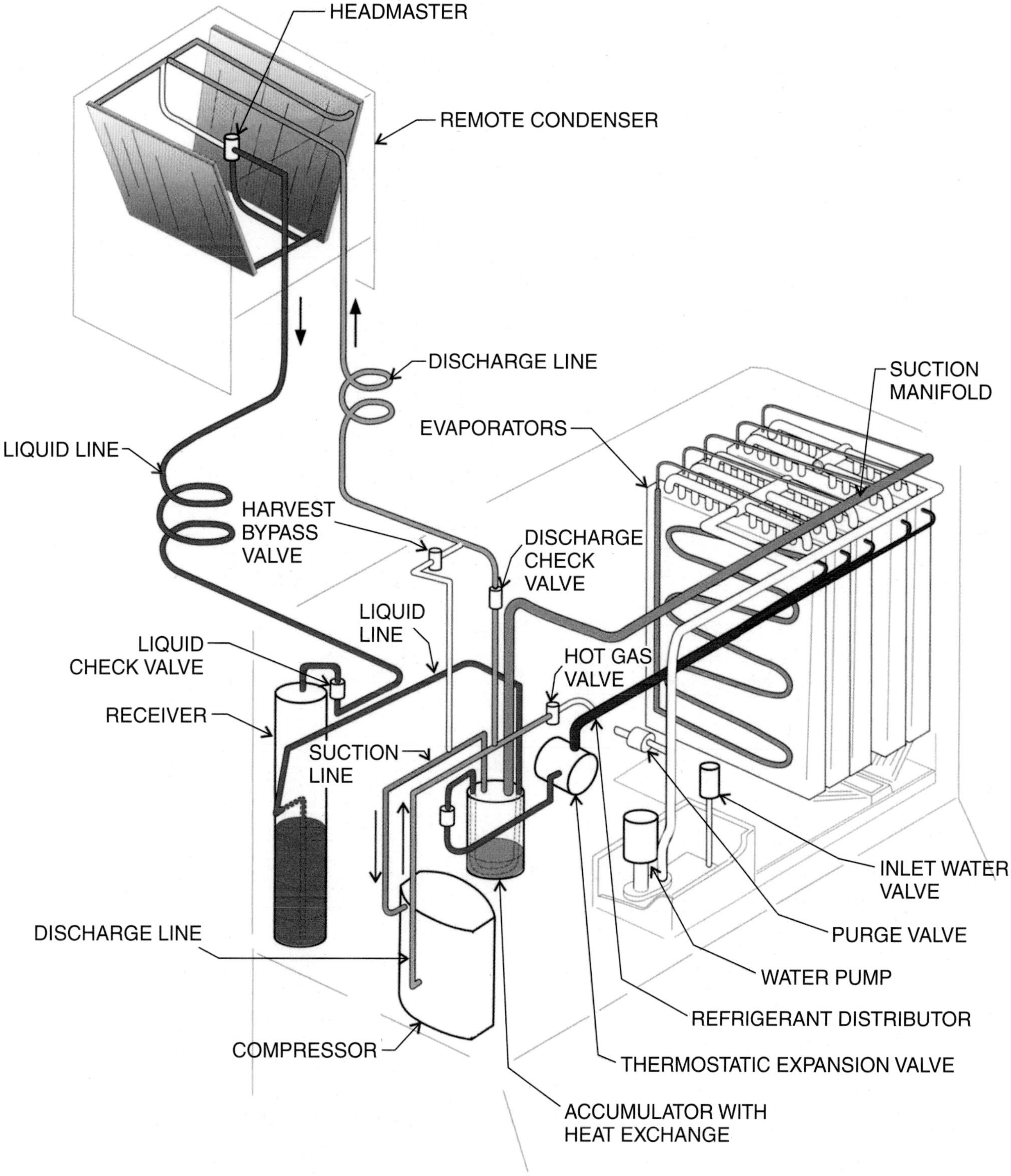

(C)

Figure 27.25 (*Continued*) (D) An ice machine during the harvest cycle. Courtesy Scotsman Ice Systems

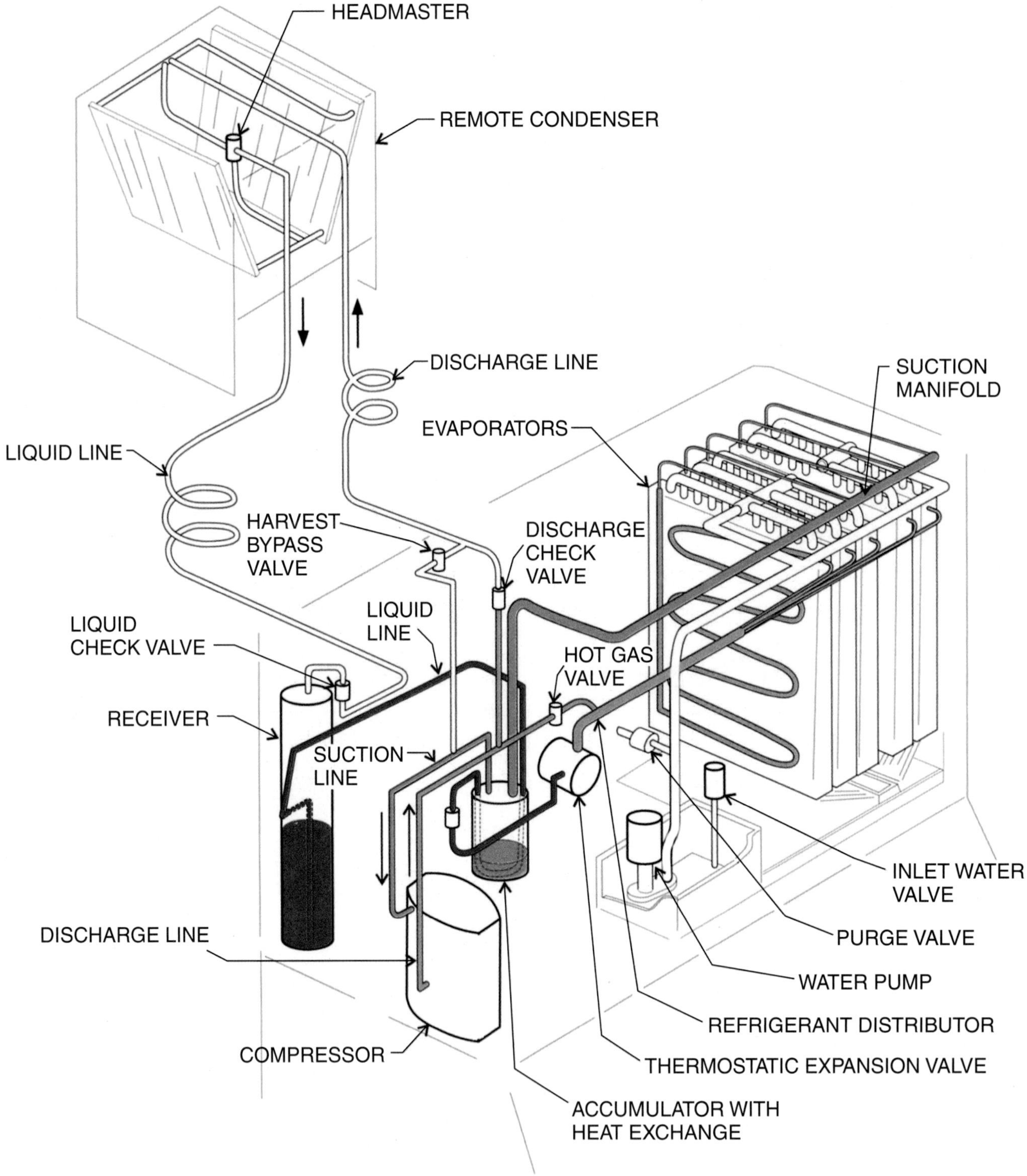

(D)

Figure 27.26 A refrigeration system diagram of a single-evaporator cell-type ice maker. Based on Manitowoc Ice, Inc.

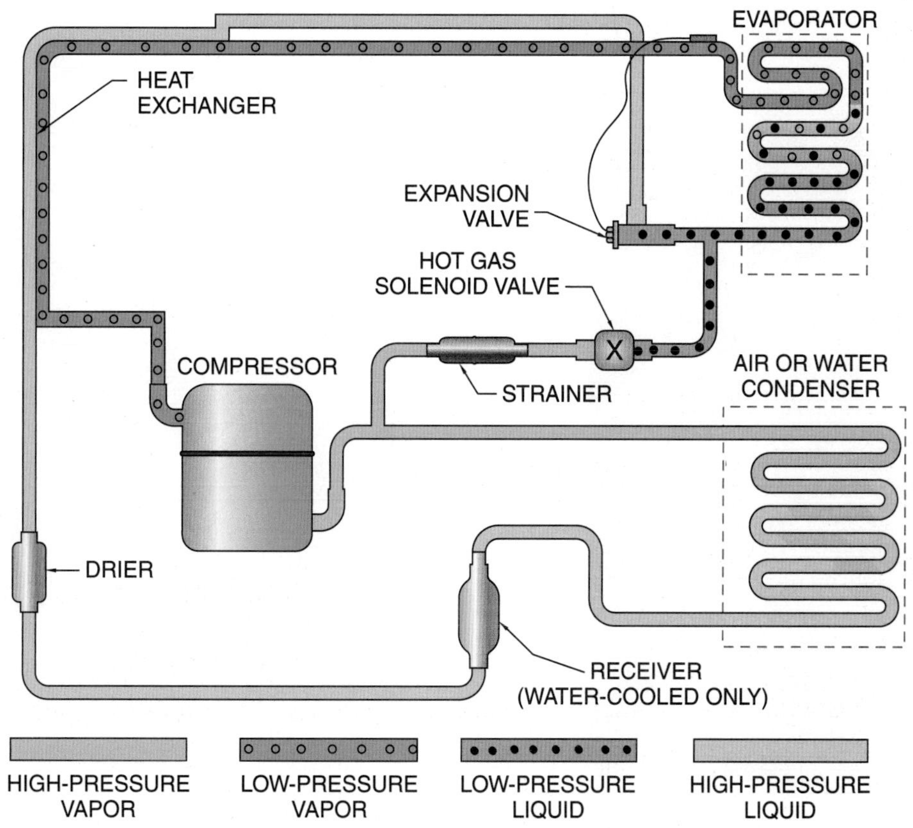

Figure 27.27 A system diagram during the harvest cycle with the hot gas valve energized. Based on Manitowoc Ice, Inc.

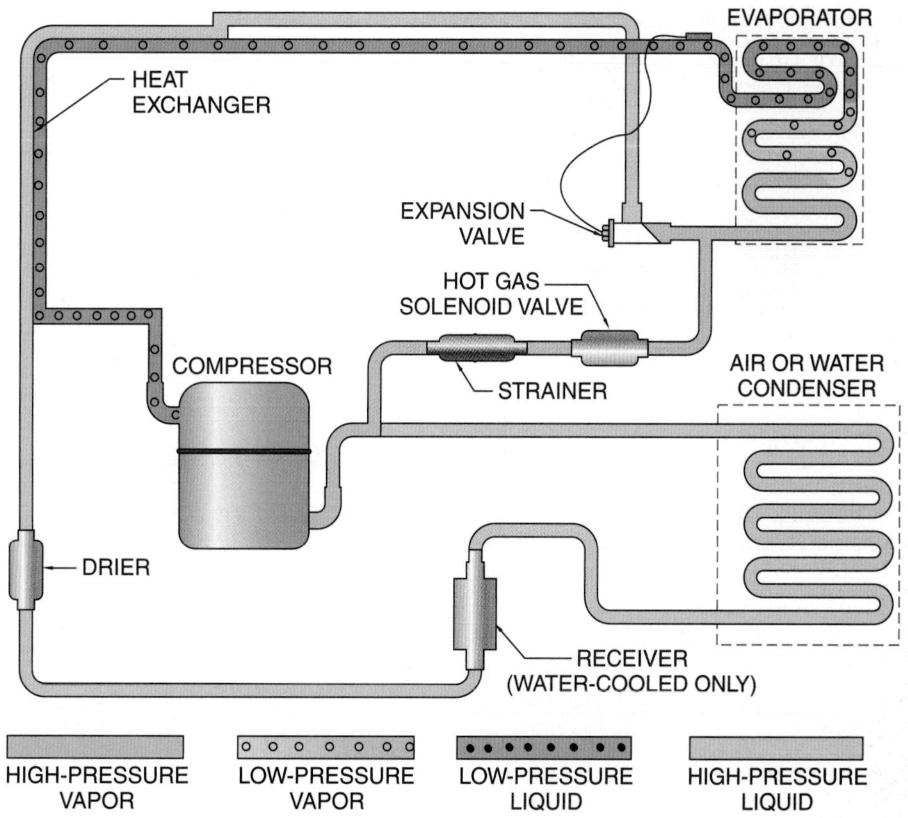

Figure 27.28 (A) The hot gas solenoid valve located on the compressor's discharge line. Courtesy Ferris State University. Photo by John Tomczyk. (B) A harvest bypass valve that is energized only for a few seconds and controls the necessary amount of refrigerant for an ice harvest. Courtesy Scotsman Ice Systems

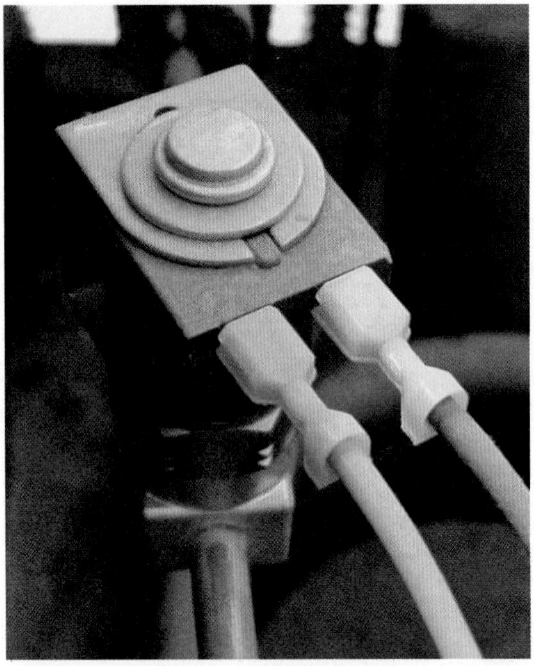

(A)

(B)

for a specific time period, usually 45 sec, **Figure 27.30(A)**. This gets rid of the old, mineral-laden water. One manufacturer uses a purge valve located just above the water reservoir for controlled draining of water from the reservoir, **Figure 27.30(B)**. When it is open and the water pump is on, water flows through this valve to the ice machine's drain. This dilutes the amount of mineral scale in the reservoir. The amount of water purged is adjustable to accommodate local water conditions. *The solenoid-operated hot gas valve is also energized for a specific time period during this start-up. This allows the system refrigerant pressures to equalize between the high and low side for an easier compressor start-up.*

Figure 27.29 An electric, solenoid-operated water dump valve. Courtesy Ferris State University. Photo by John Tomczyk

Figure 27.30 (A) Sump water being routed down a drain during harvest. Based on Manitowoc Ice, Inc.

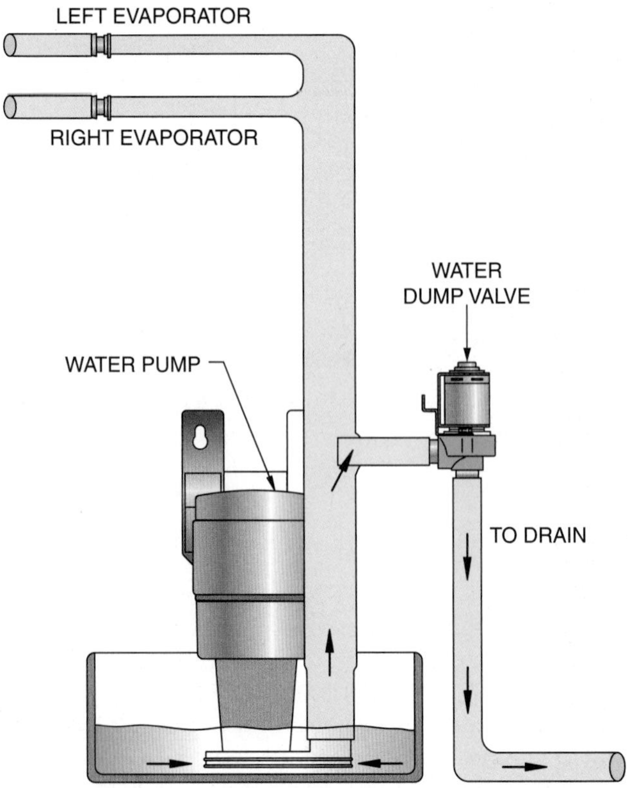

(A)

Figure 27.30 (*Continued*) (B) A purge valve located just above the water reservoir for controlled draining of water from the reservoir. Courtesy Scotsman Ice Systems

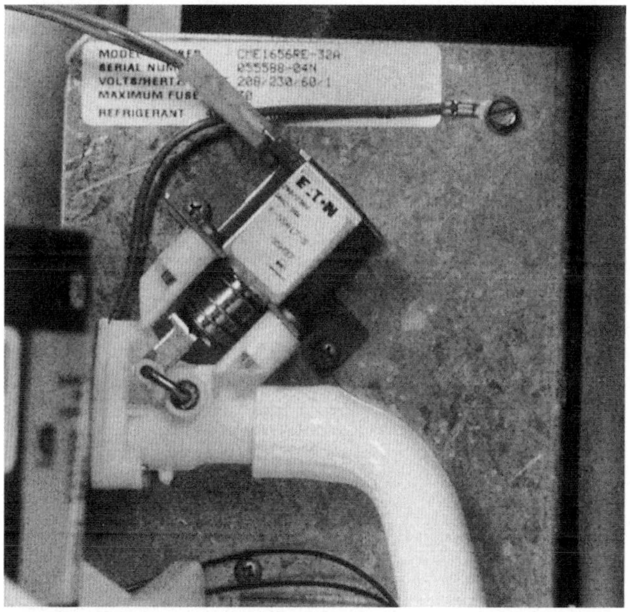

(B)

Figure 27.31 A full ice storage bin prevents the ice slab from closing the bin switch. Courtesy Ferris State University. Photo by John Tomczyk

The compressor starts after the 45-sec water purge and remains on during the freeze and harvest cycles. The compressor being on during the harvest cycle ensures quality superheated gas for melting the ice slab from the evaporator. The water fill valve is also energized at the same time the compressor is on. The water level sensor shuts off the water fill valve when the water level is satisfied. The electronic controller will not allow the water fill valve to be on longer than 6 min to prevent flooding. On air-cooled units, the condenser fan motor is energized with the compressor. **Figure 27.32** is an electrical system diagram shown in the freeze cycle.

Many manufacturers supply electrical diagrams with electrical conductor lines in bold print to illustrate which components are energized during a particular mode of operation, **Figure 27.33**. This greatly assists the service technician with any systematic troubleshooting problems that may arise with the ice machine. An energized parts chart may also be available from the ice machine manufacturer to assist the service technician in systematic troubleshooting, **Figure 27.34**. The energized parts chart tells which control board relay and contactor is energized, when, and how long. These charts, along with the electrical diagrams, can make the ice machine's sequence of operation easier to understand.

ICE HARVESTS

Most ice makers use a hot gas harvest to melt an ice sheet off the evaporator. **Figure 27.28** is a system diagram of an ice maker in the hot gas harvest mode. Notice that superheated gas is routed through the evaporator. As the ice sheet absorbs heat from the superheated gas, the gas becomes saturated. As more heat is absorbed from the saturated gas, it can eventually form a saturated liquid that can become a subcooled liquid. This liquid will travel to the suction line and eventually to the compressor. *Some manufacturers install a suction-line accumulator just before the compressor in the suction line to vaporize this liquid before it reaches the compressor.* However, if the system has no suction-line accumulator, liquid refrigerant can reach the compressor crankcase and dilute the oil in the crankcase. Liquid refrigerant returning to the compressor while the compressor is operating is called **liquid floodback**. Liquid floodback is covered in detail in Unit 25, "Special Refrigeration System Components." Liquid floodback can cause serious compressor problems because of the diluted oil in the crankcase. Bearings and other moving parts in the crankcase can be scored from the inadequate lubrication of diluted oil. The returning liquid refrigerant eventually boils off in the compressor's hot crankcase, which causes oil foaming. A cold and sweating crankcase will result. This foaming of oil and boiling refrigerant can get sucked up into the compression chamber of the compressor. Extremely high pressures can result, and valve and valve plate damage can occur. If any liquid refrigerant gets compressed, *wet compression* will result. Wet compression occurs when the compression stroke of the compressor vaporizes any liquid in the cylinder. The vaporizing refrigerant in the cylinder absorbs a lot of the heat of compression, leaving a cool cylinder. The discharge temperature will also be much cooler.

Many manufacturers install a discharge-line thermistor, **Figure 27.35**, to sense the discharge-line temperature. If it falls below a certain temperature—for example, 90°F—during a defrost period, severe liquid floodback is occurring. Since the discharge thermistor is an input to the microprocessor, the ice machine will automatically be taken out of defrost by deenergizing the hot gas solenoid. In the case of

Figure 27.32 An electrical diagram of a cell-type ice maker shown in the freeze cycle. Based on Manitowoc Ice, Inc.

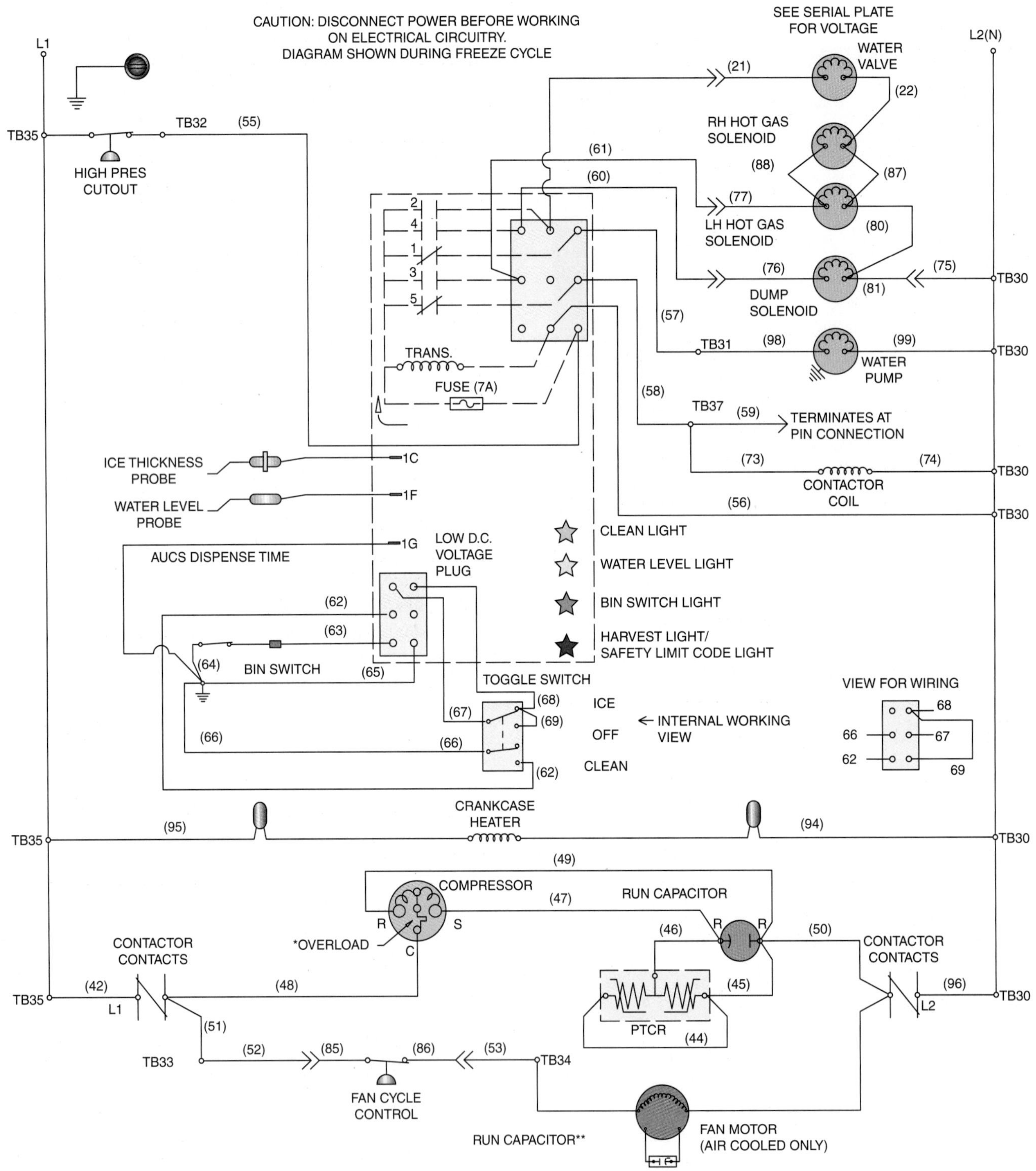

dual evaporators, just one evaporator will be taken out of defrost. Once the discharge-line temperature increases, the evaporators will be put back into defrost. This cycling of the hot gas solenoid will repeat until all of the ice has been harvested off the evaporators.

As mentioned earlier, once the ice slab grows to a certain thickness, some means of initiating defrost will put the system into a hot gas defrost. If an ice thickness probe or feeler contacts are used, the water flowing over the ice that contacts the probe must have some electrical conductivity for defrost

Figure 27.33 An electrical diagram with bold electric conductor lines showing energized loads. Based on Manitowoc Ice, Inc.

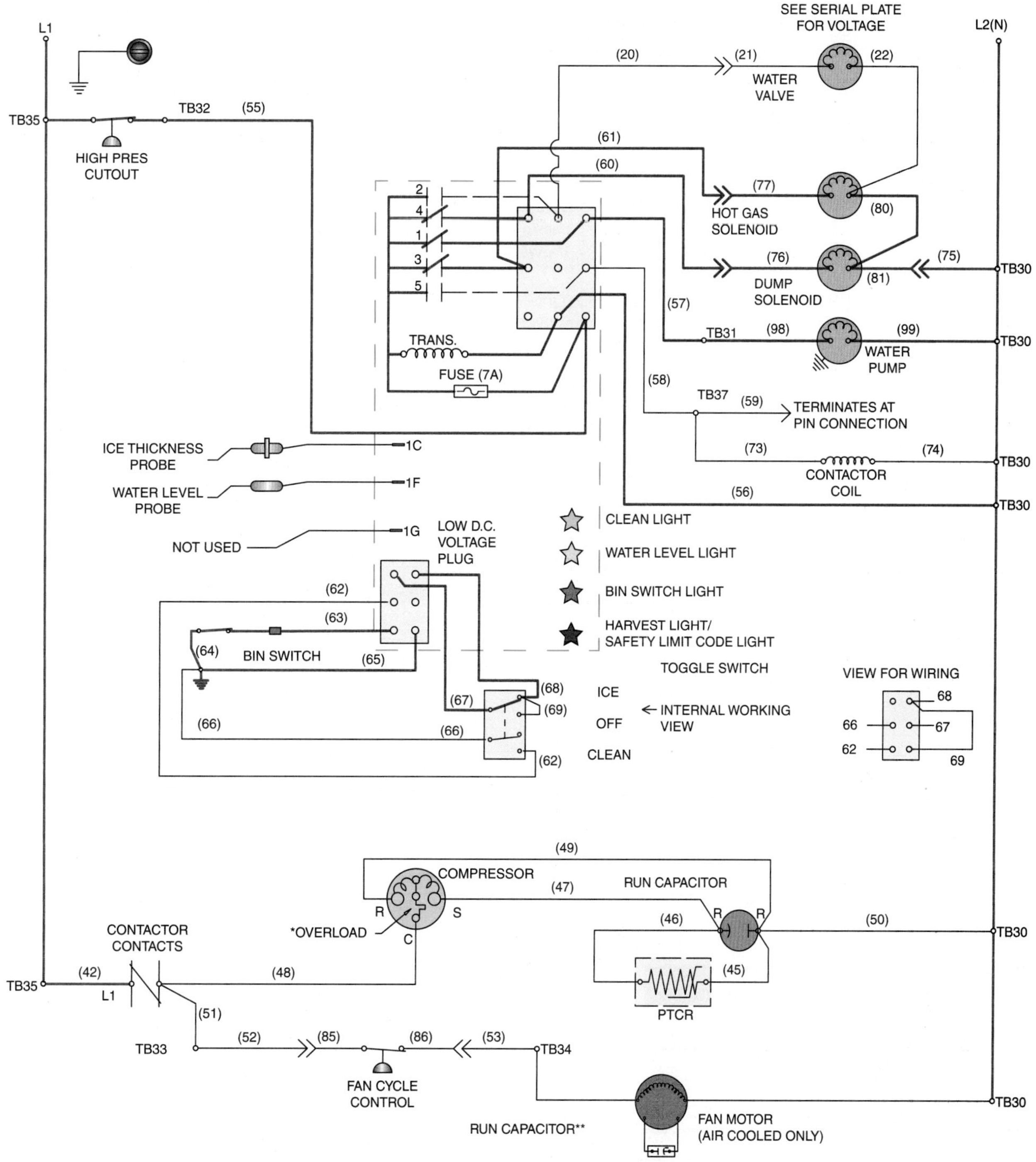

to start. If the water is too pure because too many minerals have been removed, the conductivity will not be correct and a defrost will not be initiated. A pinch of salt may be added to the circulating water to provide conductivity to initiate a defrost when the evaporator is frozen solid. If a defrost cycle occurs after adding salt, it is an indication that the water is too pure. Special controls can be obtained from the ice maker manufacturer so that the defrost will occur under pure water conditions.

Once hot gas loosens the ice slab from the evaporator by melting some of the ice next to the evaporator surface, a suction or attraction force is formed between the ice slab and the evaporator. Air must be introduced between the ice slab

Figure 27.34 An energized parts chart for systematic troubleshooting. Based on Manitowoc Ice, Inc.

ELECTRICAL SYSTEM ENERGIZED PARTS CHARTS SELF-CONTAINED AIR- AND WATER-COOLED MODELS								
	Control Board Relays					Contactor		
	1	2	3	4	5	5A	5B	
Ice Making Sequence of Operation	Water Pump	Water Fill Valve	Hot Gas Valve(s)	Water Dump Valve	Contactor Coil	Compressor	Condenser Fan Motor	Length of Time
Start-Up[1] 1. Water Purge	On	Off	On	On	Off	Off	Off	45 Seconds
2. Refrigeration System Start-Up	Off	On	On	Off	On	On	May cycle On/Off	5 Seconds
Freeze Sequence 3. Pre-Chill	Off	May cycle On/Off during first 45 sec.	Off	Off	On	On	May cycle On/Off	30 Seconds
4. Freeze	On	Cycles On, then Off 1 more time	Off	Off	On	On	May cycle On/Off	Until 7 sec. water contact with ice thickness probe
Harvest Sequence 5. Water Purge	On	30 sec. Off, 15 sec. On	On	On	On	On	May cycle On/Off	Factory-set at 45 Seconds
6. Harvest	Off	Off	On	Off	On	On	May cycle On/Off	Bin switch activation
7. Automatic Shut-Off	Off	Off	Off	Off	Off	Off	Off	Until bin switch re-closes

[1]Initial Start-Up or Start-Up after Automatic Shut-Off

Figure 27.35 A thermistor on the discharge line of an ice maker. Courtesy Ferris State University. Photo by John Tomczyk

and the evaporator to release this suction force. Weep holes between the evaporator cells allow air entering from the edges of the ice to travel along the entire ice slab and relieve the suction force. This releases the ice from the evaporator. The weep holes are shown in **Figure 27.36(A).** One manufacturer uses an air pump to force air between the ice slab and the evaporator, **Figure 27.36(B).**

During an ice harvest, the water pump continues to run and the electric, solenoid-operated water dump valve is energized for a specified amount of time. This action purges the water from the water sump either by gravity or water pump pressure. This removes any residual minerals from the previous freeze cycle, which is very important for clear, hard ice. Some older ice machines do not have water dump valves. When a harvest cycle occurs, the water pump turns off. The excess water from the pumping system that is caught in the piping and water distribution header causes an overflow in the water sump. It also causes a siphon effect in the water sump from the momentum of the overflowing water. This siphon effect is planned so that

Figure 27.36 (A) Weep holes between cells on the evaporator. Courtesy Ferris State University. Photo by John Tomczyk. (B) An air pump to force air in back of the ice sheet to assist defrost. Courtesy Ferris State University. Photo by John Tomczyk

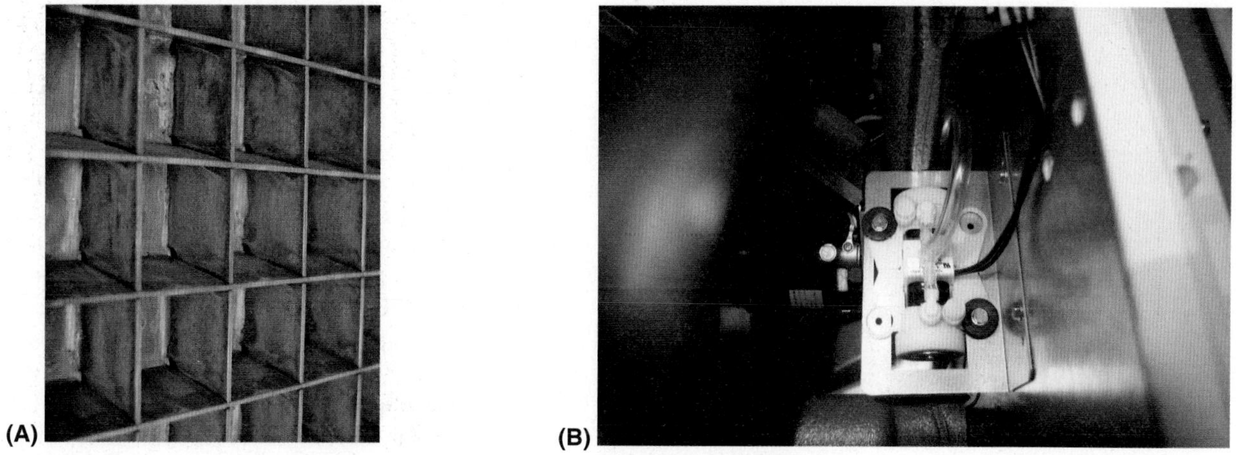

some of the mineral buildup in the sump water moves down the drain. The siphon effect continues until the water sump is emptied, allowing air to break the siphon. This allows the sump to drain every cycle by gravity and prevents any mineral buildup. The float mechanism then lets fresh water into the water sump. Plumbing the water system to allow the overflow and siphon to occur is very critical on older machines.

Once the ice falls off the face of the evaporator, something has to sense the falling ice and terminate the harvest cycle. Most of these devices are digital inputs to a controller or microprocessor. **Figure 27.20** shows a magnetic bin reed switch that is activated by the ice curtain to tell the microprocessor that the ice has fallen from the evaporator. **Figure 27.37** shows an ice curtain switch, and **Figure 27.38** shows a paddle switch activated by a pushrod and ice curtain (the pushrod and ice curtain are not shown in this figure).

Often, a pushrod or probe controlled by a harvest motor, clutch assembly, and cam switch will push the ice off the evaporator during a harvest cycle, **Figure 27.39**. This harvest assist assembly is made up of four major parts:

- Motor
- Clutch assembly
- Probe
- Cam switch

Figure 27.37 (A) An ice curtain switch arm for sensing fallen ice. Courtesy Ferris State University. Photo by John Tomczyk. (B) An ice curtain switch. Courtesy Ferris State University. Photo by John Tomczyk

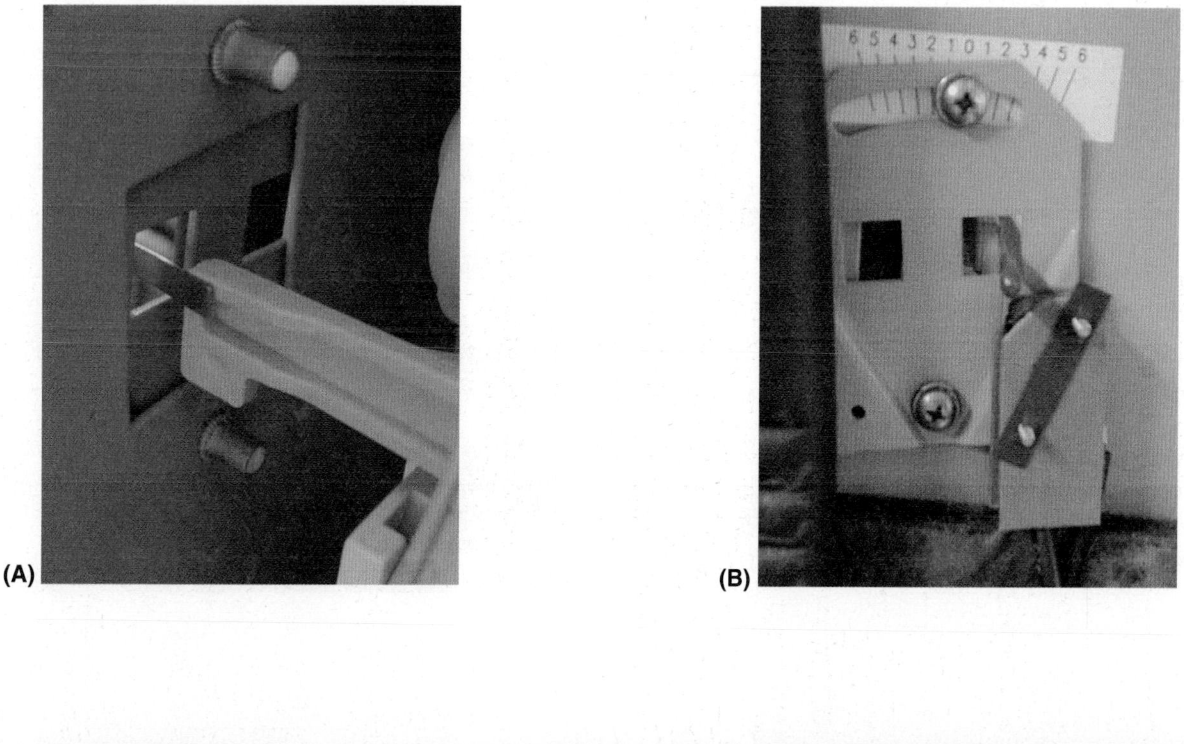

Figure 27.38 A paddle switch used to sense ice falling off an evaporator. Courtesy Ferris State University. Photo by John Tomczyk

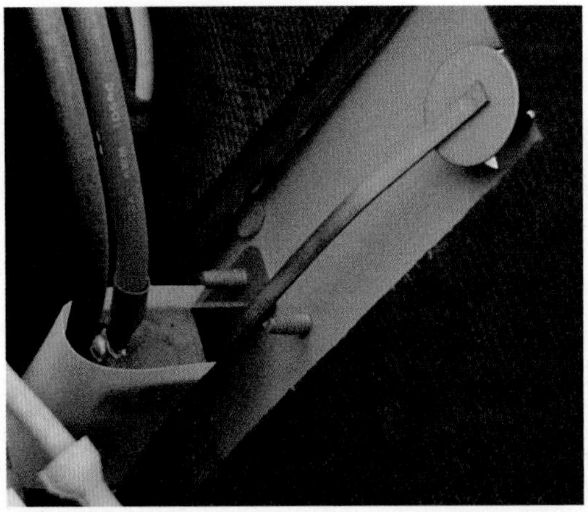

Figure 27.39 (A) A harvest assist clutch assembly showing motor, probe, cam switch, and frame. Based on Ice-O-Matic. (B) Cam switch. Based on Ice-O-Matic.

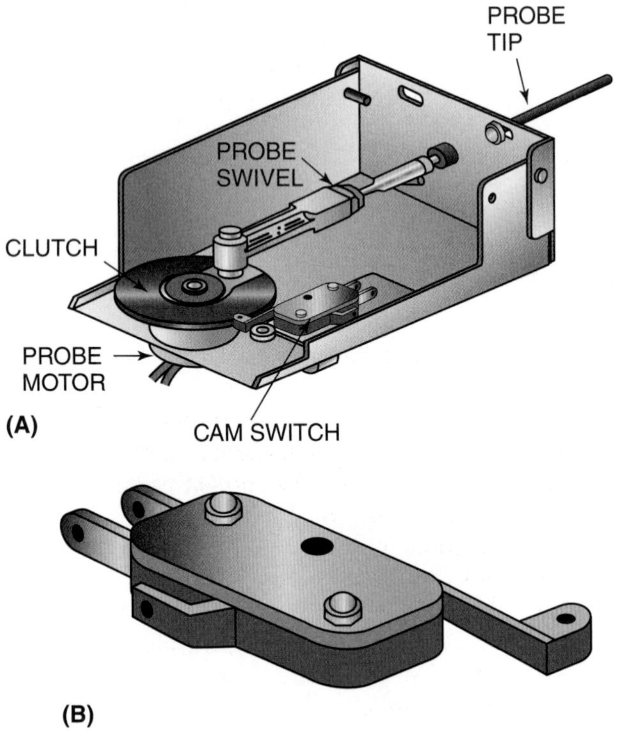

The motor drives the harvest-assist clutch assembly during the harvest cycle. The clutch assembly provides about 6 oz of force against the ice slab. It is a brass cam designed to slip as it pushes against the back of the ice slab during a harvest cycle. The upper half of the clutch slips against the lower half until the ice starts to release off the evaporator plate. The clutch then engages, which allows the probe to push the ice off the evaporator into the holding bin.

The probe is a stainless-steel rod, which is driven through a hole in the evaporator by the motor when the clutch engages. To prevent the probe from freezing into the ice during the freeze cycle, it is recessed about 1/8 in. from the back of the evaporator when not in use. The probe is adjusted by loosening a nut and screwing the probe either in or out. A ball joint between the probe and the clutch allows the probe to pivot while it is being pushed against the ice.

The cam switch rides on the clutch assembly. It is a single-pole–double-throw switch. The switch operates many ice machine components. The normally closed contacts allow the water pump and water purge valve to operate during the ice-making cycle and sometimes during the first part of the harvest cycle. The normally open contacts of the switch power the harvest-assist motors and hot gas solenoid during a harvest cycle. When the cam's arm starts to ride the lower side of the cam during a harvest, the ice machine starts a freeze cycle again.

CLEAN CYCLE

Most ice machines have an ICE/OFF/CLEAN toggle switch. It is usually a double-pole–double-throw switch that allows the owner or service technician to toggle the ice machine between either a freeze, off, or clean mode. In most cases, the clean mode energizes the water pump only, allowing the service technician or owner to add the proper chemicals for circulation through the system. Many ice makers have digital touchpad switches.

ICE MACHINE PRESSURE CONTROLS

Even though the microprocessor controls most of the operations of today's ice machines, it is not the only control device. There are other safety and component- cycling controls that are used to enhance system operation. For example, **Figure 27.40** shows a modular, nonadjustable, fan-cycling

Figure 27.40 Modular, nonadjustable, fan-cycling, and high-pressure controls. Courtesy Ferris State University. Photo by John Tomczyk

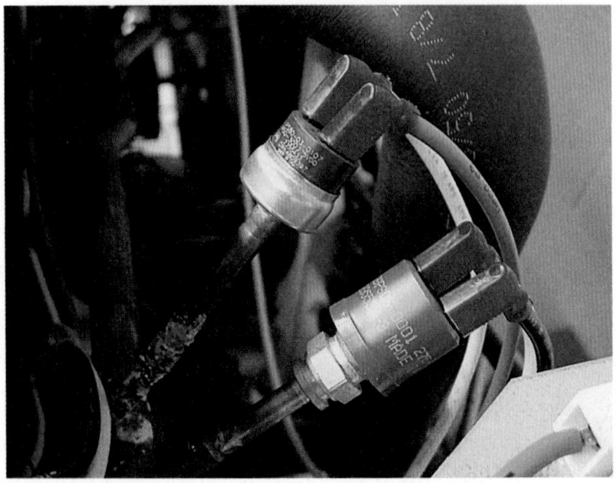

Figure 27.41 A manually reset high-pressure control. Courtesy Ferris State University. Photo by John Tomczyk

control and a high-pressure control. *The fan-cycling control turns the condenser fans on once the ice machine reaches a certain condensing pressure. It also cycles the fan off if the condensing pressure drops below a set point. This ensures the proper pressure drop across the metering device as quickly as possible.* The high-pressure control is a safety device that protects the system from excessively high operating pressures. If the high-pressure cut-out switch opens, the machine will stop immediately. It will automatically reset when the pressure falls below its cut-in point. Low-pressure controls are also available in this modular style. If the low-pressure control opens, the machine will shut off. It will also automatically reset when the pressure rises above its cut-in point. Both of the control settings are dependent on what refrigerant is being used and the ice machine's application. **Figure 27.41** shows a manually resettable high-pressure control. If the machine cycles off on high head pressure, it will not continue cycling on high head pressure. A service technician must manually reset the control for the machine's operation to resume.

Some ice machines use a reverse-acting low-pressure control to terminate the freeze cycle. As ice forms on the evaporator, the suction pressure drops from a reduced heat load on the system. A reverse-acting low-pressure control closes its contacts when the suction pressure drops to a preset level. This action terminates the freeze cycle and initiates a harvest cycle.

27.4 MICROPROCESSORS

Service technicians in the HVACR field will encounter microprocessors in their daily service work, as all manufacturers of ice machines and ice flakers incorporate

Figure 27.42 A microprocessor used in a commercial ice machine. Courtesy Ferris State University. Photo by John Tomczyk

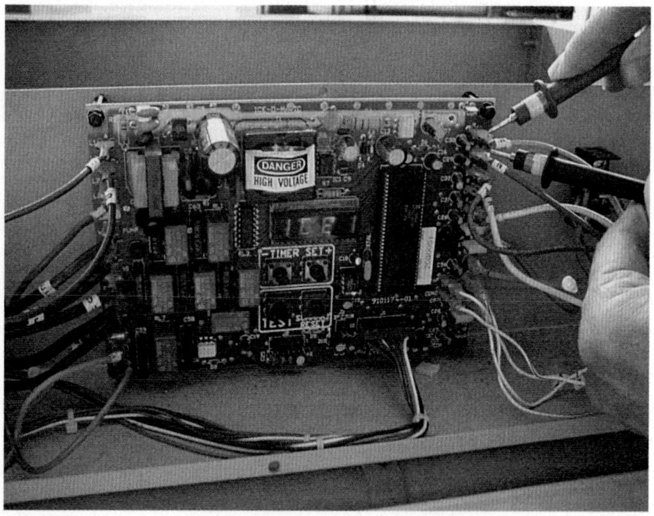

microprocessors into their units, **Figure 27.42**. Microprocessors are simply small computers that contain stored programs or algorithms and often display system status as well as operating pressures and temperatures. The sequence of operation is stored in the software of the microprocessor. **Figure 27.43** shows a small microprocessor or controller with an interfacing ice sensor and water-level controller used to control a modern ice cube machine. The controller operates the ice machine according to instructions, or programs, that reside in its internal memory. Each model made by this manufacturer operates a bit differently. For maximum efficiency, individual ice machine models have different programs called algorithms stored in their electrical erasable programmable read-only memory (EEPROM). These controllers allow the ice machine to optimize efficiency of operations, recall errors that occur during running, have self-diagnostic functions, and customize programs for individual models. Even the water fill, which takes in water in batches, is a controller function. In fact, the amount of time it takes to fill the water reservoir is measured every cycle to determine the fill time needed to maintain the correct amount of water rinse. **Figure 27.44** shows a service controller that can be custom programmed for a particular model of ice machine by rotating a selector switch.

Microprocessors are often referred to as integrated circuit controllers, electronic controllers, or just controllers. *HVACR equipment manufacturers use microprocessors to make their products more reliable, less complicated, and easier to troubleshoot by including self-diagnostics. The microprocessor has cut down on the amount of hard wiring that goes into the control circuit, and it has taken the place of a lot of the hard wiring that used to be inside an ice machine.* Many years ago, there were so many wires going in and out of controls that the entire control circuitry looked like a plate of spaghetti. Now, most

Figure 27.43 A controller, ice sensor, and water-level sensor used on an ice cube machine. Courtesy Scotsman Ice Systems

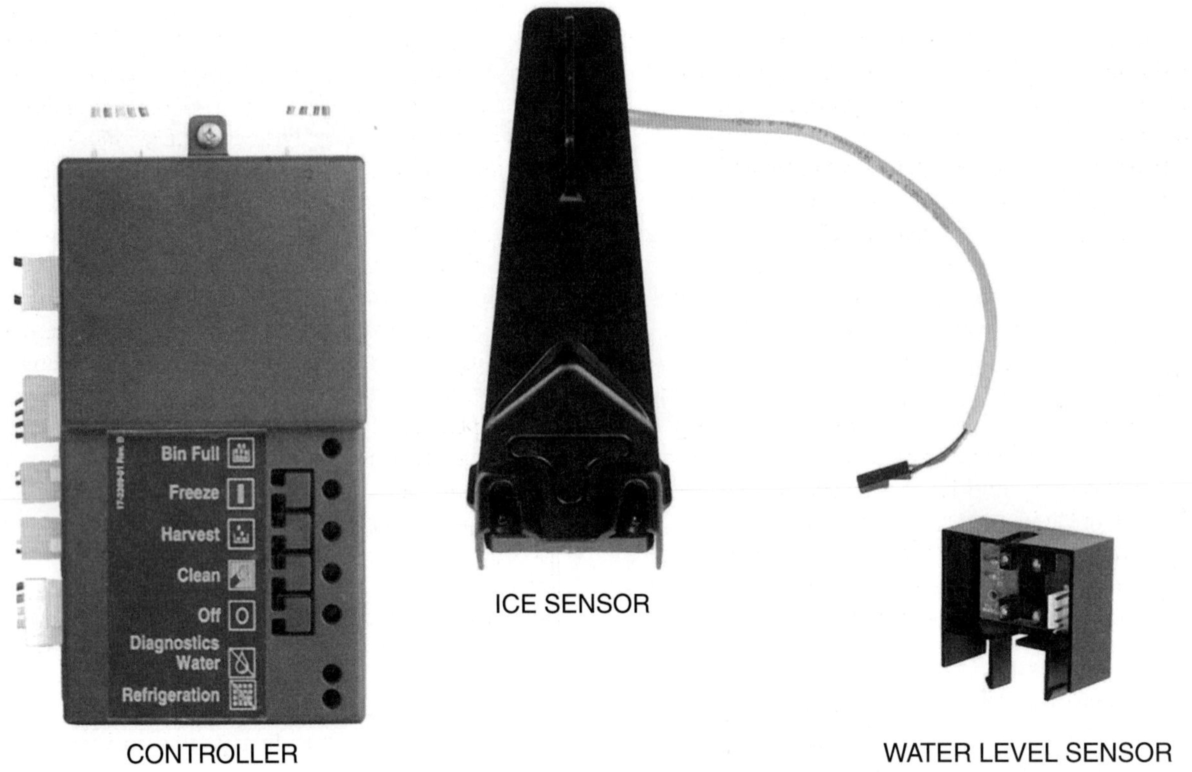

Figure 27.44 A controller that can replace many existing controllers by rotating a selector switch for a desired stored program. Courtesy Scotsman Ice Systems

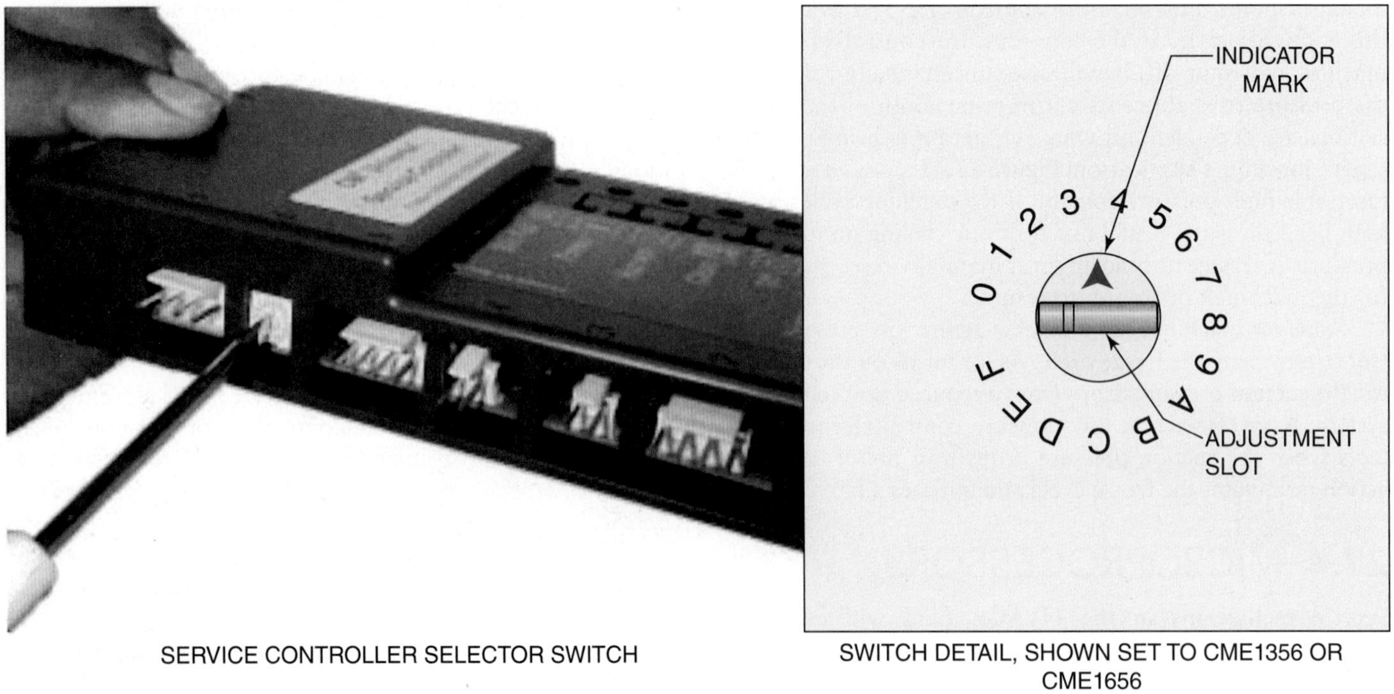

modern control circuits consist of a microprocessor; a few starting components like a start and run capacitor, a starting relay, and a compressor contactor; and the main loads or power-consuming devices. Most service technicians are not afraid of the main loads or power-consuming devices when troubleshooting, but do fear the microprocessor. However, the microprocessor can be the simplest component to troubleshoot with a little patience, understanding, and practice.

Although microprocessors have many sophisticated solid-state devices in their internal circuitry that are needed for their proper operation, service technicians do not need to understand how each solid-state device operates in order to tell if the microprocessor is good or bad. What the service technician does have to know is the microprocessor's sequence of operation, its self-diagnostic functions, and how to input/output (I/O) troubleshoot the microprocessor using its external terminals. The microprocessor's external terminals are shown at the right- and left-hand sides of **Figure 27.42**, where most of the wires enter and exit the microprocessor. These wires are the inputs and outputs of the microprocessor. Troubleshooting the external terminals of a microprocessor will be covered in detail later in this unit.

MICROPROCESSOR SEQUENCE OF EVENTS AND SELF-DIAGNOSTICS

The sequence of events of a microprocessor is usually found in the service manual. If a service manual cannot be found on-site with the refrigeration equipment, the owner or manager can be contacted to see if it has been filed away in some safe location. If the manual still cannot be located, an online search is often helpful. Usually, a model and serial number of the refrigeration machine is all that is needed to locate the manual. The service technician simply needs the pages from the manual that include the sequence of events of the microprocessor and how to initiate the self-diagnostics of the machine, if it has any. With today's modern technology, this information can be in the hands of the service technician in minutes. **Figure 27.45** shows a portion of a service manual for an ice maker that includes the sequence of operation and the component test of the microprocessor, respectively. Simply pushing a button or moving a switch initiates the tests that the microprocessor performs. Notice that the component test walks the technician through voltmeter checks of the water pump, purge solenoid, hot gas solenoid, condenser fan, harvest probe motors, and compressor. It even instructs the service technician where to place the voltmeter and when to check the component. This gives the technician the knowledge of what the ice machine components are supposed to do at a certain time, temperature, or test mode.

INPUT/OUTPUT TROUBLESHOOTING

Even if the manufacturer of the ice machine does not include a component test mode within the programming of the microprocessor, the service technician can still troubleshoot the

Figure 27.45 (A) A microprocessor sequence of operation. Based on Ice-O-Matic.

ICE MAKING PROGRAM					
When the selector switch is turned to the ICE position, in addition to displaying the evaporator and condenser temperatures, the computer will also display the part of the program that the computer is in.					
ICE 1	ICE 2	ICE 3	ICE 4	ICE 5	ICE 0
ICE 1 is the untimed portion of the freeze cycle. In this part of the program, the computer is waiting for the evaporator temperature to reach 14°F (−10°C) before it starts the timer.	ICE 2 is the timed portion of the freeze cycle. The machine is still in freeze, but the timer in the computer is now running.	ICE 3 occurs during the last 20 seconds (12 seconds on single evaporator units) of the timed portion of the freeze cycle. During this time the purge valve is energized.	Once the amount of time set on the timer has passed, the program enters ICE 4 for 20 seconds. The WATER PUMP is now switched off and the HOT GAS VALVE is energized (open).	During ICE 5 the harvest assist motor(s) are now switched on and remain energized until the cam switch(es) return to the N.O. position, at which time the program returns to ICE 1 or ICE 0.	If the splash curtain(s) are held open when the cam switch(es) opens at the end of ICE 5, the machine will shut off and ICE 0 will display. When the curtain(s) closes the program returns to ICE 1.

Figure 27.45 (*Continued*) (B) A microprocessor component test. Based on Ice-O-Matic.

ELECTRONIC CONTROLLER OPERATION
COMPONENT TEST

All high-voltage components are tested in this function. This test allows the technician to distinguish between a defective computer and a defective component or component circuitry. With the machine in the test mode, the component indicated on the display should energize and operate.

To use this test, bring up the display that describes the component to be tested. If the component is indicated on the LED, but the component does not energize, check the output at the high-voltage (left) side of the computer. This is done by checking voltage between the top terminal, marked L1, and the terminal marked with the component description being tested. If the computer output is the same as the supply voltage, the problem is in the component, or the wiring to that component (i.e., a broken wire). If the computer output is different from the supply voltage, the computer is defective and must be replaced.

H2OP	PrG	GAS	FAn	HP-2	HP-1	CnnP
The water pump is switched on. Check voltage across the terminals labeled L1 and WATER PUMP.	The purge solenoid is switched on. Check voltage across the terminals labeled L1 and PURGE.	The hot gas solenoid is switched on. Check voltage across the terminals labeled L1 and HOT GAS.	The condenser fan motor(s) is switched on. Check voltage across the terminals labeled L1 and FAN 1 and L1 and FAN 2.	Harvest probe motor 2 is switched on. Check voltage across the terminals labeled L1 and MOTOR 2.	Harvest probe motor 1 is switched on. Check voltage across the terminals labeled L1 and MOTOR 1.	The compressor is switched on. Check voltage across the terminals labeled L1 and COMP.

microprocessor. Input/output troubleshooting is actually an easy method if the technician uses some common sense. All that is needed is a voltmeter and knowledge of the sequence of events from the service manual. An ohmmeter should not be used directly on a microprocessor because the ohmmeter's battery can cause damage to the board or its components. However, if certain components (inputs or outputs) to the microprocessor are detached, an ohmmeter can safely be used on these components.

Microprocessors are fed with information from their input devices to their input terminals. Input devices can be analog, as with a thermistor (variable resistor), or digital (on/off), as with a switch. **Figure 27.46** shows a thermistor connected to the midpoint of the condenser. It is acting as an analog input device to the microprocessor for displaying the condensing temperature on an LED.

After processing the data from the inputs, an output signal is sent. The output signal can be read with a meter from the output terminals of the microprocessor. Input and output terminals of a microprocessor can be seen in **Figure 27.47**. Notice that both the input and output terminals are clearly labeled for the service technician to troubleshoot. It is from these terminals that most troubleshooting can be accomplished using a meter. Most of the time, but not always, the input signals are low-voltage (AC or DC) or resistance signals, and the output signals are of higher voltage (usually AC)

Figure 27.46 The condenser thermistor is an analog input to a microprocessor. Courtesy Ferris State University. Photo by John Tomczyk

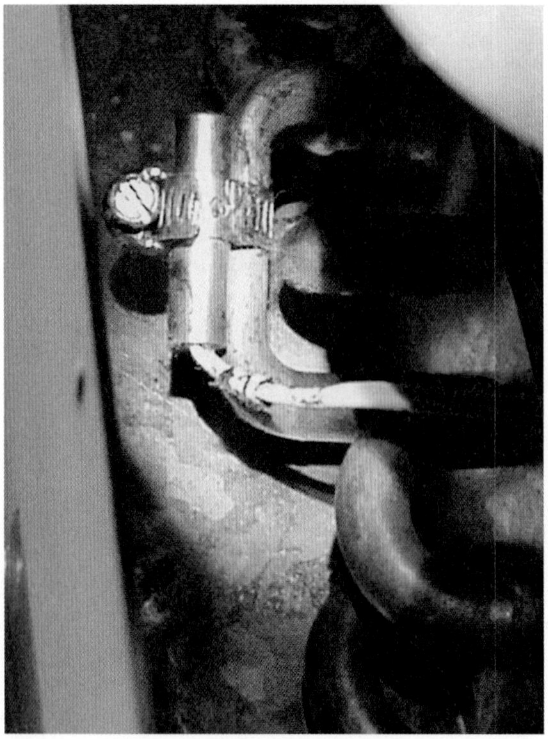

Figure 27.47 A microprocessor showing components that include the input and output terminals. Based on Ice-O-Matic.

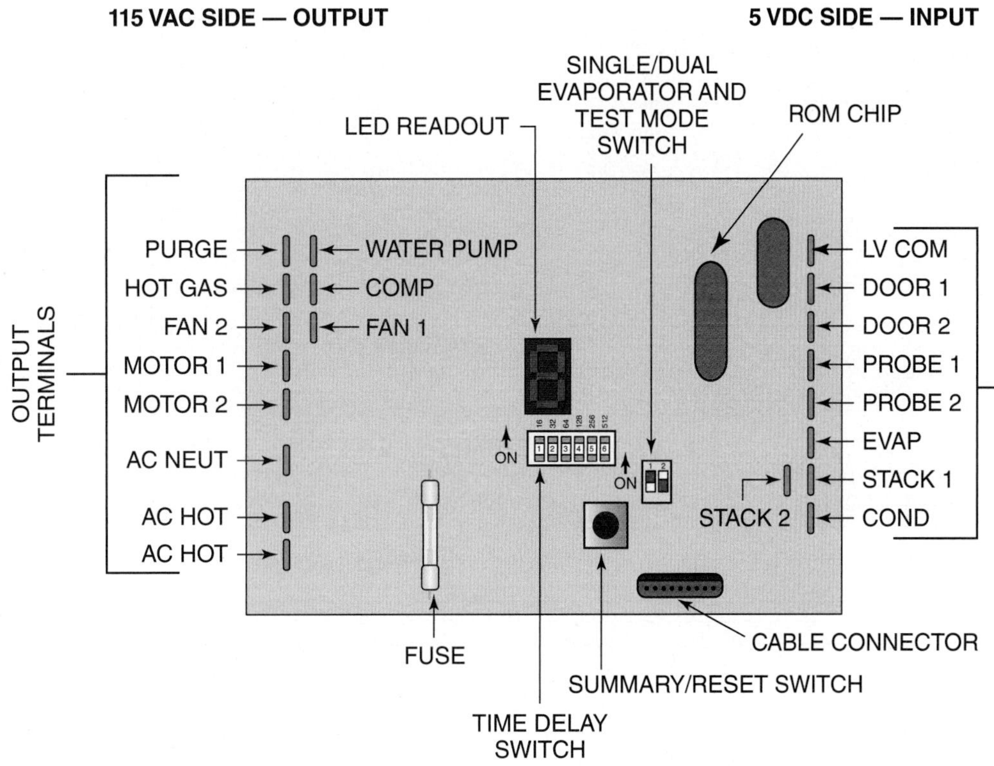

going to the power-consuming devices. Always refer to the service manual for specifics. Input devices are frequently used for digital readouts of a condensing temperature, an evaporating temperature, or a specific sequence mode. **Figure 27.42** shows a service technician measuring the input voltage from a microprocessor. It is important to use as short a measuring probe as possible to avoid shorts to other terminals of the microprocessor. A direct short could ruin any microprocessor. Taping the voltmeter probes halfway up will help prevent electrical shorts when measuring inputs or outputs from a microprocessor.

MICROPROCESSOR SELF-DIAGNOSTICS AND ERROR CODES

Many microprocessors come with a self-diagnostic mode. The microprocessor continually monitors all functions of the ice maker. If a malfunction should occur, the microprocessor will associate the error or malfunction with an error code in its memory. The error code will then be indicated on the LED display. **Figure 27.48** shows some error code descriptions of one manufacturer's ice machine. The service technician in **Figure 27.49** is using a voltmeter to troubleshoot the hot gas high-voltage circuit. Notice that the error code 12 (EC 12) is displayed on the LED of the microprocessor. By referring to the service manual, **Figure 27.48**, the technician finds that this error code indicates that the evaporator temperature exceeded 150°F.

Error codes are a valuable and time-saving troubleshooting tool. However not all problems can be diagnosed with error codes. Always check to see if the fuse on the microprocessor is good before condemning the processor. Most microprocessors have these fuses.

MICROPROCESSOR HISTORY OR SUMMARY

Microprocessors can often record an ice machine's history. This summary of the ice machine's past operations can be retrieved from memory and reviewed at any time by using switches or buttons on the microprocessor. This history may also be erased once it is recorded by the service technician. Important information that happened over long periods of time, such as the number of harvests, error codes stored, and average cycle times, can be recalled, **Figure 27.50**.

The microprocessor is not as complicated as it may seem. All the service technician really has to know is I/O troubleshooting, the sequence of operation, and how to find and follow an instruction manual.

Figure 27.48 Error code descriptions from a microprocessor. Based on Ice-O-Matic.

ERROR CODES

The computer constantly monitors the ice-making cycles to determine if everything is operating correctly. Should the computer detect a malfunction, it will record the associated error code in its memory. If the error code is a "FATAL ERROR," the computer will shut the machine down, light the front panel "SERVICE LIGHT," and indicate the error code on the LED display. NOTE: On machines manufactured after August, 1991, error codes 2, 5, 7, and 12 will automatically make 4 attempts (one attempt every 15 minutes) to reset before the machine is shut down and the service fault light illuminates. The same error code must occur 4 consecutive times before the machine shuts down. All computers incorporating this change can be identified by a version number of 0612 or higher. See version number identification, page H-2. Error codes that do not cause a machine shut-down (EC 1 and EC 3) will ONLY display on the LED when in the "SUMMARY MODE."

Below is a description of each Error code.

EC 1	EC 2
Freeze time is greater than 50 minutes or the temperature of the evaporator is above 40°F at 6 minutes into the freeze cycle.	Cam switch(es) did not cycle closed and open within 45 minutes of initiating the harvest sequence (ICE 5). See note below.

NOTE: On computers with version number 0612 or higher, time of failure for EC 2 has been changed from 45 minutes to 15 minutes.

EC 3	EC 4	EC 5	EC 6	EC 7	EC 8	EC 9
Curtain switch(es) did not close within 5 minutes of initiating the harvest sequence (ICE 5).	The microprocessor chip has a defect. The computer must be replaced.	Condenser thermistor circuit is either open or shorted.	Evaporator thermistor circuit is either open or shorted.	Condenser temperature exceeds 150°F (65°C).	Freeze time is greater than 80 minutes.	Evaporator temperature has reached 14°F (−10°C) in LESS than 90 seconds for FOUR CONSECUTIVE ice-making cycles.

NOTE: On computers with version number 0612 or higher, EC 9 has been eliminated.

EC 12
Evaporator temperature exceeds 150°F (65°C).

If a machine is shut down on a "FATAL" error code, the machine must be reset following the procedure before it can be returned to the ice making mode. Error codes will remain in the memory even after a power interruption.

ICE PRODUCTION AND PERFORMANCE DATA

Ice production, cycle times, and system operating pressures for an air-cooled ice machine mainly depend on two factors:

- Temperature of the air entering the condenser
- Temperature of the water entering the ice machine

Assuming a constant water temperature entering the ice machine, the ice production decreases as the air temperature increases through an air-cooled condenser. Notice in **Figure 27.51** that if the water temperature entering the ice machine stays constant at 50°F, the 24-hour ice production is 1880 lb at the 70°F temperature of the air entering the condenser. However, if the air entering the condenser increases in temperature to 100°F, the 24-hour ice production decreases to 1550 lb. This happens because of the inefficiencies associated with a high compression ratio caused by the increase in condensing temperature with warmer air entering the condenser.

Figure 27.49 A service technician measuring output voltage from a microprocessor. Courtesy Ferris State University. Photo by John Tomczyk

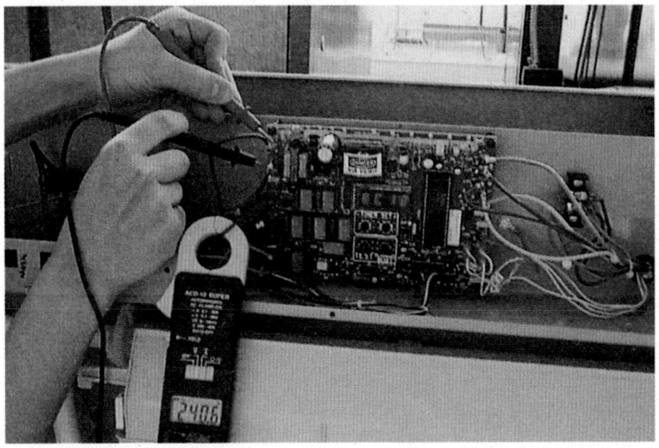

As the air entering the condenser stays constant at 70°F, the 24-hour ice production decreases from 1880 lb to 1640 lb as the water temperature entering the ice machine increases from 50°F to 90°F. This happens because it takes more refrigeration effect, and thus more time, to freeze the warmer 90°F water into ice. Notice that the ice machine has longer freeze times as the water temperature to the ice machine and the air temperature through the air-cooled condenser increase.

For both the freeze and harvest cycles, the discharge and suction pressures increase as the air entering the air-cooled condenser increases. This happens because of higher compression ratios associated with a higher condensing temperature caused by the warmer air passing through the condenser. Knowing what the discharge and suction pressures should be for different air and water temperatures is a valuable tool for the service technician.

Figure 27.50 A summary or history of the ice machine can be recalled by the microprocessor. Based on Ice-O-Matic.

SUMMARY

The machine history is recorded by the computer. This summary of past operation can be recalled and reviewed any time that the selector switch is in the ICE or OFF position. The history may only be erased when the selector switch is in the OFF position, see page H-15. The summary is not affected by power loss.

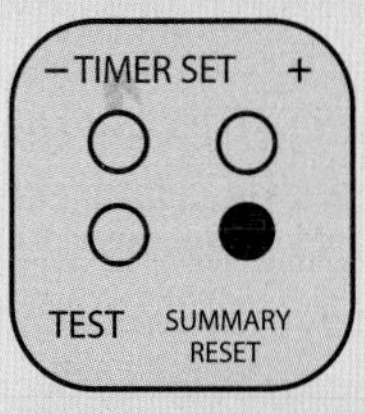

To recall the history, press the SUMMARY/RESET button momentarily. A summary of past operation will now be displayed.

H254

The drawing to the left gives an example of the first part of the summary, which is the harvest count. The H shown on the display indicates the harvest count in thousands, the number that follows (254) indicates the machine has cycled 254,000 times since last being reset.

The remaining history will be displayed in the order given below.

h	EC	rrrr	r1	15	r2
The number that appears in front of h indicates the number of harvests UP TO ONE THOUSAND since last being reset.	The number that appears in front of EC indicates any error code that has been stored in the memory and not yet erased.	This tells the service tech that the computer is now going to display information or "reviews" of prior ice-making cycles. A review is the average cycle time of a group of 10 ice-making cycles.	The computer is now going to display the first review.	The number displayed after r 1 has been displayed is the average cycle time of the 10 most recent ice-making cycles. The example to the left shows a 15-minute review.	The computer is now going to display the second review. The number displayed after r 2 has been displayed is the next 10 most recent ice-making cycles.

Note: The maximum number of reviews is 10 (a total history of 100 cycles). If the computer has not operated for a minimum of 10 cycles, there will be no reviews.

Figure 27.51 Cycle times, ice production, and operating pressures for an air-cooled system. Based on Manitowoc Ice, Inc.

CYCLE TIMES				
Freeze Time + Harvest Time = Cycle Time				
Air Temp. Entering Condenser °F/°C	Freeze Time			Harvest Time
	Water Temperature °F/°C			
	50/10.0	70/21.1	90/32.2	
70/21.1	8.5–9.3	9.4–10.3	9.9–10.9	1–2.5
80/26.7	9.0–9.9	9.8–10.8	10.5–11.5	
90/32.2	9.6–10.5	10.4–11.5	11.1–12.2	
100/37.8	10.6–11.6	11.5–12.6	12.4–13.6	

[1]Times in minutes

24 HOUR ICE PRODUCTION			
Air Temp. Entering Condenser °F/°C	Water Temperature °F/°C		
	50/10.0	70/21.1	90/32.2
70/21.1	1880	1720	1640
80/26.7	1780	1650	1560
90/32.2	1690	1570	1480
100/37.8	1550	1440	1350

[1]Based on average ice slab weight of 13.0–14.12 lb
[2]Regular cube derate is 7%

OPERATING PRESSURES				
Air Temp. Entering Condenser °F/°C	Freeze Cycle		Harvest Cycle	
	Discharge Pressure PSIG	Suction Pressure PSIG	Discharge Pressure PSIG	Suction Pressure PSIG
50/10.0	220–280	40–20	155–190	60–80
70/21.1	220–280	40–20	160–190	65–80
80/26.7	230–290	42–20	160–190	65–80
90/32.2	260–320	44–22	185–205	70–90
100/37.8	300–360	46–24	210–225	75–100
110/43.3	320–400	48–26	215–240	80–100

[1]Suction pressure drops gradually throughout the freeze cycle

NOTE: These characteristics may vary depending on operating conditions.

27.5 WATER AND ICE QUALITY

An ice machine is an important piece of equipment to any business where it is installed and is vital to a food-service customer. Most of these businesses depend heavily on ice. However, many companies ignore their ice machine and fail to consider that it is a self-contained "mini-ice factory." An ice machine takes a raw material and manufactures a consumable product. This product is then stored and provided to the customer for whatever purpose it is needed.

WATER QUALITY

Ice is considered a food source, and the single component material is water. The quality of the ice product depends on the quality of the water supplied to the unit. Good water quality will produce a crystal-clear, hard cube that provides excellent cooling capacity and lasts a long time. Poor water quality will produce ice that can be soft and cloudy and will result in less Btu cooling capacity, ice bridging in the storage container, or both.

Water can contain many different minerals at various levels, and the quality is constantly changing. In fact, water quality varies throughout the United States and may vary across a town or even across a street. Water quality can be improved through filtration and treatment. Technical experts know about the chemical makeup of water and how to treat specific problems, helping to ensure that high-quality ice is produced. Although few service technicians have this level of expertise when it comes to resolving water-quality issues, it is important to have a basic understanding of water quality in order to service an ice machine. The typical ice machine service technician is called on from time to time to recommend water filtration or treatment for ice-making equipment. Of course, the best source of information concerning filtration and treatment is a water filter manufacturer. These manufacturers will work with you and make recommendations to resolve your water-quality issues. Many websites provide educational information on water quality and water filtration and treatment. Although it is impossible to cover every aspect of water quality here, a service technician should have a basic knowledge of key water treatment terminology and how it applies to an ice machine. The following terms are used when dealing with water quality in ice machines:

Acidic water: Water containing dissolved carbon dioxide that forms carbonic acid. Acidic water is considered aggressive to most components it comes in contact with and is represented by a low pH reading.

Alkaline water: Water that has a surplus of hydroxyl ions, which results in a pH over 7. It often contains concentrations or a mixture of soluble salts or alkali that are picked up from basic minerals in the earth, such as calcium, magnesium, and silica. These are scale-causing minerals.

Chloramines: Disinfectants used to treat municipal water systems for bacteria. Chloramines are mixtures of chlorine and ammonia. A higher concentration of charcoal is generally needed to reduce the amount of chloramines in the water.

Chlorine: A disinfectant used to treat municipal water systems for bacteria. Chlorine can be corrosive to metals. Charcoal filtration is generally used to reduce the amount of chlorine in the water to improve offensive chlorine taste and odor.

Filtration: The process of removing contaminants from water. A device is installed in the water supply inlet of the ice machine to remove contaminants from the water. Coarse filtration removes the "rocks and boulders." Fine filtration removes smaller particles.

Flow rate: The volume of flow through a device measured in gallons per minute (GPM). It is important to size a filter system properly to supply the correct flow rate to the unit as per the manufacturer's required specifications.

Hardness: The amount of calcium and magnesium measured in grains per gallon. See the following standards chart from the U.S. Bureau of Standards:

Water Hardness

Description	Grains per Gallon	Parts per Million (Total Dissolved Solids)
Soft Water	Less than 1.0	Less than 17.1
Slightly Hard	1.0 to 3.5	17.1 to 60.0
Moderately Hard	3.5 to 7.0	60.0 to 120.0
Hard Water	7.0 to 10.5	120.0 to 180.0
Very Hard	10.5 and over	180.0 and over

Note: 7000 grains = 1 pound.

Iron: A compound or bacteria found generally in well water. Iron compounds cause a rust-colored scale. Iron bacteria show up as rust-colored slime.

pH: The term used to express the level of acidity or alkalinity in water or solutions. The scale for pH ranges from 0 to 14. A reading of 7 is considered a neutral pH and is the desired condition for ice machine applications. A reading of below 7 is acidic, and above 7 is basic or alkaline.

Reverse osmosis (RO): The process of removing minute contaminant particles. An RO device forces water, under pressure, against a fine membrane to remove these minute particles of contaminants.

Scale: A buildup of minerals that forms a flaky coating on the surfaces of the evaporator and water system. Scale buildup restricts water flow and reduces heat transfer in an ice machine. This buildup affects unit efficiency and usually causes service concerns.

Scale inhibitor: A material or element that slows the ability of scale to stick to a surface. Many filter manufacturers use **polyphosphate** to inhibit calcium or lime scale buildup.

Sediment: Larger particles of dirt, trash, sand, minerals, and similar substances. A sediment filter is generally installed prior to the filter/treatment device to remove larger contaminants. This allows for longer life of the filter/treatment device by reducing rapid plugging of the filter medium.

Total dissolved solids (TDS): The level of dissolved minerals in the water measured in parts per million (ppm). Higher concentrations of TDS can increase mineral buildup.

Treatment: Using a device that adds chemicals or minerals to the water to change water quality. Treatment is generally required to address scale and bacteria concerns.

Turbidity: The degree of water cloudiness generally caused by high mineral saturation or the presence of air bubbles in the water.

Water softener: A device that uses salt to provide an ion exchange that reduces the hardness of water, making it less likely to produce scale.

One last term that is usually understood by the service technician but seldom considered by the customer is ***preventive maintenance.*** *This refers to the regularly scheduled maintenance performed on a unit, including inspecting, cleaning, sanitizing, and servicing of the ice machine and external water filtration/treatment system. Preventive maintenance pays big benefits. Every equipment manufacturer provides recommendations and instructions that detail the steps necessary to perform preventive maintenance. This process, along with providing proper water filtration and treatment at the time of initial installation, is vital for protecting the customer's investment and maintaining maximum efficiency—as well as ensuring a longer life for the ice machine.* A basic understanding of these terms will help a service technician communicate any water-quality concerns to a filter manufacturer and to the customer. This will definitely help in resolving ice machine water-quality issues.

ICE QUALITY

Ice is a consumable product. Most beverages taste better over ice, especially on a hot summer day. Placing some ice in your cooler keeps your picnic foods and beverages fresh and cool. Whether you crunch it, chew it, or use it as a means of cooling something, ice is an important commodity.

The amount of refrigeration effect that you get from the ice depends on the quality of the ice. Ice quality is measured in percent of hardness. Hardness is a measurement that represents the thermal cooling capacity. The higher the percent of hardness, the more cooling ability the ice possesses. The harder the ice, the denser it is and the longer it will last in

a glass or a cooler. Do not confuse ice hardness with hard water. If you have hard water, you will have reduced ice hardness because of the minerals contained in the ice.

Ice hardness is calculated by conducting a calorimeter test. This is a specific test used by the Air-Conditioning, Heating and Refrigeration Institute (AHRI) to rate ice quality or cooling ability. The American National Standards Institute (ANSI) and the American Society of Heating, Refrigerating, and Air-Conditioning Engineers (ASHRAE) have standards for test procedures. ANSI/ASHRAE Standard 29 lists the procedure for conducting this calorimeter test. The test requires an insulated container with a specific amount of water at a specific temperature. The ambient conditions for the test are also controlled. An exact amount of ice is stirred into the container and timed until all the ice melts. The water temperature before and after the ice is added and the actual melting time is listed on a data sheet. This information is then used to calculate the ice hardness.

Water purity definitely has an effect on ice quality. Pure ice will be crystal clear. Cloudy ice usually contains air or minerals, which will limit hardness. Ice is normally made available in blocks, cubes, chunks, flakes, and cubelets or nuggets. Customer preference and use dictate which type of ice is needed. As block ice is frozen, pure water freezes first. Any minerals or trapped air in the cold water push toward the center of the block and finally freeze. This leaves a cloud in the center of the clear block of ice. You may have noticed this if you have ever used ice made in an ice tray in your home refrigerator. This type of ice falls in the 95% to 100% hardness range, depending on the mineral content.

Flaked ice falls in the 70% hardness range. This is due to the freezing process used in the ice flake machine. Water is frozen on a cylinder wall, broken away, and extruded into the storage bin. Generally, all the water that goes into the flaker comes out as ice. Any minerals in the water will be frozen into the ice. Because of the nature of flaked ice, more air is present. This also reduces the overall hardness and cooling effect of flaked ice.

As mentioned earlier, cubelets or nuggets are made of compressed flakes. The flakes of ice are squeezed or extruded through a smaller opening to remove more water. This extruded ice is then broken into chunks. These chunks fall in the 80% to 90% hardness range. Since cubelets or nuggets of ice are easy to chew, they are preferred by the health care industry.

Cubes fall in the 95% to 100% hardness range, depending on the evaporator style and ice-making process. Commercial ice cube machines circulate water into, across, or over an evaporator plate. As the water circulates and is cooled, pure water freezes first. The minerals tend to "wash out" during this freezing process. The result is a purer and thus harder cube of ice. The ice hardness will vary with the style of evaporator and materials used. Any restrictions to the water path on the evaporator surface can slow down the water flow and allow minerals to freeze into the cube. Such restrictions can occur when the water has to flow over a separator on the evaporator or into the cavity of a grid cell. These cubes may be cloudy and fall in the lower hardness range. **Figure 27.52** shows the difference between hard and soft ice. The ice sheet on the left is cloudy and soft. It contains a lot of minerals. The ice sheet on the right is clearer, denser, and will have a higher hardness percentage with more cooling capacity. **Figure 27.53(A)** shows the water sump full of a fine, white, talcy powder of undissolved minerals. **Figure 27.53(B)** shows a service technician taking a sample of the white, powdered minerals in his fingers.

Figure 27.52 Soft, mineral-laden ice versus clear, hard ice. Courtesy Ferris State University. Photo by John Tomczyk

Regardless of the type of ice or its use, a cleaner ice machine and purer water will produce harder ice. Some ice machine manufacturers provide designs that include built-in flushing or purging capabilities. This helps to eliminate minerals, thus keeping the ice machine cleaner and improving ice quality. Regularly scheduled preventive maintenance checks including cleaning and sanitizing of the ice machine water system are a must. Always use the correct cleaner and follow the manufacturer's recommendations when performing the cleaning and sanitizing procedure.

SANITIZING AND CLEANING

The sanitizing process helps to eliminate harmful bacteria that tend to thrive in the cool ice-making environment. Remember that ice is consumed as a food and comes in direct contact with the food or beverage it cools. It is used to cool down beverages and ice down fruits, vegetables, fish, poultry, and meats. Because of these uses, ice is considered a food product. *This is why ice machines must be cleaned and sanitized on a regular basis. Using a sanitizer is an important step in the preventive maintenance process. An acid-based cleaner removes scale. Sanitizing kills harmful bacteria, viruses, and protozoa.*

Bacteria, viruses, and protozoa can adhere to moist areas on the inside of the evaporator compartment. Some of these bacteria and viruses can make people very sick. These microscopic organisms can be either airborne or waterborne.

Figure 27.53 (A) Water sump with mineral deposits appearing as a white, talcy powder. Courtesy Ferris State University. Photo by John Tomczyk. (B) Service technician taking a sample of the fine, white, talcy mineral powder. Courtesy Ferris State University. Photo by John Tomczyk

(A)

(B)

Municipal water systems are relatively free of harmful waterborne organisms because chlorine has been added as a disinfectant. Water filtration or treatment, which will be covered in detail shortly, can also provide protection against bacterial contamination. However, airborne organisms can still be present. Once an organism sticks to a moist surface, it begins to grow because of the cool, damp conditions inside the evaporator compartment. The result may be a mold, algae, or slime buildup. Slime is the most visible target when sanitizing an ice machine. It is usually a jellylike substance that is made up of algae, mold, and yeast spores. These spores can be either airborne or waterborne. *Lime or calcium buildup, often referred to as scale, is removed from the ice machine by using a commercial-grade ice machine cleaner.* Ice machine cleaning solutions are made of a food-grade acid that is approved for use in food applications. They come in different strengths for many different purposes. The weaker solutions are used to protect different metals. Heavy scale deposits may require soaking in the cleaning solution. A stronger cleaner can be used on stainless-steel evaporators; however, a nickel-safe cleaner must be used on plated evaporators to avoid possible flaking or peeling of the plating. Because copper and brass are more porous than stainless steel, if flaking or peeling occurs and the brass or copper is exposed, the evaporator will be more susceptible to bacterial growth. Circulating a dilute solution of acid-based cleaner through the water system will remove scale, but it does not kill bacteria. A separate procedure called *sanitizing* must be performed to address bacteria.

Sanitizing is accomplished by using a commercial ice machine sanitizer. You cannot remove scale and sanitize with the same solution. Two different solutions are needed. In fact, mixing ice machine cleaner and sanitizer together creates a strong chlorine gas that can be dangerous to breathe. The sanitizing solution must be circulated as directed by the ice machine manufacturer. Most manufacturers provide step-by-step cleaning and sanitizing instructions, which point out specific areas that must be disassembled and manually cleaned and sanitized. Sanitizing should be done at least once a month for ice machines used in the food service industry. Many service technicians prepare their own sanitizing solution by mixing 1 oz of household bleach with 2 gal of warm, potable water. The water should be in the temperature range of 95°F to 115°F. This gives the solution about 200 ppm chlorine strength. However, before using any sanitizing solution always consult the manufacturer's directions for sanitizing the ice machine.

WATER FILTRATION AND TREATMENT

As mentioned earlier, water-related problems account for over 75% of service problems in ice makers and ice flakers. However, if cleaning and sanitizing are done on a schedule that is appropriate to local water conditions, the ice machine will have a longer and more productive life dispensing quality ice. Unfortunately, ice machines are often placed out of sight and are seldom cleaned or sanitized until they have to be. This is where water treatment methods come into play. A good water treatment program will extend the cleaning and sanitizing intervals between service calls. But first, it is important to understand the local water conditions.

Water conditions can be broken down into the following three categories:

- Suspended solids
- Dissolved minerals and metals
- Chemicals

Sand and dirt are examples of suspended solids and are probably the easiest to remove. While mechanical filters in the water line can remove both sand and dirt, dissolved minerals and metals are much more expensive to capture. Their removal requires processes like reverse osmosis, distillation,

or de-ionization. Reverse osmosis uses water pressure to force water through special membranes that reject most of the dissolved minerals and organics and flush them to waste. Distillation requires water to be heated to its boiling point to create steam. The steam is then recondensed into water, leaving the minerals behind. De-ionization uses columns or beds of special filter media such as plastic beads or mineral granules to remove minerals from the water in exchange for other substances. An example of this is water softening. Water softeners are loaded up or regenerated with sodium ions from ordinary salt. The sodium ions are exchanged for calcium and magnesium, which are often referred to as the *water hardness* ions. Scale-inhibiting chemicals, which have to be food grade, can also successfully control dissolved minerals and metals.

Chemicals, on the other hand, have to be removed by carbon filtration. The most common chemicals in water are chlorine and chloramines, which are common city water disinfectants. Other types of chemicals that make their way into the water system can affect the taste, odor, and color of the water.

Water filtration and treatment can be applied to a water supply to try to treat these categories of water conditions. *A common treatment for any type of ice machine is a triple-action filter system,* **Figure 27.54**. *These systems can filter out suspended solids, fight scale formation by adding chemicals, and improve the taste, odor, and color of water with carbon filtration.* Because the filters are micron filters, they do plug up and have to be changed on a regular basis. Often, a pressure gauge installed on the filter will indicate a pressure drop when the filter is starting to plug up. Any time the scale inhibitor or carbon is used up is a good time to change all three cartridges. However, water treatment in a specific geographical region is never an exact science. Many times, a trial-and-error method is used because there is never an exact quick fix when it comes to water treatment. Always consult with your local water-quality experts, ice machine distributors, or manufacturers for their professional opinions on the best water treatment for your geographical area. The water filter manufacturer can also assist you in testing the water at your site and recommending the correct water treatment combination to improve the water quality. **SAFETY PRECAUTION:** *These chemicals can be hazardous to humans. Always follow the manufacturer's instructions. The best time to circulate any chemicals is when there is no ice in the bin so they will not contaminate the stored ice. Because it is rare to find an ice maker empty, you may need to remove the ice from the bin and place it in a remote storage area. Then clean and rinse the ice bin according to the manufacturer's directions.* •

Figure 27.54 A triple-action filter system. Courtesy Scotsman Ice Systems

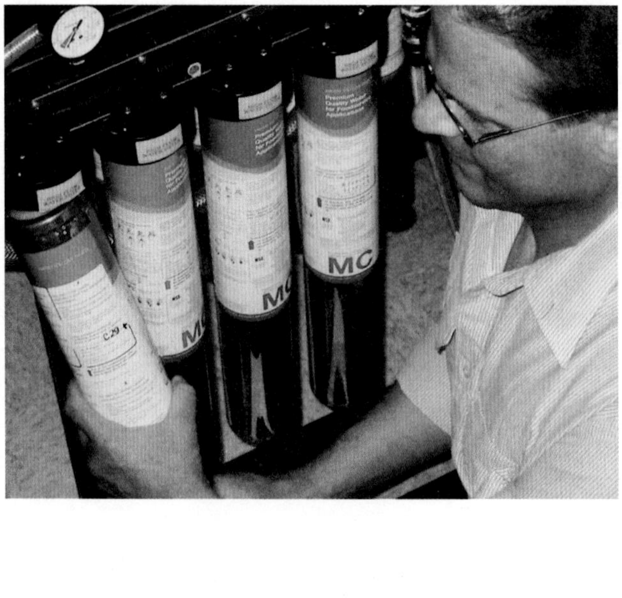

27.6 PACKAGE ICE MACHINE LOCATION

Most package ice machines are designed to be located in ambient temperatures of 40°F to 115°F. Follow the manufacturer's recommendations. If the ambient temperature is too cold for the ice maker, freezing may occur and damage the float or other components where water is left standing. If it is too hot, compressor failures and low ice capacities may occur. The ambient conditions under which the ice maker must operate are important.

WINTER OPERATION

A machine located outdoors at a motel may not make ice below about 40°F ambient temperature because the thermostat in the bin is set at about 32°F. With this setting, when ice touches the thermostat sensor, it will shut off the machine. The combination of 40°F outside and some ice in the bin will normally satisfy the thermostat. When the machine is located outside, the water should be shut off when the outside temperature approaches freezing. The water should be drained to prevent freezing and damage to the pump, the evaporator on some units, and the float assembly.

When machines are outside and the ambient temperature is below about 65°F, some method is normally used to prevent the head pressure from becoming so low that the expansion device will not feed enough refrigerant to the evaporator. Low-ambient control causing fan cycling is a common method. Mixing valves or head pressure control valves are also used to maintain both discharge- and liquid-line pressures. These valves mix hot discharge gas with liquid coming from the receiver when the outdoor ambient drops below a certain temperature. At the same time, liquid is backed up in the condenser, causing a flooded condition. This flooding of the condenser will elevate the head pressure. Head pressure control valves are covered in detail in Unit 22, "Condensers."

The correct location of an ice machine indoors is critical because of the airflow across the condenser. Air cannot be allowed to recirculate across the condenser or high head

pressures will occur. This will reduce the capacity and may in some cases harm the compressor.

27.7 TROUBLESHOOTING ICE MAKERS

When an ice maker is not functioning properly, the problem can usually be found in the water circuit, the refrigeration circuit, or the electrical circuit.

WATER CIRCUIT PROBLEMS

Water level is critical in most ice makers. The manufacturer's specifications should be followed. Most manufacturers post the water level on the float chamber or at some location close to the float. Determine what the level should be and make certain that it is set correctly, **Figure 27.55**.

The float valve has a soft seat, usually made of neoprene. This seat may become worn and a new one may be required. A float seat is inexpensive compared to an ice maker operating inefficiently for any period of time. When a float does not seal correctly, water at the supply temperature flows through the system instead of being refrigerated to the freezing temperature. It may also melt ice that has already been made. A temporary repair may be to turn the float seat over if it can be removed from its holder, **Figure 27.56**. The best time to determine whether or not a float is sealing correctly is when the ice maker has no ice. Remember, when there is ice in the bin, it is constantly melting. You cannot tell if the water in the drain is from melting ice or a leaking float. When there is no ice in the bin, the only water in the drain would be coming from a leaking float.

Water circulation over evaporators that make cube ice is accomplished by a water circulating system that is carefully designed and tested by the manufacturer. You should make every effort to understand what the manufacturer has intended before you modify anything. It is not unusual for a technician to try various methods to obtain the correct water flow over the evaporator when the manufacturer's manual would explain the correct procedure in detail. Do not overlook something as simple as a dirty evaporator. If it must be cleaned, check the manufacturer's suggestions.

Figure 27.56 Sometimes a float seat can be reversed or turned over with the good side out for a temporary repair.

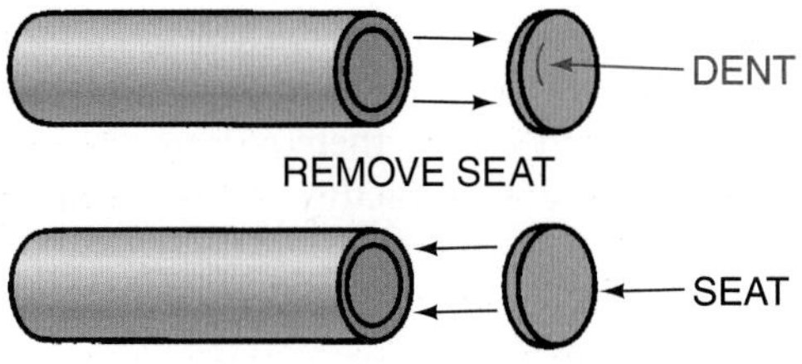

The quality of the water entering the machine must be at least as good as the manufacturer recommends. Check the filter system if the water seems to have too many minerals. A water treatment company may be contacted to help you determine the water condition and the possible correction procedures.

REFRIGERATION CIRCUIT PROBLEMS

Refrigeration circuit problems are either high-side or low-side pressure problems.

HIGH-SIDE PRESSURE PROBLEMS. High-side pressure problems can be caused by

- poor air circulation over the condenser,
- recirculated condenser air,
- a dirty condenser,
- a defective fan motor,
- overcharge of refrigerant, or
- noncondensables in the system.

Figure 27.55 Maintain the correct float level.

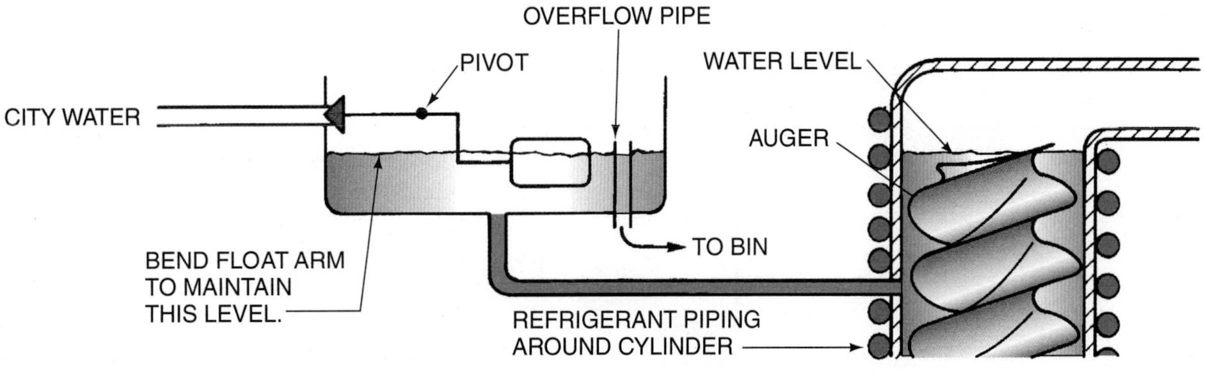

Another high-side pressure problem is a result of trying to operate a system when the ambient temperature is below the design temperature for the unit. Low head pressure can cause low suction pressure, particularly in systems with capillary tube metering devices. Owners tend to try to place the machines outside and expect them to perform all year. The design conditions the manufacturer recommends should be followed.

LOW-SIDE PRESSURE PROBLEMS. Low-side pressure problems can be caused by

- low refrigerant charge,
- incorrect water flow,
- mineral deposits on the water side of the evaporator,
- a restricted liquid line or drier,
- a defective metering device or thermostatic expansion valve,
- moisture in the refrigerant,
- an inefficient compressor, or
- a blocked metering device.

Low-side problems will show up as a pressure that is too high or too low as indicated on the refrigeration gauges. When the pressure is too low, you should suspect a starved evaporator. When the ice maker makes solid-type ice, a starved evaporator will often cause irregular ice patterns on the evaporator. The ice will become thicker where the evaporator surface is colder. When the pressure is too high, you should suspect an expansion device that is feeding too much refrigerant, a defective TXV, or a compressor that is not pumping to capacity. When the expansion valve is overfeeding, the compressor will normally be cold from the flooding refrigerant.

A stuck water float can cause excess load if the entering water is warmer than normal, **Figure 27.57**. As mentioned earlier, it is hard to detect a leaking float. Look for the correct level in the float chamber. If adjusting the float does not help, the seat is probably defective.

The troubleshooting checklist from a manufacturer's manual in **Figure 27.58** is a general guide for the ice machine owner or service technician to follow. Similar checklists are available from the manufacturer of the particular ice machine in question. As mentioned earlier, the Internet is an excellent source of quick troubleshooting information that can assist the service technician in systematically troubleshooting an ice machine.

ELECTRICAL PROBLEMS

A defrost problem may appear to be a low-pressure problem if the unit fails to defrost at the correct time. Large amounts of ice may build up. Remember, defrost is only used on machines that make cube ice. When you find heavy amounts of ice on a machine, determine what the defrost method is. The machine may have a forced defrost cycle that will allow you to clear the evaporator and start again, but some manufacturers may suggest not forcing the machine to defrost or shutting the machine off but waiting for the normal defrost cycle. Waiting for the normal cycle will help determine what type of problem caused the condition. If the evaporator is frozen solid and a defrost is forced, you may not find the problem because it may not happen again in the next cycle.

Line-voltage problems may be detected with a voltmeter. The unit nameplate will tell you what the operating voltage for the machine should be. Make sure the machine is within the correct voltage range. The plug on the end of the power cord may have become damaged or overheated due to a poor connection. Examine it while the machine is running by touching it. If it is hot to the touch, it is not properly conducting power to the machine. When the power is wired to the junction box in the ice maker, it goes to the control circuit that operates the machine. Larger machines may use relays to energize contactors for starting the compressor.

Figure 27.57 A float that does not seal will reduce the usable ice a machine can make.

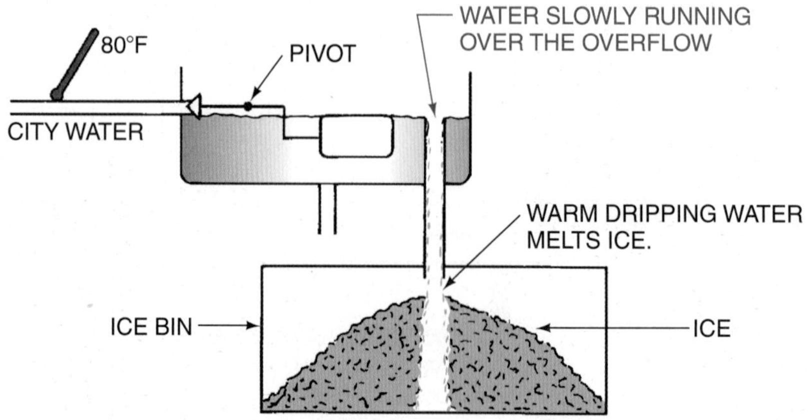

Figure 27.58 A basic service checklist for an ice machine. (Page numbers do not refer to this text.) Courtesy Manitowoc Ice, Inc.

PROBLEM	POSSIBLE CAUSE	TO CORRECT
Ice machine does not operate.	No electrical power to the ice machine.	Replace the fuse/reset the breaker/turn on the main switch.
	High pressure cutout tripping.	Clean dirty air filter and/or condenser coil (See page 4-1).
	ICE/OFF/CLEAN toggle switch set improperly.	Move the toggle switch to the ICE position.
	Water curtain stuck open.	Water curtain must be installed and swinging freely. See page 4-10.
	Remote receiver valve closed.	Open the valve.
Ice machine stops and can be restarted by moving the toggle switch to OFF and back to ICE.	Safety limit feature stopping the ice machine.	Refer to "Safety Limit Feature" on the next page.
Ice machine does not release ice or is slow to harvest.	Ice machine is dirty.	Clean and sanitize the ice machine. See pages 4-3 and 4-4.
	Ice machine is not level.	Level the ice machine. See page 2-6.
	Low air temperature around ice machine (air-cooled models).	Air temperature must be at least 35°F (1.6°C).
	Water regulating valve leaks in harvest mode (water-cooled models).	Refer to "Water-Cooled Condenser" on page 4-2.
Ice machine does not cycle into harvest mode.	The six-minute freeze time lock-in has not expired yet.	Wait for freeze lock-in to expire.
	Ice thickness probe is dirty.	Clean and sanitize the ice machine. See pages 4-3 and 4-4.
	Ice thickness probe wire is disconnected.	Connect the wire.
	Ice thickness probe is out of adjustment.	Adjust the ice thickness probe. See page 3-4.
	Uneven ice fill (thin at top of evaporator).	See "Shallow or Incomplete Cubes" on the next page.
Ice quality is poor (soft or not clear).	Poor incoming water quality.	Contact a qualified service company to test the quality of the incoming water and make appropriate filter recommendations.
	Water filtration is poor.	Replace the filter.
	Ice machine is dirty.	Clean and sanitize the ice machine. See pages 4-3 and 4-4.
	Water dump valve is not working.	Disassemble and clean the water dump valve. See page 4-7.
	Water softener is working improperly (if applicable).	Repair the water softener.
Ice machine produces shallow or incomplete cubes, or the ice fill pattern on the evaporator is incomplete.	Ice thickness probe is out of adjustment.	Adjust the ice thickness probe. See page 3-4.
	Water trough level is too high or too low.	Check the water level probe for damage. See page 3-4.
	Water inlet valve filter screen is dirty.	Remove the water inlet valve and clean the filter screen. See page 4-9.
	Water filtration is poor.	Replace the filter.
	Hot incoming water.	Connect the ice machine to a cold water supply. See page 2-10.
	Water inlet valve is not working.	Remove the water inlet valve and clean it. See page 4-9.
	Incorrect incoming water pressure.	Water pressure must be 20-80 psi (137.9–551.5 kPA).
	Ice machine is not level.	Level the ice machine. See page 2-6.
Low ice capacity.	Water inlet valve filter screen is dirty.	Remove the water inlet valve and clean the filter screen. See page 4-9.
	Incoming water supply is shut off.	Open the water service valve.
	Water inlet valve is stuck open or leaking.	Remove the water inlet valve and clean it. See page 4-9.
	The condenser is dirty.	Clean the condenser. See page 4-1.
	High air temperature around ice machine (air-cooled models).	Air temperature must not exceed 110°F (43.3°C).
	Inadequate clearance around the ice machine.	Provide adequate clearance. See page 2-5.
	Objects stacked on or around ice machine, blocking air flow to condenser (air-cooled models).	Remove items blocking air flow.
	Air baffle is not installed (air-cooled models).	Install air baffle. See page 2-12.

27.8 SERVICE TECHNICIAN CALLS

The following service calls cover some of the most common problems experienced with ice machines.

SERVICE CALL 1

A service technician is servicing a flake ice machine. After taking the covers off the machine, the technician does not notice anything wrong. The bin is half full of ice and the machine seems to be working properly. The technician then decides to talk to the restaurant manager. The manager says that the ice flake machine makes loud, intermittent screeching noises. The technician returns to the machine and does hear some noise coming from the evaporator and auger assembly. The noise gets progressively louder for about 10 min before stopping. Ten minutes later, the noise starts again. *The problem is mineral buildup in the evaporator freezing cylinder and auger compartment. This causes the auger to drag on the mineral deposits, making a* sound like *two pieces of Styrofoam being rubbed together. Sometimes the mineral deposits break loose and are delivered with the ice.*

The service technician then tells the restaurant manager that the noise he described was occurring again. By doing this, the technician lets the restaurant manager feel that he helped to find the solution to the problem. The service technician examines the ice flakes for quality and finds that they are very cloudy. This is caused by minerals trapped in the flakes. The technician determines that the ice machine needs a cleaning to rid it of the mineral deposits. The technician also decides to sanitize the machine after the cleaning to rid the machine of any harmful bacteria, viruses, or protozoa. Emptying the ice bin of all ice flakes is the first step the technician takes.

The technician follows the manufacturer's instructions in the service and maintenance manual, acquired from an Internet search, on cleaning and sanitizing the entire ice machine and bin. (Often, instructions for cleaning and sanitizing are located on a label on the ice machine itself.) After restarting the ice flake machine, the technician finds that it runs quietly and that the ice has a much better quality than before the cleaning and sanitizing procedure.

The service technician stays for about 45 min, cleaning up and talking to the restaurant manager about a preventive maintenance and water treatment program for all the ice machines in the restaurant. The technician then returns to the ice flake machine to check for screeching sounds, but hears none. It appears that the problem has been remedied. Again, the technician explains what the problem was, and then thanks the restaurant manager for his help in solving the problem. The restaurant manager is then given the billing information with a detailed description of the work completed and associated costs. Handshakes are exchanged and the service technician leaves the job site.

SERVICE CALL 2

A service technician is called to service a cell-type ice cube machine that will not harvest the ice from the evaporator. The sheet of ice grows much too thick and actually freezes the ice thickness probe into the ice sheet. *The ice thickness probe's electrodes are scaled with mineral buildup. This prevents them from completing a circuit when the thin water film next to the ice sheet contacts them. (Evaporator damage from too thick an ice sheet can occur. Also, low suction pressure from a reduced evaporator load, which may lead to compressor flooding, can occur if the problem is not fixed.)*

The service technician bypasses the freeze time lock-in feature by moving the ice/clean/off switch to *off* and back to *ice*. The technician waits for water to start running over the evaporator and then clips a jumper wire lead to the ice thickness probe and cabinet ground, **Figure 27.59**. The harvest light on the control board comes on. Eight seconds later, the ice machine goes into a harvest cycle. The technician can hear the hot gas solenoid valves energize and hot gas pass into the evaporators. The ice thickness probe is taken off the evaporator and cleaned with the proper ice machine cleaner to get rid of any mineral deposits. While cleaning up, the technician watches the ice machine through three freeze and harvest cycles and determines that the ice machine is working as designed. After talking with the store owner about preventive maintenance and water treatment, the technician leaves the job site.

SERVICE CALL 3

A service technician is called to a restaurant to service an ice machine that makes cube type ice. The restaurant owner is complaining about cloudy ice that melts very fast. Customers are complaining that there is flaky sediment in the bottom of their glasses after the ice melts. Before the service technician enters the restaurant, he makes sure that there is no dirt or grease on his face, clothes, or shoes from a previous job that day. The technician also makes sure his hands are cleaned and sanitized before working on the ice machine, knowing that ice is a food product and can be contaminated very easily. Once introduced to the manager, the technician carefully listens to what he has to say while always maintaining his professionalism. It is important never to be critical of the customer, even in a joking manner, no matter what he has to say about the service call. Always try to be sympathetic, listen carefully, and try to determine why the customer is so upset. Once the customer finishes, the technician is led to the ice machine to begin work. *The problem is a defective water dump solenoid valve* **Figure 27.30(A).** *If the old water from the water reservoir is not dumped or purged every cycle, the same water is being used and the mineral content becomes very high. Excess minerals that are frozen in the ice cause it to become soft, cloudy, and dirty.* **Figure 27.52** *shows the difference between soft, cloudy ice and hard, clear ice.*

Figure 27.59 An ice thickness probe and control board. (A) Before jumper. Based on Manitowoc Ice, Inc. (B) After jumper. Based on Manitowoc Ice, Inc.

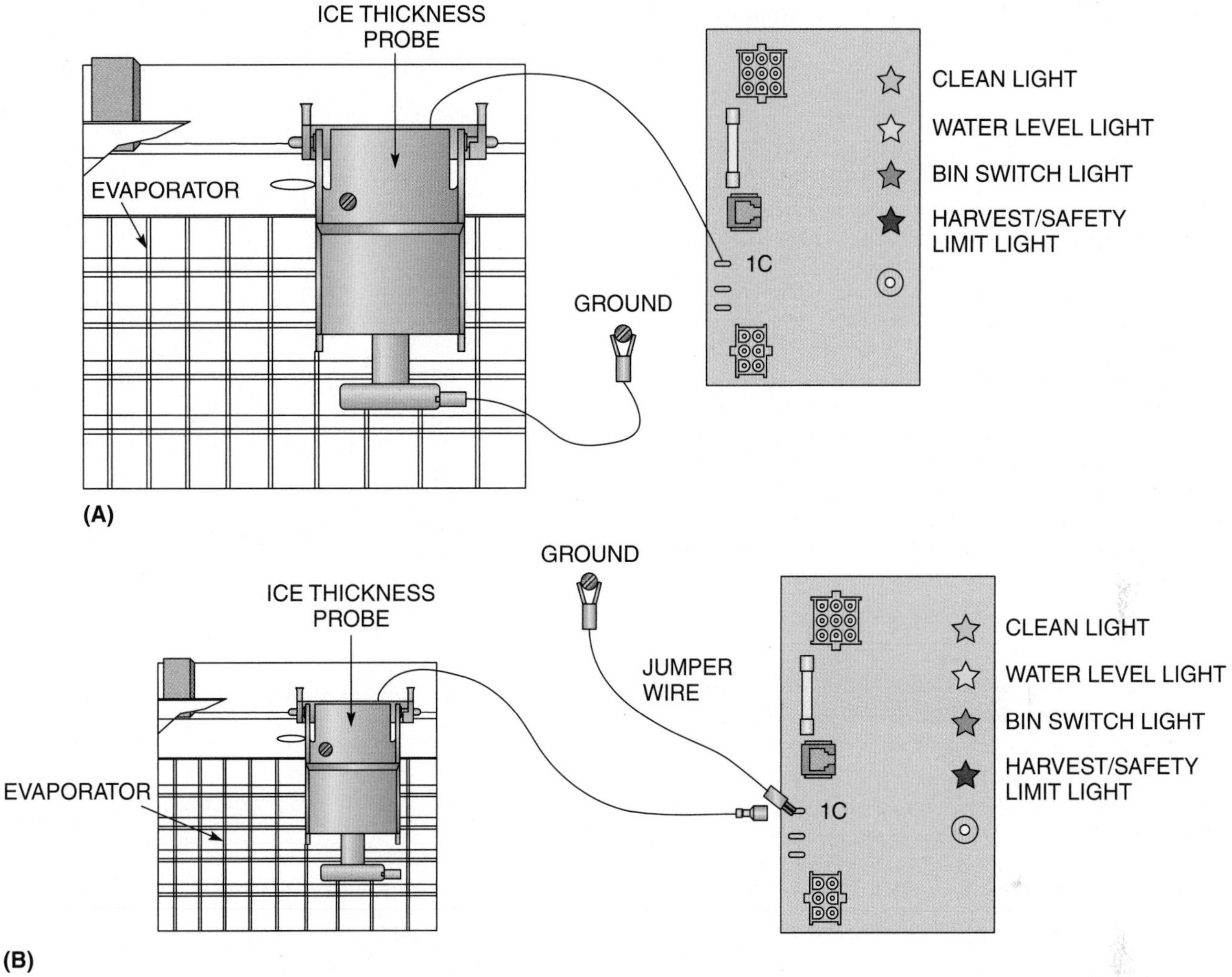

The service technician inspects some ice in the ice bin and sees that it is soft, cloudy, and dirty with minerals. After observing the freeze and harvest cycles and determining that there is no purging of water during the harvest cycle, the technician turns off the ice machine and unhooks the purge solenoid from the control board. An ohmmeter is used to check resistance of the purge solenoid's coil. The resistance checks to infinite, indicating an open solenoid coil. A new coil is put in place and is hooked back up to its control board terminals. The ice machine is turned back on and a cycle is again observed. During a harvest cycle, the control board energizes the water purge solenoid and water freely flows down the drain. This action empties the water reservoir of old, mineral-laden water, and fresh water now fills the reservoir. The technician catches the ice before it falls to the bin and observes it. The ice is crystal clear, hard, and clean. The service technician then talks to the restaurant owner about a preventive maintenance and water treatment program, cleans up around the ice machine, gives the customer the billing information and explains each step of the billing in detail, and leaves the job site.

SERVICE CALL 4

A service technician is servicing an ice maker and has determined that the ice machine will not come out of a hot gas defrost mode as quickly as it should. Too long a defrost will cook off the metal coating on the evaporator. Instead of the harvest motor's probe pushing the ice off the coil, which will terminate defrost with a curtain switch, the evaporator gets very hot from the prolonged hot gas defrost. In other words, a "cookout" occurs. The ice slowly falls off the evaporator in chunks by gravity. Sometimes, one of these chunks will trip the curtain switch and bring the machine out of defrost, and

other times it will not. *The problem is a defective harvest motor. Ice cannot be forced off the evaporator during harvest with a defective harvest motor.*

The service technician gets out a voltmeter and measures the output terminals of the microprocessor labeled motor 1 and motor 2 while in defrost, **Figure 27.49**. The technician measures 115 V AC at both terminals and determines that this is the correct voltage. This tells the technician that the microprocessor is doing its job and is sending an output signal to the harvest motor, so the microprocessor is not to blame. The problem must lie in the harvest motor or the wires leading to the harvest motor. The service technician then checks voltage at the harvest motor itself and gets 115 V AC. This rules out the problem being in the wiring between the microprocessor and the harvest motor. The problem must lie in the harvest motor itself. (If the correct voltage is going to a power-consuming device and it is not operating, the problem lies within the power-consuming device.)

The service technician retrieves a new harvest motor from the service van and installs it. After watching the ice machine for two complete cycles, the technician determines that everything is working as it should.

SERVICE CALL 5

A service technician servicing an ice maker that has a microprocessor observes that the ice machine will not come out of ICE 1 mode. The technician consults the service manual and finds that the evaporator must reach 14°F before an internal timer will start. The technician wonders if it is the microprocessor's fault or if the problem is in one of the inputs to the microprocessor. *The problem is a defective thermistor on the evaporator. With a defective thermistor, the microprocessor does not know the evaporator's temperature and cannot advance the ice machine to the next mode (ICE 2).*

The technician reads further in the manual and discovers that the microprocessor is fed an input signal from a thermistor connected to the evaporator. A thermistor is nothing but a resistor that varies its resistance as the temperature changes. This thermistor is what signals the digital LED display what mode the ice maker is in. It also tells the microprocessor when the evaporator has reached 14°F. The manufacturer has supplied a resistance versus temperature relationship table for checking thermistors. The only sure temperature that the technician can simulate accurately is 32°F. This is accomplished by placing some crushed ice in a small amount of water and stirring. The thermistor lug is then taken off the evaporator and is also electrically disconnected from the microprocessor's evaporator input terminal. The thermistor lug is given a chance to stabilize its temperature at 32°F in the ice water bath. A resistance reading taken with an ohmmeter is 5.50 thousand ohms, or 5.50 Kohms. According to the temperature versus resistance chart for the thermistor, the resistance should be 9.842 Kohms. This indicates a defective thermistor and explains why the microprocessor is not going out of ICE 1 mode. A new thermistor is ordered from the manufacturer. The customer is politely informed of the problem and the part ordered. The work site and the technician's hands are then cleaned, and handshakes are exchanged with the customer.

The technician could have measured the signal coming into the microprocessor's input terminal labeled EVAP, but that would require knowing what it was supposed to be. Most of the time these signals are a low AC or DC voltage, but they can be a resistance signal. The service technician must read the manual to find out. The service technician has a service trainee with him and was very careful to explain each step in detail while training. The technician told the trainee never to argue about a service problem in front of a customer, but to wait until they are alone in the service van before voicing a difference in opinion.

REVIEW QUESTIONS

1. Which type of ice maker has a continuous harvest cycle?
   A. Cube ice
   B. Flake ice
   C. Block ice
   D. Cylindrical ice
2. Which type of ice is made using a gear motor and auger?
   A. Cube ice
   B. Cylinder ice
   C. Flake ice
   D. Block ice
3. What would be the symptoms of a defective float on an ice maker?
   A. Excessive load on the system
   B. Increased ice production
   C. Lower operating pressures
   D. Increased system efficiency
4. What are the two functions of the dual float switch in an ice flake machine?
   A. Maintains proper ice level and provides low water level protection.
   B. Maintains proper ice level and provides high water level protection.
   C. Maintains proper water level and provides low water level protection.
   D. Maintains proper water level and provides high water level protection.
5. Which part of an ice flake machine is most prone to failure?
   A. The gear motor
   B. The float mechanism
   C. The ice chute
   D. The compressor

6. List two causes that may lead to auger wobble.
   A. Loss of grease and low water level
   B. Excessive grease and high water level
   C. Improper bearing alignment and low water level
   D. Loss of grease and improper bearing alignment
7. What is the function of a flush cycle for an ice flake machine?
   A. Increase mineral concentration
   B. Decrease mineral concentration
   C. Create softer ice
   D. Create cloudier ice
8. Name three types of ice bin controls in ice flake machines.
   A. Float, thermostat, and infrared electric eye
   B. Thermostat, float, and liquid-filled bulb
   C. Sonar, mechanical flapper, and sail switch
   D. Mechanical flapper, infrared electric eye, and thermostat
9. Ice production in an ice flake machine depends mainly on two factors. Name these two factors.
   A. Water outlet temperature and ambient temperature
   B. Water outlet temperature and water level
   C. Water inlet temperature and ambient temperature
   D. Water inlet temperature and water level
10. The purpose of weep holes on the evaporator of a cube ice machine is to
    A. ensure the ice sticks to the evaporator.
    B. allow the ice to separate from the evaporator during harvest.
    C. allow water to flow freely through the evaporator cells.
    D. ensure that minerals are not frozen in the ice cubes.
11. What material is often used for evaporators due to its durability and resistance to corrosion?
    A. Copper
    B. Brass
    C. Aluminum
    D. Stainless Steel
12. An ice bridge is best described as the ice that
    A. joins ice cubes together.
    B. melts during the harvest cycle.
    C. connects the ice to the evaporator coil.
    D. freezes at temperatures higher than 32°F.
13. What type of pump is typically used to deliver water to the evaporator coil in an ice maker?
    A. Centrifugal pump
    B. Propeller pump
    C. Orbital pump
    D. Concentric pump
14. Which of the following is used to initiate the harvest cycle on a cube ice maker?
    A. Auger assembly
    B. Ice thickness probe
    C. Float switch
    D. Water dump valve
15. Minerals are removed from the ice maker when which of the following components is energized?
    A. Float switch
    B. Auger assembly
    C. Water dump solenoid
    D. Ice bridge sensor
16. What is the most common type of defrost during an ice maker harvest cycle?
    A. Electric
    B. Natural
    C. Hot gas
    D. Air
17. Condensing hot gas that can return to the compressor during harvest is called
    A. floodback.
    B. flowback.
    C. flowthrough.
    D. floodthrough.
18. Periodic ice machine cleaning accomplishes which of the following?
    A. Making clearer ice
    B. Making harder ice
    C. Making both clearer and harder ice
    D. Making neither clearer nor harder ice
19. Which of the following is a visual indication that sanitizing of an ice maker is needed?
    A. Presence of slime
    B. Presence of bacteria
    C. Presence of protozoa
    D. Presence of viruses
20. Why can't an ice maker be cleaned and sanitized at the same time?
    A. It will take too long to do so.
    B. They can. Cleaning and sanitizing are the same thing.
    C. One requires the unit to be operating, while the other requires the unit to be off.
    D. Dangerous chlorine gas will form, making it difficult to breathe.
21. What is a common troubleshooting technique used on microprocessors?
    A. Input/output troubleshooting
    B. Ohmmeter
    C. Ammeter
    D. Guesswork
22. One of the quickest and most thorough way of obtaining technical service information for an ice machine is through the
    A. manufacturer.
    B. wholesale store.
    C. Internet.
    D. telephone.
23. Most of the time, the input signals to a microprocessor are
    A. high voltage.
    B. low voltage.
    C. steady.
    D. high frequency.

24. A way in which a microprocessor communicates to a service technician about a certain malfunction in the ice machine is a(n)
    A. summary.
    B. error code.
    C. history.
    D. data link.
25. The amount of ice that is produced in a 24-hour period will increase if the
    A. inlet water temperature increases.
    B. inlet water temperature decreases.
    C. ambient temperature increases.
    D. system's suction pressure increases.
26. A measurement that represents the thermal cooling capacity of ice and also represents the amount of calcium and magnesium in water is
    A. hardness.
    B. turbidity.
    C. alkalinity.
    D. scale.
27. A buildup of minerals that forms a flaky coating on the surface of the evaporator and water system is called
    A. sediment.
    B. scale.
    C. hardness.
    D. turbidity.
28. A method for killing bacteria, viruses, and protozoa in ice machines is
    A. sanitizing.
    B. cleaning.
    C. rinsing.
    D. brushing.
29. Scale is most effectively removed from an ice maker by
    A. scraping the scale from the evaporator surface with a sharp blade or knife.
    B. using a heat source, such as a torch, to burn the scale from the ice maker.
    C. using an alcohol-based, general-purpose cleaner to remove the scale from the machine.
    D. using a manufacturer-approved cleaner expressly intended to remove scale.
30. Water impurities are commonly classified as
    A. chemicals, dissolved minerals, and suspended metals.
    B. chemicals, suspended minerals, and dissolved metals.
    C. dissolved metals, suspended solids, and suspended minerals.
    D. chemicals, suspended solids, and dissolved minerals and metals.
31. A triple-action carbon filtration system has the ability to accomplish all of the following except:
    A. filter solids from the water.
    B. improve the taste of the water.
    C. increase the scale formation rate in the water.
    D. reduce the odor of the water.
32. The most common chemicals in water are
    A. chlorine and chloramines.
    B. fluorine and fluoromines.
    C. chloramines and fluoromines.
    D. chlorine and fluorine.
33. Usually, sand and dirt in water are easily removed by
    A. filters.
    B. reverse osmosis.
    C. distillation.
    D. de-ionization.
34. Air-cooled, package-type ice makers should be positioned so that
    A. there is adequate airflow through the condenser that does not recirculate.
    B. the condenser air can be used to defrost the evaporator coil.
    C. the discharge condenser air is directed toward the ice storage bin's door.
    D. the field-installed refrigerant lines are at the back of the unit.
35. What portion of the service manual provides information about what each system component does and when?
    A. Table of contents
    B. Sequence of operation
    C. Installation guidelines
    D. Unit's parts list

UNIT 28

SPECIAL REFRIGERATION APPLICATIONS

OBJECTIVES

After studying this unit, you should be able to

- list four common methods by which refrigerated products are transported long distances.
- describe methods used for refrigerating trucks.
- discuss phase-change plates used in truck refrigeration.
- describe two methods used to haul refrigerated freight using the railroad.
- discuss the basics of blast cooling for refrigeration.
- describe cascade refrigeration.
- describe basic methods used for ship refrigeration.
- describe methods used to transport refrigerated products by air.

SAFETY CHECKLIST

- ✓ Ensure that nitrogen and carbon dioxide systems are off before entering an enclosed area where they have been in use. The area should also be well ventilated so that oxygen is available for breathing and has not been replaced by these gases.
- ✓ Liquid nitrogen and liquid carbon dioxide are very cold when released to the atmosphere. Take precautions to avoid getting them on your skin. Wear gloves and goggles if you are working in an area where one of these gases may escape.
- ✓ Ammonia is toxic. Wear protective clothing, goggles, gloves, and breathing apparatus when servicing ammonia refrigeration systems.

28.1 TRANSPORT REFRIGERATION

Transport refrigeration is the process of refrigerating products while they are being transported from one place to another—by truck, rail, air, or water. For example, vegetables may be harvested in California and shipped to New York City for consumption. These products may need refrigeration along the way. Many products must be shipped in different ways. Some are shipped in the fresh state, such as lettuce and celery, which are shipped at about 34°F. Some are shipped in the frozen state, such as ice cream, which is shipped at about −10°F. Fresh vegetables may be shipped through climates that are very hot, such as through Arizona and New Mexico. They may also be shipped through very cold climates, such as when fresh vegetables are shipped through Minnesota in the winter. Different situations require different types of systems. For example, refrigeration may be needed for a load of lettuce when leaving California, but when it goes to the northern states, the load may need to be heated to prevent it from freezing. Products that fall into the preceding categories are carried on trucks, trains, airplanes, and ships. The refrigeration system is responsible for getting the load to the destination in good condition.

28.2 TRUCK REFRIGERATION SYSTEMS

Several methods are used to refrigerate trucks. One of the most basic still being used is to merely pack the product in ice that is made at an ice plant. This is successful in a well-insulated truck for short periods of time. However, the melting ice produces water that must be handled. You can usually spot these trucks because of the trail of water left behind, **Figure 28.1**. Packing in ice can hold the load temperature only to above freezing.

Figure 28.1 When ice is used for preserving products, it melts and often leaves a trail of water on the highway or a puddle in a parking area.

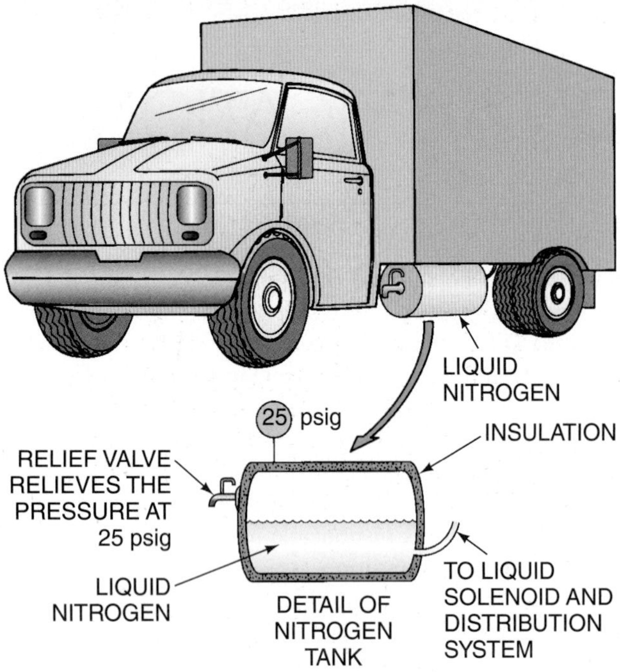

Figure 28.2 A nitrogen cylinder maintained at about 25 psig.

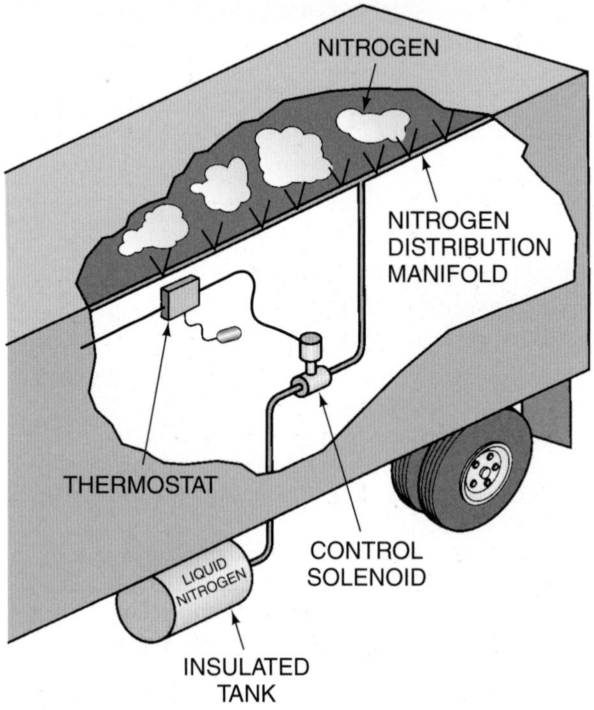

Figure 28.3 A control for a liquid nitrogen cooling system.

Dry ice, which is solidified and compressed carbon dioxide, is also used for some short-haul loads. It is somewhat harder to hold the load temperature at the correct level using dry ice because it is so cold. Dry ice changes state directly from a solid to a vapor in a process called sublimation, at −109°F. As a result, food products can be reduced to very low temperatures. But dehydration of food products that are not in airtight containers can become a problem when dry ice is used because the very low temperature of −109°F attracts moisture from the food.

Liquid nitrogen or liquid carbon dioxide (CO_2) may also be used to refrigerate products by refrigerating and storing them as a liquid in a cylinder at a low pressure. A relief valve on the liquid nitrogen cylinder releases some of the vapor to the atmosphere if the pressure in the cylinder rises above the set point of about 25 psig. This boils some of the remaining liquid and drops the cylinder temperature and pressure, **Figure 28.2**. The liquid is piped to a manifold where the refrigerated product is located and released to cool the air in the truck and the product. A thermostat located in the food space controls a solenoid valve in the liquid nitrogen line, stopping and starting the flow of liquid, **Figure 28.3**. Fans may be used to disperse the cold vapor evenly over the product.

Air in the atmosphere is 78% nitrogen; it is the source for the nitrogen used in refrigeration. It is not toxic, and it can be released to the atmosphere with confidence that no harm to the environmental will result.

SAFETY PRECAUTION: *Nitrogen has no oxygen in it, so it is not suitable for breathing. Special safety interlocks shut the refrigeration off if the door to the refrigerated truck is opened to prevent personnel from being smothered by lack of oxygen. Nitrogen is also very cold; it evaporates at −320°F at atmospheric pressure. Any contact with the skin would cause instant freezing. Never allow liquid nitrogen to get on your skin.•*

Liquid CO_2 can be used in the same manner as liquid nitrogen to refrigerate products. The system works much the same as a liquid nitrogen system, only CO_2 boils at −109°F. Even though liquid nitrogen and CO_2 are both very cold, they can be used to refrigerate medium-temperature loads at 35°F with proper controls. These controls would include a distribution system for the refrigerant that does not impinge on the product and a space-temperature thermostat that accurately senses the refrigerated space temperature. Both of these liquid injection methods have been used for many years to replace mechanical refrigeration for permanent installations and for emergency use. They are typically more expensive in the long run than a permanent installation, but they also provide some valuable benefits. For example, when used to refrigerate fresh foods, the presence of nitrogen or CO_2 vapor displaces any oxygen and helps preserve the food. The system is simple and easy to control because only a solenoid and a distribution manifold need to be maintained.

Many truck bodies are equipped with refrigerated plates (called cold plates) that contain a phase-change solution

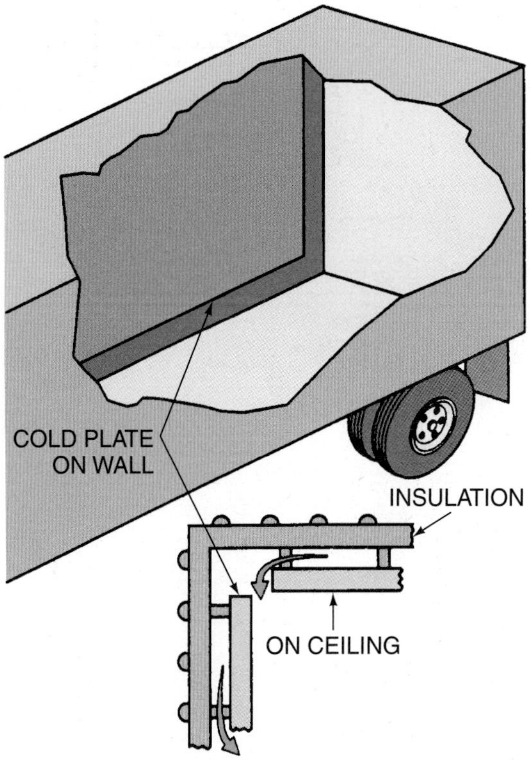

Figure 28.4 Cold plates mounted in a truck body.

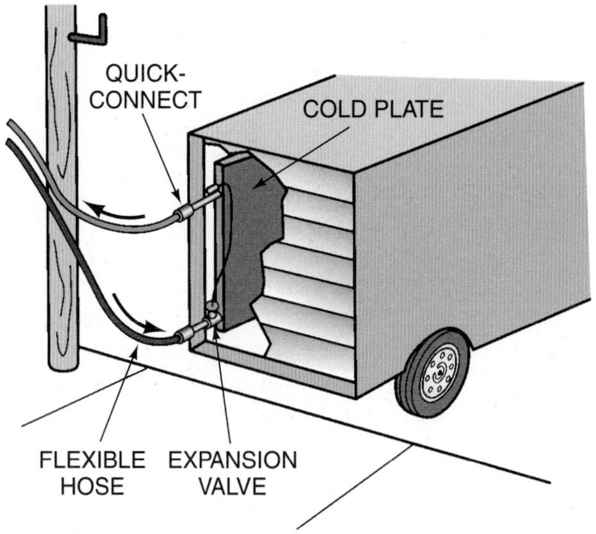

Figure 28.5 A cold plate connected to the refrigeration system at the dock.

called a eutectic solution. *Phase change* means a change in the state of matter, much like the change of state from ice to water, except at a different temperature. When the change of state occurs, much more heat is absorbed per pound of material. The eutectic solution has the capability of changing state at many different temperatures, depending on the composition of the solution. The eutectic solution in the cold plate is brine, a form of saltwater made up of either sodium chloride, a form of table salt, or calcium chloride. These brines are very corrosive and must be handled with care. Various strengths of brine solution may be used to achieve the desirable melting temperature below freezing, called the eutectic temperature. Actually, the brine does not freeze solid; salt crystals form, and the remaining solution is liquid. Different strengths of brine will cause the crystals to form at different temperatures.

The brine solution is contained in plates that are from 1-in. to 3-in. thick mounted on the wall or ceiling of a truck in such a manner that allows air to flow all around them, **Figure 28.4**. The air in the enclosed truck body transfers heat to the plates. The brine solution then changes from a solid to a liquid while absorbing heat. These plates will hold a refrigerated load at the correct temperature (in a medium- or low-temperature application) to allow an all-day delivery route to be run. For example, a truck loaded with ice cream may be loaded at night for delivery the next day. The ice cream must be at its desired temperature, and the truck's cold plates must be at the design temperature. For ice cream, this may be −10°F or lower. When the truck comes back at night, it may still be −10°F, but most of the brine solution will have changed back to a liquid and need regenerating to the solid crystal state. The heat absorbed during the day is in the brine solution and must be removed. To recharge the plates, they must be cooled to the point that the brine changes state back to solid crystals.

The plates are recharged in several ways. Some systems have a direct expansion coil located in the cold plate. This coil is connected to the refrigeration system at the loading dock, **Figure 28.5**. The refrigeration system may use one of the typical refrigerants—R-134a, R-404A, or R-717, ammonia, or an environmentally friendly alternative refrigerant. When the truck backs up to the loading dock, quick-connect hoses are connected from the central refrigeration plant to the truck. The expansion valve may be located on the loading dock or at the actual coil in the eutectic cold plate. The central refrigeration system will reduce the pressure in the evaporator in the cold plate to the point that the brine solution will change back to solid crystals in a few hours. The truck may then be loaded with product to start another trip.

Another method of recharging the cold plate is to circulate refrigerated brine solution through a coil located in the cold plate. Brine that is refrigerated at a lower temperature than the cold plate will then recharge or freeze the brine in the plate, but the circulating brine must be much colder than the solution in the plate. Connecting this system to the truck also involves quick-connect fittings at the loading dock. There is always the possibility that some brine

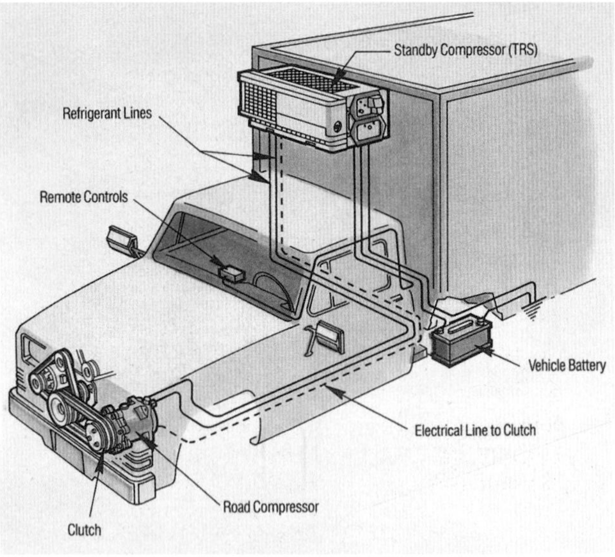

Figure 28.6 This truck refrigeration is driven from the truck's engine and may be cycled by means of an electric clutch. Source: Carrier Corporation

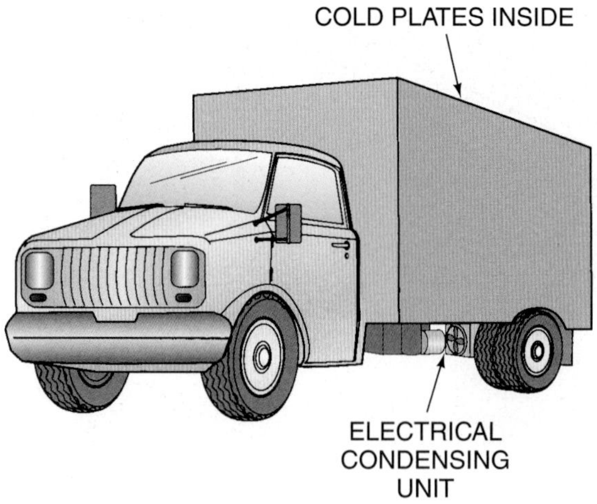

Figure 28.7 A small refrigeration system on the truck recharges the cold plates while the truck is parked at night.

will be lost during the connecting and disconnecting of the quick-connect fittings. This brine must be cleaned up, or corrosion will occur.

Trucks may also be refrigerated using a refrigeration system mounted on the trucks. Whatever the refrigeration source, it must have a power supply. This power supply may be the truck's power supply, a diesel or gas engine-driven compressor, a land line (electric power from a building), or an electric compressor operated by a motor generator. Van-type delivery trucks often use a refrigeration compressor mounted under the hood that is turned by the truck engine. The refrigeration may be controlled by cycling the compressor using an electric clutch that is thermostatically controlled, **Figure 28.6**. This system works well as long as the truck engine is running. Because the truck is usually moving all day during delivery, refrigeration can easily be maintained. Some of these trucks have an auxiliary electric motor-driven compressor that may be plugged in at the loading dock while the truck engine is not running.

Small condensing units under the truck body are also used to refrigerate the load. These units are often used to recharge cold plates at night when the truck is at the dock or other power supply; the cold plates hold the load during the day, **Figure 28.7**. The advantage of this system is that it is self-contained and does not require a connection to a central refrigeration system at the loading dock. The condensing units may be powered while on the road by a small motor generator unit mounted under the chassis. In this case, there is no need for cold plates because the generator operates on the road and the compressor is plugged in at night.

Motor generators have either a gas or diesel engine that turns the generator to generate 230-V, 60-cycle current to operate the compressor. Diesel is the preferred fuel because diesel engines will operate longer with less fuel and with less maintenance. Diesel engines cost more but seem to be less expensive in the long run. In some cases, a generator producing 230 V from the truck engine has been used. It must have a rather sophisticated control because the truck engine turns at so many different speeds. The stationary generator has only one speed. Different speeds would affect the power supplied to the compressor. As the engine turns faster, the voltage and the cycles per second vary; 60 cps is normal for alternating current electric motors. Electronic circuits may be used to maintain a constant voltage.

Larger trucks use either nose-mount or under-belly units that may be operated by either a diesel-engine-driven compressor or a motor generator driven by a diesel engine, **Figure 28.8**. An engine-driven compressor has a governor to regulate the speed of the compressor, which may have two speeds, high and low. The compressor may be controlled by means of cylinder unloading. For example, a four-cylinder compressor with a capacity of 1 ton per cylinder may be operated from 1 ton to 4 tons by unloading cylinders, **Figure 28.9**. A minimum load of 1 ton must be maintained at all times, or the diesel engine will have to be cycled off and on, so it is most desirable to keep the engine running. However, controls for diesel engines have been developed that will automatically start and stop the engine, **Figure 28.10**.

The evaporator for truck refrigeration units is typically located at the front of the truck and blows air toward the back of the truck, **Figure 28.11**. The fan is usually a type of centrifugal fan that distributes a high volume of air at a high velocity so that the air will reach the back. The air is then returned to the inlet of the air handler over the product load.

Figure 28.8 (A) A nose-mount refrigerated truck. Source: Carrier Corporation (B) A belly-mount refrigerated truck. Source: Carrier Corporation

(A)

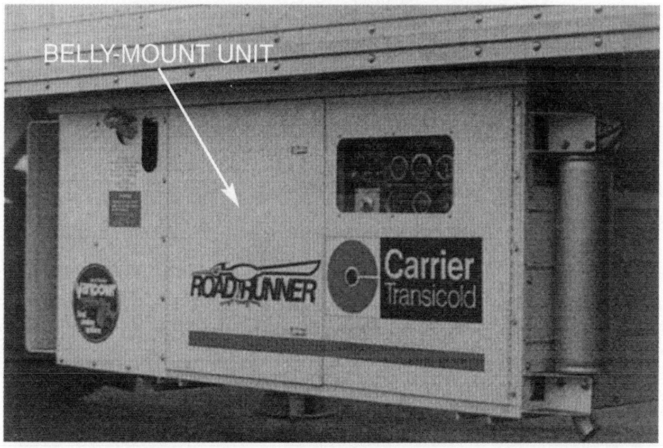

(B)

Figure 28.9 A capacity control for a compressor.

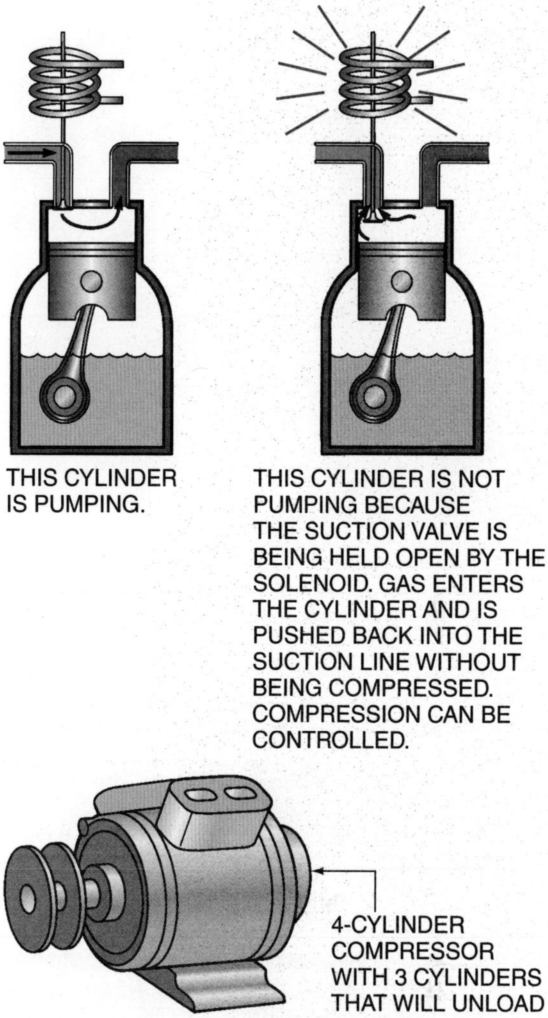

The fan is driven with belts or gearboxes from the engine that drives the compressor. These components are normally located just inside the access door to the unit, where they can be checked and serviced easily from outside the refrigerated space, **Figure 28.12**. The condenser fan is often driven from the same drive mechanism and is typically a prop-type fan, **Figure 28.13**.

The evaporator and condenser coils are usually made of copper with aluminum fins because truck systems must be designed and built to be both rugged and lightweight given that each truck has a maximum load limit. Any weight added to the refrigeration unit is weight that cannot be hauled as a paying load. Some other system components, such as compressors, are made of aluminum to help reduce the weight of the system.

Truck refrigeration systems must be designed to haul either low- or medium-temperature loads or even carry both low- and medium-temperature loads at the same time. This is accomplished using one refrigeration condensing

Figure 28.10 The air distribution system for a truck and location of the control panel. Source: Carrier Corporation

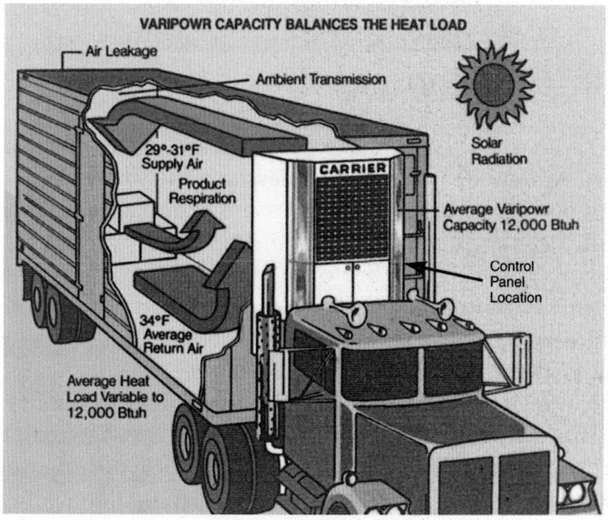

Figure 28.11 The evaporator for a nose-mount refrigeration unit.

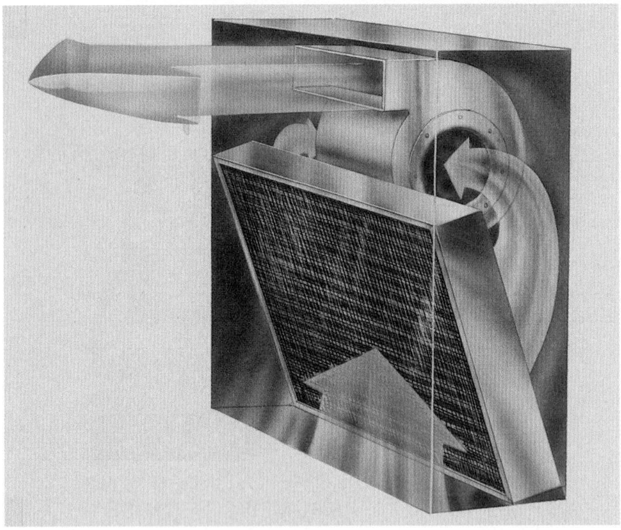

Figure 28.12 Access for service to a nose-mount unit. Source: Thermo King Corp.

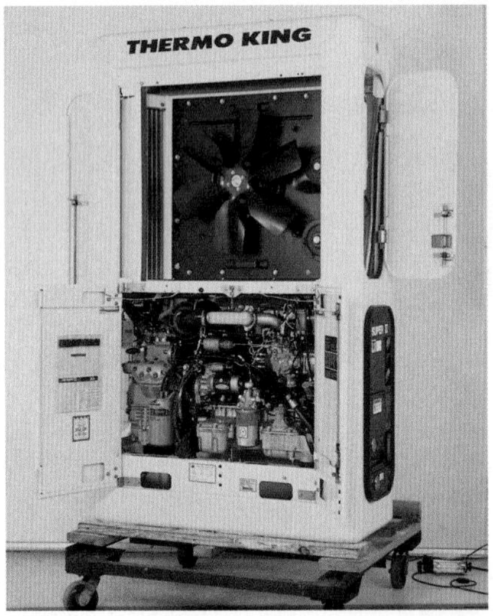

Figure 28.13 (A) Fan drive mechanism. Source: Thermo King Corp. (B) Fan drive mechanism. Source: Carrier Corporation

(A)

(B)

unit with more than one evaporator. Part of the time, trucks may be hauling fresh vegetables; at other times the load may be frozen foods. The space temperature is controlled with a thermostat that starts out at full load; when it is within approximately 2°F of the set point, the system compressor starts to unload. If the space temperature continues to drop, the unit will shut down. The air distribution system is designed to hold the load at the specific temperature, not to pull the load temperature down. As mentioned earlier, the food should be at the desired storage temperature before being loaded because the truck does not have much reserve capacity for pulling the load temperature down, especially if the interior is warmed up very much. If the refrigeration system for a medium-temperature load of food has to reduce the load temperature, the air distribution system may overcool some of the highest parts of the load because the truck body is so small that the cold air from the evaporator, being directed across the top of the refrigerated compartment, may impinge directly on the top product, **Figure 28.14**. If the load is frozen food and has warmed, it may be partially thawed. The truck unit may not have the capacity to refreeze the product, and if it does, it may be so slow that product damage may occur. Various alarm methods are used to alert the driver of problems with the load, **Figure 28.15**. The driver should stop and check the refrigeration at the first sign of problems. It may not be easy to locate a technician when the truck is away from metropolitan areas.

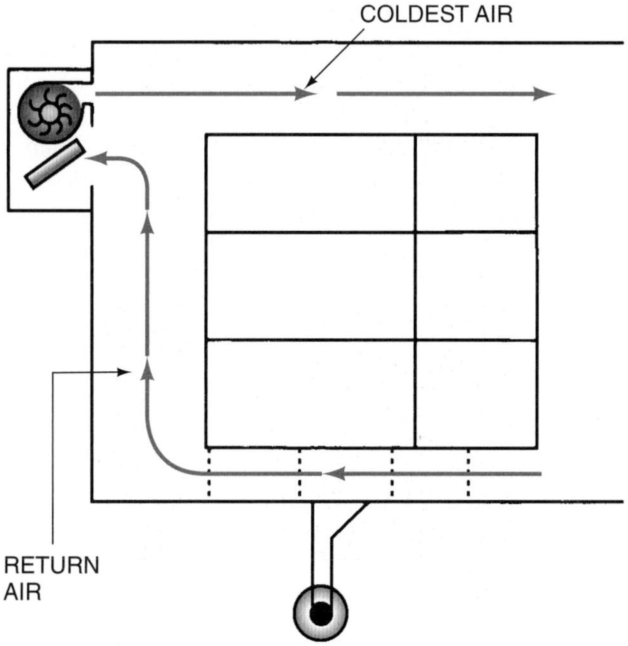

Figure 28.14 Due to air distribution, cold air may overcool food on top of the load.

Figure 28.15 An alarm may alert the driver if the load's condition is not being maintained. Source: Thermo King Corp.

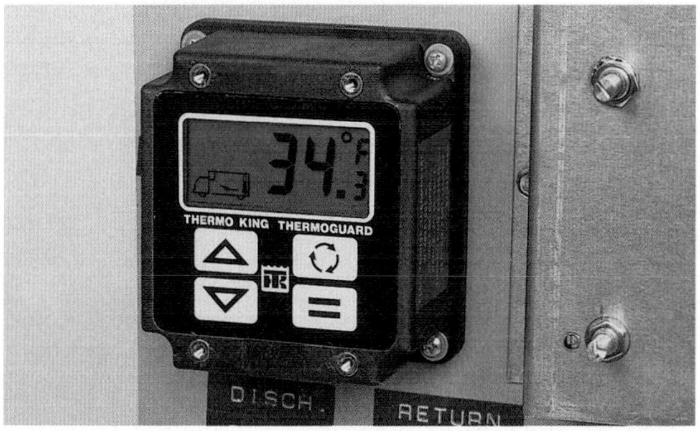

Figure 28.16 Sandwich construction for truck walls.

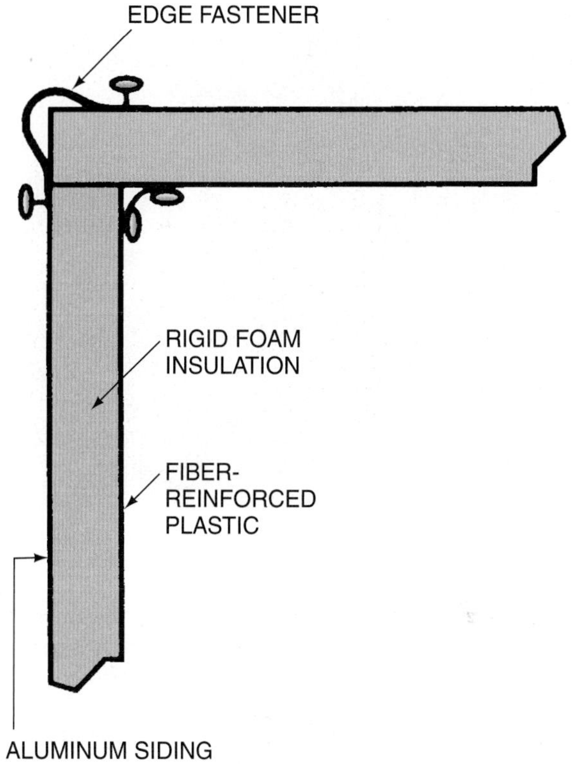

Truck bodies are insulated to keep the heat out, usually by foam insulation sprayed onto the walls with wallboard fastened to the foam. This makes a sandwich type of wall that is very strong, **Figure 28.16**. The walls must be rugged to withstand the loading and unloading of product. Aluminum- and fiber-reinforced plastic products are often used to cover the walls. The floor must be strong enough to support the weight of the load while traveling on a rough highway, and it must also be able to support a forklift that may be used for loading. The interior of a refrigerated truck must be manufactured in such a manner that it can be kept clean because food is being handled. The doors are normally in the back so that the truck may be loaded with a forklift; however, some trucks have side doors and compartments for different types of product. The doors must have a good gasket seal to prevent infiltration of the outside air. The doors must also have a secure lock to prevent the load from being stolen.

Truck transport refrigeration technicians should be diesel technicians as well as refrigeration technicians. Trained in both fields, they usually do not work on other types of refrigeration systems. Often, they are factory trained to service a particular type of refrigeration system and engine combination. A good knowledge of basic refrigeration would be good preparation for truck transportation systems, and knowledge regarding diesel engines can be added.

28.3 RAILWAY REFRIGERATION

Products shipped and refrigerated by rail are mechanically refrigerated either by a self-contained unit located at one end of the refrigerated car or by an axle-driven unit. The axle unit is an older system and not used to any extent today. The railcar using axle-driven units would have refrigeration only while the

Figure 28.17 This motor generator is used to power a conventional compressor. Source: Fruit Growers Express

train is in motion. The self-contained unit is the most popular. It is generally powered by a diesel-driven motor generator, similar to truck refrigeration, **Figure 28.17**. The compressor is a standard voltage and may be powered with a power cord while the car is in a station or on a rail siding for long periods. The evaporator blower is mounted at the end of the railcar in a compartment, along with the motor generator. All of the serviceable components can be accessed from this compartment, **Figure 28.18**.

Two-speed engines are often used when the generator has an output frequency of 60 cps at full speed and 40 cps at reduced load. The compressor electric motor must be able to operate at the different frequencies. The diesel engine speed is adjusted by the space-temperature thermostat and operates at a much lower fuel cost at low speed. When low speed is reached, the compressor may be unloaded using cylinder unloading. These systems have a rather wide range of capacities to suit the application. Like truck systems, they are capable of reducing the space temperature to 0°F or below for holding frozen foods, or they may be operated at medium temperature for fresh foods. Railcars have an air distribution system much like trucks and are designed to hold the refrigerated load, not reduce the temperature. They do not have much reserve capacity.

Refrigerated railcars are made of steel on the outside and typically have a rigid foam insulation in the walls, **Figure 28.19**. The walls are covered with various types of wallboard that can be easily cleaned. If railroad cars did not have a sophisticated suspension system for en route travel and to protect the load from the rough coupling of cars when the train is assembled, they would ride rough, with metal wheels on metal rails, and have a lot of impact vibration when being connected and disconnected. They have wide doors for forklift access, and the doors must have a very good gasket to prevent air leakage. Typically,

Figure 28.18 A railroad car refrigeration unit. Source: Fruit Growers Express

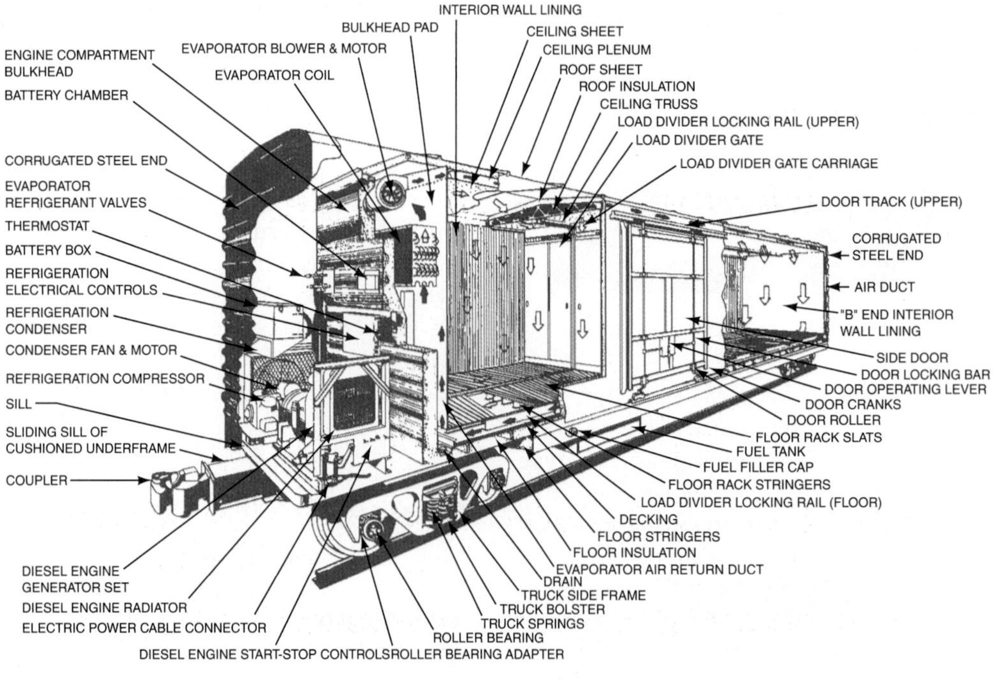

Figure 28.19 The wall construction of a railcar.

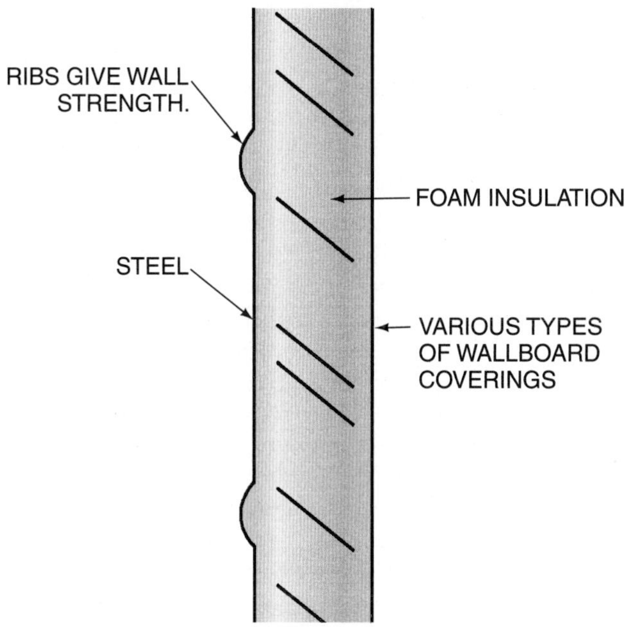

the railcar is operated under a positive air pressure and must have good door seals.

Railroads also ship refrigerated trucks in a system known as "piggyback." The trucks are loaded on the railcars and shipped across country, **Figure 28.20**, a popular system because the refrigerated load often would otherwise have to be off-loaded to a refrigerated truck in the locality where the food is to be consumed or processed. With piggy-back cars, the refrigerated load arrives on the truck for hauling to the final destination.

Figure 28.20 Piggyback railcar refrigeration.

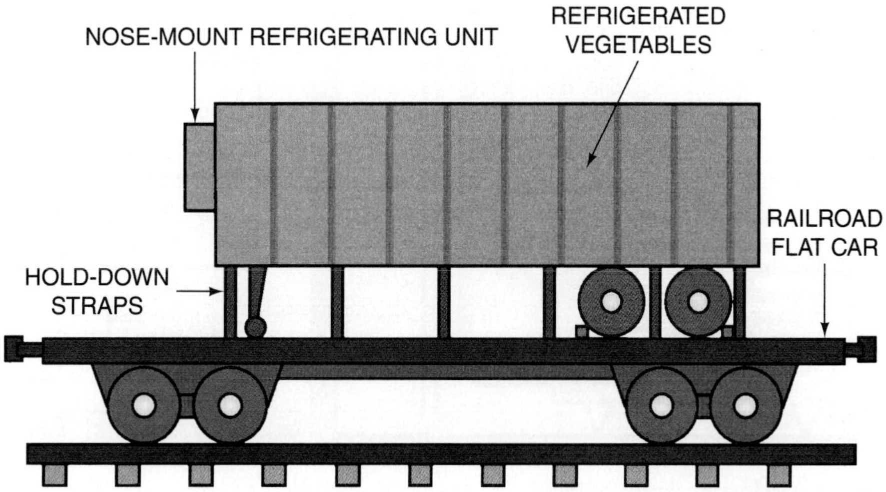

28.4 EXTRA-LOW-TEMPERATURE REFRIGERATION

This text defines extra-low temperature as below −10°F. Temperatures below that are not used to any great extent for storing food; however, they are used to fast-freeze foods commercially. These temperatures may be in the vicinity of −50°F.

When the product temperature needs to be reduced lower than −50°F, a two-stage system of freezing may be used. At extremely low evaporator temperatures, the compression ratio for the compressor becomes too great for single-stage refrigeration. The two stages of compression are accomplished by using a first-stage compressor (or cylinder) pumping into a second-stage compressor (or cylinder), **Figure 28.21**. The compression ratio is thus divided

Figure 28.21 Two-stage refrigeration.

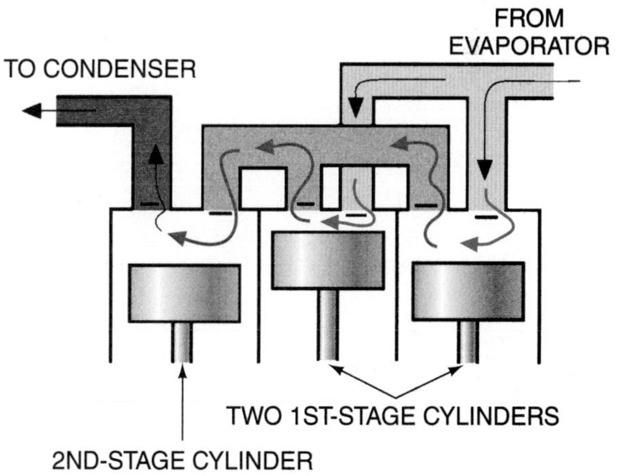

between two stages of compression. If a single-stage system were operating at a room temperature of −50°F, it would have a boiling refrigerant temperature of about −60°F, using a coil-to-air-temperature difference of 10°F, **Figure 28.22**. This would make the evaporator pressure 6.6 in. Hg (11.45 psia) for R-404A. If the condensing temperature were to be 115°F, the head pressure would be 291 psig (305.7 psia). The compression ratio would be

$$CR = \text{Absolute Discharge Pressure} \div \text{Absolute Suction Pressure}$$
$$CR = 305.7 \div 11.45$$
$$CR = 26.7{:}1$$

This is not a usable compression ratio. It can be reduced with a two-stage compressor, **Figure 28.23**. The first stage of compression has a suction pressure of 11.45 psia and a head pressure of 40 psig (54.7 psia). The compression ratio would be

$$CR = 54.7 \div 11.45$$
$$CR = 4.78{:}1$$

The second-stage compressor would have suction pressure of 40 psig (54.7 psia) and a discharge pressure of 291 psig (305.7 psia).

$$CR = 305.7 \div 54.7$$
$$CR = 5.59{:}1$$

With a lower compression ratio, the trapped refrigerant gas in the clearance volume at the top of the piston's stroke will go through less reexpansion. This increases the efficiency of the first and second stages of compression. The second-stage compressor then experiences a higher suction pressure, which also will aid in giving it a higher efficiency.

Manufacturers would always like to have the system operate in a positive pressure. The system just described operates in a vacuum, and if a leak occurs on the low-pressure side, atmosphere would enter the system. This system may not be very practical, but it is very possible. The lower the compression ratio, the better the compressor efficiency. This is particularly true for the first-stage compressor. Designers will pick the best operating conditions for the system, combining several compressors until they find the highest efficiency.

Figure 28.22 A two-stage refrigeration system in operation.

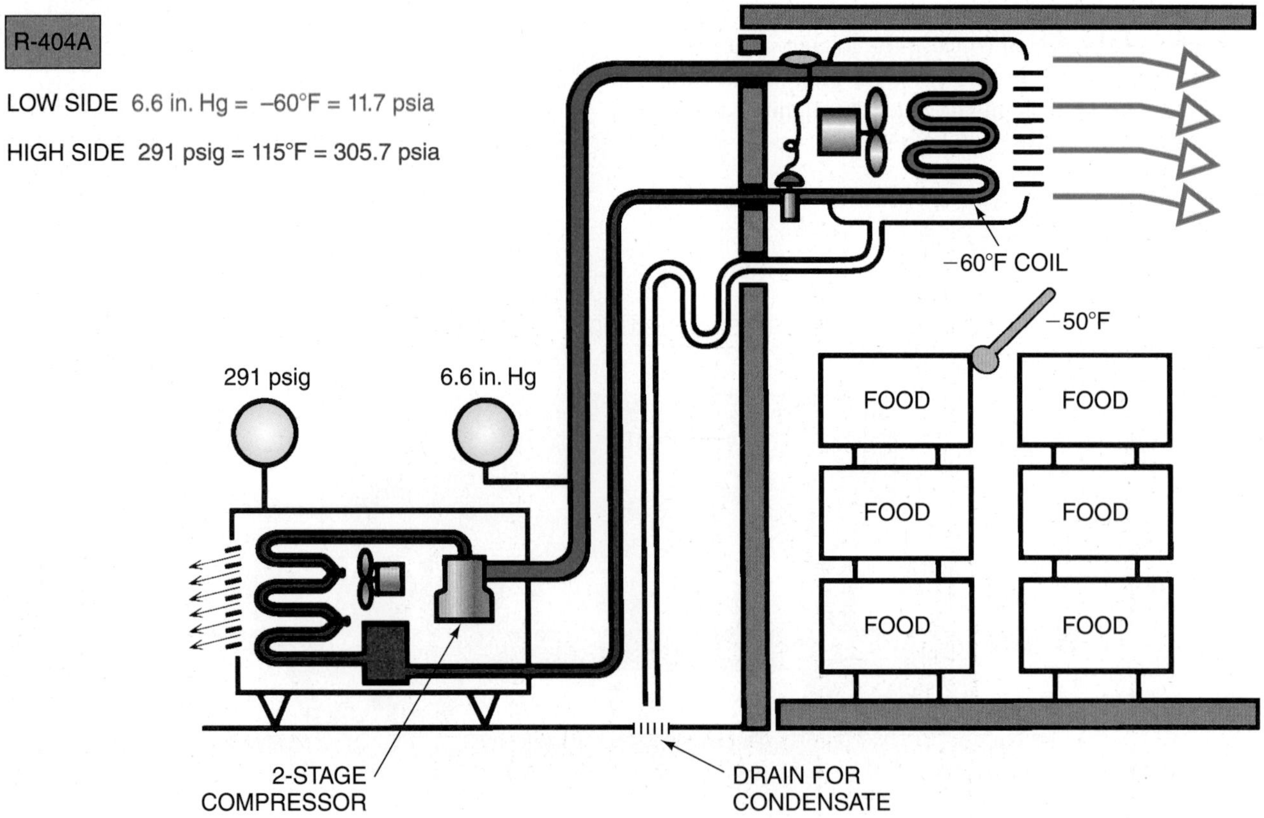

Figure 28.23 The compression ratio and two-stage refrigeration.

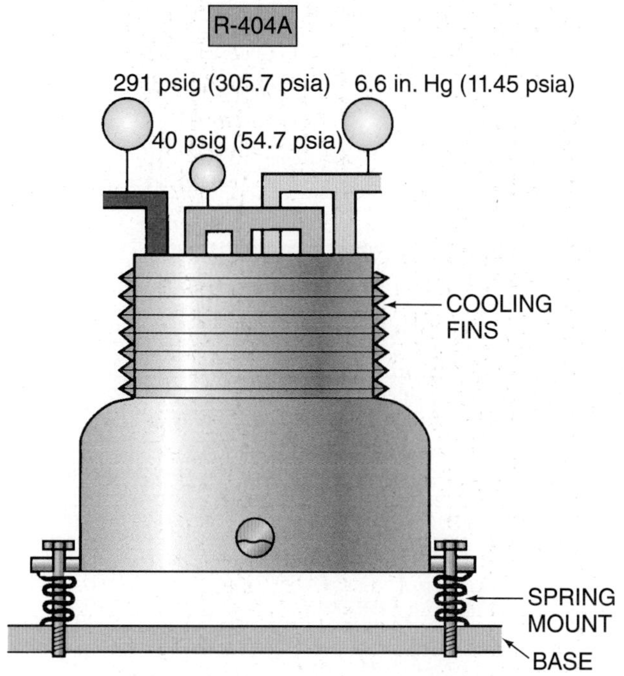

COMPRESSION RATIO

USING 1 STAGE OF COMPRESSION

$$CR = \frac{\text{ABSOLUTE DISCHARGE}}{\text{ABSOLUTE SUCTION}}$$

$$= \frac{\text{291 psig + 14.7 (ATMOSPHERE)}}{\text{11.45 psia}}$$

$$= \frac{\text{305.7 psia}}{\text{11.45 psia}}$$

$$= \text{26.7 TO 1}$$

USING 2 STAGES OF COMPRESSION

$$\text{CR (1ST STAGE)} = \frac{\text{ABSOLUTE DISCHARGE}}{\text{ABSOLUTE SUCTION}}$$

$$= \frac{\text{54.7 (40 psig + 14.7 ATM)}}{\text{11.45}}$$

$$= 4.78 : 1$$

$$\text{CR (2ND STAGE)} = \frac{\text{ABSOLUTE DISCHARGE}}{\text{ABSOLUTE SUCTION}}$$

$$= \frac{\text{305.7 psia (291 psig + 14.7 ATM)}}{\text{54.7 psia}}$$

$$= \text{5.59 TO 1}$$

28.5 CASCADE SYSTEMS

Systems with operating temperatures down to about −160°F may use what is called cascade refrigeration. Cascade refrigeration involves two or more stages, depending on the system configuration. Each stage of refrigeration is produced by a separate, complete refrigeration system. Although the stages are *separate* from each other, they are *dependent* on each other. For example, in a two-stage cascade system, the evaporator of one stage is used to condense the refrigerant in the second. If one stage or the other fails to operate correctly, the other system stages will also malfunction. The purpose of using multiple stages is to reduce the system's compression ratio, increase its coefficient of performance and as a result, increase the overall efficiency of the system.

Consider the example of a single R-134a system that operates with a condenser saturation temperature of 120°F and an evaporator saturation temperature of −40°F. Assuming an evaporator superheat of 20°F, a condenser subcooing of 20°F, and zero degrees of suction-line superheat, a simplified pressure enthalpy plot of the system is provided in **Figure 28.24**. From the values obtained from the plot, it can be determined that the compression ratio is 26.43:1 (185 psia ÷ 7 psia), and the system's coefficient of performance (COP) is 1.35. [COP = NRE ÷ HOC = (95 Btu/lb − 45 Btu/lb) ÷ (132 Btu/lb − 95 Btu/lb)]. The compression ratio is much higher than desired and the coefficient of performance is much lower than desired.

To increase the efficiency of the system, a two-stage cascade system will be used in this example, **Figure 28.25**. Stage A is operating with a condenser saturation temperature of 120°F and an evaporator saturation temperature of 20°F. Stage B is operating with a condenser saturation temperature of 40°F and an evaporator saturation temperature of −40°F. It can be seen that, in the figure, the plots for the two stages overlap each other. It is where the overlap occurs that the evaporator of stage A, at 20°F, is being used to condense the refrigerant in stage B at 40°F.

From the values obtained from the plot, it can be determined that the compression ratio for stage A is 5.6:1 (185 psia ÷ 33 psia), and the compression ratio for stage B is 7.14:1 (50 psia ÷ 7 psia). It can also be determined that the coefficient of performance (COP) for stage A is 4.33. [COP = NRE ÷ HOC = (110 Btu/lb − 45 Btu/lb) ÷ (125 Btu/lb − 110 Btu/lb)]. Similarly, the coefficient of performance (COP) for stage B can be calculated to be 3.2. [COP = NRE ÷ HOC = (95 Btu/lb – 18 Btu/lb) ÷ (119 Btu/lb − 95 Btu/lb)]. So, although two complete refrigeration systems have been used in this two-stage example, the compression ratios and COP values are much better and result in a more efficient system. In most cases, two different refrigerants are used for each of the stages in a cascade system. A single refrigerant was used for this example to clarify the concept of how the evaporator of one stage can be used to condense refrigerant in another stage. Refer to Unit 26, "Applications of Refrigeration Systems," for more detailed information on cascade refrigeration systems that utilize R-404A and carbon dioxide, R-744.

28.6 QUICK-FREEZING METHODS

Food that is frozen for later use must be frozen correctly or it will deteriorate, the quality will be reduced, or the color may be affected. When food is frozen, the faster the temperature is lowered, the higher the quality of the food. This can be done in several ways using a refrigeration system at very low temperatures. This method is often called "quick-freezing" or "flash-freezing."

Figure 28.24 Simplified pressure-enthalpy plot of a single-stage, R-134a system with a condenser saturation temperature of 120°F and an evaporator saturation temperature of –40°F.

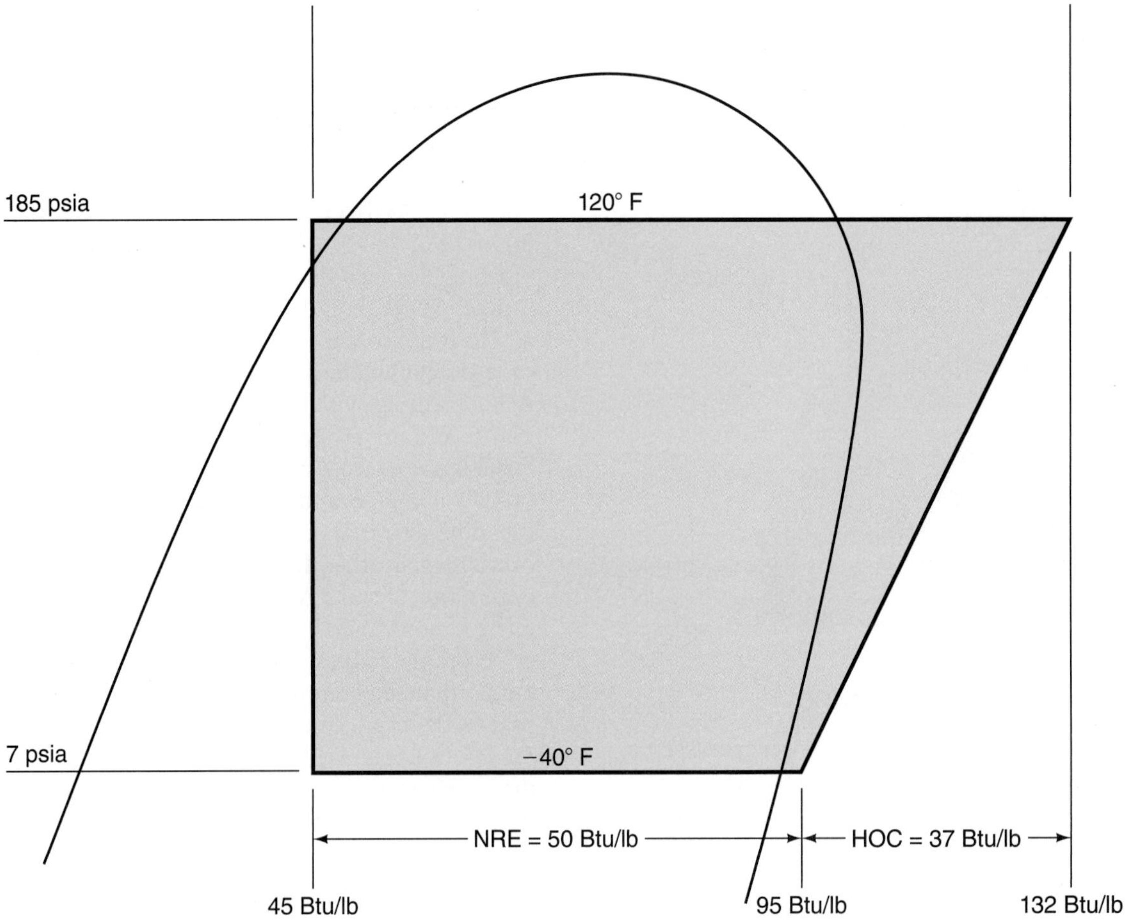

Generally, the meats that you buy at the grocery store and freeze at home do not have the same quality as the meats you buy already frozen. The reason for this is the home freezer does not freeze the meat fast enough. When meat is frozen slowly, ice crystals form in the cells and often puncture them, which is why you often see that a puddle of liquid forms when meat is thawed. When foods are frozen commercially, they are frozen quickly by moving very cold air across the product at a high velocity to remove the heat quickly. When most fresh food is frozen very quickly and thawed, it should still have the quality of fresh food.

Different foods have different requirements for freezing. For example, some poultry must be frozen very quickly to preserve the color. One method of accomplishing this is to seal the meat in an airtight plastic bag and then dip it in a very cold solution of brine, which causes the surface to freeze. The bag is then washed and the poultry removed to another area for freezing the core of the meat at a slower rate.

Foods are often frozen in bulk in packages to be thawed later for processing or cooking. When foods are frozen in this manner, the pieces may stick together, but it makes no difference because they are going to be used in bulk. Many foods are frozen in individual pieces. For example, shrimp may be frozen individually so that they can be sold by the pound while frozen. This may be accomplished by a process called blast freezing, which uses trays in a blast tunnel or on a conveyor system in a blast freezer room, **Figure 28.26.** The blast freeze method is done at very low temperatures and a high air velocity. If the conveyor is placed in a room where the temperature is −40°F or lower and the air is moving fast, by the time small food pieces travel the length of the conveyor they are frozen hard. When the food is placed on the conveyor separated from other pieces, it will be frozen individually, **Figure 28.27**. It may then be packaged and dispensed individually. Larger parcels of food, such as a carcass of beef, take longer to freeze. These would be precooled to very close to freezing and then moved to

Figure 28.25 Simplified pressure-enthalpy plot of a two-stage, R-134a cascade system. Stage A has a condenser saturation temperature of 120°F and an evaporator saturation temperature of 20°F. Stage B has a condenser saturation temperature of 40°F and an evaporator saturation temperature of –40°F.

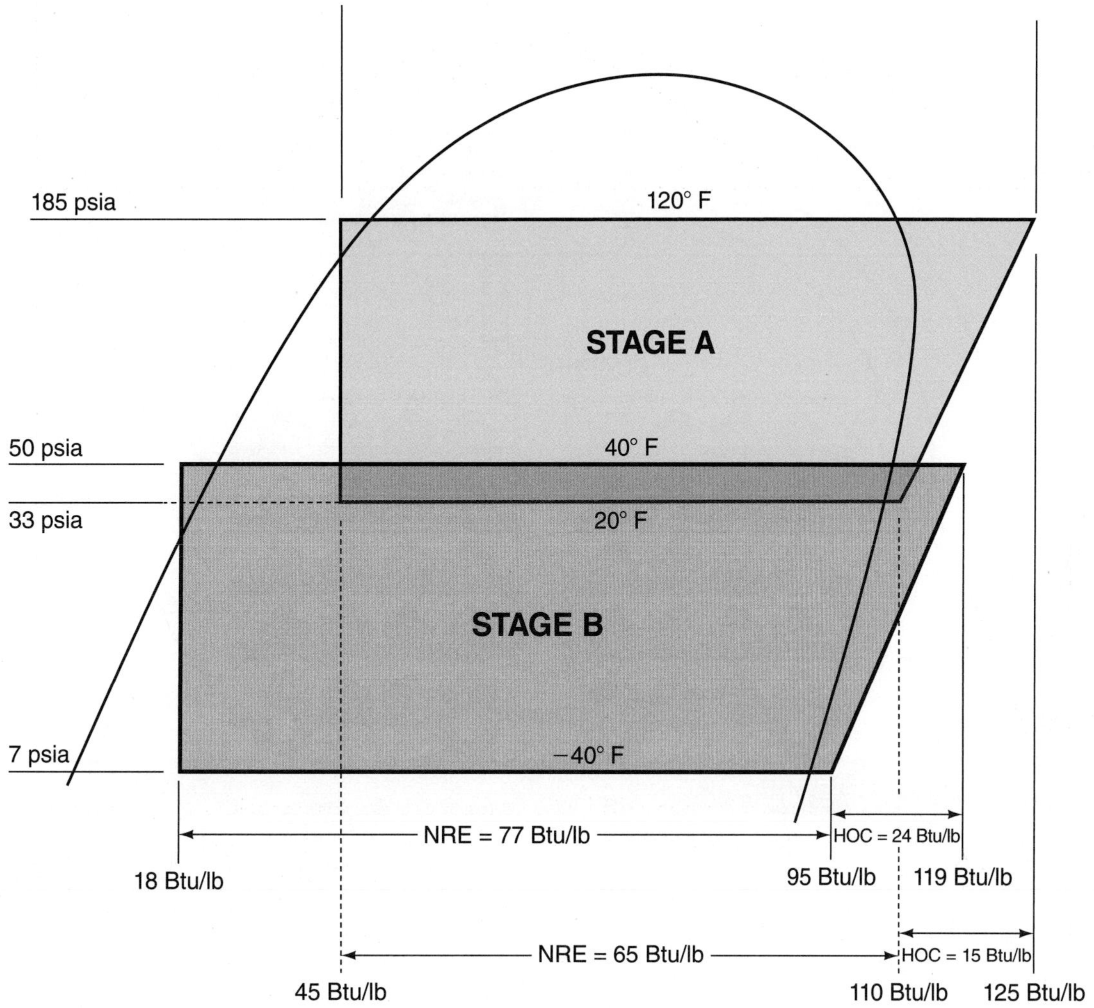

the blast freeze room for quick freezing. Different weights of food may be placed on longer or shorter conveyor belts or the belt speed may be adjusted to the correct time for freezing the food.

Food trays and tunnels may also be used for fast-freezing food. Food is placed on trays with plenty of space between the pieces. The trays are then loaded on carts and moved into a tunnel, **Figure 28.28**. When the tunnel is full and a cart is pushed in, a cart is pushed out the other end with frozen food, ready to package. This method requires more labor but is less expensive to build from a first-cost standpoint than a conveyor system, which is more automated. Quick freezing may be accomplished using standard refrigeration equipment. The larger refrigeration systems may use ammonia or newer, environmentally friendly, alternative refrigerants.

The chlorofluorocarbon (CFC)/ozone depletion issue has led to a comeback of ammonia systems, which are efficient and do not deplete the ozone layer.

28.7 MARINE REFRIGERATION

Ship refrigeration may be required to transport product in several different ways. For example, a ship may pick up a perishable product requiring medium-temperature refrigeration (34°F) in South America and deliver it to New York City. The ship may then pick up a load of frozen food to be delivered to Spain at −15°F, a low-temperature application. The ship's refrigeration system must be able to operate at either temperature range. Another application for ship refrigeration may be

Figure 28.26 A blast freezing tunnel.

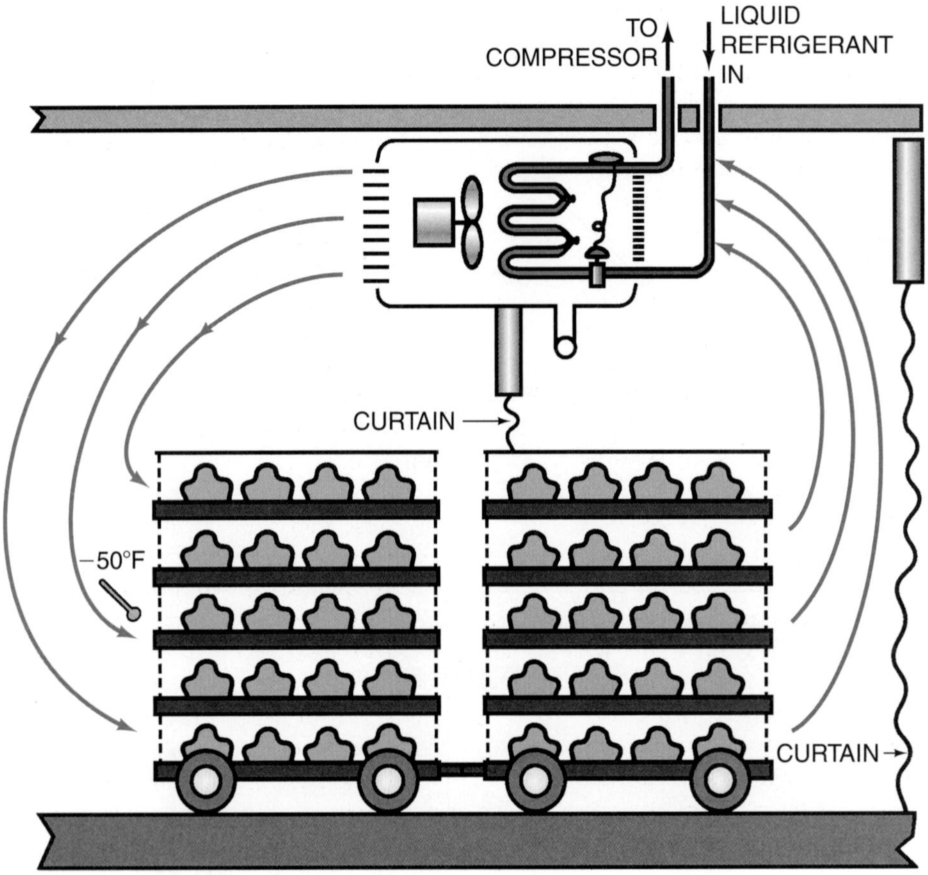

Figure 28.27 The conveyor system for a blast cooler.

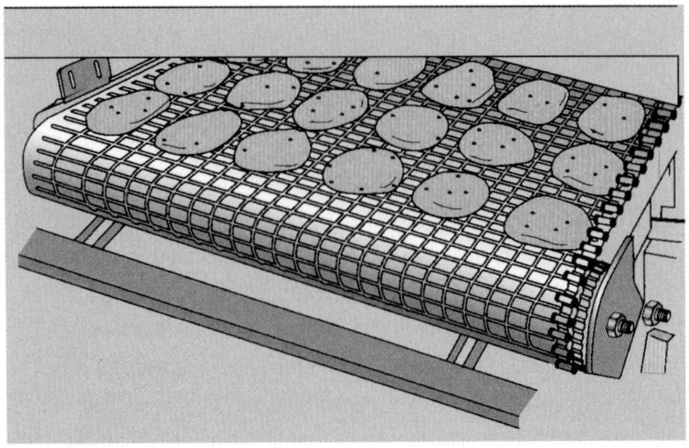

Figure 28.28 Spacing of food for tunnel quick cooling.

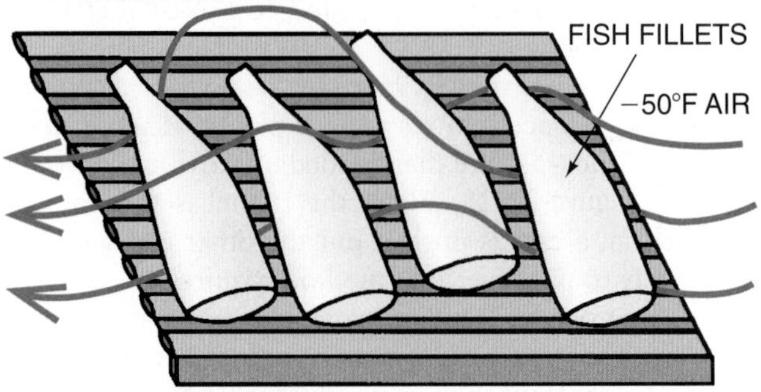

preserving fish caught at sea. If the ship is only out to sea for a few days, it may use ice carried from shore. This would be the case for a shrimp boat operating off the coast. To prevent excessive melting, auxiliary refrigeration may also be used to aid the ice in refrigerating the load. If the ship is to be at sea for long periods, the load must be frozen for preservation. It is not unusual for a ship to stay at sea for several months gathering a load of fish. Often, a fleet of small boats catch the fish and deliver it to a large, refrigerated processing ship called a mother ship. In this case, the refrigeration system may have to be very large, possibly several hundred tons. These ships must have a method for quick freezing fish to preserve the flavor and color.

Another method is to haul refrigerated freight in containerized compartments that are loaded onto the ship. These containers may have their own electric refrigeration system that may be operated at the dock and can be tied into the ship's electrical system, **Figure 28.29**. These units must be located in such a manner that air can circulate over the condensers or they will have head pressure problems. It is typical for them to be located on the deck of an open ship. The ships that have large refrigeration systems onboard may use any of the refrigerants listed in this text, including the newer alternative refrigerants and blends, depending on the age and the design of the refrigeration system. Ammonia refrigeration systems are also used for ships. This refrigerant is not considered dangerous to the environment, but is extremely toxic to humans and, therefore, must be used with care.

⚠ SAFETY PRECAUTION: *It is sometimes necessary for technicians to wear breathing apparatus as well as protective clothing, gloves, and goggles when working with ammonia systems.•*

Large-system compressors may be reciprocating, screw, or centrifugal when low-pressure refrigerants are used. Older generation, low-pressure systems operated with R-11, a CFC refrigerant, but R-123 (HCFC-123) is a popular replacement for R-11 (CFC-11). The compressors may be driven using belts or direct drive by the ship's electrical system if it has a large enough generating plant. A large system may also be driven by steam turbines using the ship's steam system. The evaporators may be large plates in the individual refrigerated rooms or bare pipes, **Figure 28.30**. There are also systems using forced air, **Figure 28.31**. These systems are direct expansion, using thermostatic expansion valves. The ship may have several refrigerated compartments so that all the refrigeration is not in one large area.

The evaporators must have some means of defrosting because they will normally be operating below freezing. If the evaporator is close enough to the compressor, this may be accomplished with hot gas. If not, seawater may be used. If the water is too cold, it may be heated and sprayed over the evaporator when the fan is not running on a fan coil unit and the coil is not refrigerating during defrost time. By stopping the fan and closing the liquid-line solenoid valve during defrost the compressor will pump the refrigerant out of the evaporator during defrost. Defrost can be terminated using coil temperature or time. The refrigerant that is pumped out of the coil will be in the receiver during defrost, and the

Figure 28.29 Containerized freight for ships.

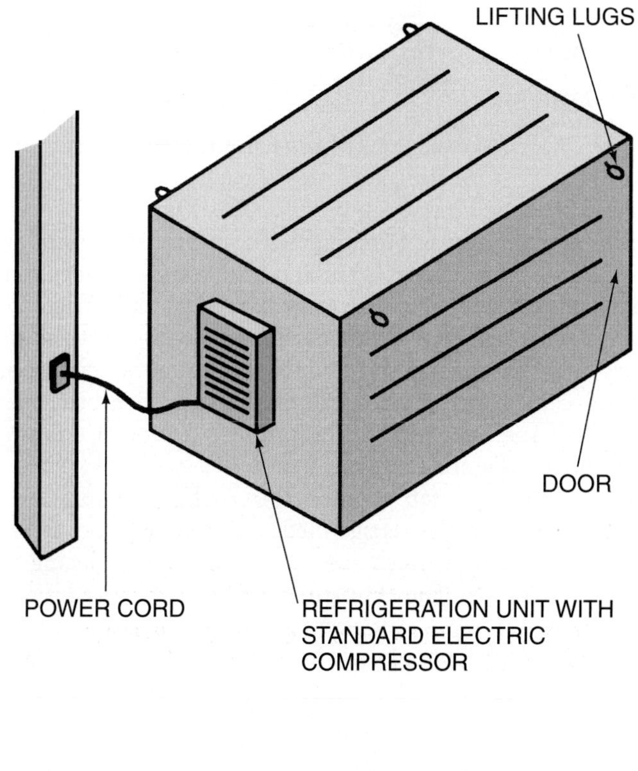

Figure 28.30 Plate evaporators for a ship.

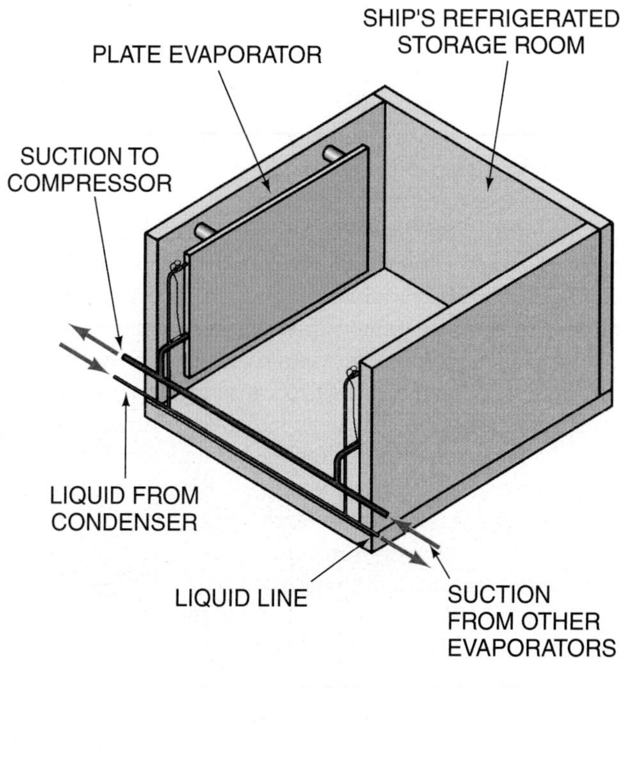

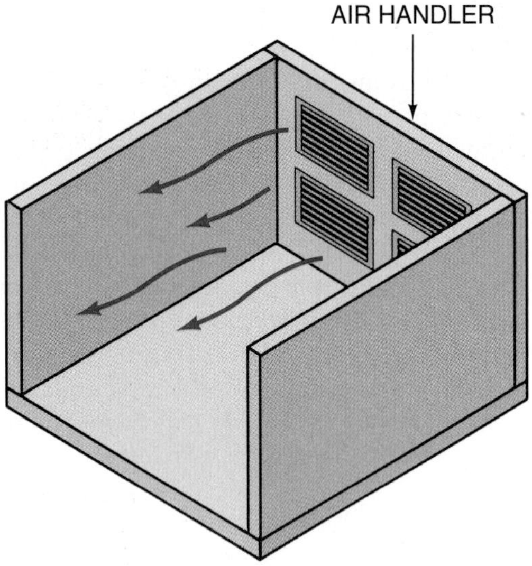

Figure 28.31 Forced-air refrigeration for a ship.

compressor can continue to run to other refrigerated compartments on the ship. Defrost can be staggered so that all coils will not call for defrost at the same time.

The condensers use seawater, which is readily available everywhere the ship goes. Seawater contains debris and must be filtered with filters that are easy to access and clean, which is commonly done with valves that allow the pumps to reverse the flow through the filters and piping to backwash the system, **Figure 28.32**. The refrigerant condensers for the typical refrigerants as well as newer alternative refrigerants and blends must be designed and built for easy cleaning because they will become dirty. These condensers have removable water box covers, called marine water boxes, and the tubes can be cleaned using a brush, **Figure 28.33**. The tubes are made of cupronickel (copper and nickel), a tough metal that does not corrode easily. The tube sheet that holds the tubes is typically made of steel, and the water box is made of brass where possible, **Figure 28.34**.

The American Bureau of Shipping recommends reserve capacity refrigeration and a spare parts inventory for making repairs at sea. It also recommends two compressor systems, either one of which should have enough capacity to maintain the load, running continuously while in tropical waters. This provides 100% compressor standby capacity, which is the most likely part of the system to cause a problem. Therefore, marine systems can become quite large. There must be someone onboard the vessel who is experienced and familiar with the system to make any necessary repairs. Repairs at sea are not uncommon. It can take days to reach a port with a repair facility.

The individual evaporators can be located far from the ship's condensers and compressors requiring long refrigerant lines to be run. This may lead to oil return problems. Oil separators may be used at the compressor to keep most of the

Figure 28.32 (A)–(B) Flushing the ship's seawater-cooled condensers.

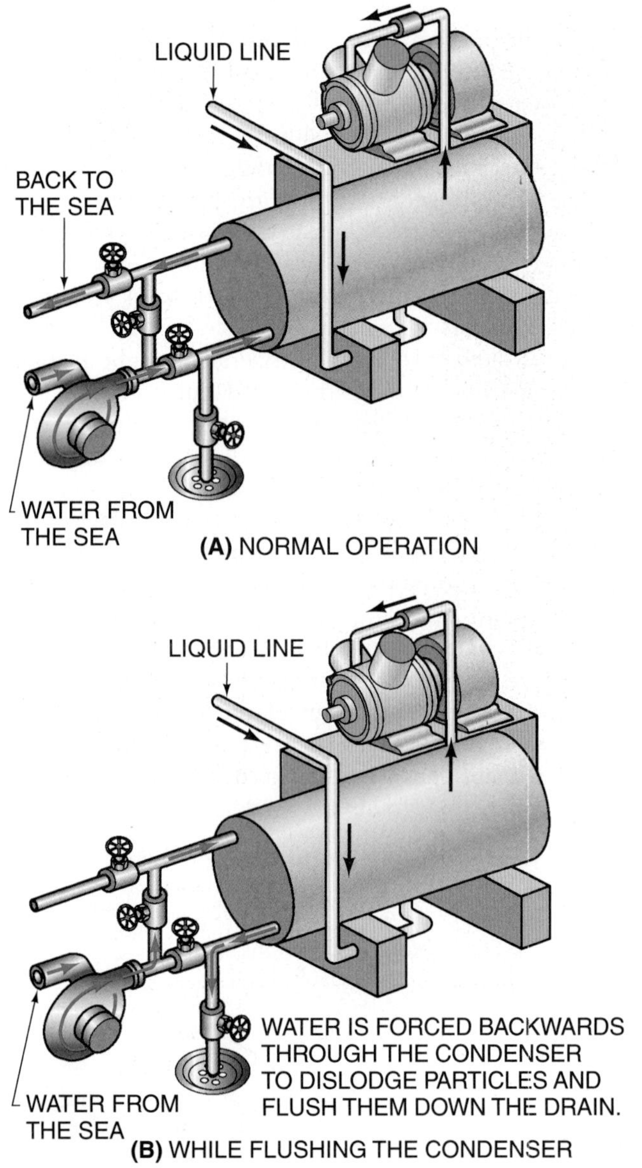

oil in the compressor, but some will get into the system and will have to be returned. Circulating brine systems have some advantages over individual evaporators in the refrigerated spaces. The primary refrigeration components and the primary refrigerant can all be contained in the equipment room with a central compressor or compressors. The brine may be chilled and circulated to the various locations for desired applications, **Figure 28.35**.

As with truck refrigeration, cargo ship refrigeration is designed to hold the refrigerated load temperature. It usually does not have much reserve capacity for reducing the load temperature. The ship storage rooms are not designed or built for the type of air circulation necessary to reduce the temperature of the load even if reserve capacity is available.

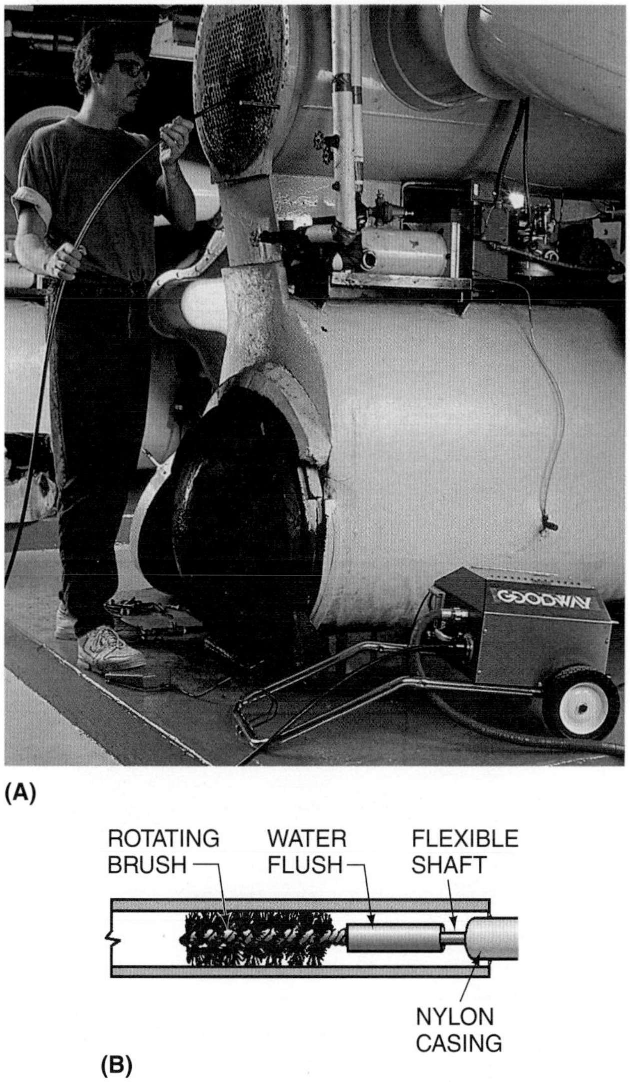

Figure 28.33 (A) Cleaning a condenser using a brush. Source: Tools Corporation (B) A cutaway view of a tube showing a rotating brush.

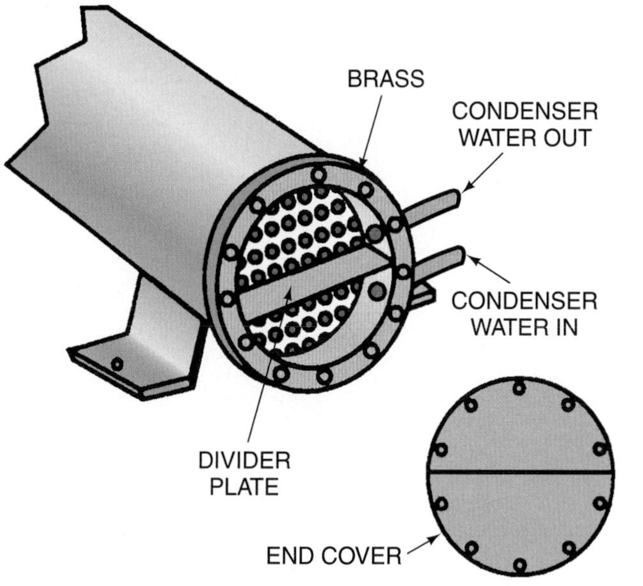

Figure 28.34 Tubes and water boxes for a condenser.

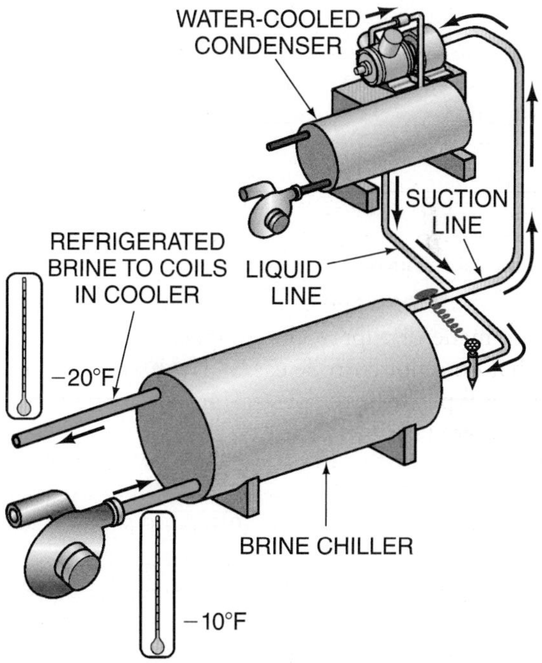

Figure 28.35 The brine circulation system for a central refrigeration system.

When the ship is loaded, every square foot of storage counts. The ship must be designed to accept the load and store it for the trip even if the seas are rough. The placement of the refrigerated load will have much to do with the successful delivery of that load. Correct circulation of air is important when the load is fresh food, such as fruit or vegetables. These foods give off heat, called heat of respiration, during storage, **Figure 28.36**. This is not at all like storing a load of frozen food that only has to be protected from heat that may enter from the outside, such as with a walk-in cooler. The heat is generated by the load and must be taken into account when calculating the size of the refrigeration equipment.

All ships must have refrigeration for the storage of foods for the crew and passengers. Refrigeration for the ship stores may be drawn from the main refrigeration system as long as it is running when the ship hauls refrigerated cargo. When the ship is not hauling refrigerated cargo, auxiliary systems may be used or the ship may have a separate system for the crew. Many of the same appliances may be used on a ship as in a supermarket. Some of these units may be individual package units with the compressor condenser built into the unit, or central refrigeration rooms may be used.

Figure 28.36 The heat of respiration for some fresh vegetables. Source: The 2014 ASHRAE Refrigeration Handbook.

THERMAL PROPERTIES OF FOOD					
Heat of Respiration (Btu/day/ton) – Approximate Values					
Food	**32°F**	**41°F**	**50°F**	**59°F**	**68°F**
Lima Beans (Unshelled)	4970	6125	15,450	24,750	29, 250
Lima Beans (Shelled)	5800	9925	24,300	38,685	53,050
Snap Beans	5215	7620	12,430	19,650	27,360
Red Beets	1390	2055	2790	4415	6050
Sweet Beets	1050	2595	4150	7710	11,275
Sweet Corn (With Husks)	9375	17,110	24,700	35,900	63,550
Cucumbers	2740	4250	5730	6300	8720

28.8 AIR CARGO HAULING

Products are often hauled by air to hasten the trip to market. Flowers are frequently shipped by air from places like Hawaii to the rest of the United States. Onboard refrigeration is out of the question because of the weight of the refrigeration systems. When the product needs to be cooled during flight, ice or dry ice is used. Specially designed containers, made of lightweight materials, are built to match the shape of the plane and slide easily into the cargo compartment of the aircraft.

REVIEW QUESTIONS

1. CO_2 is commonly known as
   A. carbon dioxide.
   B. carbon monoxide.
   C. chlorine monoxide.
   D. chlorine dioxide.
2. CO_2 in the solid form is referred to as
   A. liquid nitrogen.
   B. liquid ice.
   C. dry ice.
   D. dry nitrogen.
3. Which of the following is considered to be a major challenge associated with transport refrigeration?
   A. The transport vessels are often too small to carry the load.
   B. The load on the system changes as the vessel travels through different climates.
   C. All items being transported must be at a temperature lower than 32°F.
   D. Only fresh foods can be transported long distances without spoiling.
4. The process of sublimation is most commonly associated with
   A. dry ice.
   B. dry nitrogen.
   C. liquid nitrogen.
   D. dry nitrogen.
5. For safety purposes, the maximum pressure of a liquid nitrogen tank used in conjunction with truck refrigeration systems is
   A. 5 psig.
   B. 15 psig.
   C. 25 psig.
   D. 35 psig.
6. A eutectic or brine solution can best be described as a(n)
   A. salt/water solution.
   B. ice/water mixture.
   C. water/ammonia solution.
   D. ammonia/salt mixture.
7. Transport refrigeration encompasses movement of refrigerated products by
   A. cars, buses, trains, and boats.
   B. trains, cars, buses, and boats.
   C. planes, cars, buses, and trains.
   D. planes, trains, trucks, and boats.
8. In a truck refrigeration system, the air coming off the evaporator is typically directed
   A. down the side walls of the refrigerated space.
   B. up the side walls of the refrigerated space.
   C. across the top of the refrigerated space.
   D. across the bottom of the refrigerated space.

9. The motor that operates the compressor in a railway refrigeration transport system must
   A. be a three-phase motor.
   B. be able to operate at different frequencies.
   C. be able to operate at different voltages.
   D. be a capacitor-start, capacitor-run motor.
10. The process of transporting a refrigerated truck on a railroad flat car is called
   A. jumpstarting.
   B. hopscotching.
   C. leapfrogging.
   D. piggybacking.
11. Flash-freezing is accomplished at temperatures of approximately
   A. 0°F.
   B. −10°F.
   C. −30°F.
   D. −50°F.
12. A two-stage compressor is used to
   A. increase the compression ratio as compared to a single-stage compressor.
   B. increase the efficiency of the compression process.
   C. increase both the compression ratio and efficiency of the system.
   D. increase neither the compression ration nor the efficiency of the system.
13. In a two-stage cascade system, the condenser of the lower temperature stage is used to
   A. absorb heat from the product being cooled.
   B. evaporate the refrigerant in the higher-temperature stage.
   C. condense the refrigerant in the higher-temperature stage.
   D. absorb heat from the ambient air.
14. When food products are frozen slowly
   A. ice crystals will form, decreasing the overall quality of the product.
   B. ice crystals will form, increasing the overall quality of the product.
   C. ice crystals will not form, decreasing the overall quality of the product.
   D. ice crystals will not form, increasing the overall quality of the product.
15. The freezing method that makes it easier to sell frozen product by the piece or by the pound is
   A. bulk blast freezing.
   B. liquid nitrogen immersion.
   C. individual blast freezing.
   D. carbon dioxide immersion.
16. Which of the following makes up an accurate list of some of the components found in a two-stage, air-cooled cascade system?
   A. Two compressors, two air-cooled condensers, one tube-in-tube heat exchanger and a cooling tower
   B. One compressor, two water-cooled condensers, one tube-in-tube heat exchanger
   C. One compressor, one air-cooled condenser, one tube-in-tube heat exchanger and two evaporators
   D. Two compressors, one air-cooled condenser, one tube-in-tube heat exchanger
17. The tubes on heat exchangers that utilize sea water as their condensing medium are most often made of
   A. aluminum.
   B. copper.
   C. tin.
   D. cupronickel.
18. It is recommended that larger ships have backup refrigeration systems that are capable of providing 100% reserve capacity. The primary reason for this is in case the
   A. fishermen catch more fish than they expected.
   B. supply of condenser water is exhausted.
   C. primary refrigeration system fails.
   D. the ambient temperature drops too low.
19. Which of the following is true regarding the containers used to ship refrigerated products on large planes?
   A. They are constructed from steel.
   B. They are constructed from stainless steel.
   C. They are shaped to match the shape of the plane's cargo compartment.
   D. They are equipped with self-contained mechanical refrigeration systems.
20. The removable ends on shell and tube condensers are called
   A. marine water boxes.
   B. removable water boxes.
   C. threaded end caps.
   D. serviceable end caps.

UNIT 29

TROUBLESHOOTING AND TYPICAL OPERATING CONDITIONS FOR COMMERCIAL REFRIGERATION

OBJECTIVES

After studying this unit, you should be able to

- explain troubleshooting procedures as they relate to high-, medium-, and low-temperature refrigeration systems.
- discuss typical operating conditions for air-cooled equipment.
- discuss typical operating conditions for water-cooled equipment.
- describe the symptoms of a system that does not have the correct charge.
- describe the symptoms of a system that is operating with a restriction.
- describe the symptoms of a system operating with an inefficient system component.
- explain two methods for bench-testing the pumping capacity of a compressor.
- explain two methods for field-testing the pumping capacity of a compressor.
- discuss the diagnosis and repair of typical refrigeration system problems.

SAFETYCHECKLIST

- ✓ Wear goggles and gloves when attaching and removing gauges and when transferring refrigerant or checking pressures.
- ✓ Wear warm clothing when working in a cooler or freezer. A cold technician does not make good decisions.
- ✓ Observe all electrical safety precautions. Be careful at all times and use common sense.
- ✓ Disconnect electrical power whenever possible when working on a system. Lock and tag the panel and keep the key on your person.
- ✓ Wear rubber gloves when using chemicals to clean an evaporator or condenser. Use only approved chemicals. This is particularly important when cleaning an evaporator in a food storage unit.
- ✓ Do not attempt to perform a compressor vacuum or closed-loop test without the close supervision of an instructor or other experienced person.

29.1 ORGANIZED TROUBLESHOOTING

To begin troubleshooting any area of a refrigeration system, you need some idea of what the typical conditions should be. In commercial refrigeration you may be dealing with air temperatures inside a refrigerated box or outside at the condenser. You may need to know the current draw of the compressor or fan motors. The pressures inside the system may be important for troubleshooting and diagnoses. There are many conditions both outside and within the system that are important to know and understand.

When a piece of equipment has been running well for a period of time without noticeable problems, normally a single problem will trigger any sequence of events that may be encountered. For instance, it is common for the technician to think that two or three components of a piece of equipment have failed at one time. The chance of two parts failing at once is remote, unless one part causes the other to fail. These failures can almost always be traced to one original cause.

Knowing how the equipment is supposed to be functioning helps. You should know how it is supposed to sound, where it should be cool or hot, and when a particular fan is supposed to be operating. Knowing the correct operating pressures for a typical system can help you get started. **Before you do anything, look over the whole system for obvious problems.**

Troubleshooting procedures are divided into high-, medium-, and low-temperature ranges and some typical pressures for each. Each system has its own temperature range, depending on what it is supposed to be refrigerating. The table in **Figure 29.1** shows the temperature ranges for some refrigerated storage applications. This is only a partial table from the ASHRAE *Fundamental Guide*; refer to this guide for a complete table for all foods. Notice that for almost every food there is a different temperature requirement under more than one situation. For example, there may be a different temperature requirement for long-term storage than for short-term storage. The temperatures and pressures on the low-pressure side of

Figure 29.1 (A) This chart shows some storage requirements for typical products. Source: American Society of Heating, Refrigeration, and Air Conditioning Engineers, Inc. (B) A differences chart for coil-to-air temperature for selecting desired relative humidity. Source: American Society of Heating, Refrigeration, and Air Conditioning Engineers, Inc.

COMMODITY	STORAGE TEMP. °F	RELATIVE HUMIDITY %	APPROXIMATE STORAGE LIFE
Carrots			
Prepackaged	32	80–90	3–4 weeks
Topped	32	90–95	4–5 months
Celery	31–32	90–95	2–4 months
Cucumbers	45–50	90–95	10–14 days
Dairy products			
Cheese	30–45	65–70	
Butter	32–40	80–85	2 months
Butter	0 to −10	80–85	1 year
Cream (sweetened)	−15	–	several months
Ice cream	−15	–	several months
Milk, fluid whole			
Pasteurized Grade A	33	–	7 days
Condensed, sweetened	40	–	several months
Evaporated	Room Temp	–	1 year, plus
Milk, dried			
Whole milk	45–55	low	few months
Non-fat	45–55	low	several months
Eggs			
Shell	29–31	80–85	6–9 months
Shell, farm cooler	50–55	70–75	
Frozen, whole	0 or below	–	1 year, plus
Frozen, yolk	0 or below	–	1 year, plus
Frozen, white	0 or below	–	1 year, plus
Oranges	32–34	85–90	8–12 weeks
Potatoes			
Early crop	50–55	85–90	–
Late crop	38–50	85–90	–

(A)

COIL-TO-AIR TEMPERATURE DIFFERENCES TO MAINTAIN PROPER BOX HUMIDITY		
TEMPERATURE RANGE	DESIRED RELATIVE HUMIDITY	TD (REFRIGERANT TO AIR)
25° F to 45° F	90%	8° F to 12° F
25° F to 45° F	85%	10° F to 14° F
25° F to 45° F	80%	12° F to 16° F
25° F to 45° F	75%	16° F to 22° F
10° F and Below	–	15° F or less

(B)

the system are dependent on the product load. **An evaporator acts the same whether there is a single compressor close to the evaporator or a large compressor in the equipment room.** The examples given relate to evaporators. The function of the compressor is to lower the pressure in the evaporator to the correct boiling point to achieve the desired condition in the evaporator.

The service technician will have to deal with many numbers to address the conditions in various applications. In the past, basically one refrigerant was used for medium- and high-temperature applications and one was used for low-temperature applications. On routine service calls, the technician needed to think only of pressures that were common to these two refrigerants for the specific application. There are now many refrigerants on the market, which adds to the pressures to be considered. The technician should instead concentrate on the recommended temperatures for the particular applications. The temperatures can readily be converted to the correct pressures that should be displayed on the gauges. This text makes many references to R-134a and R-404A because many refrigeration systems are using these refrigerants. R-134a is the refrigerant replacement for R-12, and R-404A is the refrigerant replacement for R-502 in many applications. However, R-134a and R-404A are *not* drop-in replacements for any refrigerant. Because of global warming concerns, the Environmental Protection Agency's Significant New Alternatives Policy (SNAP) program is the governing body that can make changes, after a public comment period, as to what refrigerants will be acceptable for refrigeration and air-conditioning systems. At the time of this writing, there is more of a move away from HCFC and HFC refrigerants, and more of a move toward HC, HFO, carbon dioxide, and other natural refrigerants. Many refrigerants and refrigerant blends that have high global warming potentials (GWPs) will no longer be allowed to be used, even as retrofit refrigerants. It is the service technician's responsibility to keep abreast of the most recent laws and rules concerning refrigerant use and service. Refer to Unit 9, "Refrigerant and Oil Chemistry and Management—Recovery, Recycling, Reclaiming, and Retrofitting," for more detailed information on refrigerants. Always consult with the refrigeration system's manufacturer before retrofitting any system's refrigerant.

29.2 TROUBLESHOOTING HIGH-TEMPERATURE APPLICATIONS

High-temperature refrigerated box temperatures start at about 45°F and go up to about 60°F. Normally, a product temperature will be at one end of the range or the other depending on the product (flowers may be at 60°F and candy at 45°F). The low- and high-temperature and pressure conditions for high-temperature refrigeration systems must be therefore maintained. In these conditions, the coil temperature is normally 10°F to 20°F cooler than the box temperature, a difference referred to as temperature difference (TD). This means that the coil will normally be operating at 22 psig (45°F − 20°F = 25°F coil temperature) at the lowest temperature for an R-134a system. For the highest temperature of a high-temperature application, the pressure would be 57 psig (60°F coil temperature) at the end of the cycle when the compressor is off. **Figure 29.2** is a

Figure 29.2 (A) These charts can be used to determine typical operating temperatures and pressures for refrigerating systems using forced-air evaporators.

This table is intended to show the typical operating pressures for commercial refrigeration systems. Column 1 is with the compressor off and the box temperature just at the cut-in point. Column 2 is just after the compressor comes on. Column 3 is just before the compressor cuts off. This is not intended to be the only operating pressures, but the upper and lower limits as applied to high-, medium-, and low-temperature typical applications. A coil-to-inlet air temperature of 20° F (TD) was used as an average. Many systems will use 10° F or 20° F TD.

R-134a			
	Column 1 Compressor Off	Column 2 Compressor On	Column 3 Compressor On
HIGH–TEMPERATURE			
Box Temperature	60° F	60° F	45° F
Coil Temperature	60° F	40° F	25° F
Temperature Difference	0° F	20° F	20° F
Suction Pressure	57 psig	35.1 psig	22 psig
MED-TEMPERATURE			
Box Temperature	45° F	45° F	30° F
Coil Temperature	45° F	25° F	10° F
Temperature Difference	0° F	20° F	20° F
Suction Pressure	40 psig	22 psig	11.9 psig
LOW-TEMPERATURE			
Box Temperature	5° F	5° F	−20° F
Coil Temperature	5° F	−15° F	−40° F
Temperature Difference	0° F	2° F	20° F
Suction Pressure	9.1 psig	0 psig	14.7 in. Hg
R-404A			
LOW-TEMPERATURE			
Box Temperature	5° F	5° F	−20° F
Coil Temperature	5° F	−15° F	−40° F
Temperature Difference	0° F	20° F	20° F
Suction Pressure	38 psig	20.5 psig	4.8 psig
R-404A			

ICE MAKING
Ice begins to form at 20° F coil temperature.
Suction pressure 56 psig
End of cycle, ice about to harvest at 5° F coil temperature.
Suction pressure 38 psig

When a service technician installs the gauge manifold, the readings for a typical installation should not exceed or go below these readings. For instance, if the gauges were applied to a medium-temperature vegetable box, the highest suction pressure that should be encountered on a typical box should be 22 psig with the compressor running and a low of 11.9 psig at the end of the cycle. Any readings above or below these will be out of range for a box that has a 20° F temperature difference between the air temperature entering the coil and the coil's refrigerant boiling temperature.

(A)

Figure 29.2 (B) Typical low-, medium-, and high-temperature and pressure ranges for the low-pressure side of the refrigeration system.

TEMPERATURE °F	12	22	134a	502	R-404A	R-410A	TEMPERATURE AND PRESSURE RANGES FOR THE LOW-PRESSURE SIDE OF THE SYSTEM
−60	19.0	12.0		7.2	6.6	0.3	
−55	17.3	9.2		3.8	3.1	2.6	
−50	15.4	6.2		0.2	0.8	5.0	
−45	13.3	2.7		**1.9**	2.5	7.8	
−40	11.0	**0.5**	14.7	**4.1**	4.8	9.8	
−35	8.4	**2.6**	12.4	**6.5**	7.4	14.2	
−30	5.5	**4.9**	9.7	**9.2**	10.2	17.9	
−25	2.3	**7.4**	6.8	**12.1**	13.3	21.9	← LOW TEMPERATURE
−20	**0.6**	**10.1**	3.6	**15.3**	16.7	26.4	
−18	**1.3**	**11.3**	2.2	**16.7**	18.7	28.2	
−16	**2.0**	**12.5**	0.7	**18.1**	19.6	30.2	
−14	**2.8**	**13.8**	**0.3**	**19.5**	21.1	32.2	
−12	**3.6**	**15.1**	**1.2**	**21.0**	22.7	34.3	
−10	**4.5**	**16.5**	**2.0**	**22.6**	24.3	36.4	
−8	**5.4**	**17.9**	**2.8**	**24.2**	26.0	38.7	
−6	**6.3**	**19.3**	**3.7**	**25.8**	27.8	40.9	
−4	**7.2**	**20.8**	**4.6**	**27.5**	30.0	42.3	
−2	**8.2**	**22.4**	**5.5**	**29.3**	31.4	45.8	
0	**9.2**	**24.0**	**6.5**	**31.1**	33.3	48.3	
1	**9.7**	**24.8**	**7.0**	**32.0**	34.3	49.6	
2	**10.2**	**25.6**	**7.5**	**32.9**	35.3	50.9	
3	**10.7**	**26.4**	**8.0**	**33.9**	36.4	52.3	
4	**11.2**	**27.3**	**8.6**	**34.9**	37.4	53.6	
5	**11.8**	**28.2**	**9.1**	**35.8**	38.4	55.0	
6	**12.3**	**29.1**	**9.7**	**36.8**	39.5	56.4	
7	**12.9**	**30.0**	**10.2**	**37.9**	40.6	57.8	
8	**13.5**	**30.9**	**10.8**	**38.9**	41.7	59.3	
9	**14.0**	**31.8**	**11.4**	**39.9**	42.8	60.7	
10	**14.6**	**32.8**	**11.9**	**41.0**	43.9	62.2	
11	**15.2**	**33.7**	**12.5**	**42.1**	45.0	63.7	
12	**15.8**	**34.7**	**13.2**	**43.2**	46.2	65.3	
13	**16.4**	**35.7**	**13.8**	**44.3**	47.4	66.8	
14	**17.1**	**36.7**	**14.4**	**45.4**	48.6	68.4	
15	**17.7**	**37.7**	**15.1**	**46.5**	49.8	70.0	
16	**18.4**	**38.7**	**15.7**	**47.7**	51.0	71.6	
17	**19.0**	**39.8**	**16.4**	**48.8**	52.3	73.2	← MEDIUM TEMPERATURE
18	**19.7**	**40.8**	**17.1**	**50.0**	53.5	75.0	
19	**20.4**	**41.9**	**17.7**	**51.2**	54.8	76.7	
20	**21.0**	**43.0**	**18.4**	**52.4**	56.1	78.4	
21	**21.7**	**44.1**	**19.2**	**53.7**	57.4	80.1	
22	**22.4**	**45.3**	**19.9**	**54.9**	58.8	81.9	
23	**23.2**	**46.4**	**20.6**	**56.2**	60.1	83.7	
24	**23.9**	**47.6**	**21.4**	**57.5**	61.5	85.5	
25	**24.6**	**48.8**	**22.0**	**58.8**	62.9	87.3	
26	**25.4**	**49.9**	**22.9**	**60.1**	64.3	90.2	
27	**26.1**	**51.2**	**23.7**	**61.5**	65.8	91.1	
28	**26.9**	**52.4**	**24.5**	**62.8**	67.2	93.0	
29	**27.7**	**53.6**	**25.3**	**64.2**	68.7	95.0	
30	**28.4**	**54.9**	**26.1**	**65.6**	70.2	97.0	
31	**29.2**	**56.2**	**26.9**	**67.0**	71.7	99.0	
32	**30.1**	**57.2**	**27.8**	**68.4**	73.2	101.0	← HIGH TEMPERATURE
33	**30.9**	**58.8**	**28.7**	**69.9**	74.8	103.1	
34	**31.7**	**60.1**	**29.5**	**71.3**	76.4	105.1	
35	**32.6**	**61.5**	**30.4**	**72.8**	78.0	107.3	
36	**33.4**	**62.8**	**31.3**	**74.3**	79.6	108.4	
37	**34.3**	**64.2**	**32.2**	**75.8**	81.2	111.6	
38	**35.2**	**62.6**	**33.2**	**77.4**	82.9	113.8	
39	**36.1**	**67.1**	**34.1**	**79.0**	84.6	116.9	
40	**37.0**	**68.5**	**35.1**	**80.5**	86.3	118.3	
41	**37.9**	**70.0**	**36.0**	**82.1**	88.0	120.5	

REFRIGERANT

Vacuum (IN HG)—Red Figures
Gauge Pressure (psig)—Bold Figures

(B)

chart of some typical operating conditions for high-, medium-, and low-temperatures for typical applications in which R-134a and R-404A are used. For any other refrigerant or application, the temperature and pressure readings may be converted.

When the technician suspects problems, the following pressure gauge readings should be obtained on a 20°F TD high-temperature system using R-134a:

1. With the compressor running just before the cut-out point, a reading below 22 psig would be considered too low.
2. With the compressor running just after start-up with a normal box temperature, 60°F (60°F − 20°F = 40°F coil temperature), a reading above 35.1 psig would be considered too high.
3. When the compressor is off and the box temperature is at the highest point, the coil pressure would correspond to the air temperature in the box.

If a 60°F box is involved, the suction pressure could be 57 psig just before the compressor starts in a normal operation. These pressures are fairly far apart but serve as reference points for most high-temperature applications. The preceding illustration is for a 20°F TD. There will be times when a 10°F TD is normal for a particular application. The best procedure is to consult the chart in **Figure 29.1(A)** for the application and determine what the pressures should be by converting the box temperatures to pressure. This chart shows what the storage temperatures are.

The correct box humidity conditions may also be found by studying **Figure 29.1(B)**. The box humidity conditions will determine the coil-to-air temperature difference for a particular application. For example, you may want to store cucumbers in a cooler. The chart shows that cucumbers should be stored at a temperature of 45°F to 50°F and a humidity of 90% to 95% to prevent dehydration. Examining the second part of the chart indicates that to maintain these conditions the coil-to-air temperature difference should be 8°F to 12°F; 10°F is most commonly used. Typically, the higher the coil-to-air temperature difference, the more humidity is removed from the space.

The pressures from one piece of equipment to another may vary slightly. *Do not forget that any time the compressor is running, the coil should be 10°F to 20°F cooler than the fixture air temperature.*

29.3 TROUBLESHOOTING MEDIUM-TEMPERATURE APPLICATIONS

Medium-temperature refrigerated box temperatures range from about 30°F to 45°F in the refrigerated space. Many products will not freeze when the box temperature is as low as 30°F. Using the same methods as in the high-temperature examples to find the low- and high-pressure readings that would be considered normal, we find the following. The lowest pressures that would normally be encountered would be at the lowest box temperatures of 30°F. If the coil were to be boiling the refrigerant at 20°F below the inlet air temperature, the

refrigerant would be boiling at 10°F (30°F − 20°F = 10°F). This corresponds to a pressure for R-134a of 11.9 psig. If the service technician encounters pressures below 11.9 psig for an R-134a medium-temperature system, there is most likely a problem. The highest pressures that would normally be encountered while the system is running would be at the boiling temperature just after the compressor starts at 45°F. This is 45°F − 20°F = 25°F, or a corresponding pressure of 22 psig, **Figure 29.2**. The pressure in the low side while the compressor is off would correspond to the air temperature inside the cooler. When 45°F is considered the highest temperature before the compressor comes back on, the pressure would be 40 psig. Consult the chart in **Figure 29.1** for any specific application, because each is different. Use the chart as a guideline for the product type. The equipment operating conditions may vary to some extent from one manufacturer to another.

29.4 TROUBLESHOOTING LOW-TEMPERATURE APPLICATIONS

Low-temperature applications start at freezing and go down in temperature, although not many are designed for just below freezing. The first significant application is for the making of ice, which normally occurs from 20°F evaporator temperature to about 5°F evaporator temperature. Ice making has many different variables, which are associated with the type of ice being made. Flake ice is a continuous process, so the pressures are about the same regardless of the equipment manufacturer, although the manufacturer should be consulted for exact temperatures and pressures for flake ice machines.

For cube ice makers using R-404A, a suction pressure of 36 psig (20°F) at the beginning of the cycle will begin to make ice. The ice will normally be harvested before the suction pressure reaches the pressure corresponding to 5°F, or 38 psig, **Figure 29.2**. If the suction pressure for an R-404A machine does not go below 56 psig, suspect a problem. If it goes below 38 psig, also suspect a problem. These pressures indicate the normal running cycle.

The low temperatures for frozen foods are generally considered to start at 5°F space temperature and go down from there. Different foods require different temperatures to be frozen hard; some are hard at 5°F, others at −10°F. The designer or operator of the equipment will use the most economical design or operation. For instance, ice cream may be frozen hard at −10°F, but it may be maintained at −30°F without harm. The economics of cooling it to a cooler temperature may not be cost-effective. As the temperature goes down, the compressor has less capacity because the suction gas becomes less dense. The compressor may not even cut off at the very low temperatures if the box thermostat is set that low.

Using the same guidelines as for high- and medium-temperature applications, the highest normal refrigerant boiling temperature for low-temperature refrigeration applications are 5°F − 20°F = −15°F for R-134a. The refrigerant boils at 20°F below the space temperature because the compressor is running and creating a suction pressure of 0 psig. The lowest suction pressure may be anything down to the evaporator pressure corresponding to −20°F space temperature. This would be −20°F − 20°F = −40°F, or 14.7 in. Hg vacuum for an R-134a system, **Figure 29.2(B)**. With the compressor off just at the time the thermostat calls for cooling for the highest temperature, the evaporator would be 5°F, the same as the space temperature, or 9.1 psig.

A different refrigerant for low-temperature applications may be used to keep the pressures positive, above atmospheric. Using R-404A would give a new set of figures: 5°F corresponds to 38 psig when the space temperature is 5°F with the compressor off. When the space temperature is 5°F and the compressor is running, the refrigerant will be boiling at about 20°F below the space temperature, or 5°F −20°F = −15°F. This −15°F corresponds to a pressure of 20.3 psig. The lowest temperature normally encountered in low-temperature refrigeration is a space temperature of about −20°F, which makes the suction pressure about −20°F − 20°F = −40°F, or 4.8 psig, **Figure 29.2(B)**. By using R-404A in this application the system will operate well above a vacuum. R-404A boils at −50°F at 0 psig, so the cooler would go down to −30°F air temperature before the suction would go down to the atmospheric pressure of 0 psig with a 20°F TD. Temperatures below −20°F for the space or product are obtainable and are used.

The chart in **Figure 29.2(A)** can be used as a guideline for typical low-side operating conditions for the aforementioned systems. Its intent is to illustrate the average high and low temperatures and pressures for each application.

29.5 TYPICAL AIR-COOLED CONDENSER OPERATING CONDITIONS

Air-cooled condensers have an operating temperature range from cold to hot as they are exposed to outside temperatures. The equipment may even be located inside a conditioned building. The condenser has to maintain a pressure that will create enough pressure drop across the expansion device for it to feed correctly. This pressure drop across the expansion device will push enough refrigerant through the device for proper refrigerant levels in the evaporator. However, some modern thermostatic expansion valves (TXVs) can operate with much less pressure drop as long as 100% liquid is supplied to them. The balanced-port TXV is noted for its low-pressure drop performance. Expansion valves for R-134a are normally sized for a 75- to 100-psig pressure drop. For example, when a manufacturer indicates that a valve has a capacity of 1 ton at an evaporator temperature of 20°F, it is typically designed for a pressure drop of 80 psig. The same valve will have less capacity if the pressure drop falls to 60 psig. Refer to Unit 24, "Expansion Devices," for more detailed information on thermostatic expansion devices.

Figure 29.3 A refrigeration system shows the relationship of the air-cooled condenser to the entering air temperature on a cool day. This unit has head pressure control for mild weather. The outside air entering the condenser is 50°F. This would normally create a head pressure of about 85 psig. This unit has a fan cycle device to hold the head pressure up to a minimum of 125 psig.

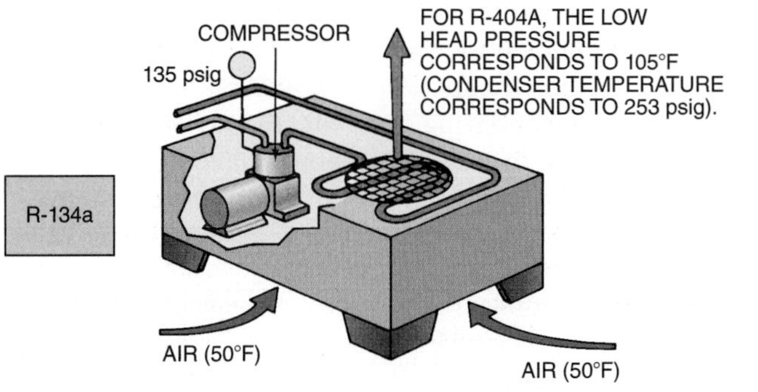

Because any piece of equipment located outside must have head pressure control, the minimum values at which the control is set would be for the minimum expected head pressure. For R-134a, this is normally about 135 psig for air-cooled equipment, **Figure 29.3**; this corresponds to 105°F condensing temperature. R-404A has a head pressure of 253 psig at 105°F. When the discharge pressure is less than 135 psig for an R-134a system, the pressure drop across the expansion valve will decrease, causing erratic operation and a drop in capacity. An exception to this is when the manufacturer *floats* the head pressure with the ambient temperature. Head pressure controls can still be used, but they are usually set at much lower settings to take advantage of the cooler ambient temperatures. Floating head pressures will increase efficiencies by enabling the compressor to run at lower compression ratios. Refer to Unit 22, "Condensers," for more in-depth information on floating the head pressure.

29.6 CALCULATING THE CORRECT HEAD PRESSURE FOR AIR-COOLED EQUIPMENT

The maximum normal high-side pressure would correspond to the maximum ambient temperature in which a condenser is operated. Most air-cooled condensers will condense the refrigerant at a temperature of about 30°F above the ambient temperature when the ambient is above 70°F. **Figure 29.4** presents an example of a refrigeration system operating under low-ambient conditions without head pressure control. When the ambient temperature drops below 70°F, the condensing relationship changes to a lower temperature. If the condenser is a higher-efficiency condenser or oversized, the condenser may condense at 25°F above the ambient. *For energy-efficiency purposes, many manufacturers are designing condensers with improved heat transfer materials and surfaces. Condenser sizes are also increasing to keep head pressures lower and compression ratios lower for better efficiencies. This will cause some condensing temperatures to be only 10°F to 15°F above the ambient. Refer to Unit 22, "Condensers," for more detailed information on high efficiency condensers.* Using the 30°F figure, if a unit were located outside and the temperature were 95°F, the condensing temperature would be 95°F + 30°F = 125°F, **Figure 29.5**. *Many newer, more environmentally friendly replacement refrigerants are being introduced into the HVACR industry. These refrigerants include hydrocarbon (HC), hydrofluoro-olefin (HFO), carbon dioxide, and other natural refrigerants with lower GWP and ODP values. Refer to Unit 9, "Refrigerant and Oil Chemistry and Management-Recovery, Recycling, Reclaiming, and Retrofitting," for more detailed information on newer environmentally friendly refrigerants.* The corresponding pressure for R-134a would be 185 psig,

Figure 29.4 A refrigeration system operating under a low-ambient condition. The expansion valve is not feeding correctly because there is not enough head pressure to push the liquid refrigerant through the valve.

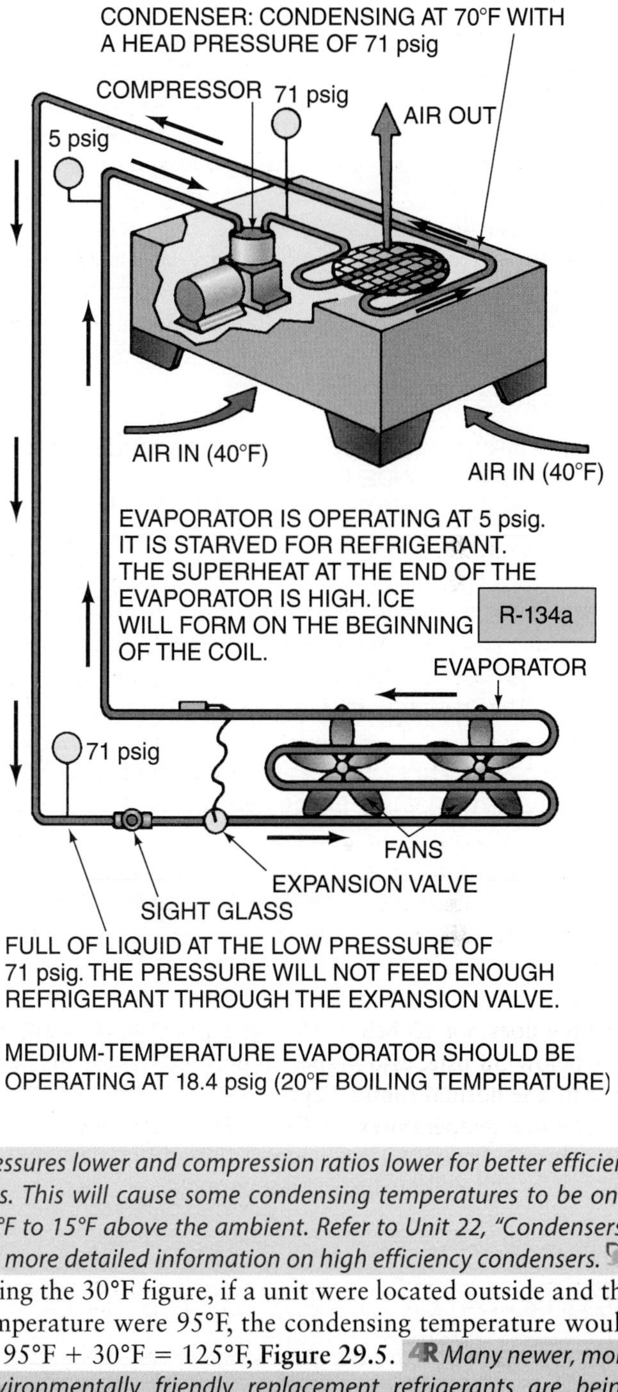

Figure 29.5 An air-cooled refrigeration system with a condenser outside. Typical pressures are shown.

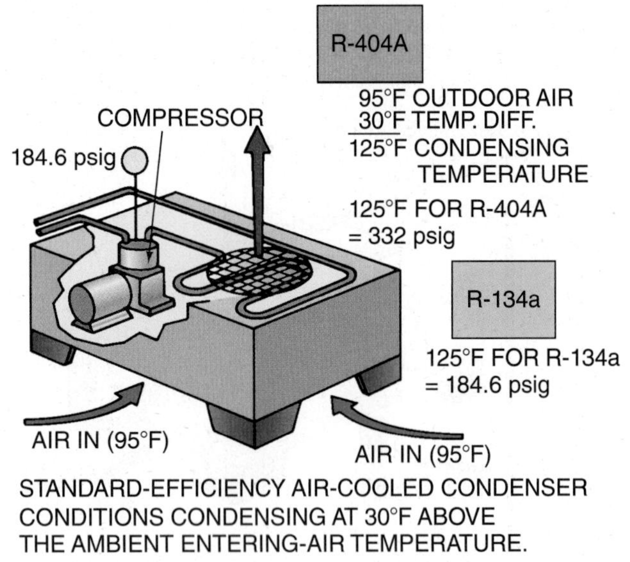

for R-404A it would be 332 psig, and for R-22 it would be 278 psig at 125°F condensing temperature.

29.7 TYPICAL OPERATING CONDITIONS FOR WATER-COOLED EQUIPMENT

Water-cooled condensers are used in many systems in one of two ways. Some use a wastewater system and some recirculate the same water and extract the heat with a cooling tower. Typically, a water-cooled condenser that uses freshwater, such as city water or well water, uses about 1.5 gal of water per minute per ton with a designed 20° temperature difference between the water inlet and the water outlet. A system using a cooling tower circulates about 3 gpm/ton with a designed 10° temperature difference between the water inlet and the water outlet. These two applications have different operating conditions and will be discussed separately.

TYPICAL OPERATING CONDITIONS FOR WASTEWATER CONDENSER SYSTEMS

Wastewater systems use the same condensers as the cooling tower systems, but the water is wasted down the drain. A water-regulating valve is normally used to regulate the water flow for economy and to regulate the head pressure. The condensers are either cleanable or a coil type that cannot be cleaned mechanically. With either type of condenser, it is advantageous to know from the outside how the condenser is performing on the inside.

When a water-regulating valve is used to control the water flow, the flow will be greater if the condenser is not performing correctly. When the head pressure goes up, the water will start to flow faster to compensate. This will take place until the capacity of the valve opening is reached. Then the head pressure will increase with maximum water flow. Sometimes the inlet water is colder, such as in winter. It would not be unusual for the inlet water to be 45°F in the winter. If the water travels through a hot ceiling in the summer, 90°F may be the high value, **Figure 29.6**. Because of the variable flow and variable temperature of water flowing through a wastewater system during different seasons of the year, there is no refrigerant-to-water temperature relationship or rule of thumb for the technician to go by for these systems. The technician must simply rely on the water-regulating valve to

Figure 29.6 (A) A water-cooled wastewater refrigeration system. The water flow is adjusted by the water-regulating valve, which keeps too much water from flowing, maintains a constant head pressure, and shuts the water off at the end of the cycle. (B) Water flow is greater with a 90°F inlet water temperature.

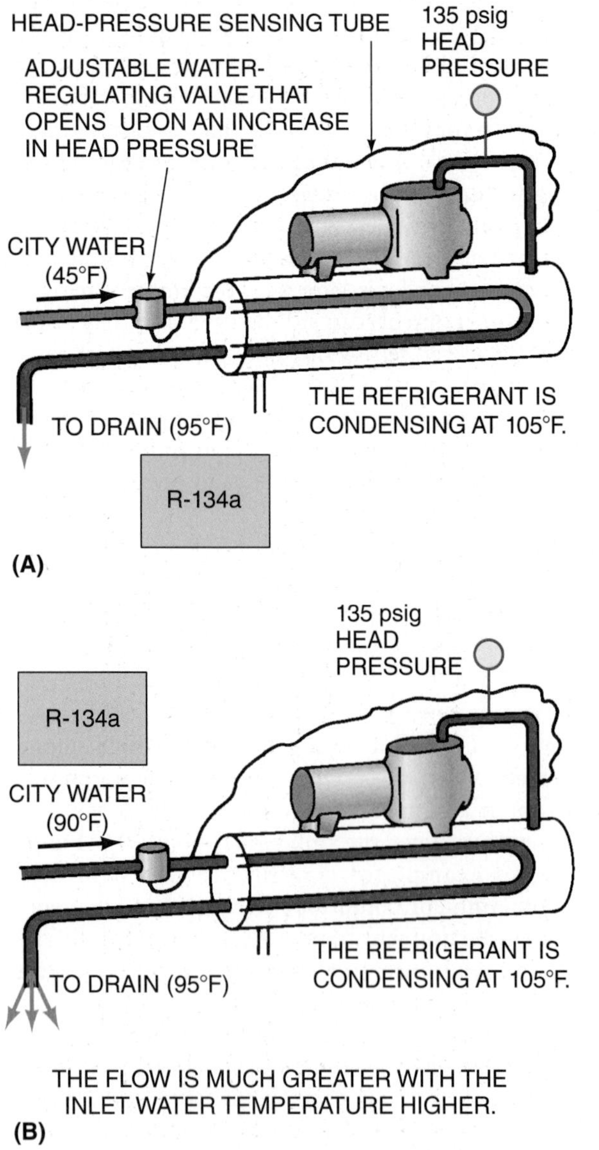

Figure 29.7 A water-cooled refrigeration system with a dirty condenser. The system has high head pressure, high water flow, and a low water temperature difference.

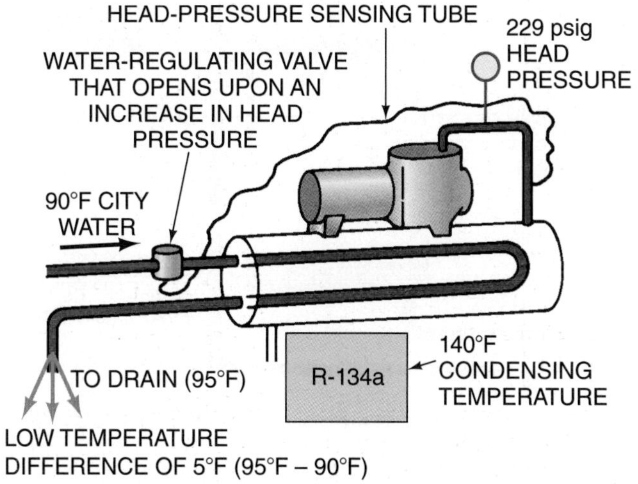

throttle more or less water at different temperatures through the condenser to maintain a constant head pressure. For more detailed information on condensers and water-regulating valves, refer to Unit 22, "Condensers."

If the condensing temperature of a water-cooled system were to rise much above 105°F, suspect a dirty or clogged condenser. For example, a system with a 140°F condensing temperature has a water inlet temperature of 90°F and an outlet temperature of 95°F with a high water flow rate, **Figure 29.7**. This would indicate that the condenser is not removing heat and that the coil is dirty. The low water temperature difference of 5°F (95°F – 90°F), high condensing temperature (140°F), and high water flow rate indicate that little to no heat transfer is taking place. Whenever there is a noticeable increase in water flow, a low temperature difference between the water in and out of the condenser, and a high head pressure, a dirty condenser should be suspected. When there is a noticeable decrease in water flow, a high water temperature difference and high head pressure, a clogged water regulating valve or clogged condenser should be suspected.

TYPICAL OPERATING CONDITIONS FOR RECIRCULATED WATER SYSTEMS

Water-cooled condensers that have a cooling tower to remove the heat from the water usually do not use water-regulating valves to control the water flow. Normally, a constant volume of water is circulated by a pump. The volume is typically designed into the system at the time of installation and can be verified by checking the pressure drop across the water circuit from the condenser inlet to the outlet. There has to be some pressure drop for there to be water flow. The original specifications for the system should include the engineer's intent with regard to the water flow, but this information may not be available on an old installation.

Figure 29.8 A water-cooled refrigeration system reusing the water after the heat is rejected to the atmosphere. There is a constant bleed of fresh water to dilute the minerals left behind by the evaporating water. The temperature difference between the incoming water and the outgoing water is 10°F. This is the typical temperature rise across a cooling tower system.

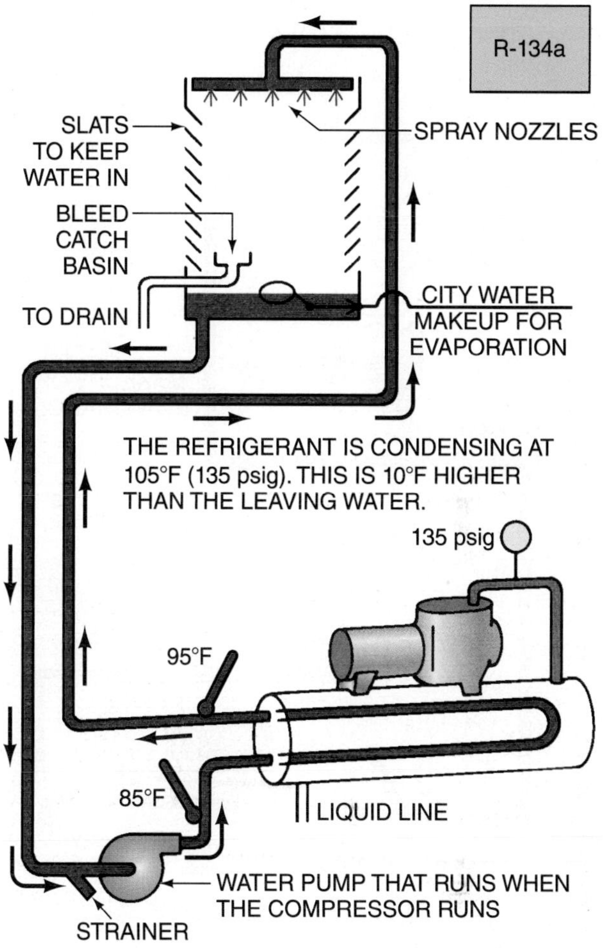

Most systems reusing the same water by means of a cooling tower have a standard 10°F water temperature rise across the condenser and a refrigerant condensing temperature that is typically 10°F above the water outlet temperature. For example, if the water from the tower were to be 85°F entering the condenser, the water leaving the condenser should be 95°F and the condensing temperature should be about 105°F, **Figure 29.8**. If the difference were to be 15°F or 20°F, the condenser appears to be doing its job of removing the heat from the refrigerant, but the water flow is insufficient, **Figure 29.9**. If the water temperature entering the condenser were to be 85°F and the leaving water were to be 90°F, it may be that there is too much water flow. If the head pressure is not high, the condenser is removing the heat and there is too much water flow. If the head pressure is high and there is the right amount of water flow, the condenser is dirty. See **Figure 29.10** for an example of a dirty condenser. **Figure 29.11** shows a 10°F water temperature rise across the condenser and related

Figure 29.9 This system has a high water temperature difference, indicating there is not enough water flow. The condensing temperature is higher than normal, which causes the head pressure to rise. A decrease in water flow may be detected by water pressure drop if the design pressure drop is known.

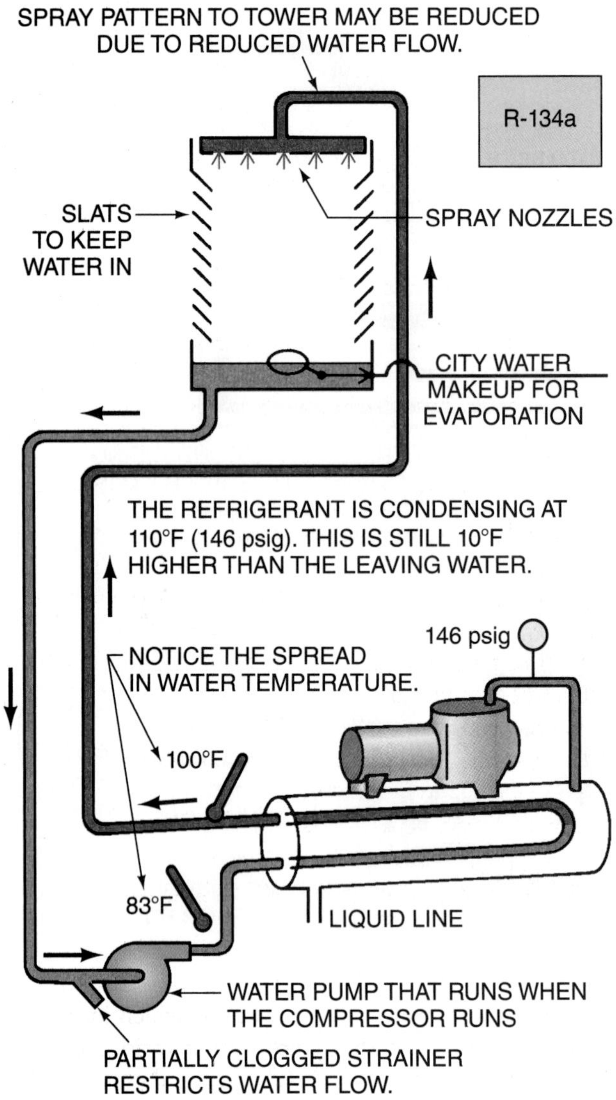

Figure 29.10 Typical temperatures and pressures of a recirculated water system operating with a dirty condenser.

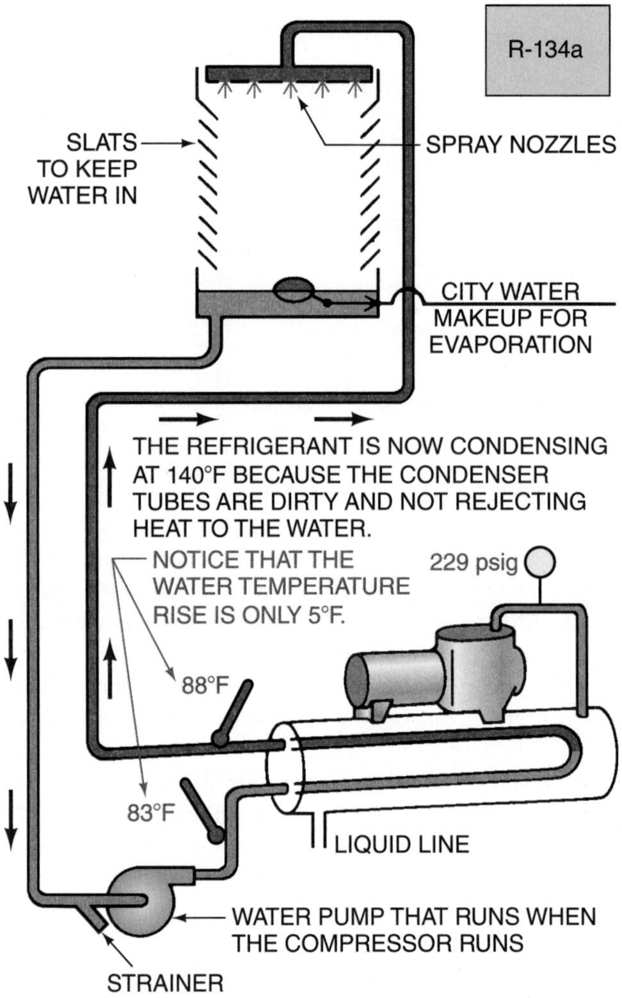

Figure 29.11 A cooling tower system showing a 10°F water temperature rise across the condenser and related condensing pressures for a 105°F condensing temperature when R-134a, R-22, and R-404A are used as the refrigerants.

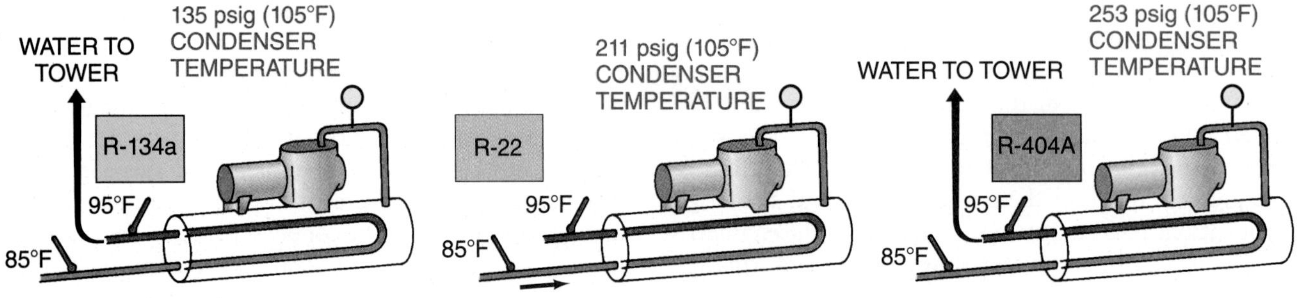

Figure 29.12 This cooling tower has a temperature relationship with the air that is cooling the water. Most cooling towers can cool the water that goes back to the condenser to within 7°F of the wet-bulb temperature of the ambient air. For example, if the wet-bulb temperature is 78°F, the tower should be able to cool the water to 85°F. If it does not, a problem with the tower should be suspected.

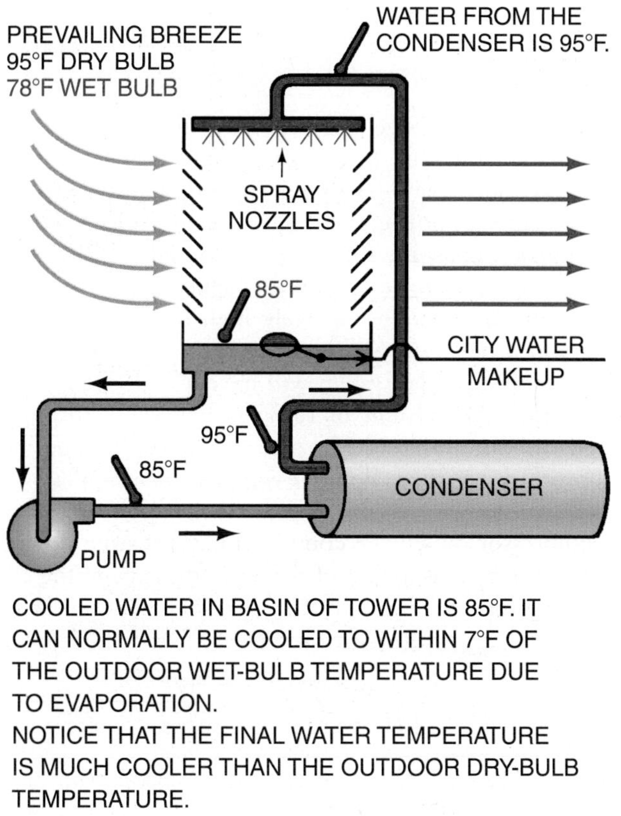

condensing pressures for a 105°F condensing temperature when R-134a, R-22, and R-404A are used as the refrigerants.

In this type of system the condenser is getting its inlet water from the cooling tower. A cooling tower has a heat exchange relationship with the ambient air. Usually, the cooling tower is located outside and may have a natural draft or forced draft (where a fan forces the air through the tower). Either type of tower will be able to supply water temperatures according to the humidity or moisture content in the outside air. The cooling tower can normally cool the water to within 7°F of the wet-bulb temperature of the outside air, **Figure 29.12**. If the outside wet-bulb temperature (taken with a psychrometer) is 78°F, the leaving water will be about 85°F if the tower is performing correctly. Wet-bulb temperature relates to the moisture content in air; it is discussed in more detail in Unit 35, "Comfort and Psychrometrics."

29.8 REFRIGERANT UNDERCHARGE

A low refrigerant charge affects most systems in about the same way, depending on the correct amount of the refrigerant needed. The normal symptoms of this problem are high superheat and low capacity. The system has a starved evaporator and cannot absorb the rated amount of Btu or heat. The suction gauge and the discharge pressures will be lower than normal. The exception to this is a system with an automatic expansion valve, which will adjust itself to maintain a constant evaporator pressure.

Figure 29.13 A sight glass to indicate when pure liquid is in the liquid line. This sight glass also has a moisture indicator in its center. Courtesy Sporlan Division, Parker Hannifin Corporation.

If the system has a sight glass, it will have bubbles in it that look like air but are actually vapor refrigerant. **Figure 29.13** shows a typical liquid-line sight glass. Remember, vapor or liquid may look the same in a sight glass. If there is only vapor in the glass, a slight film of oil may be present. This is a good indicator that only vapor is present. A system with a sight glass will generally have a TXV or automatic expansion valve and a receiver. These valves will hiss when a partial vapor–partial liquid mixture goes through the valve. A system with a capillary tube will probably not have a sight glass. The technician should know how the system feels to the touch at different points to determine the gas charge level without using gauges.

A refrigerant undercharge affects the compressor by not supplying the cool suction vapor that cools the motor. Low charges also cause low suction pressures, which increase the compression ratio. A higher compression ratio causes inefficiencies along with higher discharge temperatures. Most compressors are suction-cooled, so the result is a hot compressor motor. The motor may even overheat and be turned off by the motor-winding thermostat, **Figure 29.14.** If the compressor is air-cooled, the suction line coming back to the compressor will not need to be as cool as for a suction-cooled compressor. It may be warm by comparison.

An undercharge of refrigerant will produce low compressor amperage because the higher superheat coming back to the compressor inlet will cause the inlet vapors to expand, which decreases their density. Low-density vapors entering the compressor will mean low refrigerant flow through the compressor. This will cause low amperage (amp) draw, because the compressor does not have to work as hard compressing the low-density vapors. This low refrigerant flow may also cause refrigerant-cooled compressors to overheat. Due to the

Figure 29.14 A system with a suction-cooled hermetic compressor. The compressor is hot enough to cause it to cut off because of motor temperature. It is hard to cool off a hot hermetic compressor from the outside because the motor is suspended inside a vapor atmosphere.

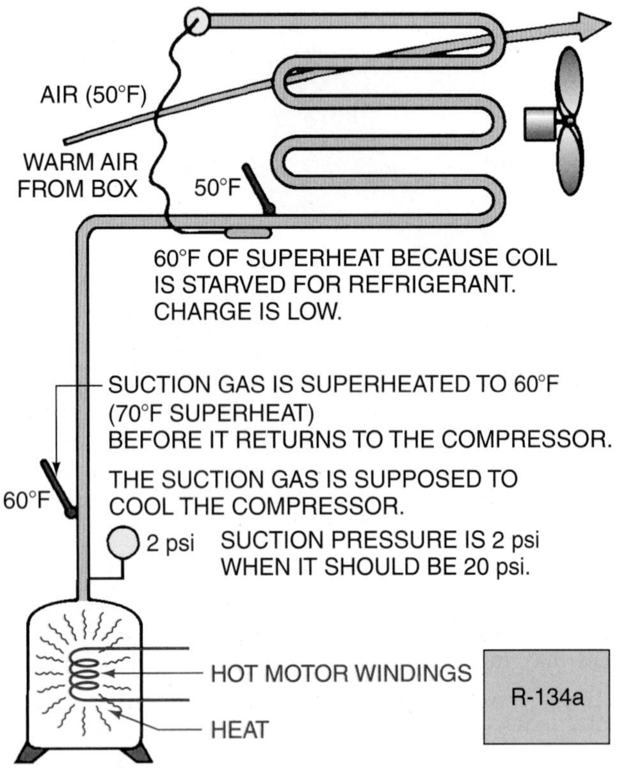

low refrigerant flow rate through the compressor and system that occurs with an undercharge of refrigerant, the 100% saturated-vapor point in the condenser will be very low, causing low condenser subcooling. The condenser will not receive enough refrigerant vapors to condense them to a liquid and feed the receiver (if the system has one). Condenser subcooling is a good indication of how much refrigerant charge is in the system. Low condenser subcooling may mean a low charge, whereas high condenser subcooling may mean an overcharge.

Capillary tube systems usually have no receiver so a capillary tube system can run high subcooling simply from a restriction in the capillary tube or liquid line. The excess refrigerant will accumulate in the condenser, causing high subcooling and high head pressures. If the liquid line of a TXV/receiver system is restricted, some refrigerant will accumulate in the condenser, but most will collect in the receiver. Remember, a receiver is designed to handle the system's entire refrigerant charge and still have a 25% vapor head for safety. This will cause low condenser subcooling and a low head pressure.

In summary, the symptoms of a refrigerant undercharge are

- low system capacity,
- starved evaporator (high superheat),
- low suction pressure,
- low discharge pressure,
- high compression ratio,
- bubbles in the sight glass,
- low compressor amperage (amp) draws, and
- low condenser subcooling.

29.9 REFRIGERANT OVERCHARGE

A refrigerant overcharge also acts much the same way from system to system. The discharge pressure is high, and the suction pressure may be high. The automatic expansion valve system will not have a high suction pressure because it maintains a constant suction pressure, **Figure 29.15**. The TXV system may have a slightly higher suction pressure if the head pressure is excessively high because the system capacity may be down, **Figure 29.16**.

The capillary tube system will have a high suction pressure because the amount of refrigerant flowing through it depends on the difference in pressure across it. The more head pressure, the more liquid it will pass. The capillary tube will allow enough refrigerant to pass to allow liquid into the compressor. When the compressor is sweating down its side or all over, it is a sign of liquid refrigerant in the compressor, **Figures 29.17** and **29.18**.

Figure 29.15 The system has an automatic expansion valve metering device. This device is designed to maintain a constant suction pressure. In an overcharged condition, the head pressure is higher than normal, but the suction pressure remains the same.

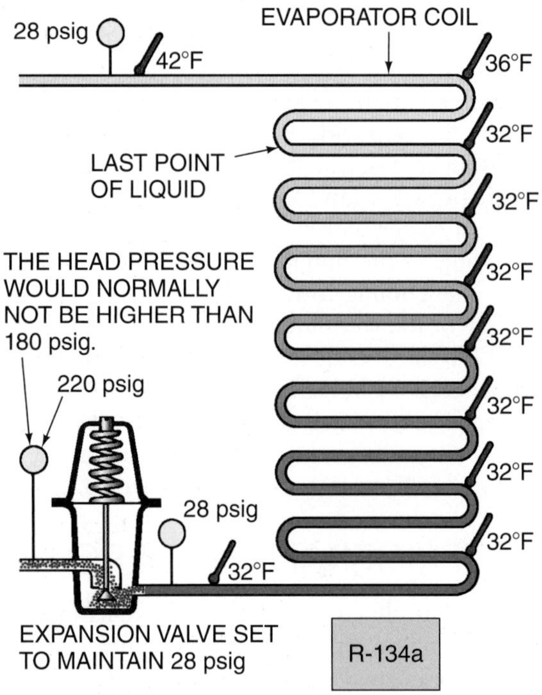

Figure 29.16 A TXV system using R-134a, which has a higher-than-normal head pressure in an overcharged condition.

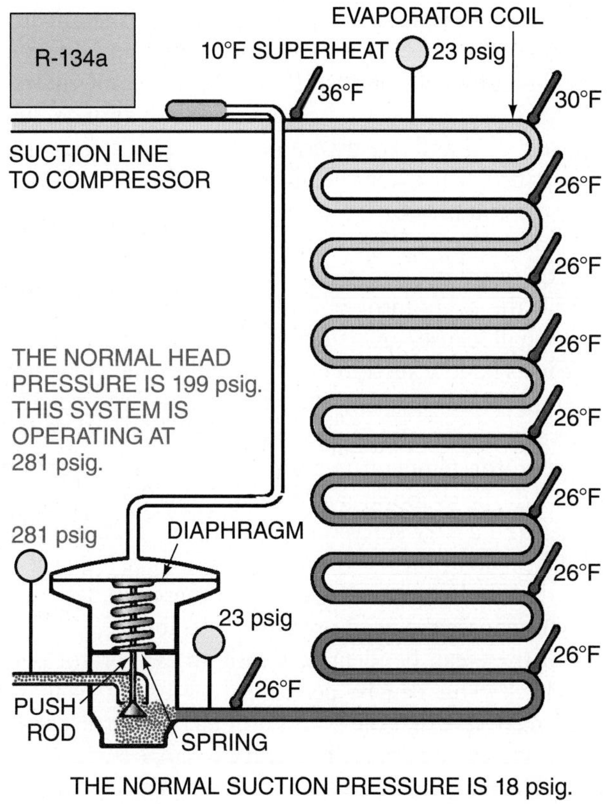

Figure 29.17 The capillary tube system using R-134a has an overcharge of refrigerant, and the head pressure is higher than normal. This tends to push more refrigerant than normal through the metering device.

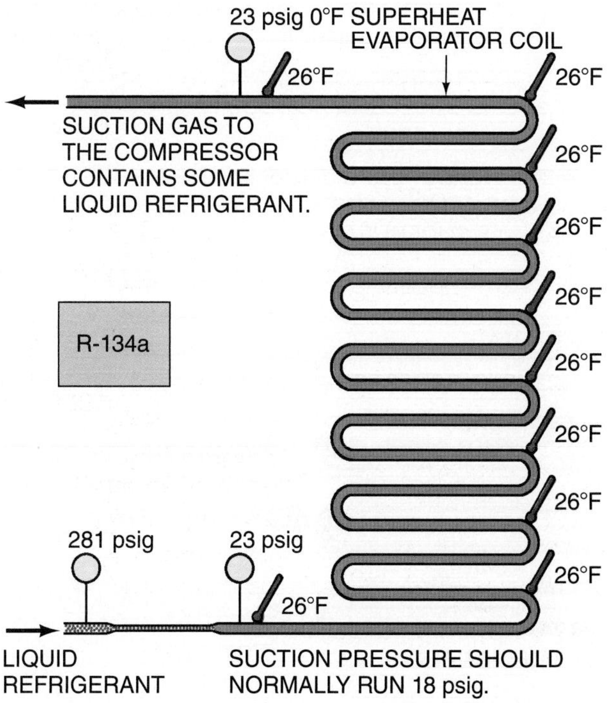

Figure 29.18 When liquid refrigerant gets back to the compressor, the latent heat that is left in the refrigerant will cause the compressor to sweat more than normal. When vapor only gets back to the compressor, the vapor changes in temperature quickly and the compressor does not sweat much.

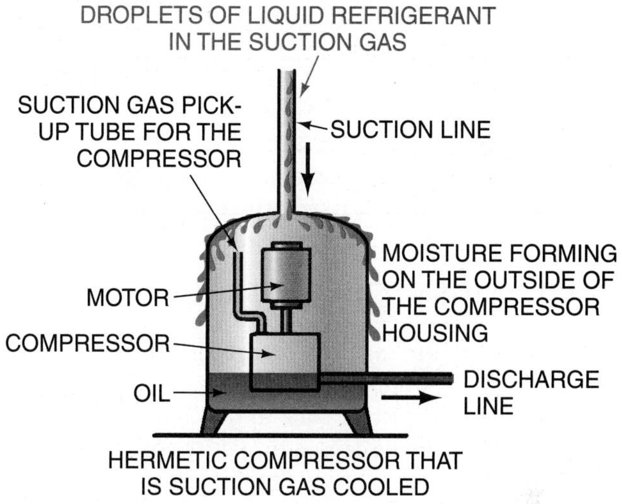

Critically charging a capillary tube system may help reduce the severity of these flooding problems. Often, in high head-pressure situations in which the system is critically charged with refrigerant, there may not be enough refrigerant in the system to adversely affect the compressor when flooding occurs. The manufacturer will determine the amount of critical charge and stamp it on the unit's nameplate. If it is not stamped on the unit, contact the manufacturer to determine the correct charge.

Only vapor should enter the compressor. Vapor will rise in temperature as soon as it touches the compressor shell, but when liquid is present, it will not rise in temperature and will cool the compressor shell. **Liquid refrigerant still has its latent heat absorption capability and will absorb a great amount of heat without changing temperature. A vapor absorbs only sensible heat and will change in temperature quickly, Figure 29.18.** Another reason that liquid may get back to the compressor in a capillary tube system is poor heat exchange in the evaporator. If liquid is getting back to the compressor, the evaporator heat exchange should be checked before removing refrigerant.

As mentioned earlier, when a system is overcharged with refrigerant, the discharge pressure and suction pressure will be higher. The condenser will be flooded with subcooled liquid, causing the discharge pressure to be excessively high. Remember, any liquid in the condenser at temperatures lower than the condensing temperature is considered to be subcooled. Excessive condenser subcooling will cause a high compression ratio, low efficiency, and low refrigerant flow rates and capacities. Forced-air condensers should have at least 6 to 8 degrees of liquid subcooling; however, amounts depend on system piping configurations and liquid-line static and friction pressure drops.

If the system has a TXV as a metering device, the TXV will try to maintain evaporator superheat even with an excessive overcharge. The TXV may overfeed slightly during its opening strokes, but it should stabilize if still in its pressure operating range.

In summary, the symptoms of a refrigerant overcharge are

- high discharge pressures,
- high suction pressures,
- high compression ratios,
- high condenser subcooling, and
- normal evaporator superheats with a TXV; low superheats with a capillary tube or fixed orifice (piston).

29.10 INEFFICIENT EVAPORATOR

An inefficient evaporator does not absorb the heat into the system and will have a low suction pressure. The suction line may be sweating or frosting back to the compressor. An inefficient evaporator can be caused by a dirty coil, a fan running too slowly, an expansion valve flooding or starving the coil, recirculated air, ice buildup, or interference from product blocking the airflow.

Whenever there is reduced airflow across the face of the evaporator, there is a reduced heat load on the coil. This reduced airflow and heat load will cause the refrigerant in the evaporator coil to remain a liquid and not vaporize. This causes low suction pressure because there is no vapor pressure being generated from vaporizing refrigerant. This liquid refrigerant will travel past the evaporator coil and eventually reach the compressor and cause compressor damage from flooding and slugging. Often, the system will have no evaporator or compressor superheat. The refrigerant in the suction line is what causes the suction line and compressor to get cold and sweat or frost. If liquid refrigerant enters the crankcase, it will mix with the oil and often vaporize, which will cause a cold and sweaty crankcase. Liquid refrigerant or a dense vapor will enter the suction valves—depending on whether the compressor is air- or refrigerant-cooled. In either case, the amp draw of the compressor will be high.

If a thermostatic expansion valve is starving the evaporator coil, the evaporator will also be very inefficient. This situation will cause low suction pressure with very high superheats. The low suction pressure will cause a low evaporator temperature—causing the suction line and compressor to frost and sweat also. Even though the suction pressure and evaporator temperatures are low, the refrigerant flow rate through the evaporator is also very low. This causes low capacities and a very starved and inefficient evaporator. The reduced heat load on the evaporator causes a reduced heat load on the condenser, and it will have, therefore, less heat to reject. This causes the head pressure to be low.

All of these can be identified with an evaporator performance check. This can be performed by using a superheat check to make sure that the evaporator has the correct amount of refrigerant, **Figure 29.19**. The heat exchange surface should be clean. The fans should be blowing enough air and not recirculating it from the discharge to the inlet of the coil.

Figure 29.19 Analyzing a coil for efficiency. This requires temperature and pressure checks at the outlet of the evaporator to calculate the superheat and air-to-refrigerant temperature difference. When a coil has the correct refrigerant level, it will absorb the correct amount of heat. The correct heat exchange is taking place if the refrigerant is boiling at 10°F to 20°F cooler than the entering air. Note that this occurs when the cooler temperature is within the design range, for example, 30°F to 45°F for medium temperature. Every circuit on coils with multiple circuits should be checked for even distribution.

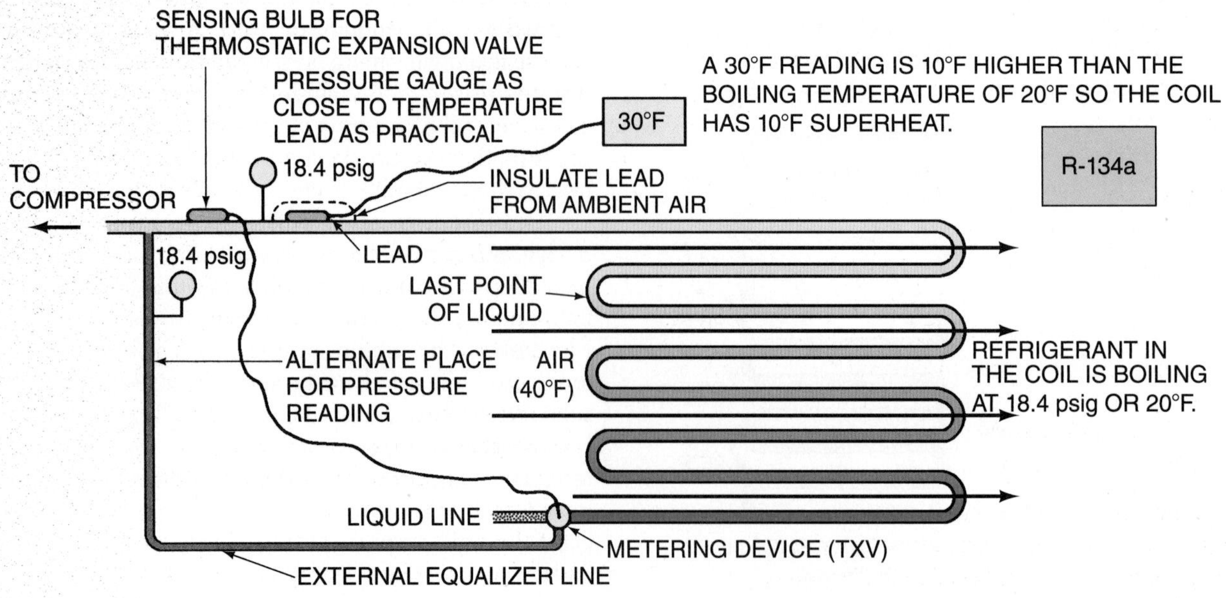

Figure 29.20 Water coils can be analyzed in much the same way as air coils. The refrigerant temperature and pressure at the end of the evaporator will indicate the refrigerant level. The refrigerant boiling temperature should not be more than 10°F cooler than the leaving water.

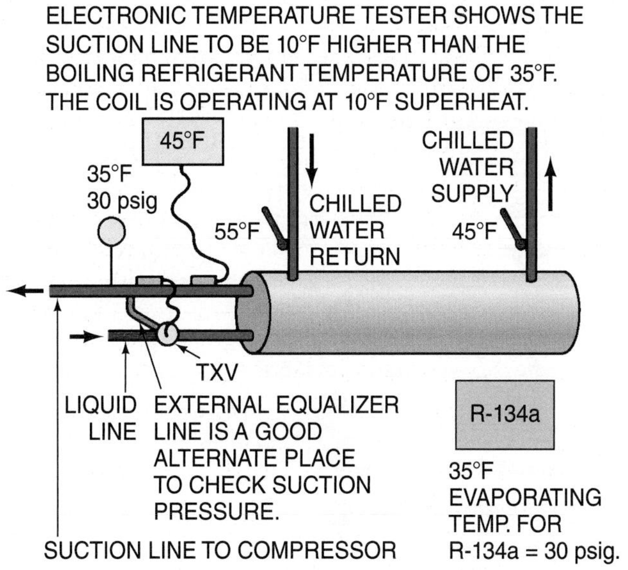

The refrigerant boiling temperature should not be more than 20°F colder than the entering air on an air evaporator coil, **Figure 29.2(A)**. A water coil should have no more than a 10°F TD in the boiling refrigerant and the leaving water, **Figure 29.20**. When the boiling refrigerant relationship to the medium being cooled starts increasing, the heat exchange is decreasing. The temperature difference between the evaporating temperature and the leaving-water temperature is referred to as the approach. The more efficient the heat exchange between the evaporating refrigerant and the water, the smaller the approach will be.

In summary, the symptoms of an inefficient evaporator are

- low suction pressure,
- sweaty or frosted suction line to the compressor,
- low head pressure,
- low superheat,
- a cold compressor crankcase, and
- high compressor amp draw.

29.11 INEFFICIENT CONDENSER

An inefficient condenser acts the same whether it is water-cooled or air-cooled. If the condenser cannot remove the heat from the refrigerant, the head pressure will go up. The condenser does three things and has to be able to do them correctly, or excessive pressures will occur.

1. Desuperheat the hot gas from the compressor. This gas may be 200°F or hotter on a hot day on an air-cooled system. Desuperheating is accomplished in the beginning of the coil.
2. Condense the refrigerant. This is done in the middle of the coil, the only place that the coil temperature will correspond to the head pressure. You can check the temperature against the head pressure if a correct temperature reading could be taken, but the fins are usually in the way.
3. Subcool the refrigerant before it leaves the coil. This subcooling is cooling the refrigerant to a point below the actual condensing temperature. A subcooling of 5°F to 20°F is typical, depending on the system. The subcooling can be checked just like the superheat, only the temperature is checked at the liquid line and compared with the high-side pressure converted to condensing temperature, **Figure 29.21**.

The condenser must have the correct amount of cooling medium (air or water). This medium must *not* be recirculated (mixed with the incoming medium) without being cooled. Ensure that air-cooled equipment is not located so that the air leaving the condenser circulates back into the inlet. This air is hot and will cause the head pressure to go up in proportion to the amount of recirculation. An air-cooled condenser located on a roof should be sufficiently elevated above the roof, even though the air comes in through the side. The temperature of the air coming directly off the roof is significantly higher than the ambient air above the roof. A clearance of about 18 in. will give better condenser performance, **Figure 29.22**. An air-cooled condenser that has a vertical coil may be influenced by prevailing winds. If the fan is trying to discharge its air into a 20-mph wind, it may not move the correct amount of air, and a high head pressure may occur, **Figure 29.23**.

Dirty condenser coils often result in inefficiency. Once the coil gets dirty, the condenser has a hard time rejecting heat. Since heat from the evaporator, suction line, compressor motor, and compression is rejected in the condenser, the condenser coil must be kept clean with the proper amount of airflow through it. If the condenser cannot reject heat fast enough, it will accumulate heat until its temperature and pressure are high enough to reject the heat to the surrounding ambient. The system is now operating at an elevated condensing temperature and pressure and causing the unwanted inefficiencies that result from high compression ratios.

Now that the system is running at elevated head pressures with unwanted inefficiencies, the suction pressure will be a bit high. Higher-than-normal suction pressures are caused by low refrigerant flow rates, and these are caused by the low volumetric efficiencies resulting from the high condensing pressures. The evaporator may not be able to keep up with the heat load, which creates higher-than-normal suction pressures. High head pressures cause high compression ratios and low volumetric efficiencies. High compression ratios will cause a greater pressure range in which the suction vapors can be compressed, requiring more work for the compressor and increasing the amp draw.

If the head pressure gets too high, the TXV may operate outside of its pressure ranges. The TXV may let too much

Figure 29.21 Checking the subcooling on a condenser. The condenser does three jobs: (1) It desuperheats the hot gas in the first part of the condenser, (2) it condenses the vapor refrigerant to a liquid in the middle of the condenser, and (3) it subcools the refrigerant at the end of the condenser. The condensing temperature corresponds to the head pressure. Subcooling is the temperature of the liquid line subtracted from the condensing temperature. A typical condenser can subcool the liquid refrigerant between 5°F and 20°F cooler than the condensing temperature.

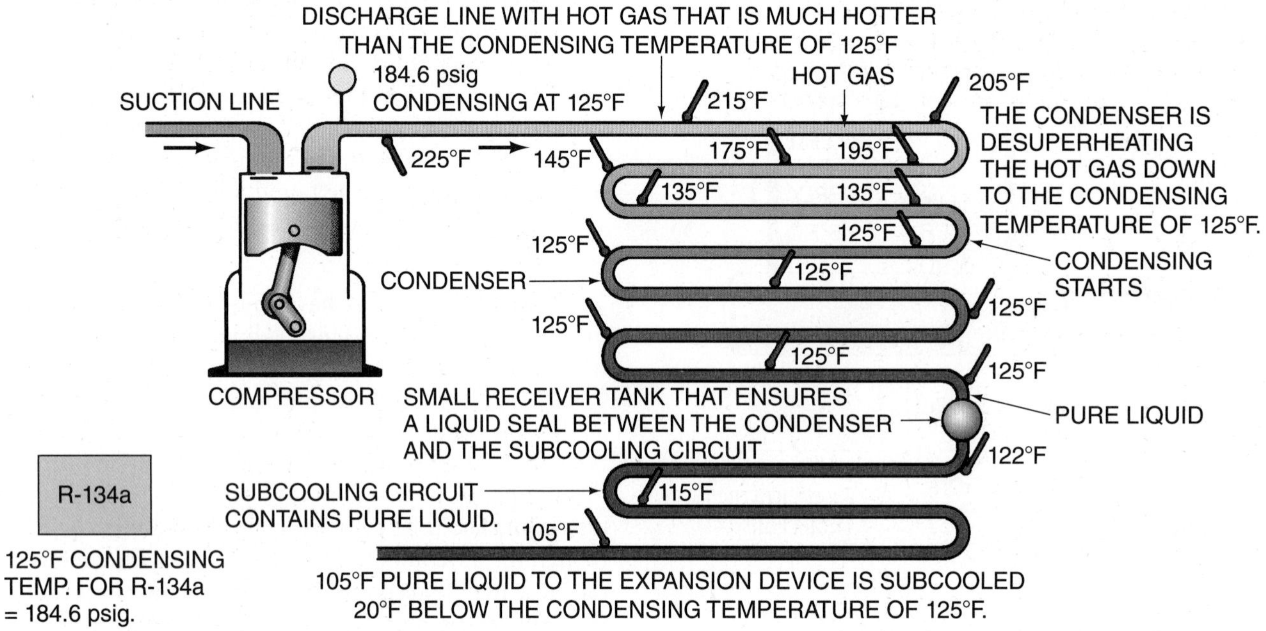

Figure 29.22 An air-cooled condenser should not sit directly on a roof. The temperature of the air coming directly off the roof is much warmer than the ambient air because the roof acts like a solar collector.

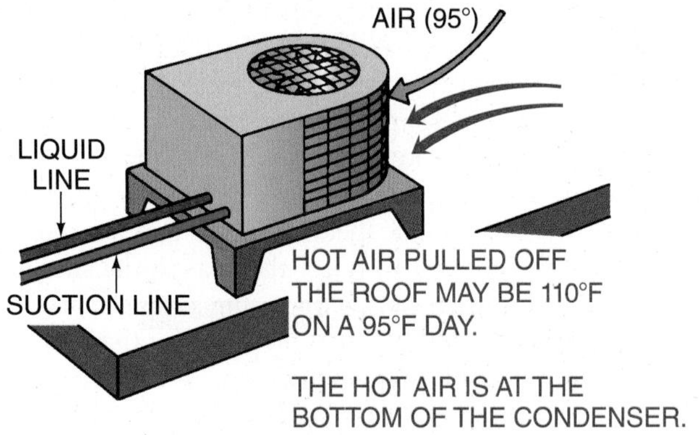

Figure 29.23 A condenser that is located in such a manner that it is discharging its air into a strong prevailing wind may not get enough air across the coil.

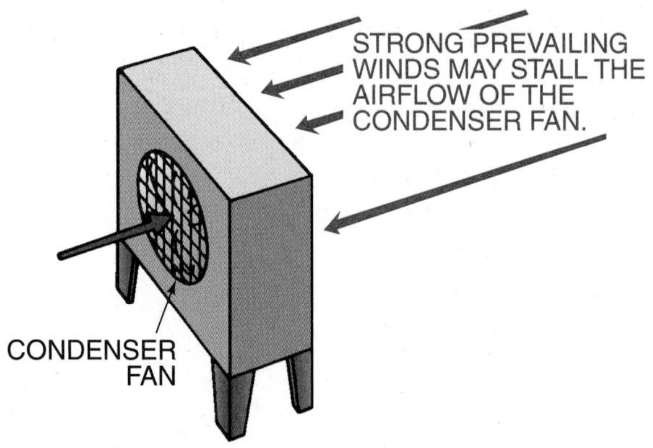

refrigerant into the evaporator during its opening stroke from the higher head pressures, causing higher-than-normal suction pressures. This action may also cause lower-than-normal evaporator superheats. Otherwise, because the TXV is maintaining superheat and doing its job, there may be normal evaporator pressures, depending on the severity of the condenser's condition.

What liquid subcooling is formed in the condenser will be at an elevated temperature and will reject heat to the ambient faster. Because of this faster heat rejection, the liquid in the condenser will cool faster and have a greater temperature difference as compared with the condensing temperature. Therefore, the condenser subcooling may be a bit high when the condensing temperature is high.

In summary, the symptoms of an inefficient condenser are

- high head pressure,
- normal to slightly high suction pressures,
- normal to slightly low evaporator superheats,
- a high compression ratio,
- high compressor amp draws, and
- normal-to-high condenser subcooling.

29.12 REFRIGERANT FLOW RESTRICTIONS

Restrictions that occur in the refrigeration circuit are either partial or full. A partial restriction may be in the vapor or the liquid line. A restriction always causes pressure drop at the point of the restriction. Different conditions will occur, depending on where the restriction is. Pressure drop can always be detected with gauges, but the gauges cannot always indicate the correct location of the restriction. Gauge ports may need to be installed for pressure testing.

If a restriction occurs due to something outside the system, it is usually physical damage, such as flattened or bent tubing. These can be hard to find if they are in hidden places, such as under the insulation or behind a fixture. If a partial restriction occurs in a liquid line, it will be evident because the refrigerant will undergo a pressure drop and will start to act like an expansion device at the point of the restriction. **When there is a pressure drop in a liquid line, there is always a temperature change.** A temperature check on each side of a restriction will pinpoint its location. Sometimes when the drop is across a drier, the temperature difference from one side to the other may not be enough to feel with bare hands, but a thermometer will detect it, **Figure 29.24**. A system that has been running for a long time may have physical damage from a restriction. If the restriction occurs soon after start-up, a filter or drier may be plugging up. When this occurs in the liquid-line drier, bubbles will appear in the sight glass when the drier is located before the sight glass, **Figure 29.25**.

Another cause of a partial restriction could be valves that do not open all the way. Normally, the TXV either works or it does not. It will function correctly, but if it loses its charge in the thermal bulb, it will close and cause a full restriction, **Figure 29.26**. There is a strainer at the inlet to most expansion devices to trap particles, and it may stop up slowly. If the device is a valve that can be removed, it can be inspected and

Figure 29.24 Each of the driers has a restriction. One of them is very slight. Where there is pressure drop in a liquid line, there is temperature drop. If the temperature drop is very slight, it can be detected with a thermometer. Sometimes gauges are not easy to install on each side of a drier to check for pressure drop. If possible, use the same gauge when checking the pressure drop through a drier to avoid inaccuracies.

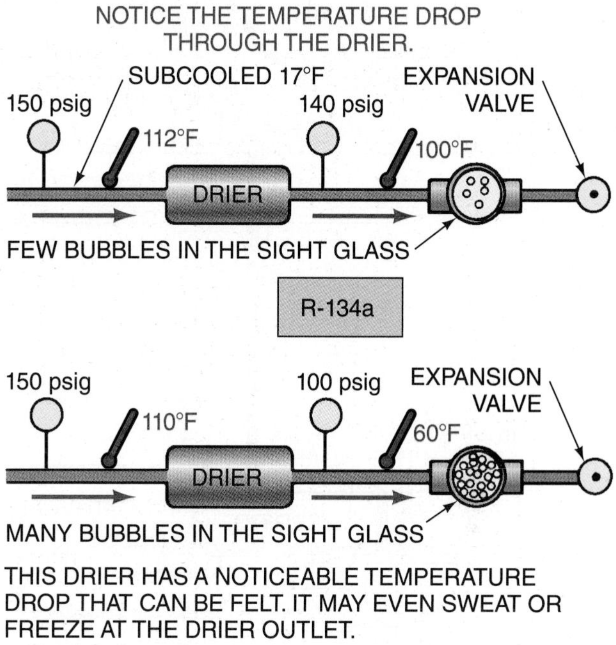

Figure 29.25 The restriction in a system may occur shortly after start-up. This indicates that solid contaminants from installing the system must be in the drier. If the restriction occurs after the system has been running for a long time, the restriction may be physical damage, such as a bent pipe. Normally, loose contamination will make its way through the system to the drier in a matter of hours.

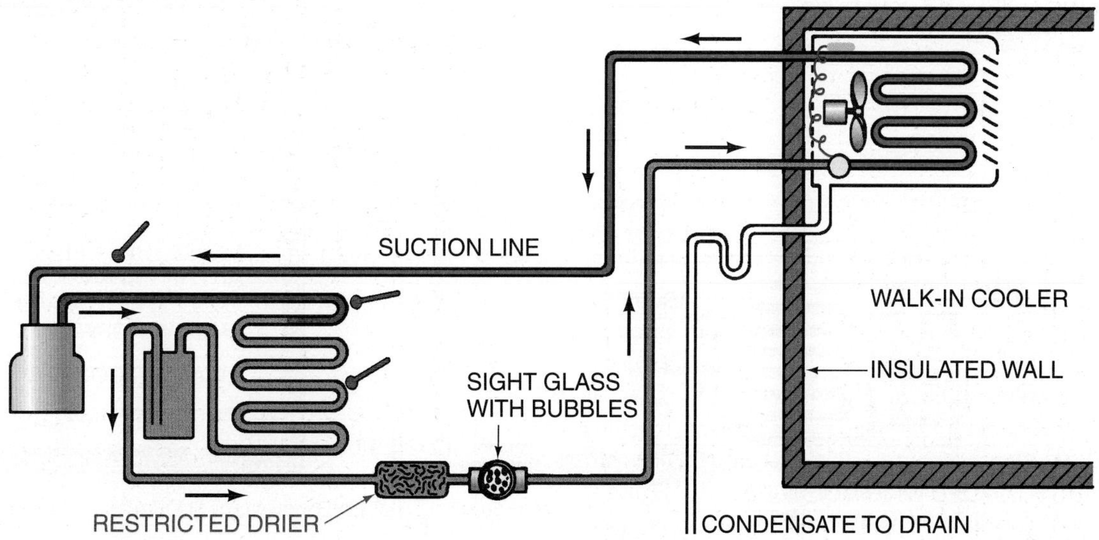

Figure 29.26 The charge in the bulb of a TXV is the only force that opens the valve, so when it loses its pressure the valve closes. The system can go into a vacuum if there is no low-pressure control to stop the compressor. A partial restriction can occur if the bulb loses part of its charge or if the inlet strainer stops up.

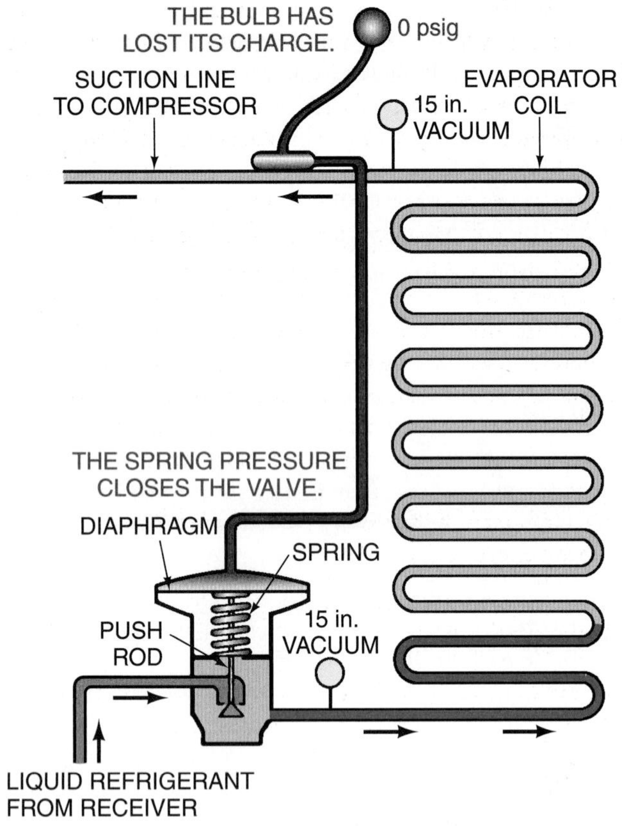

Figure 29.27 A capillary tube metering device with the strainer at the inlet. This strainer is soldered into the line and is not easy to service. Courtesy Parker Hannifin Corporation.

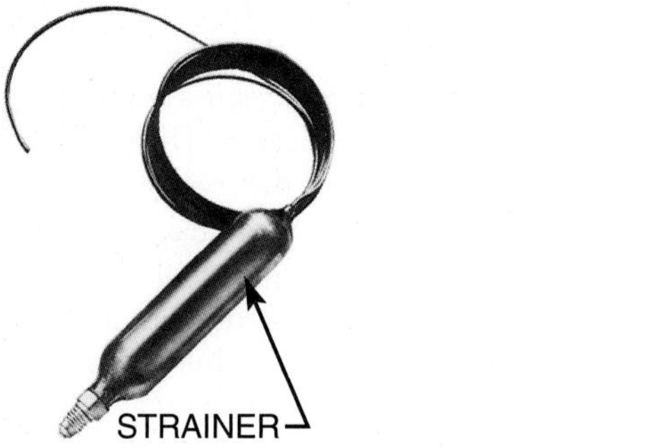

cleaned if necessary. If it is a capillary tube, it is brazed into the line and not easy to inspect, **Figure 29.27**.

Water circulating in any system that operates below freezing will freeze at the first cold place it passes through. This would be in the expansion device. One drop of water can stop a refrigeration system. Sometimes a piece of equipment that has just been serviced will show signs of moisture on the first hot day, **Figure 29.28**. This is because the drier in the liquid line will have more capacity to hold moisture when it is cool. On the first hot day, the drier may turn a drop or two of water loose, and it will freeze in the expansion device, **Figure 29.29**. When you suspect this, apply heat to thaw the ice to a liquid. **SAFETY PRECAUTION:** *Care must be used when applying heat. A hot, wet cloth is a good source.•* If applying a hot cloth to the metering device causes the system to start functioning properly, the problem is free water in the system. *Recover the refrigerant, change the drier, and evacuate the system. Recharge the system with the appropriate refrigerant.*

Figure 29.28 This system has had some moisture in it. The drier has all the moisture it can hold at mild temperatures. When the weather gets warm, the drier cannot hold all of the moisture and turns some of it loose. The moisture will freeze at the expansion device where the first refrigeration is experienced if the system operates below freezing.

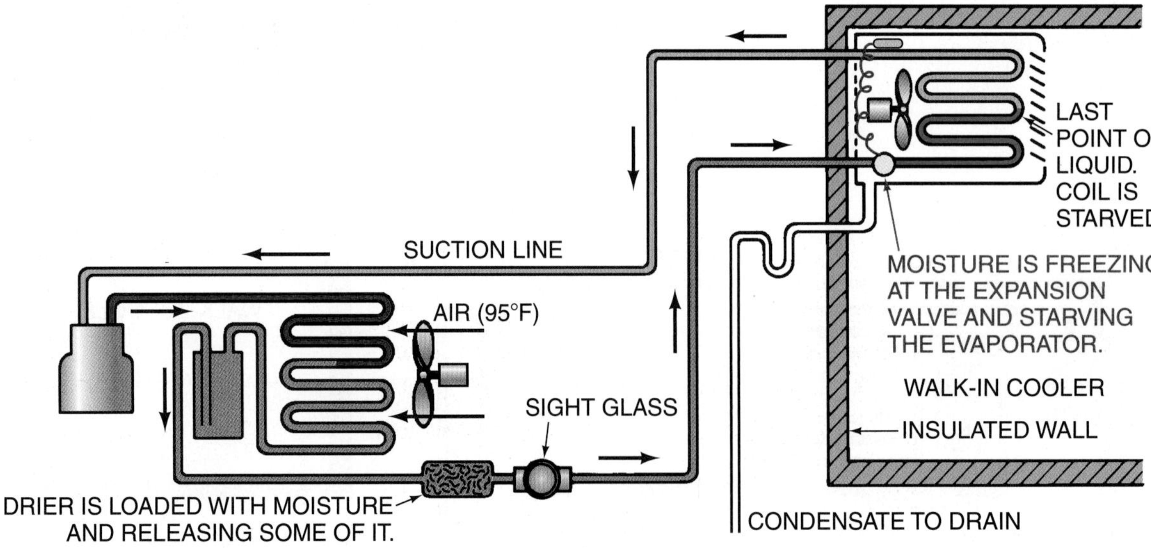

Figure 29.29 If it is suspected that moisture is frozen in the metering device, mild heat can be added. Heat from a hot, wet cloth will normally thaw the moisture out, and the system will start refrigerating again.

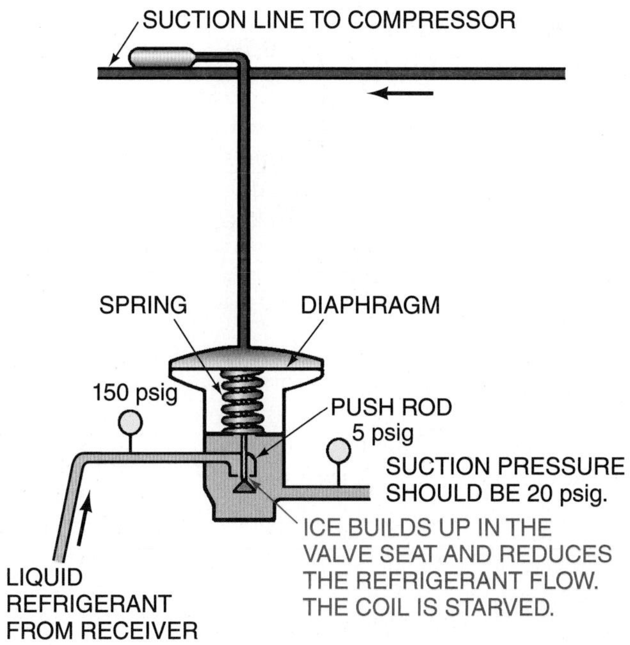

Other components that may cause restrictions are automatic valves in the lines, such as the liquid-line solenoid, crankcase pressure regulator, or the evaporator pressure regulator. These valves may easily be checked with gauges applied to both sides where pressure taps are provided, **Figure 29.30**. Restrictions in the liquid line usually occur after the receiver and will cause the evaporator to be starved of refrigerant, resulting in a low suction pressure. High superheats will also be experienced because of the starved evaporator. Because the evaporator is starved of refrigerant, there will not be much heat transferred to it from its surroundings. Because of the highly superheated vapors coming into it, the compressor will also be starved of refrigerant and will run hot and have a low amp draw. Highly superheated vapors mean low-density vapors for the compressor to compress.

Because the evaporator and compressor are starved of refrigerant, the condenser will also be starved. The condenser will not have a lot of heat to reject because the evaporator did not absorb much heat. This will cause a low condensing (head) pressure. If the restriction is ahead of the sight glass, the sight glass may bubble, depending on the severity of the restriction.

In summary, the symptoms of a restricted liquid line after the receiver are

- low suction pressures,
- high superheats,

Figure 29.30 (A–H) Valve components that can close are all capable of causing a restriction. (A, D, E, G, H) Courtesy Sporlan Division, Parker Hannifin Corporation. (C, F) Courtesy Ferris State University. Photos by John Tomczyk.

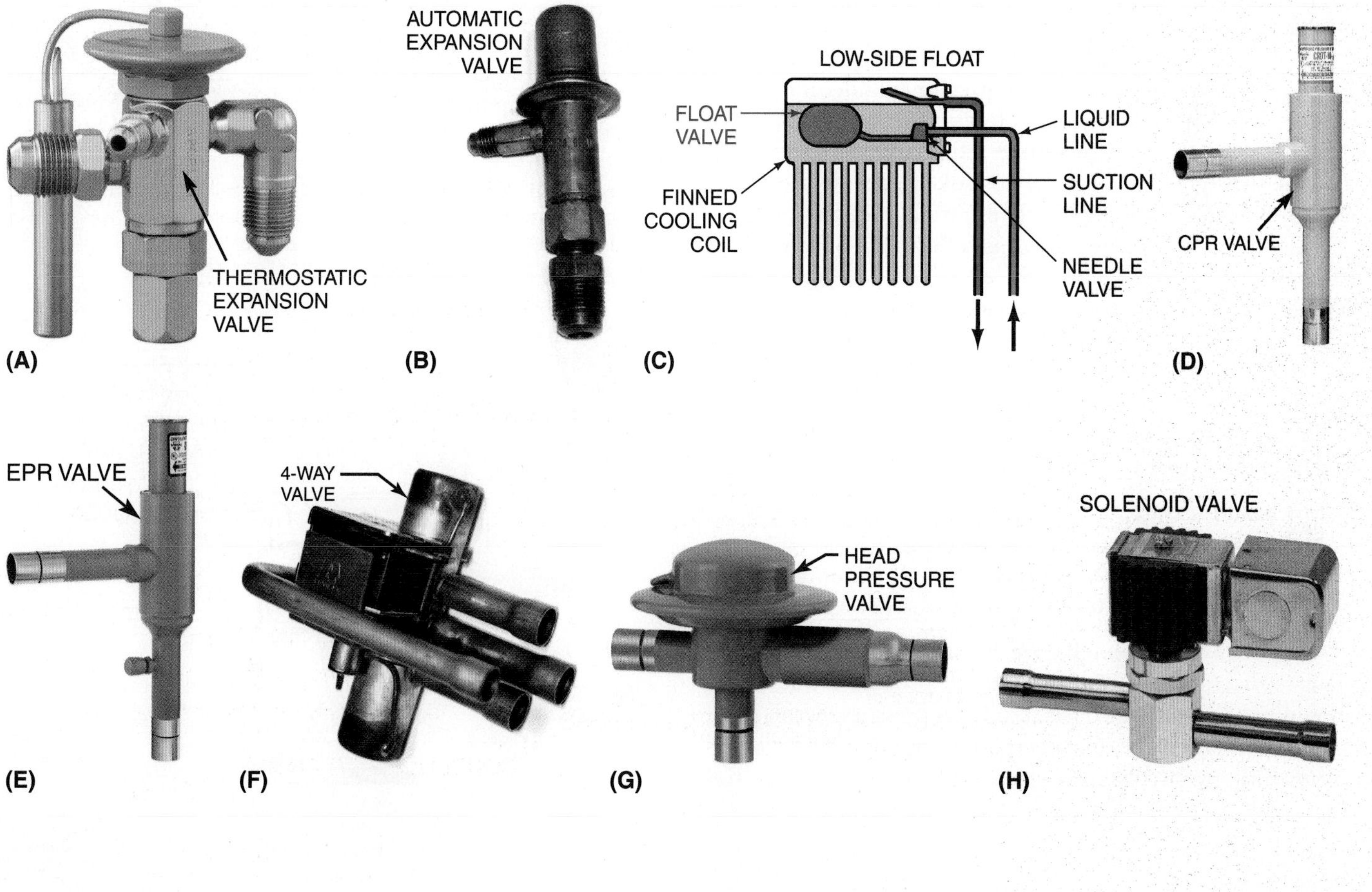

- an overheated compressor,
- low amp draw,
- low head pressure,
- bubbles in the sight glass, and
- a temperature drop in the liquid line across the restriction.

29.13 INEFFICIENT COMPRESSOR

Inefficient compressor operation can be one of the most difficult problems to find. When a compressor will not run, it is evident where the problem is. Motor troubleshooting procedures are covered in Unit 20, "Troubleshooting Electric Motors." When a compressor is pumping at slightly less than capacity, it is hard to determine the problem. It helps at this point to remember that a compressor is a vapor pump. It should be able to create a pressure from the low side of the system to the high side under design conditions.

If a reciprocating compressor is inefficient because its valves are bad, the suction pressure will be high and the head pressure will be low. There is no other situation that will give a system both low head and high suction pressures at the same time, other than worn piston rings causing blow-by of gases around the rings. If a discharge valve is not seating properly, a low head pressure will result. Refrigerant vapor will be forced out of the cylinder and into the discharge line during the upstroke of the compressor. On the downstroke, this same refrigerant that is in the discharge line will be drawn back into the cylinder because the discharge valve is not seating properly. This short cycling of refrigerant will cause high discharge temperatures out of the compressor and low head (condensing) pressures. Refrigerant vapors will be drawn from the suction line into the compressor cylinder during the downstroke of the compressor. However, if the suction valve is not seating properly, this same refrigerant may leak back into the suction line during the upstroke, reducing the refrigerant flow and causing higher-than-normal suction pressures.

Bad valves will cause a low mass flow rate of refrigerant through the system, which will usually cause higher-than-normal superheats on the low side of the system. System capacities will be low—with higher-than-normal box or conditioned space temperatures. Because of the reduced refrigerant flow rates, compressor amp draws will be lower.

In summary, the symptoms of an inefficient compressor due to bad valves are

- low head pressure,
- high suction pressure,
- higher-than-normal superheats, and
- lower-than-normal compressor amp draws.

The following methods are all used by service technicians to diagnose compressor pumping problems. These tests should not be performed on scroll compressors. Scroll compressors should not be operated under a vacuum, they will fail to lubricate properly and overheat very quickly. Most scroll compressors have some internal mechanism, low-pressure cut out or bypass, to prevent them from operating in a vacuum.

29.14 COMPRESSOR VACUUM TEST

The compressor vacuum test is usually performed on a test bench with the reciprocating compressor removed from the system. This test may be performed within the system when the system has service valves. **NOTE:** *Care should be taken not to pull air into the system while in a vacuum. Make sure that there is no residual refrigerant mixed in the oil that will affect the outcome of the test. Do this by running the system for 30 min or leaving the crankcase heater on for several hours.* •

All reciprocating compressors should immediately go into a vacuum if the suction line is valved off when the compressor is running. This test proves that the suction valves are seating correctly on at least one cylinder. This test is *not satisfactory* on a multicylinder compressor. If one cylinder pumps correctly, a vacuum will be pulled. A reciprocating compressor should pull 26 in. to 28 in. Hg vacuum with the atmosphere as the discharge pressure, **Figure 29.31**. The compressor should pull about 24 in. Hg vacuum against 100 psig discharge pressure, **Figure 29.32**. When the compressor has pumped a differential pressure and is stopped, the pressures should not equalize. For example, when a compressor that has been operated until the suction pressure is 24 in. Hg vacuum and the head pressure is 100 psig is stopped, it should have the same pressures while it is off. When refrigerant is used for this pumping test, the 100 psig will drop some because of the condensing refrigerant. Nitrogen is a better choice to pump in the test because the pressure will not drop.

Figure 29.31 A compressor pulling a vacuum with the atmosphere as the head pressure. Most reciprocating compressors can pull 26 in. to 28 in. Hg vacuum with the atmosphere as the discharge pressure.

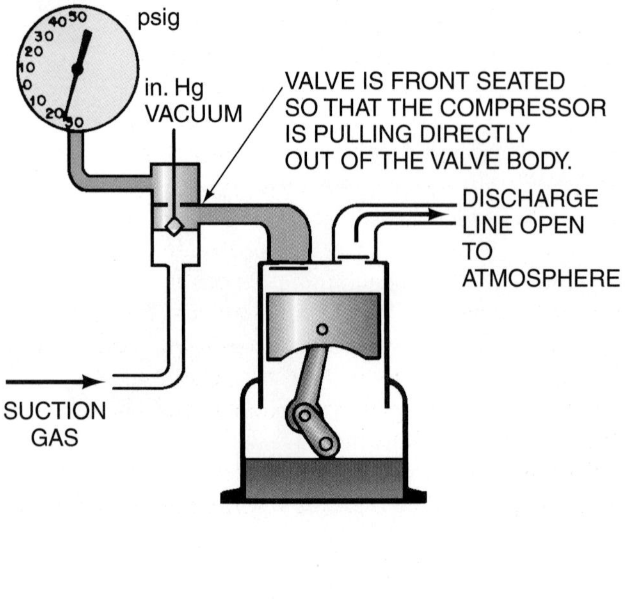

Figure 29.32 Perform a compressor check by pulling a vacuum while pumping against a head pressure. A reciprocating compressor normally can pull 24 in. Hg vacuum against 100 psig head pressure.

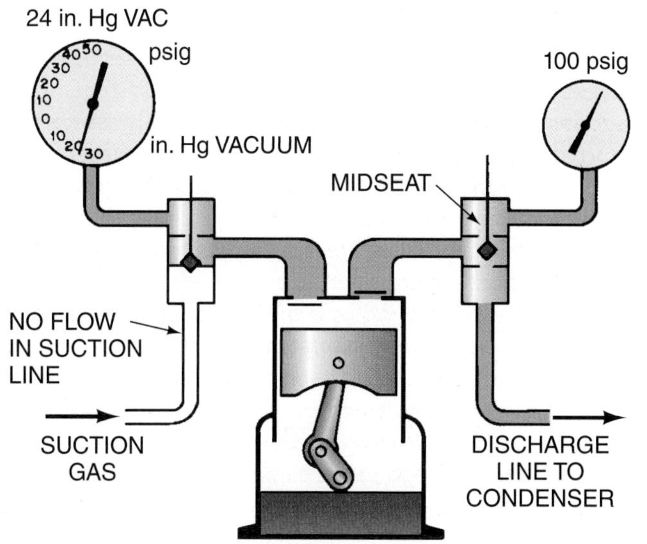

SAFETY PRECAUTION: *Care should be taken when operating a hermetic compressor in a deep vacuum (below 1000 microns) because the motor is subject to damage. This low vacuum cannot be obtained with a reciprocating compressor. Also, most compressor motors are cooled with suction gas and will get hot if operated for any length of time while performing these tests. This vacuum test should not take more than 3 to 5 min. The motor will not overheat in this period of time. The test should be performed only by experienced technicians.•*

29.15 CLOSED-LOOP COMPRESSOR RUNNING BENCH TEST

Doing a running bench test on a reciprocating compressor can be accomplished by connecting a line from the discharge to the suction of the compressor and operating the compressor in a closed loop. A difference in pressure can be obtained with a valve arrangement or gauge manifold. This will prove the compressor can pump. **NOTE:** *When the compressor is hermetic, it should operate at close to full-load current in the closed loop when design pressures are duplicated.•* Nitrogen or refrigerant can be used as the gas to be compressed. Typical operating pressures will have to be duplicated for the compressor to operate at near the full-load current rating. For example, for a medium-temperature compressor, a suction pressure of 18 psig and a head pressure of 184.6 psig will duplicate a typical condition for R-134a on a hot day. The compressor should operate close to nameplate full-load current when the design voltage is supplied to the motor.

The following is a step-by-step procedure for performing this test. Use this procedure with the information in **Figure 29.33**.

SAFETY PRECAUTION: *This test should be performed only under the close supervision of an instructor or by a qualified person and on equipment under 3 hp. Safety goggles must be worn. This test should not be performed on scroll compressors.•*

Steps for performing a closed-loop test with the reciprocating compressor removed from the system:

1. Use the gauge manifold and fasten the suction line to the low-pressure gauge line in such a manner that the compressor is pumping only from the gauge line.
2. Fasten the discharge gauge line to the discharge valve port in such a manner that the compressor is pumping only into the gauge manifold.

Figure 29.33 Checking a compressor's pumping capacity by using a closed loop. The test is accomplished by routing the discharge gas back into the suction port with a piping loop. The discharge gauge manifold valve is gradually throttled toward closed (do not entirely close). It can never be closed because tremendous pressure will occur. When the design suction and discharge pressures are reached, the compressor should be pulling close to nameplate full-load amperage. *A suction-cooled compressor cannot be run for long in this manner or the motor will get hot.* **SAFETY PRECAUTION:** *This test should only be performed by experienced technicians.•*

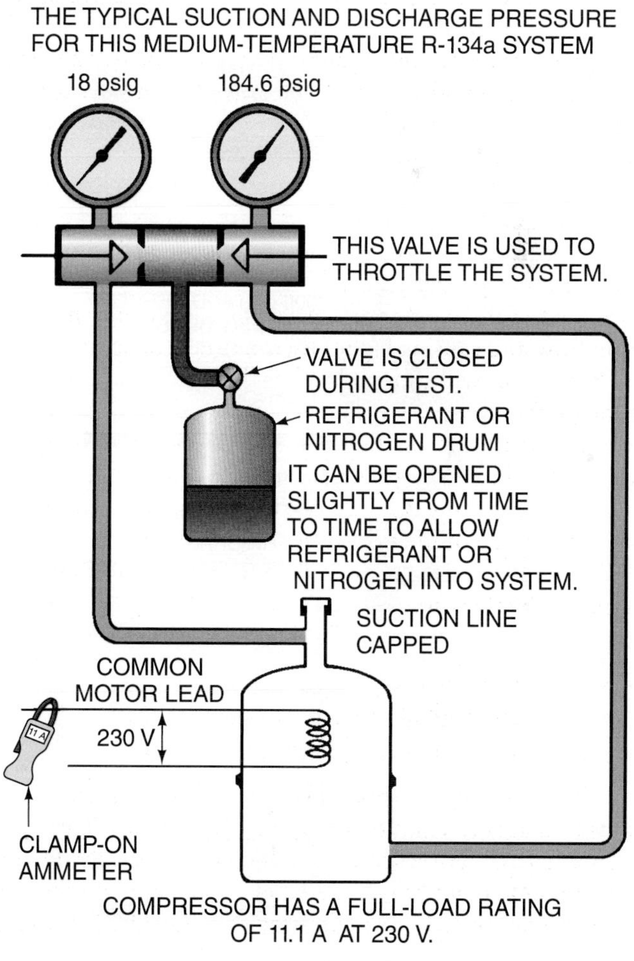

3. Plug the center line of the gauge manifold.
4. **⚠ SAFETY PRECAUTION:** *Both gauge manifold valves should be wide open, counterclockwise.•*
5. **⚠ SAFETY PRECAUTION:** *Start the compressor; keep your hand on the off switch.•*
6. The compressor should now be pumping out the discharge line and back into the suction line. The discharge gauge manifold valve may be slowly throttled (*do not close entirely*) toward closed until the discharge pressure rises to the design level and the suction pressure drops to the design level.
7. When the desired pressures are reached, the amperage reading on the compressor motor should compare closely to the full-load amperage of the compressor. If the correct pressures cannot be obtained with the amount of gas in the compressor, a small amount of gas may be added to the loop system by attaching the center line to a nitrogen cylinder and slightly opening the cylinder valve.
8. The amperage may vary slightly from full load because of the input voltage. For example, a voltage above the nameplate will cause an amperage below full load, and vice versa.

Figure 29.34 illustrates a situation in which a technician accidentally closed the valve completely.

Figure 29.34 **⚠ SAFETY PRECAUTION:** *What happens if a compressor is started up in a closed loop with no place for the discharge gas to go (discharge gauge manifold closed by accident)? One cylinder full of gas could be enough to build tremendous pressures.•*

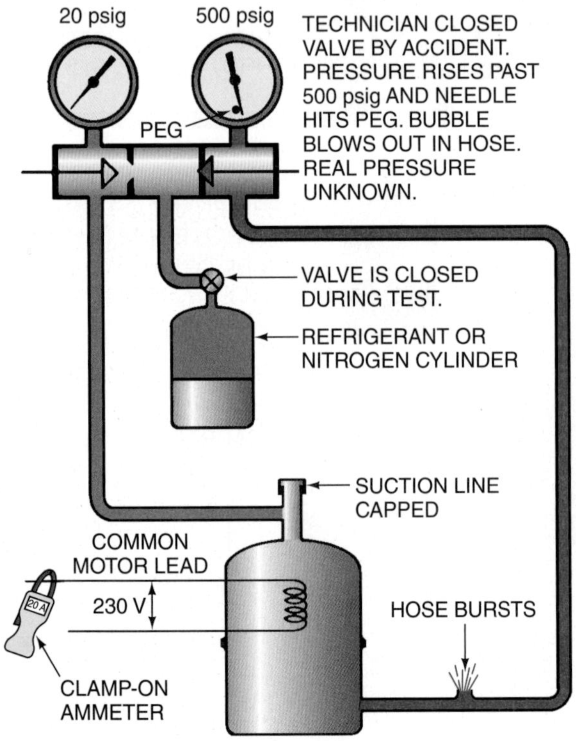

29.16 CLOSED-LOOP COMPRESSOR RUNNING FIELD TEST

When a compressor has suction and discharge service valves, this test can be performed in place in the system using a gauge manifold as the loop, **Figure 29.35**. The compressor is started with the compressor service valves turned all the way to the front seat and the gauge manifold valves open. The center line on the gauge manifold must be plugged. **⚠ SAFETY PRECAUTION:** *This can be a dangerous start-up and should be performed only under supervision of an experienced person. The compressor has to have a place to pump the discharge gas because reciprocating compressors are positive displacement pumps. When the compressor cylinder is full of gas, it is going to pump it somewhere or stall. In this test the gas goes around through the gauge manifold and back into the suction. Should the compressor be started before the gauge manifold is open for the escape route, tremendous pressures will result before you can stop the compressor.*

Figure 29.35 A compressor test being performed in a system using a gauge manifold as a closed loop. Notice that this test can be performed in the system because the compressor has service valves. Do not perform this test without experienced supervision.

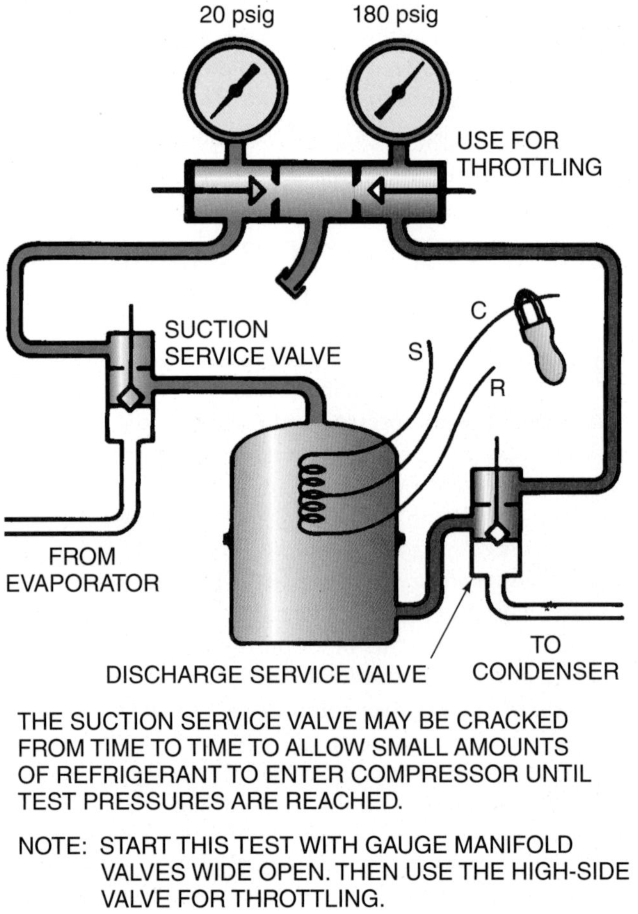

It takes only one cylinder full of vapor to fill the gauge manifold to more than capacity. This test should only be performed on compressors under 3 hp. Safety goggles must be worn.•

Steps for performing the closed-loop compressor test with the compressor in the system:

1. Turn the unit OFF and fasten the suction line of the gauge manifold to the suction service valve. Plug the center line of the gauge manifold.
2. Turn the suction service valve stem to the front seat.
3. Fasten the discharge line to the discharge valve and turn the discharge service valve stem to the front seat.
4. **SAFETY PRECAUTION:** *Open both gauge manifold valves all the way, counterclockwise.•*
5. **SAFETY PRECAUTION:** *Start the compressor and keep your hand on the switch.•*
6. The compressor should now be pumping out through the discharge port, through the gauge manifold, and back into the suction port.
7. The discharge gauge manifold valve may be throttled toward the seat to restrict the flow of refrigerant and create a differential in pressure. Throttle (*do not entirely close*) the valve until the design head and suction pressure for the system is attained (the compressor should then be pulling near to full-load amperage). As in the bench test, the amperage may vary slightly because of the line voltage.

29.17 COMPRESSOR RUNNING TEST IN THE SYSTEM

A running test in the system can be performed by creating typical design conditions in the system. Typically, a compressor will operate at a *high suction* and a *low head pressure* when it is not pumping to capacity. This will cause the compressor to operate at a low current. When the technician is called to a job, the conditions are not usually at the design level. The fixture is usually not refrigerating correctly—this is what instigated the call to begin with. The technician may not be able to create design conditions, but the following approach should be tried if the compressor capacity is suspected:

1. Install the high- and low-side gauges.
2. Make sure that the charge is correct (not over or under), using manufacturer's recommendations.
3. Check the compressor current and compare to full load.
4. Block the condenser airflow and build up the head pressure.

If the compressor will not pump the head pressure up to the equivalent of a 95°F day (95°F + 30°F = 125°F condensing temperature, or 185 psig for R-134a, 278 psig for R-22, and 332 psig for R-404A) and draw close to the nameplate full-load current, the compressor is not pumping. When the compressor is a sealed hermetic type, it may whistle when it is shut down. This whistle is evidence of an internal leak from the high side to the low side.

If the compressor has service valves, a closed-loop test can be performed on the compressor using the methods we have explained while it is in the system. **NOTE:** *Make sure no air is drawn into the system while it is in a vacuum.•* **SAFETY PRECAUTION:** *The compressor temperature should be monitored any time these tests are being conducted. If the compressor gets too warm, it should be stopped and allowed to cool. If the compressor has internal motor-temperature safety controls, do not operate it when the control is trying to stop the compressor.•*

HVACR GOLDEN RULES

When making a service call to a business:

- Never park your truck or van in a space reserved for customers.
- Look professional and be professional.
- Before starting troubleshooting procedures, get all the information you can regarding the problem.
- Be extremely careful not to scratch tile floors or to soil carpeting with your tools or by moving equipment.
- Be sure to practice good sanitary and hygiene habits when working in a food preparation area.
- Keep your tools and equipment out of the customers' and employees' way if the equipment you are servicing is located in a normal traffic area.
- Be prepared with the correct tools and ensure that they are in good condition.
- Always clean up after you have finished. Try to provide a little extra service by cleaning filters, oiling motors, or providing some other service that will impress the customer.
- Always discuss the results of your service call with the owner or representative of the company. Try to persuade the owner to call if there are any questions as a result of the service call.

PREVENTIVE MAINTENANCE FOR REFRIGERATION

PACKAGED EQUIPMENT. Packaged equipment is built and designed for minimum maintenance because the owner may be the person that takes care of it until a breakdown occurs. Most of the fan motors are permanently lubricated and will run until they quit, at which time they are replaced with new ones.

The owners should be educated to keep the condensers clean and not to stack inventory so close as to block the condenser airflow. When the unit is a reach-in cooler, the owner should be cautioned to follow the manufacturer's directions in loading the box. The load line on the inside should be observed for proper air distribution.

The owner or manager should examine each refrigerated unit frequently and be aware of any peculiar noises or actions from the box. Each box should have a thermometer that should be monitored each day by management. Any rise in temperature that does not reverse itself should alert the manager that a problem is occurring. This can prevent unnecessary loss of perishable foods.

The electrical service for all package equipment should be visually inspected for frayed wires and overheating. Power cord connections may become loose and start to build up heat. If the end that plugs into the wall receptacle becomes hot, the machine should be shut down until the problem is found and corrected. **⚠ SAFETY PRECAUTION:** *If only the plug is replaced, the wall receptacle may still be a problem. The whole connection, the wall receptacle and the cord plug, should be inspected and repaired.*•

Ice machines require special attention if they are to be reliable. Management should know what the quality of the ice should be when the machine is operating correctly. When the quality begins to deteriorate, the reason should be found. This can be as simple as looking for the correct water flow on a cube maker or the correct ice-cutting pattern on another type of cube maker. The ice falling into the bin of a flake maker should have the correct quality, neither soft and mushy nor brittle and hard. The level of the ice in the bin the morning after the machine has been left to run all night can reveal if the machine is making enough ice. For example, if it is observed that the bin is full every morning and the bin thermostat is satisfied but then one morning the bin is only half full, trouble can be suspected. Keep a close watch on the machine.

Drains and drain lines for all refrigeration equipment should be maintained and kept clean and free. This can normally be accomplished when the unit is cleaned and sanitized on the inside. When the cleaning water is flushed down the drain, the speed with which the drain moves the water should indicate if the drain is partially plugged or draining freely.

SPLIT-SYSTEM REFRIGERATION EQUIPMENT, EVAPORATOR SECTION, CONDENSER SECTION, AND INTERCONNECTING PIPING. In refrigeration split systems, the evaporator section is located in one place and the condensing unit in another. The interconnecting piping makes the system complete. The evaporator section contains the evaporator, metering device, fan motor, defrost heaters, drain pan, and drain line. Refrigeration evaporators do not need cleaning often but do need to be inspected and cleaned regularly. The technician cannot always tell when a coil is dirty by looking at the evaporator. Grease or dirt may be in the core of the coil. Routine cleaning of the evaporator once a year will usually keep the coil clean. **⚠ SAFETY PRECAUTION:** *Use only approved cleaning compounds where food is present. Turn off the power before cleaning any system. Cover the fan motors and all electrical connections when cleaning to prevent water and detergent from getting into them.*•

The motors in the evaporator unit are usually sealed and permanently lubricated. If not, they should be lubricated at recommended intervals, which are often marked on the motor. Observe the fan blade for alignment and look for bearing wear. This may be found by lifting the motor shaft. Most small motors have considerable end play. Do not mistake this for bearing wear. You must lift the shaft to discover bearing wear. The fan blade may also gather weight from dirt and become out of balance.

All wiring in the evaporator section should be visually inspected. If it is cracked or frayed, shut off the power and replace it. Do not forget the defrost heaters and any heater tapes that may be used to keep the drain line warm during cold weather.

The entire case should be cleaned and sanitized at regular intervals. You may need to be the judge of these intervals. Do not let the unit become dirty, including the floor and storage racks in walk-in coolers. **⚠ SAFETY PRECAUTION:** *Keep all ice off the floor of walk-in coolers, or an accident may occur.*•

The condensing unit may be located inside an equipment room in the store or it may be outside. When the condenser is located inside an equipment room, it must have proper ventilation. Most equipment rooms have automatic exhaust systems that turn fans on when the equipment room reaches a certain temperature. In large equipment rooms, the temperature will stay warm enough for the exhaust fan to run all the time.

Condensers in equipment rooms become dirty just as they do outdoors and must be cleaned. Most system condensers should be cleaned at least once a year to ensure the best efficiency and to prevent problems. The condensers may be cleaned with an approved condenser cleaner and then washed. **⚠ SAFETY PRECAUTION:** *Turn off the power before cleaning any system. All motors and wiring must be covered to prevent them from getting wet. Watch the equipment room floor. If oil is present and water gets on it, it will become slippery.*•

The technician can identify many future problems by close inspection of the equipment room. Leaks may be found by observing oil spots on piping. Touch testing the various components will often tell the technician that a compressor is operating too hot or too cold. Tape-on temperature indicators may be fastened to discharge lines to indicate the highest temperature the discharge line has ever reached. For example, most manufacturers would consider that oil inside the compressor will begin to break down at discharge temperatures above 250°F. A tape-on temperature indicator will tell if high temperatures are occurring when no one is around, such as a defrost problem in the middle of the night. This can lead to a search for a problem that is not apparent.

(Continued)

PREVENTIVE MAINTENANCE FOR REFRIGERATION (*Continued*)

The crankcase of a compressor should not be cold to the touch below the oil level. This is a sure sign that liquid refrigerant is in the refrigerant oil on some systems, not necessarily in great amounts, but enough to cause diluted refrigerant oil. Diluted refrigerant oil causes marginal lubrication and bearing wear.

Fan motors may be inspected for bearing wear. Again, do not mistake end play for bearing wear. Some fan motors require lubrication at the correct intervals.

A general cleaning of the equipment room at regular intervals will ensure the technician good working conditions when a failure occurs.

The interconnecting piping should be inspected for loose insulation and oil spots (indicating a leak) and to ensure the pipe is secure. Some piping is in trenches in the floor. A trench that is full of water because of a plugged floor drain will hurt the capacity of the equipment because the insulation value of the suction-line insulation is not as great when wet. A heat exchange between the liquid lines and suction lines or the ground and suction lines will occur. The pipe trenches should be kept as clean as practical.

29.18 DIAGNOSTIC CHART FOR COMMERCIAL REFRIGERATION

Commercial refrigeration equipment that controls the storage temperatures for perishable products must be kept operating properly to prevent spoilage. The technician must restore the system to good working order as quickly as possible. Orderly troubleshooting is the most effective way to arrive at sound conclusions in a short time. Following are some recommendations for troubleshooting commercial refrigeration systems:

1. Listen to the customer's description of the system's symptoms. Most commercial refrigeration equipment customers monitor the system and the temperatures carefully. Question the customer about sounds, cycle times, and temperatures on normal days, and ask what is different now.
2. Observation may be one of the most important steps you can take. Inspect the complete system before making any adjustments or repairs. Check the condenser for dirt, leaves, or trash in the coil or fan. Look for blocked airflow. Inspect the wiring for discoloration of connections. Look for loose or missing insulation on the suction line or kinks or flattened tubing. Check the evaporator for air blockage due to frost, ice, or dirt. Make sure the fan is circulating air.
3. Once the customer has been consulted and the system has been observed, determine what is functioning and what is not functioning. The following chart will help to determine possible causes of the problem.

Problem	Possible Cause	Possible Repair
Compressor will not start or attempt to start; makes no sound	Open disconnect switch	Close disconnect switch.
	Open fuse or breaker	Replace fuse or reset breaker and determine why it opened.
	Tripped overload	Reset and determine why it tripped.
	Pressure control stuck in open position	Reset or replace and determine why it opened.
	Open internal or external overload	Allow to cool down.
	Open coil on motor starter	Check coil and replace if needed.
	Faulty wiring	Repair or replace faulty wiring connectors.
Compressor will not start; hums and trips on overload	Incorrect wiring, **Single Phase**	Check common, run, and start connections.
	High discharge pressure	Overcharge of refrigerant, not enough condenser cooling, closed discharge service valve.
	Incorrect wiring, **Three Phase**	Check all three phases, phase-to-phase.
	Start relay wired wrong or defective, **Single Phase**	Rewire relay if needed or replace if defective.
	Start or run capacitor defective	Replace capacitor.
	Compressor winding open or shorted	Replace compressor.
	Internal compressor problems, stuck or tight	Replace compressor.

(*Continued*)

Problem	Possible Cause	Possible Repair
Compressor starts but stays in start, **Single Phase**	Wired incorrectly Low-voltage supply Defective start relay Tight or binding compressor	Check wiring as compared with diagram. Correct low voltage. Replace relay. Replace compressor.
Compressor starts and runs for a short time and then shuts off due to overload	Defective overload Excess current in overload circuit Low voltage Unbalanced voltage, **Three Phase** Defective run capacitor Excessive load on compressor A. HIGH SUCTION PRESSURE: Defrost heater on all the time B. Hot food placed in cooler C. HIGH DISCHARGE PRESSURE: Recirculating condenser air Reduced airflow Dirty condenser Compressor windings shorted	Check actual load on overload and replace if it trips below specifications. Check circuit for added load, such as fans or pumps. Determine reason and correct. Correct unbalance; possible redistribution of loads. Replace. Check and repair defrost control circuit. Operator education. Move object causing recirculation. Remove blockage, make sure all fans are on. Clean condenser. Replace compressor.
Compressor starts but short cycles	Overload protector shutting compressor off Thermostat High-pressure condition Low-pressure condition with pumpdown control A. Solenoid valve leaking B. Compressor discharge valve leaking during off cycle (will also leak while running) With or without pumpdown control A. Low refrigerant charge B. Restriction in system such as expansion valve, drier, or crimped pipe	See category above. Differential too close; thermostat setting; bulb in cold airstream, move location. Check airflow or water flow. Air blockage or recirculating; correct. Overcharge; remove refrigerant. Air in system; remove. Replace solenoid. Repair valve or replace compressor. Repair leak and add refrigerant. Repair or replace.
Compressor runs continuously or space temperature is too high	Low refrigerant charge Operating control (thermostat or low-pressure control) contacts stuck closed Excessive load A. Unit undersized B. Door of cooler open too often C. Product too hot when placed in cooler D. Poor door gaskets or infiltration E. Cooler setting in too hot a location Evaporator coil not defrosting Restriction in refrigerant circuit, drier, metering device, crimped pipe Poor condenser performance A. Restricted airflow B. Recirculated air C. Dirty condenser	Repair leak; add refrigerant. Replace control. A. Reduce load or replace unit. B. Train operator. C. Precool food in air-conditioned space. D. Seal air leaks. E. Move or reduce heat. Check defrost method and repair. Find and repair. Restore airflow. Move objects causing recirculation. Clean condenser.

29.19 SERVICE TECHNICIAN CALLS

In addition to the six problems typically encountered in refrigeration systems, many more problems are not so typical. The following service situations will help you understand troubleshooting. Most of these service situations have already been described, although not as actual troubleshooting procedures. Sometimes the symptoms do not describe the problem and a wrong diagnosis is made. Do not draw any conclusion until the whole system has been examined. Become a system doctor. Examine the system and say, "This needs further examination," and then do it. Refrigeration systems often cool large amounts of food so they are slow to respond. The temperature may drop very slowly on a hot or warm pulldown.

SERVICE CALL 1

A customer calls and complains that the compressor for a medium-temperature walk-in cooler with a remote condensing unit is short cycling and not cooling correctly. *The evaporator has two fans, and one is burned. The unit is short cycling on the low-pressure control because there is not enough load on the coil.*

On the way to the job the technician goes over the possible problems. This is where it helps to have some familiarity with the setup. The technician remembers that the unit has a low-pressure control, a high-pressure control, a thermostat, an overload, and an oil safety control. Defrost is accomplished with an off cycle using the air in the refrigerated space with the fans running. By the process of elimination the technician decides that the thermostat and the oil safety control are not at fault. The thermostat has a 10°F differential, and the cooler temperature should not vary 10°F in a short cycle. The oil safety control is a manual reset and will not short cycle. This narrows the possibilities down to the motor protection device and the high- or low-pressure controls.

On arrival, the technician parks so that his service vehicle does not block the driveway and makes sure his uniform is neat, clean, and tucked in. He uses a clean rag to remove some grease smudges from his shoes from a previous job that day.

The service technician then looks over the whole system before doing anything and notices that one evaporator fan is not running and that the coil is iced, so the suction pressure is too low, **Figure 29.36**. This causes the low-pressure control to shut off the compressor. After completing the examination, he replaces the fan motor and puts the system back in operation with the compressor shut off and allows the fans to run long enough to defrost the coil. The temperature inside the cooler is 50°F; it will take a long while for the cooler to pull down to the cut-out point of 35°F because of the amount of food. The technician cautions the owner to watch the thermometer in the cooler to make sure that the temperature is going down. The technician then politely gives the customer the billing information, work description, and cost of the work performed. A call later in the day will confirm that the unit is working properly.

Figure 29.36 Symptoms of a medium-temperature system with one of the two evaporator fans burned out. This system has a TXV.

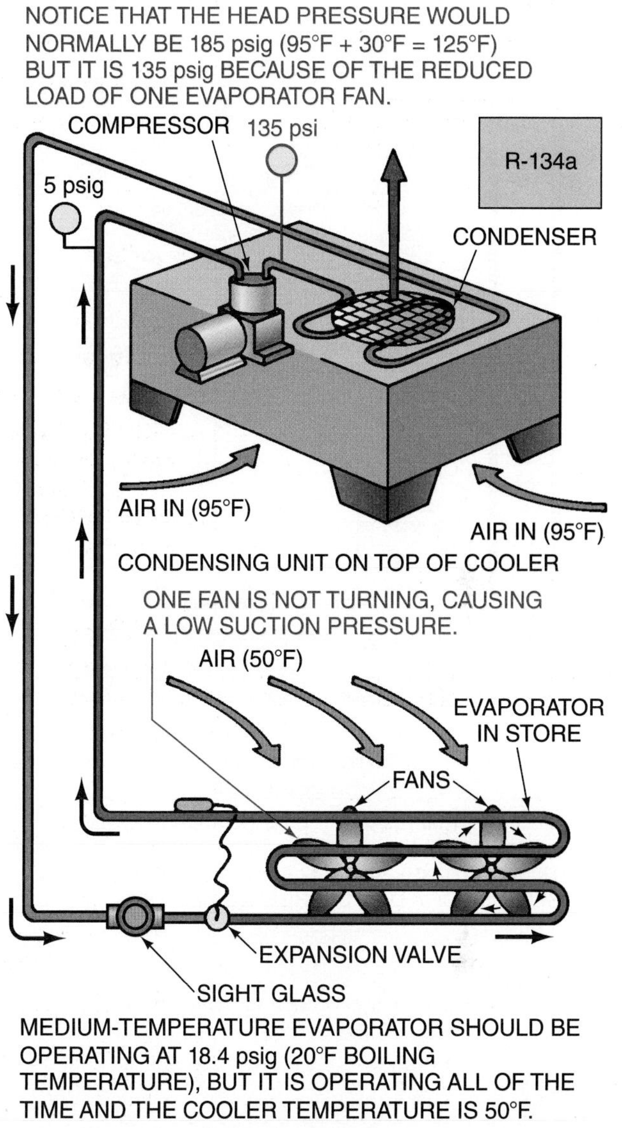

SERVICE CALL 2

A medium-temperature reach-in cooler is not cooling, and the compressor is cutting on and off. In this system the condensing unit is on top. *The unit has a TXV and a low charge. Two tubes have rubbed together and created a leak.*

The technician finds the unit is short cycling on the low-pressure control. The sight glass has bubbles in it, indicating a low charge, **Figure 29.37**. The small tube leading to the low-pressure control has rubbed a hole in the suction line, and there is oil around the point of the leak. The refrigerant is recovered using a recovery machine, and the leak is repaired. The system is pressurized with nitrogen and leak checked. The system is then evacuated to the proper micron level. Once evacuation is complete and the vacuum pump has been isolated from the system, the system is charged to the weight specified on the unit nameplate.

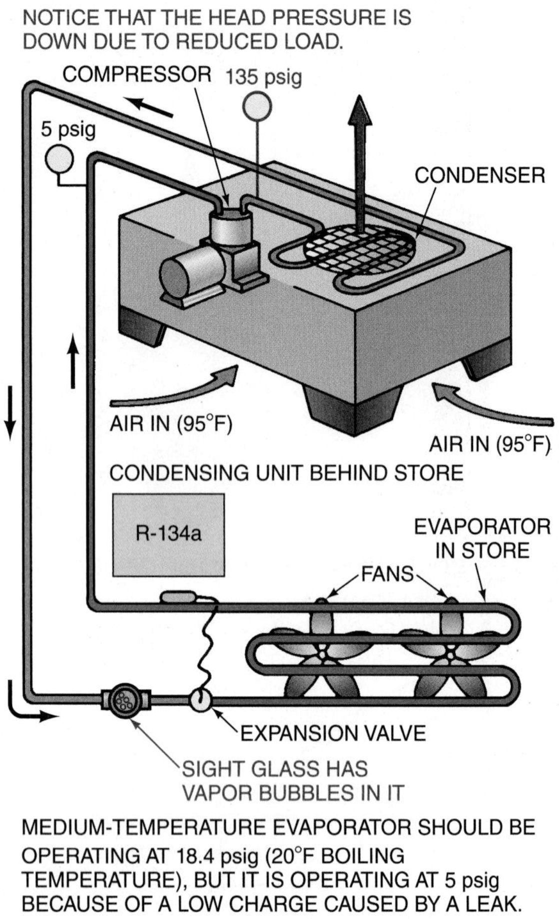

Figure 29.37 Symptoms of a medium-temperature system operating in a low-charge situation. This system has a TXV.

After the repair a leak check is performed, the technician starts the system and charges it to the correct charge. A call later in the day verifies that the system is functioning correctly.

SERVICE CALL 3

A factory cafeteria manager reports that the reach-in cooler in the lunchroom is not cooling and is running all the time. It is a medium-temperature cooler with the condensing unit at the bottom of the fixture that has not had a service call in a long time. *The evaporator is dirty and icing over,* **Figure 29.38**. *The unit has a TXV and there is no off time for defrost because the box does not get cool enough.*

The technician sees that the system is iced. The compressor is sweating down the side, and liquid is slowly coming back to the compressor—not enough to cause a noise from slugging. The TXV should maintain a constant superheat, but it may lose control if the pressure drops too low. The first thing that has to be done is to defrost the evaporator by stopping the compressor and using a heat gun (like a high-powered hair drier). The evaporator has a lot of dirt on it. **SAFETY PRECAUTION:** *The evaporator is cleaned with a coil cleaner that is approved for use in food-handling areas.•* After the evaporator is cleaned, the system is started. The unit is now operating with a full sight glass, and the suction line leaving the evaporator is cold. From this point it will take time for the unit to pull the cooler temperature down. While maintaining eye contact with the customer and remembering the customer's name, the technician politely explains how the problem was corrected. After giving the customer billing information, the work site is cleaned and the work double-checked for completeness. The service technician leaves and calls later in the day to confirm that the repair has solved the problem.

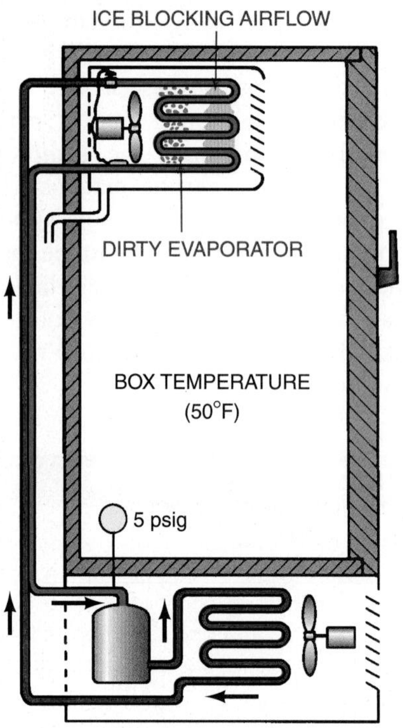

Figure 29.38 Symptoms of a medium-temperature cooler with a dirty evaporator coil. This system has a TXV.

SERVICE CALL 4

A convenience store manager reports that the reach-in cooler that stores the dairy products is not cooling properly. *The temperature is 55°F. It had been cooling well until early that morning. There is a leak due to a stress crack (caused by age and vibration) in the suction line near the compressor. The compressor is vibrating because when the customer moved the unit, the condensing unit fell down in the frame. This system has a capillary tube metering device,* **Figure 29.39**.

Figure 29.39 Symptoms of a medium-temperature system with a low charge. This system has a capillary tube.

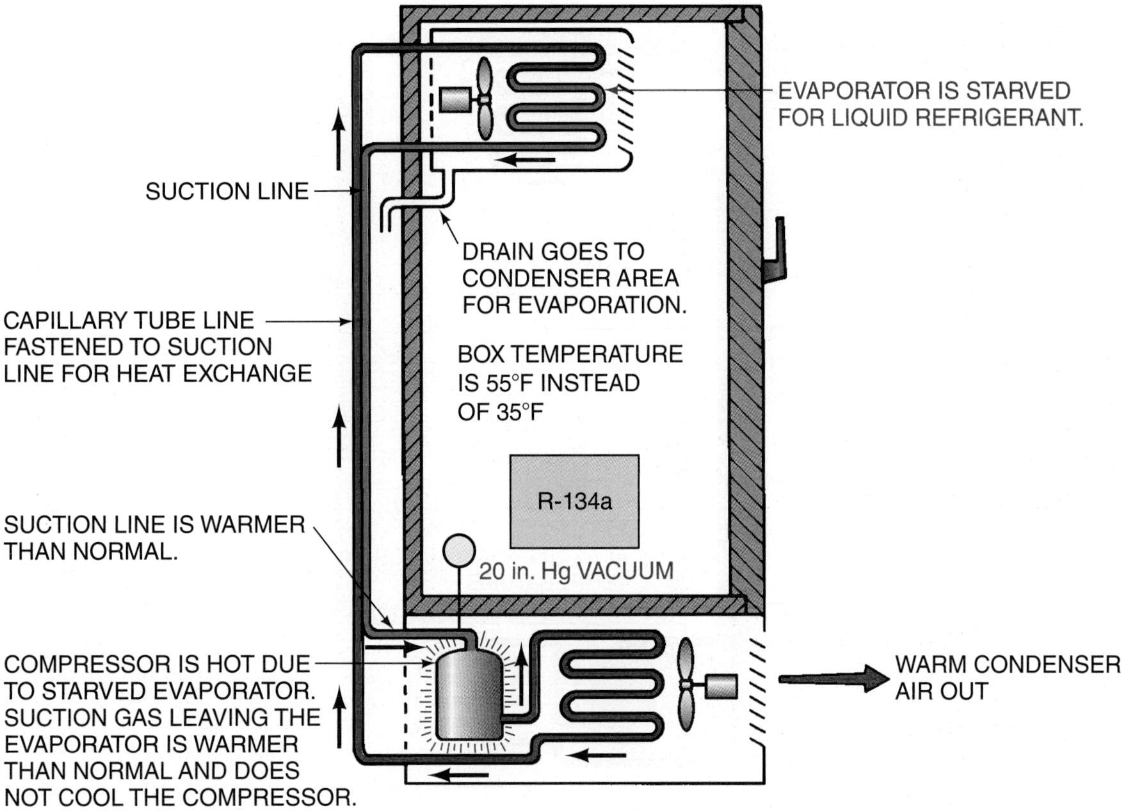

The technician's examination discloses that the compressor is vibrating because the condensing unit is not sitting properly in the frame. While securing the condensing unit, the technician notices an oil spot on the bottom of the suction line and that the compressor shell is hot. When gauges are installed, he discovers that the suction pressure is operating in a vacuum. **NOTE:** *Care must be used when installing gauges if a vacuum is suspected or air may be drawn into the system.* • The compressor is stopped and the low-side pressure is allowed to rise. There is not enough pressure in the system to accomplish a good leak test, so refrigerant is added. A leak check in the vicinity of the suction line reveals a leak, which appears to be a stress crack due to the vibration. 4R *The refrigerant is recovered from the system to repair the leak. Air must have been pulled into the system while it was operating in a vacuum, so the refrigerant charge must be recovered.* 4R **This system does not have a low-pressure control.** A short length of pipe is installed where the stress crack was found, and a new liquid-line drier is installed because the old one may not have any capacity left. The system is leak checked, triple evacuated, and charged, using a measured charge that the manufacturer recommends. The technician calls the manager later in the day and learns that the unit is functioning correctly.

SERVICE CALL 5

A restaurant manager calls to indicate that the reach-in freezer used for ice storage is running all the time. A competitor worked on this system in the early spring. A leak was found, the refrigerant was recovered, and the leak was repaired. A deep vacuum was then pulled and the unit was recharged. Hot weather is here, and the unit is running all the time. *This system has a capillary tube metering device, and the system has an overcharge of refrigerant. The other service technician did not measure the charge into the system,* **Figure 29.40**.

After parking the service van toward the back of the restaurant and away from the general public, the technician quietly knocks at the back door of the restaurant. Once invited in, the technician politely introduces himself and asks the restaurant manager pertinent questions without dominating the conversation or making the manager feel stupid about any technical topic.

After shaking hands, the technician starts work. Upon examining the fixture, the technician notices that the compressor is sweating down the side. The condenser feels hot for the first few rows and then warm. This appears to be an overcharge of refrigerant. The condenser should be warm near the bottom where the condensing is occurring. The evaporator fan is running

Figure 29.40 (A) A reach-in freezer under normal conditions. (B) Symptoms of a reach-in freezer with an overcharge of refrigerant. This system has a capillary tube.

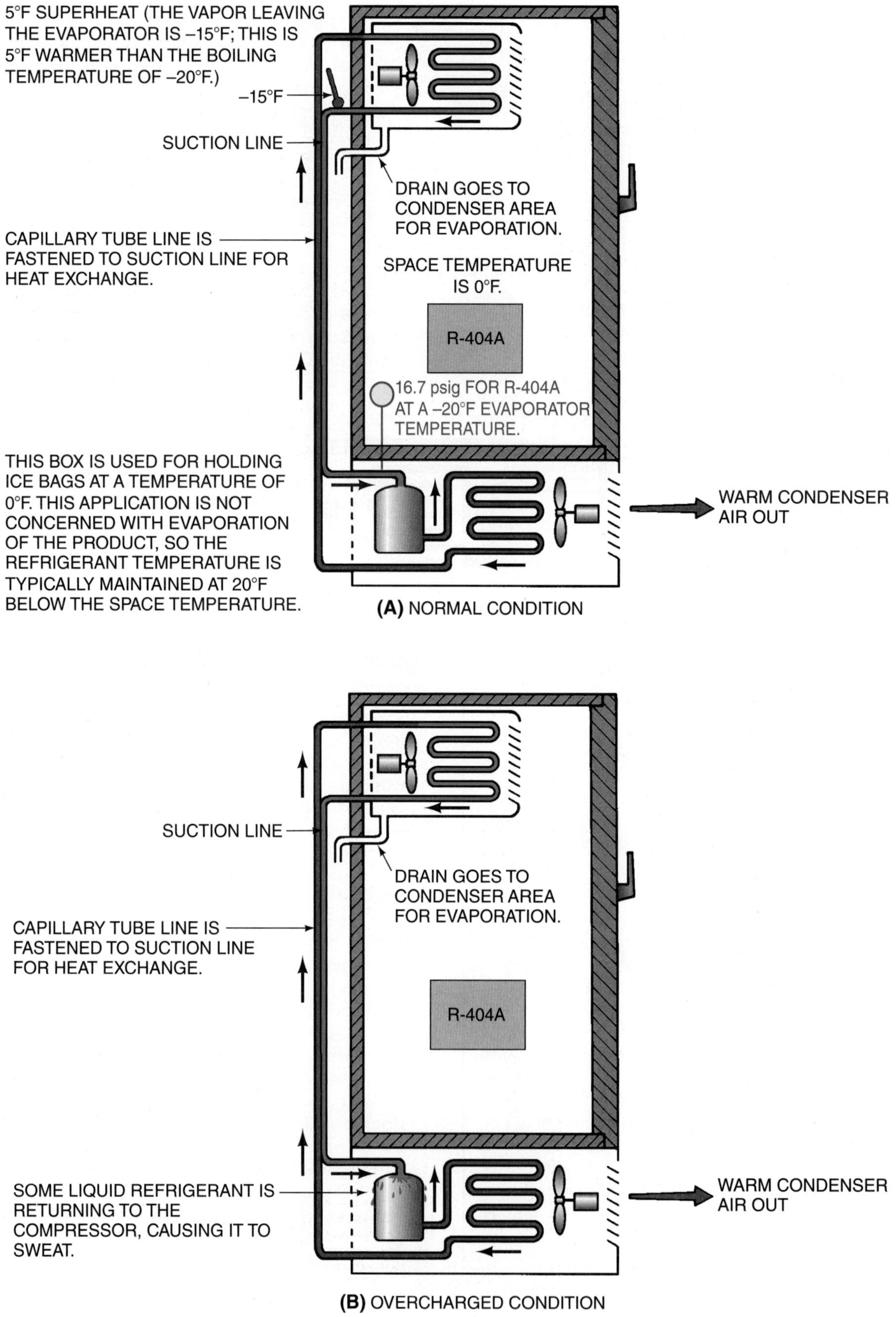

and the evaporator looks clean, so the evaporator must be doing its job. Gauges are installed, revealing that the head pressure is 445 psig with an outside temperature of 95°F; the refrigerant is R-404A. The head pressure should be no more than 332 psig on a 95°F day (95°F + 30°F = 125°F condensing temperature, or 332 psig). This system calls for a measured charge of 2 lb 8 oz. 4R *There are two possible approaches: (1) Alter the existing charge, or (2) recover the charge and measure a new charge into the unit while the unit is in a deep vacuum. It is a time-consuming process to recover the charge and evacuate the system.* 4R

The technician chooses to alter the existing charge. This is a plain capillary tube system with a heat exchanger (the capillary tube is soldered to the suction line after it leaves the evaporator). A thermometer lead is fastened to the suction line **after the evaporator** but **before the heat exchanger.** The suction pressure is checked for the boiling point of the refrigerant and compared with the suction-line temperature. The superheat is 0°F with the existing charge of refrigerant. 4R *Refrigerant is removed to an approved recovery cylinder until the superheat is 5°F at this point.* 4R **NOTE:** *The heat exchanger will allow a lower superheat than normal.* • When the system charge is balanced, the technician leaves the job and will call the manager later in the day for a report on how the system is functioning.

SERVICE CALL 6

An office manager calls and reports that the reach-in medium-temperature beverage cooler in the employee cafeteria is running all the time. The suction line is covered with frost back to the compressor. *A small liquid-line drier is stopped up with sludge from a compressor changeout after a motor burnout. A suction-line drier should have been installed,* **Figure 29.41**.

After looking the system over closely, the technician sees that the suction line is frosting and that the compressor has a frost patch on it. A first glance indicates that the fan at the evaporator is off and that the coil is dirty. Further examination, however, shows that the liquid line is frosting starting at the outlet of the drier. This means the drier is partially stopped up. The pressure drop across the drier makes the drier act like an expansion device. This effectively means there are two expansion devices in series because the drier is feeding the capillary tube. 4R *The refrigerant charge must be recovered and the drier replaced.* 4R After the refrigerant has been recovered, the drier is installed on the liquid line. There are no service valves. While the system is open, the technician solders a suction-line drier in the suction line close to the compressor. A deep vacuum is then pulled and the system is recharged.

Figure 29.41 Symptoms of a medium-temperature reach-in cooler with a partially clogged liquid-line drier.

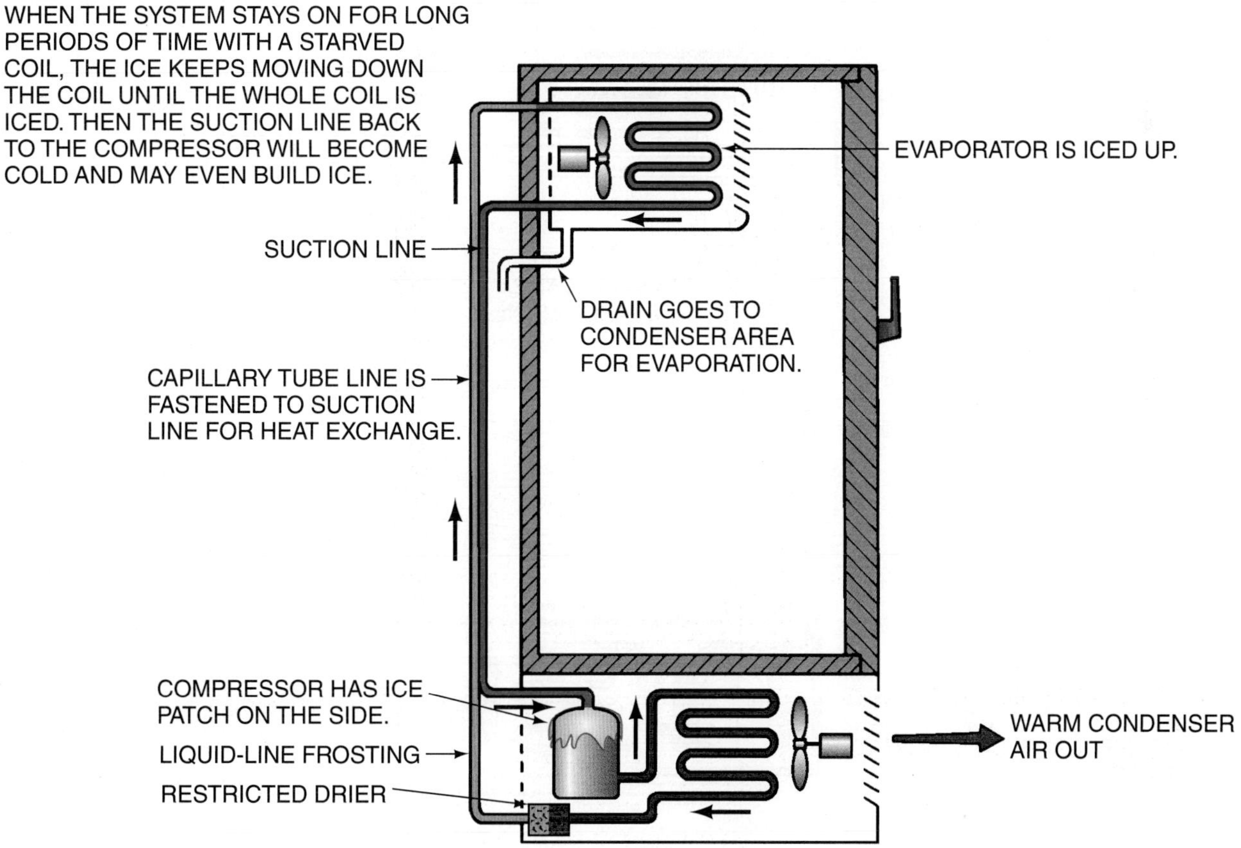

The reason the unit was frosting instead of sweating is that the evaporator was starved for liquid refrigerant. The suction pressure went down below freezing. The unit's capacity was reduced to the point where it was running constantly and had no defrost. The frost will become more dense, blocking the air through the coil and acting as an insulator. This will cause the coil to get even colder with more frost. This condition will continue with the frost line moving on to the compressor. The ice or frost acts as an insulator and blocks air. Air has to circulate across the coil for the unit to produce at its rated capacity.

SERVICE CALL 7

A restaurant owner calls indicating that the reach-in freezer for ice cream is running but the temperature is rising. *The system has a capillary tube metering device and the evaporator fan is defective. The customer hears the compressor running and thinks the whole unit is running,* **Figure 29.42**.

The technician has never been to this installation and has to examine the system thoroughly. He discovers that the frost on the suction line goes all the way to and down the side of the compressor. The first thought is that the system is not going through the proper defrost cycle. The coil is iced and has to be defrosted before anything can be done. The defrost cycle clears the coil of ice, and when defrost is over, the fan does not restart as it should. The technician takes off the panel to the fan compartment and checks the voltage. There is voltage, but the fan will not run. When the motor is checked for resistance through the windings, it is discovered that the fan motor winding is open. The fan motor is changed, and the system is started again. It will take several hours for the fixture to pull back down to the normal running temperature of −10°F.

A call later in the day verifies that the freezer temperature is going down. The service technician then returns to the restaurant and clearly and concisely explains to the customer exactly what was done to remedy the problem. The service technician then gives the manager the billing information with a work description and costs. The technician then shakes the manager's hand and gives him a business card while maintaining eye contact. Before leaving, the technician makes a quick walk-around to make sure the work site has been cleaned and the unit is still pulling down to temperature.

Figure 29.42 Symptoms of a reach-in freezer with a defective evaporator fan motor. This system has a capillary tube.

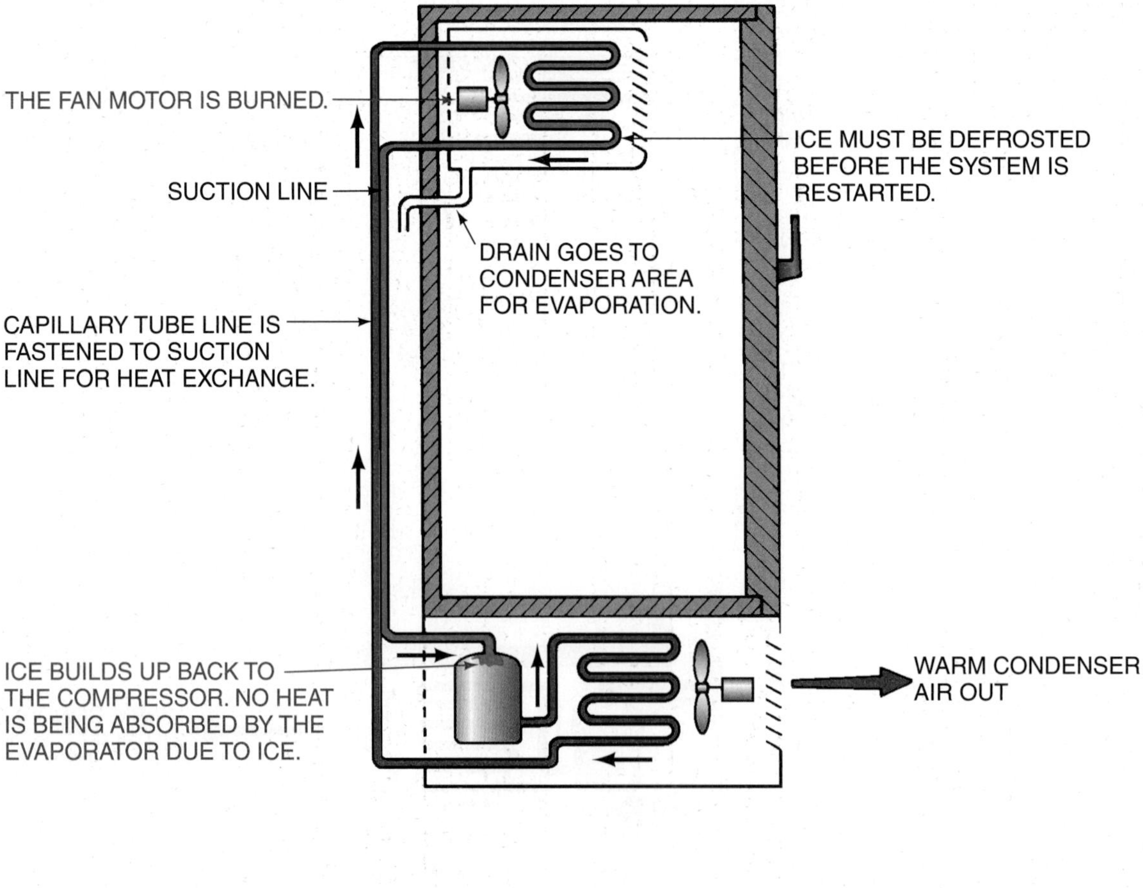

SERVICE CALL 8

A call is received indicating that the customer's reach-in dairy case used for milk storage is running constantly and the temperature is 48°F. This unit has not had a service call in 10 years. *The evaporator is dirty. This unit has an automatic expansion valve. The coils never have had filters, and years of dust have accumulated,* **Figure 29.43**.

An automatic expansion valve maintains a constant pressure. It responds in reverse to a load change. If a load of additional product is added to the cooler, the rise in suction pressure will cause the automatic expansion valve to throttle back and slightly starve the evaporator. If the load is reduced, such as with a dirty coil, the valve will overfeed to keep the refrigerant pressure up.

The technician finds that the suction line is very cold and the compressor is sweating. This is evidence of an overcharge, or that the refrigerant is not boiling to a vapor in the evaporator. A close examination of the evaporator indicates that it is not exchanging heat with the air in the cooler. **SAFETY PRECAUTION:** *The evaporator is cleaned with a special detergent approved for evaporator cleaning and for use in food-handling areas. Areas that have dairy products are particularly difficult because these products absorb odors easily.* • The system is started, the sweat gradually disappears from the compressor, and the suction line feels normally cool. The technician leaves the job and will call back later to see how the unit is performing.

Figure 29.43 Symptoms of a reach-in dairy case with a dirty evaporator coil. This system has an automatic expansion valve.

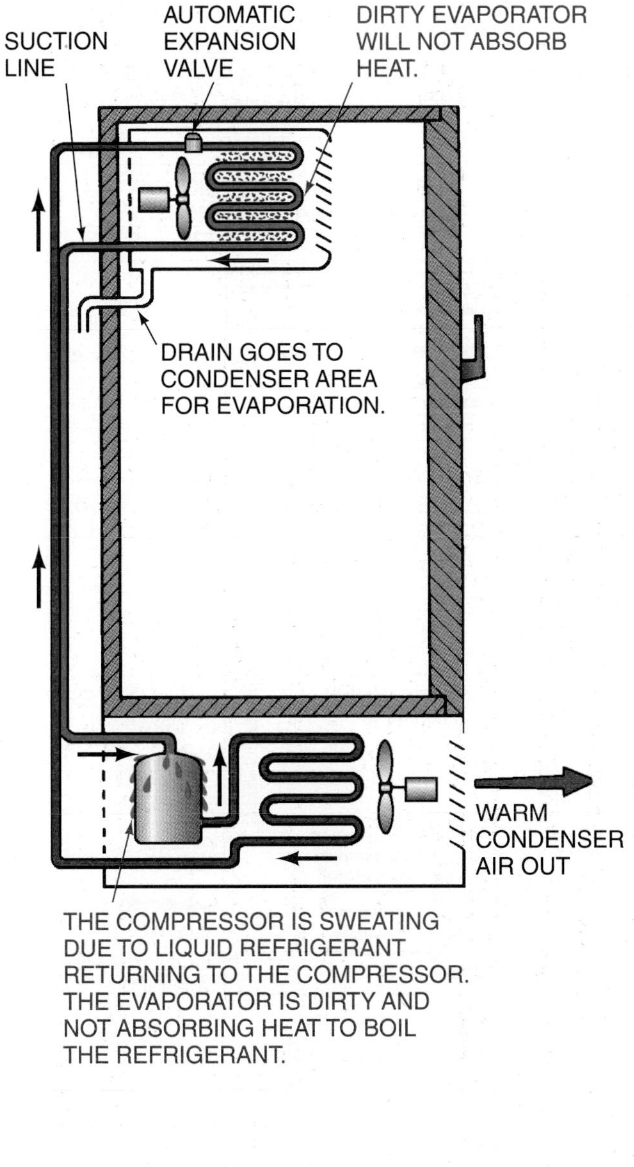

SERVICE CALL 9

A golf club restaurant manager calls to say that a small beverage cooler is not cooling drinks to the correct temperature. *A small leak at the flare nut on the outlet to the automatic expansion valve has caused a partial loss of refrigerant,* **Figure 29.44**.

Since this is an exclusive golf club and restaurant, the technician makes sure that the appearance of both the service van and himself are clean and presentable. The technician also takes care not to interfere with any golfers or restaurant customers. After ringing the bell at the service entrance and being invited in, the technician makes sure to maintain a comfortable distance between himself and the customer, politely shaking hands and maintaining eye contact while handing out a business card. After formally introducing himself and the company he works for, the technician carefully and politely listens to the customer about the service call and asks pertinent questions.

The technician begins work. He hears the expansion valve hissing, which means the valve is passing vapor along with the liquid it is supposed to pass. The sight glass shows some bubbles. The gauges that are installed indicate that the suction pressure is normal and the head pressure is low. The suction pressure is 15 psig; this corresponds to 15°F boiling temperature for R-134a. The boiling refrigerant normally is about 20°F cooler than the beverages to be cooled. The liquid beverage in the cooler is 50°F and should be 35°F. The head pressure should be about 135 psig, the pressure corresponding to 105°F for R-134a. The ambient is 75°F, and the condensing temperature should be 105°F (75°F + 30°F = 105°F). The head pressure is 100 psig. All of these signs point to an undercharge. The technician turns the cooler off and allows the low-side pressure to rise so that there will be a better chance of detecting a leak. A leak is found at the flare nut leaving the expansion valve. The nut is tightened but that does not stop the refrigerant from leaking. The flare connection must be defective.

4R *The charge is recovered, and the flare nut is removed.* 4R There is a crack in the tubing at the base of the flare nut. The flare is repaired, and the system is leak checked and evacuated. The system is charged to the weight specified on the unit nameplate. This gives the most accurate operating charge. The system is started. The technician will call back later to ensure that the unit is operating properly.

Figure 29.44 Symptoms of a reach-in beverage cooler with a low charge. This system has an automatic expansion valve.

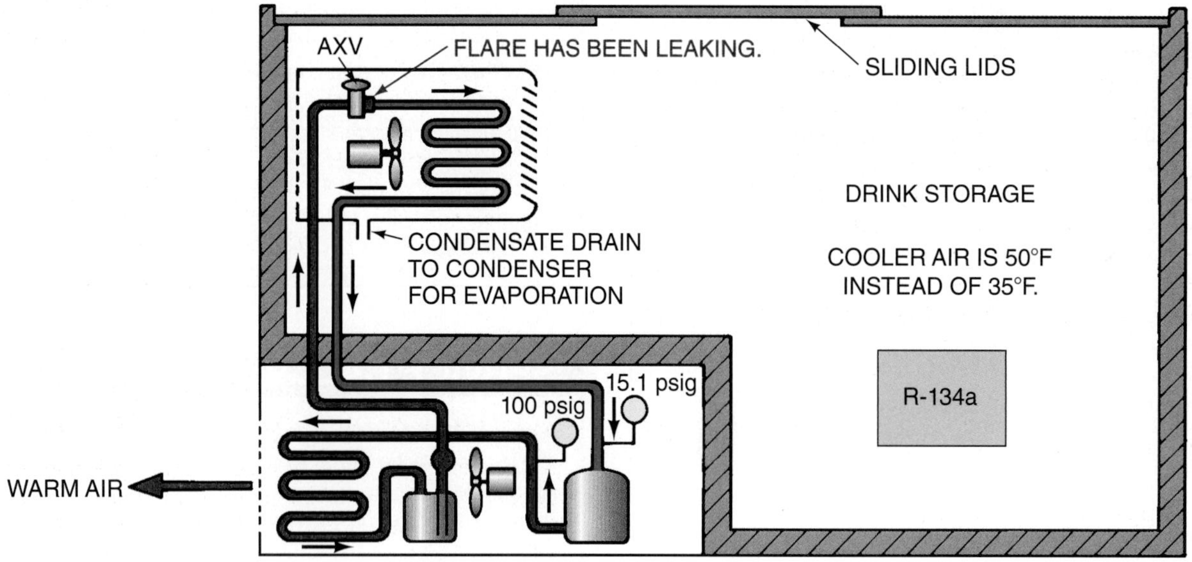

SERVICE CALL 10

A restaurant manager reports that the compressor in the pie case is cutting off and on and sounds like it is straining while it is running, **Figure 29.45**. This unit was charged after a leak and then shut down and put in storage. It has not been operated in several months. *The unit has an automatic expansion valve and has an overcharge of refrigerant. When the unit was charged it was in the back room in the winter, and it was cold in the room. The unit was charged to a full sight glass with a cold condenser.*

The service technician remembers that the unit was started and charged in a cold ambient and assumes that it may have an overcharge of refrigerant. When the technician arrives, the unit is started in the location where it is going to stay; the ambient is warm. Gauges are installed, and the head pressure is 280 psig. The pressure should be 146 psig at the highest because the ambient temperature is 80°F. This should create a condensing temperature of no more than 110°F (80°F + 30°F = 110°F). The compressor is cutting off due to the high pressure at 280 psig. 4R *Refrigerant is removed from the machine to an approved recovery cylinder until the head pressure is down to 146 psig, the correct head pressure for the ambient temperature.* 4R The sight glass is still full. A call back to the owner later in the day verifies that the system is working correctly.

Figure 29.45 Symptoms of a reach-in pie case with an overcharge of refrigerant. This system has an automatic expansion valve.

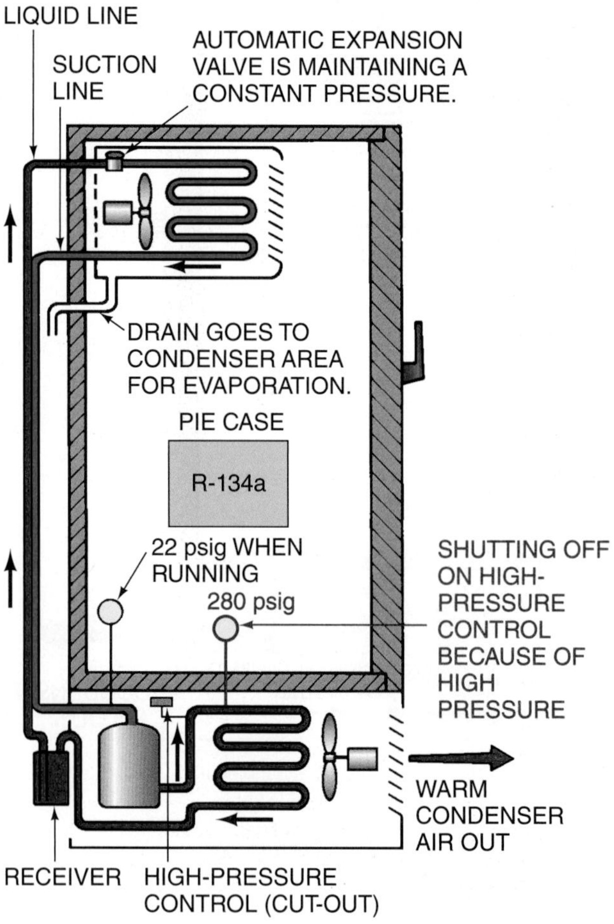

SERVICE CALL 11

A store manager calls to say that a reach-in cooler is rising in temperature and the unit is running all the time. It is a medium-temperature cooler and should be operating between 35°F and 45°F. It has an automatic expansion valve. The unit has not had a service call in 3 years. *The evaporator fan motor is not running. This system has off-cycle defrost,* **Figure 29.46**.

The service technician examines the system. The compressor is on top of the fixture and is sweating down the side. Because this unit has an automatic expansion valve, the sweating is a sign that the evaporator is not boiling the refrigerant to a vapor. The evaporator could be dirty or the fan may not be moving enough air. After removing the fan panel, the technician sees that the evaporator fan motor is not running. The blades are hard to turn, indicating that bearings are tight. The bearings are lubricated and the fan is started and seems to run like it should. This is a non-standard motor and cannot be purchased locally. The system is left in running condition until a new fan motor can be obtained.

The technician returns in a week with the correct motor and exchanges it for the old one. The system has been working correctly all week. The old motor would fail again because the bearings are scored, which is the reason for exchanging a working motor for a new one.

Figure 29.46 Symptoms of a reach-in cooler with a defective fan motor. This system has an automatic expansion valve.

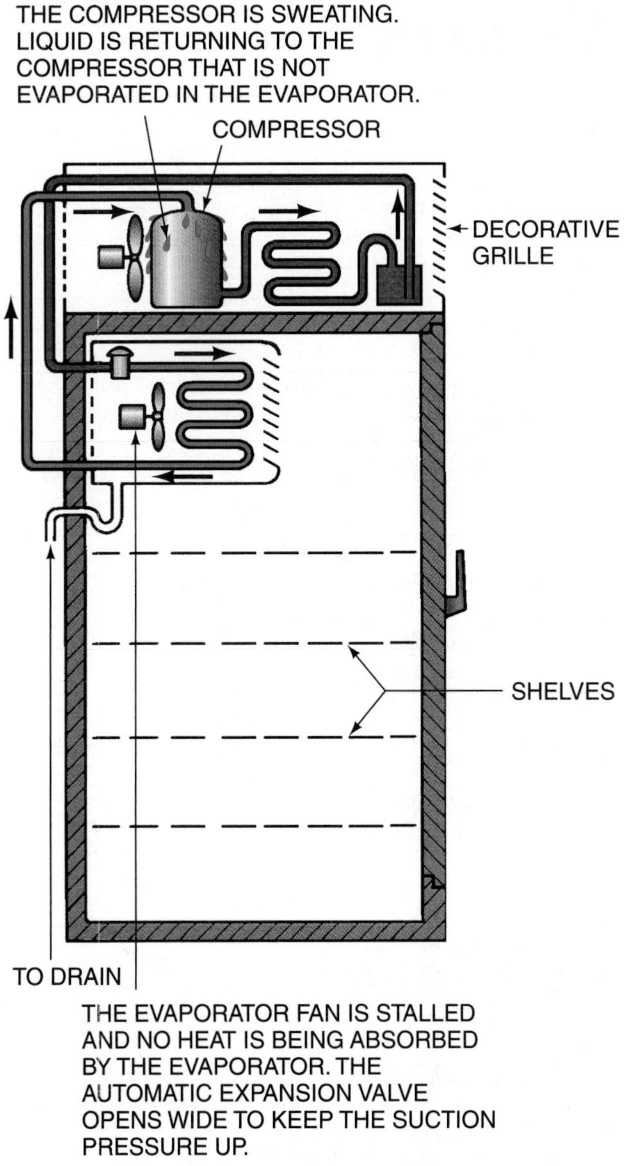

SERVICE CALL 12

A store assistant manager reports that a reach-in medium-temperature cooler is rising in temperature. *This unit has been performing satisfactorily for several years. A new stock clerk has loaded the product too high, and the product is interfering with the airflow of the evaporator fan. The system has an automatic expansion valve,* **Figure 29.47**.

The technician immediately notices that the product is too high in the product area. The extra product is moved to another cooler. The stock clerk is shown the load level lines in the cooler without making the new stock clerk or the store's assistant manager feel uncomfortable or stupid when telling them what the problem really was. Technicians must always be courteous and show respect for the customer and never brag about their training or skills. A call back later in the day indicates that the cooler is now working correctly.

SERVICE CALL 13

A customer reports that the condensing unit on a medium-temperature walk-in cooler is running all the time and the cooler is rising in temperature. *This is the first hot day of the season, and some boxes have been stacked too close to the condenser outlet. The hot air is leaving the condenser and recirculating back into the fan inlet,* **Figure 29.48**.

The service technician looks over the system. The temperature is 52°F inside the cooler. The liquid-line temperature is hot instead of warm. This is a TXV system, and the sight glass is clear. The unit appears to have a full charge with a high head pressure. Gauges are installed, which reveal that the head pressure is 229 psig. The system is using R-134a as the refrigerant. The head pressure should be no more than 171 psig, which is a condensing temperature of 120°F; the ambient temperature is 90°F (90°F + 30°F = 120°F, or 171 psig). Further examination indicates that air is leaving the condenser and recirculating back to the condenser inlet. Several boxes have been stored in front of the condenser. These are moved, and the head pressure drops to 171 psig. The reason there has been no complaint up to now is that the air has been cold enough in the previous mild weather

Figure 29.47 Symptoms of a reach-in medium-temperature cooler where the product is interfering with the air pattern. This system has an automatic expansion valve.

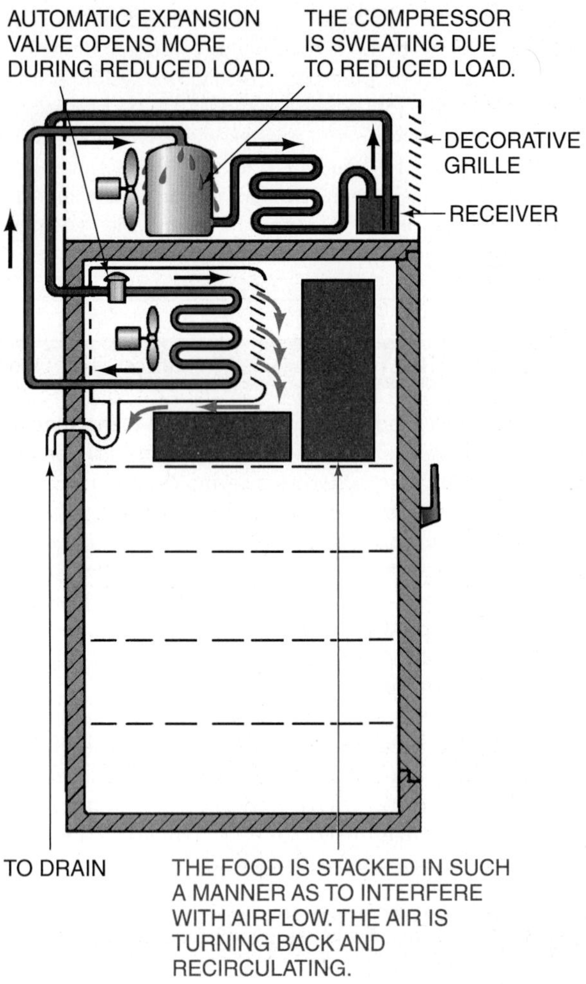

to keep the head pressure down even with the recirculation problem. A call back later in the day verifies that the cooler temperature has returned to normal.

SERVICE CALL 14

A store frozen-food manager calls to say that the temperature is going up in the low-temperature walk-in freezer. *The defrost time clock motor is burned, and the system will not go into defrost,* **Figure 29.49**.

The service technician examines the system and sees that the evaporator is coated with thick ice. It is evident that it has not been defrosting. The first thing to do is to force a defrost. The technician examines the clock dial on the timer and finds that the indicator says 4:00 AM, but the time is 2:00 PM. Either the power has been off or the timer motor is not advancing the timer. The technician advances the timer by hand until the defrost cycle starts and then marks the time. After the defrost is terminated by the temperature-sensing device, the system goes into normal operation. The technician waits about one-half hour and sees that the clock has not moved. A new time clock is installed.

The customer is cautioned to look out for heavy ice buildup. A call the next day verifies that the system is working correctly. The service technician then stops by the store at the end of the day and politely introduces himself again. After addressing the store manager by his first name (always maintaining eye contact), he gives the billing information to the customer, which includes a work description and cost of the service. He gives the customer a firm handshake and thanks him for the business. The technician then looks around quickly to make sure everything is in good working order and that the work space has been cleaned.

Figure 29.48 Symptoms of an air-cooled condenser with the hot discharge air recirculating back into the condenser inlet.

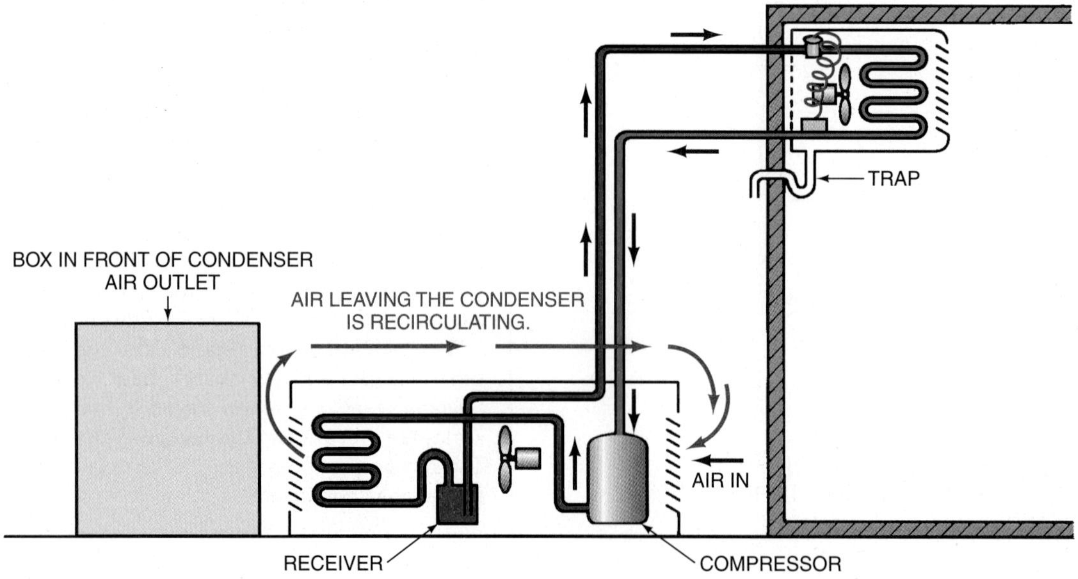

Figure 29.49 Symptoms of a defrost problem with a defective time clock.

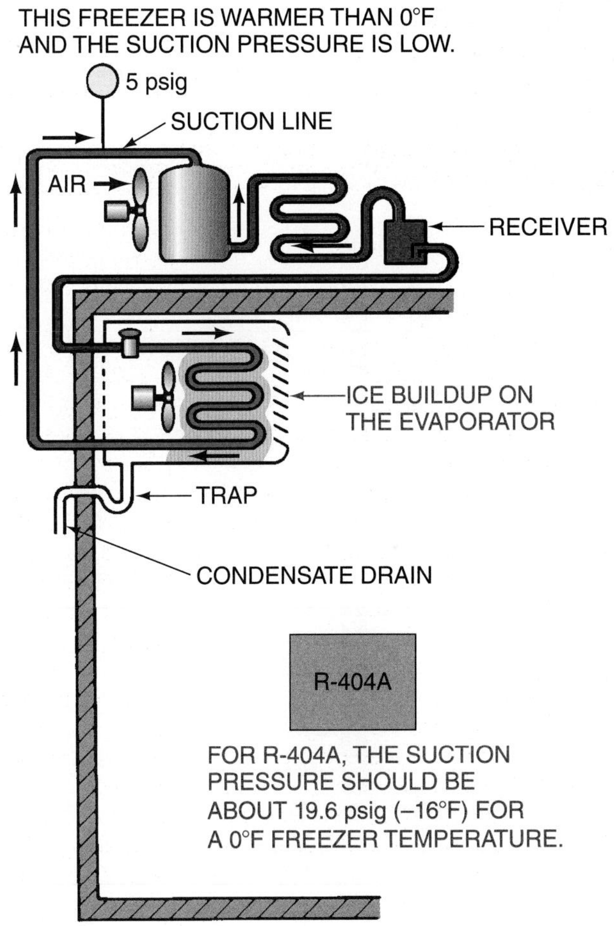

SERVICE CALL 15

A customer states that the defrost seems to be lasting too long in a low-temperature walk-in cooler. It used to defrost, and the fans would start back up in about 10 min. It is now taking 30 min. *The defrost termination switch is not terminating defrost with the temperature setting. This is an electric defrost system. The system is staying in defrost until the fail safe setting in the timer takes it out of defrost. This also causes the compressor to operate at too high a current because the fans should not start until the coil cools to below 30°F. With the defrost termination switch stuck in the cold position, the fans will come on when the timer terminates defrost. The compressor will be overloaded by the heat left in the coil from the defrost heaters,* **Figure 29.50**.

The technician examines the system. The coil is free of ice, so defrost has been working. The cooler is cold, −5°F. The technician advances the time clock to the point that defrost starts. The termination setting on the defrost timer is 30 min. This may be a little long unless the cooler is defrosted only once a day. The timer is set to defrost twice a day, at 2:00 PM and 2:00 AM. The coils are defrosted in approximately 5 min, but the defrost continues. The technician allows defrost to continue for 15 min to allow the coil to get to maximum temperature so that the defrost termination switch has a chance to close. After 15 min, the system is still in defrost, so the timer is advanced to the end of the timed cycle. When the system goes out of defrost, the fans start. They should not start until the coil gets down to below 30°F (around 25°F).

The defrost termination switch is removed. The wires on its three terminals are marked so that it can easily be replaced with a new control. After a new control is acquired and installed, the system is started. The owner called the next day to report a normal defrost and that the system was acting normally. The old defrost termination switch is checked with a digital multimeter after having been at room temperature for several hours and the cold contacts are still closed. The control is definitely defective.

SERVICE CALL 16

The frozen-food manager in a supermarket calls indicating that the walls between the doors on a reach-in freezer are sweating. *This has never happened. The freezer is about 15 years old. The mullion heater in the wall of the cooler that keeps the panel above the dew point temperature of the room is not functioning,* **Figure 29.51**.

The technician first talks to the store manager and carefully listens to what he has to say, without interrupting. The technician then asks the manager a few questions, letting him feel as though he is contributing to the solution of the problem by answering the questions. The technician then starts to work.

The technician examines the fixture to determine the reason the mullion heater is not getting hot. If the heater is burned out, the panel needs to be removed. He traces the circuit to the back of the cooler and finds the wires to the heater. A check with a multimeter shows that the heater has a measurable resistance, but a voltage check shows there is no voltage going to the heater. After further tracing, the technician locates a loose connection in a junction box. The wires are properly connected and a current check is performed to prove the heater is working. The service technician then politely explains to the store manager how he found the problem, thanks the customer for the business, and lets him know the store is a valued customer. He explains the billing information and exchanges handshakes.

Figure 29.50 Symptoms of a defective defrost termination thermostat (stuck with cold contacts closed) in a low-temperature freezer.

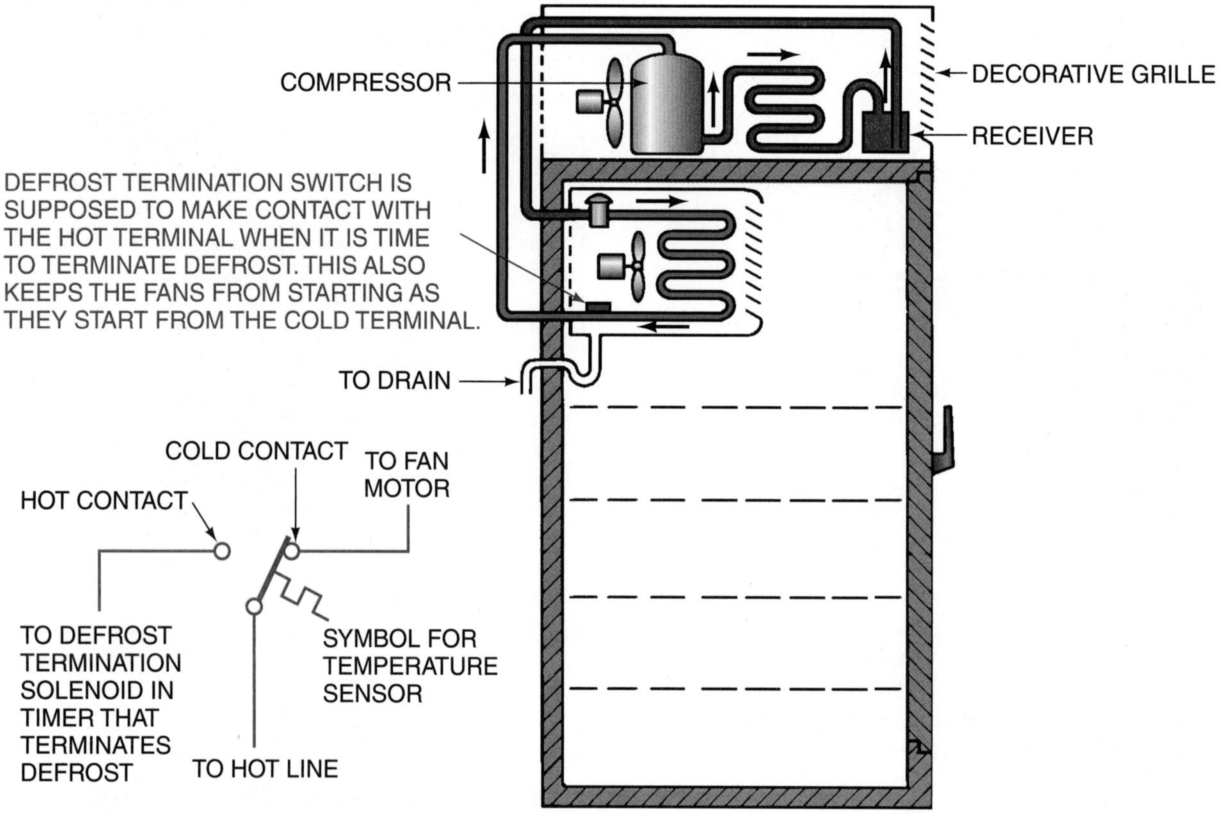

Figure 29.51 Testing a mullion heater on a low-temperature freezer.

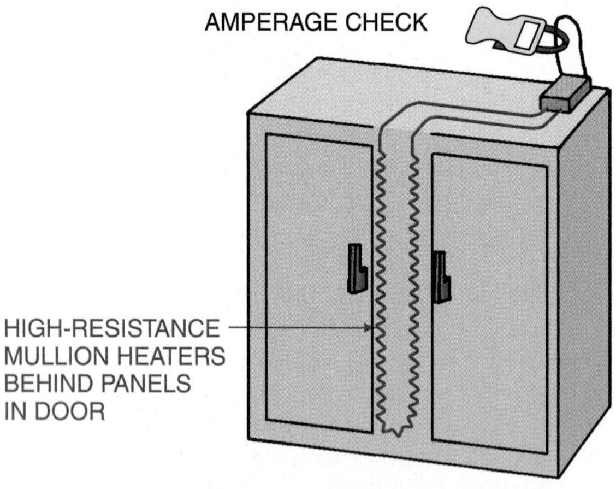

SERVICE CALL 17

A restaurant maintenance person calls to say that the medium-temperature walk-in cooler is off. The breaker was tripped and was reset, but it tripped again. *The compressor motor is burned,* **Figure 29.52**.

The technician examines the system. The food is warming up, and the cooler is up to 55°F. Before resetting the breaker, the technician uses a digital multimeter and finds the compressor has an open circuit through the run winding. Nothing can be done except to change the compressor. An acid test is performed to determine the severity of the motor burnout. The refrigerant has a high acid odor, indicating a bad motor burn. 4R *The technician uses a recovery machine to recover the refrigerant. The refrigerant will be turned in to be reclaimed.* 4R

It will be several hours before the system can be put back into service, so the food is moved to another cooler. The technician gets the required materials from the supply house, including a suction-line filter drier with high acid-removing qualities. This drier has removable cores for each changeout. The compressor is changed and the suction-line drier is installed. The system is purged with dry nitrogen to push as much of the free contaminants as possible out of the system. The system is leak checked, and a deep vacuum is pulled. The system is then charged with new refrigerant and started. After the compressor runs for an hour, the crankcase oil is tested for acid. It shows a slight acid count. The unit is run for another 4 hours, and an acid check shows even less acid. The system is pumped down, and the cores are changed in

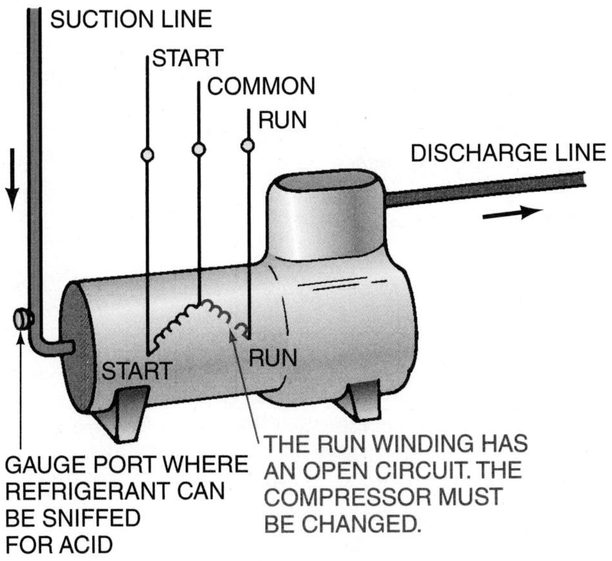

Figure 29.52 A system with a burned compressor motor.

the suction-line drier. The liquid-line drier is changed, and the system is allowed to operate overnight.

The technician returns the next day and makes another acid check. This check shows no sign of acid. The system is left to run in this condition with the drier cores still in place in case some acid becomes loose at a later date. *The motor burn is attributed to a random motor failure.*

SERVICE CALL 18

A medium-temperature reach-in cooler is not cooling properly. The temperature is 55°F inside, and it should be no higher than 45°F. The compressor sounds like it is trying to start but then cuts off. *The start capacitor is defective.*

The technician arrives at the job in time to hear the compressor try to start. Several things can keep the compressor from starting. It is best to give the compressor the benefit of the doubt and assume it is good. Check the starting components first and then check the compressor. The start capacitor is removed from the circuit for testing, **Figure 29.53**. There is discoloration around the capacitor vent at the top, and the vent is pushed upward. These are signs the capacitor is defective. After bleeding the capacitor with a 20,000-Ω, 5-W resistor, a test is performed with a capacitor analyzer which shows the capacitor is open. The capacitor is replaced with a new one, and the compressor is started. The compressor is allowed to run for several minutes to allow the suction gas to cool the motor; then it is stopped and restarted to make sure that it is operating correctly. A call back the next day indicates that the compressor is still stopping and starting correctly.

Figure 29.53 (A) Discharging a start capacitor with a 20,000-Ohm 5-Watt resistor. Photo by Jason Obrzut. (B) A capacitor analyzer is used to test the start capacitor. Photo by Jason Obrzut. (C) The vent on a start capacitor may have discoloration or bubble out when the capacitor is bad. Photo by Jason Obrzut.

(A)

(B)

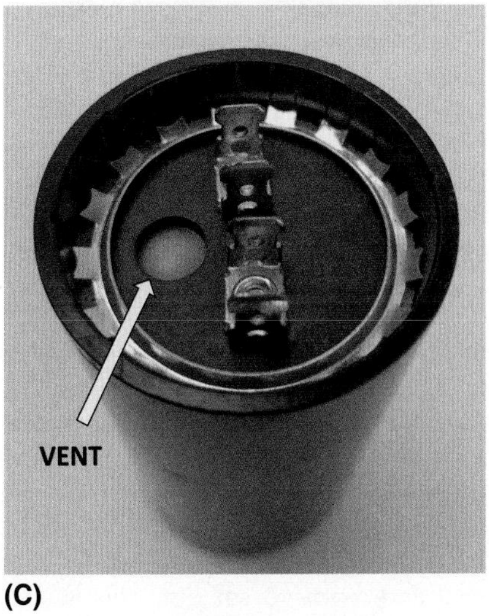

(C)

SERVICE CALL 19

The compressor in the low-temperature walk-in freezer of a supermarket is not starting. The box temperature is 0°F, and it normally operates at −10°F. *The compressor is locked. This is a multiple-evaporator installation with four evaporators piped into one suction line. An expansion valve on one of the evaporators has been allowing a small amount of liquid refrigerant to get back to the compressor. This has caused marginal lubrication, and the compressor has scored bearings that are bad enough to lock it up.*

After finishing a previous service call and before arriving at the supermarket, the service technician notices that the service van is very muddy from a rainstorm earlier that morning and decides to run the van through an automatic wash about a mile away. Once the van is clean, the technician spends some time wiping down his shoes from the previous job. He also needs to change his uniform because of grease stains on the pants and shirt sleeves. The technician then washes his hands with waterless hand cleaner and tucks in the new uniform. A quick look in the mirror verifies that the technician's face and hair are presentable.

The service technician examines the system at the supermarket. The compressor is a three-phase compressor and does not have a starting relay or starting capacitor. It is important that the compressor be started within 5 hours, or the food must be moved. The technician turns off the power to the compressor starter then removes the leads from the load side of the starter. The motor is checked for resistance using a digital multimeter, **Figure 29.54(A)**. The motor windings appear to be normal. **NOTE:** *Some compressor manufacturers furnish data that tell what the motor-winding resistances should be.* • The multimeter is then turned to the AC voltage function to test the supply voltage. The starter is then energized to test its load side to verify that each of the three phases has the correct power. This is a no-load test. The leads have to be disconnected from the motor terminals and the voltage applied to the leads while the motor is trying to start before the technician knows for sure there is a full 230 V under the starting load, **Figure 29.54(B)**.

When the load is connected, the technician connects up the motor leads and gets two more meters for checking voltage. This allows the voltage to be checked from phase 1 to phase 2, phase 1 to phase 3, and phase 2 to phase 3 at the same time the motor is being started. When the power is applied to the motor, there is 230 V from each phase to the other, but the motor will not start, **Figure 29.54(C)**. **This is conclusive—the motor is locked**. The motor can be reversed (by changing any two leads, such as L1 and L2) and it may start, but it is not likely that it will run for long.

The technician now has 4 hours to either change the compressor or move the food products. A call to the local compressor supplier shows that a compressor is in stock in town. The technician calls the shop for help, another technician is dispatched to pick up the new compressor, and a crane is ordered to set the new compressor in and take the old one out. The 30-hp compressor weighs about 800 lb.

When the crane gets to the job, the old compressor is disconnected and ready to set out. This is accomplished by front seating the suction and discharge service valves to isolate the compressor and then recovering the compressor's refrigerant charge. When the new compressor is installed, all that will need to be done after evacuating the compressor is to open the service valves and start it. The original charge is still in the system. By the time the old compressor is removed, the new compressor arrives and is set in place; the service valves are connected. While a vacuum is being pulled, the motor leads are connected. In an hour the new compressor has been evacuated and is ready to start.

When the new compressor is started, the technician notices that the suction line is frosting on the side of the compressor. Liquid is getting back to the compressor. A thermometer lead fastened to each of the evaporators at the evaporator outlet reveals that one of the evaporator suction lines is much colder than the others. The expansion valve bulb is examined and is found to be loose and not sensing the suction-line temperature. Someone has taken the screws out of the mounting strap.

The expansion valve bulb is then secured to the suction line, and the system is allowed to operate. The frost line moves back from the side of the compressor, and the system begins to function normally. After double-checking his work, the technician thoroughly cleans the job site.

The next day the technician stops by the store and determines that the system is functioning properly. He then presents billing information to the store manager with a complete list of parts required. The technician then spends time with the manager to explain in everyday language how the problems were corrected. Handshakes are exchanged and the technician leaves the job site.

SERVICE CALL 20

A restaurant owner with a cube ice-making machine calls to say that the evaporator is frozen solid and no ice is dropping into the bin. *The problem is that the ice thickness probe is dirty. This probe senses the ice thickness by creating a circuit through the ice that starts the harvest cycle when the ice reaches a predetermined thickness. The probe is monitored by an electronic circuit that measures the conductivity of the ice to determine when the correct thickness has been reached.*

The technician examines the entire system carefully. He has worked on this type of ice maker before and knows that the manufacturer recommends the machine not be shut down until the problem is found. If the technician were to shut the machine off and defrost the evaporator, the condition may not happen again. The technician knows this machine is older and suspects a dirty ice thickness probe. The manufacturer has supplied a special jumper cord with a fixed resistor to measure the value of the resistance through ice. The jumper can be used to jump the probe and determine if the sensor is defective. The technician jumps the probe connection and the unit starts a harvest cycle. This is a sure sign that the sensor is dirty where it contacts the ice.

Figure 29.54 (A–C) Symptoms of a system with a three-phase compressor that is locked.

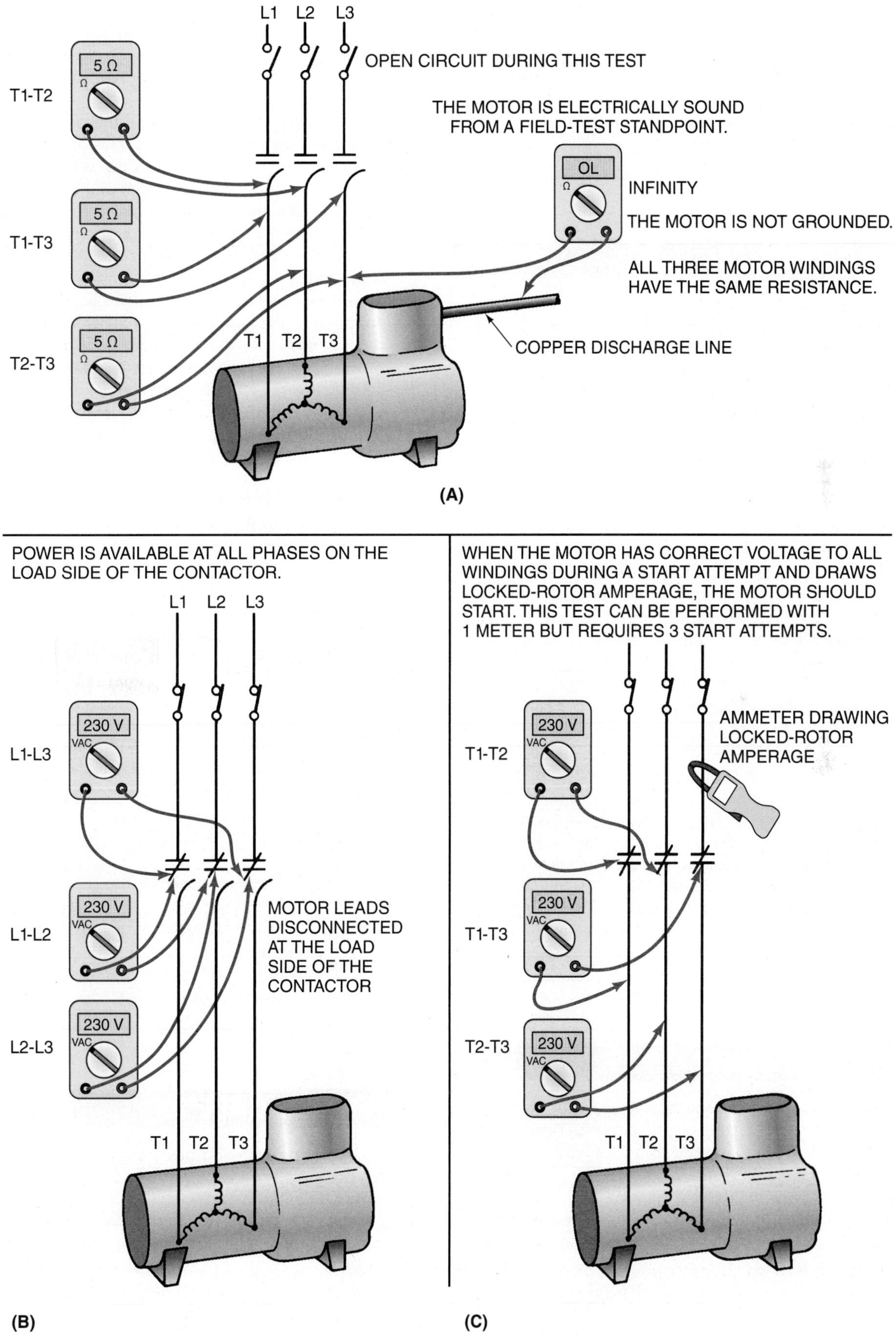

The technician waits until the unit completes the harvest cycle and the ice drops to the storage bin, then shuts the machine off and turns the power off. The probe is removed and cleaned. It is reinstalled, the machine power is restored, and the machine is started. About 25 min later, the machine starts another harvest cycle on its own. The technician leaves the job, telling the owner to watch the machine for the rest of the day for correct operation.

SERVICE CALL 21

The owner of a restaurant calls and states that one of the ice machines is not dropping new ice to the bin. This is a new installation in a new restaurant. *The problem is so many minerals have been removed from the water by filtering that the water does not have enough conductivity to allow the electronic circuit to determine when the ice is touching the probes.*

The technician surveys the situation. The evaporator is frozen solid. This is a new machine, with the best water filtration system. There is a chance the filtration is so effective that the conductivity of the water has been reduced to a point where the harvest sensor probe will not sense the conductivity. The technician remembers the last service school she attended where these symptoms were described and goes to the shelf for a box of salt. A pinch of salt (about 1/10 teaspoon) is placed in the water sump where the pump will pick it up and circulate the slightly salted water over the probes. Within a very few minutes, the machine starts a harvest cycle.

The technician knows the owner cannot put salt in the filtered water every day, so she consults the manual. A resistor is available from the manufacturer that can be placed across certain terminals on the circuit board to allow the probe and electronic circuit board to sense the conductivity of water that has been filtered. Normal harvest can then be expected.

The technician obtains the resistor from the supplier and installs it across the appropriate terminals. The machine is started; the technician watches it through two complete cycles and is then satisfied the machine will operate satisfactorily.

SERVICE CALL 22

A restaurant owner calls to request a checkup for the cube-type ice maker in the kitchen. This maker has been in the kitchen for several years and the owner cannot remember when it was last serviced. *There is no particular problem.*

The technician examines the machine thoroughly. The condenser appears to be dirty, and the evaporator section needs cleaning. The machine is full of ice. The technician decides to remove the ice and clean the bin at the same time. Before he gets started on the cleaning, the technician politely explains to the store owner, in plain English, the advantages of cleaning an ice machine on a regular basis—how efficiency will improve and also the health benefits.

The machine is turned off and the power supply is shut off, locked, and tagged. The technician starts transferring the ice to trays to be moved to the walk-in cooler. As the technician gets toward the bottom of the bin, he sees that it is dirty. So cleaning the bin is a good idea.

The panels to the condenser are removed and it is vacuumed to remove lint and light dust. The condenser fan motor and terminal box are covered with a plastic bag to prevent moisture from entering. An approved coil cleaner is applied to the condenser and allowed to set on the coil so it can soak into the dirty surface. While the condenser coil cleaner is working, the bin is cleaned and sanitized. When the coil cleaner has been on the condenser for about 30 min, the technician flushes the cleaner off the condenser with a water hose and down a nearby drain. The restaurant owner is surprised at the amount of dirt washed out of the condenser.

The water that drains off the machine is mopped up from the floor around it. The plastic is removed from the motor and the unit is dried with towels. This unit has a clean cycle for cleaning the evaporator. In the clean cycle only the water pump circulates and chemicals may be circulated with the water for a period of time. The technician pours the recommended amount of chemicals in the sump and starts the pump. The chemicals circulate for the prescribed time; the technician dumps them down the drain and flushes freshwater through the system. The evaporator section is then sanitized using the clean cycle and the water is dumped. A hose is used to wash the bin area one more time to be sure no chemicals are still present and the machine is started. The technician allows the machine to make two harvests of ice and dumps them down the drain. This ensures there are no residual chemicals and that the machine is making ice properly.

The machine is then put back into service. The technician politely thanks the store owner for his business and lets him know he is a valued customer. While giving billing information with a work description and costs to the store owner, the technician calls him by name, makes eye contact, shakes his hand, and quietly leaves the job site.

SERVICE CALL 23

The owner of a small grocery store calls a service technician complaining of high water bills and warm refrigeration temperatures for a few coolers and freezers. *The problem is that two of the helical tube-within-a-tube water-cooled condensers are fouled with mineral deposits and are not*

transferring the condenser's rejected heat to the water. This causes the condensing pressure to increase, unwanted inefficiencies in the refrigeration systems, and warmer-than-normal refrigerated-box temperatures. Because the water-regulating valve reacts to head pressure increases, the water-regulating valve remains open all the time—wasting water down the drain.

The service technician locates all the refrigeration equipment's condensing units and observes that five of the six condensing units used to keep the dairy and meats cool are water-cooled. The water-cooled units use water-regulating valves to supply water to the five tube-within-a-tube condensers. City water is used for cooling and the used water is simply dumped down the drain after it exits the condenser. The customer explains to the technician that the water bill has always been somewhat high because of his use of water-cooled condensers, but it has gradually been increasing in the last three months. Also, refrigerated-box temperatures have been warmer than usual.

The service technician politely asks the customer some pertinent questions about the equipment in everyday language that the customer can understand and is careful not to interrupt while the customer is answering the questions. The technician wants to make the customer feel that he is helping to find a solution to the problem. After listening to the customer's account of the problem and answering any questions, the technician clearly explains what work has to be done and why it must be done. The technician explains any work options and the advantages and/or disadvantages of each option.

The technician realizes that two of the water-cooled condensers are dumping water down the floor drain very fast. He then measures the temperature difference between the water going into the condenser and the water going out of the condenser on these two units and finds that there is only a 1-degree temperature difference. This indicates that both of the tube-within-a-tube condensers are fouled with mineral deposits and are preventing heat from being transferred from the condenser to the water. Because of the severity of the fouling and because helical tube-within-a-tube water-cooled condensers are not cleanable, the service technician tells the customer that he will have to order two new condensers and then provides the customer an estimate and work orders in writing. A week later, the service technician recovers the entire refrigerant charge and replaces both of the tube-within-a-tube water-cooled condensers with more easily cleanable shell-and-tube condensers of the same capacity, **Figures 22.10(G)** and **22.11**. This will allow the service technician to more easily clean the condensers the next time they become severely fouled with minerals.

The service technician now double-checks the work done, cleans the work site, and explains how the problem was corrected. He then gives the final billing information with work description and costs to the customer. After making sure his hands are clean, the technician shakes the customer's hand, politely thanks the customer for the business, and quietly leaves the work site.

SERVICE CALL 24

A customer complains that a walk-in cooler designed to keep meats at 36°F is short cycling and not keeping the refrigerated space below 45°F. The R-134a unit has an outdoor, air-cooled condensing unit. *The problem is that the outdoor ambient is 10°F and that the low-ambient head pressure control mixing valve is defective. The valve is not allowing compressor's discharge gas into the receiver to keep the receiver pressure up to feed the metering device.* The evaporator is thus being starved of refrigerant and is running a low evaporator pressure with high superheats. This causes the low-pressure control to shut the unit off prematurely and short cycle.

After installing gauges on the unit, the service technician realizes that the head pressure is really low (50 psig) and that the evaporator superheat is really high (30°F). While the machine is still running, the technician feels the line between the head pressure control valve and the receiver and finds it to be very cold. This indicates that the head pressure control valve is defective and not allowing the compressor's discharge gas into the receiver to keep the receiver's pressure up in order to feed the metering device. This line should be warm at any ambient below 70°F and get warmer as the outdoor temperature drops.

When the temperature of the condenser is above 70°F, the refrigerant flow from the compressor is directed by the mixing valve through the condenser and into the receiver. If the outside ambient temperature drops below 70°F, the condensing pressure will fall and the pressure of the liquid coming from the condenser will also fall to a point below that of the bellows in the dome of the valve. This causes a piston to move in the valve and partially restrict the flow of refrigerant leaving the condenser. The condenser will now partially flood with refrigerant to maintain a certain condensing pressure. At the same time, discharge gas will bypass the condenser and flow directly to the receiver. This hot, superheated discharge gas going to the receiver now mixes with liquid refrigerant coming from the condenser. This keeps the receiver pressure up and helps keep metering devices fed with refrigerant.

The service technician decides to replace the valve by following these steps:

1. Recover the refrigerant.
2. Cut the process tube on the valve's charged dome to remove the pressure charge.
3. With a torch, heat the valve until the solder liquefies, then remove the valve.
4. Wrap the valve's body with a heat sink to prevent damage and silver-solder a new valve in place.
5. Install a new filter drier.
6. Leak check the system and evacuate to a deep vacuum (500 microns).
7. Charge the system and leak check again.

When the system is restarted, normal evaporator superheat and system pressures are realized. The refrigerated space pulls down in a matter of 1 hour. The technician feels

the line between the low-ambient control mixing valve and the receiver and realizes that it is warm. This means the valve is allowing the compressor's discharge gas into the receiver to keep the receiver's pressure up in order to feed the metering device. At the same time, the condenser will be partially flooded with liquid refrigerant to control head pressure.

Solutions for the following service calls are not provided here. Please discuss the solutions with your instructor.

SERVICE CALL 25

A store manager reports that the food is thawing out in the low-temperature walk-in freezer. During the winter, this freezer lost its charge, the leak was repaired, and the unit recharged. The service technician remembers that this unit was serviced by a new technician when the loss-of-charge incident happened. The technician suspects that the unit may be off because of the manual reset high-pressure control. This is the only control on the unit that will keep it off, unless there is a power problem.

After arriving at the job the technician finds that the unit is off because of the high-pressure control. Before resetting the control, the technician installs a gauge manifold so that the pressures can be observed. When the high-pressure control is reset, the compressor starts but the head pressure rises to 420 psig and shuts the compressor off. The system uses R-404A. The ambient temperature is 90°F, so the head pressure should not be more than the pressure corresponding to 120°F. This was arrived at by adding 30°F to the ambient temperature: 30°F + 90°F = 120°F. The head pressure corresponding to 120°F is 312 psig. It is obvious the head pressure is too high. The condenser seems clean enough, and air is not recirculating back to the condenser inlet.

What is the likely problem and the recommended solution?

SERVICE CALL 26

A customer calls to indicate that the medium-temperature walk-in cooler unit that was just installed is running all the time and not cooling the box.

The technician looks the job over. This is a new installation of a walk-in medium-temperature cooler that was just started up yesterday. The hermetic compressor is sweating down the side of the housing. The evaporator fan is operating, and the box temperature is not down to the cut-off point, so there is enough load to boil the refrigerant to a vapor.

What is the likely problem and the recommended solution?

SERVICE CALL 27

A customer reports that the low-temperature reach-in cooler that had a burned compressor last week is not cooling properly. The food is beginning to thaw.

Before arriving on the job site, the technician calls the customer on his cell phone and asks for an account of the problem. The technician carefully listens to the details, making sure not to interrupt the customer. Once at the job site, the technician parks his van out of sight of the store's customers and knocks on the service entrance door. Once invited inside, the technician introduces himself and hands the customer his business card. The technician then offers the customer his hand to shake. The customer then shows the technician the same cooler that was serviced last week. The technician examines the whole system. The evaporator seems to be starving for refrigerant. When the technician approaches the condensing unit in the back of the store, he notices that the liquid line is sweating where it leaves the drier. A suction-line drier was not installed at the time of the motor burn.

What is the likely problem and the recommended solution?

SERVICE CALL 28

A customer calls on the first hot day in the spring and says the walk-in freezer is rising in temperature. The food will soon start to thaw if something is not done.

The service technician examines the whole system and notices that the liquid line is hot, not warm. This is a TXV system and has a sight glass. It is full, indicating a full charge of refrigerant. The compressor is cutting off because of high pressure and then restarting periodically. This system uses R-404A, and the head of the compressor is painted orange to signify the refrigerant type. Gauges are installed, and the head pressure starts out at 365 psig and rises to the cut-out point, or 400 psig. The condenser is located in an equipment room where the ambient temperature is consistently 90°F.

What is the likely problem and the recommended solution?

SERVICE CALL 29

A restaurant manager indicates that the compressor on a water-cooled unit is cutting off from time to time *and the walk-in cooler is losing temperature. This is a medium-temperature cooler, using R-134a.*

The service technician arrives to find a wastewater condenser (the water is regulated by a water-regulating valve to control the head pressure and then goes down the drain). The water is coming out of the condenser and going

down the drain at a rapid rate. The liquid line is hot, not warm. The water inlet temperature is measured at 75°F and the outlet temperature is 80°F. Gauges installed at the compressor indicate the head pressure is 223 psig. The compressor is cycling off and on because of the high-pressure control. The head pressure should be about 135 psig with a condensing temperature of 105°F.

What is the likely problem and the recommended solution?

SERVICE CALL 30

A customer reports that the temperature is rising in a frozen-food walk-in freezer. Although it tries, the compressor will not start.

Because it is the first service call of the day, the service technician makes sure he takes off his rings before starting the work day. Rings can get caught in mechanical and/or electrical devices and are a serious safety concern. A quick look in the vehicle's mirror tells the technician that his hair needs to be combed before entering the market where the service call is to be made. The technician then notices that his identification badge with his company's name and logo is still on the dash of the service van. He quickly pins the badge on his clean uniform and knocks on the door of the service entrance to the market where his services are needed. Once invited in, the technician introduces himself and carefully listens to the customer. The technician then politely answers the customer's questions, follows the customer to the walk-in freezer, and starts work.

The service technician listens to the compressor try to start and hears the overload cut it off, making a clicking sound. The compressor is hot, too hot to start many more times before it will overheat to the point of being off for a long time. Every start the motor goes through makes it hotter. The technician needs to make sure that it starts and stays on line the next time it starts. This is a single-phase compressor and has a starting relay and start capacitor. The problem could be in several places: (1) the start capacitor, (2) the starting relay, or (3) the compressor. The technician uses a digital multimeter to test the resistance in the windings of the compressor, from the common terminal to the run terminal, and from the common terminal to the start terminal. The compressor has continuity and should start, provided it is not locked. The microfarad function of a digital multimeter is used to test the start capacitor. It shows that it is within the manufacturer's specified range. The starting relay is checked for continuity from terminal 1 to terminal 2. This is the circuit through which the starting capacitor is energized. The circuit shows zero (0) resistance; the switch is closed and the contacts are good.

What are the likely problems and the recommended solutions?

SERVICE CALL 31

A motel maintenance person calls to report that the ice maker on the second floor of the motel is not making ice. This ice maker is a flake ice machine. It sits outside and the temperature is hot.

The technician arrives and parks his service van out of sight of the motel customers, being especially careful not to block any driveway exits. The technician then puts on shoe covers to protect the motel's carpets from dirt and grease that might be on his shoes from previous job sites. After politely introducing himself to the motel owner and again carefully listening to the service call complaint, the technician asks to be led to the ice machine.

The technician notices right away there is no ice in the bin. The compressor is running and hot air is coming from the condenser. Therefore, it must be refrigerating. The suction line leaving the evaporator is cold. Gauges are fastened to the compressor and reveal that the suction pressure is high, but so is the discharge pressure. The system uses R-404A and the suction pressure is 69 psig (30°F boiling temperature). No ice can be made under these conditions. The discharge pressure is 379 psig (135°F condensing temperature). It is a hot day, 97°F, but not hot enough to cause a head pressure this high. The technician wonders if the unit is overcharged or if the condenser is dirty. This could cause high head pressure, which would then cause high suction pressure and low or no ice production.

After sitting down and thinking for a minute, the technician thinks of the water circuit, then feels the water line. It is cooler than hand temperature (which is about 90°F). This is not the answer. The technician then looks at the float chamber and discovers that water is continuously running into the chamber.

What is the likely problem and the recommended solution?

REVIEW QUESTIONS

1. What is the first thing a technician should know before beginning troubleshooting?
   A. The typical operating conditions for the equipment.
   B. Type of oil that is in the system.
   C. The make and model of the equipment being serviced.
   D. Which company serviced the unit last.
2. What is the coil-to-air temperature relationship for a refrigeration system designed for minimum food dehydration?
   A. 12°F to 15°F
   B. 10°F to 15°F
   C. 8°F to 20°F
   D. 8°F to 12°F
3. When the desired relative humidity is 75%, the coil-to-air temperature relationship should be
   A. 8°F to 20°F.
   B. 10°F to 15°F.
   C. 12°F to 15°F.
   D. 16°F to 22°F.
4. High-temperature refrigeration is typically used for
   A. meat and poultry.
   B. flowers and candy.
   C. fruits and vegetables.
   D. ice storage.
5. Low-temperature refrigeration is typically used for
   A. meat and poultry.
   B. comfort cooling.
   C. ice making and ice storage.
   D. fruits and vegetables.
6. At what evaporator temperature does an ice maker usually begin to make ice?
   A. 35°F
   B. 32°F
   C. 25°F
   D. 20°F
7. When an air-cooled condensing unit is located outdoors, it requires
   A. head pressure controls.
   B. hot gas defrost.
   C. a suction-line filter/drier.
   D. an evaporator pressure regulator.
8. Most standard-efficiency air-cooled condensers will condense the refrigerant at a temperature of about ______°F above the ambient temperature.
   A. 40
   B. 30
   C. 20
   D. 10
9. A water-cooled condenser that uses freshwater, such as city water or well water, uses about ______ gal of water per minute per ton with a 20°F water temperature difference.
   A. 1.5
   B. 3
   C. 6
   D. 10
10. A water-cooled system that has a cooling tower and recirculates the water, uses about ______ gal of water per minute per ton with a 10°F water temperature difference.
    A. 1.5
    B. 3
    C. 6
    D. 10
11. A wastewater system that is experiencing high head pressure, a high water flow rate, and a low water temperature difference most likely has
    A. an undercharge.
    B. an overcharge.
    C. a dirty condenser coil.
    D. a system leak.
12. Most recirculated water systems will have a refrigerant condensing temperature that is
    A. 10°F above the water inlet temperature.
    B. 10°F above the water outlet temperature
    C. 30°F above the ambient air temperature
    D. within 7° of the ambient air wet-bulb temperature.
13. A TXV system responds to an overcharge of refrigerant with
    A. low suction pressure.
    B. high head pressure.
    C. a flooded compressor.
    D. high superheat.
14. An automatic expansion valve system responds to an overcharge of refrigerant with
    A. low suction pressure.
    B. high head pressure.
    C. a flooded compressor.
    D. a flooded evaporator.
15. A capillary tube system responds to an overcharge of refrigerant with
    A. high suction pressure and high head pressure.
    B. high head pressure and low suction pressure.
    C. a flooded evaporator and flooded compressor.
    D. high superheat and high subcooling.
16. Which of the following can cause an inefficient evaporator?
    A. A clogged expansion device or dirty condenser coil.
    B. A clogged suction line drier, an undersized receiver, or a defective EPR.
    C. A dirty coil, low airflow, or an over-feeding expansion valve.
    D. A dirty coil with proper airflow or an under-feeding expansion valve.
17. An inefficient condenser will cause
    A. a low compression ratio.
    B. low compressor amp draw.
    C. the head pressure to fall.
    D. the head pressure to rise.

18. Symptoms of an inefficient compressor are
    A. high head pressure and low suction pressure.
    B. high head pressure and high suction pressure.
    C. low head pressure and low suction pressure.
    D. low head pressure and high suction pressure.
19. A restriction always causes a(n) _______ at the point of restriction.
    A. oil trap
    B. pinhole leak
    C. pressure drop
    D. rupture
20. A reciprocating compressor should be able to pull a vacuum of _______ when being bench-tested against atmospheric pressure.
    A. 26 in. to 28 in. Hg
    B. 20 in. to 24 in. Hg
    C. 24 in. Hg
    D. 1000 microns
21. A reciprocating compressor should be able to pull a vacuum of _______ when being bench-tested against a head pressure of 100 psig.
    A. 26 in. to 28 in. Hg
    B. 20 in. to 24 in. Hg
    C. 24 in. Hg
    D. 1000 microns
22. A closed-loop compressor test can be performed with the compressor in the system when
    A. the refrigerant is too contaminated to recover.
    B. the compressor has suction and discharge service valves.
    C. the system is equipped with both a liquid- and a suction-line drier.
    D. the compressor is a hermetic scroll.
23. To test compressor capacity in the field when the system is not at design conditions, a technician can
    A. block the condenser airflow and watch for the head pressure to rise.
    B. block the evaporator airflow and watch for the suction pressure to rise.
    C. heat the TXV sensing bulb and watch for the system pressures to equalize.
    D. put the system into defrost and monitor the compressor amp draw.
24. Explain what the symptoms would be on a reach-in cooler with a dirty evaporator. The system has TXV metering device and uses off-cycle defrost.
25. Explain what the symptoms would be on a low-temperature walk-in freezer with a defective defrost time clock motor. The system has a TXV metering device and uses resistive heat for defrost.

SECTION 6

AIR-CONDITIONING (HEATING AND HUMIDIFICATION)

UNITS

UNIT 30

ELECTRIC HEAT

OBJECTIVES

After studying this unit, you should be able to

- discuss the efficiency and relative operating costs of electric heat.
- discuss the properties and operation of portable and radiant electric heating devices and panels.
- discuss the control and operation of electric baseboard, unit and wall heaters.
- discuss three reasons why electric boilers are desirable in some geographic regions.
- describe the operation of an electric force-air furnace.
- describe how thermostats, sequencers, and contactors operate in electric forced-air furnaces.
- describe two commonly used types of wiring diagrams.
- explain how power-passing and power-consuming devices work together to control the operation of an electric forced-air furnace.
- trace the circuitry in a diagram of an electric forced-air furnace.
- calculate airflow, temperature difference, and heat output of an electric forced-air furnace using the sensible heat formula.
- describe typical preventive maintenance procedures used in electric heating units and systems.
- perform basic tests in troubleshooting electrical problems in an electric forced-air furnace.

SAFETY CHECKLIST

- ☑ Be careful that nails or other objects driven into or mounted on radiant heating panels do not damage the electrical circuits.
- ☑ Any heater designed to have air forced across the element should not be operated when the blower is not operating.
- ☑ Always observe all electrical safety precautions.

30.1 INTRODUCTION

Electric heat is produced by converting electrical energy into heat energy. This is done by placing a known resistance of a particular material in an electric circuit. The resistance has relatively few free electrons and does not conduct electricity well. The resistance to electron flow produces heat. One type of material commonly used in electric heating is **nichrome**, which is short for nickel chromium. Wire made from nichrome is used in the majority of electric heaters.

Electric heat is very efficient but can be more expensive to operate compared to other sources of heat. It is efficient because very little electrical energy is lost from the meter to the heating element and there are no chimney losses as in the case of fossil-fuel heating systems. It can be expensive because it takes large amounts of electrical energy to produce the heat, and the cost of electrical energy in most areas of the country is expensive when compared to fossil fuels (coal, oil, and gas). It is, therefore, important to understand that efficiency and cost are not the same.

The purchase price of electrical heating systems is usually lower than that of other systems. The installation and maintenance is also usually easier and, therefore, less expensive. This makes electric heating systems attractive to many purchasers, especially those who live in areas where electricity is more reasonably priced. When electric heat is used as the primary heat source, insulation with a high R-value is normally used throughout the structure to lower the amount of heat required. Although less expensive from a first-cost standpoint, electric heating systems can be very expensive to operate.

This unit briefly describes several types of electric heating devices. Emphasis, however, is placed on central forced-air electric heat because, in some areas of the country, it is frequently serviced by HVACR technicians.

30.2 PORTABLE ELECTRIC HEATING DEVICES

Portable or small space heaters are sold in many retail stores and by many industrial distributors and manufacturers, **Figure 30.1**. Some have glowing coils (due to the resistance of the wire to electron flow). These coils transfer heat by radiation (infrared rays) to the solid objects in front of the heater. The radiant heat travels in a straight line and is absorbed by solid objects that warm the space around them. Radiant heat also provides comfort heating to individuals. The heat concentration decreases by the square of the distance and is soon dissipated into the space, **Figure 1.12.** Quartz and glass-panel heaters are also used to heat small spaces by radiation, **Figure 30.2**. Some space heaters use fans or small blowers to move air over the heating elements and into the space. This is called forced-convection heat because the heat-laden air is moved mechanically. The units may be designed to move enough air across the heating elements so that they do not glow.

Radiant spot heating can be effectively used at doorways, warehouses, work areas, and even outdoors. The effect is

Figure 30.1 A portable electric space heater. Courtesy Fostoria Industries, Inc.

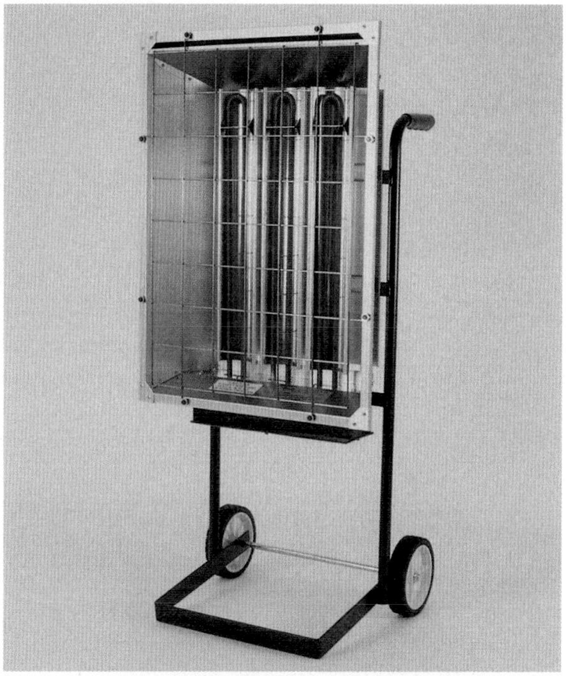

Figure 30.2 A quartz heater. Courtesy Fostoria Industries, Inc.

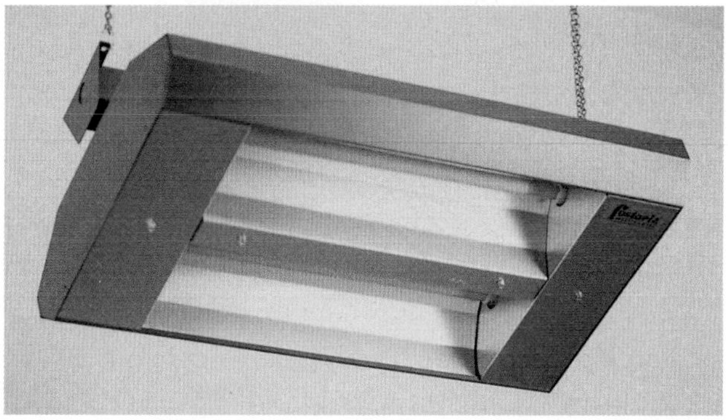

Figure 30.3 A radiant ceiling heating panel.

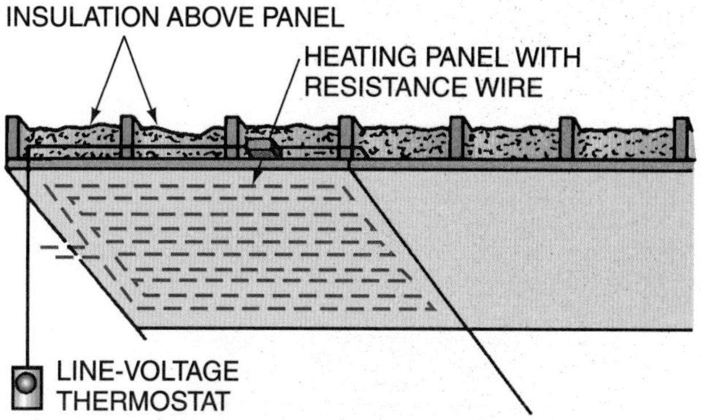

Figure 30.4 A roll of electric radiant heating element.

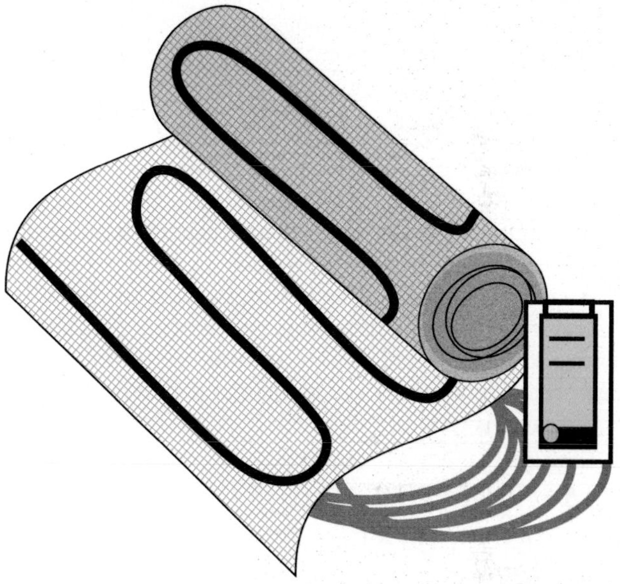

much like a sunlamp pointing at the heated area. The distance between the heater and the objects being heated has a great bearing on the effectiveness of radiant heating. Refer back to Unit 1 for more information on how distance affects heat transfer by radiation.

30.3 RADIANT HEATING PANELS

Radiant electric heating can be used in residential and light commercial buildings and can be located in either the ceilings or floors of the structure. Panels used for ceiling installation are often made of gypsum board with wire heating circuits running throughout, **Figure 30.3**. When electric radiant heat is to be installed in floors, it is often supplied in roll form, **Figure 30.4**. This type of product is typically less than 1/8-in. thick, so it has little if any effect on finished floor height. It can be installed under the thinset concrete that is typically used for ceramic or porcelain tile installations, **Figure 30.5**. Electric radiant heating elements are often controlled with individual room thermostats, providing individual room or zone control. The heat produced is even, is easy to control, and tends to keep the mass of the room warm by its radiation. This makes items in the room pleasingly warm to the touch. **SAFETY PRECAUTION:** *Be careful that nails or other objects driven into or mounted on the panels do not damage the electrical circuits. Be particularly careful when installing a cooling system where electric radiant heat is provided if the supply registers or grilles are to penetrate the ceiling or floor where the radiant wires are located.•* Electric radiant heaters must have good insulation behind (or above) to keep heat from escaping. When using the floor-heating roll-type material, it is recommended that insulation be placed under the heater to help prevent heat loss.

Figure 30.5 The installation of a rolled radiant heating element.

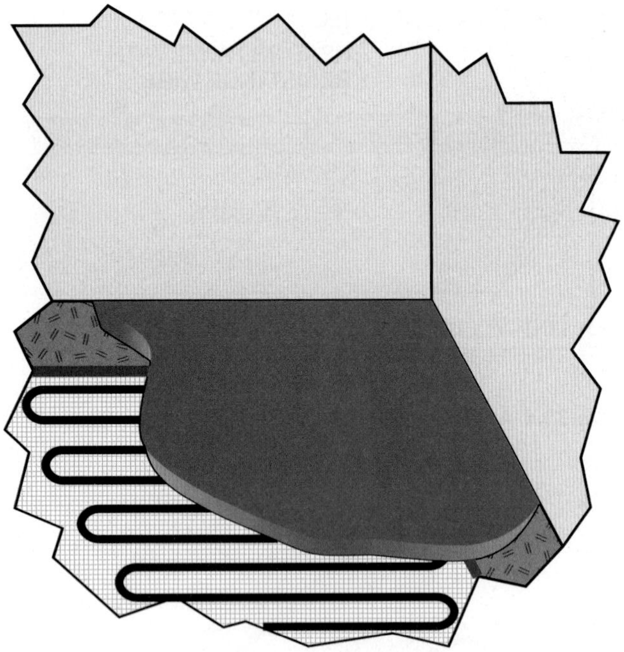

Although not required to install electric radiant heating panels, an electrician's license may be required to tie the heater circuitry into the distribution panel. This depends on the type of electric radiant heating panels being used and how they are to be powered. Before starting any installation, be sure to check local codes and guidelines to determine what, if any, licensing is required.

30.4 ELECTRIC BASEBOARD, UNIT AND WALL HEATERS

Baseboard heaters are popular convection heaters used for whole-house, spot, or individual room heating, **Figure 30.6**. They are economical to install and can be controlled by individual room thermostats, which are normally line-voltage devices that are mounted on the individual heaters themselves. Unused rooms can be closed off and the heat turned down, which makes baseboard heaters economical in certain applications.

Baseboard units are mounted on the wall just above the floor or carpet, usually on outside walls. The heater is a natural-draft unit. Air enters near the bottom, passes over the electric element, is heated, and rises in the room. As the air cools, it settles, setting up a natural-convection air current. Refer back to Unit 1 for more information on natural convection currents. Baseboard heat is easy to control, safe, quiet, and evenly distributed throughout the house.

Unit and wall heaters are typically mounted on ceilings or walls and use a fan or blower to force the air across the elements into the space to be heated, **Figure 30.7.** These heaters are controlled with line-voltage thermostats. Some heaters are recessed into the wall and all that protrudes into the occupied space is a decorative cover. Some new-generation wall heaters come with programmable touch controls, **Figure 30.8.** **SAFETY PRECAUTION:** *Any heater designed to have air forced across the element should not be operated unless proper airflow is established.*•

Figure 30.6 An electric baseboard heater.

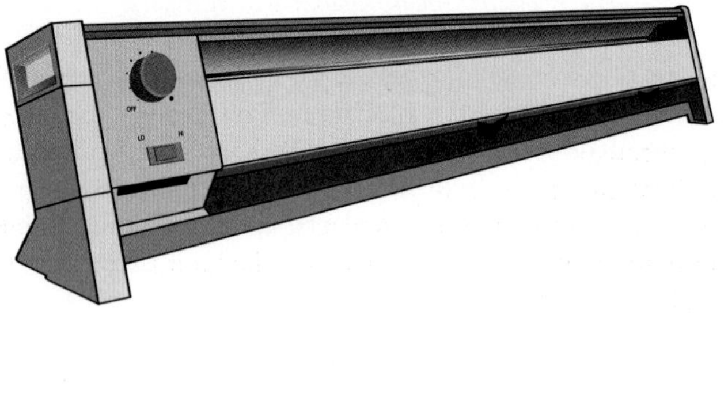

Figure 30.7 (A) Ceiling- and (B) wall-mounted unit heaters.

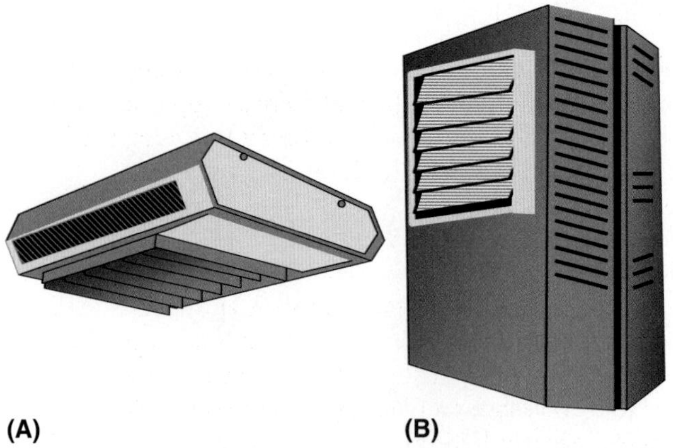

Figure 30.8 Recessed wall heater with programmable touch controls.

30.5 ELECTRIC HYDRONIC BOILERS

The electric hydronic (hot water) boiler system is used for some residential and light commercial applications. Except for the control arrangement and safety devices, the boiler is somewhat like an electric domestic water heater. It is also connected to a closed loop of piping and requires a pump to move water through the loop, which consists of the boiler and the terminal heating units, **Figure 30.9.** The electric boiler is small and compact for easy location and installation. Since there is no combustion in this type of appliance, there is no need for any flue or vent piping to be installed. It is relatively easy to troubleshoot and repair. As with other electric heating systems, electric hydronic boilers are a viable option when the cost of electricity is low.

Figure 30.9 Electric boiler. Courtesy Slant/Fin Corp.

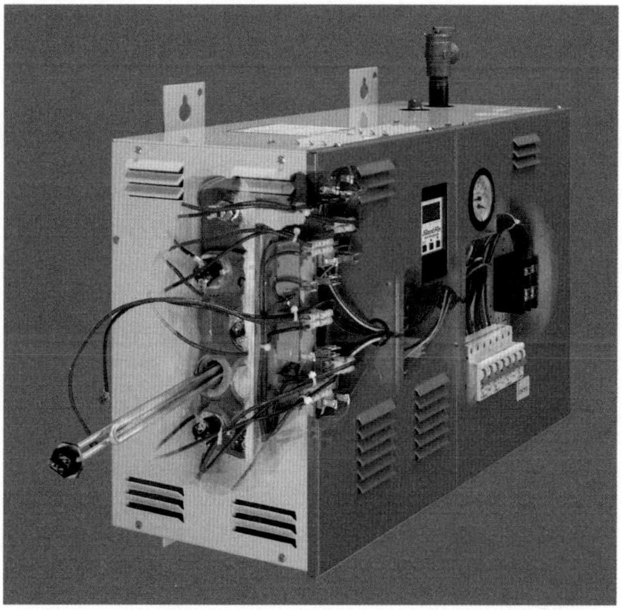

Any boiler handling water is subject to all the problems of a water-circulating system. For instance, if the boiler shuts down and the room temperature drops below freezing, pipes will burst, and the boiler will be damaged; water treatment against scale and corrosion is also required. The boiler itself is efficient because it converts virtually 100% of its electrical energy input into heat energy and transfers it to the water. When the boiler is located in the conditioned space, any heat loss through the boiler's walls or fittings is not lost from the structure. The system is quiet and reliable, but cooling and humidification systems cannot easily be added.

Electric boilers are often equipped with the following components:

- Contactors to energize and deenergize the heating elements
- Sequencers to ensure that all of the heating elements are not energized at the same time
- Relays to control the operation of the circulator pumps
- A pressure relief valve to prevent system pressure from getting too high
- Aquastats to control and maintain the temperature of the water and prevent the boiler water from getting too hot
- Circuit breaker protection to protect the circuits from an overcurrent condition

When dealing with electric heating equipment, it is important that both operational and safety devices be incorporated into the system to ensure the safety of those in the structure. Quite often, redundant, or backup, controls are used as well in the event one or more of the safeties fail.

30.6 CENTRAL FORCED-AIR ELECTRIC FURNACES

Central forced-air furnaces utilize ductwork to distribute the heated air to remotely located rooms or spaces. The heating elements are factory installed in the furnace along with the blower, or they can be purchased as duct heaters and installed within the ductwork, **Figures 30.10** and **30.11**. The heating elements are made of nichrome resistance wire mounted on ceramic or mica insulators. They are enclosed in the ductwork or in the housing of an electric furnace.

Central heaters are usually controlled by a single thermostat, resulting in one control point—there is no individual room or zone control. An advantage of duct heaters is that temperature in individual rooms or zones can be controlled by placing multiple heaters in the ductwork system, **Figure 30.12.** However, an interlock system must be incorporated. The interlock system will not allow power to reach the heating element unless the blower is operating, which keeps the heating elements from overheating and possibly causing a fire. Air-conditioning (cooling and humidification) can usually be added to this system because of the air-handling feature. **SAFETY PRECAUTION:** *Be careful when servicing these systems because many exposed electrical connections are behind the inspection panels. Electric shock can be fatal.*•

30.7 AUTOMATIC CONTROLS FOR FORCED-AIR ELECTRIC FURNACES

Automatic controls are used to maintain temperature at desired levels in given spaces and to protect the equipment and occupant. Three common controls used in electric heat applications are **thermostats**, **sequencers** (discussed later in this unit), and contactors (or relays). Safety devices commonly found on electric furnaces are the limit switch and the fusible link.

Figure 30.10 This electric duct heater raises the temperature of the air in the duct.

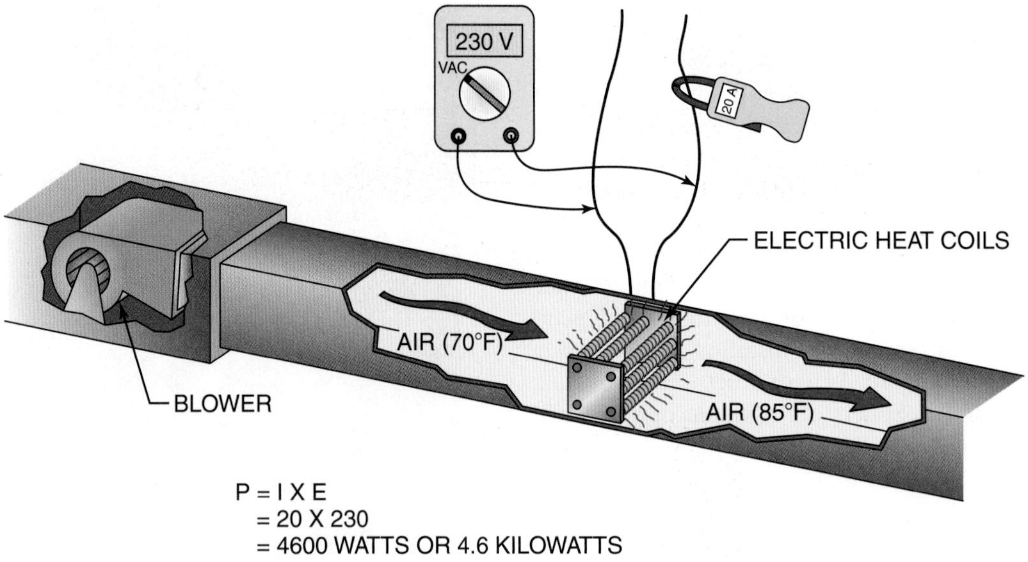

Figure 30.11 A central forced-air electric furnace.

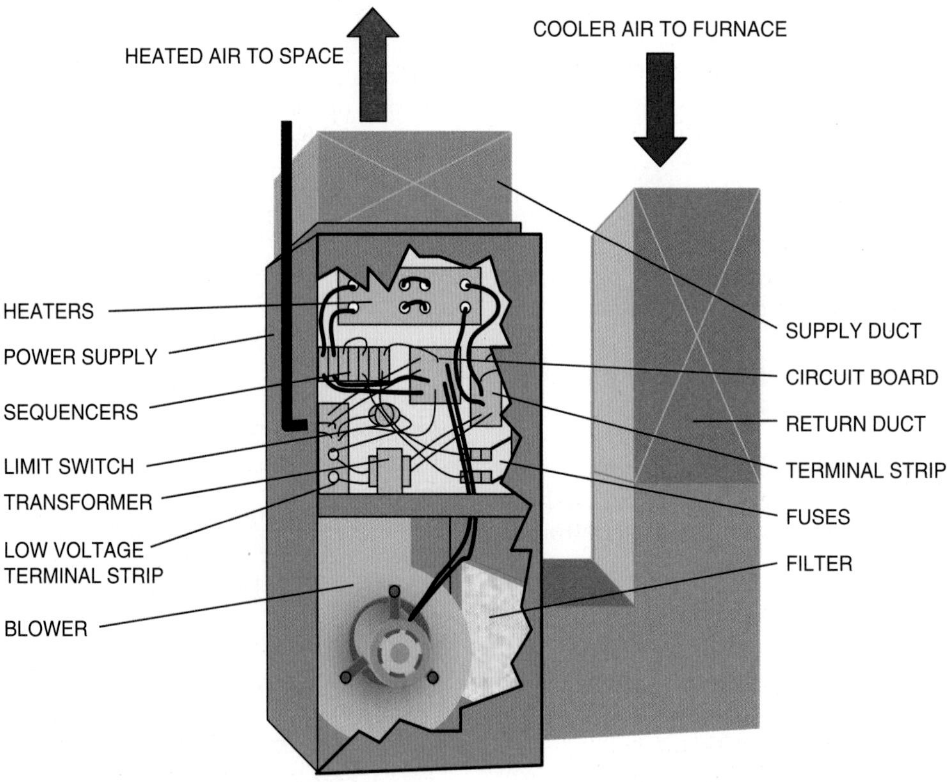

THE LOW-VOLTAGE THERMOSTAT

The low-voltage thermostat is used to control sequencers and contactors because it is compact, quick to respond to load changes, safe, and easy to install and troubleshoot. In many localities the low-voltage wiring may be installed without an electrical license. **Figure 30.13** shows a wiring diagram of a typical low-voltage thermostat that is used in conjunction with electric forced-air furnaces.

The wiring diagram in **Figure 30.13** can be used with a single transformer to control the heating and cooling modes of operation, or can be used with two separate transformers. If

Figure 30.12 An air distribution system with multiple electric heaters to serve different zones.

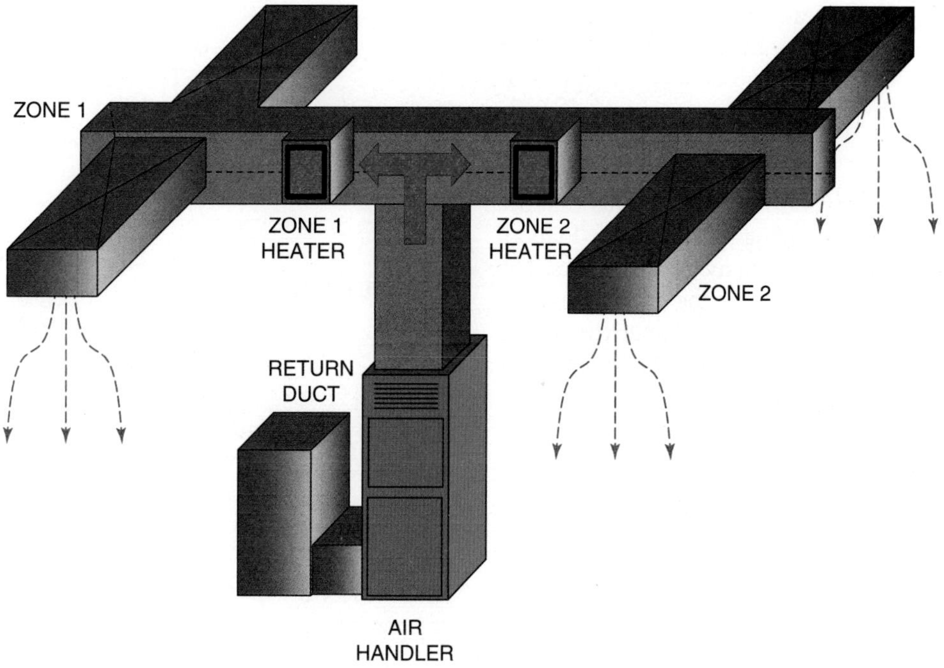

Figure 30.13 A wiring diagram of a low-voltage thermostat. Some manufacturers use different terminal designations.

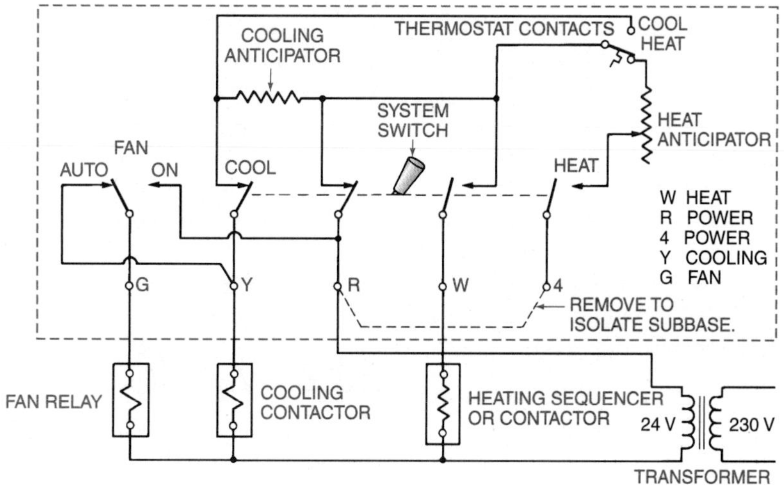

two transformers are used, the jumper wire between terminal "R" and terminal "4" is removed. Removing this jumper creates two isolated control circuits; one for the heating mode and one for the cooling mode. Thermostat subbases that have the capability of isolating the heating and cooling control circuits are referred to as isolating, or isolated, subbases. This isolated subbase may be needed when air-conditioning is added after the furnace is installed. The furnace will have its low-voltage power supply, and the air-conditioning unit may have its own low-voltage power supply. If one power supply is used for both, the jumper between terminals "R" and "4" will remain in place. The R terminal is the cooling terminal, and the 4 terminal is assigned to heating if the subbase has two power supplies. **NOTE:** *Many thermostat manufacturers identify the power terminals for the heating and cooling circuits as RH and RC. As mentioned, if only one transformer is used, the RH and RC terminals can be jumped together.* •

The heat anticipator used in a thermostat with electric heat may need to be set at the time of the installation. Heat anticipators are adjustable devices that are wired in series with the thermostat's heating contacts. The setting is determined by the total current draw in the 24-V heating control circuit that

passes through the thermostat and heat anticipator. When this current is determined, it is then set with the indicator on the heat anticipator. Digital thermostats have heat anticipator circuitry built into them and automatically adjust themselves, so field setting and adjustments are not needed.

SEQUENCERS

Most electric furnaces have several heating elements that are activated in stages to avoid energizing a high-power load all at once. Some furnaces may have as many as six heaters. The *sequencer*, **Figure 30.14**, is used to bring the heaters into the active electric circuit at the right time to prevent the underheating or overheating of the space. The sequencer uses low-voltage control power to start and stop the electric heaters. A mechanical sequencer uses a bimetal strip with a low-voltage wire wrapped around it. When the thermostat calls for heat, the low-voltage wire heats the strip and warps it out of shape in a known direction for a known distance. As it warps or bends, it closes electrical contacts to the electric heat circuit. This bending takes time, and each set of contacts closes in a certain order with a time delay between the closings. When the heat requirement has been met, the thermostat opens the low-voltage circuit, and the steps are reversed.

See **Figure 30.15** for a diagram of a package mechanical sequencer. This sequencer can start or stop three stages of strip heat. Some sequencers have five circuits: three heat circuits, a fan circuit, and a circuit to pass low-voltage power to another sequencer for three more stages of heat.

Another sequencer design, called an *individual sequencer*, has only a single circuit that can be used for starting and stopping an electric heat element, but it could have two other circuits: one to energize another sequencer through a set of low-voltage contacts and one for starting the blower motor. Several stages of electric heat may be controlled with several of these sequencers, one for each heat strip. New generation sequencers use electronic circuitry to control the opening and closing of the electrical contacts for the individual heating stages.

Figure 30.14 A multiple-type sequencer. Photo by Bill Johnson

Figure 30.15 A sequencer with three contacts. (A) OFF position. (B) ON position.

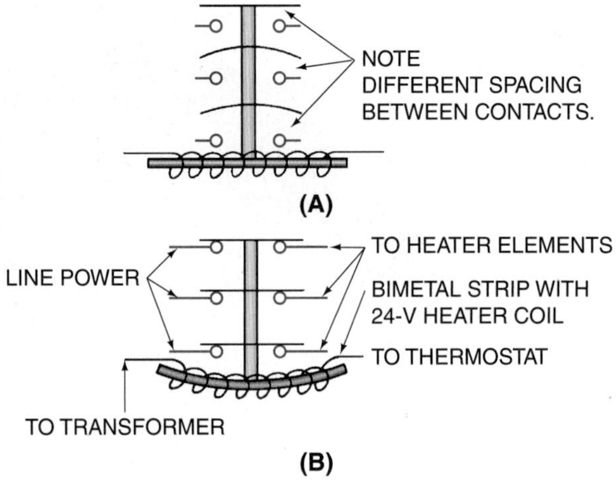

CONTACTORS FOR CONTROLLING ELECTRIC FURNACES

Using contactors or relays to control electric heat means that all of the heat will be started at one time unless separate relays, controlled by some type of time-delay or other staggering sequence, are used. Contactors are magnetic and make noise. They are normally used on commercial systems where they may not be directly attached to the ductwork. This prevents the noise from traveling through the system, **Figure 30.16**. The contactor magnetic coils are 24 V and 230 V. The 24-V coils are controlled by the room thermostat, and they energize the 230-V coil. There are four contactors and their coils in this diagram for the electric heat and one for the blower. The 24-V coils are energized at the same time and pull in together.

30.8 WIRING DIAGRAMS

Individual manufacturers vary in how they illustrate electrical circuits and components. Some use pictorial diagrams, some use schematics, and others use both types of diagrams. The *pictorial* diagram shows the location of each component as it actually appears to the person installing or servicing the equipment. When the panel door is opened, the components inside the control box are in the same location as on the pictorial diagram. The schematic wiring diagram (sometimes called the line wiring diagram or ladder diagram) shows the current path to the components. Schematic diagrams help the technician to understand and follow the intent of the design engineer.

Figure 30.17 contains a legend of electrical symbols for an electric heating circuit. Such a legend is vital in following a wiring diagram. Many symbols are standardized throughout the industry, even throughout the world.

Figure 30.16 (A) The wiring diagram for an electric furnace using contactors. (B) A line or ladder diagram.

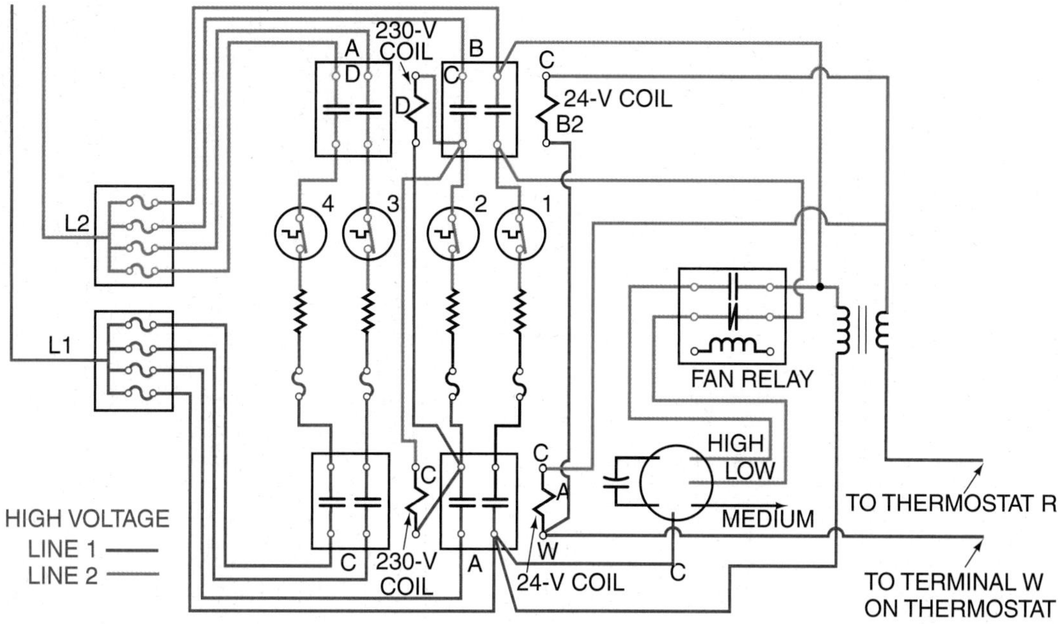

(A)

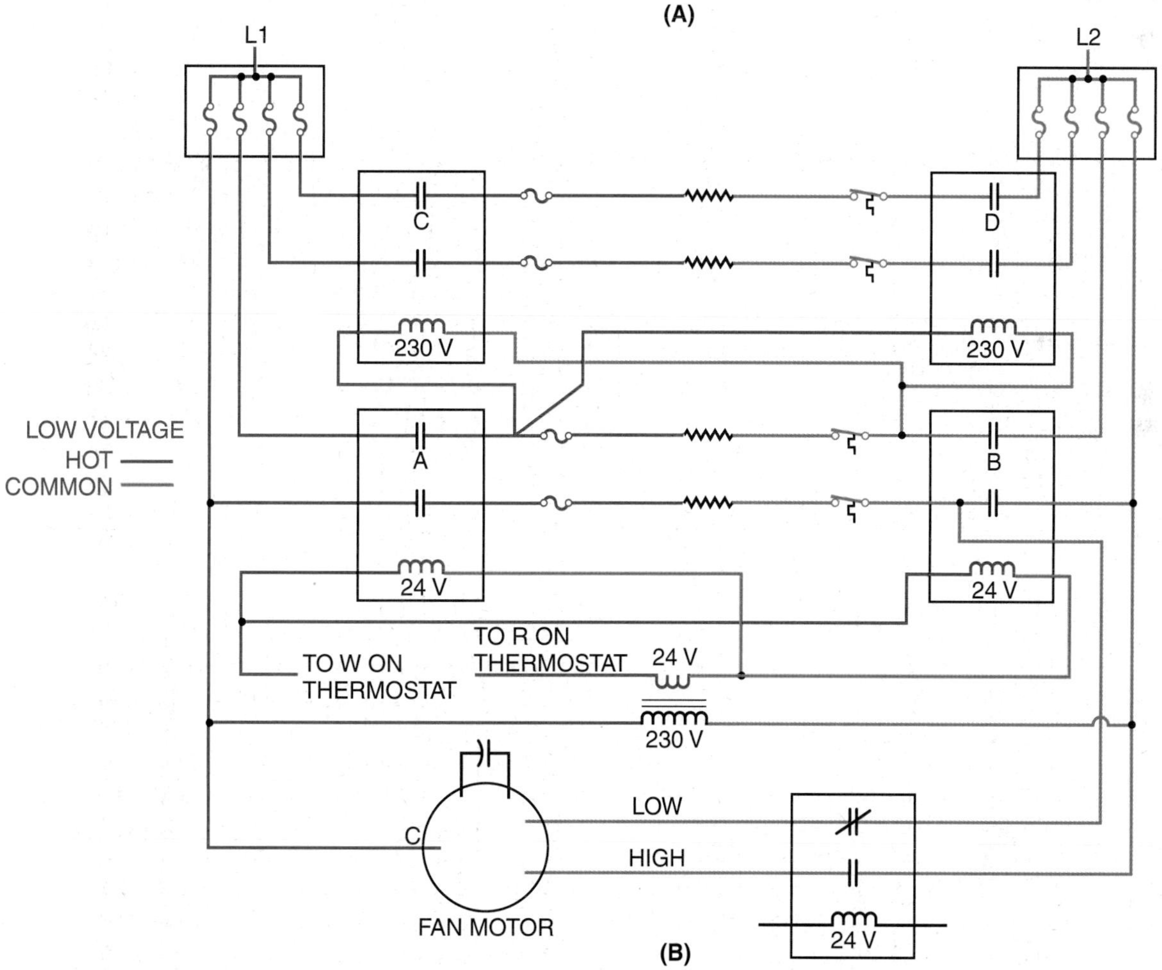

(B)

Figure 30.17 Components of a pictorial diagram.

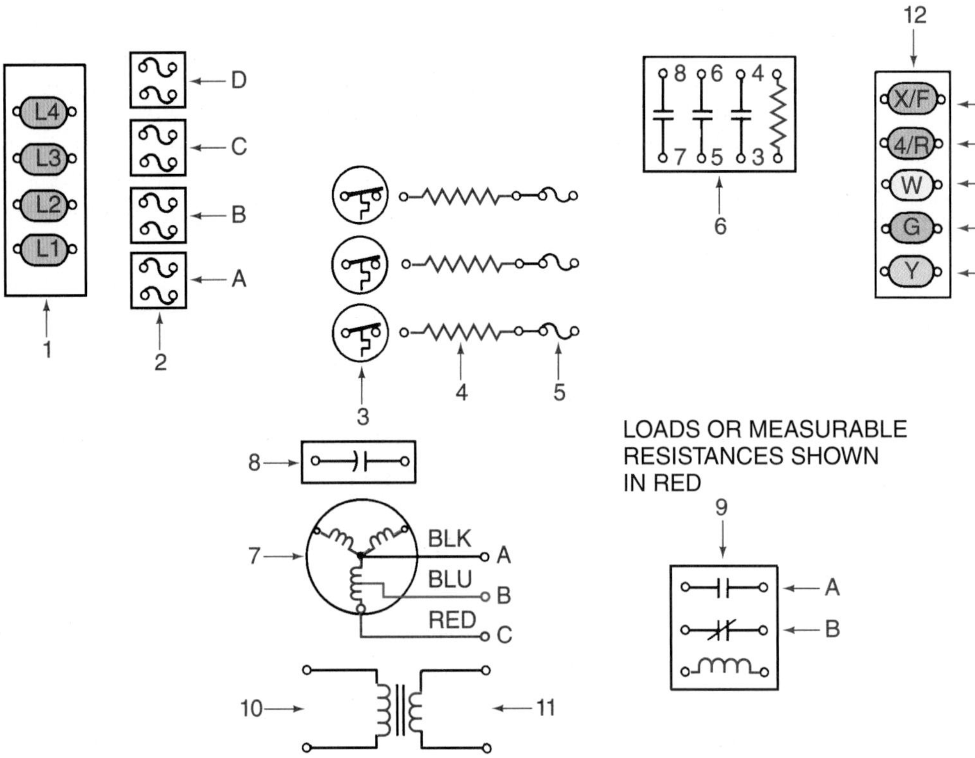

1. LINE-VOLTAGE TERMINAL BLOCK FROM DISCONNECT
2. FUSE BLOCK
   A. L1 FUSE BLOCK
   B. L2 FUSE BLOCK
   C. L3 FUSE BLOCK
   D. L4 FUSE BLOCK
3. AUTOMATIC RESET HIGH LIMIT SWITCH
4. ELECTRIC HEAT ELEMENT
5. FUSIBLE LINK
6. SEQUENCER
7. FAN MOTOR
   A. HIGH-SPEED MOTOR LEAD
   B. MEDIUM-SPEED MOTOR LEAD
   C. LOW-SPEED MOTOR LEAD
8. FAN MOTOR CAPACITOR
9. FAN RELAY (DOUBLE-POLE, SINGLE-THROW)
   A. NORMALLY OPEN CONTACTS
   B. NORMALLY CLOSED CONTACTS
10. PRIMARY SIDE OF TRANSFORMER (LINE VOLTAGE)
11. SECONDARY SIDE OF TRANSFORMER (LOW VOLTAGE—24 V)
12. LOW-VOLTAGE TERMINAL BLOCK
    A. COMMON TERMINAL FROM TRANSFORMER
    B. TERMINAL FOR HOT LEAD FROM TRANSFORMER
    C. HEATING
    D. FAN RELAY FOR COOLING
    E. COOLING

30.9 CONTROL CIRCUITS FOR FORCED-AIR ELECTRIC FURNACES

The low-voltage control circuit safely and effectively controls the heating elements that do the work. The circuit contains devices for both safety and for control. For example, the limit switch, a safety device, shuts off the unit if high temperature occurs; the room thermostat, an operational control device, stops and starts the heating system based on the difference between the desired room temperature and the actual temperature in the occupied space.

Safety and operating devices can consume power and do work or pass power to a power-consuming device. For example, the sequencer heater coil that operates the bimetal and a magnetic solenoid that moves the armature in a contactor are power-consuming devices in the low-voltage circuit. The contacts of a contactor or a limit control pass power to the power-consuming devices and are in the line-voltage circuit. **Power-passing devices are wired in series with the power-consuming devices. Power-consuming devices are wired in parallel with each other.**

Figure 30.18 is a diagram of a low-voltage control circuit. By tracing this circuit, you can see that when the thermostat contacts are closed, a circuit is completed. The transformer is the power source. The common leg of power in the diagram is blue and furnishes power to the top side of the sequencer coil. The red leg furnishes power to the thermostat and on to the bottom of the sequencer coil. When the room is colder

Figure 30.18 A diagram of a low-voltage control circuit.

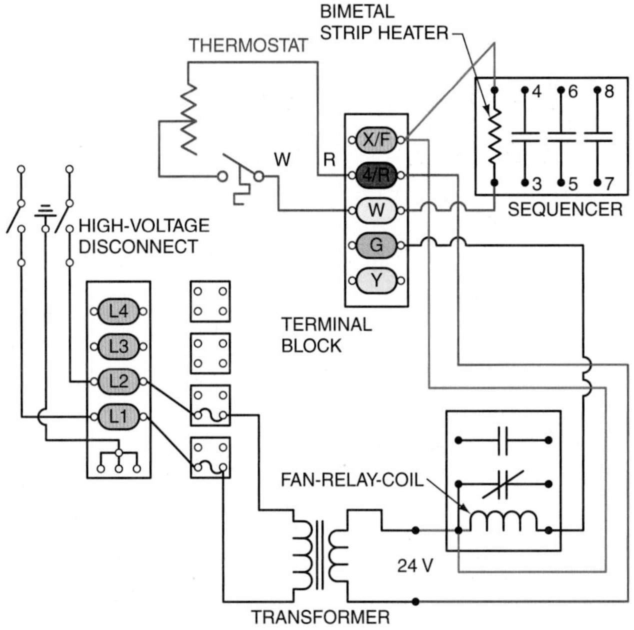

than the temperature setting on the thermostat, the thermostat contacts close. This energizes the coil on the sequencer, which causes the contacts on the sequencer to start closing. The closing contacts energize the heating elements in the line-voltage circuit. The line to the G terminal is used for the cooling circuit or when the fan switch on the thermostat is set to the ON position. The heating elements and the fan circuit have been omitted to simplify the circuit.

Control of a single heating element is illustrated in **Figure 30.19.** The electrical current from the L1 fuse block goes directly to the limit switch "A", which is a temperature-actuated switch that opens under excessive heat and provides

Figure 30.19 Circuit for a single heating element with sequencer (line voltage).

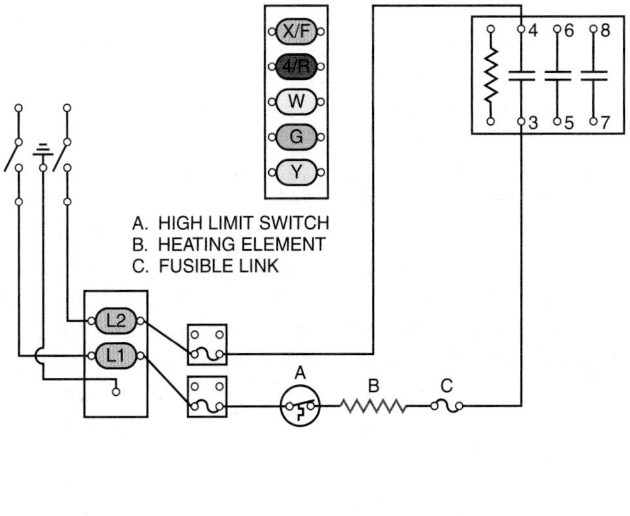

protection to the furnace. It is usually an automatic reset switch that closes when the temperature drops. It is wired to the heating element, "B", and to a fusible link, "C", that provides additional protection from overheating. When unsafe higher temperatures are present, the fusible link melts and must be replaced. The link is wired to terminal 3 of the sequencer with the circuit completed through L2 when the sequencer contact is closed. **NOTE:** *The fusible link is set to melt at a temperature that is higher than the opening temperature of the high-limit switch. So if the fusible link is found to have melted, the high-limit switch has failed and must be replaced as well.* •

Figure 30.20 presents a diagram of a furnace with two heating elements. **Figure 30.21** is an example of one with three elements. Each of these examples uses the same package sequencer shown in **Figure 30.14.** On this sequencer, contacts 3-4 close first, followed by the closing of contacts 5-6 if there is still a call for heat. After a predetermined time period contacts 7-8 will close if heating is still needed.

Figure 30.20 Circuits for two heating elements.

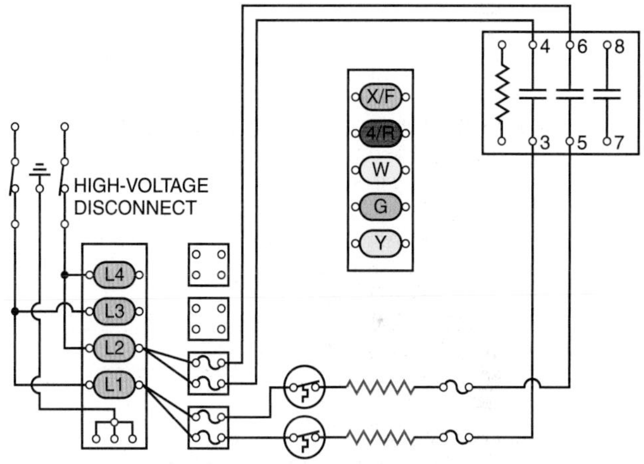

Figure 30.21 Circuits for three heating elements.

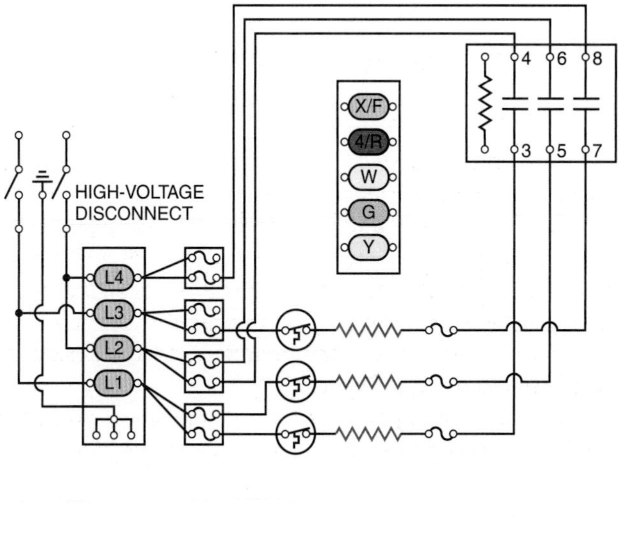

30.10 BLOWER MOTOR CIRCUITS

The blower motor, a power-consuming device, moves air over the electric heat elements and must be started and stopped at the correct time. **Figure 30.22** is an example of the wiring circuit for the blower motor. **NOTE:** *The blower must cycle on before the furnace gets too hot and continue to run until the furnace cools down.* • Note that the L1 terminal is wired directly to the blower motor and that the L2 terminal is wired directly to terminal 4 on the sequencer and to one side of the normally open (NO) set of contacts on the fan relay. If the coil on the relay is energized, the normally open contacts on the relay will close and the blower will operate at high speed. This will occur when the system is operating in the cooling mode or if the fan switch is moved to the "ON" position.

Power is passed through terminal 4 to terminal 3 when the sequencer is energized long enough for the bimetal to bend and close the contacts. The power then passes to the normally closed (NC) terminal on the fan relay and on to the low-speed winding required for heating. **Figure 30.23** combines **Figures 30.21** and **30.22** into a line diagram.

Remember that the sequencer contacts and the relay contacts *pass* power and do not consume power. **Figure 30.24** is an example of a legend for terminal designations used on electric forced-air furnaces that use multiple sequencers instead of one package sequencer. **Figure 30.25** is a wiring

Figure 30.22 The fan wiring circuit.

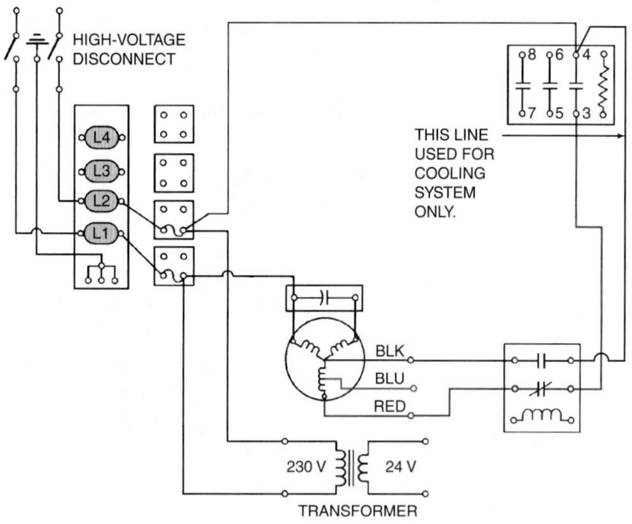

Figure 30.23 A simplified line drawing combining components in **Figure 30.21** and **Figure 30.22**.

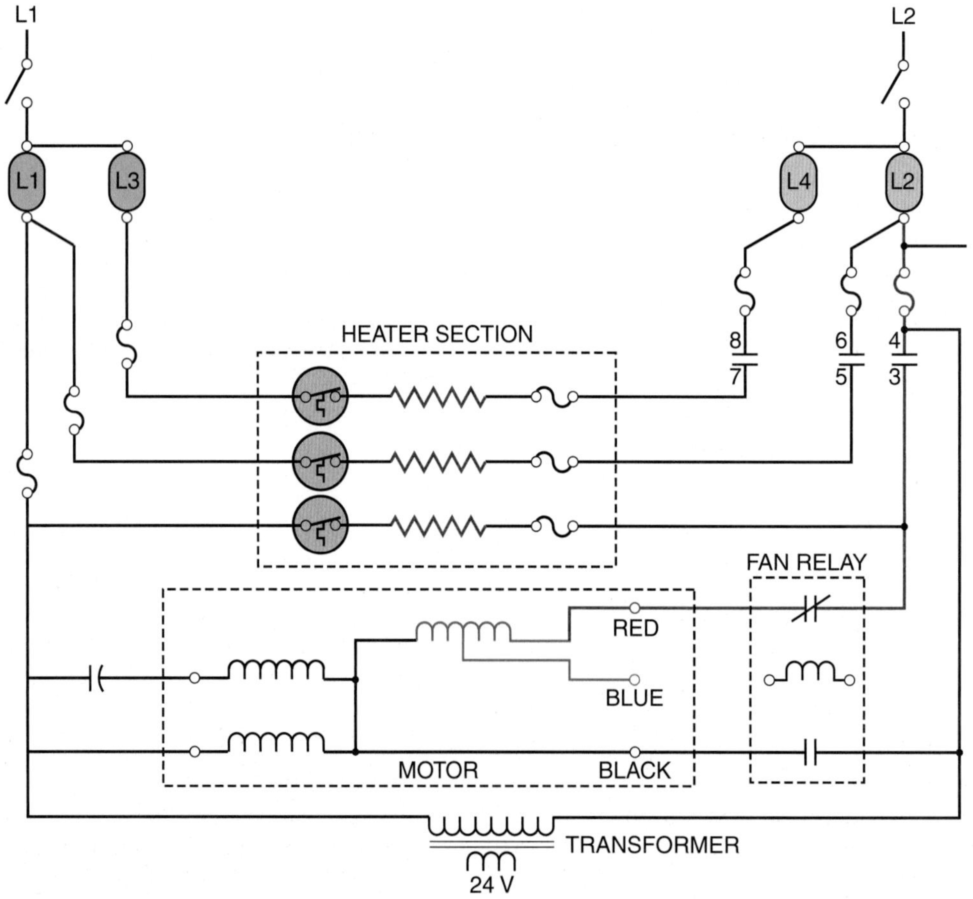

Figure 30.24 The electrical legend for **Figure 30.25**.

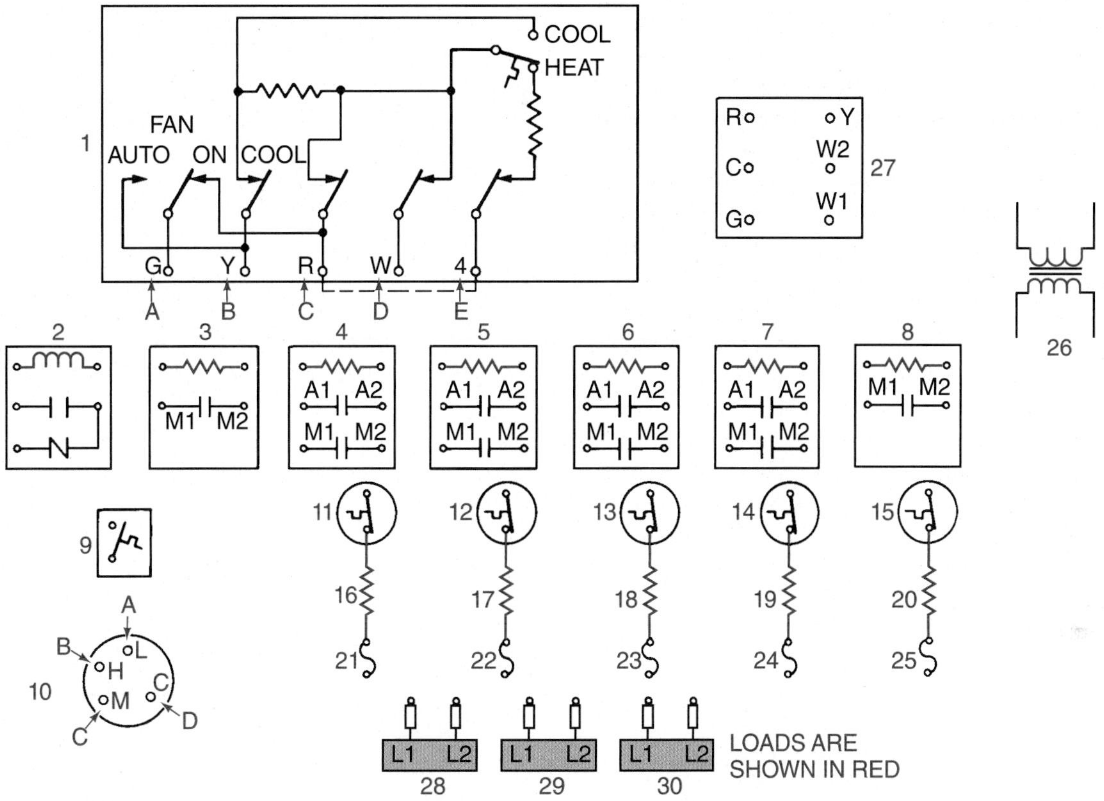

1. THERMOSTAT
   A. G TERMINAL (TO FAN RELAY)
   B. Y TERMINAL (TO COOLING CONTACTOR)
   C. R TERMINAL (FROM TRANSFORMER SECONDARY)
   D. W TERMINAL (TO HEATING SEQUENCER)
   E. 4 TERMINAL (RED OR JUMPED)
2. FAN RELAY (FOR COOLING)
3. TIMED FAN CONTROL
4. HEAT SEQUENCER I
5. HEAT SEQUENCER II
6. HEAT SEQUENCER III
7. HEAT SEQUENCER IV
8. HEAT SEQUENCER V
   (A1 AND A2 TERMINALS FOR 24 VOLTS ONLY. THESE ARE FOR AUXILIARY SEQUENCER HEATER STRIPS THAT ENERGIZE NEXT SEQUENCER. M1 AND M2 ARE TERMINALS FOR HEATING ELEMENTS.)
9. TEMPERATURE-ACTUATED FAN CONTROL
10. FAN MOTOR
    A. LOW SPEED
    B. HIGH SPEED
    C. MEDIUM SPEED
    D. COMMON

11–15. TEMPERATURE-ACTUATED LIMIT SWITCHES (AUTOMATIC RESET)
16–20. ELECTRIC HEATING ELEMENTS
21–25. FUSIBLE LINKS
26. PRIMARY SIDE OF TRANSFORMER
27. LOW-VOLTAGE TERMINAL BOARD
28–30. LINE TERMINAL BLOCKS

diagram for a forced-air furnace with individual sequencers. Note the low-voltage circuit wiring sequence:

1. Power is wired directly from the transformer to terminal C and on to the coils of all the sequencers and blower control devices.
2. The other side of the transformer is wired directly to terminal R and then on to the R terminal on the room thermostat. The room thermostat passes power to the respective power-consuming device circuits on a call for HEAT, COOL, or FAN ON.
3. On a call for heat, the thermostat energizes the W terminal and passes power to the timed fan control and the first sequencer. The first sequencer passes power to the next sequencer heater coil and then to the next.

See **Figure 30.25** for the high-voltage sequence:

1. Although there are six fuses, they are labeled either L1 or L2, meaning that there are only two lines supplying power to the unit. Three fuses are connected to L1, while the other three fuses are connected to L2. The six fuses

Figure 30.25 (A) The wiring diagram for an electric furnace. (B) The line diagram.

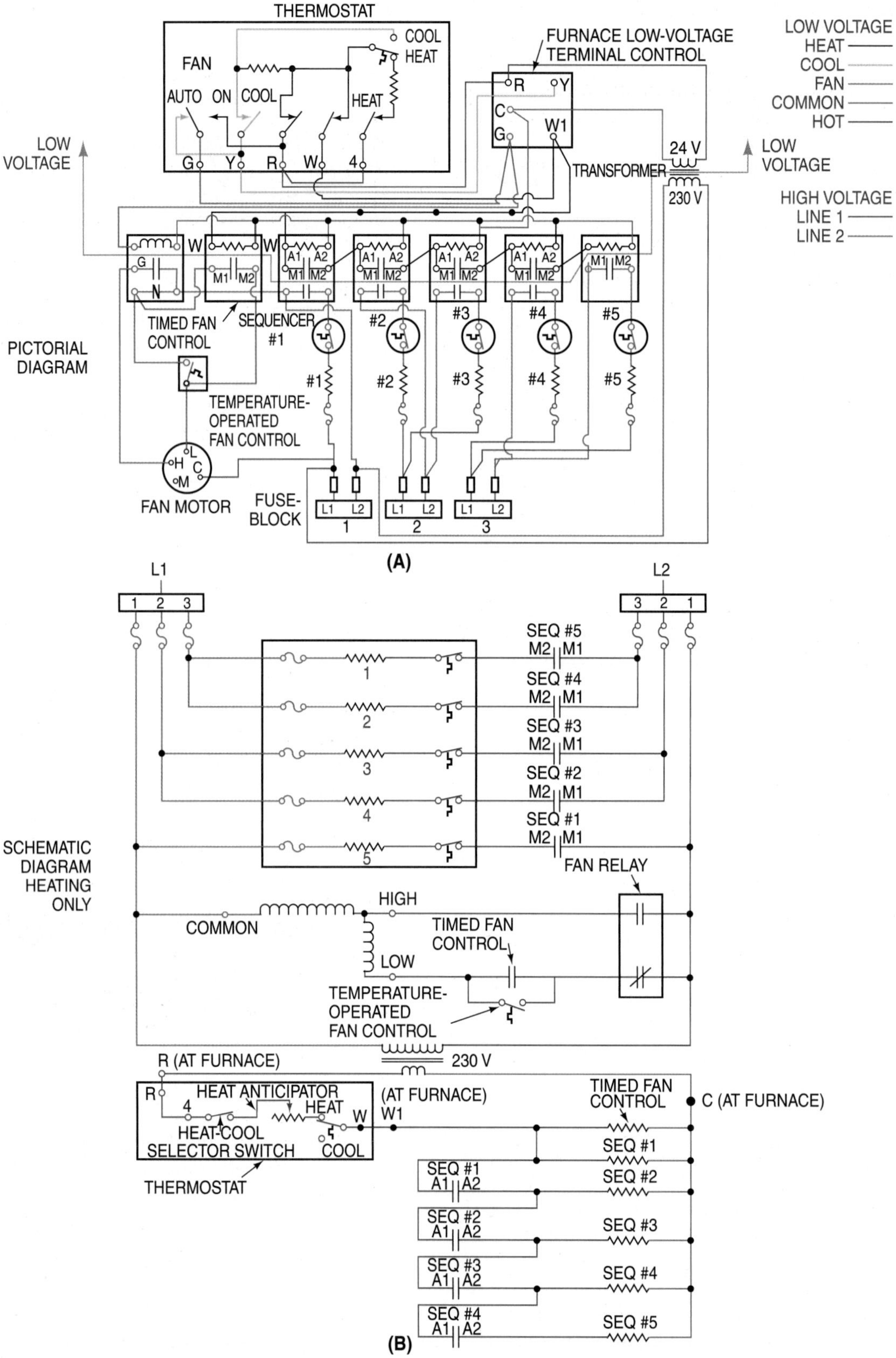

break the power-consuming heaters and fan motor down into separate circuits, each with an amperage rating that is lower than that of the entire appliance.

2. On a call for heat, the contacts in sequencer 1 close and pass power from L2 to the limit switch and on to heater 1. At the same time, the timed fan control is energized and its contacts close, starting the blower at low speed.
3. When the sequencer contacts close for the high-voltage load, the low-voltage contacts (A1 to A2) also close and pass power in the low-voltage circuit to the next sequencer's operating coil, which sets up the same sequence for the second heater. This continues down the line until all sequencer contacts are closed and all the heaters are energized.

NOTE: *This sequence may take 20 seconds or more per sequencer, and five heat sequencers are involved for the heat to be completely on. When the room thermostat is satisfied, it takes only about 20 seconds to turn off the heat because the power to all sequencers is interrupted by the room thermostat. The temperature-operated fan switch takes command of the blower at this time and keeps the blower on until the heaters cool.* •

The fan circuit is a circuit by itself. The blower will start at low speed if the fan relay is not energized. If the blower is running in low speed and someone turns the fan switch to ON, the blower will switch to high speed. It cannot run in both speeds at the same time due to the relay with NO and NC contacts.

30.11 AIRFLOW IN ELECTRIC FURNACES

The airflow in cubic feet per minute (cfm) may be verified in an electric heating system using the following formula and example. This formula is often called the sensible heat formula and is restated as it is seen in many other text calculations as

$$Q_s = 1.08 \times \text{cfm} \times \text{TD}$$

where: Q_s = Sensible heat in Btu/h
1.08 = A constant
cfm = Cubic feet per minute
TD = Temperature difference across furnace (also often denoted as ΔT)

The formula is restated and solved for cfm:

$$\text{cfm} = \frac{Q_s}{1.08 \times \text{TD}}$$

The total amount of heat energy added to the airstream may be calculated by first calculating the power consumed by the appliance. This is done by taking the total amperage to the electric furnace and multiplying it by the applied voltage. For example, if the amperage draw of a furnace is 85 amperes and the applied voltage is 208 V, the power in watts is calculated as:

$$\text{Watts} = \text{Amperes} \times \text{Volts}$$
$$\text{Watts} = 85 \times 208$$
$$\text{Watts} = 17{,}680$$

NOTE: *This calculation includes the blower motor, which is treated like a resistive load. This will give a slight error in the calculation. The motor would actually consume about 200 W less power if measured with a wattmeter. Because this is a fraction of a percent error, it is ignored.* •

To determine the Qs, the wattage must be converted to Btu/h by multiplying it by 3.413. (3.413 Btu/h of heat energy are generated for each watt of electrical power consumed.)

$$\text{Btu/h} = \text{Watts} \times 3.413$$
$$\text{Btu/h} = 17{,}680 \times 3.413$$
$$\text{Btu/h} = 60{,}341.8$$ (**Figure 30.26**)

If the temperature rise across the furnace is 50°F, we can now use the sensible heat formula to find the cfm:

$$\text{cfm} = \frac{\text{Qs}}{1.08 \times \text{TD}}$$
$$\text{cfm} = \frac{60{,}341.8}{1.08 \times 50}$$
$$\text{cfm} = 1117.4$$

Figure 30.26 An electric furnace airflow calculation.

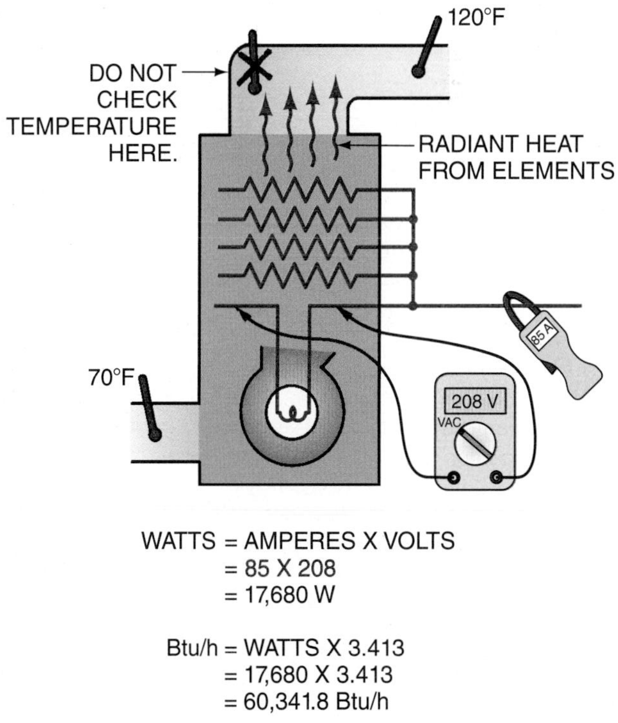

The technician may use the electric heat section of a system to establish the correct cfm for the cooling system. Control the blower speed by setting the fan switch on the thermostat to the ON position. This will ensure that the blower is running at the same speed for cooling. Reduced airflow can cause major problems with cooling systems and is covered later in the book.

HVACR GOLDEN RULES

When making a service call to a residence:

- Arrive at the job on time. If you must be late, be sure to call the customer and let them know beforehand.
- Keep your service truck neat and clean, both inside and out.
- Try to park your vehicle so that you do not block the customer's driveway.
- Look professional and be professional. Clean clothes and shoes must be worn. Wear coveralls for under-the-house work, and remove them before entering the house.
- Ask the customer about the problem. The customer can often help you solve the problem and save you time.
- Check humidifiers where applicable. Annual service is highly recommended for health reasons.

Added Value to the Customer

Here are some simple, inexpensive procedures that may be included with the basic service call:

- Repair or replace all frayed wires.
- Clean or replace dirty air filters.
- Lubricate fan motors when needed.
- Fasten all access panels securely, and replace any missing screws.
- Ask customers if they know how to program the thermostat and offer to help if they do not know how.

PREVENTIVE MAINTENANCE

Maintenance of electrical heating equipment may involve small appliances or central heating systems. Electric heat appliances, units, or systems all consume large amounts of power. This power is routed to the units by way of wires, wall plugs (appliance heaters), and connectors. These wires, plugs, and connectors carry the load and must be maintained.

ELECTRICAL HEATING APPLIANCES. The power cord at the plug-in connector is usually the first place that trouble may occur on a heating appliance. The heater should be taken out of service at the first sign of excess heat at the wall plug. It may become warm to the touch, but if it becomes too hot to hold comfortably, shut it off and check the plug and the wall receptacle. Fire may start in the vicinity of the plug if the situation is not corrected. The situation will become worse every time the heater is operated.

An electric appliance heater should be observed at all times to ensure safe operation. Do not locate the heater too close to combustible materials. Only operate the heater in the upright position.

CENTRAL ELECTRICAL HEATING SYSTEMS. Large electrical heating systems may be the panel type, baseboard type, or forced-air system. The panel and baseboard systems do not require much maintenance when properly installed. Baseboard heat has more air circulation over the elements than panel systems, so the units must be kept dust-free. Some areas will have more dust and the baseboards will require cleaning more often. Both panel and baseboard systems use line-voltage thermostats. These thermostats contain the only moving part and will normally be the first to fail. The thermostats will collect some dust and must be cleaned inside at some point. Be sure to turn the power off before servicing.

Central forced-air systems have a blower to move the air and consequently require filter maintenance. Filters should be changed or cleaned on a regular basis, depending on the rate of dust accumulation.

Some motor and blower bearings require lubrication at regular intervals. Shafts with sleeve bearings must be kept in good working order because these bearings will not take much abuse. Lift the motor or fan shaft and look for movement. Do not confuse shaft end-play with bearing wear. Many motors have up to 1/8-in. end-play.

Some systems may have belt drives and require belt tension adjustment or belt changing. Frayed or broken belts must be replaced to ensure safe and trouble-free operation. Belts should be checked at least once a year. It is very dangerous to operate electric heating systems with little or no airflow.

Filters should be checked every 30 days unless it is determined that longer periods may be allowed. Use the filters that

(continued)

PREVENTIVE MAINTENANCE (*Continued*)

fit the holder and be sure to follow the direction arrows on the filter for the correct airflow direction. When the filter becomes clogged on a forced-air system, the airflow is reduced and causes overheating of the elements. The automatic reset limit switches should be the first to trip and then reset. If the condition continues, these switches may fail, and the fusible link in the heater may open the circuit. When this happens, the link must be replaced by a qualified technician. This requires a service call that could have been prevented. If the fusible link fails, the limit switch should be replaced as well.

The electric heating elements are usually controlled by low-voltage thermostats that open or close the contacts on sequencers or contactors. The contacts in the sequencers or contactors have large amounts of electric current flowing through them. For this reason, the contacts may very well be the first component to fail in a properly maintained system. You may visually check the contacts on a contactor, but the contacts on sequencers are concealed. When the contacts on a contactor become pitted, change the contacts on the contactor. The condition will only become worse and could then cause other problems, such as terminal and wire damage to the circuit.

You can check a sequencer by visually checking the wires for discoloration and measuring the current flow through each circuit with the sequencer energized. If power will not pass through the sequencer circuit, the sequencer may be defective. Do not forget that the heater must have a path through which electrical current can flow. Many sequencers have been declared defective only to find later that the fusible links in the heater or the heater itself were defective.

Line-voltage electrical connections in the heater junction box should be examined closely. If there is any wire discoloration or connectors that appear to have been hot, these components must be changed. If the wire is discolored, and it is long enough, it may be cut back until the conductor is its normal color.

Fuse holders may lose their tension and not hold the fuses tightly. This is caused by heating and will be apparent because the fuses will blow when no overload exists and for no other apparent reason. These fuses and holders must be changed for the unit to perform safely and correctly.

30.12 DIAGNOSTIC CHART FOR ELECTRIC HEAT

Electric heat may involve either boiler systems or forced-air systems. The discussion in this text has primarily been concerned with forced-air systems as they are the most common. These forced-air systems may be the primary heating system or the secondary heating system when used with heat pumps. The technician should always listen to the customer for symptoms. The customer can often lead you to the cause of the problem with a simple description of how the system is performing.

Problem	Possible Cause	Possible Repair
No heat—thermostat calling for heat	Open disconnect switch	Close disconnect switch.
	Open fuse or breaker	Replace fuse or reset breaker and determine why it opened.
	High-temperature fuse link open circuit	Tighten loose connection at fuse link causing heat.
	Faulty high-voltage wiring or connections	Repair or replace faulty wiring or connections.
	Control-voltage power supply off	Check control-voltage fuses and safety devices.
	Faulty control-voltage wiring or connections	Repair or replace faulty wiring or connections.
	Heating element burned, open circuit	Replace heating element—check airflow.
Insufficient heat	Only some of the heating elements are energized	See possible repair options for a no-heat situation.
	Low voltage	Correct voltage.

30.13 SERVICE TECHNICIAN CALLS

SERVICE CALL 1

A customer with a new 20-kW electric furnace calls the service company and informs the dispatcher that she has no heat. The technician arrives at the job, rings the doorbell, and introduces himself. The customer tells the technician that the *blower comes on, but the heat does not. This is a system that has individual sequencers with a fan-starting sequencer. The first-stage sequencer has a burned-out coil and will not close its contacts. The first stage starts the rest of the heat, so there is no heat produced by the furnace.*

The thermostat is set to call for heat. On approaching the electric furnace, the blower can be heard running. The technician realized that, because the fan is operating, the 24-V power supply is functioning. Removing the panel covering the electric heat elements and sequencers, the technician uses the ammeter to check whether any of the heaters is energized; none are. **SAFETY PRECAUTION:** *The technician observes electrical safety precautions.*• The voltage is checked at the coils of all sequencers. See **Figure 30.25** for a diagram. 24 V is present at the timed fan control and the first-stage sequencer. The voltage across the normally open contacts is 24 V, indicating that these contacts are in the open position. These contacts should be closed because the sequencer coil is energized. The electrical disconnect is turned off and the wires on the sequencer are marked and removed from the device. Continuity through the first-stage sequencer coil is checked. There is no continuity through the coil, indicating that the coil is open. When the sequencer is changed and power is turned on, the electric heat comes on.

Since this is a relatively new installation and still under warranty, the technician simply replaced the sequencer without getting approval from the customer. Had this been a time-and-material service call, the technician would have informed the customer of the problem, provided information regarding the cost of the repair, and obtained approval to perform the work prior to doing so.

SERVICE CALL 2

A residential customer calls. *There is no heat, but the customer smells smoke. The company dispatcher advises that the customer should turn the system off until the technician arrives. The blower motor has an open circuit and will not run. The smoke smell is coming from the heating unit cycling on the limit control.*

The technician arrives on the job and notices that the front door is opened slightly. Instead of simply walking in, the technician rings the bell and waits for the homeowner to come to the door. The homeowner invites the technician in, explains the problem, and shows him to the thermostat. At the thermostat, the technician turns the system to heat. This is a system that does not have cooling, or the technician would have turned the thermostat to FAN ON to see whether the blower will start. If it did not start, this would have helped to solve the problem. When the technician goes to the electric furnace in the hall closet, a smoke smell is present. The technician removes the service panel from the front of the furnace and uses an ammeter to check the current at the electric heater. The heater is drawing 40 A, but the blower is not running. **SAFETY PRECAUTION:** *The technician observes electrical safety precautions.*•

The technician looks at the wiring diagram and discovers that the blower should start with the first-stage heat from a set of contacts on the sequencer. The blower motor is getting voltage but is not running. The unit power is turned off, locked, and tagged, and a continuity check of the motor proves that the motor winding is open.

The technician takes down all of the pertinent system and motor information and calls the shop to check on the cost and availability of the motor. Once the technician has the information, he informs the customer about the problem with the system, the cost of the repair, and when the repair can be made. Approval for the repair is given and the motor is replaced early the next morning. Once the motor is replaced, the technician checks the operation of the system to make certain that everything is okay.

SERVICE CALL 3

A customer calls at the end of the heating season and explains that the blower on his electric furnace is starting and stopping from time to time. He has the thermostat set to OFF, so there should be no activity.

The technician arrives at the job, introduces himself to the homeowner, and asks about the problem and the location of the furnace. The owner informs him about the blower problem and leads him to the furnace, which is located in the basement. When the technician touches the cabinet of the furnace, he can feel that it is warm; it shouldn't be—it should be room temperature. He removes the cabinet cover, checks the current at each of the three heating elements with an ammeter, and discovers that one of the elements is drawing 8 A. This heater should not be drawing any current since the system is set to OFF. The technician turns off the power, removes that heating element, and finds that one of the elements has been burned in two and that one of the coils is touching the frame. The element is grounded and drawing current, **Figure 30.27**. (**NOTE:** *The sequencer has no control over this grounded element. The fuse did not open because the current flow was not high enough—but there is enough current flow to create heat and activate the temperature control for the blower. The running blower will cool off the element quickly and turn off, only to turn on again.*)• **SAFETY PRECAUTION:** *It is a good electrical safety practice to check from the frame of any appliance to ground to make sure there is no danger. Do this before touching the appliance! This is particularly important when you are lying on the ground under a house checking equipment.*•

Figure 30.27 One side of this heating element is grounded to the frame of the furnace and is heating all of the time. Technicians should always check an appliance to ground for safety. This furnace is grounded and safe. If the ground were not correct, the technician could be injured.

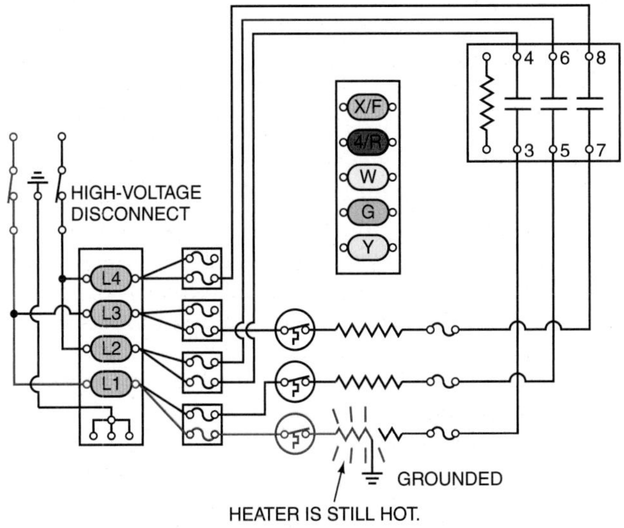

The technician shows the element to the homeowner, calls the office for the repair cost, and gets approval for the repair. The technician leaves for the supply house to get a replacement element. When the new element is installed, the unit operates correctly.

Solutions for the following service call is not provided here. Please discuss the solutions with your instructor.

SERVICE CALL 4

The owner of a small business calls. The heating system is not putting out enough heat. This is the first day of very cold weather, and the space temperature is only getting up to 65°F with the thermostat set at 72°F. The system is a package air-conditioning unit with 30 kW (six stages of 5 kW each) of strip heat located on the roof.

The technician arrives, goes to the room thermostat, checks the setting of the thermostat, and finds that the space temperature is 5°F lower than the setting. The technician goes to the roof with a volt-ohmmeter and an ammeter. **SAFETY PRECAUTION:** *After removing the panels on the side of the unit where the strip heat is located, the amperage is carefully checked.*• Two of the six stages are not pulling any current. It looks like a sequencer problem at first. A voltage check of the individual sequencer coils shows that all of the sequencers should have their contacts closed; there is 24 V at each coil. A voltage check at each heater terminal shows that all stages have voltage but are not drawing any current.

What is the likely problem and the recommended solution?

REVIEW QUESTIONS

1. Electric heating equipment is typically
   A. less expensive to purchase and less expensive to operate than fossil fuel heating equipment.
   B. more expensive to purchase and more expensive to operate than fossil fuel heating equipment.
   C. more expensive to purchase and less expensive to operate than fossil fuel heating equipment.
   D. less expensive to purchase and more expensive to operate than fossil fuel heating equipment.
2. The material often used to manufacture electric strip heater element wire is
   A. nickel cadmium (nicad).
   B. nickel chromium (nichrome).
   C. cadmium sulfate (calfate).
   D. cadmium dioxide (cadoxide).
3. Portable or small space heaters with glowing coils transfer heat by
   A. conduction.
   B. convection.
   C. evaporation.
   D. radiation.
4. Ceiling-mounted radiant heating panels are typically constructed of
   A. ¾-in. plywood.
   B. gypsum board
   C. ½-in. balsa wood.
   D. select pine board.
5. If the distance between a heat source and the heat target is doubled, the amount of sensed heat at the target location will be
   A. equal to the amount of heat sensed at the initial location.
   B. one-half of the amount of heat sensed at the initial location.
   C. one-quarter of the amount of heat sensed at the initial location.
   D. twice the amount of heat sensed at the initial location.
6. Electric radiant heating elements
   A. are often controlled by individual thermostats.
   B. are often installed as part of a ducted, forced-air heating system.
   C. must be installed in the walls of a structure.
   D. must be installed by a licensed electrician in all jurisdictions.

7. Baseboard heaters transfer heat by
   A. conduction.
   B. convection.
   C. evaporation.
   D. radiation.
8. Electric baseboard heaters are typically controlled by
   A. remotely located line-voltage thermostats.
   B. heater-mounted line-voltage thermostats.
   C. remotely located low-voltage thermostats.
   D. heater-mounted low-voltage thermostats.
9. From an installation standpoint, electric boilers have the benefit of not requiring
   A. flue or vent piping.
   B. pressure relief valves.
   C. aquastats.
   D. limit controls.
10. What is the approximate efficiency of an electric furnace with respect to its ability to convert electrical energy to heat energy?
    A. 25%
    B. 50%
    C. 75%
    D. 100%
11. The electric resistance strip heaters in an electric furnace are insulated from the rest of the appliance using which materials?
    A. Ceramic or iron
    B. Iron or copper
    C. Mica or ceramic
    D. Aluminum or mica
12. What is the purpose of an interlock circuit on an electric forced-air furnace?
    A. It prevents the blower from coming on if the heaters are not energized.
    B. It prevents the heaters from coming on if the blower is not energized.
    C. It prevents the equipment owner from receiving an electric shock.
    D. It prevents all of the system's heaters from turning on at the same time.
13. A disadvantage of a central forced-air electric furnace is that
    A. there is no zone control for individual rooms.
    B. a thermostat must be installed in each room.
    C. only line-voltage thermostats can be used.
    D. it is difficult to add air-conditioning (cooling).
14. Three common controls used in central electric heat applications are
    A. capacitors, thermostats, and contactors.
    B. thermostats, contactors and sequencers.
    C. cool anticipators, relays, and contactors.
    D. capacitors, cool anticipators, and sequencers.
15. The fusible link is best described as a(n)
    A. one-time, safety device.
    B. one-time, operational device.
    C. resettable safety device.
    D. resettable operational device.
16. Which of the following statements is true regarding the limit switch and the fusible link?
    A. The fusible link melts at a temperature that is higher than the temperature at which the limit switch opens.
    B. The fusible link melts at a temperature that is lower than the temperature at which the limit switch opens.
    C. The fusible link melts at the same temperature at which the limit switch opens.
    D. There is no direct relationship between the opening temperature of the limit switch and the melting temperature of a fusible link.
17. If a fusible link is found to be defective, which other system component must be replaced as well?
    A. The space thermostat
    B. The furnace blower motor
    C. The heating sequencer
    D. The limit control
18. The low-voltage thermostat is often used to control an electric heating system because it is compact, quick to respond to load changes, and it
    A. provides fuse protection.
    B. acts as a limit switch.
    C. is safe.
    D. acts as a fusible link.
19. An isolating thermostat subbase is easily identified because it has which of the following terminals?
    A. W, Y_C, and C
    B. R, W_C, Y, and C
    C. R, C, and G_H
    D. R_H and R_C
20. The heat anticipator is a(n)
    A. adjustable device that is wired in series with the thermostat's heating contacts.
    B. non-adjustable device that is wired in series with the thermostat's heating contacts.
    C. adjustable device that is wired in parallel with the thermostat's heating contacts.
    D. non-adjustable device that is wired in parallel with the thermostat's heating contacts.
21. The sequencer is responsible for
    A. cycling the furnace blower on and off during a call for heat.
    B. staggering the starting and stopping of multiple stages of electric heat.
    C. protecting the limit switch and fusible links.
    D. ensuring that the space temperature remains within 1 degree of the thermostat's set point.

22. The wiring diagram that shows the location of each system component is the
    A. ladder diagram.
    B. line diagram.
    C. schematic diagram.
    D. pictorial diagram.
23. Referring to **Figure 30.22**, the indoor fan motor will operate at low speed
    A. whenever the high-voltage disconnect switch is closed.
    B. all the time.
    C. whenever the high-voltage disconnect switch is closed and sequencer contacts 3-4 are closed.
    D. whenever the primary winding of the transformer is energized.
24. Referring to **Figure 30.23**, each electric strip heater is protected by how many safety devices?
    A. 1
    B. 2
    C. 3
    D. 4
25. How much heat is being put out by an electric furnace if the airflow is 1000 cfm and the temperature difference across the furnace is 50°F?
    A. 54,000 Btu/h
    B. 40,500 Btu/h
    C. 27,000 Btu/h
    D. 13,500 Btu/h
26. What is the temperature difference across a properly operating electric furnace that is using a 10-kW electric heater and is moving 900 cfm of air?
    A. 30.2°F
    B. 35.1°F
    C. 40.4°F
    D. 45.6°F
27. Dirty air filters on an electric forced-air furnace can cause
    A. reduced airflow and low internal appliance temperature.
    B. increased airflow and increased internal appliance temperature.
    C. reduced airflow and increased internal appliance temperature.
    D. increased airflow and low internal appliance temperature.
28. On furnaces that are equipped with belt-driven blower assemblies, which of the following can result if the belt is too loose?
    A. Reduced airflow and low internal appliance temperature
    B. Increased airflow and increased internal appliance temperature
    C. Reduced airflow and increased internal appliance temperature
    D. Increased airflow and low internal appliance temperature
29. If the proper voltage is supplied to an electric heating element and it does not get hot,
    A. the circuit breaker needs to be reset.
    B. the fusible link has melted.
    C. the heater is defective.
    D. the limit switch is open.
30. If the thermostat is calling for heat, but the appliance is providing insufficient heat, which of the following is the most likely cause for system malfunction?
    A. The main disconnect switch is off.
    B. The supply voltage to the appliance is too low.
    C. The thermostat is defective.
    D. The control transformer is defective.

UNIT 31

GAS HEAT

OBJECTIVES

After studying this unit, you should be able to

- list three common furnace configurations.
- explain the function of flame rollout switches, auxiliary limit switches, and draft safeguard switches.
- list two fuels burned in gas furnaces and describe characteristics of each.
- discuss gas combustion.
- explain the function of gas pressure regulators.
- discuss gas pressure measurement in inches of water column.
- describe the difference in operation between gas valves and automatic combination gas valves.
- discuss the operation of a redundant gas valve.
- discuss four types of gas burners and explain their function.
- discuss primary air and secondary air and the role they play in gas combustion.
- describe the function of a heat exchanger in a gas furnace.
- explain the operation of a fan switch, a limit switch, and a fan/limit switch.
- list three types of pilot safeties and describe their function.
- describe the difference between intermittent ignition and direct ignition.
- discuss flame rectification and explain how it is measured.
- explain dew point temperature as it applies to a high-efficiency condensing furnace.
- discuss the operation of standard-, mid-, and high-efficiency gas furnaces.
- discuss the characteristics and operation of the integrated and non-integrated ignition controls used in gas furnaces.
- describe a two-stage furnace, a modulating gas furnace, and a variable-output thermostat.
- discuss excess air and dilution air.
- describe flue gas venting systems and their categories.
- explain gas piping as it applies to gas furnaces.
- discuss the use of gas furnace wiring diagrams and troubleshooting flowcharts.
- discuss the factors that affect combustion efficiency.
- describe the troubleshooting process for standing pilot, intermittent ignition, and direct ignition systems.
- describe typical preventive maintenance procedures.

SAFETY CHECKLIST

- ☑ Fuel gases are dangerous in poorly vented areas. They can replace oxygen in the air and cause suffocation. They are also explosive. If a gas fuel leak is suspected, avoid ignition by preventing any open flame or spark. ⚠ **SAFETY PRECAUTION:** *A spark from a flashlight switch can ignite gas. Vent all enclosed areas adequately before working in them.*•
- ☑ Always shut off the gas supply when installing or servicing a gas furnace.
- ☑ A gas furnace that is not functioning properly can produce carbon monoxide, a poisonous gas. It is absolutely necessary that the production of carbon monoxide be avoided.
- ☑ Yellow tips indicate an air-starved flame emitting poisonous carbon monoxide.
- ☑ It is essential that the furnace be vented properly so that all flue gases will be dissipated into the atmosphere.
- ☑ When taking flue-gas samples, do not touch the hot vent pipe.
- ☑ A furnace with a defective heat exchanger (one with a hole or crack) must not be allowed to operate because the flue gases can mix with the air being distributed throughout the building.
- ☑ Wear goggles when using compressed air for cleaning parts.
- ☑ The limit switch must operate correctly to keep the furnace from overheating due to a restriction in the airflow or other furnace malfunction.
- ☑ Proper replacement combustion air must be available in the furnace area.
- ☑ All new gas piping assemblies should be tested for leaks.
- ☑ Gas piping systems should be purged only in well-ventilated areas.
- ☑ Follow all safety procedures when troubleshooting electrical systems. Electric shock can be fatal.
- ☑ Turn off the power before attempting to replace any electrical component. Lock and tag the disconnect panel. There should be only one key. Keep this key on your person while making any repairs.
- ☑ Perform a gas leak check after completing an installation or repair.

31.1 INTRODUCTION TO GAS-FIRED, FORCED-HOT-AIR FURNACES

Gas-fired, forced-hot-air furnaces have a heat-producing system and a heated-air distribution system. The heat-producing system includes the manifold and controls, burners, the heat exchanger, and the venting system. The heated-air distribution system consists of the blower that moves the air throughout the ductwork and the ductwork assembly. See **Figure 31.1** for a photo of a high-efficiency condensing gas furnace.

The manifold and controls meter the gas to the burners where the gas is burned, which creates flue gases in the heat exchanger and heats the air in and surrounding the heat exchanger. The

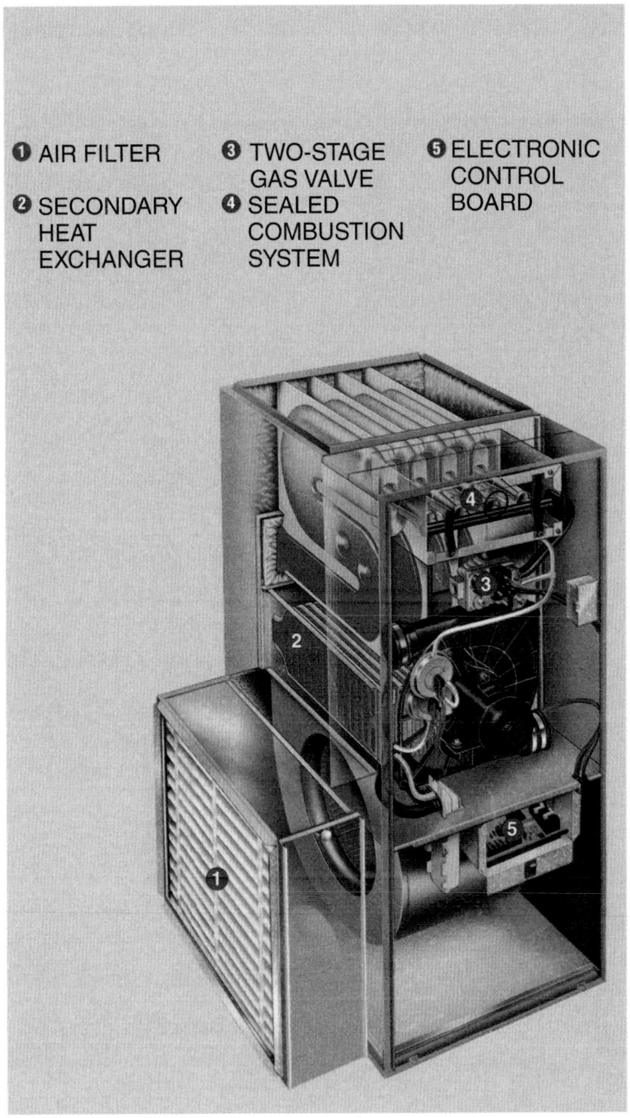

Figure 31.1 A high-efficiency condensing gas furnace. Source: Bryant Heating and Cooling Systems.

venting system allows the flue gases to be exhausted into the atmosphere. The blower distributes the heated air through ductwork to the areas where the heat is wanted.

31.2 TYPES OF FURNACES

UPFLOW

The *upflow* furnace stands vertically and needs headroom. It is designed for applications where the warm-air ductwork is above the furnace. For example: a first-floor installation with the ductwork in the attic or a basement installation with the ductwork between or under the first-floor joists. The furnace takes in cool air from the bottom, or sides near the bottom. It discharges hot air out the top, **Figure 31.2**.

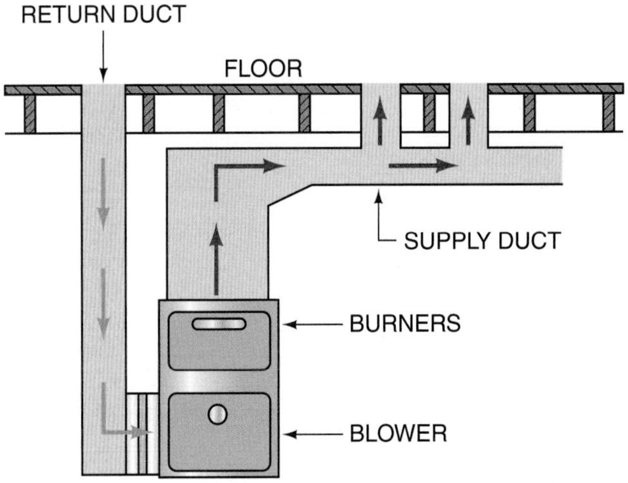

Figure 31.2 The airflow for an upflow gas furnace.

DOWNFLOW

The *downflow* furnace, sometimes referred to as a *counterflow* furnace, looks like the upflow furnace. It is designed for applications where the warm-air ductwork is below the furnace. The ductwork may be in a concrete slab floor or in a crawl space under the house. The air intake is at the top, and the discharge is at the bottom, **Figure 31.3**.

HORIZONTAL

The *horizontal* furnace is positioned on its side. It is installed in crawl spaces, in attics, or suspended from floor joists in basements. In these installations it takes no floor space. The air intake is at one end; the discharge is at the other, **Figure 31.4**.

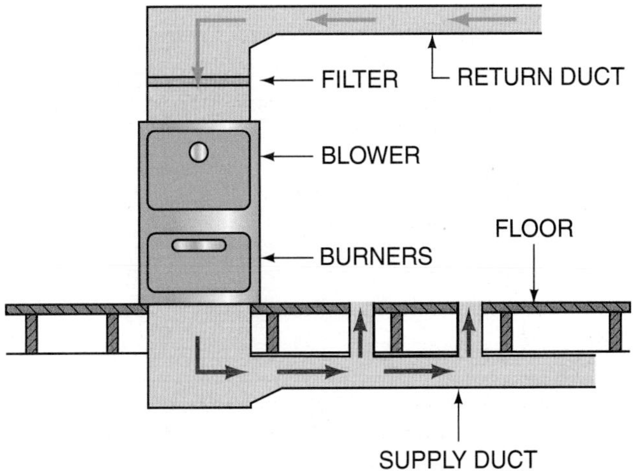

Figure 31.3 The airflow for a downflow, or counterflow, gas furnace.

Figure 31.4 The airflow for a horizontal gas furnace.

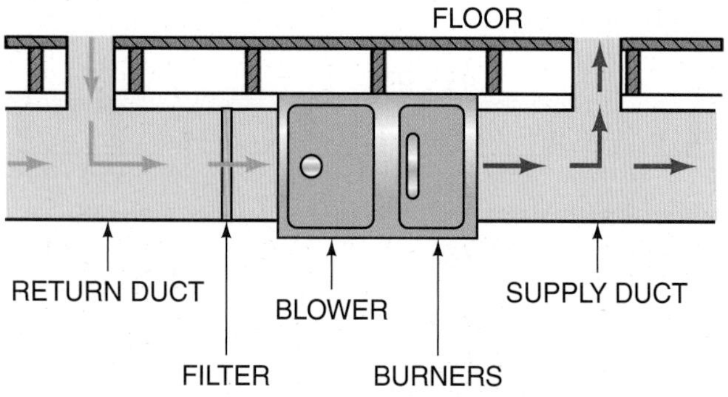

MULTIPOISE OR MULTIPOSITIONAL

A multipoise furnace can be installed in any position. This adds versatility to the furnace, especially for the installation contractor. Multipoise furnaces can be upflow, downflow, horizontal right or horizontal left air discharge, **Figure 31.5**. Many times, these furnaces come from the factory configured for the more common upflow applications but can be changed in the field with very little modification. Often, the combustion air pipe and exhaust pipe can be attached to either side of the furnace. Condensate drain lines can also be attached to either side of the furnace as long as the one not in use is capped. **Figure 31.6**(A) shows a high-efficiency condensing furnace with two combustion air and condensate drain line connection options. Many multipoise furnaces come with vent and drain extension tubes for ease in retrofitting. Multipoise furnaces look like any other furnace from the outside; however, the inside may have extra condensate drain line connections, more than one combustion air connection and vent pipe or condensate drain line extensions, and plugs for lines not in use during certain position configurations.

Because multipoise furnaces can be placed in four different configurations, multiple safety controls must be designed into the furnace. **Figure 31.6(B)** shows multiple bimetallic flame rollout or overtemperature limits very near to one another. This protects the furnace no matter what configuration it is

Figure 31.5 Four different positions for a multipoise furnace.

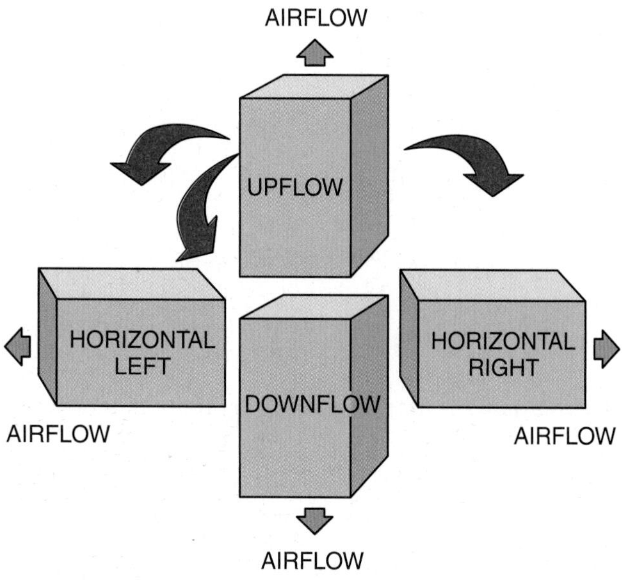

Figure 31.6 (A) A multipositional furnace showing two exhaust air location options and two condensate drain location options. Courtesy Ferris State University. Photo by John Tomczyk. (B) A multipoise furnace exhibiting multiple flame rollout switches. Courtesy Ferris State University. Photo by John Tomczyk

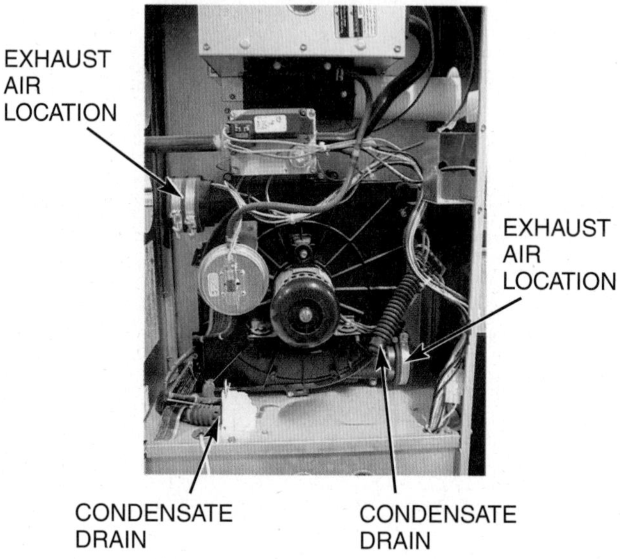

(A)

(B)

Figure 31.6 *(Continued)* (C) An auxiliary limit switch typically found on horizontal and counterflow furnaces. The switches are located on the warm-air blower housing and monitor return air temperature. Photo by Jason Obrzut. (D) The draft safeguard switch that opens the electrical circuit when there is any spillage of flue gas. Courtesy Ferris State University. Photo by John Tomczyk.

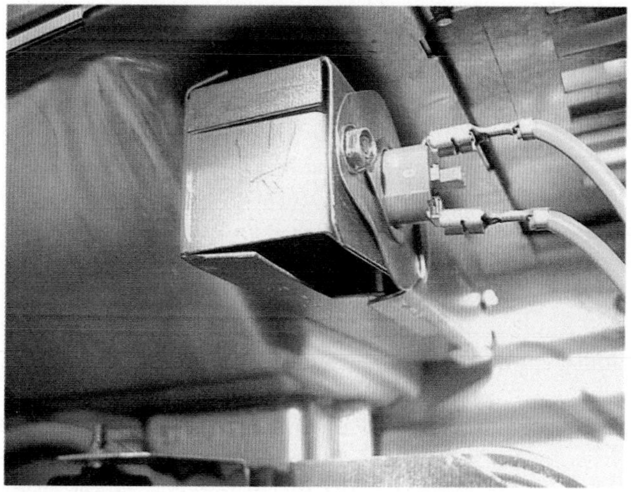

(D)

in. Normally closed (NC) flame rollout switches are usually located near the top of the burner box, where the flame is most likely to roll out. When a furnace is installed horizontally, additional rollout switches are wired in series to provide extra flame rollout protection. Control schemes on some furnaces allow for the combustion blower to operate continuously when a rollout switch has opened. Flame rollout switches may have to be manually reset by pushing a button on top of the switch.

⚠ SAFETY PRECAUTION: *Never use an automatic reset flame rollout switch to replace a damaged or inoperative manual reset flame rollout switch.•*

Auxiliary limit switches are typically used for counterflow and horizontal furnaces. They are usually bimetallic, similar to flame rollout switches, but are mounted on the warm-air blower housing, **Figure 31.6(C)**. They monitor room return air temperature and interrupt burner operation when the temperature gets too hot. This protects the furnace's blower motor from exceeding its maximum allowable temperature. Usually, the combustion blower or warm-air blower will continue to run if any auxiliary limit switch has opened. Most of these controls are automatically reset.

Vent limit or draft safeguard switches are NC switches usually mounted on the draft hood if the furnace is equipped with one. They are typically bimetallic switches similar to flame rollout and auxiliary limit switches. They monitor the draft hood temperature and open when there is any spillage of flue gas, **Figure 31.6(D)**.

31.3 GAS FUELS

Natural gas, manufactured gas, and liquefied petroleum (LP) are commonly used in gas furnaces.

NATURAL GAS

Natural gas has been forming along with oil from dead plants and animals for millions of years. The accumulated organic material gradually was washed or deposited to great depths in hollow spots in the earth, **Figure 31.7**, causing tremendous pressure and high temperature from its own weight. The pressure and temperature caused a chemical reaction, changing the organic material to oil and gas, which accumulated in pockets and porous rocks deep in the earth.

Natural gas is composed of 90% to 95% methane and other hydrocarbons, almost all of which are combustible, which makes this gas efficient and clean-burning. Natural gas has an average specific gravity of 0.60 and dry air has a specific gravity of 1.0, so natural gas weighs only 60% as much as dry air. Therefore, natural gas rises when discharged into

Figure 31.7 Gas and oil deposits deep in the earth.

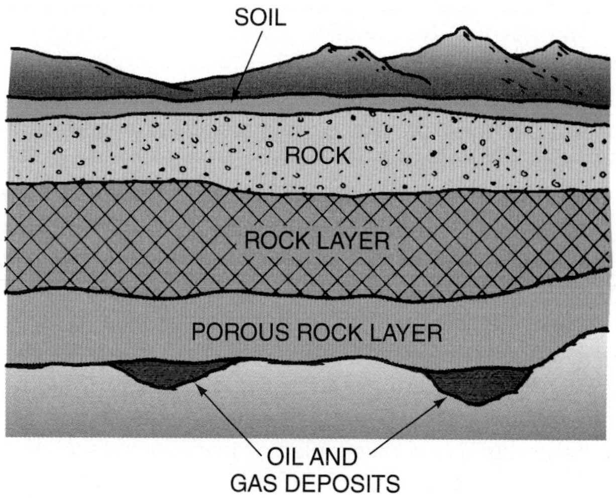

the air. When burned with air, 1 ft^3 of natural gas can produce between 900 Btu and 1100 Btu of heat energy. 1000 Btu is the average number used for calculations. Gas from these pockets deep in the earth (more generally called wells) may contain moisture and other gases that must be removed before distribution. The characteristics of natural gas will vary somewhat from one location to another. Local utility companies can provide specific characteristics for the gas in their area.

Natural gas by itself has neither odor nor color and is not poisonous. **SAFETY PRECAUTION:** *But it is dangerous because it can displace oxygen in the air, causing suffocation, and when it accumulates, it can explode.•* Sulfur compounds, called *odorants,* which have a garlic smell, are added to the gas to make leak detection easier. No odor is left, however, when the gas has been burned.

LIQUEFIED PETROLEUM

Liquefied petroleum (LP) is liquefied propane, butane, or a combination of propane and butane. When in the vapor state, these gases may be burned as a mixture of one or both of these gases and air. The gas is liquefied by keeping it under pressure until ready for use. A regulator at the tank reduces the pressure as the gas leaves the tank. As the vapor leaves the tank, it is replaced within by means of a slight boiling or vaporizing of the remaining liquid. LP gas is obtained from natural gas or as a by-product of the oil-refining process. The boiling point of propane at atmospheric pressure is −44°F, which makes it feasible to store in tanks for use in low temperatures during northern winters.

Butane in the liquid state, however, has a high boiling point. At any temperature below 31°F a tank of liquid butane would be at a negative pressure, so it must be buried or be supplied with a small heater to keep the fuel pressure high enough to move the gas to the main burner. For these reasons, butane is not very popular. It has a very high heat content but is hard to work with because of the pressures.

LP gas contains more atoms of hydrogen and carbon than natural gas does, which causes the Btu output to be greater. When burned, 1 ft^3 of *propane* produces about 2500 Btu of heat. However, it requires 24 ft^3 of air to support this combustion. The specific gravity of propane gas is 1.52, which means that it is 1½ times heavier than air, so it sinks when released in air. **SAFETY PRECAUTION:** *This is dangerous because it will replace oxygen and cause suffocation; it will also collect in low places and create pockets of highly explosive gas.•* *Butane* produces approximately 3200 Btu/ft^3 of gas when burned.

LP gas is used primarily where natural gas is not available. Occasionally, butane and propane are mixed with air to alter the characteristics of LP gas in order to closely duplicate natural gas. It is then called propane or butane air gas. The mixture is accomplished with blending equipment under pressure. Mixing with air reduces the Btu/ft^3 and the specific gravity and allows the mixture to be substituted and burned in natural gas burners without adjustment for standby purposes.

Figure 31.8 A spud showing the orifice through which the gas enters the burner.

The specific gravity of a gas is important because it affects the gas flow through the piping and through the orifice of the furnace. The orifice is a small hole in a fitting (called a *spud*) through which the gas must flow to the burners, **Figure 31.8.** The gas flow rate also depends on the pressure and on the size of the orifice. At the same pressure, more light gas than heavy gas will flow through a given orifice. When the proper mixture of LP gas and air is used, it will operate satisfactorily in natural gas furnaces because the mixture will match the orifice size and the burning rate will be the same as for natural gas. **SAFETY PRECAUTION:** *LP gas alone must not be used in a furnace that is set up for natural gas because the orifice will be too large. This results in overfiring and soot buildup.•*

MANIFOLD PRESSURES

The manifold pressure at the furnace should be set according to the manufacturer's specifications because they relate to the characteristics of the gas to be burned. The manifold gas pressure is much lower than 1 psi and is expressed in inches of water column (in. WC). *One inch of water column* pressure is the pressure required to push a column of water up 1 in. One psi of pressure will support a column of water 27.7 in. high, or 1 psi = 27.7 in. WC.

Some Common Manifold Pressures	
Natural gas and propane/air mixture	3 to 3½ in. WC
Manufactured gas (below 800 Btu quality)	2½ in. WC
LP gas	10 to 11 in. WC

A water manometer can be used to measure gas pressure. It is a tube of glass or plastic formed into a U and about half full of plain water, **Figure 31.9.** The standard manometer used in domestic and light commercial installations is graduated in in. WC. The pressure is determined

Figure 31.9 A manometer used for measuring pressure in inches of water column. Courtesy Ferris State University. Photo by John Tomczyk

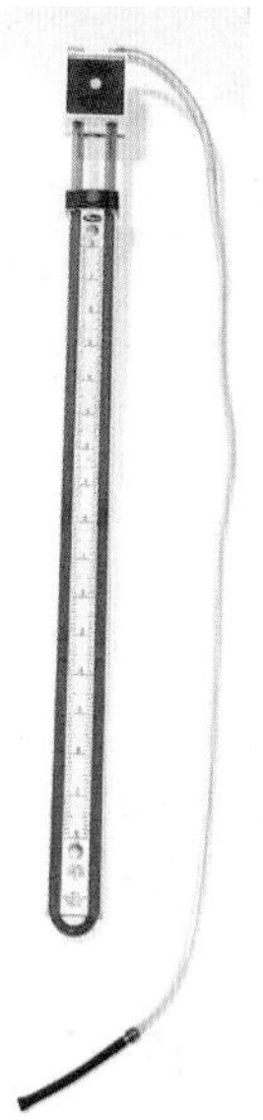

Figure 31.10 A gauge calibrated in inches of water column. Photo by Bill Johnson

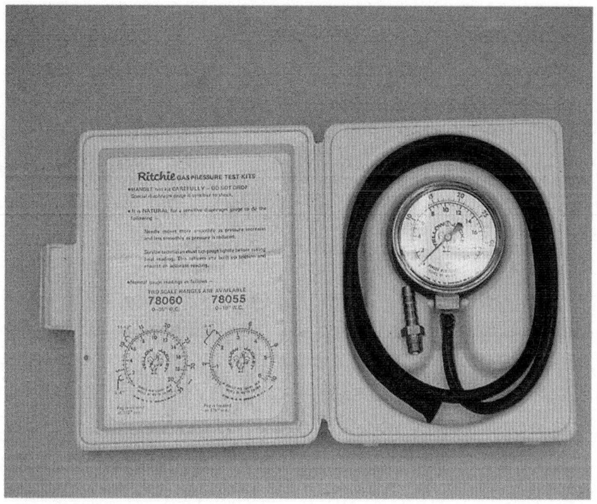

by the difference in levels between the two columns. Gas pressure is piped to one side of the tube, and the other side is open to the atmosphere to allow the water to rise, **Figure 31.10**. The instrument often has a sliding scale so it can be adjusted for easier reading. Some common pressure conversions follow:

Pressures Per Square Inch
1 lb/in^2 = 27.71 in. WC
1 oz/in^2 = 1.732 in. WC
2.02 oz/in^2 = 3.5 in. WC (standard pressure for natural gas)
6.35 oz/in^2 = 11 in. WC (standard pressure for LP gas)

Figure 31.11 A digital manometer being used to measure gas pressure in inches of water column. Courtesy Ferris State University. Photo by John Tomczyk.

A digital manometer is often used to measure gas pressures, **Figure 31.11**, and can measure both positive and negative pressures in inches of water column. One advantage of a digital manometer is that there is no fluid to spill. Another is that the pressure reading does not require the service technician to measure manually the difference in height of the water column. However, the digital manometer does require batteries. **SAFETY PRECAUTION:** *Turn the gas off while connecting any manometer.•*

31.4 GAS COMBUSTION

Heating systems must be installed to operate safely and efficiently. To properly install and service gas heating systems, a technician must know the fundamentals of combustion. Combustion needs fuel, oxygen, and heat. A reaction between the fuel, oxygen, and heat is known as rapid oxidation, or the process of burning. The fuel in this case is the natural gas (methane), propane, or butane; oxygen comes from the air; and the heat comes from a pilot flame or an electronic igniter. The fuels contain hydrocarbons that combine with the oxygen when heated to the ignition temperature. Air contains approximately 21% oxygen; enough air must be supplied to furnish the proper amount of oxygen for the combustion process. The ignition temperature for natural gas is 1100°F to 1200°F. The pilot flame or electronic igniter must provide this heat, which is the minimum temperature for burning. The chemical formula for the burning process is

$$CH_4 \text{ (methane)} + 2\,O_2 \text{ (oxygen)} \rightarrow CO_2 \text{ (carbon dioxide)} + 2\,H_2O \text{ (water vapor)} + \text{heat}$$

Figure 31.12 (A) Primary air is induced into the air shutter by the velocity of the gas stream from the orifice. (B) Ignition of the gas is on top of the burner. (C) Incomplete combustion yields yellow "lazy" flame. Any orange color indicates dust particles drawn in with the primary air. (D) A proper flame color.

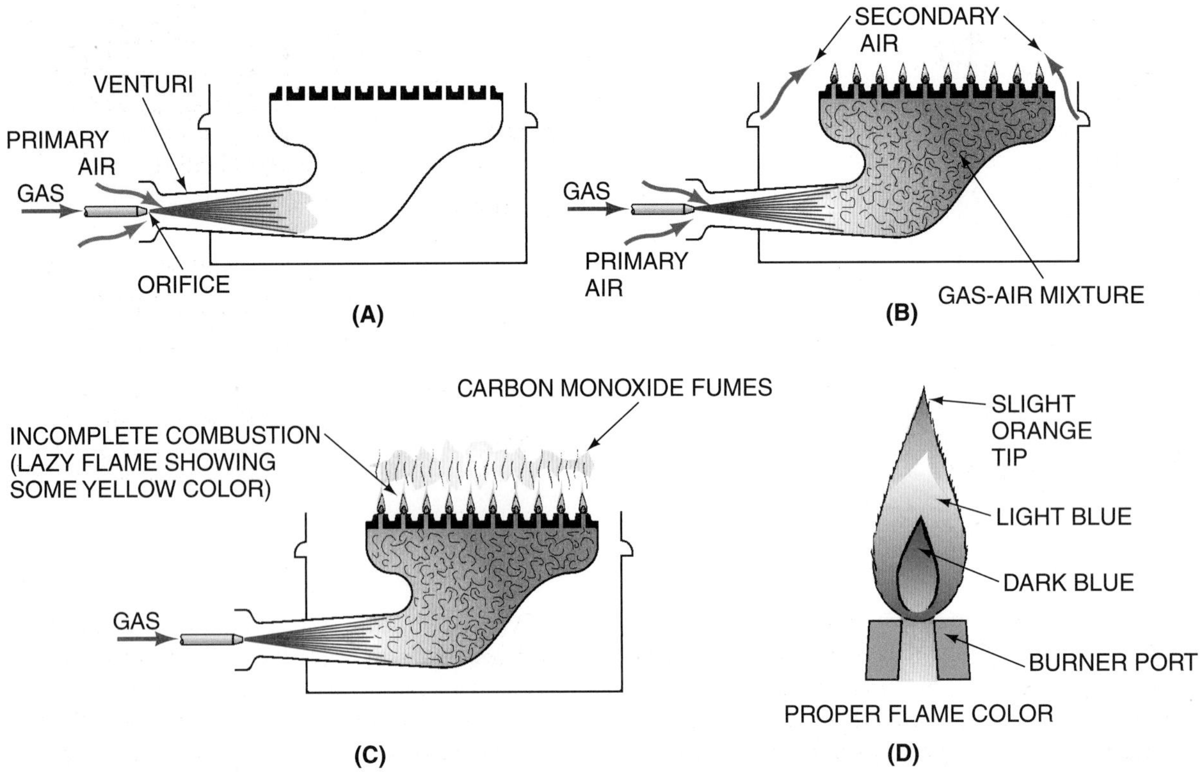

This formula represents perfect combustion. The process produces carbon dioxide, water vapor, and heat. Although perfect combustion seldom occurs, slight variations create no danger. **SAFETY PRECAUTION:** *The technician must see to it that combustion is as close to perfect as possible. By-products of poor combustion are carbon monoxide—a poisonous gas—soot, and minute amounts of other products. It should be obvious that carbon monoxide production must be avoided, as it is a colorless, odorless, poisonous gas.•* Soot lowers furnace efficiency because it collects on the heat exchanger and acts as an insulator, so it also must be kept at a minimum.

Some gas furnaces and appliances use an atmospheric burner because the gas and air mixtures are at atmospheric pressure during burning. The gas is metered to the burner through the orifice. The velocity of the gas pulls in the primary air around the orifice, **Figure 31.12(A)**.

Where the gas passes through the burner tube, the diameter is reduced to induce the air. This reduction in burner diameter is called the venturi, **Figure 31.12(A)**. The gas-air mixture moves on through the venturi into the mixing tube, where the gas is forced by its own velocity through the burner ports or slots; it is ignited as it leaves the port, **Figure 31.12(B)**. When the gas is ignited at the burner port, secondary air is drawn in around the ports to support combustion. The flame should be a well-defined blue with slightly orange, *not yellow,* tips, **Figure 31.12(C)**.

SAFETY PRECAUTION: *Yellow tips indicate an air-starved flame emitting poisonous carbon monoxide. Orange streaks in the flame are not to be confused with yellow. Orange streaks are dust particles burning.•*

Other than adjusting the gas pressure at the orifice, changes to the primary air supply are the only adjustments that can be made to the flame. Modern furnaces can only be adjusted slightly; this is intentional in order to maintain minimum standards of flame quality. The only thing, aside from gas pressure, that will change the furnace combustion characteristics to any extent is dirt or lint drawn in with the primary air. If the flame begins to burn *yellow,* primary air restriction should be suspected. For a burner to operate efficiently, the gas flow rate must be correct, and the proper quantity of air must be supplied. Modern furnaces use a primary and a secondary air supply, and the gas supply or flow rate is determined by the orifice size and the gas pressure. A normal pressure at the manifold for natural gas is 3½ in. WC. It is important, however, to use the manufacturer's specifications when adjusting the gas pressure.

A gas-air mixture that is too lean (not enough gas) or too rich (too much gas) will not burn. If the mixture of air and natural gas contains 0% to 5% natural gas, it will not burn. If the mixture contains from 5% to 15% gas, it will burn. **SAFETY PRECAUTION:** *However, if the mixture is allowed to accumulate, it will explode when ignited.•* If the gas

Figure 31.13 The quantity of air for the combustion of natural gas which produces approximately 1000 Btu.

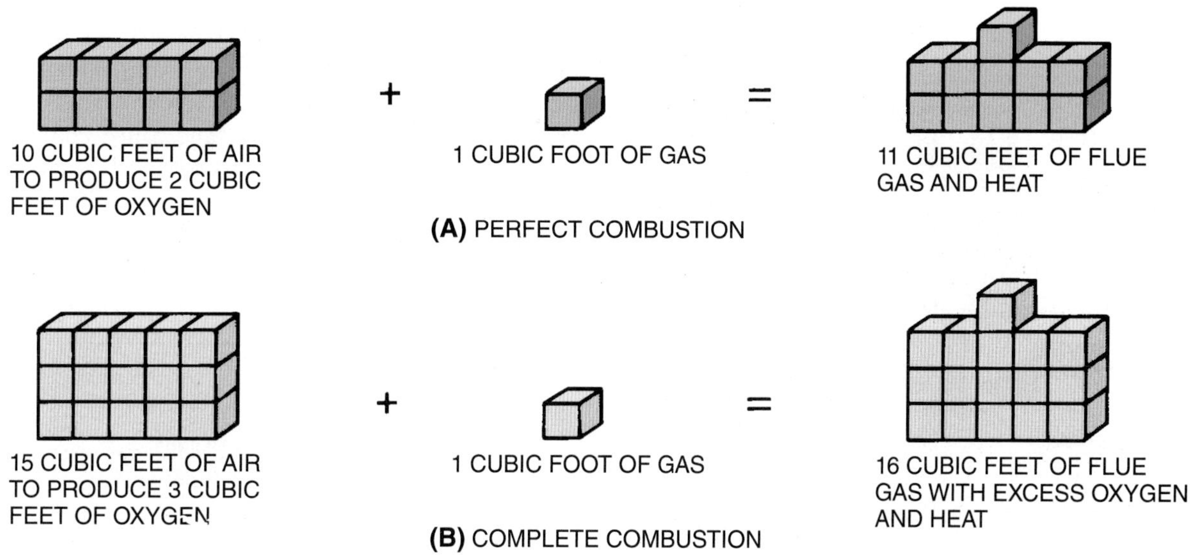

in the mixture is 15% to 100%, the mixture will not burn or explode. The burning mixtures are known as the *limits of flammability* and are different for different gases.

The rate at which the flame travels through the gas and air mixture is known as the *ignition* or *flame velocity* and is determined by the type of gas and the air-to-gas ratio. Natural gas (CH_4) has a lower burning rate than butane or propane because it has fewer hydrogen atoms per molecule. Butane (C_4H_{10}) has 10 hydrogen atoms, whereas natural gas has 4. The speed is also increased in higher gas-air mixtures within the limits of flammability; therefore, the burning speed can be changed by adjusting the airflow.

Perfect combustion requires two parts oxygen to one part methane. The atmosphere consists of approximately one-fifth oxygen and four-fifths nitrogen with very small quantities of other gases. Approximately 10 ft^3 of air is necessary to obtain 2 ft^3 of oxygen to mix with 1 ft^3 of methane. This produces about 1000 Btu and approximately 11 ft^3 of flue gases, **Figure 31.13**.

The air and gas never mix completely in the mixing tube chamber. Extra primary air is always supplied so that the methane will find enough oxygen to burn completely. This excess air is normally supplied at 50% over what would be needed if the air and gas were mixed thoroughly. This means that to burn 1 ft^3 of methane, 15 ft^3 of air needs to be supplied. This is referred to as complete combustion. Additional air also will be added at the draft hood. This is discussed later. There will be 16 ft^3 of flue gases to be vented from the combustion area, **Figure 31.13**. Flue gases contain approximately 1 ft^3 of oxygen (O_2), 12 ft^3 of nitrogen (N_2), 1 ft^3 of carbon dioxide (CO_2), and 2 ft^3 of water vapor (H_2O). **SAFETY PRECAUTION:** *It is essential that the furnace be vented properly so that these gases will all be dissipated into the atmosphere.•*

A cooler flame results in inefficient combustion. A flame is cooled when it strikes the sides of the combustion chamber (due to burner misalignment), called *flame impingement*. When the flame strikes the cooler metal of the chamber, the temperature of that part of the flame is lowered below the ignition temperature. This results in poor combustion and produces carbon monoxide and soot.

The percentages and quantities indicated here are for the burning of natural gas only. These figures vary considerably when propane, butane, propane-air, or butane-air are used. When making adjustments, always use the manufacturer's specifications for each furnace. **SAFETY PRECAUTION:** *While taking flue-gas samples, be careful not to touch the hot vent pipe.•*

31.5 GAS REGULATORS

Natural gas pressure in the supply line does not remain constant and is always at a much higher pressure than required at the manifold. The *gas regulator* drops the pressure to the proper level (in. WC) and maintains this constant pressure at the outlet where the gas is fed to the gas valve. Many regulators can be adjusted over a pressure range of several inches of water column, **Figure 31.14**. The pressure is increased when the adjusting screw is turned clockwise; it is decreased when the screw is turned counterclockwise. Some regulators have limited adjustment capabilities; others have none. Such regulators are either fixed permanently or sealed so that an adjustment cannot be made in the field. The natural gas utility company should be contacted to determine the proper setting of the regulator. In most modern furnaces, the regulator is built into the gas valve, and the manifold pressure is factory set at the most common pressure: 3½ in. WC for natural gas.

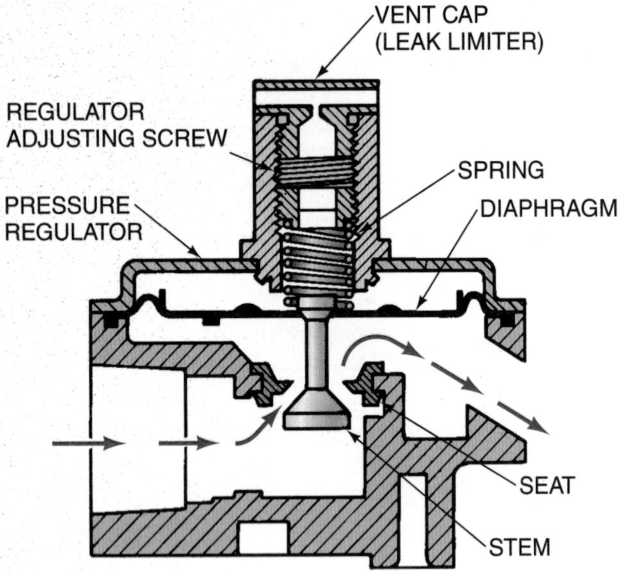

Figure 31.14 The diagram of a standard gas pressure regulator.

LP gas regulators are located at the supply tank and are furnished by the gas supplier. Check with the supplier to determine the proper pressure at the outlet from the regulator. The outlet pressure range is normally 10 to 11 in. WC. The distributor usually adjusts this regulator; in some localities the pressure is higher. The installer then provides a regulator at the appliance and sets the water column pressure to the manufacturer's specifications, usually 11 in. WC.

Combination gas valves have built-in regulators in their bodies. LP gas installations normally *do not* use gas valves with built-in regulators. Manufacturers usually will provide some sort of retrofit kit for converting from natural gas to LP gas. Many times, a higher-tension spring will have to be installed in the regulator at the gas valve to hold the valve open all the way. Some valves incorporate a push-pull knob. Pushing the knob down into the regulator will open the regulator all the way, which allows the regulator at the tank or the regulator at the side of the house to take control and regulate at 10½ to 11 in. WC. The regulators in gas valves sold as LP valves may have already had this done in some fashion. Always consult the manufacturer's specifications when setting regulators for any type of gas.

Converting from natural gas to LP gas involves many steps that may differ from manufacturer to manufacturer because of design differences in gas valves. For example, converting from natural gas to LP gas will include changing the regulator pressure from 3.5 in. WC by some method, installing smaller main burner orifices, and installing a new pilot orifice. This process also may include changing the ignition/control module to a 100% shutoff module because LP gas is heavier than air. **SAFETY PRECAUTION:** *Only experienced technicians should adjust gas pressures.•*

31.6 GAS VALVE

From the regulator the gas is piped to the **gas valve** at the manifold, **Figure 31.15**. While this valve and piping arrangement are no longer common, there are still systems in service that utilize it. There are several types of gas valves. Many are combined with pilot valves and called *combination* gas valves, **Figure 31.16**. We will first consider the gas valve separately and then in combination. Valves are generally classified as solenoid, diaphragm, and heat motor.

Figure 31.15 Natural gas installation where the gas passes through a separate regulator to the gas valve and then to the manifold.

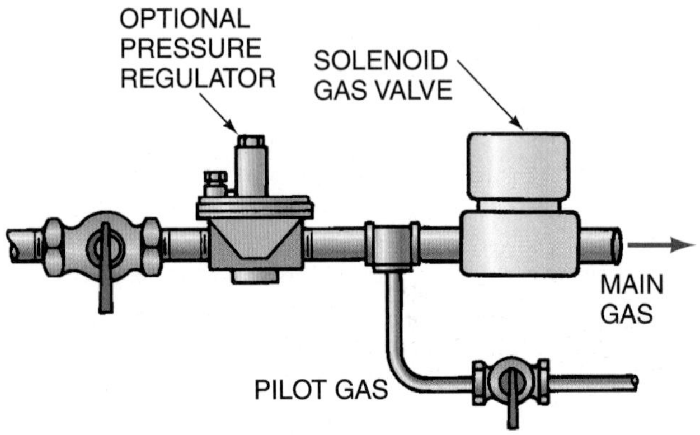

Figure 31.16 Gas valve with internal pressure regulator. Photo by Jason Obrzut

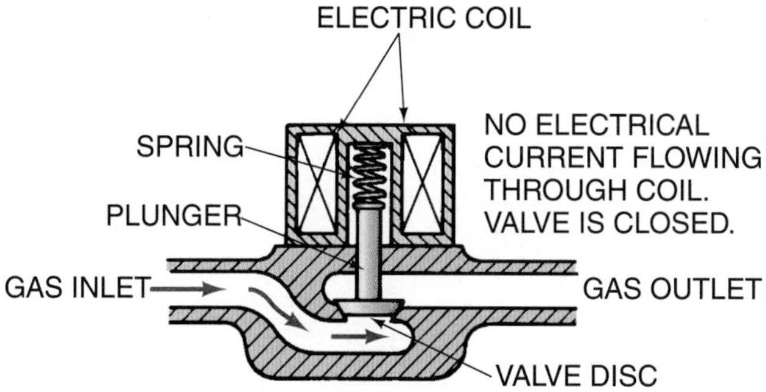

Figure 31.17 A solenoid gas valve in its normally closed position.

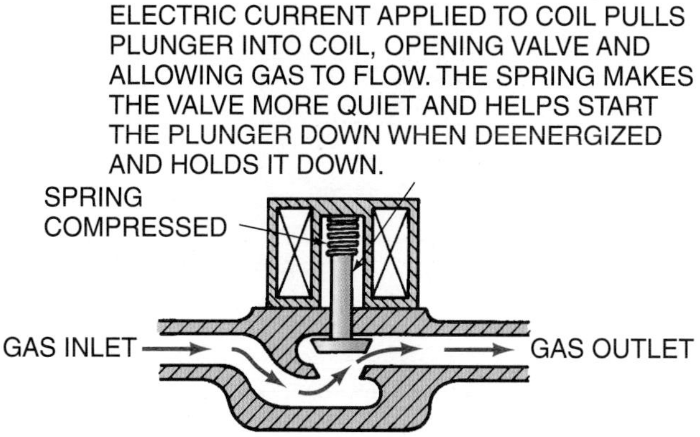

Figure 31.18 A solenoid valve in the open position.

SOLENOID VALVE

The gas *solenoid* valve is an NC valve, **Figure 31.17**. The plunger in the solenoid is attached to the valve or is in the valve. When an electric current is applied to the coil, the plunger is pulled into the coil. This opens the valve. The plunger is spring loaded so that when the current is turned off, the spring forces the plunger to its NC position, shutting off the gas, **Figure 31.18**.

DIAPHRAGM VALVE

The *diaphragm* valve uses gas pressure on one side of the diaphragm to open the valve. When there is gas pressure above the diaphragm and atmospheric pressure below it, the diaphragm will be pushed down and the main valve port will be closed, **Figure 31.19**. When the gas is removed from above the diaphragm, the pressure from below will push the diaphragm up and open the main valve port, **Figure 31.19**. This is done by a very small valve, called a *pilot-operated* valve because of its small size. It has two ports—one open while the other is closed. When the port to the upper chamber is closed and not allowing gas into the chamber above the diaphragm, the port to the atmosphere is opened. The gas already in this chamber is vented or bled to the pilot, where it is burned. The valve controlling the gas into this upper chamber is operated electrically by a small magnetic coil, **Figure 31.20**.

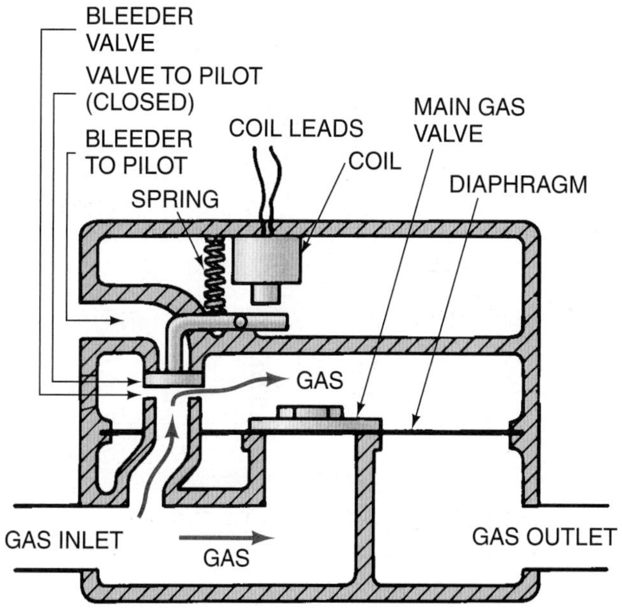

Figure 31.19 An electrically operated magnetic diaphragm valve.

Figure 31.20 When an electric current is applied to the coil, the valve to the upper chamber is closed as the lever is attracted to the coil. The gas in the upper chamber bleeds off to the pilot, reducing the pressure in this chamber. The gas pressure from below the diaphragm pushes the valve open.

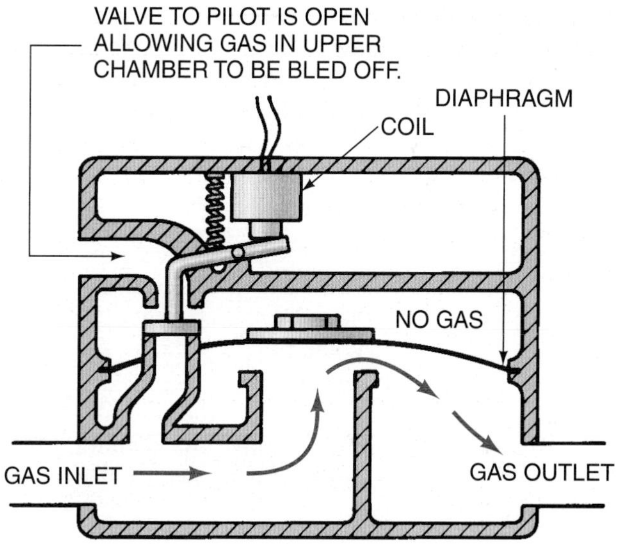

Figure 31.21 The thermally operated diaphragm gas valve.

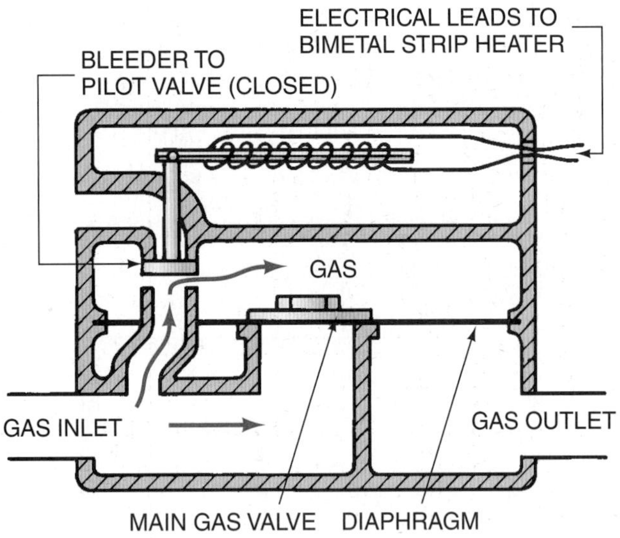

Figure 31.22 When an electric current is applied to the leads of the bimetal strip heater, the bimetal warps, closing the valve to the upper chamber and opening the valve to bleed the gas from the upper chamber. The gas pressure is then greater below the diaphragm, pushing the valve open.

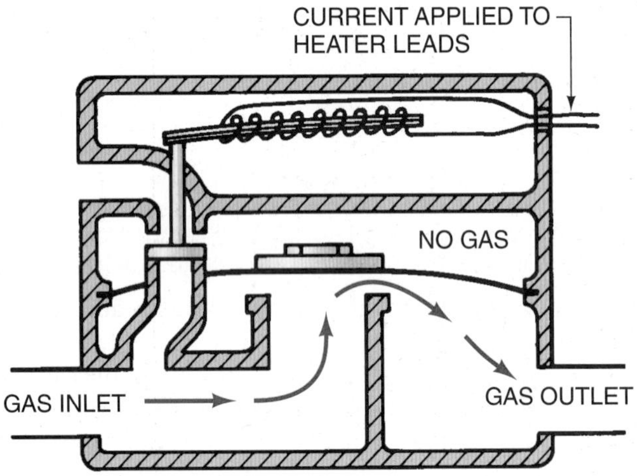

When the thermostat calls for heat in the thermally operated valve, a bimetal strip is heated and caused to warp by a small heater attached to the strip or a resistance wire wound around it, **Figure 31.21**. When the strip warps, it closes the valve to the upper chamber and opens the bleed valve. The gas in the upper chamber is bled to the pilot, where it is burned, reducing the pressure above the diaphragm. The gas pressure below the diaphragm pushes the valve open, **Figure 31.22**.

Figure 31.23 A heat motor–operated valve: (A) closed and (B) open.

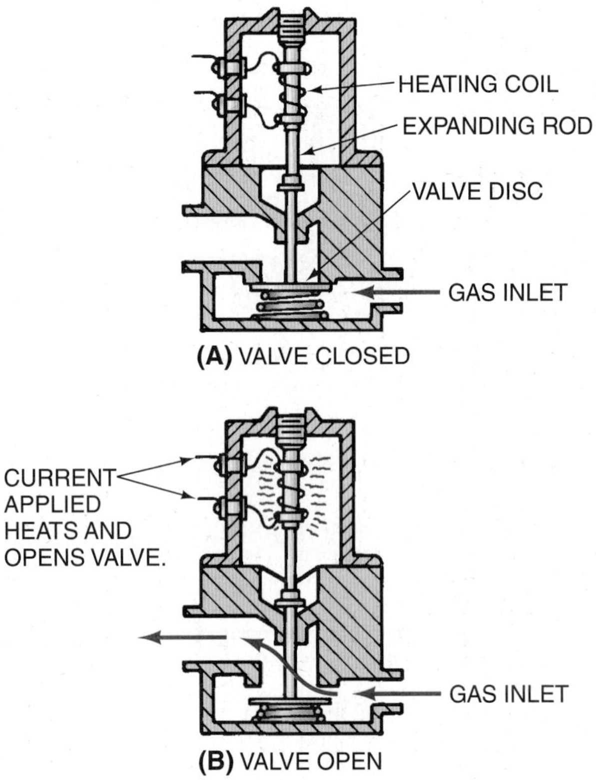

HEAT MOTOR–CONTROLLED VALVE

In a heat motor–controlled valve an electric heating element or resistance wire is wound around a rod attached to the valve, **Figure 31.23**. When the thermostat calls for heat, this heating coil or wire is energized and produces heat, which expands, or elongates, the rod. When elongated, the rod opens the valve, allowing the gas to flow. As long as heat is applied to the rod, the valve remains open. When the heating coil is deenergized by the thermostat, the rod contracts. A spring will close the valve. The time it takes time for the rod to elongate and then contract varies with the particular model, but the average is 20 sec to open the valve and 40 sec to close it.

SAFETY PRECAUTION: *Be careful while working with heat motor gas valves because of the time delay. Because there is no audible click, you cannot determine the valve's position. Gas may be escaping without your knowledge.*•

AUTOMATIC COMBINATION GAS VALVE

Many furnaces designed for residential and light commercial installations use an automatic combination gas valve (ACGV), **Figure 31.24**, also called the redundant gas valve.

Figure 31.24 An automatic combination gas valve. Source: Honeywell, Inc., Residential Division.

These valves incorporate a manual control, the gas supply for the pilot when used, the adjustment and safety shutoff features for the pilot, the pressure regulator, and the controls to operate the main gas valve. They often have dual shutoff seats for extra safety protection. These valves also combine the features described earlier relating to the control and safety shutoff of the gas. Many combination gas valves also may include programmed safe lighting features, servo pressure regulation, choices of different valve operators, and installation aids.

STANDING PILOT AUTOMATIC GAS VALVES. The internals of a typical combination gas valve for a standing pilot system are shown in **Figure 31.25**. A standing pilot system has a pilot burner that is lit all the time. The operation of the valve is described in the following paragraphs.

The power unit consists of a low-resistance coil and an iron core. When the power unit is energized, it has enough power to hold the safety shutoff valve open against its own spring pressure. If the pilot flame is ever lost, the power unit will be deenergized because the thermocouple will quickly stop generating a millivoltage as it cools. This action will stop all gas flow.

The power unit does not have enough power to pull the assembly open once closed because it is simply a holding coil. Once closed, pushing the red reset button on the combination gas valve will manually open the pilot valve. The red reset button can only be pushed down when it is in the pilot position. **Figure 31.26** shows how a safety shutoff mechanism assembly works, along with its shutoff circuit.

Once the gas passes the safety shutoff valve, it comes to the first automatic valve. The first automatic valve is controlled by the first automatic valve solenoid coil. The closing of the room thermostat energizes its solenoid. The valve is spring loaded and opens against flowing gas pressure. When the solenoid is deenergized, the spring closes the valve. This is one of the valve's built-in safeties.

Once the room thermostat calls for heat, the first automatic valve is opened. The closing of the room thermostat also energizes the second automatic valve solenoid, which raises the second automatic operator valve disc. This in turn controls the position of the second automatic valve diaphragm. The second automatic valve diaphragm is a servo-operated valve, which means that its position is controlled by the pressure of gas coming into a chamber beneath its diaphragm. This modulates the valve diaphragm between the open and closed position. Putting more pressure under the valve diaphragm modulates the valve closed, and less pressure modulates it more to the open position.

The servo pressure regulator, an outlet or working pressure regulator, is an integral part of the second automatic gas valve. During the furnace's running cycle, the servo pressure regulator closely monitors the outlet or working pressure of the gas valve, even when inlet pressures and flow rates vary widely. This is quite different from a standard gas pressure regulator, which monitors the gas valve's inlet pressure, **Figure 31.27**. Any gas outlet pressure change is instantly sensed and reflected back to the diaphragm in the servo pressure regulator. The servo pressure regulator then repositions its diaphragm and disc to change the flow rate through the servo pressure regulator valve. This changes the pressure under the second automatic valve diaphragm and thus the position of the diaphragm.

If the working pressure or outlet pressure of the gas valve rises, the servo pressure regulator diaphragm moves slightly higher. This allows less gas to bleed out of the servo pressure regulator to the gas valve outlet. However, it increases the pressure under the second automatic valve diaphragm, causing the valve to close. This action reduces gas valve outlet pressure.

If the working pressure or outlet pressure of the gas valve drops, the servo pressure regulator diaphragm moves slightly lower. This allows more gas to bleed out of the servo pressure regulator to the gas valve outlet. This decreases the pressure under the second automatic valve diaphragm, causing the valve to open. This action increases gas valve outlet pressure.

The servo pressure regulation system is a self-balancing system that works off gas valve outlet or working pressure

Figure 31.25 A standing pilot combination gas valve with servo pressure regulation. The gas inlet is on the left of the valve. A screen or filter is usually installed at the factory to keep the valve free from dirt and miscellaneous debris. The gas then encounters the safety shutoff valve. Pushing the red reset button can manually open this valve. This is how the pilot burner is lit. Pushing the red reset button allows gas to flow to the pilot gas outlet.

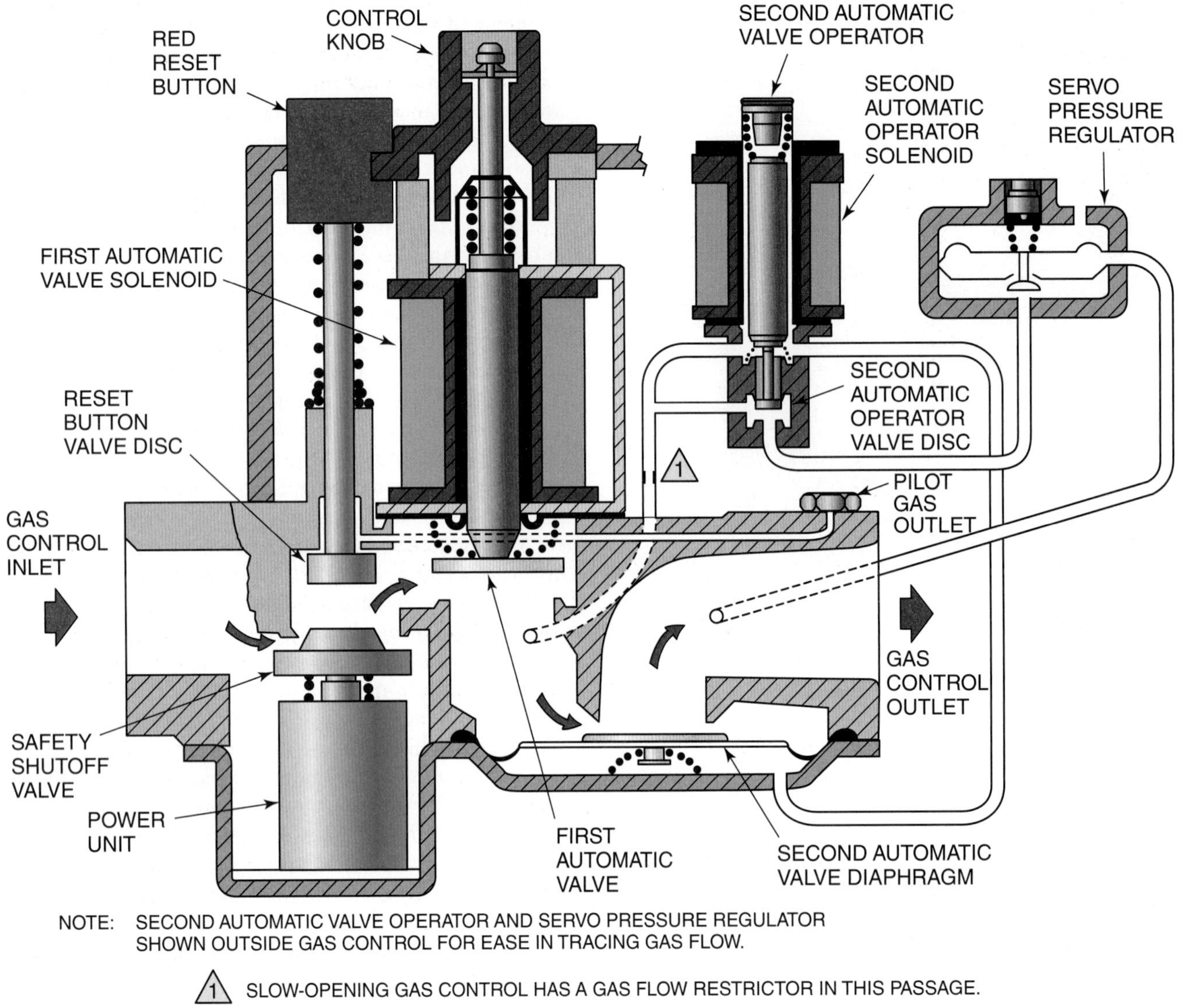

and is independent of flow rate. It can operate over a wide range of gas flow rates. The gas valve's outlet pressure is adjusted at the top of the servo pressure regulator's body. Putting more or less tension on a spring will change the outlet or working gas valve pressure.

INTERMITTENT PILOT AUTOMATIC GAS VALVES. In an intermittent pilot (IP) automatic gas valve system, the pilot is lit every time there is a call for heat by the thermostat. Once the pilot is lit and proved, it lights the main burner. When the call for heat ends, the pilot is extinguished and does not relight until the next call for heat. In other words, there is automatic and independent control of the gas flow for both the pilot and the main burner.

Figure 31.28 shows the internals of a typical combination gas valve for an IP system. Note that there is no power unit or reset button. In fact, there is no thermocouple circuit either. There are simply two automatic valves. The first valve has an electric solenoid operator, and the second valve is servo operated. The passage for the pilot gas is located between these two main valves.

The first automatic valve solenoid is energized by an electronic module or integrated furnace controller (IFC) when there is a call for heat from the thermostat. This opens the passage for pilot gas to flow. The electronic module or IFC has about 90 sec to light the pilot gas. Four electronic modules and an IFC are shown in **Figure 31.29**. Once the pilot is lit and proved, usually through flame rectification, the

Figure 31.26 (A) A safety shutoff mechanism. Source: Honeywell, Inc., Residential Division. (B) A safety shutoff circuit. Source: Honeywell, Inc., Residential Division.

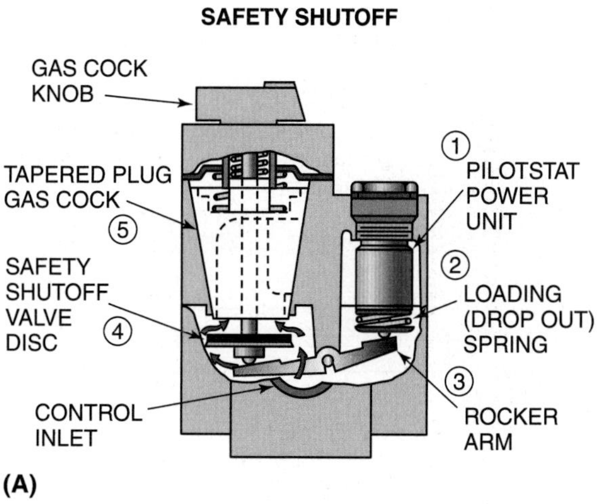

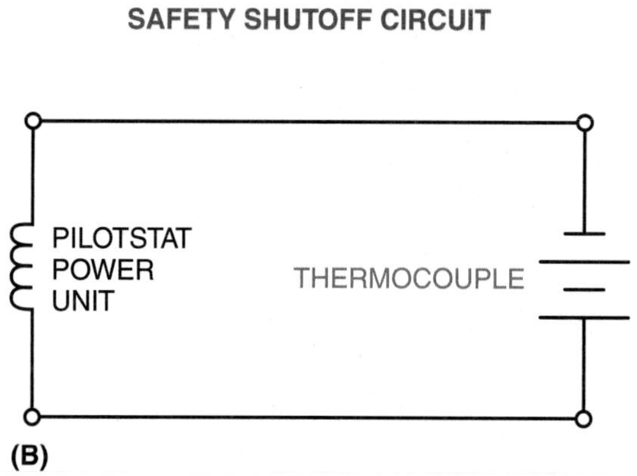

Figure 31.27 A standard pressure regulator that monitors gas inlet pressure. Source: Honeywell, Inc., Residential Division.

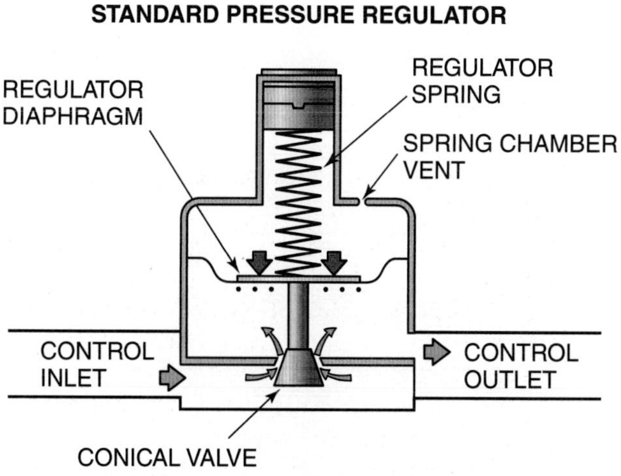

ignition module or IFC energizes the second automatic valve solenoid. Flame rectification will be covered in detail later in this unit.

Because the second automatic valve is servo operated, the second main valve will open and the main burner will be lit. Its servo operator will operate the same way as described earlier for the standing pilot automatic gas valve. The main burner and the pilot burner will remain lit until the room thermostat opens and the call for heat ends. Both automatic valves will close once the call for heat ends. This action will shut off the main burner and pilot burner until the room thermostat initiates another call for heat.

DIRECT IGNITION AUTOMATIC GAS VALVES. In a direct ignition automatic gas valve system, the electronic module or IFC lights the main burner directly without a pilot flame. The most common electronic igniters are the spark igniter and the hot surface igniter. These types of ignition will be covered in detail later in this unit. **Figure 31.30** shows the internals of a typical combination gas valve for a direct ignition system. Notice that there is no power unit or reset button. There are two automatic valves, and one is servo operated. The servo-operated valve will operate the same way as described for both the standing pilot and intermittent pilot combination gas valves.

In this system, when the room thermostat calls for heat, an electronic module or IFC energizes both the first and second automatic valve solenoids at the same time along with the ignition source. Remember, there is no pilot in a direct ignition system. Gas flows to the main burner and must be ignited within a short period of time, usually 4 to 12 sec. The reason for the short time period is safety, because of the large flow of gas to the main burners compared to the small flow of gas to a pilot burner. Once the main burner is lit and proved, usually through flame rectification, the furnace will run until the room thermostat opens and the call for heat ends. If the main burner flame is not proved, gas flow will be shut off to the main burner through the electronic module or IFC.

Some combination gas valves have a built-in slow-opening feature. The second automatic valve opens slowly, letting the flame grow slowly in the main burner. This eliminates noise and concussion in the heat exchanger by giving the main burner a slow and controlled ignition. Slow opening may be accomplished through a restriction in the passage to the second automatic valve operator. This restriction controls and limits the rate at which gas pressure builds in the servo pressure regulator system. The restriction is shown in **Figures 31.25, 31.28,** and **31.30.**

All three types of gas valves—standing pilot, intermittent pilot, and direct ignition—are classified as redundant gas valves. A redundant gas valve has two or three valve operators in series physically but wired in parallel with one another. This built-in safety feature allows any operator (pilot or main) to block gas from getting to the main burner. **Figures 31.25, 31.28,** and **31.30** are all examples of redundant gas valves. Most gas valves manufactured today are redundant gas valves.

Figure 31.28 An intermittent pilot combination gas valve with servo pressure regulation. Source: Honeywell, Inc., Residential Division.

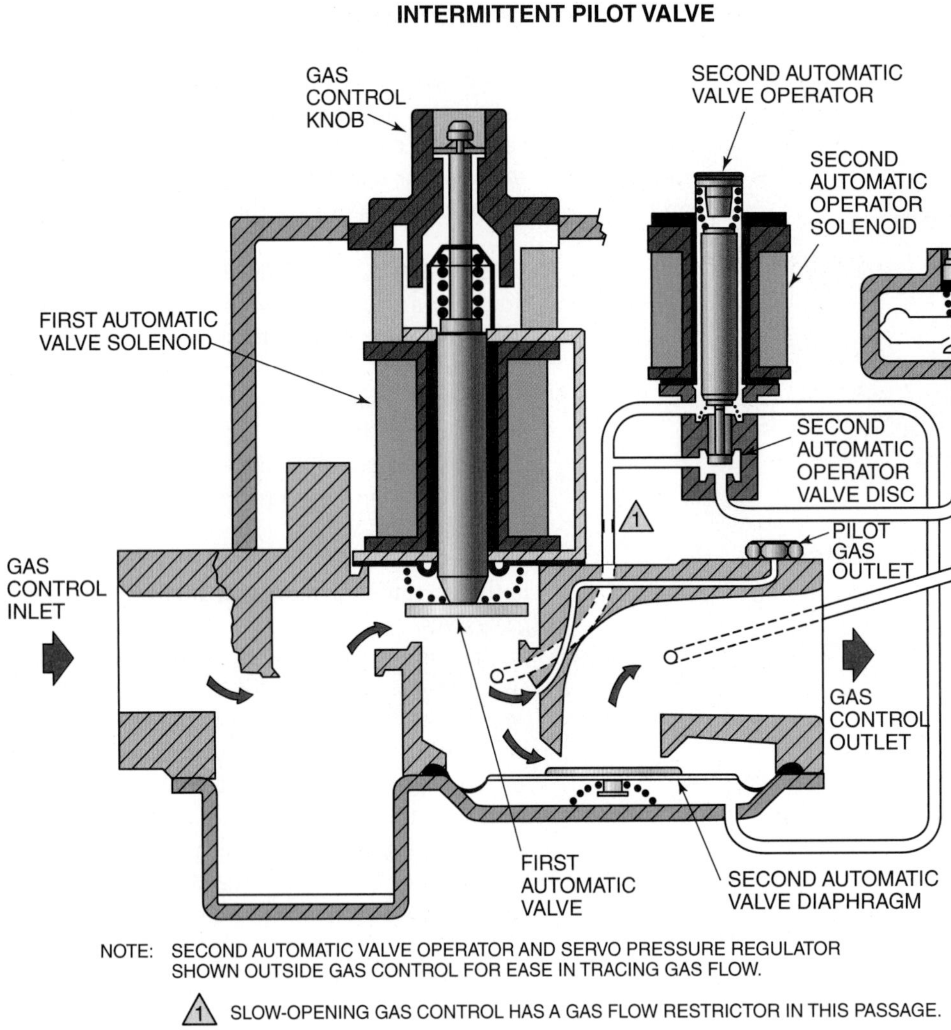

Figure 31.29 (A) An integrated furnace controller (IFC). Courtesy Ferris State University. Photo by John Tomczyk. (B) Four electronic modules for gas furnaces. Courtesy Ferris State University. Photo by John Tomczyk.

(A)

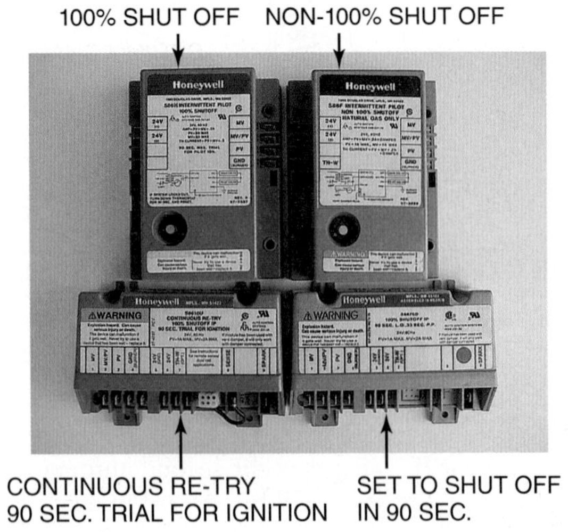

(B)

Figure 31.30 A direct burner combination gas valve with servo pressure regulation.

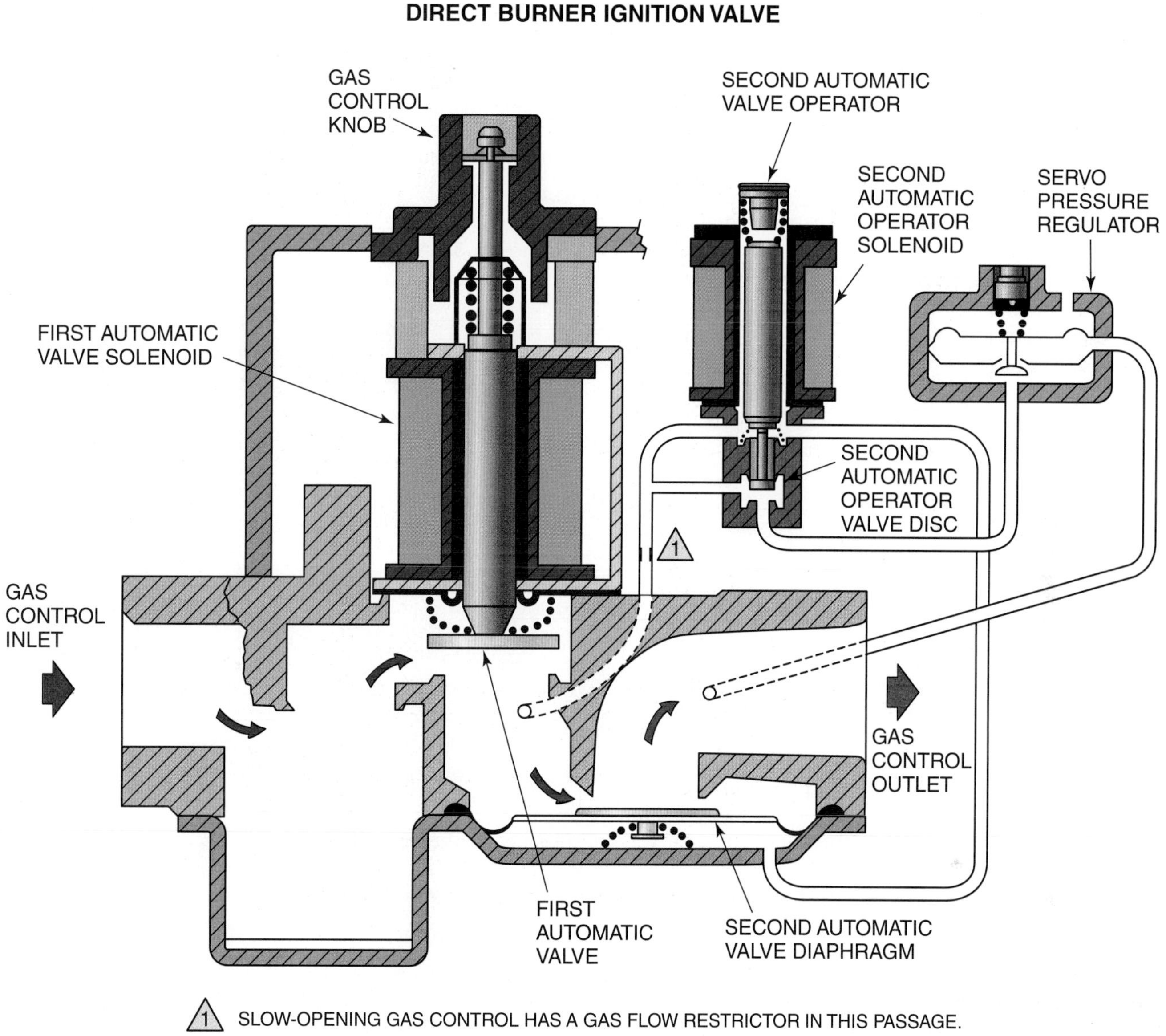

SAFETY PRECAUTION: *Most combination gas valves look alike. Their model number may be the only way to differentiate them. The correct combination gas valve must be used in each application or serious injury can occur. Always contact the furnace or gas valve manufacturer if you are unsure of what gas valve to use.•*

31.7 FUEL DELIVERY COMPONENTS

MANIFOLD

The *manifold* in the gas furnace is a pipe through which the gas flows to the burners and on which the burners are mounted. The gas orifices are drilled into a spud, which is threaded into the manifold, and direct the gas into the venturi in the burner. The manifold is piped to the outlet of the gas valve, **Figure 31.31**.

Figure 31.31 The gas valve feeds gas to the manifold.

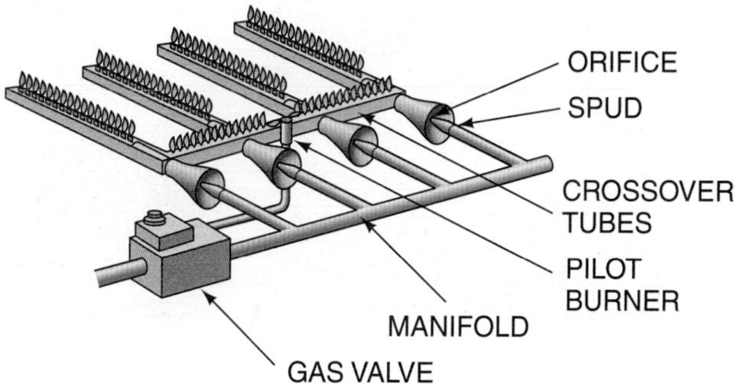

ORIFICE

The *orifice* is a precisely sized hole through which the gas flows from the manifold to the burners. The angle and depth of the orifice varies from one model to the next. The orifice is located in the spud, **Figure 31.8**. The spud is screwed into the manifold. The orifice allows the correct amount of gas into the burner.

BURNERS

Gas combustion takes place at the burners. Combustion requires primary and secondary air. Primary air enters the burner from near the orifice, **Figure 31.12**. The gas leaves the orifice with enough velocity to create a low-pressure area around it. The primary air is forced into this low-pressure area and enters the burner with the gas. The procedure for adjusting the amount of primary air entering the burner is explained later in this unit.

The primary air is not sufficient to sustain combustion. Additional air, called secondary air, is pulled into the combustion area. Both primary and secondary air must be available in the correct quantities for proper combustion. The gas is ignited at the burner by the pilot flame or electronic igniter.

The drilled-port burner is generally made of cast iron with the ports drilled. The slotted-port burner is similar to the drilled port except that the ports are slots, **Figures 31.32** and **31.33**. The ribbon burner produces a solid flame down the top of the burner, **Figure 31.34**. The single-port burner is the simplest and has, as the name implies, one port. This is often called the *inshot* or *upshot* burner, **Figure 31.35**. Most inshot burners are made of aluminized steel that has cross-over porting for good flame retention, proper flame carryover to each main burner, and excellent flame stability. Inshot burners have no primary or secondary air adjustments. The burners simply slip over the orifice spuds and are screwed in place. **Figure 31.36** shows both inshot and upshot burners. Inshot burners are generally used with

Figure 31.32 A cast-iron burner. Photo by Jason Obrzut

Figure 31.33 A stamped-steel slotted burner. Photo by Jason Obrzut

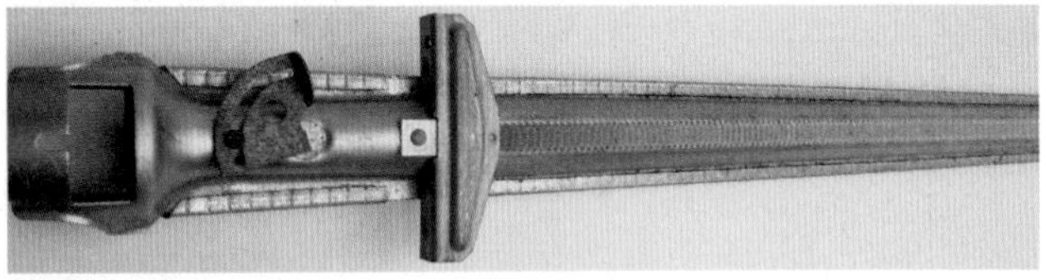

Figure 31.34 A ribbon burner. Photo by Jason Obrzut

Figure 31.35 (A) An upshot burner. Based on Carrier Corporation. (B) An inshot burner. Based on Carrier Corporation.

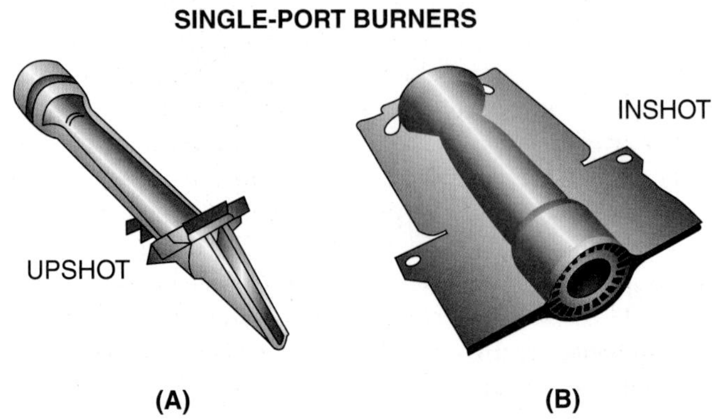

Figure 31.36 (A) Upshot burners. Courtesy Ferris State University. Photo by John Tomczyk. (B) Inshot burners. Courtesy Ferris State University. Photo by John Tomczyk.

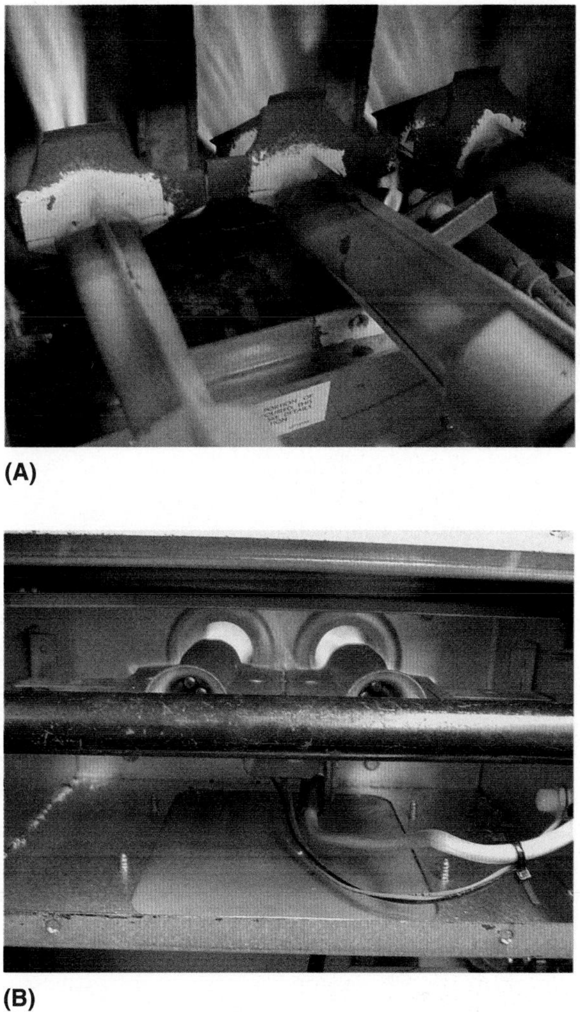

induced-draft systems, which pull or suck combustion gases through the heat exchanger.

All of these burners are known as *atmospheric* burners because the air for the burning process is at atmospheric pressure. Most modern gas furnaces use either induced- or forced-draft systems. Induced-draft systems have a combustion blower motor located on the outlet of the heat exchanger. They pull or suck combustion gases through the heat exchanger, usually causing a slight negative pressure in the heat exchanger itself. Forced-draft systems have the combustion blower motor on the inlet of the heat exchanger. They push or blow combustion gases through the heat exchanger and cause a positive pressure in the heat exchanger. **Figure 31.37** shows a combustion blower motor assembly. Both induced- and forced-draft systems will be covered in detail later in this unit.

Figure 31.37 A combustion blower assembly. Courtesy Ferris State University. Photo by John Tomczyk

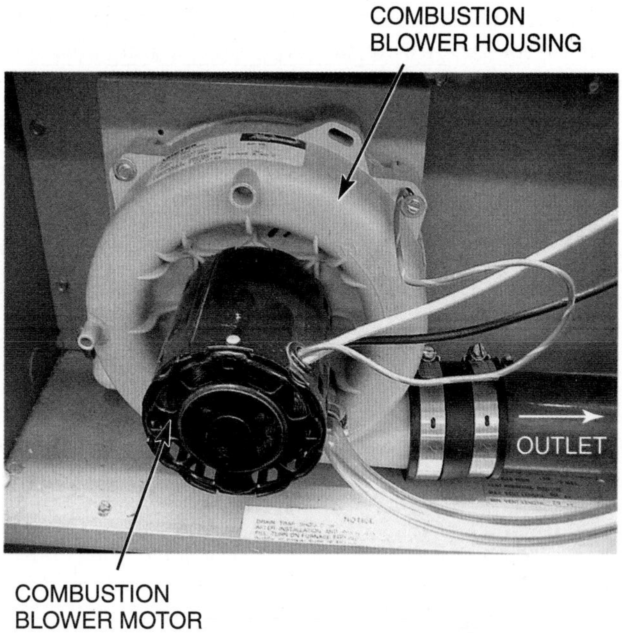

31.8 HEAT EXCHANGERS

Heat exchangers are made of sheet steel or aluminized steel and are designed to provide rapid transfer of heat from the hot combustion gases through the steel to the air that will be distributed to the space to be heated. Many burners are located at the bottom of the heat exchanger, **Figure 31.38.**

Figure 31.38 A heat exchanger with four sections.

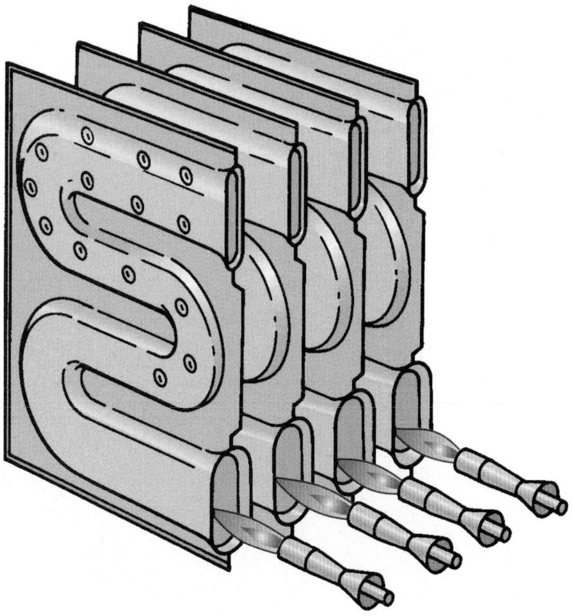

Figure 31.39 Checking the air temperature rise across a gas furnace.

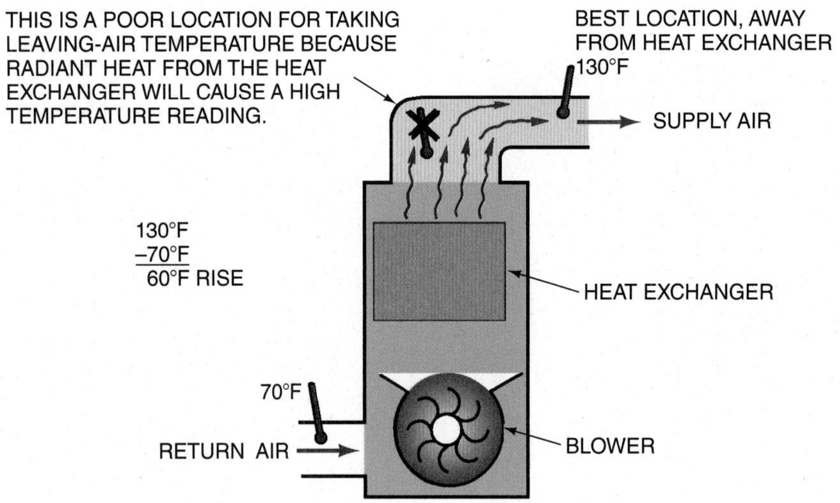

However, high-efficiency gas furnaces have their burners located at the top of the furnace in a sealed combustion chamber, **Figure 31.42**. The combustion gases are pulled or sucked through the heat exchanger. The combustion gases soon reach dew point, and liquid condensate forms and drains by gravity to a condensate sump or drain trap. The burners are located on top to allow the draining of condensate by gravity. High-efficiency furnaces will be covered in detail later in this unit. The *heat exchanger* is divided into sections with a burner in each section.

The airflow across heat exchangers must be correct or problems will occur. If too much air flows across a heat exchanger, the flue gas will become too cool and the products of combustion may condense and run down the flue pipe. These products are slightly acid and will deteriorate the flue system if not properly removed. However, high-efficiency condensing gas furnaces and their venting systems are equipped for the condensation of flue gases. Condensing furnaces will be covered in detail later in this unit.

If there is not enough airflow, the combustion chamber will overheat, and stress will occur. Furnace manufacturers print the recommended air temperature rise for most furnaces on the furnace nameplate. Temperature rise is dependent on whether the furnace is standard-, mid-, or high-efficiency. The temperature rise of the furnace decreases as the efficiency increases. This can be verified by taking the return air temperature and subtracting it from the leaving-air temperature, **Figure 31.39**. Be sure to use the correct temperature probe locations. Radiant heat from the heat exchangers can affect the temperature readings.

The following formula may be used to calculate the exact airflow through a gas furnace:

$$\text{cfm} = \frac{Q_s}{1.08 \times \Delta T}$$

where Q_s = Sensible heat in Btu/h
cfm = Cubic feet of air per minute
1.08 = Constant used to change cfm to pounds of air per minute for the formula
ΔT = Temperature difference between supply and return air

This requires some additional calculations because gas furnaces are rated by the input. If a gas furnace is 80% efficient during steady-state operation, you may multiply the input times the efficiency to find the output. For example, if a gas furnace has an input rating of 80,000 Btu/h and a temperature rise of 55°F, how much air is moving across the heat exchanger in cfm? The first thing we must do is find out what the actual heat output to the airstream is: 80,000 × 0.80 (80% estimated furnace efficiency) = 64,000 Btu/h. Use this formula:

$$\text{cfm} = \frac{Q_x}{1.08 \times \Delta T}$$

$$\text{cfm} = \frac{64{,}000}{1.08 \times 55}$$

$$\text{cfm} = 1077.4$$

There is some room for error in the calculation unless you take several temperature readings at both the supply and return ducts and average them. The more readings you take, the more accurate your answer. Also, the more accurate your thermometer is, the more accurate the results. There is also the heat added by the fan motor to consider. This will be about 300 W for a furnace this size. If you will multiply 300 W × 3.413 Btu/W, you will find that some extra heat is added to the air: 300 × 3.413 = 1023.9 Btu. For an even more accurate reading, you may take a watt reading of the motor or multiply the amperage reading times the applied voltage for a watt reading approximation:

$$\text{Watts} = \text{Amperage} \times \text{Applied Voltage}$$

Figure 31.40 (A) A modern heat exchanger showing L and S tubes, a transition box, and a collection box. Courtesy Rheem Manufacturing Company. (B) A close-up of S tubes, a transition box, and a collection box. Courtesy Rheem Manufacturing Company.

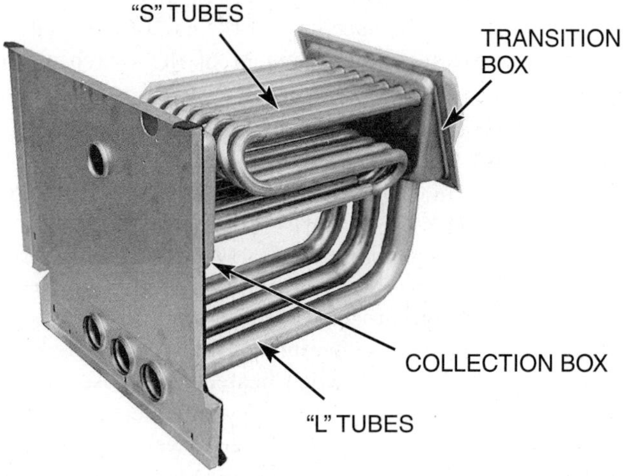

(A)

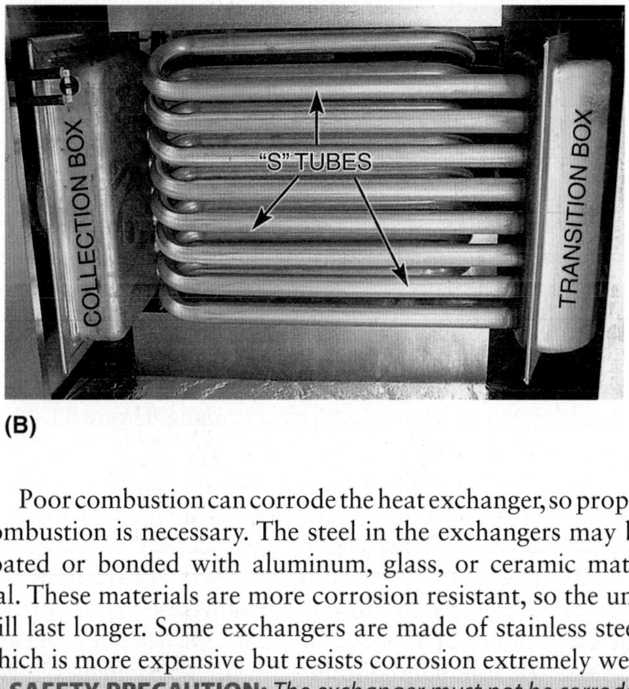

(B)

Poor combustion can corrode the heat exchanger, so proper combustion is necessary. The steel in the exchangers may be coated or bonded with aluminum, glass, or ceramic material. These materials are more corrosion resistant, so the unit will last longer. Some exchangers are made of stainless steel, which is more expensive but resists corrosion extremely well. **SAFETY PRECAUTION:** *The exchanger must not be corroded or pitted enough to leak since one of its functions is to keep combustion gases separated from the air to be heated and circulated throughout the building.•*

Today, heat exchangers come in many sizes, shapes, and materials. **Figure 31.40** shows a heat exchanger with both L-shaped and S-shaped tubes made of aluminized steel. The S tubes are smaller than the L tubes. The combustion gases travel a serpentine path. The L tubes carry the hot combustion gases to a transition box. The S tubes then carry the combustion gases from the transition box to a collection box. The collection box directs the combustion gases to an induced-draft centrifugal fan and then to the furnace's vent. The transition box is usually made of a highly corrosion-resistant stainless steel because it may see some condensation. The tubes are often swaged to the connection and transition boxes. The collection and transition box flanges are crimped for a tight fit.

Condensing furnaces may have two or three sets of heat exchangers. The secondary or tertiary heat exchanger, often referred to as a condensing heat exchanger, is typically made of stainless steel or aluminum to resist corrosion from the slightly acidic condensate, **Figure 31.41**. It may also have extended surfaces or fins for better heat conduction to the heating air. This condensing heat exchanger is designed to intentionally handle condensate and eventually drain it away from the furnace. One side of the heat exchanger has a plastic condensate drain header. **Figure 31.42** shows a high-efficiency condensing gas furnace with only two heat exchangers. The primary heat exchanger is made of two pieces of metal that are rolled or pressed together to form the serpentine path. This type of exchanger is often referred to as "clam-shell-type." The polypropylene plastic laminate coating of the second heat exchanger resists corrosion from the slightly acidic condensate.

Because of the serpentine path of combustion gases in most modern heat exchangers, they do not lend themselves well to the ribbon and slotted types of burners found in most traditional standard- and mid-efficiency furnaces, **Figures 31.33** and **31.34**. One of the main reasons for this is the height of the flame in the ribbon and slotted-port burners. Flame impingement would occur in the tubelike passages of these modern heat exchanger surfaces if ribbon

Figure 31.41 A stainless-steel condensing heat exchanger with aluminum fins. Notice the plastic condensate drain header on the left end. Courtesy Ferris State University. Photo by John Tomczyk.

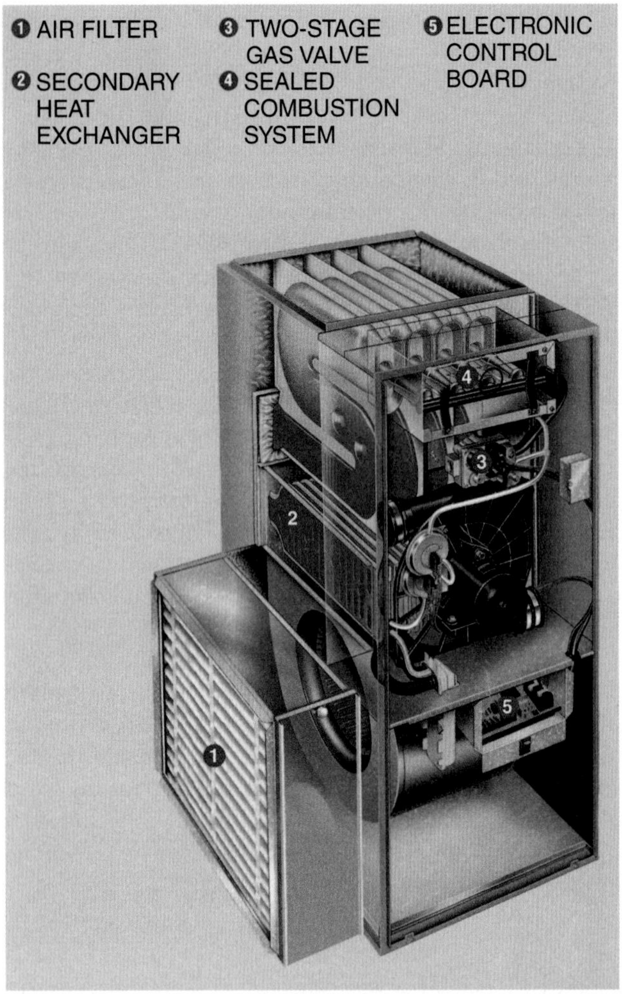

Figure 31.42 A high-efficiency condensing gas furnace with a second heat exchanger. Source: Bryant Heating and Cooling Systems.

or slotted-port burners were used. Also, if ribbon or slotted burners were used with serpentine heat exchangers, the secondary combustion air would flow at right angles to the vertical burner flame. This would cause the flame to be pulled off the ribbon or slotted-port burner. It is for these reasons that the inshot burner is used with serpentine-style heat exchangers. The inshot burner, sometimes called a *jet burner*, fires a horizontal flame straight out of its end, making it ideal for serpentine heat exchangers, **Figure 31.36(B)**.

31.9 FAN AND LIMIT CONTROLS

The fan side of the fan/limit switch automatically turns the blower on and off. The blower circulates the heated air to the conditioned space. The switch can be temperature controlled or on a time delay. In either instance, the heat exchanger is given a chance to heat up before the fan is turned on. This delay keeps cold air from being circulated through the duct system at the beginning of the cycle before the heat exchanger gets hot. The heat exchanger must be hot in a conventional furnace because the heat provides a good draft for properly venting the combustion gases. There is also a delay in shutting off the fan. This allows the heat exchanger time to cool off and dissipate the furnace heat at the end of the cycle.

The temperature-sensing element of the switch, usually a bimetal helix, is located in the airstream near the heat exchanger, **Figure 31.43**. When the furnace comes on, the air is heated, which expands the bimetal and closes the fan contacts, activating the blower motor. This is called a *temperature on–temperature off* fan switch.

In some models, the fan switch could activate the blower with a time delay and shut it off with a temperature-sensing device. When the thermostat calls for heat and the furnace starts, a small resistance heating device is activated. This heats a bimetal strip that, when heated, will close electrical contacts to the blower. This provides a time delay and allows the furnace to heat the air before the blower comes on. The bimetal helix in the airstream keeps the contacts closed even after the room thermostat is satisfied and the burner flame goes out. When the furnace shuts down, the bimetal cools at the same time as the heat exchanger and turns off the blower. This fan switch provides a positive starting of the fan and stops it by temperature and is called a *time on–temperature off* switch.

A third type of switch, called a *time on–time off* switch, uses a small heating device, such as the one used in the time on–temperature off switch. The difference is that this switch is not mounted in such a way that the heat exchanger heat will influence it. The time delay is designed into the switch and is not adjustable. Most models of the other two types of switches are adjustable.

Many gas furnaces use electronic modules or IFCs to control the fan or blower's off–delay timing. **Figure 31.44(A)** shows an electronic module that uses dual in-line pair (DIP) switches to control the fan's off–delay time. The on–delay timing is set at 30 sec and is nonadjustable. The service technician cannot control when the fan will come on once the burners light, which for the heating cycle will always be within 30 sec. However, the technician can adjust the fan or blower's off–delay timing when the fan shuts off once the burners shut off. The four configurations for the DIP switches in **Figure 31.44(B)** let the fan have a 60-, 100-, 140-, or 180-sec on–time period after the burners shut off to cool down the heat exchanger. This electronic module also controls the self-diagnostic functions of the gas furnace. IFCs often have nonadjustable fixed timing periods for blower on and off delays, which are programmed into their software, **Figure 31.29(A)**. Because there are many manufacturers of gas furnaces, the service technician must understand the furnace's sequence of operation, including the blower's on and off times and delays.

The *limit switch* is a safety device. If the fan does not come on or if there is another problem causing the heat

Figure 31.43 (A) A temperature on–temperature off fan/limit switch. Photo by Jason Obrzut. (B) Cutaway view of a fan/limit switch. (C) Helical bimetallic fan and limit controller located in close proximity to the heat exchanger. Courtesy Ferris State University. Photo by John Tomczyk.

(A)

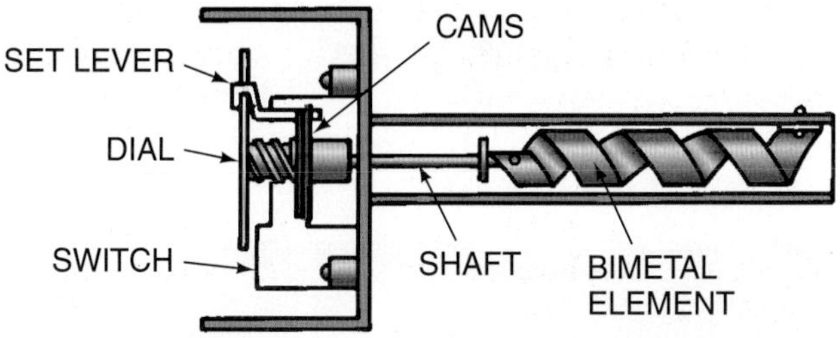

(B)

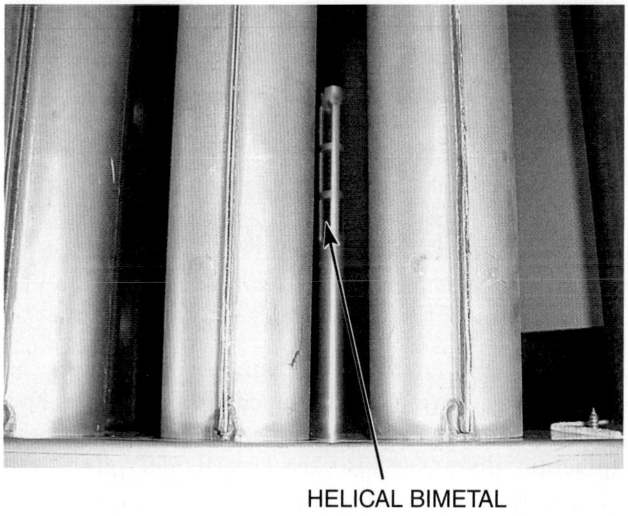

(C)

Figure 31.44 (A) An electronic module to control the blower's off–delay timing. Courtesy Ferris State University. Photo by John Tomczyk. (B) The dual in-line pair switches for controlling blower off–delay timing. Courtesy Ferris State University. Photo by John Tomczyk.

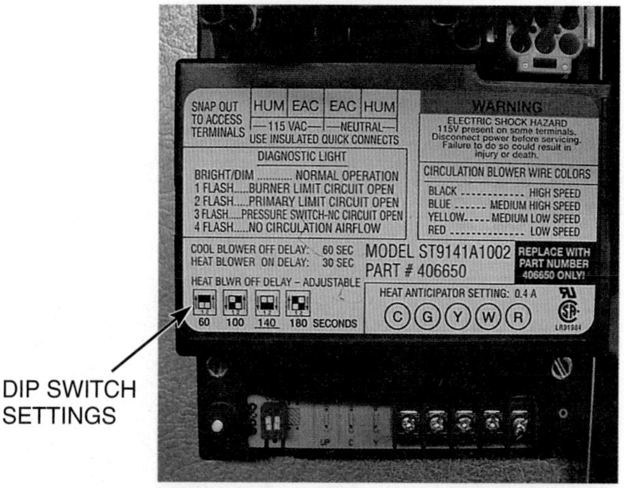

(A)

(B)

exchanger to overheat, the limit switch will open its contacts, which closes the gas valve. **⚠ SAFETY PRECAUTION:** *Almost any circumstance causing a restriction in the airflow to the conditioned space can make the furnace overheat, for example, dirty filters, a blocked duct, closed dampers, fan malfunctions, or a loose or broken fan belt. The furnace may also be overfired due to an improper setting or malfunctioning of the gas valve. It is extremely important that the limit switch operate as it is designed to do.•*

The limit switch has a heat-sensing element. When the furnace overheats, this element opens contacts, thus directly or indirectly closing the main gas valve. This switch can be combined with a fan switch, **Figure 31.45(A)**. Different state codes have different high-limit cut-out requirements, often around 200°F to 250°F. The combination fan and high-limit control in **Figure 31.45(B)** can be a high-voltage, low-voltage, or combination high- and low-voltage control. The actions of both the fan and the high-limit contacts are controlled by the same helical bimetal positioned next to the heat exchanger, **Figure 31.43(C)**. The combination control can be easily retrofitted for different voltages in the

Figure 31.45 (A) Fan/limit switch settings. Photo by Jason Obrzut. (B) A fan/limit switch showing the location of the break-away jumper.

(A)

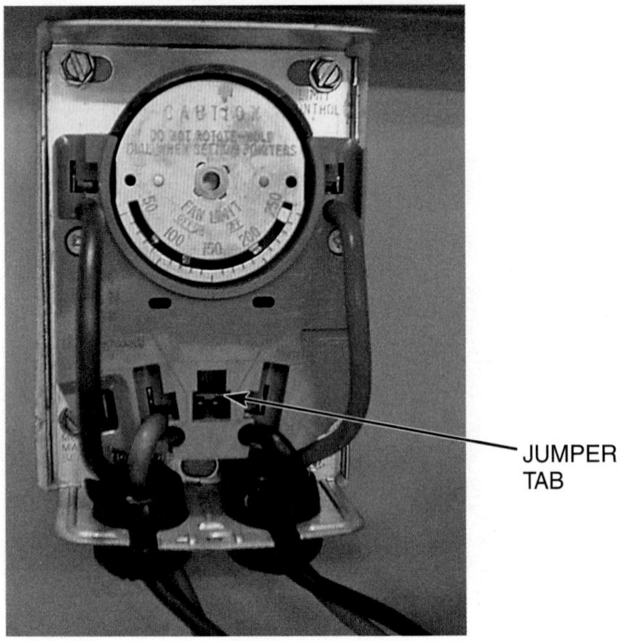

(B)

Figure 31.46 (A) A snap-action, bimetallic, high-limit control. The control is non-adjustable. Courtesy Ferris State University. Photo by John Tomczyk. (B) A snap-action, bimetallic, high-limit control positioned close to the heat exchanger. Courtesy Ferris State University. Photo by John Tomczyk.

(A)

(B)

field. If the furnace's circuit calls for both the high-limit and fan control contacts to be high voltage, a jumper tab must remain in place in the fan/limit switch control, **Figure 31.45(B)**. However, if the furnace's circuit calls for a high-voltage fan control and a low-voltage limit control, the jumper tab must be removed to separate the different voltages. This makes the control more versatile for circuit wiring purposes. **SAFETY PRECAUTION:** *Always read the manufacturer's literature for the contact's voltage rating before wiring or retrofitting any furnace control. Failure to do so may lead to a fire or explosion.*•

Many gas furnaces use snap-action, bimetallic, high-limit controls located in close proximity to the heat exchanger, **Figure 31.46**. These disks are set to snap open at a certain temperature dictated by the furnace manufacturer. Most high-limit controls will automatically reset back to the closed position when cooled to a certain temperature. However, some must be manually reset. Bimetallic, snap-action disks are also used as flame rollout safety controls, but they usually have to be manually reset. An example is shown in **Figure 31.47**. Notice that it is wired in series with the high-limit control. If either of these controls overheats and opens its contacts, the circuit to the main gas valve will be interrupted. The stem coming out of the disk must be pushed in once the control has cooled down. This resets the control. Manual reset requires a service call by a technician to troubleshoot why the furnace has flame rollout problems.

Figure 31.47 (A) A manually reset flame rollout safety switch. Courtesy Ferris State University. Photo by John Tomczyk. (B) A manual resettable flame rollout safety switch. Courtesy Ferris State University. Photo by John Tomczyk.

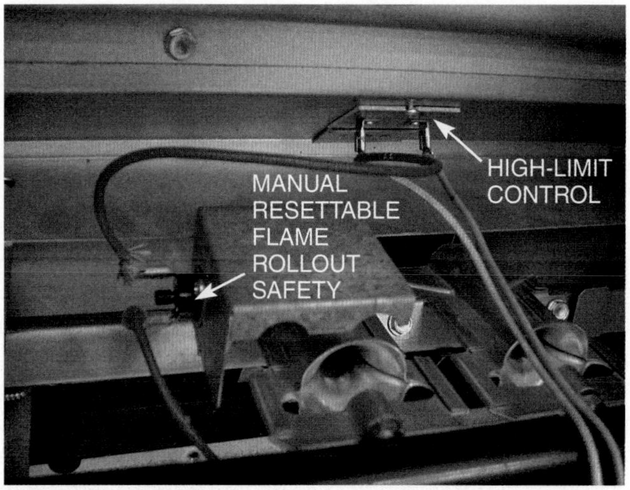

(A)

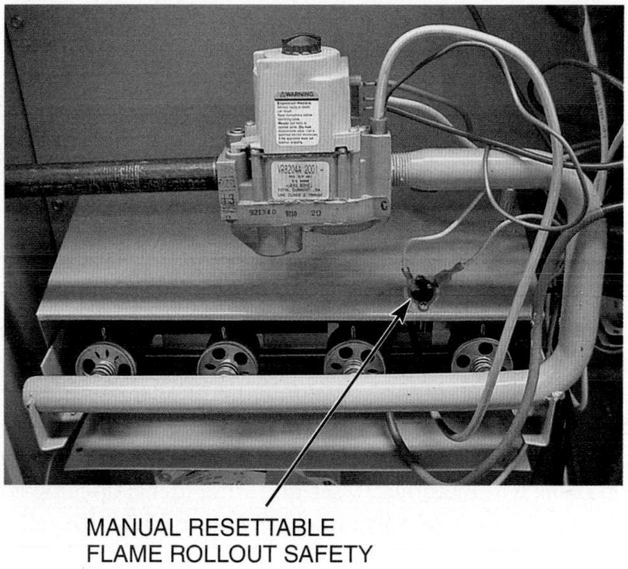

(B)

Figure 31.48 (A) The aerated pilot. (B) Cutaway view. (C) Clogged and starved for air.

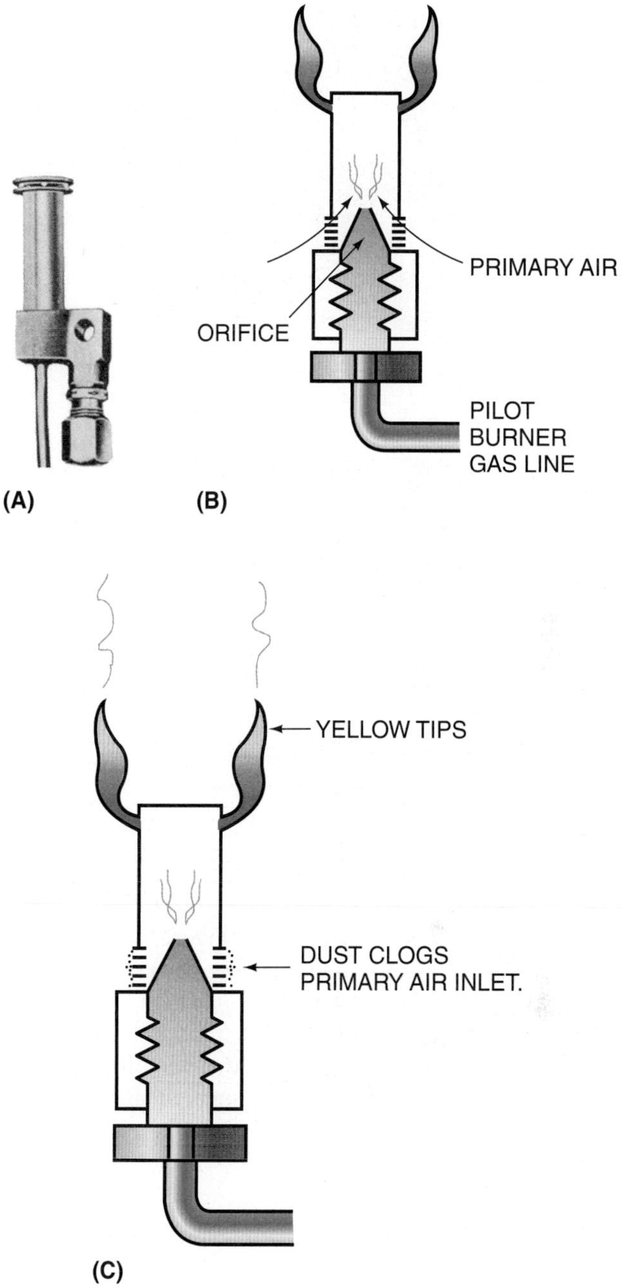

31.10 PILOTS

Pilot flames are used to ignite the gas at the burner on most conventional gas furnaces. Pilot burners can be aerated or nonaerated. In the aerated pilot, the air is mixed with the gas before it enters the pilot burner, **Figure 31.48**. However, the air openings often clog and require periodic cleaning if there is dust or lint in the air. Nonaerated pilots use only secondary air at the point where combustion occurs. They need little maintenance, so most furnaces are equipped with nonaerated pilots, **Figure 31.49**.

The pilot is actually a small burner, **Figure 31.50**. It has an orifice, similar to the main burner, through which the gas passes. If the pilot goes out or does not perform properly, a safety device will stop the gas flow. *Standing* pilots burn continuously; other pilots are ignited by an electric spark or hot surface igniter when the thermostat calls for heat. In furnaces without pilots, the ignition system ignites the gas at the burner. In furnaces with pilots, the pilot must be ignited and proved to be burning before the gas valve to the main burner will open. The pilot burner must direct the flame so that there

(A)

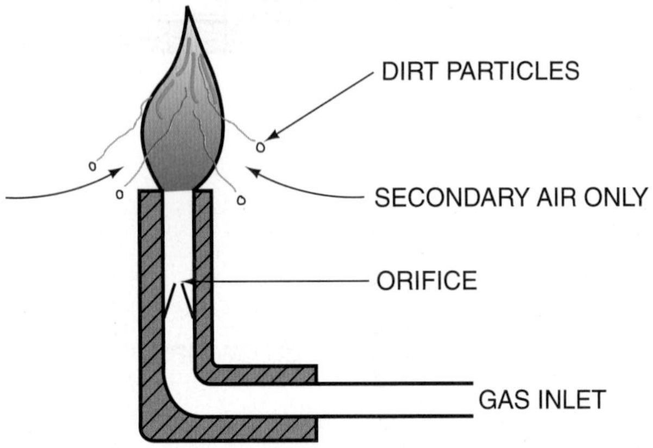

(B)

Figure 31.49 (A) A nonaerated pilot. Photo by Jason Obrzut (B) Cutaway view. Notice how dust particles burn away.

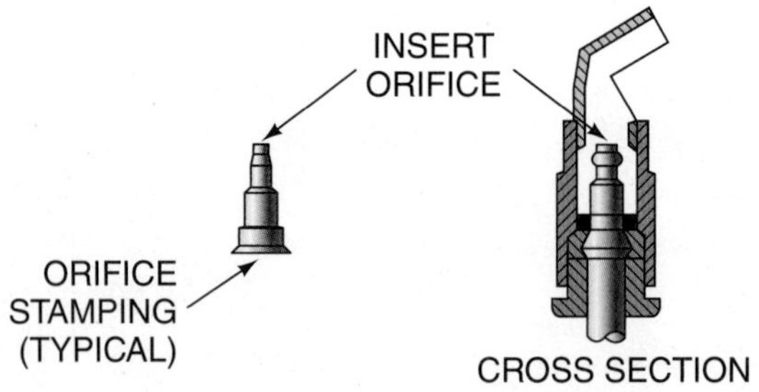

Figure 31.50 A nonaerated pilot burner showing an orifice.

will be ignition at the main burners, **Figure 31.51**. The pilot flame also provides heat for the safety device that shuts off the gas flow if the pilot flame goes out. There are three types of safety, or flame-proving devices, that prevent gas from flowing through the main valve if the pilot light goes out. These three devices are the thermocouple, the bimetallic strip, and the liquid-filled remote bulb.

THERMOCOUPLES AND THERMOPILES

The *thermocouple* consists of two dissimilar metals welded together at one end, **Figure 31.52**, called the "hot junction." When this junction is heated, it generates a small DC voltage (approximately 15 mVDC with load; 30 mVDC without load) across the two wires or metals at the other end, called the "cold junction." The thermocouple is connected to a shutoff valve, **Figure 31.53**. As long as the electrical current in the thermocouple energizes a coil, the gas can flow. If the flame goes out, the thermocouple will cool off in about 30 to 120 sec, the voltage will drop to zero and the gas valve will close. A *thermopile* consists of several thermocouples wired in series to increase the voltage. Typical voltages range from 500 mVDC to 750 mVDC, depending on the model. A thermopile performs the same function as the thermocouple, **Figure 31.54**.

BIMETALLIC SAFETY DEVICE

In the *bimetallic* safety device the pilot heats the bimetal strip, which closes electrical contacts wired to the gas safety valve, **Figure 31.55(A)**. As long as the pilot flame heats the bimetal, the gas safety valve remains energized and gas will flow when called for. When the pilot goes out, the bimetal strip cools within about 30 sec and straightens, opening contacts and causing the valve to close, **Figure 31.55(B)**.

Figure 31.51 (A)–(C) The pilot flame must be directed at the burners and adjusted to the proper height.

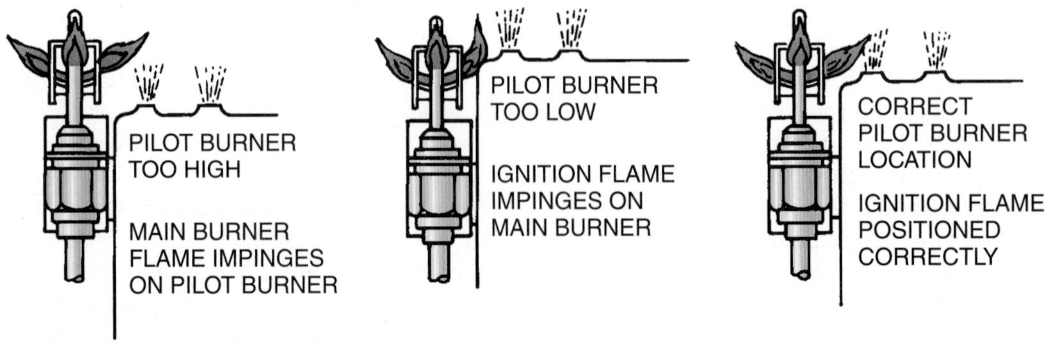

Figure 31.52 (A)–(C) The thermocouple and its functions.

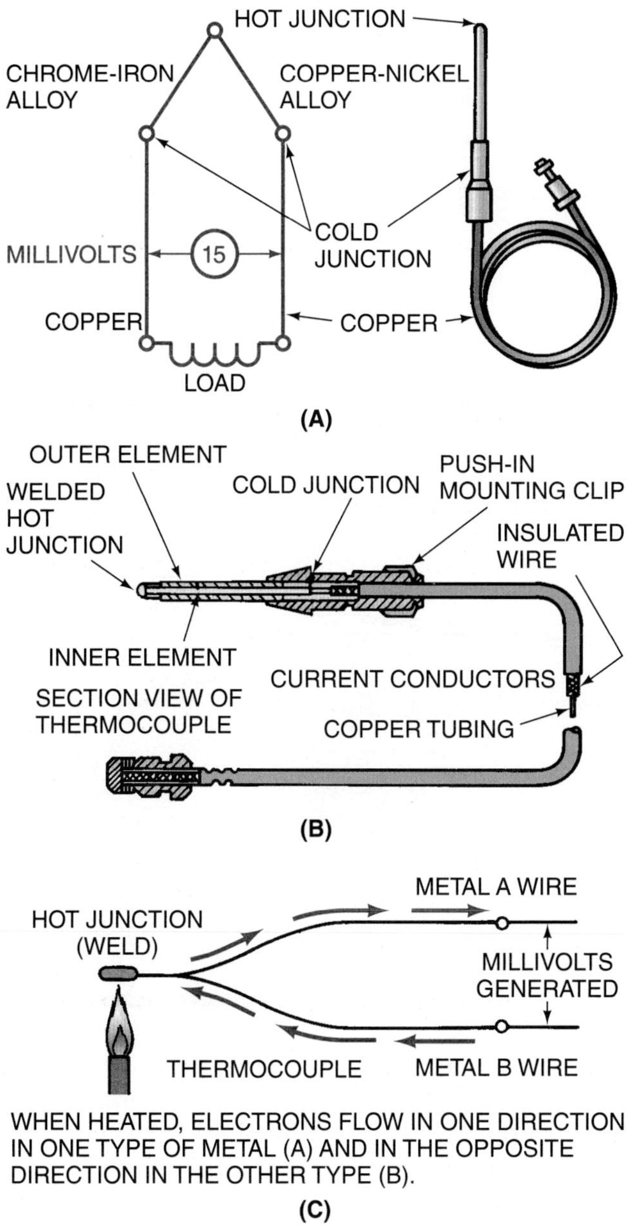

Figure 31.53 (A) and (B) The thermocouple generates electrical current when heated by the pilot flame. This induces a magnetic field in the coil of the safety valve holding it open. If flame is not present, the coil will deenergize, closing the valve, and the gas will not flow. (B) Photo by Bill Johnson.

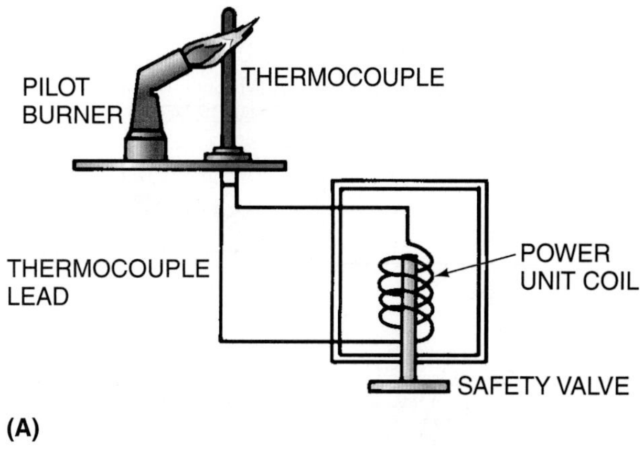

(B)

LIQUID-FILLED REMOTE BULB

The *liquid-filled remote bulb* includes a diaphragm, a tube, and a bulb, all filled with a liquid, usually mercury. The remote bulb is positioned so as to be heated by the pilot flame, **Figure 31.56.** The pilot flame heats the liquid and the liquid expands, causing the diaphragm to expand, which closes contacts wired to the gas safety valve. As long as the pilot flame is lit, the liquid is heated and the valve is open, allowing gas to flow. If the pilot flame goes out, the liquid cools in about 30 sec and contracts, opening the electrical contacts and closing the gas safety valve.

Figure 31.54 A thermopile consists of a series of thermocouples in one housing.

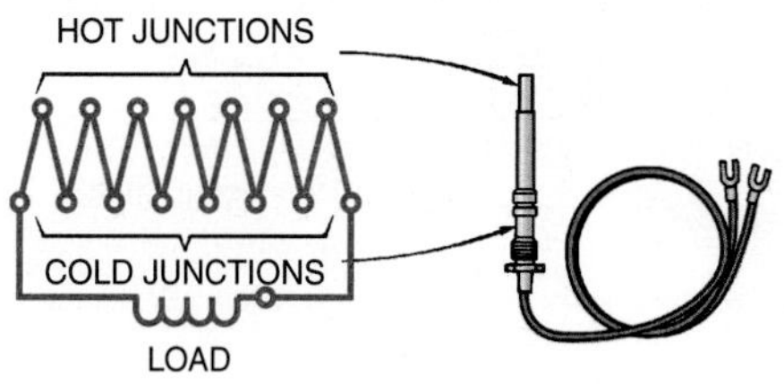

Figure 31.55 (A) When the pilot is lit, the bimetal warps, causing the contacts to close. The coil is energized, pulling the plunger into the coil, opening the valve, and allowing the gas to flow to the main gas valve. (B) When the pilot is out, the bimetal straightens, opening the contacts. The safety valve closes. No gas flows to the furnace burners. (C) A bimetallic safety device assembly. Courtesy Ferris State University. Photo by John Tomczyk.

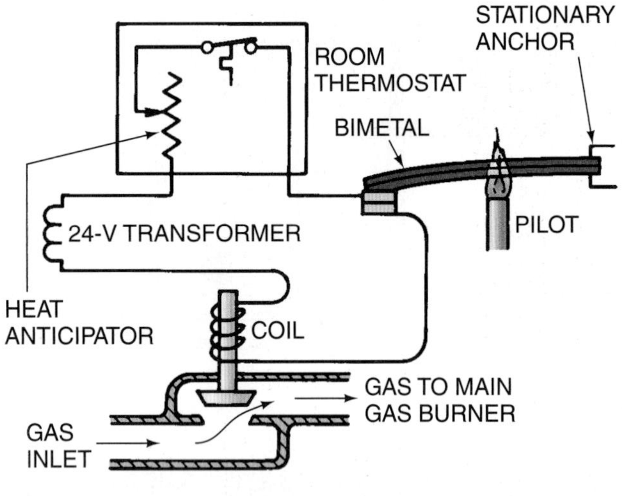

(A)

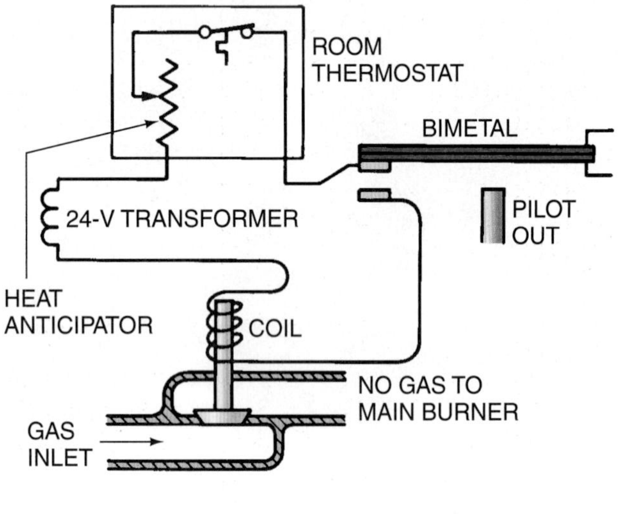

(B)

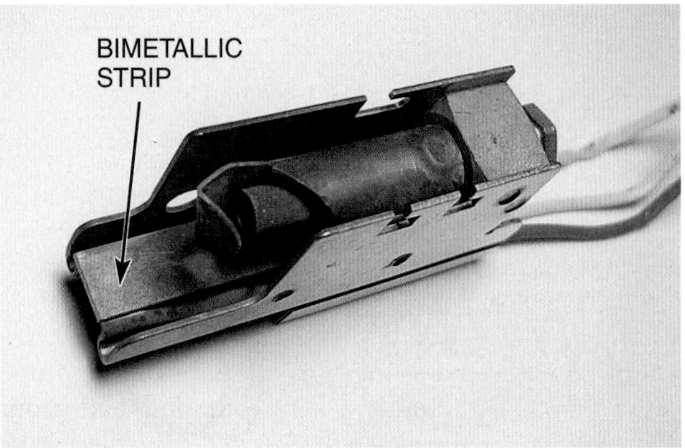

(C)

Figure 31.56 (A) A liquid-filled remote bulb. (B) A liquid-filled remote bulb assembly. Courtesy Ferris State University. Photo by John Tomczyk.

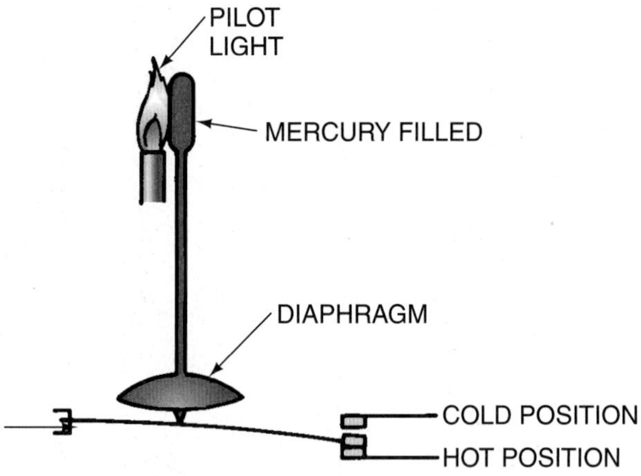

(A)

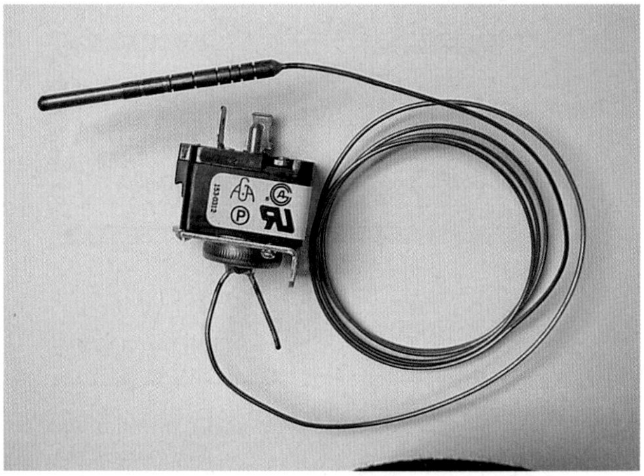

(B)

31.11 IGNITION SYSTEMS

Ignition systems can ignite either a pilot or the main burner. They can be divided into three categories:

- Intermittent pilot (IP)
- Direct-spark ignition (DSI)
- Hot surface ignition (HSI)

INTERMITTENT PILOT (IP) IGNITION

In the intermittent pilot ignition system, a spark igniter or 24-V hot surface igniter (HSI) light the pilot, which ignites the main gas burners, **Figure 31.57**. The pilot burns only when the thermostat calls for heat. This system is popular because fuel is not wasted by the pilot burning when not needed. Two types of control schemes are used with this system. One, used a lot with natural gas, is not considered a 100% shutoff system. If the pilot does not ignite, the pilot

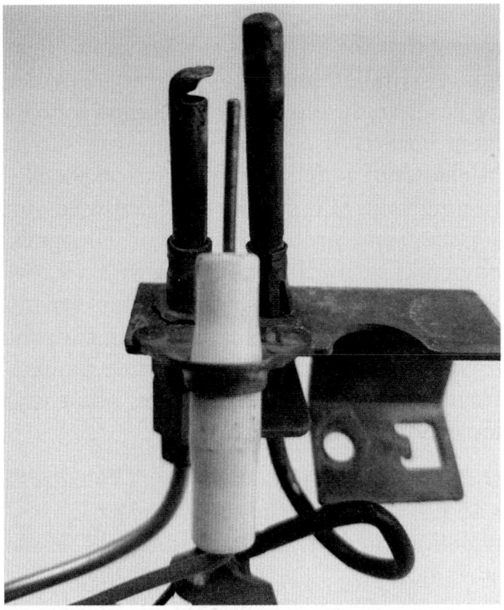

Figure 31.57 A spark-to-pilot ignition assembly. Photo by Jason Obrzut.

valve will remain open, the spark will continue, and the main gas valve will not open until the pilot is lit and proved. The other type of system is used with LP gas and some natural gas applications and is a 100% shutoff system. If there is no pilot ignition, the pilot gas valve will close and may go into safety lockout after approximately 90 sec, and the spark will stop. Safety lockout systems will be covered in detail later in this unit. This system must be manually reset, usually at the thermostat or power switch.

The 100% shutoff control system is necessary for any gas fuel heavier than air, because the fuel will accumulate in low places. Even the small amount of gas from a pilot light could be dangerous over time. Natural gas rises up through the flue system and is not dangerous in small quantities, such as the volume from a pilot light. Even though natural gas is lighter than air, some codes do call for 100% shutoff control. All technicians who service fuel gas products should be aware of the local codes. Both 100% shutoff and less than 100% shutoff systems will be covered in detail later in this unit.

When there is a call for heat in a natural gas system, contacts will close in the thermostat, providing 24 V to the ignition module, which will send power to the pilot igniter and to the pilot valve coil. The coil opens the pilot valve, and the electronic igniter lights the pilot. Once the pilot flame is lit, it must be proved. Flame rectification is the fastest, safest, and most reliable method for proving a flame.

In the *flame rectification* system, the pilot flame changes the normal alternating current to direct current. The electronic components in the system will energize and open the main gas valve only when a direct current, measured in microamps, is present. It is important that the pilot flame quality be correct to ensure proper operation. Consequently, the main gas valve will open only when a pilot flame is present. Flame rectification will be covered in detail later in this unit.

Figure 31.58 A direct-spark ignition assembly near the gas burner. Courtesy Ferris State University. Photo by Jason Obrzut

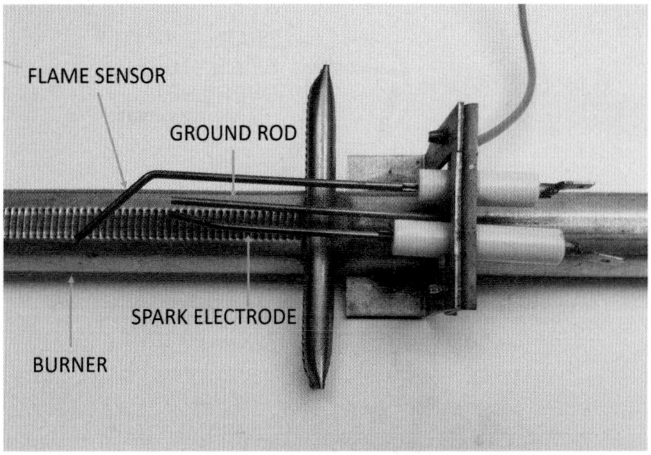

In a spark-to-pilot system, the spark is intermittent and arcs approximately 100 times per minute. It must be a high-quality arc, or the pilot will not ignite. The arc comes from the control module and is very high voltage. The voltage can reach 10,000 V in some systems but is usually very low amperage. A direct path to earth ground must be provided because the arc actually arcs to ground. A ground strap near the pilot assembly or the pilot hood often acts as ground, **Figure 31.58**.

DIRECT-SPARK IGNITION (DSI)

Direct spark ignition furnaces are designed with a spark ignition direct to the main burner or a ground strap, **Figure 31.58**. No pilot is used in this system. System components are the igniter/sensor assembly and the DSI module. The flame sensor verifies that the furnace has fired and sends a microamp signal through flame rectification to the DSI module confirming ignition. The furnace will then continue to operate. This system goes into a "safety lockout" if the flame is not established within the "trial for ignition" period (approximately 4 to 11 sec). Gas is being furnished to the main burner, so there cannot be as much time delay as there is in the 90-sec ignition trial for spark-to-pilot systems. The system can then only be reset by turning the power off to the system control and reapplying the power. This is a "typical" system. The technician should follow the wiring diagram and manufacturer's instructions for the specific furnace being installed or serviced.

Most ignition problems are caused from improperly adjusted spark gap, igniter positioning, and bad grounding, **Figure 31.59**. The igniter is centered over the left port. Most manufacturers also provide specific troubleshooting

Figure 31.59 The spark gap and igniter position for a DSI system.

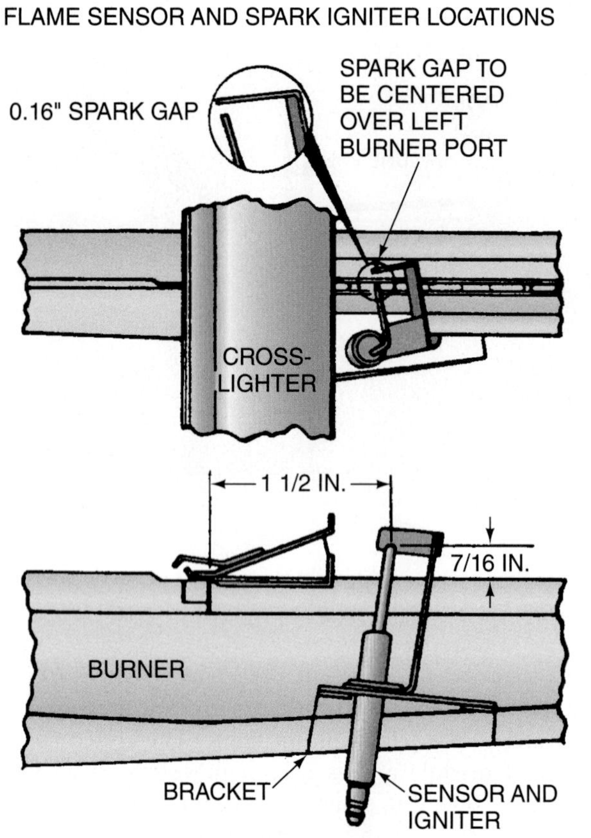

instructions. Once the main burner flame is proved by the flame rectification system, the sparking will stop. If the sparking does not stop once the main burner is lit, the electronic DSI module could be defective. The continual sparking may not be harmful to the system, but it is noisy and often can be heard in the living or working area. Read and follow the manufacturer's instructions to repair the system.

HOT SURFACE IGNITION (HSI)

The hot surface ignition (HSI) system uses a special product called silicon carbide that offers a high resistance to electrical current flow but is very tough and will not burn up. This substance is placed in the gas stream and is allowed to get very hot before the gas is allowed to impinge on the glowing hot surface. Immediate ignition should occur when the gas valve opens.

The hot surface igniter is usually operated off of 120 V, the line voltage to the furnace, and draws considerable current when energized. It is energized for only a short period of time during start-up of the furnace, so this current is not present for more than a very few minutes per day. **Figures 31.60(A)**, **(B)**, and **(C)** shows the hot surface igniter. Intermittent HSI systems operate on 24 V. These systems are hot surface to pilot, meaning that the 24-V hot surface igniter lights a pilot flame, **Figures 31.60(D)** and **31.60(E)**. For many decades the pilot flame has been the most reliable ignition source for lighting the main burner. Also, a low-voltage hot surface igniter lowers the chance of a high-voltage electrical shock.

Even though the HSI system has been used successfully for many years, some problems do occur. The igniter is very brittle and breaks easily. If you bump it with a screwdriver, it will break. Silicon nitride hot surface igniters are more durable and will not break as easily. However, care should still be taken when working around hot surface igniters because of their fragile nature. If repeated failures occur with an HSI system, look for the following:

- Higher-than-normal applied voltage. Voltage that is in excess of 125 V can shorten the igniter's life.
- Accumulation of drywall dust, fiberglass insulation, or sealant residue.
- Stress caused by the small explosion due to delayed ignition.
- Overfiring condition.
- Furnace short cycling as with a dirty filter, which may cause the furnace to cycle on high limit.

The hot surface igniter may be used for lighting the pilot in an intermittent ignition system or for direct ignition of the main furnace burner. When it is used as the direct igniter for the burner, it will have very little time, usually 4 to 11 sec, to ignite the burner; then a safety lockout will occur to prevent too much gas from escaping.

31.12 FLAME RECTIFICATION

In the flame rectification system, the pilot flame, or main flame, can conduct electricity because it contains ionized combustion gases, which are made up of positively and negatively charged particles. The flame is located between two electrodes of different sizes. The electrodes are fed with an alternating-current (AC) signal from the furnace's electronic module. Due to the different size of the electrodes, current will flow better in one direction than in the other, **Figure 31.61(A)**. The flame actually acts as a switch. If there is no flame, the switch is open and electricity will not flow. When the flame is present, the switch is closed and will conduct electricity.

The two electrodes in a spark-to-pilot (intermittent pilot) ignition system are the pilot hood (earth ground) and a flame rod or flame sensor, **Figure 31.62**. In a DSI system, the two electrodes can be the main burners (earth ground) and a flame rod or sensor. These types of systems are often referred to as dual-rod, or remote, sensing systems; one rod senses the flame and the other is used for ignition. However, the flame rod or sensor can also be part of the ignition system and is referred to as a combination sensor and igniter, **Figure 31.59**.

Figure 31.60 (A) A hot surface igniter deenergized. (B) The hot surface igniter system energized before main gas is allowed to flow. (C) The hot surface igniter system deenergized during the furnace heating cycle. (D) A 24-V intermittent hot surface ignition pilot assembly with a flame sensor. (E) The 24-V hot surface igniter energized. All photos: Courtesy Ferris State University. Photo by John Tomczyk.

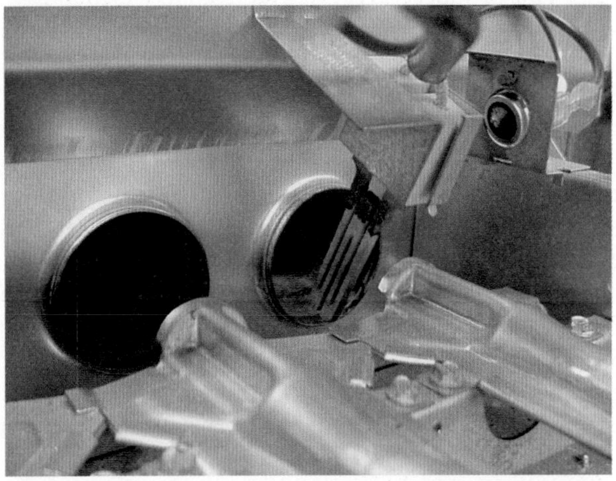

(A)

(B)

(C)

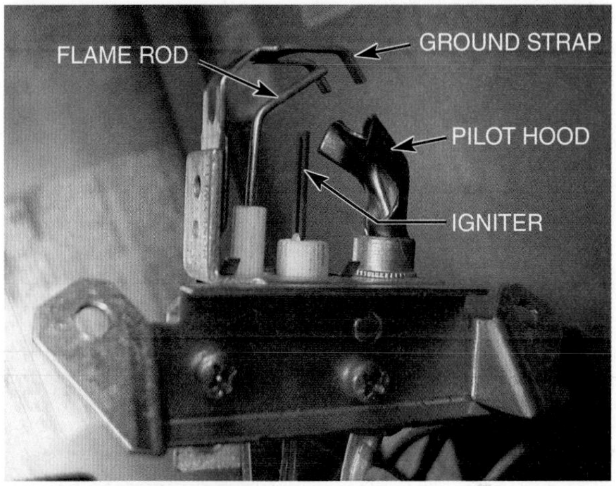

(D)

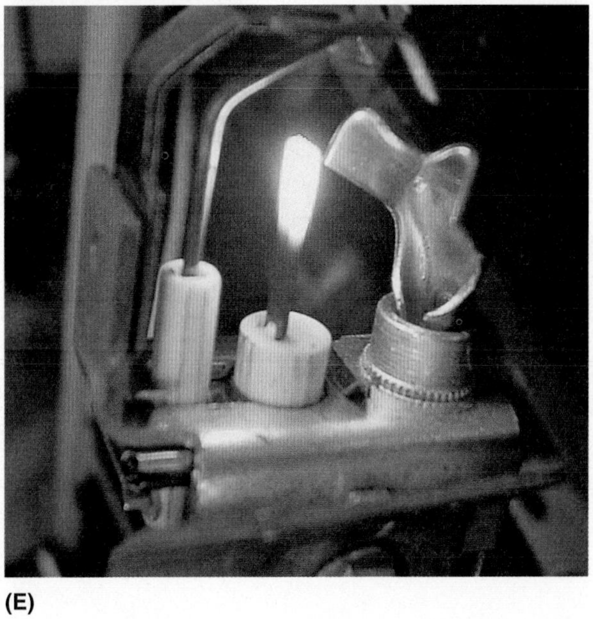

(E)

Figure 31.61 (A) The different sizes of the electrodes cause current to flow better in one direction than the other. (B) A pulsating direct-current (DC) signal that can be measured in microamps with a microammeter in series with one of the electrodes.

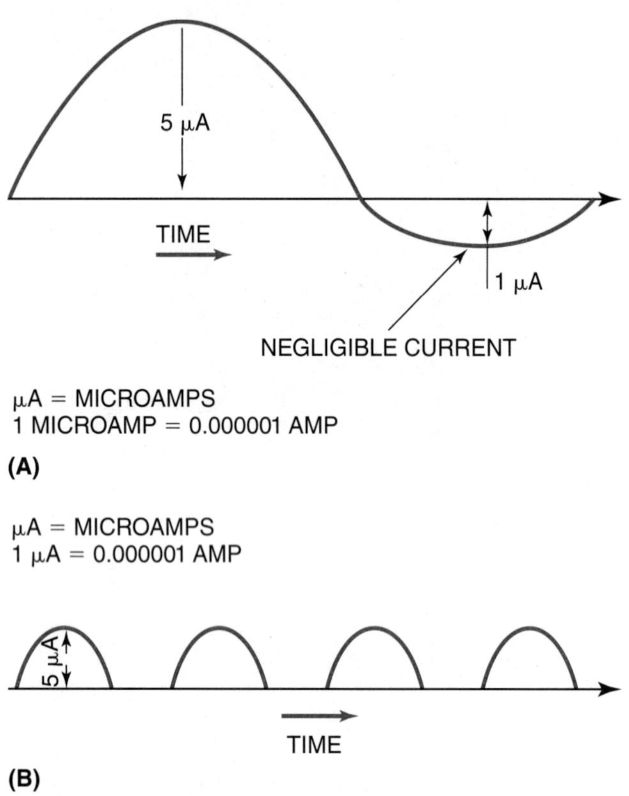

This type of system is referred to as a single-rod, or local, sensing system because it has a combination sensor and igniter in one rod.

As alternating current switches the polarity of the electrodes 60 times each second, the positive ions and negative electrons move at various speeds to the different-size electrodes. This produces a rectified AC signal, which looks much like a pulsating DC signal, because the positively charged particles (ions) are relatively larger, heavier, and slower than the negatively charged particles (electrons), **Figure 31.61(B)**. In fact, the electrons are about 100,000 times lighter than the ions. This DC signal is the only signal the electronics of the furnace recognize. It is proof that a flame exists, rather than a short resulting from humidity or direct contact between the electrodes. The DC signal can be measured with a microammeter in series with one of the electrode leads. It usually ranges between 1 and 25 microamps. The magnitude of the microamp signal may depend on the quality, size, and stability of the flame and on the electronic module and electrode design. Always consult with the furnace manufacturer for specific information on flame rectification measurements.

Humidity or direct contact between the electrodes would cause an AC current to flow, which would not be recognized by the furnace's electronic module. An AC signal thus prevents the main gas valve from opening in a spark-to-pilot system. It shuts down the main gas in a DSI system once the time period for ignition has elapsed and may put the furnace in a lockout mode.

Single-rod, or local, sensing consists of an igniter and a sensor all in one rod, often referred to as a one-rod pilot. Single-rod flame rectification systems accomplish both ignition and sensing. There is only one wire, which is the large ignition wire, running from the pilot assembly to the ignition module, **Figure 31.63**. This high-voltage lead does both the sparking and the sensing. The ignition pulse occurs during one half cycle and the sensing pulse current

Figure 31.62 (A) An intermittent pilot dual-rod system showing the pilot hood and the flame sensor rod as the two electrodes for flame rectification. Dual-rod systems are often referred to as remote sensing systems. (B) The flame rectification circuit.

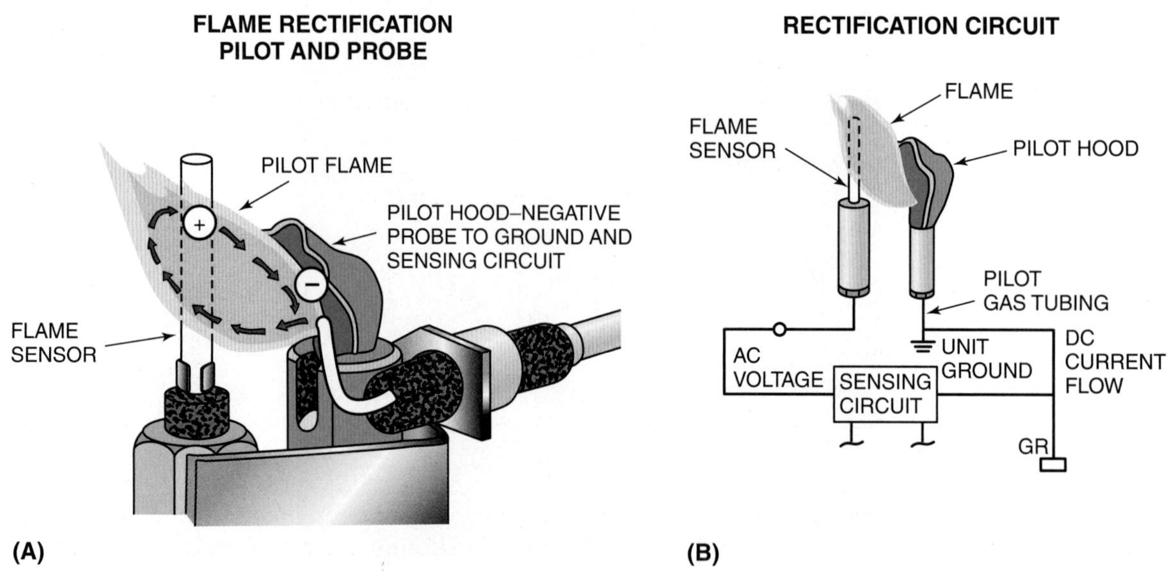

Figure 31.63 A single-rod, or local, sensing system connected to a furnace electronic module. Source: Honeywell.

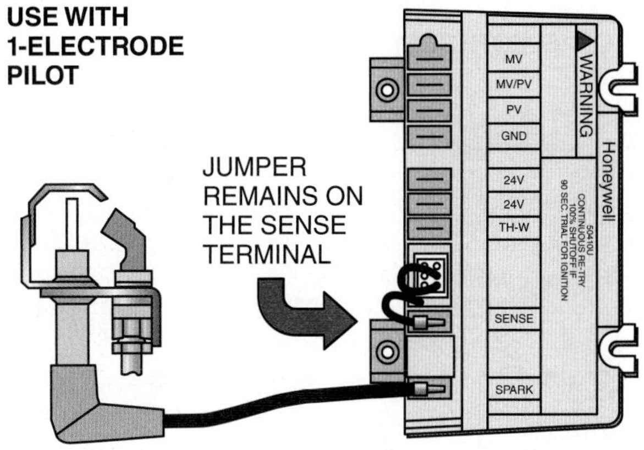

Figure 31.64 Single-rod, or local, flame rectification sensing system senses flame current during one half cycle and the ignition pulse during the other half cycle. Source: Honeywell.

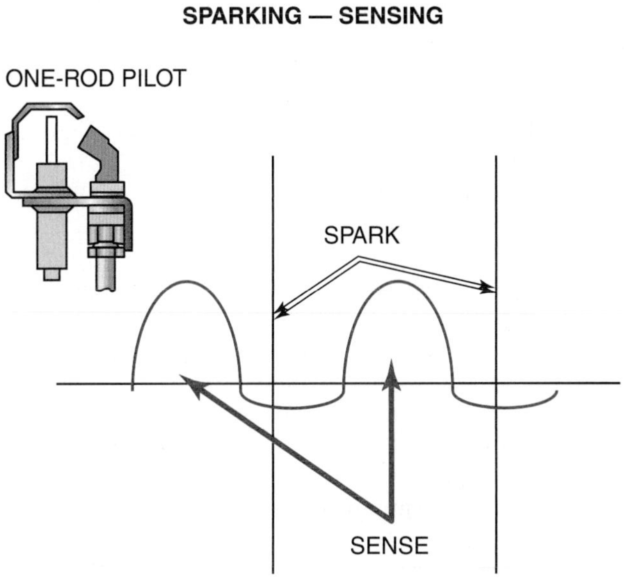

occurs on the other half cycle, **Figure 31.64**. Signal analyzers within the electronic module can differentiate between the two signals.

Dual-rod, or remote, sensing uses separate igniter and flame-sensing rods or electrodes, **Figure 31.65**. The sparking is accomplished with one rod and the sensing is done with the other rod. There are two wires running from the pilot assembly to the module. The spark electrode wire goes to the spark terminal on the module. The flame rod or sensing wire goes to the sensing terminal. A jumper wire will have to be cut off on some models, **Figure 31.66**. The flame sensor works with the grounded burner hood and the ground strap to achieve flame rectification.

Figure 31.65 A dual-rod, or remote, flame rectification sensing system showing a separate igniter spark electrode and a flame sensing rod. Source: Honeywell.

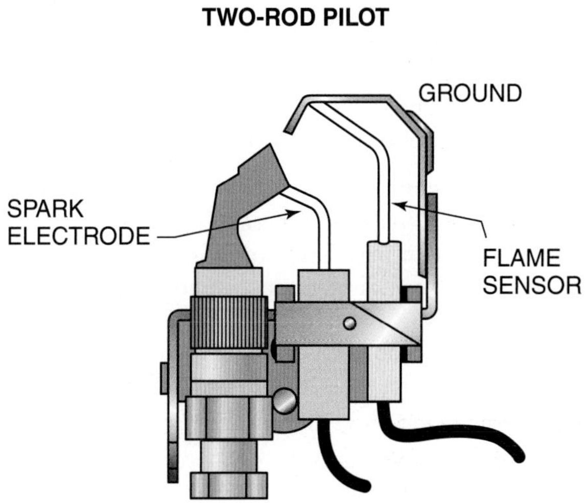

Figure 31.66 A dual-rod, or remote, flame rectification sensing system showing two separate wires going to the electronic furnace module. Source: Honeywell.

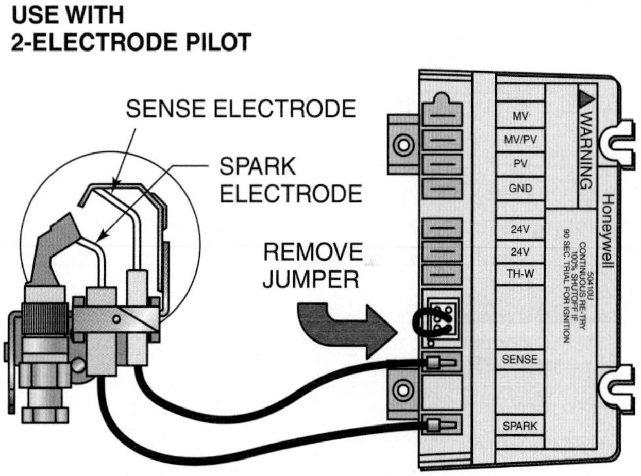

The preceding examples describe only a few of the furnace modules seen in the field today. Many newer furnaces have much more sophisticated IFCs. However, the principles of flame rectification still hold true.

TROUBLESHOOTING FLAME RECTIFICATION SYSTEMS

Troubleshooting a flame rectification system should be handled in steps. If the pilot flame or main flame does not light during a normal call for heat, the problem is not in the flame rectification system. It would be in the ignition system, because there must be a flame for there to be a problem in the flame rectification system. If a flame is present but goes out

because the system shuts off the gas valve, there could be one of several problems with the flame-sensing system:

- The flame electrodes could be dirty or corroded.
- A wire may be disconnected.
- The flame rod may not be installed or positioned properly.
- The flame signal amplifier (module or IFC) may have failed.
- There may be a poor ground wire connection.
- There may be a poor or weak flame.

The service technician must understand the ignition control sequence to determine whether the ignition source is working. The system may have either a spark or hot surface ignition element, so it is of utmost importance to know the sequence of events of the particular ignition module or IFC. If the ignition sequence seems to be fine but the pilot or main burner does not ignite, the problem is not flame rectification. It could be one of the following:

- Misadjusted pilot burner
- Out-of-place ignition source
- Low or no gas pressure
- Lack of power to the system
- Insufficient spark or insufficient power to the hot surface element
- Broken or disconnected wires or cables
- Broken electrode or HSI element

FLAME RECTIFICATION MAINTENANCE

Several factors can affect the very small flame current. Remember that this current is measured in DC microamps; 1 microamp equals 0.000001 amp. The following are some factors to consider in basic maintenance:

- Wires should be replaced every 3 to 5 years.
- Wiring between the flame electrodes and the ignition module or IFC should be as short as possible.
- Electrodes must be replaced if there are signs of bending, warping, or deterioration.
- Burner gas pressure must be correct. Poor flame contact with the metal burner components can cause failure.
- Over time, silicon contamination from the outgassing of the ignition cable or ordinary room dust, which is typically high in silicon, can cause an insulating material to build up on the sensing electrode or ground terminal.
- Aluminum oxides can also build up on the surface of the sensing electrode. The source of the aluminum can be the sensing electrode itself. Most flame rods are made of a Kanthol element, which contains aluminum and gives the rod the ability to withstand high temperatures.

MEASURING FLAME CURRENT

A microammeter placed in series with the flame rod or system ground can measure the small current produced by the flame. It is usually easier to place the meter in series with the wire coming from the burner ground terminal to the module or IFC, **Figure 31.67**.

Figure 31.67 A microammeter placed in series with the flame rod or system ground will measure DC microamps of flame rectification.

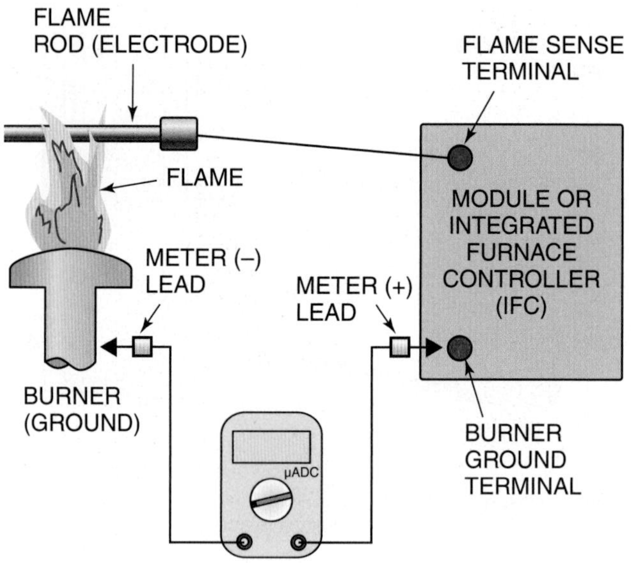

31.13 HIGH-EFFICIENCY GAS FURNACES

In addition to their efficiency, the lack of sidewall venting material for mid-efficiency furnaces has made high-efficiency gas furnaces popular. Stack temperatures are the limiting factor for sidewall venting. Sidewall venting in high-efficiency furnaces involves inexpensive materials like polyvinyl chloride (PVC). In addition, high-efficiency furnaces are quiet and energy-efficient due to their sealed combustion chamber, shown in **Figure 31.68**. These furnaces also lower gas bills. Because the sealed combustion chamber uses outside air and never receives conditioned room air, there is less outdoor air infiltration into the indoor heated space. The heated, conditioned room air is not pulled into a slight vacuum—that is, depressurized—by being used as combustion air, which would allow cold outdoor air to infiltrate the heated space.

HIGH-EFFICIENCY GAS FURNACE DESIGN

High-efficiency gas furnaces have some unique characteristics compared to conventional and mid-efficiency gas furnaces:

- A "recuperative" or "secondary" (and sometimes "tertiary") heat exchanger with connecting piping for flue-gas passages from the primary heat exchanger
- A flue-gas condensate disposal system, which includes drain, taps, drain lines, and drain traps, **Figure 31.69(A)**

Figure 31.68 A sealed combustion chamber on a high-efficiency furnace. Courtesy Ferris State University. Photo by John Tomczyk.

- A venting system that is drafted by either induced or forced power, **Figure 31.69(B)**

A direct-vented, high-efficiency gas furnace has a sealed combustion chamber. This means that the combustion air is brought in from the outside through an air pipe, which is usually made of PVC plastic, **Figure 31.70**. *The combustion chamber never sees conditioned room air. Since conditioned room air is not used for combustion, it prevents the conditioned room from being depressurized or pulled into a slight vacuum. This cuts down on the amount of infiltration entering the room, and less cold, infiltrated air has to be reheated.*

Direct venting also minimizes the interaction of the furnace with other combustion sources. Depressurization of a building or home may lead to products of combustion entering the building through the vent systems of water heaters and clothes driers. (*Depressurization* is the term for rooms that have lost pressure and are in a slight vacuum.) Even toxic fumes and particulates from a connecting or attached garage or building can infiltrate a conditioned space when it is depressurized. A high-efficiency furnace that is non-direct-vented uses indoor conditioned air for combustion. A room can become depressurized using such a system. These are reasons why the direct venting of a furnace is an added safety feature. Both direct-vented and non-direct-vented systems can be positive pressure systems and vented vertically or through the sidewall. The flue pipe in a positive pressure system has positive pressure its entire distance to its terminal end. However, not all direct-vented systems are positive pressure systems.

Some direct-vented designs need stronger flue fans or combustion blowers to overcome the additional pressure drop created by the sealed combustion chamber of the high-efficiency furnace and the additional combustion air pipe. In fact, the combustion blower must overcome three system pressure drops:

- Pressure drops from the heat exchanger system
- The pressure drop from the vent system (vent and combustion air pipe)
- Vent termination wind resistance pressure

Combustion blower motors use an air-proving switch or pressure switch to prove that the mechanical draft has been established before allowing the main burner to light, **Figure 31.69(B)**. The pressure switches are connected to the combustion blower housing or heat exchanger (burner box) by rubber tubing. These pressure switches have diaphragms that are sensitive to pressure differentials. Some furnaces use a single-port pressure switch that senses the negative pressure inside the heat exchanger caused by an induced-draft combustion blower. Others use a dual-port, or differential, pressure switch that senses the difference in pressure between the combustion blower and the burner box, **Figures 31.69(C)** and **31.69(D)**.

When the venting system is operating properly, sufficient negative pressure is created by the combustion blower to keep the pressure switch closed and the furnace operating. However, if the heat exchanger has a leak or the furnace vent pipe is blocked, the pressure switch will open from insufficient negative pressure. This will interrupt the operation of the main burners. Different furnace manufacturers use different settings for their pressure switches. The settings are usually measured in inches of water column and are sometimes indicated on the pressure switch housing. Pressures can be measured using a digital manometer.

SAFETY PRECAUTION: *Never replace a pressure switch with another switch without knowing its operating pressures. Always consult with the furnace manufacturer or use a factory-authorized substitute pressure switch.*•

DEW POINT TEMPERATURE

When natural gas is burned, water vapor is one of the by-products of combustion. More than 2 lbs of water vapor is produced for every 1 lb of natural gas burned. This water vapor contains heat. If it can be condensed, it will give off latent heat. The dew point temperature (DPT) is the temperature at which the condensation process begins. In flue gas, the DPT varies depending on the composition of the gas and the amount of excess air (combustion air and dilution air).

For every 1 lb of water vapor that is condensed into liquid, about 970 Btu are given off. This is the latent heat of condensation of water vapor at atmospheric pressure. The latent heat adds significantly to the efficiency of the condensing furnace. The condensate can be somewhat acidic and corrosive, however. The flue gas is carried away with PVC piping, and the liquid condensate should be piped down an appropriate indoor drain, **Figure 31.69(A)**.

Figure 31.69 (A) A flue-gas condensate disposal system in a high-efficiency condensing furnace. Courtesy Rheem Manufacturing Company (B) A combustion blower motor assembly with a pressure switch and diaphragm used for safely venting a furnace. Courtesy Ferris State University. Photo by John Tomczyk (C) A single-port pressure switch. Source: International Comfort Products Corporation LLC (D) A dual-port, or differential, pressure switch. Source: International Comfort Products Corporation LLC.

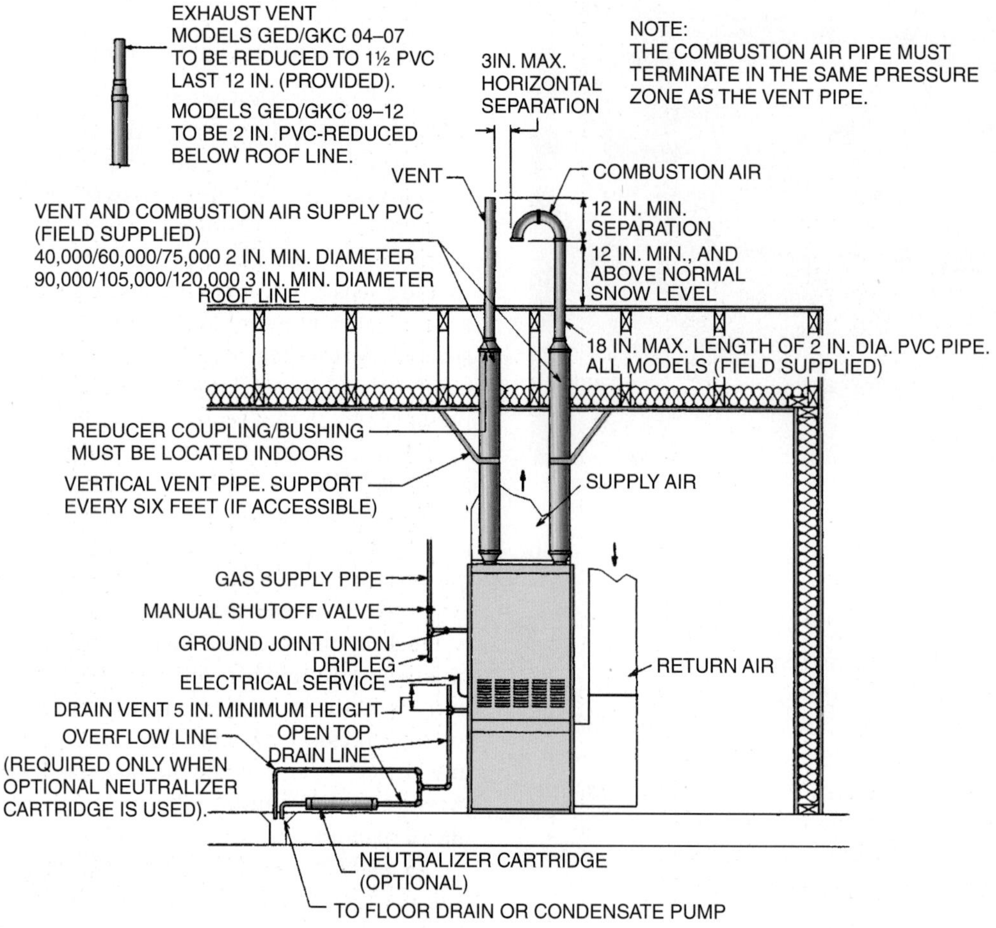

(A)

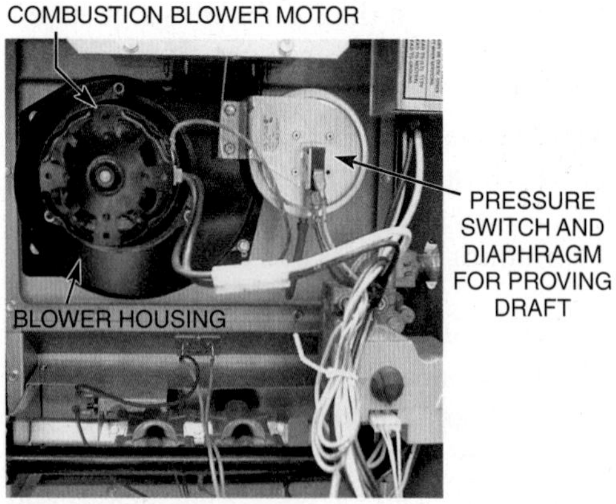

(B)

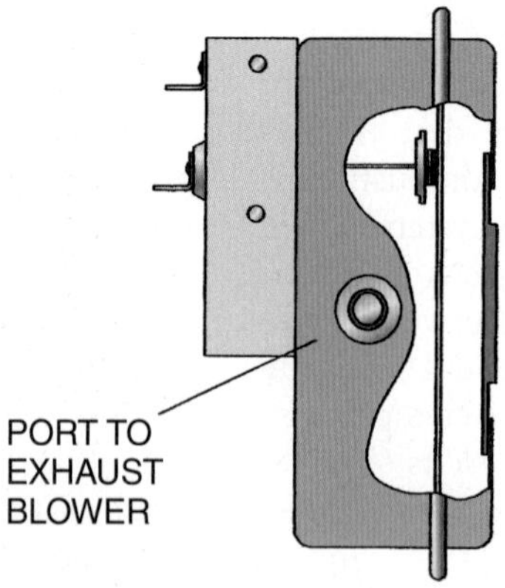

(C)

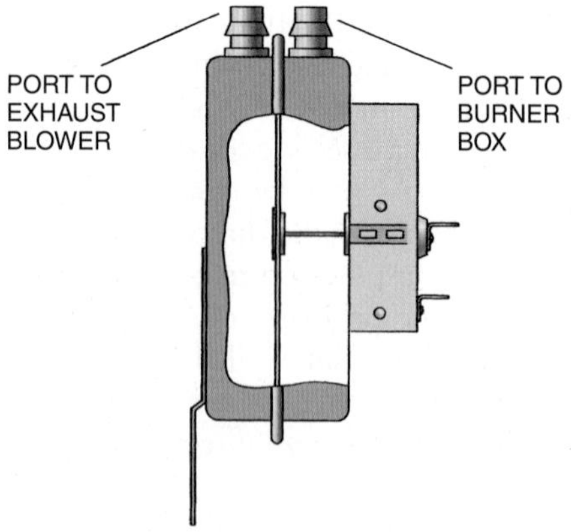

(D)

Figure 31.70 An example of a direct-vented, high-efficiency gas furnace. Courtesy Rheem Manufacturing Company.

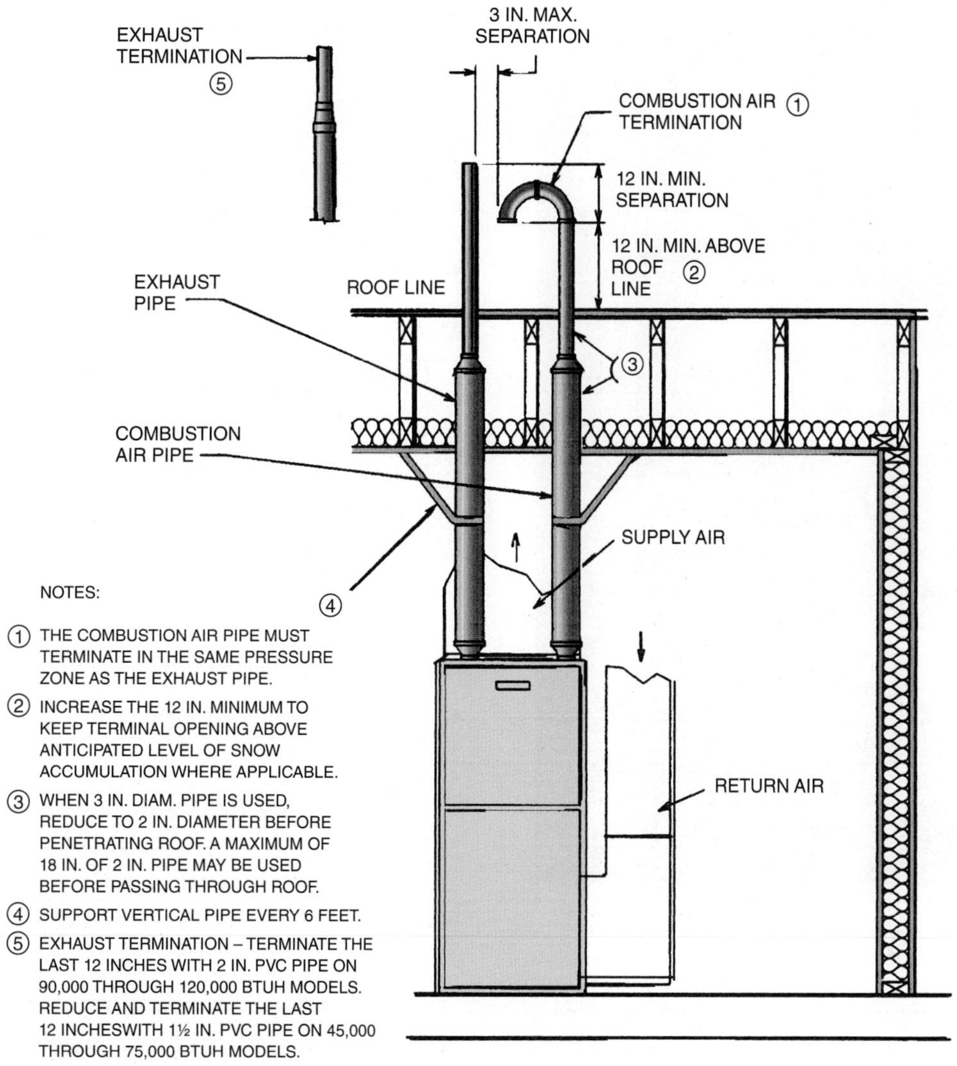

EXCESS AIR

Excess air consists of combustion air and dilution air. Combustion air is primary and/or secondary air. Primary air enters the chamber before combustion takes place. It mixes with raw gas in the burner before a flame has been established. Secondary air is air present after combustion and supports combustion. Dilution air is excess air after combustion and usually enters at the end of the heat exchanger. It is brought in by the draft hood of the furnace, **Figure 31.71**. Uncontrolled excess air can enter the combustion zone through cracks, openings, or faulty gaskets. This air, referred to as tramp air, can lower system efficiency.

As excess air decreases, the DPT increases. The DPT for undiluted flue gas is around 140°F; in diluted flue gas, the temperature may drop to about 105°F. This is the main reason high-efficiency condensing furnaces require very little excess air for combustion. High-efficiency condensing furnaces operate with stack temperatures below the DPT of the combustion gases to allow condensation to happen. The flue gas is intentionally cooled below the DPT to promote condensation.

Gas furnaces with high efficiency ratings are being installed in many homes and businesses. The U.S. Federal Trade Commission requires furnace manufacturers to provide an annual fuel utilization efficiency rating (AFUE). This

Figure 31.71 A standard-efficiency furnace showing dilution air entering a draft hood. Courtesy Ferris State University. Photo by John Tomczyk.

rating allows the consumer to compare furnace performances before buying. These annual efficiency ratings have increased in some instances from 65% to 97% or higher, partly by keeping excessive heat from being vented to the atmosphere. In conventional furnaces, stack temperatures are kept high to provide a good draft for proper venting. This also prevents condensation and therefore corrosion in the vent.

FURNACE EFFICIENCY RATINGS

All forced-air gas furnaces have efficiency ratings. The ratings are determined by the amount of heat that is transferred to the heated medium, which is usually water or air. The following are some factors that determine the efficiency of a furnace:

- Type of draft (natural atmospheric injection, induced, or forced)
- Amount of excess air used in the combustion chamber
- Delta-T, or temperature difference, of the air or water entering versus leaving the heating medium side of the heat exchanger
- Flue stack temperature

The following are general furnace classifications and approximate AFUE ratings:

Conventional, or Standard-Efficiency, Furnace (78% to 80% AFUE), Figure 31.71

- Atmospheric injection or draft hood
- 40% to 50% excess air usage
- Delta-T of 70°F to 100°F (lowest cfm airflow of furnaces)
- Stack temperatures of 350°F to 450°F
- Noncondensing
- One heat exchanger

Mid-Efficiency Furnace (78% to 83% AFUE), Figure 31.72

- Usually induced or forced draft
- No draft hoods
- 20% to 30% excess air
- Delta-T of 45°F to 75°F (more cfm airflow than conventional units)
- Stack temperatures of 275°F to 300°F
- Noncondensing
- One heat exchanger

High-Efficiency Furnace (87% to 97% AFUE), Figure 31.42

- Usually induced or forced draft
- No draft hoods
- 10% excess air
- Delta-T of 35°F to 65°F (highest cfm of all furnaces)
- Stack temperatures of 110°F to 120°F

Figure 31.72 A mid-efficiency furnace.

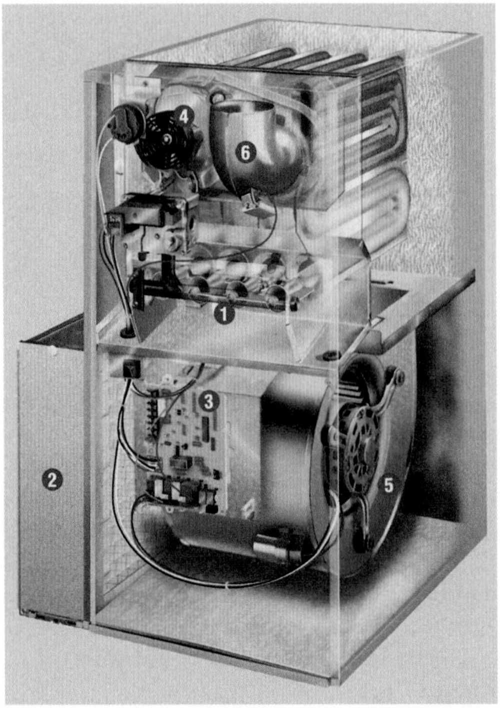

- Two or three heat exchangers
- Condensing furnace (reclaims the latent heat of condensation by condensing flue-gas water vapors to liquid water)
- Uses PVC pipe to avoid corrosion

High-efficiency furnaces have a low delta-T due to the increased amount of air (cfm) they receive through their heat exchangers, which keeps the air in contact with the heat exchanger for a shorter period of time.

Annual Fuel Utilization Efficiency (AFUE) is calculated by the manufacturer of fuel burning combustion equipment. The higher the AFUE, the more efficient the combustion equipment is at converting fuel to useful heat energy. An 85% AFUE value means that for every Btu of fuel burned, 85% of the fuel is used for useful energy within the building and 15% is wasted. The AFUE compares the amount of (seasonal) energy used in Btus to the (seasonal) output of Btus and expresses the ratio as a percentage.

Another efficiency rating is the Heating Seasonal Performance Factor (HSPF). This efficiency factor is applied to heat pumps in the heating mode and takes into consideration the energy used during the entire heating season. It is very similar to the Seasonal Energy Efficiency Ratio (SEER) of heat pumps, but it is only applied to heat pump systems in the heating mode. SEER applies to cooling equipment and heat pumps only in the cooling mode. The HSPF efficiency rates both the compressor and the electric strip heaters. HSPF is a ratio of the seasonal heat output in Btus to the seasonal watt-hours used. The higher the HSPF, the more efficient the unit is in moving Btus of heat energy per Btu of energy consumed.

Seasonal Energy Efficiency Ratio (SEER) is calculated on the basis of the total amount of cooling (in Btu) the system will provide over the entire season, divided by the total number of watt-hours it will consume. Higher SEER ratings reflect a more efficient cooling system.

31.14 ELECTRONIC IGNITION MODULES AND INTEGRATED FURNACE CONTROLLERS

Non-integrated ignition modules and integrated furnace controls (IFCs) are responsible for the ignition and sequence of operations in most gas furnaces. Non-integrated ignition controls are typically only responsible for the pilot and/or burner functions. Integrated ignition controls are responsible for the entire sequence of operations. **Figures 31.73 (A) and (B)** show examples of non-integrated electronic ignition modules and an IFC. Electronic modules usually indicate on their faceplate what basic sequence of events or operations they include within their electronics. Some of their most common control sequences and control definitions follow.

Figure 31.73 (A) Four electronic ignition modules for gas furnaces. Courtesy Ferris State University. Photo by John Tomczyk. (B) An integrated furnace controller (IFC). Courtesy Ferris State University. Photo by John Tomczyk. (C) A high-efficiency pulse furnace. Courtesy of Lennox International Inc.

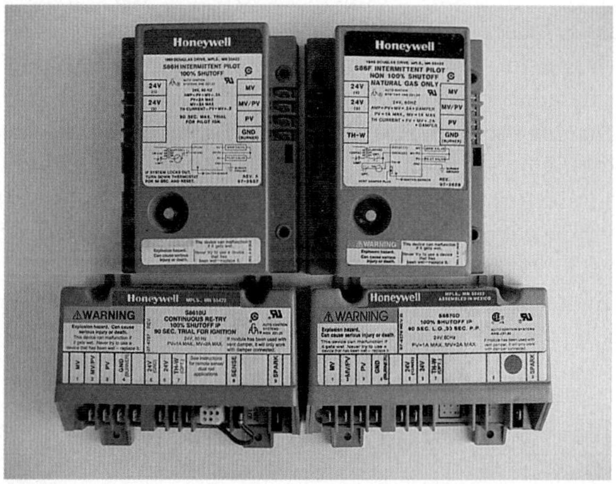

(A)

(B)

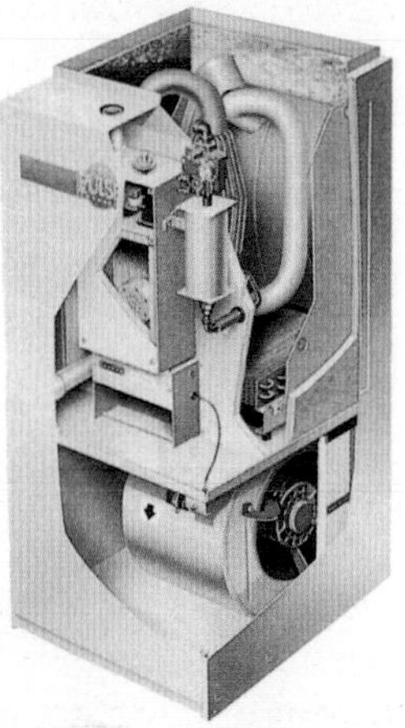

(C)

100% SHUTOFF SYSTEM

A system is 100% shut off when there is a failure in the flame-proving device system and both the pilot valve and main gas valve are shut down (closed). In the past, these systems had to be used on liquefied petroleum gas systems since LP gas is heavier than air and will not vent naturally up the flue. It was thought that letting the pilot gas bleed could create a dangerous situation. This is no longer true. Today, LP gas systems use non-100% shutoff systems in a safe manner by employing short shutdown or soft lockout times between ignition retries to let any residual gas dissipate. Lockout timing will be covered later. A 100% shutoff module is shown in **Figure 31.74**. Notice that it is designed for an IP system and usually has a 90-sec maximum trial for pilot ignition before going into a shutoff.

NON-100% SHUTOFF SYSTEM

A system is considered non-100% shutoff when the pilot valve continues to bleed gas after a failure in the flame-proving device system and the main gas valve shuts down. These systems are used where there is a need to continually try to relight the pilot to avoid nuisance shutdowns. Many commercial rooftop heating units are non-100% shutdown, which prevents nuisance shutdowns on windy days. Many newer ignition control modules have ignition retry sequences for both natural and LP gas. Many modules have soft shutdown periods of 5 min or more to allow any accumulation of fuels to dissipate. The system must have a pilot mechanism to be classified as either 100% or non-100% shutoff. A non-100% shutoff module is shown in **Figure 31.75**.

Figure 31.74 A 100% shutoff electronic ignition module. Courtesy Ferris State University. Photo by John Tomczyk.

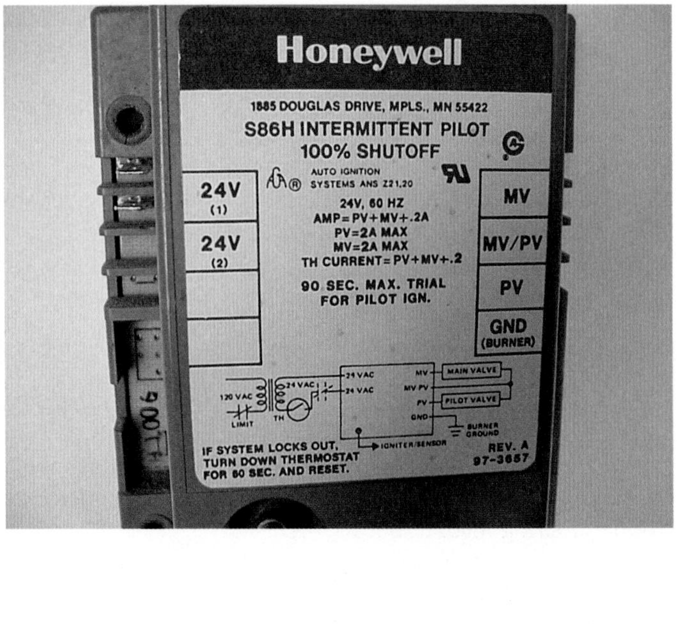

Figure 31.75 A non-100% shutoff ignition module. Courtesy Ferris State University. Photo by John Tomczyk.

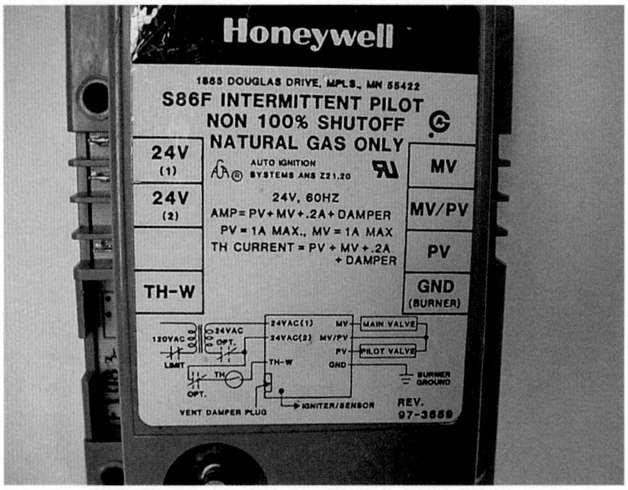

CONTINUOUS RETRY WITH 100% SHUTOFF

A module with a continuous retry control scheme is shown in **Figure 31.76**. It has 100% shutoff with a 90-sec trial for ignition. However, it will continuously try to relight the pilot after the shutoff period. Once in the shutoff period, there will be no bleeding of pilot gas.

LOCKOUT

A lockout is a time period built into the module that allows a specific amount of time for lighting or relighting the pilot or main burner. If this time period is exceeded, the module will go into either a soft lockout or a hard

Figure 31.76 A continuous retry 100% shutoff electronic ignition module. Courtesy Ferris State University. Photo by John Tomczyk.

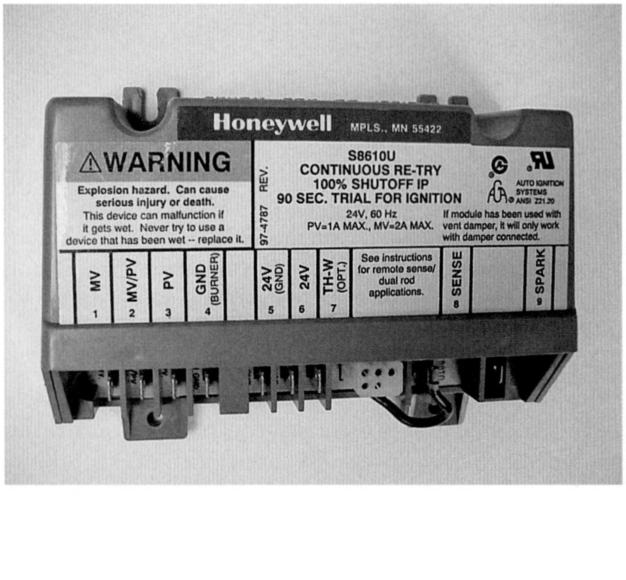

lockout. Some ignition control modules will allow for a specific number of unsuccessful trials for ignition before going into a lockout.

SOFT LOCKOUT. A soft lockout allows a certain time for the lighting or relighting of the pilot or main burner. If this time is exceeded, the module will go into a semishutdown for a certain period but will eventually keep trying to relight the system. This system may go into a hard lockout once several soft lockouts have been attempted and still no flame has been established. Soft lockouts can range from 5 min to 1 hour and so allow time to pass with the hope that a change in the local ambient or outside environment will allow the unit to light normally. They also allow for any power interruptions to clear themselves without going into a hard lockout.

HARD LOCKOUT. A hard lockout allows for a specific number of ignition trials or period of time for the lighting or relighting of the pilot or main burner. If these are exceeded, the module will go into a hard lockout or shutdown. To reset the module after this shutdown, power has to be interrupted to the module and then turned back on. This usually requires a service call and costs the building owner money.

PURGING

PREPURGE. Shown in **Figure 31.77** is a 100% shutoff module for an IP system. There is a 90-sec lockout (LO) with a 30-sec prepurge (PP). Before each heating cycle, the PP control scheme allows the combustion blower motor to run for 30 sec to clear, or PP, the heat exchanger of any unwanted flue gases, household fumes, or dust that may have accumulated during the last heating cycle.

INTERPURGE. Interpurge allows the combustion blower to operate for a certain time period between ignition tries.

Figure 31.77 A 100% shutoff module with a 90-sec lockout and a 30-sec prepurge. Courtesy Ferris State University. Photo by John Tomczyk.

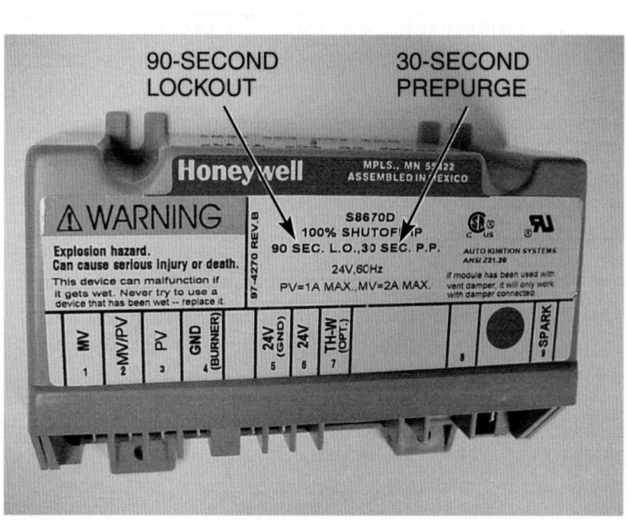

Interpurge is generally used when the furnace does not light or does not sense a flame. The heat exchanger must be purged of residual unburned gas or combustion by-products from the previous unsuccessful ignition attempt.

POSTPURGE. A postpurge allows the combustion blower motor to operate for a certain time period after each heating cycle. This control strategy, along with a PP, is double insurance that the heating cycle will start with only air in the heat exchanger, allowing a quick and safe ignition. Both prepurge and postpurge control strategies can be used alone or with one another.

The IFC provides all of the ignition, safety, and sequence of operation or control schemes for the furnace. Many furnace manufacturers use these controllers because of their safe, reliable operation and versatility. The controller carefully monitors all thermostat, temperature, resistance, and current inputs to ensure stable and safe operation of the combustion blower, gas valve, ignition, flame rectification system, and warm-air blower. They often have many DIP switches that allow the controller to be programmed for different timing functions and control scheme changes. **Figure 31.78(A)** shows a section of an IFC with DIP switches and status lights. DIP switches 1 and 2 are for setting the warm-air blower's off–delay timing in seconds, **Figure 31.78(B)**. DIP switch 3 is for furnace twinning applications. DIP switch 4 is not used in this application.

Twinning involves the side-by-side operation of two furnaces connected by a common ducting system. High-voltage power is supplied from a common source. Transformer-supplied low-voltage power is controlled by one common thermostat for both furnaces. Status lights will usually blink a certain number of times to tell the service technician that the furnace is in the twinning mode. Twinning is used when a heating capacity greater than that of the largest-capacity furnace manufactured is needed. Twinning is also used when the airflow of one furnace, for air-conditioning purposes, is greater than a single furnace can provide.

The status lights monitor incoming power, furnace setup modes, diagnostic status for troubleshooting, and the flame rectification signal. The furnace status check can be a self-diagnostic flashing light with a code to help the service technician systematically troubleshoot the furnace. **Figure 31.79** shows a furnace controller with a self-diagnostic light and flashing codes for some common furnace faults. The diagnostic light is located on the bottom left corner of the controller. Care must be taken when troubleshooting an IFC because static electricity can damage the sensitive hardware and software. Always make sure you are touching a well-grounded surface before attempting to troubleshoot an IFC board. The small voltage of an ohmmeter can also damage an IFC board. Never put an ohmmeter on an IFC board unless instructed to do so by the furnace manufacturer's troubleshooting procedures. Always disconnect the wires of a power-consuming device or sensor from the IFC before using an ohmmeter in troubleshooting these devices.

Figure 31.78 (A) An integrated furnace controller showing DIP switches and status lights. Courtesy Ferris State University. Photo by John Tomczyk. (B) DIP switches for setting the warm-air blower off–delay timing. Courtesy Rheem Manufacturing Company.

(A)

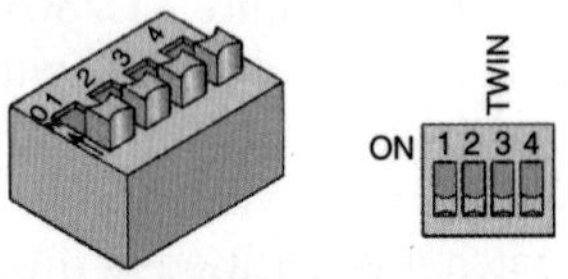

BLOWER OFF TIMES (SECONDS)				
LOW FIRE	HIGH FIRE	COOLING	SWITCH 1	SWITCH 2
90	60	30	OFF	ON
120	90	45	OFF	OFF
160	130	60	ON	OFF
180	150	90	ON	ON

(B)

Figure 31.79 An integrated furnace controller (IFC) with self-diagnostic lights and flashing fault codes. Courtesy Ferris State University. Photo by John Tomczyk.

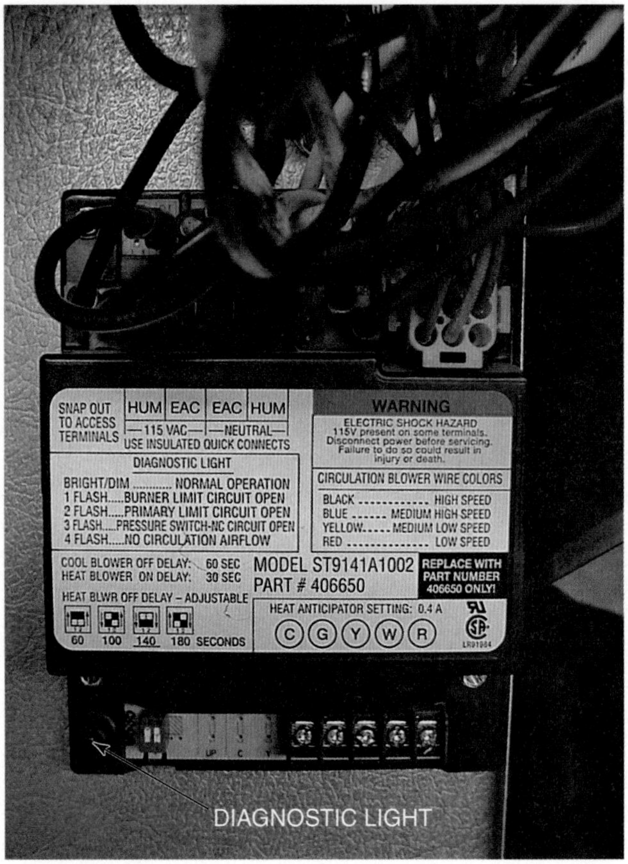

Following is a typical sequence of operation or control scheme for a residential direct-ignition, natural gas furnace employing an IFC and a single-stage thermostat. With the proper settings, either a single- or dual-stage thermostat can be used on this system. This is a two-stage furnace, meaning the gas valve has two heat stages, accomplished by gas pressure regulation and solenoid valves. The combustion blower also has two speeds. The sequence includes a PP, interpurge, and postpurge with a soft lockout. The control board or IFC used on this furnace is shown in **Figure 31.73(B)**.

1. The W terminal on the thermostat is connected to the W2 terminal on the IFC board.
2. When there is a call for heat, the R and W2 contacts close and the IFC runs a self-check routine to verify that the draft-proving pressure switch contacts are open. The limit switch contacts are constantly monitored.
3. The induced-draft motor starts on high for a few seconds, to ensure that the low-pressure switch contacts close, and then changes to low speed. After a 30-sec prepurge, the spark indicator is energized and the low stage of the gas valve opens, lighting the burners.
4. After the gas valve opens, the remote flame sensor must prove ignition for 1 sec using the process of flame rectification. If the burners do not light, the system goes through another ignition sequence. It does this up to four times (two tries on low fire and two tries on high fire).
5. The main blower starts on low speed 30 sec after the burners light. The furnace will operate on low fire for 12 min. Then, if the thermostat is not satisfied, it will shift to high fire, causing the draft inducer to go to heat speed. (If a two-stage thermostat is used, the second stage of the thermostat, W2, will initiate high fire.) The draft inducer will continue running for a 5-sec (low-speed) or 10-sec (high-speed) postpurge.

The main blower will run for 90 sec on high speed or 120 sec on low speed. (This timing is field adjustable for 60, 90, 120, or 150 sec.)

The following is the sequence if the system does not light or does not sense flame:

1. If a flame is not sensed within 8 sec after the gas valve opens, the valve closes and the igniter is deenergized. The induced-draft motor will run for 60 sec on low, stop momentarily, and then restart. The ignition process will go through one more try on low fire. If this fails, there will be two tries on high fire with a 30-sec interpurge between trials. If there is no ignition after the second trial on high fire, the furnace will go into soft lockout for 1 hour.
2. The sequence will repeat after a 1-hour delay and continue repeating until ignition is successful or the call for heat is terminated.
3. To reset the lockout, make and break power either at the thermostat or at the unit disconnect switch for 5 to 10 sec. The furnace will then go through another set of trials for ignition.

ELECTRONIC IGNITION MODULE WIRING AND TERMINAL NAMING

Most terminals in electronic modules are marked with abbreviations, **Figure 31.73(A)**. Some electronic modules include a wiring diagram on their faceplate, **Figures 31.74** and **31.75**. This can be helpful to the service technician in wiring and troubleshooting the furnace's electrical and mechanical systems. The following are important terminal abbreviations and names:

MV	Main valve (gas valve)
PV	Pilot valve
MV/PV	Common terminal for main and pilot valves
GND	Burner ground
24V	24-V source out of the transformer
24V(GND)	Common or ground out of the 24-V transformer
TH-W	Thermostat lead to the module
IGNITER/SENSOR	High-voltage igniter and flame rectification rod or sensor for local sensing (single-rod systems)
SENSE	Flame rectification rod or sensor for remote sensing (dual-rod systems)
SPARK	High-voltage igniter

SMART VALVE

The smart valve is a gas valve and an electronic control module enclosed in one package, **Figure 31.80**. It combines the features of an IP system and an HSI system in one.

Figure 31.80 (A) A smart valve gas control system. Courtesy Ferris State University. Photos by John Tomczyk. (B) Internal view of a smart valve showing the electronic module and gas valve. Courtesy Ferris State University. Photos by John Tomczyk.

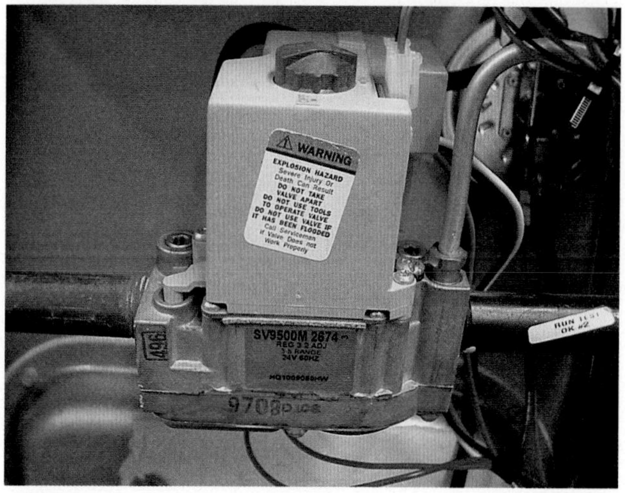

(A)

(B)

The smart valve is a continuous pilot trial system and is designed for both natural and LP gas furnaces. The pilot flame is probably one of the most reliable ignition sources in the industry. The valve's logic lights the pilot with a 24-V hot surface igniter system, **Figure 31.81**. Then, as in any other intermittent ignition system, the pilot lights the main burners. Using a 24-V HSI system instead of a high-voltage spark to light the pilot simplifies the electronics, and because there is no need for a noisy spark generator or high-voltage ignition transformer, it provides a quiet ignition system.

The 24-V hot surface igniter in the smart valve is made of a material different from that used for the conventional

Figure 31.81 A 24-V hot surface igniter that lights a pilot flame. Courtesy Ferris State University. Photo by John Tomczyk.

silicon carbide line-voltage igniters. It is designed to be stronger and less brittle. In fact, it is well protected by the pilot burner and grounding strap. The ground strap is there to provide a ground electrode for the flame rectification system. The other electrode is the flame rod, **Figure 31.81**. Once the pilot flame is proved, the 24-V hot surface igniter is deenergized and the electronic fan timer within the module is energized. The electronic fan timer tells when to start timing the fan–on delay and the fan–off delay.

The smart valve system is easy to troubleshoot because of the simplified wiring and modular electrical connections. A voltmeter can be used at the valve and electrical plug connections, **Figure 31.82**. **Figure 31.83(A)** is a sequence of operation for a smart valve control. **Figure 31.83(B)** is a systematic troubleshooting chart. **SAFETY PRECAUTION:** *Because there are different models and makes of gas valves, electronic modules, and IFCs, always consult with the valve and*

Figure 31.82 (A)–(B) The smart valve's modular electrical connections. (C)–(D) A voltmeter being used to troubleshoot the smart valve system. Source: Resideo.

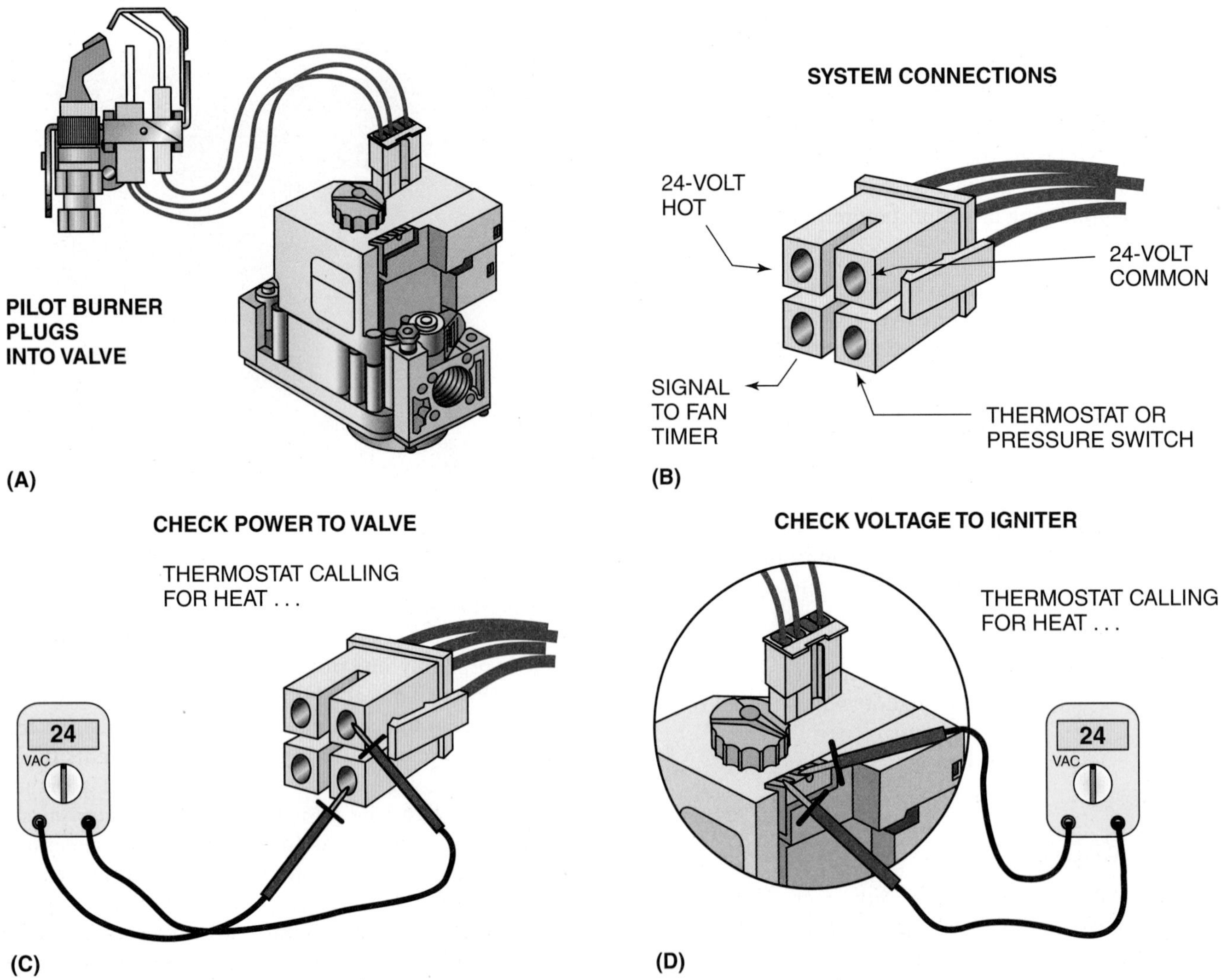

Figure 31.83 (A) The smart valve sequence of operation. (B) A smart valve troubleshooting chart. Source: Honeywell.

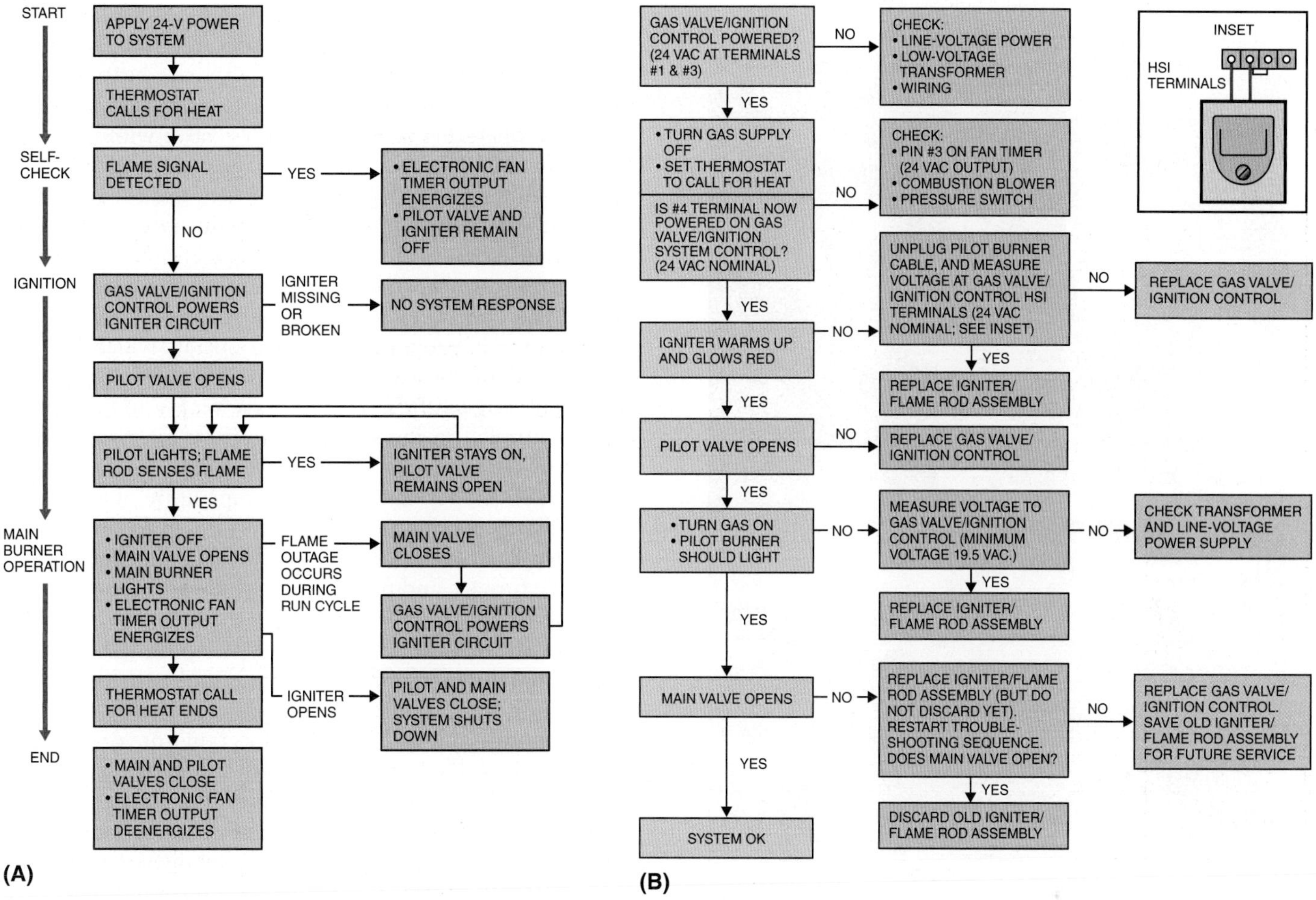

module or IFC manufacturer for specific information on systematic troubleshooting and sequence of operation to avoid personal injury.•

PULSE FURNACE

The pulse furnace, **Figure 31.73(C)**, ignites minute quantities of gas 60 to 70 times per second in a closed combustion chamber. The process begins when small amounts of fuel gas and air enter the chamber. This mixture is ignited with a spark igniter and is forced down a tailpipe to an exhaust decoupler. The pulse is reflected back to the combustion chamber, igniting another gas and air mixture, and the process is repeated. Once this process is started, the spark igniter can be turned off because the pulsing will continue on its own 60 to 70 times per second. This furnace also uses heat exchangers to absorb much of the heat produced. Air is circulated over the exchanger and then to the space to be heated.

31.15 TWO-STAGE AND MODULATING GAS FURNACES

Two-stage gas furnaces use a two-stage gas valve and a two-speed combustion blower motor with two pressure switches to prove the draft. Most two-stage furnaces are controlled by an IFC. **Figure 31.73(B)** shows an IFC for a two-stage gas furnace. **Figure 31.84** shows the two-stage gas valve, two-speed combustion blower motor, and dual pressure switches to prove the draft. One pressure switch is a low-pressure switch and the other is a high-pressure switch to prove the low and high speeds on the combustion blower motor. The first stage of heating operates with a typical manifold pressure of 1.75 in. WC by energizing the first solenoid valve in the gas valve. The first stage of heating usually operates at about 50% to 70% of the total furnace heating output. The second stage of heating puts out a manifold pressure of 3.5 in. WC by energizing the second solenoid valve in the gas valve.

Figure 31.84 A two-stage gas furnace showing the two-stage gas valve, the two-stage combustion blower motor, and dual pressure controls. Courtesy Ferris State University. Photo by John Tomczyk.

This is 100% of the total furnace output. Two-stage gas furnaces control the temperature in the conditioned space much more closely to the set point of the thermostat. Because of this, room temperature swings are smaller and comfort levels are increased.

The industry has always needed a furnace that adjusts to the heat loss of a structure. For example, a house that requires a 100,000-Btu/h furnace on the coldest day of winter, which may be 0°F, may require only 60,000 Btu/h of heat when it is 20°F. The furnace begins to cycle off and then back on as the weather warms. The warmer the weather, the more cycle times per day. These additional cycle times cause the furnace to be less efficient. An oversized furnace with a steady-state efficiency of 80% is estimated to have a seasonal efficiency as low as 50%. When cycling can be reduced or eliminated, the efficiency of the furnace becomes greater. *A modulating gas furnace can adjust to the heat loss of a structure and have longer run times, thus less cycling. This makes it a very efficient furnace.*

Most modulating gas furnaces use a modulating gas valve instead of a staged gas valve. They also come with a variable-speed warm-air blower to adjust the speed and amount of warm air being circulated. Two-stage combustion blower motors are standard on most modulating gas furnaces, and an advanced intelligent IFC monitors and controls the furnace operations to vary the heating input and warm airflow. Some modulating gas furnaces also have a variable operating solenoid or servo valve within the gas valve that is fed with a varying or proportional milliamp signal from the IFC. This servo valve acts as a capacity controller. It varies the input rate of the furnace proportionally to the signal of the controller. This allows the furnace to operate over a heating output range between 40% and 100%. The controller gets its proportional signal from a variable-output programmable thermostat, which will be covered in detail below.

Modulating gas furnaces may have supply and return air sensors installed in the supply and return air system. Usually, these sensors are thermistors that send an analog signal to the IFC. They can sense the temperature of the return and discharge air and relay this information to the furnace controller. The controller can then calculate the temperature rise of the furnace and proportionally adjust the modulating gas valve to maintain a comfortable discharge air temperature. The IFC monitors the thermostat inputs and all other temperature inputs from the furnace and provides stable operation of the variable input gas valve, variable warm air blower motor, and two-speed inducer motor. A variable-speed warm-air blower provides varied airflow through the furnace's ductwork to meet the conditioned space demands. A brushless, permanent-magnet, internally commutated motor (ICM) with variable-speed capabilities can deliver airflows as low as 300 cfm. The modulating gas furnace also has the ability to adjust itself to changes in gas heating values and air density.

The variable-output programmable thermostat is not a conventional on–off thermostat. It sends a varying or proportional signal to the furnace controller. It can sense the exact heating or cooling requirements of the conditioned space by a "fuzzy logic" control routine resident in its software. These routines calculate the room's load by evaluating recent room conditions from a sensor within the thermostat or remote to the thermostat and reactions to supplying heating and cooling. The thermostat and IFC then calculate a heating or cooling load factor. This load factor is used to determine when, for how long, and at what firing rate and airflow the furnace should be activated. This control routine optimizes furnace efficiency and reduces temperature swing in the controlled space. Modulating furnaces will have longer run times, allowing greater comfort levels in the conditioned space. Temperature swings in the conditioned space will not vary by more than one-half degree from the thermostat set point. **Figure 31.85** shows the internals of a variable-output programmable thermostat used in a modulating furnace.

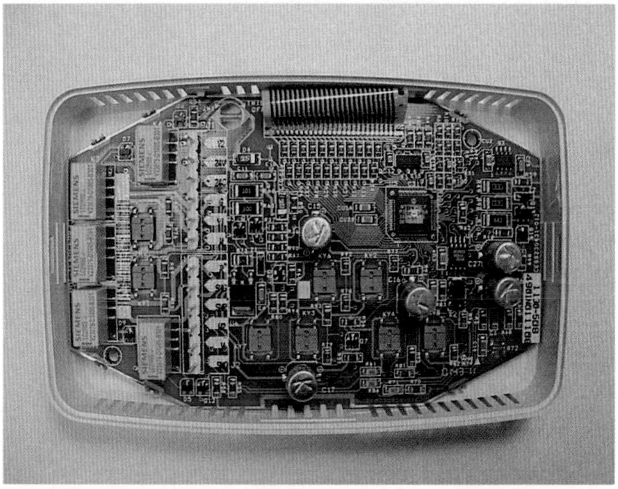

Figure 31.85 Internal view of a variable-output programmable thermostat for a modulating furnace. Courtesy Ferris State University. Photo by John Tomczyk.

31.16 VENTING

Conventional gas furnaces use flue gas temperature and natural convection to vent the products of combustion (flue gases). The hot gas is vented quickly, primarily to prevent cooling of the gas, which produces condensation and other corrosive actions. However, these furnaces lose some efficiency because considerable heat is lost up the flue. High-efficiency furnaces (90% and higher) recirculate the flue gases through a special extra heat exchanger to keep more of the heat available for space heating in the building. The gases are then pushed out of the flue by a small fan. Plastic pipe is used to handle the resulting condensation and corrosion in the flue because it is not damaged by the corrosive materials. High-efficiency furnaces may have efficiencies of up to 97%. **SAFETY PRECAUTION:** *Regardless of the type, a gas furnace must be properly vented.*•

Venting must provide a safe and effective means of moving the flue gases to the outside air. In a conventional furnace, it is important for the flue gases to be vented as quickly as possible. Conventional furnaces are equipped with a draft hood that blends some room air with the rising flue gases, **Figure 31.86(A)**. The products of combustion enter the draft hood and are mixed with air from the area around the furnace. Approximately 100% additional air (called *dilution* air) enters the draft hood at a lower temperature than the flue gases. The heated gases rise rapidly and create a draft, bringing the dilution air in to mix and move up the vent.

All furnaces must have the correct amount of excess air. Excess air is made up of combustion air and dilution air. Combustion air consists of both primary and secondary air. The amount of air required for gas-burning appliances is determined by the appliance size and is

Figure 31.86 (A) and (B) Draft hood operation.

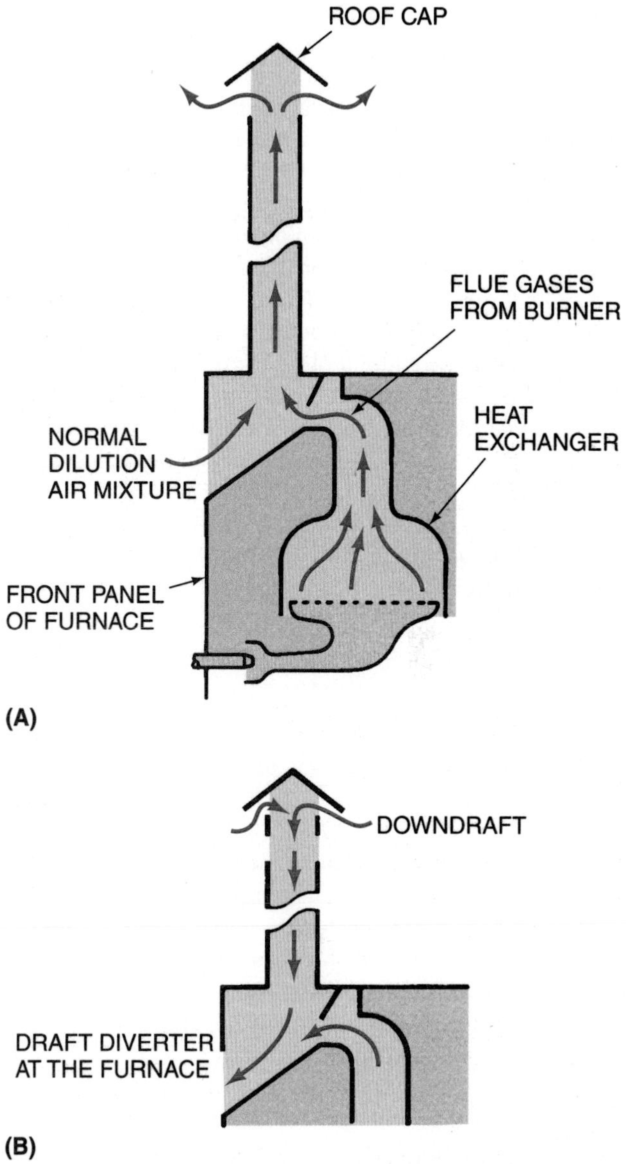

mandated by local or statewide codes. Most of these codes are based on the International Fuel Gas Code (IFGC) or the National Fuel Gas Code. The IFGC is published by the International Code Council. The National Fuel Gas Code is also considered the standard by many states. Obviously, the larger the furnace, the more excess air is needed. One rule of thumb is that a grill with a free area of 1 in^2 per 1000 Btu/h must be provided below the furnace (if the structure is over a crawl space) and above the furnace (to the attic space) to allow adequate air for combustion. For a 100,000-Btu/h furnace, this would be 100 in^2 of free area. This grill must be larger than 10 in. × 10 in. = 100 in^2 to allow for the space the louvres take up (approximately 30% of the grill area is louvres), so the grill would have to be 100/0.70 = 142.9 in^2 or 12 in. × 12 in. The vent above the furnace is for dilution

air; the vent below is for combustion air. This example of figuring out the ventilation air for this situation is only one possibility.

In the past, excess air was merely extracted from the volume of air in the structure. Structures today are built to allow less infiltration of air around the windows and doors, so makeup air for the furnace must be considered. Even old structures are being tightened up by adding storm windows and weather stripping around the doors, so there is less infiltration. A furnace installed in an older structure that has been tightened may have venting problems, as evidenced by flue gases spilling out of the draft diverter rather than exiting out the flue. If wind conditions produce a downdraft, the opening in the draft hood provides a place for the gases and air to go, diverting them away from the pilot and main burner flame. The *draft diverter* helps reduce the chance of the pilot flame being blown out and the main burner flame being altered, **Figure 31.86(B)**.

Venting categories were created when high-efficiency furnaces were introduced. Following are the four venting categories listed by the American National Standards Institute (ANSI). Based on ANSI Standard Z21.47A-2016, they set required temperature and pressure venting standards for the proper venting of gas furnaces.

- Category I—Furnace has a nonpositive vent pressure and operates with a vent gas temperature at least 140°F above its dew point (conventional, or standard-efficiency, furnace).
- Category II—Furnace has a nonpositive vent pressure and operates with a vent gas temperature less than 140°F above its dew point (no longer manufactured because of condensate problems at start-up).
- Category III—Furnace has a positive vent pressure and operates with a vent gas temperature at least 140°F above its dew point (mid-efficiency furnace).
- Category IV—Furnace has a positive vent pressure and operates with a vent gas temperature less than 140°F above its dew point (high-efficiency condensing furnace).

SAFETY PRECAUTION: *Make sure replacement air is available in the furnace area. Remember it takes 10 ft^3 of air for each 1 ft^3 of natural gas to support combustion. To this is added 5 ft^3 more to ensure enough oxygen. Another 15 ft^3 is added at the draft hood, making a total of 30 ft^3 of air for each 1 ft^3 of gas. A 100,000-Btu/h furnace would require 3000 ft^3/h of fresh air (100,000 ÷ 1000 Btu/h × 30 ft^3 = 3000 ft^3/h). All of this air must be replaced in the furnace area and also must be vented as flue gases rise up the vent. If the air is not replaced in the furnace area, it will become a negative pressure area, and air will be pulled down the flue. Products of combustion will then fill the area.*•

Type B vent or approved masonry materials are required for conventional gas furnace installations. *Type B* venting systems consist of metal vent pipe of the proper thickness that is approved by a recognized testing laboratory, **Figure 31.87**. The venting system must be continuous from the furnace to the proper height above the roof. Type B vent pipe is usually of a double-wall construction with air space between. The inner wall may be made of aluminum and the outer wall may be constructed of steel or aluminum.

Figure 31.87 A section of type B gas vent. Photo by Jason Obrzut.

The vent pipe should be at least the same diameter as the vent opening at the furnace. The horizontal run should be as short as possible. Horizontal runs should always be sloped upward as the vent pipe leaves the furnace. A slope of ¼ in. per foot of run is the minimum recommended, **Figure 31.88**. Use as few elbows as possible. Long runs lower the temperature of the flue gases before they reach the vertical vent or chimney and they reduce the draft. Do not insert the vent connector beyond the inside of the chimney wall and make sure that there are no obstructions when the gases are vented into a masonry chimney, **Figure 31.89**. If two or more gas appliances are vented into a common flue, the common flue size should be calculated based on recommendations from the National Fire Protection Association (NFPA) or local codes. The top of each gas furnace vent should have an approved vent cap to prevent the entrance of rain and debris and to help prevent wind from blowing the gases back into the building through the draft hood, **Figure 31.90**.

Vertical vents can be masonry with lining and glazing or prefabricated metal chimneys approved by a recognized

Figure 31.88 The minimum rise should be ¼ in. per foot of horizontal run.

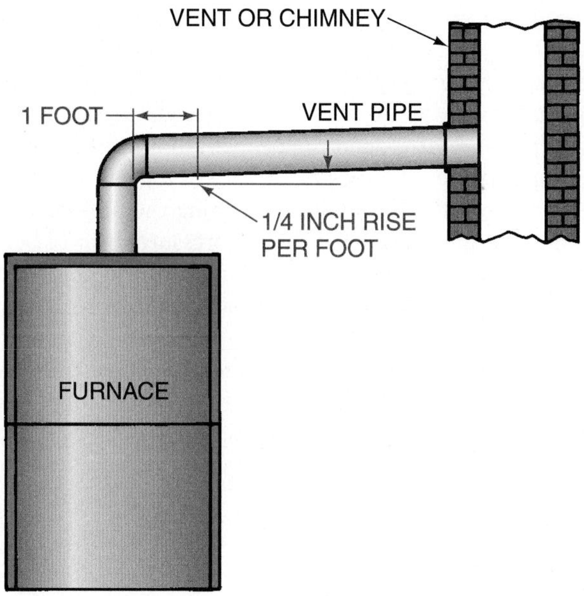

Figure 31.89 The vent pipe should not extend beyond the inside of the flue lining.

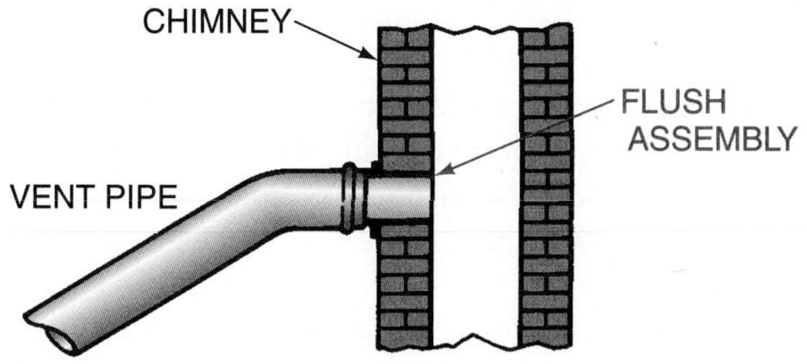

Figure 31.90 A gas vent cap. Photo by Jason Obrzut.

Figure 31.91 A flexible chimney liner. Photo by Jason Obrzut.

testing laboratory. Metal vents reach operating temperatures quicker than masonry vents do. It takes a long running cycle to heat a heavy chimney. The warmer the gases, the faster they rise and the less damage there is from condensation and other corrosive actions. Masonry chimneys tend to cool the flue gases and must be lined with a glazed-type tile or have a vent pipe installed inside the chimney. The corrosive materials in the flue gas that come from the condensation that occurs at start-up and from short running times in mild weather will destroy the mortar joints of an unlined chimney. When an unlined chimney must be used, a special corrosion-resistant flexible liner can often be installed, **Figure 31.91**. It is worked down into the chimney and connected at the bottom to the furnace vent. This provides a corrosion-proof vent system.

When the furnace is off, heated air can leave the structure through the draft hood. Automatic vent dampers that close when the furnace is off and open when the furnace is started can be placed in the vent to prevent air loss, **Figure 31.92**. Heated room air will only be vented if the furnace is non-direct-vented because these furnaces get their combustion air from the conditioned space. However, if the furnace is direct-vented, it will have a sealed combustion chamber and will get its combustion air from the outside through an air pipe usually made of PVC plastic, **Figure 31.70**. There is no possible way for conditioned room air to be vented

Figure 31.92 A thermal damper and a motorized vent damper. Photo by Jason Obrzut

because the sealed combustion chamber never receives conditioned room air. Since conditioned room air is not used for combustion, the conditioned room is not depressurized, or pulled into a slight vacuum by the furnace. This cuts down on the amount of outside air infiltrating the room, and so less cold, infiltrated air will have to be reheated. Direct venting also minimizes the interaction of the furnace with other combustion appliances.

As mentioned earlier in this unit, combustion blowers can be either forced- or induced-draft systems. In an induced-draft system, the combustion blower is at the outlet of the heat exchangers, producing a negative pressure that actually sucks the combustion gases out of the heat exchanger. In a forced-draft system, the combustion blower is before the heat exchangers and causes a positive pressure that blows the combustion gases through the heat exchanger. This blower will vent gases at a lower temperature than is possible with natural draft, allowing the heat exchanger to absorb more of the heat from the burner for distribution to the conditioned space. When combustion blowers are used for furnace venting, the room thermostat normally energizes the blower.

31.17 GAS PIPING

Installing or replacing gas piping can be an important part of a technician's job. The first thing a technician should do is to become familiar with the national and local codes governing gas piping. Local codes may vary from national codes. A technician should also be familiar with the characteristics of natural gas and LP gas in the particular area in which he or she works. Pipe sizing and furnace Btu ratings will vary from area to area due to varying gas characteristics.

The piping should be kept as simple as possible and the pipe should run as directly as possible. Distributing or piping gas is similar to distributing electricity. The piping must be large enough and there should be as few fittings as possible in the system because each fitting creates a resistance. Pipe that is too small will, along with other resistances in the system, cause a pressure drop, and the proper amount of gas will not reach the furnace. The specific gravity of the gas must also be taken into consideration when designing the piping. Systems should be designed for a maximum pressure drop of 0.35 in. WC. The amount of gas to be consumed by the furnace must also be determined. The gas company should be contacted to determine the heating value of the gas in that area. It can also furnish pipe sizing tables and helpful suggestions. Most natural gas will supply approximately 1000 Btu/ft^3 of gas. To determine the gas to be consumed in 1 hour by a typical natural gas furnace, use the following formula:

$$\frac{\text{Furnace Btu Input per Hour}}{\text{Gas Heating Rating in Btu/ft}^3} = \text{Cubic Feet of Gas Needed per Hour}$$

Suppose the Btu rating of the furnace were 100,000 and the gas heating rating were 1000. Then

$$\frac{100{,}000}{1000} = 100\ \text{ft}^3\ \text{of Natural Gas Needed per Hour}$$

Tables give the size (diameter) of pipe needed for the length of the pipe needed to provide the proper amount of gas for the furnace. The designer of the piping system would have to know whether natural gas or LP gas were going to be used because the pipe sizing would be different due to the differences in specific gravities of the gases. Steel or wrought iron pipe, copper, and corrugated stainless steel tubing are some of the common gas piping materials. Always refer to the authority having jurisdiction for what type of piping is acceptable.

SAFETY PRECAUTION: *Ensure that all piping is free of burrs and that threads are not damaged,* **Figure 31.93**. *Typical threaded pipe should have 7 to 8 complete threads, depending*

Figure 31.93 When working with threaded pipe, it should be deburred, scale and/or dirt should be removed from the inside, and the threads should be inspected for defects.

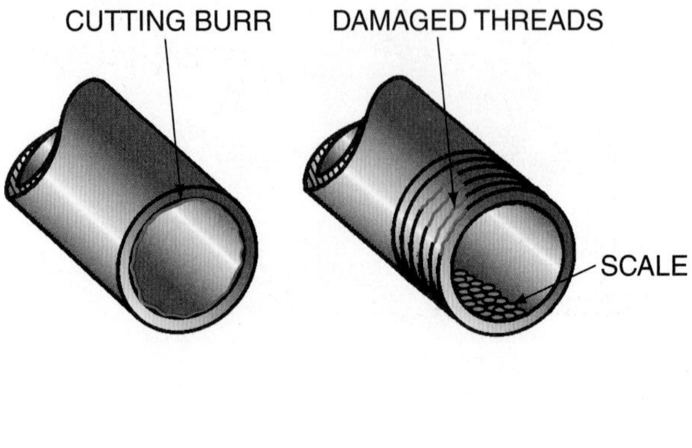

Figure 31.94 Pipe dope should be applied to male threads, but do not apply it to the last two threads at the end of the pipe.

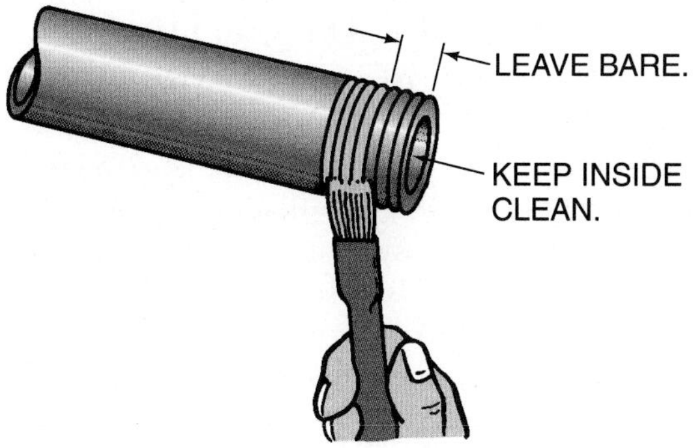

Figure 31.95 A shutoff valve, drip trap, and union should be installed ahead of the gas valve.

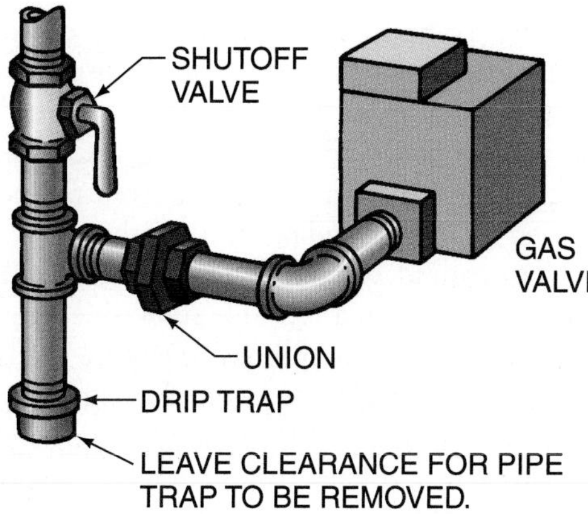

Figure 31.96 Piping from the furnace should have a rise of ¼ in. for each 15 ft of horizontal run.

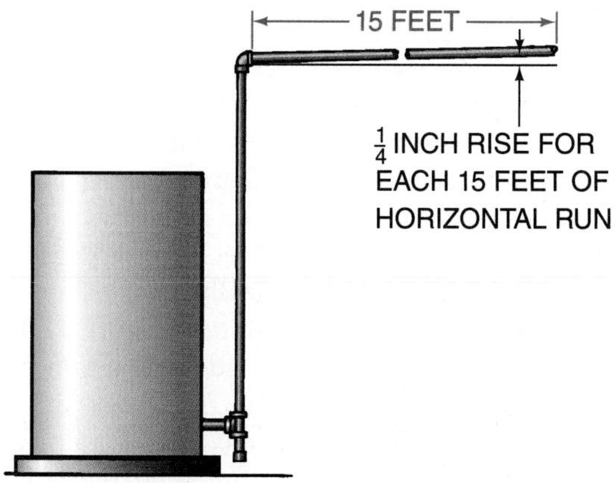

Figure 31.97 Piping should be supported adequately with pipe hooks or straps.

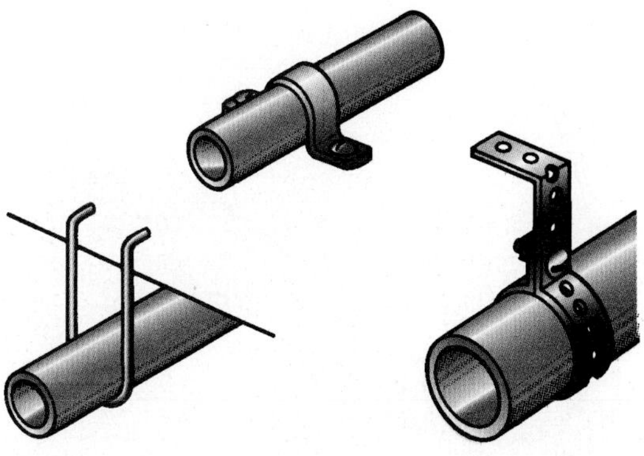

on the diameter. All scale, dirt, or other loose material should be cleaned from pipe threads and from the inside of the pipe. Any loose particles can move through the pipe to the gas valve and keep it from closing properly. It may also stop up the small pilot light orifice.•

When assembling threaded pipe and fittings, use a joint compound, commonly called *pipe dope*. Do not apply this compound on the first two threads at the end of the pipe because it could get into the pipe and plug the orifice or prevent the gas valve from closing, **Figure 31.94**. Also be careful when using Teflon tape to seal the pipe threads because it can also get into the pipe and cause similar problems.

At the furnace the piping should provide for a drip trap, a shutoff valve, and a union, **Figure 31.95**. A manual shutoff valve within 2 ft of the furnace is required by most local codes. The drip trap is installed to catch dirt, scale, or condensate (moisture) from the supply line. A union between the tee and the gas valve allows the gas valve or the entire furnace to be removed without disassembling other piping. Many jurisdictions require that the union be located outside the furnace cabinet. Piping should be installed with a pitch of ¼ in. for every 15 ft of run in the direction of flow. This will prevent trapping moisture that could block the gas flow, **Figure 31.96**. Small amounts of moisture will move to the drip leg and slowly evaporate. Use pipe hooks or straps to support the piping adequately, **Figure 31.97**.

SAFETY PRECAUTION: *When completed, test the piping assembly for leaks. There are several methods for doing this, one being to use the gas in the system. Do not use other gases, and especially do not use oxygen.*• Turn off the manual shutoff to the furnace. Make sure all joints are secure. Turn the gas on in the system. Watch the gas meter dial to see if it moves, indicating that gas is passing through the system. A check for 5 to 10 min will indicate a large leak if the meter dial

Figure 31.98 An electronic leak detector used to check for gas leaks. Photo by Jason Obrzut.

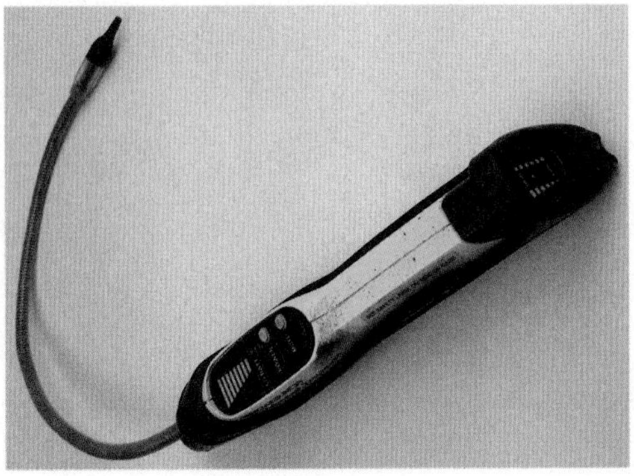

continues to show gas flow. An overnight standing check is better. If the dial moves, indicating a leak, check each joint with a bubble solution or an electronic leak detector, **Figure 31.98**. When you find the leak and repair it, repeat the same procedure to ensure that the leak has been repaired and that there are no other leaks.

Leaks can also be determined with a manometer. Install the manometer in the system to measure the gas pressure. Turn the gas on and read the pressure in in. WC on the instrument. Then turn the gas off. There should be no loss of pressure. Some technicians use an overnight standing period with no pressure drop as their standard. **NOTE:** *If a standing pilot system that is not 100% shut off is being checked, the pilot light valve must be closed because it will indicate a leak.* • A high-pressure test is more efficient. For a high-pressure leak check, use air pressure from a bicycle tire pump. Ten pounds of pressure for 10 min with no leakdown will prove there are no leaks. **SAFETY PRECAUTION:** *Do not let this pressure reach any appliance gas valve. The manual valve must be closed during the high-pressure test.*•

If there are no gas leaks, the system must be *purged,* that is, the air must be bled off to rid the system of air or other gases. **SAFETY PRECAUTION:** *Purging should be done in a well-ventilated area. It can often be performed by disconnecting the pilot tubing or loosening the union in the piping between the manual shutoff and the gas valve. Do not purge where the gas can collect in the combustion area.*• After the system has been purged, allow it to set for at least 15 min to allow any accumulated gas to dissipate. If you are concerned about gas collecting in the area, wait for a longer period. Moving the air with a hand-operated fan can speed this up. When it is evident that no gas has accumulated, the pilot can be lit. **SAFETY PRECAUTION:** *Be extremely careful, because gas is explosive. The standing leak test must be performed and all conditions satisfied before a gas system is put in operation. Be aware of and practice any local code requirements for gas piping.*•

31.18 GAS FURNACE WIRING DIAGRAMS AND TROUBLESHOOTING FLOWCHARTS

A wiring diagram in both schematic and pictorial form is shown in **Figure 31.99**. The diagram is for an intermittent hot surface ignition system with an integrated furnace controller. Notice that in the schematic diagram the IFC is shown as an empty box; the pictorial diagram shows the IFC wiring terminals. It is beyond the scope of this text to explain the internal operations of the IFC. Although service technicians do not troubleshoot the circuits within the IFC board, they do troubleshoot circuits coming to and from the IFC board by following the input/output troubleshooting procedure. Troubleshooting guides can be followed more easily as the technician comes to understand the IFC's sequence of operation, **Figure 31.106**.

The IFC is a complicated microprocessor or small computer and must be handled with care. A small static discharge can damage the IFC. Service technicians must make sure they are physically grounded before attempting IFC troubleshooting. To prevent electrical static discharge from the body, some technicians even wear a grounding strap on their wrist, which is securely wired to ground, when they are replacing an IFC board. An IFC board is shipped in a reusable static shielding bag. The service technician must be careful not to expose the IFC board to a static shock once it is removed from the bag. As mentioned, the service technician must be securely grounded before handling the board. When voltage-troubleshooting the IFC board with a voltmeter, make sure the voltmeter leads do not touch other parts of the board. This can lead to short circuits and damage the board. The voltmeter leads should be kept as short as possible. Applying some electrical tape to the bottom part of the voltmeter leads can shorten their conducting length.

The first step in systematically troubleshooting a furnace that has an IFC is to obtain the electrical diagram of the furnace and the sequence of operation of the IFC. Then the service technician should obtain a systematic troubleshooting guide or flowchart like the one in **Figure 31.100 (B) and (C)**. The sequence of operation, the electrical diagrams, and the systematic troubleshooting guide are the technician's most important tools for furnace troubleshooting.

Figure 100(A) is an example of both pictorial (connection) and schematic (ladder) diagrams for a gas furnace incorporating a furnace and fan control module. **Figures 31.100 (B) and (C)** show a flowchart for systematically troubleshooting a gas furnace that has an IFC and DSI system. **Figure 31.101(A)** is an example of a connection and ladder diagram for a gas furnace having both an IFC and a 115-V hot surface ignition system. **Figures 31.101 (B) and (C)** show a flowchart for systematically troubleshooting a different furnace that has an IFC and a hot surface ignition system.

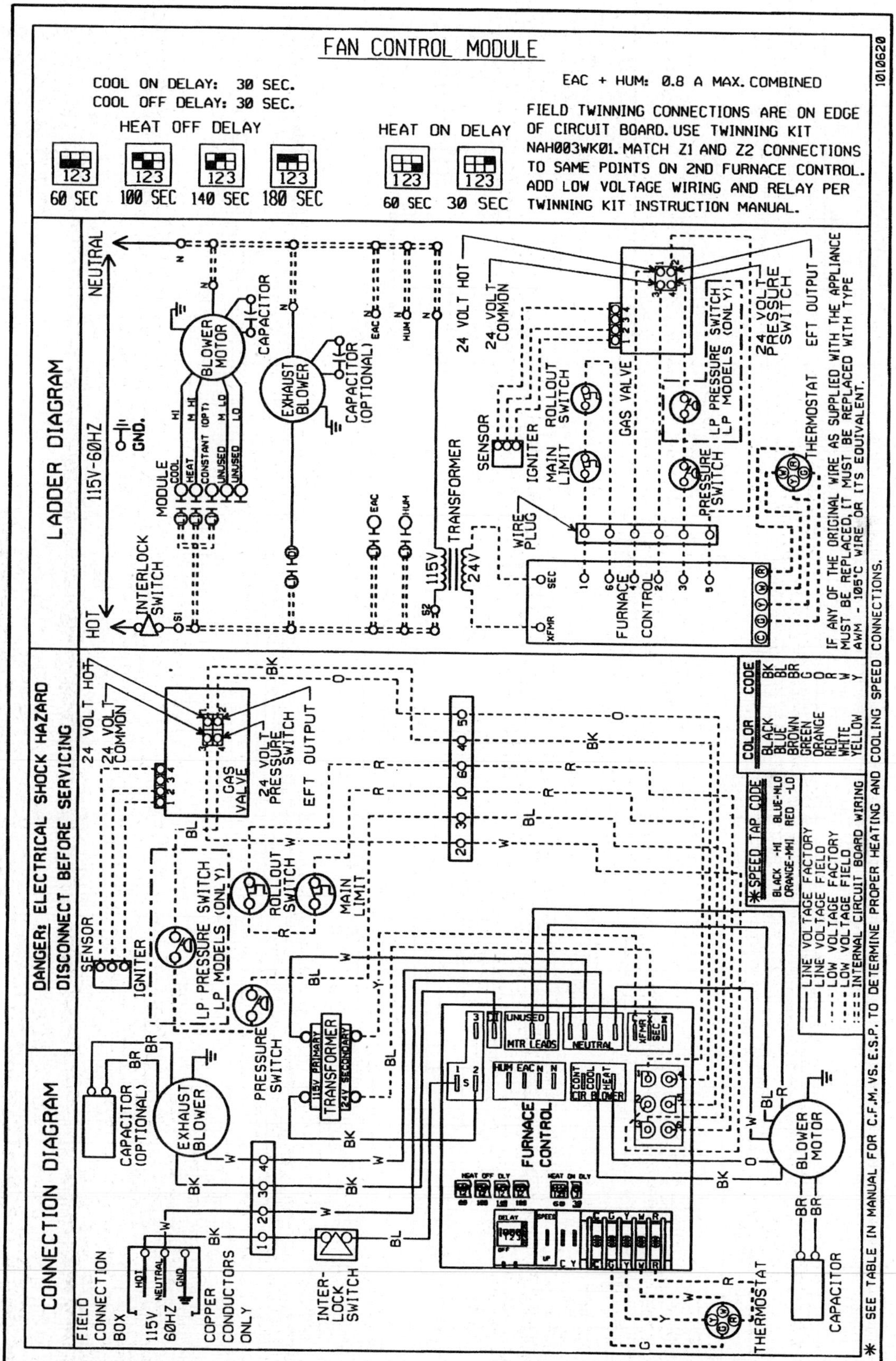

(A)

Figure 31.99 An electrical diagram for an intermittent hot surface ignition gas furnace with a furnace and fan control module. Source: International Comfort Products Corporation LLC

Figure 31.100 (A) The schematic wiring diagram for a direct spark ignition gas furnace. Courtesy Rheem Manufacturing Company.

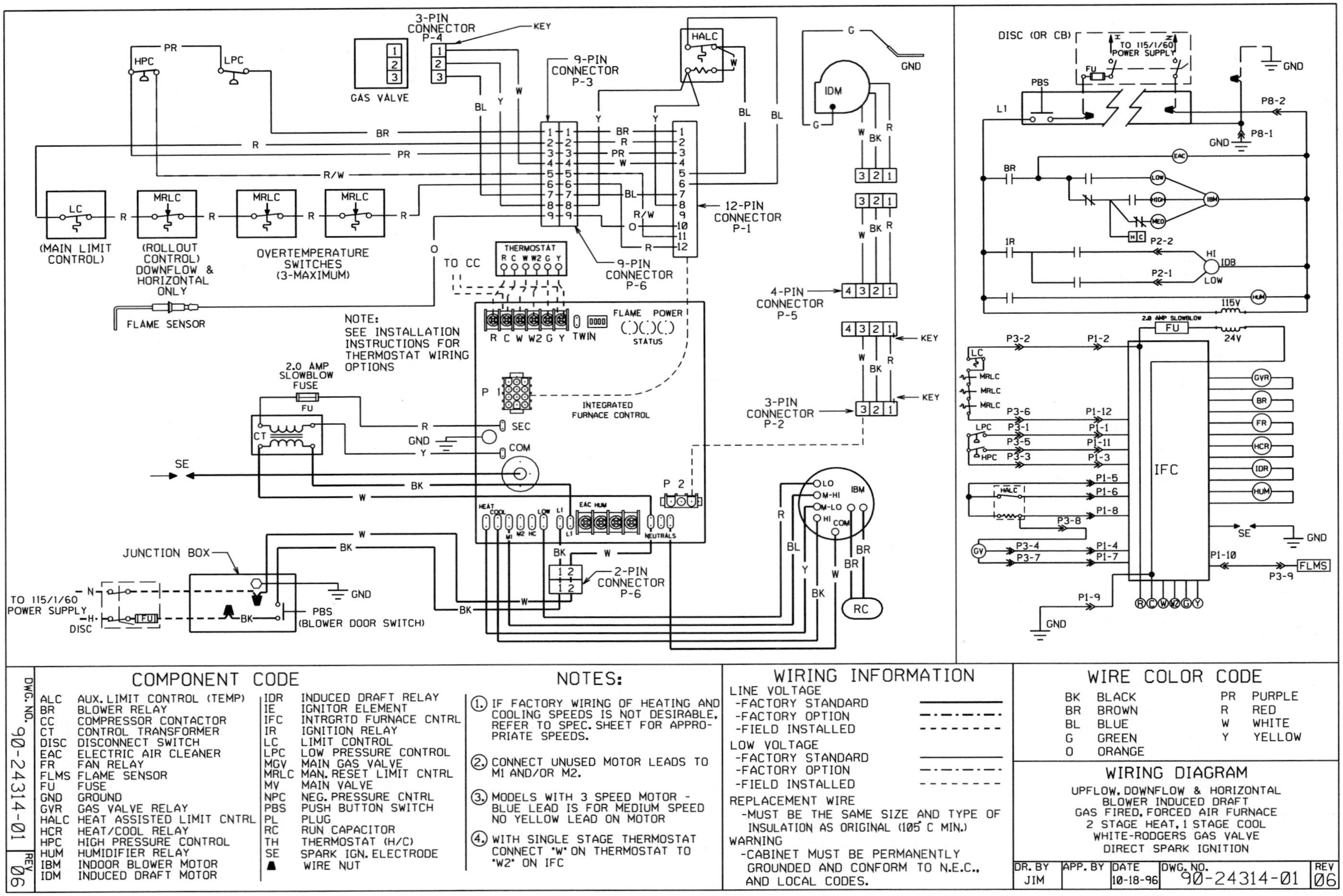

(A)

Figure 31.100 (B) (*Continued*) A troubleshooting guide for a direct spark ignition system with an integrated furnace controller (IFC). Courtesy Rheem Manufacturing Company.

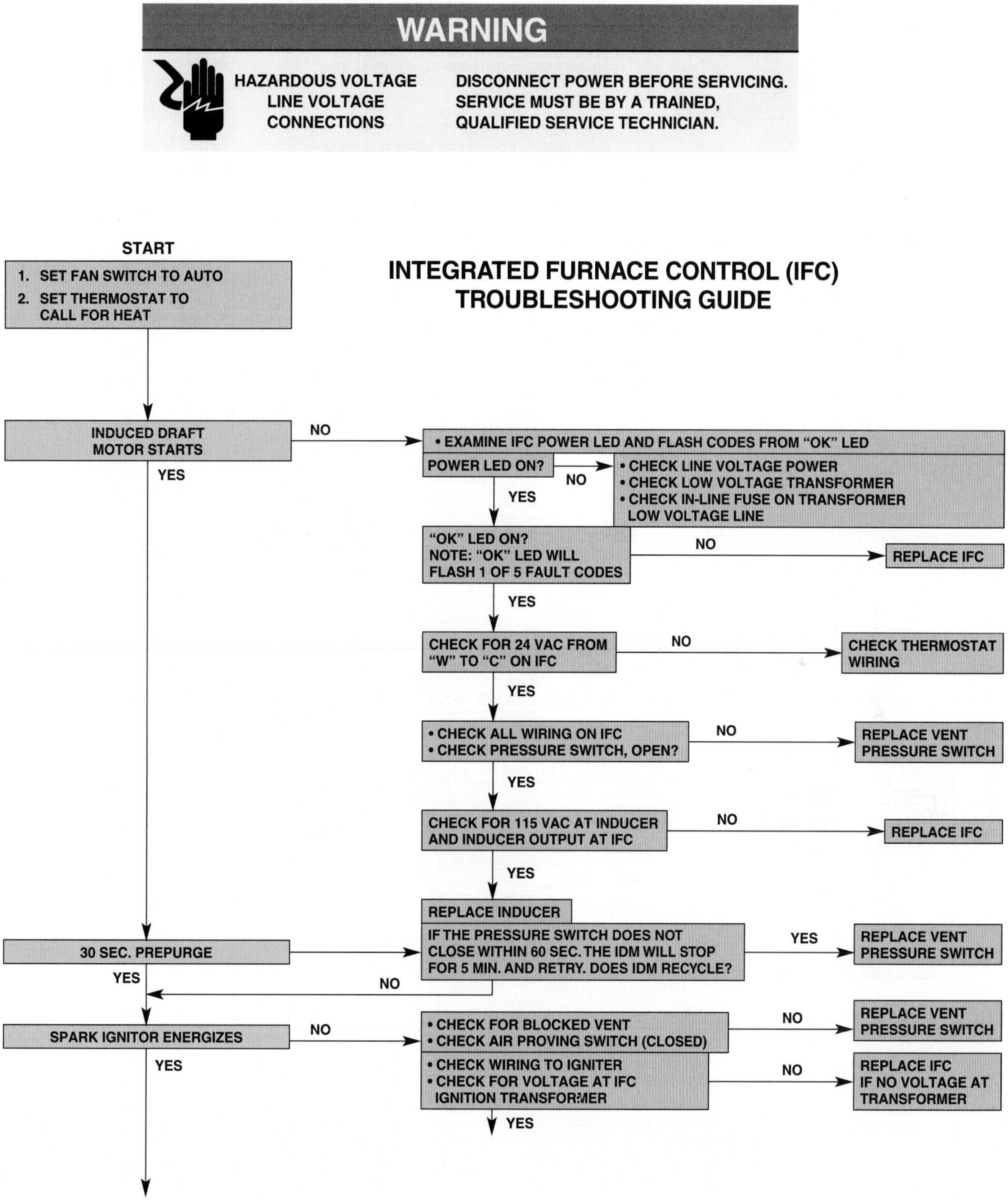

(B)

Figure 31.100 (C) (*Continued*) A troubleshooting guide for a direct spark ignition system with an integrated furnace controller (IFC). Courtesy Rheem Manufacturing Company.

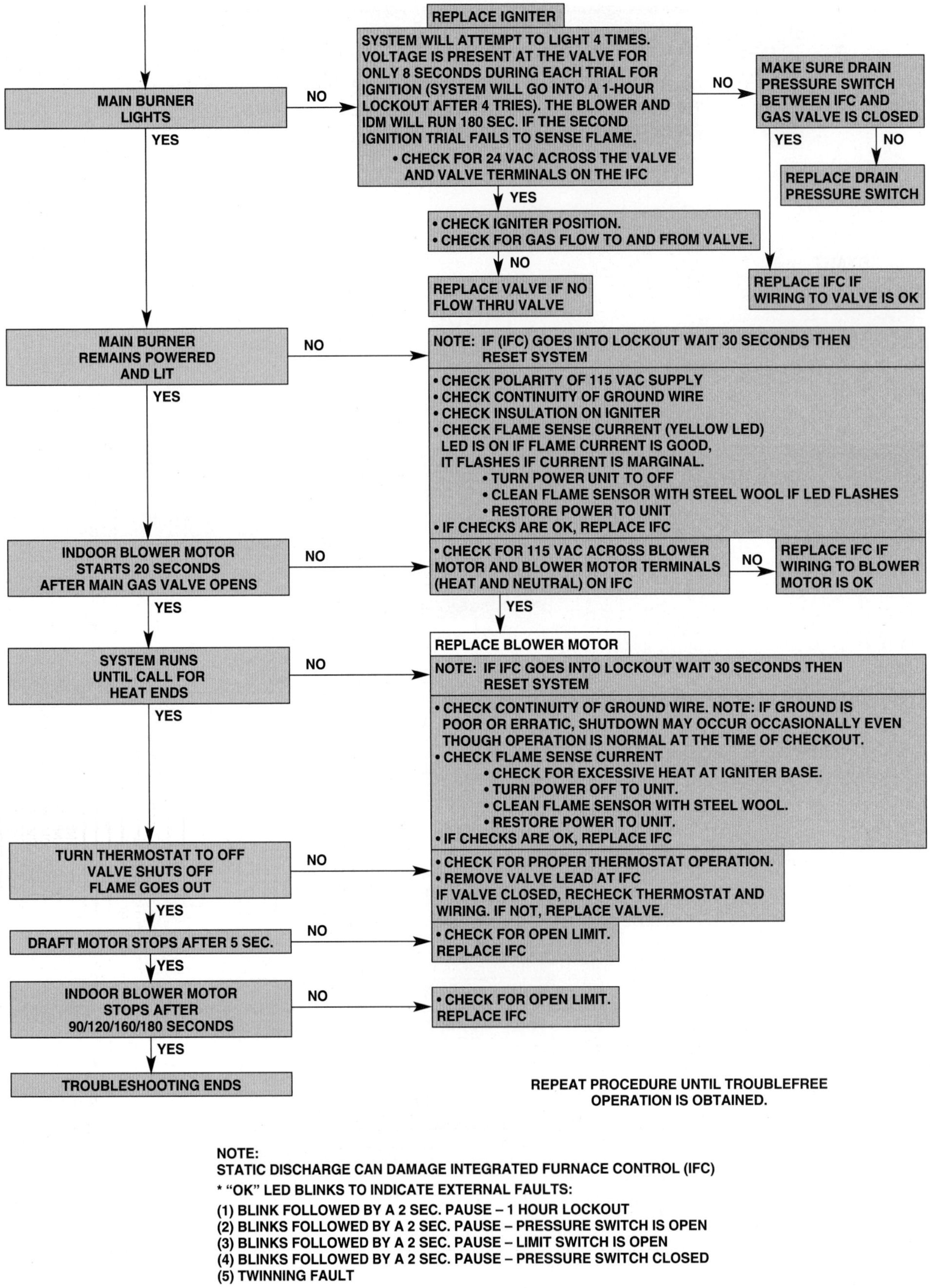

(C)

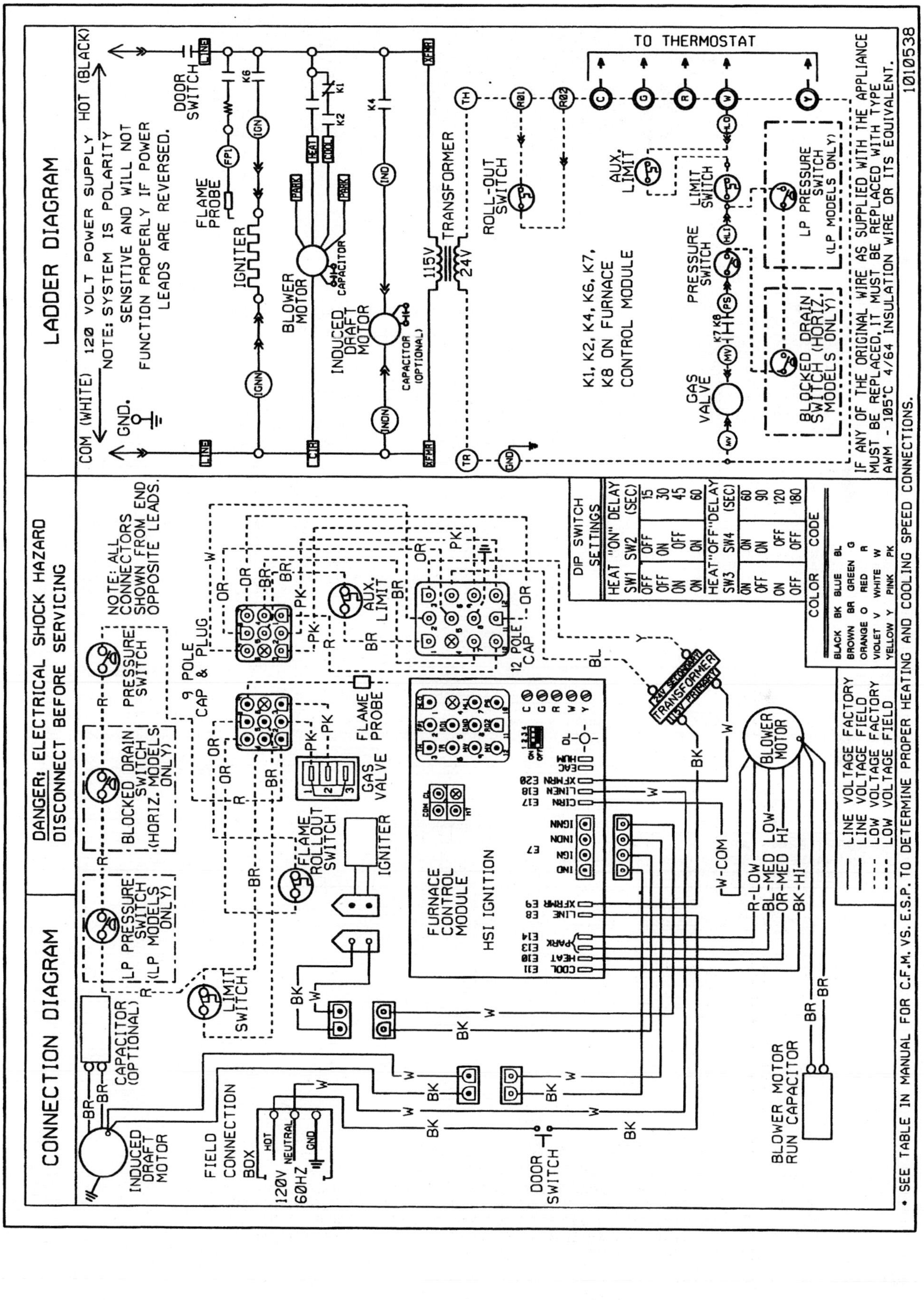

(A)

Figure 31.101 (A) A gas furnace connection and ladder diagram for direct hot surface ignition system with and IFC. Source: International Comfort Products Corporation LLC.

Figure 31.101 (B) (*Continued*) The troubleshooting guide for a direct hot surface ignition system with an integrated furnace controller (IFC). Courtesy Rheem Manufacturing Company.

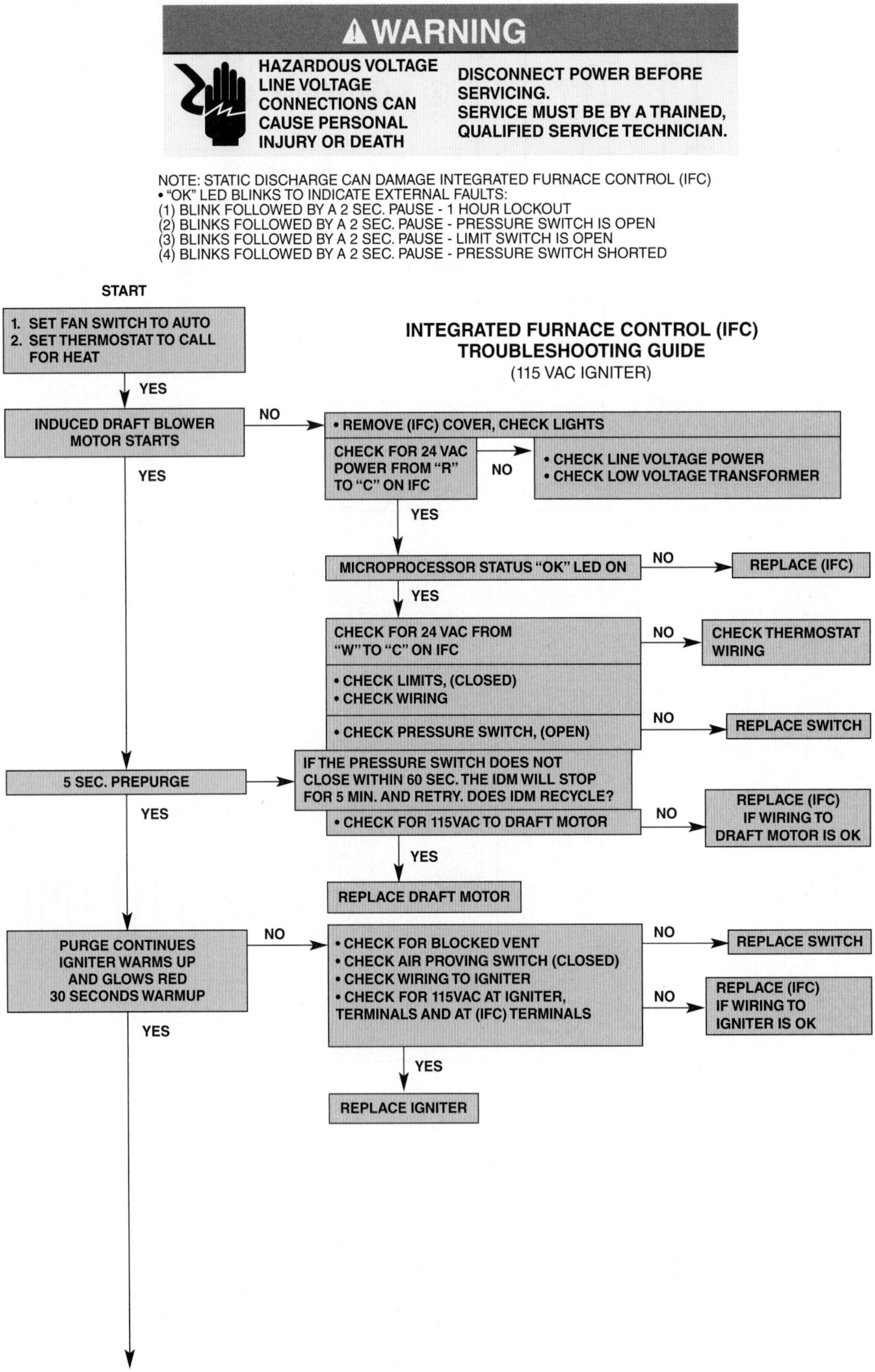

(B)

Figure 31.101 (C) (*Continued*) The troubleshooting guide for a direct hot surface ignition system with an integrated furnace controller (IFC). Courtesy Rheem Manufacturing Company.

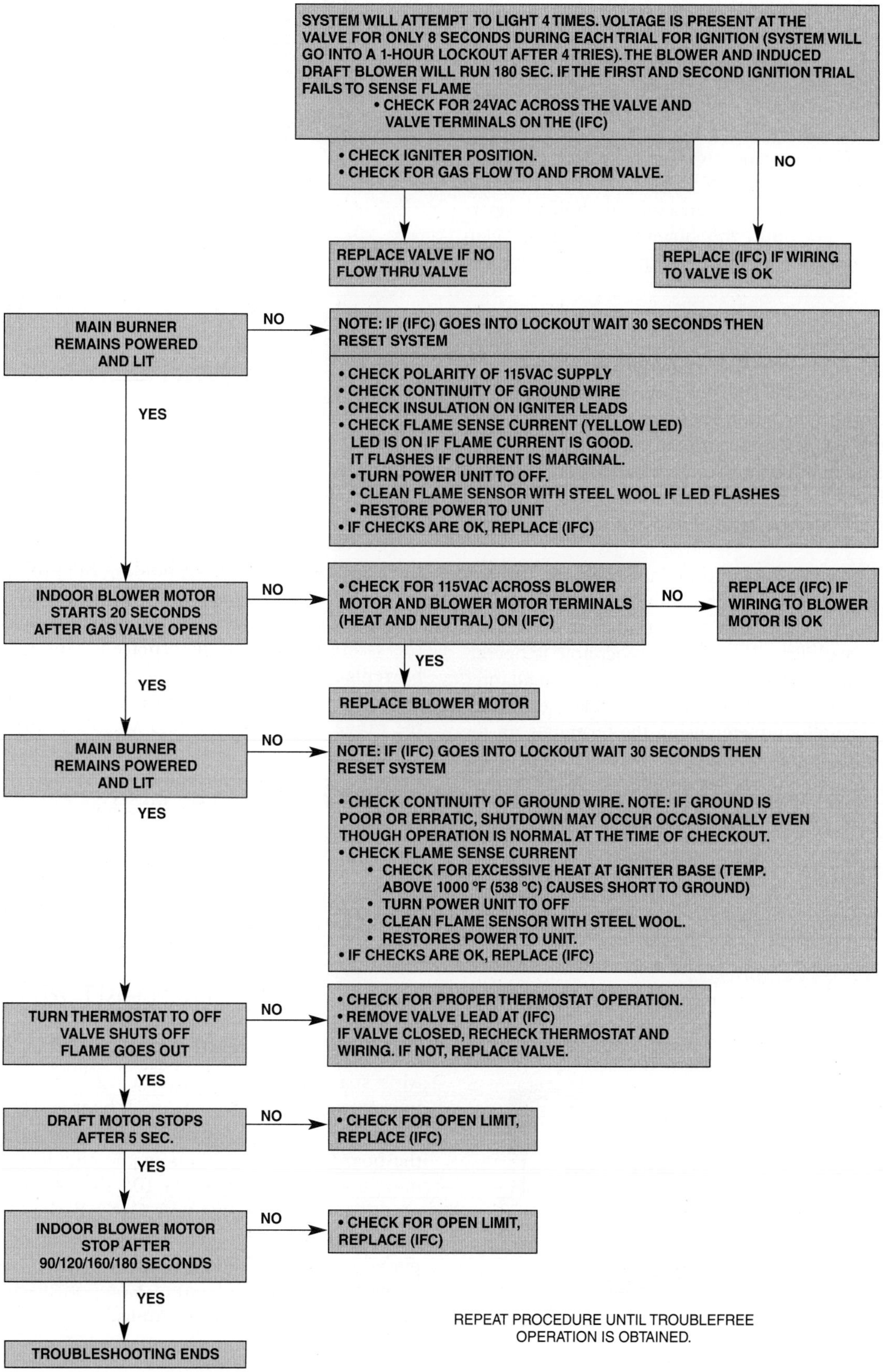

(C)

31.19 TROUBLESHOOTING THE SAFETY PILOT-PROVING DEVICE—THE THERMOCOUPLE

The thermocouple generates a small electrical current when one end is heated. When the pilot flame is lit and heating the thermocouple, it generates a current that energizes a coil holding a safety valve open. This valve is manually opened when lighting the pilot; the coil only holds it open. If the pilot flame goes out, the current is no longer generated and the safety valve closes, shutting off the supply of gas.

To light the pilot, turn the gas valve control to PILOT position. Depress the control knob and light the pilot. Hold the knob down for approximately 45 sec, release, and turn the valve to the ON position. The thermocouple may be defective if the pilot goes out when the knob is released. **Figure 31.102** describes the no-load test for a thermocouple. It is important to remember that the thermocouple only generates enough current to *hold in* the coil armature. The armature is pushed in when the valve handle is depressed.

To check under load conditions, unscrew the thermocouple from the gas valve and insert the thermocouple testing adapter into the valve. The testing adapter allows you to take voltage readings while the thermocouple is operating. Screw the thermocouple into the top of the adapter, **Figure 31.103**, and relight the pilot. Check the voltage produced with a millivoltmeter using the terminals on the adapter. The pilot flame must cover the entire top of the thermocouple rod. If the thermocouple produces at least 9 mVDC under load (while connected to its coil), it is good; otherwise, replace it.

Figure 31.103 The thermocouple test under load conditions. Disconnect the thermocouple from the gas valve. Screw the thermocouple test adapter into the valve and screw the thermocouple into the test adapter. Light the pilot and main burner and allow them to operate 5 min. Attach one lead from the millivoltmeter to either connecting post of the adapter and the other lead to the thermocouple tubing.

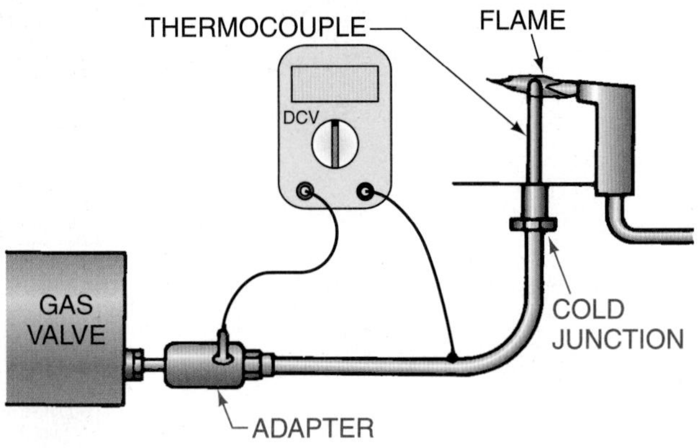

Check the manufacturer's specifications to determine the acceptable voltage for holding the different valve coils. If the thermocouple functions properly but the flame will not continue to burn when the knob is released, the coil in the safety valve must be replaced. This is normally done by replacing the entire gas valve. Thermopiles can generate much more voltage than thermocouples, sometimes in the range of 500 mVDC (0.5 V) to 750 mVDC (0.750 V). When more power is required to operate the circuit, the thermopile is a good choice. **Figure 31.104** is a systematic troubleshooting guide for a standing pilot gas furnace system.

Figure 31.102 The thermocouple no-load test. Disconnect the thermocouple from the gas valve while still holding the knob down to maintain the flame on the thermocouple. Attach the leads from the millivoltmeter to the thermocouple. Any reading below 18 DC mV indicates a defective thermocouple or poor pilot flame.

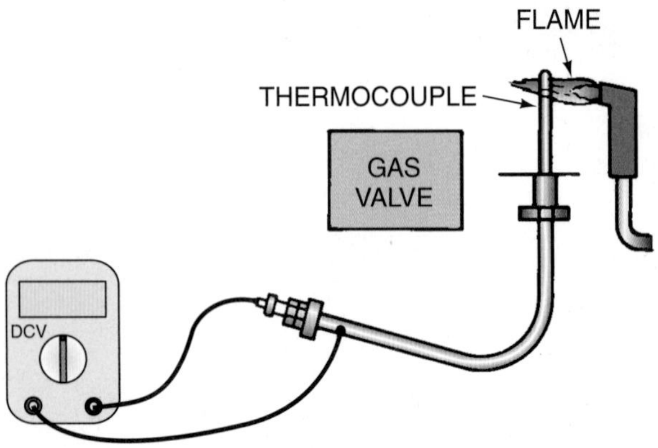

31.20 TROUBLESHOOTING SPARK IGNITION AND INTERMITTENT PILOT SYSTEMS

Most spark ignition or intermittent pilot light assemblies have internal circuits or printed circuit boards. The technician can troubleshoot only the circuit to the board, not circuits within the board. If the trouble is in the circuit board, the board must be changed. **Figure 31.105** shows how one manufacturer diagrams spark ignition. One is a pictorial diagram and the other is a line drawing. Notice that the manufacturer has placed the terminal board at the bottom of the pictorial and to the side of the line diagram. Components are not placed in the same position in the diagrams.

Figure 31.104 A guide for systematically troubleshooting a standing pilot furnace.

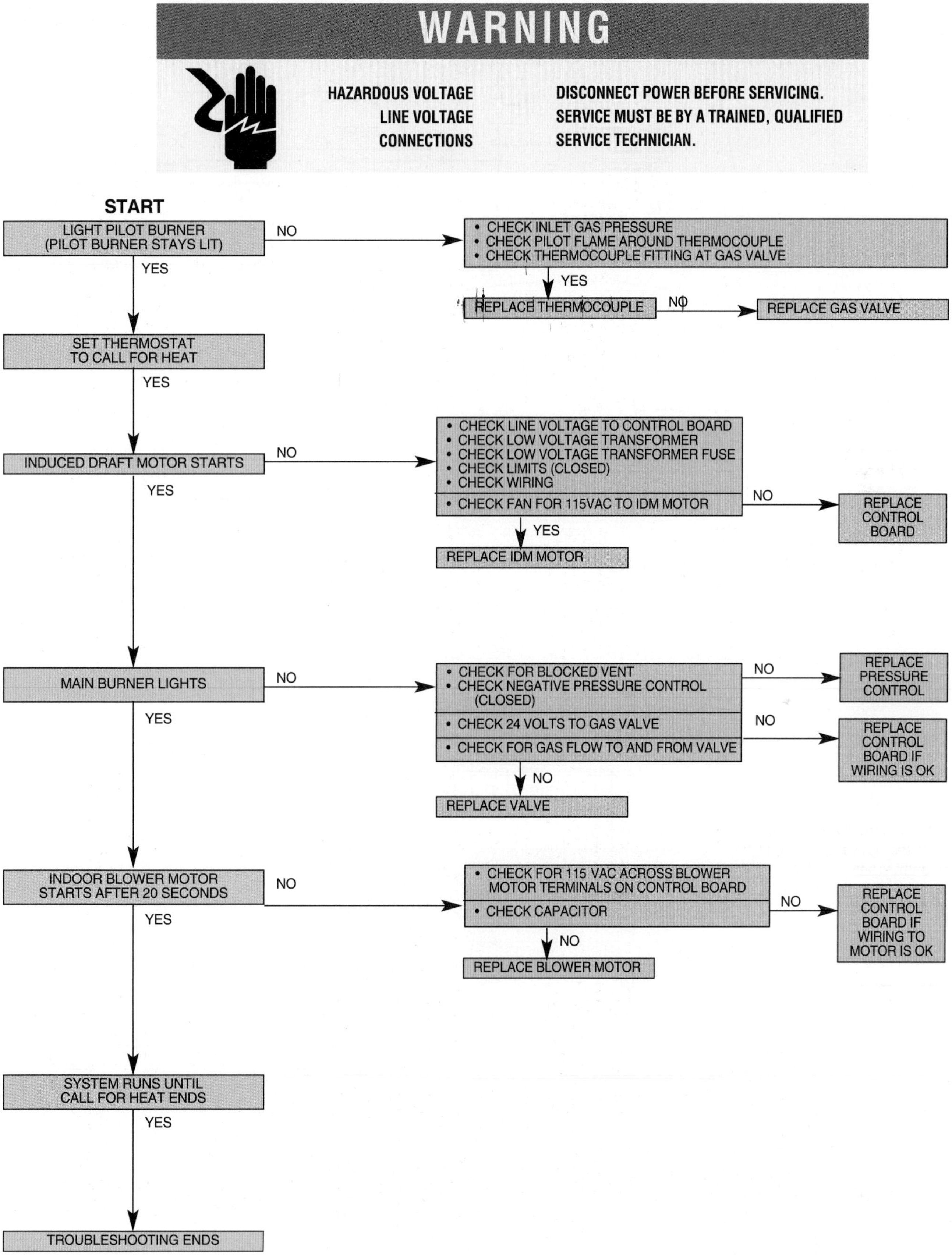

Figure 31.105 Intermittent spark ignition schematic. Note the voltages indicated.

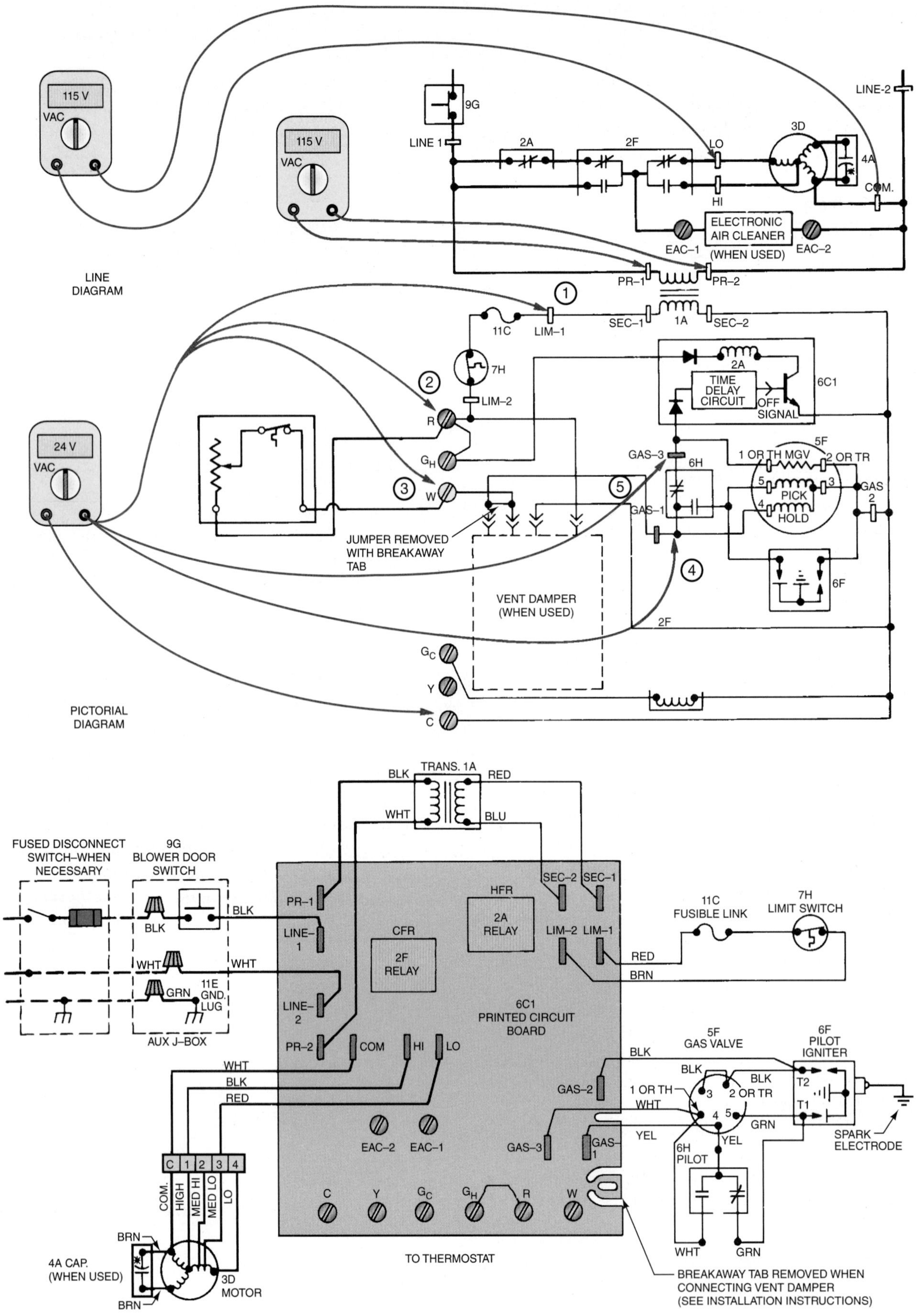

The spark ignition board in the previous example has terminals that provide for adding other components, such as an electronic air cleaner and a vent damper shutoff motor. This is handy because the installing technician only has to follow the directions to wire these components when they are used in conjunction with this furnace. The electronic air cleaner is properly interlocked to operate only when the fan is running. This component is not always easy to wire into the circuit to achieve the proper sequence, particularly in two-speed fan applications.

The following reference points can be used for the circuit board illustrated:

1. **SAFETY PRECAUTION:** *When the front panel is removed, all procedures come to a stop until the door interlock switch is closed. This should only be done by a qualified service technician.*•
2. The fan motor is started through the 2A contacts in low speed.
3. The 2A contacts are held open during the off cycle by energizing the 2A coil. This coil is deenergized during the on cycle.
4. If the control transformer is not functioning, the fan will run all the time. If the fan runs constantly but nothing else operates, suspect a bad control transformer, open limit, or open fusible link.
5. The fan starts and stops through the time delay in the heating cycle.

The following description of the electrical circuit can be used to perform an orderly troubleshooting procedure. See **Figure 31.105** for the voltages in the pictorial and line diagrams.

1. Line power should be established at the primary of the control transformer, terminals PR-1 and PR-2.
2. 24 V should be detected from the C terminal (common to all power-consuming devices) to the LIM-1 terminal. See the meter lead on the 24-V meter that is labeled 1.
3. Leaving the probe on the C terminal, move the other probe to positions 2, 3, and 4. 24 V should be detected at each terminal, or there is no need to proceed. There may be problem with the limit switch, fusible link, or the thermostat. Each component should be checked to verify that it is closed.
4. The key to this circuit is to have 24 V between the C terminal and the GAS-1 terminal. At this time, the pilot should be trying to light. The spark at the pilot should be noticeable. It makes a ticking sound with about one-half second between sounds.
5. If the pilot light is lit and the ticking has stopped, the 6H relay should have changed over. The NC contacts in the 6H relay should be open and the NO contacts should be closed. This means that there should be 24 V between the C terminal, which is the same as the GAS-2 terminal, and the GAS-3 terminal. The 6H relay is actually a bimetallic pilot-proving relay, **Figure 31.55(C)**, and heat from the pilot changes its position. If it will not pass power, it must be replaced.
6. When the circuit board will pass power to the GAS-3 terminal (probe position 5) and the burner will not light, the gas valve is the problem. Before changing it, be sure that its valve handle is turned in the correct direction and the interconnecting wire is good. The procedure of leaving one probe on the C terminal and moving the other probe can be done quickly when a furnace has a circuit board with terminals, which makes troubleshooting easier.
7. If the R terminal has 24 V and the W terminal does not, the thermostat and the interconnecting wiring are the problem. A jumper can be placed from R to W for a moment; if the spark starts to arc, the problem is the wiring or the thermostat.
8. The fan is started through the circuit board and is also part of the sequence of operation. The fan is started when the 2A relay is deenergized through the time-delay circuit. This is accomplished when the 6H relay in the circuit board changes position. When 24 V is detected from terminal C to terminal GAS-3, the time-delay circuit starts timing. When the time is up, the contacts open and deenergize the fan relay coil; the contacts close, and the fan starts.
9. The voltmeter can be relocated to the COM terminal and the LO terminal. When relay 2A is passing power, 115 V should be detected here. If there is power to the fan motor and it will not start, open up the fan section and check the motor for resistance. Check the capacitor. See if the motor is hot. If the motor cannot be started with a good capacitor, replace the motor. **Figure 31.106** shows a flowchart for systematically troubleshooting an intermittent pilot module. **SAFETY PRECAUTION:** *Not all ignition modules are alike. The service technician must consult with the ignition module manufacturer for specific systematic troubleshooting charts or procedures.*•

31.21 COMBUSTION EFFICIENCY

Given the high and ever-increasing cost of fuel, it is essential that adjustments be made to produce the most efficient but safe combustion possible. It is also necessary to ensure that an overcorrection that would produce carbon monoxide is not made.

Atmospheric gas burners use primary air, which is sucked in with the gas, and secondary air, which is pulled into the combustion area by the draft produced when ignition

Figure 31.106 A chart for systematically troubleshooting an intermittent pilot module. Source: Honeywell.

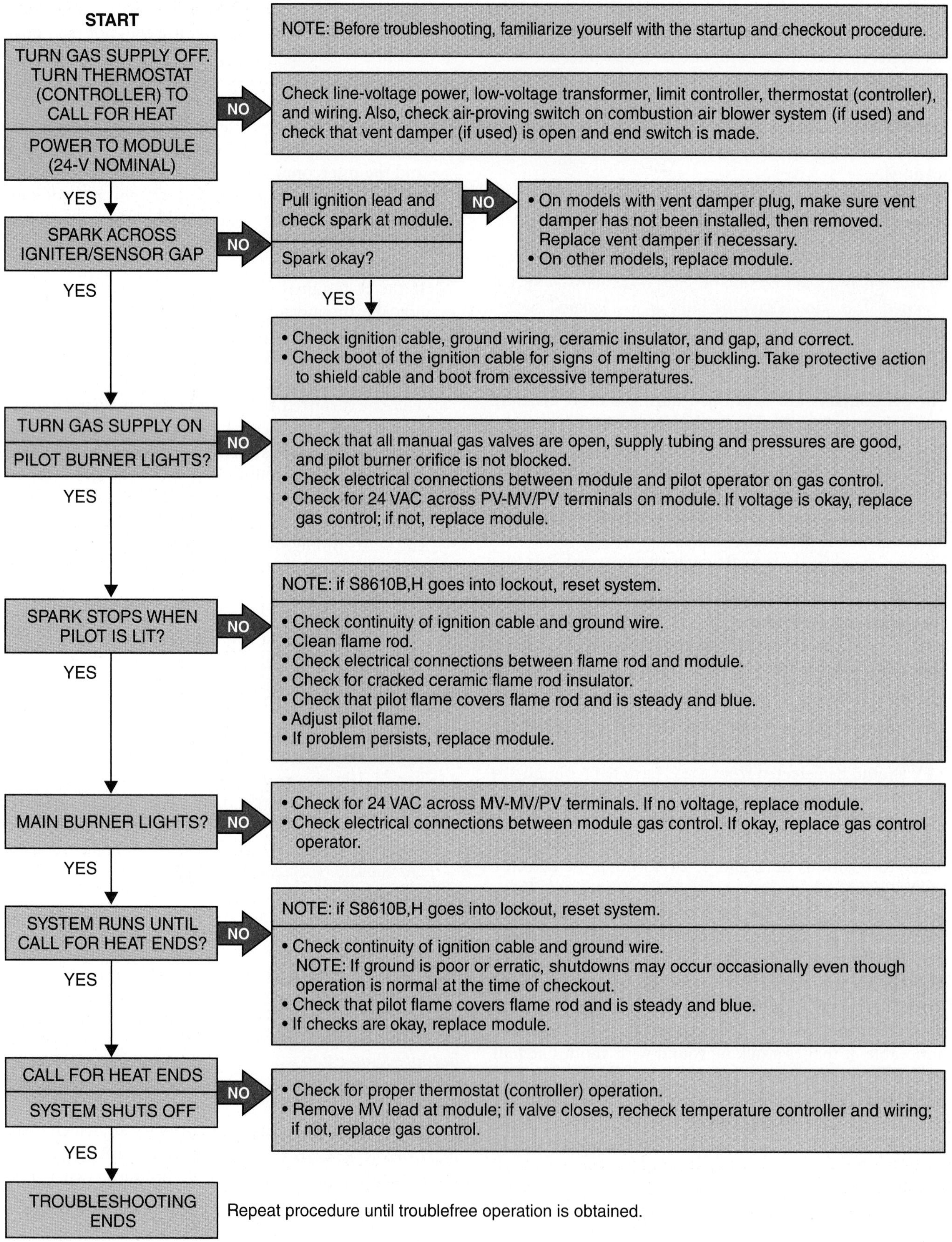

Figure 31.107 Main burner flame adjustment. The air shutter is turned to produce the proper flame. To check the flame, turn the furnace on at the thermostat. Wait a few minutes to ensure that any dust or other particles have been burned and no longer have an effect on the flame. The flames should be stable, quiet, soft, and blue with slightly orange tips. They should not be yellow. They should extend directly upward from the burner ports without curling downward, floating, or lifting off the ports, and they should not touch the side of the heat exchanger.

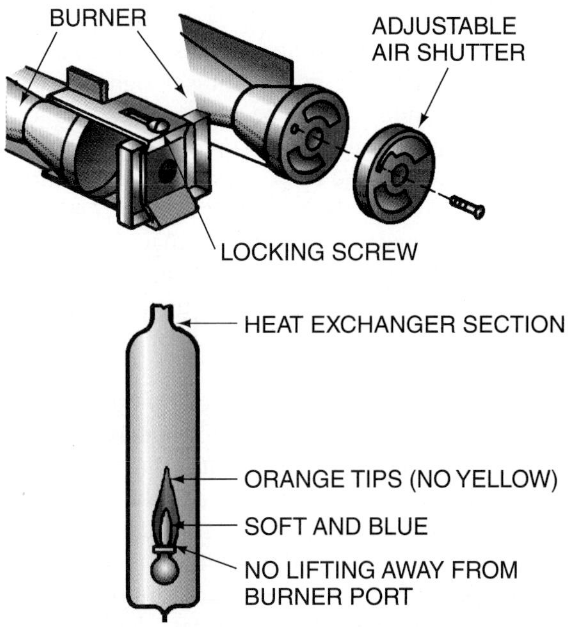

takes place. **SAFETY PRECAUTION:** *Incomplete combustion produces carbon monoxide. Enough secondary air must be supplied so that carbon monoxide will not be produced.* *Primary air intake can be adjusted by turning the shutters near the orifice. This adjustment is made so that the flame has the characteristics specified by the manufacturer. Generally, however, the flame is blue with a small orange tip when it is burning efficiently,* **Figure 31.107.**

Modern furnaces often do not have an air adjustment because of the types of burners used. With the inshot burner, air adjustment is not as critical. Conversion burners used to convert coal or oil furnaces or boilers to gas-fired normally have secondary air adjustments, and these must be adjusted in the field for maximum efficiency. Combustion analysis is used to get the necessary information from the flue gases to make this adjustment. Carbon dioxide in the flue gases is actually a measurement of the excess secondary air supplied. As the secondary air is decreased, the percentage of carbon dioxide in the flue gases increases, and the chance of producing carbon monoxide becomes greater. Other factors that may affect combustion efficiency may be gas pressure, the amount of air available for combustion, and furnace temperature rise.

If the correct amount of air is being supplied and perfect combustion is taking place, a specific amount of carbon dioxide by volume will be present in the flue gases. This is called the *ultimate carbon dioxide content*. The following table indicates the ultimate percentage of carbon dioxide content for various gases (these figures assume perfect combustion and should not be used as the carbon dioxide reading for a gas-burning appliance):

For Perfect Combustion

Natural gas	11.7%–12.2%
Butane gas	14.0%
Propane gas	13.7%

SAFETY PRECAUTION: *A gas furnace is never adjusted to produce the ultimate carbon dioxide content because there is too much danger of producing carbon monoxide. Therefore, the accepted practice is to adjust the secondary air to produce excess air—resulting in less carbon dioxide by volume in the flue gases.*

HVACR GOLDEN RULES

When making a service call to a residence:

- Always carry the proper tools and avoid extra trips through the customer's house.
- Make and keep firm appointments. Call ahead if you are delayed. The customer's time is valuable also.
- Keep customers informed if you must leave the job for parts or other reasons. Customers should not be upset when you inform them of your return schedule if it is reasonable.

Here are some simple, inexpensive procedures that may be included in the basic service call:

- Leave the furnace and area clean.
- Clean or replace dirty filters.
- Lubricate motor bearings when applicable.
- Check the burner compartment to ensure burners are clean and burning correctly.
- Replace all panel covers.

31.22 PREVENTATIVE MAINTENANCE

Gas equipment preventive maintenance consists of servicing the blower motor, belt and drive, and the combustion components. Air-side components are the same for gas as for electric forced-air heat, so only the burner section will be discussed here. Refer to Unit 30, "Electric Heat," for information on preventive maintenance practices for air-side components.

The combustion components consist of the burner, the heat exchanger, and the venting system for the products of combustion. The burner section of the modern gas furnace does not need adjustment from season to season as most burners no longer allow for air adjustments. These burners should burn efficiently and be reliable as long as they are kept clean.

Rust, dust, or scale must be kept from accumulating on top of the burner where the secondary air supports combustion or in the actual burner tube where primary air is induced into the burner. When a burner is located in a clean atmosphere (no heavy particles in the air as in a manufacturing area), most of the dust particles in the air will burn and exit through the venting system. The burner may not need cleaning for many years. A vacuum cleaner with a crevice tool may be used to remove small amounts of scale or rust deposits.

The burner and manifold may be removed for more extensive cleaning inside the burner tube where deposits may be located. Each burner may be removed and tapped lightly to break the scale loose. An air hose should be used to blow down through the burner ports while the burner is out of the system. The particles will blow back out of the primary air shutter and out of the system. **SAFETY PRECAUTION:** *Wear goggles when using compressed air for cleaning. This cleaning function cannot be accomplished while the burners are in the furnace.* •

Burner alignment should be checked to make sure that no flame impingement on the heat exchanger is occurring. The burners should sit straight and feel secure while in place. Large deposits of rust scale may cause flame impingement and must be removed. All gas burners should burn with a clear blue flame, with only orange streaks in the flame. **There should never be any yellow or yellow tips on the flame. If yellow tips cannot be removed from the flame by adjustments, the burner contains dirt or trash and should be removed for cleaning. In extreme cases, the burners may need to be replaced.**

Observe the burner for correct ignition as it is lit. Often, the crossover tubes that carry the burner flame from one burner to the next get out of alignment and the burner may light with an irregular pattern. Sometimes, a burner will "puff" if slow ignition occurs. The spuds should also be checked for obstructions.

On systems with a pilot, the pilot flame should be observed for proper characteristics. It should not have any yellow tips and it should not impinge on the burner. It should impinge on the thermocouple only. If the pilot flame characteristics are not correct, the pilot assembly may have to be removed and cleaned. Compressed air will usually clear any obstructions from the pilot parts. **SAFETY PRECAUTION:** *Do not use a needle to make the orifice holes larger. If they are too small, obtain the correct size from a supply house.* •

The venting system should be examined for obstructions. Birds may nest in the flue pipe, or the vent cap may become damaged. Visual inspection is the first procedure. Then, with the furnace operating, the technician should strike a match in the vicinity of the draft diverter and look for the flame to pull into the diverter. This is a sign that the furnace is drafting correctly.

Observe the burner flame with the furnace fan running. A flame that wavers with the fan running may indicate a crack in the heat exchanger. This would warrant further examination of the heat exchanger. It is difficult to find a small crack in a heat exchanger and one may only be seen with the heat exchanger out of the furnace housing. Extensive tests have been developed for proving whether or not a gas furnace heat exchanger has a crack. These tests may include the use of an inspection camera or combustion analyzer and should be performed carefully by experienced technicians.

High-efficiency condensing furnaces require maintenance of the combustion blower motor if it is designed to be lubricated. A digital manometer can be used to check induced draft motors for proper negative pressure. The condensate drain system should be checked to be sure it is free of obstructions and that it drains correctly. Manufacturers of condensing furnaces locate the condensing heat exchanger in the return air section where all air must move through the coil. This is a finned coil and will collect dirt if the filters are not cleaned. Routine inspection of this condensing coil for obstruction is a good idea.

All installations should be checked by observing the furnace as it goes through its sequence of operations. When the furnace is operating at a steady state, the temperature rise through the furnace should be verified. The airflow and gas pressure should be adjusted to the manufacturer's recommended settings. Some technicians disconnect the circulating fan motor and allow the furnace to operate until the high-limit control shuts it off. This proves the safety device will shut off the furnace.

Technicians will often perform a combustion analysis test to check furnace efficiency and then make adjustments. The adjustments should produce from 65% to 80% of the ultimate carbon dioxide content, which would produce 8% to 9.5% carbon dioxide in the flue. It is important that manufacturers' specifications be followed when making any adjustments.

The combustion efficiency percentage is an index of the useful heat obtained. It is not possible to obtain 100% efficiency because, except on direct-vented furnaces, heated air does leave the building through the flue. Most combustion analyzers will automatically calculate the combustion efficiency, carbon dioxide percentage, flue temperature, and undiluted carbon monoxide level, **Figure 31.108**.

When testing the level of carbon monoxide in the flue gas, an undiluted or air-free reading is used. Excess and dilution air can dilute the actual level of CO produced during combustion by the time it is measured in the flue pipe. Most combustion analyzers will calculate a carbon monoxide air-free (COAF) reading, which must be less than 400 parts per million (0.04%). Undiluted CO levels in excess of 50 parts per million should be investigated further. Carbon monoxide can be produced by flame impingement on a cool surface or by insufficient primary air (yellow flame) as well as insufficient secondary air.

Figure 31.108 A digital combustion analyzer used to test flue gas samples. Instrumentation provided by Test Products International Inc. Photo by Jason Obrzut.

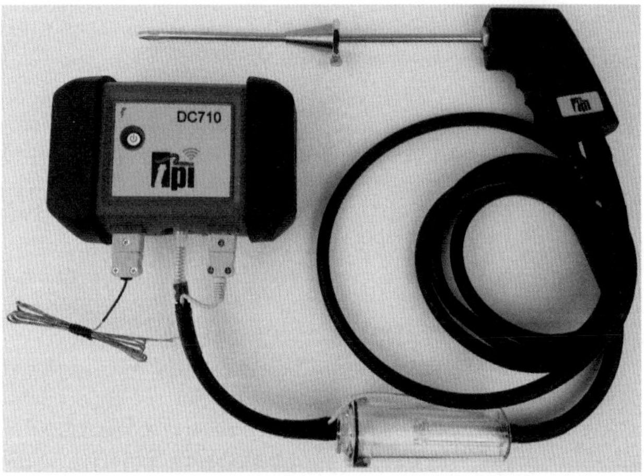

31.23 SERVICE TECHNICIAN CALLS

SERVICE CALL 1

The manager of a retail store calls. *There is no heat. This is an upflow gas furnace with a standing pilot and air-conditioning. The low-voltage transformer is burned out because the gas valve coil is shorted. This causes excess current for the transformer. The furnace is located in the stockroom.*

The technician arrives at the job and knocks on the store's service entrance door. After being invited in, the technician then asks the manager his account of the problem and listens very carefully, while maintaining a comfortable distance from the manager. After the customer is finished talking and all of his questions are answered, the technician starts work on the furnace.

The technician goes to the room thermostat and turns the fan switch to ON to see whether the indoor fan will start. It will not, so the technician suspects that the low-voltage power supply is not working. The thermostat is set to call for heat. The technician goes to the stockroom where the furnace is located. The voltage checked at the transformer secondary indicates 0 V. The technician turns the power off and uses the ohmmeter to check the resistance of the transformer. The secondary (low-voltage) coil has an open circuit, so the transformer is changed. Before connecting the secondary wires, the technician checks the resistance of the low-voltage circuit. There is only 2 Ω of resistance in the gas valve coil. This is so low that it will cause a high current flow in the secondary circuit. (The resistance is typically at least 20 Ω.) The technician goes to the truck and checks the resistance of another gas valve and finds it to be 50 Ω, so the gas valve is also changed.

The system is started with the new gas valve and transformer. A current check of the gas valve shows that it is pulling only 0.5 A, which is not overloaded. If the technician had just installed the transformer and turned the power on, it is likely that another transformer failure would have occurred quickly.

The technician changes the air filters and calls the store manager back to the stockroom for a conference before leaving. The store manager is informed that the boxes of inventory must be kept away from the furnace because they may present a fire hazard. The technician makes a quick check to make sure the job site is clean before leaving. Billing information and a work description with costs are then politely presented to the manager. After the technician makes sure his hands are clean, handshakes are exchanged and the technician leaves the job site.

SERVICE CALL 2

A customer in a residence reports that *fumes can be smelled and are probably coming from the furnace in the hall closet. The furnace has not been operated for the past two weeks because the weather has been mild. Because the fumes might be harmful, the dispatcher tells the customer to shut the furnace off until a technician arrives. The flue is stopped up with a shingle laid on top of the brick chimney by a roofer who had been making repairs.*

The technician arrives and starts the furnace. A match is held at the draft diverter to see whether the flue has a negative pressure. The flue-gas fumes are not rising up the flue for some reason; the match flame blows away from the flue pipe and draft diverter. The technician turns off the burner and examines it and the heat exchanger area with a flashlight. There is no soot that would indicate the burner had been burning incorrectly. The technician goes to the roof to check the flue and notices a shingle on top of the chimney. After the shingle is lifted off, heat rises out of the chimney.

The technician goes back to the hall closet and starts the furnace again. A match is held at the draft diverter, and the flame is drawn toward the flue. The furnace is now operating correctly. The technician changes the furnace filter and oils the fan motor before leaving the job.

SERVICE CALL 3

A customer requests that an efficiency check be performed on a furnace. *This customer thinks that the gas bill is too high for the conditions and wants the system completely checked out.*

The customer wants to watch the complete procedure of the service call. Before getting started, the technician asks the customer many pertinent questions. While the customer is talking, the technician shows respect by keeping eye contact and being a courteous listener. The technician is careful not to interrupt the customer and never

lets the customer feel stupid about a technical topic. When talking about a technical topic, the technician makes sure to use clear, concise, everyday language the customer can understand. The technician will run an efficiency test on the furnace burners at the beginning and then at the end, so the customer can see the difference.

The furnace is in the basement and easily accessible. The technician turns the thermostat to 10°F above the room temperature setting to ensure that the furnace will not shut off in the middle of the test. He then goes to the truck and gets the combustion analyzer kit, **Figure 31.108**. **SAFETY PRECAUTION:** *Be careful not to touch the stack; the stack temperature is taken at the heat exchanger side of the draft diverter.*• A sample of the flue gas is drawn into the sample chamber after the flue-gas temperature has stopped rising and the indoor fan has been running for about 5 min. This ensures that the furnace is up to maximum temperature.

The flue-gas reading shows that the furnace is operating at 80% efficiency. This is normal for a standard-efficiency furnace. After the test is completed and recorded, the technician shuts off the gas to the burner and allows the fan to run to cool down the furnace.

When the furnace has cooled enough to allow the burners to be handled, the technician takes them out. The burners are easily removed in this furnace by removing the burner shield and pushing the burners forward one at a time while raising up the back. They will then clear the gas manifold and can be removed. There is a small amount of rust, which is normal for a furnace in a basement. The draft diverter and the flue pipe are removed so that the technician and customer can see the top of the heat exchanger. All is normal—no rust or scale.

With a vacuum cleaner the technician removes the small amount of dirt and loose rust from the heat exchanger and burner area. The burners are taken outside where they are blown out with a compressed air tank at the truck. **SAFETY PRECAUTION:** *The technician wears eye protection.*• After the cleaning is complete, the technician assembles the furnace and tells the customer that it was in good condition and that there will be no difference in efficiency. Modern gas burners do not stop up as badly as the older ones, and the air adjustments will not allow the burner to get out of adjustment more than 2% or 3% at the most.

The technician oils the fan motor and changes the filter. The system is started and allowed to get up to normal operating temperature. While the furnace is heating to that temperature, the technician checks the fan current and finds it to be running at full load. The fan is doing all the work that it can. The efficiency check is run again, and the furnace is still operating at about 80%. It is now clean, and the customer has peace of mind. The technician thoroughly cleans the work site and washes his hands before presenting the customer with the billing information. The customer and technician shake hands and the technician thanks him for his business, letting the customer know that his company values the business. The thermostat is set before leaving.

SERVICE CALL 4

A customer indicates that *the pilot light will not stay lit. The pilot light goes out after it has been lit a few minutes. The heat exchanger has a hole in it very close to the pilot light. The pilot light will light, but when the fan starts it blows out the pilot light.*

The technician arrives, checks the room thermostat to see that it is set above the room temperature, and then goes to the basement. The standing pilot light is not lit, so the technician lights it and holds the button down until the thermocouple will keep the pilot light lit. When the gas valve is turned to ON, the main burner lights. Everything is normal until the fan starts, then the pilot light flame starts to wave around and after a short period it goes out. The thermocouple cools, and the gas to the main burner shuts off. Using an inspection camera, the technician shows the customer the hole in the heat exchanger. He turns the gas valve off and explains to the customer that the furnace cannot be operated in this condition because of the potential danger of gas fumes.

Given that this furnace is 18 years old, the technician advises that the customer really should consider getting a new one. The heat exchanger may be changed, but that requires considerable labor plus the price of the heat exchanger. The customer decides to replace the furnace.

SERVICE CALL 5

A residential customer calls. *There is no heat. The main gas valve coil is shorted, and the low-voltage transformer is burned. This symptom in an electronic intermittent ignition system,* ***Figure 31.105****, can lead the technician to believe that the printed circuit board is defective.*

When the technician arrives he notices that the indoor fan is running. It cannot be stopped by turning the room thermostat to OFF. **NOTE:** *(The low-voltage circuit holds the fan motor off in this system.)* • A low-voltage problem is suspected. The technician goes to the furnace and checks the output of the low-voltage transformer; the meter reads 0 V at the transformer. The transformer is burned. **SAFETY PRECAUTION:** *The technician turns off the power and replaces the transformer.*• When the power is restored, the electronic ignition circuit lights the pilot light. Everything seems normal, but the transformer becomes extra hot and smells like it is burning.

The technician turns off the power, changes the electronic circuit board, and adds a 3-A fuse to the low-voltage circuit. When power is restored, the pilot lights and everything is normal for a moment, but the transformer is still getting hot. The technician turns off the power and fastens a clamp-on ammeter to the low-voltage system. When power is restored, the pilot lights again and the amperage is normal (about 0.5 A). When the pilot proves and it is time for the main gas valve to be energized, the

current goes up to 3 A and the transformer begins to heat up. The 3-A fuse blows, protecting the transformer from burning.

The technician turns off the power and uses the ohmmeter to check the resistance of all power-consuming components in the circuit. When the main gas valve coil in the gas valve is checked, it shows 2 Ω of resistance. The gas valve is replaced. Power is restored. The pilot light is lit and proved. The main gas valve opens and the burner ignites. The system operates normally.

The technician turns off the power, and reinserts the original circuit board, and then restores the power. The furnace goes through a normal start-up.

SERVICE CALL 6

A residential customer indicates that although the furnace is heating the house, it is blowing cold air just before the fan goes off at the end of each heating cycle. *The warm-air fan's off–delay timing is set too long. The furnace has an IFC with programmable DIP switches for the fan's off–delay timing,* **Figures 31.44(A)** *and* **(B)**. *These switches have to be repositioned to shorten the fan's programmed off–delay timing. This will shut the fan off sooner after each heating cycle and prevent the cooler air from being blown into the conditioned space.*

Before arriving at the residence, the service technician stops to refuel the service van and clean up from a previous service call. The technician uses the gas station's bathroom to change into a clean uniform. After washing his hands and face, the technician looks in the mirror to make sure his hair is still combed neatly.

After knocking and politely introducing himself, the technician waits to be invited in before entering the house. After politely listening to the customer, the technician begins work. He takes the front panel off the furnace, putting some duct tape on the door switch to keep the furnace running while the front panel is removed. He then turns the thermostat 10°F higher than the room temperature to initiate a heating cycle. Once the furnace has run for 5 min, the thermostat is turned down. As usual, the flame goes out but the warm-air fan remains running. The technician times the fan's run time with a wristwatch once the flame goes out and finds that the off–delay timing for the fan is 180 sec. The technician then checks the DIP switches on the IFC and finds that switch 1 is in the up position and switch 2 is in the down position, **Figure 31.44(B)**. According to the faceplate of the IFC, the DIP switches in this position will give a 180-sec off–delay time for the warm-air fan. The service technician then changes the switches so that both switch 1 and switch 2 are in the up position, **Figure 31.44(A)**. This will give the warm-air fan a 60-sec off–delay time.

The technician then cycles the furnace through a complete cycle and checks the fan's off–delay timing again. The timing is now 60 sec as planned. The technician notices that the air coming from the furnace just before it shuts off is much warmer than it was with the longer off–delay timing. The technician explains to the homeowner what has been adjusted and how it pertains to the warm-air fan's off–delay timing, hands the customer the billing information with all costs explained in detail, thanks the customer, and sets the thermostat to the proper temperature before leaving the house.

SERVICE CALL 7

A residential customer says that the furnace is not heating the house. No airflow is coming from any of the warm-air ducts. *The combustion blower motor is burned out and the draft pressure switch is not closing, preventing ignition sequence from continuing,* **Figure 31.69(B)**.

Before taking the front cover off the furnace, the service technician notices a self-diagnostic light flashing through a clear plastic inspection hole on the front panel of the furnace, **Figure 31.109**. A technician should always look into this inspection hole before removing the front cover. Taking the front cover off the furnace will many times take power away from the IFC board and reset the self-diagnostic flashing lights. The technician makes note of the flashing intervals. The front cover is then removed, which interrupts power to the IFC and resets the self-diagnostics. The furnace door switch is taped closed to keep the furnace running while the door is off. The diagnostic light is blinking in intervals of three flashes, indicating that the draft pressure switch is open, **Figures 31.79** and **31.44(B)**.

The service technician turns power to the furnace off to make sure the IFC is reset, then turns the power back on. The combustion blower motor does not come on, which prevents the draft pressure switch from closing. The technician then shuts off power to the furnace and unplugs the wire leads going to the combustion blower motor from the IFC. An ohmmeter reading of infinity across the combustion blower motor leads proves to the technician that the combustion blower motor is open. A new combustion blower motor is installed and plugged into the IFC. The technician must take care that he is grounded so as not to cause a static shock, which can easily damage the delicate IFC.

Power is turned back on. The combustion blower motor is operating, the pressure switch has closed, and the hot surface igniter is glowing red. The burners light in about 5 sec and, a short time later, the blower motor is energized. The technician politely explains to the customer what was wrong with the furnace and that the problem has been remedied. The technician then makes sure the furnace operates successfully through three complete heating cycles.

Figure 31.109 A furnace with a clear plastic inspection hole on the front panel for observing the self-diagnostic flashing light. Courtesy Ferris State University. Photo by John Tomczyk.

SERVICE CALL 8

A residential customer complains that there is no heat in the house but the warm-air blower runs continuously. *The furnace high-limit switch has opened due to a dirty return air filter.*

Once arriving at the customer's residence, the service technician parks the service van on the street so as not to block the driveway. The technician maintains a comfortable distance from the door while waiting for the customer to answer the bell. After introducing himself and his business and waiting until the customer invites him in, the technician politely listens to the customer's account of the problem. After answering the customer's questions and asking a few questions of the customer, the technician asks to see the furnace. The customer then leads the technician to the basement and work is started.

The service technician notices that the self-diagnostic status light is blinking with only one flash, **Figure 31.78(A)**. The self-diagnostic code on the faceplate of the IFC indicates one flash is for an open high limit switch. Because the warm-air fan is running continuously, the technician pulls out the furnace filter. The filter is completely plugged. No or low airflow will heat up the heat exchanger until the furnace's high-limit control opens, **Figure 31.46(A)**. The technician then interrupts power to reset the IFC. When power is resumed, the furnace runs fine. He installs a new filter and the furnace continues to run correctly. The technician politely advises the customer to replace the furnace filter at least monthly, and sometimes even more often, to avoid having this problem again.

The service technician then hands the customer the billing information and they exchange handshakes. The technician leaves the customer his company's business card and thanks the customer for their valued business.

Solutions for the following service calls are not provided here. Please discuss the solutions with your instructor.

SERVICE CALL 9

A residential customer calls to report that *the furnace stopped in the middle of the night. The furnace is old and has a thermocouple for the pilot safety.*

The technician arrives and goes directly to the thermostat. The thermostat is set correctly for heating, but the thermometer shows a full 10°F below the thermostat setting. This unit has air-conditioning, so the technician turns the fan switch to ON to see if the indoor fan will start; it does.

The furnace is in an upstairs closet. The technician sees that the pilot light is not burning. This system has 100% shut-off, so there is no gas to the pilot unless the thermocouple holds the pilot-valve solenoid open. The technician positions the main gas valve to the PILOT position and presses the red button to allow gas to the pilot light. The pilot light burns when a lit match on an extender is placed next to it. The technician then holds the red button down for 30 sec and slowly releases it. The pilot light goes out.

What is the likely problem and the recommended solution?

SERVICE CALL 10

A residential customer calls. *The furnace is not heating the house. The furnace is located in the basement and is very hot, but no heat is moving into the house. The dispatcher tells the customer to turn the furnace off until a service technician can get to the job.*

The technician has some idea of what the problem is from the symptoms and knows what kind of furnace it is from previous service calls, so he brings some parts along. When the technician arrives, the room thermostat is set to call for heat. From the service request it is obvious that the low-voltage circuit is working because the customer says the burner will come on. The technician then goes to the basement and hears the burner operating. When the furnace has had enough time to get warm and the fan has not started, the technician takes the front control panel off the furnace and notices that the temperature-operated fan/limit switch dial (circular dial type, **Figure 31.45**) has rotated as if it were sensing heat. ⚠ **SAFETY PRECAUTION:** *The technician carefully checks the voltage entering and leaving the fan side of the switch.*•

What is the likely problem and the recommended solution?

SERVICE CALL 11

A residential customer calls. *There is no heat. The furnace is in the basement, and there is a sound like a clock ticking.*

Making sure not to block the driveway, the service technician puts on protective shoe coverings, rings the doorbell, and then steps back to a comfortable distance from the door. When the customer comes to the door, the technician politely introduces himself while handing out his business card. Once invited in, the technician politely listens to the customer and asks where the furnace is located. The service technician then goes to the basement, locates the furnace, and removes the furnace door. The technician hears the arcing sound. On removing the shield in front of the burner, the technician sees the arc. He lights a match on an extender and places it near the pilot to see whether there is gas at the pilot and whether it will light. It does. The arcing stops, as it should. The burner lights after the proper time delay.

What is the likely problem and the recommended solution?

REVIEW QUESTIONS

1. The most common furnace configurations are
   A. upflow, bi-flow, and downflow.
   B. upflow, horizontal, and bi-flow.
   C. downflow, upflow, horizontal, and multipoise.
   D. downflow, counterflow, bi-flow, and upflow.
2. A multipoise furnace
   A. requires an auxiliary vent fan motor when installed in the upflow position.
   B. may require some components to be relocated depending on the installation position.
   C. is no longer manufactured and should be replaced when encountered in the field.
   D. may require the spuds(orifices) to be changed when installed in the downflow position.
3. The specific gravity of natural gas is
   A. 0.08.
   B. 1.00.
   C. 0.42.
   D. 0.60.
4. Auxiliary limit switches are typically located
   A. in the return air stream.
   B. in the supply air stream.
   C. around the burners.
   D. near the draft hood.
5. A tool, which measures in inches of water column, that can be used to test gas pressure is a
   A. combustion analyzer.
   B. microammeter.
   C. digital manometer.
   D. digital multimeter.
6. The pressure at the manifold for natural gas is typically
   A. 10.5 to 11 in. WC.
   B. 8 in. WC.
   C. 3.5 psig.
   D. 3.5 in. WC.
7. The three things needed for combustion are
   A. air, fuel, and oxygen.
   B. fuel, air, and heat.
   C. fuel, air, and water.
   D. air, carbon, and fuel.
8. The pressure at the manifold for LP is typically
   A. 10.5 to 11 in. WC.
   B. 10.5 to 11 psig.
   C. 8 in. WC.
   D. 3.5 in. WC.
9. Burning one cubic foot of natural gas will produce approximately
   A. 1000 Btus.
   B. 1500 Btus.
   C. 2500 Btus
   D. 3200 Btus.
10. Burning one cubic foot of propane will produce approximately
    A. 1000 Btus.
    B. 1500 Btus.
    C. 2500 Btus
    D. 3200 Btus.

11. When working on a pressure regulator, turning the adjustment screw counterclockwise will
    A. increase the inlet pressure.
    B. decrease the inlet pressure.
    C. increase the outlet pressure.
    D. decrease the outlet pressure.
12. Combination gas valves have
    A. 120-V pilot solenoids.
    B. built-in pressure regulators.
    C. external pressure regulators.
    D. dual gas outlets.
13. Solenoid, diaphragm, and heat motor are all types of
    A. gas valves.
    B. pilot safety devices.
    C. flame rectification methods.
    D. pressure regulators.
14. In an intermittent pilot (IP) ignition system
    A. the pilot must be manually lit and will remain lit indefinitely.
    B. the pilot is lit electronically each time there is a call for heat.
    C. the pilot must be lit manually each time there is a call for heat.
    D. the pilot is used to heat the bimetal fan switch.
15. A servo pressure regulator reacts to changes in
    A. the temperature and pressure in the vent system.
    B. the voltage to pilot valve coil.
    C. the valve's inlet gas pressure.
    D. the valve's outlet gas pressure.
16. A gas valve that may incorporate a manual control, gas supply for a pilot, pilot safety features, a pressure regulator, and operating controls is referred to as a(n)
    A. solenoid valve.
    B. combination gas valve.
    C. automatic combination gas valve.
    D. heat motor–controlled valve.
17. A direct ignition furnace
    A. electronically lights the main burner.
    B. electronically lights a pilot each time there is a call for heat.
    C. requires the pilot to be lit manually.
    D. automatically relights the pilot in the event of a pilot outage.
18. Air that is pulled into the burner to mix with the gas prior to combustion is called
    A. primary air.
    B. secondary air.
    C. excess air.
    D. dilution air.
19. Furnaces with serpentine-type heat exchangers typically use
    A. ribbon burners.
    B. slotted burners.
    C. cast-iron burners.
    D. inshot or upshot burners.
20. Air that is pulled into combustion area to support combustion is called
    A. primary air.
    B. secondary air.
    C. tramp air.
    D. dilution air.
21. The component on which the spuds and burners are mounted to is the
    A. manifold.
    B. gas valve.
    C. heat exchanger.
    D. draft diverter.
22. The component that separates the flue gas from the medium being heated is the
    A. burner.
    B. flue pipe.
    C. heat exchanger.
    D. draft diverter.
23. The three adjustable pointers on a fan/limit switch are used to set the temperatures for
    A. igniter on, igniter off, and high limit.
    B. combustion blower on, fan on, and high limit.
    C. gas valve on, fan on, and high limit.
    D. fan on, fan off, and high limit.
24. A bimetal fan/limit switch cycles the blower using
    A. time on–time off.
    B. time on–temperature off.
    C. temperature on–time off.
    D. temperature on–temperature off.
25. A component that consists of two dissimilar metals that are fused together at one end that will generate a small voltage when heated is called a
    A. flame sensor.
    B. thermocouple.
    C. bimetallic safety device.
    D. liquid-filled remote bulb pilot safety device.
26. The component that consists of several thermocouples wired in series to increase voltage output is the
    A. flame sensor.
    B. thermocouple.
    C. bimetallic safety device.
    D. thermopile.
27. The silicon carbide hot surface igniter, used in many direct ignition systems, typically operates at
    A. 18 to 30 mVDC.
    B. 750 mVDC.
    C. 24 VAC.
    D. 120 VAC.
28. In an intermittent ignition system, the flame sensor proves
    A. the pilot flame.
    B. the main burner flame.
    C. both the pilot flame and main burner flame.
    D. the high voltage spark.

29. In the liquid-filled remote bulb pilot safety device, the liquid is usually
    A. water.
    B. mercury.
    C. refrigerant.
    D. liquefied propane.
30. A common problem in direct spark ignition systems is
    A. a dirty pilot orifice.
    B. improper pilot positioning.
    C. an improper spark gap.
    D. a defective bimetallic pilot safety device.
31. To test the flame rectification circuit, a digital multimeter should be placed
    A. across the flame sensor terminals to test microfarads.
    B. around the flame sensor wire to test for amps.
    C. in series with the flame sensor to test microamps.
    D. in series with the flame sensor to test for DCmV.
32. In a direct ignition system, the flame sensor proves
    A. the pilot flame.
    B. the main burner flame.
    C. both the pilot flame and main burner flame.
    D. the high voltage spark.
33. The principal of flame rectification is based on the fact that
    A. a flame can conduct electricity.
    B. flames are always grounded.
    C. a flame can generate small amounts of voltage.
    D. flame has a very low resistance.
34. Which of the following would cause a problem with the flame rectification circuit?
    A. An overfiring condition
    B. Furnace short cycling
    C. An improperly positioned thermostat
    D. A poor ground wire connection
35. High-efficiency furnaces are designed with which of the following?
    A. A draft diverter
    B. Ribbon or slotted burners
    C. A condensing heat exchanger
    D. A natural draft heat exchanger
36. For every 1 lb of water vapor that is condensed into liquid
    A. 970 Btus are released.
    B. 1050 Btus are released.
    C. 2500 Btus are released.
    D. 3200 Btus are released.
37. As excess air decreases,
    A. dew point decreases.
    B. dew point increases.
    C. gas pressure increases.
    D. gas pressure decreases.
38. Furnace efficiency is measured by
    A. Energy Efficiency Ratio (EER).
    B. Season Energy Efficiency Ratio (SEER).
    C. Heating Seasonal Performance Factor (HSPF).
    D. Annual Fuel Utilization Efficiency (AFUE).
39. A control's response to a specific number of unsuccessful trials for ignition is a(n)
    A. prepurge cycle.
    B. interpurge cycle.
    C. lockout.
    D. timed fan delay.
40. To cycle the blower motor during the heating sequence, an integrated furnace control (IFC) uses
    A. time on–time off.
    B. time on–temperature off.
    C. temperature on–time off.
    D. temperature on–temperature off.
41. In a two-stage furnace, the typical manifold pressure in low-fire is
    A. 3.5 in. WC.
    B. 1.75 in. WC.
    C. 10.5 in. WC.
    D. 1.0 in. WC.
42. Modulating furnaces typically have
    A. long run cycles and lower combustion efficiencies.
    B. short run cycles and high temperature swings.
    C. short run cycles and very low temperature swings.
    D. longer run cycles and very low temperature swings.
43. A high-efficiency condensing furnace falls into which venting category?
    A. Category I
    B. Category II
    C. Category III
    D. Category IV
44. The horizontal run of the flue pipe on gas-burning furnace should be sloped
    A. ¼ in. rise per foot.
    B. ¼ in. rise per 10 feet.
    C. ½ in. rise per foot.
    D. ½ in. rise per 10 feet.
45. The horizontal run of gas piping should be sloped in the direction of flow. The recommended slope is
    A. ¼ in. rise per foot.
    B. ¼ in. rise per 10 feet.
    C. ¼ in. rise per 15 feet.
    D. ¼ in. rise per 20 feet.
46. After a gas piping system is assembled, it should be
    A. pressurized with oxygen and leak tested with a digital leak detector or bubble solution.
    B. pressurized with nitrogen to 150 psig and left for 24 hours to test for leaks.
    C. pressurized with gas and leak tested with a digital leak detector or bubble solution.
    D. pressurized with gas and leak tested with a small open flame.
47. The first step in systematically troubleshooting a furnace with an integrated furnace control (IFC) is to
    A. turn off the power and gas supply to the furnace.
    B. obtain the wiring diagram for the furnace being serviced.
    C. rewire the furnace from scratch to ensure proper electrical connections.
    D. try to determine the age of the system.

48. A thermopile typically generates
    A. 24 to 120 VAC.
    B. 18 to 30 mVDC.
    C. 1 to 25 DC microamps.
    D. 500 to 750 mVDC.
49. During perfect combustion of natural gas, the carbon dioxide percentage produced is
    A. 11.7% to 12.2%.
    B. 14.0% to 14.3%.
    C. 8.0% to 9.5%.
    D. 13.2% to 13.7%.
50. The maximum allowable undiluted carbon monoxide level in the flue gas of a furnace is
    A. 50 PPM.
    B. 400 PPM.
    C. 1000 PPM.
    D. 2500 PPM.
51. What is the difference between a direct-vent and a non-direct vent furnace?
52. Describe the difference between a single-rod (local-sensing) and a dual-rod (remote-sensing) system as they pertain to flame rectification.
53. What is the difference between a hard and a soft lockout?
54. Describe the difference between a forced-draft system and an induced-draft system.
55. Explain the difference between a 100% shutoff system and a non-100% shutoff system.
56. What is the difference between an integrated control and a non-integrated control?
57. Why is excess air supplied to all gas-burning appliance burners?
58. What is a redundant gas valve?
59. Explain how a temperature rise test is done.
60. Describe the differences between intermittent ignition and direct ignition systems.
61. List some things that may be done as preventive maintenance on furnaces.

UNIT 32

OIL HEAT

OBJECTIVES

After studying this unit, you should be able to

- describe the two main systems that make up an oil-fired, warm-air furnace.
- explain how the two main furnace systems work together to provide comfort heating.
- describe four common configurations for oil-fired furnaces.
- explain how fuel oils are classified and numbered.
- describe different methods for storing fuel oil.
- explain how oil tanks are sized.
- explain the importance of periodic oil tank inspection.
- describe two common piping methods that are used to deliver oil to the heating appliance.
- describe how fuel oil and air are prepared and mixed in the oil burner unit for combustion.
- explain the combustion process for oil.
- list products produced as a result of combustion of fuel oil.
- list the component parts of a gun-type oil burner.
- sketch wiring diagrams of the oil burner primary control system and the fan circuit.
- explain the relationship that exists between combustion chamber configuration and a nozzle's spray angle.
- explain the function of a heat exchanger on an oil-fired furnace.
- explain how condensing furnaces differ from conventional furnaces.
- describe basic service procedures for oil burner components.
- describe preventive maintenance procedures.
- state tests used to determine oil burner efficiency.
- explain proper procedures to diagnose and repair oil-fired furnaces.

SAFETY CHECKLIST

- ✓ Do not reset any primary control too many times because unburned oil may accumulate after each reset. If a puddle of oil is ignited, it will burn intensely. If this should happen, stop the burner motor but allow the furnace blower to run. The air shutter should be shut off to reduce the air to the burner. Notify the fire department. Do not try to open the inspection door to put the fire out; let it burn itself out with reduced air.
- ✓ Do not start a burner if the heat exchanger is cracked or otherwise defective. The flue gases from combustion should never mix with the circulating air.
- ✓ Do not start the burner with the fuel pump bypass plug in place unless the oil delivery system is configured as a two-pipe system with both a supply and return line. Without the return line, there is no place for the excess oil to go except to possibly rupture the shaft seal.
- ✓ When conducting flue-gas efficiency tests, avoid burns by not touching the hot flue-gas pipe.
- ✓ Ignition is produced by an arc of 10,000 V or more. Keep your distance from this arc.
- ✓ Observe all electrical safety precautions.
- ✓ Inspect oil burners and lines for leaks.
- ✓ When installing oil tanks, make certain that all codes are followed.
- ✓ Oil tanks need to be inspected periodically to minimize damage to the environment in the event of tank leakage.

32.1 INTRODUCTION TO OIL-FIRED, FORCED-WARM-AIR FURNACES

The oil-fired, forced-warm-air furnace has two main systems: a heat-producing system and a heat-distributing system. The *heat-producing system* consists of the oil burner, fuel supply components, combustion chamber, and heat exchanger. The *heat-distributing system* is composed of the blower, which moves the air through the ductwork, and other related components, **Figure 32.1**.

Following is a brief description of how such a furnace operates. When the thermostat calls for heat, the ignition

Figure 32.1 An oil-fired furnace. Photo by Eugene Silberstein

system is powered and the oil burner motor starts, which turns the fuel pump and combustion air blower wheel. An oil-air mixture is sprayed into the combustion area, where it is ignited. After the flame has been produced and the heat exchanger is up to temperature, the furnace blower comes on and distributes the heated air through the ductwork to the space to be heated. When this space has reached a predetermined temperature, the thermostat will cause the burner to shut down. The furnace blower will continue to run until the heat exchanger cools to a set temperature.

32.2 PHYSICAL CHARACTERISTICS

The physical appearance and characteristics of forced-air furnaces vary to some extent. The low-boy furnace is often used when there is not much headroom, **Figure 32.2**. Low-boys may have a cooling coil on top, if space permits, to provide air-conditioning. An upflow furnace is a vertical furnace in which the air is taken in at the bottom and is forced across the heat exchanger and out the top, **Figure 32.1**. This furnace is installed with the ductwork above it and is often located in a hallway closet or in a basement area. A downflow furnace looks similar to an upflow except that the air is drawn in from the top and forced out the bottom. In the downflow configuration, the supply ductwork that carries the heated air to the occupied space is located just below or in the floor below the furnace. The cooler air coming from the occupied space to the appliance enters at the top of the furnace. A *horizontal* furnace is usually installed in a crawl space under a house or in an attic, **Figure 32.3**. These units are available with right-to-left or left-to-right airflow.

Oil-fired forced-air heating is a popular method of heating residences and light commercial buildings. Millions of furnaces of this type are being used in the United States at the present time. It is essential that the air-conditioning and heating technician understand proper installation and servicing techniques involved with oil burners. Safety is always a concern of technicians, but with the high cost of energy, efficiency in the combustion process is also extremely important.

Figure 32.3 A horizontal, oil-fired forced-air furnace. Courtesy Williamson-Thermoflo

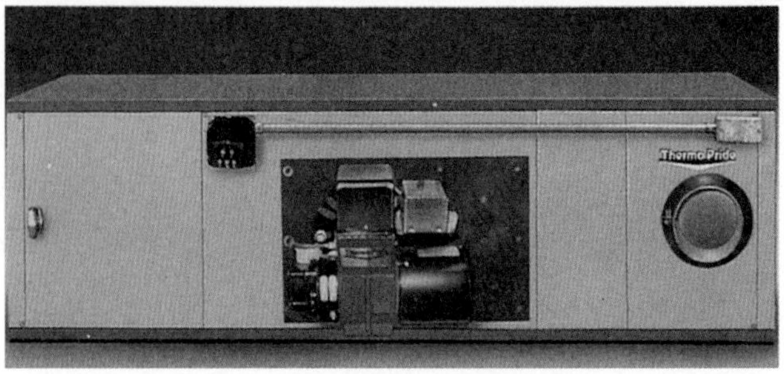

Figure 32.2 A low-boy oil-fired furnace. Courtesy Thermo Pride

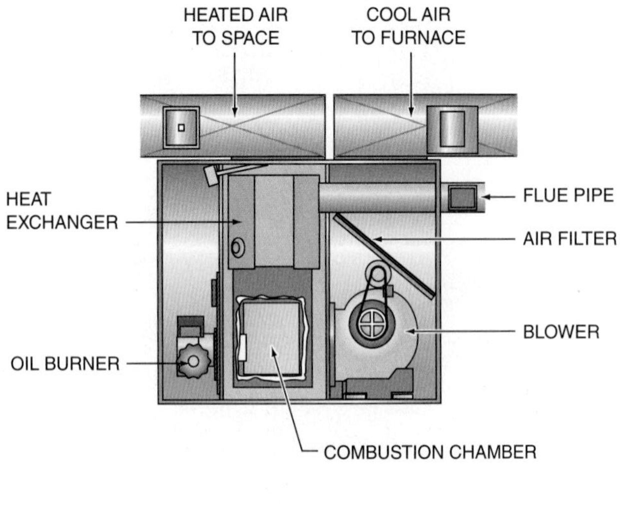

32.3 FUEL OILS

Fuel oil as we know it is a fossil fuel, as is natural gas, discussed in Unit 31. Fuel oil is a combination of liquid hydrocarbons containing hydrogen and carbon in chemical combination. Some of these hydrocarbons in the fuel oil are light, and others are heavy. Combustion occurs when the hydrocarbons combine rapidly with oxygen (O_2). Heat, carbon dioxide (CO_2), and water vapor are produced when the combustion occurs.

Fuel oil is delivered to the customer in liquid form and stored in tanks, either aboveground or underground, **Figure 32.4**. The oil used in residential and most commercial systems in the United States is basically diesel fuel that has been

Figure 32.4 A local oil dealer delivers the fuel oil in small trucks to aboveground or underground storage tanks.

dyed red for tax purposes. Taxes are higher when diesel fuel is used for transportation. By dyeing the fuel, authorities can identify the cheaper, lower-taxed product. If the dyed fuel is found in the tanks of motor vehicles or trucks, heavy fines can be levied.

Fuel oil is derived from crude oil by a process called distillation. There are six grades of fuel oil, and a numbering system, 1 through 6, is used to identify each grade. The lower-numbered oils are called *light oils* because they weigh less per gallon than the higher-numbered oils. The reason for this is the carbon content. Oils are hydrocarbons made up of carbon and hydrogen, and since carbon is heavier than hydrogen, it can be concluded that the higher-numbered oils contain more carbon than the lower-numbered oils. The higher-numbered fuel oils also have higher heat content than the lower-numbered oils. The lightest of the fuel oils, Number 1, is most commonly known as kerosene.

Since fuel oil is made up of carbon and hydrogen, it is classified as a hydrocarbon. Recall from earlier units that other pure hydrocarbons, such as methane, ethane, propane, and butane, are very good refrigerants but are typically not used because of their high flammability. **Figure 32.5** shows the molecular structure of some very common hydrocarbons. Methane, **Figure 32.5(A)**, is the base molecule for refrigerants such as R-22 and R-12. Ethane, **Figure 32.5(B)**, is the base molecule for refrigerants such as R-123 and R-134a.

The most common fuel oil in use is Number 2. It is used in most residences and commercial installations. The heavier oils are used in large commercial and industrial installations. Number 2 fuel oil has a heat content of 137,000 to 141,800 Btu/gal and weighs 6.960 to 7.296 lb/gal. (This text will use the average of 140,000 Btu/gal, which is an industry standard.) Number 6 fuel oil has a heat content of about 153,000 Btu/gal and weighs 8.053 to 8.448 lb/gal. We will use Number 2 fuel oil in all of our examples because it is the most common.

The American Society for Testing and Materials (ASTM) publishes standards for a number of industries, including the oil heat industry, and it has established minimum standards for the production and refinement of oil used for heating purposes. The fuel oil standards set acceptable ranges for fuel oil characteristics, which include

- flash point,
- ignition point,
- viscosity,
- carbon residue,
- water and sediment content,
- pour point,
- ash content, and
- distillation quality.

Figure 32.5 Chemical composition of some common hydrocarbons.

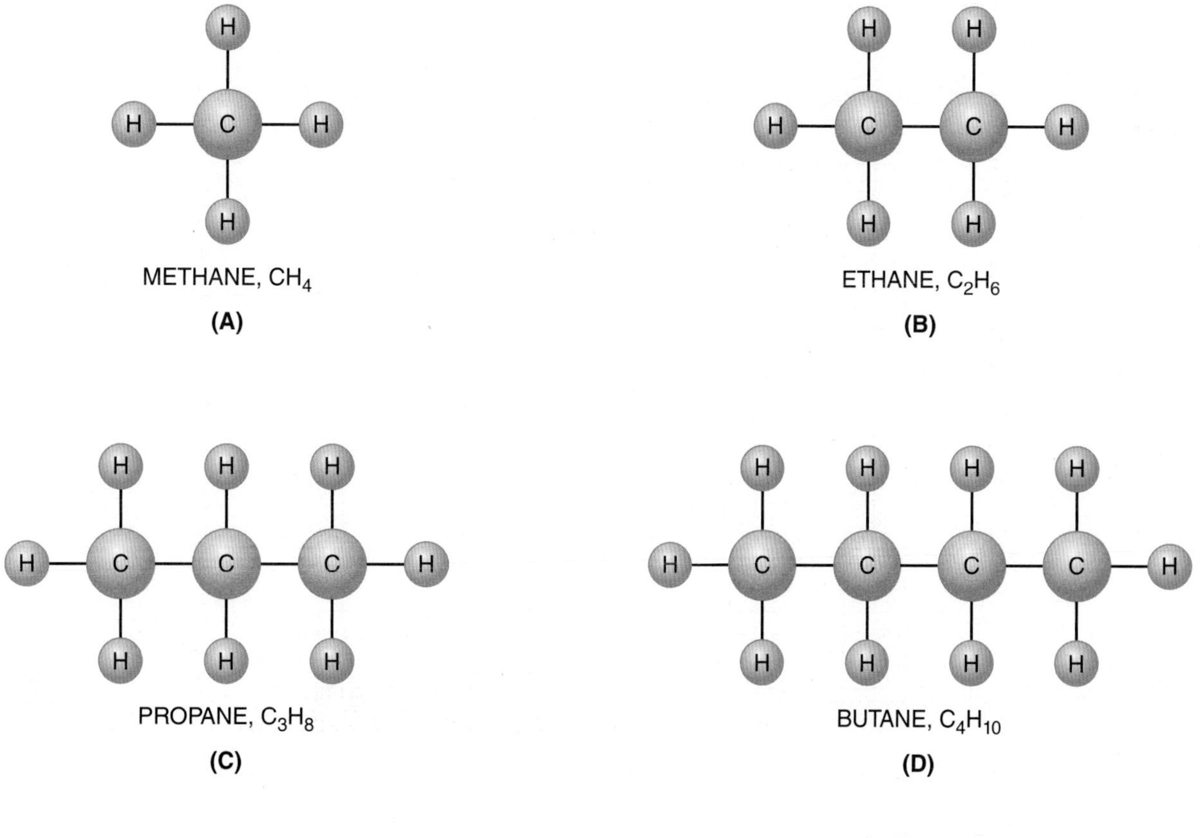

FLASH POINT

The *flash point* has to do with the maximum safe storage and handling temperature for the fuel. It is the lowest temperature at which vapors above the liquid fuel oil ignite for a short period of time when exposed to a flame. The lighter the oil, the lower the flash point. The minimum flash point for Number 2 oil is about 100°F.

IGNITION POINT

The *ignition point* is significantly higher than the flash point. The ignition point for Number 2 fuel oil is approximately 500°F, which is about 400 degrees higher than its flash point. Once the ignition point is reached, the fuel will continue to burn because vapors will continue to rise from the liquid.

VISCOSITY

Viscosity is the thickness of the oil under normal temperatures. This thickness can be compared with other oils and fluids. The heavier oils are thicker at the same temperatures. Viscosity is expressed in Saybolt Seconds Universal (SSU), which describes how much oil will drip through a calibrated hole at a certain temperature. As the temperature becomes colder, the same oil will have a reduced flow through the calibrated hole. **Figure 32.6** shows the viscosity of various oils. Warmer oil has a lower viscosity and is therefore desirable to ensure better flow through the oil lines as well as through the system filters, pumps, and nozzles.

CARBON RESIDUE

Carbon residue is the amount of carbon left in a sample of oil after it is changed from a liquid to a vapor by being boiled in an oxygen-free atmosphere. Properly burned oil does not have any appreciable carbon residue.

WATER AND SEDIMENT CONTENT

Ideally, the process of oil refining and delivery should occur in such a manner that no water or sediment is present in the oil once it arrives at its final destination. In the real world, however, this is impossible to expect. At the very least, a strong effort should be made to ensure that the water, sediment, and

Figure 32.6 This graph shows the viscosities of the different grades of fuel oil. Courtesy American Society of Heating, Refrigerating, and Air-Conditioning Engineers (ASHRAE)

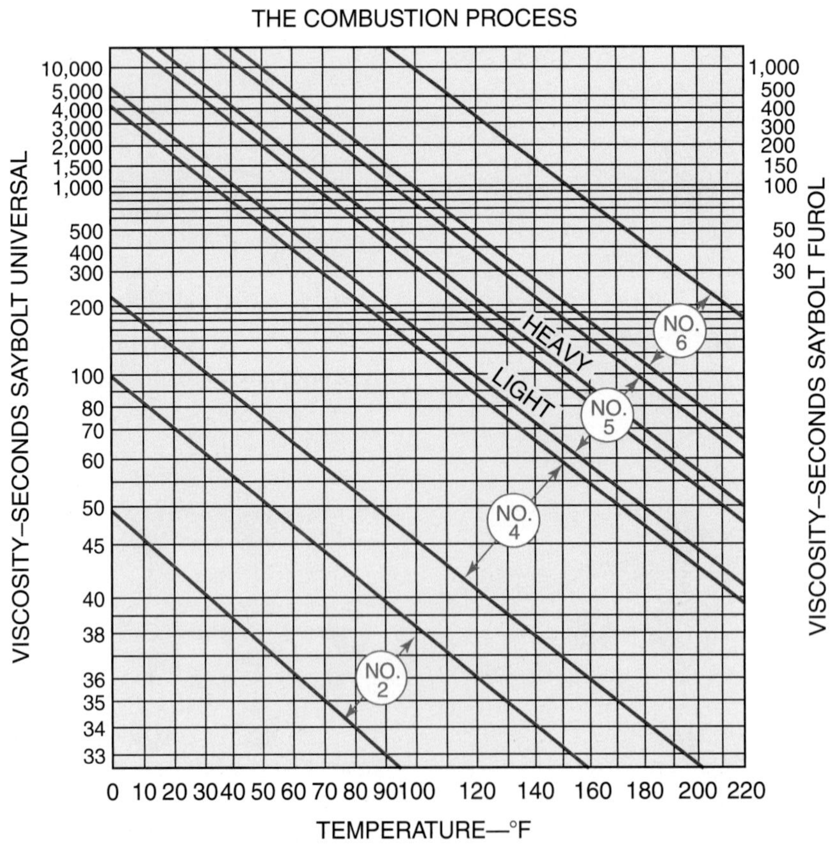

other contaminant levels are as low as possible under the given set of conditions.

Sediment can form as a result of rust forming on the internal pipe and tank surfaces, which may be made of iron or steel. In addition, sludge can form in the tank when the water, in the form of condensation, reacts with the fuel oil. Higher condensation concentrations are present when the oil tank is not filled to capacity and the bare steel walls are exposed to the moisture/humidity in the air above the oil level. Sludge and sediment can lead to clogged oil lines, nozzles, and filters. Oil impurities and fuel-line restrictions can cause poor flame characteristics and inefficient burning of the fuel. Following proper tank installation and maintenance procedures will help minimize fuel-related problems that occur after product delivery.

POUR POINT

Pour point is the lowest temperature at which the fuel can be stored and handled. Number 2 fuel oil is one of the lower pour-point fuels. The oil does not freeze, but there are waxes in the oil that become thick so that it will not flow correctly. Combustion will stop, and a no-heat call will result. Number 2 fuel oil can be successfully stored and used down to about 20°F. If the storage temperature is expected to be colder than 20°F, blending the fuel oil with 25% kerosene, known as fuel oil Number 1, will lower the pour point of the fuel. Other commercially available pour-point suppressants are available as well. Storing the oil in a basement or underground tank will often keep the oil warm enough for adequate flow. The heavier oils may have to be heated to facilitate flow when low temperatures are encountered.

ASH CONTENT

The *ash content* indicates the amount of noncombustible materials contained in the fuel oil. These materials pass through the flame without burning and are contaminants. They can also be abrasive and wear down burner components. The refinery is responsible for keeping the ash content within the required tolerances. Many states have inspection departments that regularly check oil that is distributed within the state.

DISTILLATION QUALITY

Oil must be turned into a vapor before it can be burned. *Distillation quality* describes the ability of the oil to become vaporized. The lighter oils turn into a vapor more easily than the heavier oils. However, there are many reasons to use the heavier oils. They are less expensive and have more heat content per gallon. Heavier oils are less volatile, do not ignite as easily, and are therefore safer in some situations, for example, on board a ship and in many large buildings and commercial establishments that must store the oil on the premises.

32.4 OIL STORAGE

Once oil is delivered to the customer, it must be stored until it is needed. For this reason, heating with fuel oil requires the installation of a storage tank on the property. Fuel oil storage tanks are available in a variety of shapes and sizes and can be installed either indoors or outside. Outdoor oil tanks can be installed either aboveground or underground. Whether an aboveground or an underground tank should be used is determined mainly by the ground conditions at a particular location.

When possible, tanks should be installed aboveground for environmental reasons. Aboveground tanks can be inspected more easily, and if a leak should be found, it can be repaired easily without major environmental impact. In addition, aboveground tank installations are less expensive than underground installations. When installing an oil tank, be sure to follow all local codes. Some of the codes, regulations, and organizations that govern the installation of oil tanks are the following:

- International Code Council (ICC) represents three of the legacy organizations that formed it, International Conference of Building Officials (ICBO) that also included the International Fire Code Institute (IFCI), Building Officials and Code Administrators International (BOCA) and Southern Building Code Congress International (SBCCI).
- National Fire Protection Association 30 (NFPA 30), the code that covers flammable and combustible liquids including those in aboveground oil tanks that contain more than 660 gallons
- National Fire Protection Association 31 (NFPA 31), the code that covers the installation of oil-burning equipment and aboveground and underground oil tanks that contain less than 660 gallons

Always check the local codes in your area, since they may differ from those set forth by these organizations. Also, be sure to comply with the clearances required between oil storage tanks and property lines, electric meters, combustion sources, open flames, windows, doors, air intakes and other relevant property elements. And when replacing older furnaces and boilers, consider the age of the oil tank as well. Old oil tanks should be replaced. You wouldn't take the 30-year-old gas tank from your old clunker and put it in your brand-new car, would you?

SIZING OIL TANKS

Oil storage tanks are available in a wide range of sizes, from about 100 gallons to over 1000 gallons, and selecting the right size tank for a particular job or application can be a confusing process. A delicate balance of concerns need to be

addressed when selecting the proper tank size. Two of the main issues are the frequency of oil deliveries and the quality of oil with respect to the storage time.

Larger oil tanks obviously hold more oil, so the oil company will have to make fewer deliveries to structures that have larger oil tanks. Although this may seem to be a benefit, it has a major drawback. Larger amounts of oil will sit in the tank longer and can degrade. Moisture, heat, and oxygen all contribute to the formation of sludge in the tank, which ultimately clogs the filters, nozzles, piping, and strainers on the equipment. So, in order to benefit from a larger tank without sacrificing the quality or integrity of the oil, the oil heat industry has recommended that the tank be sized to hold approximately one-third of the annual oil consumption. For example, if a house is expected to use 1000 gallons of oil a year, the tank should be sized to hold about 333 gallons. So, in this case, the oil tank used should be sized to hold 275 or 330 gallons.

UNDERGROUND TANKS

Tanks that are to be installed underground must be protected against corrosion. The two most popular types of tanks for in-ground residential fuel storage are the STI-P3 tanks and fiberglass tanks. Fiberglass tanks, **Figure 32.7**, are not subject to corrosion, but steel tanks are. What follows is a brief description of the steel tank and the methods by which this type of tank is protected from corrosion.

The acronym STI-P3 stands for "Steel Tank Institute—Protected 3 Ways." These steel tanks, **Figure 32.8**, are manufactured with an external corrosion-resistant coating that prevents the rapid deterioration of the base metal. In addition, these tanks also have dielectric bushings and sacrificial anodes. Dielectric bushings provide a noncorrosive barrier at the point where dissimilar metals are joined. It is typically at this point in any piping circuit that oxidation and corrosion take place. The sacrificial anodes are rods usually manufactured from either zinc or magnesium. The rods are secured to the base metal of the tank and serve to protect it from corrosion. The anode material oxidizes and corrodes before the base metal does and since the anode rods are in contact with the steel tank, the steel will not corrode. However, once the anodes have completely corroded, the steel tank will begin to corrode. Therefore, it is beneficial to replace the anodes prior to this point. Field technicians can perform tests to determine the condition of the anodes even though the tank is buried in the ground. Replacing the anodes, however, involves exposing the tank, which can be a very complicated process. To illustrate this, refer to the two angled "bags" at the bottom of the underground tank in **Figure 32.8**. These are the anodes. Accessing them once the tank has been installed requires a great deal of digging.

Figure 32.7 An underground fiberglass oil storage tank. Courtesy Oilheat Associates, Inc. Photo by John Levey

Figure 32.8 An STI-P3 steel oil storage tank. Courtesy Oilheat Associates, Inc. Photo by John Levey

ABOVEGROUND TANKS

For many years, aboveground oil storage tanks were constructed of either 12- or 14-gauge steel. Over the past decade, technological advances have paved the way for a new generation of tanks that are made of steel, fiberglass, or a combination of steel and plastic. Modern, residential aboveground tanks are typically fabricated from 12-gauge steel and hold between 275 and 330 gallons of oil. Smaller tanks are available and some applications call for the twinning, or joining, of smaller tanks to create larger storage capacity. The capacities of residential fiberglass tanks typically range from 240 to 300 gallons. Polyethylene/steel tanks are constructed as a tank within a tank, **Figure 32.9**. The inner oil-holding tank is made of polyethylene and the outer galvanized-steel tank is intended to protect the inner tank. The outer tank provides secondary containment in the event that the inner tank develops a leak.

Figure 32.9 A polyethylene/steel oil storage tank. Photo by Eugene Silberstein

INDOOR OIL STORAGE

When oil tanks are located outside, the temperature of the fuel oil will change as the outside ambient temperature changes. This can lead to oil flow issues as well as condensation problems. When the option presents itself, indoor oil storage is preferable. By storing the oil inside, there will be less temperature fluctuation of the fuel and there will also be less moisture and condensation on the interior surfaces of the tank. Condensation in the tank increases the rate of sludge formation, leading to tank corrosion. In addition, warmer fuel oil has better flow characteristics and helps the oil burner provide more even combustion and more even heating.

OIL TANK INSPECTION AND MAINTENANCE

For the most part, oil storage tanks do not require much maintenance. What follows in this section are some key points and suggestions that should be followed to help maintain both the integrity of the tank itself and the valuable oil contained within it.

- Aboveground tanks should be painted periodically and inspected on a regular basis. Deficiencies should be addressed promptly to reduce the possibility of premature tank failure.

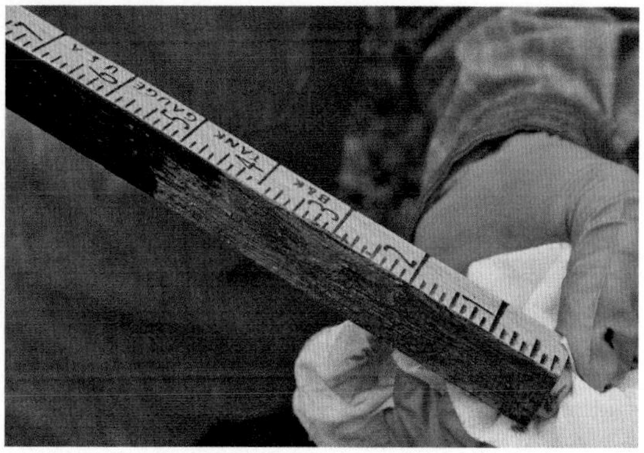

Figure 32.10 Water-sensing paste changes color when water is present in an oil tank. Courtesy Oilheat Associates, Inc. Photo by John Levey

- STI-P3 tanks should be checked periodically to ensure that the protective anodes are still in good working order. Remember that these anodes are *designed* to deteriorate in order to protect the integrity of the tank itself.
- Tank piping should be inspected regularly for imperfections and leaks. Always check for loose or missing vent and fill caps.
- Tanks should be checked for water at least once a year. If water is present in the tank, the source should be identified and the situation corrected. Tanks are tested for water content by placing a water-sensing paste on the oil tank stick. In the presence of water, the paste changes color, **Figure 32.10**. There are different brands of water-sensing pastes on the market and the initial and final colors can vary from one product to another. The important thing is to pay attention to the color change, not the color.
- Visually inspect aboveground tanks for damage to the fill and vent pipes, caps, oil lines, tank legs, and oil tank gauges. Do not fill a tank that has any deficiencies.
- Check aboveground tanks for traces of oil under the tank. Check the tank for any corrosion if oil is found. Also check the drain on the tank for potential leaks.

32.5 FUEL OIL SUPPLY SYSTEMS

In order to get the fuel oil from the storage location to the oil burner itself, oil lines must be run between these two components. Two common piping configurations are the one-pipe system and the two-pipe system. In a one-pipe system, **Figure 32.11**, only one pipe runs between the oil storage tank and the oil burner. As oil is needed, it flows from the tank and through the oil line to the burner—where it is ultimately atomized, mixed with air, and ignited. Since there is only one pipe, there is a lower potential for oil line leaks.

In a two-pipe setup, **Figure 32.12**, one pipe is the supply and the other acts as the return. Oil flows at maximum

Figure 32.11 A one-pipe system connecting the oil tank to the burner.

ITEM	DESCRIPTION
1	OIL BURNER
2	ANTIPULSATION LOOP
3	OIL FILTER
4	SHUTOFF VALVE (FUSIBLE)
5	INSTRUCTION CARD
6	OIL GAUGE
7	VENT ALARM
8	VENT
9	FILL LINE COVER

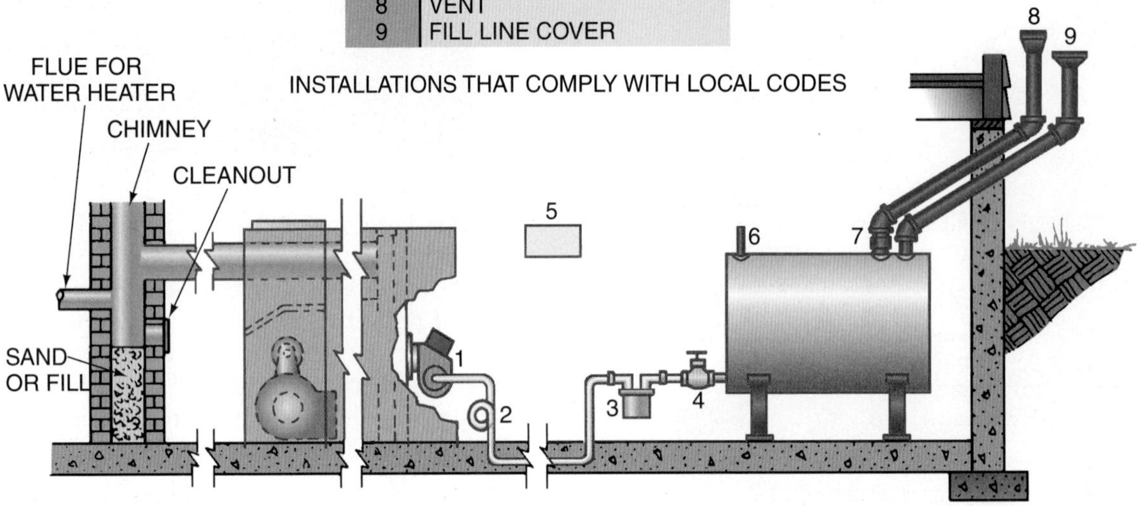

Figure 32.12 A two-pipe system connecting the oil tank to the burner.

ITEM	DESCRIPTION
1	OIL BURNER
2	OIL FILTER
3	SHUTOFF VALVE (FUSIBLE)
4	SHUTOFF VALVE (NOT FUSIBLE)
5	ANTIPULSATION LOOP
6	INSTRUCTION CARD
7	OIL GAUGE
8	VENT
9	SWING JOINT
10	FLUSH BOX WITH COVER
11	SLIP FITTING
12	VENT ALARM
13	SUPPLY LINE
14	RETURN LINE

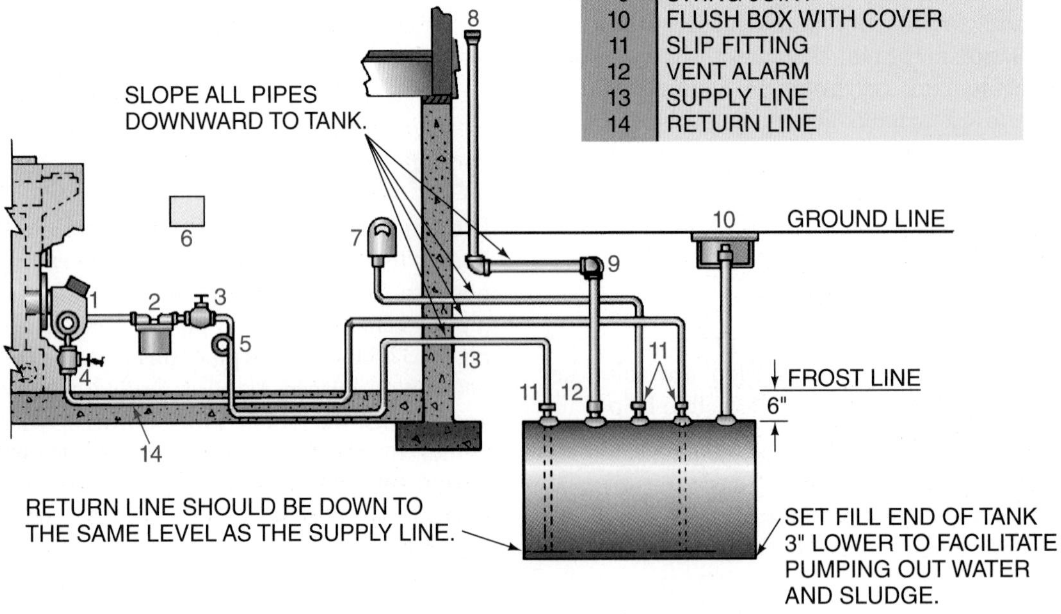

pump capacity from the tank to the burner, and whatever oil the system needs for combustion, it takes. Any oil that is not needed is pumped back to the tank via the return line.

The location of the fuel storage tank in relation to the burner determines, for the most part, whether a one-pipe or a two-pipe system will be used. If the oil tank is located above the oil burner, a one-pipe system may be used. The oil will flow by gravity from the tank, through a fuel filter, **Figure 32.11**, and then on to the burner itself. When the oil storage tank is located below the oil burner, a two-pipe system is sometimes used.

Whenever oil is supplied to the heating appliance by means of a one-pipe system, it is important that all of the air in the connecting fuel line be bled out to ensure proper fuel flow. Bleeding the line must also be done if the oil tank has been allowed to run dry. Air bleeding is not an issue with a two-pipe system, as air is eliminated automatically through the return pipe. Here are some key points to keep in mind regarding fuel oil lines and other tank-related piping:

- Oil lines are typically made of copper and should be sized at a minimum of 3/8 in. OD.
- Never use PVC pipe for oil lines.
- Use only flare connections to join copper oil lines, **Figure 32.13**. Do not use compression fittings.
- When needed, use only a nonhardening, oil-resistant pipe joint compound.
- When installing oil lines, make the runs as short as possible, using as few fittings as possible. Avoid kinking the lines.
- Be sure that all oil lines are secured and otherwise protected from damage.
- When oil lines are run outside, make certain that they are well insulated.
- Underground oil lines should be installed with secondary containment. Oil lines that come in contact with the earth or concrete must be protected, **Figure 32.13**.

Figure 32.13 A sleeve over oil lines and flare connections. Photo by Eugene Silberstein

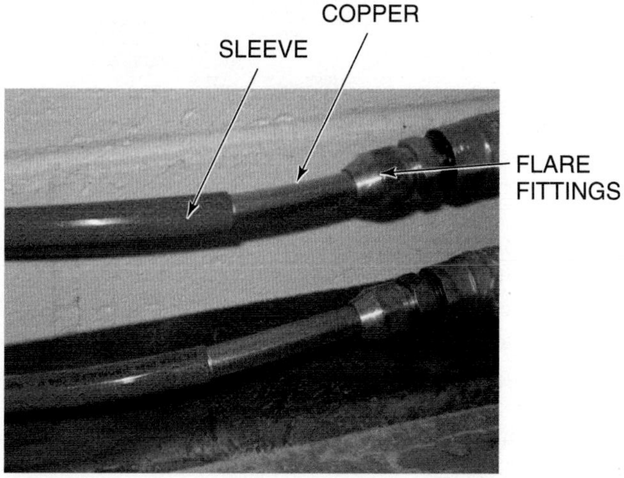

Figure 32.14 A vent alarm. Photo by Eugene Silberstein

- There should be a shutoff valve on the suction line at the oil burner.
- Fill and vent pipes should be pitched toward the tank opening to prevent the formation of oil traps.
- A vent alarm, **Figure 32.14**, should be installed in all tanks at the vent pipe. While the oil tank is being filled, the vent alarm will whistle. As the tank nears capacity, the whistling sound will stop, alerting the oil delivery person that the tank is nearly full. The National Oilheat Research Alliance (NORA) recommends a "No whistle, no fill" policy. If there is no whistle from the alarm, it is possible that the tank has been removed but the fill line is still in place. Following this policy will prevent filling a structure with oil.

OIL DEAERATORS

The oil deaerator, **Figure 32.15**, is an oil piping component that allows a one-pipe system to function as a two-pipe system. This is accomplished by connecting two pipes between the deaerator and the fuel pump on the oil burner, but only one pipe between the deaerator and the oil storage tank. This helps eliminate the air problems that are typically found in one-pipe oil supply systems.

In operation, the fuel travels from the oil tank, through the oil filter, and into the deaerator. It is in the deaerator that any air that is mixed with the oil is separated from the fuel. The fuel then passes on to the fuel pump. At the fuel pump, the desired amount of oil will be pushed through the burner for combustion, while the remaining oil is pumped back to the deaerator instead of to the oil storage tank, **Figure 32.16**.

Figure 32.15 An oil deaerator.

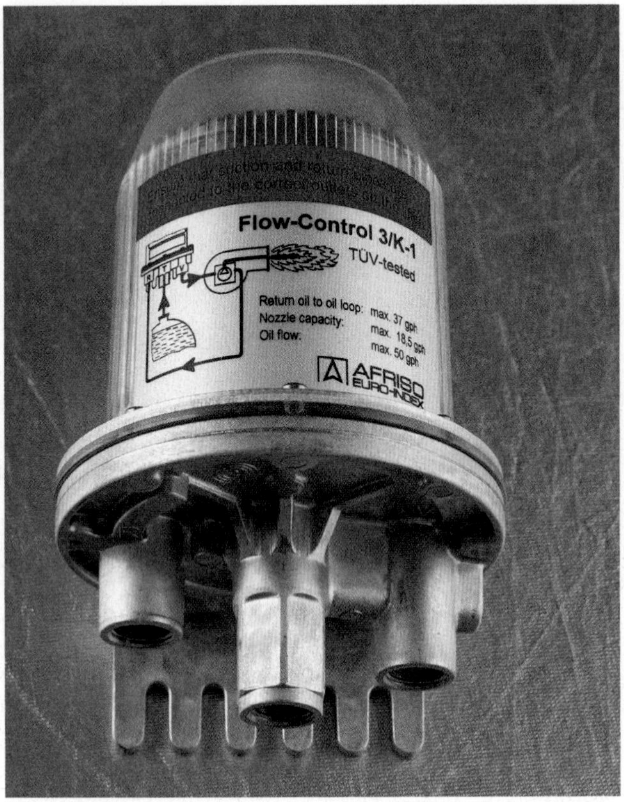

Figure 32.16 Piping diagram of the oil deaerator.

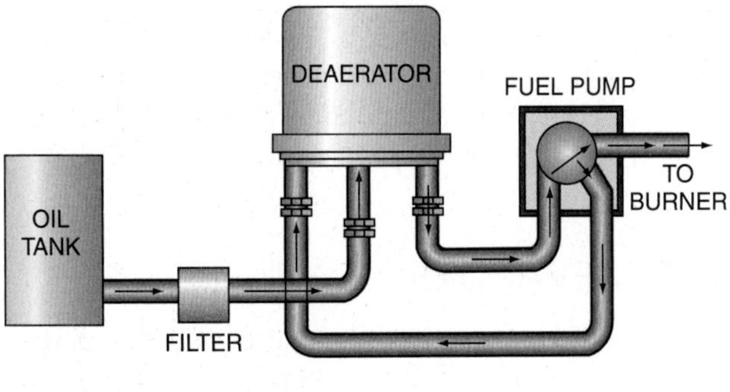

AUXILIARY FUEL SUPPLY SYSTEMS

In some installations, oil burner units will be located above the fuel oil supply tank at a height that will be beyond the lift capabilities of the burner pump. This occurs primarily in commercial applications. Heights exceeding 15 ft from the bottom of the tank are beyond the capabilities of even the two-stage systems. In such cases, an auxiliary fuel system must be used to get the fuel supply to the burner.

A fuel oil booster pump is used to pump the oil to an accumulator or reservoir tank, **Figure 32.17**. The booster pump, wired separately from the burner unit, pumps the fuel oil through its own piping system. The accumulator tank is kept full, and check valves are used to maintain the prime of the booster pump when it is not operating. An adjustable pressure regulator valve is a part of the fuel oil booster pump, and it must be set to maintain a fuel oil pressure of 5 psi, measured at the accumulator tank. Failure to properly adjust this regulator could result in seal damage to the fuel unit.

Figure 32.17 (A) A diagram of a system using an auxiliary booster pump and reservoir tank. (B) A booster pump. Courtesy Suntec Industries Inc., Rockford, Illinois

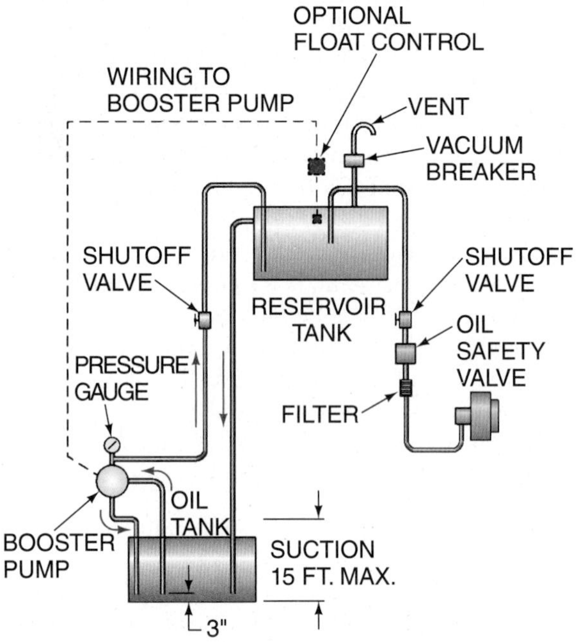

(A)

(B)

FUEL-LINE FILTERS

A filter should be located between the tank and the pump to remove any fine, solid impurities from the fuel oil before it reaches the pump, **Figure 32.18.** Filters may also be located in the fuel line at the pump outlet to further reduce the impurities that may reach the tiny oil passages in the nozzle, **Figure 32.19.** Different types of filter media are available. **Figure 32.20** shows a cutaway view of one type of oil filter.

NOTE: *Be sure to replace the gaskets on the fuel filter when changing the filter cartridge,* **Figure 32.21.** •

In many cases, systems are equipped with dual in-line oil filtration systems, **Figure 32.22.** This helps ensure that the oil

Figure 32.18 An oil filter.

Figure 32.19 (A) A fuel line filter. Courtesy Delavan Corp. (B) Location of the fuel line filter.

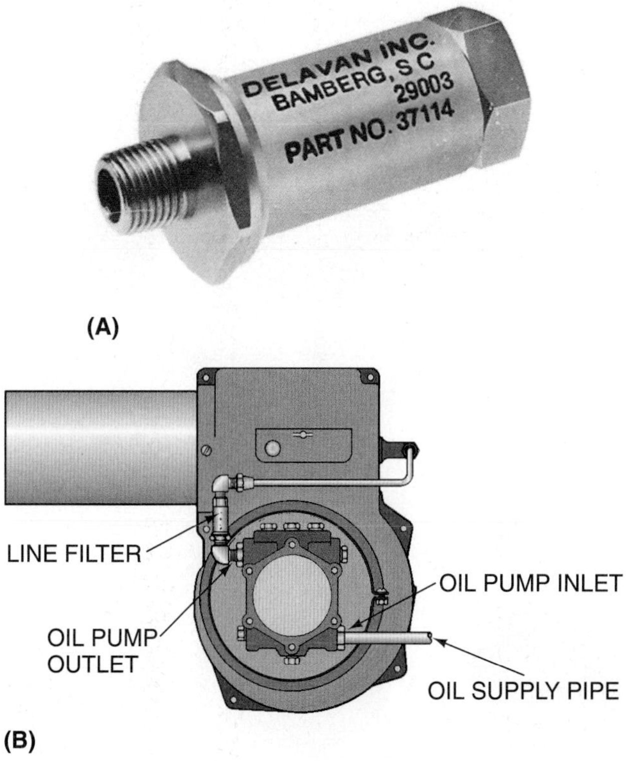

Figure 32.20 Cutaway view of an oil filter. Courtesy Oilheat Associates, Inc. Photo by John Levey

Figure 32.21 Oil filter media and gaskets. Courtesy Oilheat Associates, Inc. Photo by John Levey

Figure 32.22 Dual in-line oil filtration. Courtesy Oilheat Associates, Inc. Photo by John Levey

reaching the appliance is as clean as possible. Make certain that all filters are installed with the arrow on the device pointing in the direction of fuel flow, **Figure 32.23**. Sometimes a sludge/water isolator is installed in the fuel line as well. This device is located before the filter and is nothing more than an empty canister, **Figure 32.24**, that has a drain plug at the bottom. The fuel oil enters from the side and leaves from the top. Since water and sediment are heavier than the fuel oil, these substances fall to the bottom of the canister. Periodically, the drain plug is removed and the water and sediment are removed.

OIL SAFETY VALVES (OSV)

Another system component commonly found in the piping arrangement of an oil-fired heating appliance is the oil safety valve. Located in the suction line, the oil safety valve (OSV) is a device that stops oil flow in the event of a leak in the suction line, which is the line that carries oil from the tank to the oil burner. When the pump on the oil burner operates, a vacuum is created in the oil line, causing oil to flow from the tank to the pump. The OSV senses this vacuum and opens to allow oil flow. In the event of a suction-line leak, the vacuum will be lost and the valve will close.

Figure 32.23 Make certain that the arrows on oil filters point in the direction of oil flow. Courtesy Oilheat Associates, Inc. Photo by John Levey

Figure 32.24 Diagram of a sludge/water isolator.

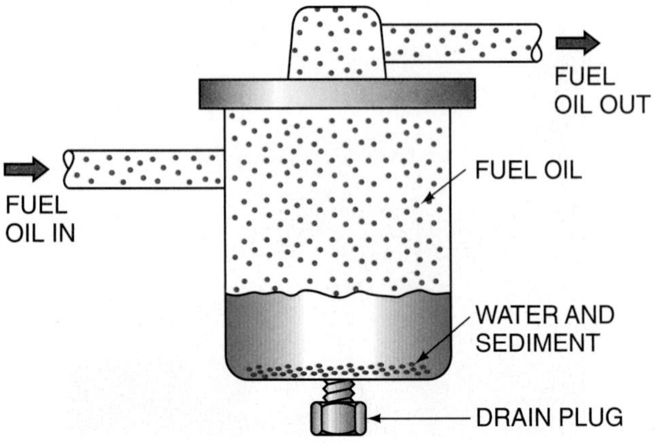

32.6 COMBUSTION

Combustion is the burning of fuel to generate heat. The combustion process is the same regardless of the fuel being burned. In order for combustion to take place, there must be ample supplies of fuel, heat, and oxygen. If any of these three components is missing, the process cannot occur. If, during combustion, the supply of one or more of these three components is depleted, the process will stop.

During combustion, the oxygen reacts with fuel oil to produce heat, carbon dioxide, and water vapor, **Figure 32.25**. If only these three by-products are produced, perfect combustion will have occurred. Perfect combustion exists only in theory—achieving the delicate balance required among the components is not possible in the real world. When combustion is incomplete, carbon monoxide, soot or free carbon, and other undesirable substances form in addition to the heat, carbon dioxide, and water vapor, **Figure 32.26**. Carbon monoxide is a colorless and odorless gas. Even in small quantities, it is highly toxic and dangerous. Fortunately, with the proper use of combustion analysis equipment, **Figure 32.27**, the carbon monoxide level can be measured. In addition, combustion testing provides important data to aid in troubleshooting.

Figure 32.25 Diagram representing complete combustion.

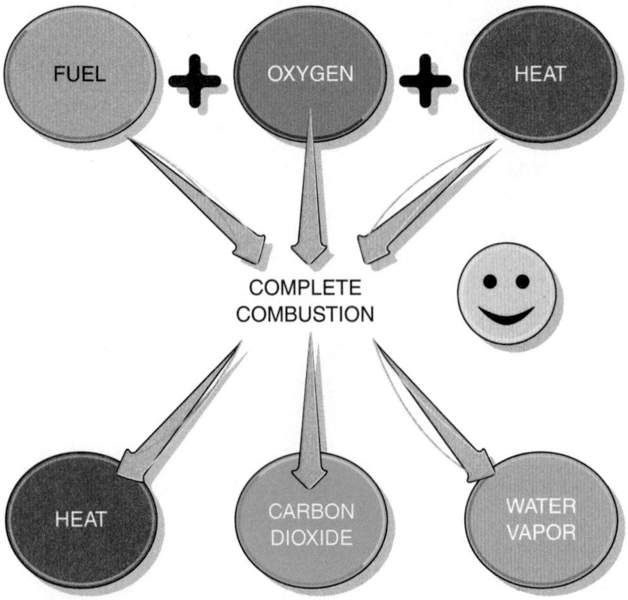

Figure 32.26 Diagram representing incomplete combustion.

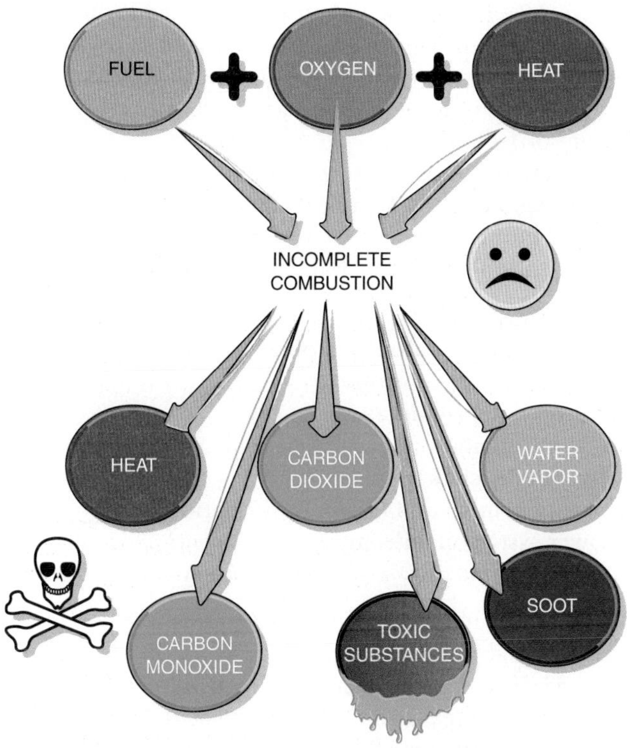

Figure 32.27 A combustion test kit. Courtesy Bacharach

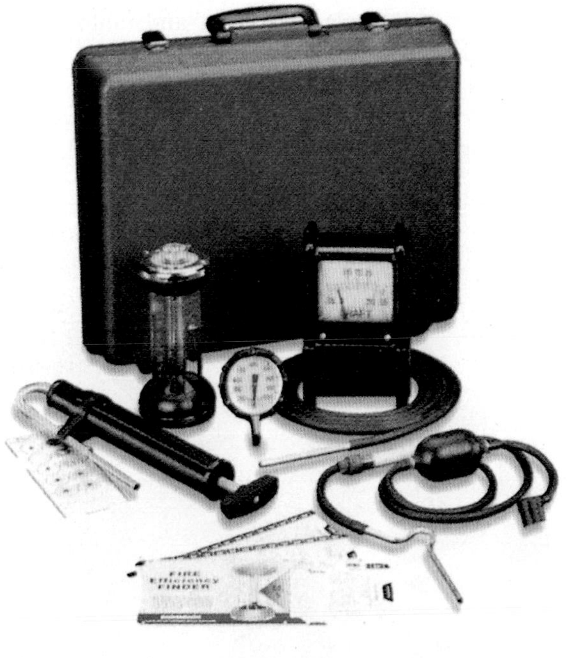

When combustion takes place in the presence of nitrogen, nitric oxide (NO) and nitrogen dioxide (NO_2) can form. These are classified as mono-nitrogen oxides and are collectively referred to as NO_X gases. These gases are pollutants and can react to form smog (tropospheric ozone) and, in some cases, toxic compounds. If there is significant exposure to NO_X gases, individuals can experience reduced lung function that, in the case of those who already suffer from asthma or other breathing ailments, can lead to death.

32.7 PREPARATION OF FUEL OIL FOR COMBUSTION

Fuel oil must be prepared for combustion. It must first be converted to a gaseous state by forcing it, under pressure, through a nozzle. The nozzle breaks the fuel oil up into tiny droplets, a process called **atomization**, **Figure 32.28**. The oil droplets are then mixed with air, which contains oxygen. The lighter hydrocarbons form envelopes of gas around the droplets. Heat is introduced at this point with a spark. The vapor ignites (combustion takes place) and the temperature rises, causing the droplets to vaporize and burn.

High-pressure gun-type oil burners are generally used to achieve combustion. Oil is fed under pressure to a nozzle, and air is forced through a tube that surrounds this nozzle. Usually the air is swirled in one direction and the oil in the opposite direction. The oil forms into tiny droplets and combines with the air. The ignition transformer or ignitor sends a high-voltage spark between two electrodes located near the front of the nozzle, and combustion occurs.

Figure 32.28 Atomized fuel oil droplets and air leaving the burner nozzle. Courtesy Delavan Corp.

32.8 BY-PRODUCTS OF COMBUSTION

The correct ratio of fuel oil to air must be maintained for efficient combustion. The fuel to be burned is made of carbon and hydrogen. To burn 1 lb of fuel oil, it must be mixed with about 3 lb of oxygen for complete combustion to take place. Since air is made up of approximately 79% nitrogen and 21% oxygen, 14.4 lb of air would be needed to give 3 lb of oxygen (3 lb of oxygen/0.21 lb of oxygen per lb of air = 14.4 lb of air). Since 1 lb of air takes up 13.33 ft^3 at standard atmospheric conditions, about 192 ft^3 (13.33 ft^3 of air per pound × 14.4 lb of air = 192 ft^3) of air would be needed to burn a single pound of fuel oil. This is the correct amount of air needed for perfect and complete combustion, but, as with fuel gases, it would be dangerous to set up a burner for exactly the correct amount of air. If the air were restricted for some reason, incomplete combustion would occur.

Practical combustion requires excess air to ensure that all of the carbon and hydrogen particles come in contact with enough oxygen for complete combustion. The air supplied to an oil burner is normally in excess of the exact amount needed. Most oil burners are set up to burn with approximately 50% excess air. When 1 lb of oil with 50% excess air is burned, 21.6 lb, or 288 ft^3, of air must be furnished (21.6 lb × 13.33 ft^3/lb of standard air = 287.9 ft^3 of air). One gallon of Number 2 fuel oil weighs approximately 7 lb, so 1 lb of oil is one-seventh of a gallon; this is 0.88 pint, or nearly the contents of a soda can. **Figure 32.28** is a picture of the process of atomizing oil to mix with air.

When the correct amounts of oil and air are mixed and the mixture is ignited, heat is given off by the appliance for use in heating. The typical oil-burning appliance produces nearly 85% of usable heat energy, with approximately 15% escaping with the flue gases. Since a gallon of fuel oil contains about 140,000 Btu of heat, 119,000 Btu are used for heating and the remaining 21,000 Btu push the by-products of combustion up the flue (140,000 × 0.85 = 119,000 and 140,000 × 0.15 = 21,000).

The by-products of combustion result from the chemical reaction during which the stored energy in the fuel is released as heat. The nitrogen in the air just heated and passes through the combustion process. Since air contains about 79% nitrogen, 227.5 ft^3 of heated nitrogen rises up the flue (288 × 0.79 = 227.5 ft^3 of nitrogen).The oxygen in the excess air also just passes through the combustion process and is not combined with the carbon and hydrogen particles of the fuel oil and burned. Of the 288 ft^3 of air, the combustion process uses only 192 ft^3, so 96 ft^3 pass through and are only heated. This air contains 21%, or 20 ft^3, of oxygen (96 × 0.21 = 20 ft^3). The other by-products of combustion are CO_2 and water vapor. The flue gases contain 27 ft^3 of CO_2 and more than a pound of water vapor per pound of fuel. The flue gases must remain hot enough both to prevent the water vapor from condensing and to rise up the flue to the outside.

32.9 GUN-TYPE OIL BURNERS

In a *gun-type* burner, oil and air are prepared for burning by being forced into the burner head for mixing and ignition. Modern oil burners are high-pressure burners, **Figure 32.29**, because the oil is delivered through the burner at a minimum pressure of 100 psi.

All new oil burners utilize a high-speed flame-retention design. This design is a radical improvement over the conventional burner design of a few decades ago. The benefits of flame-retention burners include the following:

- Faster motor speeds (3450 rpm versus 1725 rpm) for better mixing of the atomized oil and the combustion air
- Higher air static pressure
- Improved air-oil mixing
- Lower emissions
- Higher efficiency
- Cleaner combustion (less soot formation)
- Lower system maintenance requirements
- A hotter flame (300°F to 500°F hotter)

As part of the improved burner design, the air-oil mixture is swirled at high pressure to produce a tightly formed flame. **Figure 32.30** compares a conventional burner flame and a burner with a flame-retention design. The flame created by a conventional oil burner is erratic, so there are large temperature fluctuations in the flame and it is common for unburned fuel to remain behind in the combustion area. The flame of the flame-retention burner is tighter and has a more uniform temperature. Since the flame is tight and uniform, there is substantially less unburned fuel left behind.

The main parts of a gun-type oil burner are the burner motor, blower or fan wheel, pump, nozzle, air tube, electrodes, transformer/igniter, and primary control, **Figure 32.31**.

Figure 32.29 High-pressure oil burner. Courtesy R. W. Beckett Corp.

Figure 32.30 Diagram that compares the flames created by conventional and flame-retention oil burners.

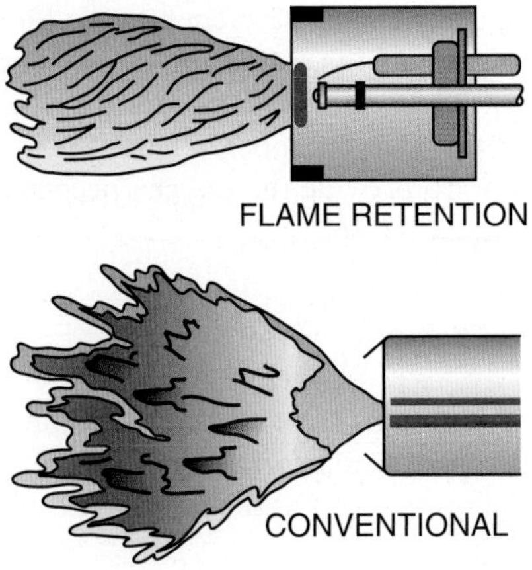

BURNER MOTOR

The oil burner *motor,* **Figure 32.32**, provides power for both the fan and the fuel pump. A flexible coupling, **Figure 32.33**, is used to connect the shaft of the motor to the shaft of the pump, **Figure 32.34**. The motor speeds may be either 1750 or 3450 rpm. The fuel pump rpm should always match the motor rpm. Most flame-retention burners have motors that operate at 3450 rpm; most conventional burners are equipped with motors that operate at 1725 rpm.

The most common type of motor in modern oil burners is the permanent split-capacitor (PSC) motor. The PSC motor is manufactured and designed to operate with a start winding, a run winding, and a run capacitor. By design, both the start

Figure 32.32 An oil burner motor. Courtesy R. W. Beckett Corp.

Figure 32.33 Flexible couplers.

Figure 32.31 Exploded view of an oil burner assembly.

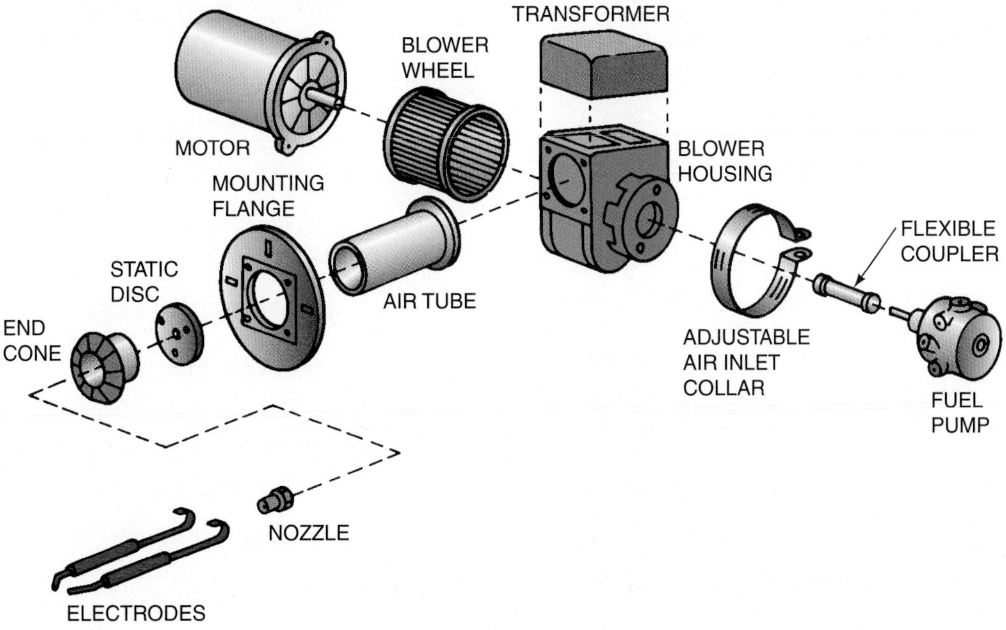

and run windings are energized whenever the motor is operating. These motors have relatively low starting torque but, since they are operating under a relatively low load, they are perfect for this application. The main benefits of this type of motor are that they have excellent running efficiency and have no external or mechanical control relays or contacts to wear out. The wiring diagram for a simple, single-speed PSC motor is shown in **Figure 32.35**. Refer to Unit 17, "Types of Electric Motors," for more information regarding electric motors.

Figure 32.34 Oil burner motor, pump, connector, and blower assembly.

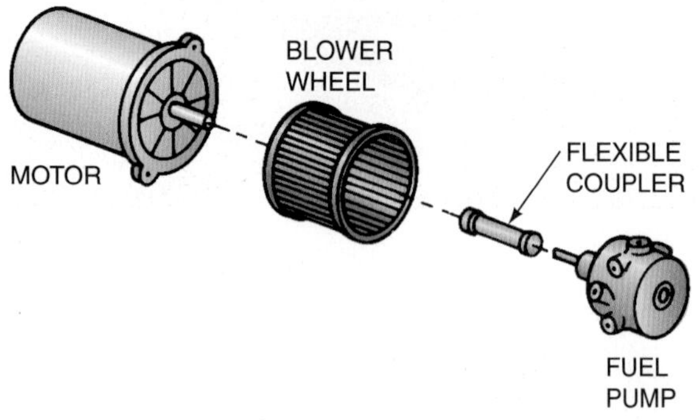

Figure 32.35 Simplified wiring diagram of a single-speed permanent split-capacitor (PSC) motor.

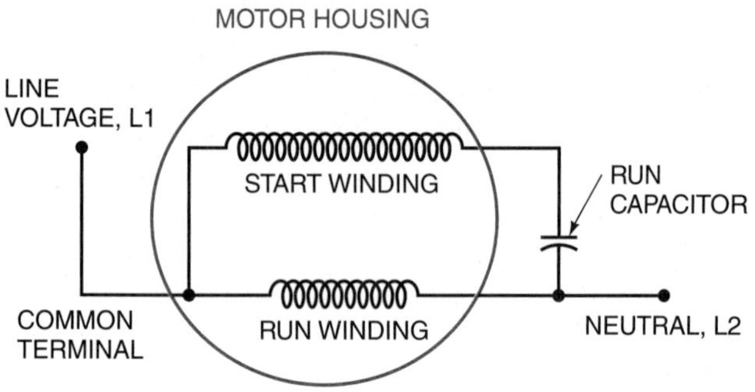

BURNER BLOWER

The burner blower is a squirrel-cage design with adjustable air inlet openings in a collar attached to the blower housing. These provide a means for regulating the volume of air being drawn into the blower. The blower forces air through the air tube to the combustion chamber, where it is mixed with the atomized fuel oil to provide the oxygen necessary to support combustion, **Figure 32.36**.

FUEL OIL PUMPS

Just like the burner fan or blower, the fuel pump, **Figure 32.37**, is connected to the shaft of the burner motor and turns at the same speed as the motor. The pump performs

Figure 32.37 An oil pump. Courtesy R. W. Beckett Corp.

Figure 32.36 Top-view diagram of a typical gun-type oil burner. Courtesy Honeywell

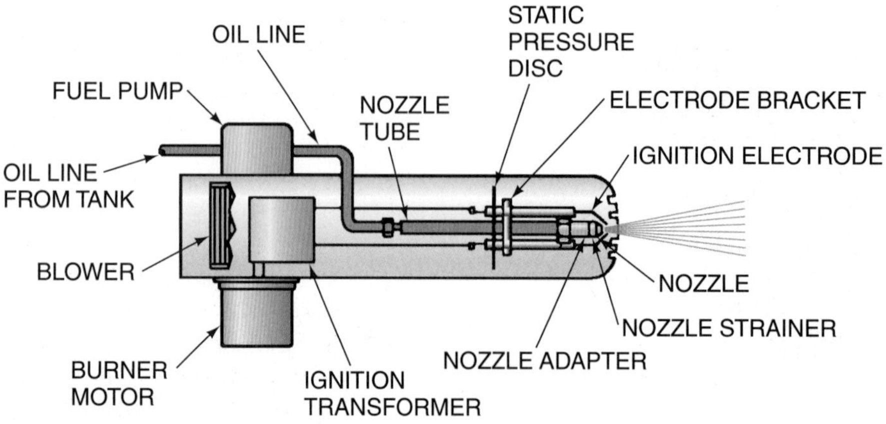

three major functions in the system. First, it moves the fuel oil from the storage tank to the burner for ignition. Second, it filters or pulverizes particulate matter that may have found its way past the filter. Third, the pump regulates the pressure at which the oil is fed through the burner into the combustion area. Oil pumps are typically set at the factory to deliver oil at a minimum pressure of 100 psig. Nozzles are rated at 100 psig but, most of the time, have oil at a higher pressure flowing through them.

Two types of oil pumps are mainly used on gun-type burners: the single-stage pump and the two-stage pump, **Figure 32.38**. Fuel pumps are manufactured with gear sets that facilitate the movement of oil through the system, **Figure 32.39**. Pumps that have one set of gears to pull oil from the storage tank are referred to as one-stage, or single-stage, pumps, **Figure 32.40**. Single-stage pumps can be used in conjunction with one-pipe or two-pipe systems so long as the vacuum necessary for operation does not exceed 12 in. Two-stage pumps, **Figure 32.41**, have two sets of gears: one set to pull oil from the oil storage tank and one set to push oil onto the nozzle. The additional gear set, called the suction pump gear, allows for a greater vacuum to lift the oil farther than a single-stage gear set can. Two-stage pumps can operate at vacuum levels as high as 17 in. Two-stage pumps are highly recommended for high-lift applications. Refer to the next section for more information regarding pump pressures and vacuums.

Figure 32.38 (A) A single-stage fuel oil pump. (B) A two-stage fuel oil pump. (A) Courtesy Webster Electric Company. (B) Courtesy Suntec Industries Inc., Rockford, Illinois

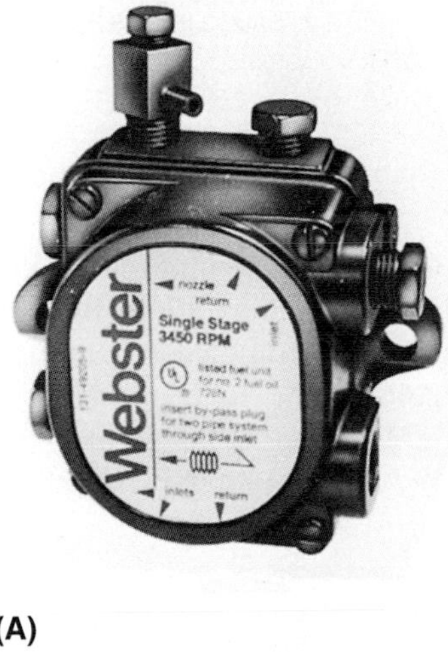

(A)

(B)

Figure 32.39 Gear set on an oil pump. Courtesy Oilheat Associates, Inc. Photo by John Levey

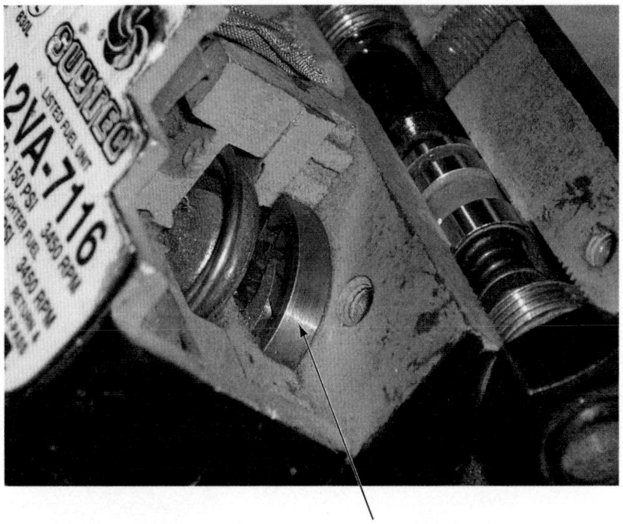

Oil pumps are equipped with a bypass plug, **Figure 32.42**. This plug will either be left in place or removed, depending on the type of piping configuration used. For a one-pipe installation, **Figure 32.11**, the bypass plug needs to be removed so that any unused oil can be recirculated back to the inlet of the pump. In a two-pipe configuration, **Figure 32.12**, or in a system piped with a deaerator, the bypass plug is left in place.

Built into each pump is a pressure-regulating valve, **Figure 32.40**. The pump provides more pressure than needed. The pressure-regulating valve can be set so that the fuel oil being delivered to the nozzle is under a specific pressure. Although nozzles are rated at 100 psig, most of today's burners operate with oil pressures of 140 psig and above. Higher nozzle pressures yield more oil flow in a nozzle. The chart in **Figure 32.43** shows nozzle capacities at various pressures. As mentioned previously, the rate of fuel flow through the oil pump is greater than the flow of fuel through the nozzle line and nozzle. The excess oil is diverted back to the inlet side of the pump in one-pipe systems and to the storage tank in two-pipe systems. **NOTE:** *Fuel pumps are equipped with screens at the inlet,* **Figure 32.44**, *to prevent particulate matter from entering the pump. Before replacing what is thought to be a defective pump, be sure to check this screen. If the screen is indeed clogged, replace it—as well as the gasket,* **Figure 32.45.** •

Figure 32.40 A single-stage fuel oil pump showing the flow of oil. Courtesy Suntec Industries Inc., Rockford, Illinois

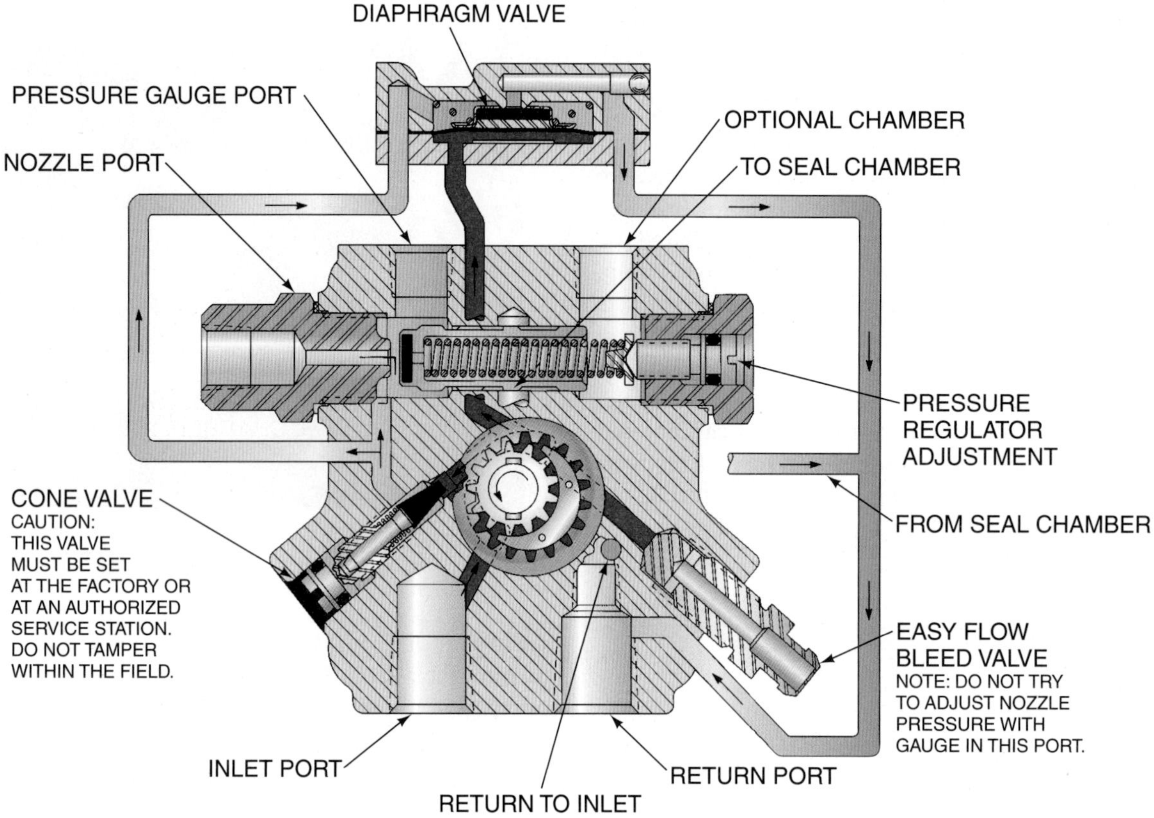

PUMP PRESSURES. Oil pumps generate a vacuum during the process of pulling oil from the storage tank. This vacuum is measured in inches of mercury. Field technicians can estimate the expected vacuum by using the following guidelines:

- Estimate 1 in. of vacuum for each foot of vertical lift between the oil tank and the oil burner.
- Estimate 1 in. of vacuum for every 10 ft of horizontal run between the oil tank and the oil burner.
- Estimate 1 in. of vacuum for each oil filter.
- Estimate 3 in. of vacuum for an oil safety valve (OSV).

So, a system that has the oil burner 5 ft above the tank at a distance of 40 ft will have an approximate vacuum at the oil burner of 9 in. of mercury, not including the filters and other components. Adding an oil filter and an OSV valve will increase the expected vacuum to about 13 in. of mercury vacuum.

Excessive vacuum at the inlet of the pump can cause cavitation (oil changing to a vapor) and oil foaming. This will have a negative effect on system operation and efficiency. During normal system operation when the pump is operating within its design vacuum range, there should be no bubbling or foaming of the oil, **Figure 32.46(A)**. As the vacuum level in the oil supply line increases, the amount of foaming at the inlet of the pump increases and the amount of oil that is being pumped decreases, **Figure 32.46(B)**. At extremely high vacuums, the oil is nearly completely foam upon entering the pump, **Figure 32.46(C)**. **NOTE:** *If there is an exceptionally high vacuum at the inlet of the pump, the technician should check for crimped or kinked oil supply lines or a dirty oil filter.* •

NOZZLE

The *nozzle* prepares the fuel oil for combustion by atomizing, metering, and patterning the fuel oil. The process of atomization is best described as breaking the fuel oil down into tiny droplets. To give some idea of how small these droplets are, 1 gallon of fuel oil, when atomized, creates approximately 55,000,000,000 (55 billion) droplets of oil. By atomizing the oil, the surface area of the oil is increased, leading to its rapid vaporization and ignition. It should be noted that fuel oil will not burn in the liquid state, so atomization is necessary in order for the combustion process to occur. The smallest of the atomized droplets will be ignited first. The larger droplets (there are more of these) provide more heat transfer to the heat exchanger when they are ignited. Atomization of fuel oil is a complex process. The straight lateral

Figure 32.41 A two-stage fuel oil pump. Courtesy Suntec Industries Inc., Rockford, Illinois

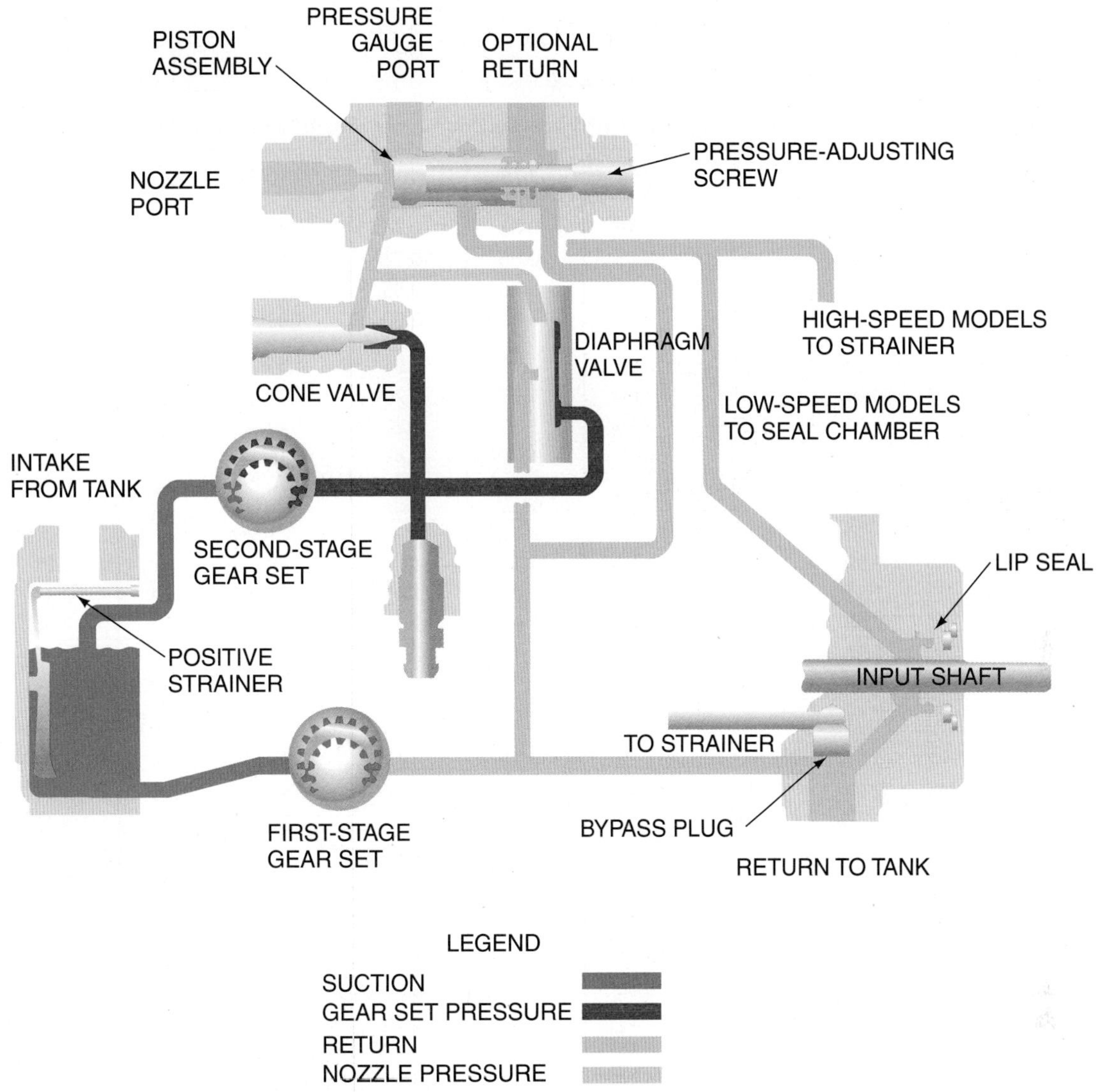

Figure 32.42 The bypass plug location. Courtesy Webster Electric Company

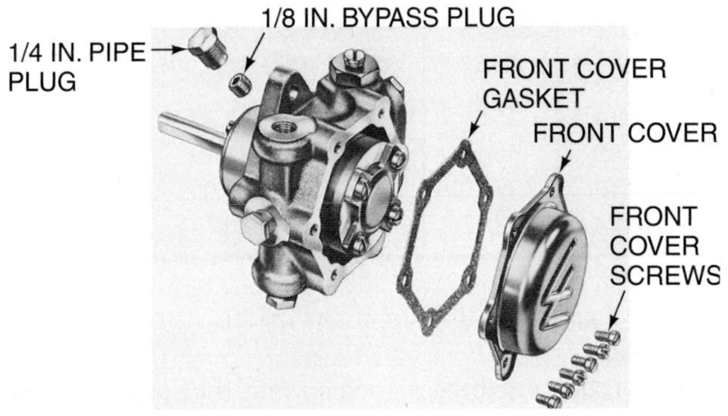

movement of the flowing fuel oil must be changed to a circular motion. This is accomplished in the swirl chamber of the nozzle. **Figure 32.47** shows a cutaway view of a nozzle—as well as a diagram showing some of the parts, including the swirl chamber, that make up this precision element of the oil burner.

The orifice, or bore size, of the nozzle is designed to allow a certain amount of fuel through at a given pressure to produce the amount of heat desired in Btu/h. Each nozzle is marked as to the amount of fuel it will deliver. A nozzle marked 1.00 will deliver 1 gallon of fuel oil per hour (gph) if the pressure at the inlet of the nozzle is 100 psig. With Number 2 fuel oil at a temperature of 60°F, approximately 140,000 Btu would be produced each hour. A 0.8 nozzle would deliver 0.8 gph, and so on. By increasing

Figure 32.43 Nozzle flow rates at different pressures.

NOZZLE CAPACITIES U.S. Gallons per Hour No. 2 Fuel Oil								
Rate gph @ 100 psi	Operating Pressure: pounds per square inch							
	125	140	150	175	200	250	275	300
0.40	0.45	0.47	0.49	0.53	0.56	0.63	0.66	0.69
0.50	0.56	0.59	0.61	0.66	0.71	0.79	0.83	0.87
0.60	0.67	0.71	0.74	0.79	0.85	0.95	1.00	1.04
0.65	0.73	0.77	0.80	0.86	0.92	1.03	1.08	1.13
0.75	0.84	0.89	0.92	0.99	1.06	1.19	1.24	1.30
0.85	0.95	1.01	1.04	1.13	1.20	1.34	1.41	1.47
0.90	1.01	1.07	1.10	1.19	1.27	1.42	1.49	1.56
1.00	1.12	1.18	1.23	1.32	1.41	1.58	1.66	1.73
1.10	1.23	1.30	1.35	1.46	1.56	1.74	1.82	1.91
1.20	1.34	1.42	1.47	1.59	1.70	1.90	1.99	2.08
1.25	1.39	1.48	1.53	1.65	1.77	1.98	2.07	2.17
1.35	1.51	1.60	1.65	1.79	1.91	2.14	2.24	2.34
1.50	1.68	1.77	1.84	1.98	2.12	2.37	2.49	2.60
1.65	1.84	1.95	2.02	2.18	2.33	2.61	2.73	2.86
1.75	1.96	2.07	2.14	2.32	2.48	2.77	2.90	3.03
2.00	2.24	2.37	2.45	2.65	2.83	3.16	3.32	3.46
2.25	2.52	2.66	2.76	2.98	3.18	3.56	3.73	3.90
2.50	2.80	2.96	3.06	3.31	3.54	3.95	4.15	4.33
2.75	3.07	3.25	3.37	3.64	3.90	4.35	4.56	4.76
3.00	3.35	3.55	3.67	3.97	4.24	4.75	4.97	5.20
3.25	3.63	3.85	3.98	4.30	4.60	5.14	5.39	5.63
3.50	3.91	4.14	4.29	4.63	4.95	5.53	5.80	6.06
3.75	4.19	4.44	4.59	4.96	5.30	5.93	6.22	6.50
4.00	4.47	4.73	4.90	5.29	5.66	6.32	6.63	6.93
4.50	5.40	5.32	5.51	5.95	6.36	7.11	7.46	7.79
5.00	5.59	5.92	6.12	6.61	7.07	7.91	8.29	8.66
5.50	6.15	6.51	6.74	7.27	7.78	8.70	9.12	9.53
6.00	6.71	7.10	7.35	7.94	8.49	9.49	9.95	10.39
6.50	7.26	7.69	7.96	8.60	9.19	10.28	10.78	11.26
7.00	7.82	8.28	8.57	9.25	9.90	11.07	11.61	12.12
7.50	8.38	8.87	9.19	9.91	10.61	11.86	12.44	12.99
8.00	8.94	9.47	9.80	10.58	11.31	12.65	13.27	13.86
8.50	9.50	10.06	10.41	11.27	12.02	13.44	14.10	14.72
9.00	10.06	10.65	11.02	11.91	12.73	14.23	14.93	15.59
9.50	10.60	11.24	11.64	12.60	13.44	15.02	15.75	16.45
10.00	11.18	11.83	12.25	13.23	14.14	15.81	16.58	17.32
10.50	11.74	12.42	12.86	13.89	14.85	16.60	17.41	18.19
11.00	12.30	13.02	13.47	14.55	15.56	17.39	18.24	19.05
12.00	13.42	14.20	14.70	15.88	16.97	18.97	19.90	20.79

the pressure of the fuel oil, the quantity of oil that the nozzle allows to flow through it will increase as well. For example, refer to **Figure 32.43** to see that a nozzle rated to deliver 0.75 gallons of fuel per hour at 100 psig will deliver about 0.92 gallons per hour if the oil pressure is increased to 150 psig.

The nozzle must be designed so that the spray will ignite smoothly and provide a steady, quiet fire that will burn

Figure 32.44 Cutaway view of an oil pump showing the screen. Courtesy Oilheat Associates, Inc. Photo by John Levey

Figure 32.45 Oil pump screen and gaskets. Courtesy Oilheat Associates, Inc. Photo by John Levey

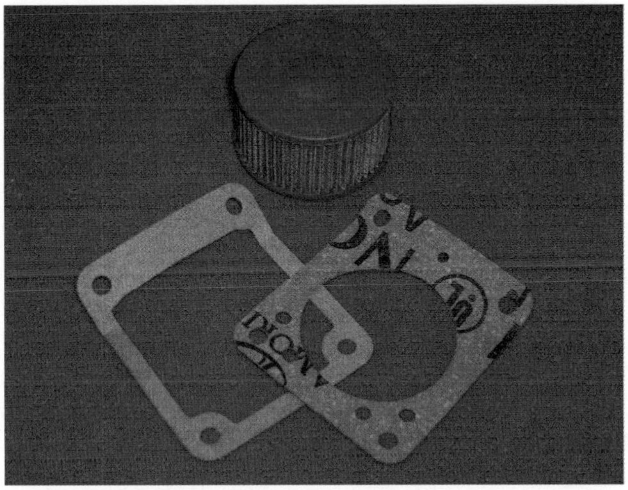

Figure 32.46 (A) No foaming at normal vacuum level. (B) Foaming begins to occur as vacuum level increases. (C) High vacuum levels at the inlet of the oil pump cause excessive oil foaming. Courtesy Oilheat Associate, Inc. Photos by John Levey

(A)

(B)

(C)

cleanly and efficiently. It must provide a uniform spray pattern and angle that is best suited to the requirements of the specific appliance. There are three basic spray patterns—hollow, semisolid, and solid—with angles from 30° to 90°, **Figure 32.48.**

Hollow cone nozzles generally produce a more stable spray angle and pattern than do solid cone nozzles of the same flow rate. These nozzles are often used where the flow rate is under 1 gph. *Solid cone* nozzles distribute droplets fairly evenly throughout the pattern. These nozzles are often used in larger burners. *Semisolid cone* nozzles are often used in place of the hollow or solid cone nozzles. The higher flow rates tend to produce a more solid spray pattern, and the lower flow rates tend to produce a more hollow spray pattern. Nozzle manufacturers often offer nozzles with hybrid spray patterns that are a combination of the three just

Figure 32.47 An oil burner nozzle. Courtesy Delavan Corp.

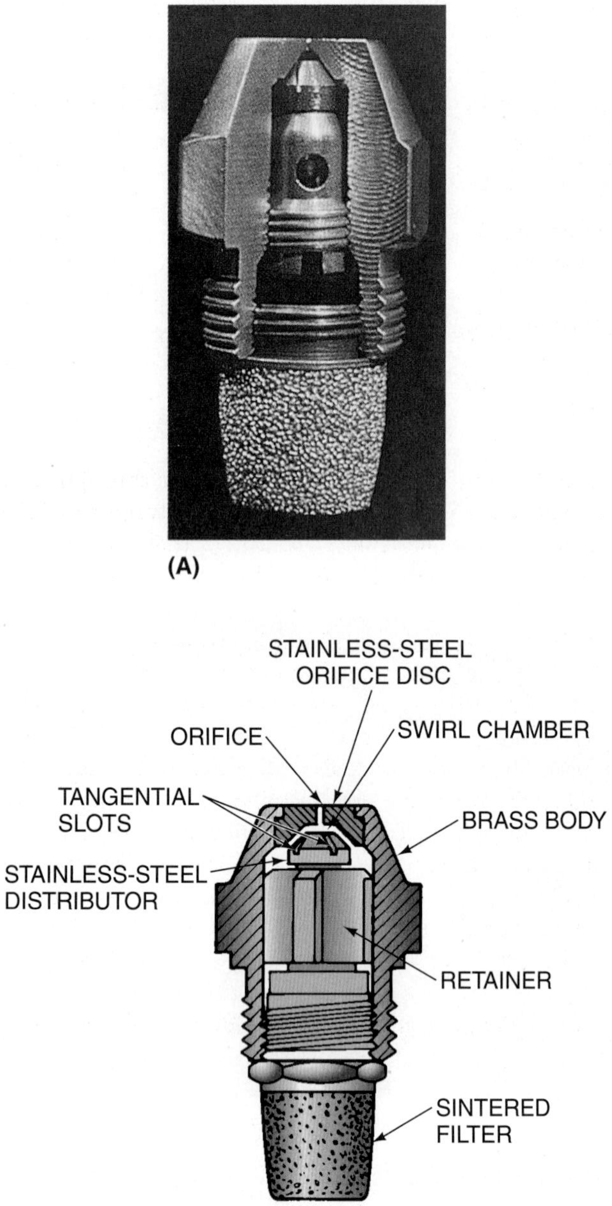

(B)

Figure 32.48 Nozzle spray patterns. Courtesy Delavan Corp.

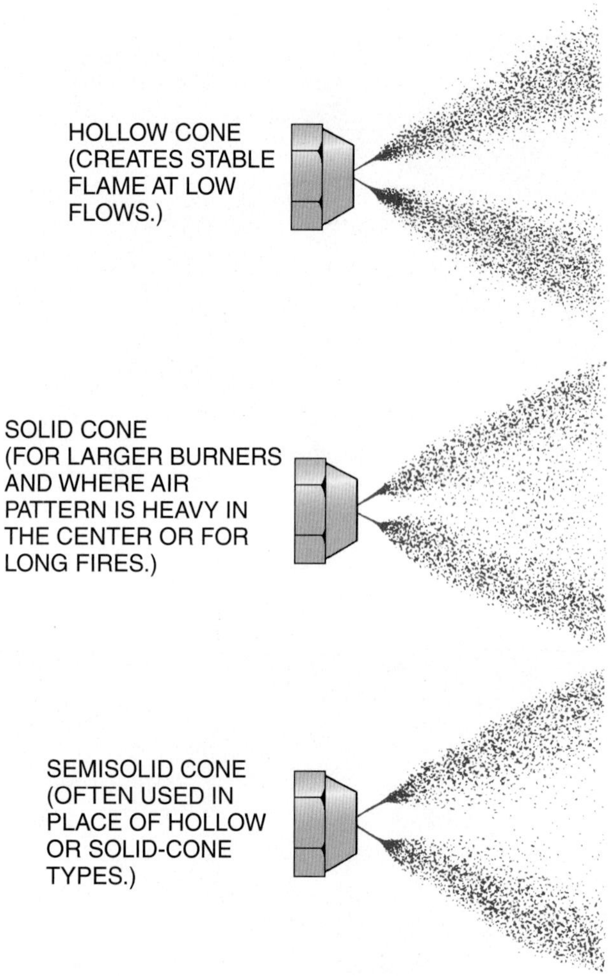

Figure 32.49 Additional spray patterns. Courtesy Delavan Corp.

mentioned. **Figure 32.49** shows some different spray patterns that are available.

The angle of the spray should be selected by following the manufacturer's recommendations. Oil burner and nozzle combinations will vary depending on the appliance in which the burner is installed. If the manufacturer's literature is not available, the following guidelines can be followed. If the combustion chamber is round or square, a nozzle with a larger spray angle is desired, **Figure 32.50(A)**. If the combustion chamber is deep, a nozzle with a smaller spray angle is desired, **Figure 32.50(B)**.

The high-pressure oil burner nozzle is a precision device that must be handled carefully. **SAFETY PRECAUTION:** *Do not attempt to clean these nozzles. Metal brushes or other cleaning devices will distort or otherwise damage the precision-machined surfaces. When the nozzle is not performing properly, replace it. Use a nozzle changer designed specifically for this purpose to remove and install a nozzle,* **Figure 32.51.**•

Figure 32.50 (A) Large spray angle used for round, square, and more shallow combustion chambers. (B) Small spray angle used on deeper combustion chambers.

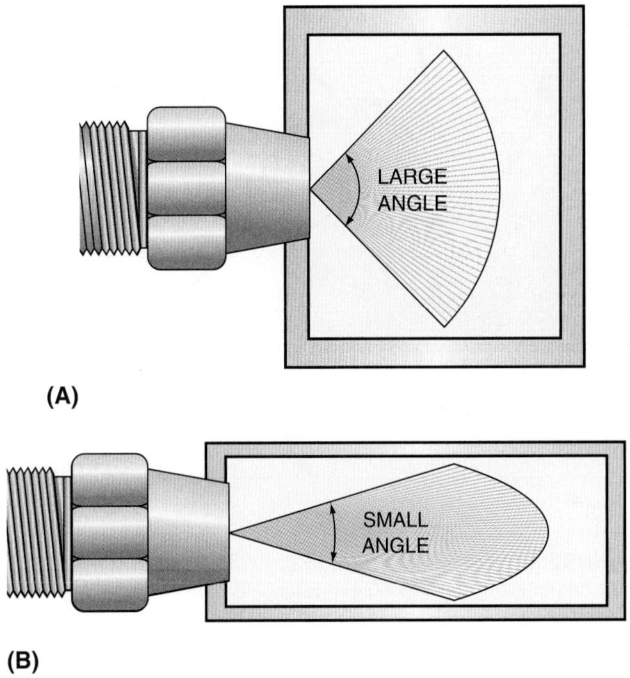

Figure 32.51 A nozzle wrench. Photo by Bill Johnson

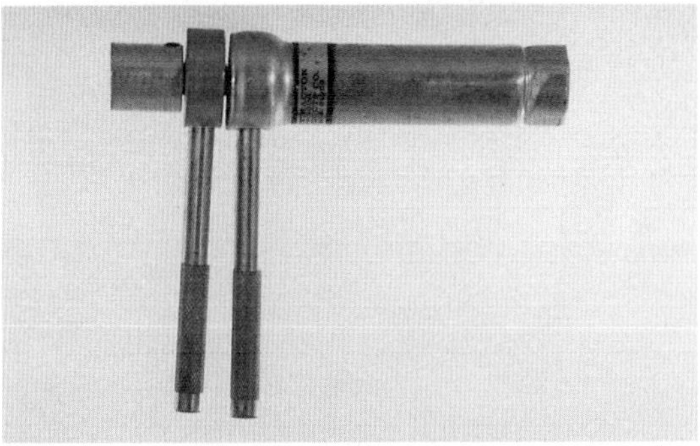

Certain conditions can cause the nozzle to drip after the burner has been shut off. Nozzle dripping at the end of the cycle is the result of expanding air bubbles in the nozzle line. When the nozzle line contains high pressure oil, any air bubbles in the line are small in size, **Figure 32.52** (top). At the end of the cycle, as the pressure in the nozzle line decreases, the air bubbles expand and push oil through the nozzle, **Figure 32.52** (bottom). This can cause a rumble at the end of the cycle, which is actually an ignition of the dripped oil. Components within the oil pump should eliminate this, but they sometimes fail to work correctly. To avoid an after-fire drip at the nozzle, a solenoid-type cut-off valve can be installed, **Figure 32.53**.

AIR TUBE

Air is blown into the combustion chamber through the *air tube,* **Figure 32.54**. Within this tube the straight or lateral movement of the air is changed to a circular motion that is opposite that of the circular motion of the fuel oil pattern leaving the nozzle. The air moving in a circular motion in one direction mixes with the fuel oil moving in a circular motion in the opposite direction in the combustion chamber. The air tube of some burners contains a stationary disc, **Figure 32.55**, referred to as an end cone. The end cone increases the static air pressure and reduces the air volume. The increase in static pressure causes an increase in air velocity to create a better air-oil mixture. The circular air pattern is achieved by vanes located in the retention head. The head of the air tube chokes down the air and the fuel oil mixture, causing a higher velocity, **Figure 32.56**. End cones are often referred to as flame-retention heads, as they are designed to retain the flame on the end of the air tube. This creates more turbulence within the air-oil mixture, resulting in greater burning efficiency. It also provides for

Figure 32.52 (top) Air bubbles are small when under pressure. (bottom) When the pressure in the nozzle line drops, the bubbles expand and oil is pushed from the nozzle.

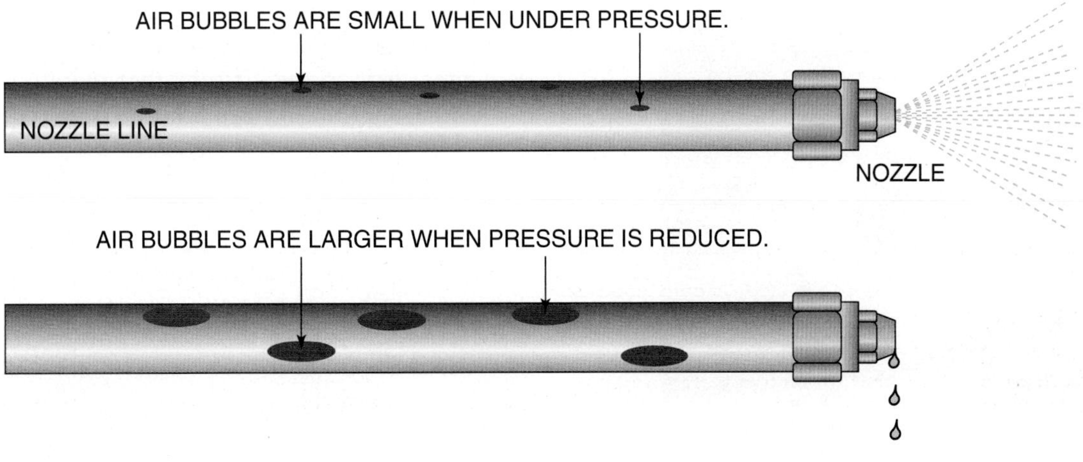

Figure 32.53 Solenoid installed on an oil burner. Courtesy Oilheat Associates, Inc. Photo by John Levey

Figure 32.54 Air tube on an oil burner. Courtesy Oilheat Associates, Inc. Photo by John Levey

Figure 32.55 End cone on a flame-retention oil burner. Courtesy Oilheat Associates, Inc. Photo by John Levey

Figure 32.56 The end cone swirls the air into the swirling oil to achieve the best air-oil mixture.

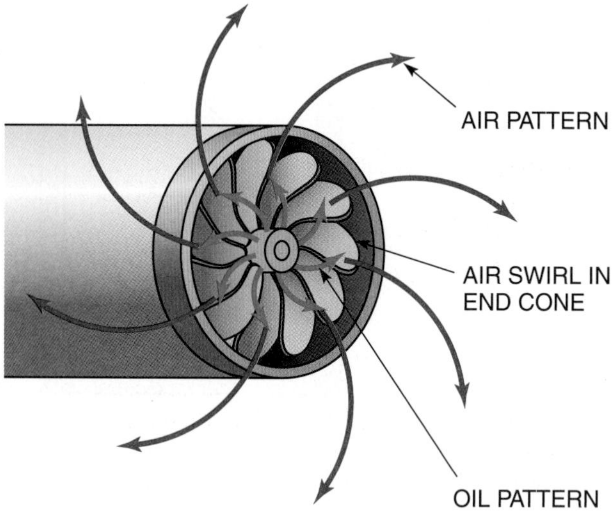

Figure 32.57 Hollow air tube showing placement of the nozzle and electrodes in the tube. The hollow tube is for illustrative purposes only. Courtesy Oilheat Associates, Inc. Photo by John Levey

more complete burning once the atomized air-oil mixture is ignited. It is in the air tube that the atomized oil from the nozzle mixes with the air. The nozzle, therefore, is positioned inside the air tube, **Figure 32.57**.

The amount of air required for the fuel oil preignition treatment is greater than the amount required for combustion. Prior to ignition, approximately 2000 ft^3 of air is needed per gallon of fuel oil, so the air blown into the combustion chamber is greater than what would be needed for the theoretical perfect combustion. Should the amount of air exceed the 2000 ft^3, the burner flame will be long and narrow and possibly impinge on or strike the rear of the

combustion chamber. Airflow adjustments should be made only after analyzing the products of combustion, the flue gases. This is discussed in detail in Section 32.18, "Combustion Efficiency."

ELECTRODES

Two electrodes are positioned close to the orifice of the nozzle, **Figure 32.58**. The electrodes are metal rods insulated with ceramic material to prevent an electrical ground. The rear portions of the electrodes are made of metal and are designed to come in contact with a high-voltage power source, **Figure 32.59**, which is either an ignition transformer or a solid-state igniter. The electrodes provide the heat necessary to both vaporize the atomized oil droplets and ignite the vaporized air-oil mixture. A high-voltage spark is generated between the electrodes to facilitate these processes. The exact position of the electrode tips varies from manufacturer to manufacturer, so be sure to check the literature that accompanies the oil burner for specific information regarding adjusting the electrodes. Refer to Section 32.17, "Service Procedures," for more information regarding electrode positioning. **Figure 32.57** shows the nozzle assembly and the electrodes in position within the air tube.

Figure 32.58 Electrode location with respect to the nozzle. Courtesy Oilheat Associates, Inc. Photo by John Levey

Figure 32.59 An electrode assembly.

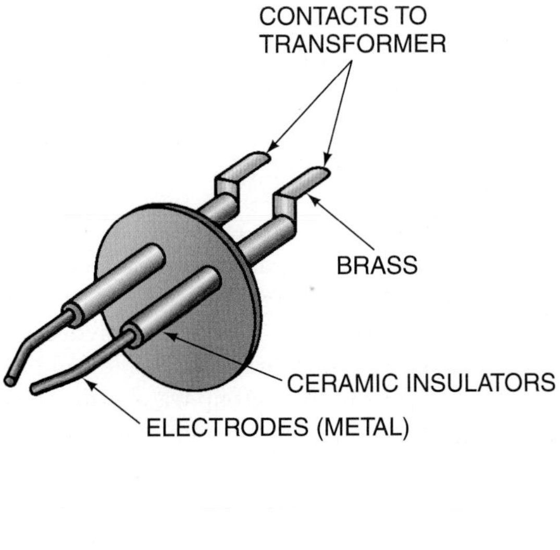

The type of primary control used determines the length of time that the electrodes will provide a spark. The two most commonly encountered ignition strategies are called **interrupted ignition** and **intermittent ignition**. Interrupted ignition employs an electrode spark for a brief period at the beginning of the burner's run cycle. This spark is intended to facilitate the lighting of the burner; then, once the burner flame has been established, the spark stops. On systems with intermittent ignition, the spark is generated continuously during the burner's run cycle. Intermittent ignition used to be referred to as constant ignition, but it isn't any more. Of the two types, interrupted ignition is a better option for a number of reasons. Here are a few:

- Because the spark is only generated for a short period at the beginning of the cycle, transformer/igniter life is extended.
- The elimination of the constant spark reduces burner operation noise.
- Combustion problems become more evident.
- The useful life of the electrodes is lengthened because sparks are generated for shorter periods of time.
- NO_x emissions are reduced. (Refer to Section 32.6 for more information on NO_x gases.)

IGNITION TRANSFORMER

Some oil burners use a *step-up transformer* to provide high voltage to the electrodes that produce the spark for ignition. The transformer is typically located on top of the oil burner, **Figure 32.60**, but can be found on the side of the burner, depending on the model of the unit. Step-up transformers are constructed with more turns in the secondary winding than in the primary. Notice in **Figure 32.61** that the size of the secondary winding is substantially larger than the primary winding. The ignition transformer is supplied with 115 V at its primary winding; the voltage output of the ignition transformer, at the secondary winding, is between 10,000 V and 14,000 V. The voltage at the secondary winding of the transformer is directly related to the voltage supplied to the device.

The ignition transformer has springs, **Figure 32.62**, that allow the secondary winding of the transformer to come in contact with the back, or flat portion, of the electrodes, **Figure 32.63**. The transformer is hinged, so the springs and the back end of the electrode rods can be accessed. Ignition transformers are not intended to be serviced in the field. When there is voltage present at the primary winding and no

Figure 32.60 Ignition transformer on an oil burner. Courtesy American Oil Burner Corp.

Figure 32.61 Clear transformer shows the size difference between the primary and the secondary windings on the ignition transformer. Courtesy Oilheat Associates, Inc. Photo by John Levey

Figure 32.62 Springs on the secondary (high-voltage) winding of the ignition transformer. Photo by Bill Johnson

Figure 32.63 Ignition transformer springs in contact with the electrode rods. Courtesy Oilheat Associates, Inc. Photo by John Levey

voltage present at the secondary winding, the ignition transformer needs to be replaced.

Service technicians can test the operation of the ignition transformer by using a tool designed for such purposes, **Figure 32.64**. The ignition transformer tester is connected to the primary winding of the ignition transformer. The display on the tester will indicate if the transformer is good or in need of replacement, **Figure 32.65**. ⚠ **SAFETY PRECAUTION:** *Be very careful around the high-voltage ignition transformer parts. The arc is much like the arc on a lawn mower or automobile spark plug. It normally will not hurt you, but you may react in such a way that you hurt yourself.*•

SOLID-STATE IGNITER

Newer oil burners do not utilize the ignition transformer discussed in the previous paragraphs. Instead, they are manufactured with solid-state igniters, **Figure 32.66**. These devices perform the same function as their older predecessors, but there are some differences. For instance, the solid-state version is much lighter than the step-up transformer, as the output voltage is generated by solid-state circuitry as opposed

Figure 32.64 Ignition transformer tester connections. Courtesy Oilheat Associates, Inc. Photo by John Levey

Figure 32.65 Display of an ignition transformer tester. Courtesy Oilheat Associates, Inc. Photo by John Levey

Figure 32.66 A solid-state igniter on an oil burner. Photo by Bill Johnson

to multiple coils of wire. The voltage output of a solid-state igniter is typically higher than that of the conventional step-up transformer: in the range of 14,000 V to 20,000 V. Some of the benefits of the solid-state igniter are as follows:

- Higher voltage at the output results in faster fuel vaporization and ignition.
- Solid-state igniters are not greatly affected by a drop in supply voltage.
- Solid-state igniters use less electricity than ignition transformers do.

Solid-state igniters can be field tested with the aid of an igniter tester, **Figure 32.67**. Be sure to follow the directions on the device to ensure proper use. **NOTE:** *Quite often, a solid-state igniter tester can be used to field test both solid-state igniters and ignition transformers. An ignition transformer tester, on the other hand, cannot be used to field test a solid-state igniter. When using a solid-state igniter tester, make certain that you follow all instructions and that you only use the tester in conjunction with the ignition devices designated by the manufacturer.* •

PRIMARY CONTROL UNIT

The oil burner primary control provides a means for operating the burner and shutting the burner down on safety in the event that combustion does not occur. As long as line voltage is supplied to the primary control, meaning that the limit controls are all closed, the primary control will turn the burner on and off in response to the low-voltage operating controls (thermostat). When the thermostat closes its contacts, low voltage is supplied to a coil that energizes and closes a switch by magnetic force. This switch transfers line voltage to the burner, and it begins to operate. If the burner does not ignite or if it flames out during the combustion cycle, the safety function of the primary control must shut down the burner

Figure 32.67 A solid-state igniter tester. Courtesy Oilheat Associates, Inc. Photo by John Levey

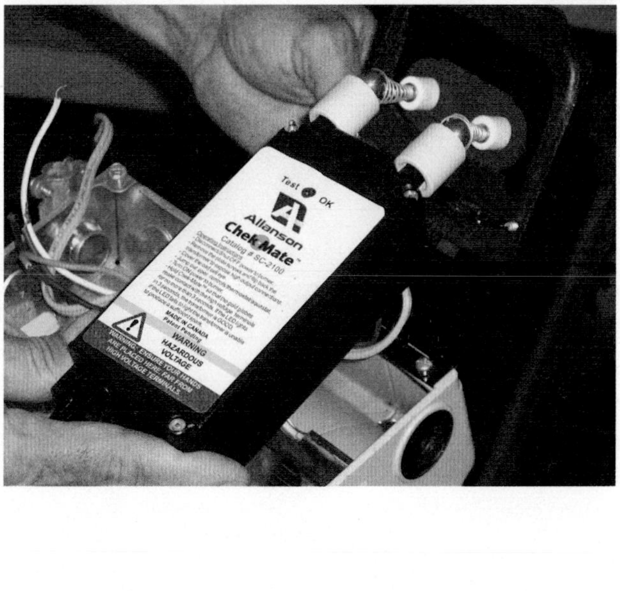

to prevent large quantities of unburned oil from accumulating in the combustion chamber.

STACK RELAY. Older burner installations may have a bimetal-actuated switch, called a *stack switch* or *stack relay*, **Figure 32.68**, as a safety feature. This switch is installed in the stack (flue) between the heat exchanger and the draft damper. This device is a heat-sensing component and is wired to shut down the burner if it does not detect heat in the stack. Although this type of primary control is no longer used on new installations, technicians will encounter the stack relay on some service calls, so it is important to understand how it operates.

Constructed with a bimetal strip that extends into the flue pipe, the stack relay senses the temperature of the flue gases to determine whether or not the burner is firing. When the burner is firing, the bimetal strip senses the rise in temperature in the flue pipe and the strip warps. This warping action causes some electrical contacts in the relay to open and others to close, **Figure 32.69**. The contacts that are in the closed position when the flue pipe is cold are referred to as the cold contacts. Those that are in the closed position when the flue pipe is hot are referred to as the hot contacts. So when the burner is firing, the hot contacts are closed and the cold contacts are open, **Figure 32.70(A)**. When the burner is not firing, the cold contacts are closed and the hot contacts are in the open position, **Figure 32.70(B)**. The wiring diagram for a stack relay is covered later in this unit.

CAD CELL RELAY. Although there are still many stack relays in operation, all new oil burners are equipped with cad cell

Figure 32.68 A stack relay.

Figure 32.69 Hot and cold contacts on a stack relay. Courtesy Oilheat Associates, Inc. Photo by John Levey

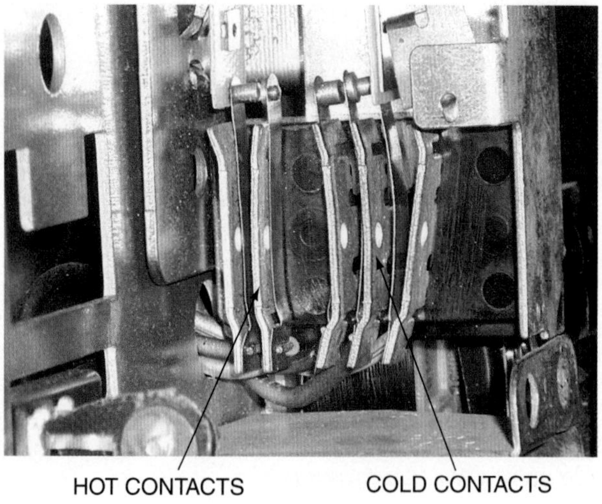

Figure 32.70 (A) The hot contacts are closed and the cold contacts are open. High temperature is being sensed in the flue pipe. (B) The hot contacts are open and the cold contacts are closed. Low temperature is being sensed in the flue pipe.

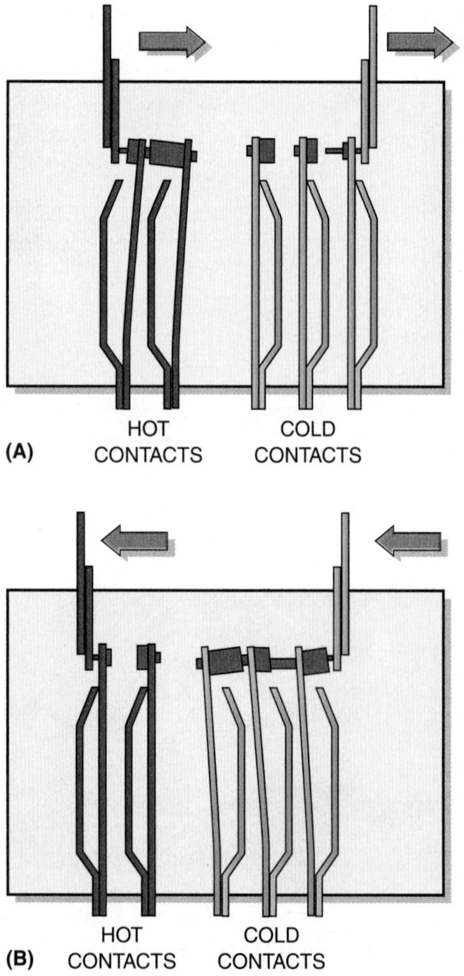

Figure 32.71 A cad cell. Photo by Eugene Silberstein

relays. The main component of this relay is the cad cell, **Figure 32.71**, which is made from a chemical compound called cadmium sulfide that changes its resistance in response to the amount of light the material is exposed to. The cad cell is made up of the cadmium sulfide, a ceramic disc, a conductive grid, and electrodes is encased in glass, **Figure 32.72**. It is positioned in the oil burner so that it can sense the light that is produced when the air-oil mixture is ignited, **Figure 32.73**.

When the oil burner is off, no light is being generated and the resistance of the cad cell is very high, about 100,000 Ω. When the burner is firing, the resistance of the cad cell typically drops to under 1000 Ω. In order to properly sense the light, the cad cell must be kept clean. A dirty cad cell may lead to operational problems with the burner. The cad cell relay circuitry is discussed later on in this unit.

Older cad cell primary controls, such as the one shown in **Figure 32.74**, are designed to operate with intermittent duty

Figure 32.72 A diagram of the face and side of a cad cell.

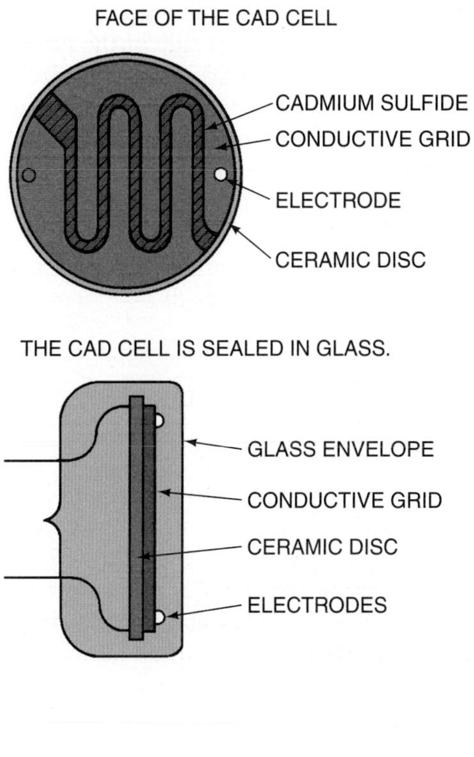

Figure 32.73 Typical cad cell location in an oil burner.

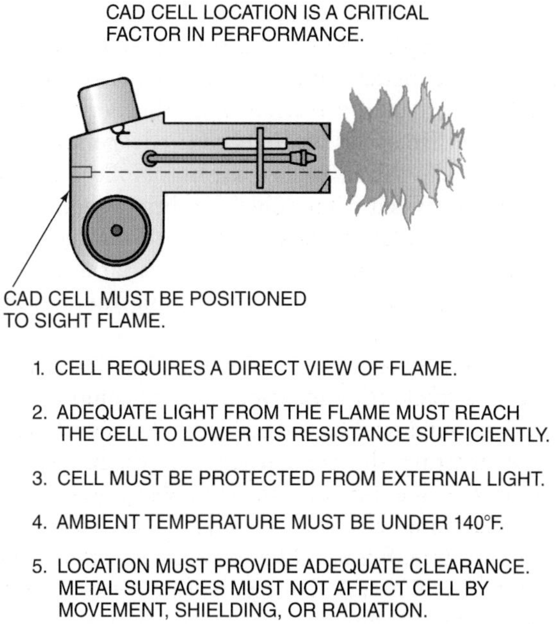

Figure 32.74 A cad cell relay. Photo by Bill Johnson

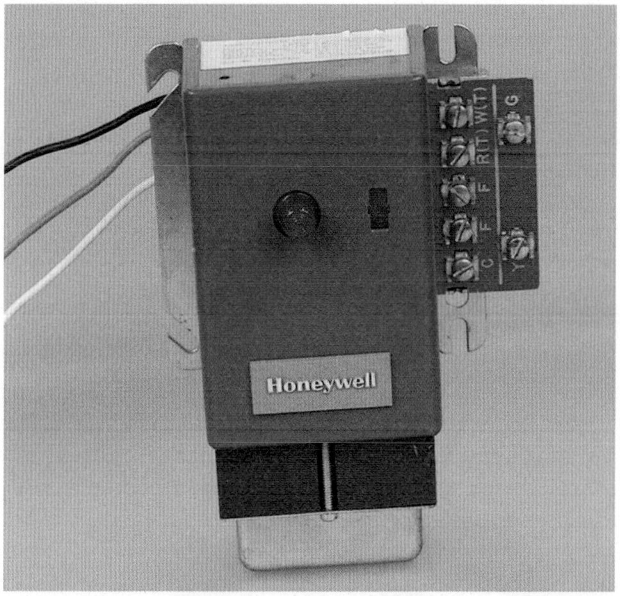

ignition. As mentioned earlier in this unit, with intermittent ignition the spark is produced the entire time the burner is operating. When configured in this manner, the same wire powers the burner motor and the ignition. For the cad cell relay in **Figure 32.74**, the wire that powers the burner motor and the ignition is the orange wire, while the black and white wire are L1 and Neutral wires, respectively. Cad cell relays that are configured like this cannot be used in conjunction with an interrupted ignition system. This is because there is

no way to separately power the burner motor and ignition circuits.

Newer generation cad cell primary controls, **Figure 32.75**, are designed to operate with interrupted duty ignition. The interrupted ignition control strategy produces an ignition spark for a short period of time when burner operation is called for and then cycles off. In order to deenergize the ignition process while still energizing the burner motor, these two circuits must be wired separately. In addition, there are other features such as "valve-delay-on" and "motor-delay-off" that are incorporated into these controls that allow for the burner's blower to operate without having fuel be pumped through the nozzle into the heat exchanger. As can be seen in **Figure 32.75**, there are many more wires that are connected to this device than the older, intermittent ignition cad cell relay shown in **Figure 32.74**. These additional wires are used to independently energize and deenergize the burner motor, ignition, and oil valve circuits at the desired times to enhance system performance.

The operational feature known as "valve-delay-on" is sometimes referred to as prepurge. During the valve-delay-on period, the burner motor and the fuel pump are operating. Air is being moved through the combustion chamber by the blower, but no fuel is being pumped to the nozzle. This is because the oil valve, **Figure 32.53**, is not energized at this point. After the valve-delay-on time period has elapsed, the solenoid on the oil valve is energized and pressurized fuel is delivered to the nozzle. The purpose of the valve-delay-on feature is to help ensure a clean start by allowing the burner motor to reach its full speed before introducing atomized fuel to the combustion chamber. The valve-delay-on feature helps to improve the fuel-air mixture, which will help reduce the amount of soot, or unburned carbon, buildup.

Figure 32.75 A cad cell primary control that is used for interrupted duty ignition. Photo by Eugene Silberstein

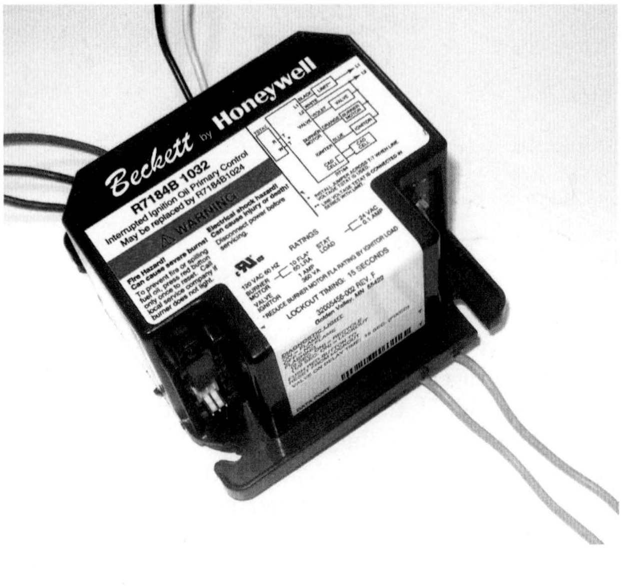

Figure 32.76 The diagnostic LED on a newer generation interrupted duty primary control. Photo by Eugene Silberstein

The operational feature known as "motor-delay-off" is sometimes referred to in our industry as postpurge. During the motor-delay-off period, the solenoid on the oil valve closes at the end of the cycle but the burner motor and the fuel pump continue to operate. Since no oil is being introduced to the chamber, combustion stops, but air is still moving through the combustion chamber. This process allows for a cleaner shutdown at the end of the cycle.

NOTE: *The valve-delay-on and motor-delay-off features cannot be accomplished with older cad cell relays that do not have the capability to energize and deenergize the burner motor, the oil valve, and the ignition circuits separately.* •

Newer generation cad cell primary controls are also typically equipped with diagnostic LEDs, **Figure 32.76**. These low-powered indicator lights provide information to the service technician which can be used to effectively troubleshoot the oil burner. As can be seen in the figure, the manufacturer has provided information on the side of the primary control that is used to interpret the LED's code.

Unlike older generation primary controls, newer devices are also equipped with a feature called **limited recycle.** Limited recycle limits the number of times that an oil burner will try to restart if the flame is lost during the operating cycle. This will help keep the appliance cleaner since less soot will be produced. If the number of attempted restarts is reached, the burner will go into lockout mode.

32.10 OIL FURNACE WIRING DIAGRAMS

Figure 32.77 is a wiring diagram for a typical forced-warm-air oil furnace. The diagram is designed to be used with an air-conditioning system as well. It includes wiring for the fan circuit for air distribution, the oil burner primary control circuit, and the 24-V control circuit. Note that the diagram illustrates the circuitry for a multispeed blower motor.

Figure 32.77 The primary control has a hot wire (black) and a neutral wire (white) feeding into it. When there is a call for heat from the thermostat, the orange wire is energized. This starts the burner motor and the sparking at the electrodes.

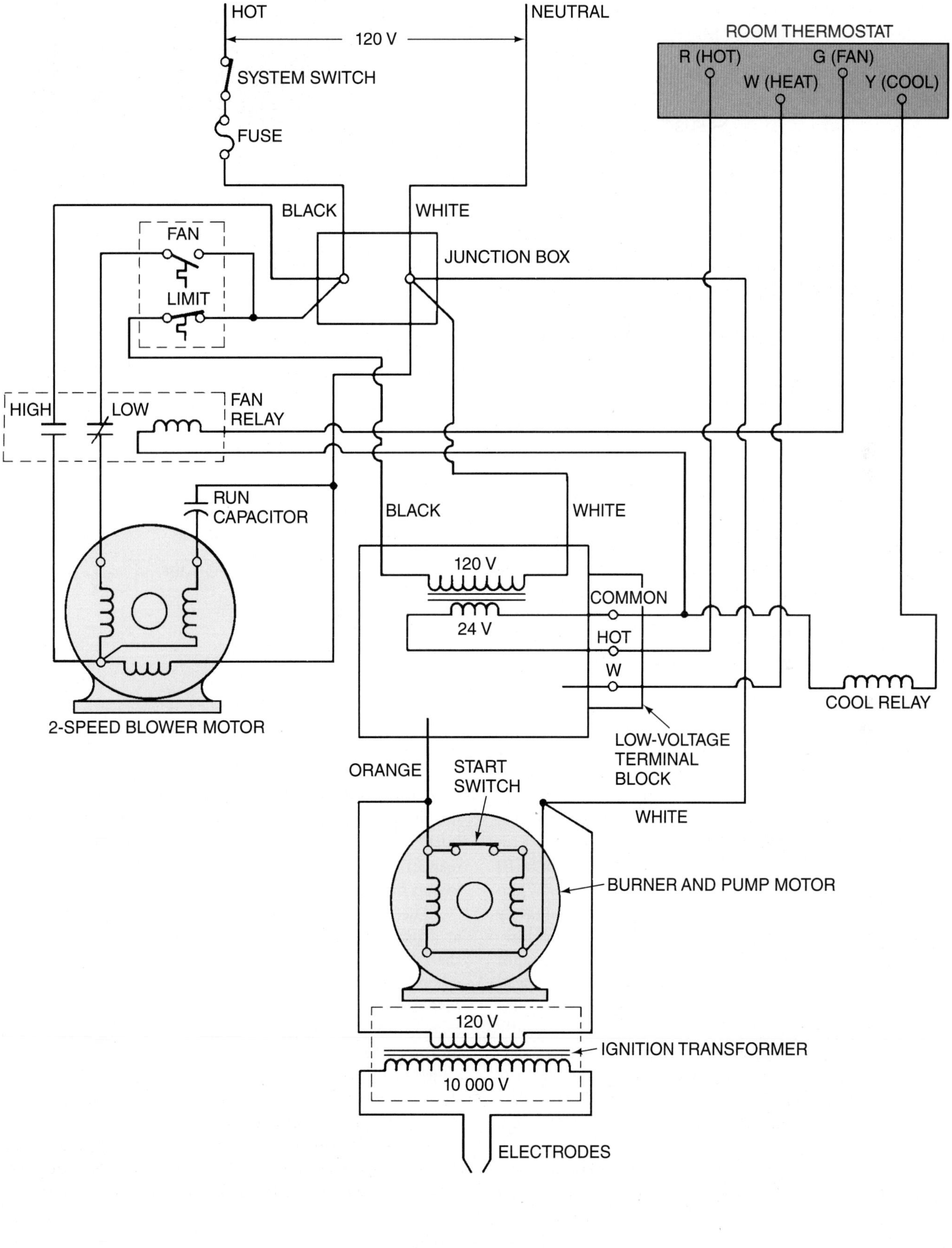

Figure 32.78 The furnace blower or fan runs using the thermally (temperature) activated fan switch in the heating cycle. If the coil on the fan relay were to be energized while in the HEAT mode, the fan would switch from low speed to high speed.

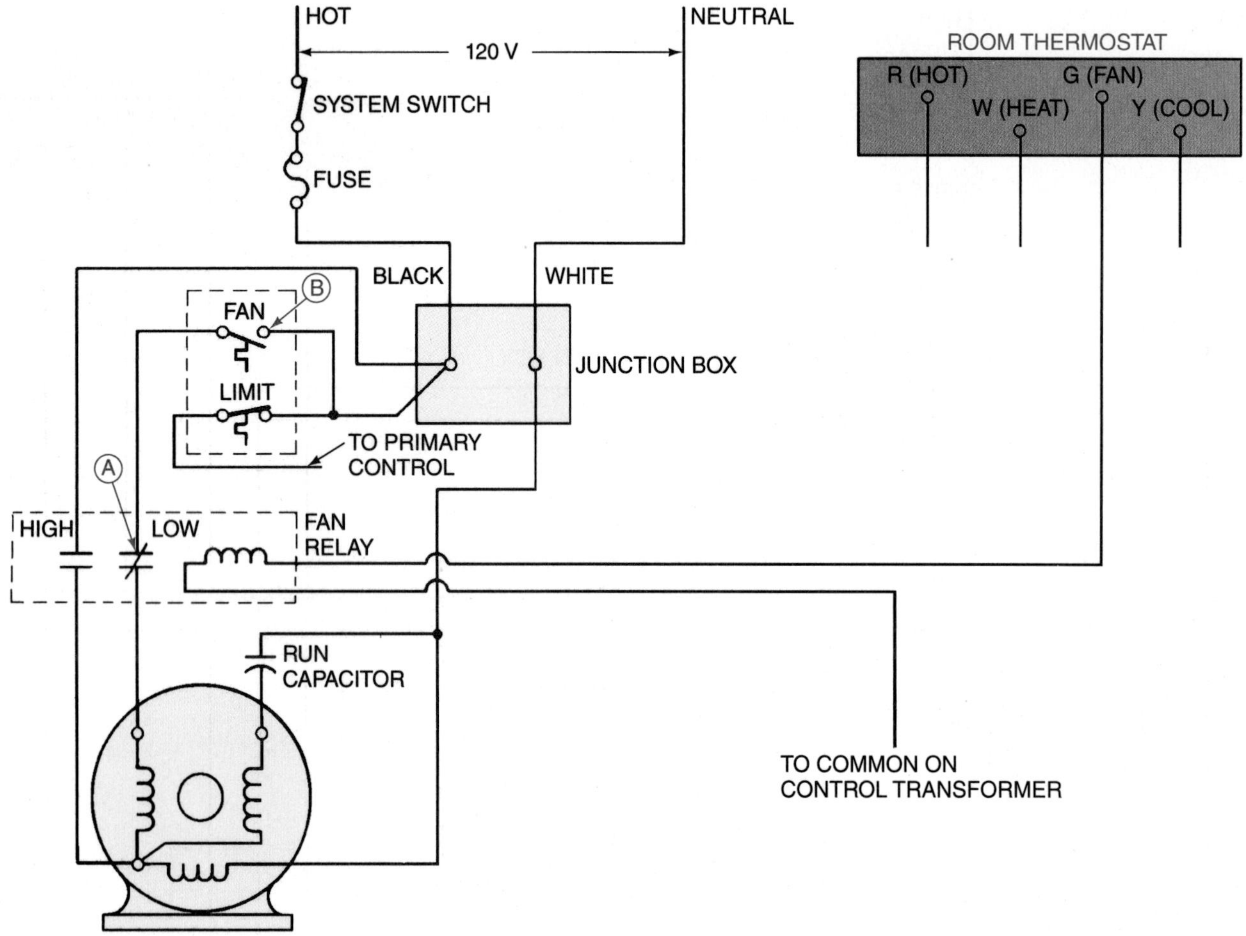

Figure 32.78 illustrates the wiring for the operation of the blower fan motor only. The fan relay (A) protects the blower motor from current flowing in the high- and low-speed circuits at the same time. If the normally open (NO) contacts on the high-speed circuit close, the NC contacts on the low-speed circuit will open. The NO fan limit switch (B) is operated by a temperature-sensing bimetal device. When the temperature at the heat exchanger reaches approximately 140°F, these contacts close, providing current to the low-speed terminal on the blower motor. The high-speed circuit can be activated only from the thermostat. The high speed on the blower motor is *not* used for heating. It will operate only when the thermostat is set to FAN ON manually or for cooling (air-conditioning). The fan circuit for heating would be from the power source (hot leg) through the temperature-actuated fan limit switch, to the NC fan relay, to the low-speed blower motor terminal, and back through the neutral wire.

Figure 32.79 shows the oil burner wiring diagram. The burner is wired through an NC limit switch (A). This limit switch is temperature actuated to protect the furnace and the building. If there is an excessive temperature buildup from the furnace overheating, this switch will open, causing the burner to shut down. Overheating could result from the burner being overfired, from a fan problem, from airflow restriction, or from similar problems. The wiring can be followed from the limit switch to the primary control (B).

The black wire and orange wire at the primary are both at a potential of 120 V, but the orange wire is powered by the 24-V control circuit (C), while the black wire is powered when limit switch (A), the system switch, is closed and the fuse is good. When the thermostat calls for heat, these contacts close, providing current to the ignition system, burner motor, and fuel valve, if one is used. The return is through the neutral (white wire) to the power source. The transformer in the primary control reduces the 115-V line voltage to 24 V for the thermostat control circuit.

If there is a problem in the start-up of the burner, the safety device (stack switch or cad cell) will shut it down. **Figure 32.80** illustrates the wiring for the 24-V thermostat control circuit.

Figure 32.79 The power to operate the burner motor and transformer passes through the NC limit switch to the primary. On a call for heat, the thermostat energizes an internal 24-V relay in the primary control. If the primary safety circuit is satisfied, power will pass through the orange wire to the burner motor and ignition transformer.

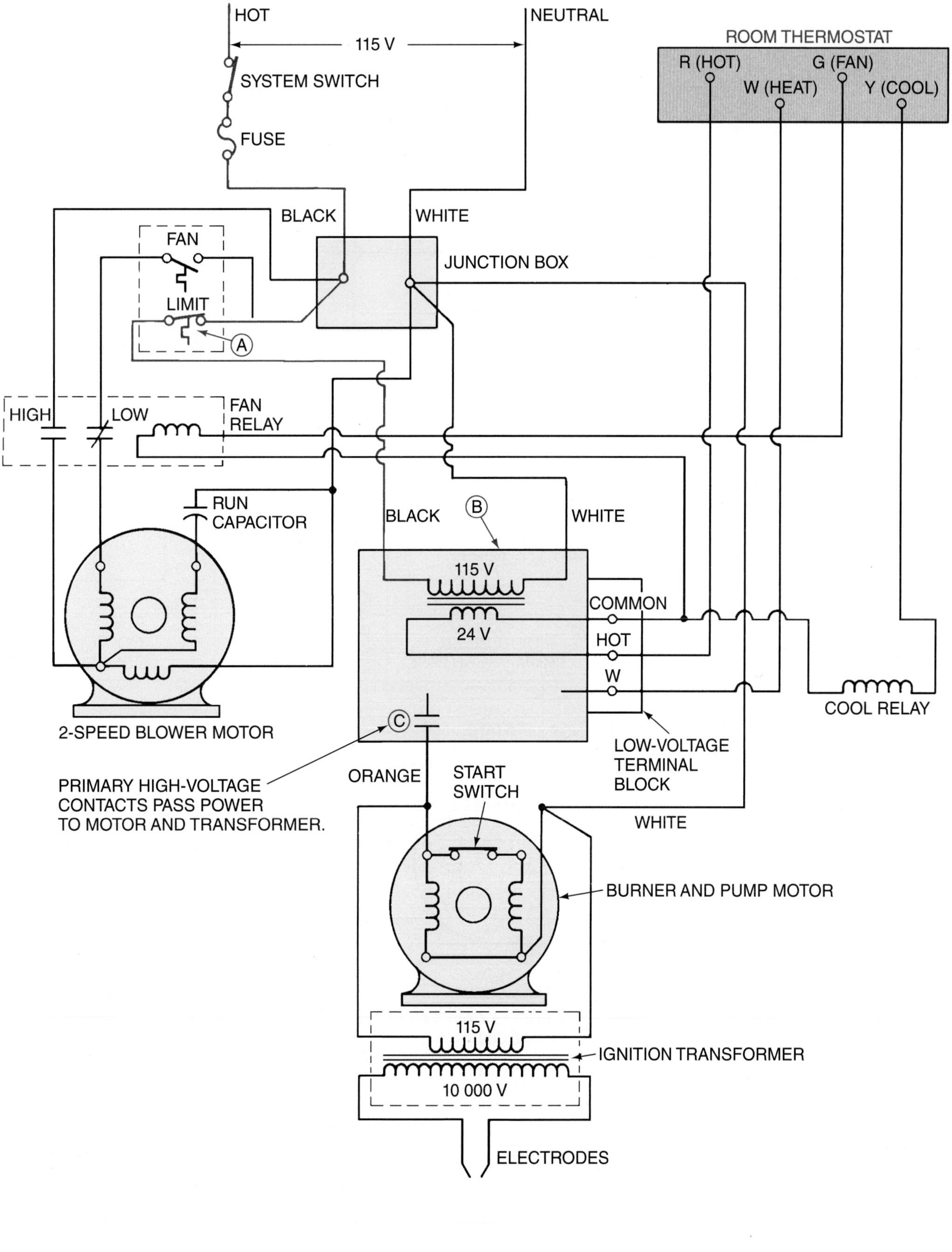

Figure 32.80 The common wire in the primary transformer does not extend into the field circuit. The room thermostat has a split subbase with the hot wire from two different transformers feeding it. These two hot circuits do not come in contact with each other because of the split, or isolating, subbase.

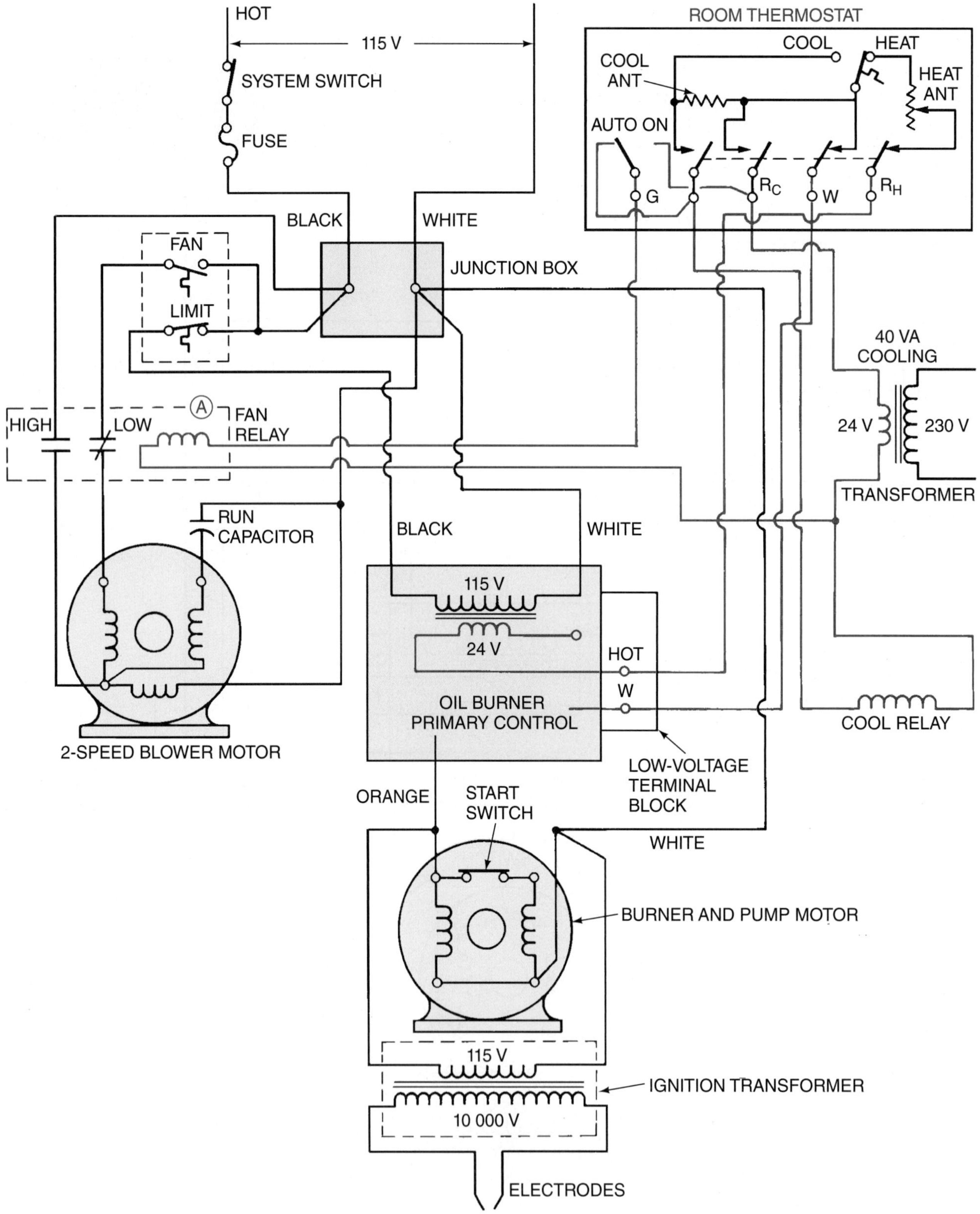

The letter designations on the diagram indicate what the wire colors would normally be: R (red), "hot" leg from the transformer; W (white), heat; Y (yellow), cooling; G (green), manual fan relay.

It will occasionally be necessary to convert a system from one designed for heating only to one that will also accommodate air-conditioning. A heating-cooling system requires a larger 24-V control transformer than the one normally supplied with a heating-only system. To make this conversion, a 40-VA transformer is required in addition to the one supplied in the normal primary control circuit. When adding a cooling package, the larger 40-VA transformer is required because of the added cooling relay and fan-relay load in the 24-V control circuit. The thermostat in such a conversion should also have an isolating subbase so that one transformer is not connected electrically with the other. A fan relay (A) is also needed in the fan circuit to activate the high-speed fan.

Power-consuming devices must be connected to both legs (i.e., connected in parallel). Power-passing devices will be wired in series through the hot leg.

Figure 32.81 (A) An illustration of a stack switch or relay. (B) A side view of an actual stack switch. (A) Courtesy Honeywell. (B) Photo by Bill Johnson.

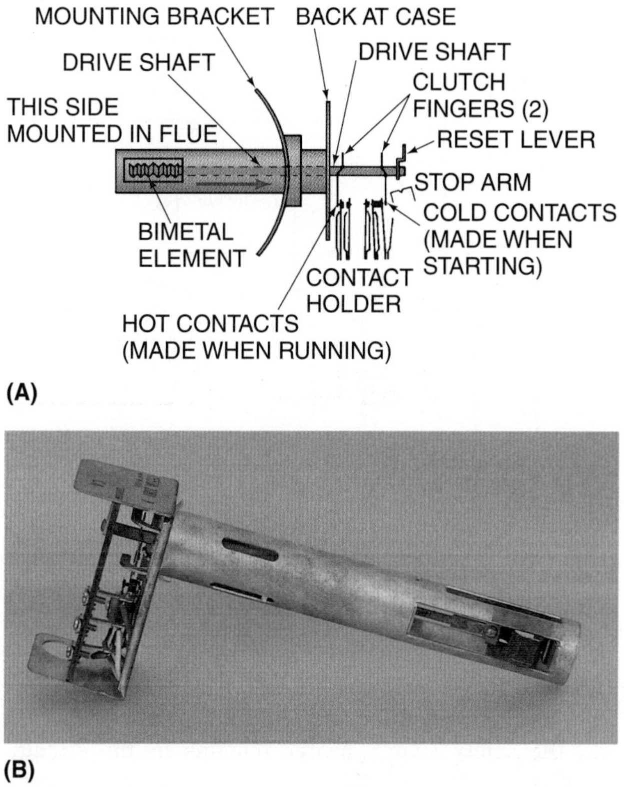

32.11 WIRING DIAGRAM FOR THE STACK SWITCH SAFETY CONTROL

The bimetal element of the stack switch is positioned in the flue pipe. When the bimetal element is heated, proving that there is ignition, it expands and pushes the drive shaft in the direction of the arrow, **Figure 32.81**. This closes the hot contacts and opens the cold contacts.

Figure 32.82 is a wiring diagram showing the current flow during the initial start-up of the oil burner. The 24-V room thermostat calls for heat. This will energize the 1K coil, closing the 1K1 and 1K2 contacts. Current will flow through the safety switch heater and cold contacts. The hot contacts remain open. The closing of the 1K1 contacts provides current to the oil burner motor, oil valve, and ignition transformer. Under normal conditions ignition and heat will be produced from the combustion of the air-oil mixture. This will provide heat to the stack switch in the flue pipe, causing

Figure 32.82 The stack switch circuit when the burner first starts. Note that current flows through the safety switch heater.

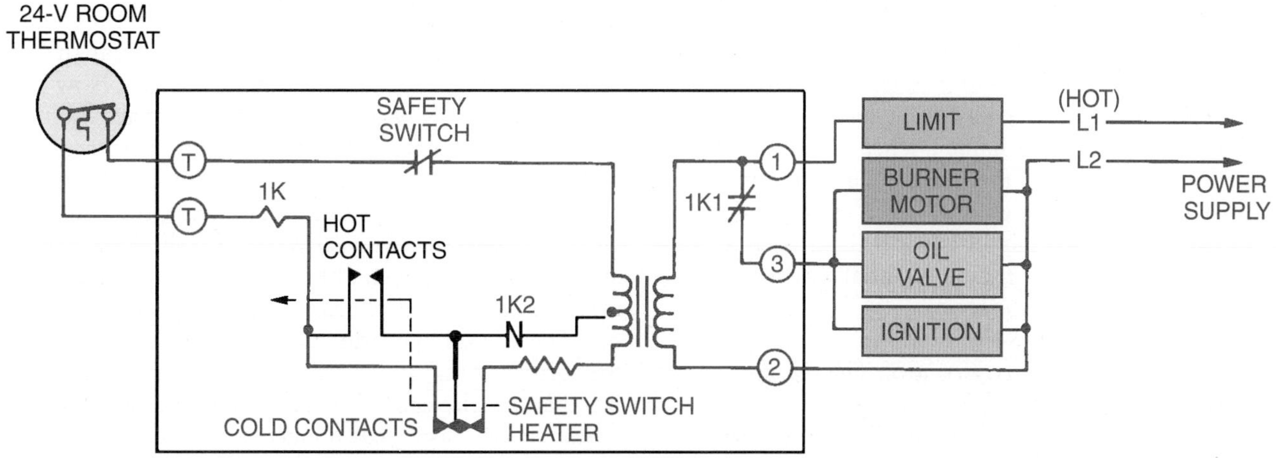

Figure 32.83 Typical stack switch circuitry with flame on, hot contacts closed, and cold contacts open. The safety switch heater is not in the circuit.

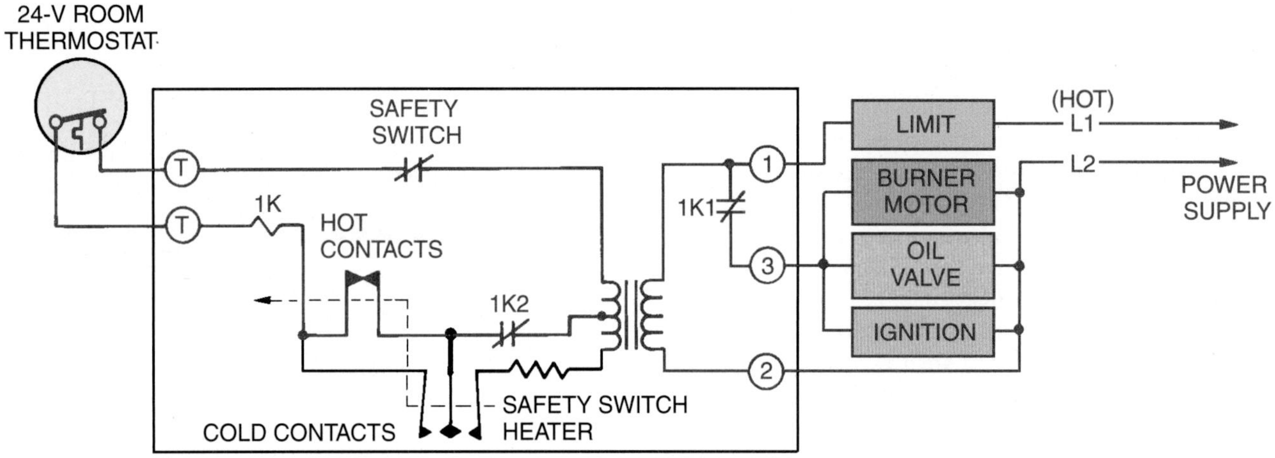

the hot contacts to close and the cold contacts to open. The current flow is shown in **Figure 32.83**. The circuit is completed through the 1K2 contact, the hot contacts to the 1K coil. The 1K1 contact remains closed and the furnace continues to run as in a normal safe start-up. The safety switch heater is no longer in the circuit.

If there is no ignition and consequently no heat in the stack, the safety switch heater remains in the circuit. In approximately 90 sec the safety switch heater opens the safety switch, which causes the burner to shut down. To start the cycle again, depress the manual reset button. Allow approximately 2 min for the safety switch heater to cool before attempting to restart.

32.12 WIRING DIAGRAM FOR CAD CELL PRIMARY CONTROL WITH INTERMITTENT IGNITION

In early generation primary control circuits, the cad cell discussed earlier may be coupled with a triac, which is a form of a solid-state device designed to pass current when the cad cell circuit resistance is high (no flame). When the burner fires, the cad cell senses light from the burner's flame and its resistance drops. This causes the triac to open, stopping current flow through it. When the burner is not firing, the resistance of the cad cell is high and the triac is essentially closed, allowing current to pass through the triac.

The circuit in **Figure 32.84** shows how the triac and the cad cell are wired in the primary control. The red lines indicate the electrical paths through which electric current is flowing. The burner represented by this diagram is currently firing and the resistance of the cad cell is low. This means that the resistance between the two F terminals is low. This low resistance causes the triac to stop passing electric current through the safety switch heater. If the safety switch heater remains in the circuit too long, the burner will shut down. It should be clear from this diagram that the burner motor, the oil valve, and the ignition are all powered by the same orange wire. Since these three elements will all be energized and deenergized at the same time, this type of primary control allows for only intermittent duty ignition operation. The ignition spark will be produced during the entire operating cycle.

In **Figure 32.85** there is no flame and a high resistance in the cad cell. By design, the triac will then conduct current that will pass through the safety switch heater. If this current continues through the heater, it will open the safety switch and cause the burner to shut down. To recycle, press the reset button.

The technician must be able to check the cad cell. It connects to the primary control on the outside at terminals F and F. To check the resistance reading of the cad cell, disconnect the two yellow wires at the F and F terminal block. With the furnace off, it should be dark inside the combustion chamber. Set the ohmmeter to check out 1000 Ω. Calibrate the meter to 0 Ω with the meter leads touching each other. Now place the meter leads on the two yellow wires. The meter should read a very high resistance, maybe even infinity, as if the leads were held apart in air. Now jump out the T to T terminals and the furnace should fire. As the furnace fires, the resistance will become less on the two yellow wires. It should be somewhere around 600 to 1000 Ω. The cad cell is good if this test proves out. The furnace should run for a short while and then lock out. This is a safety feature to prevent someone from running the furnace without the cad cell protection. In most modern installations the cad cell is preferred over the thermal stack switch. It acts faster, has no mechanical moving parts, and is consequently considered more dependable.

Figure 32.84 Cad cell circuitry when the cad cell "sights" flame. The safety switch heater is not in the circuit.

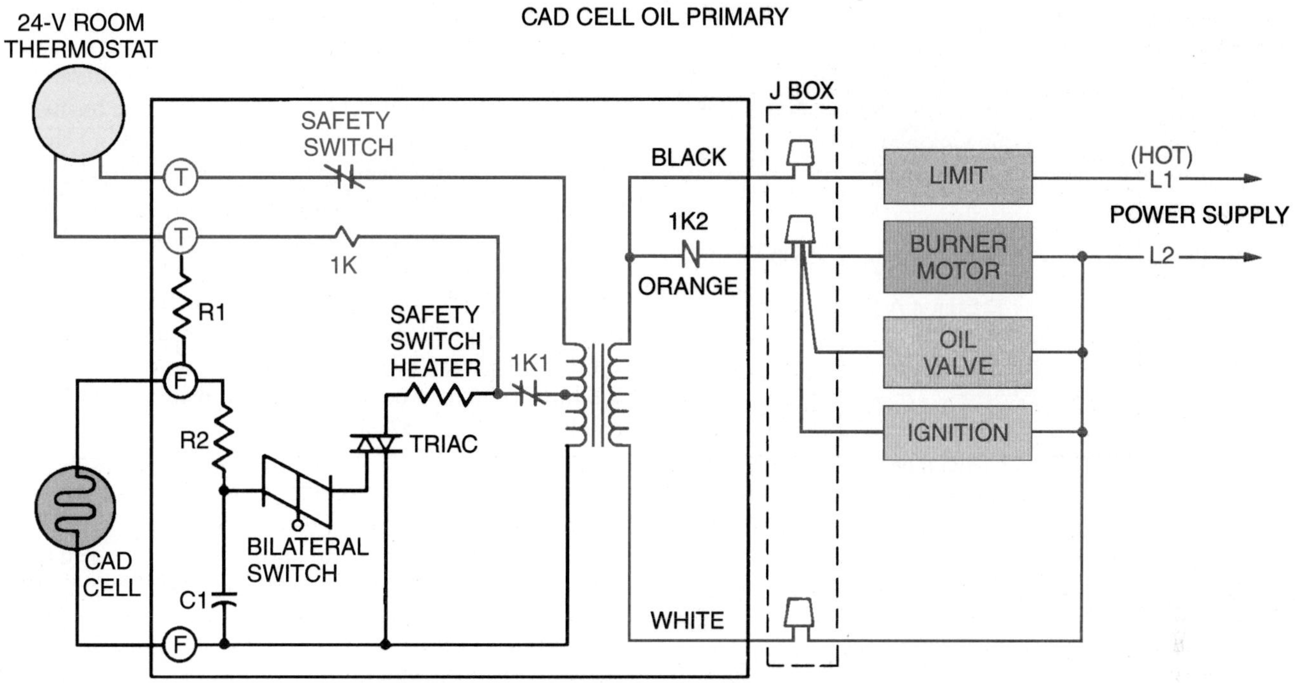

Figure 32.85 Cad cell circuitry when the cad cell does not "sight" flame. The safety switch heater will heat up and open the safety switch, shutting the burner down.

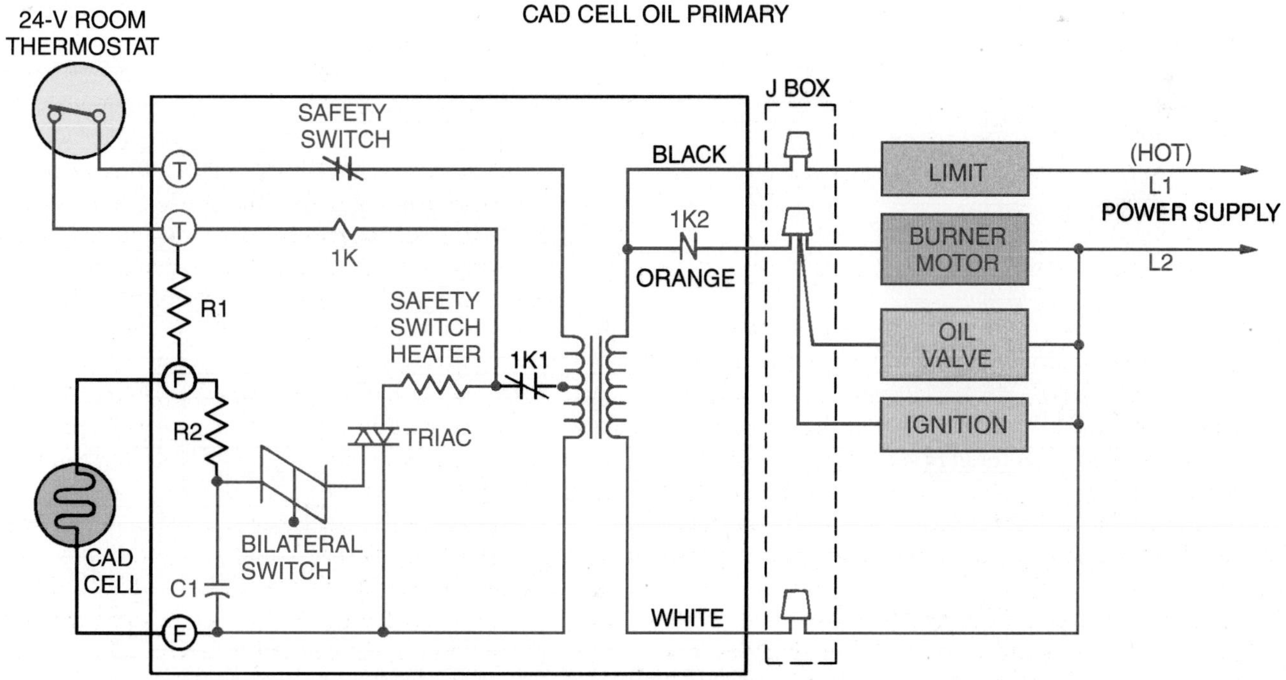

SAFETY PRECAUTION: *Care should be taken not to reset any primary control too many times because unburned oil may accumulate after each reset. Once ignited, a puddle of oil will burn intensely. If this should happen, the technician should stop the burner motor but should allow the furnace fan to run. The air shutter should be shut off to reduce the air to the burner. Notify the fire department. Do not try to open the inspection door to put the fire out. Let the fire burn itself out with reduced air.•*

32.13 WIRING DIAGRAM FOR CAD CELL PRIMARY CONTROL WITH INTERRUPTED IGNITION

The cad cell shown in **Figure 32.75** can be used for interrupted duty ignition. This type of primary control has separate wires to control the operation of the burner motor, the oil valve, and the ignition, **Figure 32.86**. This type of control is powered constantly through L1 and has another power lead that is interrupted if the limit switch is open. There are three color-coded wires that feed power to the burner motor, the ignition, and the oil valve. Be sure to check the color-coding for the particular primary control you are working on, as different primary controls may be color-coded differently. For example, some manufacturers color-code the constant L1 power lead red, while others color code the equivalent wire black.

The diagram of the primary control shown in **Figure 32.86** shows a set of dry alarm contacts. In the event the primary control goes into lockout or latch-up mode, the dry alarm contacts on the primary control close. When these contacts close, an auto-dialer, if connected and used, has the capability to send a signal to the homeowner, alarm company, and/or oil company, alerting them that a problem with the oil burner is present. Lockout occurs when the burner cannot successfully start. If the control goes into lockout three times on a single call for heat, the control goes into the restricted lockout mode called latch-up.

A sample wiring diagram for an oil-fired furnace using an interrupted ignition-duty primary control is shown in **Figure 32.87**. In this diagram it can be seen that an uninterrupted 120-V power supply is reaching the primary control via the red wire, while a switched 120-V power lead, through the limit switch, is connected to the primary control via the black wire, **Figure 32.88**.

If the limit switch is closed, on a call for heat, the space thermostat closes its contacts and power is passed to the coil on the burner relay, **Figure 32.89**. The energizing of the burner relay coil causes the contacts on the burner relay to close, making the connection between the T-T contacts on the primary control, **Figure 32.90**.

Once the T-T contacts have closed, the oil burner motor and the ignition circuits are energized, allowing the motor to come up to speed. This valve-delay-on period varies from primary control to primary control, but is often in the 10-sec range, **Figure 32.91**. Once the valve-delay-on period has elapsed, the oil valve circuit is energized and pressurized oil is delivered to the nozzle for atomization and combustion, **Figure 32.92**.

Figure 32.86 Typical wiring connections on an interrupted duty cad cell primary control.

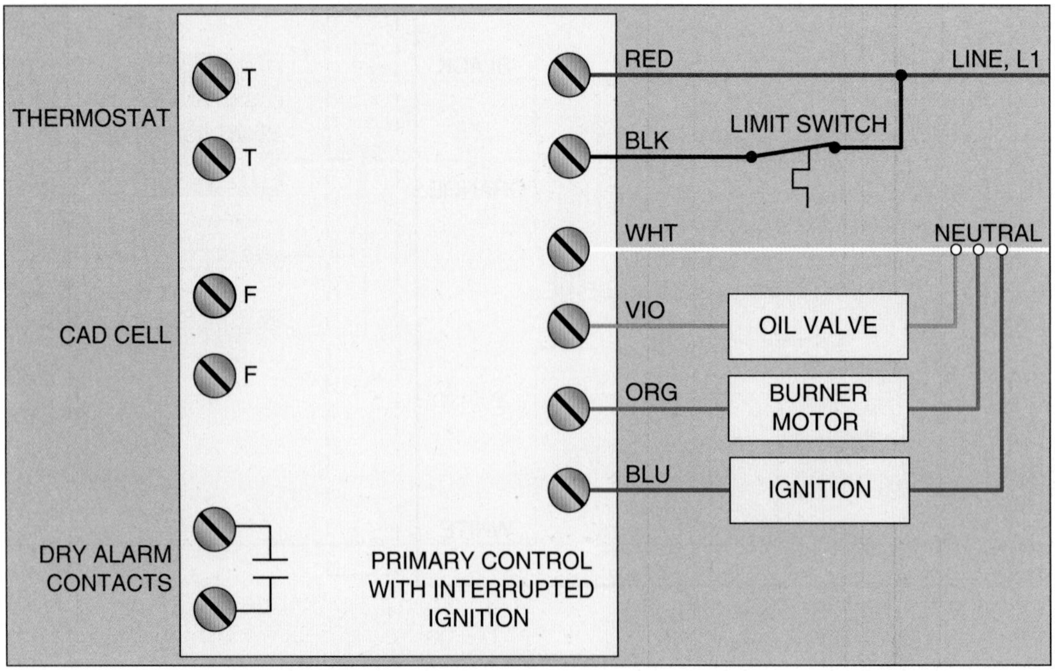

Figure 32.87 Typical wiring diagram of a furnace using an interrupted duty cad cell primary control.

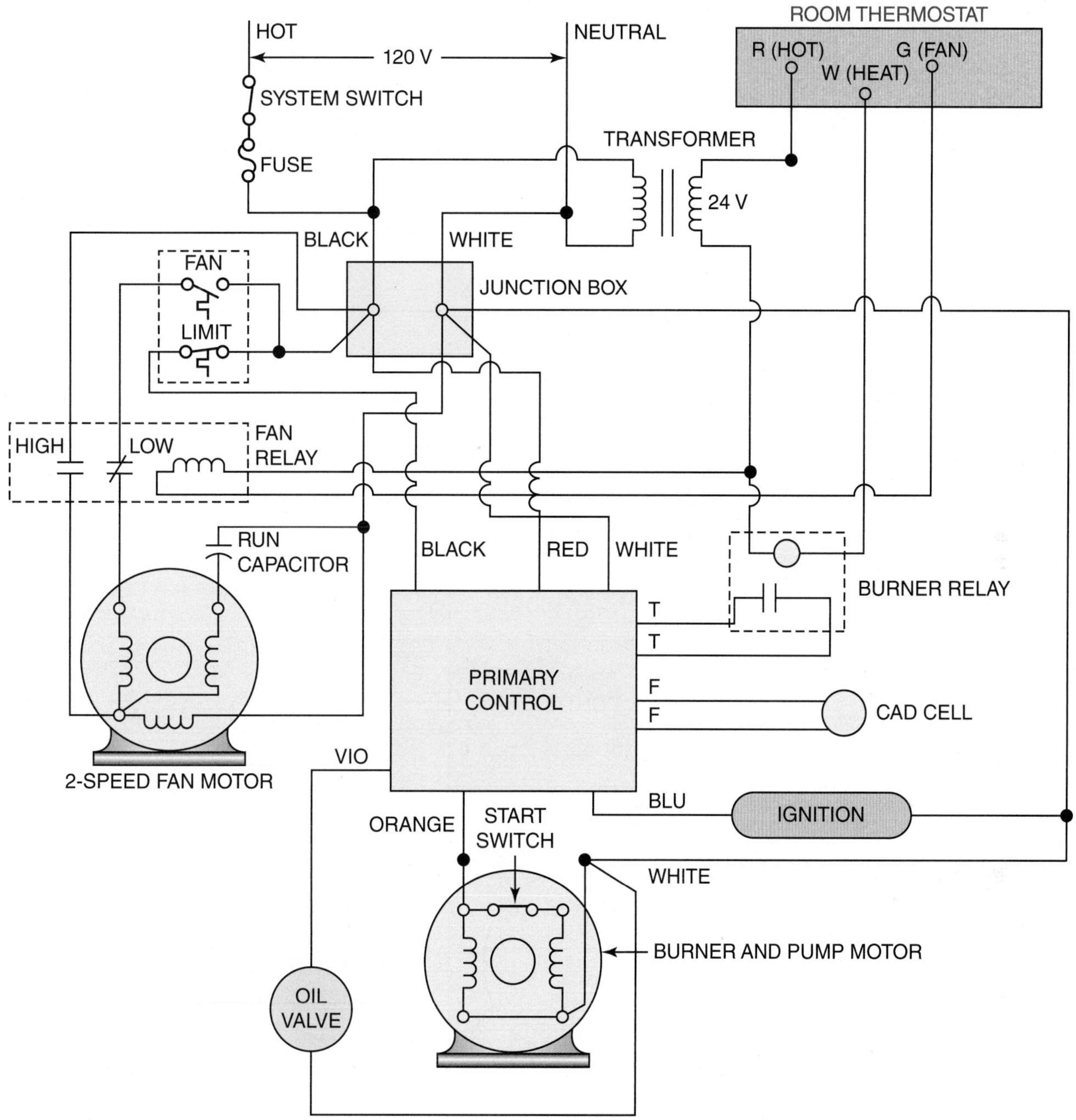

32.14 COMBUSTION CHAMBER

The atomized oil and air mixture blown from the air tube into the combustion chamber is ignited by a spark from the electrodes. The atomized oil must be burned in suspension; that is, it must be burned while in the air in the combustion chamber. If the flame hits the wall of the combustion chamber before it is totally ignited, the cooler wall will cause the oil vapor to condense, and efficient combustion will not occur. Combustion chambers should be designed to avoid this condition. They may be built of steel or a refractory material. **Figure 32.93** shows a flame in a combustion chamber.

The combustion chamber is supposed to be matched to the burner flame at the factory. It would be a rare case where the flame and combustion chamber were not a fit from the beginning. On older systems, the technician may encounter several replacement nozzles lying around the furnace. It is always best to follow the manufacturer's recommendations regarding nozzle size.

Figure 32.88 When the limit switch is closed, line voltage is supplied to the primary control at two points.

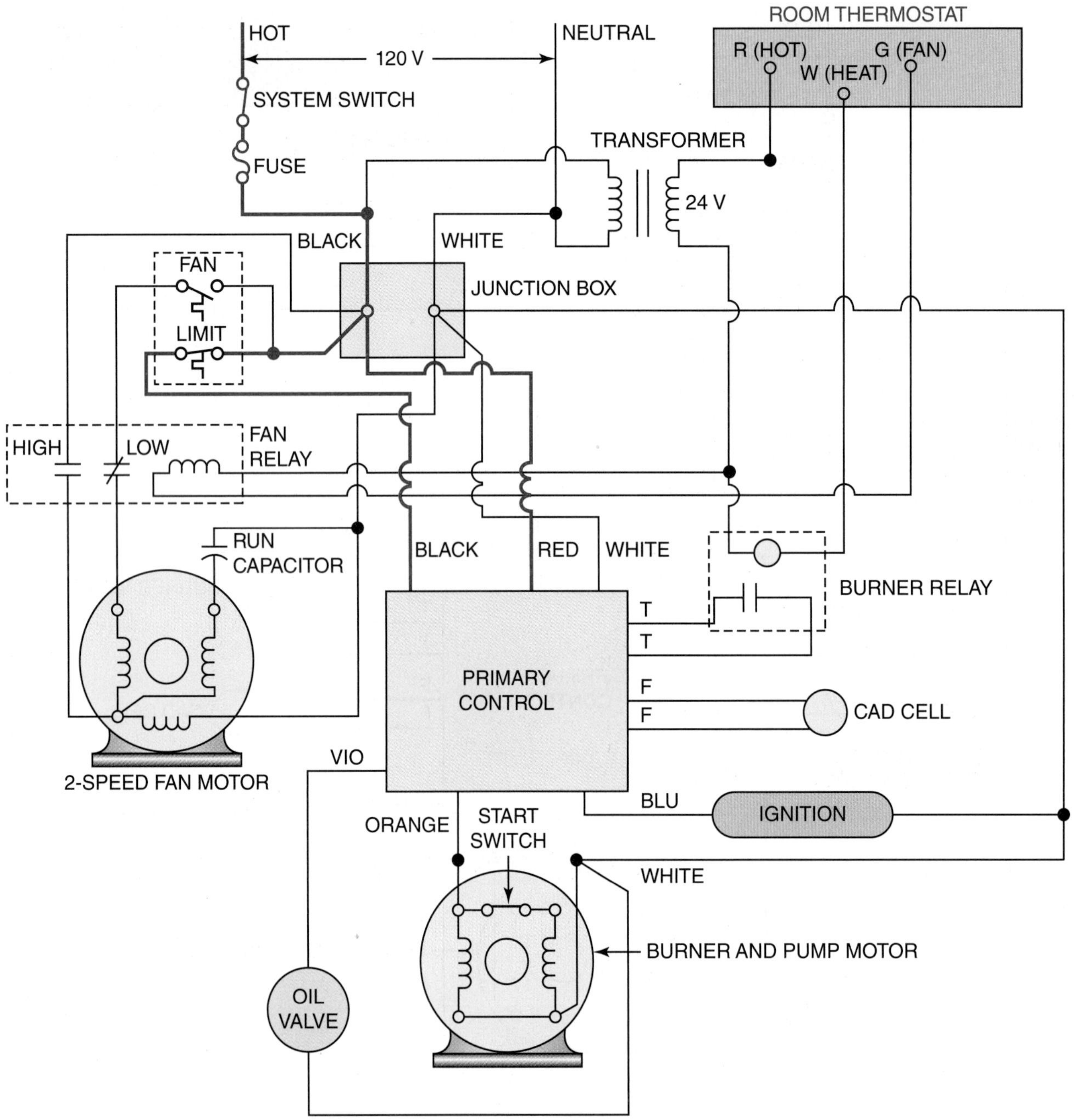

Recall from our earlier discussion on nozzles that the spray angle of the nozzle is, at least in part, determined by the shape and configuration of the combustion chamber. Nozzles with smaller spray angles are intended for use in conjunction with deeper combustion chambers; nozzles with large spray angles are intended for use with square or round combustion chambers. Oil burner manufacturers often can supply original equipment manufacturer (OEM) guides that provide information regarding the original nozzle size intended for use on a specific furnace with a specific model of oil burner. This information is extremely valuable to the service technician.

The combustion chamber is merely a firebox, or a box to contain the fire. They sometimes develop cracks. There is no connection between the combustion chamber and the recirculating air, so the chamber oftentimes may be repaired. The repair kit may consist of a wet blanket material that can be installed through the hole after the burner is removed. The wet blanket is placed over the crack and the furnace is reassembled and fired. The blanket will

Figure 32.89 On a call for heat, the low-voltage control circuit energizes the coil on the burner relay.

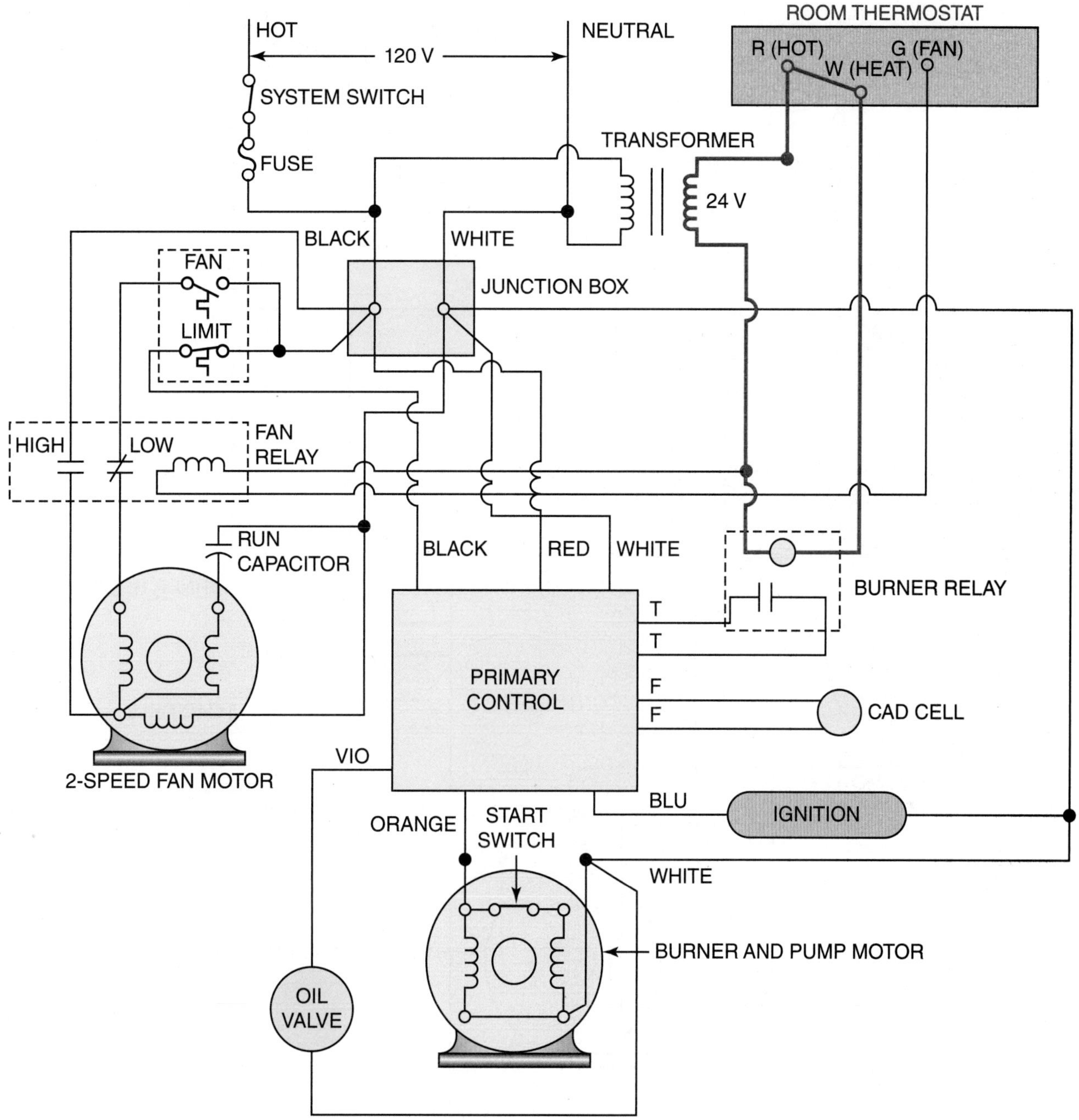

become hard and is fireproof. The other option is to replace the combustion chamber.

32.15 HEAT EXCHANGER

A heat exchanger in a forced-air system takes heat from the combustion in the furnace and transfers it to the air that is circulated to heat the building, **Figure 32.94**. The heat exchanger is made of material that will cause a rapid transfer of heat. Modern exchangers, particularly in residential furnaces, are made of sheet steel, which is frequently coated with special substances to resist the corrosion caused by the acids produced by combustion.

The heat exchanger is also designed to keep flue gases and other combustion materials separate from the air circulated through the building. **SAFETY PRECAUTION:** *The flue gases from combustion should never mix with the circulating air. A technician who discovers a crack in a heat exchanger may be tempted to repair it by welding it closed. Many state codes prohibit this type of repair.*•

Figure 32.90 When the coil on the burner relay is energized, the contacts on the relay close, completing the T-T circuit.

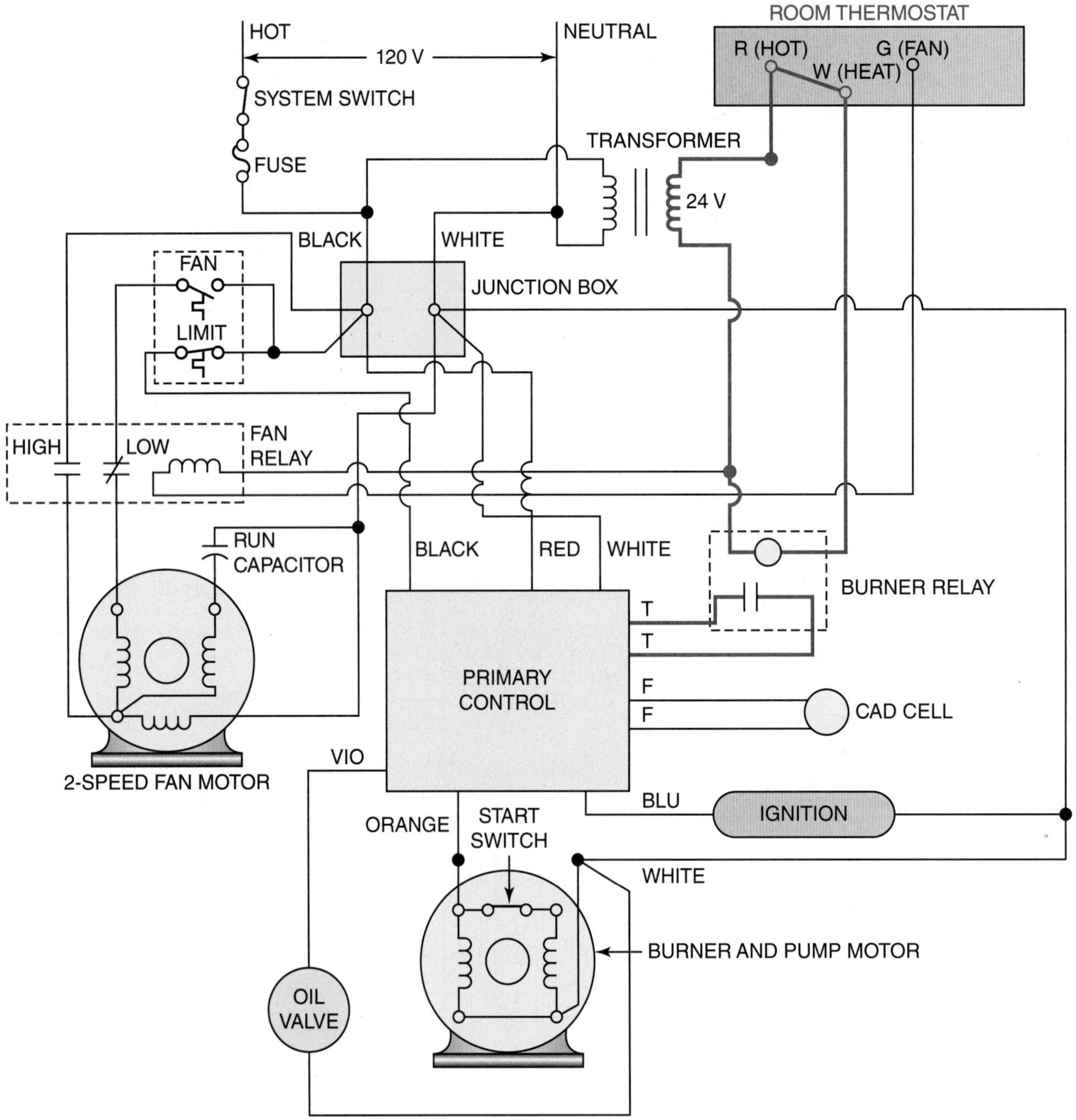

A technician often inspects heat exchangers and combustion chambers. A mirror made of chrome-plated steel is an ideal tool for this job, **Figure 32.95**, and new products on the market allow technicians to inspect heat exchangers with more accuracy and reliability than ever before. These products include a small video camera on the end of a long, snakelike, flexible rod that can be inserted into the furnace. The tip of the rod contains a light so the technician can inspect the heat exchanger by monitoring the screen on the device, **Figure 32.96**.

One way a technician can determine if there is a crack in the heat exchanger is to insert the probe from the combustion analyzer into the supply air duct far from the furnace location. If there is a measurable quantity of carbon monoxide in the supply airstream, a leak in the heat exchanger should be suspected. Damage to the heat exchanger can be the result, among other things, of insufficient airflow through the appliance or an oversized furnace that cycles on and off too often.

Figure 32.91 Closing the T-T contacts initiates burner operation by powering the burner motor and the ignition circuit.

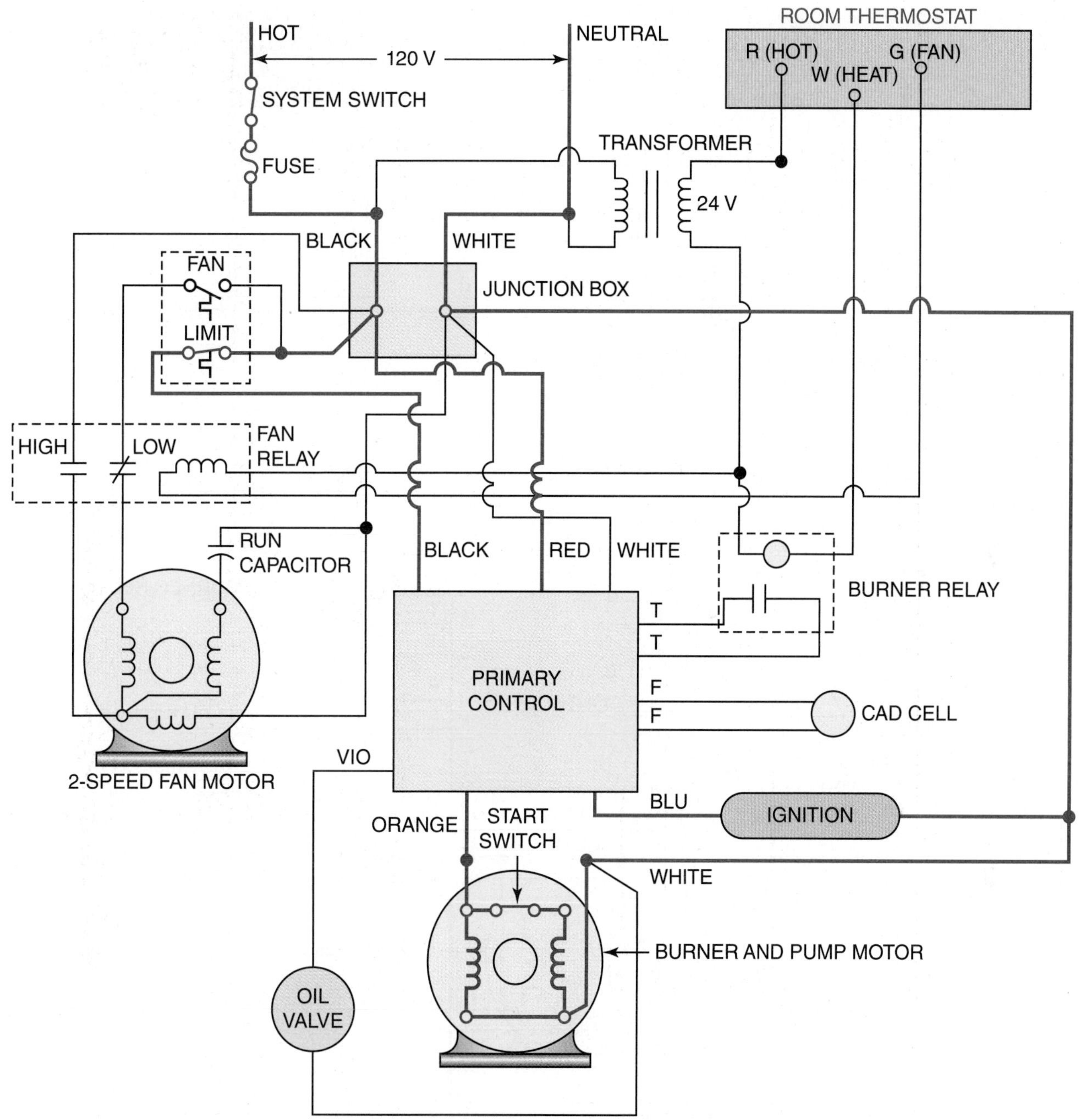

Oil furnaces should have the correct airflow across the heat exchanger. A check of the rise in air temperature may be performed if the installation is new or if it is suspected that the airflow is not correct. Often, for air-conditioning equipment the quantity of airflow must be known. The unit capacity may be found on the unit nameplate or calculated from the oil burner nozzle size. For example, the unit may have a 0.75-gph nozzle and a 65°F temperature rise. What should the cfm be?

Since one gallon of Number 2 fuel oil can produce about 140,000 Btu of heat energy per hour, a 0.75-gph nozzle will produce approximately 105,000 Btu per hour:

$$140{,}000 \text{ Btu/gal} \times 0.75 \text{ gal/h} = 105{,}000 \text{ Btu/h}$$

The 105,000 Btu/h is the input of the furnace. There will be a certain percentage of this heat that is wasted up the chimney. If we assume that the heating appliance is 80% efficient, the heat output will be

$$105{,}000 \text{ Btu/h} \times 0.80 = 84{,}000 \text{ Btu/h}$$

Figure 32.92 After the valve-delay-on period has elapsed, the oil valve's solenoid coil is energized and pressurized oil is pumped to the nozzle.

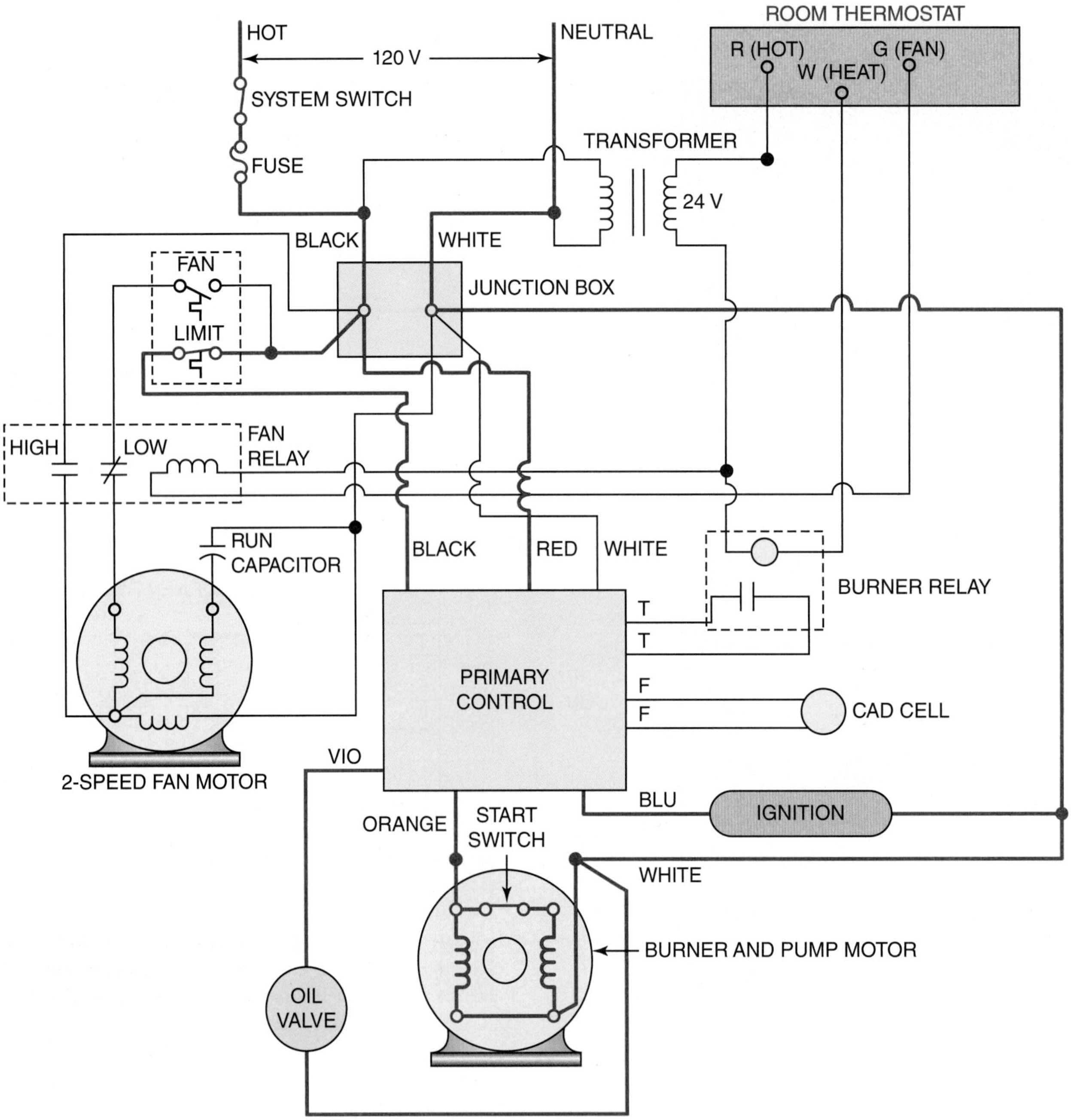

This result represents the amount of heat that is being added to the air in the structure. The following formula calculates the air volume (in cubic feet per minute) that is required to transfer this amount of heat:

$$Q_S = \text{cfm} \times 1.08 \times \Delta T$$

where Q_S = The quantity of heat in Btu/h
cfm = The air volume in cubic feet per minute
1.08 = A constant
ΔT = The temperature rise across the furnace

In this example, the Q_S is 84,000 Btu/h and the ΔT is 65°F. Plugging these values into the formula gives

$$84{,}000 \text{ Btu/h} = \text{cfm} \times 1.08 \times 65$$
$$84{,}000 \text{ Btu/h} = \text{cfm} \times 70.2$$
$$\text{cfm} = 84{,}000 \div 70.2$$
$$\text{cfm} = 1196.6$$

So, the required airflow is about 1200 cfm.

Figure 32.93 Inside a combustion chamber. Note the retention ring with flame locked to it. Courtesy R. W. Beckett Corp.

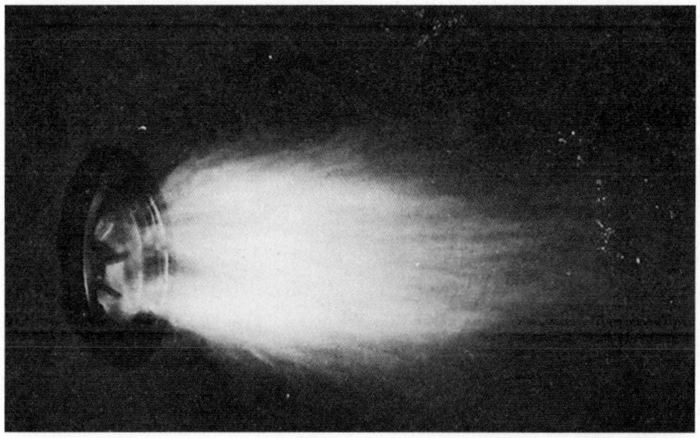

Figure 32.94 A heat exchanger in an oil-fired forced-air furnace. Courtesy The Williamson-Thermoflo Company

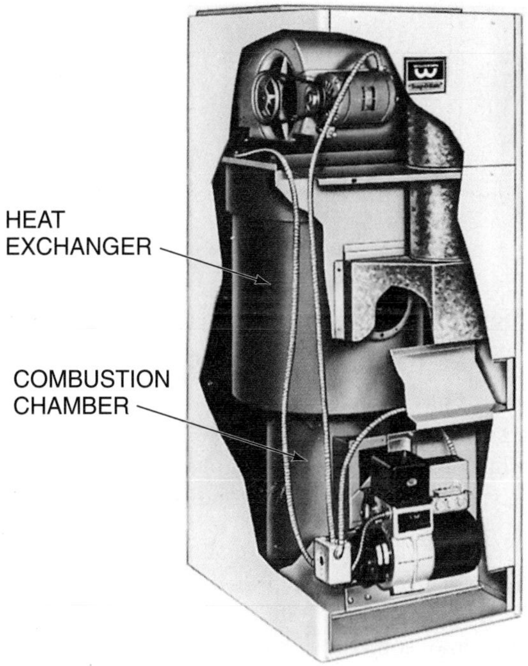

Figure 32.95 A mirror used by the heating technician to inspect the heat exchanger, the inside of the combustion chamber, and the flame. The handles on some models telescope, and the mirror angle is adjustable. Photo by Bill Johnson

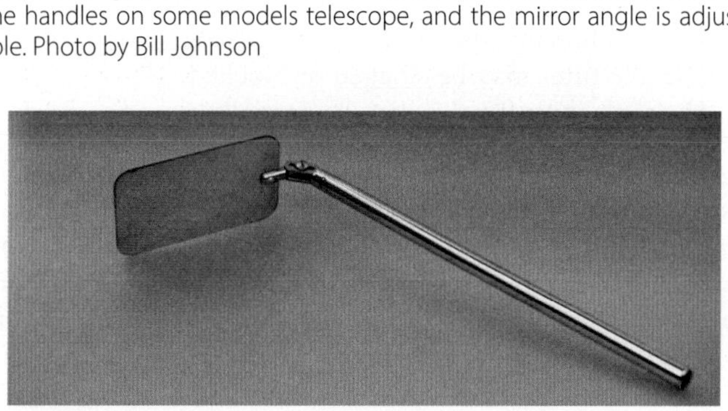

Figure 32.96 Small video cameras can be used to inspect heat exchangers and other furnace components. Photo by Eugene Silberstein

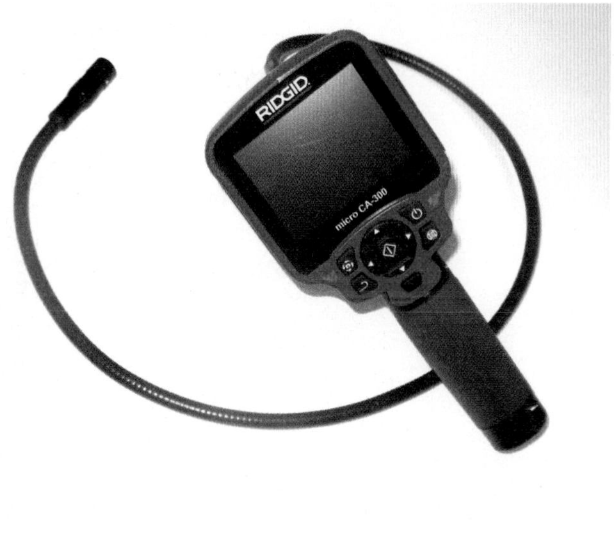

32.16 CONDENSING OIL FURNACE

Since energy prices have increased and the need for fossil-fuel conservation has become so evident, some manufacturers have worked to design more efficient oil furnaces. The condensing oil furnace is one of the product designs intended to achieve increased efficiency.

The condensing oil furnace, like the more conventional furnace, has two systems: the combustion or heat-producing system and the heated-air circulation system. The combustion system includes the burner and its related components: the combustion chamber, as many as three heat exchangers, and a vent fan and pipe. The heated-air circulation system includes the blower fan, housing, motor, plenum, and duct system. The following describes the operation of a condensing oil furnace manufactured by Yukon Energy Corporation, **Figure 32.97.**

The burner forces air and fuel oil into the combustion chamber where they are mixed and ignited. Much of this combustion heat is transferred to the main heat exchanger, which surrounds the combustion chamber. Combustion gases still containing heat are forced through a second heat exchanger where additional heat is transferred. Gases still containing some heat are forced through a third heat exchanger, where nearly all the remaining heat is removed. The third heat exchanger is a coil-type exchanger in which the temperature is reduced below the dew point, low enough for a change of state to occur, which results in moisture condensing from the flue gases. This is a latent-heat, high-efficiency exchange. About 1000 Btu are transferred to the airstream for each pound of moisture condensed (1 lb water = about 1 pint), so this can result in considerable savings. The remaining exhaust gases are vented to the outside, normally forced

Figure 32.97 A condensing oil furnace. Courtesy Yukon Energy Corp.

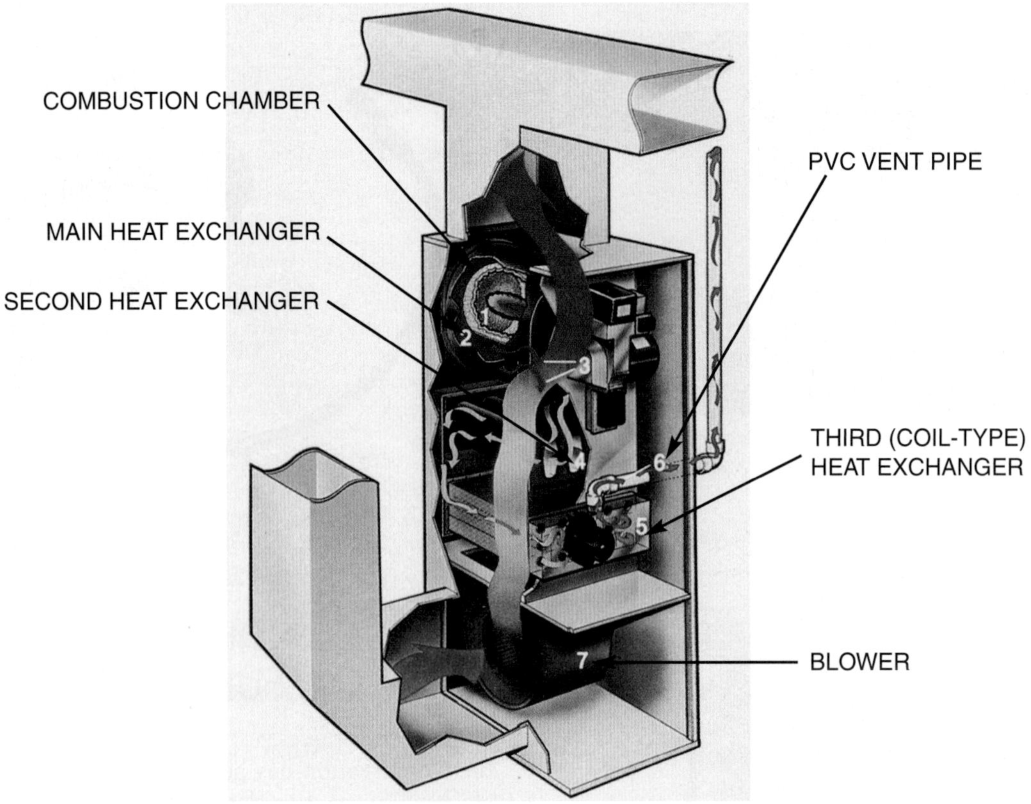

out by a blower through polyvinyl chloride (PVC) pipe. The condensing coil must be made of stainless steel to resist the acid in the flue gases. A drain is located at the condensing coil to remove the condensate to a container or to a suitable area or drain outside.

In the heated-air circulation system, the blower draws air through the return air duct and moves it across the condensing heat exchanger where the air is preheated by removing the sensible and latent heat from the coil. The air then passes around the second heat exchanger where additional heat is added. The circulating air then moves over the main heat exchanger, after which it is circulated to the rooms to be heated through the duct system. The leaving air is approximately 60°F warmer than when it entered the furnace through the return air duct.

Condensing oil furnaces have a 90+ AFUE. They also can be installed without a chimney because they are vented through a PVC pipe.

32.17 SERVICE PROCEDURES

PUMPS

The performance of residential and commercial fuel oil systems without booster pumps from tank to nozzle can be determined with a vacuum gauge and a pressure gauge. Before the vacuum and pressure checks are made, however, the following should be determined:

1. The tank has sufficient fuel oil.
2. The tank shutoff valve is open.
3. The tank location is noted (above or below burner).
4. The type of system is noted (one- or two-pipe).

Connect the vacuum and pressure gauges to the fuel oil unit as shown in **Figure 32.98**. If the tank is above the level of the burner, the supply system will typically be a one-pipe system, but may sometimes be a two-pipe system. When the tank is above the burner and the unit is operating, the vacuum gauge should read 0 in. Hg. If the gauge indicates a vacuum, one or more of the following may be the problem:

- The fuel line may be kinked.
- The line filter may be clogged or blocked.
- The tank shutoff valve may be partially closed.

If the supply tank is below the burner level, the vacuum gauge should indicate a reading: generally, 1 in. Hg for every foot of vertical lift and 1 in. Hg for every 10 ft of horizontal run. The lift and horizontal run combination should not total more than 17 in. Hg. (Remember, the lift capability of a dual-stage pump is approximately 15 ft.) Vacuum readings in excess of this calculated value may indicate a tubing size too

Figure 32.98 A pump with a pressure gauge and a vacuum gauge in place for tests and servicing. Courtesy Webster Electric Company

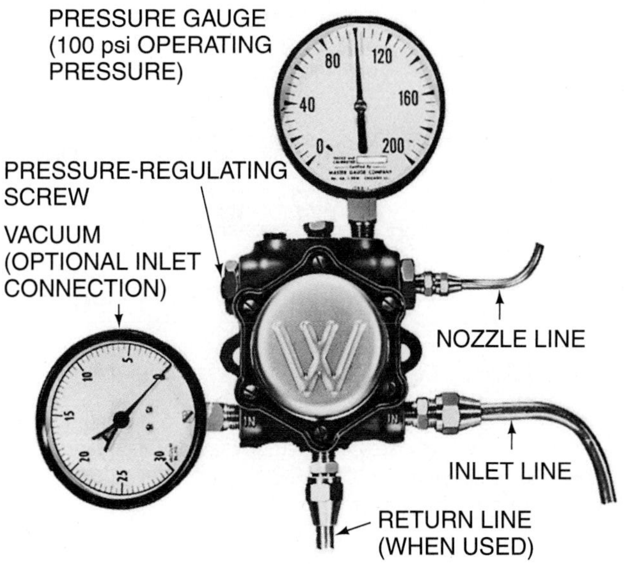

small for the run, but may also be an indication of a kinked or partially blocked line. This may cause the oil to foam.

The pressure gauge measures the performance of the pump and its ability to supply a steady, even pressure to the nozzle. The pressure regulator should be adjusted to the manufacturer's specifications, which is always above 100 psi. See **Figure 32.99** for an illustration of a pressure-adjusting valve. To adjust the regulator, with the pressure gauge in place, turn the adjusting screw until the gauge indicates 100 psi. With the pump shown in **Figure 32.99**, the valve-screw cover screw should be removed and the valve screw turned with a 1/8-in. Allen wrench. Some pumps require a screwdriver instead of an Allen wrench to adjust the regulator. With the regulator set and the unit in operation, the pressure gauge should give a steady reading. If the gauge pulsates, one or more of the following may be the problem:

- Partially clogged supply filter element
- Partially clogged unit filter or screen
- Air leak in fuel oil supply line
- Air leak in fuel oil pump cover

A pressure check of the pump cutoff is necessary if oil enters the combustion chamber after the unit has been shut down. To make this check, insert the pressure gauge in the outlet for the nozzle line, **Figure 32.100.** With the gauge in place and the unit running, record the pressure. When the unit is shut down, the pressure gauge should drop about 20% and hold. If the gauge drops to 0 psi, the fuel oil cutoff inside the pump is defective, and the pump may need to be replaced or an external solenoid installed, **Figure 32.53.** If the pressure gauge holds but there is still an indication of improper fuel oil shutoff, the problem may be one of the following:

- Possible air leak in fuel unit (pump)
- Possible air leak in fuel oil supply system
- Air trapped in the nozzle line

BURNER MOTOR

If the burner motor does not operate:

1. Press the reset button if there is one. Newer oil burners are equipped with permanent split-capacitor motors

Figure 32.99 A fuel pump and a pressure-adjusting valve. Courtesy Webster Electric Company

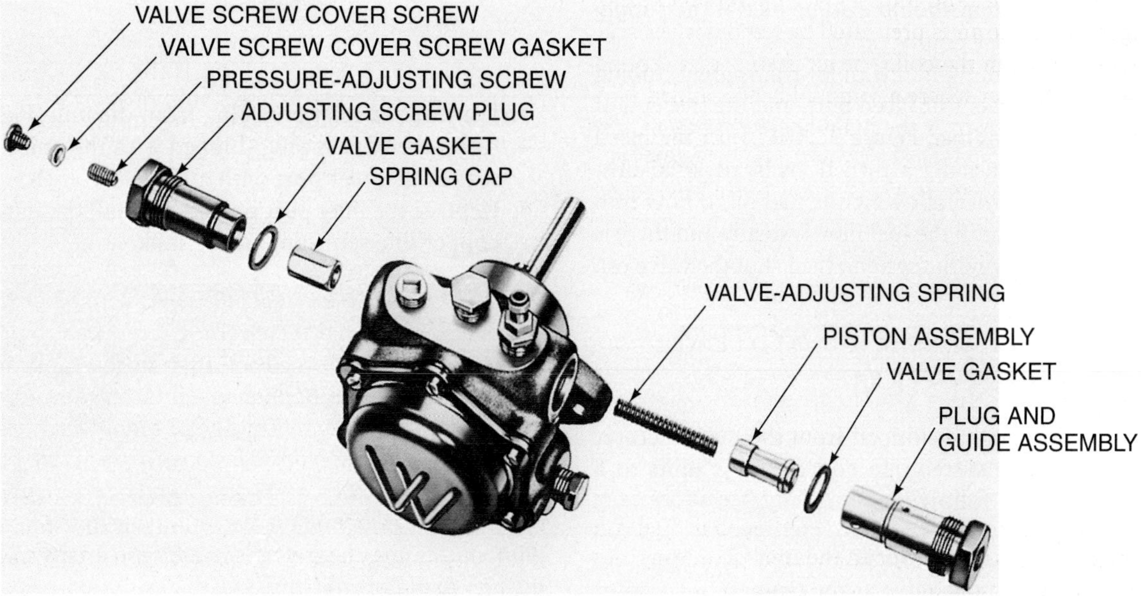

Figure 32.100 A pressure gauge in place to check the valve differential. Courtesy Webster Electric Company

Figure 32.101 A diagram of a fuel oil pump. Courtesy Webster Electric Company

that do not have reset buttons—they may just reset automatically from the internal overload. **NOTE:** *Reset may not function if the motor is too hot. Wait for the motor to cool down.* •

2. Check the electrical power to the motor. This can be done with a voltmeter across the orange and white or black and white leads to the motor. If there is power to the motor and nothing is binding it physically and it still does not operate, it should be replaced.

BLEEDING A ONE-PIPE SYSTEM

When starting an oil burner with a one-pipe system for the first time or whenever the fuel-line filter or pump is serviced, air will get into the system piping. The system must be bled to remove this air. The system should also be bled if the supply tank is allowed to become empty.

To bleed the system, ¼-in. flexible, transparent tubing should be placed on the vent or bleed port, and the free end of the tube placed in a container, **Figure 32.101**. Turn the bleed port counterclockwise (usually a turn from ⅛ to ¼ is sufficient). Then start the burner, allowing the fuel oil to flow into the container. Continue until the fuel flow is steady and there is no further evidence of air in the system; then shut the valve off.

CONVERTING A ONE-PIPE SYSTEM TO A TWO-PIPE SYSTEM

Most residential burners are shipped from the manufacturer for use with a one-pipe system. To convert these units to a two-pipe system, do the following:

1. Shut down all electrical power to the unit.
2. Close the fuel oil supply valve at the tank.

Figure 32.102 Installation of the bypass plug for conversion to a two-pipe system. Courtesy Webster Electric Company

3. Remove the inlet port plug from the unit, **Figure 32.102**.
4. Insert the bypass plug shipped with the unit into the deep seat of the inlet port with an Allen wrench.
5. Replace the inlet port plug and install the flare fitting and copper line returning to the tank.

NOTE: *Make certain that the specific manufacturer's instructions are followed when converting a one-pipe system to a two-pipe system.* **The procedure just outlined is for Webster pumps only**. *Pumps manufactured by Danfoss, for example, do not use a bypass. In addition, pumps manufactured by Suntec require that the bypass plug be installed in the return port.* •

SAFETY PRECAUTION: *Do not start the pump with the bypass plug in place unless the return line to the tank is installed. Without this line, there is no place for the excess oil to go, except possibly to rupture the shaft seal.* •

NOZZLES

Nozzle problems may be discovered by observing the condition of the flame in the combustion chamber, taking readings on a pressure gauge, and analyzing the flue gases. (Flue-gas analysis is discussed in Section 32.18, "Combustion Efficiency.") The following are common conditions that may relate to the nozzle:

- Smoke
- Flame changing in size and shape
- Flame impinging (striking) on the sides of the combustion chamber
- Sparks in the flame
- Low carbon dioxide reading in the flue gases (less than 8%)
- Delayed ignition
- Odors

When nozzle problems are apparent, the nozzle should be replaced. The nozzle is a delicate, finely machined component, and wire brushes and other cleaning tools should not be used on it. Generally, a nozzle should be replaced annually as a normal servicing procedure. **SAFETY PRECAUTION:** *Nozzles should be removed and replaced with a special nozzle wrench. Never use adjustable pliers or a pipe wrench.•* When the nozzle has been removed, keep the oil from running from the tube by plugging the end of the nozzle assembly. If oil leaves this tube, air will enter and cause an erratic flame when the burner is put back into operation. It may also cause an afterdrip or afterfire when the burner is shut down.

NOTE: *Do not attempt to clean the nozzle strainer. If the nozzle strainer is clogged, the cause should be determined. Clean the fuel oil nozzle line and the fuel unit (pump) strainer. Change the supply filter element. There may also be water in the supply tank, and the fuel oil may otherwise be contaminated.•*

It is important that nozzles do not overheat. Overheating causes the oil within the swirl chamber to break down into a varnish-like substance, which will build up. The following are major causes of overheating and varnish buildup:

- Fire burning too close to the nozzle
- Firebox too small
- Nozzle too far forward
- Inadequate air-handling components on burner
- Overfired burner

SAFETY PRECAUTION: *Do not allow an afterdrip or afterfire to go unchecked. Any leakage or afterfire will result in carbon formation and clogging of the nozzle. Check for an oil leak, air in the line, or a defective cutoff valve, if there is one. An air bubble behind the nozzle will not always be pushed out during burning. It may expand when the pump is shut off and cause an afterdrip.•*

IGNITION SYSTEM

Of all phases of oil burner operation and servicing, ignition problems are among the easiest to recognize and solve. Ignition service mostly consists of making the proper gap adjustment, cleaning the electrodes, and making all connections secure.

To check the ignition transformer, do the following:

1. Turn off power to the oil burner unit.
2. Swing back the ignition transformer.
3. Shut off the fuel supply or disconnect the burner motor lead.
4. Disconnect the wires from the primary winding of the ignition transformer.
5. Connect the ignition transformer tester, as shown in **Figure 32.64**.
6. Turn the ignition transformer tester on.
7. Observe the display on the tester to determine whether the transformer is good. **SAFETY PRECAUTION:** *Follow the manufacturer's instructions and keep your distance from the 10,000-V leads.•*

To check the electrodes, do the following:

1. Ensure that the three spark gap settings of the electrodes are set properly. Check **Figure 32.103** for the settings of

Figure 32.103 Electrode adjustments. Electrodes cannot be closer than ¼ in. to any metal part.

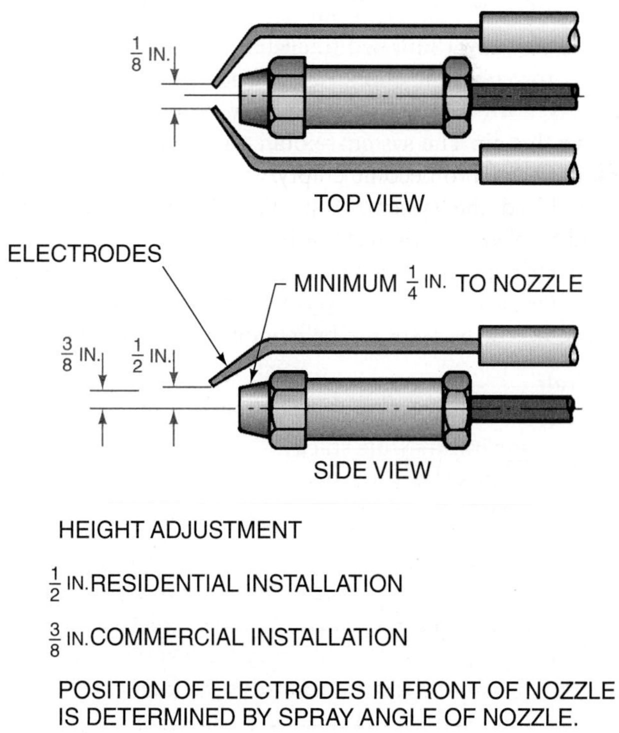

the gap, height above the center of the nozzle, and distance of the electrode tips forward from the nozzle center.

2. Make sure that the tips of the electrodes are not in the oil spray. **NOTE:** *Oil burner manufacturers have special gauges that help technicians properly position the electrode tips with respect to the nozzle orifice and to each other. When they are available, it is advisable to use these gauges, which can be easily obtained by calling the burner manufacturer.* •

ELECTRODE INSULATORS. Wipe the insulators clean with a cloth moistened with a solvent. If they are cracked or remain discolored after a good cleaning, they should be replaced.

32.18 COMBUSTION EFFICIENCY

Until a few years ago, the heating technician was primarily concerned with ensuring that the heating equipment operated cleanly and safely. Technicians made adjustments by using their eyes and ears. The high cost of fuel, however, now requires the technician to make adjustments using test equipment to ensure efficient combustion.

Fuels consist mainly of hydrocarbons in various amounts. In the combustion process, new compounds are formed and heat is released. The following are simplified formulas showing what happens during a perfect combustion process:

$$C + O_2 \rightarrow CO_2 + \text{heat}$$

$$H_2 + \tfrac{1}{2} O_2 \rightarrow H_2O + \text{heat}$$

During a normal combustion process, carbon monoxide, soot, smoke, and other impurities are produced along with heat. Excess air is supplied to ensure that the oil has enough oxygen for complete combustion, but even then the air and fuel may not be mixed perfectly. The technician must make adjustments to achieve near-perfect combustion, producing the most heat while reducing the quantity of unwanted impurities. On the other hand, too much excess air will absorb heat, which will be lost in the stack (flue gas) and reduce efficiency. Air contains only about 21% oxygen, so the remaining air does not contribute to the heating process.

The following tests can be made for proper combustion:

1. Draft
2. Smoke
3. Net temperature (flue stack)
4. Carbon dioxide

Technicians develop their own procedures for combustion testing. Making an adjustment to help correct one problem will often correct or help to correct others. Compromises have to be made also. When correcting one problem, another may be created. In some instances, problems may be caused by the furnace design. Some technicians use individual testing devices for each test, **Figure 32.104**.

Figure 32.104 Combustion efficiency testing equipment. (A) A draft gauge, nozzle gauge, and CO_2 tester. (B) A smoke tester. Photos by Bill Johnson

(A)

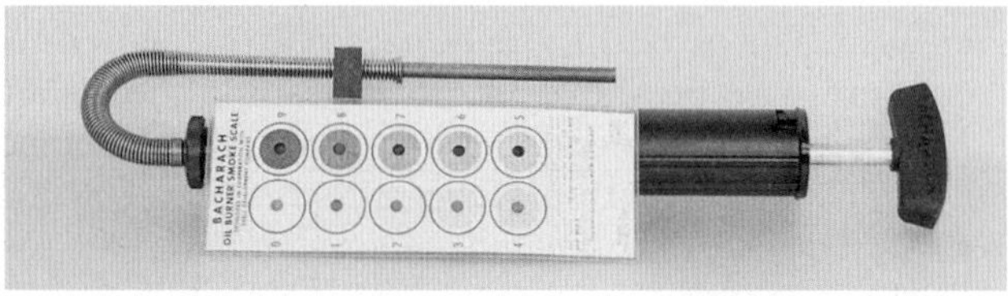

(B)

Figure 32.105 A combustion efficiency analyzer. Courtesy Bacharach, Inc., Pittsburgh, PA, USA

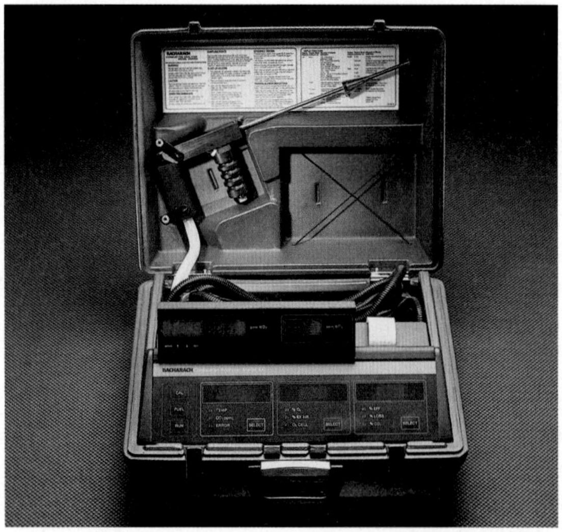

Others use electronic combustion analyzers with digital readouts, **Figure 32.105**. To make these tests with the individual testing devices, a hole must be drilled or punched in the flue pipe 12 in. from the furnace breaching, on the furnace side of the draft regulator, and at least 6 in. away from it. The hole should be of the correct size so that the stem or sampling tube of the instrument can be inserted into it.

⚠ **SAFETY PRECAUTION:** *The manufacturer's instructions furnished with the instruments should be followed carefully. The procedures indicated here are very general and should not take the place of the manufacturer's instructions.*•

DRAFT TEST

Correct draft is essential for efficient burner operation. The draft determines the rate at which combustion gases pass through the furnace, and it governs the amount of air supplied for combustion. The draft is created by the differences in temperature of the hot flue gases; it is negative pressure in relation to the atmosphere. Excessive draft can increase the stack temperature and reduce the amount of carbon dioxide in the flue gases. Insufficient draft may cause pressure in the combustion chamber, resulting in smoke and odor around the furnace. To obtain maximum efficiency, adjust the draft before you make other adjustments.

To perform the test, do the following:

1. Drill a hole into the combustion area for the draft tube. (On some furnaces, a bolt may be removed and the bolt hole used for access.)
2. Place the draft gauge on a level surface near the furnace and adjust to 0 in.
3. Turn the burner on and let it run for at least 5 min.
4. Insert the draft tube into the combustion area to check the overfire draft, **Figure 32.106.**
5. Insert the draft tube into the flue pipe to check the flue draft.

The overfire draft should be set to the manufacturer's specifications, typically, −0.02 in. WC. The flue draft should be adjusted with the draft regulator to maintain the proper overfire draft. Most residential oil burners require a flue draft of −0.04 in. to −0.06 in. WC. These are updrafts and are negative in relationship to atmospheric pressure. Longer flue passages require a higher flue draft than shorter flue passages.

Figure 32.106 Checking overfire draft. Photo by Bill Johnson

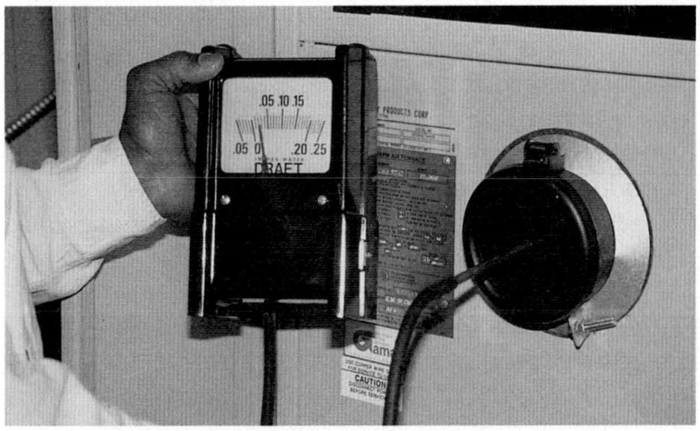

SMOKE TEST

Excessive smoke is evidence of incomplete combustion. Incomplete combustion will result in wasted fuel and higher energy bills. Excessive smoke also results in soot buildup on the heat exchanger and other heat-absorbing areas of the furnace. Soot is an insulator, so a buildup results in less heat being absorbed by the heat exchanger and increased heat loss to the flue. A 1/16-in. layer of soot can cause a 4.5% increase in fuel consumption, **Figure 32.107.**

The smoke test is accomplished by drawing a prescribed number of cubic inches of smoke-laden flue products through a specific area of filter paper. The residue on the filter paper is then compared to a scale furnished with the testing device. The degree of sooting can be read off the scale. A smoke tester such as the one illustrated in **Figure 32.108** may be used.

To perform the smoke test, **Figure 32.109**, do the following:

1. Turn on the burner and let it run for at least 5 min or until the stack thermometer stops rising.
2. Ensure that filter paper has been inserted in tester.
3. Insert the sampling tube of the test instrument into the hole in the flue.
4. Pull the tester handle the number of times indicated by the manufacturer's instructions.

Figure 32.107 The effect of soot on fuel consumption. Courtesy Bacharach, Inc., Pittsburgh, PA, USA

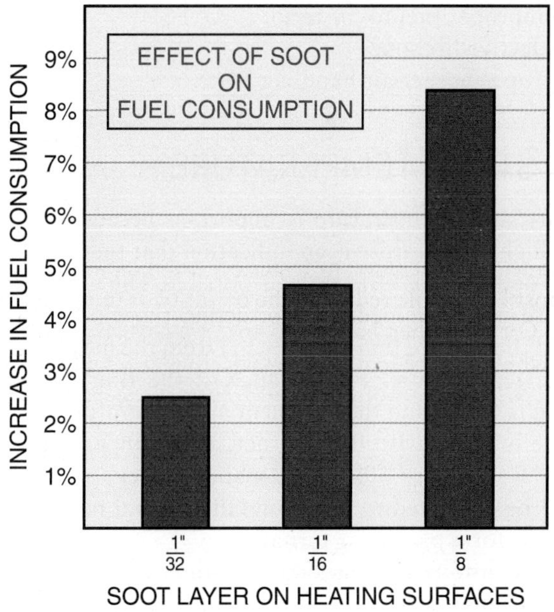

Figure 32.108 A smoke tester. Photo by Bill Johnson

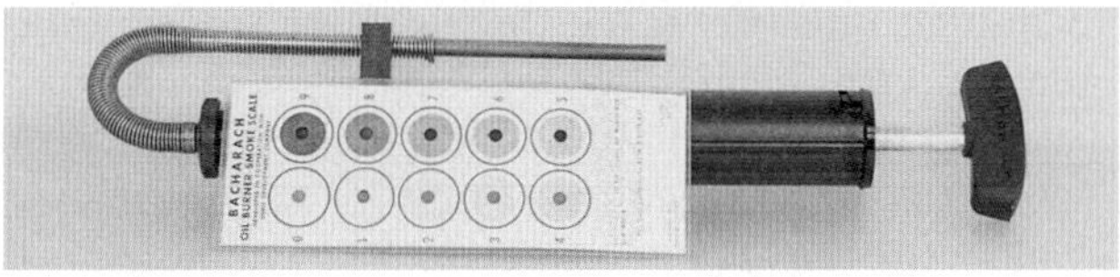

Figure 32.109 Making a smoke test. Photo by Bill Johnson

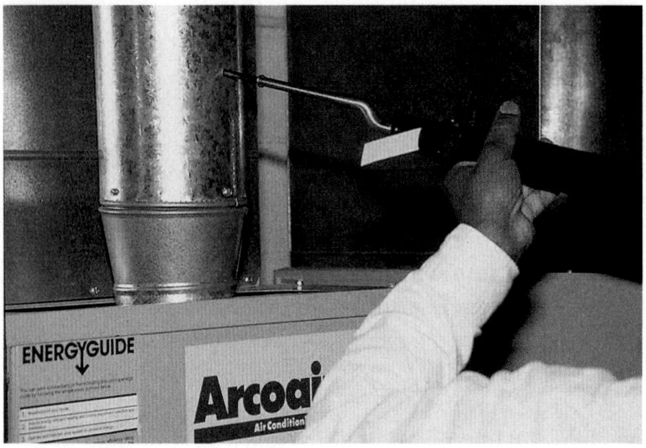

5. Remove the filter paper and compare it with the scale furnished with the instrument.

Excessive smoke can be caused by

- improper fan collar setting (burner air adjustment).
- improper draft adjustment (draft regulator may be required or need adjustment).
- a poor fuel supply (pressure).
- the oil pump not functioning properly.
- a defective or incorrect nozzle.
- excessive air leaks in the furnace (air diluting flame).
- an improper fuel-to-air ratio.
- a defective firebox.
- improper burner air-handling parts.

NET STACK TEMPERATURE

The net stack temperature is important because an abnormally high temperature is an indication that the furnace may not be operating as efficiently as possible. The net stack temperature is determined by subtracting the air temperature around the furnace (burner inlet air) from the measured stack or flue temperature. For instance, if the flue temperature reading is 650°F and the basement air temperature where the furnace is located is 60°F, the net stack temperature would be 650°F − 60°F = 590°F. The manufacturer's specifications should be consulted to determine the normal net stack temperatures for a particular furnace.

To determine the stack temperature, **Figure 32.110**, do the following:

1. Insert the thermometer stem into the hole in the flue.
2. Turn on the burner and allow it to run for at least 5 min or until the stack thermometer stops rising in temperature.
3. Subtract the basement or ambient temperature from the stack temperature reading.
4. Compare the temperature to the manufacturer's specifications.

Figure 32.110 Making a stack gas temperature test. Photo by Bill Johnson

A high stack temperature may be caused by

- excessive draft through the combustion chamber.
- a dirty or soot-covered heat exchanger.
- lack of baffling.
- an incorrect or defective combustion chamber.
- overfiring (check nozzle size and pressure).

CARBON DIOXIDE TEST

The carbon dioxide test is an important combustion efficiency test. A high carbon dioxide reading is good. If the test reading is low, it indicates that the fuel oil has not burned efficiently or completely. The reading should be considered along with all of the other test readings. Under most normal conditions, a carbon dioxide reading greater than 10% should be obtained. If problems exist that would be very difficult to correct but the furnace is considered safe to operate and has a net stack temperature of 400°F or less, a carbon dioxide reading of 8% could be acceptable. However, if the net stack temperature is over 500°F, a reading of at least 9% carbon dioxide should be obtained. **⚠ SAFETY PRECAUTION:** *A CO_2 reading over 12.5% can be dangerous.•*

To perform the carbon dioxide test, **Figure 32.111**, do the following:

1. Turn on the burner and operate it for at least 5 min.
2. Insert the thermometer and wait for the temperature to stop rising.
3. Insert the sampling tube into the hole previously made in the flue pipe.
4. Remove a test sampling, using the procedure provided by the manufacturer of the test instrument.
5. Mix the fluid in the test instrument with the sample gases from the flue according to instructions.
6. Read the percentage of carbon dioxide from the scale on the instrument.

A low percentage carbon dioxide reading may be caused by one or more of the following:

- High draft or draft regulator not working properly
- Excess combustion air

Figure 32.111 Performing a carbon dioxide test. Photo by Bill Johnson

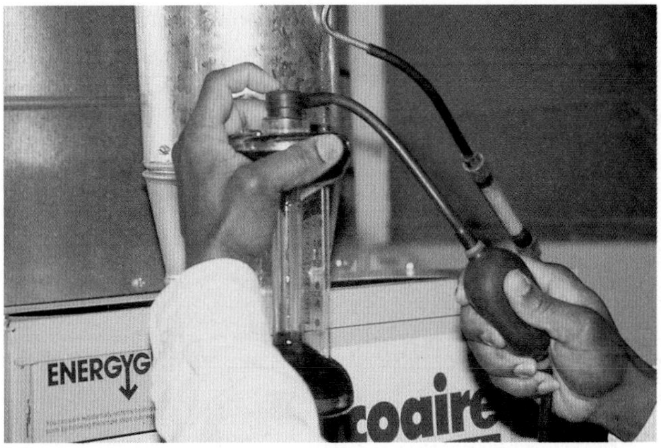

- Air leakage into the combustion chamber
- Poor oil atomization
- A worn, clogged, or incorrect nozzle
- Oil-pressure regulator set incorrectly

ELECTRONIC COMBUSTION ANALYZERS

Electronic combustion analyzers provide information similar to that provided by the test instruments already described. Electronic analyzers may measure carbon monoxide and oxygen concentrations rather than carbon dioxide percentages, but from the manufacturer's instructions the technician can determine acceptable levels and necessary corrections to be made on the furnace. Generally, the tests are made from samplings from the same position in the flue as used with the other instruments. Electronic instruments are easier to use and save time. The readings appear on a convenient digital readout system, **Figure 32.105**.

HVACR GOLDEN RULES

When making a service call to a residence, do the following:

- Ask the customer about the problem. The customer can often help you solve the problem and save you time.
- Check the humidifiers, where applicable. Annual service is highly recommended for health reasons.
- Be prepared and do not let any fuel oil escape on the ground or floor around the furnace. The customer will notice the odor.

Added Value to the Customer

Here are some simple, inexpensive procedures that may be included in the basic service call.

- Check flame characteristics; a smoke test is good insurance and may lead to extra paying service work.
- Replace all panels with the correct fasteners.
- Inspect heat exchanger for soot buildup.

32.19 PREVENTIVE MAINTENANCE

Oil heat, like gas heat, has air-side and burner-side components. Because the air-side components have been discussed in the electric heat unit, only the burner section will be discussed here.

Oil heat requires more consistent regular service than any other heating system discussed in this book. The reason is that the unit uses fuel oil that must be properly metered and burned for the best efficiency. Burning efficiency is accomplished by ensuring the correct fuel supply and properly adjusting and maintaining the burner section. This includes the combustion chamber, the heat exchanger, and the flue system. They must all be working perfectly for proper burning efficiency. Improper combustion will cause soot to form on the heat exchanger, which will reduce the rate of heat transfer in the appliance. The condition will deteriorate and cause combustion efficiency to worsen. It will pay the customer to have the system serviced each year before the season starts, using the appropriate portions of the following procedures. These procedures may look like a long service call for each furnace every year, but for a typical furnace a competent service technician should be able to accomplish the maintenance portions in an hour. Corrective actions are not considered as part of maintenance. We will start discussion of the preventive maintenance procedures at the tank and move to the flue.

Every year, the service technician should check the tank for water accumulation in the bottom. A commercial paste that changes color when in contact with water may be spread on the oil-tank measuring stick and inserted into the tank. If there is water in the bottom of the tank, when the stick is pulled out of the oil the paste will show the level of the water. If the tank is leaking, it should be replaced. Often, an aboveground tank may be substituted. Aboveground tanks are easier to inspect and, in the event of a leak, pose less of a danger to the environment.

The oil lines should be inspected for bends, dents, and rust where they leave the ground and enter the house. A small pinhole in a supply line will cause air to be pulled into the system if the tank is below the burner. This will cause an erratic flame at the burner. The burner may also have an afterburn when the unit stops, caused by the expanding air

in the system. Any flame problems, problems priming the pump, or afterburn problems should prompt you to check the supply line for leaks.

A vacuum test may be performed by disconnecting the oil supply line from the tank and letting the system pump the oil supply line into a vacuum and shutting the pump off. The line should stay in a vacuum if it is leak-free. **NOTE:** *Oil and air may leak backward through the pump. This may be prevented by disconnecting the oil return line and holding your finger over it once the pump is stopped.* •

The gun burner should be removed from the unit and the combustion chamber inspected to make sure the refractory is in good condition. Look for signs of soot, and vacuum it out as needed. While the burner is out, change the oil nozzle; be sure to replace the old one with the correct nozzle and use only a nozzle wrench. Check the burner head for overheating and cracks. Examine the static tube for signs of oil or soot, and clean as needed. Set the electrodes using a gauge, and examine the insulators for cracks and soot deposits. Clean as needed and change if cracked. Examine the flexible coupling for signs of wear. Make sure the cad cell is clean and aligned correctly and that the bracket is tight. Lubricate the burner motor. When returning the gun burner to the furnace, be sure to install the gasket and tighten all bolts.

Change the in-line oil filter. Change the gasket and tighten it correctly. This will help ensure a good start.

Start the furnace and while it is heating up to operating temperature, prepare to perform a combustion analysis. Check the draft.

Insert the thermometer in the flue and when it reaches the correct temperature, perform a smoke test. If the fire is smoking, adjust the air until the correct smoke spot is obtained. If you suspect any problem, check the oil pressure to be sure that you have the proper pressure at the nozzle. When the smoke is correct, perform a combustion analysis and adjust to the correct carbon dioxide reading.

When all of this is accomplished, the furnace should operate correctly for the next year.

32.20 DIAGNOSTIC CHART FOR OIL HEAT

The oil-burning appliances discussed here are the forced-air type. (Many of the same rules apply to oil boilers as far as their burner operation is concerned.) These oil-burning systems may be the primary heating systems in homes and businesses. Two basic types of oil safety controls are the cad cell and the stack switch. Both are discussed in the text. Only the basic procedures are discussed here. For more specific information about any particular step of the procedure, refer to descriptions of the particular oil heat unit as it is discussed in the text or consult the manufacturer. Always listen to customers; their input can help you to locate the problem. Remember, they are present most of the time and often pay attention to how the equipment functions.

Problem	Possible Cause	Possible Repair
Furnace will not start—no heat	Open disconnect switch	Close disconnect switch.
	Open fuse or breaker	Replace fuse or reset breaker and determine why it opened.
	Faulty wiring	Repair or replace faulty wiring or connections.
	Defective low-voltage transformer	Replace transformer and look for possible overload condition.
	Primary safety control off—needs reset	Check for oil accumulation in the combustion chamber; if none, reset the control and observe fire and flame characteristics.
	Tripped burner motor reset	Press reset button, check amperage; if too much, check motor or pump for binding.
Furnace starts after reset but shuts off after 90 sec	Cad cell may be out of alignment or dirty	Check cad cell for alignment and smoke on lens—if there are smoke deposits on the lens, adjust the burner for correct fire.
	Defective cad cell	Replace cad cell and reset.
	Defective primary control	Change primary control.

(continued)

Problem	Possible Cause	Possible Repair
Furnace starts but no ignition occurs	No fuel	Fill fuel tank.
	Electrodes out of alignment	Align electrodes.
	Defective ignition transformer	Replace transformer.
	Defective oil pump	Replace oil pump.
	Restricted fuel filter	Replace filter.
	Defective coupling between pump and motor	Replace coupling.
Burner runs and ignition occurs, but fan does not start	Faulty wiring or connections in fan circuit	Repair or replace faulty wiring or connectors.
	Defective fan switch	Replace fan switch.
	Defective fan motor	Replace fan motor.
Burner has delayed ignition, makes noise on start-up	Electrode out of alignment	Align electrodes.
	Clogged nozzle	Change nozzle and line filter.
	Transformer weak	Change transformer.
	Too much or too little air	Adjust air.
	Nozzle position in burner	Adjust nozzle position.
Burner makes noise on shutdown	Fuel cutoff at fuel pump	Check fuel cutoff using gauges. If not correct, change the fuel pump or add a solenoid in the nozzle line.
High stack temperature	Too much air to burner	Correct air by adjustment.
	Too much draft	Adjust or add draft regulator.
	Overfired	Change to correct nozzle size with correct oil pressure.
Smoking	Too little air	Correct air by adjustment.
	Not enough draft	Clean flue and clean furnace heat exchanger on the oil side.
Smoke in conditioned space	Cracked heat exchanger	Replace heat exchanger.
	Smoke puffing out around burner or inspection door and pulling in through fan compartment	See preceding smoking problem.
Oil smell	Oil leak or spill during service	Clean up oil spill or leak.

32.21 SERVICE TECHNICIAN CALLS

SERVICE CALL 1

A new customer calls requesting *a complete checkup of the oil furnace, including an efficiency test. This customer will stay with the service technician and watch the complete procedure.*

The technician reads on his work order that this is a new customer. The technician parks his service truck on the street, makes certain he has his company-issued photo identification around his neck, and walks to the front door. After ringing the bell, the technician stands back and waits for the customer to open the door. The technician introduces himself and the company and hands the customer his business card.

The technician explains to the customer that the first thing to do is run the furnace and perform an efficiency test. By running a test before and after adjustment, the technician can report results. The thermostat is set to about 10°F above the room temperature to allow time for it to warm up while the technician sets up. This will help ensure that the furnace will not shut off during the test. The technician gets the proper tools and goes to the basement where the furnace is located. The customer is already there and asks whether the technician would mind an observer. The technician explains that a good technician should not mind being watched.

The technician inserts the stack thermometer in the flue and observes the temperature; it is no longer rising. A sample of the combustion gas is taken and checked for efficiency. A smoke test is also performed. The unit is operating with a slight amount of smoke, and the test shows the efficiency to be 65%. This is about 15% to 20% lower than normal.

The technician then removes a low-voltage wire from the primary, which shuts off the oil burner and allows the fan to continue to run and cool the furnace. The technician then removes the burner nozzle assembly and replaces the nozzle with the correct nozzle. Before returning the nozzle assembly to its place, the technician also sets the electrode spacing and then changes the oil filter, changes the air filter, and oils the furnace motor. The furnace is ready to start again.

The technician starts the furnace and allows it to heat up. When the stack temperature has stopped rising, the technician pulls another sample of flue gas. The furnace is now operating at about 67% efficiency and is still a little smoky in the smoke test. The technician checks the furnace nameplate and notices that this furnace needs a 0.75-gph nozzle but has a 1-gph nozzle. It is common (but poor practice) for some technicians to use what they have, even if it is not exactly correct.

This technician removes the nozzle just installed as an exact replacement and then gets more serious about the furnace because it has been mishandled. The technician installs the correct nozzle, starts the unit, and checks the oil pressure. The pressure has been reduced to 75 psig to correct for the oversized nozzle. The technician looks up the appliance information and determines that, for the particular furnace and oil burner configuration, the pump pressure should be 100 psig. He adjusts the pressure accordingly.

The air to the burner is adjusted; meanwhile, the stack temperature is reading 660°F (this is a net temperature of 600°F because the room temperature is 60°F). The technician runs another smoke test, which now shows minimum smoke. The efficiency is now 73%. This is much better.

The customer has been kept informed all through the process and is surprised to learn that oil burner service is so exact. The technician sets the room thermostat back to normal and leaves a satisfied customer. **SERVICE CALL NOTE:** *The technician used a "wet kit" to perform the combustion efficiency test. If the technician had been using an electronic test kit, the smoke test would have been performed first, to avoid potential damage to the instrument.*•

SERVICE CALL 2

A residential customer calls and reports that the *oil furnace in a basement under the family room is making a noise when it shuts off. The oil pump is not shutting off the oil fast enough.*

The technician is running a little late on his present service job and calls to alert the customer that he will be arriving about 20 min later than scheduled. The home-owner thanks the technician for the update. The technician arrives, apologizes for the delay, and asks the customer for some information regarding the problems she is having with the system. Once he has heard the system complaint, he turns up the room thermostat above the room temperature to keep the furnace running. The technician removes a low-voltage wire from the primary control of the furnace to stop the burner. The fire does not extinguish immediately but shuts down slowly with a rumble. The technician determines that the installation of a solenoid valve in the oil line will eliminate the problem. He calls the office to get the price for the repair. The technician explains the situation to the customer, along with the cost of repair, and the customer gives her approval.

The technician installs a solenoid in the small oil line that runs from the pump to the nozzle and wires the solenoid coil in parallel with the burner motor so that it will be energized only when the burner is operating. The technician disconnects the line where it goes into the burner housing and places the end in a bottle to catch any oil that may escape. The burner is then turned on for a few seconds. This clears any air that may be trapped in the solenoid out of the line leading to the nozzle.

The technician reconnects the line to the housing and starts the burner. After it has been running for a few minutes, he shuts the burner down. It has a normal shutdown. The start-up and shutdown are repeated several times. The furnace operates correctly. The technician then returns the room thermostat to the correct setting before leaving. He presents the bill to the customer, explains what was done, and demonstrates to the customer that the noises on shutdown are gone. The technician gets the customer's signature on the work order, thanks the customer for her loyalty to the company, and leaves for his next job.

SERVICE CALL 3

A customer from a duplex apartment calls. There is no heat. The customer is out of fuel. The customer had fuel delivered last week, but the driver filled the wrong tank. This system has an underground tank and it is sometimes hard to get the oil pump to prime (pull fuel to the pump). The technician is misled for some time, thinking that there is fuel and that the pump is defective.

The technician arrives to find the customer very upset that there is no heat. After expressing sympathy and assuring the customer that he will make every effort to find out what the problem is and fix it, the technician goes to the room thermostat. It is set at 10°F above the room temperature. Then he goes to the furnace, which is in the garage at the end of the apartment. The furnace is off because the primary control has tripped. The technician examines the combustion chamber with a flashlight for any oil buildup that may have accumulated if the customer has been resetting the primary. **SAFETY PRECAUTION:** *If a customer has repeatedly reset the primary to get the furnace to fire and the technician then starts the furnace, the excess oil would be dangerous.*• There is no

excess oil in the combustion chamber, so the reset button is pushed. The burner does not fire.

The technician suspects that the electrodes are not sparking or that the pump is not pumping. The first thing to do is to fasten a piece of flexible tubing to the bleed port side of the oil pump. The end of the hose is placed in a bottle to catch any fuel that may escape, then the bleed port is opened and the primary control is reset. The burner and pump motor start. No oil is coming out of the tubing, so the pump must be defective. Before changing the pump, the technician decides to make sure there is oil in the line and removes the line entering the pump. There is no oil in the line. The technician opens the filter housing and finds very little oil in the filter. The technician now decides to check the tank, and he determines that there is no oil in the tank.

When the technician tells the customer that the oil tank is empty, the customer shows the receipt for the delivery. The technician calls his supervisor and, after some investigation, finds out that the oil truck driver delivered oil to the wrong tank, even though the truck had been sent out to fill the correct tank. While waiting for the oil truck, the technician apologizes for the confusion.

After the oil is delivered, technician starts the furnace, bleeds the pump until a full line of liquid oil flows, and then closes the bleed port. The burner ignites and goes through a normal cycle. The furnace is not under contract for maintenance, so the technician suggests to the occupant that a complete service call including a nozzle change, electrode adjustment, and filter change be arranged. The customer agrees. The technician completes the service and turns the thermostat to the normal setting before leaving. Upon returning to the office, the technician suggests to his service manager that the oil tanks be labeled with the apartment number to avoid this confusion in the future.

Solutions for the following service calls are not provided here. Please discuss the solutions with your instructor.

SERVICE CALL 4

A customer calls reporting a smell of smoke when the furnace starts up. This customer does not have the furnace serviced each year.

The technician goes to the furnace in the basement, followed by the customer. The technician turns the system switch off and goes back upstairs and sets the thermostat 5°F above the room temperature so the furnace can be started from the basement. When the technician gets back to the basement, he starts the furnace and observes a puff of smoke. The technician inserts a draft gauge in the burner door port; the draft is +0.01 in. WC positive pressure.

What is the likely problem and the recommended solution?

SERVICE CALL 5

A retail store manager calls stating that there is no heat in the small store.

When the technician arrives, he discovers the room thermostat is set at 10°F higher than the room temperature. The thermostat is calling for heat. The technician goes to the furnace in the basement and discovers it needs resetting in order to run. The technician examines the combustion chamber with a flashlight and finds no oil accumulation. He then resets the primary control. The burner motor starts, and the fuel ignites. The burner runs for 90 sec and then shuts down.

What is the likely problem and the recommended solution?

SERVICE CALL 6

A customer reports that there is no heat and the furnace will not start when the reset button is pushed.

The technician arrives and checks to see that the thermostat is calling for heat. The set point is much higher than the room temperature. The technician goes to the garage where the furnace is located and presses the reset; nothing happens. The primary is carefully checked to see whether there is power to the primary control. It shows 120 V. Next, the circuit leaving the primary, the orange wire, is checked. (The technician realizes that power comes into the primary on the white and black wires, the white being neutral, and leaves the orange wire on.) There is power on the white to black but none on white to orange, **Figure 32.80**.

What is the likely problem and the recommended solution?

REVIEW QUESTIONS

1. All of the following are parts of the heat-producing portion of an oil-fired, warm-air furnace except the
   A. oil burner.
   B. fuel supply components.
   C. blower assembly.
   D. combustion chamber.
2. Which of the following is true regarding the sequence of operations of an oil-fired, warm-air furnace?
   A. The furnace blower cycles on before the oil burner fires.
   B. When the oil burner shuts off at the end of a heating cycle, the furnace blower continues to operate.
   C. When the oil burner shuts off at the end of a heating cycle, the furnace blower shuts off at the same time as the burner.
   D. Air mixes with 100% atomized fuel oil only after the oil has ignited.
3. The most common furnace configuration for use in a crawl space is the
   A. horizontal furnace.
   B. vertical furnace.
   C. lowboy furnace.
   D. downflow furnace.
4. The furnace configuration that is often used in spaces where there is little headroom is the
   A. horizontal furnace.
   B. vertical furnace.
   C. lowboy furnace.
   D. downflow furnace.
5. How many grades of fuel oil are normally considered as heating oils?
   A. 3
   B. 4
   C. 5
   D. 6
6. Fuel oils are classified as
   A. hydrocarbons (HC).
   B. hydrofluorocarbons (HFC).
   C. chlorofluorocarbons (CFC).
   D. hydrochlorofluorocarbons (HCFC).
7. Perfect combustion produces which of the following?
   A. Heat, oxygen, and carbon dioxide
   B. Heat, carbon dioxide, and water vapor
   C. Oxygen, water vapor, and heat
   D. Oxygen, carbon dioxide, and heat
8. Combustion occurs when a substance rapidly unites with
   A. carbon.
   B. hydrogen.
   C. helium.
   D. oxygen.
9. The heat content of Number 2 fuel oil is about
   A. 70,000 Btu/gal.
   B. 70,000 Btu/h.
   C. 140,000 Btu/gal.
   D. 140,000 Btu/h.
10. If a residential oil customer uses, on average, 900 gallons of oil per year, what size oil tank is recommended?
    A. 125 gallons
    B. 275 gallons
    C. 450 gallons
    D. 900 gallons
11. One of the main benefits of indoor oil storage is that
    A. it is easier for the oil company to make deliveries.
    B. indoor oil tanks take up less space than buried tanks.
    C. there is less fluctuation in the fuel temperature.
    D. there is more condensation and moisture in indoor tanks.
12. How often should tanks be tested for water?
    A. Weekly
    B. Monthly
    C. Every 6 months
    D. Annually
13. Which of the following is a drawback of using a two-pipe oil delivery system?
    A. The oil filter will require more frequent changing.
    B. They are ineffective when the oil storage tank is located below the oil burner.
    C. The oil lines must be bled of air periodically.
    D. The amount of oil moved by the oil pump is greatly reduced
14. Where should the vent alarm be installed on an oil delivery system?
    A. Immediately downstream of the oil filter
    B. In the vent pipe
    C. Between the oil filter and the oil burner
    D. In the oil tank's fill line
15. An oil deaerator performs which function?
    A. It allows a two-pipe system to function as a one-pipe system.
    B. It allows a one-pipe system to function as a two-pipe system.
    C. It allows for the introduction of excess air into the oil-air mixture.
    D. It allows the fuel pump to run slower, pumping less oil than a one-pipe system.
16. In oil heating, the acronym OSV stands for
    A. oil's standard viscosity.
    B. oil supply vacuum.
    C. oil safety valve.
    D. optimal shutoff valve.

17. How many pounds of oxygen are required to completely burn 1 pound of Number 2 fuel oil?
   A. 1.6
   B. 3.0
   C. 5.9
   D. 14.4
18. Most oil burners are set up to burn oil with how much excess air?
   A. 10%
   B. 30%
   C. 50%
   D. 100%
19. Which three components are required for combustion?
   A. Air, oxygen, and heat
   B. Oxygen, fuel, and air
   C. Air, oxygen, and heat
   D. Oxygen, fuel, and heat
20. Incomplete combustion produces which of the following toxic substances?
   A. Soot
   B. Carbon monoxide
   C. Carbon dioxide
   D. Water vapor
21. Before fuel oil can burn it must be
   A. atomized.
   B. solidified.
   C. liquefied.
   D. carbonized.
22. If an oil-fired, residential furnace produces 150,000 Btu/h of heat, how many gallons of oil will have burned if the appliance operates at this capacity for 3 continuous hours?
   A. 3.2 gallons
   B. 4.8 gallons
   C. 6.4 gallons
   D. 8.0 gallons
23. The motor on an oil burner is responsible for turning the
   A. burner fan.
   B. burner fan and the fuel pump.
   C. fuel pump.
   D. fuel pump and the vent blower.
24. The oil burner component that is directly connected to the fuel pump is the
   A. burner motor.
   B. blower wheel.
   C. air tube.
   D. flexible coupler.
25. Oil burner nozzles are rated at what pressure?
   A. 50 psig
   B. 75 psig
   C. 100 psig
   D. 150 psig
26. A system that has the oil burner located 30 feet away from, and 3 feet above the oil tank, will operate with an approximate vacuum of how many inches if there is an oil filter and OSV installed in the oil line?
   A. 7 in. of mercury vacuum
   B. 10 in. of mercury vacuum
   C. 13 in. of mercury vacuum
   D. 16 in. of mercury vacuum
27. According to the chart provided in Figure 32.43, how oil will be delivered by a nozzle that is rated at 1.0 gph if the operating pressure is 150 psig?
   A. 1.0 gph
   B. 1.12 gph
   C. 1.18 gph
   D. 1.23 gph
28. Which three terms are most commonly associated with an oil burner nozzle's function?
   A. Atomize, angle, and pattern
   B. Pump, pressure, and angle
   C. Atomize, pump, and pressure
   D. Pattern, pressure, and angle
29. The voltage provided by a solid-state igniter typically ranges from
   A. 5000 V to 8 000 V.
   B. 7000 V to 11,000 V.
   C. 10,000 V to 15,000 V.
   D. 14,000 V to 20,000 V.
30. The primary control monitors/controls the operation of all of the following except
   A. the safe starting and stopping of the oil burner.
   B. the ignition of the oil-air mixture.
   C. the speed at which the oil burner motor rotates.
   D. the opening and closing of the oil valve.
31. Which of the following statements is true regarding the FAN/LIMIT switch shown in Figure 32.77?
   A. Both the fan and limit switches open on a rise in temperature.
   B. Both the fan and limit switches close on a rise in temperature.
   C. The fan switch closes on a rise in temperature and the limit switch opens on a rise in temperature.
   D. The limit switch closes on a rise in temperature and the fan switch opens on a rise in temperature.
32. According to the wiring diagram in Figure 32.80, 115 V will be supplied to the primary control when which of the following sets of conditions is met?
   A. System switch open, limit switch closed, fuse good, 115 V at distribution panel
   B. System switch closed, limit switch closed, fuse good, 115 V at distribution panel
   C. System switch closed, limit switch open, fuse good, 115 V at distribution panel
   D. System switch open, limit switch open, fuse good, 115 V at distribution panel

33. On a stack relay,
    A. the hot and cold contacts open and close at the same time.
    B. the hot contacts open when the flue pipe is hot.
    C. the hot contacts close when the flue pipe is hot.
    D. the cold contacts open when the flue pipe is cold.
34. Which of the following statements is true regarding the cad cell?
    A. The cad cell is a component part of a stack relay primary control.
    B. The resistance of a cad cell decreases when it senses light.
    C. The resistance of a cad cell increases when it senses light.
    D. The resistance of a cad cell decreases when no light is sensed.
35. A primary control with an intermittent ignition control strategy, such as the one shown in Figure 32.84,
    A. energizes the burner motor, the oil valve, and ignition during a call for heat.
    B. energizes the ignition only at the beginning of a heating cycle.
    C. does not allow for a pre-purge or post-purge cycle.
    D. energizes the oil valve a few seconds after the burner motor is energized.
36. A primary control with an interrupted ignition control strategy, such as the one shown in Figure 32.86,
    A. must energize the burner motor, the oil valve, and ignition at the same time.
    B. can energize the ignition for a short time at the beginning of a heating cycle.
    C. does not allow for a pre-purge or post-purge cycle.
    D. does not allow the oil burner motor operate unless the oil burner is firing.
37. To prevent flame impingement in the combustion chamber,
    A. the combustion chamber and the burner flame should be factory-matched.
    B. a nozzle with a small spray angle should be used on all oil-fired appliances.
    C. a nozzle with a large spray angle should be used on all oil-fired appliances.
    D. the oil pressure should be set as high as possible.
38. If an oil-fired furnace is operating with a 1.25 gph nozzle and an oil pressure is 100 psig, what is the heat output of the appliance if it is operating at an efficiency of 80%?
    A. 140,000 Btu/h
    B. 175,000 Btu/h
    C. 210,000 Btu/h
    D. 245,000 Btu/h
39. The furnace component that is intended to prevent flue gases from mixing with the air from the conditioned space is the
    A. backdraft damper.
    B. draft diverter.
    C. heat exchanger.
    D. burner blower.
40. Condensing furnaces have an Annual Fuel Utilization Efficiency (AFUE) of
    A. 60+.
    B. 70+.
    C. 90+.
    D. 100+.
41. Condensing furnaces are vented with piping made from
    A. PVC.
    B. aluminum.
    C. copper.
    D. steel.
42. Which of the following is true regarding condensing furnaces?
    A. The flue gases are much hotter than on conventional furnaces.
    B. The efficiencies of condensing furnaces are lower than those of conventional furnaces
    C. Condensing furnaces must tie into the chimney of the structure.
    D. Condensing furnaces have multiple heat exchangers.
43. Typically, what is the lowest vacuum that a fuel pump should operate at?
    A. 5 in. of mercury
    B. 10 in. of mercury
    C. 13 in. of mercury
    D. 17 in. of mercury
44. If the vacuum level on an oil delivery system is too low, what can happen to the oil?
    A. It can decompose
    B. It can foam
    C. It can vaporize
    D. It can freeze
45. When connecting an oil pump to operate on a two-pipe system
    A. the pump must operate at 1725 rpm.
    B. the pump must operate at 3450 rpm.
    C. the bypass plug on the pump must be installed.
    D. the bypass plug on the pump must be removed.
46. What should be done if it is suspected that the nozzle is clogged or otherwise defective?
    A. The nozzle can be cleaned with a fine brush.
    B. The nozzle should be replaced.
    C. The nozzle should be cleaned with mineral spirits.
    D. The nozzle should be blown out with dry nitrogen.

47. All of the following tests should be made for proper combustion testing except
    A. draft and smoke.
    B. net flue stack temperature.
    C. carbon dioxide.
    D. water vapor.
48. Which of the following is the correct formula for calculating net stack temperature?
    A. Net Stack Temperature = Flue Temperature + Indoor Ambient Air Temperature
    B. Net Stack Temperature = Flue Temperature × Indoor Ambient Air Temperature
    C. Net Stack Temperature = Flue Temperature − Indoor Ambient Air Temperature
    D. Net Stack Temperature = Flue Temperature ÷ Indoor Ambient Air Temperature
49. Which of the following is true regarding preventive maintenance on oil-fired heating equipment?
    A. Oil-fired heating systems do not require preventive maintenance.
    B. Oil-fired heating system typically requires more preventive maintenance than other types of heating appliances.
    C. Once buried, oil storage tanks do not need to be inspected or checked.
    D. Signs of soot in the air tube and/or heat exchanger is an indication that the appliance is operating at its maximum efficiency.
50. If an oil burner fires, but shuts off after about 90 sec, the problem is most likely related to the
    A. cad cell being dirty and not sensing the flame.
    B. solid-state igniter being defective and not producing the required spark.
    C. fuel filter being clogged and not allowing oil to pass through it.
    D. heat exchanger being cracked and not properly venting the flue gases.

UNIT 33

HYDRONIC HEAT

OBJECTIVES

After studying this unit, you should be able to

- describe the basic operation of a hydronic heating system.
- describe reasons why it is desirable for a hydronic system to have more than one zone.
- list four heat sources commonly used in hydronic heating systems.
- explain the difference between conventional and high-efficiency condensing boilers.
- explain the difference between a wet-base and a dry-base boiler.
- explain the difference between a high-mass boiler and a low-mass boiler.
- explain why a boiler is constructed in sections or tubes.
- discuss the reasons why air should be eliminated from hydronic heating systems.
- describe the function of the air cushion or expansion tank.
- explain the operation of circulator pumps as they apply to hydronic heating systems.
- describe the importance of the "point of no pressure change."
- describe the purpose of limit controls and low-water cutoff devices.
- state the purpose of a pressure relief valve.
- state the purpose of a zone valve.
- explain the difference between a two-port zone valve and a three-port zone valve.
- explain how "outdoor reset" can be used to increase system efficiency.
- describe the characteristics of a series-loop hydronic heating system.
- describe the characteristics of a one-pipe hydronic heating system.
- explain the function of the diverter tee.
- explain the differences between a two-pipe, direct-return hydronic heating system and a two-pipe, reverse-return hydronic heating system.
- explain the application that requires the use of a balancing valve.
- describe a primary–secondary piping system.
- describe the operation and function of mixing valves.
- describe the differences between radiant and conventional hydronic heating systems.
- explain how a hydronic system can incorporate both high-temperature and low-temperature heating circuits.
- describe a tankless domestic hot water heater used with a hydronic space-heating system.
- describe how solar energy can be used to heat water for heating applications.
- explain the importance of preventive maintenance on hydronic heating systems.

SAFETY CHECKLIST

- ✔ When checking a motor pump coupling, turn off the power and lock and tag the disconnect before removing the cover to the coupling.
- ✔ If you are checking the electrical service to a pump that has a contactor, shut off the power and remove the cover to the contactor. Do not touch any electrical connections with meter leads until you have placed one voltmeter lead on a ground source such as a conduit. Then touch the other meter lead to every electrical connection in the contactor box. The reason for this is that some pumps are interlocked electrically with other disconnects and a circuit from another source may be in the box.
- ✔ Ensure that all controls are working properly in a hot water system, because an overheated boiler has great potential for explosion.
- ✔ If an antifreeze solution is used in a solar domestic hot water system, a double-walled heat exchanger must be used.

33.1 INTRODUCTION TO HYDRONIC HEATING

Hydronic heating systems are systems in which water or steam carries the heat through pipes to the areas to be heated. This text covers only hot water systems. In hot water hydronic systems, water is heated and circulated through pipes to a heat transfer component called a **terminal unit**, such as a radiator or finned-tube baseboard unit. It is from the terminal units that heat is given off to the air in the room. The cooler water is then returned to the heat source to be reheated. In most residential installations, the airflow across the radiators or terminal units is due to natural convection currents. Hot water stays in the tubing and heating units even when the boiler is not running, so there are no sensations of rapid cooling or heating, which might occur with forced-warm-air systems. Air-conditioning cannot readily be added to a structure that has hydronic heat, so a separate ducted air-conditioning system must be added.

Hydronic systems are designed with one or more zones. Generally speaking, zones are groupings of one or more rooms in a structure that have similar heat gain, heat loss, and occupancy patterns. A small home, for example, might be treated as a single zone, where all rooms are heated whenever the heating system is operating. Larger homes, on the other hand, might be divided into numerous zones. For example, a two-story home with the bedrooms on the second floor and the living area on the first floor can be easily divided into a first floor zone and a

Figure 33.1 A two-zone hydronic heating system.

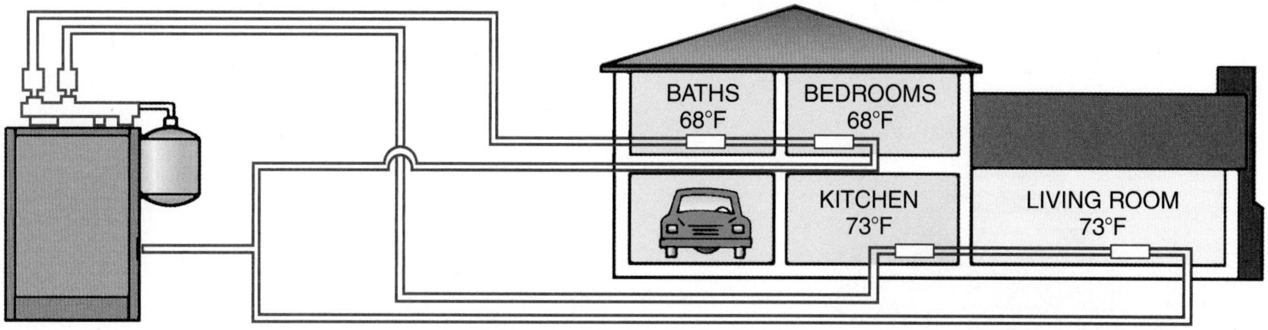

second floor zone, **Figure 33.1**. Zoning is desirable for a number of reasons, including comfort and energy usage. Since each zone has its own thermostat, it is possible to maintain each zone at its own unique temperature. Since unoccupied spaces can be maintained at lower temperatures than those that are occupied, significant energy savings can result.

The water in hydronic systems can be heated using a number of heat sources such as oil, gas, and electric. Sensing elements start and stop the heat source according to the water temperature. The water is circulated through the system with one or more centrifugal pumps. Thermostatically controlled zone valves can be used to control the flow of heated water to the zone needing the heat. Flow can also be controlled by separate, thermostatically controlled circulator pumps located in each zone's piping circuit. Most residential installations use finned-tube baseboard units to transfer heat from the water to the air.

33.2 THE HEAT SOURCE

A major part of the hydronic heating system is the component that heats the water. The piping circuits and strategies that follow later on in this unit are for the most part independent of the heat source—because the strategies involve what is done with the water once it is heated. The heat used in hydronic systems can be generated by a number of different methods, including the burning of fossil fuels, the collection of solar energy, the use of electric heaters, and the use of heat pump technology.

For the most part, the heated water used in hydronic systems comes from a boiler, but geothermal heat pump systems are becoming more and more popular for low-temperature (radiant) applications. Boilers can be used for both high- and low-temperature applications, but accommodations must be made to keep the temperature of the water that returns to the boiler hot enough to prevent the flue gases from condensing—unless the boiler is of the condensing variety.

THE BOILER

A **boiler** is, in its simplest form, an appliance that heats water, using oil, gas, or electricity as the heat source. Some larger

Figure 33.2 A gas-fired boiler. Courtesy Weil-McLain Corp.

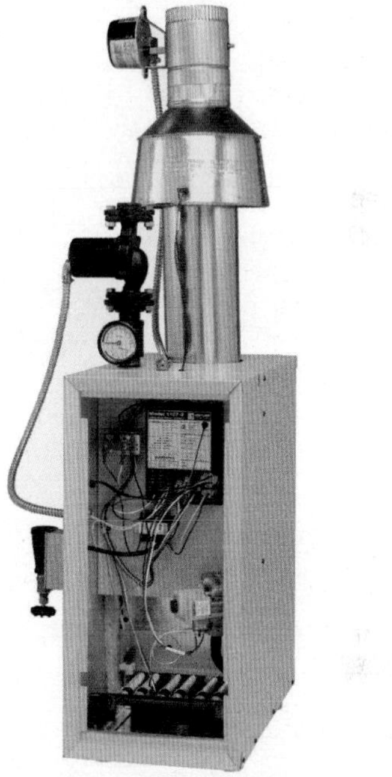

commercial boilers use a combination of two fuels. When one fuel is more readily available than another, the boiler can easily be changed over. **Figure 33.2** shows a gas-fired boiler, **Figure 33.3** shows an oil-fired boiler, and **Figure 33.4** shows an electric boiler.

Boilers can be used to supply water at various temperatures to the areas being heated, but the most common temperature to which water is heated in older, conventional boilers, is about 180°F. Newer technology, however, provides for a wide range of attainable boiler water temperatures that are based on the heating requirements of the space and the outside ambient temperature. Refer to the "Outdoor Reset Control" section in this unit for more on this very important topic. Hydronic systems can be designed to operate with water temperatures ranging from 90°F to over 200°F.

The temperature of the water returning to the boiler affects the operation of the boiler as well as the process of removing flue gases from the structure. As the temperature of the return

Figure 33.3 An oil-fired boiler. Courtesy Weil-McLain Corp.

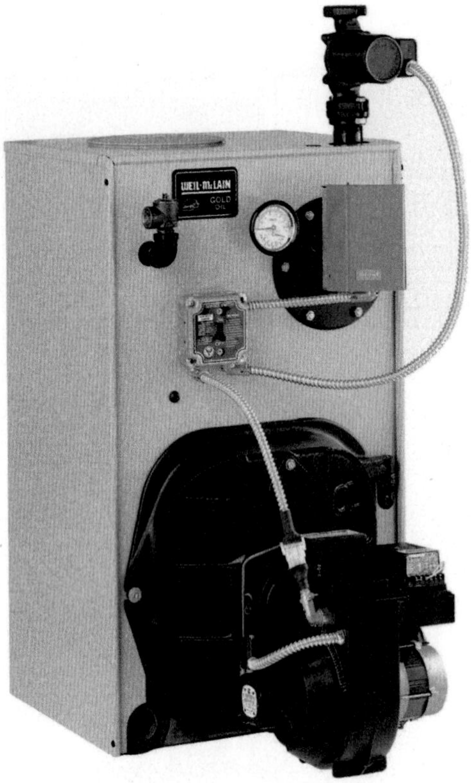

Figure 33.4 An electric boiler. Courtesy Slant/Fin Corp.

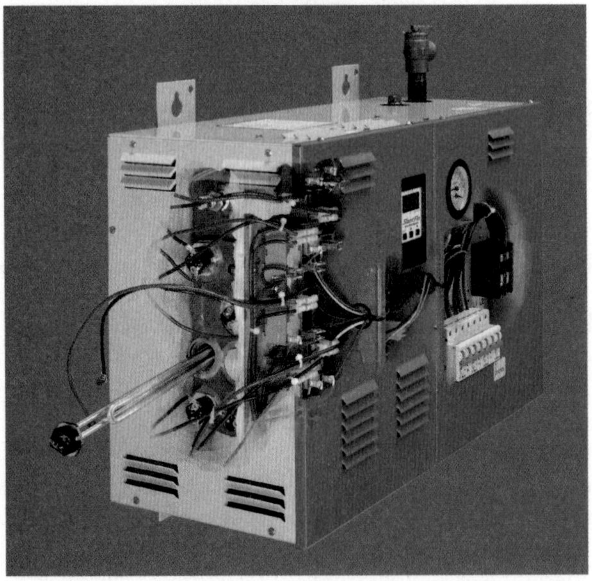

water to the boiler decreases, the temperature of the flue gases leaving the combustion chamber will also decrease. This is because the temperature of the heat exchanger will be lower and more heat will be transferred from the hot flue gases to the cooler heat exchanger. If the temperature of the flue gases

Figure 33.5 Efficiency of a boiler in both condensing and non-condensing modes.

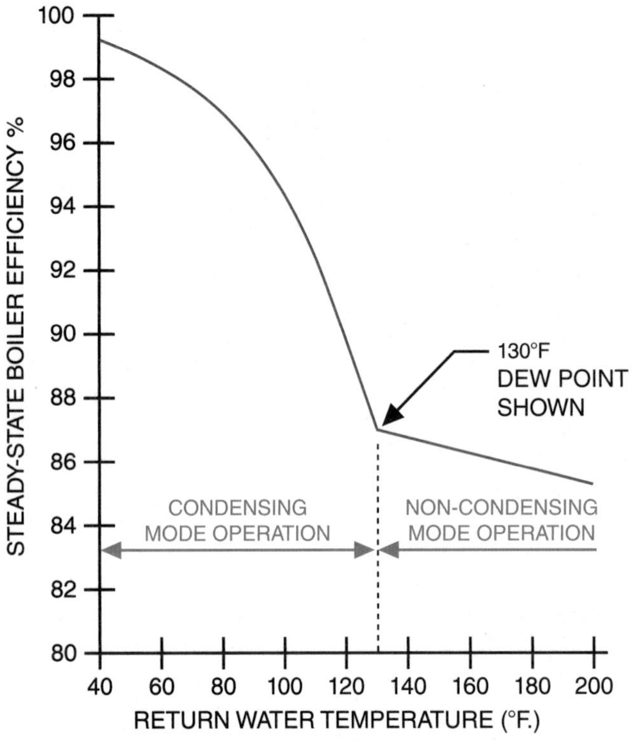

is permitted to drop below about 130°F, the flue gases may begin to condense into a liquid. Most boilers, classified as conventional boilers, are designed to operate with the flue-gas temperature above the 130°F dew point temperature to prevent condensation. Condensing flue gases are highly corrosive and can damage the chimney and the heating equipment. Non-condensing boilers are designed with efficiencies up to about 86%. **Figure 33.5** shows the relationship among boiler return-water temperature, system efficiency, and dew point temperature for the flue gases.

In order to increase the efficiency of boilers, additional and/or more efficient heat exchangers have been incorporated. These improved heat transfer surfaces remove more heat from the flue gases, thereby lowering the temperature of the gases. Condensing flue gases result if the flue gases cool to a temperature below their dew point temperature. But, since some systems are designed to condense the flue gases, equipment design engineers are able to plan for and design around the issue. These systems are referred to as condensing boilers, **Figure 33.6**. Condensing boilers are classified as 90+ because their efficiencies are over 90%. Condensing boilers are available in both gas- and oil-fired models.

CAST-IRON BOILERS. One type of boiler commonly encountered in residential and light commercial applications is the cast-iron boiler. It is often made up of individual sections. Two classifications of boiler section assemblies are the

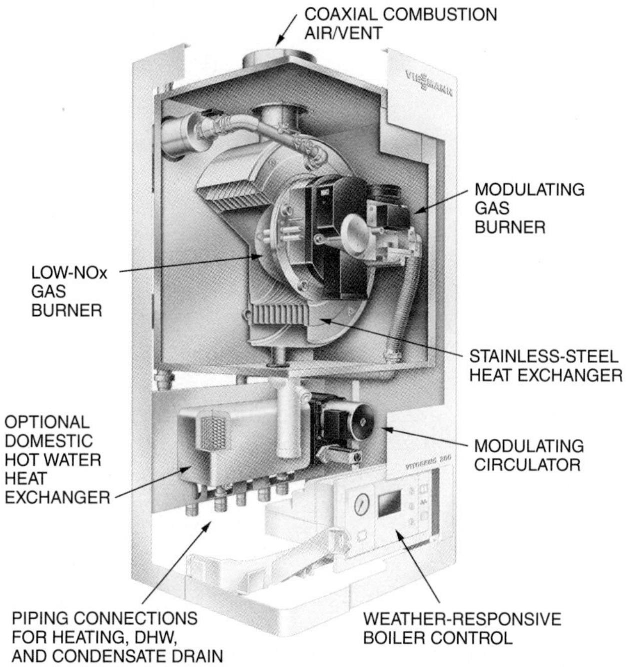

Figure 33.6 Example of a wall-hung condensing boiler. Courtesy Viessmann Manufacturing

dry-base type and the wet-base type. **Figure 33.7(A)** shows a section of a dry-base boiler. The term **dry-base** means that the area under the combustion chamber is dry—there is no water there. **Figure 33.7(B)** shows a section of a wet-base boiler. In a **wet-base boiler** setup, the water being heated is located both above and below the combustion area. Multiple sections are bolted together to form the complete heat exchange surface between the burning fuel and the water being heated, **Figure 33.8**. The more sections a boiler has, the higher its capacity will be. The following are some key characteristics associated with the cast-iron boiler:

- They are very heavy appliances.
- They hold between 15 and 30 gallons of water.
- They are classified as high-mass boilers.
- They take a long time to heat up.
- They retain their heat for a long period of time.
- They have longer run times and longer off cycles.
- They are used on closed-loop systems.
- Air in the system can result in boiler corrosion.

STEEL BOILERS. The water being heated within steel boilers, **Figure 33.9**, surrounds a bundle of steel tubes through which the combustion gases pass on their way to the chimney. These tubes have a series of baffles located inside to increase the heat transfer surface area and to slow the speed of the flue gases as they pass through the tubes. Without the baffles, the hot flue gases would flow out of the appliance very quickly at a hotter temperature. By slowing the rate of flue-gas flow, more heat can be extracted from the gases, increasing the efficiency of the boiler.

Figure 33.7 (A) Cast-iron dry-base boiler section. (B) Wet-base boiler section. Courtesy Weil-McLain Corp.

(A)

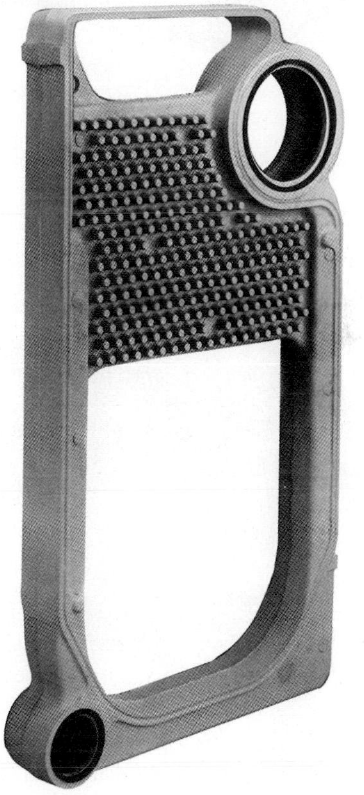

(B)

COPPER WATER-TUBE BOILERS. The heat exchanger in a copper-tube boiler is constructed as a series of finned copper tubes, **Figure 33.10**, that are heated by the flue gases in the combustion chamber of the boiler, **Figure 33.11**. The heat

Figure 33.8 Boiler block assembly for a wet-base boiler. Courtesy Weil-McLain Corp.

Figure 33.9 Cutaway view of a vertical fire-tube steel boiler. Courtesy Columbia Boiler Company.

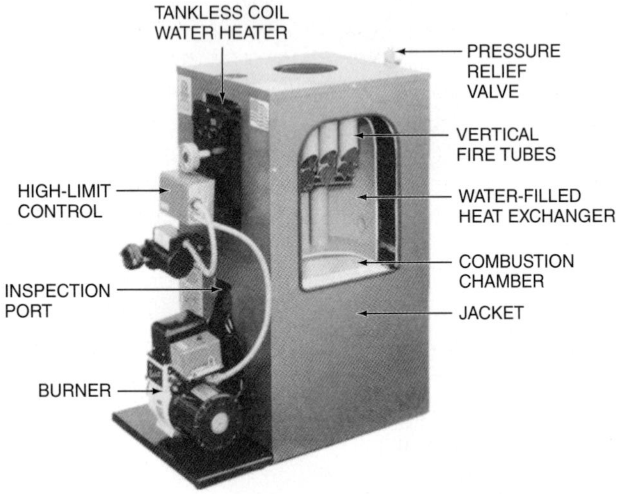

exchangers on copper-tube boilers have much less surface area than cast-iron or steel boilers do but since copper is such a good conductor of heat, they are still able to heat water quickly and efficiently.

Some key characteristics of copper-tube boilers are:

- They are much lighter than similarly sized cast-iron boilers.
- They hold only a few gallons of water.

Figure 33.10 Finned copper-tube heat exchanger assembly.

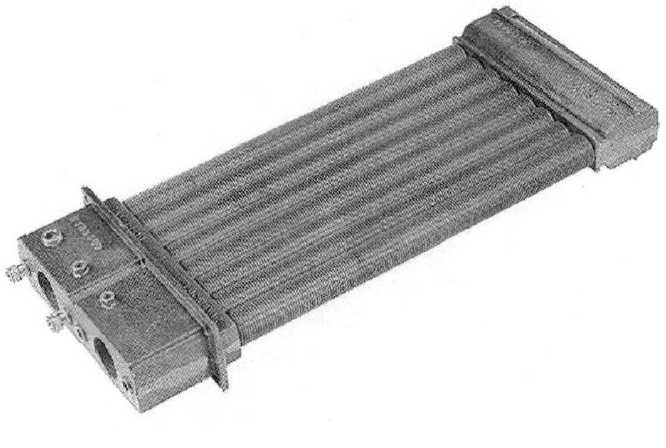

Figure 33.11 Cutaway view of a copper water-tube boiler. Courtesy Lochinvar Corp.

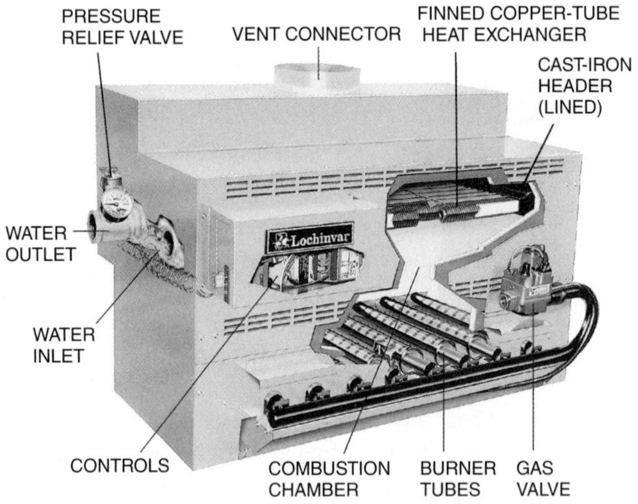

- They are classified as low-mass boilers.
- They heat up quickly.
- They cool down quickly.
- They have shorter run times and shorter off times.
- They are used on closed-loop systems.
- They are less susceptible to corrosion when compared to cast-iron boilers.

THE GEOTHERMAL HEAT PUMP

One heat source that is becoming popular for use in conjunction with hydronic heating systems is the geothermal heat pump. To provide heat to the occupied space, the geothermal heat pump transfers heat from the earth to a water/water-antifreeze mixture. This heat-laden water mixture then transfers its heat to the space to be heated. Geothermal heat pump systems can heat system water to a temperature as high as 130°F, so it is a good option for low-temperature, radiant,

hydronic heating systems. Refer to Unit 44, "Geothermal Heat Pumps," for more information on this topic. Radiant heating systems will be discussed later on in this unit.

33.3 THE BASIC HYDRONIC SYSTEM

Although there are many individual parts that make up a hydronic heating system, we will begin our discussion by examining two system components to see how they function in a completely sealed water loop. This will give us a basic understanding of some of the properties of water and of the function of the circulator pump in the hydronic system. As the unit progresses, more components will be discussed to help broaden your understanding of these popular systems.

EXPANSION TANK

Hot water hydronic systems are closed-loop systems and are, ideally, air-free. As water is heated, it expands, and if the system is totally air-free, the pressure in the piping circuit will increase as the water is heated and expands. This excess pressure in the system will cause the pressure relief valve to open every time the water is heated. To prevent this from happening, an extra volume or space is provided to accommodate the additional volume of the heated water. The expansion tank, **Figure 33.12**, provides this extra volume.

Figure 33.12 An expansion tank. Photo by Eugene Silberstein

Figure 33.13 A steel expansion tank.

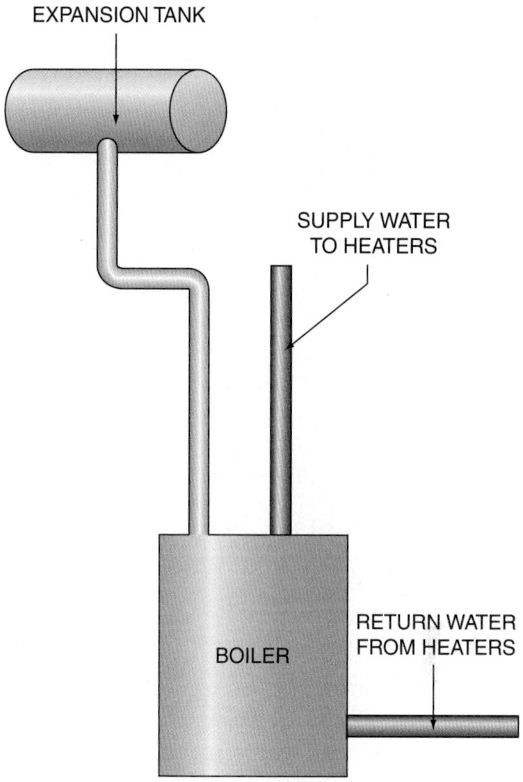

There are two types of expansion tanks. One type is called the compression tank, or standard expansion tank, **Figure 33.13**. This tank is nothing more than a steel tank located above the boiler. Upon installation, the tank is filled with air at atmospheric pressure. As water is added to the boiler, the air that is being displaced is pushed up into the expansion tank, causing the pressure in the tank to increase, **Figure 33.14**. This creates a pressurized air cushion within the shell of the tank. As the water in the system is heated, the volume of water in the system increases and further compresses the air in the tank. Over time, however, the tank can become completely filled with water, thereby removing the air cushion. This situation can be easily identified because the pressure relief valve will open each time the water in the system is heated.

A more common type of expansion tank is the diaphragm-type expansion tank. This tank has a rubber, semipermeable membrane within its shell, **Figure 33.15**. One side of the tank contains pressurized air, whereas the other side of the tank is open to the system water. The air portion of the tank is pressurized at the factory, but it is a good idea to check the pressure with a tire gauge prior to installing the tank in a system. For residential applications, these tanks often come precharged with 12 psi of air. The tank pressure is noted on the nameplate of the tank itself, **Figure 33.16**.

To determine the actual pressure required for the expansion tank, it is necessary to measure the vertical distance

Figure 33.14 Pressure in the tank increases as water is added to the system.

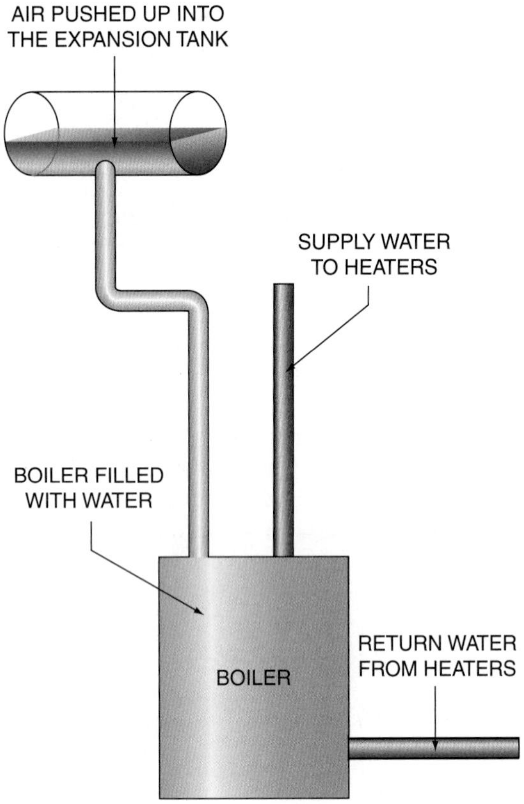

Figure 33.15 Cutaway view of a diaphragm-type expansion tank. Photo by Eugene Silberstein

(in feet) between the highest pipe in the system and the inlet of the expansion tank and then use the following formula:

$$P_{tank} = (H \div 2.31) + 5$$

where P_{tank} = The pressure in the expansion tank in psi
H = The vertical distance between the highest pipe in the system and the inlet of the expansion tank
2.31 = The amount of vertical lift (head) for each pound of pressure

Figure 33.16 Expansion tank data tag. Photo by Eugene Silberstein

5 = The additional pressure added to the system to ensure that there is a positive pressure at the highest pipe in the system

Consider the following residential application:

- The boiler is located in the basement.
- The opening of the expansion tank is 4 ft from the basement ceiling.
- The home has two floors, each with 9-ft ceilings.
- The second floor has baseboard heating and the highest pipe is 2 ft from the floor.

From this we can conclude that the "*H*" measurement is 15 ft (9 ft + 2 ft + 4 ft) and that therefore the required tank pressure is

$$P_{tank} = (H \div 2.31) + 5$$

$$P_{tank} = (15 \div 2.31) + 5$$

$$P_{tank} = 6.49 + 5$$

$$P_{tank} = 11.49 \text{ psi}$$

Since most residential applications will provide similar results, it is easy to see why the tanks are supplied by the manufacturer with a pressure of 12 psi.

CIRCULATOR/CENTRIFUGAL PUMPS

Centrifugal pumps, also called *circulators,* **Figure 33.17**, force the hot water from the heat source through the piping to the heat transfer units and back to the boiler. These pumps rely on centrifugal force to circulate the water through the system. Centrifugal force is generated whenever an object is rotated around a central axis. The object or matter being rotated tends to fly away from the center due to its velocity. This force increases proportionately with the speed of the rotation, **Figure 33.18**.

The impeller, **Figure 33.19**, is the part of the pump that spins and forces water through the system. It is essential

Figure 33.17 A centrifugal pump. Courtesy Ferris State University. Photo by John Tomczyk

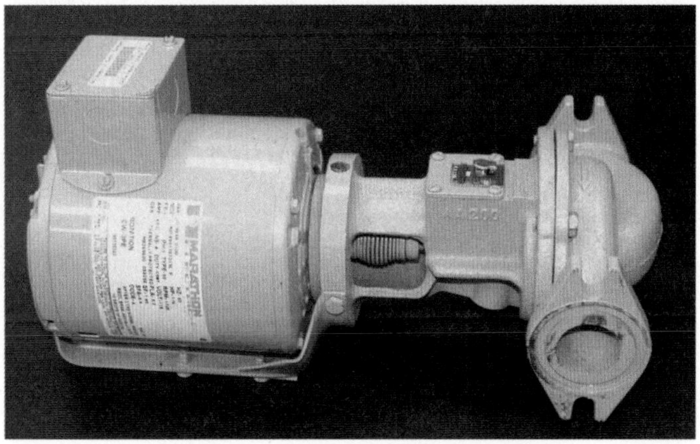

Figure 33.18 When the impeller is rotated, it "throws" the water away from the center of the pump and out through the opening.

Figure 33.19 A pump impeller. Note the direction of the rotation as indicated by the arrow. This is an example of a closed impeller.

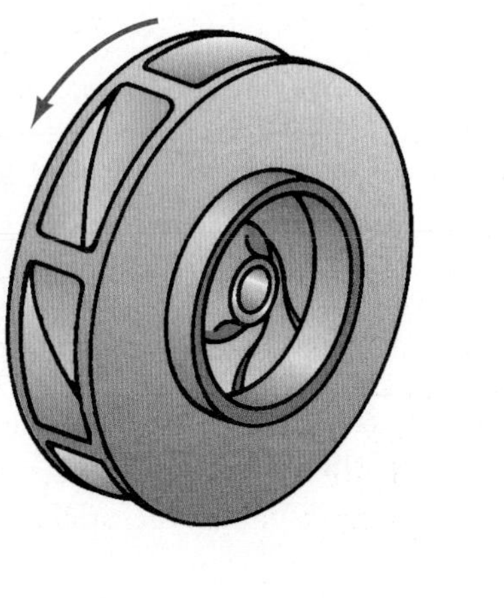

Figure 33.20 Cutaway view of a centrifugal pump. Courtesy Bell and Gossett

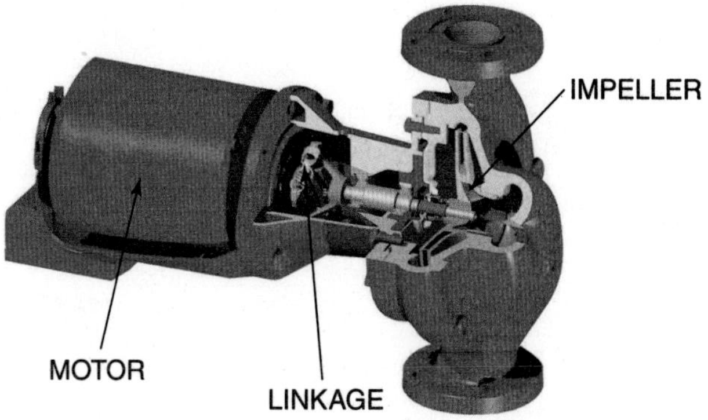

that the impeller rotate in the proper direction. The vanes or blades in the impeller must "slap" and then throw the water. The vanes of the impellers used in circulating pumps in hot water heating systems are usually enclosed, and are referred to as *closed* impellers. A cutaway of a centrifugal pump is shown in **Figure 33.20**. Pumps generally have a cast-iron body with the impeller and other moving parts made of bronze or nonferrous metals. Other pumps have stainless-steel or all bronze parts.

Centrifugal water pumps are not positive displacement pumps, as are most compressors. The term *circulating pump* is used for them because these pumps do not add much pressure to the water from the inlet to the outlet in small systems. It is important to understand how a centrifugal pump responds to pressure and load changes. The centrifugal pump is responsible for creating a pressure difference between the water at its inlet and the water at the outlet of the pump. It is this pressure differential that makes the water flow through the piping circuit. However, before water can flow through the circuit, the pressure difference generated by the pump must overcome the resistance of the piping circuit itself. Consider a pump that is operating in a system that offers no resistance to flow. If this pump were supplied with an unlimited amount of water to pump—and this water were immediately pumped to atmospheric pressure once it left the pump—it is safe to say that the amount of water moved by the pump would be the pump's maximum capacity. As resistance to flow is added, the volume of water moved by the pump will decrease, reducing the pump's capacity. A point will eventually be reached where the amount of water pumped will reach zero because the resistance of the piping arrangement is too much for the pump to overcome.

The following progressive example will demonstrate how a performance curve for a centrifugal pump is developed. See **Figure 33.21(A)–(J)**.

A. A pump is connected to a reservoir that maintains the water level just above the pump inlet. The pump meets no resistance to the flow, and maximum flow exists. The water flow is 80 gpm.

Figure 33.21 (A)–(F) Developing a centrifugal pump performance curve.

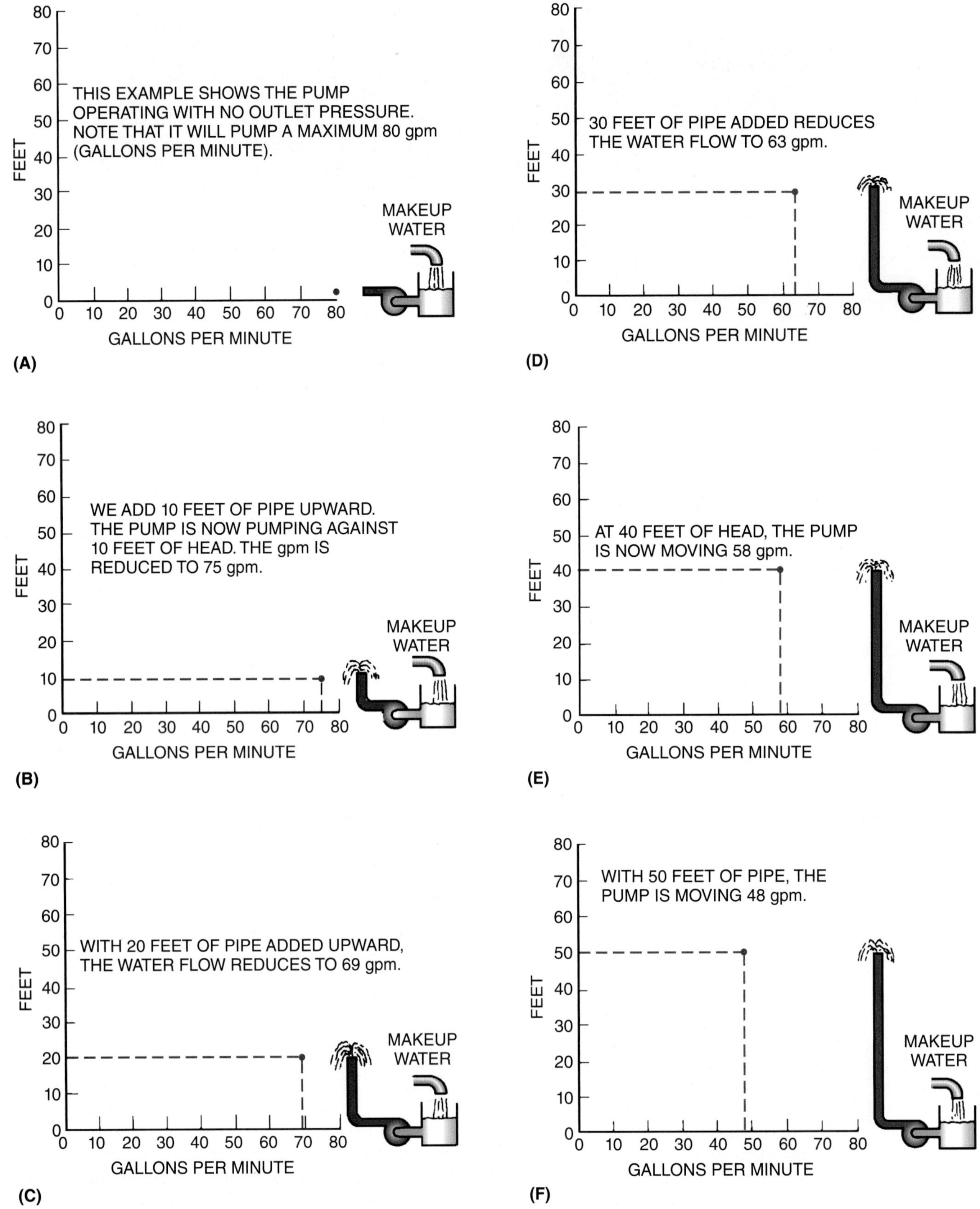

Figure 33.21 (*Continued*) (G)–(J) Developing a centrifugal pump performance curve.

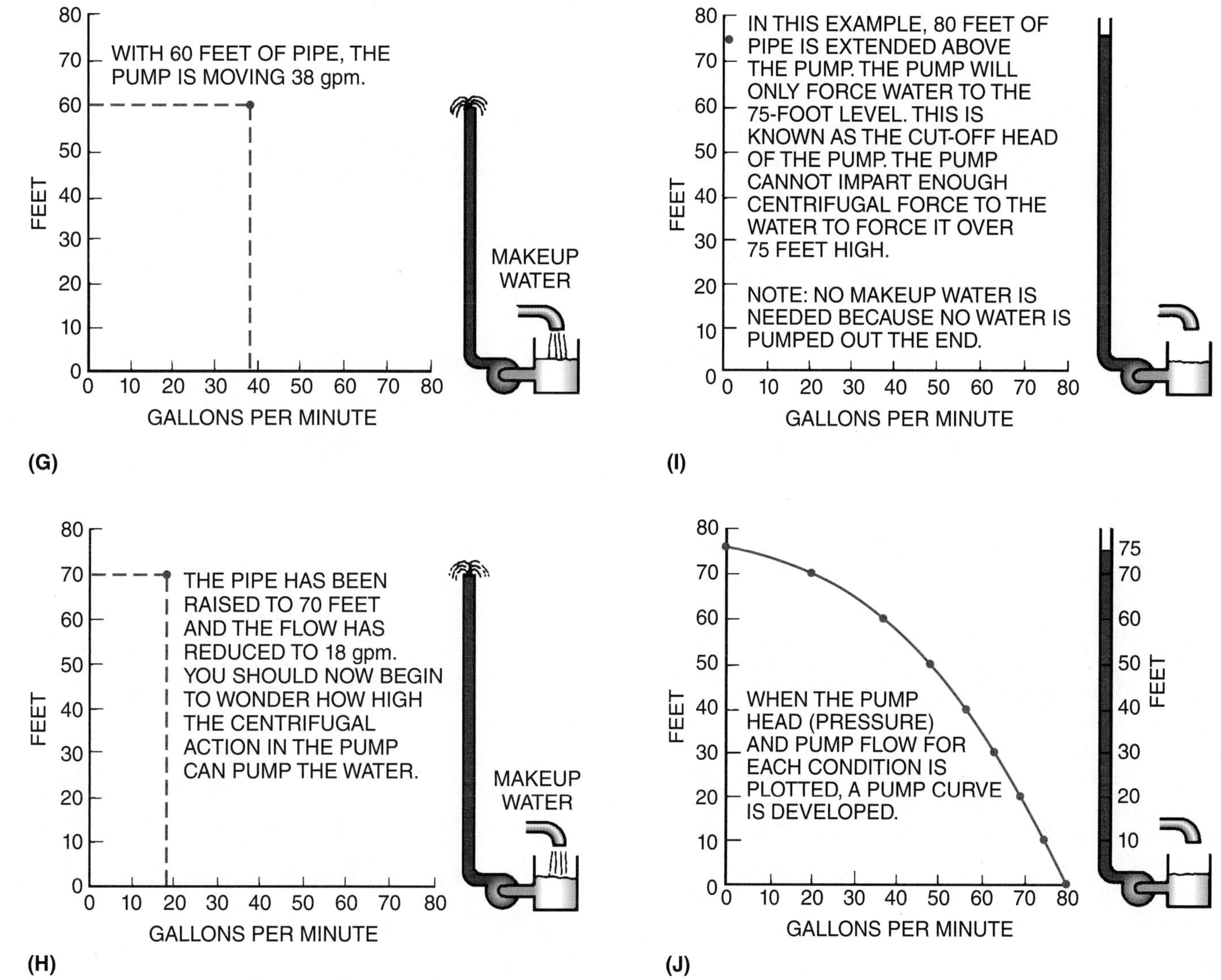

B. A pipe is extended 10 ft high (known as feet of head, or pressure), and the pump begins to meet some pumping resistance. The water flow reduces to 75 gpm.
C. The pipe is extended to 20 ft, and the flow reduces to 69 gpm.
D. When 30 ft of pipe is added, the water flow reduces to 63 gpm.
E. The pipe is extended to 40 ft. The pump is meeting more and more resistance to flow. The water flow is now 58 gpm.
F. At 50 ft, the flow slows to 48 gpm.
G. At 60 ft, the flow slows to 38 gpm.
H. At 70 ft, the water flow has been reduced to 18 gpm. The pump has almost reached its pumping limit. The pump can impart only a fixed amount of centrifugal force.
I. When 80 ft of pipe is added to the pump outlet, the water will rise to the 75-ft level. If you could look down into the pipe, you would be able to see the water gently moving up and down. The pump has reached its pumping pressure head capacity. This is known as the shutoff point of the pump.
J. This is a manufacturer's pump curve; the preceding examples show how the manufacturer arrives at the pump curve. It is important to note that each pump has its own unique pump curve.

It would be awkward for manufacturers to add vertical pipe to all pump sizes for testing purposes, so they use valves to simulate vertical head. To use the valve system of measurement, liquid water vertical head must be converted to gauge readings. Gauges read in psig. When a gauge reads 1 psig, it is the same pressure that a column of water 2.309 ft (27.7 in.) high exerts at the bottom, **Figure 33.22**. This would be the same as saying that a column of water 1 ft high (feet

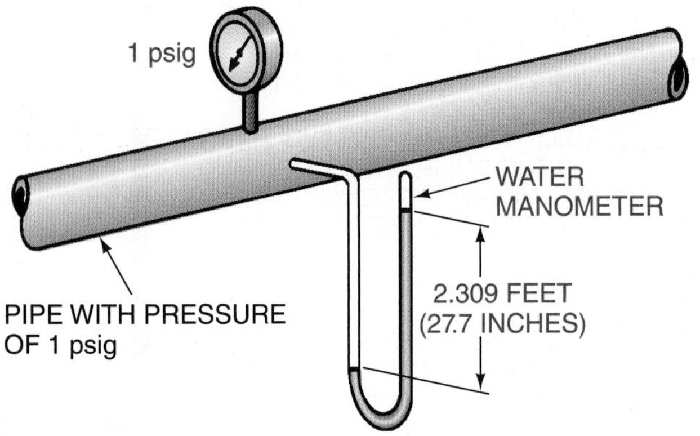

Figure 33.22 A gauge pressure reading of 1 psi is equal to 2.309 ft of water.

of head) is equal to 0.433 psig, because 1 psig divided by 2.3093 ft = 0.433. Some examples of water columns and pressure are shown in **Figure 33.23**. A pump manufacturer may use gauges to establish pump head (pressure) by placing a gauge at the inlet and a gauge at the outlet and reading the difference. This is known as psi difference. Psi difference is then converted to feet of head, **Figure 33.24**.

The power consumed by the motor driving a centrifugal pump is proportional to the quantity of water the pump circulates. For example, the pump operating under the conditions shown in **Figure 33.21(A)** requires more horsepower than when operating under the conditions shown in **Figure 33.21** (I). This might sound counter-intuitive. When a valve is installed in a piping circuit and closed, it would seem that the power consumption should rise due to an increase in pump head, but it does not. As the discharge valve is closed, the water begins to recirculate in the pump and the power consumption reduces. This can be demonstrated by shutting the valve at the outlet of a centrifugal pump and monitoring the amperage, **Figure 33.25**.

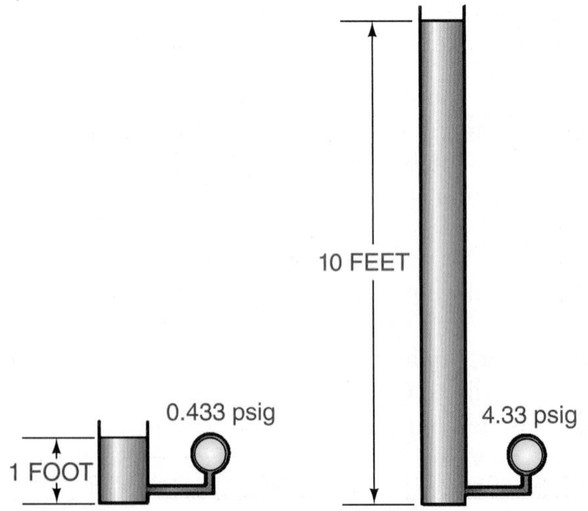

Figure 33.23 This figure shows that different heights of water columns exert different pressures. Each time you increase the water column 1 ft, the pressure rises 0.433 psi.

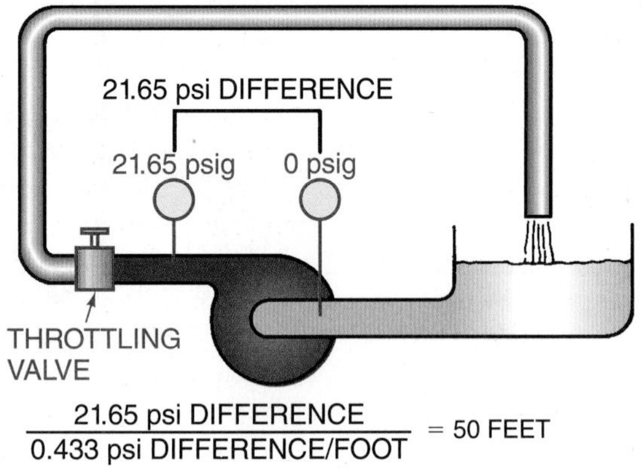

Figure 33.24 This figure illustrates how a pump manufacturer may test a pump for pumping head. More elaborate systems that measure gallons per minute at the same time are used to establish a pump curve.

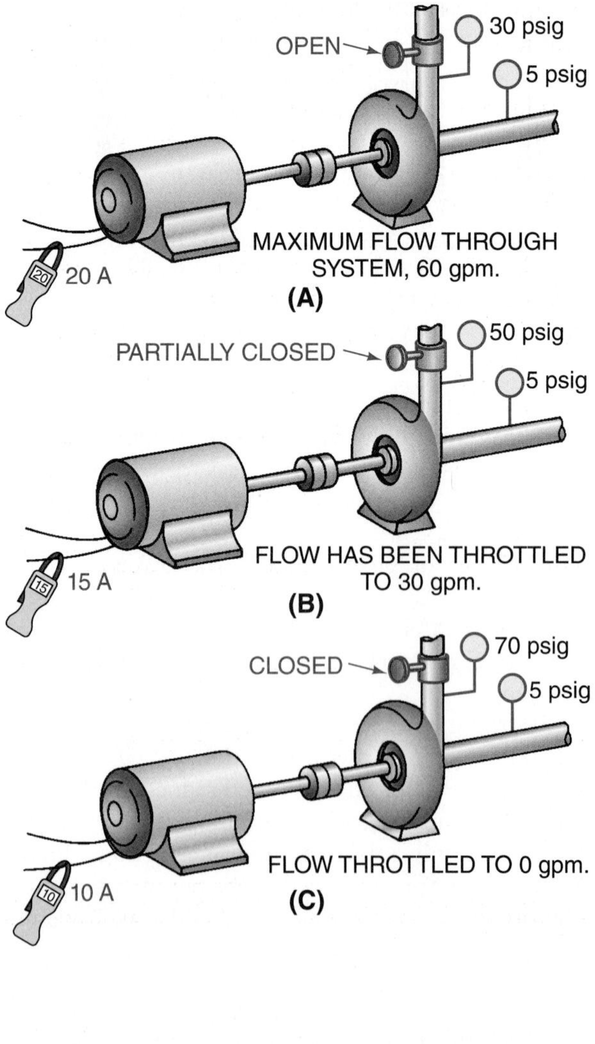

Figure 33.25 (A) Maximum flow. (B) As the flow is reduced by throttling the outlet valve, the power to drive the motor reduces the current required to turn the pump impeller in the water. (C) Throttled to 0 gpm.

33.4 THE POINT OF NO PRESSURE CHANGE

Probably one of the most important and busy locations in the hydronic hot water system is the "point of no pressure change." As its name implies, this is the point in the system where the pressure, no matter what the system is doing, will remain the same. This is a great reference point for evaluating a system and also provides a location for multiple system-component connections.

Consider a piping arrangement that contains only the two components discussed so far in this unit: the expansion tank and the circulator pump, **Figure 33.26**. Notice that, in the figure, the pump is discharging in a direction pointing toward the expansion tank. Assume that the pressure in the expansion tank prior to installation was 10 psi and that the pump generates a pressure difference of 15 psig between its inlet and outlet. When the piping loop is filled, the pressure will be 10 psi, as determined by the pressure in the expansion tank. Since the loop is completely filled and the expansion tank has a 10-psi cushion of air inside it, what will happen when the pump is turned on? Will the pump push water into the expansion tank? Will the pump pull water out of the expansion tank?

The answer is that water will neither be pushed into nor pulled out of the expansion tank and the pressure in the circuit at the inlet of the expansion tank will remain unchanged at 10 psi. Water cannot be transferred from the loop to the expansion tank because there would be nothing to replace the water that leaves the loop. Also, water cannot be relocated from the expansion tank to the loop because there is no room in the loop to accept any additional water. So, if water cannot enter or leave the expansion tank, the pressure in the loop at the inlet of the expansion tank will remain constant. This will change slightly once the water is actually heated, as the water will expand and some will be pushed into the expansion tank when heating occurs.

Figure 33.26 Simple loop with only an expansion tank and a circulator.

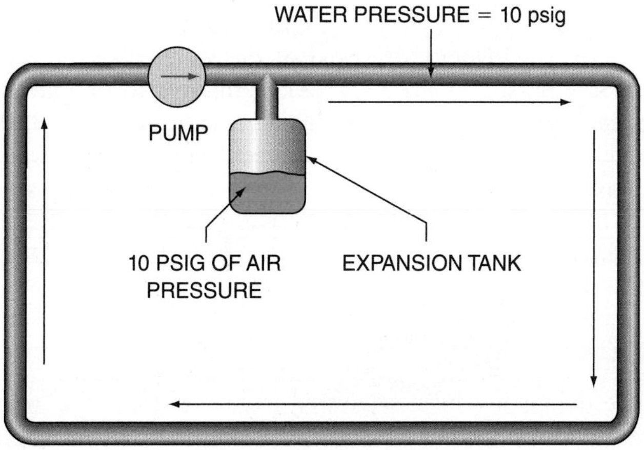

Figure 33.27 Pumping toward the expansion tank can result in a vacuum in the piping circuit.

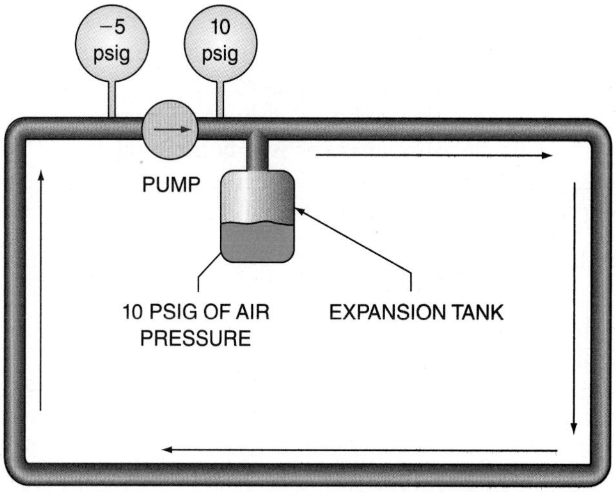

So, if the pump creates a pressure difference of 15 psi and the pressure at the outlet of the pump cannot change, the pressure at the inlet of the pump will have to drop to account for the 15-psi drop across the pump, **Figure 33.27**. This will create a vacuum in the piping circuit. If a leak should develop (at the pump's mounting flange, for example), air will be pulled into the system, creating problems. Therefore, the operating pressure range for this system will be between −5 psi and +10 psi.

By moving the pump to the other side of the expansion tank, the vacuum situation is eliminated. The pressure at the inlet of the circulator will be 10 psi and the pressure at the outlet of the pump will be 25 psig (10 psi + 15 psi), **Figure 33.28**. Relocating the pump in this manner minimizes the possibility of the system pulling into a vacuum and also reduces the effects of air in the system. We will be addressing

Figure 33.28 Point of no pressure change.

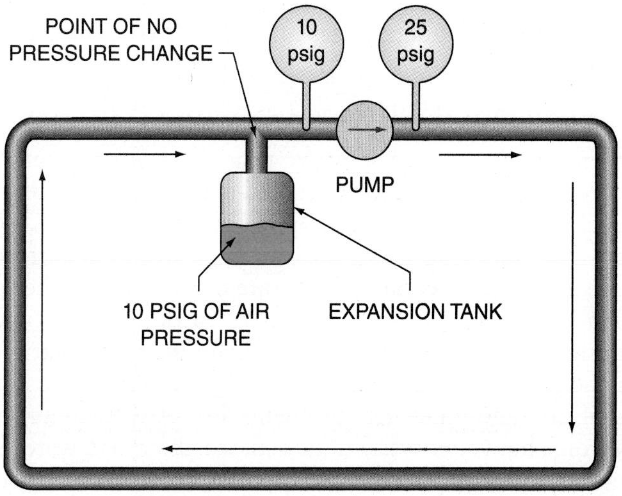

air in the next few sections, as it is an important aspect of hot water systems, but one issue needs to be mentioned here. At higher system pressures, air bubbles are smaller and are more likely to remain in solution. Air bubbles that remain in solution with the water are far less likely to have a large negative effect on system operation and performance.

As we move through this unit, the importance of the "point of no pressure change" will become more and more evident. This point in the system is the desired connection point for the following system components:

- Inlet of the circulator pump
- Inlet of the expansion tank
- Air separator
- Air vent
- Outlet of the pressure-reducing (water-regulating) valve

33.5 OTHER HYDRONIC SYSTEM COMPONENTS

The expansion tank and the circulator are the two components needed to make up even the most basic of hydronic systems. A system such as this would not, however, function effectively because a number of important operational and safety issues would not be accounted for. In addition to simply moving the water and allowing for expansion, it must also do the following:

- Separate air from the water
- Remove the air from the system
- Provide for makeup water to the system
- Provide safety controls in the event of system overpressurization
- Provide for the desired flow control
- Provide for the desired temperature control
- Allow for heat transfer from the system to the occupied space

The system components discussed in the next sections allow the system to operate as needed and desired while simultaneously ensuring that it operates in a safe and efficient manner.

AIR SEPARATOR AND AIR SCOOP

When working with hot water systems, it is important to remove any air that may be trapped in the piping circuit, the boiler, or any other system component. Air, as mentioned earlier in the unit, can lead to excessive corrosion in cast-iron and steel boilers. As the temperature of the water in the system increases, the rate of corrosion increases. Corrosion takes time to occur, so the effects of air on the structural integrity of the boiler may not be readily seen. However, there are more significant and immediate problems associated with air in a hydronic hot water system. Since air is lighter than water, air pockets will form at the highest points in the system. These trapped air pockets prevent water flow and can result in insufficient heating. In addition, trapped air can lead to noisy pipes and noisy system operation.

Figure 33.29 An air separator. Courtesy Bell and Gossett

The air separator, **Figure 33.29**, is the hydronic system component that separates air from the water as it flows through the system. It is important to note that the air separator does not remove air from the system—it simply separates the air from the water. Another system component, the air vent, is responsible for venting, or removing, the air from the system. Air separators are capable of removing small "microbubbles" from the system. The new generation of air separators operates on the concept of "collision and adhesion." As water passes through the separator, it comes into contact with a mesh screen or similar material, **Figure 33.30**. The air bubbles collide with the screen and adhere to it. As more and more air bubbles adhere to the mesh, the bubbles get larger, break loose, and travel up into the air vent, where they are vented from the system.

Other air separation devices, known as air scoops, **Figure 33.31**, are installed on horizontal runs of straight pipe, which cause the air in the pipe to separate from the water and rise to the top of the pipe. Using a baffle, the air scoops then scoop the air out of the pipe, **Figure 33.32**. In order to work properly, the velocity or speed of the water must be lower than 4 ft per sec, **Figure 33.32(A)**, or the smaller bubbles will not be able to rise to the top of the device, **Figure 33.32(B)**. At the top, the trapped air is directed to the air vent for removal from the system. Air scoops are directional, so be sure to install them in the system with the arrow pointing in the direction of water flow.

AIR VENT

Once the air has been separated from the circulating water, the air vent is responsible for removing it from the system.

Figure 33.30 (A) Cutaway of an air separator. (B) Mesh screen from an air separator. Photos by Eugene Silberstein

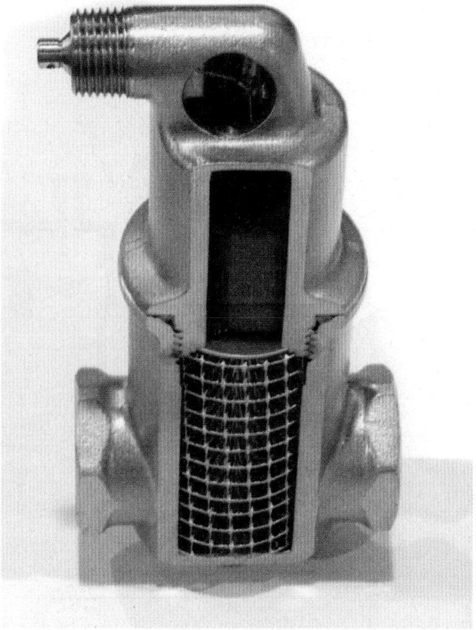

(A)

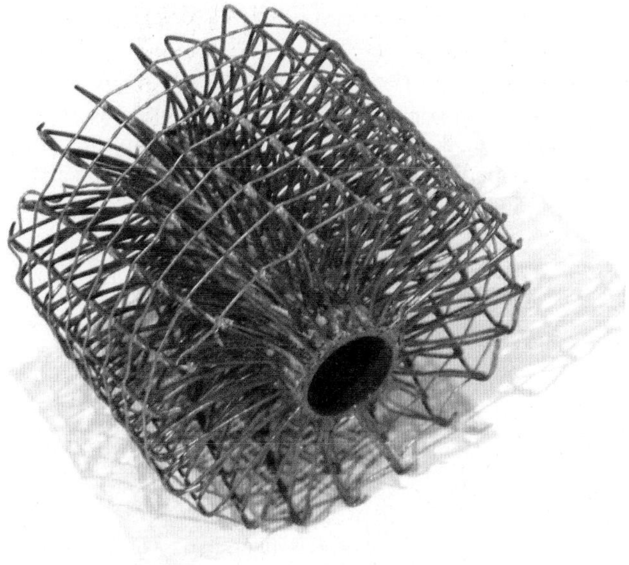

(B)

Figure 33.31 A cast-iron air scoop. Courtesy Watts Regulator Co.

Figure 33.32 (A) When the flow velocity is lower than 4 ft per sec, the vanes will deflect the air bubbles into the upper chamber. (B) If the velocity of the water is more than 4 ft per sec, the bubbles remain trapped in the water and cannot be scooped by the baffle.

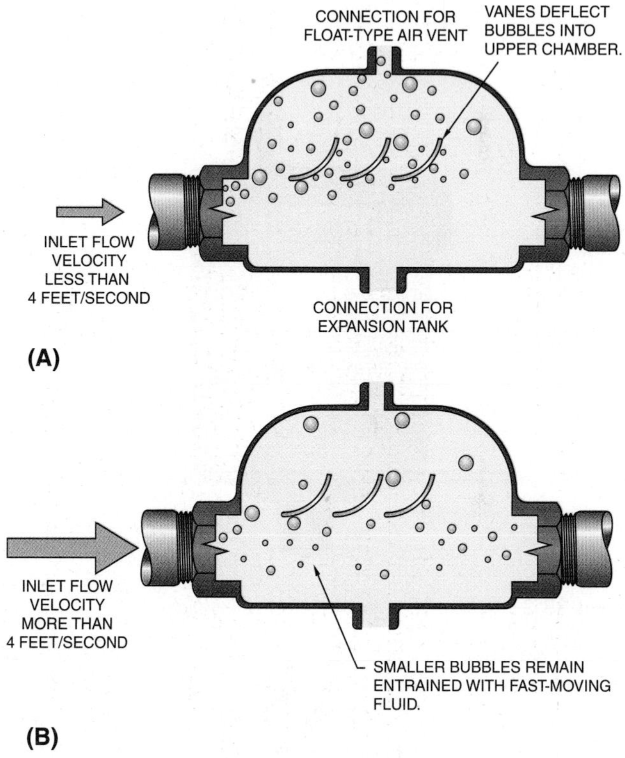

Air vents can be manually operated or automatic devices. They are typically located at high points in the hydronic system as well as directly above air separators and air scoops, as these are the devices that accumulate air that is separated from the water in the system. Manual air vents are the simplest of the vents, **Figure 33.33**. They are often referred to as "coin-operated" air vents or "bleeders." As the name implies, these vents are opened and closed manually as the need arises. Automatic air vents, **Figure 33.34**, rely on an internal disc that swells when water is sensed. In operation, if there is no air at the vent location, the disc will be wet and swollen, which will seal off the air vent port. As air accumulates, the disc will dry out and shrink, allowing the air to pass through to the vent port. As the air is removed and replaced with water, the disc will swell, closing off the port. Since it takes time to fully close, there is often a small amount of water loss when the device vents.

Another type of automatic air vent utilizes a float mechanism, **Figure 33.35**. The position of the float is determined by

Figure 33.33 A "coin-operated" air vent. Courtesy Bell and Gossett

Figure 33.34 An automatic air vent. Courtesy Bell and Gossett.

Figure 33.35 A float-type air vent. Courtesy Maid O' Mist.

Figure 33.36 (A) The float closes the valve when there is no air in the air vent. (B) When air is present, the float falls and the valve opens to vent the air.

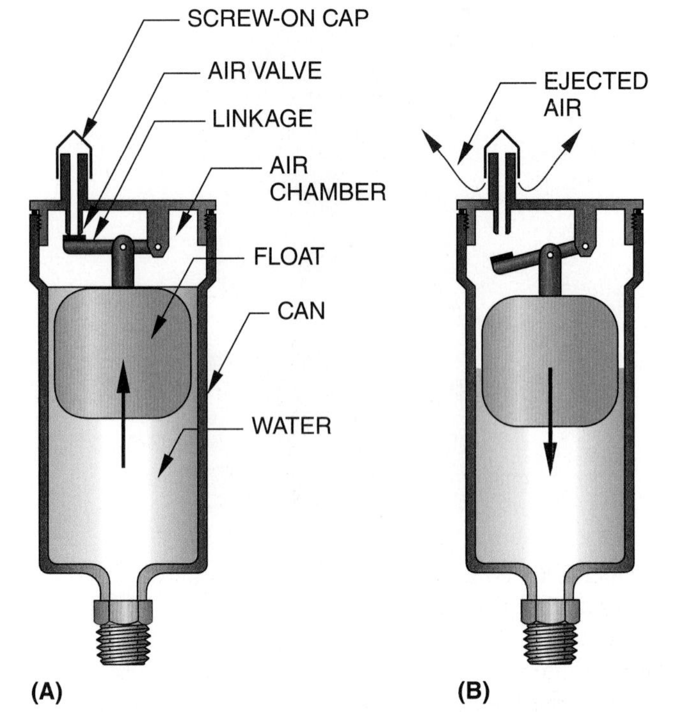

the amount of air in the device. When there is no air in the air vent, the float is at the top of the vent and the vent port is closed, **Figure 33.36(A)**. As air accumulates in the vent, the float drops, opening the vent port and removing the air from the system, **Figure 33.36(B)**.

TEMPERATURE-LIMITING CONTROL (AQUASTAT)

The temperature-limiting control, more commonly known as the aquastat, is a temperature-controlled switch that is intended to maintain the temperature of the water in a hydronic heating system, **Figure 33.37**. If a heating system is to maintain a temperature of 180°F with a 10° differential, the boiler will cycle off when the water is heated to 180°F. When the water cools to 170°F (180°F − 10°F), the boiler cycles back on. The water temperature will, therefore, fluctuate between 170°F and 180°F. The aquastat is commonly equipped with a remote-sensing bulb that is positioned in the boiler to provide accurate control of the boiler water temperature. It is an operational control that opens and closes its contacts to maintain the desired water temperature.

HIGH-LIMIT CONTROL

The high-limit control, like the aquastat, is a device that opens and closes its contacts in response to boiler water temperature.

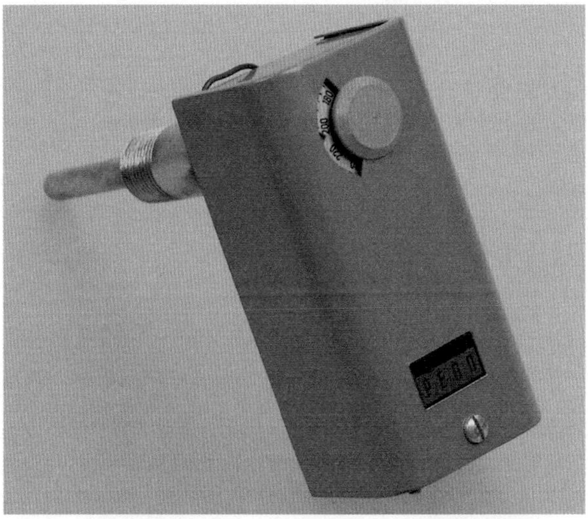

Figure 33.37 A temperature-limiting control, or aquastat. This is the system component responsible for maintaining the boiler water at the desired temperature. Photo by Bill Johnson

The high-limit control is not, however, an operational control. It is a safety device that will open its contacts to deenergize the heat source if the boiler water temperature rises too high. If the maximum desired boiler water temperature is 180°F, for example, the high-limit control may open its contacts when the boiler water temperature rises to 200°F. Since 200°F boiler water is above the desired level, a system problem is likely and the high-limit control will interrupt power to the heat source.

WATER-REGULATING VALVE (PRESSURE-REDUCING VALVE)

Water heating systems should have an automatic method of adding water back into the system if water is lost due to leaks. The source of this water may be at a pressure too great for the system. A *water-regulating valve,* **Figure 33.38**, is therefore installed in the water makeup line leading to the boiler and is set to maintain the pressure on its leaving side (entering the boiler) at less than the relief valve on the boiler. Low-pressure water boilers typically have a relief valve pressure setting of 30 psig.

The pressure-reducing valve should be connected to the point of no pressure change to ensure that the system is filled to the desired pressure, **Figure 33.39**. Since the pressure will change at every other point in the system as it cycles on and off, the system fill pressure will also change unless the point of no pressure change is the system connection point.

PRESSURE RELIEF VALVE

The American Society of Mechanical Engineers (ASME) Boiler and Pressure Vessel Codes require that each hot water heating boiler have at least one officially rated *pressure relief valve* set to relieve at or below the maximum allowable

Figure 33.38 A pressure-reducing valve. Courtesy Bell and Gossett

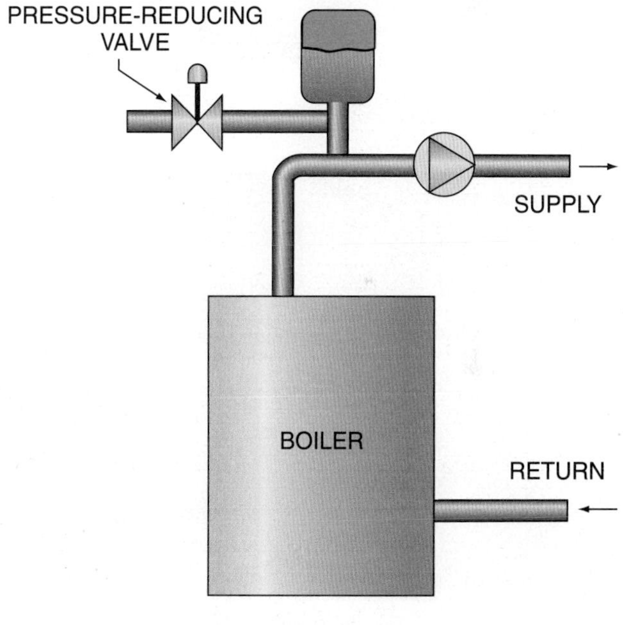

Figure 33.39 Proper location for the pressure-reducing valve.

working pressure of the low-pressure boiler. This would be 30 psig. This valve discharges excessive water when pressure is created by expansion. It also releases excessive pressure if there is a runaway overfiring emergency, **Figure 33.40**. It is important that pressure relief valves be inspected periodically. If a relief valve has rusted, the valve should be replaced.

LOW-WATER CUTOFF

The low-water cutoff valve, **Figure 33.41**, is responsible for deenergizing the system in the event the level of water in the system drops below the desired point. In some states,

Figure 33.40 A pressure relief valve. Courtesy Bell and Gossett

Figure 33.41 A low-water cutoff. Photo by Eugene Silberstein

low-water cutoffs are required by law, but even in states where they are not required, it is good field practice to equip all systems with one. This is especially true on radiant heating systems, as the tubing is usually below the level of the heat source and often buried in concrete. A leak in the tubing circuit can cause the system to drain itself.

ZONE VALVES

Zone control valves, or simply zone valves, are thermostatically controlled valves that regulate water flow to the various zones in a hydronic heating system. Some zone valves are equipped with gear motor actuators, whereas others have heat motor actuators, **Figure 33.42**. When the temperature in a zone falls below the thermostat's set point in that area, the space thermostat closes its contacts, energizing the zone valve, causing it to open. On the gear motor variety of zone valve, when potential is applied

Figure 33.42 Two examples of hydronic zone valves. (A) Zone valve with a gear motor actuator. (B) Zone valve with a heat motor actuator. Courtesy Sparco, Inc.

(A)

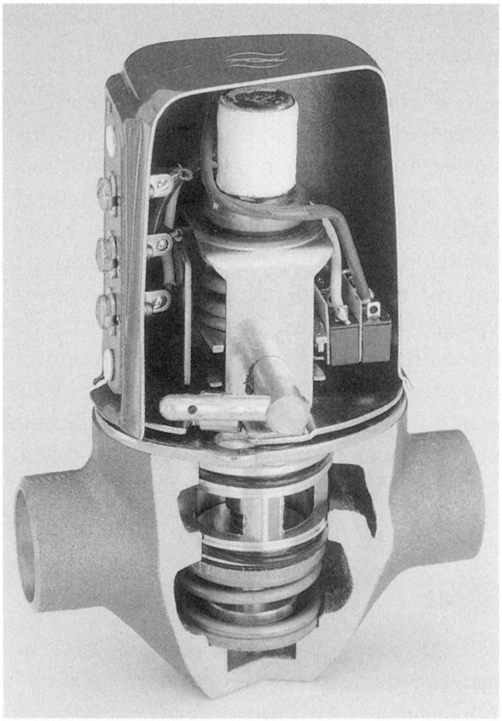

(B)

to the valve, a motor-driven gear assembly slowly opens the valve. When potential is applied to the heat motor variety of zone valve, electrical current generates heat in the valve and a bimetal strip warps to open the valve. Zone valves are often equipped with a setting that allows the valve to be opened manually. This position can be used to help troubleshoot the heating system and, in the event of zone-valve malfunction, provide temporary heat in the conditioned space.

Although there are many different system control strategies and options, here is one application that would employ zone valves. When an individual zone calls for heat, the zone valve for that zone opens. When the zone valve has completely opened, an end switch on the valve closes to start the circulator pump. This will circulate heated water from the boiler to the occupied space until the thermostat is satisfied. When the space reaches the desired temperature, the zone valve closes and the circulator is deenergized. **Figure 33.43** shows two zones being supplied with hot water by a single circulator. When either zone is calling for heat, the circulator is energized.

Zone valves are available in both two-port and three-port varieties. The piping circuit in **Figure 33.44** shows an application that uses two-port zone valves. If only one of the three heating zones is calling for heat, the other two supply circuits will be closed off and the pressure in the supply header will likely increase. Refer to the section "Pressure Differential Bypass Valve," for a discussion of valves that help alleviate the problem associated with increased system pressure due to closed zone valves.

Figure 33.43 Gear motor zone valves on a multiple-zone loop. Photo by Eugene Silberstein

With three-port zone valves, the water flow rate remains constant because water flows through all valves, whether or not all of them are calling for heat, as long as at least one zone is calling for heat. **Figure 33.45** shows a piping circuit that uses three-port zone valves. It can be seen that Zone 2 is the only zone calling for heat. When hot water enters zone valve #2, it leaves and flows through the terminal or heating units in that zone before returning to the boiler. When hot water flows through zone valve #1 and zone valve #3, water leaving the valves is directed right back to the boiler without flowing through the terminal units in those zones.

BALANCING VALVES

When designing a hot water heating system, consideration must be given to the flow rate of the water and the friction in the system. Friction is caused by the resistance of the water flowing through the piping, valves, and fittings in the system. The flow rate is the number of gallons of water flowing each minute through the system. A system is considered to be in balance when the resistance to the water flow is the same in each flow path. A means for balancing the system should be provided in all installations. One method is to install a *balancing valve,* **Figure 33.46,** in each heating circuit branch. This valve is adjustable.

The balancing valve has pressure taps on either side where a very accurate pressure gauge can be fastened to check the pressure drop across the valve. The gauge is calibrated in in. WC. The pressure drop then can be plotted on a chart for that specific valve to show the gallons of water per minute (gpm) and then compared with the system specifications. The valve then can be adjusted to the correct flow rate. **NOTE:** *The valve adjustment is generally determined by the system designer, and the installer sets them accordingly. Be sure to read the manufacturer's instructions for the correct procedure for setting the valve.* •

PRESSURE DIFFERENTIAL BYPASS VALVE

On systems that use two-port zone valves, excessive pressure in the supply header may result when multiple zone valves close. The piping circuit may become excessively noisy as a result of the increased water velocity and water banging within the piping circuit. The pressure differential bypass valve, **Figure 33.47**, is often used to help alleviate this situation. The bypass valve is located between the supply-and-return headers of the system, **Figure 33.48.** When all, or most, of the zone valves are open, the bypass valve is in the closed position and the system operates normally.

Figure 33.44 Hydronic piping circuit using two-port zone valves.

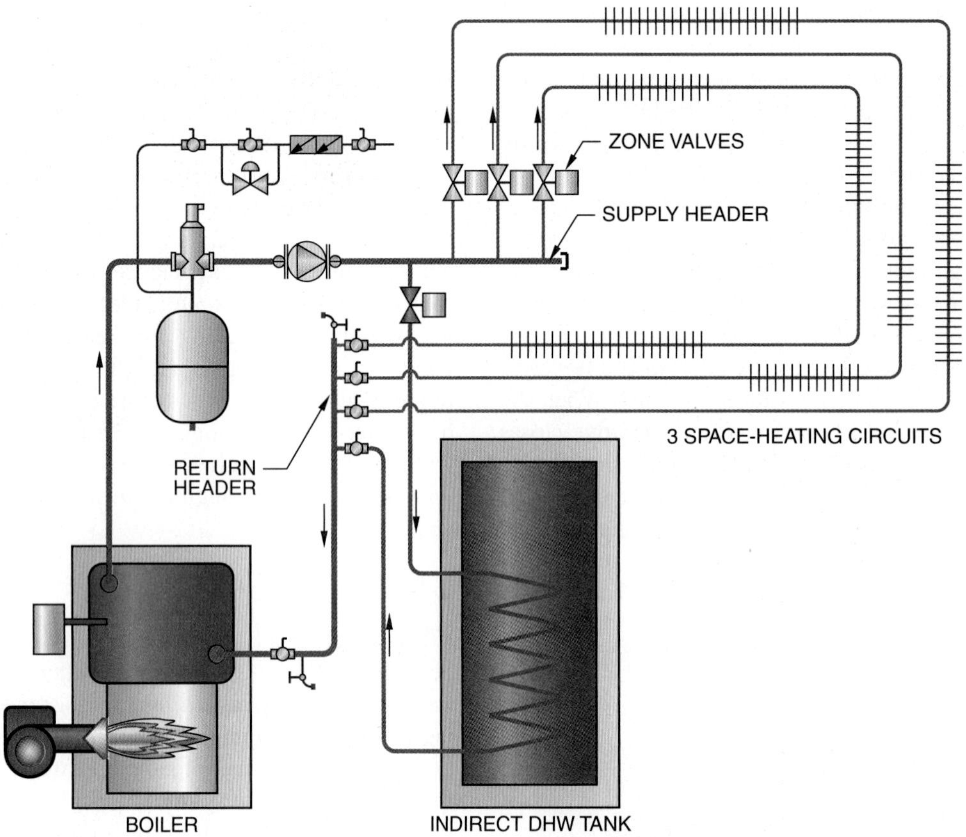

Figure 33.45 Hydronic piping circuit using three-port zone valves.

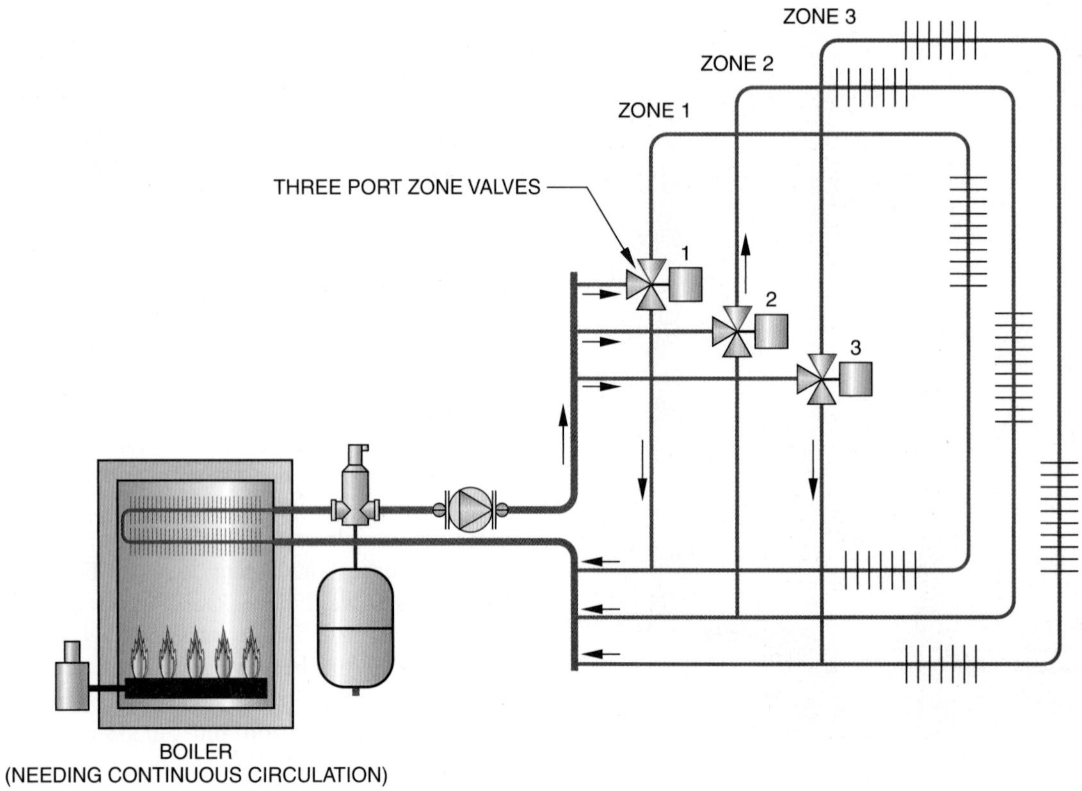

Figure 33.46 A balancing valve. Photo by Bill Johnson

Figure 33.47 A pressure differential bypass valve.

Figure 33.48 Location of the pressure differential bypass valve on systems with zone valves.

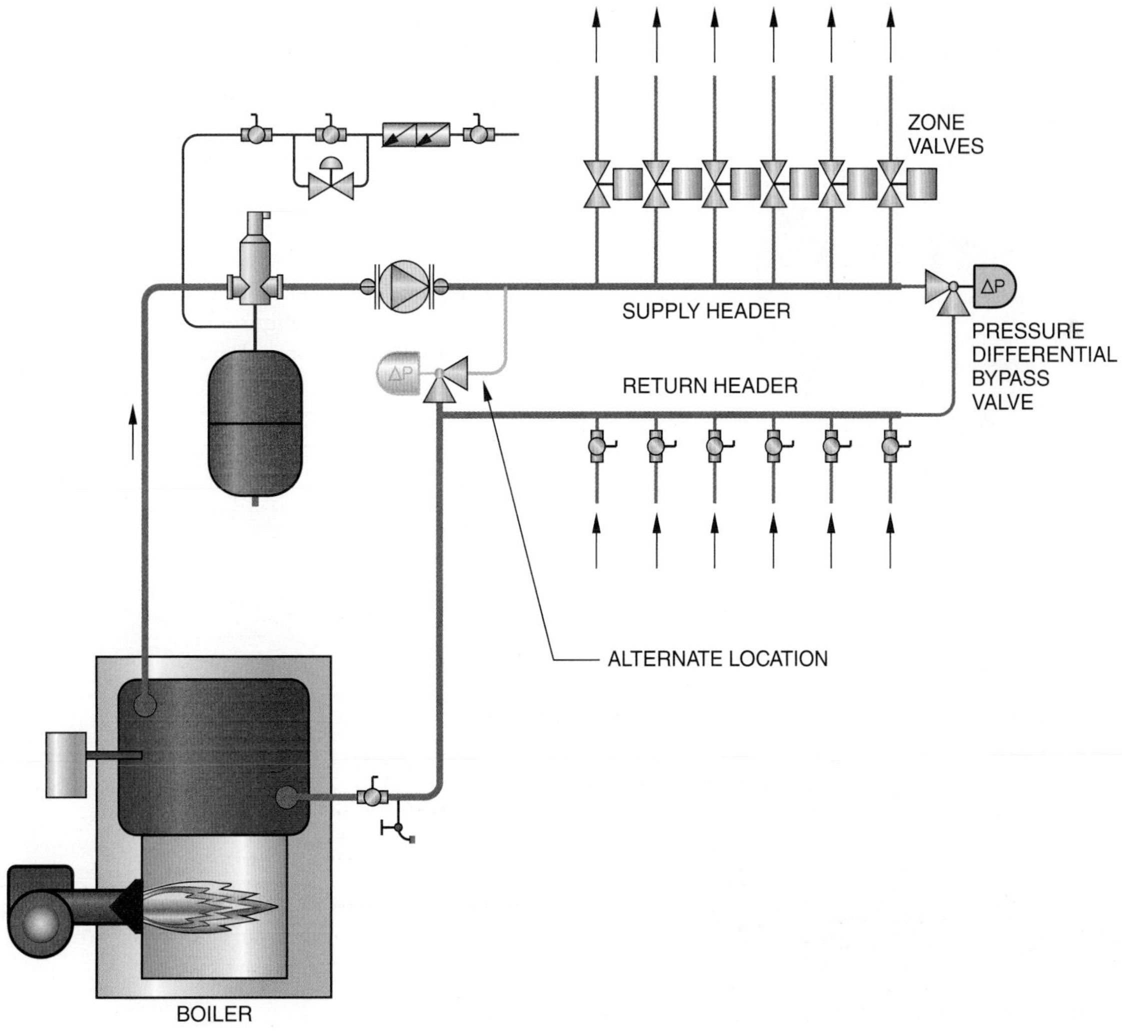

As zone valves close, the pressure in the supply header increases and the bypass valve begins to open. The partially open valve allows the circulator to move a fixed amount of water (although only a portion of the circulated water flows through the heating zones) while the remaining hot water recirculates back to the boiler.

FLOW CONTROL VALVES

The flow control valve, **Figure 33.49**, is actually a flow check valve that prevents hot water in the system from flowing through a heating loop when no flow through that particular loop is desired—often referred to as *ghost flow*. A system using zone valves to provide a positive shutoff between the heat source and the heating loops does not need flow control valves. However, when the system utilizes only circulators, flow control valves are desirable. Some circulator manufacturers are now installing flow control valves right on the circulators themselves to ease the installation process, **Figure 33.50**.

Figure 33.49 A flow control valve. Courtesy Bell and Gossett

Figure 33.50 Example of a circulator pump with an integral flow check valve. Courtesy Taco, Inc.

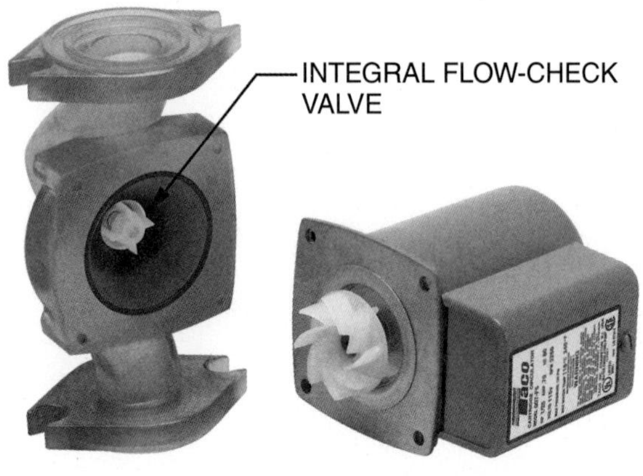

OUTDOOR RESET CONTROL

The outdoor reset control is the device that senses the outside ambient temperature and adjusts the boiler water temperature accordingly. In theory, as the outside ambient temperature increases, the occupied space will require less heating, so cooler water can be used. Conversely, when the outside ambient temperature drops, hotter water will be needed to heat the space. When cooler water can be used, energy is saved.

The reset control measures two temperatures: the outside ambient temperature and the temperature of the water in the boiler. At the time of initial installation, the control must be set up manually to select which set of "commands" the control will obey. **Figure 33.51** shows a set of control options for a particular reset control. In this case, a reset control setting of 1.0 will cause the boiler to maintain a water temperature of 160°F if the outside ambient temperature is 20°F. The same control setting will cause the boiler water temperature to drop to 140°F if the outside ambient temperature rises to 40°F. *As of September 2012, all newly installed boilers are required to have some type of reset control strategy to automatically adjust the boiler water temperature.*

THERMOSTATIC RADIATOR VALVES

On hydronic systems with multiple zones, one common method for controlling the temperature in each individual zone is to use thermostatic radiator valves, **Figure 33.52**. These valves can be used on many different types of hydronic piping circuits, with the exception of the series loop system, which will be discussed later on in this unit. It should be noted that when the thermostatic radiator valve is installed, it should be protected from the heat that will rise from the terminal heater unit, **Figure 33.53**. If the sensing element is located above the terminal unit, the radiator valve will close before the space reaches the desired temperature.

Figure 33.51 Residential outdoor reset control curve.

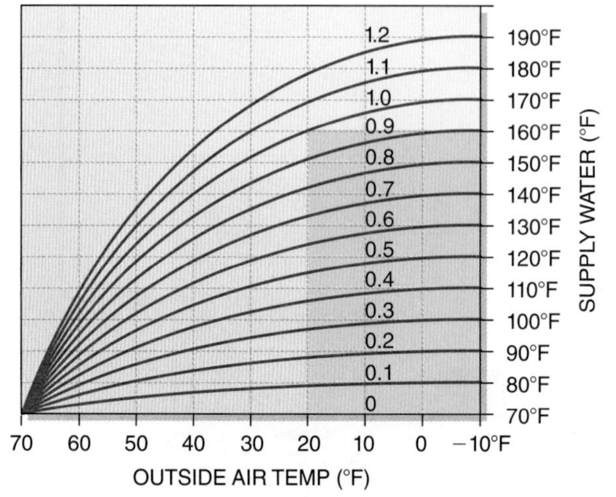

Figure 33.52 Thermostatic radiator valve. Courtesy Danfoss, Inc.

FINNED-TUBE BASEBOARD UNITS

Most residences with modern hydronic heating systems use finned-tube baseboard heat transfer for terminal heating units, **Figure 33.54**. Air enters the bottom of these units and passes over the hot fins. Heat is given off to the cooler air, causing it to rise by convection. The heated air leaves the unit through the damper area. The damper can be adjusted to regulate the heat flow. Baseboard units are generally available in lengths from 2 ft to 8 ft and are relatively easy to install following a manufacturer's instructions. **Figure 33.55** shows a radiator and fan coil, which are other types of terminal units.

Finned-tube heating units are rated in Btuh/ft of pipe in two different flow rates, approximately 500 lb/h and 2000 lb/h. Five hundred lb/h is the equivalent of 1 gal of water flow per minute, and 4 gal/min equals 2000 lb/h.

$$1 \text{ gal/min} = 8.33 \text{ lb/min}$$

$$8.33 \text{ lb/min} \times 60 \text{ min/h} = 499.8 \text{ lb/h (Rounded to 500 lb/h)}$$

$$8.33 \text{ lb/gal} \times 4 \text{ gal/min} = 33.32 \text{ lb/min}$$

$$33.32 \text{ lb/min} \times 60 \text{ min/h} = 1999.2 \text{ lb/h}$$
$$\text{(Rounded to 2000 lb/h)}$$

The manufacturer will rate the finned-tube radiation at so many Btuh/ft at different temperatures, **Figure 33.56**.

These same convectors are available in ¾-in. ID tubing. This will give you some idea of how the convectors are chosen. If a room needs 16,000 Btuh and the flow rate has been

Figure 33.53 Proper and improper mounting locations for the operator on the thermostatic radiator valve.

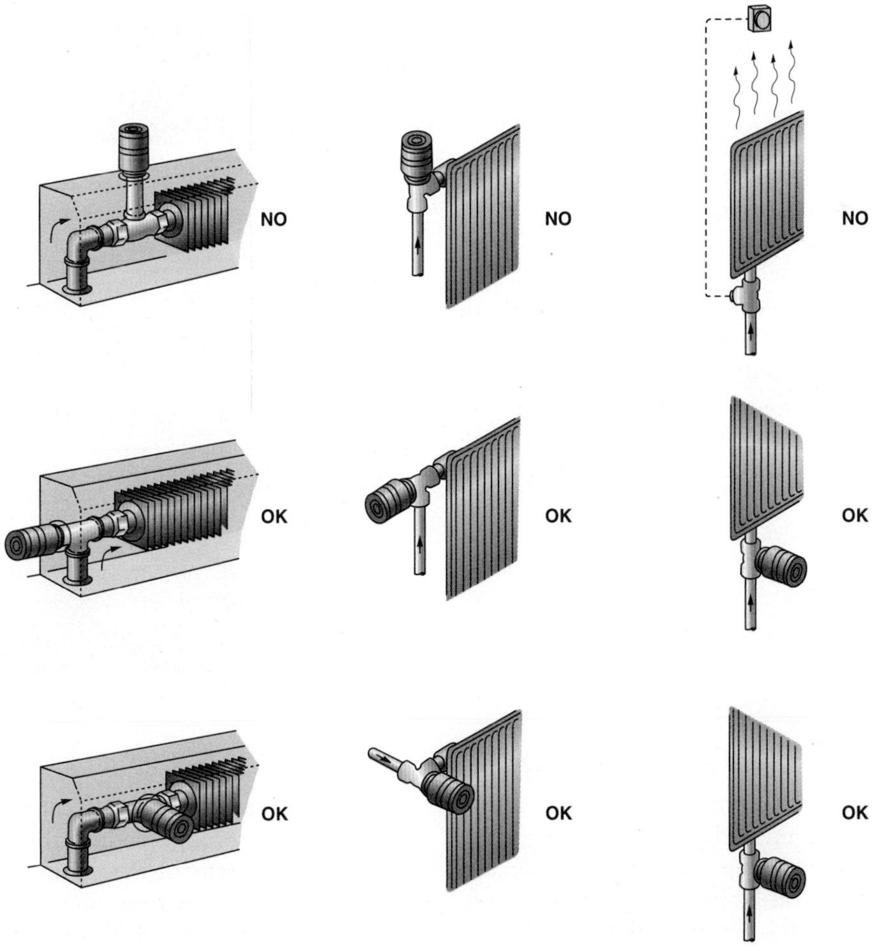

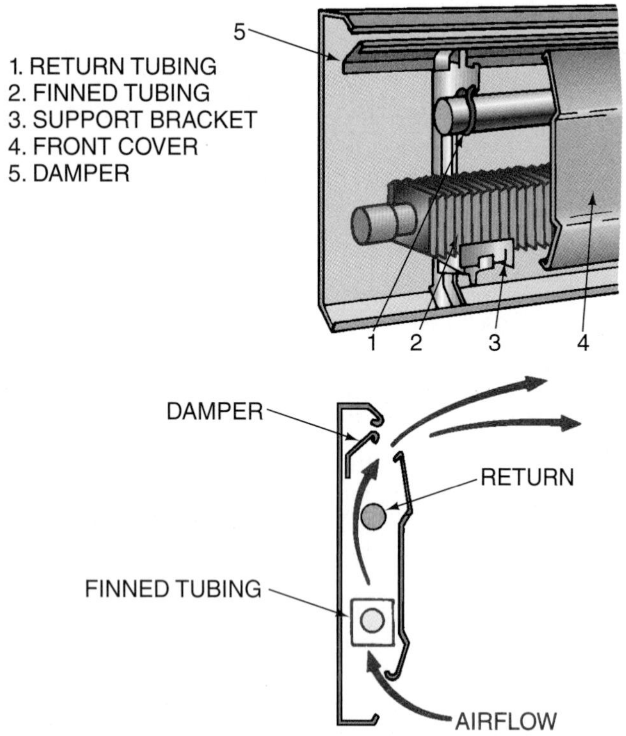

Figure 33.54 A finned-tube baseboard unit.

determined to be 1 gal/min and the system's average water temperature is 180°F, the convector length would be 27.6 ft:

$$16,000 \text{ Btuh} \div 580 \text{ Btuh/ft} = 27.6 \text{ ft}$$

Other types of terminal heating units are selected from manufacturers' literature in much the same manner.

When installing baseboard terminal units, provide for expansion due to heat. The hot water passing through the unit makes it expand. Expansion occurs toward both ends of the unit, so the ends should not be restricted. It is a good practice to install expansion joints in longer units, **Figure 33.57**. If using one expansion joint, place it in the center. If using two, space them evenly at intervals one-third the length of the unit. When piping drops below the floor level, the vertical risers should each be at least 1 ft long. The table in **Figure 33.58** shows the diameter of the holes needed through the floor for the various pipe sizes. The table in **Figure 33.59** lists the maximum lengths that can be installed when the water temperature is raised from 70°F to the temperature shown.

Figure 33.55 Terminal hydronic heating units. (A) Water in the radiator radiates heat out into the room. (B) The fan coil unit is similar to baseboard units but has a blower that circulates air over the hot surface of the terminal unit. (A) Photo by Bill Johnson. (B) Courtesy Ferris State University. Photo by John Tomczyk

(A)

(B)

33.6 HIGH-TEMPERATURE HYDRONIC PIPING SYSTEMS

Up to this point in the unit, we have discussed the components that are used to make up hydronic heating systems. Some of these components are connected together and make up what is known as the "near-boiler piping." A common near-boiler piping configuration is shown in **Figure 33.60**. We will now start putting these pieces together to form a number of different types of heating systems. As you will see, each of these systems will have advantages and disadvantages. It is important to evaluate each project separately, as the priorities and concerns of the equipment owner and operator will undoubtedly vary from one job to the next. The systems that will be discussed in the following sections are basic high-temperature applications, but variations or combinations of

Figure 33.56 Example of a thermal rating table for a finned-tube baseboard. Courtesy Slant/Fin Corp.

RATINGS: Fine/Line 15 Series
(Hot water ratings, BTU/HR per linear ft. with 65°F (18.3°C) entering air)

ELEMENT	WATER FLOW	PRESSURE DROP*	140°F†	150°F	160°F	170°F	180°F	190°F	200°F	210°F	215°F	220°F
No. 15-75E Baseboard with ¾" element	1 GPM	47	290	350	420	480	550	620	680	750	780	820
	4 GPM	525	310	370	440	510	580	660	720	790	820	870
No. 15-50 Baseboard with ½" element	1 GPM	260	310	370	430	490	550	610	680	740	770	800
	4 GPM	2880	330	390	450	520	580	640	720	780	810	850

* Millinches per foot.
† Ratings at 140°F determined by multiplying 150°F rating by the I=B=R conversion multiplier of 84.

NOTE: Ratings are for element installed with damper open, with expansion cradles. Ratings are based on active finned length (5" to 6" less than overall length) and include 15% heating effect factor. Use 4 gpm ratings only when flow is known to be equal to or greater than 4 gpm; otherwise, 1 gpm ratings must be used.

Figure 33.57 An expansion joint. Courtesy Edwards Engineering Corp.

Figure 33.58 This table indicates the diameter of the holes required for each pipe size to allow for expansion.

TUBE SIZE		RECOMMENDED MINIMUM HOLE (INCHES)
NOMINAL (INCHES)	O. D. (INCHES)	
$\frac{1}{2}$	$\frac{5}{8}$	1
$\frac{3}{4}$	$\frac{7}{8}$	$1\frac{1}{4}$

Figure 33.59 The maximum lengths of a baseboard unit that can be installed when the temperature is raised from 70°F to the temperature indicated in the table.

AVERAGE WATER TEMPERATURE °F	MAXIMUM LENGTH OF STRAIGHT RUN (FEET)
220	26
210	28
200	30
190	33
180	35
170	39
160	42
150	47

these configurations are possible, depending on the design requirements of the system.

THE SERIES LOOP SYSTEM

The series loop system is the hydronic system that is found most often, primarily because of the low installation costs involved in executing this piping configuration. Similar to a series electric circuit, all of the terminal units are piped in series with each other so that the outlet of one heat emitter is the inlet of the next, **Figure 33.61**. The main drawback of the series loop system is that it is not possible to have individual temperature control for each area being heated. This is because there is only one path for the hot water to take, and stopping the water flow through one terminal unit will stop water flow in the entire system. Another drawback is that the first terminal units being supplied will be warmer than the later ones. This is because, as the water travels through the loop and heat is transferred from the hot water to the occupied space, the temperature of the water in the loop will drop.

Consider the series loop system in **Figure 33.62**. This system has three terminal units that are designed to provide 20,000 Btuh, 30,000 Btuh, and 50,000 Btuh, respectively. We will also assume that the water supplied by the boiler is 180°F and that the temperature of the water returning to the boiler is 160°F. From this information, we can determine the amount of water flow by using the following formula:

$$\text{Water flow (gpm)} = \frac{Q_T}{500 \times \Delta T}$$

$$\text{Water flow (gpm)} = [(20{,}000\text{ Btuh} + 30{,}000\text{ Btuh} + 50{,}0000\text{ Btuh})] \div [500 \times (180°\text{F} - 160°\text{F})]$$

$$\text{Water flow (gpm)} = 100{,}000 \div 10{,}000$$

$$\text{Water flow (gpm)} = 10\text{ gpm}$$

We now know that there will be 10 gpm of water flow through the piping circuit when the circulator is operating.

Figure 33.60 Hot water supply manifold. Courtesy Bell and Gossett

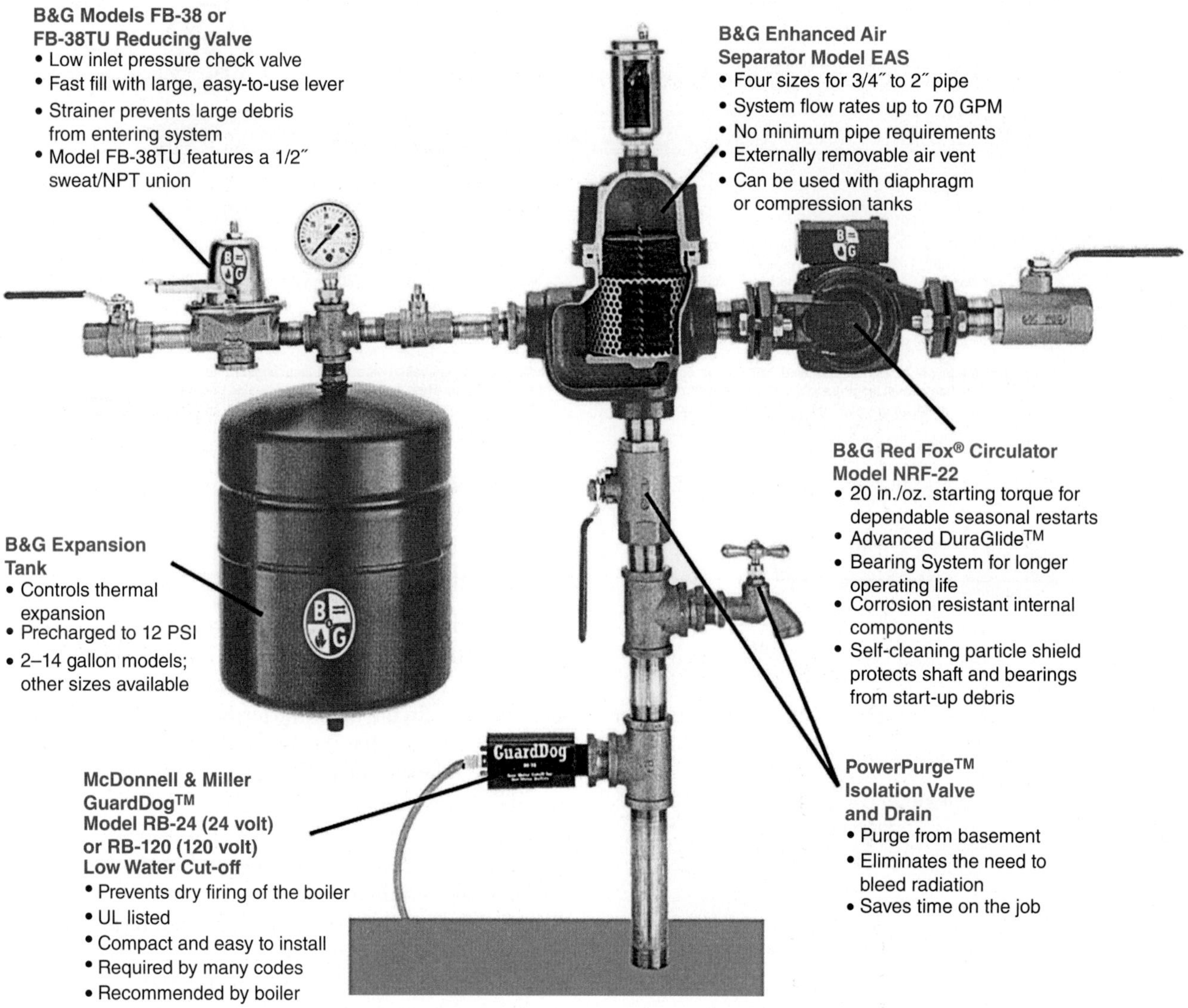

We can determine the temperature of the water at the outlet of the first terminal unit by using the following variation on the previous formula:

$$\Delta T = Q_1 \div (500 \times \text{gpm})$$
$$\Delta T = 20{,}000 \div (500 \times 10\text{ gpm})$$
$$\Delta T = 20{,}000 \div (500 \times 10\text{ gpm})$$
$$\Delta T = 4°F$$

So, if the temperature of the water entering the first terminal unit is 180°F, the temperature of the water at the outlet of the first terminal unit will be 4° less than that, or 176°F, **Figure 33.63**. The temperature of the water at the outlet of the second terminal unit can be calculated as follows:

$$\Delta T = Q_2 \div (500 \times \text{gpm})$$
$$\Delta T = 30{,}000 \div (500 \times 10\text{ gpm})$$
$$\Delta T = 30{,}000 \div (500 \times 10\text{ gpm})$$
$$\Delta T = 6°F$$

Since the temperature of the water entering the second terminal unit is 176°F, the temperature at the outlet of the second terminal unit is 170°F (176°F − 6°F), **Figure 33.63.** Finally, let's calculate the temperature at the outlet of the third terminal:

$$\Delta T = Q_3 \div (500 \times \text{gpm})$$
$$\Delta T = 50{,}000 \div (500 \times 10\text{ gpm})$$
$$\Delta T = 50{,}000 \div (500 \times 10\text{ gpm})$$
$$\Delta T = 10°F$$

Since the temperature of the water entering the third terminal unit is 170°F, the temperature at the outlet of the third terminal unit is 160°F (170°F − 10°F), **Figure 33.63.**

Using the formula a little differently, we can determine the output of a particular terminal unit if we know the temperature difference across it. Consider a terminal unit that has 8 gpm of hot water flowing through it. The temperature at the inlet of the terminal unit is 180°F and the temperature

Figure 33.61 Series loop piping circuit.

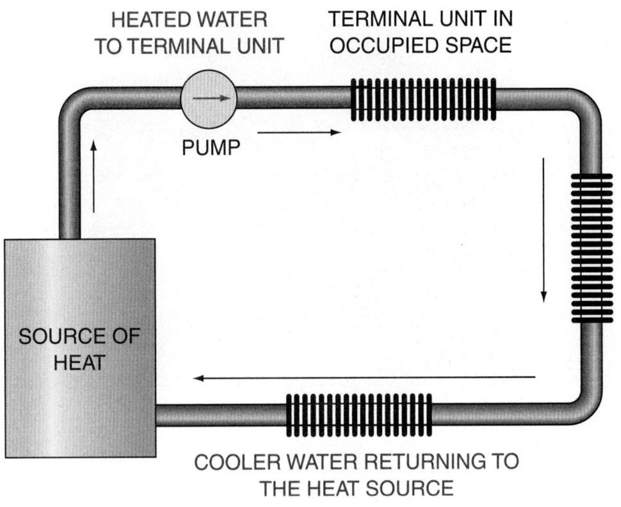

Figure 33.62 A series loop piping system with a 100,000 Btu/h output and a ΔT of 20°F across the boiler.

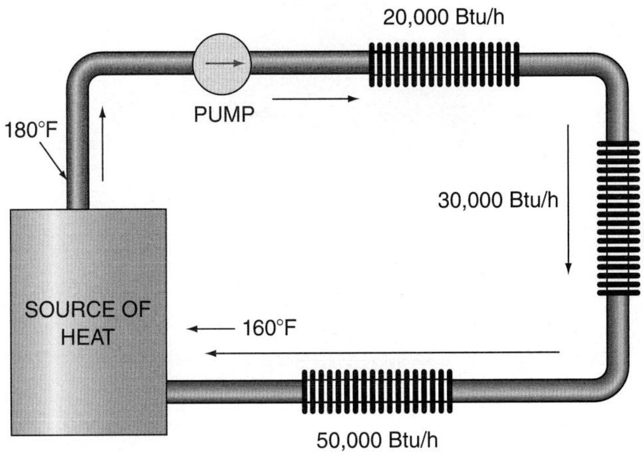

at the outlet of the terminal unit is 178°F. The capacity of the terminal unit is as follows:

$$Q = \text{Water flow (gpm)} \times 500 \times \Delta T$$
$$Q = 8\text{ gpm} \times 500 \times (180°F - 178°F)$$
$$Q = 8\text{ gpm} \times 500 \times 2$$
$$Q = 8000\text{ Btu/h}$$

THE ONE-PIPE SYSTEM

In a one-pipe hydronic system, one main piping loop extends around the occupied space and connects the outlet of the boiler back to the return of the boiler. Each individual terminal unit is connected to this main loop with two tees, **Figure 33.64**, one or more of which may be specially designed for use on one-pipe systems. The proper operation

Figure 33.63 The water temperature at the outlet of each terminal unit. The ΔT is determined by taking both the Btu output of each terminal unit and the water flow through each terminal unit into account.

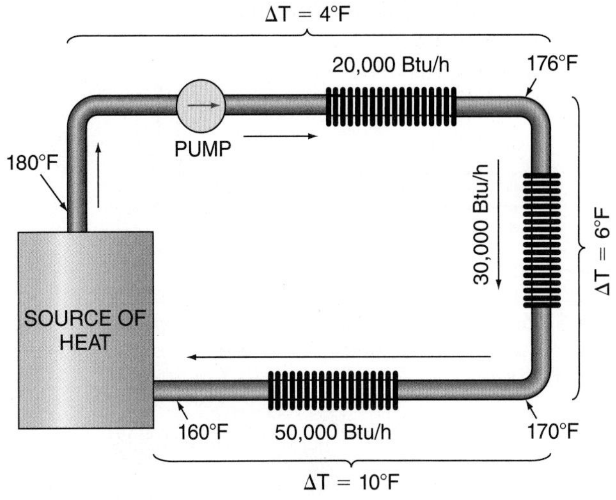

Figure 33.64 Simplified one-pipe hydronic system.

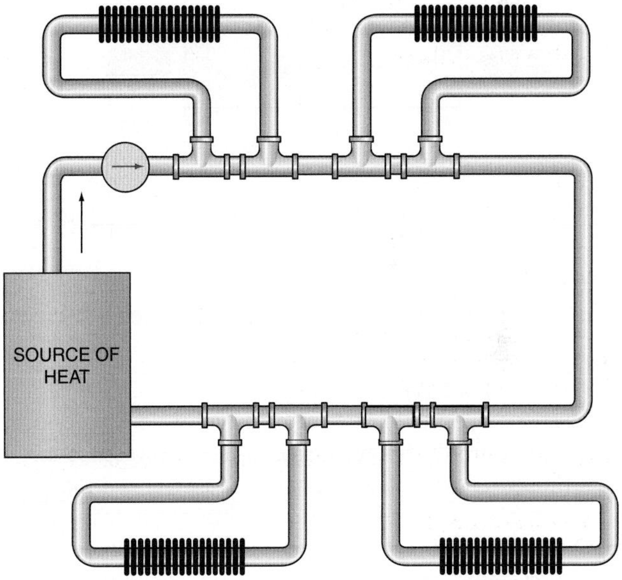

of a one-pipe hydronic system relies on the proper ratios of resistance between the terminal unit branch and the resistance to flow in the section of pipe between the two tees. Consider the setup in **Figure 33.65**. If there is a water flow of 4 gpm entering the tee at point A, 4 gpm must leave the tee. How much water will flow through the terminal unit and how much water will bypass the terminal unit depends on the resistances of each branch. If the resistance of the terminal unit branch is three times the resistance of the bypass branch, the flow through the bypass branch will be three times the flow through the terminal unit, **Figure 33.66**.

Figure 33.65 Common piping between the tees. In this application, the diverter tee is located at the return side of the terminal unit.

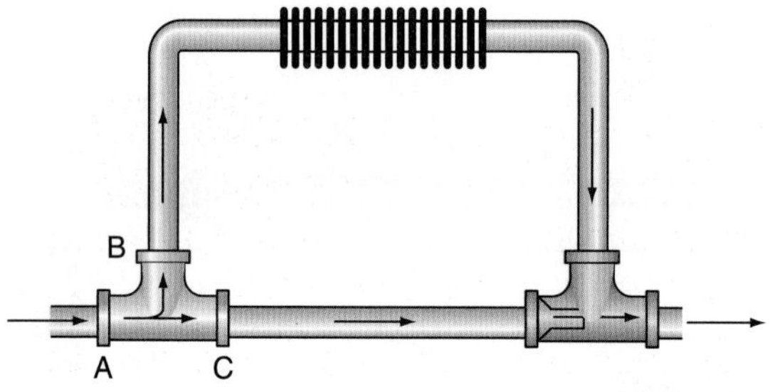

Figure 33.66 The high resistance at point B lowers the flow through that branch, while the lower resistance at point C results in greater flow through that branch.

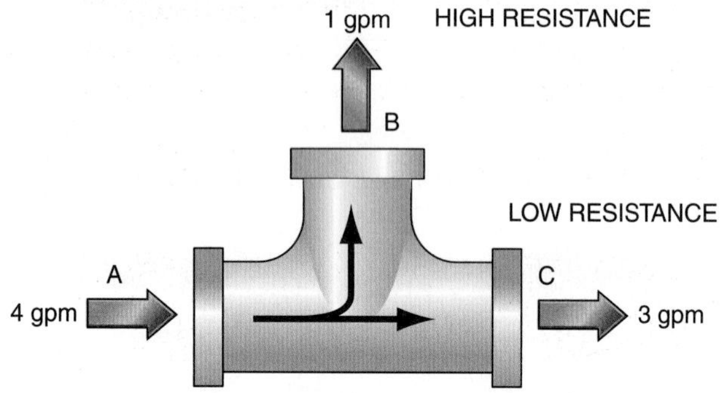

For this reason, a number of factors must be considered when laying out, evaluating, or installing a one-pipe system. These factors include the following:

- The length of the terminal unit branch circuit
- The distance between the tees
- The size of the piping in the branch circuit
- The size of the piping between the tees
- The location of the terminal unit with respect to the main loop

THE DIVERTER TEE. The special tees used in one-pipe hydronic systems are called **diverter tees** or *Monoflo tees,* **Figure 33.67.** Diverter tees are designed to increase the resistance in the main-loop pipe section between the two tees. By increasing the resistance in the main loop, more water will be directed through the terminal heating units. The diverter tee is constructed with an interior cone, **Figure 33.68**, that reduces the diameter of the pipe, thereby increasing resistance to water flow. When working with diverter tees, keep the following in mind:

- If the terminal unit is located above the hot water main and the length of the terminal unit branch is not excessive, one diverter tee should be used on the return side of the terminal unit, **Figure 33.65.**

Figure 33.67 Diverter tee for use on one-pipe systems.

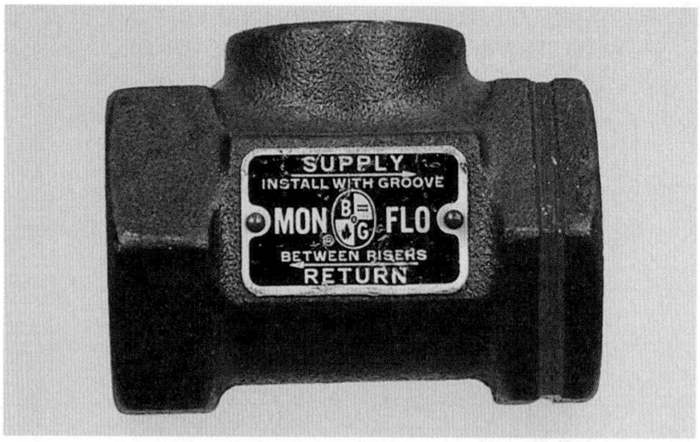

Figure 33.68 Cross-sectional view of a diverter tee.

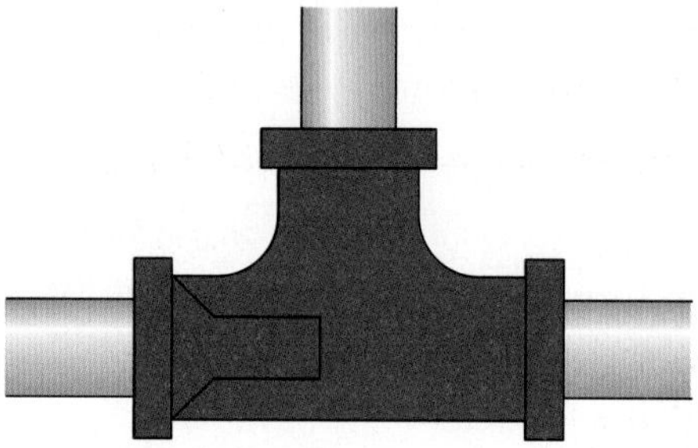

Figure 33.69 Two diverter tees are recommended if the radiator is located above the main loop and there is significant resistance in that branch.

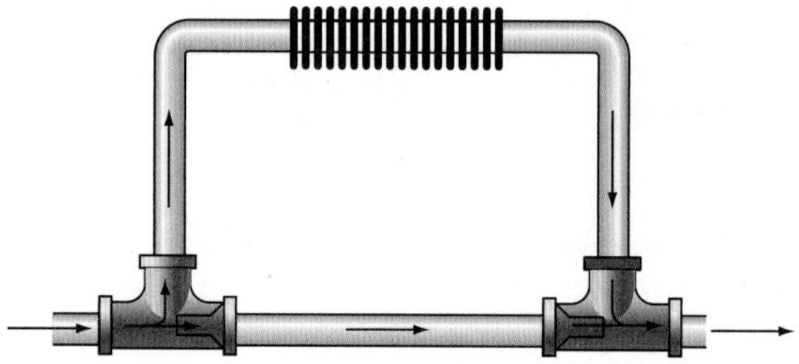

- If the terminal unit is located above the hot water main and the length of the terminal unit branch is very long, two diverter tees should be used—one on the supply side of the terminal unit and one on the return side of the terminal unit, **Figure 33.69.**
- If the terminal unit is located below the hot water main, two diverter tees should be used, **Figure 33.70.**
- If the terminal unit is located above the main hot water loop, the tees should be no closer than 6 or 12 in., depending on the manufacturer of the diverter tee.
- If the terminal unit is located below the main hot water loop, the tees should be as far apart as the ends of the

Figure 33.70 Two diverter tees are recommended if the radiator is located below the main loop.

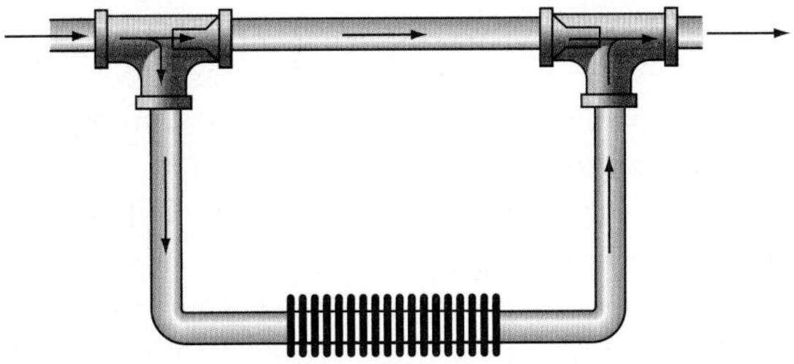

terminal unit. If the terminal unit is 2 ft long, the tees should be 2 ft apart.

- If there are multiple terminal units, and some are above and some below the main hot water loop, it is recommended that the tees be staggered, **Figure 33.71**.
- Always check the installation literature that comes with the diverter tees to ensure that they are installed correctly and that they are pointing in the right direction.

The above information is summarized in **Figure 33.72**.

THE HEAT FORMULA AND THE ONE-PIPE HYDRONIC SYSTEM. Consider a portion of the one-pipe system shown in **Figure 33.73**. Let's assume we have a one-pipe hydronic system that has a total heating load of 100,000 Btu/h. If there is a ΔT from boiler outlet to boiler return of 20°F, the water flow through the main hot water loop will be as follows:

$$\text{Water flow (gpm)} = \frac{Q_T}{500 \times \Delta T}$$

$$\text{Water flow (gpm)} = 100{,}000 \div (500 \times 20)$$

$$\text{Water flow (gpm)} = 100{,}000 \div 10{,}000$$

$$\text{Water flow (gpm)} = 10 \text{ gpm}$$

Figure 33.71 Alternate the diverter tees if there are terminal units both above and below the main loop.

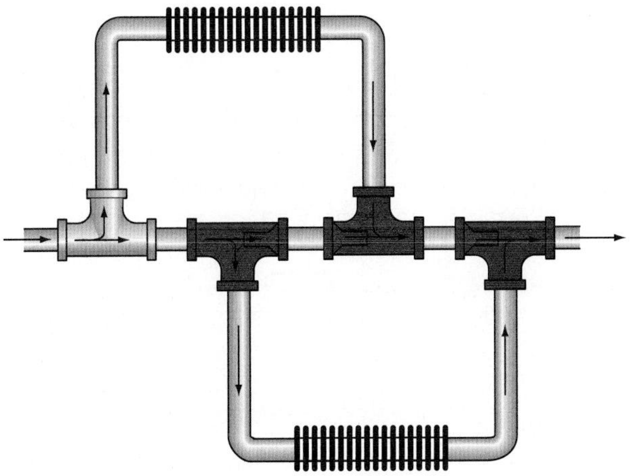

Figure 33.72 Partial water flow is diverted through the tee to and from the terminal unit. Courtesy ITT Fluid Handling Division

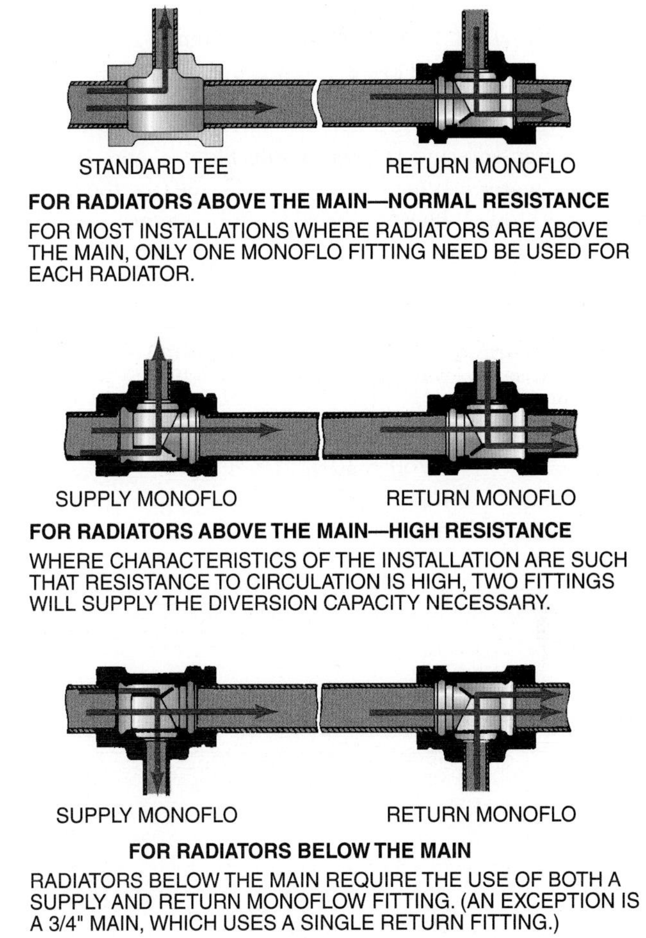

Figure 33.73 The flow rate through this terminal unit is 2 gpm.

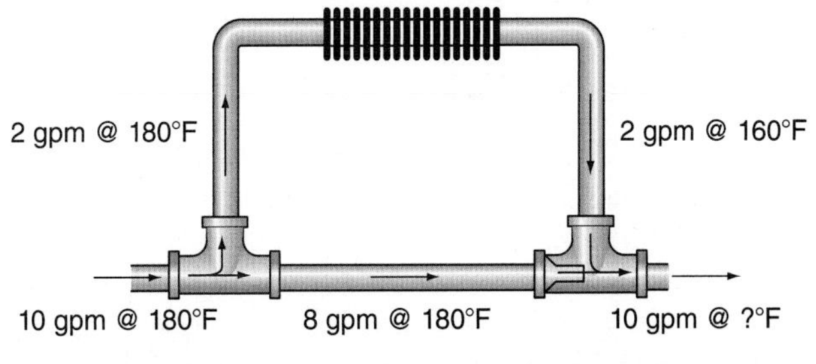

If the first terminal unit is to provide 20,000 Btu of heating, this will require that we have 2 gpm (20,000 ÷ 10,000) of hot water flowing through the terminal unit. The selected diverter tee should be capable of providing the desired flow. Since 2 gpm of water will flow through the terminal unit, 8 gpm will continue to flow through the main hot water loop. At the second tee, there will be 8 gpm of 180°F water entering

from the main loop and 2 gpm of 160°F water entering from the terminal unit. We can determine the mixed water temperature by using the following formula:

$$(\text{Flow 1} \times \text{Temp. 1}) + (\text{Flow 2} \times \text{Temp. 2}) = (\text{Flow 3} \times \text{Temp. 3})$$

In the above formula, Flow 1, Flow 2, Temp. 1, and Temp. 2 refer to water entering the tee, while Flow 3 and Temp. 3 refer to the water leaving the tee. Flow 3 must be the sum of Flow 1 and Flow 2. Here's how the numbers work out:

$$(\text{Flow 1} \times \text{Temp. 1}) + (\text{Flow 2} \times \text{Temp. 2}) = (\text{Flow 3} \times \text{Temp. 3})$$

$$(8\text{ gpm} \times 180°\text{F}) + (2\text{ gpm} \times 160°\text{F}) = (10\text{ gpm} \times \text{Temp. 3})$$

$$1440 + 320 = 10\text{ gpm} \times \text{Temp. 3}$$
$$1760 = 10\text{ gpm} \times \text{Temp. 3}$$
$$\text{Temp. 3} = 1760 \div 10$$
$$\text{Temp. 3} = 176°\text{F}$$

From this we can conclude that the temperature of the water entering the second terminal unit will be 176°F, **Figure 33.74**. A sample one-pipe system using thermostatic radiator valves is shown in **Figure 33.75**.

THE TWO-PIPE DIRECT-RETURN SYSTEM

Two-pipe hydronic systems use two pipes. One pipe is used to carry water to the terminal units, while the other pipe is used to carry water from the terminal units back to the boiler, **Figure 33.76**. In a two-pipe direct-return system, the terminal unit closest to the boiler will have the shortest piping run, while the terminal unit farthest from the boiler will have

Figure 33.74 The mixed water temperature at the outlet of the terminal unit is 176°F.

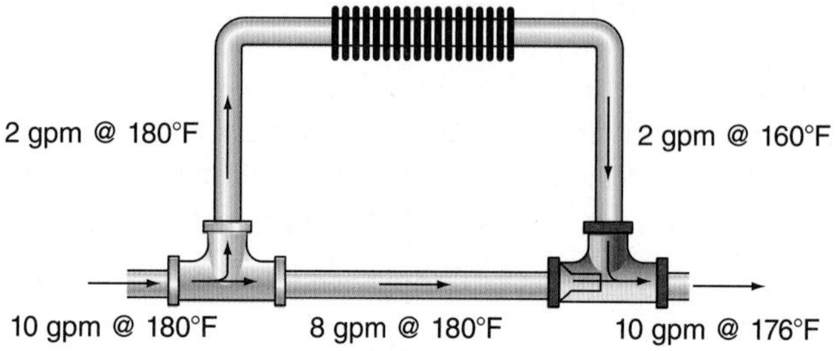

Figure 33.75 One-pipe hydronic system utilizing thermostatic radiator valves.

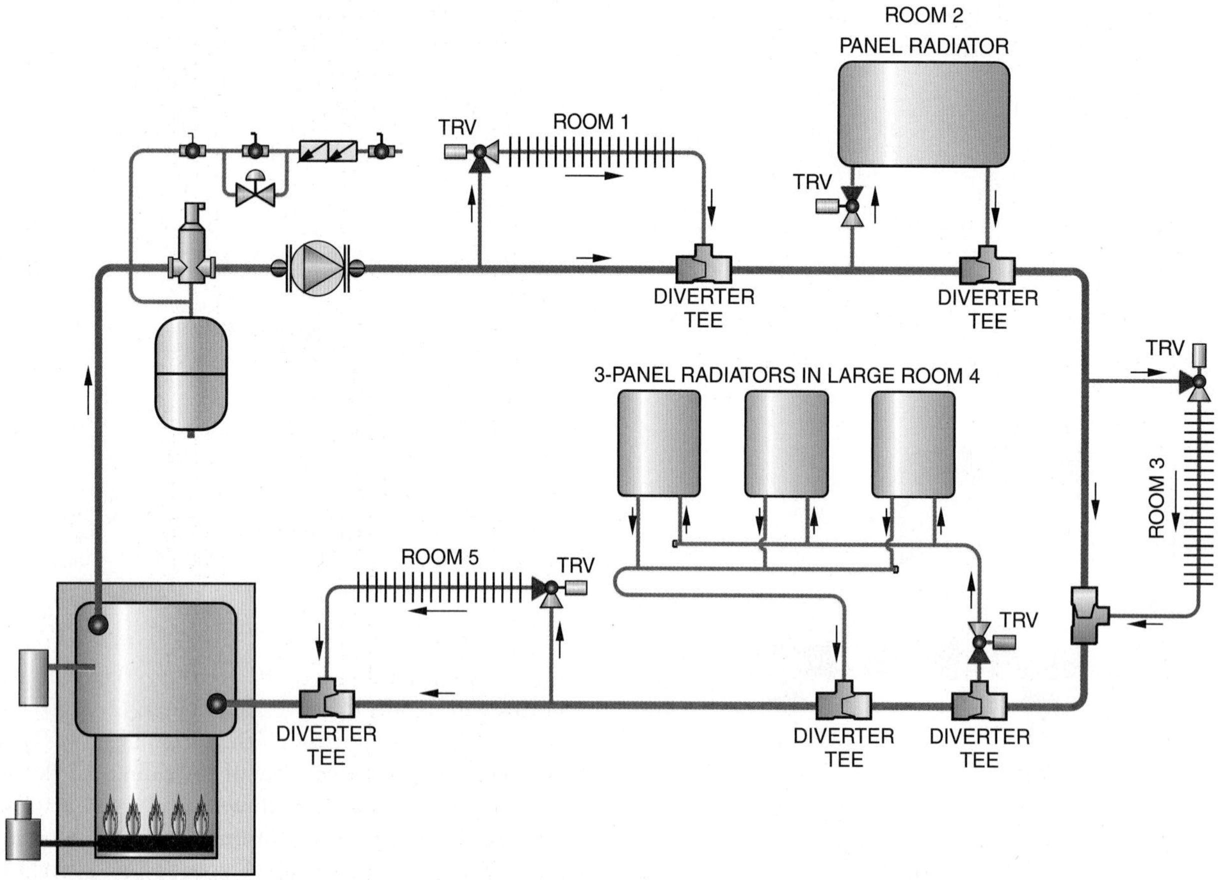

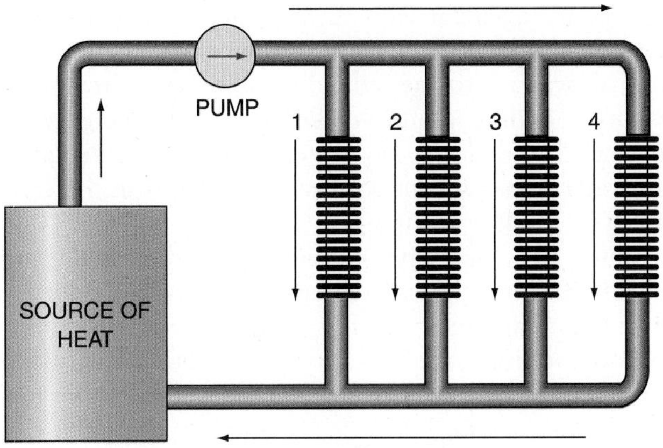

Figure 33.76 Simplified two-pipe direct-return piping arrangement.

the longest piping run. This means that the resistance of the closer terminal unit circuits is lower than that of the terminal units located far from the boiler. Consider the piping circuit shown in **Figure 33.77**. This circuit is a two-pipe direct-return system that has four terminal units numbered 1 through 4. For water to flow from the boiler through terminal unit 1 and back to the boiler, the water must travel 24 ft (3 + 9 + 5 + 7). For water to make a complete path through terminal unit 4, the water must travel 30 ft. Because the piping circuit through the more distant terminal unit location is longer, less water will flow to this location. In order to even out the water flow, balancing valves, **Figure 33.46**, are used to increase the resistance of the shorter piping loops, **Figure 33.78**.

THE TWO-PIPE REVERSE-RETURN SYSTEM

In the two-pipe reverse-return system, the first terminal unit that is supplied with water is the last to have its water return to the boiler. In other words, the terminal unit with the shortest supply pipe will have the longest return pipe and vice versa, **Figure 33.79**. By configuring the piping in this manner, the distance traveled by the water will be the same for all terminal units in the piping arrangements. It is left as an exercise for the reader to confirm that water must flow 31 ft to get from the boiler through any of the terminal units and back to the boiler in the piping arrangement shown in **Figure 33.80**. Two-pipe reverse-return systems do not require the use of balancing valves.

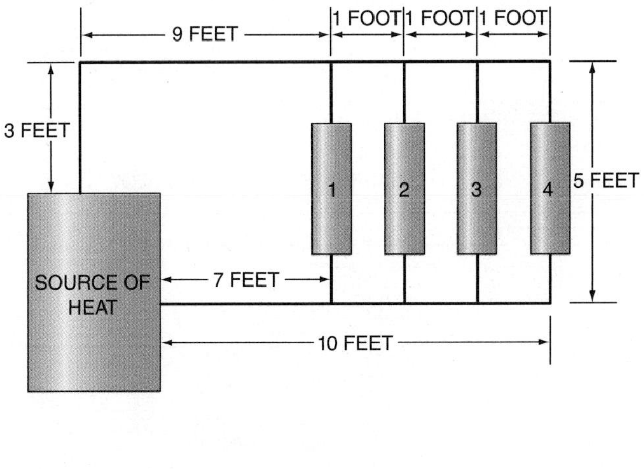

Figure 33.77 Water flowing through terminal unit 1 must travel 24 ft, while water flowing through terminal unit 4 must travel 30 ft.

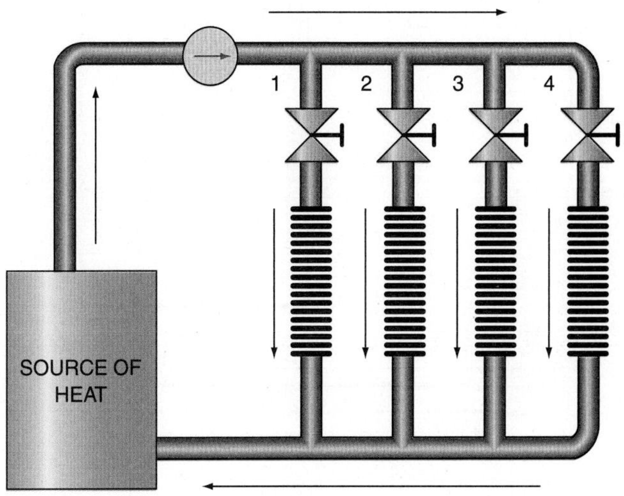

Figure 33.78 Balancing valves are used to even the resistance in each branch of a two-pipe direct-return system.

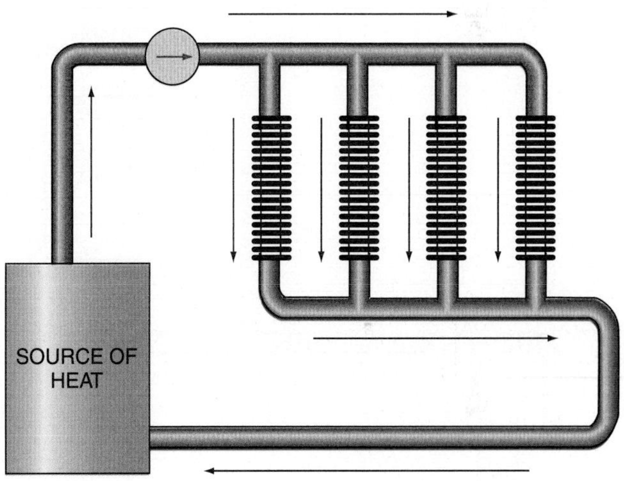

Figure 33.79 A two-pipe reverse-return piping arrangement.

PRIMARY–SECONDARY PUMPING

Primary–secondary pumping involves at least two separate piping circuits between the boiler and the terminal-unit circuits. One circuit path flows from the boiler supply back to the boiler return. Flow through this path is made possible by using a circulator pump. The secondary circuit(s) is(are) connected to this main/primary circuit and shares a portion of the piping with that circuit, **Figure 33.81**. **Figure 33.81(A)** shows

Figure 33.80 The length of the piping through any one branch is exactly the same as the length through any other branch.

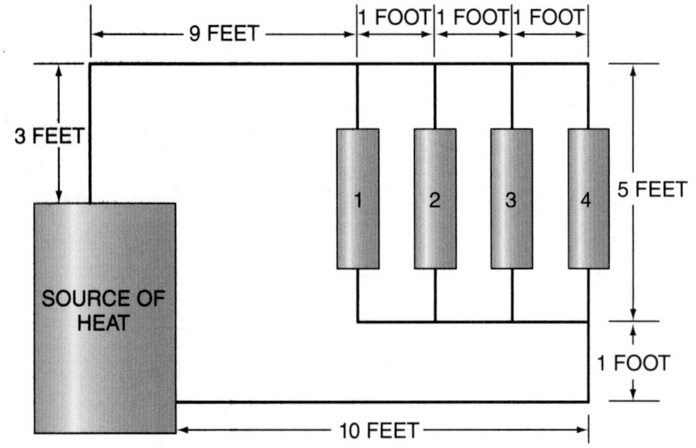

Figure 33.81 (A) Primary–secondary piping arrangement where one secondary loop is feeding three terminal units. (B) Primary–secondary piping arrangement where three secondary loops are each supplying hot water to one terminal unit.

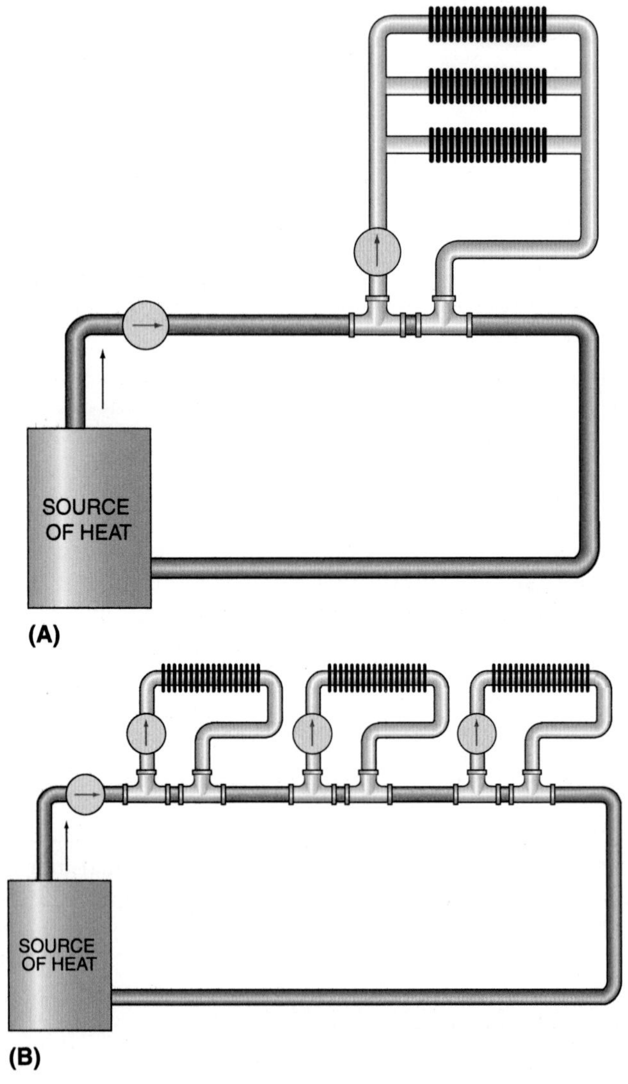

a primary–secondary pumping arrangement with one secondary circuit; **Figure 33.81(B)** shows a primary–secondary pumping arrangement with three secondary circuits. Notice that each secondary circuit has its own circulator pump that will push water into the individual secondary circuits when the corresponding pump is energized. The industry term that is used to describe any system that uses a circulator to push water into a secondary heating loop (as in the case of primary–secondary pumping) is **injection**.

Unlike the one-pipe configuration discussed earlier, the tees that are used to create a primary–secondary system are standard tees and not of the diverter or Monoflo variety. In addition, the tees are to be positioned as close to each other as possible. The space in between the two tees is referred to as *common piping*. By closely spacing these tees, the resistance of the common piping will be very low. This is a desirable condition, as we do not want water to flow through the secondary loops when there is no call for heat in them. The lower the resistance of the common piping is, the more likely it is that the water will bypass (as opposed to flow into) the secondary loops, **Figure 33.82.**

CIRCULATOR PUMPS IN PRIMARY–SECONDARY SYSTEMS. It can be seen in **Figure 33.81(B)** that primary–secondary pumping involves the use of multiple circulator pumps. One circulator pump is located in the primary loop and is typically the largest pump in the system. It is this pump that is responsible for making the heated water available to the secondary loops. When no remote areas or zones are calling for heat, all system pumps are off but, depending on system design, the boiler may be cycling on and off to maintain the water in the

Figure 33.82 More resistance between the tees results in more water flow through the terminal unit connected to the tees. Less resistance between the tees results in less water flow through the terminal unit.

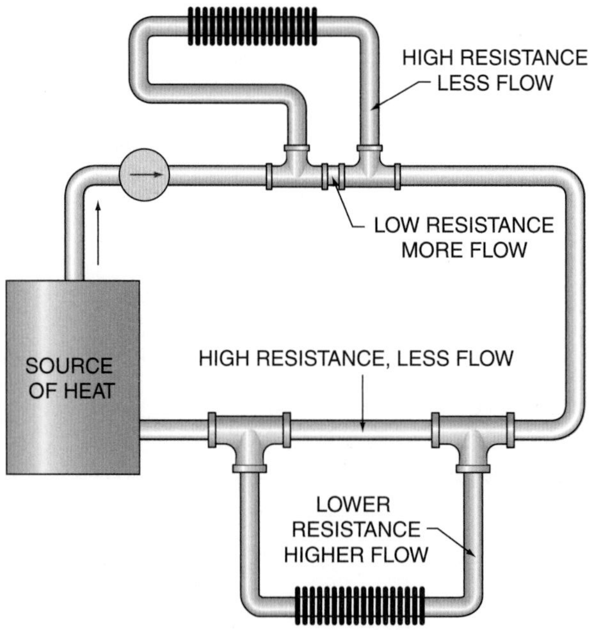

Figure 33.83 Sample one-pipe system with three terminal units providing 200,000 Btu/h of heating to the occupied space.

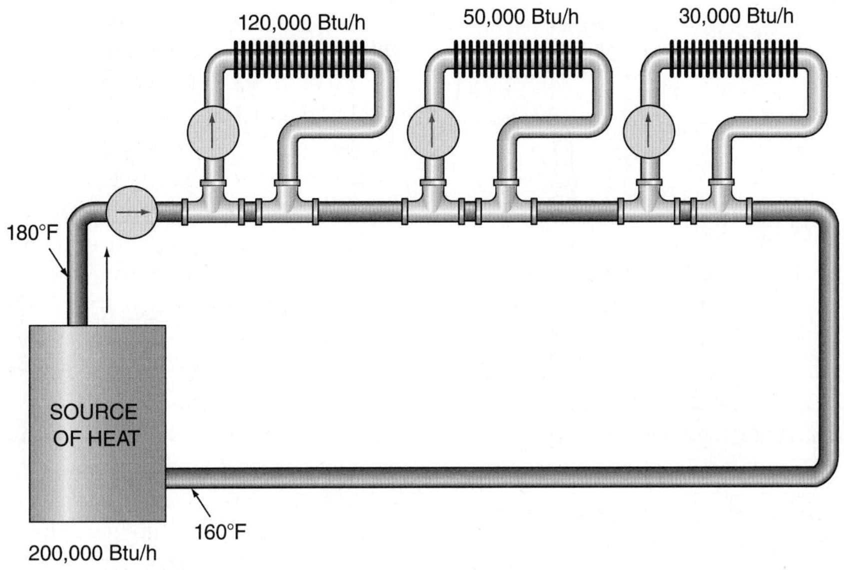

boiler at the desired operating temperature. When an individual zone calls for heat, both the secondary pump for that loop and the primary pump will cycle on. The primary pump will bring the heated water to the common piping section and the secondary pump will push heated water into the secondary loop to satisfy the heating requirements of the space.

Consider the primary–secondary setup in **Figure 33.83**. The system has a capacity of 200,000 Btu/h and the boiler operates with a ΔT of 20°F. There are three secondary loops with the following heating capacities:

- Zone 1: 120,000 Btu/h
- Zone 2: 50,000 Btu/h
- Zone 3: 30,000 Btu/h

The total water flow that should be made available by the system is as follows:

$$\text{Water flow (gpm)} = \frac{Q_T}{500 \times \Delta T}$$

$$\text{Water flow (gpm)} = 200{,}000 \div (500 \times 20)$$

$$\text{Water flow (gpm)} = 200{,}000 \div 10{,}000$$

$$\text{Water flow (gpm)} = 20 \text{ gpm}$$

If all three secondary loops are calling for heat, the water temperature at the outlet of each secondary loop is as shown **Figure 33.84**. Notice that the temperature of the water at the outlet of the third secondary loop is the temperature of the boiler supply water (180°F) less the 20°F that was indicated

Figure 33.84 Water temperatures at various points in the system when all three zones are calling for heat.

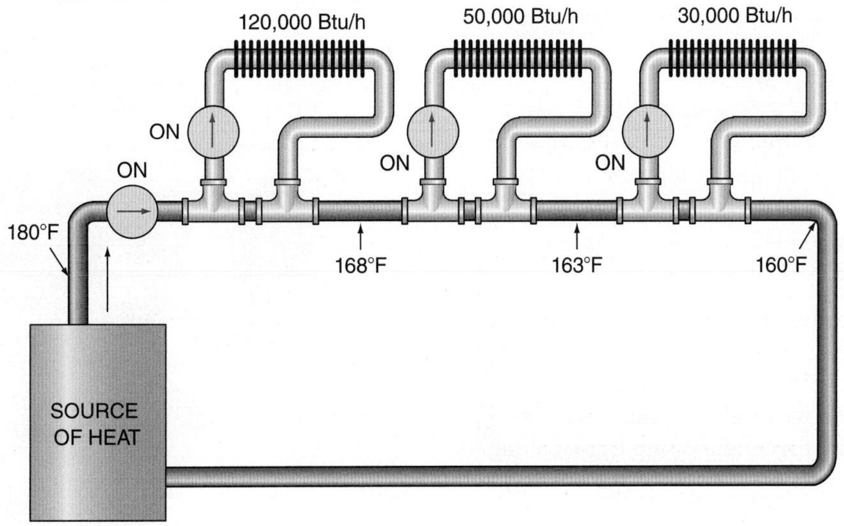

Figure 33.85 Water temperatures at various points in the system when only the second and third zones are calling for heat.

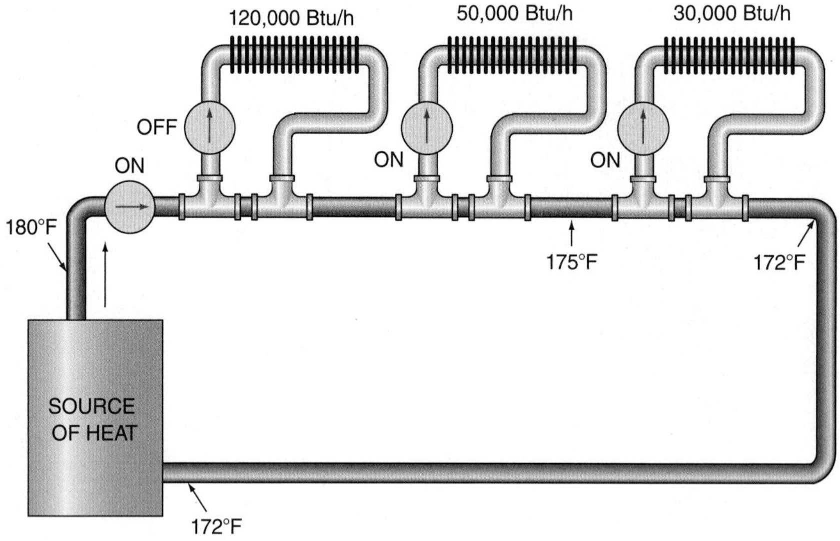

as the ΔT of the boiler. Let's now assume that the thermostat for the first secondary loop is satisfied. In that case, the corresponding secondary pump will cycle off and there will be no water flow into that loop. When the first loop is no longer active, water at a temperature of 180°F is fed to the second loop. The temperature of the water at the outlet of the second loop, **Figure 33.85**, is calculated as follows:

$$(\text{Flow 1} \times \text{Temp. 1}) + (\text{Flow 2} \times \text{Temp. 2}) = (\text{Flow 3} \times \text{Temp. 3})$$

$$(15\text{ gpm} \times 180°\text{F}) + (5\text{ gpm} \times 160°\text{F}) = (20\text{ gpm} \times \text{Temp. 3})$$

$$2700 + 800 = 20\text{ gpm} \times \text{Temp. 3}$$

$$3500 = 20\text{ gpm} \times \text{Temp. 3}$$

$$\text{Temp. 3} = 3500 \div 20$$

$$\text{Temp. 3} = 175°\text{F}$$

The water temperature at the outlet of the third secondary loop will be as follows:

$$(\text{Flow 1} \times \text{Temp. 1}) + (\text{Flow 2} \times \text{Temp. 2}) = (\text{Flow 3} \times \text{Temp. 3})$$

$$(17\text{ gpm} \times 175°\text{F}) + (3\text{ gpm} \times 155°\text{F}) = (20\text{ gpm} \times \text{Temp. 3})$$

$$2975 + 465 = 20\text{ gpm} \times \text{Temp. 3}$$

$$3440 = 20\text{ gpm} \times \text{Temp. 3}$$

$$\text{Temp. 3} = 3440 \div 20$$

$$\text{Temp. 3} = 172°\text{F}$$

The temperature of the water returning to the boiler will be 172°F, **Figure 33.85**. Notice that the water left the boiler at 180°F and returned at 172°F. This represents a ΔT of only 8°F across the boiler, when we originally anticipated a ΔT of 20°F. This is because the first secondary loop is not calling for heat and this loop represents 60% (0.6) of the total system capacity (120,000 Btu/h ÷ 200,000). Since 60% of our 20°F ΔT is 12°, we only see the portion of the system (20°F − 12°F = 8°F) that is actively being heated. Notice that when all three secondary loops are actively heating, the return temperature at the boiler reflects the ΔT of 20°F.

MIXING VALVES IN PRIMARY–SECONDARY SYSTEMS. Mixing valves are used to combine two water streams of different temperatures to produce water flow that is at a temperature between the two entering temperatures. Common types of mixing valves include three-way valves, **Figure 33.86**, and four-way valves, **Figure 33.87**. Three-way valves have two

Figure 33.86 A three-way mixing valve with a manual adjustment knob. Courtesy Danfoss, Inc.

Figure 33.87 Four-way mixing valves. Courtesy Tekmar Control Systems

inlet ports and one outlet port, whereas four-way valves have two inlet ports and two outlet ports.

Mixing valves can be manually set, **Figure 33.86**, or they can be thermostatically controlled, **Figure 33.88**. A manually set mixing valve is set to the desired position on the basis of the percentages of water flow desired from each inlet. At either extreme, the water at the outlet of the valve will be 100% from the hot water inlet or 100% from the cold water inlet. At any point in between, the outlet water will be a mixture of the two, **Figure 33.89**. Four-way mixing valves allow some of the water leaving the boiler to return to the boiler to keep the return-water temperature high enough to prevent flue-gas condensation, **Figure 33.90**.

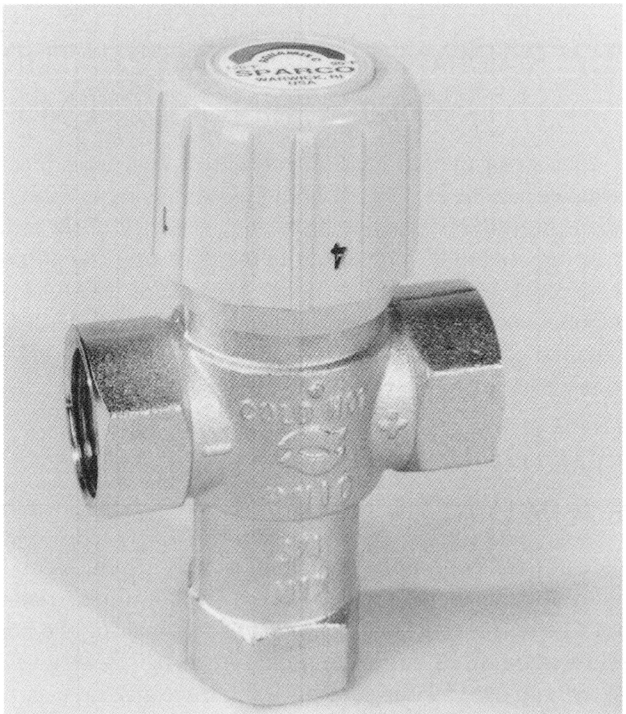

Figure 33.88 Three-way thermostatic mixing valve. Courtesy Sparco, Inc.

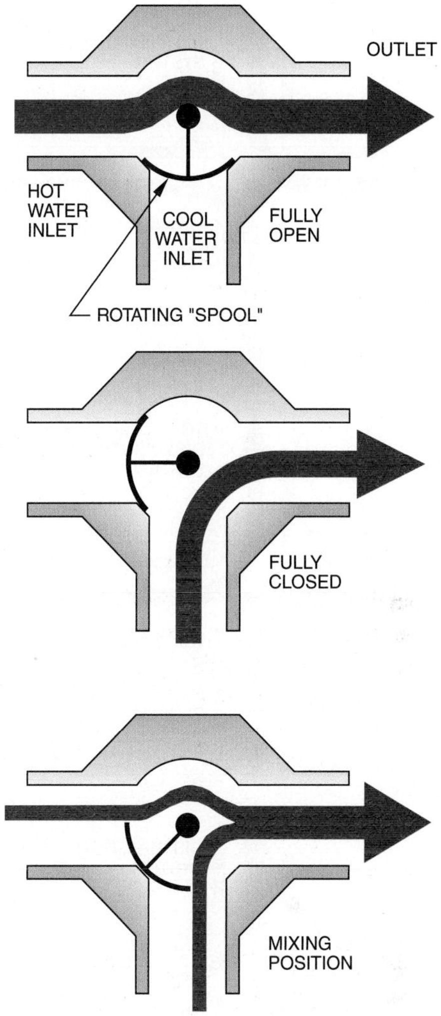

Figure 33.89 Cutaway illustration showing internal construction of a three-way valve.

One of the many benefits of primary–secondary pumping is that each individual loop can be used to supply water at different temperatures to the terminal units in that loop. For example, one portion of the structure may have baseboard heating elements installed, while another portion of the structure may have radiant heat. The area with baseboard heating elements will require higher-temperature water than will the area with radiant heat. By using primary–secondary pumping in conjunction with a mixing valve, both conditions can be met. Multitemperature combination systems will be discussed later on in this unit.

EXPANSION TANKS IN PRIMARY–SECONDARY SYSTEMS. Earlier in this unit, we discussed the location of the expansion tank with respect to the circulator pump. We concluded

Figure 33.90 Cutaway illustration showing internal construction of a four-way valve.

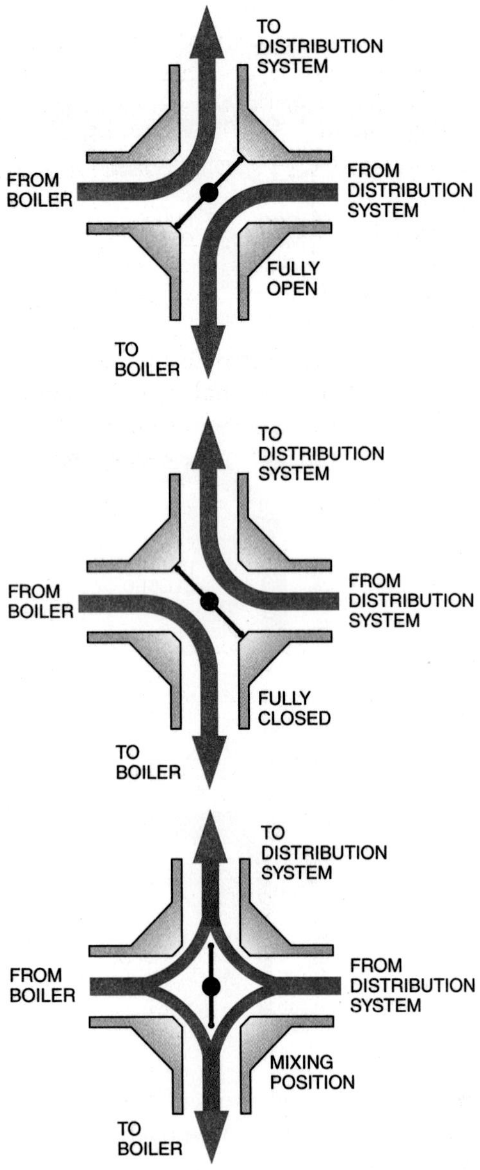

that the best place for the circulator was on the supply piping coming from the boiler, pumping away from the expansion tank. This was referred to as the "point of no pressure change." Since primary–secondary systems have additional circulator pumps, it may seem logical that there should be expansion tanks in each of the secondary loops to provide for the expansion of heated water in these locations. The truth of the matter is that the common piping between the primary and secondary loops serves as the expansion tank for that loop. As water is heated, it will move from the secondary loop into the primary loop and then into the expansion tank located in the primary loop. Since we want to pump away from the point of no pressure change, the circulators in the secondary loops should be located on the pipe closest to the primary pump—and should be directed to pump into the secondary circuit and away from the common piping.

Regardless of the type of piping layout, the estimator or engineer will pay close attention to the sizing of the pipe for any system. Pipe sizing and pump size determine how much water will flow in each circuit. If the pipe is too large, it will be expensive and it may not meet the low-limit flow requirements for convectors. For example, the minimum design flow rates for convector pipes of different diameters are:

½ in. ID	0.3 gpm
¾ in. ID	0.5 gpm
1 in. ID	0.9 gpm
1¼ in. ID	1.6 gpm

If the flow rates below these are used at the convectors, the heat exchange between the water and the air will be less. The velocity of the water in the convector helps the heat exchange. Flow that is too slow is called *laminar flow.*

If a system is designed with too fast a flow, the pipe will be too small and the velocity noise of the water will be objectionable. The designer tries to reach a balance between oversized and undersized pipe. The following table lists some water velocities in ft per minute (fpm) recommended for hydronic piping systems.

Pump discharge	8 to 12 fpm
Pump suction	4 to 7 fpm
Drain lines	4 to 7 fpm
Header	5 to 15 fpm
Riser	3 to 10 fpm
General service	5 to 10 fpm
City water lines	3 to 7 fpm

33.7 RADIANT, LOW-TEMPERATURE HYDRONIC PIPING SYSTEMS

The concept of radiant heat differs a great deal from that of convective heating as it has been discussed thus far in this unit. Radiant heating systems rely on heating the shell of the structure as opposed to heating the air in the structure. The purpose of convective heating is to determine the rate at which heat is lost from the room by conduction, convection, and radiation at the desired conditions. Radiant heating involves regulating the rate at which the human body loses heat.

HOW THE BODY FUNCTIONS AS A RADIATOR

Under normal conditions, the human body produces about 500 Btu/h. The body requires about 100 Btu/h to remain alive. The rest of this heat must be rejected, making the body act as a radiator. By controlling the rate at which we give up heat, we can control our comfort level. If we give up heat too fast, we will feel cold. If we do not give up heat fast enough,

we will feel warm and uncomfortable. Typically, a room at a temperature of 68°F allows us to shed at an acceptable rate the extra 400 Btu/h that our body generates. For this reason, it is desirable that the occupied space be about 68°F—with the exception of the areas close to the ceiling and floor. Since our heads are not close to the ceiling, there is no need to keep that area heated, so it can be somewhat cooler than 68°F. In addition, we lose a lot of heat from our feet, so the floor should be somewhat warmer than 68°F. **Figure 33.91** shows room temperature changes as we go from the floor to the ceiling. This represents an ideal temperature pattern, one which radiant heating systems closely follow. **Figure 33.92** shows the same ideal temperature pattern as compared with the temperature pattern created by forced-air heating systems.

THE RADIANT SYSTEM

Like hot water hydronic systems, radiant systems rely on the circulation of hot water through a series of piping circuits. There are three main differences between radiant heating systems and conventional hot water hydronic systems. These differences are as follows:

- Radiant heating systems are nearly invisible as compared with conventional hydronic systems. The piping arrangements are concealed in the floors, walls, and sometimes ceilings of the structure.
- The piping material used in radiant heating systems is different. Instead of copper piping materials, radiant heating systems often use *polyethylene tubing,* more commonly known as PEX tubing, **Figure 33.93**. PEX is installed using a minimum of pipe fittings, **Figure 33.94**.
- The temperature of the water in radiant heating systems is considerably lower than the temperature of the water utilized by conventional hydronic systems. Typical hydronic systems use water as hot as 180°F, whereas radiant systems use water at an average temperature of about 110°F.

Figure 33.91 Ideal comfort chart compared with that of a radiant heating system. Courtesy Uponor

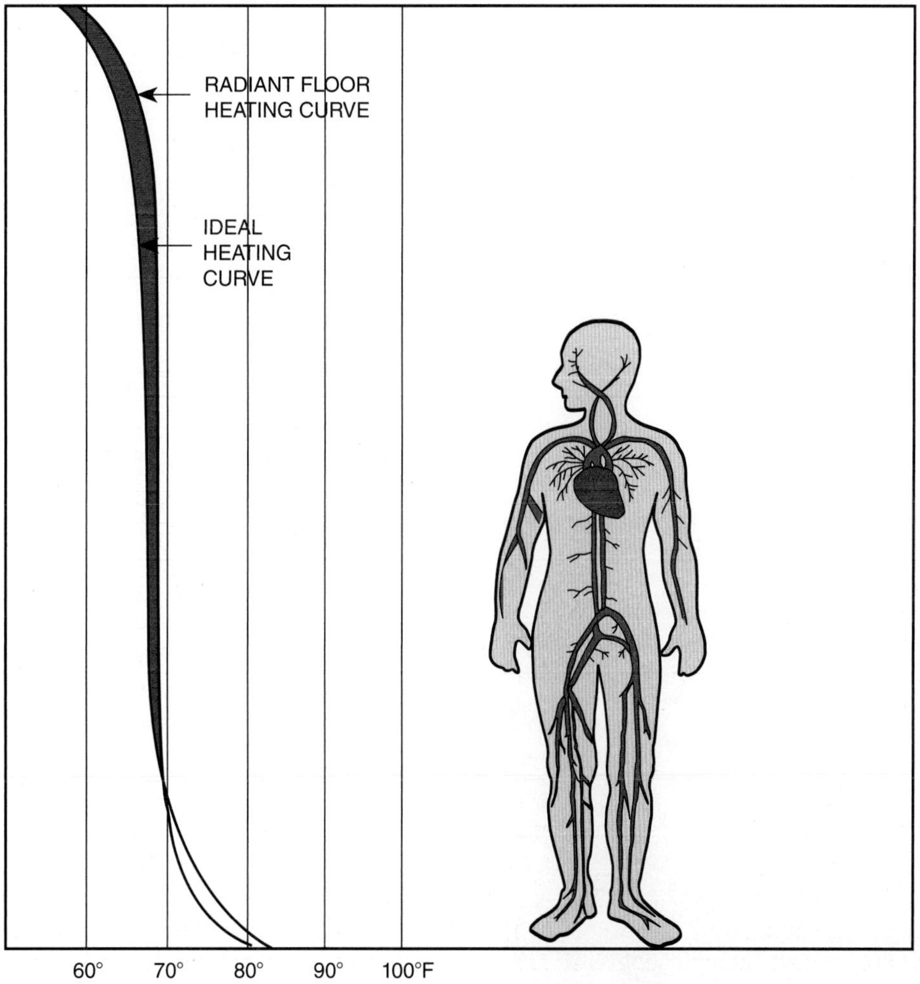

Figure 33.92 Ideal comfort chart compared with that of a forced-air heating system. Courtesy Uponor

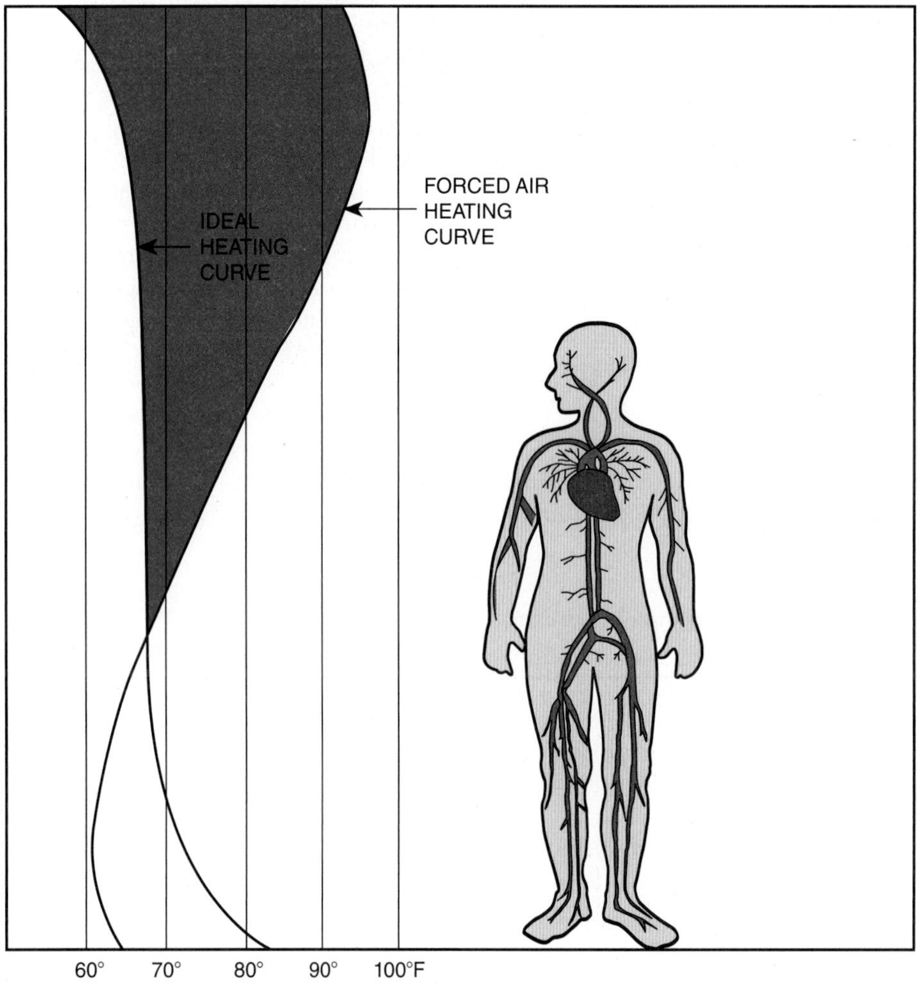

Figure 33.93 PEX tubing. Courtesy Uponor

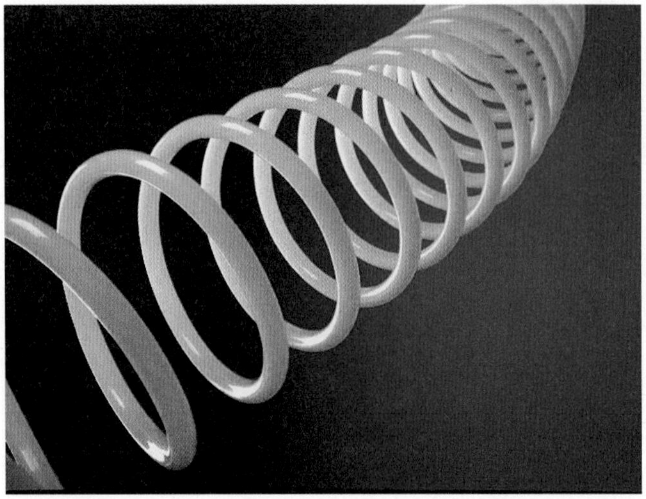

HEAT SOURCES FOR RADIANT HEATING SYSTEMS

As mentioned at the beginning of this unit, the heat source for hydronic systems is quite often a boiler. It was also mentioned that in the case of low-temperature systems such as radiant applications, geothermal heat pumps are becoming more and more popular. Here are some quick notes and suggestions regarding the heat sources used for radiant heating systems:

- When using a geothermal heat pump system for radiant heating, be sure to use a buffer tank. This will prevent water flow restrictions on the condenser water side, which can lead to high-pressure problems in the refrigerant circuit.
- The buffer tank serves as a storage tank for the heated water; the heat pump will cycle on and off to maintain the desired water temperature in the tank.

Figure 33.94 (A)–(C) Fittings used to connect PEX tubing. Courtesy Uponor

(A)

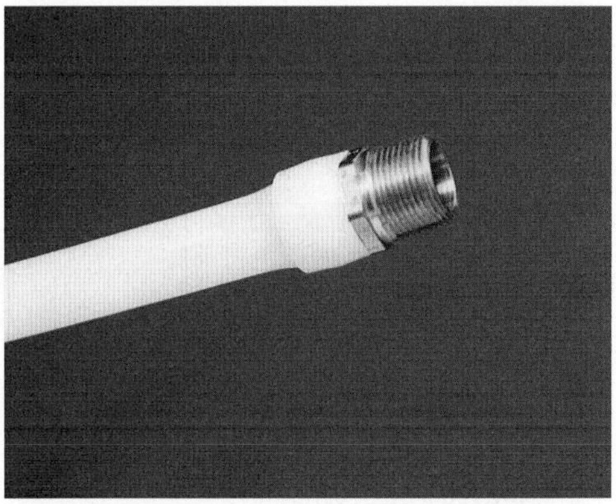

(B)

(C)

- Larger temperature differential settings on the buffer tank increase the run time of the heat pump, but they also increase the off time of the system, thereby increasing the efficiency of the system.
- It is a good idea to use a buffer tank as well when using a copper-tube, low-mass boiler for radiant applications, This will prevent short cycling of the boiler, **Figure 33.95**.
- If direct piping is being used, **Figure 33.96**, make sure that the boiler is designed to handle the potentially low temperature of the water returning to the appliance.
- When the boiler return-water temperature may be a problem, make sure to provide a means for keeping the return-water temperature high enough to prevent flue-gas condensation, **Figure 33.97**.

RADIANT HEATING PIPING

Given the versatility of radiant heating systems, the possibilities are boundless when it comes to installation variations and custom applications. The most common types of radiant heating system applications are the following:

- Slab on grade
- Thin slab
- Dry or concrete-free applications

Figure 33.95 A buffer tank used in conjunction with a low-mass boiler.

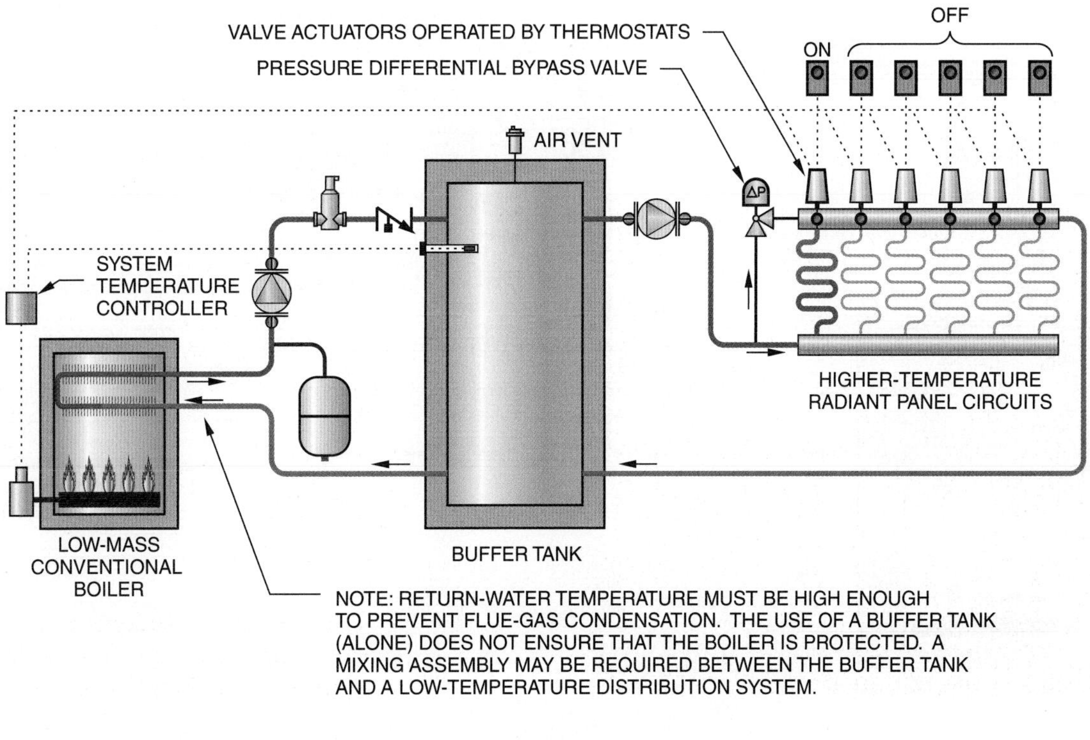

Figure 33.96 Direct piping from the heat source to the heating circuits. There is no mixing taking place in this piping arrangement.

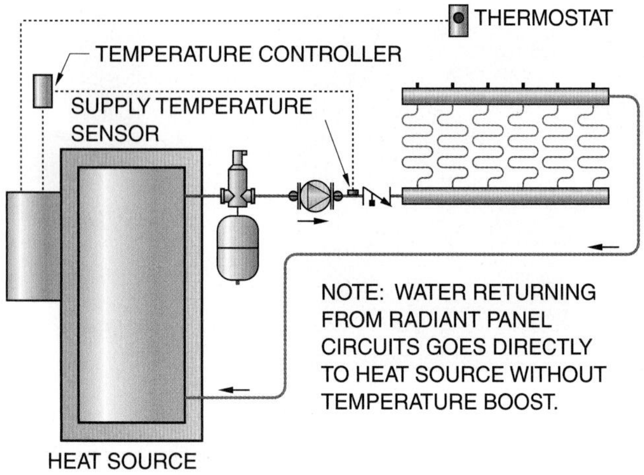

SLAB-ON-GRADE. The slab-on-grade application is very popular, especially for new construction and projects. Since the concrete for these applications has yet to be poured, it is a relatively simple task to position the PEX tubing loops right in the concrete slab. The tubing material is secured to the steel reinforcement mesh in the floor using straps or clips specially designed for that use, **Figure 33.98.** The spacing between the PEX tubes is determined by the depth of the tubes below the concrete surface, the size of the tubes, the type of floor being installed, and the location of the tubes with respect to the outside walls of the structure. For higher system efficiency, insulation should be placed below the concrete slab as well as below a vapor barrier to prevent the ground from robbing the slab of its heat.

Figure 33.98 Cutaway view of a slab-on-grade radiant piping layout. Courtesy Uponor

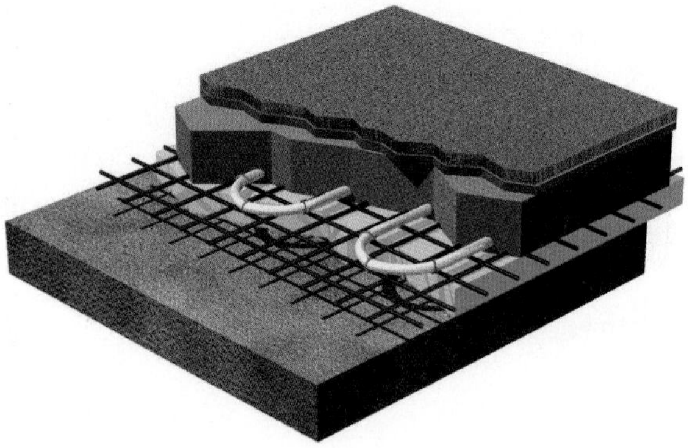

THIN SLAB. When there is an existing floor in place, thin-slab installations are common. The PEX tubing is stapled in place to the top of the frame floor and concrete is then poured over the existing floor. The thickness of the concrete can range from 1.5 in. to 2 in. and should completely

Figure 33.97 Boiler piping allows for control of boiler return-water temperature.

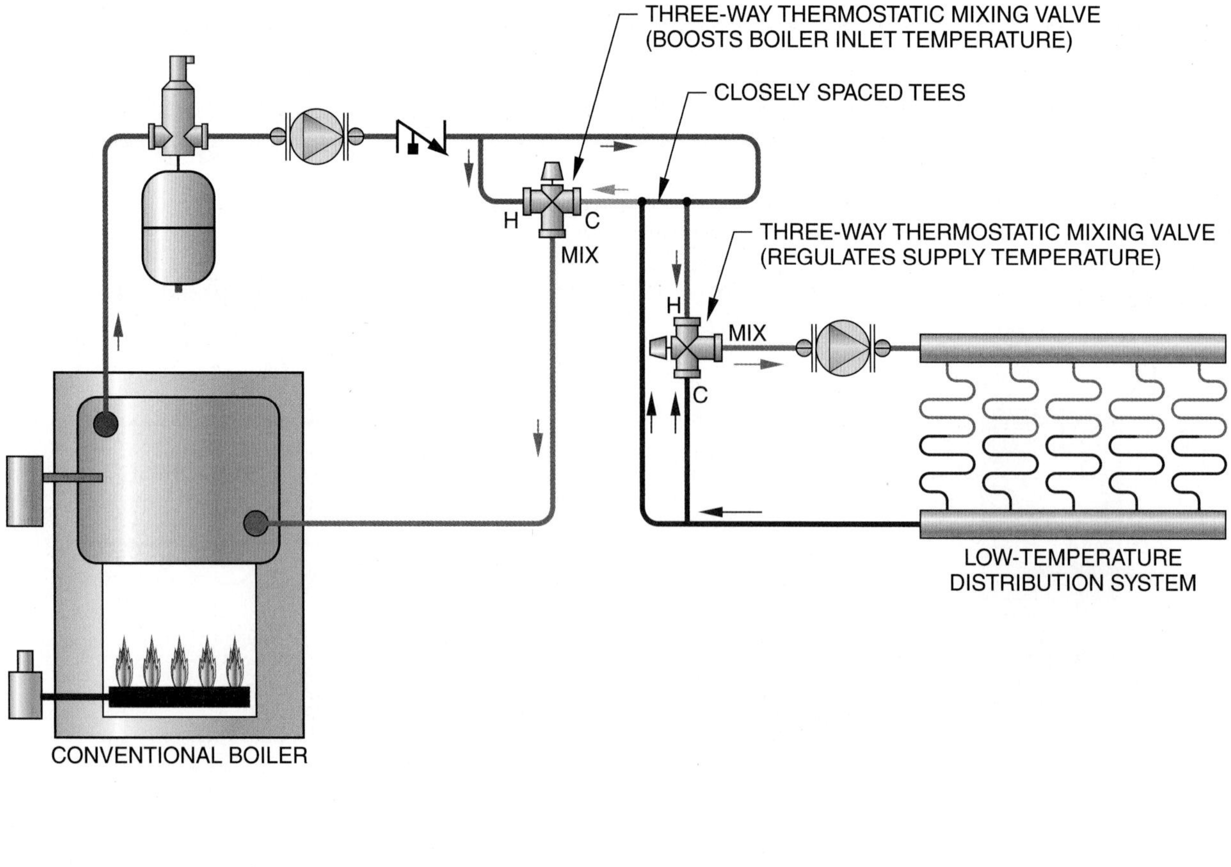

Figure 33.99 Cutaway view of a thin-slab radiant piping layout. Courtesy Uponor

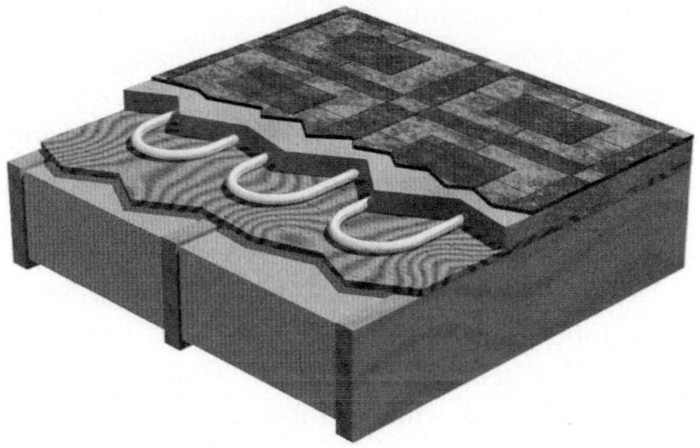

cover the PEX tubing with at least ¾ in. of poured material, **Figure 33.99**. To ensure that the existing floor will be able to handle the weight of the concrete, be sure to consult an engineer or architect when installing a thin-slab system.

DRY APPLICATIONS. Another option for installing radiant systems is to staple the tubing to the bottom of the flooring material. This is referred to as a staple-up job, **Figure 33.100**. When installing this type of system, it is important to space the staples closely, as the tubing should be in loose contact with the floor material for effective heat transfer. Insulation should be positioned under the stapled-up tubing as illustrated in the figure.

Another option for a concrete-free radiant heating system installation is the use of aluminum "wings" or fins, **Figure 33.101**, that can be attached to PEX tubing that is fed through the floor joists (as opposed to being stapled up to the underside of the floor). These fins are installed in sets of two and are designed to completely surround the PEX tubing, **Figure 33.102.** Installation is relatively simple, given that holes are predrilled in the material. The completed installation is illustrated in **Figure 33.103.**

Figure 33.100 Cutaway view of a staple-up radiant piping layout. Courtesy Uponor

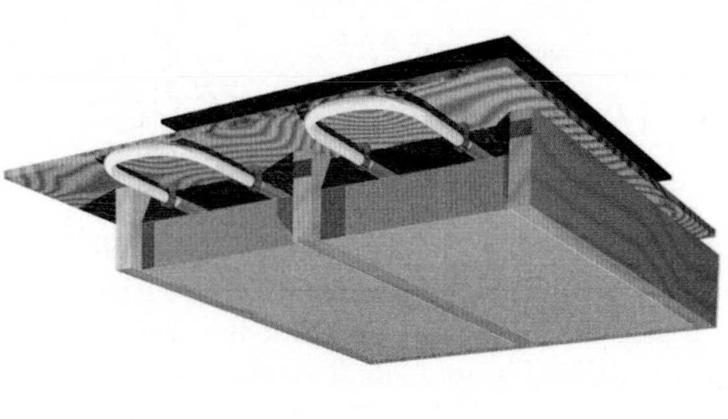

Figure 33.101 Aluminum "fins" used to increase the heat transfer surface of the PEX tubing in a dry, concrete-free radiant application. Photo by Eugene Silberstein

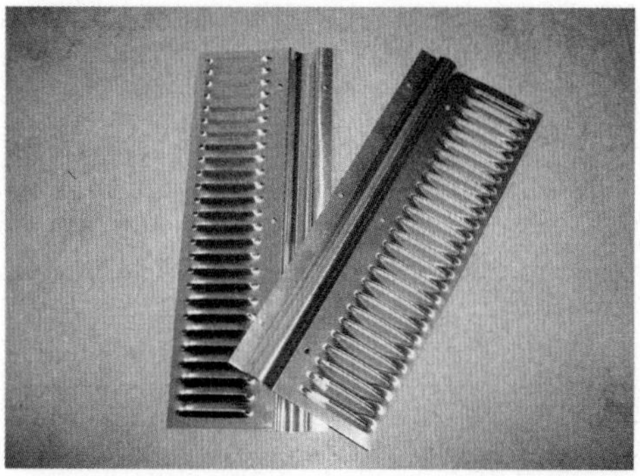

Figure 33.102 Predrilled holes ease the installation process. Photo by Eugene Silberstein

Figure 33.103 Completed installation.

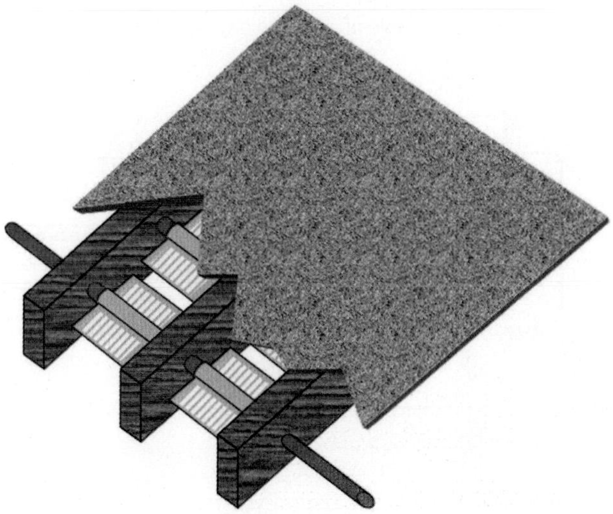

PIPING ARRANGEMENTS

Many different configurations of piping can be utilized on radiant heating systems. What follows is intended to provide merely a small sampling of the options available—attempting to discuss all installation and piping schemes here would be beyond the scope of this text.

DIRECT PIPING. By far, the simplest configuration of piping in a radiant system is direct piping. In a direct-piping scheme, the water supplied to the boiler is pumped into the radiant heating circuits, **Figure 33.96.** This type of piping configuration is more system-friendly when the heat source is a buffer tank as opposed to a boiler. If the heat source is a boiler and the desired water temperature at the radiant heating circuits is low, there is a great chance that the flue gases in the boiler will condense, because the return-water temperature will be low as well. If a buffer tank is used, the water that returns to the boiler comes from the buffer tank, not the outlet of the radiant heating circuits.

MANUALLY SET MIXING VALVES. As mentioned earlier in the radiant section, manually set mixing valves do not respond to changes in water temperature or to changing water flow rates. For this reason, these valves are not the best option, as they do not prevent low-temperature water from returning to the boiler, **Figure 33.104**.

THERMOSTATIC MIXING VALVES. Thermostatic mixing valves will adjust their internal settings to supply water at the desired temperature to the radiant heating loops. Consider the system shown in **Figure 33.105**. As a point of reference, let's assume that the entire system is completely filled with 180°F water and the desired water temperature in the radiant loops is 110°F. The radiant loops operate with a ΔT of 20°F. Since the water temperature at the inlet of the radiant loops is 70°F higher (180°F–110°F) than desired, the hot water inlet port of the thermostatic mixing valve is completely closed. The 180°F water enters the radiant loops and leaves at 160°F. All of the 160°F water flows back into the cold inlet port of the mixing valve, **Figure 33.106**. **NOTE:** *No water from the radiant loop can return to the boiler because no water has entered the radiant loop from the hot water inlet port. Because the radiant loops are completely filled, we cannot add water without removing water, and we cannot remove water without adding water.•*

Figure 33.104 An example of a hydronic system using a manually set three-way valve.

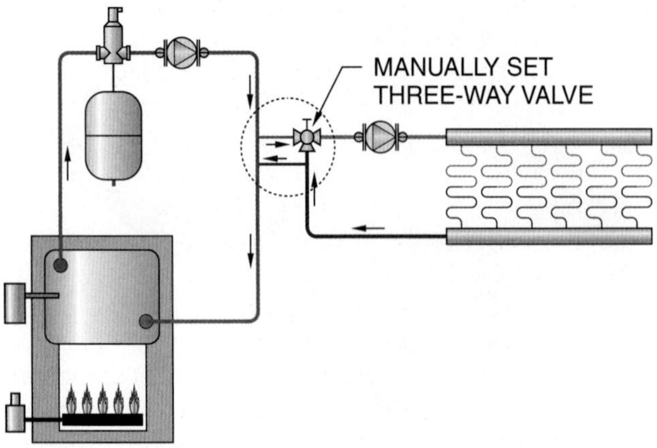

The water enters the cold inlet port of the mixing valve at 160°F and, because the water is still too hot, the hot water inlet port will remain closed. The 160°F water enters the radiant loops, and the water leaves the loops at a temperature of 140°F. This process continues until the water leaving the radiant loops is at a temperature of 100°F. Since the water enters the cold water inlet port of the mixing valve at 100°F, and since the desired water temperature is 110°F, the hot water inlet port will open slightly to mix some 180°F water with the 100°F water to provide 110°F water at the outlet of the mixing valve, **Figure 33.107**. Because some water enters the mixing valve through the hot water inlet port, the same amount of water will return to the boiler from the outlet of the radiant heating circuits.

Another application for thermostatic mixing valves is to ensure that the temperature of the return water to the boiler is high enough to prevent flue-gas condensation. The system in **Figure 33.108** utilizes a mixing valve to regulate water supplied to the heating circuits and also uses another mixing valve to regulate the temperature of the water returning to the boiler.

33.8 COMBINATION (HIGH- AND LOW-TEMPERATURE) PIPING SYSTEMS

In a case where a structure has both high- and low-temperature heating circuits, the same boiler can serve both applications by using a primary–secondary pumping arrangement, **Figure 33.109**. In the figure we can see that water is being supplied to the first two secondary loops at a temperature close to 180°F. The water in the main loop after these two secondary loops will be about 165°F. This water can then return to the boiler, while a portion of the water will mix with the water in the radiant loop to maintain the desired water temperature in that circuit. The temperature of the water returning to the boiler will be about 160°F. When both high- and low-temperature water circuits are to be used, thermostatic mixing valves are used to lower and maintain the water at the desired temperature in the low-temperature loops.

33.9 TANKLESS DOMESTIC HOT WATER HEATERS

Most hot water heating boilers (oil- and gas-fired) can be furnished with a domestic hot water heater consisting of a coil inserted into the boiler. The domestic hot water is contained within the coil and heated, on demand, by the boiler quickly.

Figure 33.105 Low-temperature hydronic system using a three-way thermostatic mixing valve. Since the water in the heating loop is warmer than desired, the hot water inlet port on the mixing valve is in the closed position. Under these conditions, no water can enter or leave the low-temperature heating circuit.

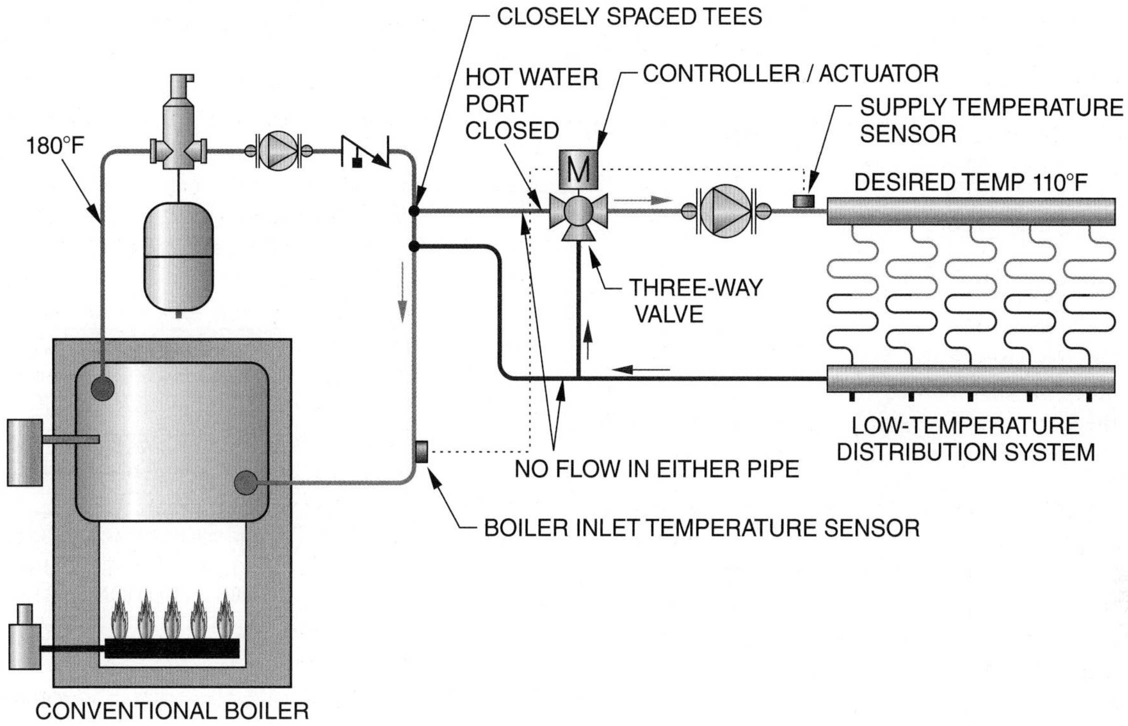

Figure 33.106 Water that is too hot for the heating loop will continue to circulate without mixing until the temperature at the outlet of the heating loop (cold water inlet) falls below the desired water temperature.

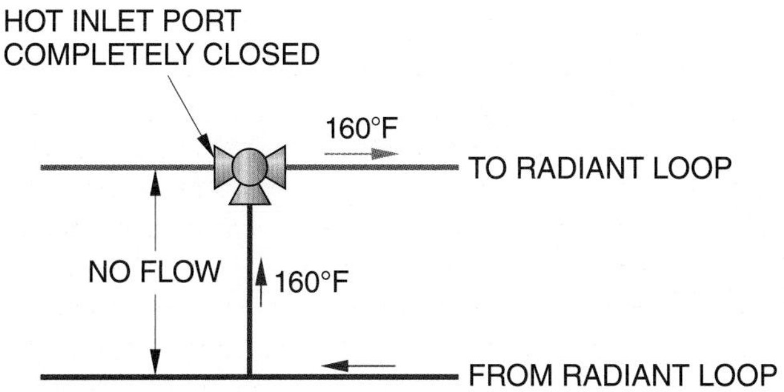

Figure 33.107 Since the water at the outlet of the heating loop is at a temperature that is lower than desired, the hot water inlet port will open slightly. This will allow hot water to enter the mixing valve—and also allow cool water from the heating loop to leave the loop and return to the boiler.

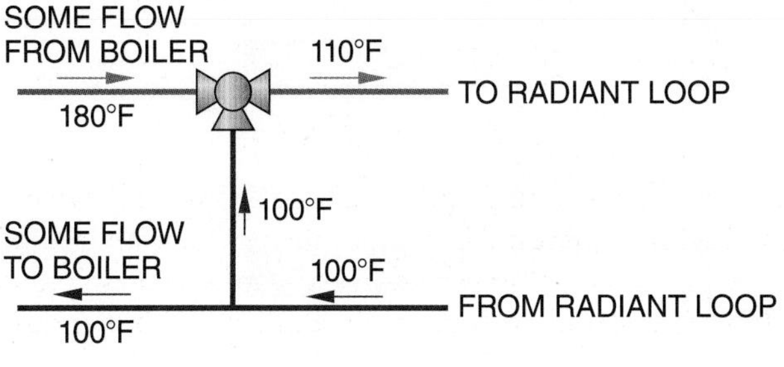

This eliminates the need for a storage tank. This system can meet most domestic hot water needs without using a separate tank, **Figure 33.110**. Because there is no storage tank, there are no standby losses, making this an efficient way to produce hot water. Standby losses are the heat and energy losses associated with heating water that is not immediately used. When water is heated and not used immediately, it will cool down and will need to be heated again.

33.10 SOLAR HEATING AS A SUPPLEMENTAL HEAT SOURCE

The sun furnishes the earth with tremendous amounts of direct energy each day. It is estimated that 2 weeks of the sun's energy reaching the earth is equal to all of the known deposits of coal, gas, and oil. The challenge facing scientists, engineers, and technicians is to better harness and use this energy. We know that the sun heats the earth, which heats the air immediately above the earth. One of the challenges is to learn how to collect, store, and distribute this heat to provide heat and hot water for homes and businesses. Many advances have been made, but the design and installation of solar systems has progressed very slowly. It is assumed, however, that as the fossil-fuel resources are further depleted and as economics or political actions cause energy crises, it will be only a matter of time before the direct energy of the sun is used extensively.

Figure 33.108 This system uses two thermostatic mixing valves: one to control the temperature of the heating loop, and one to control the temperature of the water returning to the boiler.

Figure 33.109 A primary–secondary piping arrangement that incorporates both high- and low-temperature heating circuits.

Figure 33.110 A tankless domestic water heater installed in a boiler. Courtesy Weil McLain Corp.

PASSIVE SOLAR DESIGN

Many structures being built presently are using passive solar designs. These designs use nonmoving parts of a building or structure to help provide heat or cooling, or they eliminate certain parts of a building that help cause inefficient heating or cooling. Some examples follow:

- In areas where there are harsh winters, more windows can be placed on the east, south, or west sides of homes and fewer on the north side. This allows warming from the morning sun in the east and from the sun throughout the rest of the day from the south and west. By eliminating windows on the north side, the coldest side of the house can be better insulated.
- Place greenhouses, usually on the south side, to collect heat from the sun to help heat the house, **Figure 33.111.**
- Design roof overhangs to shade windows from the sun in the summer but allow sun to shine through in the winter, **Figure 33.112.**

Figure 33.111 A greenhouse will allow the sun to shine into the house. It may also have a masonry floor, or barrels of water may be located in the greenhouse, to help store the heat until evening when the sun goes down. This heat then may be circulated throughout the house.

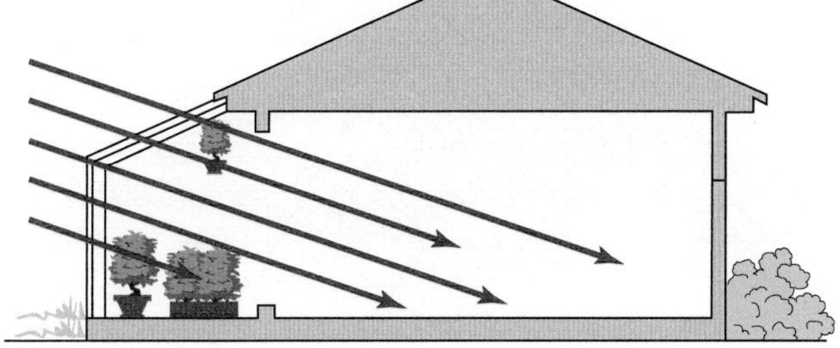

- Provide a large mass, such as a concrete or brick wall, to absorb heat from the sun and temper the inside environment naturally.
- Planting trees around the structure provides benefits in both the summer and the winter. In the summer, the shade reduces the heat load on the structure. In the winter, when the trees have no leaves, the sun helps create a solar load on the structure.
- Place latent-heat storage tubes containing phase-change materials where they can collect heat to be released at a later time, **Figure 33.113.**

Figure 33.112 An overhang may be constructed to allow the sun to shine into the house in the winter when it is lower in the sky and yet shade the window in the summer.

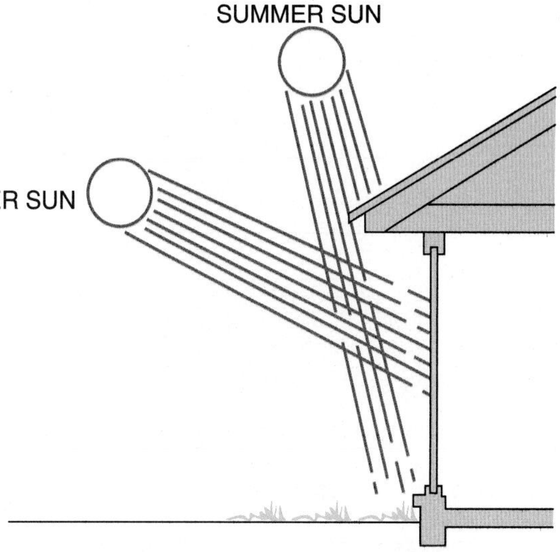

Figure 33.113 Latent-heat storage tubes may be placed in windows to store heat when the sun is shining.

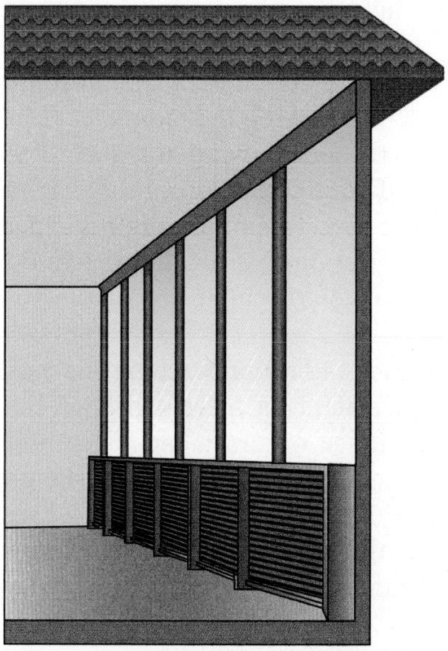

Figure 33.114 Radiation striking the earth directly from the sun is considered direct radiation. If it reaches the earth after it has been deflected by clouds or dust particles, it is called diffuse radiation.

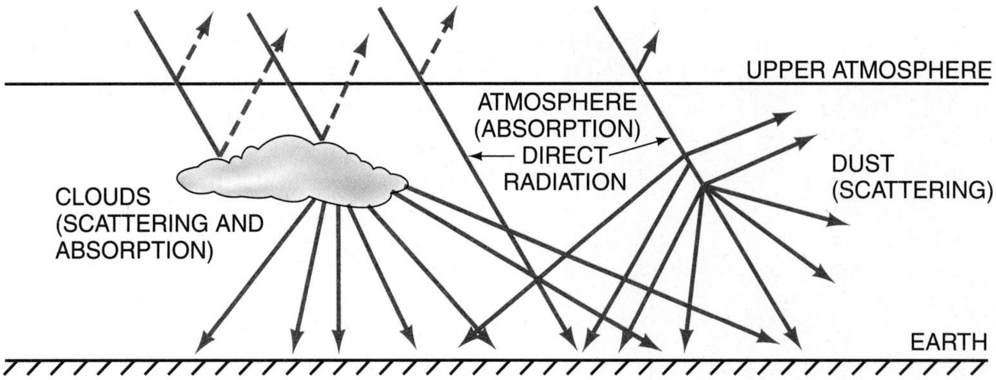

The remainder of this unit is limited to a discussion of active solar systems. These systems use electrical or mechanical devices to help collect, store, and distribute the sun's energy. The distribution of this heat to the conditioned space is accomplished by means of the same type of equipment used in fossil-fuel furnaces.

DIRECT AND DIFFUSE RADIATION. Only a very small amount, about 0.000000045%, of the sun's energy reaches the earth. Much of the energy that does reach the earth's atmosphere is reflected into space or absorbed by moisture and pollutants before reaching the earth. The energy reaching the earth directly is direct radiation. The reflected or scattered energy is diffuse radiation, **Figure 33.114**.

SOLAR CONSTANT AND DECLINATION ANGLE. The rate of solar energy reaching the outer limits of the earth's atmosphere is the same at all times. It has been determined that the radiation from the sun at these outer limits produces 429 Btu/ft^2/h on a surface perpendicular to the direction of the sun's rays. This is known as the solar constant. The energy from the sun that has been received is often called insolation.

The earth revolves once each day around an axis that passes through the north and south poles. This axis is tilted 23.5°, so the intensity of the sun's energy reaching the northern and southern atmospheres varies as the earth orbits around the sun. This tilt or angle is called the declination angle and is responsible for differences during the year in the distribution of the intensity of the solar radiation, **Figure 33.115**.

The amount of radiation reaching the earth also varies according to the distance it travels through the atmosphere. The shortest distance occurs when the sun is perpendicular to a particular surface and is when the greatest energy reaches that section of the earth, **Figure 33.116**. The angle of the sun's rays with regard to a particular place on the earth plays an important part in the collection of the sun's energy.

Figure 33.115 The angle of declination (23.5°) tilts the earth so that the angle from the sun north of the equator is greater in the winter. The sun's rays are not as direct, and the normal temperatures are colder than during the summer.

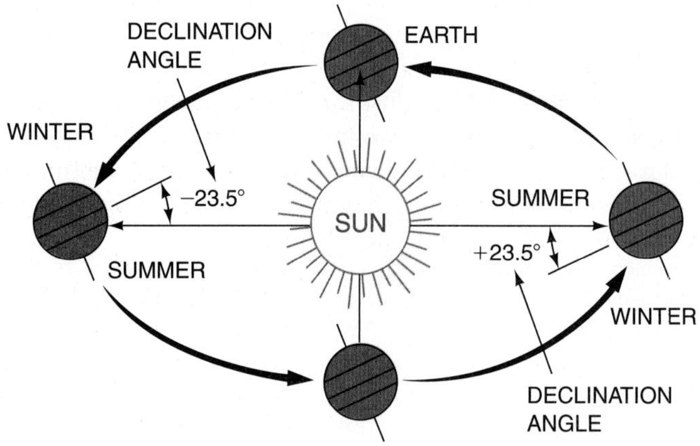

Figure 33.116 Radiation from the sun will be warmer at point 1, because that is the shortest distance it has traveled through the atmosphere.

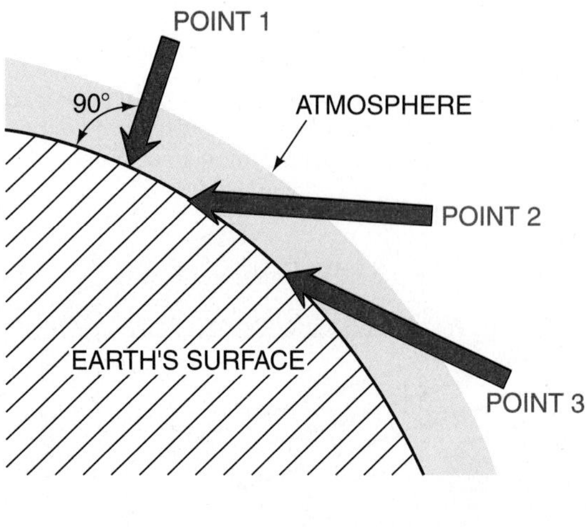

ACTIVE SOLAR DESIGN

The systems discussed and illustrated in this unit are basic systems with a minimum of controls, check valves, pressure relief valves, and other safety protection. Manufacturers' instructions and local and state codes must be followed when installing this equipment. Active solar design utilizes collectors, storage systems, distribution devices such as pumps and fans, and control systems. These solar systems are used primarily for space heating, domestic hot water, and swimming pool heating. One useful resource that has collected and organized all of the solar energy provisions of the International Code Council codes is the ISEP (International Solar Energy provisions), the latest edition being the 2018 editions and it has a companion commentary volume as well.

LIQUID SOLAR FORCED-AIR SPACE-HEATING SYSTEMS. **Figure 33.117** illustrates a basic water drain-down liquid space-heating system, which also includes a domestic hot water tank. Drain-down systems are a type of system used in areas where water to and from and within the collectors would freeze if left in the system. When collectors are producing heat at a higher temperature than the storage water, circulator pump A moves water from storage to be heated and from the collectors back to storage. When the house thermostat calls for heat, circulator pump B moves water from storage to the heat exchanger in the furnace duct. The furnace blower moves air across the heat exchanger, warming the air, and distributes it to the house. Also shown is a domestic hot water tank heated through a heat exchanger in the water storage. When circulator pump A shuts down, the collector water will all drain to a heated area to keep from freezing in cold weather. Note that the pipe from the collector into the storage tank does not extend into the water. This provides venting and allows the collectors and piping to drain. If this space is not provided, other venting in the collector system is necessary.

An alternative design that is often preferred, particularly in colder climates, requires a closed-collector piping system using an antifreeze solution instead of water, **Figure 33.118**. The figure illustrates a closed liquid collection water storage system with air distribution to the house. The liquid in the collector system is an antifreeze and water solution. It is heated at the collectors and pumped through the coil at the heat exchanger and back to the collectors. This continues as long as the collectors are absorbing heat at a predetermined temperature.

When the room thermostat is calling for heat, three-way valve 1 allows the water to be circulated by pump B through the liquid-to-air heat exchanger and back to the liquid-to-liquid heat exchanger, where it absorbs more heat. If the room thermostat is not calling for heat, three-way valve 1 diverts the water to storage. If the storage temperature has reached a predetermined temperature, the collector solution is circulated through the purge coil through three-way valve 2. (The heat is dissipated into the air outside through this coil.) This is a necessary safety precaution because it is possible for the collector system to overheat and damage the equipment. Note that a liquid-to-liquid heat exchanger is used to heat the storage water. The collector antifreeze fluid must be kept separated from the storage water. This system is less efficient than the drain-down system, which does not require the extra heat exchanger.

In most areas, an auxiliary heating system is required to provide comfort heating at all times. A conventional hot air

Figure 33.117 Collectors use water as the liquid. This is a drain-down system because the water in the collectors would freeze on cold nights.

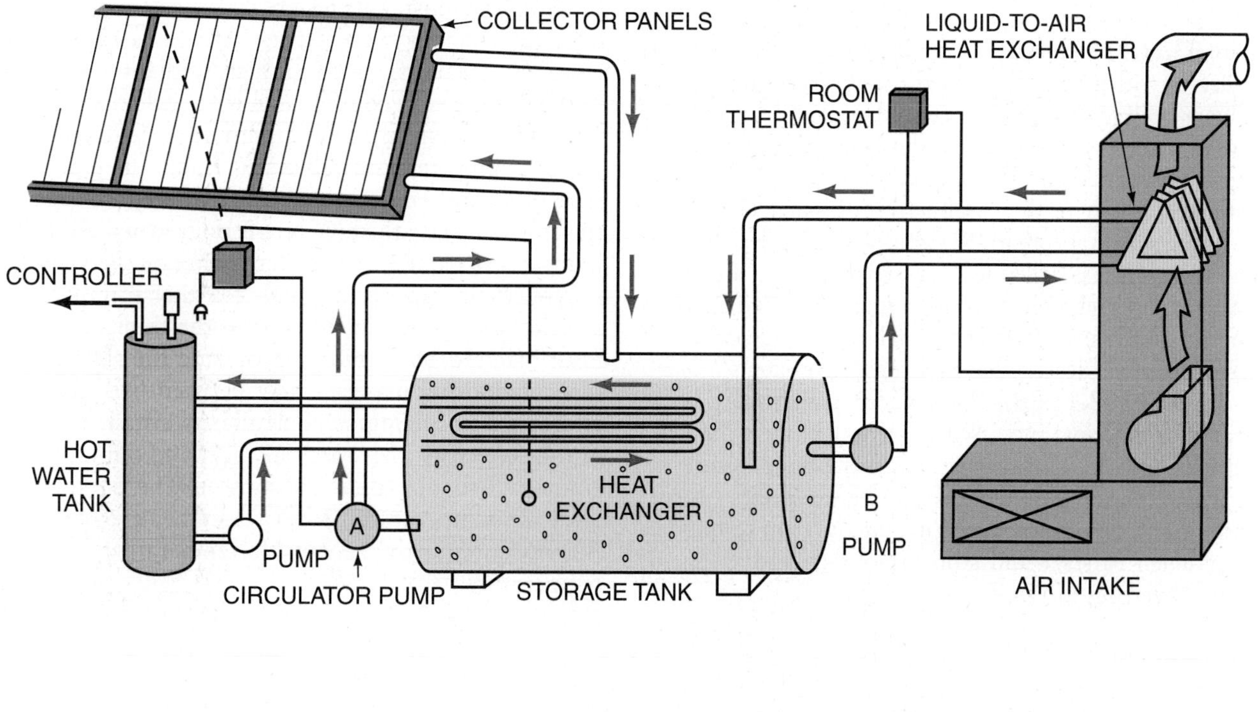

Figure 33.118 A closed liquid collection water storage system with air distribution to the house.

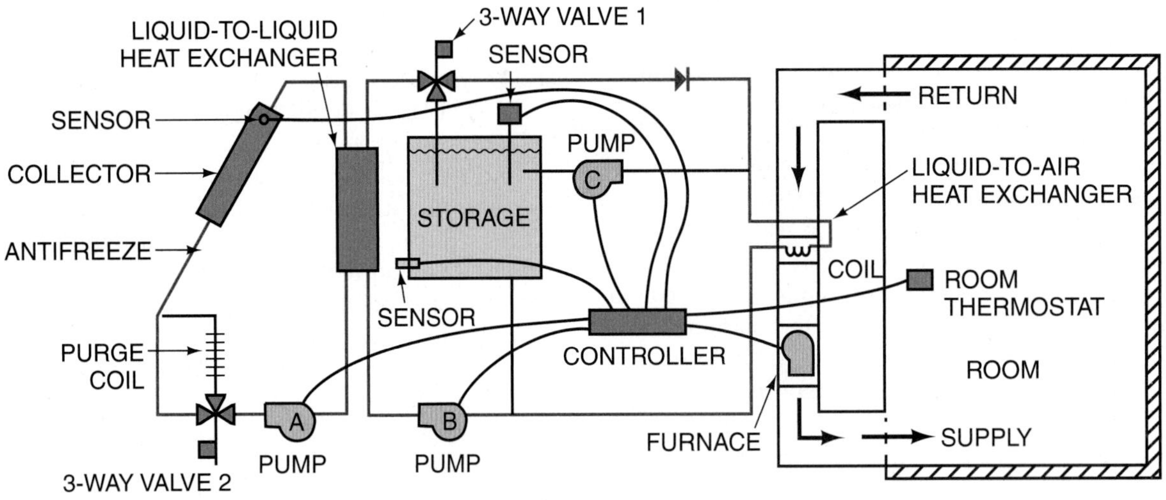

heating system can use the same ducts used for the solar system. In each system room thermostats may be used to provide steady, comfortable heat.

LIQUID COLLECTION/WATER STORAGE/AUXILIARY CONVENTIONAL HOT WATER BOILER. A liquid-based solar collector system may be paired with a hot water finned-tube convector heating furnace. The collector system can be of the same design as an air distribution system. It can be a drain-down water design or one that uses antifreeze in the collectors with a heat exchanger in the storage unit. Solar collector and storage temperatures may be lower than those produced by a hot water boiler. This can be compensated for—to some extent—by using more fin tubing than would be used with a conventional hot water system.

Figure 33.119(A) illustrates a design using a liquid-based solar collector with a hot water boiler system. This design uses the auxiliary boiler when the storage water is not hot enough. If the solar storage is hot enough, the boiler does not operate. Water is pumped from the storage tank and bypasses the boiler. It goes directly to the baseboard fin tubing. The three-way diverting valve automatically causes this boiler bypass. When auxiliary heat is called for, the two three-way diverting valves cause the water to bypass the storage. It would be very inefficient to pump heated water through the storage tank.

A system can be designed as two separate systems in parallel, as in **Figure 33.119(B)**. The separate systems can both be operated at the same time. With this design, the solar system can help the auxiliary system when the temperature of the storage water is below what would be satisfactory for it to operate alone. Sensors are used in either system to tell the controller when to start and stop the pumps and when to control the diverting valves.

SOLAR RADIANT HEAT

Figure 33.120 illustrates a water storage, radiant heating system with an auxiliary hot water boiler. Either water or an antifreeze solution is piped through a collector system and, in the case of an antifreeze solution, through a heat exchanger in the water storage. The heating coils may be embedded in concrete in the floor or in plaster in ceilings or walls. The normal surface temperature for floor heating is 85°F, and for wall or ceiling panels, 120°F.

Floor installations are most common. The coils are embedded in concrete below the surface. If a concrete slab on grade, it should be insulated underneath. In ceiling and wall applications, there should be coils for each room to be heated. When installations are on outside walls, they should be insulated very well between the coils and outside surfaces to prevent extreme heat loss. Radiant heating installed in the floor is very comfortable with little temperature variation from room to room. **Figure 33.121** illustrates a polybutylene pipe being installed in a floor. The pipe is tied to a wire mesh and propped on blocks to keep it near the top of the concrete slab; the concrete is poured around it, covering it anywhere from 1 to 4 in. depending on the system design. Polybutylene pipe is popular for this installation. There should be no joints in the pipe in the floor. Any joints necessary should be manifolded outside the slab. Almost any type of flooring material may be used over the slab—tile, wood, and carpeting. A controller is installed with sensors at the collectors, the water storage, and outdoors. It is also connected to the room thermostat. The controller modulates the temperature of the water being circulated in the floor according to the outside temperature and the room temperature.

Figure 33.119 (A) A liquid-to-liquid solar space-heating system. (B) Both solar and auxiliary heat can be used at the same time.

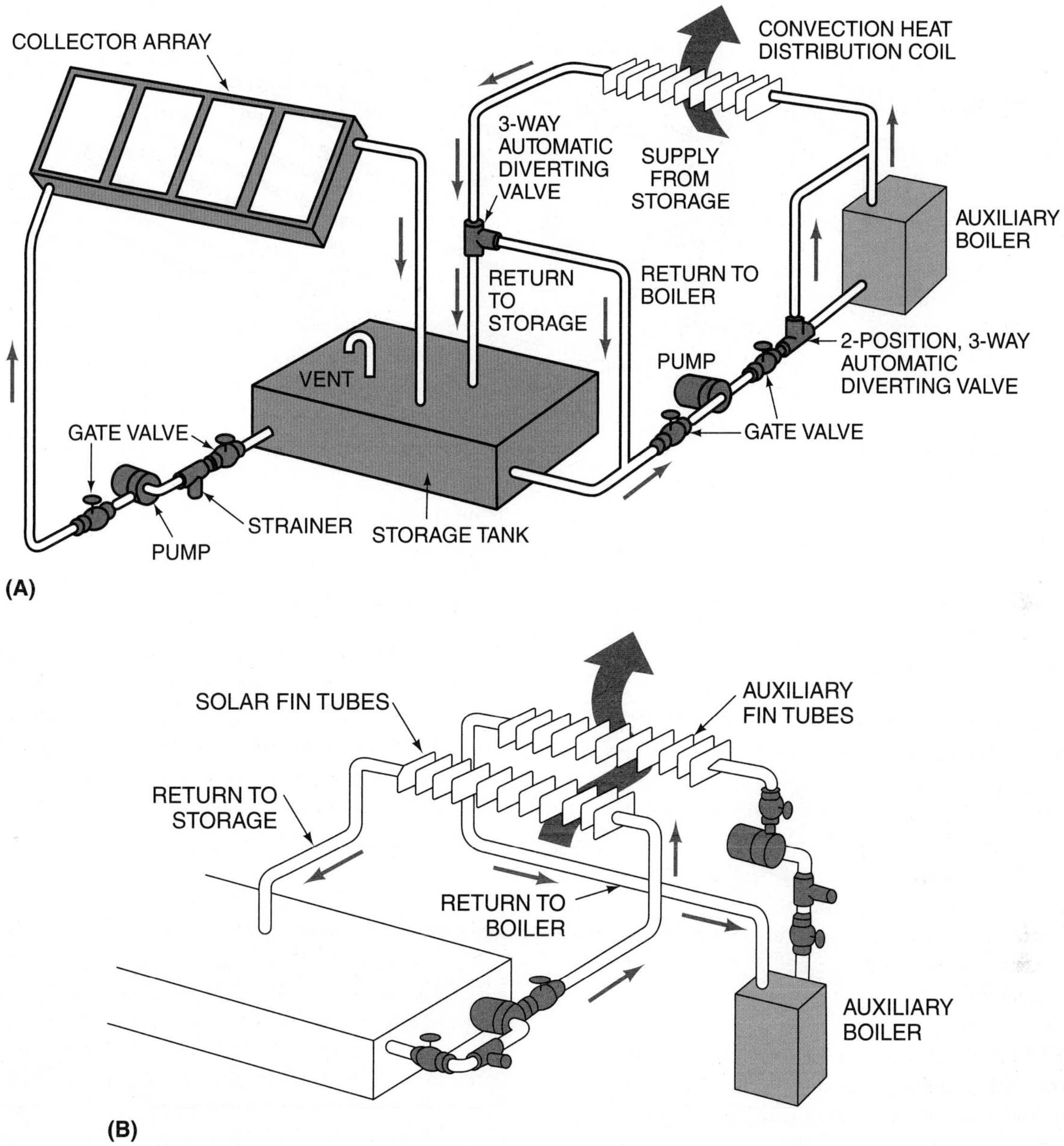

SOLAR-HEATED DOMESTIC HOT WATER

Many solar systems are designed and installed specifically to heat or to assist in heating domestic hot water. These systems may use components similar to those used in space-heating systems, **Figure 33.122**. The heated collector water or antifreeze is pumped through the heat exchanger where it provides heat to the water in the tank. This tank must have conventional heating components also because the solar heating probably will not meet the demand at all times. If an antifreeze solution is used, the heat exchanger must have a double wall to prevent the toxic solution from mixing with the water used in the home or business, **Figure 33.123**. Many systems are designed to use a solar preheat or storage tank in conjunction with a conventional hot water tank. This gives additional solar-heated water to the system. Conventional heating (electric, gas, oil) will not have to be used as much with this extra storage, **Figure 33.124**.

⚠ SAFETY PRECAUTION: *Ethylene glycol may be used as an antifreeze if allowed by local or state codes. It is toxic and must be used only in a double-wall heat exchanger. Propylene glycol is a "food grade" antifreeze and is recommended for use in a domestic hot water system.*•

A space-heating system may include domestic hot water heating as well. This can be designed in several ways,

Figure 33.120 A solar space-heating radiant system. The radiant heating coils are normally embedded in a concrete floor.

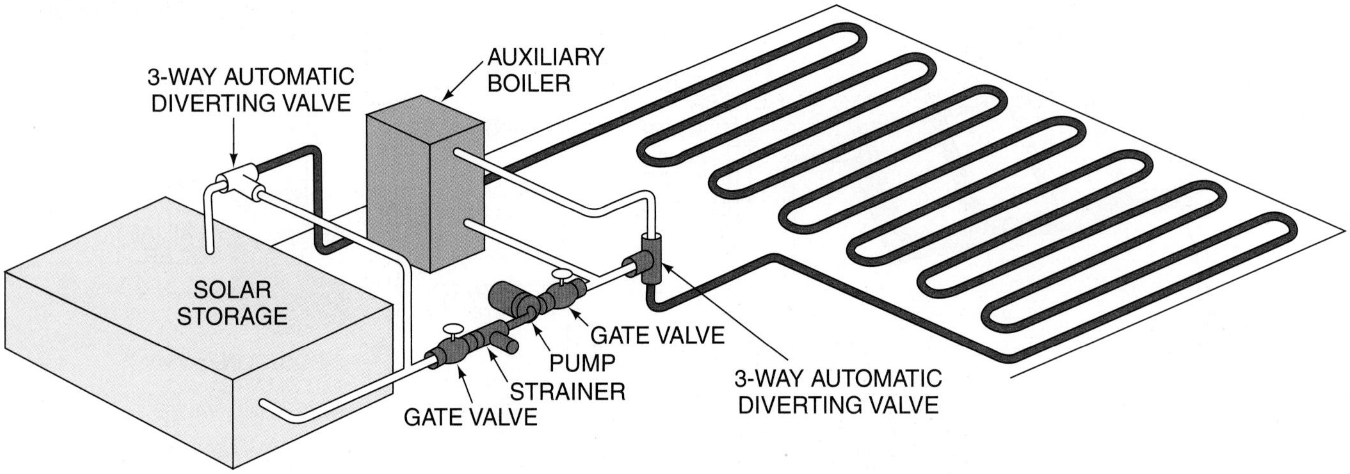

Figure 33.121 A polybutylene pipe being installed in a concrete slab to be used in a solar radiant heating system.

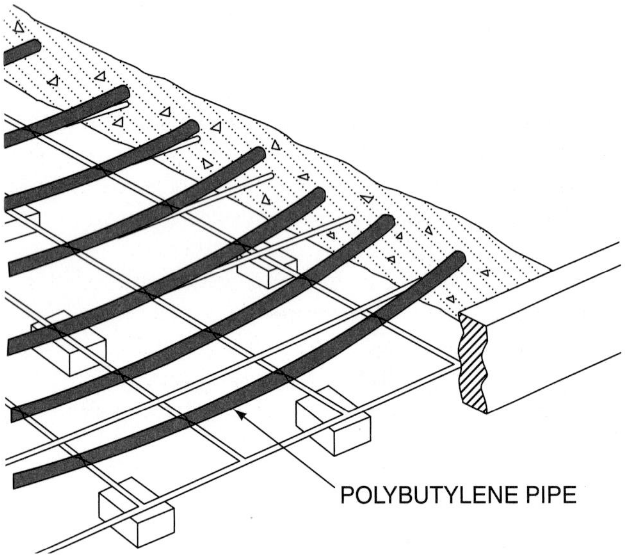

Figure 33.117. The cold makeup water is passed through the heat exchanger in the storage tank. It absorbs whatever heat is available before entering the hot water tank, thus reducing the amount of heat that must be supplied by conventional means.

LIQUID COLLECTOR DESIGN. Liquid solar collectors designed for space heating and domestic hot water systems are usually flat-plate collectors. As indicated earlier, a liquid (either water or an antifreeze solution) is passed through the collector in copper tubing, **Figure 33.125**. The liquid is heated as it passes through the collector. These collectors may use one or two panels of glass. The glass may vary in quality and design, but a low-iron tempered glass is often used. The absorber plate is usually made of copper and formed tightly around the copper tubing and/or may consist of fins welded to the tubing. These assemblies are painted black, using a paint designed for this purpose, to absorb the maximum heat from the sun.

SOLAR POOL HEATING

Many solar pool heating systems simply use the pump that circulates the pool water through the filter. The pool water is pumped through the filtering device(s) and then is diverted

HVACR GOLDEN RULES

When making a service call to a residence:

- Try to park your vehicle so that you do not block the customer's driveway.
- Look professional and be professional. Wear clean clothes and shoes. Wear coveralls for under-the-house work and remove them before entering the house.
- Check humidifiers, where applicable. Frequent service is highly recommended for health reasons.

Added Value to the Customer

Here are some simple, inexpensive procedures that may be added to the basic service call.

- Do not leave water leaks.
- Check the boiler for service needs; this may lead to more paying service work, depending on the type. Oil-heated boilers require the most extra work.
- Check for the correct water treatment.
- Lubricate all bearings on pumps and fans as needed.

Figure 33.122 A solar domestic hot water system.

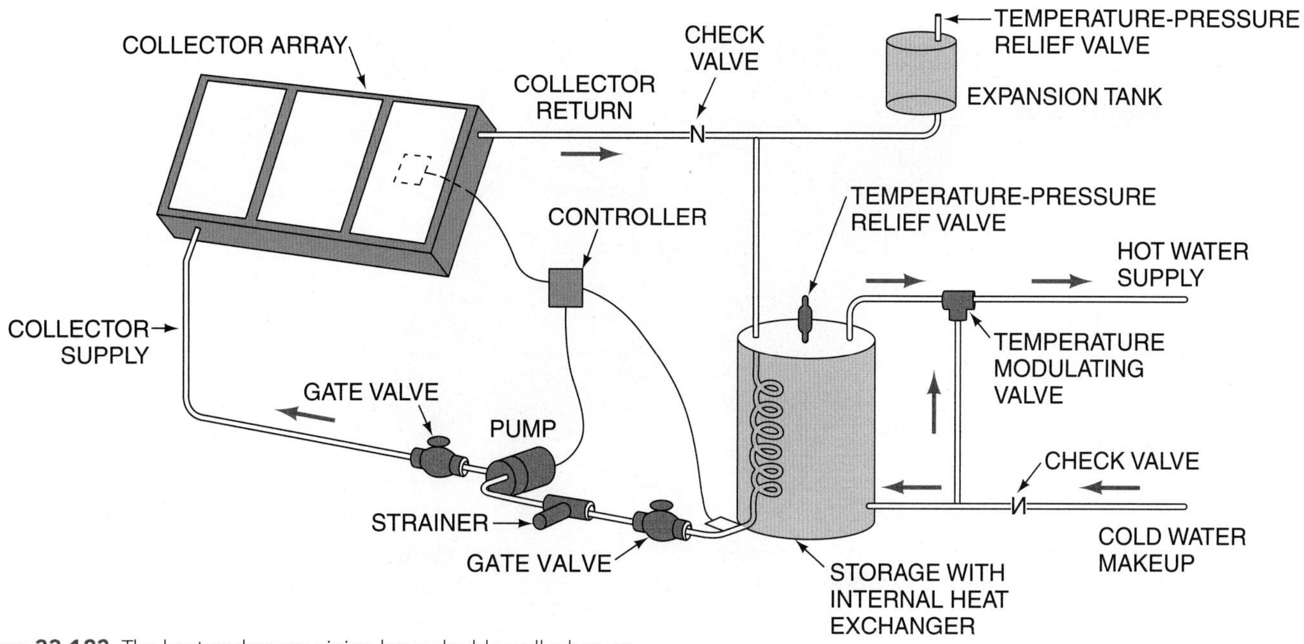

Figure 33.123 The heat exchanger piping has a double wall when an antifreeze solution is used in the collectors in a domestic hot water system.

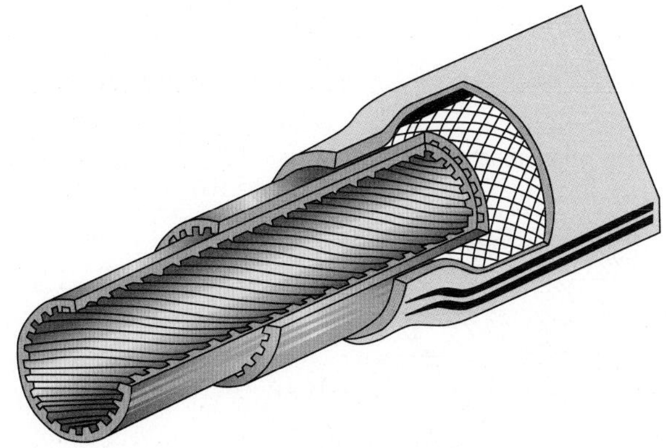

to the solar collectors, where it is heated. The water then flows back to the pool, **Figure 33.126**. This continuous flow of water through the collectors and back to the pool gradually heats the water. When the pool has reached the desired temperature level, the water bypasses the collectors and is returned directly to the pool from the filters.

Collectors for pool systems are often made of extruded plastic, **Figure 33.127**. A cross section of a collector produced by one manufacturer is shown in **Figure 33.128**. As this illustration shows, the collector receives diffuse radiation and direct radiation for optimum heating of the water passing through.

Figure 33.124 A solar preheat tank provides additional storage for a domestic hot water system.

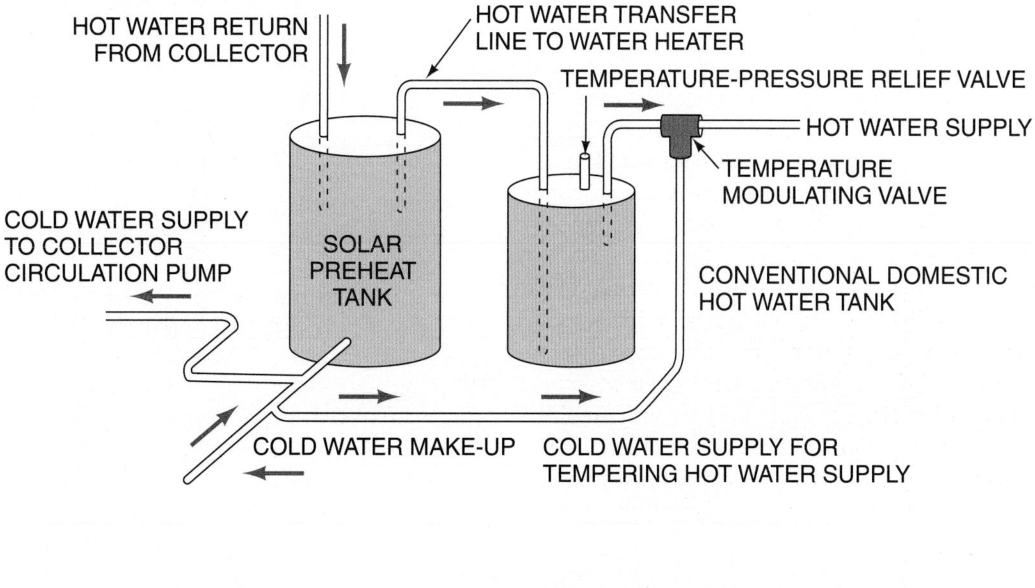

Figure 33.125 A solar collector.

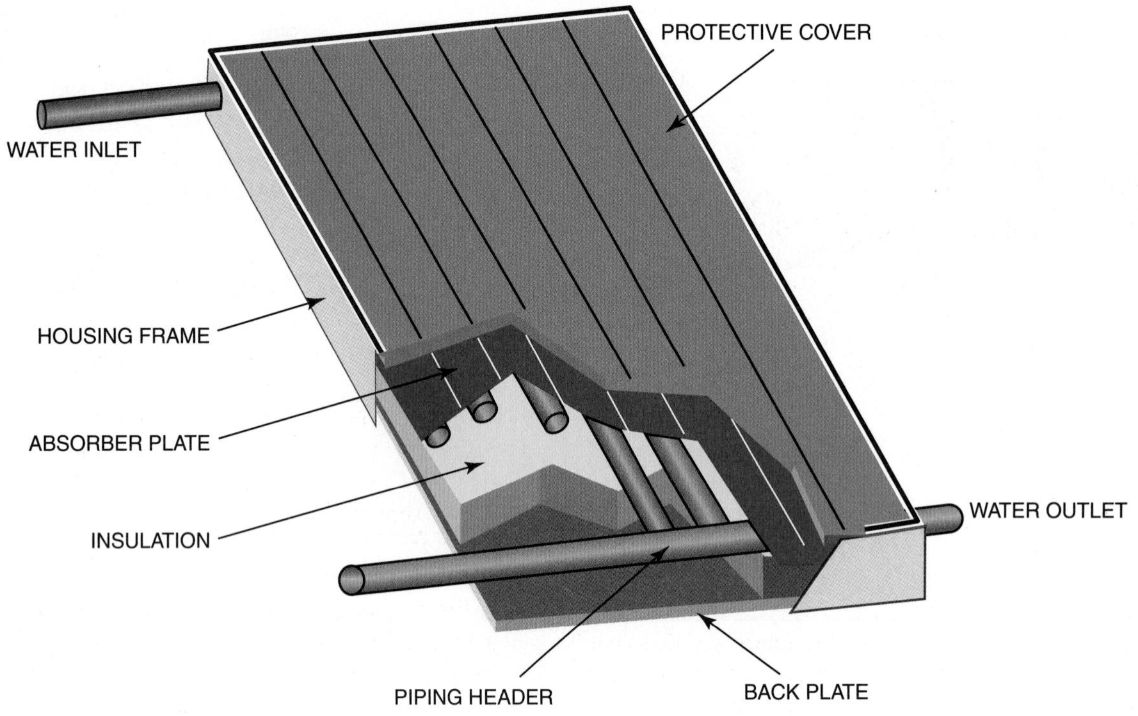

Figure 33.126 A swimming pool solar heating system. Courtesy Aquatherm Industries/Solar Industries

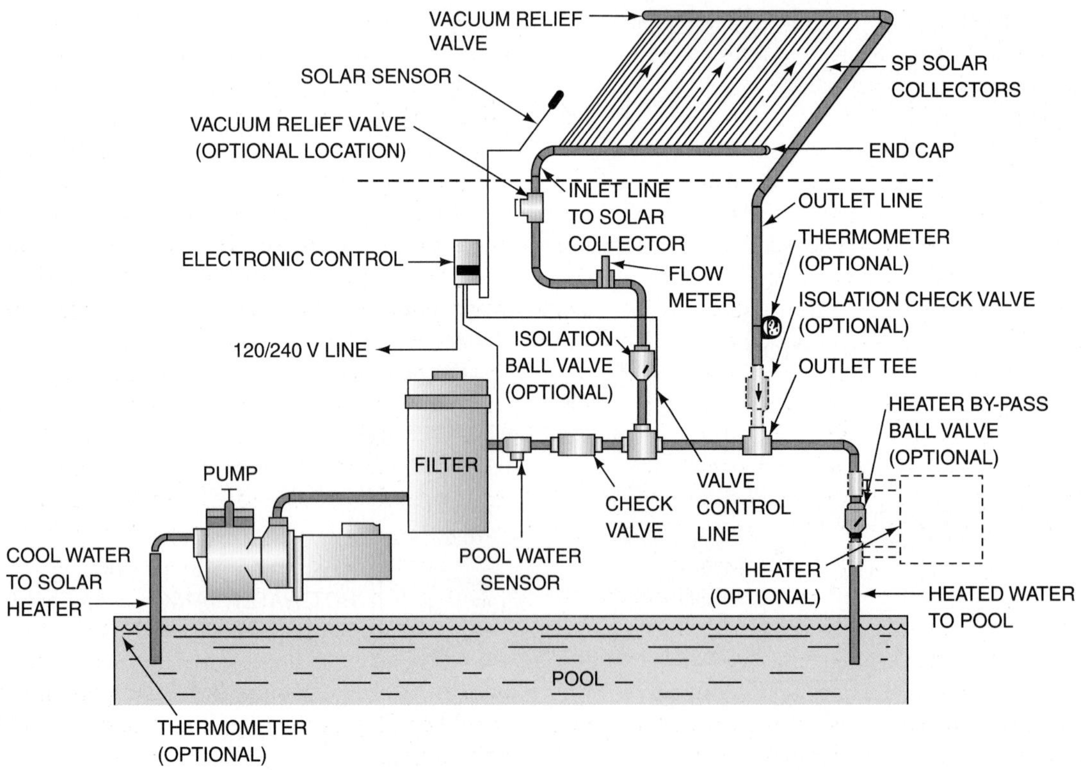

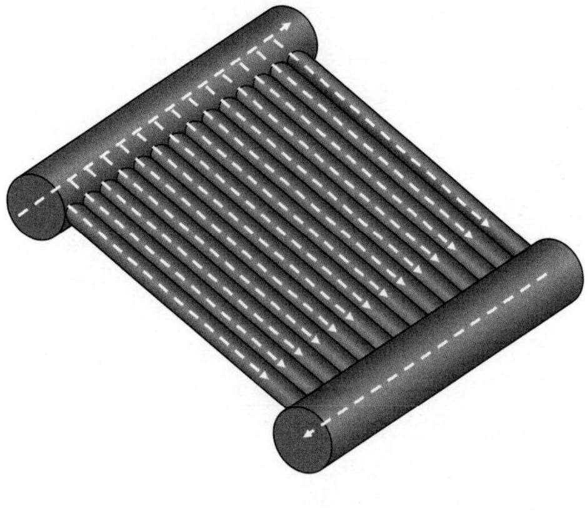

Figure 33.127 The solar collector for a swimming pool solar heating system.

33.11 PREVENTIVE MAINTENANCE

The air side of hydronic systems can be either natural draft or forced draft. If the system is natural draft, the convectors will be next to the floor and will be subjected to dust on the bottom side where the air enters. There are no filters on these natural-draft systems and consequently there is no filter maintenance. The coil becomes the filter, but the air moves so slowly that not much dust is picked up. The cover should be removed after several years and the coil vacuumed clean.

If the system is forced air, routine filter maintenance will be required, as described in Unit 30. Coils may become dirty after many years of good filter maintenance because all particles of dust are not filtered and some will deposit on the coil. When the coil becomes dirty, capacity will be reduced. Most coils are oversized and the capacity reduction may not be noticed for a long time.

Cleaning a coil is accomplished with approved liquid detergents mixed with water and applied to the coil. If the coil is in an equipment room, it may be cleaned in place. If the coil is in the attic or above a ceiling, it may have to be removed from the air handler for cleaning.

The fan blades may also have built-up dust deposits and need cleaning. The fan wheel may be cleaned in the fan if the motor is not in the way. This may be accomplished by scrapping the dust buildup with a screwdriver blade and using a vacuum cleaner to remove it. If the fan wheel is very dirty, it may be best to remove the fan section and clean it outdoors. **NOTE:** *Do not allow water to enter the bearings or motor. Plastic bags may be used to protect them.•* If the bearings get soaked, purge them with oil or grease to remove the water. If the motor gets wet, set it in the sun and use an ohmmeter to verify that the motor is not grounded. Setting the motor on an operating air-conditioning condensing unit fan outlet will quickly dry it as the warm, dry air passes over it.

The water side of a hydronic system requires some maintenance for the system to give long, trouble-free performance.

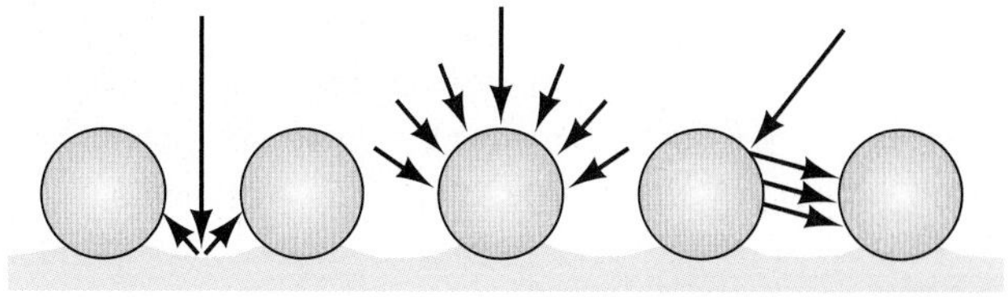

Figure 33.128 A cross-sectional view of a swimming pool solar collector. Courtesy Aquatherm Industries/Solar Industries

Fresh water contains oxygen, which causes corrosion in the pipes. Corrosion is a form of rust. The proper procedure for starting up a hydronic system is to wash the system inside with an approved solution to remove any oils and construction debris. This may or may not have been performed on any system you may encounter. When the system is cleaned initially, it is supposed to be filled with fresh water and chemicals that will treat the initial charge of water for corrosion. Again, this may or may not have been done. If the system is several years old and no water treatment has been used, the system will surely need maintenance. It should be drained and flushed. Water treatment should be added when the system is filled with fresh water. Be sure to follow the water treatment company's instructions. While inspecting the water side of a hydronic heating system, be sure to inspect the pressure relief valves. They should not be leaking and they should definitely not be rusted. Rusted pressure relief valves should be replaced immediately. Rusted valves may not operate properly when unsafe system pressures are present and can cause severe personal injury.

Systems with leaks have a continuous water makeup of fresh water, causing continuing corrosion because the fresh water contains oxygen. Water leaks must be repaired. It is common for water from leaks in coils to be carried away in drain lines and not be discovered. When a system has an automatic air bleed, air will not build up.

Water pumps must be maintained. They have lubrication ports for oil or grease. Keep the oil reservoir at the correct level on the pump with the correct oil. The motor will not have an oil reservoir but will require a certain amount of oil. Do not exceed the recommended amount.

When a motor has grease fittings, proper procedures must be used when applying grease. You cannot continue to pump grease into the bearings without some relief. The seal will rupture. **NOTE:** *Unscrew the relief fitting on the bottom of the bearing and allow grease to escape through the relief hole. Grease may be added until fresh grease is forced out the relief opening. •*

Pumps also have couplings that should be inspected. **SAFETY PRECAUTION:** *Turn off the power and lock and tag the disconnect before removing the cover to the coupling.•* Check the coupling for filings caused by wear that may be in the housing. Check for loose bolts and play in the coupling if it is rubber. Play in the coupling is natural for the spring type of coupling.

The electrical service should be checked if the pump has a contactor. ⚠ **SAFETY PRECAUTION:** *Shut off the power and remove the cover to the contactor. Do not touch any electrical connections with meter leads until you have placed one voltmeter lead on a ground source—a conduit pipe will do—and touch the other meter lead to every electrical connection in the contactor box. The reason for this is that some pumps are interlocked electrically with other disconnects and there may be a circuit from another source in the box. Use all precautions.•*

When you find burned contactor contacts, replace them; if you find frayed or burned wire, repair or otherwise correct the situation.

33.12 SERVICE TECHNICIAN CALLS

SERVICE CALL 1

A motel manager calls indicating that there is no heat at the motel. *The problem is on the top two floors of a four-story building. The system has just been started for the first time this season, and it has air in the circulating hot water. The automatic vent valves have rust and scale in them due to lack of water treatment and proper maintenance.*

Arriving at the motel, the technician asks at the front desk to speak with the manager. The technician introduces himself to the manager, who thanks the technician for arriving so promptly, and asks him about the problem with the heat. The manager brings the technician up-to-date and then shows him where the boiler room is located. The boiler is hot, so the heat source is working. The water pump is running because the technician can hear water circulating. The technician goes to the top floor to one of the room units and removes the cover; the coil is at room temperature. The technician listens closely to the coil and can hear nothing circulating. There may be air in the system, and the air vents, sometimes called bleed ports, must be located. The technician knows they must be at the high point in the system and normally would be on the top of the water risers from the basement.

The technician goes to the basement where the water lines start up through the building and finds a reference point. He then goes to the top floor and finds the top portions of these pipes. The pipes rise through the building next to the elevator shaft. The technician goes to the top floor to the approximate spot and finds a service panel in the hall ceiling. Using a ladder, he finds the automatic bleed port for the water supply over the panel. This is the pump discharge line. (See **Figure 33.34** for an example of an automatic air vent or bleed port.)

The technician removes the rubber line from the top of the automatic vent. The rubber line carries any water that bleeds to a drain. There is a valve stem in the automatic vent (as in an automobile tire). The technician presses the stem but nothing escapes. There is a hand valve under the automatic vent, so he closes the valve and removes the automatic vent. The technician then carefully opens the hand valve and air begins to flow out. Air is allowed to bleed out until water starts to run. The water is very dirty. ⚠ **SAFETY PRECAUTION:** *Care should be used around hot water.•*

The technician then bleeds the return pipe in the same manner until all the air is out and water only is at the top of both pipes. The technician then goes to the heating coil where the cover was previously removed. The coil now has hot water circulating.

The technician takes the motel manager to the basement, drains some water from the system, and shows the manager the dirty water. The water should be clear or slightly colored with water treatment. The technician suggests that the manager call a water treatment company that specializes in boiler water treatment, telling him that the system will have some major troubles in the future.

The technician then replaces the two automatic vents with new ones and opens the hand valves so that they can operate correctly.

SERVICE CALL 2

A customer in a small building calls. *There is no heat in one section of the building although the system is heating well in most of the building. The building has four hot water circulating pumps, and one of them is locked up—the bearings are seized. These pumps are each 230 V, three-phase, and 2 hp.*

The technician consults the customer at the building. They go to the part of the building where there is no heat. The thermostat is calling for heat. They go to the basement where the boiler and pumps are located. They can tell from examination that the number 3 pump, which serves the part of the building that is cool, is not turning. The technician carefully touches the pump motor; it is cool and has not been trying to run. Either the motor or the pump has problems.

The technician chooses to check the voltage to the motor first. He goes to the disconnect switch on the wall and measures 230 V across phases 1 to 2, 2 to 3, and 1 to 3. This is a 230-V three-phase motor with a motor starter. The motor overload protection is in the starter, and it is tripped.

The technician still does not know whether the motor or the pump is the problem. All three windings have equal resistance, and there is no ground circuit in the motor. He turns the power on, clamps an ammeter on one of the motor leads, and presses the overload reset. The technician can hear the motor try to start, and it is pulling locked-rotor current. He pulls the disconnect to stop the power from continuing to the motor. It is evident that either the pump or the motor is stuck.

The technician turns off the power, locks it out, and then returns to the pump. It is on the floor next to the others. Removing the guard over the pump shaft coupling, the technician tries to turn the pump over by hand—remember, the power is off. The pump is very hard to turn. The question now is whether it is the pump or the motor that is tight.

The technician disassembles the pump coupling and tries to turn the motor by hand. It is free. The technician then tries the pump; it is too tight. The pump bearings must be defective. The technician gets permission from the owner to disassemble the pump. He shuts off the valves at the pump inlet and outlet, removes the bolts around the pump housing, and then removes the impeller and housing from the main pump body. The technician takes the pump impeller housing and bearings to the shop and replaces them, installs a new shaft seal, and returns to the job.

While the technician is assembling the pump, the customer asks why the whole pump was not taken instead of just the impeller and housing assembly. The technician shows the customer where the pump housing is fastened to the floor with bolts and dowel pins. The relationship of the pump shaft and motor shaft has been maintained by not removing the entire pump, so they will not have to be realigned after they are reassembled. The pump is manufactured to be rebuilt in place to avoid this.

After completing the pump assembly, the technician turns it over by hand, assembles the pump coupling, and affirms that all fasteners are right. Then the pump is started. It runs and the amperage is normal. The technician turns off the power, locks it out, and assembles the pump coupling guard. The power is turned on and the technician leaves.

SERVICE CALL 3

An apartment house manager reports that *there is no heat in one of the apartments. This apartment house has 25 fan coil units with a zone control valve on each and a central boiler. One of the zone control valves is defective.*

The technician checks in with the property manager on arrival. The manager informs the technician about the one apartment with no heat and that there have been no other complaints from the other occupants regarding a lack of heat. Since the occupant is not home, the technician is brought up to the apartment by one of the building's service personnel. Once in the apartment, the technician notices that the room thermostat is set above the room temperature. The technician goes to the fan coil unit located in the hallway ceiling and opens the fan coil compartment door, which drops down on hinges. Standing on a short ladder to reach the controls, the technician finds there is no heat in the coil and there is no water flowing.

This unit has a zone control valve with a small heat motor. The technician checks for power to the valve's coil; it has 24 V, as it should. (See **Figure 33.42** for an example of this type of valve.) This valve has a manual open feature, so he opens the valve by hand. The technician can hear the water start to flow and can feel the coil get hot. The valve heat motor or valve assembly must be changed. The technician chooses to change the assembly because the valve and its valve seat are old.

The technician obtains a valve assembly from the stock of parts at the apartment and proceeds to change it. *The technician shuts off the power and the water to the valve and water coil. He then spreads a plastic drop cloth to catch any water that may fall and places a bucket under the valve before removing the assembly.* The old valve is removed and the new valve installed. The electrical connections are made, and the system valves are opened, allowing water back into the coil.

When the technician turns on the power he can see the valve begin to move after a few seconds (this is a heat motor valve, and it responds slowly). The technician can now feel heat in the coil. He closes the compartment door, removes the plastic drop cloth, sets the room thermostat, and leaves the apartment for the manager's office. The technician discusses the repair with the manager, completes his paperwork, gets his work order signed, and leaves for his next job.

Solutions for the following service calls are not provided here. Please discuss the solutions with your instructor.

SERVICE CALL 4

A customer calls and indicates that *one of the pumps beside the boiler is making a noise.* This is a large home with three pumps, one for each zone.

The technician arrives, goes to the basement, hears the coupling, and notices there are metal filings around the pump shaft.

What is the problem and the recommended solution?

SERVICE CALL 5

A homeowner calls and states that *there is water on the floor around the boiler in the basement. The relief valve is relieving water.*

The technician goes to the basement and sees the boiler relief valve seeping. The boiler gauge reads 30 psig, which is the rating of the relief valve. The system normally operates at 20 psig. The technician looks at the flame in the burner section and sees the burner burning at a low fire. A voltmeter check shows there is no voltage to the gas valve operating coil.

What is the problem and the recommended solution?

REVIEW QUESTIONS

1. A hydronic system provides comfort heating by producing
   A. heated hot water.
   B. steam.
   C. heated hot water or steam.
   D. neither heated hot water nor steam.
2. Zoning a hydronic heating system is beneficial because this strategy saves energy by
   A. overheating the areas of the structure that are not occupied.
   B. only heating the spaces that need to be heated at any given time.
   C. underheating the areas of the structure that are occupied.
   D. preventing the boiler from operating when the occupied space gets too cold.
3. Which of the following is most likely true with regards to rooms that are zoned together?
   A. A living room and a bedroom will very likely be zoned together.
   B. A garage, kitchen, and dining room will very likely be zoned together.
   C. A garage, dining room, and bedroom will very likely be zoned together.
   D. All of the bedrooms in a house will very likely be zoned together.
4. Boilers that operate with efficiencies over 90% are classified as
   A. conventional boilers.
   B. condensing boilers.
   C. conversion boilers.
   D. compensating boilers.
5. A high-mass boiler
   A. heats up quickly and cools down slowly.
   B. heats up quickly and cools down quickly.
   C. heats up slowly and cools down slowly.
   D. heats up slowly and cools down quickly.
6. A low-mass boiler
   A. heats up quickly and cools down slowly.
   B. heats up quickly and cools down quickly.
   C. heats up slowly and cools down slowly.
   D. heats up slowly and cools down quickly.
7. The term "wet-base boiler" refers to a hydronic appliance that
   A. leaks water at a controlled rate to keep the base wet at all times.
   B. allows for the water being heated to be both above and below the combustion chamber.
   C. only uses steam as the heat transfer medium to provide comfort heating.
   D. is configured so the water being heated is always below the combustion chamber.
8. Air in a cast iron boiler can cause appliance damage resulting from
   A. corrosion.
   B. combustion.
   C. condensation.
   D. cavitation.
9. In a dry-base boiler,
   A. the combustion chamber is always above the water being heated.
   B. the combustion chamber is always below the water being heated.
   C. the combustion chamber and the heat exchanger are side-by-side.
   D. the heat exchanger is mounted below the combustion chamber.
10. An air cushion is a necessary component of a hot water hydronic heating system because
   A. water expands when it is heated.
   B. air contracts when it is heated.
   C. air condenses when it is heated.
   D. water evaporates when it is cooled.
11. An example of a low-mass boiler is a
   A. cast-iron boiler.
   B. copper water-tube boiler.
   C. wet-base boiler.
   D. steel boiler.
12. The most common type of pump found on a hot water hydronic heating system is the
   A. reciprocating.
   B. pulsating.
   C. scroll.
   D. centrifugal.
13. Another term that is commonly used in conjunction with pump pressure is
   A. inches of water column.
   B. inches of head.
   C. feet of water column.
   D. feet of head.
14. If the highest pipe in a hot water hydronic heating system is 12 feet above the inlet of the expansion tank, what is the desired pressure in the expansion tank?
   A. 5.2 psig
   B. 10.2 psig
   C. 15.4 psig
   D. 20.6 psig
15. On a hot water hydronic heating system, the "point of no pressure change" is located at the
   A. outlet of the circulator pump.
   B. inlet of the pressure-reducing valve.
   C. inlet of the expansion tank.
   D. outlet of the pressure relief valve.

16. The purpose of a limit control in a boiler is to
    A. vent air from the heating system.
    B. cycle the boiler on and off to maintain water at the desired temperature.
    C. shut the boiler off before the near-boiler piping freezes.
    D. limit the amount of water that is fed into the boiler.
17. A relief valve's pressure setting on a low-pressure boiler is approximately
    A. 10 psig.
    B. 20 psig.
    C. 30 psig.
    D. 40 psig.
18. The opening and closing of a zone valve is controlled by the temperature of the
    A. air in the occupied space.
    B. water returning to the boiler.
    C. water leaving the boiler.
    D. air in the expansion tank.
19. In an outdoor reset control strategy,
    A. as the outdoor air temperature rises, the target boiler water temperature rises.
    B. as the outdoor air temperature rises, the target boiler water temperature drops.
    C. as the boiler water temperature rises, the outdoor air temperature rises.
    D. as the boiler water temperature rises, the outdoor temperature drops.
20. A series loop, hot water hydronic heating system is characterized by
    A. higher system installation costs when compared to most other piping configurations.
    B. its ability to provide individual temperature control in each of the structure's rooms.
    C. warmer terminal units at the beginning of the loop and cooler ones at the end of the loop.
    D. the need for excessive amounts of redundant piping to ensure proper system operation.
21. Which of the following statements is true regarding a series loop, hot water hydronic heating system?
    A. The heat output of each of the terminal units is the same.
    B. The temperature of the water entering each of the terminal units is the same.
    C. The temperature difference across each of the terminal units is the same.
    D. The rate of water flow through all of the terminal units is the same.
22. In a one-pipe hot water hydronic heating system, all of the terminal units
    A. are piped in parallel with each other.
    B. are piped in series with each other.
    C. are piped in parallel with the main hot water loop.
    D. are piped in series with the main hot water loop.
23. The diverter tee is found on which hot water piping arrangement?
    A. Series loop system
    B. One-pipe system
    C. Two-pipe, direct-return system
    D. Two-pipe, reverse-return system
24. Balancing valves are most commonly found on which hot water piping arrangement?
    A. Series loop system
    B. One-pipe system
    C. Two-pipe, direct-return system
    D. Two-pipe, reverse-return system
25. In a primary-secondary piping arrangement,
    A. multiple circulator pumps are used.
    B. all of the system water passes through all of the terminal units.
    C. each terminal unit must have its own circulator pump.
    D. the temperature of the water entering each of the terminal units is the same.
26. The temperature of the water at the outlet of a three-way mixing valve will most likely be
    A. equal to the temperature of the water at the hot water inlet.
    B. lower than the temperature of the water at the cold water inlet.
    C. higher than the temperature of the water at the hot water inlet.
    D. higher than the temperature of the water at the cold water inlet.
27. Which of the following is correct when comparing a radiant heating system to a conventional hot water heating system?
    A. The water temperature in a radiant heating system is lower than that in a conventional hot water heating system.
    B. The water temperature in a radiant heating system is higher than that in a conventional hot water heating system.
    C. Conventional hot water heating systems use primarily polyethylene, cross-linked tubing.
    D. Conventional hot water heating systems have piping circuits that are typically concealed in floors.
28. All of the following are common types of radiant heating system except
    A. slab on grade.
    B. thick slab.
    C. thin slab.
    D. staple-up.
29. When a hydronic heating system is to use a combination of high- and low-temperature heating circuits, which of the following components must be used?
    A. Thermostatic expansion valve
    B. Thermostatic diverter valve
    C. Thermostatic mixing valve
    D. Thermostatic combination valve

30. A tankless, domestic water heater
    A. heats domestic water as needed, but experiences large standby losses.
    B. heats domestic water as needed, greatly reducing system standby losses.
    C. continuously heats domestic water and stores it in the boiler until needed.
    D. continuously heats domestic water, greatly reducing system standby losses.
31. Which of the following is true regarding the configuration of a tankless, domestic water heater?
    A. A water coil is mounted in the boiler, keeping the domestic water separate.
    B. Domestic water is continuously mixed with the boiler water.
    C. Boiler water continually flows through the domestic water coil.
    D. A circulator pump continually circulates domestic hot water through the boiler.
32. What percentage of the sun's energy reaches the earth?
    A. Between 0% and 1%
    B. Between 1% and 3%
    C. Between 4% and 6%
    D. Between 7% and 9%
33. The varying intensity of the sun's energy that reaches the earth is due in part to the
    A. solar constant.
    B. declination angle.
    C. lunar constant.
    D. solar design strategy.
34. Drain down solar systems are best suited for use in
    A. colder climates.
    B. warmer climates.
    C. climates with very low humidity.
    D. climates with very high humidity.
35. All of the following are recommended procedures for maintaining hydronic heating systems except
    A. when greasing circulator pumps, be sure to remove the pump's relief fitting first.
    B. inspect the couplings on system pumps for wear and damage.
    C. inspect pressure relief valves and replace those that are rusted.
    D. boilers should be drained and refilled every time the system is serviced.
36. List four heat sources that are commonly used in hydronic heating systems.
37. Explain why a boiler's heat exchangers are made up of sections.
38. Describe the function of a low-water cutoff.
39. Explain the difference between a two-port zone valve and a three-port zone valve.
40. Explain the differences between a two-pipe, direct-return system and a two-pipe, reverse-return system.

UNIT 34

INDOOR AIR QUALITY

OBJECTIVES

After studying this unit, you should be able to

- explain why good indoor air quality is an important factor in helping to keep people healthy.
- state the industry standard that establishes acceptable indoor air-quality guidelines.
- explain how indoor air contamination is controlled.
- explain how building construction affects indoor air quality.
- list five sources or types of indoor air pollution that are directly related to HVACR equipment.
- describe six common air pollutants.
- describe the phenomenon known as sick building syndrome.
- explain the importance of proper mold remediation.
- explain how carbon dioxide is formed.
- describe the dangers of common pollutants, such as carbon monoxide and asbestos.
- explain the importance of providing proper ventilation.
- explain the difference between a heat recovery ventilator and an energy recovery ventilator.
- describe the MERV rating for filters.
- describe six types of air filters.
- describe the term face-loading as it applies to air filters.
- list the four main parts of an electronic air cleaner.
- explain how ion generators and ultraviolet light are used to help improve indoor air quality.
- explain the difference between absorption and adsorption as it applies to air cleaning.
- explain how cleaning the air distribution system can help improve indoor air quality.
- explain the term relative humidity as it applies to comfort.
- explain why it is important to provide humidification in winter months.
- describe four common types of humidifiers.
- list general factors used when sizing humidifiers.
- troubleshoot basic problems with filtration and humidification systems.

SAFETY CHECKLIST

- ☑ Do not disturb materials containing asbestos.
- ☑ Use all electrical safety precautions when servicing or troubleshooting electrical or electronic circuits. Use common sense.
- ☑ When replacing or servicing line-voltage components, ensure that power is off, that the panel is locked and tagged, and that you have the only key.
- ☑ Handle ultraviolet (UV) lights carefully as they contain mercury.
- ☑ Dispose of non-working ultraviolet light bulbs in accordance with the manufacturer's guidelines as well as any local disposal laws.
- ☑ Do not look directly at energized ultraviolet lights.

34.1 INTRODUCTION

Scientific evidence suggests that the air in homes and other buildings is becoming increasingly polluted. In many cases, evidence indicates that indoor air is much more polluted than the outdoor air. Since many people spend as much as 90% of their time indoors, the chance of being exposed to poor quality air for extended periods of time is great. Certain groups of people, such as the young, elderly, and chronically ill, typically spend much more time indoors than other groups and are, therefore, more susceptible to the effects of indoor air pollution. These people face a serious risk from the cumulative effects of this indoor air pollution. ASHRAE Standard 62.1, "Ventilation for Acceptable Indoor Air Quality," is the standard generally used by engineers and technicians in this industry to determine acceptable indoor air quality.

34.2 SOURCES OF INDOOR AIR POLLUTION

Some potential sources of indoor air pollution include:

Adhesives	Mold
Air freshener	Moth repellents
Asbestos	Paint supplies
Car exhaust	Paneling
Carbon monoxide	Personal care products
Carpets and drapes	Pesticides
Cleaning supplies	Pet waste, dander and hair
Dry-cleaned goods	Pet products
Fireplaces	Pressed wood furniture
Hobby supplies	Pressed wood products
House dust mites	Tobacco smoke
Household chemicals	Radon
Humidifiers	Stored fuels
Lead-based paints	Unvented appliances
Moisture	Wood stoves

Many of the items listed above, such as cleaning supplies, paint, pesticides, and adhesives, contain organic compounds that are released when the products are used. These compounds may also be released to the air if the products are not properly stored. Always follow the manufacturer's recommendations for proper usage, storage, and disposal of items to ensure that their effect on indoor air quality is minimal.

Figure 34.1 illustrates some sources and causes of indoor pollution. In older homes, outdoor air may enter through

Figure 34.1 Sources and causes of indoor pollution. Courtesy Aerotech Laboratories, Inc., www.aerotechpk.com

openings, joints, and spaces around windows and doors. Most modern homes and office buildings, however, are designed and constructed more tightly to help keep the outside air out in order to maximize the efficiency of heating and cooling systems. This may result in an accumulation of pollution inside. To offset the negative effects of tight construction, *mechanical ventilation* may be required.

Inadequate ventilation can increase the effects of indoor air pollution. Outdoor air in sufficient quantities must be brought inside to mix with and dilute the pollutant emissions, and the indoor air must be exhausted to the outside to carry out pollutants. Without this ventilation, the pollutants may accumulate to the point where they can cause health and comfort problems. Mechanical ventilation can include exhaust fans in a single room, such as in a bathroom or kitchen, or an air-handling system to intermittently or continuously remove indoor air and replace it with filtered and often conditioned air from outside. Results of inadequate ventilation may include moisture condensation on windows or walls, smelly or stuffy air, dirty central heating and cooling equipment, and areas in which books, shoes, or other items become moldy. In addition to providing adequate ventilation, other methods can be used to control indoor air contamination. These methods include eliminating the source of the contamination and providing a means for cleaning the air.

34.3 COMMON POLLUTANTS

Pollutants can be found in many different places and their presence can be traced back to different sources. Some pollutants are naturally occurring elements, such as radon and lead, while others are naturally occurring compounds, such as formaldehyde and asbestos. Some contaminants are produced during processes such as combustion. Combustion can refer to a number of activities including the burning of fuel for home heating or the burning of a cigarette. Pollutants are often classified as being biological or non-biological, where the term biological refers to being organic, or derived from living organisms.

There are many sources of biological contaminants. Central air-handling systems can grow mold, mildew, and other contaminants. Viruses can be transmitted by people and animals. Dust mites, a source of biological allergens, can be found in carpeting, bedding, and other fabrics. Many allergens can become airborne under certain conditions. Water-damaged materials and wet surfaces provide opportunities for molds, mildew, and bacteria to grow.

Wood-burning or gas stoves and fireplaces, and unvented kerosene and gas space heaters may be the source of pollutants from combustion products. The pollutants released may be carbon monoxide, nitrogen dioxide, sulfur dioxide, and solid particulate matter suspended in the air. The combustion gases may come from improperly sealed stove doors, from chimneys and flues that are improperly installed or maintained, or from cracked heat exchangers. The combustion process also produces water vapor. Although not a pollutant in itself, water vapor can result in high humidity and wet surfaces, which can encourage the growth of biological pollutants. The following paragraphs discuss some commonly encountered pollutants.

RADON

Radon is a colorless, odorless, and radioactive gas. The main health risk of breathing air polluted with radon is lung cancer. Consuming water that is contaminated with high radon levels may also cause health risks, but these risks are believed to be much lower than breathing air that contains radon gas. The most common source of radon is uranium in the soil or rock on which homes are built. As uranium breaks down, it releases radon gas, which can enter buildings through cracks in concrete floors and walls, floor drains, and sumps. The presence of some types of radon contamination can be identified by using a home radon test kit. Testing equipment should be approved by the Environmental Protection Agency (EPA) or be state certified. The test kit comes with detailed instructions on how to perform the test and, once completed, the test canister is sent to a laboratory for analysis. If the radon level, whether in the air or water, is determined to be above the acceptable limits, corrective action should be taken to eliminate the source. Procedures should be carried out by contractors or individuals who have satisfactorily completed the EPA Radon Contractor Proficiency Test or who are otherwise certified to perform this type of remediation.

ENVIRONMENTAL TOBACCO SMOKE

Environmental tobacco smoke (ETS) is the mixture of smoke that comes from the burning end of a cigarette, pipe, or cigar and from smoke exhaled by the smoker. It consists of more than 4000 compounds, more than 40 of which are known to cause cancer in humans or animals and many of which are strong irritants. ETS is often referred to as "secondhand smoke." To reduce exposure to environmental tobacco smoke, smoking should not be permitted indoors. Separating smokers and nonsmokers in different rooms in the same house will do very little to eliminate exposure to ETS.

DUST MITES

Dust mites are microscopic, spiderlike insects usually found indoors. These biological contaminants thrive in warm, humid conditions, where they produce waste pellets. Their bodies disintegrate when they die, producing very small particulates that mix with household dust. Dust mites and their remains are often thought to be a primary irritant to

people sensitive to dust irritants. Carpeting should be vacuumed often, but vacuum cleaners have the potential to stir up the dust in the carpeting and often raise dust levels in the air. Vacuum cleaners with high-efficiency particulate arrestor (HEPA) filters or central vacuuming systems that discharge air outside the building can be used to help control this dust. Replacing carpeting with hard surface floors is helpful in many instances, but some manufactured flooring materials contribute negatively to the quality of indoor air. Smaller area rugs can be used if cleaned regularly. Bedding should be washed often with hot water. Mattresses and pillows can be covered with plastic in order to more easily remove the dust.

MOLDS

Molds are simple organisms and, like dust mites, are classified as biological contaminants. Molds are commonly found both outdoors and indoors. Molds found outside the house break down many organic substances necessary to plant, animal, and human life. Mold, a fungus, releases spores into the air. Many of the spores found indoors come from outside. However, quantities of mold growing indoors can produce many spores that are harmful to humans. Mold is usually found in areas where there is high humidity. **Figure 34.2** shows one kind of mold as seen under a microscope. Estimates indicate that there may be over 100,000 different types of mold.

To grow, molds need moisture and food, which includes paper, textiles, dirt, leaves, drywall, or wood. The molds digest this material and gradually destroy it. As long as there is moisture and food, mold will continue to thrive. **Figures 34.3(A)** and **(B)** show examples of extreme cases of indoor mold. Molds can grow on the surface of damp inorganic matter such as glass and concrete when there is a thin, invisible layer of organic material that acts as food for the mold. Left unchecked, mold can continue to grow and cause health problems for those who are sensitive to mold as well as those who suffer from breathing disorders such as chronic obstructive pulmonary disease (COPD) or asthma. Mold can usually be seen, or it may be smelled. It has a musty odor. It may grow in wall cavities, above ceilings, under carpets, and behind wallpaper. Because air-conditioning and heating systems are responsible for controlling humidity inside structures, the technicians in the HVACR field, who work with these systems, should be able to recognize mold. There are many different types, each with its unique look and description. A moisture meter, **Figure 34.4**, may be used to help detect moisture.

Figure 34.2 Photo of *Stachybotrys* mold under a microscope. Courtesy Aerotech Laboratories, Inc. www.aerotechpk.com

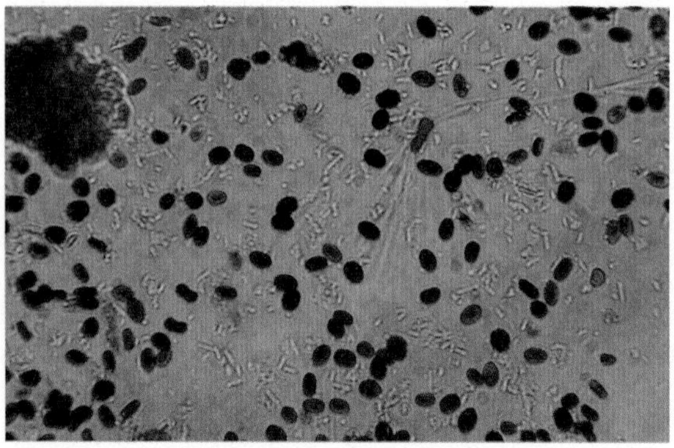

Figure 34.3 (A) A mold, *Stachybotrys*, colonized on a ceiling and ceiling cavity. (B) A mixture of *Cladosporium*, *Aspergillus*, and *Penicillium* mold on a painted drywall surface. Courtesy Aerotech Laboratories, Inc. www.aerotechpk.com

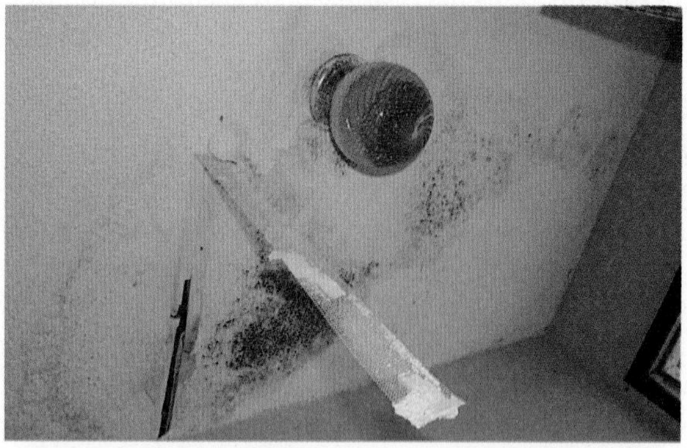

(A)

(B)

Molds consist of multicelled filaments called hyphae. These filaments can unite to form balls or mats and may penetrate the food material, making removal and cleanup difficult. Molds reproduce through the production of spores. Spores are released into the air after they are produced and dry out. They move with the air currents, land, germinate, and grow. Spores can survive for long periods of time in both hot and cold environments. They require only moisture and food to germinate and grow. Spores contain allergens, which have the potential to cause allergic reactions in humans. Individuals may ingest, inhale, or be exposed to spores through skin

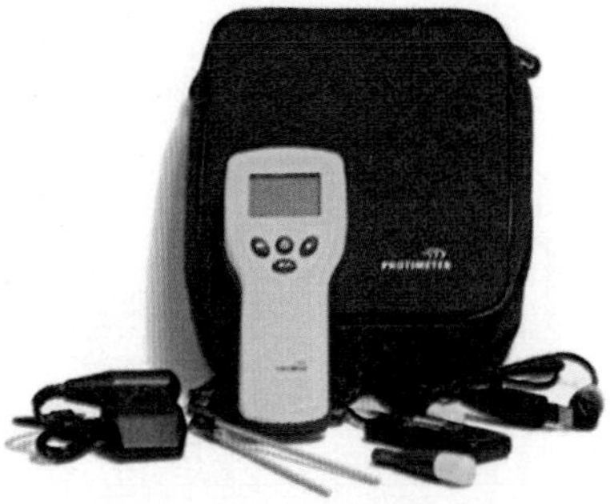

Figure 34.4 A digital moisture meter. This meter has a pointed probe that can be inserted into carpet or wood to measure the moisture level. Courtesy Aerotech Laboratories, Inc. www.aerotechpk.com

Figure 34.5 A testing device that will measure ppm of carbon monoxide. Courtesy Bacharach, Inc., Bacharach Institute of Technical Training

contact. Individuals may experience a temporary effect from this exposure or, in some cases, the effects can be long-lasting.

Molds can produce a variety of health problems for humans. The severity of the problems are determined, in part, by the type and quantity of the mold as well as the sensitivity of the individual. Molds produce volatile organic compounds (VOCs). Exposure to a large quantity of these VOCs can irritate mucous membranes and produce headaches, inability to concentrate, and dizziness. Some molds may have the potential to produce mycotoxins, which may be toxic to humans.

CARBON MONOXIDE

Carbon monoxide (CO) is a very poisonous, lighter-than-air gas. It is produced by incomplete combustion in oil or gas-fired appliances. Incomplete combustion may be a result of the lack of oxygen or the incomplete mixing of oxygen and fuel. Excess combustion air or an insufficient fuel supply can also result in the production of CO. When a flame temperature is lower than 1128°F, CO will be produced. CO combines with hemoglobin in the bloodstream, gradually replacing the oxygen. Its concentration is cumulative to the point that, when there is too little oxygen in the bloodstream, life will no longer be supported. Carbon monoxide especially affects fetuses, infants, and people with anemia or a history of heart disease. Low levels of exposure can cause fatigue and increased chest pain in people with chronic heart disease. Higher exposure levels may cause headaches, dizziness, and weakness in healthy people as well as sleepiness, nausea, vomiting, confusion, disorientation, loss of consciousness, and possibly death.

CO is very dangerous because it is colorless, odorless, tasteless, and nonirritating. It can only be measured with a testing device, **Figure 34.5**. However, alarms can be purchased that, if properly installed, will respond when CO levels become dangerous. CO detectors are presently required to "alarm" at CO levels set by the Underwriters Laboratory (UL Standard 2034). However, those who are sick or prone to illness, including infants and the aged, should use detectors that are designed to alarm at lower levels of carbon monoxide. Typically, lower-priced detectors initiate an alarm at higher levels of CO. Carbon monoxide in the ambient air is measured by the number of CO molecules present in a million molecules of air, referred to as parts per million (ppm). See **Figure 34.6** for concentrations and the symptoms they develop. Carbon monoxide monitors or alarms should be installed in all areas of a building where a combustion process is being used and in other parts of a building where CO may collect.

Auto exhaust is considered to be the leading cause of carbon monoxide alarm responses. When individuals let their cars warm up or otherwise let them run in a garage that is attached to or below the house, the CO may become trapped and infiltrate the living areas of the house, producing a dangerous condition and setting off the alarm. Unvented, poorly installed, and improperly maintained oil and gas appliances account for a large percentage of CO alarms.

Another frequent cause of CO alarms is the backdrafting of natural-draft combustion appliances. These backdrafts may be caused by blockages in the vent, or by pressure differences. As the exhaust systems in the bathrooms, kitchen, attic, and other areas of the house exhaust air from the structure, the house will have a net negative pressure with respect to the outside air. This tends to push vented air back into the house. **Figure 34.7** shows air being exhausted from within the house and other air returning through the venting or chimneys to make up for the exhausted air. This backdrafting air may contain carbon monoxide. Combustion appliances must be properly installed and adjusted, and a proper supply of fresh air, that includes oxygen, must be available. In all vented appliances a proper flue passageway must be available to vent the flue gases to the atmosphere.

The other pollutants released from combustion appliances include **nitrogen dioxide**, which causes irritation of the respiratory tract and can cause shortness of breath, **sulfur dioxide**, which can cause eye, nose, and respiratory tract irritation and breathing problems, and *particles from combustion*. These

Figure 34.6 This chart indicates carbon monoxide concentrations and symptoms that may develop. Courtesy Bacharach, Inc., Bacharach Institute of Technical Training

CO CONCENTRATIONS & SYMPTOMS DEVELOPED	
Concentrations of CO in the air	Inhalation time and toxic symptoms developed.
9 ppm (0.0009%)	The maximum allowable concentration for short-term exposure in a living area, according to ASHRAE.
50 ppm (0.0050%)	The maximum allowable concentration for continuous exposure in any 8-hour period, according to federal law.
200 ppm (0.02%)	Slight headache, tiredness, dizziness, and nausea after 2–3 hours.
400 ppm (0.04%)	Frontal headaches within 1–2 hours, life-threatening after 3 hours, also maximum parts per million in flue gas (on an air-free basis), according to EPA and AGA.
800 ppm (0.08%)	Dizziness, nausea, and convulsions within 45 minutes. Unconsciousness within 2 hours. Death within 2–3 hours.
1,600 ppm (0.16%)	Headache, dizziness, and nausea within 20 minutes. Death within 1 hour.
3,200 ppm (0.32%)	Headache, dizziness, and nausea within 5–10 minutes. Death within 30 minutes.
6,400 ppm (0.64%)	Headache, dizziness, and nausea within 1–2 minutes. Death within 10–15 minutes.
12,800 ppm (1.28%)	Death within 1–3 minutes.

particles are suspended in the air and can cause eye, nose, throat, and lung irritation. Certain chemicals attached to the particles can cause lung cancer. All of these risks depend on the length of exposure, the size of the particles, and the chemical makeup of the contaminants. Wood smoke, for example, contains both polycyclic organic matter (POM) and non-polycyclic organic matter, both of which are considered to be health hazards, and are by-products of wood combustion.

It is important to keep wood stove emissions to a minimum and to make sure that stove doors close tightly. All joints in stove pipe and/or fittings should be tight or sealed. EPA-certified wood stoves should be used whenever possible. Central air-handling systems, including furnaces, flues, and chimneys, need to be inspected regularly. Cracked heat exchangers should be replaced immediately. Cracked or damaged flues and chimneys should be repaired. Ductwork

Figure 34.7 Air exhausted from a house causes a negative pressure within the house. This interferes with the venting of combustion appliances, causing a backdraft that forces some of the flue gases back down the chimneys and vents into the living areas. Courtesy Bacharach, Inc., Bacharach Institute of Technical Training

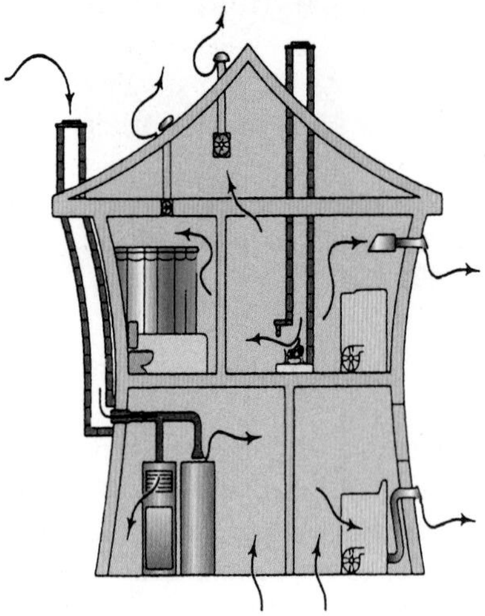

in central air-conditioning and heating systems should be cleaned periodically (See Section 34.8, "Duct Cleaning.") Exhaust fans should be installed over all cooking stoves. These exhaust fans should be vented to a location outside the structure. Failure to do so would simply result in the moving of the cooking fumes from one location within the structure to another. This would be the same as sweeping dirt from one room of the house to another without picking it up.

FORMALDEHYDE

Formaldehyde is a naturally occurring organic compound that is chemically known as methanal, CH_2O. Formaldehyde is commonly used in the manufacture of some building materials and certain household products. It is also contained in combustion gases and may exist in the gases from unvented fuel-burning appliances such as unvented gas stoves or kerosene space heaters. Pressed wood products that have adhesives containing urea-formaldehyde (UF) resins include particleboard, which may be used as subflooring and shelving, hardwood plywood paneling, medium-density fiberboard, cabinets, and some furniture tops. Pressed wood, such as softwood plywood and flake board, is intended for exterior use and contains phenolformaldehyde (PF) resin. Formaldehyde has a strong odor, can cause cancer, and can be fatal if ingested. Brief exposure to formaldehyde can be irritating to the eyes, nose, and throat.

PESTICIDES

Pesticides may be used in and/or around a residence to control insects, termites, rodents, or fungi. Most disinfectants

contain pesticides. Weed killers, some pool chemicals, and kitchen sanitizers, for example, contain pesticides. Research has indicated that a high percentage of exposure to pesticides occurs indoors. Pesticides should only be used in accordance with the manufacturer's instructions. As with other household chemicals, make certain that all containers are properly sealed prior to storage. Since there are many different types of pesticides, the health effects due to exposure vary greatly. The mildest reactions to pesticide exposure include nose, throat, and eye irritation. Contact with pesticides can also result in more severe reactions and, depending on the exposure time, can cause various cancers and reproductive problems.

Under the Food Quality Protection Act (FQPA), the EPA is in charge of evaluating pesticide levels in food to ensure that the products are safe to consume. Many people assume that simply rinsing foods will wash the pesticides away, but this is not true. In order to minimize pesticide exposure from food, it is a good practice to wash, trim, and peel all food products that have been exposed to pesticides.

ASBESTOS

Asbestos is a naturally occurring mineral substance that can be found in older homes and buildings. It was once used in the manufacturing of construction materials as a strengthening agent. It is no longer used in many countries, as it has been proven to be a carcinogen. It may be found in pipe and furnace insulation materials, shingles, textured paints, and floor tiles. Concentrations of airborne asbestos can be found when asbestos-containing materials have been disturbed by cutting, sanding, or removal activities. The most dangerous asbestos fibers are too small to be seen, can accumulate in the lungs, and can cause lung cancer and other diseases. Before a renovation project is started, asbestos experts should be brought in to evaluate the construction materials used on the job to determine if asbestos is present. If its presence is detected, a qualified remediation crew will have to be hired to remove the contaminants. The process of removing asbestos is time-consuming and costly. ⚠ **SAFETY PRECAUTION:** *If there is any possibility that deteriorating, damaged, or disturbed materials contain asbestos, they should be examined and tested by a person certified to do so. If the materials do contain asbestos, they must be removed by contractors licensed and equipped to do so. No other persons should handle the material.* •

LEAD

Humans may be exposed to *lead* through air, drinking water, food, contaminated soil, deteriorating paint, and dust. Old lead-based paint is probably the greatest source of exposure. In 1978, lead paints were banned by the federal government. Lead particles can come from outdoor sources of lead dust and from indoor activities using lead solder. Lead pipes and soldering materials were used until the mid-1980s. The Safe Drinking Water Act, last updated in 2017, set new, stricter guidelines for the allowable lead content in piping materials. The present limit is 0.25%. ⚠ **SAFETY PRECAUTION:** *Deteriorating, damaged, or disturbed materials that contain lead, particularly lead paint, should be removed by contractors licensed and equipped to do so. No other persons should handle the material.* •

34.4 SICK BUILDING SYNDROME

Sick building syndrome (SBS) is a term given to situations when people who spend considerable time in a particular building experience health problems. These illnesses may be caused by poor ventilation or biological contaminants such as molds or bacteria. Mold in buildings has become a significant issue in the United States. Over the past two decades, several attempts have been made to introduce legislation at the national level intended to protect home buyers and consumers from moving into mold-infested homes. Proposed legislation also intended to provide legal protection to home buyers and renters who are exposed to dangerous levels of indoor mold. In addition, the proposed legislation was expected to establish national standards, including insurance provisions, health benefits, and education programs. Although no federal laws were enacted with regards to these important issues, some states have passed their own legislation to protect individuals from the dangers of indoor air pollution. Check with your local government for more information about IAQ laws and guidelines in your state.

Mold has many legal ramifications. Contractors can be sued. Insurance companies may or may not cover expenses and liability costs when presented with claims involving mold. Some sources suggest that HVACR and other contractors should obtain legal advice to determine their liability and that contractors should meet with their insurance representatives to ensure they have adequate coverage. To help minimize liability, it is important that HVACR service technicians receive approved mold-related training. This training will help enable the technician to identify potential problem areas before they become full-blown mold conditions.

34.5 MOLD REMEDIATION

The following general recommendations regarding mold remediation relate primarily to small areas of mold growth. Larger areas should be cleaned and/or materials should be removed and disposed of by contractors certified to do so.

Mold contamination is classified by levels, which are determined by the square footage of contamination. These levels are:

- Level I: contamination areas of less than 10 square feet
- Level II: contamination areas from 10 to 30 square feet

- Level III: contamination areas from 30 to 100 square feet
- Level IV: contamination areas over 100 square feet

For Level I and II mold contamination, maintenance staff members are permitted to perform the cleanup as long as they are properly trained under the Occupational Safety and Health Administration's (OSHA) Hazard Communication Standard. For Level III contamination, those performing the remediation should be trained to handle hazardous materials and should be outfitted with a respirator, gloves, and eye protection. Level IV remediators must also receive hazardous materials training. Level IV mold remediation requires the use of a full-face HEPA filter respirator and the wearing of disposable clothing that covers all parts of the body.

SAFETY PRECAUTION: *Technicians involved with removing mold should wear outer clothing designed for this purpose, including gloves, hat, and respirator or an approved alternative.* •

Removing the source of the moisture should be the first step in mold cleanup, and it should be done quickly. All leaks should be repaired, and all damp areas should be thoroughly dried and cleaned up. The humidity level may need to be decreased. Because molds can often penetrate the surfaces they are feeding on, merely cleaning the surface may not be enough. Using a household bleach or other chemicals available for that purpose may be helpful in eradicating mold growing on hard surfaces. Carpets with mold growing within the fabric or underneath should be removed, professionally cleaned, and dried. The floor beneath the carpet must be cleaned, ensuring that all mold is removed, and any moisture problems must be resolved. Wood, wallboard, or wallpaper that has been heavily contaminated with mold may need to be discarded. This should be done by contractors and technicians certified to do so. Dead molds can still cause problems to human health, so complete removal is necessary.

Air-conditioning (cooling) equipment should not be oversized. Sometimes when a house is being constructed, owners want the equipment to be oversized because they are planning on extending the house at some point in the future. However, oversized air-conditioning equipment may not run long enough to sufficiently dehumidify the space. When oversized equipment satisfies the cooling requirement, the unit shuts down, and the evaporator coil warms up. A warm evaporator coil cannot remove moisture from the air passing through it, as the coil will be at a temperature that is higher than the dew point temperature of the air stream. If this is the case, separate dehumidifiers may be needed. (Refer to Unit 42 for more on the importance of properly sized heating and cooling equipment.) In addition, the ground around the foundation should slope away from the building to ensure that water does not collect around or under the house.

Air-conditioning condensate pans and drain lines should be cleaned regularly. This will help ensure that the condensate will drain from the equipment properly. Humidification equipment should also be cleaned regularly. If mold is or has been present within a house, the HVACR duct system may need to be inspected. If mold is found in the ducts, they should be cleaned or replaced by qualified technicians. Clothes driers and stoves should be vented to the outside. Exhaust fans should be used and vented to the outside when bathing or showering, when cooking, or when running the dishwasher. An architect or engineer may be consulted to determine whether whole-house ventilation is needed.

34.6 VENTILATION

Ventilation is the process of supplying air by natural or mechanical means to a space. Ventilation air is part of the air supplied from outdoors plus any recirculated air that has been cleaned or treated. Ventilation air should be supplied to the occupants' breathing space. Indoor air quality can be improved when proper ventilation procedures are followed. These procedures remove some of the polluted air and dilute the remaining pollutants with outside air. Outside air and/or cleaned filtered air does not benefit building occupants if it short-cycles from the supply diffuser immediately back to the return grille.

Ventilation has a negative effect on the efficiency of heating and cooling systems. The air that is brought in from the outside must be heated or cooled to maintain the desired temperature in the occupied space. See Unit 35, "Comfort and Psychrometrics," for how to calculate the heating or cooling capacity that must be added to account for outdoor air from ventilation. ASHRAE Standard 62 prescribes ventilation and other measures to ensure acceptable air quality.

Commercial buildings are often designed with the appropriate mechanical ventilation, which takes into account the expected number of occupants and any special requirements arising from the activities taking place in the building. Few homes, however, have heating, cooling, or ventilation systems that mechanically bring outside air inside. Therefore, opening windows and doors or operating window or attic fans becomes necessary. Bathroom or kitchen fans may be used to exhaust, or remove air from the structure, but this air must be replaced to prevent creating a negative pressure in the structure. This replacement air may infiltrate, or leak into, the house, or a door or window may be opened for short periods of time. Mechanical ventilation methods can also be used.

HEAT RECOVERY VENTILATORS (HRVs)

In an effort to increase the efficiency of air-conditioning and heating systems, the structural integrity of homes and other structures has become much tighter. Partly because of more energy-efficient structural elements, such as windows and doors, and improved insulation techniques and products available, the amount of air leakage, both into and out of a

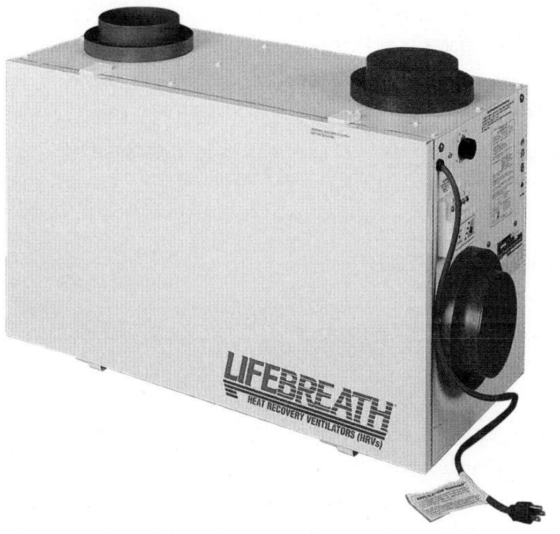

Figure 34.8 *A heat recovery ventilator.* Courtesy Nutech Energy Systems, Inc. (Life Breath)

structure, has been dramatically reduced. Increased structural tightness has benefits, such as reduced cooling and heating costs and reduced drafts, but there is a major drawback to tighter construction.

The major problem associated with tight construction is that the same air remains in the structure for a longer period of time, increasing the concentration of air contaminants and pollutants such as smoke, carbon dioxide, and other compounds found in cleaning supplies and construction materials. In order to help ensure that there is ample fresh air and ventilation, many structures are equipped with heat recovery ventilators, or HRVs, **Figure 34.8**. HRVs perform a number of functions that include:

- Removing stagnant air from the structure at a controlled rate.
- Introducing outside (fresh) air into the structure at the same rate at which air is removed.
- Recovering heat from the air being removed and transferring this heat to the air entering the structure. (Heating Mode)
- Transferring heat from the intake air to the exhaust air. (Cooling Mode)

Heat recovery ventilators are referred to as balanced systems, meaning that the amount of air removed from the structure and the amount of air into the structure are the same. If the system was unbalanced and the amount of air introduced and removed were different, a number of problems could result. An excess of air removal could result in a negative pressure in the space. This can cause improper venting of fossil-fuel heating system flue gases, introducing carbon dioxide to the occupied space. Introducing too much air into the structure could result in a positive pressure in the space. This pressure can push moisture into the walls and

Figure 34.9 Airflow patterns through a heat recovery ventilator.

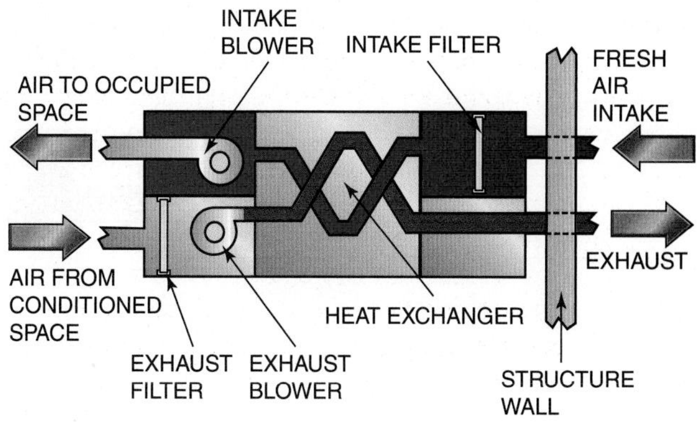

ceilings, resulting in the slow deterioration of the construction materials.

A simplified diagram of a heat recovery ventilator is shown in **Figure 34.9**. The HRV has four duct sections connected directly to it. Two of the ducts are for the fresh air intake and two are for the exhaust air. In the center of the unit, there is a heat exchange surface that facilitates the transfer of heat between the intake and exhaust air streams. Both of the airstreams are filtered before reaching the heat exchanger in an effort to keep the heat exchanger clean and keep the system operating as efficiently as possible.

ENERGY RECOVERY VENTILATORS (ERVs)

Energy recovery ventilators (ERVs) are commonly used to reduce the negative effects on system efficiency by transferring both latent and sensible heat between the entering and leaving air streams. One design for mechanical equipment that provides air dilution consists of two insulated ducts run to an outside wall. One duct allows outside air to enter the building, and the other exhausts air to the outside. In one version of this design, the ventilation and exhaust ducts are configured to allow sensible heat to transfer between the two air streams. In addition, a desiccant-coated disc rotates between the two air-streams to allow for latent heat to transfer between them.

In the colder winter months, the ERV recovers both latent and sensible heat from the air stream being exhausted from the space. Consider the system shown in **Figure 34-10**. The air in the occupied space is at a temperature of 70°F with 50% R.H. (1). The outside air is at a temperature of 20°F with 30% R.H. (2). The exhaust (red) air stream and the ventilation (blue) air stream are in contact with each other to allow for heat transfer. It can be seen that the temperature of the exhaust air has dropped from 70°F (1) to 40°F (3), while the temperature of the ventilation air has increased

Figure 34.10 Airflow patterns through an energy recovery ventilator.

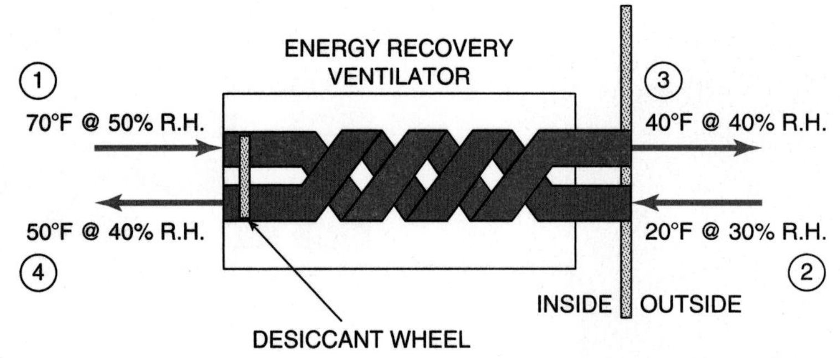

from 20°F (2) to 50°F (4). In addition, it can be seen that the relative humidity of the exhaust air stream has dropped, while the relative humidity of the ventilation air stream has increased. This latent heat transfer is a result of the rotating desiccant wheel between the two air streams. Were it not for the ERV, the ventilation air stream (2) would be brought directly into the structure and the heating equipment would have to treat this air, reducing the effectiveness and efficiency of the system. A similar situation arises in the warmer summer months. In the summer months, the exhausted cooler air is used to cool and dehumidify the ventilation air before it enters the structure.

SECURITY

Security is an important topic to address, especially as it relates to acts of war, terrorism, or other accidents and natural disasters. Building owners, managers, and technicians should meet and assess all risks and dangers that could possibly be present and develop a plan for each of these potential events. Universal plans are seldom useful due to the differences in building architecture, HVACR systems, and building occupancy and use.

Outdoor air intakes for buildings should be located away from public areas when possible. An elevated position or a roof location is the best in most instances. Under normal conditions these locations may also produce cleaner air. If an event does occur that results in an air pollutant, some recommend that ventilation systems not be shut down until the source of the pollutant is determined. Of course, the source should be determined as quickly as possible. If it is outside, the ventilation system should be shut down. If it is inside, the building will need all the outside ventilation it can get to dilute the pollutant. If the ventilation system is shut down, the outside louvers should be closed and sealed if they are movable.

Filtering systems with the highest physically and economically possible minimum efficiency reporting value (MERV) ratings should be used in buildings. MERV ratings range from 1 to 16, with the upper levels providing the most filtering. For example, a filter with a MERV rating of 7 is about 30% efficient, while a filter with a MERV rating of 13 is about 80% efficient. Filters should be chosen not only for their capacity to arrest particles but also according to the amount of pressure drop they would cause.

Higher efficiency filters will operate with larger pressure drops across them. Using filters that are too efficient, therefore, can result in improper airflow through the appliance. In air-conditioning systems, inadequate airflow can result in evaporator coil freeze-ups and liquid floodback to the system compressor. On heating systems, inadequate airflow through a furnace can cause excessively high temperature rises across furnaces, possibly leading to cracked heat exchangers and a dangerous carbon monoxide condition.

The filter retainer frame should be inspected regularly to ensure that it has not come loose or deteriorated. Clips should hold the frame tight. For the best seal, gaskets should be used. Considerable amounts of air can pass around the filter frames if they are not installed and maintained properly. Air filters are only effective if air is directed through them. Filters should be sized so that they completely fill the opening in the duct or filter grill or rack. Under no circumstances should air be permitted to bypass the filter medium.

A qualified and knowledgeable technician can be very valuable in preparing a security plan and maintaining building equipment. The National Energy Management Institute (NEMI) has developed the Testing, Adjusting and Balancing Bureau (TABB) certification, which technicians can obtain.

34.7 AIR CLEANING

Along with source control and ventilation, air cleaning is another method of providing acceptable indoor quality. *There are four general types of air cleaners: mechanical filters, electronic air cleaners, ion generators, and ultraviolet-C light. Some systems are considered "hybrid" devices because they may contain two or more particle-removing devices. For instance, a mechanical filter may be combined with an electrostatic precipitator or an*

Figure 34.11 Filter media in rolls and pads. Courtesy Aerostar Filtration Group

Figure 34.12 A section of fiberglass medium showing that the medium becomes denser as the air passes through it. This adds to the capacity of the filter to filter particles because the larger particles are filtered first in the less dense parts of the filter.

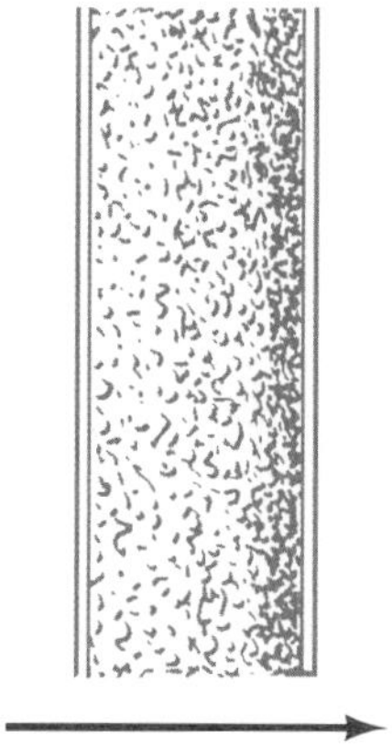

ion generator. Air cleaners may contain absorbent or adsorbent materials to remove gaseous materials. Absorption is the process of one substance being absorbed by another. For instance, carbon dioxide and carbon monoxide can be absorbed by either solid or liquid-spray filters. Adsorption occurs when a thin film of a liquid or gas adheres to the surface of a solid substance. Charcoal is used extensively as an adsorber.

FILTERS

HIGH-LOFT POLYESTER FILTER MEDIA. These filters are spray-bonded and can be purchased in bulk, in media rolls, or in cut pads, **Figure 34.11**. Bonded filters are often made up of different layers of materials, each with a different medium concentration, making the filter more dense as air passes through them, **Figure 34.12**.

FIBERGLASS THROWAWAY FILTERS. These filters are constructed of continuous glass fibers bonded together with a thermosetting agent. They are coated with a dust-holding adhesive that helps prevent particulates from migrating downstream, **Figure 34.13**. The filter material is designed to get progressively denser as the air passes through it.

PLEATED FILTERS. Pleated filters have much more surface area than non-pleated filters and typically require less frequent changing. *These filters can accumulate heavier, more restrictive particles at the bottom of the pleats, leaving the sides open longer for effective filtration,* **Figure 34.14**.

CUBE FILTERS. *These filters have a very high dust-holding capacity and are designed to operate effectively in high humidity applications. Three filter layers are heat sealed together. The first*

Figure 34.13 Fiberglass throwaway filters. Courtesy Aerostar Filtration Group

Figure 34.14 *A pleated filter. These filters have more filter surface area per front area of the filter.* Courtesy Aerostar Filtration Group

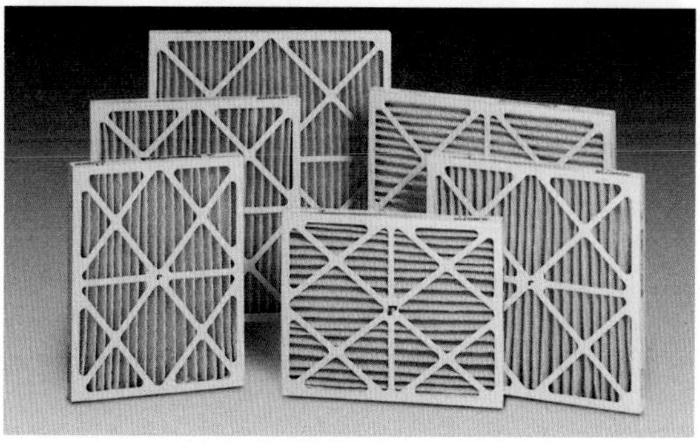

Figure 34.15 *A cube filter. This filter has a lot of surface area and requires a deep filter case.* Courtesy Aerostar Filtration Group

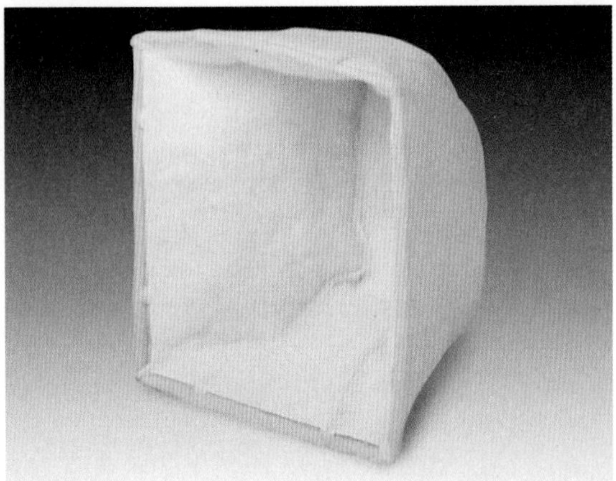

layer acts as a high-porosity impingement filter, which captures large particulate matter and helps prevent a condition known as face-loading. Face-loading occurs when the lead, or front, surface of the filter becomes clogged with dirt and prevents the smaller particulate matter from reaching the deeper, more dense, layers of the filter. The second and third layers of the filter capture smaller and smaller particulate matter. Cube filters, **Figure 34.15**, *are available in a number of configurations, including single, double, and triple pocket arrangements.*

POCKET FILTERS. *Vertical separators for each pocket channel air throughout the medium to prevent turbulence and allow even contaminant loading,* **Figure 34.16**. *Pocket filters are often used as either a final air filter for commercial applications or as a prefilter to a High Efficiency Particulate Arrestor (HEPA filter) when used in sterile or hospital applications.*

Figure 34.16 *A pocket filter.* Courtesy Aerostar Filtration Group

HEPA FILTERS. *This type of filter is used where the highest degree of filtration is desired or required. Many dust particles are in the 20 to 30 micron (μm) size range. (The symbol μm stands for micrometer, often called a micron.) Other particles are much smaller and require a much more efficient filter. HEPA filters are highly efficient—they can capture particles down to 0.3 μm or smaller. These minute particles can contain mold, dust mites, bacteria, pollen, and other allergens that, if breathed in, may remain in the lungs.*

Figure 34.17 *is an example of a HEPA filter. This particular filter can remove from the air 99.99% of particulates having a minimum size of 0.3 μm. Many HEPA filters remove 99.97% of particulates having a minimum size of 0.3 μm. For comparison, a human hair ranges from 70 to 100 μm in diameter, and an object 300 μm in diameter would fit easily through the eye of a sewing needle.*

SAFETY PRECAUTION (FILTER APPLICATION WARNING): *Common sense dictates that individuals would naturally want to use the most effective filtration system possible, but caution must be used when selecting a filter type or when changing from one filter type to another. Filter enclosures or racks have fixed dimensions, and the new filters are often selected based on the size and configuration of these racks. Should the filter configuration change, the racks would also need to be changed. If a technician replaces a low-efficiency or low-density filter a high-density filter, it is very likely that the airflow will be reduced. The blower will have more difficulty moving air through the higher density filter medium. If the airflow is reduced too much, there will be problems with the heat exchanger. In summer, there may be coil freezing problems. In winter, the heating system may overheat, so the technician must be aware of airflow requirements. Filters are rated in cfm at a prescribed pressure drop expressed in inches of water column. Be sure to match the new filter with the old filter airflow to reduce the chances of system performance problems.* •

Figure 34.17 *A HEPA filter.* Courtesy Aerostar Filtration Group

ELECTROSTATIC PRECIPITATORS OR ELECTRONIC AIR CLEANERS

There are several types of **electrostatic precipitators**. *These cleaners are designed to be mounted (1) at the furnace, (2) within the throwaway filter frame, or (3) within duct systems. They may also be stand-alone portable systems.* **Figure 34.18A** *shows a duct-mounted electronic air cleaner. These systems generally have a prefilter section that filters out the larger airborne particles and an ionizing or charging section,* **Figure 34.18B**. *Some also have a charcoal filter,* **Figure 34.18C**. *The prefilter traps larger particles and airborne contaminants. In the ionizing or charging section, particles are charged with a positive charge. These particles then pass through a negatively charged collection plate. The positively charged contaminants are attracted to this collection plate. The air then passes through a charcoal filter, on those systems that have them, to remove many of the odors and gases by adsorption.*

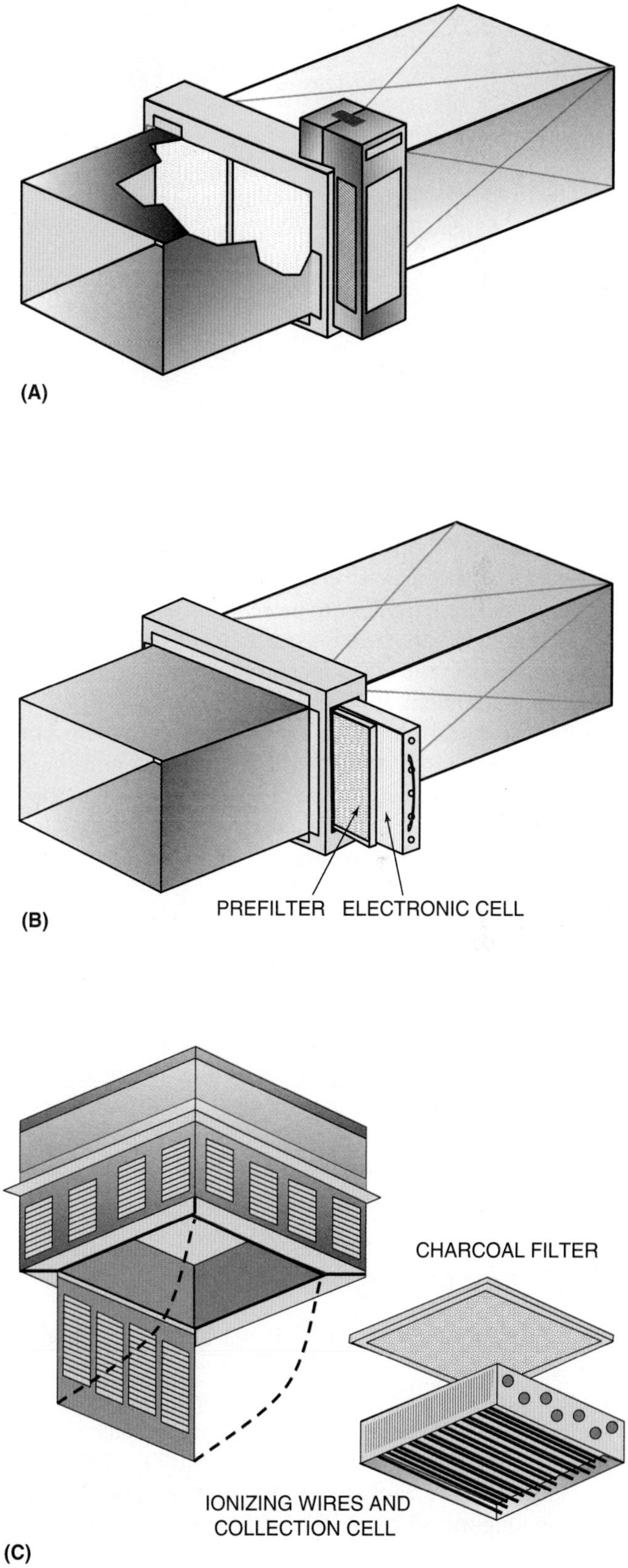

Figure 34.18 *Electrostatic precipitators.* (A) Duct-mounted air cleaner. (B) Prefilter and electronic cells on a duct-mounted air cleaner. (C) Ceiling-mounted, self-contained air cleaner.

ACTIVATED CHARCOAL AIR PURIFIER. *Activated charcoal, also called activated carbon, is manufactured in pellet form from coal or coconut shells. The material is treated during the manufacturing process so that the pellets produced have a very large internal surface area. Activated charcoal is especially effective when used to adsorb solvents, other organic materials, and odors. The purifier,* **Figure 34.19**, *is mounted in the return air duct of a heating and air-conditioning system. The velocity of the air within the duct forces the gaseous material to impinge on the activated charcoal, which is held in tubes suspended in the airstream. The gaseous material is attracted to the carbon particles, and eventually it condenses into a liquid and is held in place. These gases and/or odors are held in the carbon media. Periodically, the carbon media must be replaced. Some typical gases that can be removed by this type of purifier are*

- *compounds, such as alcohols, aldehydes, and acids from household products, construction materials, carpeting, furniture, or other household items.*
- *chlorinated hydrocarbons.*
- *inorganic compounds, such as halogens, phosgene, and sulfur compounds.*
- *a variety of odors from cooking, waste storage, pets, and humans.*

BI-POLAR IONIZATION AND POLARIZED FILTRATION

Ions are commonly found in clean environments, such as areas with low populations and low levels of industrial contamination. In overpopulated areas like city centers, ions are almost nonexistent. The concentration of ions is also very low in indoor environments. This is partially due to the fact that houses and other structures are being constructed more tightly, reducing the amount of entering natural ventilation.

Figure 34.19 *An activated carbon adsorber removes many gaseous contaminants found in the home.* Courtesy General Filters, Inc.

Ions help to reduce air contamination because they cause contaminants to bond to each other, making them heavier and larger. Larger particles are more likely to get trapped in filters and heavier particles are more likely to fall to the floor. Two technologies are being used as ion generators, which charge particles in the air to increase the number of ions and reduce air contamination levels.

Polarized filtration uses a high voltage at low current to put an electric charge on both the filter media and the particles that are present in the air. This causes the particles to become attracted to the filter media since the charges on the filter and on the contaminants are opposite. Charging the particles also causes them to bond to each other, creating larger and heavier particles. These larger particles are more likely to get trapped by the charged filter.

Bi-polar ionization creates air ions but does not charge the filter media. The ions of contaminants such as mold and dust bond together and get trapped in the filter. One of the benefits of bi-polar ionization is that it does not require additional filtration, so the addition of this type of unit does not further restrict airflow. In addition, as in the case of polarized filtration, the heavier particles fall to the floor and pose less of a problem than do airborne contaminants.

ULTRAVIOLET LIGHT

The use of ultraviolet (UV) light for providing cleaner indoor air has been studied.

UV light occupies a specific range in the light spectrum. Its wavelength ranges from 200 to 400 nanometers (nm), a unit that represents a length equal to 0.000000001 meter. Just below the UV range is the microwave portion of the spectrum (**Figure 34.20**); the visible light spectrum ranges

Figure 34.20 Diagram showing the wavelength of UV light with respect to that of visible light.

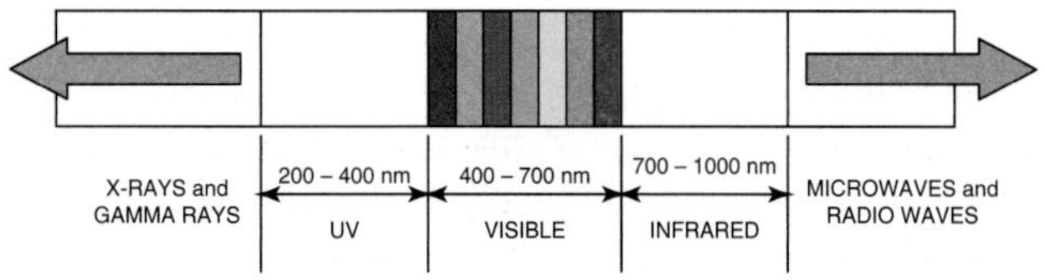

from 400 nm to 700 nm, and the infrared spectrum starts at 700 nm and goes up from there. Within the UV range there are three frequency bands:

- Long wave: 315 to 400 nm
- Midrange: 280 to 315 nm
- Low range: 200 to 280 nm

The long wave band represents what is called UV-A and the midrange is referred to as UV-B. The lower end of the UV spectrum represents light waves referred to as UV-C waves, and it is these light waves that help clean the air in forced-air heating and cooling systems. The UV-C (germicidal) range has been studied for use in destroying microbes that may cause unhealthy air in homes, offices, or other buildings, including hospitals and schools. The UV-C light has the ability to penetrate the walls of microbes that can have negative effects on indoor air quality. Once penetrated, the UV-C light can damage the DNA of the microbe, killing smaller and weaker ones, and preventing larger ones from reproducing.

Ultraviolet air treatment systems (**Figure 34.21**) are used on forced-air conditioning or heating systems and

Figure 34.21 An ultraviolet air cleaner with a photocatalytic oxidation (PCO) filter.

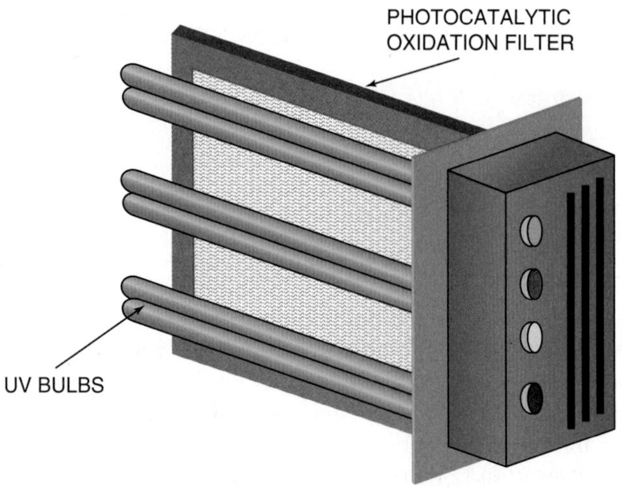

are most effective when used in conjunction with high-efficiency air filtration media. These units are easily installed in duct systems in much the same way that electric duct heaters are installed. They can be part of a new installation or added to existing systems. The location of the UV light is very important and the manufacturer's guidelines should be followed. These devices are typically positioned either close to the air filter or the evaporator coil, as both of these are prime locations for microbes such as mold to accumulate. Some manufacturers recommend installing UV devices in both the supply and return ducts, depending on the needs of the system. Most UV systems are available in different sizes for residential, commercial and industrial installations.

UV bulbs contain mercury, which has the potential to cause health issues as well as environmental damage if not handled and disposed of properly.

Many UV air purifiers come equipped with a supplemental photocatalytic oxidation filter, commonly known as a PCO filter. These filters are coated with titanium oxide (TiO_2) to facilitate the photocatalytic oxidation process. This process creates hydroxyl radicals and superoxide ions, which are highly reactive electrons. These electrons combine with pollutants in the air, such as various forms of bacteria and volatile organic compounds, or VOCs, and break them down into carbon dioxide and water. (VOC is an umbrella term used to describe a wide range of substances, such as formaldehyde, acetaldehyde, ammonia, and various acids that are found in building materials and household cleaning products.)

It is important to select the right UV light for the system. If the airflow through the PCO filter is not sufficient, the process of breaking down the VOCs may not be completed, resulting in the possible formation of even more VOCs. The type of volatile organic compounds present also has an effect on the time needed by the air treatment system to break them down. For example, cigarette smoke contains hundreds of different chemicals. Laboratory studies have indicated that, when air that is heavily contaminated with cigarette smoke is treated with a PCO filter, formaldehyde is formed as a by-product of the process.

SAFETY PRECAUTION: *Although the UV light may not be visible, do not look at it when it is activated. The unit should be placed where technicians or building occupants will not be exposed to it accidentally. Other safety issues include:*

- *Do not turn the light on before installing it.*
- *Place warning signs for building occupants and future technicians.*
- *Check the duct for possible light leaks, and seal them if any are found.*
- *Turn off the light before performing any maintenance or service on the HVACR system.*
- *Replace the light element annually.* •

34.8 DUCT CLEANING

Dirty air ducts may be a possible source of pollutants in homes. Homeowners should consider having the air ducts cleaned if

- there is visible mold growth present on hard-surface (sheet metal) ducts or on other components of the heating and cooling system.
- ducts are infested with vermin (rodents or insects).
- ducts are clogged with excessive amounts of dust and debris and/or particles are actually released into the home from supply registers.

Duct cleaning generally refers to the cleaning of the interior of ductwork and the cleaning of various heating and cooling system components of forced-air systems. This includes the supply and return air ducts and registers, grilles and diffusers, heat exchangers, heating and cooling coils, condensate drain pans, fan motor and fan housing, any humidification components within the system, and the air-handling unit housing. If the duct system is to be cleaned, all of the listed components should be cleaned. Failure to clean all units may result in recontamination of the entire system. If there is one inlet to the return duct system, having the filter placed there would help keep the return duct system clean.

SAFETY PRECAUTION: *The entire system should be inspected before cleaning to make sure that no asbestos-containing materials are in the heating and cooling system. If asbestos is found, specialized procedures are required. The asbestos should not be disturbed or removed except by specially trained and equipped contractors.* •

The vacuum system used should exhaust particles outdoors or high-efficiency particle air-vacuuming equipment should be used if the exhaust is inside the building. All carpeting and household furnishings should be protected. To dislodge and contain dust and other particles, equipment that allows controlled brushing of duct surfaces and contact vacuuming should be used, **Figure 34.22**. Soft-bristled brushes should be used on fiberglass duct board and sheet metal ducts lined internally with fiberglass. Some newer duct-cleaning systems, like the one shown in **Figure 34.22**, are equipped with variable speed controls, so low speed can be used on flexible duct runs and high speed can be used on rigid metal duct runs. When cleaning is completed, all access holes and/or access joints should be resealed and reinsulated, when appropriate, so that they are airtight. If moisture and/or mold are found, the source of this problem should be determined and the problem should be corrected, or it will reoccur.

Duct cleaning technicians should be familiar with and follow the National Air Duct Cleaning Association's standards, and if the ducts are constructed of flex duct, duct board, or lined with fiberglass, guidelines of the North American Insulation Manufacturers Association (NAIMA) should be followed.

Figure 34.22 *Duct cleaning equipment.* Courtesy G.I. Industries

34.9 AIR HUMIDIFICATION

In fall and winter, homes often feel dry because cold air from the outside infiltrates the conditioned space. The air in the home feels dry when it is heated because it expands, spreading out the moisture. The amount of moisture in the air is often expressed as relative humidity, which is the percentage of moisture in the air compared to the air's ability to hold moisture. In other words, if the relative humidity is 50%, each cubic foot of air is holding one-half the moisture it is capable of holding. The relative humidity of the air decreases as the temperature increases, because air with higher temperatures can hold more moisture. When a cubic foot of 20°F outside air at 50% relative humidity is heated to room temperature (75°F), the relative humidity of that air drops.

For comfort, the "dry" air should be replenished with moisture. The recommended relative humidity for a home is between 40% and 60%. When the relative humidity goes above these limits, bacteria, viruses, fungi, and other organisms become more active. In conditioned spaces with lower relative humidity, the dry warm air draws moisture from everything in the conditioned space, including carpets, furniture, woodwork, plants, and people. Furniture joints loosen, nasal and throat passages dry out, and skin becomes dry. Dry air causes more energy consumption than necessary because the air gets moisture from the human body through evaporation from the skin. The person then feels cold and sets the thermostat a few degrees higher to become comfortable. With more humidity in the air, a person can be more comfortable at a lower temperature. Static electricity is also much greater in dry air. A person also may receive a small electrical shock when touching something after having walked across the room.

Humidifiers are commonly used to add moisture to the air, by the process of evaporation, to increase its relative humidity. The evaporation process is speeded up by using power or heat or by passing air over large areas of water. The area of the water can be increased by spreading it over pads or by atomizing it.

HUMIDIFIERS

Evaporative humidifiers work on the principle of providing moisture to a surface called a medium and exposing it to the dry air. This is normally done by forcing the air through or around the medium; the air picks up the moisture from the medium as a vapor or a gas. There are several types of evaporative humidifiers.

Bypass humidifiers rely on the difference in pressure between the supply (warm) side of the furnace and the return (cool) side. It may be mounted either on the supply plenum or duct or the cold air return plenum or duct. Piping must be run from the plenum or duct where the humidifier is mounted to the other plenum or duct. If mounted in the supply duct, piping must be run to the cold air return, **Figure 34.23**. The difference in pressure between the two plenums draws some heated air through the humidifier to the return duct.

Plenum-mount humidifiers are mounted in the supply plenum or the return air plenum. The furnace blower forces heated air through the medium, where it picks up moisture. The air and moisture are then distributed throughout the conditioned space, **Figure 34.24**.

Under-duct humidifiers are mounted on the underside of the supply duct so that the medium is extending into the heated airflow, where moisture is picked up in the airstream. **Figure 34.25** illustrates an under-duct humidifier.

HUMIDIFIER MEDIA

Humidifiers are available in several designs with various kinds of media. **Figure 34.25** is a photo of a humidifier with drum-shaped medium. A motor turns the drum, which picks up moisture from the reservoir. The moisture then is evaporated from the drum into the moving airstream. The drums can be screens or sponges.

The screen-type humidifier medium resembles the drum medium but uses disc screens instead. The discs are mounted

Figure 34.23 (A) A bypass humidifier. (B) Cutaway view showing airflow from plenum through media to the return. (C) Typical installations. Courtesy AutoFlo Company

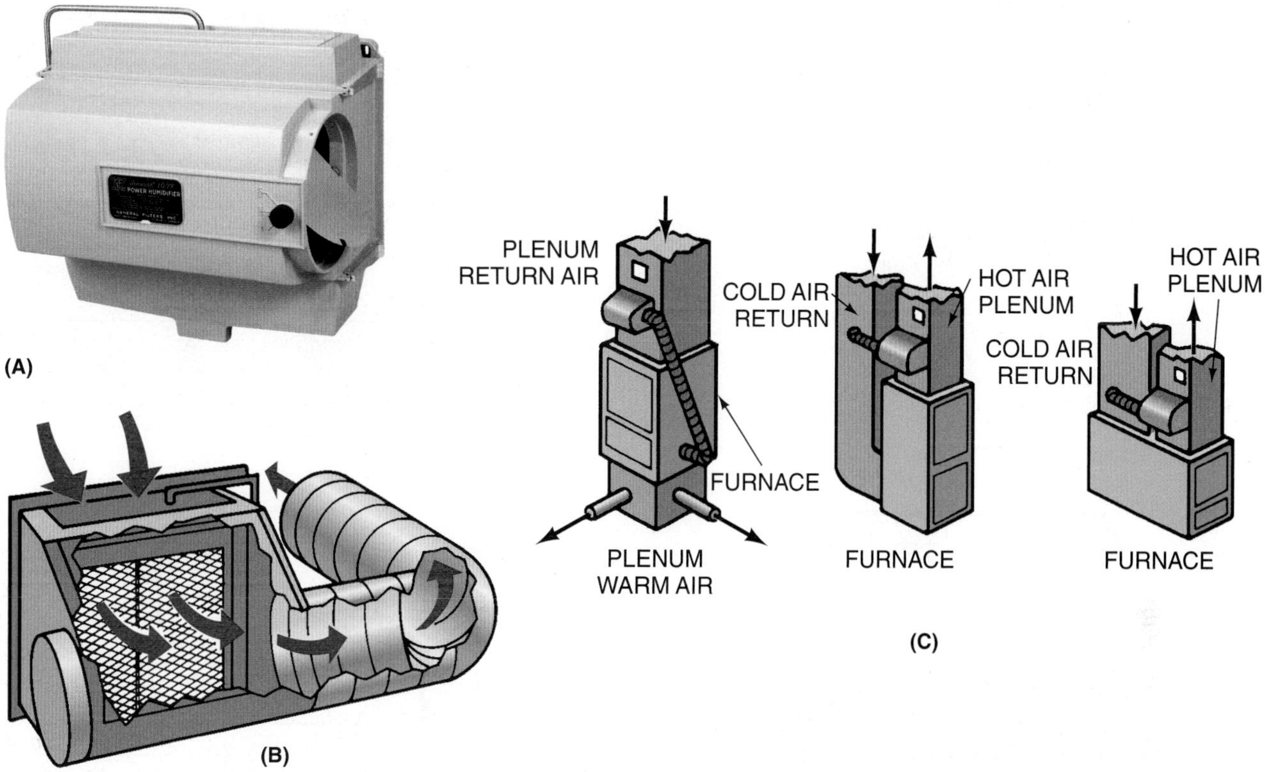

Figure 34.24 A plenum-mounted humidifier with a plate-type medium that absorbs water from the reservoir and evaporates it into the air in the plenum.

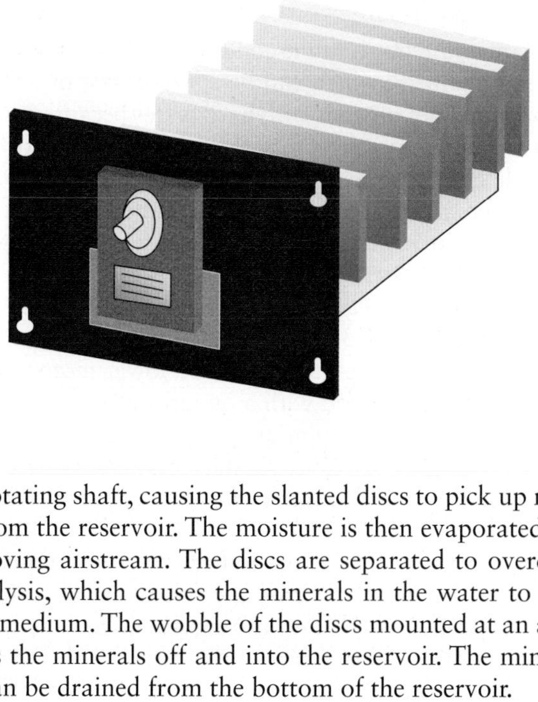

Figure 34.25 An under-duct humidifier using drum-style media.

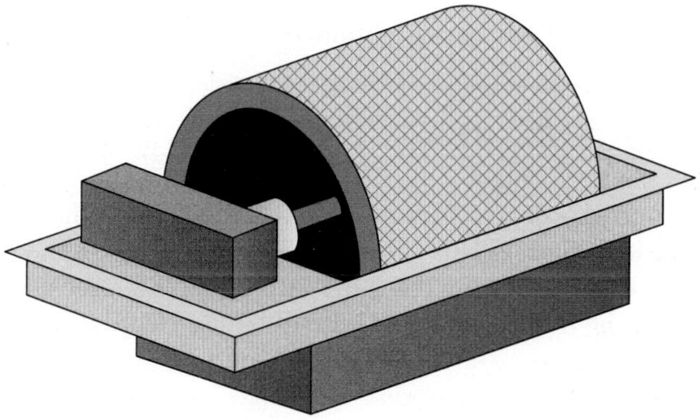

on a rotating shaft, causing the slanted discs to pick up moisture from the reservoir. The moisture is then evaporated into the moving airstream. The discs are separated to overcome electrolysis, which causes the minerals in the water to form on the medium. The wobble of the discs mounted at an angle washes the minerals off and into the reservoir. The minerals then can be drained from the bottom of the reservoir.

A plate- or pad-type medium is shown in **Figure 34.24**. The plates form a wick that absorbs water from the reservoir. The airstream in the duct or plenum causes the water to evaporate from the wicks or plates.

Figure 34.26 shows an electrically heated water humidifier. In electric furnace and heat pump installations, the temperature in the duct is not as high as in other types of hot-air furnaces. Media evaporation is not as easy as with lower temperatures. The electrically heated humidifier heats the water with an electric element, causing it to evaporate

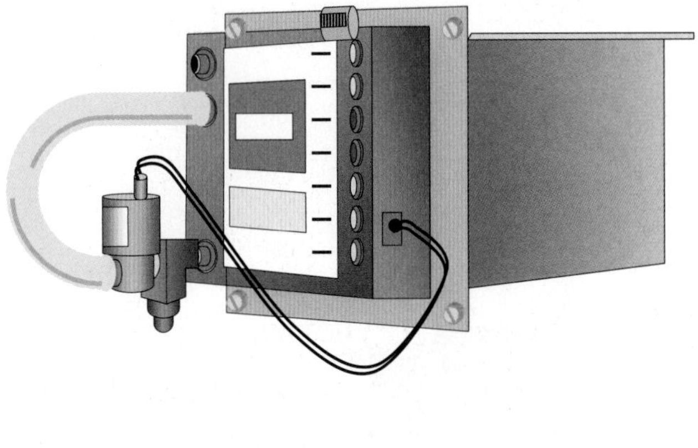

Figure 34.26 A humidifier with electric heating elements.

Figure 34.27 Operational diagram of an infrared humidifier.

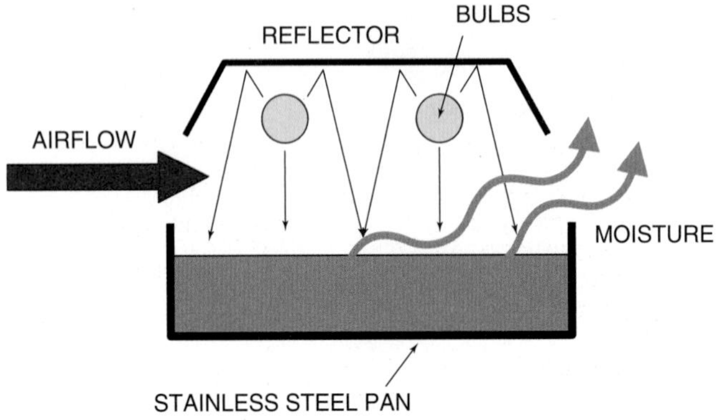

Figure 34.28 (A) A humidistat. (B) A wiring diagram with humidistat wired in. Courtesy of Honeywell, Inc., Residential Division

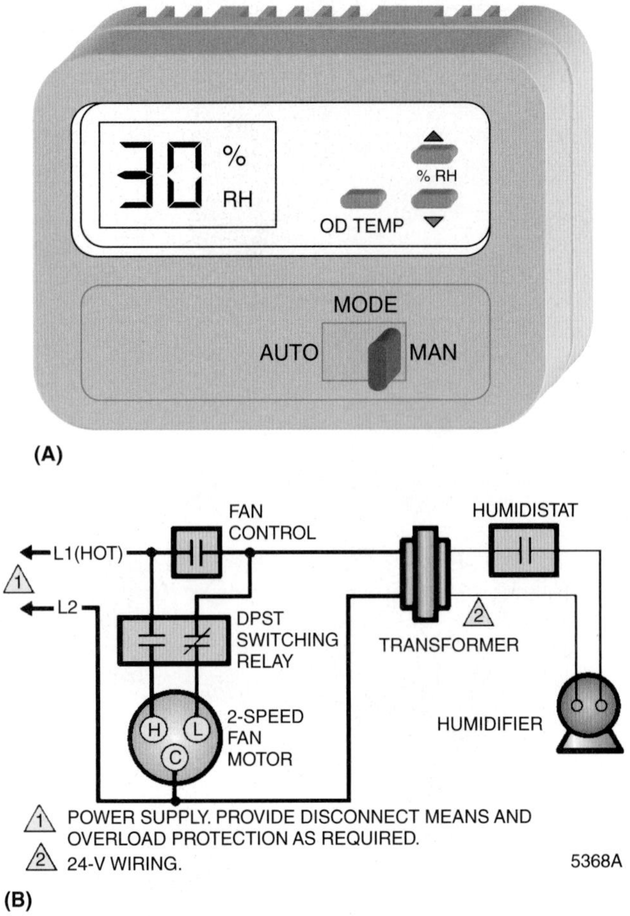

and be carried into the conditioned space by the airstream in the duct.

Infrared humidifiers are mounted in the duct and have infrared lamps with reflectors that reflect the infrared energy onto the water, **Figure 34.27**. The water thus evaporates rapidly into the duct airstream and is carried throughout the conditioned space. This is similar to the sun's rays shining on a lake and evaporating the water into the air. Both the electrical and the infrared types of humidifier use electrical energy.

Humidifiers are often controlled by a **humidistat**, **Figure 34.28**. The humidistat controls the motor and the heating elements and has a moisture-sensitive element, often made of hair or nylon ribbon. This material is wound around two or more bobbins and shrinks or expands, depending on the humidity. Dry air causes the element to shrink, which activates a snap-action switch and starts the humidifier. Many other control devices are used, including solid-state electronic components that vary in resistance with the humidity.

ATOMIZING HUMIDIFIERS

Atomizing humidifiers discharge tiny water droplets (mist) into the air, which evaporate very rapidly into the duct airstream or directly into the conditioned space, **Figure 34.29**. These humidifiers, which can be *spray-nozzle* or *centrifugal* types, should not be used with hard water because it contains minerals (lime, iron, etc.) that leave the water vapor as dust and will be distributed throughout the house or building. Eight to ten grains of water hardness is the maximum recommended for atomizing humidifiers. Some models are operated by a thermostat that controls a solenoid valve which turns the unit on and off. The furnace must be on and heating before this type will operate. Others, wired in parallel with the blower motor, operate when the blower motor operates. Most are also controlled with a humidistat.

The spray-nozzle humidifier sprays water through a metered bore of a nozzle into the duct airstream, where it is distributed to the occupied space. Another type sprays the water onto an evaporative media where it is absorbed by the airstream as a vapor. These humidifiers can be mounted in the plenum, under the duct, or on the side of the duct. It is generally recommended that atomizing humidifiers be mounted on the hot air or supply side of the furnace. The *centrifugal atomizing humidifier* uses an impeller or slinger to throw the

Figure 34.29 An atomizing humidifier. Courtesy AutoFlo Company

Figure 34.30 A centrifugal humidifier. Courtesy of Air System Components, Inc. Trion Division

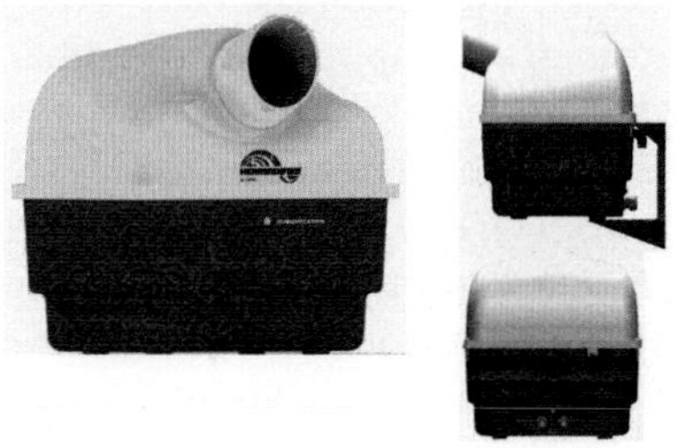

water and break it into particles that are evaporated in the airstream, **Figure 34.30.** **SAFETY PRECAUTION:** *Atomizing humidifiers should operate only when the furnace is operating or moisture will accumulate and cause corrosion, mildew, and a major moisture problem where the humidifier is located.* •

SELF-CONTAINED HUMIDIFIERS

Many residences and light-commercial buildings do not have heating equipment with ductwork through which the heated air is distributed. Hydronic heating systems, electric baseboards, or unit heaters, for example, do not use ductwork.

To provide humidification in these systems, *self-contained humidifiers* may be installed. These generally operate in the same way as those used with forced-air furnaces, using the evaporative, atomizing, or infrared processes. They may include an electric heating device to heat the water, or the water may be distributed over an evaporative media. A fan must be incorporated in the unit to distribute the moisture throughout the room or area. **Figure 34.31** illustrates a drum-type self-contained humidifier. A design using steam

Figure 34.31 A drum-type self-contained humidifier.

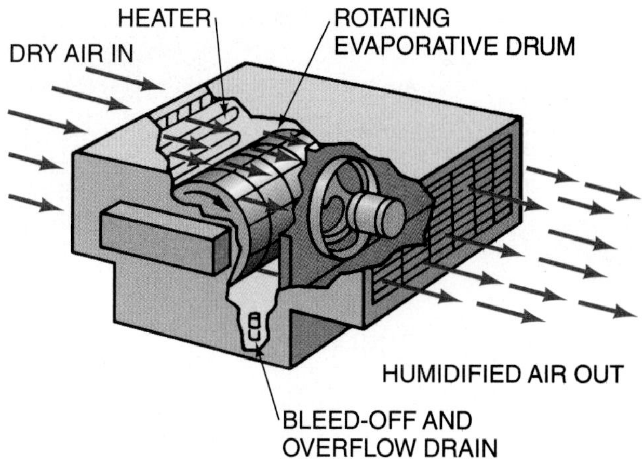

Figure 34.32 A self-contained steam humidifier.

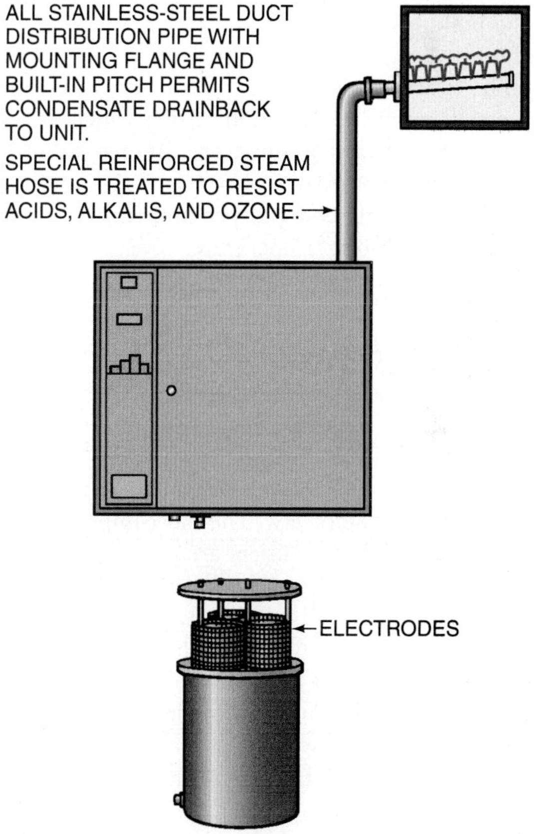

is shown in **Figure 34.32.** In this system the electrodes heat the water, converting it to steam. The steam passes through a hose to a stainless steel duct. Steam humidification is also used in large industrial applications where steam boilers are available. The steam is distributed through a duct system or directly into the air.

PNEUMATIC ATOMIZING SYSTEMS. *Pneumatic atomizing* systems use air pressure to break up and disperse the water in a mist of tiny droplets. **⚠ SAFETY PRECAUTION:** *These systems as well as other atomizing systems should be installed only where the atmosphere does not have to be kept clean or where the water has a very low mineral content, because the minerals in the water are also dispersed in the mist throughout the air. The minerals fall out and accumulate on surfaces. These systems are often used in manufacturing establishments, such as textile mills.* •

HUMIDIFIER SIZING

A humidifier of the proper size should be installed. This text emphasizes installation and service, so details for determining the size or capacity of humidifiers will not be covered. However, the technician should be aware of some general factors involved in the sizing process:

1. The number of cubic feet of space to be humidified. This is determined by taking the number of heated square feet of the building and multiplying it by the ceiling height. A 1500-ft^2 house with an 8-ft ceiling height would have 1500 $ft^2 \times 8$ ft = 12,000 ft^3.
2. The construction of the building. This includes quality of insulation, storm windows, fireplaces, building "tightness," and so on.
3. The number of air changes per hour and the approximate lowest outdoor temperature.
4. The level of relative humidity desired.

HUMIDIFIER INSTALLATION

Plan the installation carefully, including the location of the humidifier and the wiring and plumbing (with drain). A licensed electrician or plumber must provide these services where required by code or law. The most important facts to know about installing humidifiers are provided in the manufacturer's instructions. Evaporative humidifiers often are operated independently of the furnace. It is normally recommended that they be controlled by a humidistat, but it does no real harm for them to operate continuously, even when the furnace is not operating. Atomizing humidifiers, however, should not operate when the furnace and blower are not. Moisture will accumulate in the duct if they are allowed to do so.

Particular attention should be given to clearances within the duct or plenum. The humidifier should not exhaust directly onto air-conditioning coils, air filters, electronic air cleaners, blowers, or turns in the duct. If mounting on a supply duct, choose one that serves the largest space in the house. The humid air will spread throughout the house, but the process will be more efficient when given the largest distribution possible.

HUMIDIFIER SERVICE, TROUBLESHOOTING, AND PREVENTIVE MAINTENANCE

Proper service, troubleshooting, and preventive maintenance play a big part in keeping humidifying equipment operating efficiently. Cleaning the components that are in contact with the water is the most important factor. The frequency of cleaning depends on the hardness of the water: the harder the water, the more minerals in the water. In evaporative systems, these minerals collect on the media, on other moving parts, and in the reservoir. In addition, mold, algae, bacteria, and virus growth can cause problems, even to the extent of blocking the output of the humidifier. Algaecides can be used to help neutralize algae growth. The reservoir should be drained regularly if possible, and components, particularly the media, should be cleaned periodically.

Indoor air quality can be adversely affected by the humidifier because it uses water and any mold, mildew, or algae can be distributed by the moving airstream. Mold, mildew, or algae may cause allergic reactions or upper respiratory problems. One of the maintenance factors that must be attended to regularly is the cleanliness of the water reservoir and evaporative surface when an evaporative humidifier is used. These surfaces must be cleaned and sanitized on a regular basis to prevent mold and algae from accumulating. Some manufacturers have used UV lights to kill these growing molds and algae and to prevent them from contaminating the area. Follow the manufacturer's recommendations for cleaning a particular unit. If the humidifier is an atomizing type, look for mold or mildew in the duct downstream from the nozzle.

HUMIDIFIER NOT RUNNING

When the humidifier does not run, the problem is usually electrical or a component is bound tight or locked due to a mineral buildup. A locked condition may cause a thermal overload protector to open. Using typical troubleshooting techniques, check overload protection, circuit breakers, humidistat, and low-voltage controls if there are any. Check the motor to see whether it is burned out. Clean all components and disinfect if appropriate.

EXCESSIVE DUST

If it is the humidifier that is causing excessive dust, the dust will be white due to mineral buildup on the media. Clean or replace the media. If excessive dust occurs in an atomizing humidifier, the wrong equipment has been installed.

WATER OVERFLOW

Water overflow indicates a defective float valve assembly. It may need cleaning, adjusting, or replacing.

MOISTURE IN OR AROUND DUCTS

It is primarily atomizing humidifiers that cause moisture in ducts. **Remember, this equipment should operate only when the furnace operates.** Check the control to see whether it operates at other times, such as in the COOL or FAN ON modes. A restricted airflow may also cause this problem.

LOW OR HIGH LEVELS OF HUMIDITY

If the humidity level is too high or too low, check the calibration of the humidistat by using a sling psychrometer. If it is out of calibration, it may be possible to adjust it. The humidistat may need to be relocated if it is too close to a window or door. Ensure that the humidifier is clean and operating properly.

34.10 DIAGNOSTIC CHART FOR FILTRATION AND HUMIDIFICATION SYSTEMS

All forced-air systems have a filtration system of some sort. These vary from the basic air filter to the more complex electronic air filter. Most of the problems with basic air filters are caused by reduced airflow due to lack of maintenance. The problems that occur with electronic air-filter systems are more complex.

Humidification systems all contain water, and water can cause problems if the system is not properly cared for. The following chart lists some of the common problems of filtration and humidification systems.

Problem	Possible Cause	Possible Repair
Filters, Media type		
Restricted	Filters are not changed often enough	Change on a more regular schedule.
Dust entering conditioned space, possibly caused by dirty evaporator coil and fan blades	Filter media too coarse	Change to finer filter media.
	Leak in return duct after the filter	Repair leaks in return duct.
	Air bypassing filter	Install filters of the correct size.
Electronic Filter		
No power to filter	Filter interlock with fan	Establish correct fan interlock circuit.
Power to filter but filter not operating	Filter grounded in element	Clean filter.
	Contacts in filter element not making contact	Clean contacts and reassemble.
	Defective power assembly	Change power assembly.
Filter makes cracking sound too often	High dust content in air	Change to finer prefilter.
Humidifiers		
Unit not operating—depends on electricity	Faulty electrical circuits	Change fuse or reset breaker.
	Humidistat not calling for humidity	Repair faulty electrical circuits or connections.
	Interlock with fan circuit	Check humidistat calibration; adjust or change humidistat.
		Determine what type of interlock and correct.
Unit operating, but humidity is low	No water supply	Reestablish water supply.
	Defective float	Change float.
	Evaporation media saturated with minerals and will not absorb water	Change media.
	Power assembly that turns evaporation media not turning	Make sure that power is available; if it is, change unit.

REVIEW QUESTIONS

1. Which of the following is generally true regarding the amount of time people spend indoors and outdoors?
   A. People spend as much as 90% of their time indoors
   B. People spend as much as 60% of their time indoors
   C. People spend as much as 90% of their time outdoors
   D. People spend as much as 60% of their time outdoors
2. Of the following, which groups of individuals are most likely to suffer from poor indoor air quality related illnesses?
   A. Athletes and infants
   B. Chronically ill people and the elderly
   C. The elderly and construction workers
   D. Construction workers and infants
3. All of the following are methods used to control indoor contamination except
   A. cleaning the air.
   B. removing the source of contamination.
   C. providing adequate ventilation.
   D. sealing all leaks in the building envelope.
4. The ASHRAE standard that is used to determine acceptable indoor air quality is
   A. ASHRAE Standard 5.1.
   B. ASHRAE Standard 13.1.
   C. ASHRAE Standard 26.1.
   D. ASHRAE Standard 62.1.
5. The most common source of radon is
   A. plutonium.
   B. uranium.
   C. germanium.
   D. cranium.
6. How many compounds are found in tobacco smoke?
   A. 0 to 100
   B. 200 to 500
   C. 700 to 900
   D. Over 1000
7. Dust mites and mold fall under which classification?
   A. Biological contaminants
   B. Inorganic contaminants
   C. Metallic contaminants
   D. Herbicidal contaminants
8. The effects of tight building construction can be offset by
   A. installing additional weather-stripping.
   B. providing mechanical ventilation.
   C. utilizing self-contained humidifiers.
   D. keeping all interior doors open.
9. Sick Building Syndrome is a blanket term used to describe
   A. a building that has experienced severe damage to its thermal envelope.
   B. a building that is being operated/managed without a certificate of occupancy.
   C. the cause for health problems associated with poor indoor air quality.
   D. the effect that results when sick people transmit their illnesses to others.
10. For which level of mold contamination must individuals performing the remediation be trained in the handling of hazardous materials?
    A. Level I only
    B. Level I and Level II
    C. Level III only
    D. Level III and Level IV
11. For which level of mold contamination are trained maintenance staff members permitted to perform mold cleanups?
    A. Level I only
    B. Level I and Level II
    C. Level III only
    D. Level III and Level IV
12. Which of the following is true regarding mold?
    A. Only living molds can cause human health problems.
    B. Living and dead molds can cause human health problems.
    C. Oversized air-conditioning system help prevent mold growth.
    D. Molds grow only on surfaces that are visible to the naked eye.
13. Carbon monoxide is commonly formed when
    A. incomplete combustion takes place.
    B. a carbon dioxide molecule gains an additional oxygen atom.
    C. an ozone molecule looses an electron.
    D. complete combustion takes place.
14. Death from carbon monoxide can result if continued exposure exceeds
    A. 800 parts per million.
    B. 800 parts per billion.
    C. 200 parts per million.
    D. 200 parts per billion.
15. Asbestos is known to cause which of the following health problems?
    A. High blood pressure
    B. Diabetes
    C. Lung cancer
    D. Skin cancer
16. Generally speaking, ventilation has a
    A. positive effect on the efficiency of heating systems.
    B. positive effect on the efficiency of both heating and cooling systems.
    C. negative effect on the efficiency of cooling systems.
    D. negative effect on the efficiency of both heating and cooling systems.

17. A heat recovery ventilator transfers
    A. latent heat from the intake air to the exhaust air in the cooling mode.
    B. latent heat from the exhaust air to the intake air in the cooling mode.
    C. sensible heat from the intake air to the exhaust air in the cooling mode.
    D. sensible heat from the exhaust air to the intake air in the cooling mode.
18. The main difference between an HRV and an ERV is that
    A. an HRV has the ability to recover both latent and sensible heat.
    B. an ERV has the ability to recover both latent and sensible heat.
    C. an HRV has the ability to recover only latent heat.
    D. an ERV has the ability to recover only sensible heat.
19. The main drawback of using a high MERV filter is that
    A. they can cause system condenser pressure to rise.
    B. the pressure drop across them is too high.
    C. they can cause system evaporator pressure to rise.
    D. the airflow through them will be higher than desired.
20. The "V" in a filter's MERV rating stands for
    A. value.
    B. velocity.
    C. viscosity.
    D. volume.
21. The efficiency of a MERV 7 filter is approximately
    A. 7%. C. 50%.
    B. 30%. D. 70%.
22. HEPA filters can capture particles in the air that are as small as
    A. 0.3 microns in size.
    B. 3 microns in size.
    C. 30 microns in size.
    D. 300 microns in size.
23. Pleated filters are desirable because they
    A. are self-cleaning.
    B. have greater surface area than non-pleated filters.
    C. are far less expensive than non-pleated filters.
    D. can be installed in any direction.
24. The filter type that is best-suited for high humidity applications is the
    A. pocket filter.
    B. HEPA filter.
    C. high-loft polyester filter.
    D. cube filter.
25. With regards to air filtration, the term *face-loading* occurs when the
    A. back side of the filter becomes clogged with dirt.
    B. filter is installed with the arrow pointing in the wrong direction.
    C. front side of the filter becomes clogged with dirt.
    D. dirt bypasses the filter and clogs the evaporator coil.
26. Which part of the electronic air cleaner is responsible for charging the dust and dirt particles?
    A. The prefilter
    B. The post-filter
    C. The ionizing section
    D. The collection plate
27. Ion generators are used to promote clean environments by causing contaminants to
    A. bond to each other, making them larger and easier to capture in air filters.
    B. repel each other, making them easier to capture in air filters.
    C. stick to the interior surfaces of the return duct, removing them from the airstream.
    D. disintegrate, reducing their ability to negatively affect indoor air quality.
28. Air cleaning systems that utilize ultraviolet light target which type of UV?
    A. UV-A
    B. UV-B
    C. UV-C
    D. UV-D
29. Care should be taken when handling UV lamps because they contain
    A. cyanide.
    B. mercury.
    C. acid.
    D. sulfur dioxide.
30. To help keep the entire return duct clean and dirt free,
    A. the air filter should be as close to the air handler as possible.
    B. the air filter should be as far from the air handler as possible.
    C. the air filter should be smaller than the filter rack.
    D. the air filter should be larger than the filter rack.
31. An electronic precipitator is the same as a(n)
    A. extended surface air filter.
    B. fiberglass media filter.
    C. electronic air cleaner.
    D. washable steel air filter.
32. The recommended relative humidity for a home is between
    A. 50% and 70%.
    B. 20% and 40%.
    C. 30% and 50%.
    D. 40% and 60%.
33. The plenum-mount humidifier is mounted
    A. only in the supply plenum.
    B. only in the return plenum.
    C. in either the supply plenum or the return plenum.
    D. in neither the supply plenum nor the return plenum.

34. What is the driving force that moves air through a bypass humidifier?
    A. The pressure difference between the supply and return airstreams
    B. The propeller-type fan in the humidifier
    C. The forward-curved centrifugal blower in the humidifier
    D. The blower in the air handler or furnace
35. All of the following should be taken into account when sizing humidifiers except
    A. volume of the structure.
    B. building construction.
    C. desired indoor air temperature.
    D. the number of air changes per hour in the structure.
36. If an air filter is installed in a ceiling-mounted return grille and the evaporator coil is dirty, which of the following is not a possible cause?
    A. Air leaks in the return duct
    B. Air leaks in the supply duct
    C. Improperly sized filter
    D. Poor filter quality/application
37. In the winter months, humidification is often needed because
    A. the heating system removes moisture from the air in the heated space.
    B. high levels of humidity will make the occupants of the space colder.
    C. the relative humidity of air decreases as it is heated.
    D. the absolute humidity of the air increases as it is heated.
38. List the system components that should be cleaned in conjunction with a complete "duct-cleaning" process.
39. Briefly explain the difference between *absorption* and *adsorption*.
40. Briefly explain how ventilation helps to improve indoor air quality.
41. List five sources or types of indoor air pollution that can be directly related to HVACR equipment.
42. List two possible causes for dust entering the conditioned space.
43. List five possible causes for the relative humidity to be too low in a heated space, even if the humidifier is operating.

SECTION 7

AIR-CONDITIONING (COOLING)

UNITS

UNIT 35

COMFORT AND PSYCHROMETRICS

OBJECTIVES

After studying this unit, you should be able to

- recognize the four factors that affect comfort.
- explain how the body absorbs and rejects heat.
- explain how comfort guidelines change with the seasons.
- describe what is meant by a comfort chart.
- explain the concepts of density and specific volume as they apply to the study of air.
- describe two common types of humidity and how each of them is expressed.
- discuss the concept of superheated gases as it applies to the primary elements in air.
- define wet-bulb and dry-bulb temperature.
- define dew point temperature.
- define enthalpy as it applies to the study of air.
- Identify the lines and scales on a psychrometric chart.
- plot air conditions on a psychrometric chart.
- plot an air-conditioning process on a psychrometric chart.
- explain the difference between infiltration and ventilation.
- use the sensible heat, latent heat, and total heat formulas to evaluate air-conditioning and ventilation processes.

35.1 COMFORT

Comfort describes a delicate balance of pleasant feeling in the body. A comfortable atmosphere describes our surroundings when we are not aware of discomfort. Providing a comfortable atmosphere for people is the job of the heating and air-conditioning profession. Comfort involves four things: (1) temperature, (2) humidity, (3) air movement, and (4) air cleanliness, **Figure 35.1**.

The human body has a sophisticated control system for both protection and comfort. When we move from a warm house to the cold outdoors, the body begins to adjust to the changing conditions. If we move from a cool house to the warm outdoors on a hot summer day, the body will also adjust itself to keep comfortable and prevent us from overheating. Body adjustments are accomplished by the circulatory and respiratory systems. When the body gets too warm, the vessels next to the skin dilate to get the blood closer to the surrounding air in an effort to increase the heat exchange with the air. This is why you turn red when you become hot. If this does not cool the body, you will perspire. When the perspiration evaporates, it takes heat from the body and cools it. Excess perspiration in hot weather explains why you must drink more fluids.

Figure 35.1 Comfort is described as a balance of air temperature, humidity, air movement, and air cleanliness.

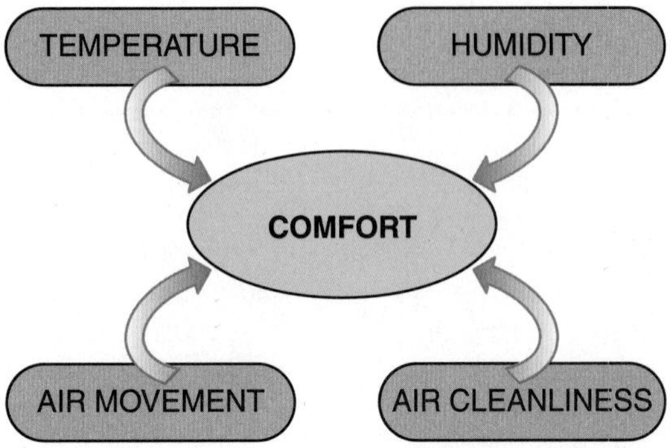

35.2 FOOD ENERGY AND THE BODY

The human body may be compared to a coal-burning boiler. Coal is burned in the boiler to create heat energy. Supplying food to the human body can be equated to adding coal to the boiler. When the coal is burned and converted to heat, some of this energy goes up the flue or chimney, some escapes to the surroundings, and some is carried away in the ashes. If fuel is added to the fire and the heat cannot be dissipated, the boiler will overheat.

The body uses food to produce energy. Some energy is stored as fatty tissue, some leaves as waste, some leaves as heat, and some is used as energy to keep the body functioning. If the body needs to dissipate some of its heat to the surroundings and cannot, it will overheat. The average core body temperature is 98.6°F. The surface temperature of the body, or the temperature of our skin, is considerably lower than 98.6°F and varies according to the temperature of its surroundings. So, if someone is standing outside in 100°F air, the body will not be able to reject heat to its surroundings, **Figure 35.2**. Instead, the body will actually absorb

Figure 35.2 The body will absorb heat if the surroundings are warmer than the body.

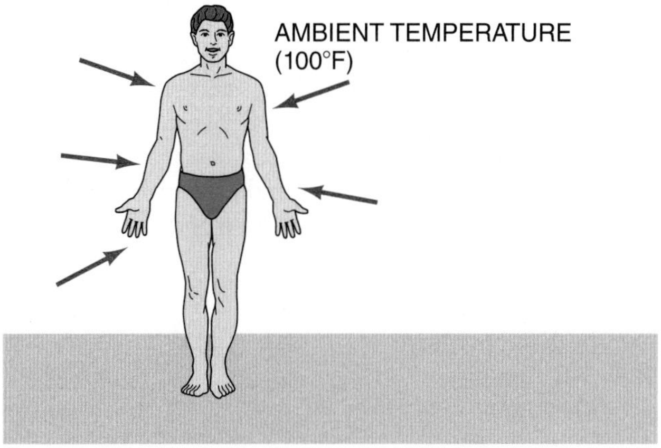

Figure 35.3 Average body temperature is 98.6°F.

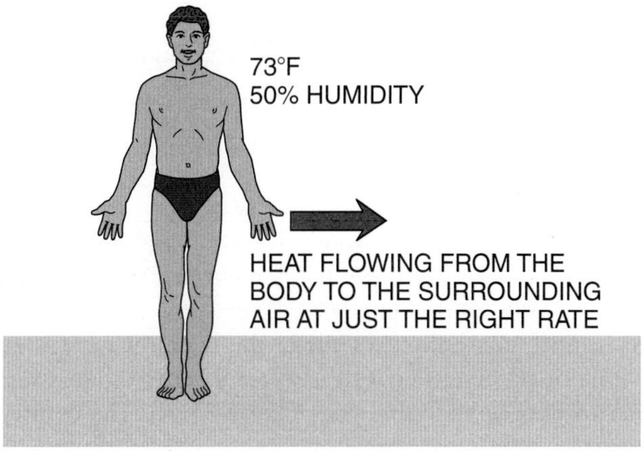

heat. When such is the case, the body's internal temperature-control system will cause perspiration to form on the surface of the skin in an attempt to cool the body. As the surrounding temperature drops, the body will reject heat naturally, as a warmer substance transfers heat to cooler ones, **Figure 35.3**. We are comfortable when our body transfers heat to its surroundings at the correct rate.

35.3 HEAT TRANSFER TO AND FROM THE BODY

The body gives off and absorbs heat by the three methods: conduction, convection, and radiation, **Figure 35.4**. The evaporation of perspiration could be considered a fourth way, **Figure 35.5**. When the surroundings are at a particular comfort condition, the body is giving up heat at a steady rate that is comfortable. The surroundings must be cooler than

Figure 35.4 The three direct ways the human body gives off heat. (A) Conduction. (B) Convection. (C) Radiation.

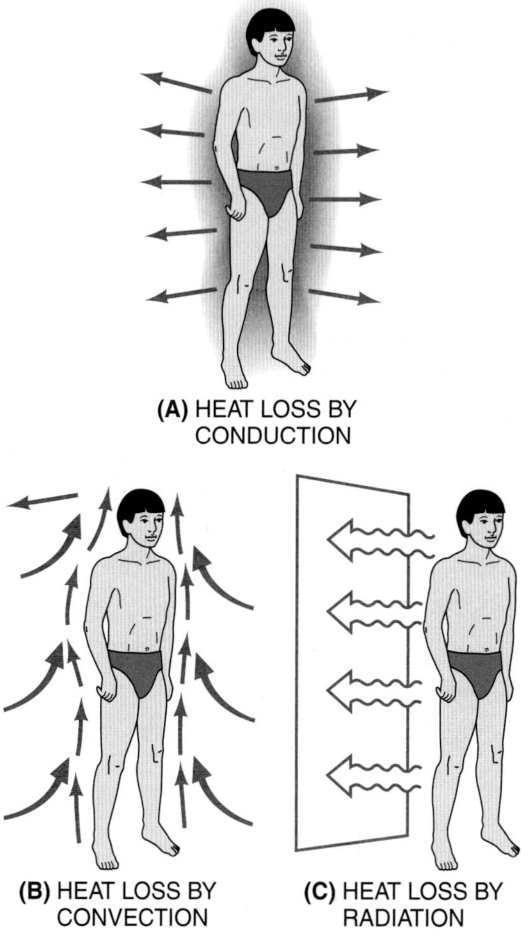

Figure 35.5 A fourth way the body gives off heat—the evaporation of perspiration.

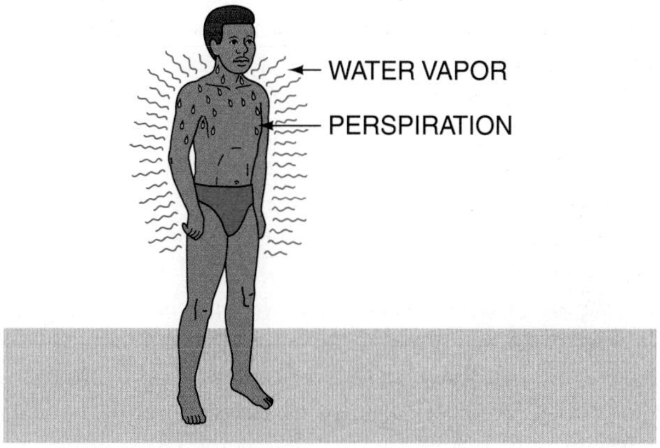

the body for the body to be comfortable. Typically, when the body is at rest (sitting) and in surroundings of 75°F and 50% humidity with a slight air movement, the body is close to

being comfortable. In cooler weather, a different set of conditions applies (e.g., we wear more clothing). The following statements can be used as guidelines for comfort.

1. In winter:
   A. The effects of lower temperatures can be offset by a higher relative humidity.
   B. The lower the relative humidity is, the higher the temperature must be.
   C. Air movement is more noticeable.
2. In summer:
   A. Air movement helps reduce the negative effects of high relative humidity.
   B. The effects of higher temperatures can be offset with a lower relative humidity.
3. The comfort conditions in winter and in summer are different.
4. Styles of clothes in different parts of the country make a difference in the conditioned space-temperature requirements for comfort. For example, in Maine, the clothing worn during the winter months would be warmer than the clothing worn during the winter in Georgia, so the temperatures of similar interior spaces in each location will be somewhat different.
5. Body metabolism varies from person to person, based on a number of factors including gender and age. Women, for example, are not as warm-bodied as men and, as a result, typically have less tolerance than men do for large temperature swings. The circulatory system generally does not function as well in older people as compared to younger people.

35.4 THE COMFORT CHART

The chart in **Figure 35.6**, often called a *generalized comfort chart*, can be used to compare one situation with another. **Figure 35.6(A)** represents the comfort chart for the summer months. In the center of this chart there is a bold rectangular area that represents the temperature and humidity combinations at which most people will be comfortable. The air temperatures are located along the bottom of the chart, and the (relative) humidity percentages are shown by the curved lines as indicated in **Figure 35.7**.

From the comfort chart for the summer months, it can be determined that the average individual will be comfortable if the surrounding air is 75°F with a relative humidity of 50%, **Figure 35.8**. This is because the 75°F line and the 50% relative-humidity line cross at a point that is located within the "comfort" zone. It can also be determined that the average person will *not* be comfortable if the surrounding air is 85°F and the relative humidity is 50%, **Figure 35.9**. This is because these two lines intersect at a point that is outside the comfort region. The same concept applies to the comfort chart for the winter months, as shown in **Figure 35.6(B)**.

A technician may use charts like those shown in **Figure 35.6** to plot and compare comfort conditions in an occupied space. In general, the closer the plot falls to the middle of the chart, the more people would be comfortable at those conditions. Notice that there is a different chart for summer and winter conditions. The charts shown in **Figures 35.8** and **35.9** show only the relationship between air temperature and relative humidity. There are many other properties of air that affect comfort levels. The complete graph is referred to as a psychrometric chart and compiles information about the following air characteristics:

- Dry-bulb temperature
- Wet-bulb temperature
- Relative humidity
- Specific volume
- Moisture content
- Heat content
- Dew point temperature

Each of these items will be discussed in more detail in this unit. Section 35.12, "Plotting on the Psychrometric Chart," will explain how to use the chart to determine the various characteristics of air at different sets of conditions.

35.5 PSYCHROMETRICS

The study of air and its properties is called psychrometrics. When we move through a room, we are not aware of the air inside the room, but the air has weight and occupies space like water in a swimming pool. The water in a swimming pool is more dense than the air in the room; it weighs more per unit of volume. Perfectly dry air weighs 0.075 lb/ft^3 at standard atmospheric pressure, **Figure 35.10**. The density of water is 62.4 lb/ft^3.

Air, like water, offers resistance to movement. To prove this, take a large piece of cardboard and try to swing it around with the flat side moving through the air, **Figure 35.11**. It is hard to do because of the resistance of the air. The larger the area of the cardboard, the more resistance there will be. For example, if you took a large piece of cardboard outside on a windy day, the wind will try to take the cardboard from you. The cardboard acts like a sail on a boat, **Figure 35.12**. Air not only offers resistance to movement, it also takes up space. If you invert an empty glass and push it down in water, you can observe that the air completely fills the glass and, by creating resistance, prevents the water from flowing into the glass, **Figure 35.13**.

The weight of air in a room can be calculated by multiplying the room volume by the weight of a cubic foot of air. In a room 10 ft × 10 ft × 10 ft, the volume is 1000 ft^3, so the room air weighs 1000 $ft^3 \times 0.075\ lb/ft^3 = 75$ lb, **Figure 35.14**. It is important to remember that the 0.075 lb/ft^3 that is used in this example assumes that the air is at standard atmospheric conditions and is *dry*. Deviations from these

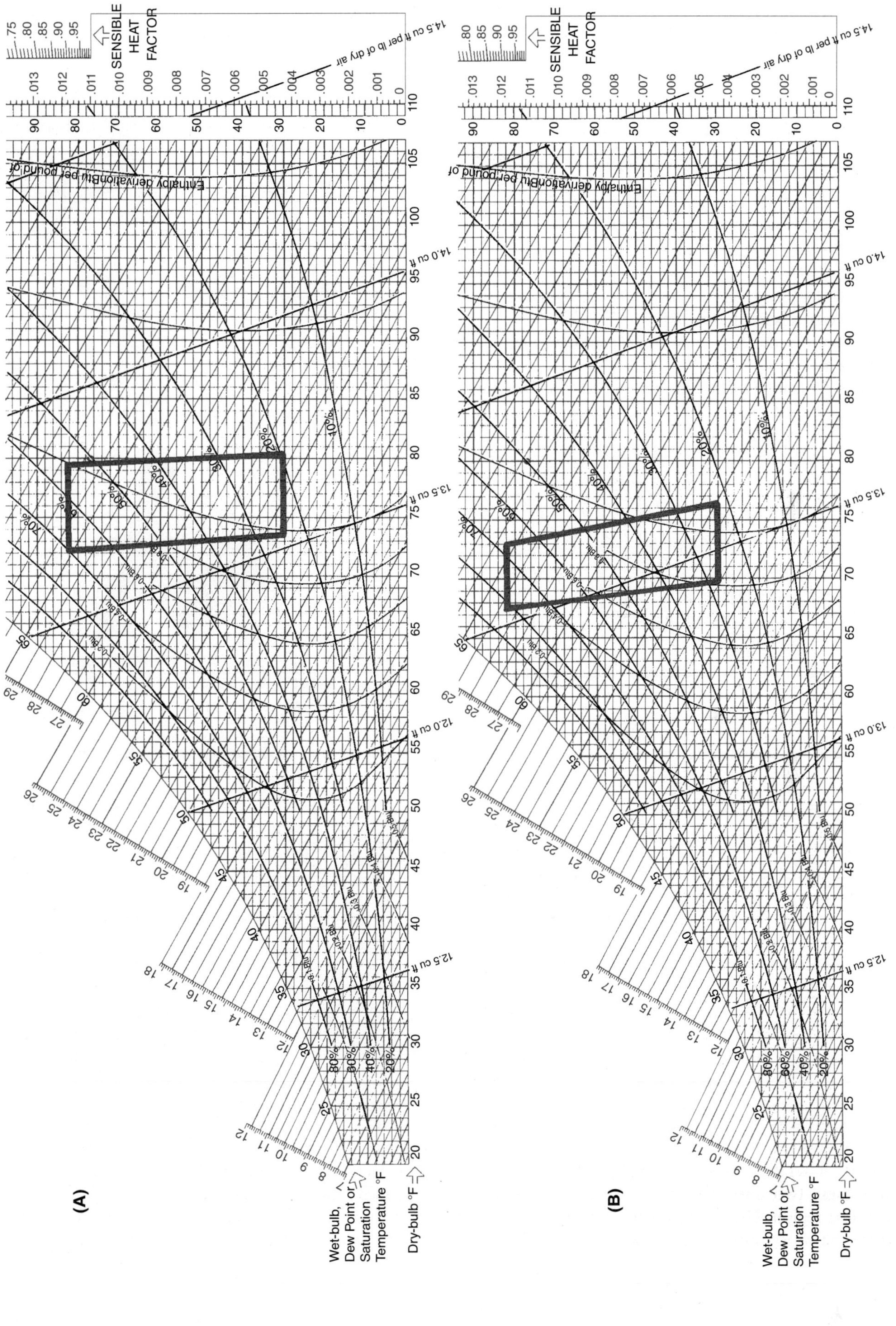

Figure 35.6 Comfort charts for (A) summer and (B) winter. Adapted from Carrier Corp. Psychrometric Chart

Figure 35.7 Combinations of temperature and humidity make up the boundaries of the comfort charts.

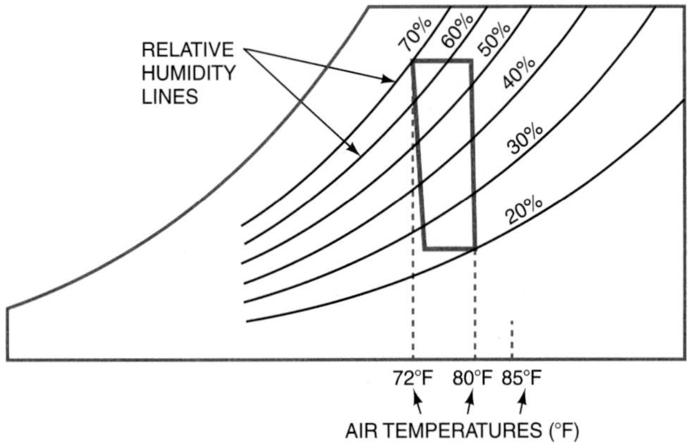

Figure 35.8 This combination of temperature and humidity lies within the boundaries of the comfort chart.

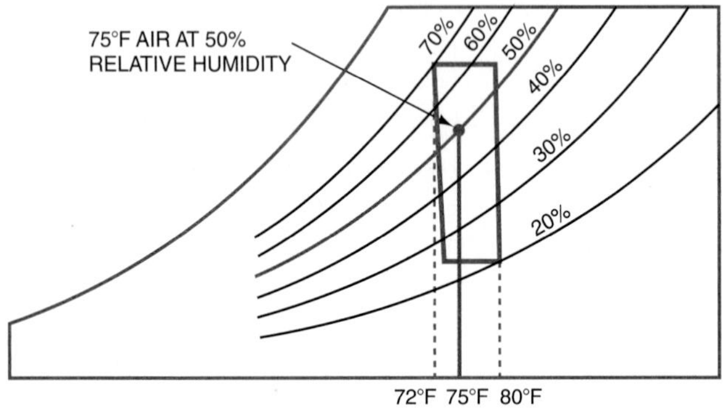

Figure 35.9 This combination of temperature and humidity lies outside the boundaries of the comfort chart.

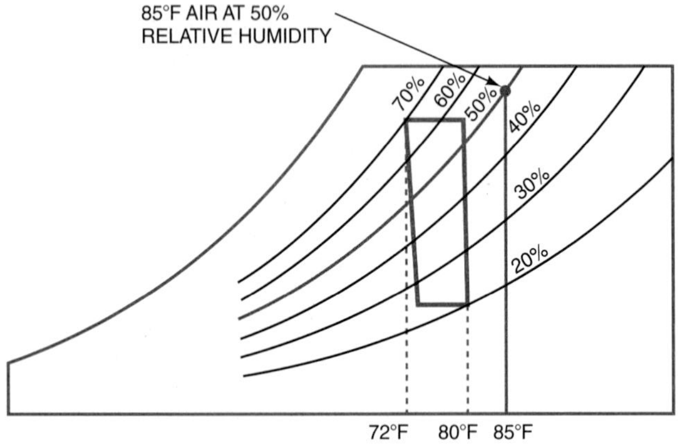

conditions will have an effect on the weight of the air. The number of cubic feet of air to make a pound of air can be obtained by taking the reciprocal of the density. The reciprocal of a number is 1 divided by that number. The reciprocal of the density of air from the previous example is 1 divided by 0.075, or 13.33 ft³/lb of air, **Figure 35.15**. The reciprocal

Figure 35.10 The difference between the weight of air and the weight of water.

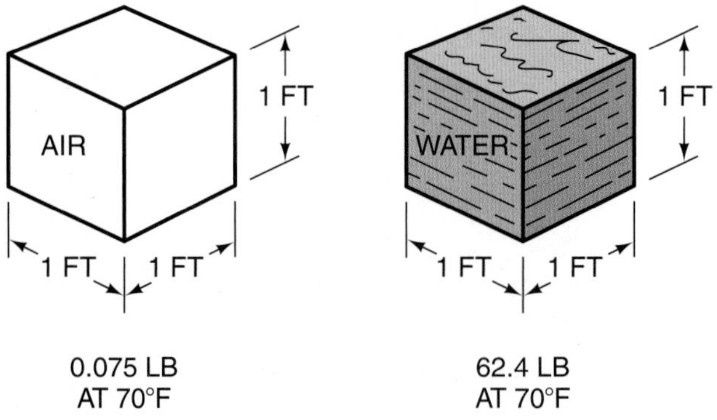

Figure 35.11 This man is having a hard time swinging a piece of cardboard around because the air around him is taking up space and causing resistance to the movement of the cardboard.

Figure 35.12 This man is walking in a breeze. He is pushed along by the breeze against the cardboard because the air has weight and takes up space.

Figure 35.13 Air creating resistance that prevents water from entering the glass.

Figure 35.14 This room is 10 ft × 10 ft × 10 ft. The volume of the room is 1000 ft³ of air that weighs 0.075 lb/ft³. 1000 ft³ × 0.075 lb = 75 lb of air in the room.

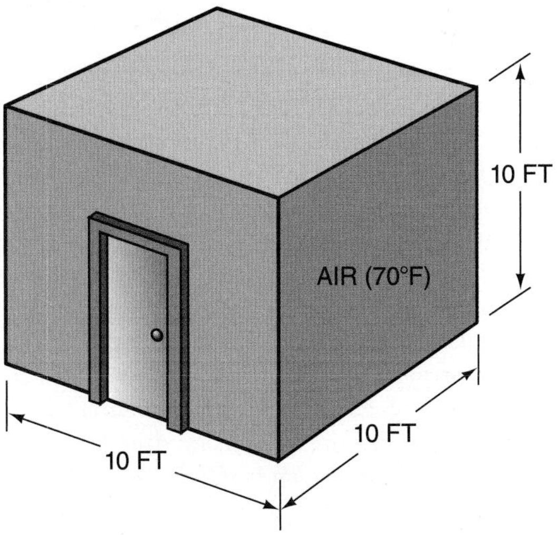

of density is known as the specific volume, or the volume that 1 lb of a substance occupies. Both density and specific volume values are used in many calculations. Multiplying the density of a gas and its specific volume together will always produce a result of 1.

35.6 ABSOLUTE AND RELATIVE HUMIDITY

Air consists of approximately 78% nitrogen, 21% oxygen, and 1% other gases, **Figure 35.16**. But air is not totally dry. Surface water and rain keep moisture in the atmosphere everywhere at all times, even in a desert. Two-thirds of the earth's surface is covered with water. Water, in the form of low-pressure vapor, is suspended in the air and is called *humidity,* **Figure 35.17**. Two types of humidity that are studied are absolute humidity and relative humidity.

Figure 35.15 At standard atmospheric pressure, it will take 13.33 cubic feet of air to make up 1 lb of air.

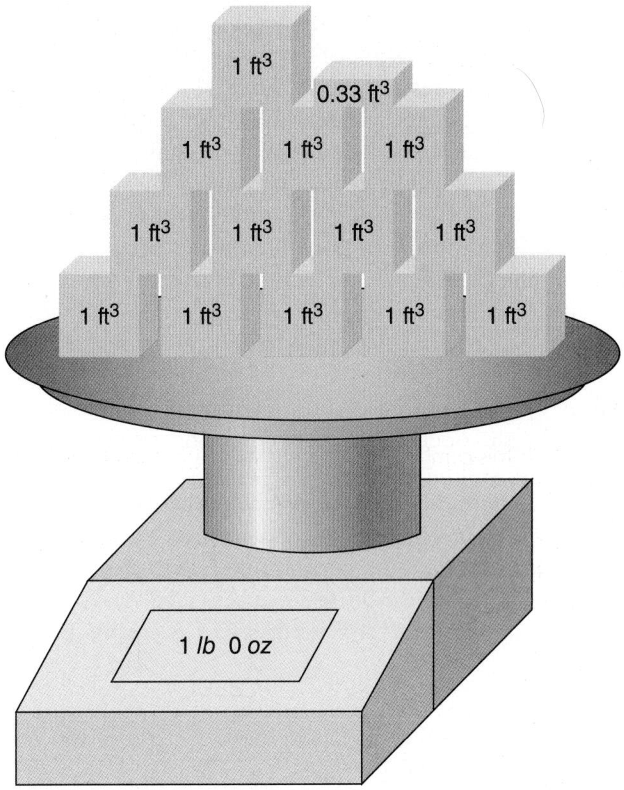

Figure 35.16 Substances that make up air.

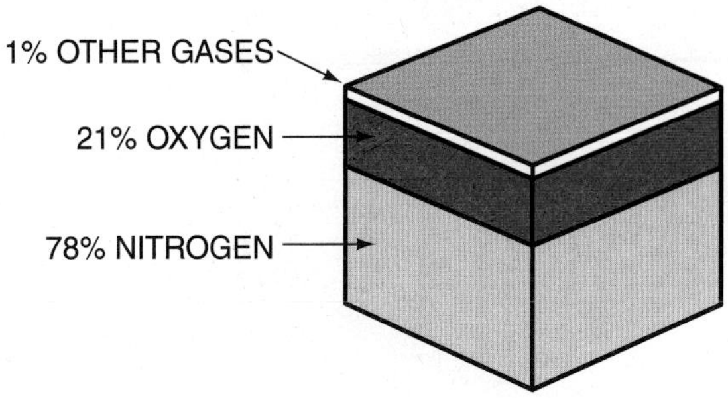

Figure 35.17 Moisture suspended in air.

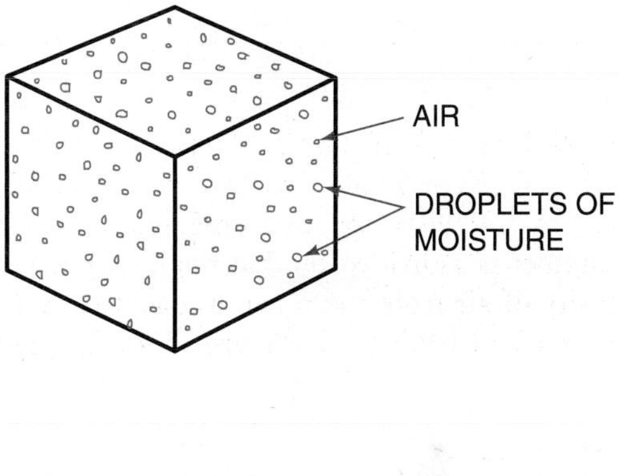

The moisture content in air (humidity) is measured by weight and expressed in pounds or grains (7000 grains = 1 pound) per pound of air. Air can hold very little water vapor. Each pound of air at sea level (standard atmospheric pressure) at a temperature of 70°F can hold a maximum of 110.5 grains of moisture, or 0.01578 (110.5/7000) pounds of water. The amount of water vapor that an air sample can hold will change as the air conditions change. The actual amount of water vapor that is contained in an air sample is referred to as the *absolute* humidity and is expressed in either grains per pound or pounds of water per pound of air. For example, a particular air sample that contains 10 grains of moisture per pound of air has an absolute humidity of 10 grains per pound. Absolute humidity is not commonly used for field calculations. Relative humidity, on the other hand, is very useful and is often used to determine air conditions and the operating effectiveness and efficiency of air-conditioning equipment.

Relative humidity is the relationship between the weight of water vapor in a pound of air compared to the weight of water vapor that a pound of air could hold at a 100% saturated condition, **Figure 35.18. Figure 35.18(A)** shows that 100% saturated air at 70°F holds 110.5 grains of moisture per pound, and **Figure 35.18(B)** shows that 50% saturated air will hold 55.25 grains at the same temperature. The air in **Figure 35.18(A)** has a relative humidity of 100%, while the air sample in **Figure 35.18(B)** has a relative humidity of 50%.

Figure 35.18 (A) This air sample is 100% saturated, so the relative humidity is 100%. (B) This air sample is 50% saturated, so the relative humidity is 50%.

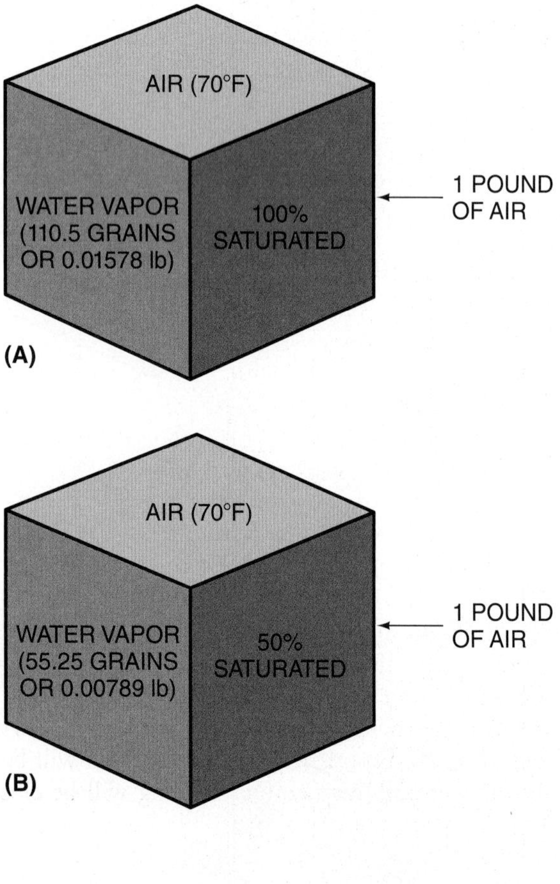

Large buildings typically have a set of specifications that tell the technician what the design conditions are for the occupied space. The HVACR system should be able to maintain these conditions when the system and controls are set and operating correctly. When there are complaints in certain areas of the building, the technician may be called on to determine whether these spaces are within the correct temperature and humidity ranges. Often, the only way to settle a dispute is to plot the air conditions on a comfort chart or compare the readings with the building's specifications. The technician must be prepared to accurately measure and record the conditions that are present in order to help determine what the problem is.

35.7 SUPERHEATED GASES IN AIR

Because air is made up of several different gases, it is not a pure element or gas. As we have mentioned, air is made up of nitrogen (78%), oxygen (21%), and approximately 1% of other gases, **Figure 35.16.** These gases are highly superheated. Nitrogen, for instance, boils at −319°F, and oxygen boils at −297°F at atmospheric pressure, **Figure 35.19.** Hence, nitrogen and oxygen in the atmosphere are *superheated gases*—they are superheated several hundred degrees above absolute zero (0° Rankine). Each gas exerts pressure according to Dalton's law of partial pressures. This law states that each gas in a mixture of gases acts independently of the other gases and the total pressure of a gas mixture is the sum of the pressures of each gas in the mixture.

Water vapor suspended in air is a gas that exerts its own pressure and occupies space with the other gases. This

Figure 35.19 If you place a container of liquid nitrogen and a container of liquid oxygen in a room, the nitrogen and oxygen will start to evaporate or boil.

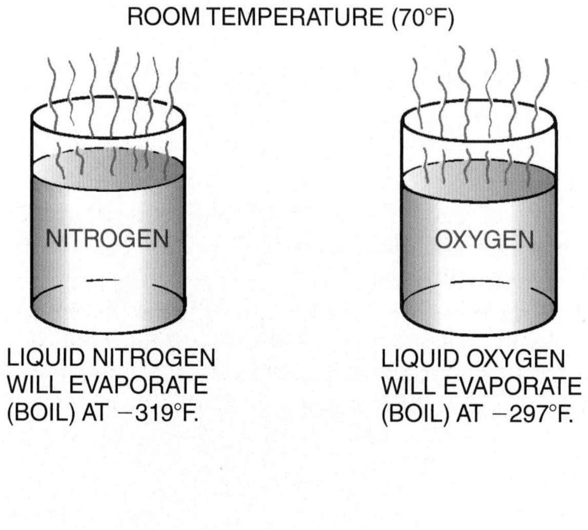

pressure, known as the water vapor pressure, changes with temperature. As the temperature increases, the water vapor pressure increases. Consider the situation where a dish of water is resting on a table. If the water vapor pressure in the air is lower than the water vapor pressure in the dish, the water in the dish will evaporate slowly to the lower pressure area of the water vapor in the air. Values for the vapor pressure of moisture in air can be found in some psychrometric charts and in saturated water tables. If the water vapor pressure in the dish is lower than the pressure of the vapor in the air, water vapor from the air will condense into the water in the dish. This can occur if the water in the dish has ice added to it and the vapor pressure is lowered as a result of the reduced water temperature.

When water vapor is suspended in the air, the air is sometimes called "wet air." If the air has a large amount of moisture, the moisture can be seen (e.g., fog or a cloud). Actually, the air is not wet because the moisture is suspended in the air. This could more accurately be called a nitrogen, oxygen, and water vapor mixture.

Figure 35.20 Wet- and dry-bulb thermometer bulbs.

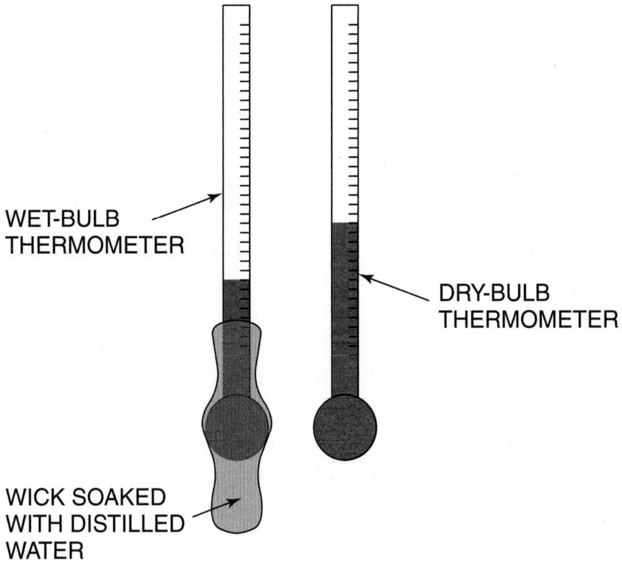

Figure 35.21 Sling psychrometer.

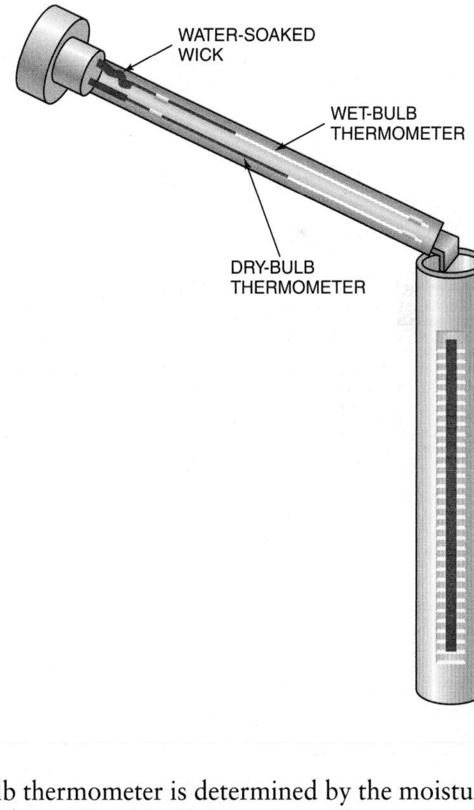

35.8 DRY-BULB AND WET-BULB TEMPERATURES

Up to this point in the unit, we have made several references to the temperature of air. When doing so, we have been referring to the level of heat intensity that is obtained when measured by a standard thermometer. In actuality, there are two types of temperature readings that are very useful to the field technician. These are the *dry-bulb* temperature and the *wet-bulb* temperature.

The moisture content of air can be checked by using a combination of dry-bulb and wet-bulb temperatures. *Dry-bulb* temperature is the sensible-heat level of air and is taken with an ordinary thermometer. Wet-bulb temperature is measured by a thermometer with a wick on the end that is soaked with distilled water, **Figure 35.20**. Both the wet-bulb and dry-bulb temperature readings are obtained at the same time by using an instrument referred to as a psychrometer. Some psychrometers are referred to as sling psychrometers, **Figure 35.21**, that require that the thermometers be spun in an airstream. Digital or electronic psychrometers, **Figure 35.22**, can be positioned in a moving airstream and do not need to be spun manually. The reading from a wet-bulb thermometer takes into account the moisture content of the air and reflects the total heat content of air.

To take accurate wet- and dry-bulb temperature readings, the sling psychrometer is spun in the air. The wet-bulb thermometer will get cooler than the dry-bulb thermometer because of the evaporation of the distilled water. Distilled water is used because some water has undesirable mineral deposits. Some minerals will change the boiling temperature. The rate at which water will evaporate from the wick on the wet-bulb thermometer is determined by the moisture content of the surrounding air. If the surrounding air is very dry, the moisture will evaporate quickly, causing the wet-bulb temperature to drop lower. If the surrounding air is very wet (high relative humidity) the rate of evaporation will be very low and the wet-bulb temperature reading will be closer to

Figure 35.22 A digital psychrometer. Photo by Eugene Silberstein.

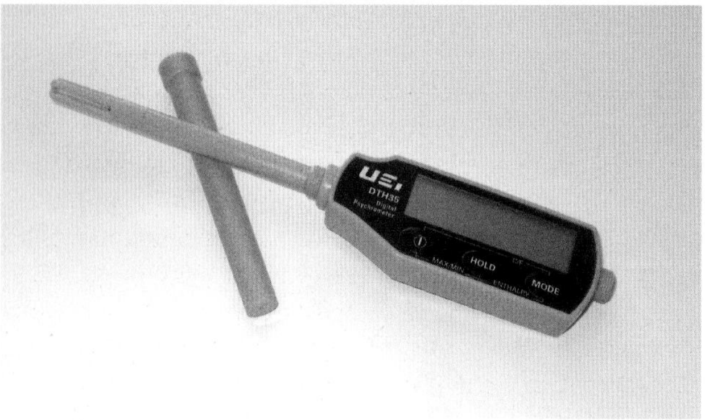

the dry-bulb temperature reading. The closer the wet-bulb and dry-bulb temperatures are to each other, the higher the relative humidity. For example, if the wet-bulb and dry-bulb temperature readings are both 78°F, the relative humidity will be 100%. This is true because, if the air is saturated with moisture, no moisture can evaporate from the wetted wick on the wet-bulb thermometer and no reduction in wet-bulb temperature will result. The wet-bulb temperature can never be higher than the dry-bulb temperature. If it ever measures higher, there is a mistake.

The difference between the dry-bulb reading and the wet-bulb reading is called the *wet-bulb depression.* **Figure 35.23** shows a wet-bulb depression chart. As the amount of moisture suspended in the air decreases, moisture from the wet-bulb thermometer's wick will evaporate and the wet-bulb temperature reading will drop. This will cause the wet-bulb depression to increase. As the moisture content in the air increases, the rate of evaporation from the wick will decrease and the wet-bulb depression will be low. For example, a room with a dry-bulb temperature of 76°F and a wet-bulb temperature of 64°F has a wet-bulb depression of 12°F (76 degrees − 64 degrees) and a relative humidity of 52%. This is found by locating the point where the 76-degree dry-bulb temperature line (left side of the table in **Figure 35.23**) and the 12-degree wet-bulb depression line (across the top of the chart in **Figure 35.23**) cross. If the 76°F dry-bulb temperature is maintained and moisture is added to the room so that the wet-bulb temperature rises to 74°F, the relative humidity increases to 91% and the new wet-bulb depression is 2°F. If the wet-bulb depression is allowed to go to 0°F (e.g., 76°F dry bulb and 76°F wet bulb), the relative humidity will be 100%. The air is holding all of the moisture it can—it is *saturated* with moisture.

35.9 DEW POINT TEMPERATURE

The dew point temperature is the temperature at which moisture begins to condense out of the air. If a glass of ice water is placed on a table in a 75°F room that has a relative humidity of 50%, moisture will begin to form on the surface of the glass, **Figure 35.24**. Moisture from the room will also collect in the water in the glass and its level will begin to rise. It is

Figure 35.23 A wet-bulb depression chart.

DB	WB DEPRESSION																													
TEMP.	1	2	3	4	5	6	7	8	9	10	11	12	13	14	15	16	17	18	19	20	21	22	23	24	25	26	27	28	29	30
32	90	79	69	60	50	41	31	22	13	4																				
36	91	82	73	65	56	48	39	31	23	14	6																			
40	92	84	76	68	61	53	46	38	31	23	16	9	2																	
44	93	85	78	71	64	57	51	44	37	31	24	18	12	5																
48	93	87	80	73	67	60	54	48	42	36	34	25	19	14	8															
52	94	88	81	75	69	63	58	52	46	41	36	30	25	20	15	10	6	0												
56	94	88	82	77	71	66	61	55	50	45	40	35	34	26	24	17	12	8	4											
60	94	89	84	78	73	68	63	58	53	49	44	40	35	31	27	22	18	14	6	2										
64	95	90	85	79	75	70	66	61	56	52	48	43	39	35	34	27	23	20	16	12	9									
68	95	90	85	81	76	72	67	63	59	55	51	47	43	39	35	31	28	24	21	17	14									
72	95	91	86	82	78	73	69	65	61	57	53	49	46	42	39	35	32	28	25	22	19									
76	96	91	87	83	78	74	70	67	63	59	55	52	48	45	42	38	35	32	29	26	23									
80	96	91	87	83	79	76	72	68	64	61	57	54	51	47	44	41	38	35	32	29	27	24	21	18	16	13	11	8	6	1
84	96	92	88	84	80	77	73	70	66	63	59	56	53	50	47	44	41	38	35	32	30	27	25	22	20	17	15	12	10	8
88	96	92	88	85	81	78	74	71	67	64	61	58	55	52	49	46	43	41	38	35	33	30	28	25	23	21	18	16	14	12
92	96	92	89	85	82	78	75	72	69	65	62	59	57	54	51	48	45	43	40	38	35	33	30	28	26	24	22	19	17	15
96	96	93	89	86	82	79	76	73	70	67	64	61	58	55	53	50	47	45	42	40	37	35	33	31	29	26	24	22	20	18
100	96	93	90	86	83	80	77	74	71	68	65	62	59	57	54	52	49	47	44	42	40	37	35	33	31	29	27	25	23	21
104	97	93	90	87	84	80	77	74	72	69	66	63	61	58	56	53	51	48	46	44	41	39	37	35	33	31	29	27	25	24
108	97	93	90	87	84	81	78	75	73	70	67	64	62	59	57	54	52	50	47	45	43	41	39	37	35	33	31	29	28	26

left as a student exercise to verify, by using the psychrometric chart, that the dew point temperature of air at 75°F and 50% relative humidity is 55.5°F. Air can be dehumidified by passing it over a surface that is below its dew point temperature; moisture will collect on the cold surface, for example, an air-conditioning coil, **Figure 35.25**. The condensed moisture is drained, and this is the moisture that you see running out of the condensate line of an air conditioner, **Figure 35.26**.

35.10 ENTHALPY

As air is heated or cooled, the heat content of the air changes. When cooled, the air's heat content decreases, and when heated, the air's heat content increases. The enthalpy of air is the measure of heat content and is expressed in Btu per pound of air. Enthalpy is the total heat content of air and has both sensible- and latent-heat components.

Figure 35.24 The glass was gradually cooled until beads of water began to form on the outside of the glass.

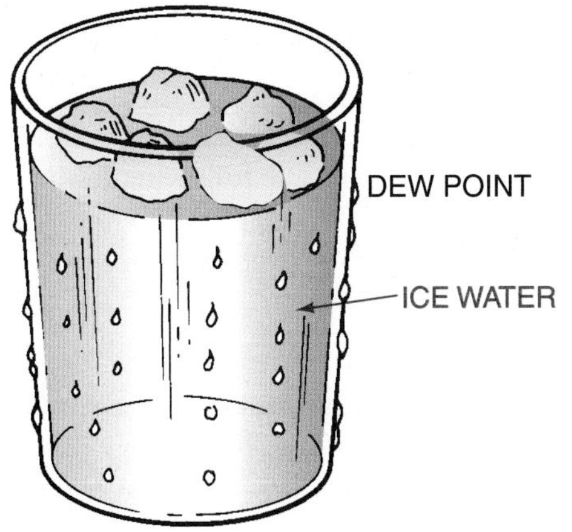

Given the conditions of an air sample, the heat content of the sample can be determined by using the psychrometric chart, which will be discussed in the next section. If the conditions of an airstream are known both before and after the air passes through the heating or cooling equipment, we can easily determine how much heat has been either added or removed from the air as it passes through the heat-transfer surface. This information, together with some other basic calculations, will enable the service technician to verify the capacity and operating efficiency of the system. As we will soon see, enthalpy and wet-bulb temperatures are closely related. So, by measuring the wet-bulb temperature of the air before and after it moves through a heat exchanger, we can get a good idea of how much heat energy, in Btu/lb, is being transferred. If the amount of air moving through the

Figure 35.26 The water dripping from the back of this window air conditioner is moisture that was collected from the room onto the cooling coil of the unit.

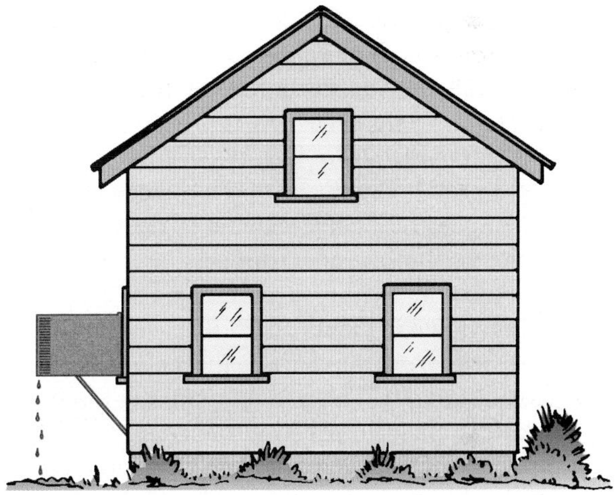

Figure 35.25 The cold surface of the evaporator coil causes moisture to condense from the air as the air passes over the surface of the coil.

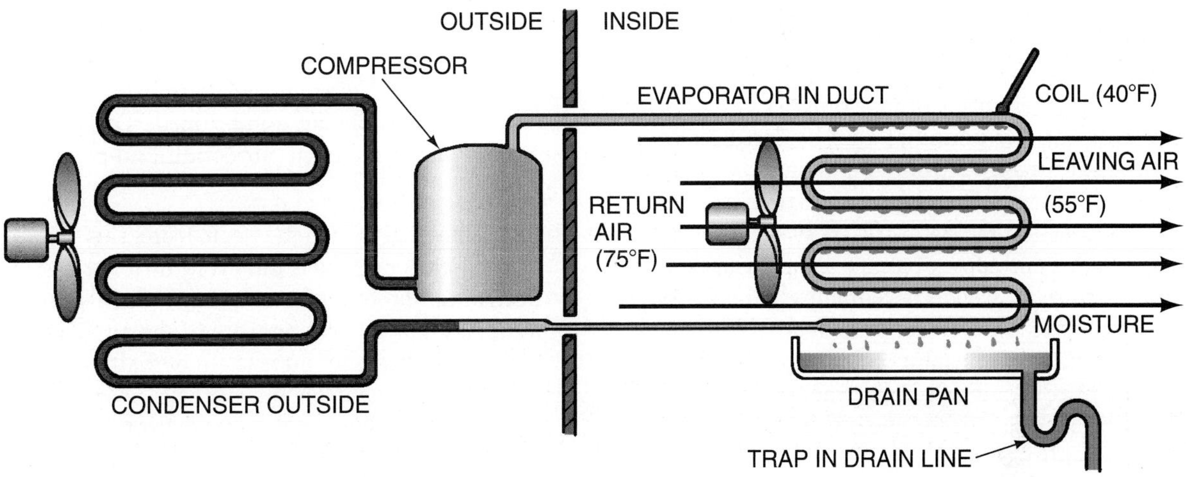

heat exchanger, in cubic feet per minute, is known, we can formulate a fairly accurate estimate of the performance of the heat exchanger.

35.11 THE PSYCHROMETRIC CHART

All of the properties of air are compiled on the psychrometric chart, **Figure 35.27**. The bottom scales on the chart represent the sensible-heat portion, and the vertical scales on the right-hand side of the chart represent the latent-heat portion. Just to recap, sensible heat is the term used to describe a heat transfer that can be measured with a thermometer, while a latent-heat transfer is one that represents a change in state of a substance with no change in temperature. Since enthalpy is the total heat, the enthalpy scale, which is located on the curved top portion of the chart, is a combination of the latent-heat (vertical) and sensible-heat (horizontal) properties of the air. The chart looks complicated, but a clear plastic straightedge and a pencil will help you understand it.

The vertical lines extending from the bottom of the chart to the top represent lines of constant dry-bulb temperature, **Figure 35.28**. The dry-bulb temperature scale is found across the bottom of the chart. As a process moves up and down a dry-bulb temperature line, the dry-bulb temperature remains the same. Moving to the right indicates an increase in dry-bulb temperature, while moving to the left indicates a decrease in the dry-bulb temperature.

The wet-bulb temperature is represented by a set of gently sloped lines, as shown in **Figure 35.29**. The wet-bulb temperature values are read from the chart on the scale indicated along the left edge of the chart. The relative humidity is represented by the upward sloping lines within the chart, **Figure 35.30**. The bottom, horizontal line on the chart represents 0% relative humidity, while the left-hand border of the chart represents 100% relative humidity. Within the chart itself, there are nine other lines, each of which represents a 10% increment of relative humidity.

The moisture content, or absolute humidity, is represented by the horizontal lines on the chart, **Figure 35.31**. The absolute humidity can be read from the scales on the right side of the chart in either grains of moisture per pound of air or pounds of moisture per pound of air. The dew point temperature is also indicated by the horizontal lines, but the dew point temperature values are read from the same scale that is used to read the wet-bulb temperature.

The total heat, or enthalpy, of an air sample can be measured from the chart as well. The enthalpy scale is found above and to the left of the wet-bulb temperature scale, **Figure 35.32**. The lines of constant enthalpy are nearly identical to the lines of constant wet-bulb temperature. So, if the wet-bulb temperature of an air sample is known, the enthalpy of the air sample is also known.

The steeply sloped lines on the chart represent the specific volume of an air sample, **Figure 35.33**. These lines represent how many cubic feet of air will be needed to make up one pound. Remember, the specific volume is the inverse of density.

If you know any of the two conditions previously mentioned, you can easily determine any of the other conditions. The easiest conditions to determine from room air are the wet-bulb and the dry-bulb temperatures. For example, you can use a digital, multi-probe thermometer as a wet-bulb/dry-bulb thermometer if you do not have a psychrometer. Take two leads and tape them together with one lead about 2 in. below the other, **Figure 35.34**. A simple wick can be made from a piece of white cotton. Wet the lower bulb (with the wick attached) with distilled water that is warmer than the room air. Tap water can be used but may result in slightly incorrect results due to the water's impurities. Hold the leads about 3 ft back from the element on the end and slowly spin them in the air. The wet lead will drop to a colder temperature than the dry lead. Keep spinning them until the lower lead stops dropping in temperature but is still damp. The selector switch on the thermometer allows you to determine which probe's temperature value is being displayed. Quickly read the wet-bulb and dry-bulb temperatures without touching the bulbs. Suppose the reading is 75°F DB (dry bulb) and 62.5°F WB (wet bulb). Put your pencil point at this place on the psychrometric chart, **Figure 35.35**, and make a dot. Draw a light circle around it so you can find the dot again. The following information can be determined from this point:

1.	Dry-bulb temperature	**75°F**
2.	Wet-bulb temperature	**62.5°F**
3.	Dew point temperature	**55.5°F**
4.	Total heat content of 1 lb of air	**28.2 Btu**
5.	Moisture content of 1 lb of air	**65 gr**
6.	Relative humidity	**50%**
7.	Specific volume of air	**13.7 ft^3/lb**

35.12 PLOTTING ON THE PSYCHROMETRIC CHART

The conditions of air can be plotted on the psychrometric chart as the air is being conditioned. The following examples show how different air-conditioning applications are plotted:

- Air is heated. It can be seen from this diagram that a heating-only process is indicated by an arrow moving in a horizontal direction from left to right, **Figure 35.36**. This indicates that the temperature of the air is increasing. This will also indicate an increase in enthalpy. Since there is no vertical movement on the system's process line, moisture is not being added or removed from the air, so there is no latent activity taking place. The absolute humidity will

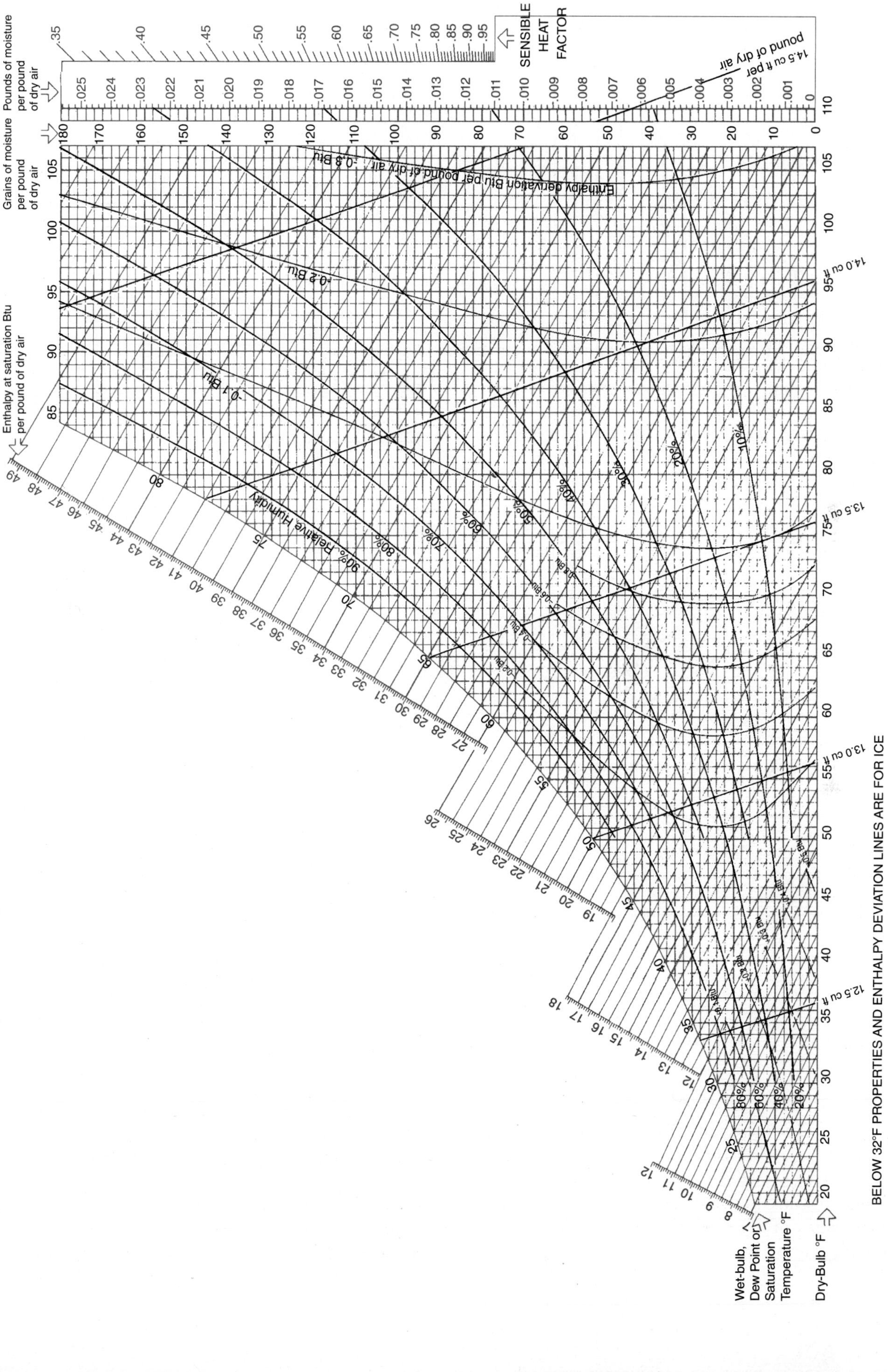

Figure 35.27 A psychrometric chart. Reproduced courtesy of Carrier Corp.

Figure 35.28 A skeleton chart showing the dry-bulb temperature.

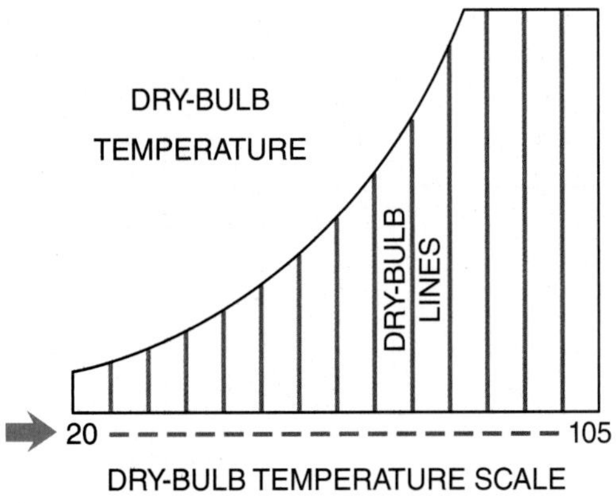

Figure 35.29 A skeleton chart showing the wet-bulb temperature.

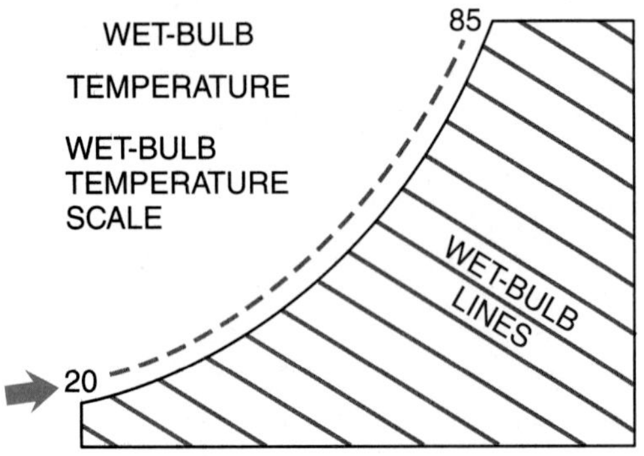

Figure 35.30 A skeleton chart showing the relative humidity lines.

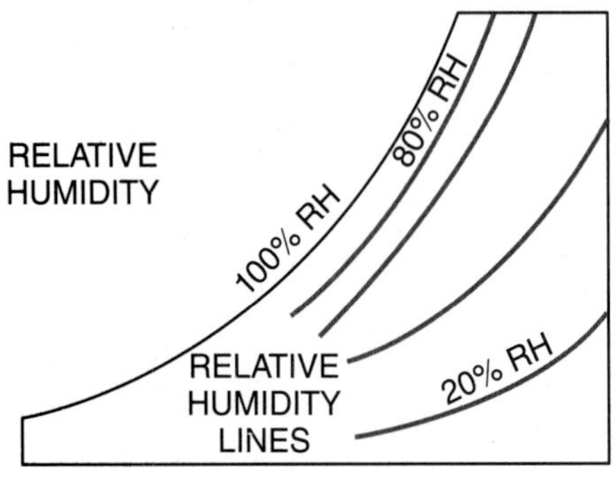

Figure 35.31 A skeleton chart showing the moisture content of air expressed in grains per pound of air.

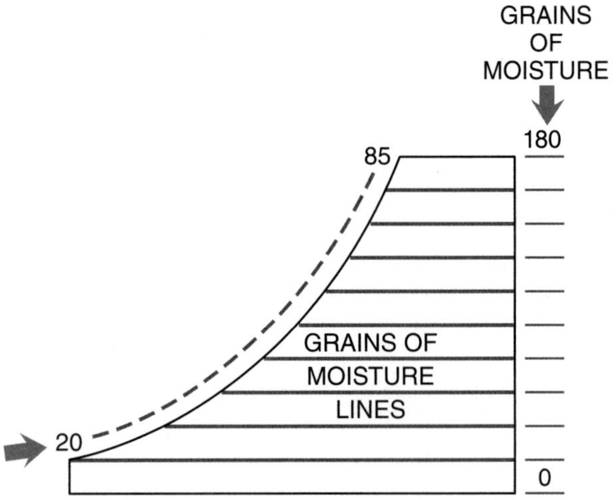

Figure 35.32 A skeleton chart showing the total heat content of the air in Btu/lb. These lines are almost parallel to the wet-bulb lines.

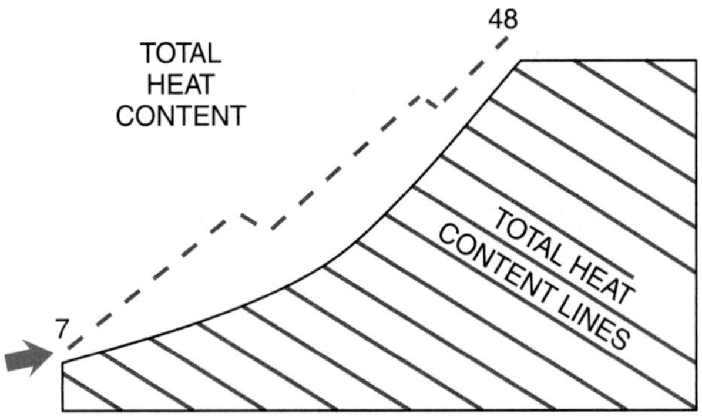

remain the same, but the relative humidity will decrease. This is because the air's ability to hold moisture increases as the temperature of the air increases.

- Air is cooled. A cooling-only process is indicated by a horizontal arrow moving from right to left, **Figure 35.37**. Since the air is not being humidified or dehumidified, there is no latent-heat activity in this process. The temperature of the air is dropping, so the enthalpy of the air is also dropping. In addition, the relative humidity of the air is increasing because, as air is cooled, its ability to hold moisture decreases.
- Air is humidified. When air is humidified, there is no change in the dry-bulb temperature of the air sample. The humidification process is indicated by an upward-pointing arrow, **Figure 35.38**. The enthalpy of the air sample will increase, along with the wet-bulb temperature of the air sample.

Figure 35.33 A skeleton chart showing the specific volume of air at different conditions.

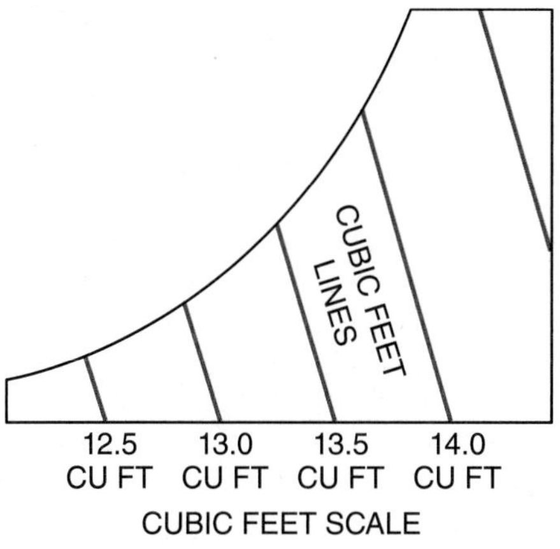

Figure 35.34 How to make a simple sling psychrometer.

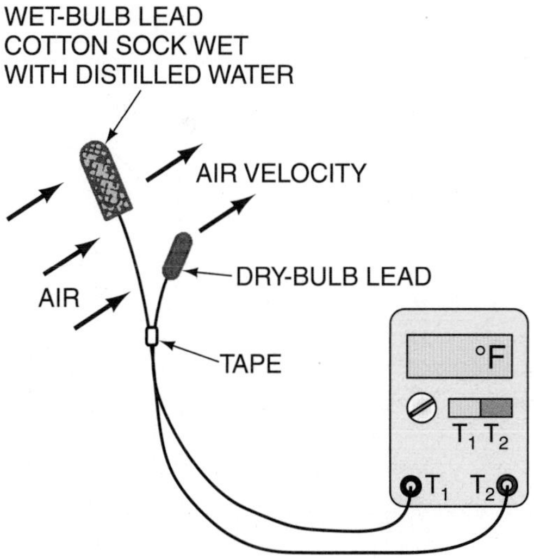

- Air is dehumidified. When air is dehumidified, there is no change in the dry-bulb temperature of the air sample. The dehumidification process is indicated by a downward-pointing arrow, **Figure 35.39**. The enthalpy of the air sample will decrease, along with the wet-bulb temperature of the air sample.
- Air is cooled and humidified using an evaporative cooler. During this process, the dry-bulb temperature decreases, while the relative humidity increases, **Figure 35.40**. This is popular in hot, dry climates.

When an airstream enters air-conditioning equipment, the airstream can be heated, cooled, humidified, or dehumidified, **Figure 35.41**. Some equipment will both add heat and moisture or cool and remove moisture. The following examples will show what happens in the most common heating and cooling systems:

- The most common winter application is to heat and humidify. This process will produce both a rise in temperature and an increase in moisture and dew point temperature, **Figure 35.42**.
- The most common summer application is to cool and dehumidify air. A decrease in temperature, moisture content, and dew point will occur, **Figure 35.43**.

35.13 FRESH AIR, INFILTRATION, AND VENTILATION

For us to be comfortable, the air that surrounds us has to be maintained at the correct conditions. The air in our homes is treated by heating it, cooling it, dehumidifying it, humidifying it, and cleaning it so that our bodies will give off the correct amount of heat for comfort. A small amount of air is induced from the outside into the conditioner to keep the air from becoming oxygen-starved and stagnant. This is called fresh air intake, or *ventilation*. If a system has no ventilation, it is relying on air infiltrating the structure around doors and windows.

Modern energy-efficient homes can be built so tightly that infiltration does not provide enough fresh air. Recent studies indicate that indoor pollution in homes and buildings has increased as a result of an increase in energy conservation in heating and air-conditioning systems and the structures in which they are used. People are more energy conscious now than in the past, and modern buildings are constructed to allow much less air infiltration from the outside. A typical homeowner may take the following measures to prevent outside air from entering the structure:

- Installing storm windows
- Installing storm doors
- Caulking around windows and doors
- Installing dampers on exhaust fans and dryer vents

All of these will reduce the amount of outside air that leaks into the structure and will reduce energy costs. This may not be all good because of the following indoor pollution sources:

- Chemicals in new carpets, drapes, and upholstered furniture
- Cooking odors
- Vapors from cleaning chemicals
- Bathroom odors
- Vapors from freshly painted rooms
- Vapors from aerosol cans, hair sprays, and room deodorizers
- Vapors from particle-board epoxy resins
- Pets and their upkeep
- Radon gas leaking into the structure from the soil

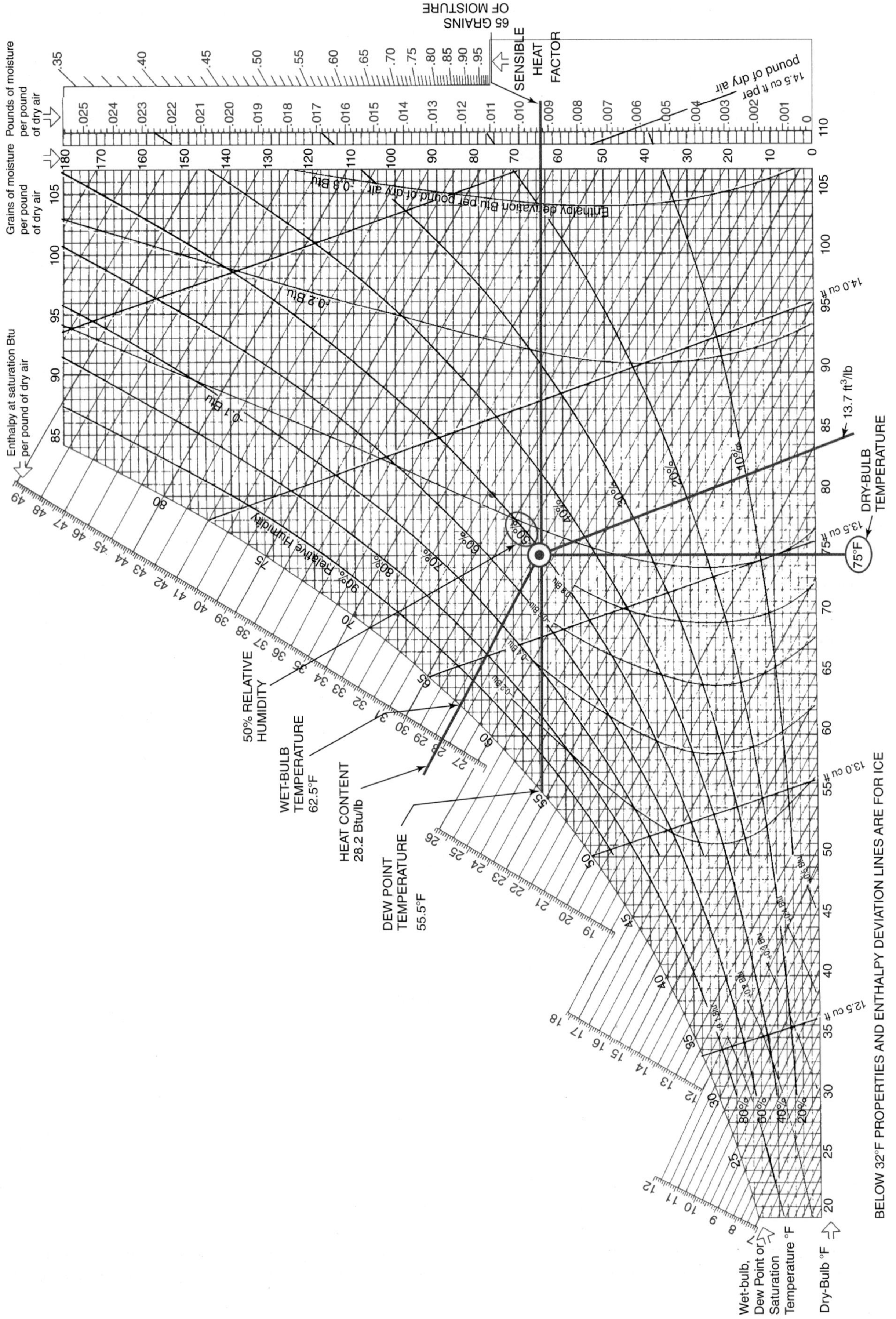

Figure 35.35 Psychrometric chart plotting example. Adapted from Carrier Corp. Psychrometric Chart

Figure 35.36 Air passing through a sensible-heat exchange furnace.

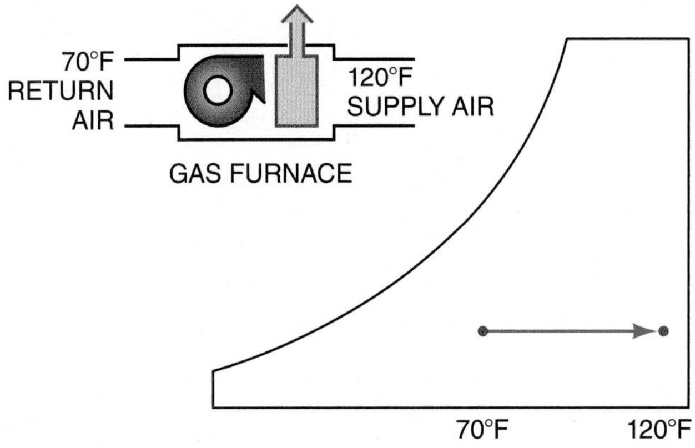

Figure 35.37 Air is cooled with a dry evaporator coil operating above the dew point temperature of the air. No moisture is removed. This is not a typical situation.

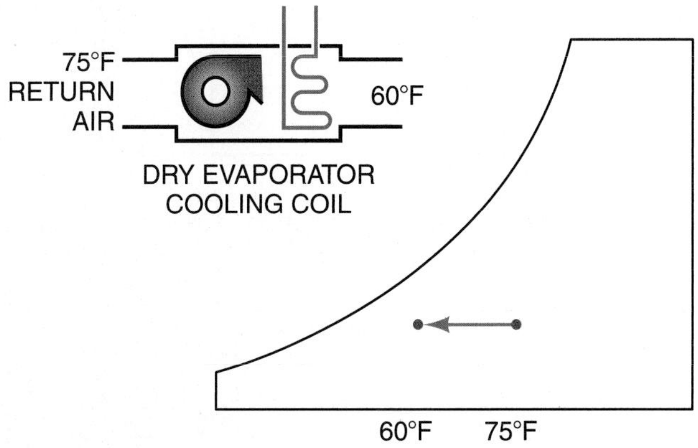

Figure 35.39 Moisture is removed from the air. This is not a typical application and is shown only as an example.

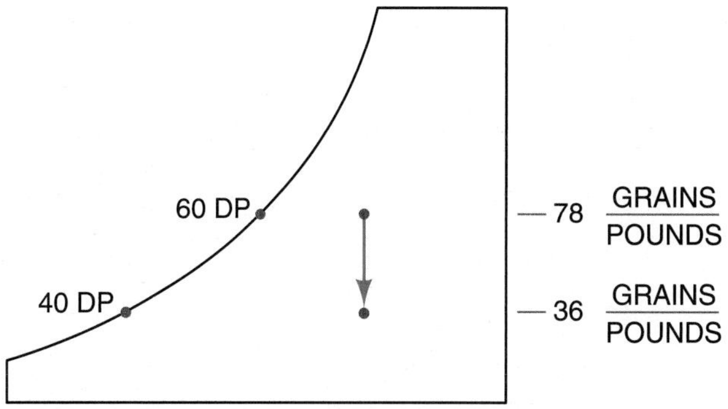

Figure 35.40 Hot, dry air passes through the water circuit of the evaporative cooler. Heat is given up to the cooler water, and the air entering the house is cooled and humidified; water evaporates to cool the air.

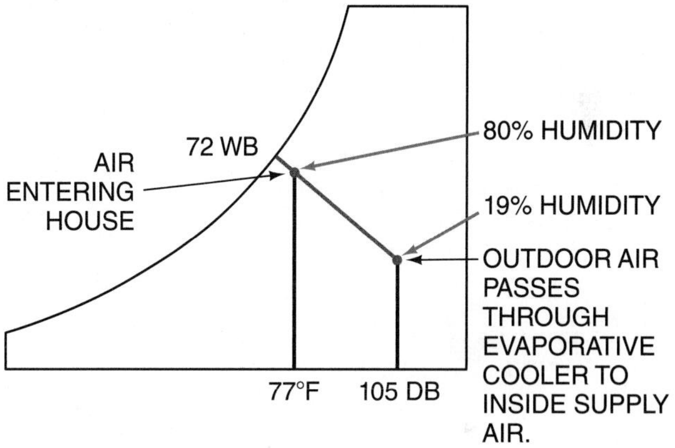

Figure 35.38 Spray atomizers are used to add moisture to the air. The dew point temperature and moisture content both increase.

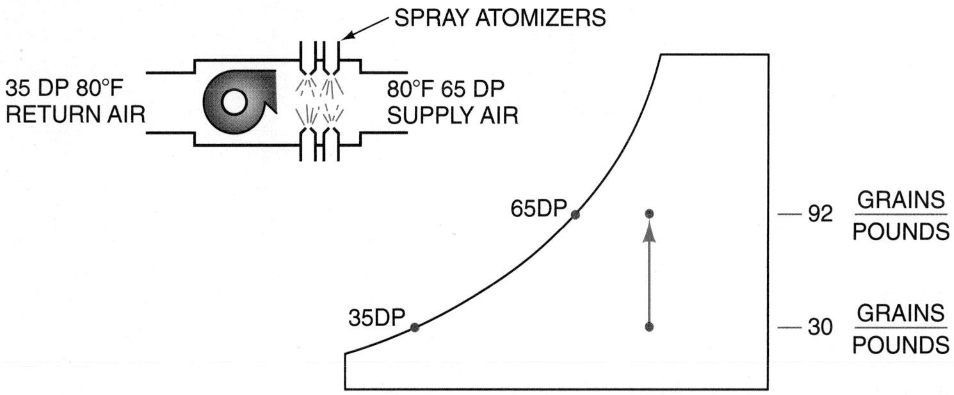

These indoor pollutants may be diluted with outdoor air in the form of ventilation. *Infiltration* is the term used for random air that leaks into a structure; ventilation is planned, fresh air added to the structure. When air is introduced into the system before the heating or air-conditioning system, it is called ventilation. This may be accomplished by using a duct from the outside to the return-air side of the equipment, **Figure 35.44.**

There is some discussion as to how much ventilation air should be introduced to a residential structure, but it

Figure 35.41 A summation of sensible and latent heat.

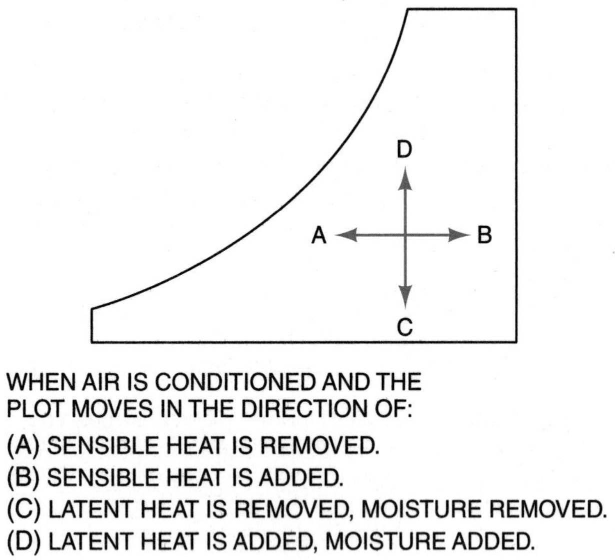

Figure 35.42 Sensible heat raises the temperature of the air from 70°F to 120°F. Moisture is evaporated, and latent heat is added to the air.

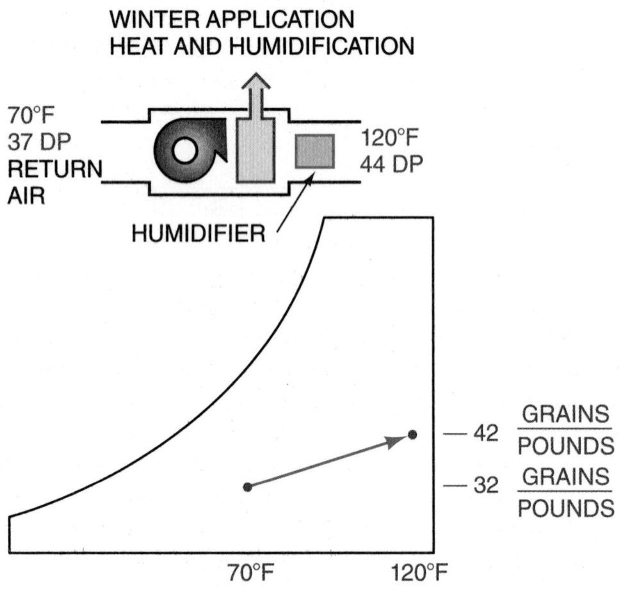

Figure 35.43 Removal of sensible heat cools the air. Removal of latent heat removes moisture from the air.

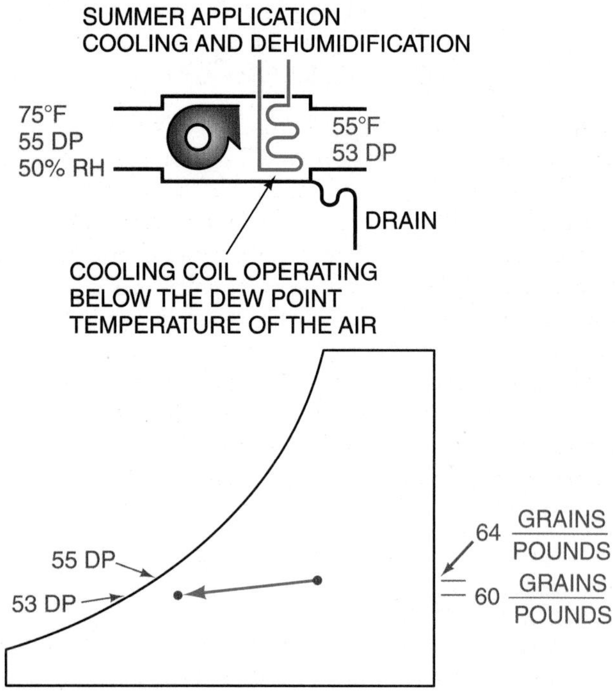

Figure 35.44 Fresh air is drawn into the return-air duct to improve air quality inside the house.

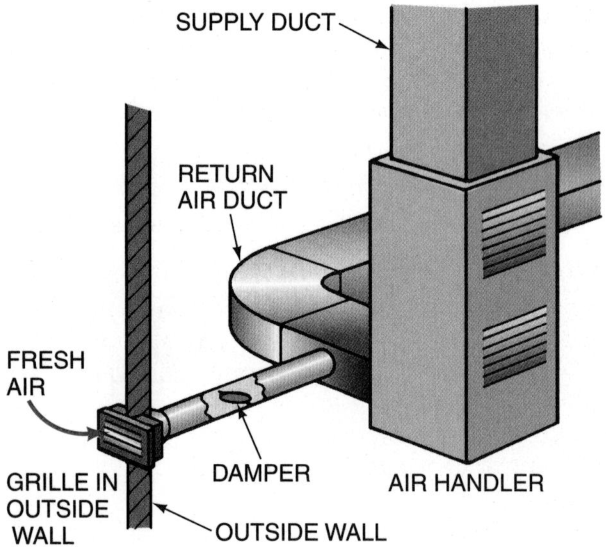

is generally agreed that changing the air in the space once every 4 hours or so is desirable. This means that 25% of the indoor air is pushed out by inducing air into the system. For example, suppose a 2000-ft^2 home with 8-ft ceilings needs ventilation because the home is very tight. How many cubic feet of air per minute must be introduced to change 25% of the air per hour?

$$2000 \text{ ft}^2 \times 8\text{-ft ceiling} \times 0.25 = 4000 \text{ ft}^3/\text{h}$$

$$\frac{4000 \text{ ft}^3/\text{h}}{60 \text{ min/h}} = 67 \text{ cfm}$$

This ventilation air has the potential to add a considerable load to the equipment. For example, suppose the house is located in Atlanta, Georgia, where the outdoor design temperature is 17°F in the winter. (See **Figure 36.15**, Design Dry-Bulb column 99%.) If the home is to be maintained at 70°F indoors when the outdoor temperature is 17°F, there is

a 53°F temperature difference (TD). The load on the heating equipment due to ventilation alone will be

$$Q_s = 1.08 \times \text{cfm} \times \text{TD}$$
$$Q_s = 1.08 \times 67 \times 53$$
$$Q_s = 3835 \text{ Btu/h}$$

This formula is explained in detail in Unit 30, "Electric Heat." An outside temperature of 17°F is only for 1% of the year; the fresh air will be warmer than 17°F the other 99% of the year. This problem depicts the worst possible case. Many system designers will use the 97.5% column for designing a system and may select smaller equipment because of the warmer design temperatures. This may be taking a chance in the event of a cold winter.

The summertime calculation is made in much the same manner, except a different formula is used. Sensible and latent heat must be considered. In the summer, the design temperatures are 95°F dry-bulb and 74°F wet-bulb. The total heat formula may be used for this calculation:

$$Q_t = 4.5 \times \text{cfm} \times \Delta h$$

where Q_t = total heat
4.5 = a constant used to change pounds of air to cfm
Δh = difference in total heat (enthalpy) indoors and outdoors

To solve the problem of fresh air, plot the indoor air and the outdoor air on the psychrometric chart.

Outdoors: 95°F dry-bulb and 74°F wet-bulb = 37.68 Btu/lb
Indoors: 75°F dry-bulb and 50% relative humidity indoors = 28.2 Btu/lb, **Figure 35.45.**
Total heat difference = 37.68 − 28.2 = 9.48 Btu/lb

$$Q_t = 4.5 \times \text{cfm} \times \text{Total Heat Difference}$$
$$Q_t = 4.5 \times 67 \times 9.48$$
$$Q_t = 2858 \text{ Btu/h total heat added due to ventilation}$$

NOTE: *The 4.5 that is used in the total heat calculation is only an estimate and assumes that the average specific volume of the air is 13.33 ft³/lb. Since the conversion involves changing time units (hours to minutes) as well as taking the specific volume into account, the 60 min/h ÷ 13.33 ft³/lb calculation will yield a result of 4.5. However, if the average specific volume differs greatly from the 13.33 value, there will be small calculation errors down the road.* •

System designers would require this total heat calculation to be broken down into the sensible-heat gain and the latent-heat gain. Equipment must be selected for the correct sensible- and latent-heat capacities or space humidity will not be correct. This calculation is done by using the sensible-heat and the latent-heat formulas as separate calculations. You will not arrive at the exact same total heat, because you cannot see the lines on the psychrometric chart closely enough for total accuracy. (In addition, refer to the calculation note above.)

$$Q_s = 1.08 \times \text{cfm} \times \text{TD}$$
plus
$$Q_l = 0.68 \times \text{cfm} \times \text{Grains Difference}$$

where Q_l = latent heat
0.68 = a constant used to change cfm to pounds of air and grains per pound
Grains difference = the difference in the grains per pound of air for the indoor air and the outdoor air

Example:

$$Q_s = 1.08 \times \text{cfm} \times \text{TD}$$
$$Q_s = 1.08 \times 67 \times 20$$
$$Q_s = 1447 \text{ Btu/h sensible heat}$$

From the psychrometric chart points plotted earlier, you will find that the outdoor air contains 92 gr/lb and the indoor air contains 64 gr/lb for a difference of 28 gr/lb of air.

$$Q_l = 0.68 \times \text{cfm} \times \text{grains difference}$$
$$Q_l = 0.68 \times 67 \times 28$$
$$Q_l = 1276$$
$$\text{Total Heat} = Q_s + Q_l$$
$$\text{Total Heat} = 1447 + 1276$$
$$\text{Total Heat} = 2723 \text{ Btu/h}$$

Notice the small difference between the 2858 Btu/h and the 2723 Btu/h results. This difference is not large enough to cause a problem in the evaluation of the total system performance.

Office buildings have the same pollution problems, except they have more sources of pollution because there are more people and different and varied activities. Office buildings have a tendency to be remodeled more often, which leads to more construction-type pollution. Also, copying machines that use liquid copy methods give off vapors, and there are more people per square foot in office buildings. The national code requirements for fresh air in buildings where the public works are based on the number of people in the building and the type of use. Different buildings may have different indoor pollution rates because of the activity in the building. For example, buildings that have a lot of copying machines or blueprint machines may have more indoor pollution than ones with only electronic processors, such as computers. Department stores have different requirements than restaurants do. **Figure 35.46** shows some of the fresh-air requirements for different applications recommended by ASHRAE. These figures are constantly changing as new research is done and are for comparison only.

The design engineer is responsible for choosing the correct fresh-air makeup for a building, and the service technician is responsible for regulating the airflow in the field. The technician may be given a set of specifications for a building and told to regulate the outdoor air dampers to induce the correct amount of outdoor air into the building.

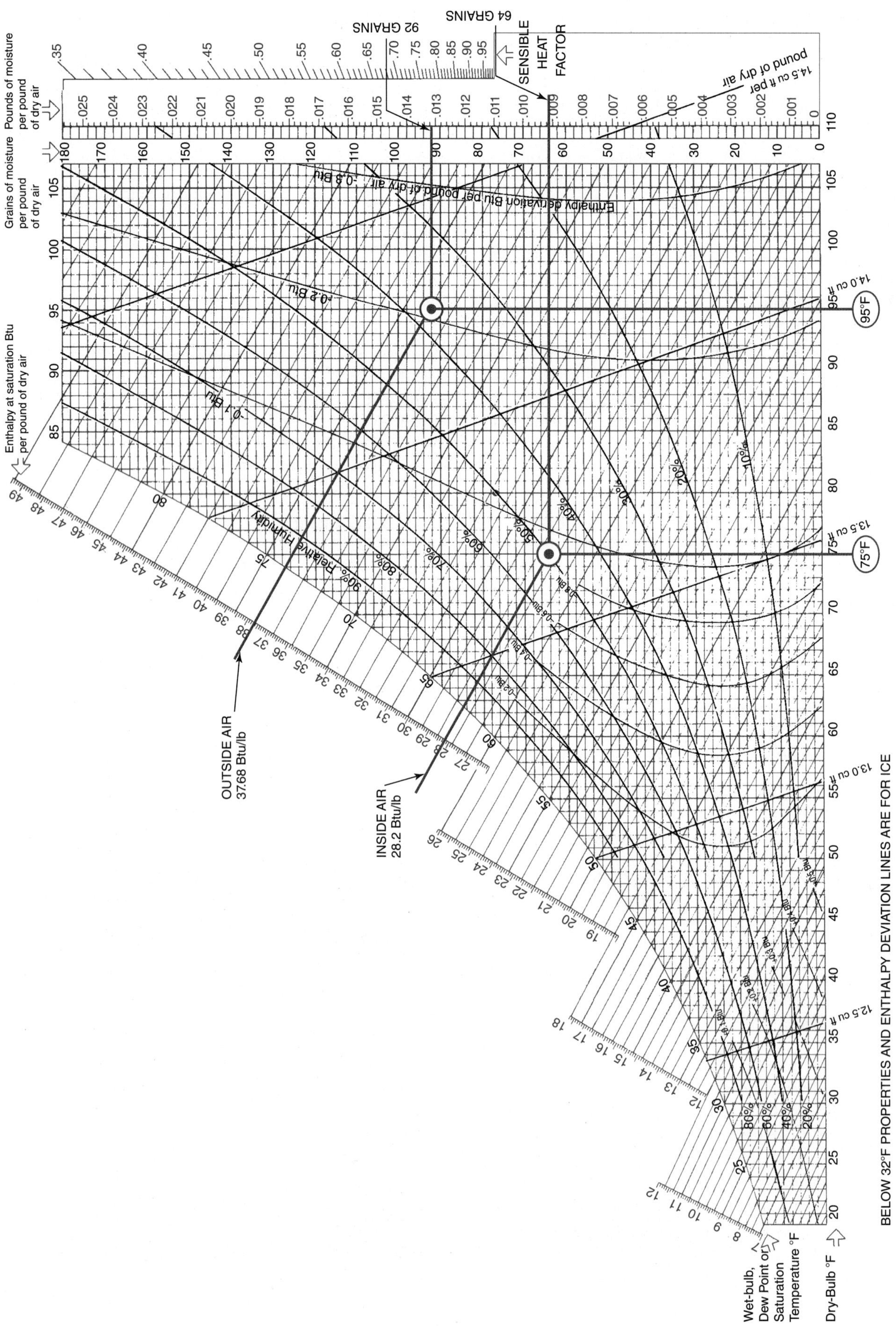

Figure 35.45 Graphic representation of ventilation example. Adapted from Carrier Corp. Psychrometric Chart

Figure 35.46 Fresh-air requirements for some typical applications. Reprinted with permission from ASHRAE, Inc.—Standard 62

RECOMMENDED OUTDOOR AIR VENTILATING RATES Abbreviated from ASHRAE Standard 62	
	cfm or see footnote
Dining rooms	20
Bars & cocktail lounges	30
Hotel conference rooms	20
Office spaces	20
Office conference rooms	35
Retail stores	.02 to .03(a)
Beauty shops	25
Ballrooms & discos	25
Spectator areas	15
Theater auditoriums	15
Transportation waiting rooms	15
Classrooms	15
Hospital patient rooms	25
Residences	.35(b)
Smoking lounges	60

a cfm per square feet of space
b air changes per hour

For the purpose of setting the outside air dampers, the mixed-air condition may be determined by calculation. The condition may be plotted on the psychrometric chart for ease of understanding. The dampers may then be adjusted for the correct conditions. For example, suppose a building with a return air condition of 73°F dry-bulb and 60°F wet-bulb is taking in and mixing air from the outside of 90°F dry-bulb and 75°F wet-bulb. When these two conditions are plotted on the psychrometric chart and a line is drawn between them, the condition of the mixture air will fall on the line, **Figure 35.47**. If the mixture were half outdoor air and half indoor air, the condition of the mixture air would be midway between points A and B.

The point on the line between A and B may be calculated for any air mixture if the percentage of either air source, outdoor or indoor, is known. For example, when the point for 25% outdoor air and 75% indoor air mixture was calculated, the following calculation was used to find the dry-bulb point on the line. When the dry-bulb point is known and it falls on the line, you have a coordinate to find any other information for this air condition.

25% outdoor air: 0.25 × 90°F = 22.50°F
75% indoor air: 0.75 × 73°F = 54.75°F
The mixture (dry-bulb) = 77.25°F

Figure 35.47 Point A is the condition of the indoor air, and point B is the condition of the outdoor air. When the two are mixed, the mixed air condition will fall somewhere on the line that connects point A to point B. Adapted from Carrier Corp. Psychrometric Chart

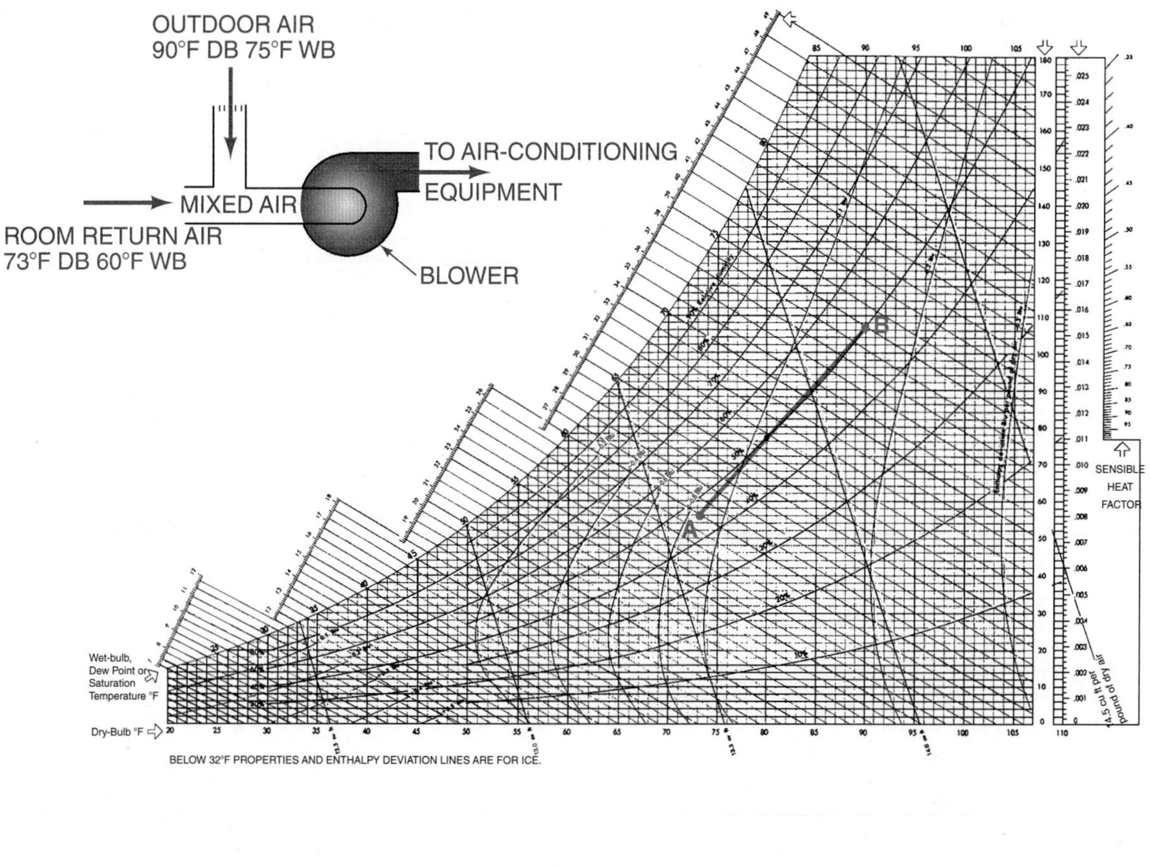

When you look on the chart in **Figure 35.48**, you will find the point (C). Note that the point always falls closest to the air temperature with the highest percentage of air.

An actual problem may be something like the following:

BUILDING AIR-CONDITIONING SPECIFICATIONS

- Total building air = 50,000 cfm
- Recirculated air = 37,500 cfm, 75% return air
- Makeup air = 12,500 cfm, 25% outside air
- Indoor design temperature = 75°F dry-bulb, 50% relative humidity (62.5 wet-bulb)
- Outdoor design temperature (Atlanta, Georgia) = 94°F dry-bulb, 74°F wet-bulb

A technician checking the outdoor air percentage finds that the outdoor temperature is not at the design temperature. It is 93°F dry-bulb/75°F wet-bulb and the indoor conditions are 73°F dry-bulb/59°F wet-bulb. What should the mixture air temperature be?

A calculation for the mixture air is made. The actual conditions are plotted in **Figure 35.49**.

$$93°F \times 0.25 = 23.25°F$$
$$73°F \times 0.75 = 54.75°F$$

The mixture (dry-bulb) = 23.25°F + 54.75°F = 78°F

This calculation tells the technician that 25% of the mixed air dry-bulb temperature comes from the outside air, which is at 93°F, and therefore constitutes 23.25°F; 75% of the temperature is from the 73°F inside air dry-bulb temperature, which constitutes 54.75°F. Adding them together gives the mixed air dry-bulb temperature of 78°F (23.25 + 54.75 = 78). The technician can use dry-bulb temperatures alone to calculate air mixture percentages.

The technician would now check the mixture air temperature. If it is too high, he would slightly close the outdoor air dampers until the mixture air temperature stabilizes at the correct mixture temperature. If the mixture temperature is too low, he would slightly open the outdoor dampers until the mixture air temperature is stable and correct. **NOTE:** *The outdoor and indoor conditions should be rechecked after adjustment because of changes in weather and indoor condition. These calculations and proper damper settings also depend on correct instrumentation while checking the air temperatures.* •

Figure 35.48 Mixture of different percentages of outside and inside air. Adapted from Carrier Corp. Psychrometric Chart

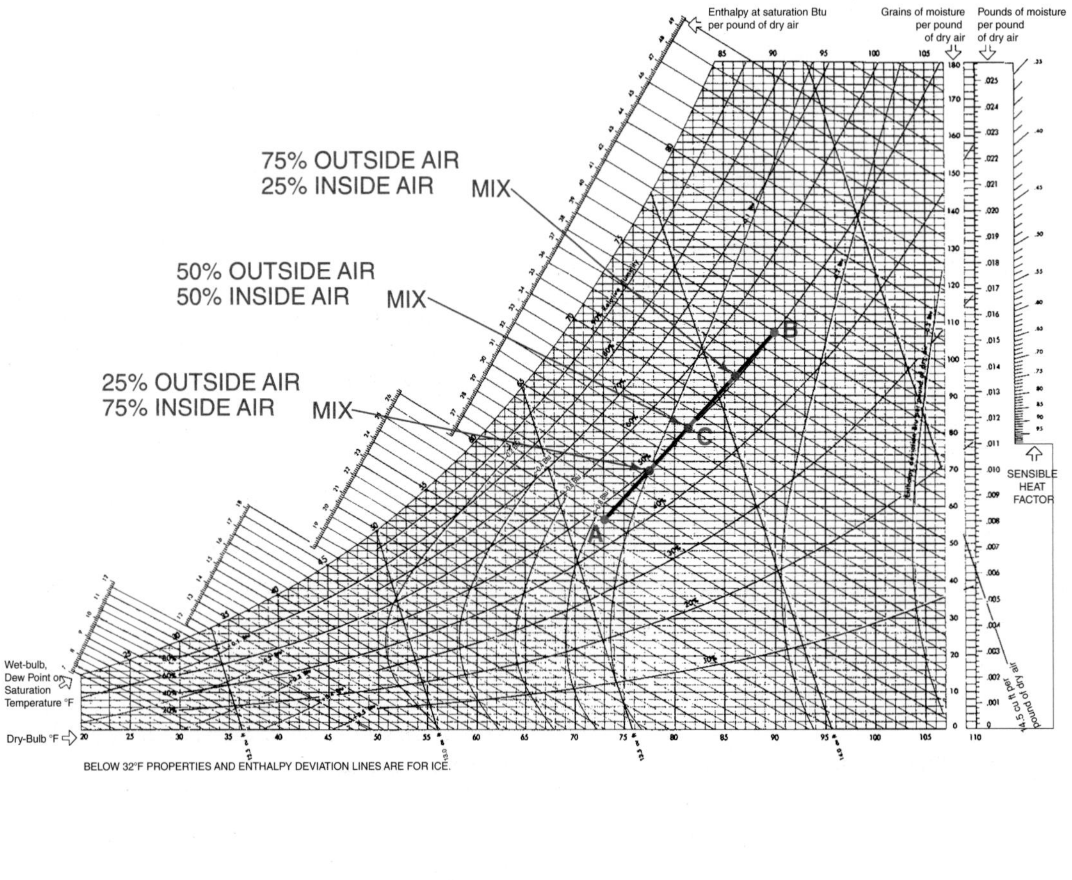

Figure 35.49 This practical problem of finding the mixed air condition is shown plotted on the chart. Point C is the mixed air point. Adapted from Carrier Corp. Psychrometric Chart

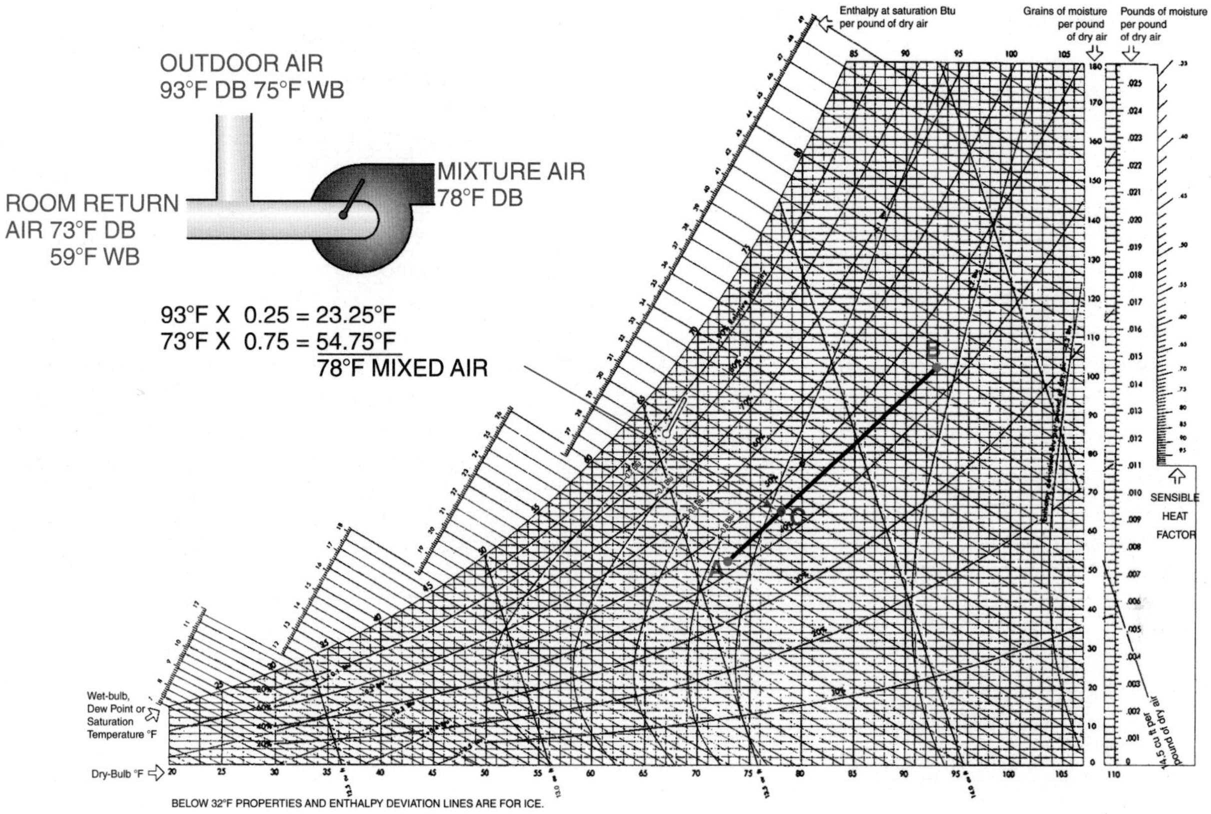

A technician may also use the psychrometric chart for field checking the capacity of a piece of air-conditioning equipment. For example, suppose the capacity of a 5-ton (60,000 Btu/h) unit is in question and the unit uses gas as the heat source. The heat source is used to determine the cfm of the equipment.

A technician may want to perform the cfm part of this problem first thing in the morning because it will involve operating the heating system long enough to arrive at the cfm. The unit heat is turned on and an accurate temperature rise across the gas heat exchanger is taken at 69.5°F. **NOTE:** *The technician was careful not to let radiant heat from the heat exchanger influence the thermometer lead.•*

The unit nameplate shows the unit to have 187,500 Btu/h input. The output would be 80% of the input, or 187,500 Btu/h × 0.80 = 150,000 Btu/h output to the airstream. The fan motor would add about 600 W × 3.431 Btuh/W = 2048 Btuh, for a total of 152,048 Btu/h. **NOTE:** *The blower is operated using the FAN ON position on the room thermostat. This ensures the same blower speed in heating and cooling.•*

Using the sensible-heat formula, the technician can find the cfm.

$$Q_s = 1.08 \times \text{cfm} \times \text{TD}$$

Solved for cfm:

$$\text{cfm} = \frac{Q_s}{1.08 \times \text{TD}}$$

$$\text{cfm} = \frac{152{,}048}{1.08 \times 69.5}$$

$$\text{cfm} = 2026\ \text{ft}^3/\text{min}$$

The technician then would turn the unit to cooling and check the wet-bulb temperature in the entering and leaving airstream. Suppose that the entering wet-bulb is 62°F and the leaving wet-bulb is 53°F, **Figure 35.50**.

Total heat at 62°F wet-bulb is 27.85
Total heat at 53°F wet-bulb is 22.00
Total heat difference = 5.85 Btu/lb of air

Using the total heat formula:

$$Q_t = 4.5 \times \text{cfm} \times \text{Total Heat Difference}$$

$$Q_t = 4.5 \times 2026 \times 5.85$$

$$Q_t = 53{,}334\ \text{Btu/h Total Heat}$$

The unit has a capacity of 60,000 Btu/h, so it is operating very close to capacity. This is about as close as a technician can expect to get.

Figure 35.50 Capacity check for cooling mode. Adapted from Carrier Corp. Psychrometric Chart

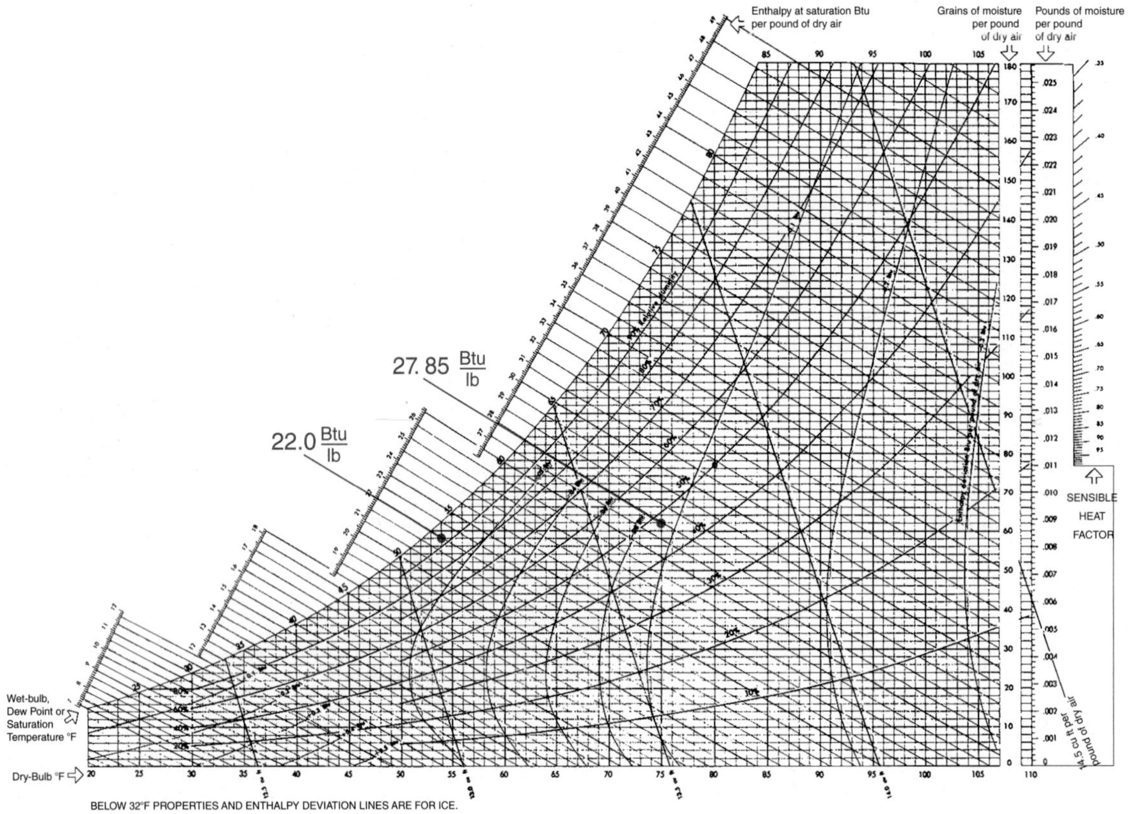

REVIEW QUESTIONS

1. The body gives off heat in all of the following ways except
   A. conduction.
   B. convection.
   C. radiation.
   D. condensation.
2. If the human body is exposed to very warm temperatures, which of the following is true?
   A. Perspiration will form to warm the body.
   B. Perspiration will form to cool the body.
   C. Respiration will form to warm the body.
   D. Respiration will form to cool the body.
3. In the winter, lower room temperatures can be offset by
   A. lowering the absolute humidity.
   B. avoiding strenuous activities.
   C. increasing the relative humidity.
   D. opening the structure's doors and windows.
4. Relative humidity is measured using which of the following instruments?
   A. Psychrometer
   B. Thermotyporiter
   C. Volt-ohm-milliammeter
   D. Dry-bulb thermometer
5. According to the summer comfort chart, Figure 35.6(A), people will generally be comfortable when exposed to which of the following sets of conditions?
   A. 70°F @ 50% relative humidity
   B. 75°F @ 50% relative humidity
   C. 80°F @ 50% relative humidity
   D. 85°F @ 50% relative humidity
6. According to the winter comfort chart, Figure 35.6(B), people will generally be comfortable when exposed to which of the following sets of conditions?
   A. 70°F @ 50% relative humidity
   B. 75°F @ 50% relative humidity
   C. 80°F @ 50% relative humidity
   D. 85°F @ 50% relative humidity
7. At standard atmospheric pressure, air has a density of
   A. 0.075 lb/ft3.
   B. 0.75 lb/ft3.
   C. 1.33 lb/ft3.
   D. 13.33 lb/ft^3.
8. If the density of a gas is 0.5 lb/ft^3, what is its specific volume?
   A. 1 ft^3/lb
   B. 2 ft^3/lb
   C. 4 ft^3/lb
   D. 8 ft^3/lb

9. Multiplying the density of a gas and its specific volume together will always produce a result equal to
   A. 4.
   B. 3.
   C. 2.
   D. 1.
10. Air is primarily made up of
   A. 78% oxygen, 21% nitrogen.
   B. 25% nitrogen, 75% oxygen.
   C. 50% oxygen, 50% nitrogen.
   D. 78% nitrogen, 21% oxygen.
11. In order for an evaporator to remove moisture from the airstream passing through it, the evaporator coil must be at a temperature that is lower than the air's
   A. wet-bulb temperature.
   B. dry-bulb temperature.
   C. dew point temperature.
   D. critical temperature.
12. Air that contains all of the moisture that it can hold is referred to as being
   A. subcooled.
   B. superheated.
   C. sublimated.
   D. saturated.
13. If an air sample holds 60 grains of moisture (per pound) and has the ability to hold a maximum of 90 grains of moisture (per pound), what is the relative humidity of the air sample?
   A. 33%
   B. 67%
   C. 100%
   D. 150%
14. Which of the following statements is true regarding the heating of an air sample?
   A. The dry-bulb temperature will rise and the wet-bulb temperature will drop.
   B. The relative humidity will drop and the absolute humidity will remain the same.
   C. The dry-bulb temperature will drop and the wet-bulb temperature will rise.
   D. The relative humidity will rise and the absolute humidity will remain the same.
15. The nitrogen and oxygen in our air are classified as
   A. inert gases.
   B. subcooled gases.
   C. superheated gases.
   D. saturated gases.
16. If the dry-bulb temperature is 75°F and the wet-bulb temperature is 60°F, the wet-bulb depression is
   A. 15°F.
   B. 30°F.
   C. 60°F.
   D. 75°F.
17. If the wet-bulb depression is 0°F, what is the relative humidity of the air sample?
   A. 25%
   B. 50%
   C. 75%
   D. 100%
18. Which of the following is true regarding wet-bulb and dry-bulb temperatures?
   A. The wet-bulb temperature reading is never lower than the dry-bulb temperature reading.
   B. The wet-bulb temperature reading is always lower than the dry-bulb temperature reading.
   C. The wet-bulb temperature reading can be equal to the dry-bulb temperature reading.
   D. The wet-bulb temperature reading is always equal to the dry-bulb temperature reading.
19. What is the dew point temperature of a 75°F air sample with a relative humidity of 60%?
   A. 58°F
   B. 60°F
   C. 62°F
   D. 64°F
20. Condensate dripping from an air-conditioning system is an indication that the evaporator coil temperature is
   A. cooler than the air's dew point temperature.
   B. warmer than the air's dew point temperature.
   C. cooler than the air's dry-bulb temperature.
   D. warmer than the air's dry-bulb temperature.
21. The term enthalpy applies to
   A. a combination of hidden heat and latent heat.
   B. a combination of latent heat and sensible heat.
   C. the latent heat portion of an air-conditioning process.
   D. the sensible heat portion of an air-conditioning process.
22. Enthalpy is expressed in which of the following units?
   A. ft^3/lb
   B. Btu/lb
   C. lb/ft^3
   D. lb/Btu
23. On the psychrometric chart, the scale along the bottom of the chart represents
   A. wet-bulb temperature.
   B. dew point temperature.
   C. enthalpy.
   D. dry-bulb temperature.
24. If only the absolute humidity of an air sample is known, which of the following values can be determined from the psychrometric chart?
   A. Wet-bulb temperature
   B. Dew point temperature
   C. Enthalpy
   D. Dry-bulb temperature

25. Of the following sets of air properties, which two can be identified by following the same line?
    A. Wet-bulb temperature and dry-bulb temperature
    B. Relative humidity and absolute humidity
    C. Enthalpy and wet-bulb temperature
    D. Absolute humidity and dry-bulb temperature
26. In a cooling-only process, all of the following are true except
    A. the dry-bulb temperature drops.
    B. the wet bulb temperature drops.
    C. the specific volume of the air drops.
    D. the absolute humidity drops.
27. The term used to describe the random air that leaks into a structure is
    A. ventilation.
    B. infiltration.
    C. exhaust.
    D. exfiltration.
28. Which of the following is a correct formula for total heat?
    A. $Q_T = Q_S - Q_L$
    B. $Q_T = Q_S + Q_L$
    C. $Q_T = Q_S \times Q_L$
    D. $Q_T = Q_S \div Q_L$
29. Name the four factors that affect comfort.
30. Determine the dew point temperature of an air sample if the sample's dry-bulb temperature is 70°F and its wet-bulb temperature is 61°F.
31. Calculate the volume of a house that is 3500 ft^2 with a 9-foot ceiling.
32. Determine the total heat content of an outdoor air sample if the sample's dry-bulb temperature is 93°F and its wet-bulb temperature is 74°F.
33. Determine the airflow through a furnace that is producing 60,000 Btu/h of heat when the return air temperature is 72°F and the supply air temperature is 130°F.
34. Consider two airstreams. Airstream 1 is at a temperature of 90°F and airstream 2 is at a temperature of 50°F. Assuming that the mixed air is made up of 20% airstream 1 and 80% airstream 2, calculate the temperature of the mixed air.
35. Explain the concepts of relative and absolute humidity.

UNIT 36

REFRIGERATION APPLIED TO AIR-CONDITIONING

OBJECTIVES

After studying this unit, you should be able to

- explain the term air-conditioning as it applies to refrigeration.
- list three ways that heat leaks into a structure.
- list two design temperature readings that must be taken into account when sizing and selecting air-conditioning equipment.
- explain the process of evaporative cooling.
- explain why ductless split air-conditioning systems can help to reduce energy consumption.
- describe package air-conditioning equipment.
- describe split-system equipment.
- describe the function of an air-conditioning evaporator.
- describe three common configurations of an air-conditioning evaporator.
- obtain design temperature information from published industry charts and tables.
- list three types of compressors used in air-conditioning systems.
- describe the operation of a reciprocating compressor.
- explain how the operating temperatures of a compressor are affected.
- describe the operation of a rotary compressor.
- describe the operation of a scroll compressor.
- describe the function of an air-conditioning condenser.
- explain how "high efficiency" is accomplished.
- describe the function of an air-conditioning metering device.
- explain the importance of a properly installed and insulated air distribution system.
- identify the field-installed refrigerant lines on a split-type air-conditioning system.
- state the operating conditions that might be observed on a properly operating R-410A air-conditioning system.

SAFETY CHECKLIST

- ✔ Use caution when loading or unloading air-conditioning equipment from the truck. Wear an approved back brace when lifting and use your legs, keeping your back straight and upright. Trucks with lift gates may be used to lower the equipment to the ground.
- ✔ When installing equipment in attics and under buildings, be alert for stinging insects. Avoid being scratched or cut by exposed nails and other sharp objects in attics and in crawl spaces under buildings. Be sure to treat flesh wounds quickly to prevent infection.
- ✔ Wear goggles and gloves when installing charged line sets or gauges because escaping refrigerant can cause frostbite. Treat frostbite just like a burn. Warm the affected body part back to body temperature and apply burn ointments or creams to the affected areas.

36.1 AIR-CONDITIONING

The term *air-conditioning* refers to a refrigeration process that is applied to comfort cooling. Comfort cooling systems are intended to cool and maintain the temperature of buildings and structures during periods when the temperature within the structure is higher than desired. Although this typically occurs in the warmer summer months, there are many applications and situations where comfort cooling may be required year round.

Air-conditioning and comfort-cooling processes are classified as high-temperature refrigeration systems, as they operate with evaporator saturation temperatures above freezing and also operate to maintain the conditioned space at temperatures above freezing. Air-conditioning systems are similar to other refrigeration systems in that the same four major system components (compressor, condenser, metering device, evaporator) are used to cool the space. The basics of refrigeration and its components were initially discussed in Unit 3, "Refrigeration and Refrigerants." Refer back to that unit as needed.

In operation, the air-conditioning system absorbs heat from within the structure and deposits this heat to the outside. Some of the heat absorbed into the air-conditioning system likely leaked into the structure from the outside, while some was likely generated inside the structure itself. People are sources of internally-generated structural heat, as are cooking equipment, computers, televisions, and other appliances.

36.2 STRUCTURAL HEAT GAIN

Heat leaks into a structure by conduction, infiltration, and radiation (the sun's rays or solar load). In the northern hemisphere, the summer *solar load* on a structure is greater on the east and west sides of the structure. This is because the sun shines for longer periods of time on these parts of the

Figure 36.1 The solar load on a home.

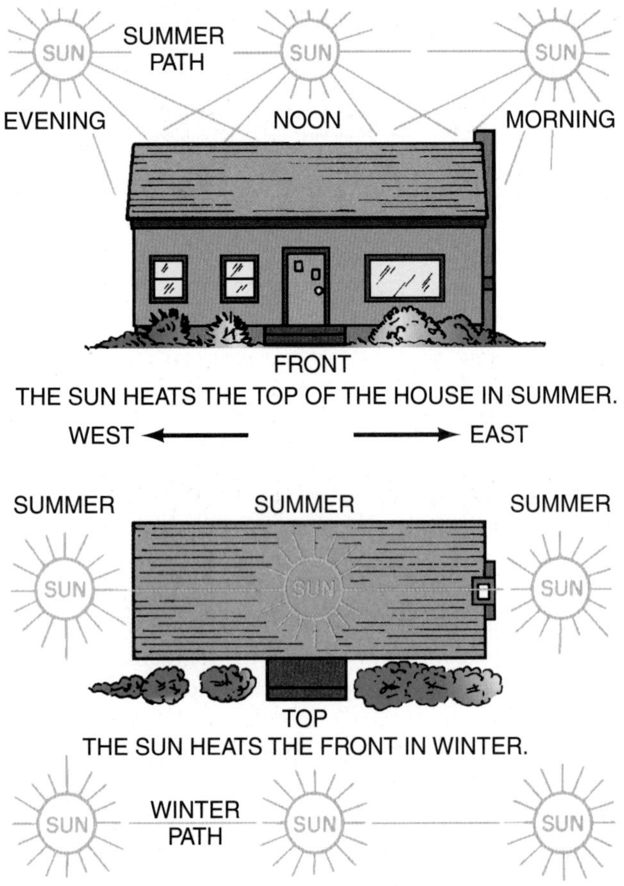

Figure 36.2 A ventilated attic helps keep the solar heat from the ceiling of the house.

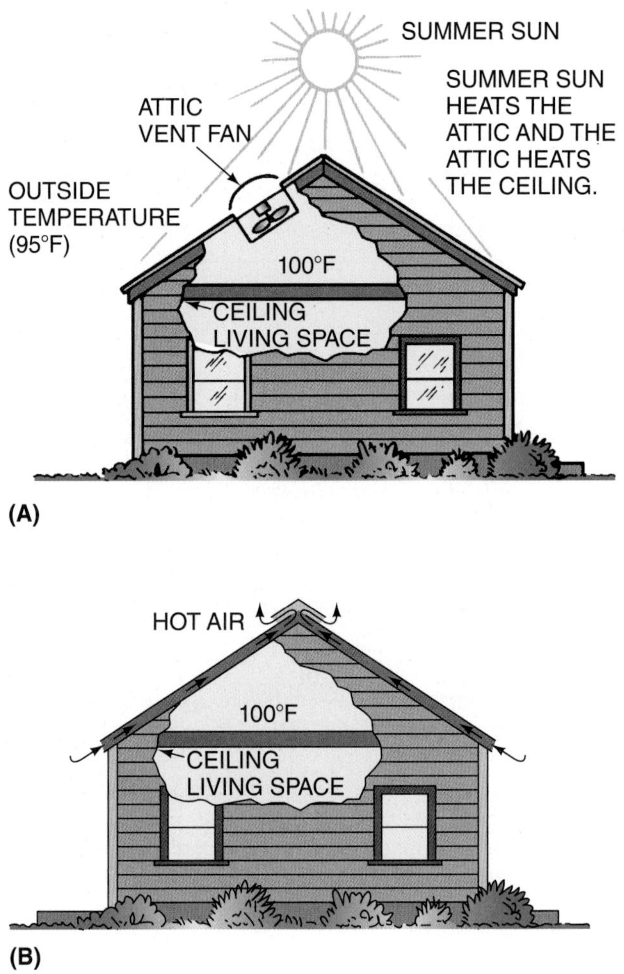

structure, **Figure 36.1**. If a building has an attic, ventilating this space will help reduce the heat gain on the structure. Attic ventilation is commonly accomplished with either powered or natural-draft ventilation systems. **Figure 36.2(A)** shows a power ventilator. It is a thermostatically controlled fan that cycles on when the attic temperature reaches a predetermined temperature, typically 85°F. This fan must have a high-limit control to shut the fan off in case of an attic fire. The power ventilator would just feed air to a fire. **Figure 36.2(B)** shows a natural-draft ventilation system that is used more extensively. It is called a ridge vent system. The ridge vent is cut along the top of the entire roof ridge. Another vent, called a soffit vent, typically runs along the bottom of the roof overhang. These two vents facilitate air circulation along the entire underside of the roof surface. In some instances, solar-powered ventilation fans can be used. If the structure has no attic, the roof of the house is also the ceiling of the living space. There should be sufficient insulation in the roof to minimize the heat gain in the occupied space, **Figure 36.3**.

Heat enters the structure through walls, closed windows, and closed doors by conduction. The rate of heat

Figure 36.3 This house has no attic. The sun shines directly on the ceiling of the living space.

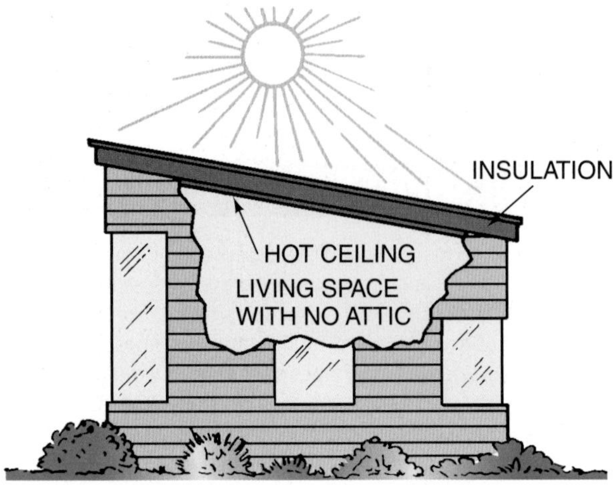

Figure 36.4 The difference between the inside and outside temperatures of a home in Augusta, Georgia, and a home in Phoenix, Arizona.

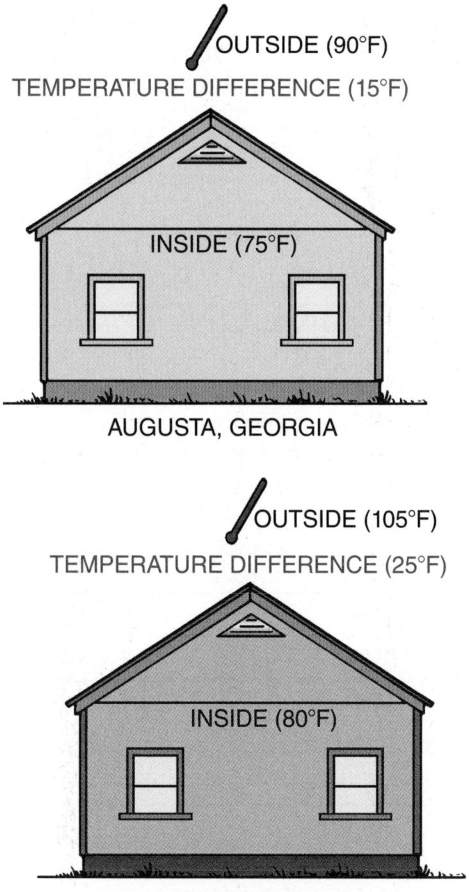

transfer into the structure depends on, among other things, the temperature difference between the inside and the outside of the house, **Figure 36.4**. Some of the warm air that gets into the structure infiltrates through the cracks around the windows and doors. Air also leaks in when the doors or windows are opened. This infiltrated air has different characteristics in different parts of the country. Using the example in **Figure 36.4**, the typical design condition in Phoenix, Arizona, is 105°F dry-bulb and 71°F wet-bulb. In Augusta, Georgia, the air may be 90°F dry-bulb and 73°F wet-bulb. When the air leaks into a structure in Phoenix, it is cooled to the space temperature. The air that leaks into a structure also contains moisture. In Augusta, the infiltration air will have a higher moisture content than the infiltration air in Phoenix because Phoenix has a drier climate when compared to Augusta, Georgia.

The moisture content of the air in different locations must be taken into account when selecting air-conditioning equipment. Equipment selected for parts of the country that are humid must be sized properly to remove moisture from the conditioned space.

36.3 EVAPORATIVE COOLING

Air can be conditioned in more than one way to achieve comfort. In climates where the humidity is low, a device called an **evaporative cooler**, **Figure 36.5**, *is commonly used. Evaporative cooler are also commonly referred to as swamp coolers. Water slowly running down a fiber pad mounted in a frame is the cooling media. Fresh air is drawn through the water-soaked fiber pad and cooled by evaporation to a point close to the wet-bulb temperature of the air. The air entering the structure is very humid but cooler than the dry-bulb temperature. For example, in Phoenix, Arizona, the design dry-bulb temperature in summer is 105°F. At the same time the dry-bulb is 105°F, the wet-bulb temperature may be 70°F. An evaporative cooler may lower the air temperature entering the room to 80°F dry-bulb, which is cool compared to 105°F, even if the humidity is high.*

Evaporative cooling units use 100% outdoor air. Since outside air is being pushed into the structure, there must be an outlet through which air in the structure can leave. This will prevent the structure from becoming pressurized. This outlet could very well be a partially open window at the far side of the structure.

Evaporative coolers come in three common configurations: the drip-type swamp cooler, the spray or slinger-type swamp cooler, and the rotary-type swamp cooler. The drip-type evaporative cooler, **Figure 36.6(A)**, uses a pump to lift a column of water to the top of the evaporative pad, where the water drips down and wets the surface of the pad. As the hot, dry outside air passes through the wet pad, heat causes some of the moisture from the pad to evaporate. The evaporating

Figure 36.5 An evaporative cooler.

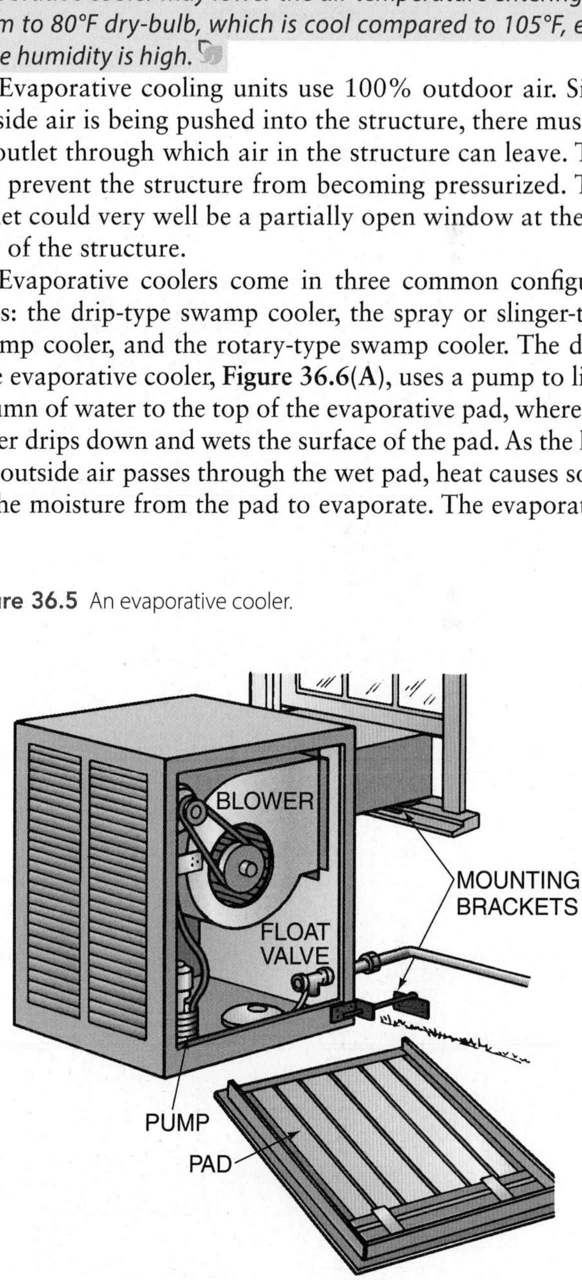

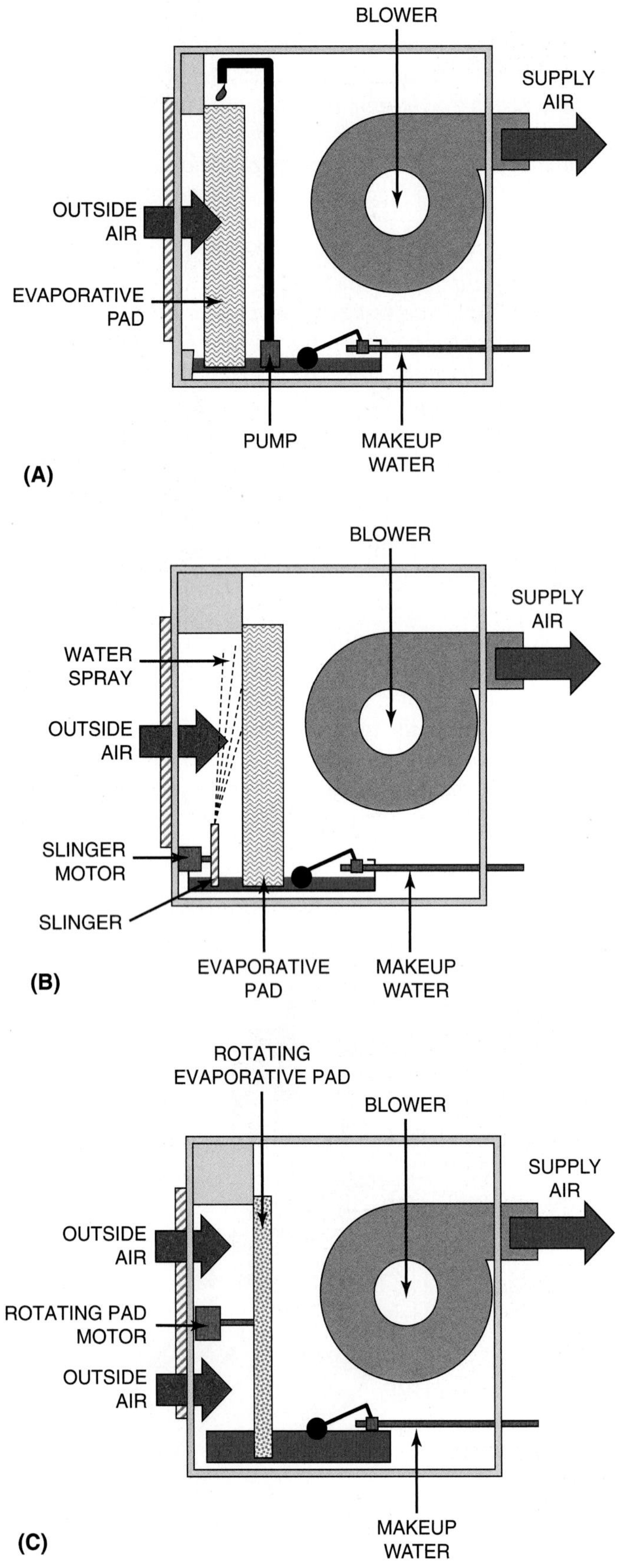

Figure 36.6 (A) A drip-type evaporative cooler. (B) A spray or slinger-type evaporative cooler. (C) A rotary-type evaporative cooler.

moisture absorbs heat from the airstream passing through the pad, and the airstream is cooled. The spray or slinger-type evaporative cooler uses a pump to spray water onto the evaporative pad, **Figure 36.6(B)**. Just as with the drip-type swamp cooler, the moisture from the pad evaporates, cooling the airstream. The rotary-type evaporative cooler uses a motor to rotate the evaporative pad in the water sump, **Figure 36.6(C)**. As the pad rotates, it becomes wet and functions in a manner similar to the other two types just discussed.

36.4 PACKAGE-TYPE AND SPLIT-TYPE AIR-CONDITIONING SYSTEMS

Air-conditioning, comfort cooling, or high-temperature refrigeration systems operate with the same four main system components that are found in the typical refrigeration system. The system components can be configured differently depending on the application for which the system is to be used. Common system configurations are the package-type system and the split-type system.

PACKAGE AIR-CONDITIONING

With package equipment all of the components are built into one cabinet much like a window unit. It is also referred to as *self-contained* equipment. Package-type air-conditioning units have all of the refrigerant piping contained within the unit's cabinet and come charged from the factory. There is no need for field charging of these systems. Air is carried between the unit and the conditioned space by two dedicated runs of ductwork, one for the supply air and one for the return air. Package equipment may be located beside the structure or on top of it. In some instances, the heating equipment is built into the same cabinet. Package units have three main heating options: natural gas heating, electric heating, and heat pump—the latter using the reverse cycle of refrigeration absorbing heat from outdoors during winter and rejecting the heat to the conditioned space. Package units come in three styles: ground level, rooftop, and multi position. Ground level package units are used when the roof cannot support the weight of the unit or when servicing the unit becomes a problem. Rooftop package units are the most common in the field. They provide safe operation away from the occupants while hiding the big bulky unit from sight. Rooftop units are the preferred method of installation due to location, which creates perfect conditions for optimum air circulation—hot air rising while cold air falls. Multi position package units have the ability to be used for either rooftop application or ground level application, and can save storage space for a wholesaler or contractor. Package units will be discussed in greater detail in Unit 50, "Commercial, Packaged Rooftop, Variable Refrigerant Flow, and Variable Air Volume Systems."

SPLIT-SYSTEM AIR-CONDITIONING

In **split-system** air-conditioning, the condenser and compressor are located outside, remote from the evaporator, and uses interconnecting refrigerant lines known in the field as the line set. In cooling-only applications, the outdoor portion of the system is referred to as the condensing unit, and is made up of the compressor and condenser coil. The evaporator may be located in the attic or a crawl space in the horizontal application, or a closet for upflow or downflow applications. The blower that moves air across the evaporator may be included in the furnace, or the air handler may be used for the air-conditioning system, **Figure 36.7**.

Interconnecting refrigerant lines must be field installed on split-type air-conditioning systems. The lines that are typically field installed are the liquid line and the suction line. The liquid line connects the outlet of the condenser at the outdoor unit to the inlet of the metering device at the indoor unit. The suction line connects the outlet of the evaporator at the indoor unit to the inlet of the compressor at the outdoor unit. The suction line, the larger of the two, will need to be insulated to prevent sweating. **Figure 36.8** shows the indoor and outdoor units of a split-type air-conditioning system.

One type of split system that is becoming very popular is the ductless split system. Ductless split systems consist of a matching outdoor unit, **Figure 36.9**, connected to one or more indoor fan coil units, **Figure 36.10**. Each room in the structure

Figure 36.7 A split air-conditioning system. Courtesy Climate Control

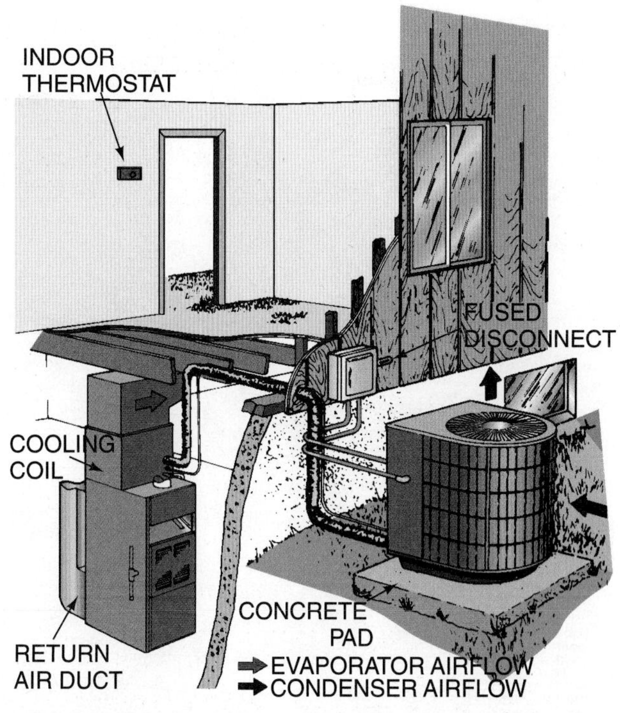

Figure 36.8 The operating conditions of an evaporator and a condenser. Notice that the cooling coil is in an enclosure to prevent sweating on the outside.

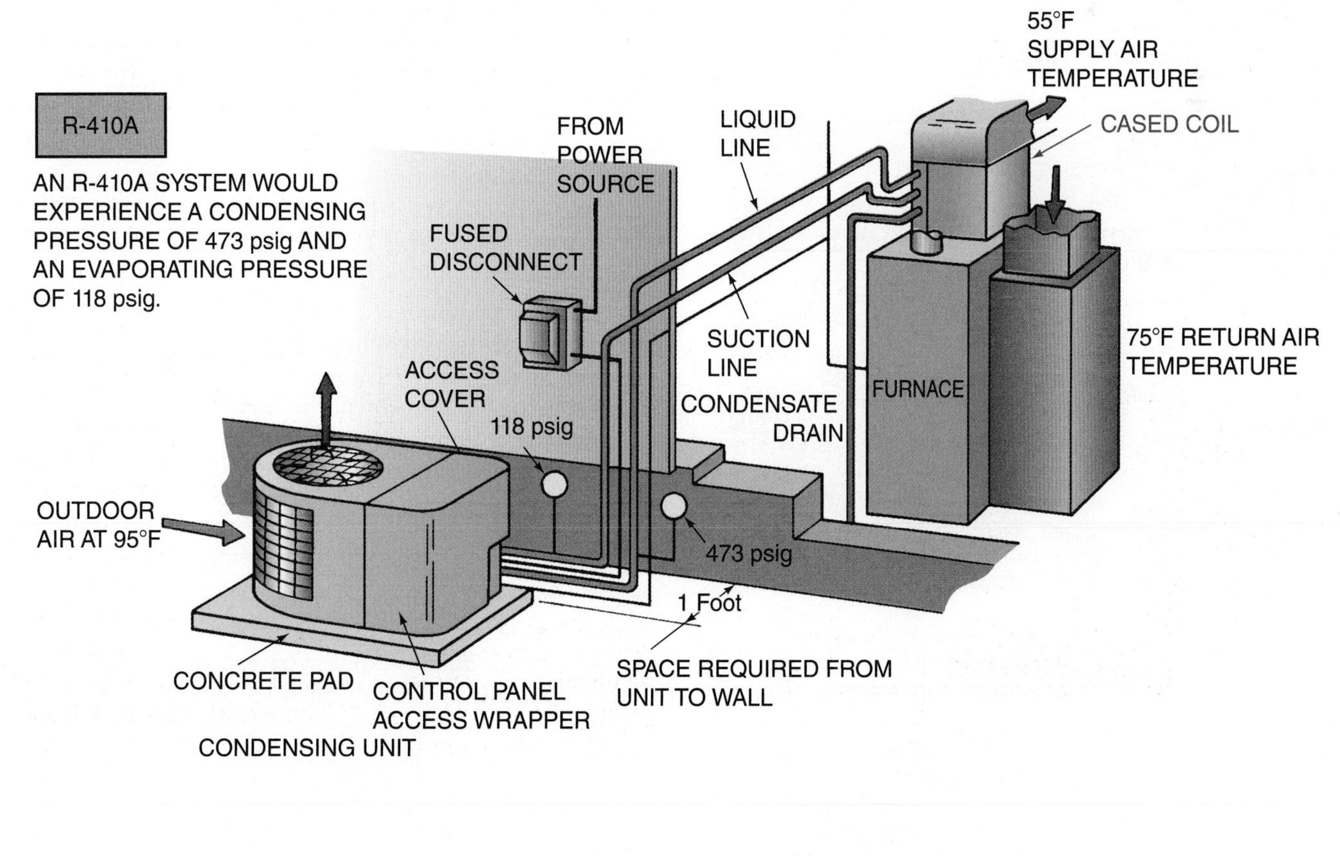

Figure 36.9 The outdoor unit of a ductless split-type air-conditioning system. Courtesy Daikin AC.

Figure 36.10 The indoor unit of a ductless split-type air-conditioning system. Photo by Eugene Silberstein.

has its own indoor unit. Ductless split systems with one outdoor unit and multiple indoor units, or heads, are often referred to as multihead systems. They have grown in popularity in the United States over the last 30 years due to their many possible configurations, which require no ductwork and can achieve high seasonal energy efficiency ratings. Most ductless split systems are equipped with remote controls, some wired and some wireless, to give the equipment user the ability to control the operation of the unit from a distance.

The outdoor portion of a ductless split system is referred to as the outdoor unit, not the condensing unit. This is because the vast majority of ductless split systems are heat pump systems, which have the ability to provide both comfort heating and comfort cooling. When these heat pump systems operate in the heating mode, the outdoor coil functions as the evaporator, so the outdoor coil is not responsible for rejecting system heat. In fact, when operating in the heating mode, the outdoor coil, is responsible for absorbing heat from the outdoor air. Refer to Unit 43, "Air Source Heat Pumps," for more information on heat pump theory.

As with traditional split-type air-conditioning systems, the installation of interconnecting refrigerant lines is required. On ductless split systems, however, the two lines that are field installed must both be insulated. This is because, when the system operates in the cooling mode, both refrigerant lines will carry low-pressure, low-temperature refrigerant, and will likely sweat.

Ductless split systems come in many styles such as compact ceiling cassette, ceiling cassette, high wall, under ceiling, slim duct, and floor mounted. These systems are used in many residential applications such as additions, renovations, home preservation projects, garages, attic conversations, sun rooms, and finished basements. Commercial applications include computer rooms, churches, bars, and restaurants, to list a few.

The biggest advantage of the ductless split system over conventional split systems is the duct-free design. Conventional split systems with ductwork have the potential, according to the DOE, to lose as much as 30% of their airflow due to air leaks; not so with ductless split systems.

Ductless split systems are able to achieve the desired space temperature faster with reduced air volume losses using inverter-controlled compressors, motors, and capacity controls. Refer to Unit 17, "Types of Electric Motors," for more information on inverter technology and inverter-controlled motors. Ductless split systems also offer smaller temperature swings in the conditioned space, making them a highly efficient option, while providing even heating and cooling in the occupied space. Moreover, individual room zoning provides the means for the homeowner to control the temperature in each and every room, via individual fan coil units. This zoning option allows for greater comfort and savings.

These systems can be installed in just about any application where heating and cooling is desirable, which makes them extremely marketable options for both homeowners and contractors. Contractors are gravitating to ductless split systems for their clients due to their ease of installation.

Heat pump ductless split systems give the advantage of both heating and cooling throughout the year. The operation is similar to conventional heat pump systems, which absorb heat, both sensible and latent, from the conditioned space in

the cooling mode, while rejecting heat into the conditioned space in the heating mode.

Appearance is a major drawback with ductless systems, so manufacturers continue to seek other options. Many manufacturers of ductless systems have tried to improve their fan coil unit appearance by making them look less obtrusive in the living space. These systems require that precise heat gain and heat loss calculations be performed to ensure proper system sizing.

36.5 THE EVAPORATOR

The evaporator is a heat exchange surface that takes the heat from the room air and transfers it to the refrigerant. Two kinds of heat must be transferred: sensible heat and latent heat. Sensible heat lowers the air temperature, and latent heat changes the water vapor in the air to condensate. On average, about one quart of condensate will be produced per ton of system capacity per hour. The condensate collects on the coil and runs through the drain pan to a p-trap (to stop air from pulling into the drain), and then it is normally piped to a drain.

Typically, room air may be 75°F dry-bulb and have a humidity of 50%, which is 62.5°F wet-bulb. The evaporator coil generally operates at a refrigerant temperature that is about 35°F below the temperature of the return air. At design conditions, this equates to an evaporator saturation temperature of about 40°F. At this temperature, the coil can remove the required amount of sensible heat (to lower the air temperature) and latent heat (to remove the correct amount of moisture). The air leaving the coil is approximately 55°F dry-bulb with a relative humidity of about 95%, which is 54°F wet-bulb. These points have been plotted on the psychrometric chart in **Figure 36.11**. The return air conditions have been plotted in red, while the supply air conditions have been plotted in blue. Notice that the relative humidity of the air leaving the cooling coil is much higher than the relative humidity of the return air. This is because the air has been cooled. However, the absolute humidity of the supply air is 60 grains per pound of air, as opposed to the absolute humidity of the return air, which is 65 grains per pound of air. The result is dehumidification. When the supply air mixes with the air in the room, it heats up and the relative humidity begins to drop.

These conditions are average for a climate with high humidity. If the humidity is very high, such as in coastal locations, the coil temperature may be a little lower to remove more humidity. If the system is in a locality where the humidity is very low, such as in desert areas, the coil temperature

Figure 36.11 Supply and return air conditions plotted on a psychrometric chart. The absolute humidity drops as a result of the reduced moisture content of the cooled, supply airstream.

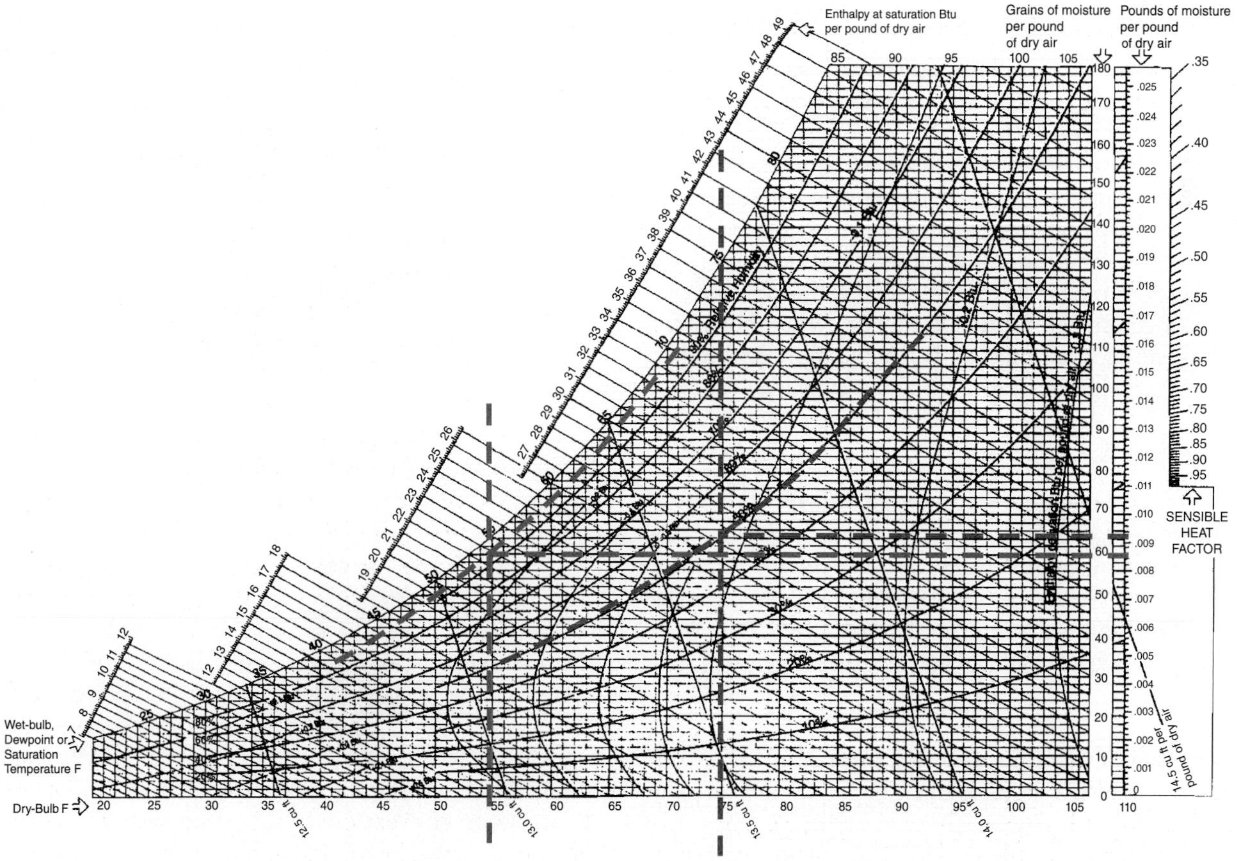

may be a little higher than 40°F. The coil temperature is controlled by the airflow across the evaporator coil. More airflow will cause higher coil temperatures, and less airflow will cause lower coil temperatures. The same equipment can be sold in all parts of the country because the airflow can be varied to accomplish the proper evaporator temperature for the proper humidity. Fin spacing on the evaporator may also affect the leaving air's humidity levels. Closer fin spacing will take more humidity out of the air. In an effort to provide greater comfort to the condition space, manufacturers are using variable frequency drive motors that modulate air volume to provide the greatest amount of evaporator efficiency.

36.6 EVAPORATOR COIL CONFIGURATIONS

The evaporator is the air-conditioning system component that absorbs heat into the refrigeration system. It is a refrigeration coil typically made of aluminum or copper with aluminum fins attached to the coil of either type to give more surface area for better heat exchange. There are several designs for airflow through the evaporator coil and for draining the condensate water from the coil, depending on the installation. Common evaporator coil configurations are the *A* coil, the *slant* coil, and the *H* coil.

THE A COIL

The *A coil* is used for upflow, downflow, and horizontal flow applications. It consists of two coils with side-by-side circuits that are spread apart at the bottom in the shape of the letter A, **Figure 36.12.** In upflow or downflow applications, the condensate pan is at the bottom of the A pattern. When used for horizontal flow, a pan is placed at the bottom of the coil and the coil is turned on its side. With the two coils in series the airflow in an A coil is through the core of the coil; it cannot be from side to side. When horizontal airflow is required, slant or H coils may be more desirable.

THE SLANT COIL

The *slant coil* is a one-piece coil mounted in the duct on an angle (usually 60°), or slant, to give the coil more surface area. The slant of the coil causes the condensate water to drain to the condensate pan located at the bottom of the slant. The coil can be used for upflow, downflow, or horizontal flow when designed for these applications, **Figure 36.13.**

THE H COIL

The *H coil* is normally used in horizontal applications where space is limited, such as crawl space, attics, basements, and above drop ceilings. They can be adapted to vertical applications by using special drain pan configurations. The condensate drains are normally at the bottom of the H pattern, **Figure 36.14,** to allow for proper drainage.

Figure 36.12 An A coil. Notice that two coils are standing at an angle and touching at the top—shaped like an A.

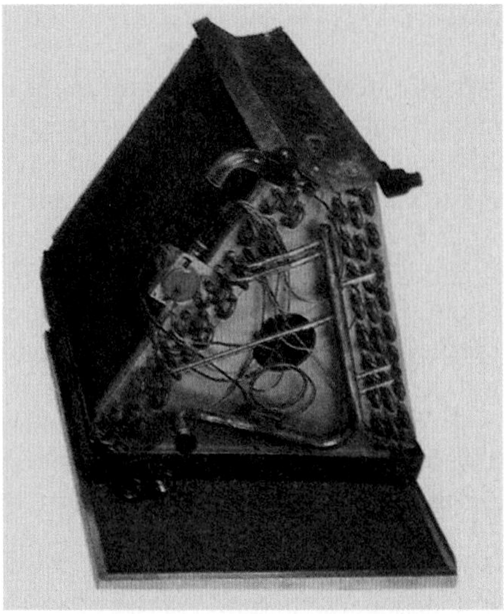

Figure 36.13 A slant coil. Courtesy BDP Company

Figure 36.14 An H coil. Courtesy BDP Company

Figure 36.15 A multi-circuit coil. Notice that the expansion valve has four outlet tubes and that the suction line has four connections. There are four completely different circuits.

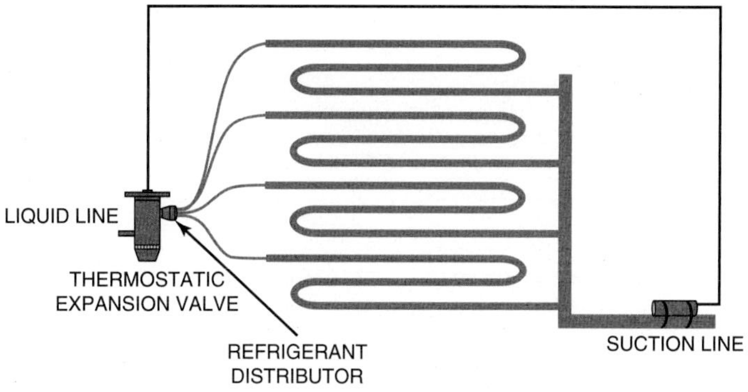

COIL CIRCUITS

All of the aforementioned coils may have more than one refrigeration circuit for the refrigerant to travel. As was indicated in Unit 21, "Evaporators and the Refrigeration System," when a coil becomes too long and an excessive pressure drop occurs, it is advisable to configure the coil with multiple parallel circuits, **Figure 36.15**. Coils configured in this manner are referred to as multicircuit coils. The coil may have as many circuits as necessary to reduce the pressure drop through the coil while maintaining the proper amount of superheated vapor at the outlet of the evaporator. When there is more than one circuit, a distributor must be used to distribute the correct amount of refrigerant to the individual circuits, **Figure 36.16**. The distributor is located at the outlet of the metering device to ensure that refrigerant is fed evenly to each of the circuits in the coil. These circuits travel through the coil to the suction manifold, where superheated suction gas makes its way through the suction line.

MICROCHANNEL (FLAT-PLATE-AND-FIN) TECHNOLOGY

Efficient heat exchange between the refrigerant and air has been improved in the past few years by means of different coil designs. In the past, all coils have been copper or aluminum tubes with fins attached to the tubes. Coil sizes were limited to the tube sizes, typically 3/8 in. and 5/16 in. New coil designs have been developed that resemble automobile radiators, **Figure 36.17**. Small passages with fins, called microchannel or flat-plate-and-fin coils, have made it possible to build a smaller coil with more heat exchange surface that requires less refrigerant charge. Many manufacturers use these coils for evaporators and condensers. Remember, when smaller, more efficient equipment can be built, it is a savings in materials, manufacturing time, storage space, and shipping costs and helps all the way from material purchasing to installation. These coils are discussed in detail in Unit 21, "Evaporators and the Refrigeration System."

Figure 36.16 A refrigerant distributor. It is responsible for equally distributing refrigerant to each of the three circuits in this coil.

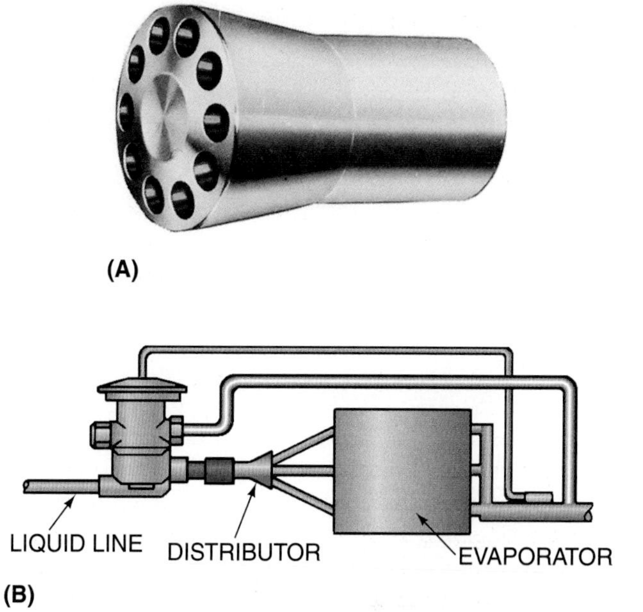

Figure 36.17 Cutaway of a microchannel coil. Courtesy Modine Manufacturing Co., Racine, WI.

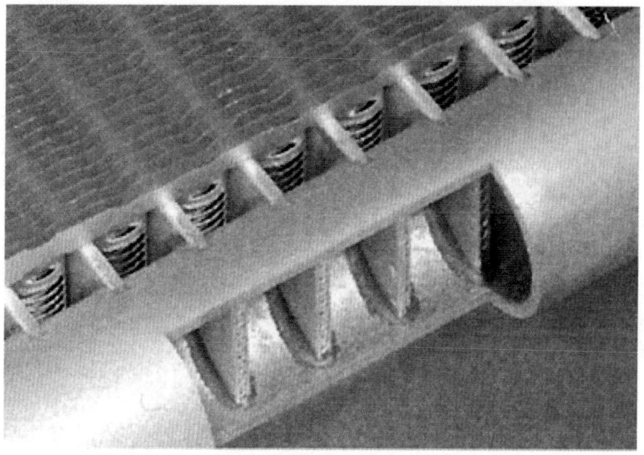

36.7 DESIGN CONDITIONS

An air-conditioning system designer or engineer must be familiar with local design temperatures and conditions. This information is published in tables such as the one shown in **Figure 36.18**. Comparing columns 6 and 8 in **Figure 36.18**,

Figure 36.18 This excerpt from a table shows different design conditions for various parts of the United States. Reprinted with permission from American Society of Heating, Refrigerating, and Air-Conditioning Engineers

CLIMATIC CONDITIONS FOR THE UNITED STATES[a]

Col. 1	Col. 2		Col. 3		Col. 4	Col. 5 (Winter[d])		Col. 6 (Summer[e])			Col. 7	Col. 8		
State and Station	Latitude[b]		Longitude[b]		Elevation[c]	Design Dry-Bulb		Design Dry-Bulb and Mean Coincident Wet-Bulb			Mean Daily Range	Design Wet-Bulb		
	°	′	°	′	Ft	99%	97.5%	1%	2.5%	5%		1%	2.5%	5%
ARIZONA														
Douglas AP	31	3	109	3	4098	27	31	98/63	95/63	93/63	31	70	69	68
Flagstaff AP	35	1	111	4	6973	−2	4	84/55	82/55	80/54	31	61	60	59
Fort Huachuca AP (S)	31	3	110	2	4664	24	28	95/62	92/62	90/62	27	69	68	67
Kingman AP	35	2	114	0	3446	18	25	103/65	100/64	97/64	30	70	69	69
Nogales	31	2	111	0	3800	28	32	99/64	96/64	94/64	31	71	70	69
Phoenix AP (S)	33	3	112	0	1117	31	34	109/71	107/71	105/71	27	76	75	75
Prescott AP	34	4	112	3	5014	4	9	96/61	94/60	92/60	30	66	65	64
Tucson AP (S)	32	1	111	0	2584	28	32	104/66	102/66	100/66	26	72	71	71
Winslow AP	35	0	110	4	4880	5	10	97/61	95/60	93/60	32	66	65	64
Yuma AP	32	4	114	4	199	36	39	111/72	109/72	107/71	27	79	78	77
CALIFORNIA														
Bakersfield AP	35	2	119	0	495	30	32	104/70	101/69	98/68	32	73	71	70
Barstow AP	34	5	116	5	2142	26	29	106/68	104/68	102/67	37	73	71	70
Blythe AP	33	4	114	3	390	30	33	112/71	110/71	108/70	28	75	75	74
Burbank AP	34	1	118	2	699	37	39	95/68	91/68	88/67	25	71	70	69
Chico	39	5	121	5	205	28	30	103/69	101/68	98/67	36	71	70	68
Los Angeles AP (S)	34	0	118	2	99	41	43	83/68	80/68	77/67	15	70	69	68
Los Angeles CO (S)	34	0	118	1	312	37	40	93/70	89/70	86/69	20	72	71	70
Merced-Castle AFB	37	2	120	3	178	29	31	102/70	99/69	96/68	36	72	71	70
Modesto	37	4	121	0	91	28	30	101/69	98/68	95/67	36	71	70	69
Monterey	36	4	121	5	38	35	38	75/63	71/61	68/61	20	64	62	61
Napa	38	2	122	2	16	30	32	100/69	96/68	92/67	30	71	69	68
Needles AP	34	5	114	4	913	30	33	112/71	110/71	108/70	27	75	75	74
Oakland AP	37	4	122	1	3	34	36	85/64	80/63	75/62	19	66	64	63
Oceanside	33	1	117	2	30	41	43	83/68	80/68	77/67	13	70	69	68
Ontario	34	0	117	36	995	31	33	102/70	99/69	96/67	36	74	72	71
GEORGIA														
Albany, Turner AFB	31	3	84	1	224	25	29	97/77	95/76	93/76	20	80	79	78
Americus	32	0	84	2	476	21	25	97/77	94/76	92/75	20	79	78	77
Athens	34	0	83	2	700	18	22	94/74	92/74	90/74	21	78	77	76
Atlanta AP (S)	33	4	84	3	1005	17	22	94/74	92/74	90/73	19	77	76	75
Augusta AP	33	2	82	0	143	20	23	97/77	95/76	93/76	19	80	79	78
Brunswick	31	1	81	3	14	29	32	92/78	89/78	87/78	18	80	79	79
Columbus, Lawson AFB	32	3	85	0	242	21	24	95/76	93/76	91/75	21	79	78	77
Dalton	34	5	85	0	720	17	22	94/76	93/76	91/76	22	79	78	77
Dublin	32	3	83	0	215	21	25	96/77	93/76	91/75	20	79	78	77
Gainesville	34	2	83	5	1254	16	21	93/74	91/74	89/73	21	77	76	75
NEW YORK														
Albany AP (S)	42	5	73	5	277	−6	−1	91/73	88/72	85/70	23	75	74	72
Albany CO	42	5	73	5	19	−4	1	91/73	88/72	85/70	20	75	74	72
Auburn	43	0	76	3	715	−3	2	90/73	87/71	84/70	22	75	73	72
Batavia	43	0	78	1	900	1	5	90/72	87/71	84/70	22	75	73	72
Binghamton AP	42	1	76	0	1590	−2	1	86/71	83/69	81/68	20	73	72	70
Buffalo AP	43	0	78	4	705r	2	6	88/71	85/70	83/69	21	74	73	72
Cortland	42	4	76	1	1129	−5	0	88/71	85/71	82/70	23	74	73	71
Dunkirk	42	3	79	2	590	4	9	88/73	85/72	83/71	18	75	74	72
Elmira AP	42	1	76	5	860	−4	1	89/71	86/71	83/70	24	74	73	71
Geneva (S)	42	5	77	0	590	−3	2	90/73	87/71	84/70	22	75	73	72

[a]Table 1 was prepared by ASHRAE Technical Committee 4.2, Weather Data, from data compiled from official weather stations where hourly weather observations are made by trained observers, See also Ref 1, 2, 3, 5 and 6.

[b]Latitude, for use in calculating solar loads, and longitude are given to the nearest 10 minutes. For example, the latitude and longitude for Anniston, Alabama are given as 33 34 and 85 55 respectively, or 33° 40, and 85° 50.

[c]Elevations are ground elevations for each station. Temperature readings are generally made at an elevation of 5 ft above ground, except for locations marked r, indicating roof exposure of thermometer.

[d]Percentage of winter design data shows the percent of the 3-month period, December through February.

[e]Percentage of summer design data shows the percent of the 4-month period, June through September.

it can be seen that an air-conditioning system for a house built in Augusta, Georgia will be designed using a dry-bulb temperature of 97°F (from the left portion of column 6 in **Figure 36.18**) and a dry-bulb temperature of 80°F (from the left portion of column 8 in **Figure 36.18**). A similar house that is to be built in Phoenix, Arizona will be designed using a dry-bulb temperature of 109°F (from the left portion of column 6 in **Figure 36.18**) and a dry-bulb temperature of 76°F (from the left portion of column 8 in **Figure 36.18**). The design dry-bulb temperature in Phoenix is higher than that in Augusta, but the design wet-bulb temperature in Phoenix is lower than that in Augusta. The airflow across the same coil in the two different parts of the country may be varied to account and compensate for the different air conditions. People who choose equipment for conditioning homes and buildings either use the chart shown in **Figure 36.18** or some other chart that has been derived from this chart. The estimator, either a practical-design person or an engineer, will want the equipment to be sized for the best efficiency for the owner. The technician should also be familiar with this chart as a point of reference for understanding how designers choose equipment.

Examining the summer conditions of the two cities Augusta, Georgia, and Phoenix, Arizona, let's explore the different possibilities. There are three columns for the design dry-bulb (DB) and wet-bulb (WB) conditions. The 1% column for Augusta lists 97 DB/77 WB. This means that, on average, from records taken over time, Augusta conditions will be at 97 DB/77 WB 1% of the time during the summer. The 2.5% column shows 95/76 and the 5% column shows 93/76. The person selecting the equipment may use any of these figures. It is important to note that the average wet-bulb temperatures shown in column 6 are not the design wet-bulb temperatures. The design wet-bulb temperatures are provided in column 8 of **Figure 36.18**.

Single-speed, non-inverter-controlled equipment will perform at its best efficiency when running at full load. The starting and stopping of the system reduces the overall efficiency and life-expectancy of the system. Therefore, many designers try to size equipment so it will run as much as possible. If the equipment is selected for the 1% conditions, it will run at full load only 1% of the time. The rest of the time, it will be stopping and starting. If the equipment is selected for 5% conditions, it will be stopping and starting at conditions that fall below the 5% conditions. The problem with selecting equipment for 5% is that when the 1% conditions occur, the equipment will run all of the time and will not always maintain the conditions at a comfortable level for the occupants. This would also mean that the equipment would have no reserve capacity to lower the structure temperature in case of equipment downtime. For example, if the equipment is shut off for any reason for a period of time and the space conditions rise, it may not be able to reduce the space conditions until after an overnight run time. Suppose that a storm causes a power failure for several hours during the day—the unit will probably not catch up until the middle of the night when it can run at reduced load. In order to alleviate difficulties such as this, many designers compromise by selecting equipment for the 2.5% conditions.

Another strategy that is becoming very popular to effectively match the system to the required capacity is to use variable refrigerant flow systems. Variable capacity equipment utilizes inverter-controlled compressor motors and fan motors to modulate both the flow of air and the flow of refrigerant through the system to adjust the capacity of the unit to fit the design conditions at the time. The variable-capacity units all use some form of electronics to control capacity. They can control the space temperature and humidity, yet are designed so that you do not have to be an electronics technician to troubleshoot the systems.

36.8 THE COMPRESSOR

The following types of compressors are used in air-conditioning systems: reciprocating, rotary, scroll, centrifugal, and screw. These are of the same design as similar types used in commercial refrigeration. The centrifugal and screw compressors, described briefly in Unit 23, "Compressors," are used primarily in large commercial and industrial applications and will be discussed further later in this text.

Compressors are vapor pumps that pump heat-laden refrigerant vapor from the low-pressure side of the system, the evaporator side, to the high-pressure, or condenser, side. On a properly operating system, the refrigerant that enters and leaves the compressor is a superheated vapor. To do this they compress the vapor from the low side, increasing the pressure and raising the temperature.

Older small and medium-size air-conditioning compressors were single-speed pumps and turned at the standard motor speeds of 3450 rpm or 1725 rpm. Early compressors were 1750 rpm, used R-12, and were large and heavy. Newer single-speed compressors operated at the faster speed and used more efficient refrigerants like R-22, R-410A, and R-407C. Most systems that are now being manufactured use variable-speed electronics to vary the compressor motor speed to control and adjust capacity. This is referred to as variable refrigerant flow technology. For more detailed information on variable refrigerant flow systems, refer to Unit 50, "Commercial, Packaged Rooftop, Variable Refrigerant Flow, and Variable Air Volume Systems."

36.9 THE RECIPROCATING COMPRESSOR

The reciprocating compressors used for residential and light commercial air-conditioning are similar to those discussed in Unit 23, "Compressors." Reciprocating compressors used in air-conditioning are mainly cooled by suction gas and air passing over the shell of the compressor. If suction-gas-cooled, the suction line is piped to the motor end of the compressor, **Figure 36.19**. The maximum suction-gas temperature is usually about 70°F. Reciprocating compressors rely on pistons that

Figure 36.19 A serviceable hermetic compressor. Courtesy Copeland Corporation

move up and down in cylinders to perform the compression process. These compressors are not typically found in newly manufactured systems, as reciprocating compressors are not as efficient as other compressor types.

Reciprocating compressors in residential systems are usually the fully hermetic, suction-gas-cooled pumps, **Figure 36.20**. They are positive-displacement compressors using R-22, R-410A, or R-407C refrigerants. Because R-22 is an HCFC refrigerant, which contains chlorine, it was phased out of production in January 2020. R-410A and R-407C are two popular refrigerant blend replacements for R-22 in residential and commercial air-conditioning applications. R-410A is designed for new applications and R-407C can be used in new and retrofit applications.

Figure 36.20 A suction-gas-cooled compressor. Courtesy Tecumseh Products Company

36.10 COOLING THE COMPRESSOR AND MOTOR

All compressors must be cooled because a significant amount of heat energy is generated during the compression process. Without cooling, the compressor motor would burn out and the oil would become hot enough to break down and form carbon. Hermetic compressors have always been suction-gas-cooled. Rotary compressors are cooled with the discharge gas leaving the compressor. Discharge gas cools the compressor motor due to the fact the motor is much hotter than the discharge gas. However, the discharge gas temperature is influenced by the suction gas temperature. If the correct superheat is not maintained at the compressor, it will overheat because the discharge gas temperature will rise. So in the end the compressor motor temperature is still controlled by the suction gas temperature.

Newer compressor motors are able to run at higher temperatures because the motor windings are made of better materials and lubricating oils are of higher quality. Better insulation materials are used to help ensure that the windings will not overheat and break down while running under high-heat conditions. **NOTE:** *When dealing with a discharge-cooled compressor, the technician should determine that it is not running too hot.•* The housing of a discharge-cooled compressor is hotter than that of a suction-cooled compressor. Checking the discharge line temperature is the sure test. Most manufacturers agree that the discharge line leaving the compressor should not exceed 225°F. The discharge line temperature on R-410A systems is considerably lower, with an average temperature in the 175°F range.

36.11 THE ROTARY COMPRESSOR

The rotary compressor is small and light. It is sometimes cooled with compressor discharge gas, which makes the compressor appear to be running too hot. A warning about the hot temperature may be posted in the compressor compartment. The compressor and motor are pressed into the shell of a rotary compressor, and there is no vapor space between the compressor and the shell. This is another

Figure 36.21 A rotary compressor (A)–(B). Courtesy Motors and Armatures, Inc.

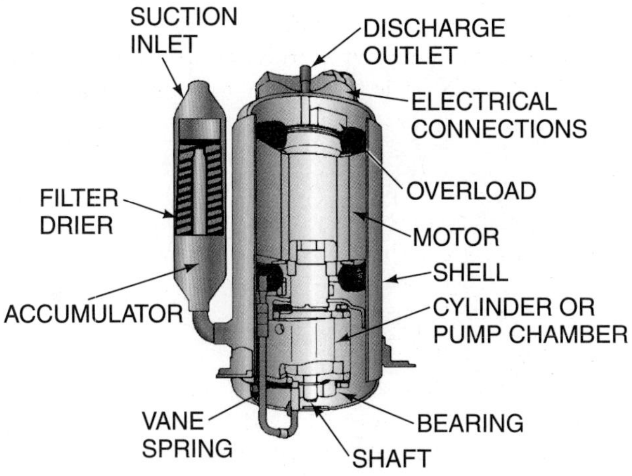

(A)

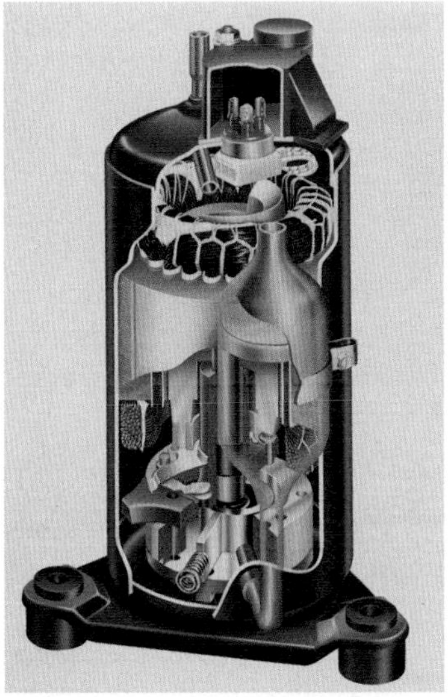

(B)

reason why the rotary compressor is smaller than a reciprocating compressor. The rotary compressor, which is used in small- to medium-sized systems, is more efficient than the reciprocating compressor, **Figure 36.21**. The primary reason the rotary compressor is highly efficient is that all the refrigerant that enters the intake port is discharged through the exhaust or discharge port. There is no clearance volume as there is in the reciprocating compressor. Rotary compressors are manufactured in two basic design types: the stationary vane and the rotary vane.

STATIONARY VANE ROTARY COMPRESSOR

The components of the stationary vane compressor are the housing, a blade or vane, a shaft with an off-center (eccentric) rotor, and a discharge valve. The shaft turns the off-center rotor so that it "rolls" around the cylinder, **Figure 36.22**. The blade or vane keeps the intake and compression chambers of the cylinder separate. The tolerances between the rotor and cylinder must be very close, and the vane must be machined so that gas does not escape from the discharge side to the intake. As the rotor turns, the vane slides in and out to remain tight against the rotor. The valve at the discharge keeps the compressed gas from leaking back into the chamber and into the suction side during the off cycle.

As the shaft turns, the rotor rolls around the cylinder, allowing suction gas to enter through the intake and compresses the gas on the compression side. This is a continuous process as long as the compressor is running. The results are similar to those with a reciprocating compressor. Low-pressure suction gas enters the cylinder, is compressed (which also causes its temperature to rise), and leaves through the discharge opening.

ROTARY VANE ROTARY COMPRESSOR

The rotary vane compressor has a rotor fitted to the center of the shaft, and the rotor and the shaft have the same center. The rotor has two or more vanes that slide in and out and trap and compress the gas, **Figure 36.23**. The shaft and rotor are positioned off-center in the cylinder. As a vane passes by the suction intake opening, low-pressure gas follows it. This gas continues to enter and is trapped by the next vane, which compresses it and pushes it out the discharge opening. Notice that the intake opening is much larger than the discharge opening. This is to allow the low-pressure gas to enter more readily. The process of intake and compression is continuous as long as the compressor is running.

36.12 THE SCROLL COMPRESSOR

Compression in a scroll compressor, **Figure 36.24**, takes place between two spiral-shaped, nested scrolls. One of the scrolls is stationary, while the other operates with an orbiting action within it, **Figure 36.25**. The orbiting movement draws gas into a pocket between the two spirals. As this action continues, the gas opening is sealed off and the gas is forced into a smaller pocket at the center, **Figure 36.26**. The figure illustrates only one pocket of gas, but actually, most of the scroll or spiral is filled with gas and compression or "squeezing" occurs in all the pockets. Several stages

Figure 36.22 Operation of a stationary vane rotary compressor.

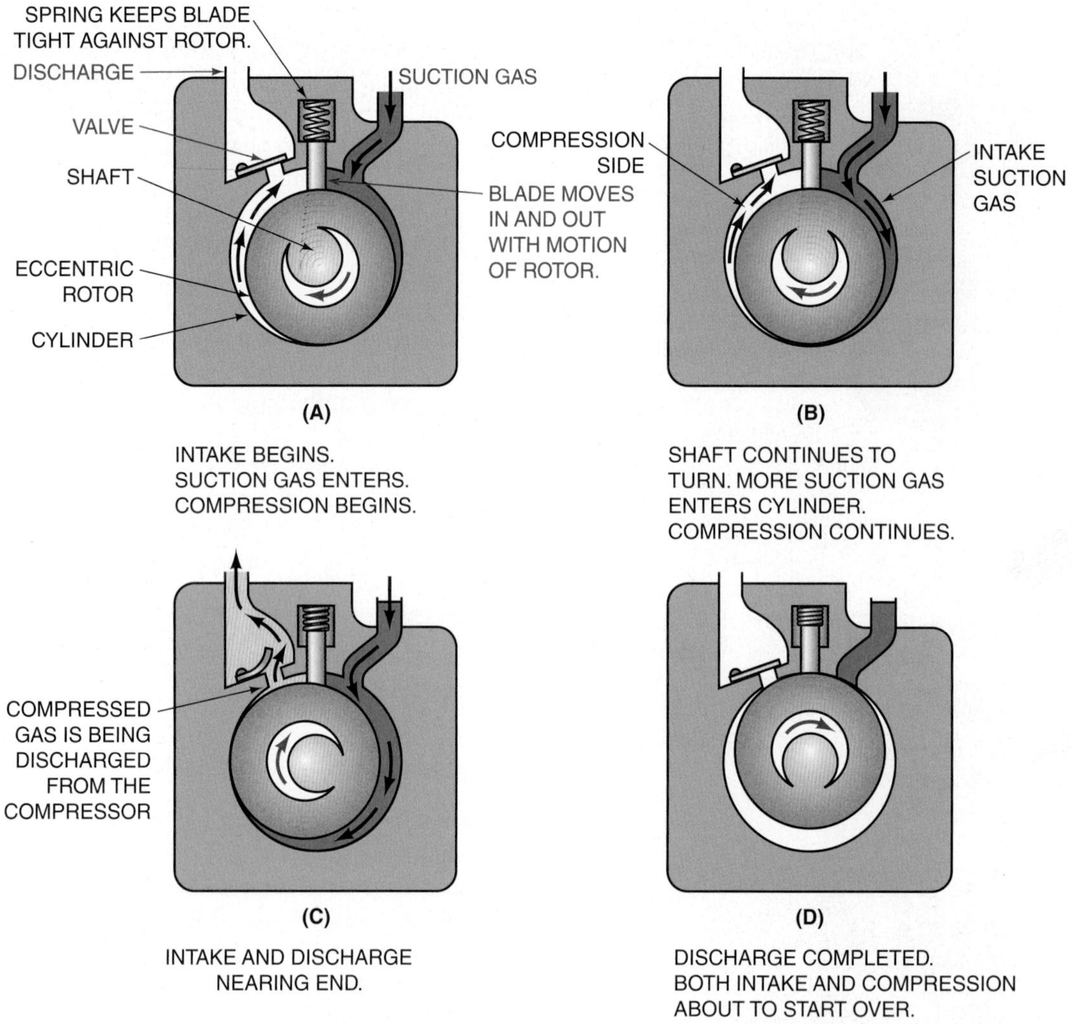

Figure 36.23 Operation of a rotary vane rotary compressor.

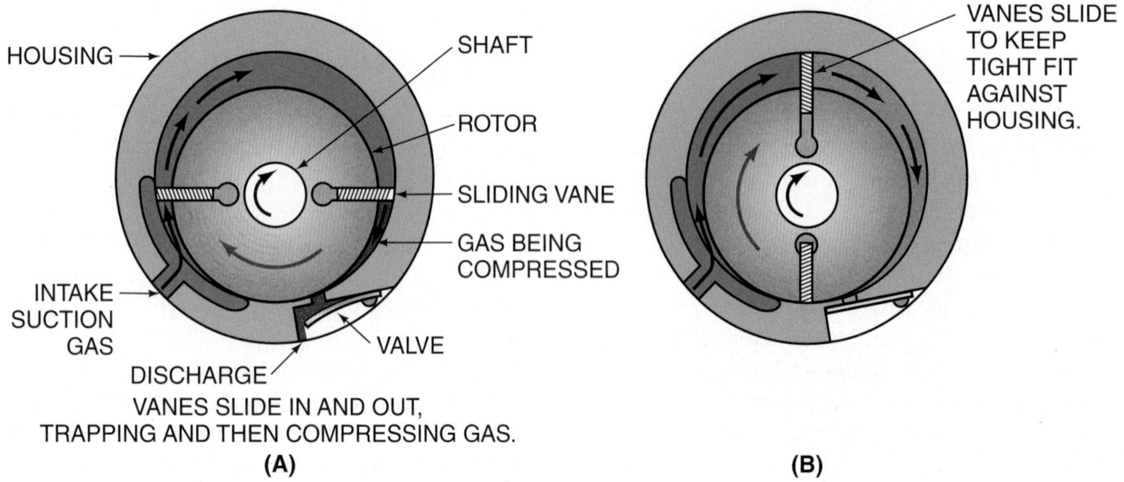

Figure 36.24 Cutaway of a scroll compressor. Courtesy Hampden Engineering.

Figure 36.25 Orbiting action between the scrolls in a scroll compressor.

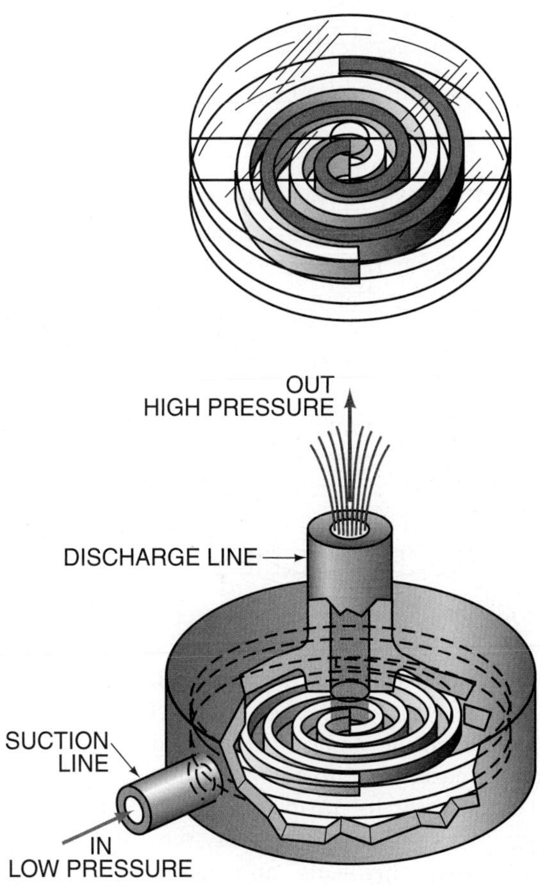

Figure 36.26 Compression of one pocket of refrigerant in a scroll compressor. Courtesy Copeland Corporation

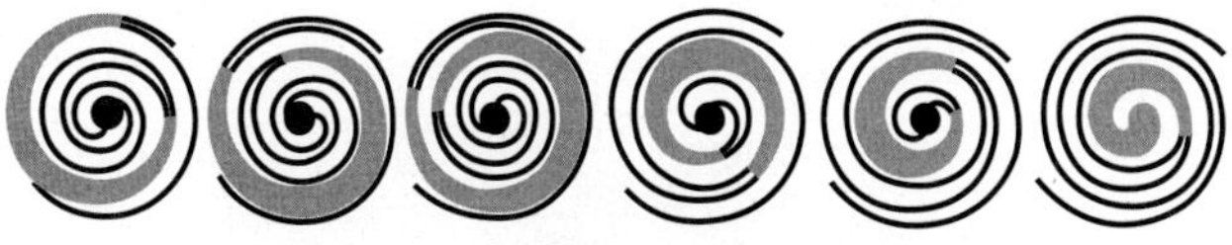

of compression are occurring at the same time, which acts to balance the compression action. The scroll compressor is quiet compared to reciprocating compressors of the same size. In addition, scroll compressors are more efficient than reciprocating compressors, making them the compressor of choice for most newly manufactured air-conditioning equipment.

Contact between the spirals that seal off the pockets is achieved using centrifugal force. This minimizes gas leakage. However, should liquid refrigerant or debris be present in the system, the scrolls will separate without damage or stress to the compressor. The high performance of the scroll design depends on good axial and radial sealing. Axial sealing is usually accomplished with floating seals. Radial sealing is usually accomplished by only a thin film of oil. Therefore, there is little wear.

The scroll compressor is highly efficient due to the following factors:

- It requires no valves; consequently, there are no valve compression losses.
- The suction and discharge locations are separate, which reduces the heat transfer between the suction and discharge gas, so there is less difference in temperature between the various stages of compression occurring in the compressor.
- There is no clearance volume providing reexpansion gas as in a reciprocating compressor.
- Over time, the nested scrolls tend to "wear in," creating a better seal between them.

36.13 THE CONDENSER

In air-conditioning equipment the condenser is the component that rejects the heat from the system. Most equipment is air-cooled and rejects heat to the air. The coils are copper or aluminum and have aluminum fins to increase the heat exchange surface area, **Figure 36.27**. As mentioned earlier, new coil designs are being used for some condensers to make them smaller and more efficient. The condenser is part of the condensing unit, which is the piece of equipment that contains the compressor, the condenser coil, the condenser fan, and the outdoor unit controls. Refrigerant enters the condenser as a high-pressure, high-temperature vapor and leaves the condenser as a high-pressure, high-temperature subcooled liquid.

Figure 36.27 Condensers have either copper or aluminum tubes with aluminum fins. Courtesy Carrier Corporation

Figure 36.29 Equipment that discharges air from the top of the cabinet. Courtesy York Corporation

SIDE-AIR-DISCHARGE CONDENSING UNITS

Air-cooled condensers all discharge hot air, which contains heat absorbed from inside the structure. Some condensing units discharge the air out the side and are called side discharge units. The advantage of this equipment is that the fan and motor are located under the top panel, **Figure 36.28**. However, any noise generated inside the cabinet is discharged into the leaving airstream and may be clearly heard in a neighbor's yard. **NOTE:** *The heat from the condenser coil can be hot enough to kill plants that it blows on. These condensers are still in use.•* Ductless split systems utilize side-discharge outdoor units. The condensing unit shown in **Figure 36.28** is a side discharge unit. Since ductless split systems utilize inverter-controlled compressors and fan motors, the noise that is generated by these systems is very low.

Figure 36.28 The fan motor and all components are located under the top panel. Courtesy Carrier Corporation

TOP-AIR-DISCHARGE CONDENSING UNITS

The most popular configuration for condensing units in residential equipment is the top-discharge design, **Figure 36.29**. In this type of unit, the hot air and noise are discharged from the top of the unit into the air. This is advantageous as far as air and noise are concerned, but because the fan and motor are on top, rain, snow, and leaves can fall directly into the unit. The fan motor needs to be protected with a rain shield, **Figure 36.30**. The fan motor bearings are in a vertical position, which means there is more thrust on the end of the bearing. In this type of application, therefore, the bearing needs a thrust surface, **Figure 36.31**.

CONDENSER COIL DESIGN

Some coil surfaces are positioned vertically, which allows grass and dirt to easily get into the bottom of the coil. Coils must be clean for the condenser to operate efficiently. Condenser coils should be cleaned every 3 to 5 years. Caution should be taken to read cleaning products' instructions carefully. Most coil cleaning products must be diluted with water for proper cleaning. Not all condenser coil chemicals can be used on every type of condenser coil. Improperly used chemicals can cause refrigerant leakages.

Figure 36.30 In units that discharge air from the top, the fan is usually on the top.

Figure 36.31 Top-air-discharge units place an additional load on the bearings. The fan is trying to fly down while it is running and pushing air out the top. During the starting and stopping of the fan, the fan blade and shaft rest on the bottom bearing. This bearing must have a thrust surface.

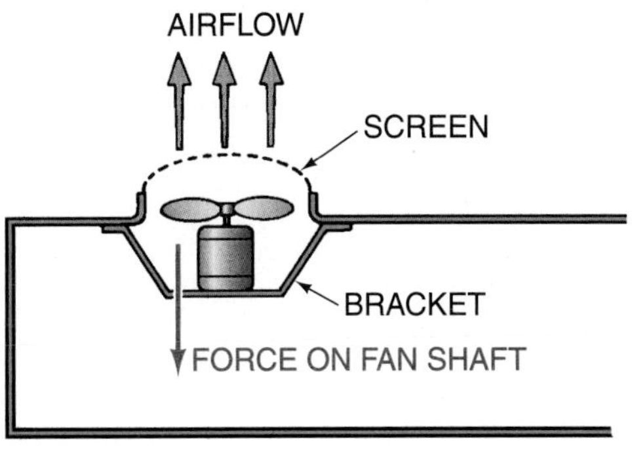

The bottom few rows of the coil in some equipment are used as a subcooling circuit to lower the condensed refrigerant temperature below the condensing temperature. It is common for the liquid line to be 10°F to 15°F cooler than the condensing temperature. Each degree of subcooling will add approximately 1% to the efficiency of the system. If the subcooling circuit is dirty, it could lower capacity by 10% to 15%. A decrease in capacity may affect the system's ability to satisfy the cooling needs of the structure.

HIGH-EFFICIENCY CONDENSERS

The HVACR industry as a whole and the federal government have demanded that air-conditioning equipment become more efficient. Probably the best way to improve efficiency is to lower the head pressure so that the compressor does not work as hard due to a lower compression ratio. The most common way to accomplish this is to increase the physical size of the condenser coils. More surface area in the condenser reduces the compressor head pressure even in the hottest weather. This means lower compressor current and less power consumed for the same amount of air-conditioning. Refer to Unit 22, "Condensers," for more detailed information on high-efficiency condensers, condenser splitting, condenser fan cycling, variable-speed condenser fan controls, variable-frequency drives, and head pressure control devices.

CONDENSER CABINET DESIGN

The condenser cabinet is usually located outside, so it needs to be weatherproof. Most cabinets are galvanized and painted to give them more years of life without rusting. Some cabinets are made of aluminum, which is lighter but may not last as long in a salty environment, such as coastal areas. Most small equipment is assembled with self-tapping sheet metal screws. These screws should be made of weather-resistant material that will last for years out in the weather. In the salt air of coastal areas, stainless steel is a good choice for sheet metal screws. When equipment that is assembled with drill screws is installed in the field, all of the screws should be fastened back into the cabinet tightly, or the unit may rattle. After being threaded many times, the screw holes may become oversized; if so, use the next half-size screw to tighten the cabinet panels.

36.14 EXPANSION DEVICES

The expansion device controls the flow of refrigerant into the evaporator. The TXV and the fixed-bore metering device (either a capillary tube or an orifice) are the types most often encountered on air-conditioning systems, but the TXV is almost exclusively used on all newly manufactured equipment. Electronic expansion valves (EEVs) are used on almost all ductless split systems. The TXVs are the same types as those described in Unit 24, "Expansion Devices," except that they have a different temperature range. Air-conditioning expansion devices are in the high-temperature range. TXVs are more efficient than fixed-bore devices because they allow the evaporator to reach peak performance faster. They allow more refrigerant into the evaporator coil during a hot pulldown—when the conditioned space is allowed to get too warm before the unit is started. Since the TXV maintains a constant evaporator superheat,

an increase in system load will increase the evaporator superheat, causing the TXV to open.

With a TXV, the refrigerant pressures do not equalize during the off cycle unless the valve is equipped with a bleed port. When the pressures do not equalize during the off cycle, a compressor with a high-starting-torque motor must be used. This means that the compressor must have a start capacitor and starting relay for start-up after the system has been off. Some expansion valve manufacturers make a bleed port that always allows a small amount of refrigerant to bleed through. During the off cycle, this valve will allow pressures to equalize and so a compressor with a low-starting-torque motor can be used. Low-starting-torque motors are less expensive and have fewer parts that may malfunction.

36.15 AIR-SIDE COMPONENTS

The air side of an air-conditioning system consists of the indoor blower as well as the supply air and the return air ducts. The airflow in an air-conditioning system is normally in the range of 400 cfm/ton in average humidity climates, but can vary significantly since many newer systems operate with variable speed blowers. A different airflow may be used for other climates. In humid coastal areas, airflow may be 350 cfm/ton, and 450 cfm/ton may be used in desert areas. The air leaving the air handler could be expected to be about 55°F in the average system operating at or near design conditions. The ductwork carrying this air must be insulated when it runs through an unconditioned space or it will sweat and gain unwanted heat from the surroundings. In the case when the duct work carries heated air, uninsulated ducts will result in the transfer of heat from the air to the unconditioned air surrounding the ducts. The insulation should have a vapor barrier to prevent the moisture from penetrating and collecting on the metal duct. All seams in the insulation must be sealed securely. If an air distribution system is located within a conditioned space, there will be no duct heat losses. This is because the air in the duct and the air surrounding the duct will ultimately be mixed together anyway.

When operating under near-design conditions, the return air will normally be about 75°F. If the return air duct is run through an unconditioned space, it may not need insulation. Even if a small amount of heat did exchange, the cost of insulating versus allowing the small heat exchange should be evaluated to see which is more economical. If the duct is run through a hot attic, the duct must be insulated.

Cool air will distribute better from supply registers located high in the room. Ideally, the cool air will flow across the ceiling and then fall slowly through the room. The final distribution point is where the cool air is mixed with the room air to arrive at a comfort condition. **Figure 36.32** shows some examples of air-conditioning diffusers and registers.

Figure 36.32 The final task of the air-conditioning system is to get the refrigerated air properly distributed in the conditioned space.

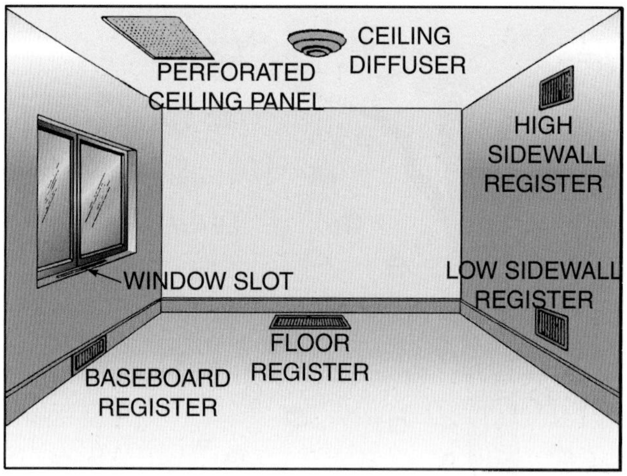

36.16 INSTALLATION PROCEDURES

Package air-conditioning systems were described earlier as systems in which the entire air conditioner is built into one cabinet. A window unit is a package air conditioner designed to blow freely into the conditioned space. Most package air conditioners described in this text have two fan motors, one for the evaporator and one for the condenser, **Figure 36.33**. A window unit, however, typically uses one fan motor with a double shaft that drives both fans. The outdoor fan is not designed to push air through a duct, but the indoor blower is. The package air conditioner has the advantage that all equipment is located outside the structure, so all service can be performed from outside the structure. *This equipment is more efficient than the split system because it is completely factory*

Figure 36.33 A package air conditioner with two fan motors—one for the evaporator coil and one for the condenser. Courtesy Climate Control

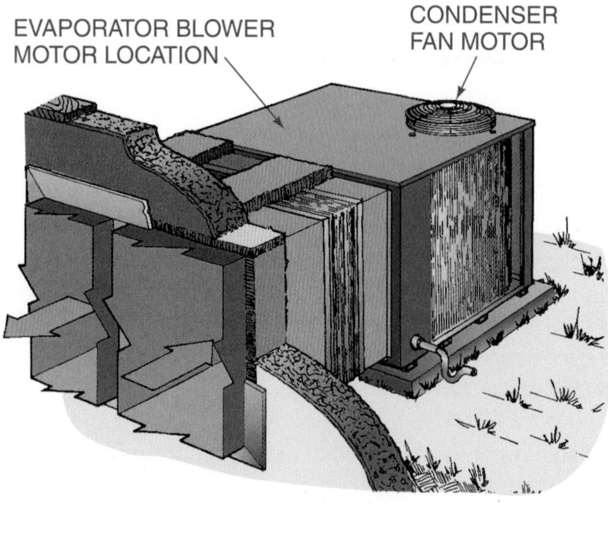

Figure 36.34 Package unit installations.

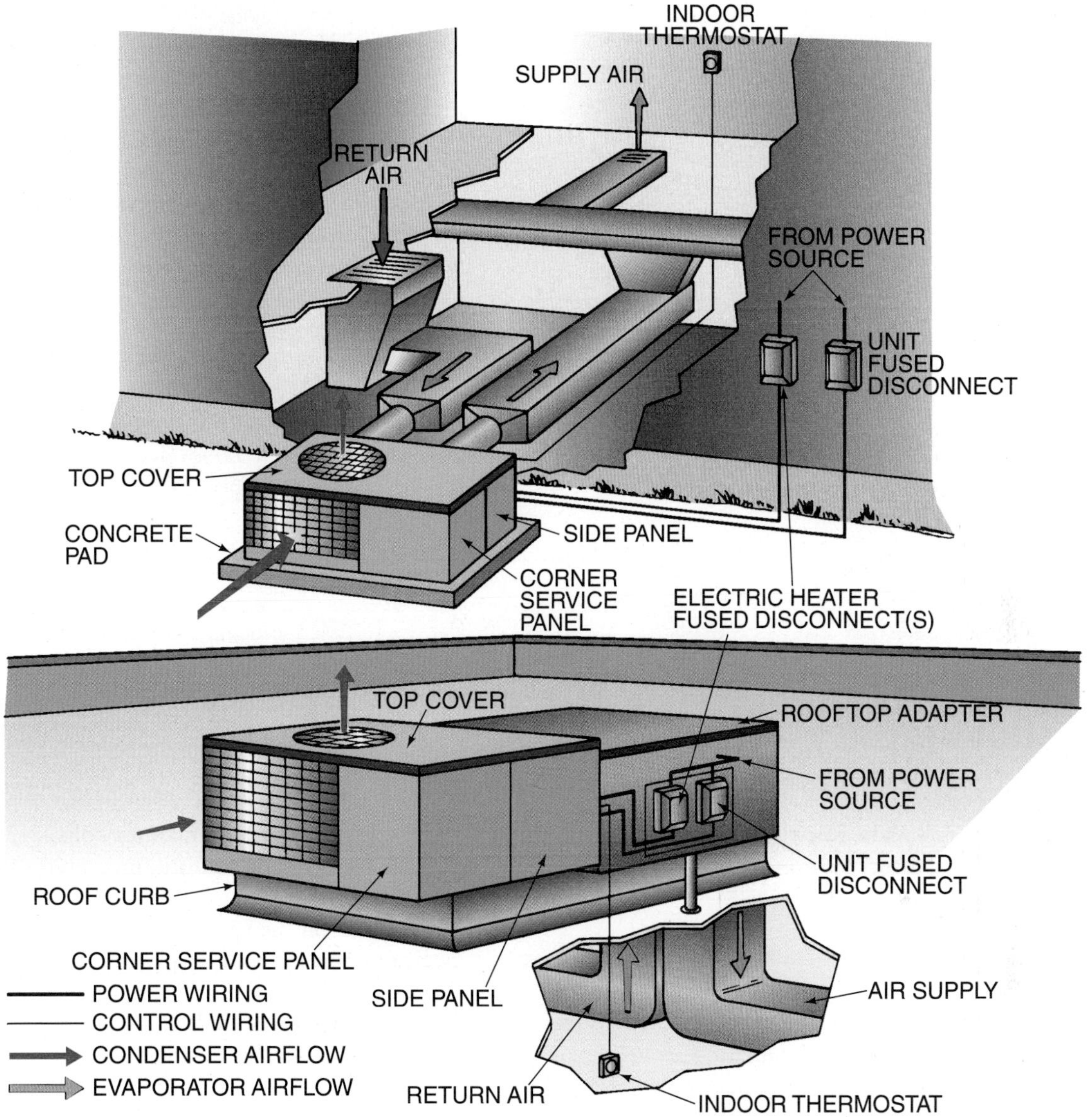

assembled and charged with refrigerant. The refrigerant lines are short, so there is less efficiency loss due to the refrigeration lines.

The installation of a package system consists of mounting the unit on a firm foundation or roof curb, fastening the package unit to the ductwork, and connecting the electrical service to the unit. These units can be installed easily in some locations where the ductwork can be readily attached to the unit. Some common installation types are on the rooftop, ground level beside the structure, and in the eaves of structures, **Figure 36.34**.

The split air-conditioning system is used when the condensing unit must be located away from the air handler. There must be interconnecting piping, often supplied in the form of a line set, which is furnished by the installing contractor or the equipment supplier. The condensing unit should be located as close as possible to the air handler to keep the line set as short as possible. The shorter the line set, the more efficient the system will operate. The tubing consists of a cool gas line and a liquid line. The cool gas line is insulated and is the larger of the two lines. The foam insulation on the suction line keeps unwanted heat from conducting into the line and keeps the line from sweating and dripping.

A typical installation and its pressures and temperatures are shown in **Figure 36.35**, which provides some

Figure 36.35 An installation showing the pressures and temperatures for a system in the humid southern part of the United States.

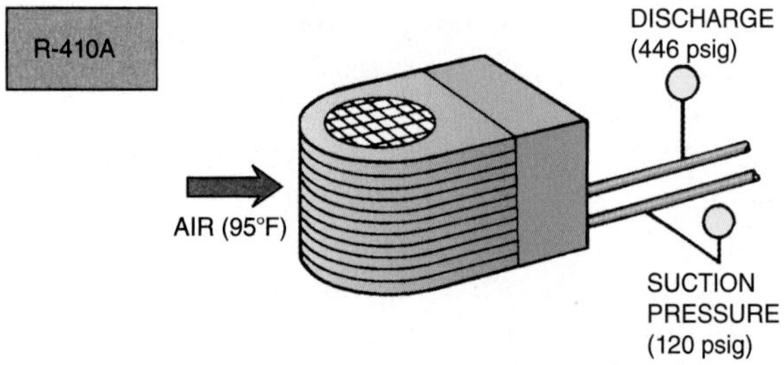

guidelines about the operating characteristics of such a system.

SAFETY PRECAUTION: *Installation of equipment may require the technician to unload the equipment from a truck or trailer and move it to various locations. The condensing unit may be located on the other side of the structure, far away from the driveway, or it may even be located on a rooftop. Proper care should be taken in handling the equipment. Lift-gate trucks may be used to set the equipment down to the ground.*

When installing the equipment, care should be taken while working under structures and in attics. Spiders and other types of stinging insects are often found in these places, and sharp objects such as nails are often left uncovered. When working in an attic, be careful not to step through a ceiling.

Care should also be taken while handling the line sets during connection if they contain refrigerant. Liquid R-22 boils at −41°F and R-410A boils at −60°F at atmospheric pressure and can inflict serious frostbite to the hands and eyes. Goggles and gloves must be worn when connecting line sets.•

REVIEW QUESTIONS

1. Air-conditioning is often referred to as
   A. comfort cooling.
   B. comfort heating.
   C. mechanical exhausting.
   D. mechanical ventilating.
2. Air-conditioning is classified as which type of refrigeration?
   A. Ultra low temperature
   B. Low temperature
   C. Medium temperature
   D. High temperature
3. Heat leaks into a structure by which three methods?
   A. Conduction, convection, and radiation
   B. Infiltration, convection, and radiation
   C. Conduction, convection, and radiation
   D. Conduction, infiltration, and radiation
4. When sizing and selecting air-conditioning equipment, which of the following must be taken into account?
   A. Dirt content
   B. Moisture content
   C. Smoke content
   D. Pollen content
5. Evaporative cooling systems are best-suited for which type of climate?
   A. Hot, humid climate
   B. Hot, dry climate
   C. Cool, humid climate
   D. Cool, dry climate
6. Which of the following best describes the air that enters the structure after being treated by an evaporative cooler?
   A. Warm and humid
   B. Warm and dry
   C. Cool and humid
   D. Cool and dry
7. The condensing unit of a split-type air-conditioning system is made up of the
   A. metering device and evaporator coil.
   B. compressor and metering device.
   C. condenser coil and compressor.
   D. evaporator and condenser coils.
8. Package-type air-conditioning units are intended to be located
   A. either on the roof of a structure or at ground level.
   B. only on the roof of a structure.
   C. only in the attic of a residence.
   D. only at ground level.
9. A typical multihead, ductless split system is configured with
   A. multiple outdoor units and one indoor unit.
   B. multiple indoor units and one outdoor unit.
   C. multiple outdoor units and multiple indoor units.
   D. one indoor unit and one outdoor unit.
10. What refrigerant lines are field installed on the typical split-type air-conditioning system?
    A. Liquid line and suction line
    B. Discharge line and liquid line
    C. Expansion line and discharge line
    D. Suction line and expansion line
11. What refrigerant lines are field installed on the typical package-type air-conditioning unit?
    A. Liquid line and suction line
    B. Discharge line and liquid line
    C. Discharge and suction line
    D. Package-type equipment has no field-installed refrigerant lines.

12. Three common evaporator coil configurations found on air-conditioning systems are
    A. slant, A, and V.
    B. A, V, and H.
    C. slant, A, and H.
    D. A, H, and V.
13. Multicircuit evaporators are equipped with which of the following components?
    A. A refrigerant distributor located at the inlet of the metering device.
    B. A refrigerant distributor located at the outlet of the metering device.
    C. A refrigerant receiver located at the outlet of the metering device.
    D. A refrigerant receiver located at the outlet of the evaporator coil.
14. On average, each ton of system capacity has the ability to remove approximately how much moisture from the air?
    A. One cup
    B. One pint
    C. One quart
    D. One gallon
15. The two types of heat transfer that are facilitated by the evaporator are
    A. superheat and subcooling.
    B. sensible heat and superheat.
    C. subcooling and latent heat.
    D. sensible heat and latent heat.
16. The temperature of the evaporator coil is affected
    A. only by the amount of air passing through the coil.
    B. only by the temperature of the air passing through the coil.
    C. by both the temperature and amount of air passing through the coil
    D. by neither the temperature nor the amount of air passing through the coil.
17. Referring to Figure 36.18, the 1% design wet-bulb and dry-bulb temperatures for Napa, California in the summer are
    A. 100°F dry-bulb, 71°F wet-bulb.
    B. 100°F dry-bulb, 69°F wet-bulb.
    C. 96°F dry-bulb, 69°F wet-bulb.
    D. 96°F dry-bulb, 67°F wet-bulb.
18. The refrigerant that both enters and leaves a properly operating compressor can best be described as
    A. low- pressure vapor.
    B. high-pressure vapor.
    C. superheated vapor.
    D. subcooled vapor.
19. As refrigerant undergoes the compression process
    A. the pressure of the refrigerant increases, while its temperature decreases.
    B. the pressure and temperature of the refrigerant increase.
    C. the pressure of the refrigerant decreases, while its temperature increases.
    D. the pressure and temperature of the refrigerant decrease.
20. Reciprocating compressors used in residential applications are most likely to be
    A. fully hermetic, suction-gas-cooled pumps.
    B. semihermetic, suction-gas-cooled pumps.
    C. fully hermetic, discharge-gas-cooled pumps.
    D. semihermetic, discharge-gas-cooled pumps.
21. The compressor type that relies on pistons moving up and down within cylinders is the
    A. stationary-vane rotary compressor.
    B. rotary-vane rotary compressor.
    C. reciprocating compressor.
    D. scroll compressor.
22. The maximum desired discharge line temperature for an R-410A compressor is
    A. 100°F.
    B. 125°F.
    C. 150°F.
    D. 175°F.
23. Discharge gas temperature is affected most by
    A. the temperature of the suction gas.
    B. the type of metering device used.
    C. the type of condenser used.
    D. the temperature of the refrigerant oil.
24. Which of the following is true regarding the rotary compressor?
    A. There is no clearance volume on these compressors.
    B. They are less efficient than reciprocating compressors.
    C. They are larger than reciprocating compressors.
    D. They rely on pistons and cylinders to compress refrigerant.
25. Rotary compressors are built with
    A. both suction and discharge valves.
    B. neither suction nor discharge valves.
    C. only discharge valves.
    D. only suction valves.
26. All of the following are true regarding the scroll compressor except
    A. there are no valves in the scroll compressor.
    B. the scrolls tend to "wear in" over time.
    C. there is a clearance volume that traps some refrigerant.
    D. they are more efficient than reciprocating compressors.
27. Which of the following best describes the scrolls in a scroll compressor?
    A. One scroll is stationary and one scroll orbits.
    B. Both scrolls are stationary.
    C. Both scrolls orbit clockwise.
    D. Both scrolls orbit counterclockwise.
28. Radial sealing on a scroll compressor is typically accomplished with
    A. floating seals.
    B. oil.
    C. mechanical connections.
    D. grease fittings.

29. Which of the following best describes the refrigerant at the outlet of the condenser?
    A. High-pressure vapor
    B. Low-pressure vapor
    C. Low-pressure liquid
    D. High-pressure liquid
30. Each degree of condenser subcooling will increase system efficiency by approximately how much?
    A. 1%
    B. 3%
    C. 5%
    D. 8%
31. What feature on a thermostatic expansion valve allows the system pressures to equalize during the off-cycle?
    A. Thermal bulb
    B. External equalizer line
    C. Bleed port
    D. Needle and seat assembly
32. Thermostatic expansion valves are designed to maintain constant
    A. evaporator pressure.
    B. evaporator superheat.
    C. condenser pressure.
    D. condenser subcooling.
33. If a duct system is uninsulated and runs through the conditioned space
    A. there will be a net duct loss while the system is operating in the cooling mode only.
    B. there will be a net duct loss while the system is operating in the heating mode only.
    C. there will be a net duct loss while the system is operating in both the cooling and heating mode.
    D. there will be no net duct losses while the system is operating in either the cooling or heating mode.
34. At or near design conditions the supply and return temperatures for an air-conditioning system are closest to
    A. 75°F supply air, 55°F return air.
    B. 70°F supply air, 60°F return air.
    C. 60°F supply air, 65°F return air.
    D. 55°F supply air, 75°F return air.
35. The suction line is insulated to
    A. prevent the liquid refrigerant in the line from rejecting additional, unwanted heat.
    B. prevent the liquid refrigerant in the line from absorbing additional, unwanted heat.
    C. prevent the vapor refrigerant in the line from rejecting additional, unwanted heat.
    D. prevent the vapor refrigerant in the line from absorbing additional, unwanted heat.
36. Which of the following is a possible condenser saturation temperature for an R-410A air-conditioning system that is operating with an outside ambient temperature of 90°F?
    A. 30°F
    B. 80°F
    C. 90°F
    D. 120°F
37. List the four major components of an air-conditioning system.
38. List three common types of compressors that are used in air-conditioning systems.
39. Explain how manufacturers are changing condenser coil design to increase overall system efficiency.
40. Explain why systems with TXVs may require compressors with high-starting-torque motors.

UNIT 37

AIR DISTRIBUTION AND BALANCE

OBJECTIVES

After studying this unit, you should be able to

- list two ways that fresh air enters a structure.
- explain the difference between a forced-air system and a natural-draft system.
- explain why some rooms in a structure require more airflow than other rooms in the same structure.
- list four desirable characteristics of a forced-air system.
- explain the units that are used to measure the pressures in air distribution systems.
- describe the three pressures that act on a duct system.
- describe the instrument that is commonly used to measure the pressures in a duct system.
- describe the function of an anemometer.
- describe the construction and use of a pitot tube.
- explain the relationship that exists between air velocity and air volume.
- explain the difference between a propeller-type fan and a squirrel cage blower.
- differentiate between direct-drive and belt-drive blower assemblies.
- explain the importance of a properly sized duct system.
- describe the properties of four common duct system configurations.
- explain how duct sections are joined together.
- list four common materials from which air distribution systems are fabricated.
- explain the importance of addressing the issue of inertia as it applies to air movement through a duct system.
- explain the purpose and location of balancing dampers.
- explain the concept of zoning.
- explain why it is often desirable for air ducts to be insulated.
- explain how conditioned air is blended with the air in the occupied space.
- list two common types of return air system.
- explain how friction losses affect the performance of an air distribution system.
- explain how to accurately determine the average velocity of air flowing through a duct system.
- explain the purpose of an air friction chart.
- explain how to troubleshoot air distribution system problems.

SAFETY CHECKLIST

✔ Use a grounded, double-insulated, or cordless portable electric drill when installing or drilling into metal duct.

✔ Sheet metal can cause serious cuts. Wear gloves when installing or working with sheet metal.

✔ Wear goggles whenever working on ductwork if the system blower is operating.

37.1 CONDITIONING EQUIPMENT

Generally speaking, air has to be conditioned for people to be comfortable. One way to condition air is to use a blower to move the air through the conditioning equipment. This equipment may consist of a cooling coil, a heating device, a humidifier, or a device to clean the air. The forced-air system typically uses the same room air over and over again. Air from the room enters the system, is conditioned, and is returned to the room. Fresh air enters the structure either by infiltration around the windows and doors or by ventilation from a fresh-air inlet connected to the outside, **Figure 37.1**. Blower-assisted airflow is referred to as forced convection. Some specialized air-conditioning systems, such as those used in hospital operating rooms, are classified as 100% outside air systems. These systems do not recirculate the air from the occupied space back to the air-conditioning equipment. Instead, the air from the space is exhausted outdoors.

The forced-air system is different from a natural-draft system, where the air passes naturally over the conditioning equipment. Hot water baseboard heat is an example of natural-draft heat.

Figure 37.1 Ventilation using fresh air.

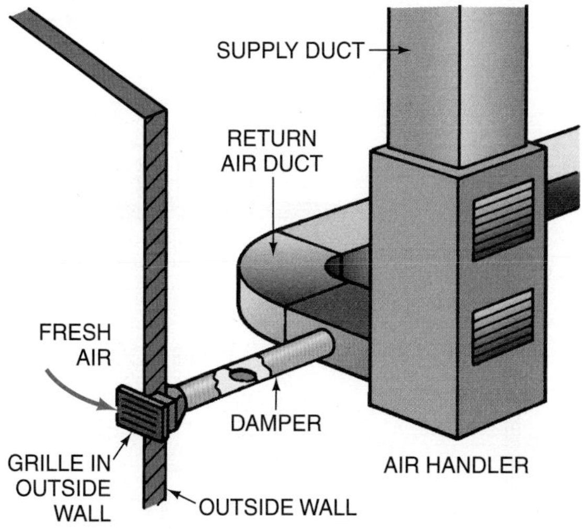

Figure 37.2 Air near the heated baseboard will become heated; it will expand and then rise.

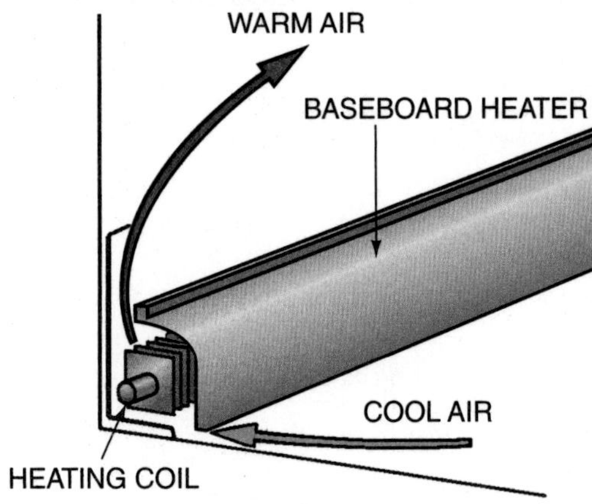

The warmer water in the pipe heats the air in the vicinity of the pipe. The warmer air is less dense than cooler air, causing it to rise. New, cooler air from the floor takes the place of the heated air, **Figure 37.2**.

37.2 CORRECT AIR QUANTITY

The object of the forced-air system is to deliver the correct quantity of conditioned air to the occupied space. When this occurs, the air mixes with the room air and creates a comfortable atmosphere in that space. Delivering the correct quantity of filtered air to the space also increases the efficiency of the system and overall indoor air quality. Different spaces have different air quantity requirements, and the same structure may have several different cooling requirements. For example, a house has rooms of different sizes with different requirements. A bedroom requires less heat and cooling than does a large living room. Different amounts of air need to be delivered to these rooms to maintain comfortable conditions, **Figures 37.3** and **37.4**. Another example is a small office building with a high cooling requirement in the lobby and a low cooling requirement in the individual offices. One factor that is taken into account is the windows in each particular room or area in the structure. The number, type, and direction of the windows in the structure will have a significant effect on the amount of conditioned air that is required for each individual space. The correct amount of air must be delivered to each part of the building so that some areas will not be overcooled while other areas are undercooled.

37.3 THE FORCED-AIR SYSTEM

The components that make up the *forced-air system* are the *blower* (fan), the *air distribution system*, the *return-air system*, and the grilles and registers where the circulated air enters the room and returns to the conditioning equipment. See **Figure 37.5** for an example of duct fittings. When these

Figure 37.3 This floor plan has the heating and cooling requirements for each room indicated.

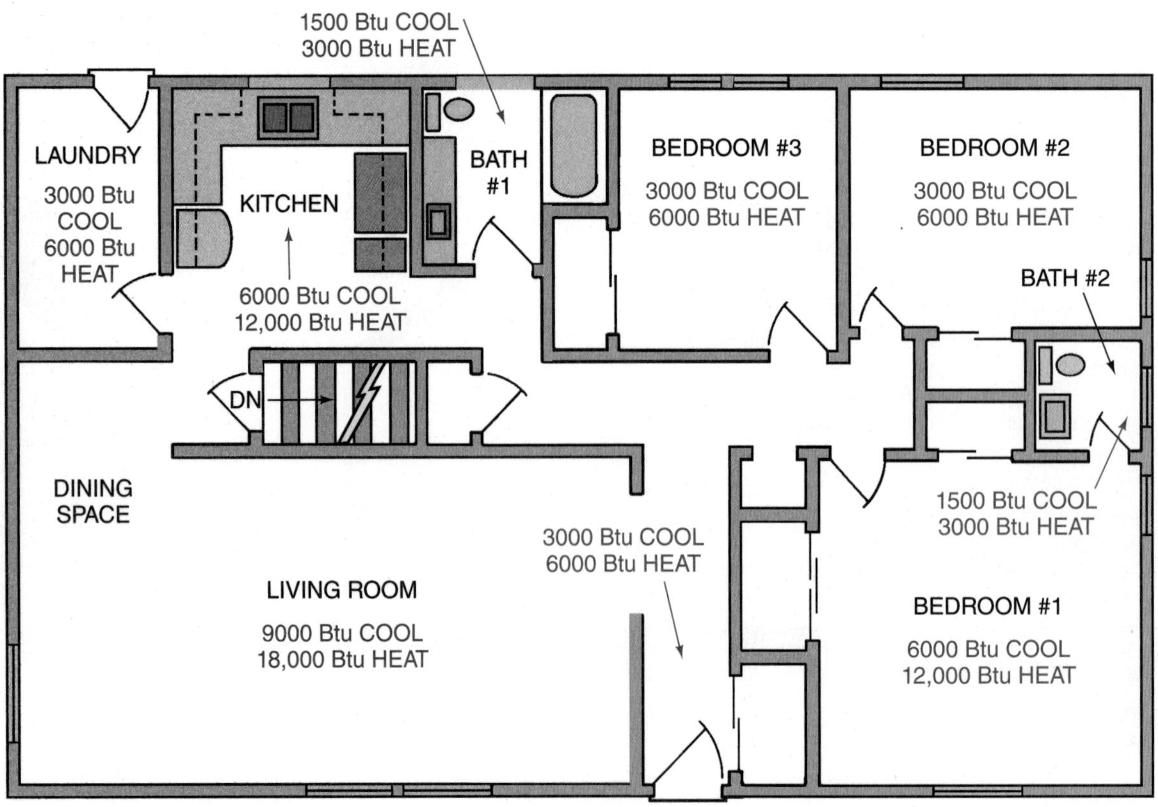

Figure 37.4 This is the same floor plan as that shown in **Figure 37.3**. The quantity of air, in cfm, to be delivered to each room is indicated.

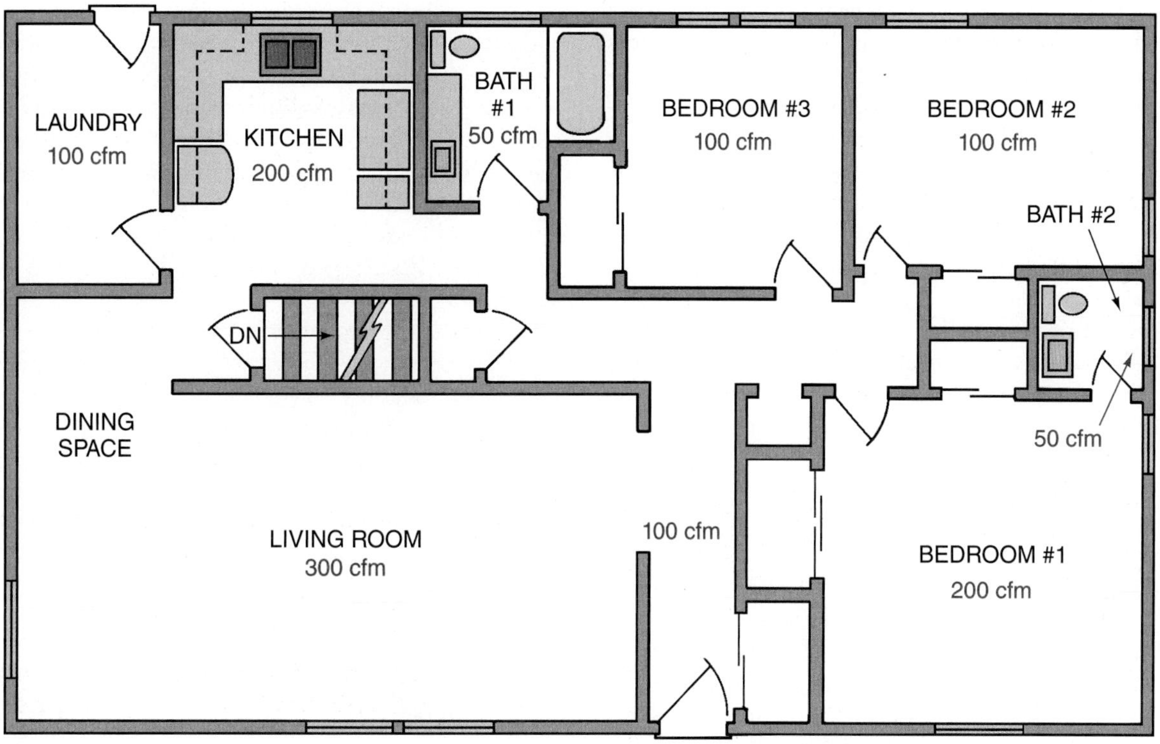

components are chosen correctly, they work together as a system with the following characteristics:

1. No air movement will be felt in the conditioned space.
2. No air noise will be noticed in the conditioned space.
3. No temperature swings will be felt by the occupants.
4. The occupants will not be aware that the system is on or off unless it stops for a long time and the temperature changes.

To summarize, if the air-conditioning system is operating properly, the occupants of the space should not be aware of the system.

Figure 37.5 Duct fittings.

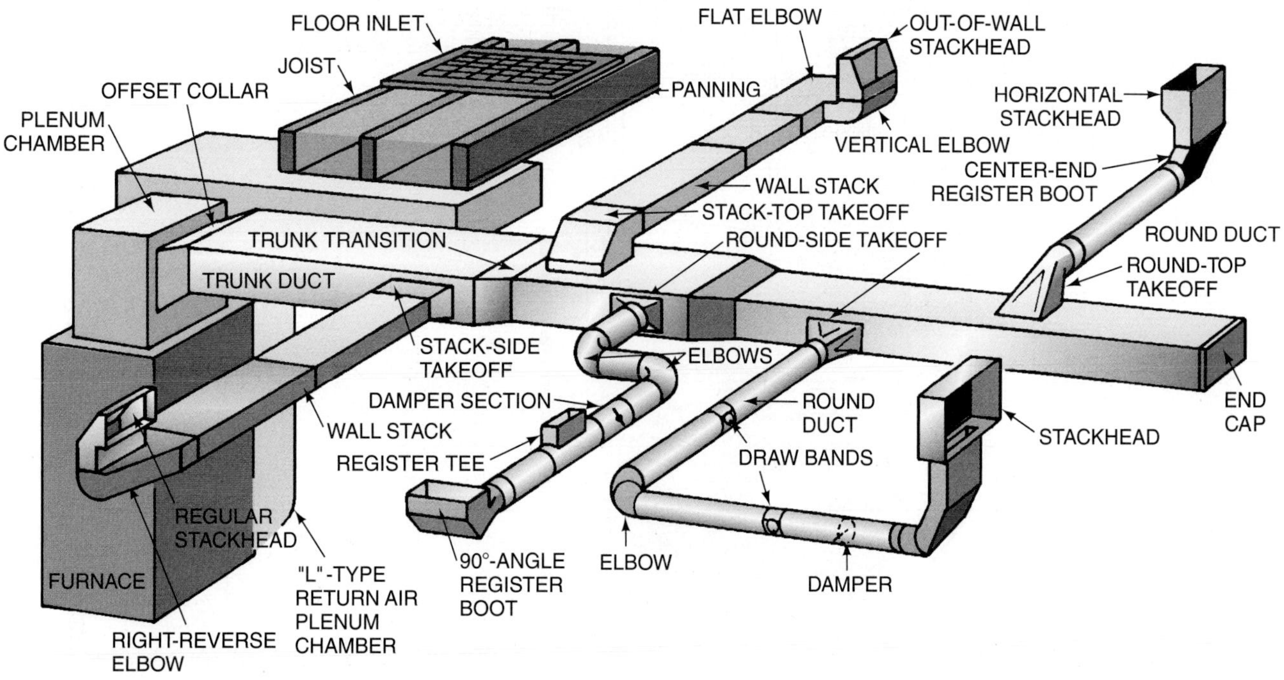

37.4 THE BLOWER

The *blower* provides the pressure difference needed to force the air into the duct system, through the heat-transfer surfaces in the conditioning equipment, through the grilles and registers, and into the room, **Figure 37.6**. Air has weight and creates resistance to movement. This means that it takes energy to move the airstream to and from the conditioned space. If the blower on a 3-ton air-conditioning system is required to move 1200 cubic feet of air through the system per minute (400 cfm/ton), this is about 90 pounds of air moving through the system every minute (1200 ft^3/min × 0.075 lb/ft^3 = 90.02 lb/min). Ninety pounds of air moving through the system every minute equates to 5400 pounds of air every hour, or 129,600 pounds of per day. In order to move the required air quantities, the blower must overcome the resistance and friction created by other components located in the air distribution system: the return duct, the supply duct, the air filter, the cooling coil, the heat exchanger, the supply registers, and the return-air grilles.

The pressure in a duct system for a residence or a small office building is too small to be measured in psi. Instead, it is measured in units called inches of water column. Inches of water column is abbreviated as IWC or in. WC, where 27.7 IWC = 1 psi. Inches of water column was introduced earlier in the text in the gas heat unit where the operating pressures of natural gas and liquefied petroleum were discussed. A pressure of 1 in. WC is the pressure necessary to raise a column of water 1 in. Air pressure in a duct system is measured with a manometer, which is an instrument that accurately measures these very low pressures, **Figure 37.7**.

Figure 37.7 Digital electronic manometer. Courtesy Ritchie Engineering Company, Inc.

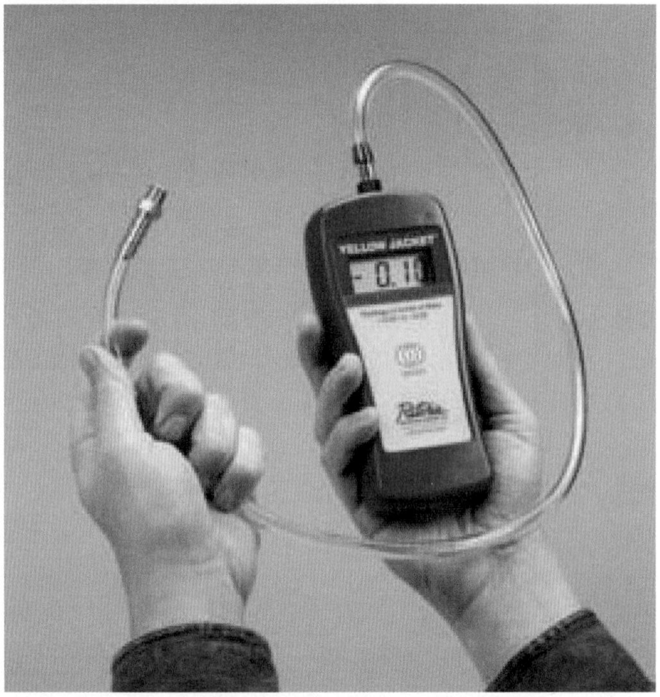

Since 1 psi = 27.7 in., or 2.31 feet, as represented in **Figure 37.8**, we can conclude that, at standard conditions, atmospheric pressure will support a column of water approximately 34 feet high. This is because the atmosphere exerts a

Figure 37.6 A blower and motor for moving air.

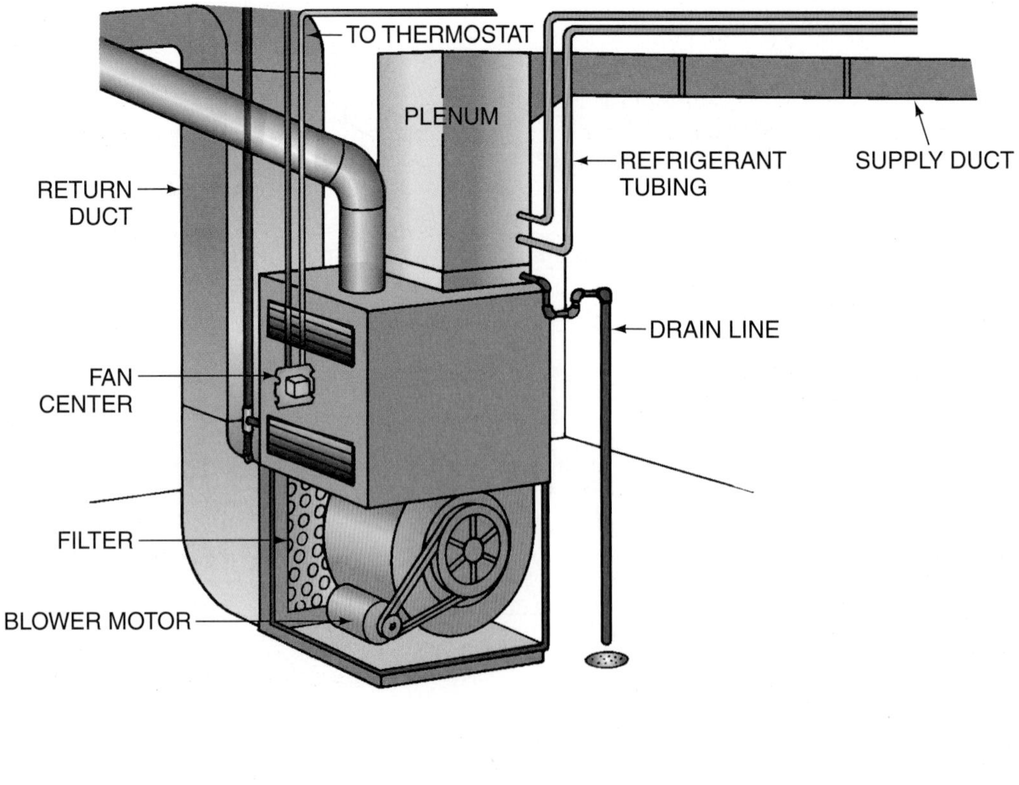

Figure 37.8 A vessel with a pressure of 1 psi inside. The water manometer has a column 27.7 in. high (2.31 ft).

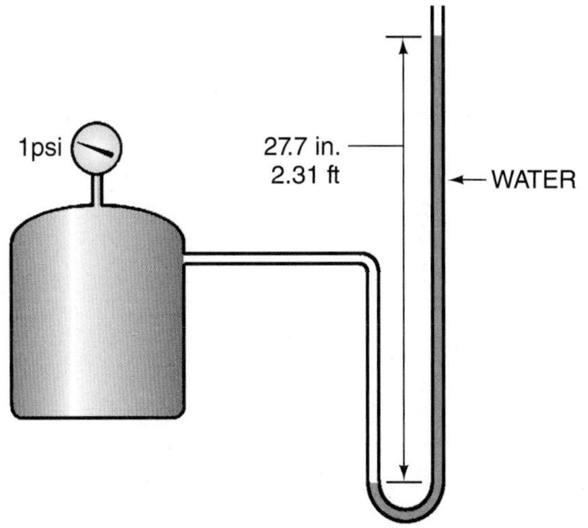

pressure of 14.696 psi at standard atmospheric conditions (2.31 × 14.696 = 34 feet). The average duct system will not exceed a pressure of 1 in. WC, which equates to a pressure of about 0.036 psig. This means that a pressure of 0.036 psig will support a column of water about 1-in. high. Airflow pressure in ductwork is measured in inches of water column.

37.5 SYSTEM PRESSURES

A duct system is pressurized by three pressures—static pressure, velocity pressure, and total pressure. Static pressure + velocity pressure = total pressure. Static pressure is the same as the pressure on an enclosed vessel and can be thought of as the pressure of the air pushing against the walls of the duct sections. This is like the pressure of the refrigerant in a cylinder that is pushing outward. The probe on the manometer is positioned perpendicular to the walls of the duct and therefore senses only the outward, or static pressure, in the duct, **Figure 37.9.**

Figure 37.9 A manometer connected to measure the static pressure.

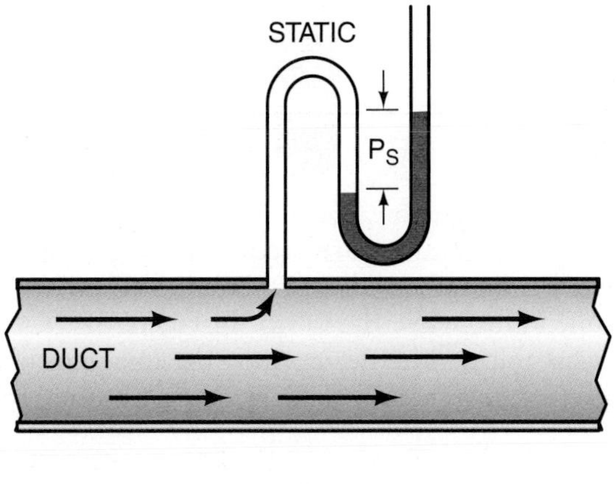

The air moving in a duct system flows in a path that is parallel to the walls of the duct section. The velocity and weight of the air create what is called velocity pressure. **Figure 37.10** shows how a manometer is connected to measure velocity pressure in an air duct. Notice the position of the sensing probes. The air flows straight into the inlet of the probe, which registers both velocity and static pressure. The manometer determines the velocity pressure by canceling the static pressure with the second probe. We have velocity pressure and static pressure pushing on the left side of the manometer and only the static pressure pushing on the right side of the manometer.

The total pressure of a duct can be measured by using the manometer a little differently. In this case, only the probe that is placed in the airstream is needed, as it senses both the velocity and static pressures in the duct. This is the duct system's total pressure, **Figure 37.11**. The main difference between the manometer connections in **Figures 37.10** and **37.11** is that there is no second connection to cancel out the static pressure from the measurement.

Figure 37.10 A manometer connected to measure the velocity pressure of the air moving in the duct.

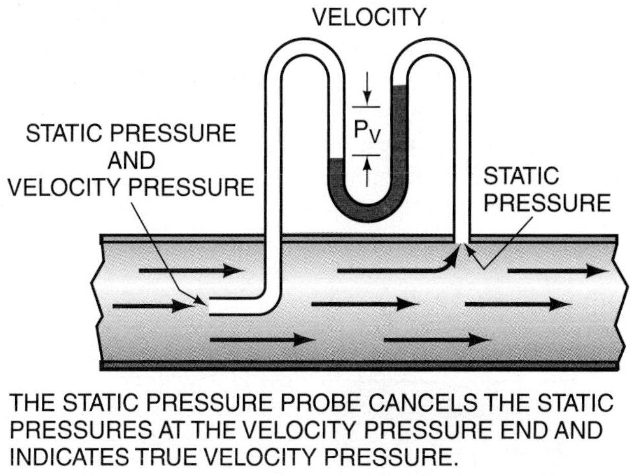

Figure 37.11 A manometer connected to measure the total air pressure in the duct.

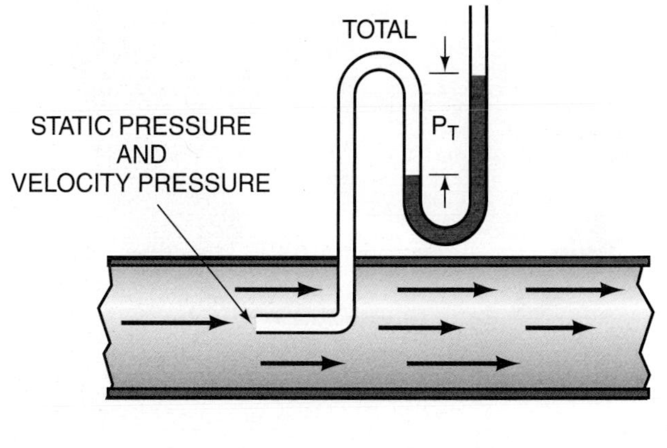

37.6 AIR-MEASURING INSTRUMENTS FOR DUCT SYSTEMS

The manometer that was just discussed is an instrument that measures air pressure. An instrument that is commonly used on air distribution systems is the anemometer, or velometer, which is used to measure air velocity, or speed, **Figure 37.12**. The velometer measures how fast air is moving past a particular point in an air duct. Once the average air speed, in ft/min, has been determined using an instrument similar to that shown in **Figure 37.12**, the air volume that is moving in the duct can be easily determined, as long as the cross-sectional area of the duct section, in ft^2, is known. For example, if the average air speed through a 9 in. $\times$ 8 in. rectangular duct section is 700 ft/min, we can determine the air volume as follows:

$$\text{Air Volume (cfm)} = \text{Air Velocity (ft/min)} \times \text{Cross-sectional Area (ft}^2\text{)}$$
$$\text{Air Volume (cfm)} = 700 \text{ ft/min} \times [(9 \text{ in.} \times 8 \text{ in.}) \div 144 \text{ in}^2/\text{ft}^2]$$
$$\text{Air Volume (cfm)} = 700 \text{ ft/min} \times (72 \text{ in}^2 \div 144 \text{ in}^2/\text{ft}^2)$$
$$\text{Air volume} = 350 \text{ cfm}$$

These calculations and concepts are covered in more detail in Section 37.18, "Measuring Air Movement for Balancing."

A number of additional functions have been incorporated into the design of new-generation air-measuring instruments. The instrument in **Figure 37.13**, for example, has the capability to measure the temperature, dew point temperature, wet-bulb temperature, and relative humidity of the air as it passes through the device. Other new-generation air-measuring instruments are wireless. The instrument shown in **Figure 37.14** has the ability to measure airflow, air velocity, and temperature, while the device in **Figure 37.15** can also measure humidity. These devices have the ability to communicate with electronic hand-held devices, such as smart phones,

Figure 37.12 A digital anemometer. Courtesy Dwyer Instruments, Inc.

Figure 37.13 An airflow instrument that has the ability to measure dry-bulb and wet-bulb temperatures in addition to relative humidity and dew point temperature. Courtesy Dwyer Instruments, Inc.

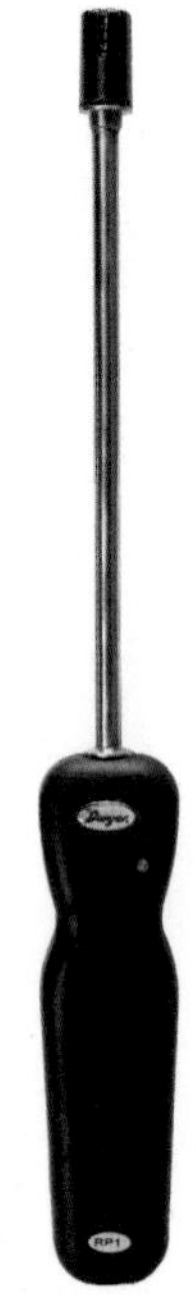

Figure 37.14 An airflow instrument that has the ability to measure airflow, air velocity, and temperature. Courtesy Dwyer Instruments, Inc.

when the appropriate application is installed, **Figure 37.16.** Some other instruments have a modified keypad so that the service technician can enter the duct measurements into the device. This enables the instrument to display the actual cfm of

Figure 37.15 This device has the ability to measure relative humidity in addition to other airflow properties. Courtesy Dwyer Instruments, Inc.

Figure 37.16 New generation airflow instruments are wireless and have the ability to transmit data to technicians via their smart phones and other hand-held electronic devices through proprietary applications. Courtesy Dwyer Instruments, Inc.

Figure 37.17 A telescopic pitot tube. Courtesy Dwyer Instruments, Inc.

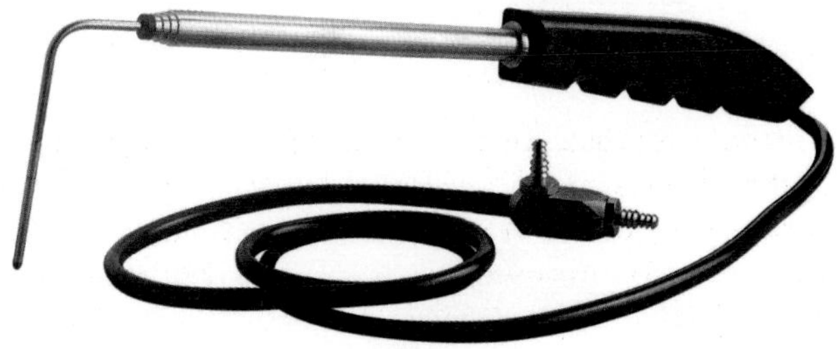

Figure 37.18 How a pitot tube arrangement is used to measure velocity pressure.

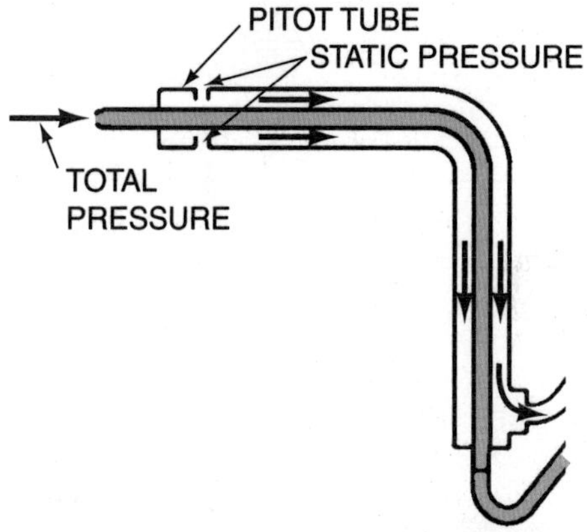

the air, saving the technician time on the job and reducing the number of manual calculations that need to be performed.

A special device called a pitot tube, developed many years ago, is used to check duct pressures, **Figure 37.17.** The pitot tube is constructed as two concentric tubes and is bent into a 90-degree angle. The pitot tube is inserted into the duct section and directed into the stream so the air flowing in the duct will enter the tube. The small holes in the outer tube are perpendicular to the airstream's direction of flow. The airflow into the center tube represents the total pressure, while the pressure at the perpendicular holes represents the static pressure, **Figure 37.18. Figure 37.19** shows such a setup, where PT is the total pressure, PV is the velocity pressure, and PS is the static pressure. The velocity pressure reading is obtained by subtracting the static pressure from the total pressure. Modules connected to the pitot tube sense both the static and velocity pressures and determine the pressure difference internally, **Figure 37.20.**

37.7 TYPES OF FANS AND BLOWERS

Fans and blowers are devices that are used to move air through, among other things, heat exchangers and duct systems. The two types of air-moving devices that are discussed in this

Figure 37.19 Pressures exerted on the pitot tube.

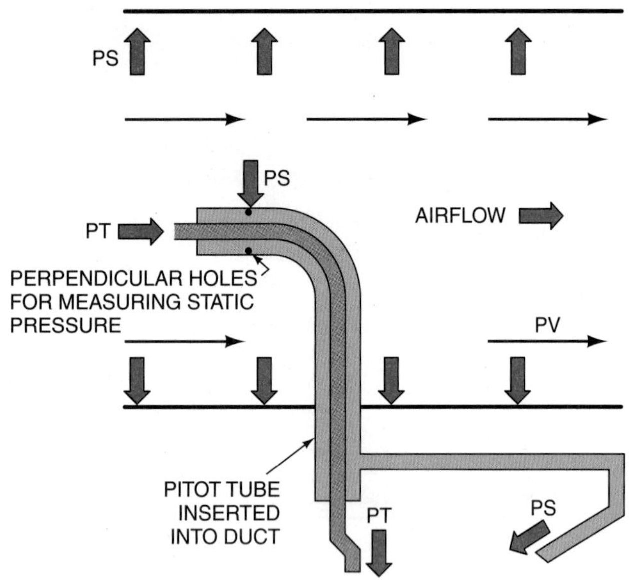

Figure 37.20 The pitot tube can connect to modules that determine the pressure differential automatically. Courtesy Dwyer Instruments, Inc.

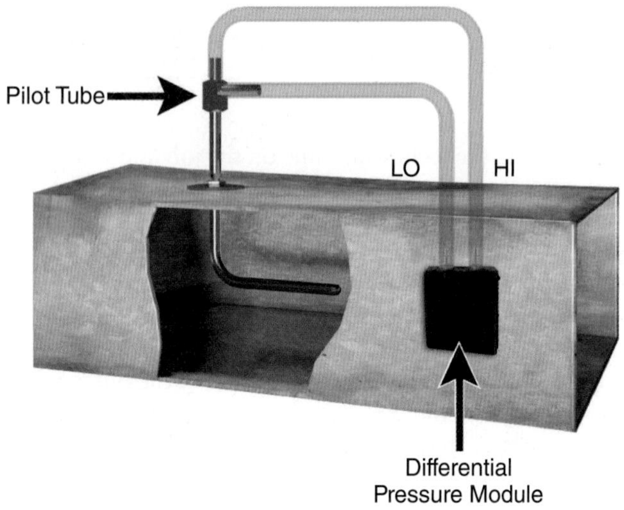

text are the propeller fan and the *forward curved centrifugal blower,* also called the squirrel cage blower or *blower wheel.*

The propeller-type fan is often used as an exhaust fan or a condenser fan. Propeller-type fans can move large volumes of air at low pressure differentials. The propeller can be cast iron, aluminum, or stamped steel and is set into a housing called a *venturi* to help ensure that air flows in a straight line from one side of the fan to the other, **Figure 37.21**. Propeller-type fans make more noise than centrifugal blowers so they are normally used where noise is not a factor.

The squirrel cage or centrifugal blower creates more pressure from the inlet to the outlet and moves more air against more pressure. For this reason, these blowers are desirable for use in ducted air distribution systems. This blower has a forward curved blade and a cutoff to prevent the air from spinning around the blower wheel within the housing. Air enters the blower from the center and is then thrown by centrifugal action to the outer perimeter of the blower housing. Some of the air would keep going around with the blower wheel if it were not for the shear that cuts off the air and sends it out the blower outlet, **Figure 37.22**. The centrifugal

Figure 37.21 A propeller-type fan.

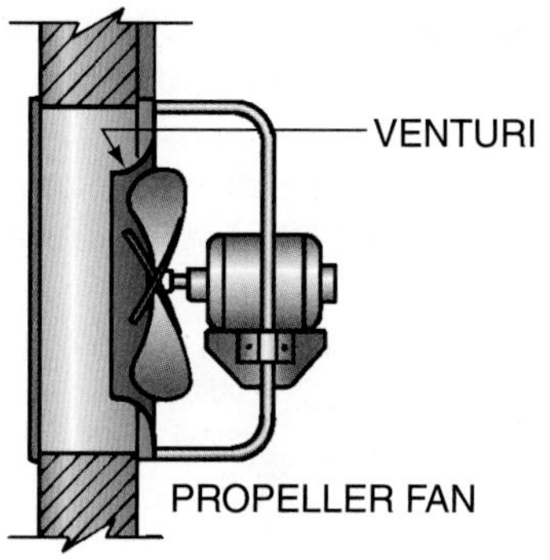

Figure 37.22 A centrifugal blower.

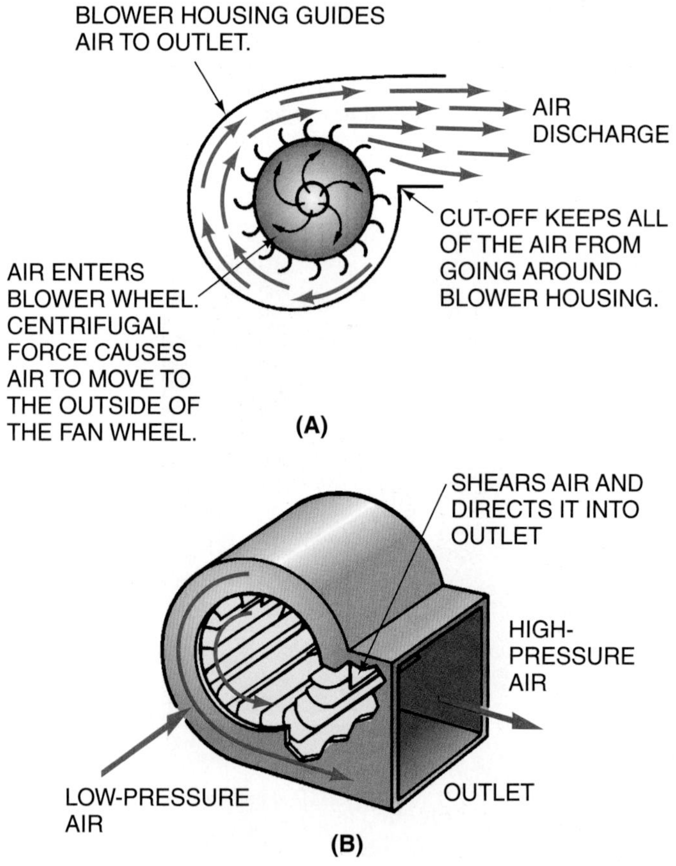

blower is very quiet when compared to the noise generated by propeller-type fans. It meets all the requirements for very large, high-pressure duct systems. High-pressure systems have pressures of 1 in. WC or higher. The rotation of a centrifugal blower is critical to its ability to move air. The rotation can be determined by looking at the blower from the side and finding the blower outlet. The blower wheel must turn so that the top of the wheel is rotating toward the blower outlet.

One characteristic that makes troubleshooting the centrifugal blower easier is the air volume-to-horsepower requirement. This blower uses energy at the rate at which it moves air through the ductwork. The current draw of the fan motor is proportional to the number of pounds of air it moves. For example, a fan motor that draws full-load amperage at the rated blower capacity will draw less than full-load amperage when operating under any conditions that are lower than the motor's maximum capacity. If the blower inlet is blocked, the suction side of the blower will be starved for air, and the current draw of the motor will go down. If the discharge side of the blower is blocked, the pressure will go up in the discharge side and the current draw will also go down because the blower is not moving as much air, **Figure 37.23**. The air is merely spinning around in the blower and housing and not being forced into the ductwork.

The airflow through this particular type of blower can be checked with an ammeter when making simple field measurements. If the airflow is reduced, the current draw will drop. If the airflow is increased, the current draw will increase. For example, if the door on the blower compartment is removed or not secured properly, the motor current will go up. This is

Figure 37.23 (A) When the airflow through the air handler is correct, the amperage draw of the motor will be correct as well. (*Continued*)

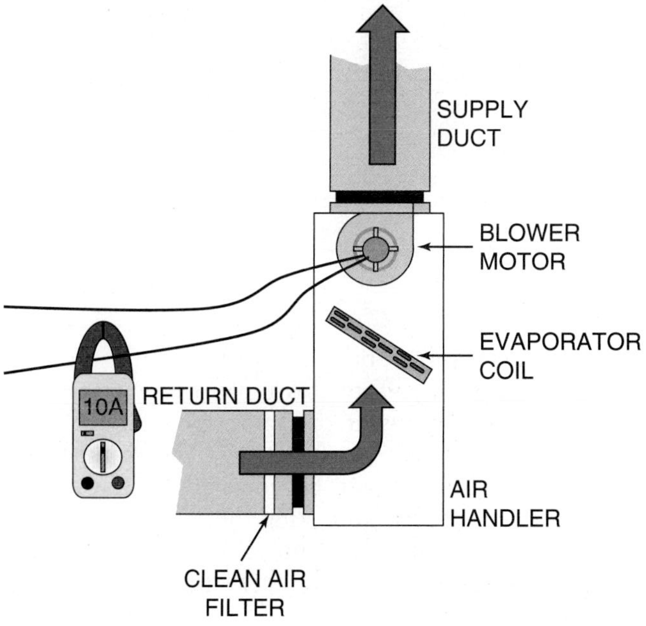

(A)

Figure 37.23 (*Continued*) (B) If the air filter is clogged, airflow will be reduced and the amperage draw of the motor will go down. (C) If the supply ductwork is restricted for any reason, airflow will be reduced and the amperage draw of the motor will go down.

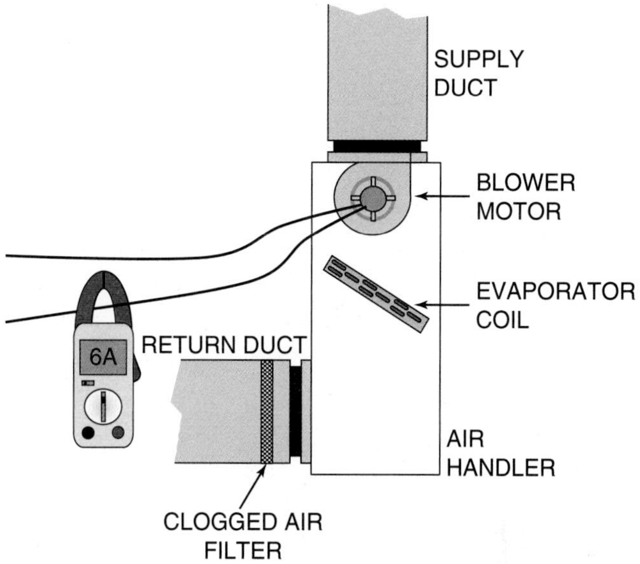

(B)

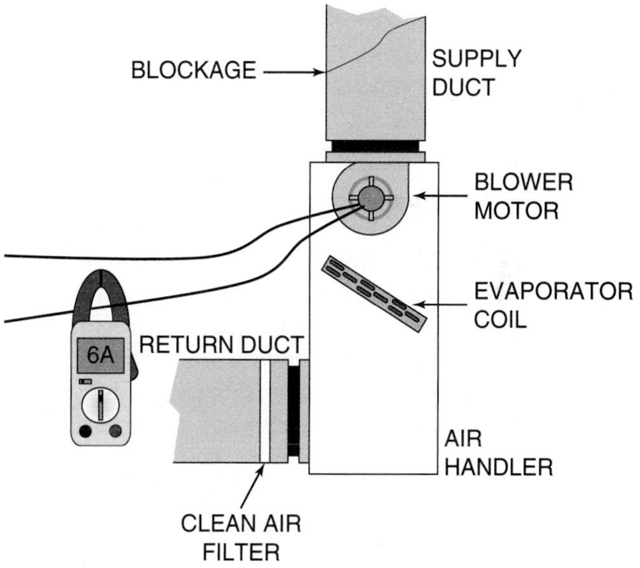

(C)

because the restrictions created by the heat exchanger, return duct, and air filter have been removed and more air will be moved by the blower, **Figure 37.24**. In addition, the amperage reading can be a useful tool for properly setting up belt-driven blower assemblies. If the driven pulley is too small, the blower will turn too fast and the amperage draw of the motor will increase. The same condition will be present if the drive pulley is too large. This is a relatively common situation given that the drive pulley is often a variable-pitch component. Some technicians will use the variable-pitch feature on the drive pulley to adjust the tension on the belt. This practice

Figure 37.24 (A) When service panels are properly secured and the airflow through the air handler is correct, the amperage draw of the motor will be correct as well. (B) If the service panel is removed, airflow through the air handler will increase and the amperage draw of the motor will go up.

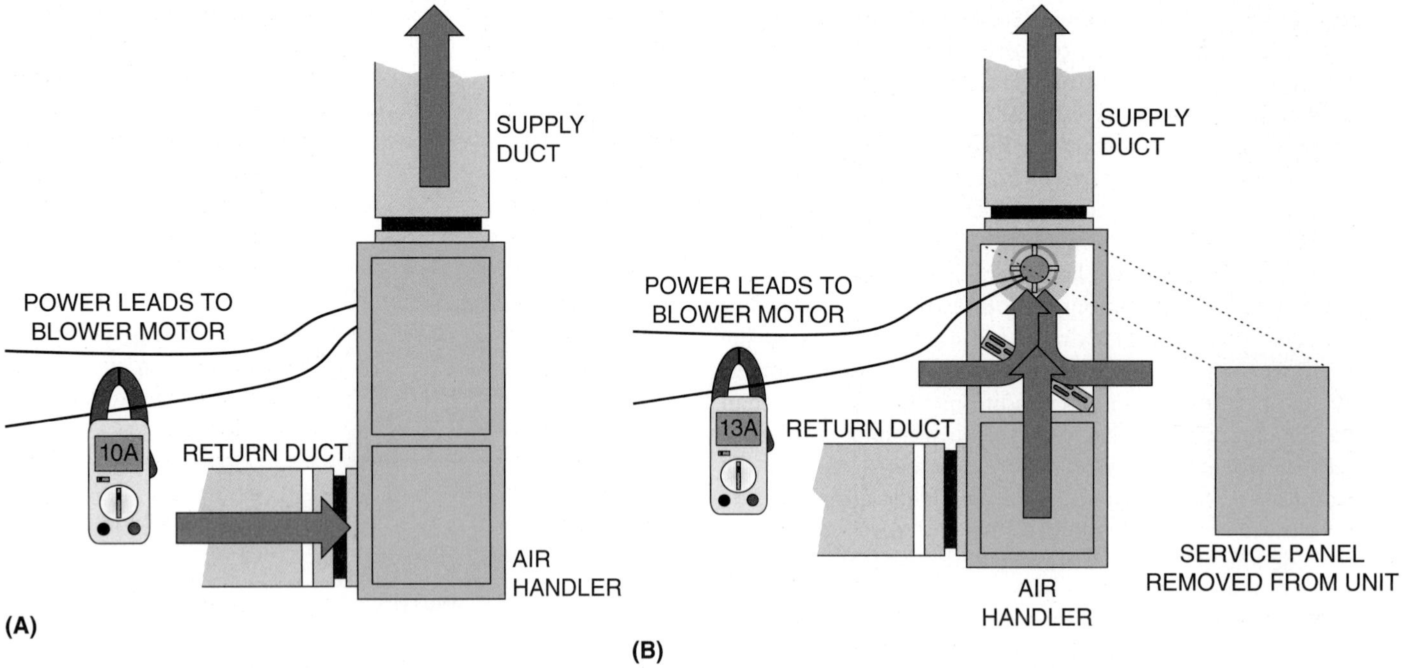

is not acceptable as doing so will alter the rate at which air is moving through the system and will also affect the amperage draw of the motor.

37.8 TYPES OF DRIVE ASSEMBLIES

There are two types of drive assemblies that are commonly used to turn fans and blowers: the belt drive and the direct drive.

Belt-driven assemblies have two bearings on the blower shaft and two bearings on the motor shaft. Sleeve bearings and resilient (rubber) mountings are used to keep bearing noise out of the blower section. Sometimes these bearings are permanently lubricated by the manufacturer, so a technician cannot oil them. The drive pulley on the motor, the driven pulley on the fan shaft, and the belt all must be maintained. Because the pulleys can be adjusted or changed to change the blower speed, this motor and blower combination has many uses, **Figure 37.25**.

The direct-drive blower assembly is commonly used on cooling equipment up to 5 tons. The motor is mounted directly to the blower housing, usually with rubber mounts, with the motor shaft extending into the blower wheel, **Figure 37.26**. The direct-drive blower is very quiet and does not have belts and pulleys that can wear out or need to be adjusted.

With direct-drive blower assemblies, the number of bearing surfaces is reduced from four to two. The bearings may be permanently lubricated at the factory. The front bearing, which is located inside the blower wheel, may be difficult to access for lubrication purposes if a special oil port is not furnished. With this type of blower assembly, the blower wheel turns at the same speed as the motor, so multispeed motors are often used. The volume of air moved by the blower may be adjusted by the different motor speeds instead of adjusting the pitch or size of the pulley. The motor may have up to four different speeds that can be changed by switching wires at the motor terminal box. Common speeds are from about 1500 rpm down to about 800 rpm. The motor, for example, can be

Figure 37.25 This blower is driven with a motor using a belt and two pulleys to transfer the motor energy to the fan wheel.

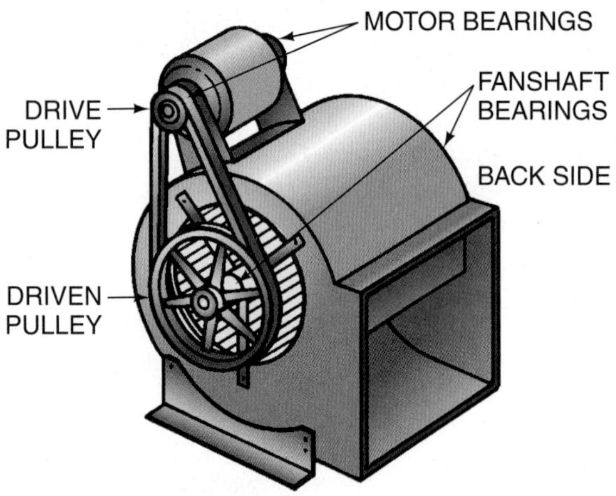

Figure 37.26 A direct-drive blower assembly. Photo by Eugene Silberstein

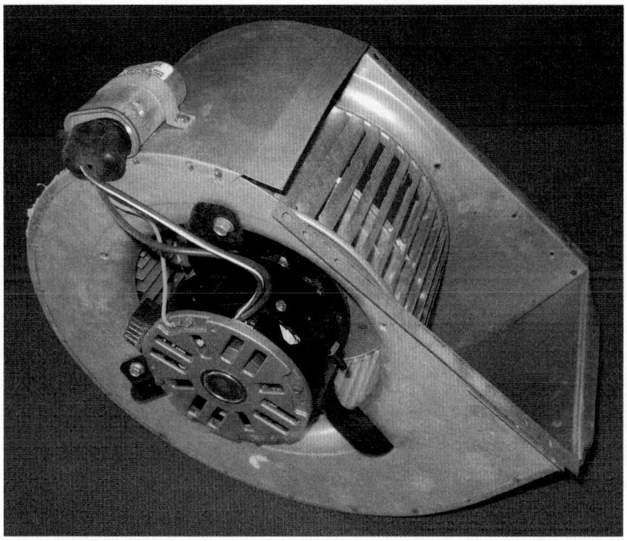

operated at a faster speed in the summer for more airflow in the cooling mode, **Figure 37.27**. The changing of the motor speed is accomplished by the fan relay as shown in the diagram. In the heating mode, the normally closed contacts on the fan relay allow the blower to operate at low speed when the heat exchanger is warm and the fan switch contacts are closed. In the cooling mode, the G terminal on the thermostat is powered, and the fan relay coil is energized. This closes the normally open contacts on the fan relay, allowing the motor to operate at high speed.

New-generation blower assemblies utilize electronically commutated motors (ECMs). These motors are programmed to deliver the desired amount of airflow under a wide range of conditions and do not require the use of multiple wire taps to adjust the speed of the motor.

Figure 37.27 The wiring diagram of a multiple-speed motor.

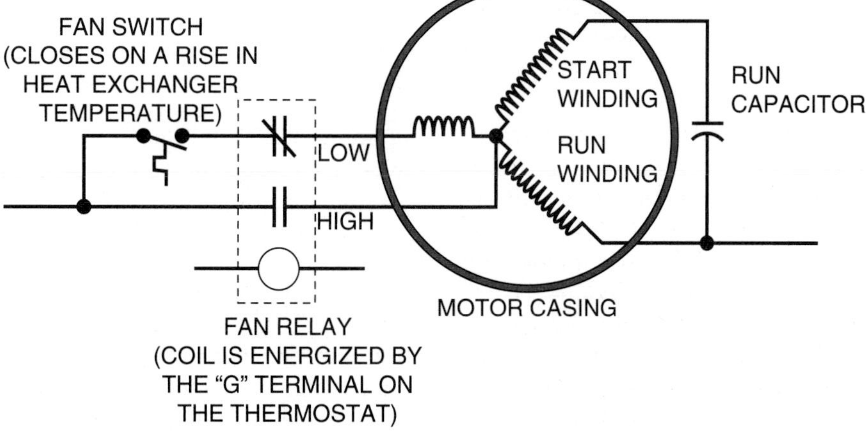

37.9 THE SUPPLY DUCT SYSTEM

The supply duct system distributes air to the terminal units, registers, grilles, or diffusers in the conditioned space. Starting at the blower outlet, the duct can be fastened to the blower or blower housing directly or by means of a fireproof vibration eliminator, **Figure 37.28**, between the blower and the ductwork. This vibration eliminator is often referred to as a canvas connector or collar, although these are very rarely made of canvas anymore. Flexible rubber or a similar material is more commonly used. The flexible rubber material is attached to two wide strips of sheet metal that ultimately connect to the duct system and the air handler, **Figure 37.29**. The vibration eliminator is recommended on all installations to prevent noise transmission through the duct system. Given the relatively low cost of the vibration eliminator material, it is wise to incorporate this component into all air distribution systems.

The supply duct system must be designed so that it allows the air to move toward the conditioned space as freely as possible. It must be sized properly: neither undersized nor

Figure 37.28 A canvas connector. Photo by Eugene Silberstein

Figure 37.29 Material used to fabricate canvas connectors. Photo by Eugene Silberstein

oversized. Oversized duct systems cost more to fabricate and install and will also result in the slowing of the air as it flows through the duct. Recall from earlier in this unit that the volume of airflow through a duct section is calculated by multiplying the velocity of the air by the cross-sectional area of the duct. Consider an air duct that is designed to move 350 cfm and has a cross-sectional area of one-half of a square foot (72 in^2). The velocity of the air is determined as follows:

Air Volume (cfm) = Air Velocity (ft/min) × Cross-sectional Area (ft^2)
Air Velocity (ft/min) = Air Volume (cfm) ÷ Cross-sectional Area (ft^2)
Air Velocity (ft/min) = 350 cfm ÷ 0.5 ft^2
Air Velocity (ft/min) = 700 ft/min

Let's now assume that the ductwork was oversized by 50% for this application. The desired airflow is still 350 cfm, but the cross-sectional area of the duct is now 0.75 ft^2. The velocity of the air is now as follows:

Air Velocity (ft/min) = Air Volume (cfm) ÷ Cross-sectional Area (ft^2)
Air Velocity (ft/min) = 350 cfm ÷ 0.75 ft^2
Air Velocity (ft/min) = 467 ft/min

By reducing the velocity of the air, the path the air takes as it leaves the supply register will also be changed. The air will not come out with the force that was originally planned for at the design stage. In addition, undersizing a duct will result in excessive air velocity, which will increase the noise levels of the system and, once again, alter the desired air patterns.

Duct systems typically follow one of four common configurations. Duct systems can be plenum, *extended plenum*, *reducing plenum*, or *perimeter loop*, **Figure 37.30**. Each system has its advantages and disadvantages.

THE PLENUM SYSTEM

The plenum system is well suited for a job in which the room outlets are all close to the unit. This system is economical from a first-cost standpoint and can be installed easily by an installer with a minimum of training and experience.

Plenum systems work better with fossil-fuel (coal, oil, or gas) systems than with heat pumps because the leaving-air temperatures are much warmer in fossil-fuel systems. The plenum system is desirable when the duct runs are short, as in the case of apartment installations. The return-air system can be a single return located at the air handler, which makes materials economical. The single-return system will be discussed in more detail later. **Figure 37.31** shows an example of a plenum system with registers in the ceiling.

THE EXTENDED PLENUM SYSTEM

The extended plenum system is a popular choice for use in a long structure such as the ranch-style house. This system extends the main plenum closer to the farthest supply register. The extended plenum is called the *trunk duct* and can be round, square, or rectangular, **Figure 37.32**. Small ducts called *branches* complete the connection to the terminal units. These small ducts also can be round, square, or rectangular. Smaller branch ducts are usually round because round ducts in smaller diameters are relatively inexpensive, commercially available, and do not need to be specially fabricated for the project. Average homes typically use 6-in. round duct for the branches.

The main drawback of the extended plenum system is that the velocity of the air is lower as the air reaches the last takeoffs on the extended plenum. Consider the portion of the air distribution system shown in **Figure 37.33**. An airflow of 1000 cfm is moving through the duct at a velocity of 700 ft/min. It can be determined that the cross-sectional area of the duct is 1.43 ft^2 (1000 cfm ÷ 700 ft/min). After the first takeoff, where 200 cfm is directed to an area in the conditioned space, there are 800 cfm of air remaining. Because this duct system is an extended plenum system, the duct size remains the same. This means that the velocity of the air after the first 200-cfm takeoff will be only 560 ft/min (800 cfm ÷ 1.43 ft^2). After the second 200-cfm takeoff, the air velocity will be only 420 ft/min.

THE REDUCING PLENUM SYSTEM

The reducing plenum system reduces the size of the trunk duct as the volume of air in the duct decreases. This reduction in duct size is accomplished with duct fittings called

Figure 37.30 (A) A plenum system. (B) An extended plenum system. (C) A reducing extended plenum system. (D) A perimeter loop system.

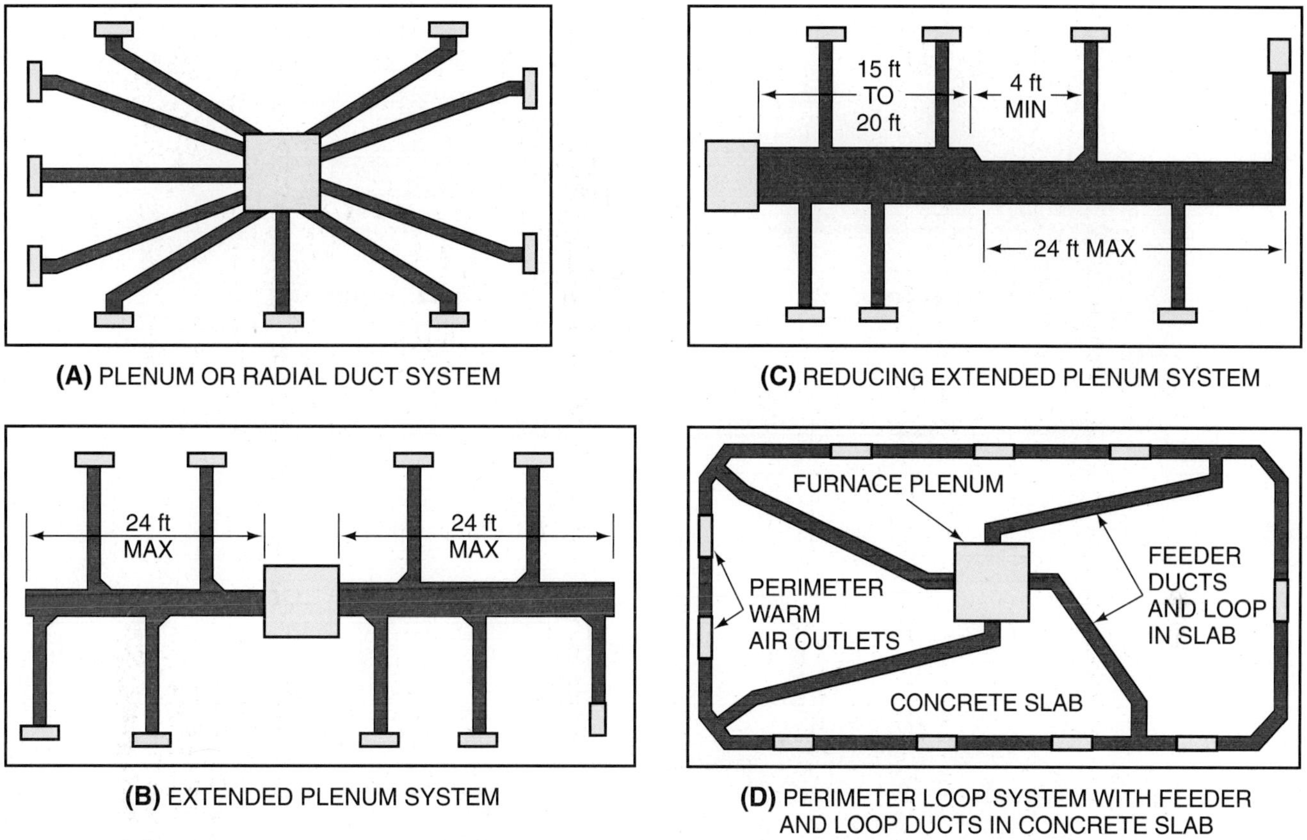

Figure 37.31 A plenum system.

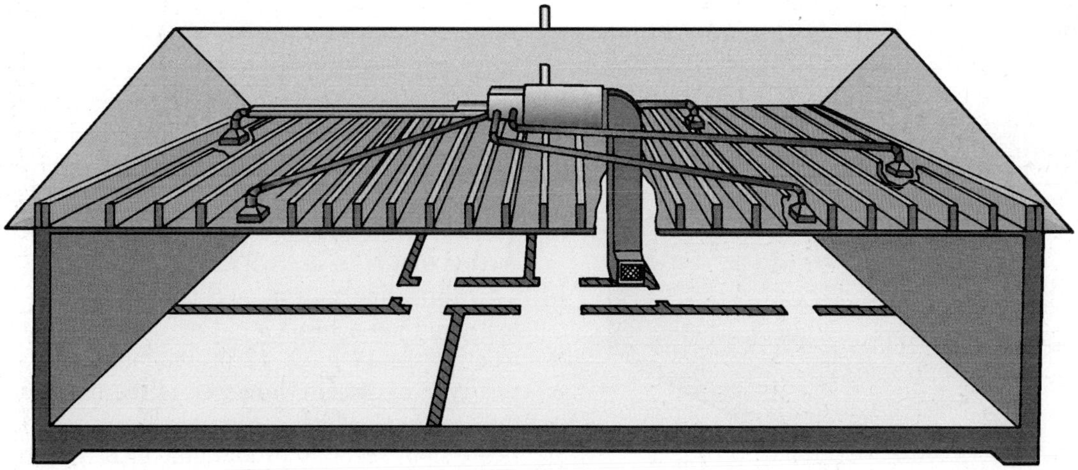

transitions, **Figure 37.34**. Transitions are used to connect duct sections of different sizes, **Figure 37.35**. This system has the advantage of conserving materials and being able to maintain a constant air pressure and velocity from one end of the duct system to the other, when properly sized. This layout also ensures that approximately the same pressure and velocity are pushing air into each branch duct.

Let's take a look at the same duct system originally shown in **Figure 37.33**. That system will now be shown as a reducing plenum system, **Figure 37.36**. The initial volume of air is

Figure 37.32 An extended plenum system.

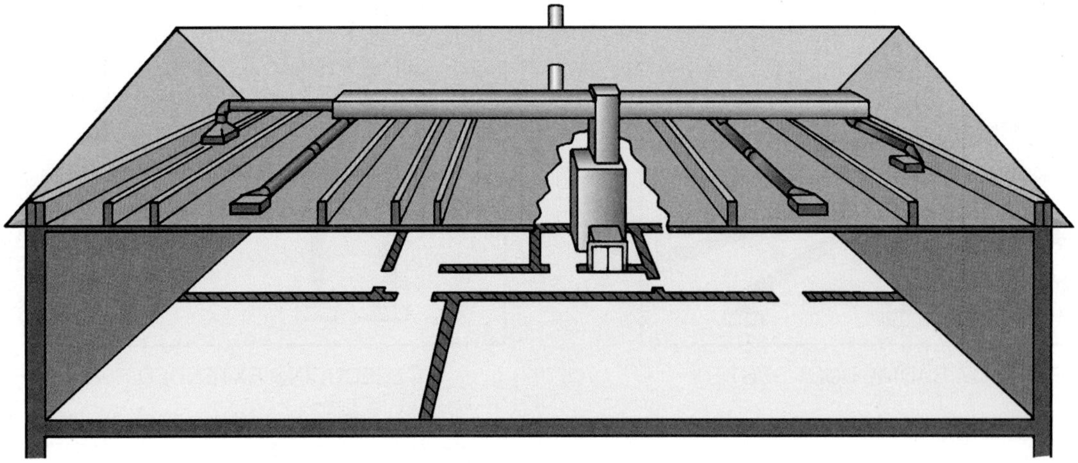

Figure 37.33 In an extended plenum system, the velocity of the air decreases as the end of the trunk line is reached.

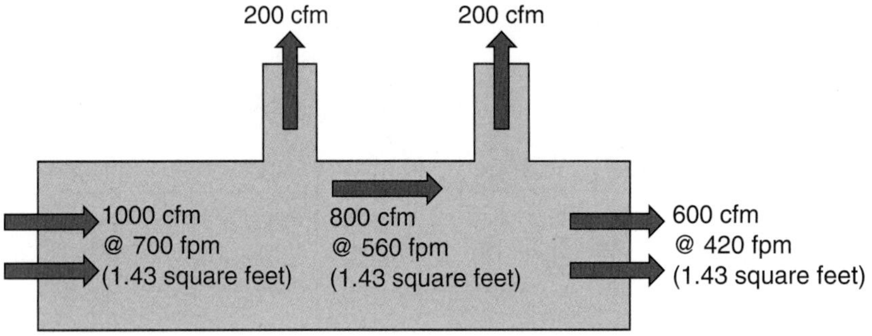

Figure 37.34 A transition fitting.

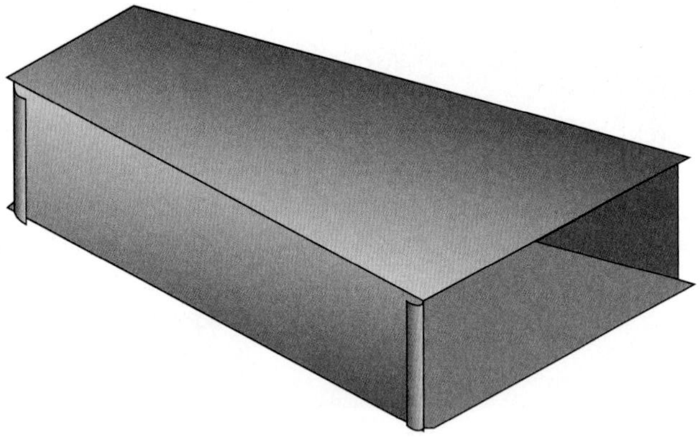

the same (1000 cfm) and the two takeoffs are still designed to carry 200 cfm each. The 1.43 ft^2 cross-sectional area is equivalent to a 14 in. × 14 in. square duct section. After the first takeoff, there are 800 cfm of air left in the main trunk line. To keep the velocity of the air the same, the cross-sectional area of the duct will have to be reduced to 1.14 ft^2 (800 cfm ÷ 700 ft/min). This is equivalent to a rectangular duct section that is about 14 in. × 12 in. After the second takeoff, there are 600 cfm of air left in the main trunk line. To keep the velocity of the air the same, the cross-sectional area of the duct will have to be reduced to 0.86 ft^2 (600 cfm ÷ 700 ft/min). This last duct section is equivalent to a rectangular duct section that is 14 in. × 9 in.

Fabricating a 14 in. × 9 in. duct section will require less material than a 14 in. × 14 in. duct section. The same holds true for the 14 in. × 12 in. section. The main downside to this type of system, however, is the need for the transition duct sections, which are more difficult and expensive to fabricate than sections of straight duct. Most of the time, the air distribution system is designed so that the different-sized duct sections will have one measurement in common to make the fabrication of transition fittings easier. We can see from the previous example that the duct measurements changed from 14 in. × 14 in. to 14 in. × 12 in. to 14 in. × 9 in. To further ease the process, one side of the transition fitting is often at a 90-degree angle to the edges of the fitting, **Figure 37.37(A)**. When the air distribution system is located

Figure 37.35 A reducing plenum system. Courtesy Climate Control

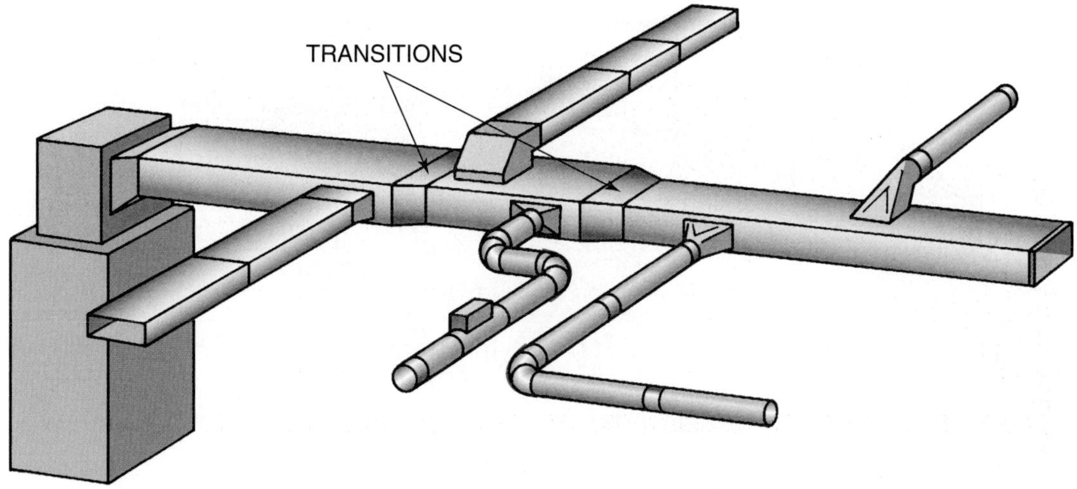

Figure 37.36 The velocity of the air that reaches the end of the trunk line can be maintained by reducing the size of the trunk line.

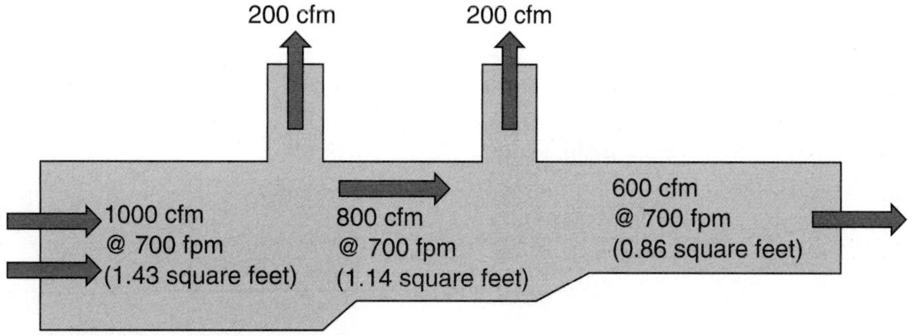

Figure 37.37 (A) Common transition fitting layout. (B) Symmetrical transition fitting.

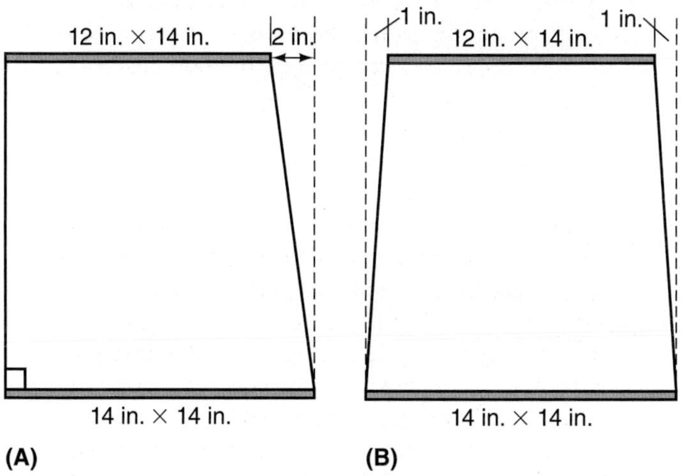

within the occupied space and is visible to the occupants—or for other design considerations—the transition fittings may have a symmetrical layout, **Figure 37.37(B)**. Laying out and fabricating symmetric transition fittings is more complicated than fabricating non-symmetric fittings.

THE PERIMETER LOOP SYSTEM

The perimeter loop duct system is particularly well suited for installation in a concrete floor in a colder climate. The loop can be run under the slab close to the outer walls with the outlets next to the wall. Warm air is in the whole loop when the furnace blower is running, and this keeps the slab at a more even temperature. The loop has a constant pressure around the system and provides the same pressure to all outlets, **Figure 37.38**.

37.10 DUCT MATERIALS

The ductwork that carries the air from the blower to the conditioned space can be made of different materials. For many years galvanized sheet metal was used exclusively, but galvanized duct sections are expensive to fabricate and require highly skilled individuals to properly install them. Galvanized metal is by far the most durable material.

Figure 37.38 A perimeter loop system.

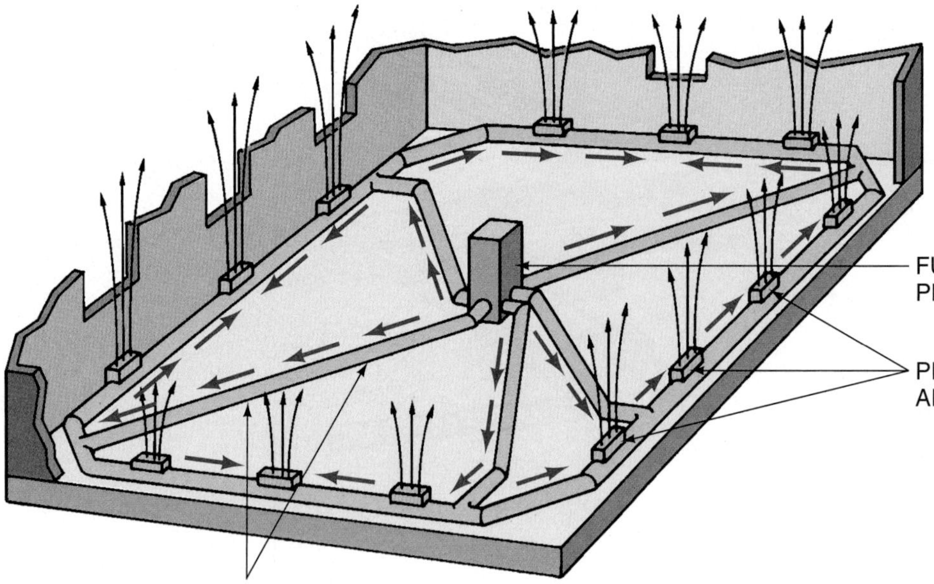

It can be used in walls where easy access for servicing cannot be made available. Aluminum, fiberglass ductboard, spiral metal duct, and flexible duct have also been used successfully. **SAFETY PRECAUTION:** *The duct material selected for a particular project must meet local building and fire protection codes. such as International Mechanical Code (IMC), International Building Code (IBC) and International Energy Conservation Code (IECC)•*

GALVANIZED-STEEL DUCT

Aluminum duct follows the same guidelines and has the same characteristics as galvanized duct. The cost of aluminum prevents it from being used for many applications. Many air distribution systems are fabricated from a material known as galvanized-steel sheet metal. Galvanized steel is steel sheet metal that has been treated with zinc to form a coating on the surface of the metal. The zinc coating will ultimately form zinc carbonate on the surface of the metal, which acts to prevent corrosion of the steel underneath. Galvanized steel is easily identified by the crystal-like patterning on the surface of the metal, often referred to as spangle, **Figure 37.39**.

Galvanized sheet metal comes in several different thicknesses, called the *gauge* of the metal. As the gauge number gets larger, the thickness of the sheet metal gets thinner. For example, the thickness of 30-gauge galvanized sheet metal is 0.0157 in., while the thickness of 22-gauge galvanized sheet metal is 0.0336 in. Thus, 22-gauge galvanized sheet metal is about twice as thick as 30 gauge. **Figure 37.40** shows the thicknesses of some commonly used gauges of steel, galvanized-steel, and aluminum sheet metal.

Figure 37.39 Spangle on galvanized sheet metal. Photo by Eugene Silberstein

Figure 37.40 Common thicknesses for steel, galvanized-steel, and aluminum sheet metal.

SHEET THICKNESS IN INCHES			
GAUGE	STEEL	GALVANIZED STEEL	ALUMINUM
22	0.0299	0.0336	0.0253
24	0.0239	0.0276	0.0201
26	0.0179	0.0217	0.0159
28	0.0149	0.0187	0.0126
30	0.012	0.0157	0.0100

Figure 37.41 A table of recommended metal thickness for different sizes of duct.

GAUGES OF METAL DUCTS AND PLENUMS USED FOR COMFORT HEATING OR COOLING FOR A SINGLE DWELLING UNIT				
	COMFORT HEATING OR COOLING			Comfort Heating Only
	Galvanized Steel			
	Nominal Thickness (In Inches)	Equivalent Galvanized Sheet Gauge No.	Approximate Aluminum B & S Gauge	Minimum Weight Tin-Plate Pounds Per Base Box
Round Ducts and Enclosed Rectangular Ducts				
14 in. or less	0.016	30	26	135
Over 14 in.	0.019	28	24	—
Exposed Rectangular Ducts				
14 in. or less	0.019	28	24	—
Over 14 in.	0.022	26	23	—

The thickness of a section of sheet metal is directly related to its weight. The Manufacturer's Standard Gauge for Sheet Metal (sheet steel) is 41.82 lb/ft². This is a constant factor and, when multiplied by the thickness of the steel sheet metal, will provide the weight of 1 ft² of the material. For example, the thickness of 24-gauge sheet steel is 0.0239 in. (from **Figure 37.40**). The weight of 1 ft² of this material is 1 lb (0.0239 in. × 41.82 lb/ft²). It is left as an exercise for the reader to confirm that 1 ft² of 30-gauge steel sheet metal weighs 0.5 lb. The thickness of the metal used to fabricate smaller duct sections can be thinner than the metal used to fabricate larger sections of ductwork. When the ductwork is larger, it must be more rigid or it will swell and make noises when the blower starts or stops. **Figure 37.41** shows a table to be used as a guideline for choosing duct metal thickness. Quite often the duct manufacturer will cross-break or make a slight bend from corner to corner on large fittings to make the duct more rigid, **Figure 37.42.**

Figure 37.42 Cross-breaks on ductwork helps to keep the duct section rigid.

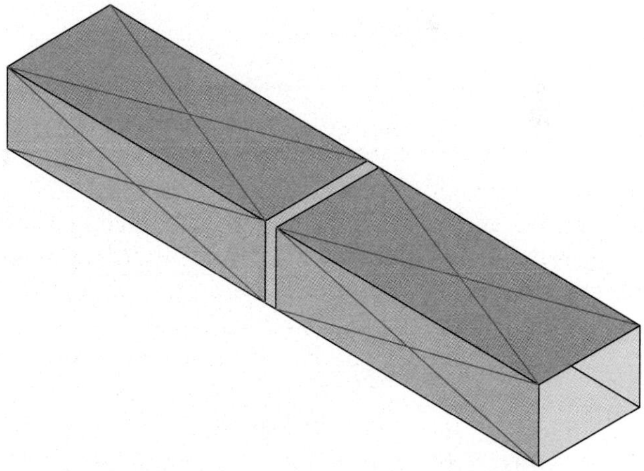

Metal duct is normally fabricated in 4-foot lengths and can be round, square, or rectangular. Smaller round duct can be purchased in lengths up to 10 ft. Duct lengths can be fastened together with special fasteners (called S fasteners or slip fasteners) and drive cleats if the duct is square or rectangular, or fastened with self-tapping sheet metal screws if the duct is round. These fasteners make a secure connection, **Figure 37.43**. If there is any question of air leaking out, special tape or sealant can be applied to the connections. **Figure 37.44** shows a duct section being connected to a furnace using self-tapping sheet metal screws. **Figure 37.45** shows a connection that has been taped after being fastened. A special product called mastic is used to help make airtight duct connections. Mastic is a very thick, puttylike substance that is applied with a stiff bristle paintbrush, **Figure 37.46.** It dries to a very hard, but slightly flexible, sealant. Most mastics are water soluble, so they are easy to clean up after use. All duct systems should be properly sealed to minimize air leakage.

Figure 37.43 (A)–(E) Fasteners for square and round duct for low-pressure systems only.

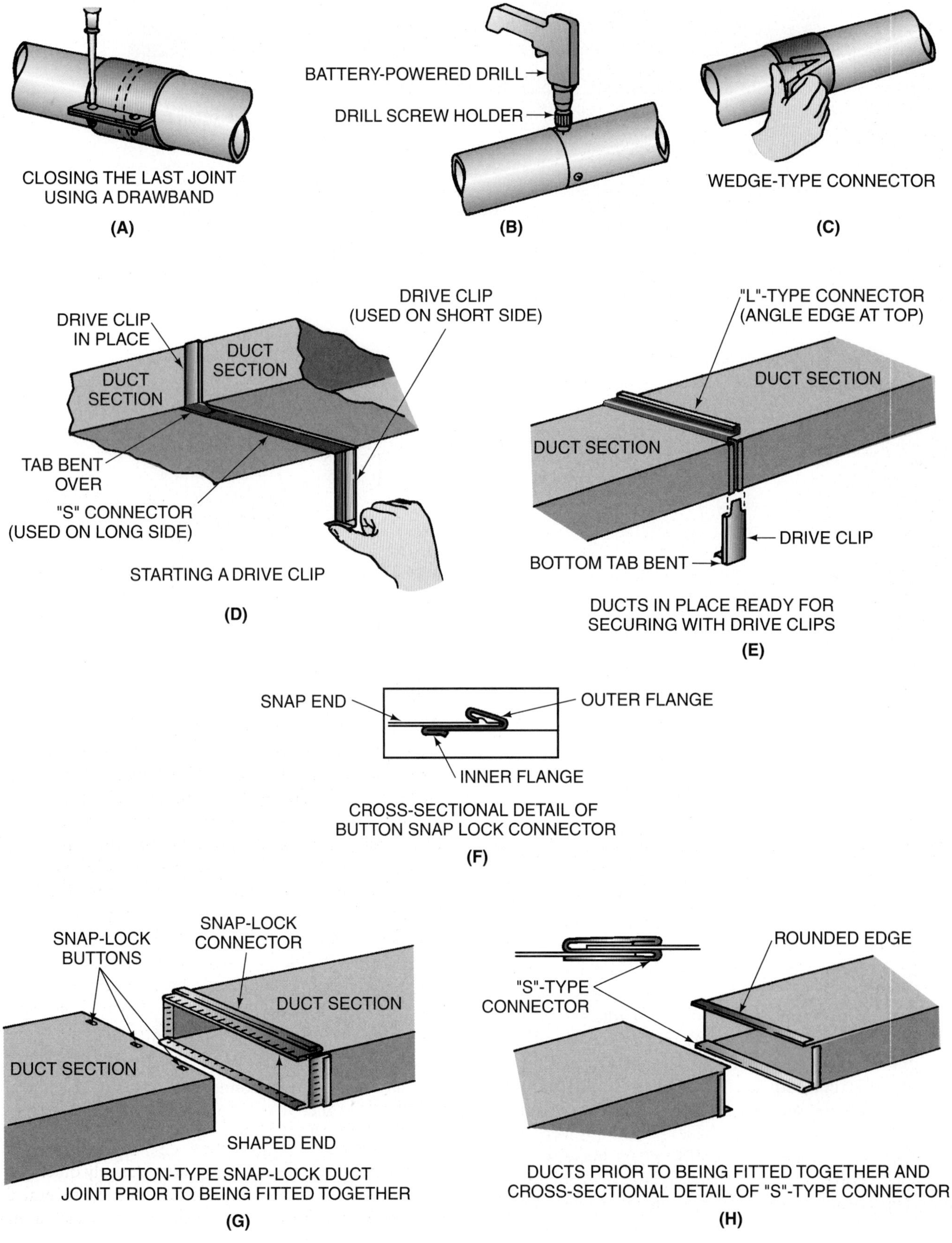

Figure 37.44 Fastening duct using a portable electric drill and sheet metal screws. Photo by Bill Johnson

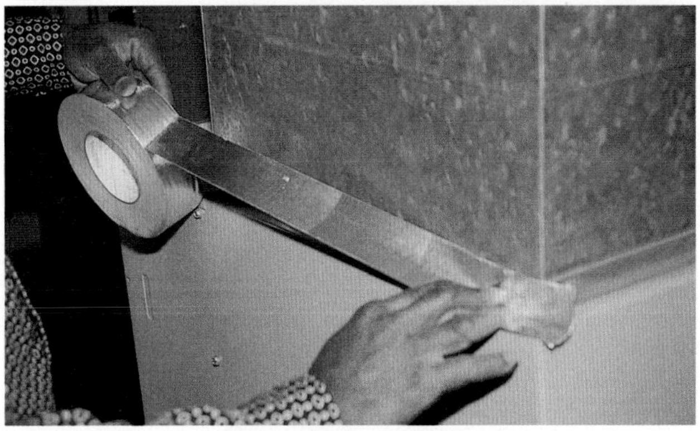

Figure 37.45 A taped duct connection. Photo by Bill Johnson

Figure 37.46 Mastic being applied to a duct connection.

Figure 37.47 (A) Rigid fiberglass ductboard. (B) Prefabricated rigid round fiberglass duct sections.

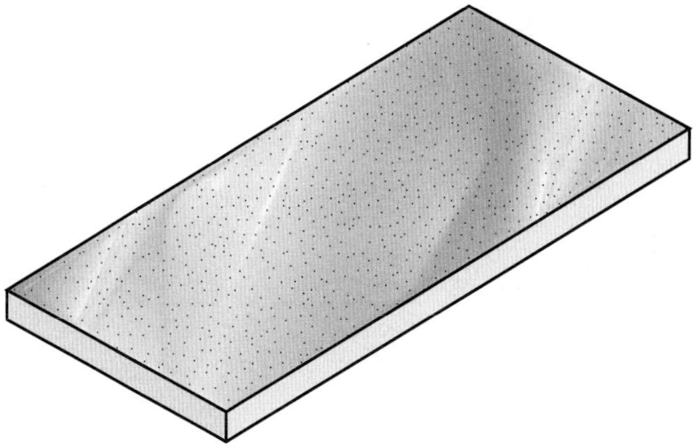

(A)

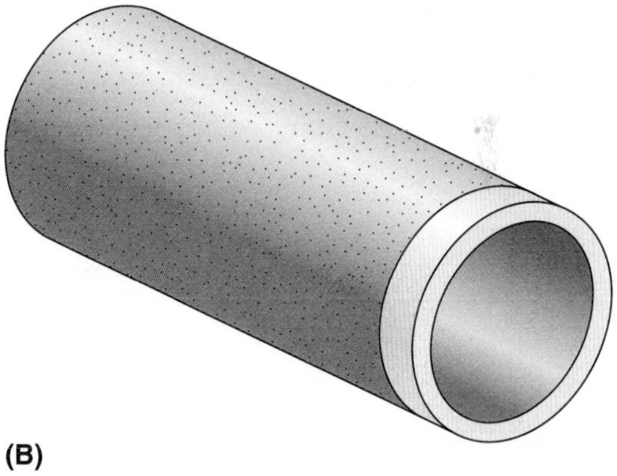

(B)

FIBERGLASS DUCT

Fiberglass duct is furnished in two styles: flat sheets for fabricating square and rectangular duct sections and round prefabricated duct sections, **Figure 37.47.** Fiberglass duct is normally 1 or 1½ in. thick with an aluminum foil backing, **Figure 37.48.** The foil backing contains fiber reinforcement to make it strong. When a duct is fabricated, portions of the fiberglass material is cut away, while the reinforced foil backing is left intact to support the connection and provide an air barrier, **Figure 37.49.** The duct sections are easily transported and connected in the field. The rough inside lining of fiberglass duct acts as a sound dampener. As a result, these systems are very quiet.

Figure 37.50 shows a layout of a fiberglass ductboard system. Main trunk lines, offsets, and takeoffs can be fabricated on the job. Special round takeoffs are available that can be inserted into the main trunk line and connected to prefabricated round duct sections for supplying air to the remote locations. One of the main benefits of using the fiberboard duct system is that the sections can be fabricated or modified on the job. In the event that a section has not been fabricated properly, the job can continue because the piece can be remade relatively quickly.

The ductboard can be made into duct in several different ways. Special knives that cut the board in such a manner

Figure 37.48 Foil backing on the fiberboard material.

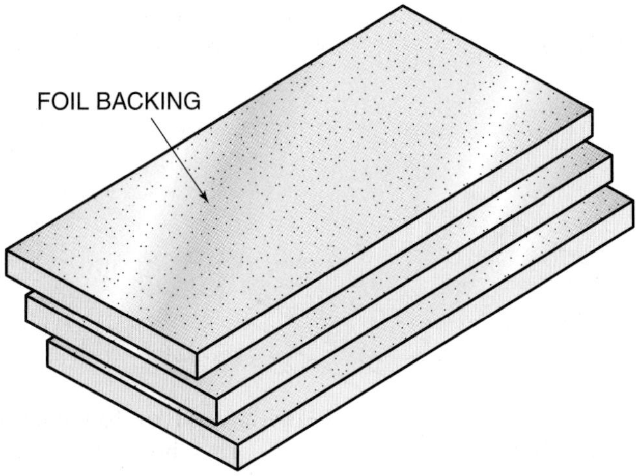

as to produce overlapping connections can be used in the field by placing the ductboard on any flat surface and using a straightedge to guide the knife, **Figure 37.51**. When fabricating a duct section, the fiberboard is positioned on the work surface with the fiberglass facing up and the foil backing against the work surface. Using the appropriate knife, cuts are made in the material, **Figure 37.49**. Notice that only the fiberglass is cut away and that the foil backing is left intact. Once all of the cuts have been made, the duct section is folded, **Figure 37.52(A)**. After folding, the flap of foil backing will serve as on overlap for completing the connection, **Figure 37.52(B)**. To finish the section, the flap is stapled and taped to seal the section, **Figure 37.52(C)**. When the duct is made in the shop, it can be cut and left flat and stored in the original boxes. This makes transportation easy. The pieces can be marked for easy assembly at the job site.

Figure 37.49 (A) Ductboard is positioned with the foil backing down. (B) Some of the fiberglass is cut away, leaving the foil backing intact.

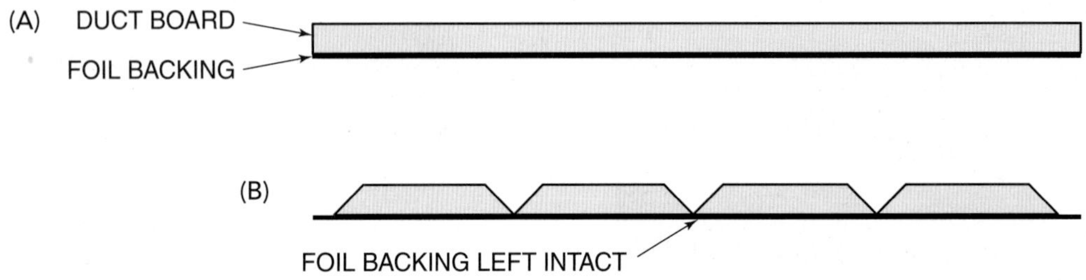

Figure 37.50 Fiberboard duct system layout.

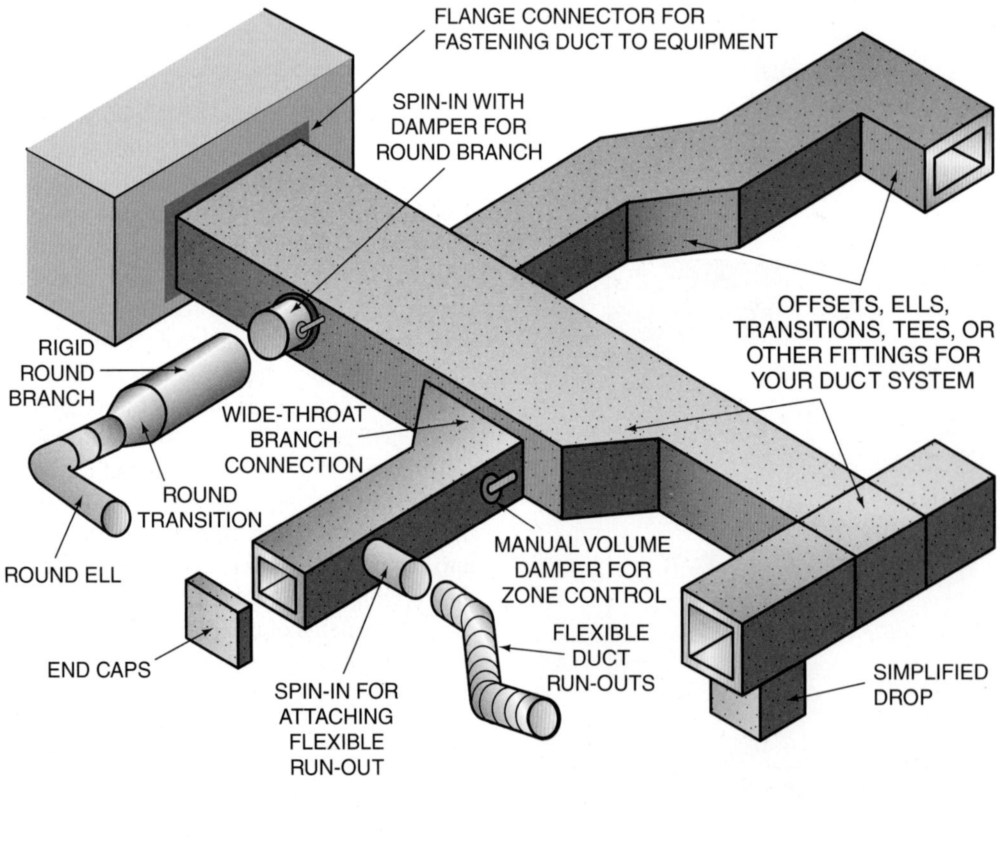

Figure 37.51 Special knives cut various shapes into the fiberboard panels for easy duct fabrication and assembly.

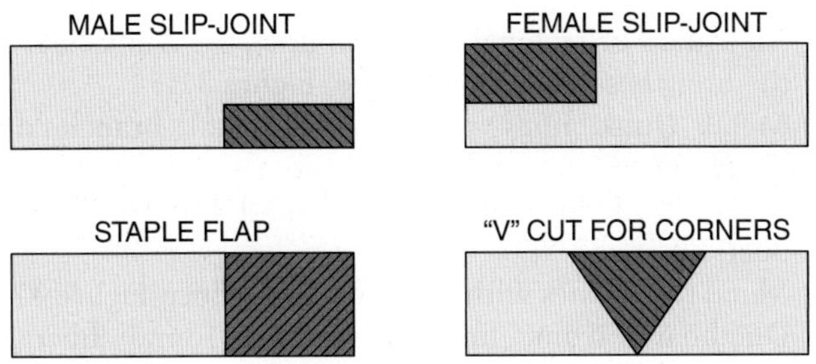

SPIRAL METAL DUCT

Spiral metal duct is more commonly found on larger systems. Spiral metal duct can be purchased in various diameters from 8 in. to over 36 in., but can also be fabricated on the job site. On-site fabrication is expensive, as special machinery is required, but is a viable option for contractors that do a large amount of this type of work. The material used to fabricate these duct sections comes in rolls of flat, narrow metal. The machine winds the metal off the spool and folds a seam into it. One benefit of fabricating spiral metal duct sections on the job is that very long pieces of duct can be fabricated on site that would otherwise have to be made in much smaller lengths and then joined together.

FLEXIBLE DUCT

Flexible round duct, **Figure 37.53**, comes in sizes up to about 24 in. in diameter. Flexible duct material can be either insulated or uninsulated. Without the insulation, the duct looks like a coil spring wrapped with a flexible foil or plastic backing. Installing flexible duct is easy because the runs can be aligned fairly loosely, unlike rigid duct, which must be perfectly aligned.

Some flexible duct comes with vinyl or foil backing with insulation, in lengths of 25 ft to a box. One benefit of flexible duct material is that it is easy to route around corners or other obstructions. The duct runs should be kept as short as practical and not allowed to make tight turns that may cause the flexible duct to collapse. Flexible duct restricts airflow more than metal duct does, but it also serves as a sound attenuator that reduces the transmission of blower noise. For best airflow, flexible duct should be

Figure 37.52 (A) Once the fiberglass material has been cut away, the duct section can be folded. (B) A flap of foil backing material is used to fold over the last corner. (C) The foil backing flap is then taped to complete the section.

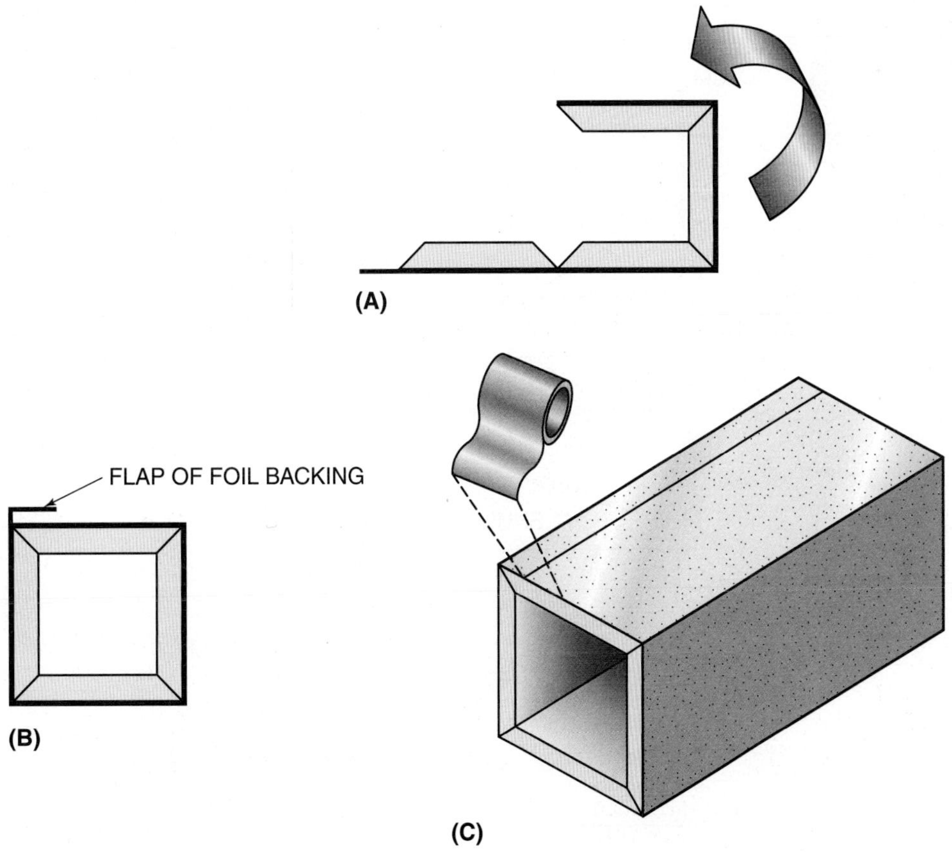

stretched as tightly as is practical, **Figure 37.54**, to reduce the resistance to airflow, and properly supported to prevent sagging. In addition, flexible duct should be used for only very short runs, such as when connecting a duct takeoff to a supply boot.

Figure 37.53 Flexible duct.

Figure 37.54 Flexible duct should be stretched for proper installation.

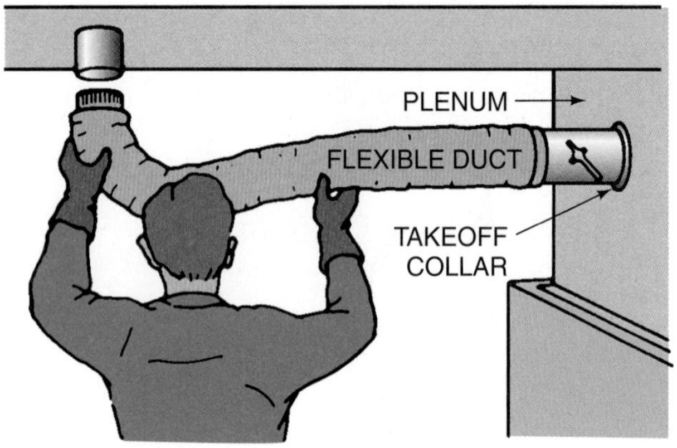

COMBINATION DUCT SYSTEMS

Duct systems can be combined in various ways. For example:

1. All square or rectangular metal duct, the trunk line, and the branches of the same shape, **Figure 37.55**
2. Metal trunk lines with round metal branch duct, **Figure 37.56**
3. Metal trunk lines with rigid round fiberglass branch duct, **Figure 37.57**
4. Metal trunk lines with flexible branch duct, **Figure 37.58**
5. Ductboard trunk lines with rigid round fiberglass branches, **Figure 37.59**
6. Ductboard trunk lines with round metal branches, **Figure 37.60**
7. Ductboard trunk lines with flexible branches, **Figure 37.61**
8. Round metal duct with round metal branch ducts, **Figure 37.62**
9. Round metal trunk lines with flexible branch ducts, **Figure 37.63**

Figure 37.55 Square or rectangular duct.

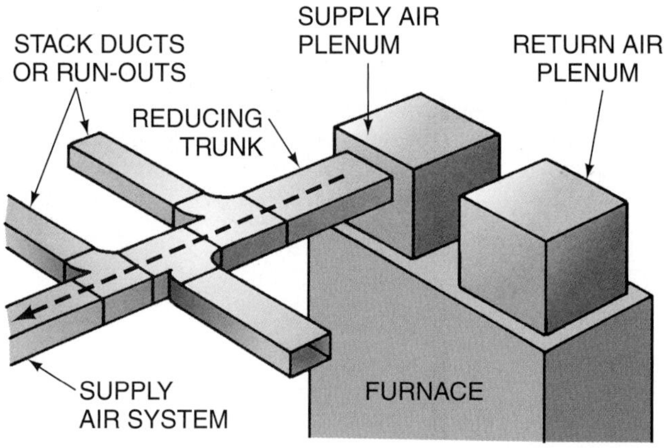

Figure 37.56 Rectangular metal trunk duct with round metal branch ducts.

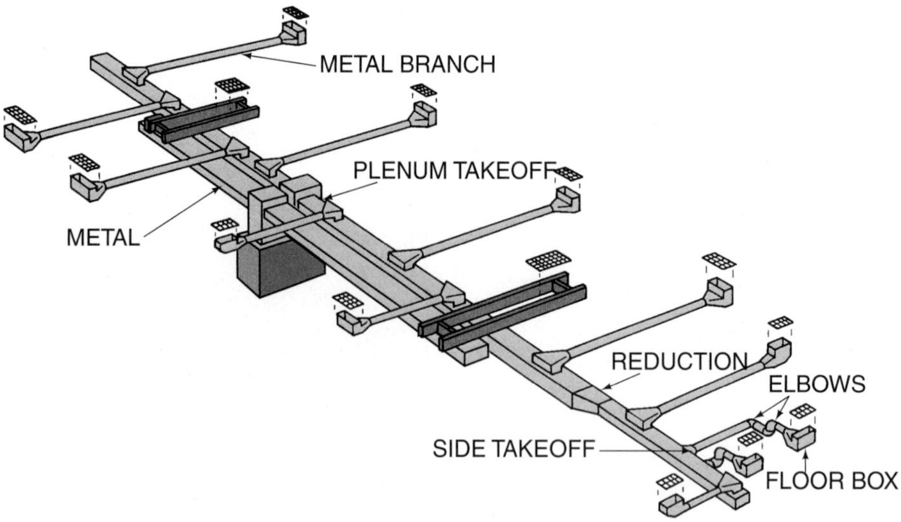

Figure 37.57 Metal trunk duct with round fiberglass branch ducts.

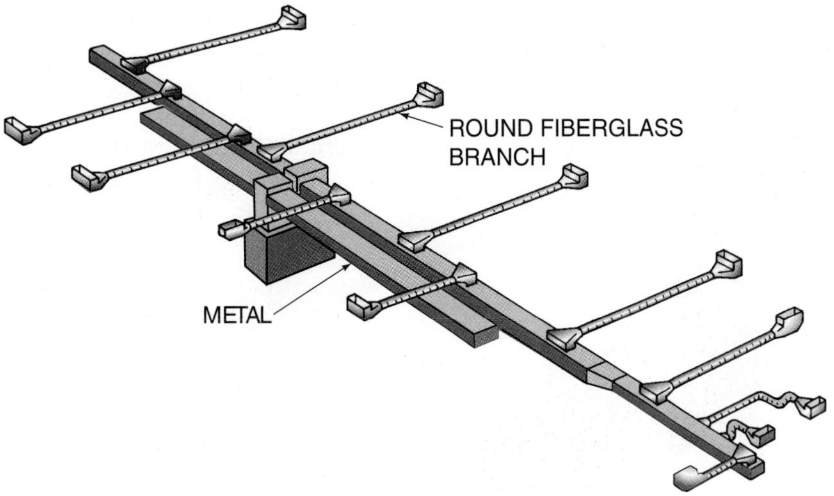

Figure 37.58 Metal trunk duct with flexible branch ducts.

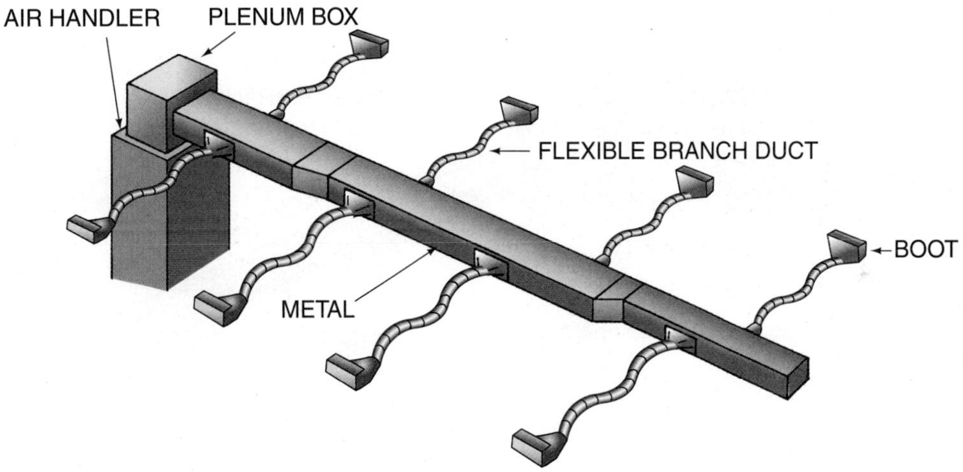

Figure 37.59 Fiberglass duct with round fiberglass ducts.

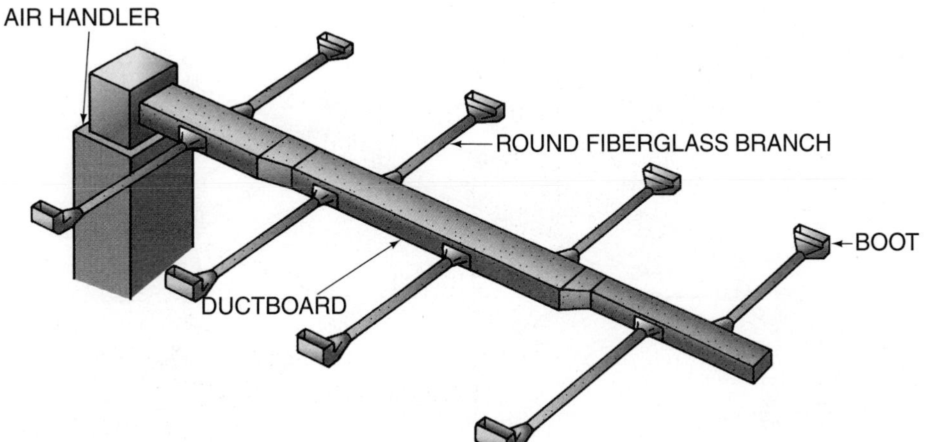

Figure 37.60 Fiberglass ductboard trunk and round metal branches.

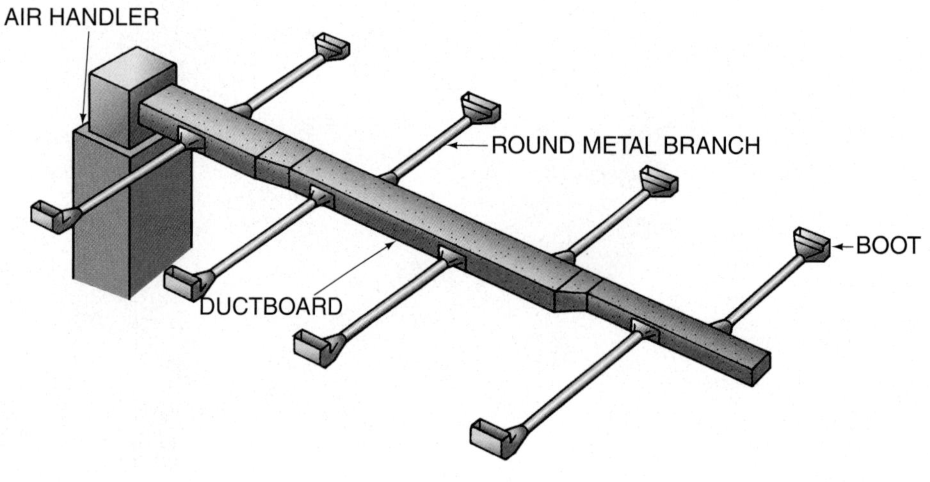

Figure 37.61 Fiberglass ductboard trunk and flexible branch ducts.

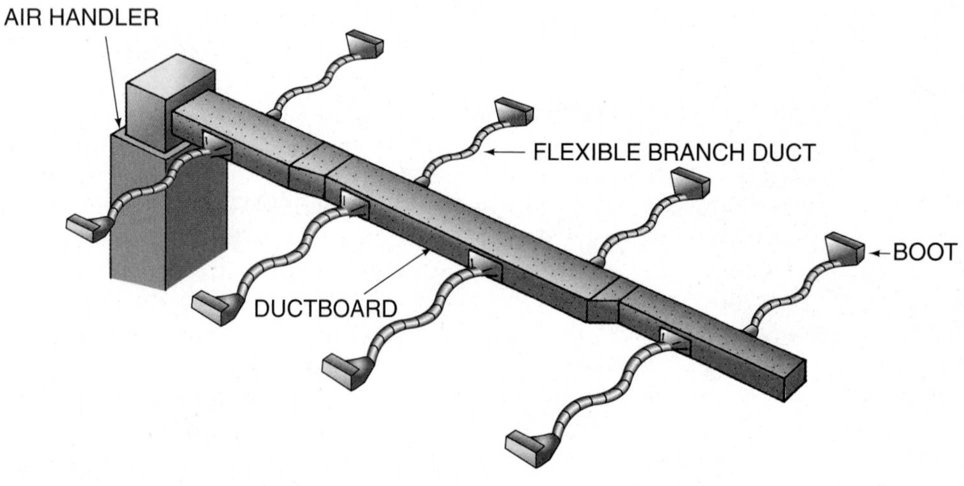

Figure 37.62 Round metal duct.

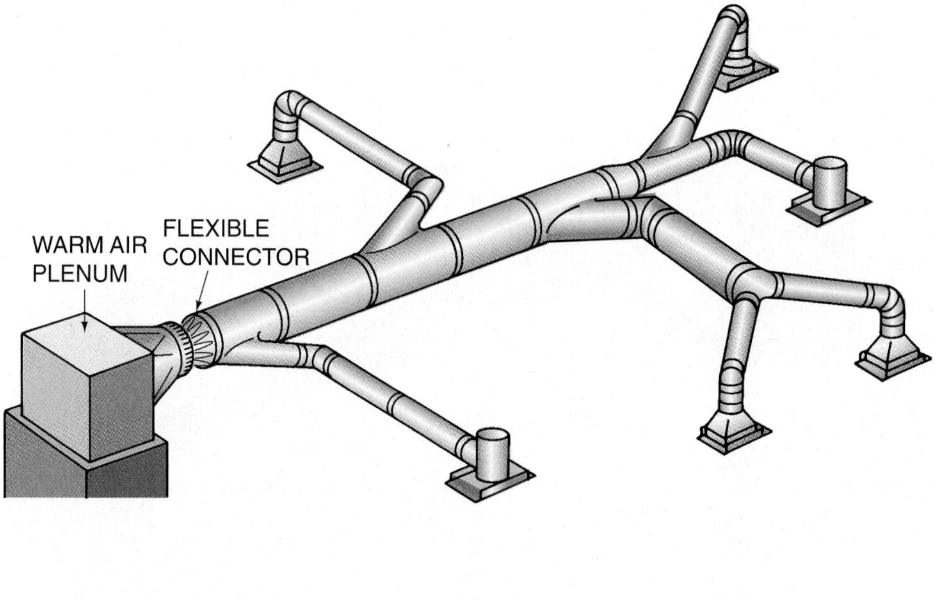

Figure 37.63 Round metal trunk line with flexible branch lines.

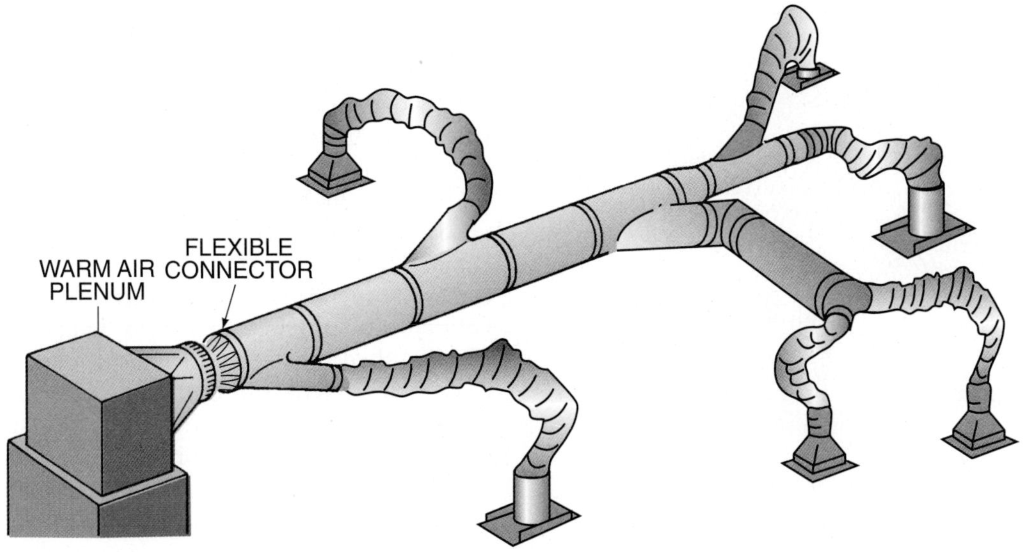

37.11 DUCT AIR MOVEMENT

Special attention should be given to the point where the branch duct leaves the main trunk duct in order to get the correct amount of air into the branch duct. The branch duct must be fastened to the main trunk line with a *takeoff fitting*. The takeoff is the connecting point between the main trunk line and the branch duct. The connection at the main trunk line is the throat portion of the takeoff. The other side of the takeoff is connected to the runout, or branch, duct. Often, an air distribution system installation will utilize takeoffs that have the same throat and runout measurements, **Figure 37.64(A)**. Utilizing such takeoffs may very well not be the best option. The takeoff that has a larger throat area than the runout duct will allow the air to leave the trunk duct with a minimum of effort. This could be called a streamlined takeoff, **Figure 37.64(B)**. The streamlined takeoff encourages the air moving down the main duct to enter the takeoff to the branch duct.

Air moving in a duct has *inertia*—it wants to continue moving in a straight line. If sudden or sharp turns are made in the duct run, there will be a great deal of turbulence created, **Figure 37.65**. For example, a square-throated elbow offers more resistance to airflow than a round-throated elbow. If the duct is rectangular or square, turning vanes, **Figure 37.66**, will improve the airflow around a corner. The turning vanes separate the air into individual streams and direct them around the bend in a duct in a manner similar to the white dashed lines on a highway that help to keep vehicular traffic flowing properly around bends and turns in the road, **Figure 37.67**.

Figure 37.64 (A) A standard takeoff fitting. (B) A streamlined takeoff fitting.

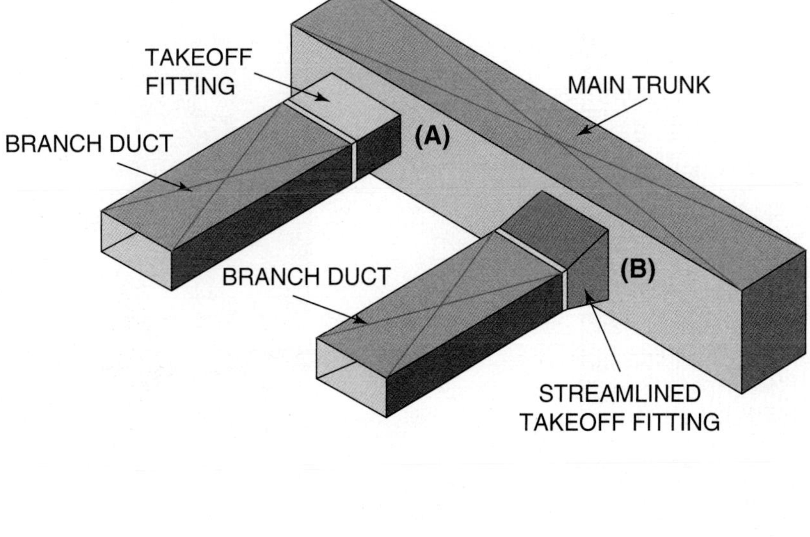

Figure 37.65 Turbulence in a duct when air tries to go around a corner.

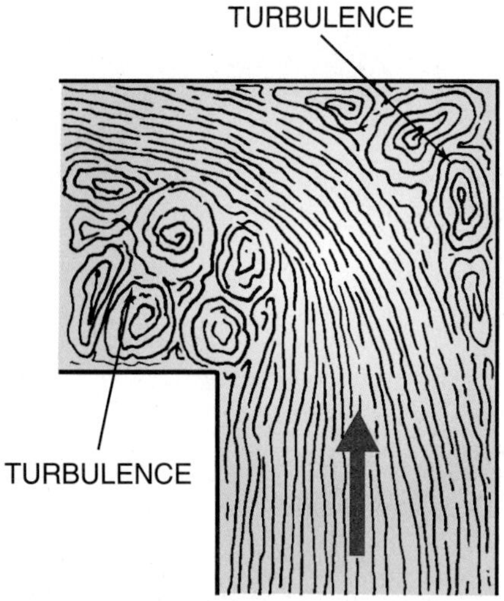

Figure 37.66 Turning vanes. Photo by Eugene Silberstein

Figure 37.67 A square elbow with turning vanes.

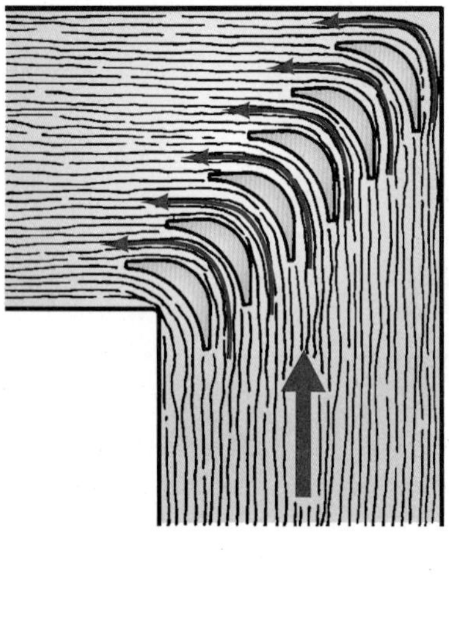

37.12 BALANCING DAMPERS

A well-designed system will have *balancing dampers,* **Figure 37.68**, in each of the branch ducts to balance the flow of air in the various parts of the system. Balancing the airflow with dampers enables the technician to direct the correct volume of air to each room or area for better temperature control. The dampers should be located as close as practical to the trunk line, with the damper handles accessible and uncovered if the duct is insulated. The best place at which to balance airflow is as close to the trunk as possible. This way, if there is any air velocity noise its effects will be minimized before the air enters the room. A damper consists of a piece of metal shaped like the inside of the duct with a handle protruding through the side of the duct to the outside. The handle allows the damper to be turned at an angle to the airstream to slow the air down. Damper handles are directional. When the handle is positioned perpendicular to the duct, the damper is closed, **Figure 37.69**. When the handle is positioned parallel to the direction of airflow, the damper is open, **Figure 37.70**.

Figure 37.68 Balancing damper. Photo by Eugene Silberstein

Figure 37.69 Damper in the closed position. Photo by Eugene Silberstein

Figure 37.70 Damper in the open position. Photo by Eugene Silberstein

Figure 37.71 A zone damper.

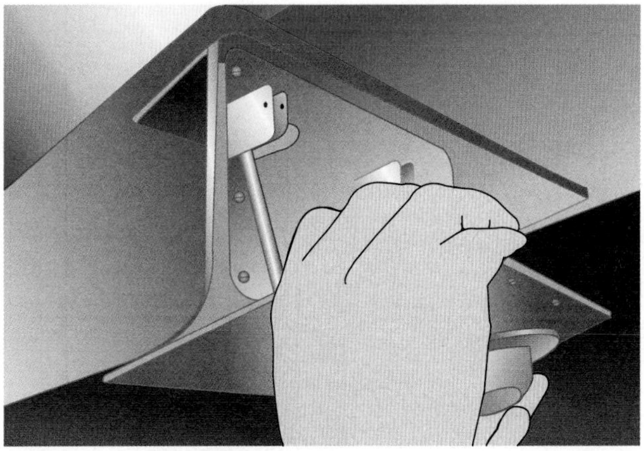

37.13 ZONING

As we mentioned in the previous section, balancing dampers are used to ensure that the proper amount of air is delivered to each area in the conditioned space. Balancing the airflow in the system helps ensure that the space is heated and/or cooled evenly. Once the system is properly balanced, the balancing dampers should never have to be touched again. Another type of damper that is found in an air distribution system is the zone damper. Here are some characteristics that will differentiate between the two types:

- Balancing dampers are set to the desired position and locked in place at the time of initial system start-up or shortly thereafter.
- Zone dampers are opened and closed in response to changes in room temperature.
- Balancing dampers are manually opened and closed.
- Zone dampers open and close automatically in response to temperature changes. These dampers can be pneumatically (air) or electrically controlled.
- Balancing dampers are used to set airflow at some point between minimum and maximum on the basis of the size of the duct and the velocity of the air moving through the duct.
- Zone dampers will move from the fully open to the fully closed position depending on whether or not the temperature requirements for the zone are satisfied.

The zoning process utilizes zone dampers, ***Figure 37.71****, and enables a single forced-air heating and/or cooling system to maintain different temperatures in different rooms or areas of the conditioned space. By zoning a system, efficiency is increased.* Consider, for example, a single central air-conditioning system that is being used to cool a two-story house. Typically, the second floor of the structure will be considerably warmer than the lower level. This temperature difference can be as high as 10 degrees. If the thermostat is located on the lower level, the upper floor will never get cool. If the thermostat is located upstairs, the lower level will be too cool. The same condition can occur in some ranch-style homes. In this case, one side of the house may be warmer than the other. Depending on the location of the thermostat, one side of the home will be comfortable, while the other will be either too hot or too cold.

Zoning an air-conditioning system involves installing a thermostat and a zone damper for each room or zone in the structure. These thermostats and dampers are wired to a central control panel that will control the operation of the system on the basis of the actual conditions in the space. The complexity of these systems varies greatly depending on what features and tasks the system is able to perform.

A very basic zoning system is shown in **Figure 37.72**. This system has two separate zones, each with its own zone damper and thermostat. In a system similar to the one shown in this figure, it is easy to divide the structure into two separate areas, as the main plenum branches off in two distinct directions. For this example, we will assume that each of the two zones has the same airflow, cfm, and requirements. When configured in this manner (as two otherwise equal zones), the ductwork serving each zone should be oversized about 50%. If the total system were to provide 2000 cfm, this would mean that each trunk should be able to carry 1000 cfm of air. For zoning purposes, each trunk should be sized to carry approximately 1500 cfm, or 75%, of the total airflow requirement for the system. The wiring/control system for the zoned system shown in **Figure 37.72** is comprised of a central control panel, two damper motors, and two thermostats. One thermostat is centrally located in each of the two zones.

Of course there are more intricate and complicated zoning systems available. Typically, the two-zone system just discussed can set up zones in one of the following ways:

- Zone 1: upstairs; Zone 2: downstairs
- Zone 1: living room and kitchen; Zone 2: bedrooms and bathrooms

Figure 37.72 A common two-zone system layout.

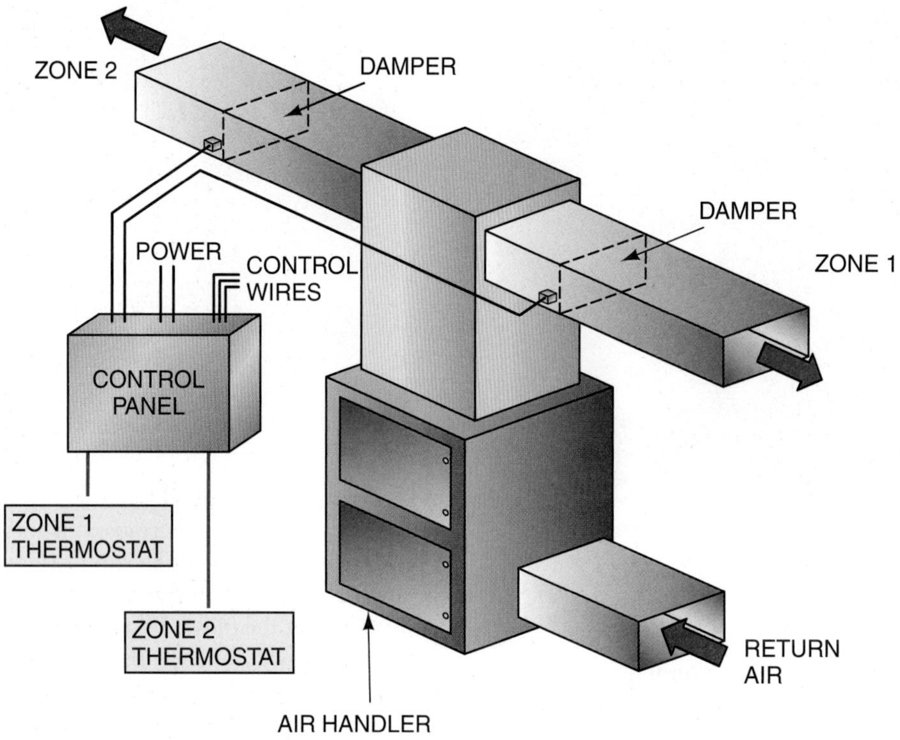

As homes get larger, the zoning requirements change as well. For example, the master bedroom suite in some higher-end homes has its own zone. The possibilities for zoning are limited only by the system designer's imagination. It is not unreasonable to have a separate zone for every room in a home. Obviously, systems like this are very costly to design and install and are more complicated to troubleshoot.

ZONING WITH A SINGLE-SPEED BLOWER MOTOR

The opening and closing of the zone dampers will have an effect on the velocity and volume of the air moving through the system. This is not an action that should be undertaken lightly, as a reduction in airflow through a cooling coil can, for example, cause the coil to freeze, further reducing airflow and causing a major reduction in cooling. In a two-zone system, as in the situation discussed in the previous section, oversizing the duct system by approximately 50% will, for the most part, alleviate this problem.

Let's take a closer look. If the system were designed to move 2000 cfm at a velocity of 700 ft/min, the supply plenum would have to have a cross-sectional area of 2.86 ft^2 (2000 cfm ÷ 700 ft/min). If the air is to be divided between the two zones as shown in **Figure 37.73**, each zone will have to carry 1000 cfm at the same 700 ft/min. This means that the cross-sectional

Figure 37.73 (A) Both zones are calling for cooling, so each zone is being supplied with 1000 cfm of cooled air. (B) Only Zone 1 is calling for cooling, so this zone is receiving all of the air.

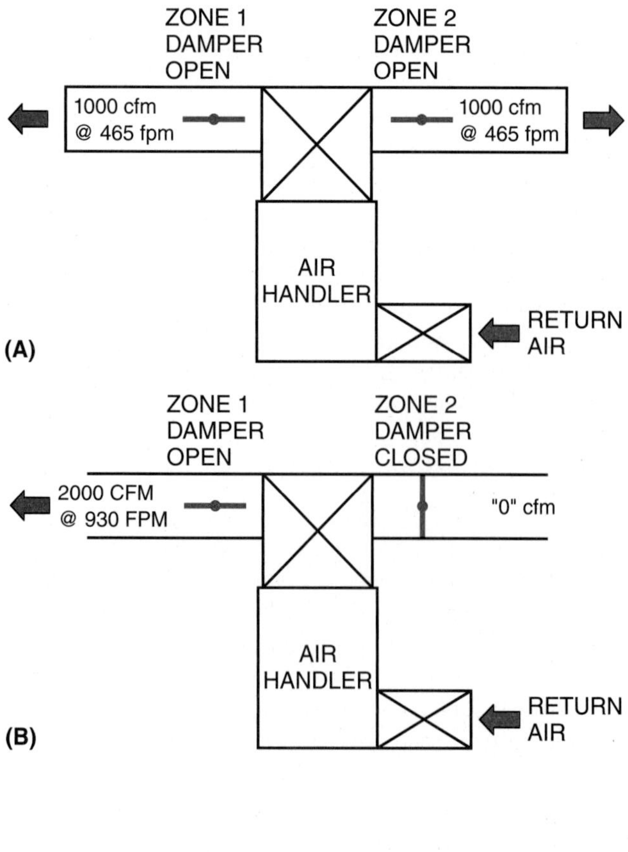

area of the duct is going to be 1.43 ft². By oversizing this zone duct by 50%, the cross-sectional area will be increased to 2.15 ft². The duct will still be carrying the 1000 cfm, so the velocity of the air will be reduced to 465 ft/min (1000 cfm ÷ 2.15 ft²), which will be the velocity when both zones are calling for cooling, **Figure 37.73(A)**. When only one zone is calling for cooling, the entire 2000 cfm will flow through the duct feeding that zone. The speed of the air flowing through that duct will be 930 ft/min (2000 cfm ÷ 2.15 ft²), **Figure 37.73(B)**.

In systems with three or more zones, oversizing the duct to each zone by 50% is not practical. In such cases, oversizing the ducts by 10% to 15% will often suffice. In addition to this minimal oversizing of the duct, a bypass duct can be installed in the system, **Figure 37.74**. The bypass duct allows the blower to move the required amount of air even if a number of zones are not calling for conditioning. The bypass duct should be equipped with a damper so that the airflow can be adjusted. This damper can be manually or automatically controlled. When the damper is manual, it should be set for the "worst-case scenario," which is when only the smallest zone is calling for conditioning.

Consider the system shown in **Figure 37.75(A)**. This is an air-conditioning system with four zones. In this case, all four zones are calling for cooling, so there is no problem with excess or unwanted air volume. In **Figure 37.75(B)**, only the smallest zone is calling for cooling, so all zone dampers are closed with the exception of the one serving Zone 4. The blower will still deliver 2000 cfm, but only 400 cfm are needed. This is the worst-case scenario, as only the smallest zone is calling for cooling. The bypass duct should be sized to handle enough air for this condition, which, in this case, is 1600 cfm (2000 cfm − 400 cfm). When the damper in the bypass zone is controlled automatically, a pressure switch is often used to open and close the damper. The pressure switch senses the pressure in the duct and, as the pressure in the duct rises, causes the bypass damper to open.

When using the bypass-duct method, it is important to sense the temperature of the air as it moves through the system. By having conditioned air fed back through the system, the air can become overheated in the heating mode or overcooled in the cooling mode. This can damage the equipment. For example, if the return air is too cool, it may very well cause the evaporator coil to freeze. This can lead to liquid floodback to the compressor, resulting in major system damage. The sensor should send a signal to the central control panel to cycle the compressor off when such conditions are present. Different manufacturers tackle this situation differently, so be sure to follow the recommendations and instructions provided with each specific zoning product.

One alternative to using the bypass duct is to incorporate a dump zone into the system. Instead of recirculating the air back through the equipment, the excess air is directed to an

Figure 37.74 A bypass duct.

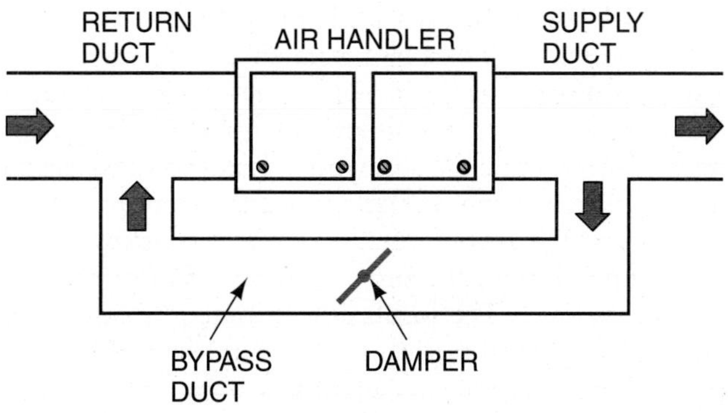

Figure 37.75 (A) All four zones are calling for system operation. (B) This is the worst-case scenario, as only the smallest zone is calling for system operation. The bypass damper must be able to handle enough air for this situation.

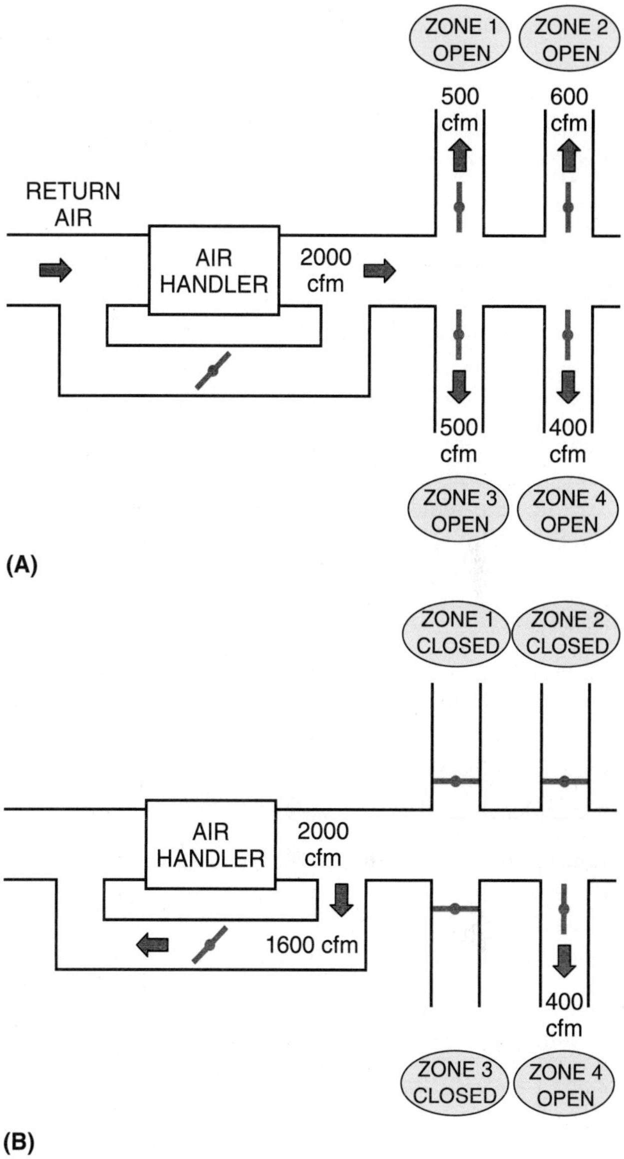

area in the structure that does not have critical temperature requirements. Such areas include basements, common hallways, or similar spaces. Since this may be objectionable to the occupants of the structure, options should be discussed prior to the design and installation stages of the job.

ZONING WITH A MULTISPEED COMPRESSOR AND VARIABLE-SPEED BLOWER

Some of the most desirable zoning systems utilize multispeed compressors and variable-speed blower motors. These systems use computer-controlled panels that store information regarding the air requirements for each zone. Some of these zoning systems use the following pieces of data:

- Return-air temperature
- Supply-air temperature
- Airflow requirements for each zone
- Percentage of airflow for each zone as a percentage of total system airflow

More advanced zoning systems have the ability to

- provide simultaneous heating and cooling by zone.
- provide continuous fan operation options by zone.
- control individual zone temperature.
- set heating or cooling priority by zone.

Consider the following system setup:

- Zone 1: 200 cfm—den and bedroom 4
- Zone 2: 450 cfm—bedrooms 2 and 3
- Zone 3: 500 cfm—master bedroom suite
- Zone 4: 450 cfm—living room and dining room area

The total cfm requirement for this home is 1600 cfm, so the percentage of air going to each zone is as follows:

- Zone 1 $\rightarrow$ (200 cfm $\div$ 1600 cfm) $\times$ 100% = 13%
- Zone 2 $\rightarrow$ (450 cfm $\div$ 1600 cfm) $\times$ 100% = 28%
- Zone 3 $\rightarrow$ (500 cfm $\div$ 1600 cfm) $\times$ 100% = 31%
- Zone 4 $\rightarrow$ (450 cfm $\div$ 1600 cfm) $\times$ 100% = 28%

Notice that the total air volume is equal to 100%. These air volume percentages are rounded off and will get programmed into the central control panel. The average air volumes for Zones 1 through 4 will then be estimated as 10%, 30%, 30%, and 30%, respectively. On the basis of this information and the zones that are calling for conditioning, the compressor and blower speeds will be selected automatically by the system. For example, if Zones 3 and 4 are calling for cooling, the blower will operate at a speed that will deliver 960 cfm of air (60% of 1600 cfm). The compressor speed will be selected on the basis of this airflow as well as the data gathered from the return- and/or supply-air sensors. If another zone now calls for cooling, the blower will ramp up its speed to deliver more air volume.

If the system is equipped to provide both heating and cooling, the zoning equipment may have the capability of selecting a priority mode of operation. For example, if a heating priority is programmed into the system and there are simultaneous calls for heating and cooling, the system will satisfy the heating demand first and then switch over to satisfy the cooling demand. Similarly, if there is a cooling priority programmed into the system, the system will satisfy the cooling demand first and then switch over to the heating mode for those zones that require it. There may also be a priority referred to as *zone priority.* Zone priority gives preference to whatever mode of operation is called for by the priority zone. For example, some homeowners set the master bedroom suite as the priority zone. So if that zone is calling for cooling and the rest of the home is calling for heating, the cooling demand will be satisfied first.

ADDING ZONING TO AN EXISTING SYSTEM

Although it is easier to zone a system when it is initially designed and installed, retrofitting existing systems has become a very popular option. Adding zones to an existing system has become an easier task as well. Compact, easy-to-install dampers are readily available in both round, square, and rectangular configurations to fit into existing duct sections. The manufacturers of these products provide templates along with their products to further ease the installation process. **Figure 37.76(A)–(D)** shows the installation of round and square zone dampers.

37.14 DUCT INSULATION

When ductwork passes through an unconditioned space, heat transfer may take place between the air in the duct and the air in the unconditioned space. If the heat exchange adds or removes very much heat from the conditioned air, insulation should be applied to the ductwork. A 15°F temperature difference from the inside to the outside of the duct is considered the maximum difference allowable before insulation is necessary.

Insulation is built into a fiberglass duct by the manufacturer. Metal duct can be insulated in two ways: on the outside or on the inside. When applied to the outside, the insulation is usually a foil- or vinyl-backed fiberglass. It comes in several thicknesses, with a 2-in. thickness the most common. The backing creates a moisture vapor barrier. This is important where the duct may operate below the dew point temperature of the surroundings and moisture would form on the duct. The insulation is joined by lapping and stapling it. It is then taped to prevent moisture from entering the seams. External insulation can be added after the

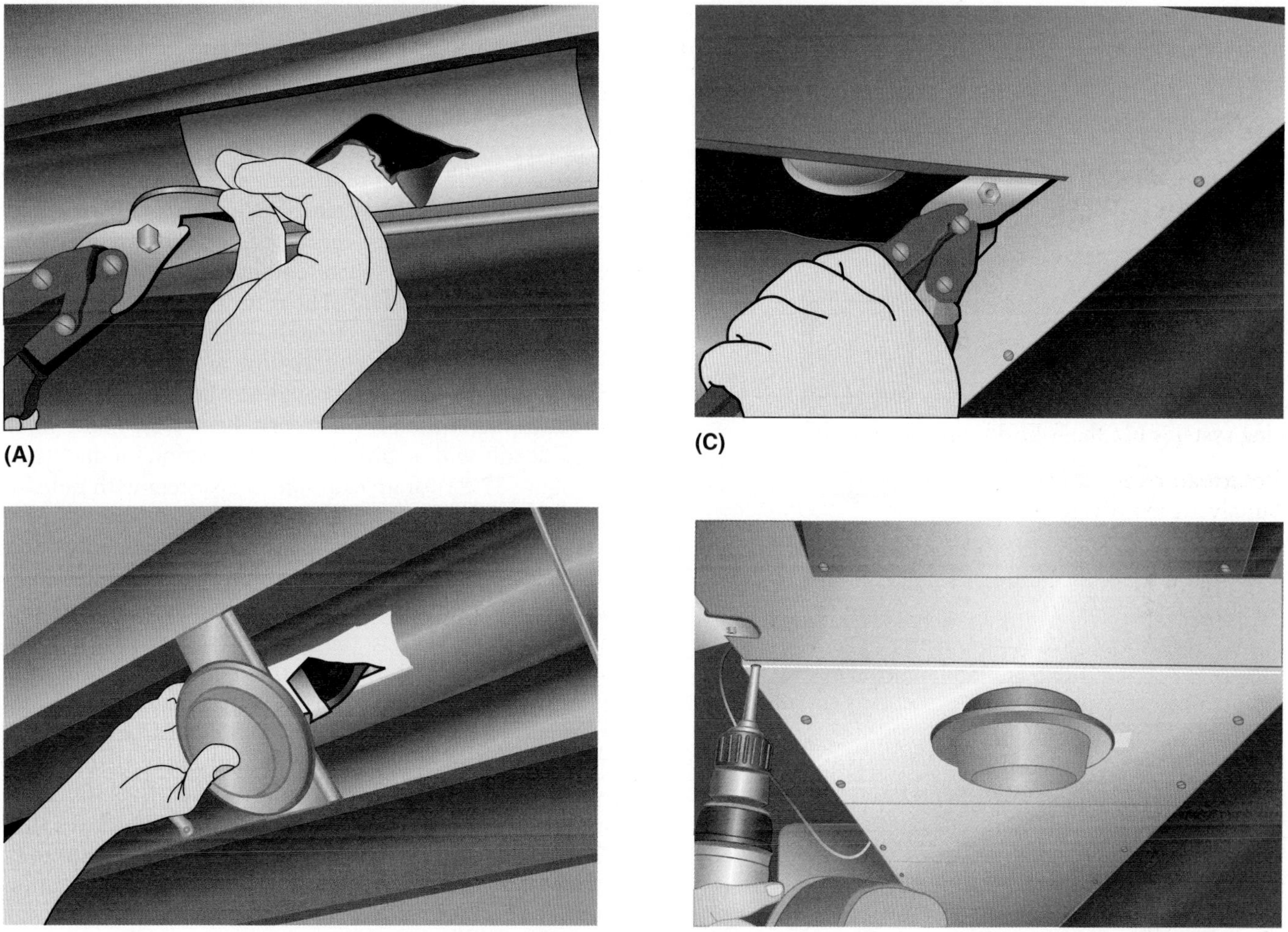

(A) (B) (C) (D)

Figure 37.76 Installing a zone damper in a round duct (A)–(B). Installing a zone damper in a square duct (C)–(D).

duct has been installed if the duct has enough clearance all around. When applied to the inside of the duct, the insulation is either glued or fastened to tabs mounted on the duct by spot weld or glue. This insulation must be applied when the duct is being manufactured.

37.15 BLENDING THE CONDITIONED AIR WITH ROOM AIR

When air reaches the conditioned space, it must be properly distributed into the room so that the room will be comfortable without anyone being aware that a conditioning system is operating. This means that the final components in the system must place the air in the proper area of the conditioned space for proper air blending. Following are guidelines that can be used for room air distribution.

1. When possible, air should be directed onto the walls. They are the load where the heat exchange occurs. For example, in winter air can be directed to the outside walls to cancel the load (cold wall) and keep the wall warmer. This will keep the wall from absorbing heat from the room air. In summer the same distribution will work; it will keep the wall cool and keep room air from absorbing heat from the wall. The diffuser spreads the air in the desired air pattern, **Figure 37.77**.
2. Warm air for heating distributes better from the floor because warm air tends to rise, **Figure 37.78**.
3. Cool air distributes better from the ceiling because cool air tends to fall, **Figure 37.79**.
4. The most modern concept for both heating and cooling is to place the diffusers next to the outside walls to accomplish this load-canceling effect.
5. The amount of throw (how far the air from the diffuser will blow into the room from the diffuser) depends on the air pressure behind the diffuser and the style of the diffuser

blades. Air pressure in the duct behind the diffuser creates the velocity for the air leaving the diffuser. **Figure 37.80** shows some air registers and diffusers. The various types can be used for low-side wall, high-side wall, floor, ceiling, or baseboard.

Figure 37.77 Diffusers spread and distribute the air.

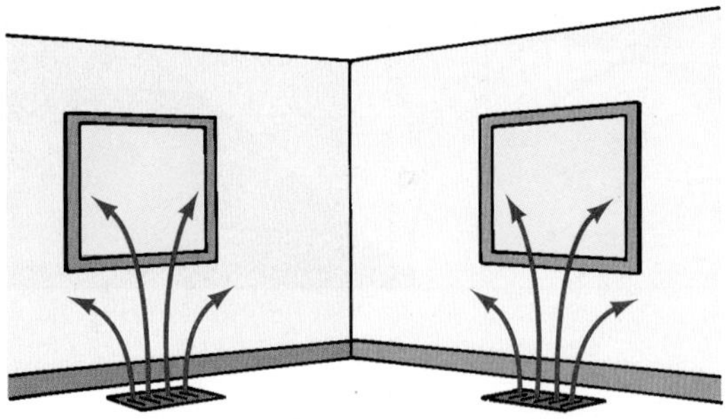

Figure 37.78 Warm air distribution.

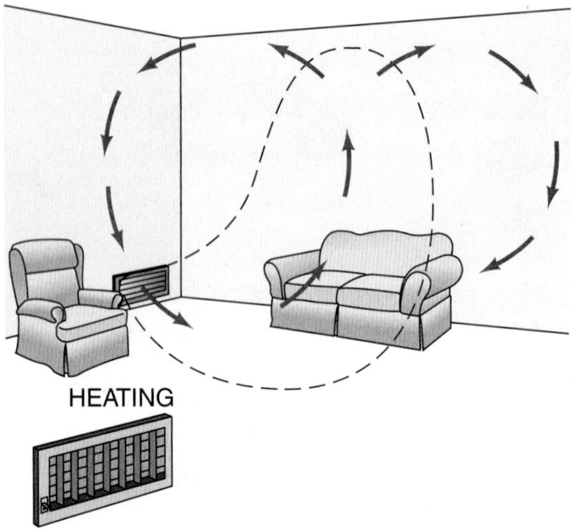

Figure 37.79 Cool air distribution.

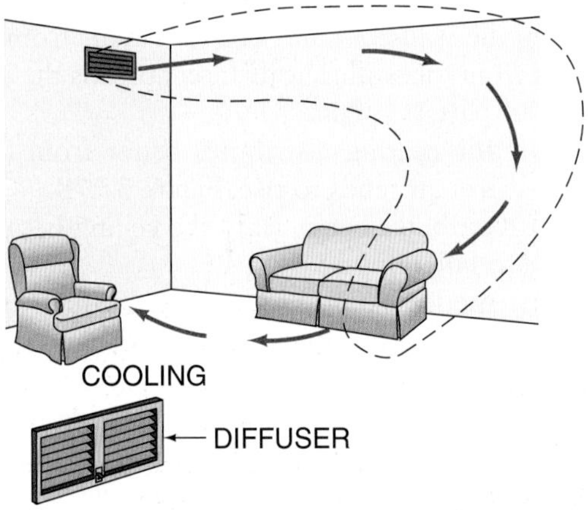

37.16 THE RETURN-AIR DUCT SYSTEM

The return-air duct system is constructed in much the same manner as the supply duct system except that many installations utilize central returns instead of individual room returns. Individual return-air systems have a return-air grille in each room that has a supply diffuser (with the exception of bathrooms and kitchens). Individual return systems provide the best return air, but they are much more expensive to install when compared to similarly sized, central return air systems. The return-air duct normally has a slightly larger cross-sectional area than the supply duct, so there is less resistance to airflow in the return system. In addition, the air flowing through the larger return duct will be much quieter than if the return ducts were made smaller. (more details will be provided in the section on duct sizing). See **Figure 37.81** for an example of a system with individual room returns.

A central return system is usually satisfactory for a one-level residence. Larger return-air grilles are typically centrally located in a hallway so that air from the individual rooms can easily move back to the common return. In order for air to return to the central return, there must be a path, such as doors with grille work, open doorways, and undercut doors in common hallways. However, these open areas in the doors can reduce privacy. In a structure with more than one floor level, a return should be installed on each level. Remember, cold air is more dense than warmer air and moves downward. Warm air is less dense than colder air and tends to flow upward naturally. **Figure 37.82** illustrates air stratification in a two-level house.

A properly constructed central return air system helps eliminate blower noise in the conditioned space. The return-air grille should not be mounted directly to the furnace or air handler because the noise of the blower will be noticeable several feet away. There should be no line of sight between the return-air grille and the inlet of the blower. There should be at least one elbow or bend between the return-air grille and the blower to help keep noise levels low. If this cannot be done, the return-air plenum can be insulated on the inside to help deaden the blower noise. Return-air grilles are normally large and meant to be somewhat decorative. They do not have another function unless they house a filter. They are usually made of stamped metal or have a metal frame with grille work, **Figure 37.83**.

37.17 SIZING DUCT FOR MOVING AIR

To move air takes energy because (1) air has weight; (2) the air tumbles down the duct, rubbing against itself and the ductwork; and (3) fittings create resistance to the airflow.

Figure 37.80 Air registers and diffusers.

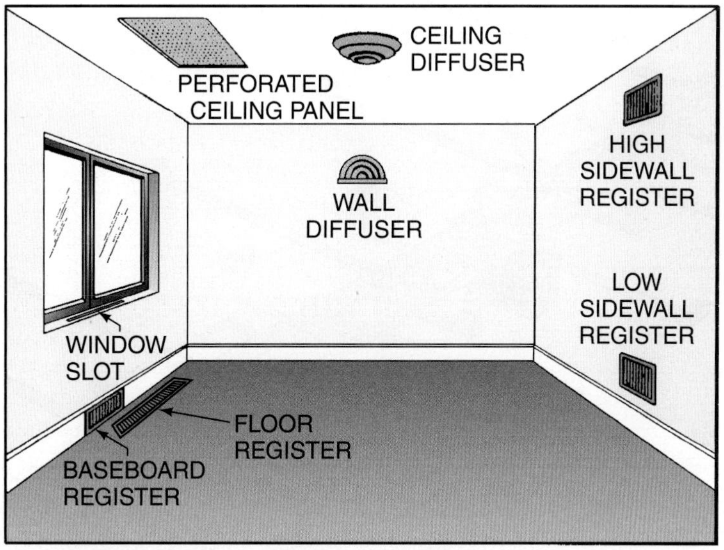

	 FLOOR	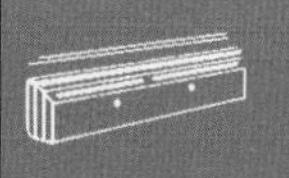 BASEBOARD	 LOW SIDEWALL	 HIGH SIDEWALL	 CEILING
COOLING PERFORMANCE	Excellent	Excellent if used with perimeter systems	Excellent if designed to discharge upward	Good	Good
HEATING PERFORMANCE	Excellent	Excellent if used with perimeter systems	Excellent if used with perimeter systems	Fair–should not be used to heat slab houses in northern climates	Good–should not be used to heat slab houses in northern climates
INTERFERENCE WITH DECOR	Easily concealed because it fits flush with the floor and can be painted to match	Not quite so easy to conceal because it projects from the baseboard	Hard to conceal because it is usually in a flat wall	Impossible to conceal because it is above furniture and in a flat wall	Impossible to conceal but special decorative types are available
INTERFERENCE WITH FURNITURE PLACEMENT	No interference–located at outside wall under a window	No interference–located at outside wall under a window	Can interfere because air discharge is not vertical	No interference	No interference
INTERFERENCE WITH FULL-LENGTH DRAPES	No interference–located 6 or 7 inches from the wall	When drapes are closed they will cover the outlet	When located under a window, drapes will close over it	No interference	No interference
INTERFERENCE WITH WALL-TO-WALL CARPETING	Carpeting must be cut	Carpeting must be notched	No interference	No interference	No interference
OUTLET COST	Low	Medium	Low to medium, depending on the type selected	Low	Low to high—wide variety of types are available
INSTALLATION COST	Low because the sill need not be cut	Low when fed from below–sill need not be cut	Medium–requires wall stack and cutting of plates	Low on furred-ceiling system; high when using under-floor system	High because attic ducts require insulation

Friction losses in ductwork are due to the actual rubbing action of the air against the side of the duct and the turbulence of the air rubbing against itself while moving down the duct. Friction due to air rubbing against the walls of the ductwork cannot be eliminated but can be minimized by good design practices. Proper duct sizing for the amount of airflow helps maximize system performance. The smoother the interior duct surface, the less friction there is. The slower the air is moving, the less friction there will be. Although it is beyond the scope of this text to go into details of duct design, the following information can be used as basic guidelines for a typical residential installation. Each foot of duct offers a known resistance to airflow. This is called friction loss. It can be determined from tables and special slide calculators designed for this purpose. The following example explains friction loss in a duct system, **Figure 37.84.**

Figure 37.81 A duct plan of individual room return-air inlets. Reproduced courtesy of Carrier Corporation.

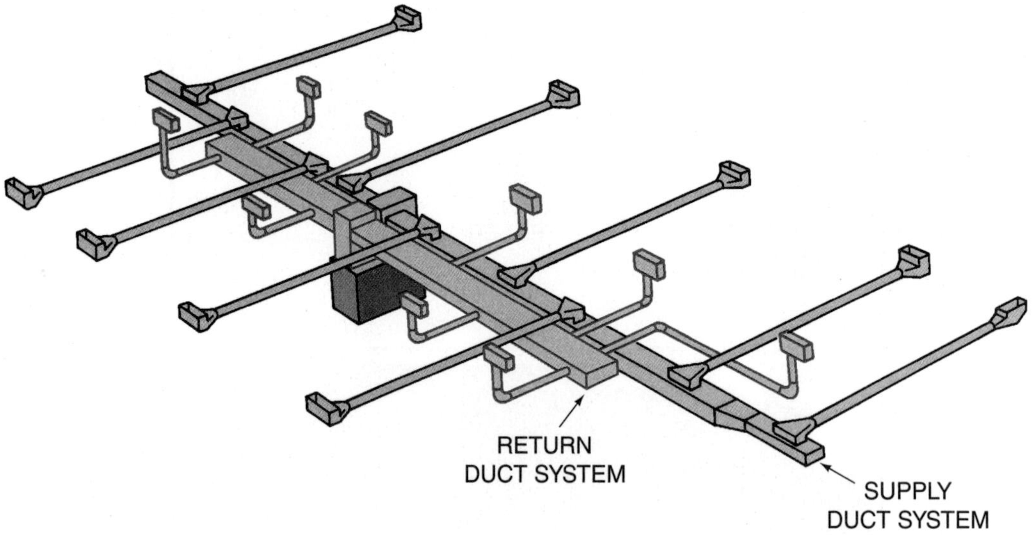

Figure 37.82 Air stratifies even when distributed because warm air rises and cold air falls.

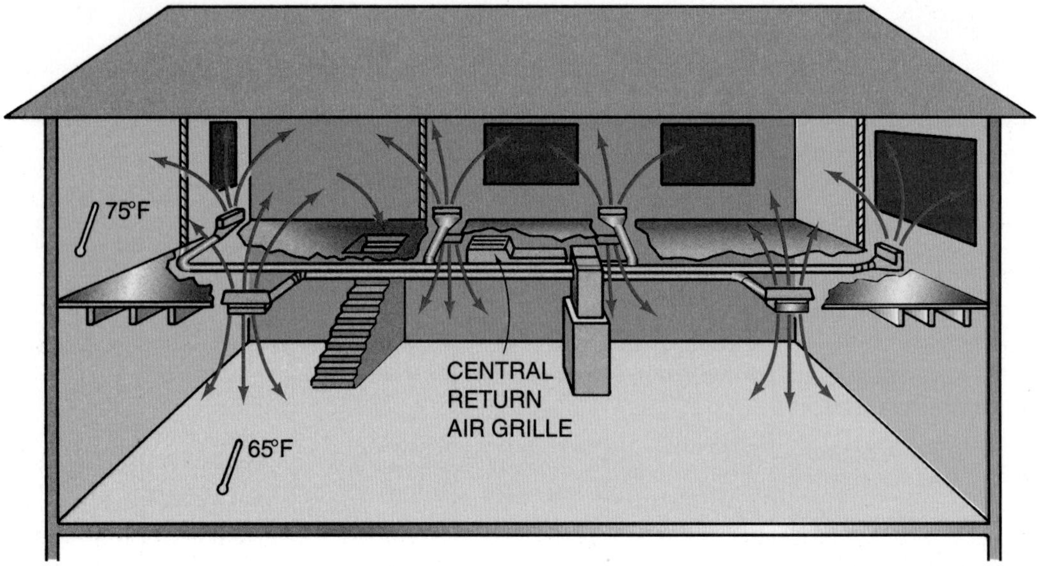

Figure 37.83 Return-air grilles.

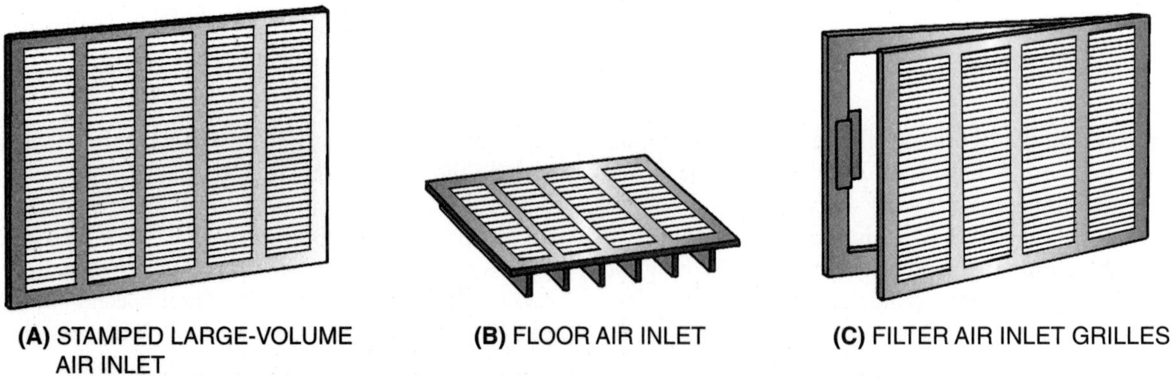

Figure 37.84 A duct system.

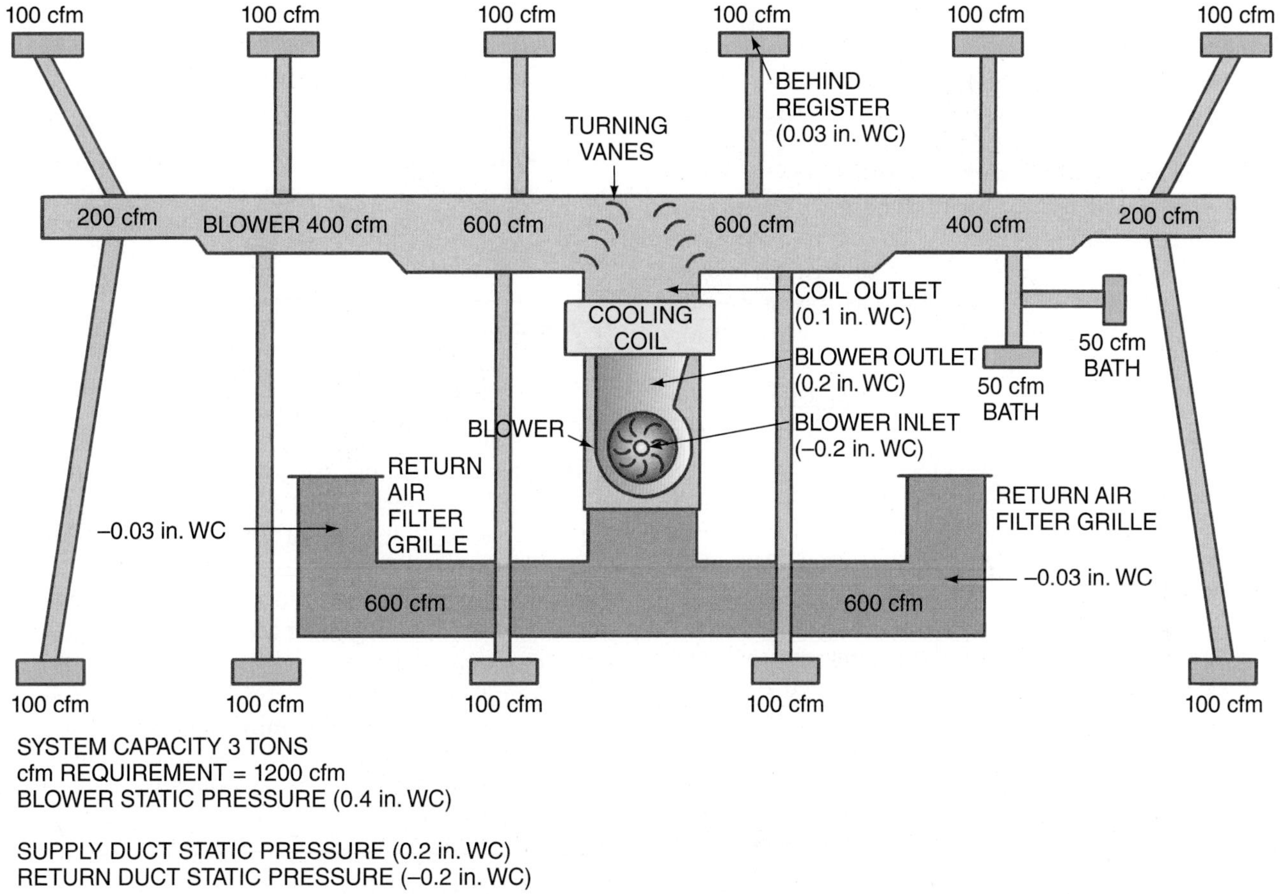

1. Ranch-style home requires 3 tons of cooling.
2. Cooling provided by a 3-ton cooling coil in the ductwork.
3. The heat is provided by a 100,000 Btu/h furnace input, 80,000 Btu/h output (80% efficiency).
4. The blower has a capacity of 1360 cfm of air while operating against 0.40 in. WC static pressure with the system blower operating at medium-high speed, **Figure 37.85**. This system requires only 1200 cfm of air in the cooling mode. The system blower will easily be able to achieve this rate of airflow.
5. The system has 11 outlets, each requiring 100 cfm in the main part of the house, and 2 outlets, each requiring 50 cfm, located in the bathrooms. Most of these outlets are on the exterior walls of the house and distribute the conditioned air to the outside walls.
6. The return air is taken into the system from a common hallway, one return at each end of the hall.
7. While reviewing this system, think of the entire house as the system. The supply air must leave the supply registers and sweep the walls. It then makes its way through the

Figure 37.85 A manufacturer's table for furnace airflow characteristics. Courtesy BDP Company

SIZE	Blower Motor HP	Speed	External Static Pressure in. W.C.							
			0.1	0.2	0.3	0.4	0.5	0.6	0.7	0.8
048100	1/2 PSC	High	1750	1750	1720	1685	1610	1530	1430	—
		Med-High	1360	1370	1370	1360	1340	1315	—	—
		Med-Low	1090	1120	1140	1130	1100	—	—	—
		Low	930	960	980	980	965	945	—	—

rooms to the door adjacent to the hall. The air then flows under the door and makes its way to the return-air grille.

8. The return-air grille is where the duct system starts. A slight negative pressure (in relation to the room pressure) at the grille gives the air the incentive to enter the system. The filters are located in the return-air grilles. The pressure on the blower side of the filter will be −0.03 in. WC, which is less than the pressure in the room, so the room pressure pushes the air through the filter into the return duct.
9. As the air proceeds down the duct toward the blower, the pressure continues to decrease. The lowest pressure in the system is at the blower's inlet, −0.20 in. WC below the room pressure.
10. The air is forced through the blower and the pressure of the air increases. The greatest pressure in the system is at the blower outlet, 0.20 in. WC above the room pressure. The pressure difference between the inlet and the outlet of the blower is 0.40 in. WC.
11. The air is then pushed through the heat exchanger in the furnace, where the air experiences an additional pressure drop.
12. The air then moves through the cooling coil, where it enters the supply duct system at a pressure of 0.10 in. WC.
13. The air will undergo a slight pressure drop as it passes through the tee that splits the duct into two reducing extended plenums, one for each end of the house. This tee in the duct has turning vanes to help reduce the pressure drop as the air passes through the fitting.
14. The first section of each reducing trunk has to handle an equal amount of air. One half of the total cfm, 600 cfm, will flow to the left, while the other 600 cfm will flow to the right. The first two branches on each side will deliver 100 cfm each. This reduces the capacity of the trunk to 400 cfm on each side. A smaller trunk can be used at this point, and materials can be conserved while maintaining the correct air velocity.
15. The duct is reduced to a smaller size to handle 400 cfm on each side. Because another 200 cfm of air is distributed to the conditioned space, another reduction can be made.
16. The last part of the reducing trunk on each side of the system needs to handle only 200 cfm for each side of the system.

This supply duct system will distribute the air for this house with minimal noise and maximum comfort. The pressure in the duct will be about the same all along the duct because the air was distributed off the trunk line and the duct size was reduced to keep the pressure inside the duct at the prescribed value.

At each branch, dampers should be installed in the duct to balance the system air supply to each room. The system will furnish 100 cfm to each outlet, but if a room does not need that much air, the dampers can be adjusted. The branch to each bathroom will need to be adjusted to 50 cfm each.

The return-air system is the same size on each side of the system. It returns 600 cfm per side with the filters located in the return air grilles in the halls. The furnace blower is located far enough from the grilles so that it will not be heard.

Entire books have been written on duct sizing and air distribution system design. Manufacturers' representatives may also be helpful for specific applications. Manufacturers also offer classes in duct sizing that use their methods and techniques.

37.18 MEASURING AIR MOVEMENT FOR BALANCING

Air balancing is sometimes accomplished by measuring the air leaving each supply register. When one outlet has too much air, the damper in that run is throttled to slow down the air. This, of course, redistributes air to the other outlets and will increase their flow.

The air quantity of an individual duct can be measured in the field to some degree of accuracy by using instruments to determine the average velocity of the air in the duct. To accomplish this, a velometer is used. Multiple velocity readings must be taken in a uniform cross-sectional pattern of the duct. The average of the readings taken is used for the calculation. This process is called *traversing* the duct. The average velocity, along with the cross-sectional area of the duct, is used to determine the airflow through a duct, expressed in cfm. For example, if the air in a 1-ft^2 duct (12 in. × 12 in. = 144 in^2) is traveling at a velocity of 1 ft/min, the volume of air passing a point in the duct is 1 cfm. If the velocity is 100 ft/min, the volume of air passing the same point is 100 cfm. The cross-sectional area of the duct is multiplied by the average velocity of the air to determine the volume of the moving air, **Figure 37.86. Figure 37.87** shows the patterns used while traversing the duct.

When you know the average velocity of the air in any duct, you can determine the cfm using the following formula:

$$\text{cfm} = \text{Area (ft}^2) \times \text{Velocity (fpm)}$$

For example, suppose a duct is 20 in. × 30 in. and the average velocity is 850 fpm. The cfm can be found by first finding the area in square feet:

$$\text{Area} = \text{Width} \times \text{Height}$$

$$\text{Area} = 20 \text{ in.} \times 30 \text{ in.}$$

$$\text{Area} = \frac{600 \text{ in}^2}{144 \text{ in}^2/\text{ft}^2}$$

$$\text{Area} = 4.2 \text{ ft}^2$$

$$\text{cfm} = \text{Area} \times \text{Velocity}$$

$$\text{cfm} = 4.2 \text{ ft}^2 \times 850 \text{ fpm}$$

$$\text{cfm} = 3570 \text{ ft}^3 \text{ of air per minute}$$

Figure 37.86 This cross section of duct with airflow shows how to measure duct area and air velocity.

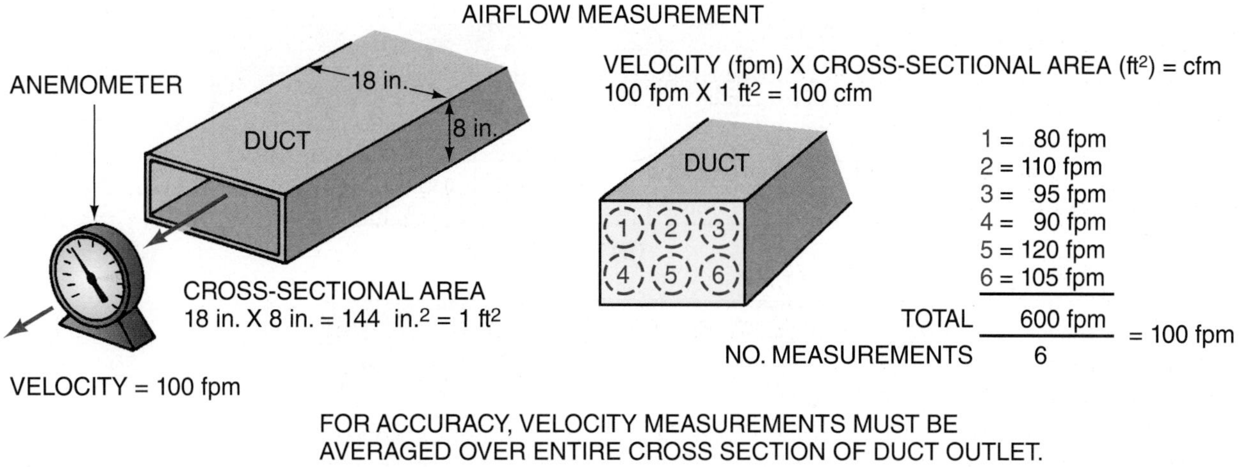

Figure 37.87 This figure shows the pattern that should be used for traversing square, rectangular, and round duct for the most accurate average reading (A). (*Continued*)

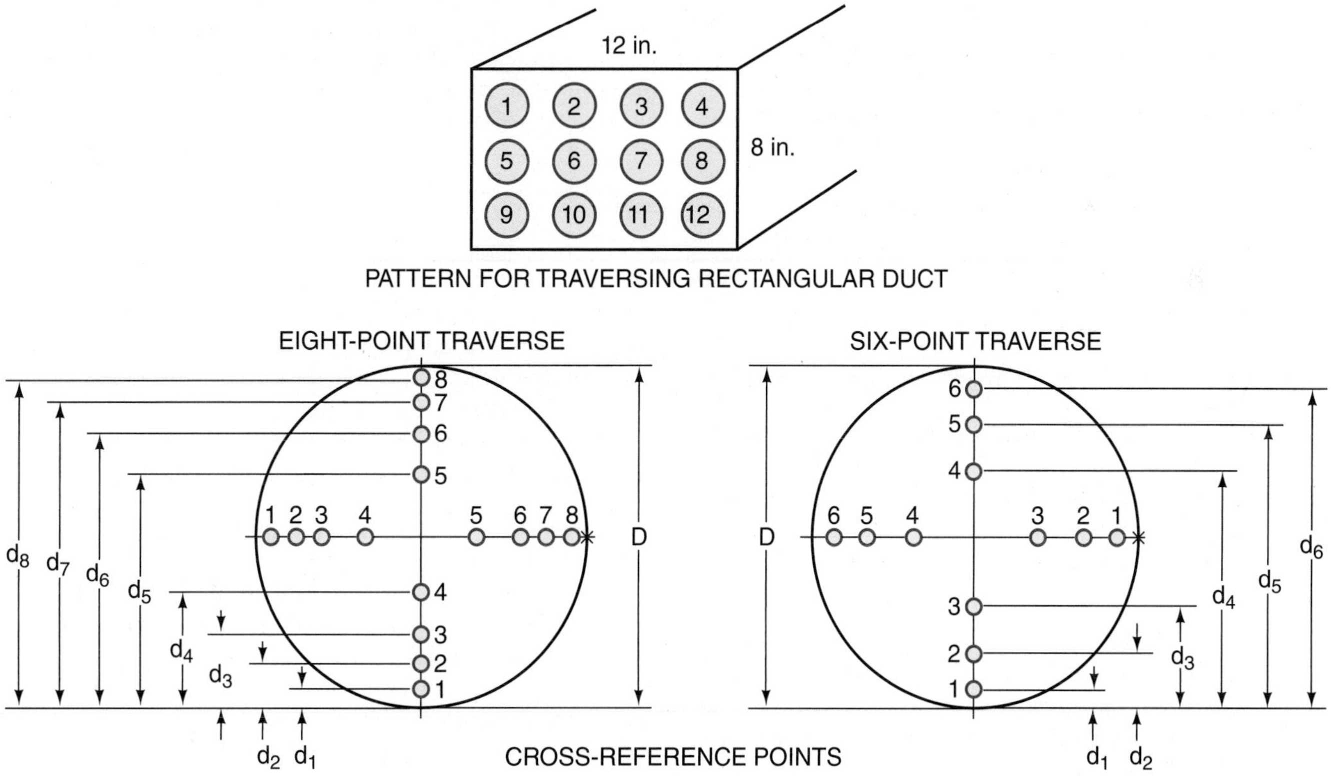

(A)

Figure 37.87 *(Continued)* This figure shows the pattern that should be used for traversing square, rectangular, and round duct for the most accurate average reading (B)–(D).

TRAVERSE METHOD	PROBE IMMERSION EXPRESSED IN DUCT DIAMETERS							
	d_1	d_2	d_3	d_4	d_5	d_6	d_7	d_8
6 POINT	0.043	0.147	0.296	0.704	0.853	0.957	–	–
8 POINT	0.032	0.105	0.194	0.323	0.677	0.806	0.895	0.968

(B)

DUCT. DIA. (IN.)	PROBE IMMERSION FOR 6 PT. TRAVERSE					
	d_1	d_2	d_3	d_4	d_5	d_6
10	3/8	1-1/2	3	7	8-1/2	9-5/8
12	1/2	1-3/4	3-1/2	8-1/2	10-1/4	11-1/2
14	5/8	2	4-1/8	9-7/8	12	13-3/8
16	3/4	2-3/8	4-3/4	11-1/4	13-5/8	15-1/4
18	3/4	2-5/8	5-3/8	12-5/8	15-3/8	17-1/4
20	7/8	3	6	14	17	19-1/8
22	1	3-1/4	6-1/2	15-1/2	18-3/4	21
24	1	3-1/2	7-1/8	16-7/8	20-1/2	23

(C)

DUCT. DIA. (IN.)	PROBE IMMERSION FOR 8 PT. TRAVERSE							
	d_1	d_2	d_3	d_4	d_5	d_6	d_7	d_8
10	5/16	1	2	3-1/4	6-3/4	8	9	9-5/8
12	3/8	1-1/4	2-3/8	3-7/8	8-1/8	9-5/8	10-3/4	11-1/2
14	7/16	1-1/2	2-3/4	4-1/2	9-1/2	11-1/4	12-1/2	13-1/2
16	1/2	1-5/8	3-1/8	5-1/8	10-7/8	12-7/8	14-3/8	15-1/2
18	9/16	1-7/8	3-1/2	5-7/8	12-1/4	13-1/2	16-1/8	17-1/2
20	5/8	2-1/8	3-7/8	6-1/2	18-1/2	16-1/8	17-7/8	19-3/8
22	11/16	2-3/8	4-1/4	7-1/8	14-7/8	17-3/4	19-3/4	21-1/4
24	3/4	2-1/2	4-5/8	7-3/4	16-1/4	19-1/2	21-1/2	23-1/4

(D)

Suppose the duct were round and had a diameter of 12 in. with an average velocity of 900 fpm.

$$\text{Area} = \pi \times \text{Radius squared}$$

$$\text{Area} = 3.14 \times 6 \times 6$$

$$\text{Area} = \frac{113\ \text{in}^2}{144\ \text{in}^2/\text{ft}^2}$$

$$\text{Area} = 0.78\ \text{ft2}$$

$$\text{cfm} = \text{Area} \times \text{Velocity}$$

$$\text{cfm} = 0.78\ \text{ft}^2 \times 900$$

$$\text{cfm} = 702\ \text{ft}^3 \text{ of air per minute}$$

Special techniques and good instrumentation must be used to find the correct average velocity in a duct. For example, readings should not be taken within 10 duct diameters of the nearest fitting. This means that long, straight runs are desirable for taking accurate readings. This is not always possible because of the fittings and takeoffs in a typical system, particularly residential systems. The proper duct traverse must be used and it is different for round duct than for square or rectangular duct. **Figure 37.87** shows the pattern that should be used for traversing square, rectangular, and round duct for the most accurate average reading. See Section 37.21 for some practical tips for measuring airflow.

37.19 THE AIR FRICTION CHART

The system described in the previous section can be plotted on the friction chart, **Figure 37.88**. This chart is for volumes of air up to 2000 cfm. **Figure 37.89** shows a chart for larger

Figure 37.88 An air friction chart. Used with permission from ASHRAE, Inc.

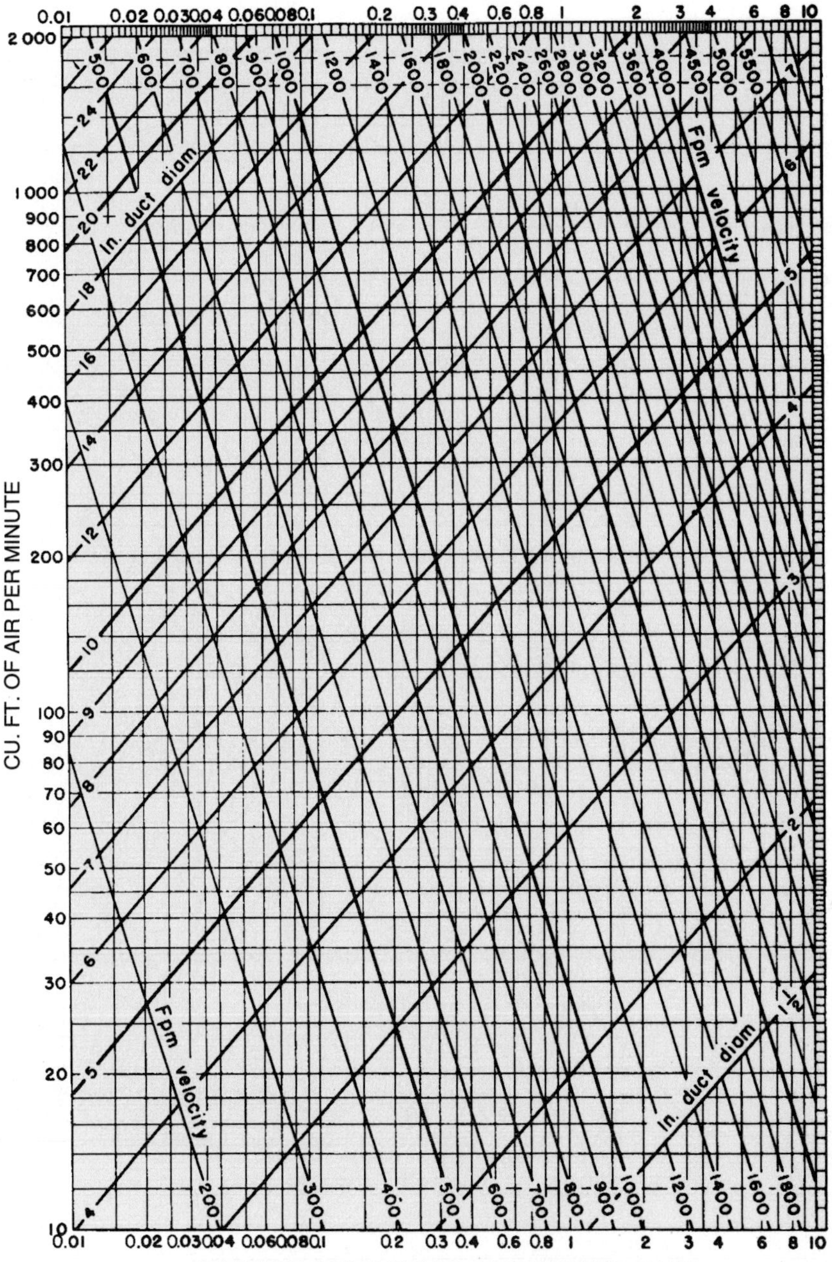

systems, up to 100,000 cfm. The horizontal lines on the chart represent airflow in cfm, as indicated by the scale on the left side of the chart. Round duct diameters, expressed in inches, are represented by the lines that slope from the bottom-left to the top-right of the chart. The velocity of the airflow is represented by the lines that run from the top-left to the bottom-right of the chart. The vertical lines on the chart represent the friction rate.

The duct sizes are rated in round-pipe sizes on the chart and can be converted to square or rectangular duct by using the table in **Figure 37.90**. The round-pipe sizes are for galvanized steel duct material and assume that the air is at standard conditions and has a density of 0.075 lb/ft^3. Other charts are available for ductboard and flexible duct. The friction loss is expressed in inches of water column per 100 ft of duct, and the scale is shown along the top and bottom of the chart. For example, using this chart, a run of pipe that carries 100 cfm (on the left of the chart) can be plotted over the intersection of the 6-in. pipe line. When this pipe is carrying 100 cfm of air, it has a velocity of just over 500 ft/min and a pressure drop of 0.085 in. WC/100 ft. A 50-ft run would have half the pressure drop—0.0425 in. WC.

The friction chart can be used by the designer to size the duct system before the job price is quoted. This duct sizing

Figure 37.89 A friction chart for larger duct systems. Reprinted with permission from American Society of Heating, Refrigerating, and Air-Conditioning Engineers, Inc.

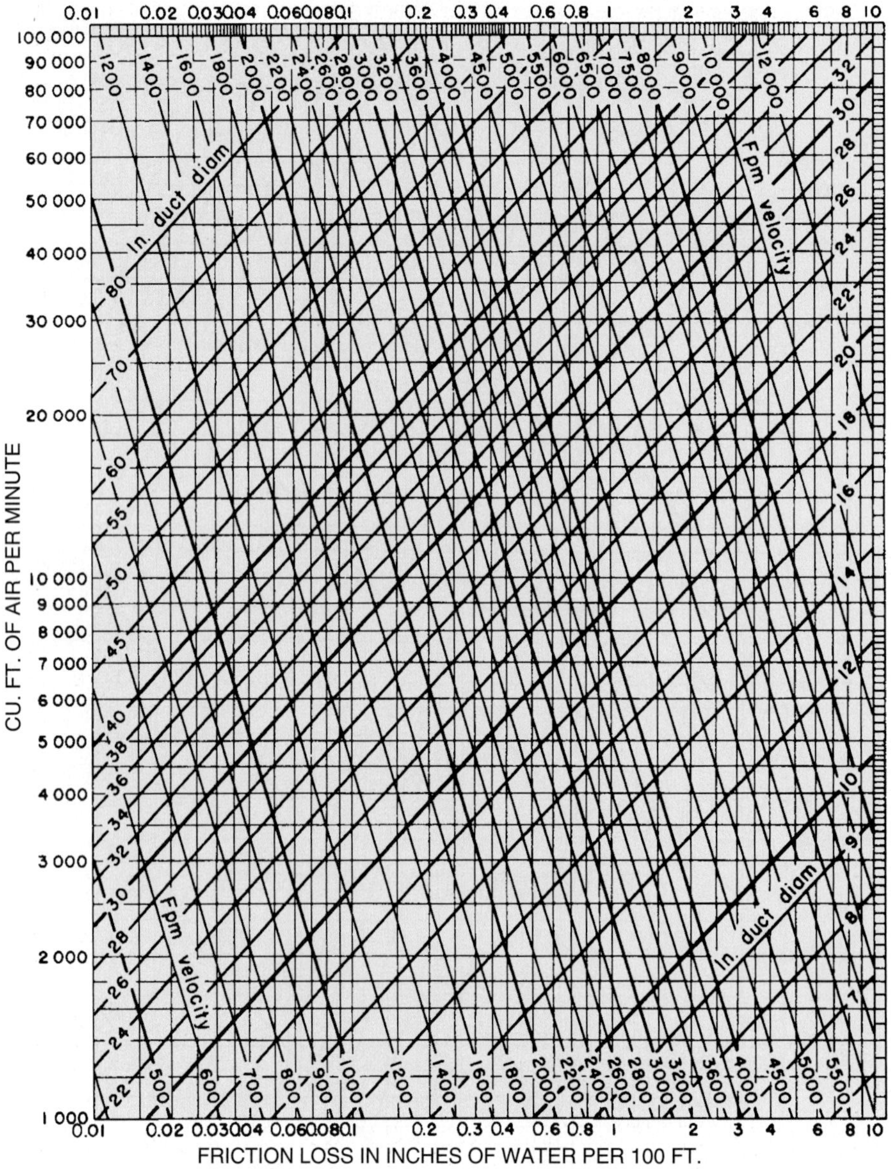

provides the sizes that will be used for figuring the duct materials. The duct should be sized using the chart in **Figure 37.88** and the recommended velocities from the table in **Figure 37.91**. In the previous example, a 4-in. pipe could have been used, but the velocity would have been nearly 1200 ft/min. This would be noisy, and the fan may not have enough capacity to push sufficient air through the duct.

High-velocity systems have been designed and used successfully in small applications. Such systems normally have a high air velocity in the trunk and branch ducts; the velocity is then reduced at the register to avoid drafts from the high-velocity air. If the air velocity is not reduced at the register, the register has a streamlined effect and is normally located in the corners of the room where someone is not likely to walk under it.

The friction chart also can be used by the field technician to troubleshoot airflow problems. Airflow problems come from system design, installation, and owner problems. System design problems can be a result of a poor choice of blower or incorrect duct sizes. They may result from not understanding the fact that the air friction chart is for 100 ft of duct only. The technician should have an understanding of how to size ductwork to determine whether or not a particular duct is sized correctly.

Figure 37.90 A chart used to convert from round to square or rectangular duct. Reprinted with permission from American Society of Heating, Refrigerating, and Air-Conditioning Engineers, Inc.

Lgth Adj.[b]	Length of One Side of Rectangular Duct (a), in.																				Lgth Adj.[b]
	6	7	8	9	10	11	12	13	14	15	16	17	18	19	20	22	24	26	28	30	
6	6.6																				6
7	7.1	7.7																			7
8	7.6	8.2	8.7																		8
9	8.0	8.7	9.3	9.8																	9
10	8.4	9.1	9.8	10.4	10.9																10
11	8.8	9.5	10.2	10.9	11.5	12.0															11
12	9.1	9.9	10.7	11.3	12.0	12.6	13.1														12
13	9.5	10.3	11.1	11.8	12.4	13.1	13.7	14.2													13
14	9.8	10.7	11.5	12.2	12.9	13.5	14.2	14.7	15.3												14
15	10.1	11.0	11.8	12.6	13.3	14.0	14.6	15.3	15.8	16.4											15
16	10.4	11.3	12.2	13.0	13.7	14.4	15.1	15.7	16.4	16.9	17.5										16
17	10.7	11.6	12.5	13.4	14.1	14.9	15.6	16.2	16.8	17.4	18.0	18.6									17
18	11.0	11.9	12.9	13.7	14.5	15.3	16.0	16.7	17.3	17.9	18.5	19.1	19.7								18
19	11.2	12.2	13.2	14.1	14.9	15.7	16.4	17.1	17.8	18.4	19.0	19.6	20.2	20.8							19
20	11.5	12.5	13.5	14.4	15.2	16.0	16.8	17.5	18.2	18.9	19.5	20.1	20.7	21.3	21.9						20
22	12.0	13.0	14.1	15.0	15.9	16.8	17.6	18.3	19.1	19.8	20.4	21.1	21.7	22.3	22.9	24.0					22
24	12.4	13.5	14.6	15.6	16.5	17.4	18.3	19.1	19.9	20.6	21.3	22.0	22.7	23.3	23.9	25.1	26.2				24
26	12.8	14.0	15.1	16.2	17.1	18.1	19.0	19.8	20.6	21.4	22.1	22.9	23.5	24.2	24.9	26.1	27.3	28.4			26
28	13.2	14.5	15.6	16.7	17.7	18.7	19.6	20.5	21.3	22.1	22.9	23.7	24.4	25.1	25.8	27.1	28.3	29.5	30.6		28
30	13.6	14.9	16.1	17.2	18.3	19.3	20.2	21.1	22.0	22.9	23.7	24.4	25.2	25.9	26.6	28.0	29.3	30.5	31.7	32.8	30
32	14.0	15.3	16.5	17.7	18.8	19.8	20.8	21.8	22.7	23.5	24.4	25.2	26.0	26.7	27.5	28.9	30.2	31.5	32.7	33.9	32
34	14.4	15.7	17.0	18.2	19.3	20.4	21.4	22.4	23.3	24.2	25.1	25.9	26.7	27.5	28.3	29.7	31.0	32.4	33.7	34.9	34
36	14.7	16.1	17.4	18.6	19.8	20.9	21.9	22.9	23.9	24.8	25.7	26.6	27.4	28.2	29.0	30.5	32.0	33.3	34.6	35.9	36
38	15.0	16.5	17.8	19.0	20.2	21.4	22.4	23.5	24.5	25.4	26.4	27.2	28.1	28.9	29.8	31.3	32.8	34.2	35.6	36.8	38
40	15.3	16.8	18.2	19.5	20.7	21.8	22.9	24.0	25.0	26.0	27.0	27.9	28.8	29.6	30.5	32.1	33.6	35.1	36.4	37.8	40
42	15.6	17.1	18.5	19.9	21.1	22.3	23.4	24.5	25.6	26.6	27.6	28.5	29.4	30.3	31.2	32.8	34.4	35.9	37.3	38.7	42
44	15.9	17.5	18.9	20.3	21.5	22.7	23.9	25.0	26.1	27.1	28.1	29.1	30.0	30.9	31.8	33.5	35.1	36.7	38.1	39.5	44
46	16.2	17.8	19.3	20.6	21.9	23.2	24.4	25.5	26.6	27.7	28.7	29.7	30.6	31.6	32.5	34.2	35.9	37.4	38.9	40.4	46
48	16.5	18.1	19.6	21.0	22.3	23.6	24.8	26.0	27.1	28.2	29.2	30.2	31.2	32.2	33.1	34.9	36.6	38.2	39.7	41.2	48
50	16.8	18.4	19.9	21.4	22.7	24.0	25.2	26.4	27.6	28.7	29.8	30.8	31.8	32.8	33.7	35.5	37.2	38.9	40.5	42.0	50
52	17.1	18.7	20.2	21.7	23.1	24.4	25.7	26.9	28.0	29.2	30.3	31.3	32.3	33.3	34.3	36.2	37.9	39.6	41.2	42.8	52
54	17.3	19.0	20.6	22.0	23.5	24.8	26.1	27.3	28.5	29.7	30.8	31.8	32.9	33.9	34.9	36.8	38.6	40.3	41.9	43.5	54
56	17.6	19.3	20.9	22.4	23.8	25.2	26.5	27.7	28.9	30.1	31.2	32.3	33.4	34.4	35.4	37.4	39.2	41.0	42.7	44.3	56
58	17.8	19.5	21.2	22.7	24.2	25.5	26.9	28.2	29.4	30.6	31.7	32.8	33.9	35.0	36.0	38.0	39.8	41.6	43.3	45.0	58
60	18.1	19.8	21.5	23.0	24.5	25.9	27.3	28.6	29.8	31.0	32.2	33.3	34.4	35.5	36.5	38.5	40.4	42.3	44.0	45.7	60
62		20.1	21.7	23.3	24.8	26.3	27.6	28.9	30.2	31.5	32.6	33.8	34.9	36.0	37.1	39.1	41.0	42.9	44.7	46.4	62
64		20.3	22.0	23.6	25.1	26.6	28.0	29.3	30.6	31.9	33.1	34.3	35.4	36.5	37.6	39.6	41.6	43.5	45.3	47.1	64
66		20.6	22.3	23.9	25.5	26.9	28.4	29.7	31.0	32.3	33.5	34.7	35.9	37.0	38.1	40.2	42.2	44.1	46.0	47.7	66
68		20.8	22.6	24.2	25.8	27.3	28.7	30.1	31.4	32.7	33.9	35.2	36.3	37.5	38.6	40.7	42.8	44.7	46.6	48.4	68
70		21.1	22.8	24.5	26.1	27.6	29.1	30.4	31.8	33.1	34.4	35.6	36.8	37.9	39.1	41.2	43.3	45.3	47.2	49.0	70
72			23.1	24.8	26.4	27.9	29.4	30.8	32.2	33.5	34.8	36.0	37.2	38.4	39.5	41.7	43.8	45.8	47.8	49.6	72
74			23.3	25.1	26.7	28.2	29.7	31.2	32.5	33.9	35.2	36.4	37.7	38.8	40.0	42.2	44.4	46.4	48.4	50.3	74
76			23.6	25.3	27.0	28.5	30.0	31.5	32.9	34.3	35.6	36.8	38.1	39.3	40.5	42.7	44.9	47.0	48.9	50.9	76
78			23.8	25.6	27.3	28.8	30.4	31.8	33.3	34.6	36.0	37.2	38.5	39.7	40.9	43.2	45.4	47.5	49.5	51.4	78
80			24.1	25.8	27.5	29.1	30.7	32.2	33.6	35.0	36.3	37.6	38.9	40.2	41.4	43.7	45.9	48.0	50.1	52.0	80
82				26.1	27.8	29.4	31.0	32.5	34.0	35.4	36.7	38.0	39.3	40.6	41.8	44.1	46.4	48.5	50.6	52.6	82
84				26.4	28.1	29.7	31.3	32.8	34.3	35.7	37.1	38.4	39.7	41.0	42.2	44.6	46.9	49.0	51.1	53.2	84
86				26.6	28.3	30.0	31.6	33.1	34.6	36.1	37.4	38.8	40.1	41.4	42.6	45.0	47.3	49.6	51.7	53.7	86
88				26.9	28.6	30.3	31.9	33.4	34.9	36.4	37.8	39.2	40.5	41.8	43.1	45.5	47.8	50.0	52.2	54.3	88
90				27.1	28.9	30.6	32.2	33.8	35.3	36.7	38.2	39.5	40.9	42.2	43.5	45.9	48.3	50.5	52.7	54.8	90
92					29.1	30.8	32.5	34.1	35.6	37.1	38.5	39.9	41.3	42.6	43.9	46.4	48.7	51.0	53.2	55.3	92
96					29.6	31.4	33.0	34.7	36.2	37.7	39.2	40.6	42.0	43.3	44.7	47.2	49.6	52.0	54.2	56.4	96

To size duct properly, the technician must understand the calculation of pressure drop in fittings and pipe. The pressure drop in duct is calculated and read on the friction chart in feet of duct, from the air handler to the terminal outlet. This is a straightforward measurement. The fittings are calculated in a different manner. Pressure drop across different types of fittings in in. WC has been determined by laboratory experiment and is known. This pressure drop has been converted to equivalent feet of duct for the convenience of the designer and technician. The technician

Figure 37.91 A chart of recommended velocities for different duct designs.

Structure	Supply Outlet	Return Openings	Main Supply	Branch Supply	Main Return	Branch Return
Residential	500–750	500	1000	600	800	600
Apartments, Hotel Bedrooms, Hospital Bedrooms	500–750	500	1200	800	1000	800
Private Offices, Churches, Libraries, Schools	500–1000	600	1500	1200	1200	1000
General Offices, Deluxe Restaurants, Deluxe Stores, Banks	1200–1500	700	1700	1600	1500	1200
Average Stores, Cafeterias	1500	800	2000	1600	1500	1200

must know the type of fittings used in the system and then consult a chart for the equivalent feet of duct:

- **Figure 37.92** shows some sample supply plenum fittings (Group 1) and their equivalent lengths.
- **Figure 37.93** shows some sample branch supply trunk takeoff fittings (Group 2) and their equivalent lengths.
- **Figure 37.94** shows some sample reducing trunk takeoff fittings (Group 3) and their equivalent lengths.
- **Figure 37.95** shows some sample supply air boot and stack head fittings (Group 4) and their equivalent lengths.

The equivalent feet of duct for all of the fittings for a particular run must be added together and then added to the

Figure 37.92 Various supply plenum fittings (Group 1) and their equivalent length, expressed in feet.

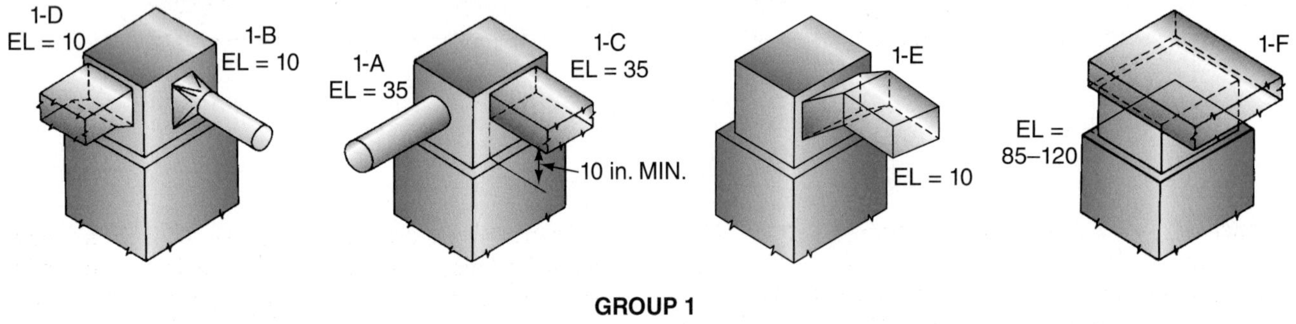

Figure 37.93 Various branch supply trunk takeoff fittings (Group 2) and their equivalent length, expressed in feet.

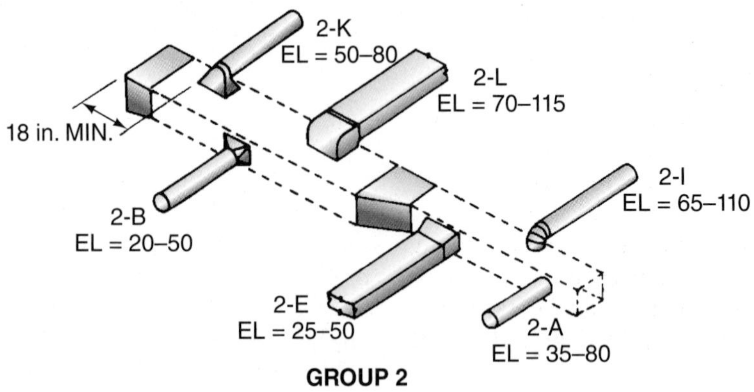

Figure 37.94 Various reducing trunk takeoff fittings (Group 3) and their equivalent length, expressed in feet.

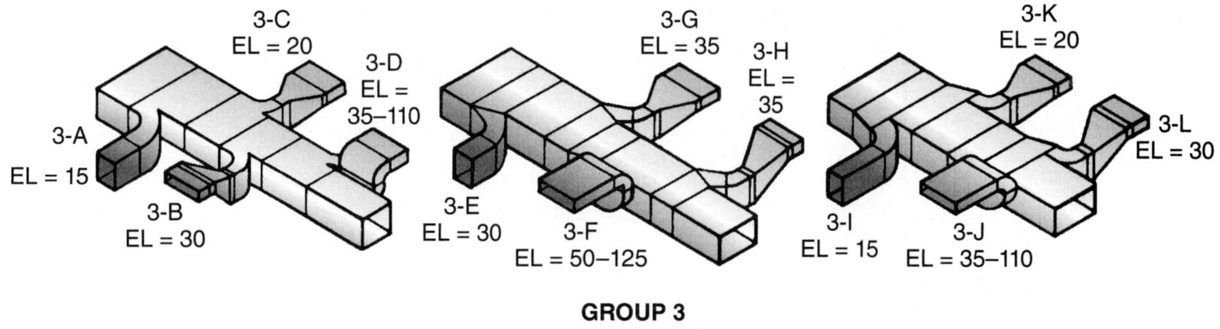

Figure 37.95 Various supply air boot and stack head fittings (Group 4) and their equivalent length, expressed in feet.

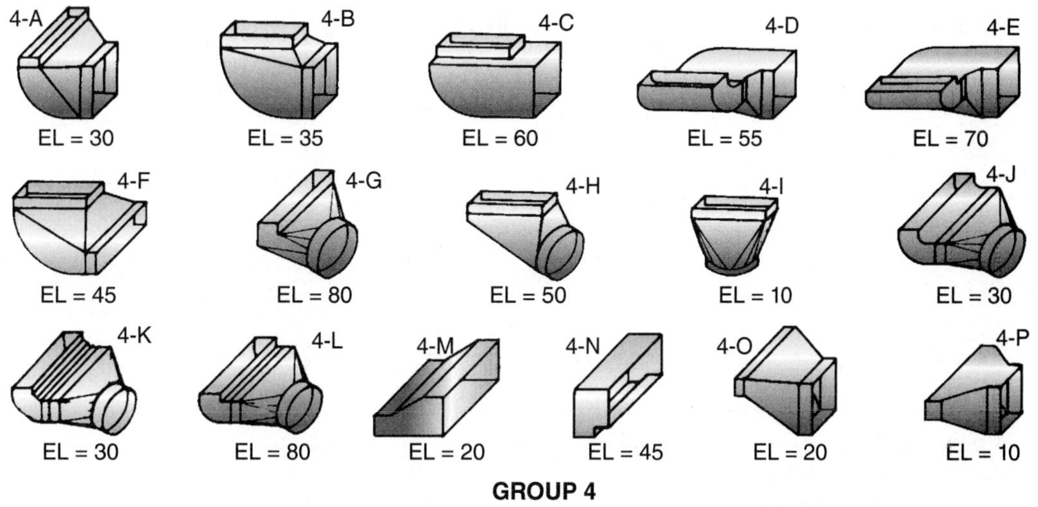

actual duct length to determine the proper duct size for a particular run.

When a duct run is under 100 ft, a corrected friction factor must be used or the duct will not have enough resistance to airflow and too much flow will occur. The corrected friction factor may be determined by multiplying the design friction loss in the duct system times 100 (the friction loss used on the chart) and dividing by the actual length of run. For example, a system may have a duct run handling 300 cfm of air to an inside room. This duct is very close to the main duct and air handler. There are 60 effective feet, including the friction loss in the fittings and the duct length, **Figure 37.96.** The blower may be capable of moving the correct airflow in the supply duct with a friction loss of 0.36 in. WC. The cooling coil may have a loss of 0.24 and the supply registers a loss of 0.03. When the coil and register loss are subtracted from the friction loss for the supply duct, the duct must be designed for a loss of 0.09 in. WC (0.36 − 0.27 = 0.09).

The following describes a mistake the designer made and how the technician discovered it. The designer chose 9-in. duct from the friction chart at the junction of 0.09 in. WC and 300 cfm, from **Figure 37.88,** without making a correction for the fact that the duct is only 60 ft long. The technician corrected the static pressure loss of 60 ft of duct using the formula:

$$\text{Adjusted Static} = \frac{\text{Design Static} \times 100}{\text{Total Equivalent Length of Duct}}$$

$$= \frac{0.09 \times 100}{60}$$

$$= 0.15$$

Using this adjusted static pressure, the technician referred to the friction chart and plotted at 300 cfm, then moved to the right to 0.15 adjusted static. The duct size now became 8-in. duct. This is not a significant savings in materials, but it does now reflect accurate friction rate and velocity information. An alternative is to use a throttling damper where the branch line

Figure 37.96 This duct has an equivalent length of less than 100 feet.

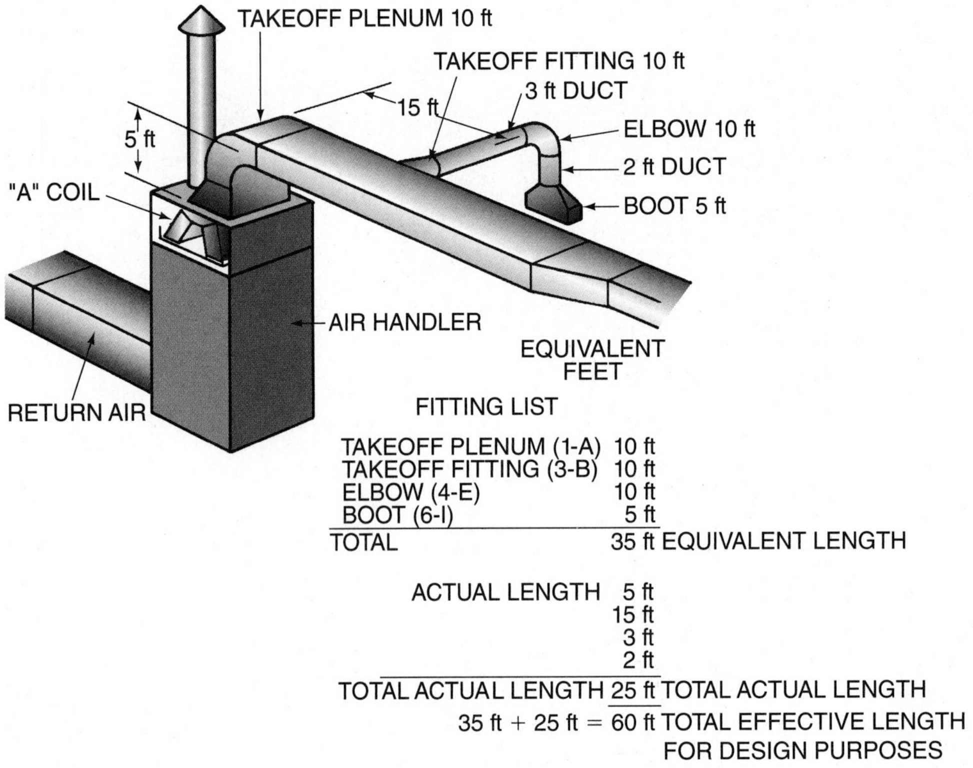

leaves the trunk duct. **The key to this problem is correcting the static pressure to an adjusted value.** This causes the chart to undersize the duct for the purpose of reducing the airflow to this run because it is less than 100 ft long. Otherwise, the air will have so little resistance that too much air will flow from this run, starving other runs. One run that is oversized will not make too much difference, but a whole system sized using the same technique will experience problems.

This problem illustrates a run that is under 100 ft. The same technique applies to a run that is more than 100 ft, except that the run will be undersized and starved for air. For example, suppose that another run off the same system requires 300 cfm of air and is a total of 170 ft (fittings and actual duct length), **Figure 37.97**. The designer sized the duct for 300 cfm at a static pressure loss of 0.09 and chose a duct size of 9-in. round. The technician finds the airflow is too low and checks the figures using the formula:

$$\text{Adjusted Static} = \frac{\text{Design Static} \times 100}{\text{Total Equivalent Length of Duct}}$$

$$= \frac{0.09 \times 100}{170}$$

$$= 0.053$$

When the duct is sized using the adjusted static, the chart reading oversizes the duct, and it moves the correct amount of air. In this case, the technician enters the chart at 300 cfm, moves to the right to 0.053, and finds the correct duct size to be 10 in. instead of 9 in. This is the reason this system is starved for air. All duct is first sized in round, then converted to square or rectangular. **NOTE:** *All runs of duct are measured from air handler to the end of the run for sizing.*•

As the previous examples show, when some duct is undersized and some oversized the result will probably produce problems. The technician should pay close attention to any job with airflow problems and try to find the original duct-sizing calculations. This will make possible a full analysis of a problem duct installation. Any equipment supplier will help in selecting the correct blower and equipment. Suppliers may even help with the duct design.

37.20 PRACTICAL TROUBLESHOOTING TECHNIQUES

Most duct systems are either residential or commercial installations. Residential systems will be discussed first. Generally, residential systems have less than 5 tons of cooling and 200,000 Btu/h of heating. These systems are often installed based on experience rather than on the use of

Figure 37.97 This duct has an equivalent length of more than 100 feet.

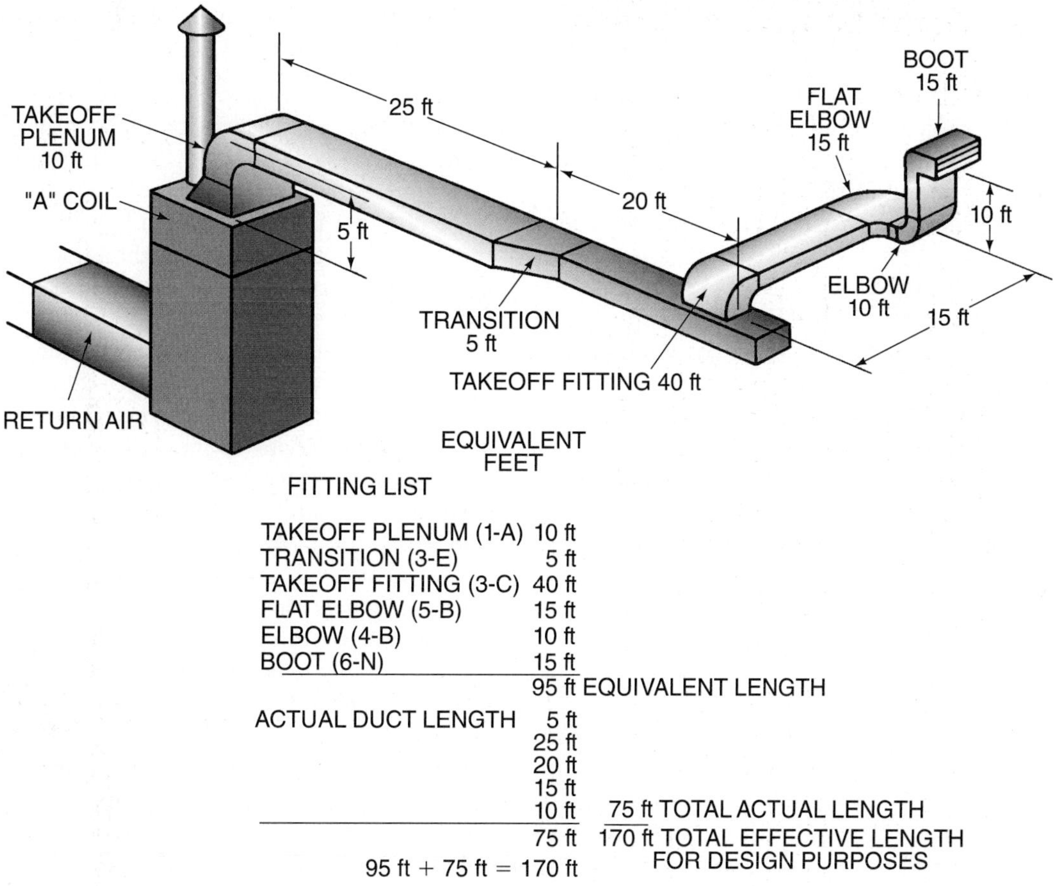

industry-approved design techniques. Many contractors estimate how much air should be delivered to a particular room in a home using rules of thumb. Since most areas in residential structures are open to each other, deficiencies in the system design are often not noticed. It is strongly recommended, however, that nationally recognized methods, such as ACCA Manual D, be used to properly design and lay out residential air distribution systems. One characteristic that is common to most residential air distribution systems is that they typically have short duct runs. As explained earlier, for instruments to work correctly, measurements must be made in long runs of duct. Therefore, instruments may not be necessary for locating and repairing airflow problems. Most testing of residential duct systems is performed at the registers and grilles because of this lack of long duct runs. Commercial systems are much different and will require the use of instruments.

RESIDENTIAL DUCT SYSTEM PROBLEMS

Many contractors size the branch duct system in residences using two sizes of duct for economy of the installation. The duct to the rooms is often sized as 6-in. or 8-in. round (which may be converted to square or rectangular duct) with dampers in each run to adjust the airflow. A 6-in. round duct will move 100 cfm of air with a friction loss of about 0.08 in. WC pressure drop per 100 ft. An 8-in. round duct will move 200 cfm of air with a friction loss of about 0.075 in. WC. Many contractors divide the house into 100- and 200-cfm zones and use dampers to compensate for any airflow unbalance. The trunk duct that furnishes the air to the branch ducts is often sized by simply choosing a size that will furnish the required number of branch ducts off the main. Larger rooms will have several outlet registers. Although this may seem careless, it is simple and practical. When a system is designed in this manner, the technician must know if there is adequate airflow at the terminal units and the supply registers.

The simple way to check for airflow is to see if you can feel the airflow at a distance of about 2 to 3 ft from each register. If the registers are floor registers and it is the cooling season, you can walk to each register, one at a time, while holding your hand down at your side. This can often be done while touring the home with the homeowner, who may not even be aware of what you are doing. If you find one that has

a very low flow, follow the duct from the trunk to the register. The following are some problems that you may encounter:

- Flex duct runs may be too long. Sometimes the installing contractor will use an entire 25-ft box of duct for a 15-ft run and just leave the excess in the attic. Some installers simply do not want to cut off the extra duct.
- When the system is installed in a new home, construction debris can often be found in the duct. The register holes should be covered during construction when floor registers are used.
- Runs may have been disconnected. Duct systems that are run where people, such as cable TV or telephone repair technicians, crawl over or step on them can become disconnected or crushed.
- The damper on one or more of the branch ducts may be closed. Look for the damper handle that is supposed to be installed where you can see it. Even if the duct is insulated on the outside, the handle should protrude out of the insulation. If the damper handle is under the insulation, it can often be found by feel.
- Flexible duct will collapse inside if it is installed with tight bends. This may not be easy to see from the outside.
- When the duct is lined on the inside, it is hard to determine where the obstruction may be. The sure sign of obstruction is reduced airflow at the supply registers for that trunk run.
- Look for furniture that is placed over registers. In some cases homeowners will block registers and grilles because they feel a draft. Floor registers or grilles can be covered with area rugs, **Figure 37.98**.
- The wrong filter media may have been chosen. Many companies advertise better filters for residential systems. A better filter will have a tighter weave and offer more air restriction. If the filter rack was designed for a typical media filter and a high-density filter is used, it is likely that the airflow is reduced to the point of low flow. This is because the high-density filters require more area for the same airflow. Take the filter out and recheck the airflow. Refer to Unit 34, "Indoor Air Quality," for more information on air filters.

After checking each individual register for noticeable airflow, look for signs of airflow into the return-air grilles. If the return-air grille is oversized, you may not notice airflow. A piece of paper placed close to the grille will be sucked onto the grille surface with only a slight amount of airflow velocity. Oversized return-air grilles are often desirable because they reduce the air noise.

To verify airflow for floor and sidewall registers, tie a piece of ribbon about 12 in. long to the end of a stick or piece of wire. Hold it over the register. The ribbon will show the air pattern by waving in the air; if there is no airflow, it will not wave. This does not work as well with ceiling registers because the airflow is usually down, just as the thread or ribbon would normally hang.

If you are using instruments to measure air velocity instruments, readings can be taken at the various registers. The industry standard for the maximum recommended air velocity for a supply register in a home is 750 fpm, **Figure 37.91**. You should be able to feel this airflow.

The airflow from a register is not the same across the entire surface of the register. Although it may seem that the static pressure in a duct would provide even pressure to all surfaces of the register, there are other considerations. For example, when a register is mounted on the duct itself, the air turning into the register can create a low-pressure area at one end, **Figure 37.99**. This situation may be remedied by the use of turning vanes at the register. Manufacturers of registers and grilles provide tables that show what the airflow through each device is at a particular velocity. A major consideration is that the velocity is often not a constant across the entire device.

Figure 37.98 Obstructed airflow. Reproduced courtesy of Carrier Corporation

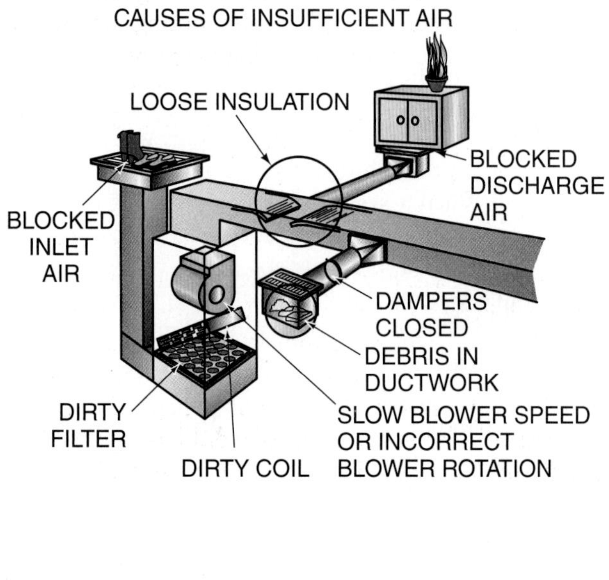

Figure 37.99 This register is installed just after a reducing transition and will need turning vanes to turn the air into the register.

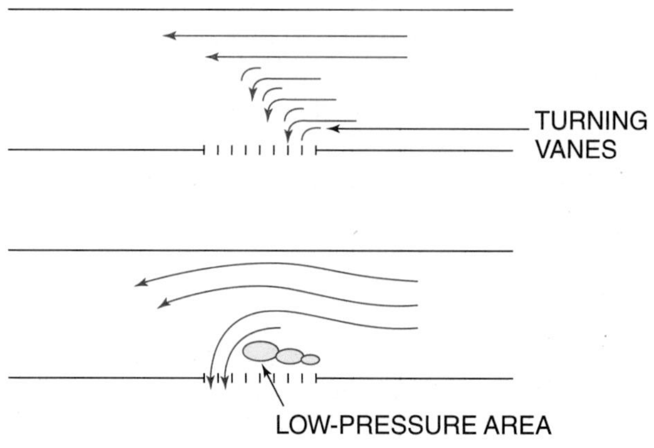

If you think there is an overall low airflow problem because there does not seem to be enough air at all of the registers, do a total system examination, starting with the filter. The air handler is often shipped with a filter already installed in the cabinet. The installer who installs a return-air filter grille may forget to remove the filter shipped with the unit. Even if the common return filter is faithfully changed, the filter at the air handler will still eventually become obstructed. An overall airflow reduction would result in low suction pressure in the cooling mode and a high temperature rise across the furnace in the heating mode. Reduced airflow can cause the furnace to trip off on a high limit control because of high heat exchanger temperatures. The technician can check the recommended air temperature rise in most furnaces. If the rise is above the maximum, there is not enough air. For example, if a gas furnace manufacturer recommends an air temperature rise of between 45°F and 65°F and the rise is 80°F, there is not enough airflow to satisfy the furnace manufacturer's specifications. It is easy for the technician to measure the temperature rise across the appliance by measuring the difference between the supply and return air temperatures.

An amperage reading of the fan motor may provide a clue to overall reduced airflow. If the fan is supposed to have a draw of 7 A and it is drawing only 4 A, suspect an obstruction or even an undersized duct. When a new system is started up for the first time and the airflow is established, the technician should record the fan motor amperage on the unit to use later for comparison testing.

When a system has one or two common return-air grilles, the average air velocity can be taken at the grilles and calculated to find the overall airflow. The free or usable area of the grille is first measured, then the average velocity across the grille is multiplied by the area [(cubic feet per minute = velocity (feet per minute) × area (square feet)]. The effective area of the grille is not the overall area of the grille because of the grille's louvers. The effective area of the grille is about 70% of the measured area in square feet. For example, suppose that the average velocity of a return-air grille is 350 fpm and the measured size of the grille opening is 14 in. × 20 in. The total area would be 1.944 ft^2 (14 × 20/144 = 1.944 ft^2). Since only 70% of the grille area is usable because of the louvers, the actual usable area is 1.944 × 0.7, or 1.44 ft^2. The volume of air would be cfm = velocity × area, 350 × 1.44 = 504 cfm. This is approximately the amount of airflow required for a 1½-ton system for cooling. If the return-air grille is too small, it will normally make a whistling sound as air passes through it. Be sure to check the manufacturer's specifications for specific grilles, as different grilles have different effective areas.

Other methods may be used for determining the flow for a residential system. These are discussed in the individual units that apply to gas, oil, or electric heat. These methods determine the airflow by measuring the energy input to the airstream and calculating the airflow.

No matter what method you use, the accuracy will probably be within ±20%. This is about as close as you can get for residential systems, assuming they are not leaking air at the duct connections. **NOTE:** *All air measurements for residential systems should be performed at maximum airflow. Any system that is both heating and cooling and that has a multiple-speed motor may have a different airflow for heating versus cooling. Switch the room thermostat to FAN ON for all testing. This will be the maximum airflow.*•

If you discover there is not enough air for the total system and there are no physical barriers, the next step is to see if the air volume delivered by the blower can be increased. If the system has a belt-drive blower, check the amperage as compared with the motor full-load amperage. If the blower motor amperage is low, the pulleys might be able to be adjusted to increase the airflow rate through the duct system.

COMMERCIAL DUCT SYSTEMS

Commercial duct systems vary greatly depending on the size of the buildings. Commercial duct systems should always have a blueprint of the duct layout along with a set of specifications that tell how much air should be flowing at any particular part of the system. This makes it easier to use the proper instruments and identify problems.

When a large multistory building is erected, a certified testing and balancing company should be present at the building system's start-up to verify that the airflow is correct before the building is turned over to the owner-operator. The company will verify the total building airflow, including the flow in all of the trunk ducts, in all of the branch ducts, and at the terminal units. These airflows are supposed to match the building specifications within an acceptable range, which is usually about 10% for comfort cooling applications. Upon completion, a balancing report is completed, verifying that all airflow rates are within the acceptable range.

An advantage of commercial high-rise buildings is that they usually have a separate air distribution system for each floor and the individual floors of the building are often similar in layout to each other. Troubleshooting airflow problems in large systems involves reading the prints to determine what the airflow is supposed to be under the conditions you are trying to check. If the system is a variable-air-volume (VAV) system, the airflow may be reduced because of the thermostat setting. If it is a constant flow system, the flow should be known. In either case, when the correct air volume is known, measure the airflow with quality instruments.

There should be test ports in the duct where the testing and balancing company took its readings from. If not, you will have to make your own. Be sure to use the correct drill size and wear eye protection when making these test holes. Make certain the blower is off when drilling airflow measurement holes in the duct to avoid having metal shavings blow in your eyes. Testing will involve taking traverse

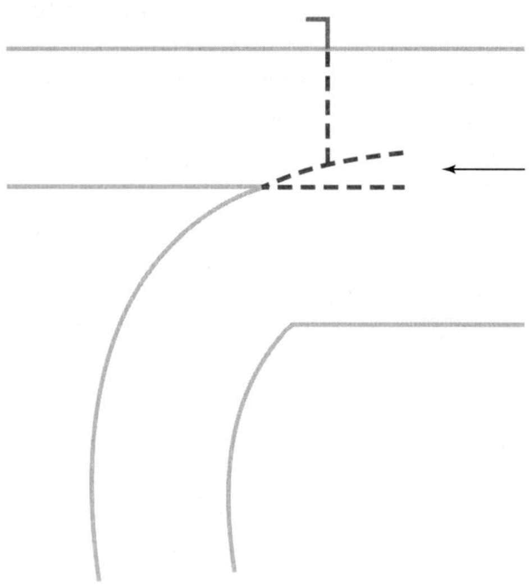

Figure 37.100 This splitter, or diverter, damper can help control how much airflow is in the two different ducts.

Figure 37.101 This hood fits directly over ceiling registers up to about 2 ft × 2 ft. The hood forces the air from the register through the meter and displays the airflow in cfm. Courtesy Alnor Instrument Company

readings of the duct, determining the average velocity of the air, and then using the airflow formula to determine the cfm of air. Commercial duct systems often have *dampers* that divert or divide the airflow in ductwork. These dampers can be adjusted to cause more or less air to flow into branch ducts, **Figure 37.100**. The system prints and drawings should provide information regarding the desired cfm in each duct run. If there are no prints available at the building, you may be able to get a copy of the original drawings from the installing contractor, the engineering firm, or the building architect. Be sure to plug the test ports when you are finished. Tapered rubber stoppers are commonly used for this.

Flow hoods are available for measuring the airflow at terminal units. These flow hoods fit over the register, sealing off the air, **Figure 37.101**. Because they read out directly in cfm, the readings can be easily compared to the job specifications.

Total airflow of commercial systems can be verified by checking the main duct for the correct airflow. If the airflow in the main supply duct is not correct, the technician should check for any problems that could reduce the airflow. Filters should be checked first. Dampers may be blocking airflow if they are not positioned correctly. All large blowers are powered by three-phase power, and fan motors should be checked for proper rotation. It is not uncommon for a motor to be running backward because some technician reversed two wires while servicing the electrical system. A squirrel cage blower will run just as well in either direction, but it will only discharge a fraction of the correct volume if the blower is rotating in the wrong direction. This reduction in airflow may not be noticed until peak season on some systems. **SAFETY PRECAUTION:** *The technician is responsible for obtaining air pressure and velocity readings in ductwork. This involves drilling and punching holes in metal duct. Use a grounded drill cord or a cordless drill and be careful with drill bits and the rotating drill. Sheet metal can cause serious cuts, and air blowing from the duct can blow chips into your eyes, so be sure to deenergize the blower before making any test holes in the duct. Protective eyewear must be worn.*•

REVIEW QUESTIONS

1. Air that enters a structure from leaks around windows and doors is best described as
   A. exhaust air.
   B. ventilation air.
   C. infiltration air.
   D. exfiltration air.

2. An example of a natural-draft system is
   A. hot water baseboard heating.
   B. radiant heating.
   C. reverse-cycle (heat pump) refrigeration.
   D. central air-conditioning.

3. All of the following are desirable characteristics of a forced-air system except
   A. air movement is not felt by the occupants of the space.
   B. air noise is not noticed by the occupants of the space.
   C. temperature differences between rooms are minimal.
   D. equipment cycling is noticeable by the occupants of the structure.
4. The pressures in a duct system are measured in which units?
   A. Feet of water column (FWC)
   B. Feet of air column (FAC)
   C. Inches of water column (IWC)
   D. Inches of air column (IAC)
5. At standard conditions, atmospheric pressure can support a column of water that is approximately how high?
   A. 3.4 feet
   B. 8.5 feet
   C. 17 feet
   D. 34 feet
6. The pressure that represents the force pushing outward against the walls of the duct system is referred to as the
   A. viscosity pressure.
   B. static pressure.
   C. velocity pressure.
   D. total pressure.
7. Which of the following is an accurate expression that relates the three duct system pressures?
   A. Total Pressure = Viscosity Pressure − Static Pressure
   B. Velocity Pressure = Total Pressure − Static Pressure
   C. Static Pressure − Velocity Pressure = Viscosity Pressure
   D. Total Pressure = Velocity Pressure ÷ Static Pressure
8. The instrument that is used to measure duct system pressures is the
   A. velometer.
   B. speedometer.
   C. anemometer.
   D. manometer.
9. An anemometer is used to directly measure the
   A. volume of air flowing through a duct.
   B. speed of air flowing through a duct.
   C. density of air flowing through a duct.
   D. moisture content of air flowing through a duct.
10. How much air is flowing through a 10 in. × 12 in. duct if the air is traveling at a speed of 700 feet per minute?
    A. 122 cfm
    B. 351 cfm
    C. 583 cfm
    D. 827 cfm
11. Which of the following sets of instruments are most similar to each other?
    A. Anemometer and velometer
    B. Potentiometer and anemometer
    C. Manometer and velometer
    D. Potentiometer and manometer
12. Squirrel-cage-type blowers are used in conjunction with duct system because they
    A. are less expensive than propeller-type fans for this application.
    B. create less of a pressure differential between their inlet and outlet.
    C. are able to overcome the pressures that are present in the duct system.
    D. are easier to mount in duct systems than propeller-type fans.
13. Which of the following is true regarding the amperage draw of a blower motor being used to move air through a duct system?
    A. Motor amperage will go up if the blower's inlet is blocked.
    B. Motor amperage will go up if the blower's outlet is blocked.
    C. Motor amperage will go down if the door on the blower compartment is removed.
    D. Motor amperage will go up if the door on the blower compartment is removed.
14. The type of blower assembly that utilizes both motor shaft and blower shaft bearings is the
    A. combination-drive blower assembly.
    B. direct-drive blower assembly.
    C. belt-drive blower assembly.
    D. isolation-drive blower assembly.
15. The type of blower assembly that utilizes only motor shaft bearings is the
    A. combination-drive blower assembly.
    B. direct-drive blower assembly.
    C. belt-drive blower assembly.
    D. isolation-drive blower assembly.
16. Which of the following blower assemblies is used in conjunction with changeable or adjustable pulleys that can vary the blower speed?
    A. Combination-drive blower assembly
    B. Direct-drive blower assembly
    C. Belt-drive blower assembly
    D. Isolation-drive blower assembly
17. Which of the following is false regarding a duct system that is oversized?
    A. Increased fabrication costs
    B. Increased installation costs
    C. Increase noise levels
    D. Decreased air velocity
18. Which of the following is a benefit of the reducing plenum air distribution system?
    A. The velocity of the air is maintained throughout the system
    B. It is the least expensive of all duct system options
    C. It requires only a minimal amount of rigid ductwork
    D. The air pressures in the duct system vary greatly from inlet to outlet

19. Which of the following air distribution systems is commonly installed in a poured slab floor?
    A. Plenum system
    B. Extended plenum system
    C. Reducing plenum system
    D. Perimeter loop system
20. Rectangular metal duct sections are commonly joined together with
    A. duct tape.
    B. slips and drives.
    C. mastic.
    D. self-tapping sheet metal screws.
21. Galvanized sheet metal is treated/coated with which substance to slow the corrosion process?
    A. Zinc
    B. Tin
    C. Aluminum
    D. Steel
22. Sheet metal duct sections are often fabricated with cross-breaks to
    A. reduce air turbulence in the duct.
    B. prevent the duct from sweating.
    C. make the duct section more rigid.
    D. improve the appearance of duct.
23. Flexible duct
    A. is desirable because it has a very low resistance to airflow.
    B. Is ideally used for very short, direct duct runs.
    C. Is available in sizes up to 8 in. in diameter.
    D. is desirable because it is less likely to kink than rigid duct.
24. The branch duct is connected to the main trunk line with a
    A. boot fitting.
    B. takeoff fitting.
    C. transition fitting.
    D. offset fitting.
25. A streamlined takeoff fitting is designed with
    A. a throat that is larger than the runout duct it is connected to.
    B. a throat that is smaller than the runout duct it is connected to.
    C. a throat that is the same size as the runout duct it is connected to.
    D. a throat that is the same size as the main trunk line it is connected to.
26. Balancing dampers are ideally located
    A. as close to the supply register location as possible.
    B. as far from the main supply trunk as possible.
    C. as close as possible to the main supply trunk.
    D. as close as possible to the air handler in the main supply duct.
27. Which of the following statements is accurate with regards to simple system zoning?
    A. Zone dampers are manually operated based on space temperature.
    B. Zone dampers are manually operated based on space occupancy.
    C. Zone dampers are automatically operated based on space temperature.
    D. Zone dampers are automatically operated based on space occupancy.
28. If an air-conditioning system has three independent zones, how many space thermostats will the system have?
    A. 1
    B. 2
    C. 3
    D. 4
29. Which of the following are challenges that need to be addressed when a single speed blower motor is used on a zoned air-conditioning system?
    A. Reduced airflow, increased noise, evaporator coil freeze-ups
    B. Increased airflow, reduced air velocity, high system operating pressures
    C. Increased airflow, increased noise, evaporator coil freeze-ups
    D. Reduced airflow, lower system operating pressures, condenser coil freeze-ups
30. Ducts are insulated to
    A. prevent duct sweating and increase the heat transfer rate between the air in and around the duct.
    B. encourage duct sweating and increase the heat transfer rate between the air in and around the duct.
    C. encourage duct sweating and decrease the heat transfer rate between the air in and around the duct.
    D. prevent duct sweating and decrease the heat transfer rate between the air in and around the duct
31. Supply registers should be positioned and configured so that
    A. the supply air is directed to the center of the room.
    B. cooling air is introduced to the occupied space via floor registers.
    C. the supply air is directed onto the walls of the room.
    D. heating air is introduced to the occupied space via ceiling registers.
32. Duct systems must be properly sized and designed to reduce and overcome the effects of
    A. resistance to flow at the supply registers.
    B. friction within the duct system.
    C. resistance to flow at the return grilles.
    D. excessive airflow in the main supply trunk.

33. A technician is traversing a duct and obtains the following 8 velocity readings: 500 fpm, 400 fpm, 375 fpm, 625 fpm, 275 fpm, 360 fpm, 420 fpm, and 575 fpm. What is the velocity value that will be used for system calculations?
    A. 350 fpm
    B. 440 fpm
    C. 530 fpm
    D. 620 fpm
34. Using the chart in Figure 37.88, what is the velocity and friction loss of a 400 cfm airstream that is flowing through a 9-in. round duct?
    A. 90 fpm with a friction loss of 1.5 IWC/100 feet
    B. 900 fpm with a friction loss of 1.5 IWC/100 feet
    C. 90 fpm with a friction loss of 0.15 IWC/100 feet
    D. 900 fpm with a friction loss of 0.15 IWC/100 feet
35. Referring to Figure 37.90, a rectangular duct with cross-sectional measurements of 20 in. × 11 in. will be the equivalent of a round duct with a diameter of
    A. 13 in.
    B. 14 in.
    C. 15 in.
    D. 16 in.
36. If there seems to be reduced airflow at all supply register locations, which of the following is the most likely problem with the system?
    A. One or more supply registers are closed.
    B. One or more balancing dampers are closed.
    C. The system air filter is dirty.
    D. A branch duct has come loose from the main trunk line.
37. If a gas-fired furnace is constantly tripping on a high limit safety control, which of the following is likely NOT a cause for this condition?
    A. Dirty air filter
    B. Improperly balanced air distribution system
    C. Defective furnace blower motor
    D. Duct lining has come loose from the main trunk line
38. Explain why some areas in a structure might be too cool, while others are too warm.
39. Describe the construction of a pitot tube.
40. Explain the purpose of canvas connectors and where they are located in an air distribution system.
41. Explain why the extended plenum system is desirable for longer structures.
42. Explain how the gauge of sheet metal is related to the thickness of the metal.
43. Explain how turning vanes help to reduce turbulence in an air duct.
44. List two types of return air systems commonly found on residential air-conditioning systems.

UNIT 38

INSTALLATION

OBJECTIVES

After completing this unit, you should be able to

- list three crafts involved in air-conditioning installation.
- identify three common types of duct system installations.
- describe the installation of square or rectangular metal duct.
- describe the installation of round metal duct.
- explain how and why metal ducts are insulated and/or lined.
- describe the installation of ductboard systems.
- describe the installation of flexible duct.
- explain how to calculate the acceptable range of voltage supplied to air-conditioning equipment.
- describe good installation practices for package air-conditioning equipment.
- discuss different connections for package air-conditioning equipment.
- describe the split air-conditioning system installation.
- describe best practices for installing the condensing unit of a split air-conditioning system.
- describe correct refrigeration piping practices.
- state start-up procedures for air-conditioning equipment.

SAFETY CHECKLIST

- ✔ Be careful when handling sharp tools, fasteners, and metal duct. Wear gloves whenever practicable.
- ✔ Use only grounded, double-insulated, or cordless electrical tools.
- ✔ Wear gloves, goggles, and clothing that will protect your skin when working with or around fiberglass insulation.
- ✔ Extreme care should be used when working with any electrical circuit. All safety rules must be adhered to. Particular care should be taken while working with any primary power supply to a building because the fuse that protects that circuit may be on the power pole outside. Keep the power off whenever possible when installing or servicing equipment. Lock and tag the panel. There should be only one key. Keep this on your person at all times.
- ✔ Although low-voltage circuits are generally harmless, you can experience shock if you are wet and touch live wires. Exercise caution when working on or around any electric circuit.
- ✔ Before starting a system for the first time, examine all electrical connections and moving parts. Connections, pulleys, or other moving parts may be loose.
- ✔ Follow all manufacturer's instructions before and during initial start-up.
- ✔ Before starting a system, make certain that all fans and blowers spin freely and that all compressor shipping blocks have been removed.

38.1 INTRODUCTION TO EQUIPMENT INSTALLATION

Installing air-conditioning equipment requires three crafts: duct, electrical, and mechanical, which includes refrigeration. Some contractors use separate crews to carry out the different tasks. Others may perform the duties of two of the crafts within their own company and subcontract the third to a more specialized contractor. Some small contractors do all three jobs with a few highly skilled people. The three job disciplines are often licensed at local and state levels and may be licensed by different departments.

Included in the following sections are duct installation procedures for (1) all-metal square or rectangular duct, (2) all-metal round duct, (3) fiberglass ductboard, and (4) flexible duct. **NOTE:** *Local codes must be followed while performing all work.* • Since codes can vary greatly from one area to another, be sure to be aware of, and adhere to, the codes that are in effect in the work is being performed.

38.2 INSTALLING SQUARE AND RECTANGULAR DUCT

Square metal duct is fabricated in a sheet metal shop by qualified sheet metal layout and fabrication personnel. The duct is then moved to the job site and assembled to create the air distribution system. Because the all-metal duct system is rigid, all dimensions have to be precise or the job will not fit together properly. Runs of ductwork must sometimes rise over or go beneath objects and still measure out correctly so that the takeoff reaches the correct location for the branch duct connection. The branch duct must be accurately measured and fabricated so that it reaches the terminal point, which is the boot for the room register. **Figure 38.1** illustrates an example of a duct system layout. The measuring, fabricating, and installing of the duct system must all be done properly in order to ensure a proper installation. If the system is not properly measured, there is no way that the fabricators will be able to create the correct duct sections. Improperly fabricated duct sections cannot be installed without having to make modifications to the sections or having to remake some sections. This results in lost time and money for the installation company.

Figure 38.1 A duct system.

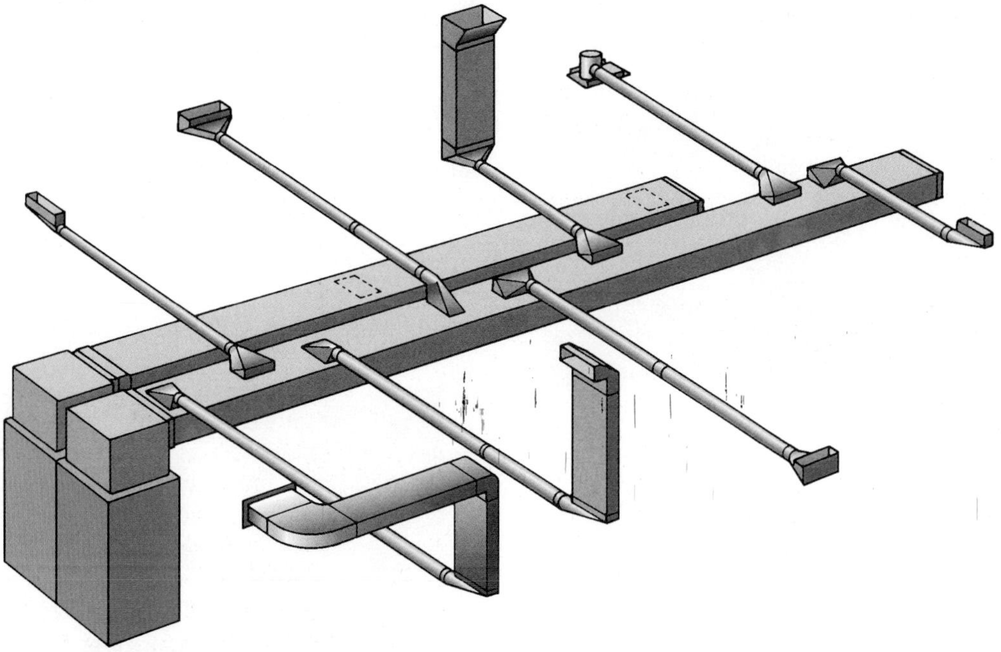

Square duct or rectangular duct is assembled with S fasteners, often referred to as slips, and drive clips. To seal the duct sections, a special product called mastic may be applied to the duct joints with a stiff paintbrush, **Figure 38.2**. It dries hard but is slightly pliable. Mastic helps create an airtight seal, minimizing air leaks in the duct system.

While the duct is being assembled, it has to be fastened to the structure for support. This can be accomplished in several ways. In an attic installation it may be laid flat or be hung by hanger straps. The duct should be supported so that it will be steady when the blower starts and will not transmit noise to the structure. The rush of air (which has weight) down the duct will move the duct if it is not fastened. Vibration eliminators next to the blower section will prevent the transmission of noise down the duct. See **Figure 38.3** for a flexible duct connector that can be installed between the blower and the metal duct. When installed, the metal portions of the connector should not touch each other to help ensure that sound does not transmit, **Figure 38.4**. Flexible connectors are always recommended, but not always used.

Metal duct can be purchased in the popular sizes for small and medium systems from some supply houses. This makes metal duct systems available to the small contractor who may not have a sheet metal shop. The standard duct sizes can be assembled with an assortment of standard fittings to build a system that appears to be custom-made for the job, **Figure 38.5**.

Figure 38.2 Technician applying mastic to a joint between two duct sections.

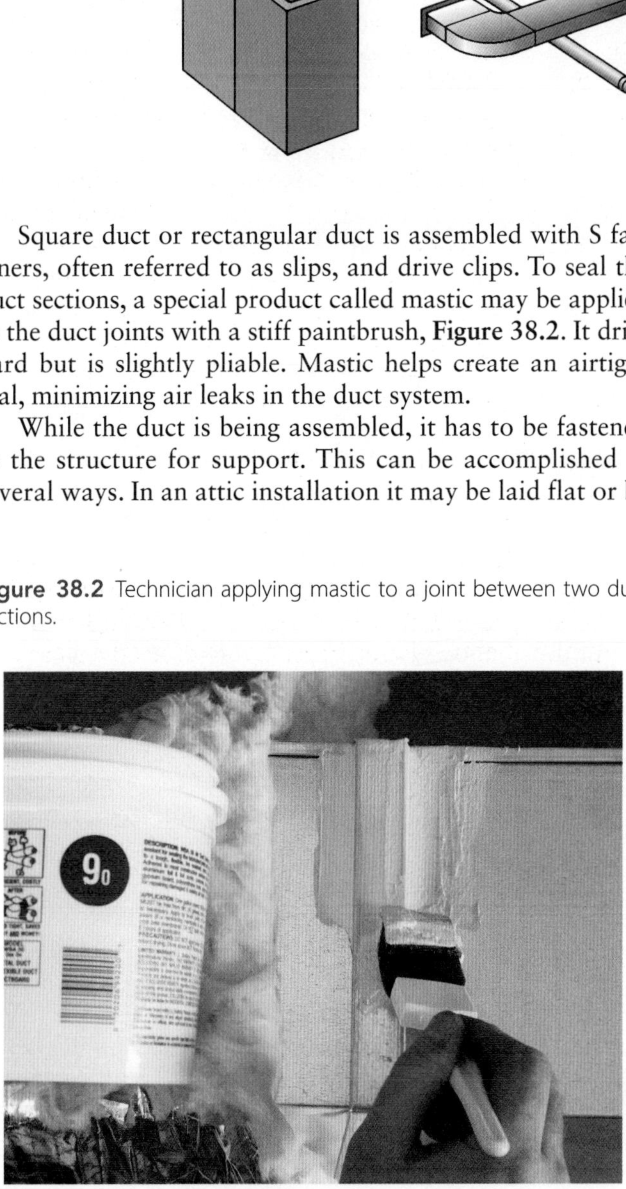

Figure 38.3 A flexible duct connector.

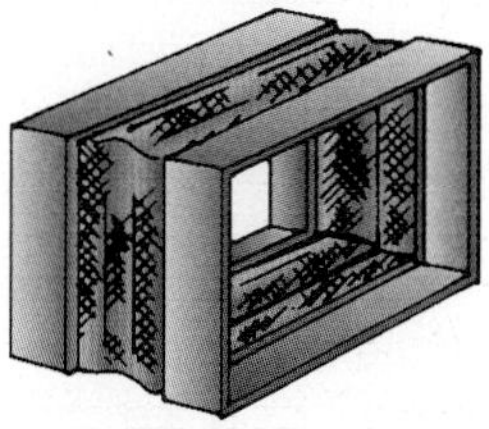

Figure 38.4 The metal portions of the flexible connector should not come in contact with each other. Photo by Eugene Silberstein

38.3 INSTALLING ROUND METAL DUCT SYSTEMS

Round metal duct systems are easy to install and are available from some supply houses in standard sizes for small and medium systems. These systems, which use reducing fittings from a main trunk line, may be assembled in the field. They must be fastened together at all connections. Self-tapping sheet metal screws are popular fasteners. The screws are held by a magnetized screw holder while an electric drill turns and starts the screw, **Figure 38.6**. Each connection should have a minimum of three screws spaced evenly to keep the duct steady. A good installer can fasten the connections as fast as the screws can be placed in the screw holder, **Figure 38.7**. A reversible, cordless, variable-speed drill is the ideal tool to use.

Round metal duct requires more clearance space than square or rectangular duct. It must be supported and

Figure 38.6 A battery-powered hand drill. Photo by Eugene Silberstein

Figure 38.5 Standard duct fittings.

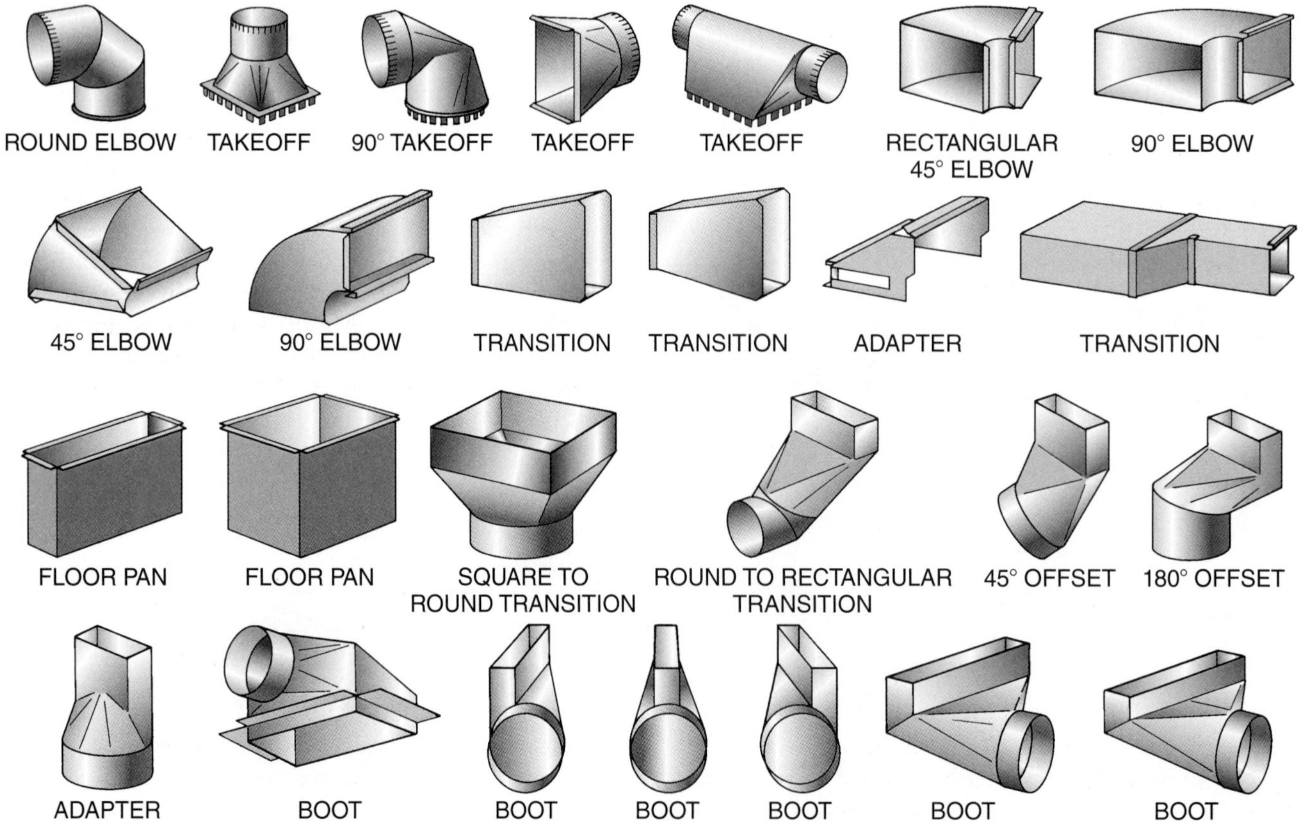

Figure 38.7 A self-tapping sheet metal screw with a magnetic screw holder. Photo by Bill Johnson

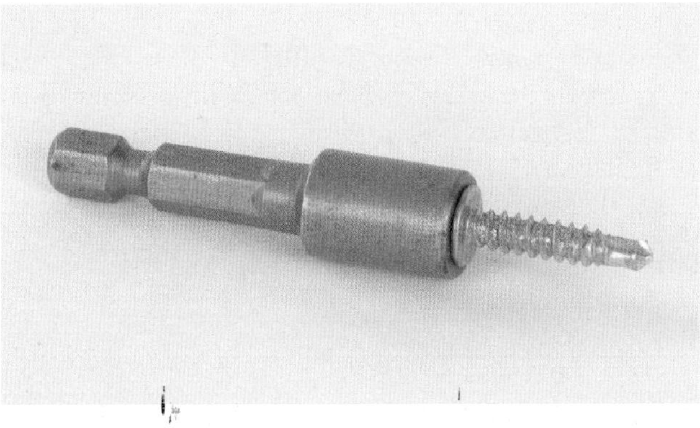

Figure 38.8 A tab that holds the fiberglass duct liner to the inside of the duct.

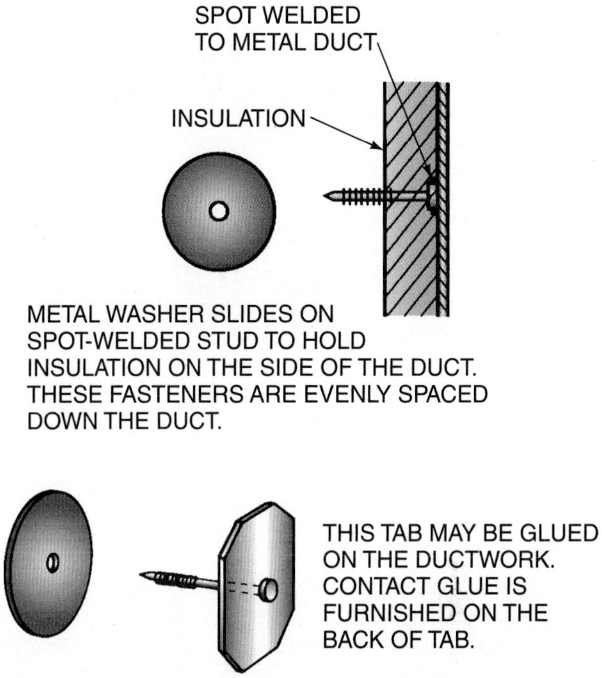

mounted at the correct intervals to keep it straight. Exposed round metal duct does not look as good as square duct and rectangular duct, so it is often used in places that are out of sight. **SAFETY PRECAUTION:** *Be careful when handling the sharp tools and fasteners. Use grounded, double-insulated, or cordless electrical tools.* •

38.4 INSULATION AND ACOUSTICAL LINING FOR METAL DUCT

Insulation and acoustical lining for metal duct can be applied to either the inside or the outside of the duct. Insulation is typically wrapped around the duct, and acoustical lining is applied to the interior of the duct. When lining is to be applied to the inside, it is usually done in the shop when the duct is being fabricated. The lining can be fastened with tabs, glue, or both. The tabs, which are fastened to the inside of the duct, have a shaft that looks like a nail protruding from the duct wall. The liner and a washer are pushed over the tab shaft. The tab base, **Figure 38.8**, is fastened to the inside of the duct by glue or spot welding. Spot welding is the most permanent method, but it is difficult to do and expensive. Liner that is glued to the duct may come loose and block airflow, perhaps years after the initial installation date. This problem is difficult to find and repair, particularly if the duct is in a wall or framed in by the building structure. Using tabs with glue is a more permanent method.

When acoustical lining is to be applied to the interior surfaces of the ductwork, adjustments must be made to the duct's cross-sectional dimensions to ensure proper airflow through the duct. Since the acoustical lining is typically 1-in. thick, the amount of air the duct can carry can be greatly reduced, especially in smaller ductwork. Consider a section of 10 in. × 10 in. duct work that has a cross-sectional area of 100 in^2. If 1-in. lining is installed in the duct, the interior measurements become 8 in. × 8 in. and the cross-sectional area drops to 64 in^2, **Figure 38.9**. This is a 36% reduction in duct area!

The liner is normally fiberglass and is coated on the air side to keep the fibers from being pulled from the material and allowed to enter the airstream. An additional coating is often added to the liner to reduce the potential for microbial growth on the insulation. Over time, some dust will accumulate on the surface of the duct liner, which can support mold growth should the liner become damp for any reason.

Many pounds of air pass through a duct each season. An average air-conditioning system moves approximately 400 cfm/ton of air. A 3-ton system moves approximately 1200 cfm or 72,000 cfh (1200 cfm × 60 min/h = 72,000 cfh); 72,000 cfh ÷ 13.33 ft^3/lb = 5401 lb of air per hour. This is more than 2½ tons of air per hour.

SAFETY PRECAUTION: *Fiberglass insulation may irritate the skin. Gloves and goggles should be worn while handling it.* •

38.5 INSTALLING DUCTBOARD SYSTEMS

Fiberglass ductboard is popular among many contractors because it requires little special training to construct a system, **Figure 38.10**. Special knives can be used to fabricate the duct in the field. If the knives are not available, simple cuts and joints can be made with a utility or contractor's knife.

Figure 38.9 Acoustical lining reduces the cross-sectional area of the duct.

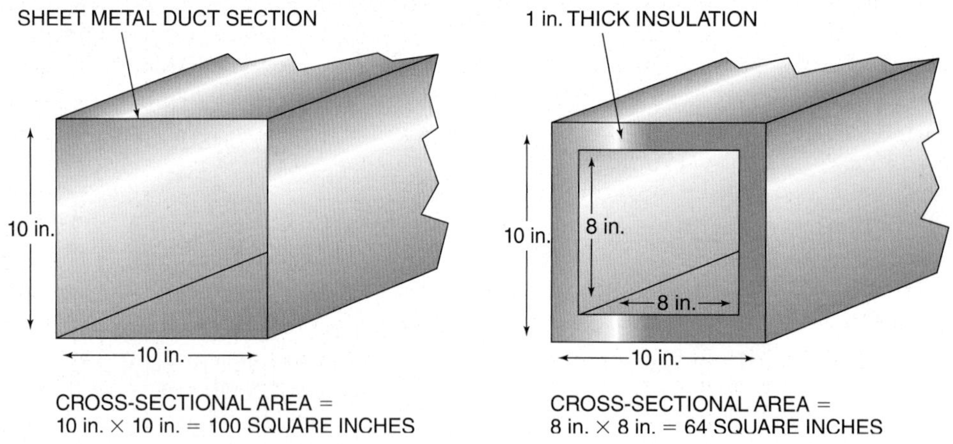

Figure 38.10 Ducts manufactured from compressed fiberglass with foil backing. Courtesy Johns Manville

Fiberglass ductboard comes with insulation already attached to the outside skin. The skin is made of foil with a fiber running through it to give it strength. When assembling this duct, it is important to cut some of the insulation away so that the outer skin can overlap the surface that it is joining. The two skins are strengthened by stapling them together and taping to make them airtight, **Figure 38.11**.

Fiberglass ductboard can be made into almost all of the configurations that metal duct can. It is lightweight and easy to transport because the duct can be cut, laid out flat, and assembled at the job site. The original shipping boxes can be used to transport the ductboard and to keep it dry. Metal duct fittings, on the other hand, are large and take up a lot of space in a truck. Round fiberglass duct is as easy to install as flat ductboard because it can also be cut with a knife.

Figure 38.11 The outer skin is lapped over, and the backing of one piece is stapled to the backing of the other piece.

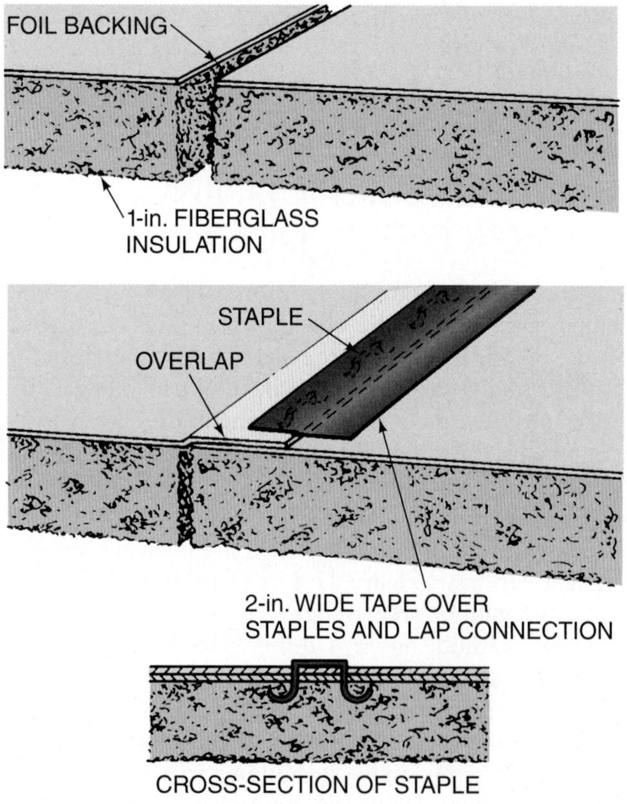

Fiberglass duct must be supported like metal duct to keep it straight. The weight of the ductboard itself will cause it to sag over long spans. A broad hanger that will not cut through the foil backing of the ductboard is necessary. Fiberglass duct deadens sound because the inside of the duct has a coating that helps to greatly reduce any air or fan noise that may be transmitted into the duct.

38.6 INSTALLING FLEXIBLE DUCT

Flexible duct has a flexible liner and may have a fiberglass outer jacket for insulation if needed, for example, where heat exchange takes place between the air in the duct and the space where the duct is routed. The outer jacket is held by a moisture-resistant cover made of fiber-reinforced foil or vinyl. The duct may be used for the supply or return and should be run in a direct path to keep bends from narrowing the duct. Sharp bends can cause the duct to kink and greatly reduce the airflow, therefore, sharp bends and turns should be avoided.

In the supply system, flexible duct may be used to connect the main trunk to the boot. (The boot is the fitting that connects the duct to the supply register. In this case, it has a round connection on one end for the flexible duct and a rectangular connection on the other end where the floor register fastens to it.) The flexibility of the duct makes it valuable as a connector for metal duct systems. The metal duct can be installed close to the boot and the flexible duct can be used to make the final connection. Flexible duct used at the end of a metal duct run will help reduce any noise that may be traveling through the duct. This can help reduce the noise in the conditioned space if the blower or heater is noisy. Long runs of flexible duct are not recommended.

Flexible duct must be properly supported. A support band of 1.5 in. or more in width is the best method of keeping the duct from collapsing and reducing its inside dimension. Tight turns should be avoided to keep the duct from collapsing on the inside and reducing the inside dimension. Flexible duct should be stretched tight to keep the liner from sagging, kinking, and creating friction loss.

38.7 ELECTRICAL INSTALLATION

The air-conditioning technician should be familiar with certain guidelines regarding electrical installation to make sure that the unit has the correct power supply and that the power supply is safe for the equipment, the service technician, and the owner. The control voltage for the space thermostat is often installed by the air-conditioning contractor even if the line-voltage power supply is installed by a licensed electrical technician.

The power supply must include the correct voltage and wiring practices, including wire size. The law requires the manufacturer to provide a nameplate with each electrical device that gives the voltage requirements and the current the unit will draw, **Figure 38.12**. The applied voltage (the voltage that the unit will actually be using) should be within the range of $\pm 10\%$ of the rated voltage of the unit. That is, if the rated voltage of a unit is 230 V, the maximum operating voltage at which the unit should be allowed to operate would be 10% over 230 V, or 253 V ($230 \times 1.10 = 253$). The minimum voltage would be 10% below 230 V, or 207 V ($230 \times 0.90 = 207$). Some motors are rated at 208/230 V. This is a versatile rating because the motor will operate safely within a large voltage range. The lower limit would be 187 V ($208 \times 0.90 = 187$) and the upper limit would be 253 V ($230 \times 1.10 = 253$). **NOTE:** *If the unit operates for a long time beyond these limits, the motors and its controls can become damaged.* •

Package-type equipment has one power supply for the unit. If the system is a split system there will be two power supplies: one for the indoor unit and one for the outdoor unit. Both power supplies will go back to a main panel, but there will be a separate circuit breaker for each at the main panel, **Figure 38.13. SAFETY PRECAUTION:** *A disconnect or cutoff switch should be within 25 ft of each unit and within sight of the unit.* • If the disconnect is located remotely, someone may mistakenly turn the unit on while the technician is working on the electrical system. **SAFETY PRECAUTION:** *All safety rules must be adhered to while working on any electrical circuit.* •

Wire-sizing tables specify the wire size that each circuit will require. The *National Electrical Code* (*NEC*) provides installation standards, including workmanship, wire sizes, methods of routing wires, and types of enclosures for wiring

Figure 38.12 An air-conditioner unit nameplate. Photo by Bill Johnson

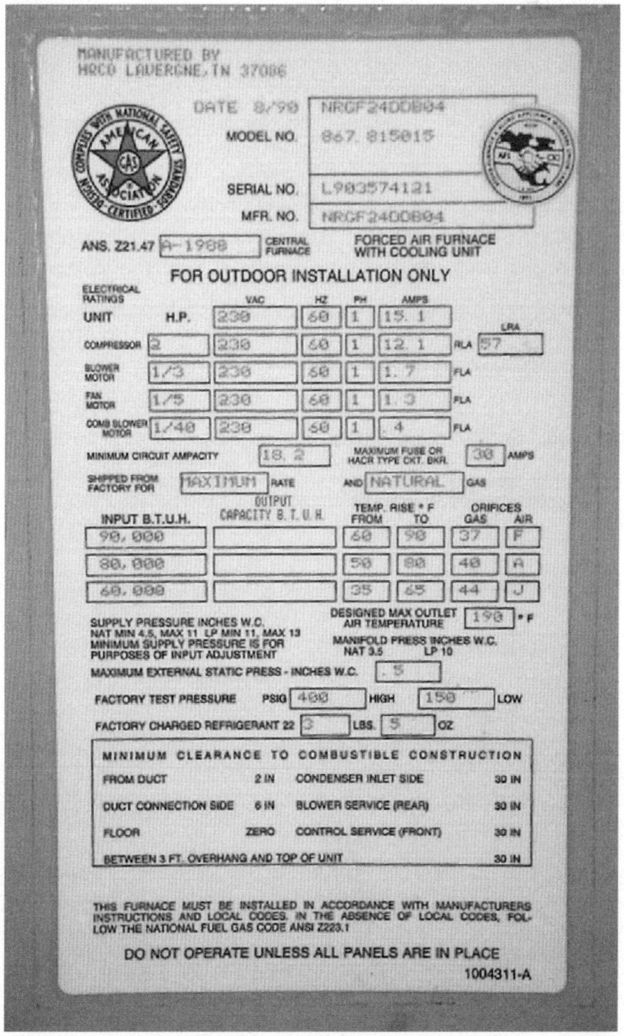

Figure 38.13 Wiring connecting indoor and outdoor units to the main power service.

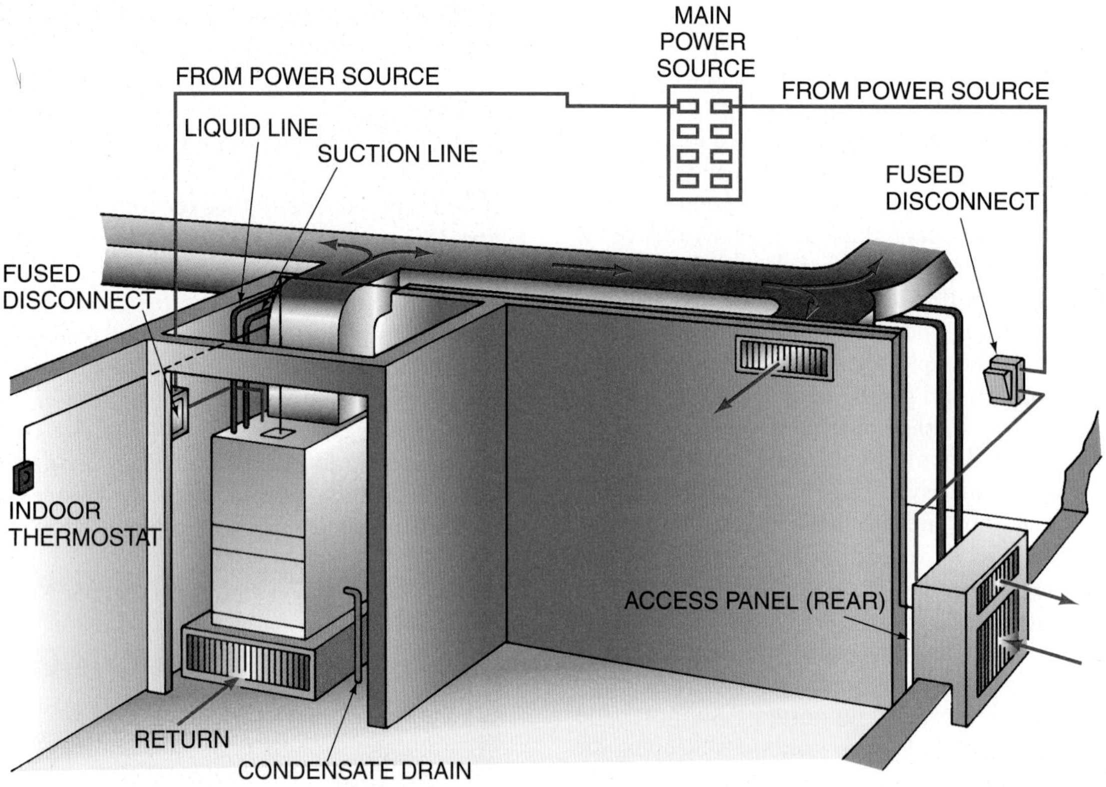

and disconnects. Use it for all electrical wiring installation unless a local code prevails. The control-voltage wiring in air-conditioning and heating equipment is the line voltage that is reduced through a step-down transformer. It is installed with color-coded or numbered wires so that the circuit can be followed through the various components. For example, the 230-V power supply is often at the air handler, which may be in a closet, an attic, a basement, or a crawl space. The interconnecting low-voltage wiring may have to leave the air handler and go to the room thermostat and the condensing unit. The air handler is often used as the junction for these connections, **Figure 38.14**.

Figure 38.14 This pictorial diagram shows the relative position of the wiring as it is routed to the room thermostat, the air handler, and the condensing unit.

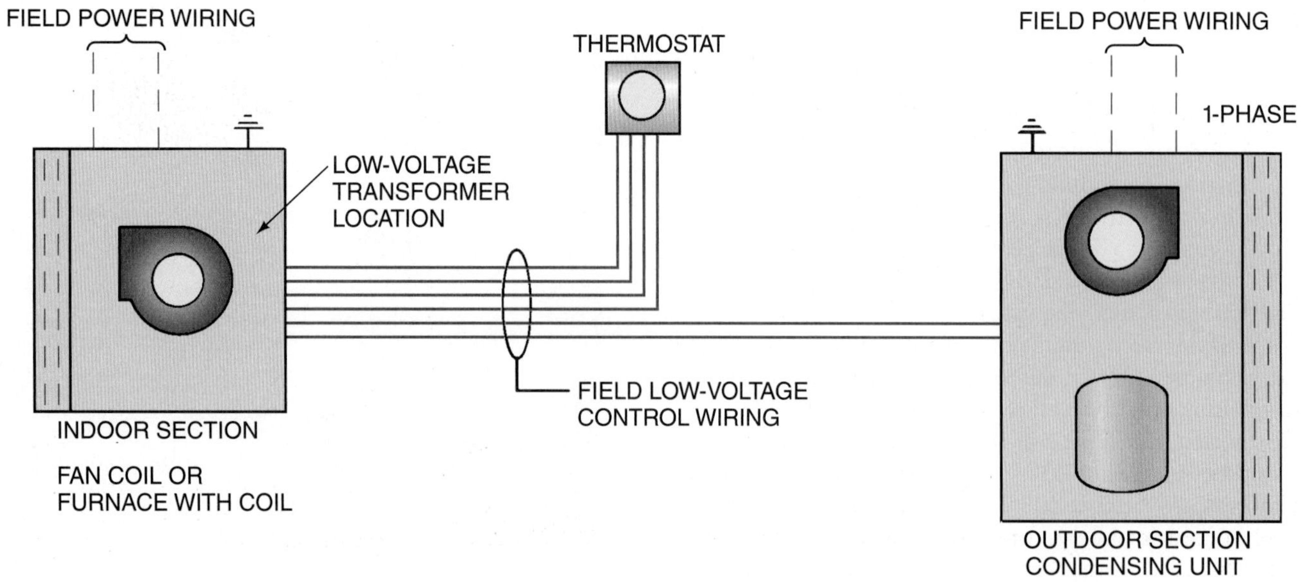

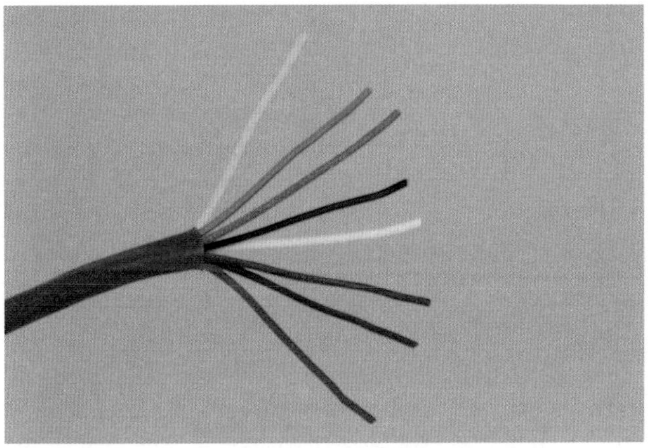

Figure 38.15 This is called 18–8 thermostat wire. Its size is 18 gauge and there are 8 wires in the cable. Photo by Bill Johnson

The control wire is a light-duty wire because it carries a low voltage and current. The standard wire size is 18 gauge. A standard air-conditioning cable has four wires each of 18 gauge in the same plastic-coated sheath and is called 18–4 (18 gauge, 4 wires). Red, white, yellow (or blue), and green wires are in the colored wires commonly found in a 4-conductor, low-voltage cable. Some wires may have eight conductors and are called 18–8, **Figure 38.15**. When conductors are bundled, such as shown in **Figure 38.15**, the bundle is called a cable. Low-voltage control wiring can be installed by the air-conditioning contractor in most areas because it is low voltage and electrician's license is not required. Be sure to check the licensure requirements in your area to determine whether or not a license is required to install low-voltage control wiring. **⚠ SAFETY PRECAUTION:** *Although low-voltage circuits are generally harmless, you can be hurt if you are wet and touch live wires. Use precautions for all electrical circuits.* •

38.8 INSTALLING PACKAGE SYSTEMS

Package or self-contained equipment includes all components in one housing. The window air conditioner is, since all of its components are contained in a single cabinet, considered to be a small package unit. Larger package systems may have 100 tons or more of air-conditioning capacity. The package unit is available in several different configurations for different applications, some of which are described here.

1. An air-to-air application is similar to a window air conditioner except that it has two motors. Because this unit is the most common, it will be discussed here. The term *air-to-air* refers to the fact that the refrigeration unit absorbs its heat from the air and rejects it into the air, **Figure 38.16**.
2. **Figure 38.17** shows an air-to-water unit. This system absorbs heat from the conditioned space air and rejects the heat into water. The water is wasted or passed

Figure 38.16 An air-to-air unit. Courtesy Heil-Quaker Corp.

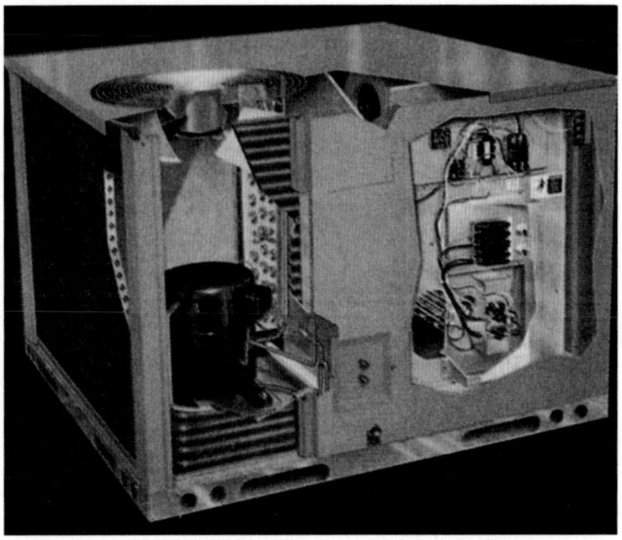

Figure 38.17 An air-to-water package unit. Reproduced courtesy of Carrier Corp.

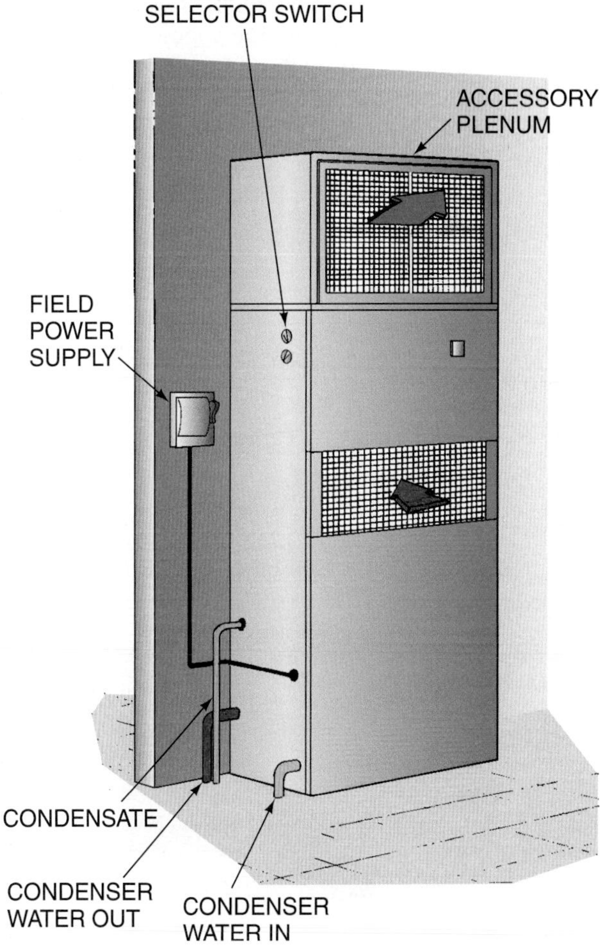

through a cooling tower to reject the heat to the atmosphere. This system is sometimes referred to as a water-cooled package unit.

3. Water-to-water equipment, **Figure 38.18**, has two water-to-refrigerant heat exchangers and is used in large commercial systems. Water is cooled in the system's chiller barrel and then circulated through the building to absorb the structure's heat, **Figure 38.19**. This system uses two pumps and two water circuits in addition to the fans to circulate the air in the conditioned space. To properly maintain this system, service technicians need to be able to service pumps and water circuits.
4. Water-to-air equipment, **Figure 38.20**, absorbs heat from the water circuit and rejects the heat directly into the atmosphere. This equipment is used for larger commercial systems.

Air-cooled air-conditioning systems are by far the most common in residential and light commercial installations. These units must be set on a firm foundation. For rooftop installations the unit may be furnished with a roof curb. The roof curb raises the unit off the roof and provides waterproof duct connections to the conditioned space below. In new construction, the roof curb can be shipped separately and a roofer can install it, **Figure 38.21**. The air-conditioning contractor can then set the package unit on the roof curb for a watertight installation. The foundation for another type of an installation located beside the conditioned space may have a foundation made of high-impact plastic, concrete, or a metal frame, **Figure 38.22**. The unit should be positioned high enough so accumulating water will not be able to reach, and potentially damage, the equipment.

Figure 38.18 A water-to-water package unit. Reproduced: courtesy of Carrier Corp.

Figure 38.19 The total water-to-water system must have two pumps to move the water in addition to the fans that move the air across the coils.

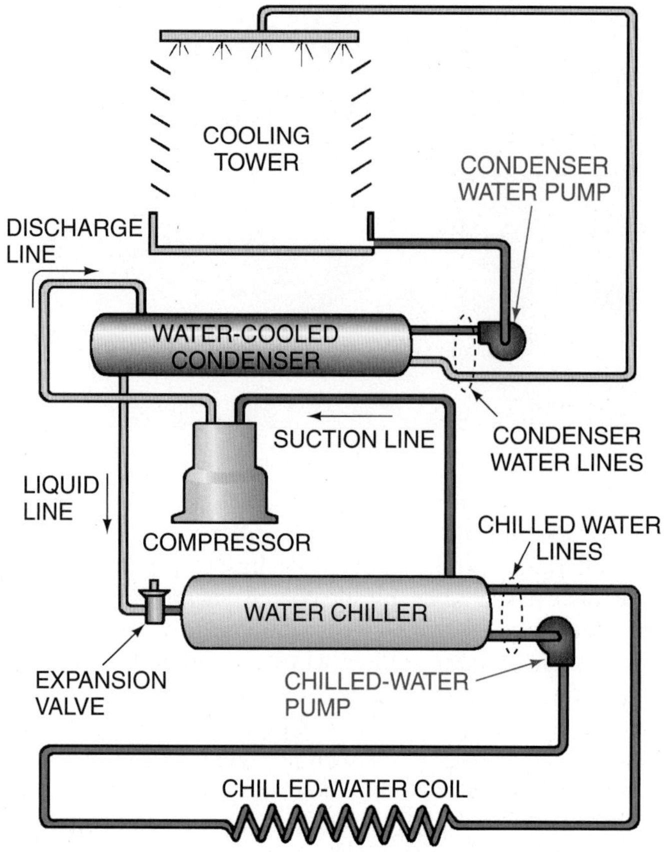

Figure 38.20 A water-to-air package unit. Courtesy York International Corp.

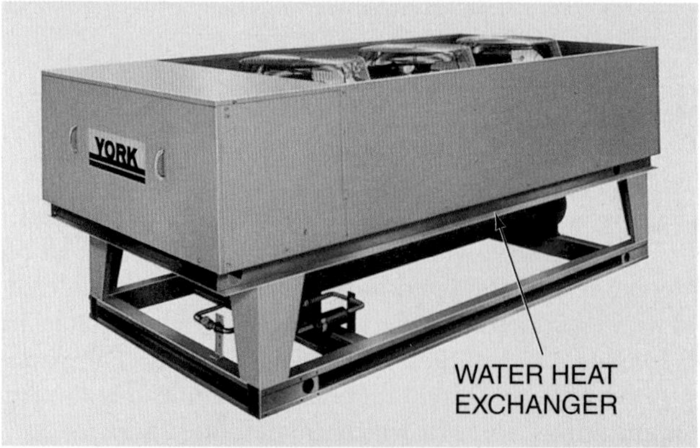

VIBRATION ISOLATION

The foundation of the unit should be placed in such a manner that vibration from the unit is not transmitted into the building. Some system for isolating vibration from the

Figure 38.21 This unit will sit on the roof curb. Courtesy Heil Heating and Cooling Products

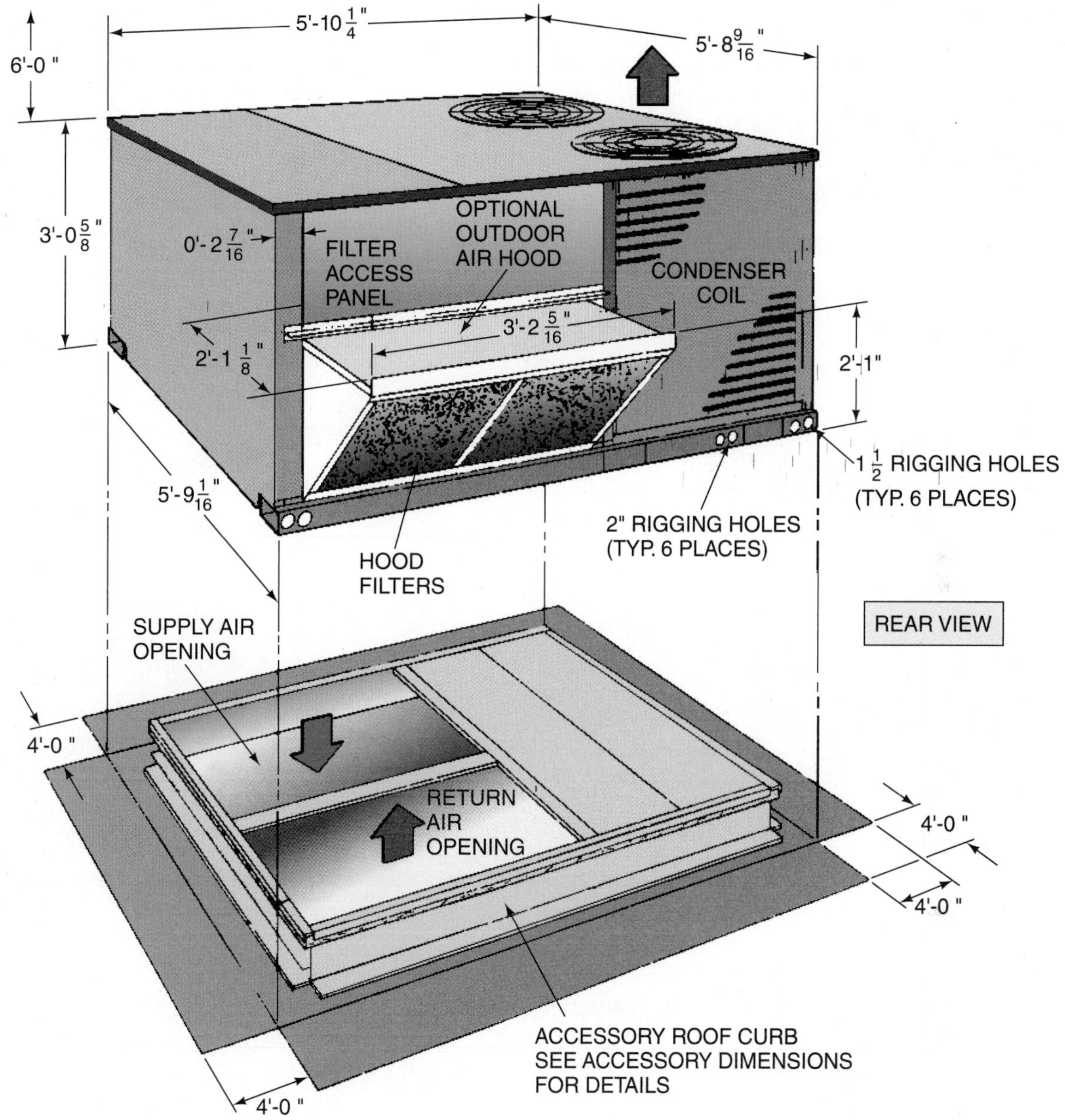

building structure may be needed. Two common methods are rubber and cork pads and spring isolators. Rubber and cork pads are the simplest and least expensive method for simple installations, **Figure 38.23**. The pads come in sheets that can be cut to the desired size. They are placed under the unit at the point where the unit rests on its foundation. If there are raised areas on the bottom of the unit that make contact, the pads may be placed there. Spring isolators may also be used and need to be matched to the weight of the unit. The springs will compress and lose their effectiveness under too heavy a load. They also may be too stiff for a light load if not chosen correctly.

DUCT CONNECTIONS FOR PACKAGE EQUIPMENT

The ductwork that enters the structure may be furnished by the manufacturer or fabricated by the contractor. Manufacturers furnish a variety of connections for rooftop installations but do not normally provide many factory-made fittings for those going through a wall. See **Figure 38.24** for a rooftop installation.

Package equipment comes in two different duct connection configurations relative to the placement of the return and supply ducts, either side-by-side or over-and-under. In

Figure 38.22 Various types of unit pads or foundations (A)–(C).

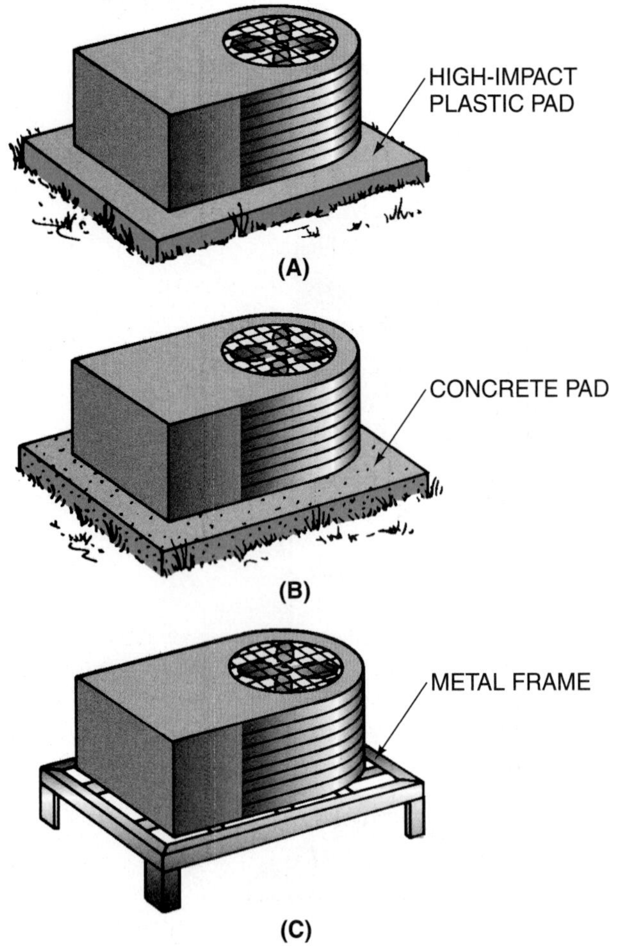

the *side-by-side* pattern, the return duct connection is beside the supply connection, **Figure 38.25**, and the connections are almost the same size. However, there is a large difference in the connection sizes on an over-and-under unit. If the unit is installed in a crawl space, the return duct must run on the correct side of the supply duct if the crawl space is low. It is difficult to cross a return duct over or under a supply duct if there is not enough room. The side-by-side duct design makes it easier to connect the duct to the unit because of the sizes of the connections on the unit. They are closer to square in dimension than are the over-and-under duct connections. If the duct has to rise or fall going into the crawl space, a standard duct size can be used.

The *over-and-under* connection is more difficult to fasten ductwork because the connections are more oblong, **Figure 38.26**. They are wide because they extend from one side of the unit to the other and shallow because one is on top of the other. A duct transition fitting is almost always necessary to connect this unit to the ductwork. The transition is normally from wide shallow duct to almost square.

Figure 38.23 (A) Rubber and cork pads may be placed under the unit to reduce vibration. (B) Some smaller pads are constructed of a single cork layer. (B) Photo by Eugene Silberstein

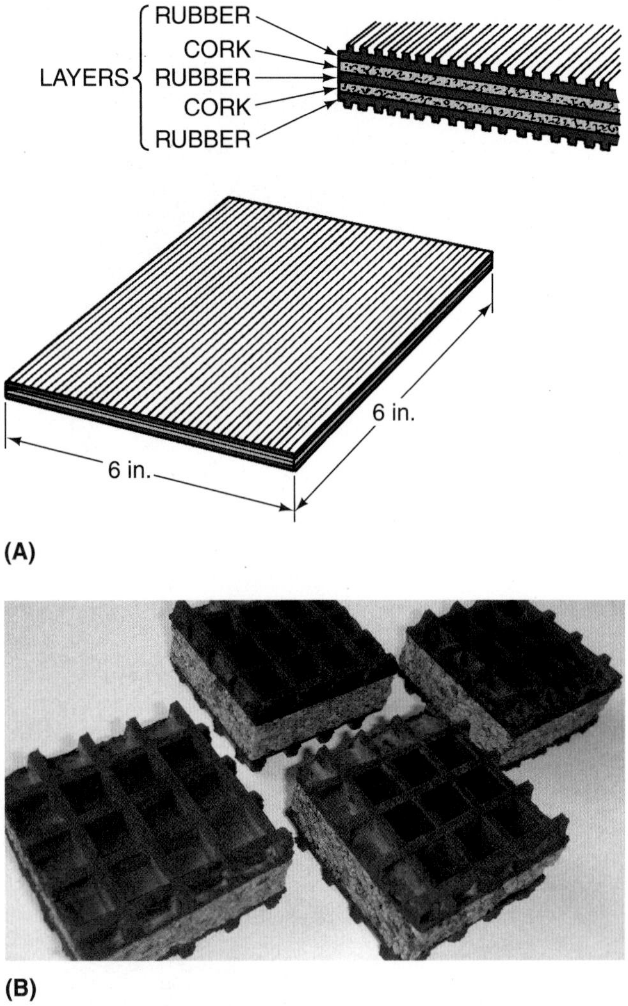

Figure 38.24 Duct connector for a rooftop installation. Courtesy Climate Control

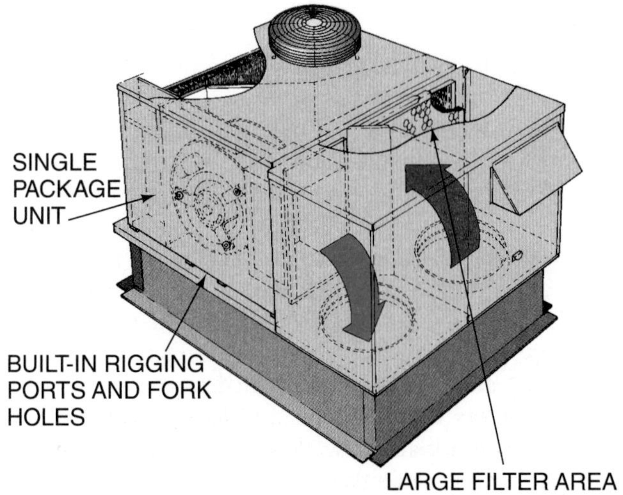

Figure 38.25 This air-to-air package unit has side-by-side duct connections. Courtesy Climate Control

Figure 38.26 Over-and-under duct connections.

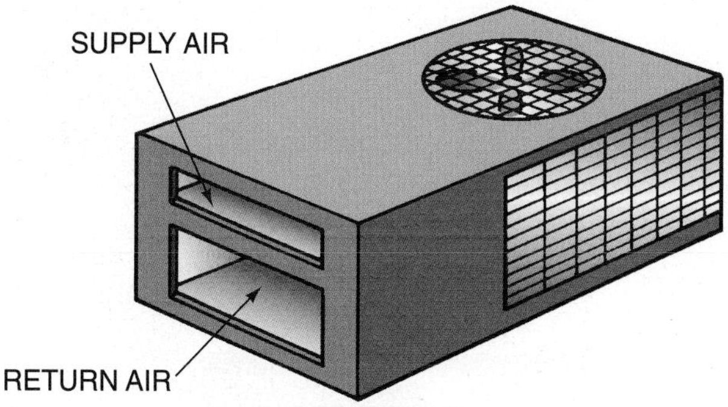

When over-and-under systems are used, the duct system may be designed to be shallow and wide to keep from complicating the duct transition. The duct connection in either style must be watertight and insulated. Some contractors cover the duct with a weatherproof hood.

At some point in time, package equipment will need replacing. The duct system usually outlasts two or three units. Manufacturers may change their equipment style from over and under to side by side. This complicates choosing new equipment because when ducting already exists for an over-and-under connection, an over-and-under unit may not be as easy to find as a side-by-side unit, **Figure 38.27**.

Package air conditioners have no refrigerant piping that needs to be field installed. The piping is assembled within the unit at the factory. Package units are charged with refrigerant at the factory and, as such, the refrigerant charge is included in the price of the equipment. Except for electrical and duct

Figure 38.27 An air-cooled package unit with over-and-under duct connections. The manufacturer has changed the design to a side-by-side duct connection, so the installing contractor has a problem.

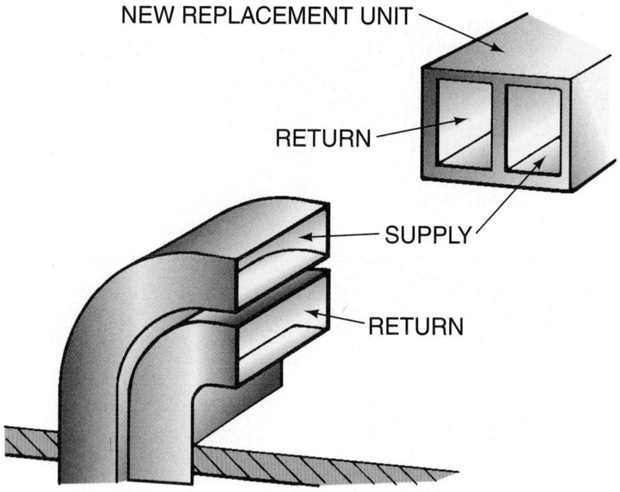

connections, the equipment is ready to start. One precaution must be observed, however. Most of these systems use refrigerants that have an affinity (attraction) for the oil in the system. The refrigerant will all be in the compressor if it is not forced out by a crankcase heater. **NOTE:** *All manufacturers that use crankcase heaters on their compressors issue a warning to leave the crankcase heater energized for some time, perhaps as long as 24 hours, before starting the compressor.* • The installing contractor must coordinate the electrical connection and start-up times closely because the unit should not be started immediately when the power is connected.

38.9 INSTALLING SPLIT-SYSTEM AIR CONDITIONERS

In split-system units, the evaporator is normally located close to the blower section regardless of whether the blower is in a furnace or in a special air handler. The air handler (blower section) and coil must be located on a solid base or suspended from a strong support. The base for the air handler in upflow and downflow equipment often may be fireproof and rigid. Some vertical-mount air-handler units have a wall-hanging support.

When a unit is installed horizontally, it may rest on the ceiling joists in an attic, **Figure 38.28**, or on a foundation of blocks or concrete in a crawl space. Be sure to check local codes about resting the horizontal unit on the ceiling joists, as not all jurisdictions allow this practice both from a mechanical code perspective and a ceiling joist supporting the additional load. The air handler may be hung from the floor or ceiling joists in different ways, **Figure 38.29**. If the air handler is hung from above, vibration isolators are often located under it to

Figure 38.28 A furnace located in an attic crawl space. See the manufacturer's instructions for suggestions. The unit may require a fireproofed base.

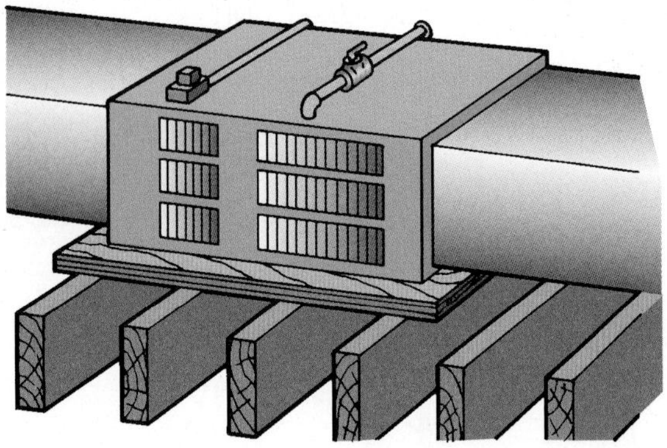

Figure 38.29 The air handler is hung from above, and the ductwork is connected and hung at the same height.

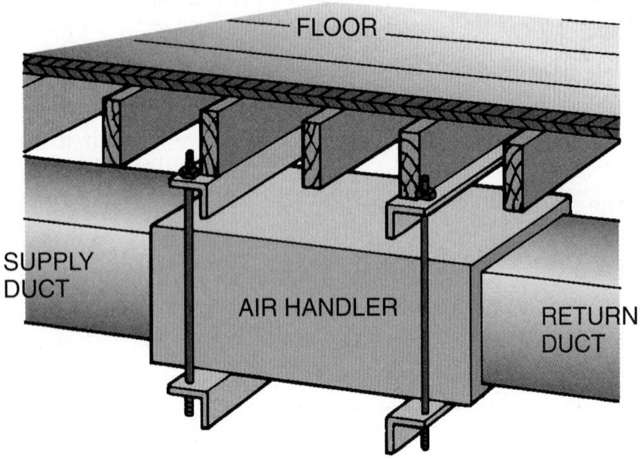

keep blower noise or vibration from being transmitted into the structure. **Figure 38.30** is a trapeze hanger using vibration isolation pads.

The air handler (fan section) should always be installed so that it will be easily accessible for future service. The air handler will always contain the blower and sometimes the controls and heat exchanger. Most manufacturers have designed their air handlers or furnaces so that, when installed vertically, they are totally accessible from the front. This works well for closet installations where there is insufficient room between the closet walls and the sidewalls of the furnace or air handler. **Figure 38.31** is an electric furnace used as an air-conditioning air handler. When a furnace or air handler is located in a crawl space, side access is important, so the air handler should be located off the ground next to the floor joists. The ductwork also will be installed off the ground, **Figure 38.32**. If the air handler is top-access, it will have to

Figure 38.30 A trapeze hanger with vibration isolation pads.

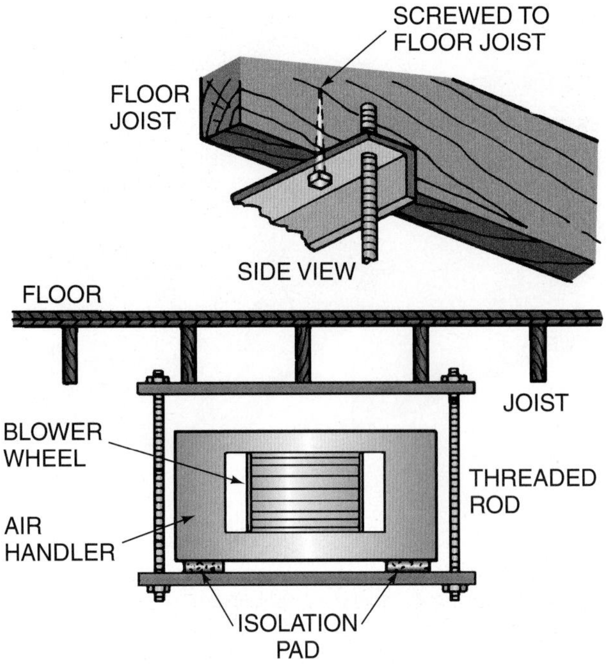

Figure 38.31 An electric furnace with an air-conditioning coil placed in the duct. Courtesy Climate Control

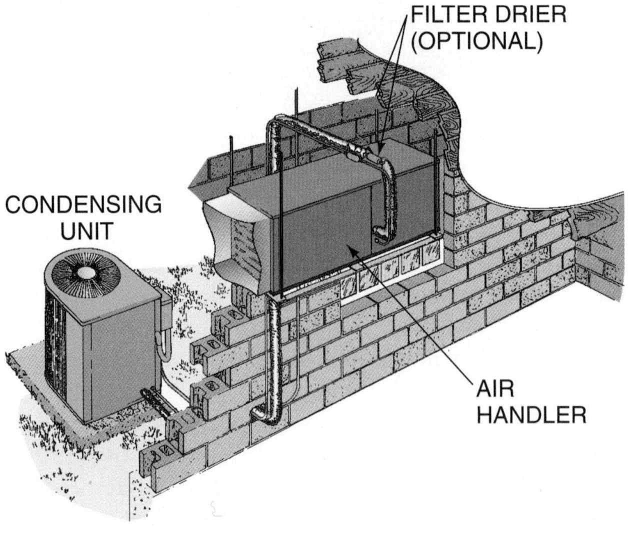

be located lower than the duct and the duct must make a transition downward to the air handler on the supply and return ends, **Figure 38.33**. The best location for a top-access air handler is in an attic crawl space where the air handler can be set on the joists. The technician can work on the unit from above, **Figure 38.34**.

Figure 38.32 A side-access air handler in a crawl space.

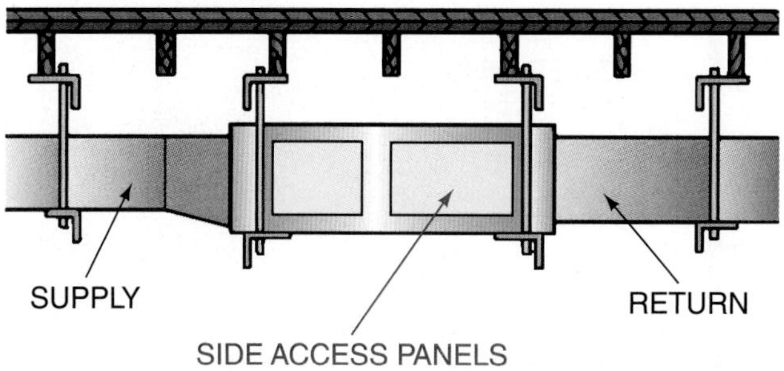

Figure 38.33 A top-access air handler.

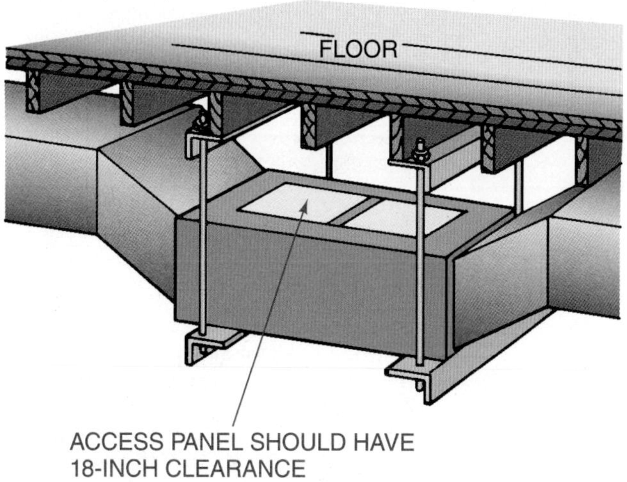

Figure 38.34 An evaporator installed in an attic crawl space.

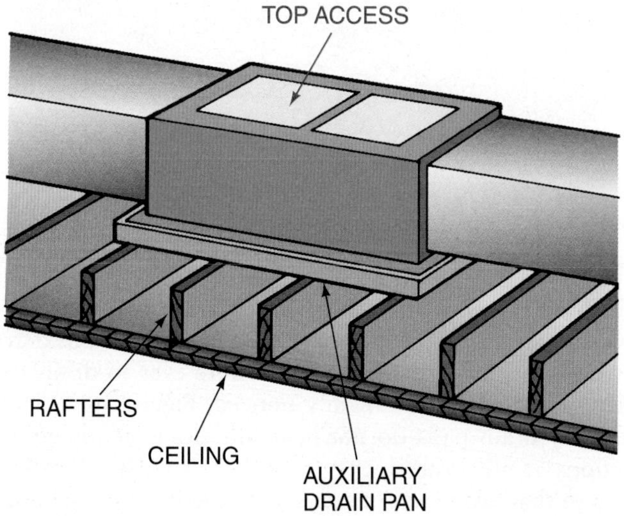

CONDENSATE DRAIN PIPING

When installing the evaporator, provisions must be made for the condensate that will be collected in the air-conditioning cycle. An air conditioner in a climate with average humidity will collect about 1 quart of condensate per hour of operation for each ton of air-conditioning. A 3-ton system would, therefore, produce about 3 quarts of condensate per hour of operation. This is approximately 18 gal in a 24-hour operating period. This can add up to a great deal of water over a period of time. If the unit is near a drain that is below the drain pan, simply pipe the condensate to the drain, **Figure 38.35**. A trap in the drain line will hold some water and keep air from pulling into the unit from the termination point of the drain. The drain may terminate in an area where foreign particles may be pulled into the drain pan. The trap will prevent this, **Figure 38.36**. If no drain is close to the unit, the condensate must be drained or pumped to another location, **Figure 38.37**. Some locations call for the condensate to be piped to a dry well. A dry well is a hole in the ground filled with stones and gravel. The condensate is drained into the well and absorbed into the ground, **Figure 38.38**. For this to be successful, the soil must be able to absorb the amount of water that the unit will collect.

When the evaporator and drain are located above the conditioned space, an auxiliary drain pan under the unit is recommended and often required, **Figure 38.39**. Airborne particles, such as dust and pollen, can get into the drains. Algae will also grow in the water in the lines, traps, and pans and may eventually plug the drain. If the drain system

Figure 38.35 Condensate piped to a drain below the evaporator drain pan.

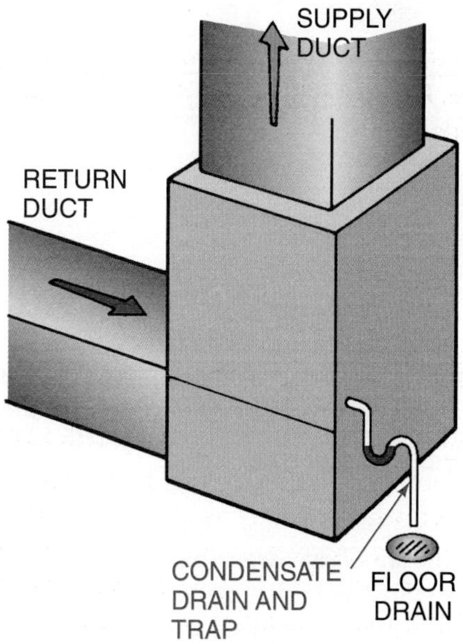

Figure 38.36 Cutaway view of a drain trap.

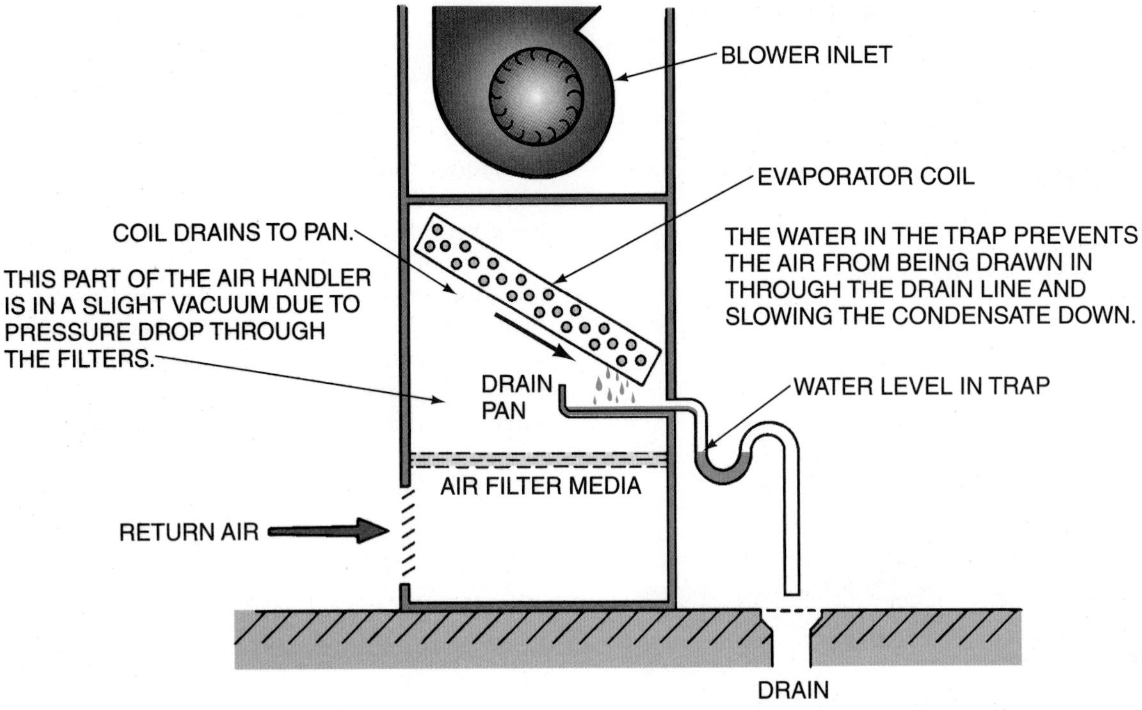

Figure 38.37 The drain in this installation is above the drain connection on the evaporator, and the condensate must be pumped to a drain at a higher level.

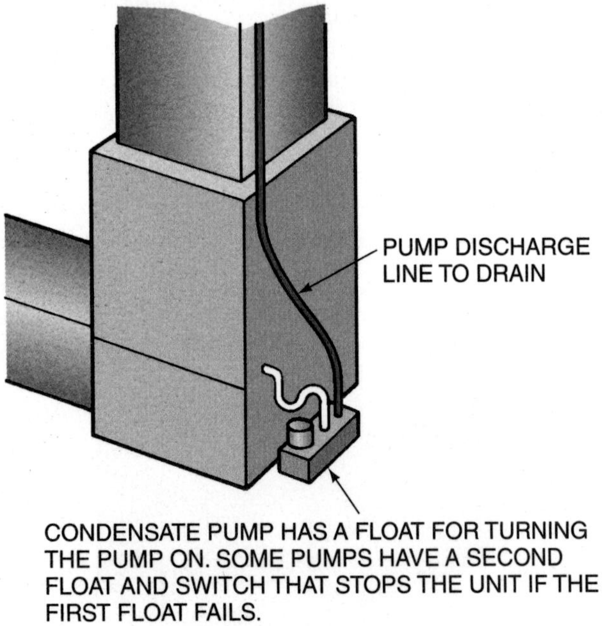

Figure 38.38 A dry well for condensate.

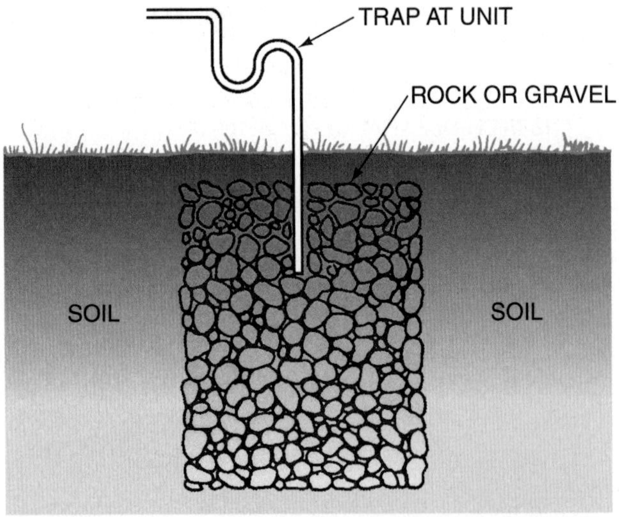

is plugged, the auxiliary drain pan will catch the overflow and keep the water from damaging whatever is below it. This auxiliary drain should be piped to a conspicuous place. The owner should be warned that if the water ever comes from this drain line, a service call is necessary. Some contractors pipe this drain to the end of the house and out next to a driveway or patio so that if water were ever to drain from this point it would be readily noticed, **Figure 38.40**. Some auxiliary drain pans do not have any drain or piping connections at all. Instead, they have float switches located in the pan that are wired in series with the low-voltage power supply. If water accumulates in the auxiliary drain pan,

Figure 38.39 An auxiliary drain-pan installation.

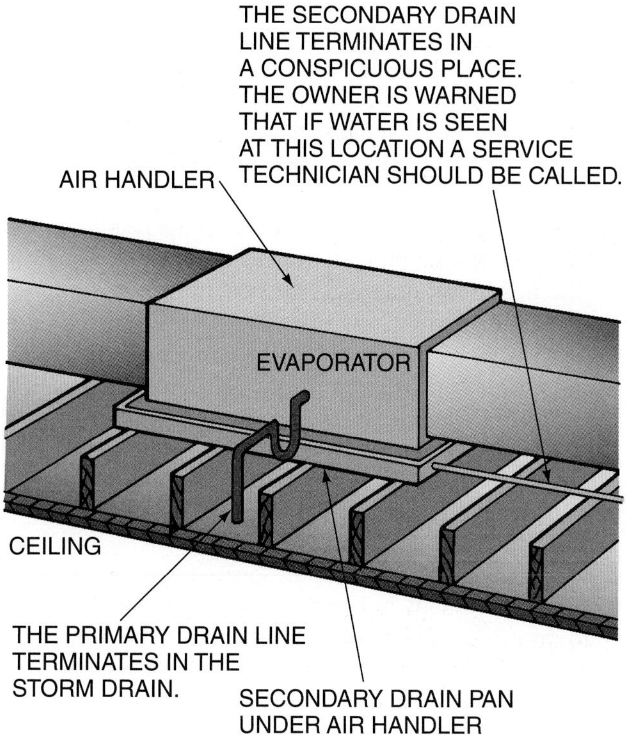

Figure 38.40 An auxiliary drain piped to the end of a house. If water is noticed at this location, the owner should alert the service technician.

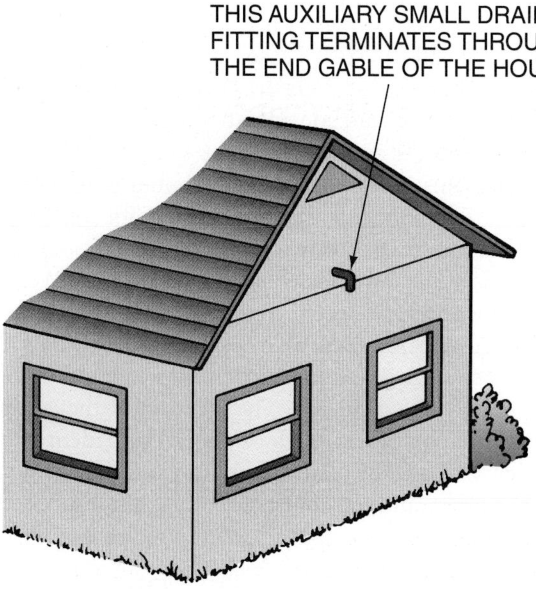

the float will rise and open the control circuit. The system will not operate until the service technician responds to the no-cooling call that will result.

38.10 THE SPLIT-SYSTEM CONDENSING UNIT

The condensing unit location is separate from that of the evaporator. The following must be considered carefully when placing a condensing unit:

1. Proper air circulation
2. Convenience for piping and electrical service
3. Future service
4. Natural water and roof drainage
5. Solar influence
6. Appearance

AIR CIRCULATION AND INSTALLATION

The unit must have adequate air circulation. The air discharge from the condenser may be from the side or from the top. Discharged air must not hit an object and circulate back through the condenser. The air leaving the condenser is warm or hot and will create high head pressure and poor operating efficiency if it passes through the condenser again. Follow the manufacturer's literature for minimum clearances for the unit, **Figure 38.41**.

ELECTRICAL AND PIPING CONSIDERATIONS

The refrigerant piping and the electrical service must be connected to the condensing unit. Piping is discussed later in this unit, but, for now, realize that the piping will be routed between the evaporator and the condenser. When the condenser is located next to a house, the piping must be routed between the house and the unit, usually behind or beside the house with the piping routed next to the ground. If the unit is placed too far from the house, the piping and electrical service become natural obstructions between the house and the unit.

Figure 38.41 The condensing unit for a split system located so that it has adequate airflow and service room. Courtesy Carrier Corp.

Children may jump on the piping and electrical conduit. If the unit is placed too close to the house, it may be difficult to remove the service panels. Before placing the unit, study the electrical and the refrigeration line connections and make them as short as practical, leaving adequate room for service.

SERVICE ACCESSIBILITY

Unit placement may determine whether a service technician gives good service or barely adequate service. The technician must be able to see what is being worked on. A unit is often placed so that the technician can touch a particular component but is unable to see it, or, by shifting positions, can see the component but not touch it. The technician should be able to both see and touch the work at the same time.

WATER DRAINAGE FROM NATURAL SOURCES

The natural drainage of the groundwater and roof water should be considered in unit placement. ⚠ **SAFETY PRECAUTION:** *The unit should not be located in a low place where groundwater will rise in the unit. If this happens, the controls may short to ground, and the wiring will be ruined.* • All units should be placed on a base pad of some sort. Concrete and high-impact plastic are commonly used. Metal frames can be used to raise the unit when needed.

The roof drainage from a structure may run off into gutters. The condensing unit should not be located where the gutter drain or roof drainage will pour down onto it. Condensing units are made to withstand rainwater but not large volumes of drainage. If it is a top-discharge unit, the drainage from above will be directly into the fan motor, **Figure 38.42**.

Figure 38.42 This unit is located improperly and allows the roof drain to pour down into the top of the unit.

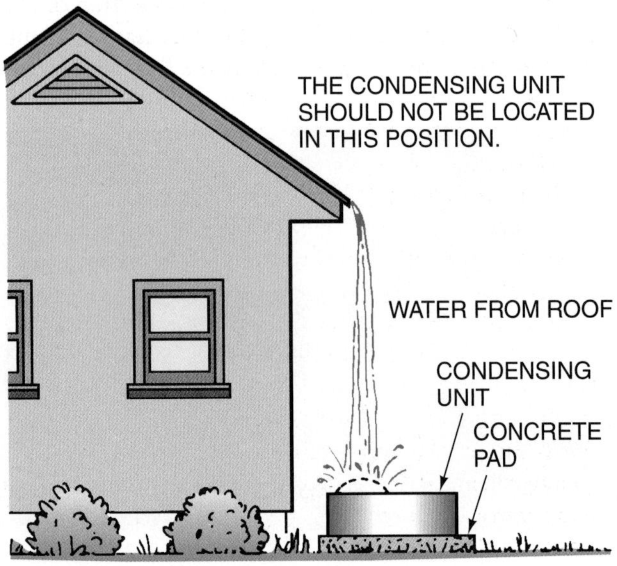

SOLAR INFLUENCE

If possible, put the condensing unit on the shady side of the house because the sun shining on the panels and coils will lower the efficiency. However, the difference is not crucial, and it would not pay to pipe the refrigerant tubing and electrical lines long distances just to keep the sun from shining on the unit. Shade helps cool the unit, but it may also cause problems. Some trees have small leaves, sap, berries, or flowers that may harm the finish of the unit. Pine needles may fall into the unit, and the pine pitch that falls on the cabinet may harm the finish and outweigh the benefit of placing the unit in the shade of many types of trees. In addition, leaves and sap may restrict airflow through the condenser coil, limiting its ability to reject heat from the air-conditioning system.

POSITIONING A CONDENSING UNIT FOR BEST APPEARANCE

The condensing unit should be located where it will not be noticeable or will not make objectionable noise. When located on the side of a house, the unit may be hidden from the street with a low shrub. If the unit is a side-discharge type, the fan discharge must be away from the shrub or the shrub may not survive. A top-discharge unit will not affect plants, but the noise of the unit will be higher, which may be objectionable in a bedroom located above the unit.

Locating the condensing unit at the back of the house may place it closer to the evaporator and would mean shorter piping, but the back of the house is the usual location of patios and porches, and the homeowner may not want to hear the unit while sitting outside. In such a case, a side location at the end of the house where there are no bedrooms may be the best choice.

Each location has its considerations. The salesperson and the technician should consult with the owner about locating the various components. The salesperson should be familiar with all local code requirements. Local codes provide guidelines, for example, with respect to how close a condensing unit can be to the property line. Issues such as this need to be investigated before the location is selected to avoid having to relocate the unit later on. A floor plan of the structure may help. Some companies use large graph paper to draw a rough floor plan to scale when estimating the job. The location of the equipment can be indicated on the rough floor plan to help identify potential problems. The floor plan can be shown to homeowners to help them understand the contractor's suggestions.

38.11 INSTALLING REFRIGERANT PIPING ON SPLIT-SYSTEMS

The refrigerant piping is always a big consideration when installing a split-system air conditioner. The choice of the piping system may make a difference in the start-up time. The piping should always be kept as short as practical. For

an air-conditioning installation, three methods are used to connect the evaporator and the condensing unit on almost all equipment under 5 tons (refrigeration systems for commercial refrigeration are not the same): (1) contractor-furnished piping; (2) flare fittings, with the manufacturer furnishing the tubing (called a line set); and (3) precharged tubing with sealed quick-connect fittings, called a precharged line set. **In all of these systems the operating charge for the system is shipped in the equipment.**

THE REFRIGERANT CHARGE

Regardless of the source of the connecting tubing, an initial refrigerant charge for the system is normally furnished by the manufacturer. The charge is shipped in the condensing unit. The manufacturer furnishes enough charge to operate the unit with a predetermined line length, typically 25 ft. Be sure to check the installation and start-up instructions that come with the equipment to learn more about the refrigerant charge and its associated refrigerant line length. The manufacturer holds and stores the refrigerant charge in the condensing unit with service valves if the tubing uses flare or compression fittings. The most common method to connect field-installed refrigerant lines is to braze them in place. Some condensing units are equipped with service valves that can accept copper tubing for brazing. Other service valves are configured so refrigerant lines can be brazed into them or, with the addition of a special fitting, be used in conjunction with flare nuts, **Figure 38.43.** *It is not legal to manufacture or import air-conditioning equipment that is precharged with either R-22 or R-142b. This law covers equipment components such as line sets, condensing units, and evaporator coils.*

THE LINE SET

Manufacturer-furnished tubing is called a *line set.* *The suction line is insulated,* and the tubing may be charged with nitrogen, contained by rubber plugs in the tube ends. When

Figure 38.43 Some valves can be used with either brazed or flared connections. Courtesy Aeroquip Corp.

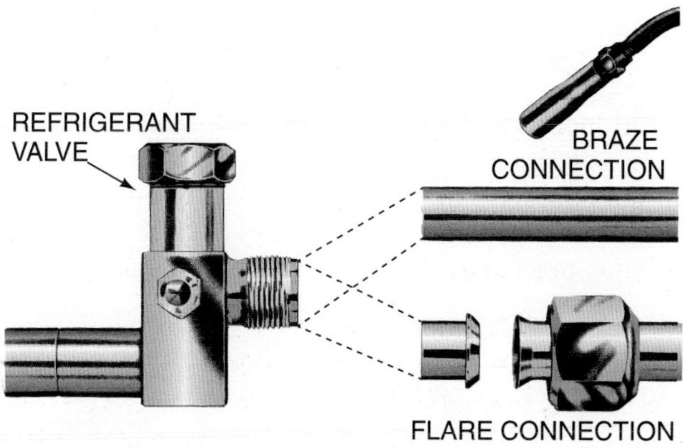

Figure 38.44 Uncoiling tubing. Photo by Bill Johnson

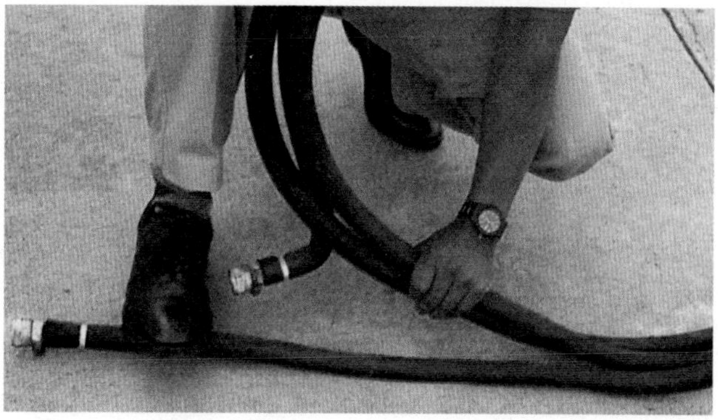

the rubber plugs are removed, the nitrogen rushes out with a loud hiss. This indicates that the tubing is not leaking. Make sure you know the type of system you are working on when installing the line set. Some ductless split systems, for example, have the metering device in the outdoor unit, so both field-installed refrigerant lines are on the low-pressure side of the system and both will need to be insulated. Uncoil tubing from the end of the coil. Place one end of the coil on the ground and unroll it while keeping your foot on the tubing end on the ground, **Figure 38.44.** **NOTE:** *Be careful not to kink the tubing when going around corners, or it may collapse. Because of the insulation on the pipe, you may not even see the kink.* •

TUBING LEAK TEST AND EVACUATION

When the piping is routed and in place, the following procedure is commonly used to make the final connections:

1. The tubing is connected to the air handler or indoor unit. Many tubing sets offer either flare fittings or a braze connection. Brazed connections are the connection of choice when done correctly. When the tubing has flare nuts, a drop of oil applied to the back of the flare will help prevent the flare nut from turning the tubing when the flare connection is tightened. The suction line may be as large as 1⅛-in. OD tubing. Large tubing connections will have to be made very tight. Two adjustable wrenches are recommended. The liquid line will not be any smaller than ¼-in. OD tubing, and it may be as large as ½-in. OD on larger systems.
2. The smaller (liquid) line is connected to the liquid-line service valve at the outdoor unit.
3. The larger (suction) line is connected to the suction-line service valve at the outdoor unit.
4. With the service valves still closed, introduce dry nitrogen to the line set to bring the pressure in the line set and evaporator up to the test pressure, as indicated by the equipment manufacturer.

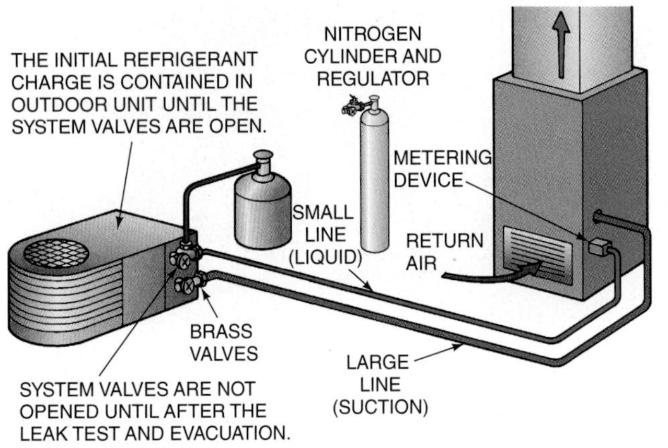

Figure 38.45 After all line set connections are made tight, the line set is leak checked using refrigerant and nitrogen. The line set is then evacuated, and the system valves are opened.

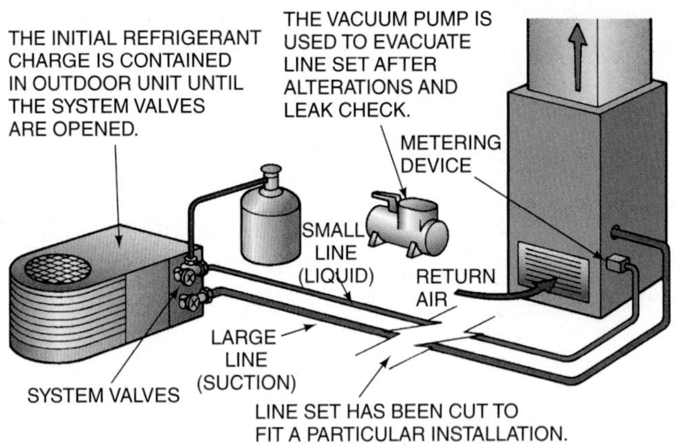

Figure 38.46 Altering the line set. After the line set has been altered, it is connected as in **Figure 38.45** and charged with nitrogen and a trace gas of system refrigerant for leak-checking purposes. After checking for leaks, the refrigerant trace gas and nitrogen are released to the atmosphere. The line set and evaporator are then evacuated. After evacuation, the system valves are opened and the system is ready to run.

5. After the leak check, you are allowed to release the nitrogen to the atmosphere. Connect the vacuum pump to the line set and evaporator portion of the system and evacuate it to a deep vacuum, **Figure 38.45**. 4R *EPA rules allow a trace of the system refrigerant to be mixed with the dry nitrogen for leak detection purposes. The refrigerant/nitrogen mixture can be released to the atmosphere as this is referred to as a de minimis release.* 4R
6. Once the system has reached and held the required vacuum level, open the valves at the condensing unit. This will release the initial factory refrigerant charge into the system. Check the manufacturer's installation manual for the correct charge for the length of line set. Follow the instructions to properly adjust the system's refrigerant charge.

Line sets come in standard lengths of 25 or 50 ft. If lengths other than these are needed, the manufacturer should be consulted. If the line length has to be changed from the standard length, the manufacturer will also recommend how to adjust the unit charge for the new line length. Most units are shipped with a charge for 25 ft of line. If the line is shortened, refrigerant must be recovered. If the line is longer than 25 ft, refrigerant must be added.

ALTERED LINE-SET LENGTHS

When line sets must be altered, they may be treated as a self-contained system of their own. The following procedures should be followed:

1. Alter the line sets as needed for the proper length. If the lines are pressurized with dry nitrogen, the nitrogen charge will escape during this alteration. **Do not remove the rubber plugs** from the ends of the lines.
2. When the alterations are complete, the lines may be pressured to about 25 psig to leak check any connections you have made. The rubber plugs should stay in place. If you desire to pressure test with higher pressures, remove the rubber plugs and connect the lines to the indoor and outdoor units. You may then safely pressurize the lines, and the indoor coil, to the factory test-pressure as indicated on the coil. The line's service port is common to the line side of the system. If the valves are not opened at this time, the line set and indoor coil may be thought of as a sealed system for pressure testing and evacuation, **Figure 38.46**.
3. After the leak test has been completed, evacuate the line set (and indoor coil, if connected). When all connections are made and checked, the system valves may be opened and the system started.

PRECHARGED LINE SETS (QUICK-CONNECT LINE SETS)

Precharged refrigerant line sets are typically encountered in refrigeration and ductless split systems. These line sets are manufactured with quick-connect fittings and are available in most of the standard lengths. The difference between line sets and precharged line sets is that the correct refrigerant operating charge is shipped in the precharged line set. Refrigerant does not have to be added or taken out unless the line set is altered. The following procedures are recommended for connecting precharged line sets with quick-connect fittings:

1. Roll the tubing out straight.
2. Determine the routing of the tubing from the evaporator to the condenser and put it in place.

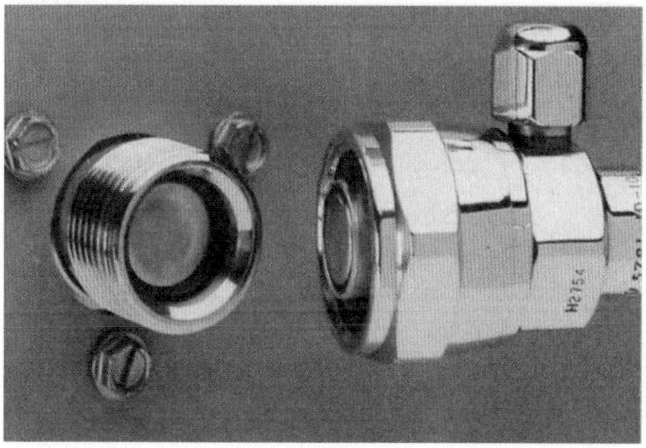

Figure 38.47 Precharged line set fittings. Courtesy Aeroquip Corp.

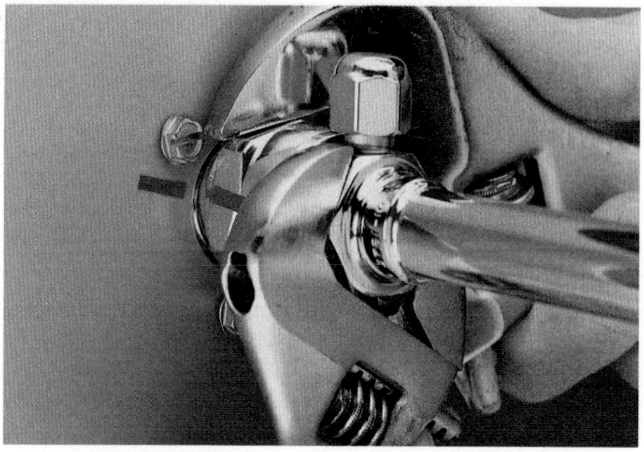

Figure 38.48 Using two wrenches to tighten fittings—one to hold the nut with the service tap on it and one to turn the movable nut next to the cabinet. Courtesy Aeroquip Corp.

3. Remove any protective plastic caps on the evaporator and line set fittings, **Figure 38.47**. Place a drop of refrigerant oil (this is sometimes furnished with the tubing set) on the neoprene O-rings of each line fitting. The O-ring is used to prevent refrigerant from leaking out while the fitting is being connected; it serves no purpose after the connection is made tight. It may tear while making the connection if it is not lubricated.
4. Start the threaded fitting and tighten hand tight. Making sure that several threads can be tightened by hand will ensure that the fitting is not cross-threaded.
5. When the fitting is tightened hand tight, finish tightening the connection with a wrench. You may hear a slight escaping of vapor while making the fitting tight; this purges the fitting of any air that may have been in the fitting. The O-ring should be seating at this time. Once you have started tightening the fitting, do not stop until you are finished. Two adjustable wrenches are recommended, **Figure 38.48**. If you stop in the middle, some of the system charge may be lost.
6. Tighten all fittings as indicated.
7. After all connections have been tightened, leak check all connections that you have made.

ALTERED PRECHARGED LINE SETS

When the lengths of quick-connect line sets are altered, the charge will also have to be altered. The line set may be treated as a self-contained system for alterations. The following is the recommended procedure:

1. 4R *Before cutting or altering the length of the line set, recover the refrigerant. Do not just cut the line because both the liquid and suction lines contain refrigerant. Recovering the refrigerant is required.* 4R
2. Cut the line set and alter the length as needed.
3. Pressure test the line set.
4. Evacuate the line set to a deep vacuum of at least 500 microns. Make certain that the deep vacuum level is reached and maintained.
5. Valve off the vacuum pump and pressurize the line set to about 10 psig (using refrigerant that is the same as the refrigerant in the system) and connect it to the system using the procedures already given.
6. Read the manufacturer's recommendation as to the amount of charge for the new line lengths and add this much refrigerant to the system. If no recommendations are provided, see **Figure 38.49** for a table of liquid-line and suction-line capacities for some commonly encountered refrigerant capacities. Of the two field-installed refrigerant lines, the liquid line contains the most refrigerant, and the proper charge should be added to that line to make up for what was lost when the line was cut. If an R-410A air-conditioning system is being installed with a 40-ft line set (½-in. liquid line and 1⅛-in. suction line) and the system's holding charge is sufficient for a 25-ft refrigerant line set, refrigerant will need to be added to the system. The amount of refrigerant to be added will be for the additional 15 ft of refrigerant line set. The amount of refrigerant will be, from **Figure 38.49**, 15 ft × (0.035 + 5.35), or 80.8 ounces of R-410A.

PIPING ADVICE

The piping practices that have been described for air-conditioning equipment are typical of most manufacturers' recommendations. **The manufacturer's recommendation should always be followed.** Each manufacturer ships installation and start-up literature inside the shipping crate. If it is not there, request it from the manufacturer or obtain it online at the manufacturer's website.

Figure 38.49 A table of liquid-line capacities.

Line Size (OD)	R-22 (Ounces per foot)		R-134a (Ounces per foot)		R-410A (Ounces per foot)		R-407C (Ounces per foot)	
L-Type	Liquid (105°F)	Suction (40°F)	Liquid (105°F)	Suction (40°F)	Liquid (105°F)	Suction (40°F)	Liquid (105°F)	Suction (40°F)
¼″	0.20	0.004	0.195	0.003	0.16	0.006	0.18	0.005
3/8″	0.56	0.012	0.57	0.009	0.47	0.018	0.52	0.014
½″	1.11	0.024	1.13	0.018	0.94	0.035	1.04	0.027
5/8″	1.78	0.040	1.81	0.027	1.51	0.058	1.67	0.043
7/8″	3.70	0.082	3.76	0.056	3.14	0.119	3.48	0.091
11/8″	6.30	0.139	6.41	0.096	5.35	0.202	5.93	0.155
13/8″	9.60	0.213	9.77	0.147	8.15	0.309	9.03	0.237
15/8″	13.58	0.301	13.82	0.208	11.54	0.435	12.78	0.334
21/8″	29.82	0.661	30.34	0.456	25.33	0.957	28.05	0.734

38.12 EQUIPMENT START-UP

The final step in installation is starting the equipment. The manufacturer will furnish start-up instructions. After the unit or units are in place, are leak checked, have the correct charge, and are wired for line and control voltage, follow these guidelines:

1. Test the condensate drain lines for proper drainage and leaks by pouring water into them. Be sure to double check the installation instructions and confirm that the condensate lines were installed according to the manufacturer's guidelines.
2. The line voltage must be connected to the unit disconnect panel. This should be done by a licensed electrician. Disconnect a low-voltage wire, such as the Y wire at the condensing unit, to prevent the compressor from starting. Turn on the line voltage. The line voltage allows the crankcase heater to heat the compressor crankcase. **NOTE:** *This applies to any unit with crankcase heaters. Heat must be applied to the compressor crankcase for the amount of time the manufacturer recommends (usually not more than 24 h). Heating the crankcase boils out any refrigerant before the compressor is started. If the compressor is started with liquid in the crankcase, some of the liquid will reach the compressor cylinders and may cause damage. The oil also will foam and provide only marginal lubrication until the system has run for some time.* •
3. It is normally a good idea to plan on energizing the crankcase heater in the afternoon and starting the unit the next day.
4. If the system has service valves, open them. **NOTE:** *Some units have valves that do not have back seats. Do not try to back seat these valves, or you may damage the valve.* • Open the valve until resistance is felt, then stop, **Figure 38.50**.

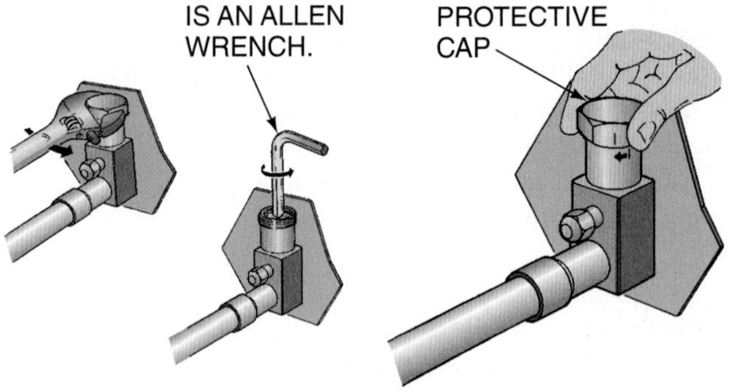

Figure 38.50 Service valves furnished with equipment may not have a "back seat." Do not try to turn the valve system all the way out. Courtesy Aeroquip Corp.

5. Check the power at the indoor and outdoor units and make certain that the voltage is within the recommended limits.
6. Check all electrical connections, including those made at the factory, and ensure that they are tight and secure. **SAFETY PRECAUTION:** *Turn the power off when checking electrical connections.* •
7. Set the fan switch on the room thermostat to the ON position and check the indoor blower for proper operation and current draw. You should feel air at all registers. Make sure that no air blockages exist at the supply and return openings.
8. Turn the fan switch to the AUTO position and, with the HEAT-OFF-COOL selector switch to the OFF position, replace the Y wire connection at the condensing unit. **SAFETY PRECAUTION:** *The power at the unit is turned off and the panel is locked and tagged while making this connection.* •

9. Place your clamp-on ammeter around the common wire to the compressor. Move the HEAT-OFF-COOL selector switch on the space thermostat to COOL and adjust the temperature setting to a temperature that is lower than the temperature in the conditioned space. The compressor should start.
10. In most cases, the manufacturer will provide detailed information regarding the procedures to follow to ensure that the system is properly charged. Refer to Unit 10, "System Charging," for details regarding proper system charging methods. Regardless of the manufacturer, evaluating the refrigerant charge in an air-conditioning system relies on a number of factors that might include:

    - Outside ambient temperature
    - Return air temperature (wet-bulb and/or dry-bulb)
    - Condenser saturation temperature
    - Evaporator saturation temperature
    - Evaporator superheat
    - Condenser subcooling

New generation field tools and instruments make the taking and evaluating of system readings much easier. For example, a set of sensors, **Figure 38.51**, can be used to monitor system pressures and saturation temperatures, as well as evaporator superheat and condenser subcooling. This particular set contains two pressure sensors, one for the high side and one for the low side, and two temperature sensors, one for the high side and one for the low side. The pressure sensors are connected to the service ports on the system, **Figure 38.52**, and the temperature sensors are simply clamped onto the suction and liquid lines of the system, **Figure 38.53**.

Once the four sensors have been secured, the system's operating pressures and temperatures, as well as superheat and subcooling, can be obtained from a smart phone or other compatible device, **Figure 38.54**. Once the application has

Figure 38.51 New generation, wireless pressure and temperature sensors. Instrumentation provided by Sporlan. Photo by Eugene Silberstein

Figure 38.52 Pressure sensors connected to system. Instrumentation provided by Sporlan. Photo by Eugene Silberstein

Figure 38.53 Pressure and temperature sensors connected to system. Instrumentation provided by Sporlan. Photo by Eugene Silberstein

Figure 38.54 Using the smart phone app, a technician can obtain system pressure and temperature information, in addition to evaporator superheat and condenser subcooling. Instrumentation provided by Sporlan. Photo by Eugene Silberstein

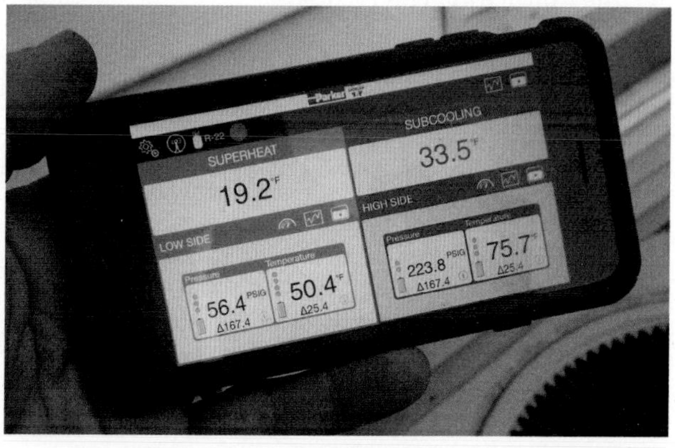

been launched, the technician can select the correct refrigerant for the system in order to ensure proper readings, **Figure 38.55.**

The amperage of the compressor is a good indicator of system performance. If the outside temperature is hot and there is a significant heat load on the evaporator, the compressor will be pumping at near capacity. The amperage draw of the compressor motor will be near nameplate amperage. It is rare for the compressor amperage to be higher than the nameplate rating due only to the system load. It is also normal for the compressor amperage to be slightly below the nameplate rating.

⚠ SAFETY PRECAUTION: *Before starting the system, examine all electrical connections and moving parts. The equipment may have faults and defects that can be harmful (e.g., a loose pulley may fly off when the motor starts to turn). Remember that vessels and hoses are under pressure. Always be mindful of potential electrical shock.* •

Figure 38.55 Selecting the proper refrigerant for the system being worked on. Instrumentation provided by Sporlan. Photo by Eugene Silberstein

REVIEW QUESTIONS

1. Ductboard duct sections are fabricated from which material?
   A. Fiberglass
   B. Plastic
   C. Sheet metal
   D. Aluminum
2. Square duct or rectangular duct is assembled with
   A. mastic and drive clips.
   B. foil tape and mastic.
   C. "S" fasteners and drive clips.
   D. "S" fasteners and zip ties.
3. To help create an airtight seal between duct sections
   A. mastic may be applied.
   B. self-tapping sheet metal screws should be used.
   C. duct insulation can be used.
   D. additional duct supports should be installed.
4. __________ installed next to the blower section will help prevent the transmission of noise down the duct.
   A. Duct insulation
   B. Vibration eliminators
   C. Threaded rods
   D. Mastic
5. When connecting sections of round duct
   A. canvas collars should be used.
   B. flexible connectors are used.
   C. "S" fasteners and drive clips are used.
   D. a minimum of three screws should be used.
6. Acoustical lining for metal ducts is normally made of
   A. aluminum.
   B. mastic or adhesive.
   C. coated fiberglass.
   D. a flexible rubber or canvas material.
7. When joining two sections of fiberglass ductboard
   A. "S" fasteners and drive clips are used.
   B. staples and tape are used.
   C. a minimum of three screws should be used.
   D. glue or spot welds are used.
8. Flexible duct used at the end of a metal duct run will
   A. help reduce noise transmission in the ducts.
   B. greatly reduce airflow and should be avoided.
   C. cause excessive friction losses.
   D. cause the duct to collapse and reduce its inner diameter.
9. The applied voltage (the voltage that the unit will actually be using) should be within ±_____% of the rated voltage of the unit.
   A. 3
   B. 5
   C. 10
   D. 15
10. The standard wire size for light-duty low-voltage wire is
    A. 20 gauge.
    B. 18 gauge.
    C. 16 gauge.
    D. 14 gauge.

11. The ________ provides installation standards, wire sizes, methods of routing wires, and types of enclosures for wiring and disconnects.
   A. National Fire Protection Association (NFPA)
   B. International Fuel Gas Code (IFGC)
   C. U.S. Department of Energy (DOE)
   D. National Electrical Code (NEC)
12. Two common methods for preventing the vibration of a package system from being transferred to the building or structure are
   A. rubber and cork pads and spring isolators.
   B. using fiberglass ductboard and flex duct.
   C. canvas connectors and duct insulation.
   D. a roof curb and rigging holes.
13. Package equipment comes in two different duct connection configurations,
   A. upflow and downflow.
   B. metal connection and flex connection.
   C. side-by-side and over-and-under.
   D. rooftop connection and sidewall connection.
14. On a new package system, the refrigerant charge is
   A. added in the field, after installation.
   B. very large compared to a spilt-system.
   C. calculated based the length of the field-installed piping.
   D. added to the system at the factory.
15. When insulation on the inside of the duct comes loose it can cause
   A. the system to shut off.
   B. high head pressure.
   C. increased airflow.
   D. reduced airflow.
16. An air conditioner in a climate with average humidity will collect about ______ of condensate per hour of operation for each ton of air-conditioning.
   A. 1 quart
   B. 3 gallons
   C. 9 pints
   D. 24 gallons
17. A float switch, located in an auxiliary drain pan, is typically wired to
   A. cycle both the evaporator and condenser fans.
   B. open the low-voltage control circuit.
   C. control a condensate pump.
   D. open the line voltage circuit to the air handler.
18. If the hot air being discharged from the condenser is recirculated back through the condenser
   A. the condenser coil will require more frequent cleaning.
   B. the condensing unit will not drain properly.
   C. high head pressure and poor operating efficiency will result.
   D. low suction pressure and higher system capacity will result.
19. When installing the condensing unit of a split-system, it should be
   A. located out of sight, such as under a deck.
   B. set on a concrete or high-impact plastic pad.
   C. lifted off the ground 24 in. to 48 in.
   D. located near a municipal water source for cleaning.
20. On a line set used to install a split-system, the ______ line is insulated.
   A. smaller
   B. compressor discharge
   C. liquid
   D. suction
21. The condenser of a split-system typically contains enough refrigerant for ____ feet of line set.
   A. 10
   B. 25
   C. 35
   D. 50
22. After installing the refrigerant lines on a split-system, the next step would be to
   A. pressurize the system with nitrogen for a leak test.
   B. pull the system down into a vacuum.
   C. add or remove refrigerant based on the length of the line set.
   D. open the service valves to allow the stored refrigerant into the system.
23. Before a newly installed system is started up, it is recommended that
   A. the factory Schrader pins be replaced with new ones.
   B. the condenser be cleaned with a foaming cleanser.
   C. the crankcase heater be allowed to operate for a specified period of time.
   D. the service valves are returned to their closed position.
24. Line sets typically come in standard lengths of
   A. 10 and 20 feet.
   B. 25 and 35 feet.
   C. 15 and 25 feet.
   D. 25 and 50 feet.
25. What are the differences between a package system and a split system?
26. Name at least three criteria for a good location for an air-cooled condensing unit.
27. When evaluating the refrigerant charge in a system, what factors would a technician take into account?
28. List the three crafts that are involved in the installation of an air-conditioning system.

UNIT 39

RESIDENTIAL ENERGY AUDITING

OBJECTIVES

After studying this unit, you should be able to

- list four organizations that offer energy audit certifications.
- describe the three broad categories of a residential energy audit.
- discuss the meaning of thermal boundaries, thermal bypass, and intermediate zones.
- list and briefly explain the five diagnostic tests of a residential energy audit.
- discuss health and safety issues as they relate to indoor air quality for energy audits.
- describe infrared scanning using a thermal imaging camera as it applies to auditing.
- discuss how sealing air leaks in a residential structure will help save energy.
- discuss how a duct leakage test is performed.
- briefly describe the differences between total duct air leakage tests and duct leakage tests to the outdoors.
- explain what is meant by a blower door test.
- explain what is meant by perfect, complete, and incomplete combustion and discuss the characteristics of each.
- discuss combustion analysis and how it is related to combustion efficiency.
- explain the importance of draft as it applies to combustion appliances.
- discuss the importance and safety issues of properly venting combustion appliances.
- describe three different flame safeguard controls and their importance for the health and safety of the home's residents.
- discuss standard-, mid-, and high-efficiency furnace design and operation.
- describe routine maintenance typically performed on HVACR systems.
- state the importance of numerical analysis and reporting for residential energy audits.
- describe the difference between base load and seasonal usage of energy as it applies to a home.
- discuss what is meant by a home's energy index.

SAFETY CHECKLIST

✔ Blower door testing may cause some safety problems within the structure especially with combustion appliances. The negative pressure or vacuum of 50 Pascals may cause combustion appliances to backdraft. It is important to shut off all combustion appliances when performing a blower door test.

✔ The auditor must also identify all thermal boundaries, conditioned zones, and combustion appliance zones (CAZ) in the house, close all exterior windows and doors, and open all interior doors before performing a blower door test.

✔ Return duct air leaks within a combustion appliance zone (CAZ) need to be sealed. If not sealed, the CAZ can become depressurized and cause the combustion appliances to backdraft. Backdrafting can cause harmful combustion by-products to enter the conditioned space and be potentially dangerous to humans.

✔ The energy auditor must anticipate any safety issues before performing a blower door test by turning off all combustion appliances, inspecting all sources of pollutants like fireplaces or wood stoves, and inspecting for any large air leaks that may prevent the blower door from achieving the desired vacuum level. The auditor must plug these openings temporarily.

✔ Because it may contain asbestos, vermiculite insulation poses a health hazard when performing a blower door test on a structure. It is strongly recommended not to perform a blower door test that draws a negative pressure on the home if this type of insulation insulates any part of the structure. Blower door tests that put a positive pressure on the structure to check for air exfiltration can be used.

✔ Recessed light fixtures (can lights) can be a major source of heat loss, but you need to be careful how close you place insulation next to a fixture unless it is marked (IC)—designed for direct insulation contact. Otherwise, a fire hazard exists. Recessed light fixtures are also rated for insulation contact and air tightness (ICAT). Check your local building codes for recommendations.

39.1 INTRODUCTION

Residential (home) energy audits not only provide homeowners with an opportunity to lower their monthly utility bills, they also can help our nation conserve energy. The conservation of natural resources is an added attraction for many homeowners to have an energy audit performed on their home. Homeowners can cut utility bills by 15% or more by following energy audit recommendations. However, before energy conservation and savings can occur, energy audit recommendations must be followed. It is important that we all do our part to make this world a better world to live in.

According to the United States Department of Energy (DOE), 55% of the energy used by the average home goes toward space heating and space cooling. The U.S. Department of Energy also says that a home's heating and cooling costs can be reduced by as much as 30% through proper insulation and air-sealing techniques. The DOE further

states that 12% of a home's total energy bill usually goes toward heating water for household use. Federal help to pay for these modifications may be available through state and local utilities or from low-income housing programs. Check with local utilities or city or the DOE-sponsored Database of State Incentives for Renewables and Efficiency (DSIRE).

Much research has been conducted by the U.S. Department of Energy's Building America and Home Performance with ENERGY STAR, sponsored by the U.S Environmental Protection Agency (EPA) and the DOE. Home Performance with ENERGY STAR offers a comprehensive, whole-house approach to improving the energy efficiency and comfort of existing homes and the testing of combustion products. The DOE's Building America program has worked with some of the nation's leading building scientists and, at the time of this writing, more than 1.5 million homes have earned the ENERGY STAR label. Building America research applies building science to the goal of achieving efficient, comfortable, healthy, and durable homes.

There are a number of organizations that offer certifications for home auditors and contractors including:

1. Association for Energy Engineers (AEE)
2. Building Performance Institute (BPI)
3. Green Mechanical Council
4. Residential Energy Services Network (RESNET)

BPI offers the Building Analyst certification and RESNET offers the HERS Rater certification. Historically, BPI certification has focused on understanding the building science of retrofitting existing homes; RESNET has focused on building science in new home construction. BPI is a nonprofit organization that accredits auditors, contractors, and other building professionals. Auditors and building analysts specialize in evaluating building systems and potential energy savings in homes.

A certified BPI Building Analyst energy auditor has passed both written and field exams, and must recertify every three years. Contractors learn about building systems and are trained to install energy-efficient measures. The Green Mechanical Council offers a certification series which includes two written exams and a field or performance based examination. The written examinations includes a Load Calculation Analyst exam as well as a Residential Energy Analyst exam. The performance or field exam requires observation and validation that the candidate can apply their knowledge of energy audit tasks in the field. The Green Mechanical Certification is most often utilized in conjunction with a training curriculum developed by HVACR Excellence. A certified RESNET energy auditor is called a HERS Rater. HERS is an acronym for the Home Energy Rating System. A written exam and field observation along with continuing education is required to become a HERS Rater. HERS provides a mile-per-gallon type of rating for expected energy consumption in homes based on computer models. Each home receives a score that can be compared with other new or existing homes. The lower the score, the more energy efficient the home is. An easy way to find an energy contractor is through a national or regional retrofit program. One such program is Home Performance with ENERGY STAR. This program is a national program from the EPA and the DOE that promotes a comprehensive, whole-house approach to energy-efficiency improvements.

Green awareness, climate change, energy efficiency, energy savings, sustainability, and high-performance buildings are on the environmental and energy forefront. Energy audits of residential, commercial, or industrial buildings have become an integral part of evaluating and assessing an existing building's energy performance. Higher efficiency standards have been established for the energy performance of new buildings, and higher levels of training and certification have been developed for auditors, contractors, and HVACR technicians to meet the needs of more sophisticated, energy-efficient buildings and HVACR equipment. This unit will cover residential energy audits, which are also referred to as home energy audits.

39.2 RESIDENTIAL (HOME) ENERGY AUDITING

More HVACR contractors are offering home auditing services because of its direct impact on a home's heating and cooling costs. Once the home audit is completed, they perform weatherization techniques to improve the energy efficiency of the residential structure. The terms *residential energy audit* and *home energy audit* are often used interchangeably. A residential energy audit can be defined as the steps taken to increase the homeowner's comfort, safety, and health and the life of the home, while conserving energy and saving the homeowner money. Audits also are environmentally beneficial because they decrease environmental pollution. The home energy audit can be divided into three broad categories:

1. Screening, visual inspections, and measurements
2. Diagnostic testing
3. Numerical analysis with reporting

The depth of the energy audit will depend on the homeowner's commitment to energy and cost savings, the qualifications and capabilities of the energy auditor, and what program, if any, is requiring certain energy-saving parameters to be tested, quantified, and reported. However, a comprehensive residential energy audit generally consists of the following:

1. Screening is a preliminary evaluation of the homeowner's needs. It can involve gathering information by phone calls or completing an Internet-based pre-audit by homeowners themselves. During this phase, a walk-through visual inspection of the residential structure to diagram and record the square footages and volumes of all rooms is performed. Visual inspections should include both the interior and exterior of the residence. HVACR systems should also be visually inspected for proper installation

and safety concerns. The walk-through visual phase should identify any health and safety issues relating to asbestos, fiberglass, radon, mold, venting, ventilation, or the like. (Please refer to Unit 34, "Indoor Air Quality," for more detailed information on indoor air pollutants and safety.) Especially when the house and its HVACR duct system may be pressurized and/or depressurized when performing parts of the testing phase of the energy audit, safety issues are of major concern. The testing phase of the energy audit can also alter the operation of the HVACR equipment. Moisture levels, ventilation rates, and indoor air quality within a residence can also be altered if proposed changes or energy conservation measures (ECMs) recommended by the auditor's report are adopted.

2. Diagnostic testing and evaluation of the home and its systems is perhaps the most important phase of a residential audit. Systems include the building envelope, lighting, indoor air quality, water heating, and the HVACR systems. The inspection and evaluation should be done by a qualified, reputable expert in residential energy auditing. A blower door test on the entire structure in conjunction with an infrared inspection using a thermal imaging camera is often performed in this phase of the energy audit. Also employed are duct pressurization tests with a duct blower to test for leaky air ducts. Combustion efficiency tests for safety and energy savings are also performed at this diagnostic phase, along with HVACR diagnostics, which include refrigerant charge checks, duct airflow and pressure measurements, and specific temperature measurements.
3. Numerical analysis and review of the home's energy-consumption history, which will include both electrical and heating fuel bills, is an important part of a residential audit, as well as comparing the energy-consumption history of the audited home to other homes of similar size. Looking for any patterns of energy consumption and any opportunities to improve energy efficiencies is very important.

The findings should be analyzed and reported both manually and with reputable computer analysis simulation software in a detailed written report to the homeowner. The report should include a numerical analysis on where energy is being used, lost, or wasted and how to fix any energy-related problems. The report should also suggest any human behavioral changes that will help with energy conservation measures (ECMs). In addition, the report should include recommendations for fixes or changes that will benefit the homeowner by saving the most money and/or energy. The report should also educate the homeowner on how to maximize the payoffs from the energy audit's findings by tapping potential financial support from federal, state, and local sources and/or utility companies. It should include information on tax incentives, grants, and rebates that may help cover the cost of the audit.

39.3 PERFORMING A HOME ENERGY AUDIT

SCREENING, VISUAL INSPECTIONS, AND MEASUREMENT

Performing a home energy audit starts with a visual inspection of a home and its energy systems, which include:

- the building envelope,
- HVACR systems,
- base load inspection, and
- indoor air quality.

VISUAL INSPECTION OF THE BUILDING

The age and type of construction of the residential structure must be identified. A visual inspection of the home's exterior should be the energy auditor's first task. The exterior visual inspection gives the auditor a general idea of the condition of the home's exterior and what construction materials make up its shell, or envelope. How much heat flows through the shell of the house will depend on three factors:

1. Any solar heat gains depend on the area and color of the shell's external surfaces and the solar transmittance and area of the windows.
2. It is the thermal resistance of the insulation and the surface area of the shell of the house that will determine how much heat is transmitted through the shell.
3. Air leakage through the shell will be determined by the pressure difference between the inside and outside of the shell and the size of the holes or cracks where the air is leaking.

Sealing and insulating the envelope or shell of a home is often the most cost-effective way to improve energy efficiency and comfort. Air leaks and insulation issues often go unnoticed simply because they cannot be seen unless a thermal imaging camera and chemical smoke generators are used. Thermal imaging photography and the use of chemical smoke generators will be covered in detail later in this unit.

The condition and construction materials of the doors, windows, roof, siding, and any overhangs should be inspected. Modern multipane windows with vinyl, wood, or fiberglass frames far outperform old metal-frame, single-pane models in terms of air leakage. All windows should be inspected for type and condition. Check to see if they are shaded by trees, overhangs, or awnings that may decrease the solar gain on the structure. Windows should be inspected for low emittance (low-e) factors. (Low emittance is also referred to as low emissivity factor.) Any shading factors on the roof should be observed as well as any reflectivity factors and insulation qualities. (Please refer to Unit 42, "Heat Gains and Heat Losses in Structures," for more detailed information on

heat gains and heat loss through residential structures.) Site issues are also noted in this phase of the audit, such as roofs and windows shaded by trees, awnings, overhangs, or other structures. What direction each side of the home faces should be noted, because that may affect solar heat gains. The foundation of the house should also be visually inspected in this phase of the audit—its condition, how much of the foundation is exposed directly to the outside air, and any drainage problems or existing moisture stains. The condition of all chimneys and exhaust vents should also be inspected.

Once the exterior of the home has been inspected, the auditor moves to the interior of the residential structure. Measurements made with a laser distance meter are used to determine square footage and volume calculations of each room in the house. All walls, ceilings, attics, floors, and foundations are inspected for insulation, and the type and thickness of the insulation are observed. *First, the insulation in the attic, ceilings, exterior and basement walls, floors, and crawl spaces is checked to see if it meets the levels recommended for the area. Insulation is measured in R-values—the higher the R-value, the better the walls and roof will resist the transfer of heat.* DOE recommends ranges of R-values based on local heating and cooling costs and climate conditions in different areas of the nation. The map and chart in **Figures 39.1** and **39.2**, respectively, show the DOE recommendations for different geographical areas. State and local code minimum insulation requirements may be less than the DOE recommendations, which are based on cost-effectiveness. For more customized insulation recommendations, a Zip Code Insulation Calculator tool can provide insulation levels for new or existing homes based on zip code and other basic information about the home.

INSULATION TIPS

Although insulation can be made from a variety of materials, it usually comes in four types; each type has different characteristics.

1. **Rolls and batts, or blankets**—flexible products made from mineral fibers, such as fiberglass and rock wool. They are available in widths suited to standard spacings

Figure 39.1 U.S. Department of Energy's specific zone maps of the United States used in conjunction with Figure 39.2 for recommended total R-values for new wood-framed houses in various zones. Courtesy U.S. Department of Energy

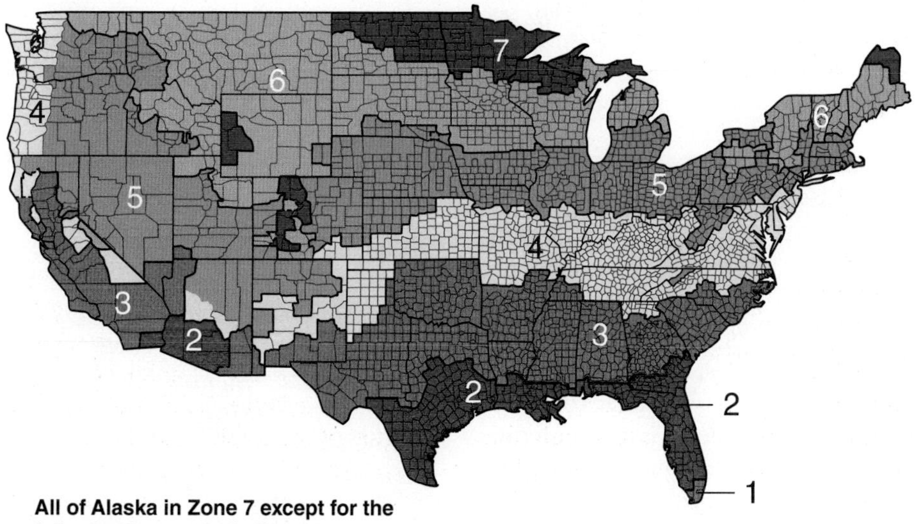

All of Alaska in Zone 7 except for the following boroughs in Zone 8:

- Bethel
- Northwest Arctic
- Dellingham
- Southeast Fairbanks
- Fairbanks N. Star
- Wade Hampton
- Nome
- Yukon-Koyukuk
- North Slope

Zone 1 includes

- Hawaii
- Guam
- Puerto Rico
- Virgin Islands

Figure 39.2 U.S. Department of Energy's recommended total R-values for new wood-framed houses in specific zones and recommended combinations of sheathing and cavity insulation. Courtesy U.S. Department of Energy

ZONE	GAS	HEAT PUMP	FUEL OIL	ELECTRIC FURNACE	ATTIC	CATHEDRAL CEILING	WALL		FLOOR
							CAVITY	INSULATION SHEATHING	
1	✓	✓	✓	✓	R30 to R49	R22 to R38	R13 to R15	None	R13
2	✓	✓	✓		R30 to R60	R22 to R38	R13 to R15	None	R13
2				✓	R30 to R60	R22 to R98	R13 to R15	None	R19 to R25
3	✓	✓	✓		R30 to R60	R22 to R38	R13 to R15	None	R25
3				✓	R30 to R60	R22 to R38	R13 to R15	R2.5 to R5	R25
4	✓	✓	✓		R38 to R60	R30 to R38	R13 to R15	R2.5 to R6	R25-R30
4				✓	R38 to R60	R30 to R38	R13 to R15	R5 to R6	R25-R30
5	✓	✓	✓		R38 to R60	R30 to R38	R13 to R15	R2.5 to R6	R25-R30
5				✓	R38 to R60	R30 to R60	R13 to R21	R5 to R6	R25-R30
6	✓	✓	✓	✓	R49 to R60	R30 to R60	R13 to R21	R5 to R6	R25-R30
7	✓	✓	✓	✓	R49 to R60	R30 to R60	R13 to R21	R5 to R6	R25-R30
8	✓	✓	✓	✓	R49 to R60	R30 to R60	R13 to R21	R5 to R6	R25-R30

of wall studs and attic or floor joists: 2 × 4 walls can hold R-13 or R-15 batts; 2 × 6 walls can have R-19 or R-21 products.

2. **Loose-fill insulation**—usually made of fiberglass, rock wool, or cellulose in the form of loose fibers or fiber pellets; it should be blown into spaces using special pneumatic equipment. The blown-in material conforms readily to building cavities and attics. Therefore, loose-fill insulation is well suited for places where it is difficult to install other types of insulation.
3. **Rigid foam insulation**—typically more expensive than fiber insulation, but very effective in buildings with space limitations and where higher R-values are needed. Foam insulation R-values range from R-4 to R-6.5 per inch of thickness, which is up to 2 times greater than most other insulating materials of the same thickness.
4. **Foam-in-place insulation**—can be blown into walls and reduces air leakage, if blown into cracks, such as around window and door frames.

SAFETY PRECAUTION: *Because some vermiculite insulation may contain asbestos, it may pose a health hazard.*•

- *Consider factors such as the climate, building design, and budget when selecting insulation R-values for a home.*
- *Use higher-density insulation, such as rigid foam boards, in cathedral ceilings and on exterior walls.*
- *Ventilation helps with moisture control and with reducing summer cooling bills. Attic vents can be installed along the entire ceiling cavity to help ensure proper airflow from the soffit to the attic to make a home more comfortable and energy efficient. Do not ventilate an attic if there is insulation on the underside of the roof. Check with a qualified contractor.*
- *Recessed light fixtures (can lights) can be a major source of heat loss, but care needs to be taken when placing insulation next to a fixture unless it is marked IC—designed for direct insulation contact. Another option is to replace the recessed fixtures with ones that are rated for insulation contact and air tightness (ICAT),* **Figure 39.3(A)**. *Local building codes should be checked for recommendations.*
- *One of the most cost-effective ways to make a home more comfortable year-round is to add insulation to the attic,* **Figure 39.3(B)**.

Figure 39.3 (A) Replace old, leaky can fixtures with insulated, airtight recessed light fixtures and caulk them where the house meets the drywall. Courtesy U.S. Department of Energy

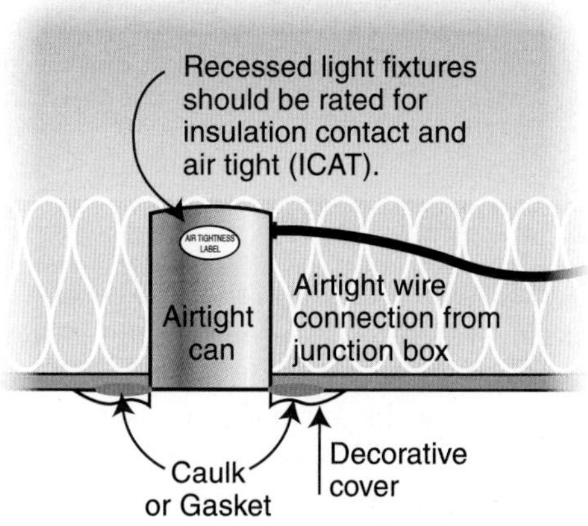

Figure 39.3 (B) A well-insulated attic. Courtesy McIntyre Builders, Rockford, MI

SAFETY PRECAUTION: *Follow the installation instructions as specified on the product packaging and wear the proper protective gear when installing insulation.•*

Adding insulation to an attic is relatively easy and very cost-effective. To find out if there is enough attic insulation, measure its thickness. If it is less than R-30 (11 in. of fiberglass or rock wool or 8 in. of cellulose), there is probably a benefit from adding more. Most U.S. homes should have between R-30 and R-60 insulation in the attic, including the attic trap or access door. If the attic has enough insulation and the home still feels drafty and cold in the winter or too warm in the summer, chances are insulation needs to be added to the exterior walls as well. This is a more expensive measure that usually requires a contractor, but it may be worth the cost if in a very hot or cold climate. When the exterior siding on a home is replaced, it is a good time to consider adding insulation.

Insulation may also need to be added to a crawl space or basement. Check with a professional contractor. For new homes in most climates, money and energy will be saved if a combination of cavity insulation and insulative sheathing is installed. Cavity insulation can be installed at levels up to R-15 in a 2 in. × 4 in. wall and up to R-21 in a 2 in. × 6 in. wall. The insulative sheathing, used in addition to cavity insulation, helps to reduce the energy that would otherwise be lost through the wood frame. **Figure 39.2** shows the recommended combinations. For example, in Zone 5, either a 2 × 4 wall could be insulated with R-13 or a 2 × 6 wall with R-21. For either of those two walls, an inch of insulative sheathing that has an R-value of R-5 or R-6 should also be used.

Today, new products on the market provide both insulation and structural support and they should be considered for new home construction or additions. Structural insulated panels, known as SIPs, and masonry products like insulating concrete forms are among these. Some homebuilders are even using an old technique borrowed from the pioneers: building walls using straw bales. Radiant barriers (in hot climates), reflective insulation, and foundation insulation should all be considered for new home construction. Check with your contractor for more information about these options.

Look and smell for any moisture problems which have caused stains or mold issues. Make sure crawl spaces are well ventilated and covered with a sealed ground moisture barrier to control ground moisture. Look in the attic for any moisture, mold, or large air leakage problems. Keep an eye out for any health and safety issues that the attic may hide. Inspect the wiring in the attic and look to make sure all lighting that protrudes through the ceiling is insulated properly.

Identify the thermal boundaries and the components that make up the thermal boundary in the residence. Thermal boundaries consist of insulation and an air barrier that surrounds a conditioned space. Thermal bypasses are areas in the thermal boundary where air can either infiltrate or exfiltrate. Conditioned spaces include the areas of the residence that are heated or cooled. Unconditioned spaces are areas of the residence that are not heated or cooled. Examples of unconditioned spaces are crawlspaces, attached garages, and attics. Some houses contain unintentionally conditioned spaces. These spaces are usually warmed by waste heat and include boiler rooms or furnace rooms. They are also referred to as intermediate zones. All conditioned spaces should have a thermal boundary surrounding them made up of insulation and an air space barrier.

VISUAL INSPECTION OF THE COOLING SYSTEM

The condenser fan in the outdoor condensing unit should be operational and the condenser coil clean and free of debris. The condenser's airflow should not be short-cycled or restricted. Often, awnings, overhangs, or bushes can rebound or short-cycle hot air back to the condensing coil causing high condensing pressures and unwanted inefficiencies. The extended surfaces or fins on the condensing coil may need straightening to allow more air flow to pass through for better heat transfer.

The suction line of the air-conditioning unit or gas line of a heat pump should have good insulation. Uninsulated suction lines can lower vapor densities coming back to the compressor, which causes system inefficiencies. The condensing unit should be located on the proper side of the house, away from mid-afternoon solar gains.

The air conditioner's evaporator coil and filter should be visually inspected, if easily accessible. They should not be clogged with debris, algae, or mold and the evaporator coil drain should not be clogged and backed up with condensate. Standing water can cause algae, bacteria, and viruses to be present around the evaporator coil and drain pan. If it is the cooling season, a reading of the return-air and cool-air discharge temperature and relative humidity should be taken and a note of its temperature for later analysis.

VISUAL INSPECTION OF THE HEATING SYSTEM

The age, condition, and efficiency of the heating equipment and whether older units can be upgraded to newer, higher-efficiency models should be determined. Any heating system problem found during the inspection can be documented and addressed later in the audit.

The flue gases should be inspected to determine the combustion efficiency of the heating equipment, as should the condition of air filters, ducts, piping, and radiators. Any blockage of supply-air outlets and leaking ductwork should be noted. The thermostat should also be programmed to a reasonable, yet economical, heating setting.

Safety issues relating to venting of appliances should be identified. The flame in the furnace and water heater should be watched for signs of roll-out, which may indicate **backdrafting**. The auditor should be alert to signs of combustion by-product spillage at the water heater and furnace, make sure the furnace and water heater vents are properly sloped and sized for their input ratings, and inspect how far any chimney or vent connector is from a combustible material.

Visual signs of corrosion on water heaters and boilers should be noted. If the furnace is a forced-air furnace, the combustion chamber should be inspected for rust, corrosion or signs of deterioration. Water stains on the floor near the boiler that lead toward the drain could mean the boiler is running too hot and the pressure relief valve is opening. The zone valves should be inspected for proper operation and to make sure they have not been manually overridden to stay in the open position.

BASE LOAD INSPECTION

Base loads are consistent, everyday consumers of energy within a residential structure. It is important that these energy-consuming loads be evaluated in a residential energy audit. Together they can make a significant contribution to daily energy usage in a home. These everyday loads are often referred to by energy auditors as **LAMEL** loads. LAMEL is an acronym for lighting, appliance, and miscellaneous electrical loads. Typical residential base loads or LAMELs are listed below:

- Lights
- Refrigerators
- Freezers
- Water heaters
- Clothes dryers
- Computers
- Printers
- Televisions
- Cell phone or smart device chargers

LAMEL loads include what used to be referred to as phantom, vampire, or plug loads. The auditor should inspect lighting to see if there are appropriate or too many lighting levels in the residence and make note if the lighting is incandescent, compact fluorescent, or light emitting diode (LED). LED fixtures are much more energy-efficient.

The exterior (condenser) coils of all refrigerators and/or freezers should be inspected to make sure that there is proper air circulation through them and that they are not clogged with dirt, dust, lint, or fuzz. Any of these materials clogging the condenser coil will impede heat transfer from the coil. This will raise the condensing pressure and cause unwanted inefficiencies and higher power usage. The refrigerator/freezer electricity (power) consumption should be measured. Refrigerator temperatures should range from 35°F to 39°F. Freezer compartments should range from 0°F to −10°F, depending on the particular product frozen and the length of time the homeowner plans to keep the frozen goods.

The water heater temperature should be below 120°F. Water leakage from the water heater should be noted and what type of energy (natural gas, propane, fuel oil, or electricity) is heating the water in the water heater. Whether the water heater has an insulating blanket on its exterior should be noted. If the water heater is a combustion appliance, it should be inspected for any backdrafting problems.

The interior drum of the clothes dryer should be inspected, as should the vent to the outdoors and the lint screen for excessive lint build up. What type of energy is used in drying the clothes should be noted, and the age of the clothes dryer.

INDOOR AIR QUALITY

Energy audits and weatherization often identify and interact with a home's indoor air quality. Health and safety issues relating to the following list are identified and dealt with during an energy audit and weatherization process:

- Asbestos
- Fiberglass
- Radon
- Mold
- Pollen
- Venting issues
- Moisture issues
- Backdrafting
- Ventilation
- Cigarette smoke
- Cooking odors
- Carpet offgassing
- Paints
- Finishes
- Cleaning chemicals
- Pesticides
- Pet dander
- Dust mites
- Home electronics

Safety issues are of major concern especially when the house and its HVACR duct system may be pressurized and/or depressurized when performing parts of the testing phase of the energy audit. The testing phase of the audit can also alter the operation of the HVACR equipment. Moisture levels, ventilation rates, and indoor air quality within a residence can also be altered if proposed changes recommended by the auditor's report are adopted. Also, once the testing phase is over and the home is sealed and weatherized, its structure will be much tighter and combustion appliance backdrafting issues may have to be dealt with, **Figure 39.4**. Often, because homes are so "airtight" after sealing and weatherization, an energy recovery ventilator may have to be installed in the home for safety reasons and moisture management, **Figure 39.5**. Energy recovery ventilators bring in fresh air from the outside of the home which experiences a heat exchange with warmer, stale exhaust air being exhausted from the inside of the home. The auditor may also recommend that air from the outside be brought into the combustion appliance zone (CAZ) to supply combustion air for all gas-burning appliances. (For a more detailed analysis on indoor air quality in houses and energy recovery ventilators, please refer to Unit 34, "Indoor Air Quality.")

Figure 39.4 Air exhausted from a house creates a negative pressure within the house, causing backdrafting of combustion appliances. Courtesy Bacharach, Inc., Pittsburgh, PA, USA

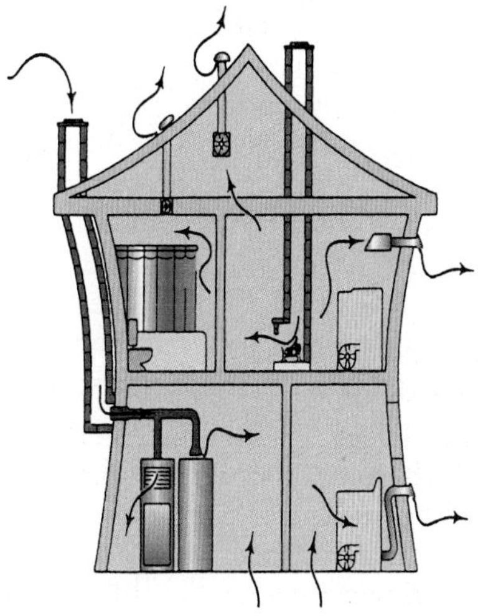

Figure 39.5 An energy recovery ventilator. Source: Honeywell, Inc.

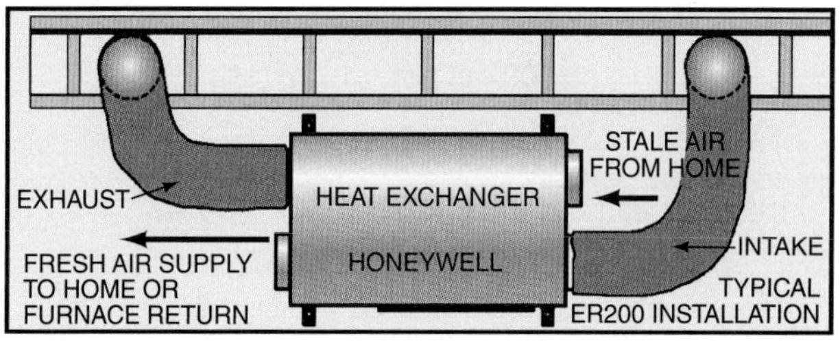

39.4 DIAGNOSTIC TESTING

Diagnostic testing and evaluation of the home and its systems that consume energy is probably the most important phase of the energy audit. *These diagnostic tests bring out many of the building's "hidden unknowns," such as structural and ductwork air leakages, moisture problems, combustion efficiencies, refrigerant charges, and wasted heat leakages, which can be quantified by diagnostic testing.* The most common diagnostic tests performed during a residential energy audit are:

- Blower door testing
- Infrared scanning testing (by thermal imaging camera)
- Duct leakage testing
- Combustion efficiency and safety testing
- HVACR system testing

39.5 BLOWER DOOR TESTING

Blower door tests give the auditor insight on how tight or leaky the home's building envelope is overall. Blower doors consist of a frame and shroud that fit inside the door frame of a house. The shroud or membrane forms an airtight seal on an outside doorway of the residence. Mounted in each blower door is a variable-speed fan that allows the fan to suck air out of the house to create a low-level vacuum (negative pressure) on the inside of the house, **Figure 39.6.** Blower door instruments include gauges like manometers and anemometers that can measure the flow of air through the blower door's fan at a set pressure differential between the inside and outside of the house, **Figure 39.7.** The set pressure differential is 50 Pascals for air-tightness testing. The slight vacuum on the inside of the house of 50 Pascals is measured with reference to (WRT) the outdoor pressure. However, if a thermal imaging camera is used in conjunction with a blower door test for air infiltration and exfiltration inspections, 20 Pascals is used for the pressure differential.

The Pascal (Pa) is the Systems International (SI) or metric unit for measuring pressure. One Pascal is a very small amount of pressure.

$$1 \text{ Pascal (Pa)} = 0.004 \text{ in. of water column (in. WC)} = 0.000145 \text{ pounds/in}^2 \text{ (psi)}$$

or

$$1 \text{ Pa} = 0.004 \text{ in. WC} = 0.000145 \text{ psi}$$

Figure 39.6 A blower door assembly. Courtesy The Energy Conservatory

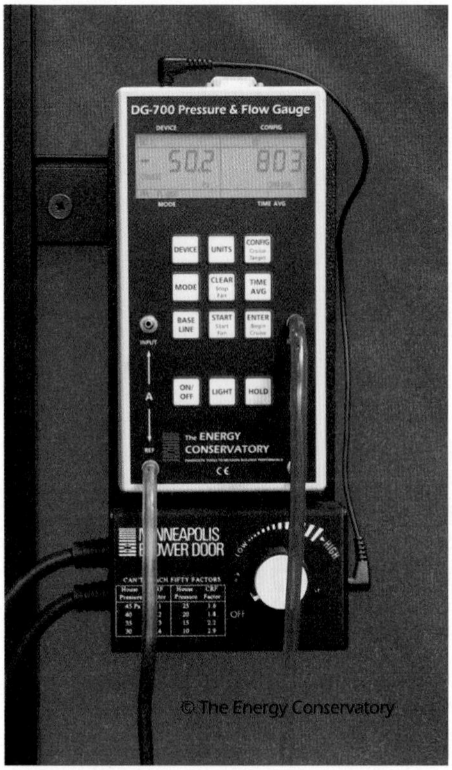

Figure 39.7 Blower door instrumentation, which includes digital gauges like manometers and anemometers that measure the flow of air through the blower door's fan at a set pressure differential between the inside and outside of the house. Courtesy The Energy Conservatory

For comparison purposes, if a person were to put a straw in a glass of water and suck the water up 1 in. in the straw, the negative pressure needed to do this would be 1 in. of water column, or approximately 250 Pascals. Wind stack effects on buildings usually generate air pressures across the building's shell that range from 0.05 to 10 Pascals. Bathroom exhaust fans also generate negative pressures that fall between 1.0 and 10 Pascals.

Instrumentation that accompanies a blower door inspection includes pressure gauges with which a technician can measure the flow of air through the fan as well as the pressure differential between the living space and the outdoors. The greater the airflow in CFM through the blower door's fan required to reach a certain pressure differential (50 Pascals), the leakier the house is. The blower door test alone will give a general idea of the relative "leakiness" of a home. **Figure 39.8** shows air leakages into a residential structure using a blower door test. **Figure 39.9** shows common areas that leak air in and out of a residential structure.

It takes a certain amount of airflow through the blower door to maintain a constant vacuum of 50 Pascals inside a home. This airflow is measured in cubic feet per minute (CFM) and is referred to as CFM_{50}. For example, a house that measures 1700 CFM_{50} on the blower door's digital manometer means that the house has 1700 CFM of air leakage while the blower door is connected and pulling 50 Pascals of vacuum on the

Figure 39.8 Air leakages into a residential structure using a blower door test. Courtesy The Energy Conservatory

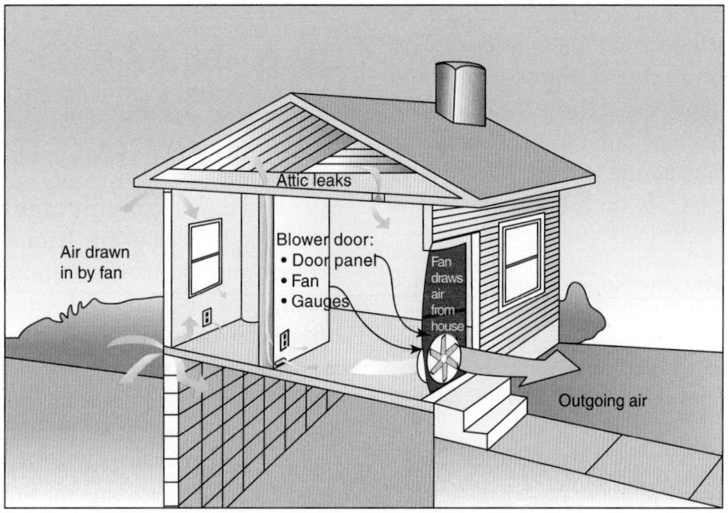

Figure 39.9 Common areas that leak air in and out of a residential structure. Source: Environmental Protection Agency (EPA)

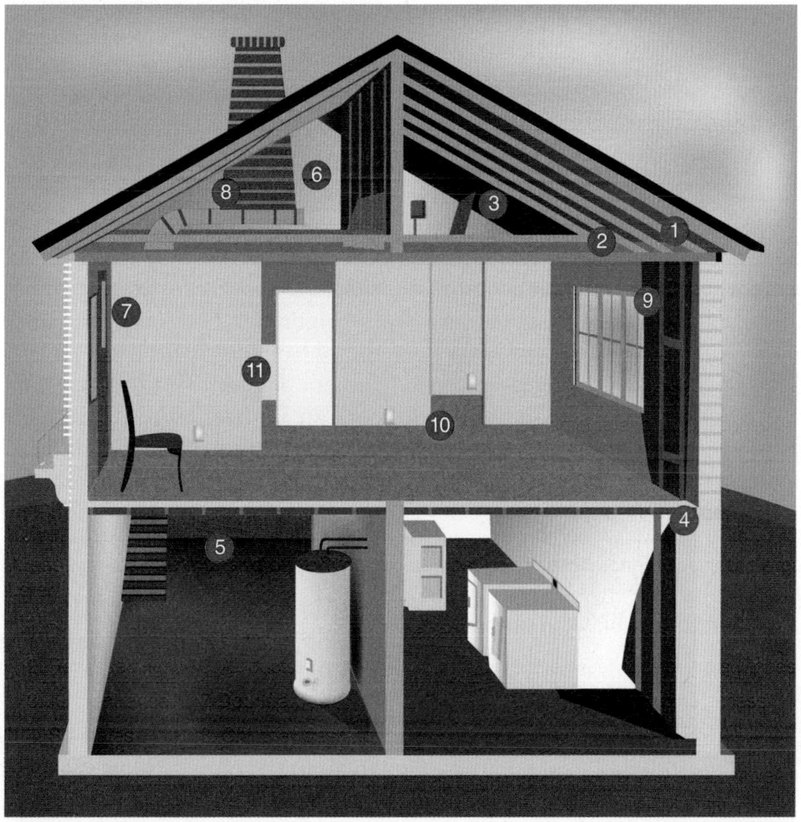

Areas that leak air into and out of your home cost you lots of money. Check the areas listed below.

1. Dropped ceiling
2. Recessed light
3. Attic entrance
4. Sill plates
5. Water heater and furnace flues
6. All ducts
7. Door frames
8. Chimney flashing
9. Window frames
10. Electrical outlets and switches
11. Plumbing and utility access

house. By comparing that flow to the volume (ft^3) of air in the building, an auditor can determine the number of air changes per hour (ACH_{50}) that occur in the structure at a 50 Pascal pressure differential between the house and the outdoors. This will tell the auditor the air leakage rate of the structure and how well the structure is sealed. This leakage rate measurement can be used to compare to the leakage rate measured after the house is air-sealed. (Air-sealing a structure will be covered later in this unit.) To figure out the ACH_{50} for the house, auditors use the following equation:

$$ACH_{50} = (CFM_{50} \times 60) \text{ divided by the (House's Conditioned Volume)}$$

where

ACH_{50} = Air changes per hour
60 = 60 min/h
House Conditioned Volume = Total Volume of the Conditioned House in Cubic Feet (ft^3)

Suppose a house had a conditioned volume of 15,660 ft^3 and the blower door test measured 600 CFM_{50}. The ACH_{50} of the house would be:

$$ACH_{50} = (600\ CFM_{50} \times 60\ \text{min/h}) \text{ divided by } (15{,}660\ ft^3)$$
$$ACH_{50} = 2.29$$

The standard for an Energy Star dwelling is 4 ACH at 50 Pascals for Department of Energy climate zones 5, 6, 7, see **Figure 39.1**. For comparison purposes, some leaky or drafty homes can test at 20 ACH at 50 Pascals.

Blower door testing may cause some safety problems within the structure, especially with combustion appliances. The negative pressure or vacuum of 50 Pascals may cause combustion appliances to backdraft. **SAFETY PRECAUTION:** *Even fireplaces or woodstoves may backdraft and cause ash to enter the structure during the blower door testing procedures.•* If combustion appliances are left on, by-products of combustion or other pollutants may enter the structure during testing. The auditor must anticipate any safety issues before performing a blower door test by:

1. turning off all combustion appliances,
2. inspecting all sources of pollutants like fireplaces or wood stoves,
3. inspecting for any large air leakages that may prevent the blower door from achieving the desired vacuum level and plugging these openings temporarily,
4. identifying all thermal boundaries and conditioned zones in the house,
5. closing all windows and doors, and
6. opening all interior doors.

Blower doors can also be used to create a positive pressure in a house. In these cases, the blower door is often used in conjunction with a duct blower for testing air lost to the outdoors in ductwork systems. The duct blower used in conjunction with the blower door will be covered later in the unit under the heading "Duct Leakage to the Outdoors."

39.6 INFRARED SCANNING USING A THERMAL IMAGING CAMERA

Standard digital cameras capture images of objects that reflect visible light. The visible light spectrum consists of white light broken into its many component colors (red, orange, yellow, green, blue, indigo, and violet). Thermal imaging cameras, often referred to as infrared cameras, create pictures by detecting infrared energy or heat. Infrared energy has a lower frequency than visible light. The thermal imaging camera then assigns colors based on the temperature differences it detects. **Figure 39.10(A)** is a photo taken by a thermal imaging camera that shows heat escaping the building envelope. **Figure 39.10(B)** shows a thermal imaging camera. **Figure 39.10(C)** shows an energy auditor taking a thermal image on the interior of a window and the surrounding window framing.

Using a thermal imaging camera in conjunction with a blower door test is strongly recommended for best results in testing for air infiltration and exfiltration. Testing with both a blower door and a thermal imaging camera can show the "exact" location of the building envelope problems. With a blower door in operation, a technician using an infrared thermal imaging camera and a pressure gauge (to verify the pressure differential in various parts of the house) can find areas that contribute to the loss of conditioned air by convection. These losses include heated and humidified air in the winter and cooled and dehumidified air in the summer caused by either infiltration or exfiltration.

Figure 39.11 shows cold air infiltrating into the house at the bottom sill plate through the eyes of a thermal imaging camera. **Figures 39.12** through **Figure 39.14** are examples of cool air infiltration into a residential structure through light fixtures. **Figure 39.14(C)** illustrates a canned (recessed) light losing heat to an unconditioned attic space. **Figure 39.15(A)** shows cool air infiltrating around the attic door. **Figure 39.15(B)** shows heat escaping through an improperly sealed or insulated attic. **Figure 39.16** shows a residential ceiling diffuser. **Figure 39.17** shows cool air infiltrating around the top plate of a residential structure. **Figure 39.18** through **Figure 39.20** show cold air infiltration around windows. **Figure 39.21** through **Figure 39.23** show cold air infiltration around doors. **Figure 39.24** shows cool air infiltration around an electrical chase and wall switch plate. **Figure 39.25** shows cool air infiltrating around an electrical outlet and the floor. **Figure 39.26** through **Figure 39.28** show warm air exfiltrating from a residential structure. The most effective thermal imaging work in an energy audit will occur indoors when a living space is being heated in the winter or cooled in the summer, not in the spring or fall. This happens because there are greater temperature differences between the inside and outside of the house in the summer and winter months.

Figure 39.10 (A) A thermal imaging camera showing heat escaping a building envelope. Courtesy Fluke Corporation

Figure 39.10 (B) A thermal imaging camera. Courtesy Fluke Corporation

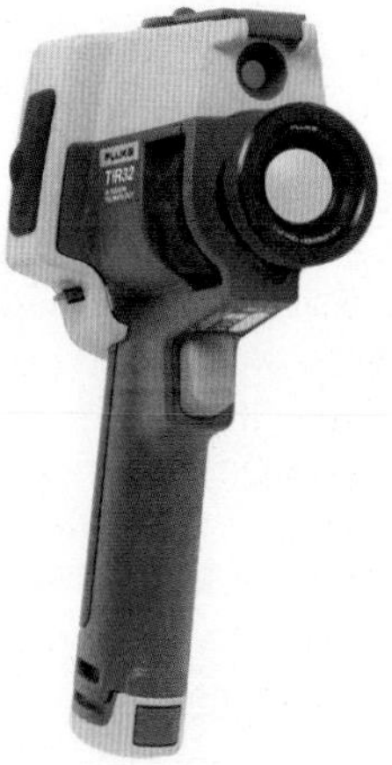

Figure 39.10 (C) An energy auditor taking a thermal image on the interior of a window and the surrounding window framing. Courtesy Fluke Corporation

Figure 39.11 Cold air infiltrating into the house at the bottom sill plate through the eyes of a thermal imaging camera. Courtesy Fluke Corporation

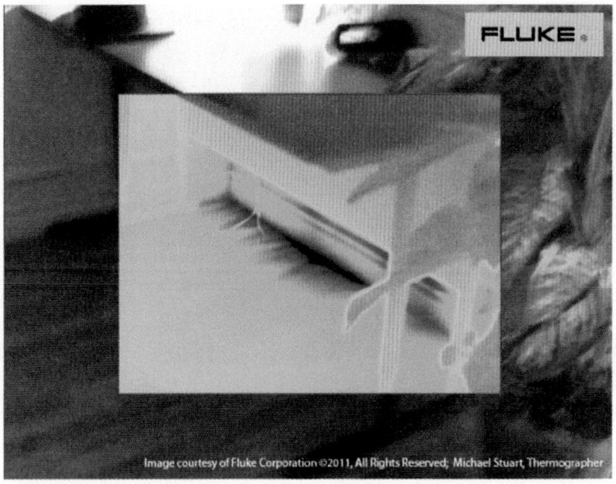

Figure 39.12 Cool air infiltration into a residential structure through ceiling canned light fixtures. Courtesy Fluke Corporation

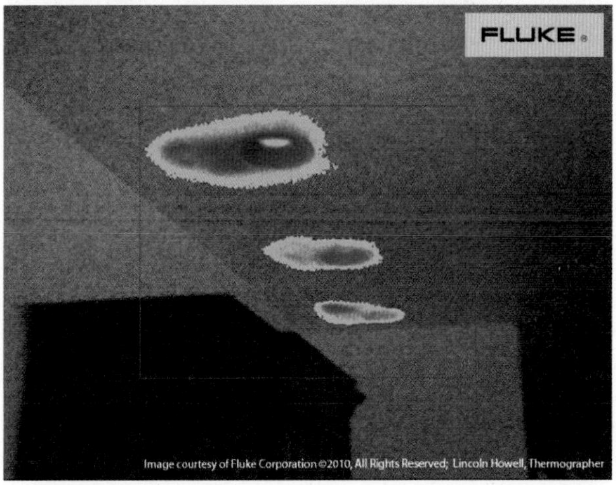

Figure 39.13 Cool air infiltration into a residential structure through a ceiling canned light. Courtesy Fluke Corporation

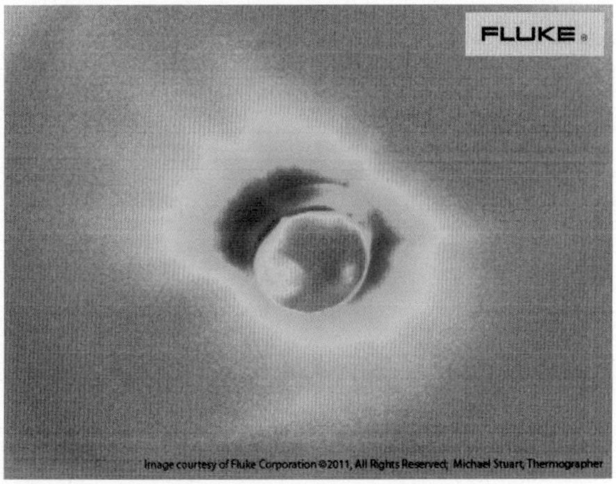

Figure 39.14 (A) An auditor aiming a thermal imaging camera at a light fixture. Courtesy Fluke Corporation

Figure 39.14 (B) Cool air infiltration into a residential structure through a ceiling light fixture. Courtesy Fluke Corporation

Figure 39.14 (C) A thermal imaging camera showing a canned (recessed) light losing heat to an unconditioned attic space. Courtesy McIntyre Builders, Rockford, MI

Figure 39.15 (A) A thermal imaging camera showing cool air infiltrating around an attic door. Courtesy Fluke Corporation

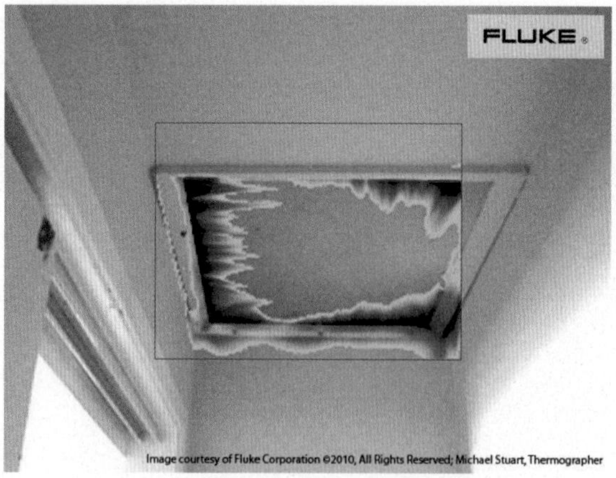

Figure 39.15 (B) Heat escaping through an improperly sealed or insulated attic. Photos Courtesy Fluke Corporation

Figure 39.16 A thermal imaging camera showing a ceiling diffuser. Courtesy Fluke Corporation

Figure 39.17 A thermal imaging camera showing cool air infiltrating around the top plate of a residential structure. Courtesy Fluke Corporation

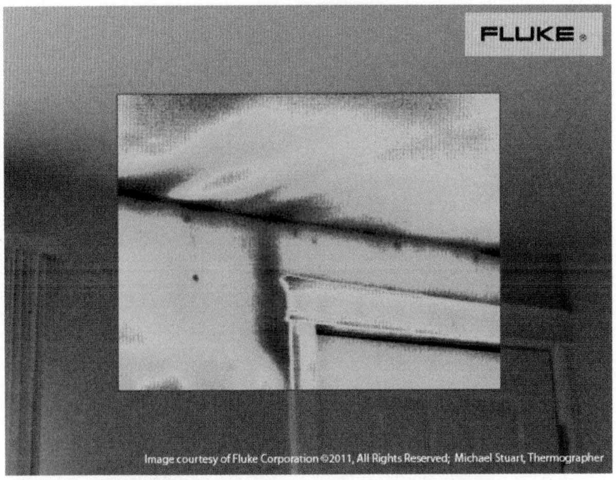

Figure 39.18 A thermal imaging camera showing cold air infiltration around a window. Courtesy Fluke Corporation

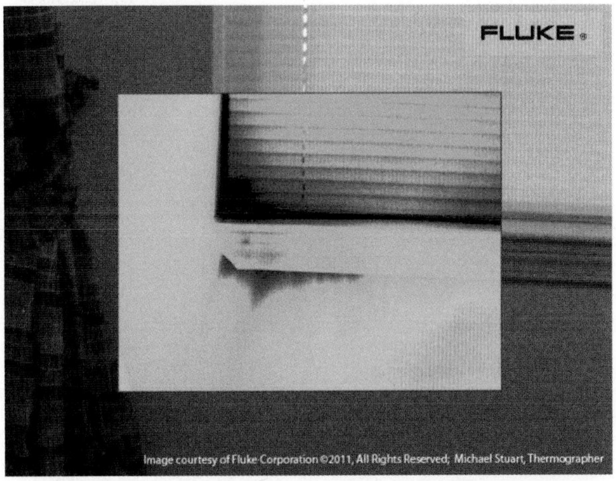

Figure 39.19 A thermal imaging camera showing cold air infiltration around a window. Courtesy Fluke Corporation

Figure 39.20 A thermal imaging camera showing cold air infiltration around a window. Courtesy Fluke Corporation

Figure 39.21 A thermal imaging camera showing cold air infiltration around the bottom of a door. Courtesy Fluke Corporation

Figure 39.22 A thermal imaging camera showing cold air infiltration around the top of a door. Courtesy Fluke Corporation

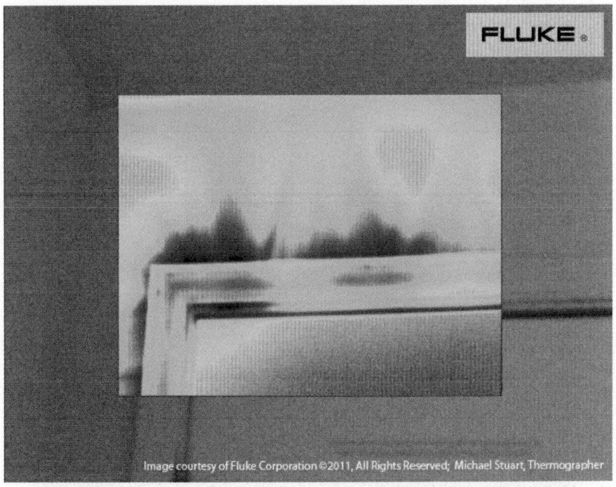

Figure 39.23 A thermal imaging camera showing cold air infiltration around the bottom of a door. Courtesy Fluke Corporation

Figure 39.24 A thermal imaging camera showing cool air infiltration around an electrical chase and wall switch plate. Courtesy Fluke Corporation

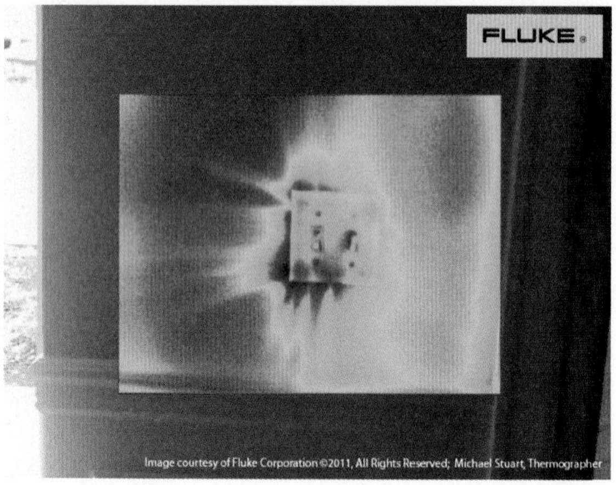

Figure 39.25 A thermal imaging camera showing cool air infiltrating around an electrical outlet and the floor. Courtesy Fluke Corporation

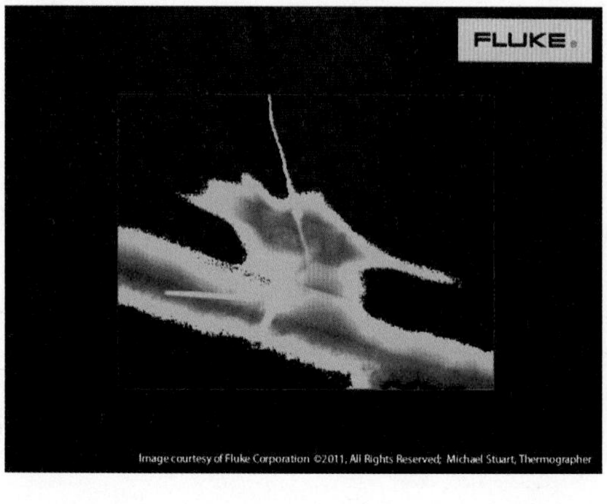

Figure 39.26 A thermal imaging camera showing warm air exfiltrating around the side of a window. Courtesy Fluke Corporation

Figure 39.27 A thermal imaging camera showing warm air exfiltrating above a door. Courtesy Fluke Corporation

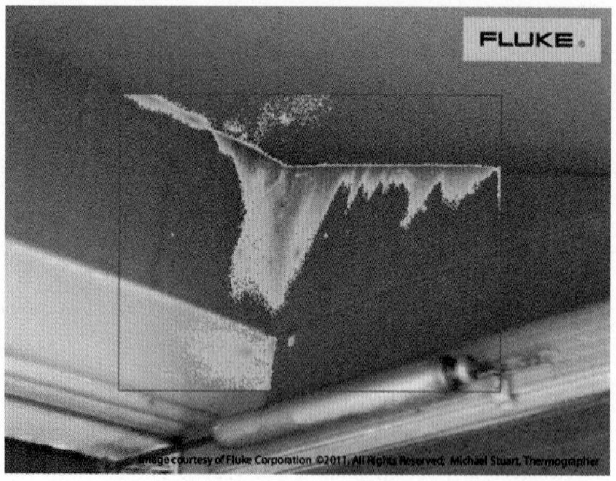

Figure 39.28 A thermal imaging camera showing warm air exfiltrating at the top of a window. Courtesy Fluke Corporation

Heating and cooling losses can occur by both conduction and convection. Conduction heat losses can also happen along with convective losses when insulation is missing from an outside wall, **Figure 39.15(B)**. Conductive heat losses and gains can also occur through a process called thermal bridging. Thermal bridging is rapid heat being conducted through the residential structure's shell by materials in direct contact with one another that are very thermally conductive. An example of thermal bridging is a cold aluminum or steel window frame in contact with the window's glass, which will rapidly conduct heat through the building's shell to the outdoors in the winter. The result is rapid heat loss to the outside and a cold interior surface on the window frame and glass for condensate to form. **Figure 39.29(A)** shows a window where condensate that has ran down the glass and rotted the bottom structure of the window because of thermal bridging. *Thermal bridging can be lessened by installing a gasket material made of a poor thermal conductive material between the window frame and the glass.* The gasket material is often referred to as a thermal break, **Figure 39.29(B)**.

For best results when performing thermal image testing, make sure there is a temperature difference of approximately 18°F from the interior to the exterior surface. For air leakage inspection work, the temperature difference should be at least 9°F. Make sure these temperature differences have stabilized for a period of 4 hours or more before starting the test. Record the interior and exterior air temperature and relative humidity. Make notes as to wind direction, sun position, precipitation, and any other environmental factor that may affect test results. Close all outside doors, open all inside doors, and make sure the HVACR system has been off for at least 15 minutes.

Install the blower door on an exterior door. The blower door will induce or draw a negative pressure on the inside of the house. As mentioned earlier, this negative pressure should be 20 Pascals for infiltration and exfiltration inspections with a thermal imaging camera. This will pull outside air to the inside of the house through every crack and orifice because of the pressure difference caused by the blower door. The thermal imaging camera can spot these leaks and show them as thermal images of different colors. The thermal imager cannot see the incoming air, but it can detect cool or warm surfaces created as air leaks across them. Some of the more common places for incoming air leakages are

- window sills,
- electrical fixtures,
- windows and doors,
- penetrations in the wall, roof, and floor caused by plumbing, lighting, and electrical outlets,
- joints around chimneys,
- joints between walls,
- joints between ceilings and floors, and
- joints between top plates and sill plates.

Research has shown that the most serious air leaks occur at the top and bottom of the conditioned envelope, like attics and basements. Large gaps for air to travel are often found around

- plumbing pipes,
- recessed lighting fixtures,
- chimneys,
- eave soffits,
- chaseways,
- basement rim joists where the house foundation meets the wood framing,
- top and bottom of plumbing runs and vents especially in attics and basements,
- utilities such as electrical conduit and TV cables entering the conditioned air space,
- cold floors in the winter, especially above crawl spaces and garages, and
- floor slabs that extend outdoors and have no thermal barrier between the indoor and outdoor portion, which can account for large conduction heat losses.

Figure 39.29 (A) A window where condensate has run down the glass and rotted the bottom structure of the window because of thermal bridging. Courtesy Ferris State University. Photo by John Tomczyk

Figure 39.29 (B) Gasket material between the window frame and the glass of a double-pane window. Thermal bridging can be lessened by installing a gasket material made of a poor thermal conductive material between the window frame and the glass. The gasket material is often referred to as a thermal break. Courtesy Fluke Corporation

Figure 39.30(A) illustrates common air leakages within a residential structure. Blue arrows show air leaking into the house and red arrows show air leaking out of the house. **Figure 39.30(B)** is a pie chart showing common air leak sources with their respective percentages.

A blower door can also pressurize a house. Once the house is pressurized, a thermal imaging camera can capture where air exfiltrates from a home. Often, air exits from different places than where it enters a home, **Figures 39.26, 39.27,** and **39.28** show air exfiltration leakages.

Thermal imaging cameras can reveal the presence of moisture and the problems it creates. Thermal imaging cameras can also be used to validate the effectiveness of repairs and improvements, such as caulking, filling voids with spray foams, and adding insulation. **Figures 39.31(A)** and **39.31(B)** show moisture on the floor and wall of a home. Moisture can be caused by leaks, condensate, or flooding. Moisture can be found in walls, floors, ceilings, and roofs. The resistance value (R-value) of insulation and building materials will significantly decrease when they are wet. Moisture and condensation often go hand-in-hand with air leaks in a structure because air can provide a means for moisture to travel. *Moisture, if not properly remedied, can lead to building damage, reduced insulation effectiveness, and mold growth.* Mold and mildew will start to grow where moisture is present, causing other health-related problems. The effects of evaporative cooling (usually a 2° to 5°F surface temperature difference) helps reveal the extent of moisture damage, even when the surface feels dry to the touch. However, all suspected moisture problems should be validated with a moisture meter. Please refer to Unit 34, "Indoor Air Quality," for more detailed information on mold, moisture, indoor air pollutants, and safety.

Figure 39.30 (A) Common air leakages within a residential structure. Source: Environmental Protection Agency (EPA)

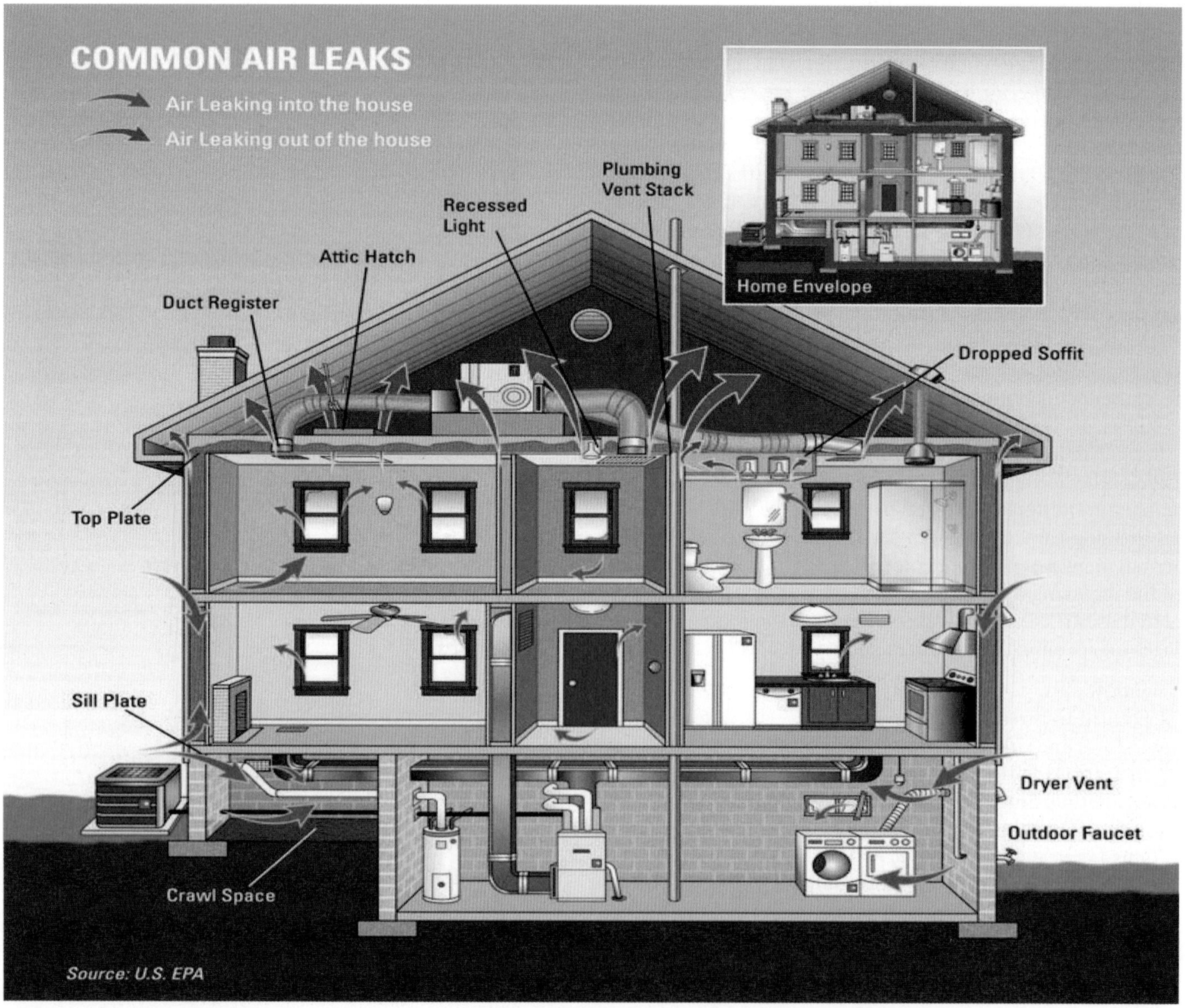

Figure 39.30 (B) Pie chart showing common air leak sources with their respective percentages for residential structures. Source: Environmental Protection Agency (EPA)

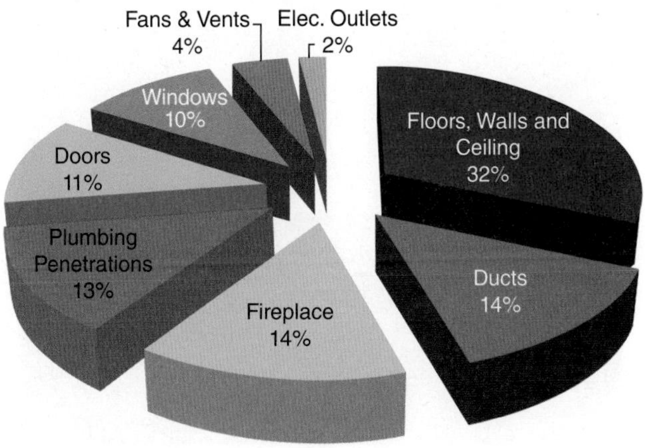

Figure 39.31 (A) A thermal imaging camera showing moisture on the floor and wall of a home. Courtesy Fluke Corporation

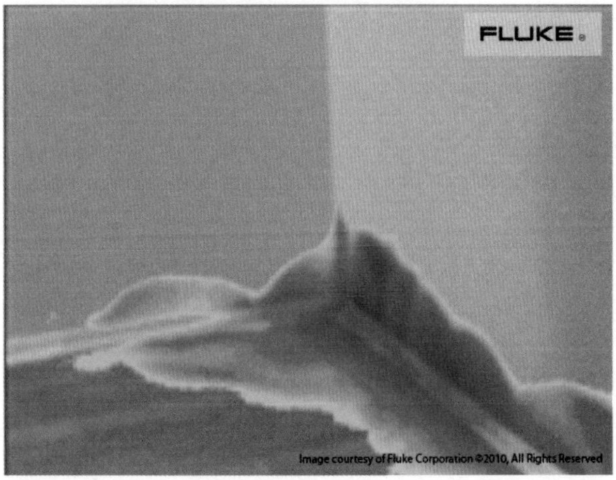

Figure 39.31 (B) A thermal imaging camera showing moisture on the floor and wall of a home. Courtesy Fluke Corporation

39.7 SEALING AIR LEAKS

Warm air leaking into a home during the summer and out of a home during the winter can waste a lot of energy dollars. One of the quickest dollar-saving tasks a homeowner can do is caulk, seal, and weather-strip all seams, cracks, and openings to the outside. Savings on heating and cooling bills can be achieved by reducing the air leaks in a home. Air infiltrates into and out of a home through every hole and crack. About one-third of this air infiltrates through openings in ceilings, walls, and floors. Some main reasons why a homeowner would air-seal their home are listed below:

- Damage control for frozen pipes which may burst and cause water damage
- Preventing pest entry and damage
- Ice dam prevention
- Comfort of the entire home by preventing drafts and cold spots
- Preventing water damage in walls and other building materials
- Money savings on heating and cooling
- Reducing outdoor noise and indoor noises between rooms
- Improving indoor air quality by sealing out outside air

First, test the home for air tightness. A certified energy auditor can use a blower door test. The homeowner can perform a simple test themselves using a smoke pen, lit ensconce stick, or other nontoxic chemical smoke-generating device. On a windy day, a lit incense stick or a smoke generator is carefully held next to windows, doors, electrical boxes, plumbing fixtures, electrical outlets, ceiling fixtures, attic hatches, and other locations where there is a possible air path to the outside, **Figure 39.32**. A smoke stream that travels horizontally locates an air leak that may need caulking, sealing, or weather-stripping.

The 2018 International Energy Conservation Code (IECC) and the 2018 International Residential Code (IRC) have several mandatory requirements for air sealing in new construction and additions, adapted by local jurisdiction. In general, the requirements do not apply to retrofit projects unless the project adds living space to the building or changes the building's energy load. Listed below are important places to seal air leakages.

Figure 39.32 An auditor holding a chemical smoke generator up to a recessed light to find out if there is a possible air path to the outside.

- *Caulk and weather-strip doors and windows that leak air,* **Figures 39.33(A)** *and* **(B)**.
- *Caulk and seal air leaks where plumbing, ducting, or electrical wiring penetrates through walls, floors, ceilings, and soffits over cabinets,* **Figures 39.34(A)** *and* **(B)**.

Figure 39.33 (A) Weather-strip on doors to prevent air leakage. Courtesy Fluke Corporation

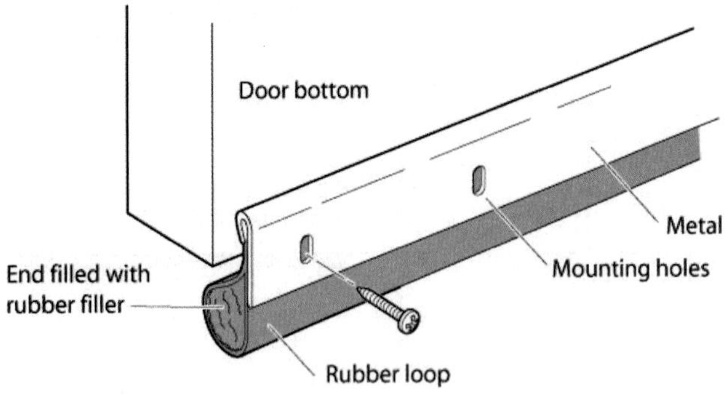

Figure 39.33 (B) Caulking around windows to prevent air leakage. Photos Courtesy Fluke Corporation

Figure 39.34 (A) Insulating foam sealant sealing air leaks where plumbing fixtures penetrate through floors. Courtesy Fluke Corporation

Figure 39.34 (B) Insulating foam sealant sealing air leaks where electrical wiring penetrates through walls. Courtesy U.S. Department of Energy

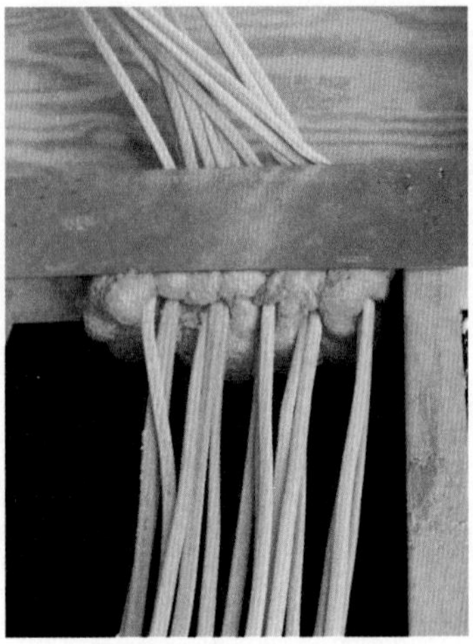

Figure 39.35 (A) Use caulk or spray foam to air seal where the foundation wall meets the sill plate, where the sill plate meets the rim joist, and where the rim joist meets the subfloor. Courtesy U.S. Department of Energy

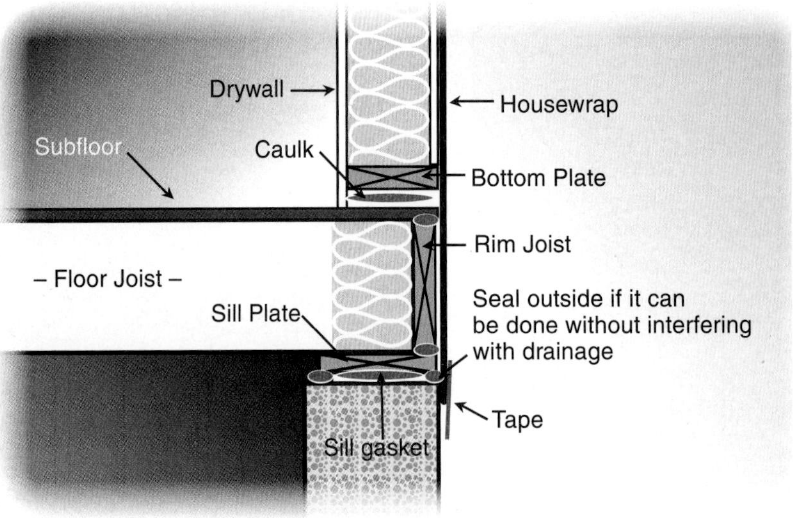

Figure 39.35 (B) Spray foam along the basement rim joist to provide a complete air barrier connecting the foundation wall, sill plate, rim joist, and subfloor. Photos Courtesy U.S. Department of Energy

- *Install foam gaskets behind outlet and switch plates on walls.*
- *Seal all joints, seams, and penetrations.*
- *Seal rim joists, sill plates, foundations, and floors,* **Figures 39.35(A)** *and* **(B)**.
- *Seal fireplace walls,* **Figure 39.36**.
- *Look for dirty spots in your insulation, which often indicate holes where air leaks into and out of the house. The holes can be sealed with low-expansion spray foam made for this purpose.*
- *Look for dirty spots on the ceiling paint and carpet, which may indicate air leaks at interior wall/ceiling joints and wall/floor joists. These joints can be caulked.*
- *Install storm windows over single-pane windows or replace them with more efficient windows, such as double-pane.*
- *Seal openings between window and door assemblies and their respective jambs and framing.*
- *When the fireplace is not in use, keep the flue damper tightly closed. A chimney is designed specifically for smoke to escape, so until it is closed, warm air escapes—24 hours a day!*

Figure 39.36 Air seal the enclosure and flue. Courtesy U.S. Department of Energy

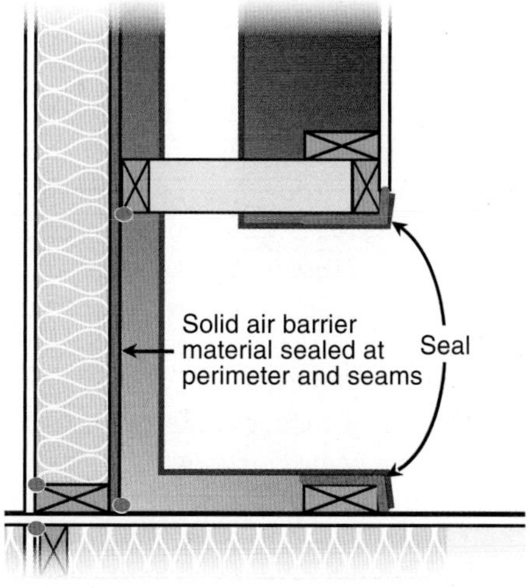

Figure 39.37 Finish the walls that separate the garage from the rest of the home with drywall that is sealed to the top and bottom plate with a bead of caulk. Courtesy U.S. Department of Energy

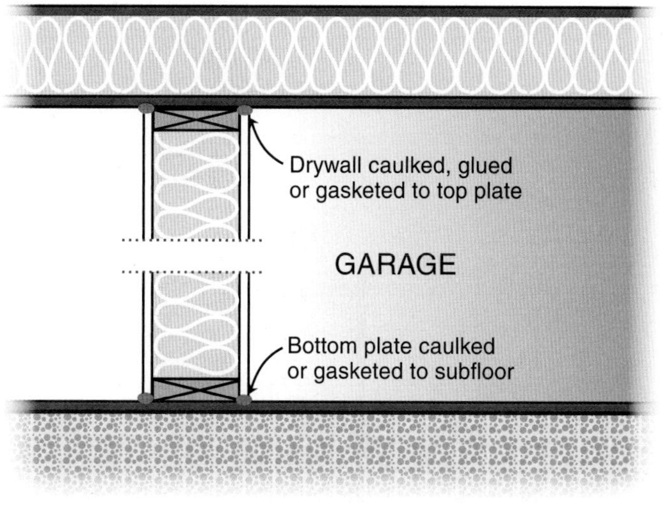

- *For new construction, reduce exterior wall leaks by installing house wrap, taping the joints of exterior sheathing, and comprehensively caulking and sealing the exterior walls.*
- *Seal utility penetrations.*
- *Seal garage/living space walls,* **Figure 39.37**.
- *Seal attic knee walls and attic air sealing,* **Figures 39.38(A)** *and* **(B)**.
- *Use foam sealant around larger gaps around windows, baseboards, and other places where warm air may be leaking out.*
- *Install kitchen exhaust fan covers to keep air from leaking in when the exhaust fan is not in use. The covers typically attach via magnets for ease of replacement.*
- *Replace existing door bottoms and thresholds with ones that have pliable sealing gaskets to prevent conditioned air leaking out from underneath the doors.*
- *Check fireplace flues. Fireplace flues are made of metal, and over time repeated heating and cooling can cause the metal to warp or break, permitting hot or cold air loss.*
- *Seal dropped ceilings/soffit or chases adjacent to the thermal envelope,* **Figure 39.39**.

Figure 39.38 (A) Insulate and air seal the knee wall itself, or along the roof line. Courtesy U.S. Department of Energy

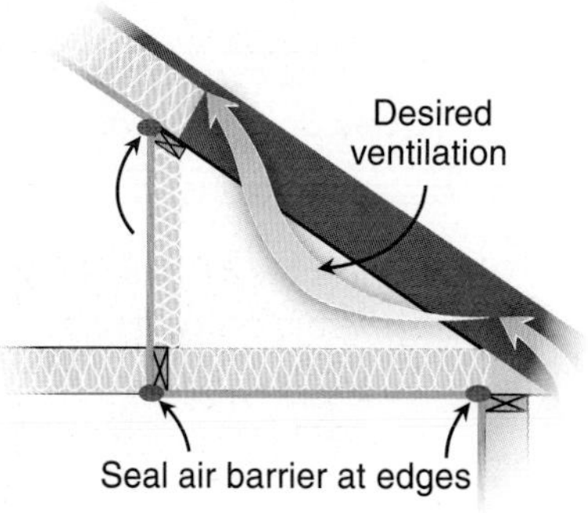

Figure 39.38 (B) Seal the drywall to the top plate and ceiling drywall. Pull back insulation to seal drywall to the top plate with spray foam, caulk, or other sealer. Photos Courtesy U.S. Department of Energy

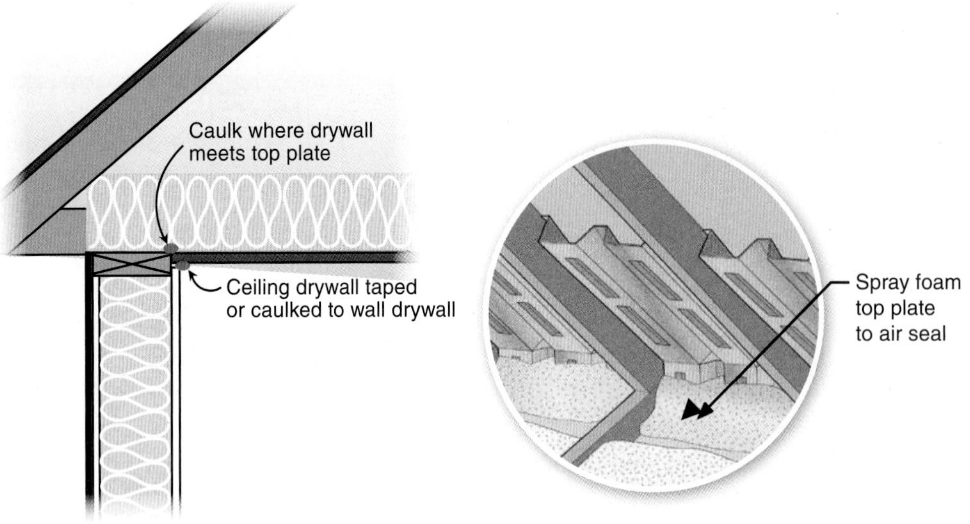

Figure 39.39 Place a solid air barrier over soffits. Courtesy U.S. Department of Energy

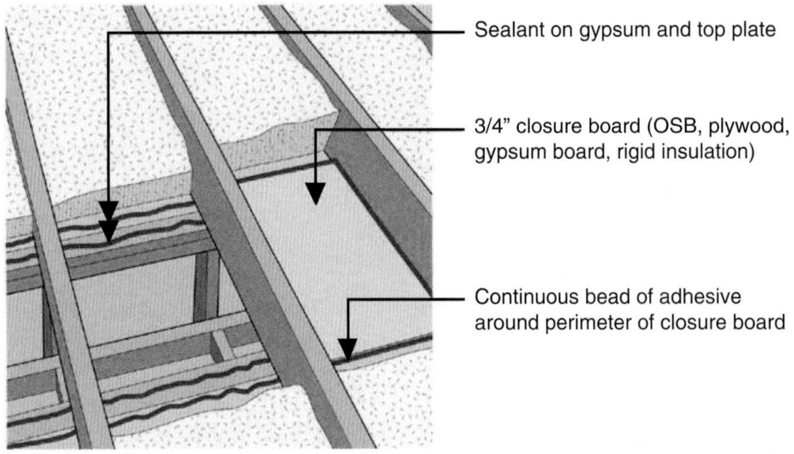

- *Seal duct shafts/pipe shafts and penetrations,* **Figure 39.40**.
- *Seal staircase framing at exterior wall/attic,* **Figure 39.41**.
- *Seal walls and ceilings separating a garage from conditioned spaces.*
- *Seal exterior walls, electrical outlets, exhaust fans, plumbing penetrations, and tub and shower,* **Figure 39.42**.
- *Install inflatable chimney balloons beneath a fireplace flue during periods of nonuse. They are made from several layers of durable plastic and can be removed easily and reused hundreds of times. Should the balloon remain in place accidentally before making a fire, the balloon will automatically deflate within seconds of coming into contact with heat.*
- *Seal rim joist junctions.*
- *Seal attic access or pull-down stairs,* **Figure 39.43**.
- *Seal the flue or chimney shaft.*
- *Air-seal recessed lights (can lights) by installing a box around them and air-sealing the box by using caulk or a gasket under the decorative cover,* **Figure 39.3(A)**. *Another option is to replace the recessed fixtures with ones that are rated for insulation contact and air tightness (ICAT).*
- *Seal attic access openings.*
- *Seal return and supply ducts,* **Figure 39.44(A)** *and* **(B)**.
- *Seal drywall and top plate junctions.*

Figure 39.40 Seal attic and wall penetrations associated with mechanical ventilation systems, electrical chase openings, and dropped soffits. Courtesy U.S. Department of Energy

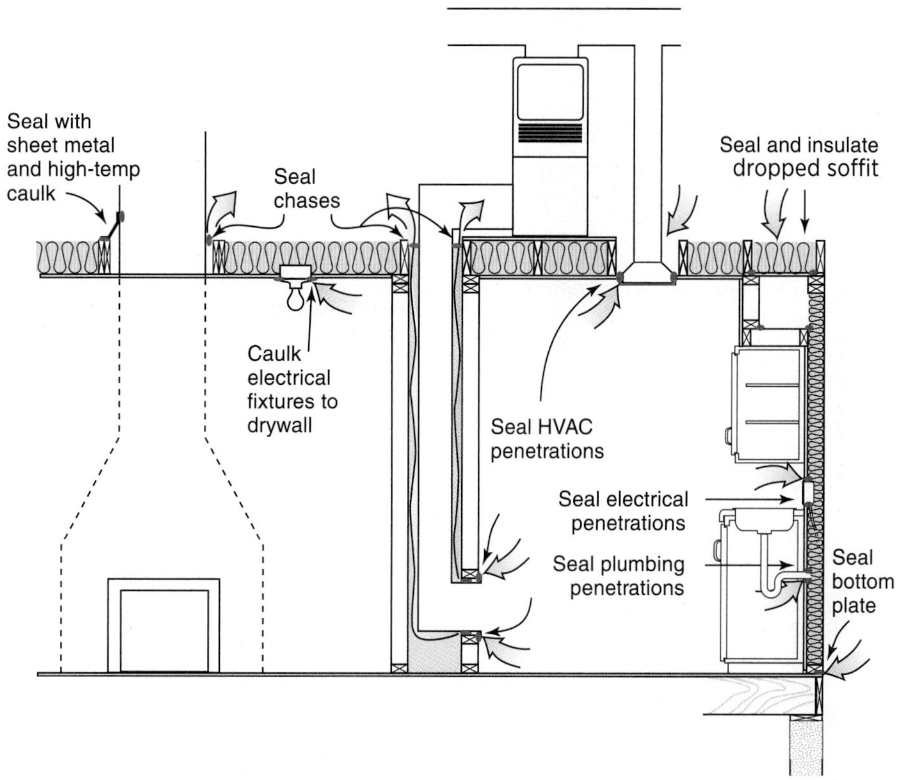

Figure 39.41 Install an air barrier and air sealing on exterior walls behind stairs. Courtesy U.S. Department of Energy

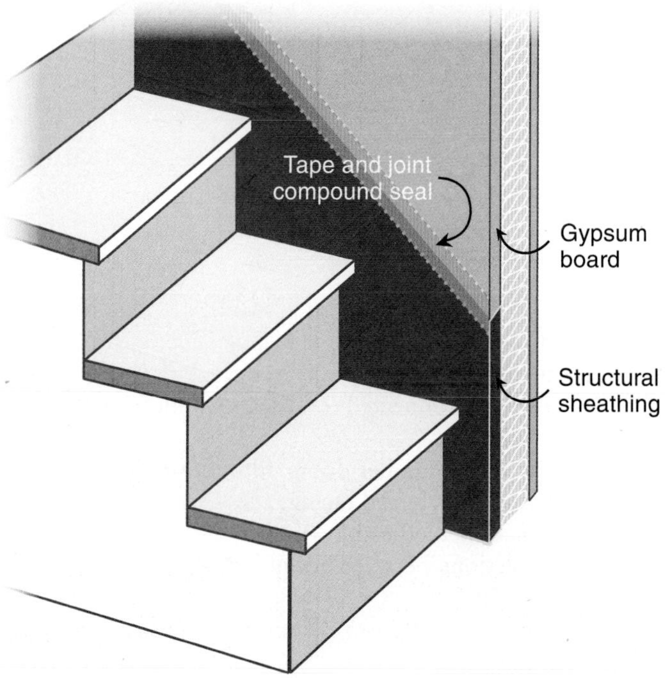

39.8 DUCT LEAKAGE TESTING

In both new and existing homes, as much as 25% of the total energy losses can often be blamed on air leakage in forced-air duct systems. *Testing and sealing leaky air distribution systems is one of the most cost-effective energy improvements for homeowners.* Because air is invisible, most duct leaks go unnoticed by HVACR contractors and homeowners. Often, ducts are in attics, crawl spaces, or other hard-to-access places. It is usually these hard-to-find-and-access leaky ducts that are the important ones to fix because of their being connected to hot attics or humid crawl spaces. Listed below are a few problems caused by leaky ductwork:

1. Leaks in return ducts draw air into the building from crawl spaces, garages, and attics, bringing with it dust, mold spores, pollen, insulation fibers, and other contaminants.
2. Leaks in supply ducts cause expensive conditioned air to be dumped directly outside, in the attic or crawl space, rather than delivering it to the building.
3. Duct leakage greatly increases the use of electric strip heaters in heat pumps during the heating season.
4. In humid climates, return-air ducts drawing in moist air through leaks can overwhelm air-conditioning humidity loads, causing occupants to feel clammy.

Figure 39.42 Seal all plumbing penetrations, electrical outlets, exhaust fans, and tub and showers. Courtesy U.S. Department of Energy

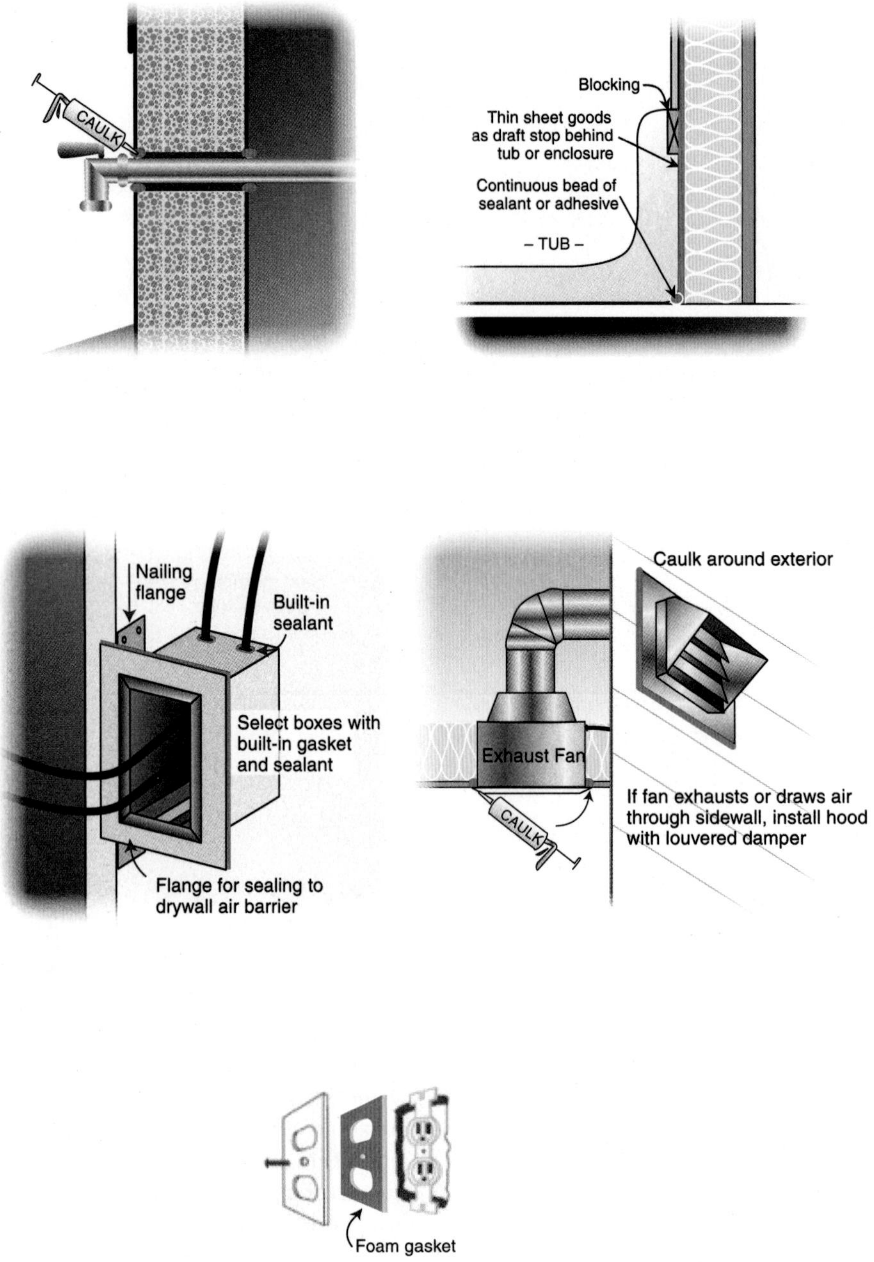

5. Unconditioned air pulled in by leaky return-air ducts will reduce both efficiency and capacity of the air-conditioning system.

Duct leaks can be caused by the initial installation, by general wear and tear on the HVACR system over time, and by equipment or material failures, which may include:

- failed taped duct joints,
- holes in the duct run,
- plenum joints at the air handler,
- tabbed sleeves,
- panned floor joists,
- poorly fitting seams and joints in ductwork,
- return plenums connected to unsealed building cavities,
- any duct joints,
- poorly fitting air handler doors, filter doors, and cabinets,
- deteriorating duct board facing,
- joints in sectioned elbows,
- poor connections between room air registers and the air register boots,
- poor connections between the room air register's boot and the floor or wall,

Figure 39.43 Insulate and air seal the pull-down attic stairs. Courtesy U.S. Department of Energy

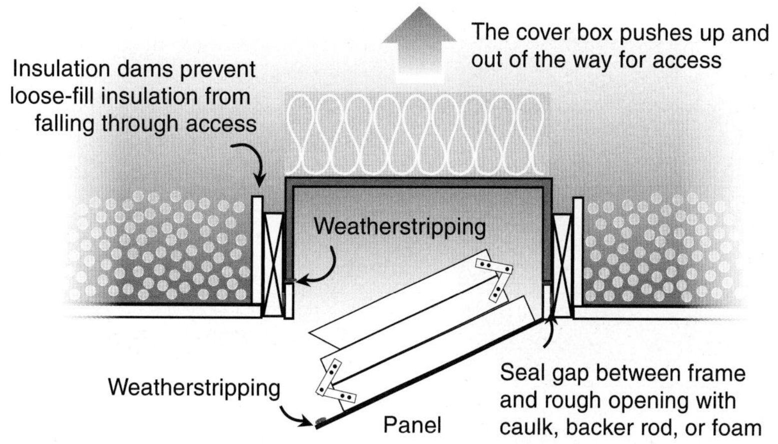

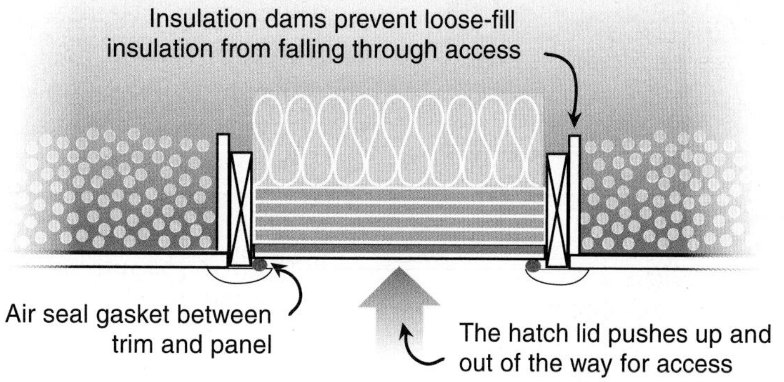

Figure 39.44 (A) Mastic seal all supply- and return-air ducts. Courtesy U.S. Department of Energy

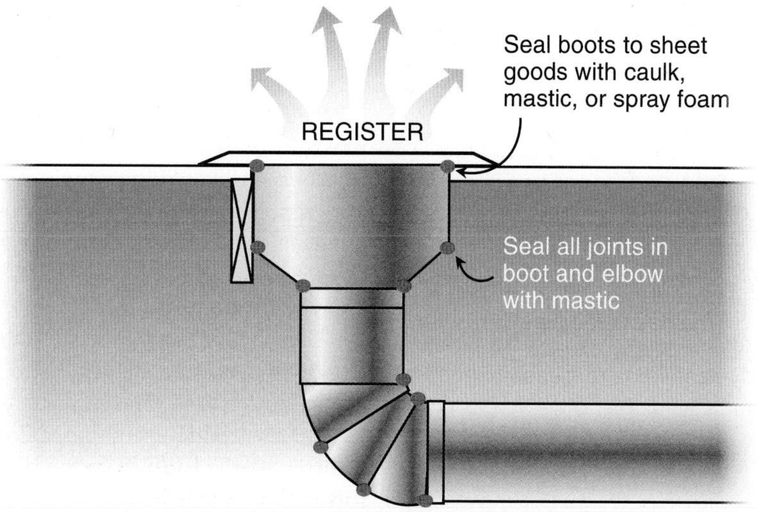

Figure 39.44 (B) Seal all return-air cavities. Courtesy The Energy Conservatory

- joints at branch takeoffs,
- supply and return registers and flooring,
- disconnected or partially disconnected air register boot connections,
- flexible duct to metal duct connections,
- improperly sealed building cavities for supply or return ducts,
- connections on main duct trunks that usually occur every 48 in.,
- collars attached to the main duct trunk,
- leaky duct intersections, or
- leaks at adjustable joints.

What effect duct leakage will have on a home will depend on:

1. whether the leak is connected to an outdoor (unconditioned) space or indoor (conditioned) space.
2. location of the duct leak in relation to the air handler.
3. the size of the duct leak.

Duct leakage to the outside of the conditioned space has a much larger impact on HVACR system performance than duct leakage inside the conditioned space. Duct leakage to the outside usually occurs when leaky duct systems run through attics, garages, or crawl spaces. However, even if leaky ducts are located inside the building envelope in a conditioned space, air leakage to the outside can still occur. This happens when leaky ducts passing through walls or floor cavities create a pressure difference between the cavity the ductwork is in and other building cavities indirectly connected to the outside. *All framed cavities should be lined with OSB, plywood, duct board, drywall, or sheet metal and then sealed with mastic at every joint.* When the air handler blower is operating, air can be forced through these leaks. Some building cavities themselves are used as supply or return ducts. Also, the nearer the leak is to the air handler; the higher the pressure will be in the ductwork. This is why leaks closer to the air handler are more important than leaks farther away.

Duct leakage to a conditioned space inside the house does not have as much impact on system performance and has a smaller energy penalty compared to duct leakage to the outside. If the leaky ductwork is located completely within the conditioned living space of a home, leaking air simply enters the already conditioned space. However, air-balance problems may result from too much or too little air released to a conditioned space because of duct leakage. Duct leakages to conditioned spaces may cause air pressure differences that create air leakages through the house's shell.

Many times, houses have basement return-air systems where panned underfloor joists (joist bays) are used as return ductwork. It is hard to completely seal this type of return-air system, but if the basement is conditioned, the energy loss penalty is very small. However, this may cause the basement to depressurize to the point where combustion products from the water heater and furnace spill into the basement. Moisture leakage and radon gas can also easily enter a house with a depressurized basement. If a basement is conditioned in the winter, it usually is not conditioned in the summer. Supply-duct air leakage during the air-conditioning season will leak cooling air to the unconditioned basement even though the basement has no cooling load.

39.9 DUCT PRESSURIZATION TEST FOR TOTAL AIR LEAKAGE

Measuring the total leakage of air from ducts is accomplished with a duct blower test. This test measures the air leakage rate in cfm of the entire air handling system, which includes both the supply- and return-air ducts, along with the air handler cabinet. A duct blower is shown in **Figure 39.45**. This test gives the energy auditor the air leakage rate once the air handling system is subjected to a uniform test pressure of +25 Pascals of positive pressure using a duct blower. The test pressure of 25 Pascals has been adopted by the majority of residential duct testing programs in the United States because it is representative of typical duct pressures in a residential environment. The total air leakage test measures air leakage rates to both the outside (attics, crawl spaces, and garages) and the inside of the structure.

Using a duct blower to test the air tightness of duct systems is one of the most accurate and common methods for measuring air leakage rates in duct systems. The duct blower consists of a variable-speed fan, a digital manometer, and a set of reducer plates for measuring different air leakage rates. Following the directions in the operations manual of whichever

Figure 39.45 A duct blower. Courtesy The Energy Conservatory

duct blower is purchased is strongly recommended. Digital manometers need a warm-up period and calibration before each use. They also have to be set to the correct range, fan type, and mode. Below are some general guidelines or steps for measuring "total" (indoors and outdoors) duct leakage of an air handling system using a duct blower:

- Open all windows, doors, and vents of the house, including intermediate zones, to the outdoors. This allows all zones (indoor, outdoor, and intermediate) to be open to each other. (This step assures the test is measuring only the duct system tightness, not the tightness of the ducts plus other air barriers, such as foundations, roofs, garages, and walls.)
- Install the duct blower in the air handler with tape and cardboard, **Figure 39.45**. The duct blower can also be connected to a large return-air register using its connector duct.
- Remove the air filters from the air handling and duct system.
- Tape off all supply and return registers with a duct sealing material such as Duct Mast or a high-quality painter's masking tape and masking paper, **Figures 39.46(A)** and **39.46(B)**.
- Drill a ¼-in. hole in the supply duct a short distance away from the air handler and insert a manometer hose. When the manometer is connected, this hose will measure the duct pressure WRT outdoor pressure.
- Connect an airflow manometer hose near the duct blower's fan airflow in cfm.
- Pressurize the duct system to 25 Pascals and record the duct blower's airflow in cfm, **Figure 39.47**.

Figure 39.46 (A) Tape off all supply registers with a duct sealing material such as Duct Mast or a high-quality painter's masking tape and masking paper. Courtesy Ferris State University. Photo by John Tomczyk

Figure 39.46 (B) Tape off all return registers with a duct sealing material such as Duct Mast or a high-quality painter's masking tape and masking paper. Courtesy Ferris State University. Photo by John Tomczyk

- While the ducts and air handler cabinet are pressurized to 25 Pascals, the auditor or technician can check for leaks in the ducts, air handler, and duct branches.
- Once leaks are found and testing is complete, seal the leaks, replace all filters, and remove all tape and paper seals from air registers. Sealing leaks will be covered later in this unit.

39.10 DUCT LEAKAGE TO THE OUTDOORS

As mentioned earlier, duct leakage to the outside of the conditioned space has a much larger impact on HVACR system performance than duct leakage to the inside of the conditioned space. Duct leakage to the outside of the structure's conditioned space usually occurs when leaky duct systems run through attics, garages, or crawl spaces. A duct leakage test to the outdoors is a much more meaningful test to the

Figure 39.47 Pressurize the duct system to 25 Pascals and record the duct blower's airflow in cfm. Courtesy The Energy Conservatory

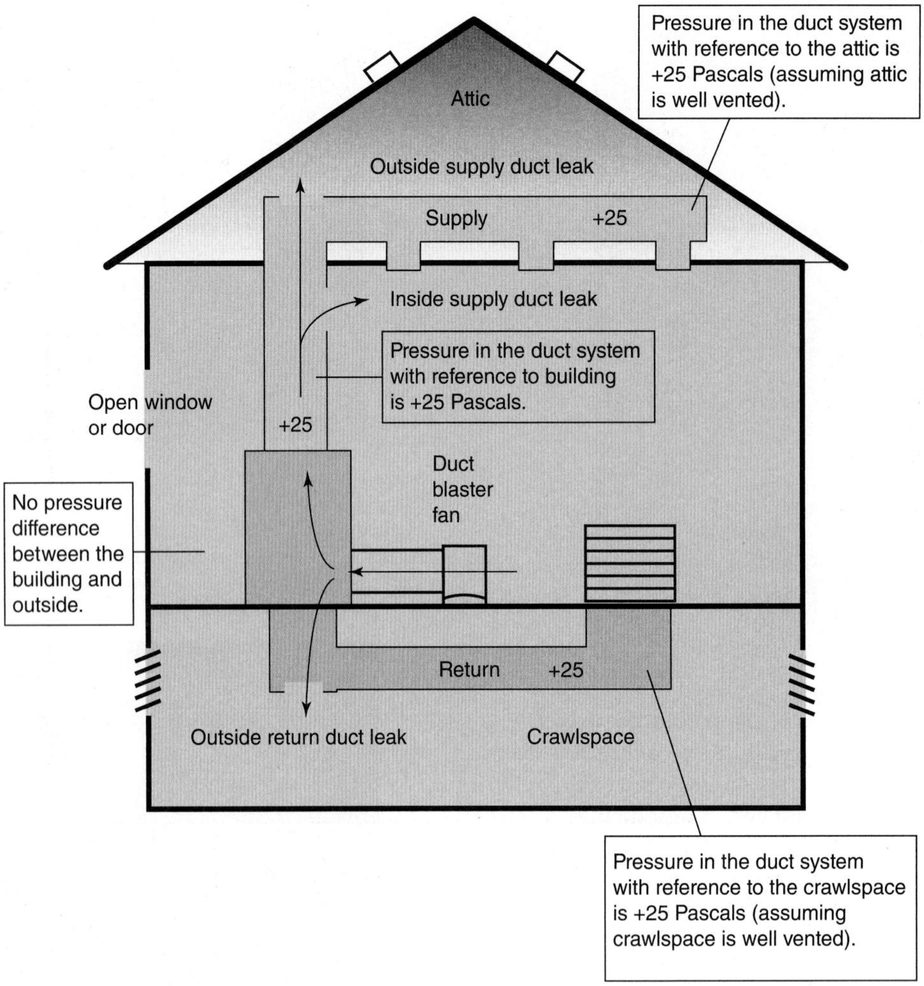

energy auditor than a total duct leakage test because the leakage is directly related to wasted energy and so has much more potential for energy savings than a total duct leakage test.

This test requires the use of both a blower door and a duct blower. Again, following the directions in the operations manual of whichever blower door and duct blower is purchased is strongly recommended. And remember, digital manometers need a warm-up period and calibration before each use. Below are steps or general guidelines for measuring duct air leakage to the outdoors using both a blower door and a duct blower:

- Close all windows and outside doors. Make sure all indoor conditioned areas are opened to one another. This simulates a typical heating or cooling mode inside the home.
- Install a blower door in a configuration to pressurize the home, **Figure 39.48**.
- Connect the duct blower to the air handler or to a return air register, **Figures 39.45** and **39.48**.
- With the duct blower, pressurize the duct system to 25 Pascals. The variable-speed fan on the duct blower can be adjusted until 25 Pascals is reached.
- Pressurize the house with the blower door until the pressure difference between the house and the ductwork is 0.0 Pascals. This is referred to as the pressure of the house WRT ducts.
- The duct blower will give a digital readout of the airflow in cfm into the duct with pressures being equal (25 Pascals).

Using this configuration, all the air that the duct blower is pushing into the ducts will leak outside of the conditioned space. This will give an accurate calculation of the duct leakage to the outdoors. The number is usually given either as the CFM_{25} (cubic feet per minute at 25 Pascals) or as a ratio of the CFM_{25} to the area of the conditioned space.

39.11 COMBUSTION EFFICIENCY AND SAFETY TESTING

It is important that energy auditors, contractors, and HVACR service technicians know the fundamentals of combustion. They will be checking combustion appliances such as furnaces, wood stoves, water heaters, and fireplaces for carbon

Figure 39.48 Install a blower door in a configuration to pressurize the home, then connect the duct blower to the air handler or to a return-air register. Courtesy The Energy Conservatory

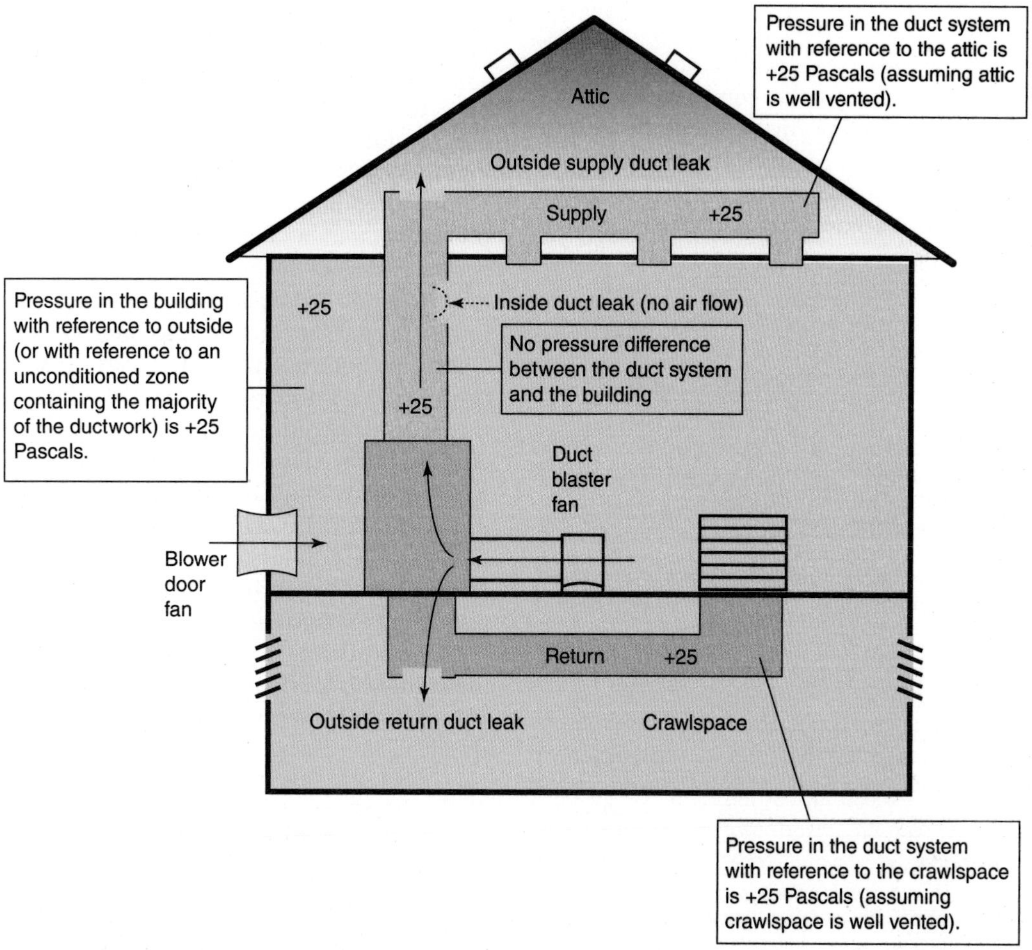

monoxide levels, backdrafting, and other safety hazards such as cracked heat exchangers and gas leakage.

Combustion needs fuel, oxygen, and heat. The reaction of fuel, oxygen, and heat is known as rapid oxidation, or the process of burning. The fuel in this case is natural gas (methane), propane, butane, or fuel oil. The oxygen for combustion comes from the air, which contains approximately 21% oxygen and 79% nitrogen. It takes approximately 5 cubic feet of air to produce 1 cubic foot of oxygen. The heat comes from some type of ignition system, which can be a spark or a hot surface igniter. The fuels contain hydrocarbons that unite with the oxygen when heated to the ignition temperature. Natural gas, propane, butane, and fuel oil are all considered hydrocarbons (HC). Enough air must be supplied to furnish the proper amount of oxygen for the combustion process. For perfect combustion, 2 cubic feet of oxygen (10 cubic feet of air) are needed for each cubic foot of gas. To ensure complete combustion, 3 cubic feet of oxygen (15 cubic feet of air) are supplied for each cubic foot of gas.

The ignition temperature for natural gas is 1100°F to 1200°F. The ignition source must provide this heat, which is the minimum temperature for burning. When the burning process involves methane, the chemical formula is

Perfect combustion:

$$CH_4 + 2O_2 \rightarrow CO_2 + 2H_2O + \text{Heat}$$

where

CH_4 = methane
O_2 = oxygen
CO_2 = carbon dioxide
H_2O = water vapor

This formula represents perfect combustion. This process produces carbon dioxide, water vapor, and heat. Although perfect combustion does not occur in gas-burning appliances, slight variations create no danger. However, large variations away from perfect combustion can be hazardous to human health.

Since perfect combustion does not occur in gas-burning appliances, additional, or excess air, is supplied to ensure complete combustion. The typical amount of excess air for

most applications is 50%. With the excess air being added to the process, the balanced chemical formula for complete combustion is expressed as:

$$CH_4 + 3O_2 \rightarrow 2H_2O + CO_2 + O_2 + \text{Heat}$$

The oxygen (O_2) at the end of the equation is excess oxygen supplied to the process to ensure complete combustion. It is this percentage of excess oxygen that informs us about the fuel–air mixture and directly relates to how complete the combustion is.

Incomplete combustion, in addition to water vapor and carbon dioxide, produces carbon monoxide and possibly unburned carbon. A sample balanced chemical formula for incomplete combustion would be

Incomplete combustion:

$$4CH_4 + 6O_2 \rightarrow CO_2 + 8H_2O + 2CO + C + \text{Heat}$$

where

CO = carbon monoxide
C = unburned carbon or soot

One of the primary causes of incomplete combustion is lack of combustion air, as can be seen in the previous chemical formula. The (C) at the end of the equation is unburned carbon or soot.

SAFETY PRECAUTION: *The technician must see to it that the combustion is as close to perfect as possible. By-products of poor or imperfect combustion are carbon monoxide—a poisonous gas—soot, and minute amounts of other products like nitrous oxides (NO_x) and sulfur dioxide (SO_2). NO_x is a toxic gas that is formed in the hottest part of the flame. It should be obvious that carbon monoxide production must be avoided. Carbon monoxide (CO) is a very toxic gas that can go unnoticed because it is odorless, colorless, and tasteless. It is deadly at high levels, and at low levels it will make people very sick, cause nagging illnesses that can hide behind other symptoms of illness. Technicians can measure the exhaust gases of combustion appliances with very accurate, sophisticated digital testing instrumentation,* ***Figures 39.50, 39.53(B),*** *and* ***39.55****, these instruments can measure carbon monoxide in parts per million (ppm). It is very important for homeowners to install carbon monoxide detectors to protect them from the dangers of carbon monoxide poisoning. Refer to Unit 34, "Indoor Air Quality," for more detailed information on carbon monoxide and other indoor pollutants.*•

39.12 COMBUSTION ANALYSIS

Digital combustion analyzers can measure:

1. carbon monoxide (CO),
2. oxygen (O_2),
3. draft, and
4. flue gas temperature.

Digital combustion analyzers can calculate:

1. carbon dioxide (CO_2)
2. carbon monoxide air-free (COAF)
3. net stack temperature
4. combustion efficiency

These digital flue gas analyzers will calculate combustion efficiency in a percentage and in steady-state efficiency (SSE). The term *steady-state efficiency* refers to combustion efficiency when the flue gas temperature (stack temperature) has stabilized and the flue gas samples reach equilibrium. Combustion efficiency is the ability of the combustion appliance to convert fuel into heat energy. Today, technicians and/or auditors can quickly test the efficiency of furnaces, **Figure 39.49.**

Some common efficiency problems are low fuel input pressure and high O_2 (excess air). This will result in very poor heat transfer. Only by performing a flue gas analysis with a combustion analyzer can an auditor or technician detect this problem. The remedy is to optimize the fuel–air mixture, which can increase the SSE up to 10%. Another common problem is a lower than normal flue gas temperature. Low flue gas temperature is usually an indicator of good efficiency. However, if flue gas temperature is too low, especially in older conventional furnaces, condensation will form in the vent pipe. This happens because the flue gas has cooled below its dew point temperature (DPT). This condensation has been known to be a bit acidic and can rust metal vent pipes and damage masonry chimneys.

Certain furnace parameters are very important for indicating how efficiently the furnace is performing. These parameters also tell the auditor or technician how safely the furnace is operating.

Figure 39.49 A technician performing a combustion analysis on a furnace. Photo by Jason Obrzut

- **Percent oxygen (O_2)**—Indicates whether the fuel–air mixture is correct. Percent O_2 indicates the percentage of excess air and is directly related to furnace efficiency. As the percent O_2 increases, the furnace efficiency decreases along with the percentage of carbon dioxide.
- **Stack (flue gas) temperature**—Stack temperature or flue gas temperature is indirectly related to furnace efficiency. As flue gas temperature increases, furnace efficiency decreases. Stack temperature is a major indicator of combustion efficiency; it reflects the heat that did not transfer from the combustion gases to the heating medium (air, water, or steam). A high flue gas temperature means a lot of heat is escaping out the vent of the furnace, making it less efficient. A too low flue gas temperature also may cause inadequate drafting of the furnace. Low flue gas temperatures can also cause condensation in the vent pipe when the flue gas cools enough to reach its dew point temperature. Stack temperature must be measured before the introduction of dilution air, which means before the draft diverter or draft hood of the combustion appliance, if one is used. Net stack temperature is the stack temperature minus the combustion air temperature. For example, if the actual measured stack temperature is 300°F and the ambient around the furnace is 70°F, then the net stack temperature is 230°F (300°F − 70°F). Modern combustion analyzers usually measure the room's ambient temperature during the warm-up period. Always follow the combustion analyzer's instructions for proper operating procedures when combustion air is piped from the outdoors directly to the combustion appliance.
- **Carbon monoxide (CO)**—Incomplete combustion will create CO. By the time it reaches the flue pipe, the carbon monoxide has been diluted with excess or dilution air. When measuring carbon monoxide in the flue gas, an undiluted, or air-free reading, is necessary. Most combustion analyzers will calculate a carbon monoxide air-free (COAF) reading. A properly operating furnace or boiler will not typically exceed 100 ppm (undiluted) at steady-state operation. Adjustments to the primary air, gas pressure, and burner maintenance can usually lower the level of CO, **Figure 39.50**.

Figure 39.50 A technician testing the level of carbon monoxide (CO) in a furnace's flue gases. Photo by Jason Obrzut

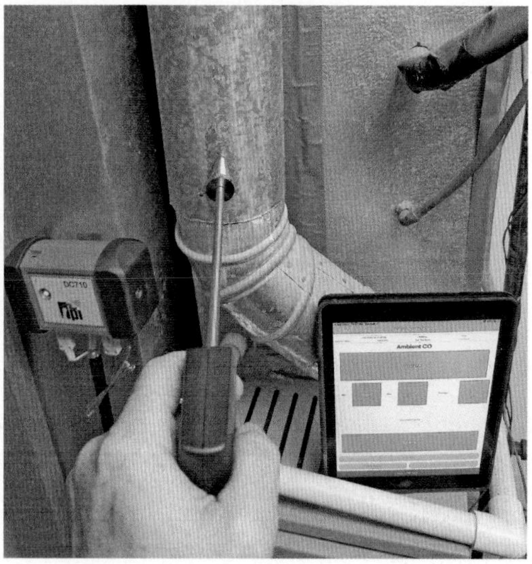

- **Airflow**—Airflow through the furnace is directly related to the temperature rise through the furnace. The more airflow in cubic feet per minute (cfm), the lower the temperature rise. The temperature rise is a measurement of the supply-air temperature minus the return-air temperature. **Figure 39.51** illustrates how to measure the temperature rise across a furnace. Airflow in cfm has a large impact on flue gas temperature, temperature rise, and furnace output in Btu/h. The higher the cfm, the lower the flue gas temperature and temperature rise, but the higher the capacity of the furnace in Btu/h.

Figure 39.51 How to measure the temperature rise through a furnace.

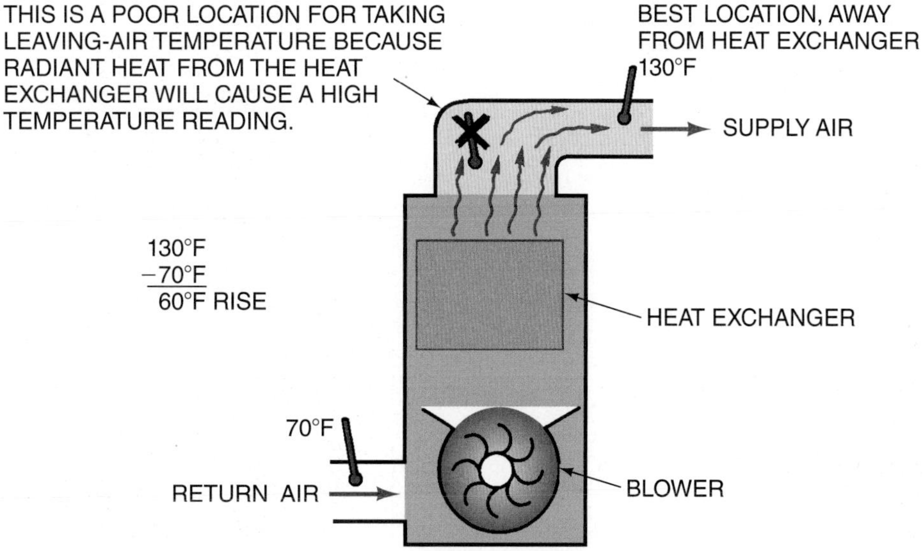

Figure 39.52 General combustion guideline standards for gas furnaces.

PERFORMANCE INDICATORS	SSE 70+	SSE 80+	SSE 90+
Carbon Monoxide (CO) ppm	Less than 100 ppm	Less than 100 ppm	Less than 100 ppm
Temperature Rise (°F)	60°–90° *	45°–75° *	35°–65° *
Stack Temperature (°F)	350°–450°	275°–350°	110°–120°
Oxygen (% O_2)	5–10%	4–9%	4–9%
Gas Pressure (in. WC)	3–4 *	3–4 *	3–4 *
Draft in. WC (Pascals) Pa	−0.02 to −0.04 in. WC (−5 to −10 Pa)	−0.02 to −0.04 in. WC (−5 to −10 Pa)	 +20 to 60 Pa
Steady State Efficiency (SSE) %	73–78%	79–82%	91–97%

*Follow manufacturer's instructions.

As mentioned earlier, CO_2 is a product of complete combustion. The higher the percentage of CO_2, the greater the steady-state efficiency (SSE). CO_2 will range usually between 3% and 14%. Oxygen (O_2) is an indicator of the amount of excess air in the combustion gases. As the O_2 gets higher, the SSE will decrease. Oxygen will also range from 3% to 10%. Generally, the SSE will decrease as the stack temperature increases. However, to determine SSE, the stack temperature must be compared and analyzed with the O_2 and/or the CO_2 levels. Refer to **Figure 39.52** for general guidelines on combustion standards for gas furnaces.

Furnace heat exchangers are made of sheet steel or aluminized steel and are designed to provide rapid transfer of heat from the hot combustion products through the steel to the air that will be distributed to the space to be heated. Heat exchangers must have the correct airflow across them or problems will occur. If too much air flows across a heat exchanger, the flue gas will become too cool and the products of combustion may condense and run down the flue pipe. These products are slightly acidic and will deteriorate the flue system if not properly removed. However, high-efficiency condensing gas furnaces and their venting systems are equipped to handle the condensation of flue gases. If there is not enough airflow, the combustion chamber will overheat, and stress will occur.

Generally, too much airflow will decrease the temperature rise across the furnace. This happens because the air is in contact with the heat exchanger for only a short time. Too little airflow will increase the temperature rise across the furnace because the air is in contact with the heat exchanger for a longer time period. Furnace manufacturers print the recommended air temperature rise for most furnaces on the furnace nameplate. However, the nameplate-rated temperature rise is dependent on whether the furnace is standard-, mid-, or high-efficiency. (This will be explained later in this unit.) The temperature rise of the furnace decreases as the efficiency increases. This can be verified by taking the return-air temperature and subtracting it from the leaving-air temperature, **Figure 39.51**. Be sure to use the correct temperature probe locations. Radiant heat can affect the temperature readings.

The following formula may be used to calculate the exact airflow across a gas furnace:

$$\text{cfm} = Q_s \div (1.08 \times \Delta T)$$

where

Q_s = Sensible heat in Btu/h
cfm = Cubic feet of air per minute
1.08 = A constant
ΔT = Temperature difference between supply and return air

This requires some additional calculations because gas furnaces are rated by the input. If a gas furnace is 80% efficient during steady-state operation, multiply the input times the efficiency to find the output. For example, if a gas furnace has an input rating of 80,000 Btu/h and a temperature rise of 55°F, how much air is moving across the heat exchanger in cfm? The first thing to do is to find out what the actual heat output to the airstream is: 80,000 × 0.80 (80% estimated furnace efficiency) = 64,000 Btu/h. Use this formula:

$$\text{cfm} = Q_s \div (1.08 \times \Delta T)$$
$$\text{cfm} = 64{,}000 \div (1.08 \times 55)$$
$$\text{cfm} = 1077.4$$

The answer's accuracy will be increased by taking several temperature readings at both the supply- and return-air ducts and averaging them. The more readings taken, the more accurate the answer will be. A good digital thermometer will also increase accuracy.

39.13 FURNACE EFFICIENCY RATINGS

Furnace efficiency ratings are determined by the amount of heat that is transferred to the heated medium, which is usually water or air. The following factors determine the efficiency of the furnace:

1. Type of draft (natural atmospheric injection, induced, or forced)
2. Amount of excess air used for combustion
3. Delta-T, or temperature difference, of the air or water entering versus leaving the heating medium side of the heat exchanger
4. Flue stack temperature

The following are general furnace classifications and approximate Annual Fuel Utilization Efficiency (AFUE) ratings:

Conventional, or Standard-Efficiency, Furnace (78% to 80% AFUE)

- Atmospheric injection or draft hood
- 40% to 50% excess air usage
- Delta-T of 70°F to 100°F (lowest cfm airflow of furnaces)
- Stack temperatures of 350°F to 450°F
- Noncondensing
- One heat exchanger

Mid-Efficiency Furnace (78% to 83% AFUE)

- Usually induced or forced draft
- No draft hoods
- 20% to 30% excess air
- Delta-T of 45°F to 75°F (more cfm airflow than conventional units)
- Stack temperatures of 275°F to 300°F
- Noncondensing
- One heat exchanger

High-Efficiency Furnace (87% to 97% AFUE)

- Usually induced or forced draft
- No draft hoods
- 10% excess air
- Delta-T of 35°F to 65°F (highest cfm of all furnaces)
- Stack temperatures of 110°F to 120°F
- Two or three heat exchangers
- Condensing furnace (reclaims the latent heat of condensation by condensing flue-gas water vapors to liquid water)
- Use of PVC pipe for appliance venting to avoid corrosion

Gas furnaces with high-efficiency ratings are being installed in many homes and businesses. The U.S. Federal Trade Commission requires manufacturers to provide an Annual Fuel Utilization Efficiency (AFUE) rating for their furnaces. This rating allows the consumer to compare furnace performance before buying. Annual efficiency ratings have been increased in some instances from 65% to 97% or higher. A part of this increase in efficiency is accomplished by transferring more heat from the flue gases to the conditioned space. As a result, less heat is lost to flue pipe and the venting process. In conventional furnaces, stack temperatures are kept high to provide a good draft for proper venting. High temperatures also prevent condensation and therefore corrosion in the vent. But because of the high stack temperatures, excess heat is allowed to escape to the atmosphere, reducing efficiency. The reason for the low Delta-T on high-efficiency furnaces is the increased amount of air (cfm) through their heat exchangers. This keeps the air in contact with the heat exchanger for a shorter period of time.

39.14 FLAME COLOR

A good flame has an inner and outer mantle and is hard and blue. Orange streaks in the flame are not to be confused with yellow streaks. Orange streaks are dust particles burning. Other than pressure to the gas orifice, the primary air is often the only adjustment that can be made to the flame. Modern furnaces have very little adjustability. This is intentional in order to maintain minimum standards of flame quality. The only thing that will change the furnace combustion characteristics to any extent, other than primary air and gas pressure adjustments, is dirt or lint drawn in with the primary air. If the flame begins to burn yellow, primary air restriction should be suspected. For a burner to operate efficiently, the gas flow rate must be correct, and the proper quantity of air must be supplied.

⚠SAFETY PRECAUTION: *Yellow tips indicate an air-starved flame emitting poisonous carbon monoxide.*•

Cooling the flame will cause inefficient combustion. This happens when the flame strikes the sides of the combustion chamber (due to burner misalignment), which is called flame impingement. When the flame strikes the cooler metal of the chamber, the temperature of that part of the flame is lowered below the ignition temperature. This results in poor combustion, which produces carbon monoxide and soot. Soot, which is basically unburned carbon, lowers furnace efficiency because it collects on the heat exchanger and acts as an insulator, so it also must be kept at a minimum. Refer to Unit 31, "Gas Heat," for more detailed information on furnace efficiency ratings and correct flame color.

39.15 SPILLAGE AND BACKDRAFTING

As mentioned earlier, carbon monoxide (CO) is a colorless and odorless gas which can sicken or kill persons residing in the home. Carbon monoxide can enter the homes living spaces by

- backdrafting,
- spillage, or
- cracked heat exchangers.

Backdrafting can be caused by pressure differences. When the pressure is lower in the living space of the home with respect to the pressure in the vent or chimney for the combustion appliances, backdrafting will occur. The pressure difference will draw combustion by-products down the chimney or vent pipe and often over the flame, causing a suffocated flame and carbon monoxide, **Figure 39.56(C)**. **Figure 39.53(A)** shows a technician testing for backdrafting or downdrafting with a chemical smoke generator at the draft diverter of a residential furnace. **Figure 39.53(B)** shows a technician testing for backdrafting on a water heater by checking for the presence of carbon monoxide (CO) with a digital meter. In **Figure 39.53(C)**, a technician is checking for backdrafting in a residential fireplace using a chemical smoke generator. **Figure 39.53(D)** shows a

Figure 39.53 (A) A technician using a chemical smoke generator in the vicinity of the draft diverter on a standard-efficiency furnace to check for proper draft. Courtesy Ferris State University. Photo by John Tomczyk

Figure 39.53 (B) A technician checking a water heater with a digital carbon monoxide sensing meter. The meter will sense ppm CO coming into the home from backdrafting of the combustion appliance. Courtesy Fluke Corporation

technician measuring the airflow from an exhaust fan using an exhaust fan flow meter and digital gauge. Exhaust fans can cause a negative pressure in the home because they exhaust inside conditioned air, which must be made up or a negative pressure will result. Backdrafting can be caused by

- high winds,
- combustion using inside air,
- open fireplaces,

Figure 39.53 (C) A technician checking for backdrafting in a residential fireplace using a chemical smoke generator. Courtesy Ferris State University. Photo by John Tomczyk

Figure 39.53 (D) A technician measuring the airflow from an exhaust fan using an exhaust fan flow meter and digital gauge. Courtesy The Energy Conservatory

- clothes dryers,
- exhaust fans,
- leaky return-air ducts in the combustion appliance zone (CAZ),
- a poorly designed chimney, or
- a poorly designed venting system.

Some remedies for backdrafting problems include sealing all leaky ducts, especially return-air ducts in the CAZ; eliminating all sources of depressurization (clothes dryers and exhaust fans) near the vent or chimney; and providing outdoor air for all combustion appliances. Large, open fireplaces are especially a hazard because their draft is so powerful it will bring the inside living space of a residence into

a negative pressure quickly. Make sure all fireplaces have sealed combustion chambers where outside air is used for the combustion process. This will prevent depressurization of the inside of the home and backdrafting of other combustion appliances. In all cases, bring outdoor air into the CAZ and replace open-combustion, atmospheric (non-direct-vented) appliances with sealed-combustion (direct-vented) power-drafted, high-efficiency appliances, **Figure 39.54**. The Building Performance Institute (BPI) considers naturally drafting appliances and venting systems antiquated and both unsafe and inefficient. It recommends replacing them with direct-vented power-draft-assisted appliances.

Spillage is not as serious as backdrafting, but it is an unhealthy situation and can cause sickness in households if not controlled. Spillage is a temporary flow of combustion by-products into the living space. Spillage usually happens when a furnace (combustion appliance) has just started up and the chimney or vent has a weak draft. The weak draft is caused by a cool chimney or vent pipe. Spillage usually stops once the chimney or vent warms up. **Figure 39.55** shows a technician checking a furnace for CO caused by spillage.

Figure 39.54 A sealed combustion chamber with a flame inspection port on a high-efficiency furnace. Courtesy Ferris State University. Photo by John Tomczyk

Figure 39.55 A technician checking a furnace for ppm carbon monoxide (CO) caused by spillage. Courtesy Fluke Corporation

Cracked heat exchangers can allow combustion products to mix with the heating air that enters the living spaces of the house and cause serious health effects. Heat exchangers can become cracked by

- failed weld joints,
- rust from condensation,
- stress cracks from excessive heat, or
- old age.

Heat exchangers can be inspected for cracks by probing the combustion chamber with an inspection camera. Using lights, chemicals, and smoke generators in the combustion chamber can also search out leaky heat exchangers. Detection devices placed at the supply registers in living areas can detect the chemicals or smoke. A light placed in the heat exchanger can often be seen from the outside if there is a crack. If there is any variation in color, size, or motion of the flame when the furnace's blower turns on, the heat exchanger is probable cracked.

39.16 FLAME SAFEGUARD CONTROLS

Flame safeguard controls protect against unignited fuel collecting in the combustion chamber. The build up of such fuel can quickly become a serious fire and explosion hazard. The three most common types of flame safeguard controls are

1. thermocouples,
2. flame rectification, and
3. a cadmium sulfide (Cad) cell or photo cell.

Thermocouples are found mainly on older gas furnaces. This control consists of two dissimilar metals welded together at one end, called the "hot junction." When this junction is heated, it generates a small voltage (approximately 15 mVDC with load; 30 mVDC without load) across two wires at the other end. As long as the electrical current in the thermocouple energizes a coil, the gas can flow. If the flame goes out, the thermocouple will cool off in about 30 to 120 sec, no current will flow, and the gas valve will close. Most older gas furnaces use thermocouple systems, but they are gradually being phased out and replaced with more efficient furnace models incorporating flame rectification safeguard controls. However, many water heaters still use thermocouples as flame safeguard controls.

Flame rectification is based on the fact that the pilot flame or main flame can conduct electricity because it contains ionized combustion gases, which are made up of positively and negatively charged particles. The flame is located between two electrodes of different sizes. The electrodes are fed with an alternating current (AC) signal from

the furnace's electronic module. Because the electrodes are of different sizes, current will flow better in one direction than in the other with the flame acting as a switch. If there is no flame, the "switch" is open and electricity will not flow. When the flame is present, the switch is closed and will conduct electricity. The two electrodes in a spark-to-pilot (intermittent pilot) ignition system are the pilot hood (earth ground) and a flame rod or flame sensor.

As AC switches the polarity of the electrodes 60 times each second, the positive ions and negative electrons move to the different-size electrodes. The particles move at different speeds because the positively charged particles (ions) are relatively larger, heavier, and slower than the negatively charged particles (electrons). In fact, the electrons are about 100,000 times lighter than the ions. This produces a rectified AC signal, which looks much like a pulsating DC signal. It is the DC signal that the electronics of the furnace recognize. It is proof that a flame exists and is not a short resulting from humidity or direct contact between the electrodes. Humidity or direct contact would cause an AC current to flow, which the furnace's electronic module would not recognize. An AC signal would prevent the main gas valve from opening in a spark-to-pilot system. In a DSI system it would shut down the main gas once the trial for ignition timing had elapsed and might put the furnace in a lockout mode. The DC signal can be measured with a microammeter in series with one of the electrode leads and usually ranges between 1 and 25 microamps. The magnitude of the microamp signal may depend on the quality, size, and stability of the flame and on the electronic module and electrode design. Refer to Unit 31, "Gas Heat," for more detailed information on thermocouple and flame rectification operation.

Cad cells, or photo cells, are usually used in conjunction with a primary control and are most often found on oil-fired furnaces. The main component of the relay is the cad cell, which is made from a chemical compound called cadmium sulfide that changes its resistance in response to the amount of light the material is exposed to. The cad cell is made up of cadmium sulfide, a ceramic disc, a conductive grid, and electrodes. The entire assembly is encased in glass. The cell is positioned in the oil burner so that it can sense the light produced when the air-oil mixture is ignited. When the burner is off, no light is being generated and the resistance of the cad cell is very high, about 100,000 ohms. When the burner is firing, the resistance of the cad cell drops to under 1000 ohms. The primary control will shut the burner down in a preset amount of time if no light is sensed by the cad cell and its resistance remains high. Refer to Unit 32, "Oil Heat," for more detailed information on cad cell operation.

39.17 EXCESS AIR

Excess air consists of (1) combustion air and (2) dilution air. Combustion air is primary and/or secondary air. Primary air enters before combustion takes place. It mixes with raw gas in the burner before a flame has been established. Too much primary air can cause the flame to "lift off" the burner and be noisy or unsettled. Too little primary air can cause a very lazy flame that looks like it is searching for air. A good flame has an inner and outer mantle and is hard and blue. Secondary air enters after combustion and supports it. Dilution air is excess air that enters after combustion, usually at the end of the heat exchanger. It is brought in by the draft hood of the furnace, **Figure 39.56(A)**.

As excess air decreases, the dew point temperature (DPT) increases. The DPT for undiluted flue gas is around 140°F. In diluted flue gas, the temperature may drop to about 105°F. This is the main reason high-efficiency condensing furnaces require very little excess air for combustion. They operate with stack temperatures below the DPT of the combustion gases to promote condensation. Refer to Unit 31, "Gas Heat," for more information on excess air.

39.18 VENTING

Conventional gas furnaces use high flue gas temperatures and natural convection to vent the products of combustion (flue gases). The hot gas is vented quickly, primarily to prevent cooling, which produces condensation and other corrosive activity. As mentioned, conventional non-direct-vented furnaces lose some efficiency because considerable heat is lost up the flue. High-efficiency furnaces (90% and higher) recirculate the flue gases through a condensing heat exchanger to keep more of the heat available for space heating in the building. The gases are then pushed out of the flue by a small fan. There is some condensation and thus corrosion in the flue, so plastic pipe is used because it is not damaged by the corrosive materials. **⚠ SAFETY PRECAUTION:** *Regardless of the type, combustion appliances must be properly vented so that the by-products of combustion will all be dissipated into the atmosphere.•*

A venting system must provide a safe and effective means of moving the flue gases to the outside air. With a conventional furnace, it is important for the flue gases to be vented as quickly as possible. They are equipped with a draft hood that blends some room air with the rising flue gases, **Figures 39.56(A)** and **56(B)**. The products of combustion enter the draft hood and are mixed with air from the area around the furnace. Approximately 100% additional air (called dilution air) enters the draft hood at a lower temperature than the flue gases. The heated gases rise rapidly

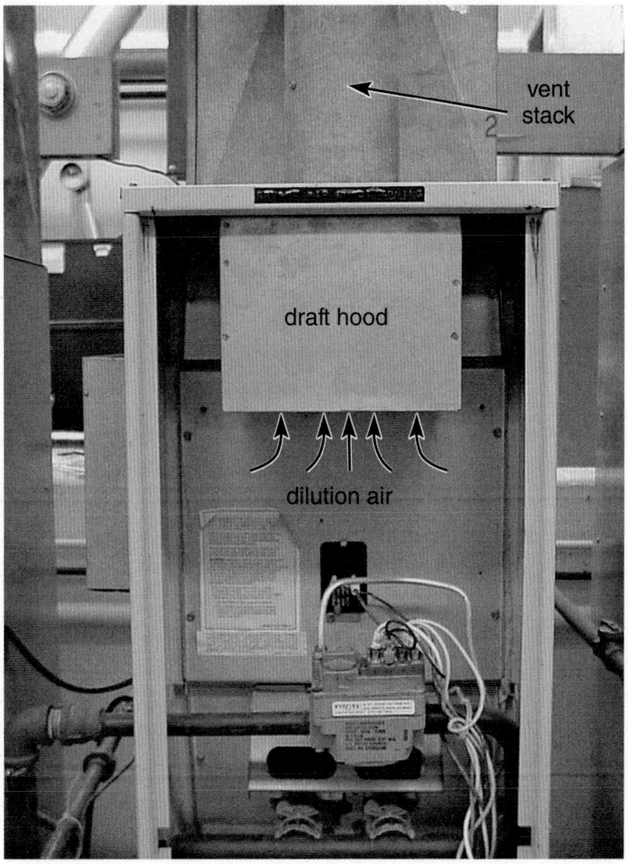

Figure 39.56 (A) Dilution air is excess air after combustion and usually enters at the end of the heat exchanger. It is brought in by the draft hood of the furnace. Courtesy Ferris State University. Photo by John Tomczyk

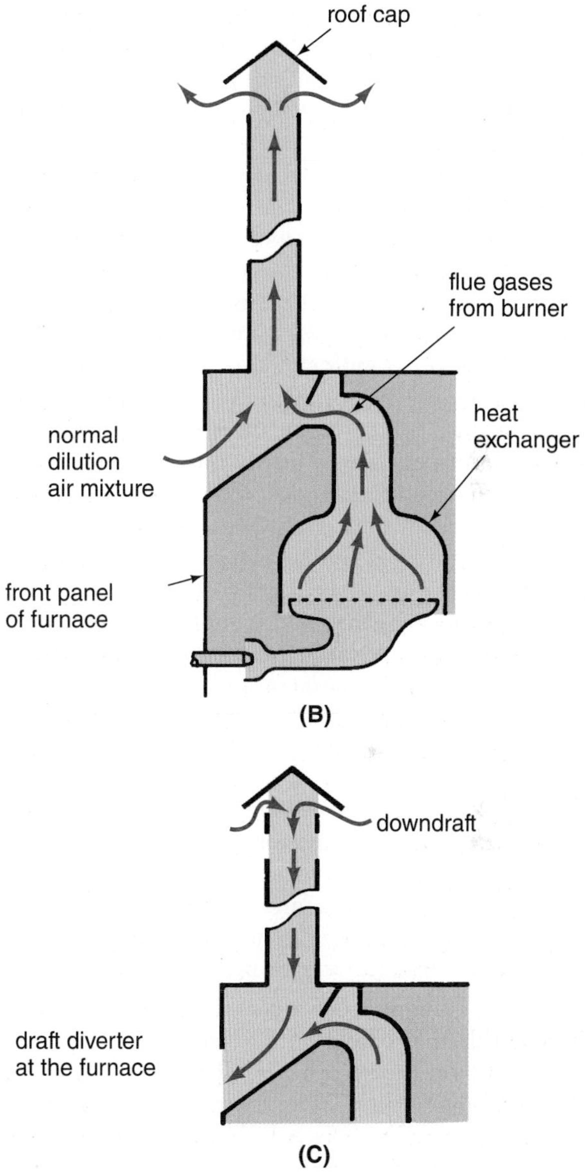

Figure 39.56 (B) and (C) Conventional furnaces are equipped with a draft diverter or draft hood that blends some room air with the rising flue gases. The diverter or hood also prevents downdraft air from blowing out the burner flame and pilot light, if the furnace has one.

and create a draft, bringing the dilution air in to mix as they move up the vent.

All furnaces must be supplied with the correct amount of excess air. As mentioned earlier, excess air consists of combustion air and dilution air. Combustion air consists of both primary and secondary air. This amount of air for gas-burning appliances is determined by the appliance size and is mandated by local codes or the National Fuel Gas code, which is considered the standard by many states and the International Fuel Code which is the code most jurisdictions adopt. Combustion air can be brought into the CAZ from adjacent rooms or from the outdoors. According to the guidelines set forth by the NFPA, if indoor air from an adjacent space is to be used, two openings, each with a minimum free area of 1 in^2 per 1000 Btu/h are required. One of the openings is to be located high in the space, closer to the ceiling, while the other is to be located low in the space, closer to the floor. If combustion air is to be brought in directly from outdoors, only one opening, located high in the space, with a free area of 1 in^2 per 3000 Btu/h is required. The term free area refers to the portion of a grille through which combustion air can flow. Grilles are the decorative, and often functional, covers that conceal the openings.

The louvers of the grille take up space. Metal grilles have approximately 75% free area, while wood grilles have approximately 25%. For example, a 12 in. × 12 in. metal grille has a surface area of 144 in^2. The free area of the

grille would be approximately 75% of 144 in^2 (144 × 0.75), or 108 in^2. For a 100,000 Btu/h furnace, two openings of 100 in^2 each would be needed to bring in enough indoor combustion air. Therefore, two openings of 12 in. × 12 in. with a metal grille would be sufficient for this particular application.

This example of figuring out what adequate ventilation air should be is only one possibility. In the past, excess air was merely extracted from the volume of air in the structure. Structures today are built tighter with less infiltration air entering around the windows and doors, so makeup air for the furnace must be considered. Even old structures are being tightened by the use of storm windows, weather stripping, and other weatherization methods, so there is less air infiltration and exfiltration. A furnace installed in an older structure that has been tightened may have venting problems as evidenced by flue gases spilling out of the draft diverter rather than exiting out the flue. If wind conditions produce a downdraft, the opening in the draft hood provides a place for the gases and air to go, diverting it away from the pilot and main burner flame. The draft diverter helps reduce the chance of the pilot flame being blown out and the main burner flame altered, **Figure 39.56(B)**.

Venting categories were created when high-efficiency furnaces were introduced. Following are the four venting categories listed by the American National Standards Institute (ANSI). Based on ANSI Standard Z21.47A-2016, they set venting standards for temperature and pressure for the proper venting of gas furnaces.

- **Category I**—Furnace has a nonpositive vent pressure and operates with a vent gas temperature at least 140°F above its dew point (conventional or standard-efficiency furnace).
- **Category II**—Furnace has a nonpositive vent pressure and operates with a vent gas temperature less than 140°F above its dew point (no longer manufactured because of condensate problems at start-up).
- **Category III**—Furnace has a positive vent pressure and operates with a vent gas temperature at least 140°F above its dew point (mid-efficiency furnace).
- **Category IV**—Furnace has a positive vent pressure and operates with a vent gas temperature less than 140°F above its dew point (high-efficiency condensing furnace).

It takes 10 ft^3 of air to support combustion for each 1 ft^3 of natural gas. To this is added 5 ft^3 of excess air to ensure enough oxygen. For induced- or forced-draft furnaces, at total of 15 ft^3 of air for each 1 ft^3 of gas is required. A 100,000 Btu/h furnace would require 1500 ft^3/h of fresh air for proper combustion: (100,000 Btu/h ÷ 1000 Btu/ft^3) × 15 ft^3/h = 1500 ft^3/h.

For natural-draft appliances, another 15 ft^3 of dilution air is added at the draft hood, making a total of 30 ft^3 of air for each 1 ft^3 of gas. A 100,000 Btu/h furnace would require 3000 ft^3/h of fresh air: (100,000 Btu/h ÷ 1000 Btu/ft^3) × 30 ft^3/h = 3000 ft^3/h. All of this air must be replaced in the furnace area and also must be vented as flue gases rise up the vent. If the air is not replaced, the furnace area will become a negative pressure area and air will be pulled down the flue. Products of combustion will then fill the area.

A venting system must be continuous from the furnace to the proper height above the roof. Type B vent pipe is usually of a double-wall construction with air space between. The inner wall may be made of aluminum and the outer wall may be constructed of steel or aluminum. The vent pipe should be at least the same diameter as the vent opening at the furnace. The horizontal run should be as short as possible; long runs lower the temperature of the flue gases before they reach the vertical vent or chimney and they reduce the draft. The run should always be sloped upward as the vent pipe leaves the furnace. A slope of ¼ in. per foot of run is the minimum recommended, **Figure 39.57**. As few elbows as possible should be used. The vent connector should not be inserted beyond the inside of the chimney wall, and there should be no obstructions in the chimney when venting into a masonry-type chimney, **Figure 39.58**. If two or more gas appliances are vented into a common flue, the common flue size should be calculated based on recommendations from the National Fire Protection Association (NFPA) or local codes. The top of each gas furnace vent should have an approved vent cap to prevent the entrance of rain and debris and to help prevent wind from blowing the gases back into the building through the draft hood.

Figure 39.57 The horizontal run should be as short as possible. Horizontal runs should always be sloped upward as the vent pipe leaves the furnace. A slope of ¼ in. per foot of run is the minimum recommended.

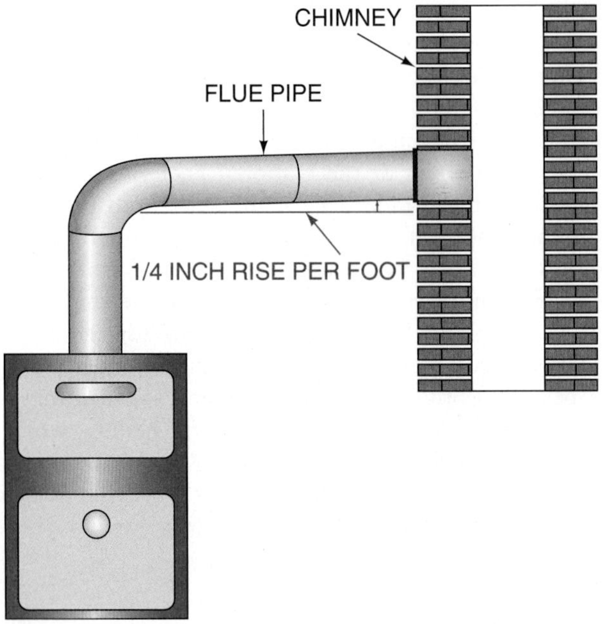

Figure 39.58 The vent connector should not be inserted beyond the inside of the wall of the chimney, and there should be no obstructions in the chimney when venting into a masonry-type chimney.

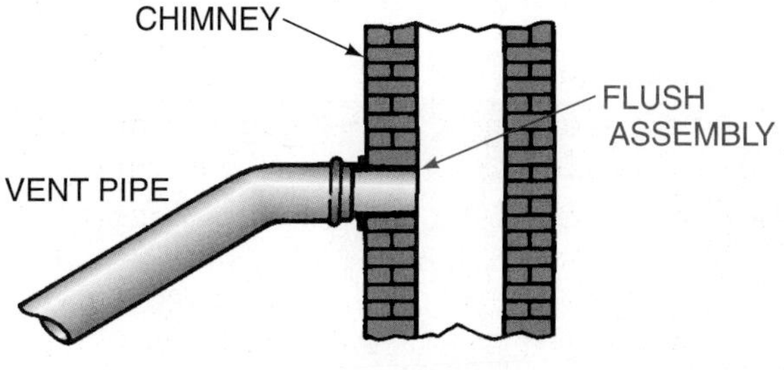

Metal vents reach operating temperatures quicker than masonry vents do. Masonry chimneys tend to cool the flue gases. They must be lined with a glazed-type tile or vent pipe. If an unlined chimney is used, the corrosive materials from the flue gas will destroy the mortar joints. It takes a long running cycle to heat a heavy chimney. Condensation occurs at start-up and during short running times in mild weather. The warmer the gases, the faster they rise and the less damage there is from condensation and other corrosive actions. Vertical vents can be lined and glazed masonry or prefabricated metal chimneys approved by a recognized testing laboratory. When an unlined chimney must be used, a special corrosion-resistant flexible liner often can be installed. It is worked down into the chimney and connected at the bottom to the furnace vent. This provides a corrosion-proof vent system.

Figure 39.59 (A) The proper location for the installation of a vent damper. (B) A thermal vent damper and a motorized vent damper. Photo by Jason Obrzut

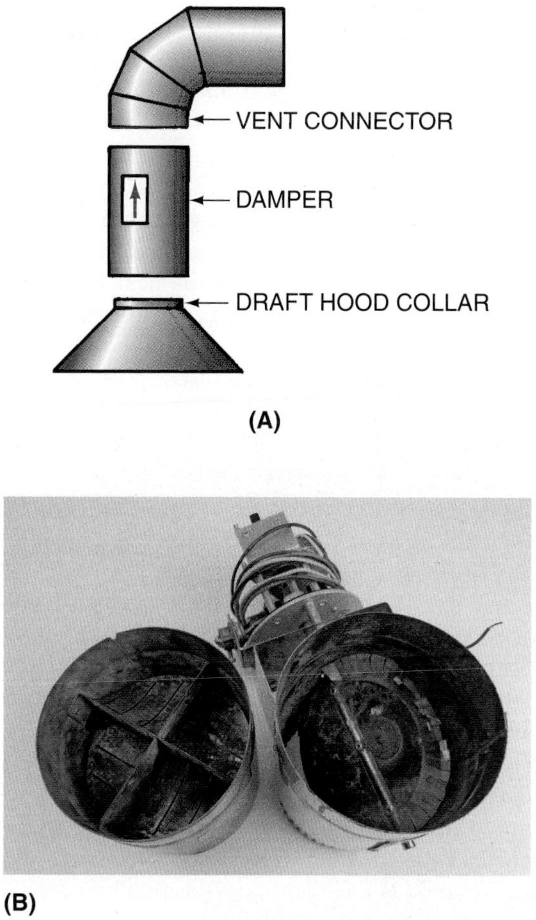

When the furnace is off, heated air can leave the structure through the draft hood. Automatic vent dampers that close when the furnace is off and open when the furnace is started can be placed in the vent to prevent air loss. Several types of dampers are available. A thermal type uses damper blades constructed of bimetal. When the system is off, the damper is closed. When the vent is heated by the flue gases, the bimetal action on the damper blades opens them, **Figure 39.59**. Other automatic vent dampers are controlled by a small electric motor. For safety reasons, the motors are electrically linked to the burner controls, which prevent the furnace from operating when the damper is closed and causing an unwanted health hazard from combustion by-product spillage into the structure.

39.19 DRAFT

Draft is the force that allows air to travel into the combustion chamber of a combustion appliance and pulls the combustion gases out through a vent or chimney. Correct draft is essential for efficient burner operation. The draft determines the rate at which combustion gases pass through the furnace. Draft also governs the amount of air supplied for combustion. *The correct draft is essential for efficient operation of the furnace and for the safety of the people living in the residence or structure.* Relative to the atmosphere, correct draft is a negative pressure and is referred to as an updraft because it travels up the vent or chimney. Insufficient draft may cause pressure in the combustion chamber, resulting in smoke and odor around the furnace (combustion appliance). Excessive draft can increase the stack temperature and reduce the amount of carbon dioxide in the flue gases.

Atmospheric-draft heating appliances are often referred to as open combustion devices. They draw combustion air from the surrounding room. Some older models draw indoor air into their chimney or vents through a dilution air device. In gas furnaces and boilers, the dilution air device is often referred to as a draft diverter or draft hood. In oil-fired furnaces and boilers, the dilution device is referred to as a barometric damper. Barometric dampers are used to maintain constant over-the-fire drafts regardless of chimney or vent system draft conditions, **Figure 39.60**. Double-acting barometric damper controls allow the control damper to swing open to introduce air into the vent system when the

Figure 39.60 Barometric dampers are used to maintain constant over-the-fire drafts regardless of chimney or vent system draft conditions. Courtesy Ferris State University. Photo by John Tomczyk

draft is too high. They also allow the damper to swing open to relieve downdrafts. Draft diverters, draft hoods, and barometric dampers are also used to provide draft control and introduce dilution air into the venting system to reduce the possibility of condensation in the vent system. High-efficiency combustion appliances (Category IV) do not use draft controls. Their vent pressures are positive all the way to the terminal end of the vent.

In natural- or atmospheric-draft appliances, the combustion gases are less dense than the air around the furnace and in the chimney or vent piping. Because of this, the hot combustion gases are more buoyant. The added buoyancy causes the gases to rise and induce a draft. Factors that determine the strength of the draft are

- temperature of the flue gases,
- chimney or vent height, or
- chimney or vent cross-sectional area.

The hotter the flue gases, the stronger the draft. Also, the higher the chimney or vent piping, the stronger the draft will be. However, as the cross-sectional area of the chimney or vent increases, the draft will decrease. The draft will be stronger with a narrower vent or chimney as long as the vent or chimney is not undersized for the heating appliance. It will take a longer time to heat up a chimney or vent that has a large cross-sectional area and induce a good draft compared to one with a smaller cross-sectional area. Atmospheric-draft heating appliances are those in which the air for the combustion process is at atmospheric pressure; they rely solely on the buoyancy of their combustion gases for draft, **Figure 39.56(A)**.

Today, most furnaces are power drafted using a combustion blower motor and a fan. These systems are sometimes referred to as fan-assisted draft. Power drafting can be either

Figure 39.61 Both induced-draft and forced-draft systems have a combustion blower motor, fan, and housing. Courtesy Ferris State University. Photo by John Tomczyk

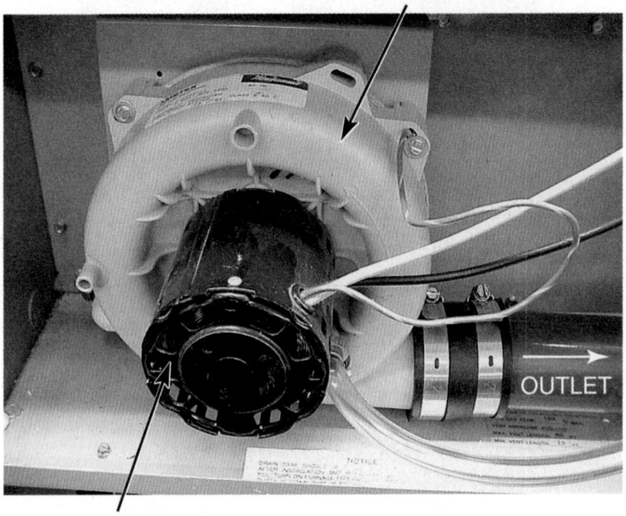

(1) induced draft or (2) forced draft, **Figures 39.61** and **39.63**. Neither system has a device that allows dilution air into the venting system. This is because their fans control the strength of the over-the-fire draft and create a regulated flow of combustion air and combustion gases, which minimizes the use of excess air. Induced-draft systems have a combustion blower motor located at the outlet of the heat exchanger. They pull or suck combustion gases through the heat exchanger, usually causing a slight negative pressure in the heat exchanger itself. Forced-draft systems push or blow combustion gases through the heat exchanger and cause a positive pressure in the exchanger. Forced-draft systems have the combustion blower motor at the inlet of the heat exchanger.

⚠ SAFETY PRECAUTION: *The most dangerous hazards in terms of air pollution and fire danger in residential homes are combustion appliances. It is of utmost importance that the combustion appliance is installed and vented correctly.•*

The burners for condensing high-efficiency combustion appliances are located at the top of the furnace in a sealed combustion chamber. The combustion gases are pulled or sucked through the heat exchanger and when they reach the dew point, liquid condensate forms and drains by gravity to a condensate sump or drain trap. The burners are located on top to allow the draining of condensate by gravity. High-efficiency furnaces will be covered in detail later in this unit.

Draft can be divided into two categories: (1) over-the-fire draft and (2) stack (chimney) draft. Over-the-fire draft is the draft over the fire and will be a lower reading—lower

Figure 39.62 A technician taking an over-the-fire draft reading. Courtesy Ferris State University. Photo by John Tomczyk

than chimney draft. If the combustion appliance allows this draft to be read, there will be a place to take the reading located just over the burners, **Figure 39.62**. Over-the-fire draft does not apply to atmospheric combustion devices.

The reading for chimney or stack draft is taken in the vent stack between the combustion device and the draft diverter, **Figure 39.49**. Stack or chimney draft is often referred to as flue draft. The stack draft is typically between −0.02 in. WC and −0.04 in. WC. Category I appliances with power burners usually have over-the-fire drafts of −0.02 in. WC, and the stack draft is usually between −0.02 in. WC and −0.04 in. WC. Category IV appliances will have a positive pressure stack draft. **SAFETY PRECAUTION:** *Always consult the combustion appliance manufacturer's specifications or the unit nameplate for specific draft requirements.*•

39.20 HIGH-EFFICIENCY GAS FURNACE DESIGN

Unlike conventional and mid-efficiency gas furnaces, high-efficiency gas furnaces have some unique characteristics:

- A "recuperative" or "secondary" (and sometimes "tertiary") heat exchanger with connecting piping for flue-gas passage from the primary heat exchanger.
- A flue-gas condensate disposal system, which includes a drain, taps, drain lines, and drain traps.
- A direct-vent configuration with a separate combustion air pipe to provide outdoor air for combustion.

A direct-vented high-efficiency gas furnace has a sealed combustion chamber. This means that the combustion air is brought in from the outside through an air pipe, which is usually made of PVC plastic. The combustion chamber never receives conditioned room air. Combustion gases are then exhausted to the outside through a separate, dedicated vent pipe. The exhaust gases are sealed within the system to prevent backdrafting—exhaust fumes coming back down the flue into the living space. Since conditioned room air is not used for combustion, it does not have to be replaced. Using outside air instead of room air prevents the conditioned room from being depressurized or pulled into a slight vacuum. This cuts down on the amount of infiltration entering the room and of cold, infiltrated air that has to be reheated.

Direct venting also minimizes the interaction of the furnace with other combustion appliances. Depressurization of a building or home may lead to combustion products entering from the vent systems of water heaters and clothes driers and from exhaust products in attached garages. When a conditioned space is depressurized, even toxic fumes and particulates from a connecting or attached garage or building can infiltrate.

As mentioned earlier in this unit, combustion blowers can be forced- or induced-draft. An induced-draft system places the combustion blower at the outlet of the heat exchangers, causing a negative pressure that actually sucks the combustion gases out of the exchangers. In a forced-draft system, the combustion blower is before the heat exchangers and creates a positive pressure that blows the combustion gases through the exchangers. Forced-draft blowers will vent gases at a lower temperature than is possible with a natural draft, allowing the heat exchanger to absorb more of the heat from the burner for distribution to the conditioned space. Some direct-vented designs need stronger combustion blower systems to overcome the additional pressure drop created by the sealed combustion chamber and added combustion air pipe. In fact, the combustion blower must overcome three system pressure drops:

1. The heat exchangers' system pressure drops
2. The vent system (vent and combustion air pipe) pressure drop
3. Vent termination wind resistance pressure

Combustion blower motors use an air-proving switch or pressure switch to prove that the mechanical draft has been established before allowing the main burner to light, **Figure 39.63.** The pressure switches are connected to the combustion blower housing or heat exchanger (burner box) by rubber tubing. The pressure switches have diaphragms that are sensitive to pressure differentials. Some furnaces use a single-port pressure switch that senses the negative pressure inside the heat exchanger caused by an induced-draft combustion blower. Others use a dual-port, or differential, pressure switch that senses the difference in pressure between the combustion blower and the burner box. When the venting system is operating properly, sufficient negative pressure is created by the combustion blower to keep the pressure switch closed and the furnace operating. However, if the heat exchanger has a leak or the furnace vent pipe is blocked, the pressure switch will open due to insufficient

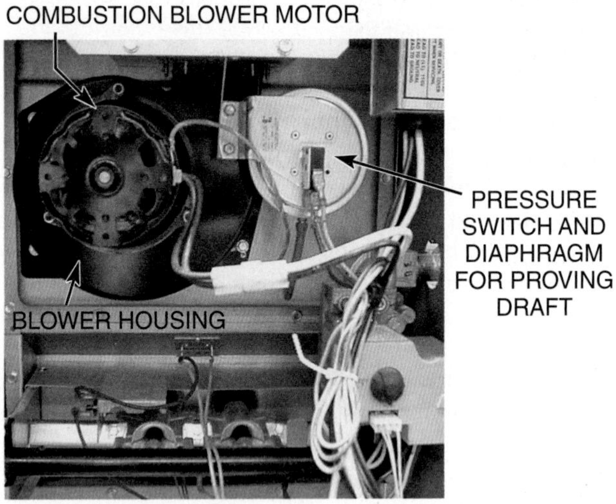

Figure 39.63 Combustion blower motors use an air-proving pressure switch to prove that the mechanical draft has been established before allowing the main burner to light. Courtesy Ferris State University. Photo by John Tomczyk

negative pressure and interrupt the operation of the main burners. Different furnace manufacturers use various pressure settings for their pressure switches. The settings are usually measured in inches of water column and are sometimes indicated on the pressure switch housing. The pressures can be measured using a digital or liquid-filled manometer. **SAFETY PRECAUTION:** *Never replace a pressure switch with another pressure switch without knowing its operating pressures. Always consult with the furnace manufacturer or use a factory-authorized substitute pressure switch.*•

39.21 HVACR SYSTEM TESTING

The proper and efficient operation of the heating, cooling, refrigeration, water heating, and humidification systems all greatly affect the energy usage of a residence. *Routine maintenance and sporadic testing of this equipment can assure the homeowner that the equipment is working efficiently. HVACR diagnostics include refrigerant charge checks, equipment and duct airflow and pressure measurements, along with specific temperature measurements.*

Listed here are some routine maintenance and tests for cooling equipment, some of which can be performed by the homeowner and some by service technicians and/or residential energy auditors.

- Clean the condenser and evaporator coil annually.
- Clean the blower fan blades.
- Check airflow quantities (cfm), **Figure 39.64**.
- Straighten fins on condenser and evaporator.
- Replace air handler filters often.
- Keep all supply and return registers clear and clean.

Figure 39.64 A residential energy auditor using a flow hood to measure the airflow (cfm) out of a ceiling diffuser in a residence. Courtesy Ferris State University. Photo by John Tomczyk

- Check low-side superheat and condenser subcooling for proper charge.
- Replace compressor contactor every 3 to 4 years.
- Make sure the suction line is insulated.
- Make sure air ducts do not leak.

Listed here are some routine maintenance and tests for heating equipment. Some can be performed by the homeowner and some by a service technician.

- Replace air handler filters often.
- Keep all supply and return registers clear and clean.
- Clean the blower fan blades.
- Check flame characteristics.
- Perform a combustion analysis annually.
- Inspect thermostat and heat anticipator settings.
- Test for carbon monoxide.
- Check the heat exchanger for cracks.
- Check for and seal any duct leaks.
- Test for temperature rise.
- Test for proper flue draft.
- Test high-limit control.
- Inspect burners for dirt, rust, and other debris.
- Check manifold gas pressure and adjust as needed.
- Look for any vent or chimney blockages.
- Walk the appliance through the sequence of operation.
- Set the controls for highest efficiency.

Listed next are some routine maintenance and tests for oil-fired combustion appliances that should be performed by a service technician.

- Test the oil pump pressure.
- Clean the oil strainer.
- Perform a combustion analysis annually.
- Replace the oil nozzle annually.

- Test the cad cell flame sensor.
- Test the high-limit control.
- Check electrode placement and gap.
- Adjust the air shutter for best efficiency.
- Adjust the barometric damper.
- Test for carbon monoxide.
- Check flame characteristics.
- Check the temperature rise.
- Check the heat exchanger for cracks.
- Clean the blast tube and blower.
- Check for oil leaks.

For more detailed information on testing, routine maintenance, venting, installation, typical operating conditions, troubleshooting, or charging of specific HVACR systems, refer to the appropriate chapter in this text.

39.22 NUMERICAL ANALYSIS AND REPORTING

Energy auditors use powerful computer programs and other electronic tools to model energy usage within a residence. Computer programs containing energy algorithms help auditors produce analytical reports and report summaries. Worksheets and spreadsheets also assist both the energy auditor and the homeowner in the actual audit and with their decision making after the audit is completed and a summary report is generated. Remember, one of the most important purposes of an energy audit is to determine where energy is used and/or wasted. Homes with the highest energy usage have the greatest potential for energy savings. Decision making encompasses what energy conservation measure (ECM) to choose and apply to the residence, usually according to their potential for saving energy. This decision-making process is also dependent on a solid understanding on how residential structures use energy.

An energy auditor will analyze and report the findings both manually and with reputable simulation software. The detailed written report of the findings prepared and presented to the homeowner should include a numerical analysis of where energy is being used, lost, or wasted and how to fix any energy-related problems. The report should also suggest any human behavioral changes that will help with energy conservation measures. In addition, the report should include recommendations on what fixes or changes will be of benefit by saving the most money and/or energy. The report should also educate the homeowner on how to maximize the payoffs from the energy audit's findings by tapping potential financial support from federal, state, and local sources and/or utility companies. It should include information on tax incentives, grants, and rebates that may help cover the cost of the audit.

WORKSHEETS FOR REPORTING

Following are examples of some common topics used for compiling worksheet data in home energy audits:

1. Building Occupant and Information Sheet
2. Health Safety and Structural Worksheet
3. Heating/Air Handler Equipment Worksheet
4. Cooling and Water Heating Equipment Worksheet
5. CAZ and Blower Door Testing
6. Duct Testing and Audit Diagnostics
7. House Diagrams
8. Post-inspection Worksheet
9. Post-inspection (Infiltration, Health, Safety) Worksheet

REPORTING BASE LOAD VERSUS SEASONAL USE

As mentioned earlier in this unit, LAMEL is an acronym for lighting, appliance, and miscellaneous electrical loads. Energy used in a residential structure can be categorized into two types: (1) base loads (LAMELs) and (2) seasonal energy usages. Refer back to "Base Load Inspection" for more information on LAMELs.

Seasonal energy usages include only heating and cooling energy. It is important that the energy auditor understand whether base loads or seasonal energy usages are the dominant energy usage in the home. Often, homes are serviced with both combustion fuel and electricity. Electricity is usually too expensive to handle both base loads and seasonal energy usages, but in some regions of the country nothing else is available. Auditors must know which energy loads in the home are serviced by electricity, by combustion fuels, or both.

Figures 39.65(A) and **(B)** are examples of seasonal-dominated and base load-dominated energy usage. Notice in **Figure 39.65(A)** that the home has a small energy base load but high seasonal energy usage. This is typical of homes with poorly insulated structures or shells, defined as the floors, roof, and walls of a building, or, in other words, the exterior envelope of the building. The home could also be located in a very cold climate, or both. **Figure 39.65(B)** illustrates a home with a large energy base load usage, but smaller seasonal energy usage. Homes like these are either geographically located where weather conditions are less severe, or constructed much more efficiently.

ENERGY INDEX REPORTING

A way to compare one home to another and to characterize their energy efficiency is through an energy index. One of the simplest methods of determining an energy index is to divide the home's energy usage in BTUs/h by the square

Figure 39.65 (A) Heavily seasonal-weighted energy usage. (B) Heavily baseload-weighted energy usage.

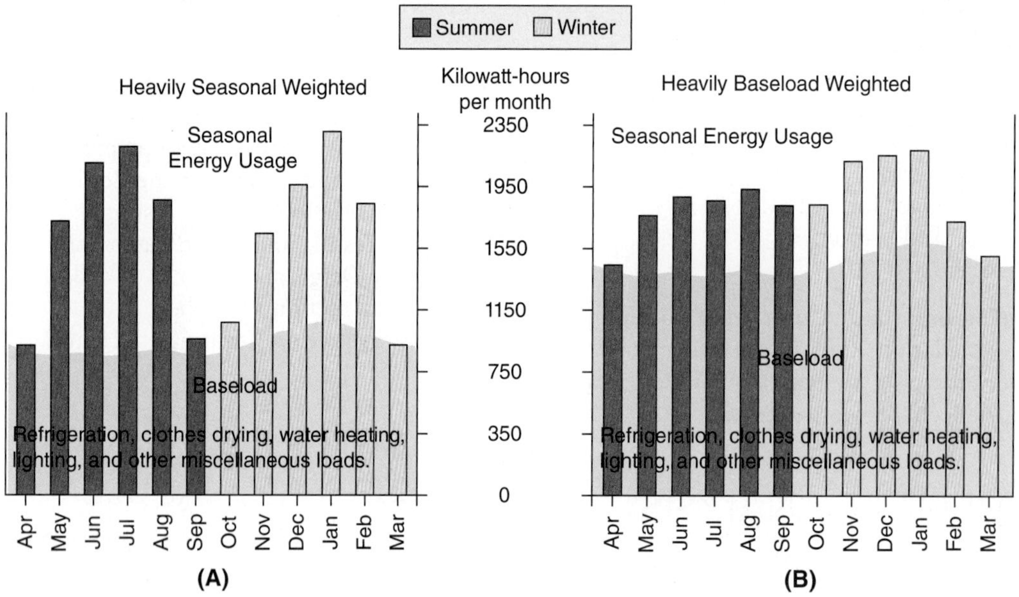

footage of the conditioned space. Another way is to divide the energy usage in kilowatt-hours by the square footage of the conditioned space. An energy index that takes into account the severity of the climate where the house is located also uses the heating degree days. The energy usage in BTUs/h is divided by both the square footage of the conditioned space and the heating degree days for that geographical location.

The most popular energy index is the **HERS index**. HERS is an acronym for Home Energy Rating Systems, a standardized rating procedure administered by the **Residential Energy Services Network (RESNET)**. The HERS index score is determined by computer modeling of the home with **Residential Energy Analysis and Rating Software (REM/Rate)**. The score provides a "miles-per-gallon" type of rating for expected energy consumption. Each home receives a score that can be compared with those of other new or existing homes. The lower the HERS score, the more efficient the home. **Figure 39.66** is an example of a HERS Home Energy Rating Certificate. This home has a HERS index of 60. An index of 85 or lower earns the home an Energy Star label. An index of 100 is a reference home built to the International Energy Conservation Code (IECC) standards. An index can be over 100. An index of 0.0 (zero) would mean the home was a "net-zero" energy home. This means the home uses energy but also either generates and/or captures energy to negate its energy usage. The capture and/or generation of energy could be through wind power, solar heat capture for heating household air or water, or photovoltaic electricity generation, to mention a few.

The Home Energy Rating Certificate is a summary of

1. general information,
2. mechanical system features,
3. building shell features, including infiltration rates (ACH) for heating and cooling using the blower door test,
4. lights and appliance features, and
5. estimated energy costs.

Notice in **Figure 39.66** that the estimated annual energy cost category consists of

- heating,
- cooling,
- hot water,
- lights/appliances,
- photovoltaics, and
- service charges.

As mentioned earlier, residential energy audits not only provide homeowners with an opportunity to lower their monthly utility bills, they also can help our nation conserve energy. The ability to conserve natural resources is an added attraction to many homeowners for getting an energy audit performed on their home. It is important that we all do our part to make this world a better place to live.

Figure 39.66 (A) An example of a HERS Home Energy Rating Certificate. Based on Home Energy Rating System (HERS).

WISCONSIN ENERGY STAR HOMES

Home Energy Rating Certificate

FERRIS STATE UNIVERSITY
Energy Center

5 Stars Plus
Confirmed Rating

Uniform Energy Rating System — **Energy Efficient**

1 Star	1 Star Plus	2 Stars	2 Stars Plus	3 Stars	3 Stars Plus	4 Stars	4 Stars Plus	5 Stars	5 Stars Plus
500-401	400-301	300-251	250-201	200-151	150-101	100-91	90-86	85-71	70 or Less

HERS Index: **60**

General Information

Conditioned Area: 1908 sq. ft.
Conditioned Volume: 15660 cubic ft.
Bedrooms: 4
HouseType: Single-family detached
Foundation: Conditioned basement

Mechanical Systems Features

Heating: Fuel-fired air distribution, Natural gas, 95.0 AFUE.
Water Heating: Conventional, Natural gas, 0.68 EF, 40.0 Gal.

Duct Leakage to Outside: RESNET/HERS default
Ventilation System: Balanced: ERV, 69 cfm, 74.5 watts.
Programmable Thermostat: Heating: Yes Cooling: No

Building Shell Features

Ceiling Flat: R-52
Vaulted Ceiling: R-38
Above Grade Walls: R-20
Foundation Walls: R-11.1, R-15.1
Slab: R-0.0 Edge, R-0.0 Under
Exposed Floor: NA
Window Type: Dbl/LoE/Arg - Vinyl****
Infiltration:
Rate: Htg: 600 Clg: 600 CFM50
Method: Blower door test

Lights and Appliance Features

Percent Fluorescent Pin-Based: 0.00
Percent Fluorescent CFL: 100.00
Refrigerator (kWh/yr): 500.00
Dishwasher Energy Factor: 0.46
Clothes Dryer Fuel: Natural gas
Range/Oven Fuel: Natural gas
Ceiling Fan (cfm/Watt): 0.00

The Home Energy Rating Standard Disclosure for this home is available from the rating provider.

REM/Rate - Residential Energy Analysis and Rating Software v12.85 Wisconsin

This information does not constitute any warranty of energy cost or savings.

Rating Number:
Certified Energy Rater:
Rating Date:
Rating Ordered For:

Estimated Annual Energy Cost
Confirmed Rating

Use	MMBtu	Cost	Percent
Heating	36.3	$307	26%
Cooling	0	$0	0%
Hot Water	20.7	$164	14%
Lights/Appliances	21.0	$461	39%
Photovoltaics	-0.0	$-0	0%
Service Charges		$239	20%
Total		**$1170**	**100%**

This home meets or exceeds the minimum criteria for all of the following:
EPA ENERGY STAR Home

(A)

Figure 39.66 (*Continued*) (B) An example of a HERS Home Energy Rating Certificate. Based on Home Energy Rating System (HERS) Based on Home Energy Rating System (HERS).

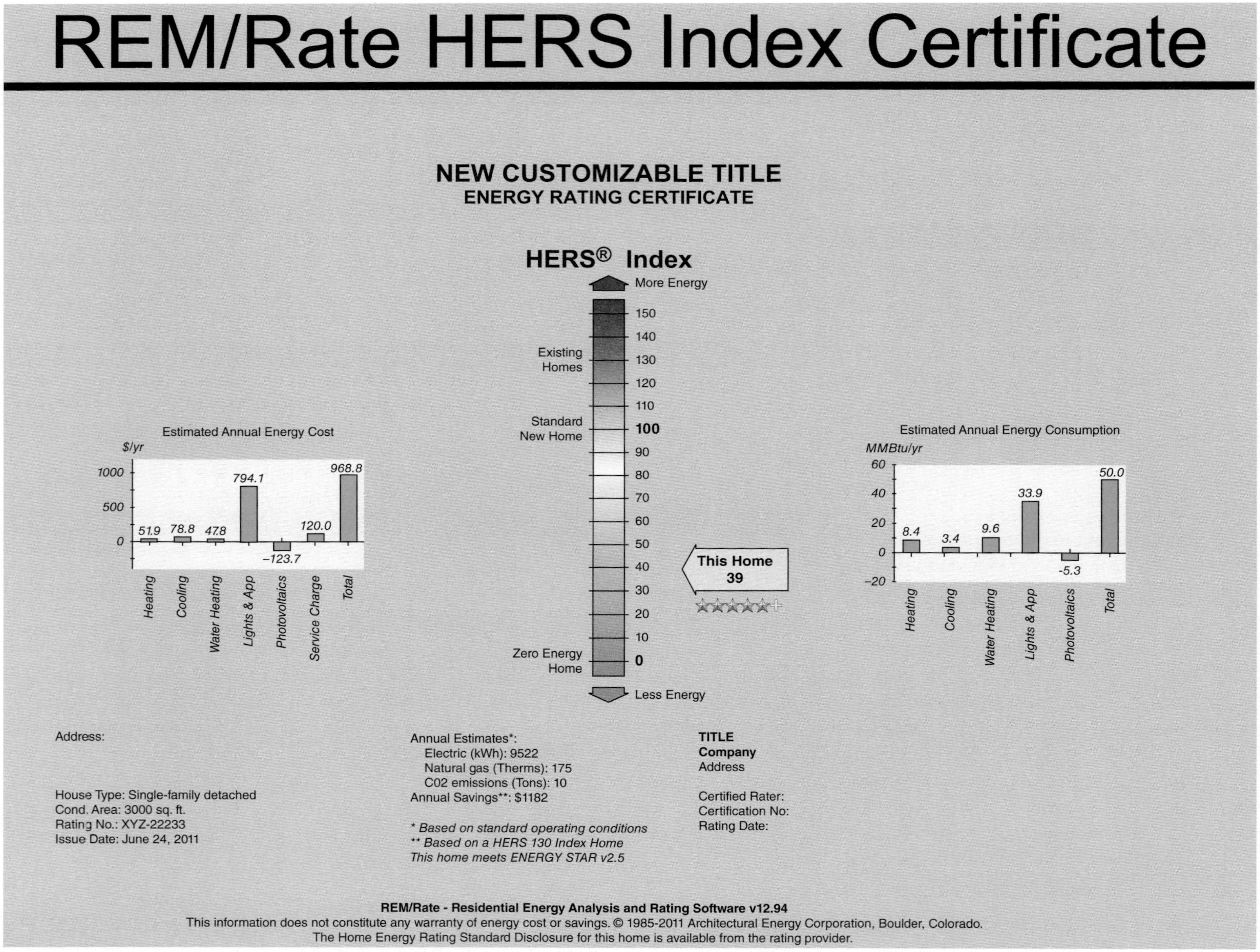

(B)

REVIEW QUESTIONS

1. The U.S. Department of Energy also says that a home's heating and cooling costs can be reduced by as much as 30% by
   A. replacing the structure's windows.
   B. planting trees and vegetation to shade the property.
   C. performing a combustion analysis on all gas-burning appliances.
   D. using proper insulation and air-sealing techniques.
2. Screening (visual inspections), diagnostic testing, and numerical analysis are three broad categories of
   A. a home energy audit.
   B. air-sealing techniques.
   C. duct testing and sealing.
   D. a combustion analysis.
3. The thermal resistance of the insulation and the surface area of the shell of the house will determine
   A. the type of heating and cooling equipment that should be installed.
   B. how much heat is transmitted through the shell.
   C. the square footage of the conditioned space being audited.
   D. the emittance factor of the windows.
4. A type of insulation that can be blown into walls to reduce air leakage, if blown into cracks, such as around windows and door frames is
   A. rigid foam insulation
   B. loose-fill insulation.
   C. foam-in-place insulation.
   D. rolls and batts insulation.
5. Thermal boundaries consist of
   A. areas where air can infiltrate or exfiltrate.
   B. insulation and an air space barrier.
   C. attics and crawl spaces.
   D. thermal bypasses and intermediate zones.
6. Which of the following are base loads or LAMELs?
   A. Refrigerators, computers, and water heaters
   B. Furnaces, boilers, and air conditioners
   C. Windows, siding, and insulation
   D. Cordless drills, cell phones, and generators
7. ________ bring out many of the buildings "hidden unknowns" such as structural and ductwork air leakages.
   A. Combustion tests C. Thermal tests
   B. Draft tests D. Diagnostic tests
8. When a blower door test is performed, the air leakage rate of the structure is determined by calculating
   A. the cfm of the blower door's fan at 20 Pascals (CFM_{20}).
   B. the emittance (low e) and reflectivity of the structure's window.
   C. the number of air changes per hour at 50 Pascals (ACH_{50}).
   D. the amount of air being passed through the shell by leaky ducts.
9. In the Department of Energy climate zones 5, 6, 7, the standard air changes per hour for an Energy Star dwelling is
   A. 4. C. 20.
   B. 10. D. 50.
10. One safety concern that may arise when performing a blower door test that a technician must address is
    A. cold air may be drawn into the structure, causing condensation.
    B. negative pressure may cause combustion appliances to backdraft.
    C. heated air may be forced from the structure, causing the furnace to run longer.
    D. negative pressure may cause the ductwork to collapse, causing airflow problems.
11. Thermal imaging cameras, often referred to as infrared cameras, create pictures by detecting
    A. reflected colors of the visible light spectrum.
    B. motion of the object being imaged.
    C. electromagnetic waves.
    D. heat energy.
12. When using a thermal imaging camera to inspect for thermal leaks, the recommended temperature difference from interior to exterior should be approximately
    A. 9°F. C. 25°F.
    B. 18°F. D. 30°F.
13. Research has shown that the most serious air leaks occur
    A. through the HVACR system.
    B. through draft hoods and fire places.
    C. at the top and bottom of the conditioned envelope.
    D. at the top of the conditioned envelope.
14. Mandatory requirements for air-sealing in new construction and additions are set by
    A. National Fire Protection Association (NFPA).
    B. Environmental Protection Agency (EPA).
    C. Building Performance Institute (BPI) and Green Mechanical Council.
    D. International Energy Conservation Code (IECC) and the International Residential Code (IRC).
15. In both new and existing homes, as much as ____% of the total energy losses can often be blamed on air leakage in forced-air duct systems.
    A. 25
    B. 35
    C. 45
    D. 55
16. Of the two types of duct leaks, the one that has a larger impact on HVACR system performance is
    A. duct leakage to the conditioned space.
    B. duct leakage to the outside.
    C. cold air infiltration in the supply duct.
    D. hot air infiltration in the supply duct.

17. Which of the following would be a cause of an air duct leak?
    A. A poor connection between a boot and the floor or wall.
    B. An unsealed wall penetration around a plumbing fixture.
    C. A floor slab that extends outdoors without a thermal barrier.
    D. A penetration in the shell where TV cables enter the conditioned space.
18. When performing a duct leakage test, what pressure should be used?
    A. 50 Pascals of positive pressure
    B. 25 psig of positive pressure
    C. 25 Pascals of positive pressure
    D. 25 Pascals of negative pressure
19. To measure the total air leakage from a duct system
    A. a blower door test is done.
    B. a smoke pen is used.
    C. a thermal imaging test is done.
    D. a duct blower test is done.
20. The leakage rate of the entire duct system is measured in
    A. cubic feet per minute (CFM).
    B. Pascals (Pa).
    C. air changes per hour (ACH).
    D. pounds per square inch (PSI).
21. When testing for duct leakage to the outdoors
    A. a duct blower should be used.
    B. both a blower door and a duct blower should be used.
    C. a duct blower and thermal imager should be used.
    D. a smoke pen and a thermal imager should be used.
22. A duct leakage test to the outdoors is a much more meaningful test to the energy auditor than a total duct leakage test because
    A. ducts that are run in enclosed spaces can't be tested.
    B. duct leakage to the conditioned space adds to the HVACR system performance.
    C. the leakage is directly related to wasted energy.
    D. total duct leakage testing does not produce reliable results.
23. The ignition temperature of natural gas is
    A. 500°F to 600°F.
    B. 700°F to 800°F.
    C. 1500°F to 1600°F.
    D. 1100°F to 1200°F.
24. Along with heat, perfect combustion of natural gas will produce
    A. carbon dioxide and carbon monoxide.
    B. carbon dioxide and water vapor.
    C. carbon (soot), carbon monoxide, and oxygen.
    D. carbon dioxide, water vapor, oxygen, and carbon (soot).
25. For complete combustion, approximately ______ ft^3 are need for each 1 ft^3 of gas.
    A. 5   C. 15
    B. 10   D. 20
26. A digital combustion analyzer *calculates* which of the following?
    A. Carbon monoxide (CO)
    B. Oxygen (O_2)
    C. Methane (CH_4)
    D. Undiluted carbon monoxide (COAF)
27. Which of the following, present in the flue gas, is an indication of complete combustion?
    A. Carbon monoxide (CO)
    B. Oxygen (O_2)
    C. Carbon dioxide (CO_2)
    D. Undiluted carbon monoxide (COAF)
28. As the percentage of O_2 (excess air) in the flue gas increases, the furnace efficiency
    A. increases.   C. stays constant.
    B. decreases.   D. is not affected.
29. Net stack temperature is calculated by
    A. subtracting the combustion air temperature from the stack temperature.
    B. adding the combustion air temperature to the stack temperature.
    C. subtracting the stack temperature from the combustion air temperature.
    D. dividing the stack temperature by the cubic feet of excess air.
30. When performing diagnostic tests on a furnace as part of a residential energy audit, the airflow can be verified by testing the
    A. manifold gas pressure.
    B. flue gas composition.
    C. draft of the vent system.
    D. temperature rise through the furnace.
31. High-efficiency, condensing, furnaces are typically rated at _____ AFUE.
    A. 65% to 75%
    B. 78% to 80%
    C. 78% to 83%
    D. 87% to 97%
32. Yellow tips of a gas flame typically indicate the production of
    A. carbon monoxide.
    B. heat.
    C. carbon dioxide.
    D. water vapor.
33. When the pressure of the combustion appliance zone (CAZ) is lower than the pressure of the vent system, ______ will occur.
    A. thermal bypass
    B. backdrafting
    C. condensation
    D. perfect combustion
34. A temporary flow of combustion by-products into the living space is called
    A. backdrafting.   C. spillage.
    B. condensing.   D. reverse draft.

35. Cad cells, flame rectification, and thermocouples are all types of
    A. combustion test methods.
    B. flame safeguard controls.
    C. limit safety controls.
    D. natural-draft vent systems.
36. When the burner is off, no light is being generated and the resistance of the cad cell is about
    A. 100,000 ohms.
    B. 10,000 ohms.
    C. 1000 ohms.
    D. 100 ohms.
37. As excess air decreases
    A. flame current will increase.
    B. gas pressure increases.
    C. dew point temperature decreases.
    D. dew point temperature increases.
38. The dew point temperature of undiluted flue gas is about
    A. 105°F.
    B. 140°F.
    C. 200°F.
    D. 400°F.
39. The recommended pitch for the horizontal run of a flue pipe is
    A. ½ in. per foot.
    B. ½ in. per 10 feet.
    C. ¼ in. per foot.
    D. ¼ in. per 10 feet.
40. An induced-draft mid-efficiency furnace would be classified as
    A. Category I.
    B. Category II.
    C. Category III.
    D. Category IV.
41. An induced-draft high-efficiency condensing furnace would be classified as
    A. Category I.
    B. Category II.
    C. Category III.
    D. Category IV.
42. The dilution air device on an oil furnace or boiler is called a
    A. power-burner.
    B. draft hood.
    C. barometric damper.
    D. combustion blower.
43. Induced-draft systems have a combustion blower motor located
    A. at the inlet of the heat exchanger.
    B. at the outlet of the heat exchanger.
    C. above the dilution air device.
    D. at the end of the flue pipe or top of the stack pipe.
44. Some direct-vented high-efficiency furnaces need stronger combustion blower systems to overcome the additional pressure drop created by
    A. the sealed combustion chamber and combustion air pipe.
    B. the secondary or condensing heat exchanger.
    C. a dual-port air-proving or pressure switch.
    D. the serpentine design of the primary and secondary heat exchangers.
45. Which of the following would be a routine maintenance task that should be performed on heating equipment?
    A. Insulate the suction line.
    B. Straighten fins on condenser and evaporator coils.
    C. Perform a combustion analysis.
    D. Caulk air leaks around window casings.
46. A way to compare one home to another and to characterize their energy efficiency is to
    A. compare conditioned square footage.
    B. use an energy index.
    C. compare geographic locations.
    D. compare heating costs.
47. Seasonal energy usages include
    A. only heating and cooling energy consumption.
    B. base or LAMEL loads.
    C. weather conditions.
    D. heating, cooling, and base load energy usage.
48. The lower the HERS score
    A. the lower the home efficiency is.
    B. the more ECMs are needed.
    C. the more humid the climate is.
    D. the more efficient the home is.
49. The most popular energy index is the
    A. EPA index.
    B. IECC index.
    C. Energy star.
    D. HERS index.
50. A home that is classified as a "net zero" home
    A. has an index score of 100.0.
    B. has an index score of 0.0.
    C. is completely off the grid.
    D. has an Energy Star rating of 85.
51. What is the difference between over-the-fire draft and stack (chimney) draft?
52. Name four nationally recognized energy certifications for home auditors and contractors.
53. What is meant by an energy conservation measure (ECM)?
54. What is the reason for performing a residential energy audit?
55. When additional combustion air is to be brought into the combustion appliance zone (CAZ) from an adjacent space, how should the openings be sized?
56. Name and briefly explain the five most common diagnostic tests performed during a residential energy audit.
57. What is the difference between a base load and seasonal energy usage as they apply to a residential home?
58. List some common issues that affect the indoor air quality of a home.
59. Describe what a typical energy audit numerical analysis report is used for.
60. List some common areas in a residential structure that are sealed to prevent air leaks.

UNIT 40

TYPICAL OPERATING CONDITIONS

OBJECTIVES

After studying this unit, you should be able to

- explain what conditions will vary the evaporator pressures and temperatures.
- describe how the various conditions in the evaporator and ambient air affect condenser performance.
- state the relationship of the evaporator to the rest of the system.
- describe the relationship of the condenser to the total system performance.
- compare high-efficiency equipment and standard-efficiency equipment.
- describe how to determine the electrical requirements for a system.
- explain how to test the electrical characteristics of an operating system.
- establish reference points when working on unfamiliar equipment to determine what the typical conditions should be.
- describe how humidity affects equipment suction and discharge pressure.
- explain three methods that manufacturers use to make air-conditioning equipment more efficient.

SAFETY CHECKLIST

✔ The technician must use caution while observing equipment operating conditions. Electrical, pressure, and temperature readings must be taken while the technician observes the equipment, and many times the readings must be taken while the equipment is in operation.

✔ Wear goggles and gloves when attaching gauges to a system.

✔ Observe all electrical safety precautions. Be careful at all times and use common sense.

Air-conditioning technicians must be able to evaluate both mechanical and electrical systems. The mechanical operating conditions are determined or evaluated with gauges and thermometers; electrical conditions are determined with electrical instruments.

40.1 MECHANICAL OPERATING CONDITIONS

Air-conditioning equipment is designed to operate at its rated capacity and efficiency at one set of design conditions. The design conditions are generally considered to be an outside temperature of 95°F and an inside temperature of 80°F with humidity at 50%. This rating is established by the Air-Conditioning, Heating and Refrigeration Institute (AHRI). Not only must equipment have a rating as a standard from which estimators and buyers can work, but it is also rated so that the buyer will have a common basis on which to compare one piece of equipment with another. All equipment in the AHRI directory is rated under the same conditions—80°F dry-bulb and 50% humidity indoors and 95°F dry-bulb outdoors.

A piece of equipment rated at 3 tons will perform at a 3-ton capacity, or 36,000 Btu/h, under the stated conditions. Under different conditions, the equipment will perform differently. Many occupants might not be comfortable at 80°F with a relative humidity of 50% and so will normally operate their system at about 75°F with the relative humidity in the conditioned space close to 50%. The equipment will not have quite the capacity at 75°F that it has at 80°F. If the designer wants the system to have a capacity of 3 tons at 75°F and 50% relative humidity, the manufacturer's literature should be consulted to choose other equipment. **The 75°F, 50% humidity condition will be used as the design condition in this unit because it is a common operating condition of equipment in the field.**

AHRI rates condensers on the basis of 95°F air passing over them. With 95°F air passing over it, a new standard-efficiency condenser will condense refrigerant at about 125°F. As the condenser ages, dirt accumulates on the outdoor coil and the efficiency decreases. The refrigerant will then condense at a higher temperature, which can easily approach 130°F, a value often found in the field. However, the unit uses a 125°F condensing temperature, assuming that equipment has been properly maintained. It is not easy to see the change in load conditions on gauges and instruments used in the field. An increase in humidity is not followed by a proportional rise in suction pressure and amperage. Perhaps the most noticeable aspect of an increase in humidity is an increase in the condensate accumulated in the condensate drain system.

The inside relative humidity adds a significant load to the evaporator coil and has to be considered as part of the total load. When conditions vary from the design conditions, the capacity of the equipment will vary. The pressures and temperatures will also change.

40.2 RELATIONSHIPS OF SYSTEM COMPONENTS UNDER LOAD CHANGES

If the outside temperature increases from 95°F to 100°F, the equipment will be operating at a higher head pressure and will not have as much capacity. The capacity also varies when the space temperature goes up or down or when the humidity varies. The relationship among the various components in the system also varies. The evaporator absorbs heat. When anything happens to increase the amount of heat absorbed into the system, the system pressures will rise. The condenser rejects heat. If anything happens to prevent the condenser from rejecting heat from the system, the system pressures will rise. The compressor pumps heat-laden vapor. Vapor at different pressure levels and saturation points (in reference to the amount of superheat) will hold different amounts of heat.

40.3 EVAPORATOR OPERATING CONDITIONS

The evaporator normally will operate at a 40°F boiling temperature when operating at the 75°F, 50% humidity condition. This will cause the suction pressure to be 70 psig for R-22. (The actual pressure corresponding to 40°F is 68.5 psig; for our purposes, we will round this off to 70 psig.) If R-410A is used as the refrigerant, the suction pressure would be 118 psig for a 40°F evaporating temperature at design conditions and under a steady-state load. Under these conditions the evaporator is boiling the refrigerant exactly as fast as the expansion device is metering it into the evaporator. As an example, suppose that the evaporator has a return-air temperature of 75°F and the air has a relative humidity of 50%. The liquid refrigerant goes nearly to the end of the coil, and the coil has a superheat of 10°F. This coil is operating as intended at this typical condition, **Figure 40.1**.

Late in the day, after the sun has been shining on the house, the heat load inside becomes greater. A new condition is being established. The example evaporator in **Figure 40.2** has a fixed-bore metering device that will feed only a certain amount of liquid refrigerant. The space temperature in the house has climbed to 77°F and is causing the liquid refrigerant in the evaporator to boil faster, **Figure 40.3**. This causes the suction pressure and the superheat to go up slightly. The new suction pressure is 73 psig, and the new superheat is 13°F. This is well within the range of typical operating conditions for an evaporator. The system actually has a little more capacity at this point if the outside temperature is not too much above design. If the head pressure goes up because the outside temperature is 100°F, for example, the suction pressure will even go higher than 73 psig, because head pressure will influence it, **Figure 40.4**.

Figure 40.1 An evaporator operating at typical conditions. The refrigerant is boiling at 40°F in the coil. This corresponds to 68.5 psig (typically rounded off to 70 psig) for R-22 and 118 psig for R-410A.

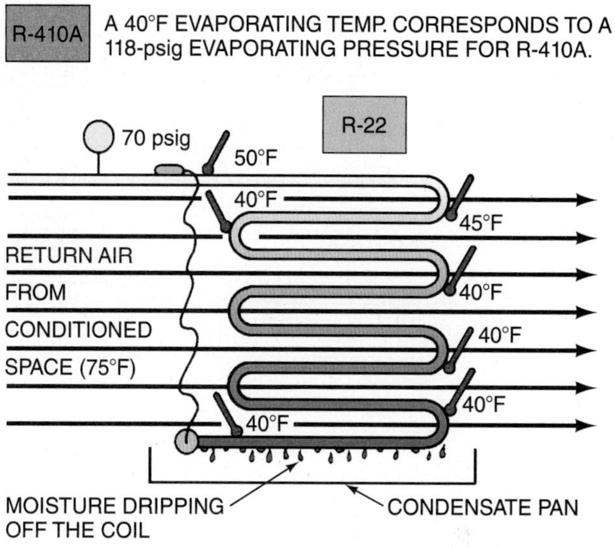

Many different conditions will affect the operating pressures and air temperatures in the conditioned space. There can actually be as many different pressures as there are different inside and outside temperature and humidity variations. This can be confusing, particularly to the new service technician. However, the service technician can use some common conditions when troubleshooting. The technician rarely has a chance to work on a piece of equipment when the conditions are perfect. Most of the time when the technician is assigned a job, the system has been off for some time or not operating correctly for long enough that the conditioned space temperature and humidity are higher than normal, **Figure 40.5**. After all, a rise in space temperature is what often prompts a customer to call for service.

40.4 HIGH EVAPORATOR LOAD AND A COOL CONDENSER

A high temperature in the conditioned space is not the only thing that will cause the pressures and capacity of a system to vary. The reverse can happen if the inside temperature is warm and the outside temperature is cooler than normal, **Figure 40.6**. For example, before going to work, a couple may turn the air conditioner off to save electricity. They may not get home until after dark to turn the air conditioner back on. By this time it may be 75°F outside but still be 80°F inside the structure. The air passing over the condenser is now cooler than the air passing over the evaporator. The evaporator may also have a large humidity load. The condenser becomes so efficient that it will hold some of the charge because it starts to condense refrigerant in the first part of the condenser. Thus, more of the refrigerant charge is in the condenser tubes, and this will slightly starve the evaporator.

Figure 40.2 This fixed-bore metering device will not vary the amount of refrigerant feeding the evaporator as much as the thermostatic expansion valve. It is operating at an efficient state of 10°F superheat.

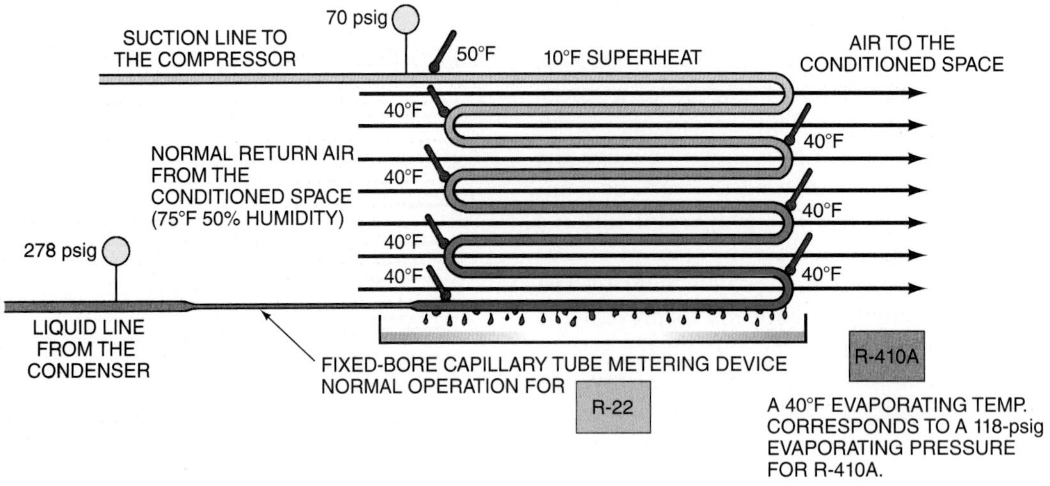

Figure 40.3 This is a fixed-bore metering device when an increase in load has caused the suction pressure to rise.

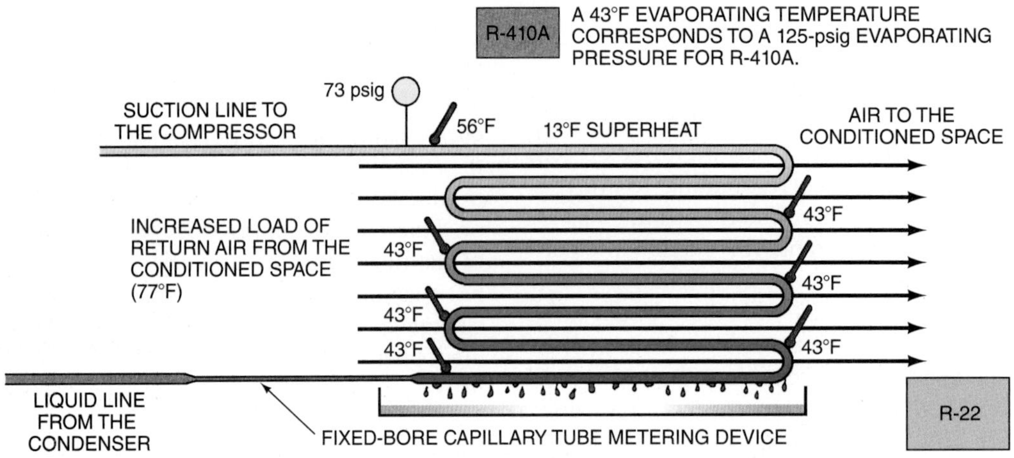

Figure 40.4 A system with a high head pressure. The increase in head pressure has increased the flow of refrigerant through the fixed-bore metering device, and the superheat is decreased. This system has a 135°F condensing temperature.

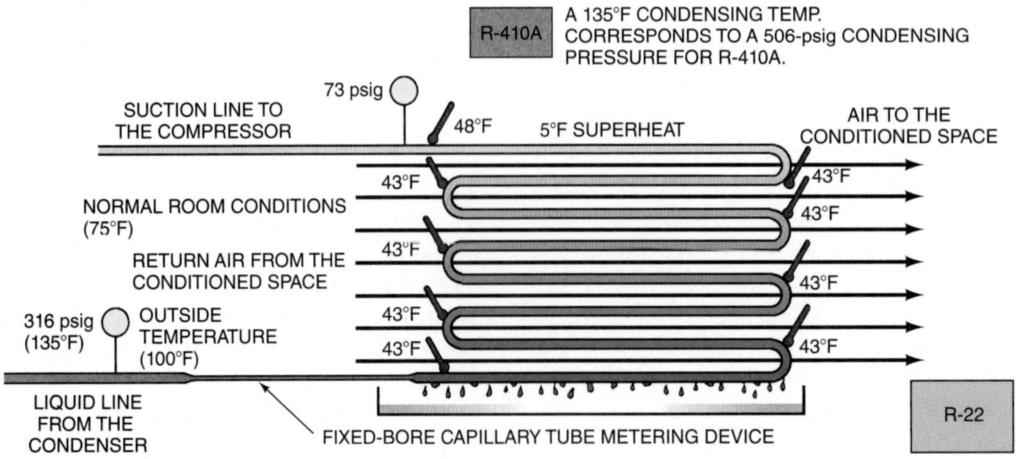

Figure 40.5 This system has been off long enough that the temperature and the humidity have gone up inside the conditioned space. Notice the excess moisture forming on the coil and going down the drain.

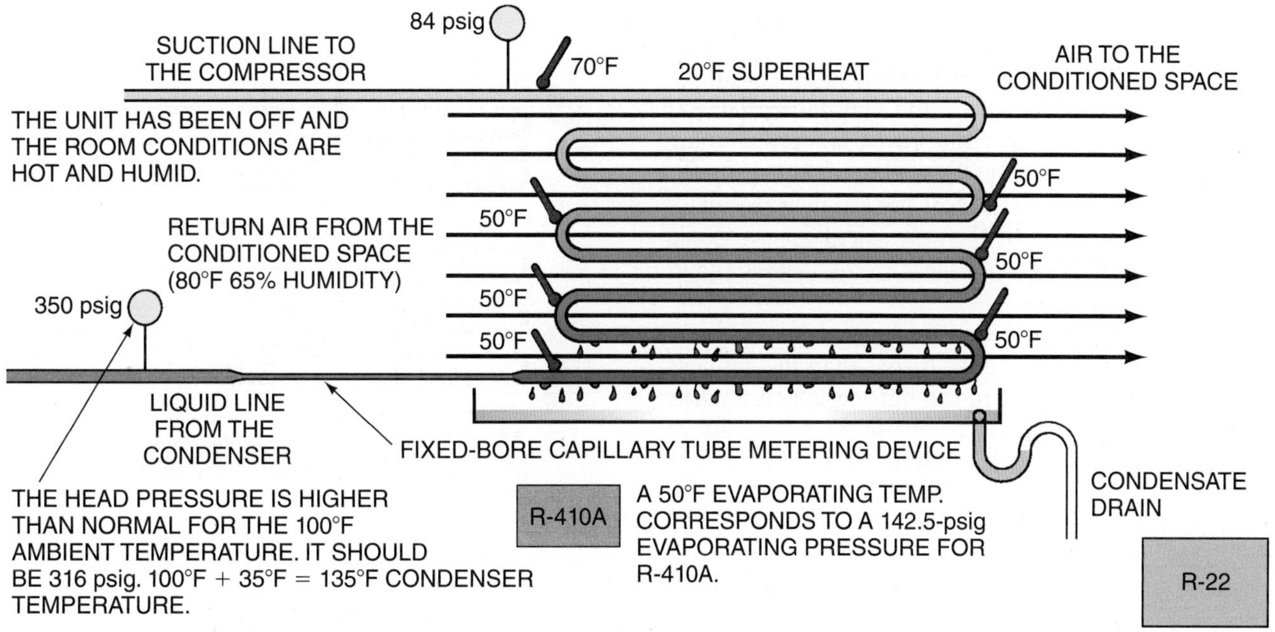

Figure 40.6 A system operating when the outside ambient air is cooler than the inside space temperature air. As the return air cools down, the suction pressure and boiling temperature will decrease.

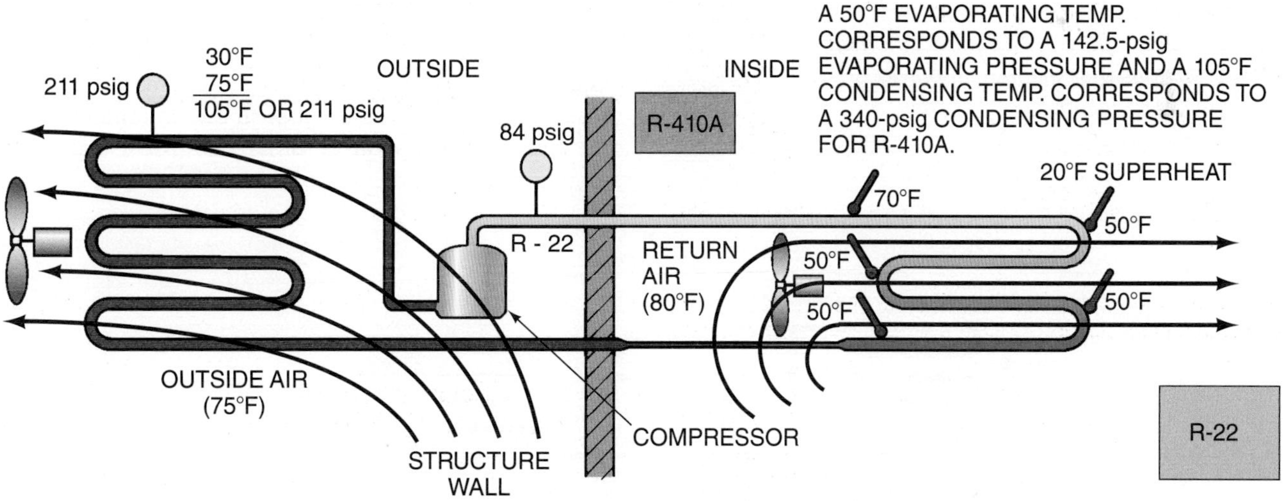

The system may not have enough capacity to cool the home for several hours. The condenser may hold back enough refrigerant to cause the evaporator to operate below freezing and freeze up before it can cool the house and satisfy the thermostat.

This condition has been improved by some manufacturers by means of a variable-speed fan operated through an inverter or by two-speed operation for the condenser fan. The fan is controlled by a modulating controller and fan motor or a single-pole–double-throw thermostat that operates the fan at both high speeds (when the outdoor temperature is high) and low speeds (in mild temperatures). Typically, the fan will run on high speed when the outdoor temperature is above 85°F and at low speed below 85°F. When the fan is operating at low speed in mild weather, the head pressure is increased, reducing the chances of the evaporator condensate freezing.

40.5 GRADES OF EQUIPMENT

Manufacturers have worked continuously to make air-conditioning equipment more efficient. There are three grades of equipment: economy, standard-efficiency, and high-efficiency grade. Some companies manufacture all three grades. Some offer only one grade and may take offense if someone calls their equipment the lower grade. Economy grade and standard-efficiency grade are about equal in efficiency, but they differ in the materials used in their manufacture and their appearance. *High-efficiency equipment may be much more efficient and will not have the same operating characteristics,* **Figure 40.7**.

A condenser normally will condense refrigerant at a temperature of about 30°F to 35°F higher than the ambient temperature. For example, when the outside temperature is 95°F, the average condenser will condense the refrigerant at 125°F to 130°F, and the head pressure for these condensing temperatures would be 278 to 297 psig for R-22 or 446 to 475 psig for R-410A. *High-efficiency air-conditioning equipment may have a much lower operating head pressure. The high efficiency is gained by using a larger condenser surface or by using more modern alloys with more extended surfaces or fins. The condensing temperature may be as low as 20°F greater than the outdoor ambient temperature. This would bring the head pressure down to a temperature corresponding to 115°F, or 243 psig for R-22 and 390 psig for R-410A. The compressor will not use as much power under this condition because of the lower compression ratio,* **Figure 40.8**. *When large condensers are used, the head pressure is reduced. Lower power requirements are a result. More efficient condensers require some means of head pressure control, such as that mentioned previously. High-efficiency equipment will also incorporate TXV metering devices to keep the evaporator active under all heat load conditions. Electrically commutated motors (ECMs) for evaporators and condensers will also usually accompany high-efficiency equipment. Refer to Unit 10, "System Charging," when charging a TXV system.*

Figure 40.7 A standard-efficiency condenser and a high-efficiency condenser. The high-efficiency condenser will have more surface area and may be made of a different metal or alloy with more extended surfaces or fins.

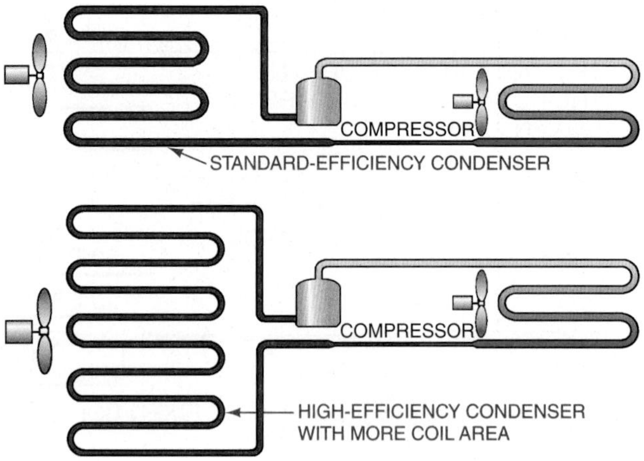

This description of condensers and their typical operating conditions holds true when the air-conditioning system is operating under a typical load. However, under high evaporator heat loads, the temperature difference between the condenser and the ambient will be much higher. The reason for this is simple. As the evaporator absorbs heat, it must reject that heat in the condenser, and as more heat is absorbed in the evaporator, more heat must be rejected in the condenser. However, a fixed-sized condenser operating under a relatively constant outside ambient cannot accomplish this amount of heat rejection at a 20°F temperature difference (for a high-efficiency

Figure 40.8 A standard-efficiency unit and a high-efficiency unit.

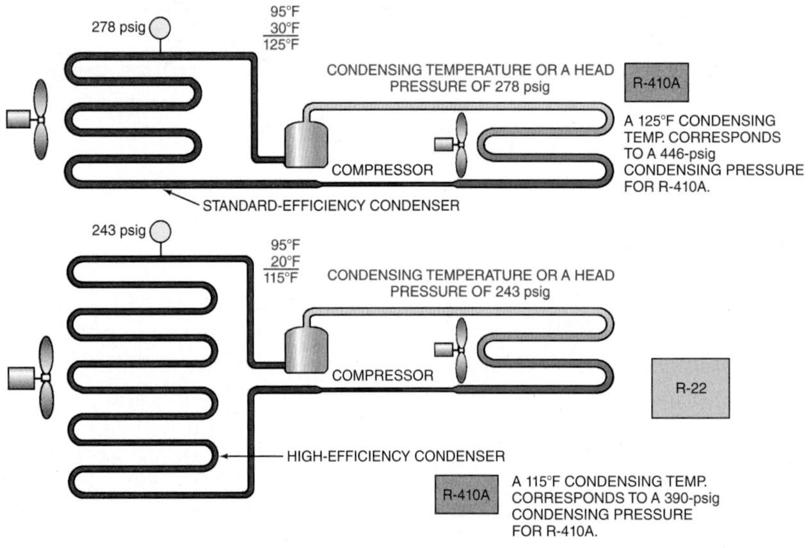

Figure 40.9 Charts are furnished by some manufacturers to monitor unit performance. Source: Carrier Corporation

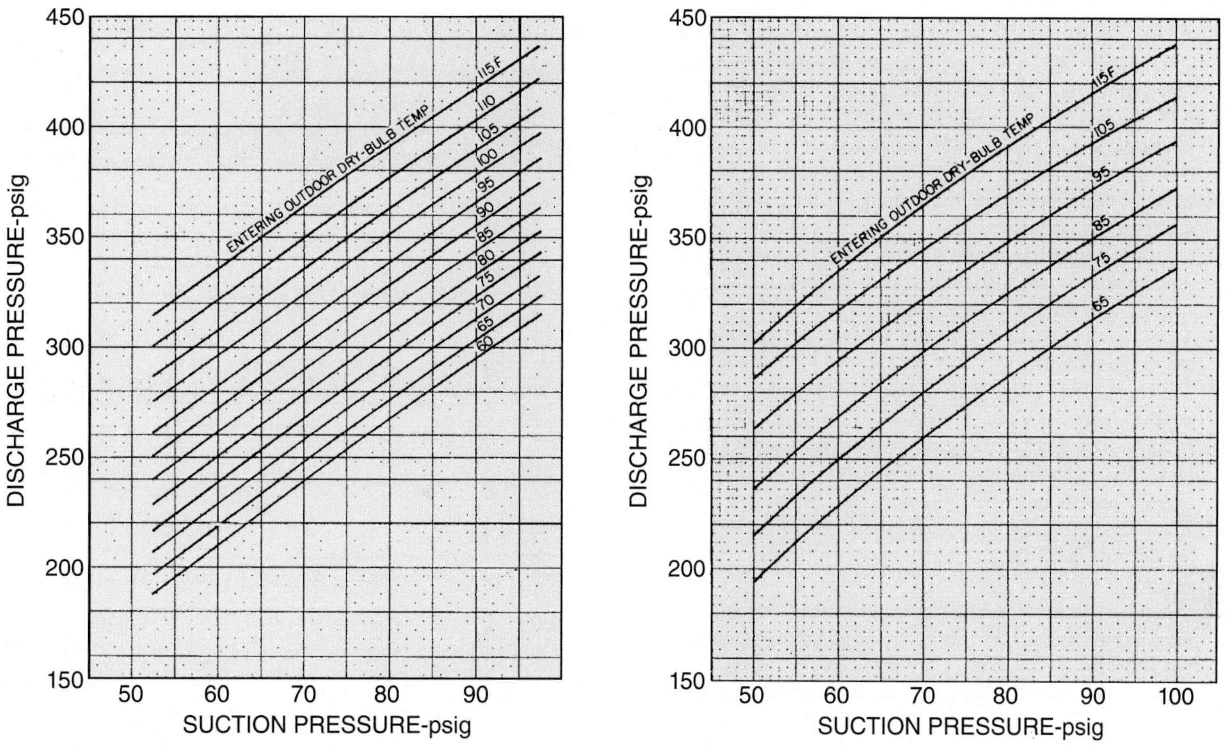

condenser) or at a 30°F temperature difference (for a standard condenser). So excess heat starts to accumulate in the condenser and the condensing temperature increases. After the condensing temperature increases, there exists a greater temperature difference between the ambient and the condensing temperature. This temperature difference increases the heat transfer between the condenser and the ambient, and the condenser will then be able to reject the higher evaporator loads. However, it will be rejecting this higher heat load at a higher condensing temperature and higher compression ratio. The same is true under low evaporator loads. However, the condenser will operate at a lower temperature difference between itself and the ambient because of the reduced load.

40.6 DOCUMENTATION WITH THE UNIT

The technician needs to know what the typical operating pressures should be under different conditions. Some manufacturers furnish a chart that tells what the suction and discharge pressures should be at various conditions, **Figures 40.9** and **40.10**. Some publish a bulletin listing all of their equipment along with the typical operating pressures and temperatures. Others furnish this information with the unit in the installation and start-up manual, **Figure 40.11**. The homeowner may have this booklet, which may be helpful to the technician.

Figure 40.10 A page from a manufacturer's bulletin explaining how to charge one model of equipment. Source: Carrier Corporation

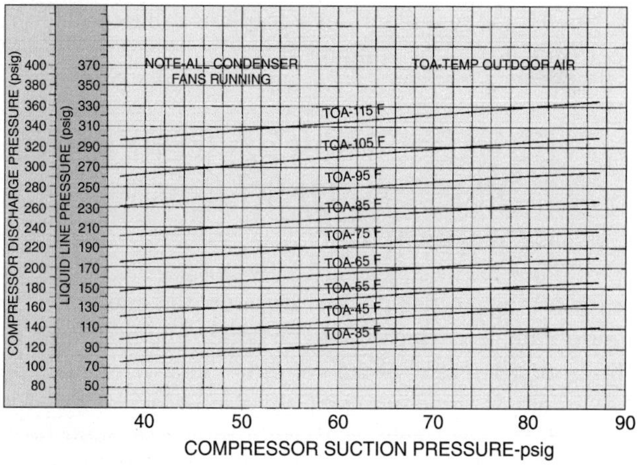

Three things must be considered along with the manufacturer's published list of typical operating conditions:

1. The load on the outdoor coil, which is influenced by the outdoor temperature and the evaporator heat loading.
2. The sensible-heat load on the indoor coil, which is influenced by the indoor dry-bulb temperature.
3. The latent-heat load on the indoor coil, which is influenced by the humidity. The humidity is determined by taking the wet-bulb temperature of the indoor air.

Figure 40.11 A chart furnished by a manufacturer in the installation and start-up literature. Source: Carrier Corporation

REFRIGERANT CHARGING

SAFETY PRECAUTION: To prevent personal injury, wear safety glasses and gloves when handling refrigerant. Do not overcharge system. This can cause compressor flooding.

1. Operate unit a minimum of 15 minutes before checking charge.
2. Measure suction pressure by attaching a gage to suction valve service port.
3. Measure suction line temperature by attaching a service thermometer to unit suction line near suction valve. Insulate thermometer for accurate readings.
4. Measure outdoor coil inlet air dry-bulb temperature with a second thermometer.
5. Measure indoor coil inlet air wet-bulb temperature with a sling psychrometer.
6. Refer to table. Find air temperature entering outdoor coil and wet-bulb temperature entering indoor coil. At this intersection note the superheat.
7. If unit has higher suction line temperature than charted temperature, add refrigerant until charted temperature is reached.
8. If unit has lower suction line temperature than charted temperature, recover refrigerant until charted temperature is reached.
9. If air temperature entering outdoor coil or pressure at suction valve changes, charge to new suction line temperature indicated on chart.
10. This procedure is valid, independent of indoor air quantity.

SUPERHEAT CHARGING TABLE (SUPERHEAT ENTERING SUCTION SERVICE VALVE)

Outdoor Temp (°F)	INDOOR COIL ENTERING AIR °F WB													
	50	52	54	56	58	60	62	64	66	68	70	72	74	76
55	9	12	14	17	20	23	26	29	32	35	37	40	42	45
60	7	10	12	15	18	21	24	27	30	33	35	38	40	43
65	—	6	10	13	16	19	21	24	27	30	33	36	38	41
70	—	—	7	10	13	16	19	21	24	27	30	33	36	39
75	—	—	—	6	9	12	15	18	21	24	28	31	34	37
80	—	—	—	—	5	8	12	15	18	21	25	28	31	35
85	—	—	—	—	—	—	8	11	15	19	22	26	30	33
90	—	—	—	—	—	—	5	9	13	16	20	24	27	31
95	—	—	—	—	—	—	—	6	10	14	18	22	25	29
100	—	—	—	—	—	—	—	—	8	12	15	20	23	27
105	—	—	—	—	—	—	—	—	5	9	13	17	22	26
110	—	—	—	—	—	—	—	—	—	6	11	15	20	25
115	—	—	—	—	—	—	—	—	—	—	8	14	18	23

The technician needs to record these temperatures and plot them against the manufacturer's graph or table to determine the performance of the equipment.

40.7 ESTABLISHING A REFERENCE POINT ON UNKNOWN EQUIPMENT

When a technician has no literature on the equipment and cannot obtain any, what should be done? The first thing is to try to establish some known condition as a reference point. For example, is the equipment standard-efficiency or high-efficiency? This will help establish a reference point for the suction and head pressures mentioned earlier. High-efficiency equipment is often larger than normal. It is not always identified as high-efficiency, and the technician may have to compare the size of the condenser to another one to determine the head pressure. It should be obvious that a larger or oversized condenser would have a lower head pressure. For example, a 3-ton compressor will have a full-load amperage (FLA) rating of about 17 A at 230 V. The amperage rating of the compressor may help to determine the rating of the equipment, **Figure 40.12**. Although a 3-ton high-efficiency piece of equipment will have a lesser amperage rating than standard equipment, the ratings will be close enough to compare and

Figure 40.12 A table of current ratings for different-sized motors at different voltages. Courtesy BDP Company

APPROXIMATE FULL-LOAD AMPERAGE VALUES FOR ALTERNATING-CURRENT MOTORS

Motor	Single Phase		3-Phase-Squirrel Cage Induction		
HP	115 V	230 V	230 V	460 V	575 V
$\frac{1}{6}$	4.4	2.2			
$\frac{1}{4}$	5.8	2.9			
$\frac{1}{3}$	7.2	3.6			
$\frac{1}{2}$	9.8	4.9	2	1.0	0.8
$\frac{3}{4}$	13.8	6.9	2.8	1.4	1.1
1	16	8	3.6	1.8	1.4
$1\frac{1}{2}$	20	10	5.2	2.6	2.1
2	24	12	6.8	3.4	2.7
3	34	17	9.6	4.8	3.9
5	56	28	15.2	7.6	6.1
$7\frac{1}{2}$			22	11.0	9.0
10			28	14.0	11.0

Does not include shaded pole.

to determine the capacity of the equipment. If the condenser is very large for an amperage rating that should be 3 tons, the equipment is probably high-efficiency and the head pressure will not be as high as in a standard piece of equipment. *In addition, high-efficiency air-conditioning equipment often uses a thermostatic expansion valve rather than a fixed-bore metering device to gain a certain amount of efficiency.* The evaporator may also be larger than normal, which will add to the efficiency of the system. It is more difficult, however, to determine the evaporator size because, being enclosed in a casing or the ductwork, it is not as easy to see.

40.8 SYSTEM PRESSURES AND TEMPERATURES FOR DIFFERENT OPERATING CONDITIONS

The operating conditions of a system can vary so much that this unit cannot cover all possibilities; however, a few general statements committed to memory may help. Standard-efficiency equipment and high-efficiency equipment will operate along the lines of the general conditions listed in the following subsections.

OPERATION OF STANDARD-EFFICIENCY EQUIPMENT AT NEAR-DESIGN CONDITIONS

1. The evaporator saturation temperature is 40°F (70 psig for R-22 and 118 psig for R-410A), **Figure 40.13**.

Figure 40.13 An evaporator operating at close-to-design conditions.

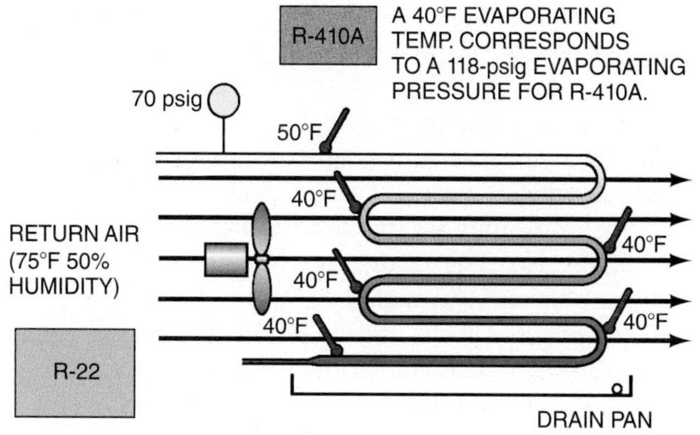

2. The head pressure should correspond to a condensing temperature of no more than 35°F above the outside ambient temperature. This would be 297 psig for R-22 and 475 psig for R-410A when the outside temperature is 95°F (95°F + 35°F = 130°F condensing temperature). A typical condition would be condensing at 30°F above the ambient, or 95°F + 30°F = 125°F, **Figure 40.14.** The head pressure would be 278 psig for R-22 and 446 psig for R-410A.

OPERATION OF STANDARD-EFFICIENCY EQUIPMENT AT HIGHER-THAN-NORMAL INDOOR TEMPERATURES

1. The suction pressure will be higher than normal. Normally, the refrigerant boiling temperature is about 35°F

Figure 40.14 The normal conditions of a standard condenser operating on a 95°F day.

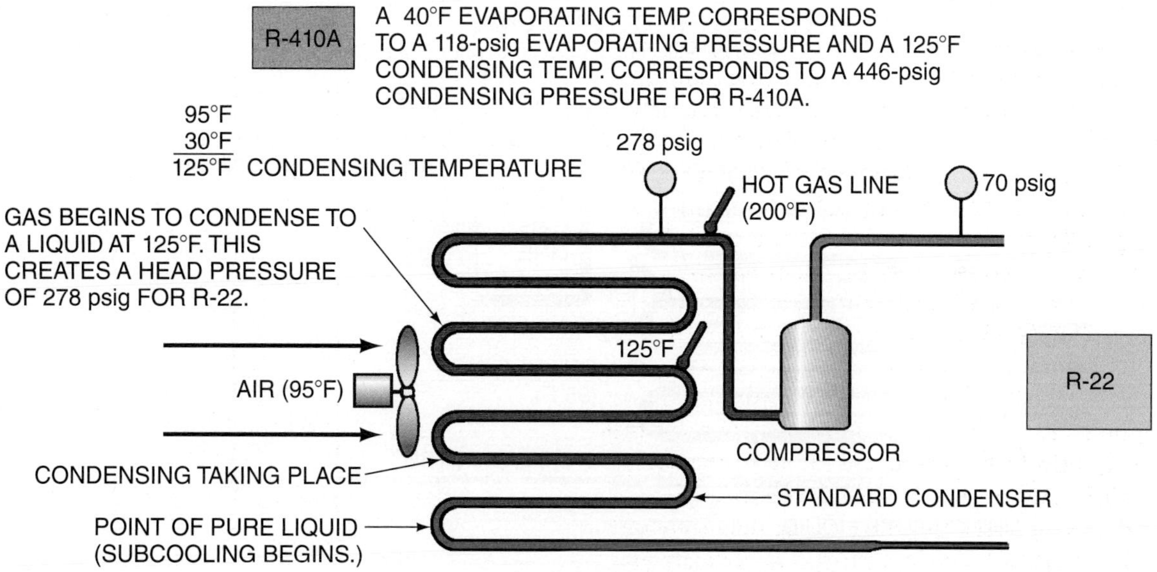

cooler than the entering-air temperature. (Recall the relationship of the evaporator to the entering-air temperature described in the refrigeration section.) When the conditions are normal, the refrigerant boiling temperature would be 40°F when the return air temperature is 75°F. This is true when the humidity is normal. When the space temperature is higher than normal because the equipment has been off for a long time, the return-air temperature may go up to 85°F and the humidity may be high also. The refrigerant boiling temperature may then go up to 50°F, with a corresponding pressure of 84 to 93 psig for R-22 and 142 to 156 psig for R-410A, **Figure 40.15**. This higher-than-normal suction pressure may cause the discharge pressure to rise also.

2. The discharge pressure is influenced by the outside temperature and the suction pressure. For example, the discharge pressure is supposed to correspond to a condensing temperature of no more than 30°F higher than the ambient temperature under normal conditions. When the suction pressure is 80 psig, the discharge pressure is going to rise accordingly. It may move up to a new condensing temperature 10°F higher than normal. This would mean that the discharge pressure for R-22 could be 317 psig (95°F + 30°F + 10°F = 135°F) or 506 psig for R-410A while the suction pressure is high, **Figure 40.16**. When the unit begins to reduce the space temperature and humidity, the evaporator pressure will begin to come down. The load on the evaporator is reduced. The head pressure will come down also.

Figure 40.15 Pressures and temperatures as they may occur with an evaporator when the space temperature and humidity are above design conditions.

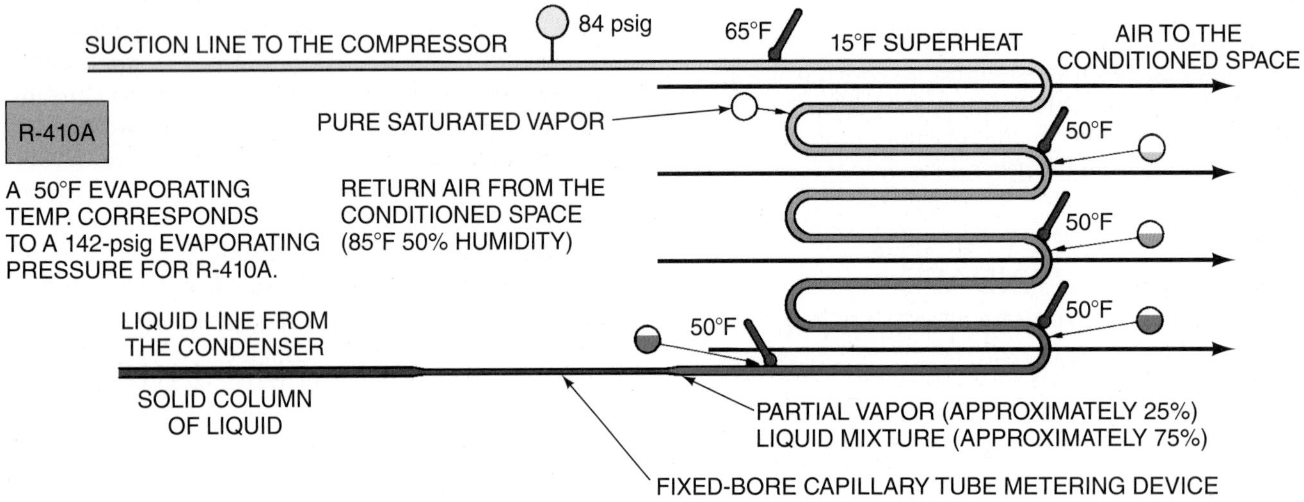

Figure 40.16 The condenser is operating at the design condition as far as the outdoor ambient air is concerned, but the pressure is high because the evaporator is under a load that is above design.

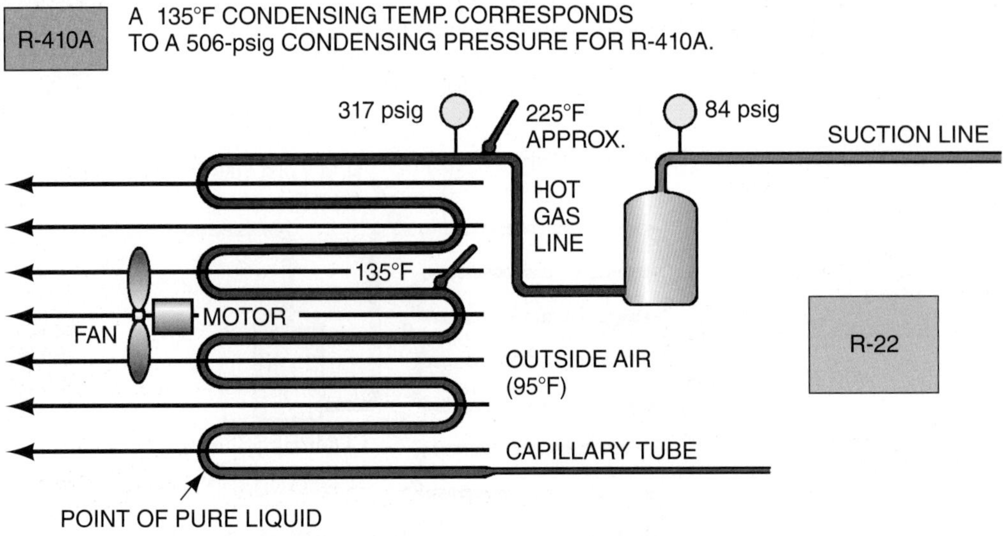

OPERATION OF HIGH-EFFICIENCY EQUIPMENT AT NEAR DESIGN CONDITIONS

1. The evaporator design temperature may in some cases operate at a slightly higher pressure and temperature on high-efficiency equipment because the evaporator is larger. The boiling refrigerant temperature may be 45°F at design conditions, and this would create a normal suction pressure of 76 psig for R-22 and 130 psig for R-410A, **Figure 40.17**. The evaporator will employ a TXV metering device to keep the entire evaporator active under all heat load conditions.
2. The refrigerant may condense at a temperature as low as 20°F higher than the outside ambient temperature. For a 95°F day with R-22, the head pressure may be as low as 243 psig, and as low as 390 psig for R-410A, **Figure 40.17**. If the condensing temperature is as high as 30°F above the ambient temperature, the technician should suspect a problem. For example, the head pressure should not be more than 277 psig on a 95°F day for R-22 and 446 psig for R-410A.

OPERATION OF HIGH-EFFICIENCY EQUIPMENT AT NON-DESIGN CONDITIONS

1. When the unit has been off long enough for the load to build up, the space temperature and humidity are above design conditions and the high-efficiency system pressures will be higher than normal, as they would in the standard-efficiency system. With standard equipment, when the return-air temperature is 75°F and the humidity is approximately 50%, the refrigerant boils at about 40°F. This is a temperature difference of 35°F. A high-efficiency evaporator is larger, and the refrigerant may boil at a temperature difference of 30°F, or around 55°F when the space temperature is 85°F, **Figure 40.18**. The exact boiling temperature relationship depends on the manufacturer's design and how much coil surface area was selected. The TXV metering device will keep the evaporator active under all heat load conditions.
2. The high-efficiency condenser, like the evaporator, will operate at a higher pressure when the load is increased. The head pressure will not be as high as it would with a standard-efficiency condenser because the condenser has extra surface area.

The capacity of high-efficiency systems is not up to the rated capacity when the outdoor temperature is much below design. The conditions in the earlier example of the couple that shut off the air-conditioning system before going to work and then turned it back on when they came home will be much worse with a high-efficiency system. The fact that the standard condenser became too efficient at night when the air was cooler will be much more evident with the larger, high-efficiency condenser. Most manufacturers produce variable-speed or two-speed condenser fans so that lower fan speeds can be used in mild weather to help compensate for this temperature difference. The service technician who tries to analyze a component of high-efficiency equipment on a mild day will find the head pressure low. This will also cause the suction pressure to be low. Using

Figure 40.17 High-efficiency system operating conditions—45°F evaporator temperature and 115°F condensing temperature.

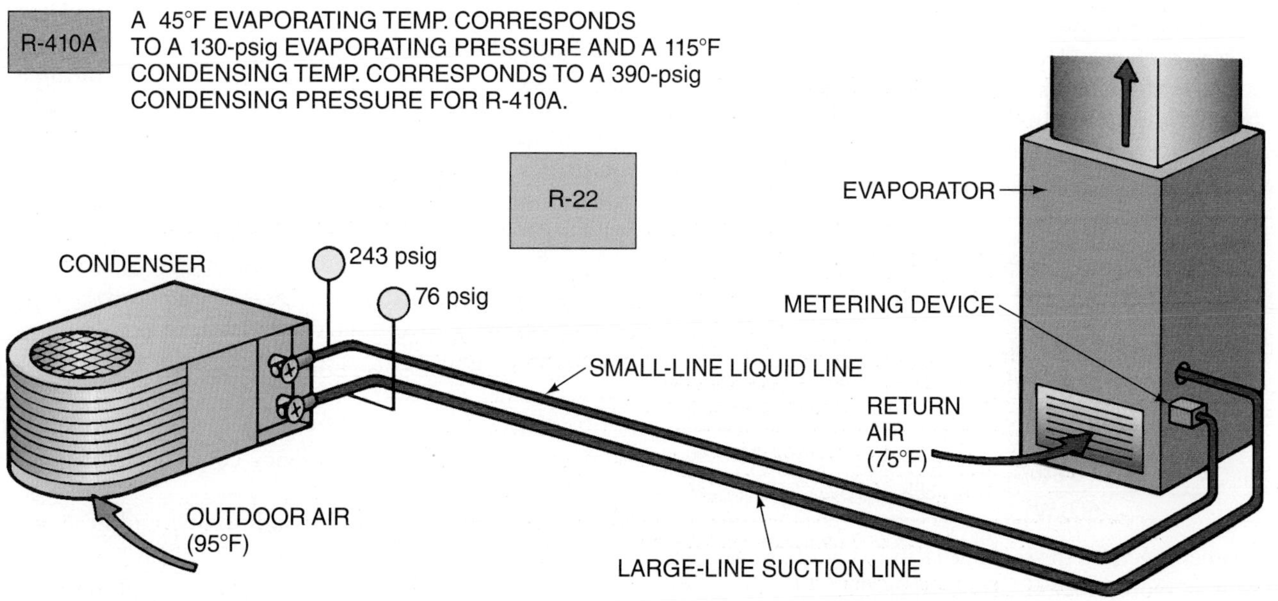

Figure 40.18 An evaporator in a high-efficiency system with a load above design conditions.

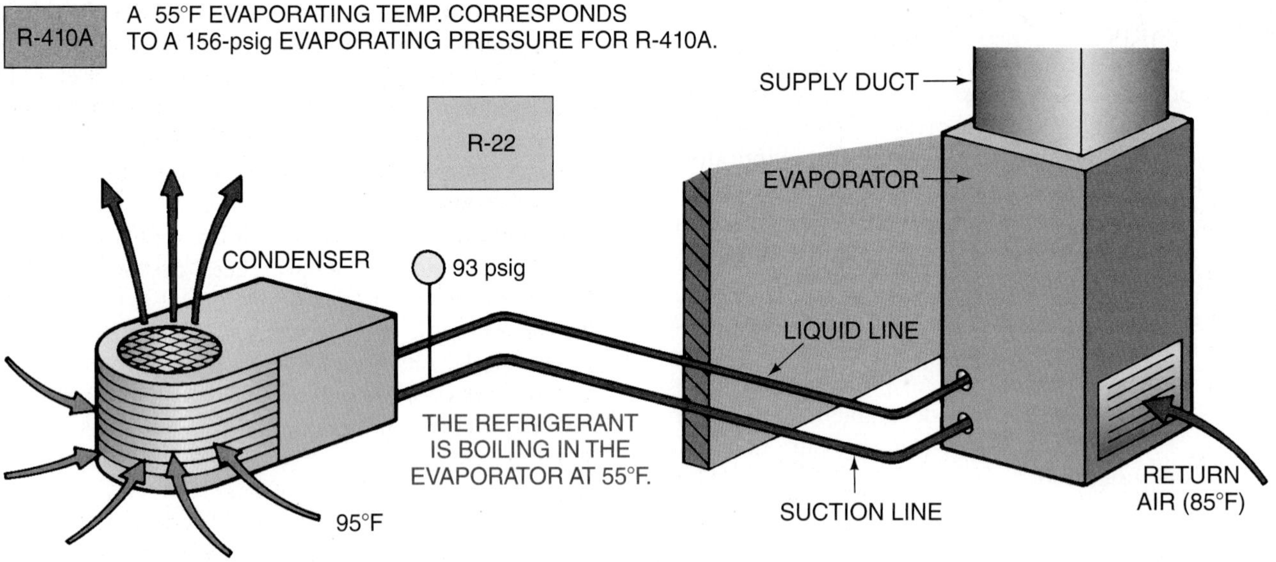

the coil-to-air relationships for the condenser and the evaporator will help determine the correct pressures and temperatures.

Remember these two statements:

1. The evaporator absorbs heat, which is related to its operating pressures and temperatures.
2. The condenser rejects heat and has a predictable relationship to the evaporator load and outdoor ambient temperature.

40.9 EQUIPMENT EFFICIENCY RATINGS

Manufacturers have a method of rating equipment so that the designer and the owner can tell high efficiency from low efficiency at a glance. This rating was originally called the EER rating, which stands for energy efficiency ratio and is actually the output in Btu/h divided by the input in watts of power used to produce the output. For example, a system may have an output of 36,000 Btu/h with an input of 4000 W:

$$\frac{36{,}000 \text{ Btu/h}}{4000 \text{ W}} = \text{an EER of } 9$$

The larger the EER rating, the more efficient the equipment. For example, suppose that the 36,000 Btu/h air conditioner required only a 3600-W input:

$$\frac{36{,}000 \text{ Btu/h}}{3600 \text{ W}} = \text{an EER of } 10$$

The customer is getting the same capacity using less power. The equipment is more efficient. **The higher the EER rating, the more efficient the equipment.**

The EER rating is a steady-state rating and does not account for the time the unit operates before reaching peak efficiency. This operating time has an unknown efficiency. It also does not account for shutting the system down at the end of the cycle (when the thermostat is satisfied), leaving a cold coil in the duct. The cold coil continues to absorb heat from the surroundings, not the conditioned space. This accounts for some of the inefficiency at the beginning of the cycle. Some manufacturers include controls that keep the fan running at the end of the cycle to take advantage of this cold-coil heat exchange. The indoor fan will run for several minutes after the compressor shuts off. This is accomplished with the electronic circuit board that controls the fan operation.

The EER rating system does not give the complete picture, so a rating of seasonal efficiency has been developed, called the seasonal energy efficiency ratio (SEER). This rating is tested and verified by a rating agency and includes the start-up and shutdown cycles. The rating agency is the Air-Conditioning Heating and Refrigeration Institute (AHRI), which publishes ratings of manufacturers' equipment that they list in their catalogs. A typical rating may look like **Figure 40.19.** *SEER ratings in the mid-20s are not uncommon today for high-efficiency air-conditioning equipment incorporating variable-frequency drives (VFDs).* High SEER equipment has a higher initial cost; not everyone will make this kind of investment, so lower-efficiency equipment is still popular. Speculative home builders have a tendency to install the lower-price and lower-efficiency equipment for quicker sale of the home.

Vapor-compression air-conditioning and heat pump systems are required to meet minimum SEER rating guidelines as set forth by the Air-Conditioning, Heating and Refrigeration Institute (AHRI). SEER is calculated by dividing

Figure 40.19 Examples of SEER ratings.

MODEL	CAPACITY (Btu/h)	SEER
A	24,000	9.00
B	24,000	10.00
C	24,000	10.50
D	24,000	11.00
E	24,000	11.50
F	24,000	12.00
G	24,000	15.00
H	24,000	17.00
I	24,000	19.00
J	24,000	22.00
K	24,000	24.00

the total amount of cooling produced, expressed in Btu/h, over an entire cooling season by the number of watt-hours of energy consumed to produce the cooling. The following national requirements are presently in place for minimum efficiency, single-phase systems with nameplate capacities of 65,000 Btu/h or less:

- Packaged Air-Conditioning Systems w/Gas Electric Heat — 14 SEER, 81% AFUE
- Packaged Heat Pump Systems/Dual-Fuel — 14 SEER, 8.0 HSPF, 81% AFUE
- Split System Heat Pumps — 14 SEER, 8.2 HSPF

Refer to Unit 31 for more information on Annual Fuel Utilization Efficiency (AFUE) and Unit 43 for more information on Heating Seasonal Performance Factor (HSPF).

The minimum SEER rating for Split Air-Conditioning Systems varies based on regional standards. States south of, and including, California, Nevada, Arizona, New Mexico, Oklahoma, Arkansas, Kentucky, Virginia, Delaware, and Maryland, are in the southern region, while all other states are in the northern region. The minimum SEER rating requirements under the regional standards are:

- Split Air-Conditioning Systems in the Southern Region — 14 SEER
- Split Air-Conditioning Systems in the Northern Region — 13 SEER

Be sure to visit the AHRI directory website (www.ahridirectory.org) for the most up-to-date information regarding the minimum SEER rating requirements for your state.

40.10 TYPICAL ELECTRICAL OPERATING CONDITIONS

The electrical operating conditions are measured using a digital multimeter with an amp clamp. Three major power-consuming devices that should be tested during a routine maintenance check are: the indoor fan motor, the outdoor fan motor, and the compressor.

The starting point for evaluating electrical operating conditions is knowing what the system supply voltage is supposed to be. For residential units, 230 V single-phase is typical. Light-commercial equipment will nearly always use 208 V or 230 V single-phase or three-phase. Single- and three-phase power may be obtained from a three-phase power supply. An equipment rating may be 208/230 V. The reason for the two different ratings is that some power companies provide 208 V and others provide 230. Some light-commercial equipment may have a supply voltage of 460 V three-phase if it is installed at a large commercial complex. For example, an office may have a 3- or 4-ton air-conditioning unit that operates separately from the main central system. If the supply voltage is 460 V three-phase, the small unit may operate from the same power supply. When 208/230-V equipment is used at a commercial installation, the compressor may be three-phase and the fan motors single-phase. The number of phases that the power company furnishes makes a difference in the method of starting the compressor. Single-phase compressors may have a start assist, such as the positive temperature coefficient device or a start relay and start capacitor. Three-phase compressors will have no start-assist accessories.

40.11 MATCHING THE UNIT TO THE CORRECT POWER SUPPLY

The typical operating voltages for any air-conditioning system must be within the manufacturer's specifications. This is typically $\pm 10\%$ of the rated voltage. For the 208/230-V motor, the minimum allowable operating voltage would be $208 \times 0.90 = 187.2$ V. The maximum allowable operating voltage would be $230 \times 1.10 = 253$ V. Notice that because this application is 208/230 V, the calculations used the 208-V rating as the base for figuring the low voltage and the 230-V rating for calculating the highest voltage. If the motor is rated at 208 V or 230 V alone, that value (208 V or 230 V) is used for evaluating the voltage.

Equipment may be started under some conditions that are beyond the rated conditions. The technician must use some judgment. For example, if 180 V is measured, the equipment should not be started because the voltage will drop further with the current draw of the motor. If the voltage reads 260 V, the equipment may be started because the voltage may drop slightly when the motor is started. If the voltage drops to within the limits, the motor is allowed to run.

40.12 STARTING THE EQUIPMENT WITH THE CORRECT DATA

When the correct rated voltage is known and the minimum and maximum voltages are determined, the equipment may be started if the voltages are within the limits. The three motors—indoor fan, outdoor fan, and the compressor—can be checked for the correct current draw.

The indoor fan builds the air pressure to move air through the ductwork, filters, and grilles to the conditioned space. By law, the voltage characteristics must be printed on the fan motor in such a way that they will not come off. However, the motor might be mounted so that the data cannot be easily seen. In some cases the motor may be inside the blower wheel. If so, removing the motor may be the only way to view the motor nameplate. If the supplier can be easily contacted, the technician may obtain the information in that way. If the electrical characteristics of the motor cannot be obtained from the motor nameplate, they might be available on the unit nameplate. The fan motor may have been changed to a larger motor for more fan capacity, however.

40.13 FINDING A POINT OF REFERENCE FOR AN UNKNOWN MOTOR RATING

When a motor is mounted so that the electrical characteristics cannot be determined, the technician must improvise. We know that air-conditioning systems normally move about 400 cfm of air per ton. This can help determine the amperage of the indoor fan motor—the fan amperage on an unknown system can be compared with the amperage of a known system. All that is needed is the approximate system capacity. As discussed earlier, compressor amperages of the unit in question can be compared to a known unit. For example, the compressor amperage of a 3-ton system is about 17 A when operating on 230 V. If the amperage of the compressor in the system being checked is 17 A, the technician can assume that the system is close to 3 tons. The indoor fan motor for a 3-ton system should be about 1/3 hp for a typical duct system. The fan motor amperage for a 1/3-hp permanent split-capacitor motor is 3.6 A at 230 V. If the fan in question is pulling 5 A, the technician should suspect a problem. Fan motors shipped in a warm-air furnace may have been changed if air-conditioning was added at a later date. In this case, the furnace nameplate may not give the correct fan motor data. The condensing unit will have a nameplate for the condenser fan motor. This motor should be sized fairly close to its actual load and should pull close to nameplate amperage.

40.14 DETERMINING THE COMPRESSOR RUNNING AMPERAGE

The compressor current draw may not be as easy to determine as the fan motor current draw because not all compressor manufacturers stamp the compressor run-load amperage (RLA) rating on the compressor nameplate. Running-load amperage is an older term that is being replaced with rated-load amperage (RLA).

Figure 40.20 A condensing unit nameplate. Photo by Bill Johnson

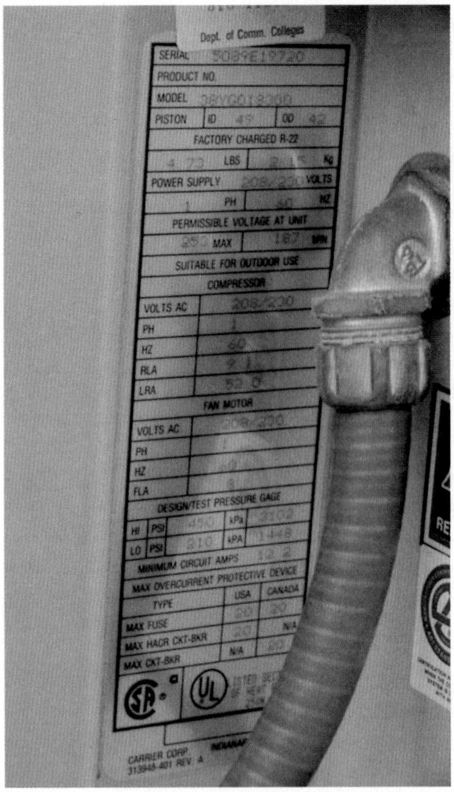

Because there are so many different compressor sizes, it is hard to state the correct full-load amperage. For example, motors normally come in the following increments: 1, 1½, 2, 3, and 5 hp. A unit rated at 34,000 Btu/h is considered to be a 3-ton unit, even though a true 3-ton unit has a capacity of 36,000 Btu/h. Ratings that are not completely accurate are known as **nominal ratings**. They are rounded off to the closest rating. A typical 3-ton air-conditioning unit may have a 3-hp compressor motor. A unit with a rating of 34,000 Btu/h does not need a full 3-hp motor, but it is supplied with one because there is no standard horsepower motor to meet its needs. If the motor amperage for a 3-hp motor were stamped on the compressor, it could cause confusion because the motor may never operate at that amperage. A unit nameplate lists electrical information, **Figure 40.20**. **NOTE:** *The manufacturer may stamp the compressor RLA on the unit nameplate. Please refer to Unit 18, "Application of Motors," when referring to full-load amperage (FLA), rated-load amperage (RLA), service-factor amperage (SFA), and lock-rotor amperage (LRA).* •

40.15 COMPRESSORS OPERATING AT FULL-LOAD CURRENT

It is rare for a compressor motor to operate at its RLA rating. If design or above-design conditions were present, the compressor would operate at close to full load. When the unit is

operating at a condition greater than design, such as when the unit has been off for some time in very hot weather, it might appear that the compressor is operating at more than RLA. However, other conditions usually keep the compressor from drawing too much current. The compressor is pumping vapor, and vapor is very light. It takes a substantial increase in pressure difference to create a significantly greater workload.

The compressor motor may be larger than needed most of the operating time and may not reach its rated horsepower until it gets to the maximum design rating for the system. It may be designed to operate at 105°F or 115°F outdoor ambient for very hot regions.

40.16 TWO-SPEED AND VARIABLE SPEED COMPRESSORS

Some air-conditioning manufacturers use two-speed compressors to achieve better seasonal efficiencies. These compressors may use a motor capable of operating as a two-pole or a four-pole motor. A four-pole motor runs at 1800 rpm; a two-pole motor runs at 3600 rpm. A motor control circuit will slow the motor to reduced speed for mild weather.

Variable speed may be obtained with electronic circuits. This greatly improves the system's efficiency because there is no need to stop the compressor and restart it at the beginning of the next cycle. However, this means more factors for which a technician must establish typical running conditions under other-than-design conditions. The equipment should perform just as typical high-efficiency equipment would when operating at design conditions.

SAFETY PRECAUTION: *The technician must use caution while observing equipment to determine the operating conditions. Electrical, pressure, and temperature readings must be taken while the technician observes the equipment. Many times, the readings must be made while the equipment is in operation.•*

REVIEW QUESTIONS

1. The standard design conditions for air-conditioning systems established by the Air-Conditioning, Heating and Refrigeration Institute (AHRI) are
   A. 95°F outside, 70°F inside, 40% relative humidity.
   B. 100°F outside, 75°F inside, 50% relative humidity.
   C. 90°F outside, 80°F inside, 30% relative humidity.
   D. 95°F outside, 80°F inside, 50% relative humidity.
2. The typical temperature relationship between a standard-efficiency air-cooled condenser and the ambient temperature is
   A. 50°F.
   B. 40°F.
   C. 30°F.
   D. 20°F.
3. The evaporator will normally operate at a ______°F boiling temperature when the indoor temperature is 75°F with a 50% humidity condition.
   A. 40
   B. 50
   C. 60
   D. 70
4. A typical temperature relationship between a high-efficiency condenser and the ambient temperature is
   A. 10°F.
   B. 20°F.
   C. 30°F.
   D. 40°F.
5. One way manufacturers gain a higher efficiency is by
   A. designing the unit to run with a higher head pressure.
   B. equipping the system with a fixed-bore metering device.
   C. using a larger condenser surface area.
   D. using a lower evaporator saturation temperature.
6. Under high evaporator heat loads, the temperature difference between the condenser and the ambient will be
   A. much higher.
   B. much lower.
   C. about the same.
   D. unpredictable.
7. The load on the outdoor coil is influenced by
   A. the indoor dry-bulb temperature.
   B. the wet-bulb temperature of the indoor air.
   C. the outdoor temperature and the evaporator heat load.
   D. the location of the indoor coil.
8. The sensible-heat load on the indoor coil is influenced by
   A. the location of the outdoor coil.
   B. the outdoor temperature and the evaporator heat load.
   C. the wet-bulb temperature of the indoor air.
   D. the indoor dry-bulb temperature.
9. The latent-heat load on the indoor coil is influenced by
   A. the outdoor temperature and the evaporator heat load.
   B. the wet-bulb temperature of the indoor air.
   C. the indoor dry-bulb temperature.
   D. the location of the outdoor coil.
10. When a technician has no literature on the equipment and cannot obtain any, what should be done?
    A. Reference a temperature/pressure chart for refrigerant information.
    B. Establish some known condition as a reference point.
    C. Calculate the system superheat and subcooling.
    D. Test the refrigerant oil for the presence of acid.

11. As the outdoor ambient temperature rises, the system capacity will
    A. decrease.
    B. increase.
    C. remain constant.
    D. fluctuate.
12. When the air passing over the condenser is cooler than the air passing over the evaporator
    A. the capacity of the condenser will greatly increase.
    B. the condenser will begin to frost up or freeze.
    C. the capacity of the condenser will be reduced.
    D. the refrigerant in the evaporator will begin to condense.
13. On a standard-efficiency system that is operating with an indoor temperature that is higher than normal
    A. the suction pressure will be lower than normal.
    B. the discharge pressure will be lower than normal.
    C. the suction and discharge pressure will both be lower than normal.
    D. the suction and discharge pressure will both be higher than normal.
14. On a high-efficiency system, if the condensing temperature is more than ______°F above the ambient temperature, the system should be checked for problems.
    A. 5
    B. 10
    C. 20
    D. 30
15. The system output in Btu/h divided by the watts of power used to produce the output is called
    A. SEER.
    B. EER.
    C. AHRI.
    D. HSPF.
16. An equipment rating system that includes the start-up and shut-down cycles as well as the steady-state operation is
    A. SEER.
    B. EER.
    C. AHRI.
    D. HSPF.
17. The higher the EER rating is
    A. the smaller the equipment is.
    B. the higher the system capacity is.
    C. the more efficient the equipment is.
    D. the lower the system efficiency is.
18. A typical analysis of a system's major power-consuming devices would include the
    A. thermostat, contactor, and the capacitor.
    B. indoor fan motor, outdoor fan motor, and the compressor.
    C. compressor, transformer, and the contactor.
    D. thermostat, compressor, contactor, and the outdoor fan motor.
19. Most manufacturers specify that the supply voltage to the equipment should be within _____% of the nameplate voltage.
    A. 5 C. 15
    B. 10 D. 20
20. The voltage being supplied to a 208/230-V motor should be between
    A. 187 and 229 volts.
    B. 207 and 253 volts.
    C. 187 and 253 volts.
    D. 207 and 229 volts.
21. In some cases, the electrical data for the indoor blower motor may be difficult to obtain because
    A. the motor may be mounted inside the blower wheel, making it hard to see the nameplate.
    B. smaller motors are not required to list electrical data on the motor nameplate.
    C. the nameplate on the furnace or air handler is not accurate.
    D. the information is only printed in the manufacturer's literature which may have been lost.
22. The amperage for a typical 1/3-hp permanent split-capacitor motor operating at 230 V is about
    A. 9.4 A. C. 10 A.
    B. 8.5 A. D. 3.6 A.
23. Ratings that are not completely accurate or that are rounded to the next closest rating are known as
    A. EER ratings.
    B. nominal ratings.
    C. SEER ratings.
    D. AHRI ratings.
24. A compressor that is operating at design or above-design conditions will
    A. overheat.
    B. operate at close to full-load amperage.
    C. operate over its RLA.
    D. operate at its service factor amperage.
25. A four-pole motor typically runs at
    A. 960 rpm.
    B. 1800 rpm.
    C. 3600 rpm.
    D. 4200 rpm.
26. A two-pole motor typically runs at
    A. 960 rpm.
    B. 1800 rpm.
    C. 3600 rpm.
    D. 4200 rpm.
27. One way manufacturers gain a higher efficiency is by
    A. designing the unit to run with a higher head pressure.
    B. equipping the system with a fixed-bore metering device.
    C. using a two-speed or variable speed compressor.
    D. using a lower evaporator saturation temperature.

UNIT 41

TROUBLESHOOTING

OBJECTIVES

After studying this unit, you should be able to

- select the correct instruments for checking an air-conditioning unit with a mechanical problem.
- calculate the correct operating suction pressures for both standard- and high-efficiency air-conditioning equipment under various conditions.
- calculate the standard operating discharge pressures at various ambient conditions.
- explain temperature difference, temperature split, and approach temperature.
- discuss charging systems in the field.
- describe basic mechanical troubleshooting procedures.
- select the correct instruments for troubleshooting electrical problems in an air-conditioning system.
- describe troubleshooting basic electrical problems in an air-conditioning system.
- use a digital multimeter to check the various components of the electrical system.

SAFETY CHECKLIST

- ✓ Be extremely careful when installing or removing gauges. Escaping refrigerant can injure your skin and eyes. Wear gloves and goggles. If possible, turn off the system to install a gauge on the high-pressure side. Most air-conditioning systems are designed so that the pressures will equalize when turned off.
- ✓ 4R *Wear goggles and gloves when transferring refrigerant from a cylinder to a system or when recovering refrigerant from a system. Do not recover refrigerant into a disposable cylinder. Use only tanks or cylinders approved by the Department of Transportation (DOT).* 4R
- ✓ Be very careful when working around a compressor discharge line. The temperature may be as high as 300°F.
- ✓ All safety practices must be observed when troubleshooting electrical systems. Often, readings must be taken while the power is on. Only let the contact tips on the insulated meter leads touch the hot terminals. Never use a screwdriver or other tools around terminals that are energized. Never use tools in a hot electrical panel.
- ✓ Turn the power off whenever possible when working on a system. Lock and tag the disconnect panel and keep the only key on your person.
- ✓ Do not stand in water when making any electrical measurements.
- ✓ Short capacitor terminals with a 20,000-Ω resistor before checking with a meter.
- ✓ Students and other inexperienced persons should perform troubleshooting tasks only under the supervision of an experienced person.

41.1 INTRODUCTION

Troubleshooting air-conditioning equipment involves troubleshooting both mechanical and electrical systems; they may have symptoms that overlap. For example, if an evaporator fan motor capacitor fails, the motor will slow down and begin to get hot. It may even get hot enough for the internal overload protector to stop it. While it is running slowly, the suction pressure will go down and give symptoms of a restriction or low charge. If the technician diagnoses the problem based on suction pressure readings only, a wrong decision may be made.

41.2 MECHANICAL TROUBLESHOOTING

Gauges and temperature-testing equipment are used when performing mechanical troubleshooting. The gauges used are those on the gauge manifold as shown in **Figures 41.1(A)** and **(B)**. The suction or low-side gauge is on the left side of the manifold, and the discharge, or the high-side, gauge is on the right side. The most common refrigerants used in air conditioners are R-22 and R-410A. However, because of ozone depletion and climate change concerns, more environmentally safe refrigerants have entered the market and may be used instead. Refer to Unit 9, "Refrigeration and Oil Chemistry and Management—Recovery, Recycling, Reclaiming, and Retrofitting," for more detailed information on newer refrigerants and their compatible oils. An R-22 temperature chart is printed on many gauges for determining the saturation temperature for the low- and high-pressure sides of the system. The production of R-22 will be banned starting in the year 2020. As of 2020, no new R-22 can be produced or imported. Because these same gauges are often used for refrigeration, temperature scales for refrigerants used in refrigeration temperature ranges are also printed on most gauges, **Figure 41.2(A)**. The R-12 scale is not used for residential air-conditioning, but was commonly used for early-generation automobiles. In 1992, the automotive industry switched to R-134a. However, because R-134a has a relatively high global warming potential, it too will soon be replaced with a more environmentally safe refrigerant. Many newer gauges are more universal and may not have a temperature relationship printed on them, or they may have a printed temperature relationship for a specific refrigerant, **Figure 41.2(B)**.

Figure 41.1 (A) This manifold has a low-side gauge on the left and a high-side gauge on the right. Courtesy Ritchie Engineering Company, Inc., Yellow Jacket Products Division

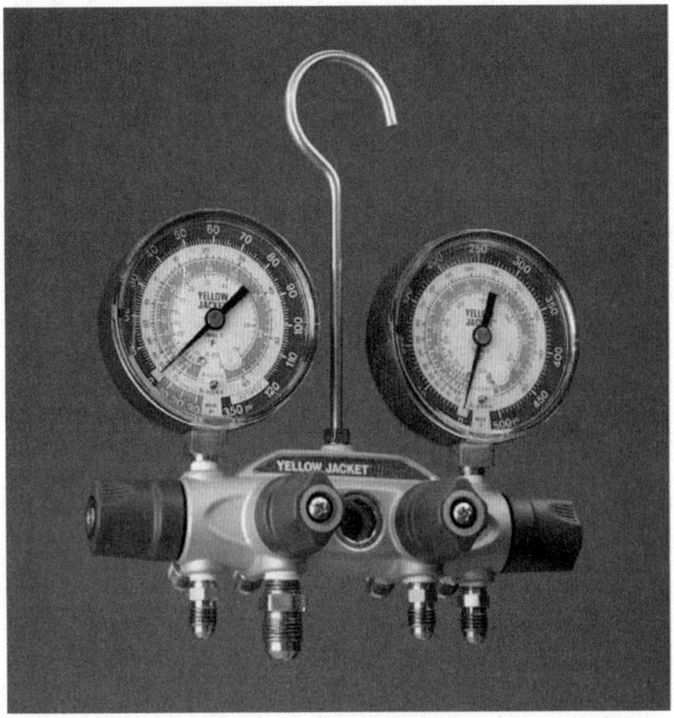

Figure 41.1 (B) A modern digital gauge manifold set with temperature measuring probes for measuring superheat and subcooling. This gauge set also has pressure/temperature relationships of different refrigerants resident in its memory. Courtesy Fieldpiece Corporation

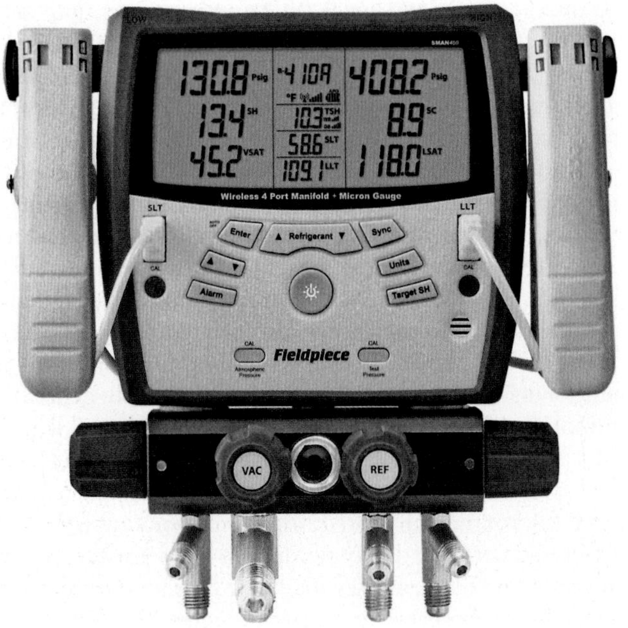

Figure 41.2 (A)–(B) Many gauges have common refrigerant temperature relationships for different refrigerants printed on them. Other gauges are universal and have no printed temperature relationships, or they may be for specific refrigerants. (A) Photo by Jason Obrzut (B) Courtesy of Robinair Division, SPX Corporation

(A)

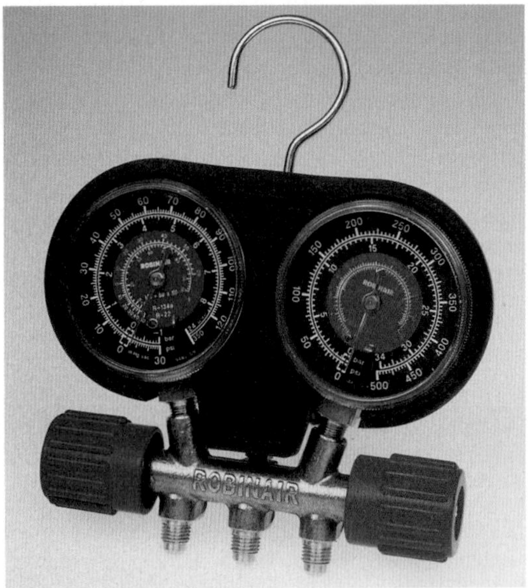

(B)

Many manufacturers use R-410A as a refrigerant. It is a popular refrigerant blend that is replacing R-22 in new equipment. This refrigerant has a much higher working pressure than R-22. Gauges have to be built to handle the higher pressures of R-410A or injury can occur. The gauge ports are the same size, 1/4 in., as for R-22, and either gauge manifold will attach to the R-410A system, but gauges rated for R-22 will be overpressurized if attached to an R-410A system. If R-22 gauges are used on an R-410A system, they will read at the highest point on the gauge. Gauges are most accurate in the middle of their scale range. The technician should make sure what the refrigerant is and apply the correct gauge manifold and hoses. **Figure 41.2(C)** illustrates a gauge built to handle both R-410(A) and R-22. As newer refrigerants and refrigerant blends enter the market, the pressure and temperature scales printed on them will differ.

Figure 41.2 (C) A gauge designed to handle the higher pressures of R-410A. Photo by Jason Obrzut

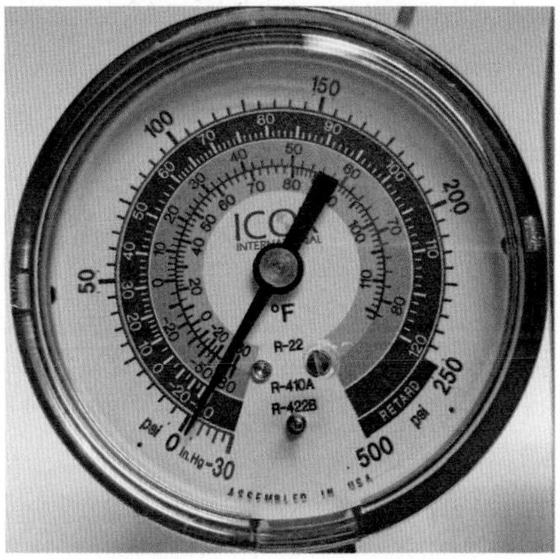

A wireless digital pressure/temperature gauge with clamp-on temperature probes can provide wireless connection to your smart device for measuring system pressures and temperatures, as well as superheat and subcooling amounts, **Figure 41.2(D)**. This wireless device uses Bluetooth technology and can be connected to HVACR systems without using a gauge manifold set or hoses, **Figure 41.2(E)**.

The device has a built-in Schrader port for the HVACR technician to add or remove refrigerant according to the superheat and subcooling amounts without the need for gauges or a manifold.

Notice in **Figure 41.2(E)** the HVACR technician is using a smart device to read, record, send, and share data to others. The smart device shows a low-side pressure of 113.4 psig for R-410A, which correlates to a low-side saturated vapor

Figure 41.2 (D) Wireless digital pressure/temperature gauges with clamp-on temperature probes can provide wireless connection to a smart device. Courtesy Ritchie Engineering Company, Inc. Yellow Jacket Products Division

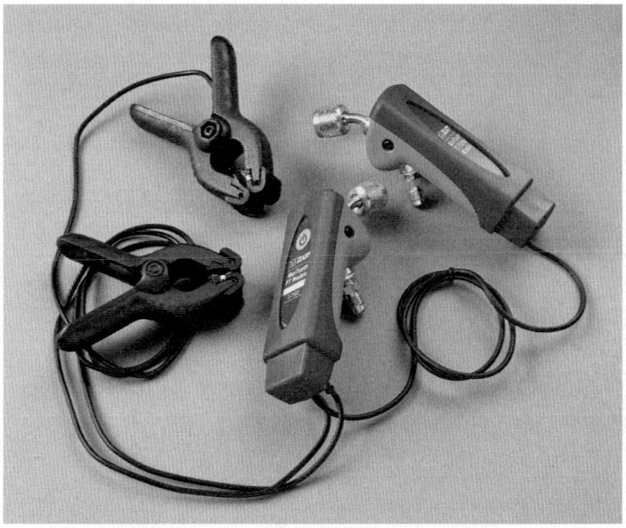

Figure 41.2 (E) A service technician using a smart device to read, record, and share data from wireless, digital pressure/temperature gauges and clamp-on temperature probes in the background. Courtesy Ritchie Engineering Company, Inc. Yellow Jacket Products Division

temperature of 37.6°F. The P/T conversion is done by the app. The clamp-on temperature probe reads a low-side compressor in temperature of 53.4°F. These temperatures and pressures are all shown on the right side of the smart device in the service technician's hand. The difference between the two temperatures of 53.4°F and 37.6°F is 15.8°F of total (compressor) superheat, which is automatically calculated by the app.

The smart device also shows a high-side pressure reading of 296.4 psig for R-410A, which correlates to a high-side saturation temperature of 94.9°F. The clamp-on probe on the liquid line at the condenser outlet reads 84.2°F. The difference between these two temperatures is 10.7°F of liquid subcooling in the condenser.

The service technician can also use a screenshot feature which allows them to save pressure, temperature, subcooling, and superheat readings via e-mail. This data can now be organized by location and unit being serviced for easy tracking and storage history. The Bluetooth radio has a range of up to 400 feet with a 10-hour rechargeable battery life. There are 98 refrigerant profiles within the app, which can be upgraded. The wireless digital pressure/temperature gauge with clamp-on temperature probe can be bought individually for individual temperature and pressure measurements, or for dual pressure and temperature readings, **Figure 41.2(D)**. Superheat and subcooling are probably two of the HVACR service technician's most important measurements and calculations when troubleshooting a system. Using the most recent state-of-the-art digital tools can save time and money, automate measurements, and permit technicians to share data.

A digital manifold that streams data to smart devices is shown in **Figure 41.2(F)**. This digital manifold technology has remote monitoring along with data reporting. The smart

Figure 41.2 (F) A digital manifold that is compatible with smart devices. Courtesy North Park Innovations

manifold wirelessly displays the system's actual pressures, temperatures, superheat, and subcooling while simultaneously calculating performance targets, **Figure 41.2(G)**. This smart manifold replaces traditional analog gauges with real-time graphical feedback. By using electronic pressure transducers and various wired or wireless probes, the manifold can collect, organize, and present system performance in a detailed and concise way. The system utilizes a smart phone or tablet to communicate system diagnostics wirelessly and capitalize on GPS, video, and communications platforms already available in smart devices. This manifold is engineered and constructed to withstand the abuse of everyday use while still holding the accuracy of measurement that is required for the industry. The app includes an easy-to-read display; target pressures calculated and displayed; over 50 refrigerants; equipment profiling; calculated target zones for superheat, subcooling, and discharge line temperature; and troubleshooting.

Service technicians often multitask and work on more than one refrigeration or air-conditioning system at a time. Many times with today's digital service tools, a technician will remotely monitor the system for the proper operating parameters like evaporator and condensing pressures, evaporator superheat, and condenser subcooling after servicing the system. **Figure 41.2(H)** shows a student service technician setting up an ammeter and Bluetooth high- and low-side pressure gauges on a compressor, after fixing a leak near the liquid line solenoid valve and charging refrigerant to a glass door reach-in freezer. **Figure 41.2(I)** shows

Figure 41.2 (G) A smart manifold wirelessly displaying the system's actual pressures, temperatures, and superheat and subcooling amounts to a smart device while simultaneously calculating performance targets. Courtesy North Park Innovations

Figure 41.2 (H) A student service technician setting up an ammeter and Bluetooth high- and low-side pressure gauges on a compressor. Courtesy Ferris State University. Photo by John Tomczyk

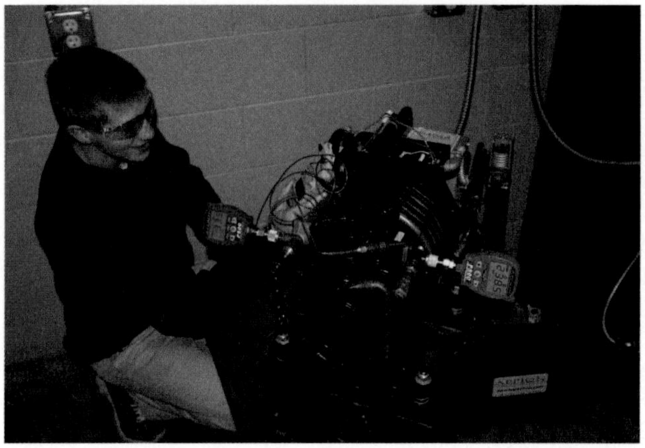

Figure 41.2 (I) A service technician remotely monitoring system pressures with a tablet and appropriate software from an outside unit 2 hours after adding refrigerant. Courtesy Ferris State University. Photo by John Tomczyk

the service technician's partner remotely monitoring system pressures with a tablet and appropriate downloaded application or "App" from an outside unit 2 hours after refrigerant was added and freezing temperatures were achieved. This same service technician is currently working on an outdoor heat pump that is servicing a computer room for the university. The Bluetooth gauge technology allows this kind of multitasking of serviced units without going in and out of the building, saving time and money. Many times in the hot summer months, a service technician can remotely monitor system pressures from an air-conditioned service van while simultaneously doing needed paperwork for their company and the customer.

Unlike radios and mobile phones, which send information wirelessly over many miles, Bluetooth technology exchanges data wirelessly over short distances (10 meters) using radio waves. Bluetooth technology is built into many products like computers, mobile phones, and medical devices. Bluetooth technology was created as an open standard to allow connectivity and collaboration between different products within different industries. Bluetooth technology is a combination of hardware and software. Bluetooth products contain a computer chip containing a Bluetooth radio, **Figure 41.2(J)**. However, software (an application or App) is also needed to connect or interface to other products using Bluetooth wireless technology. Many of these Apps can be downloaded for free. Bluetooth technology is bringing everyday devices into the digital world, which allows them to connect to one another without wires. Today, just about all top-selling tablets, personal computers, headsets, computer mice, keyboards, and smart phones incorporate Bluetooth technology.

Figure 41.2 (J) A Bluetooth product containing a computer chip containing a Bluetooth radio. Courtesy Appion Inc.

41.3 TEMPERATURE DIFFERENCE, TEMPERATURE SPLIT, AND APPROACH TEMPERATURE

Many technicians use the terms approach temperature, temperature split, or temperature difference to describe the difference in temperature between heat exchange mediums. For example, the difference in temperature between the indoor coil and the return air for a central air-conditioning system is 35°F for standard air-conditioning equipment, **Figure 41.3**. The condenser also has temperature difference, or temperature split, **Figure 41.4**. These temperatures are used by the technician to determine when a heat exchanger is functioning correctly. For example, when an air-cooled condenser temperature split begins to increase, it is a sign that the condenser is not condensing efficiently. This tells the technician that the coil is dirty or that there is not enough air passing over the coil.

Delta T (ΔT) is another term that is used by technicians, but it has a different meaning. Delta T stands for change in temperature and it applies to the inlet versus the outlet temperature of the same medium. For example, the air entering an evaporator may be 75°F and the leaving-air temperature may be 55°F; the difference is 20°F. The term applies only to the difference in the temperature of the air. Air-conditioning applications typically operate with an evaporator temperature between 40°F and 45°F. As result, these systems do not require a defrost system.

Figure 41.3 The difference between the entering-air temperature and the refrigerant boiling temperature is known as the approach temperature, temperature difference, or temperature split.

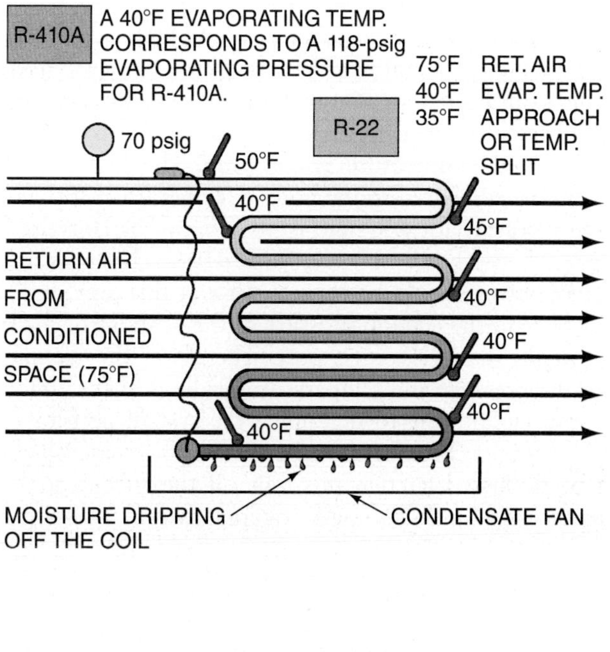

Figure 41.4 The temperature difference, or temperature split, works the same for the condenser.

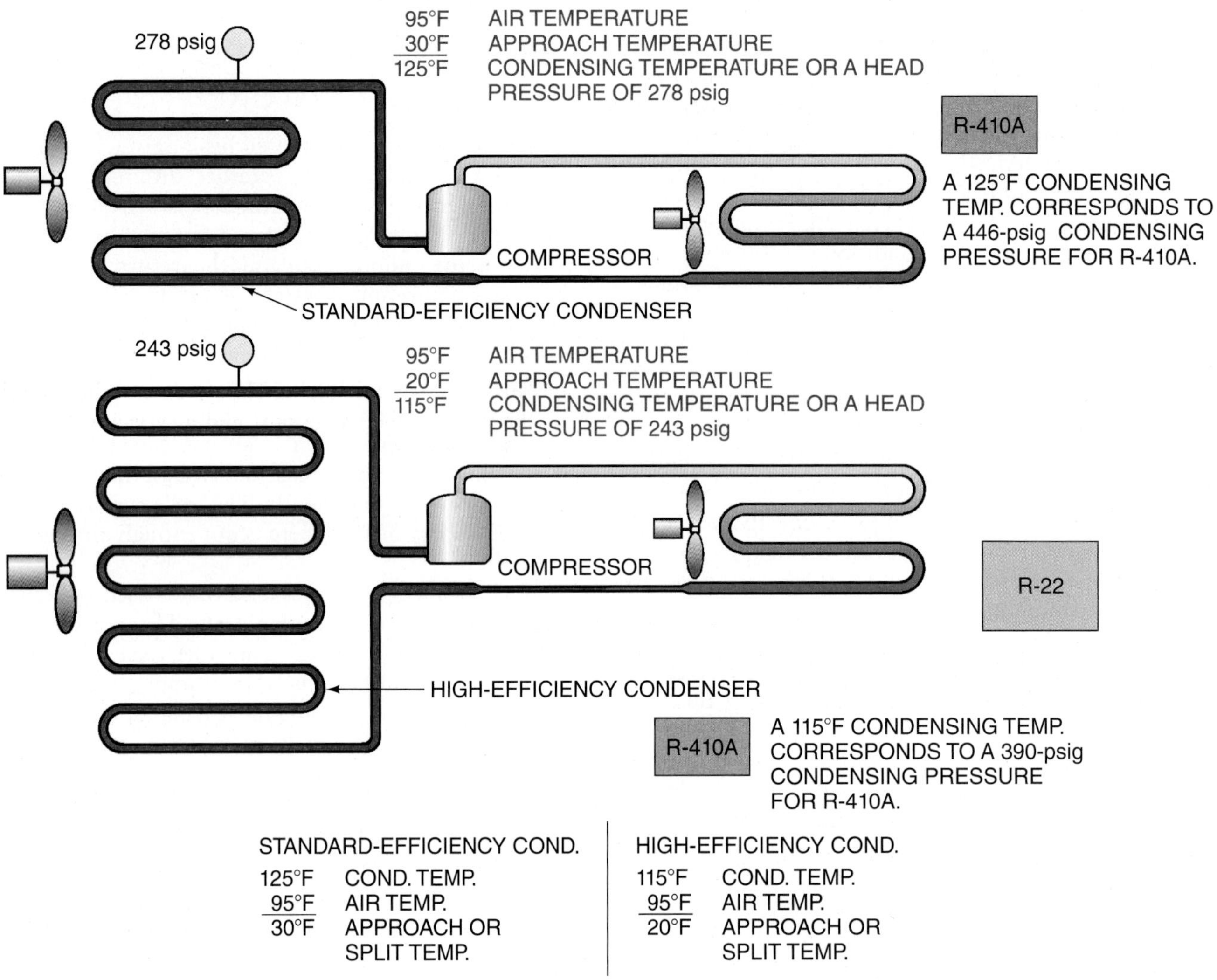

41.4 GAUGE MANIFOLD USAGE

The gauge manifold displays the low- and high-side pressures while the unit is operating. These pressures can be converted to the saturation temperatures for the evaporating (boiling) refrigerant and the condensing refrigerant by using the pressure and temperature relationship. The low-side pressure can be converted to the boiling temperature. If the boiling temperature for a system is 40°F, it can be converted to 70 psig for R-22 and 118 psig for R-410A. The superheat at the evaporator should be close to 10°F at this time. It is difficult to read the suction pressure at the evaporator because it normally has no gauge port. Therefore, the technician takes the pressure and temperature readings at the condensing unit suction line to determine the system performance. (Guidelines for checking the superheat at the condensing unit will be discussed later in this unit.) If the suction pressure were 48 psig for R-22, the refrigerant would be boiling at about 24°F, which is cold enough to freeze the condensate on the evaporator coil and too low for continuous operation. The probable causes of low boiling temperature are low charge or restricted airflow, **Figure 41.5**.

⚠ SAFETY PRECAUTION: *The technician must be extremely careful while attaching manifold gauges. High-pressure refrigerant will injure your skin and eyes. Wearing goggles and gloves is necessary. The danger from attaching the high-pressure gauges can be reduced by shutting off the unit and allowing the pressures to equalize to a lower pressure. Liquid R-22 can cause serious frostbite because it boils at −42°F at atmospheric pressure, while R-410A boils at −60°F.* •

The high-side gauge may be used to convert pressures to condensing temperatures. For example, if the high-side gauge reads 278 psig for R-22 and the outside ambient temperature is 80°F, the head pressure may seem too high. The gauge manifold chart shows that the refrigerant is condensing at 125°F. However, the refrigerant should not condense at a temperature more than 30°F higher than the ambient temperature. Adding 30°F to the ambient temperature of 80°F = 110°F, so

Figure 41.5 This coil was operated below freezing until the condensate on the coil froze.

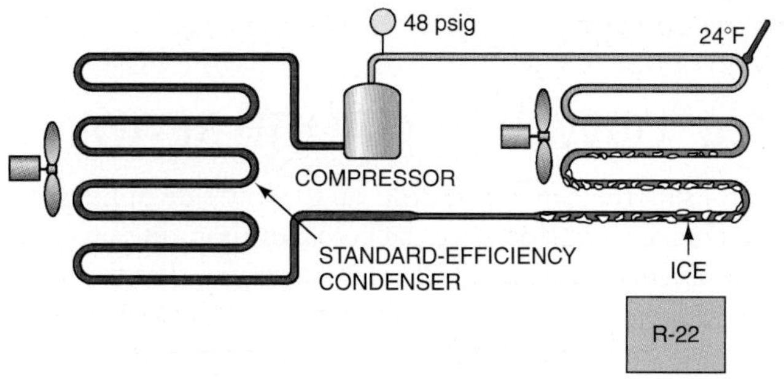

the condensing temperature is actually 15°F too high. Probable causes are a dirty condenser, an overcharge of refrigerant, or noncondensables in the system.

The gauge manifold is used whenever the pressures need to be known. Two types of pressure connections are used with air-conditioning equipment: the Schrader valve and the service valve, **Figure 41.6**. The **Schrader valve** is a pressure connection only. The **service valve** can be used to isolate the system for service. The system may have a Schrader connection for the gauge port and a service valve for isolation, **Figure 41.7**.

Figure 41.6 Two valves commonly used when taking pressure readings on modern air-conditioning equipment. (A) A Schrader valve. Courtesy J/B Industries. (B) Two service valves. Courtesy Ferris State University. Photo by John Tomczyk

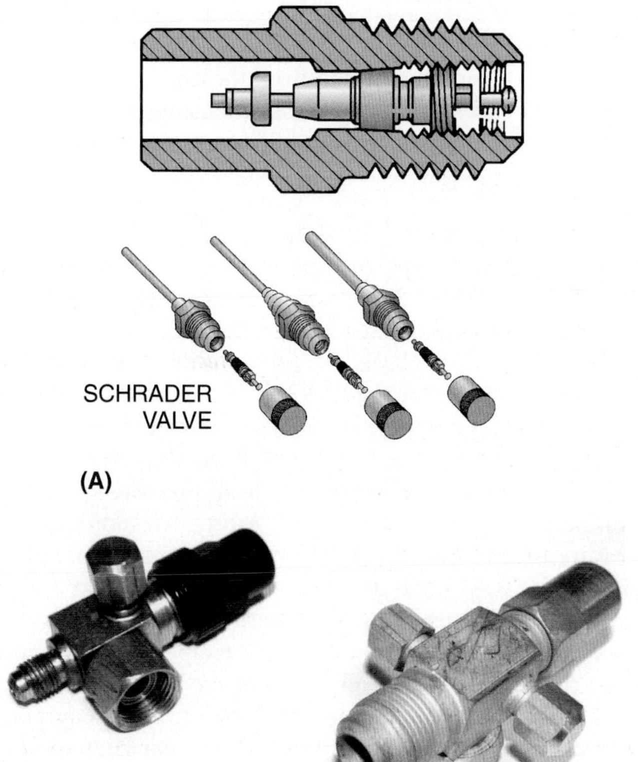

Figure 41.7 This service valve has a Schrader port instead of a back seat like a refrigeration service valve. Photo by Jason Obrzut

41.5 WHEN TO CONNECT THE GAUGES

For small systems, the technician should not connect a gauge manifold every time the system is serviced. A small amount of refrigerant escapes each time a gauge is connected, and some residential and small commercial systems have a critical refrigerant charge. When the high-side line is connected, high-pressure refrigerant will condense in the gauge line. The refrigerant will escape when the gauge line is disconnected from the Schrader connector. A gauge line full of liquid refrigerant lost while checking pressures may be enough to affect the system charge. A short gauge line connector for the high side will help prevent refrigerant loss, **Figure 41.8**. This can be used only to check pressure. It cannot be used to transfer refrigerant out of the system because it is not a manifold. Turning the system off before connecting the high-side gauge will reduce the pressure when connecting the gauge.

Another method of connecting the gauge manifold to the Schrader valve service port is with a small core depressor, which has a mechanism for depressing the stem in the Schrader valve, **Figure 41.9(A)**. The core depressor can be used to keep the refrigerant in the system when disconnecting the gauges. Follow these steps: (1) Turn the stem on the core depressor; this allows the Schrader valve to close; (2) make sure there is a plug in the center line of the gauge manifold; (3) open the gauge manifold valves. This procedure will cause the gauge lines to equalize to the low side of the system. The liquid refrigerant will move from the high-pressure line to the low side of the system. The only refrigerant that will be lost is an insignificant amount of vapor in the three gauge lines. This vapor is at the suction pressure while the system is running and should be about 70 psig for R-22 air-conditioning systems. A core removal tool, **Figure 41.9(B)**, can be used to remove Schrader valves while pulling a vacuum on a system.

Figure 41.8 This short service connection is used on the high-side gauge port to keep too much refrigerant from condensing in the gauge line. Courtesy Ferris State University. Photo by John Tomczyk

Figure 41.9 (A) A core depressor used to press the Schrader valve stem and control the pressure. Photo by Jason Obrzut. (B) A core removal tool can be used to back the valve stem out. Backing the valve stem out while pulling a vacuum on a system causes less system restriction and a faster and more effective evacuation procedure. Courtesy Ritchie Engineering Company, Inc., Yellow Jacket Products Division

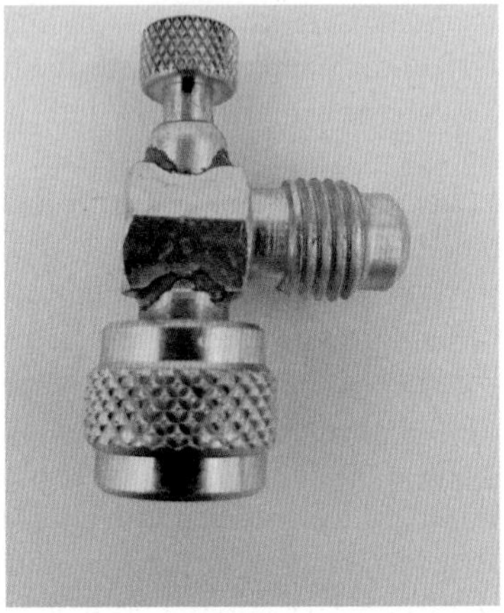

(A)

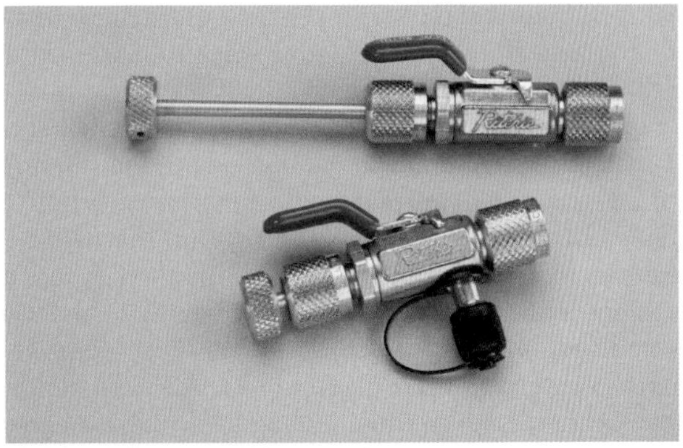

(B)

Removing the Schrader valve while pulling a vacuum causes less restriction and a faster and more effective evacuation process. **NOTE:** *This procedure should only be done with a clean and purged gauge manifold,* **Figure 41.10.** •

41.6 LOW-SIDE GAUGE READINGS

When using the gauge manifold on the low side of the system, the technician can compare the actual evaporating pressure to the normal evaporating pressure. This verifies that the refrigerant is boiling at the correct temperature for the low side of the system at some load condition. High-efficiency systems often have oversized evaporators. This makes the suction pressure slightly higher than normal. A standard-efficiency system usually has a refrigerant boiling temperature of about 35°F cooler than the entering-air temperature at the standard operating condition of 75°F return air with a 50% humidity. This results in a typical evaporator temperature of 40°F.

If the space temperature is 85°F and the humidity is 70%, the evaporator has an oversized load. It is absorbing an extra heat load, both sensible heat and latent heat, from the moisture in the air. The technician needs to wait a sufficient time for the system to reduce the load to determine whether the equipment is functioning correctly. Gauge readings before this time will not reveal the kind of information that will verify the system performance unless a manufacturer's performance chart is available.

It is good practice to start the unit and check for problems. If none are apparent, the technician can let the unit run for the rest of the day and come back the next morning before evaluating system performance. There may be a larger-than-normal volume of condensate pouring from the drain line. The moisture content in the form of condensate must be removed before the actual air temperature will begin to drop significantly.

41.7 HIGH-SIDE GAUGE READINGS

Gauge readings obtained on the high-pressure side of the system are used to check the relationship of the condensing refrigerant to the ambient air temperature. Standard air-conditioning equipment condenses the refrigerant at no more than 30°F higher than the ambient temperature. For a 95°F entering-air temperature, the head pressure should correspond to 95°F + 30°F = 125°F for a corresponding head pressure of 278 psig for R-22; the head pressure should be 445 psig for R-410A. If the head pressure shows that the condensing temperature is higher than this, something is wrong.

When checking the condenser air temperature, be sure to check the actual temperature. Do not use the weather report as the ambient temperature. Air-conditioning equipment located on a black roof is influenced by solar heating of the air being pulled across the roof, **Figure 41.11.** If the condenser is located close to any obstacles, such as below a sundeck, air may be circulated back through the condenser and cause the head pressure to be higher than normal, **Figure 41.12.**

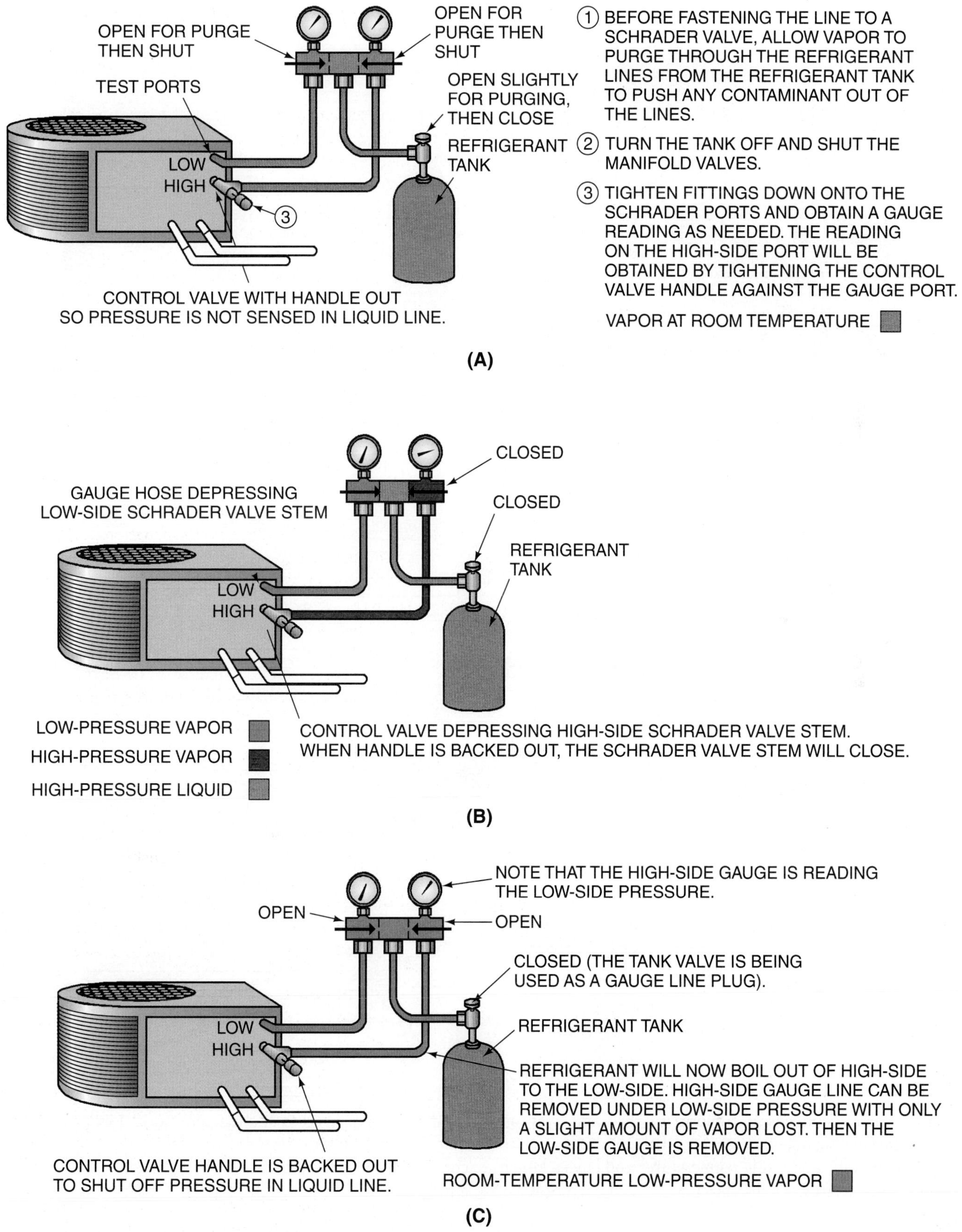

Figure 41.10 A method of pumping refrigerant condensed in the high-pressure line over into the suction line. (A) Purging a manifold. (B) Taking pressure readings. (C) Removing liquid refrigerant from the high-side gauge line.

Figure 41.11 A condenser located low on a roof. Hot air from the roof level enters it. A better installation would be to have the condenser mounted up about 20 in. so that air could at least mix with the roof ambient air. The ambient air could be 95°F and the air off the roof is 105°F.

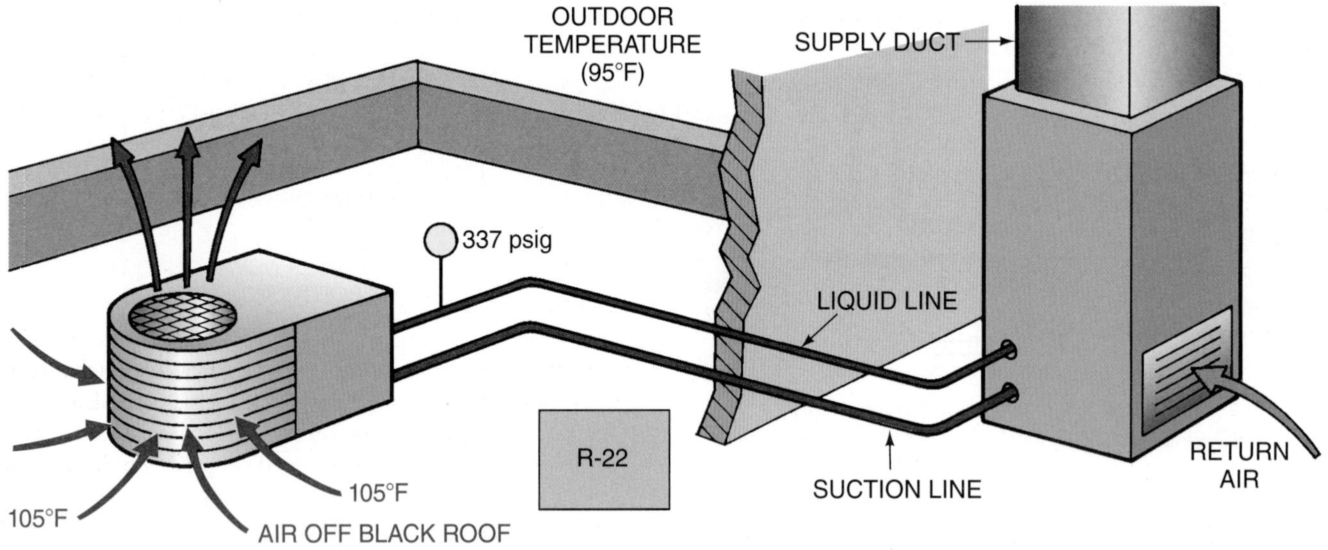

Figure 41.12 A condenser installed so that the outlet air is hitting a barrier in front of it and recirculating to the condenser inlet.

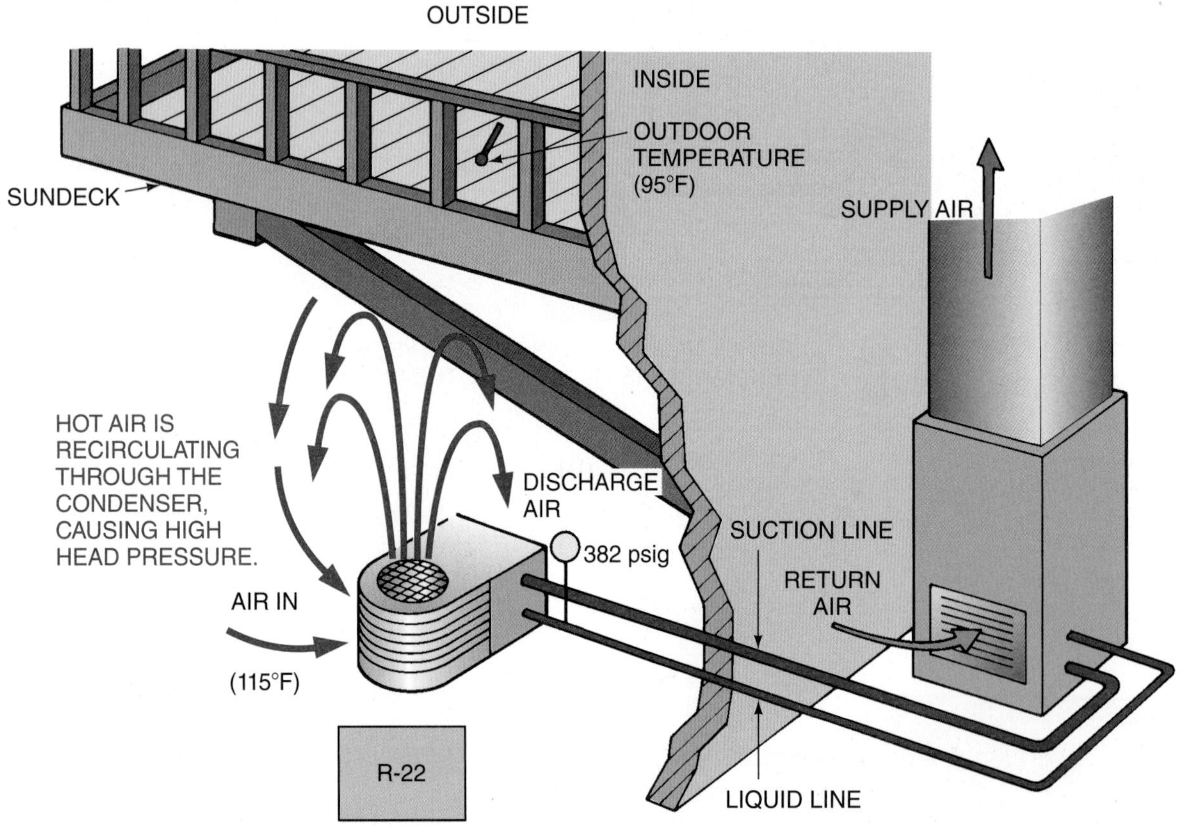

High-efficiency condensers perform in the same way as standard-efficiency condensers except that they operate at lower pressures and condensing temperatures and have a larger condenser surface area. High-efficiency condensers normally condense the refrigerant at a temperature as low as 20°F higher than the ambient temperature. On a 95°F day, the head pressure corresponds to a temperature of 95°F + 20°F = 115°F, which is a pressure of 243 psig for R-22 and 390 psig for R-410A.

41.8 TEMPERATURE READINGS

Temperature readings also can be useful. An electronic thermometer performs very well as a temperature-reading instrument, **Figure 41.13**. It has small temperature leads that respond quickly to temperature changes. The leads can be attached easily to the refrigerant piping with electrical tape and insulated from the ambient temperature with short pieces of foam line insulation, **Figure 41.14**. It is important that the lead be insulated from the ambient if the ambient temperature is different from the line temperature. The temperature lead is better and easier to use than the glass thermometers used in the past. It was almost impossible to get a true line reading by strapping a glass thermometer to a copper line, **Figure 41.15**.

Figure 41.13 (A) and (B) Electronic thermometers. (A) Photo by Bill Johnson. (B) Photo by Jason Obrzut

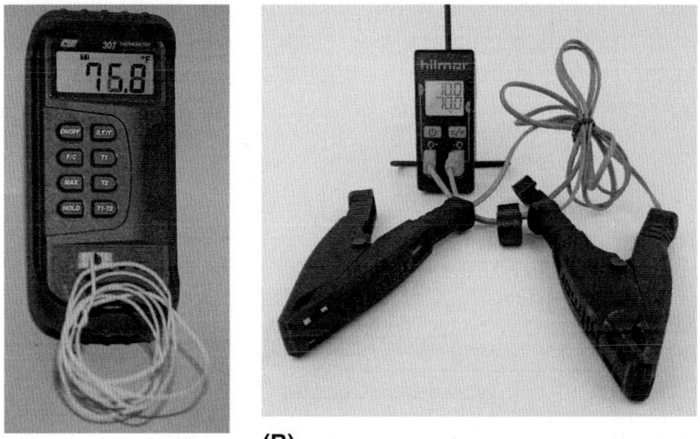

(A) (B)

Figure 41.14 A temperature lead attached to a refrigerant line in the correct manner. It must also be insulated. Photo by Bill Johnson

Figure 41.15 A glass thermometer strapped to a refrigerant line. Photo by Bill Johnson

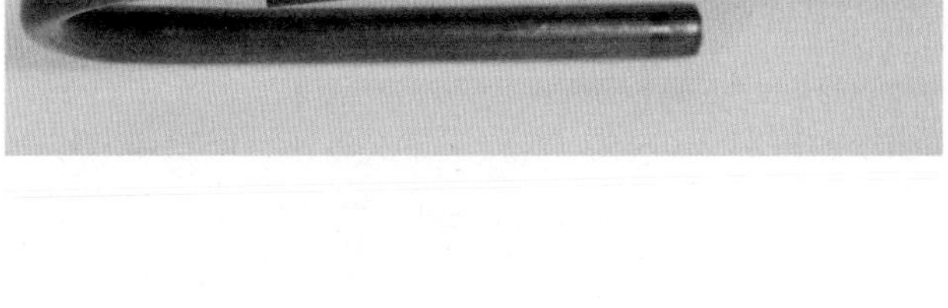

Temperatures vary from system to system. The technician must be prepared to accurately record these temperatures to evaluate the various types of equipment. Some technicians record temperature readings under different conditions for future reference. The common temperatures used are inlet air wet- and dry-bulb, outdoor air dry-bulb, and the liquid- and suction-line temperatures. Sometimes, the compressor discharge temperature needs to be known. A thermometer with a range from −50°F to +250°F is commonly used for all of these tests.

INLET AIR TEMPERATURES

It may be necessary to know the inlet air temperature to the evaporator for a complete analysis of a system. A wet-bulb reading to determine the humidity may be necessary. Such a reading can be obtained by using one of the temperature leads with a cotton sock that is saturated with pure water. A wet-bulb and a dry-bulb reading may be obtained by placing a dry-bulb temperature lead next to a wet-bulb temperature lead in the return airstream, **Figure 41.16 (A)**. The velocity of the return air will be enough to accomplish the evaporation for the wet-bulb reading. A digital psychrometer could also be used, **Figure 41.16(B)**.

EVAPORATOR OUTLET TEMPERATURES

The evaporator outlet air temperature is seldom important. It may be obtained, however, in the same manner as the inlet air temperature. The outlet air dry-bulb temperature will normally be about 20°F less than the inlet air temperature.

Figure 41.16 (A) This electronic thermometer temperature lead has a damp cotton cloth wrapped around it to convert it to a wet-bulb lead. Photo by Bill Johnson. (B) A digital psychrometer. Photo by Jason Obrzut

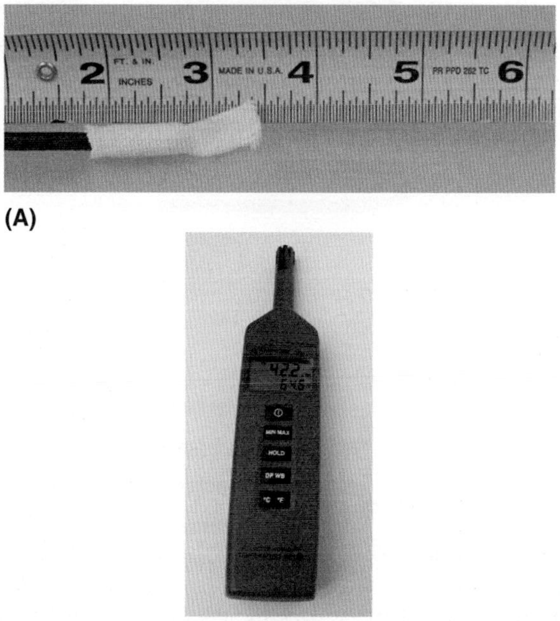

(A)

(B)

The temperature drop across an evaporator coil is about 20°F when it is operating at typical operating conditions of 75°F and 50% relative humidity return air. If the conditioned space temperature is high with a high humidity, the temperature drop across the same coil will be much less because of the latent-heat load of the moisture in the air.

If a wet-bulb reading is taken, there will be approximately a 10°F wet-bulb drop from the inlet to the outlet during standard operating conditions. The outlet humidity will be almost 90%. This air is going to mix with the room air and will soon drop in humidity because it will expand to the room air temperature. It is very humid because it has been contracted or shrunk during the cooling process. **SAFETY PRECAUTION:** *The temperature lead must not be allowed to touch the moving fan while taking air temperature readings.* •

SUCTION-LINE TEMPERATURES

The temperature of the suction line returning to the compressor and the suction pressure will help the technician understand the characteristics of the suction gas. The suction gas may be part liquid if the return-air filters are clogged or if the evaporator coil is dirty, **Figure 41.17**. The suction gas may have a high superheat if the unit has a low charge or if there is a refrigerant restriction, **Figure 41.18**. The combination of suction-line temperature and pressure will help the service technician decide whether the system has a low charge or a dirty air filter in the air handler. For example, if the suction pressure is too low and the suction line is warm, the evaporator is starved, **Figure 41.19**. If the suction-line temperature is cold and the pressure indicates that the refrigerant is boiling at a low temperature, the coil is not absorbing heat as it should. The coil may be dirty or the airflow may be insufficient, **Figure 41.20**. The cold suction line indicates that the unit has enough charge because the evaporator must be full for the refrigerant to get back to the suction line, **Figure 41.21**.

DISCHARGE-LINE TEMPERATURES

The temperature of the discharge line may tell the technician that something is wrong inside the compressor. If there is an internal refrigerant leak from the high-pressure side to the low-pressure side, the discharge gas temperature will

Figure 41.17 An evaporator flooded with refrigerant.

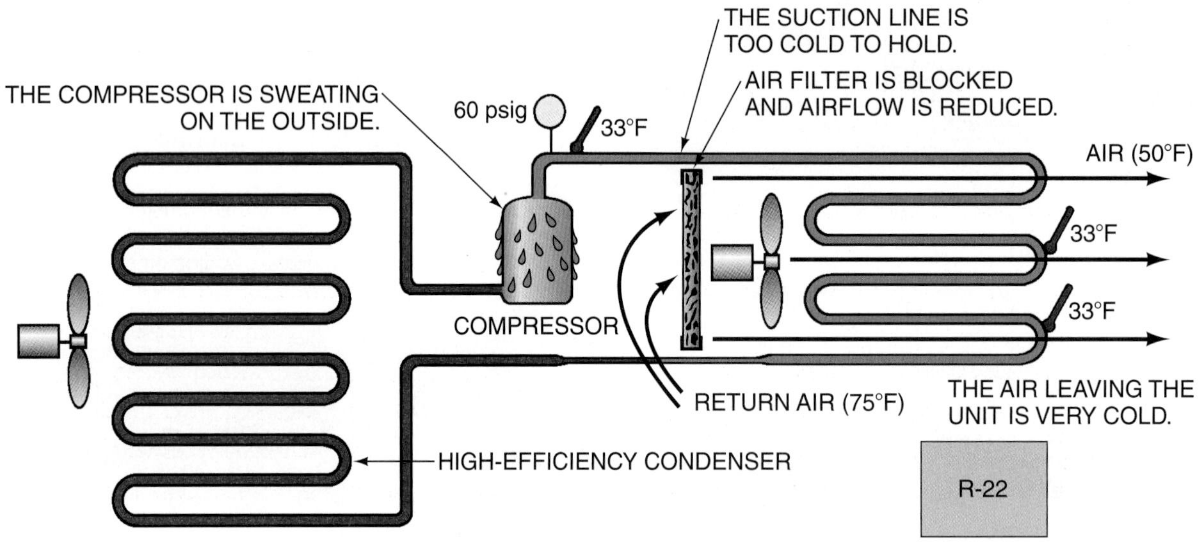

Figure 41.18 The evaporator is starved for refrigerant because of a low refrigerant charge.

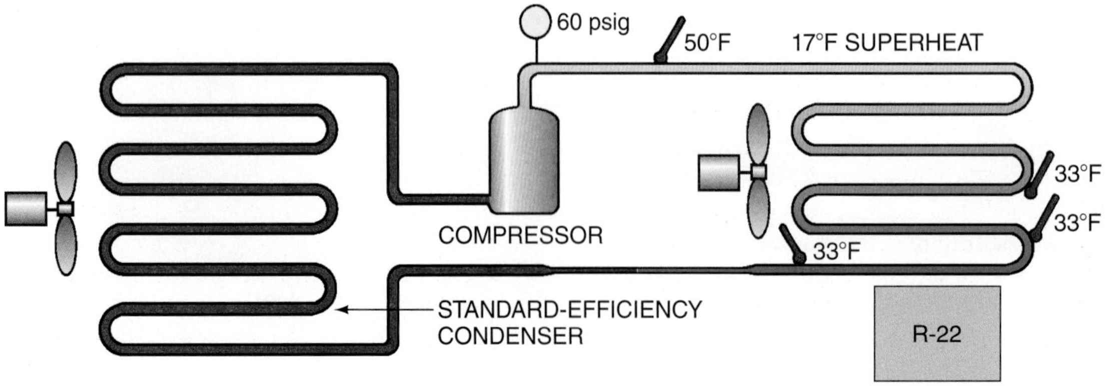

Figure 41.19 If the suction pressure is too low, and the suction line is not as cool as normal, the evaporator is starved for refrigerant. The unit refrigerant charge may be low.

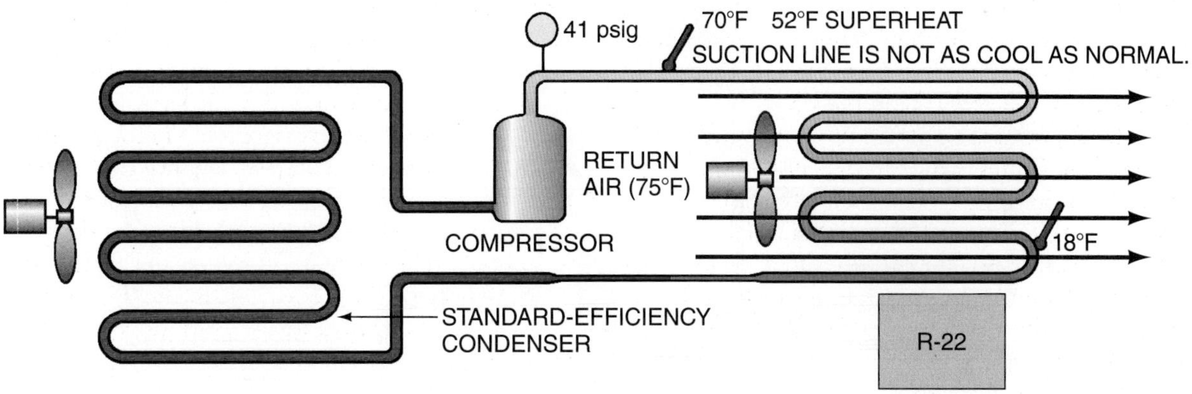

Figure 41.20 When the suction pressure is too low and the superheat is low, the unit is not boiling the refrigerant in the evaporator. The coil is flooded with refrigerant.

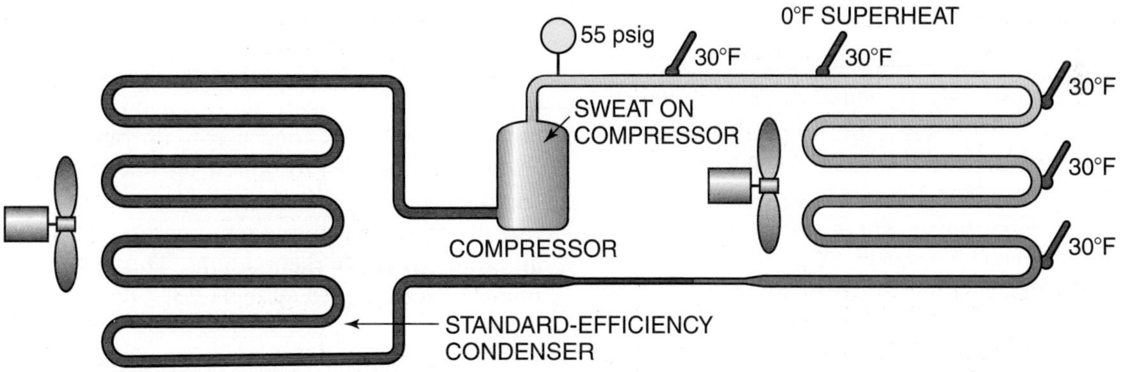

Figure 41.21 The evaporator is full of refrigerant.

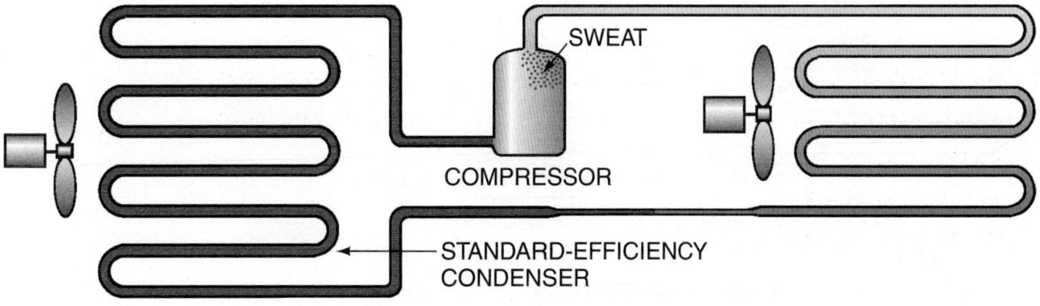

go up, **Figure 41.22.** Normally the discharge-line temperature at the compressor would not exceed 220°F for an air-conditioning application even in very hot weather. When a high discharge-line temperature is discovered, the probable cause is an internal leak. The technician can prove this by building up the head pressure as high as 300 psig for R-22 and then shutting off the unit. If there is an internal leak, this pressure difference between the high and the low sides can often be heard (as a whistle) equalizing through the compressor. If the suction line at the compressor shell starts to warm up immediately, the heat is coming from the discharge of the compressor.

High discharge temperatures also can be created by high compression ratios. High compression ratios can be the result of low suction pressures, high condensing pressures, or a combination of both. High discharge temperatures also can be caused from high superheats coming back to the compressor, which is the result of an undercharge of refrigerant, a restricted filter drier, or a metering device starving the system. **⚠ SAFETY PRECAUTION:** *The discharge line of a compressor may be as hot as 220°F for R-22 systems under normal conditions, so be careful while working on these compressors or when attaching a temperature lead to this line. Although the compressor discharge line on R-410A systems will be somewhat cooler, in the 175°F range, technicians should still be careful when working on these compressors.* •

Figure 41.22 A thermometer is attached to the discharge line on this compressor. If the compressor has an internal leak, the hot discharge gas will circulate back through the compressor. The discharge gas will be abnormally hot. When a compressor is cooled by suction gas, a high superheat will cause the compressor discharge gas to be extra hot.

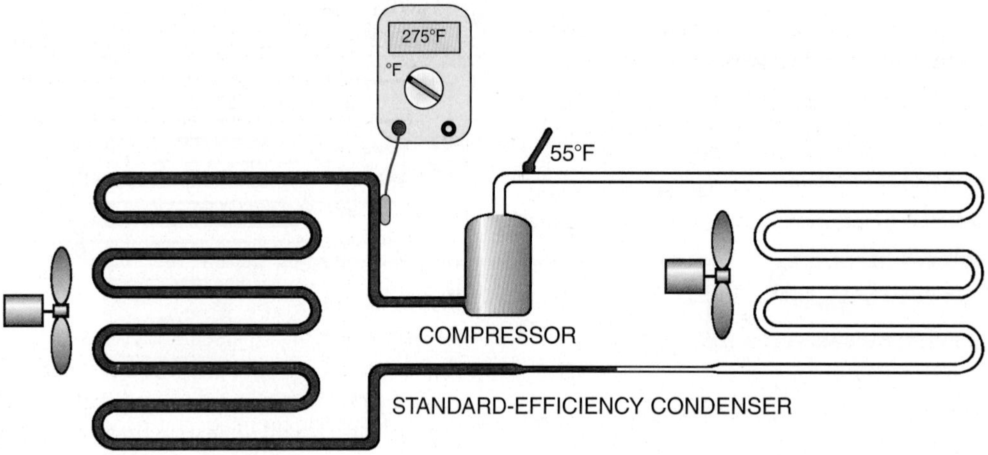

LIQUID-LINE TEMPERATURES

Liquid-line temperature may be used to check the subcooling efficiency of a condenser. Most condensers will subcool the refrigerant to between 10°F and 20°F below the condensing temperature of the refrigerant. If the condensing temperature is 125°F on a 95°F day, the liquid line leaving the condenser may be 105°F to 115°F when the system is operating normally. If there is a slight low charge, there will not be as much subcooling, and the system efficiency therefore will not be as good. The condenser performs three functions: (1) it removes the superheat from the discharge gas, (2) condenses the refrigerant to a liquid, and (3) subcools the liquid refrigerant below the condensing temperature. All three of these functions must be successfully accomplished for the condenser to operate at its rated capacity.

It is a good idea for a new technician to take the time to completely check out a working system operating at the correct pressures. Apply the temperature probes and gauges to all points to verify the readings. This will provide reference points to remember, **Figure 41.23**.

Figure 41.23 This condenser is operating at normal conditions. A technician can use these readings as points of reference for a condenser operating at full-load normal conditions.

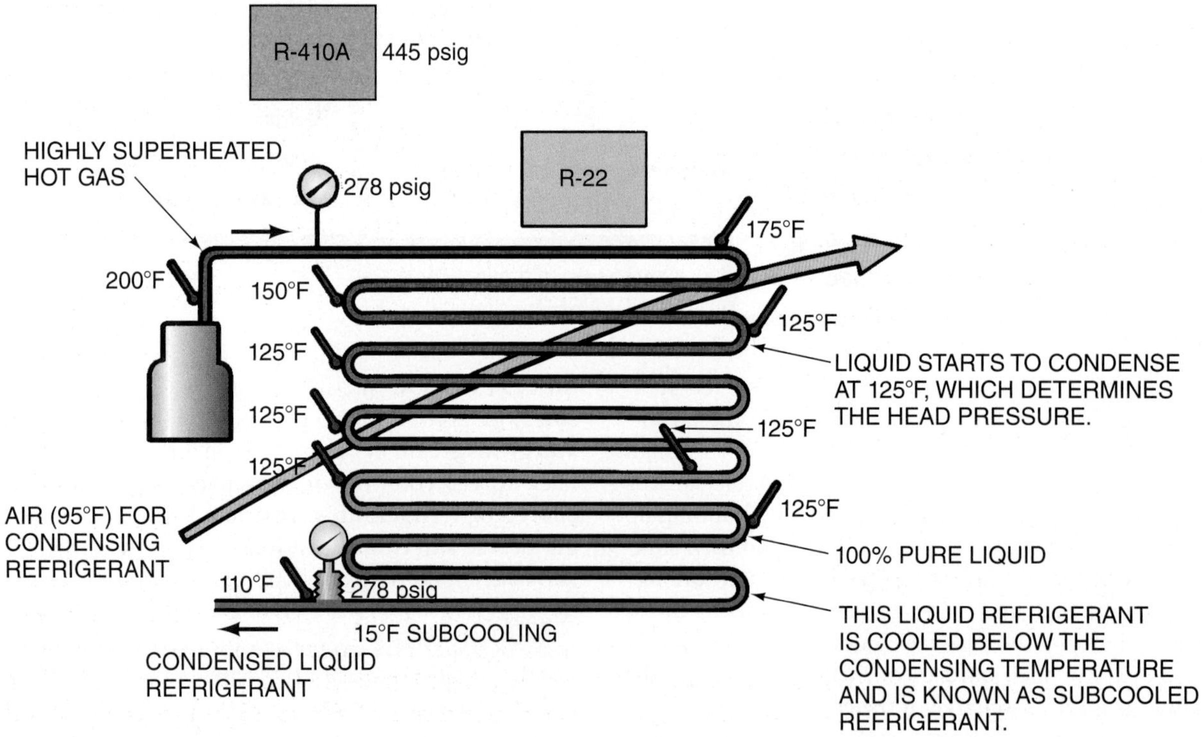

41.9 CHARGING PROCEDURES IN THE FIELD

While performing charging procedures in the field, the technician should keep in mind what the designers of the equipment intended and how the equipment should perform. The charge consists of the correct amount of refrigerant in the evaporator, the liquid line, the discharge line between the compressor and the condenser, and the suction line. The discharge line and the suction line do not hold as much refrigerant as the liquid line because the refrigerant is in the vapor state. Actually, the liquid line is the only interconnecting line that contains much refrigerant. When a system is operating correctly, under design conditions, a prescribed amount of refrigerant should be in the condenser, the evaporator, and the liquid line. Understanding the following statements is important for the technician in the field:

1. The amount of refrigerant in the evaporator can be measured by the superheat method.
2. The amount of refrigerant in the condenser can be measured by the subcooling method.
3. The amount of refrigerant in the liquid line may be determined by measuring the length and calculating the refrigerant charge. However, in field service work, if the evaporator is performing correctly, the liquid line should have the correct charge.

When the preceding statements are understood, the technician can check for the correct refrigerant level to determine the correct charge.

A field charging procedure may be used to check the charge of most typical systems. The technician sometimes needs typical reference points when adding small amounts of vapor refrigerant to adjust the amount of refrigerant in equipment that has no charging directions. **NOTE:** *Remember, if the refrigerant is a zeotropic blend that may fractionate when vapor-charged, the refrigerant must be throttled in the low side of the system as a liquid to avoid fractionation.* • Often the technician finds that the system charge needs adjusting. This can occur due to an over- or undercharge in the factory or from a previous technician's work. It can also occur because of system leaks. *System leaks should always be repaired to prevent further loss of charge to the atmosphere.*

The technician must establish charging procedures to use in the field for all types of equipment. These procedures will help get the system back on line under emergency situations. Following are some methods used for different types of equipment.

FIXED-BORE METERING DEVICES—CAPILLARY TUBE AND ORIFICE TYPE

Fixed-bore metering devices like the capillary tube do not throttle the refrigerant as the *thermostatic expansion valve* (TXV) does. They allow refrigerant flow based on the difference in the inlet and the outlet pressures. The one time when the system can be checked for the correct charge and produce normal readings is at the typical operating condition of 75°F and 50% humidity return air and 95°F outside ambient air. Other conditions will produce different pressures and different superheat readings. The condition that most affects the readings is the outside ambient temperature. When it is lower than normal, the condenser will become more efficient and will condense the refrigerant sooner in the coil. This will have the effect of partially starving the evaporator of refrigerant. Refrigerant in the condenser that should be in the evaporator starves the evaporator.

When a technician needs to check the system for correct charge or to add refrigerant, the best method to follow is the manufacturer's instructions. Please refer to Unit 10, "System Charging," for detailed instructions on charging refrigerant into air-conditioning systems with different types of metering devices. If they are not available, the typical operating condition may be simulated by reducing the airflow across the condenser to cause the head pressure to rise. On a 95°F day, the highest condenser head pressure is usually 278 psig for R-22 (95°F ambient + 30°F added condensing temperature difference = 125°F condensing temperature, or 278 psig). The pressure for R-410A would be 445 psig. Because the high pressure pushes the refrigerant through the metering device, no refrigerant is held back in the condenser when the head pressure is up to the high-normal end of the operating conditions.

When the condenser is pushing the refrigerant through the metering device at the correct rate, the remainder of the charge must be in the evaporator. A superheat check at the evaporator is not always easy with a split air-conditioning system; instead, a superheat check at the condensing unit can be done. The suction line from the evaporator to the condensing unit may be long or short. For a test comparison, we will use two different lengths: up to 30 ft and from 30 to 50 ft. When the system is correctly charged, the superheat may be 10°F to 15°F at the condensing unit with a line length of 10 to 30 ft. The superheat may be 15°F to 18°F when the line is 30 to 50 ft long. Both of these conditions are with a head pressure of 278 psig for R-22 and 445 psig for R-410A ± 10 psig. At these conditions the actual superheat at the evaporator will be close to the correct superheat. When using this method, be sure to allow enough time for the system to stabilize after adding refrigerant before drawing any conclusions, **Figure 41.24**.

Superheat temperatures taken at the compressor will vary when typical operating conditions are not met. Factors that cause this variation are outside ambient temperatures and inside dry- and wet-bulb temperatures. An air-conditioning system may be operating normally at 30°F superheat at the compressor if the evaporator is seeing a high latent- and sensible-heat load. On the other hand, superheats of 8°F to 15°F may be normal if outside

Figure 41.24 (A) This system has a 25-ft line set and is charged for 10 to 15 degrees of superheat. (B) This line set is 45 ft and is charged for 15 to 18 degrees of superheat.

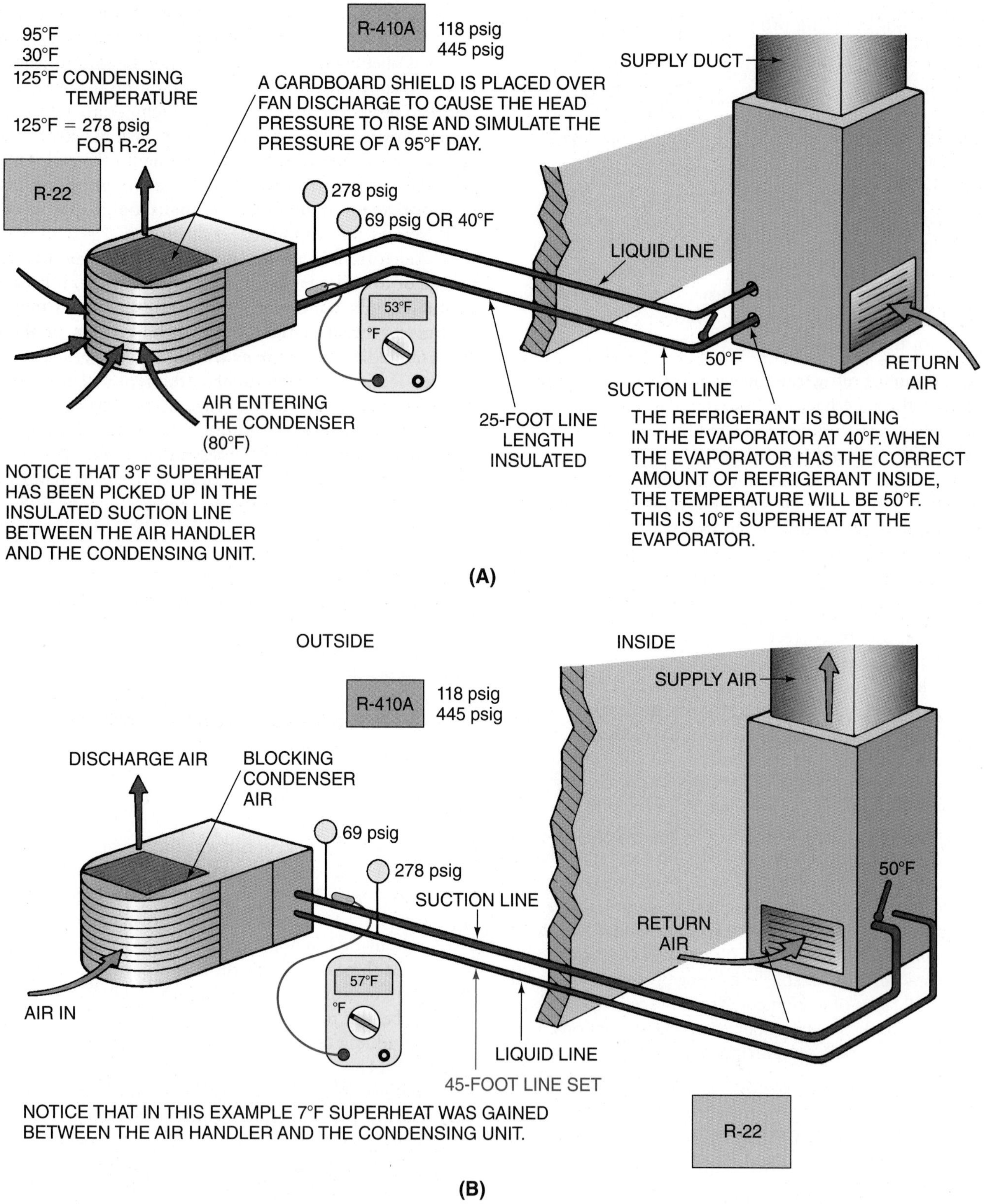

ambient temperatures are high. This is because refrigerant is being forced through the capillary tube at a higher rate. The manufacturer's superheat charging information should be used, if available. It will take into account all of the varying conditions that affect superheat at the compressor.

Orifice-type metering devices have a tendency to hunt while reading steady-rate operation. This hunting can be observed by watching the suction pressure rise and fall accompanied by the superheat as the suction-line temperature changes. When this occurs, the technician will have to use averaging to arrive at the proper superheat for the coil. For example, if it is varying between 6°F and 14°F, the average value of 10°F superheat may be used.

FIELD CHARGING THE TXV SYSTEM

The TXV system can be charged in much the same way as the fixed-bore system, with some modifications. Please refer to Unit 10, "System Charging," for detailed instructions on charging refrigerant into air-conditioning systems with different types of metering devices. The condenser on a TXV system will also hold refrigerant back in mild ambient conditions. This system always has a refrigerant reservoir or receiver to store refrigerant and will not be affected as much as the capillary tube by lower ambient conditions. To check the charge, restrict the airflow across the condenser until the head pressure simulates a 95°F ambient, 278 psig head pressure for R-22 or 445 psig for R-410A. Using the superheat method will not work for this valve because if there is an overcharge, the superheat will remain the same. Superheats of 15°F to 18°F measured at the compressor's inlet are not uncommon for air-conditioning systems incorporating TXVs as metering devices. If the sight glass is full, the unit has at least enough refrigerant, but it may have an overcharge. If the unit does not have a sight glass, a measure of the subcooling of the condenser may reveal the amount of charge. For example, a typical subcooling circuit will subcool the liquid refrigerant from 10°F to 20°F cooler than the condensing temperature. A temperature lead attached to the liquid line should read 115°F to 105°F, or 10°F to 20°F cooler than the condensing temperature of 125°F, **Figure 41.25**. If the subcooling temperature is 20°F to 25°F cooler than the condensing temperature, the unit may have an overcharge of refrigerant, and the bottom of the condenser is acting as a large subcooling surface.

The charging procedures just described will also work for high-efficiency equipment. The head pressure does not need to be operated quite as high. A head pressure of 250 psig will be sufficient for an R-22 system, and 390 psig for R-410A, when charging. **SAFETY PRECAUTION:** *Refrigerant in cylinders and in the system is under great pressure (as high as 300 psi for R-22 and 500 psig for R-410A). Use proper safety precautions when transferring refrigerant, and be careful not to overfill tanks or cylinders when recovering refrigerant. Do not use disposable cylinders by refilling them. Use only*

Figure 41.25 A unit with a TXV cannot be charged using the superheat method because the TXV controls superheat. The head pressure is raised to simulate a 95°F day, and the temperature of the liquid line is checked for the subcooling level. A typical system will have 10°F to 20°F of subcooling when the condenser contains the correct charge.

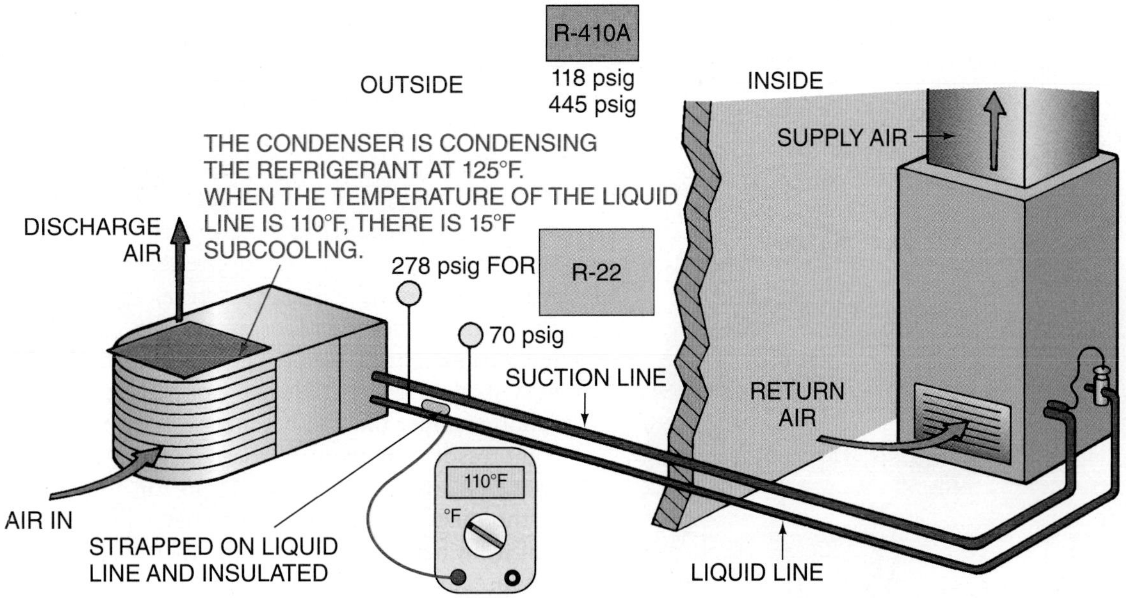

DOT-approved tanks or cylinders. The high-side pressure may be reduced for attaching and removing gauge lines by shutting off the unit and allowing the unit pressures to equalize. •

41.10 ELECTRICAL TROUBLESHOOTING

Electrical troubleshooting is often required at the same time as mechanical troubleshooting. The digital multimeter equipped with an amp clamp is the primary instrument used, **Figure 41.26**. Knowledge of what the readings should be is required to know whether the existing readings on a particular unit are correct. This is often not easy to determine because the desired reading may not be furnished. **Figure 41.27** shows typical horsepower-to-amperage ratings. It is a valuable tool for determining the correct amperage for a particular motor.

For a residence or small commercial building, one main power panel will normally serve the building. This panel is divided into many circuits. For a split-system type of cooling system, the main panel usually has separate breakers (or fuses) for the air handler or furnace for the indoor unit and for the outdoor unit. For a package or self-contained system, usually one breaker (or fuse) serves the unit, **Figure 41.28**. The power supply voltage is stepped down by the control transformer to the control voltage of 24 V.

Begin any electrical troubleshooting by verifying that the power supply is energized and that the voltage is correct. One way to do this is to see whether the indoor fan will start with the room thermostat FAN ON switch. See **Figure 41.29** for a wiring diagram of a typical split-system air conditioner. The air handler or furnace, where the low-voltage transformer is located, can be under the house or in the attic. This quick check with the FAN ON switch can save a trip under the house or to the attic. If the fan will start with the fan relay, several things are apparent: (1) the indoor fan operates, (2) there is control voltage, and (3) there is line voltage to the unit because the fan will run. When taking a service call over the phone, ask the homeowner whether the indoor fan will run. If it does not, bring a transformer. This could save a trip to the supply house or the shop. **SAFETY PRECAUTION:** *All safety practices must be observed while troubleshooting electrical systems. Many times, the system must be inspected while power is on. Let only the insulated meter lead contact probes touch the hot terminals. Special care should be taken while troubleshooting the main power supply because the fuses may be correctly sized large enough to allow great amounts of current to flow before they blow (e.g., when a screwdriver slips in the panel and shorts across hot terminals). Never use a screwdriver in a hot panel. Capacitors should be discharged using a 20,000-ohm, 5-watt resistor before being tested with a meter.* •

If the power supply voltages are correct, move on to the various components. The path to the load may be the next item to check. When trying to get the compressor to run, remember that the compressor motor is operated by the compressor contactor. Is the contactor energized? Are the contacts closed? See **Figure 41.30** for a diagram. Note that in the diagram the only thing that will keep the contactor coil from being energized is the thermostat, the path, or the low-pressure control. If the outdoor fan is operating and the compressor is not, the contactor is energized because it also

Figure 41.26 Instruments used to troubleshoot the electrical part of an air-conditioning system. Photo by Jason Obrzut

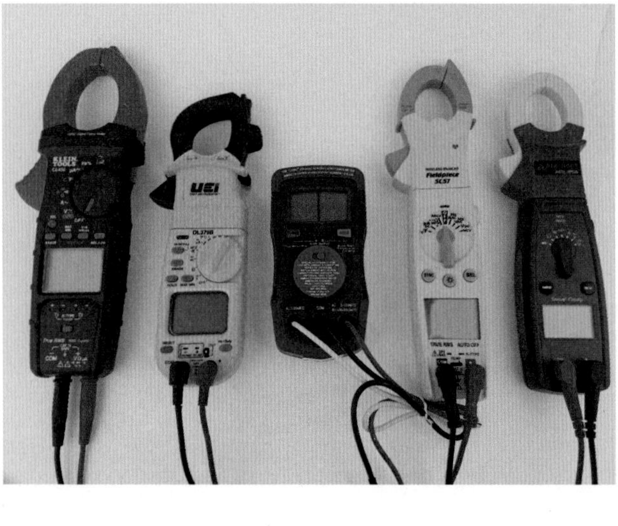

Figure 41.27 This chart shows approximate full-load amperage values. Courtesy BDP Company

APPROXIMATE FULL-LOAD AMPERAGE VALUES FOR ALTERNATING CURRENT MOTORS

Motor	Single Phase		3-Phase Squirrel Cage Induction		
HP	120V	230V	230V	460V	575V
1/6	4.4	2.2			
1/4	5.8	2.9			
1/3	7.2	3.6			
1/2	9.8	4.9	2	1.0	0.8
3/4	13.8	6.9	2.8	1.4	1.1
1	16	8	3.6	1.8	1.4
$1\frac{1}{2}$	20	10	5.2	2.6	2.1
2	24	12	6.8	3.4	2.7
3	34	17	9.6	4.8	3.9
5	56	28	15.2	7.6	6.1
$7\frac{1}{2}$			22	11.0	9.0
10			28	14.0	11.0

Note: Does not include shaded pole.

Figure 41.28 A typical package and split-system installation.

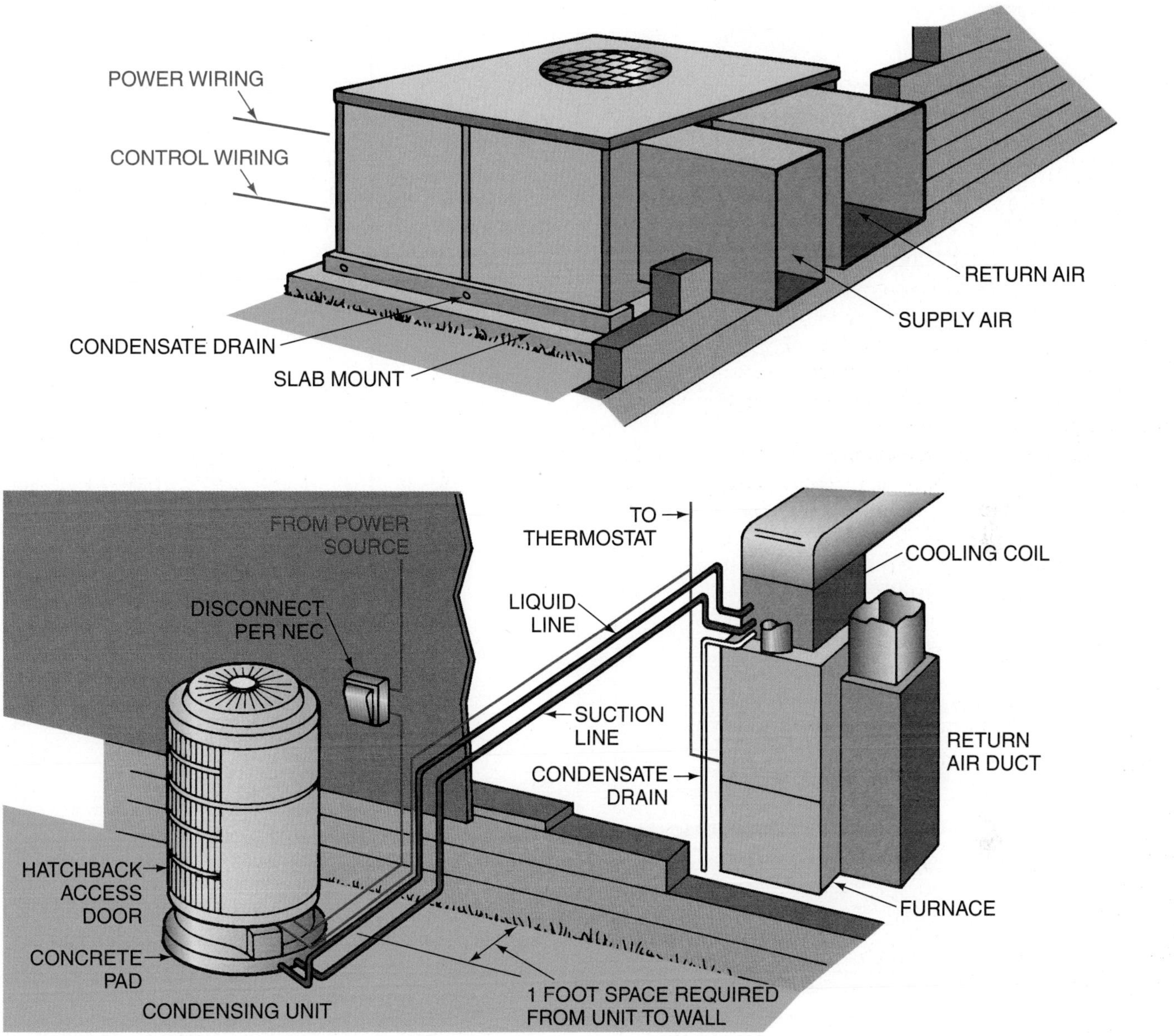

starts the fan. If the fan is running and the compressor is not, either the path (wiring or terminals) is not making good contact or the compressor internal overload protector is open.

41.11 COMPRESSOR OVERLOAD PROBLEMS

When the compressor overload protector is open, if you cannot hold your hand on the compressor shell, the motor is too hot. Ask these questions: Can the charge be low (this compressor is suction-gas-cooled)? Can the start-assist circuit not be working and the compressor not starting?

Allow the compressor to cool before restarting it. It is best to set the unit so that it will not come back on for several hours. The best way to do this is to remove a low-voltage wire. If the disconnect switch is pulled, by the next day the refrigerant charge may have migrated to the crankcase because there was no crankcase heat. To start the unit within the hour rather than waiting, pull the disconnect switch and run a small amount of water through a hose over the compressor. **SAFETY PRECAUTION:** *The standing water poses a potential electrical hazard. Be sure that all electrical components are protected with plastic or other waterproof covering. Do not come in contact with the water or electrical current when working around live electricity.* • It will take about 30 min to cool.

Figure 41.29 The wiring diagram of a split-system summer air conditioner. Based on Climate Control

AFS	Airflow Switch
CC	Compressor Contactor (Located Outdoors)
CC1, CC2	Compressor Contactor Contacts (Located Outdoors)
CFM	Condenser Fan Motor (Located Outdoors)
COMP	Compressor (Located outdoors)
F	Fuse
G	Thermostat Fan Terminal
HB	Blower Relay Coil (Heating)
HB1	Blower Relay Contacts (Heating, Located Indoors)
HPS	High-Pressure Switch
HR	Heating Relay/Sequencer (Located Indoors)
HR1, HR2	Heating Relay Contacts
HTR	Heater (Located Indoors)
IBM	Indoor Blower Motor (Located Indoors)
IBR	Indoor Blower Relay Coil (Located Indoors)
IBR1, IBR2	Indoor Blower Relay Contacts
L1, L2	Incoming Power Lines
LA	Low Ambient Control
LIM	High-Limit Switch (Located Outdoors)
LPS	Low-Pressure Switch (Located Outdoors)
OL	Compressor Overload (Located Outdoors)
R	Thermostat Power Terminal
RC	Run Capacitor
T-STAT	Space Thermostat (Located in the Conditioned Space)
TR	Control Transformer (Located Indoors)
W	Thermostat Heating Terminal
Y	Thermostat Cooling Terminal

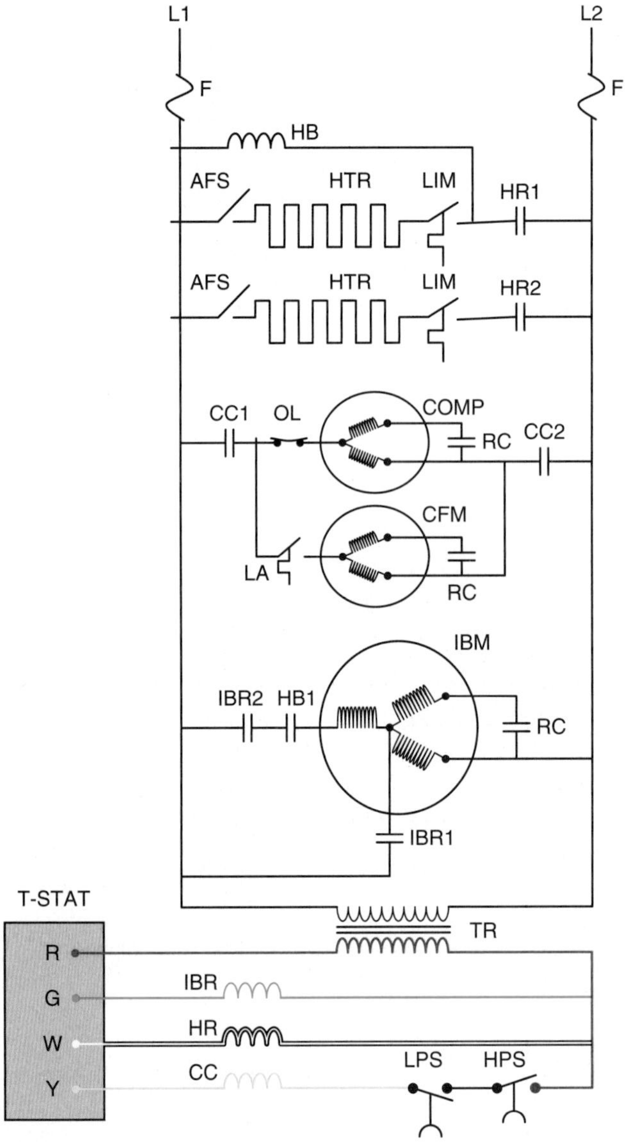

Have the gauges and a cylinder of refrigerant connected on the unit because when the compressor is started up by closing the disconnect, it may need refrigerant. If the system has a low charge, taking the time to set up to charge after starting the system may allow the compressor to cut off again from overheating before the gauges are connected.

41.12 COMPRESSOR ELECTRICAL CHECKUP

Manufacturers report that a fairly high percentage of compressors returned to them under warranty are actually not defective, but were reported by the technician as defective. The technician should use great care when condemning a compressor, as it costs a lot of money in labor and materials to reclaim the refrigerant, ship in a new compressor, and ship the defective compressor back to the manufacturer for in-warranty work. When the compressor is out of warranty, the owner must bear that burden. The technician must be correct with the diagnosis.

When a compressor will not start, the problem can be either electrical or mechanical. The electrical portion of the problem is rather easy to prove. When the technician can prove the compressor is electrically sound and it will not start, the only other conclusion would be mechanical. The steps for checking a single-phase compressor electrically are as follows:

1. Check the compressor from winding to ground using either an ohmmeter or a megohmmeter. A megohmmeter will detect much smaller ground circuits and is

Figure 41.30 The wiring diagram of basic components in a control-and-compressor circuit.

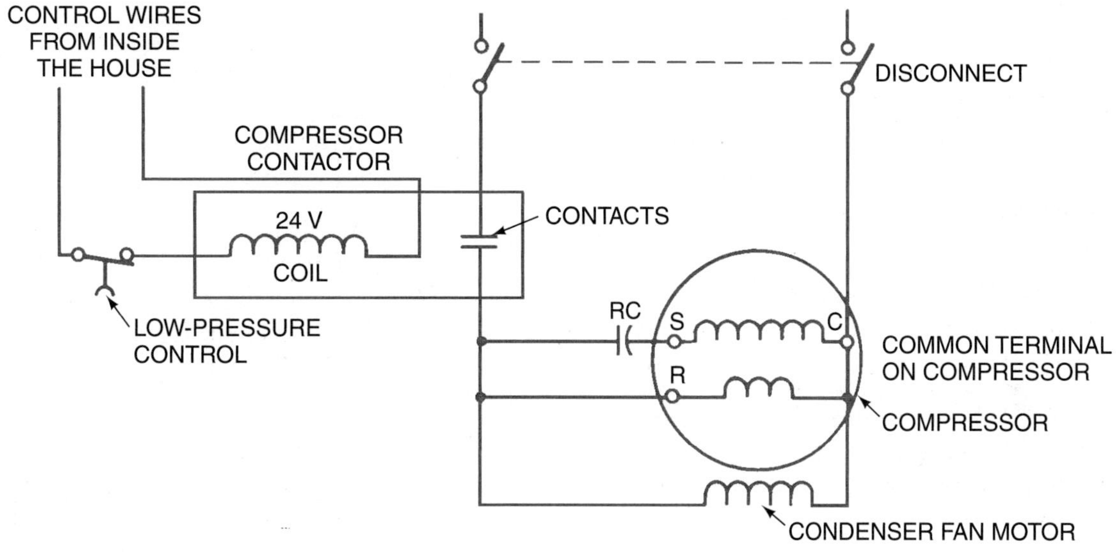

used on larger equipment exclusively. If there is a circuit showing to ground, remove all wires from the compressor terminals and check just the terminals. If the ground is no longer there, the ground is somewhere in the circuit wiring, not the compressor. If a ground is still indicated, clean the compressor terminal block if there is a possibility that current can leak to ground due to dirt. If there is still a ground, condemn the compressor as grounded. **⚠ SAFETY PRECAUTION:** *If a compressor has a large amount of liquid refrigerant in the compressor housing, such as when the crankcase heater has been off for a long period, a circuit to ground may be detected. If the compressor crankcase heat has been off and the compressor housing is cool, it is advisable to let it heat for about 24 hours and check it again. The reason for this condition is probably that the refrigerant has floated up into the motor windings some system contamination that is normally dormant. After the 24-hour heating period, the system should operate correctly. Also, dirt around the motor terminals can cause a slight circuit. If there is a question, clean them with a degreasing solvent and try again.* •

2. Check the start winding for correct resistance from common to start.
3. Check the run winding for correct resistance from common to run. If the correct resistances are not readily available, make sure that the start winding and run winding have different resistances. The start winding should have much more resistance than the run winding. Either winding can have some shorted turns in the motor. **Figure 41.31** shows a start winding with *shorted,* sometimes called *shunted,* winding. **Figure 41.32** shows a start winding that has an open circuit.
4. Check for resistance from the run to the common terminal. If this circuit is open, the winding thermostat may be open. If the compressor is hot to the touch, let it cool for several hours or cool it with water, **Figure 41.33.** **⚠ SAFETY PRECAUTION:** *Make sure all power is off and do not stand in water when electrically checking any circuit.* •

Figure 41.31 A compressor with a shorted, or shunted, winding.

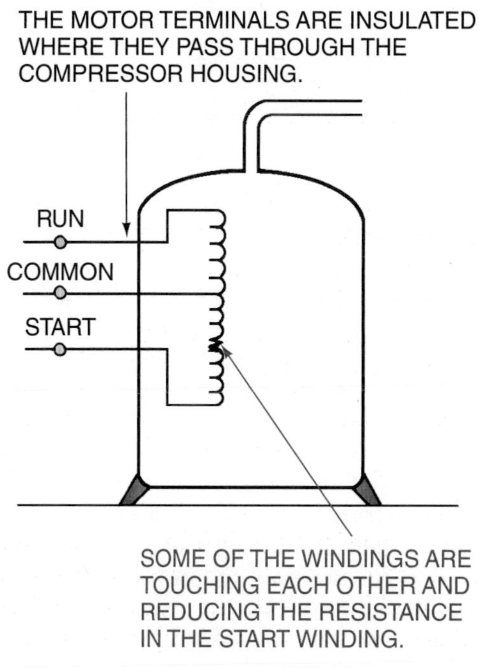

Figure 41.32 A compressor with an open start winding.

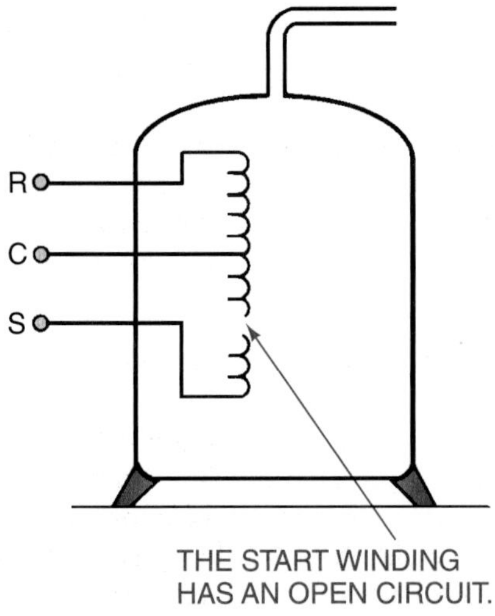

Figure 41.33 Cooling a compressor with water.

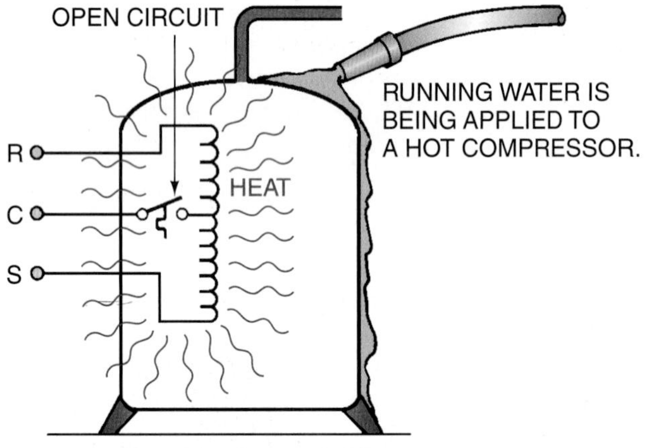

5. Check the voltage from common to run and from common to start using two different voltmeters while starting the motor. This voltage must be within ±10% of the motor's ratings. For example, many motors are rated at 208/230. The minimum voltage would be 187.2 V ($208 \times 0.90 = 187.2$) and the maximum voltage would be 253 V ($230 \times 1.10 = 253$). This voltage must be verified as close to the compressor terminals as possible, but do not use the terminals themselves. Follow the wiring back to the first junction, such as the load side of the compressor contactor. It is not unusual for a loose connection to create a low-line-voltage problem at the compressor. If there is any discoloration at the motor terminals, suspect a loose connection. **SAFETY PRECAUTION:** *Never start a compressor with the motor terminal cover box removed. If a motor terminal were to blow out while starting, there is the potential for a high-velocity blowtorch-type of fire. The electrical arc may ignite the escaping oil with the refrigerant, propelling it outward. The box and cover are built to enclose this type of malfunction.* •

The steps for checking a three-phase compressor electrically are as follows:

1. Check the compressor from all windings to ground using either an ohmmeter or a megohmmeter as described previously in step 1. Pay attention to the Safety Precautions.
2. Check each set of windings from winding to winding—T1 to T2, T1 to T3, and T2 to T3—on a three-lead motor. Some larger motors may be dual motors in the same housing; check both motors. There should be the same resistance across each motor winding on a three-phase motor. There should be no circuit from motor to motor in a dual motor application.

After completing these tests, and the motor passes them successfully, all that can be done in the field to prove that the motor is sound has been done. One final option on single-phase compressors may be a hard start capacitor. When a compressor locks up, adding a hard start capacitor to the circuit may be able to get it going. This is typically a temporary fix as the cause of the compressor lock-up has not been remedied. It may be able to keep the compressor going until a new compressor can be obtained.

Mechanically checking the compressor comes next. When there is full voltage to the compressor terminals and it will not start, the compressor could be stuck. To test for this in single-phase compressors, use two voltmeters and check from common to run and common to start at the same time. For a three-phase compressor, use three voltmeters and check all three windings at the same time. With three-phase compressors, try reversing any two of the main leads to the contactor and start it again. If the motor has two contactors, it is a part winding-start compressor and it has two motors. Be sure to reverse the leads that feed both contactors. This will reverse the motor. If it will start backward, it may run long enough to allow for planning to change the compressor at a more convenient time, but it should not be relied on to run normally for an extended period. Internally, the compressor may be binding, and the bind will reoccur. Changing the compressor is the most permanent solution.

The most difficult problem to troubleshoot is checking the capacity of a compressor. The complaint will probably be that the compressor runs all the time but the space temperature is not cool enough. This is a common complaint for both cooling

and refrigeration systems. Many compressors will have two, three, or four cylinders with one or more of them not functioning. The smaller the compressor, the more difficult this is to prove. For any compressor that does not have the capacity to unload, or to run at reduced load or speed, follow these guidelines:

1. Try to get the system to operate as close to design conditions as possible. The amperage is the key to this procedure. If the amperage is considerably lower than the rated-load amperage (RLA) and the voltage is correct, the compressor is not pumping to capacity.
2. If there are gauges on the system, the suction pressure may be a little high with the discharge pressure a little low. This is a sign that the compressor is not doing maximum work. **Figure 41.34** shows an example of a residential system that is not operating at the proper gauge readings. Notice that the outdoor temperature is hot and the indoor unit is operating at the inside design temperature. This system will not remove humidity because the coil temperature is well above the dew point temperature of the indoor air. The indoor air may be cool, but it will feel warm because of the humidity and the compressor will run all the time. This could very well be first thing in the morning after the unit has run all night, because as the day goes by, the indoor air temperature will rise.
3. This is probably the only time that the suction pressure rises, the head pressure drops, and the amperage drops.
4. One possibility is to see if the head pressure will rise when the condenser airflow is partially blocked. Usually, the head pressure will not rise to the head pressure of a design day; if it does, the suction pressure will rise accordingly.

The other mechanical problem that can occur with a compressor is an internal leak, such as when a discharge valve reed becomes split or a head gasket leaks through. To determine what the problem is, install gauges and an ammeter. The unit may appear to be running correctly, but the compressor will be hot and will likely shut off from time to time from high temperature. Try to raise the head pressure to about 300 psig for R-22 and 475 psig for R-410A, and then shut the system off. Be prepared to listen closely to the compressor discharge line. A long screwdriver or stick with one end placed on the discharge line and the other end to your ear is a rudimentary "stethoscope" that works well for listening to internal sounds. **SAFETY PRECAUTION:** *Never use an ohmmeter to check a live circuit.* •

Figure 41.34 This system compressor is not functioning up to capacity. The head pressure is low, the suction pressure is high, and the amperage is low. The compressor is defective.

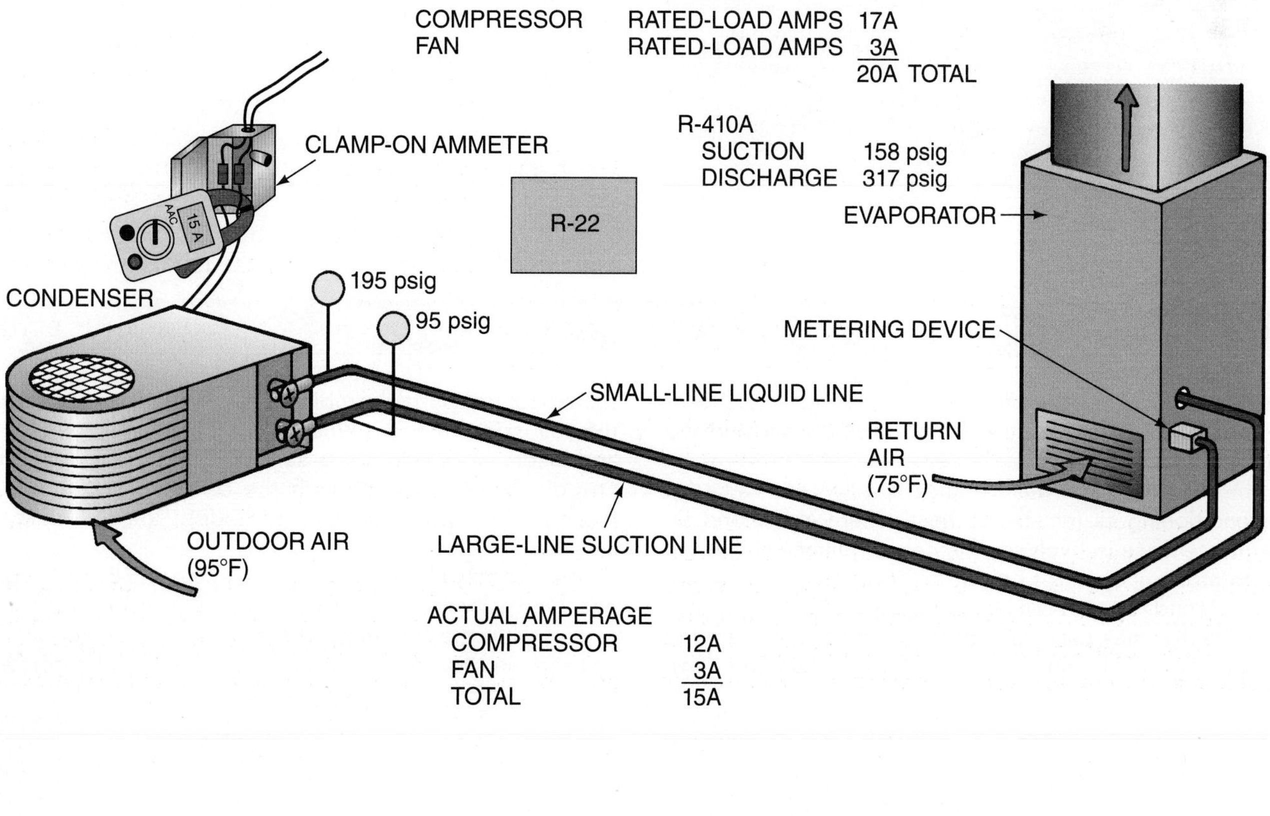

41.13 TROUBLESHOOTING THE CIRCUIT ELECTRICAL PROTECTORS—FUSES AND BREAKERS

One service call that must be treated cautiously is one involving a circuit protector, such as a fuse or breaker, opening the circuit. The compressor and fan motors have protection that will normally guard them from minor problems. The breaker or fuse is for large current surges in the circuit. When one is tripped, do not simply reset it. Perform a resistance check of the compressor section, including the fan motor. **SAFETY PRECAUTION:** *The compressor may be grounded (may have a circuit to the case of the compressor), and it could be dangerous to try to start it.* • **Be sure to isolate the compressor circuit before condemning the compressor.** Take the motor leads off the compressor to check for a ground circuit in the compressor, **Figure 41.35.**

Figure 41.35 A compressor with winding shorted to the casing.

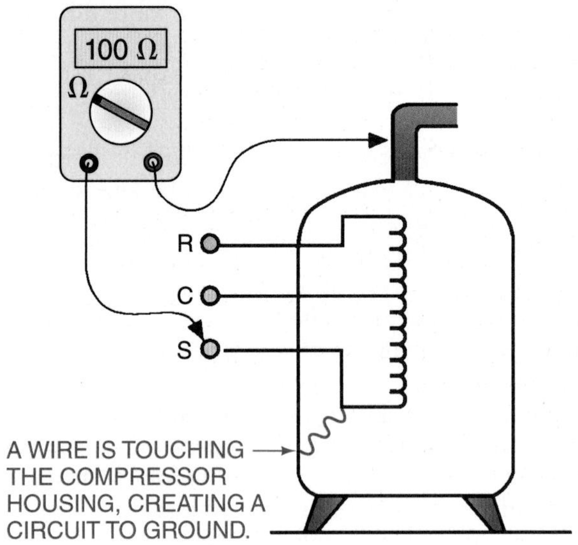

HVACR GOLDEN RULES

- Always carry the proper tools and avoid extra trips through the customer's house.
- Make and keep firm appointments. Call ahead if delayed. The customer's time is valuable, too.
- If possible, do not block the customer's driveway.

Added Value to the Customer

Here are some simple, inexpensive procedures that may be included in the basic service call.

- Touch-test the suction and liquid lines to determine possible over- or undercharge.
- Clean or change filters as needed.
- Lubricate all bearings as needed.
- Replace all panels with the correct fasteners.
- Inspect all contactors and wiring. This may lead to more paying service.
- Make sure that the condensate line is draining properly. Clean if needed. Algaecide tablets may be necessary.
- Check evaporator and condenser coils for blockage or dirt.

PREVENTIVE MAINTENANCE

Preventive maintenance for air-conditioning equipment involves the indoor airside, the outdoor airside (air-cooled and water-cooled), and electrical circuits.

The indoor airside maintenance is much the same for air-conditioning as for electric heat, where motor and filter maintenance is involved. The only difference is that the evaporator coil operates below the dew point temperature of the air and is wet. It will become a superfilter for any dust particles that may pass through the filter or leak in around loose panel compartment doors. Many air handlers have draw-through coil-fan combinations. The air passes through the coil before the fan. When the fan blades become dirty and loaded with dirt, it is a sure sign that the coil is dirty. The dirt has to pass through the coil first, and it is wet much of the time. When the fan is dirty, the coil must be cleaned.

The coil may be cleaned in place in the unit by two methods. One is by the use of a special detergent manufactured to work while the coil is wet. This detergent is sprayed on the coil and into the core of the coil with a hand-pump sprayer,

(continued)

PREVENTIVE MAINTENANCE (*Continued*)

similar to a garden sprayer. When the unit is started, the condensate will carry the dirt down the coil and down the condensate drain line. This type of cleaner is for light-duty cleaning. Care must be taken that the condensate drain line does not become clogged with the dirt from the coil and pan.

The coil may also be cleaned by shutting the unit down and applying a more powerful detergent to the coil, forcing it into the coil core. After the detergent has had time to work, the coil is then sprayed with a water source, such as a water hose. Care must be taken not to force too much water into the coil, causing the drain pan to overflow. The water must not be applied faster than the drain can accept it.

Special pressure cleaners may be used. These units have a low water flow, about 5 gal/min, and will not overflow an adequate drain system. The nozzle pressure enables the water to clean the core of the coil.

It is always best to "backwash" a coil when cleaning with water. For backwashing, the water is forced through the coil in the opposite direction to the airflow. Most of the dirt will accumulate on the inlet of the coil and become progressively less as the air moves through the coil. If the coil is not backwashed, the accumulation of dirt may only be driven to the center of the coil. **SAFETY PRECAUTION:** *Never use hot water or a steam cleaner on refrigeration equipment if refrigerant is in the unit because the pressure will rise high enough to burst the unit at the weakest point. This may be the compressor shell. A refrigerant system must be open to the atmosphere before hot water or steam is used for cleaning.* •

In some cases, the coil cannot be cleaned in the unit. If the unit has service valves, the refrigerant may be pumped into the condenser-receiver and the coil is removed from the unit. The coil will not contain any refrigerant and may be cleaned with approved detergent and hot water or with a steam cleaner. Approved cleaners may be purchased at any air-conditioning supply house. **SAFETY PRECAUTION:** *Follow the directions. Make sure no water can enter the coil through the piping connections.* •

The outdoor unit may be either air-cooled or water-cooled. Air-cooled units have fan motors that must be lubricated. Some motors require lubrication only after several years of operation. At that time, a recommended amount of approved oil is added to the oil cup. Some motors require more frequent lubrication.

The fan blades should be checked to make sure that they are secure on the shaft. At the same time, the rain shield, if there is one, on top of the fan motor should be checked. On upflow units, the shield should not allow water to enter the motor where the motor is out in the open.

Coils may become dirty from the dust that is being pulled through. It is hard to determine just by looking at a coil how much dirt is in the core. The coil should be cleaned at the first sign of dirt buildup or high operating head pressure.

Condenser coils may be easier to clean because they are on the outside. **SAFETY PRECAUTION:** *Make sure that all power is turned off, locked, and tagged so that someone cannot accidentally turn it on. The fan motor must be covered and care should be taken that water does not enter the controls. They may not all be in the control cabinet. Apply an approved detergent on the coil using a hand-pump sprayer. Soak the coil to the middle. Let the detergent soak for 15 to 30 min, then backwash the coil using a garden hose or spray cleaner.* •

Water-cooled equipment must be maintained like air-cooled equipment. Two things are done as part of the maintenance program to minimize the mineral concentrations in the water. One is to make sure the cooling tower has an adequate water bleed system. This is a measured amount of water that goes down the drain and is added back to the system with the supply water makeup line. Owners may suggest that the bleed system be shut off because all they see is water going down the drain. They may not realize that the purpose of this is to dilute the minerals in the water.

The other maintenance procedure is to use the correct water treatment. A water treatment specialist should supervise the setup of the treatment procedures. This will ensure the best-quality water in the system.

Electrical preventive maintenance consists of examining any contactors and relays for frayed wire and pitted contacts. When these are found, they should be taken care of before leaving the job. There can be no certainty about how long they will last. Single-phase compressors will just stop running if a wire burns in two or a contact burns away. A three-phase compressor will try to run on two phases, which is very hard on the motor. Wires should be examined to make sure they are not rubbing on the frame and wearing the insulation off the wire. Wires resting on copper lines may rub a hole in the line, resulting in a refrigerant leak.

41.14 DIAGNOSTIC CHART FOR AIR-CONDITIONING (COOLING) SYSTEMS

Many homes and businesses have air-conditioning (cooling). These systems operate in the spring, summer, and fall for the purpose of maintaining comfort for the occupants. The systems are often connected to the heating system. Air-conditioning may also be a part of a heat pump system. The air-conditioning system controls will be discussed here. Always listen to the owner to help to identify the potential problem.

Problem	Possible Cause	Possible Repair
No cooling, outdoor unit not running, indoor fan running	Open outdoor disconnect switch Open fuse or breaker Faulty wiring	Close disconnect switch. Replace fuse, reset breaker, and determine problem. Repair or replace faulty wiring or connections.
No cooling; indoor fan and outdoor unit will not run	Low-voltage control problem A. Thermostat B. Interconnecting wiring or connections C. Transformer	Repair loose connections or replace thermostat and/or subbase. Repair or replace wiring or connections. Replace if defective. **SAFETY PRECAUTION:** *Look out for too much current draw due to a ground circuit or shorted coil.* •
No cooling; indoor and outdoor fan running, but compressor not running	Tripped compressor internal overload A. Low line voltage B. High head pressure 1. Dirty outdoor (condenser) 2. Condenser fan not running all the time 3. Condenser air recirculating 4. Overcharge of refrigerant C. Low charge, motor not being properly cooled	Correct low voltage, call power company, repair loose connections. Clean condenser. Check condenser fan motor and capacitor. Correct recirculation problem. Correct charge. Correct charge; if due to leak, repair leak.
Indoor coil freezing	Restricted airflow Low charge Metering device Restricted filter drier Operating unit during low ambient conditions without proper head pressure control	Change filters. Open all supply register dampers. Clear return air blockage. Clean fan blades. Speed fan up to higher speed. Adjust unit charge—repair leak if refrigerant has been lost. Change or clean metering device. Change filter drier. Add proper low-ambient head pressure control.

41.15 SERVICE TECHNICIAN CALLS

Troubleshooting can take many forms and cover many situations. The following actual troubleshooting situations help show what the service technician does to solve actual problems.

SERVICE CALL 1

A residential customer reports that the central air conditioner at a residence is not cooling enough and runs continuously. *The problem is a low R-410A refrigerant charge.*

When the technician arrives, the unit is running. The temperature indicator on the room thermostat shows the thermostat is set at 72°F, and the thermometer on the thermostat indicates that the space temperature is 80°F. The air feels very humid. The technician notices that the indoor fan motor is running. The velocity of the air coming out of the registers seems adequate, so the filters are not stopped up.

The technician goes to the condensing unit and hears the fan running. The air coming out of the fan is not warm. This indicates that the compressor is not running. Removing the door to the compressor compartment, the technician notices that the compressor is hot to the touch. This is an indication that the compressor has been trying to run. Gauges installed on the service ports both read the same because the system has been off. The readings are 235 psig, which corresponds to 80°F. The residence is 80°F inside, and the ambient is 90°F, so it can be assumed that the unit has some liquid refrigerant in the system because the pressure corresponds so closely to the chart. If the system pressure were 170 psig, it would be obvious that little liquid refrigerant is left in the system. A large leak would be suspected.

The technician decides that the unit must have a low charge and that the compressor is off because of the internal overload protector. The technician pulls the electrical disconnect and takes a resistance reading across the compressor terminals with the motor leads removed. The meter shows an open circuit from common to run and common to start and a measurable resistance between run and start.

SERVICE CALL 3

A commercial customer calls to report that the air-conditioning unit at a small office building is not cooling. The unit was operating and cooling yesterday afternoon when the office closed for the day. *The problem is that someone turned the thermostat down to 55°F late yesterday afternoon, and the condensate on the evaporator froze solid overnight trying to pull the space temperature down to 55°F.*

The technician notices that the thermostat is set at 55°F when he arrives at the site. An inquiry reveals that one of the employees felt too warm in his office at the end of the building and turned down the thermostat to cool that office. The technician notices no air is coming out of the registers. The air handler is located in a closet at the front of the building. Examination shows that the fan is running and the suction line is frozen solid at the air handler.

The technician stops the compressor and leaves the evaporator fan operating by turning the heat-cool selector switch to OFF and the fan switch to ON. It is going to take a long time to thaw the evaporator, probably an hour. The technician leaves the following directions with the office manager:

1. Let the fan run until air comes out of the registers, which will verify that airflow has started through the ice.
2. When air is felt at the registers, wait 30 min to ensure that all ice is melted and turn the thermostat back to COOL to start the compressor.

The technician then looks in the ceiling to check the air damper on the duct run serving the back office of the employee who was not cool enough and had turned down the thermostat. The damper is nearly closed; it probably was brushed against by a telephone technician who had been working in the ceiling. The damper is reopened. A call later in the day indicates that the system is working again.

SERVICE CALL 4

A residential customer reports that the air-conditioning unit is not cooling correctly. *The problem is a dirty condenser.* The unit is in a residence and next to the side yard. When the homeowner mows the grass, the lawnmower throws grass on the condenser. This is the first hot day of summer, and the unit had been cooling the house until the weather became hot.

The technician notices that the thermostat is set for 75°F but the space temperature is 80°F. Plenty of slightly cool air is coming out of the registers. The technician goes to the side of the house where the unit is located. The suction line feels cool, but the liquid line is very hot. An examination of the condenser coil shows that the coil is clogged with grass and is dirty. **SAFETY PRECAUTION:** *The technician shuts off the unit at the disconnect and locks it out, then takes enough panels off to be able to spray coil cleaner on the coil.* • A high-detergent coil cleaner is applied to the coil and allowed to stand for about 15 min to soak into the coil dirt. While this is occurring, he oils the motors and changes the filters.

A water hose with a nozzle to concentrate the water stream is used to wash the coil in the direction opposite to the airflow. One washing is not enough, so the coil cleaner is applied to the coil again and allowed to set for another 15 min. The coil is washed again and is now clean.

The unit is assembled and started. The suction line is cool, and the liquid line is warm to the touch. The technician decides not to put gauges on the system and leaves. A call back later in the day to the homeowner indicates that the system is operating correctly.

SERVICE CALL 5

A homeowner reports that the air-conditioning unit is not cooling. This is a residential high-efficiency unit. *The problem is that the TXV is defective.*

The service technician finds the house warm upon arrival. The thermostat is set at 74°F, and the house temperature is 82°F. The indoor fan is running, and plenty of air is coming out of the registers.

Before leaving the house to observe the condensing unit, the technician informs the homeowner why he is leaving and going outside. At the condensing unit the technician finds that the fan and compressor are running, but they quickly stop. The suction line is not cool. After gauges are attached to the gauge ports, the low-side pressure reads 25 psig, and the head pressure reads 170 psig; this corresponds closely to the ambient air temperature. The unit is off because of the low-pressure control. When the pressure rises, the compressor restarts but stops in about 15 sec. The liquid-line sight glass is full of refrigerant. The technician correctly concludes that there is a restriction in the system. The TXV is a good place to start when there is an almost complete blockage. 4R *The system does not have service valves, so the charge has to be removed and recovered in an approved cylinder.* 4R

Before entering the house, the technician makes sure his shoes are clean and he puts the covers back on to protect flooring. He installs a new TXV into the system, which takes an hour to complete. After the valve is changed, the system is leak checked and evacuated to a deep vacuum. A charge is measured into the system, and the system is started. It is evident from the beginning that the valve change has repaired the unit.

After cleaning the work site, the technician explains to the homeowner how the problem was corrected. He gives billing information while letting the homeowner know that the business is valued. While always making eye contact and remembering the customer's name, the technician shakes the customer's hand with a clean hand.

SERVICE CALL 6

An office manager calls to report that the air-conditioning unit in a small office building is not cooling. The unit was cooling correctly yesterday afternoon when the office closed for the day. *The problem is that at night an electrical storm tripped a breaker on the air handler.* The power supply is at the air handler.

The technician first goes to the space thermostat. It is set at 73°F, and the temperature is 78°F. There is no air coming out of the registers. The fan switch is turned to ON, and the fan still does not start. The technician checks the low-voltage power supply, which is in the attic at the air handler, and finds the tripped breaker. **SAFETY PRECAUTION:** *Before resetting it, the technician decides to check the unit electrically.* •

A resistance check of the fan circuit and the low-voltage control transformer proves there is a measurable resistance. The circuit seems to be safe. The circuit breaker is reset and stays in. The thermostat is set at COOL, and the system starts. A call later indicates that the system is operating properly.

SERVICE CALL 7

A residential customer calls. The homeowner found the breaker at the condensing unit tripped and reset it several times, but it did not stay set. *The problem is that the compressor motor winding is grounded to the compressor shell and tripping the breaker.* **SAFETY PRECAUTION:** *The homeowner should be warned to reset a breaker no more than once.* •

The technician goes straight to the condensing unit with electrical test equipment. The breaker is in the tripped position and is moved to the OFF position. **SAFETY PRECAUTION:** *The voltage is checked at the load side of the breaker to ensure that there is no voltage. The breaker has been reset several times and is not to be trusted.* • When the technician determines that the power is definitely off, he connects the ohmmeter to the load side of the breaker. The ohmmeter reads infinity, meaning no circuit. The compressor contactor is not energized, so the fan and compressor are not included in the reading. The technician pushes in the armature of the compressor contactor to make the contacts close. The ohmmeter now reads 0, or no resistance, indicating that a short exists. It cannot be determined whether the fan or the compressor is the problem, so the compressor wires are disconnected from the bottom of the contactor. The short does not exist when the compressor is disconnected. This verifies that the short is in the compressor circuit, possibly the wiring. The meter is moved to the compressor terminal box and the wiring is disconnected. The ohmmeter is attached to the motor terminals at the compressor. The short is still there, so the compressor motor is condemned.

The technician must return the next day to change the compressor. Before he leaves, he disconnects the control wiring and insulates the disconnected compressor wiring so that power can be restored. This must be done to keep crankcase heat on the unit until the following day. If it is not done, most of the refrigerant charge will be in the compressor crankcase and too much oil will move with the refrigerant when it is recovered. The technician performs one more task before leaving the job. The Schrader valve fitting at the compressor is slightly depressed to determine the smell of the refrigerant. The refrigerant has a strong acid odor. A suction-line filter drier with high acid removal capacity needs to be installed along with the normal liquid-line drier.

4R *The technician returns on the following day, recovers the refrigerant from the system, and then changes the compressor, adding the suction- and liquid-line driers.* 4R The unit is leak checked, evacuated, and the charge measured into the system with a set of accurate scales. The system is started, and the technician asks the customer to leave the air conditioner running, even though the weather is mild, to keep refrigerant circulating through the driers to clean it in case any acid is left in the system.

The technician returns on the fourth day and measures the pressure drop across the suction-line drier to make sure that it is not restricted with acid from the burned-out compressor. It is well within the manufacturer's specification.

SERVICE CALL 8

A residential customer calls to say that he can hear the indoor fan motor running for a short time; then it stops. This is happening repeatedly. *The problem is that the indoor fan motor is cycling on and off on its internal thermal overload protector because the fan capacitor is defective.* The fan will start and run slowly then stop.

The technician knows from the work order to go straight to the indoor fan section. It is in the crawl space under the house, so he carries electrical instruments on the first trip. The fan has been off long enough to cause the suction pressure to go so low that the evaporator coil is frozen. The breaker is turned off during the check. The technician suspects the fan capacitor or the bearings, so he also carries along a fan capacitor. After bleeding the charge from the fan capacitor, it is checked with a meter to see if it is within the manufacturer's specified range. **SAFETY PRECAUTION:** *Ensure that no electrical charge is in the capacitor before checking it with a meter. A 20,000-Ω resistor should be used between the two terminals to bleed off any charge.* • The meter shows that the capacitor is open and must be replaced.

The technician oils the fan motor and starts it from under the house with the breaker. The fan motor is drawing the correct amperage. The coil is still frozen, and the technician must set the space thermostat to operate only

the fan until the homeowner feels air coming out of the registers. Then the fan only should be operated for one-half hour (to melt the rest of the ice from the coil) before the compressor is started.

The service technician then clearly explains to the customer in everyday language what went wrong with the system and the work that was accomplished and politely asks the customer if there are any questions. Billing information is given to the customer, the customer is again thanked for the business, and the technician leaves the work site.

Solutions for the following service calls are not provided here. Please discuss the solutions with your instructor.

SERVICE CALL 9

A commercial customer reports that an air-conditioning unit in a small office building is not cooling on the first warm day.

The technician first goes to the thermostat and finds that it is set at 75°F, and the space temperature is 82°F. The air feels very humid. No air is coming from the registers. The air handler is in the attic. An examination of the condensing unit shows that the suction line has ice all the way back to the compressor. The compressor is still running.

What is the likely problem and the recommended solution?

SERVICE CALL 10

A residential customer reports the unit in the residence was worked on in the early spring under the service contract. The earlier work order shows that the unit is an R-22 system and that refrigerant was added by a technician new to the company.

The technician finds that the thermostat is set at 73°F, and the house is 77°F. The unit is running, and cold air is coming out of the registers. Everything seems normal until the condensing unit is examined. The suction line is cold and sweating, but the liquid line is only warm.

Gauges are fastened to the gauge ports; the suction pressure is 85 psig, far from the correct pressure of 74 psig (77°F − 35°F = 42°F, which corresponds to about 74 psig). The head pressure is supposed to be 243 psig to correspond to a condensing temperature of 115°F (85°F + 30°F = 115°F, or 243 psig). The head pressure is 350 psig. The unit shuts off after about 10 min running time.

What is the likely problem and the recommended solution?

SERVICE CALL 11

A residential customer calls to report that an R-22 air-conditioning unit is not cooling enough to reduce the space temperature to the thermostat setting. It is running continuously on this first hot day.

The technician finds the thermostat set at 73°F, and the house temperature is 78°F. Air is coming out of the registers, so the filters must be clean. The air is cool but not as cold as it should be. The air temperature should be about 55°F, and it is 63°F.

At the condensing unit the technician finds that the suction line is cool but not cold. The liquid line seems extra cool. It is 90°F outside, and the condensing unit should be condensing at about 120°F. If the unit had 15°F of subcooling, the liquid line should be warmer than hand temperature, yet it is not.

Gauges are fastened to the service ports to check the suction pressure. The suction pressure is 95 psig, and the discharge pressure is 225 psig. The airflow is restricted to the condenser, and the head pressure gradually climbs to 250 psig; the suction pressure goes up to 110 psig. The compressor should have a current draw of 27 A, but it only draws 15 A.

What is the likely problem and the recommended solution?

SERVICE CALL 12

A commercial customer calls to indicate that the air-conditioning unit in a small office building was running but suddenly shut off.

The technician introduces himself to the customer and carefully listens to what he has to say. Before being invited inside, the technician puts protective covers over his shoes. The technician then goes to the space thermostat and finds it set at 74°F; the space temperature is 77°F. The fan is not running. The fan switch is moved to ON, but the fan motor does not start. It is decided that the control voltage power supply should be checked first. The power supply is in the roof condensing unit on this particular installation. The technician explains to the customer the reason for leaving the building and proceeds outside.

⚠ **SAFETY PRECAUTION:** *The technician places a ladder against the building away from the power line entrance.* • Electrical test instruments are taken to the roof along with tools to remove the panels. The breakers are checked and seem to be in the correct ON position. The panel is removed where the low-voltage terminal block is mounted. A voltmeter check shows there is line voltage but no control voltage.

What is the likely problem and the recommended solution?

REVIEW QUESTIONS

1. Before installing a gauge on the high-pressure side of a system, you should do which of the following?
   A. Turn on the system
   B. Turn off the system
   C. Adjust the system to a neutral position
   D. Have a cylinder approved by the DOT nearby
2. Capacitors should be shorted with a ______ before checking with an ohmmeter.
   A. 20,000-ohm, 5-watt resistor
   B. screwdriver
   C. digital multimeter
   D. capacitor analyzer
3. The high- and low-side pressures on an operating air-conditioning system can be converted to the ______ temperatures by using a temperature/pressure relationship chart.
   A. dew point and wet-bulb
   B. dry-bulb and wet-bulb
   C. latent and sensible
   D. condensing and boiling
4. Why do most air-conditioning systems not need a defrost system?
   A. They typically operate at temperatures below freezing.
   B. Heat from the heating system is used to save energy.
   C. They do not typically operate at temperatures below freezing.
   D. There is not enough condensate to cause frost on the evaporator.
5. A typical standard-efficiency evaporator under standard conditions will operate at a temperature of
   A. 30°F.
   B. 40°F.
   C. 50°F.
   D. 60°F.
6. The typical temperature difference between the entering air and the boiling refrigerant on a standard air-conditioning evaporator is ______°F.
   A. 10
   B. 15
   C. 25
   D. 35
7. When the ambient outside air temperature is 95°F and the unit is a high-efficiency unit, the lowest condensing refrigerant temperature when the unit is operating properly is
   A. 110°F.
   B. 115°F.
   C. 120°F.
   D. 125°F.
8. When the outside ambient air temperature is 90°F, the temperature of the condensing refrigerant in a standard-efficiency unit should be approximately
   A. 120°F.
   B. 130°F.
   C. 140°F.
   D. 150°F.
9. If a condensing pressure is 260 psig for R-22, the condensing temperature will be
   A. 110°F.
   B. 120°F.
   C. 130°F.
   D. 140°F.
10. How is high efficiency accomplished in an air-conditioning condensing unit?
    A. They use blended refrigerants with mineral oil.
    B. They incorporate multiple filter driers.
    C. They use a larger condenser coil surface area.
    D. They use small, single-speed compressors.
11. When troubleshooting a small system
    A. the gauge manifold should not be connected every time.
    B. turn the unit off and check the wiring.
    C. always attach the high- and low-pressure gauges.
    D. recover the refrigerant and recharge with new refrigerant.
12. The suction gas may have a high superheat if the unit
    A. is undercharged.
    B. is overcharged.
    C. has low evaporator airflow.
    D. has dirty evaporator and condenser coils.
13. If the suction pressure is low and the suction line is warm, the system has a ______ evaporator.
    A. flooded
    B. starved
    C. full
    D. defective
14. Which function on a digital multimeter is used to measure resistance in a compressor winding?
    A. Microfarads
    B. Volts
    C. Amperage
    D. Ohms
15. Which function on a digital multimeter is used to measure the current draw of a compressor?
    A. Microfarads
    B. Volts
    C. Amperage
    D. Ohms

16. Most condensers will subcool the refrigerant to between ______°F below the condensing temperature of the refrigerant.
    A. 0 and 5
    B. 5 and 10
    C. 10 and 20
    D. 20 and 30
17. What are two types of fixed-bore metering devices?
    A. Capillary tube and orifice (piston).
    B. TXV and capillary tube.
    C. EEV and orifice (piston).
    D. TXV and EEV.
18. If no crankcase heat is at the compressor, what may happen to the refrigerant charge?
    A. It might be vented to the atmosphere.
    B. It can migrate to the compressor crankcase.
    C. It can breakdown or fractionate.
    D. It can mix with the system oil to create acid.
19. The primary tools used when troubleshooting the mechanical side of a system are
    A. gauges and temperature-testing equipment.
    B. a digital multimeter and a digital thermometer.
    C. a vacuum pump and recovery machine.
    D. a core removal tool and a set of gauges.
20. If a condenser circuit breaker is tripped, you should
    A. reset it.
    B. check the suction pressure and discharge pressure.
    C. check the low-voltage circuit.
    D. perform a resistance check of the compressor section.

SECTION 8

ALL-WEATHER SYSTEMS

UNITS

UNIT 42

HEAT GAINS AND HEAT LOSSES IN STRUCTURES

OBJECTIVES

After studying this unit, you should be able to

- explain the concepts of heat gain and heat loss.
- explain the importance of determining the heat gain and heat loss of a structure.
- list the four items that should be considered when performing a heat gain/heat loss calculation.
- explain the difference between an interior wall and an exterior wall.
- use appropriate charts and tables to determine the design heating and cooling temperatures for a particular city.
- explain how cooling and heating temperature differentials are determined using design indoor and outdoor temperatures.
- explain what is meant by a U-value and an R-value.
- explain how to use charts and tables to determine the U-value for particular building construction elements.
- calculate the heat transfer through construction members.
- determine the factors that go into a heat loss calculation.
- explain the difference between net wall area and gross wall area.
- determine the factors that go into a heat gain calculation.
- explain how the heat load on windows is affected by their orientation.

42.1 INTRODUCTION TO HEAT GAIN AND HEAT LOSS

With many things in life, bigger is better. We want bigger houses, bigger boats, bigger paychecks, and bigger families. But, as we get bigger and bigger, the problems associated with these things get bigger as well. Air-conditioning and heating equipment is no exception. When it comes to sizing air-conditioning and heating equipment, however, bigger is definitely not better. When we refer to "bigger," we are referring to the capacity of the system to move Btus either into or out of a structure. The use of oversized equipment has many negative effects that more than outweigh the perceived benefits of larger equipment. Larger equipment, in general,

- costs more to manufacture, purchase, install, and maintain,
- costs more to operate because of larger motors and other electrical loads,
- requires larger air distribution systems which cost more than smaller ones, and
- will cool or heat the space quicker, resulting in more frequent system starts and stops.

Oversized air-conditioning equipment will cool the space faster. This results in more system off-time, decreasing the dehumidification effect that air-conditioning systems have on the occupied space. Decreased dehumidification leads to higher relative humidity in the occupied space. This leads to an increase in the possibility of mold growth in the space. Mold growth in the space puts the contractor at increased risk of litigation as a result of mold-related illness.

Undersizing equipment is not desirable either. The obvious result of an undersized heating or cooling system is that the equipment will not be able to maintain the space at the desired temperature at certain times of the season. If, during the summer months, the temperatures are much lower than average, an undersized air-conditioning system will be able to cool the space down to acceptable levels. If, however, the temperatures are higher than average, the system will not be able to handle the heat load on the structure. A similar situation arises with undersized heating equipment.

So, in order to make certain that the heating and cooling equipment that is selected for use in a structure is the right size, we need to know how much heat the structure loses in the cooler heating months and how much heat the structure gains in the warmer cooling months. In order to do this, a series of calculations must be made. These calculations are called heat gain and heat loss calculations. These calculations determine the amount of heat lost and gained by means of the envelope or shell of the structure, **Figure 42.1**. In simple terms, the envelope separates the inside of the structure from the outside. In the summer months heat leaks into the house, while in the winter, heat leaks out of the house.

Heat gain and heat loss calculations are important parts of the system design process. When these calculations are performed accurately, the equipment capacities selected will properly match the structure, helping to ensure that

Figure 42.1 (A) In the warmer summer months, heat leaks into the structure. (B) In the colder winter months, heat leaks out of the structure.

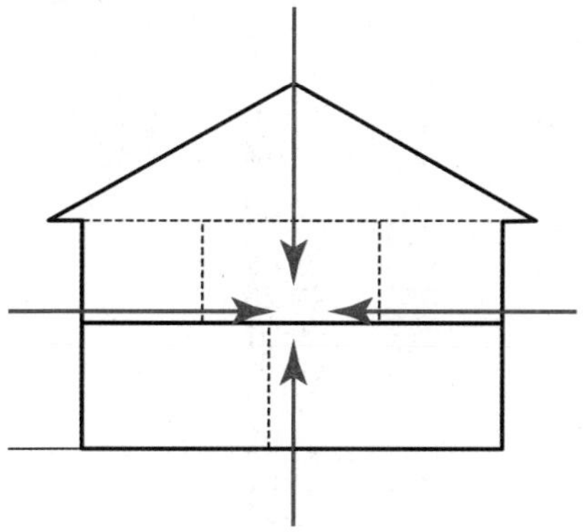

(A)

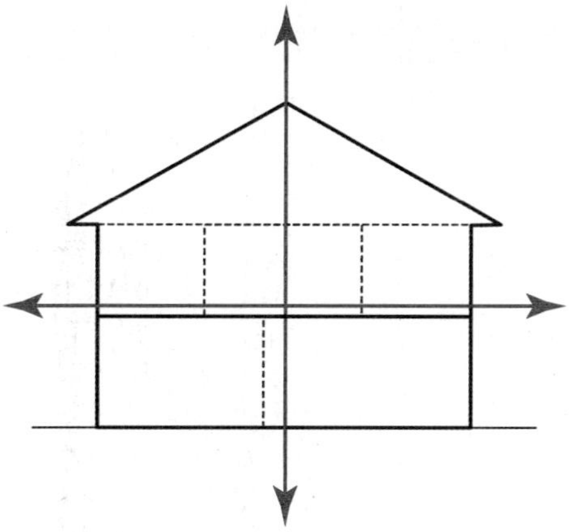

(B)

the desired levels in the structure will be maintained over a large range of operating conditions. As was mentioned in Unit 35, "Comfort and Psychrometrics," comfort is the delicate balance among air temperature, relative humidity, air cleanliness, and air movement. Properly selected and installed equipment is, therefore, the basis for providing comfort. Quite often, comfort-related issues can be traced back to the system, so it's important to start out on the right foot with the right equipment. *Governmental agencies have stressed the importance and the environmental benefits of properly sized heating and cooling equipment. In order to qualify for many government rebates for new heating and cooling equipment, proof of an accepted heat gain and/or heat loss calculation must be provided.* Refer to the next paragraph for more about "acceptable" methods for estimating the heat gain and heat loss of a structure. It is very important to note that the material in this unit is not intended to replace industry-accepted methods for performing these calculations, but it will provide the reader with the concepts and methods involved in making them.

42.2 METHODS TO DETERMINE THE HEAT GAIN AND HEAT LOSS OF A STRUCTURE

There are many methods that are commonly used for estimating the heat gain and heat loss of a structure. Most of these methods rely on rules of thumb that provide very quick results and, as a result, do not provide very accurate estimations of the needs of the particular structure. When performing a heat gain or heat loss calculation, a few things must be taken into account:

- How accurate do the results have to be?
- How critical does the temperature control need to be?
- How critical does the humidity control need to be?
- Is an equipment rebate being applied for?

The most basic heat gain and heat loss calculations are based solely on the area or volume of the space being conditioned. For example, an air-conditioning system might be sized at so many Btus per square foot or so many Btus per cubic foot, **Figure 42.2**. These down-and-dirty, cookie-cutter methods do not take into account the geographic location of the structure; the construction materials used; the slope of the roof; the number of windows, doors, and other construction elements; or the orientation (north, south, east, or west) of the structure.

Some equipment manufacturers provide single-page forms that can be used to provide a general idea of the equipment size needed. These forms are typically much easier to complete than are other industry-accepted methods. If rebates are being applied for, it is important to check with the agency for a list of calculation methods that are acceptable.

The most popular method for determining the heat gain and heat loss of a residential structure is to use Manual J, **Figure 42.3**. Manual J is prepared by the Air Conditioning Contractors of America, ACCA, and is among the most comprehensive methods of determining the heating and cooling needs of residential structures. Manual J is available in both

Figure 42.2 Cookie-cutter methods of performing heat gain and heat loss calculations are not acceptable. Photo by Eugene Silberstein

Figure 42.3 The heat gain and heat loss methods used should be nationally recognized and accepted. Photo by Eugene Silberstein

abridged and unabridged versions. The abridged version can be used for most residential applications without sacrificing accuracy.

Whatever method is ultimately used to determine the heat gain and heat loss values for the project, it should be nationally recognized and approved. It might be necessary for a contractor or system designer to justify or defend the equipment selection process in court. If it is suspected that the equipment is responsible for structural damage, mold-related issues, or health-related problems, having a properly completed, industry-accepted heat gain or heat loss report will prove invaluable.

42.3 INDOOR AND OUTDOOR DESIGN CONDITIONS FOR HEATING AND COOLING

One important factor in determining the heat gain and heat loss of a structure is knowing the temperature difference between the inside and the outside of the structure. As this temperature differential, TD, increases, the rate of heat transfer from one side of the structure to the other will increase. Similarly, as this temperature differential decreases, the rate of heat transfer from one side of the structure to the other will decrease.

Consider the internal wall of a house that separates the hallway from a bedroom. This wall or panel has 70°F air on both sides. The heat transfer through the wall will be zero, since there is no temperature difference across the panel, **Figure 42.4**. Now consider an exterior wall of a house in the summer. Exterior walls separate conditioned interior spaces within a structure from the outside. The indoor house temperature is 75°F and the outside air temperature is 90°F. There will obviously be heat flowing through the wall since there is a 15°F TD across the panel, **Figure 42.5**. Now consider the same wall in the middle of the winter. The indoor house temperature is 70°F and the outside air temperature is 0°F. Now, the rate of heat transfer through the wall will be much higher, because of the 70°F TD. In addition, the heat will be moving in the opposite direction, **Figure 42.6**.

Figure 42.4 No heat transfer takes place through a wall or panel that has no temperature differential across it.

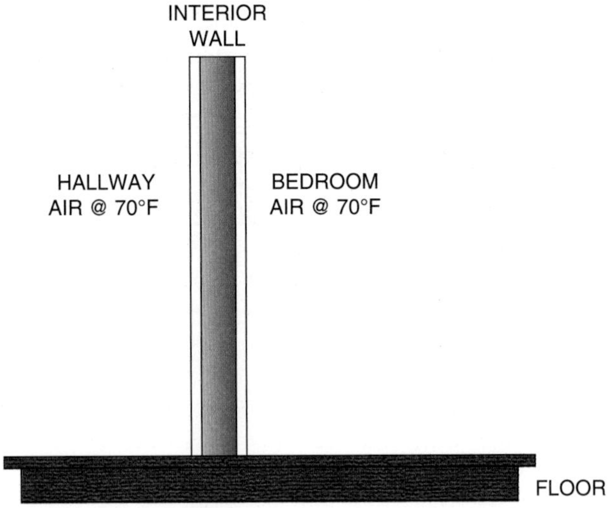

Figure 42.5 This wall has a temperature difference of 15 degrees across it, so heat will enter the structure since the outside air temperature is higher than the indoor temperature.

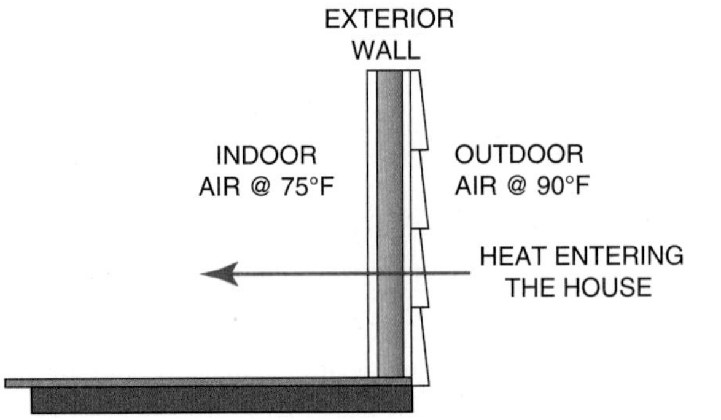

Figure 42.6 This wall has a temperature difference of 70 degrees across it, so heat will leak from the structure since the indoor air temperature is higher than the outside temperature.

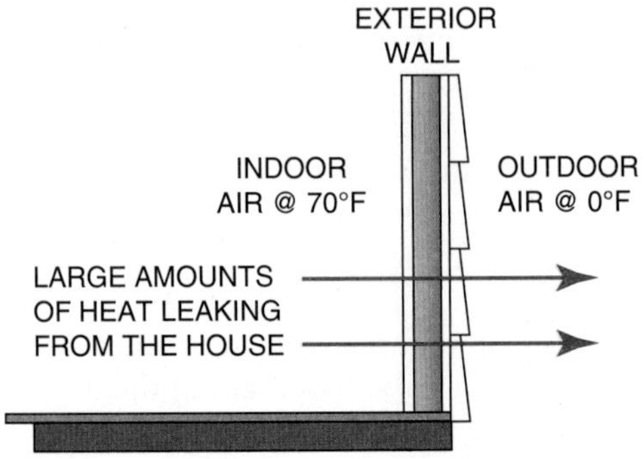

Since the temperature on both sides of an exterior wall or panel is needed to help determine how much heat is entering or leaving the structure, we need to know something about the expected indoor temperatures as well as the expected outdoor temperatures in different geographic regions at different times of the year. Typically, the design indoor temperatures used for heat gain and heat loss calculations are as follows:

- Indoor design temperature for the heating season: 70°F
- Indoor design temperature for the cooling season: 75°F

Determining the design outdoor temperature requires the use of a table similar to that shown in **Figure 36.13** in Unit 36, "Refrigeration Applied to Air-Conditioning." Although tables that are prepared by different organizations will present data differently, the information, for the most part, is the same. For example, when determining the design outdoor temperature for cooling, ACCA Manual J provides one column, the 1% column, while the American Society of Heating, Refrigerating and Air-Conditioning Engineers (ASHRAE) table provides three temperature columns: the 1% column, the 2.5% column, and the 5% column. Since most system designers use the ASHRAE 2.5% column for sizing cooling equipment, the ASHRAE 2.5% information is the same data that are included in the ACCA Manual J 1% column. To clarify this idea, a simplified comparison for Nogales, Arizona, and Modesto, California, is shown in **Figure 42.7**. Notice that the ACCA 99%–1% values correspond to the ASHRAE 97.5%–2.5% values.

To determine the temperature differentials for heating and cooling equipment in Nogales, Arizona, we would use the data from the table and the indoor design temperatures to get:

Cooling Design TD = Outdoor Design Temperature − Indoor Design Temperature

Cooling Design TD = 96°F − 75°F

Cooling Design TD = 21°F

Heating Design TD = Indoor Design Temperature − Outdoor Design Temperature

Heating Design TD = 70°F − 32°F

Heating Design TD = 38°F

For systems that will be installed in Modesto, California, we would use the data from the table and the indoor design temperatures to get:

Cooling Design TD = Outdoor Design Temperature − Indoor Design Temperature

Cooling Design TD = 98°F − 75°F

Cooling Design TD = 23°F

Heating Design TD = Indoor Design Temperature − Outdoor Design Temperature

Heating Design TD = 70°F − 30°F

Heating Design TD = 40°F

For a system that is being installed in Central Islip (Long Island), New York, with design cooling and heating temperatures of 85°F and 15°F, respectively, the cooling design TD would be 10°F (85°F − 75°F), while the heating design TD would be 55°F (70°F − 15°F). In a colder climate such as Big Rapids, Michigan, which has design cooling and heating temperatures of 87°F and 4°F, respectively, the cooling design TD would be 12°F (87°F − 75°F), while the heating design TD would be 66°F (70°F − 4°F).

42.4 U-VALUES AND R-VALUES

In addition to knowing the design indoor and outdoor temperatures for the structure, information about the construction materials used is also needed. The construction materials will have a major effect on the rate of heat transfer through the structural envelope. One of the most commonly used characteristics of building construction materials is the R-value. The R-value is typically used to rate the thermal resistance of insulation materials. The higher the R-value, the harder it is for heat to pass through the material. The R-values for some common building materials are shown in **Figure 42.8**. If the wall of a structure is made up of a number of different materials, the total R-value will be the sum of all of the individual R-values that make up the wall.

Consider a wall that is constructed from 4 in. brick, has 5/8 in. drywall on the inside surface, and has 3/4 in. plywood sheathing and insulating board-backed aluminum siding

Figure 42.7 For the most part, ACCA 99% and 1% values correspond to the ASHRAE 97.5% and 2.5% values. Table derived from data contained in ASHRAE manuals and ACCA Manual J charts.

LOCATION	ELEVATION	LATITUDE	ASHRAE HEATING 97.5% DRY-BULB	ACCA HEATING 99% DRY-BULB	ASHRAE COOLING 2.5% DRY-BULB	ACCA COOLING 1% DRY-BULB	DESIGN GRAINS @ 50% RELATIVE HUMIDITY
Nogales, Arizona	3932ft	31°	32°F	32°F	96°F	96°F	−25
Modesto, California	91ft	37°	30°F	30°F	98°F	98°F	−9
Central Islip, New York	98ft	40°	15°F	15°F	85°F	85°F	35

Figure 42.8 The R-values for some commonly used construction materials. Table derived from data contained in ASHRAE manuals and ACCA Manual J charts.

MATERIAL	R-VALUE	MATERIAL	R-VALUE
4″ Common Brick	0.80	Fiberglass Blown Insulation (attic)	2.2-4.3 per inch
4″ Brick Face	0.44	Fiberglass Blown Insulation (wall)	3.7-4.3 per inch
4″ Concrete Block	0.80	Cellulose Blown Insulation (attic)	3.13 per inch
8″ Concrete Block	1.11	Cellulose Blown Insulation (wall)	3.7 per inch
12″ Concrete Block	1.28	3/4″ Plywood Flooring	0.93
Lumber (2 x 4)	4.38	5/8″ Particleboard (Underlayment)	0.82
Lumber (2 x 6)	6.88	3/4″ Hardwood flooring	0.68
1/2″ Plywood Sheathing	0.63	Linoleum Tiles	0.05
5/8″ Plywood Sheathing	0.77	Carpet with fibrous pad	2.08
3/4″ Plywood Sheathing	0.94	Carpet with rubber pad	1.23
1/2″ Hardboard Siding	0.34	Wood Door (hollow core)	2.17
5/8″ Plywood Siding	0.77	Wood Door (1.75″ solid)	3.03
3/4″ Plywood Siding	0.93	Wood Door (2.25″ solid)	3.70
Hollow-backed aluminum siding	0.61	Single pane windows	0.91
Insulating board-backed aluminum siding	1.82	Single pane windows w/storm	2.00
Insulating board-backed aluminum siding (foil-backed)	2.96	Double pane windows with 0.25″ air space	1.69
1/2″ Drywall	0.45	Double pane windows with 0.50″ air space	2.04
5/8″ Drywall	0.56	Triple pane windows with 0.25″ air space	2.56
Fiberglass Batt Insulation	3.14 – 4.3 per inch	Asphalt Shingles	0.44
Rock Wool Batt Insulation	3.14 – 4.0 per inch	Wood Shingles	0.97

(foil-backed) on the outside, **Figure 42.9**. We can calculate the R-value for the wall by adding up the R-values of the materials:

4 in. Brick:	R-value = 0.80
5/8 in. Drywall:	R-value = 0.56
3/4 in. Plywood sheathing:	R-value = 0.94
Hollow-backed aluminum siding:	R-value = 2.96
	Total R-value = 5.26

Another characteristic of building construction materials is the U-value. The U-value describes the thermal conductivity of a material. The higher the U-value, the easier it is for heat to pass through the material. When designing a house, for example, it would be a good idea to use construction materials with high R-values and low U-values. The R-value and the U-value are inverses of each other. If a material has an R-value of 30, then the corresponding U-value will be 0.033, which is 1/30. In the example provided above, the U-value of the wall would be 0.19, which is 1/5.26.

The units of the U-value are Btu/h/°F/ft^2, or Btu per hour per degree Fahrenheit per square foot. This means that if the U-value of a material is 0.19, it will let 0.19 Btu of heat energy pass through each square foot of the material if the TD across the material is 1°F. The next section will provide some more numerical examples. **NOTE:** *The published R-value for insulation is very rarely indicative of the material's actual ability to restrict heat transfer. As the material is compressed, the R-value of the material decreases. R-19 batt-type cavity insulation, for example, will have an R-value of 19 at a thickness of 6.5 in. Since the space in a 2 × 6 wall cavity is only 5.5 in. the insulation will*

Figure 42.9 Illustration of a sample wall composition.

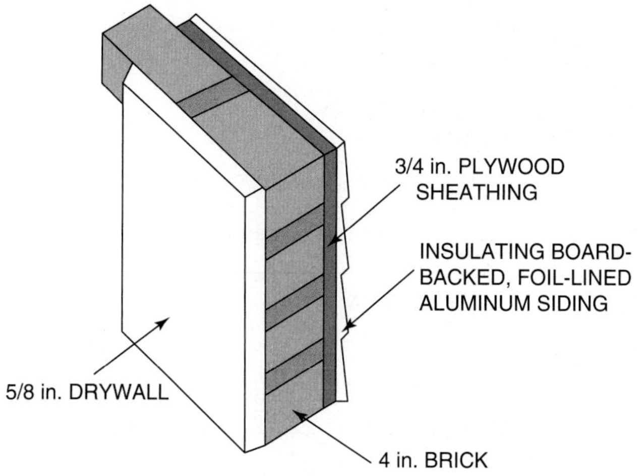

be compressed somewhat in order to fit in the wall cavity and the R-value will drop to roughly 16. •

42.5 INTRODUCTION TO HEAT GAIN AND HEAT LOSS CALCULATIONS

Up to this point, we have discussed the basic concepts that go into a heat gain or heat loss calculation. We will now take a look at some examples to provide some more insight into the calculation process.

Example 1: Consider the wall from **Figure 42.9**. If the wall is 8 ft high and 20 ft long, how much heat will pass through the wall if the TD is 20 degrees?

Solution: In general, the rate of heat transfer through a wall will be determined by

$$Q = \text{U-value} \times \text{TD} \times \text{Area}$$

where

Q is the quantity of heat, in Btu/h
U-value is the U-value of the wall
TD is the temperature differential across the wall
Area is the area of the wall

By plugging the values into the formula, we will get:

$$Q = \text{U-value} \times \text{TD} \times \text{Area}$$
$$Q = 0.19 \times 20°\text{F} \times (8 \times 20)\ \text{ft}^2$$
$$Q = 0.19 \times 20°\text{F} \times 160\ \text{ft}^2$$
$$Q = 608\ \text{Btu/h}$$

Example 2: Consider a single-pane window with a storm. If the window is 4 ft high and 6 ft long, how much heat will pass through the window if the TD is 30 degrees?

Solution: The U-value of the window can be found by first looking up the R-value of the window in **Figure 42.8**. Since the R-value is 2.00, the U-value is 0.5.

By plugging the values into the formula, we will get:

$$Q = \text{U-value} \times \text{TD} \times \text{Area}$$
$$Q = 0.5 \times 30°\text{F} \times (4 \times 6)\ \text{ft}^2$$
$$Q = 0.5 \times 30°\text{F} \times 24\ \text{ft}^2$$
$$Q = 360\ \text{Btu/h}$$

42.6 ELEMENTS OF STRUCTURAL HEAT LOSS (HEATING MODE)

When calculating the heat loss of a structure, it is necessary to take into account the entire envelope and all the contact the envelope has with the outside. Heat losses from the conditioned space can occur in a number of ways, including:

- Through exposed (exterior) walls
- Through windows and doors
- Through roofs
- Through ceilings to unconditioned attics or ceiling cavities
- Through partition panels
- Through below-grade (underground) walls
- Through slab and basement floors
- Through floors located above unconditioned spaces
- Through cracks, openings, and deficiencies in structure
- Duct losses to attics, ceiling cavities, and other unconditioned spaces

The total heat loss for a structure is simply the sum of the heat losses experienced by all of the construction elements that make up a particular structure.

HEAT LOSS THROUGH EXPOSED WALLS

The walls of a structure make up a large portion of the structural envelope. Walls account for a large percentage of the total heat loss of a structure, so it is very important to have accurate information regarding the construction materials used to fabricate them. Frame houses commonly have either 2 × 4 or 2 × 6 construction with insulation in between the wall studs. When calculating heat loss, it is important to know not only what lumber the house was framed with but also the R-value of the insulation. The size of the lumber can be determined in a number of different ways, including:

- Blueprint information
- Measuring the thickness of the wall
- Inspection of the attic or basement
- Measuring through an electrical outlet or switch opening

Probably the easiest way to determine what size lumber was used to frame a house is to inspect the blueprint, if available. If the blueprint is not handy, an open window and a tape measure will help. The thickness of the wall will be the width of the window minus any moldings and other construction elements, such as drywall, paneling, shingles, and sheathing.

It might be possible to visually inspect the lumber from the attic or basement, depending on the construction of the house. Another possible method for determining the size of the lumber used for framing is to access the wall cavity through an electrical switch or outlet box. Be sure to disconnect the power to the circuit before removing the switch or receptacle cover. Gently insert a blunt, nonmetallic object such as a paint stirrer or unsharpened pencil in the gap between the electric box and the drywall until it reaches the back end of the wall cavity.

The R-value of the insulation used can be determined by visual inspection or by determining its value based on the size of the studs that were used for construction. For example, if the house was constructed from 2 × 4 lumber and is insulated with fiberglass batt insulation, the R-value of the insulation be estimated to be between 10.15 and 13.3. This is because a 2 × 4 is 3.5 in. wide and fiberglass batt-type insulation has an R-value that ranges from 2.9 to 3.8 per inch. From this information, it is very likely that the insulation in the wall has an R-value of R-13, since R-13 is among the most common insulation values. Similarly, a wall framed with 2 × 6 lumber will likely have R-19 insulation if fiberglass batting has been used.

The insulation and studs are not the only components that make up a wall. In addition to these items, sheathing, siding, vapor barriers, and drywall are also used to create the final product. To make it easier to determine the values needed for calculations, tables are available that allow the user to simply look up the wall type and construction and obtain the needed values, **Figure 42.10**. So, if a house is framed with 2 × 6 wood studs and has board insulation with an R-value of R-6 and insulation with an R-value of R-19, the U-value for the wall will be 0.049.

GROSS WALL AREA. *Gross wall area* is a term that is used to describe the area of a wall before any doors or windows are cut into it. The gross wall area of the wall in Example 1 in Section 42.4 is 160 ft², which is simply the height of the wall multiplied by the width of the wall. Consider the house in **Figure 42.11**, which has no windows or doors. The gross wall area of Wall A is 250 ft², while the gross wall area of Wall B is 160 ft² (ignoring the triangular portion of the wall in the attic).

NET WALL AREA. *Net wall area* is a term that is used to describe the area of a wall after the doors and windows are cut into it. The net wall area can be expressed mathematically as:

Net Wall Area = Gross Wall Area − Door Area − Window Area − Area of Other Components

Let's take a look at the wall in **Figure 42.12**. This wall has a gross wall area of 300 ft², which is simply the height of the wall multiplied by the width of the wall. To find the net wall

Figure 42.11 Gross wall area is determined by multiplying the length and the width of the wall. The gross wall area of Wall A is 250 ft² and the gross wall area of Wall B is 160 ft².

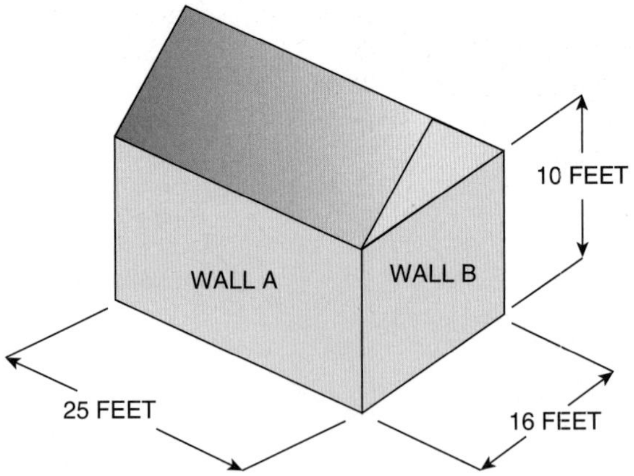

Figure 42.10 The U-values for some exterior walls. Table derived from data contained in ASHRAE manuals and ACCA Manual J charts.

WALL CONSTRUCTION	STUD SIZE (WOOD)	BOARD INSULATION	INSUALATION R-VALUE	U-VALUE
Frame	2 x 4	R-2	R-11	0.086
Frame	2 x 4	R-2	R-13	0.081
Frame	2 x 4	R-4	R-11	0.073
Frame	2 x 4	R-4	R-13	0.069
Frame	2 x 4	R-6	R-11	0.064
Frame	2 x 4	R-6	R-13	0.060
Frame	2 x 6	R-2	R-15	0.077
Frame	2 x 6	R-2	R-19	0.063
Frame	2 x 6	R-4	R-15	0.066
Frame	2 x 6	R-4	R-19	0.055
Frame	2 x 6	R-6	R-15	0.058
Frame	2 x 6	R-6	R-19	0.049

Figure 42.12 The net wall area is the gross wall area minus the area of all windows and doors in the wall. The net wall area for this wall is 261 ft².

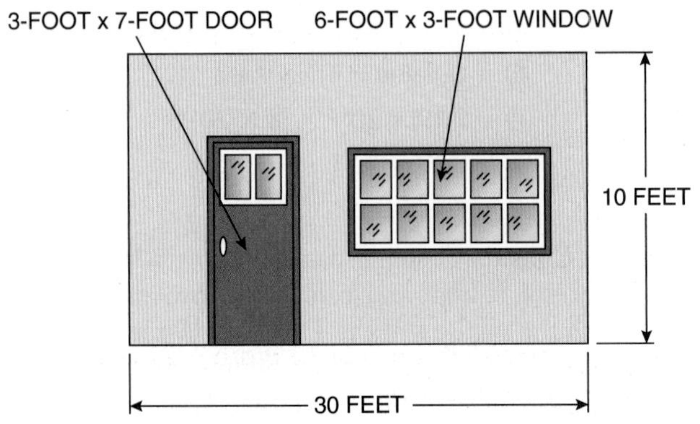

area, we would subtract the area of the window and door. Using the formula for net wall area, we get:

Net Wall Area = Gross Wall Area − Door Area − Window Area
Net Wall Area = 300 ft^2 − Door Area − Window Area
Net Wall Area = 300 ft^2 − (3' × 7') − (6' − 3')
Net Wall Area = 300 ft^2 − 21 ft^2 − 18 ft^2
Net Wall Area = 300 ft^2 − 39 ft^2
Net Wall Area = 261 ft^2

HEAT LOSS THROUGH WINDOWS AND DOORS

The method used to determine the heat loss through windows and doors is basically the same as that used to determine the heat loss through the wall discussed earlier. For the wall in **Figure 42.12**, the construction materials are as follows:

Wall:	2 × 6 studs, R-19 cavity insulation, R-4 board insulation
Window:	Double pane with 0.50in. of air space
Door:	2.25in. solid wood
TD across panel:	40°F

To determine the heat loss via the construction materials, the U-values are needed. These values are obtained as follows:

Wall U-value:	0.055 (from **Figure 42.10**)
Window U-value:	0.49 (from **Figure 42.8** and using U = 1/R, 1/2.04 = 0.49)
Door U-value:	0.27 (from **Figure 42.8** and using U = 1/R, 1/3.70 = 0.27)

The heat transfer rate through the window is found as follows:

$$Q = \text{U-value} \times TD \times \text{Area}$$
$$Q = 0.49 \times 40°F \times (3 \times 6)\ ft^2$$
$$Q = 0.49 \times 40°F \times 18\ ft^2$$
$$Q = 352.8\ Btu/h$$

Similarly, the heat transfer rate through the door is found as follows:

$$Q = \text{U-value} \times TD \times \text{Area}$$
$$Q = 0.27 \times 40°F \times (3 \times 7)\ ft^2$$
$$Q = 0.27 \times 40°F \times 21\ ft^2$$
$$Q = 226.8\ Btu/h$$

We can then determine the heat loss for the entire wall by using the following:

$$Q_{Entire\ Wall} = Q_{Net\ Wall} + Q_{Window} + Q_{Door}$$
$$Q_{Entire\ Wall} = [0.055 \times 261\ ft^2 \times 40°F] + Q_{Window} + Q_{Door}$$
$$Q_{Entire\ Wall} = 574.2\ Btu/h + 352.8\ Btu/h + 226.8\ Btu/h$$
$$Q_{Entire\ Wall} = 1153.8\ Btu/h$$

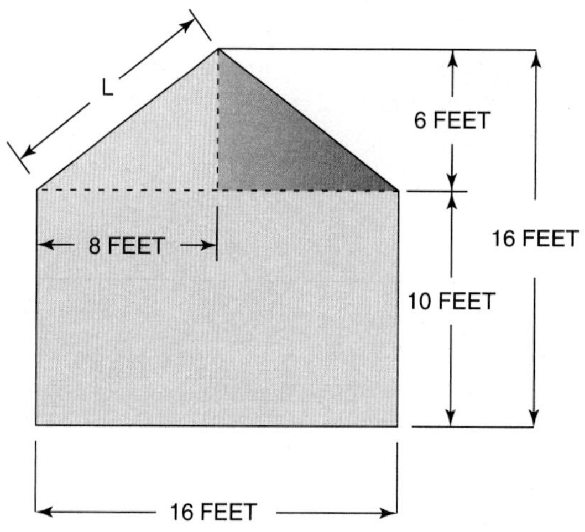

Figure 42.13 Geometry showing how to calculate the "L" dimension of this roof.

HEAT LOSS THROUGH ROOFS

A great deal of heat is lost through the roof. The calculation to determine the heat loss through the roof is similar to the method used earlier, but calculating the area of the roof involves a little geometry. If the roof in **Figure 42.11** has its peak at a height of 16 ft from the ground, we can calculate the length, L, of the sloped portions of the roof, **Figure 42.13**, by using Pythagorean Theorem as follows:

Height of triangle:	6 ft (16-ft peak height − 10-ft wall height)
Base of triangle:	8 ft (one half of the width of the house)

So, using the Pythagorean Theorem, we get:

$$L^2 = \text{Height}^2 + \text{Base}^2$$

or

$$L = \sqrt{\text{Height}^2 + \text{Base}^2}$$
$$L = \sqrt{6^2 + 8^2}$$
$$L = \sqrt{36 + 64}$$
$$L = \sqrt{100}$$
$$L = 10\ ft$$

NOTE: *It could also be noted that the length "L" is the hypotenuse of a 3-4-5 triangle, where one side is 6 (3 × 2) and the second is 8 (4 × 2). The hypotenuse, L, is therefore 10 (5 × 2).* • Since the house is 25 ft long, each half of the roof will be 10 ft × 25 ft, or 250 ft^2. So, the total area of the roof will be 500 ft^2. If the roof, for example has a U-value of 0.25 and the TD across the roof is 60°F, the heat loss attributed to the roof is

$$Q_{Roof} = \text{U-value} \times TD \times \text{Area}$$
$$Q_{Roof} = 0.25 \times 60°F \times 500\ ft^2$$
$$Q_{Roof} = 7500\ Btu/h$$

Figure 42.14 This attic is separated from the occupied space by drywall and R-30 insulation.

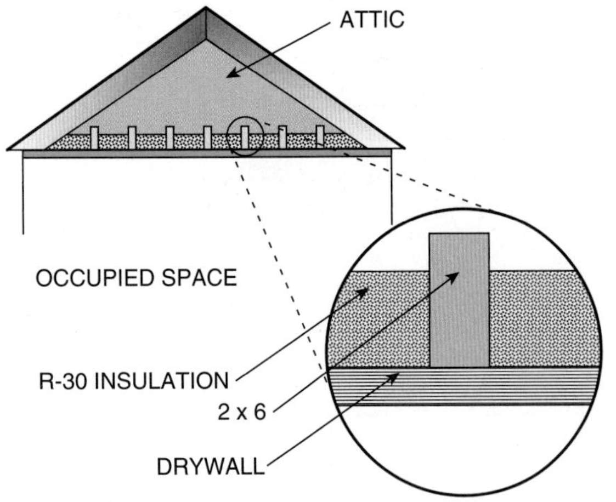

NOTE: *This example assumes that the roof separates the outside from the occupied space. If there is a flat ceiling located under the roof, the heat loss is calculated from the occupied space to the attic.* •

HEAT LOSS THROUGH CEILINGS TO UNCONDITIONED ATTICS OR CEILING CAVITIES

Since the basics are known, the best way to cover this is by example. Consider a house that has a footprint of 400 ft^2 and a ceiling is separating the occupied space from the unconditioned attic. The ceiling is insulated with R-30 insulation and the TD between the space and the attic is 60°F, **Figure 42.14**. The U-value of the ceiling will be 0.033, since the U-value = 1/R. The heat loss from the structure via the ceiling can be found by:

$$Q_{Ceiling} = \text{U-value} \times TD \times Area$$
$$Q_{Ceiling} = 0.033 \times 60°F \times 400\ ft^2$$
$$Q_{Ceiling} = 800\ Btu/h$$

This example has been simplified a great deal, as the actual calculation is much more complicated. The drywall under the insulation was not accounted for, nor was the 2 × 6 lumber. Both of these construction elements, along with the R-30 insulation, make up the barrier that separates the attic from the living space. In actuality, the barrier can be broken up into two different sections. One portion is made up of the R-30 insulation and the 5/8 in. drywall, and the other is made up of the 2 × 6 lumber and the 5/8 in. drywall. Since the width of a 2 × 6 is 1.5 in., and the beams are typically 16-in. apart from each other on center, there is a 14.5-in. gap between the beams, **Figure 42.15**.

The R-value for the lumber/drywall portion of the barrier is 7.44 (6.88 + 0.56), and the R-value for the insulation/drywall portion of the barrier is 30.56 (30 + 0.56). In any 16-in. portion of the barrier, 1.5 in. will be the lumber/drywall combination and 14.5 in. will be the insulation/drywall combination. This means that the barrier is made up of 9.4% (1.5/16) of the lumber/drywall combination and 90.6% (14.5/16) of the insulation/drywall combination. Mathematically, the R-value for the barrier can be calculated as:

Figure 42.15 Attic barrier made up of insulation, beams, and drywall. Typical beams are positioned 16-in. apart, on center.

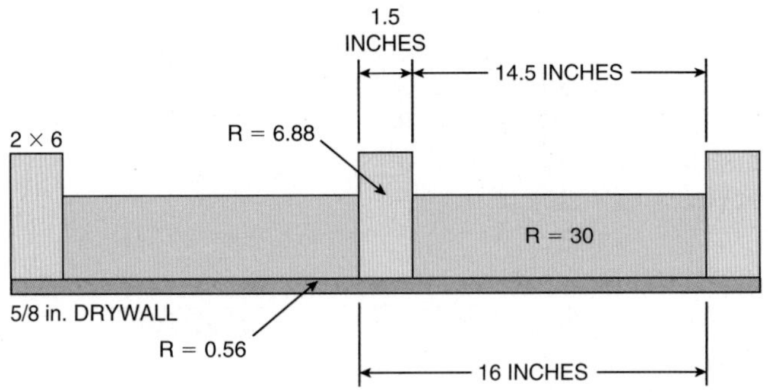

R-value = 9.4% of the lumber/drywall combo + 90.6% of the insulation/drywall combo

R-value = 0.094 (7.44) + 0.906 (30.56)

R-value = 0.094 (7.44) + 0.906 (30.56)

R-value = 0.70 + 27.69

R-value = 28.39

The R-value for the barrier can then be converted to a U-value of 0.035. Using the same TD of 60°F and the area of 400 ft^2, the heat loss from the structure via the ceiling can be calculated as:

$$Q_{Ceiling} = \text{U-value} \times TD \times Area$$
$$Q_{Ceiling} = 0.035 \times 60°F \times 400\ ft^2$$
$$Q_{Ceiling} = 840\ Btu/h$$

It can be seen that the difference between the original calculation and this, more complicated version, is approximately 5%.

HEAT LOSS THROUGH PARTITION WALLS

Partition walls are walls that separate a conditioned interior space from an unconditioned interior space. The heat that is lost from the structure is determined using the same formula that has been utilized for most of this chapter, but the TD that is used is not the difference between the indoor and outdoor temperatures. Instead, the TD is the difference between the design indoor air temperature and the temperature of the air in the unconditioned space. If the outside temperature is 0°F, the indoor temperature is 70°F, and the temperature of an unheated garage is 40°F, the TD across the partition wall that separates the garage from the conditioned space is 30°F, **Figure 42.16**.

Figure 42.16 Partition walls separate a conditioned interior space from an unconditioned interior space.

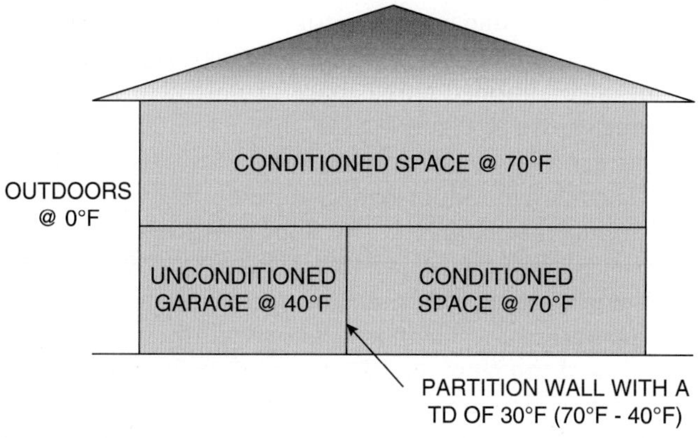

Figure 42.17 Some walls are partially above grade and partially below grade.

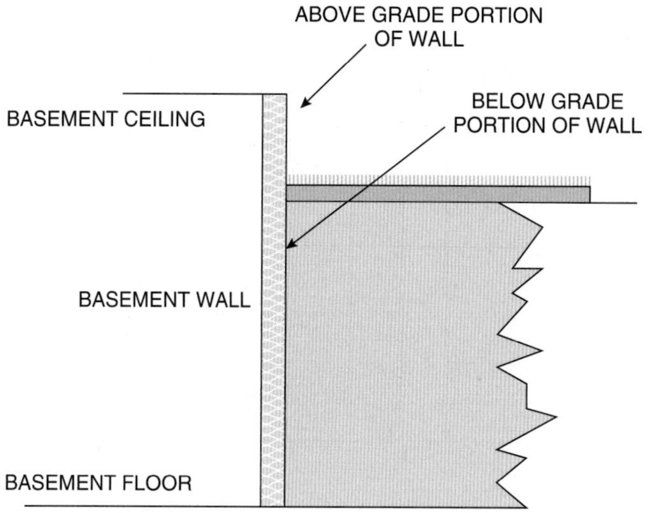

HEAT LOSS THROUGH BELOW-GRADE WALLS

The term *below grade* is used to describe walls or portions of walls that are below ground level. Some walls are partially above grade and partially below grade, **Figure 42.17**. When such is the case, the heat loss through the wall must be calculated as the sum of the heat loss from the above-grade portion of the wall and the heat loss from the below-grade portion of the wall. The U-values for the above-grade and below-grade portions of the wall vary as well.

The two important factors that affect the U-value for the below-grade portion of the wall are the materials used to construct the wall and how far below grade the floor is. The rate of heat loss through a below-grade wall decreases as the floor gets deeper into the ground. Let us take a look at a sample chart that might be used for basement walls that are made up of 8 in. of brick or stone and are framed with wood studs, **Figure 42.18**.

Let us consider a house that has a 40' × 60' footprint and a basement with a full 8′ ceiling. The basement floor is 6′ below grade and the design TD for the structure is 70°F. The basement walls are constructed of 8 in. of brick with 2 × 6 framing and R-19 insulation, **Figure 42.19**.

The heat loss through the basement walls is divided between the above-grade and below-grade portions of the wall. The above-grade portion of the walls will have a U-value of 0.069 and the below-grade portion of the walls will have a U-value of 0.043, from **Figure 42.18**. The area of the above-grade walls can be found by:

$$\text{Above-Grade Wall Area} = \text{Height of Above-Grade Wall} \times \text{Perimeter of Above-Grade Wall}$$
$$\text{Above-Grade Wall Area} = 2' \times (40' + 60' + 40' + 60')$$
$$\text{Above-Grade Wall Area} = 2' \times 200'$$
$$\text{Above-Grade Wall Area} = 400\ \text{ft}^2$$

The heat loss through the above-grade portion of the wall is found by:

$$Q_{\text{Above-Grade Walls}} = \text{U-value} \times \text{TD} \times \text{Area}$$
$$Q_{\text{Above-Grade Walls}} = 0.069 \times 70°\text{F} \times 400\ \text{ft}^2$$
$$Q_{\text{Above-Grade Walls}} = 1932\ \text{Btu/h}$$

Figure 42.18 The U-values for some exterior walls. Notice that this chart shows values for both the above-grade and below-grade portions of the wall. Table derived from data contained in ASHRAE manuals and ACCA Manual J charts

WALL CONSTRUCTION	STUD SIZE (WOOD)	INSUALATION R-VALUE (TO FLOOR)	U-VALUE (BASED ON FLOOR DEPTH FOR BELOW GRADE WALLS) 2′	4′	6′	8′	10′	U-VALUE (ABOVE GRADE WALLS)
8 inches (Brick or Stone)	2 x 4	R-11	0.071	0.062	0.056	0.051	0.047	0.099
8 inches (Brick or Stone)	2 x 4	R-13	0.068	0.059	0.053	0.049	0.045	0.093
8 inches (Brick or Stone)	2 x 4	R-15	0.065	0.057	0.051	0.047	0.043	0.088
8 inches (Brick or Stone)	2 x 6	R-19	0.052	0.047	0.043	0.040	0.037	0.069
8 inches (Brick or Stone)	2 x 6	R-21	0.051	0.046	0.042	0.039	0.036	0.067

Figure 42.19 This 8-ft wall has a 2-ft portion above grade and a 6-ft portion below grade.

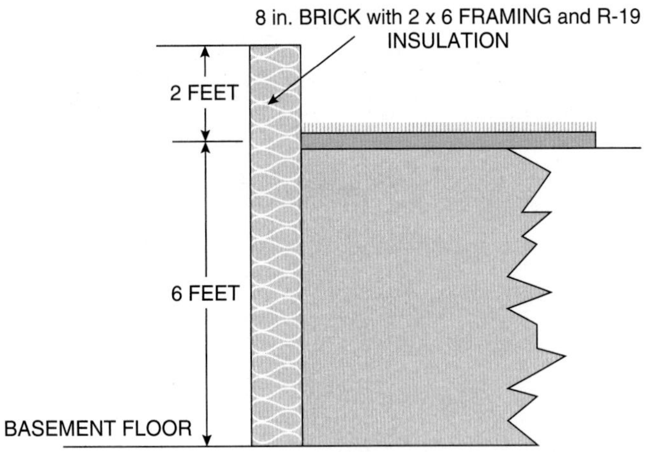

The heat loss through the below-grade portion of the wall is found by:

$$Q_{\text{Below-Grade Walls}} = \text{U-value} \times \text{TD} \times \text{Area}$$
$$Q_{\text{Below-Grade Walls}} = 0.043 \times 70°F \times (6 \times 200)\ ft^2$$
$$Q_{\text{Below-Grade Walls}} = 0.043 \times 70°F \times 1200\ ft^2$$
$$Q_{\text{Below-Grade Walls}} = 3612\ \text{Btu/h}$$

The total heat loss for the basement walls is the sum of the heat losses for the below-grade and above-grade walls as follows:

$$Q_{\text{Basement Walls}} = Q_{\text{Above-Grade Walls}} + Q_{\text{Below-Grade Walls}}$$
$$Q_{\text{Basement Walls}} = 1932\ \text{Btu/h} + 3612\ \text{Btu/h}$$
$$Q_{\text{Basement Walls}} = 5544\ \text{Btu/h}$$

HEAT LOSS THROUGH FLOORS LOCATED ABOVE UNCONDITIONED SPACES

Heat loss calculations performed on houses that have unconditioned basements or crawl spaces need to take into account the heat that is lost between the conditioned and unconditioned areas. Just as we addressed the concept of heat losses through partition walls a little earlier in this unit, the floors that separate conditioned from unconditioned areas act in much the same way as partition walls. The tables that are used to calculate the heat loss through a floor located over an unconditioned crawl space or basement are similar to the one shown in **Figure 42.20**.

Consider a basement that has a 40' × 60' footprint. The house has a design indoor temperature of 70°F and a design outside temperature of 10°F. The floor is insulated with R-19 blanket insulation. The calculated TD is 60°F (70°F − 10°F). Since the unconditioned basement is not as cold as the outside temperature, the TD that will be used for the calculation will be found in **Figure 42.20**. For the floor in our example, the TD that will be used is 29.1°F, which is the value that is found at the point where the R-19 floor insulation row intersects with the 60°F TD column. It can also be found from **Figure 42.20** that the U-value for the floor is 0.049. The floor loss calculation is as follows:

$$Q_{\text{Floor Over Unconditioned Basement}} = \text{U-value} \times \text{TD} \times \text{Area}$$
$$Q_{\text{Floor Over Unconditioned Basement}} = 0.049 \times 29.1°F \times 2400\ ft^2$$
$$Q_{\text{Floor Over Unconditioned Basement}} = 3422.16\ \text{Btu/h}$$

HEAT LOSS THROUGH SLAB FLOORS

Quite often, houses are built on slabs of concrete that are at ground level. These are referred to as slab-on-grade floors. A great deal of heat is lost from the edges of the slab and the heat loss is calculated using the running feet of exposed slab edge, not the area of the floor. The running feet of exposed edge is the perimeter of the slab. Insulation for the slab can be vertical or horizontal, or can be vertical insulation that turns under the slab, **Figure 42.21**. This type of construction element utilizes an F-value instead of a U-value. F-values are expressed in Btuh per running foot per degree Fahrenheit (Btuh/ft/°F) of the perimeter of a slab floor, as opposed to U-values, which are expressed in Btuh per square foot per degree Fahrenheit (Btuh/ft²/°F). An F-value table for various types of slab insulation would look similar to the table in **Figure 42.22**.

Figure 42.20 Some U-values and TD values for floors that are located above unconditioned crawl spaces or basements. Table derived from data contained in ASHRAE manuals and ACCA Manual J charts.

FLOOR INSULATION	FLOOR U-VALUE	FLOOR TEMPERATURE DIFFERENCE (FOR SPECIFIC HEATING TEMPERATURE DIFERENTIAL, TD)								
		30°F	35°F	40°F	45°F	50°F	55°F	60°F	65°F	70°F
R-2 Board	0.206	5.4	6.4	7.3	8.2	9.1	10.0	10.9	11.8	12.7
R-5 Board	0.125	8.0	9.4	10.7	12.0	13.4	14.7	16.1	17.4	18.7
R-11 Blanket	0.073	11.5	13.4	15.4	17.3	19.2	21.1	23.0	25.0	26.9
R-19 Blanket	0.049	14.6	17.0	19.4	21.8	24.3	26.7	29.1	31.5	34.0
R-30 Blanket	0.034	17.1	20.0	22.8	25.7	28.6	31.4	34.3	37.1	40.0

Figure 42.21 (A) Vertical insulation along the perimeter of a slab. (B) Horizontal insulation along the perimeter of a slab. (C) Vertical insulation along the perimeter of a slab that has a turn under.

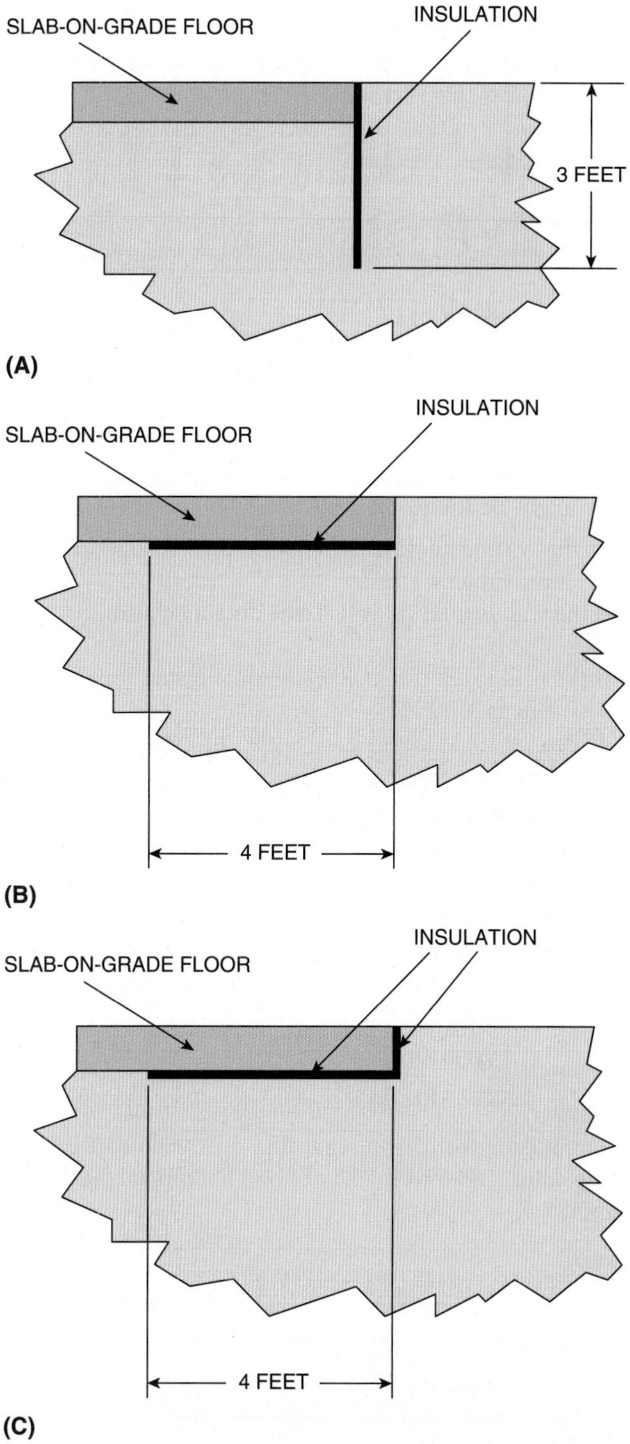

Consider a slab-on-grade floor that has a footprint of 40' × 70'. This slab is located in light, dry soil and has vertical, R-15 insulation. The TD that will be used is 60°F. The heat loss through this slab can be found by:

Figure 42.22 F-values for various types of slab insulation. Table derived from data contained in ASHRAE manuals and ACCA Manual J charts.

INSULATION CONFIGURATION	INSULATION R-VALUE	F-VALUE FOR HEAVY MOIST SOIL	F-VALUE FOR LIGHT DRY SOIL
No Insulation	R-0	1.358	0.989
Vertical	R-5	0.589	0.289
Vertical	R-15	0.432	0.178
Horizontal	R-5	1.266	0.98
Horizontal	R-15	1.194	0.967
Vertical with Turn Under	R-5	0.574	0.287
Vertical with Turn Under	R-15	0.401	0.174

$$Q_{Slab} = \text{F-value} \times \text{TD} \times \text{Running Feet of Exposed Edge}$$
$$Q_{Slab} = 0.178 \times \text{TD} \times \text{Running Feet of Exposed Edge}$$
$$Q_{Slab} = 0.178 \times 60°F \times (40' + 70' + 40' + 70')$$
$$Q_{Slab} = 0.178 \times 60°F \times 220'$$
$$Q_{Slab} = 2349.6 \text{ Btu/h}$$

HEAT LOSS THROUGH CRACKS, OPENINGS, AND DEFICIENCIES IN STRUCTURE

No structure is completely airtight, so losses due to infiltration must be accounted for. The sensible heat formula, from Unit 35, "Comfort and Psychrometrics," is used to calculate these losses:

$$Q_{Infiltration} = 1.08 \times \text{cfm} \times \text{TD}$$

In this formula, 1.08 is a constant and the TD is the design temperature differential that has been used throughout this unit. The cfm value is an estimate of how many cubic feet of air are entering the structure each minute. This can be estimated by determining how many air changes occur each hour, abbreviated as ACH, and the internal volume of the house. The number of air changes per hour is determined from a table similar to that shown in **Figure 42.23**.

To calculate the cfm, we can use the following formula:

$$\text{cfm (infiltration)} = (\text{ACH} \times \text{HOUSE VOLUME}) \div 60 \text{ min/h}$$

Figure 42.23 Table used to determine the number of air changes per hour for heat loss calculations. Table derived from data contained in ASHRAE manuals and ACCA Manual J charts.

AIR CHANGES PER HOUR (FOR HEATING)				
	FLOOR AREA OF CONDITIONED SPACE			
CONSTRUCTION	900–1500 ft²	1500–2000 ft²	2000–3000 ft²	3000 and up ft²
TIGHT	0.30	0.25	0.20	0.20
AVERAGE	0.75	0.60	0.50	0.45
LEAKY	1.70	1.45	1.15	1.00

Consider a house with the following characteristics:

- Average construction
- 2500 ft^2
- Single-story house with 8-ft ceilings
- Design TD is 60°F

From this information, we can determine that the ACH is 0.50 (from **Figure 42.23**), and that the volume of the house is 20,000 ft^3 (2500 ft^2 × 8 ft). We can then calculate the cfm of the infiltration by:

$$\text{cfm (infiltration)} = (\text{ACH} \times \text{HOUSE VOLUME}) \div 60 \text{ min/h}$$

$$\text{cfm (infiltration)} = (0.50 \times 20{,}000) \div 60$$

$$\text{cfm (infiltration)} = 167 \text{ cfm}$$

We can then determine the heat loss due to infiltration by:

$$Q_{\text{Infiltration}} = 1.08 \times \text{cfm} \times \text{TD}$$

$$Q_{\text{Infiltration}} = 1.08 \times 167 \times 60°\text{F}$$

$$Q_{\text{Infiltration}} = 10{,}821.6 \text{ Btu/h}$$

DUCT LOSSES TO ATTICS, CEILING CAVITIES, AND OTHER UNCONDITIONED SPACES

When heated air passes through ducts that are located in unconditioned spaces, heat is lost to the unconditioned space, **Figure 42.24.** The rate at which heat is lost depends on a number of factors, including the temperature of the air in the duct, the temperature of the air in the unconditioned space, and the amount of insulation on the duct. The duct losses are calculated as a percentage of the envelope load, but the percentage changes based on the conditions and location of the system. In general, duct losses are determined by the following:

$$Q_{\text{Duct Losses}} = \text{Envelope Load} \times \text{Loss Factor} \times \text{Duct Insulation Correction Value}$$

The envelope load is simply the sum of the individual heat losses of the structure. The heat loss factor is determined by a table similar to the one shown in **Figure 42.25**. The duct wall insulation correction value is determined by the duct wall insulation, as well as the type of duct system and its installation location. For example, the following correction values might be found for a particular duct system type:

If the wall insulation is R-2, the duct insulation correction value is 1.48.
If the wall insulation is R-4, the duct insulation correction value is 1.15.
If the wall insulation is R-6, the duct insulation correction value is 1.00.
If the wall insulation is R-8, the duct insulation correction value is 0.93.

Consider a 2000 ft^2 house that has a total envelope loss of 15,000 Btu/h. The design outdoor air temperature is 0°F and the duct wall insulation is rated at R-4. The design supply air temperature to the occupied space is 110°F. The duct losses can be calculated as:

$$Q_{\text{Duct Losses}} = \text{Envelope Load} \times \text{Loss Factor} \times \text{Duct Insulation Correction Value}$$

$$Q_{\text{Duct Losses}} = 15{,}000 \text{ Btu/h} \times 0.25 \times 1.15$$

$$Q_{\text{Duct Losses}} = 4312.5 \text{ Btu/h}$$

For complete information regarding duct wall insulation correction values, refer to the Duct Load Factors tables in ACCA Manual J.

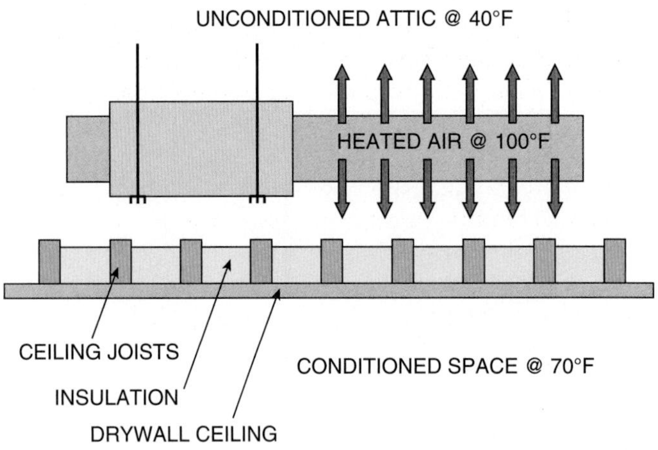

Figure 42.24 Heated air passing through ducts that are located in unconditioned spaces gives up heat to the unconditioned space.

Figure 42.25 Tables used to determine heat loss factors for duct loss calculations. Table derived from data contained in ASHRAE manuals and ACCA Manual J charts.

HEAT LOSS FACTOR									
DISCHARGE AIR TEMPEATURE = 100°F					DISCHARGE AIR TEMPEATURE = 110°F				
DESIGN OUTSIDE AIR TEMP	FLOOR AREA OF CONDITIONED SPACE				DESIGN OUTSIDE AIR TEMP	FLOOR AREA OF CONDITIONED SPACE			
	1000 ft²	1500 ft²	2000 ft²	2500 ft²		1000 ft²	1500 ft²	2000 ft²	2500 ft²
0°F	0.21	0.24	0.29	0.34	0°F	0.18	0.21	0.25	0.26
10°F	0.18	0.22	0.26	0.29	10°F	0.16	0.18	0.22	0.24
20°F	0.16	0.19	0.22	0.25	20°F	0.14	0.17	0.18	0.21
30°F	0.14	0.17	0.20	0.21	30°F	0.12	0.15	0.17	0.19

42.7 ELEMENTS OF STRUCTURAL HEAT GAIN (COOLING MODE)

When calculating the heat gain of a structure, it is necessary to take into account the entire envelope and all the contact the envelope has with the outside. In addition, the occupants of the structure as well as any appliances, lighting, plants, fish tanks, and other heat/moisture-generating elements must be factored into the calculations. Since the sun creates a varying heat load on the structure as it moves across the sky based on the direction the house is facing, the orientation of the house is also a major consideration. Heat gains in the conditioned space can occur in a number of ways that include:

- Through exposed (exterior) walls
- Through doors
- Through roofs, walls, floors, and ceilings
- Through windows
- From people, appliances, lighting, and other elements
- Through cracks, openings, and deficiencies in structure
- Through duct runs through attics, ceiling cavities, and other unconditioned spaces

In summary, the heat gain on a structure is the sum of the conduction loads, the internal loads, the solar loads, and the infiltration loads. It is important to note that there is no heat gain through below-grade walls, slabs, and basement floors, so these construction elements do not need to be accounted for.

There are different methods of calculating the heat gain on a structure based on the type of house and the number of wall exposures it has. What is presented in the following sections is a general guideline to the methods and calculations used to determine the heat gain of a structure and is not intended to be a comprehensive, step-by-step procedure for obtaining an accurate heat gain calculation.

HEAT GAIN THROUGH EXPOSED (EXTERIOR) WALLS

Just as with the heat loss calculations, the heat gain through exterior walls is determined by the following formula:

$$Q = \text{U-value} \times \text{TD} \times \text{Area}$$

Since we are now dealing with a heat gain, we will add a subscript to the TD to clarify that this is the TD used to calculate the heat gain for cooling purposes:

$$Q = \text{U-value} \times \text{TD}_{\text{Cool}} \times \text{Area}$$

The TD_{Cool} is *not* necessarily the design TD that was found by determining the difference between the design outside temperature and the design indoor temperature of 75°F (Section 42.3). To determine the TD_{Cool} that will be used, a two-part chart is often needed, **Figure 42.26.**

Figure 42.26 Tables used to obtain the temperature differentials that are used to calculate the heat gains through walls and panels. The U-value is obtained from the first chart along with the group number. The group number is used to obtain the TD value to be used for the calculations. Table derived from data contained in ASHRAE manuals and ACCA Manual J charts.

WOOD STUD SIZE	CAVITY INSULATION	BOARD INSULATION	U-VALUE	GROUP
2 x 4	R-11	R-3	0.079	D
2 x 4	R-11	R-5	0.068	E
2 x 4	R-13	R-3	0.075	E
2 x 4	R-13	R-5	0.064	F
2 x 4	R-15	R-3	0.071	F
2 x 4	R-15	R-5	0.062	H
2 x 6	R-19	R-3	0.059	G
2 x 6	R-19	R-5	0.052	H
2 x 6	R-21	R-3	0.058	H
2 x 6	R-21	R-5	0.050	I

		DESIGN COOLING TD				
GROUP	CONSTRUCTION	10°F	15°F	20°F	25°F	30°F
D	WALL	14.5	19.4	24.5	29.5	29.5
D	PARTITION	5.4	10.4	15.4	20.4	20.4
E	WALL	12.4	17.4	22.4	27.4	27.4
E	PARTITION	4.0	9.0	14.0	19.0	19.0
F	WALL	10.3	15.3	20.3	25.3	25.3
F	PARTITION	2.9	7.9	12.9	17.9	17.9
G	WALL	8.3	13.3	18.3	23.3	23.3
G	PARTITION	1.6	6.6	11.6	16.6	16.6
H	WALL	7.7	12.7	17.7	22.7	22.7
H	PARTITION	1.3	6.3	11.3	16.3	16.3
I	WALL	7.0	12.0	17.0	22.0	22.0
I	PARTITION	0.9	5.9	10.9	15.9	15.9

Figure 42.27 Table to obtain the U-values and temperature differentials that are used to calculate the heat gains through doors. Table derived from data contained in ASHRAE manuals and ACCA Manual J charts.

		DESIGN COOLING TD				
CONSTRUCTION	U-VALUE	10°F	15°F	20°F	25°F	30°F
Hollow Core Wood	0.47	21.0	26.0	31.0	36.0	36.0
Hollow Core Wood with Storm	0.31	21.0	26.0	31.0	36.0	36.0
Solid Core Wood	0.39	21.0	26.0	31.0	36.0	36.0
Solid Core Wood with Storm	0.27	21.0	26.0	31.0	36.0	36.0
Fiberglass Core Metal	0.60	21.0	26.0	31.0	36.0	36.0
Fiberglass Core Metal with Storm	0.36	21.0	26.0	31.0	36.0	36.0

Consider a 10' × 50' wall that has the following characteristics:

Wall: 2 × 6 studs, R-19 cavity insulation, R-5 board insulation

TD across panel: Design TD of 20°F

From this information, we can determine that this particular wall has a U-value of 0.052 and is in Group H. We can then look up the TD_{Cool} to be used by finding the intersection point of the Group H wall and the 20°F TD to get a TD_{Cool} of 17.7°F.

The heat gain attributed to this wall is then found as follows:

$$Q_{Wall} = \text{U-value} \times TD_{Cool} \times \text{Area}$$
$$Q_{Wall} = 0.052 \times 17.7°F \times (10 \times 50)\ ft^2$$
$$Q_{Wall} = 0.052 \times 17.7°F \times 500\ ft^2$$
$$Q_{Wall} = 460.2\ Btu/h$$

HEAT GAIN THROUGH DOORS

The same idea is applied to doors. Although there are no group numbers associated with doors, the TD_{Cool} concept is used here. The table that is used to find door-related information is similar to **Figure 42.27**.

Let's take a look a 3' × 7' solid-core wood door without a storm that is exposed to a design cooling TD of 20°F. This door has a U-value of 0.39 and a TD_{Cool} of 31°F. We can find the heat gain through this door as follows:

$$Q_{Door} = \text{U-value} \times TD_{Cool} \times \text{Area}$$
$$Q_{Door} = 0.39 \times 31°F \times (3 \times 7)\ ft^2$$
$$Q_{Door} = 0.39 \times 31°F \times 21\ ft^2$$
$$Q_{Door} = 253.89\ Btu/h$$

HEAT GAIN THROUGH ROOFS, WALLS, FLOORS, AND CEILINGS

The same method that was used to calculate the heat gain through a door is applied to roof, walls, floors, and ceilings. There are no group numbers associated with these construction elements, but the TD_{Cool} concept is used here as well. The table that is used to find ceiling information is similar to **Figure 42.28**, which shows some information for insulated ceilings under an attic. Probably one of the most important things to keep in mind when performing a heat gain calculation on a structure is to use the proper charts and tables. Many charts and tables look similar to each other but have significant differences that can dramatically alter the results. It is also important to consider the fact that roofs and ceilings may not have the same area as the footprint of the structure if they are not parallel to the floor. As in the earlier heat loss example of a sloped roof, some geometry is needed to determine the actual area of the panel.

Let's take a look at a 25' × 40' ceiling that is located under an attic. The ceiling has R-19 insulation and a design cooling TD of 15°F. This ceiling has a U-value of 0.049 and a TD_{Cool} of 40°F. We can find the heat gain through this ceiling as follows:

$$Q_{Ceiling} = \text{U-value} \times TD_{Cool} \times \text{Area}$$
$$Q_{Ceiling} = 0.049 \times 40°F \times (25 \times 40)\ ft^2$$
$$Q_{Ceiling} = 0.049 \times 40°F \times 1000\ ft^2$$
$$Q_{Ceiling} = 1960\ Btu/h$$

Figure 42.28 Table to obtain the U-values and temperature differentials that are used to calculate the heat gains through ceilings. Table derived from data contained in ASHRAE manuals and ACCA Manual J charts.

		DESIGN COOLING TD				
CEILING INSULATION	U-VALUE	10°F	15°F	20°F	25°F	30°F
R-7	0.112	35.0	40.0	45.0	50.0	50.0
R-11	0.081	35.0	40.0	45.0	50.0	50.0
R-15	0.061	35.0	40.0	45.0	50.0	50.0
R-19	0.049	35.0	40.0	45.0	50.0	50.0
R-21	0.044	35.0	40.0	45.0	50.0	50.0
R-30	0.032	35.0	40.0	45.0	50.0	50.0

HEAT GAIN THROUGH WINDOWS

One of the major sources for heat gain on a structure is the windows. Not only is there heat transfer through the glass, there is also the possibility for a major solar load, depending on the direction the window faces. The amount of shading, from structural overhangs, on the window affects the solar load, as does the type and position of the internal shades, curtains, and blinds. In this text, we will be assuming that there is no structural shade on the windows and that all internal blinds, curtains, and shades are opened halfway. Since most new systems are designed long before people occupy the house, it is difficult to determine what type of window coverings will ultimately be used.

As was mentioned in the ceiling and roof section, many of the charts and tables that are used to determine the heat gain through windows look similar to each other but have significant differences that can dramatically alter the results. It is also important to make certain that the correct orientation (north, south, east, west, northwest, northeast, southeast, or southwest) for each window is used.

The tables used to calculate the heat gain of a window have a slightly different configuration, as they provide a multiplier that gives the heat gain per square foot for the window. All that is needed once the multiplier is determined is the area of the window. A table for window heat gain might look something like the table in **Figure 42.29**. This table provides information for single- and double-pane windows that have the drapes/curtains half open.

Consider a single-pane, 4' × 6' window that is exposed to a design cooling TD of 15°F. This window faces north. The shades on the window are half open. The heat gain through this window is determined by the following:

$$Q_{Window} = MULTIPLIER_{Window} \times Area$$
$$Q_{Window} = 22\ Btu/h/ft^2 \times Area$$
$$Q_{Window} = 22\ Btu/h/ft^2 \times 24\ ft^2$$
$$Q_{Window} = 528\ Btu/h\ \text{(north-facing window)}$$

If we repeat the same calculation for the exact same window that is facing east, we will get a very different result:

$$Q_{Window} = MULTIPLIER_{Window} \times Area$$
$$Q_{Window} = 63\ Btu/h/ft^2 \times Area$$
$$Q_{Window} = 63\ Btu/h/ft^2 \times 24\ ft^2$$
$$Q_{Window} = 1512\ Btu/h\ \text{(east-facing window)}$$

HEAT GAIN FROM PEOPLE, APPLIANCES, LIGHTING, AND OTHER ELEMENTS

The occupants of a house add to the heat gain of the structure. Since the human body acts as a radiator and gives off heat to the space, it contributes to the sensible heat load. In addition, the body produces a latent load since moisture evaporates from the surface of the body. The number of occupants in

Figure 42.29 Table to obtain the multiplier (based on design cooling TD) that is used to determine the heat gain through a window. Table derived from data contained in ASHRAE manuals and ACCA Manual J charts.

WINDOW TYPE/COVERING	ORIENTATION	DESIGN COOLING TD 10°F	15°F	20°F	25°F	30°F
Single pane with drapes half open	North	17	22	27	32	37
Single pane with drapes half open	NE or NW	40	45	50	55	60
Single pane with drapes half open	East or West	58	63	68	73	78
Single pane with drapes half open	SE or SW	50	54	59	64	69
Single pane with drapes half open	South	28	32	37	42	47
Double pane with drapes half open	North	12	14	17	20	23
Double pane with drapes half open	NE or NW	32	34	37	40	43
Double pane with drapes half open	East or West	47	50	53	55	58
Double pane with drapes half open	SE or SW	40	43	45	48	51
Double pane with drapes half open	South	21	24	26	29	32

a particular house is estimated as the number of bedrooms plus one. So, a three-bedroom house would count four people when determining the heat gain. The estimated heat gain for each full-time occupant of the house is about 430 Btu/h, which breaks down to 230 Btu/h of sensible heat and 200 Btu/h of latent heat. So, a two-bedroom house would have three occupants and a heat gain component of 1290 Btu/h attributed to the occupants.

As far as appliance loads go, those performing a heat gain calculation resort to using default values based on the appliances present in the house. Typically, the following guidelines are used:

- 1200 Btu/h: House with a refrigerator and a vented cooking surface
- 2400 Btu/h: House with a refrigerator, vented cooking surface, dishwasher, clothes washer, vented clothes dryer, and miscellaneous electronic devices
- 3600 Btu/h: House with two refrigerators, vented cooking surface, dishwasher, clothes washer, vented clothes dryer, and miscellaneous electronic devices

HEAT GAIN THROUGH CRACKS, OPENINGS, AND DEFICIENCIES IN STRUCTURE

The infiltration loads are found in a manner similar to the method used when calculating the heat loss in a structure. The heat gain due to infiltration has both a sensible portion and a latent portion. The formulas for each are shown here:

$$Q_{\text{Infiltration Sensible}} = 1.08 \times \text{cfm} \times \text{TD}$$
$$Q_{\text{Infiltration Latent}} = 0.68 \times \text{cfm} \times \text{DG}$$

where

cfm is the amount of infiltration in cubic feet per minute
TD is the design cooling temperature differential
DG is the grains of moisture figure at design conditions and design relative humidity (refer to **Figure 42.7**)

Consider a house with the following characteristics:

- Single-story, average construction
- 10-ft ceilings
- 40' × 60' footprint (2400 ft^2)
- Design cooling TD: 20°F
- Design grains of moisture @ 50% relative humidity: 30 grains

The cfm is found by the following:

$$\text{cfm (infiltration)} = (\text{ACH} \times \text{HOUSE VOLUME}) \div 60 \text{ min/h}$$

Figure 42.30 Table used to determine the number of air changes per hour for heat gain calculations. Table derived from data contained in ASHRAE manuals and ACCA Manual J charts.

AIR CHANGES PER HOUR (FOR COOLING)				
	FLOOR AREA OF CONDITIONED SPACE			
CONSTRUCTION	900–1500 ft^2	1500–2000 ft^2	2000–2500 ft^2	Over 2500 ft^2
TIGHT	0.15	0.10	0.10	0.10
AVERAGE	0.35	0.30	0.25	0.20
LEAKY	0.90	0.75	0.60	0.50

where the ACH, air changes per hour, for cooling is found from a table similar to that shown in **Figure 42.30**. In this case, the ACH will be 0.25.

Plugging the values into the cfm formula we get:

$$\text{cfm (infiltration)} = (\text{ACH} \times \text{HOUSE VOLUME}) \div 60 \text{ min/h}$$
$$\text{cfm (infiltration)} = (0.25 \times 24{,}000) \div 60$$
$$\text{cfm (infiltration)} = 100 \text{ cfm}$$

We can then determine the sensible heat gain due to infiltration by:

$$Q_{\text{Infiltration Sensible}} = 1.08 \times \text{cfm} \times \text{TD}$$
$$Q_{\text{Infiltration Sensible}} = 1.08 \times 100 \times 20°\text{F}$$
$$Q_{\text{Infiltration Sensible}} = 2160 \text{ Btu/h}$$

We can then determine the latent heat gain due to infiltration by:

$$Q_{\text{Infiltration Latent}} = 0.68 \times \text{cfm} \times \text{DG}$$
$$Q_{\text{Infiltration Latent}} = 0.68 \times 100 \times 30 \text{ grains}$$
$$Q_{\text{Infiltration Latent}} = 2040 \text{ Btu/h}$$

HEAT GAIN THROUGH DUCT RUNS THROUGH ATTICS, CEILING CAVITIES, AND OTHER UNCONDITIONED SPACES

The portion of the load on the system that is attributed to the duct system is expressed as a fraction of the load on the structure. The duct load, just as with the infiltration load, has both sensible and latent components. The sensible heat portion of the duct load is found by using the following:

$$Q_{\text{Sensible Duct Load}} = \text{Envelope Load} \times \text{Duct Load Factor} \times \text{Duct Insulation Correction}$$

The envelope load is the sum of all of the sensible heat gains in the structure. The duct load factor is the sensible heat factor and is found in a table or chart similar to the one shown in **Figure 42.31**. The duct insulation correction

Figure 42.31 Table used to calculate sensible and latent heat gains attributed to duct systems. Table derived from data contained in ASHRAE manuals and ACCA Manual J charts.

HEAT GAIN FACTORS									
SENSIBLE HEAT GAIN FACTOR					LATENT HEAT GAIN (Btuh)				
DESIGN OUTSIDE AIR TEMP	FLOOR AREA OF CONDITIONED SPACE				DESIGN GRAINS OF MOISTURE	FLOOR AREA OF CONDITIONED SPACE			
	1000 ft²	1500 ft²	2000 ft²	2500 ft²		1000 ft²	1500 ft²	2000 ft²	2500 ft²
85°F	0.23	0.28	0.34	0.35	+/− 10	316	521	725	929
90°F	0.25	0.28	0.36	0.42	+/− 20	495	817	1,136	1,456
95°F	0.27	0.33	0.40	0.42	+/− 30	682	1,126	1,566	2,007
100°F	0.28	0.36	0.43	0.48	+/− 40	878	1,449	2,015	2,582

factor is often provided in table form and often contains the following data:

- If the duct wall insulation is R-2, the duct insulation correction value is 1.71.
- If the duct wall insulation is R-4, the duct insulation correction value is 1.15.
- If the duct wall insulation is R-6, the duct insulation correction value is 1.00.
- If the duct wall insulation is R-8, the duct insulation correction value is 0.91.

Consider a 2500-ft² house that has a total envelope heat gain of 24,000 Btu/h. The design outdoor air temperature is 90°F and the duct wall insulation is rated at R-4. The design grains for the structure is 20 grains. The sensible duct heat gain can be calculated as:

$$Q_{\text{Sensible Duct Load}} = \text{Envelope Load} \times \text{Duct Load Factor} \times \text{Duct Insulation Correction}$$

$$Q_{\text{Sensible Duct Load}} = 24{,}000 \text{ Btu/h} \times 0.42 \times 1.15$$

$$Q_{\text{Sensible Duct Load}} = 11{,}592 \text{ Btu/h}$$

The latent duct heat gain can be determined directly from the right-hand portion of the table in **Figure 42.31**. For our 2500-ft² house that has a design grains value of 20, it can be seen that the latent duct load is 1456 Btu/h.

REVIEW QUESTIONS

1. An oversized air-conditioning system will
   A. cost more to install and operate.
   B. be more efficient than an undersized air-conditioning system.
   C. be better able to maintain the desired humidity in the conditioned space.
   D. require a smaller air distribution system than an undersized air-conditioning system.
2. Methods used to calculate the heat gain and heat loss of a structure should
   A. provide the same results for all houses with the same size footprint.
   B. provide only a very rough estimate of the needs of the structure.
   C. be nationally recognized and accepted.
   D. rely solely on rules of thumb to quickly obtain results.
3. An exterior wall
   A. separates a conditioned interior space from an unconditioned interior space.
   B. separates an interior conditioned space from the outside.
   C. separates one interior conditioned space from another conditioned interior space.
   D. separates an interior unconditioned space from the outside.
4. A partition wall
   A. separates a conditioned interior space from an unconditioned interior space.
   B. separates an interior conditioned space from the outside.
   C. separates one interior conditioned space from another conditioned interior space.
   D. separates an interior unconditioned space from the outside.

5. As the temperature differential across a panel increases,
   A. the rate of heat transfer through the panel will increase.
   B. the rate of heat transfer through the panel will decrease.
   C. the relative humidity in the occupied space will increase.
   D. the relative humidity in the occupied space will decrease.
6. The typical indoor design temperature used for heat gain calculations is
   A. 65°F.
   B. 70°F.
   C. 75°F.
   D. 80°F.
7. The typical indoor design temperature used for heat loss calculations is
   A. 65°F.
   B. 70°F.
   C. 75°F.
   D. 80°F.
8. If a heating system was being designed for use in Modesto, California, what temperature difference would be used for calculations?
   A. 30°F
   B. 40°F
   C. 50°F
   D. 60°F
9. If an air-conditioning system was being designed for use in Nogales, Arizona, what temperature difference would be used for calculations?
   A. 21°F
   B. 68°F
   C. 75°F
   D. 96°F
10. If the R-value of a construction material is 2.0, what is its U-value?
    A. 0.25
    B. 0.50
    C. 1.0
    D. 2.0
11. If a wall is made up of four different materials, all layered one on top of the other, how will the total R-value of the wall be determined?
    A. By the one material with the highest R-value
    B. By the one material with the lowest R-value
    C. By averaging the R-values of the materials
    D. By adding the R-values of the materials together
12. All of the following are taken into account on a heat loss calculation except
    A. exterior walls.
    B. interior walls.
    C. basement floors.
    D. top-floor ceilings.
13. All of the following are taken into account on a heat gain calculation except
    A. exterior walls.
    B. partition walls.
    C. basement floors.
    D. top-floor ceilings.
14. How much heat is lost through a 3' × 5' single-pane window with a storm that is exposed to a 60°F temperature differential?
    A. 450 Btu/h
    B. 900 Btu/h
    C. 1350 Btu/h
    D. 1800 Btu/h
15. Which of the following is(are) true regarding a 10' × 30' wall that has a 21-ft² door and a 20-ft² window cut into it?
    A. The gross wall area is 300 ft².
    B. The gross wall area is 259 ft².
    C. The net wall area is 300 ft².
    D. The net wall area is 341 ft².
16. Which of the following is(are) true regarding the solar gain on windows?
    A. The solar gain due to north-facing windows is greater than the solar gain due to south-facing windows.
    B. The solar gain due to east-facing windows is greater than the solar gain due to west-facing windows.
    C. The solar gain due to west-facing windows is greater than the solar gain due to east-facing windows.
    D. The solar gain due to south-facing windows is greater than the solar gain due to north-facing windows.
17. Which of the following statements best describes the reason why heat gain and heat loss calculations should be performed?
    A. To make certain that the cooling system will be able to cool the structure, even on the hottest day of the year
    B. To make certain that the heating system will be able to heat the structure, even on the coldest day of the year
    C. To properly size heating and cooling equipment based on design temperatures for the city in which the equipment will be installed.
    D. To properly match the square footage of the structure to the heating and cooling equipment's nameplate capacity
18. Which of the following statements is most likely true regarding two insulation materials, one with an R-value of 20 and the other with an R-value of 10?
    A. The R-10 material will prevent heat from passing through it better than the R-20 material.
    B. The R-20 material will prevent heat from passing through it better than the R-10 material.
    C. The R-10 material will likely be thicker than the R-20 material.
    D. The R-20 material will likely be thinner than the R-10 material.
19. Which of the following statements is true regarding the relationship between a material's R-value and its U-value?
    A. R-value × U-value = 1
    B. R-value ÷ U-value = 1
    C. R-value + U-value = 1
    D. R-value − U-value = 1

UNIT 43

AIR SOURCE HEAT PUMPS

OBJECTIVES

After studying this unit, you should be able to

- describe how an air-to-air heat pump is similar to a vapor-compression air-conditioning system.
- describe the purpose of a four-way reversing valve.
- describe the operation of a four-way reversing valve.
- list four common heat pump system configurations.
- describe the indoor and outdoor coil configurations on an air-to-air heat pump system.
- identify the state of refrigerant in the field-installed refrigerant lines for each mode of system operation.
- describe four common metering devices used on heat pump systems.
- describe the type of filter drier commonly used on heat pump systems.
- explain how an air-to-air heat pump system is able to absorb heat from the outside air in the heating mode.
- explain the term "balance point" as it applies to heat pump systems.
- explain how the term "coefficient of performance" (COP) relates to the efficiency of heat pump systems.
- discuss how the location of the indoor and outdoor coils of an air-to-air heat pump can affect system operation.
- describe the configuration of a package heat pump system.
- discuss the sequence of operations of an air-to-air heat pump system.
- explain the concept of emergency heat.
- explain the terms SEER and HSPF as they apply to heat pump systems.
- discuss why defrost is needed on heat pump systems.
- explain how the defrost cycle on a heat pump system is initiated and terminated.
- describe the term "second-stage heat" and how it applies to heat pump systems.
- explain the importance of proper heat pump system maintenance and service.
- explain how to properly evaluate the electrical circuits in a heat pump system.
- explain how to properly evaluate the mechanical portion of a heat pump system.
- explain how to properly evaluate the four-way reversing valve.
- explain how to properly evaluate the compressor in a heat pump system.
- explain when gauges should be installed on a heat pump system.
- explain how a fossil fuel heating system can be used as a second-stage heat source.
- explain why scroll compressors are a popular choice for use in heat pump systems.
- explain how variable speed motors help heat pump systems match the heating and cooling needs of the structure.
- discuss proper preventive maintenance procedures for heat pump systems.
- explain how to properly diagnose a heat pump system.

SAFETY CHECKLIST

✔ When installing insulated ductwork or otherwise working around fiberglass, wear gloves, goggles, and clothing that will cover your skin. If fiberglass particles are in the air, wear a mask that covers your nose and mouth. To avoid skin irritation, do not wear short-sleeve shirts or short pants when working with fiberglass insulation.

✔ Be sure to protect your hands when working with metal duct, panels, and fasteners.

✔ Exercise all precautions when installing units on rooftops or in other hazardous locations.

✔ Observe all safety precautions when troubleshooting the electrical components of a heat pump. When units are located in crawl spaces and you are in contact with the ground, be careful not to establish a path for current to flow through your body to ground.

✔ Troubleshoot with the electrical power on only when it is absolutely necessary. At all other times, ensure that the power is off. When the power is off, lock and tag the disconnect panel. There should be only one key. Keep it with you so that another person will not be able to turn the power on while you are working on the equipment.

✔ Exercise caution if conditions are wet when you are troubleshooting an outside unit.

✔ When installing gauges or transferring refrigerant, wear gloves, goggles, and clothing to protect your skin, because refrigerant can freeze your skin and eye tissue.

✔ The hot gas line can cause serious burns and should be avoided.

✔ If water is used to melt the ice from the coil of the outdoor unit, turn the power off and protect all electrical components. If troubleshooting later with the power on, do not stand in this water.

✔ Use common sense and caution at all times while installing, maintaining, or troubleshooting any piece of equipment. As a technician you are constantly exposed to *potential danger*. Refrigerant, electrical shock hazard, rotating equipment, hot metal, sharp metal, and lifting heavy objects are among the most common hazards you will face.

43.1 REVERSE-CYCLE REFRIGERATION

Heat pumps, like cooling-only air-conditioning systems, are refrigeration machines. Refrigeration involves the removal of heat from a place where it is not wanted and depositing it in a place where it makes little or no difference. The heat can, under certain conditions, be deposited as heat reclaim in a place where it is wanted. The main difference between a heat pump and a cooling-only air-conditioning system is that the air conditioner can pump heat only one way, but the heat pump is a refrigeration system that can pump heat two ways. Since the heat pump has the ability to pump heat into as well as out of a structure, this very special type of system can provide both heating and cooling.

All compression-cycle refrigeration systems are heat pumps in that they pump heat-laden vapor. The evaporator of a heat pump absorbs heat into the refrigeration system, and the condenser rejects the heat. The compressor pumps the heat-laden vapor. The metering device controls the refrigerant flow. These four components—the evaporator, the condenser, the compressor, and the metering device—are the basic building blocks for the vapor-compression refrigeration system.

The four-way reversing valve in a heat pump system is used to switch the unit between the heating and cooling modes of operation. By changing the mode of operation of the system, the functions of the indoor and outdoor coils change as well. In the cooling mode, the indoor coil acts as the evaporator and the outdoor coil functions as the condenser. In the heating mode, the indoor coil functions as the condenser and the outdoor coil operates as the evaporator.

43.2 HEAT SOURCES FOR WINTER

The cooling system in a typical residence absorbs heat into the refrigeration system through the evaporator and rejects this heat to the outside of the house through the condenser. The house might be 75°F inside while the outside temperature might be 95°F or higher. The air conditioner, a high-temperature refrigeration system, pumps heat from a low temperature inside the house to a higher temperature outside the house; that is, it pumps heat up the temperature scale. A low-temperature refrigeration system, such as a freezer in a supermarket, takes the heat out of ice cream at 0°F to cool it to −10°F so that it will be frozen hard. The heat removed from the freezer can be felt at the condenser as hot air. This example shows that there is usable heat in a substance, even at 0°F, **Figure 43.1**. There is heat in any substance until it is cooled down to −460°F, which is absolute zero.

If heat can be removed from 0°F ice cream, heat can also be removed from 0°F outside air. The typical heat pump does just that. It removes heat from the outside air in the winter and deposits the heat in the conditioned space to heat the house, **Figure 43.2**. Since air contains heat as long as it is above absolute zero, about 85% of usable heat will still be in air at 0°F. Since air is both the heat source and the heat sink, we refer to this type of heat pump as an *air-to-air* heat pump. In summer, the heat pump acts like a conventional air conditioner, removing heat from the house and depositing it outside. From the outside, the outdoor unit of an air-to-air heat pump looks like a central cooling air conditioner. The phrase "air-to-air" indicates the source of heat while the system is operating in the heating mode and the medium that is ultimately being treated. The first "air" represents the heat source, and the second "air" represents the medium being heated. For example, an "air-to-water" heat pump uses "air" to heat "water." Water-source heat pumps use water as the heat source in the heating mode and can be "water-to-water" or "water-to-air." The water-to-air heat pump, therefore, uses water as the heat source to heat air. The water source can be, for example, lakes or wells. Water-source heat pumps are discussed in Unit 44, "Geothermal Heat Pumps."

Figure 43.1 This low-temperature refrigerated box is removing heat from ice cream at 0°F. Part of the heat coming out of the back of the box is coming from the ice cream.

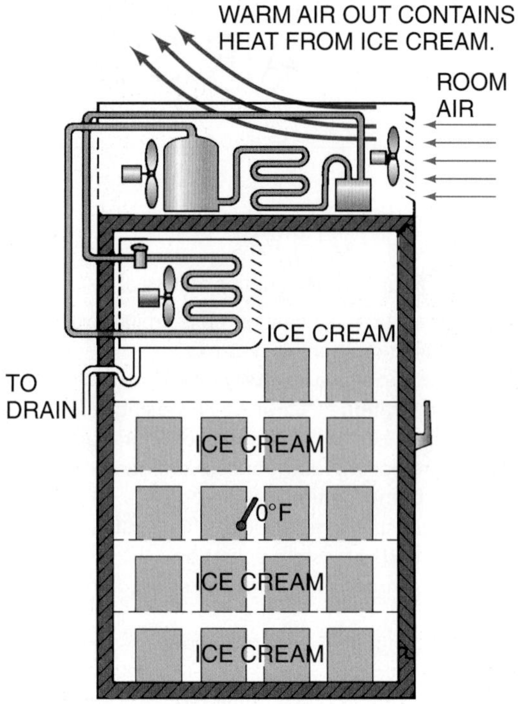

43.3 THE FOUR-WAY REVERSING VALVE

The refrigeration principles that a heat pump uses are the same as those covered earlier in the text, when traditional air-conditioning and refrigeration systems were discussed.

Figure 43.2 In the winter, the heat pump moves heat into the structure.

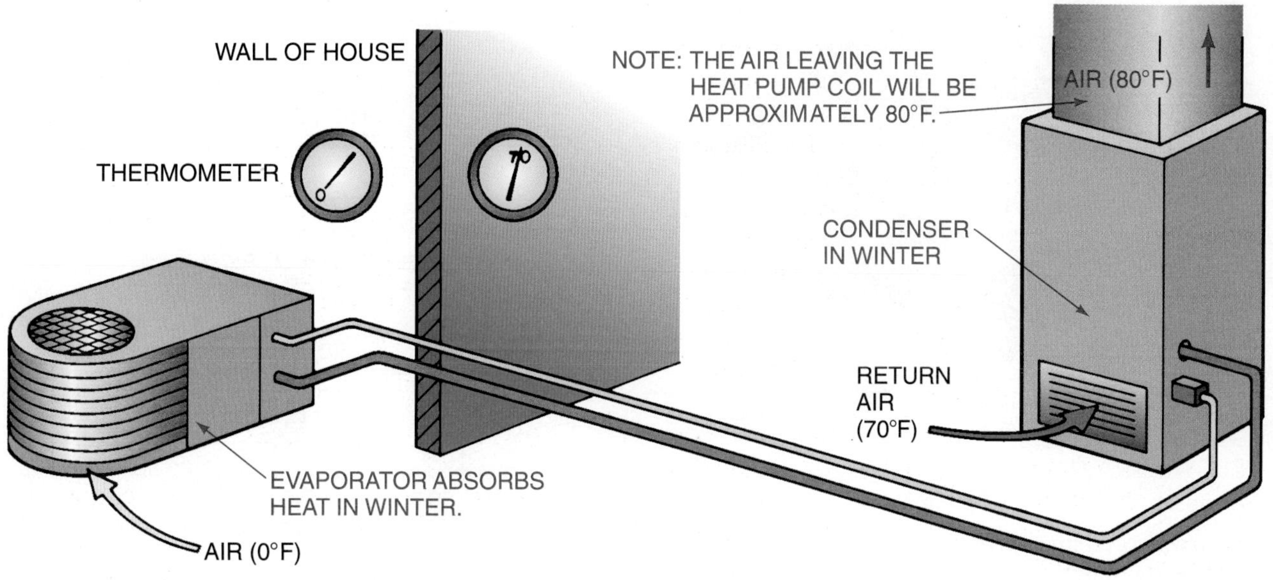

However, a new component is added to the system that allows the heat pump system to move heat either into or out of a structure by changing the direction of refrigerant flow. The air-to-air heat pump shown in **Figure 43.3** is operating in the cooling mode and is moving heat from inside the conditioned space to the outside. When operating in the heating mode, heat energy is transferred from the outside of the structure to the inside. This change of direction is accomplished with a special component called a *four-way reversing valve,* **Figure 43.4.** This valve can best be described in the following way. The system refrigerant, which contains the heat absorbed into the refrigeration system, is pumped through the system by the compressor. The discharge gas from the compressor contains the heat that the system absorbs, as well as heat that is generated during the compression process. The four-way valve diverts the heat-laden discharge gas in the proper direction to either heat or cool the conditioned space. In the heating mode, the discharge gas is directed from the compressor to the indoor coil. In the cooling mode, the discharge gas is directed from the compressor to the outdoor coil. The reversing valve's position is controlled by the space temperature thermostat, which determines the system's mode of operation.

Figure 43.3 An air-to-air heat pump moving heat from the inside of a structure to the outside.

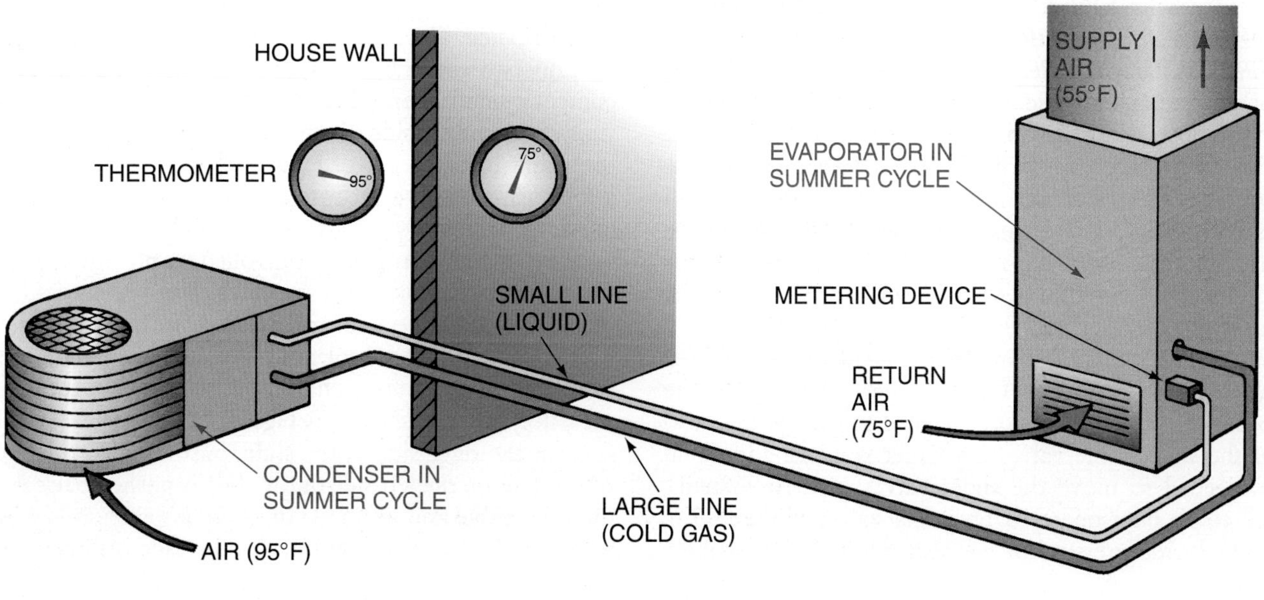

Figure 43.4 A four-way reversing valve. Photo by Eugene Silberstein

Figure 43.5 Internal slide in the four-way reversing valve. The solenoid determines the position of the slide. Most systems operate in the heating mode when the solenoid is deenergized. (A) Position of the slide when the solenoid is deenergized. (B) Position of the slide when the solenoid is energized.

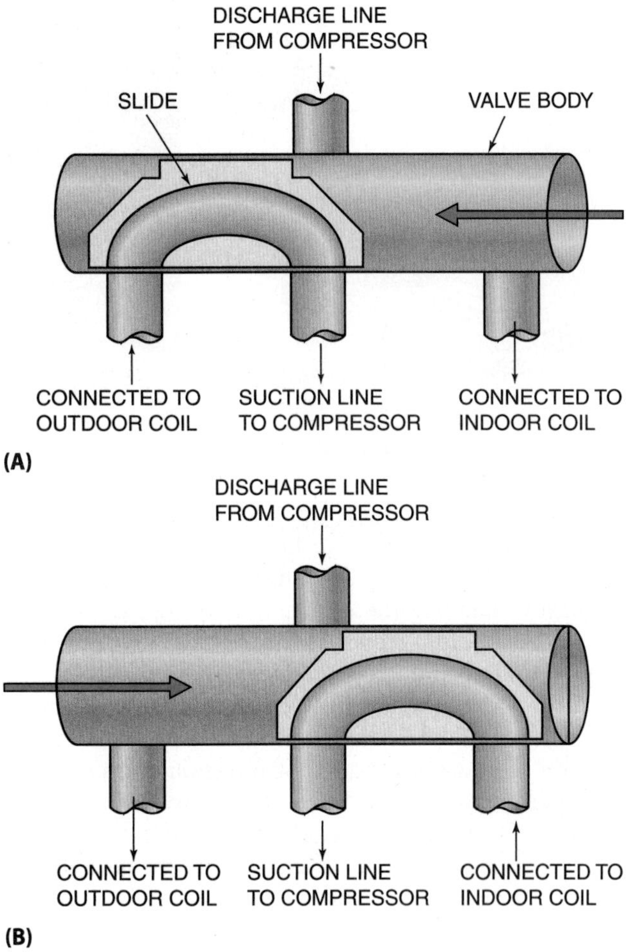

The four-way reversing valve is a four-port valve that has a slide mechanism in it, **Figure 43.5**. The position of the slide is determined by a solenoid that is energized in one mode of operation and not the other. The position of the slide also determines the mode of system operation. Most heat pump systems are designed to operate in the heating mode when the solenoid is not energized, **Figure 43.5(A)**. This is referred to as *failing in the heating mode*. For the remainder of this text it will be assumed that the heat pump system will operate this way. This may not always be the situation, so be sure to check each individual system being worked on. **Figure 43.5(B)** shows the position of the slide when the solenoid is energized and the system is operating in the cooling mode.

Typical reversing valves have one port isolated on one side and the remaining three ports on the other. The isolated port is where the hot gas from the compressor enters the valve. **Figure 43.5** shows the piping connections on the valve. Depending on the position of the slide, the hot gas will be directed to either the outdoor coil or the indoor coil. In either case, the coil that accepts the hot gas from the compressor is functioning as the condenser coil. Note that in **Figure 43.5(A)** the hot gas enters the valve at the top and is directed to the right-hand port, which is connected to the indoor coil. Since the indoor coil is receiving the hot gas from the compressor, the indoor coil is the condenser and the system is operating in the heating mode. Compare this with **Figure 43.5(B)**, where the hot gas from the compressor is being directed to the outdoor coil. Under these conditions, the system is operating in the cooling mode.

Smaller heat pump systems utilize smaller, direct-acting, four-way reversing valves. "Direct acting" implies that the solenoid itself provides the force needed to change the position of the slide in the valve. On larger systems, additional help is needed to move the slide. Valves in these systems, referred to as pilot-operated reversing valves, **Figure 43.6**, use the high-side pressure in the heat pump system to help push the slide. The pilot-operated four-way reversing valve is actually two valves built into one housing. The control wires control a small solenoid valve that has very small piping that runs along the outside of the valve, as can be seen in the bottom right-hand corner of **Figure 43.6**. This valve is called a *pilot-operated* valve because a small valve controls the flow in a large valve.

The valve movement is controlled by pressure differences within the large portion of the valve. This pressure difference pushes the slide mechanism back and forth inside the body of the valve. When the solenoid is deenergized, high pressure from the compressor discharge pushes through the tubes on the pilot valve as shown in **Figure 43.7**. This exerts a pressure on the right side of the slide, pushing it to the left. The refrigerant on the left side of the slide is pushed back through the pilot valve and into the compressor's suction line. In this position, the hot gas from the compressor is directed to the

Figure 43.6 A pilot-operated four-way reversing valve. Notice the pilot valve in the bottom right-hand corner of the figure.

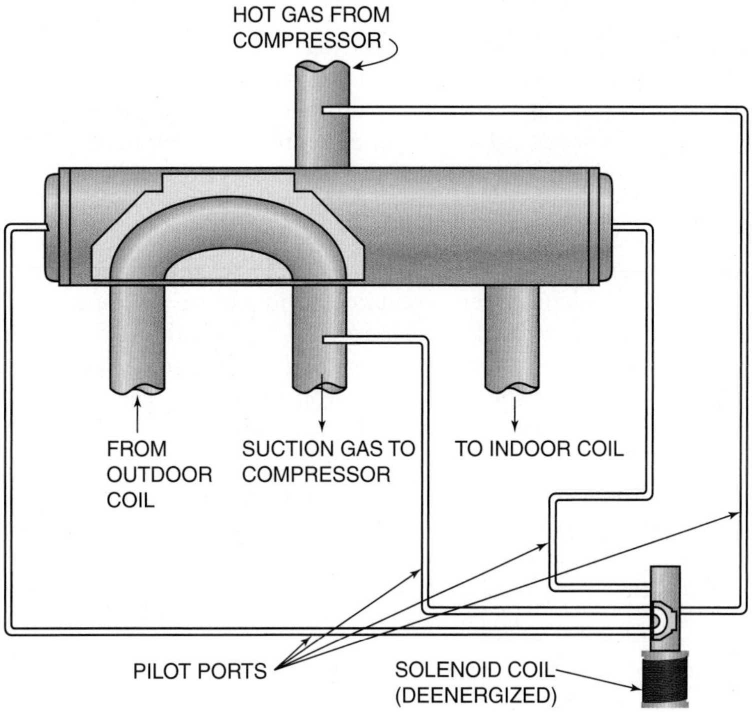

Figure 43.7 In the heating mode, the high-pressure hot gas from the compressor is directed to push the main valve's slide to the left. This directs the hot gas from the compressor to the indoor coil to provide space heating. Notice that the solenoid is deenergized.

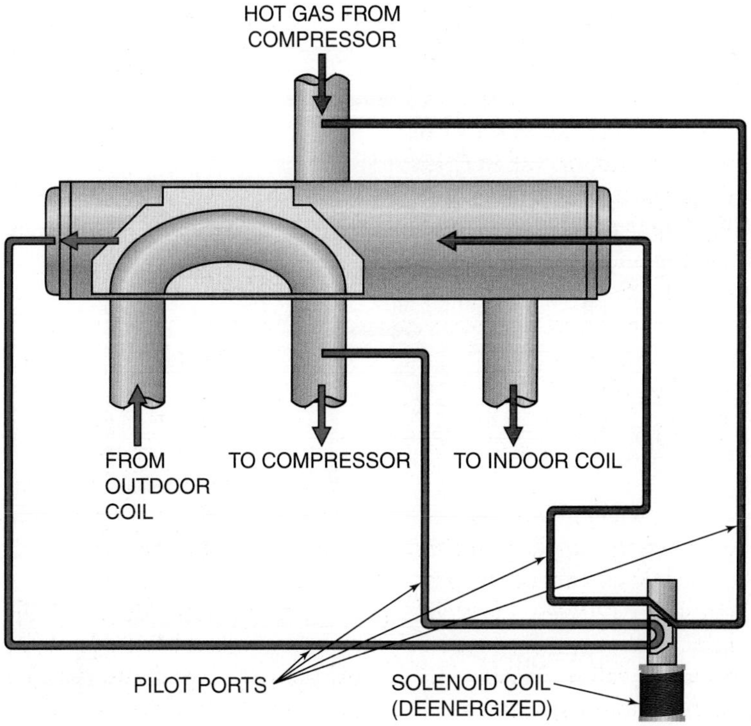

indoor coil, putting the system into the heating mode. When the solenoid valve is energized, the high pressure is directed to the other end of the valve body and the slide mechanism moves to the right, putting the system in the cooling mode, **Figure 43.8**. This is why, when starting up a heat pump, you sometimes hear the swishing of refrigerant as the valve changes position while the pressure difference builds up in the system.

In case of failure, the four-way reversing valve is somewhat difficult to change due to the three connections located on one side. These are the largest piping connections in the system and are often very close to each other. A tubing cutter cannot be used, as the pipes are too close together. A hacksaw should never be used to cut the valve out of the system because doing so will create filings that can travel through the system, potentially causing problems with the compressor, metering device, and the movement of the slide mechanism inside the four-way reversing valve. In addition, unbrazing the valve is a difficult process as three ports would need to be heated at the same time. Tool and equipment manufacturers make torch tips with multiple flame heads that are intended to make the replacing of a four-way reversing valve much easier, **Figure 43.9**.

Also remember that there is a sliding piston inside the valve body. It slides on a neoprene, teflon, or nylon pad that has a low melting temperature. The valve body must be kept as cool as possible when installing the new valve. The valve body can be kept cool enough by a heat sink paste or by a wet rag wrapped around the valve body while it is being brazed in, **Figure 43.10**. If the wet rag option is used, make certain that no water is allowed to enter the piping, as this will pose problems later on. Avoid pointing the torch directly at the valve to avoid overheating and possibly causing damage to the reversing valve.

When removing a defective four-way reversing valve, it is often better to cut the piping back far enough with a tubing cutter and just braze additional pipe stubs into the system, **Figure 43.11**. You can get a clean installation this way and the brazing of the stubs to the valve can be done in a vise, where you can control the heat much better. If the old valve is removed by using one of the specialty torch tips and without cutting the refrigerant lines back, the old stubs must be cleaned prior to the installation of the new valve. All of the old filler material must be removed or the pipe stubs will not fit into the new valve, **Figure 43.12**. This filler material is hard to remove; it must be filed off in a manner that will prevent the filings from entering the piping. It is often is difficult to clean up these stubs correctly. More information on checking the four-way reversing valve will be provided later in this unit.

Figure 43.8 In the cooling mode, the high-pressure hot gas from the compressor is directed to push the main valve's slide to the right. This directs the hot gas from the compressor to the outdoor coil. Notice that the solenoid is energized.

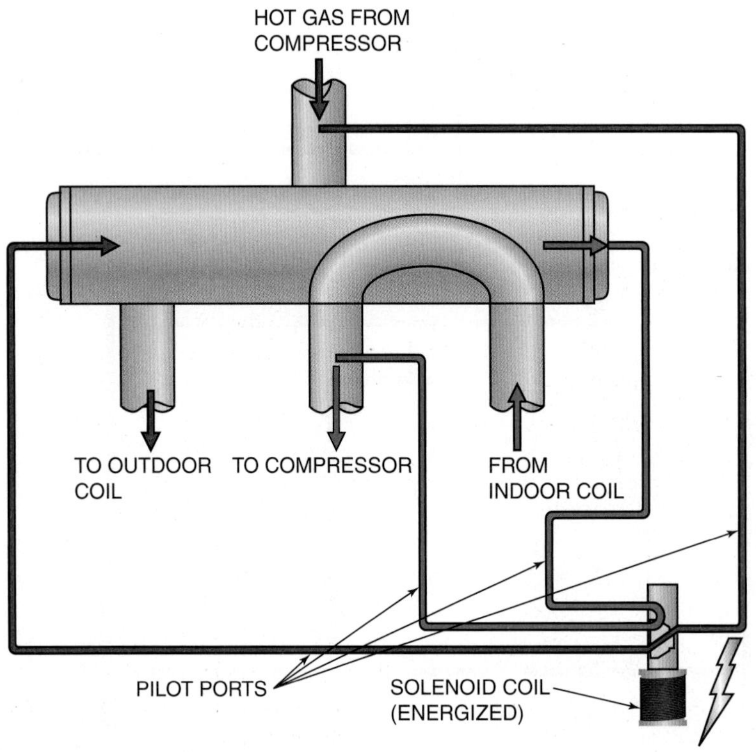

Figure 43.9 (A) Specialized torch tip designed for removing reversing valves. (B) Using the torch tip. Photos courtesy Uniweld Products, Inc.

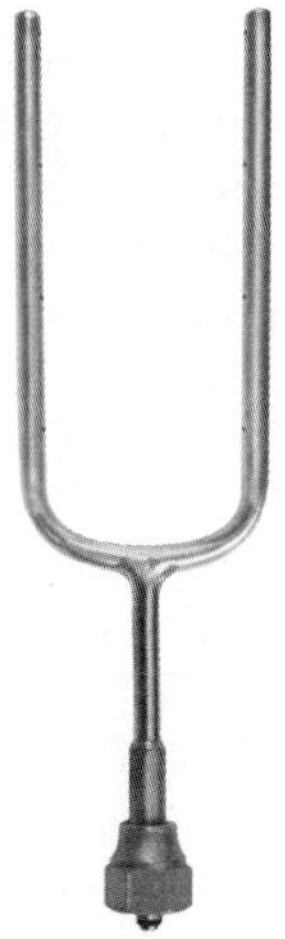

(A)

(B)

Figure 43.10 This valve body has a wet rag wrapped around it and water is dripped on the rag to keep the valve body cool. Photo by Bill Johnson

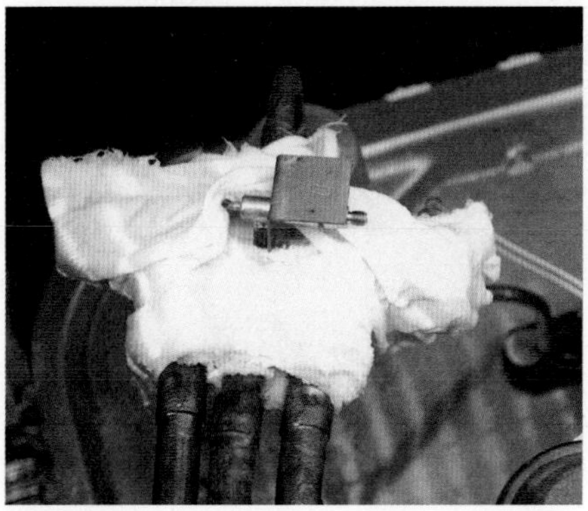

Figure 43.11 The valve and stubs are held in a vise while the additional stubs are brazed in. Remember, the valve body must be kept cool during this process. The valve and stubs can be installed in the unit by cutting the stubs to fit.

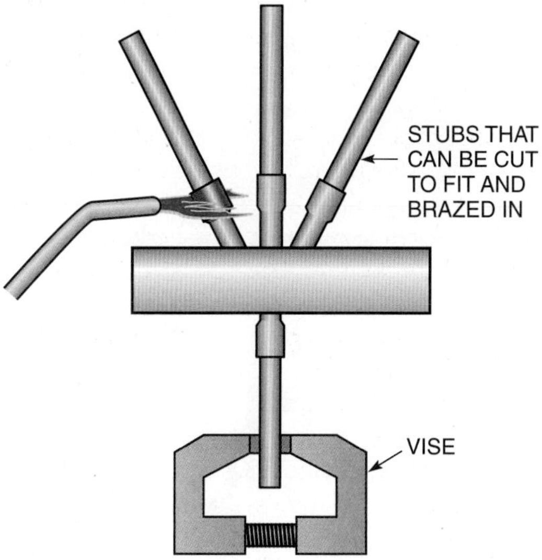

Figure 43.12 This stub has been pulled from a connection on a valve and has residue of silver brazing stuck to it. This braze material is very hard to clean off, which must be done before the stub is inserted into a new fitting.

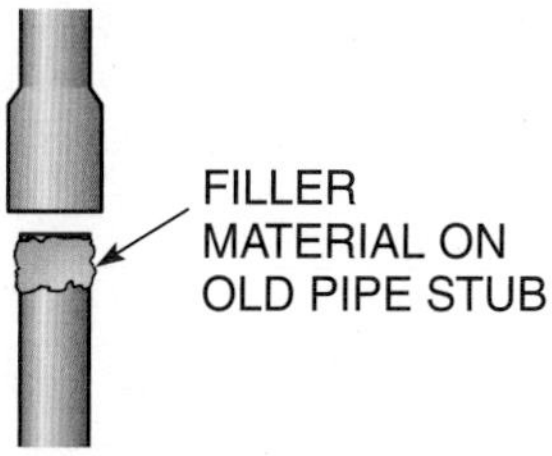

43.4 THE AIR-TO-AIR HEAT PUMP

The air-to-air heat pump resembles the central air-conditioning system. Both have indoor and outdoor system components. When discussing typical air-conditioning systems, the heat exchangers are often called the evaporator (indoor unit) and the condenser (outdoor unit), but this terminology does not work for a heat pump system. The coil that serves the inside of the house in a heat pump system is called the *indoor coil*. The coil that is located outside the house is called the *outdoor coil*. The reason for the change in terminology is that the indoor coil of the heat pump is the condenser in the heating mode and the evaporator in the cooling mode. The outdoor coil functions as the condenser in the cooling mode and the evaporator in the heating mode. The mode is determined by the direction in which the refrigerant flows. See **Figure 43.13** for an example of the direction of the gas flow.

Figure 43.13 The heat pump refrigeration cycle shows the direction of the refrigerant gas flow. Courtesy Carrier Corp.

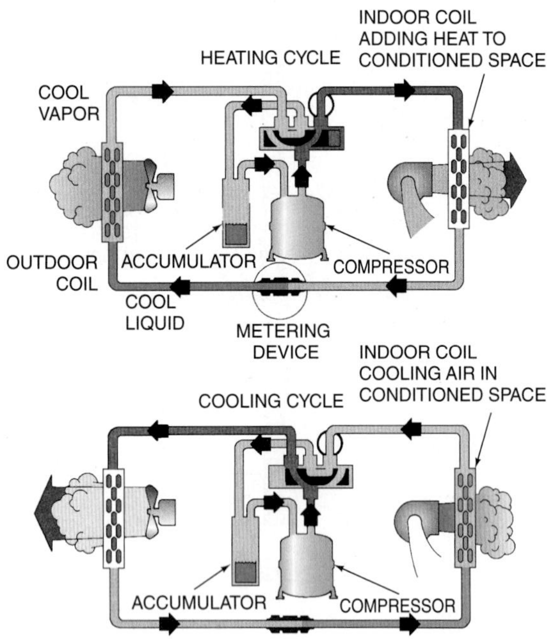

43.5 REFRIGERANT LINE IDENTIFICATION

Like cooling equipment, heat pumps are manufactured as split systems and package systems. On cooling-only, split-type systems, the field-installed refrigerant lines are typically the liquid line and the suction line. In a split-type air-to-air heat pump, the interconnecting refrigerant lines that are connected between the indoor unit and the outdoor unit are identified differently. The larger line of the two is called a *gas line* because it always carries gas. In previous units it was called a cold gas line or a suction line. In a heat pump it is always a gas line because it is the suction line when the heat pump is operating in the cooling mode and the discharge line when the system is operating in the heating mode, **Figure 43.14**. The smaller line carries liquid refrigerant in both the heating and cooling modes and is, therefore, called the liquid line, just as with cooling-only air-conditioning systems. The liquid flows toward the inside unit in summer and toward the outside unit in winter, **Figure 43.15**. The line is the same size as for cooling; the liquid direction is just reversed.

Figure 43.14 This split-system heat pump shows the interconnecting refrigerant lines.

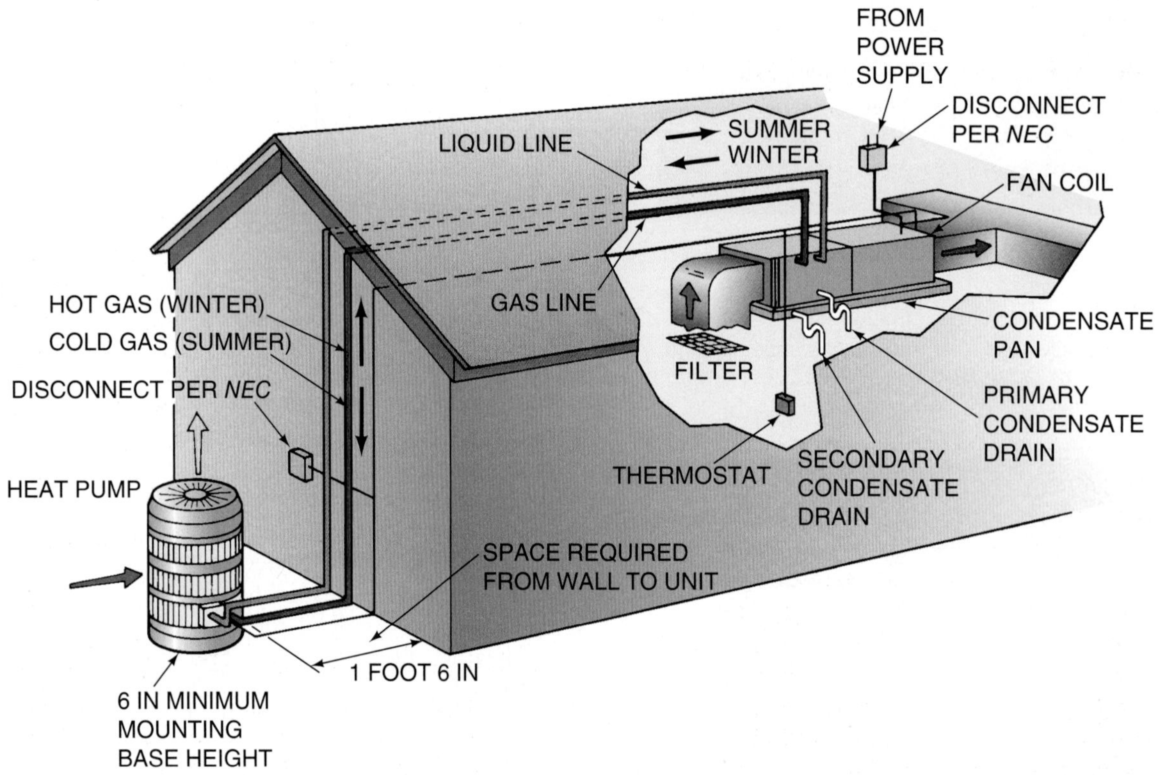

Figure 43.15 (A) Table identifying the refrigerant lines for summer and winter operation as well as the direction of refrigerant flow. (B) The small line is the liquid line for both summer and winter operation.

SUMMER OPERATION		
REFRIGERANT LINE	REFRIGERANT STATE	DIRECTION OF FLOW
Larger refrigerant line	Cold gas	Indoors → Outdoors
Smaller refrigerant line	Warm liquid	Outdoors → Indoors

WINTER OPERATION		
REFRIGERANT LINE	REFRIGERANT STATE	DIRECTION OF FLOW
Larger refrigerant line	Hot gas	Outdoors → Indoors
Smaller refrigerant line	Warm liquid	Indoors → Outdoors

(A)

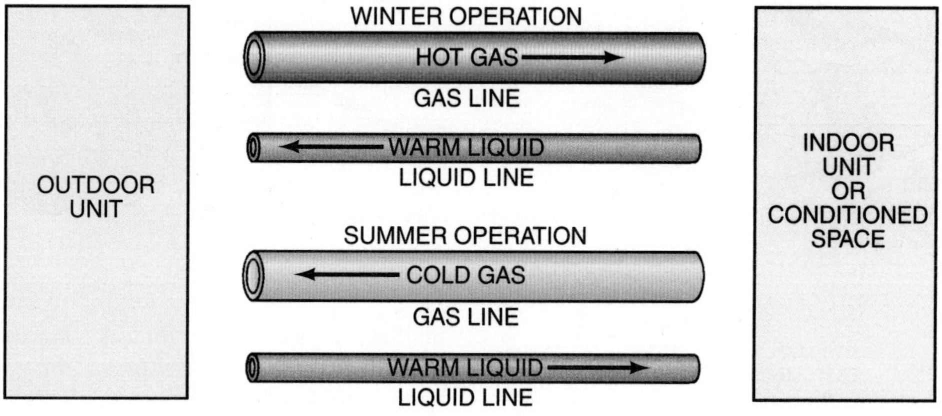

(B)

On ductless split systems, the metering device is typically located within the outdoor unit. When such is the case, when a system is operating in the cooling mode, the smaller line will carry low-pressure, low-temperature saturated refrigerant and the larger line will carry low-pressure, low-temperature superheated vapor. In the heating mode, the larger line will contain high-pressure, high-temperature superheated vapor and the smaller line will carry high-pressure, high-temperature subcooled liquid.

43.6 METERING DEVICES

Because the direction of the liquid flow is reversed as the heat pump system changes over from one mode to the other, some of the refrigeration components in heat pumps are slightly different from those in cooling-only air-conditioning systems. The metering devices are different because there must be a metering device at the indoor unit as well as at the outdoor unit. For example, when the unit is in the cooling mode, the metering device is at the indoor coil. When the system changes over to heating, a metering device must then meter refrigerant to the outdoor coil. The methods used to switch from one metering device to the other will be discussed later in this unit.

THERMOSTATIC EXPANSION VALVES

The *thermostatic expansion valve* (TXV) is the metering device that is commonly found on heat pump systems. TXVs are designed to maintain constant superheat in the evaporator. Because many TXVs allow refrigerant to flow in only one direction, a check valve needs to be piped in parallel with the valve to allow the refrigerant to bypass the valve when needed. For example, in the cooling mode, the valve needs to meter the flow of liquid refrigerant into the indoor coil. When the system reverses to the heating mode, the indoor coil becomes the condenser, and the liquid needs to be able to move freely toward the outdoor unit. The liquid flows through the check valve in the heating mode and is metered through the indoor TXV in the cooling cycle, **Figure 43.16**. **NOTE:** *The TXV used in a heat pump must be carefully chosen. The sensing bulb for the valve is on the gas line, which may become too hot for a typical TXV bulb.* • It is not unusual for the hot gas line to reach temperatures of 200°F. If the sensing bulb for a typical TXV is exposed to these temperatures, it is subject to rupture, **Figure 43.17**. A vapor-charged TXV, one with a limited charge in the bulb, is therefore usually used in a heat pump application. Remember from Unit 24,

Figure 43.16 (A) In the heating mode, hot gas from the compressor flows through the indoor coil and through the open check valve to bypass the indoor metering device. (B) In the cooling mode, high-pressure liquid refrigerant is directed through the metering device because the check valve will be in the closed position.

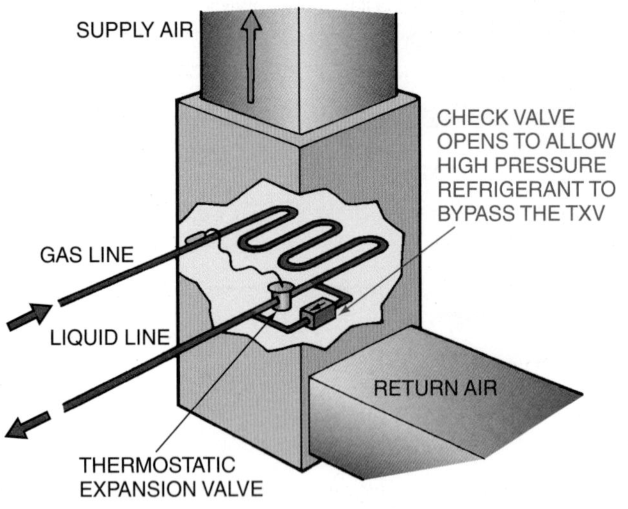

(A)

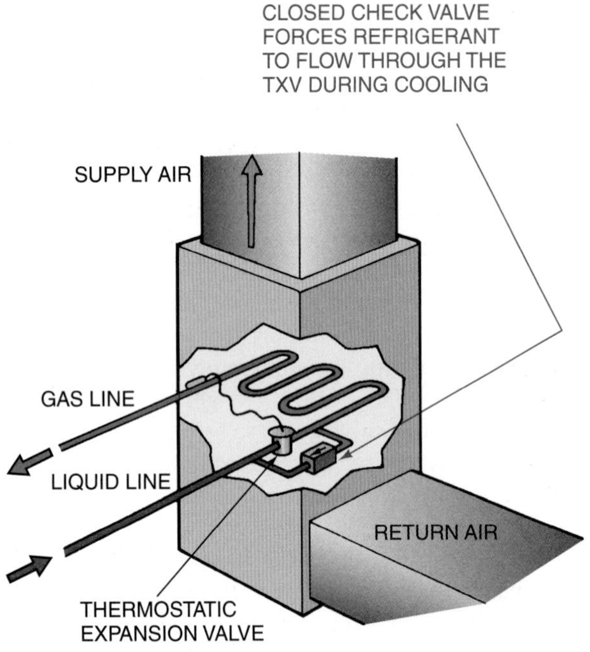

(B)

"Expansion Devices," that a limited-charge bulb does not continue to build pressure with added heat. It builds only a certain amount of pressure; then additional heat causes no appreciable increase in pressure.

The use of thermostatic expansion valves, along with filter driers and check valves, adds to the complexity of the refrigerant piping of a heat pump system. These additional components, piping, and pipe connections increase the possibility of system malfunction and refrigerant leaks. For these reasons, manufacturers offer *bidirectional thermostatic expansion valves,* which have a check valve built into the body of the TXV. The valve looks and operates in a manner similar to that of other TXVs, with the exception of the internal check valve. In operation, when refrigerant flow is desired through the expansion valve, the internal check valve will be in the closed position and the refrigerant will be directed through the needle and seat of the expansion valve. When refrigerant flows through the valve in the opposite direction, the check valve opens and the refrigerant is able to bypass the needle and seat in the valve. Be sure to check the part number of the TXV you are working with to determine whether or not it has an internal check valve.

Figure 43.17 This illustration shows that care must be taken in the choice of a TXV. If an ordinary TXV sensing bulb were to be mounted on this gas line, it would get hot enough to rupture the valve diaphragm.

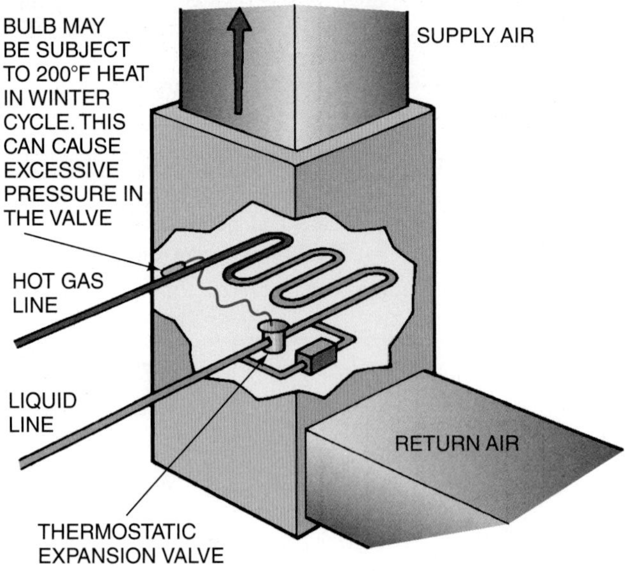

THE CAPILLARY TUBE

The capillary tube metering device is not typically used on new-generation heat pump systems, but will often be encountered on older pieces of equipment. They were a popular choice for use in equipment because they are inexpensive and have no moving parts. The capillary tube allows refrigerant flow in either direction, but this characteristic is not often used because capillary tubes of different sizes are required for the cooling and heating modes. Because of this, two capillary tubes are typically used. When used in this manner, check valves are used to reverse flow, allowing refrigerant to bypass the capillary tube that is not being used. In the heating mode of operation, the capillary tube at the indoor coil is bypassed. In the cooling mode, the capillary tube at the outdoor coil is bypassed, **Figure 43.18**.

Figure 43.18 Capillary tube metering devices used on a heat pump. Notice that there are two check valves and two different capillary tubes.

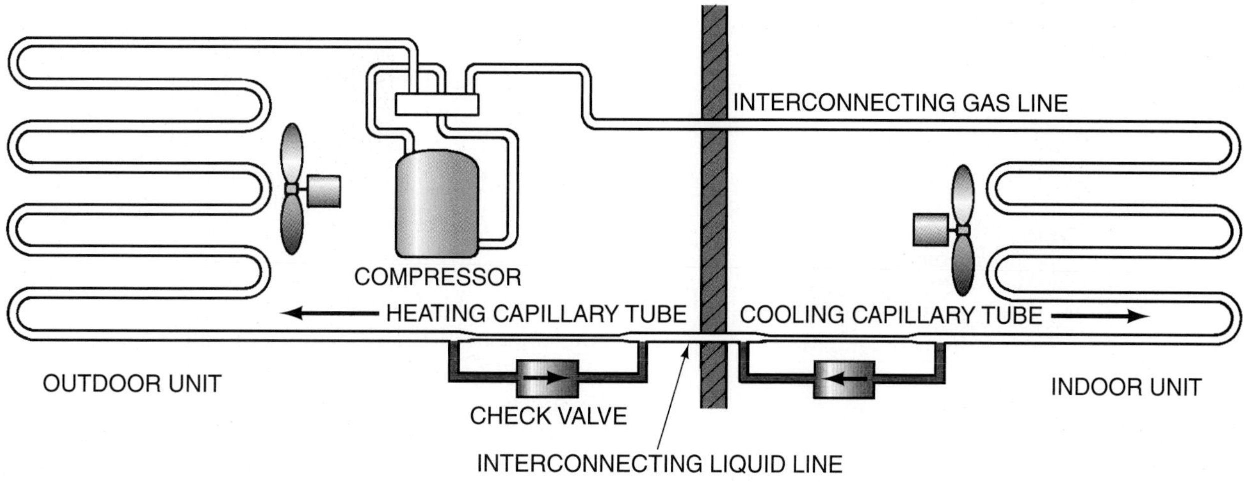

ELECTRONIC EXPANSION VALVES

The electronic expansion valve (EEV), discussed earlier in the text, is commonly found on newer heat pump systems, especially on ductless split-type heat pump systems. If the heat pump is a close-coupled unit with the indoor coil close to the outdoor unit, such as a package system (discussed in Unit 24, "Expansion Devices"), a single electronic expansion valve can be used. The valve will modulate refrigerant flow in either direction and maintain the correct superheat at the compressor's common suction line in the heating and cooling modes. The sensing element may be located on the common suction line just before the compressor to maintain refrigerant flow. On ductless split systems, the EEV is typically located in the system's outdoor unit.

ORIFICE METERING DEVICES

Another common metering device used by some manufacturers is a combination flow device and check valve. This device, referred to as a piston, allows full flow in one direction and restricted flow in the other direction, **Figure 43.19.** Two of these devices are necessary with a split-type heat pump system—one at the indoor coil and one at the outdoor coil. The metering device at the indoor coil has a larger bore than the one at the outdoor coil because the two coils have different flow characteristics. The summer cycle uses more refrigerant in normal operation. When this metering device is present, a biflow filter drier is normally used in the liquid line when field repairs are done.

COMBINATIONS OF METERING DEVICES

Sometimes two types of metering device are used in a single system. One popular combination is a capillary tube as the indoor metering device for summer operation, with a check valve piped parallel to allow flow in the other direction for winter operation. The outside unit may have a TXV with a check valve. This system is efficient because it uses the capillary tube only in the cooling mode. It is an efficient metering device for summer because the load conditions are nearly constant. The TXV used on the outdoor coil is efficient for the heating mode because winter conditions are not constant. The TXV will reach maximum efficiency sooner than the capillary tube in the same application because the TXV has the capability to open wide when needed. The capillary tube, being a fixed-bore-type metering device, cannot open or close. This allows the evaporator to become full of refrigerant earlier in the running cycle. This dual system uses each device at its best, **Figure 43.20.**

43.7 LIQUID-LINE ACCESSORIES

Liquid-line filter driers must be used in conjunction with all of the metering devices just described. Liquid-line filter driers help prevent particulate matter that may be in the refrigerant piping circuit from clogging or blocking the metering device. When standard liquid-line filter driers are installed in a system with check valves to control the flow through the metering devices, each drier is installed in series with one check valve and one expansion device. The flow direction is the same as with the metering device. If the same filter drier were installed in the common liquid line, it would filter in one direction and the particles would wash out in the other direction, **Figure 43.21.**

In the heating mode, the high-pressure refrigerant leaves the indoor coil and bypasses the cooling capillary tube by flowing through the indoor check valve and filter drier,

Figure 43.19 (A) This piston works like the ones illustrated in (B) and (C). (B) When the piston is positioned to the left, it only allows flow through the orifice, creating a metering device. (C) When the flow is reversed during the next season, the piston slides to the right and full flow is allowed. A heat pump may have two of these: one at the indoor unit for cooling and one at the outdoor coil for heating. These are very small devices and may be in a flare fitting. (A) Courtesy Carrier Corp.

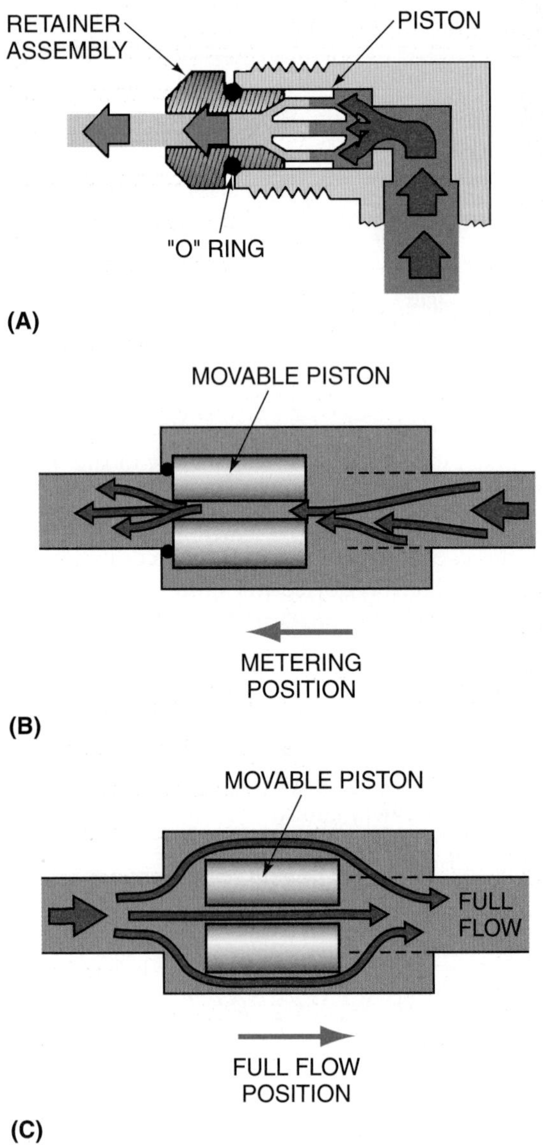

Figure 43.22. Once at the outdoor unit, the refrigerant is directed through the outdoor metering device, as the outdoor check valve is in the closed position. In the cooling mode, the outdoor heating capillary tube is bypassed and the refrigerant is directed through the cooling metering device located at the indoor coil, **Figure 43.23**. Special *biflow* filter driers are manufactured to allow flow in either direction. They are actually two driers in one shell with check valves inside the drier shell to cause the liquid to flow in the proper direction at the proper time, **Figure 43.24**. Although filter driers are commonly included in the piping when a new system is installed, it is important to refer to, and follow, the manufacturer's installation literature before doing so. Some manufacturers install filter driers at the factory, while others discourage the use of field-installed driers altogether.

43.8 APPLICATION OF THE AIR-TO-AIR HEAT PUMP

Traditionally, air-to-air heat pump systems were installed primarily in milder climates—in the "heat pump belt"—basically those parts of the United States where winter temperatures can be as low as 10°F. The reason for this geographical restriction was that the air-to-air heat pump absorbs heat from the outside air in the heating mode. As the outside air temperature drops, it was more difficult for the heat pump to absorb heat from it. For example, the evaporator must be cooler than the outside air for heat to be transferred into the evaporator. Normally, in cold weather, the heat pump evaporator will be about 20°F to 25°F cooler than the air from which it is absorbing heat. We will use a 25°F temperature difference (TD) in this text as the example, **Figure 43.25**. On a 10°F day, the heat pump evaporator will be boiling (evaporating) the liquid refrigerant at 25°F lower than the 10°F air. This means that the boiling refrigerant temperature will be −15°F. As the evaporator temperature goes down, the compressor loses capacity. Older compressors were single-speed, fixed-volume pumps and had more capacity on a 30°F day than on a 10°F day. The heat pump loses capacity as the outdoor ambient temperature drops and the heating capacity needs of the structure increase. On a 10°F day, the structure needs more heating capacity than on a 30°F day, but the traditional heat pump's capacity is lower. New generation heat pumps utilize inverter-controlled compressors and have the ability to change the speed of the compressor motor to closely match the needs of the conditioned space under a wide range of ambient conditions. For example, as the outside ambient temperature drops while the system is operating in the heating mode, an inverter-controlled compressor will speed up to lower the evaporator temperature. By lowering the temperature of the evaporator, more heat can be absorbed from the ambient air. As the compressor speeds up to lower the evaporator coil temperature, the system's head pressure rises. This results in an increase in compression ratio and a decrease in system efficiency. Even with this technology, there are times when heat pump systems need help in order to meet the heating needs of the occupied space.

AUXILIARY HEAT

In an air-to-air heat pump system, the vapor-compression portion of the system may not, under certain conditions, be able to meet the heating needs of the structure. For this

Figure 43.20 This unit uses a capillary tube for the indoor cycle and a TXV for the outdoor unit. The sensing element for this TXV is mounted on the compressor permanent suction line.

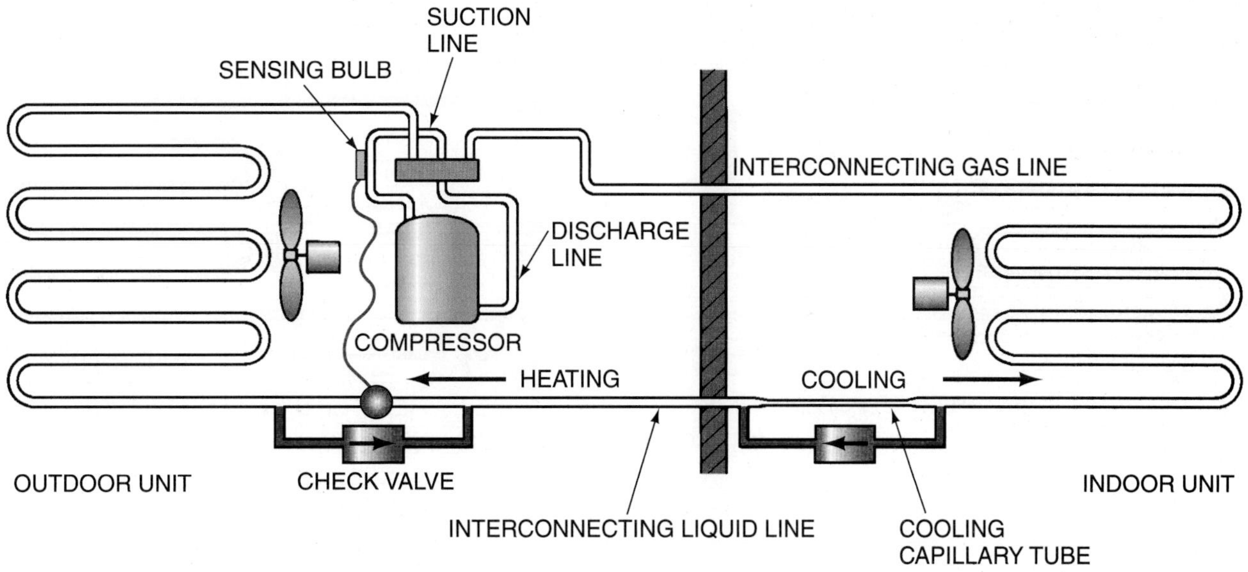

Figure 43.21 Placing a liquid-line filter drier in the heat pump refrigerant piping.

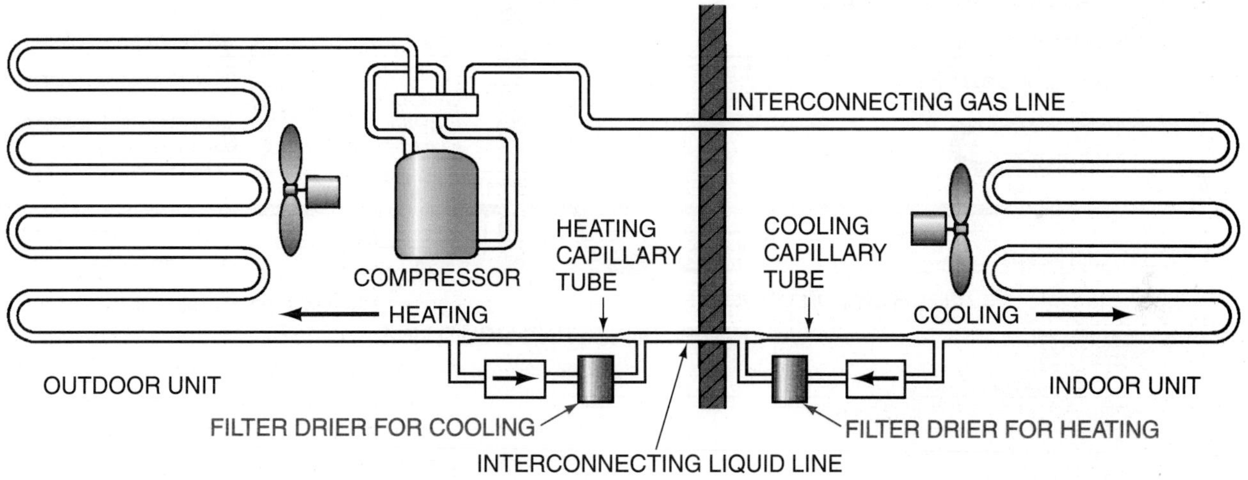

reason, some heat pump systems are equipped with second-stage heat. In many instances, this second-stage heat source is called *auxiliary heat*. The heat pump itself is the primary source of heat; the auxiliary heat source may be electric, oil, or gas. Electric auxiliary heat is the most popular because it is easy to incorporate electric heating elements into a heat pump system. Second-stage electric heat is covered in Section 43.14.

The occupied space will have different heating requirements as the outside temperature changes. The heating requirements for the structure are inversely related to the outside ambient temperature. For example, a house might require 30,000 Btu/h of heat during the day when the outside temperature is 30°F. As the outside temperature drops, the structure requires more heat. It might need 60,000 Btu/h of heat in the middle of the night when the temperature has dropped to 0°F. The heat pump could have a capacity of 30,000 Btu/h at 30°F and 20,000 Btu/h at 0°F. The difference of 40,000 Btu/h must be made up with auxiliary heat.

BALANCE POINT

The term **balance point** is used to describe the outside ambient temperature at which a heat pump can pump in exactly

Figure 43.22 In the heating mode, the cooling metering device is bypassed.

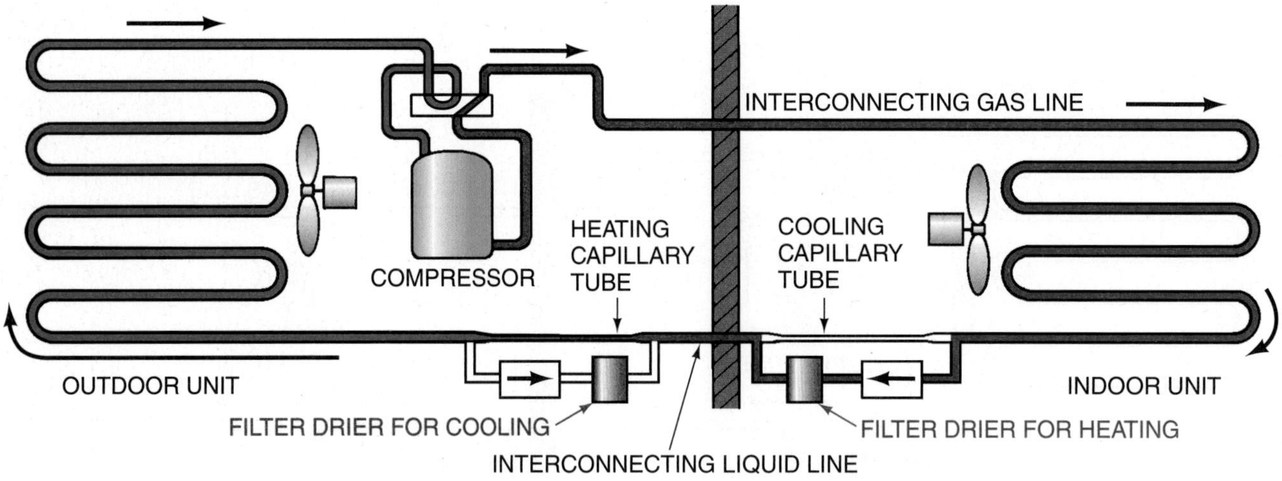

Figure 43.23 In the cooling mode, the heating metering device is bypassed.

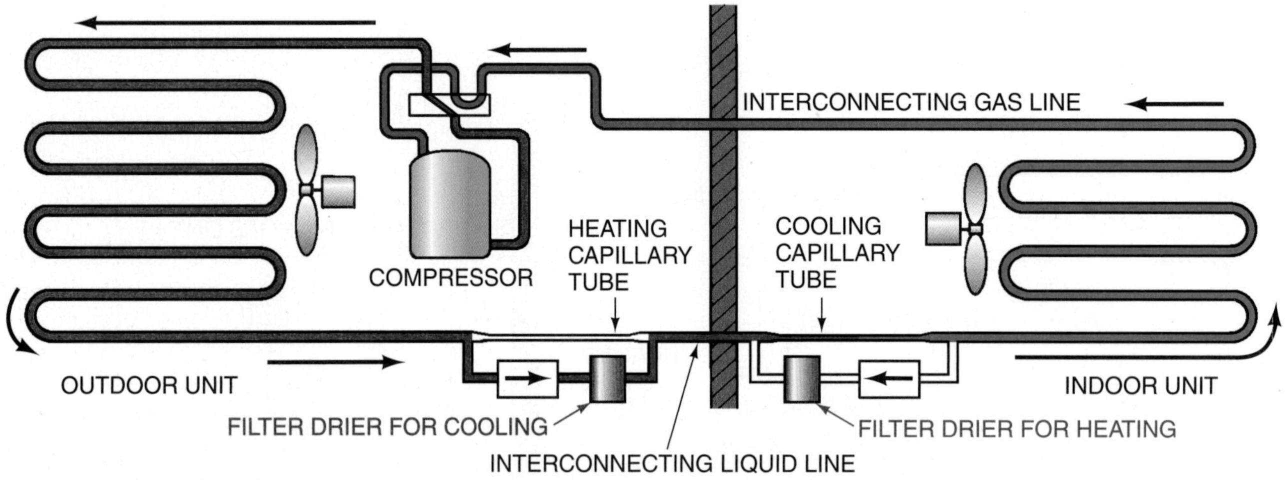

Figure 43.24 A biflow drier. Courtesy Carrier Corp.

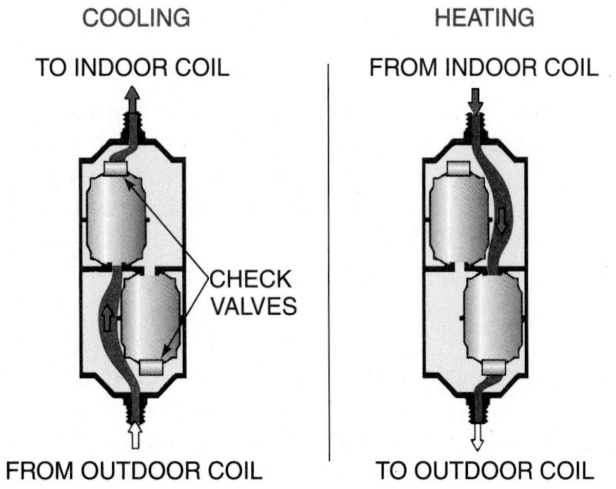

as much heat as the structure is leaking out. In more technical terms, the balance point is the lowest outside ambient temperature at which the heat pump system can satisfy the heating needs of the structure without requiring the assistance of second-stage supplementary or auxiliary heat. At this point, the heat pump will completely heat the structure by running continuously. At any temperature above the balance point, the heat pump will cycle off and on. Below this point the heat pump will run continuously but will not be able to maintain the desired temperature. Although the auxiliary heat portion of a heat pump system will be operational at some temperature above the balance point, it is impossible for a heat pump system to satisfy the heating requirements of a structure without auxiliary heat if the temperature falls below the balance point.

Figure 43.25 An R-410A heat pump outdoor coil operating in the heating mode.

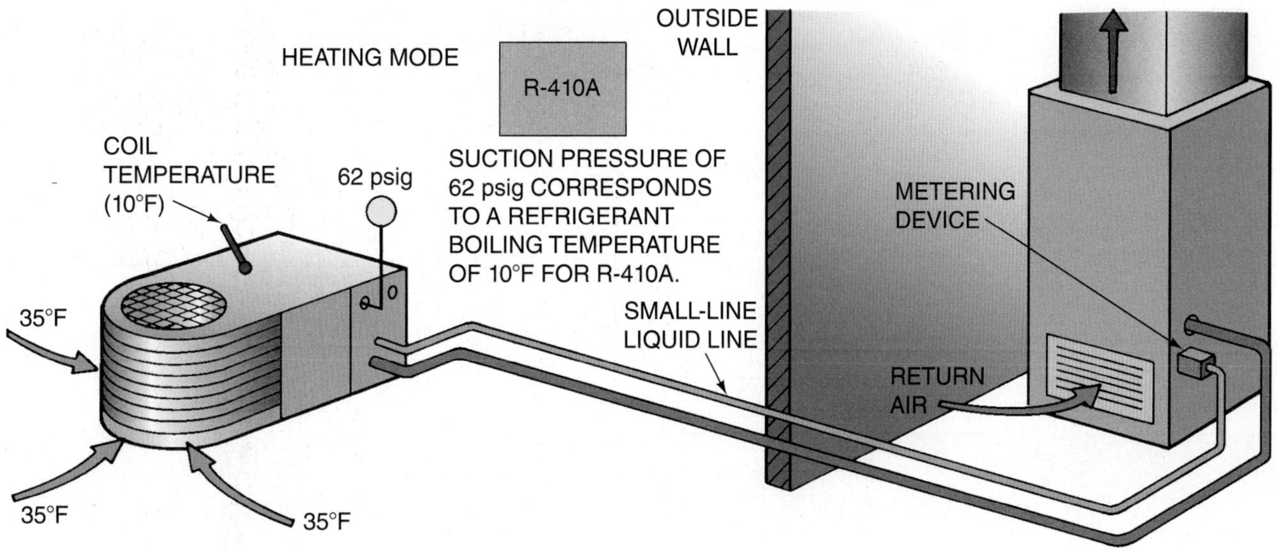

43.9 COEFFICIENT OF PERFORMANCE

If the heat pump will not heat the structure all winter, why is it so popular? This can be answered in one word—*efficiency*. To understand the efficiency of an air-to-air heat pump, you need an understanding of electric heat. A customer receives 1 W of usable heat for each watt of energy purchased from the power company while using electric resistance heat. This is called 100% efficient, or a coefficient of performance (COP) of 1:1. The output is the same as the input. With a heat pump, the efficiency may be improved to as much as 3.5:1 for an air-to-air system with typical equipment. When the 1 W of electrical energy is used in the compression cycle to absorb heat from the outside air and pump this heat into the structure, the unit could furnish 3.5 W of usable heat. Thus its COP is 3.5:1, or it can be thought of as 350% efficient. See **Figure 43.26** for a manufacturer's rating table for an air-to-air heat pump.

To provide some insight into why the COP of a heat pump system is so high, it is necessary to take a look at the pressure enthalpy chart shown in **Figure 43.27**. For a system that provides cooling, the COP is defined as

COP = Net Refrigeration Effect ÷ Heat of Compression

From the values shown in **Figure 43.27**, the net refrigeration effect (NRE) is 75 Btu/lb (115 Btu/lb − 40 Btu/lb) and the heat of compression (HOC) is 25 Btu/lb (140 Btu/lb − 115 Btu/lb). From this we can determine the COP to be 3 (75 Btu/lb ÷ 25 Btu/lb). For a system that is providing cooling, the NRE is what the occupant of the structure is benefiting from, since the refrigeration effect is what is taking place in the evaporator, or cooling, coil.

On a heat pump system that is operating in the heating mode, it is not the NRE that is providing the heat. Instead, the total heat of rejection (THOR) is what is conditioning the space. A portion of the THOR is the heat that is generated during the compression process. So, the compressor itself is contributing in two ways: It is providing the pressure difference to facilitate refrigerant flow through the system, and it is also generating heat that is directly transferred to the medium being heated. The COP for a heat pump system operating in the heating mode is given by the following formula:

COP = Total Heat of Rejection ÷ Heat of Compression

For the system shown in **Figure 43.27**, the COP is calculated as

COP = Total Heat of Rejection ÷ Heat of Compression
COP = 100 Btu/lb ÷ 25 Btu/lb
COP = 4

Notice that the COP of the heat pump system operating in the heating mode is higher than the COP of the same heat pump system operating in the cooling mode. This example is greatly simplified and intended to illustrate the contribution of the compressor heat to the efficiency of the heat pump system.

A high COP occurs only during higher outdoor winter temperatures. As the temperature falls, the COP also falls. A conventional air-to-air heat pump will have a COP of 1.5:1 at 0°F, so it is still economical to operate the compression system at these temperatures along with the auxiliary heating system. Some manufacturers include controls to shut off the compressor at low temperatures, such as 0°F to 10°F. When

Figure 43.26 Rating tables of air-to-air heat pumps. Courtesy Trane Corp.

OUTDOOR UNIT WITH HEAT PUMP COILS							
	TXC030D4	TXC031C4	TXC031D4	TXC035C4	TXC035D4	TXC036C4	TXC036D4
EXPANSION TYPE	CHG TO 59	CHG TO 59	CHG TO 59	CHG TO 59	CHG TO 59	CHG TO 59	CHG TO 59
RATINGS (COOLING) ①							
BTUH (TOTAL)	29000	29200	29200	29400	29400	29800	29800
BTUH (SENSIBLE)	22100	22200	22200	23200	23200	23900	23900
INDOOR AIRFLOW (CFM)	1000	1000	1000	1100	1100	1115	1115
SYSTEM POWER (KW)	2.70	2.70	2.70	2.75	2.75	2.77	2.77
EER / SEER (BTU/WATT-HR.)	10.75/11.75	10.80/12.00	10.80/12.00	10.70/11.75	10.70/11.75	10.75/11.75	10.75/11.75
RATINGS (HEATING) ①							
(HIGH TEMP.) BTUH	28200	28400	28400	28400	28400	28600	28600
SYSTEM POWER (KW)	2.77	2.65	2.65	2.74	2.74	2.56	2.56
COP	2.98	3.14	3.14	3.04	3.04	3.28	3.28
HSPF (BTU/WATT-HR.)	7.75	8.05	8.05	7.80	7.80	8.30	8.30

(A)

OUTDOOR UNIT WITH AIR HANDLERS							
	4TEE3F31A	4TEE3F37A	4TEE3F40A	4TEP3F24A	4TEP3F30A	4TEP3F36A	4TEP3F42A
EXPANSION TYPE	TXV-NB	TXV-NB	TXV-NB	TXV-NB	TXV-NB	TXV-NB	TXV-NB
RATINGS (COOLING) ①							
BTUH (TOTAL)	29600	30000	31400	29000	29200	29600	30200
BTUH (SENSIBLE)	23400	23200	24300	21400	22000	23700	24200
INDOOR AIRFLOW (CFM)	1060	1000	1010	900	940	1110	1125
SYSTEM POWER (KW)	2.54	2.48	2.50	2.62	2.62	2.73	2.73
EER / SEER (BTU/WATT-HR.)	11.65/14.00	12.10/14.25	12.55/14.75	11.05/13.00	11.15/13.00	10.85/12.50	11.05/13.00
RATINGS (HEATING) ①							
(HIGH TEMP.) BTUH	28400	27600	28000	28000	28000	28600	28800
SYSTEM POWER (KW)	2.49	2.42	2.23	2.72	2.66	2.60	2.54
COP	3.34	3.34	3.68	3.02	3.08	3.22	3.32
HSPF (BTU/WATT-HR.)	8.45	8.50	9.00	7.80	8.00	8.20	8.40

(B)

Figure 43.27 Pressure enthalpy chart showing the relationship between the total heat of rejection (THOR), net refrigeration effect (NRE), and the heat of compression (HOC).

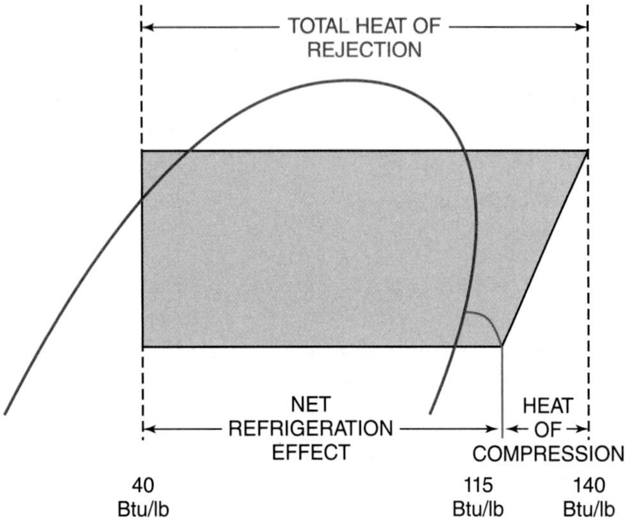

the temperature rises, the compressor comes back on. Some manufacturers do not allow the compressor shut off at all. Long running times will not hurt a compressor or wear it out to any extent. They would rather have the compressor run all the time than have it restart with a cold crankcase when the temperature rises.

The *water-to-air* heat pump might not need auxiliary heat because the heat source (the water) can be relatively constant temperature all winter. The heat loss (the heat requirement) and the heat gain (the cooling requirement) of the structure might be almost equal. Consequently, these heat pumps have COP ratings as high as 4:1. Just remember that the higher COP ratings of water-to-air heat pumps are due not to the components of the refrigeration equipment but to the temperature of the heat source. Earth and lakes have a more constant temperature than air. Such pumps are very efficient for winter heating because of the COP and for summer cooling because they operate at water-cooled air-conditioning temperatures and head pressures.

43.10 THE SPLIT-TYPE, AIR-TO-AIR HEAT PUMP

Like cooling-only air-conditioning equipment, air-to-air heat pumps come in two styles: split systems and package (self-contained) systems. The split-system air-to-air heat pump resembles the split-system cooling air-conditioning system. The components look exactly alike from the outside.

THE INDOOR UNIT

The indoor unit of an air-to-air split system may be an electric furnace with a heat pump indoor coil, **Figure 43.28**. The outdoor unit may resemble the condensing unit on a conventional split-type air-conditioning system. The indoor unit, which contains the blower and coil, is the part of the system that circulates the air in the structure. The airflow pattern may be upflow, downflow, or horizontal to serve different applications. Some manufacturers have designed their units so that one unit may be adapted to all of these flow patterns by correctly placing the pan for catching and containing condensate, **Figure 43.29**.

Coil placement in the airstream is important in the indoor unit. The electric auxiliary heat and the primary heat (the heat pump) may need to operate at the same time in weather below the system's balance point. **The refrigerant coil must be located in the airstream before the auxiliary heating coil.** Otherwise, heat from the auxiliary unit will pass through the refrigerant coil when both are operating in winter. If the auxiliary heat is operating and is located before the heat pump coil, the head pressure will be too high and could rupture the coil or burn the compressor motor. Remember, the coil is operating as a *condenser* in the heating mode. It rejects heat from the refrigeration system, so any heat added to it will cause the head pressure to rise, **Figure 43.30**.

Figure 43.28 An air-to-air heat pump indoor unit. It is an electric furnace containing a heat pump coil. Notice that the air flows through the heat pump coil and then through the heating elements. Photo by Bill Johnson

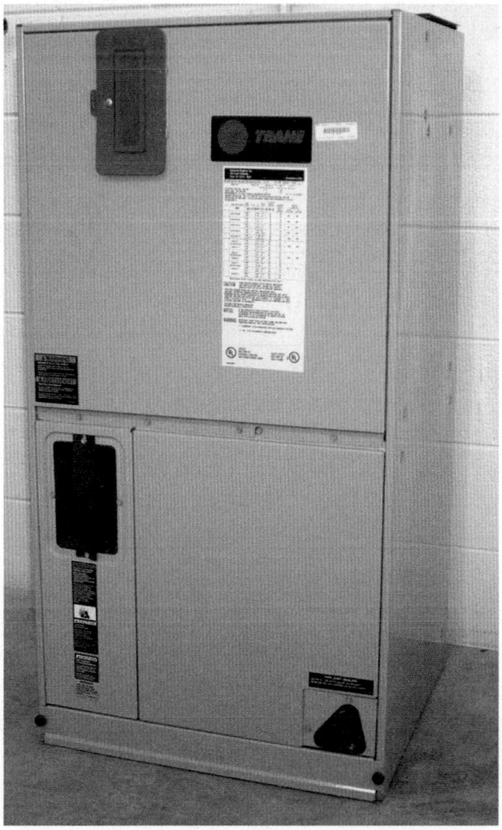

Figure 43.29 This heat pump indoor unit can be positioned either in the vertical (upflow, downflow) or horizontal position by placing the condensate pan under the coil in the respective position.

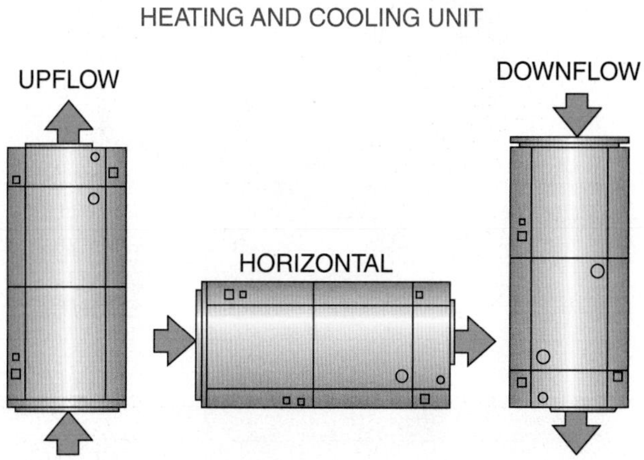

The indoor unit may be a gas or oil furnace. If this is the case, the indoor coil must be located in the outlet airstream of the furnace. When the auxiliary heat is gas or oil, the heat pump does not run while the auxiliary heat is operating. If the coil is not located after the furnace heat exchanger, it would sweat in summer, when it would be operating as an evaporator, and the outlet air temperature could be lower than the dew point temperature of the air surrounding the heat exchanger. For this reason, most local codes will not allow a gas or oil heat exchanger to be located in a cold airstream. Special control arrangements must be made when the auxiliary heat is gas or oil so that the heat pump will not operate while the gas or oil heat is operating. When a heat pump is added to an electric furnace, the indoor coil must be located after the heat strips. The same rules apply as for gas or oil.

TEMPERATURE OF THE CONDITIONED AIR

The heat pump indoor unit is installed in much the same way as a split-system cooling unit, but the air distribution system must be designed more precisely due to the temperature of the air leaving the air handler during the heating mode. Unlike the temperatures produced by gas and oil-fired systems that are normally used with electric cooling, the temperature of the air supplied by heat pumps is not as hot. The leaving-air temperatures are in the 100°F range when just the heat pump (first-stage heating) is operating. If the airflow is restricted, the air temperature will go up slightly, but the system's COP will drop. A reduced COP indicates a reduction in system efficiency.

Figure 43.30 This illustration shows the right and wrong way a heat pump coil can be installed with respect to the electric heating elements.

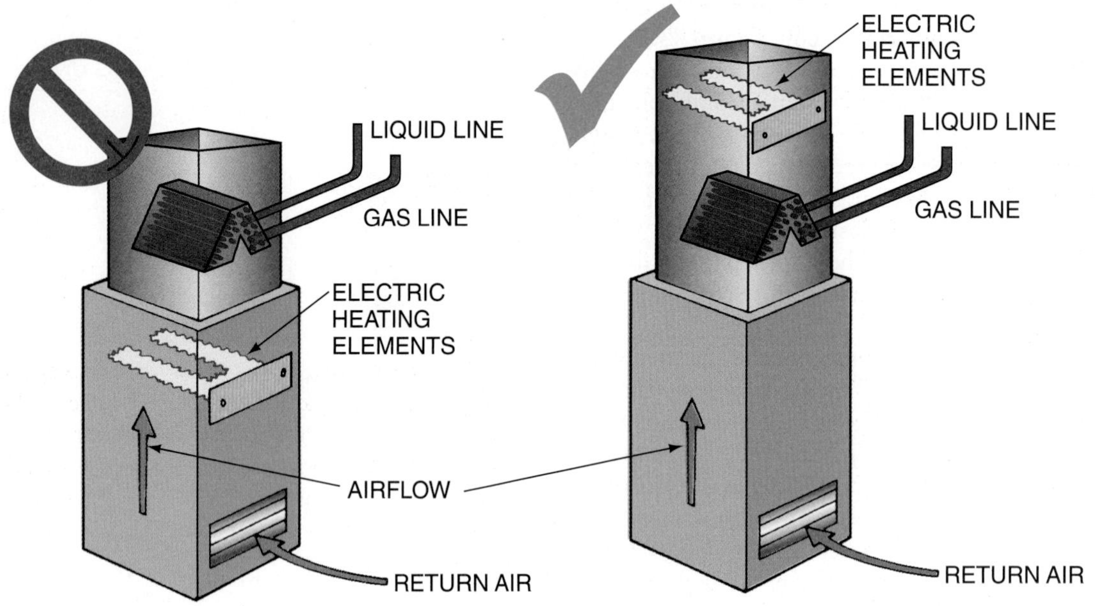

When 100°F air is distributed to a conditioned space, it must be distributed carefully or drafts will occur. Normally, the air is introduced to the structure along the outside walls. The supply registers are either in the ceiling or the floor, depending on the structure. It is common in two-story houses for the first-story registers to be in the floor with the air handler in the crawl space below the house. The upstairs unit is commonly located in the attic with the supply registers located in the ceiling near the outside walls. The outside walls are where heat leaks out of a house in the winter and leaks into the house in the summer. If these walls are slightly heated in the winter by the airstream, less heat is taken out of the air in the conditioned space and the room stays warmer, **Figure 43.31**. Air at a temperature of 100°F mixed with room air does not feel like 130°F air mixed with room air from a gas or oil furnace. It may feel like a draft.

Figure 43.31 This heat pump installation shows a good method of distributing the warm air in the heating mode.

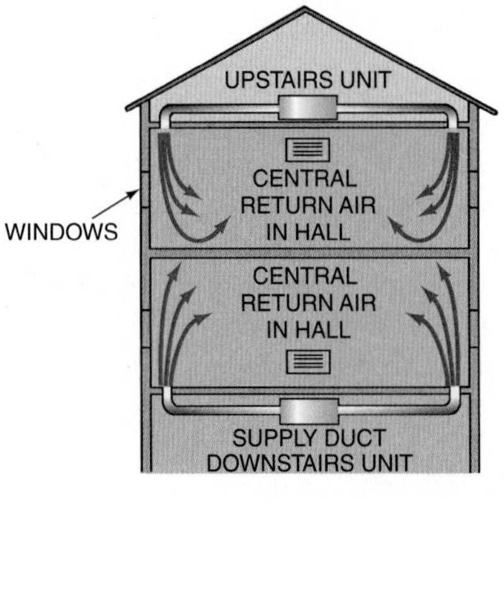

When air at 100°F enters the room from the supply register, it mixes with the air that is around the register, which may be at a temperature of 75°F. This would result in a mixed air temperature of about 80°F. If this mixed air blows on a person with a skin temperature of about 85°F, it will seem like the room is cool, not warm. Proper air distribution can help reduce the number of "draft" complaints. When 100°F air is distributed from the inside walls, such as high sidewall registers, the system may not heat satisfactorily. The air will mix with the room air and might feel drafty. The outside walls will also be cooler than if the air were directed on them, and a cold wall effect could be noticed in cold weather, **Figure 43.32**.

Figure 43.32 This air distribution system with the air being distributed from the inside using high sidewall registers shows the heat pump's 100°F air mixing with the room air of 75°F.

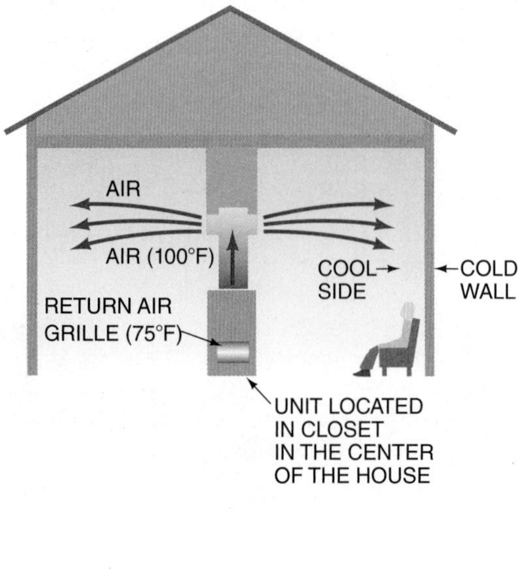

THE OUTDOOR UNIT

Manufacturers provide recommended installation procedures for their particular units that show the installer how to locate the unit for the best efficiency and water drainage. The installation of a heat pump system's outdoor unit is much like that of a traditional central air-conditioning system from an airflow standpoint. The unit must have a good air circulation around it, and the discharge air must not be allowed to recirculate. A more serious consideration is the direction of the prevailing wind in the winter, which could lower heat pump performance. If the unit is located in a prevailing north wind or a prevailing wind from a lake, system performance may be negatively affected. A prevailing north wind might cause the evaporator (outdoor coil in the heating mode) to operate at a lower-than-normal temperature. A wind blowing inland off a lake will be very humid and might cause excessive outdoor coil freezing when the system operates in the heating mode.

The outdoor unit must *not* be located where roof water will pour into it. The coil will be operating at a temperature that is often below freezing, so moisture that is not in the air itself should be kept away from the unit's coil. If it is not, excess freezing will occur. The outdoor unit is an evaporator in winter and will attract moisture from the outside air. If the coil is operating below freezing, the moisture will freeze on the coil. If the coil is above freezing, the moisture will run off the coil as it does in an air-conditioning evaporator. This moisture must have a place to go. If the unit is in a yard, the moisture will soak into the ground. **⚠ SAFETY PRECAUTION:** *If the unit is on a porch or walk, the moisture could freeze and create slippery conditions.* •

The outdoor unit is designed with drain holes or slots in the bottom of the unit to allow free movement of the water away from the coil. If they are inadequate, the coil will become a solid block of ice in cold weather. Ice accumulation on the outdoor coil acts as an insulator and will negatively affect the transfer of heat between the outside air and the refrigerant in the coil. As a result, the COP will be reduced. Generally, manufacturers recommend that the outdoor unit be placed on risers that keep the unit above the typical snow level for the region. When ice buildup is excessive, the defrost procedures for the unit may not be adequate. When this happens, a little more ice builds up each run cycle, until the normal defrost cycle will not remove the ice and other measures must be taken to get the unit back in normal operation. Defrosting methods are discussed in Section 43.13, "The Defrost Cycle."

⚠ SAFETY PRECAUTION: *Installing a heat pump involves observing the same safety precautions as when installing a "cooling only" air-conditioning system. The ductwork must be insulated, which requires working with fiberglass. Be careful when working with metal duct and fasteners and especially when installing units on rooftops and other hazardous locations.* •

43.11 PACKAGE AIR-TO-AIR HEAT PUMPS

Package air-to-air heat pumps are much like package air conditioners. They look alike and are installed in the same way. Therefore, give the same considerations to the prevailing wind and water conditions. Package heat pumps have all of the components in one housing and are easy to service, **Figure 43.33**. They have a separate compartment in which the supplementary electric strip heaters are mounted. Most package heat pump systems can accept different electric heat sizes from 5000 W (5 kW) to 25,000 W (25 kW).

The metering devices used for package air-to-air heat pumps are much like those for split systems. Some manufacturers use a common metering device for the indoor and outdoor coils because the two are so close together. When one metering device is used, it must be able to adjust for the varying refrigerant flow requirements in the heating and cooling modes of operation.

Package air-to-air heat pumps require that supply and return ductwork be connected to the unit. Depending on the location of the unit, these duct runs can be either very short, or very long, **Figure 43.34**. In some cases, the duct will need to extend all the way to an outside wall of the structure for a crawl space installation. Supply and return ducts must be insulated to prevent heat exchange between the duct and the ambient air.

Package air-to-air heat pumps have the same advantage from a service standpoint as package cooling air conditioners or package gas-burning equipment. All of the controls and components may be serviced from the outside. They also have the advantage of being factory assembled and charged. They are leak checked and bench-tested before leaving the factory.

Figure 43.33 A package heat pump. All components are contained in one housing. Courtesy Carrier Corp.

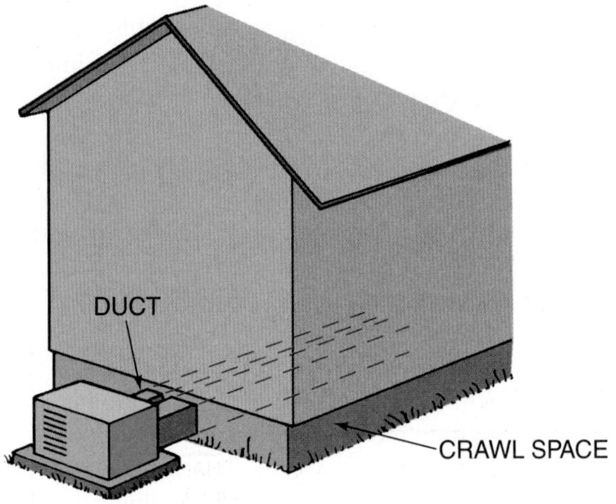

Figure 43.34 The self-contained unit contains the outdoor coil and the indoor coil. The ductwork must be routed to the unit.

43.12 CONTROLS FOR THE AIR-TO-AIR HEAT PUMP

The air-to-air heat pump is different from any other combination of heating and cooling equipment. The following items must be controlled at the same time for the heat pump to be efficient: space temperature, defrost cycle, indoor blower, the compressor, the outdoor fan, auxiliary heat, and emergency heat. Each manufacturer has its own method of controlling the heating and cooling sequence.

The space temperature for an air-to-air heat pump is not controlled in the same way as that for a typical heating and cooling system. There are actually two complete heating systems and one cooling system. The two heating systems are the reverse-cycle refrigeration cycle of the heat pump and the auxiliary heat from the supplemental heating system—electric, oil, or gas. The auxiliary heating system may be operated as a system by itself if the heat pump fails. When the auxiliary heat becomes the primary heating system because of heat pump failure, it is called *emergency heat* and is normally operated only long enough to get the heat pump repaired. The reason is that the COP of the auxiliary heating system is lower than the COP of the heat pump.

The space thermostat is the key to controlling the system. It is normally a two-stage heating and two-stage cooling thermostat that is used exclusively for heat pump applications. The number of stages of heating or cooling may vary. **Figure 43.35** shows a typical heat pump thermostat. These thermostats are easily identified because they have an EMERGENCY HEAT position on the selector switch, **Figure 43.35(B)**. The wiring diagram in **Figure 43.36** represents the control wiring of a typical heat pump system.

(A)

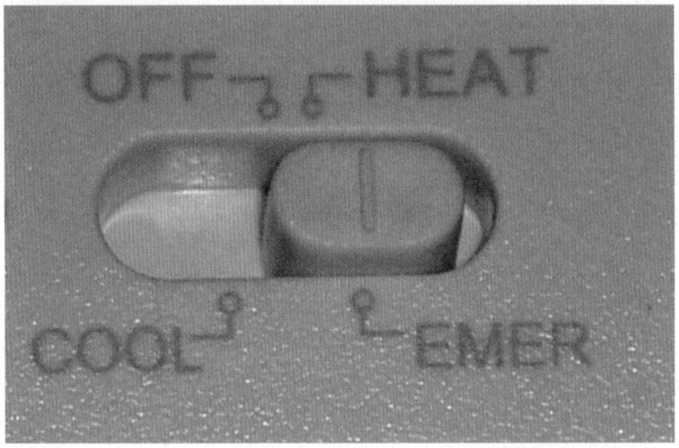

(B)

Figure 43.35 (A) A heat pump thermostat. (B) Notice the EMERGENCY HEAT option on the thermostat. Photos by Eugene Silberstein

The following sections describe the sequence of cooling and heating control of an automatic changeover thermostat—one that automatically changes from cooling to heating and back. **NOTE:** *This control sequence is only one of many. Each manufacturer has its own. The temperature-sensing element is a bimetal that controls mercury bulb contacts. Although not used in new thermostats, mercury bulbs will be encountered on older installations. The auxiliary heat is electric. The thermostat blower switch is in AUTO.* •

COOLING CYCLE CONTROL

When the first-stage bulb of the thermostat closes its contacts (on a rise in space temperature), the four-way reversing valve solenoid coil is energized, **Figure 43.37**. When the space temperature rises about 1°F, the second-stage cooling contacts of the thermostat close. The second-stage contacts energize the compressor contactor coil and the indoor blower relay coil

Figure 43.36 A heat pump thermostat with two-stage heating and two-stage cooling. Courtesy Honeywell Corp.

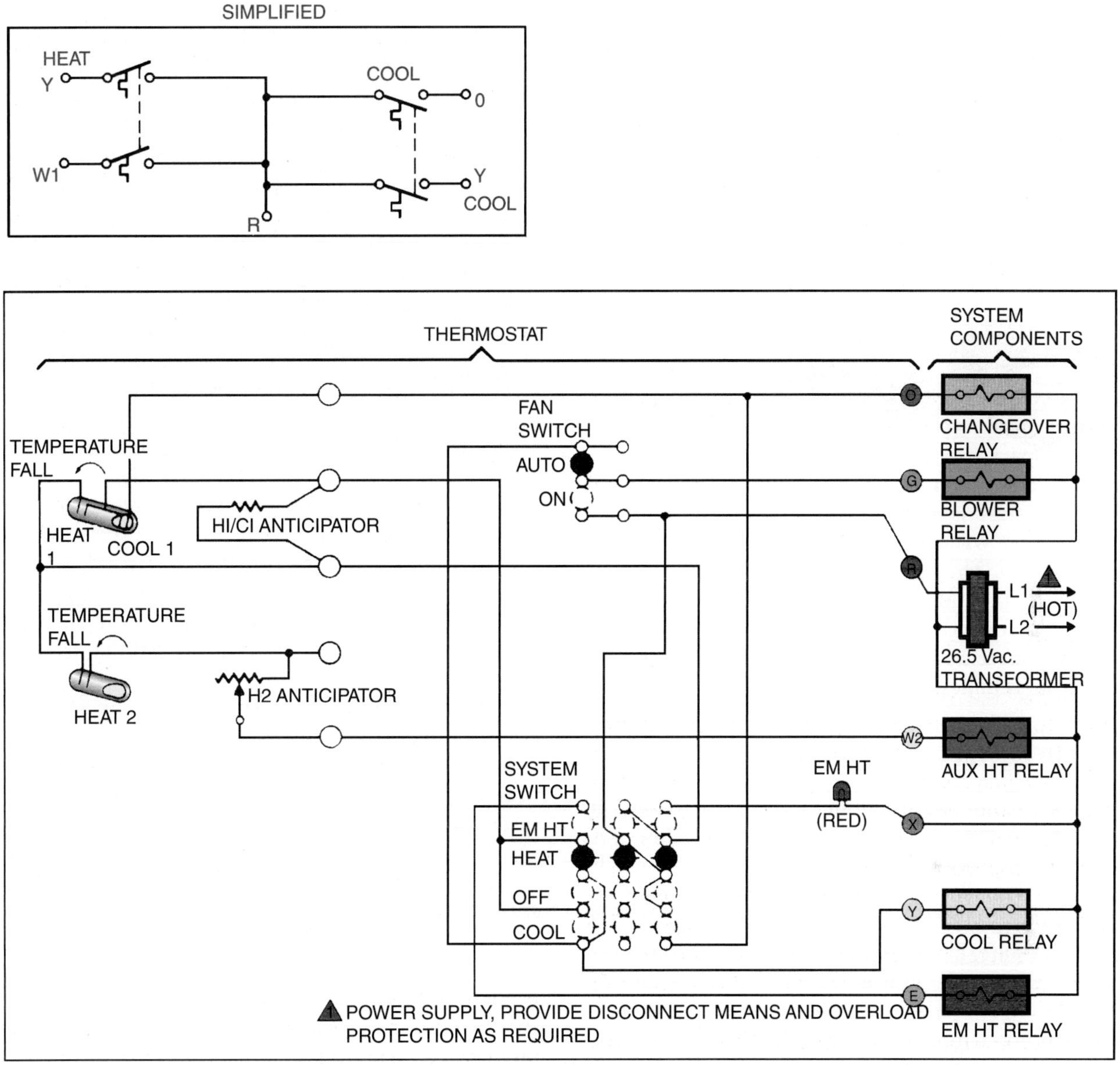

Figure 43.37 The first stage of the cooling cycle in which the four-way valve magnetic holding coil is energized.

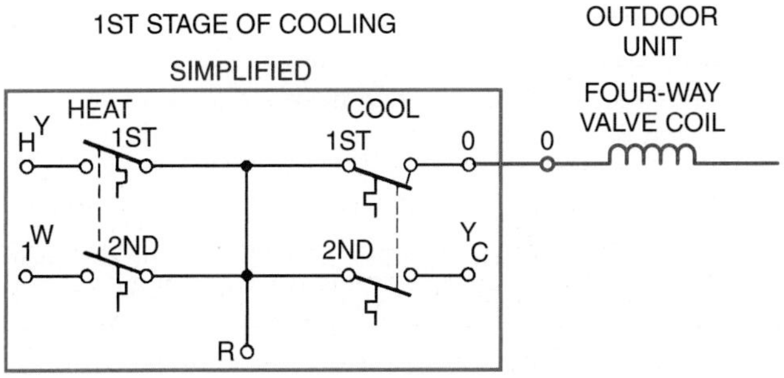

to start the compressor, outdoor fan, and the indoor blower, **Figure 43.38**. When the compressor starts the second stage of operation, it diverts the hot gas from the compressor to the outdoor coil, and the system is in the cooling cycle.

When the space temperature begins to fall and is satisfied (because the compressor is running and removing heat), the first contacts to open are the second-stage contacts. The compressor (and outdoor fan) and the indoor blower stop. The first-stage contacts remain closed, and the four-way valve remains energized. The system pressure will now equalize through the metering device for ease of compressor starting, but the four-way valve will not change position. By not having the coil on the reversing valve energize and deenergize

Figure 43.38 This diagram shows what happens with a rise in temperature. The first stage of the thermostat is already closed; then the second stage closes and starts the compressor. The compressor was the last on and will be the first off.

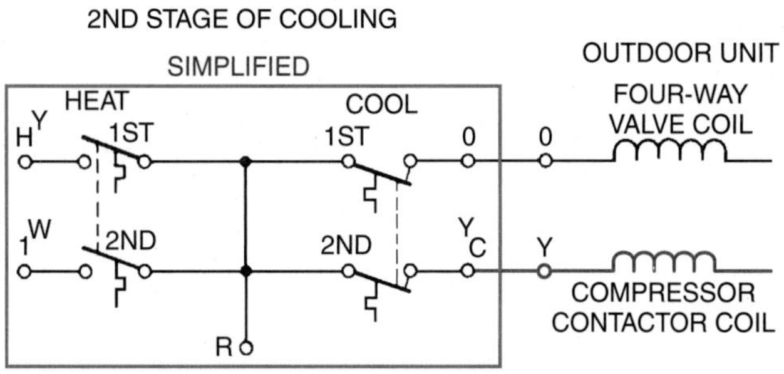

each time the compressor cycles on and off, there is much less wear-and-tear on the system.

When the space temperature rises again, the compressor (and outdoor fan) and indoor blower will start again. If the outdoor temperature is getting cooler (e.g., in autumn), the space temperature will continue to fall, and the first-stage contacts will open. This deenergizes the four-way reversing valve solenoid coil, and the unit can change over to heat when it starts up. The four-way valve is a pilot-operated valve. It will not change over until the compressor starts and a pressure differential is built up.

SPACE HEATING CONTROL

When the space temperature continues to fall due to outdoor conditions, the first-stage heating contacts close. This starts the compressor (and outdoor fan) and the indoor blower. The four-way valve is not energized, so the hot discharge gas from the compressor is directed to the indoor coil. The unit is now in the heating mode, **Figure 43.39**. When the space temperature warms, the first-stage heating contacts open, and the compressor (and outdoor fan) and indoor blower stop.

If the outdoor temperature is cold, below the balance point of the structure, the space temperature will continue to fall because the heat pump alone will not be able to provide enough heat to the structure. The second-stage heating contacts will close and start the auxiliary electric heat to help the compressor. The second-stage heat was the last component to be energized, so it will be the first component off. This is the key to how a heat pump can get the best efficiency from the refrigerated cycle. It will operate continuously because it was the first on, and the second-stage auxiliary heat will stop and start to assist the compressor, **Figure 43.40**. There are as many methods of controlling the foregoing cooling and heating sequence as there are manufacturers. One common variation for controlling the heating system is to use outdoor thermostats to control the auxiliary heat, **Figure 43.41**. This keeps all of the auxiliary heat from coming on with the second stage of heating. It will come on based on outdoor temperature by the use of outdoor thermostats. These thermostats only allow the auxiliary heat to operate when needed, using the structure balance point as a guideline. Balance point is discussed in more detail in Section 43.14.

Figure 43.39 The space temperature drops to below the cooling set points for the first stage of heating; the compressor starts. This time, the four-way valve magnetic coil is deenergized, and the unit will be in the heating mode.

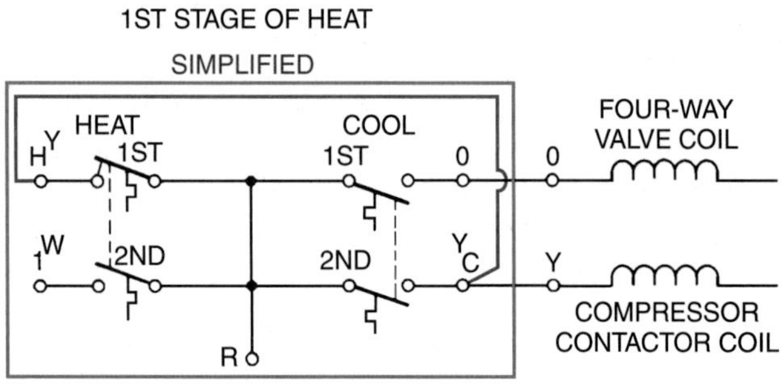

There are actually several balance points for any structure when a sophisticated heat pump system is installed. The first one mentioned previously was for the heat pump alone. Several stages of electric strip heat may be applied using balance points. It is vital that the strip heat not be operated any more than necessary. Later in this unit, we will explain why.

Consider a heat pump system that has 30 kW of total electric strip heat available in 10 kW increments that will be energized as needed. If the outdoor temperature drops down to a temperature that is just below the balance point of the system, a portion of the electric strip heaters will be energized. If after a period of time, the heat pump, along with some of the electric strip heaters, are not able to heat the space and the temperature in the structure continues to drop, additional electric strip heaters will be energized. If the space temperature still continues to drop, possibly from a drop in outside ambient temperature, still more additional electric strip heaters will be energized. The control of the supplementary electric strip heaters is accomplished by the thermostat as well as the outdoor temperature sensors or thermostats. The outdoor temperature sensors are typically low-voltage devices and are used to help anticipate the heating needs of the space, and to conserve energy. The goal is only to energize enough electric strip heaters as necessary to help the heat pump reach and maintain the desired space temperature.

One more aspect of this type of system should be discussed. Heat pump systems have an emergency heat mode that allows the owner to start all of the strip heat in the event the vapor-compression portion of the heat pump system fails. This mode of operation uses a relay that shuts off the heat pump and thermostatically controls all of the electric strip heaters. When in the emergency heat mode, an indicator light on the thermostat will illuminate to let the owner know that the system is operating in that mode. The control circuit wiring for the emergency heat mode is shown in **Figure 43.42**.

Figure 43.40 When the outside temperature continues to fall, it will pass the balance point of the heat pump. When the space temperature drops approximately 1.5°F, the second-stage contacts of the thermostat will close, and the auxiliary heat contactors will start the auxiliary heat.

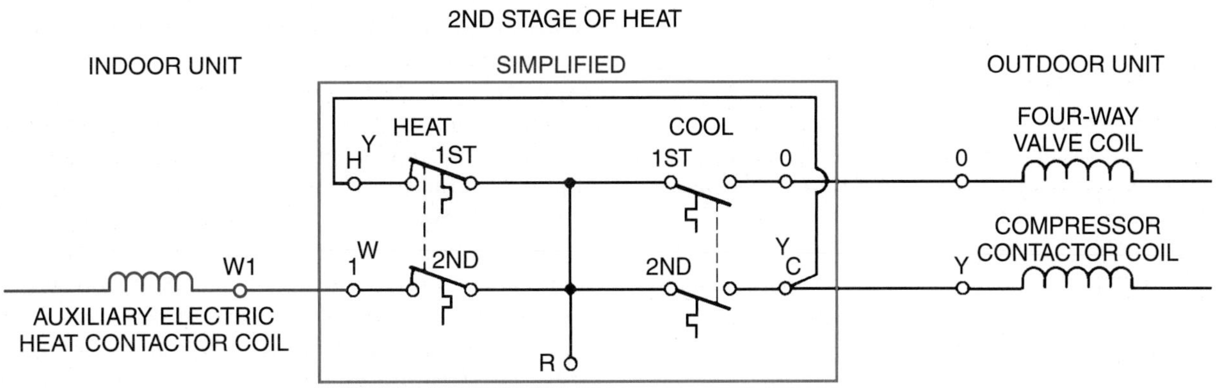

Figure 43.41 A wiring diagram with outdoor thermostats for controlling the electric auxiliary heat.

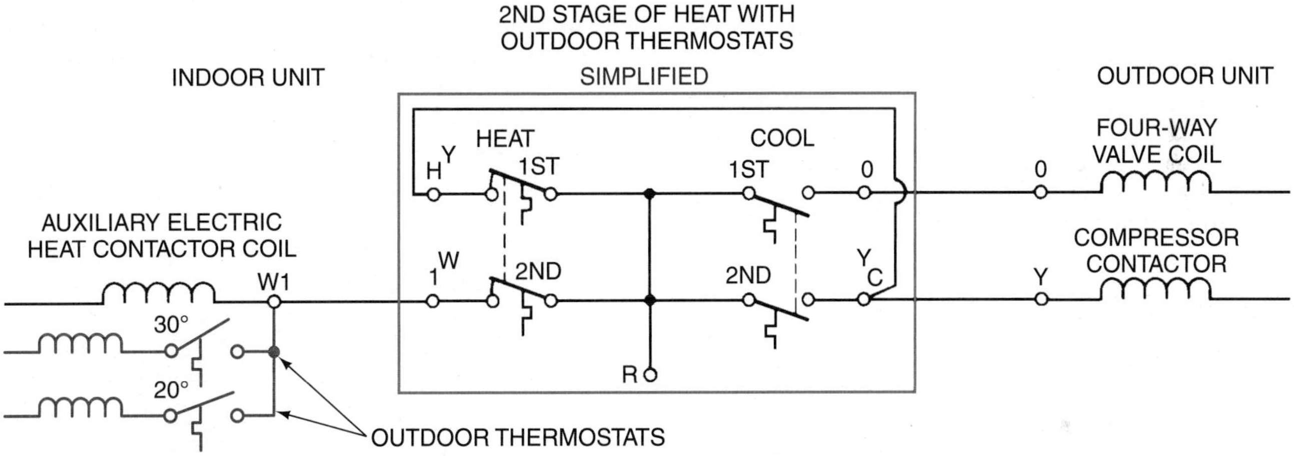

Figure 43.42 The wiring diagram of a space thermostat set to the emergency heat mode.

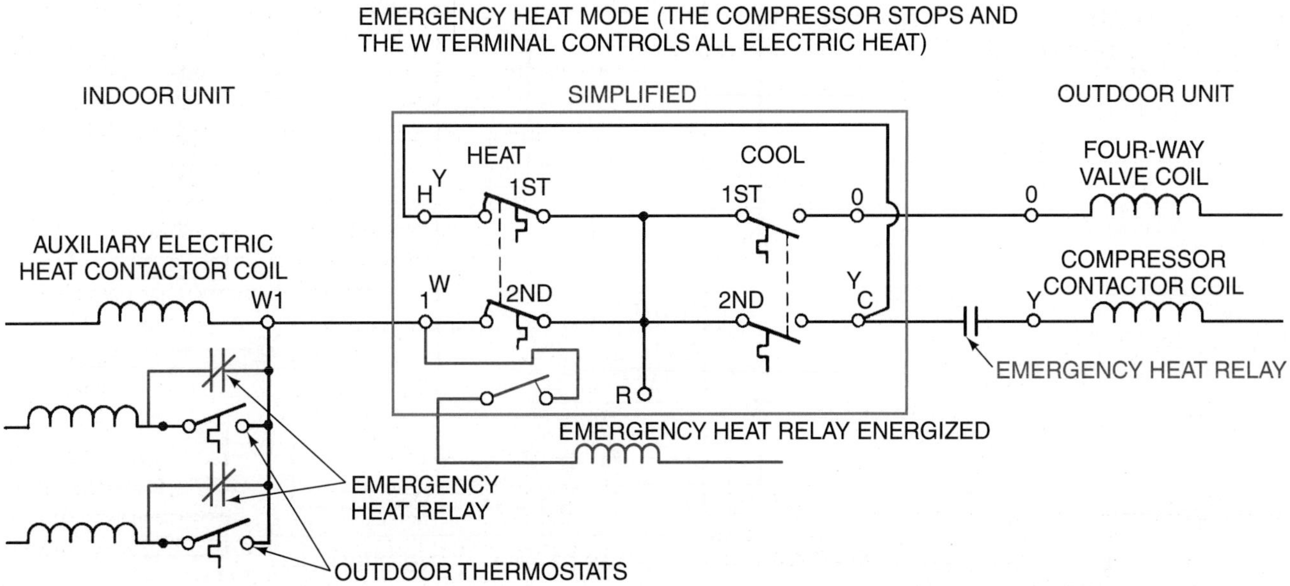

When strip heat is energized and operates along with the heat pump it is referred to as auxiliary, or supplementary, heat. When strip heat is used instead of the heat pump, in case of heat pump failure, it is called emergency heat. Many manufacturers use a signal light to alert the owner that the auxiliary heat is energized. This is typically in the form of an indicator light at the thermostat. If the indicator light is on in mild weather, the owner might assume that the heat pump itself is not working properly because the strip auxiliary heat is helping it more than usual. This should prompt a call for service. When the system is switched to emergency heat, a red light is often illuminated to remind the owner to get the system serviced soon.

The technician should be familiar with four wiring sequences. The next four figures show how they work. These illustrations are from a generic HEAT-COOL thermostat that may be used with several manufacturers' equipment. Following the arrows, notice there is a change of direction in the way the power feeds some of these devices from one mode to the other, **Figure 43.43**, while the unit is in the cooling mode. Notice what devices are energized.

The four-way reversing valve is energized by the thermostat selector switch when it is placed in the cooling mode. The other devices that must be energized are the compressor, the outdoor fan, and the indoor blower. These components must operate in a coordinated fashion; they can be checked from the thermostat subbase by removing the thermostat. The terminals shown with letter designations are on the thermostat subbase. These may be jumped one at a time to see if the component will start. For example, if the technician jumps from R, which is the hot terminal, to G, the indoor blower will start up if the relay and the wiring are good. The technician can also shut off all power and use an ohmmeter between each terminal to the R terminal to establish that the circuit is good.

Figure 43.44 shows the first stage of heat. The four-way reversing valve is not energized, so the system will start up in the heating mode with the compressor and outdoor fan starting from the cool relay. The indoor blower is running. **Figure 43.45** shows the system operating in the heat mode with the first stage of electric heat operating from the second stage of the thermostat. Notice that the indicator light on the thermostat is energized to show the homeowner that the strip heat is operating. The outdoor temperature should be below the heat pump's balance point if second-stage heat is being called for. If the system remains in second-stage heat even in mild weather, either the homeowner is turning the thermostat up to make it call or the heat pump system itself is not functioning correctly.

Figure 43.43 A generic HEAT-COOL heat pump thermostat in the cooling mode.

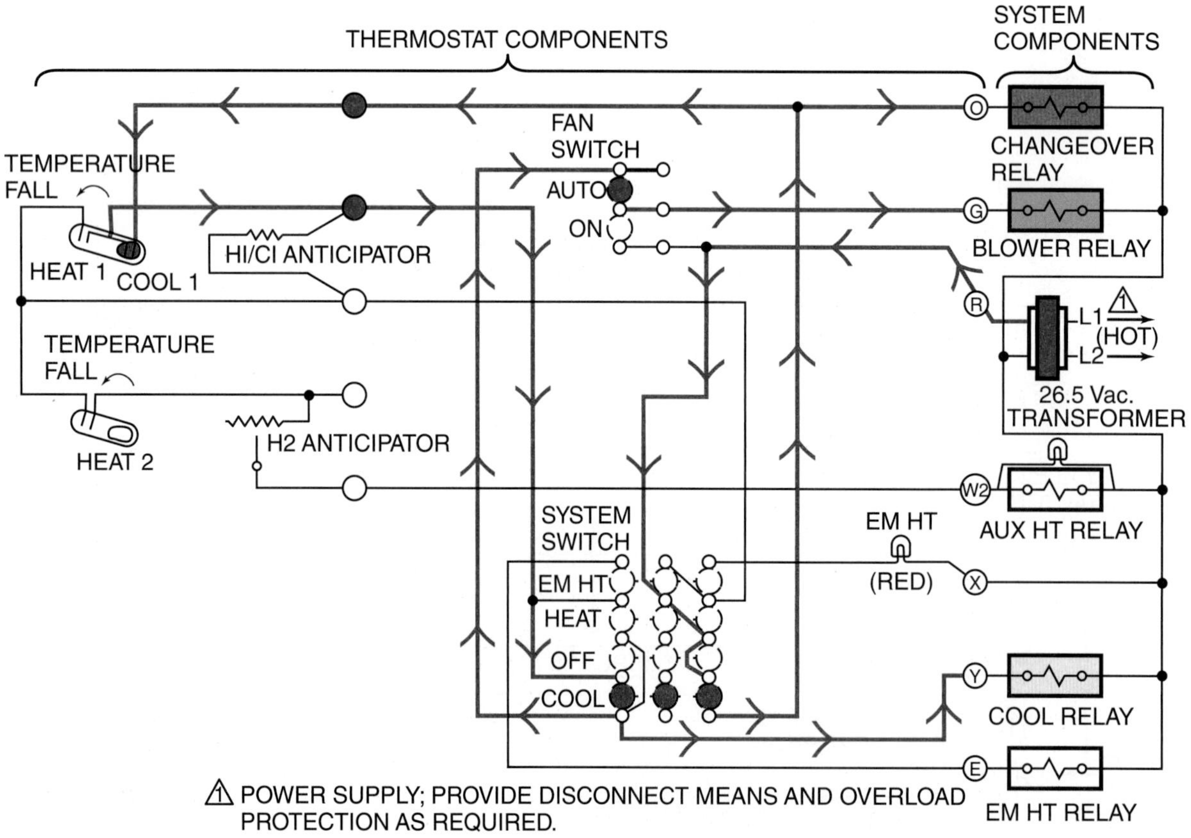

Figure 43.44 A generic HEAT-COOL heat pump thermostat in the first-stage heat mode with only the heat pump operating.

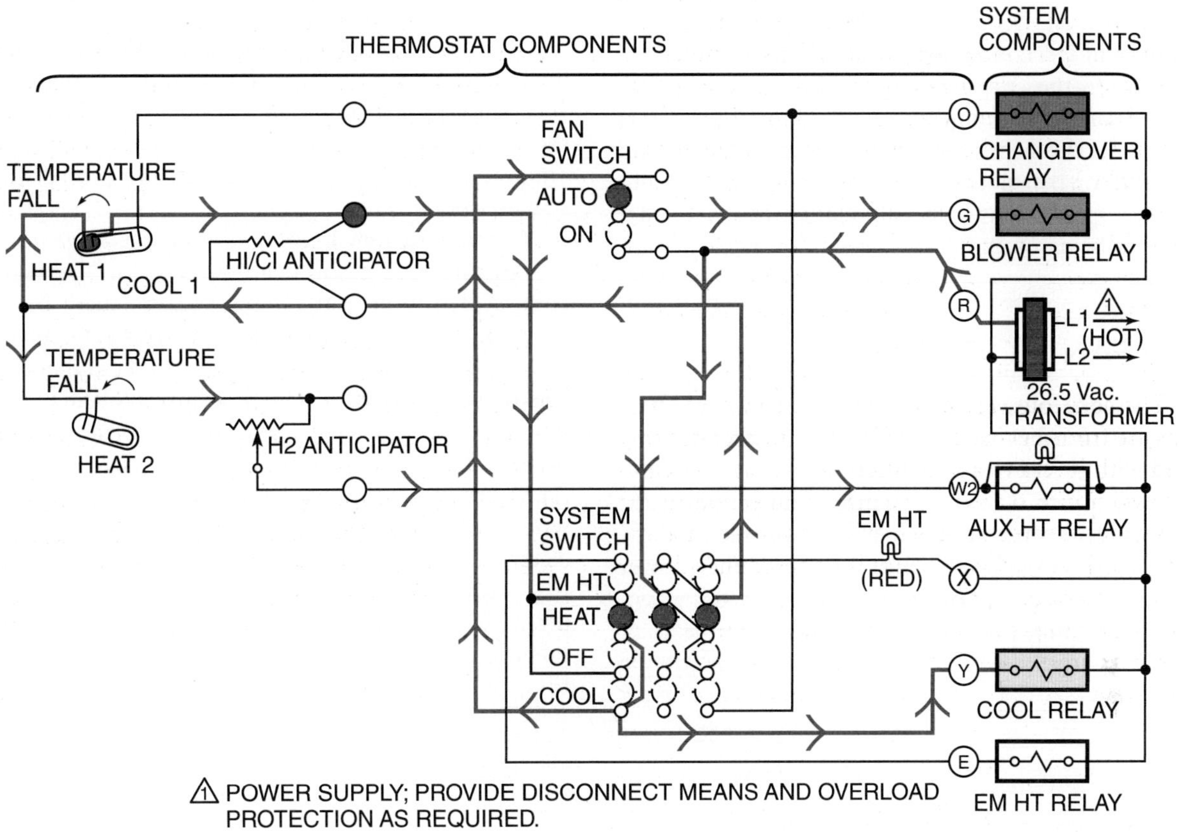

Figure 43.45 A generic HEAT-COOL heat pump thermostat operating with both the heat pump and the auxiliary heat working.

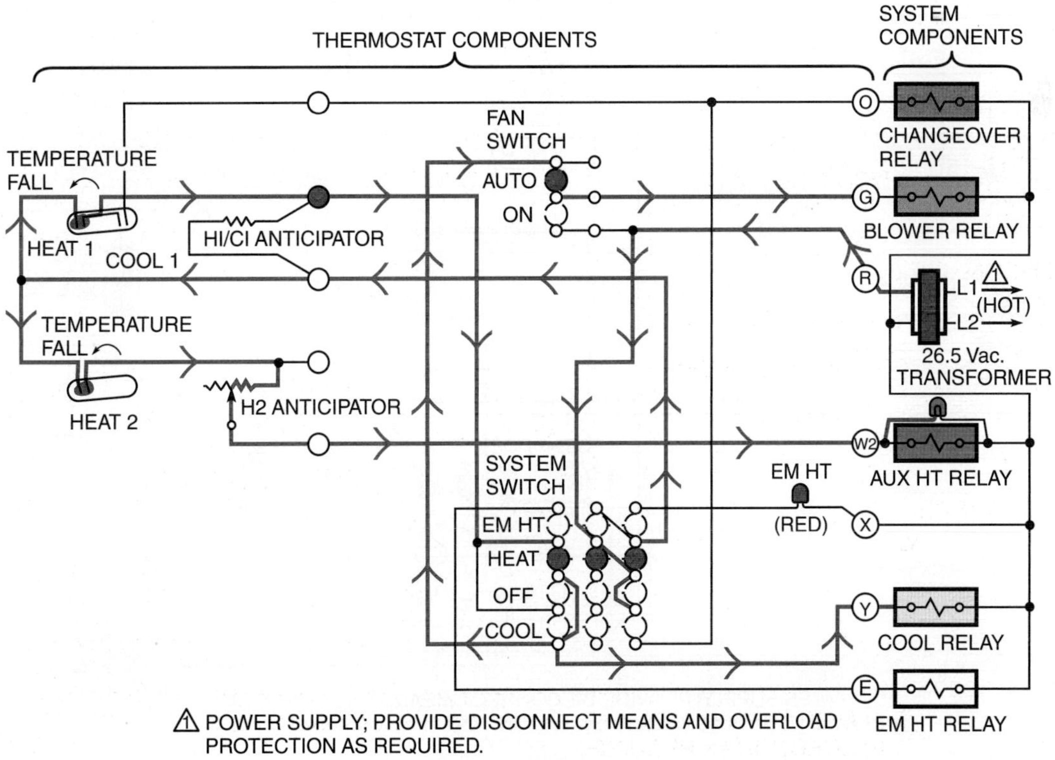

Figure 43.46 shows the system operating in the emergency heat mode. Notice there is no circuit to start the indoor blower. It must be started from an electric heat sequencer or from a thermal fan control or switch (see Unit 30, "Electric Heat"). When the emergency heat relay is energized, it will provide a circuit to jump out all outdoor thermostats and make all electric heat controllable through the second stage of the indoor thermostat. Both the auxiliary heat light and the red emergency heat light will remind the owner that excess energy is being used.

INDOOR FAN MOTOR CONTROL

Starting the indoor blower motor of a heat pump differs from starting it in other heating systems. In other heating systems, the indoor blower is started with a temperature-operated switch. With a heat pump, the blower must start at the beginning of the cycle, which is controlled by the thermostat. The indoor blower is started with the blower terminal, often labeled G. The compressor operates whenever the unit is calling for cooling or heating, so the G or fan terminal is energized whenever there is a call for cooling or heating.

HEAT ANTICIPATOR

Heat anticipators are incorporated into space thermostats and are intended to reduce temperature swing in the occupied space and overshoot. Temperature swing is the difference between the minimum and maximum temperatures in the occupied space during the heating cycle. Overshoot is the amount by which the space temperature exceeds the thermostat's set point.

Conventional heat pump thermostats have heat anticipators just as gas and oil heat systems do. They can be more complicated with heat pumps because there are often two anticipators. The thermostat in **Figure 43.36** has two anticipators. The one in the first stage is preset and not adjustable. The one in the second stage is adjustable and is coordinated with the auxiliary heat relay. Digital thermostats have anticipators, but they are self-adjusting and need not be field-set. A study of Unit 15, "Troubleshooting Basic Controls," would help to answer any heat anticipator questions. When electric heat is involved, there can be many control variables that require understanding the heat anticipator because often the auxiliary heat has several controls that must be taken into consideration while working with heat anticipation. Unit 30, "Electric Heat," contains a lengthy discussion

Figure 43.46 A generic HEAT-COOL heat pump thermostat operating in the emergency heat mode.

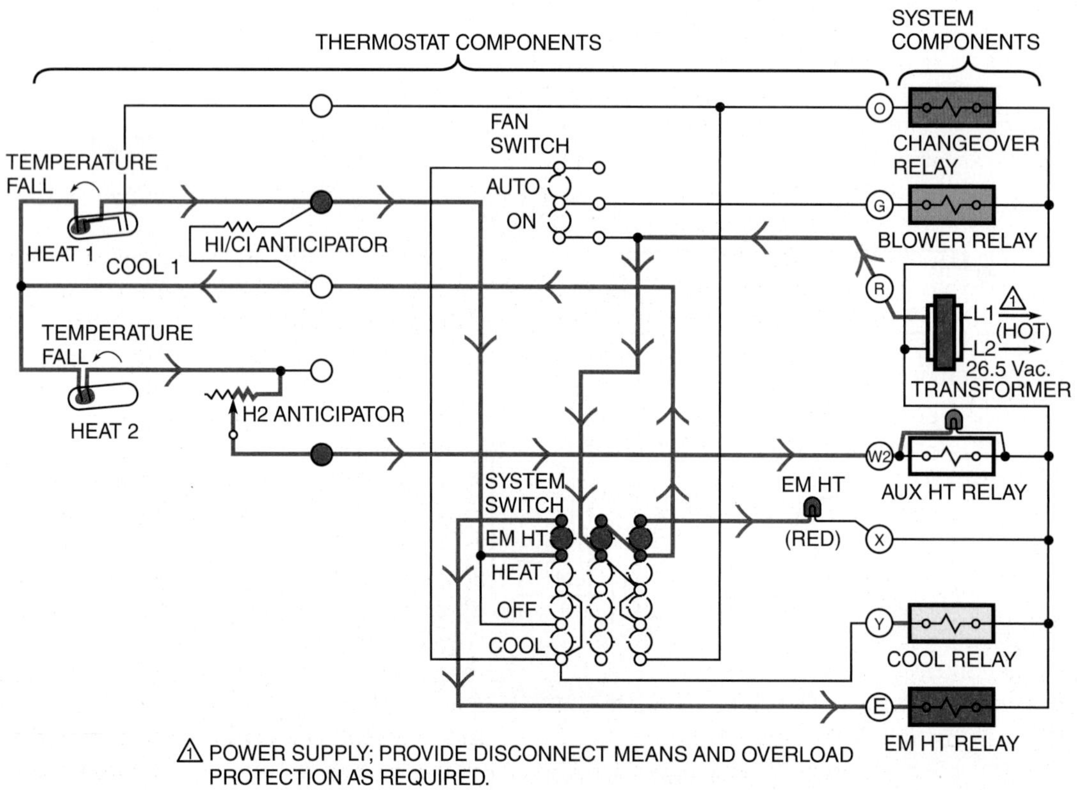

on electric heat that could also apply to auxiliary heat for a heat pump system.

ELECTRONIC THERMOSTATS

Most new thermostats use thermistors as the heat-sensing element. Remember, the thermistor is a resistor-type device that changes resistance with a change in temperature. There are several reasons why the thermistor is used. For one, mercury (Hg), which was used exclusively in older thermostats, is a poisonous, toxic, environmental hazard. There is a very small amount of mercury in each bulb of these older thermostats. A two-stage heat, two-stage cooling thermostat will often have four of these bulbs. When mercury bulbs were used, the temperature-sensing device was a bimetal strip or coil. The mass of the bimetal used to sense temperature caused rather large temperature swings. Remember, the thermostat's sensing element is located in a nearly dead air space under the cover of the thermostat. The thermistor has a very low mass and will respond much faster to temperature changes. For more information about electronic thermostats, see Unit 16, "Advanced Automatic Controls—Direct Digital Controls (DDCs) and Pneumatics."

One big difference in a system that has electronic controls as opposed to older analog controls is the number of features that can be incorporated into the system. These features often include timing functions, sensing functions, occupied and unoccupied settings, and error codes, which are explained in detail in Unit 16, "Advanced Automatic Controls—Direct Digital Controls (DDCs) and Pneumatics." Otherwise, troubleshooting the electronic thermostat is very similar to troubleshooting an older, analog, mercury bulb thermostat. If the electronic thermostat or control displays error codes, you follow them. Otherwise, treat it like a switch and jump from terminal to terminal at the subbase to determine if the thermostat has a problem. With a typical thermostat terminal designation, the R terminal is the hot terminal, the Y terminal is the compressor, the W terminal is heat, and the G terminal is the indoor blower. When the thermostat is disconnected from the subbase, a simple jumper placed from R to G should run the indoor fan. **NOTE:** *There may be a time delay built in, so the fan may not start for several minutes.* • See Units 15, "Troubleshooting Basic Controls," and 16, "Advanced Automatic Controls—Direct Digital Controls (DDCs) and Pneumatics," for information on troubleshooting thermostats and wiring circuits. **Figure 43.47** shows an electronic thermostat along with its features.

Figure 43.47 This electronic thermostat has many features. Courtesy Carrier Corp.

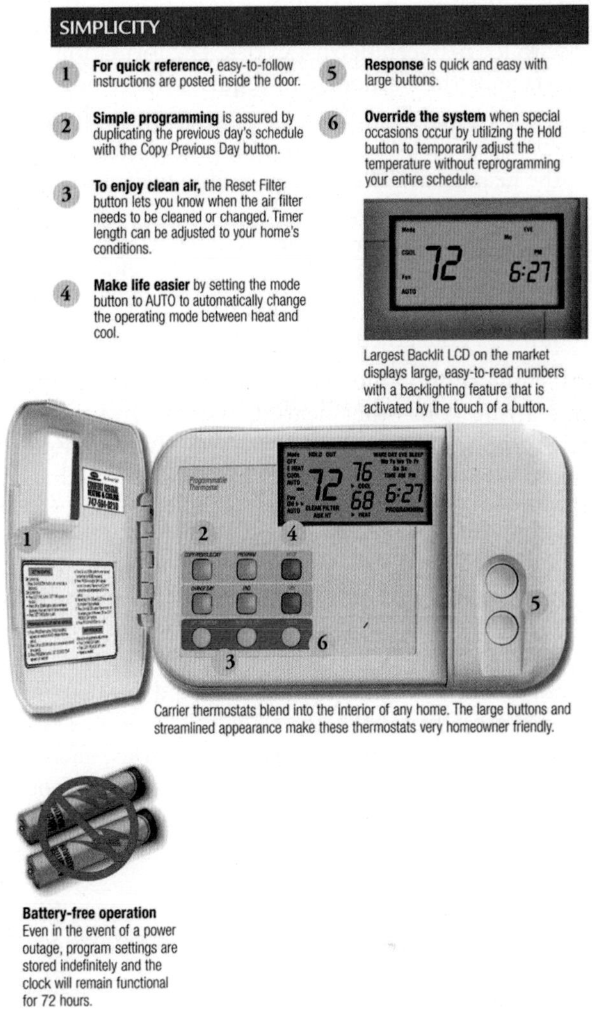

43.13 THE DEFROST CYCLE

While operating in the heating mode, the outdoor coil functions as the evaporator. When the outside ambient temperature is warm—above about 50°F—the evaporator saturation temperature will likely be above freezing and the coil will not freeze. When the outside ambient temperature drops, the evaporator saturation temperature can drop below 32°F, and the outdoor coil can begin to freeze. If the frost on the coil is allowed to accumulate, system performance will be reduced. The defrost cycle is used to defrost the ice from the outdoor coil during the heating mode.

Remember that, in the heating mode, the outdoor coil must be colder than the outside air in order for the refrigerant in the coil to absorb heat from the outside air. The need for defrost varies with the outdoor air conditions, which include temperature and relative humidity, and the running time of the heat pump. For example, if the outdoor temperature is 45°F, the heat pump will very likely be able to satisfy the heating needs of the structure and cycle off from time to time. The frost will melt from the coil during this off time. As the outside ambient temperature falls, increased system run time causes more frost to form on the outdoor coil. If the outside

air contains a great deal of moisture, the amount of ice that accumulates on the coil will increase. During cold, rainy weather, when the air temperature is 35°F to 45°F, frost will form so fast that the outdoor coil will very likely become covered with ice between defrost cycles. The formation of ice on the surface of the outdoor coil requires the removal of heat (144 Btu/lb) from the water. This heat energy is absorbed into the heat pump system. Later on, when a defrost cycle is initiated, the heat from the system is used to melt the ice that has formed on the outdoor coil.

Manufacturers have made significant advances in the managing of the defrost cycle to increase the efficiency of their equipment. Many years ago, heat pump manufacturers were not too concerned about system efficiency, but stricter government guidelines require that systems achieve higher efficiencies. Two important efficiency indicators are the seasonal energy efficiency ratio (SEER) for cooling and the heating seasonal performance factor (HSPF) for heating.

The SEER rating of a piece of equipment is the total cooling output for the system for an estimated typical cooling season divided by the actual power input for the entire season in watt-hours. Starting with the energy efficiency ratio (EER) rating where the steady-state output was divided by the power required to obtain the output, the EER was expanded to a seasonal ratio because it did not take into consideration the effects of starting and stopping the system, the defrost cycle, or different weather conditions, such as mild spring and fall seasons. Remember, a system does not get up to maximum efficiency for about 15 min after start-up. When the system cycles off, the cold cooling coil is still available to absorb more heat from the conditioned space. Many manufacturers include a time-delay start on the indoor blower to let the coil become cool before air is passed through it. They may also include a time delay incorporated into the control scheme to keep the indoor blower operating at the end of the cooling cycle. This allows the cooling coil to remove more heat from the airstream at the end of the cycle. All of this is available because of electronic control systems.

The HSPF rating describes the seasonal performance for a particular piece of equipment. In the United States, we have cold dry areas, moderate humid areas, and everything in between these extremes. The HSPF rating for each piece of equipment can be used to calculate the cost of operation in six different zones in the United States. These zones were established by the Department of Energy (DOE). This gives the potential owner or the calculating contractor a method for predicting the cost of operation for a piece of equipment and compare it to other equipment for both summer and winter operation. The defrost cycle figures into this cost analysis.

The SEER and the HSPF ratings are a result of federal energy policies and are used by the Air-Conditioning, Heating and Refrigeration Institute (AHRI) to rate equipment. AHRI is a national trade association, whose members include the majority of the equipment manufacturers in the industry. It maintains a test laboratory for verifying equipment standards and publishes the SEER and HSPF ratings for equipment. The higher the efficiency rating, either SEER or HSPF, the higher the Btu output for the same input.

HOW IS DEFROST ACCOMPLISHED?

For the heat pump system to operate in the defrost mode, a number of things must take place:

- The four-way reversing valve switches the system into the cooling mode. This involves energizing the reversing valve coil for systems that fail in the heating mode—or deenergizing the reversing valve coil for heat pump systems that fail in the cooling mode.
- The outdoor fan motor is deenergized. With the fan not operating, more heat will be concentrated in the coil, allowing the frost to melt much faster than if the fan were operating.
- The indoor blower will remain on. Since the indoor coil is functioning as the evaporator in the defrost mode, the coil can freeze if the indoor blower is not operating.
- Some auxiliary heat is energized during defrost. This will help temper the cooler air being introduced to the occupied space during defrost. During defrost, the temperature of the occupied space should not be increased. If the temperature of the space is increased, the thermostat's set point may be reached and the heating cycle will end. If this happens, the defrost cycle will end as well and the ice removal process will stop.

When the heat pump system is in defrost with the auxiliary heat energized, the system is cooling and heating at the same time and is not efficient—so the defrost time must be minimized. *Demand defrost* means defrosting only when needed. Combinations of time, temperature, and pressure drop across the outdoor coil are used to determine when defrost should be started and stopped.

The factors that vary the operating cost of a piece of equipment are the number of defrosts per season, length of defrost, and whether the manufacturer tempers the air to the conditioned space during defrost with auxiliary heat. Remember, the unit will be in the cooling mode and blowing cold air during defrost. If the manufacturer can get the equipment rated without the auxiliary heat, the rating will be better. The equipment will have a better HSPF, but the customer may not be satisfied with the cool air that blows into the space during defrost.

INITIATING THE DEFROST CYCLE

Manufacturers often design their defrost control strategies to initiate defrost shortly after the coil starts to accumulate frost in order to maximize system performance. Some manufacturers use time and temperature to start defrost. This is called *time and temperature initiated* and is performed with a timer and

Figure 43.48 A solid-state defrost control board. Photo by Eugene Silberstein

Figure 43.49 A thermistor used to sense the temperature of the outdoor coil. Photo by Eugene Silberstein

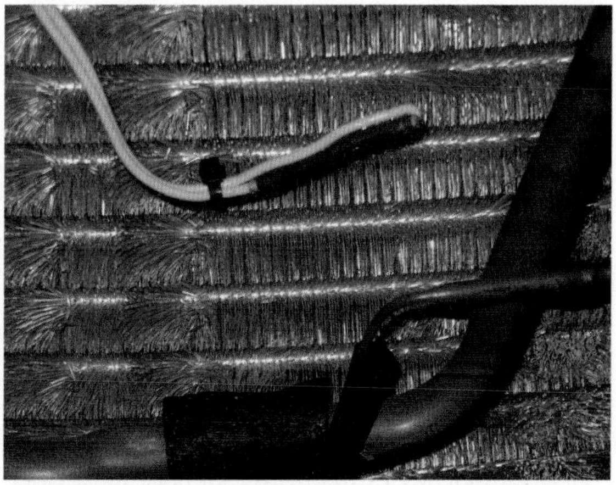

temperature-sensing device. The timer is often incorporated into a solid-state, printed circuit board, **Figure 43.48**. The temperature sensor, such as the one used to measure the temperature of the outdoor coil, is typically a thermistor, **Figure 43.49**. The timer closes a set of contacts for 10 to 20 sec for a trial defrost every 30, 60, or 90 min. The time interval between defrost trials is field set and is determined in part by climatic conditions, mainly humidity. In high-humidity climates, more frequent defrost may be needed. In very dry climates, the 90-min time interval may be used. The timer contacts are in series with the temperature sensor contacts so that both contacts must be made at the same time. This means that two conditions must be met before defrost can start, **Figure 43.50**. In operation, when the time delay period has elapsed, the timer will attempt to bring the heat pump system into defrost. If the temperature sensor determines that there is ice on the coil, defrost will start.

Another common method of starting defrost uses an air pressure switch that measures the air pressure drop across the outdoor coil. When the unit begins to accumulate ice, a pressure drop occurs and the air switch contacts close. This switch can be wired in conjunction with the timer and the temperature sensor to ensure that an actual ice buildup is on the coil. The combination of time and temperature only ensures that time has passed and the coil is cold enough to actually accumulate ice. It does not actually sense an ice thickness. It can be more efficient to defrost when an actual buildup of ice exists, **Figure 43.51**.

TERMINATING THE DEFROST CYCLE

Terminating the defrost cycle at the correct time is just as important as starting defrost only when needed. This is done

Figure 43.50 A wiring diagram of conditions to be met for defrost to start.

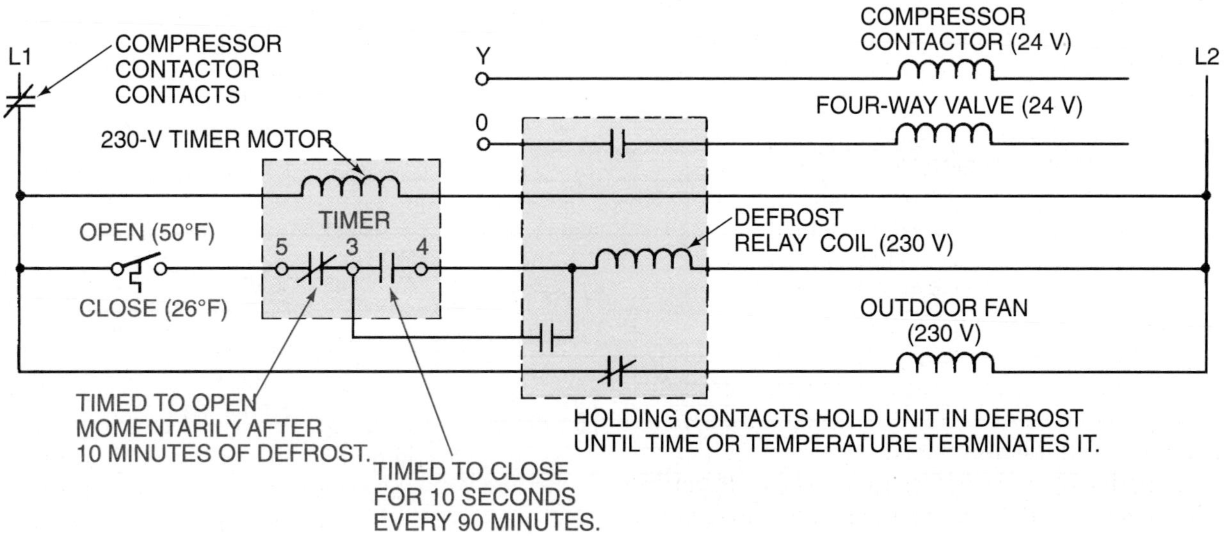

Figure 43.51 A wiring diagram taking ice buildup into consideration.

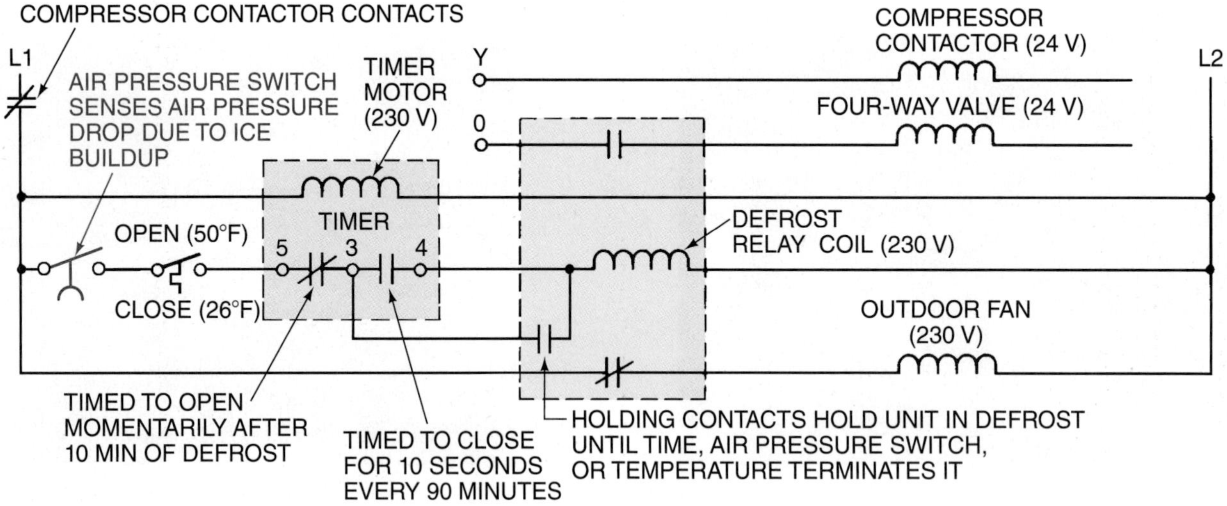

in several ways. *Time and temperature* was the terminology used to start defrost. *Time or temperature terminated* is the terminology used to terminate one type of defrost cycle. The difference is that two conditions must be satisfied to start defrost, and either one of two conditions can terminate defrost. After defrost has started, time or temperature can terminate it. For example, if the defrost temperature sensor warms up to the point that it is obvious ice is no longer on the coil and its contacts break, defrost will stop. This temperature termination is normally 50°F at the location of the temperature sensor. If the contacts do not open because it is too cold outside, the timer will terminate defrost within a predetermined time period. Normally, the maximum time the timer will allow for defrost is 10 min, **Figure 43.52**.

ELECTRONIC CONTROL OF DEFROST

Electronic circuit boards and their controls can help the manufacturer more closely control the defrost cycle. They have the ability to accurately control time and temperature with electronic timers and thermistors. Some systems may use the difference in the entering air temperature and the coil temperature to arrive at a temperature split where defrost is desired. For example, suppose the outdoor temperature is 35°F. The coil will operate at about 20°F to 25°F colder than the outdoor coil. Using a 20°F temperature split as an example, suppose the coil begins to develop ice on the surface. As ice forms on the coil, the pressure across the coil will drop, but the temperature will also drop inside the coil

Figure 43.52 A wiring diagram for a time- or temperature-terminated method of defrost.

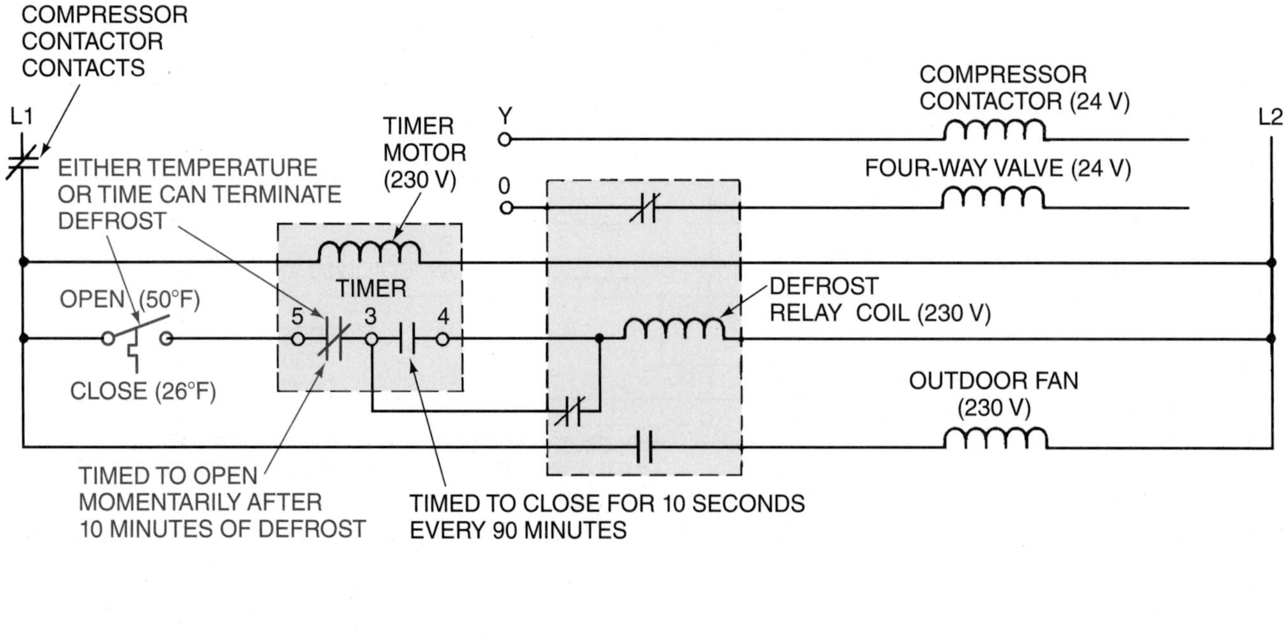

Figure 43.53 This system has ice accumulating on the outdoor coil and is in need of defrost. The manufacturer is using the difference between the temperature of the outdoor air entering the coil and the actual coil temperature, measured with a thermistor mounted on the coil.

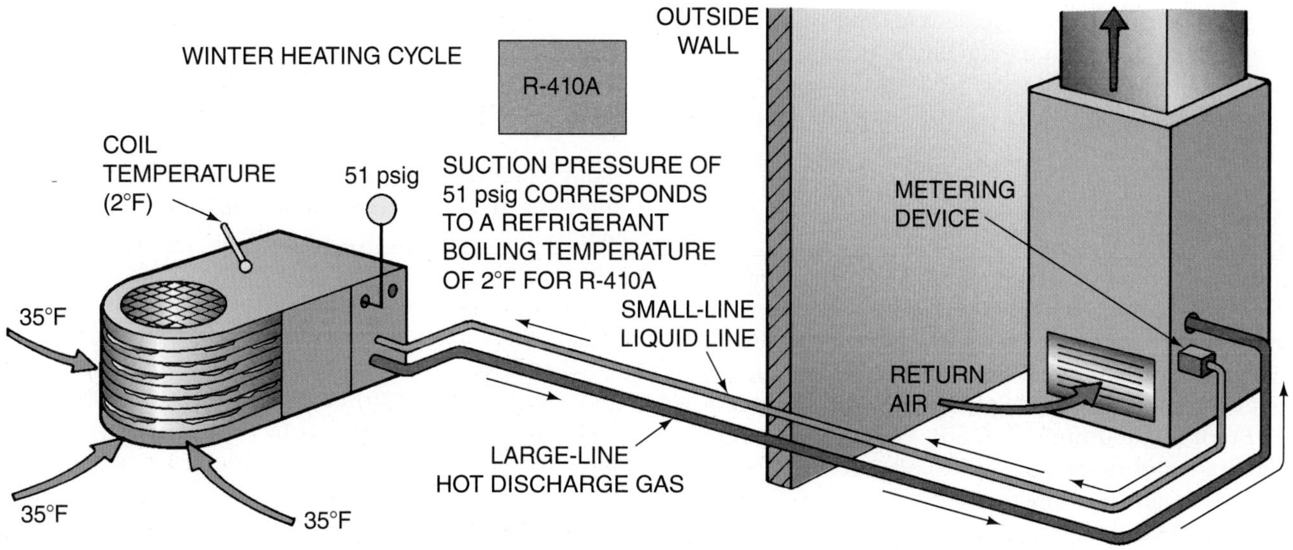

as ice develops. The manufacturer may decide that it is time for a defrost when the coil temperature has dropped to 8°F below the normal operating temperature for that condition, **Figure 43.53**. If the time portion of the defrost initiation has been satisfied, the unit will start to defrost and will be terminated by the coil temperature. All of this is accomplished with the electronic circuit board and two thermistors.

Manufacturers of heat pumps may use other combinations of conditions to operate the defrost cycle that include air pressure drop across the coil. They look for something economical, reliable, and effective for making the HSPF number higher for a better rating.

43.14 SECOND-STAGE ELECTRIC HEAT

During periods of colder weather, the air-to-air heat pump will often require auxiliary heat. This is usually accomplished with an electric furnace with a heat pump coil for cooling and heating. Electric heat is often started with a sequencer that has a time-start and time-stop feature for the electric heating element. This means that the electric heating elements are energized for a time period if the unit is shut off by the thermostat's on-off switch. When the thermostat is allowed to control the system normally, the heat pump will be the first component to come on and the last to shut off. The indoor blower will start with the compressor through the blower relay circuit. The electric strip heat will operate as the last component to come on and the first to shut off. When the owner decides to shut off the heat at the thermostat and the strip heat is energized, the heating elements will remain energized until the sequencers cycle them off. A temperature switch is often used to sense this heat and keep the blower running until the heaters cool. The electric furnaces used for heat pumps are like the ones described in Unit 30, "Electric Heat."

Electric strip heaters are used for three purposes in the heat pump system. They provide

- supplementary heat when the heating system is operating below the balance point of the system,
- emergency heat when the reverse-cycle heat pump system is not functioning, and
- air tempering when the system is operating in the defrost mode.

Depending on the function these electric strip heaters are intended to serve, the number of heaters to be energized will change. When acting to provide supplementary heat, the number of heaters energized is determined by the outside ambient temperature. When the temperature is colder outside, more heaters can be energized. When the outside temperature is warmer, fewer, if any, heaters need to be brought into operation.

When providing emergency heat, all of the available strip heaters are energized. This is because the main/primary heat source is not operating and the strip heaters are attempting to satisfy the heating requirements of the space by themselves. When used to temper the air during defrost, it is very important not to energize too many strip heaters. As mentioned earlier in this unit, the heaters used to temper the air during defrost should not raise the temperature of the occupied space. Consider a heat pump system that has just gone into defrost. The space temperature is 70°F and the thermostat is

set to 71°F. If the heaters temper the air too much, it is possible that the temperature of the space may increase by 1°F. If the temperature sensed by the thermostat reaches 71°F, the call for heat will end, as will the defrost cycle. If the defrost ends, any unmelted ice will remain on the outdoor coil when the system returns to the heating mode. If this continues, ice will continue to accumulate on the outdoor coil and system efficiency will be affected. So, to prevent this from occurring, the amount of heat added to the air during defrost should not exceed 80% of the sensible cooling capacity of the system. Consider the following example.

A heat pump system has a sensible cooling capacity of 60,000 Btu/h. During defrost, the system switches over to the cooling mode, so the air introduced to the occupied space will be cooled. We want to temper this air with enough heat to offset 80% of the cooling capacity—which, in this case, is 48,000 Btu/h (60,000 Btu/h × 0.80). The conversion factor between Btu/h and watts is 3.413 Btu/h/W, or Btu/Wh, so we will need to energize no more than 14,064 W (48,000 Btu/h ÷ 3.413 Btu/Wh) of electric heat. If the heat pump system is equipped with six 3.5-kW electric strip heaters (3500 W each), up to four heaters can be energized during defrost. These four heaters will provide 14 kW of heat (3.5 kW × 4) to the air passing over the strip heaters. Since the 14 kW (14,000 watts) is less than the maximum calculated value, the space will not be heated above its present temperature during defrost. This will allow the system to complete its defrost cycle as needed.

43.15 SERVICING THE AIR-TO-AIR HEAT PUMP

Servicing the air-to-air heat pump is much like servicing a conventional air-conditioning system. During the cooling season, the unit is operating as a high-temperature refrigeration system; and during the heating season, it is operating as a low-temperature system with planned defrost. The servicing of the system is divided into electrical and mechanical servicing.

Servicing the electrical system is much like servicing any electrical system that is used to operate and control refrigeration equipment. The components that the manufacturer furnishes are built to last for many years of normal service. One significant point about a heat pump is that it has much more running time than a cooling-only unit. A typical cooling unit in a residence in Atlanta, Georgia, will run about 120 days per year. A heat pump may run 250 days, more than twice the time. In winter the heat pump may run for days and not stop when the temperature is below the balance point of the structure.

SAFETY PRECAUTION: *The same safety precautions must be observed when troubleshooting electrical components of a heat pump as were observed when servicing an air-conditioning system. When the units are located in crawl spaces and you are in contact with the ground, be careful not to establish a path for current to flow through your body to ground.*•

43.16 TROUBLESHOOTING THE ELECTRICAL SYSTEM

Troubleshooting the electrical system of an air-to-air heat pump is similar to troubleshooting a cooling-only, air-conditioning system. Recall that an electric circuit has a power supply, a complete path for the electric current to take, and a load. The power supply provides the electrical energy to the unit, the complete path is the wiring, and the closed switches or electrical contacts and the loads are the electrical components that perform the work. Every circuit in the heat pump wiring can be reduced to these items. It is important that you think in terms of power, conductor, switch, and load, **Figure 43.54.**

It is usually easy to find electrical problems if some component does not function. For example, when a compressor contactor coil is open, it is evident that the problem is in the contactor. A voltage reading can be taken on each terminal of the contactor coil. If there is voltage and it will not energize the contacts, a resistance check of the coil will show whether there is a circuit through it. A correctly operating circuit has the correct resistance. No resistance or too little resistance will create too much current flow. Infinite resistance results in no current flow, as there is no path. It sometimes helps to be able to compare a suspected bad component (e.g., one with too little resistance) to a component of correct resistance, **Figure 43.55.**

Electrical malfunctions that cause the technician real problems are the intermittent ones, such as a unit that does not always defrost so that ice builds up on the outdoor

Figure 43.54 This is the basic example of a power supply, a path or conductor, and a load for consuming power.

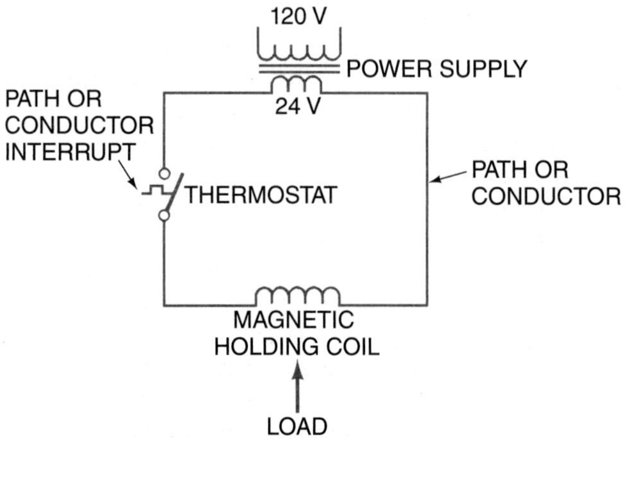

Figure 43.55 Magnetic holding coils used for the contactor. One of them has a measurable resistance of 0.5 Ω. The other one, the correct one, has a measurable resistance of 20 Ω. The one with the lower resistance is shorted. It will draw too much current and place a load on the control transformer. If the load is too great, it will overload the transformer.

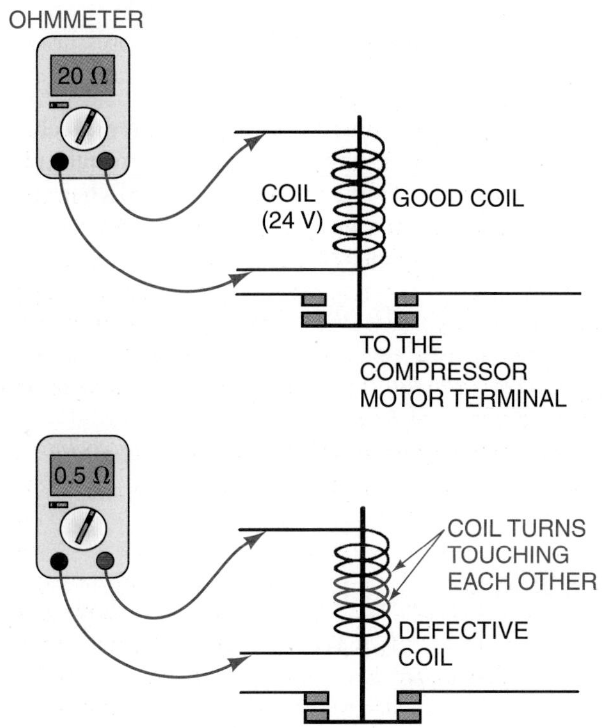

coil. The heat pump can be difficult to troubleshoot in winter because of the weather. **SAFETY PRECAUTION:** *If it is wet outside, the emergency heat feature of the heat pump should be used until the weather improves. It is hard enough to find electrical problems in cold weather; the moisture makes it dangerous.•*

TYPICAL ELECTRICAL PROBLEMS

1. Indoor blower motor and outdoor fan motor
   A. Open winding: motor will not start; may draw locked rotor amperage; may have infinite resistance across one winding, normal resistance across the other winding; motor will overheat.
   B. Shorted winding: motor may or may not start; motor will draw higher amperage.
   C. Shorted winding to ground: motor may or may not start; circuit breaker may trip or fuse may blow; motor may draw excessive current; there will be measurable resistance between the motor terminals and ground.
   D. Open run capacitor: motor will not start; motor will draw locked rotor amperage; motor will overheat.
   E. Open start capacitor: motor will likely not start; motor will draw high amperage; motor will overheat if unable to start.
   F. Shorted run capacitor: motor may or may not start; motor will draw higher amperage; motor will run slowly if it does start.
   G. Shorted start capacitor: motor will likely not start; if motor starts it will turn slowly and draw high amperage.

2. Compressor contactor, fan relays
   A. Coil winding open: device will not close NO contacts; will not open NC contacts.
   B. Shorted coil winding: low-voltage control circuit protection device may trip; control transformer may overload; device will not close NO contacts; will not open NC contacts.
   C. Pitted contacts: reduced voltage to the load being controlled; there will be a measurable voltage across closed contacts.

3. Defrost relay
   A. Coil winding open: contacts on the relay will not close or open automatically.
   B. Coil winding shorted: circuit overload.
   C. Pitted contacts: reduced voltage to the load being controlled; there will be a measurable voltage across closed contacts.

4. Compressor
   A. Open compressor motor winding: will not start; may draw locked-rotor amperage if only one winding is open.
   B. Shorted compressor motor winding: circuit breaker trips when compressor is energized.
   C. Shorted winding to ground: The breaker or fuse will open up. A short from winding to winding may allow the motor to start and run until the breaker trips or the overload opens.
   D. Open internal overload: no current draw; no compressor operation; compressor may be hot to the touch; there may be infinite resistance readings between common and run and common and start because the motor winding thermostat is open. There will be a measurable resistance between run and start because the meter is reading through the run and start windings.

5. Reversing valve solenoid
   A. Open coil: system will remain in the mode in which the system fails (usually heating); no defrost cycle.
   B. Shorted coil: fuse protection trips when system changes over.

6. Electric strip heaters
   A. Open heater: may or may not be noticeable, depending on how many heaters are affected; there may be

a reduced air-tempering effect, a reduced heating capacity (second-stage and emergency heat modes), and/or no current draw through the heater circuit. Sometimes a heater will short to ground and stay energized operating in the 115-V mode. In the cooling cycle, it will heat the structure and cause the cooling to operate to cancel the heat being introduced to the structure. In the heating mode, it will cause excessive electrical costs.
   B. Open safety switch: no current flow through the heater.
   C. Open blower-proving device: no voltage supplied to the coil of the heater control relays or sequencer; no current flow through the heaters.

7. General wiring problems
   A. Loose electrical connections: burned connections; localized heat; intermittent system problems.
   B. Short circuit: tripping breakers or blowing fuses. **SAFETY PRECAUTION:** *Breakers should not be reset repeatedly; damage can occur.•*

43.17 TROUBLESHOOTING MECHANICAL PROBLEMS

Mechanical problems can be hard to find in a heat pump, particularly in winter operation. Summer operation of a heat pump is similar to that of a conventional cooling unit. The pressures and temperatures are the same, and the weather is not as difficult to work in. Mechanical problems are solved with gauge manifolds, wet- and dry-bulb thermometers, and air-measuring instruments.

SOME TYPICAL MECHANICAL PROBLEMS

1. Indoor unit
   A. Air filter dirty (winter); high head pressure; low COP.
   B. Dirty coil (winter); high head pressure; low COP.
   C. Air filter dirty (summer); low suction pressure.
   D. Dirty coil (summer); low suction pressure.
   E. Refrigerant restriction (summer); low suction.
   F. Refrigerant restriction (winter); high head pressure.
   G. Defective TXV on the indoor coil (summer); low suction pressure.
   H. Defective TXV on the outdoor coil (winter); low suction.
   I. Leaking check valve (summer); flooded indoor coil.
2. Outdoor unit
   A. Dirty coil (winter); low suction pressure.
   B. Dirty coil (summer); high head pressure.
   C. Four-way valve will not shift; stays in same mode.
   D. Four-way valve halfway; pressures will equalize while running and appear as a compressor that will not pump.
   E. Compressor efficiency down; capacity down, suction and discharge pressures too close together. Compressor will not pump pressure differential.
   F. Defective TXVs (winter); low suction pressure.
   G. Leaking check valve (winter); flooded coil.

The technician should be aware of three mechanical problems: (1) the four-way valve leaking through, (2) the compressor not pumping to capacity, and (3) charging the heat pump. These particular problems are different enough from commercial refrigeration that they need individual attention.

SAFETY PRECAUTION: *Be aware of the high-pressure refrigerant in the system while connecting gauge lines. High-pressure refrigerant can pierce your skin and blow particles in your eyes; use eye protection. Liquid refrigerant can freeze your skin; wear gloves. The hot gas line can cause serious burns and should be avoided.•*

43.18 TROUBLESHOOTING THE FOUR-WAY REVERSING VALVE

The four-way reversing valve can cause a technician a lot of confusion when troubleshooting. Several things can cause the four-way valve to malfunction:

1. The valve may be stuck in heating or cooling.
2. The coil may be defective.
3. The valve may have an internal leak.

The sign of a stuck reversing valve may be that it will not change from heating to cooling and back. When the valve is energized in the cooling mode and is trying to operate in the heating position, the first thing the technician should do is make sure the coil is energized. The technician may do this by using a voltmeter at the coil lead wires. Another procedure is to hold a screwdriver close to the coil to see if there is a magnetic field when the valve is supposed to be energized. If the valve is getting power and will not change over while the unit is running and there is a pressure difference from the high to the low side, the valve may be stuck. Remember, the valve will not change over unless there is a pressure difference between the high- and the low-pressure sides of the system because it is a pilot-operated valve. The technician may also feel the valve coil to see if it is hot. That is a sign that it is energized.

If the reversing valve is truly stuck, the technician may take a soft-face hammer or some other soft-surface instrument and tap the valve on one end and then the other to see

if it will break loose. Something inside may be binding it. If it breaks loose, the technician should force the valve to change positions from heat to cool several times to see if it is free to operate. The valve should still be replaced, as the cause for the malfunction has not been addressed. The valve will stick again at some point in the future.

The solenoid coil on the reversing valve can have an open circuit, which will cause it to operate only in the deenergized mode. Since the coil is removable, it can be changed. It is always desirable to change the coil if that will repair the valve, rather than to change the complete valve.

A four-way reversing valve with an internal leak can easily be confused with a compressor that is not pumping to capacity. The suction pressure on a system with a leaking reversing valve will be higher than normal and the discharge pressure will be lower than normal. The capacity of the system will be reduced in both the heating and cooling modes. This is the same symptom as small amounts of hot gas leaking from the high-pressure side of the system to the low-pressure side. When a four-way reversing valve is suspected of leaking internally, a good-quality thermometer is used to check the temperature of the low-side line, the suction line from the evaporator (the indoor coil in summer or the outdoor coil in winter), and the permanent suction line between the four-way valve and the compressor. The temperature difference should not be more than about 3°F. **NOTE:** *Take these special precautions when recording temperatures: (1) Take the temperatures at least 5 in. from the valve body (to keep valve body temperature from affecting the reading); and (2) insulate the temperature lead that is fastened tightly to the refrigerant line,* **Figure 43.56.**•

REPLACING THE FOUR-WAY REVERSING VALVE

If it is determined that the four-way reversing valve needs to be replaced, a few things should be kept in mind to ensure that the system will operate as intended after the repair is made. Make certain that

- the new valve is thoroughly inspected before installation. Check the valve for scratches, dents, and other imperfections on the bodies of the main and pilot valves. Dents can prevent the internal slide from moving within the valve. If possible, inspect the valve while still at the parts depot to prevent a second trip if the valve turns out to be damaged.
- the new valve is not dropped or mishandled.
- excessive heat is avoided when installing the valve.
- the new valve is piped so that the valve will fail in the same mode of operation as the original.
- the new valve is installed horizontally with the pilot tubes on top. This will prevent oil from entering the pilot ports and causing the valve to malfunction. Always follow the manufacturer's guidelines when replacing the valve.
- an exact part replacement is selected whenever possible. If the ports on the new valve are too small, excessive pressure drops will be present across the valve, compromising system performance. If the new valve is too large, the switchover time will be increased, resulting in sluggish system operation.

43.19 TROUBLESHOOTING THE COMPRESSOR

Checking a compressor in a heat pump for pumping to capacity is much like checking a refrigeration compressor except that there are normally no service valves to work with. Some manufacturers furnish a chart with the unit to show what the compressor characteristics should be under different operating conditions, **Figure 43.57.** If such a chart is available, use it.

The following is a reliable test, using field working conditions, that will tell whether the compressor is pumping at near capacity. The test will reveal any large inefficiencies.

1. Whether summer or winter, operate the unit in the cooling mode. In winter, the four-way valve may be switched by energizing it with a jumper or deenergizing it by disconnecting a wire to switch the unit to the summer mode. This will allow the auxiliary heat to heat the structure and keep it from getting too cold.
2. Block the condenser airflow until the head pressure is 440 psig (this simulates a 95°F day) and the suction pressure is about 118 psig. This assumes that the system is operating with R-410A as its refrigerant.
3. The compressor amperage should be at close to full load. The combination of discharge pressure, suction pressure, and amperage should reveal whether the compressor is working at near full load. If the suction pressure is high and the discharge pressure is low, the amperage will be low. This indicates that the compressor is not pumping to capacity. Sometimes a whistle can be heard when the compressor is shut off under these conditions. The suction line may also get warm immediately after shutting off the unit under these conditions. Such symptoms indicate that the compressor is leaking internally. **NOTE:** *If there is any doubt as to where the leak may be, perform a temperature check on the four-way valve.*•

43.20 CHECKING THE CHARGE

Some heat pumps have a critical refrigerant charge. The tolerance could be as close as ±½ oz of refrigerant. **Therefore, do not install a standard gauge manifold each time you suspect a problem.** When a heat pump has a partial charge, it is obvious that the refrigerant leaked out of the

Figure 43.56 (A) A method for checking performance of a four-way valve using a temperature comparison. (B) A line diagram showing a defective valve in cooling and heating.

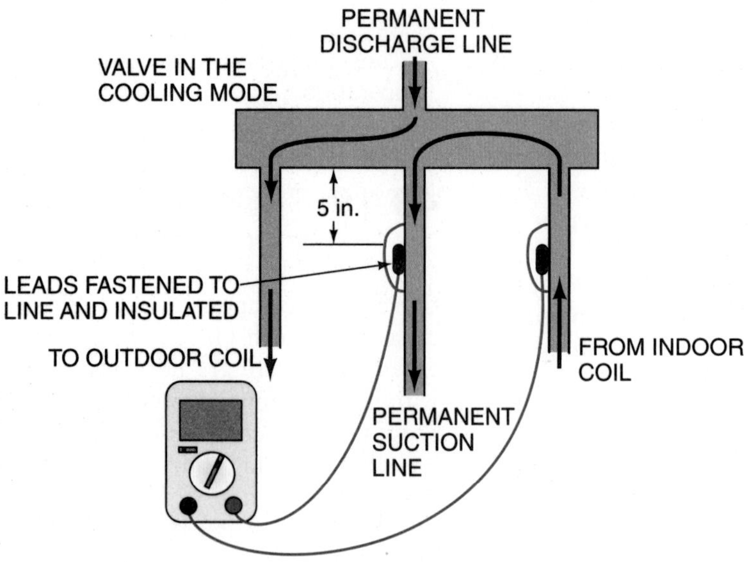

(A)

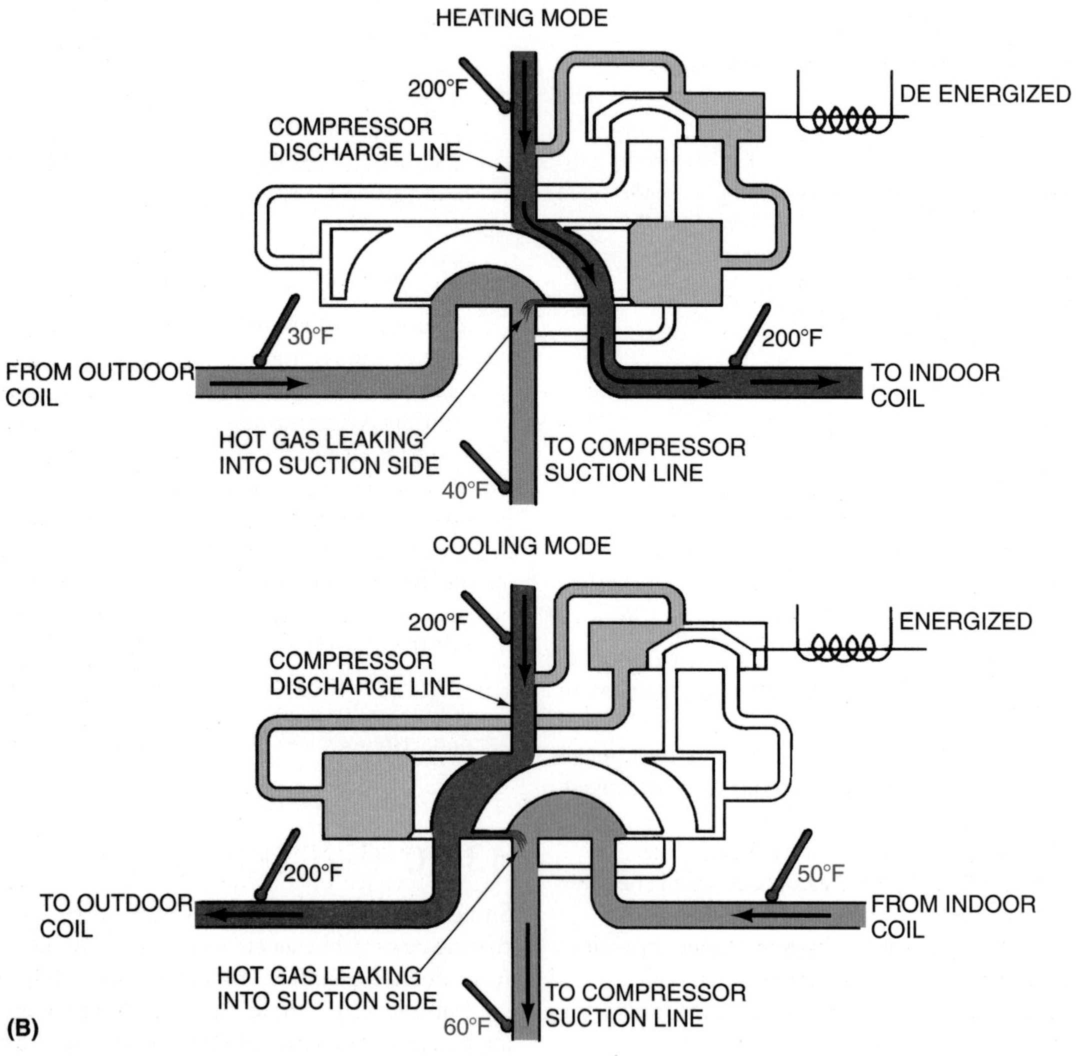

(B)

Figure 43.57 The performance chart for a 36,000-Btu/h R-410A heat pump system. Courtesy Carrier Corp.

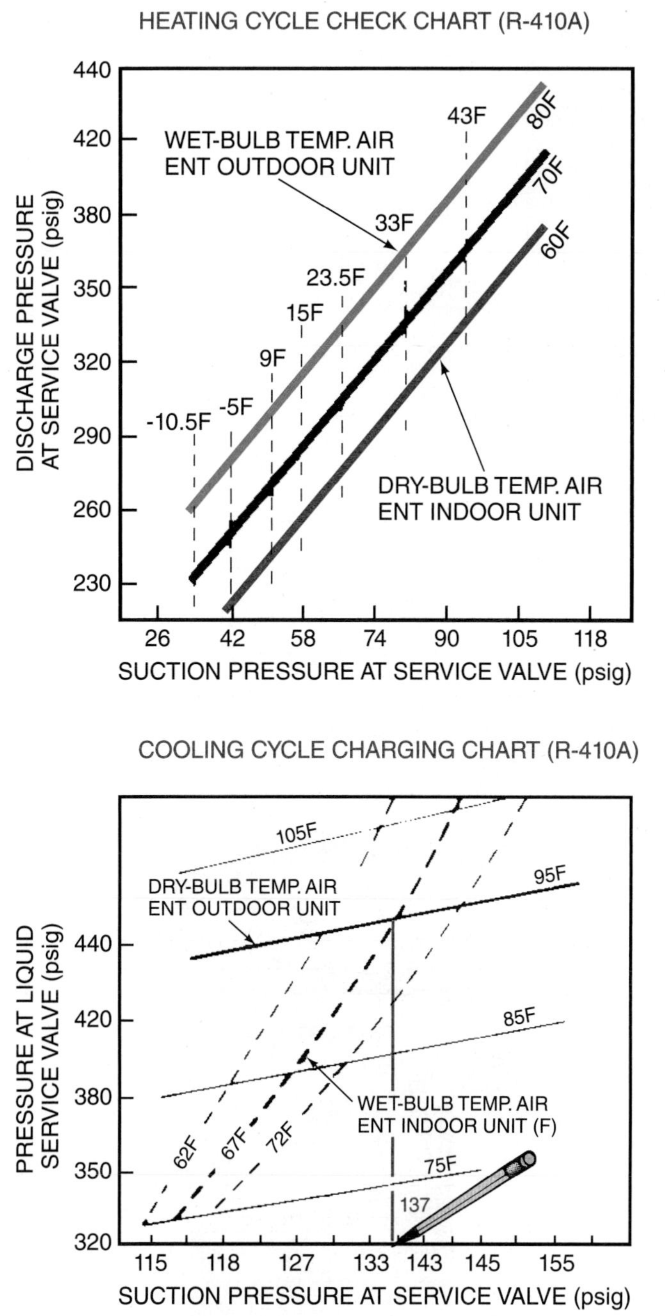

system. Leak check the system and repair the leak before charging.

When a partial charge is found in a heat pump, some manufacturers recommend that the system refrigerant be recovered, then evacuated to a deep vacuum and recharged by measuring the charge into the system. Some manufacturers furnish a charging procedure to allow adding a partial charge. It is always best to follow the manufacturer's recommendation.

43.21 SPECIAL APPLICATIONS FOR HEAT PUMPS

The use of oil or gas furnaces for auxiliary heat with heat pumps is a special application, and several manufacturers have designed such systems because they provide several advantages. Usually, natural gas is less expensive to use as the auxiliary fuel than electricity. Fuel oil may be considered if such as system is already installed. Both systems have similar control arrangements. The heat pump's indoor coil can be used in conjunction with an oil or gas furnace, but the coil must be downstream of the oil or gas heat exchanger. In other words, the air must flow through the oil or gas heat exchanger first and then through the heat pump indoor coil. **SAFETY PRECAUTION:** *The gas or oil furnace must not operate at the same time that the heat pump is operating or high head pressures will result.•*

The control function can be accomplished with an outdoor thermostat set at the balance point of the structure to change the call for heating to the oil or gas furnace and to shut off the heat pump. This allows the heat pump to operate down to the balance point, when the oil or gas furnace will take over. When defrost occurs, the oil or gas furnace will come on during defrost because the same controls are used as for a standard heat pump and a call for auxiliary heat to warm the cool air during defrost is still part of the control sequence. **SAFETY PRECAUTION:** *A high-pressure control can prevent the compressor from overloading if a defrost does not terminate due to a defective defrost control.•*

MAXIMUM HEAT PUMP RUNNING TIME

Other control modifications allow the heat pump to run below the balance point. Such arrangements have more relays and controls. The heat pump is designed to operate whenever it can. This is accomplished by using the second-stage contacts to start the oil or gas heat. This sequence stops the heat pump until the second-stage thermostat contacts satisfy. The heat pump then starts again. This sequence repeats itself each cycle. Before considering this control sequence, the system should be studied to see whether it is more economical to switch over from the heat pump to the auxiliary source or to restart the heat pump. The economic balance point of the structure may be determined by comparing the cost of fuels.

HEAT PUMP ADDED ON TO EXISTING ELECTRIC FURNACE

When a heat pump indoor coil is added to an existing electric furnace installation, the older furnace might not be set up to allow the heat pump coil to be located before the electric heating elements. If so, a wiring configuration similar to an oil or gas installation may be used, **Figure 43.58**.

Manufacturers have many different methods of building a heat pump with individual performance characteristics.

Figure 43.58 The electric furnace heat coils are below the heat pump coil and will cause a high head pressure if used incorrectly with a heat pump.

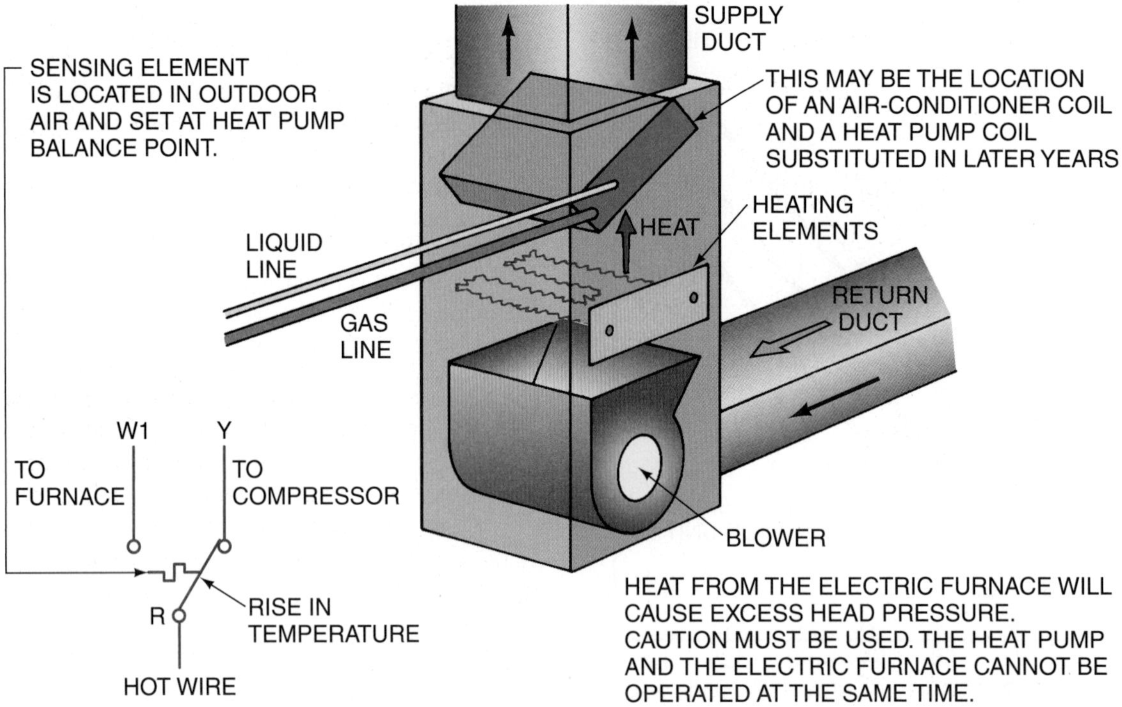

There are many different piping configurations and special heat exchangers. This text is intended to explain the typical heat pump. If the technician encounters a different system from that described here, the manufacturer should be consulted.

43.22 HEAT PUMPS USING SCROLL COMPRESSORS

Manufacturers are continually working to make heating and cooling equipment more efficient and less expensive to purchase. Some of the new techniques involve the use of improved compressors. Compressor pumping efficiency has been improved with the *scroll compressor,* **Figure 43.59**. This compressor is ideally suited for heat pump application because of its pumping characteristics. The scroll compressor does not lose as much capacity as the reciprocating compressor at the higher head pressures of summer or the lower suction pressures of winter operation. Because the compression takes place between two spiral-shaped forms, **Figure 43.60**, the scroll compressor does not have the top-of-the-stroke clearance volume loss that a reciprocating compressor has. The gas trapped in the clearance volume of a reciprocating compressor must reexpand before the cylinder starts to fill, which reduces efficiency (see Unit 23, "Compressors"). The scroll compressor is about 15% more efficient than the reciprocating compressor, and it has many of the same pumping characteristics as the rotary compressor. The scroll compressor does not require crankcase

Figure 43.59 A scroll compressor. Courtesy Copeland Corp.

Figure 43.60 The orbiting action between the scrolls in a scroll compressor provides compression.

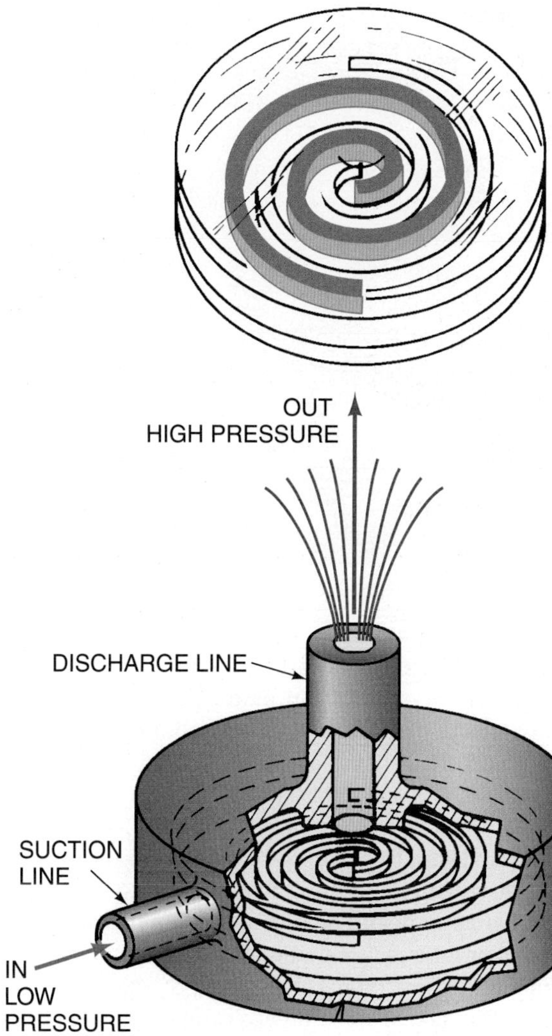

heat because it is not as sensitive to liquid as the reciprocating compressor. This results in a saving in power.

Scroll compressor pressures are about the same as those of reciprocating compressors, so the technician should not see any difference in the gauge readings. Scroll compressors are cooled by the discharge gas, so the technician will notice that the compressor shell is much hotter than that of a compressor that is cooled by suction gas. The scroll compressor runs with less vibration and a lower noise level while simultaneously operating at several stages of compression. Each pocket of gas represents a stage of compression. **Figure 43.61** illustrates one pocket.

To prevent pressures from equalizing back through the compressor during the off cycle, the scroll compressor has a check valve at its discharge. When the compressor is shut off, the gas trapped between the check valve and the compressor scroll will back up through the scroll; the compressor will make a strange sound—resembling the rustling of a handful of marbles—at the moment it is shut off until these pressures

Figure 43.61 The compression of one pocket of refrigerant in a scroll compressor. Courtesy Copeland Corp.

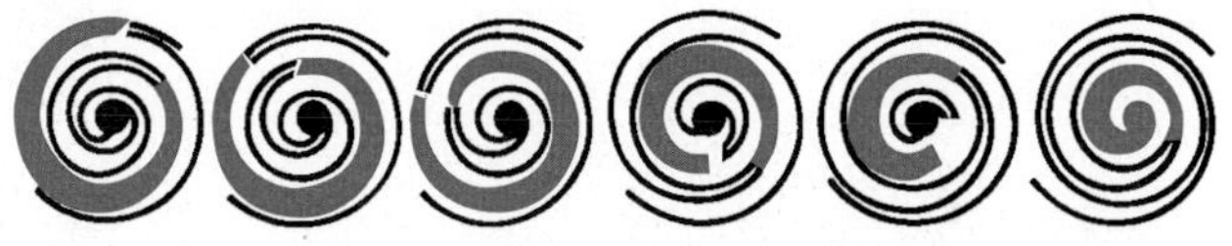

equalize. The high-side pressure will then equalize through the metering device to the low side, just as in a reciprocating compressor system. The equalizing of pressure between the check valve and the scroll allows the compressor to start up without hard-start assistance because there is no pressure differential for the first starting revolutions. Thus there are fewer start components to purchase and fewer components to cause trouble.

The scroll compressor does not normally require a suction-line accumulator, which is usually included in reciprocating compressor systems, because it is not as sensitive to liquid floodback as the reciprocating compressor. The scroll compressor is more efficient, stronger, lighter, and requires fewer accessories for good performance. Unlike reciprocating compressors, scroll compressors tend to "wear in" with time, as opposed to wearing out.

43.23 HEAT PUMP SYSTEMS WITH VARIABLE-SPEED MOTORS

Another method of improving the efficiency of heat pump systems is the use of variable-speed motors for the compressor and both fan motors. This permits the selection of a different heat pump capacity. Previously, the selection of heat pump systems had been determined by their cooling capacity, but a typical structure in the heat pump belt requires about 2 to 2½ times as much heat as cooling. For example, a home in Atlanta, Georgia, may require 30,000 Btu/h for cooling and would typically require 60,000 to 75,000 Btu/h for heating. A heat pump that provides 30,000 Btu/h requires considerable auxiliary heat for winter conditions below the balance point of the structure. If the heat pump is selected to satisfy the winter capacity need, it would be oversized in the summer, which would result in poor performance and high humidity, along with low efficiency. The variable-speed heat pump narrows the gap because a system that comes closer to satisfying the heating requirement at full load can be chosen and it can run at part load and reduced power in warmer months. Less auxiliary heat is required, so year-round efficiency is achieved. Electronics and electronically controlled motors are what provide variable speeds. Each manufacturer has its own method of control and its own checkout procedures. Repair of a unit with variable-speed motors and electronic controls should not be attempted without the manufacturer's checklist.

HVACR GOLDEN RULES

When making a service call to a residence:

- Look professional and be professional. Wear clean clothes and shoes.
- Have a neat haircut, trim facial hair, and practice good oral hygiene by brushing teeth daily. Showering and daily use of underarm deodorant are strongly recommended.
- Wear coveralls for work under the house and remove them before entering the house.
- Ask the customer about the problem. The customer can often help you solve the problem and save you time.

Added Value to the Customer

Here are some simple, inexpensive procedures that may be included in the basic service call.

- Check line temperatures by touch for the correct temperature in summer or winter cycle.
- Clean or replace air filters as needed. This is very important for heat pumps.
- Make sure that all air registers are open—again very important for heat pumps.
- Replace all panels with the correct fasteners.
- Check indoor and outdoor coils for dirt or debris.
- Ask owners whether they have noticed any ice buildup in winter operation. This may lead to checking out the defrost cycle for the system.
- Check the gas line for proper insulation. Heat pumps may lose capacity due to loss of line insulation.

43.24 PREVENTIVE MAINTENANCE

The typical heat pump requires the same maintenance as an air-cooled air-conditioning system with electric heat. It is important that the air side of the system should always handle the maximum air for which it is designed. Some homeowners shut registers off in unused rooms, but this should never be done with a heat pump. The indoor system is the condenser in the winter cycle. Any reduction in airflow will cause high head pressure and a low COP, and so higher-than-normal power bills with less heating. For the same reason, filters should be maintained more carefully in heat pump systems. Reduced airflow cannot be allowed.

The outdoor unit should be examined for a dirty coil. The insulation on the outdoor gas line should extend all the way to the unit, or heat will be lost to the outdoor air. The gas-line temperature may be as high as 200°F and will transfer a lot of heat to the cold outside air. This is heat that is paid for but lost.

The contactor should be examined at least once a year for pitting and loose connections. This contactor functions about twice as many times each year as an air-conditioning system contactor. Frayed wires should be repaired or replaced. Wires or capillary tubes rubbing on the cabinet or refrigerant lines will soon cause problems and they should be adjusted or isolated.

The fan motor may require lubrication. Follow the manufacturer's directions.

The customer should be quizzed about the winter performance of a heat pump, including the amount of ice buildup on the outdoor unit. If excess ice seems to be a problem, defrost problems may be causing the unit to stay in defrost. Defrost may be simulated even in hot weather with most heat pumps by disconnecting the outdoor fan and running the unit in heating mode. With the outdoor fan motor disconnected, the unit will not overload because of high evaporator temperatures. Most manufacturers have a recommended procedure for checking defrost. If the system has the same timer as shown in **Figure 43.49**, the unit may be run until the coil and sensor are cold and covered with frost. This should be below the make point of 26°F. At this point, carefully jump terminal 3 to 4 on the timer with an insulated jumper. The unit should immediately go into defrost if the sensor circuit is made. Should it not go into defrost at this time, wait a few more minutes and try again. The sensor is large and may take a few minutes to make. If it will not go into defrost, leave the jumper on the 3 to 4 terminals and jump the sensor. If the unit immediately goes into defrost, the sensor is either defective or not in good contact with the liquid line.

After the unit goes into defrost normally, remove the jumper from 3 to 4 and allow the unit to terminate defrost. If the ambient temperature is above freezing and there is no ice accumulation, this should take about a minute. If it takes a full 10 min, the defrost control contacts may be permanently closed. They may be checked by allowing the control to rise to ambient temperature above 26°F and using an ohmmeter to see whether the contacts remain closed. They should be open when the sensor is above 50°F. One sign of permanently closed contacts is if the unit will go into defrost every time the timer calls for it to do so, whether the sensor is cold or not, and stay in defrost the full 10 min. This is hard on the compressor if the unit does not have a high-pressure control, because high pressure will occur at this time.

When performing maintenance on systems that have electronic circuit boards, the technician will need to work with the timing circuits. If the technician has to wait for each

sequence to time out, the service will take too long. Most manufacturers have a method of advancing the timing circuits to a "quick time" that will speed up the service call and any checking procedures being performed.

The last thing the technician should do is make sure that all fasteners are in place and tight. It is common for the cabinet screws to become loose. These should be tightened to prevent the cabinet from rattling. If the screw holes are oversized due to service, use a screw one size larger. It is good practice to have a supply of rust-resistant, one-half oversize screws. Stainless screws cost more, but the heads do not rust off with time.

43.25 DIAGNOSTIC CHART FOR HEAT PUMPS IN THE HEATING MODE

The heat pump is the most complex of the heating and air-conditioning systems. They are found in many homes and businesses in the south and southwest parts of the United States, where the climate is best suited for heat pumps. These systems operate in both heating and cooling modes. The heating mode is discussed here.

Problem	Possible Cause	Possible Repair
Cooling Season	See summer air-conditioning (Unit 42)	
Heating Season		
No heat—outdoor unit will not run—indoor fan runs	Open outdoor disconnect	Close disconnect.
	Open fuse or breaker	Replace fuse, reset breaker, and determine problem.
	Faulty wiring	Repair or replace faulty wiring or connection.
No heat—indoor fan or outdoor unit will not run	Low-voltage control problem	Repair loose connections or replace thermostat and/or subbase.
	A. Thermostat B. Interconnecting—wiring or connections C. Transformer	Repair or replace wiring or connections. Replace if defective. **SAFETY PRECAUTION:** *Look out for too much current draw due to ground circuit or shorted coil.* •
No heating; indoor and outdoor fans running, but compressor not running	Compressor overload tripped	Correct low voltage, call power company, or correct loose connections.
	A. Low line voltage B. High head pressure 1. Dirty indoor (condenser) coil 2. Indoor fan not running all the time 3. Overcharge of refrigerant 4. Restricted indoor airflow	Clean indoor coil. Check indoor fan motor and capacitor. Correct charge. Change filters. Open all supply registers. Clear return air blockage.
Outdoor coil (evaporator) freezes and ice will not melt	Defrost control sequence not operating	Follow manufacturer's directions and correct defrost sequence.
	Low charge; not enough refrigerant to perform adequate defrost	Correct charge and run unit through enough defrost cycles to clear ice off; then allow unit to run normally.
Unit will not change from cooling to heating or heating to cooling	Four-way reversing valve not changing over A. Defective defrost relay or circuit board B. Four-way reversing valve stuck C. Thermostat not changing to heat in subbase	Replace relay or circuit board. Change four-way reversing valve. Repair or replace subbase.
Excessive power bill	Compressor not running; heating off or operating on auxiliary heat	Repair compressor circuit or replace compressor.
	Suction pressure too high and head pressure too low for conditions	Perform four-way reversing valve temperature check. If valve is defective, change valve. If valve is good, change compressor.

43.26 SERVICE TECHNICIAN CALLS

When reading through these service calls, keep in mind that three things are necessary for an electrical circuit to be complete and current to flow: a power supply (line voltage), a path (wire), and a load (a measurable resistance). The mechanical part of the system has four components that work together: The evaporator absorbs heat, the condenser rejects heat, the compressor pumps the heat-laden vapor, and the metering device meters the refrigerant. The four-way valve is the component added to a typical air-conditioning system that makes the system a heat pump.

SERVICE CALL 1

A homeowner calls reporting that their heat pump system is heating, even though it is set to operating in the cooling mode. This is the first time the unit has been operated in the cooling mode this season. *The problem is that the solenoid coil on the four-way reversing valve is defective and will not switch the valve over into the cooling mode.*

The service technician rings the doorbell and stands back until the customer comes to the door. He listens as the homeowner explains the problem, then turns the space thermostat to the cooling mode and starts the system. This is a split system with the air handler in a crawl space under the house. **SAFETY PRECAUTION:** *The technician goes to the outdoor unit and carefully touches the gas line (the large line); it is hot, meaning the unit is in the heating mode.•*

The technician removes the panel and checks the voltage supplied to the solenoid coil on the four-way reversing valve. A voltage of 24 V is present, so the coil should be energized. The unit is shut off, and one side of the solenoid coil is disconnected. A continuity check shows that the solenoid coil is open. The technician conveys his findings to his supervisor and, upon getting the customer's cost for the repair, discusses the problem with the customer. After receiving customer approval to make the repair, the technician obtains the part, replaces the coil and starts the unit. It now operates correctly in the cooling mode. The technician then presents the bill to the customer, reviews the repair, obtains the required signature and payment and thanks the customer for their business before leaving.

SERVICE CALL 2

A store owner calls stating that the heat pump system in the small store will not start. The owner meets the technician upon arrival and explains the problem. The technician is glad that he is well groomed, because the owner gives him a close examination when they shake hands. This owner is very careful about the people working in the store because of the valuable merchandise. The store has a split-system heat pump with the outdoor coil on the roof. *The problem is that the 24-V holding coil for the four-way valve has shorted and has overloaded the control transformer.* This has blown the fuse in the 24-V control circuit.

The technician begins by trying to start the unit at the space thermostat. The unit will not start. When the FAN ON switch is turned to FAN ON, the fan will not start. He suspects a control voltage problem. The indoor unit is in a utility closet downstairs. **SAFETY PRECAUTION:** *The technician turns the power off and removes the door to the indoor unit.•* When the power is restored, there is no power at the low-voltage transformer secondary. The low-voltage fuse is checked and found to be blown. Before installing the new fuse, the technician turns off the power and then removes one lead from the control transformer and fastens an ohmmeter to the two leads leaving the transformer on the low-voltage side. The resistance is 0.5 Ω. This seems very low and will likely blow another fuse if power is restored. A low-voltage control circuit component must have shorted. The meter should indicate a 20-Ω resistance. This will increase the current flow enough to blow the fuse.

The technician decides to check the components in the indoor unit first and save a trip onto the roof, if possible. Each component is checked, one at a time, by removing one lead and measuring its resistance with an ohmmeter. There are three sequencers for the electric heat and one fan relay. There is no problem here. **SAFETY PRECAUTION:** *The technician goes to the roof and turns off the power.•* The compressor contactor is checked. It shows a resistance of 26 Ω. The four-way valve holding coil shows a resistance of 0.5 Ω. The technician installs a new coil, which shows a resistance of 20 Ω. The problem has been found.

The unit is put into operation with a clamp-on ammeter installed around the lead leaving the control transformer to check for excessive current. The current is 1 A, which is normal. When the owner is presented with the bill, she comments on the professional look and attitude of the technician. She says that she will continue to use the company.

SERVICE CALL 3

A customer reports that a significant amount of ice has built up on the outdoor coil of the heat pump. The customer is advised to turn the thermostat to the emergency heat mode until a technician can get there. *The problem is that the defrost relay coil winding is open, and the unit will not go into defrost.*

The weather is below freezing, so there is still ice on the outdoor coil when the technician arrives. The customer says the house is getting cold and requests that the technician do a fast job. **SAFETY PRECAUTION:** *With the unit off, the technician applies a jumper wire across terminals 3 and 4 on the defrost timer,* **Figure 43.62.•** The technician then starts the unit. This should force the unit into defrost if the timer contacts are not closing, but the unit does not go into defrost, so the jumper is left in place and the defrost temperature sensor is jumped. This satisfies the

Figure 43.62 With the unit off, the technician applies a jumper wire from terminals 3 to 4 to cause a defrost to avoid having to wait for the timer. Because this is a high-voltage circuit, this exercise should not be performed unless the technician is very experienced. Courtesy Carrier Corp.

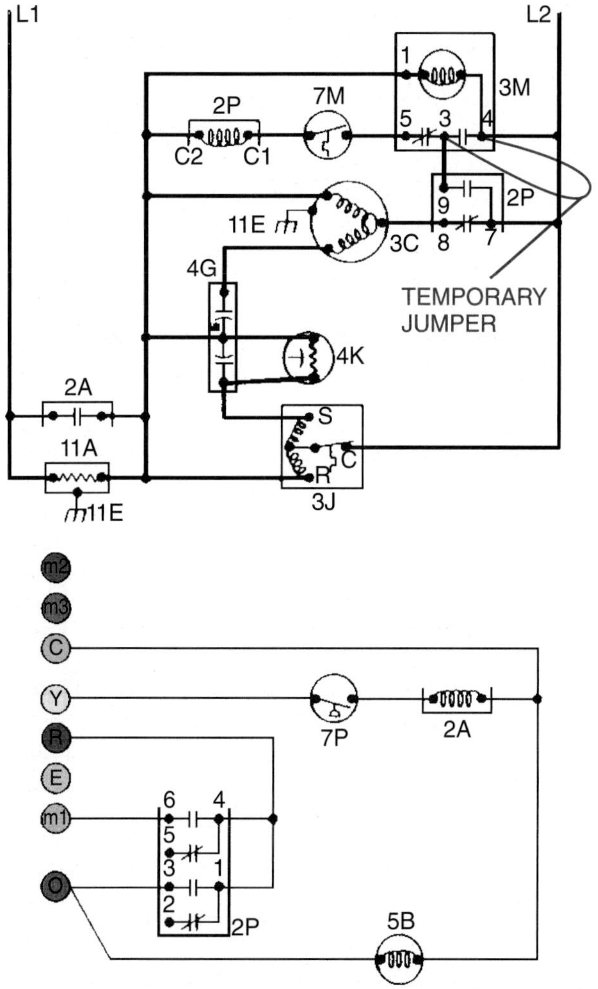

two conditions that must be met to start defrost, but nothing happens. **SAFETY PRECAUTION:** *The technician turns off the unit, turns off the power, locks and tags the panel, removes the jumpers, and checks the defrost relay.•*

The coil on the defrost relay is open. The relay is changed for a new one, and the unit is started again. When the jumper is again applied to terminals 3 and 4, the unit goes into defrost. Remember that two conditions, time and temperature, must be satisfied before defrost can start. The temperature condition is satisfied because the coil is frozen. When the connection from terminals 3 to 4 is made, defrost is started. This would have occurred normally when the defrost timer advanced to a trial defrost. By jumping from 3 to 4, the technician did not have to wait for the timer. The unit is allowed to go through defrost, but the coil is still iced.

The technician decides to shut off the unit and manually defrost the coil. **SAFETY PRECAUTION:** *The power is turned off, the panel is locked and tagged, and the electrical components are protected with plastic. A water hose is pulled over to the unit and city water is used to melt the ice. City water is about 45°F. Care is taken to not direct water onto the electrical components.•* Once the ice is melted, the unit is put back in operation. When presented with the bill, the customer thanks the technician for an efficient repair.

SERVICE CALL 4

A homeowner reports that the heat pump unit is running continuously and not heating as well as it used to. *The problem is that the outdoor fan is not running because the defrost relay has a burned set of contacts—the ones that furnish power to the outdoor fan.* These contacts are used to stop the outdoor fan in the defrost cycle. They should also start the fan at the end of the defrost cycle.

The technician arrives, goes to the outdoor unit and finds that the outdoor fan is not operating. The gas line is warm, and the outdoor coil is iced up, so the unit is in the heating cycle. The technician shuts the unit off and removes the panel to the outdoor unit's control compartment. After connecting a voltmeter to the fan motor leads, the technician starts the unit. No voltage is going to the fan motor. The wiring diagram shows that the power supply to the fan goes through the defrost relay. **SAFETY RECAUTION:** *The power is checked entering and leaving the defrost relay.•* Power is going in but not coming out. The defrost relay is changed, and the unit starts and runs as it should. The owner is happy to have the system operating again.

SERVICE CALL 5

A residential customer calls to report that the heat pump is blowing cold air for short periods of time. This is a split-system heat pump with the indoor unit in the crawl space under the house. *The problem is that the contacts in the defrost relay that bring on the auxiliary heat during defrost are open and will not allow the heaters to come on during defrost.*

The technician talks to the homeowner about the symptoms. Being familiar with the particular heat pump, the technician believes that the problem is in the defrost relay or the first-stage heat sequencer. The first thing to do is to see whether or not the sequencer is working correctly. The space thermostat is turned up until the second-stage contacts are closed. The technician goes under the house with an ammeter and verifies that the auxiliary heat will operate with the space thermostat. Checking the defrost relay is the next step. This can be done by manually energizing the relay, by waiting for a defrost cycle, or by simulating a defrost cycle. The technician decides to simulate a defrost cycle in the following manner. **SAFETY PRECAUTION:** *The power is turned off. To prevent the outdoor fan from running, one of the motor leads is removed.•* Then the unit is restarted. The system is now

operating in the heating mode with the outdoor fan motor inoperative. This causes the outdoor coil to form ice very quickly, causing the defrost temperature thermostat contacts close. When the coil has ice on it for 5 min, the technician jumps the 3 to 4 contacts on the defrost timer, and the unit goes into defrost, **Figure 43.62**.

The technician goes under the house and checks the electric strip heat; it is not operating. The problem must be the defrost relay. By the time the technician gets back to the outdoor unit, the unit is out of defrost. He decides that the defrost cycle is functioning as it should, so he carefully energizes the defrost relay with a jumper wire. The 24-V power supply feeds terminal 4 and then should go out on terminal 6 to the auxiliary heat sequencer. These contacts are not closing. The technician changes the defrost relay and energizes the new relay. Going back under the house, the technician verifies that the strip heat is working. The technician presents the bill to the owner, gets it signed, and leaves.

SERVICE CALL 6

A customer reports that a large amount of ice has built up on the outdoor unit of the heat pump. *The problem is that the changeover contacts in the defrost relay are open.* The reversing valve is not energizing in the defrost cycle.

After knocking on the door, the technician stands back to wait for the customer to open it. He asks the customer to describe the situation. She explains that after watching other heat pumps in the neighborhood, she knows hers is not defrosting correctly. The technician goes to the outdoor unit first because it is obvious there is a defrost problem. The coil looks like a large block of ice.

The technician removes the control box panel and jumps the timer contacts from 3 to 4; the outdoor unit's fan stops, but the cycle does not reverse as it should. (See the wiring diagram in **Figure 43.62**.) A voltage check at the four-way valve coil shows that no voltage is getting to the valve. The technician then jumps from the R terminal to the O terminal, and the four-way valve changes over to cooling, so he removes the jumper. The valve must not be getting voltage through the defrost relay. The defrost relay contacts are jumped from 1 to 3, and the valve reverses. He changes the defrost relay and simulates another defrost by jumping the 3 to 4 contacts. The defrost thermostat contacts are still closed, and the unit goes into defrost. The unit has so much ice accumulation that the technician uses water to defrost the coil and get it back to a normal running condition.

SERVICE CALL 7

A shop owner calls to report that the heat pump in the small shop is running constantly. The unit is blowing cold air for short periods every now and then. The power bill was excessive last month, and it was not a cold month. This is a split system with 20-kW strip heat for quick recovery so that the customer can turn the thermostat back to 60°F on the weekend and have quick recovery on Monday morning. The outdoor unit is in the rear of the shop. *The problem is that the contact in the defrost relay that changes the unit to cooling in the defrost mode is stuck closed.* The unit is running in the cooling mode, and the strip heat is heating the structure. It must also make up for the cooling effect of the cooling mode.

The technician sets the thermostat to the first stage of heating. The air at the registers is cool, not warm. The unit should have warm air in the first stage of heat (85°F to 100°F), depending on the outside air temperature. The indoor coil is in a closet in the back of the shop. The gas line is not hot but cold. This indicates that the unit is in the cooling mode, but a voltage check should be performed. (See the wiring diagram in **Figure 43.63**.)

Figure 43.63 The contacts that change the unit to cooling for defrost are stuck closed. Courtesy Carrier Corp.

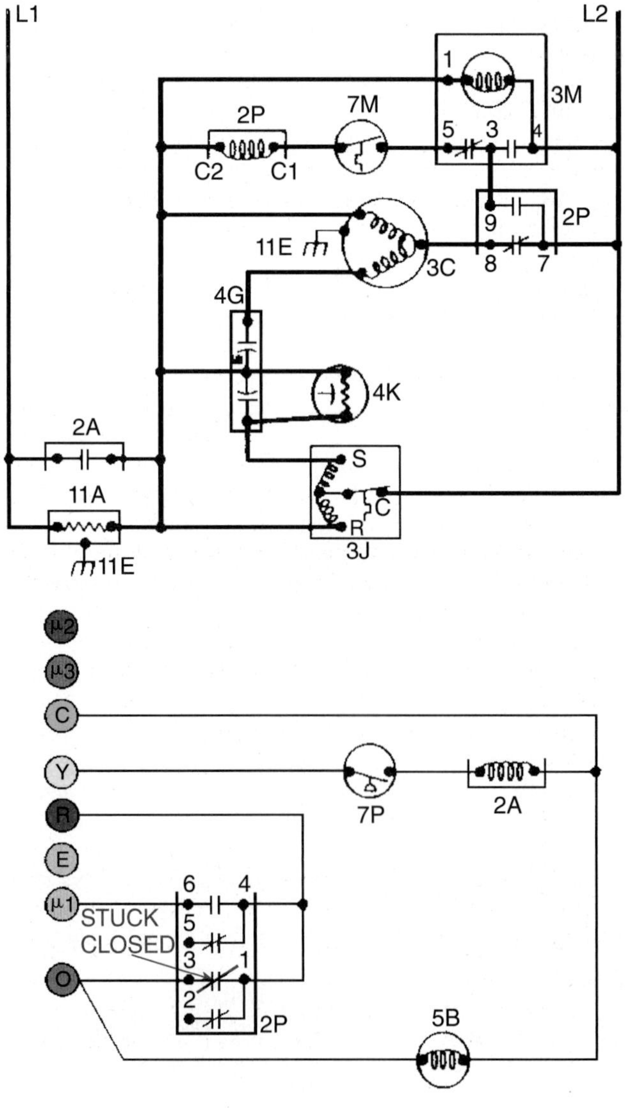

The technician removes the control voltage panel cover and finds 24 V at the C to O terminals. This means the four-way valve coil is energized, and the unit should be in the cooling mode. Now the technician must decide whether the 24 V is coming from the space thermostat or the defrost relay. He removes the field wire from the space thermostat; the valve stays energized. The voltage must be coming from the defrost relay. The wire is replaced on the O terminal, and the wire on terminal 3 on the defrost relay is carefully removed. The unit changes to heating. The defrost relay is changed, and the unit operates correctly.

The technician explains in simple terms what happened, showing the shop owner the part. The owner thanks him for the explanation and goes on to say that most technicians do not take the time to inform customers what the problem was. A callback later finds the unit to be operating as expected.

SERVICE CALL 8

While on a routine service call at an apartment house complex, the technician notices a unit go into defrost when there is no ice on the coil. The unit stays in defrost for a long time, probably the full 10 min allowable. This is a split-system heat pump and should have defrosted for only 2 or 3 min with a minimum of ice buildup. The compressor was loud, as if it were pumping against a high head pressure. *The problem is that the defrost thermostat contacts are shut.* The unit is going into defrost every 90 min and staying for the full 10 min. The timer is terminating the defrost cycle after 10 min. Toward the end of the cycle, the head pressure is very high. The technician goes to the manager and explains what the problem is and gets the go-ahead to perform the repair.

The technician shuts off the unit, removes the panel to the control box, removes one lead of the defrost thermostat, and takes a continuity check across the thermostat. The circuit should be open if the coil temperature is above 50°F but the contacts are closed. (This control should close at about 25°F and open at about 50°F. Allow plenty of time for this control to function because it is large and it takes a while for the unit's line to exchange heat from and to the control.) The technician changes the defrost thermostat and puts the unit back into operation. After the work is complete, the technician takes the defective part to the manager with an explanation and presents the bill.

SERVICE CALL 9

A residential customer reports no heat in the home. There is a smell of something hot or burning. ⚠ **SAFETY PRECAUTION:** *The customer is told to turn off the unit at the indoor breaker to stop all power to the unit.*• *The problem is that the indoor fan motor has an open circuit and the electric strip heat is coming on from time to time and then shutting off because of excessive temperature.*

Upon arrival, the technician explains to the customer that there is not likely to be any fire problem because the unit is totally enclosed. The technician goes to the space thermostat and sets it to OFF. The electrical breakers are then energized. He turns the thermostat to FAN ON. The fan relay can be heard energizing, but the fan will not start. The indoor unit is under the house in the crawl space, so the technician takes tools and electrical test equipment to the unit. ⚠ **SAFETY PRECAUTION:** *The breaker under the house is turned off and the door to the fan compartment is opened. The fan motor is checked and found to have an open winding.*• The motor is replaced with a new one, and the unit is started.

The technician uses an ammeter to check all three stages of electric heat for current draw. One stage draws no current. ⚠ **SAFETY PRECAUTION:** *The breaker is again turned off and the electric heat is checked with an ohmmeter.*• The circuit in one unit is open. This happened because the fan was not running and the unit overheated. A new fuse link is installed, and the heat is started again. This time it is drawing current. The technician puts the unit panels back together and again goes to the owner and explains what was done and that the safety controls did what they were supposed to do.

SERVICE CALL 10

A homeowner reports that the residential heat pump smells hot and not much air is coming out of the registers. ⚠ **SAFETY PRECAUTION:** *The customer is told to shut off the unit at the breaker for the indoor unit.*• *The problem is that the indoor fan capacitor is defective, and the fan motor is not running up to speed.* The fan motor is cutting off periodically from an internal overload.

When he arrives and hears the owner's explanation, the technician sets the thermostat to OFF and turns on the breaker. The FAN ON switch at the thermostat is set to ON. The fan relay can be heard energizing, and the blower starts. There does not seem to be enough air coming out of the registers. The blower motor is not running smoothly. The technician goes to the attic crawl space where the indoor unit is located, taking tools and electrical test instruments. ⚠ **SAFETY PRECAUTION:** *The breaker next to the indoor unit is shut off and the access panel to the control box is removed.*• The breaker is then turned on. The blower does not turn up to speed. The breaker is turned off again. The fan capacitor is removed and a new one is installed. The breaker is energized again, and the blower comes up to speed. A current check of the blower motor shows the fan is operating correctly. ⚠ **SAFETY PRECAUTION:** *The power is shut off, locked, and tagged, and the motor is oiled.*• The technician then replaces all panels and turns on the power. After completing the work, the technician explains to the owner what repairs were made and presents the bill.

SERVICE CALL 11

A homeowner calls indicating that his unit is blowing cool air from time to time. One of the air registers is close to his television chair, and the cool air is a bother. *The problem is that the breaker serving the outdoor unit is tripped, apparently for no reason.* The control voltage is supplied at the indoor unit, so the electric auxiliary heat is still operable.

The technician talks to the customer about the system symptoms and then turns the space temperature thermostat to the first stage of heat; he notices that the outdoor unit does not start. When the thermostat is turned to a higher setting, the auxiliary heat comes on and heat comes out of the registers. With the thermostat in the first-stage setting, only the outdoor unit should be operating. The auxiliary heat should not be operating. In the second stage, the heat pump and the auxiliary heat should both be operating.

The technician goes to the outdoor unit and finds that the breaker is tripped. **SAFETY PRECAUTION:** *There is no way of knowing how long the breaker has been off, but it is not wise to start the unit because there has been no crankcase heat on the compressor. The breaker is switched to the off position and the panel is locked and tagged.*• Using an ohmmeter, the technician checks the resistance across the contactor load-side circuit to make sure there is a measurable resistance. It is normal, about 2 Ω (there is a compressor motor, fan motor, and defrost timer motor to read resistance through). The contactor load-side terminals are checked to ground to see whether any of the power-consuming devices are grounded. The meter reads infinity, the highest resistance. Someone may have turned the space temperature thermostat off and back on before the system pressures were equalized so the compressor motor would not start. This means that the compressor tried to start in a locked-rotor condition and the breaker tripped before the motor overload protector turned off the compressor. A sudden, brief power outage then power coming back on will cause the same thing.

The technician does not reset the breaker before going to the indoor thermostat and setting it to the emergency heat mode. This prevents the compressor from starting with a cold crankcase. The technician then tells the homeowner to switch back to normal heat mode in 10 hours. This gives the crankcase heater time to warm the compressor. The homeowner is instructed to see whether the outdoor unit starts correctly or if the breaker trips. If the breaker trips, the thermostat should be set back to emergency heat, the breaker reset, and the technician called. If the technician must go back, the compressor could have a starting problem and may not start the next time. The technician explains to the customer that this procedure will save another service call and another expense. The customer does not call back, so the unit must be operating correctly.

SERVICE CALL 12

A customer says that the heat pump is running constantly in the heating mode in mild weather. It did not do this last year; it cut off and on in the same kind of weather. *The problem is that the check valve in the liquid line (parallel to the TXV) is stuck open and not forcing liquid refrigerant through the TXV.* This is overfeeding the evaporator and causing the suction pressure to be too high, hurting the efficiency of the heat pump. For example, if it is 40°F outside, the evaporator should be operating at about 15°F for the outside air to give up heat to the evaporator. If the evaporator is operating at 30°F because it is overfeeding, it will not exchange as much heat as it should. The capacity will be down.

The technician talks to the owner to get some details, then turns the room thermostat to the first stage of heating, and goes to the outdoor unit at the back of the house. The unit is running. The gas-line temperature is warm but not as hot as it should be. The compressor compartment door is removed so the compressor can be observed. It is sweating all over from liquid refrigerant returning to the compressor. This means that the TXV is not controlling the liquid refrigerant flow to the evaporator. After fastening a gauge manifold to the high and low sides of the system, the technician discovers that the suction pressure is higher than normal and the head pressure is slightly lower than normal. The head pressure is down slightly because some of the charge that would normally be in the condenser is in the evaporator. After attaching the thermometer lead to the suction line just before the four-way valve, the superheat is found to be 0°F. The outdoor coil is flooded. Not enough refrigerant is flooding to the compressor to make a noise because the liquid is not reaching the cylinders. It is boiling away in the compressor crankcase.

The technician tries adjusting the TXV but it makes no difference. The TXV is open and not closing or else the check valve is open and bypassing the TXV. The technician gets a TXV and a check valve from the supply house. 4R*The unit charge is recovered into an approved recovery cylinder.* 4R The old check valve is first removed and tested. The technician can blow nitrogen vapor through the valve in either direction, so the check valve is defective. He changes it, but leaves the TXV alone. The unit is pressurized, leak checked, evacuated, and charged. When the unit is started in the heating mode, it operates correctly. The check valve was the problem. The superheat is checked to be sure that it is normal (8°F to 12°F) and the unit panels are assembled. The technician gives an explanation to the customer about the repair. The customer thanks the technician for being professional. A call later verifies that the unit is cycling as it should.

SERVICE CALL 13

A residential customer reports that the R-410A heat pump serving the home is not performing to capacity. The power bills are much higher than last year's, although the heating season has been much the same. The customer would like a performance check of the heat pump. *The problem is that the compressor has a damaged suction valve due to liquid floodback on start-up.* The owner had shut the disconnect off at the outdoor unit last summer and left it off all summer. When the unit was started in the fall, the compressor had enough liquid in the crankcase to damage one of the suction valves. If the crankcase heat had been allowed to operate for 8 to 10 hours before starting the unit, this would not have happened. Note that if one cylinder of a two-cylinder compressor is not pumping, the compressor will pump at half capacity.

The technician meets the customer and gets an explanation of the system operation. Based on the explanation, the technician installs gauges on the system and starts it. He uses the manufacturer's performance chart to compare pressures. The unit is running a high suction pressure and a low discharge pressure. The technician switches the unit to cooling and blocks the outdoor coil airflow to build up head pressure. The head pressure goes to 320 psig and the suction pressure to 170 psig. The compressor current is below normal. The compressor is not pumping to capacity; it is defective. A temperature check of the gas lines at the four-way reversing valve proves that the reversing valve is not leaking internally.

The technician gets a new compressor from a supply house. The old compressor is not under warranty, so there is no compressor exchange. 4R *The refrigerant is recovered and the compressor is changed.* 4R A new bi-flow liquid-line filter drier is installed. The system is pressurized, leak checked, and evacuated before charging. The charge is measured into the system, the system is started, and it operates correctly. The technician explains, in simple terms, what happened: The disconnect must be turned on for at least 10 hours before starting the compressor after the power has been off for the summer. A call to the customer later in the day proves that the system is functioning correctly. The outdoor unit is satisfying the room thermostat and shutting the unit off from time to time because the weather is above the balance point of the house.

The technician talks to the customer about the heat pump performance. She then turns the room thermostat to the first stage of heating and goes to the outdoor unit. The gas line leaving the unit is warm, not hot as it should be. Gauges installed on the system show that the head pressure is low and the suction pressure is high. The compressor amperage is low. All of these are signs of a compressor that is not pumping up to capacity. The outside temperature is 40°F and the coil does not have an ice buildup, so the unit should be working at a high capacity.

The technician suspects the compressor and so performs a compressor performance test. A jumper wire is installed between the R and O terminals to energize the four-way reversing valve coil and change the unit over to the cooling mode. The thermostat is left in the first-stage heat mode. If the house begins to cool off, the second stage of heat will come on and keep the house from getting too cool. A plastic bag is wrapped around the condenser coil to slow the air entering the coil. The head pressure begins to climb, but the suction pressure rises with it. The compressor will not build a sufficient differential between the high and low sides. Generally, the unit should be able to build 500 psig of head pressure with 118 to 140 psig suction, but this unit has a 200 psig suction and a 400 psig head pressure.

Before condemning the compressor, the technician decides to check the four-way valve operation. He fastens the two leads of an electronic thermometer on the cold gas line entering and leaving the four-way valve. A difference in temperature should not be more than 3°F between these two lines, but it is 15°F. 4R *The four-way reversing valve is leaking hot gas into the suction gas. The four-way reversing valve is changed after the refrigerant is recovered.* 4R The unit is started after the pressure leak test, evacuation, and charging. It now performs to capacity with normal suction and head pressures. The technician explains the service procedures to the customer and presents the bill before leaving.

⚠ SAFETY PRECAUTION: *Use common sense and caution while troubleshooting any piece of equipment. As a technician, you are constantly exposed to potential danger. High-pressure refrigerant, electrical shock hazard, rotating equipment shafts, hot metal, sharp metal, and lifting heavy objects are among the most common hazards you will face. Experience and listening to experienced people will help to minimize the dangers.*•

SERVICE CALL 14

A customer calls the heating and cooling service company to claim that the R-410A heat pump serving the residence must not be operating efficiently because the power bill has been abnormally high for the month of January compared with last year's bill. Both years have had about the same weather pattern. *The problem is that the four-way reversing valve is stuck in mid-position and will not change all the way to heat.*

Solutions for the following service calls are not provided here. Please discuss the solutions with your instructor.

SERVICE CALL 15

A beauty shop owner calls to say that the unit serving the shop is blowing cool air from time to time.

This is a split system with the outdoor unit on the roof. When the technician turns the thermostat to the first stage of

heat, cool air comes out of the air registers. The technician goes to the outdoor unit—the outdoor fan is running, but the compressor is not. **SAFETY PRECAUTION:** *The power is turned off. The panel is locked and tagged before the technician removes the cover to the compressor compartment.* An ohm check shows that there is an open circuit between the R and C terminals and between the S and C terminals, but there is continuity between R and S. The compressor body temperature is warm, not hot.

What is the likely problem and the recommended solution?

SERVICE CALL 16

A residential customer calls to say the heat pump outdoor unit is off because of a tripped breaker. The customer reset it three times, but it would not stay reset.

The technician goes to the outdoor unit first and sees that the breaker is in the tripped position. He decides that before resetting it a motor test should be performed. **SAFETY PRECAUTION:** *A volt reading is taken at the compressor contactor line side; 50 V is present.* This must be caused by the repeated resetting of the breaker—voltage is leaking through the breaker. When the main heating breaker is turned off, the 50 V disappears. **SAFETY PRECAUTION:** *If the technician had taken a continuity reading first, the meter would have been damaged.* With the main breaker off, a resistance check on the load side of the compressor contactor shows a reading of 0 Ω to ground (the compressor suction line is a convenient ground).

What is the likely problem and the recommended solution?

SERVICE CALL 17

A call from an owner of a two-story house indicates that the unit serving the upstairs of the house is blowing cold air part of the time and not heating properly.

The technician turns the space temperature thermostat to the first stage of heating and notices that cool air is coming out of the air registers. The outdoor unit can be seen from a window and the fan is running. The technician goes to the outdoor unit and finds that the gas line leaving the unit is barely warm, not hot. The compressor is either not running, or it is running and not working. The compressor access panel is removed, and the compressor can be touched and seen. It is hot to the touch.

What is the likely problem and the recommended solution?

SERVICE CALL 18

A dress shop manager reports that the heat pump serving the shop will not cool properly. This is the first day of operation for the system this summer. The compressor was changed about two weeks ago after a bad motor burn. The system has not operated much during the two-week period because the weather has been mild.

The technician turns the thermostat to the cooling mode and starts the unit. The compressor is on the roof, and the indoor unit is in a closet. When the technician goes to the indoor unit and feels the gas line, it is hot. The unit is in the heating mode with the thermostat set in the cooling mode. The technician goes to the roof and removes the low-voltage control panel. In this unit, the four-way valve is energized in the cooling mode, so it should be energized at this time. A voltmeter check shows that power is at the C to O terminals (a common terminal and the four-way valve coil terminal). The technician removes the wire from the O terminal and hears the pilot solenoid valve change (with a click). Every time the wire is touched to the terminal, the valve can be heard to position. This means that the pilot valve is moving.

What is the likely problem and the recommended solution?

SERVICE CALL 19

A retail store owner reports that the R-410A heat pump serving the store is blowing cold air from time to time and not heating as it normally does. It runs constantly. The unit has a TXV.

The technician turns the thermostat to the first stage of the heating mode. The air coming out of the air registers is cool; it feels like the return air. The outdoor unit is on the roof, and the indoor unit is in an attic crawl space. The thermostat is set with the first stage calling for heat; the outdoor unit should be on. The technician goes to the roof. The outdoor unit is running, but the gas line is cool. A gauge manifold is installed on the high and low sides of the system. The low side is operating in a vacuum, and the high-side pressure is 202 psig, which corresponds to the inside temperature of 70°F. This indicates that liquid refrigerant is in the condenser (the indoor coil). If the condenser pressure were 105 psig, well below the 202 psig corresponding to 70°F, it would indicate that the unit is low in refrigerant charge.

What is the likely problem and the recommended solution?

REVIEW QUESTIONS

1. Which component of a heat pump is responsible for changing the direction of refrigerant flow through the system?
   A. Compressor
   B. Reversing valve
   C. Check valve
   D. Metering device
2. Which of the following statements is true regarding heat pump system operation?
   A. The indoor coil functions as the evaporator in the heating mode.
   B. The outdoor coil functions as the evaporator in the cooling mode.
   C. The indoor coil functions as the condenser in the heating mode.
   D. The outdoor coil functions as the condenser in the heating mode.
3. An air-to-water heat pump system uses heat contained in
   A. water to cool air.
   B. water to heat air.
   C. air to heat water.
   D. air to cool water.
4. What is the approximate percentage of usable heat that remains in outside air when it drops to a temperature of 0°F?
   A. 0%
   B. 23%
   C. 57%
   D. 85%
5. The majority of heat pump systems will operate in which mode when the reversing valve's solenoid is deenergized?
   A. Defrost
   B. Cooling
   C. Emergency heat
   D. Heating
6. The isolated port on a reversing valve contains high-pressure
   A. gas that is entering the valve from the compressor.
   B. gas that is leaving the valve flowing to the compressor.
   C. liquid that is entering the valve from the compressor.
   D. liquid that is leaving the valve flowing to the compressor.
7. A reversing valve that relies on the system's operating pressures to change the position of the valve is called a
   A. direct-acting reversing valve.
   B. pilot-operated reversing valve.
   C. indirectly controlled reversing valve.
   D. pressure-activated reversing valve.
8. On a split-type heat pump system, the larger field-installed line carries
   A. high-pressure liquid in the cooling mode and high-pressure gas in the heating mode.
   B. high-pressure gas in the cooling mode and low-pressure gas in the heating mode.
   C. high-pressure gas in the heating mode and low-pressure gas in the cooling mode.
   D. high-pressure liquid in the heating mode and low-pressure gas in the cooling mode.
9. On ductless split-type heat pump system, which of the following statements is true regarding the field-installed refrigerant lines if the system is operating in the heating mode?
   A. The larger line will contain low-pressure, low-temperature vapor.
   B. The larger line will contain high-pressure, high-temperature vapor.
   C. The smaller line will contain high-pressure, high-temperature vapor.
   D. The smaller line will contain low-pressure, low-temperature vapor.
10. What device is often used in conjunction with a thermostatic expansion valve that allows refrigerant to flow in the opposite direction, bypassing the valve's needle and seat assembly?
    A. Bypass valve
    B. Biflow valve
    C. Counterflow valve
    D. Check valve
11. Which of the following metering devices has no moving parts?
    A. Capillary tube
    B. Piston
    C. Electronic expansion valve
    D. Thermostatic expansion valve
12. The filter drier that operates as two filter driers in one casing to filter refrigerant flowing in either direction is called the
    A. bypass filter drier.
    B. biflow filter drier.
    C. counterflow filter drier.
    D. check-type filter drier.
13. As an inverter-controlled compressor speeds up
    A. the suction pressure and the discharge pressure drop.
    B. the suction pressure and the discharge pressure rise.
    C. the suction pressure rises and the discharge pressure drops.
    D. the suction pressure drops and the discharge pressure rises.

14. If a properly operating, conventional heat pump system is operating in the heating mode with an outside ambient temperature of 30°F, which of the following is the most likely outside ambient temperature?
    A. −10°F
    B. 10°F
    C. 25°F
    D. 35°F
15. Which of the following statements is true regarding the coefficient of performance (COP) of a heat pump system?
    A. As the compression ratio increases, the efficiency of the system increases.
    B. As the heat of compression (HOC) increases, the efficiency of the system decreases.
    C. As the compression ratio increases, the heat of compression (HOC) decreases.
    D. As the heat of compression decreases, the compression ratio increases.
16. The coefficient of performance (COP) of electric strip heaters is closest to
    A. 1:1.
    B. 2:1.
    C. 3:1.
    D. 4:1.
17. What system conditions are likely to be observed if the heat pump's A-coil is located downstream of the electric strip heaters?
    A. High operating pressures
    B. Low operating pressures
    C. High suction pressure and low discharge pressure
    D. Low suction pressure and high discharge pressure
18. Which of the following is true regarding the discharge air temperature of a heat pump system operating in first-stage heating?
    A. The discharge air temperature is higher than that of an oil-fired heating system, but lower than that of a gas-fired heating system.
    B. The discharge air temperature is higher than that of a gas-fired heating system, but lower than that of an oil-fired heating system.
    C. The discharge air temperature is higher than that of both an oil-fired heating system and a gas-fired heating system.
    D. The discharge air temperature is lower than that of both an oil-fired heating system and a gas-fired heating system.
19. An air-to-air heat pump operating in the heating mode might experience which of the following if it is installed in an area that is near a lake or other body of water?
    A. Higher than normal suction pressure
    B. Higher than normal discharge pressure
    C. Excessive ice buildup on the outdoor coil
    D. Shortened defrost cycles
20. When a single metering device is used on a package heat pump system,
    A. the capillary tube is the most popular choice.
    B. it must be able to adjust refrigerant flow for the heating and cooling modes.
    C. it must be configured to meet the flow requirements for the heating mode.
    D. it must be configured to meet the flow requirements for the cooling mode.
21. Which of the following is true regarding package-type heat pump systems?
    A. All field-installed refrigerant lines must be insulated.
    B. They are charged at the factory and bench-tested before they are shipped.
    C. They are shipped with a holding charge of dry nitrogen.
    D. Only the larger of the two field-installed refrigerant lines must be insulated.
22. How can a system be easily identified as a heat pump by simply looking at the thermostat?
    A. Heat pump thermostats have both heating and cooling settings.
    B. Heat pump thermostats have a FAN-ON, FAN-AUTO feature.
    C. Heat pump thermostats have an emergency heat setting.
    D. Heat pump thermostats have an automatic changeover feature.
23. Which of the following statements is true regarding the cooling cycle of a heat pump system?
    A. The compressor cycles on when the thermostat's selector switch is set to the cooling mode and the reversing valve's coil energizes and deenergizes depending on the space temperature.
    B. The coil on the reversing valve is energized when the thermostat's selector switch is set to the cooling mode and the compressor cycles on and off depending on the space temperature.
    C. Both the compressor and the reversing valve's coil are energized when the thermostat's selector switch is set to the cooling mode and the reversing valve's coil energizes and deenergizes depending on the space temperature.
    D. Neither the compressor nor the reversing valve's coil are energized when the thermostat's selector switch is set to the cooling mode, but they are both energized and deenergized depending on the space temperature.
24. Which of the following best describes the operation of second-stage heat on a heat pump system?
    A. First on, first off
    B. First on, last off
    C. Last on, first off
    D. Last on, last off

25. The term "balance point" is best described as
    A. the point where the heating and cooling capacities of a heat pump system are equal to each other.
    B. the lowest temperature at which a heat pump system can heat a structure without the aid of any supplementary heat.
    C. the lowest temperature at which a heat pump's electric strip heaters can heat a structure.
    D. the point where the heat pump's supplementary heating capacity exactly offsets the system's cooling capacity.
26. In the emergency heat mode, what percentage of the heat pump system's electric strip heaters will be energized?
    A. 25% C. 75%
    B. 50% D. 100%
27. System overshoot and temperature swing are controlled by a device known as a(n)
    A. anticipator.
    B. predictor.
    C. regulator.
    D. moderator.
28. Which of the following statements is true regarding SEER and HSPF?
    A. SEER is an efficiency rating for cooling, while HSPF is an efficiency rating for heating.
    B. SEER is an efficiency rating for heating, while HSPF is an efficiency rating for cooling.
    C. Both SEER and HSPF are efficiency ratings for cooling.
    D. Both SEER and HSPF are efficiency ratings for heating.
29. Why is defrost required on some heat pump systems?
    A. The outside ambient air is colder than the evaporator coil.
    B. The evaporator coil is colder than the outside ambient air.
    C. The outside ambient air is warmer than the condenser coil.
    D. The evaporator coil is warmer than the air in the occupied space.
30. During the defrost cycle of an air-to-air heat pump system
    A. the reversing valve coil is energized and the compressor cycles off.
    B. the reversing valve coil is energized and the outdoor fan cycles off.
    C. the reversing valve coil is deenergized and the compressor cycles off.
    D. the reversing valve coil is deenergized and the outdoor fan cycles off.
31. During the defrost cycle of an air-to-air heat pump system
    A. the outdoor fan remains on and some strip heaters turn on.
    B. the outdoor fan remains on and all strip heaters turn on.
    C. the outdoor fan cycles off and some strip heaters turn on.
    D. the outdoor fan cycles off and all strip heaters turn on.
32. Three factors that are commonly used to initiate the defrost cycle on a heat pump system are time, temperature, and
    A. pressure drop through the outdoor coil.
    B. pressure drop through the indoor coil.
    C. ice thickness on the outdoor coil.
    D. ice thickness on the indoor coil.
33. The defrost cycle on a heat pump system is often terminated by
    A. a combination of time and temperature.
    B. a combination of temperature and pressure.
    C. either time or temperature.
    D. either temperature or pressure.
34. During defrost, what percentage of the heat pump system's cooling capacity should be offset by the electric strip heaters?
    A. 20% C. 60%
    B. 40% D. 80%
35. At design voltage, how much heat will a 5-kW electric strip heater produce?
    A. 3413 Btu/h
    B. 6826 Btu/h
    C. 13,652 Btu/h
    D. 17,065 Btu/h
36. A typical electrical circuit is made up of
    A. a power supply, a complete path for current, switches, and a load.
    B. a power supply, a complete path for current, and a switch.
    C. a complete path for current, a switch, and a load.
    D. a current path to the load, a current path from the load, and a switch.
37. A shorted load will result in
    A. reduced resistance and increased current flow.
    B. reduced resistance and decreased current flow.
    C. increased resistance and increased current flow.
    D. increased resistance and decreased current flow.
38. If a fan relay has a defective coil, which of the following is true?
    A. The normally open contacts will close, but the normally closed contacts will stay closed.
    B. The normally closed contacts will open, but the normally open contacts will stay open.
    C. The normally open contacts will stay open, but the normally closed contacts will stay closed.
    D. The normally closed contacts will open, but the normally open contacts will stay closed.
39. If the coil on the four-way reversing valve fails,
    A. the heat pump will run in the heating mode when set to the cooling mode.
    B. the heat pump will run in the cooling mode when set to the heating mode.
    C. there will be no ice buildup on the surface of the outdoor coil.
    D. there will be excessive ice buildup on the surface of the indoor coil.

40. A dirty air filter will cause which of the following sets of conditions?
    A. Low suction pressure in heating, high discharge pressure in cooling
    B. Low suction pressure in both heating and cooling
    C. High discharge pressure in heating, low suction pressure in cooling
    D. High suction pressure in both heating and cooling
41. A dirty outdoor coil will cause which of the following sets of conditions?
    A. Low suction pressure in heating, high discharge pressure in cooling
    B. Low suction pressure in both heating and cooling
    C. High discharge pressure in heating, low suction pressure in cooling
    D. High suction pressure in both heating and cooling
42. A reversing valve that is leaking internally will
    A. cause the high-side pressure to increase and the low-side pressure to drop.
    B. cause the high-side pressure to decrease and the low-side pressure to rise.
    C. cause the compressor to run cold and pump excessive amounts of refrigerant.
    D. cause the compressor to run hot and pump excessive amounts of refrigerant.
43. Which of the following indicates that the reversing valve is not leaking internally?
    A. Two of the valve's ports are hot, one is cold and one is warm
    B. Two of the valve's ports are hot and two are cold
    C. Three of the valve's ports are hot and one is warm
    D. Three of the valve's ports are cold and one is hot
44. Referring to the heating portion of Figure 43.57, what is the expected suction pressure for an R-410A heat pump system that is operating with a 300 psig discharge pressure if the wet-bulb temperature of the air entering the outdoor coil is 9°F?
    A. 32°F
    B. 40°F
    C. 50°F
    D. 74°F
45. When servicing a heat pump system, which of the following is true regarding the installation of gauges on the system?
    A. Gauges should be installed on the system each time the system is serviced.
    B. Only the low-side gauge should be connected to the system during service.
    C. Only the high-side gauge should be connected to the system during service.
    D. Gauges should only be installed on a system if absolutely necessary.
46. When an oil-fired or gas-fired furnace is used as a second-stage heat source,
    A. the furnace will be designed to operate at the same time as the heat pump.
    B. the heat pump will operate only in the defrost mode to remove ice from the coil.
    C. the heat pump will cycle off when the furnace starts to operate.
    D. the furnace will only operate when the system is set to the emergency heat mode.
47. One benefit of using scroll compressors instead of reciprocating compressors to increase heat pump system efficiency is that scroll compressors
    A. do not have "top-of-the-stroke" clearance volume losses.
    B. are lighter than reciprocating compressors.
    C. are cooled by suction gas.
    D. do not require a suction-line accumulator.
48. What component prevents refrigerant from flowing back into the scroll compressor during its off cycle?
    A. Service valve
    B. Check valve
    C. Accumulator
    D. Receiver
49. Explain why it is important to keep all supply registers open when operating a heat pump system.
50. List four causes for high head pressure on a heat pump system operating in the heating mode.
51. Explain how a heat pump equipped with variable-speed motors is able to satisfy the heating and cooling requirements of a structure better than a single speed system.
52. Explain the conditions that would be observed if a compressor is not pumping to capacity.
53. Explain what will likely happen to a PSC motor if its run capacitor is "open."
54. Explain why heat pump system servicing is especially important as compared to the servicing of a cooling-only air-conditioning system.
55. List the three modes of heat pump operation that require the operation of some or all of the system's electric strip heaters to be in operation.
56. Explain how a heat pump system is similar to a cooling-only, vapor-compression refrigeration system.
57. List four common configurations for heat pump systems based on the source of heat and what medium the appliance ultimately treats.
58. List two features of electronic thermostats that have helped make them the choice for controlling HVACR systems.

UNIT 44

GEOTHERMAL HEAT PUMPS

OBJECTIVES

After studying this unit, you should be able to

- explain how geothermal heat pump systems utilize the heat contained in the earth.
- describe two common classifications of geothermal heat pump systems.
- describe an open-loop geothermal heat pump system.
- explain how water quality affects an open-loop geothermal heat pump.
- describe a closed-loop geothermal heat pump system.
- describe different ground-loop configurations for closed-loop geothermal heat pump systems.
- explain the advantages and disadvantages of series- and parallel-flow configurations in geothermal heat pump systems.
- explain the different system fluids and heat exchanger materials.
- describe different geothermal well types and water sources for heat pumps.
- describe the heat exchangers used on water-to-water heat pump systems.
- explain how to properly evaluate geothermal heat pump systems.
- describe a direct geothermal heat pump system.

SAFETY CHECKLIST

☑ Always wear protective clothing and eye shields when working with geothermal heat pumps. They contain high-pressure refrigerants, which can freeze skin and cause serious injury.

☑ When working around geothermal heat pumps, antifreeze solution or water leaking from the system can cause an electrical hazard. When troubleshooting with the electrical power on, extra care must be taken to prevent electrical shocks.

☑ Closed-loop geothermal heat pumps also contain a solution in the ground loop that can be harmful if not handled properly. Both protective clothing and eye protection must be worn when in contact with this fluid. Also, some antifreeze solutions used in closed-loop systems can be flammable, so never expose them to extreme heat or an open flame.

44.1 REVERSE-CYCLE REFRIGERATION

Geothermal heat pumps are refrigeration machines. They are very similar to air source heat pumps in that they can remove heat from one place and transfer it to another. However, geothermal heat pumps use the earth, or water in the earth, for their heat source and heat sink. Energy is transferred daily to and from the earth by the sun's radiation, rain, and wind. Each year, more than 6000 times the amount of energy currently used by humans is striking the earth from the sun. In fact, less than 4% of the stored energy in the earth's crust comes from its hot molten center. Heat pumps use this energy stored in the earth's crust for heating. *In the summer months, because the temperature of the earth's crust is cooler than the air just above it, summer air-conditioning heat loads can be transferred to the earth.*

Geothermal heat pumps can pump heat in two directions. Because of this, they are normally used for space conditioning: heating and cooling. The same four basic components of heat pumps mentioned in Unit 43, "Air Source Heat Pumps"—the compressor, condenser, evaporator, and metering device—operate and control geothermal heat pumps. The *four-way reversing valve* also controls the direction of refrigerant flow in geothermal heat pumps. For a review of basic heat pump theory and operation, refer back to Unit 43.

44.2 GEOTHERMAL HEAT PUMP CLASSIFICATIONS

Geothermal heat pumps are classified as either **open-loop** or **closed-loop** systems. Open-loop, or *water-source*, systems use water from the earth as the heat-transfer medium and then expel the water back to the earth in some manner. This process usually involves a well, lake, or pond. Open-loop systems need a large volume of clean water to operate properly. This same water supply also can be used for drinking and cooking.

Closed-loop, or **earth-coupled**, systems use a heat-transfer fluid, which is reused and circulated in plastic pipes buried within the earth or within a lake or pond. Closed-loop or earth-coupled systems are used where the water is rich in minerals, where local codes prohibit open-loop systems, or where not enough water exists to support an open-loop well water system. Both open-loop and closed-loop systems will be covered in detail in the following sections.

Whether the geothermal heat pump system is an open-loop, closed-loop, or earth-coupled system, water is still the source of heat when the system is operating in the heating mode. For this reason, these heat pumps are referred to as water-source heat pumps. Water-source heat pumps can be classified as

either water-to-air or water-to-water. Water-to-air heat pumps are used to heat air in the occupied space; water-to-water heat pumps are used to heat water, which would be appropriate for an application such as a radiant heating system.

44.3 OPEN-LOOP SYSTEMS

Open-loop, water-source heat pump systems involve the transfer of heat between a water source and the air or water being circulated to a conditioned space. Remember, open-loop systems use water from the earth as the heat-transfer medium and then expel the water back to the earth. During the heating mode, heat is being transferred from the water source to the conditioned space. During the cooling mode, the heat removed from the conditioned space is deposited into the water. **Figure 44.1** illustrates both a heating and a cooling application of an open-loop, water-to-air heat pump. Defrost systems are not needed in geothermal heat pump systems.

In the *heating mode*, water is supplied from a water source to a coiled, coaxial heat exchanger, **Figure 44.2**, by a circulating pump, **Figure 44.1(A)**. The heat exchange takes place between the water and the refrigerant. Refrigerant is carried in the outer section of the coaxial heat exchanger and the water flows in the inner tube, **Figure 44.2(A)**. Notice that the inner tube is ribbed to increase both the surface area and the heat-transfer rate between the fluids, **Figure 44.2(B)**. The refrigerant side of the heat exchanger is the refrigeration system's evaporator. Heat is absorbed from the water into the vaporizing refrigerant. The refrigerant vapor then travels through the reversing valve to the compressor, where it is compressed. The heat-laden hot gas from the compressor then travels to the condenser. The condenser is a refrigerant-to-air, finned-tube heat exchanger located in the ductwork. It is often referred to as the *air coil*. Heat is then rejected to the air as the refrigerant condenses. A fan delivers the heated air to the conditioned space. The condensed liquid then travels through the expansion valve and vaporizes in the evaporator, absorbing heat from the water source. The process is then repeated. High-resistance, electric strip heaters can be used for auxiliary and/or emergency heat when the heat pump needs assistance.

In the *cooling mode*, the water loop acts as the condensing medium for the refrigerant, **Figure 44.1(B)**. Discharge gas from the compressor travels through the reversing valve to the outer portion of the coaxial (tube-within-a-tube) heat exchanger of the water loop. The refrigerant side of the heat exchanger is the refrigerant system's condenser. A refrigerant-to-water heat exchange takes place. The water loop absorbs heat from the refrigerant and condenses it. Subcooled liquid refrigerant then travels to the expansion valve and on to the *air coil*, where it evaporates and absorbs heat from the air. The air is cooled and dehumidified. A fan delivers this air to the conditioned space. The heat-laden, superheated, refrigerant gas then travels from the air coil and through the reversing valve to the compressor, where the refrigerant is compressed. The process is then repeated.

For smaller residences, a single water-source heat pump is normally used, but larger houses may require more than one. For commercial applications, multiple heat pumps are combined into a system with a common water-piping loop. This loop provides a means for transferring the rejected heat from one heat pump in the cooling mode to another heat pump, which is in the heating mode. Commercial applications may also be connected to a well, a pond, a lake, or an earth-coupled closed-loop system. These systems may also contain a *boiler* and *cooling tower* to maintain desired loop temperatures, **Figure 44.3**.

Figure 44.1 An open-loop, water-source heat pump in both (A) heating and (B) cooling mode. Source: Mammoth Corporation

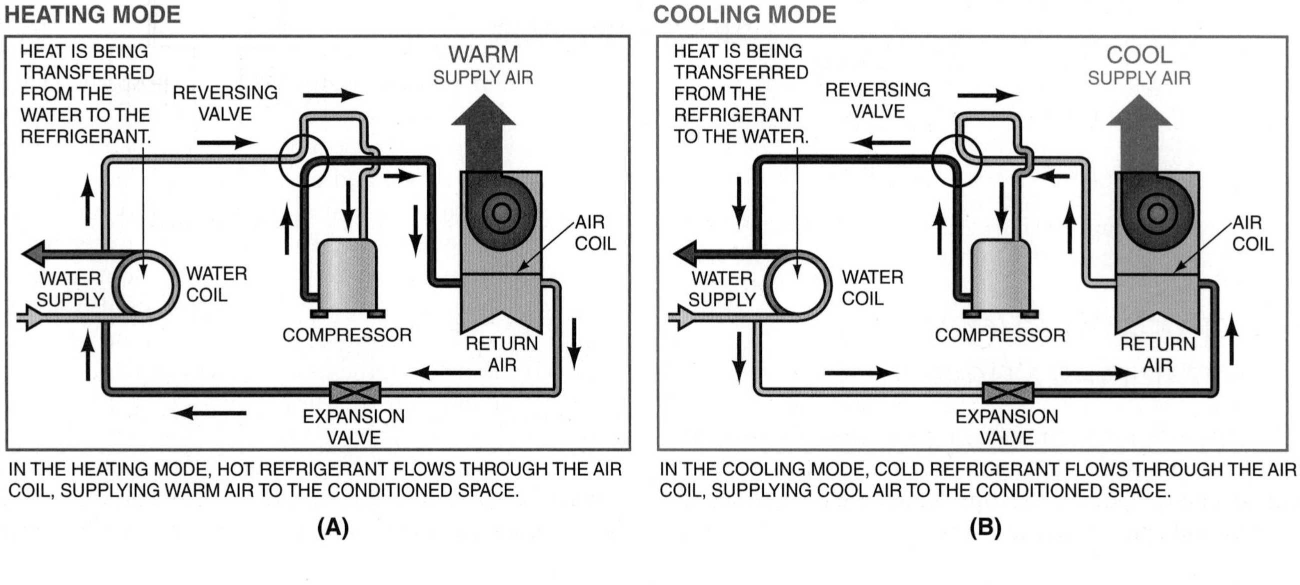

Figure 44.2 Cutaway view of a coaxial (tube-within-a-tube) heat exchanger. (A) Refrigerant flows in the outer tube while water flows in the inner tube. (B) Notice the shape of the inner tube. This increases the surface area and increases the rate of heat transfer. Photos by Eugene Silberstein

(A)

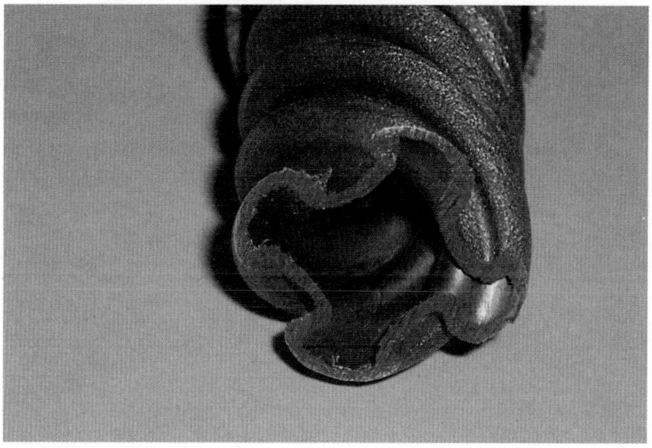

(B)

Figure 44.3 An open-loop, water-source heat pump with boiler and cooling tower to maintain the loop temperature.

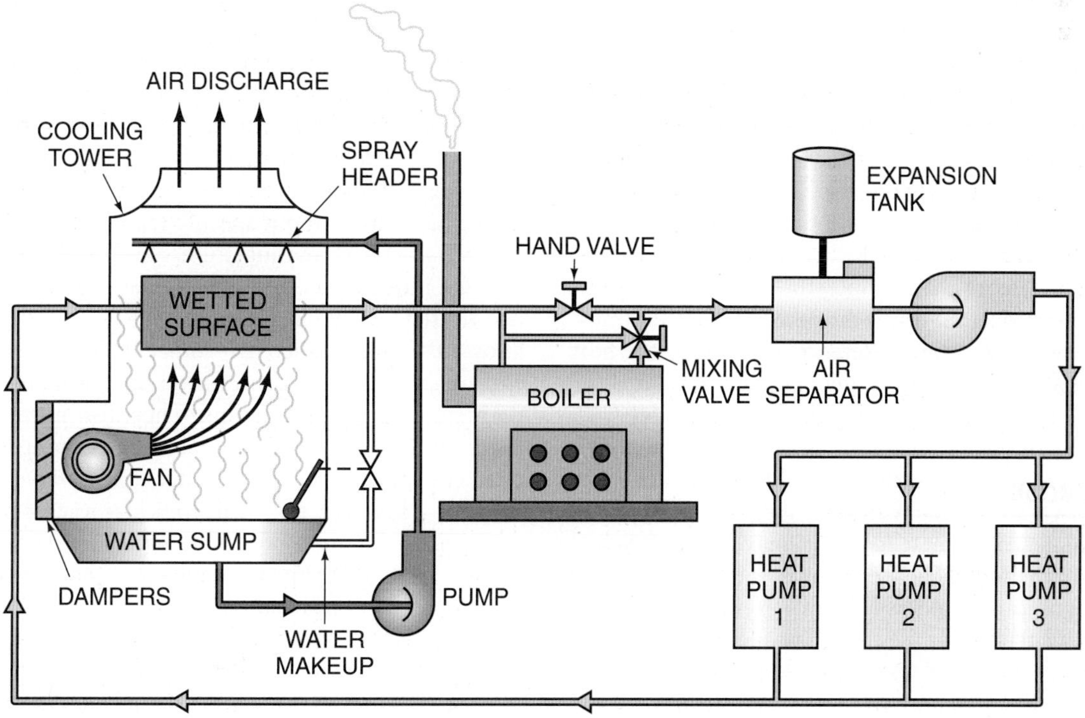

44.4 WATER QUALITY

Water quality is the most important factor to consider when dealing with open-loop systems that rely on well water. Listed here are three of the most important questions involving water quality that technicians and designers have to look at before choosing an open-loop, well-water system for heating and cooling applications.

- Will the well deliver enough water in gallons per minute (gpm) to the heat pump?
- What is the temperature of the well water?
- Is the well water clean and low in minerals?

Enough water flow in gpm has to be available to properly feed the system and handle the required heating and cooling capacities. Remember, this well water is the heat-transfer

Figure 44.4 Temperature (°F) in wells ranging from 50 ft to 150 ft deep. Source: Mammoth Corporation

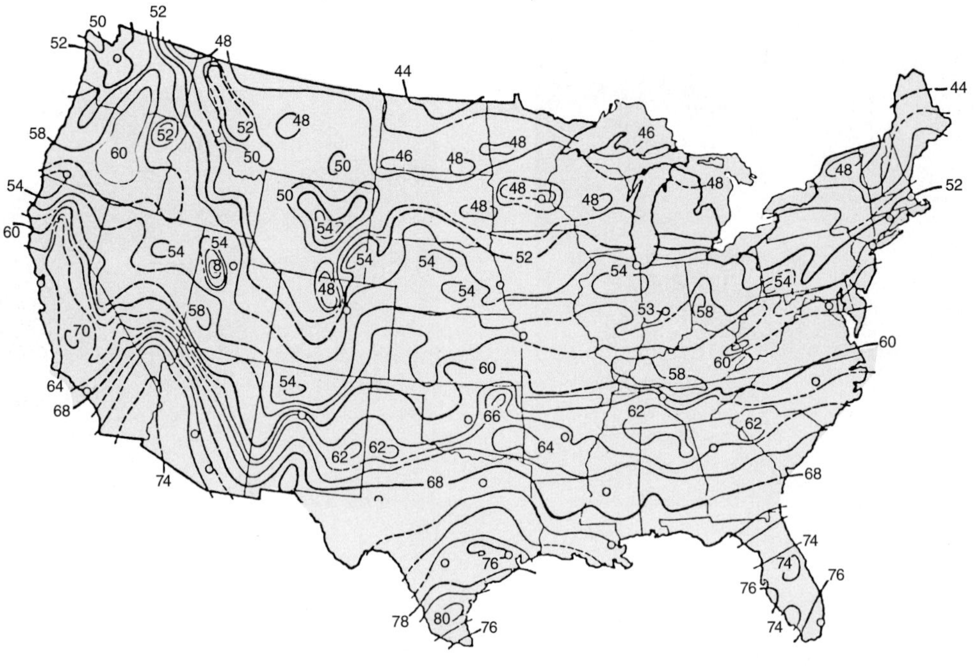

medium between the earth and the refrigerant in the heat pump. Consult with the heat pump manufacturer for specific information on water flow rates in gpm required for their heat pumps. The local well driller can determine whether enough water is in the well to properly operate the heat pump.

Well-water temperature is another important factor. It is one of the factors that determines the heat pump's capability to transfer heat energy because the temperature difference between the water and the refrigerant drives the potential for heat transfer. The larger the temperature difference, the greater the heat transfer. **Figure 44.4** illustrates water temperatures in wells ranging from 50 to 150 ft in depth within the United States.

Cleanliness of the well water is another major factor to consider when dealing with an open-loop, well-water system. The water cannot be contaminated with sand, dirt, clay, or any solid mineral particulate. All of these materials may cause a gradual wearing down of the coaxial heat exchanger in which the water flows with a significant velocity. Minerals eventually will foul the walls of the water side of the coaxial heat exchanger.

The heat exchangers of water-source heat pumps allow water to flow in one direction and refrigerant to flow in the other, called counterflowing. The water and refrigerant never come in physical contact with one another. The heat exchangers are the coiled, coaxial, tube-within-a-tube type. Water flows through the inner tube, and refrigerant flows in the outer jacket, which is made of steel. **Figure 44.5** shows the heat exchanger and a cross section. Notice that the inner tubes have extended surfaces, called *leads*, which increase the surface area for better heat transfer. The leads cause turbulence in both the water and refrigerant loops, which also increases the heat transfer between the water and the refrigerant. The temperature differential (delta-T) between the entering and leaving water in open-loop systems usually ranges between 7°F and 10°F. This means that if water is entering the heat exchanger at 55°F and leaving at 47°F during heating, the delta-T would be 8°F. This is an acceptable differential because it falls between the 7°F to 10°F range. However, always consult with the manufacturer of the heat pump about specific temperature differences; they may vary between manufacturers and geographical locations.

The inner tubes of most coaxial heat exchangers, which handle the water circuit, are made of copper. However, an alloy of copper, called *cupronickel*, is also available. Heat exchangers made of cupronickel have a higher resistance to corrosion from acid cleaning. They also have a higher resistance to abrasion than copper heat exchangers. Because of this, cupronickel heat exchangers will have a longer service life. However, they will foul just as fast as copper heat exchangers. **Figure 44.6** can be used as a guideline for determining whether a cupronickel heat exchanger should be used in an open-loop heat pump system.

44.5 CLOSED-LOOP SYSTEMS

In closed-loop heat pump systems, many yards of plastic pipe are buried in the earth. The pipe can be placed in either a horizontal or vertical configuration, depending on how much land is available and the soil composition. The loops of piping are called either ground loops or water loops, **Figure 44.7** through **Figure 44.13**. A completely sealed and pressurized loop of water, or water and antifreeze solution, is circulated by a low-wattage centrifugal pump through the pipe buried in the ground. In the heating mode, heat is transferred from the ground, through the plastic pipe, to

Figure 44.5 (A) A coiled, coaxial, tube-within-a-tube heat exchanger. The tubes are sealed in such a manner that the inside tube is separate from the outside tube. Source: Noranda Metal Industries, Inc. (B) The cross section of a counterflow, coaxial heat exchanger showing leads or extended surfaces for better heat transfer.

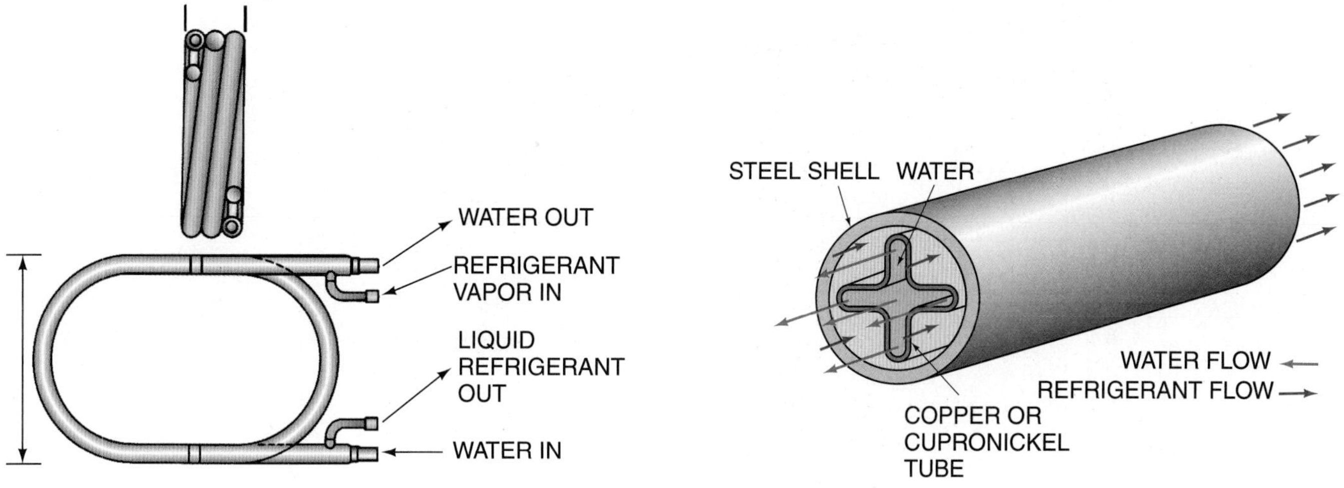

(A) (B)

Figure 44.6 A selection guide for determining whether a copper or a cupronickel heat exchanger should be used. Source: Mammoth Corporation

MAMMOTH WATER COIL SELECTION GUIDE

POTENTIAL PROBLEM	USE COPPER COIL	USE CUPRONICKEL COIL
SCALING- Calcium and magnesium salts (hardness)	Less than 350 ppm (25 grain/gallon)	More than 350 ppm (up to sea water)
Iron oxide	Low	High
CORROSION-* pH	7–9	5–7 and 8–10
Hydrogen sulfide	Less than 10 ppm	10–50 ppm
Carbon dioxide	Less than 50 ppm	50–75 ppm
Dissolved oxygen	Only with pressurized water tank	All systems
Chloride	Less than 300 ppm	300–600 ppm
Total dissolved solids	Less than 1000 ppm	1000–1500 ppm
BIOLOGICAL GROWTH- Iron bacteria	Low	High
SUSPENDED SOLIDS	Low	High

*Important—If the concentration of these corrosive exceeds the maximum tabulated in the cupronickel column, then the potential for serious corrosion problems exists. Water treatment may be required.

Figure 44.7 A ground loop showing a series-vertical configuration in the heating mode.

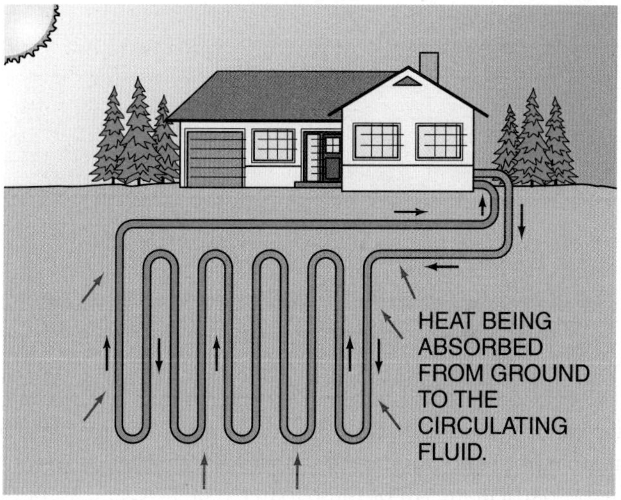

the liquid in the ground loop. During the summer or cooling mode, heat is rejected away from the circulating fluid, through the plastic pipe, and into the ground.

This fluid circulating in the plastic pipes or ground loop exchanges its energy with a refrigerant loop. The antifreeze-to-refrigerant heat exchange occurs within the heat pump's antifreeze-to-refrigerant heat exchanger, a coiled, coaxial, tube-within-a-tube exchanger contained within the cabinet of the heat pump, **Figures 44.14** and **44.15**. This heat exchanger is not subject to fouling because the closed-loop system uses the same treated water in its loop. *This is probably one of the main advantages a closed-loop system has over an open-loop system. Water-quality problems, which often foul the water side of open-loop heat exchangers, are almost nonexistent with closed-loop systems.*

An air loop is used to distribute heated or cooled air to the building. This is accomplished through a finned-coil, air-to-refrigerant heat exchanger located in the ductwork. A *squirrel cage blower* is used to move the air through the air distribution system. **Figures 44.14** and **44.15** illustrate the ground, refrigerant, and air loops in the heating and cooling modes, respectively.

Figure 44.8 A ground loop showing a parallel-vertical configuration in the cooling mode.

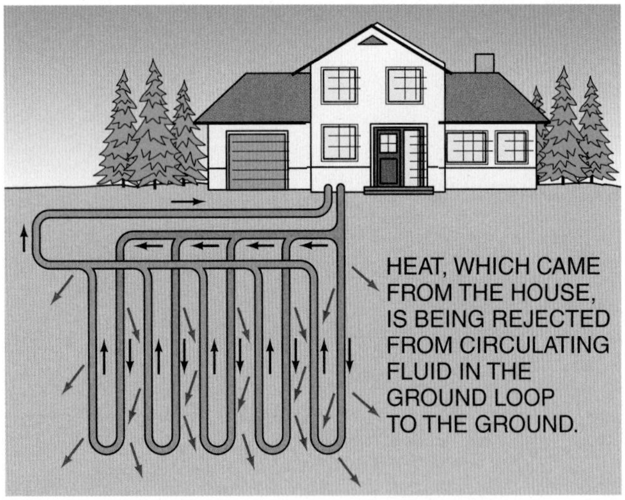

Figure 44.9 A single-layer, horizontal ground loop in the heating mode.

Figure 44.10 A two-layer, horizontal ground loop in the cooling mode.

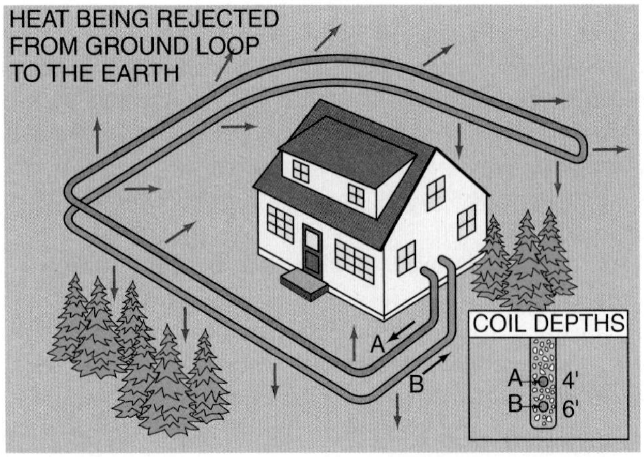

Figure 44.11 A four-pipe, horizontal ground loop.

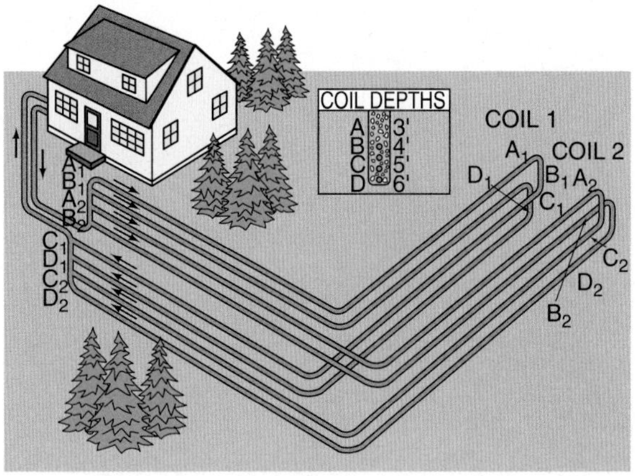

Figure 44.12 (A) A slinky ground loop.

(A)

Figure 44.12 (B) Prefabricated slinky ground loop waiting to be transported to the job site. Courtesy Patrick Pung, Ferris State University HVACR Graduate

(B)

Figure 44.12 (C) Four slinky ground loops in a freshly excavated trench waiting to be rolled out. Courtesy Patrick Pung, Ferris State University HVACR Graduate

(C)

Figure 44.12 (D) A technician in the process of rolling out a slinky ground loop. Courtesy Patrick Pung, Ferris State University HVACR Graduate

(D)

Figure 44.12 (E) Technicians rolling out the last of four slinky ground loops. Courtesy Patrick Pung, Ferris State University HVACR Graduate

(E)

Figure 44.12 (F) Freshly excavated trench showing four slinky ground loops with a large backhoe in the background. Courtesy Patrick Pung, Ferris State University HVACR Graduate

(F)

Figure 44.12 (G) A technician working on the supply and return headers for the slinky ground loops that have entered the basement of the house. Courtesy Patrick Pung, Ferris State University HVACR Graduate

(G)

Figure 44.13 A pond or lake loop.

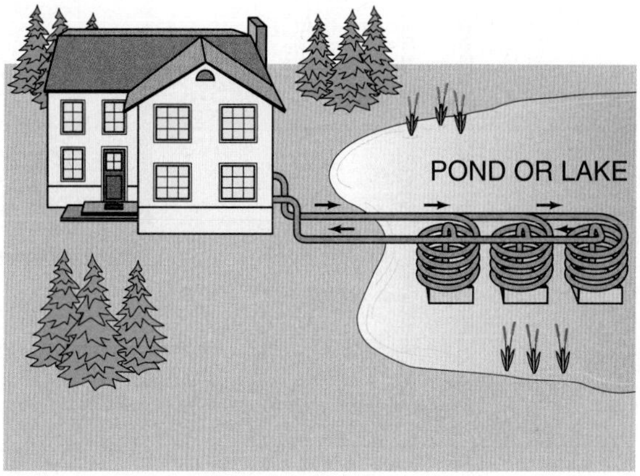

Figure 44.14 A closed-loop, water-source heat pump in the cooling mode. Source: Oklahoma State University

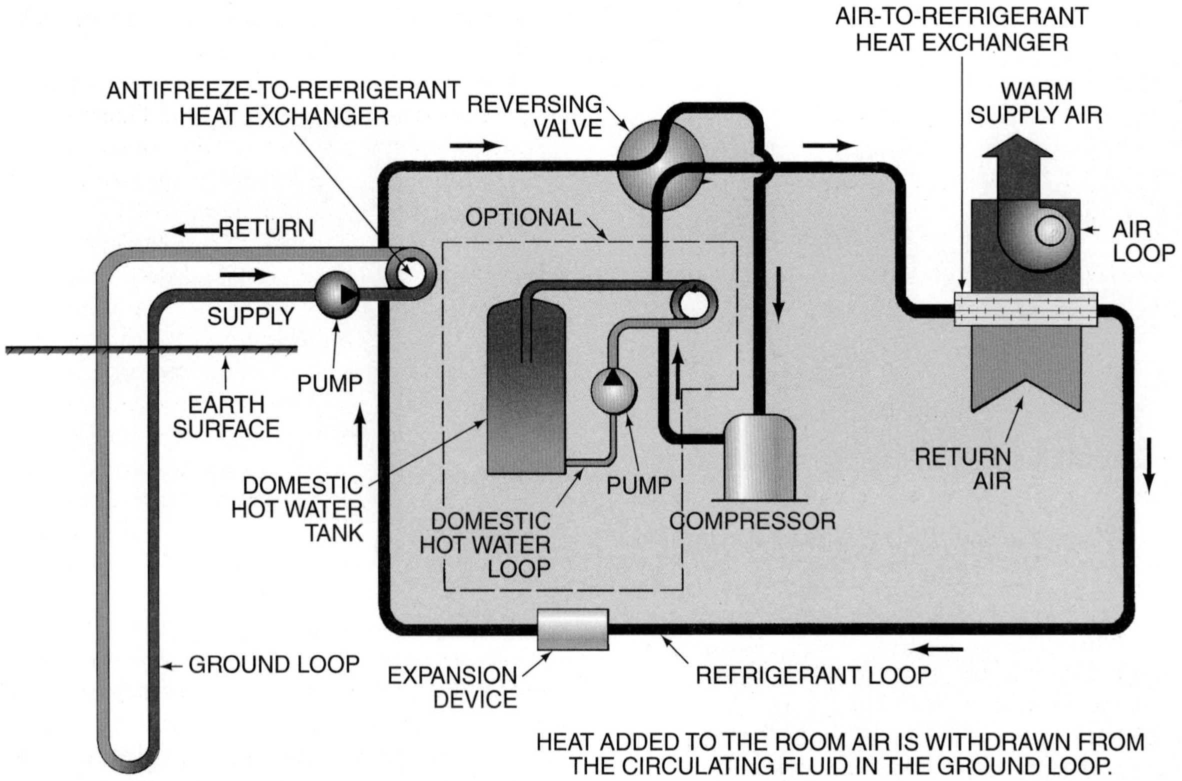

Figure 44.15 A closed-loop, water-source heat pump in the heating mode. Source: Oklahoma State University

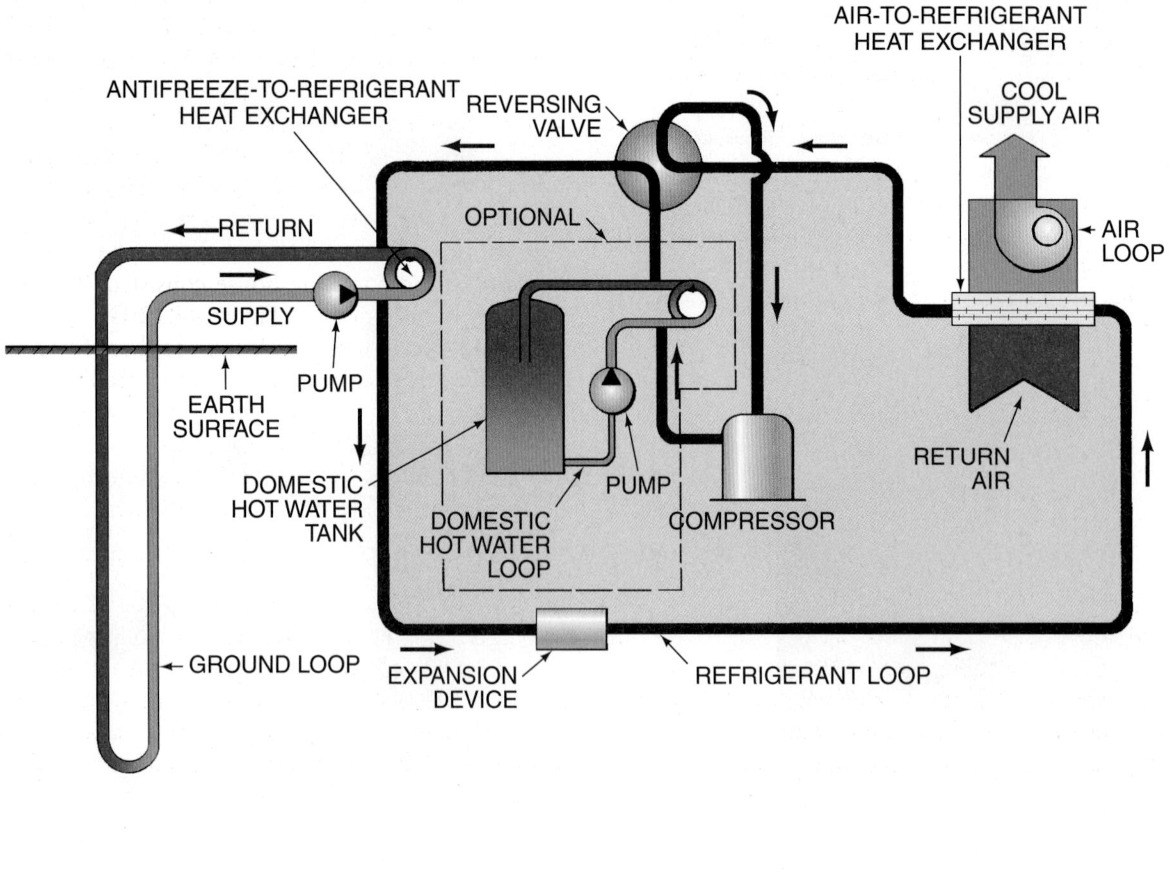

Figure 44.16 A domestic hot water loop showing a counterflow, coaxial heat exchanger. The water is heated by the refrigerant loop.

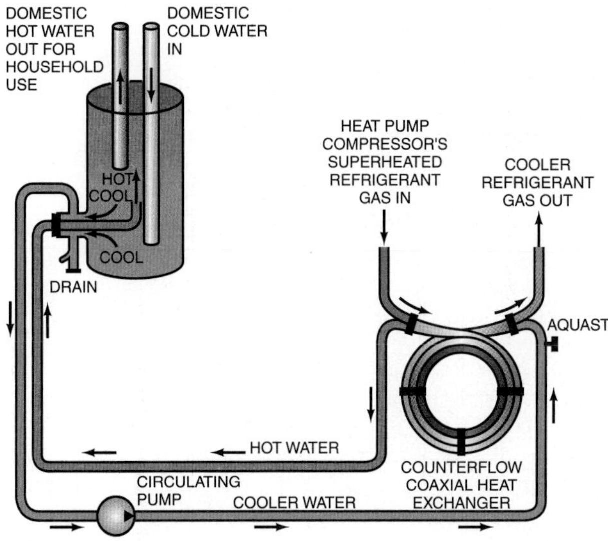

A fourth loop, or domestic hot water loop, is often used for heating domestic hot water from the hot discharge gas coming from the heat pump's refrigerant compressor. Because the discharge of the compressor is the hottest part of the refrigeration system, heat can be transferred easily from the hot refrigerant to the cool domestic water. However, a separate heat exchanger is again needed. This heat exchanger is also a coiled, coaxial, tube-within-a-tube. The domestic hot water is pressurized and circulated through the loop by a circulating pump. The refrigerant loop, which is in the outer shell of the heat exchanger, gives up heat to the cooler domestic water, which is in the inner tube of the same heat exchanger. The hot refrigerant and cool water counterflow one another, meaning they flow in opposite directions through the heat exchanger. This helps desuperheat the refrigerant and at the same time heat the domestic water. More modern systems have heat exchangers that not only desuperheat the refrigerant but also condense it for increased heat transfer. These systems can often supply 100% of domestic hot water demands. **Figure 44.16** illustrates a domestic hot water heat-exchanger loop.

44.6 GROUND-LOOP CONFIGURATIONS AND FLOWS

Ground loops can be vertical, horizontal, slinky, or of a pond/lake type. Loops can also have series or parallel fluid flows, which will be covered later in this section. **Figure 44.7** through **Figure 44.13** illustrate these loop configurations. Loop configuration choices will depend on how much land is available. The choice also depends on the type of soil found on the land and the contour of the land. Both of these factors affect excavation and drilling costs because very hilly or rocky land can become expensive to excavate or drill.

Vertical systems are used when there is a shortage of land or space restrictions, **Figure 44.7** and **Figure 44.8**. If the soil is rocky, a rock bit is used on the rotary drill. If the land available is without hard rock, a *horizontal loop* should be considered, **Figures 44.9**, **44.10**, and **44.11**.

The slinky loop is a flattened, circular coil of plastic pipe resembling a Slinky®, **Figure 44.12(A)**. Due to their configuration, slinky loops can reduce the trench length from one-third to two-thirds as compared with other loop systems. Loops can also be installed in lakes or ponds, **Figure 44.13**. A prefabricated slinky ground loop is shown rolled-up and waiting to be transported to the job site in **Figure 44.12(B)**. Slinky ground loops are very popular because their geometry allows good heat-transfer qualities between the earth and the ground loop while still maintaining a relatively small footprint. Four slinky ground loops in a freshly excavated trench are shown waiting to be rolled out against the floor of the trench in **Figure 44.12(C)**. Rolling out the loops against the floor of the excavated trench assures good thermal contact between the earth and the polyethylene piping making up each slinky loop, **Figures 44.12(D)**, **(E)**, and **(F)**.

In the residential, closed-loop, geothermal heat pump design illustrated, each loop is about 600 ft in length. The piping making up the slinky loops often averages from 600 to 700 ft; the lengths will vary depending on geographical location and soil conditions. Each slinky ground loop should ideally be about 12 to 15 ft apart horizontally. The depth that the ground loops are buried varies from 5 to 8 ft, depending on soil conditions. In the residential design shown, each slinky ground loop has the equivalent of 1 ton of capacity, giving the application illustrated a 4-ton capacity, **Figure 44.12(F)**. Also, in this residential design, the piping making up the ground loops is 0.75-in.-diameter polyethylene plastic pipe. The supply and return headers that connect the slinky loops in parallel and enter the house consist of 1.25-in.-diameter polyethylene plastic pipe, **Figure 44.12(G)**.

Plastic polyethylene tee fittings are heat-fused to the slinky loops. The supply and return headers are then heat-fused to the tee fittings. The pipe and coupling tee are heated to 500°F for approximately 15 sec with a fusion iron. After the heating of the plastic pipe and coupling tee is complete, the fusion iron is removed and the plastic pipe is pushed into the coupling tee. The plastic pipe and coupling tee must be held together until they have cooled in order to properly complete the fusion process. When the fusion process is performed properly, the two plastic pieces that have been fused will cool and harden together, creating a watertight seal for the ground loop.

Another decision that needs to be made is whether the ground heat exchanger or loop should be *series* or *parallel flow*. **Figure 44.17** illustrates both series and parallel flow paths in

Figure 44.17 Different flow paths in ground loops.

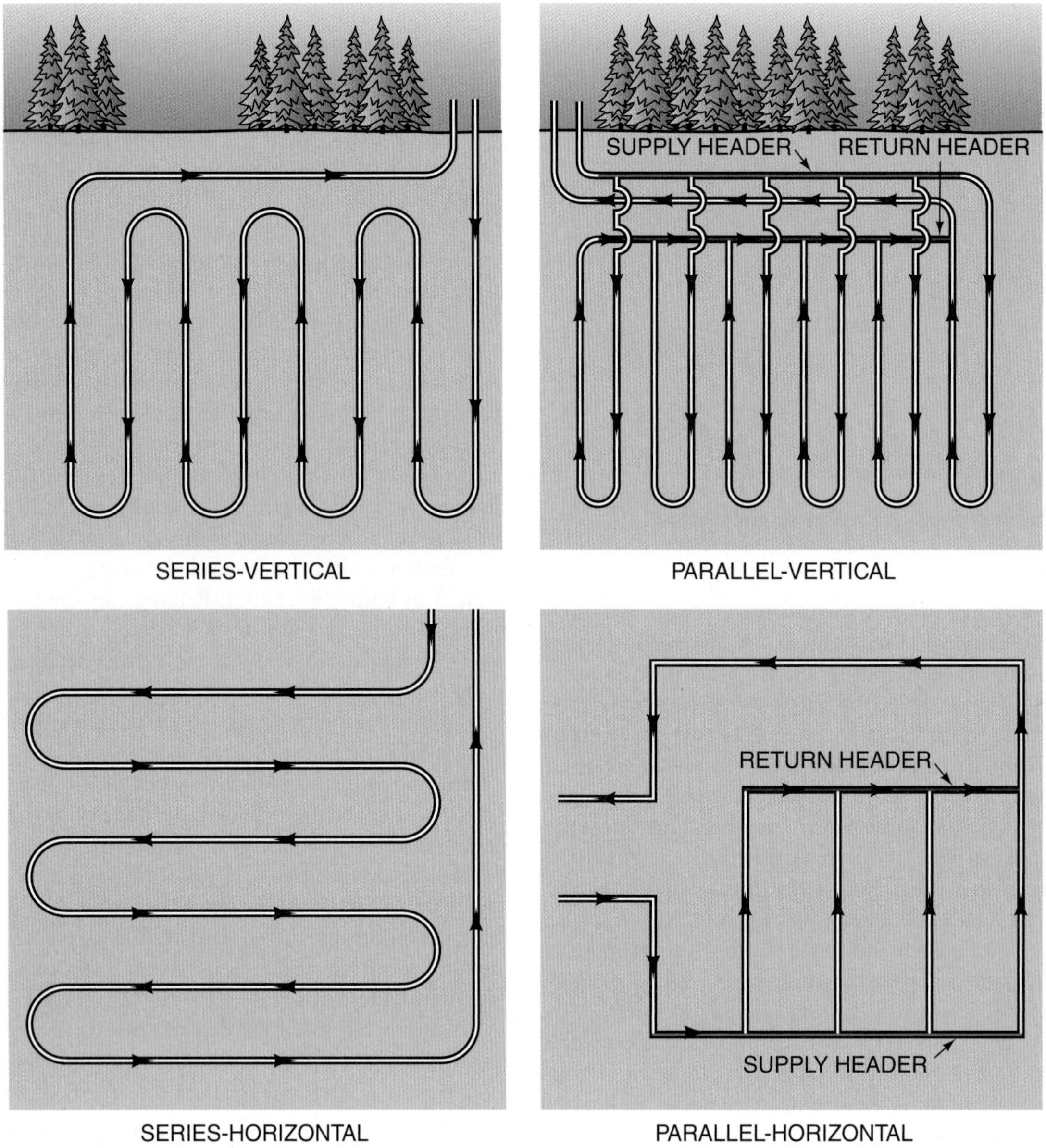

horizontal and vertical ground loops. In the series flow, only one path exists for the fluid to flow along. Air trapped in the plastic pipe is much easier to remove in a series loop, because the fluid's path is very well defined. The trapped air can be removed by power flushing, which involves connecting the heat pump and ground loop to a purging unit or pump stand that has a built-in, high-volume, high-velocity, and high-head circulating pump. The pump stand or purging unit also has a built-in air separator. The air then can be purged from the system, **Figure 44.18**. It is important that all the air be removed before the system is started, because air can erode and corrode any metal components in the water loop and can block fluid flow in the loop. Power flushing also removes any debris that may damage the bearings of the circulating pump. A system breakdown can result in either of these cases.

Listed here are advantages of a series-flow system:

- Ease of removing trapped air
- Simplified flow path
- Higher heat transfer per foot of pipe

Some disadvantages of a series-flow system are:

- Larger-diameter plastic pipe is needed, meaning more antifreeze solution
- Higher installation costs
- Higher pressure drops

Figure 44.18 A schematic of a purging unit hooked up to a heat pump for power flushing air and debris from the ground loop.

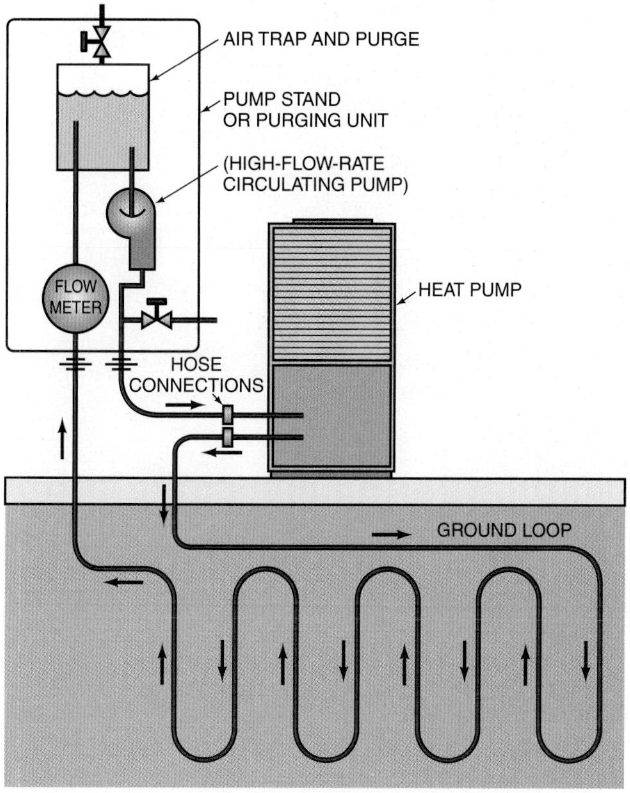

Parallel-flow systems can use smaller-diameter pipe because of their lower pressure drops. Smaller-diameter pipe has a lower cost; however, if there are air problems in the water loop, it is very difficult to remove the air with high-velocity water flushing because of the path choices. Excess air can also block the water flow in parallel systems. To make sure that pressure drops are equal in each parallel path, pipe lengths must be fairly equal. If not, the parallel path with the least resistance, or shortest length, will be fed with the most fluid, and the path with the most resistance will be partially starved of fluid. Unequal flow paths will seriously affect heat pump capacity. As a result, large-diameter headers are used at the inlet and outlets of the water loops to ensure that each loop experiences the same pressure for equal fluid flow.

Listed here are advantages of a parallel-flow system:

- Smaller-diameter and lower-cost piping
- Lower installation and labor costs
- Less antifreeze required

Some disadvantages of parallel fluid flow in ground loops are:

- Difficulty of removing air in the ground loops because of parallel paths
- Balancing problems if piping is of unequal length or unequal resistance

44.7 SYSTEM MATERIALS AND HEAT EXCHANGE FLUIDS

The buried piping or underground heat exchanger is usually made of polyethylene or polybutylene. These two materials can be bonded or joined through a heating process, which will give the piping joints a very long life. The antifreeze solutions inside the buried piping are used to prevent freezing of the heat pump heat exchanger and for good heat transfer. A lot of heat pumps in southern states use only pure water as a heat exchange fluid for the ground loop, because there is no threat of the loop freezing in the winter. Three choices for antifreeze solutions are:

- Salts—Calcium chloride and sodium chloride
- Glycols—Ethylene glycol and propylene glycol
- Alcohols—Methyl, isopropyl, and ethyl

Any fluid used must

- last a long time without breaking down.
- transfer heat readily.
- have a low cost.
- be environmentally safe.
- be noncorrosive.
- be nontoxic.

Salts are safe, nontoxic, and transfer heat well. Salts also are economical, environmentally safe, and will not break down in time. Salts are, however, corrosive when in contact with some metals, especially when mixed with air. Because of this, designers must select the proper metals to use in systems. Installers and service technicians must also make sure that the systems remain air-free if salts are to be used successfully in ground loops.

Glycols are safe, most of the time are noncorrosive, transfer heat fairly well, and also are economical. Glycols do have a track record of turning to gel at really low temperatures, which decreases their heat transfer capabilities and burdens the circulating pump responsible for their flows. Glycols have a shorter life than salts.

Alcohols can burn and are combustible if they are mixed with air. This is their major disadvantage. Alcohols are somewhat toxic, are relatively noncorrosive, transfer heat fairly well, and are relatively inexpensive to purchase. If handled properly and not mixed with air, alcohols can be a safe alternative to salts and glycols.

All system components—the coaxial heat exchangers, circulating pumps and flanges, and of course all plastic piping—must be carefully chosen when used in contact with salts, glycols, or alcohols. Sometimes, certain alcohols and salts are mixed with softened water from a domestic well-water system. They make good low-viscosity, water-soluble antifreeze solutions, but they can also be strong solvents, so strong as to dissolve some metals in the system. When some alcohols, like methanol, are mixed with high-sodium-content water from a water supply containing a water softener, they can attack the iron bodies of the

circulator pumps in the system. The iron bodies of the pumps actually become the sacrificial anode for the system because they are the least noble metals in the galvanic series. This problem can be corrected by replacing the antifreeze solution with a reputable premixed inhibited antifreeze made with deionized or distilled water. Many chemical companies market premixed geothermal loop fluids that have great antifreeze, anticorrosive, and heat-transfer properties. These premixed commercial solutions are also nontoxic. Common automotive windshield deicer or antifreeze should never be used in a geothermal loop because they often are toxic and do not have the proper inhibitors in their solutions. The flushing and fill process using geothermal loop fluids are not a homeowner's or business owner's do-it-yourself project. It should always be done by a reputable trained and licensed professional.

Historically, the refrigerant used in water-source heat pumps was R-22. However, the Clean Air Act phaseout of R-22 because of its chlorine content—and ozone-depleting potential—will soon replace R-22 with chlorine-free, environmentally friendly alternative refrigerants. R-410A is the leading alternative refrigerant for replacing R-22 in new heat pump and air-conditioning equipment. However, there are other, alternative refrigerants and refrigerant blends that are also substitutes for R-22 in new equipment and retrofitted systems. Always follow the refrigerant manufacturer's retrofit guidelines before using any alternative refrigerant to replace R-22. Refer to Unit 3, "Refrigeration and Refrigerants," and Unit 9, "Refrigerant and Oil Chemistry and Management—Recovery, Recycle, Reclaiming, and Retrofitting," for more detailed information on alternative refrigerants and refrigerant blends.

Beginning in 2015, the production and importing of any HCFC refrigerant was banned, except as refrigerant for equipment manufactured before January 1, 2020. As of January 1, 2020, HCFC-22 is completely banned, and beginning in the year 2030, there will be a total ban on all HCFC refrigerants. However, existing systems with HCFC refrigerants will still be in use and it will still be legal to use recovered and reclaimed HCFC refrigerants.

44.8 GEOTHERMAL WELLS AND WATER SOURCES FOR OPEN-LOOP SYSTEMS

Water sources for open-loop systems may be an existing well or a new well. A well pump delivers the water from the well to the heat pump. Some popular well categories are the following:

- Drilled well
- Return well
- Geothermal well
- Dry well

Figure 44.19 shows a drilled well. Notice that an electric water pump, complete with electrical lines, is encased underground within the well casing. This type of pump is often referred to as a submersible well-water pump. The water pump gets its water from an underground aquifer. Other water sources may be a pond, lake, or swimming pool. The water is then discharged into a lake, stream, or marsh. **Figure 44.20** illustrates an open-loop, geothermal heat pump system in which a water pump delivers water to the heat pump from a well and the water is then discharged to a pond.

Figure 44.19 A basic drilled well. Source: Mammoth Corporation

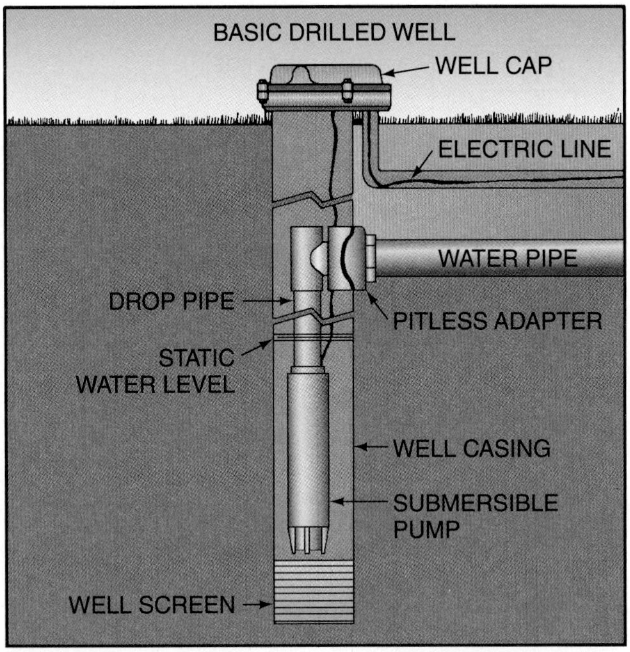

Most wells for geothermal heat pump systems are grouted. Grouting is a procedure in which a cement-like material is injected between the well casing and the hole drilled for the well. When the grout hardens, it forms a seal and prevents any contamination from other water sources in the same local area. It also prevents rain or other miscellaneous surface water from seeping into the well and polluting it. Grouting also makes the entire well structure sturdier and can prevent rusting of the well casing.

Return wells are used to discharge the water back into the ground after it has gone through the heat pump's heat exchanger, **Figure 44.21**. Return wells should be at least 100 ft from the supply well to prevent early mixing of the supply and return water. Also, the return well must be at least as large as the supply well to handle the water flow. If early mixing of supply and return water occurs, the supply-water's temperature could be increased or decreased, depending on the season. This could seriously affect the capacity of the heat pump. In some cases, the return well may discharge its water into the supply well. This ensures that a good volume of water is available to the heat pump's heat exchanger. However, adequate supplies of water must be available to ensure that supply water temperatures are not seriously affected.

Figure 44.20 Lake, stream, pond, or marsh discharge for an open-loop geothermal heat pump in the heating mode. Source: Mammoth Corporation

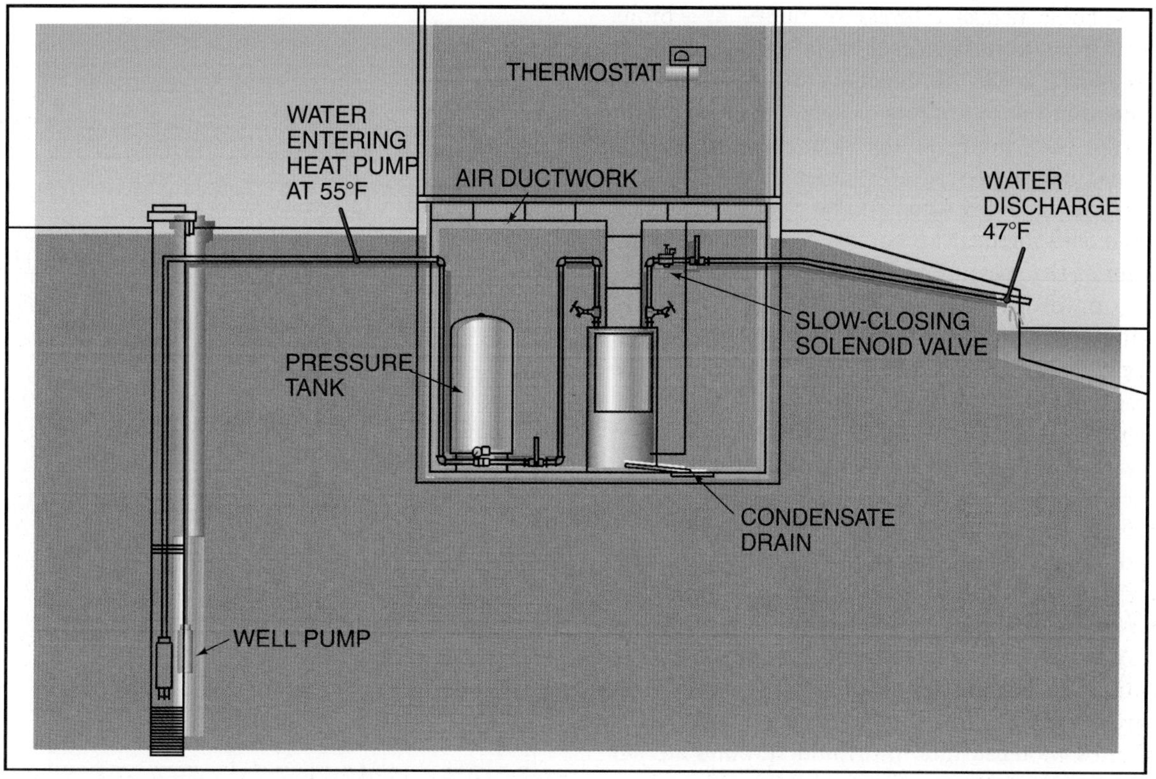

Figure 44.21 A return well system.

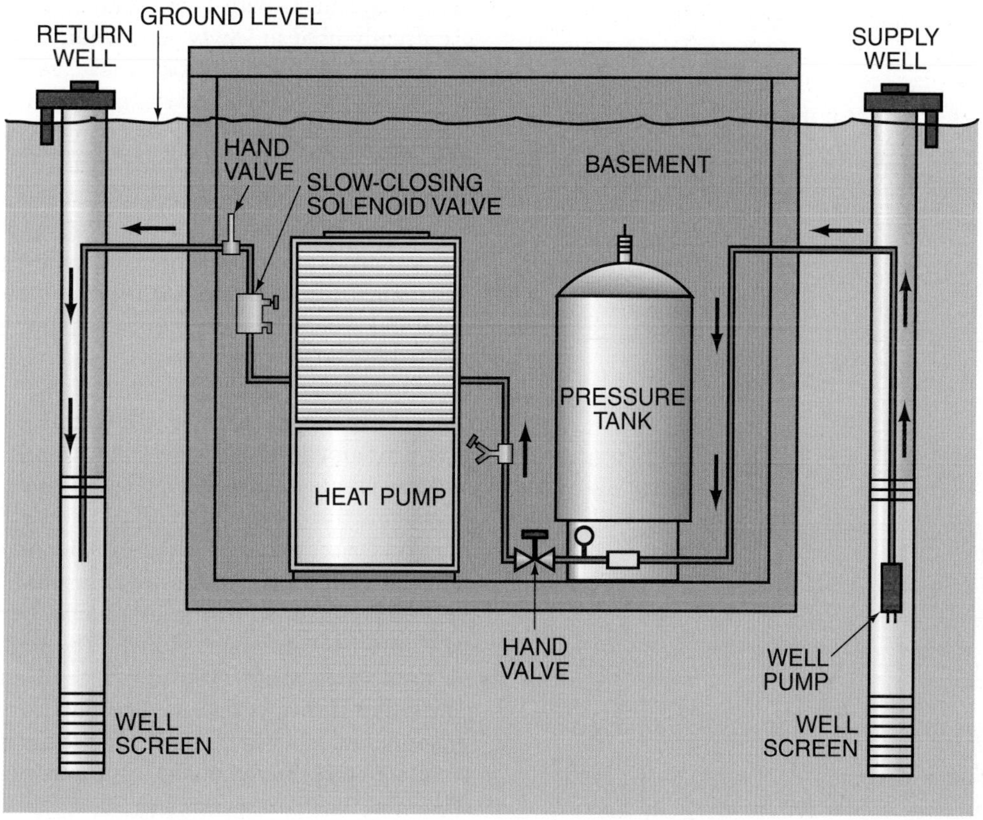

Notice in **Figure 44.21** that a slow-closing *solenoid valve* is installed in the return line. The slow closing of the solenoid valve prevents a pressure pulse, known as water hammer, from occurring each time the valve opens when the heat pump cycles on. Solenoid valves are almost always positioned in the return line to keep the coaxial heat exchanger in the heat pump at the same pressure as the pressure tank during the off cycle. Pressurization helps keep any dissolved minerals in the water so that they will not precipitate out and foul the heat exchanger. Minerals are more soluble in water as the pressure increases.

The return or drop pipe in the return well must end below the static water level in the well. Static water level is the level to which water will rise in a well as it naturally seeks its own level. The level of the return pipe keeps it free of air and helps to prevent the growth of algae and bacteria within the return line. Such a growth would restrict water flows and decrease capacities by coating the inside of the line.

Dedicated geothermal wells draw water from the top of the water column, circulate it through the heat pump where it either absorbs or rejects heat energy, and then return the water at a different temperature to the bottom of the water column, **Figure 44.22**. The water discharged to the bottom of the well will be at its normal temperature before it is drawn to the top of the well. Dedicated geothermal wells are used where there is not enough water in the underground aquifer to use other standard well systems.

A dry well is used to discharge water in an open-loop system. Dry wells are nothing more than a large reservoir in the ground filled with gravel and sand, **Figure 44.23**. They are usually used in sandy soil conditions. The used water is filtered as it seeps through the gravel and sand, and it eventually makes its way back to the underground aquifer.

Figure 44.22 A dedicated geothermal well. Source: Mammoth Corporation

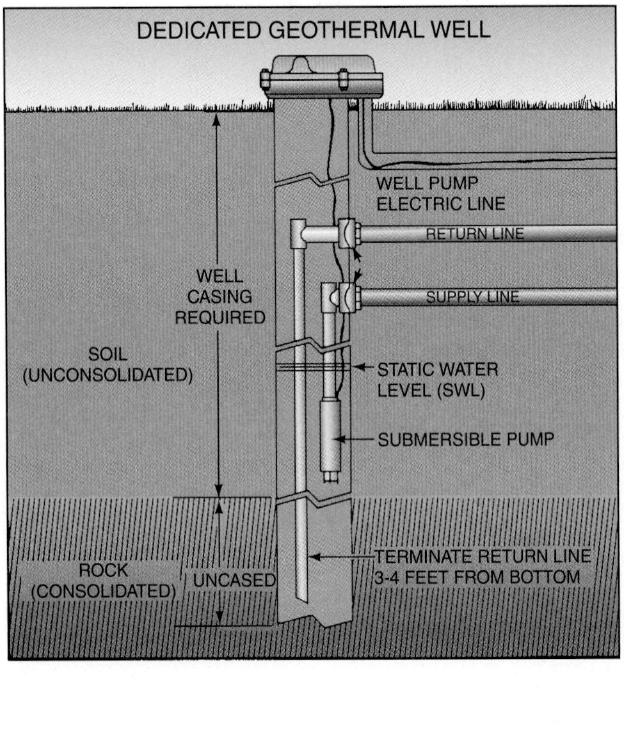

Figure 44.23 A dry well.

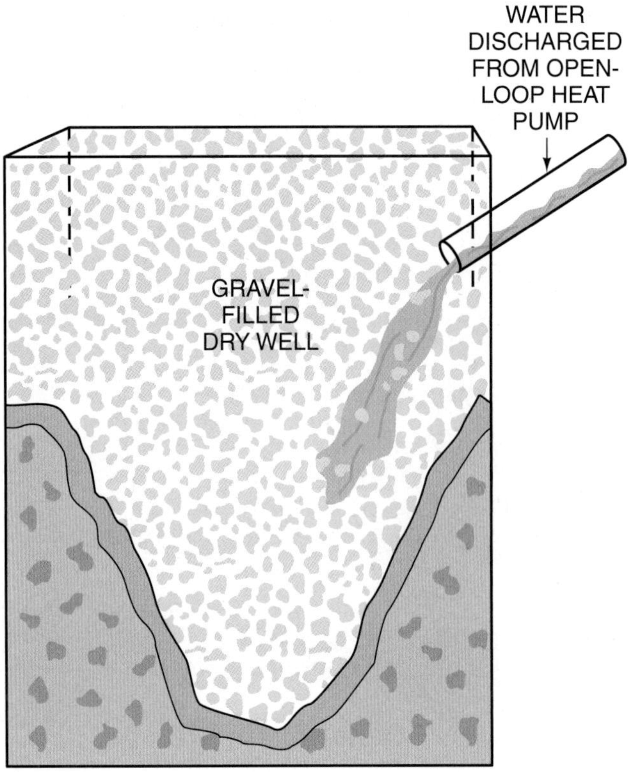

Most well systems in houses today use pressure tanks. They are also used extensively with open-loop geothermal heat pump systems. The pressure tank is nothing more than a small pressurized tank for water storage. The main purpose of the tank is to prevent the well pump from short cycling. Even when the well pump is off, the pressure tank can supply water when the open-loop heat pump calls for heating or cooling. When they come from the factory, pressure tanks have a compressed air charge in them. Some tanks have a fitting that allows the addition or subtraction of air for customization. This air charge is inserted in the tank on top of the water and is separated from the water by a rubber bladder. The well pump fills the pressure tank with water. The water displaces the bladder, creating higher pressure within the air charge. This increases the pressure in the pressure tank. When an maximum operating pressure is reached, the well pump will automatically be shut off by a pressure switch. The well pump will remain off until the pressure in the tank reaches a minimum pressure. The pump is then cycled on, and the process is repeated. The pressure tank should be sized so that the well pump does not start any more often than one time every 10 min. **Figure 44.24** illustrates the operation of a pressure tank.

Figure 44.24 The operation of a well system's pressure tank.

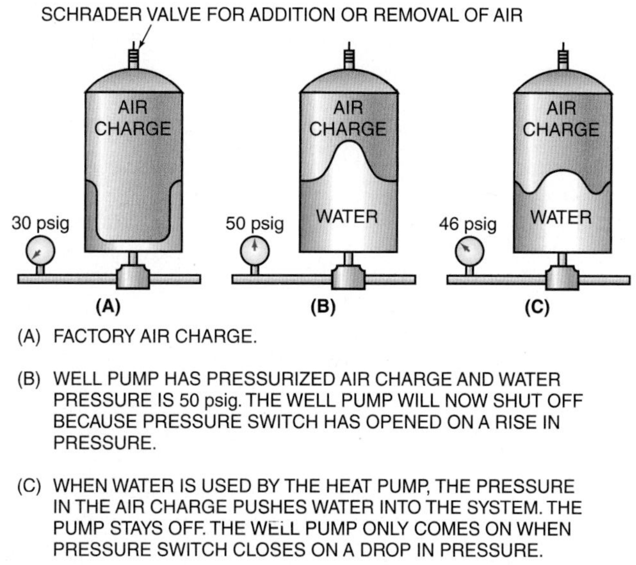

Figure 44.25 A water-to-water heat pump installed as part of a hydronic heating system.

44.9 WATER-TO-WATER HEAT PUMPS

Up to this point in this unit, our discussion has primarily focused on the water-to-air heat pump system. Now we will examine another heat pump configuration that is gaining popularity at a surprisingly fast rate: the water-to-water heat pump, **Figure 44.25**. In the past, applications for water-to-water systems were rather limited, given the fact that the water temperatures attainable were rather low for many heating situations. In an R-410A heat pump system, water can be heated to a temperature as high as 120°F. Given the increasing popularity of radiant heating systems that rely on lower-temperature water than do conventional baseboard heating systems, the water-to-water heat pump has become a viable option for providing hot water at a reasonable cost.

Water-to-water heat pumps utilize two coaxial heat exchangers, as compared to only one in the water-to-air system, **Figure 44.26**. Water-to-water systems can be configured as either open-loop or closed-loop systems. When operating in the heating mode, the coaxial heat exchanger that is linked to the ground loops is functioning as the evaporator, while the heat exchanger that is heating the water is operating as the condenser. In order to ensure continued satisfactory operation of the system, the water flow through the condenser must be unrestricted. A restricted water flow will cause the head pressure to rise, possibly causing the system to shut off on safety. This could pose a problem, especially since the heat pump system often controls more than one heating circuit. When only one or two zones are calling for heat, a high head-pressure situation may arise. For this reason, it is very common to see a buffer tank installed on the condenser water side of the water-to-water heat pump system, **Figure 44.27**.

Figure 44.26 Heat exchanger configuration on a water-to-water heat pump system.

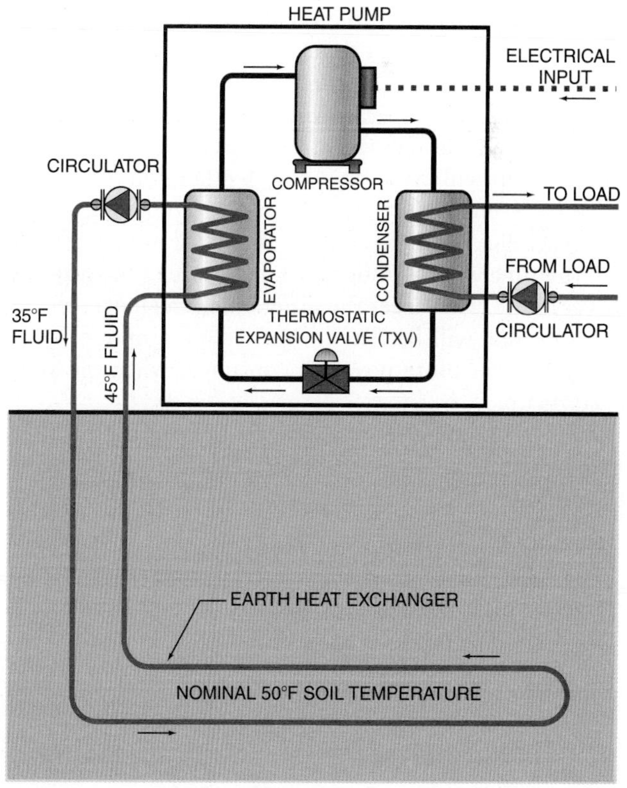

Figure 44.27 Buffer tank location on a water-to-water heat pump system.

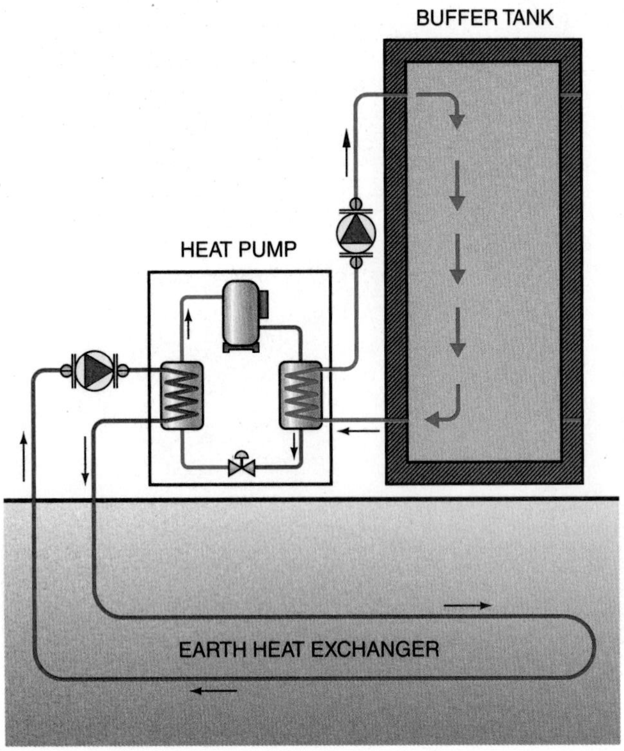

The buffer tank is nothing more than a large tank that both receives water from the condenser and supplies water to the condenser. The heat pump will cycle on and off on the basis of the temperature of the water in the tank. The cut-in temperature is the temperature at which the heat pump will cycle on; the cut-out temperature is the temperature at which the heat pump will cycle off. As with other controls, the difference between the cut-in and cut-out temperatures is the differential. The smaller the differential, the shorter the heat pump run time will be—but the temperature of the water will vary very little. The larger the differential, the longer the heat pump run time will be. The longer run-time option is more efficient because more power is consumed when the system first starts up.

The size of the buffer tank can be determined by crunching some numbers. The factors that affect the size of the tank are as follows:

- The capacity of the heat pump, $Q_{\text{Heat Pump}}$
- The desired temperature differential (ΔT) of the tank water
- The heat load on the space, Q_{Space}
- The desired run time of the heat pump, T

The formula for the buffer tank volume is

$$\text{Volume}_{\text{Buffer Tank}} = \frac{T(Q_{\text{HeatPump}} - Q_{\text{Space}})}{\Delta T \times 500}$$

Let's assume that we have a heat pump with a 24,000-Btu capacity and that we want the temperature of the water in the buffer tank to range from 95°F to 105°F. There is also a 10,000-Btu load on the space and we want the heat pump to run for 10 min. From this information, we get the following:

$$\text{Volume}_{\text{Buffer Tank}} = \frac{T(Q_{\text{HeatPump}} - Q_{\text{Space}})}{\Delta T \times 500}$$

$$\text{Volume}_{\text{Buffer Tank}} = \frac{10 \times (24,000 - 10,000)}{(105°\text{F} - 95°\text{F}) \times 500}$$

$$\text{Volume}_{\text{Buffer Tank}} = 140,000 \div 5,000$$

$$\text{Volume}_{\text{Buffer Tank}} = 28 \text{ Gallons}$$

By using a tank that is larger than calculated, the run time of the heat pump can be increased.

The buffer tank functions as the water-supply tank for the radiant heating system. One pump moves the water from the condenser to the tank and from the tank to the condenser. The second pump moves water from the buffer tank to the hot water heating circuits and from the heating circuits back to the tank, **Figure 44.28.** As long as the heating zones are calling for heat, the pump that carries the heated water into the heating zones will operate. This does not mean, however, that the heat pump is operating, as the heat pump will cycle on and off to maintain the desired water temperature in the buffer tank. As more and more zones call for heat, the temperature of the water returning to the buffer tank will be lower, so the heat pump will cycle on for a longer period of time to maintain temperature.

44.10 TROUBLESHOOTING

Troubleshooting geothermal heat pumps is much like servicing the air-to-air heat pumps covered in Unit 43, "Air Source Heat Pumps." However, one of the minor differences is how access is gained to pressures and temperatures in the water or ground loop. **Figure 44.29(A)** shows pressure and temperature port locations and mountings on the water or ground loop. **Figure 44.29(B)** illustrates a pressure and temperature port adapter used in performance testing the heat pump. The temperature probe is used for measuring the temperature difference between the inlet and outlet of the water's coaxial heat exchanger. The pressure gauge is used for determining the flow rate through the heat exchanger by acquiring the pressure drop across the heat exchanger. A curve or chart of flow in gpm versus pressure drop has to be used in conjunction with the pressure measurements.

The refrigerant system is accessed just as in any other heat pump system, and troubleshooting is very similar to troubleshooting any other refrigeration or air-conditioning system. Even the electrical system is very similar to other refrigeration and air-conditioning systems. Because of this,

Figure 44.28 Piping connections between (1) the heat pump and the buffer tank and (2) the buffer tank and the heating circuits.

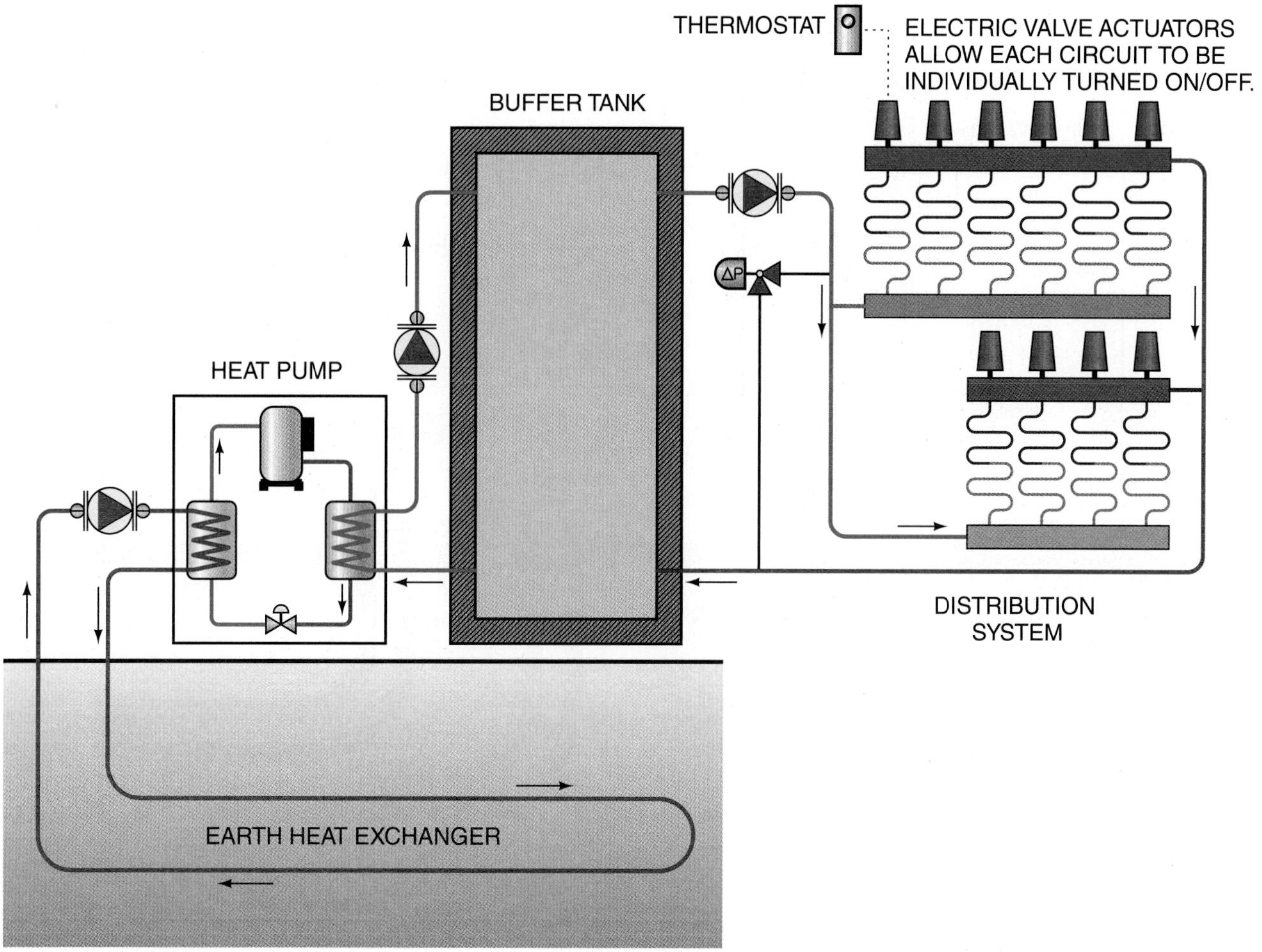

this troubleshooting section will cover mainly the water loop or ground loop of geothermal heat pump systems. If a review of the refrigeration system (refrigerant loop) or airflow system (air loop) is needed, see Unit 36, "Refrigeration Applied to Air-Conditioning"; Unit 37, "Air Distribution and Balance"; Unit 41, "Troubleshooting"; and Unit 43, "Air Source Heat Pumps."

The *ground loop* in closed-loop systems, which is sometimes referred to as the water loop, consists of the antifreeze solution, which is contained in the plastic pipes buried within the earth or lake. An open-loop system uses water from a well that is discarded back to the earth in some way. In either case, a water-to-refrigerant heat exchange takes place in the coaxial heat exchanger within the geothermal heat pump. The *governing formula* that determines the amount of heat absorbed in the heating mode or the amount of heat rejected in the cooling mode is

Heat quantity in Btu/h = (gpm) × (Temp. Diff.) × (500)

gpm = Gallons per minute of fluid circulated

Temp. Diff. = Difference in temperature between water in and out of the heat exchanger.

500 = A number that accounts for the specific heat of the fluid circulated, the weight of 1 gal of water (8.33 lb), and a time factor of 60 min per hour, to convert from gpm to Btu/h. 500 is used for pure water. 485 is usually used for an antifreeze solution.

The two variables in the preceding list are the gpm of antifreeze solution and the temperature difference of the antifreeze solution in and out of the coaxial, water-to-refrigerant heat exchanger. If either of these two variables changes, the heating and/or cooling capacity of the heat pump will be affected. Flow rates of antifreeze in the ground loop can be measured by flowmeters installed in the pipes or by the pressure drop across the coaxial heat exchanger. If the pressure drop technique is used, a curve of pressure drop versus flow rate in gpm must be used.

Figure 44.29 (A) Hookup from ground loop to heat pump. (B) Pressure- and temperature-sensing adapters for ground-loop access. Source: Oklahoma State University

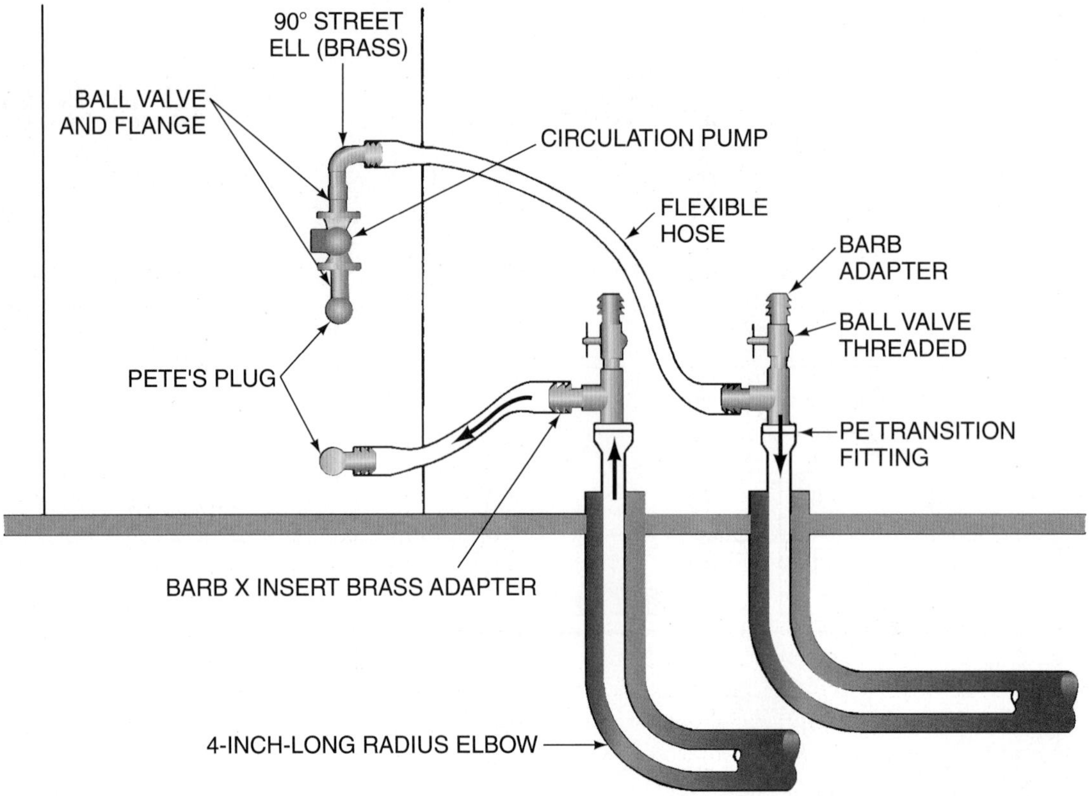

(A)

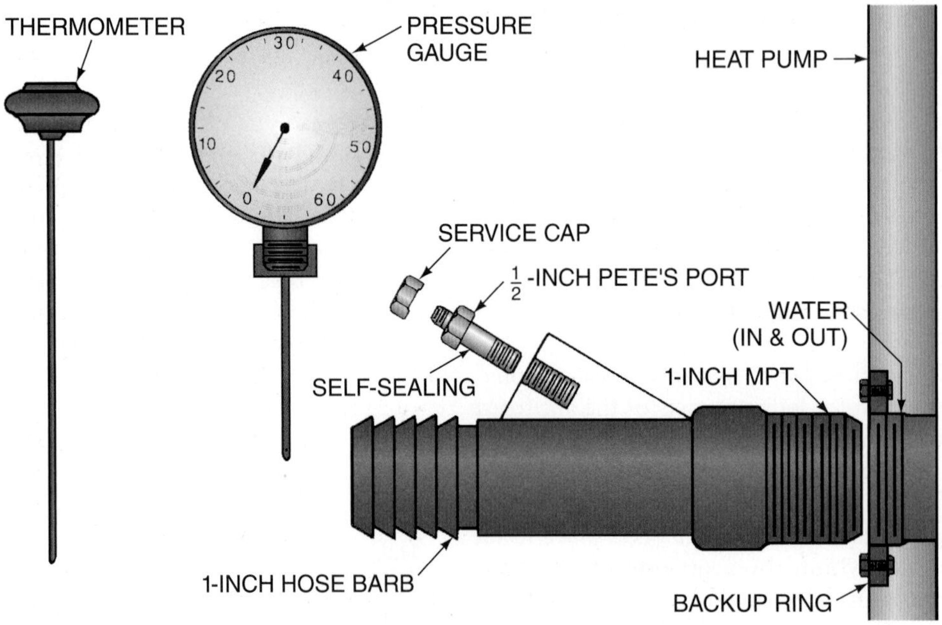

(B)

For example, a low gpm flow rate of antifreeze can be caused by a bad circulating pump, air restriction in the piping, contamination or a kinked pipe in the closed loop, or a low water-supply pressure in the open-loop system. If the heat pump is in the *heating mode,* the refrigerant side of the coaxial heat exchanger is the evaporator and low suction pressure will be noticed in the refrigeration loop. This causes low evaporating temperatures. Because of the reduced flow rates in the water side of the coaxial heat exchanger, the water will stay in contact longer with a colder coil. This will increase the temperature difference within the water. The service technician will notice a lower suction pressure and a higher temperature difference between water in and water out of the heat exchanger.

If the heat pump is in the *cooling mode,* the refrigerant side of the coaxial heat exchanger is now the condenser and the ground loop or water loop is the condensing medium. A reduced flow of water will cause high head pressures in the refrigerant loop. An increase in the temperature difference of the water in and out of the heat exchanger will also occur. This is because the water stays in contact with a hotter coil for a longer time.

If minerals in the water of an open-loop system foul and restrict the coaxial heat exchanger's surface on the water side, poor heat transfer will take place between the water and the refrigerant because the scale will act as an insulator. This will cause a low temperature difference between water in and out of the heat exchanger in both the heating and cooling modes. This results in a high head pressure in the cooling mode and low suction pressure in the heating mode. Closed-loop systems hardly ever have this problem because the same treated water/antifreeze solution is circulated throughout the loop.

If for some reason the loop has been designed too short in a closed-loop system, low water or antifreeze solution temperatures coming into the heat pump will be experienced in the heating mode. This will cause low suction pressures in the refrigerant loop. In the cooling mode, high water or antifreeze solution temperatures coming into the heat pump will be experienced, which will cause high head pressures. These same symptoms will occur if the loop is in the wrong type of soil or if there is poor thermal contact between the soil and the water loop.

44.11 DIRECT GEOTHERMAL HEAT PUMP SYSTEMS

The heat pump systems discussed previously in this unit are referred to as indirect systems because the antifreeze or water solution is a secondary refrigerant. The heat added to the conditioned space is withdrawn from the circulating fluid in the ground loop by the refrigerant in the refrigerant loop. We will now turn our attention to direct geothermal heat pump systems.

Figure 44.30 Polyethylene plastic-coated pipe being readied to be buried in a trench. Source: CoEnergies LLC, Traverse City, Michigan

Direct geothermal heat pump systems consist of small 3/8- to 7/16-in. copper pipes or polyethylene plastic-coated copper pipes buried in the ground with refrigerant flowing through them, **Figure 44.30**. The copper pipes act as the evaporator where the refrigerant experiences a phase change from liquid to vapor. This is often referred to as the *earth loop, phase-change loop,* or *refrigerant loop.* For clarification purposes, this loop will be referred to as the refrigerant loop for the rest of this unit. In the heating mode, heat is absorbed from the earth "directly" to the vaporizing refrigerant in the refrigerant loop. The refrigerant vapor then travels through the reversing valve to a refrigerant management system and then to the compressor, where it is compressed and superheated to a higher temperature. The superheated, heat-laden hot gas from the compressor then travels to the condenser. The condenser is a refrigerant-to-air, finned-tube heat exchanger located in the ductwork of the conditioned space or building. It is often referred to as the *air coil.* Heat is then rejected to the air as the refrigerant condenses at a higher temperature than the air. A blower delivers the heated air to the conditioned space. The condensed liquid then travels through the expansion valve and vaporizes in the evaporator or refrigerant earth loop buried in the ground.

The direct geothermal heat pump system can be easily retrofitted to a common home furnace gas-heating and conventional air-conditioning system. When such a retrofit is performed, a refrigerant management module is installed, **Figure 44.31**. The refrigerant management module contains a reversing valve, metering device and flow controls required to ensure that the refrigerant charge is correct in both modes of operation. Once retrofitted, the home's conventional air-conditioning outside coil and fan are not used for the heating mode. However, the outdoor coil will function as the condenser during cooling, as before the conversion, but will operate with a perfect or critical refrigerant charge because of the refrigerant management system.

Figure 44.31 A retrofitted system showing the refrigerant management module. Source: CoEnergies LLC, Traverse City, Michigan

A piping diagram for the waterless, earth-coupled, closed-loop geothermal heat pump system operating in the heating mode is shown in **Figure 44.32**. In the cooling mode, the refrigerant earth loop is not used, and the system will function as a conventional cooling-only air-conditioning system, **Figure 44.33**.

No water or antifreeze solution acts as a heat transfer fluid in these systems. The refrigerant earth loop or evaporator, which is buried in the earth, extracts heat directly from the earth. This is often referred to as a *direct* or *one-step* heat exchange. The American Society of Heating, Refrigeration, and Air-Conditioning Engineers (ASHRAE) and the Air-Conditioning, Heating and Refrigeration Institute (AHRI) refer to these types of heat pump systems as Direct GeoExchange systems. This system eliminates the coaxial heat exchanger, the liquid centrifugal pump, antifreeze or water heat-transfer fluids, and the need for system flushing common in traditional closed-loop, indirect geothermal heat pump systems. Because there is one less heat exchanger, the thermal efficiency of the heat pump is improved.

Also, by using one less heat exchanger, one less temperature difference does not have to be created and maintained within the geothermal heat transfer system. Now, there are only heat transfers from (1) the earth to the refrigerant loop and (2) the refrigerant loop to the air inside the house. The refrigerant-loop phase changes liquid refrigerant to vapor and takes advantage of the *latent heat of vaporization* of the refrigerant being used. The refrigerant loop also absorbs heat directly from the earth, which allows for higher efficiencies. In some system designs, the buried refrigerant loop is not utilized in the cooling mode, and the conventional air conditioner's outside condenser takes its place.

Home energy costs for heating and cooling can be reduced as much as 50% when the system is retrofitted to a direct geothermal system. This is especially true when fossil fuels are used exclusively. Because the system uses a precise

Figure 44.32 A heating-mode schematic. Based on CoEnergies LLC, Traverse City, Michigan

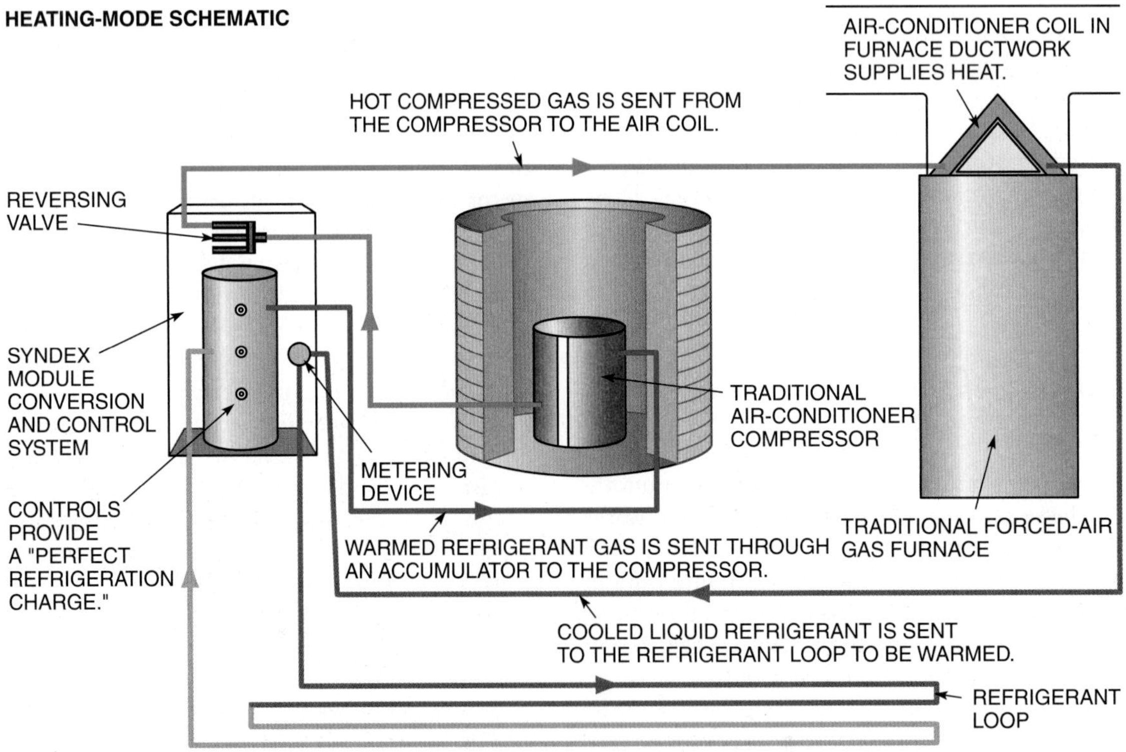

Figure 44.33 A cooling-mode schematic. Based on CoEnergies LLC, Traverse City, Michigan.

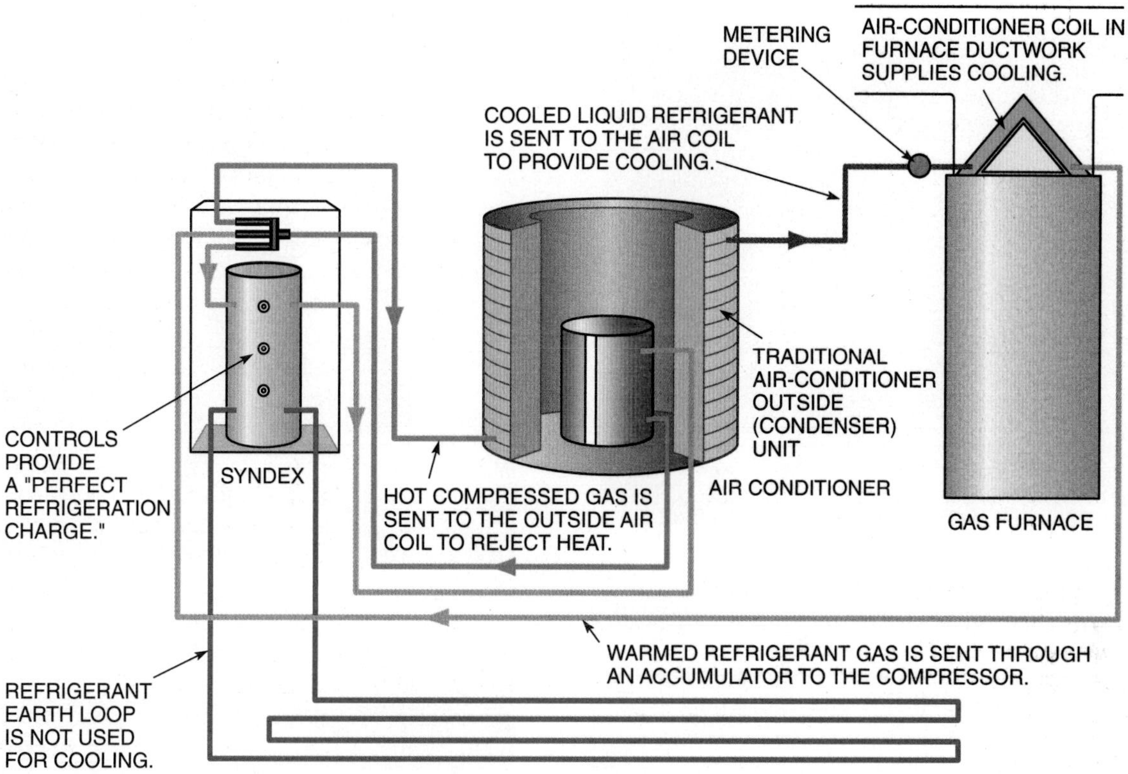

refrigerant-management module that will eliminate the common problems of an improperly charged air-conditioning system, the cooling system's performance will be greatly improved. The improved heating and cooling efficiencies result in overall energy savings for almost every geographical region of North America.

When used in conjunction with a home's conventional fossil fuel furnace, the furnace can be used for second-stage or emergency heating, with the direct geothermal heat pump providing first-stage heat. The furnace will not operate at the same time as the heat pump system because once second-stage heat is turned on, the heat pump will be locked out until the thermostat is satisfied.

INSTALLATION AND REFRIGERANT-LOOP PIPING

Installation costs for a retrofit are generally less than the installation of a comparable traditional, stand-alone, earth-coupled geothermal heating and cooling system. By using the one-step or direct heat exchange process between the phase-changing refrigerant and the earth, the homeowner's existing conventional air-conditioning compressor acts as the refrigerant pump and heat generator. All other heat energy to and from the system is directly transferred heat energy.

As mentioned earlier, the refrigerant loop buried in the ground consists of 3/8- to 7/16-in. copper pipe or copper pipe covered with polyethylene plastic. The polyethylene plastic pipe has two small vents or grooves that allow the pipe to act as a vented double-wall heat exchanger, **Figure 44.34**. Any refrigerant or oil leaks in the buried copper piping can vent to the grooves in the plastic coating, which will be vented to the surface. This surface venting allows for ease of leak detection and protects the ground from refrigerant and oil contamination, making it a more environmentally

Figure 44.34 Two small vents, or grooves, allowing the pipe to act as a vented double-wall heat exchanger.

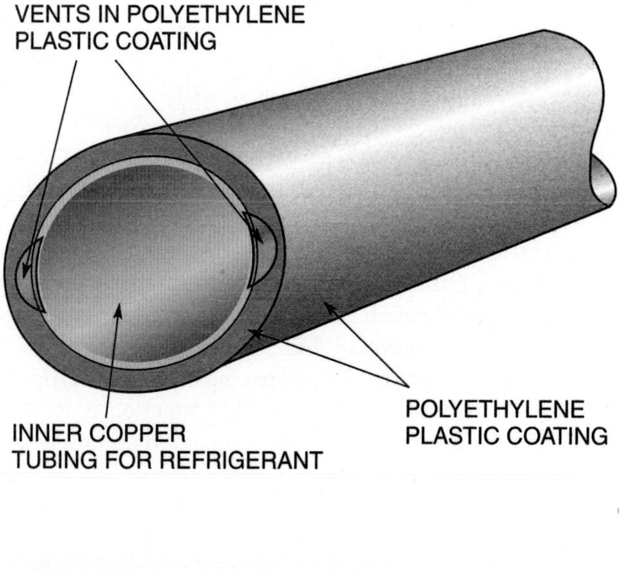

Figure 44.35 A trencher digging the trench for the refrigerant loop. Source: CoEnergies LLC, Traverse City, Michigan

Figure 44.36 (A) A copper header being brazed to supply and return loops by a technician. Source: CoEnergies LLC, Traverse City, Michigan. (B) A header box being positioned over a copper header. Source: CoEnergies LLC, Traverse City, Michigan

(A)

(B)

conscious system. Each copper pipe loop is approximately 70 to 100 ft long and is buried in trenches 3 to 4 ft deep. A trencher is used to dig the trench for the polyethylene plastic-coated copper pipe, **Figure 44.35**, and the piping is then covered with sand for good thermal efficiency. The trenches are then backfilled and restored to their natural grade. These shallow trenches are generally quicker to dig than traditional closed-loop, indirect, geothermal heat pump systems containing antifreeze or water in their ground loops. Each buried refrigerant loop is a one-piece circuit without any buried joints. The number of loops depends on the capacity of the system and the soil composition.

Each refrigerant-loop circuit's pipe supply and return ends are brazed to a copper header that is contained in an easily accessible header box, **Figure 44.36**. In the unlikely event that a leak should occur, the header is easily accessible through the header box. The individual refrigerant loops in separate trenches all meet at the header box or header pit, where the circuits are soldered together. There are no joints or connections anywhere below the ground. The two supply and return lines are then insulated and routed to the air conditioner, **Figure 44.37**. The refrigerant lines of the existing conventional air conditioner are tapped into and diverted to the header pit, where the refrigerant can be circulated through the waterless earth loop buried in the ground.

The earth loop, often referred to as the refrigerant loop, may be installed in three different configurations: (1) diagonal, (2) vertical, and (3) horizontal. All three configurations are shown in **Figure 44.38**. Diagonal drilling rigs are shown in **Figures 44.39** through **Figures 44.43**. A horizontal installation is shown in **Figure 44.44**.

The refrigerant (earth) loops are connected to a refrigerant distributor or manifold. The manifold's function is to divide the refrigerant flow equally to each loop and maintain balance within the refrigerant ground-loop system. There is a vapor and a liquid manifold for each refrigerant ground loop. The vapor manifold is the larger of the two manifolds because it transports evaporated refrigerant vapors back to the compressor. The liquid manifold has smaller tubing than the vapor manifold because of the liquid refrigerant's higher density. Most earth loops are grouted with a plasticized grout for good heat transfer with the earth and to protect any underground aquifers nearby. The grout puts a seal around the earth loops that are lowered into the small-bore holes. This grout seal

Figure 44.37 Insulated supply and return lines routed to the retrofitted air conditioner. Source: CoEnergies LLC, Traverse City, Michigan

Figure 44.38 Diagonal, vertical, and horizontal earth loop configurations. Source: Earthlinked Technologies, Inc.

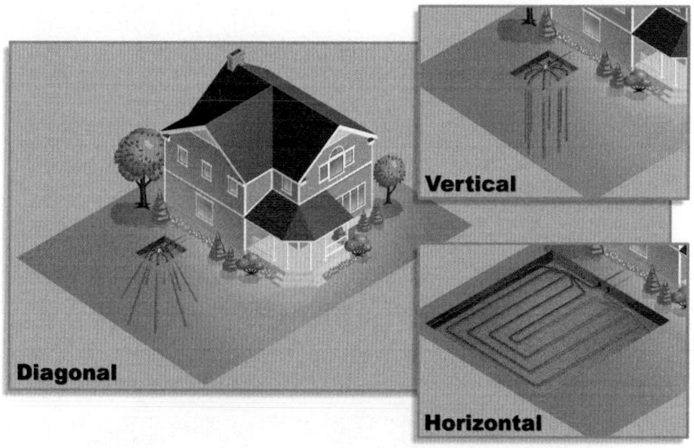

prevents any surface water from running down into an aquifer by way of the bore holes that house the earth loops.

The main ports to each manifold are connected to the heat pump's compressor unit by means of a line set comprised of one vapor and one liquid tube. **Figures 44.45** through **Figures 44.47** show some refrigerant- (earth) loop manifold systems. To ensure equal distribution of refrigerant, it is critical that the manifolds are installed vertically, as shown in **Figures 44.45** and **44.46. Figure 44.45** illustrates the interface of a line set, the liquid and vapor manifolds, and the individual refrigerant (earth) loops for a typical direct geothermal system.

There are many configurations possible for diagonal, vertical, and horizontal earth loops. The choice of the specific configuration depends on several factors involved in the

Figure 44.39 Diagonal drilling rig. Source: Earthlinked Technologies, Inc.

Figure 44.40 Diagonal drilling rig near a residence. Source: Earthlinked Technologies, Inc.

Figure 44.41 Diagonal drilling rig showing small-bore drilling equipment. Source: Earthlinked Technologies, Inc.

Figure 44.42 Diagonal drilling rig showing residence and small-bore drill. Source: Earthlinked Technologies, Inc.

Figure 44.43 Diagonal drilling rig with workers feeding copper earth loop into small-bored hole. Source: Earthlinked Technologies, Inc.

Figure 44.44 Horizontal earth loop installation. Source: Earthlinked Technologies, Inc.

Figure 44.45 Details of an earth-loop manifold. Source: Earthlinked Technologies, Inc.

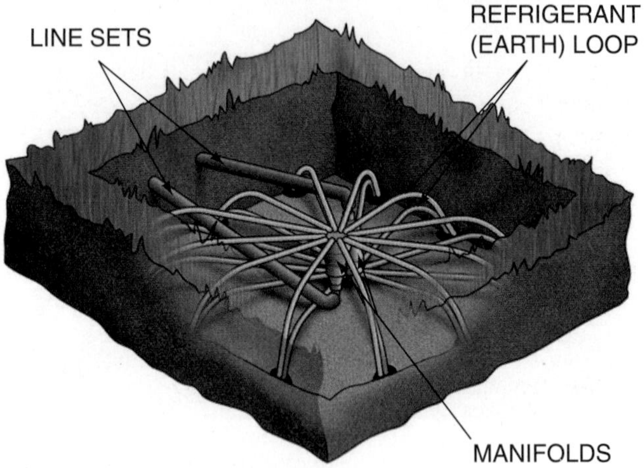

Figure 44.46 House with manifold details and a diagonal earth loop. Source: Earthlinked Technologies, Inc.

Figure 44.47 Diagonal earth-loop supply (liquid) and return (vapor) manifolds in a pit. Source: Earthlinked Technologies, Inc.

design and application of these direct geothermal systems. One such design incorporates a diagonal earth-loop configuration consisting of four 50-ft earth loops per ton of capacity. The earth loops are bored at a 45-degree angle, putting their lowest depth at about 35 ft in the earth.

The American Society of Heating, Refrigeration, and Air Conditioning Engineers (ASHRAE) and the Air-Conditioning, Heating and Refrigeration Institute (AHRI) refer to these types of heat pump systems as Direct GeoExchange systems.

NON-RETROFIT APPLICATION FOR A DIRECT GEOTHERMAL HEAT PUMP SYSTEM

When an existing air-conditioning system is not being converted to a direct geothermal heat pump system, a stand-alone direct geothermal heat pump system can be installed. This type of system uses the earth loop piping in both the heating and cooling modes of operation. The heat pump's refrigerant (earth) loop piping, air handler, and compressor unit are shown in **Figure 44.48**. The internals of the heat pump's compressor unit are shown in **Figure 44.49**.

HEATING MODE FOR A NON-RETROFIT APPLICATION

In the heating mode, liquid refrigerant from the heat pump's indoor coil (condenser) is expanded and enters the smaller-diameter tube of the refrigerant (earth) loop or evaporator. Heat is then transferred from the warmer earth into the refrigerant loop. While absorbing this heat, the refrigerant in the loop is boiled (evaporated) to a vapor as it progresses

Figure 44.48 Compressor unit, earth loop, and air handler. Source: Earthlinked Technologies, Inc.

Figure 44.49 Internals of the heat pump's compressor unit showing the active charge control with two sight glasses.

through the refrigerant (earth) loop, **Figure 44.50(A)**. The refrigerant vapor then exits the earth loop and is returned to the compressor unit. The refrigerant vapor is now compressed by the heat pump's compressor to a higher pressure and temperature. The compression process usually raises the refrigerant vapor temperature from about 40°F to about 160°F. The hot, superheated refrigerant vapor is now delivered to the indoor coil (condenser), where it gives off heat to the building's indoor air with the assistance of a blower, **Figure 44.50(B)**. As the refrigerant vapor gives off heat, it gradually condenses back to a 100% liquid state.

As mentioned earlier, the compression process raises the temperature of the refrigerant vapor to about 160°F. Because this hot, superheated vapor leaving the compressor is hotter than the home or buildings inside air temperature, heat transfers into the air as the air passes over the fan coil located in the air handler, **Figure 44.50(B)**. This warms your home or building.

COOLING MODE FOR A NON-RETROFIT APPLICATION

In the cooling mode, superheated refrigerant vapor is pumped by the heat pump's compressor into the larger-diameter (vapor) earth loop tubing. In this mode, the refrigerant temperature entering the refrigerant (earth) loop is higher than that of the surrounding earth itself. Heat will now be transferred from the refrigerant in the earth loop

Figure 44.50 (A) The earth loop showing heat being transferred from the earth into the phase-changing (evaporating) refrigerant. Source: Earthlinked Technologies, Inc.

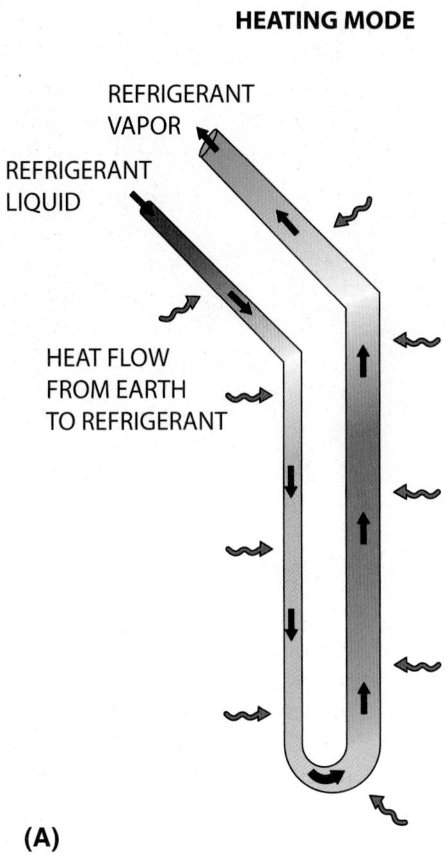

Figure 44.50 (B) System schematic in the heating mode. Source: Earthlinked Technologies, Inc.

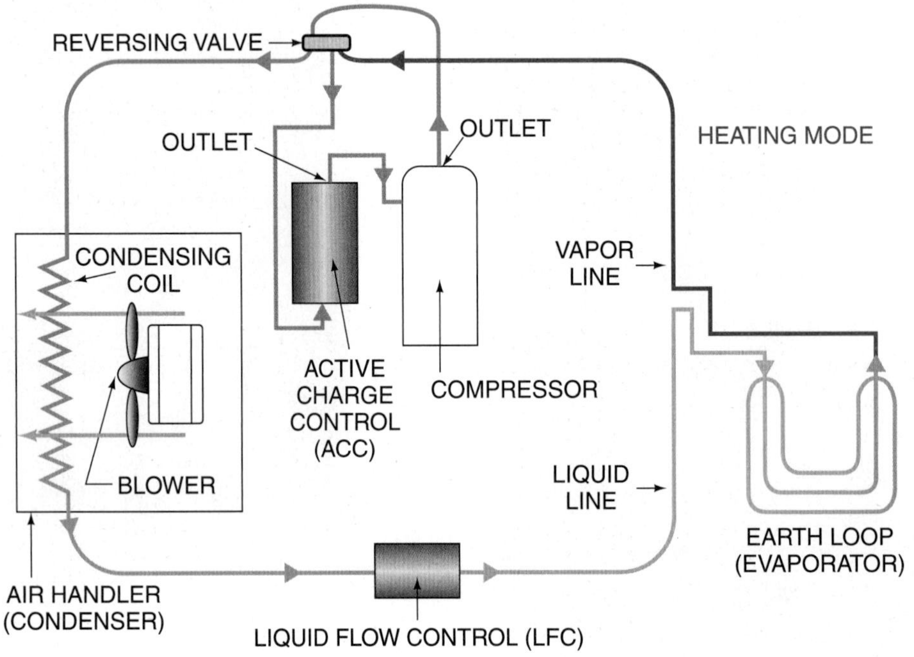

to the earth, **Figure 44.51(A)**. As this superheated refrigerant loses heat to the surrounding earth, it becomes saturated and gradually starts turning to a liquid (condensing). More and more of the vapor is condensed as it travels through the earth loop until it reaches 100% liquid. The refrigerant now exits the earth loop as a liquid. This liquid is now reexpanded to a lower pressure and temperature in the indoor coil (evaporator). Heat is absorbed into the indoor coil from the house or building with the assistance of a blower, **Figure 44.51(B)**.

REFRIGERANT-MANAGEMENT SYSTEM

The refrigerant-management system consists of two components: (1) liquid flow control and (2) active charge control. The refrigerant-management system components have three main objectives:

1. To improve the system efficiency, reliability, and serviceability
2. To continuously return lubricating oil back to the compressor without returning liquid refrigerant
3. To stabilize liquid and vapor refrigerant flow in long refrigerant (earth) loop runs, under all heat loading conditions

To operate at optimum efficiency, the three major components of all heat pumps (compressor, condenser, and evaporator) require the refrigerant to be in a particular physical state appropriate for each component.

Figure 44.51 (A) The earth loop showing heat being transferred to the earth from the phase-changing (condensing) refrigerant. Source: Earthlinked Technologies, Inc.

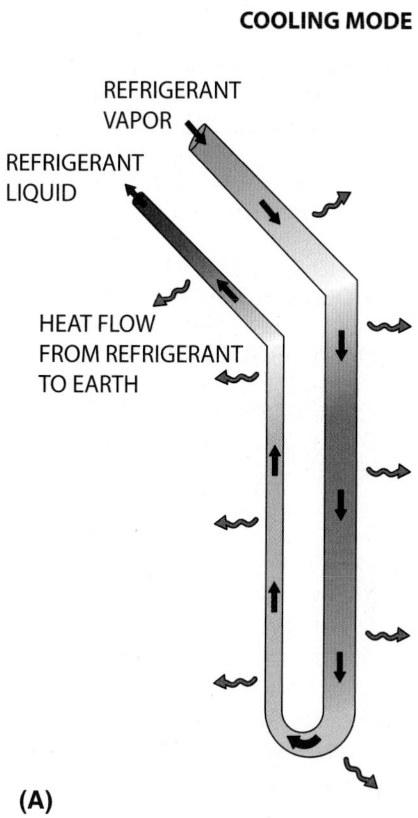

Figure 44.51 (B) System schematic in the cooling mode. Source: Earthlinked Technologies, Inc.

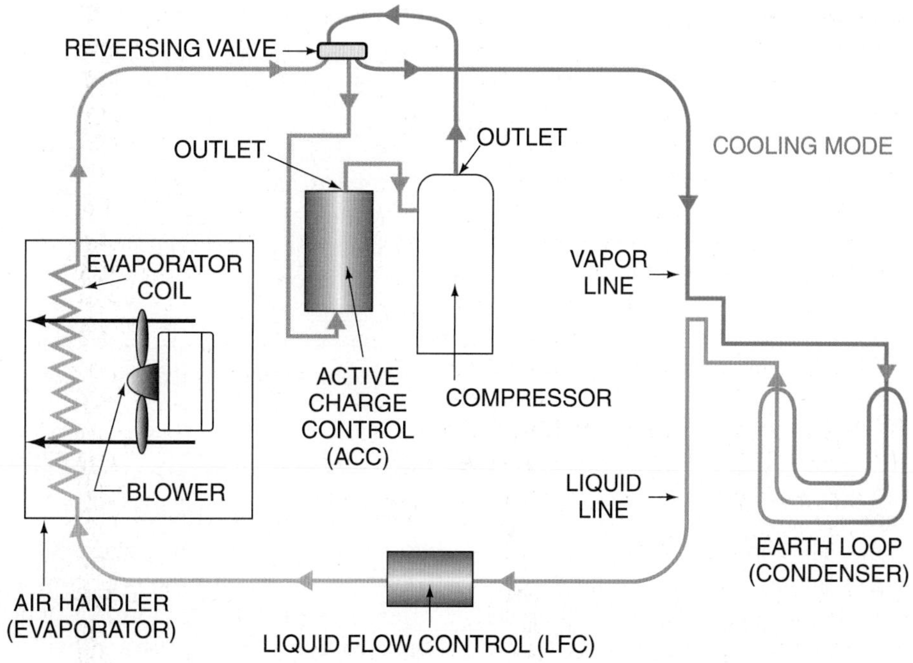

Figure 44.52 Liquid flow control. Source: Earthlinked Technologies, Inc.

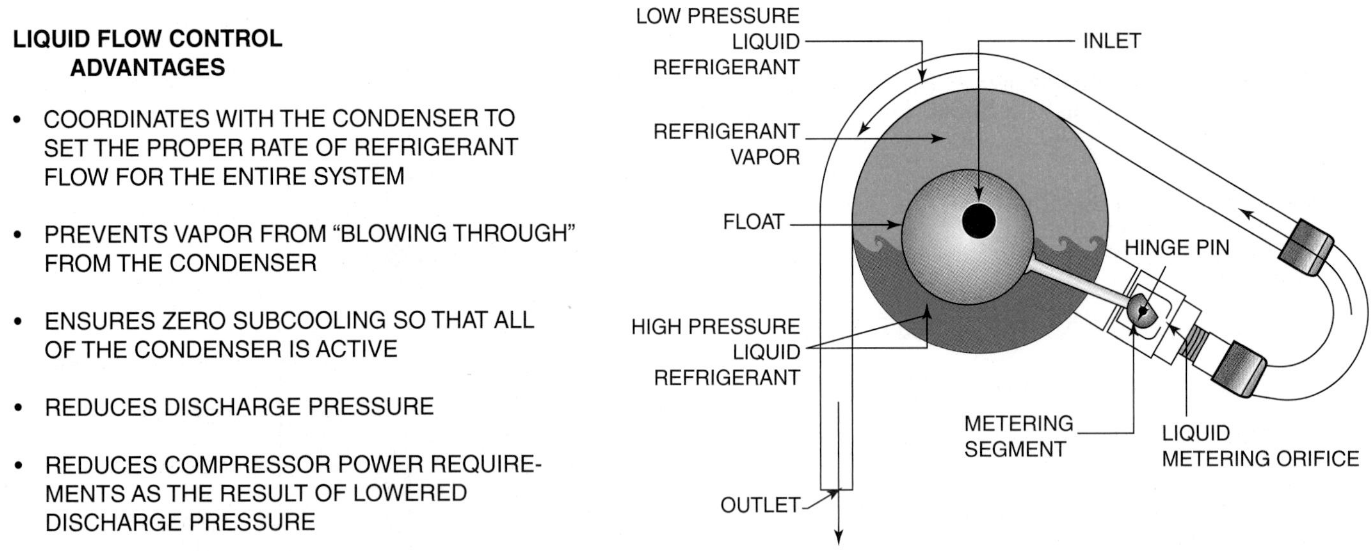

The compressor needs liquid-free refrigerant vapor from the evaporator that contains little or no superheat at its inlet. The condenser needs the saturated refrigerant vapor to be completely condensed just before it reaches the condenser outlet. The condenser's bottom passes are not used for subcooling liquid; only desuperheat and condensing are to take place in the condenser. This prevents liquid refrigerant from "backing up" in the condenser's bottom and producing subcooled liquid, which would take away valuable internal volume for condensing. The condensed liquid seal at the condenser's outlet also ensures that no uncondensed vapor passes through the condenser. The evaporator needs liquid refrigerant at its inlet. As the liquid in the evaporator evaporates, the last drop of liquid refrigerant should complete the evaporation process just as it reaches the evaporator outlet. This is the optimum "flooded" evaporator condition that produces maximum system efficiency without any evaporator superheat at the tailpipe of the evaporator. By eliminating superheat, the entire evaporator remains active to phase-changing (evaporating) refrigerant. Also, any liquid refrigerant that passes through the evaporator should not reach the compressor.

LIQUID FLOW CONTROL

Figure 44.52 shows the liquid flow control. This proprietary control replaces all other expansion devices, including thermostatic expansion valves (TXVs), electronic expansion valves (EXVs), automatic expansion valves (AXVs), fixed orifices, and capillary tubes. The control regulates the rate of liquid refrigerant flowing from the condenser to the evaporator by directly responding to the amount of vapor bubbles arriving at the control from the condenser's outlet. It positively meters the liquid refrigerant flow to the evaporator and passes liquid at the rate it is produced in the condenser. By using the vapor bubbles from the condenser's outlet as a signal for how much liquid is being produced in the condenser, the control ensures that vapor is present throughout the whole condenser, thus eliminating liquid subcooling at the condenser's outlet. Because subcooled liquid is not allowed to back up in the last passes of the condenser, the condenser will have more internal volume for desuperheating and condensing to take place. *The end result is a larger condenser with lower condensing pressures, lower compression ratios, and higher system efficiencies.* The liquid flow control also prevents vapor from blowing through from the condenser to the evaporator due to the lack of a liquid seal (subcooled liquid) at the condenser's outlet.

Some advantages of the liquid flow control are that it:

- *coordinates with the condenser to set the proper rate of refrigerant flow for the entire system,*
- *prevents refrigerant vapor from blowing through the condenser to the evaporator,*
- *ensures zero liquid subcooling at the condenser's outlet,*
- *reduces the condensing, which gives the system a lower compression ratio, higher efficiencies, and a higher mass flow rate of refrigerant, and*
- *reduces the compressor's power requirements because of lower condensing pressures and lower compression ratios.*

ACTIVE CHARGE CONTROL

The active charge control (ACC) consists of a thermally insulated reservoir that replaces the standard accumulator, **Figure 44.53**. Its purpose is to constantly deliver refrigerant vapor and oil to the compressor under optimum conditions and quantities. The control provides a means of bringing vapor from the evaporator into contact with liquid refrigerant stored within its reservoir.

Figure 44.53 Active charge control. Source: Earthlinked Technologies, Inc.

ACTIVE CHARGE CONTROL ADVANTAGES

- PREVENTS ANY LIQUID OR UNEVAPORATED REFRIGERANT FROM REACHING THE COMPRESSOR
- EVAPORATES OR TRAPS REFRIGERANT AS NECESSARY TO KEEP THE SYSTEM PROPERLY CHARGED AND THE EVAPORATOR "FLOODED" AND TO PREVENT BUILD-UP OF SUPERHEAT AT THE COMPRESSOR INLET
- INCREASES SUCTION PRESSURE BY ELIMINATING SUPERHEAT
- IMPROVES VOLUMETRIC EFFICIENCY OF THE COMPRESSOR
- CONTINUOUSLY ALLOWS THE PASSAGE OF COMPRESSOR OIL ENTRAINED IN THE REFRIGERANT
- PROVIDES A MEANS FOR QUICKLY AND EASILY DETERMINING WHETHER THE SYSTEM IS PROPERLY CHARGED
- ALLOWS THE COMPRESSOR TO RUN COOLER
- REDUCES COMPRESSOR POWER REQUIREMENT AS A RESULT OF INCREASED SUCTION PRESSURE

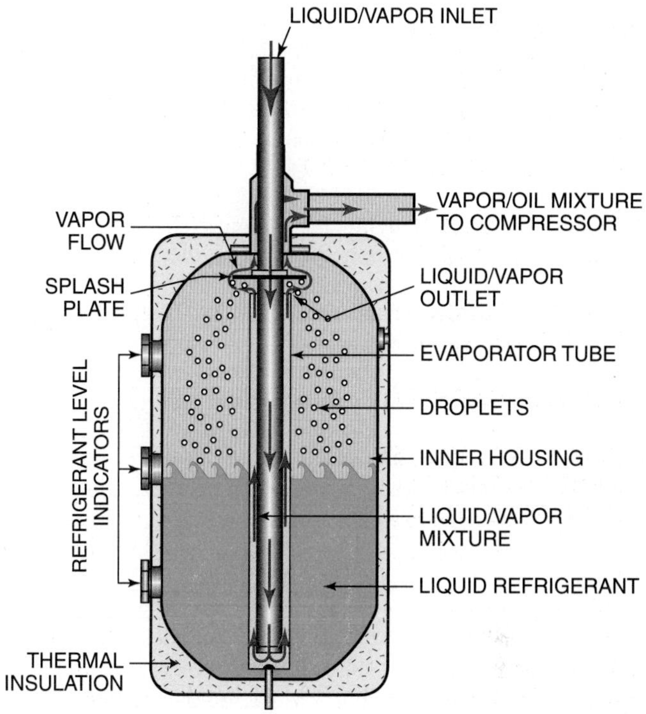

Inside this control, liquid from the reservoir flows through an orifice into an evaporator tube through which the incoming vapor passes. The liquid is entrained in the vapor stream and circulated back to the reservoir by way of a deflector plate located at the exit of the evaporator tube. If the incoming vapor is superheated, the contact with the liquid evaporates some of the stored liquid, thus reducing the superheat to near zero. Conversely, if the incoming vapor contains liquid refrigerant, the liquid is trapped in the reservoir. In either event, the refrigerant vapor leaving the active charge control has very little superheat and because of this, contains no liquid refrigerant. The vapor is almost a saturated vapor when it enters the compressor. The high vapor density entering the compressor results in higher mass flow rates of refrigerant through the compressor and system. Refrigerant coming from the evaporator and into the active charge control is continuously mixed with liquid refrigerant, which passes through an orifice from the liquid reservoir of the control.

The turbulence of the liquid/vapor mixture in the tube causes foaming and misting of the refrigerant/oil mixture. A circulator deflector plate above the inlet tube deflects the mixture radially outward, causing a vortex. This action also reduces the velocity of the mixture as it continues to move outward. The refrigerant vapor is drawn upward to the outlet at the upper portion of the tube. Oil mist entrained in the vapor mist also exits the tube at its top. Any liquid refrigerant droplets and any foam bubbles containing liquid refrigerant are too heavy to be entrained in the vapor stream and fall into the reservoir. Any liquid refrigerant arriving at the active charge control's inlet from the evaporator's outlet is deflected and trapped in the reservoir by the same action. Refrigerant must be evaporated to leave the active charge control, which ensures that no liquid refrigerant returns to the compressor.

The active charge control also acts as a reserve of refrigerant so that the refrigerant charge in the system is not critical. Its function will increase the system's active charge in circulation if the vapor reaching the ACC is superheated. The ACC will also reduce the active system charge if arriving vapor contains liquid refrigerant. Consequently, the system operates with an optimum refrigerant charge in active circulation under all heat-loading conditions, providing maximum system efficiency.

The ACC provides a means of quickly and easily determining when the system is properly charged without using gauges, wet- and dry-bulb readings, or charging charts. A service technician can determine if the system is properly charged simply by observing the liquid level through refrigerant-level indicators or sight glasses mounted on the side of the ACC, **Figures 44.49** and **44.53**.

44.12 SERVICE TECHNICIAN CALLS

The following service calls deal with both open- and closed-loop systems. They include some of the most common problems experienced with geothermal heat pump systems. Notice that a problem with one loop will usually affect the performance of the other loops in the heat pump.

SERVICE CALL 1

A rural homeowner reports that his heat pump seems to be running much too often and often shuts down on the low-temperature control. The low-temperature control consists of a thermistor located on the refrigerant line (evaporator) and connected to the control board. It is the third service call in one month to reset the low-temperature protection control. The homeowner also has noticed that his electric bills have risen steadily in the past two months. It is the heating season, and the heat pump is an open-loop system that gets its water from a drilled well located near the house. The heat pump's refrigerant loop contains R-22. *The problem is that the water side of the coaxial heat exchanger has been fouled with mineral deposits. This will not allow a good heat exchange between the water loop and the refrigerant loop.*

Arriving at the residence, the service technician makes sure his uniform is clean and his service van is not blocking the customer's driveway. After ringing the doorbell and maintaining a reasonable distance from the door, the technician introduces himself and his company when the homeowner answers. Once invited into the house, the technician slips on a pair of protective shoe covers to protect the customer's floors. After patiently listening to the customer and politely asking a few questions, the technician proceeds to the thermostat on the inside wall of the living room. He turns the space thermostat to a high setting in an attempt to keep the unit running while troubleshooting.

The technician installs gauges on the refrigerant loop and notices low suction pressure of about 48 psig. The normal suction pressure for that particular system should be about 55 psig for this geographical region with an entering-water temperature of 47°F. The technician then measures the temperature difference between the inlet and outlet of the water through the heat exchanger. There is a drop of only 4°F across the heat exchanger. The normal temperature drop for this geographical region is usually 7°F to 10°F.

The combination of the low suction pressure and a low temperature drop in the water indicates that a reduced heat transfer is taking place between the refrigerant loop and the open water loop. This could be caused by minerals in the water that have built up and fouled the heat exchanger on the water side, which would explain the high monthly electric bills. Because a good heat exchange could not take place between the water and the refrigerant, the heat pump's capacity was reduced. This forced the system to rely on electric strip heaters to meet the heating demands of the home.

The technician then tells the homeowner that the heat exchanger will have to be chemically cleaned and maybe replaced. He then clearly and concisely explains what work has to be done and why, explaining politely any options and the advantages and disadvantages of each. The customer decides to wait a few weeks for spring to break before having the system serviced. The technician then politely shakes the customer's hand and lets him know that he is a valued customer. The technician again thanks the customer and leaves the house.

SERVICE CALL 2

Homeowners are complaining that their heat pump keeps shutting down automatically. Power has to be manually switched off and then back on to reset the electrical lock-out relay. The heat pump is a closed-loop system with two circulating pumps hooked in parallel in the water loop. It is the heating season, and the heat pump's refrigeration circuit runs on R-22. *The problem is that one of the two parallel pumps in the water loop is not operating, causing low flow rates.*

The service technician turns the space thermostat to a high setting in an effort to keep the heat pump running. The pump runs for about 5 min and then shuts down on its lock-out relay. After turning the power off, the technician uses an ohmmeter and determines that the heat pump's low-pressure control (LPC) is opening, which causes the lock-out relay to hold the circuit open. The technician lets the heat pump stay off for about 20 min and analyzes the situation. He then quickly restores power to the heat pump. This resets the lock-out relay, and the heat pump runs again.

The technician then measures the temperature difference between the water in and out of the refrigerant-to-water heat exchanger. The temperature drop is 9°F instead of the normal 5°F for this closed-loop system. This indicates that the antifreeze solution in the ground loop is staying in contact with the heat exchanger longer because of the slower flow rate, causing an increased temperature drop in the solution. A lower suction pressure on the refrigerant loop is noticed after gauges are installed.

The service technician then examines the circulating pumps and determines that one of the two parallel pumps is not running. The antifreeze flowmeter indicates that there are only 7 gpm of fluid being circulated instead of the normal 14 gpm. This explains the increase in temperature difference of the antifreeze solution through the heat exchanger. The technician then applies a voltmeter to the motor terminals of the circulating pump that is not operating. The correct line voltage of 115 V is measured at the terminals, which means that the motor is defective. The technician then shuts down the heat pump and replaces the circulating pump's motor with a new one. The power is turned back on, and the system runs normally.

SERVICE CALL 3

A store owner is complaining of hardly any cooling and a reduced airflow coming from the registers in the building. The owner also has to manually reset the heat pump's electronic lock-out relay by turning the power off and then back on about eight times daily. It is the cooling season. *The problem is that the heat pump's air loop air filter is plugged with lint and dust. This prevents the correct air volume from entering the air coil.*

A service technician arrives, installs gauges on the unit, and notices that the suction pressure is 35 psig. This is a very low suction pressure for an R-22 system; the normal suction pressure should be about 60 to 65 psig for that geographical region. The technician also checks out the airflow problem and agrees with the store owner that something is restricting the airflow. A current reading of the air coil's fan motor is 3 A. The nameplate says the motor should be pulling around 9 A. This low amperage indicates that the fan is not moving enough air and the fan motor is only partially loaded. The technician then pulls out the air filter located in the return air just before the air coil and notices that it is almost completely filled with lint and dust. However, even with the air filter pulled, a restricted airflow problem and low suction pressure still exist and the fan motor continues to pull low current. The technician then decides to have a look at the air coil. He shuts off the power to the unit and removes the plenum to the heat pump. The air coil is completely covered with a blanket of ice and frost. The technician then melts the iced coil with a large blow drier while the power is turned off.

The technician then explains to the building owner that a dirty air filter has caused a restricted airflow to the air coil. This restriction in the airflow has caused a low suction pressure because of the reduced heat load from the building entering the air coil. The low suction pressure made the refrigerant flowing through the air coil's tubes colder, which finally froze the coil solid. The low suction pressure was tripping the LPC, and the unit was being locked out by its lockout relay. This is why the store owner would have to reset the unit about eight times daily. The restricted airflow also unloaded the fan motor, causing it to draw low current. The technician then explains to the store owner the importance of keeping the air coil filters clean and also tells the building owner that it is not wise to reset the heat pump any more than once before calling a service technician. After putting the plenum back on the unit and installing a clean air filter, the technician starts the heat pump. The proper airflow has been established, and the suction pressure is normal at 65 psig. The fan motor is also drawing normal current.

SERVICE CALL 4

A homeowner is complaining about his heat pump running continuously in the heating mode. Another complaint is that his winter electric bills are unusually high. When the homeowner enters the basement where the heat pump is located, a crackling sound can be heard in the water loop as it enters the circulating pump. The ground loop is a parallel-vertical bore, **Figure 44.8**. *The problem is that a flange leak at the inlet of the circulating pump has let air enter the water or ground loop and has blocked one of its parallel paths. This actually shortens the loop and causes colder return-water temperatures—which decreases the heat pump capacity.*

A service technician arrives at the scene, politely introduces himself and his company, and asks if he can be escorted to where the crackling sound is coming from. He then carefully listens to the crackling sound within the water in the ground loop as it flows through the circulating pump. The technician also notices that the electric strip heaters are on. This is what has been causing the high electric bills.

The technician carefully questions the homeowner about the system configuration of the water loop. The homeowner says that the configuration is a parallel-vertical bore system. The service technician then measures the temperature difference between the water in and the water out of the coaxial heat exchanger. The measured temperature difference is only 2°F. The temperature difference should be 4°F to 5°F for that geographical region. This indicates to the service technician that the water is coming back to the heat exchanger too cold. One loop of the parallel system is being blocked by air, thus shortening the entire loop. This severely decreases the capacity of the heat pump and explains why the electric strip heaters are coming on, causing high electric bills. This would also explain the low temperature difference across the heat exchanger and the crackling sounds at the circulating pump.

The technician determines that the ground loop needs to be power flushed to rid it of air. He then politely explains to the customer what work has to be done and why. The system's circulating pump is valved off, and a new flange is installed on the circulating pump to prevent any more air from entering the system. A portable pump cart or purging unit is brought in and hooked up to the ground loop and heat pump. The purging unit power flushes the loop and heat exchangers. The air that was trapped in the ground loop is purged from the top of the pump cart, **Figure 44.18**. The system is then hooked back up to the normal circulating pump and put back into operation.

After about 2 hours of run time, the technician measures the temperature difference across the coaxial heat exchanger. The temperature drop is a normal 5°F. No crackling sound can be heard in the water circuit. The house is up to temperature without the electric strip heaters operating. The technician is satisfied that the heat pump is operating normally. After answering the customer's questions, the technician hands the customer the billing information, work description, and costs for the services rendered. The technician shakes the customer's hand while thanking him for being a valued customer and then leaves the house.

SERVICE CALL 5

A homeowner is complaining about high electric bills in the winter. The residence has a geothermal heat pump that has a closed-loop, horizontal, two-layer ground loop, **Figure 44.10**. The system uses R-22 as a refrigerant. The owner of the house uses a setback (night/day) thermostat, which is kept at 72°F during the day and 65°F at night. *The*

problem is with the setback thermostat. At 8 AM the thermostat switches its set point from 65°F to 72°F. However, the microprocessor (smart board) on the heat pump is programmed to kick on the electric strip heat whenever more than a 2°F differential exists between the set point on the thermostat and the actual room temperature. In this case, there would be a 7°F temperature differential at 8 AM every morning. A technician is called after the homeowner has received the third high electric bill.

A service technician arrives at the residence at 10 AM and immediately enters the basement where the heat pump is housed. He installs gauges on the high and low sides of the system and temperature probes on the inlet and outlet of the coaxial heat exchanger. While the heat pump is running, the temperature difference across the heat exchanger is determined to be 5°F. This is the recommended temperature difference for a closed-loop system in that geographical region. The suction pressure also reads normal at 45 psig for an entering-water temperature of 33°F.

The technician then checks the flow rate of the antifreeze solution at the flowmeter and determines it to be normal. Just as the technician tells the homeowner that everything looks good, the unit shuts down for a normal off cycle. The technician then asks the homeowner where the thermostat is so the unit can be cycled back on and more tests can be made.

The technician realizes the thermostat is a setback thermostat. The homeowner is questioned as to the setback function on the thermostat with this heat pump. The homeowner explains how it is set back from 72°F to 65°F in the evening. The technician then explains how the heat pump's electric strip heaters will kick on if a 2°F differential is reached between the set point of the thermostat and the room temperature. The technician also explains to the homeowner that this is happening at 8 AM every morning and running up his electric bill. He then tells the homeowner that it is best not to use the setback function with this style of heat pump. The homeowner agrees and thanks the technician for the advice. The technician reprograms the thermostat to a normal operation, makes sure the heat pump is operating correctly, and then leaves.

SERVICE CALL 6

A closed-loop geothermal heat pump system keeps shutting down in the heating mode. It is the heating season and the customer has called for service five times in the last month. *The problem is that this particular system is a closed-loop system and the equipment was installed with the control board set for open-loop temperature conditions. A dual-in-line pair (DIP) switch on the control board should have been changed from open-loop to closed-loop configuration. The system was cycling off on low-temperature control.* Installers commonly overlook this switch setting or often forget to change it to the proper loop configuration. The oversight will result in high heating bills and repeated service callbacks until the setting is discovered and corrected.

The service technician politely introduces himself to the customer. He recognizes that the customer is a bit upset because of this being the fifth service call in the month. The technician is very sympathetic and listens carefully, making sure not to criticize the customer even in a joking manner. The technician also maintains his professionalism and does not take anything the customer says personally. After politely listening to the customer and asking a few questions, the technician goes to the basement and opens up the front cover of the heat pump. Having handled a problem like this before, the technician is experienced enough to check the setting on the control board. Sure enough, the technician recognizes the problem and makes a quick change.

Often, a magnifying glass has to be used to read the small print near the DIP switch on the control board. Other manufacturers may use a jumper wire, or a control wire change to a sensor, instead of a DIP switch as a method of changing the loop configuration at the control board.

The service technician then politely explains to the customer what was done and why the system acted in such a manner. The technician lets the customer know that he is a valued customer and explains that this service call and the previous service calls are covered by his service contract. Handshakes are then exchanged and the technician again thanks the customer while maintaining eye contact. The service technician then leaves the house but calls the customer back the next day to make sure the system has not cycled off.

REVIEW QUESTIONS

1. Geothermal heat pump systems rely on
   A. the energy stored in the earth's crust as a heat source.
   B. the earth to act as a heat source in the cooling mode.
   C. the earth to act as a heat sink in the heating mode.
   D. the energy stored in the earth's crust as a heat sink.

2. An earth-coupled heat pump system uses
   A. refrigerant to directly transfer heat between the earth and the system.
   B. a well system to bring water from the ground to the system.
   C. a heat-transfer fluid to transfer heat between the earth and the system.
   D. the minerals in the earth to transfer heat between the earth and the system.

3. An open-loop heat pump system uses
   A. refrigerant to directly transfer heat between the earth and the system.
   B. a well or similar system to bring water from the ground to the system.
   C. a heat-transfer fluid to transfer heat between the earth and the system.
   D. the minerals in the earth to transfer heat between the earth and the system.
4. While operating in the heating mode, the coaxial heat exchanger on an open-loop heat pump system operates as the
   A. expansion device.
   B. condenser.
   C. evaporator.
   D. refrigerant receiver.
5. The term used to describe when water and refrigerant flow in opposite directions through a heat exchanger is
   A. perpendicular flow.
   B. counterflow.
   C. universal flow.
   D. parallel flow.
6. Which of the following materials is a popular choice for heat exchangers on open-loop hat pump systems based on its durability and high resistance to corrosion?
   A. Steel
   B. Aluminum
   C. Tin
   D. Cupronickel
7. Which of the following is a typical temperature differential between the water entering and leaving the heat exchanger on an open-loop heat pump system?
   A. 0°F to 3°F
   B. 4°F to 6°F
   C. 7°F to 10°F
   D. 11°F to 15°F
8. Closed-loop, geothermal heat pump systems are a popular choice when
   A. there is no space to bury large amounts of plastic piping material.
   B. there are water-quality problems.
   C. the ground is extremely hard and rocky.
   D. there is a large quantity of clean water at a moderate temperature.
9. The inner tube of a coaxial heat exchanger on a closed-loop heat pump system carries
   A. air from the conditioned space.
   B. liquid refrigerant.
   C. vapor refrigerant.
   D. water or a water/antifreeze mixture.
10. Water can be heated for domestic use by redirecting
   A. hot discharge gas from the compressor to the water heater.
   B. hot gas from the condenser outlet to the water heater.
   C. hot liquid from the condenser outlet to the water heater.
   D. superheated vapor from the evaporator to the water heater.
11. When there are space and land restrictions, which of the following ground loop configurations would most likely be chosen for a closed-loop heat pump system installation?
   A. Vertical loop system
   B. Horizontal, single-layer system
   C. Horizontal, two-layer system
   D. Slinky configuration
12. Two common materials used to manufacture ground loop piping are
   A. monoethylene and polyethylene.
   B. polyethylene and monobutylene.
   C. polyethylene and polybutylene.
   D. monoethylene and monobutylene.
13. Salts are sometimes used as part of an antifreeze solution. One of the main drawbacks of salts, however, is that they
   A. are toxic.
   B. break down easily over time.
   C. are corrosive under certain circumstances.
   D. do not transfer heat well.
14. The major disadvantage of using alcohols as part of an antifreeze solution is that they are
   A. combustible if they mix with air.
   B. highly corrosive.
   C. highly toxic.
   D. very expensive to purchase.
15. Which of the following is the process by which a cement-like material is injected between the well casing and the hole drilled for the well?
   A. Cementing
   B. Grouting
   C. Weatherproofing
   D. Sealing
16. The type of well that draws water from the top of the water column, circulates it through the heat pump, and then returns the water to the bottom of the same water column is the
   A. basic drilled well.
   B. return well.
   C. dry well.
   D. dedicated geothermal well.
17. The recommended minimum distance between a supply well and a return well is
   A. 10 feet.
   B. 25 feet.
   C. 50 feet.
   D. 100 feet.
18. What device is used to help prevent water hammer?
   A. Slow-closing solenoid valve
   B. Fast-opening solenoid valve
   C. Pressurization tank
   D. Depressurization tank
19. The main purpose of a pressure tank is to
   A. eliminate the need for a well pump.
   B. prevent the well pump from short-cycling.
   C. increase the run time of the well pump.
   D. bypass the well pump during heat pump service.

20. Water-to-water heat pump systems are easily identified because they are configured with two
    A. air-to-refrigerant heat exchangers.
    B. water-to-refrigerant heat exchangers.
    C. water-to-water heat exchangers.
    D. water-to-air heat exchangers.
21. How much heat, in Btu/h, is transferred if water flows at a rate of 10 gpm through a heat exchanger with a temperature difference of 10°F?
    A. 20,000 Btu/h
    B. 30,000 Btu/h
    C. 40,000 Btu/h
    D. 50,000 Btu/h
22. Low water/antifreeze flow through the ground loop of a geothermal heat pump system will result in
    A. low system pressures in both the heating and cooling modes of operation.
    B. high system pressures in both the heating and cooling modes of operation.
    C. low system pressures in the heating mode and high system pressures in the cooling mode.
    D. high system pressures in the heating mode and low system pressures in the cooling mode.
23. If a water-source heat pump operates with low pressures in the heating mode and high pressures in the cooling mode, but the water flow through the ground loop is correct, what is the most likely problem with the system?
    A. The heat exchanger is fouled.
    B. The circulator pump is defective.
    C. The indoor blower motor is defective.
    D. The heat pump system is undercharged.
24. When converting a conventional air-conditioning system to a direct geothermal heat pump system, what is the function of the refrigerant earth loop during the cooling mode?
    A. The refrigerant earth loop functions as the evaporator.
    B. The refrigerant earth loop functions as the condenser.
    C. The refrigerant earth loop functions as a secondary heat exchanger.
    D. The refrigerant earth loop serves no function.
25. The refrigerant management module on a direct geothermal heat pump system performs all of the following functions except
    A. routing system refrigerant to the above ground condenser when the system is operating in the cooling mode.
    B. controlling refrigerant flow to ensure the refrigerant charge is correct in both modes of operation.
    C. bypassing the reversing valve in the cooling mode when low ambient conditions could cause coil freeze-ups.
    D. routing system refrigerant to the earth loop (evaporator) when the system is operating in the heating mode.
26. On a newly installed (non-retrofit) direct geothermal heat pump system, the refrigerant ground loop
    A. functions as the evaporator in the heating and cooling modes.
    B. functions as the condenser in the heating and cooling modes.
    C. functions as the condenser in the cooling mode and the evaporator in the heating mode.
    D. functions as the condenser in the heating mode and the evaporator in the cooling mode.
27. List the two loop classifications that are associated with geothermal heat pump systems.
28. List three factors that should be considered when dealing with open-loop heat pump systems that rely on well water.
29. List four types, other than series or parallel, of ground loop piping configurations.
30. List three advantages and three disadvantages of using a series ground loop configuration.
31. List three advantages and two disadvantages of using a parallel ground loop configuration.
32. Lists six desirable properties for a fluid that is to be used as an antifreeze in a ground-loop piping arrangement.
33. Explain the function and operation of a buffer tank on a water-to-water heat pump system.

SECTION 9

DOMESTIC APPLIANCES

UNIT 45

DOMESTIC REFRIGERATORS AND FREEZERS

OBJECTIVES

After studying this unit, you should be able to

- describe the refrigeration cycle of household refrigerators and freezers.
- explain why the doors of domestic appliances are removed for disposal.
- discuss typical refrigerator and freezer cabinet design.
- explain the purpose of mullion and panel heaters.
- explain the capacity limits of domestic refrigerators and freezers.
- describe the types, physical characteristics, and typical locations of the evaporator in a domestic appliance.
- describe the types, physical characteristics, and typical locations of the compressor in a domestic appliance.
- describe the types, physical characteristics, and typical locations of the condenser in a domestic appliance.
- describe the types, physical characteristics, and typical locations of the metering device in a domestic appliance.
- describe the typical operating conditions for domestic evaporators, condensers, and compressors.
- explain how the evaporator is defrosted.
- describe how to dispose of the condensate.
- describe the electrical controls used in household refrigerators and freezers.
- discuss ice-maker operation.
- explain how to properly service and troubleshoot domestic refrigerators and freezers.
- describe condenser efficiency relative to ambient air passing over it.

SAFETY CHECKLIST

- ☑ Technicians should wear appropriate back brace belts and use proper personal equipment when moving or lifting appliances.
- ☑ Never place your hands under a freezer that has been lifted. Use a stick or screwdriver to position a mat under the freezer feet.
- ☑ Never allow refrigerant pressures to exceed the manufacturer's recommendations.
- ☑ Always follow all electrical safety precautions.
- ☑ Never use a sharp object to remove ice from an evaporator.
- ☑ Remove refrigerator doors or latch mechanisms before disposing of a refrigerator.
- ☑ Tubing lines may contain oil that may flare up and burn when soldering. Always keep a fire extinguisher within reach when soldering.
- ☑ When accessing the refrigerant circuit on a domestic refrigerator or freezer, be sure to cut the line with a tubing cutter. Never use a torch to open the system, as pressures within the system can build to unsafe levels.
- ☑ Be sure to throw away any cardboard boxes, plastic wrapping, and other shipping/packing materials to reduce suffocation/choking potential.
- ☑ Be sure to unplug the appliance before servicing.
- ☑ Make certain the appliance is properly grounded.
- ☑ Check to make certain that all safety-related devices and controls are operational.

45.1 REFRIGERATION

The term refrigeration refers to the moving of heat from a place where it is not wanted to a place where it makes little or no difference. Domestic refrigerators and freezers are no exception to this definition. Heat enters domestic refrigerators and freezers through the walls of the box by conduction and convection, as well as when warm food is placed inside. When the food is warmer than the box temperature, the temperature in the box increases. Earlier discussions on heat transfer told us that heat travels naturally from a warm to a cold substance, **Figure 45.1**. In the case of the domestic refrigerator or freezer, the cold substance is the low-pressure, low-temperature refrigerant in the evaporator coil, while the warm substance is the product being cooled in the appliance. The refrigerator ultimately moves the heat from inside the box to the surrounding area, where it makes little or no difference, **Figure 45.2**. In order to accomplish this heat transfer, heat must be transferred from the box into the refrigeration system and then transferred from the refrigeration system to the surrounding air. The evaporator acts to absorb heat from the product being cooled as well as from the box itself into the refrigeration system, while the condenser acts to reject heat from the refrigeration system to the outside.

Domestic refrigerators and freezers are package units that are completely assembled and charged with refrigerant at the factory, so no refrigerant lines need to be run and no evacuation or field charging must be done. The air inside the box flows through or across the cold, refrigerated coil or evaporator, **Figure 45.3**, transferring both sensible and latent heat to the coil. The sensible heat transfer is responsible for lowering the air temperature in the box. The latent heat transfer results in the dehumidification of the air in the box. The product of the dehumidification process is known as condensation. Since the evaporator coil on a properly operating refrigerator or freezer operates at a temperature below 32°F, the condensate will freeze on the surface of the coil. After the air in the box has transferred heat to the coil, it circulates back to the box at a colder temperature so that it can

Figure 45.1 Heat from surrounding air seeps through the walls of the refrigerator or freezer into the box itself. Heat is also added to the box when the door is opened and when warm product is placed inside.

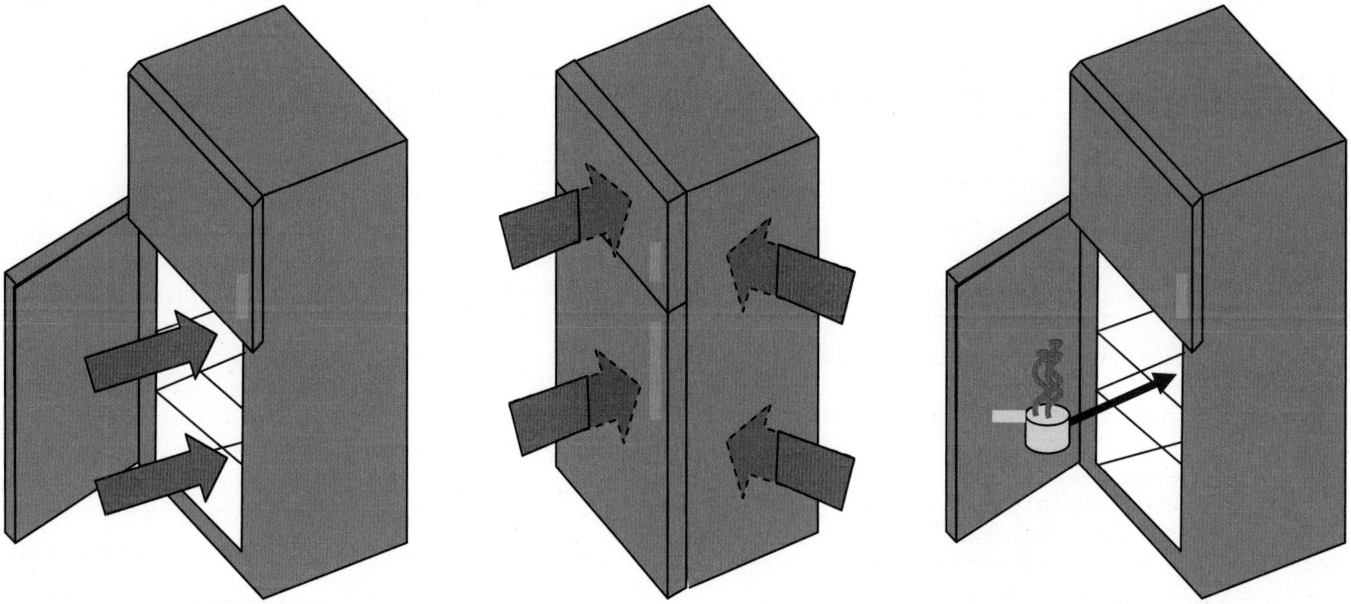

Figure 45.2 The evaporator coil absorbs heat from the box, while the condenser coil is the heat transfer surface that is responsible for rejecting heat to the surrounding air.

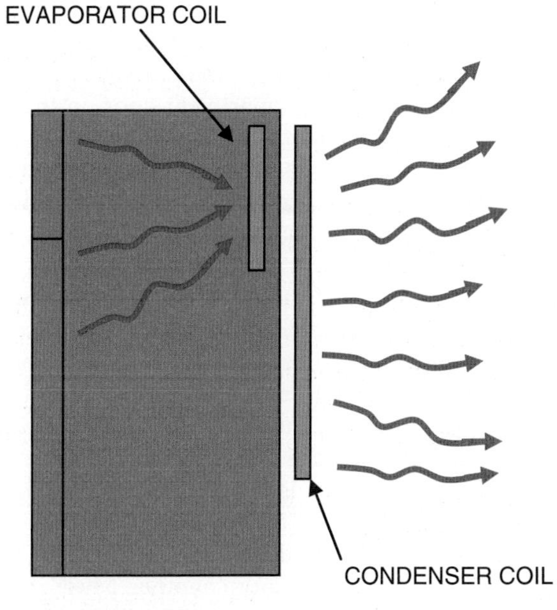

absorb more heat and humidity. This process continues until the box temperature reaches the desired level.

The typical box temperature of the refrigerator portion of the appliance is about 35°F to 40°F when the room temperature is in the 70°F to 75°F range. However, if a refrigerator is subjected to extreme temperatures, such as when the appliance is located outside in the summer or winter, **Figure 45.4**, it will not perform the same. The typical temperature for the

Figure 45.3 Air gives up heat to the cold evaporator coil.

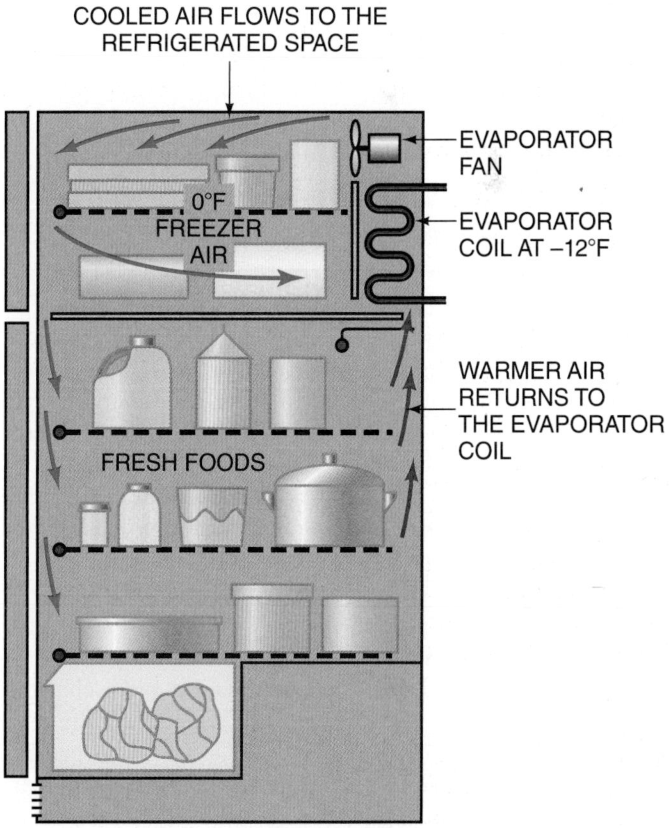

Figure 45.4 The ambient temperature for these refrigerators is not within their proper operating range.

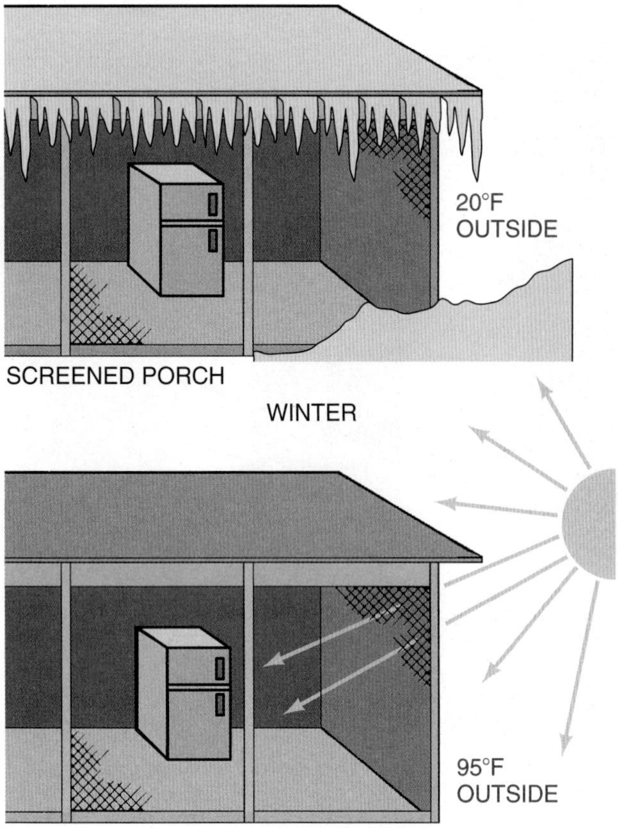

freezer portion of the system is about 0°F, given normal conditions for the air surrounding the appliance. The typical box temperature for a stand-alone freezer is also about 0°F.

PRODUCT PACKAGING

It is important that products being stored in refrigerators or freezers be packaged properly to prevent dehydration and freezer burn. With regards to fresh foods, improperly wrapped products can dehydrate and dry out much faster than otherwise intended, shortening the shelf life of the products. Freezer burn, a result of product dehydration, causes the product to look dry and burned. Although it does not completely ruin the product, freezer burn does make it look unattractive and may change the flavor of the item.

Freezer burn occurs in the vicinity of a tear in a package where moisture leaves the food and transfers to the air. The moisture is in the solid state while in the food and is changed to the vapor state in the air, where it collects on the coil. This moisture is then carried off during defrost. This process of changing from a solid to a vapor is known as sublimation. Ice cubes, for example, become smaller as a result of sublimation if stored for long periods of time in the freezing compartment. Moisture changes from a solid to a vapor and collects on the evaporator as ice. It is melted during defrost and leaves in the condensate.

THE REFRIGERATED BOX

As we have mentioned earlier in this unit, heat passes through the walls of the box and ultimately winds up inside the box. As the amount of heat that enters the box increases, the refrigeration system will have to work harder and harder. It is therefore desirable to reduce the rate at which heat enters the storage portions of refrigerators and freezers. The next two sections will discuss some features that are incorporated to increase the efficiency of the appliances.

THE DOMESTIC REFRIGERATED BOX. As kitchens become more like showplaces than simply functional, food preparation areas, box design has become an important marketing and selling tool. Once ignored by interior designers, kitchens are fast becoming the single most important room in the house from the point of view of real estate merchandising. So manufacturers are now faced with the task of creating appliances that are both energy-efficient and stylish. Not only are domestic refrigeration systems available in a wide range of colors but many come with such features as ice and water dispensers, partial access from the outside, and even computer interfaces with Internet access and monitors built right in, **Figure 45.5**.

Early-generation refrigerators were designed with one outside door and a freezer compartment door on the inside, **Figure 45.6**. The doors had gaskets with air pockets inside them that compressed when the door was closed, **Figure 45.7**. Eventually, the ability of the gasket to effectively seal would be greatly reduced as the material aged. The door would then allow room air to enter the box, which increased the load on the evaporator and caused more frost accumulation, **Figure 45.8**. Older-style refrigerator doors had mechanical door latches that could be opened only from the outside. If a child were to get inside and close the door, they could get trapped and suffocation could result. **SAFETY PRECAUTION:** *Caution must be used when one of these refrigerators is taken out of service. The door must be made safe. This is most commonly done by removing the door from the appliance.•* Be sure to check with local requirements regarding appliance disposal, as they vary from one geographic region to another. Most, if not all, guidelines require that, at a minimum, the appliance doors be removed before disposal.

The box must be airtight to prevent excess load and frost. If the door is shut and the air inside the box cools, the volume of the air will decrease and the door may be hard to open immediately after closing. *Modern refrigerators have magnetic strip gaskets all around the doors, which also have compression-type gaskets that maintain a good seal to help keep air out,* **Figure 45.9.** These magnetic gaskets are much easier to

Figure 45.5 A new, hi-tech refrigerator. Photo by Eugene Silberstein

Figure 45.6 A one-door refrigerator. Photo by Bill Johnson

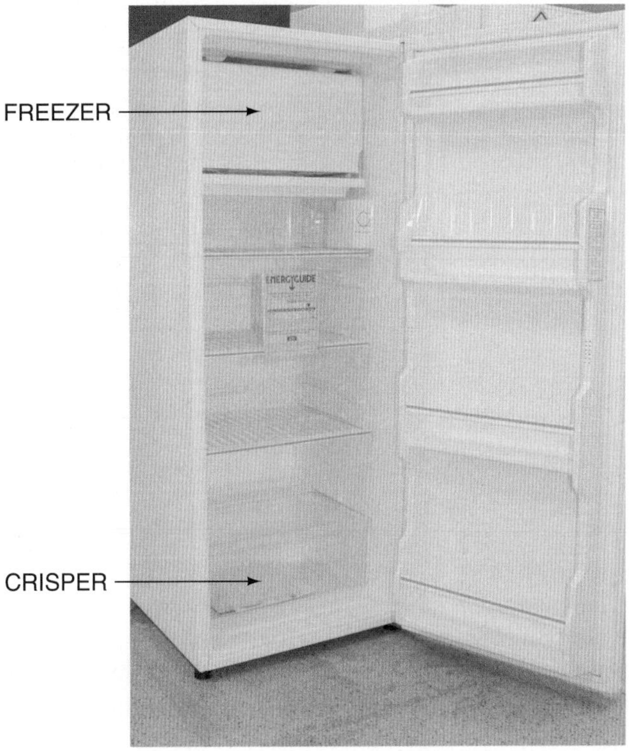

Figure 45.7 Early door gasket construction.

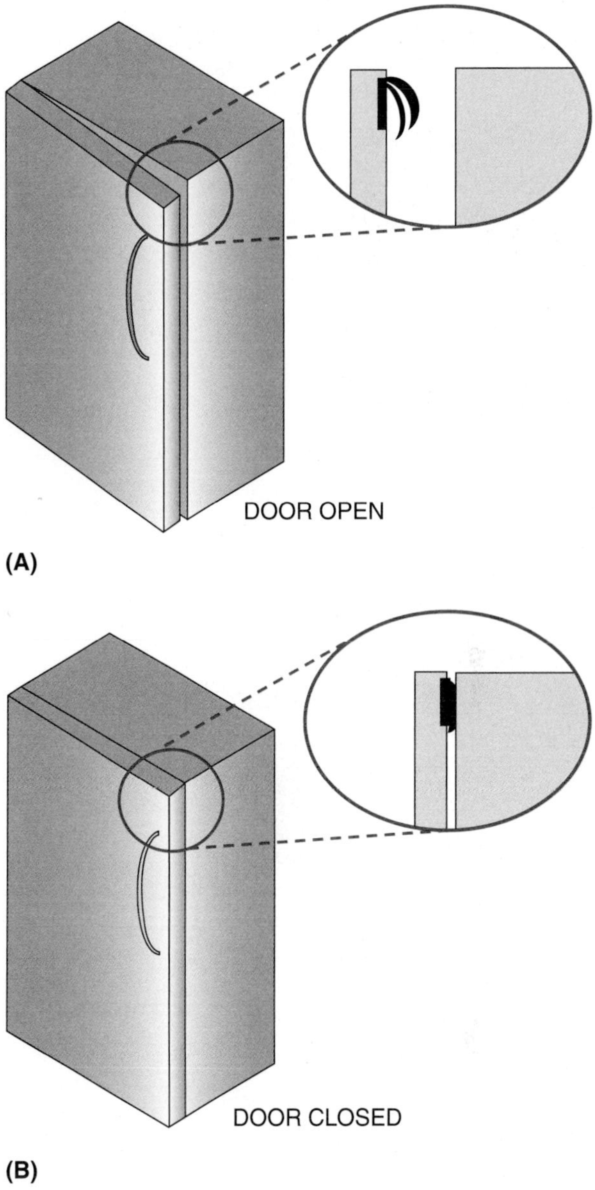

replace than the old type of gasket. The typical box has two outside doors: side by side or over and under. In the side-by-side styles, the freezer is usually on the left. The appliance shown in **Figure 45.5** is configured in this manner. Notice that the ice dispenser is in the left-hand door. Condensate is allowed to travel from the inside of the box to a collecting pan that is located under the box, where it is evaporated. A trap arrangement in the drain line prevents air from traveling into the box through this opening, **Figure 45.10**.

The refrigerator may be designed to accommodate an ice maker in the frozen-food compartment. Water piping connections and a bracket to which the ice maker can be fastened are provided. A wiring harness may also be included so the ice maker can be easily powered. This arrangement does

Figure 45.8 An older door gasket that is worn out, allowing air to leak into the box.

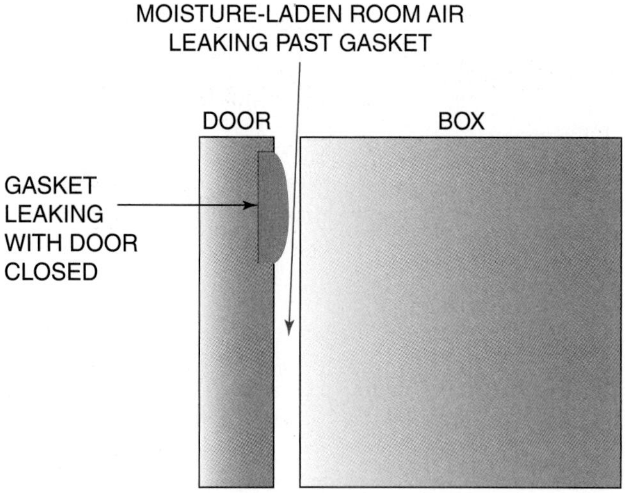

Figure 45.9 A modern refrigerator gasket with a magnet.

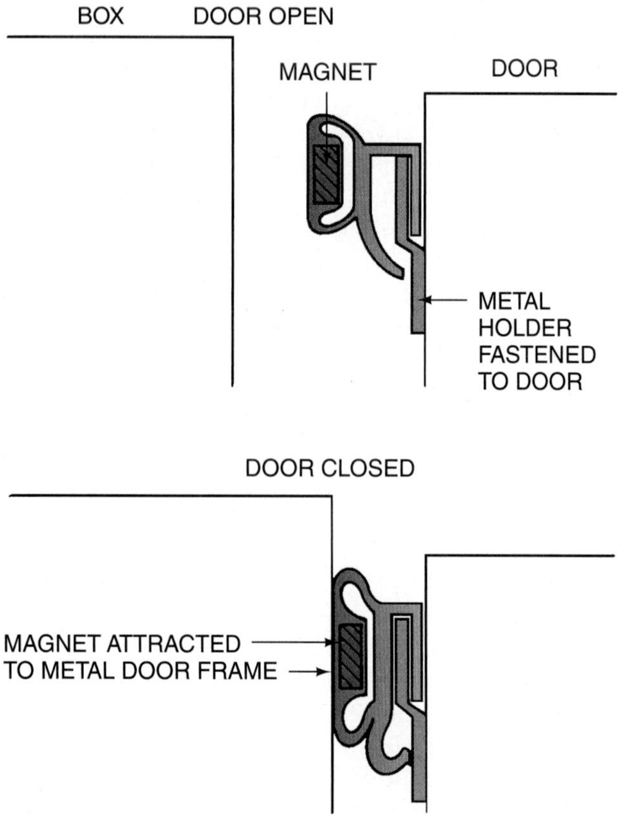

Figure 45.10 Two types of traps to allow condensate to drain and prevent air from entering.

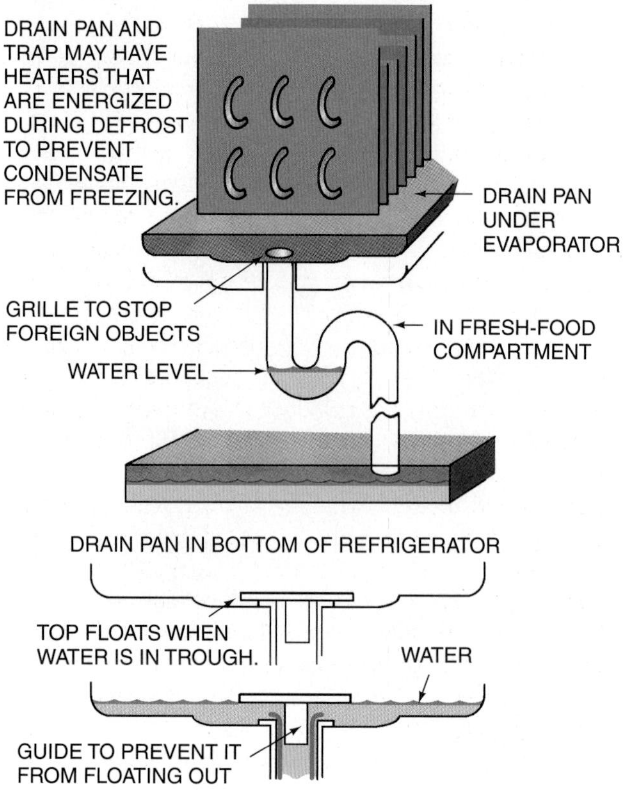

Figure 45.11 A crisper drawer.

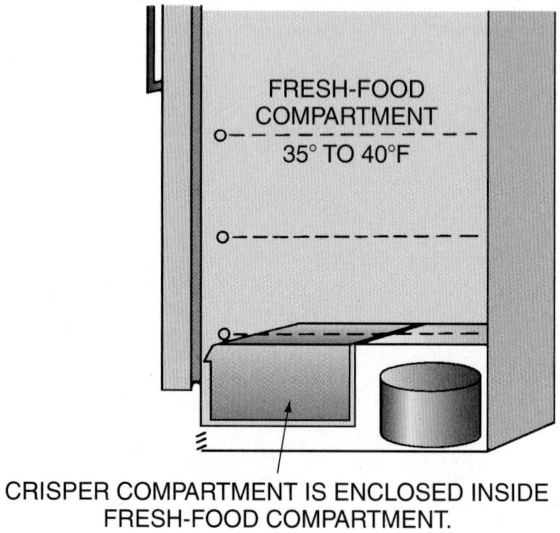

not require any wiring by the technician, since the ice maker just needs to be plugged in.

Even though the main storage area in the refrigerator maintains a temperature of about 40°F, various compartments may be maintained at different temperatures, such as the fresh vegetable crisper and the butter warmer. The conditions in the crisper are maintained by enclosing it in a drawer to keep the temperature slightly cooler and slow the rate of food dehydration, **Figure 45.11**. The butter warmer may have a small heater in a closed compartment to keep the butter at a slightly higher temperature than the space temperature so it will spread more easily.

The interior surfaces of most refrigerators or freezers are made of plastic. This surface is easy to keep clean and will last for years if not abused. The plastic in the freezing compartment is at very cold temperatures and, as a result, is quite fragile. For this reason, care must be taken not to drop frozen food on the bottom rail because it may break.

A sweating problem around the doors due to the differences in temperature in the various refrigerator compartments may occur because the gasketing material may be below the dew point temperature of the room air. Special heaters, called mullion and panel heaters, are located around the doors to keep the temperature of the box facing above the dew point temperature of the room air, **Figure 45.12**. Some units have an energy-saver switch that allows the owner to shut off some of the heaters when not needed, such as in the winter when the humidity is normally low. An explanation of the wiring for these heaters is discussed in Section 45.9, "Wiring and Controls."

Refrigerated boxes, whether vertical two-temperature appliances or chest-types freezers, should be leveled according to the manufacturer's instructions. Usually, they should be tilted slightly to the rear for proper door closing. Leveling is important, especially if the unit has an ice maker. In this case, the water may overflow the ice maker when it is automatically filled. Most refrigerators have leveling devices for the feet or rollers, **Figure 45.13**.

As mentioned, refrigerators are manufactured in over-and-under and side-by-side door configurations. The boxes that are over-and-under seem to have more room because the compartments are typically wider than those in the side-by-side models. Compartments in the side-by-side models appear narrow and deep. This does not affect the function of the box but may make service more difficult.

Figure 45.12 Mullion and panel heaters.

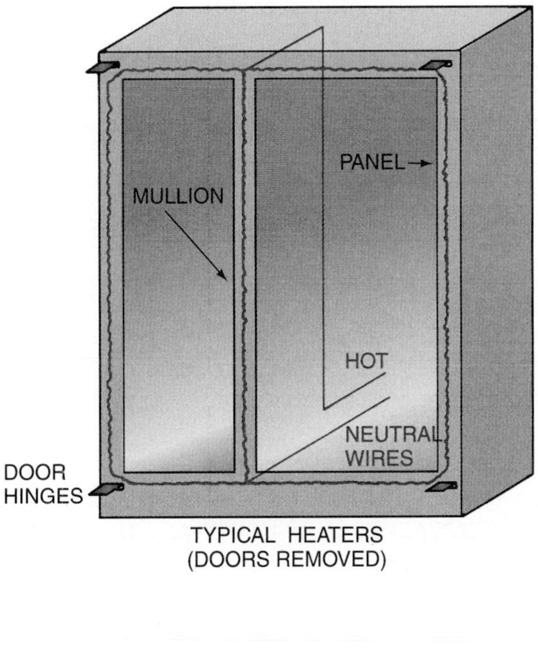

Figure 45.13 Leveling devices for feet or rollers.

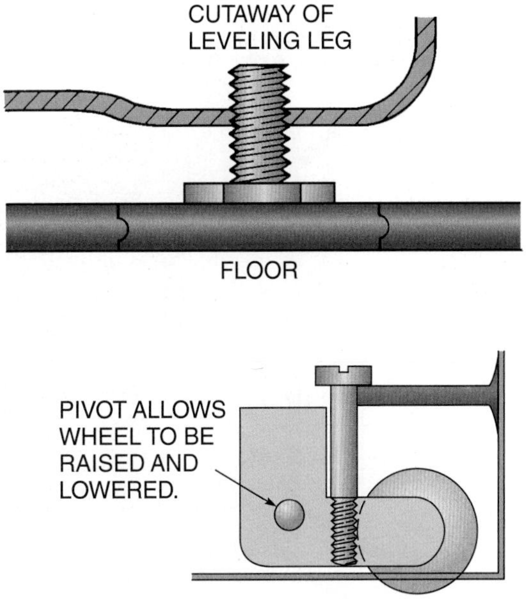

The stand-alone domestic freezer may be configured as either an upright type or a chest. The door of the upright freezer may open to the left or right, as convenient. The door of a chest freezer, known as the lid, lifts up, **Figure 45.14**. The upright box takes up much less space and is often used in the kitchen, where floor space is at a premium. It is probably not as efficient as a chest freezer, because every time the door is opened, air in the box falls out at the bottom of the opening, **Figure 45.15**. This does not change the food temperature much, just the temperature of the air, which must be recooled, thus causing increased compressor run time and wasted energy. Moisture also enters with the air and will

Figure 45.14 Chest freezer with lid open. Photo by Eugene Silberstein

Figure 45.15 Cold air falls out of an upright freezer.

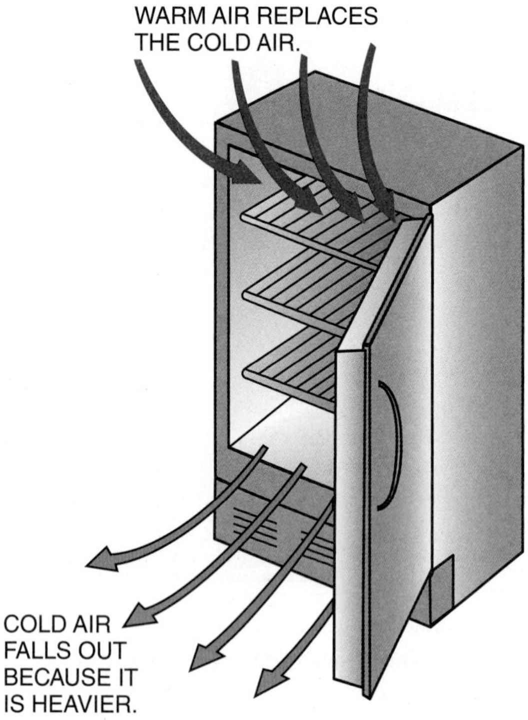

collect on the surface of the evaporator coil. The amount of time that the freezer door is open should be kept to a minimum. When a chest freezer is opened, the air stays in the box, since the cooler air in the box is heavier than the warmer surrounding air. Locating food in a chest freezer is not as easy as it is in an upright freezer because the upright freezer has multiple shelves, while the chest freezer may have lift-out baskets so food stored on the bottom may be reached.

A refrigerator may have some extra features for convenience, such as an outside ice dispenser or ice-water dispenser, **Figure 45.16.** The unit shown in **Figure 45.16** also has a digital display of the freezer and refrigerator compartment temperatures, along with a water filter alarm and a vacation mode. Appliances that have door-mounted accessories require wiring and water connections to the door, as well as a chute that connects the ice maker to the dispenser. These features add to the cost and make service more complex.

DOMESTIC FREEZER BOX INTERIOR

Many freezer interiors may have a plastic lining. The plastic is strong and easy to clean and maintain. However, it may be brittle at the low temperatures inside the box, so care should be taken not to strike it. Some boxes have metal liners and plastic trim. The metal may be either painted or coated with porcelain. The walls of the typical modern box are much like those of the refrigerator—a sandwich construction of metal or plastic on the inside, metal on the outside, and foam insulation, in between, **Figure 45.17.** Older boxes typically have fiberglass insulation within the walls.

Figure 45.16 A water and ice dispenser on the outside. Photo by Eugene Silberstein

INSIDE THE UPRIGHT BOX. A light is normally located in the freezer so that the food can be more easily seen. A door switch is used to turn the light on and off. The upright box has door storage for small packages, so the door must be strong enough to support the weight of these packages as well as the door itself. Also, it is not uncommon for a package of frozen meat to be dropped and break the bottom plastic rail of an upright freezer, **Figure 45.18.** All damage to the

Figure 45.17 Sandwich construction of a refrigerator wall.

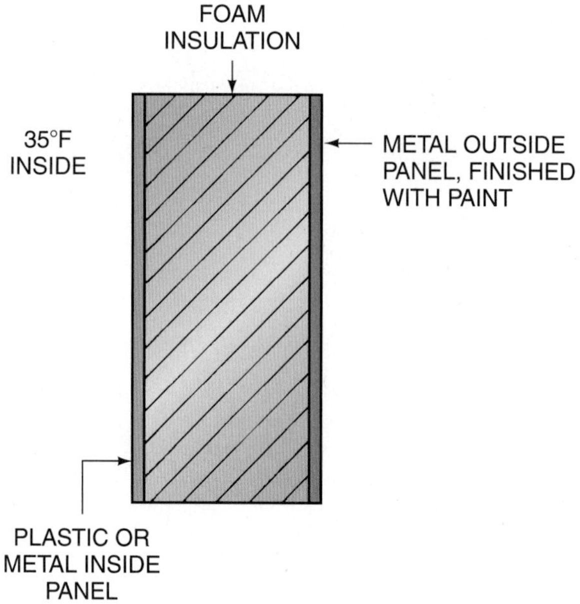

Figure 45.18 If frozen food is dropped, it may break the plastic rail in a freezer.

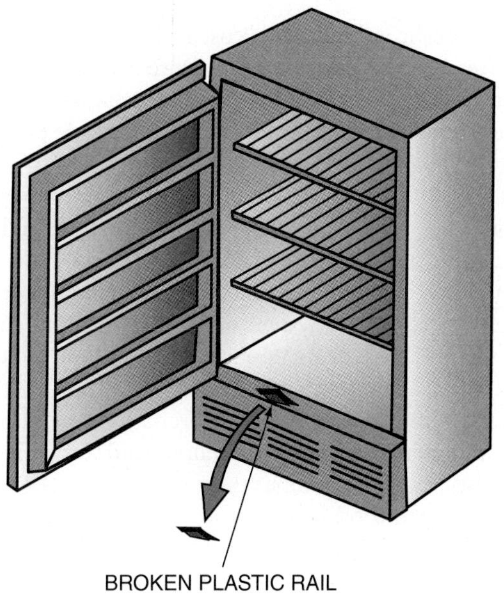

interior plastic surfaces of the box should be repaired. The plastic interior is typically repaired by refastening the broken piece with epoxy glue, **Figure 45.19**. Most adhesives cannot be applied while the surfaces being joined are cold, so the appliance should be turned off or put into defrost so the rail can warm up enough for the repair to be made.

The upright box has many of the same gasket and door alignment features as the refrigerator. The gaskets must remain tight fitting and the door must remain in alignment, or leaks will occur. Leaks will cause the compressor to run more than it was designed to and cause frost buildup. Many freezers have a manual defrost because defrosting is not needed as often as with refrigerators because the doors are not opened as frequently.

Figure 45.19 The piece may be glued back in place when the freezer is defrosted and cleaned.

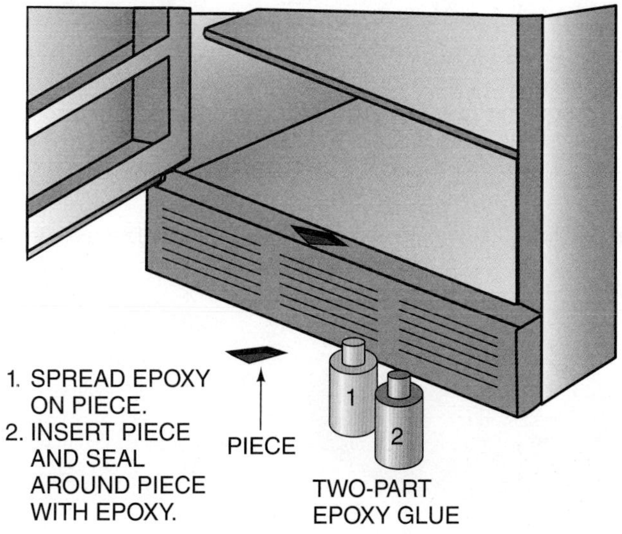

The freezer evaporator, which freezes the food inside, may be one of three types—forced-air, shelf, or a plate in the wall of the freezer, **Figure 45.20**. Plate evaporators may be tubes with wires connecting them or stamped to make a shelf, **Figure 45.21**. The fastest method of freezing food is to lay it directly on the plate or tube evaporator. However, the amount of available plate surface area is limited.

INSIDE THE FREEZER CHEST. The lid of a freezer chest is hinged at the back, and the liner may be metal or plastic. The gaskets located around the lid of the box must be in good condition, or moist air will leak in. This will cause frost buildup, normally right inside the lid where air is entering. The step at one end of the freezer compartment is where the compressor is located, **Figure 45.22**. If care is not taken when loading a chest-type freezer, food will be hard to find at the bottom. Larger parcels should be located on the bottom, and all parcels should be dated. Most freezers have baskets that may be removed and set aside while locating food that is at the bottom.

Figure 45.20 Three types of evaporators: (A) Forced-air. (B) Shelf. (C) Plate (wall).

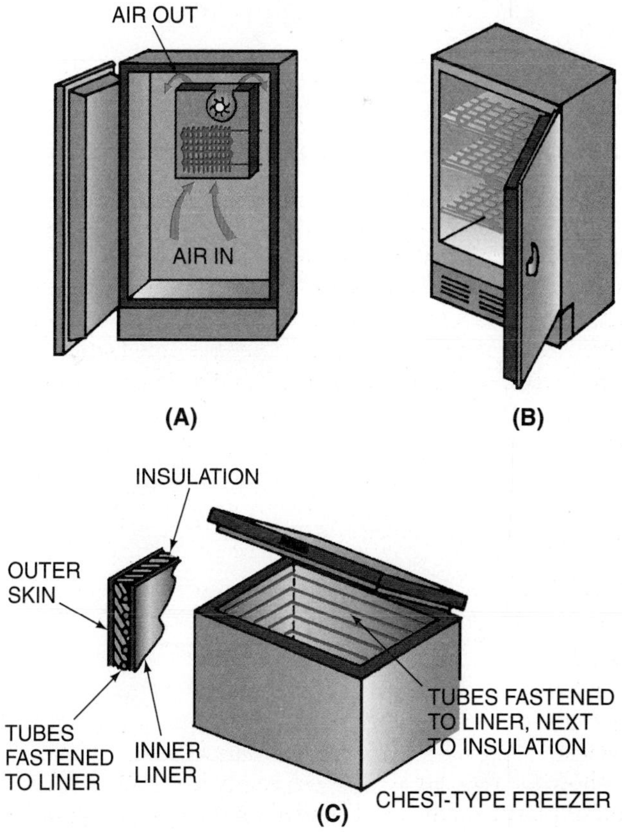

Figure 45.21 Plate evaporators—stamped plate or steel tubing with wire—may form the shelves.

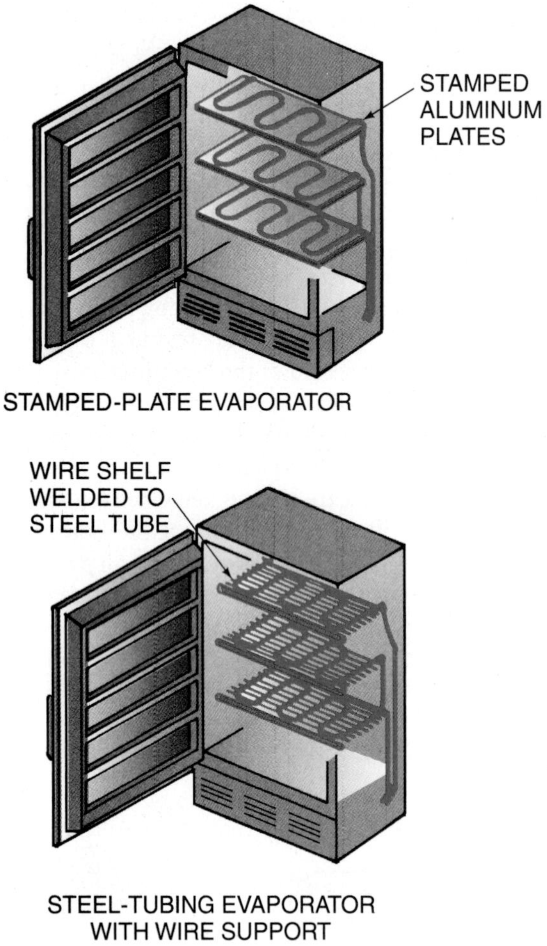

Figure 45.22 The step at the bottom of the chest freezer is where the compressor is located. Photo by Eugene Silberstein

45.2 CAPACITY OF DOMESTIC SYSTEMS

Ideally, domestic appliances are used to store and maintain perishable items at the desired temperature. Domestic freezers, for example, work more efficiently when frozen items rather than unfrozen items are placed inside them for storage purposes. The capacity of domestic freezers is not great enough to accomplish the freezing of large amounts of product at once. Manufacturers' guidelines differ as to how much product should be introduced to the refrigerator or freezer portions of the appliance at a time. For example, a manufacturer might suggest that in order to ensure proper system operation, the appliance owner should not add more than 3 pounds of fresh food per cubic foot of freezer space at one time.

Many home owners may purchase large amounts of freshly butchered meat and poultry and try to freeze the whole amount at one time in their home freezers. This causes problems. When food is frozen, ice crystals form in the product, **Figure 45.23.** When food is frozen slowly, the ice crystals that form are larger than those that form when food is frozen quickly at lower temperatures. These larger ice crystals are sharp and may puncture the cells of the product. When a steak, for example, is frozen slowly at a high temperature and then thawed out, there will be a noticeable puddle of water and blood around the steak, **Figure 45.24.** This puddle is a direct result of the cell punctures. In order to minimize the amount of cell damage to the product, it is a good idea to lower the

Figure 45.23 Food frozen slowly creates large ice crystals.

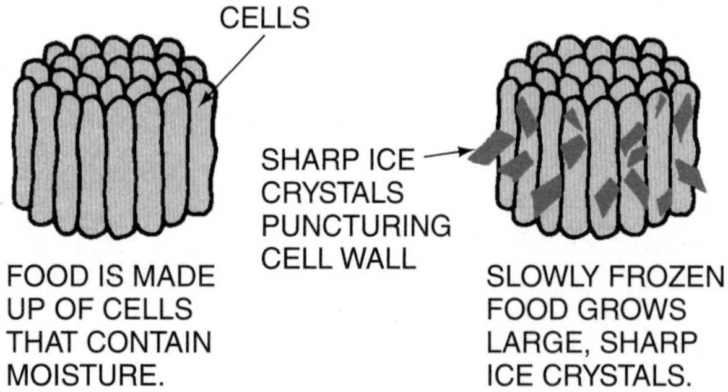

Figure 45.24 Large ice crystals may puncture the cell walls of the food.

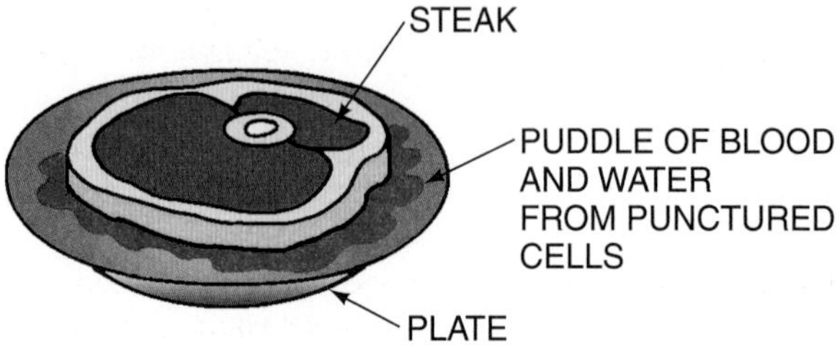

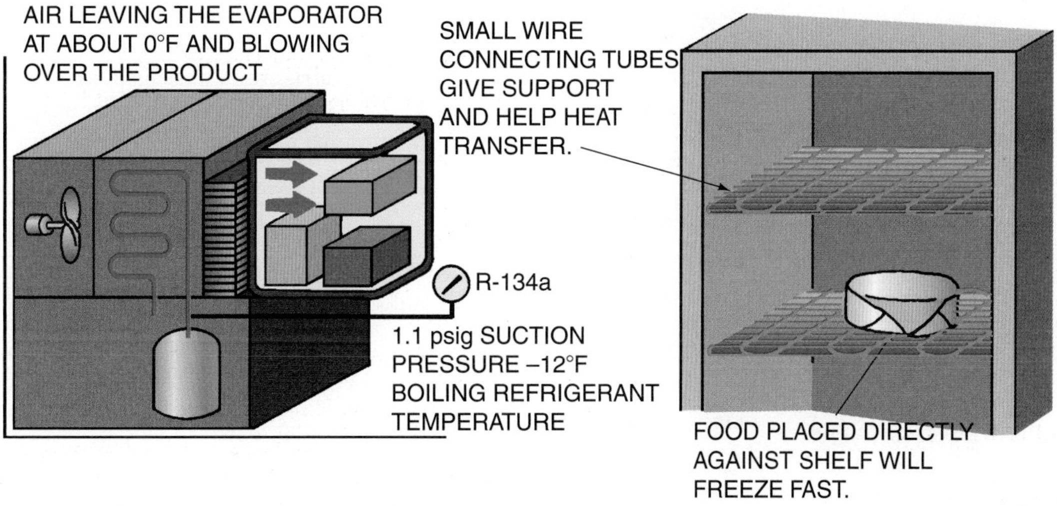

Figure 45.25 Two quick-freeze methods—using a special compartment or placing food directly on the evaporator.

temperature of the product to as close to 32°F as possible by placing it in the coldest portion of the refrigerator for several hours, then move it to the freezer and locate it in the coldest place in the freezer, which may be a quick-freeze compartment or an evaporator plate, **Figure 45.25**. The quick-freeze compartment is typically used to make ice quickly when the appliance owner is, for example, throwing a party.

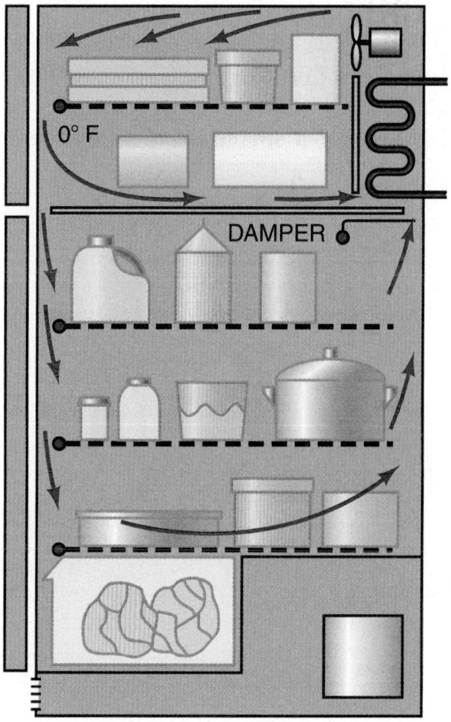

Figure 45.26 Air flows inside the refrigerated box from the low-temperature compartment to the medium-temperature compartment.

45.3 THE EVAPORATOR

The evaporator on a domestic refrigerator or freezer is the heat transfer surface that absorbs heat into the refrigeration system. To accomplish this, the surface on the evaporator coil must be cooler than the air in the refrigerated box. In a stand-alone domestic freezer, the evaporator will operate at a temperature that will maintain the entire box at the desired temperature. The household refrigerator must maintain two different compartments at two different temperatures with a single compressor. Therefore, the single compressor operates under conditions for the lowest box temperature. The freezer compartment is the cooler of the two spaces in the appliance and is typically maintained at a temperature close to 0°F. Since new domestic refrigerators and freezers operate with refrigerant R-134a, any box temperature below 0°F will cause the low-side pressure of the system to be in a vacuum. This is not desirable since any pinhole leak on the low side of the system will pull air and moisture into the system.

The evaporator in the household refrigerator typically maintains refrigerator and freezer compartments at different temperatures by one of two methods. One method is by allowing part of the air from the frozen-food compartment to flow into the fresh-food compartment, **Figure 45.26**. It may also be accomplished with two evaporators that are in series, one for the frozen-food compartment and the other for the fresh-food medium-temperature compartment, **Figure 45.27**. When configured in this manner, both evaporators are of the mechanical draft type. In either case, frost will form on the evaporator and a defrost method must be used. This is described in more detail later.

The evaporators in household refrigerators can be of two types, natural draft or mechanical draft, **Figure 45.28**. Natural-draft evaporators rely on natural convection current to

Figure 45.27 A two-evaporator box.

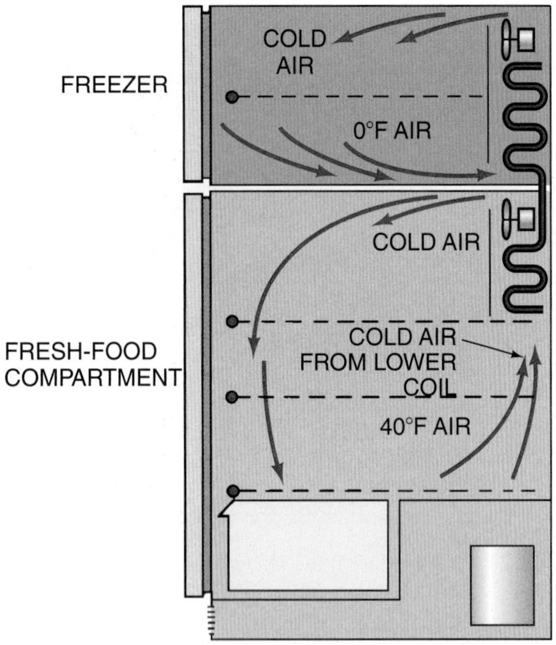

circulate air through the refrigerated space, while mechanical-draft evaporators use fans to move air across and through the evaporator coil. The fan improves the efficiency of the evaporator and allows for a smaller evaporator. Space saving is

Figure 45.28 Natural-draft and mechanical-draft evaporators.

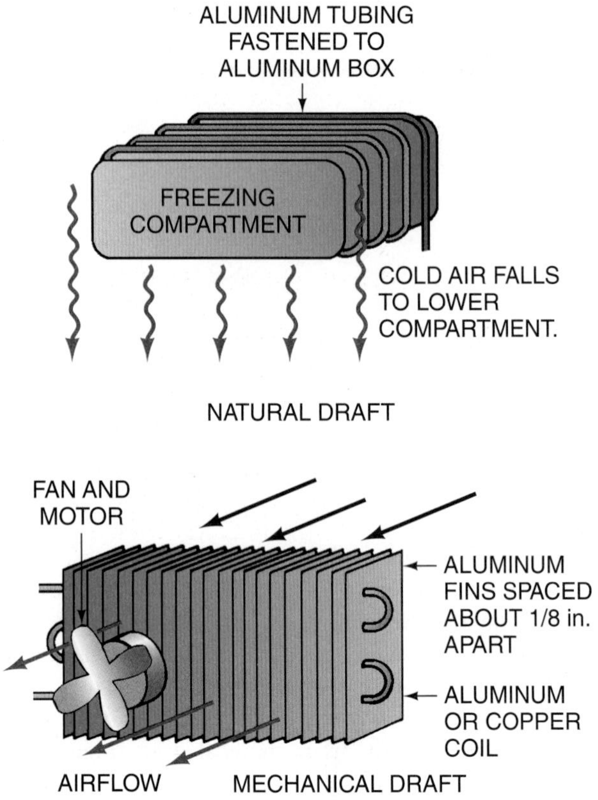

desirable in a household refrigerator, so most use mechanical-draft coils. However, other units are manufactured with natural-draft coils for economy and simplicity.

Natural-draft evaporators are normally the flat-plate type with the refrigerant passages stamped into the plate, **Figure 45.29**. They are effective from a heat-transfer standpoint and require that natural air currents be able to flow freely over them. The food in the frozen-food compartment may be in direct contact with the flat-plate evaporator, as they are quite often used as shelves on which product is placed. Air from the bottom and sides may flow to the fresh-food compartment, **Figure 45.30**.

Figure 45.29 A stamped-plate evaporator.

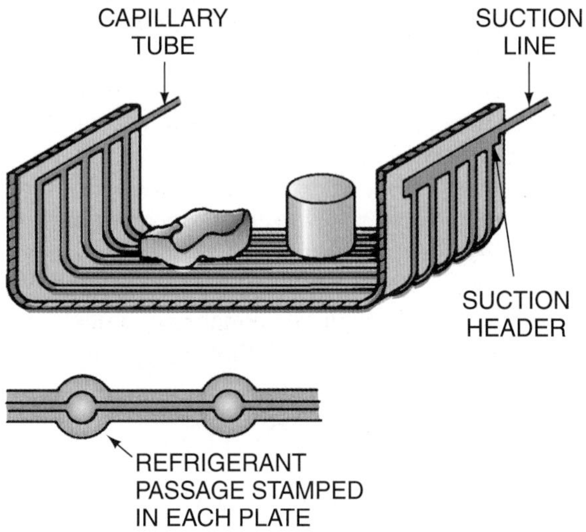

Figure 45.30 Air flows to the fresh-food compartment from the flat-plate evaporator.

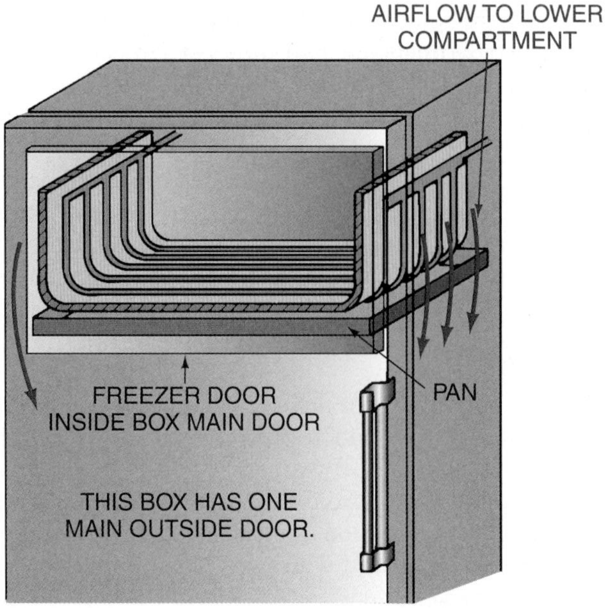

Natural-draft evaporators are more visible than forced-draft evaporators and are subject to physical abuse. Although there are automatic methods for defrosting evaporator coils on domestic refrigerators and freezers, appliances manufactured with natural-draft evaporators almost always employ the manual defrost method. A manual defrost system requires that the unit be shut off and the door to the compartment normally left open to accomplish the defrost. Frost is melted by room temperature, **Figure 45.31**. In a few instances, owners have become impatient and have used sharp objects to remove the ice. This may puncture the evaporator. **SAFETY PRECAUTION:** *Sharp objects should never be used around the evaporator,* **Figure 45.32.•** Defrost may be more quickly accomplished with a small amount of external heat, such as a hair drier, heat gun, or a small fan that blows room air into the box until the ice is melted. These methods are not recommended as the electrically powered fan or heater is put in a wet location. A pan of warm water may be placed under the coil. The melted ice normally drips into a pan below the evaporator, **Figure 45.33**.

Forced-draft evaporators are used to reduce the space the evaporator normally takes up. The amount of air moving through the evaporator coil is measured in cubic feet per minute, or cfm, and is obtained by multiplying the velocity of the air passing through the evaporator coil, in feet per minute, by the cross-sectional area of the coil, in square feet.

Figure 45.31 Manual defrost.

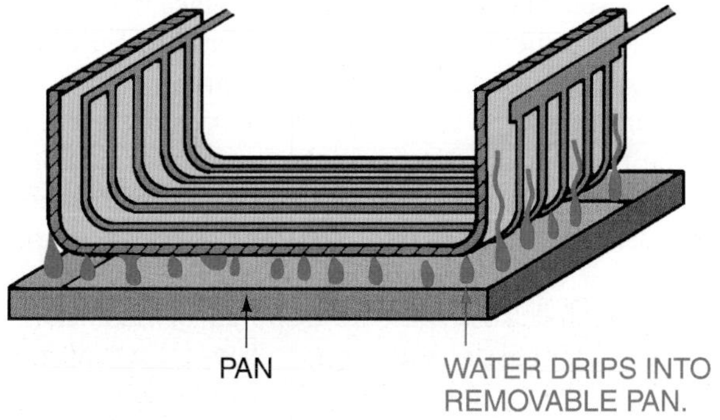

Figure 45.32 Sharp objects may puncture the evaporator. **SAFETY PRECAUTION:** *Do not use sharp objects to chip frost and ice.•*

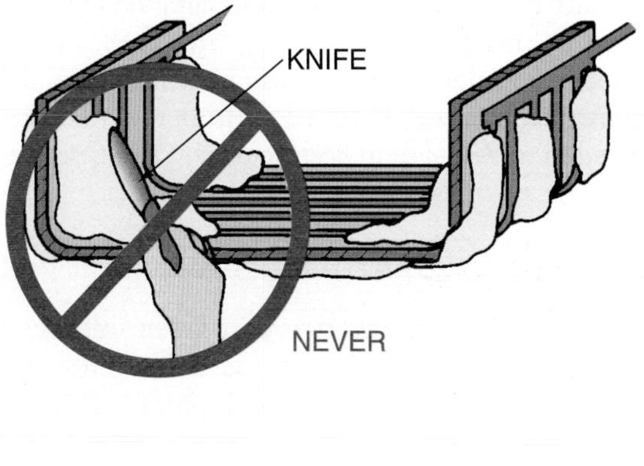

Figure 45.33 Melted ice (condensate) is caught in the pan.

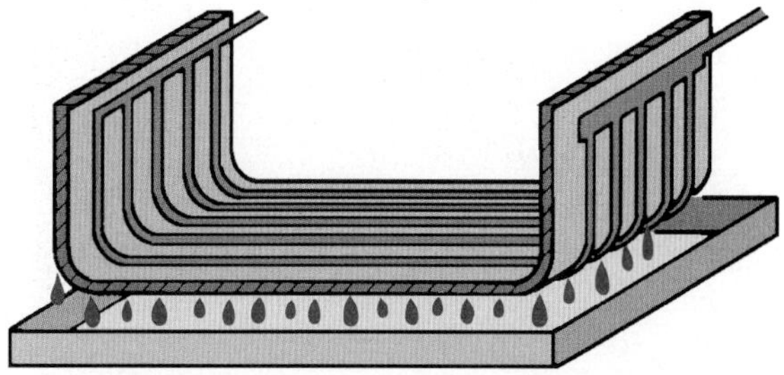

If the velocity of the air passing through the coil is increased by using a fan, the evaporator coil can be smaller and the same air volume can be maintained. The smaller the evaporator, the more internal space is available for food. The evaporator fan and coils are normally recessed in the cabinet and not exposed, **Figure 45.34.** Because the coils and fans are recessed, air ducts may provide the airflow direction; dampers may help control the volume of the air to the various compartments. Most evaporators have an accumulator installed in the suction line at the outlet of the coil. The accumulator allows the evaporator to operate as full as possible with liquid refrigerant and still protect the compressor by allowing liquid to collect and boil into a vapor before returning to the compressor.

EVAPORATOR COIL CONSTRUCTION

Evaporators may be made of aluminum tubing, regardless of whether they are of the forced-draft or natural-draft type. When the evaporator coil tubing is used for shelves, the

Figure 45.34 A forced-draft evaporator.

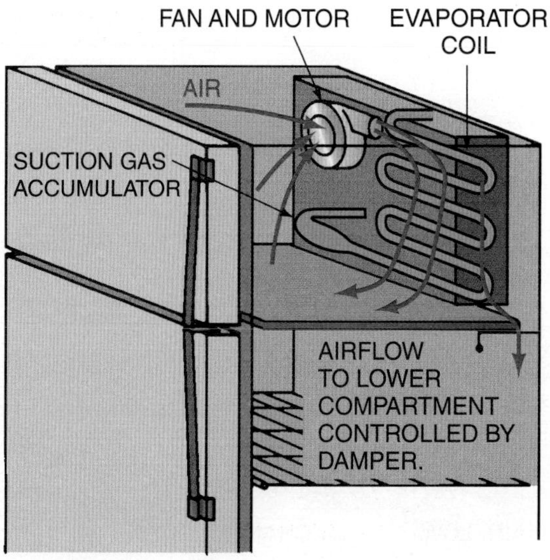

Figure 45.35 Shelf-type evaporators are normally steel tubing with wire to add support. Photo by Bill Johnson

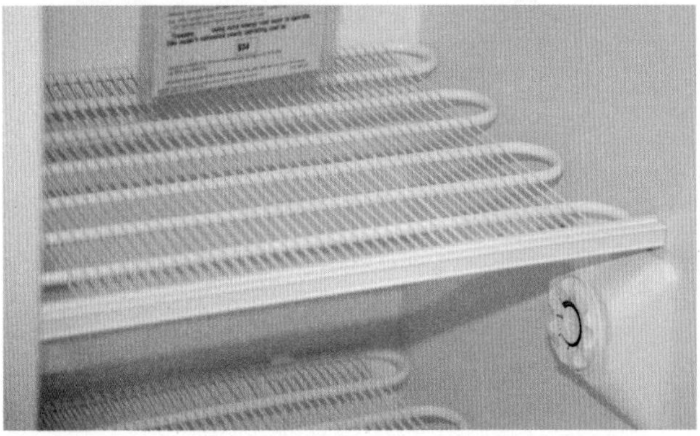

evaporator is commonly made of steel. This gives it strength to hold the food and provides a method of fastening the wire connectors to the tubes, **Figure 45.35**. One reason for using steel for the shelves is increased durability, since the products being cooled will be in contact with them and subject to user abuse. Another reason for using steel is the ease with which the support wires can be connected to the tubing.

Forced-draft evaporators are normally made of aluminum tubing with fins to increase the effective heat-transfer surface area of the coil. The fins are spaced fairly wide apart to allow for frost to build up and not block the airflow, **Figure 45.36**. The spacing of the evaporator fins on a domestic refrigerator or freezer is much wider than that found in an air-conditioning system. This is because frost does not form on the evaporator coil of a properly operating air-conditioning system, so the fins can be positioned closer together. Evaporators on domestic refrigerators and freezers do not typically require regular maintenance because air is recirculated within the refrigerated box and they do not have air filters, as is the case with an air-conditioning system.

Figure 45.36 Fin spacing on a low-temperature evaporator. Photo by Bill Johnson

Figure 45.37 Electric defrost.

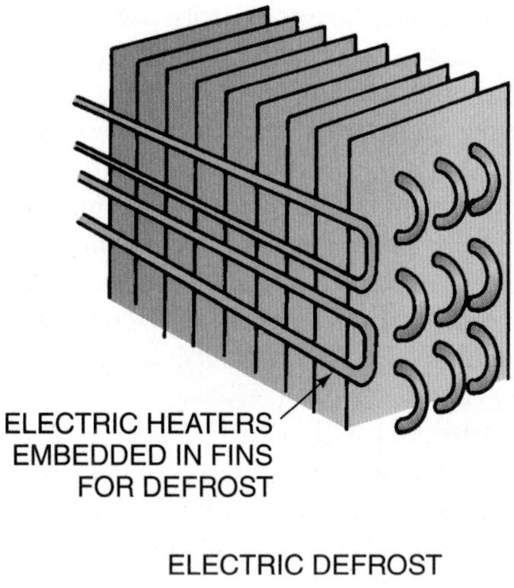

EVAPORATOR DEFROST

Even though the temperatures in the refrigerated portions of a domestic refrigerator are above freezing, it is important to remember that the evaporator coil must be at a temperature that is lower than the desired space temperature in order to absorb heat from the box and the products contained within. In the case of the domestic refrigerator or freezer, the evaporator coil, under normal conditions, will always be a temperature that is lower than 32°F. Since this is the case, moisture that condenses on the evaporator coil surface will freeze. This frost must be removed. Two common methods for defrosting evaporator coils are manual and automatic.

Manual defrost is accomplished by turning off the unit, removing the food, and using room heat, a pan of hot water, or a small heater to melt the ice that has formed on the coil. A large amount of frost may accumulate on an evaporator by the time it is defrosted. The water that forms as a result of this type of defrost must be disposed of manually. Units that require manual defrost normally have coils that food has touched. The shelves should be cleaned and sanitized when the frost is removed. Automatic defrost is accomplished by external heat from electric heating elements that are positioned close to the evaporator coil surface, **Figure 45.37**.

No matter how the defrost is accomplished, the water from the coil must be dealt with. With automatic defrost, a pint of water may be melted from the evaporator with each defrost. This water is typically evaporated using the heat from the compressor discharge line or by heated air from the condenser. This is discussed in more detail in Section 45.5, "The Condenser."

45.4 THE COMPRESSOR

The compressor circulates the heat-laden refrigerant by removing it from the evaporator as a low-pressure, low-temperature superheated vapor and pumping it into the

condenser as a high-temperature, high-pressure superheated vapor. The compressors used in domestic refrigerators and freezers are very small compared to the ones used in air-conditioning and commercial refrigeration systems. They are all in the fractional horsepower size, ranging from about 1/10 horsepower to 1/3 horsepower, depending on the size of the box. In many of these systems, it is hard to tell which is the suction and which is the discharge line, as the sizes of these lines do not differ by much.

Compressors have a suction line, a discharge line, and one or two process tubes protruding from the shell. Older compressors often had oil cooler lines connected to them as well, making the identification of the particular tubes all the more difficult. Although newer compressors do not, for the most part, have oil cooler lines, a shell diagram is required to know which line to pipe to which connection when replacing a compressor with one that is not an exact replacement, **Figure 45.38**.

The compressors used in household refrigerators are all welded, hermetically sealed types, **Figure 45.39.** These compressors are reliable and made to last many years. A typical refrigerator may be used continuously for 20 or more years.

The compressor is located at the bottom of the refrigerator and accessed from the back. The refrigerator must be moved away from the wall for service. Most modern, large refrigerators have small wheels so the box can be moved out for cleaning and to provide easy access for service, **Figure 45.40**. These wheels are often retractable. Once the appliance has been placed in its desired position, the wheels are retracted into the appliance so that the unit itself rests on the floor and will not slide back and forth. This feature is common on refrigerators and freezers that have slide-out compartments, **Figure 45.41**. So, in order to slide the unit from the wall, the wheels must first be lowered to prevent potential damage to the floor. Compressors are typically mounted on internal springs and external flexible, rubber-like feet, **Figure 45.42**. The refrigerator is located in the living area and must have a low noise level. The lines connecting the compressor to the refrigerant piping may be of copper or steel. Correct solder must be used when repairing any piping connections.

As mentioned, domestic refrigerators and freezers utilize welded hermetic compressors. Servicing them requires a skilled technician. Many units do not even have service ports for attaching gauges to get compressor suction and discharge readings. When the manufacturer does furnish a service port, it may be in the form of a special port that requires special connectors, **Figure 45.43**. When the special connectors are not furnished, line tap valves, **Figure 45.44**, are often used.

Figure 45.38 An illustration of a compressor and related piping at the refrigerator.

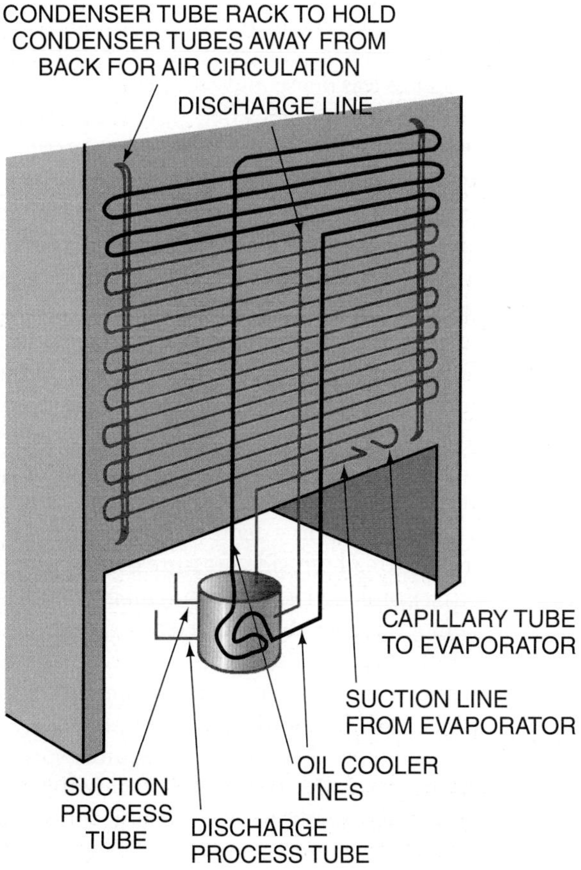

Figure 45.39 A welded, hermetically sealed compressor. Photo by Bill Johnson

Figure 45.40 Wheels on a modern refrigerator.

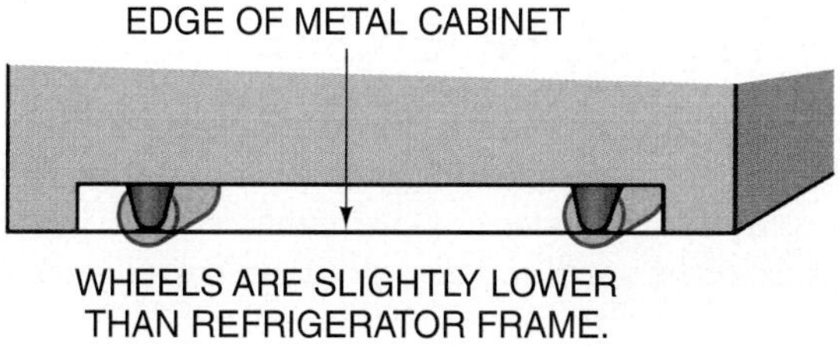

Figure 45.41 Slide-out freezer compartments on a refrigerator/freezer. Photo by Eugene Silberstein

These should be used only when necessary, and the manufacturer's instructions must be strictly followed, or the results will be unsatisfactory. Only soldered line tap valves may be left on the line. *It is a violation of the Clean Air Act of 1990 to leave nonsoldered line tap valves on a system. These valves do leak and leaving them on a system is, in effect, an example of intentional refrigerant venting. Individuals who violate the laws set forth by the EPA regarding safe and proper refrigerant handling are subject to heavy fines.*

45.5 THE CONDENSER

Domestic refrigerator condensers are all air-cooled. This permits the refrigerator to be moved to a new location or house as necessary. The condensers are either cooled by natural convection or by small induced-draft-air fans, **Figure 45.45**.

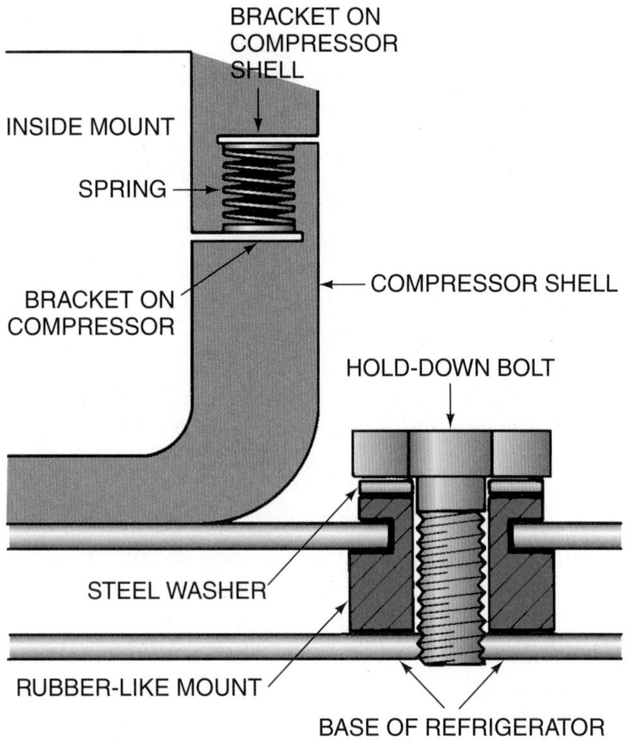

Figure 45.42 A compressor mount.

Natural convection (static) condensers were the first type used and are very simple in operation. There are no fans to move air across the surface of the coil. Natural convection currents are created as the air, as it absorbs heat from the refrigerant, becomes less dense and rises. This phenomenon is often referred to as the chimney effect, and these condensers, when located at the back of the appliance, are also referred to as chimney-type condensers. Care must be used when locating a refrigerator that has a natural convection condenser to ensure that air can flow freely over it. Poor air circulation through and across the condenser coil results in high operating pressures and reduced system capacity. Because of the reduced capacity, unit run time also increases. In severe cases, the unit may run all the time and still not keep the food compartment cool. The appliance shown in **Figure 45.46** is operating with a low-side pressure of 6.7 psig, which corresponds to an evaporator saturation temperature of about 1°F. This is much higher than desired, especially if the freezer compartment is to maintain a temperature of 0°F. This unit is also operating with a high-side pressure of 191 psig, which corresponds to a condenser saturation temperature of about 127°F. This is much higher than expected for a unit installed inside a conditioned space. This compressor is operating with high head pressure and may very well fail prematurely.

Care must be used when moving any appliance, but the external condenser type of refrigerator requires special care or the condenser coil may be damaged. If a two-wheeled hand truck is used, the belt should be placed under the condenser

Figure 45.43 A special service port valve. (B) Photo by Bill Johnson

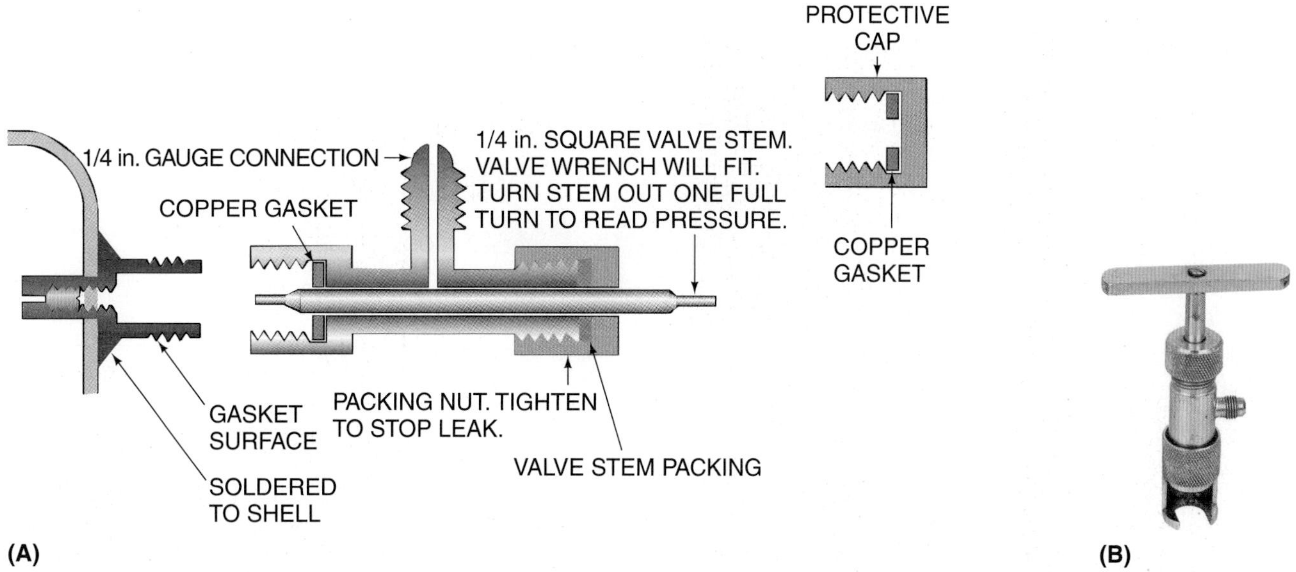

Figure 45.44 A tube-piercing valve. Photo by Eugene Silberstein

and around the unit, **Figure 45.47**. Because of the potential for damage to the exposed condenser coil, steel is commonly used in the construction of this heat-transfer surface. Some natural convection condensers have been located in the outside wall of the refrigerator by fastening the condenser tubes to the inside of the outer sheet metal shell. This unit may work well in the open, where air can circulate over it, but may not work in a tight alcove. The exterior walls of this type of unit are normally very warm, and the condensers are often referred to as hot-wall condensers.

Induced-draft condensers have solved many of the problems just described. The induced-draft condenser is located under the refrigerator and typically at the back, **Figure 45.48**. Air is taken in on one side of the bottom front of the box, pulled across the condenser coil and then discharged out the other side of the front, **Figure 45.49**. Because both inlet and outlet are close to the floor and at the front, they are not easily obstructed. Induced-draft condensers require an established air pattern to pull the air over the finned tubes. The air pattern is often maintained by cardboard partitions and a cardboard backing at the bottom of the compressor compartment, **Figure 45.50**. **NOTE:** *All partitions must be in the correct position or high head pressures and long or continuous running times will occur.•* Many service technicians and homeowners may have discarded the cardboard back cover thinking it was part of the appliance's packaging and not necessary. The refrigerator will not perform correctly without the back cover in place, and permanent damage can result.

Induced-draft condensers are typically used on units that have automatic defrost, and they may require more care. They are smaller and lint or large particles may become trapped in the fins, **Figure 45.51**. The air pulled through the condenser, in conjunction with the heat of the discharge line, may be used to evaporate any water from defrost, **Figure 45.52**. Although refrigerators are typically located in the kitchen, stand-alone freezers with induced-draft condensers are often located in remote areas of the home, such as laundry rooms. These rooms have more lint than other parts of the house and may be much warmer. These freezers must be more closely observed because of potential problems such as lint/dirt and higher-than-desirable operating temperatures and pressures. Lint will collect in the fins of the condenser and the defrost drain pan and can cause a problem.

Figure 45.45 (A) Natural-draft condenser. (B) Induced-draft condenser.

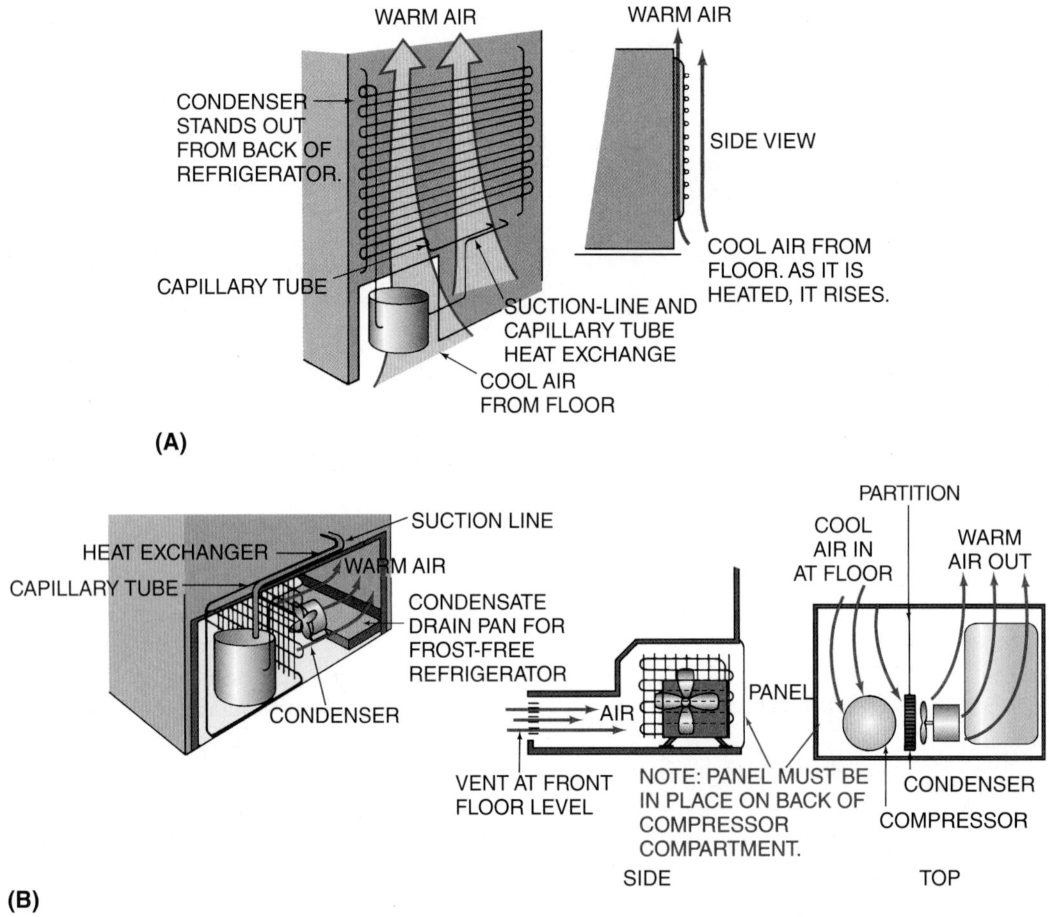

Figure 45.46 A unit running all the time due to poor location.

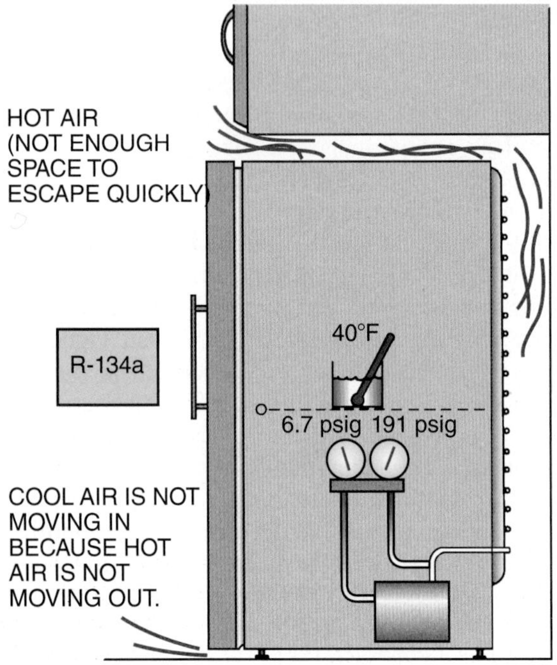

Figure 45.47 Care should be used when moving a refrigerator.

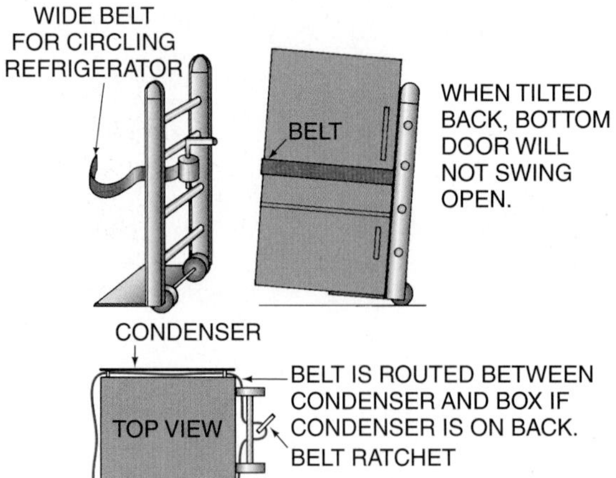

1. USE REFRIGERATOR HAND TRUCKS ONLY.
2. TRUCK REFRIGERATOR FROM SIDE SO THAT IT WILL GO THROUGH DOORS.
3. DO NOT WRAP BELT AROUND NATURAL-DRAFT CONDENSER.
4. BE AWARE OF WHICH WAY THE DOOR WILL SWING IF TILTED BACK.

Figure 45.48 Location of an induced-draft condenser.

Figure 45.49 Airflow through an induced-draft condenser.

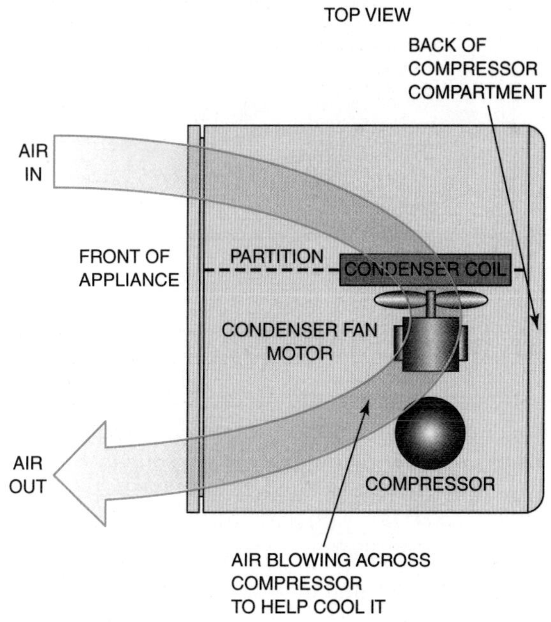

USING THE CONDENSER FOR CONDENSATE DISPOSAL

All domestic refrigerators are low-temperature appliances and accumulate frost on the evaporator. When defrost occurs, something must be done with the water. The compressor and condenser sections of the refrigerator are used to evaporate this water.

Figure 45.50 Cardboard partitions. Be sure to replace all cardboard partitions on an induced-draft condenser.

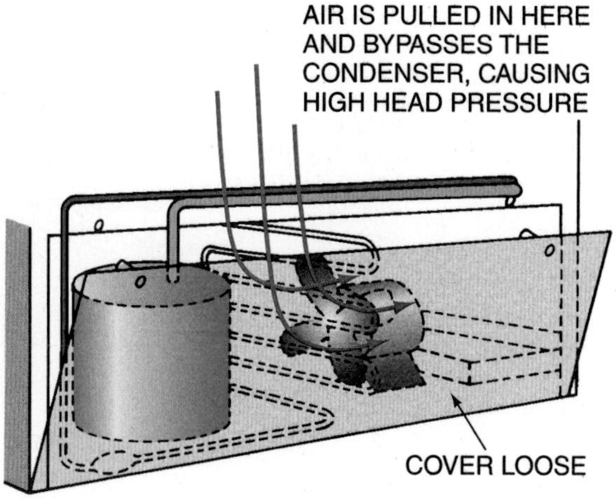

Figure 45.51 Induced-draft condenser located close to the floor collecting lint and dust. Photos by Eugene Silberstein

The heat available at the compressor discharge line can be used to evaporate the water when direct contact is made. Many units are designed so that the discharge line is passed through the pan that collects the defrost water, **Figure 45.53.**

Figure 45.52 The discharge air in conjunction with heat in the discharge line may be used to evaporate condensate from defrost.

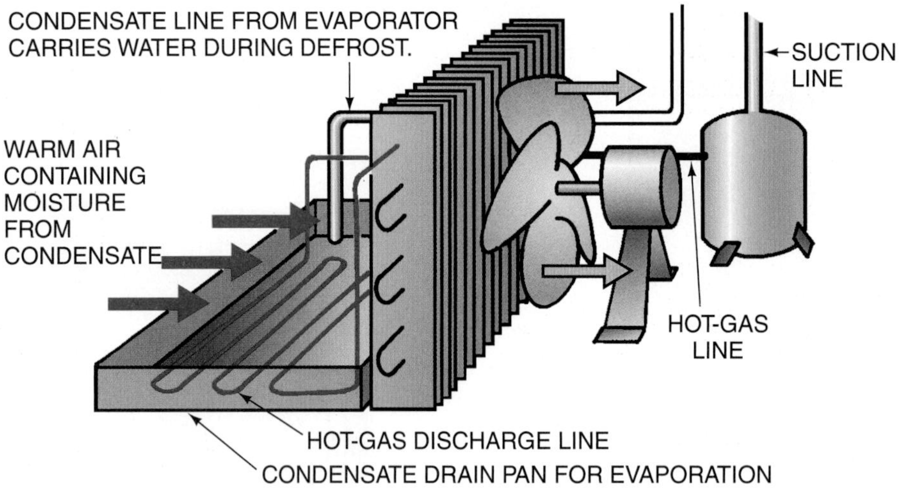

Figure 45.53 Compressor heat used to evaporate condensate.

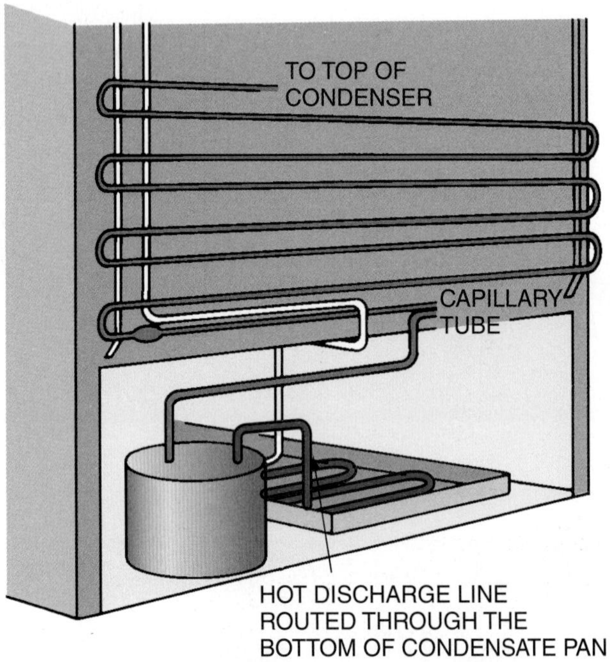

This method is used primarily on units with natural-draft condensers. When the unit has an induced-draft condenser, warm air from the condenser may be pulled over a collection pan of water for the purpose of evaporation, **Figure 45.54.** In either case, the unit has the compressor running time from one defrost cycle to the next to evaporate the water, depending on the manufacturer's design. Defrost may occur during the off cycle or as few as two or three times per 24 hours. The collection pan in the bottom of the unit is a place where lint and dirt may collect. This pan can and should be removed occasionally and cleaned.

Figure 45.54 Warm air from the condenser used to evaporate condensate.

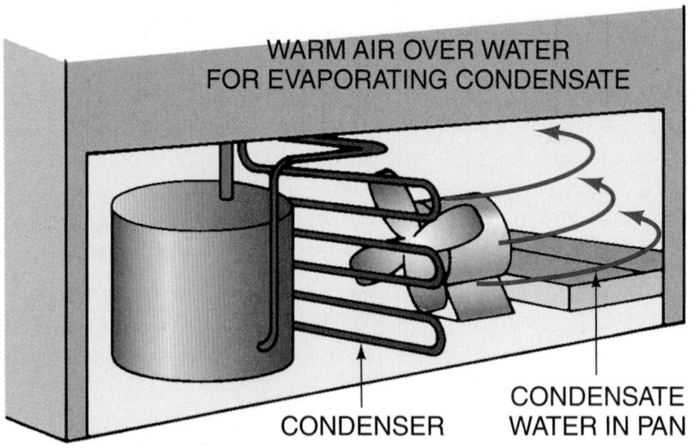

45.6 METERING DEVICE

Domestic refrigerators and freezers use the capillary tube metering device. This is a fixed-bore device, and the amount of refrigerant flow through it is determined by the bore of the tube and its length, which are predetermined by the manufacturer. The capillary tube is usually fastened for heat exchange to the suction line, **Figure 45.55**. In some cases, the capillary tube may be run inside the suction line, **Figure 45.56**. This helps prevent liquid from returning to the compressor and improves the evaporator capacity by subcooling the liquid in the capillary tube, **Figure 45.57**. The capillary tube may be serviced if necessary and may in some cases be changed for another one, but this is difficult and usually not practical.

It was mentioned earlier that a domestic refrigerator performs correctly only when located in the living space. The capillary tube metering device is one reason for this. This device is sized to pass a certain amount of liquid refrigerant at typical living condition temperatures, which

Figure 45.55 The capillary tube fastened to the suction line.

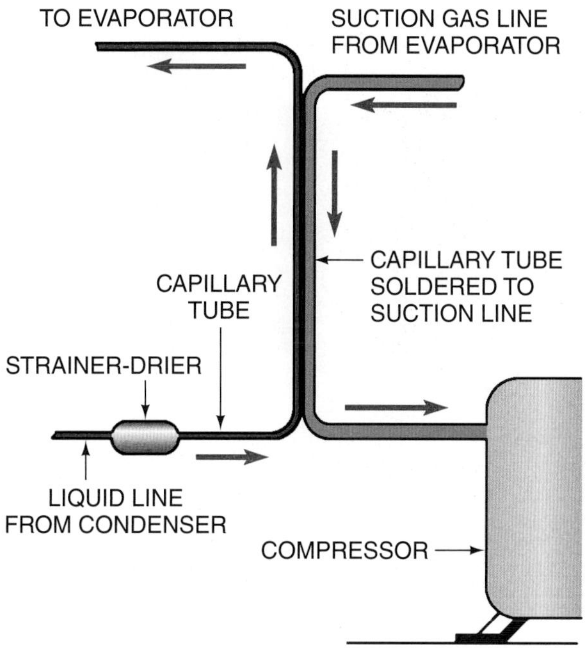

Figure 45.56 The capillary tube routed through the suction line.

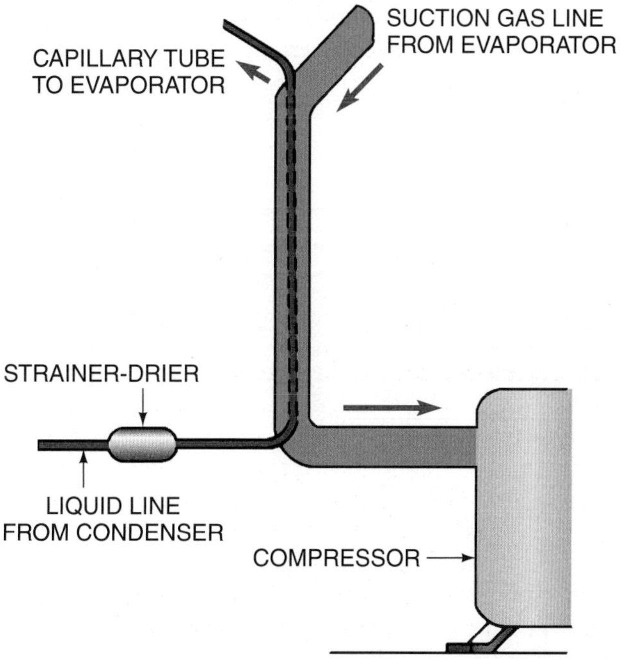

may be considered to be between 65°F and 95°F. If the room ambient temperature is greater, the head pressure will climb and push more refrigerant through the capillary tube, causing the suction pressure to rise. Capacity will suffer to some extent, **Figure 45.58**. Most residences will not exceed the extreme temperatures for long periods of time. For example, the temperature may rise to 100°F in the daytime if the house does not have air-conditioning, but it will cool down at night. The refrigerator may work all day and not cool down to the correct setting but may cycle off at night.

Figure 45.57 Heat exchange between the capillary tube and the suction line.

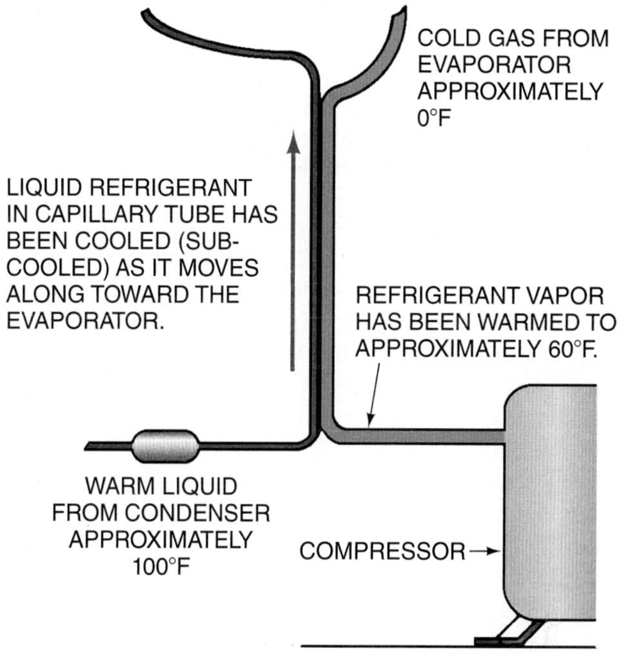

Figure 45.58 The capacity of a refrigerator is reduced by high ambient temperature.

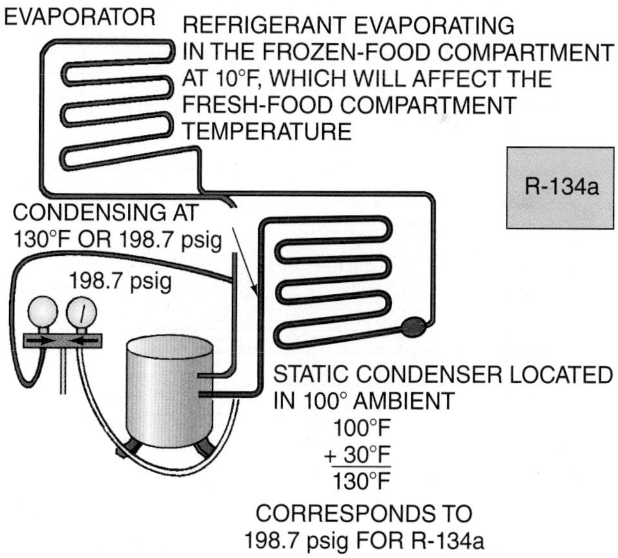

Domestic refrigerators and freezers are not intended to be placed outdoors or in buildings where the temperatures go below 65°F and rise higher than 95°F—unless the manufacturer's literature states that they can be operated at other temperatures. Poor performance and shorter life span may be expected if the manufacturer's directions are not followed.

45.7 TYPICAL OPERATING CONDITIONS

The common refrigerant found in existing, older domestic refrigerators and freezers is R-134a. However, the technician must first read the unit's nameplate to determine the type of refrigerant used and then consult a temperature/pressure chart to determine the correct pressures and temperatures for that particular system. Although R-134a is commonly found in older domestic appliances, newly manufactured appliances will likely contain new hydrocarbon (HC) refrigerants, such as R-290 (propane) and R-600a (isobutane). These refrigerants are flammable and the appliances will be clearly marked if they contain these refrigerants. Refer to Unit 3 and visit the EPA's website at www.epa.gov. for safety and other important information about these and other new refrigerants.

Ideally, a home freezer operates with a box temperature cold enough to freeze ice cream hard, that is, a temperature of about 0°F. The evaporator must operate at a temperature that will lower the air temperature to a point below 0°F. When the air temperature is 0°F, the evaporator coil temperature may be as low as −12°F. This saturation temperature corresponds to a pressure of 1.1 psig for R-134a. With a forced-draft evaporator, the coil temperature may be a little higher because air is forced over the evaporator. **Figure 45.59** provides saturation temperature and pressure information for an R-134a appliance with a static, or natural-draft, condenser. **Figure 45.60** provides typical operating pressures and temperatures for R-134a appliances operating with both forced-draft and natural-draft evaporators. These are the temperatures and pressures expected at the end of the cycle, which is about the time the compressor

Figure 45.59 Typical operating pressures and temperatures for an R-134a appliance with a static condenser.

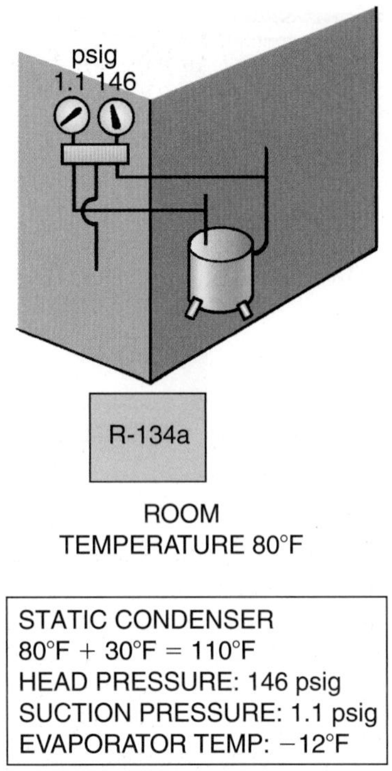

Figure 45.60 Some typical evaporator conditions for forced- and natural-draft evaporators.

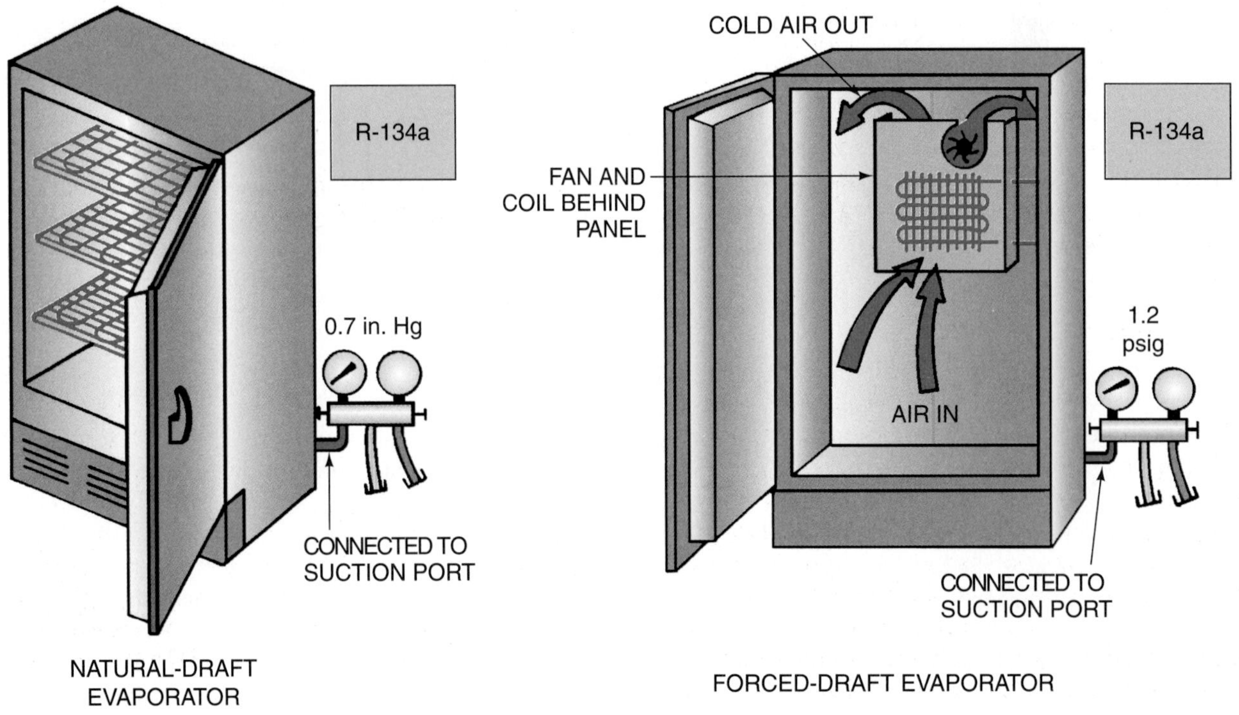

Figure 45.61 The refrigerant typically condenses at 20°F to 35°F higher than the ambient temperature.

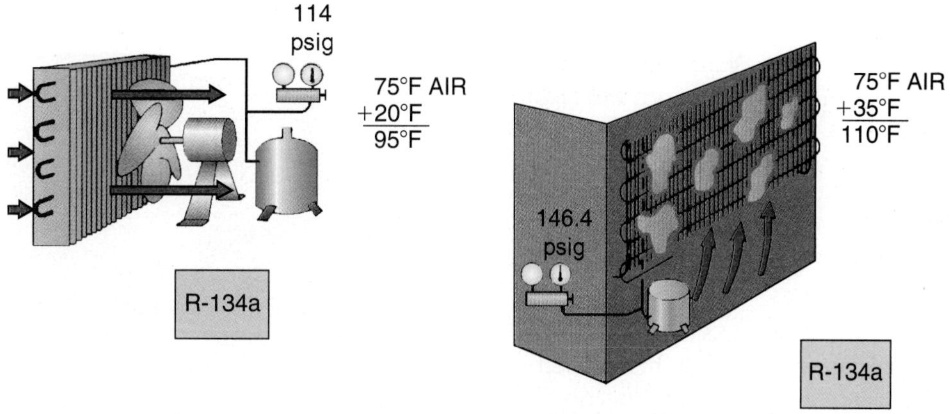

BEST CASE: CLEAN INDUCED-DRAFT CONDENSER.

WORST CASE: NATURAL-DRAFT CONDENSER THAT IS PARTIALLY DIRTY.

is going to stop. Temperatures and pressures during a pulldown will, of course, be higher, depending on the box temperature. It can be seen from the figures that evaporator saturation temperatures lower than −12°F should be avoided for R-134a to prevent the low-side pressure from pulling into a vacuum.

Operating pressures and temperatures for the condenser are determined using gauge readings. The freezer, like the refrigerator, may have no gauge ports. Gauges should not be installed on the appliance except as a last resort. The typical refrigerator or freezer is designed to be located within the living space, so 65°F to 95°F air will be passing over the condenser. Static, natural-draft, condensers should be able to condense the refrigerant at a temperature ranging from 30°F to 35°F higher than the ambient air temperature. For example, if the room temperature is 75°F, then the condenser saturation temperature should be between 105°F (75°F + 30°F) and 110°F (75°F + 35°F). Fan-assisted condensers should be able to condense the refrigerant at a temperature that is about 20°F higher than the ambient air temperature. So, if the room temperature is 75°F, then the condenser saturation temperature should be about 95°F (75°F + 20°F). The closer the condensing temperature is to the temperature of the air passing over it, the more efficient the condenser. A condenser that condenses at 25°F higher than the ambient temperature is more efficient than a condenser that condenses at 35°F higher than the ambient, **Figure 45.61**. Some typical head pressures for R-134a are shown in **Figure 45.62**.

The compressor is the heart of the system, and the technician should be able to readily recognize any compressor problems. On a freezer that is operating normally the compressor is hot to the touch. All compressors are air- or refrigerant-cooled. Compressors that are air-cooled should have fins. Refrigerant-cooled compressors may be cooled with the suction gas. The compressor should not be noisy; it is mounted on rubber feet to keep it quiet. But if the freezer is not level, it may vibrate or bind, so leveling a freezer is important. If a compressor is sitting level and steady and still vibrates, check to see whether the lines are secure and are not touching any part of the appliance. If vibration and noise persist, liquid refrigerant may be returning to the compressor. If it is, the compressor will sweat on the side, **Figure 45.63**.

Figure 45.62 Some typical R-134a head pressures.

	FORCED-AIR CONDENSER	NATURAL-DRAFT CONDENSER
COOL ROOM	65° F + 20° F = 85° F CONDENSING TEMPERATURE 95.3 psig R-134a	65° F + 30° F = 95° F CONDENSING TEMPERATURE 114 psig R-134a
NORMAL ROOM	75° F + 20° F = 95° F CONDENSING TEMPERATURE 114 psig R-134a	65° F + 30° F = 95° F CONDENSING TEMPERATURE 135 psig R-134a
HOT ROOM	95° F + 20° F = 115° F CONDENSING TEMPERATURE 158.5 psig R-134a	95° F + 30° F = 125° F CONDENSING TEMPERATURE 184.6 psig R-134a
HOT ROOM DIRTY CONDENSER	95° F + 35° F = 130° F CONDENSING TEMPERATURE 198.7 psig R-134a	95° F + 40° F = 135° F CONDENSING TEMPERATURE 213.5 psig R-134a

Figure 45.63 If liquid refrigerant is returning to the compressor, it will be sweating on the side.

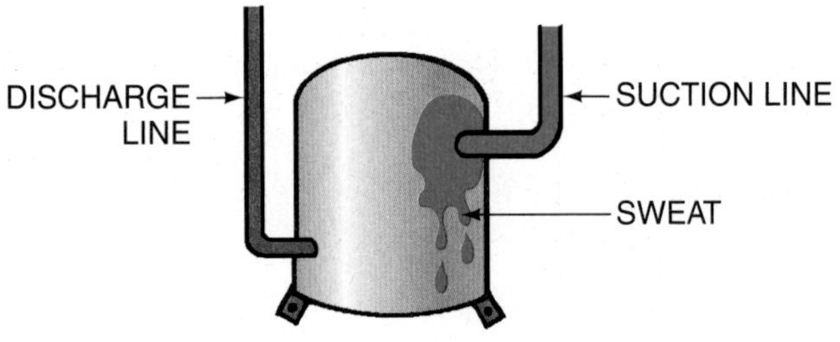

If there is a question concerning the capacity of the compressor, such as when the unit is not cold enough, first suspect an additional load on the appliance. For example, an interior light that remains on will add additional heat to the system and may cause the compressor to keep running even though the space temperature may be cold enough. A leaky gasket will also cause an extra load, or the owner may be using the box too often and opening the door too much. All conditions should be analyzed before the compressor is condemned.

45.8 ICE-MAKER OPERATION

The ice maker in a domestic refrigerator used to freeze water into ice cubes is in the low-temperature compartment of the box. Generally, a tray filled with water from the home water supply is placed in the freezing compartment. Most newer refrigerators that are equipped with ice makers and water dispensers have water filters built into the appliance to filter the water that is used for consumption and ice prior to use, **Figure 45.64.**

Domestic ice makers do not make ice like commercial ice makers, which use augers or water flowing over inverted evaporators. Domestic ice makers use time more than any other method to determine the sequence of events, but some use electronic circuitry to control the operation of the ice maker. A solenoid valve in the water line opens for a predetermined time to fill the tray. Once the tray is filled and enough time has passed for the water to freeze into cubes, the ice maker will harvest the ice. Harvesting the ice is most commonly accomplished by a rotating "rake" that pushes the cubes from the tray. A small electric resistive heater loosens the cubes for harvest. This heater is attached to the bottom of the ice tray and, when energized, causes the tray to heat up. This melts the ice closest to the tray and allows the rake to remove the finished cubes. When they turn loose, the gear motor moves them to the storage area and the process starts again. This is repeated until the ice bin is full, **Figure 45.65.**

The amount of ice a domestic ice maker produces depends on a number of factors, including the temperature of the appliance's freezer compartment. The heat load on the appliance and the number of times the door is opened also affect the rate of ice production. Keeping the freezer

Figure 45.64 (A) Water filter compartment located in the refrigerated space. (B) Compartment opens to expose the replaceable filter media cartridge. Photos by Eugene Silberstein

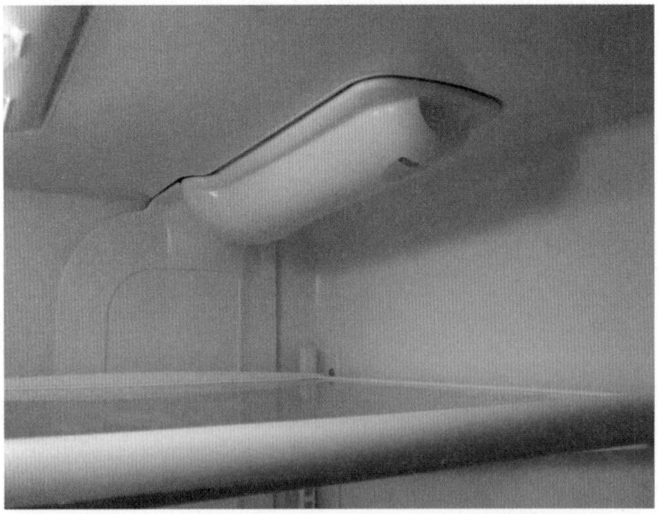

(A)

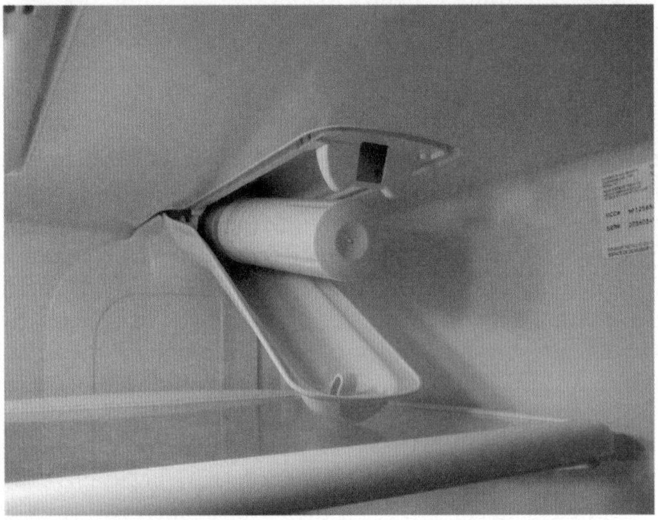

(B)

Figure 45.65 The water-filled tray freezes. When the control circuit determines that the water is frozen, the heater energizes. Then the fingers push the ice out of the tray to the storage bin. When the bin is full, the bail switch is moved to the right and will not return; production stops.

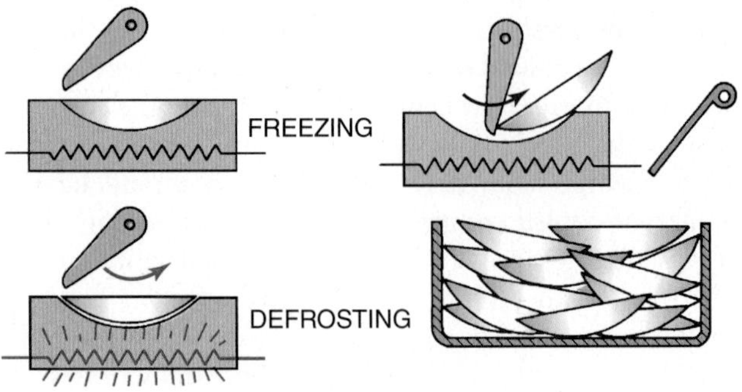

filled with frozen product keeps the temperature in the box more stable and actually helps to increase the efficiency of the appliance. A typical domestic ice maker produces between 3 and 5 pounds of ice a day, depending on the model and the conditions just mentioned. When first installed, it may take up to a full day before the unit begins to make ice.

45.9 WIRING AND CONTROLS

A wiring diagram should be permanently fastened to each refrigerator box, normally on the back. Most of the time, the diagram is placed inside a plastic sleeve that is permanently mounted to the back of the appliance, **Figure 45.66**. These wiring diagrams are usually either pictorial or linear. The line diagram, also commonly referred to as a ladder diagram, shows each separate circuit as a separate "rung" on a ladder. The pictorial diagram typically shows the location of each component in addition to all of the wires in the appliance, **Figure 45.67**. The line diagram is used to illustrate how the circuit functions by showing all power-passing and power-consuming devices between the two lines, **Figure 45.68**. It is especially useful for troubleshooting purposes, as each load can be easily identified as well as all of the controls and switches that are wired in series with it. Keep in mind that these are 115 V appliances, so there will be one hot wire and one neutral to operate the equipment and a ground (green wire) for the frame or box ground protection.

The components of the typical refrigerator that are under electrical control include the following:

- Compressor
- Defrost components
- Various heaters—butter and panel or mullion
- Lights for the interior

Figure 45.66 Wiring diagram on back of the appliance. Photo by Eugene Silberstein

Figure 45.67 A pictorial wiring diagram. Courtesy White Consolidated Industries, Inc.

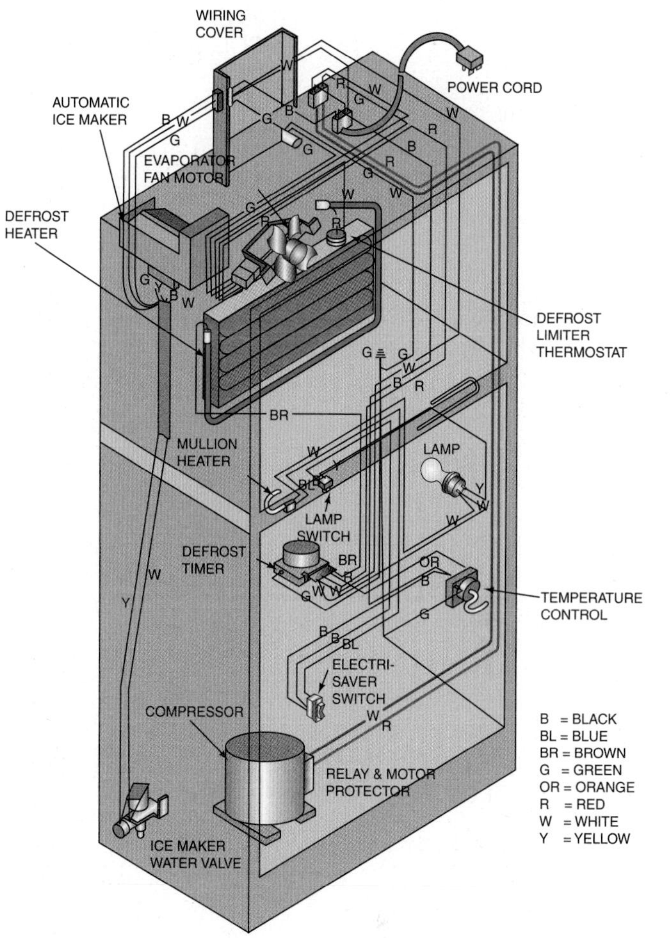

- Evaporator fan
- Ice maker
- Condenser fan

Except for automatic defrost controls, controls for freezers are simple. A typical freezer has only two: a thermostat and a door switch for turning off the light (and possibly a fan on some models). As with any control circuitry, the technician must know what the manufacturer's intent was when the system was designed.

COMPRESSOR CONTROLS

The space-temperature thermostat, also referred to as the cold control, **Figure 45.69**, controls the compressor. It is a line-voltage device that passes line power to the compressor start/run circuit. It is a make-on-a-rise-in-temperature device.

Several methods are used to control the operation of the compressor. In fully automatic defrost refrigerators, the thermostat, which controls compressor operation, is positioned so that the sensor is in the path of the freezer's supply air. The thermostat in refrigerators with two evaporators is

Figure 45.68 A line wiring diagram. Courtesy White Consolidated Industries, Inc.

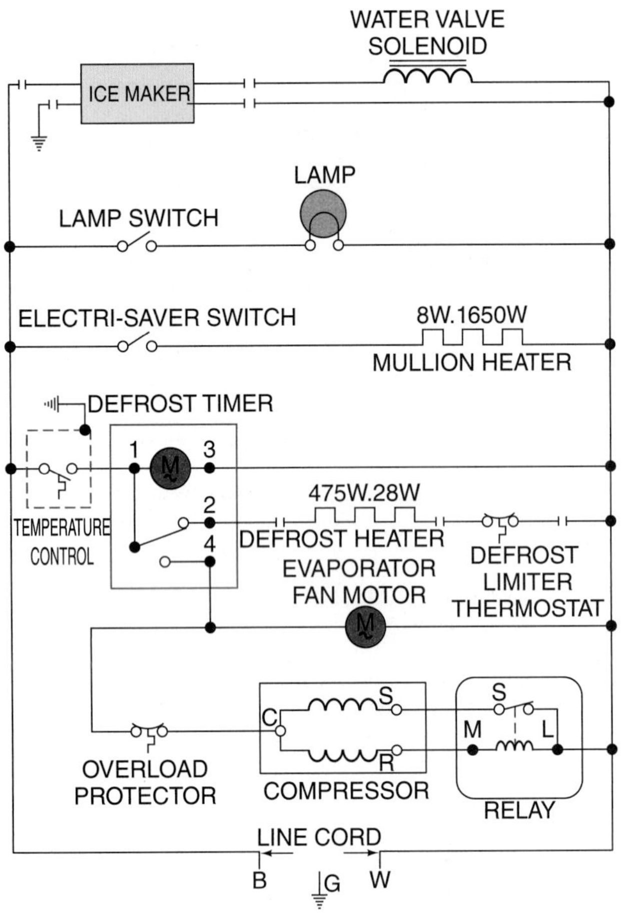

Figure 45.69 The thermostat or cold control. Photo by Bill Johnson

positioned so that it senses the temperature of the evaporator coil in the refrigerator compartment of the unit. On refrigerators that are designed for manual defrost, the thermostat senses the temperature of the evaporator coil, which is located in the freezer compartment of the unit. Refrigerator thermostats often have a large dial with graduated numbers, which typically run from 1 to 10 and have no relation to actual temperatures. The dial may bear the label "colder" with an arrow pointing in the direction where the numbers will yield colder temperatures. This control must be electrically rated and have contacts that are able to stop and start the compressor for years of service.

No matter which control method is used, the compressor is still shut off by a thermostat based on some condition inside the refrigerated box with the intent of keeping both the frozen-food compartment and the fresh-food compartment at the correct temperature. The thermostat is adjustable and can be considered a remote-bulb thermostat, since the sensing element can be uncoiled and positioned to sense the air temperature at a point that is a fair distance from the thermostat itself. Inside the sensing bulb of refrigeration thermostats is a fluid that exerts pressure against the bottom of a diaphragm or bellows, **Figure 45.70**. The atmosphere is on the other side of the bellows. If the atmospheric pressure is much lower than normal, the control setting must be calibrated for a new pressure, such as when a refrigerator is located at a high altitude. Altitude adjustment varies from one manufacturer to another, so the manual should be consulted.

COMPRESSOR START CIRCUIT. The compressor receives power from the thermostat circuit when its contacts close. A current magnetic relay, **Figure 45.71**, or a PTC device, **Figure 45.72**, are often used to assist in the starting of the compressor. Most new appliances use a PTC device to assist in compressor starting. The starting components and their circuits are typically located in the back in the vicinity of the compressor. **Figure 45.71** shows the current magnetic relay as it would be observed on the appliance.

The current magnetic relay, also referred to as a CMR, is a device that senses the amount of current flowing through the run winding of the circuit and, in response to this current, will open or close the contacts on the device. The CMR has a low-resistance coil and a normally open set of contacts. The coil is wired in series with the run winding, and the contacts are wired in series with the start winding, **Figure 45.73**.

Figure 45.70 Sketch of a thermostat.

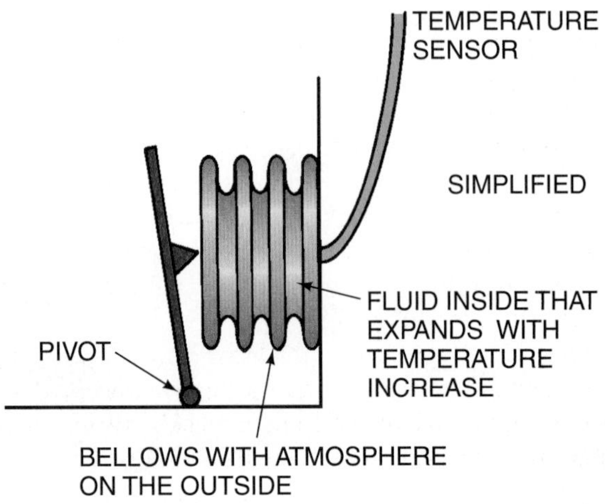

Figure 45.73 Wiring diagram of a compressor using a current relay.

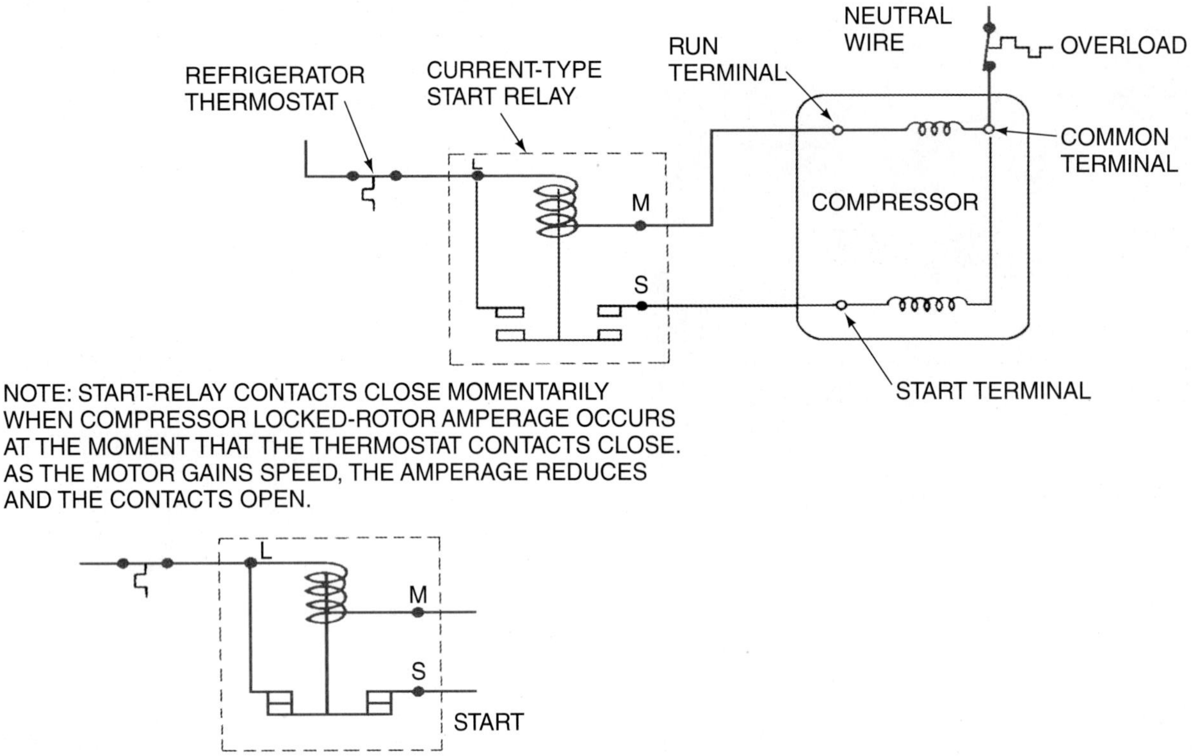

Figure 45.74 Wiring diagram of a compressor with a PTC device.

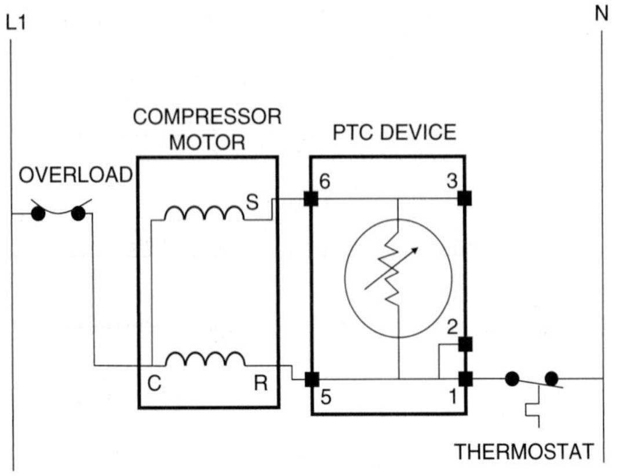

Figure 45.75 Wiring diagram of a compressor with a PTC device and a capacitor.

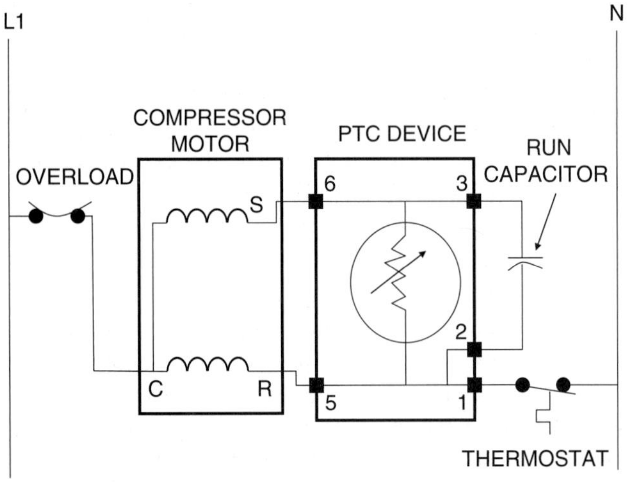

Quite often, a run capacitor will be wired in parallel with the PTC device, **Figure 45.75**. When this is the case, the PTC acts as a closed switch when the motor is initially energized and the run capacitor is bypassed. After a short time, the resistance of the PTC increases and current flow through the device stops. When the PTC "opens," current cannot bypass the run capacitor, which is now a part of the active circuit. The motor now operates as a permanent split capacitor (PSC) motor. Be sure to refer back to the units on motors and motor starting to brush up on these concepts.

COMPRESSOR WIRING DIAGRAMS. **Figure 45.76** shows a simplified ladder wiring diagram for a domestic refrigerator. Toward the top of the diagram is the compressor and the circuitry that controls its operation. From the diagram we can see that the compressor will operate only when there is power to the unit, the thermostat is calling for refrigeration, and the defrost timer is in the refrigeration mode. (The defrost timer will be discussed in more detail in the next section.)

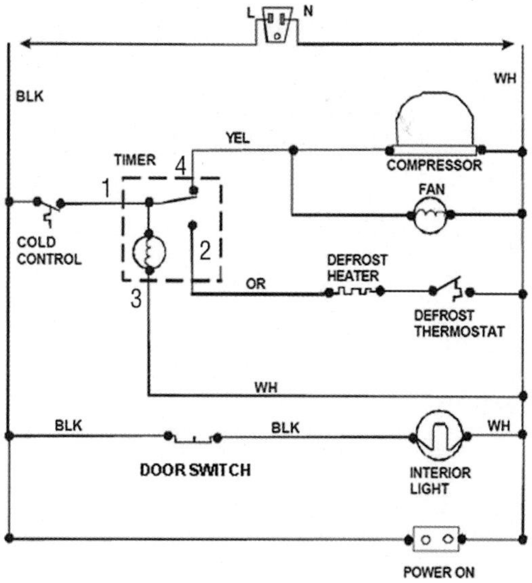

Figure 45.76 Ladder diagram of a domestic refrigerator.

Using the ladder diagram makes troubleshooting the compressor circuit much easier. For example, if the unit is being supplied with the proper line voltage and there is no voltage being supplied to the compressor, we can conclude, assuming the wiring is intact, that either the thermostat or the defrost timer is not allowing current to flow. It may be that the thermostat is not calling for refrigeration or the defrost timer is not calling for refrigeration, or it may be that one of these components is defective. Since the thermostat and the defrost timer are the only components in series with the compressor motor circuit, these are the only devices that will prevent power from reaching the compressor circuit. It is important to note that the compressor starting relays are not included in this diagram, as they are included as part of the compressor. When troubleshooting a system, be sure not to overlook these starting components.

DEFROST CYCLE

All refrigerators have freezing compartments, so they are low-temperature refrigeration systems. Low-temperature refrigeration systems operate with an evaporator coil temperature below 32°F, and for this reason, the evaporator will gather frost. This frost must be removed from the evaporator coil to maintain the coil's efficiency and also to ensure that air is able to pass through the coil. Ice acts as an insulator, so when ice forms on the evaporator coil, the rate of heat transfer between the air in the refrigerated space and the refrigerant is reduced. Automatic defrost is called *frost-free* by many manufacturers, customers, and service people. It is desirable because manual defrosting is a chore and often is not done when needed. *Automatic defrost will help the refrigeration system to operate more efficiently because the frost will be kept off the coil.*

Automatic defrost is controlled by either a mechanical defrost timer, **Figure 45.77**, or an electronic, solid-state, defrost timer, **Figure 45.78**. The mechanical defrost timer is made up of a timer motor, a normally open set of contacts, and a normally closed set of contacts, **Figure 45.79**. The defrost timer motor is often wired so that it is energized

Figure 45.77 Defrost timer.

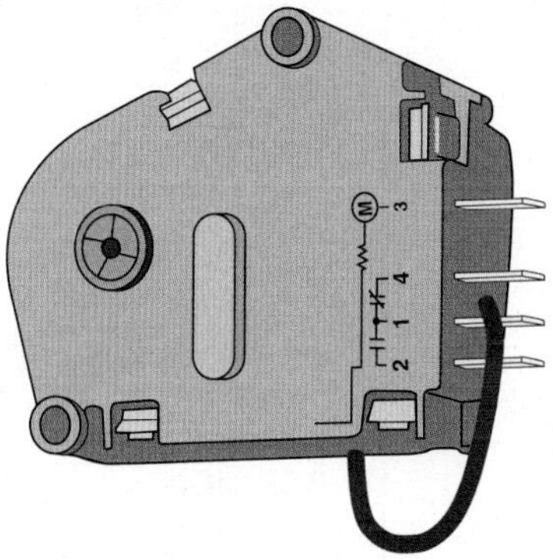

Figure 45.78 Electric, solid-state timer.

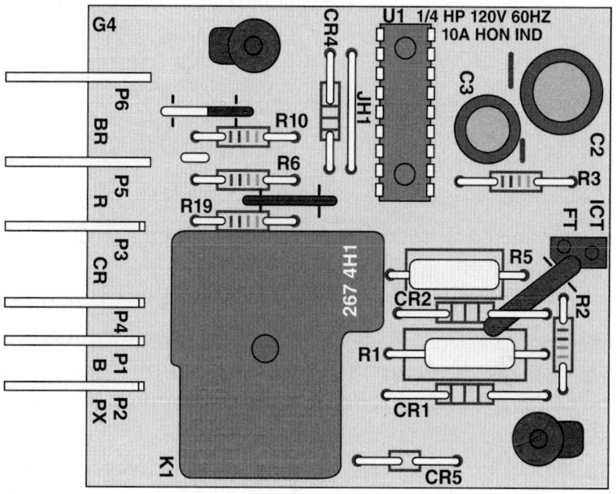

Figure 45.79 Wiring diagram of a mechanical defrost timer.

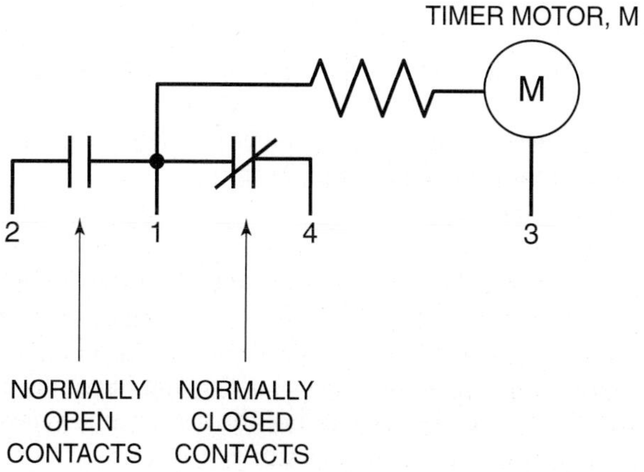

Figure 45.80 Timer wiring for refrigerator defrost.

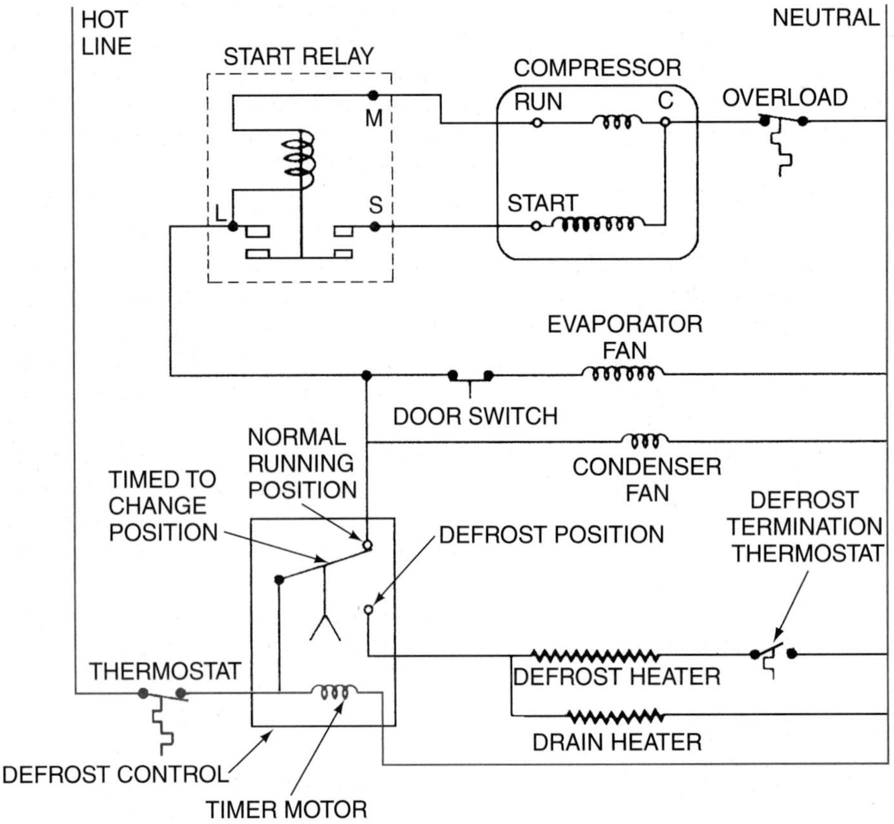

whenever the compressor is operating. When the timer motor is energized, internal gears rotate to keep track of the total run time of the compressor. After the desired run time has elapsed, the defrost cycle is initiated. The normally closed contacts are used to energize the refrigeration cycle when the refrigerator is cooling. The normally open contacts are used to energize the defrost mode of operation.

In **Figure 45.79**, we can see that the timer motor, M, is connected to terminals 1 and 3 on the timer. The timer is connected to terminal 1 by means of the jumper wire that is attached to the device. The normally open set of contacts is located between terminals 1 and 2, and the normally closed set of contacts is located between terminals 1 and 4. From the label on the timer we can see that for every 8 hours of compressor run time, the unit will go into defrost for a period that is determined by the manufacturer. The many different defrost timers available have different combinations of run and defrost times.

Among the several methods manufacturers use to start defrost is compressor running time. By wiring a defrost timer in parallel with the compressor, the time circuit only advances when the compressor runs, **Figure 45.80.** Compressor running time is directly associated with door openings, infiltration, and warm food placed in the box. In the wiring diagram in **Figure 45.80**, it should be noted that the defrost timer motor is wired in series with the refrigerator's thermostat. This means that the defrost timer motor will only be energized when the thermostat is calling for compressor operation. The defrost cycle may be terminated by two methods, time or temperature. Some units use a termination thermostat, **Figure 45.81**, that is backed up by the timer. If the thermostat does not terminate defrost, the timer will act as a safety feature.

Figure 45.81 Defrost termination thermostat. Courtesy Electrolux Home Appliances, North America.

Electric heat defrost is accomplished by electric heaters located close to the evaporator to melt the ice, **Figure 45.82**. When the unit calls for defrost, the compressor and the evaporator fan stop and the heaters are energized. This condition is maintained until the end of the defrost cycle. Drain pan heaters may be energized when electric heat defrost is used to keep the condensate from freezing as it leaves the drain pan.

DEFROST TIMER WIRING DIAGRAMS. As discussed in the previous paragraphs, the defrost timer is made up of the timer motor as well as normally open and normally closed sets of contacts. Assuming that the manufacturer intends for the appliance to go into defrost after a predetermined period of compressor run time, the timer motor will be wired in series with the cold control or refrigerator thermostat, **Figure 45.83**. It can be seen from the wiring diagram that the timer motor is identified by terminals 1 and 3 and that power cannot be supplied to the timer motor unless the cold control is in the closed position. When the cold control is in the open position, the timer motor is deenergized and the internal gears on the timer stop turning. The normally closed contacts are the contacts that energize the compressor and evaporator fan motor in the refrigeration mode and deenergize these components when the system is in defrost. These contacts are identified by terminals 1 and 4 in **Figure 45.84**.

The normally open contacts energize the defrost heater when the system goes into defrost. They also deenergize the heater when the system is operating in the refrigeration mode. These contacts are identified by terminals 1 and 2 in **Figure 45.85**. Notice that there is a defrost termination switch wired in series with the defrost heater. This component is a bimetal device that will open its contacts if it senses that there is no ice on the evaporator coil. Although the system is still technically in the defrost cycle, as indicated by the defrost timer, the defrost termination switch deenergizes the heater because its operation is not needed. This reduces the amount of unnecessary heat that is added to the box which would have to be removed later when the compressor cycles back on. This device also prevents the evaporator from overheating during defrost and causing damage to the appliance cabinet.

Figure 45.82 Electric heaters for defrost.

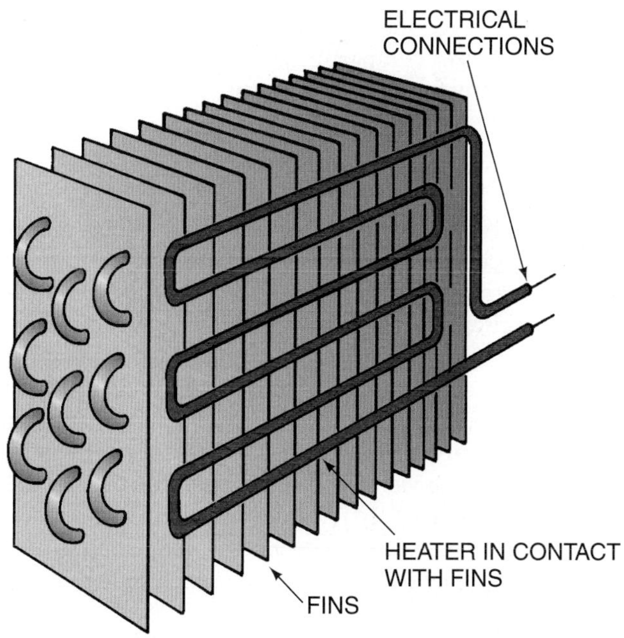

Figure 45.83 Defrost timer motor wired in series with the thermostat.

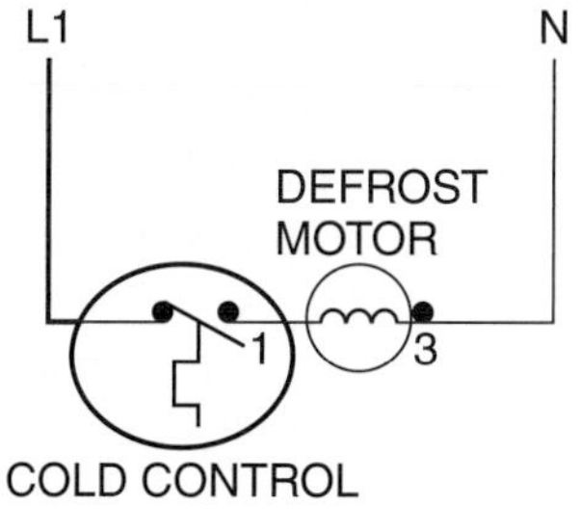

Figure 45.84 On this timer, the normally closed contacts are between terminals 1 and 4. These contacts energize the compressor circuit.

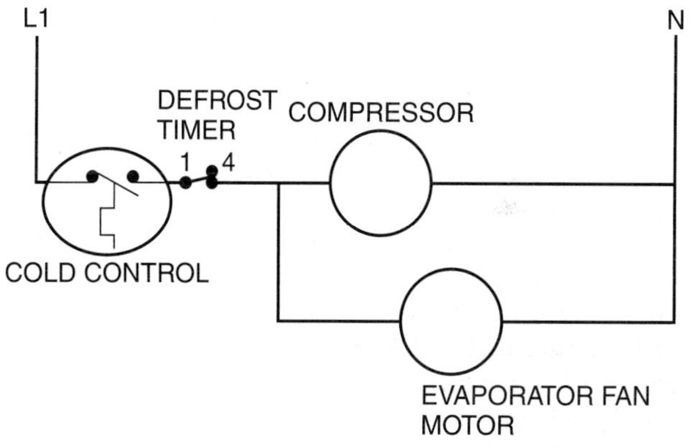

Figure 45.85 On this timer, the normally closed contacts are between terminals 1 and 2. These contacts energize the defrost heater circuit.

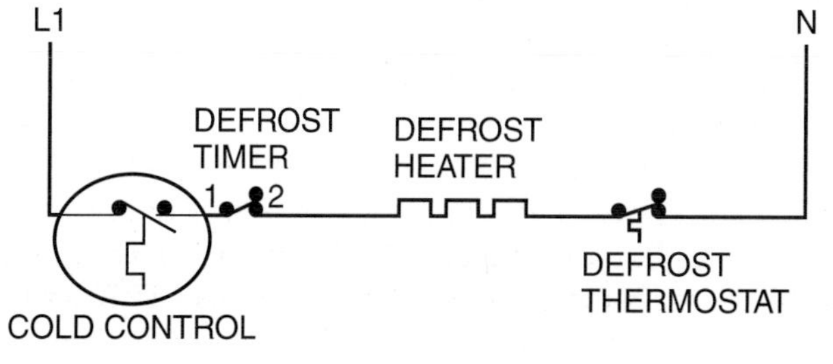

SWEAT PREVENTION HEATERS

Most sweat prevention heaters are small, electrically insulated wire heaters that are mounted against the cabinet walls at the door openings. Their function is to keep the outside cabinet temperature above the dew point temperature of the room so that the box will not sweat. *Some units may have energy-saver switches that allow the owner to switch part of the heaters off if the room humidity is low.* **Figure 45.86** shows the wiring diagram of a refrigerator/freezer with an energy-saver switch. Notice from the diagram that the mullion heater will be energized whenever the energy-saver switch is closed, regardless of whether or not the compressor is operating. If the appliance owner understands the feature and how best to use it, money can be saved.

There are, however, conflicting opinions with regard to the whether the energy-saver switch on refrigerators actually provides monetary savings over the long term. It can be argued that it costs money to keep any heater on, but because of the extremely high resistance of the heaters, the actual operating cost is very low. Most manufacturers simplify the two switch positions by giving them such labels as "HUMID" (which turns the heaters on) and "SAVE ENERGY" (which turns the heaters off). Most appliance owners are not aware of how the switch will save them money but are eager to conserve energy and money and so choose the "Save Energy" setting. Although this may be fine in the winter months when the humidity is low, problems can arise when the more humid summer months arrive. In very dry climates, the heaters may not be needed at all.

With humid air and cool refrigerator cabinet surfaces, condensate can form on the door, resulting in dripping water, which is a nuisance. The main problem is not the water dripping on the outside of the box, but the water that can build up on the interior of the refrigerator panels as well. The insulation can get wet, and over time, the wet insulation can sag, leaving bare spots with no insulation in the walls. This can lead to an increase in heat seepage through the box walls, which, in turn, can lead to increased compressor run time and an increase in energy consumption. For this reason, many industry professionals recommend that the sweat prevention heaters be powered at all times.

Figure 45.86 Wiring diagram of a refrigerator/freezer equipped with an energy-saver switch.

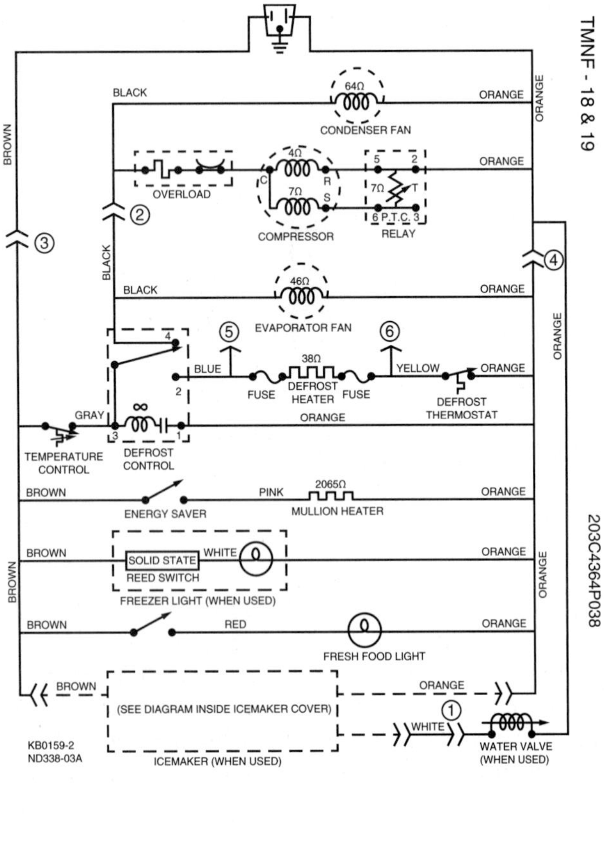

LIGHTS

Most refrigerators have lights mounted in the fresh- and frozen-food compartments. These lights are typically controlled by door switches that close a circuit when the door is opened. The wiring diagram in **Figure 45.87** shows circuits for both the freezer compartment and the fresh-food (refrigerated) compartment of the appliance.

REFRIGERATOR FAN MOTORS

In mechanical-draft refrigerators, two types of fans are used: one for the condenser and one for the evaporator. Condenser fans are usually prop types with shaded-pole motors. Small squirrel-cage centrifugal blowers may be used for the evaporator.

The evaporator fan, used on frost-free models only, may run all the time except in defrost, so in a few years it may have many operating hours, but it is typically reliable. The fan is located in the vicinity of the evaporator, usually under a panel that may be easily removed for service. The motor is often an open type with no covers over the windings. These fans have permanently lubricated bearings that require no service, **Figure 45.88**. The condenser fan is located under the refrigerator in the back and is typically a shaded-pole motor with a prop-type fan. It is permanently lubricated also but is covered, not open, **Figure 45.89**.

These small fan motors are simple to troubleshoot. If there is power to the motor leads and the motor will not turn, either the bearings are tight or the motor is defective. Most domestic refrigerator and freezer evaporator and condenser fan motors will operate whenever the compressor is operating. This means that these motors will cycle on whenever the box thermostat or cold control is calling for refrigeration and the defrost timer is not in the defrost mode. The wiring diagram in **Figure 45.87** shows that the condenser and evaporator fan motors are wired in parallel with the compressor

Figure 45.87 Wiring diagram of a refrigerator/freezer sharing circuits for refrigerator and freezer compartment lights.

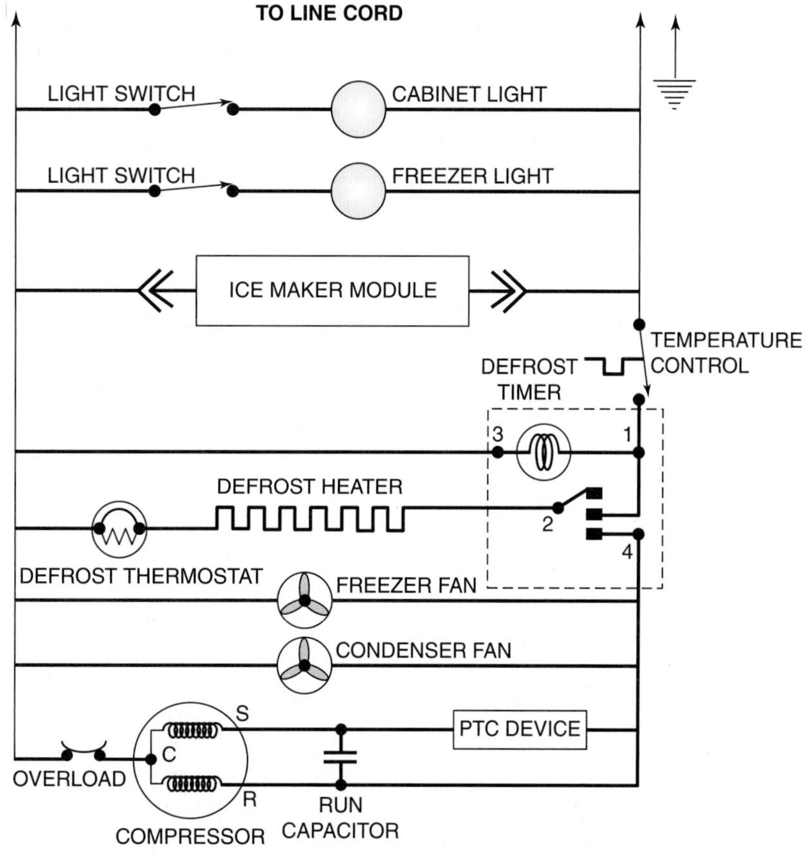

Figure 45.88 An evaporator fan motor. Photo by Eugene Silberstein

Figure 45.89 A condenser fan motor. Photo by Bill Johnson

circuit, so all three components will cycle on an off together during normal system operation. All three components will cycle "off" during the defrost cycle.

ICE MAKER WIRING

The ice makers found in domestic refrigerators are add-on options and are supplied with constant power. The ice maker has its own controls built right into it, so all that needs to be done is mount it, plug it in, and make certain that the water line is connected to the refrigerator as instructed. No field wiring is required to install the ice maker, except for plugging it into the plug provided for the wiring harness. The wiring diagram in **Figure 45.87** shows an ice maker and

where it is connected to the circuit. Notice that there are no controls or switches external to the component itself that will affect the operation of the device.

45.10 SERVICING THE APPLIANCE

The technician should make every effort to separate problems into definite categories. Some problems are electrical and some are mechanical. It is also important to know the type of appliance. For example, when servicing freezers, the technician does not have to think about the fresh-food compartment—only the low-temperature compartment. Also, the type of box determines the type of service. For example, a box that has a forced-draft evaporator will have a fan motor that can be a source of problems.

Unit 5, "Tools, Instrumentation, and Equipment," describes tools and equipment used by technicians for the servicing of refrigeration equipment. A poorly equipped technician works at a disadvantage. Lack of proper instruments may prevent discovery of the problem, and the customer's equipment or property can be damaged or the technician may be injured if, for example, the correct equipment for moving a refrigerated box is not used. A professional is well equipped. The sections that follow address a number of service-related issues regarding the domestic refrigerator and freezer.

ELECTRICAL PROBLEMS

Electrical problems in a domestic refrigerator or freezer are relatively easy to diagnose. This is because the appliance has relatively few electrical loads and switches and all of the components are located within its cabinet. In addition, a complete wiring diagram is attached to the appliance, usually on the back. Be sure to locate this very important document before attempting to repair the unit. One very important thing to know is the sequence of operations for the particular unit being worked on. Knowing what is supposed to happen and when can save a lot of time when troubleshooting the system. For example, if the compressor and condenser fan motor are wired to cycle on and off together, troubleshooting the problem will be a breeze if the condenser fan is operating and the compressor is not. The unit is being supplied with power because the condenser fan motor is operating. Also, the cold control is closed and operational because it is feeding power to the condenser fan motor. It is not difficult to conclude that the problem lies with the compressor, its wiring, or its associated starting components.

Here are a few things to keep in mind when troubleshooting the electrical circuits of a refrigerator or freezer:

- Have the wiring diagram handy.
- Observe what the appliance is doing and what it is not doing.
- Observe what appliance components are operating and those that are not.
- Know the sequence of operations of the appliance.
- Know the difference between a load (power-consuming device) and a switch (power-passing device).
- Know how to correctly use a voltmeter and ohmmeter.
- When testing resistance or continuity, make certain that the circuit power is disconnected and that the circuit or component being tested is completely removed/disconnected from the circuit to prevent false/feedback readings.
- When possible, unplug the appliance and discharge all capacitors to reduce the risk of electric shock.
- Make certain all electrical connections are tight.
- Make certain that all electrical wires are as far away as possible from wet locations, such as where condensate may drip or accumulate.
- Make certain that all electrical wires are as far away as possible from rotating components, such as fan motors.
- When replacing an electrical component, use an exact replacement whenever possible.

MECHANICAL PROBLEMS

Mechanical problems in domestic refrigerators and freezers often involve the refrigerant circuit and the compressor. Mechanical problems include system leaks, system undercharge, system overcharge, and compressor-related issues. The sections that follow are geared toward the mechanical portion of system evaluation and provide valuable service-related procedures to properly identify and remedy mechanical problems in the appliance.

CABINET PROBLEMS. In order for the refrigerant to correctly circulate through the evaporator and condenser, domestic refrigerators must be leveled according to the manufacturer's specifications. If it is not level, the ice maker may overflow when filled with water, and condensate may not completely drain during defrost. The leveling screws or wheels are on the bottom of the box or cabinet. Leveling feet may be adjusted with a wrench or a pair of adjustable pliers. Wheels have leveling adjustments to raise and lower them, **Figure 45.13**. If the floor is too low, spacers may need to be added at the lowest point so that all four feet or wheels are touching the floor or spacers. Vibration may make the refrigerator, as well as its contents (rattling glass containers, for example), noisy if all four points do not touch with equal pressure. Also, if the box is not level, the door or doors may not close correctly or may tend to swing open, **Figure 45.90**. With magnetic gaskets, it is important that the box not be sloped down and toward the front.

The door gets the most abuse of any part of the box because it is opened and closed so many times. It also may have a lot of weight in it due to food storage, so it must have strong hinges. Many refrigerator doors have bearing surfaces

Figure 45.90 If the box is not leveled per the manufacturer's instructions, the doors may not function correctly.

built into these hinges that may need changing after excessive use, **Figure 45.91**. Wires and water piping may be threaded through the door hinges to furnish power and water to circuits that may operate ice or water dispensers located in the door, **Figure 45.92**. These connections may wear with use and require service. As door gaskets age, they may need replacement. The various manufacturers use different gaskets that may require special tools that help in removing the old gaskets and installing the new ones. The manufacturer's directions should be followed when changing gaskets to ensure that the door is properly aligned and the gaskets fit properly.

MOVING THE APPLIANCE. A refrigerator or freezer is among the heaviest and most awkward types of equipment to move.

Figure 45.91 Hinges may have bearing surfaces to make door opening easier.

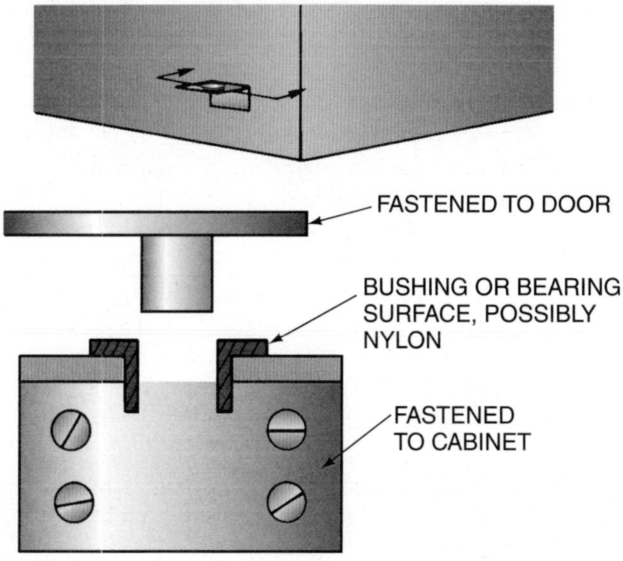

Figure 45.92 Power circuits and tubing for water may be connected to the door through the hinges. Courtesy White Consolidated Industries, Inc.

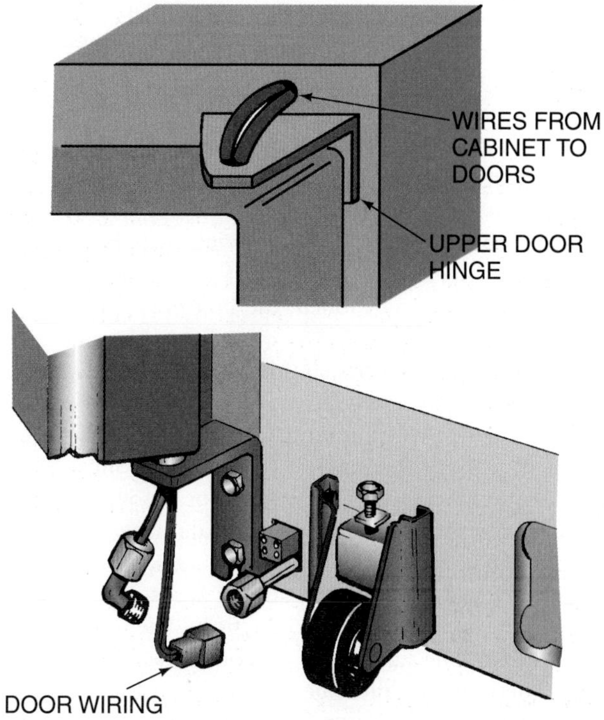

⚠SAFETY PRECAUTION: *Technicians must use great care when lifting and moving any heavy object and think out the safest method. They should use special tools and equipment at all times and wear appropriate back brace belts when moving and lifting objects,* **Figure 45.93**.• When it is necessary to lift, it should

Figure 45.93 Use a back brace to lift objects. Use legs, not the back; keep the back straight. Courtesy Wagner Products Corp.

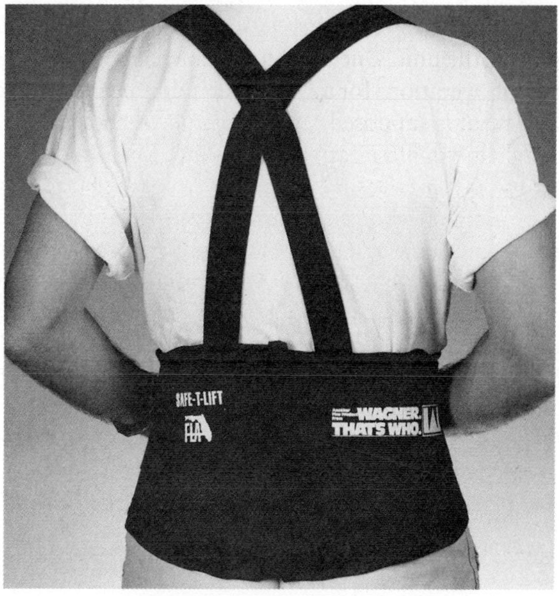

be done with the legs, not the back, and the back should be kept straight. Because freezers are frequently located in tight places and often need to be moved upstairs or downstairs, the surroundings or the freezer may be damaged when moving it. The correct moving devices must be used. **NOTE:** *Because of oil migration, never lay a freezer on its side.•*

MOVING UPRIGHT APPLIANCES. To move upright boxes, refrigerator hand trucks with wide belts are used, **Figure 45.94**. The trucks must be positioned in such a manner that they do not damage the condenser or let the door swing open, **Figure 45.95**. If possible, the box should be trucked from the side where the door closes. When repositioned, the box must be on a solid foundation.

Freezers or refrigerators should not be moved with food in them, although they may be moved far enough away from the wall for service without removing the food if this is done carefully. If the box is located on a floor that must be protected, the technician is responsible for knowing how to accomplish the move without causing damage to the floor or freezer. If the box does not have wheels, a small rug or mat inserted under the legs will slide it across a fine-finished floor without causing damage to the floor. The box should be slightly tilted to one side while someone slides the mat under the feet. Then the box must be tilted to the other side to get the mat under the other feet. This method can be used only where there is room to tilt the unit, **Figure 45.96**. If there is no tilt room, the box may have to be pulled out from the wall far enough for the mat to be placed under the front feet—and then it may be pulled out from the wall far enough for service.

Figure 45.94 Refrigerator hand trucks.

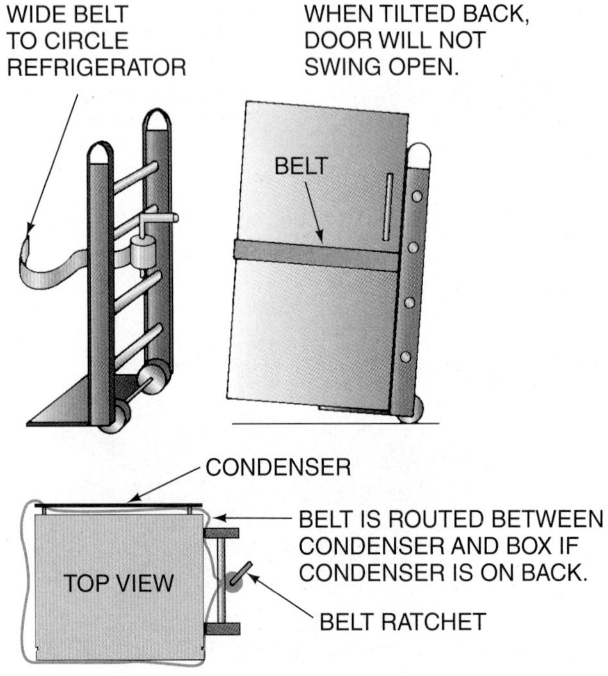

Figure 45.95 Pull the belt between the condenser and the box when the condenser is in the back.

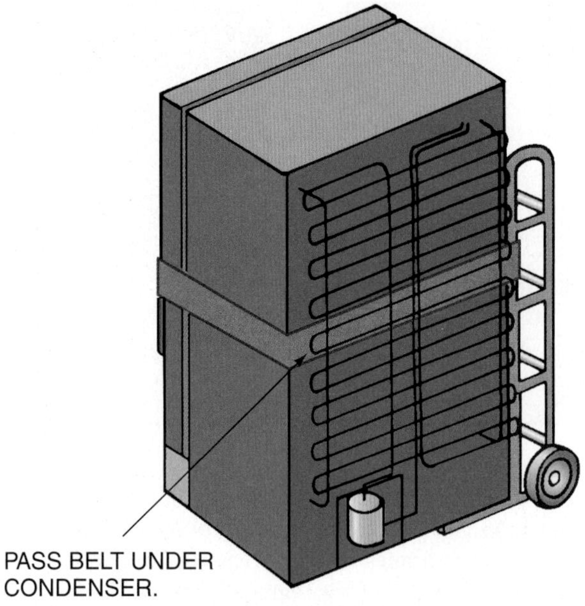

Figure 45.96 Enough room is available to tilt the box and place a mat under the feet.

The front feet may be screwed upward one at a time and a mat placed under them—and then screwed back out if no side or backward tilt is possible, **Figure 45.97**. Care should be used when placing a mat under the feet. **SAFETY PRECAUTION:** *Avoid putting hands under the box. A screwdriver or stick may be used to slide the mat under the feet,* **Figure 45.98.•**

Figure 45.97 The feet may be screwed upward, one at a time, to work the mat under the feet.

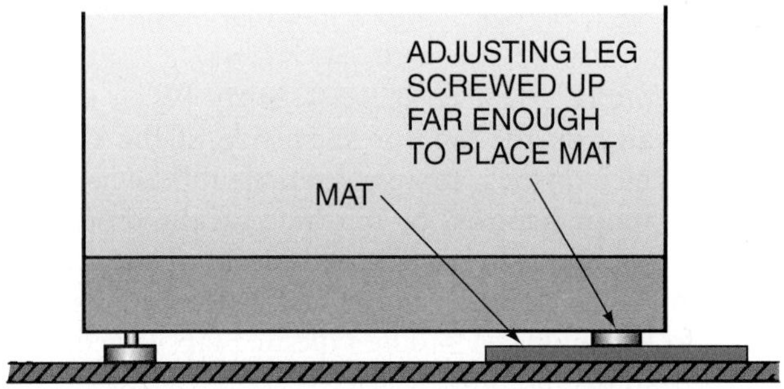

Figure 45.98 Use a small stick or long screwdriver, not hands, to place the mat.

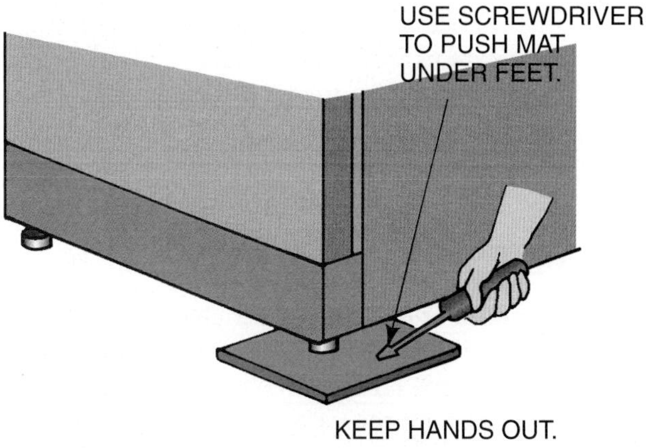

MOVING CHEST-TYPE FREEZERS. Chest-type freezers are more difficult to move than upright refrigerators or freezers. They may be larger, take up more floor space, and contain more food. Two-wheel refrigerator hand trucks will not work as well on chest freezers because they are longer than they are tall, **Figure 45.99**. When a chest freezer must be moved out from the wall for service, the same guidelines apply as for an upright refrigerator or freezer. By carefully using a pry bar resting on a piece of plywood, a mat can be placed under the feet, **Figure 45.100**. Two four-wheel dollies, **Figure 45.101**, may be used when the chest freezer is to be moved from one room to another. The freezer must carefully be raised to the level of the dollies. This requires at least two people and caution. Chest freezers for domestic use are sized so that they will fit through standard doors, but care should be used when moving the dolly wheels over the door's threshold.

Figure 45.99 Refrigerator hand trucks will not work on a chest freezer.

GAUGE CONNECTIONS

Domestic refrigerated appliances do not have factory-installed gauge ports like commercial refrigeration or air-conditioning systems. The system is hermetically sealed at the factory, and gauges may never have to be installed on it. When a system has been charged correctly, it should never have to be adjusted, as long as the appliance remains leak-free.

Leaks, field analysis requiring pressure readings, and field repair of components are the only reasons for installing gauges. Many service technicians have a tendency to routinely install gauges. This is poor practice, especially when dealing with domestic refrigeration appliances, since many have refrigerant charges amounting to only a few ounces. Taking a high-pressure reading from the liquid line on a domestic

Figure 45.100 A mat can be placed under a chest freezer by using a pry bar.

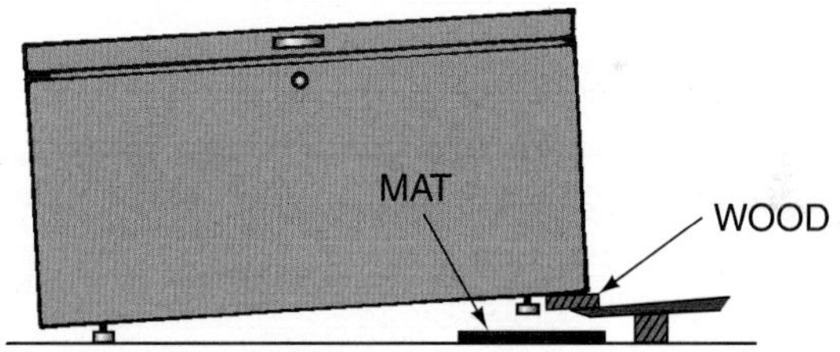

Figure 45.101 The use of two dollies with four wheels each is a good choice for moving a full chest freezer.

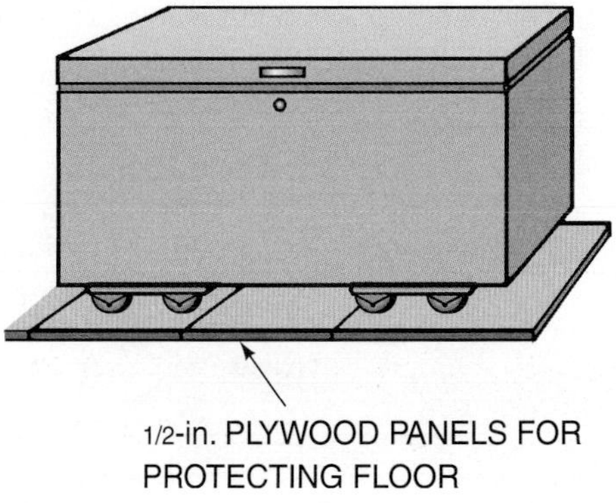

appliance can cause enough refrigerant to leave the system and enter the gauge hoses to adversely affect the operation of the unit. Many systems start with the correct charge, but as a result of a technician taking pressure readings, operate with a deficiency of refrigerant, causing system problems. Gauges should be installed on the system only as a last resort, and when installed, great care should be taken to ensure that the system remains free of contaminants. Some methods for determining system problems without the use of gauges are discussed in the next section, "Low Refrigerant Charge."

All gauge manifolds and gauge hoses must be leak-free, clean, and free from contaminants. It is good practice for a technician to have a set of gauges for each type of refrigerant commonly encountered. The gauge manifold should be stored with pressure in the hoses to reduce the possibility of manifold contamination. It may be desirable to remove the Schrader valve depressors furnished in the gauge hoses and use special adapters for depressing Schrader valve stems, **Figure 45.102**. This provides unobstructed gauge hoses for quick evacuation of a wet or damp system. Low-loss fittings must also be used on gauge hoses.

Some manufacturers provide a service port arrangement in which an attachment may be fastened to the compressor for taking gauge readings, **Figure 45.103**. Others do not furnish any service ports, so field service ports may be installed in the field in the form of line tap valves, **Figure 45.44**, following the manufacturer's instructions. Always, the correct valve size based on the line size should be used. **NOTE:** *If there is a chance that the system pressure may be in a vacuum, either purge the gauge hoses with clean refrigerant or shut the unit off and let the pressures equalize before installing a low-side line tap valve; otherwise, atmosphere will enter the system.•* When installing a line tap valve on the high-pressure side of the system, it is best to install it on the compressor process tube, where it may be soldered on or removed by pinching off the process tube between the valve and compressor housing, **Figure 45.104**. All line tap valves left in the active part of the system must be soldered to the tubing. A saddle-type line tap valve must be removed from the unit after service is complete.

If repairs are made to components in the refrigerant cycle or a refrigerant charge is lost completely, it is best to use the process tubes on the compressor for pressure readings. Fittings with Schrader valves may be soldered to the process tubes for the service work and evacuation, **Figure 45.105**. These tubes may be capped, or they may be pinched off using a special pinch-off tool, **Figure 45.106**, and soldered shut. A special pinch-off tool is necessary because refrigerant lines cannot be pinched off with general purpose pliers.

Figure 45.102 Remove the Schrader valve depressors from gauge hoses and use the special fitting on the right to depress the valve cores. Photo by Bill Johnson

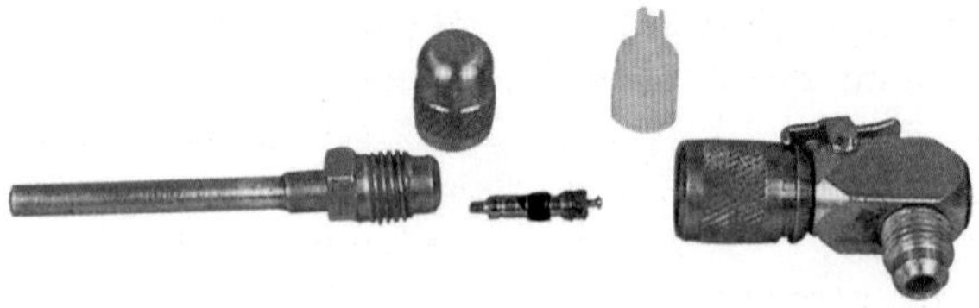

Figure 45.103 A service valve assembly. Photo by Bill Johnson

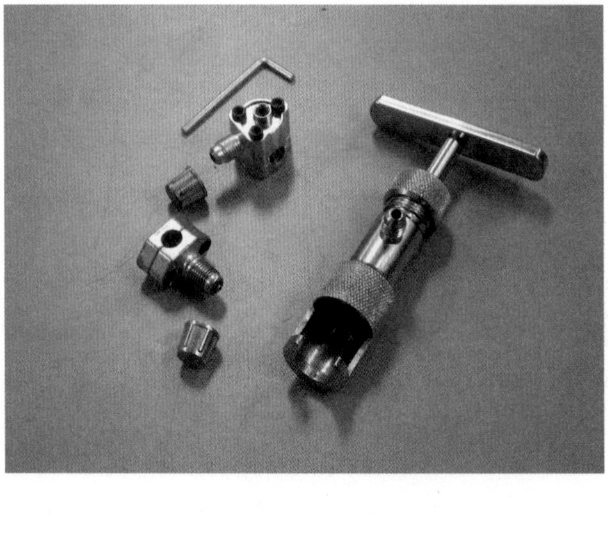

LOW REFRIGERANT CHARGE

If a refrigeration unit had the correct charge when it left the factory, it will maintain that charge until a leak develops. If a unit does have a low refrigerant charge, every effort should be made to determine the cause. A low refrigerant charge results in reduced appliance capacity, increased compressor run time, warmer-than-normal box temperatures, low operating pressures, high evaporator superheat, and low condenser subcooling.

Figure 45.104 Locate the high-pressure line tap valve on the hot gas line, not the liquid line. When finished, the line can be pinched off and soldered shut.

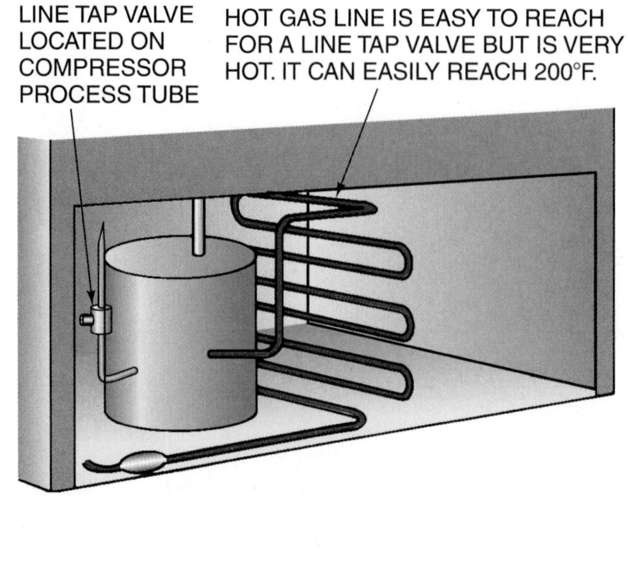

Figure 45.105 Fittings may be soldered on the process tube.

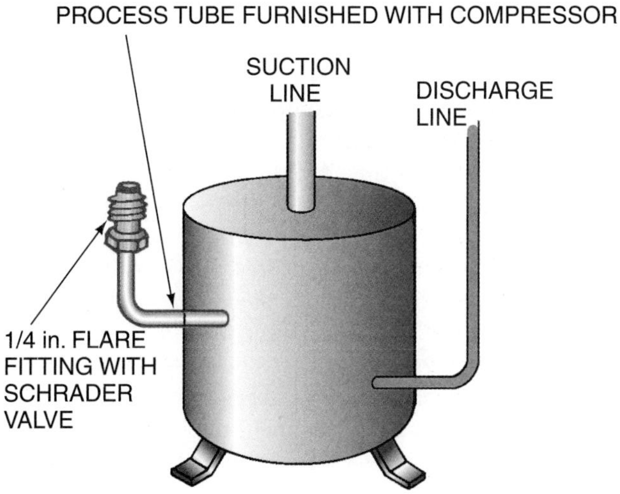

Figure 45.106 A pinch-off tool. Photo by Bill Johnson

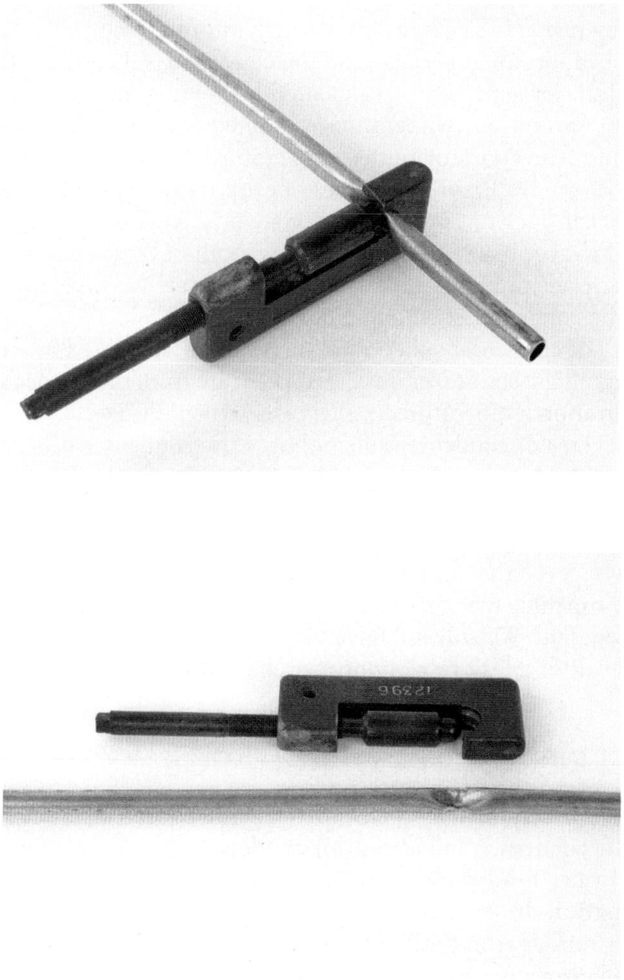

One service technique that many experienced service technicians use to determine whether the charge is approximately correct in a system (before connecting gauges) requires only that the unit be stopped and restarted. The unit is shut off and the pressures allowed to equalize, which takes about 5 min. Before the compressor is restarted, the technician places a hand on the suction line where it enters the compressor in such a manner that the line temperature may be sensed by touch. The compressor is then started. If the suction line gets cold for a short time, the chances are the refrigerant charge is correct. This occurs because a small amount of liquid is pulled from the evaporator into the suction line and the line gets cold for a short period of time. In addition, inspecting the evaporator coil can give an indication of a proper system charge. If the evaporator coil has a full frost on it, the system is very likely not undercharged. However, if the coil is only partially frosted, at the inlet of the evaporator, an undercharge should be suspected. Checking the amperage draw of the compressor, **Figure 45.107**, can also provide useful information regarding the charge. If the amperage draw of the compressor is very low as compared to the amperage rating on the compressor, an undercharge is possible.

All refrigerators currently being manufactured in the United States use R-134a. If a technician suspects a low charge, even after the items mentioned before have been checked, the compressor should be stopped and uncontaminated gauges installed on the system. It is good practice to wait 5 min before installing the gauges. This will allow the unit pressures to equalize and lower the possibility of having the low side of the system in a vacuum, which may be the case as long as the compressor is running. The compressor may then be restarted. If the system low-pressure gauge reads in a vacuum (below atmosphere) for a period of time, about 15 min, the unit is probably low on refrigerant or the capillary tube may be restricted. (Note that it is not uncommon for a refrigerator to operate in a vacuum for a short period of time after start-up.)

Figure 45.107 Taking a current reading at the compressor. Photo by Eugene Silberstein

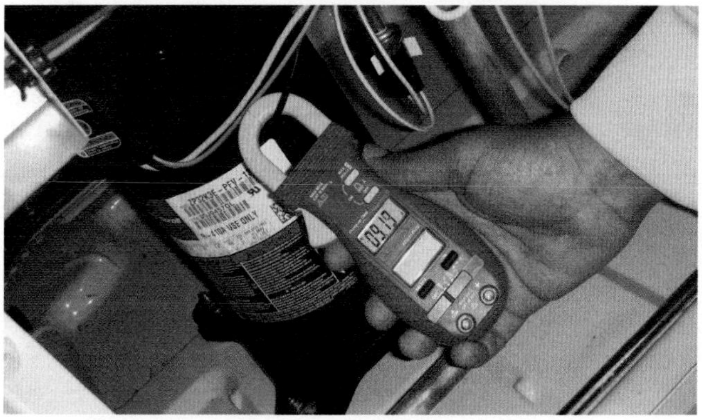

A small amount of refrigerant may be added to the system and the pressures observed again. **SAFETY PRECAUTION:** *High-pressure readings should be taken when adding refrigerant because a restricted capillary tube will cause high head pressure readings if refrigerant is added. This problem cannot be determined from a low-pressure reading only. If a high-pressure reading is not possible, attach a clamp-on ammeter to the compressor common wire and do not allow the compressor amperage to rise above the running-load amperage (RLA) rating of the compressor. If it does, shut it off.* •

4R *Manufacturers recommend that a low charge be fixed by finding the leak, removing and recovering the charge, and repairing the leak. A measured charge may then be transferred into the system.* 4R Some experienced service technicians may successfully add a partial charge by the frost-line method. This method is used to add refrigerant while the unit is operating, and it works as follows. A point on the suction line leaving the refrigerated box is located where the frost line may be observed, possibly where the suction line leaves the back of the box, **Figure 45.108**. Refrigerant is added very slowly by opening and closing the gauge manifold low-side valve until frost appears at this point. Then no more is added. The suction pressure may be 10 to 20 psig at this time and should reduce to about 2 to 5 psig just before the refrigerator thermostat shuts the compressor off. Using the frost-line method of charging a domestic refrigerator is a slow, tedious process that is not recommended unless just topping off a charge. The recommended method is to start from a deep vacuum and measure the charge into the system using either a charging cylinder or accurate scales, because the charge is critical to about ¼ ounce. 4R *If the frost line creeps toward the compressor as the box temperature reduces, refrigerant may be recovered slowly through the low side until the frost line is correct.* 4R

The typical operating conditions for the low-pressure side are fairly straightforward. The conditions are based on the coldest coil, the freezer coil. The typical low temperature for the freezer is 0°F. The refrigerant in the evaporator would typically be 12°F colder than the food temperature, so the refrigerant would boil at about −12°F. The corresponding pressures would be about 1.1 psig for R-134a when the thermostat is ready to shut the compressor off. If the box temperature were set to a lower point, the pressures would move downward. When it is time for the compressor to start, the pressures would be higher. Typically, the temperatures may fluctuate between 0°F and +5°F with corresponding pressures. **NOTE:** *Many units will operate in a vacuum for a short period of time after start-up. This condition will last until the refrigerant feeds through the capillary tube from the condenser and the system charge is in balance. This is particularly true if a unit is charged on the high-pressure side after evacuation.* •

Figure 45.108 Checking the refrigerator charge using the frost-line method.

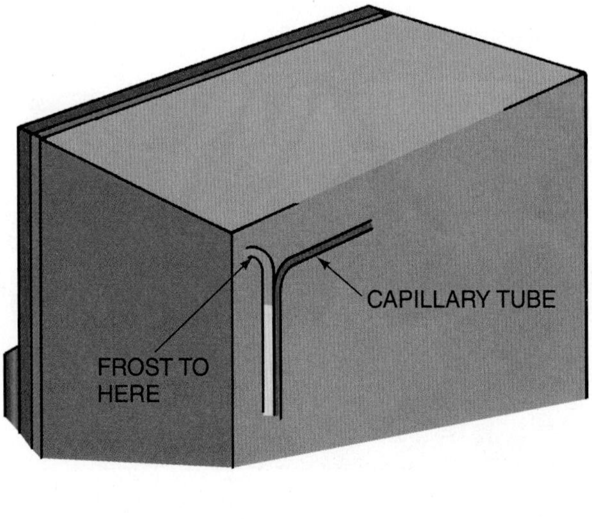

One sure-fire way to determine if the system is operating with less than a full charge of refrigerant is to evaluate the evaporator superheat and condenser subcooling of the unit. If a system is undercharged, due to either a leak or improper charging, the evaporator superheat will be high and the condenser subcooling will be low. Consider an R-134a domestic refrigerator that is operating with a low-side pressure of 9.7 in. Hg and a high-side pressure of 104.4 psig. If the evaporator outlet temperature is 10°F and the condenser outlet temperature is 85°F, we have enough information to calculate the evaporator superheat and the condenser subcooling. The evaporator superheat is calculated by subtracting the evaporator saturation temperature, obtained from the T/P chart, from the evaporator outlet temperature. In this case, we get 10°F − (−30°F), or 40 degrees of evaporator superheat. This by itself does not tell us if the system is short of refrigerant or if the capillary tube is clogged, as both of these conditions will result in higher-than-normal superheat. We take a look at the condenser subcooling next.

The condenser subcooling superheat is calculated by subtracting the condenser outlet temperature from the condenser saturation temperature. In this case, we get (90°F − 85°F), or 5 degrees of condenser subcooling. Low condenser subcooling indicates that there is a deficiency of refrigerant on the high side of the system. Since this system has high evaporator superheat, which indicates a deficiency of refrigerant on the low side of the system, and low condenser subcooling, we can conclude that the system is operating with a deficiency of refrigerant. We will still have to determine if this deficiency is a result of a leak or of improper charging.

REFRIGERANT OVERCHARGE

The refrigerator may have either a natural-draft or induced-draft condenser. Induced-draft condensers are more efficient and their head pressure will typically be lower than that of a natural-draft condenser. If too much refrigerant is added to a refrigerator, the head pressure will be too high. Typical head pressures for static condensers should correspond to

a condensing temperature that is 30°F to 35°F higher than room temperature at design operating conditions. Refrigerators that use mechanical, induced-draft condensers typically would condense refrigerant at a temperature that is about 20°F higher than room temperature.

For example, if the room temperature at the floor level is 70°F, the head pressure on a unit with a static condenser should be between 124.2 and 135 psig for R-134a, **Figure 45.109.** (Refer to the temperature/pressure chart in Unit 3, Figure 3.13.) Head pressure higher than 136 psig is too high. **NOTE:** *Be sure to check the load on the refrigerated space before drawing any final conclusions. Head pressures will be higher during a hot pulldown of the refrigerated space,* **Figure 45.110.**•

Figure 45.109 Typical head pressure for an R-134a appliance.

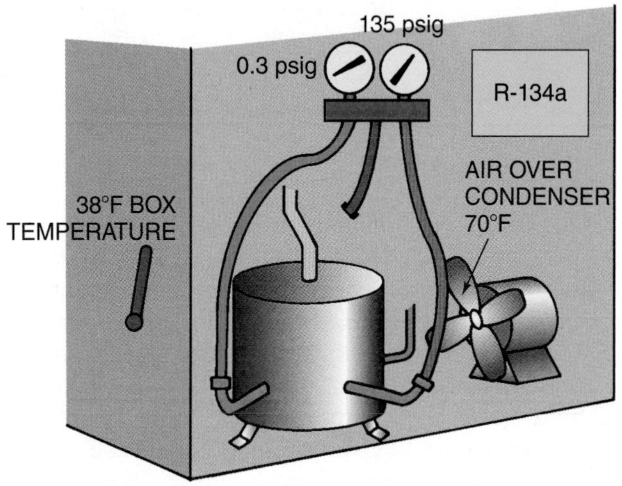

Figure 45.110 Head pressure under an abnormally high load.

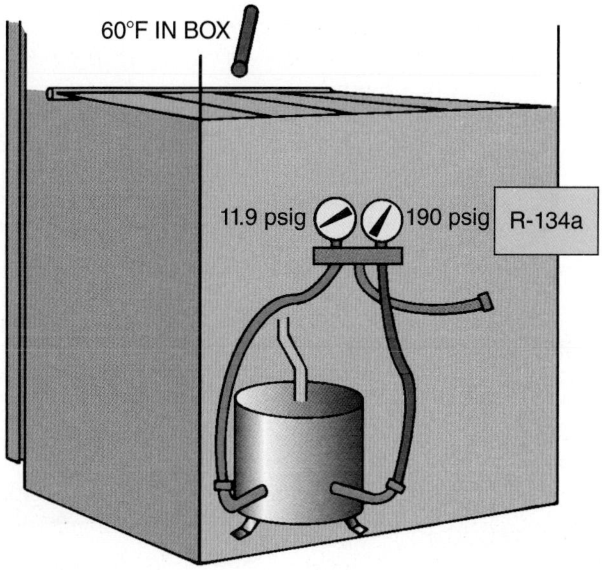

If the compressor is sweating around the suction line, there is too much refrigerant. The suction pressure will also be too high with an overcharge of refrigerant. Dirty condenser coils, defective condenser fan motors, and box lights that fail to turn off can also be the cause for high operating pressures, so it is important to investigate these as well.

A technician can determine if the system is operating with a refrigeration overcharge by evaluating the evaporator superheat and condenser subcooling of the unit, just as was done in the case of the refrigerant undercharge. If a system is overcharged, the evaporator superheat will be low and the condenser subcooling will be high. Consider an R-134a domestic refrigerator that is operating with a low-side pressure of 22 psig and a high-side pressure of 171 psig. If the evaporator outlet temperature is 28°F and the condenser outlet temperature is 80°F, we have enough information to calculate the evaporator superheat and the condenser subcooling. The evaporator superheat is calculated by subtracting the evaporator saturation temperature from the evaporator outlet temperature. In this case, we get (28°F − 25°F), or 3 degrees of evaporator superheat. Low superheat is an indication that there is too much refrigerant in the low side of the system. Since low evaporator superheat indicates that the evaporator is not absorbing heat fast enough, this condition can also be caused by a faulty defrost system or a faulty evaporator fan. In either case, the amount of heat being transferred from the air in the box to the refrigerant would be reduced, causing the low superheat condition. By evaluating the condenser subcooling, the cause for system malfunction can be narrowed down.

Condenser subcooling is calculated by subtracting the condenser outlet temperature from the condenser saturation temperature. In this case, we get (120°F − 80°F), or 40 degrees of condenser subcooling. High condenser subcooling indicates that there is an excess of refrigerant on the high side of the system. Since this system has high condenser subcooling, which indicates an excess of refrigerant on the high side of the system, and low evaporator superheat, we can conclude that the system is operating with an excess of refrigerant.

FINDING AND REPAIRING REFRIGERANT LEAKS

Because domestic refrigerators and freezers typically contain only a small amount of refrigerant, even a very small amount of refrigerant lost from a household refrigerator will affect the performance of the appliance. In some instances, a low refrigerant charge that occurred in the factory may be discovered when the refrigerator is put into service, since it is very likely that the appliance will not refrigerate properly from the start. If the appliance does function when it is initially put into service, it is very unlikely that the leak occurred at the factory. In any event, refrigerant leaks should be located and repaired for both environmental and system operational reasons.

REFRIGERANT LEAKS. When the box has run for some period of time and a leak occurs, it may be hard to locate in the field. Very small leaks may be found only with the best leak-detection equipment, such as electronic leak detectors, **Figure 45.111**. Many technicians prefer to move the unit to the shop and provide the customer with a temporary replacement. They can then repair the defective unit at their own pace when there is no food in it. However, removing the appliance and providing a loaner refrigerator involves a great deal of appliance transporting. A well-trained technician with a properly stocked service vehicle should be able to perform the required repairs on site.

In the event of a leak, there might not be enough refrigerant in the system to effectively leak check the unit. In this case, the system pressure must be increased by using nitrogen, **Figure 45.112**. **SAFETY PRECAUTION:** *Do not raise the pressure above the manufacturer's specified low-side working pressures. This is usually 150 psig.* **SAFETY PRECAUTION:** *Never operate the compressor on an appliance that is charged with nitrogen!* Once the leak has been located and repaired, it is important to properly leak check, evacuate, and recharge the system. *The EPA allows that a trace of system refrigerant be mixed with nitrogen for leak detection purposes. This nitrogen/refrigerant mixture can be released to the atmosphere as stated in Section 608 of the Clean Air Act of 1990.*

EVAPORATOR LEAKS. The use of a sharp tool when defrosting the unit manually may cause a leak in the evaporator. When this occurs in an aluminum evaporator, it may be repaired. Leaks in aluminum may be repaired with special solders. But soldering a puncture leak may be difficult because of location and the contraction and expansion of the evaporator. Leaks may also be repaired with the proper epoxy. Special epoxy products are available that are compatible with refrigerants.

After cleaning the surface according to the epoxy manufacturer's directions, the epoxy is applied to the hole while the unit is in a slight vacuum, about 5 in. Hg. This will pull a small amount of epoxy into the hole and form a mushroom-shaped mound on the inside of the pipe, **Figure 45.113**. The

Figure 45.111 An electronic leak detector. Courtesy Robinair SPX Corp.

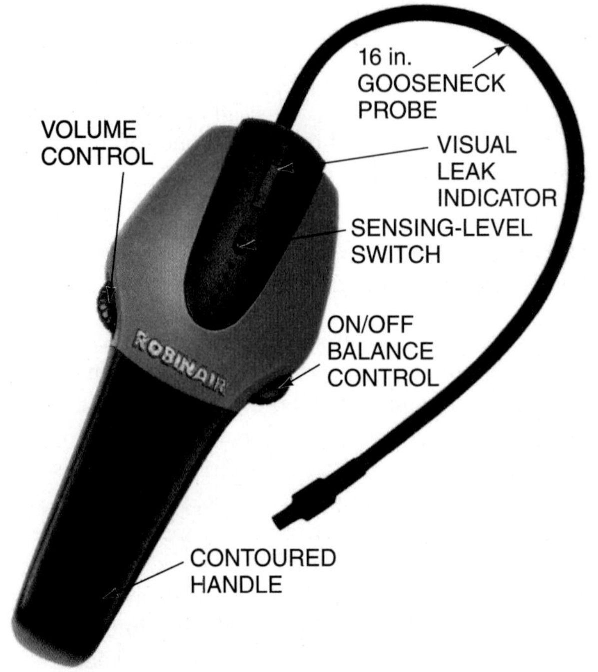

Figure 45.112 Nitrogen may be used to create more pressure for leak checking.

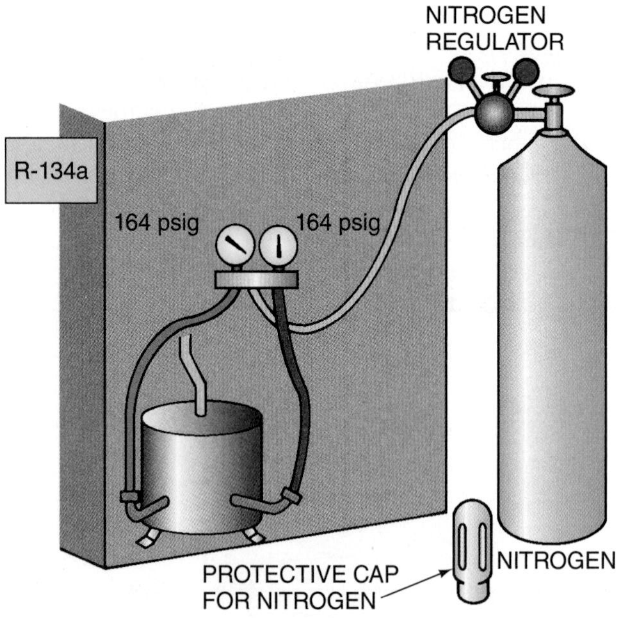

Figure 45.113 A slight vacuum may be used to pull a small amount of epoxy into the puncture.

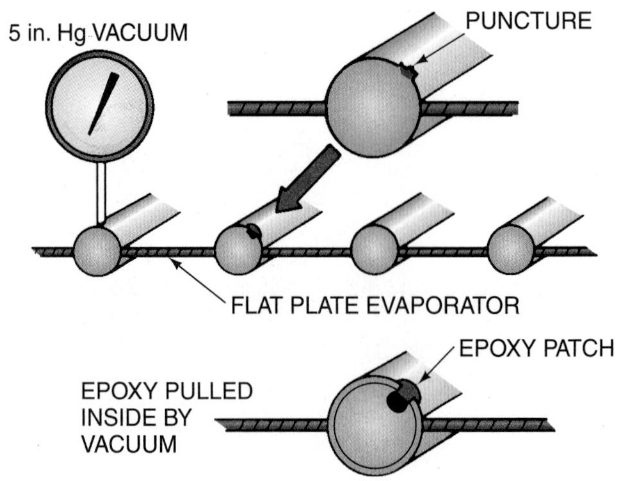

Figure 45.114 The pressure inside may push the patch off the puncture under some circumstances. This may occur during storage when the low-side pressure is high.

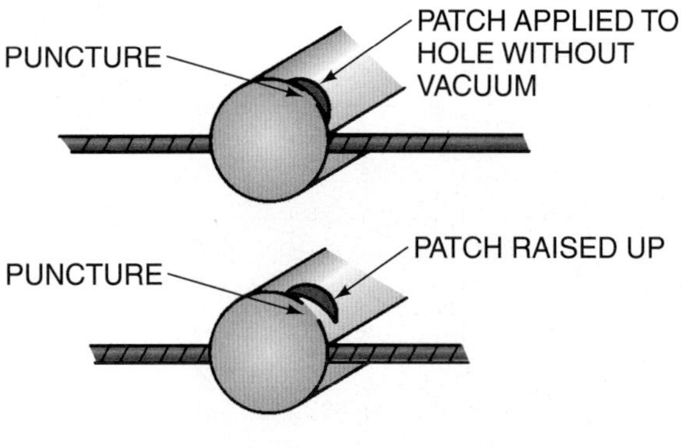

mound will prevent the patch from being pushed out when the refrigerator is unplugged and the low-side pressure rises to the pressure corresponding to the room temperature. However, the pressure inside a refrigerated box located outside where it may reach 100°F may rise to 124 psig for R-134a, which may be enough to push a plain patch off the hole, **Figure 45.114**. Care also should be taken not to allow too much epoxy to be drawn into the hole or a restriction may occur. Another method of fixing the leak may be to use a short sheet metal (self-tapping) screw and epoxy in the hole. The vessel must have enough room inside for the screw for this to work. After cleaning the puncture and the screw according to the epoxy manufacturer's recommendations, the epoxy is applied to the hole and the screw is tightened so that the head is snug against the hole to hold the epoxy under high pressure, **Figure 45.115**.

CONDENSER LEAKS. Refrigerator condensers are made of steel. Leaks usually do not occur in the middle of the tube, but at the end where connections are made or where a tube has vibrated against the cabinet, causing a hole. This part of the system operates under the high-pressure condition, and a small leak will lose refrigerant faster than the same size leak on the low-pressure side, **Figure 45.116**. Wherever the leak occurs in steel tubing, the best repair is solder. The correct solder must be used, one compatible with steel; usually it has a high silver content. When flux is used, it must be cleaned away from the connection after the repair is made. Always leak check after the repair is made. When the condenser is located under the refrigerator, it is often hard to gain access for leak repair. The box may be tilted to the side or back to do the repair, **Figure 45.117**.

Figure 45.115 A screw may be used in some cases to hold the epoxy in the puncture hole.

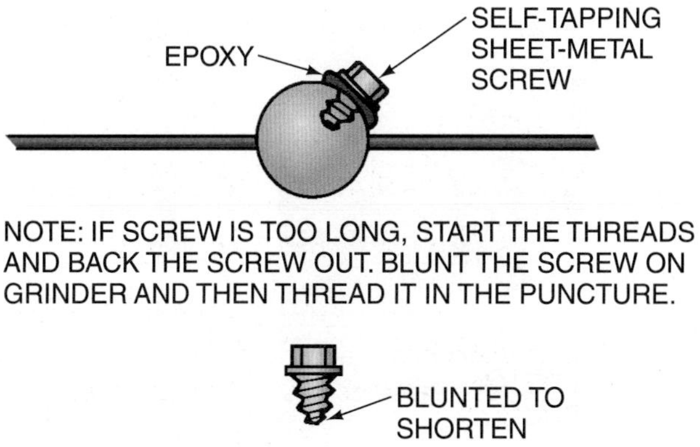

Figure 45.116 A small leak in the high-pressure side of the system will cause a greater refrigerant loss than a leak on the low-pressure side.

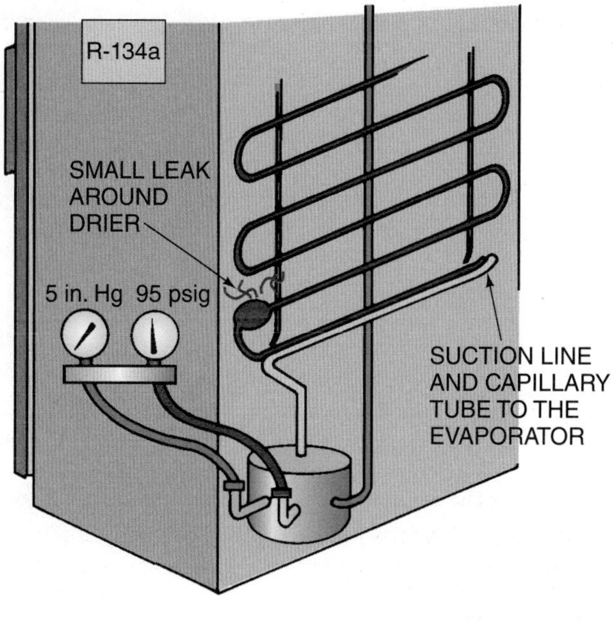

Some refrigerators and freezers are manufactured as hot-wall appliances, in which the condenser is embedded in the insulation of the box cabinet. When configured this way, locating a leak in the condenser is difficult at best, since major damage can be done to the appliance cabinet. If a hot-wall condenser is suspected of leaking, the following procedure will confirm if such is indeed the case:

1. Recover refrigerant from the system.
2. Disconnect the condenser piping from the filter drier.
3. Pinch and solder closed the open ends created in step 2.
4. Remove the discharge line connection from the compressor.
5. Seal the compressor's discharge port.
6. Connect an access port to the discharge line.
7. Using dry nitrogen, pressurize the condenser through this access port to a pressure of about 200 psig to 250 psig.
8. Monitor the pressure in the condenser. A drop in pressure indicates a leak in the condenser coil.

When a condenser leak is confirmed in a hot-wall refrigerator or freezer, it is recommended that the existing condenser be abandoned and replaced with a chimney condenser located

Figure 45.117 A refrigerator may be tilted to one side or to the back to service the condenser under the box.

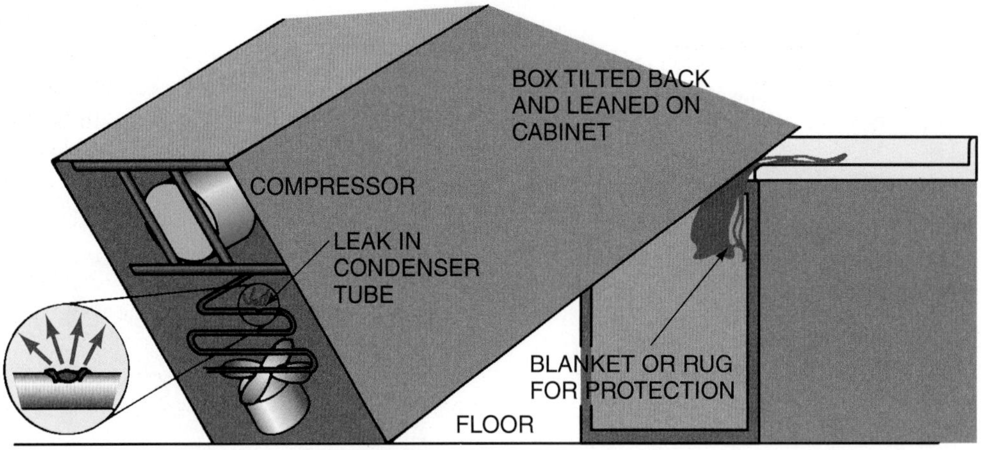

at the back of the appliance. **SERVICE NOTE:** *If a condenser replacement is needed, be sure that it is compatible with the existing appliance.•*

COMPRESSOR CHANGEOUT

4R*Compressors may be changed in a refrigerator by recovering the refrigerant charge and removing the old compressor.* 4R A new compressor should be at hand before the old one is removed. This is especially important for R-134a appliances, as the oils used in conjunction with this refrigerant are hygroscopic, meaning they attract moisture, so keeping the system open too long can contaminate the appliance. An exact replacement is the best choice but may not be available. A diagram of the tube connections and the mounting should be available to help connect it correctly. Remember, several lines, including suction, discharge, suction access, and discharge access, may all be the same size, **Figure 45.118**.

4R*The best way to remove the lines from the old compressor is to recover the charge and pinch the lines off close to the compressor,* **Figure 45.119**, *or cut them using a very small tubing cutter.* 4R If the old tubing is removed with a torch, the tubing ends should be cleaned using a file to remove excess solder, being careful not to allow filings to enter the system. **SAFETY PRECAUTION:** *Some of the lines may contain oil, which may flame up when the lines are separated with a torch. A fire extinguisher should always be at hand.•* The old tubes should be filed until they are clean and will slide inside the compressor fittings, **Figure 45.120**. Approved sand tape (sand tape with nonconducting abrasive) should be used to further clean the tubing ends. They must be perfectly clean with no dirty pits, **Figure 45.121**. Dirt trapped in pits will expand into the solder connection when heated, **Figure 45.122**, and cause leaks.

The new compressor should then be set in place in the compressor compartment to compare the connections on the box with those on the compressor. When certain that the

Figure 45.118 A compressor may have three lines coming from the shell. Photo by Bill Johnson

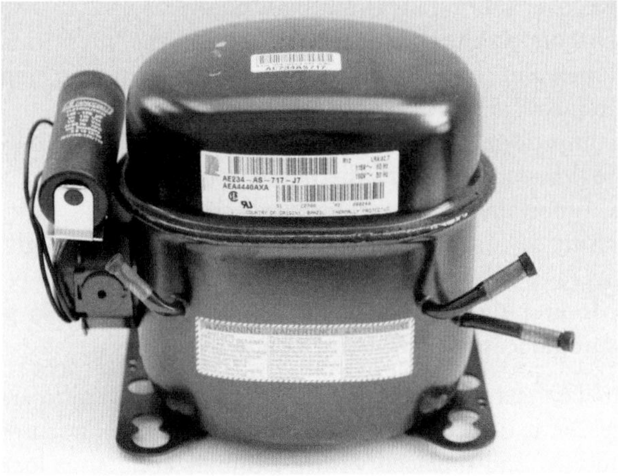

Figure 45.119 Pinch the old lines close to the compressor shell using side-cutting pliers.

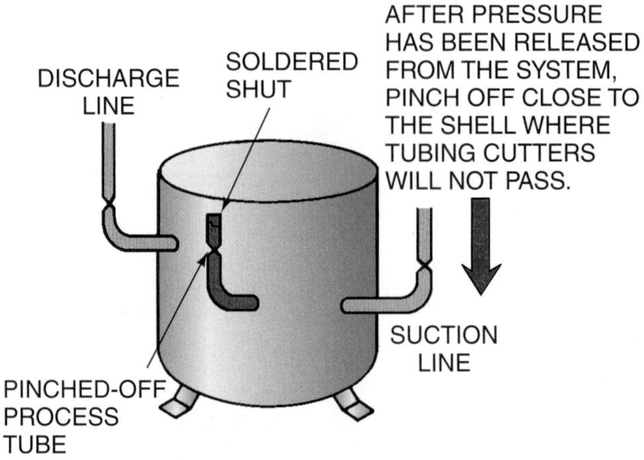

Figure 45.120 The old tubing must be cleaned of all old solder; a file may be needed.

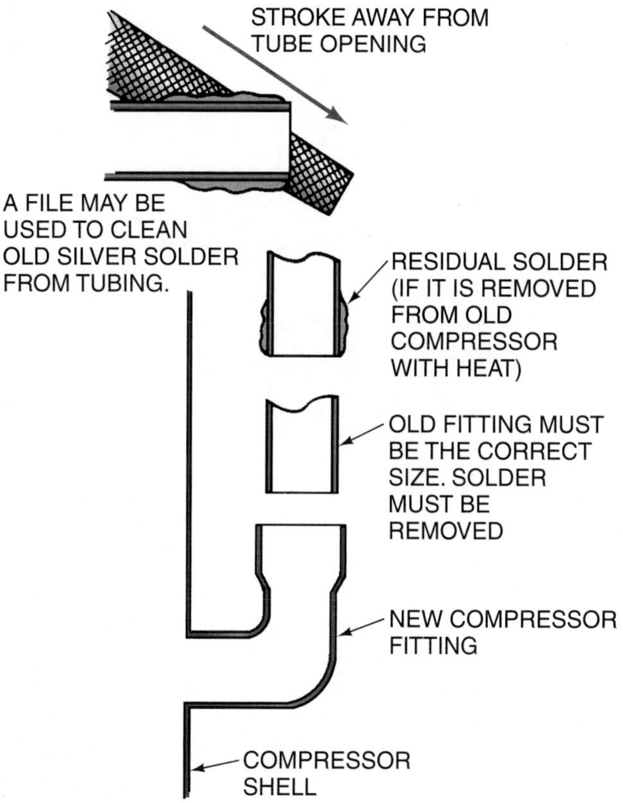

Figure 45.121 All pits in steel tubing must be cleaned.

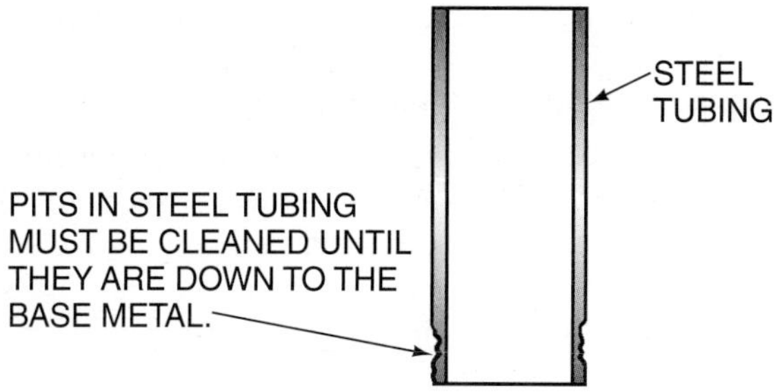

Figure 45.122 Dirt trapped in pits will expand when heated and cause leaks.

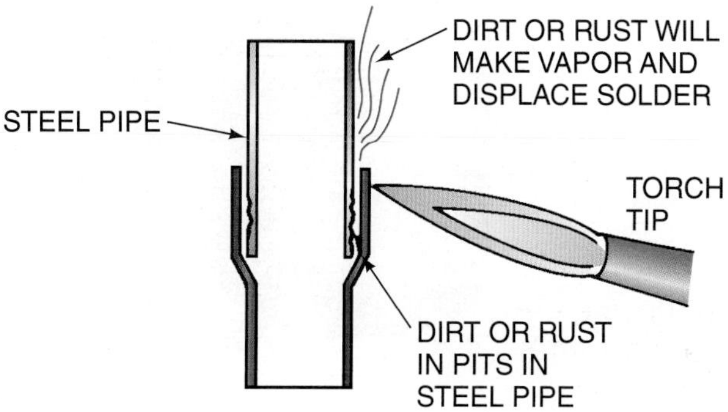

connections line up correctly, remove the new compressor from the compartment. Remove the plugs from the compressor lines and clean the tubing ends, **Figure 45.123**. **NOTE:** *Take care not to allow anything to enter the compressor lines.•* Set the new compressor in place and connect all lines. Flux may be applied if this is the last time the lines are to be connected before soldering. **SAFETY PRECAUTION:** *Solder the connections carefully, being particularly careful not to overheat the surrounding parts or cabinet. A heat shield may be used to prevent the heat of the torch flame from touching the surrounding components and cabinet.•* Use the minimum of heat recommended for the type of connection being made. While soldering the compressor to the lines, it is a good time to also solder process tubes to the compressor. Sometimes a "tee" fitting is soldered into the suction and discharge lines with a Schrader fitting, **Figure 45.124**.

Figure 45.123 Remove the plugs from the compressor lines and clean the tubing ends.

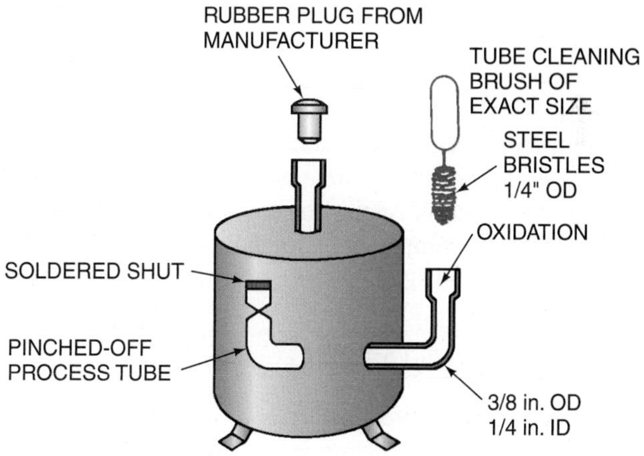

Figure 45.124 A "tee" fitting may be soldered in the suction and discharge lines for service connection while the system is open.

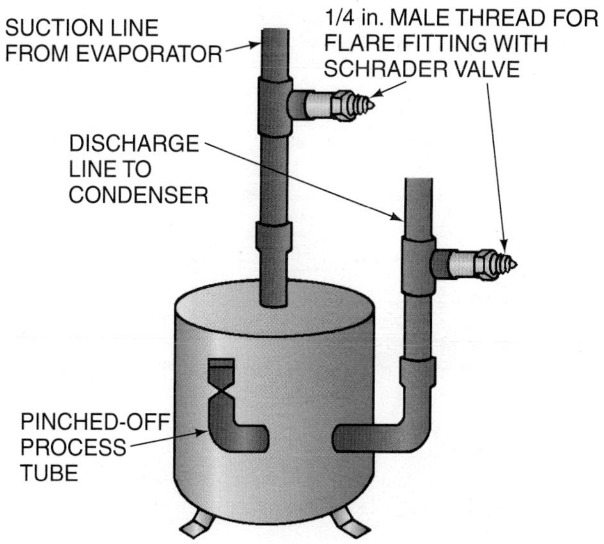

A filter drier should be added to a system open long enough to change a compressor. The refrigerator manufacturer may recommend using a liquid-line drier of the correct size. If an oversized liquid-line drier is used, additional refrigerant charge must be added. After the compressor is installed, it is recommended that the system first be swept with nitrogen and then the drier installed. This prevents contamination of the drier with whatever may be in the system. This sweep may be accomplished by cutting the liquid line or the suction line, wherever the drier is going to be installed. Connect the gauge manifold to the service ports and a cylinder of nitrogen. Allow vapor to flow first into the high side of the system. This will force pressure into the high side, then through the capillary tube, through the evaporator, and out the loose suction line (this example is for suction-line installation), **Figure 45.125**. The vapor will flow very slowly from the suction line because it must move through the capillary tube, but it will sweep the entire system, except the compressor, which is new.

Now open the whole system to the atmosphere by leaving the gauge valves open so that there will be no pressure buildup, **Figure 45.126**, and connect the drier. Close the gauge manifold valves immediately after completing the solder connection so that when the vapor in the system cools it will not shrink and draw air inside. When the new compressor is in place with process tubes, leak check the whole assembly at maximum low-side working pressure. Again, the low-side working pressure may be used as the upper limit for pressure testing. If a unit holds 150 psig of nitrogen pressure overnight, it is leak-free.

SYSTEM EVACUATION

When the system is proved leak-free, a vacuum may be pulled on the entire system with confidence. Unit 8, "System Evacuation," covers evacuation procedures. Briefly, removing the

Figure 45.125 Purging or sweeping a system.

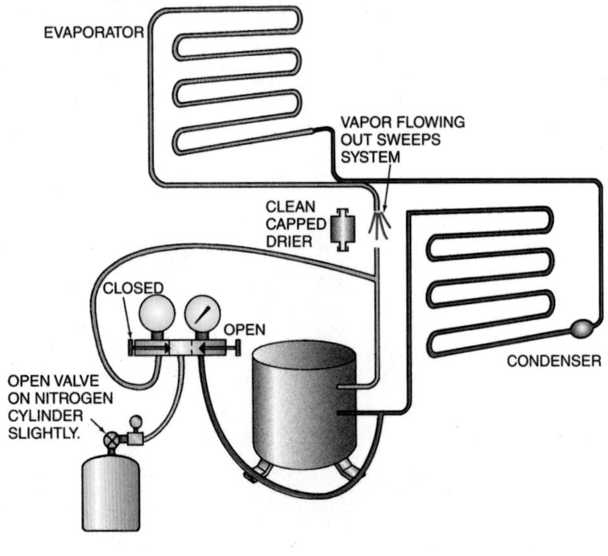

Figure 45.126 Be sure to open gauges before attempting to solder the drier in the suction line.

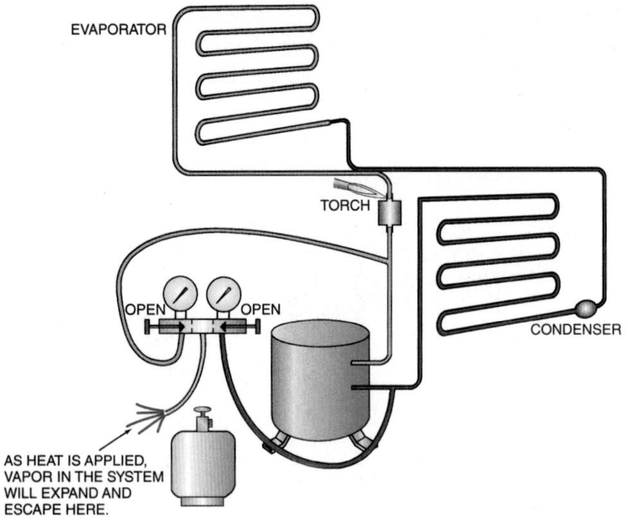

Schrader valve stem from the service stems and using gauge hoses without Schrader valve depressors will speed the vacuum. In triple evacuation, the first two vacuums may be performed and then the pressure in the system may be brought up to about 5 psig above atmospheric. The Schrader valve stems may then be installed along with Schrader valve depressors and a final evacuation performed. When a deep vacuum is achieved, the charging cylinder or accurate electronic scales may be attached and the measured charge allowed into the system.

Special evacuation procedures will be required when a system has had moisture pulled inside, because the moisture will move from the evaporator through the suction line to the compressor crankcase. When the compressor is restarted, moisture may be trapped under the oil, **Figure 45.127**. Although the triple evacuation will work to remove the moisture in a relatively short period of time, the following procedure can be used when the unit is in the shop and time is not a factor.

Install full-size gauge connections, such as "tees" in the suction and discharge lines so evacuation is from both

Figure 45.127 Moisture may be trapped under the oil in the compressor.

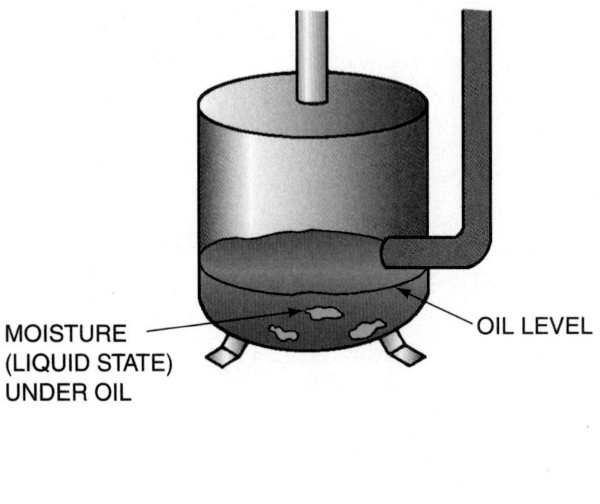

Figure 45.128 Full-size gauge connections must be installed to remove moisture.

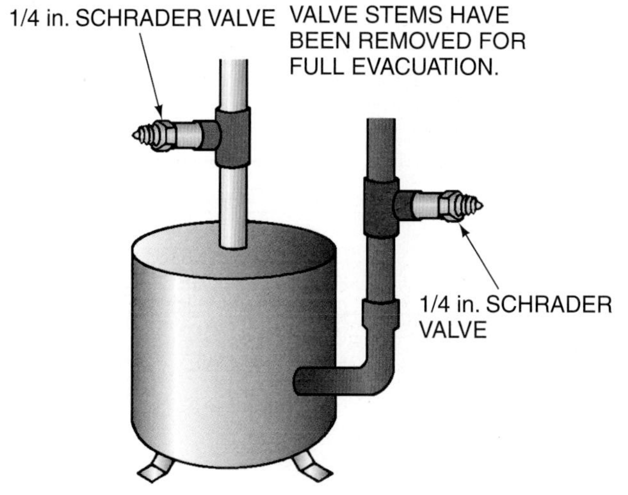

sides of the system, **Figure 45.128.** **NOTE:** *Make sure that there are no Schrader depressors in the gauge hoses or fittings on the compressor.*• Start a two-stage vacuum pump and apply heat to the compressor crankcase by placing a light bulb next to the compressor. Also place a small light bulb (about 60 W) in both the low- and medium-temperature compartments and partially shut the doors, **Figure 45.129.** **NOTE:** *Do not shut the doors all the way, or the heat will melt the plastic inside the refrigerator.*• Allow the vacuum pump to run for at least 8 hours. Break the vacuum using nitrogen

Figure 45.129 Placing light bulbs in a refrigerator to heat the refrigerant circuit but not overheat the plastic.

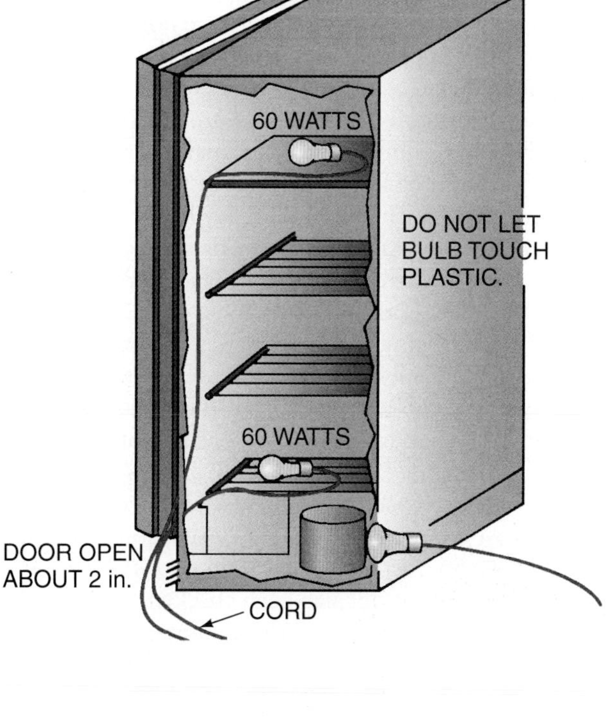

and pull another vacuum. This time, monitor it with an electronic vacuum gauge. When a deep vacuum has been achieved, such as 500 microns, break the vacuum with nitrogen to atmospheric pressure and remove the heat. Cut the suction line and install a suction-line filter drier. The liquid-line drier should also be replaced. Pull one more vacuum and charge the unit with a measured charge of refrigerant.

CAPILLARY TUBE REPAIR

Capillary tube repair may involve patching a leak in the tube because it has rubbed against some other component or the cabinet. You may need to change the drier strainer at the tube inlet, or repair may consist of clearing a partial restriction from the tube or even replacing the tube. Whatever the repair, the capillary tube must be handled with care because it is small and delicate. It can be easily pinched.

When a capillary tube must be cut for any reason, the following is recommended. Using a triangular file, cut a V-notch in the capillary tube, making certain not to cut into the bore of the tube, **Figure 45.130.** Move the file to the underside of the capillary tube and make another V-notch that lines up with the first, **Figure 45.131**, once again making certain not to cut into the bore. Hold the capillary tube with one hand on each side of the notch and gently bend the tube back and forth. The capillary tube will break at the notch and no filings will have entered the tube. **NOTE:** *The full dimension of the inside diameter of the tube must be maintained or it will cause a restriction. Proper care at this time cannot be overemphasized.*•

When it is necessary to solder a capillary tube into a fitting, such as into the end of a new strainer, do not apply flux or clean the tube all the way to the end. Allow the outside of the end of the tube to remain dirty because the solder will not flow over the end of the tube if it is not cleaned, **Figure 45.132.** A capillary tube that is broken in two may usually be repaired by cleaning both ends so that the inside dimension is maintained and then pieced together with tubing one size larger, **Figure 45.133.** When a capillary tube has a rub hole in it, it is recommended that the tube be cut in two at this point and repaired as a broken tube.

Figure 45.130 Using a triangular file, cut a V-notch in the capillary tube.

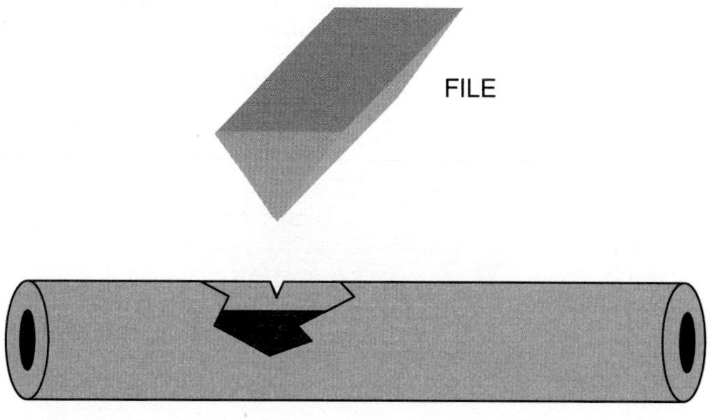

Figure 45.131 By placing the file under the notched capillary tube, another notch is made on the other side of the tube.

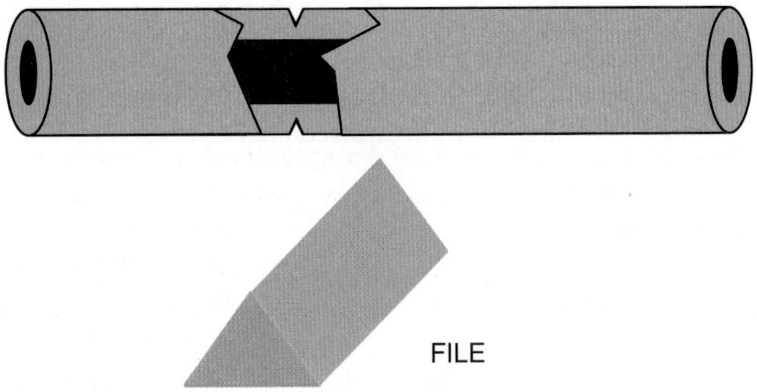

Figure 45.132 Application of flux to a capillary tube to prevent solder from entering the tube.

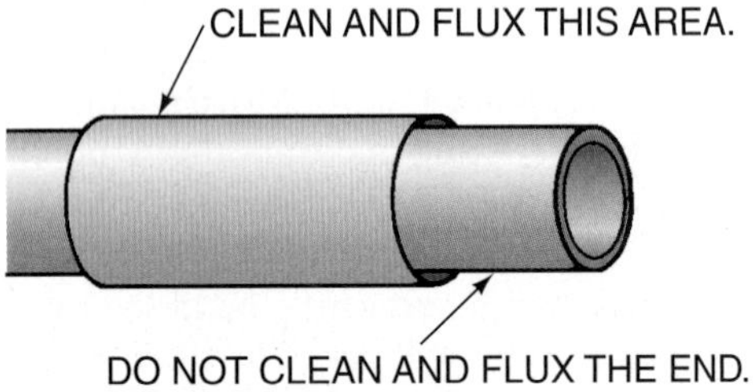

The capillary tube may be changed in some instances, but usually it is not economical because it is fastened to the suction line for a heat exchange between the capillary tube and the suction line. This connection can never be duplicated exactly, but a repair may come very close by insulating the suction line along with the capillary tube under the insulation, **Figure 45.134**.

Figure 45.133 Repairing a broken capillary tube.

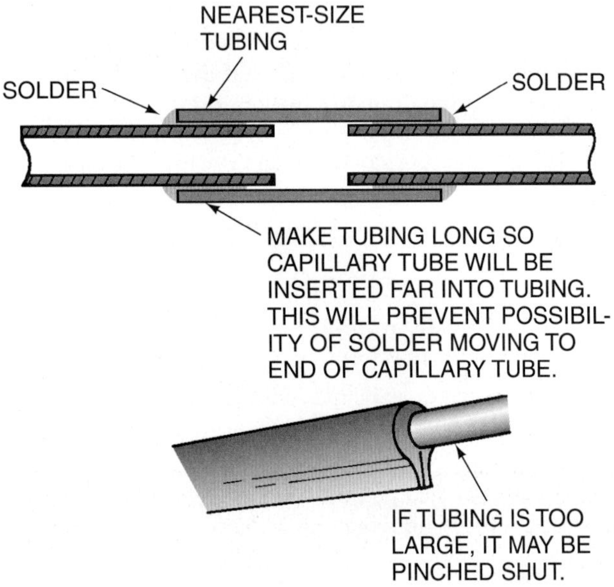

Figure 45.134 The capillary tube may pass close to the suction line under insulation to get some heat exchange.

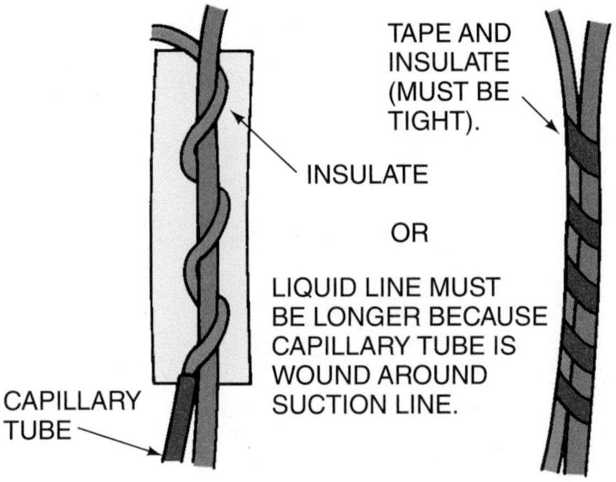

COMPRESSOR PUMPING CAPACITY

One of the most difficult problems to diagnose is a compressor that is pumping to partial capacity. The customer may complain that the unit is running all the time. It may still have enough capacity to maintain the food compartments at reasonable temperatures, but one of the first signs that the compartment temperature is not being maintained is that previously hard ice cream will be slightly soft. Liquids such as water or milk served from the fresh-food compartment also may not seem as cool. When the conditions of running all the time and not maintaining temperature are noticed, either the refrigerator has a false load or the compressor is not operating to capacity. The technician may have to decide which.

First, make sure all door gaskets are in good condition and that the doors shut tight. Then make sure there is no extra load, such as hot food being put in the refrigerator too often, **Figure 45.135**. The light bulb may be on all the time in one of the compartments. Make sure that the condenser has the proper airflow, with all baffles in place for forced draft, **Figure 45.136**, and not under a cabinet if the condenser is natural draft. Make sure that the unit is not in a location that is too hot for its capacity. Any temperature greater than 100°F will usually cause the unit to run all the time, but even so it should maintain conditions. A unit located outside, particularly where it is affected by the sun, will always be a problem because it is not designed to be outside. Another possibility is that the defrost contacts are welded closed, so the appliance is trying to defrost the evaporator and cool the box at the same time. When all of this has been checked and everything proves satisfactory, then suspect the compressor.

Figure 45.135 Extra load on a refrigerator may be caused by defective gaskets, hot food placed in the box too often, or a light bulb burning all the time.

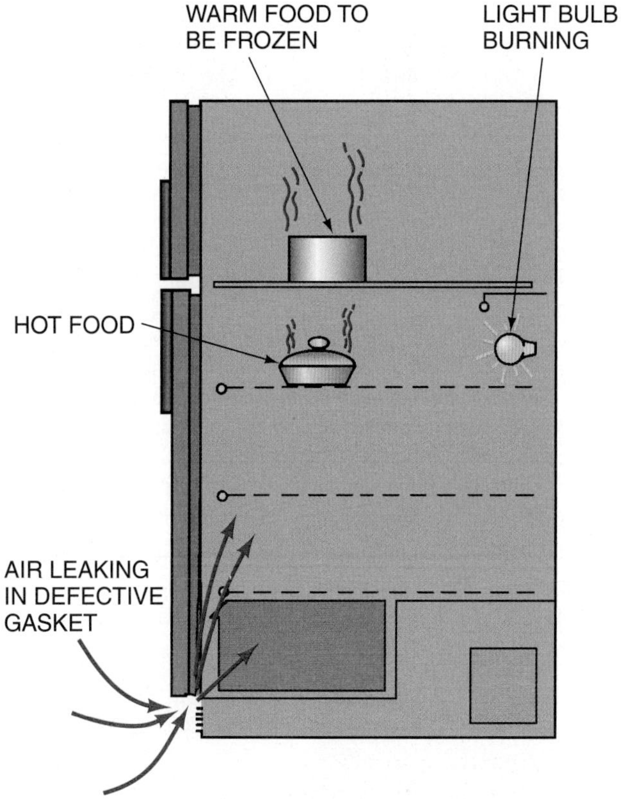

Figure 45.136 Make sure that all baffles are in place in the condenser area for correct airflow.

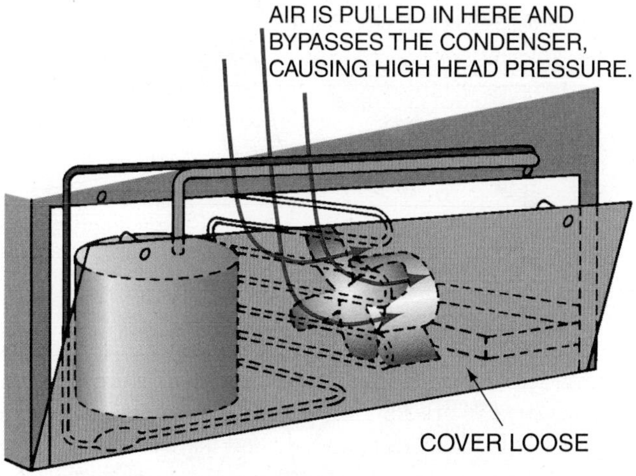

Loss of compressor pumping capacity can be caused by poor operating conditions. For example, if the box has been operating with an extremely high head pressure due to a dirty condenser or an inoperable condenser fan motor or if it has been operating in a very hot location for a long time, the compressor valves could be leaking due to valve wear. When this happens, the compressor does not pump the correct quantity of refrigerant gas and a loss of capacity occurs. Before even thinking about compressor capacity loss, always make sure that the condenser is clean and that both types of condensers have unobstructed airflow. The condenser must be able to dissipate the heat, or capacity loss is ensured. Induced-draft condensers must have all cardboard partitions in place, or air may recirculate across the condenser and cause overheating.

After operating conditions are checked, gauges should be fastened to the system to check the suction and discharge pressures. Do not forget to let the pressures equalize before attaching the gauges. If it is available, use the manufacturer's literature to check the performance, or call the distributor of the product or another technician who may have had considerable experience with the type of appliance. Declaring a compressor defective is a big decision, especially for a new technician, so it may often be best to get a second opinion from an experienced technician.

HVACR GOLDEN RULES

When making a service call to a residence:

- Make and keep firm appointments. Call ahead if you are delayed. The customer's time is valuable also.
- Keep customers informed if you must leave the job for parts or other reasons. Customers should not be upset when you inform them of your return schedule if it is reasonable.

Added Value to the Customer

Here are some simple, inexpensive procedures that may be included in the basic service call:

- Clean the condenser.
- Make sure that the condenser fan is turning freely (if applicable).
- Check the refrigerator light and replace it if needed.
- Make sure the evaporator coil has no ice buildup.
- Make sure the box is level.

45.11 SERVICE TECHNICIAN CALLS

SERVICE CALL 1

A customer calls and tells the dispatcher that a new refrigerator is sweating on the outside of the cabinet between the side-by-side doors. *The problem is a defective connection at the back of the refrigerator in a mullion heater circuit.*

Arriving at the job, the technician makes certain to park in the street, not in the customer's driveway. When the customer opens the door, the technician shakes hands and greets the customer by name. Explaining to the technician that the appliance is new, the customer is a little upset that it is not working as expected. The technician apologizes for the inconvenience and assures the homeowner that he would do whatever he could to remedy the problem.

Making his way to the refrigerator, the technician lays down a drop cloth on which he places his toolbox. He can see right away that a mullion heater is not heating. The house temperature and humidity are normal because the house has central air-conditioning. The technician dreads pulling the panels off to get to the heater between the doors if it is defective because this is a difficult, time-consuming job. In addition, the appliance is brand new and the technician does not want to void the warranty. A look at the diagram on the back of the refrigerator reveals a junction box at the back corner of the unit where the mullion heater wires are connected before running on to the front. He unplugs the refrigerator and locates the junction box and the heater wires. The connection seems loose, but the technician wants to be sure, so he takes the connection apart. An ohm check of the heater circuit proves the heater has a complete circuit. He reconnects the wires in a secure manner, plugs in and starts the refrigerator, and checks the power at the connection. Power is available from the neutral wire to the hot wire going to the heater. To make sure, the technician takes an amperage reading of the heat circuit and determines that current is flowing and feels confident the problem has been corrected.

The technician explains the situation to the customer and informs him that it is very possible that the wiring came loose during the shipping process. Before leaving the home, the technician explains some of the energy-saving features on the appliance.

SERVICE CALL 2

A homeowner who lives in a rural area calls. This customer had recently purchased a freezer and placed in it an entire butchered, quick-frozen steer. *The problem is that the condenser fan motor has failed, and the freezer compressor is running all the time. The food is still frozen hard, but the temperature is beginning to rise.*

Before arriving at the job, the technician verifies with his office that the appliance is new and still under warranty. When the customer answers the door, the technician introduces himself and presents the company's business card. The customer explains that the freezer is not working as expected. The technician apologizes for the inconvenience as the customer brings him over to the unit.

The technician notices right away that the condenser fan is not running. The thermometer in the box reads 15°F. No time can be wasted. After the unit is unplugged, the box is moved away from the wall to give access to the fan motor. The fan motor turns freely, so the technician removes one motor lead and performs an ohm check. The motor has an open circuit through the windings, **Figure 45.137**. The technician does not have a new motor with him and he must go get one. He asks the customer whether a floor fan is available that may be used to cool the condenser until he can obtain a new fan motor. She has one, so the technician places the fan where it can blow over the condenser coil and plugs it in, **Figure 45.138**. The fan is allowed to operate for a while to remove some of the excess heat from the condenser. Then the freezer is plugged in, and the compressor starts. The technician leaves to get the new fan motor.

When the technician returns to the job about 3 hours later, the freezer is still running. The thermometer inside reads 1°F, so the freezer temperature is going down. Unplugging the freezer, the technician replaces the fan motor, reinstalls all panels with the proper fasteners, and plugs the freezer in. The compressor and the new condenser fan start. The technician pushes the freezer back against the wall and cleans up any scuff marks on the floor left by the appliance. The technician once again apologizes for the problem and thanks the customer for choosing his company for service.

Figure 45.137 A motor with an open circuit through the windings.

Figure 45.138 A fan placed where it can temporarily pass air over the condenser.

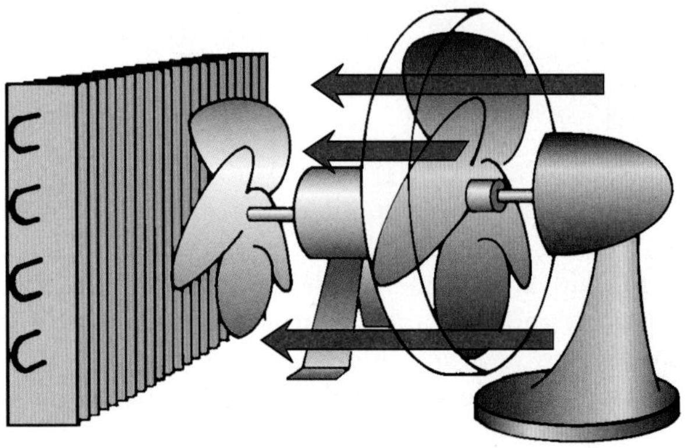

SERVICE CALL 3

A customer calls to report that the freezer is running all the time. *The problem is that this freezer has a vertical condenser on the back and it is covered with lint. The freezer is located in the laundry room. It is summer, and the laundry room is hot from the summer heat and the clothes drier.*

When the technician arrives at the job, he notices that his work shirt is soiled from his previous job. He changes into a spare, clean shirt he keeps in his truck before approaching the customer's house. Ringing the bell, he steps back and waits for the customer to answer. The customer is very happy that the technician has arrived so quickly. She leads the technician inside, shows him to the freezer, and explains the problem.

The technician checks the temperature of the box with a thermometer. It is already evident that the box is not cold enough because the ice cream is not hard. The technician places the temperature probe between two packages and closes the door. In about 5 min the instrument shows 14°F. While the temperature of the probe was pulling down, the technician had a look at the surroundings. It is evident that the condenser is dirty and the room is hot, **Figure 45.139.** He pulls the freezer out from the wall and uses a vacuum cleaner brush to remove the lint from the condenser. The technician then slides the box back to the wall.

The technician can do no more and gives the owner the following directions: "Do not dry any more clothes until the next morning." The customer purchases a small, dial-type thermometer, and places it in the food compartment. The technician calls the next morning to ask what the temperature indication is. It is −5°F, so the box is working fine. He also instructs the customer how to slide the unit from the wall and clean the condenser coil periodically and suggests that the freezer be relocated so that it is not in the same room as the washer and dryer.

SERVICE CALL 4

A customer calls to report that the refrigerator in the lunchroom is not cooling correctly. It is running all the time and the box does not seem cool enough. *The problem is that the defrost timer motor is burned out and will not advance the timer into defrost. The evaporator is frozen solid.*

When the technician arrives he opens the freezing compartment inside the top door. He hears the evaporator fan running but can feel no air coming out of the vents. A package of ice cream is soft, indicating the compartment is not cold enough. The technician removes the cover to the evaporator and finds it frozen, a sure sign of lack of defrost.

Shutting the compartment door, the technician pulls the box from the wall. The timer is in the back and has a small window for observing rotation of the timer motor. It can be seen that the timer is not turning. A voltage check of the timer terminals shows voltage to the timer motor. The winding is checked for continuity; it is open and defective.

The technician opens the refrigerator door and removes the food, placing it in another box nearby. He locates a small fan in such a manner as to blow room air into the box for rapid defrost, **Figure 45.140,** and asks the customer to put towels on the floor to absorb the water that will drip as the frost begins to melt. The manager is informed of the problem and the cost for repair. Once the repair cost is approved, the technician informs the manager that since it will take a couple of hours to melt the ice, he is going to make another service call on the way to the supply house for a defrost timer.

On returning, the technician replaces the defrost timer. The evaporator is thawed, but the pan underneath the unit is full and must be emptied. After it is emptied, the pan is cleaned, sanitized, and replaced; the evaporator cover is replaced, and the unit is started. The coil begins to cool. Since it will take some time for the box to cool down to the

Figure 45.139 A dirty chimney-type condenser.

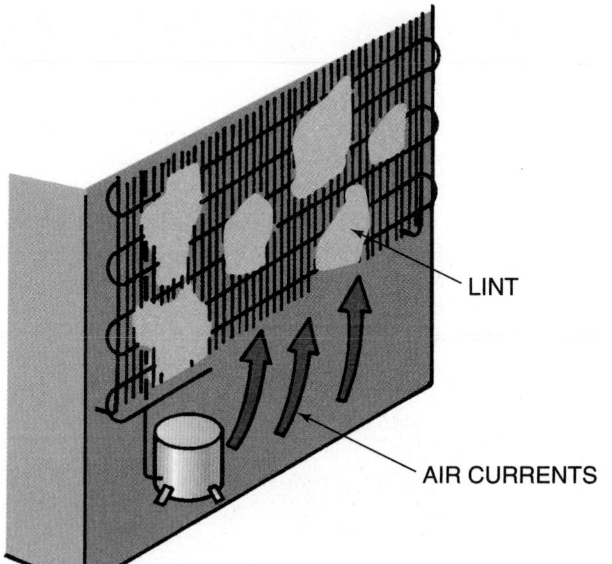

Figure 45.140 A room fan may be used for rapid defrost.

desired temperature, the technician recommends that the food be left in the other box until the next day.

Upon determining that the timer motor was defective, the technician could have simply replaced the timer and brought the system into defrost without having to manually defrost the evaporator. In addition, the motors on mechanical defrost timers rarely go bad due to the extremely low power consumption of the motor. The main reason for mechanical defrost timer failure is the jamming of the plastic gears within the timer.

SERVICE CALL 5

A homeowner reports that the refrigerator is running all the time and not shutting off. *The problem is that the door gaskets are defective and the door is out of alignment. The customer has four children.*

The customer explains the problem to the technician when he arrives and introduces himself. The technician can easily see the problem when he opens the refrigerator door. The gaskets are badly worn and daylight is visible under the door when looking from the side. The technician writes down the model number of the refrigerator and tells the owner the cost of the repair and that a trip to the supply house will have to be made to get the right gaskets. Since the supply house is about 2 hours away, they agree that the repair would be completed the next day.

The technician returns to the job with the new gaskets and replaces them following the manufacturer's recommendations. The door does not close tightly at the bottom, so the technician removes the internal shelving and adjusts the brackets in the door so that it hangs straight. Although he cannot be certain, the technician assumes that the children might have been constantly opening and closing the door or, even worse, have been swinging on the door, which would take it out of alignment. In a very professional manner, the technician informs the customer that improper use of the appliance can result in damage to the appliance or even personal injury. The technician completes his paperwork, obtains payment for the repair, and leaves for his next service call.

SERVICE CALL 6

A customer calls to say that his freezer shocked him when he touched it and the water heater located beside it at the same time. *The problem is that the freezer has a slight ground circuit due to moisture behind the box. The house is old and only has two-pronged wall outlets.* These are ungrounded, and an adapter has been used to plug the freezer into the wall outlet. The water heater is wired straight to the electrical panel, where it is grounded. The customer is advised not to touch the box. It cannot be unplugged or loss of food may result.

The technician listens attentively as the customer informs him of the situation with the appliance. He thanks the homeowner for the information and asks to be shown to the freezer. The technician uses a volt-ohmmeter to check the voltage between the freezer and water heater, duplicating the customer's experience, **Figure 45.141.** The voltmeter indicates 115 V. This could be a dangerous situation. The technician removes the fuse to the freezer circuit and pulls the freezer away from the wall. When it is unplugged, the technician discovers at least half of the problem—a two-prong plug. If the unit were connected through a three-prong plug and grounded, the ground would have flowed through the green ground wire.

The technician then fastens one lead of the ohmmeter to the hot plug and the other to the ground (green) plug for the freezer. He measures 25,000 ohms between the hot wire and the cabinet, **Figure 45.142.** If the cabinet had been properly grounded, the current flow through this circuit would have blown the fuse and would have flowed to the ground through the green ground wire, **Figure 45.143.** The technician must now find the circuit. The compressor is disconnected while the meter is still connected; the reading remains the same, **Figure 45.144**.

The technician eliminates the circuits one at a time, until he discovers that the junction box at the bottom of the unit is damp due to water dripping from a very slight leak in a water heater valve. The technician tightens the valve packing gland and the leak stops. A hair drier is used to dry the junction box, and the ground is eliminated. The technician has a talk with the homeowner about the circuit and the danger involved.

Before the technician leaves, an electrician is called to properly ground this circuit and several more that have the same potential problem. The technician has done all that can be expected. It would not have been acceptable to leave the job without knowing an electrician was already coming to make the system safe.

NOTE: *Typically, when there is a shock problem the very first thing to check is a short-to-ground in the compressor. If not that, the next most common cause is, oddly enough, that the appliance insulation got saturated over time. This is usually because the owner turned off the mullion heater trying to save energy and the wet insulation shorts out a wiring harness embedded in the walls.*•

SERVICE CALL 7

A customer reports that beef frozen two weeks ago has no flavor, is dry, and seems to be tough. *The problem is that this is a new freezer—and was sold to a person with no experience in freezing food. The customer bought a fresh side of beef and took it home to freeze it in the new freezer. Ordinarily, the technician would not get involved in this problem, but the customer is an old one and relations must be maintained.*

The night before his call, the technician asks the customer to thaw a steak overnight so that it can be checked in the morning. When he arrives, the technician sees a puddle of water and blood under the steak and begins to ask questions. The customer explains how the meat was frozen. She picked it up at a packing house and made three stops before getting it home; the weather was hot. The packing-house clerk tried to get her to let the packing house to

Figure 45.141 To determine the cause of the shock, a meter is being substituted for a person in the circuit.

freeze the meat, but she thought of her own new freezer and felt that it would do a better job.

The whole side of beef was put in the freezer at one time, about 200 lb, packaged in small parcels. The technician asks to see the owner's manual. In the manual, it plainly states that the capacity of this 15 ft^3 freezer is 40 lb. The customer therefore has a freezer full of meat that will not be of the best quality. This is partly the salesperson's fault for not explaining the capabilities of the freezer. The technician goes over the manual with the owner.

SERVICE CALL 8

A customer reports that the refrigerator is running all the time. It is cool but never shuts off. This refrigerator is under warranty. *The problem is that a piece of frozen food has fallen and knocked the door light switch plunger off, causing the light in the freezer to stay on all the time. This is enough increased load to keep the compressor continuously.*

The customer informs the technician of the problem when he arrives. After looking the refrigerator over, the

Figure 45.142 A resistance of 25,000 Ω is measured between the hot wire and the cabinet.

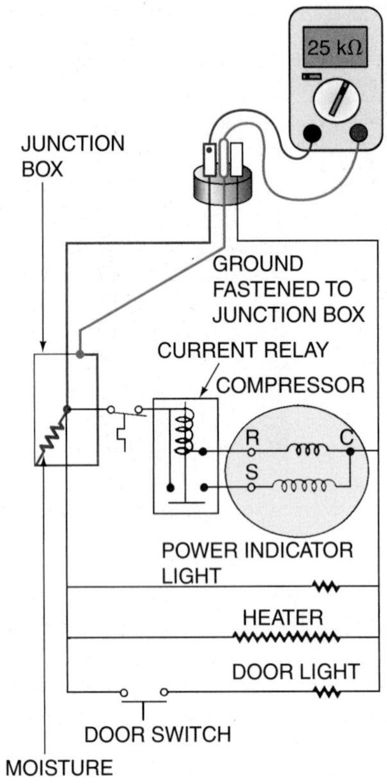

Figure 45.143 If this circuit had been properly grounded, a fuse might have blown.

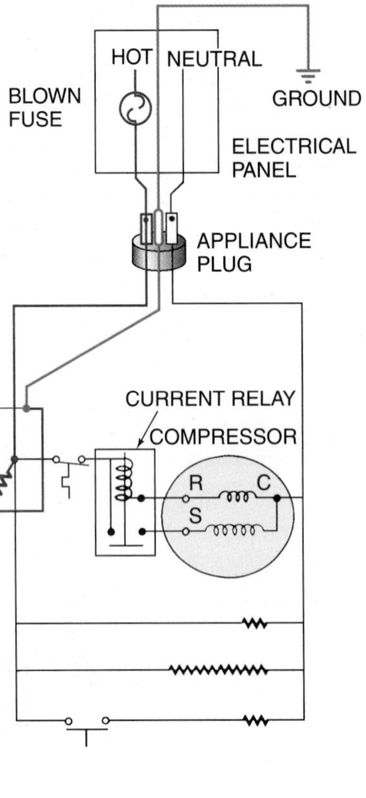

Figure 45.144 When the compressor is disconnected, the ground circuit is still present.

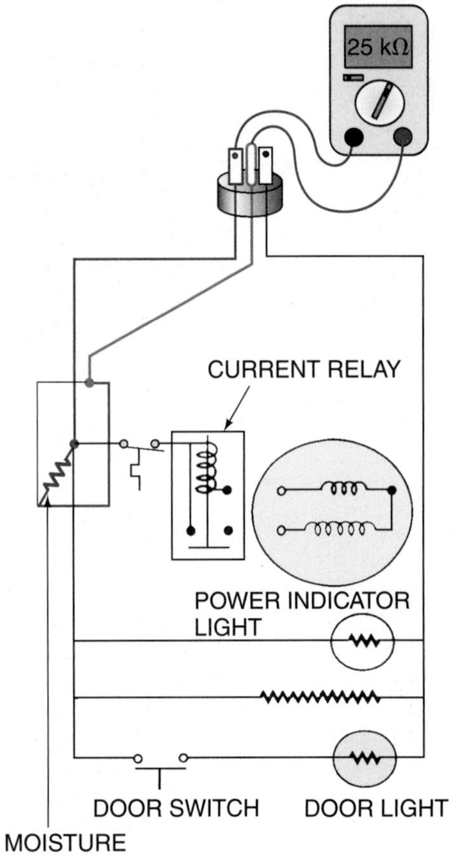

technician does not see any frost buildup on the evaporator. He questions the owner about placing hot food in the refrigerator or leaving the door open. This does not seem to be the problem. The customer says the refrigerator is always running. It should be cycling off in the morning, no matter how much load it has.

It seems like the compressor is not pumping, or maybe there is an additional load on it. The unit has electric defrost, and the evaporator coil seems to be clean of frost. The technician opens the freezer door and notices that the light switch in the freezer is missing the plunger that touches the door when it is closed, **Figure 45.145.** The light is staying on all the time.

The technician informs the customer that a switch will have to be ordered from the manufacturer. Until a switch can be obtained, the light bulb in the freezer is removed. The technician returns the next day with a switch and replaces it. The owner tells the technician that the refrigerator was not running at breakfast time, so the diagnosis was good.

Figure 45.145 The plunger is missing from the door switch.

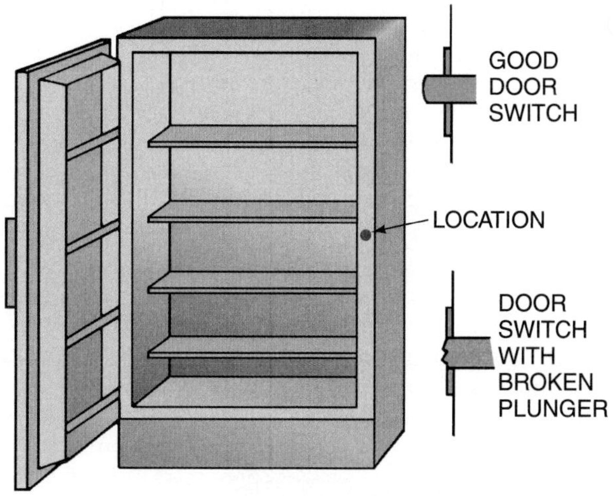

SERVICE CALL 9

An apartment-house tenant has an old refrigerator that must be defrosted manually. The ice was not melting fast enough so the tenant used an ice pick, which punctured the evaporator. She heard the hiss of the refrigerant leaking out but decided to continue anyway. *The problem is that the refrigerant leaked out, and when the refrigerator was restarted, water was pulled into the system.* (Without refrigerant the low-pressure side of the system will pull into a vacuum.) When the refrigerator would not cool, the tenant called the management office. This would normally be a throwaway situation, but the apartment house has 200 refrigerators like this one and a repair shop.

The technician takes a replacement refrigerator along after reading the service ticket. This has happened many times—so many that a definite procedure has been established. The food is transferred to the replacement box and the other one is trucked to the shop in the basement.

The technician solders two process tubes into the system, one in the suction line and the other in the discharge line, **Figure 45.146.** Pressure is added to the circuit using nitrogen, and the leak is located. It is in the evaporator plate, so the repair procedure will be to patch with epoxy. The technician allows the nitrogen pressure to escape from the system, and cleans the puncture with a nonconductive sand tape. He uses a solvent recommended by the epoxy manufacturer to remove all dirt and grease from the puncture area and then connects a vacuum pump to the service ports. **NOTE:** *The service ports have no Schrader valve plungers and the gauge hoses have no Schrader valve depressors; they have been removed for the time being.•*

The technician starts up the vacuum pump and allows it to run until 5 in. Hg is registered on the suction compound gauge and then valves off the gauges. Then mixing the epoxy (remember, this is a two-part mixture and it must be used very fast or it will harden, usually within 5 min), he spreads some over the puncture hole. The vacuum pulls a small amount into the hole to form the mushroom shape mentioned earlier in the text, **Figure 45.147.** The vacuum pulls a hole through the center of the epoxy patch, causing more epoxy to spread over the hole. The epoxy is now becoming solid. The gauges are opened to the atmosphere to equalize the pressure on each side of the patch, and the epoxy is allowed to dry.

The epoxy is allowed to dry for several hours; meanwhile, other service tasks are performed. Nitrogen is used to pressure the system to 100 psig. The epoxy patch is checked for a leak using soap bubbles. The service stems are also checked for leaks, and there are none. The tricky part of the service procedure comes next. This technician knows the procedure well and knows from experience that a step must not be skipped.

The pressure is allowed to escape from the system, and the vacuum pump is started without stems or depressors in the Schrader fittings. This allows the gauge hose to operate at full bore. A 60-W bulb is placed in the freezer compartment and one in the fresh-food compartment. **SAFETY PRECAUTION:** *The doors are shut only partially.•* A 150-W bulb is placed so that it touches the compressor crankcase, **Figure 45.148.** None of these bulbs should ever be in contact with plastic because it will melt. As part of the triple evacuation process, once the first vacuum has been reached, the technician breaks the vacuum to about 2 psig pressure using nitrogen and disconnects the vacuum pump.

Figure 45.146 Process tubes soldered into the suction and discharge lines offer the best service access.

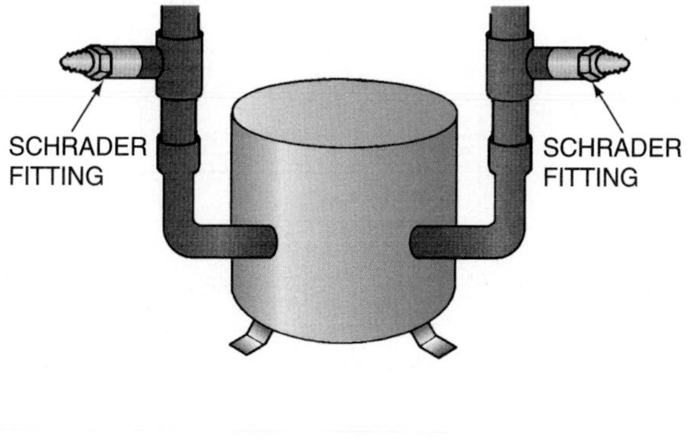

Figure 45.147 The vacuum is used to pull a small amount of epoxy into the evaporator puncture.

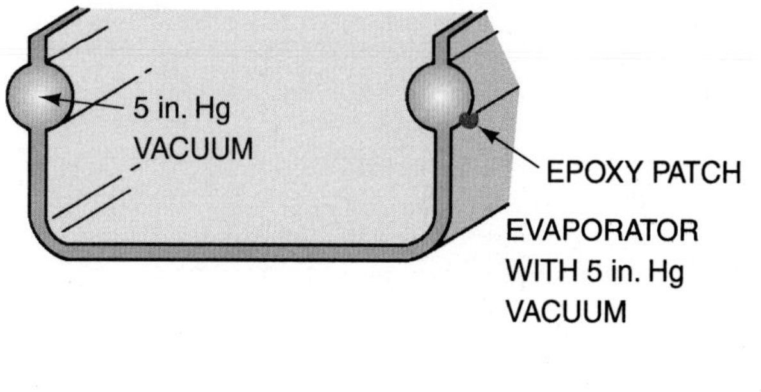

Figure 45.148 Heat is applied to the compressor to boil any moisture from under the oil.

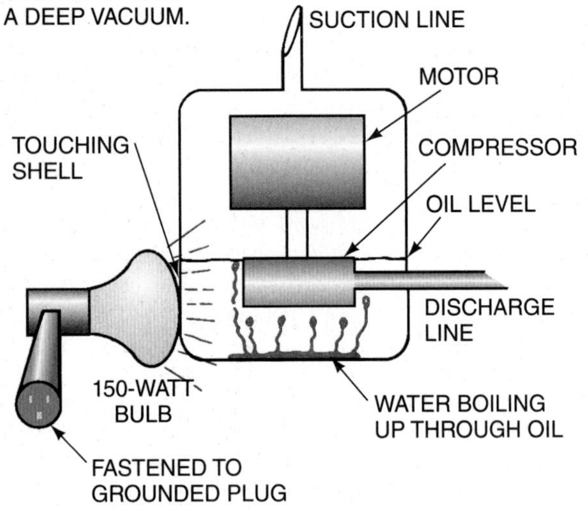

The vacuum pump is removed to the workbench and the oil is replaced. The vacuum pump is now started and allowed to run until the second stage of the evacuation is completed. Once the vacuum level has been reached again, the vacuum is again broken to about 5 psig using nitrogen. The gauge hoses are removed one at a time, and the Schrader valve stems and depressors are added back to the gauge hoses.

A suction-line filter drier is installed in the suction line and the system is pressured back to 100 psig using nitrogen to leak check the drier connection. This may seem like a long procedure, but the service technician has tried shortcuts before, and they do not work. The most obvious one would be to cut the drier into the circuit at the beginning, but it would become contaminated before the refrigerator is started if cut into a wet system. The liquid-line drier should be changed at this time, as it has become saturated. It has been partially reactivated with the evacuation, but changing it will ensure that all moisture is removed. The vacuum pump is started for the third and final evacuation and allowed to run until the required vacuum is reached and maintained as determined by a micron gauge. During these waiting periods, the technician may have other service duties to perform.

The high-side line is pinched off and soldered shut; the refrigerant is charged into the system (see Unit 10, "System Charging"). The refrigerator is started to pull the last of the charge into the system. The suction line is disconnected and a cap is placed on the Schrader valve. The discharge line port is soldered shut so the next time a gauge reading is needed, a suction-line port is available, but no discharge port. This is common; there is less chance of a leak under the low-side pressure than under the high-side pressure.

SERVICE CALL 10

A customer who has been on vacation calls. While the family was gone, the freezer had warmed and thawed. *The problem is that a compressor wire has been damaged.*

The technician arrives with a helper. It is obvious that the freezer will have to be moved. They unload it and dispose of the spoiled food. After that, the technician and helper move the freezer to the carport. They wash the box with ammonia and water, several times, and leave the door open to allow the sun to shine in.

Now the technician starts to investigate the problem. When the box is plugged in, the compressor does not start. A close examination shows that the common wire going to the compressor has been hot, **Figure 45.149.** The box is unplugged and after being examined more closely, the wire is found to be burned in two in the junction box. The wire is replaced with the proper connectors and the correct wire size. The box is plugged in again and the compressor starts. The technician can hear refrigerant circulating and feels that the box is now in good working order.

This box smells so bad that the technician advises the homeowner to try the following to remove the smell:

- Leave the box open for several days.
- Then close it and place some activated charcoal, furnished by the technician, inside and start the box.
- If odor is still present, spread several pans of ground coffee inside the box and leave the freezer running for several more days.
- Then wash the freezer inside with warm soda water—about a small package to a gallon of water—and let it run with more coffee grounds inside, for several days. If this does not remove the odor, the box may need replacing. This seems like a long procedure, but a new freezer is expensive and the old one should be saved if possible.

SERVICE CALL 11

A customer calls indicating that the freezer is not running and is making a clicking sound from time to time. *The problem is that the customer has added some appliances to the circuit*

Figure 45.149 The insulation and wire are burned.

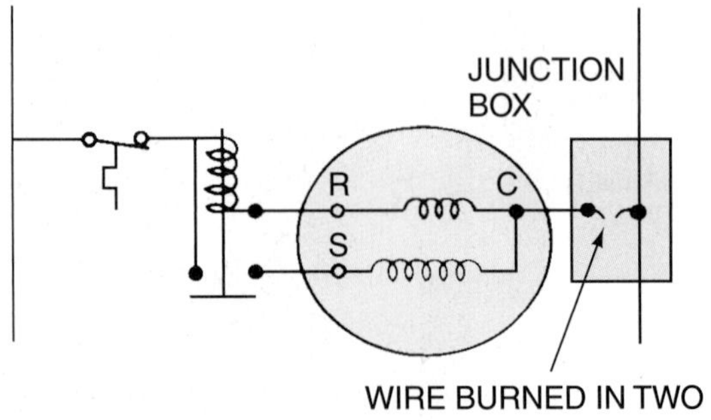

and the voltage is low. A small 115-V electric heater has been added to the mother-in-law's bedroom, and she is operating it all the time. The voltage is only 95 V at the wall outlet while the heater is operating. This is not enough voltage to start the freezer compressor. The homeowner has an old system with fuses and has replaced the 15-A fuse with a 20-A fuse. The owner is advised to shut the freezer off until the technician arrives and to keep the door to the freezer shut.

When the technician arrives, the homeowner tells him about the clicking sound and that the freezer is not running. As the customer had implied, the compressor is not operating. The technician places a temperature probe between two packages; after about 5 min, the temperature reads 15°F. There is no time to spare. The technician decides to check the amperage and the voltage at start-up. He fastens an ammeter around the common compressor wire and plugs a voltmeter into the other plug in the wall outlet. The voltage is 105 V; this is low to begin with, **Figure 45.150**. The technician plugs the box in and watches the voltmeter and the ammeter. The compressor does not start and the voltage drops to 95 V, so he unplugs the freezer before it trips the overload, **Figure 45.151.** It is time to look further into the house circuits.

The technician asks the owner whether any additional load has been added to the house; she mentions the heater in the mother-in-law's room, which turns out to be a 1500-W heater. By itself it should draw 13 A (Watts = Amperes × Volt; solved for amperes, this would be Amperes = Watts divided by Volts, or Watts = 1500 divided by 115 = 13 A). However, the house has only 110 V at the main, so 1500 W divided by 110 V = 13.6 A for the heater under these circumstances, **Figure 45.152.** The fuse being used on this circuit is a 20-A fuse. The rest of the circuits with the same wire size have 15-A fuses. A voltage check at the main electrical panel shows 110 V. With just the electric heater operating, the circuit is losing 5 V.

The technician unplugs the electric heater and measures the voltage at the freezer outlet; it is 110 V. Obviously, the electric heater must be moved to another circuit, a circuit by itself, and a smaller heater would be a better choice. Upon examining the heater, the technician discovers that it can be operated in a lower mode that consumes 1000 W, which should be enough heat for the small room. It would pull 9.1 A (1000 W divided by 110 V = 9.1 A). Every little bit of savings helps—this place is operating at near maximum capacity.

The heater is plugged into another circuit at the lower output, 1000 W. Then the freezer is plugged in and it starts; while it is running the voltage is 105 V, **Figure 45.153.** This is slightly above the −10% allowed for the motor voltage. The rated voltage is 115 V; 115 × 0.10 = 11.5. The rated voltage of 115 V minus 11.5 V is 103.5 V, which is the minimum voltage that can be used. Before, when the freezer and the heater were on the circuit at the same time, the freezer would not start.

Figure 45.150 There are only 105 V in the circuit before the freezer is plugged in. The freezer will be an additional load on the circuit.

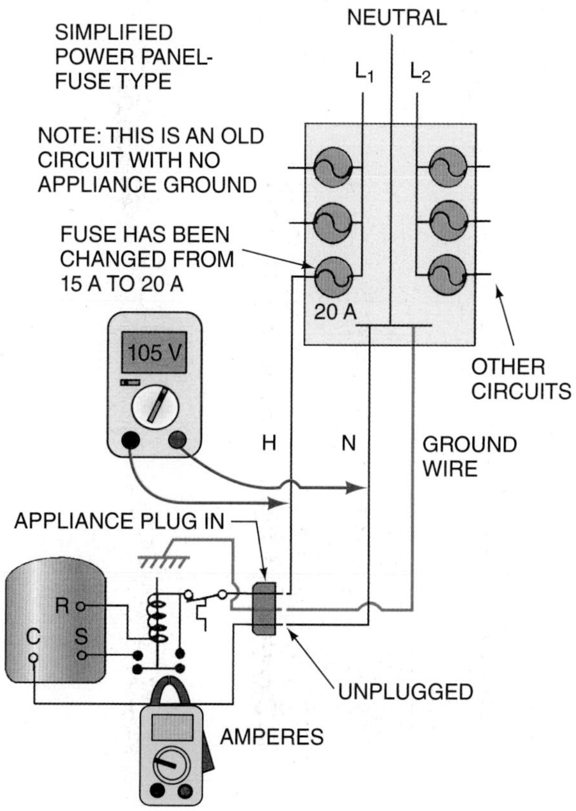

Figure 45.151 When the unit is plugged in, the voltage drops to 95 V and the compressor will not start.

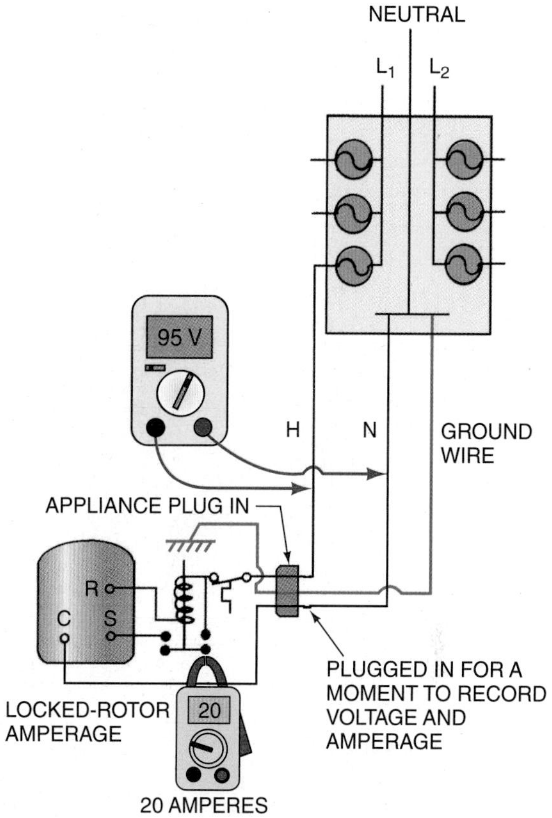

Figure 45.152 The home voltage is only 110 V; a 1500-W heater will pull 13.6 A at 110 V.

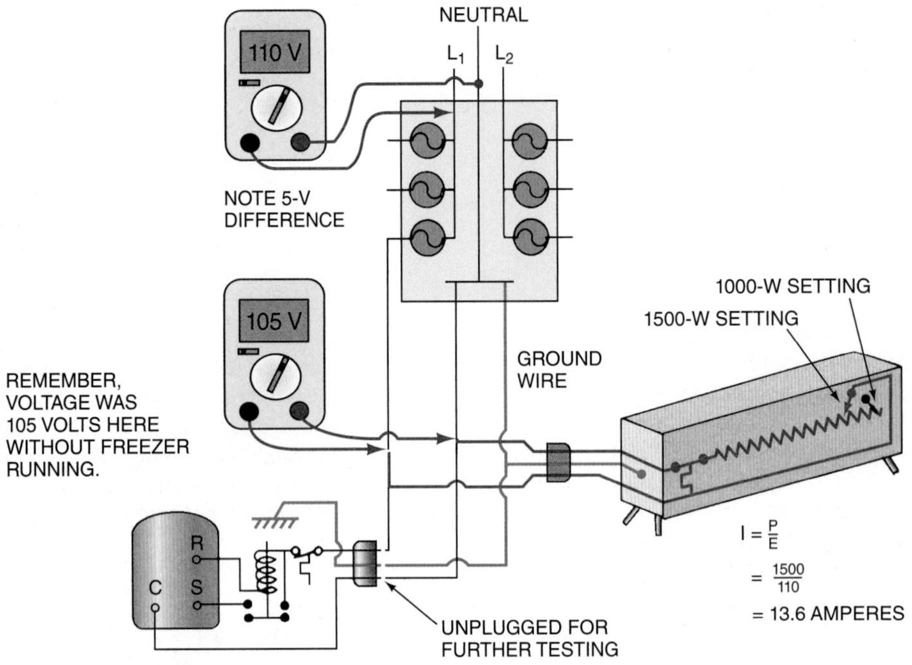

Figure 45.153 When the heater is moved to a new circuit, the freezer operates at 105 V, which is a little low but is the best that can be done here.

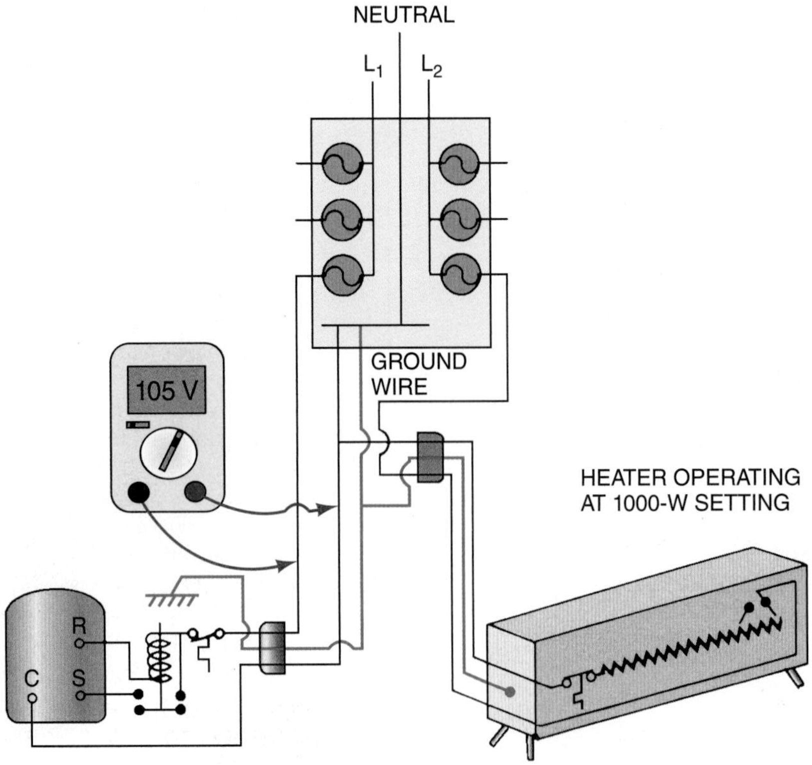

The technician replaces the 20-A fuse with a 15-A fuse and advises the owner that an electrician should be consulted to rework the electrical system in the house. The owner did not realize the circuits were operating at near capacity.

SERVICE CALL 12

A customer reports that the refrigerator has shut off; it tripped a breaker. When he tried to reset the breaker, it tripped again. The dispatcher advises the customer to unplug the refrigerator and reset the breaker because there may be something else on the circuit that he may need. *The problem is that the compressor motor is grounded and tripping the breaker.*

The technician takes with him a helper, a refrigerator to loan the customer, and a hand truck. He suspects a bad electrical problem. This unit is under warranty and will be moved to the shop if the problem is complicated.

The technician introduces himself and his helper to the customer and informs him that a loaner refrigerator is on the truck should it be needed. The customer thanks the technician for thinking of this and shows him where the unit is. Using an ohmmeter at the power cord, the technician checks for a ground. He reads 0 ohms, indicating a short circuit, when he touches one meter lead to the line prong on the plug and the other to the ground prong on the plug, **Figure 45.154**.

The refrigerator is moved away from the wall to see whether the ground can be located. It could be in the power cord and, if so, it could be repaired on the spot. Removing the leads from the compressor, the compressor terminals are checked with one lead on a compressor terminal and the other on the cabinet or one of the refrigerant lines. The compressor shows 0 Ω to ground, **Figure 45.155.** It has internal problems and must be changed.

The technician and the helper move the refrigerator to the middle of the kitchen floor and put the spare refrigerator in place. The food is transferred to the spare refrigerator, which is cold because it was plugged in at the shop. The faulty refrigerator is trucked to the shop and unloaded. The technician knows that the refrigerant in the box may be badly contaminated. *The refrigerant is properly recovered and stored in a recovery cylinder used for contaminated refrigerants.* The service technician decides to change the compressor and the liquid-line drier and add a suction-line drier.

The refrigerator is set up on a workbench for easier access. The compressor is changed and process tubes soldered in place with ¼-in. fittings on the ends. (**Schrader fittings will not be used in this example to demonstrate how a system is totally sealed after an evacuation.**) The liquid line is cut and the old drier is removed. The technician then fastens the gauge hoses to the process tubes. No Schrader stems or depressors are installed or used. Nitrogen is used to purge the system from the suction line back through the evaporator and capillary tube, which is loose at the drier. Nitrogen under pressure is blown through the high side and out of the liquid line that is attached to the liquid-line drier, **Figure 45.156.** The purge method used in this system differs because there is no moisture present, just contaminated refrigerant and oil

Figure 45.154 Checking the unit for a grounded circuit using the appliance plug.

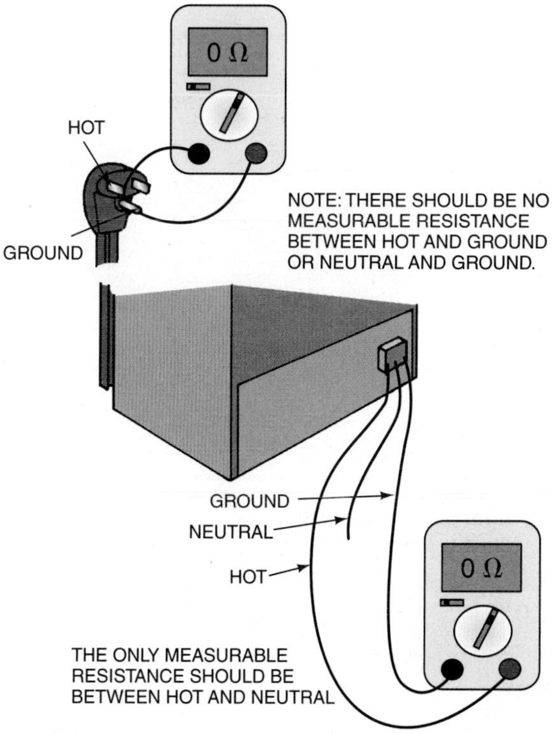

Figure 45.155 Using the compressor frame or lines to check for an electrical ground circuit.

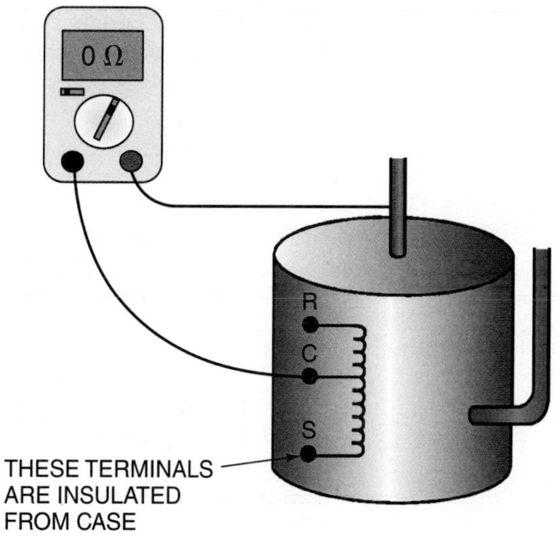

Figure 45.156 Purging the system at the suction-line drier connection before soldering the drier in the line.

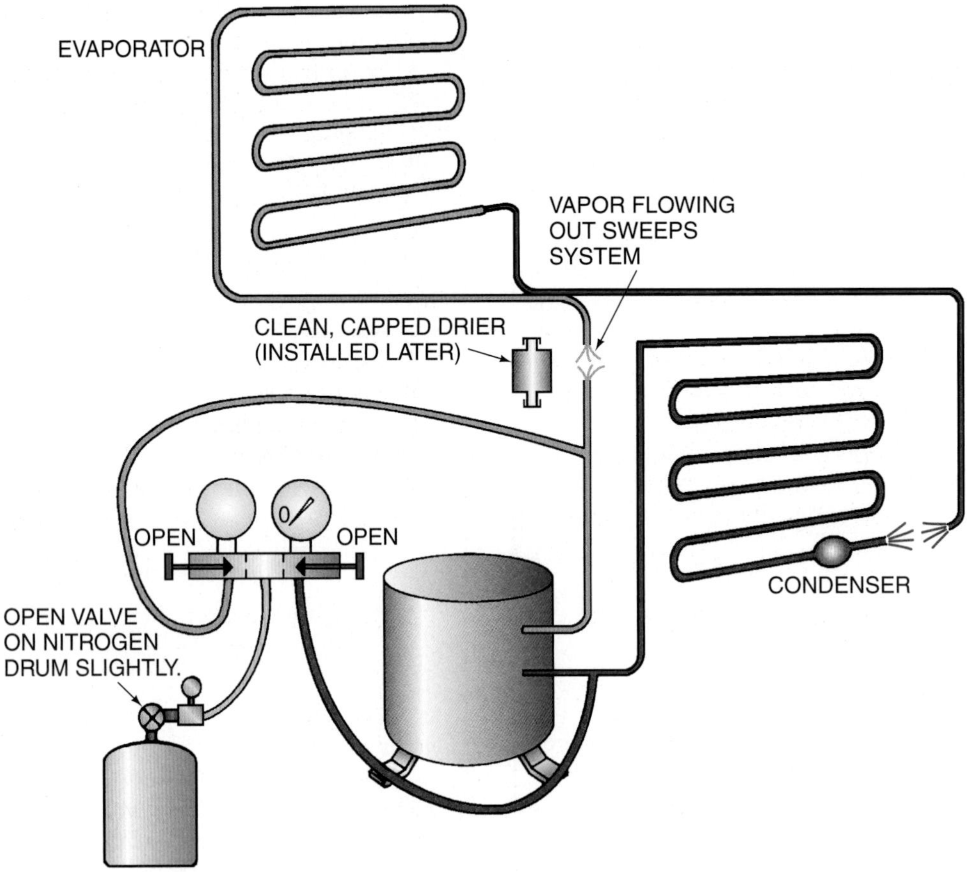

and possibly smoke from the motor ground in the system. Purging can remove the large particles, and the driers will remove the rest. After this step, the liquid-line drier and then the suction-line drier are soldered in place.

The system is leak checked and triple evacuated. At the end of the third evacuation, the discharge service tube is pinched off using a pinch-off tool. The system is charged with a measured charge and started. The low-side pressure is observed to be correct during the pulldown of the box, **Figure 45.157.** The box is allowed to run for 24 hours before it is returned to the customer.

Figure 45.157 Typical pressure while the box is pulling down.

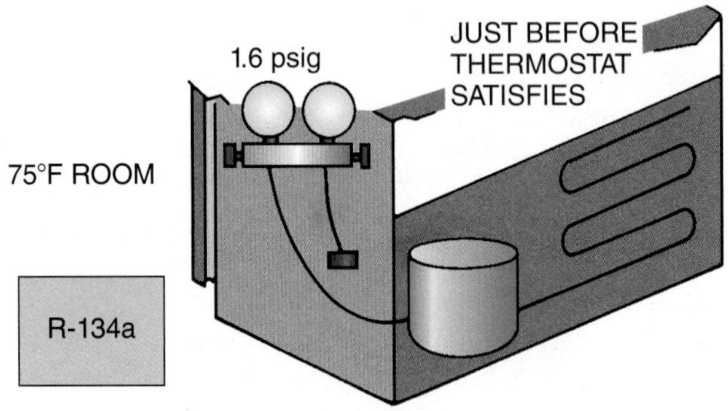

REVIEW QUESTIONS

1. The typical temperature inside the fresh-food compartment in a domestic refrigerator is
   A. 25°F to 30°F.
   B. 30°F to 35°F.
   C. 35°F to 40°F.
   D. 40°F to 45°F.

2. The operating condition for the single compressor in a household refrigerator is the lowest box temperature, which is typically
   A. 0°F
   B. −20°F
   C. 20°F
   D. 40°F

3. When the air temperature in a domestic freezer is 0°F, the plate-type evaporator temperature may be as low as
   A. −30°F.
   B. −12°F.
   C. −6°F.
   D. 0°F.
4. Typical head pressures should correspond to a condensing temperature ________°F higher than room temperature at design operating conditions.
   A. 5°F to 10°F
   B. 10°F to 15°F
   C. 20°F to 35°F
   D. 35°F to 45°F
5. Frost accumulates on the evaporators of forced-draft refrigerators because
   A. they are generally located within a house.
   B. they are generally located in high-humidity areas.
   C. they are low-temperature appliances.
   D. there is a heat exchange with the suction line.
6. When unfrozen product is placed in the freezer compartment of a domestic refrigerator/freezer, the product
   A. will freeze slowly and large ice crystals will likely form.
   B. will freeze slowly and small ice crystals will likely form.
   C. will freeze quickly and large ice crystals will likely form.
   D. will freeze quickly and small ice crystals will likely form.
7. The compressors used in domestic appliances typically range in size from
   A. 1/20 horsepower to 1/10 horsepower.
   B. 1/10 horsepower to 1/3 horsepower.
   C. 1/3 horsepower to 1/2 horsepower.
   D. 1/2 horsepower to 1 horsepower.
8. The flow through the capillary tube metering device is determined by the
   A. outside diameter of the tubing and its inside (bore) diameter.
   B. inside (bore) diameter and the appliance voltage.
   C. outside diameter of the tubing and the length of the tubing.
   D. inside (bore) diameter and the length of the tubing.
9. A manual defrost system is most likely to be found on what type of system?
   A. An appliance with an induced-draft evaporator
   B. An appliance with an induced-draft condenser
   C. An appliance with a natural-draft evaporator
   D. An appliance with two parallel compressors
10. The capillary tube metering device is often fastened where to improve heat exchange and system performance?
    A. Surface of the evaporator
    B. Compressor discharge line
    C. Compressor suction line
    D. Liquid line
11. The doors of domestic appliances are typically removed when the appliance is disposed of to prevent
    A. other individuals from using the appliance.
    B. someone from getting trapped inside the appliance.
    C. mold from forming inside the appliance.
    D. water from accumulating inside the appliance.
12. Condensers in domestic refrigerators and freezers are
    A. air-cooled.
    B. water-cooled.
    C. oil-cooled.
    D. brine-cooled.
13. What is the name given to the heaters that are located around the doors of the appliance to prevent sweating?
    A. Defrost heaters
    B. Oil heaters
    C. Condensate heaters
    D. Mullion heaters
14. Which of the following statements is true regarding the interior surfaces of most refrigerators and freezers?
    A. They are made of metal, are easy to clean, and can become brittle when cold.
    B. They are made of plastic, are easy to clean, and can become brittle when cold.
    C. They are made of metal, are difficult to clean, and can become brittle when cold.
    D. They are made of plastic, are difficult to clean, and can become brittle when cold.
15. The adjustment device used for low atmospheric pressure conditions is called the
    A. altitude thermostatic adjustment control.
    B. automatic thermostatic adjustment control.
    C. altitude electrostatic adjustment control.
    D. automatic electrostatic adjustment control.
16. The ice-making process in a domestic freezer is typically controlled by
    A. temperature.
    B. demand.
    C. time.
    D. pressure.
17. Depending on the type of condenser, the condenser saturation temperature for a domestic appliance located in a cool, 65°F room will likely be between
    A. 130°F and 140°F
    B. 115°F and 125°F
    C. 100°F and 110°F
    D. 85°F and 95°F
18. When should gauges be installed on a domestic refrigerator or freezer?
    A. Whenever the system is serviced
    B. Each time the water filter is replaced
    C. Once a year to check the refrigerant charge
    D. Only when absolutely necessary

19. A dirty condenser coil on a domestic refrigerator will cause the
    A. appliance's operating pressures to rise.
    B. appliance's operating pressures to drop.
    C. suction pressure to rise and the discharge pressure to drop.
    D. discharge pressure to rise and the suction pressure to drop.
20. Forced-draft evaporators are characterized as being
    A. physically small with small quantities of air moving through them.
    B. physically large with small quantities of air moving through them.
    C. physically small with large quantities of air moving through them.
    D. physically large with large quantities of air moving through them.
21. The type of compressor typically found on a domestic appliance is
    A. welded, semihermetic.
    B. welded, hermetic.
    C. flanged, semihermetic.
    D. flanged, hermetic.
22. On a domestic appliance with a mechanical draft condenser,
    A. the compressor cycles on and off with the evaporator fan motor.
    B. the compressor cycles on and off with the condenser fan motor.
    C. the condenser fan motor cycles on and off with the evaporator fan motor.
    D. the condenser fan motor, evaporator fan motor, and compressor cycle on and off together.
23. The door switch on a domestic appliance controls the operation of the
    A. compressor.
    B. condenser fan motor.
    C. evaporator fan motor.
    D. defrost cycle.

UNIT 46

ROOM AIR CONDITIONERS

OBJECTIVES

After studying this unit, you should be able to

- explain the purpose of window and through-the-wall room units.
- explain how to differentiate between a window unit and a through-the-wall unit.
- explain how to differentiate between a unit with a slide-out chassis and a unit with a wrap-around sleeve.
- describe the refrigerants that are found in older window units as well as newly manufactured models.
- describe the temperature range that best describes window and through-the-wall air-conditioning units.
- list the major components in the refrigeration cycle of a window cooling unit.
- describe the construction and function of the evaporator coil on a window or through-the-wall air conditioner.
- describe the heating cycle in the heat pump or reverse-cycle room air conditioner.
- explain how window air-conditioning units are installed.
- describe the electrical requirements for window units.
- describe the controls used on cooling-only room units.
- describe the controls used on heating and cooling room units.
- explain how to properly maintain window and through-the-wall units.
- explain what system problems might arise if the air filter on a window or through-the-wall unit is not kept clean.
- state the proper procedures for charging a room air conditioner.
- explain how to check the operation of the selector switch on a room air conditioner.

SAFETY CHECKLIST

- ✓ Wear back brace belts when lifting or moving any appliance.
- ✓ Lift with your legs, keeping your back straight. Whenever possible, use appropriate equipment to move heavy objects, or get help from someone else.
- ✓ Observe proper electrical safety techniques when servicing any appliance or system. Ensure that electrical power to the unit is off except when needed for troubleshooting.
- ✓ Wear goggles and gloves when transferring refrigerant into or out of a room air-conditioning unit.
- ✓ When performing service on window units, make sure that all loose clothing and hair are secured because they could become entangled in rotating fans and shafts.

46.1 AIR-CONDITIONING AND HEATING WITH ROOM UNITS

Room units are used to provide comfort cooling, and sometimes heating, to an individual room or relatively small open area in a structure. Room air-conditioning units are constructed of the same four major vapor-compression refrigeration system components that have been discussed in detail earlier in this book. When a particular component or service technique is mentioned, refer to the appropriate sections of the text for the necessary details. Systems and components characteristic of room units are discussed in this unit.

Single-room air-conditioning can be accomplished in several ways, some involving the use of package (self-contained) systems of some type. A common type of package unit is the room air-conditioning window unit for cooling only, **Figure 46.1**. Variations of this type of unit include electric strip heaters in the airstream with controls for heating and cooling, **Figure 46.2**. If there is adequate air circulation between multiple rooms, it may be possible to use a single unit to condition more than one room.

Figure 46.1 A window unit for cooling. Courtesy Friedrich Air Conditioning Co.

Figure 46.2 A window unit with heat added.

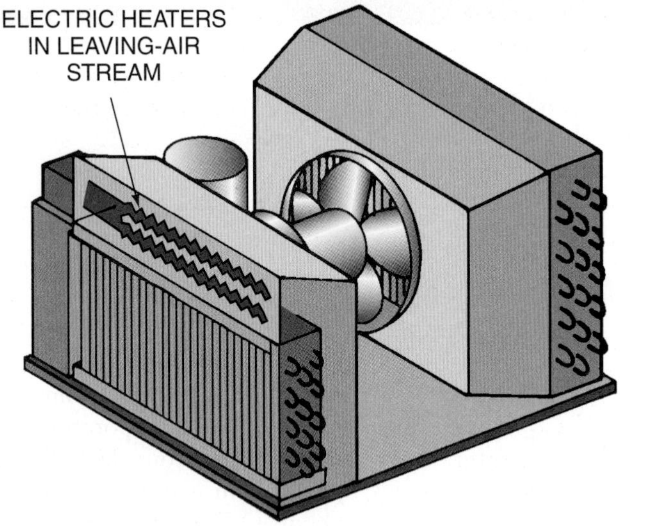

Figure 46.3 (A) Window and (B) through-the-wall-type units. Courtesy Friedrich Air Conditioning Co.

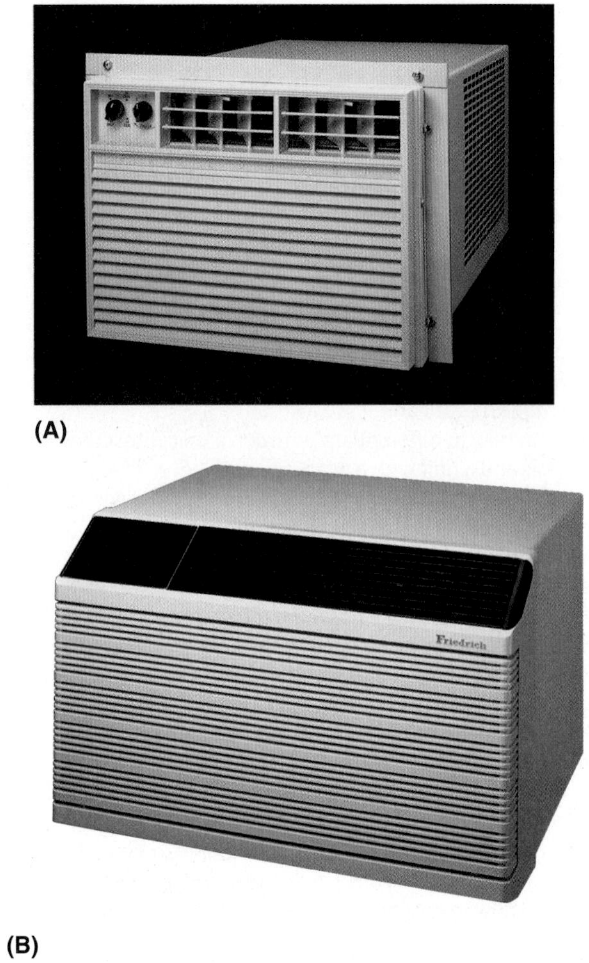

46.2 ROOM AIR-CONDITIONING—COOLING

The manufacturer's design objectives for room units are to achieve high efficiencies and low noise levels. Most units are as compact as manufacturing and design standards will allow. The intent is to get the most capacity from the smallest unit possible.

Cooling-only units may be either window or through-the-wall types, **Figure 46.3**. It is easy to distinguish between window and through-the-wall units. Window units, **Figure 46.3(A)**, have louvers on the sides, while through-the-wall units have solid sides, **Figure 46.3(B)**. Window and through-the-wall units are much the same, typically having only one double-shaft fan motor for the evaporator and the condenser. The capacity of these units may range from about 4000 Btu/h (1/3 ton) to 24,000 Btu/h (2 tons). Some units are front-air discharge and some are top-air discharge, **Figure 46.4**. Top-air discharge is a more common configuration for through-the-wall units; the controls are located on top. Frequently, these top-air discharge units are installed in motels or similar structures where individual room control is desirable. The controls are built into the unit so the appliance may be serviced in place or changed out for a spare unit and repaired in the shop.

Window and wall units are designed for easy installation and service. Window units have two types of cases or sleeves. In one type, the case or sleeve is fixed to the chassis of the unit. This is referred to as a wrap-around sleeve. In order to service this type of air conditioner, the entire unit must be removed from the window or wall. The term chassis is used to reference the assembly of all of the components that make up the air-conditioning system. The second type of window unit has a slide-out chassis. With this type of unit, the sleeve or case is secured to the window opening and the chassis slides in and out, **Figure 46.5**. The sleeve configuration is important from a service standpoint. For units in which the case is built on the chassis, the entire unit including the case must be removed for service. A slide-out chassis may be pulled out of the case for simple service.

Several companies make a roof-mount unit for travel trailers and motor homes, **Figure 46.6**. This type of unit may also be seen in gas station attendant booths. The case on this unit lifts off the top, **Figure 46.7**. The controls for this type of unit are located inside the conditioned space, **Figure 46.8**.

46.3 THE REFRIGERATION CYCLE—COOLING

The most common refrigerant used in older room units is R-22. R-22 is an HCFC refrigerant and, by the time of this printing, will have been completely phased-out. The phaseout date for R-22 is January 1, 2020, while the total

Figure 46.4 Some room units are front-air discharge and some are top-air discharge.

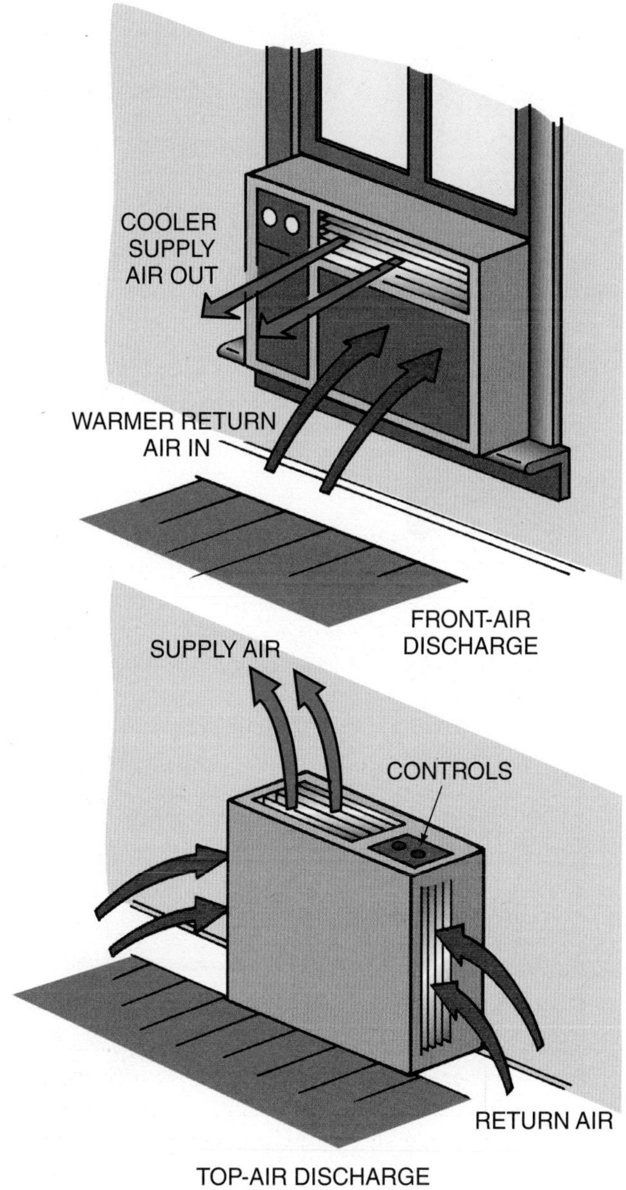

phaseout for all HCFC refrigerants is scheduled for January 1, 2030. However, newly manufactured units contain R-410A, which is an HFC refrigerant and, as such, does not contain chlorine and does not deplete ozone. R-410A does, however, have a high global warming potential, GWP, and, as a result, is classified as a super-polluter. R-410A will be the only refrigerant discussed in this Unit. When other refrigerants are encountered, the temperature/pressure chart in Unit 3, "Refrigeration and Refrigerants," may be consulted for the pressure conversions.

The basic refrigeration cycle utilized in a window or room unit consists of the same four major components as those described in Unit 3: an evaporator to absorb heat into the system, **Figure 46.9**; a compressor to pump the heat-laden refrigerant through the system, **Figure 46.10**; a condenser to reject the heat from the system, **Figure 46.11**; and an expansion device to control the flow of refrigerant, **Figure 46.12**. Although commonly found on domestic refrigerators and freezers, heat exchangers between the metering device (capillary tube) and the suction line are not typically found on window air conditioners, **Figure 46.13**. The complete refrigeration cycle is illustrated in **Figure 46.14**.

Air-conditioning units are considered high-temperature refrigeration systems, meaning the evaporator coil is intended to operate above freezing to prevent condensate from freezing on the surface of the coil. Typically, the evaporators operate with a saturation temperature of about 35°F. Central air-conditioning systems operate with a design evaporator saturation temperature of about 40°F. Room units boil the refrigerant much closer to freezing, so more care must be taken to prevent freezing of the evaporator.

The evaporator is typically made up of copper or aluminum tubing with aluminum fins. The fins may be straight or spiny, and all are in close contact with the copper tubing for the best heat exchange. The evaporator coil may be a single circuit or can be manufactured with multiple, parallel refrigerant circuits, **Figure 46.15**. Evaporators are small and designed for the maximum obtainable heat exchange. Fins that force the air to move from side to side and tubes that are staggered to force the air to pass in contact with each tube are typical, **Figure 46.16**.

The evaporator typically operates below the dew point temperature of the room air for the purpose of dehumidification, so condensate forms on the coil. The condensate falls into a pan beneath the coil, **Figure 46.17**, and is generally drained back to the condenser section where it is evaporated. **Figure 46.18** shows a typical evaporator with temperatures and pressures.

The compressor for room units is a typical hermetically sealed air-conditioning compressor, **Figure 46.19**. The compressor may be reciprocating, **Figure 46.19**, or rotary, **Figure 46.20**. Most, if not all, new room air-conditioning units use rotary compressors. Compressors are described in detail in Unit 23, "Compressors".

Similar to the evaporator, condensers are typically copper or aluminum tubes with aluminum fins, **Figure 46.21**. The condenser is responsible for rejecting heat from the system, and some of this heat can be used to evaporate the condensate from the evaporator section. The evaporation of condensate is generally accomplished by using the heat from the discharge line and a slinger ring on the condenser fan blade, **Figures 46.22** and **46.23**. *Evaporating the condensate keeps the unit from dripping and improves the efficiency of the condenser.*

Figure 46.5 Some window units are of the slide-out design, and in some, the cover is bolted to the chassis.

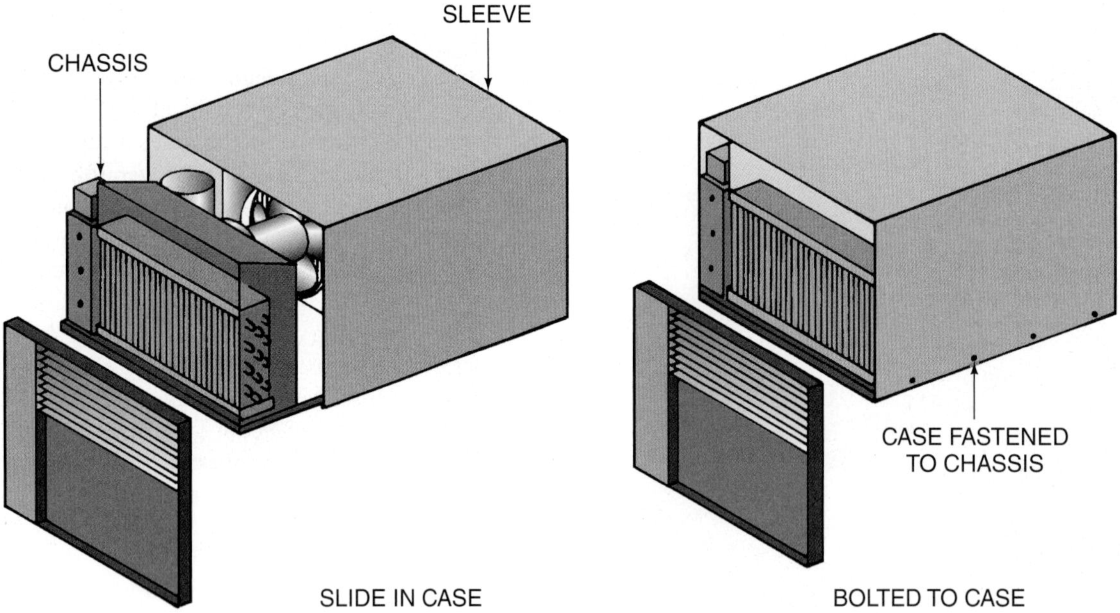

Figure 46.6 A roof-mount unit for recreational vehicles. Sometimes these are used in small stand-alone buildings. Photo by Bill Johnson

Figure 46.7 The case lifts off the roof-mount unit. Photo by Bill Johnson

Figure 46.8 Part of the control circuit is located inside, under the front cover. Photo by Bill Johnson

Figure 46.9 A room unit evaporator. Photo by Eugene Silberstein

Figure 46.10 A room unit compressor. Photo by Eugene Silberstein

Figure 46.11 A room unit condenser. Photo by Eugene Silberstein

Figure 46.12 A room unit capillary tube. Photo by Eugene Silberstein

Figure 46.13 Although found on domestic refrigerators and freezers, window air conditioners do not typically utilize suction-to-capillary-tube heat exchange.

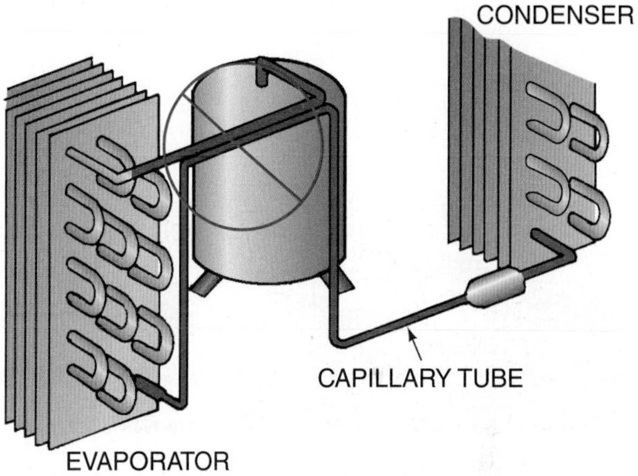

Figure 46.14 Refrigeration cycle component.

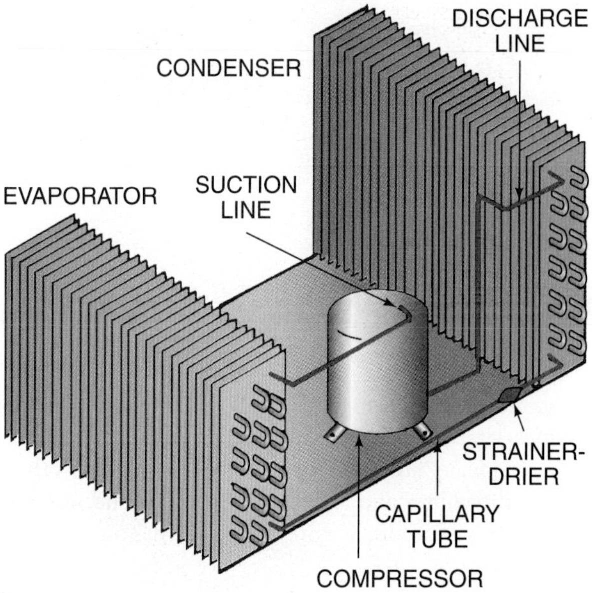

Figure 46.15 (A) An evaporator coil with three parallel refrigerant paths or circuits. (B) An evaporator coil with a single refrigerant circuit.

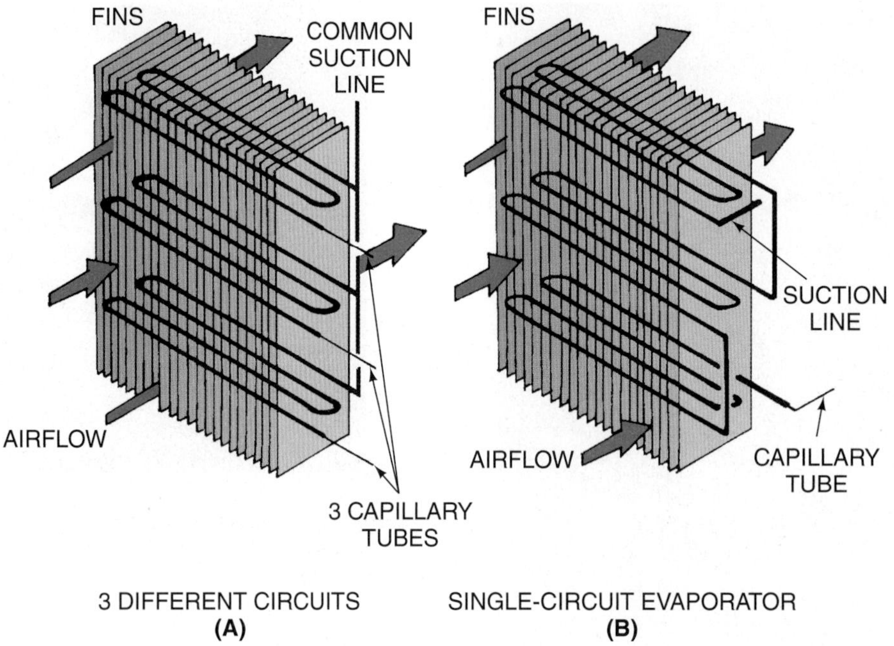

Figure 46.16 Some fins force the air to have contact with the fins and tubes.

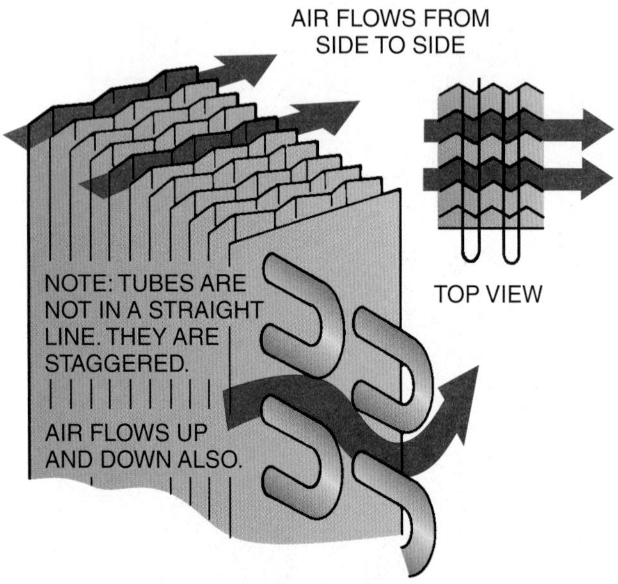

Figure 46.17 A condensate pan.

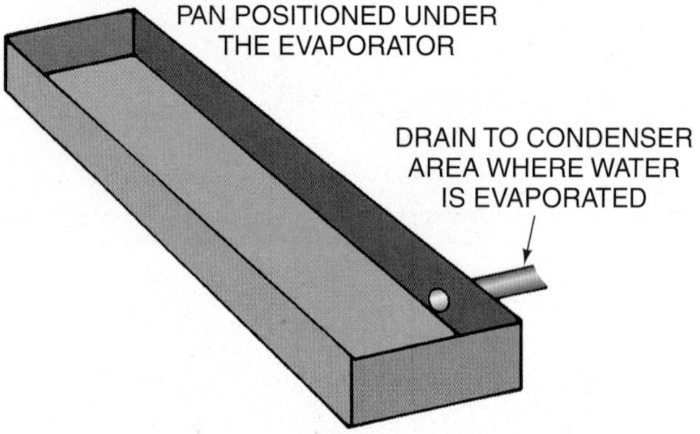

46.4 HEAT-PUMP-STYLE ROOM UNITS

Some window and through-the-wall units have reverse-cycle capabilities similar to the heat pumps discussed in Unit 43, "Air Source Heat Pumps." They can absorb heat from the outdoor air in the winter and reject this heat to the indoors. This is accomplished with a four-way reversing valve. The valve is used to redirect the suction and discharge gas at the proper time to provide heat or cooling, **Figure 46.24**. Check valves are used to ensure correct flow through the correct metering device at the proper time. A study of Unit 43 will help in understanding the basic heat-pump cycle because the same principles are involved. The following is a description of a typical cooling cycle, illustrated in **Figure 46.25**.

During the cooling cycle, the refrigerant leaves the compressor, via the discharge line, as a hot gas. The hot gas enters the four-way valve and is then directed to the outdoor coil, where heat is rejected to the outdoors. The refrigerant is condensed to a liquid, leaves the condenser, and flows to the indoor coil after flowing through the capillary tube. The refrigerant is expanded in the indoor coil, where it boils to a vapor, just as in the cooling cycle in a regular cool-only unit. The cold, heat-laden vapor leaves the evaporator and

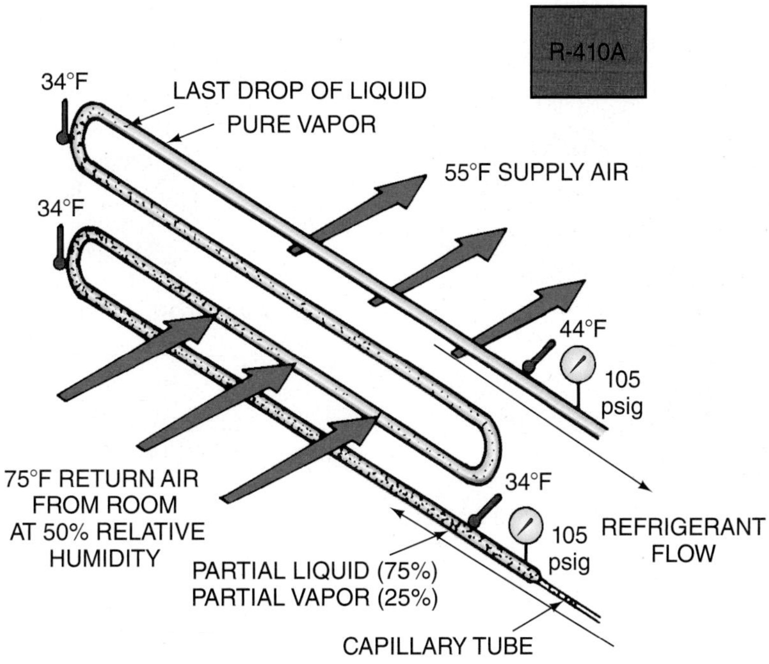

Figure 46.18 A typical R-410A evaporator with pressures and temperatures.

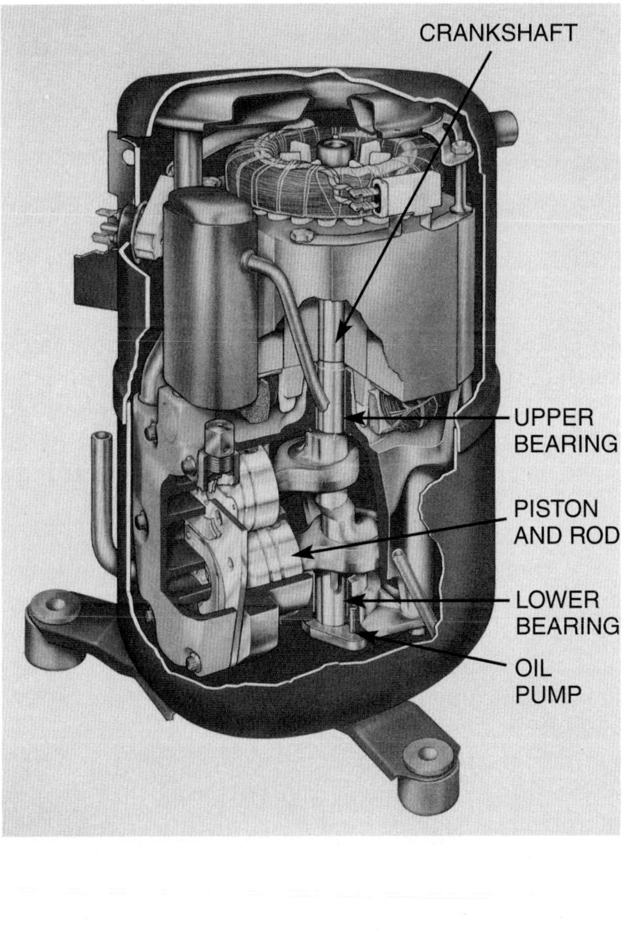

Figure 46.19 A hermetically sealed compressor. Courtesy Tecumseh Products Company.

Figure 46.20 A rotary compressor. Reprinted with permission of Motors and Armatures, Inc.

Figure 46.21 A finned-tube condenser.

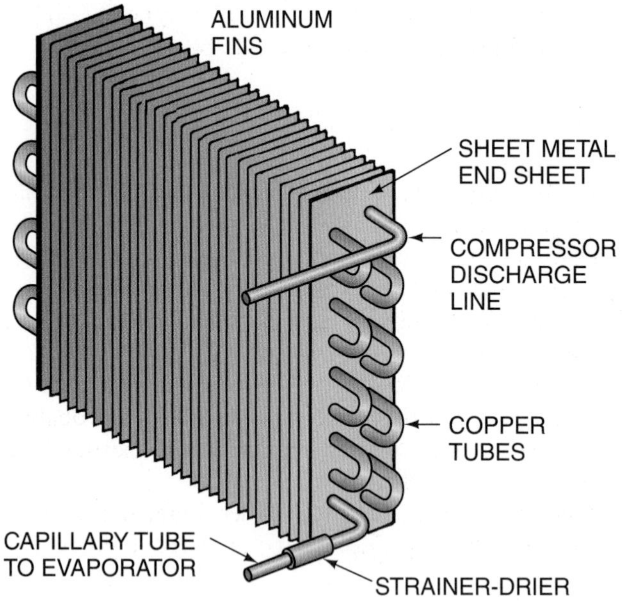

Figure 46.22 Moisture evaporated with the compressor discharge line.

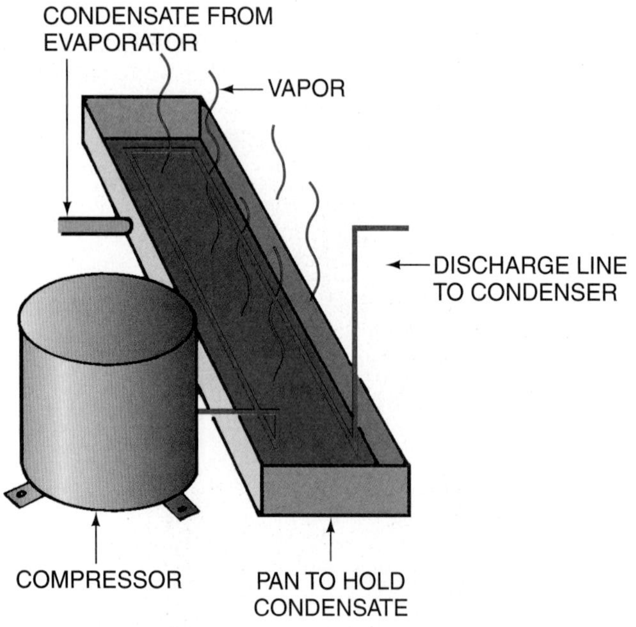

Figure 46.23 The slinger ring attached to the condenser fan blade.

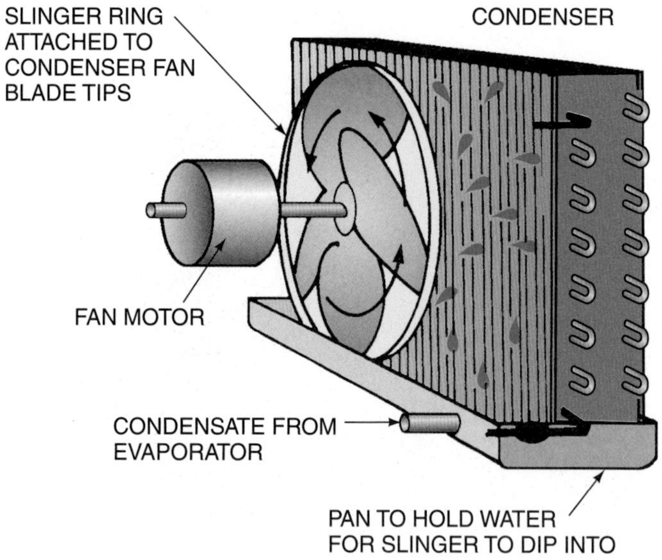

Figure 46.24 The four-way reversing valve for a heat pump. Photo by Eugene Silberstein

enters the four-way valve body. The four-way valve directs the refrigerant to the suction line, which carries the refrigerant to the compressor, where the refrigerant is compressed and the cycle repeats itself.

The heating cycle may be followed in **Figure 46.26**. The hot gas leaves the compressor and enters the four-way valve body as in the cooling cycle. However, in the heating mode, the four-way valve directs the hot gas to the indoor coil. The indoor coil functions as the condenser and rejects heat to the conditioned space. The refrigerant condenses to a liquid and flows out of the indoor coil to the outdoor coil after flowing through the capillary tube. In the outdoor coil, which is functioning as the evaporator, the refrigerant absorbs heat while boiling to a vapor. The heat-laden vapor leaves the outdoor coil and enters the four-way valve, where it is directed to the compressor suction line. The vapor is compressed and moves to the compressor discharge line to repeat the cycle.

The compressor found in a heat-pump system is not a conventional air-conditioning compressor. It has a different cylinder displacement and horsepower. The compressor must have enough pumping capacity for low-temperature operation. The unit evaporator also operates below freezing during the heating cycle, and frost will build up on it. Different manufacturers handle this frost buildup in different ways, so the technician will have to consult their literature. A defrost cycle may be used to defrost this ice, for example. Another point to remember is that the capacity of a

Figure 46.25 The refrigeration cycle and description—cooling.

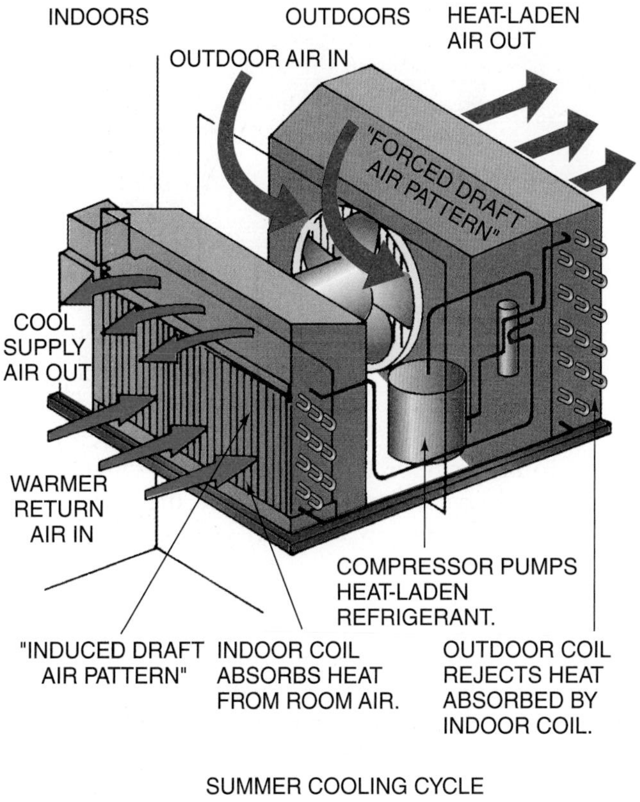

Figure 46.26 The refrigeration cycle and description—heating.

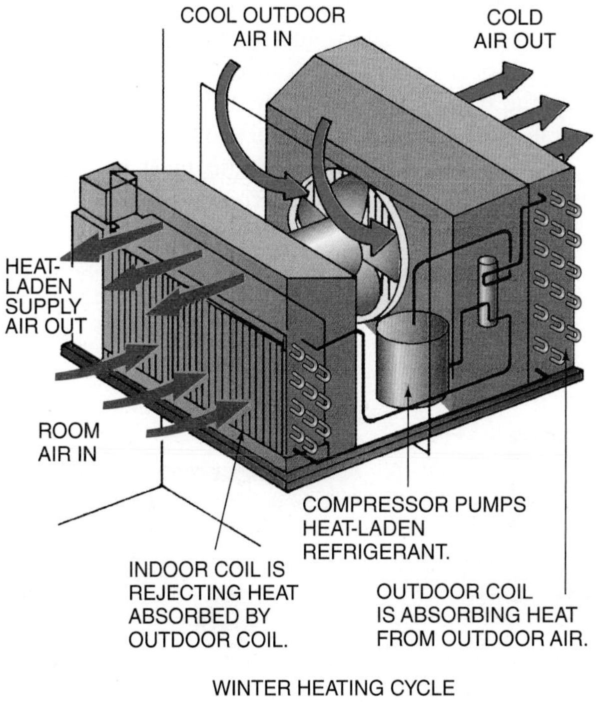

heat pump is less in colder weather, so supplemental heat—commonly electrical strip heat—will probably be provided. Supplemental heat may also be used to prevent the unit from blowing cold air during the defrost cycle.

46.5 INSTALLATION

There are two types of installation for both cooling and cooling-heating room units: window installation and wall installation. The window installation may be considered temporary because a unit may be removed and the window restored to its original use. The wall installation is permanent because a hole is cut in the wall. This hole must be repaired if the unit is removed. In a window installation, when a unit must be changed for a new one a new unit that fits the window would probably be available. A new unit that fits perfectly into an existing hole in a wall may be very difficult to obtain.

Window installations are for either double-hung or casement windows, **Figure 46.27**. Units for double-hung windows

Figure 46.27 Window units are for either (A) double-hung or (B) casement windows. Courtesy Friedrich Air Conditioning Co.

(A)

(B)

are the most popular because there are more of these types of windows. Units may be installed in other types of windows, such as a picture window or jalousie window, but doing so requires a great deal of carpentry skill, and often requires a certain amount of glasswork. A special unit is manufactured for casement window installation, **Figure 46.28**.

Some general guidelines should be followed in every installation. On the inside, a dedicated electrical outlet must be located within the length of the cord furnished with the unit. Extension cords are not recommended by manufacturers. The proper electrical outlet also must match the plug on the end of the cord, **Figure 46.29**. Window units may be either 115- or 208/230-V operation and should be on a circuit by themselves. Usually, a 115-V unit is chosen to avoid having to install an electrical circuit, but this choice is often made without investigating whether the existing circuit already has some load on it. For example, a 115-V electrical outlet may be part of the lighting circuit and a television and several lamps may be on the same circuit. The addition of a window air conditioner of any size may overload the circuit.

Figure 46.28 A window unit installed in a casement window.

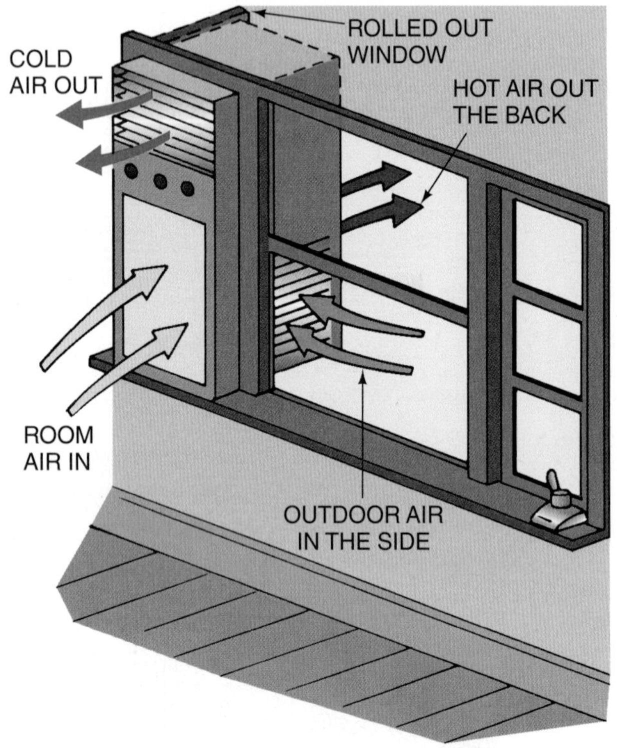

Figure 46.29 The proper outlet must be used that matches the plug.

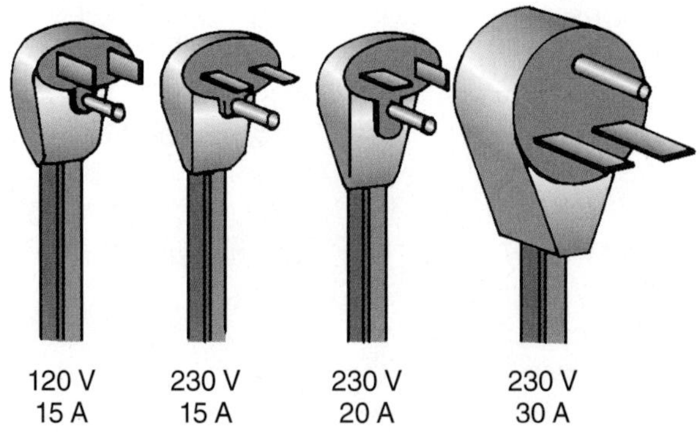

Investigating the electrical circuit to see what extra capacity is available is necessary. This may be accomplished by turning on all the lights in the room nearest the intended outlet, and possibly the lights in the adjacent room, and then turning off the breaker that controls the outlet. The current load from the lights may be totaled by adding them mathematically or using an ammeter on the wire to the circuit. If a 15-A circuit is already using 10 A, a window unit consuming 6 A will overload the circuit. A typical 5000 Btu/h unit will pull about 7 A in a 115-V circuit. An inexperienced homeowner who buys a unit at the department store and installs it may be in trouble.

Other considerations on the inside of the room are the best window to use and the air direction. A window that will give the best total room circulation may not be the best choice because it may be next to the easy chair in a den, by the bed in a bedroom, or behind the dining table, **Figure 46.30.** Some people will purchase a large window unit for several rooms. The unit must be able to circulate the air to the adjacent rooms or the thermostat will shut the unit off and cool only one room. A floor fan may be used to move the air from the room with the unit, **Figure 46.31.** Through-the-wall units in motels usually

Figure 46.30 Some bad locations for window air conditioners.

seem to be located on the outside wall, under the window and where the chair or chairs and table are located. This is the best practical place for the unit, but not necessarily the best for the occupants. However, there may be no place in a small room other than on an outside wall, **Figure 46.32**.

A unit located where air will recirculate—for example, behind drapes—will be a problem in any installation. Air will continually recirculate to the return-air grille and be cold enough to satisfy the thermostat, shutting the unit off. The airflow should usually be directed upward, as cold air will fall, **Figure 46.33**. This will also keep the supply air from mixing with the return air and freezing the evaporator or satisfying the thermostat.

Windows must be wide enough for the unit and must open enough so that the unit can be set on the window ledge, **Figure 46.34**. Smaller units are of a size that when placed on the window ledge, they will almost balance. In larger units, the compressor is located more to the rear, causing the unit to have a tendency to fall out of the window unless half of it is extended into the room. This, of course, is not desirable. Most manufacturers provide a brace kit to support the back of the unit, **46.35**. The bracket takes

Figure 46.31 A floor fan may be used to circulate air to adjacent rooms.

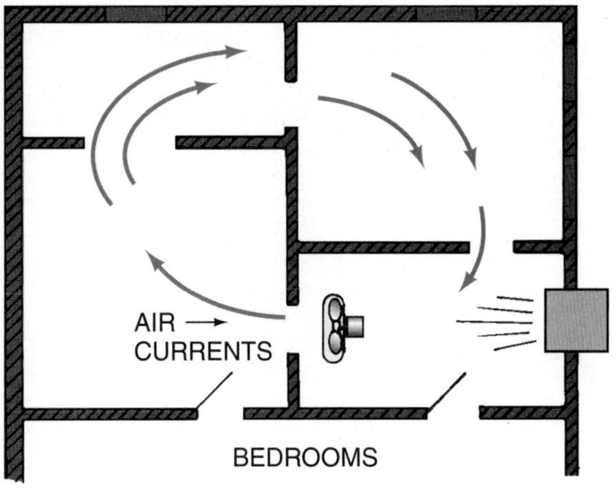

Figure 46.32 The outside wall may not be the best location for comfort, but it is the only place in most rooms.

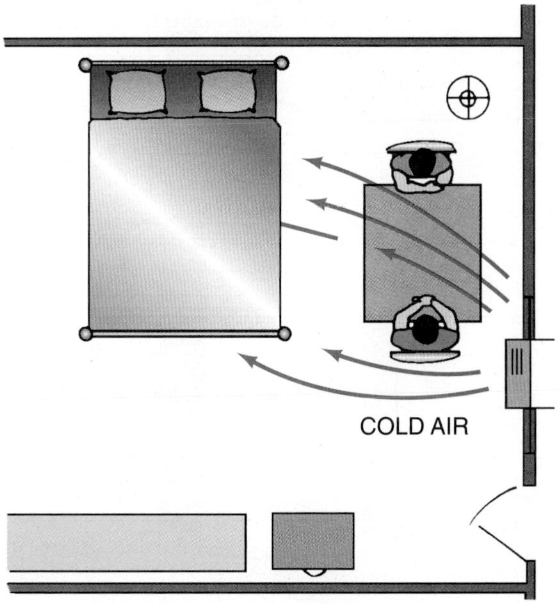

Figure 46.33 Cold air should be directed upward for best distribution.

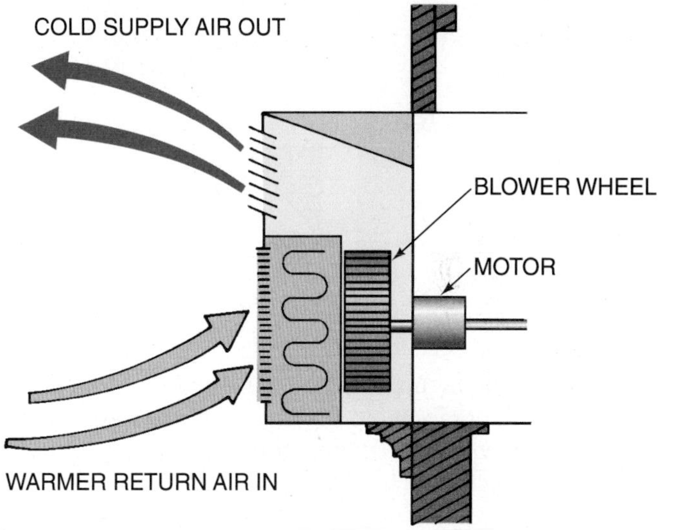

Figure 46.34 The window must open wide enough for the unit to fit.

Figure 46.35 Larger units may be located farther to the rear so that they do not protrude into the room. A brace kit should be used.

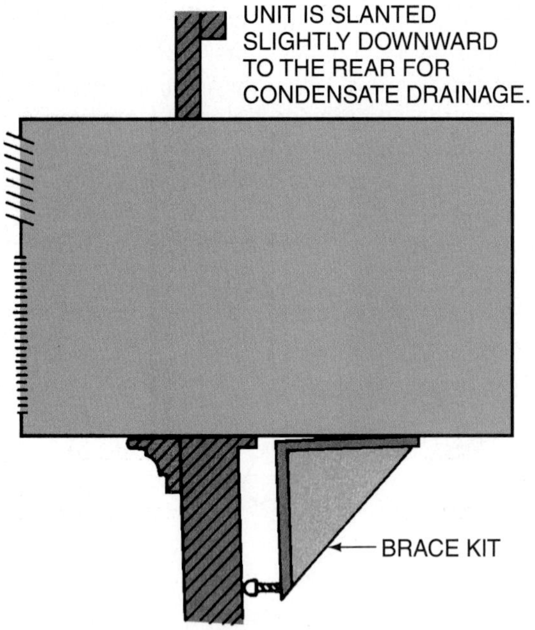

Figure 46.36 A foam strip to fill the gap between the window sections.

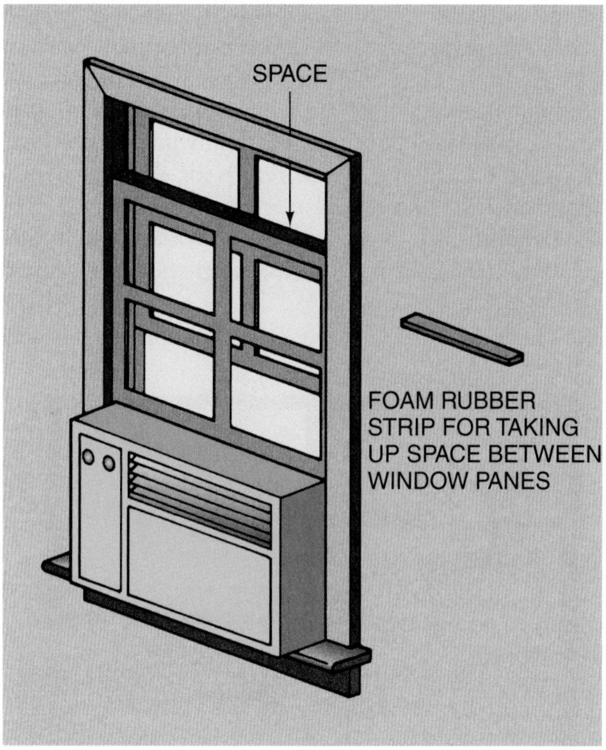

the pressure off the upper part of the window by placing the load on the bracket. A window unit never completely fills the window. Each new unit comes with a kit to aid in neatly filling the gap. The kit may contain telescoping side panels or panels matching the unit color that may be cut to fit. Also, the space between the partly raised movable window section and the upper part of the window must be insulated. Usually a foam strip is furnished for this purpose, **Figure 46.36**.

Directions are included with each new unit. Be sure to follow them and use all of the appropriate mounting hardware supplied with the unit. Units installed without the proper braces or window kits tend to vibrate, **Figure 46.37**. A properly braced unit will typically slant slightly to the rear for proper condensate drainage. When the weight is supported at the back, the pressure on the window is relieved and vibration is reduced.

Most units evaporate the condensate, but some allow it to drain out the back. This condensate should drain to a proper location, such as a flower bed, the lawn, or a gutter, **Figure 46.38**. Properly sized units operating under a normal load will be able to completely evaporate the condensate without having the unit drip. However, when there is an excessively high load on the system, even properly sized units may drip. Sometimes condensate is allowed to drip on a sidewalk. This is not good, as algae may form and the walkway could become slick, **Figure 46.39**. The unit must also be installed in such a manner that air can circulate across the

Figure 46.37 An awkward installation because the bracket kit has been lost.

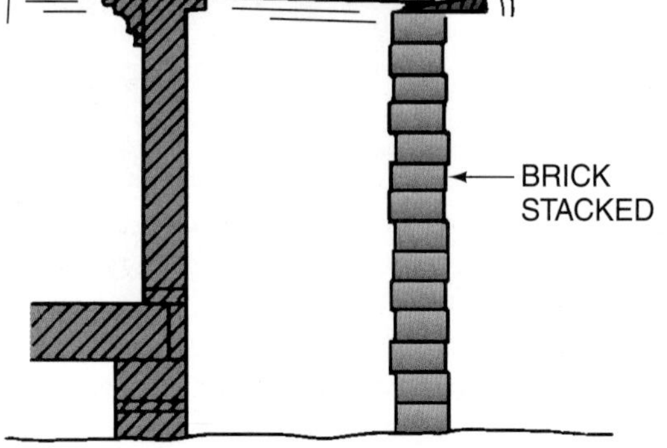

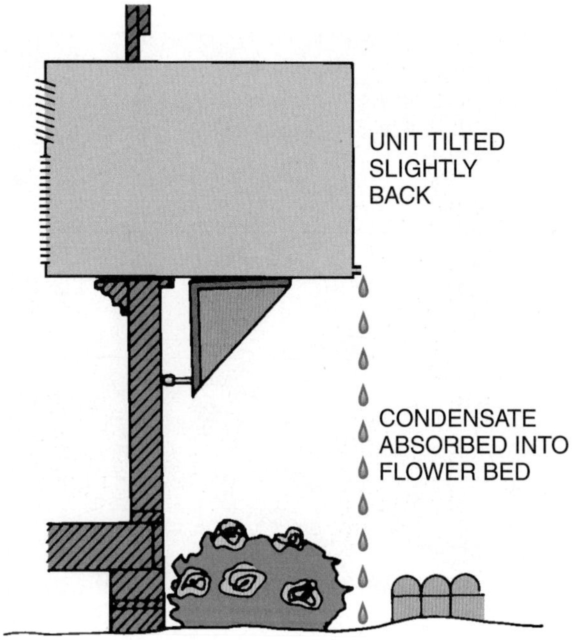

Figure 46.38 The condensate should drain to the proper place.

Figure 46.39 Condensate on a walkway can be dangerous.

condenser and not recirculate. It must be allowed to escape the vicinity and not heat the surroundings. Window units have been located with the condenser in a spare room, such as at a business location, **Figure 46.40**. This is poor practice because the unit heats the air, and high operating conditions will occur. Recirculation will cause high head pressure and high operating cost, **Figure 46.41**. In addition, there should be no obstructions located close to the air discharge.

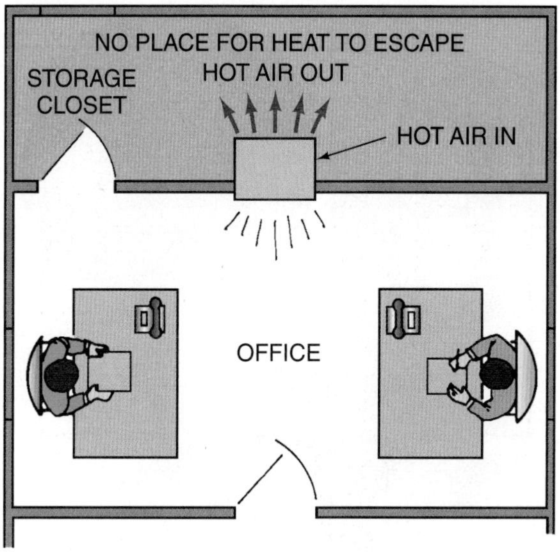

Figure 46.40 The heat must be able to escape from around the unit. Units with the condenser in an adjacent room will not operate efficiently.

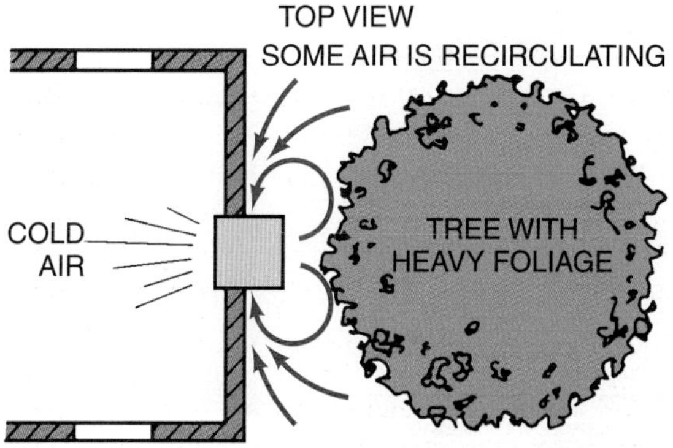

Figure 46.41 Recirculated air will cause high head pressure.

Through-the-wall units typically protrude into the room only a slight amount with the bulk of the unit on the outside, **Figure 46.42**. When these units are installed in a motel, the outdoor portion protrudes into the walkway slightly. If condensate is not completely evaporated, it will run across the walkway and cause a hazard. Through-the-wall units may be installed while the building is under construction. This is accomplished by installing wall sleeves and covering the openings until the unit is set in place, **Figure 46.43**. The wiring is run to the vicinity of the unit, and the unit is connected when installed. It is important that holes of the correct size be cut. The wall sleeve should be purchased from the manufacturer in advance to ensure this.

Figure 46.42 Most of the through-the-wall unit is located on the outside of the wall.

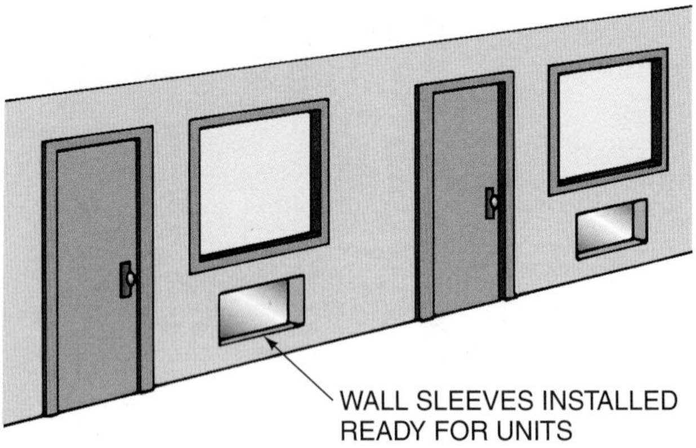

Figure 46.43 Wall sleeves may be installed in advance of the unit.

Figure 46.44 The thermostat is located in the return airstream. Photo by Eugene Silberstein

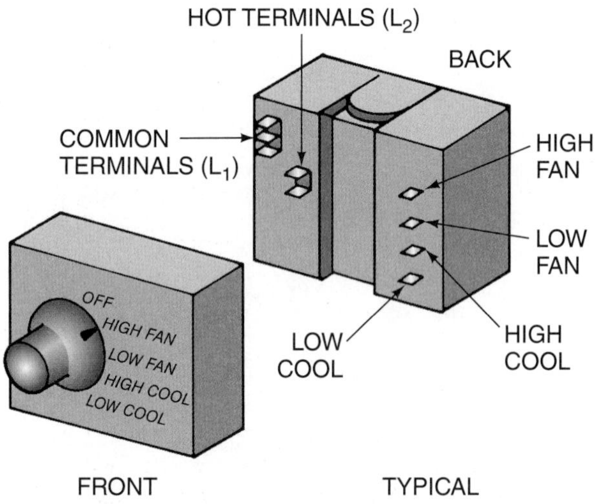

Figure 46.45 A typical selector switch for a room unit.

46.6 CONTROLS FOR COOLING-ONLY ROOM UNITS

Room units that provide only cooling are plug-in appliances furnished with a power cord. Most of the time, all of the controls are contained in the unit, as there is no wall-mounted thermostat. Many new room units, however, do come supplied with wireless remote controls so that the user can turn the unit on and off, as well as change the temperature setting in the space without having to approach the unit. In all cases, the room thermostat sensor is located in the unit's return airstream, **Figure 46.44**. A typical room unit may have a selector switch to control the fan speed and provide power to the compressor circuit, **Figure 46.45**. The selector switch may be considered a power distribution center. The power cord will usually be wired straight to the selector switch for convenience. Therefore it may contain a hot, grounded neutral and a frame safety (green) ground wire for a 115-V unit. A 208/230-V unit will have two hot wires and a ground, **Figure 46.47**. The ground or neutral wire for 115-V service is routed straight to the power-consuming device. The hot wire is connected directly to the other side of the power-consuming device.

Figure 46.46 Typical selector switch terminal designations for 120-V units.

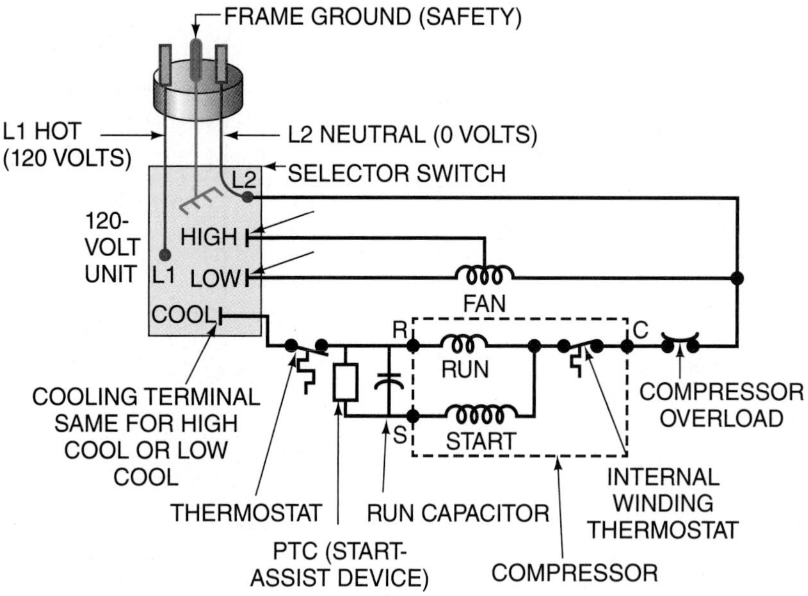

Figure 46.47 Typical selector switch terminal designations for 208/230-V units.

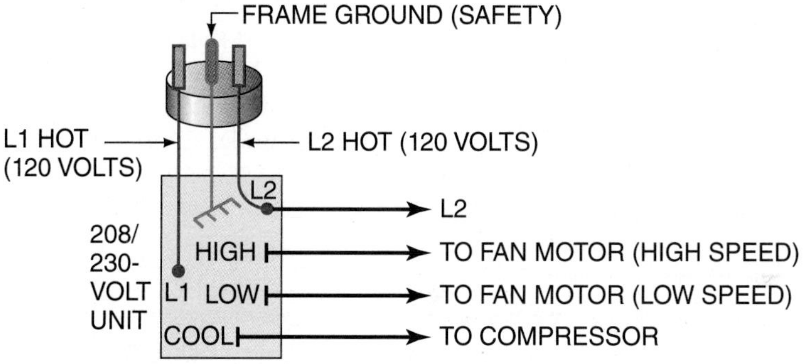

Many combinations of selector switches will allow the owner to select fan speeds, high and low cool, exhaust, and fresh air. High and low cool is accomplished with fan speed. Because the unit has only one fan motor, a reduction in fan speed slows both the indoor and the outdoor fan. This reduces the capacity and the noise level of the unit.

Some units are equipped with energy-saver switches. The purpose of this switch is to allow the user to determine if the fan motor should cycle on and off with the compressor. The basic circuit for an air conditioner equipped with this control is shown in **Figure 46.48**. In this diagram, contacts L1-C will close whenever the selector switch is set to the COOL position. With the energy-saver switch in the OFF position, power from L1 will be fed to Terminal A all the time, whether or not the unit is set to COOL or if the space thermostat is calling for compressor operation. This means

Figure 46.48 Control switch wiring for a room unit equipped with an energy-saver switch.

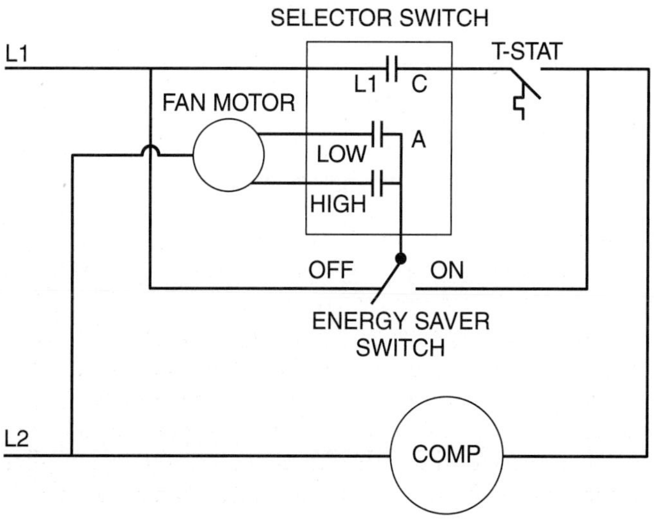

Figure 46.49 With the energy-saver switch OFF, the fan motor operates independently of the compressor.

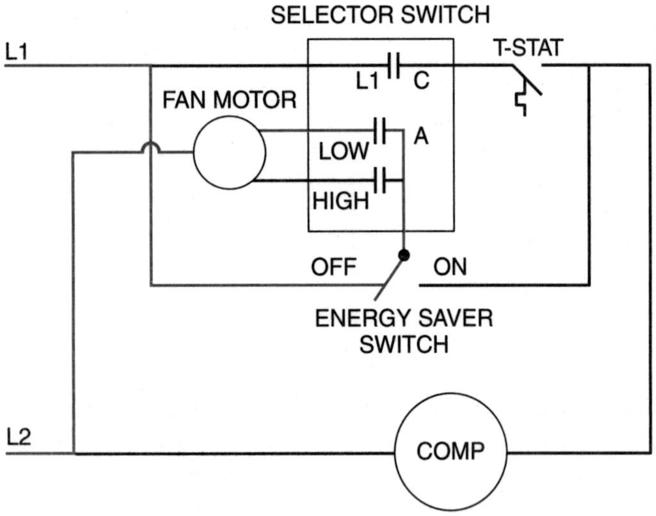

Figure 46.51 Diagram showing the path of electric current when the compressor is operating and the energy-saver switch is on.

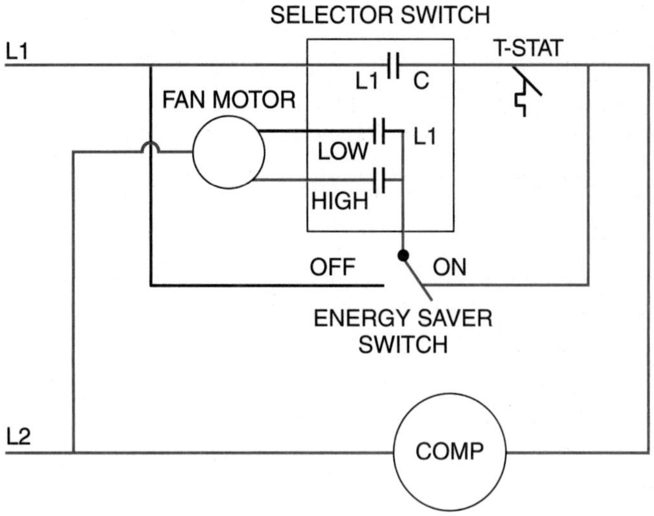

that, with the energy-saver switch in the OFF position, the fan will operate independently of the thermostat and cooling contacts on the switch, **Figure 46.49**. With the energy-saver switch in the ON position, power from L1 will be fed to Terminal A only when the selector switch is calling for cooling and the space thermostat is calling for compressor operation, **Figure 46.50**. If the energy-saver switch is in the ON position *and* the selector switch is set to COOL *and* the space thermostat is calling for cooling, Terminal A will be powered and the indoor fan will operate, **Figure 46.51**. If the thermostat is satisfied, both the compressor and the fan will cycle off. Some newer units have electronic wireless remote controls. They often have cool and fan speed options, and 24-hour ON-OFF timers. The exhaust and fresh-air control is a lever that positions a damper to bring in outdoor air or exhaust indoor air, **Figure 46.52**. The selector switch sends power to various circuits (including the compressor start circuit) to start the compressor. Motor starting was discussed in Unit 19, "Motor Controls," and will not be further discussed here. **Figure 46.53** shows some selector switch positions and various modes of operation.

Figure 46.50 With the energy-saver switch ON, the fan motor can only operate when compressor operation is called for.

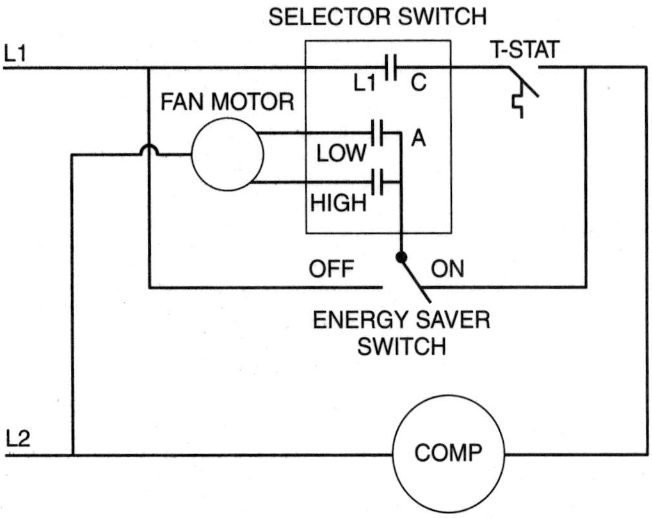

Figure 46.52 Exhaust and fresh air are accomplished with a damper operated from the front of the unit.

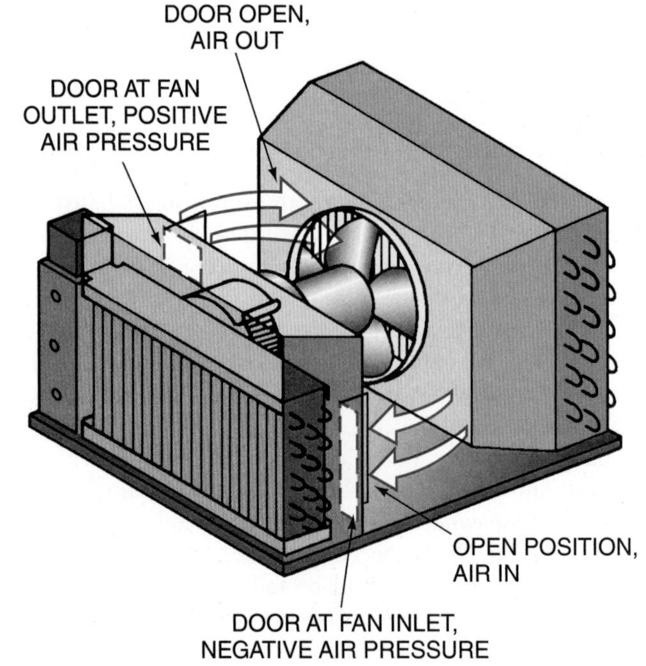

Figure 46.53 Some selector switch wiring diagrams and compressor starting diagrams. (A) In the OFF position, no electrical connections are made. (B) In the HI FAN position, an electrical connection is made between L1 and the HI FAN terminal. (C) In the LOW FAN position, an electrical connection is made between L1 and the LOW FAN terminal. (D) In the HI COOL position, electrical connections are made between L1 and the HI FAN terminal as well as between the L1 and COOL terminals. (E) In the LOW COOL position, electrical connections are made between L1 and the LOW FAN terminal as well as between the L1 and COOL terminals.

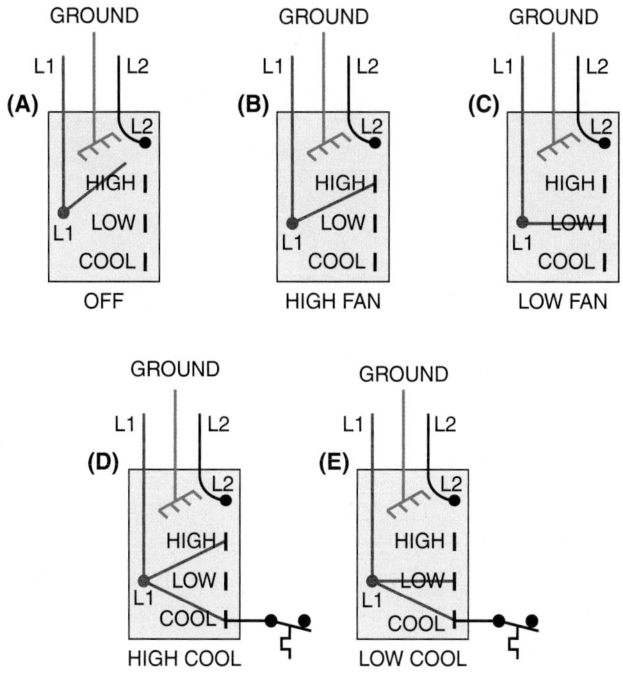

46.7 CONTROLS IN COOLING AND HEATING UNITS

The cooling controls for cooling and heating units are the same as previously discussed for the cooling cycle. Typically, a selector switch changes the unit from cooling to heating. When the unit has electric strip heat, the selector switch merely directs the power to the heating element through the thermostat, **Figure 46.54**. The heating contacts on the thermostat will close on a drop in sensed space temperature. If the unit is a heat pump, it will have a selector switch to select heating or cooling. The cooling cycle should be a typical cycle with the addition of the four-way valve. In the heating cycle, defrost may have to be considered. Many newer units with strip heat have electronic controls. In these units, a thermostat usually controls the heating elements to maintain constant room temperature, and they may have digital touch-pad controls with remotes and a 24-hour ON-OFF timer.

Figure 46.54 The selector switch directs the power to the heating or cooling circuits.

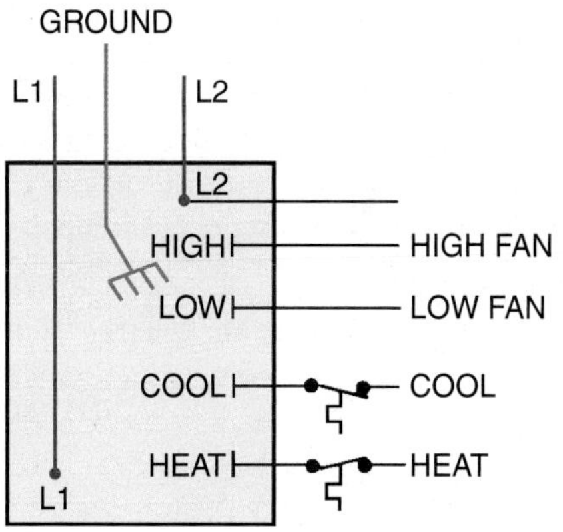

46.8 MAINTAINING AND SERVICING ROOM UNITS

Maintenance of room units basically involves keeping the filters and coils clean. Most newer motors are typically permanently lubricated. If the filter is not maintained, the indoor coil will become dirty and cause the unit to operate at low suction pressures. This will cause the evaporator coil temperature to drop. While in the cooling mode, the coil is already operating at about 35°F. If the coil temperature drops below this, ice will begin to form. Some manufacturers provide freeze protection that will shut the compressor off and allow the indoor fan to run until the threat of freezing is over. This may be done with a thermostat located on the indoor coil or a thermostat mounted on the suction line, **Figure 46.55**.

Figure 46.55 Freeze protection for a room unit.

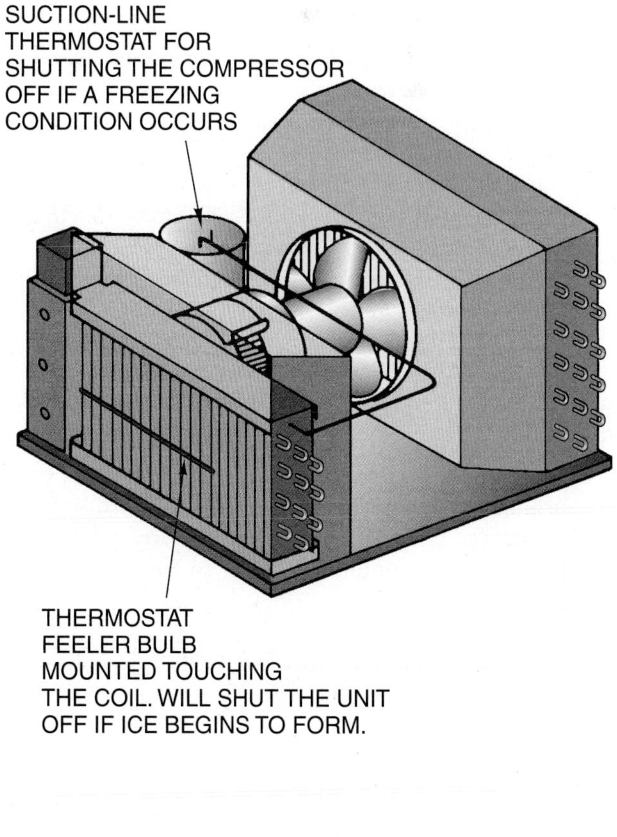

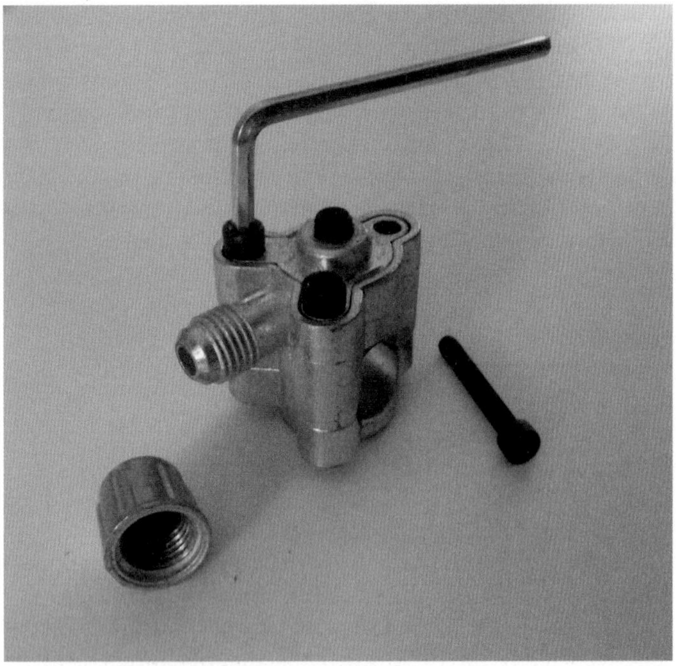

Figure 46.56 A line tap valve. Photo by Eugene Silberstein

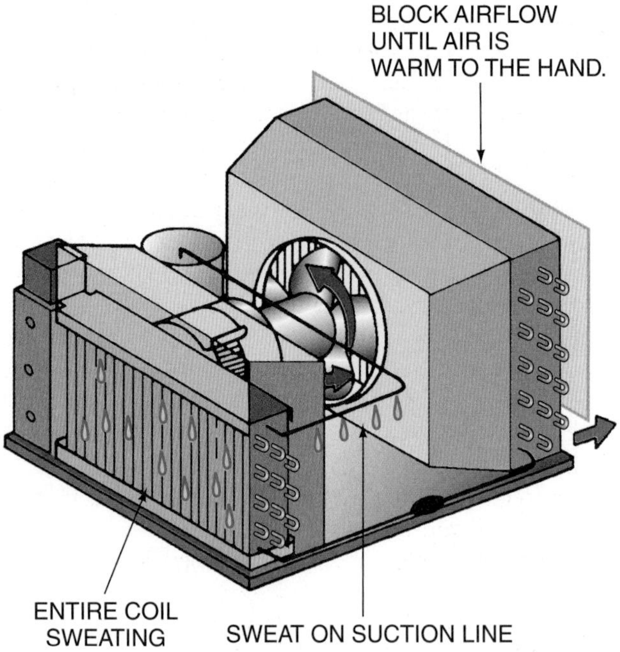

Figure 46.57 Testing for a low charge before installing gauges.

Service may involve mechanical or electrical problems. Mechanical problems usually arise in the fan motor and bearing or refrigerant circuits. All room units have a critical refrigerant charge, and gauges should be installed only after there is some proof that they are needed. The system will normally be sealed and for gauges to be installed line tap valves or the installation of service ports on the process tubes will be required. Line tap valves and their use is discussed in Unit 45, "Domestic Refrigerators and Freezers." **Figure 46.56** shows an example of a line tap valve. Line tap valves must be removed from the system after use. Since these valves leak, leaving one installed on a room unit is a violation of Section 608 of the Clean Air Act of 1990. **NOTE:** *Be sure to follow the valve manufacturer's instructions.* •

A bench test at the workshop is often performed on room units. Before installing gauges, the technician should perform the following test when a low charge is suspected. Because the condenser and the evaporator operate in air that is the same temperature, the condenser airflow may need to be reduced to move any refrigerant from the condenser due to low head pressure. A full explanation of condenser operation and charging under low ambient conditions is discussed in Unit 29, Section 29.6, "Calculating the Correct Head Pressure for Air-Cooled Equipment." Refer to **Figure 46.57** while reading the following:

- Remove the unit from its case.
- Make sure that air flows through the coils. For temporary testing, cardboard may be positioned over places where panels force air to flow through the coils.
- Start the unit in HIGH COOL.
- Let the unit run for about 5 min and observe the sweat on the suction line. It should come close to the compressor. The evaporator should be cold from bottom to top.
- Cover a portion of the condenser to force the head pressure to rise. If the room is cool, a portion of the charge may be held in the condenser. The condenser airflow may be blocked to the point that the air leaving it is hot, about 110°F.
- With the head pressure increased, allow the unit to run for 5 min; the sweat line should move to the compressor. If the humidity is too low for the line to sweat, the suction line should be cold. The evaporator should be cold from bottom to top. If only part of the evaporator is cold (frost may form), the evaporator is starved. This may be a result of a restriction or a low charge.
- If the compressor is not pumping to capacity and the charge is correct, the coil will not be cold anywhere, only cool. An ammeter may be clamped on the common wire to the compressor and the amperage compared to the full-load amperage of the compressor.

When this test indicates that the charge is low, because the suction line is not cold or sweating, gauges should be installed.

When the charge is low, the leak should be found. If the unit has never had a line tap or service port installed and has operated for a long period of time without a problem, it can be assumed that the leak has just occurred. An electronic leak detector may be used to help locate the leak. Turn the unit off and allow the pressures to equalize. Make sure there is enough refrigerant in the unit to properly leak check. If the system temperature and pressure relationship indicates there is no liquid present, there is so little refrigerant in the system that what is left could be considered trace refrigerant for leak-checking purposes. If the leak cannot be found with the unit off, dry nitrogen may be used to increase the pressure to the pressure at which the unit was leak tested at the factory. Be sure to check the nameplate on the unit for this information. Once the leak has been located, the nitrogen and refrigerant trace can be released to the atmosphere. After the leak has been repaired, the unit must be pressure tested, properly evacuated, and recharged.

4R *The correct method of charging a unit is to remove and recover the remaining charge, evacuate the unit, and measure in the correct charge printed on the nameplate.* 4R **SAFETY PRECAUTION:** *Do not start the unit with a nitrogen charge in it because high pressure will occur and the motor will quickly overheat due to lack of cooling.* • Technicians should use accurate scales similar to the one in **Figure 46.58** to measure the charge. At times, the technician may want to adjust the charge rather than start from the beginning. A charge may be adjusted by using the following method, which, together with **Figure 46.59**, illustrates the procedure:

- Install suction and discharge gauge lines.
- Purge the gauge lines and connect them to the correct refrigerant cylinder.
- Make sure that air is passing through the evaporator and the condenser coils; cardboard may be used as temporary panels.
- Start the unit in HIGH COOL.
- Watch the suction pressure and do not let it fall below atmospheric pressure into a vacuum; add refrigerant through the suction gauge line if the pressure falls too low.
- Adjust the airflow across the condenser until the head pressure is 420 psig and corresponds to a condensing temperature of 120°F for R-410A. Refrigerant may have to be added to obtain the correct head pressure.
- Add refrigerant vapor at intervals with the head pressure at 420 psig until the compressor suction line is sweating at the compressor. As refrigerant is added, the airflow will have to be adjusted over the condenser or the pressure will rise too high. Note also that the evaporator coil should become colder toward the end of the coil as refrigerant is added.

Figure 46.58 Charging scale. Photo by Eugene Silberstein

Figure 46.59 Charging a unit by sweat line.

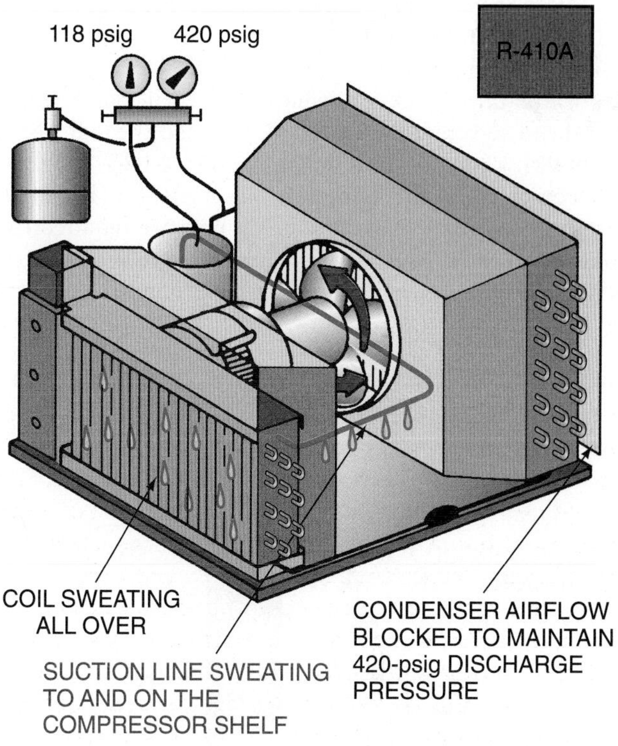

- The correct charge is verified after about 15 min of operation at 420 psig and by a cold (probably sweating) suction line at the compressor. The suction pressure should be about 118 psig and the compressor should be operating at near full-load amperage.

When a system must be evacuated, the same procedures are used as those discussed in Unit 8, "Leak Detection, System Evacuation, and System Cleanup." The larger the evacuation ports, the better. Unit 45, "Domestic Refrigerators and Freezers," has a complete evacuation procedure for refrigerators and freezers; this same procedure would apply here.

The capillary tube may be serviced in much the same manner as described in Unit 45, "Domestic Refrigerators and Freezers." If the head pressure rises and the suction pressure does not and refrigerant will not fill the evaporator coil, there is a restriction. In this case, the technician must make sure the compressor has a starting relay and capacitor or it may not start. Thermostatic and automatic expansion valves do not equalize during the off cycle. This is a workable situation for a room unit, but not a refrigerator.

It may be difficult to remove the fan motor in many units. They are so compact that the fans and motor are in close proximity to the other parts. The evaporator fan is normally a centrifugal type (squirrel cage design) and the condenser fan is typically a propeller type. The condenser fan shaft usually extends to a point close to the condenser coil. Often there is not room to slide the propeller fan to the end of the shaft, **Figure 46.60**. Either the fan motor must be raised up or the condenser coil moved, **Figure 46.61**. The evaporator fan wheel is normally locked in place with an Allen-head setscrew that must be accessed through the fan wheel with a long Allen wrench, **Figure 46.62**.

Electrical service requires the technician to be familiar with the electrical power supply. All units are required to be grounded to earth. This involves running a green or bare copper ground wire to every unit from the grounded neutral bus bar in the main control panel. Ground wires provide a safety circuit to protect the service technician and others from electrical shock hazard. The electrical service must be of the correct voltage and have adequate current-carrying capacity. The example described in Section 46.5, "Installation," describes what typically happens if the circuit was not planned to operate an air-conditioning unit.

Electrical service may involve the fan motor, the thermostat, the selector switch, the compressor, and the power cord. The fan motor will be either a shaded-pole or a permanent

Figure 46.61 The condenser may be moved back or the fan motor may be raised in order to remove the fan blade.

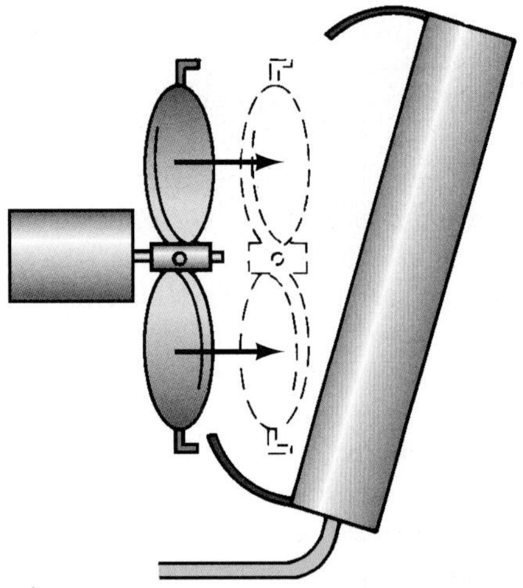

Figure 46.60 Often the condenser fan is hard to remove from the shaft because of the clearance between the condenser and fan.

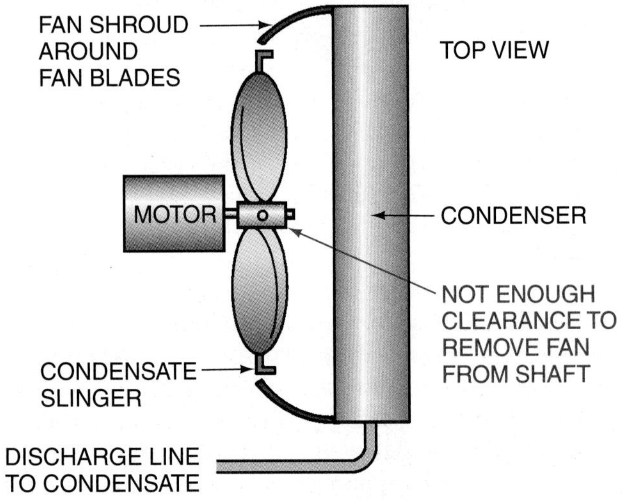

Figure 46.62 An Allen-head setscrew may be loosened by inserting a long Allen wrench through the blades.

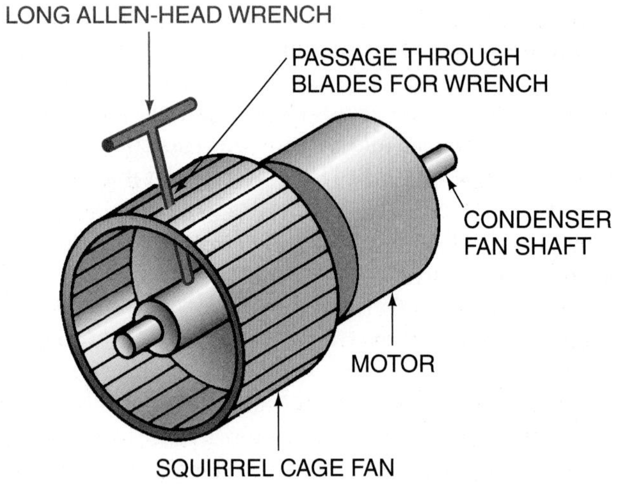

Figure 46.65 There should be a line voltage reading between the L1 and L2 Terminals on the switch.

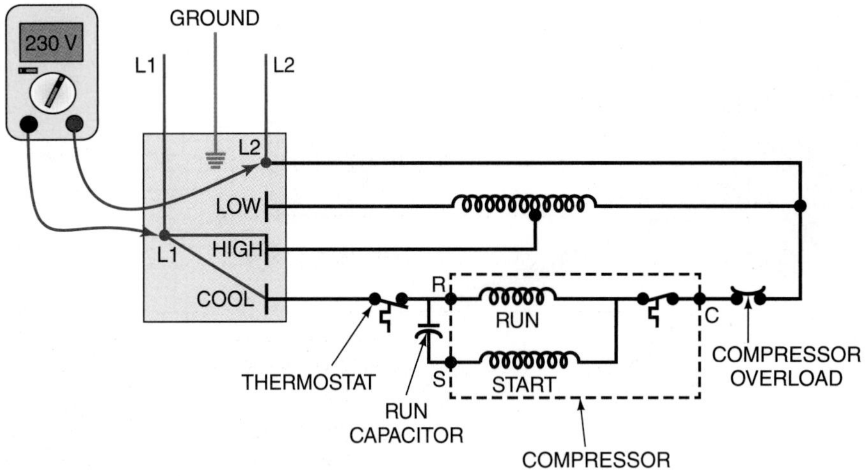

Figure 46.66 Line voltage reading between (A) the L2 Terminal and the HIGH FAN terminal and (B) the L2 Terminal and the COOL terminal. This switch is good.

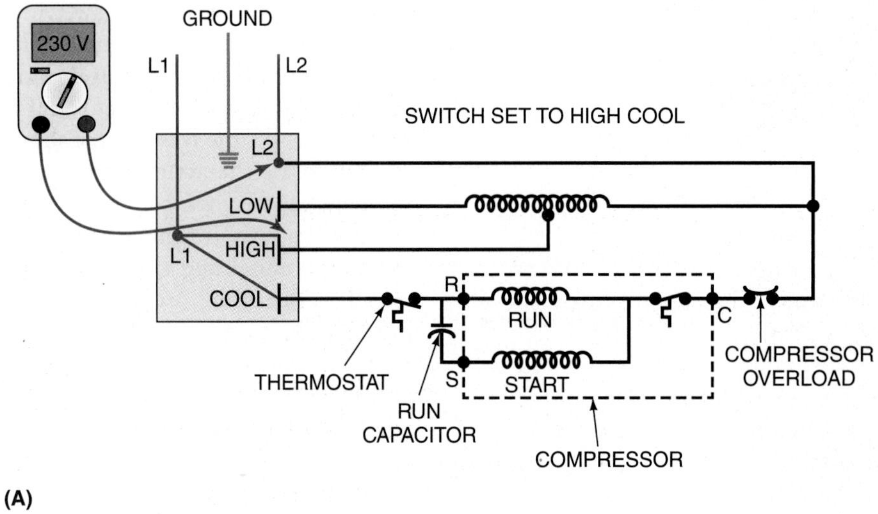

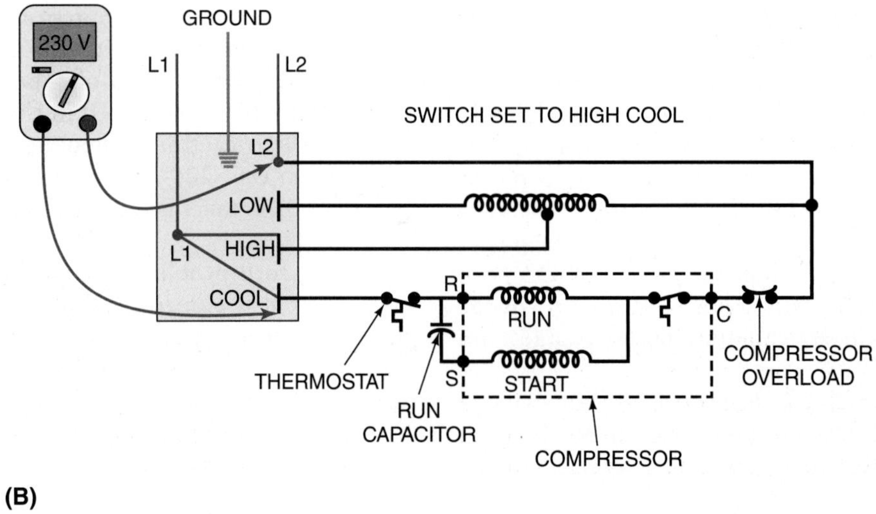

Figure 46.67 Voltage reading of 0 volts between (A) the L2 Terminal and the HIGH FAN terminal or (B) the L2 Terminal and the COOL terminal. This switch is defective.

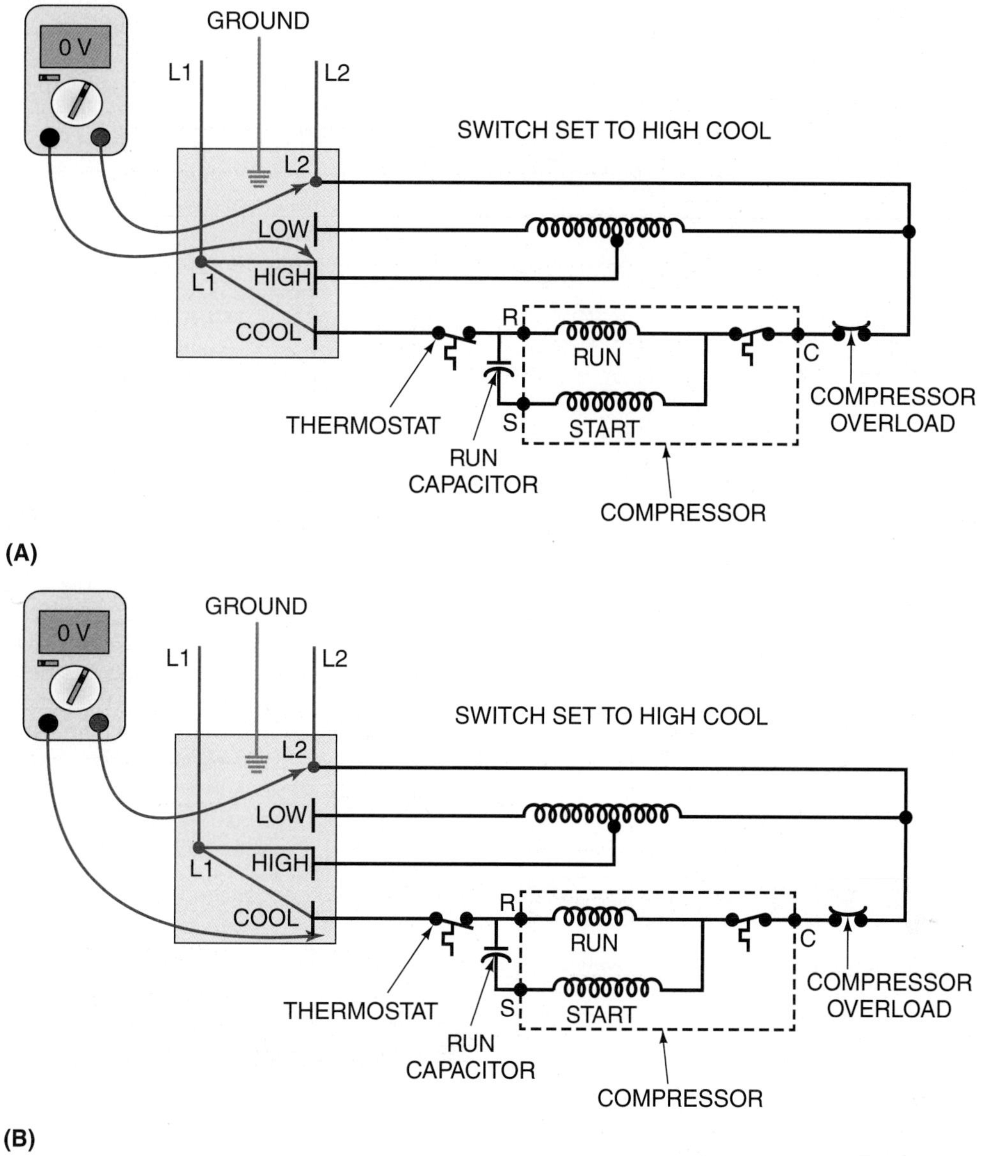

Figure 46.68 A power cord may need support.

connection, **Figure 46.68**. If the cord becomes hot to the point of discoloration or the plug swells, the plug should be changed. The wall receptacle may need inspection to make sure that damage has not occurred inside it. If the receptacle is discolored or distorted, it should be changed; installing a new plug if the receptacle is defective will not help solve the problem.

HVACR GOLDEN RULES

When making a service call to a residence:

- Try to park your vehicle so that you do not block the customer's driveway.
- Always carry the proper tools and avoid extra trips through the customer's house.
- Ask the customer about the problem. The customer can often help you solve the problem and save you time.

Added Value to the Customer

Here are some simple, inexpensive procedures that may be added to the basic service call.

- Clean or replace the air filters.
- Advise customers about proper operation; for example, tell them not to allow conditioned air to recirculate into the return air.
- If removal of the unit is necessary, obtain the proper assistance and use dropcloths to prevent dirt from the unit from falling on the floor.

46.9 SERVICE TECHNICIAN CALLS

SERVICE CALL 1

A customer in an apartment reports that the window unit in the dining room, used to serve the kitchen, dining room, and living room, is freezing solid with ice every night. The tenants both work during the day. They shut the unit off in the morning and turn it on when they get home from work. The air temperature is 80°F inside and outside by the time they turn the unit on. *The problem is that the head pressure is too low during the last part of the cycle, starving the evaporator and causing it to operate below freezing.*

The apartment house technician arrives late in the afternoon when the tenants get home. The technician asks them to describe how they are using and setting the controls on the unit to get an idea about what the problem might be. The tenants show him how the thermostat is set each night. The technician explains that the outdoor temperature is cooler than the temperature in the apartment by the time the unit is turned on and that a better approach may be to leave the unit on LOW COOL during the day and to raise the thermostat setting several degrees, letting the unit run some during the day. A call back a few days later verifies that the unit is performing well using this approach. The unit had no problem; it was lack of operator knowledge.

SERVICE CALL 2

A customer calls to indicate that the air conditioner in the guest bedroom is not cooling the room. The coil behind the filter has ice on it. This sounds like a low-charge situation. *The problem is that a drape is partially hanging in front of the unit, causing the cool supply air to recirculate. The supply air louvres are also pointing downward.*

Wiping his feet before entering the apartment, the technician makes his way to the unit. He notices that the drape hangs in front of the unit and explains that the drape will have to be kept out of the airstream. Also, the air discharge is pointed down, causing the air to move toward the return-air inlet. This is adjusted.

Just to be sure that the unit does not also have a low charge, the technician pulls it out of the case and places it on a small stool so it can be started. The technician notices that when the unit is started, air blows out of the top of the blower compartment and bypasses the indoor coil, so he places a piece of cardboard over the fan compartment to force the air through the coil. He starts the unit again and partially blocks the condenser to force the head pressure higher. After the unit has run for a few minutes, the suction line starts to sweat back to the compressor.

If the unit had a low charge, ice would form on the bottom of the coil because there would not be enough refrigerant to fill it. If the coil were dirty or the airflow blocked, the coil might ice all over, even to the compressor. The compressor might sweat excessively because the refrigerant wouldn't be boiled to a vapor and liquid would enter the compressor. This unit has never had gauges installed, so it has the original charge and it is still correct. The technician stops the unit, pushes it back into its case in the window, and cleans the filter. The unit is started and left operating normally.

SERVICE CALL 3

A customer calls to report that his room air conditioner is not cooling so it has been turned off. *The problem is a leak in the high-pressure side of the system.*

The technician asks the owner about the problem he is having with the unit, explaining that he will look at the unit and make every effort to resolve the problem. He then starts the unit. The compressor runs, but the unit does not cool. The technician turns the unit off and slides it out of its case onto a stool and restarts the compressor. He feels the discharge line and notices that it is cool to the touch. The technician tells the customer that the unit will have to be taken to the shop for further evaluation. He then brings in a hand truck as well as materials to seal the opening left by the removed air conditioner. He informs the customer that he will receive a call as soon as there is more information about the unit.

At the shop, the unit is moved to a workbench, and a line tap valve is installed on the suction process tube. When the

gauges are fastened to the line tap valve, a pressure of 0 psig registers on the gauge. The system is out of refrigerant. The technician allows a very small amount of R-410A, the refrigerant for the system, to flow into the unit and then increases the pressure to about 100 psig with dry nitrogen. No leak can be heard, so an electronic leak detector is used, which locates a leak on the discharge-line connection to the condenser. The nitrogen, along with the trace of R-410A, is released from the unit. The process tube is cut off below the line tap valve connection. A 1/4-in. flare connection is soldered to the end of the process tube and a gauge manifold connected to the low side; the discharge line leak is repaired using a high-temperature, silver-bearing brazing rod.

After removing the torch from the connection, the gauge manifold valve is closed to prevent air from being drawn into the system when the vapor inside cools and shrinks. When the connection is cooled, nitrogen is introduced to the system and the unit is leak-checked again, this time with a liquid leak detection solution. There is no leak. The nitrogen is released from the unit, and the system is ready for evacuation.

The vacuum pump is attached to the system and started. While the system is being evacuated, the technician moves the electronic scales to the system and prepares for charging. Because the service connection is on the low-pressure side of the system, the charge will have to be in the vapor state to prevent liquid refrigerant from entering the compressor. The unit is designed for 18.5 oz of refrigerant. The 30-lb cylinder should deliver enough refrigerant to keep the pressure from dropping during charging. When the appropriate level on the vacuum gauge, 500 microns, has been reached and held, the vacuum pump is disconnected from the unit and the center hose on the gauge manifold is connected to the refrigerant cylinder. The center hose on the manifold is purged of air, and the correct charge is allowed into the system.

If the charge will not move into the system, the technician may start the compressor to lower the low-side pressure where refrigerant will move from the cylinder into the system. **NOTE:** *DO NOT allow the system to operate in a vacuum. If it tries to go into a vacuum, shut the compressor off for a few minutes.* • When the correct charge is established, the unit is started. It is cooling as it should.

The technician takes the unit back to the customer and places it in the window, starts it, and makes sure that it operates correctly. Before leaving, the technician asks the customer if there are any questions about the unit.

SERVICE CALL 4

A customer in a warehouse office reports that the window unit cooling the office is not adequately cooling the space. *The problem is a slow leak.*

When the technician looks at the unit, he determines that it definitely has a low charge; the evaporator coil is only cold halfway through. He pulls the unit out of the case and sets it on a stool. The technician points out that the unit is old and needs replacing. The customer asks if there is anything that can be done to get them through the day.

The technician fastens a line tap to the suction line and connects a gauge line; the suction pressure is 90 psig. The unit has R-410A, which should yield an evaporator temperature of 26°F. The unit is showing signs of frost on the bottom of the coil. This is an old unit, so the technician decides to add refrigerant to adjust the charge. It is about 80°F in the room. The condenser airflow is blocked until the leaving air is warm to the hand. Refrigerant is added until the suction line sweats to the compressor. The gauges are removed and the unit is placed back in the case, but the leak was not found. The company requests that a new unit be installed the next day. 4R *When the old unit is discarded, the refrigerant will have to be recovered before it is scrapped.* 4R

Classroom Discussion: How would you have handled this call? What would you have done with the line tap valve? Are any laws being violated?

SERVICE CALL 5

A new unit sold last month is brought to the shop for the technician to repair. The service repair order reads, "no cooling." *The problem is that the compressor is stuck and will not start.*

The technician checks to make sure the selector switch is in the OFF position and plugs the unit into a 230-V wall plug. This workbench is set up for servicing appliances so it has an ammeter mounted on it. The technician turns the fan to HIGH FAN; the fan runs. He then turns the selector switch to LOW FAN and the fan still runs. The amperage is normal, so the technician turns the selector switch to HIGH COOL, keeping a hand on the switch. The fan speeds up but the compressor does not start. The thermostat is not calling for cooling. When the thermostat is turned to a lower setting, the compressor tries to start. The ammeter registers locked-rotor amperage and does not fall back. The technician turns the unit off before the compressor overload reacts and disconnects the power cord.

Removing the unit from the case, the technician looks for the compressor terminal box. He removes the box cover and installs a hard-start kit on the compressor. The technician tries to start the compressor with the hard-start kit. It will not start; the compressor is stuck and must be changed.

4R *The technician installs a line tap valve on the suction process tube. Before piercing the refrigerant line, the technician leak tests the line tap valve. It is leak-free, so he pierces the refrigerant line and starts to recover the refrigerant. While the refrigerant is being recovered slowly, a new compressor, a 1/4-in. flare fitting for the process tube, and a suction-line drier are brought from the supply room. The technician removes the compressor hold-down bolts. When the refrigerant has been recovered, he uses an air-acetylene torch with a high-velocity tip to remove the suction and discharge lines.* 4R

The old compressor is removed and the new one set in place. Since the compressor failed, the technician decides to test the unit for acid and also install a suction line filter drier along with the new compressor for extra protection. The acid test results showed that the system was not acidic, but the technician still opted to install a suction-line drier, just in case. The suction line is altered to get it ready for the installation of the suction-line drier. Before that is done, the compressor discharge line is brazed in place, and then the 1/4-in. flare fitting is installed on the process-tube connection (the process tube was cut off below the line tap connection). The suction-line filter drier is soldered in place and then the suction line is brazed to the suction port of the compressor. A gauge line is fastened to the 1/4-in. flare connection. A small amount of R-410A is added to the system (to 5 psig), then nitrogen is used to pressure the system to 150 psig. A leak check is performed on all fittings; the unit is leak-free.

The small amount of refrigerant vapor and nitrogen is purged from the system and the vacuum pump is fastened to the gauge manifold and started. When the vacuum pump has obtained a deep vacuum according to the micron gauge, the correct charge is weighed into the unit. The technician reconnects the compressor wiring while the vacuum pump operates so the unit will be ready to start. The technician plugs in the cord and turns the selector switch to HIGH COOL; the compressor starts. He allows the unit to run for 30 min and it appears to be a little short of refrigerant. The technician then realized that he added a suction-line filter drier to the system. After adding an additional 1.5 ounces of refrigerant to the unit, the unit is checked again and it operates satisfactorily. The process tube is pinched and cut off below the 1/4-in. flare fitting. The process tube is closed on the end and soldered shut.

SERVICE CALL 6

A customer reports that a window unit servicing the den and kitchen has tripped the breaker. This is a fairly new unit, 6 years old. The owner is told to shut the unit off and to not reset the breaker. *The problem is that the unit compressor is grounded.*

The technician introduces himself to the customer and asks about the problem with the unit and the circuit breaker. He decides to give the unit an ohm test before trying to reset the breaker and start the unit. The multimeter is set to read resistance, and the technician unplugs the unit, fastens one ohmmeter lead to each of the hot prongs (L1 and L2) on the 230-V plug, **Figure 46.69**, and turns the unit selector switch to HIGH COOL. There is a measurable resistance; all appears to be well. Before plugging the unit in, one more ohm test is necessary for the complete picture, a ground test. The multimeter is once again set to read resistance, and one test lead is connected to L2, while the other is connected to the ground prong on the plug. The meter reads 0 ohms; there is a ground in the unit somewhere, **Figure 46.70**. It could be in the fan motor or the compressor. Pulling the unit from its case, the technician disconnects the compressor terminals and touches one

Figure 46.69 The unit has a ground circuit according to the ohmmeter.

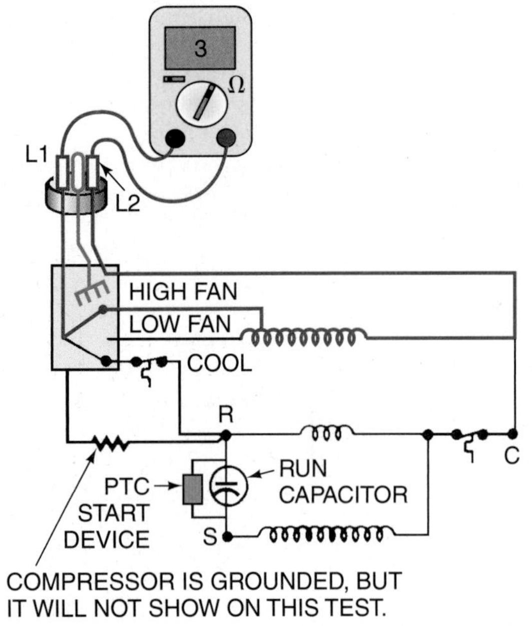

Figure 46.70 The ground circuit is through the compressor.

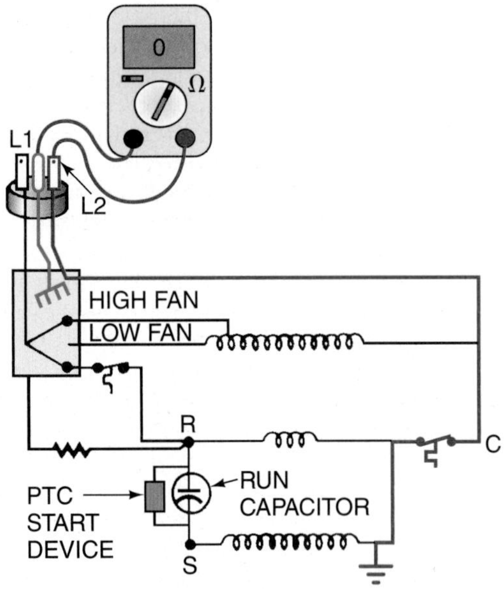

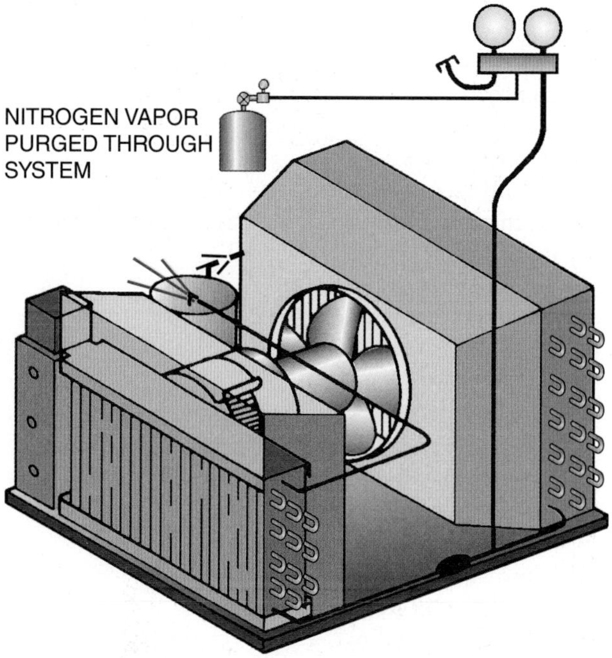

Figure 46.71 The system is purged using the "tee" fitting installed in the liquid line.

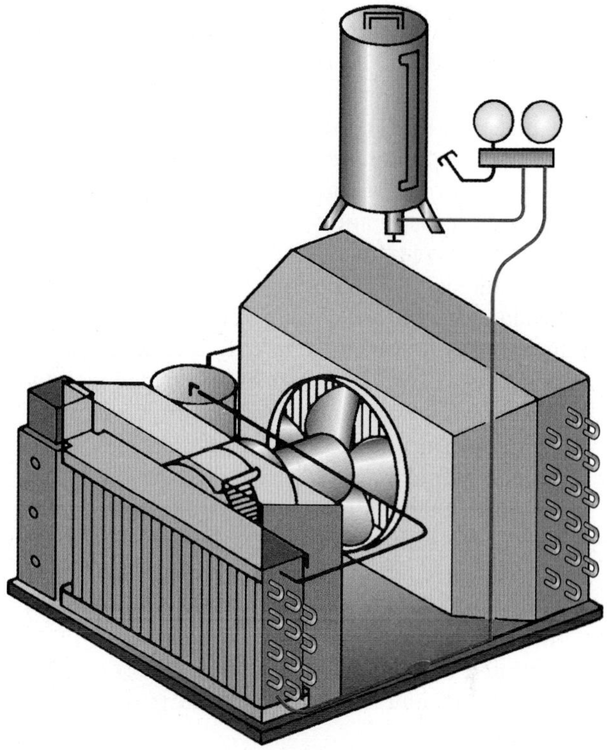

Figure 46.72 Liquid is charged into the system; this saves time.

lead of the meter to the discharge line and the other to the common terminal; the meter reads 0 ohms. The compressor has the ground circuit. The customer is informed of the problem and given the price of repairing the unit.

The technician takes the unit to the shop. The technician obtains a liquid-line drier, a suction-line drier, a 1/4-in. "tee" fitting for the liquid line, a compressor, and a 1/4-in. fitting for the process tube. After the refrigerant is recovered from the unit, the technician cuts the compressor discharge line at the compressor. He then cuts the suction line with a small tube cutter and removes the compressor from the unit. A "tee" fitting is installed in the liquid line, and a gauge line fastened to the "tee" fitting allows nitrogen to purge the unit, **Figure 46.71**. This will push much of the contamination out of the unit.

The technician sets the new compressor in place and solders the discharge line to the new compressor. The suction line is cut to the correct configuration for the suction-line drier. The liquid line is now cut, just before the strainer entering the capillary tube, and the new small drier is installed. The suction-line drier and the 1/4-in. fitting for the process tube are soldered in place. The gauges are fastened to the process port and the "tee" fitting in the liquid line. The unit is pressured with a small amount of R-410A (to 5 psig) and nitrogen is added to push the pressure up to 150 psig. The unit is then leak checked with an electronic leak detector. After the leak checking, the trace refrigerant and nitrogen are released from the system. The vacuum pump is then connected and started. While the unit is being evacuated, the technician checks the run capacitor and relay. The capacitor is good, the coil on the relay has a measurable resistance, and the contacts look good.

The charge for the unit is 18 oz; adding the capacity of the small liquid-line drier, 1.9 oz, the total charge is 19.9 oz (this information should be packed in with the drier directions but not all manufacturers do so). When a 500-micron vacuum is obtained according to the micron gauge, the proper charge is added. The charge is in a charging cylinder, as explained in Unit 9, "Refrigerant and Oil Chemistry and Management—Recovery, Recycling, Reclaiming, and Retrofitting," which has a liquid valve. The high-pressure gauge line is connected to the liquid line on the unit, and the charge is allowed to enter the system through the liquid line in the liquid state, **Figure 46.72**. When the complete charge is inside the unit, the liquid-line valve is pinched off and soldered shut.

The unit is started and allowed to run while the technician is monitoring the suction pressure and amperage. The unit is ready to be returned to the customer.

REVIEW QUESTIONS

1. Window or through-the-wall air-conditioning units are primarily used to
   A. work in conjunction with central cooling systems when additional cooling is needed.
   B. work in conjunction with central heating systems when additional heating is needed.
   C. provide cooling (and sometimes heating) to one room in a structure.
   D. provide comfort cooling and heating to multiple rooms with the aid of ductwork.
2. Which of the following is an easy way to tell the difference between a wall unit and a through-the-wall unit?
   A. Through-the-wall units have louvered sides
   B. Window units have louvered sides
   C. Window units have hard-wired power cords
   D. Through-the-wall units have hard-wired power cords
3. Which of the following is true regarding a window unit with a louvered, wrap-around sleeve?
   A. The unit can be installed though the wall without modification.
   B. The unit can be slid out of the sleeve for easy service.
   C. The entire unit must be removed from the window for service.
   D. The chassis can be accessed for service from outside the structure.
4. A newly purchased window or through-the-wall air-conditioning unit will most likely contain which of the following refrigerants?
   A. R-12
   B. R-134a
   C. R-22
   D. R-410A
5. Window and through-the-wall air-conditioning units are classified as
   A. high-temperature refrigeration systems.
   B. medium-temperature refrigeration systems.
   C. low-temperature refrigeration systems.
   D. ultra-low-temperature refrigeration systems.
6. What type of defrost control systems are commonly used on window and through-the-wall air conditioners?
   A. Hot gas defrost
   B. Electric strip heater defrost
   C. Off cycle defrost
   D. Window and through-the-wall air-conditioning units do not have defrost control systems.
7. The evaporator coil of a properly functioning window unit operates at a
   A. temperature that is lower than the dew point temperature of the air in the room.
   B. pressure that is higher than the pressure in the condenser.
   C. temperature that is higher than the dry-bulb temperature of the air in the room.
   D. pressure that is lower than atmospheric pressure.
8. Each of the following is one of the four main refrigeration system components except the
   A. evaporator.
   B. condenser.
   C. suction line.
   D. metering device.
9. Heat pump window or through-the-wall units are able to direct the refrigerant to the correct heat transfer surface by using which type of valve?
   A. Two-way reversing valve
   B. Three-way reversing valve
   C. Four-way reversing valve
   D. Five-way reversing valve
10. In the heating mode of operation, refrigerant leaves the compressor, flows through the reversing valve, and then flows through which system component?
    A. The indoor coil
    B. The outdoor coil
    C. The evaporator
    D. The metering device
11. Which of the following is a challenge regarding he replacement of a through-the-wall unit?
    A. Locating a replacement unit with the same cooling capacity as the original
    B. Locating a replacement unit with the same heating capacity as the original
    C. Locating a replacement unit with the same electrical ratings as the original
    D. Locating a replacement unit with the same physical dimensions as the original
12. The most common type of window in which an air-conditioning unit is installed is the
    A. casement window.
    B. double-hung window.
    C. triple-hung window.
    D. picture window.
13. When providing electrical service for a window unit,
    A. the unit must be hard-wired to the building's power distribution panel.
    B. an extension cord is to be used to increase resistance and reduce current draw.
    C. the receptacle must be close enough so the unit's power cord can reach it.
    D. an extension cord is to be used to decrease resistance and increase current draw.
14. The receptacle that is used to supply power to a window or through-the-wall unit must be
    A. located on the switch-side of the unit.
    B. wired to a dedicated circuit.
    C. located at least 6 feet away from the unit.
    D. able to power at least three other household appliances.

15. A properly installed window unit will allow
    A. condensate to drain to the outside of the structure.
    B. condensate to drain toward the box that contains the unit's controls.
    C. outside air to leak into the structure from around the sleeve.
    D. supply air to immediately recirculate back to the unit's return airstream.
16. On window or through-the-wall units, the thermostat's sensing bulb is located
    A. in the return air stream.
    B. in the supply air stream.
    C. in the condenser's inlet air stream.
    D. in condenser's discharge air stream.
17. On window or through-the-wall units equipped with wireless remote controls, the thermostat's sensing bulb is located
    A. in the wireless remote control.
    B. as far from the remote control as possible.
    C. as far from the unit as possible.
    D. in the return air stream.
18. On unit's equipped with an energy-saver switch
    A. the fan motor runs continuously when the energy-saver switch is on.
    B. the compressor runs continuously when the energy-saver switch is on.
    C. the compressor and fan motor cycle on and off when the energy-saver switch is on.
    D. the compressor and fan motor cycle on and off when the energy-saver switch is off.
19. A 208/230-volt window unit will have a power cord that connects to
    A. one hot leg, one neutral leg, and ground.
    B. two neutral legs and ground.
    C. two hot legs and one neutral
    D. two hot legs and ground.
20. The heating contacts on a heating/cooling window or through-the-wall unit will
    A. open and close at the same time as the cooling contacts on the thermostat.
    B. open as the sensed space temperature drops.
    C. close as the sensed space temperature rises.
    D. close as the sensed space temperature drops.
21. The most important aspect of servicing window or through-the-wall units is
    A. keeping the air filter clean.
    B. checking the refrigerant charge.
    C. oiling the permanently lubricated fan motor.
    D. cleaning the electrical contacts.
22. A dirty air filter will cause the unit's
    A. operating pressures to increase.
    B. operating pressures to decrease.
    C. suction pressure to drop and the discharge pressure to rise.
    D. suction pressure to rise and the discharge pressure to drop.
23. When should gauges be installed on a window or through-the-wall unit?
    A. Whenever the system is serviced
    B. Each time the air filter is replaced
    C. Once a year to check the refrigerant charge
    D. Only when absolutely necessary
24. If an R-410A window unit is operating with a head pressure of 420 psig, at what temperature is the refrigerant condensing in the unit?
    A. 100°F
    B. 110°F
    C. 120°F
    D. 130°F
25. A line-voltage reading across a set of contacts on a window unit's thermostat indicates that
    A. the thermostat is defective.
    B. the main power switch to the unit is off.
    C. the thermostat's contacts are open.
    D. the thermostat's contacts are closed.

SECTION 10

COMMERCIAL AIR-CONDITIONING AND CHILLED-WATER SYSTEMS

UNITS

UNIT 47

HIGH-PRESSURE, LOW-PRESSURE, AND ABSORPTION CHILLED-WATER SYSTEMS

OBJECTIVES

After studying this unit, you should be able to

- describe how chilled-water air-conditioning systems operate.
- explain why chilled water is commonly used to cool buildings.
- list the components that make up a vapor-compression, high-pressure chiller.
- describe the design conditions for a vapor-compression, high-pressure chiller.
- explain how capacity is controlled on a high-pressure chiller equipped with a reciprocating compressor.
- describe the use of scroll compressors on high-pressure chillers.
- describe the characteristics of a package-type chiller.
- explain how capacity is controlled on a high-pressure chiller equipped with a rotary screw compressor.
- describe the operation of a centrifugal compressor in a high-pressure chiller.
- explain how to control the capacity on a chiller equipped with a centrifugal compressor.
- explain how heat transfer takes place in a chiller barrel.
- explain the difference between direct-expansion and flooded chiller evaporators.
- explain the concept of approach temperature as it applies to a chiller barrel.
- explain the operation of the condenser on a high-pressure chiller.
- list the types of metering devices used in high-pressure chillers.
- describe the type of compressor found on low-pressure chillers.
- explain the function of a rupture disc on a low-pressure chiller.
- explain the purge system used on a low-pressure chiller condenser.
- explain what happens if a low-pressure chiller develops a leak.
- explain the function of the concentrator in an absorption chiller.
- state the refrigerant generally used in large absorption chillers.
- describe the solutions that are used in an absorption chiller.
- explain how capacity is controlled on an absorption chiller.
- explain the importance of maintaining an absorption chiller.
- explain how compressor motors are cooled.
- describe the types of electric motors typically used on chiller air-conditioning systems.
- explain how pitted contacts affect the operation of system motors.
- explain the concept of voltage imbalance.

SAFETY CHECKLIST

- ✔ Wear gloves and use caution when working around hot steam pipes and other heated components.
- ✔ Use caution when working around high-pressure systems. Do not attempt to loosen fittings or connections when the system is pressurized. Follow recommended procedures when using nitrogen.
- ✔ Follow all recommended safety procedures when working around electrical circuits.
- ✔ Use lockout/tagout procedures to disconnect boxes or panels when power has been turned off to work on an electrical system. There should be one key for this lock, and it should be in the technician's possession.
- ✔ Never start a motor with the door to the starter components open.
- ✔ Check pressures regularly as indicated by the manufacturer of the system.

47.1 INTRODUCTION TO CHILLED-WATER SYSTEMS

Chilled-water systems are used for larger central air-conditioning applications because of the ease with which chilled water can be circulated throughout the system. If refrigerant were piped to all floors of a multistory building, there would be too many possibilities for leaks, not to mention the expense of the refrigerant to charge such a large system. The design temperature for boiling refrigerant in a coil used for cooling air is 40°F. If water can be cooled to a low enough temperature, it can also be used to cool air,

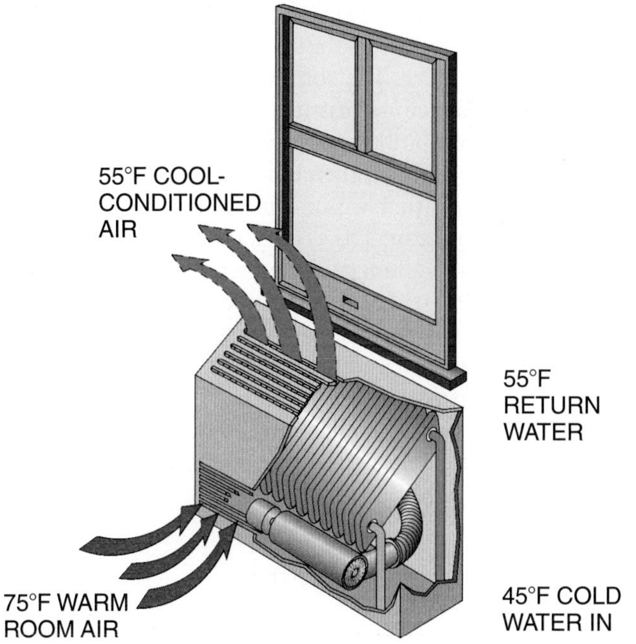

Figure 47.1 Water circulating in a fan coil unit for cooling a room.

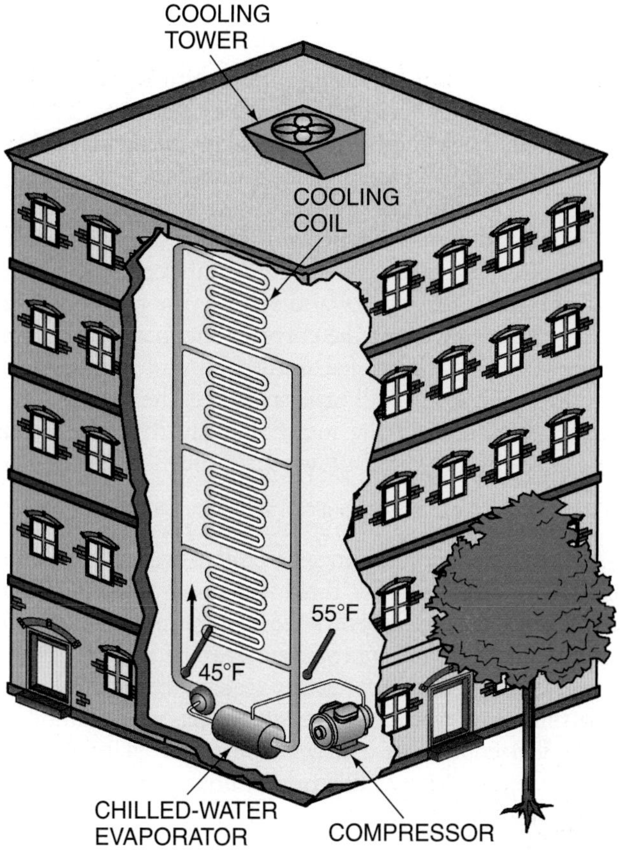

Figure 47.2 Water is circulating throughout the building to the fan coil units. It is considered to be a secondary refrigerant.

Figure 47.1. This is the logic behind circulating chilled-water systems. Water is cooled to about 45°F and circulated throughout the building to heat exchange coils that absorb heat from the building air. Water used in this way is called a secondary refrigerant. Water is much less expensive to circulate than refrigerant, **Figure 47.2.** *Lower water temperatures may be used to make the building side of the chilled-water system more efficient.* Some systems chill water to 42°F, and with the addition of electronic controls, some equipment is actually able to supply 34°F chilled water. *The colder the circulating water, the smaller the terminal equipment—the coils and fans—can be. Also, with colder water, the interconnecting piping for the entire system can be smaller, which saves on piping supplies, fittings, and insulation for the entire system as well as on labor. Also, smaller fans and pumps consume less power.*

Systems should be able to vary their capacity, as the structural heat load, which includes the load created by the occupants of the building, is not constant. In a residential system, it is typical to just turn the unit off when the desired conditions are met. In large installations, it is not wise to simply cycle the compressor or chiller on and off as the load on the structure increases and decreases. This is because much of the wear-and-tear on the equipment occurs when the chiller is started. The motor bearings are not as well lubricated at start-up as they are when the compressor is running, so the motor is stressed when the compressor is started up. In most comfort-cooling installations, the load varies during the season as well as during the day. We saw in Unit 36, "Refrigeration Applied to Air-Conditioning," that when a system is designed for very high temperature conditions, the equipment will run at full load only a small percentage of the time.

Manufacturers have developed many capacity-control methods for their equipment. Some chillers manage reduced capacity with variable-speed, or unloading, compressors. Variable-capacity control for each type of chiller will be discussed in the appropriate sections of this unit. Manufacturers usually rate chiller capacities and efficiencies at full-load operation. With energy costs rising and the increased emphasis on system efficiency and reduced environmental impact, manufacturers have implemented technologies to achieve efficient operation at part load. In the past, equipment did not typically operate efficiently at part load.

In operation, water passes through the evaporator section of the machine, often referred to as the chiller barrel, where the temperature of the water is lowered. This chilled water is then circulated throughout the building, where it picks up heat. The typical design temperatures for a circulating chilled-water system are 45°F for the water supplied to the building and 55°F for water returned to the chiller from the

building. Therefore, the building heat adds about 10°F to the water supplied by the chiller. Upon returning to the chiller, the heat is removed from the water, which is then recirculated through the building.

Cooling buildings is not the only application for chilled-water systems. Although most of this unit will cover comfort-cooling chillers—a thorough discussion of all chiller applications would be beyond the scope of this text. Many industries use chillers for process cooling, such as the manufacturing of plastic milk bottles. These systems often circulate water and glycol (antifreeze) mixtures at temperatures well below freezing to cool the hot, molten plastic that is injected into molds to cool and solidify the plastic. When a water valve is opened at the correct time in the process, the plastic milk bottle is cooled until solid and is then ejected from the mold. After the bottle is ejected, the flow of chilled water is stopped and the mold is ready to make another bottle. The chilled water allows this process to happen very quickly.

The textile industry is another example of process cooling. Mills use huge amounts of chilled water to maintain the conditions that allow thread to move quickly through the machines. If the temperature and humidity are held at the correct levels, maximum production can be accomplished. In the years before chilled water was used, mills could only operate at full capacity about 3 months of the year. With chilled water, ideal conditions can be maintained 24 hours a day, 365 days a year. Many mills run 24 hours a day year-round, with the likely exception of a one-week annual shutdown for equipment maintenance.

There are two basic categories of chillers: the compression cycle chiller and the absorption chiller. The compression chiller uses a compressor to provide the pressure differences inside the system to boil and condense refrigerant. The absorption chiller uses a salt solution and water to accomplish the same result. These chillers are very different and are discussed separately. Manufacturers of both types provide different design features, but all of them attempt to build a reliable, low-cost chiller that will last a long time.

47.2 COMPRESSION CYCLE IN HIGH-PRESSURE CHILLERS

The compression cycle chiller has the same four basic components as a typical air conditioner: a *compressor*, an *evaporator*, a *condenser*, and a *metering device*. These components are generally larger, however, in order to operate at larger capacities. In addition, they may use a refrigerant different from that used in air conditioners.

The heart of the vapor-compression refrigeration cycle is the compressor. As mentioned in Unit 3, "Refrigeration and Refrigerants," there are several types of compressors. Those common in water chillers are the *reciprocating, scroll, screw*, and *centrifugal* compressors. Photos of these can be found in Unit 23, "Compressors." The compressor is the component in the system that both lowers the suction pressure and increases the discharge pressure. The compressor can be thought of as a *vapor pump*. The compressor lowers the evaporator pressure to the desired boiling point of the refrigerant, which is about 38°F for a chiller. It then builds the pressure in the condenser to the point that vapor will condense to a liquid for reuse in the evaporator. The typical condensing temperature for a chilled-water system equipped with a water-cooled condenser is about 105°F. The technician can use temperatures to determine whether a typical chiller is operating within design parameters for pumping and compressing the vapor to meet the needs of a particular installation.

Compression cycle chillers may be classified as either *high-pressure* or *low-pressure systems*. Following is a discussion of high-pressure chilled water systems.

47.3 RECIPROCATING COMPRESSORS IN HIGH-PRESSURE CHILLERS

Large reciprocating compressors in water chillers operate in the same way as those used in any other application, with a few exceptions. Reciprocating compressors range in size from about ½ hp to approximately 150 hp, depending on the application. Many manufacturers use multiple smaller compressors instead of a single, larger pump. Reciprocating compressors are classified as positive displacement pumps and cannot pump liquid refrigerant without risk of damage. Several refrigerants are used in reciprocating compressor chillers, most commonly R-134a, R-410A, and R-22.

Large reciprocating compressors have as many as sixteen cylinders to produce the pumping capacity needed to move large amounts of refrigerant. These compressors have many moving parts and much internal friction. If one cylinder of the compressor fails, the whole system will go off-line. If one compressor fails in a multiple-compressor system, the others can carry the load, so multiple compressors provide some protection from total system failure. This is the reason, as well as for capacity control, that many manufacturers choose to use multiple smaller compressors as opposed to one large compressor in their systems.

All large chillers must have some means to control capacity or the compressor will cycle on and off. This is not satisfactory because most compressor wear occurs during start-up before adequate oil pressure is established. A better design approach is to keep the compressor running but operate it at reduced capacity. Reduced-capacity operation also smooths out temperature fluctuations that occur from shutting off the compressor and waiting for the water to warm up to bring the compressor back on.

CYLINDER UNLOADING AND VARIABLE-FREQUENCY DRIVES (VFDs)

Reduced capacity for a reciprocating compressor is accomplished through *cylinder unloading* and *variable-frequency drives* (VFDs). For example, suppose the chiller for a large office building uses a 100-ton compressor with eight cylinders and the chiller has 12.5 tons of capacity per cylinder. When all eight cylinders are pumping, the compressor has a capacity of 100 tons ($8 \times 12.5 = 100$). As the cylinders are unloaded, the capacity is reduced. For example, the cylinders may unload in pairs, which would be 25 tons per unloading step. The compressor may have four unloading steps, which would give it four different capacities: 100 tons (eight cylinders pumping), 75 tons (six cylinders pumping), 50 tons (four cylinders pumping), and 25 tons (two cylinders pumping).

In the morning when the system first starts, the building may need only 25 tons of cooling. As the temperature outside rises and as people begin to enter the building, the chiller may need more capacity and the compressor will automatically load two more cylinders for 50 tons of capacity. As the heat load on the structure continues to rise, the compressor can load up to 100% capacity, or 100 tons. If the building—a hotel, for example—stays open at night, the compressor will start to unload as the outside temperature drops. If the load on the system is too low, the chiller may cycle off.

A big advantage of cylinder unloading is that the power needed to operate the compressor is reduced as capacity is reduced. Power consumption is greatly reduced at part load. In addition to the reduction in workload by cylinder unloading, the compression ratio is reduced; when a cylinder is unloaded, the suction pressure rises slightly and the head pressure is reduced slightly. Cylinder unloading is accomplished in several ways; blocked suction unloading and suction-valve-lift unloading are the most common.

BLOCKED SUCTION UNLOADING. Blocked suction is accomplished by placing a solenoid valve in the suction passage to the cylinder being unloaded, **Figure 47.3**. If the refrigerant gas cannot reach the cylinder, no gas can be pumped. If a compressor has four cylinders and the suction gas is blocked to one of them, the capacity of the compressor is reduced by 25% and the compressor then pumps at 75% capacity. Power consumption also decreases by approximately 25%. Power consumption is related to the amperage draw of the compressor. Amperage can be measured in the field by using a clamp-on ammeter. When a compressor is running at half capacity, the amperage will be about half the full-load amperage. (The power consumption of a compressor is actually measured in watts. Using amperage as a measure of compressor capacity is close enough for field troubleshooting.)

Suction-Valve-Lift Unloading. If the suction valve is lifted off the seat of a cylinder while the compressor is pumping, the cylinder will stop pumping. Gas that enters the cylinder will be pushed back out into the suction side of the system on the upstroke. There is very little resistance to the refrigerant being pumped back into the suction side, so it requires almost no energy. As in the example of blocked suction, power consumption will be reduced. One of the advantages of lifting the suction valve is that the gas that enters the cylinder will contain oil, and good cylinder lubrication will occur even while the cylinder is not pumping.

Except for cylinder unloading, the large reciprocating chiller compressor is the same as smaller reciprocating compressors. Most compressors over 5 hp have pressure-controlled lubricating systems. The lubricating pressure is provided by an oil pump that is typically mounted on the end of, and driven by, the compressor crankshaft, **Figure 47.4**.

Figure 47.3 Blocking the flow of refrigerant to one of the cylinders in a reciprocating compressor will unload that cylinder's load from the system.

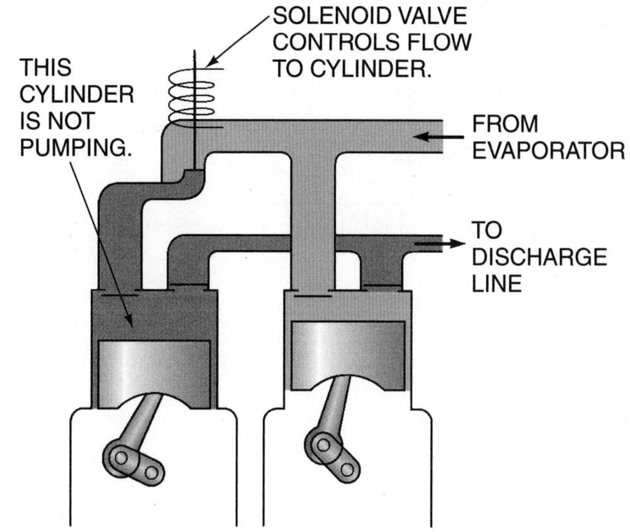

Figure 47.4 The pressure lubrication system for a reciprocating compressor. Notice that the oil pump is on the end of the shaft. The oil is picked up from the sump and pumped through the drilled passages to all moving parts. Courtesy of Trane Company.

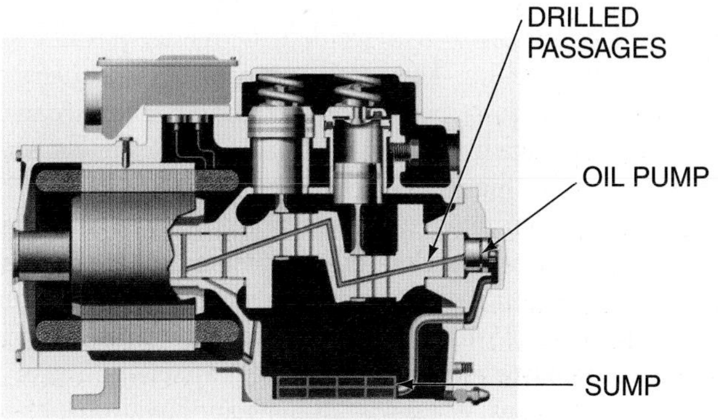

Figure 47.5 The oil pump's inlet pressure is the same as the system's suction pressure in the crankcase. The oil pump then increases the oil pressure to the pump's discharge pressure.

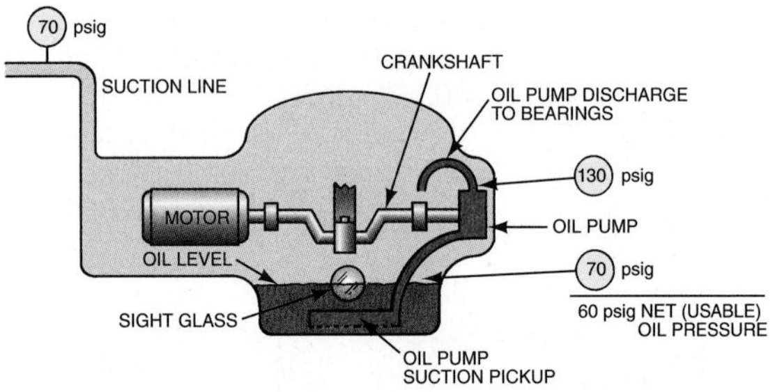

The oil pump picks up oil in the sump at evaporator pressure and delivers it to the bearings at a pressure that is about 30 to 60 psig greater than the suction pressure. The difference between the oil pump's outlet pressure and the suction pressure is called net oil pressure, **Figure 47.5**. The compressors also have an oil safety shutdown in case of oil pressure failure, which has a time delay of about 90 sec to allow the compressor to get started and to establish oil pressure before it shuts the compressor off.

Variable-Frequency Drives. Variable frequency drives, VFDs, electronically alter the frequency of alternating-current (AC) sine waves through a device called an inverter. Today, compressor motors typically operate within a wide range of speeds that can vary system capacity without shutting off. *The speeds used are in conjunction with electronic expansion valves that also have an almost unlimited ability to vary refrigerant flow to the evaporator. The condenser water pump and the chilled-water pumps may also have VFDs that can vary the water flow. The system flow rates must be synchronized to take into account the desired system conditions to prevent problems and maintain minimum and maximum flow rates for all system components being controlled.* See Unit 17, "Types of Electric Motors," for details.

47.4 SCROLL COMPRESSORS IN HIGH-PRESSURE CHILLERS

The scroll compressor is a welded, hermetic, positive displacement compressor. The scroll compressors in chillers are larger than those described earlier in this text. They are in the 10- to 25-ton range but operate the same as the smaller scroll compressors. See **Figure 47.6** for an illustration of a scroll compressor. When these compressors are used in chillers, the capacity control of the chiller is maintained by cycling the compressors off and on in increments of 10 and 25 tons. For example, a 25-ton chiller would have a 10- and a 15-ton compressor and would have capacity control of 10, 15, and 25 tons. A 60-ton chiller may have four 15-ton compressors and be able to operate at 100% (60 tons), 75% (45 tons), 50% (30 tons), and 25% (15 tons).

Figure 47.6 The check valve in a scroll compressor prevents discharge pressure from flowing back through the compressor when the compressor shuts down. Courtesy of Trane Company.

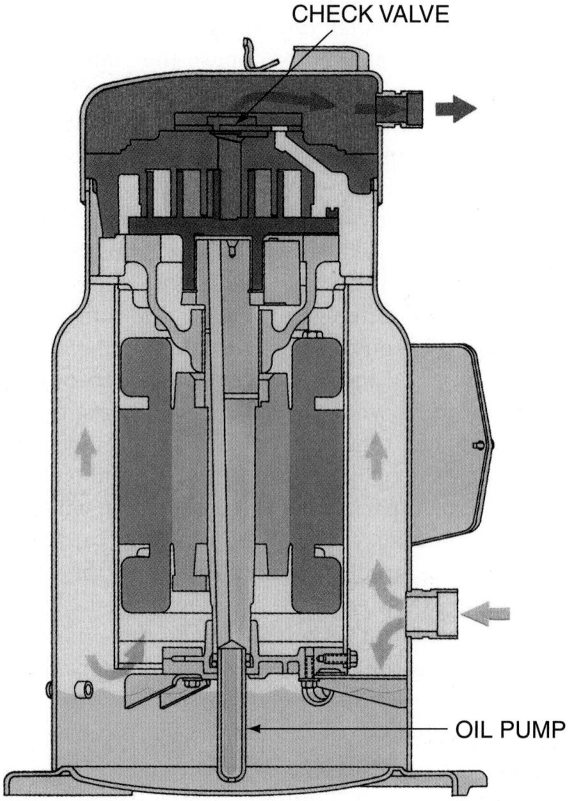

Some of the advantages of the scroll compressor include the following:

1. Efficiency
2. Quietness
3. Fewer moving parts
4. Size and weight
5. Ability to pump small amounts of liquid refrigerant without compressor damage

The scroll compressor offers little resistance to refrigerant flow from the high side of the system to the low side during the off cycle. A check valve is provided to prevent backward flow when the system is shut down, **Figure 47.6.** Lubrication of the scroll compressor is provided by an oil pump at the bottom of the crankshaft. The oil pump picks up oil and lubricates all moving parts on the shaft.

One type of chilled-water system that utilizes scroll compressors is the reverse-cycle chiller. This type of system, which is becoming more popular, has the ability to provide both heated and chilled water for many different applications, **Figure 47.7.** Uses for this type of system include

Figure 47.7 A reverse-cycle (heat pump) chilled-water system with multiple scroll compressors. Courtesy of WaterFurnace International.

- Swimming pool heating
- Commercial aquariums
- Radiant heating
- Ice/Snow-melting
- Conventional chilled-water comfort cooling

Systems such as the one shown in **Figure 47.7** have multiple scroll compressors, allowing them to function efficiently and effectively during both high- and low-load periods. These are packaged chillers, so no refrigerant lines need to be installed in the field. However, interconnecting water lines do need to be installed and all local codes must be followed when doing so. As with any chilled-water system, the water that circulates through the system's heat exchangers must be treated. This will help ensure that the amount of scaling and corrosion on the interior heat exchanger surfaces is minimized, prolonging the useful life of the equipment.

47.5 ROTARY SCREW COMPRESSORS IN HIGH-PRESSURE CHILLERS

Most of the major manufacturers are building rotary screw compressors for larger-capacity chillers, which use high-pressure refrigerants. The rotary screw compressor is capable of handling large volumes of refrigerant with few moving parts, **Figure 47.8**. This type of compressor is a positive displacement pump that is able to handle *some* liquid refrigerant without compressor damage. Rotary screw compressors are manufactured in sizes from about 50 to 700 tons of capacity and are reliable and trouble-free.

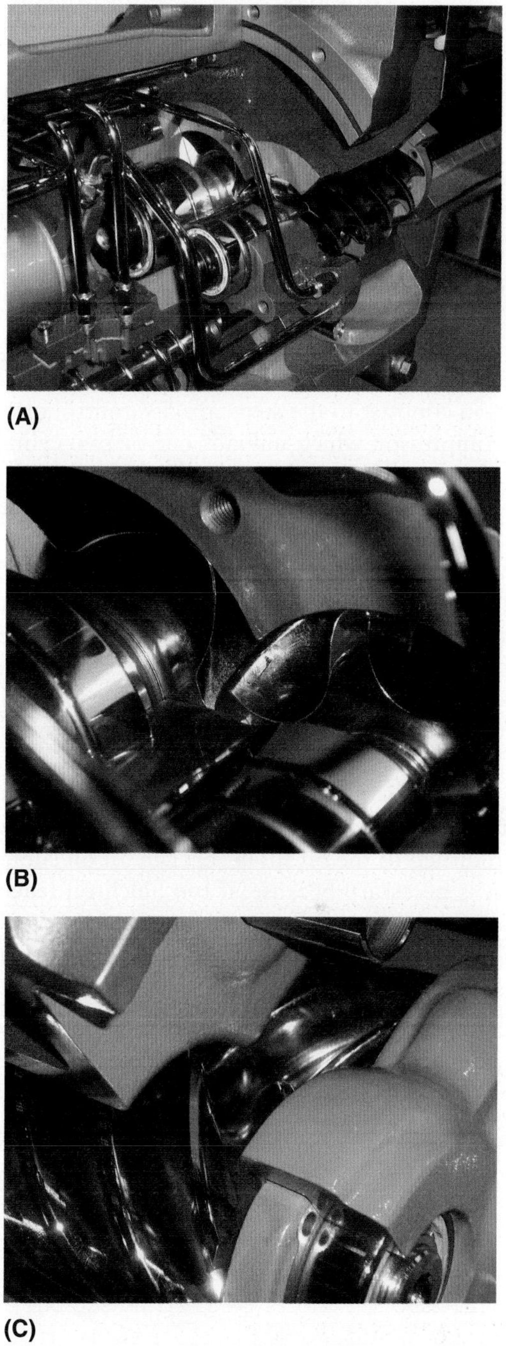

(A) (B) (C)

Figure 47.8 (A) Cutaway view of a rotary screw compressor. (B) Closeup of the nesting screws. (C) Closeup of the screws. Photos by Eugene Silberstein

Manufacturers build both *semihermetic rotary screw compressors* and *open-drive compressors*. The open-drive models are direct drive assemblies and have a shaft seal to contain the

refrigerant at the point where the rotating shaft penetrates the compressor shell. It is common practice to start up open-drive compressors regularly, especially during the off-season, to keep the shaft seal lubricated. Long periods of system downtime can cause the shaft seal to dry out and shrink somewhat, allowing refrigerant to leak from the system.

Capacity control for a rotary screw compressor may be accomplished by means of a *slide valve* that blocks the suction line, limiting the amount of refrigerant that enters the rotary screws in the compressor or by means of compressor VFD speed control. The slide valve is typically operated by system differential pressure. *The slide valve may be moved to the completely unloaded position before shutdown so that, on start-up, the compressor is unloaded, reducing the inrush current. Most of these compressors can function anywhere between 10% and 100% load because of the nature of the slide valve unloader.* This is in contrast to the step capacity control of a reciprocating compressor, which unloads one or two cylinders at a time.

Screw compressors pump a great deal of oil while compressing refrigerant, so they typically have an oil separator to return as much oil to the compressor reservoir as possible, **Figure 47.9**. The oil is moved to the rotating parts of the compressor by means of the pressure differential within the compressor. No pump is used. Oil accumulates in the oil reservoirs and is moved to the parts that need lubrication by means of gravity. The rotating screws are close together but do not touch. The gap between the screws is sealed by oil that is pumped into the rotary screws as they turn. This oil is separated from the hot gas in the discharge line and returned to the oil sump through an oil cooler. This separation is necessary because, if too much oil reaches the system, a poor heat exchange will occur in the evaporator. This will result in a loss of capacity. The natural place to separate the oil from the refrigerant is in the discharge line and manufacturers use different methods to facilitate this process.

Figure 47.9 An oil separator for a rotary screw compressor. Courtesy of Carrier Corp.

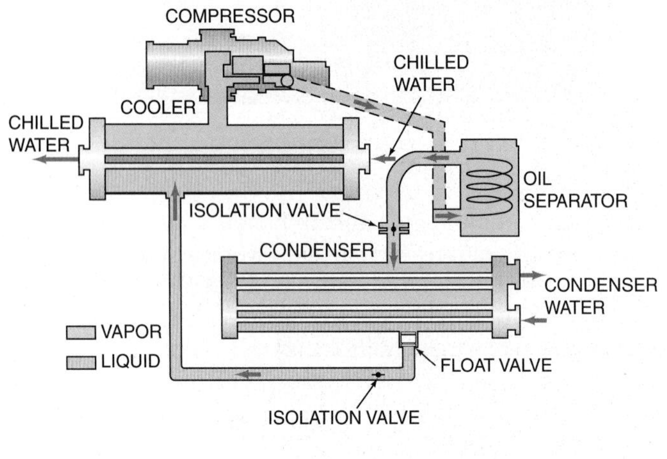

47.6 CENTRIFUGAL COMPRESSORS IN HIGH-PRESSURE CHILLERS

The centrifugal compressor applies only centrifugal force to the refrigerant to move it from the low- to the high-pressure side of the system, **Figure 47.10**. It functions like a large fan that creates a pressure difference from one side of the compressor to the other. Although it does not exert a great deal of force, it can handle a large volume of refrigerant, and some companies manufacture centrifugal compressors for use in high-pressure systems. In high-pressure systems, the compressor is turned very fast by means of a gearbox or multiple stages of compression to provide the pressure difference between the evaporator and the condenser. Typical motor speeds in direct-drive reciprocating, scroll, and rotary screw compressors are approximately 3600 rpm. In centrifugal systems, a gearbox is used to speed the compressor up to higher rpm levels when faster speeds are needed. Some single-stage centrifugal compressors operate at speeds of about 30,000 rpm. The gearbox used to obtain the high speeds of the centrifugal compressor adds some friction to the system that must be overcome; it causes a slight loss of efficiency due to the horsepower needed to turn the gears. There are typically two gears: The drive gear is the larger of the two, and the driven gear is the smaller and turns the fastest. **Figure 47.11** illustrates a gearbox.

Figure 47.10 Centrifugal action is used to compress refrigerant.

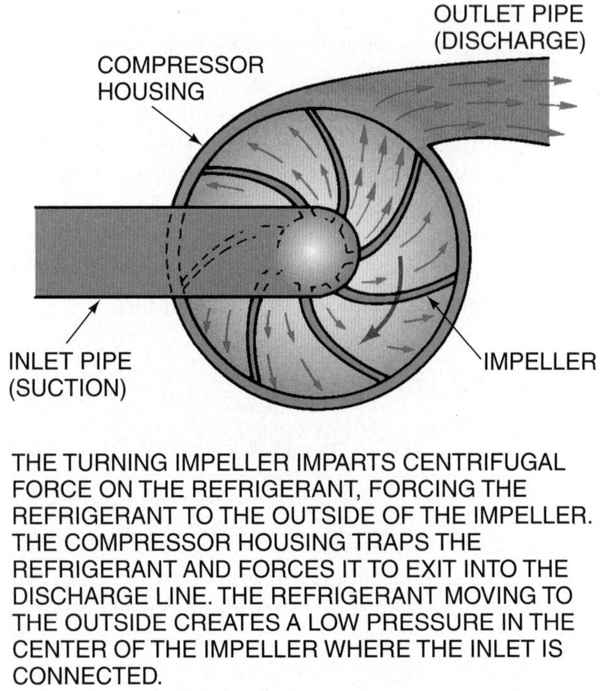

Figure 47.11 The gearbox for a high-speed compressor. Courtesy of York International.

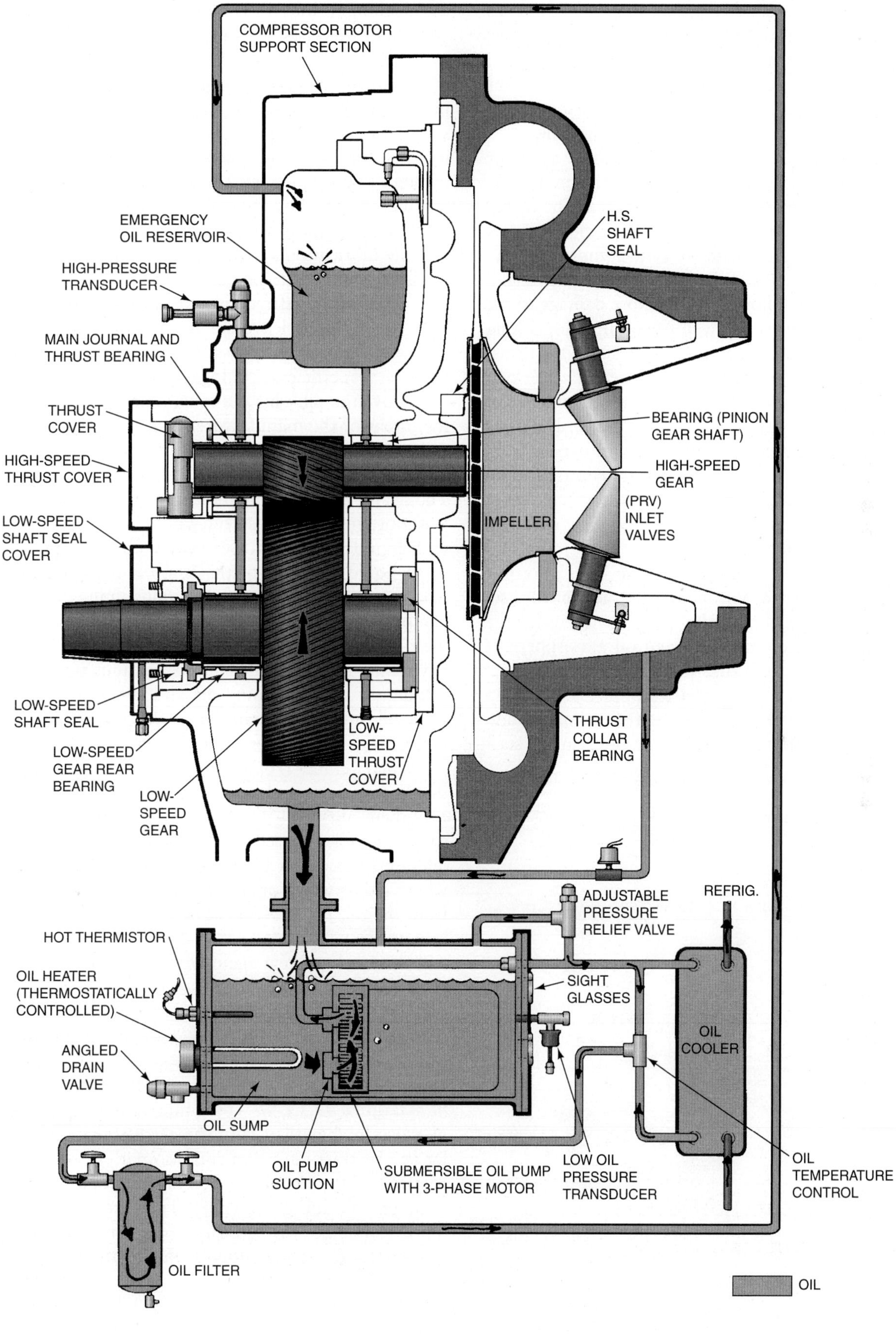

The gearbox and impeller of the centrifugal compressor have bearings that support the weight of the turning shaft. The bearings are typically made of a soft material known as babbitt. The babbitt is backed and supported by a steel sleeve. A single-stage compressor must also have a thrust bearing to counter the thrust of the refrigerant entering the compressor impeller. The thrust bearing counters the sideways movement of the shaft due to the force of the entering refrigerant.

If the head pressure becomes too high or the evaporator pressure too low, the centrifugal compressor cannot overcome the pressure difference and stops pumping. The motor and compressor still turn, but refrigerant stops moving from the low- to the high-pressure side, and the compressor may make a loud whistling sound, called a "surge." Even though this is a loud noise, it will normally not cause damage to the compressor or motor unless allowed to continue for a long time. If damage does occur due to prolonged surge, the thrust surface of the bearings or the high-speed gearbox are two common points of failure. The clearances in the impeller of a centrifugal compressor are critical, **Figure 47.12**. The parts must be very close together or refrigerant will bypass from the high-pressure side of the compressor to the low-pressure side. If such a bypass occurs, the system's low-side pressure will rise and the high pressure will drop, reducing the mass flow rate of the refrigerant.

A separate motor and oil pump provide lubrication for the centrifugal compressor. A three-phase fractional horsepower motor (usually 1/4 hp to 3/4 hp) is located inside the oil sump and runs in the refrigerant and oil atmosphere. A three-phase motor is used so that there will be no internal motor start switch, as in a single-phase motor. The oil pump is wired so that it can start before the centrifugal motor and establish lubrication before the compressor and gearbox start to turn. It also allows the oil pump to run as the centrifugal impeller and gearbox coast down at the end of a cycle when the machine is shut off. Enough oil is also present in the reservoirs to prevent bearing and gear failure during a coast down in the event of a power failure. The oil pump, usually a gear-type positive displacement pump, would normally furnish a net oil pressure of about 15 psig to the bearings and, with some compressors, to an upper sump where oil is fed by gravity to the bearings and gearbox during a coast down because of power failure, **Figure 47.11**. There are many methods of routing the oil, but the oil must have enough pressure to reach the farthest bearings and circulate around them.

Figure 47.12 The parts in a high-speed centrifugal compressor have a very close tolerance. Courtesy of Carrier Corp.

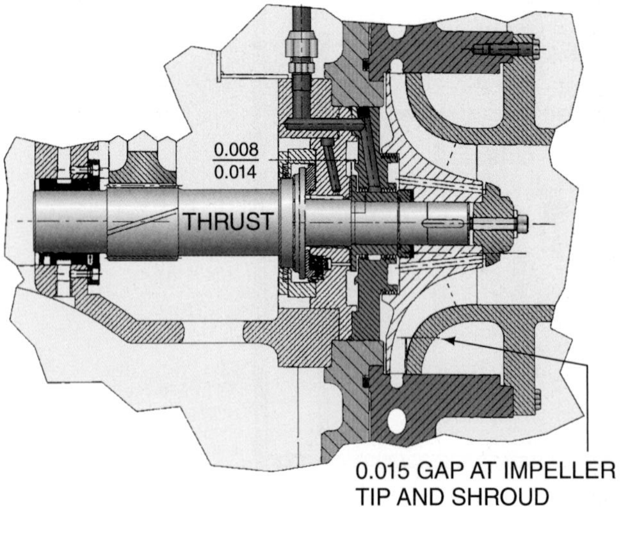

The lubrication system for a centrifugal chiller is sealed. Oil is not intended to be mixed with the refrigerant. If the oil does get into the refrigerant, it is difficult to remove because only vapor should leave the evaporator. When refrigerant in the evaporator becomes saturated with oil, heat exchange becomes less efficient. Oil-logged refrigerant must be distilled to remove the oil. This is a long, time-consuming process, so the intent of the manufacturer is to keep the two separated. A heater in the oil sump prevents liquid refrigerant from migrating to the sump when the system is off, **Figure 47.11**. The heater typically maintains the oil sump at about 140°F. Without this heater, the oil would soon become saturated with liquid refrigerant. But even with the heater operating, some liquid migrates to the sump and the oil may have a tendency to foam for a few minutes immediately after the compressor is started. The operator should keep an eye on the oil sump and run the compressor at reduced load until any foaming in the oil is reduced. Foaming oil contains liquid refrigerant that is boiling off and is not a good lubricant. Once all of the refrigerant has boiled off, the oil's maximum lubricating ability is restored.

When a chiller is operating, the oil is heated as it lubricates the moving parts and also absorbs heat from the bearings to cool them. If the oil is not cooled, it will become overheated, so an oil cooler is used. The oil is typically pumped from the oil sump through a filter and an oil cooler heat exchanger where either water or refrigerant removes some of the heat from the oil, **Figure 47.11**. The oil cooler cools the oil from a sump temperature of 140°F–160°F to about 120°F. The oil will then pick up heat again while lubricating.

Capacity control in the centrifugal compressor is accomplished by means of guide vanes at the entrance to the impeller eye. These guide vanes, usually pie-shaped devices arranged in a circle, can rotate to allow full flow or reduced flow down to about 15% or 20%, depending on the clearances in the vane mechanism, **Figure 47.13**. The guide vanes also serve another purpose in that they help the refrigerant enter the impeller eye by starting it in a rotation pattern that matches the rotation of the impeller. The guide vanes are often referred to as prerotation guide vanes. They are controlled by vane motors that vary the capacity of the centrifugal compressor. The controller sends either a pneumatic (air) or electronic

Figure 47.13 The prerotation vanes of a centrifugal compressor start the refrigerant rotating and also serve as capacity control. Courtesy of York International.

signal to the operating motor of the inlet guide vane, which is either a pneumatic or electronic motor. When the building load begins to drop, the controller tells the vane motor to start closing the vanes. The capacity control motor is controlled by the temperature of the entering or leaving chilled water, depending on the engineer's design. If the guide vanes are fully closed, the compressor pumps at only 15% to 20% of its rated capacity. When open all of the way, the compressor pumps at 100% of its rated capacity. A 1000-ton chiller can operate from 150–200 tons to 1000 tons of capacity in any number of steps. The guide vanes can also be used for two other purposes—to prevent motor overload and to start the motor at reduced capacity to reduce the system's current draw on start-up.

When a chiller is operating at a condition in which the chilled water is above the design temperature—for example, when a building is hot on a Monday morning start-up—the compressor motor would run under an overloaded condition. At that time, the return chilled water may be 75°F instead of the design temperature of 55°F. A control known as a *load limiter* on the chiller control panel will sense the increased motor amperage and partially close the guide vanes to limit compressor amperage to its full-load value. When the controls have established that the compressor motor is up to speed, the control system will take over and rotate the vanes to the correct position for the application. When the compressor starts, the prerotation guide vanes are closed and do not open until the motor is up to speed. Therefore, the compressor starts up unloaded, which reduces the power required to start the compressor.

Some centrifugal compressors are hermetically sealed and some have open drives, **Figure 47.14**. The motors of hermetically sealed compressors are refrigerant-cooled. **Figure 47.15** shows a liquid-cooled motor. Liquid refrigerant is allowed to flow around the motor housing. Systems with open-drive motors use the air in the equipment room to cool the motor. The heat from a large motor is considerable and must be exhausted from the equipment room. The rotation of the motor in many centrifugal chillers is critical. If the motor is started in the wrong direction, some chillers can be damaged, as will be discussed later in this unit.

Figure 47.14 (A) A hermetically sealed centrifugal chiller. (B) An open-drive centrifugal chiller. (A) Courtesy of Carrier Corp. (B) Courtesy of York International.

(A)

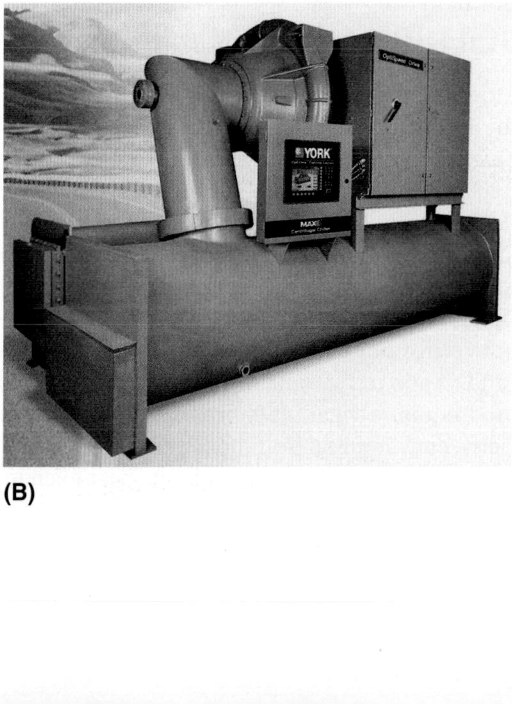

(B)

Figure 47.15 A refrigerant-cooled hermetic centrifugal compressor motor. Courtesy of Carrier Corp.

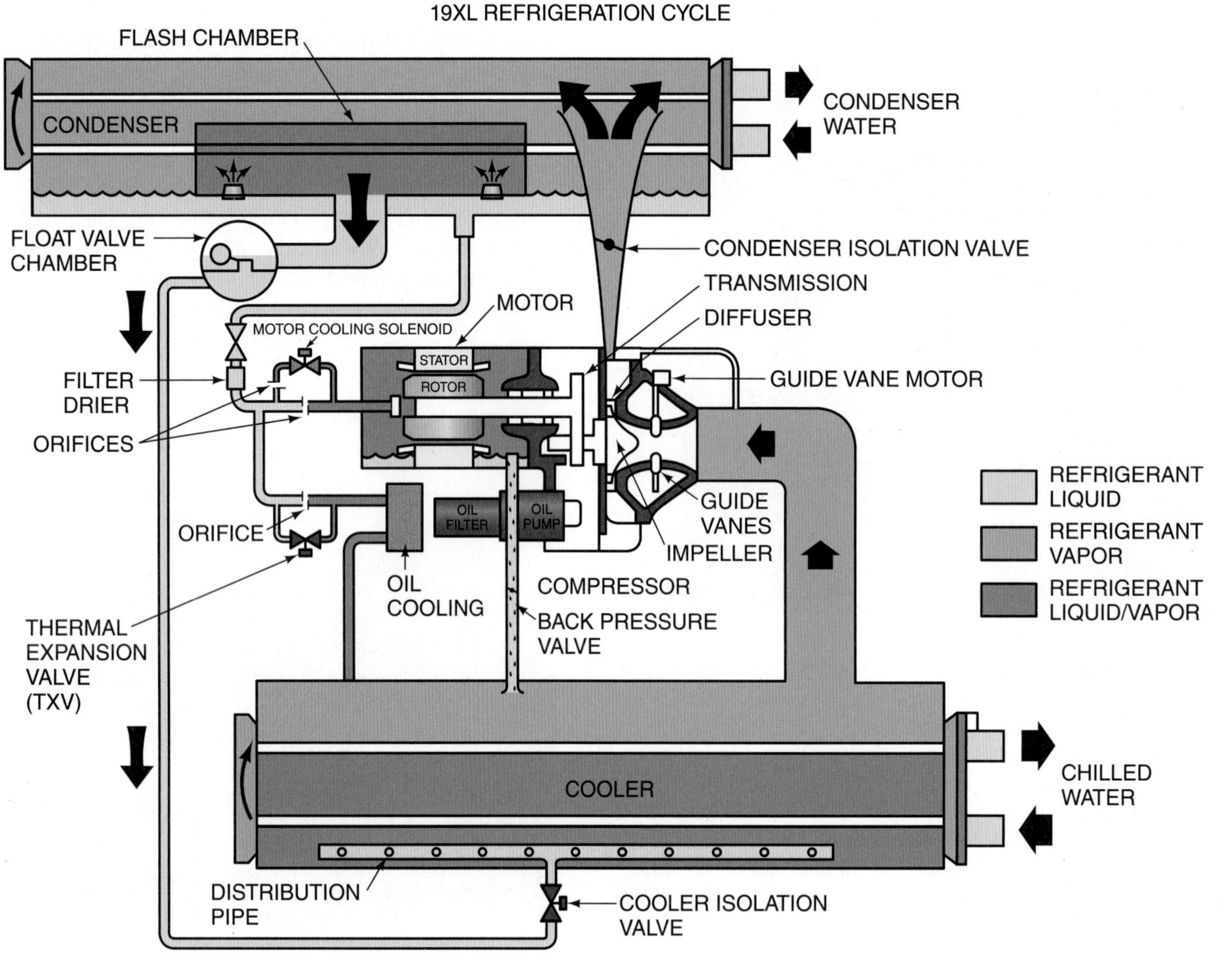

47.7 EVAPORATORS FOR HIGH-PRESSURE CHILLERS

The evaporator is the component that absorbs heat into a system. In the evaporator, liquid refrigerant is boiled to a vapor by the circulating water. This portion of a chilled-water system is often called the chiller barrel. For water chillers, the heat exchange surface is typically copper. Other materials may be used if corrosive fluids are being circulated for manufacturing processes. In typical smaller air-conditioning systems, air is on one side of the heat exchange process and liquid or vapor refrigerant on the other. The rate of exchange between air and vapor refrigerant is only fair. The exchange between air and liquid refrigerant is better, and the best heat exchange is between water and liquid refrigerant. When a liquid-to-liquid heat exchanger is being used, the heat exchange surface can be small and still produce the desired results.

DIRECT-EXPANSION EVAPORATORS

The evaporators used in high-pressure chillers are either direct-expansion evaporators or flooded evaporators. Direct-expansion evaporators are also known as dry-type evaporators, meaning they have an established superheat at the evaporator outlet, and they normally use thermostatic expansion devices to meter the refrigerant. They are used on smaller chillers, of up to approximately 150 tons for older types and to about 100 tons for more modern chillers. Refrigerant in direct-expansion chillers is introduced into the end of the chiller barrel and water is introduced into the side of the shell, **Figure 47.16**. The water is on the outside of the tubes, where baffles cause it to come in contact with as many tubes as possible for the best heat exchange. The problem with this arrangement occurs when the water side of the circuit gets fouled or dirty. The only way the chiller can be cleaned is with chemicals. It is in this way that these chillers differ from

Figure 47.16 A direct-expansion chiller with an expansion device on the end of the chiller. Courtesy of York International.

those in which the water circulates in the tubes and the refrigerant circulates around the tubes—as with flooded chillers.

FLOODED EVAPORATOR CHILLERS

Flooded chillers introduce refrigerant at the bottom of the chiller barrel and the water circulates through the tubes. There are some advantages to this arrangement because the tubes may be totally submerged in the refrigerant for the best heat exchange, **Figure 47.17**. The tubes may also be physically cleaned in several different ways. **Figure 47.18** shows tubes being cleaned with a brush. The water side of an evaporator should not become dirty with this system unless the chiller is being used in an open process, such as a manufacturing operation. Most chillers use a closed water circuit that should remain clean unless there are many leaks and water must be continually added. Flooded chillers use

Figure 47.17 Water tubes submerged under the liquid refrigerant. Courtesy of York International.

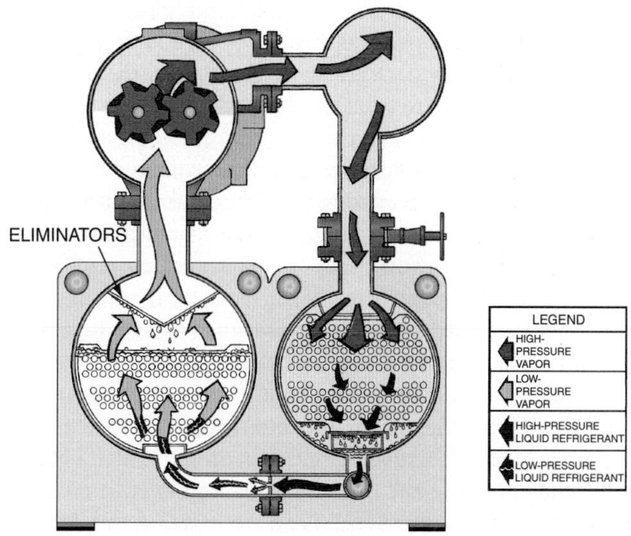

Figure 47.18 *Cleaning the tubes in a condenser by using a power brush and flushing with water.* Courtesy of Goodway Tools Corp.

much more refrigerant charge than direct-expansion chillers, so leak monitoring must be part of regular maintenance with high-pressure systems.

Water introduced into the end of the chiller is contained in water boxes. Water boxes with removable covers are known as marine water boxes, **Figure 47.19**. Water-box covers with attached piping that must be detached to remove the covers are known as standard covers, **Figure 47.20**. The water box is used to direct the water by means of partitions. For example, water may need to pass through the chiller one, two, three, or four times to achieve different purposes. Thus, a manufacturer can use one chiller with combinations of water boxes and partitions for different applications.

Figure 47.19 Marine water boxes allow the technician to clean the tubes by removing the water-box cover; piping does not have to be disconnected. Courtesy of Trane Company.

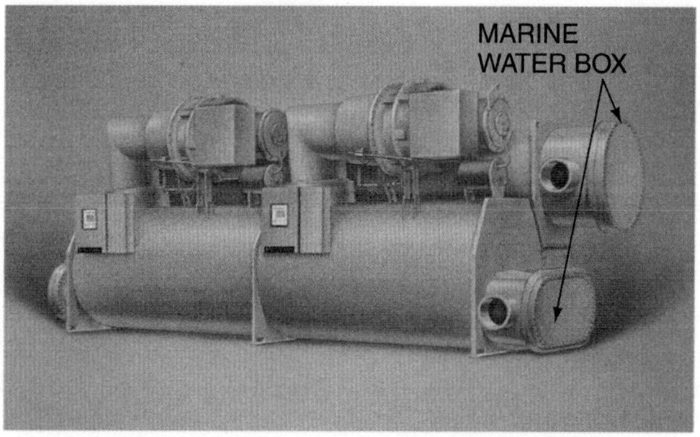

Figure 47.20 Standard water-box covers with piping connections. Courtesy of York International.

Figure 47.21 Heat exchangers for evaporators or condensers that are one-, two-, three-, and four-pass shells.

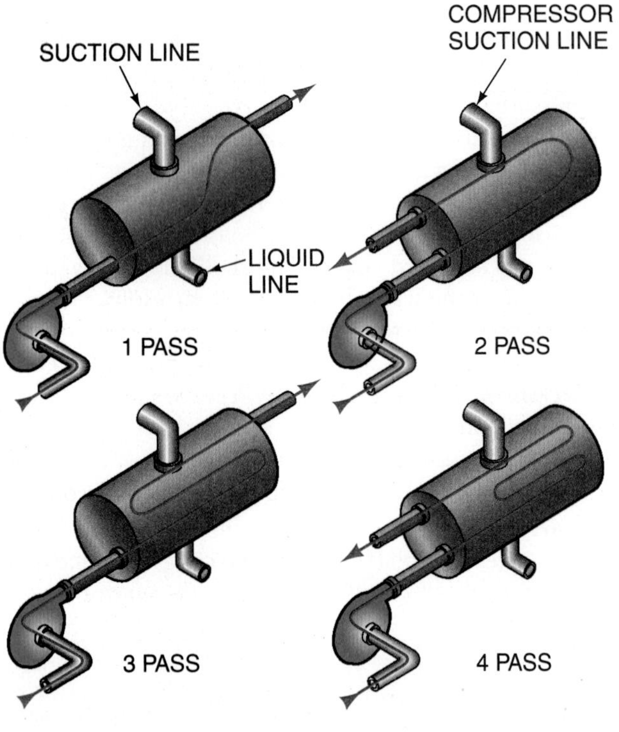

Water passing through a chiller once is in contact with the refrigerant for only a short period of time. Using two, three, or four passes keeps the water in contact longer but also creates a larger pressure drop and requires more pumping horsepower, **Figure 47.21**. In a two-pass chiller, the design water temperatures are typically 55°F for inlet water and 45°F for outlet water. The refrigerant is absorbing heat from the inlet water, and its boiling temperature is typically about 7°F cooler than that of the leaving water. The difference between

Figure 47.22 Approach temperature for a two-pass chiller.

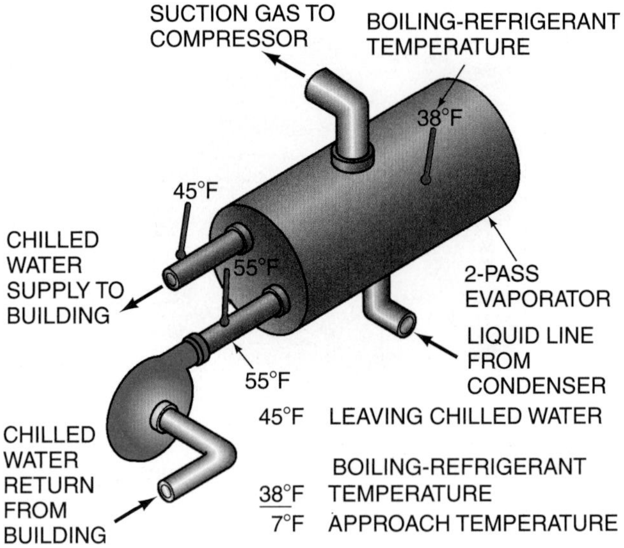

the inlet water temperature and the refrigerant's boiling, or saturation, temperature is called the approach temperature, **Figure 47.22. The approach temperature is very important to the technician when troubleshooting chiller performance.** If the chiller tubes become dirty, the approach temperature increases because the heat exchange rate between the refrigerant and the water will decrease. If the water cannot give up heat to the refrigerant, the temperature of water will be higher than desired and the saturation temperature of the refrigerant will be lower than desired.

The approach temperature is different for chillers with a different number of passes. Direct-expansion chillers have only one pass of water (with baffles to sweep the tubes), and the approach temperature may be about 8°F. Chillers with three-pass evaporators will have an approach temperature of about 5°F, and chillers with four passes may have an approach temperature as low as 3°F or 4°F. The longer the refrigerant is in contact with the water, the lower the approach temperature will be. Flooded chillers generally have a means of measuring the actual refrigerant temperature in the evaporator, either with a thermometer well or as a direct readout on the control panel. Direct-expansion chillers typically have a pressure gauge on the low-pressure side of the system; the pressure must be converted to temperature to determine the evaporating refrigerant temperature. When a chiller is first started up, a record known as an operating performance log should be kept. Data should be recorded when the chiller is at full-load conditions. Often, equipment manufacturers rather than the contractor choose to start up larger chillers because of the cost of warranty. They will update the entries in the operating log and keep it on file for future reference. **Figure 47.23** is an example of an equipment manufacturer's log sheet.

Figure 47.23 A chiller operation log sheet. Courtesy of York International.

YORK

ROTARY SCREW
LIQUID CHILLER LOG SHEET
125–400 TONS

CHILLER LOCATION ______________

SYSTEM NO. ______________

Date												
Time												
Hour Meter Reading												
O.A. Temperature D.B./W.B.			/	/	/	/	/	/	/	/	/	/
Compressor		Oil Level										
		Oil Pressure										
		Oil Temperature										
		Suction Temperature										
		Discharge Temperature										
		Filter PSID										
		Slide Valve Position %										
		Oil Added (gallons)										
Motor		Volts										
		Amps										
Cooler	Refrig.	Suction Pressure										
	Liquid	Inlet Temperature										
		Inlet Pressure										
		Outlet Temperature										
		Outlet Pressure										
		Flow Rate — GPM										
Condenser	Refrig.	Discharge Pressure										
		Corresponding Temperature										
		High Pressure Liquid Temperature										
		System Air — Degrees										
	Water	Inlet Temperature										
		Inlet Pressure										
		Outlet Temperature										
		Outlet Pressure										
		Flow Rate — GPM										

FORM 160.47-F6

When water is boiled in an open pot and the heat is turned up to the point that the water is boiling vigorously, water will splash out of the pot. Refrigerant in a flooded chiller acts much the same way; the compressor removes vapor from the chiller cavity, which will accelerate the process due to vapor velocity. Liquid eliminators are often placed above the tubes to prevent liquid from carrying over into the compressor suction intake at full load, **Figure 47.17**. In a flooded chiller, the suction gas is often saturated as it enters the suction intake of a compressor, unlike in a direct-expansion chiller where the gas will always be superheated when leaving the evaporator.

Evaporators are constructed as a shell with end sheets to hold the tubes in place. *The heat transfer efficiencies of both direct-expansion and flooded tube surfaces may be improved by machining fins on the outside of the tubes. Direct-expansion evaporator tubes may have inserts that cause the refrigerant to sweep the sides of the tube, or the inside of the tube may be rifled like the bore of a rifle,* ***Figure 47.24.*** These features increase the pressure drop through the heat exchanger slightly, which is not desirable, but the net overall positive effect on efficiency makes it worthwhile. Typically, the tubes are made leak-free in the tube sheets by a process called rolling. Grooves are cut in the hole in the tube sheet into which the tube fits. After the tube is inserted in the hole, it is expanded into the grooves by

Figure 47.24 *Fins on the outside of a chiller tube and a rifle-like bore on the inside are used to expand the surface area and increase the heat exchange per length of tube.*

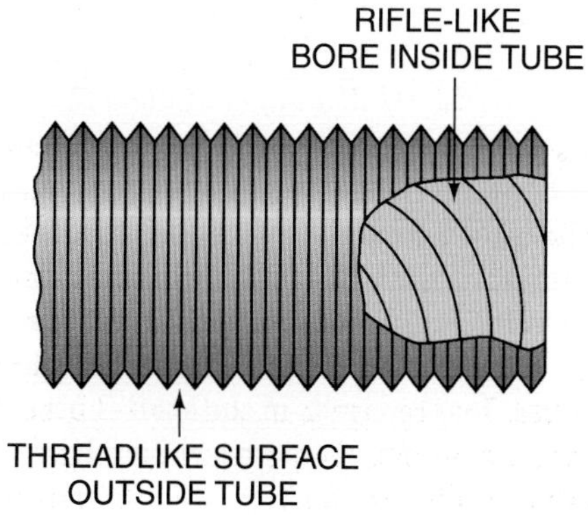

Figure 47.25 The tubes are rolled into the tube-sheet grooves to make a leak-free connection between the refrigerant side and the water side.

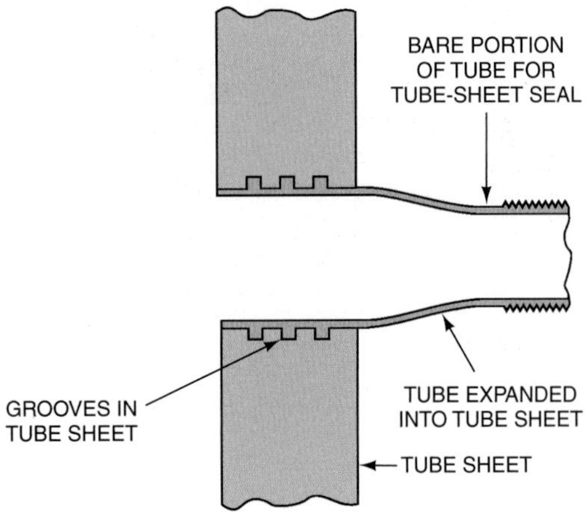

being rolled for a leak-free connection, **Figure 47.25**. Some chiller evaporators may have tubes that are silver-soldered at the tube sheet. These tubes are harder to replace than those that are rolled in place.

Evaporators have a working pressure for the refrigerant and the water circuits. High-pressure chillers must be able to accommodate the refrigerant used, generally R-22, R-410A, or R-134a. Newer, environmentally friendly, alternative refrigerants may also be used in newer equipment. The water side of the chiller may have one of two working pressures, 150 or 300 psig. The 300-psig chillers are used when the chiller is to be located on the bottom floor of a tall building where the water column will create a high pressure from the standing column of water. In Unit 37, "Air Distribution and Balance," it was discussed that 1 psi of pressure will support a column of water 2.31 ft high. A column of water 1 ft high will exert a pressure of 0.433 psig (1 psi/2.31 ft = 0.4329, or 0.433 psi/ft). A chiller with a 300-psig working pressure can have only 693 ft of water standing above it (300 psi/0.433 ft/psi = 692.8 ft). Each story of a multistory building is about 12 ft, so this chiller could serve a building that is only 58 stories high (692.8/12 = 57.5 stories). Chillers can be located on the upper floors of multistory buildings. Service becomes more difficult, but not many other options exist.

Evaporators are leak tested at the time of manufacture. If proven to be leak-free, they are then insulated. Many larger chillers are insulated with rubber foam sheeting material that is glued directly to the shell. Smaller chillers may be insulated by being placed in a skin of sheet metal with foam insulation applied between the chiller shell and the skin. A repair requires the skin to be removed and the foam cut away. The skin can be replaced later and foam insulation added back to insulate the chiller barrel. Some evaporators may be located outside, as is the case with an air-cooled package chiller, so freeze protection for the water in the shell is necessary. Resistance heaters can be added to the chiller barrel under the insulation. The heaters may be wired into the chiller control circuit and energized by a thermostat that closes its contacts if the ambient temperature approaches freezing and water is not being circulated.

47.8 CONDENSERS FOR HIGH-PRESSURE CHILLERS

The condenser is the component that transfers heat out of the system. The condenser for high-pressure chillers may be either water-cooled or air-cooled. Both will transfer heat out of the system, usually to the atmosphere. In some manufacturing processes, the rejected heat may be recovered and used for other purposes.

WATER-COOLED CONDENSERS

Water-cooled condensers used for high-pressure chillers are the shell-and-tube type with the water circulating in the tubes and the refrigerant filling the space around the tubes. The shell must have a working pressure to accommodate the refrigerant used and a water-side working pressure of 150 to 300 psig, like the evaporators mentioned earlier. The hot discharge gas from the compressor is normally discharged into the top of the condenser, **Figure 47.26**. The refrigerant is condensed to a liquid and falls to the bottom of the condenser, where it gathers and drains into the liquid line. As with evaporators, the heat exchange can be improved by use of extended surfaces on the refrigerant side of the condenser tube, **Figure 47.24**. The inside of the condenser tubes may also be grooved, as with the evaporator tubes, for improved water-side heat exchange.

Condensers must also have a method for piping the water into the shell. There are two basic types of connections to the water box that is attached to the condenser shell. In a standard water box, the piping is attached to the removable water box and some of the piping must be removed before the cover can be removed on the piping end. A marine water box has a removable cover for easy access to the piping end of the machine. Tube access is more important for the condenser than the evaporator in most installations because the water in the cooling tower is open to the atmosphere and more likely to carry dirt and other foreign matter into the condenser.

Some comfort-cooling installations serve both cooling and heating needs at the same time. While the outside, or perimeter, of the building may need heat in the winter, the core of the building and its occupants may be generating heat from lighting, equipment that needs to be removed. These installations may have what is known as double-bundled condensers, which are two condensers in one shell. *When perimeter heating is needed, the warm condenser water can be circulated in the perimeter heating circuit.* When no perimeter heating is needed and the heat needs to be transferred to the outside, the

Figure 47.26 *The subcooling circuit improves the efficiency of the chiller. Courtesy York International.*

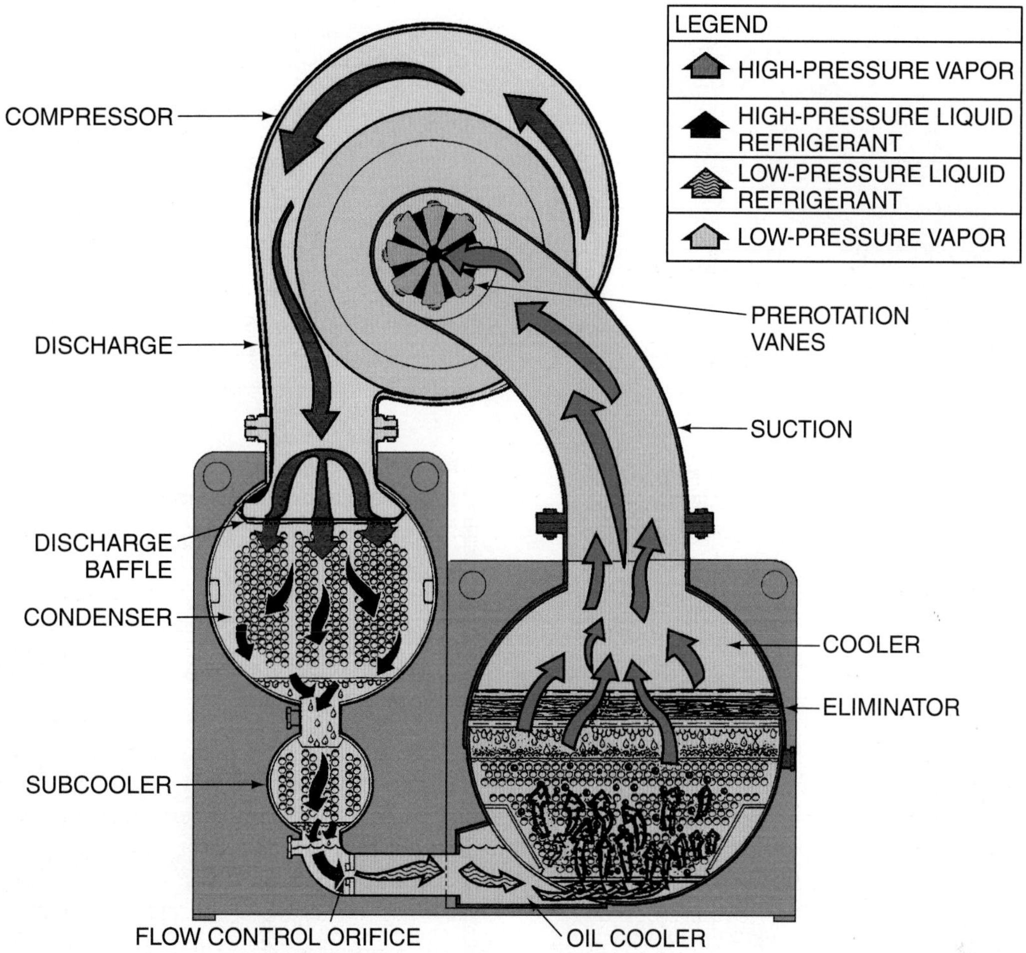

other condenser may be used and the heat transferred to the cooling tower. With this type of installation, the temperature of the water from the condenser may not be warm enough to satisfy the heating needs of the perimeter zones, so the water temperature may be increased with a boiler. *The net result is that the water entering the boiler is preheated with the heat from the building interior.*

CONDENSER SUBCOOLING

The design condensing temperature for a water-cooled condenser is around 105°F on a day when 85°F water is supplied to the condenser and the chiller is operating at full load. The saturated liquid refrigerant would be 105°F. Many condensers have subcooling circuits that reduce the liquid temperature to below 105°F. If the entering water at 85°F is allowed to exchange heat from the saturated refrigerant, the liquid-line temperature can be reduced. *If the water can be reduced to 95°F, a considerable amount of capacity can be gained for the evaporator, which can increase the machine capacity by about 1%/°F of subcooling. For a 300-ton chiller, 10°F subcooling would increase the capacity about 30 tons with very little energy use.* To subcool refrigerant, it must be isolated from the condensing process with a separate circuit, **Figure 47.26**, so the only expense for subcooling is the extra circuit in the condenser.

As in the evaporator, there is a condenser approach temperature relationship between the refrigerant's condenser saturation temperature and the temperature of the water leaving the condenser. Most water-cooled condensers are two-pass condensers and are designed for 85°F entering water and 95°F leaving water with a condensing temperature of about 105°F. The approach temperature in the condenser is calculated as the condenser saturation temperature minus the temperature of the water leaving the condenser. In this example, we have 105°F–95°F for an approach temperature of 10°F, **Figure 47.27**. Other condensers may be configured as one-, three-, or four-pass with different approach temperatures. As with evaporators, the longer the refrigerant is in contact with the water, the lower the approach temperature will be. The original operating log will reveal what the approach

Figure 47.27 The approach temperature for a two-pass condenser.

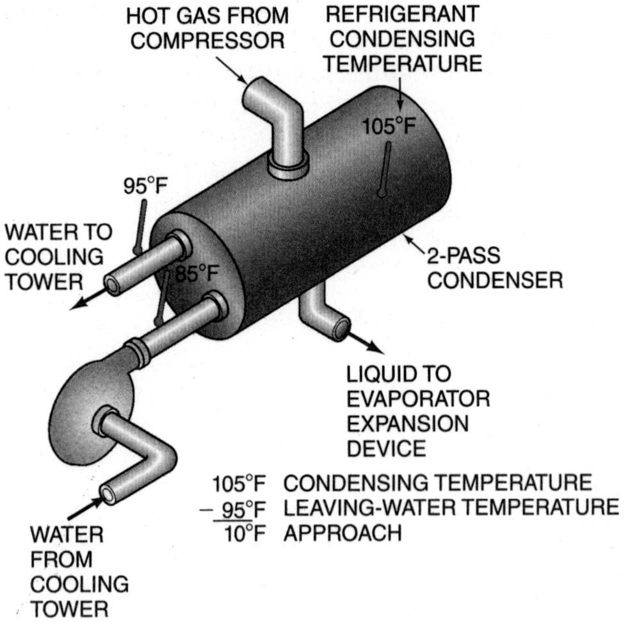

Figure 47.28 Heat moves from the refrigerant to the water and then into the atmosphere from the cooling tower.

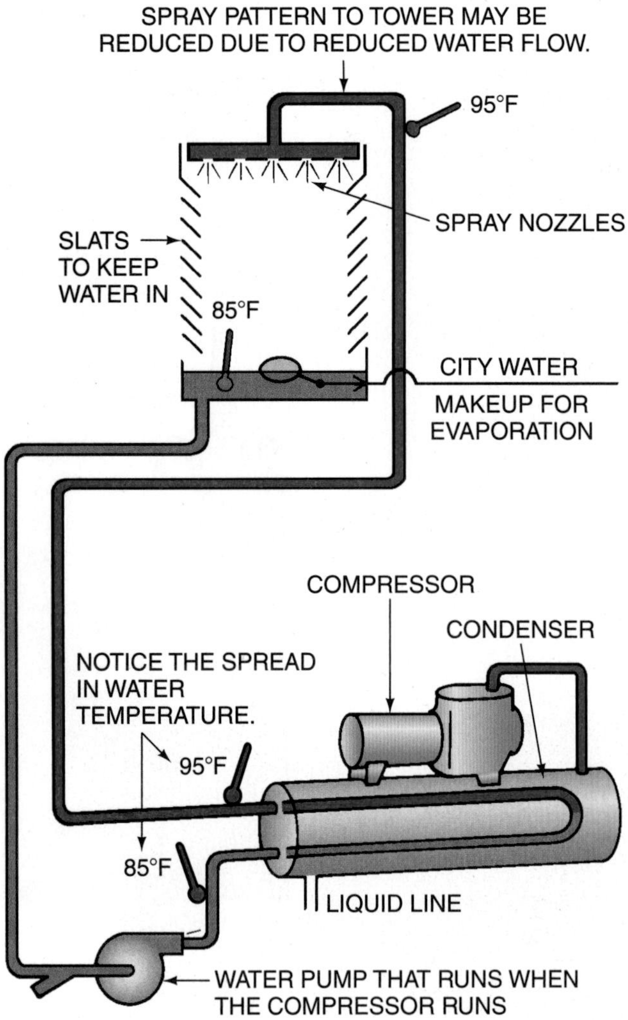

temperature was at start-up. It should be the same for similar conditions at a later date, provided the tubes are clean.

Because the condenser rejects heat from the system, its capacity to do so determines the head pressure for the refrigeration process. There must be some pressure in the condenser to push the refrigerant through the expansion device. The head pressure must be controlled. If it becomes too low, problems may occur at start-up and in very cold weather when there is still a call for air-conditioning. When there is a call to start up a water-cooled chiller and the water in the cooling tower is cold, the head pressure may become so low that the suction pressure will be low enough to trip either the low-pressure control (direct-expansion chillers) or the evaporator freeze control (flooded chillers). Continued operation will also cause oil migration in many machines. A bypass valve is typically located in the condenser circuit to bypass water during start-up to prevent nuisance shutdowns due to cold condenser water. This same bypass may be used during normal operation. (This is discussed in more detail in Unit 48, "Cooling Towers and Pumps.") The heat from the refrigerant is transferred into the water circulating in the condenser circuit and in a typical installation the heat must then be transferred to the atmosphere. This is done by means of a cooling tower, **Figure 47.28**, which are discussed in Unit 48.

AIR-COOLED CONDENSERS

The tubes in which the refrigerant circulates in air-cooled condensers are generally constructed of copper with aluminum fins on the outside of the tubes to provide a larger heat exchange surface, **Figure 47.29**. These air-cooled heat exchange surfaces have been used for many years with great success. Some manufacturers furnish copper-finned tubes or special coatings for the aluminum fins in locations where salt air may corrode the aluminum. The coils are positioned differently by manufacturers to suit their own design purposes, some horizontally and some vertically. Many air-cooled condensers use multiple fans that can be cycled on and off for head pressure control.

Air-cooled condensers range in size from very small to several hundred tons. They eliminate the need for cooling towers and the problems often associated with water-cooled equipment and are popular because they require less maintenance. Systems using air-cooled condensers do not operate at the same condensing temperatures and head pressures as water-cooled condensers. Water-cooled condensers typically condense at about 105°F. Air-cooled condensers condense

Figure 47.29 (A) An air-cooled condenser coil. (B) An air-cooled chiller. The condenser coils are at the top. (B) Courtesy of York International.

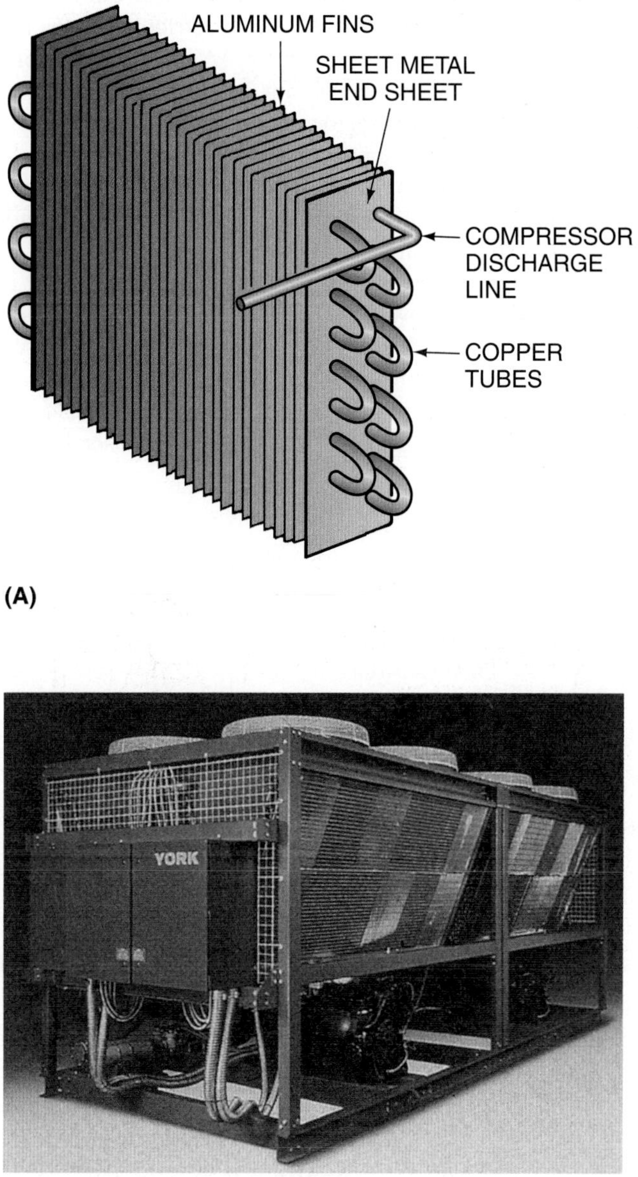

refrigerant at temperatures that are typically 20°F to 30°F higher than the entering air temperature, depending on the condenser surface area per ton. For example, an R-134a chilled-water system with a water-cooled condenser would have a head pressure of about 135 psig (105°F compares to 135 psig). An air-cooled condenser on a 95°F day would have a head pressure that could range from 157 psig (95°F + 20°F = 115°F) to 185 psig (95°F + 30°F = 125°F). This is a considerable difference in head pressure and will increase the operating cost of the chiller.

SUBCOOLING CIRCUIT

Subcooling in an air-cooled condenser is just as important as in a water-cooled unit. *It gives the unit more capacity by about 1%/°F of subcooling.* Subcooling is accomplished in an air-cooled condenser by means of a small reservoir in the condenser to separate the subcooling circuit from the main condenser, **Figure 47.30**. *A condenser operating on a 95°F day can expect to achieve 10°F to 15°F of subcooling.*

Many units accomplish the subcooling process by flooding the bottom of the condenser with refrigerant. This situation resembles an overcharge of refrigerant, except that the condenser has additional tubes to hold the extra charge. The technician must be aware of this subcooling circuit when charging the unit, or the system will not operate at its rated capacity. Often, there is a sight glass in the reservoir and refrigerant is added until the proper level is reached in the reservoir. Sometimes there is a sight glass in the condenser end sheet to show the level of liquid refrigerant in the condenser. Otherwise, the technician must use a temperature lead attached to the liquid line to determine the subcooling temperature while adding refrigerant. This is accomplished by converting head pressure to condensing temperature and adding refrigerant until the liquid line is lower in temperature than the condensing temperature. For example, the condensing temperature may be 125°F. To arrive at a subcooling of 15°F, the technician would charge the unit until the liquid-line temperature is 110°F.

Figure 47.30 *The subcooling circuit in an air-cooled condenser.*

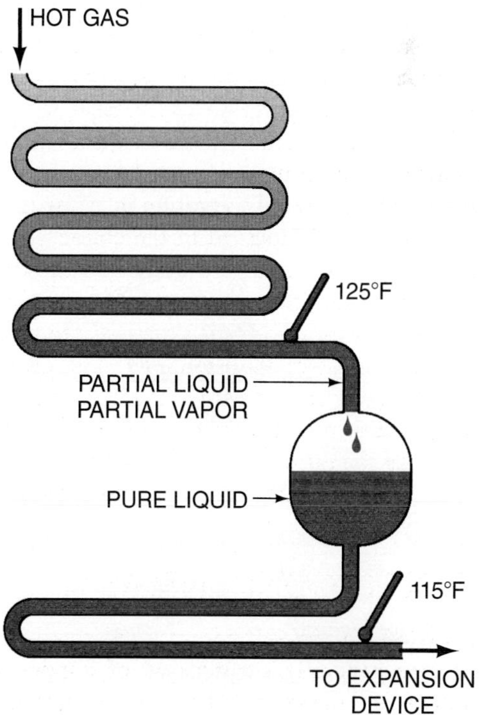

47.9 METERING DEVICES FOR HIGH-PRESSURE CHILLERS

The metering device allows refrigerant to flow into the evaporator at the correct rate. Four types of metering devices may be used for large chillers: thermostatic expansion valves, orifices, high- and low-side floats, and electronic expansion valves.

THERMOSTATIC EXPANSION VALVES

The *thermostatic expansion valve* (*TXV*) was discussed in detail in Unit 24, "Expansion Devices." The TXVs used with chillers are the same, only larger. The TXV maintains a constant superheat at the end of the evaporator; usually this is 8°F to 12°F. The more superheat, the more vapor there will be in the evaporator and the less heat exchange will occur. The more liquid the evaporator has, the better the heat exchange, so TXVs are not used except on smaller chillers, up to about 150 tons.

ORIFICES

The orifice is a fixed-bore metering device that is merely a restriction in the liquid line between the condenser and the evaporator, **Figure 47.31**. These devices are trouble-free because they have no moving parts. The flow of refrigerant through an orifice is constant at a given pressure drop. A greater pressure difference will produce more flow. Greater loads produce a higher head pressure, and when the head pressure is allowed to rise, more flow will occur. Any chiller with an orifice for a metering device has a critical refrigerant charge. The unit must not be overcharged or liquid refrigerant will enter the compressor. This could be detrimental to a reciprocating compressor, so orifices may not be used with these compressors. Small amounts of liquid refrigerant, often in the form of wet vapor, will not harm the scroll, rotary screw, or most centrifugal compressors.

Figure 47.31 An orifice metering device. Courtesy of Trane Company.

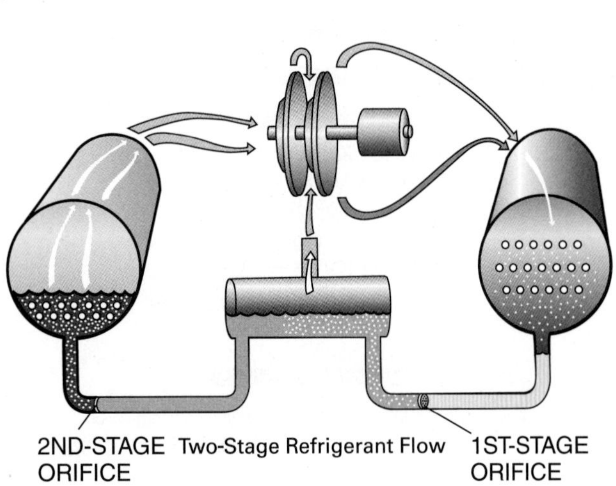

FLOAT-TYPE METERING DEVICES

There are two types of float-type metering devices: the **low-side float** and the **high-side float**. The low-side float rises to reduce refrigerant flow when the liquid refrigerant level rises in the low side of the chiller, **Figure 47.32**. It is positioned so that it maintains the refrigerant level as required by the chiller. The high-side float is located in the liquid line entering the evaporator. When the refrigerant level is higher in the liquid line than in the evaporator, the evaporator contains less refrigerant than it requires and the float rises, allowing more liquid refrigerant to enter, **Figure 47.33**.

Floats can be a problem if they rub against the side of the float chamber, which will cause resistance or rub a hole in the float. The float will sink if a hole develops. In the case of the low-side float, the line will open and allow a full flow of refrigerant to the evaporator; no refrigerant will be held back from the liquid line and the bottom of the condenser. If a hole

Figure 47.32 A low-side float.

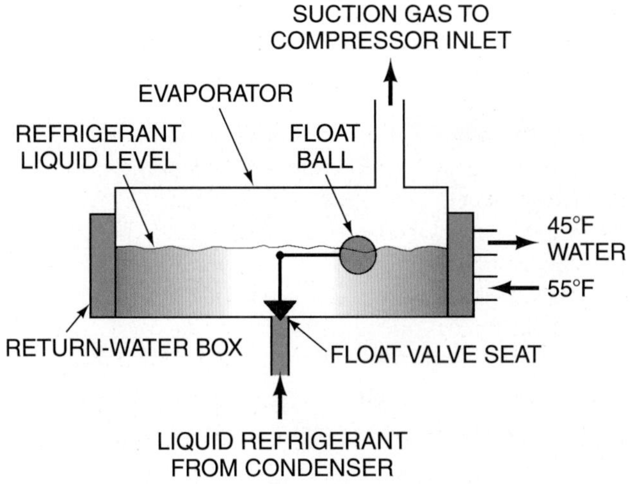

Figure 47.33 A high-side float.

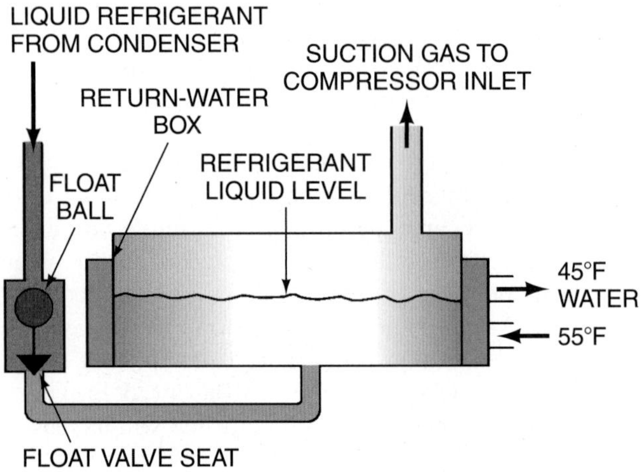

wears in the high-side float, it will sink and block the flow of liquid refrigerant to the evaporator.

The refrigerant charge is critical when either the low- or high-side floats are used. An overcharge with a low-side float will back refrigerant up into the liquid line and condenser. An overcharge of refrigerant with a high-side float will flood the evaporator.

ELECTRONIC EXPANSION VALVES

The electronic expansion valve used with large chillers is similar to the one explained in Unit 24, "Expansion Devices." In the line, it looks much like a solenoid valve, **Figure 47.34**. Expansion valves operate by means of a thermistor to monitor the refrigerant temperature. The thermistor sensor can be inserted in the evaporator piping. The liquid does not touch the sensor; an electronic circuit is used instead. The sensing element operates much like a thermometer except its readout is not temperature. A pressure transducer gives the electronic circuit a pressure reading that can be converted to temperature. The pressure reading from the evaporator converted to temperature along with the actual suction-line temperature can measure the real superheat and make close control possible.

An advantage of the electronic expansion valve is the versatility of the electronic control circuit, which makes it able to handle a wider variation in load than a typical TXV. It can also be used to allow more refrigerant to flow under low head pressure conditions during low-ambient operation. This is particularly helpful in air-cooled units when the weather is cold. The electronic expansion valve can allow maximum liquid refrigerant flow due to its large liquid-handling capacity.

The ability to allow maximum flow on a cold day gives the condenser much more capacity to pass on to the system. For example, if it is 40°F outside and the building is calling for cooling, the condenser can condense the refrigerant at about 60°F to 70°F. This is a head pressure of 58 to 71 psig for R-134a. This would not be enough head pressure for a typical TXV. But an electronic expansion valve (because of its flow capability) can use this low pressure to feed the evaporator. If the refrigerant is then subcooled 10°F more, the evaporator can be furnished with 50°F to 60°F liquid refrigerant. *The compressor is only required to pump from an evaporator pressure of approximately 33 psig (corresponding to 38°F evaporator boiling temperature for R-134a) to 71 psig maximum head pressure.* This is a big difference from a typical summer condition where the evaporator suction pressure may be 33 psig and the head pressure 185 psig (corresponding to a condensing temperature of 125°F) for R-134a.

Electronic expansion valves are often called step-control valves because they can modulate flow in so many increments. **Figure 47.35** is a photo of a step-control expansion valve

Figure 47.34 An electronic expansion valve. Courtesy of Trane Company.

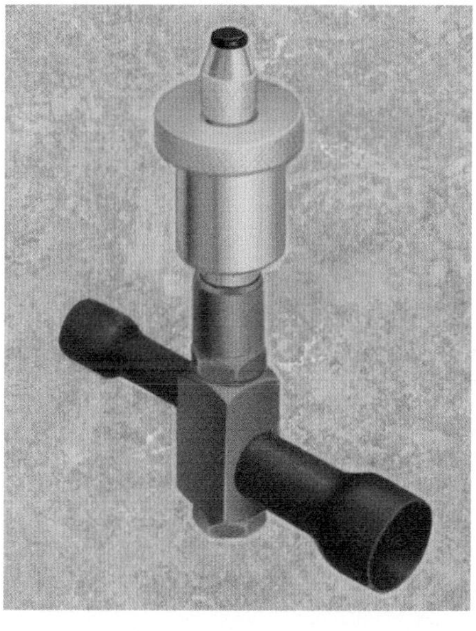

Figure 47.35 (A) A step-motor-control expansion valve for large systems. (B) This illustration shows how the needle is threaded into the seat by means of the motor. Courtesy of Sporlan Division, Parker Hannifin Corp.

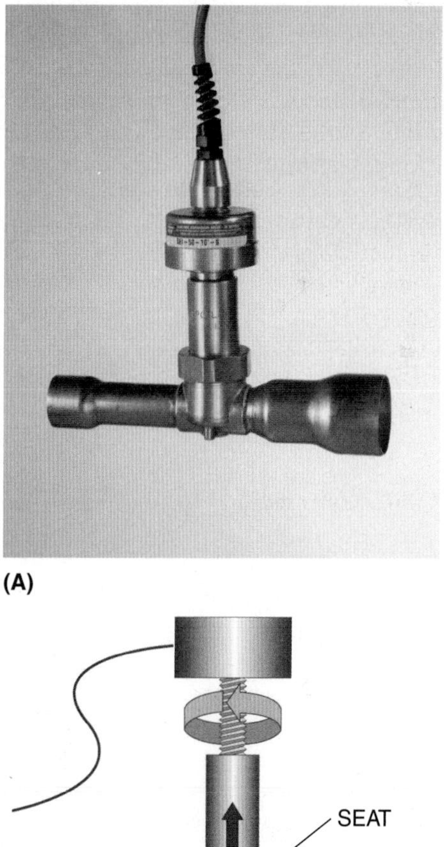

(A)

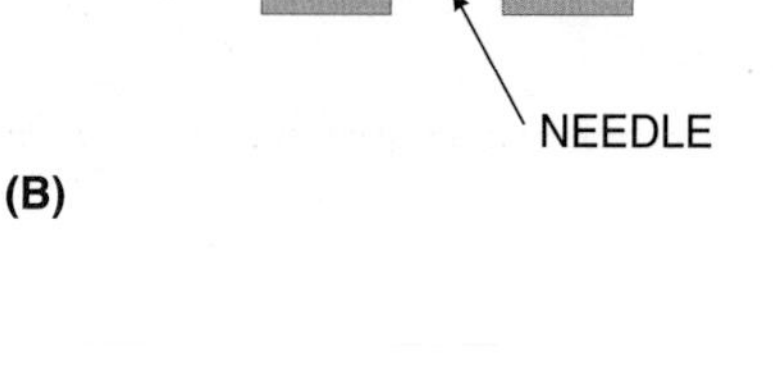

(B)

Figure 47.36 This is an exploded view of the gears and motor that accomplish step control. Courtesy of Sporlan Division, Parker Hannifin Corp.

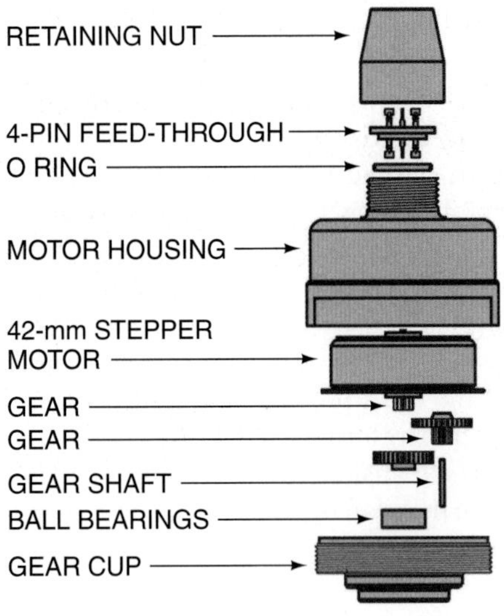

along with an illustration that shows the stepper motor turning a shaft with a needle and seat to modulate flow through the valve seat. **Figure 47.36** shows an exploded view of the valve with all of its intricate parts. **Figure 47.37** is a wiring diagram of how a valve may be wired into the circuit. Notice that there are two sensors (thermistors): one on the inlet of the coil and one on the outlet. There is no pressure connection for checking superheat; temperature difference is used to do that. The valve also has an optional pumpdown feature. It can be closed when the system thermostat is satisfied and the refrigerant can be pumped into the condenser. The valve reopens at the beginning of the next cycle to allow refrigerant to flow naturally during the ON cycle. The pumpdown cycle is used for any system that might suffer if liquid refrigerant is pumped into the compressor at the beginning of a new cycle because the refrigerant had migrated to the evaporator during the OFF cycle.

47.10 LOW-PRESSURE CHILLERS

Low-pressure chillers have all of the same components as high-pressure chillers: a compressor, an evaporator, a condenser, and a metering device. Many of the features and operations of low-pressure centrifugal chillers are the same or similar to those found on or with high-pressure units. Older low-pressure chillers typically have used R-11 or R-113, both CFCs, as refrigerants. Some newer low-pressure chillers and those that have been converted use R-123, which is an HCFC refrigerant.

COMPRESSORS FOR LOW-PRESSURE CHILLERS

All low-pressure chillers as well as some high-pressure chillers use centrifugal compressors. The speed of centrifugal compressors in low-pressure chillers ranges from a motor's

Figure 47.37 This wiring and piping diagram shows two temperature sensors and the wiring diagram that would be used for a step-control expansion valve.

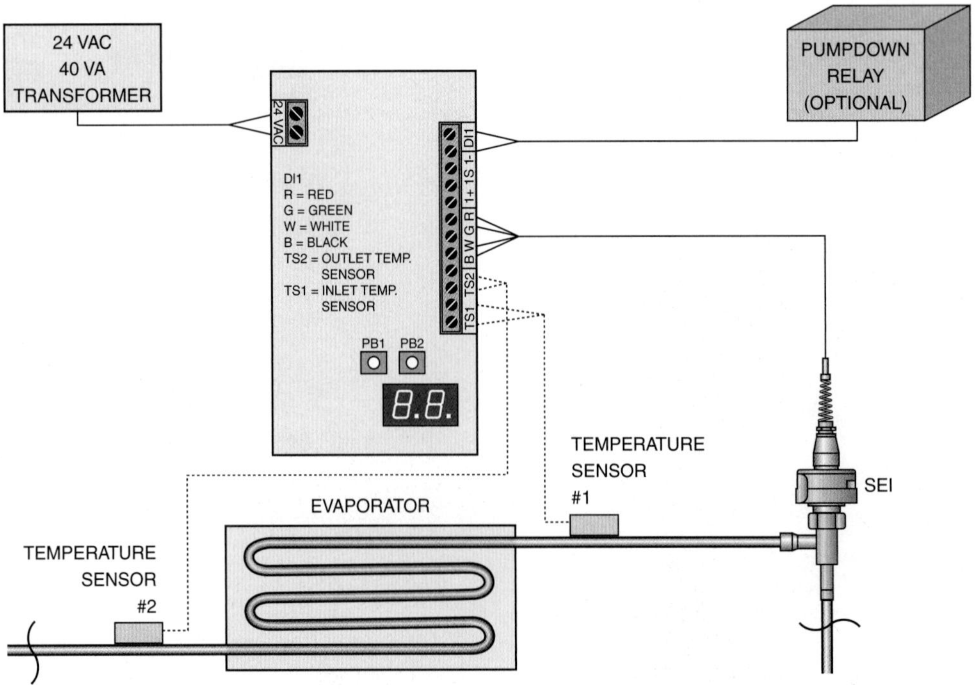

synchronous speed (3600 rpm) for direct-drive compressors to very high speeds (about 30,000 rpm) in gear-drive machines. The turning of the impeller of a centrifugal compressor creates a low-pressure area in the center where the suction line is fastened to the housing, **Figure 47.38**. The compressed refrigerant gas is trapped in the outer shell, known as the volute, and is guided down the discharge line to the condenser, **Figure 47.39**.

Systems with centrifugal compressors can be manufactured to produce more pressure by including more than one compressor and operating them in series, called multistage, or operating one stage of compression at a high speed. Single-stage compression can be accomplished using a single compressor that turns at a high speed, up to about 30,000 rpm. A gearbox is used with a 3600-rpm motor to step up the speed of the compressor, **Figure 47.11**. The advantage of this arrangement is lower size and weight.

With multistage compressors, the discharge from one compressor actually becomes the suction for the next compressor, **Figure 47.40**. It is common for compressors in air-conditioning applications to have up to three stages of

Figure 47.38 A multiple-stage compressor and chiller. Courtesy of Trane Company.

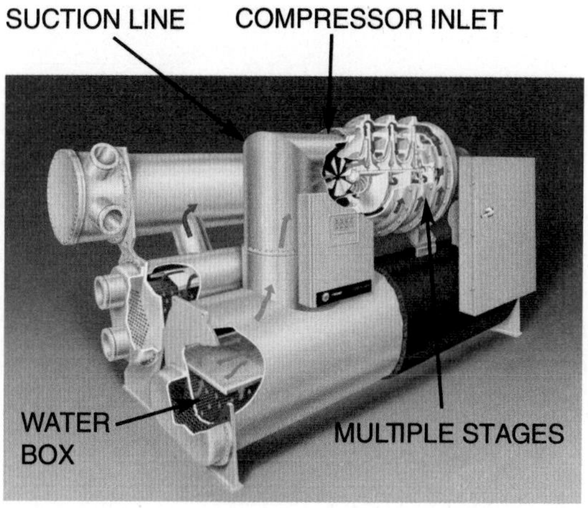

Figure 47.39 The hot gas enters the top of the condenser through the discharge line. Courtesy Trane Company

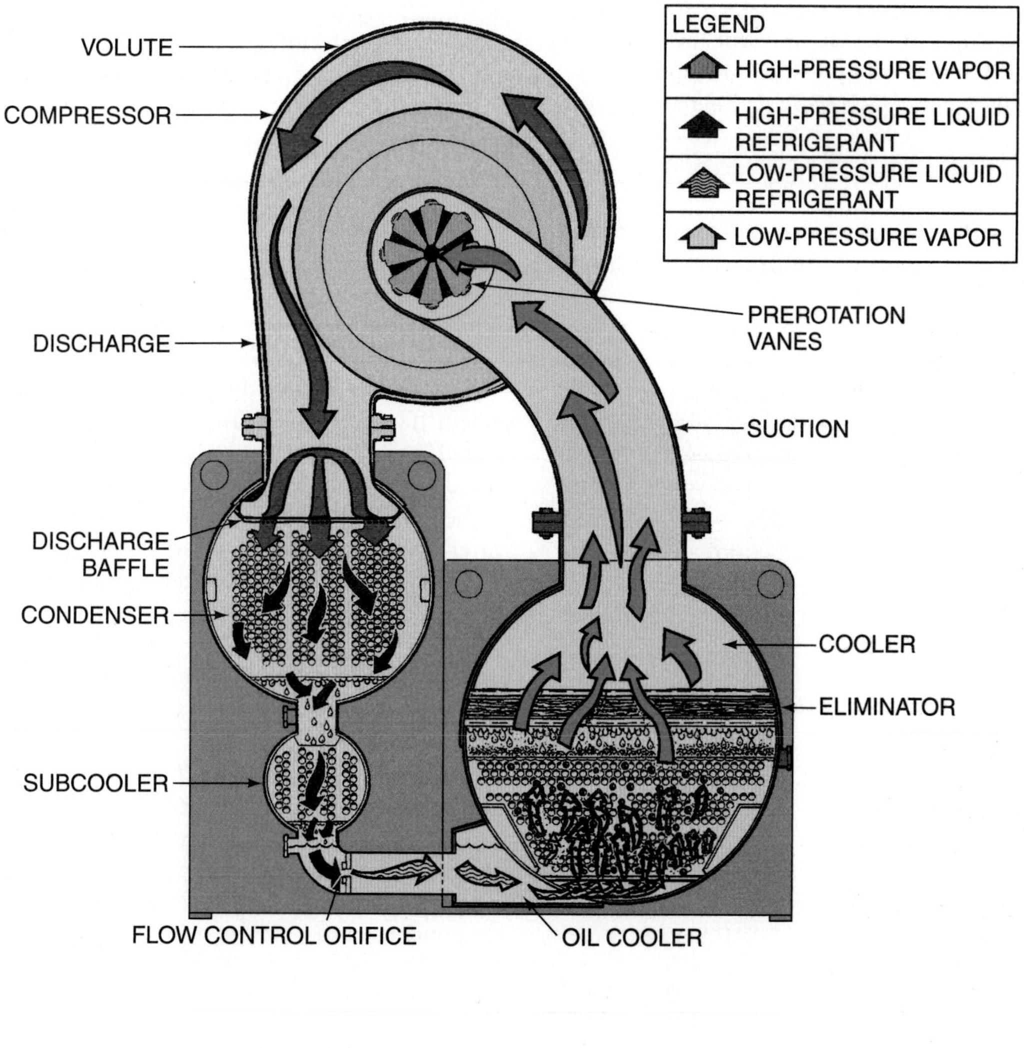

Figure 47.40 A three-stage centrifugal chiller with two economizer cycles. Courtesy of Trane Company.

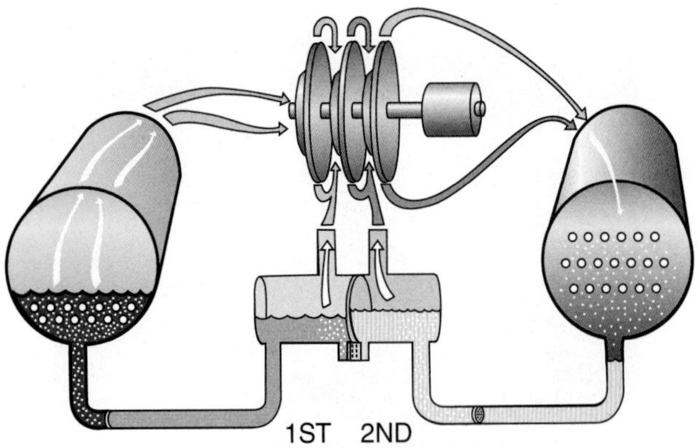

compression, **Figure 47.38**. One of the advantages of multistage operation is that the compressor can be operated at slow speed, typically the speed of the motor, about 3600 rpm. *Refrigerant vapor can also be drawn off the liquid leaving the condenser to subcool the liquid entering the evaporator metering device.* The vapor is drawn off using the intermediate compressor stages, so the liquid temperature corresponds to the pressure of the compressor intermediate suction pressure; it is less than the pressure corresponding to the condensing temperature. *The subcooling process gives the evaporator more capacity for its respective size at very little additional pumping cost. This process is called an economizer cycle,* **Figure 47.40.**

The amount of compression needed depends on the refrigerant used. For example, most centrifugal compressors in chillers use two-pass evaporators and two-pass condensers. If the evaporator has a 7°F approach temperature, the refrigerant in the evaporator would boil at 40°F, while the chilled water leaving the system would be at a temperature of 47°F. **NOTE:** *We are using 47°F water to get an evaporator temperature of 40°F for use with a typical temperature/pressure chart. If the condenser has a 10°F approach temperature, is receiving 85°F water from the tower, and has 95°F leaving water, the condenser should be condensing the refrigerant at about 105°F,* **Figure 47.27.** • The compressor must overcome the following pressures to meet the system requirements. Some examples of the pressures that must be overcome are provided here:

Refrigerant Type	Evaporator Saturation Pressure (40°F)	Condenser Saturation Pressure (105°F)
R-113	24.4 in. Hg Vac	6.4 in. Hg Vac
R-123	18.1 in. Hg Vac	8.1 psig
R-12	37 psig	126.4 psig
R-134a	35 psig	134.9 psig
R-22	68.5 psig	210.8 psig

On an R-123 chiller, the compressor only has to raise the pressure of the refrigerant from 18.1 in. Hg vacuum, which corresponds to an evaporator saturation temperature of 40°F, to 8.1 psig, which corresponds to a condenser saturation temperature of 105°F. This results in a pressure rise of about 17 psig. For comparison, an R-22 compressor that needed to pump refrigerant under the same conditions would have to raise the pressure about 142 psig, from 68.5 psig at 40°F to 210.8 psig at 105°F.

Although R-123, an HCFC, is a popular refrigerant for use in low-pressure chillers, it is no longer used in newly manufactured chillers. This restraint took effect on January 1, 2020, which is part of more sweeping restrictions on HCFCs which include their total phaseout on January 1, 2030. New, alternative refrigerants, such as HFO-1233zd and R-415A, an HFO blend, are presently the frontrunners as replacements for R-123. Refer to Unit 3, "Refrigeration and Refrigerants," for more information on HFO refrigerants.

Low-pressure chillers must be able to accommodate the refrigerant used. Because the chillers are low pressure, the refrigerant side of the shell does not have to be as strong as those of high-pressure chillers and can be much lighter in weight because of the lower pressures.

The evaporator on a low-pressure system has a working pressure for the refrigerant circuit and for the water circuit. The low-pressure system may have a refrigerant working pressure as low as 15 psig. The system has a refrigerant safety device called a rupture disc; its relief pressure is 15 psig, and it is located in the low-pressure side of the system, on the evaporator section or the suction line. This device prevents excess pressure on the shell. If a fire occurs or a technician uses too much pressure, the shell is safe from rupture.

CONDENSERS FOR LOW-PRESSURE CHILLERS

The condenser is the component that rejects heat from the system. The condensers for low-pressure chillers are water-cooled. As in high-pressure systems, the heat is usually transferred to the atmosphere. In some manufacturing processes, the heat may be recovered and used for other purposes. The water-cooled condensers used in low-pressure chillers are the shell-and-tube type; water circulates in the tubes and the refrigerant circulates around the tubes. The shell must have a working pressure to accommodate the refrigerant used and a water-side working pressure of 150 to 300 psig. The hot discharge gas enters at the top of the condenser, **Figure 47.39**.

Because it is the condenser that rejects heat from the system, its capacity to do so determines the head pressure for the refrigeration process. In low-pressure chillers, the condenser is usually located above the evaporator. The compressor lifts the refrigerant to the condenser level in the vapor state. When it is condensed to a liquid, it flows by gravity with very little

pressure difference to the evaporator through the metering device. The head pressure must be controlled so that it does not drop too low below the design pressure drop.

Water-cooled condensers may also have a subcooling circuit. The purpose of the subcooling circuit is to lower the liquid refrigerant temperature to a level below the condensing temperature. This is accomplished by means of a separate chamber in the bottom of the condenser where the entering condenser water (at about 85°F) can remove heat from the condensed liquid refrigerant (at about 105°F). This gives the system more efficiency, **Figure 47.26**.

METERING DEVICES FOR LOW-PRESSURE CHILLERS

The metering device holds liquid refrigerant back and meters the liquid into the evaporator at the correct rate. Two types of metering devices are typically used for low-pressure chillers, the orifice and the high- or low-side float, which were discussed previously in the section on high-pressure chillers.

PURGE UNITS

When a low-pressure refrigerant is used in a centrifugal system, the low-pressure side is always in a vacuum when operating. This means that any time there is a leak on the low-pressure side of the system, air will enter the machine, **Figure 47.41**. Air may cause several problems. The oxygen and moisture in the air will mix with the refrigerant, which will create a mild acid that will eventually attack the motor windings and cause damage. While the machine is running, air will move to the condenser but will not condense. This will cause the head pressure to rise. If much air enters the system, the head pressure will rise to the point that the machine will either "surge" or shut down due to high head pressure.

When air moves to the condenser, it will collect at the top and take up condenser space. This air can be collected and removed with a purge system. The purge is a separate device, often with its own compressor, that collects a sample of whatever is occupying the top portion of the condenser and tries to condense it. If the product from the top of the condenser can be condensed, it is refrigerant and it will be returned to the system's evaporator, **Figure 47.42**. If the sample cannot be condensed, it is most likely air and the purge pressure will rise. A relief valve will open and allow the sample to be exhausted to the atmosphere.

There are several types of purge systems, but they all perform the same function, some more efficiently than others. Some equipment relieves purge pressure by means of a pressure switch and a solenoid valve. A counter tells the operator when the purge is functioning. The operator can use this as a guide for determining when a system has a leak and air is entering. When too many purge reliefs occur, the system is

Figure 47.41 When a low-pressure chiller leaks while in a vacuum, atmosphere enters the chiller shell.

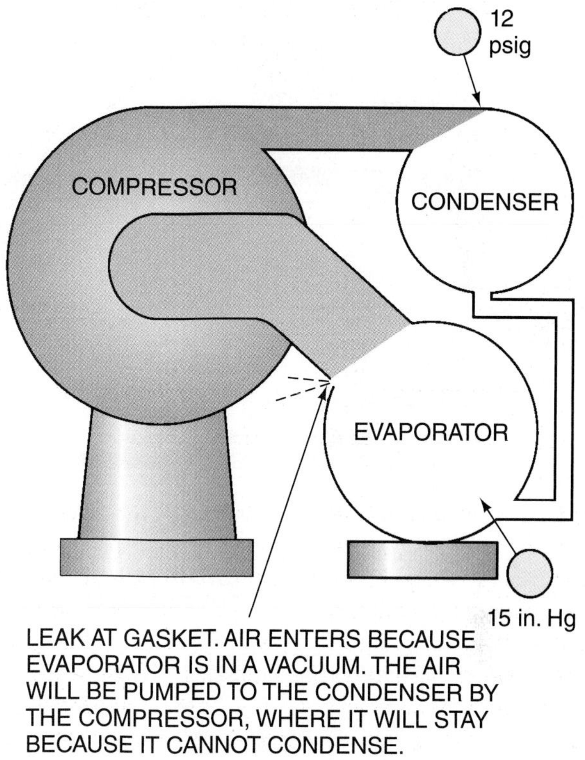

considered to have a leak. Some purge systems use the difference in pressure from the high-pressure side of the system to the low-pressure side to create the same pressure difference that a purge compressor would create—they use a condenser to condense any refrigerant from the sample. 4R *Older purge systems were not efficient, and a high percentage of the sample that was discharged to the atmosphere contained refrigerant. This is considered poor practice today because of environmental concerns and the price of lost refrigerant. Modern-day purges must be efficient and allow only a small fraction of 1% refrigerant loss during the relieving process.* 4R **Figure 47.43** shows a high-efficiency purge unit that releases only a very small amount of refrigerant when relieving the pressure. The vapor is discharged to the atmosphere only when air is present. A leak-free system will contain no air and the purge system will not be called upon to function.

47.11 ABSORPTION A[illegible]CONDITIONING C[illegible]LERS

The absorpti[illegible]igeration process is considerably different from the [illegible]ession refrigeration process discussed in the preceding [illegible]ns. The absorption process uses heat as

Figure 47.42 A simplified purge unit in operation.

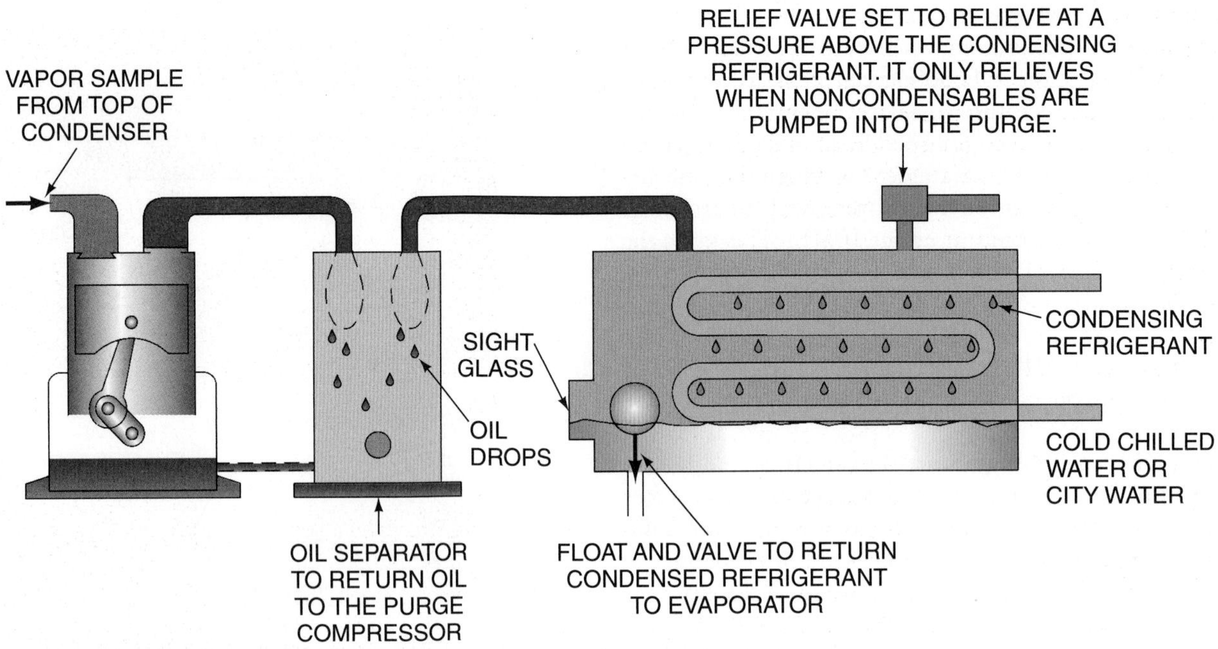

Figure 47.43 A modern purge unit with features that reduce the amount of refrigerant that escapes with the noncondensable gases. Courtesy of Carolina Products, Inc.

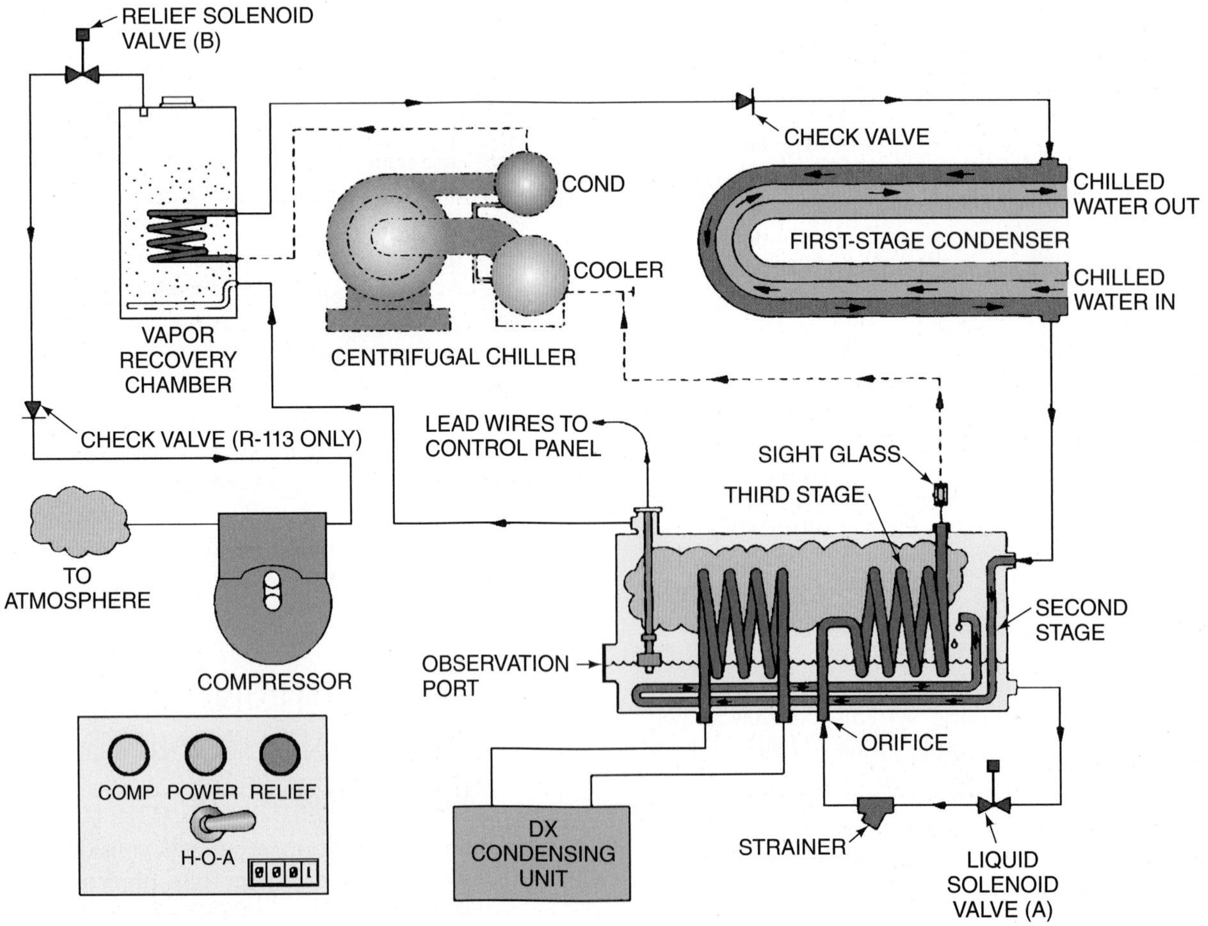

the driving force instead of a compressor. *When heat is plentiful or economical, or when it is a by-product of some other process, absorption cooling can be attractive. For example, in a manufacturing process that uses steam, which is often at 100 psig, it must be condensed before it can be reintroduced to the boiler.* In the absorption system process, condensing steam is a popular choice to drive the production of chilled water for air-conditioning. *In this case, chilled water for air-conditioning can be furnished at an economical cost, basically for the cost of operating all the related pumps, because no compressor is required.* In other cases, natural gas is an inexpensive fuel in the summer. The gas company is looking for ways to market gas because its winter quota may depend on how much gas it can sell in the summer. Some gas companies may offer incentives to install gas appliances that consume fuel in the off-peak summer months. 4R *Absorption refrigeration equipment also operates in a vacuum and contains non-CFC refrigerants.* 4R

Equipment for an absorption system looks much like a boiler except that it has chilled-water piping and condenser-water piping routed to it in addition to the piping for steam or hot water. Oil or gas burners are part of the system if the chiller is direct-fired. **Figure 47.44** shows several absorption systems from different manufacturers. The small machines are considered package machines, meaning they are completely assembled and tested at the factory and shipped as one component. They are all built in one shell or a series of chambers, so the range of sizes available may be limited because the equipment must be moved as a single component. The machines come in sizes ranging from approximately 100 to 1700 tons. The large-tonnage machines may be manufactured in sections for ease of rigging and moving into machine rooms, **Figure 47.45**. These sections are usually assembled in the factory to achieve proper alignment, except for some field welding that must be done.

The absorption process is very different in some respects from the compression cycle refrigeration process but also similar in others. It is helpful to reach an understanding of the compression cycle before trying to understand the absorption cycle. In compression cycle chillers, the boiling temperature for the refrigerant is controlled by the pressure above the boiling liquid, and it is the compressor that controls this boiling pressure. **Figure 47.46** shows a refrigeration system using water as the refrigerant. The water must be boiled at 0.248 in. Hg absolute, or 0.122 psia, to boil the water at 40°F. **Figure 47.47** shows a limited temperature/pressure chart for water. This refrigeration cycle would be workable using a compressor except that the volume of vapor rising from the boiling water is excessive. The compressor would have to pump 2444 ft^3 of water vapor for every pound of water boiled at 40°F. **Figure 47.48** shows a limited specific-volume chart for water at different temperatures. A conventional compressor is obviously impractical for use in a refrigeration system using water as the refrigerant.

Figure 47.44 Absorption chillers. (A) Courtesy of Trane Company. (B) Courtesy of Carrier Corp. (C) Courtesy of York International.

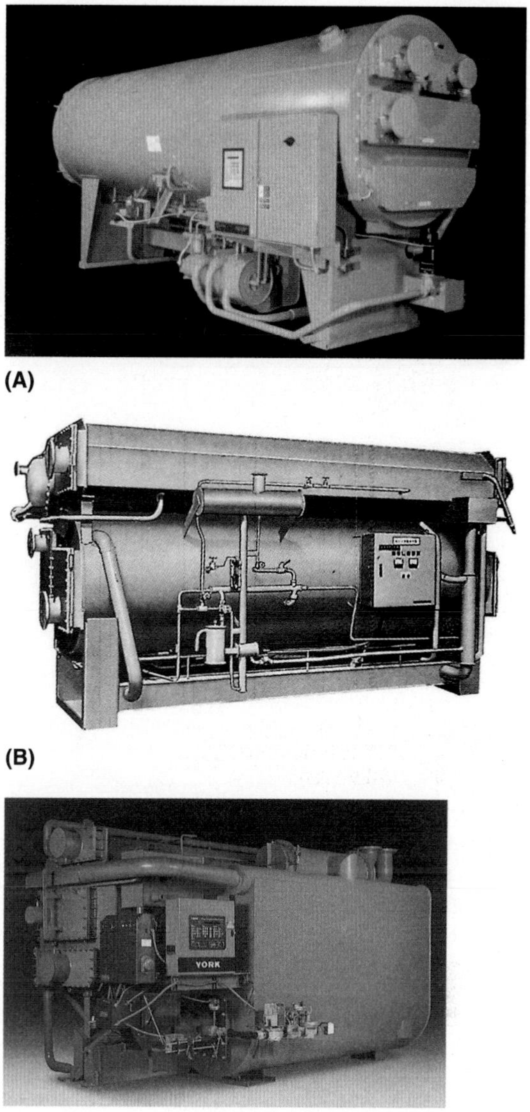

(A)

(B)

(C)

Figure 47.45 This absorption chiller is assembled after it is placed in the equipment room. Courtesy of Carrier Corp.

Figure 47.46 Water can be used as a refrigerant when the pressure is lowered enough.

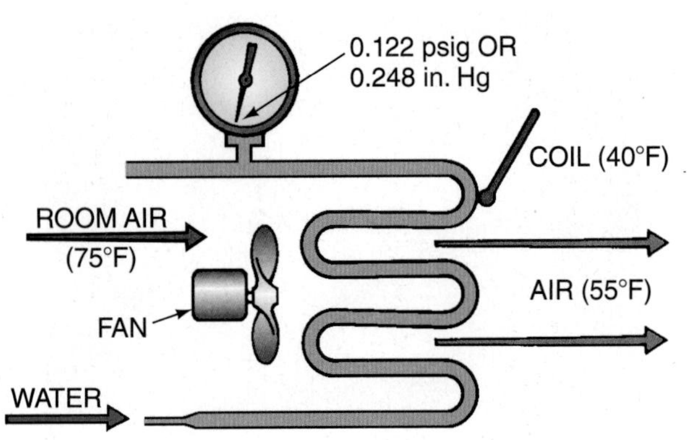

Figure 47.47 A temperature/pressure chart for water.

TEMPERATURE °F	ABSOLUTE PRESSURE lb/in²	ABSOLUTE PRESSURE in. Hg
10	0.031	0.063
20	0.050	0.103
30	0.081	0.165
32	0.089	0.180
34	0.096	0.195
36	0.104	0.212
38	0.112	0.229
40	0.122	0.248
42	0.131	0.268
44	0.142	0.289
46	0.153	0.312
48	0.165	0.336
50	0.178	0.362
60	0.256	0.522
70	0.363	0.739
80	0.507	1.032
90	0.698	1.422
100	0.950	1.933
110	1.275	2.597
120	1.693	3.448
130	2.224	4.527
140	2.890	5.881
150	3.719	7.573
160	4.742	9.656
170	5.994	12.203
180	7.512	15.295
190	9.340	19.017
200	11.526	23.468
210	14.123	28.754
212	14.696	29.921

The term *absorption* means "to attract moisture." The absorption machine uses water as the refrigerant and does not use a compressor to create the pressure difference. Instead, it uses the fact that certain salt solutions have

Figure 47.48 A specific-volume chart for water.

TEMPERATURE °C	TEMPERATURE °F	SPECIFIC VOLUME OF WATER VAPOR ft³/lb	ABSOLUTE PRESSURE lb/in²	ABSOLUTE PRESSURE kPa	ABSOLUTE PRESSURE in. Hg
−12.2	10	9054	0.031	0.214	0.063
−6.7	20	5657	0.050	0.345	0.103
−1.1	30	3606	0.081	0.558	0.165
0.0	32	3302	0.089	0.613	0.180
1.1	34	3059	0.096	0.661	0.195
2.2	36	2837	0.104	0.717	0.212
3.3	38	2632	0.112	0.772	0.229
4.4	40	2444	0.122	0.841	0.248
5.6	42	2270	0.131	0.903	0.268
6.7	44	2111	0.142	0.978	0.289
7.8	46	1964	0.153	1.054	0.312
8.9	48	1828	0.165	1.137	0.336
10.0	50	1702	0.178	1.266	0.362
15.6	60	1206	0.256	1.764	0.522
21.1	70	867	0.363	2.501	0.739
26.7	80	633	0.507	3.493	1.032
32.2	90	468	0.698	4.809	1.422
37.8	100	350	0.950	6.546	1.933
43.3	110	265	1.275	8.785	2.597
48.9	120	203	1.693	11.665	3.448
54.4	130	157	2.224	15.323	4.527
60.0	140	123	2.890	19.912	5.881
65.6	150	97	3.719	25.624	7.573
71.1	160	77	4.742	32.672	9.656
76.7	170	62	5.994	41.299	12.203
82.2	180	50	7.512	51.758	15.295
87.8	190	41	9.340	64.353	19.017
93.3	200	34	11.526	79.414	23.468
98.9	210	28	14.123	97.307	28.754
100.0	212	27	14.696	101.255	29.921

enough attraction for water that they can be used to create a pressure difference. You may have noticed that a grain of salt on a hot, humid day will attract moisture to the point that it will become wet, **Figure 47.49**. This same concept is used in an absorption refrigeration system to reduce the pressure of the water to a point where it will boil at a low temperature. A type of salt solution called lithium-bromide (Li-Br) is the absorbent (attractant) for the water in absorption cooling systems. The liquid solution of Li-Br is diluted with distilled water—actually, it is a mixture of about 60% Li-Br and 40% water.

A simplified absorption system would look like the one in **Figure 47.50**. The illustration shows a simplified cycle for ease of understanding. The following steps are a progressive description of the absorption cycle.

1. **Figure 47.51** shows the equivalent of the evaporator system. Refrigerant (water) is metered into the

Figure 47.49 A grain of salt in a humid climate will attract water.

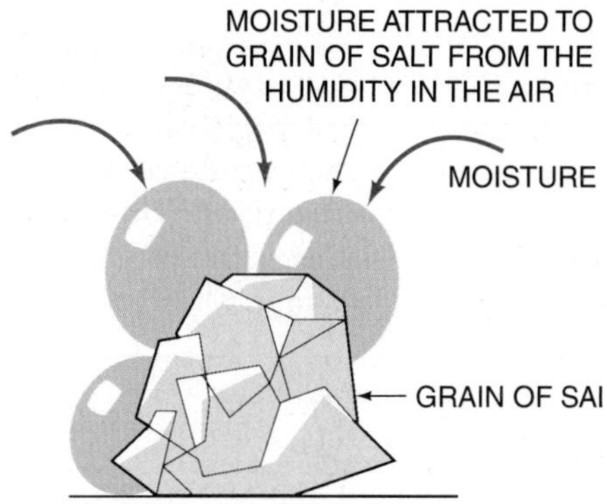

Figure 47.50 A simplified absorption refrigeration cycle.

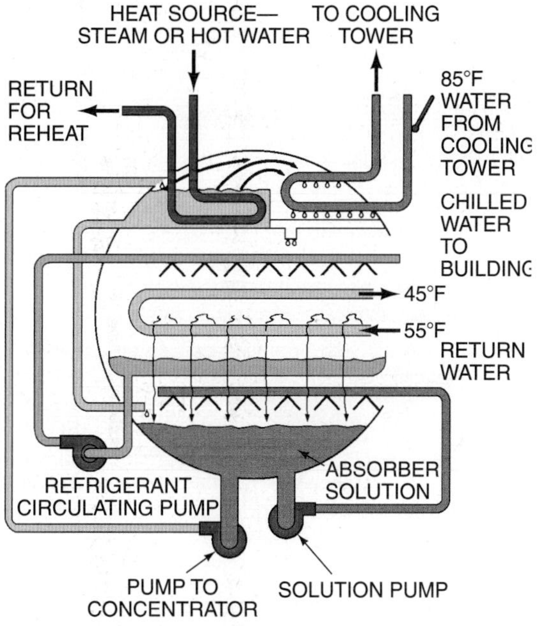

Figure 47.51 The evaporator section of an absorption machine.

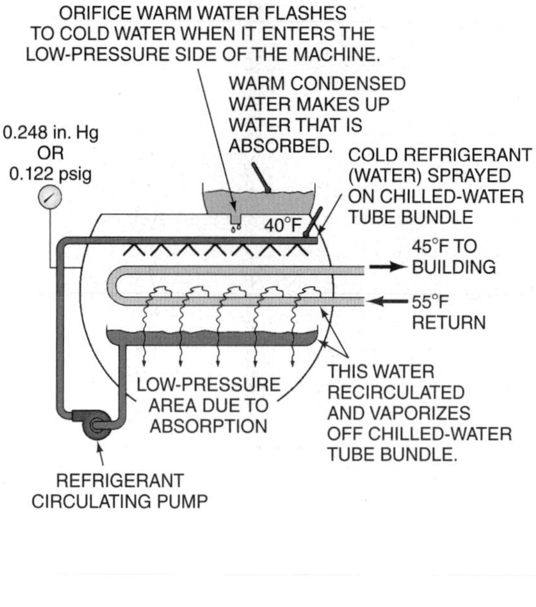

evaporator section through a restriction (orifice). It is warm until it passes through the orifice, where it flashes to a low-pressure area (about 0.248 in. Hg absolute, or 0.122 psia). The reduction in pressure also reduces the temperature of the water. The water drops to the pan below the evaporator tube bundle. A refrigerant-circulating pump circulates the water through spray heads, which spray it over the evaporator tube bundle. This wets the tube bundle through which the circulating water from the system passes. The heat from the system water within the tube bundle evaporates the refrigerant. Water is constantly being evaporated and must be made up through the orifice at the top.

2. **Figure 47.52** shows the absorber section, which would be the equivalent of the suction side of the compressor in a compression cycle system. The salt solution (Li-Br) spray is at a very low pressure and attracts the water vapor, which is readily absorbed into the spray solution. The solution is then recirculated through the spray heads to give it more surface area for attracting more water. As the solution absorbs more water, it becomes diluted. If the excess water is not removed, the solution will become so diluted that it will no longer attract water and the process will stop. To keep the process going, another pump constantly removes some of the solution and pumps it to the concentrator. This solution that is pumped to the concentrator is called the weak solution because it is diluted and contains water absorbed from the evaporator.

Figure 47.52 The absorber section of an absorption machine.

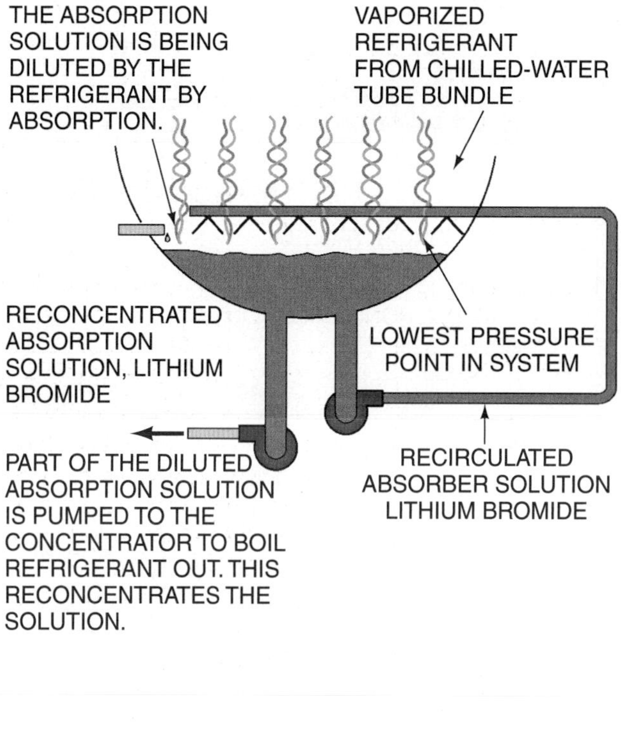

3. **Figure 47.53** shows the concentrator and condenser segment. The dilute solution is pumped to the concentrator, where it is boiled. Boiling changes the water in the solution to a vapor, and it leaves the solution, making the solution stronger and less diluted. This solution is commonly called the strong solution. The water vapor that is removed from the solution is attracted to the condenser coils. In the condenser, the water is condensed to a liquid and is gathered and metered to the evaporator section through the orifice. (In this example, the heat source for boiling the water is either steam or hot water. Direct-fired machines, discussed later, are also available.) The concentrated solution is drained back to the absorber area for circulation by the absorber pump.

As you can see from this description, the absorption process is not complicated. The only moving parts are the pump motors and impellers.

Some other features are added to the absorption cycle to make it more efficient. One of these is shown in **Figure 47.54**. It is a heat exchange between the cooling tower water and the absorber solution in the absorber. This heat exchange removes heat that is generated when the water vapor is absorbed into the solution. **Figure 47.55** shows a heat exchange between the dilute and the concentrated solution. This exchange serves two purposes: It preheats the dilute solution before it enters the concentrator and precools the concentrated solution before it enters the absorber section. It is much like the heat exchange between the suction and liquid line on a household refrigerator. Without this heat exchange, the machine would be less efficient. Some manufacturers have developed a two-stage absorption machine

Figure 47.53 The concentrator and condenser section of an absorption machine.

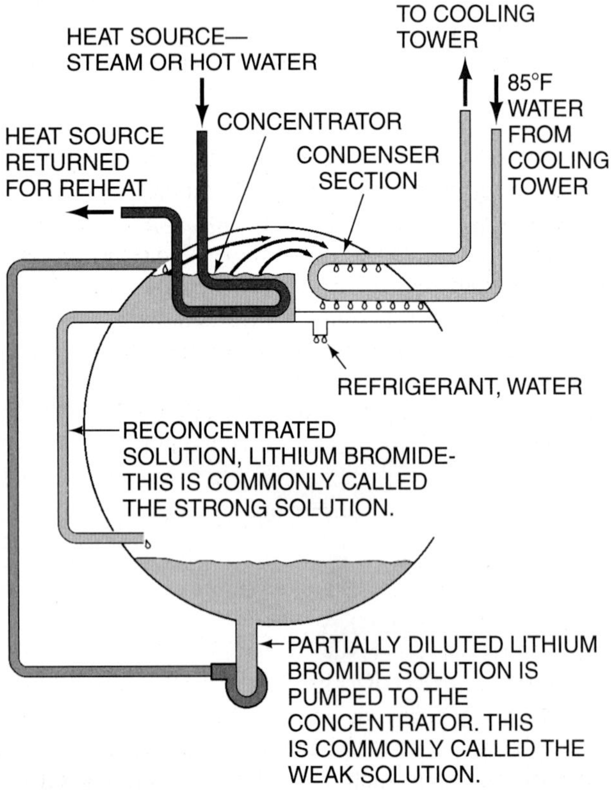

Figure 47.54 A heat exchange in the absorber to increase efficiency.

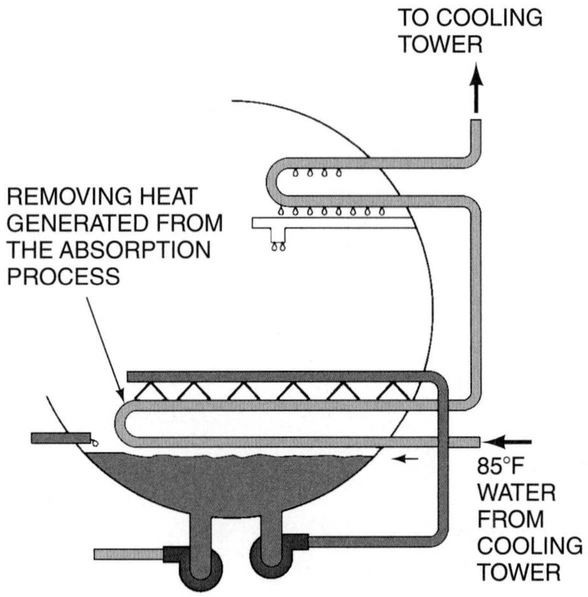

Figure 47.55 A heat exchange between the solutions.

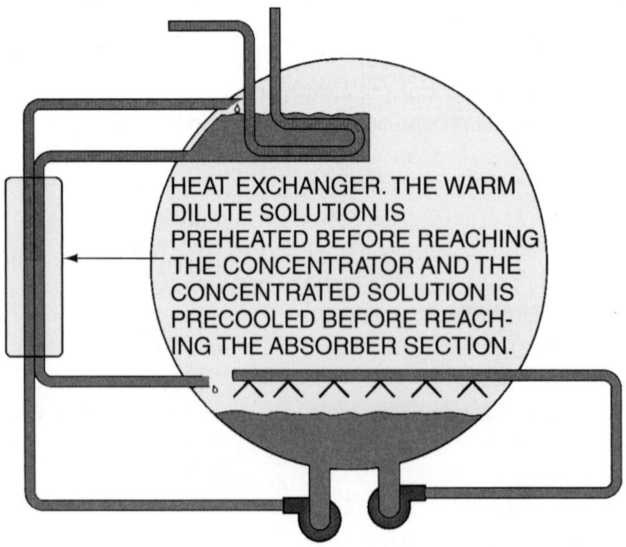

Figure 47.56 A two-stage absorption machine. Courtesy of Trane Company.

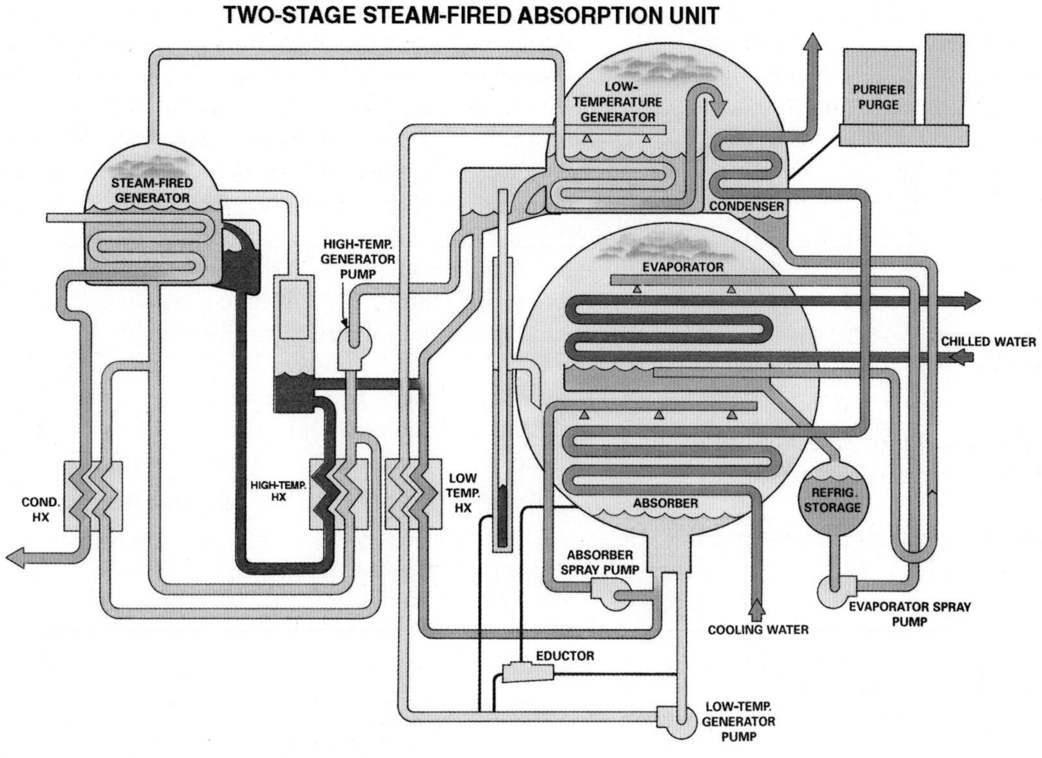

that uses higher-pressure steam or hot water, **Figure 47.56**. The steam pressure may be 115 psig, or hot water at 370°F. These machines are more efficient than the single-stage machine mentioned earlier.

SOLUTION STRENGTH

The concentration strength of the solutions determines the performance of the absorption chiller. The wider the spread between the dilute and the strong solution, the more capacity the machine has to lower the pressure to absorb the water from the evaporator. If the concentrated solution becomes too concentrated, the solution will solidify and become rock salt. If this occurs, the flow of concentrated solution may be partially or totally blocked, which causes the machine to stop cooling. The solution is said to have "become solid" or "crystallized," and service procedures must be performed in order to dissolve the hard salt crystals. The balance between the concentrations of water and Li-Br is delicate. Adjusting the strengths of these solutions is the job of the start-up technician. Some machines are shipped with no charge and the Li-Br is shipped in steel drums to be added in the field. The estimated amount of Li-Br is added first. Distilled water is used as the refrigerant charge and added next. When the charge is adjusted, the technician calls it the trim. A compression cycle system is charged and an absorption system is trimmed.

After a technician trims the machine with the approximate correct amount of Li-Br and water, the machine is started and gradually run up to full load. Full load may be determined by the temperature drop across the evaporator and full steam pressure, which is normally 12 to 14 psig (hot water or direct-fire equipment would be similar). When full load is obtained, the technician pulls a sample of dilute and strong solution and measures the specific gravity using a hydrometer, **Figure 47.57**. Specific gravity is the weight of a substance compared with the weight of an equal volume of water. The technician would know what these specific gravity readings should be from the manufacturer's literature or a Li-Br temperature/pressure chart. Water is either added or removed to obtain the correct specific gravity. This may be a relatively long process in some cases because the technician cannot always obtain full-load operation at start-up. The machine may have to be trimmed to an approximate trim and finished later when full load can be achieved. Some manufacturers furnish a trim chart for a partial load.

SOLUTIONS INSIDE THE ABSORPTION SYSTEM

The solutions inside an absorption system are corrosive. Rust, which is oxidation, occurs where corrosive materials and oxygen are present. The absorption machine is manufactured with materials such as steel and contains a saltwater

Figure 47.57 A hydrometer for measuring the specific gravity of absorption solutions.

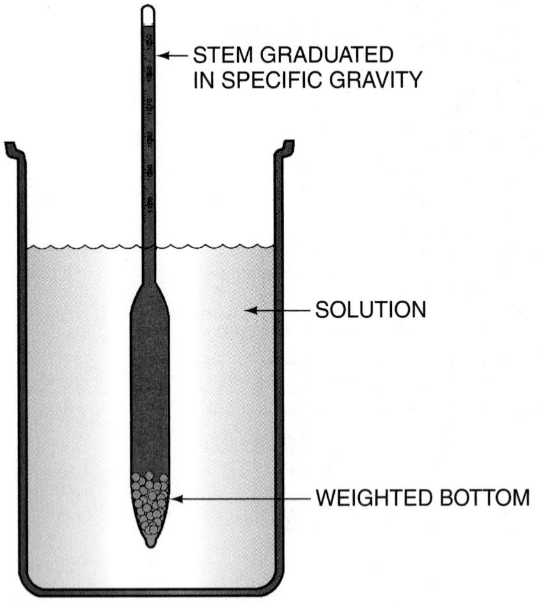

solution that will cause corrosion if air is present. It is next to impossible to manufacture and maintain a system that can keep oxygen away from its working parts.

The system must be kept as clean as possible as the solutions circulate. Some of the passages are very small, such as the spray heads in the absorber. There are many different methods of removing solid materials that may clog any small passages. Filters will stop solid particles, and magnetic devices will attract steel particles that may be in circulation. But even though the liquid circuits have filtration, the solutions in an absorption machine may become discolored. When looking at the fluids in the machine, a technician's first opinion may be that it should be scrapped and a new machine installed. However, the fluids can take on the color of rusty water and still not affect the functioning of the machine. It is normal for the fluids to look very discolored.

CIRCULATING PUMPS FOR ABSORPTION SYSTEMS

There are two distinct fluid flows for the solutions circulating throughout the various parts of the absorption system: the solution flow and the refrigerant flow. Some absorption machines use two circulating pumps, **Figure 47.58**, and some use three, depending on the manufacturer.

Whatever the number of pumps, they are similar. They are centrifugal pumps and the pump impeller and shaft must be made of a material that will not corrode or the pumps could never be serviced after they have been operated and corrosion has occurred. Special care must be taken not to let the atmosphere enter the pump motors during the pumping process. It is typical for the motors to be hermetically sealed and to operate only within the system atmosphere. The pump itself is immersed in the actual solution to be pumped, but the motor windings are sealed in their own atmosphere where no system solution can enter. Because these motors require cooling, cold refrigerant water from the evaporator or normal supply water in a closed circuit is used for this purpose.

Over time, motors must be serviced. Manufacturers recommend servicing only after many years have passed, but it will eventually be required because moving parts, such as bearings, are involved. Different manufacturers require different procedures, but it is much easier to service the motor and drive if the solution does not have to be removed. However, if it does, the process involves pressurizing the system with dry nitrogen to above atmospheric pressure and pushing the solution out. Once this is accomplished and service completed, the nitrogen must be removed, which takes a long time. Then the system must be recharged.

CAPACITY CONTROL

The capacity of a typical absorption system can be controlled by throttling back the supply of heat to the concentrator. A typical system that operates on steam uses 12 to 14 psig of steam at full load and 6 psig at half capacity. The steam can be controlled with a modulating valve, **Figure 47.59**. Hot water and direct-fired machines can be controlled in a similar manner. Some manufacturers may provide other means for capacity control, for example, controls for the internal fluid flow in the machine. Reducing the flow of weak (dilute) solution to the concentrator is a practical control because at reduced load, less solution needs to be concentrated.

CRYSTALLIZATION

Using a salt solution for absorption cooling raises the possibility of the solution becoming too concentrated and actually turning into rock salt. This is called **crystallization** and may occur if the machine is operated under the wrong conditions. For example, if the cooling tower water in some systems is allowed to become too cold while the system is operating at full load, the condenser will become too efficient and remove too much water from the concentrate. This will result in a strong solution that is too concentrated, as it does not contain enough water. When this solution passes through the heat exchanger, it will turn to crystals and restrict the flow of the solution. If left uncorrected, the machine will become completely blocked and will completely stop cooling. Because this is a difficult problem to fix when it happens, manufacturers have developed various methods to prevent it. One manufacturer, for example, installs a valve between the refrigerant circuit and the absorber fluid circuit. This valve may be opened long enough to further dilute the weak solution, which helps to break up and remove the restriction. When the situation is

Figure 47.58 A two-pump absorption machine application. Courtesy of Carrier Corp.

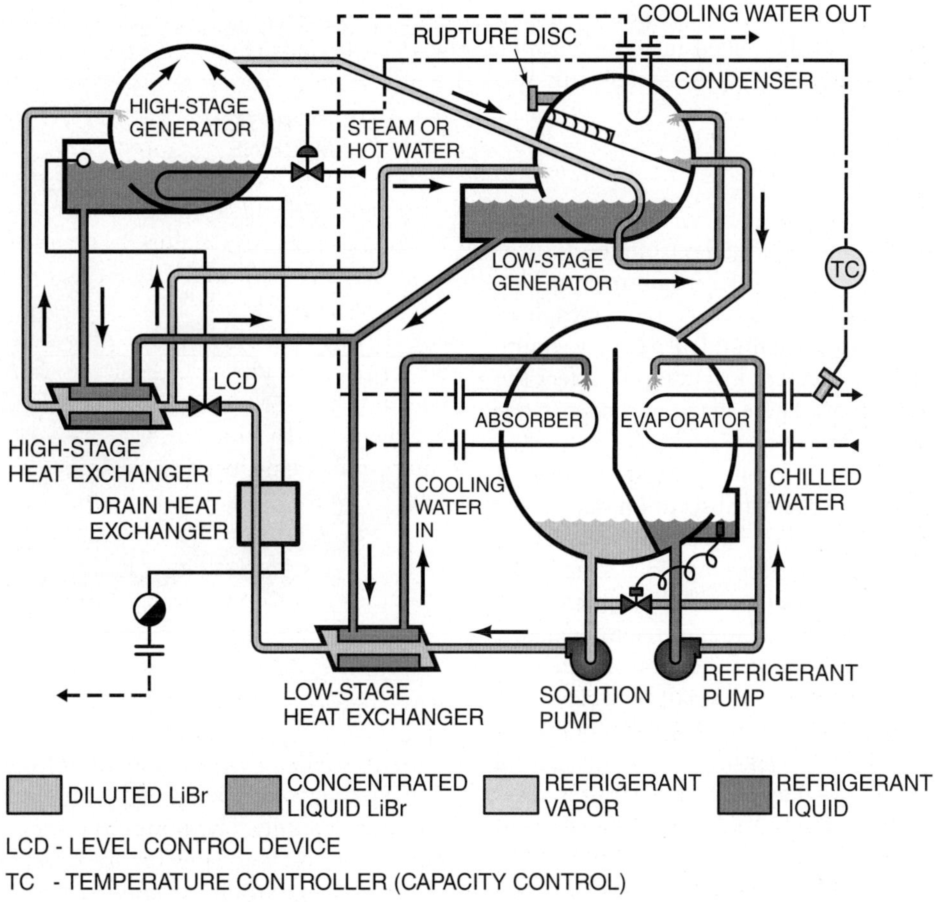

Figure 47.59 *Capacity control using a modulating steam valve.* Courtesy of York International

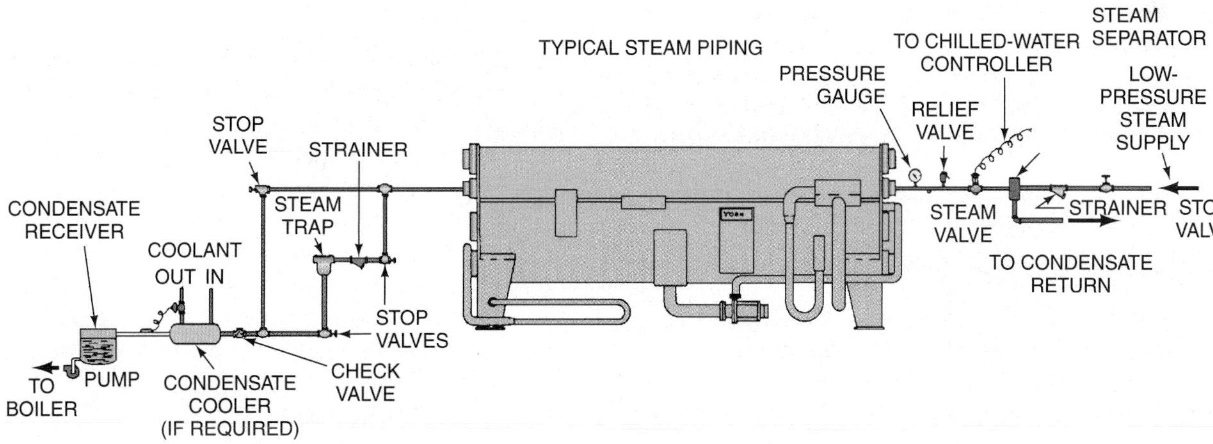

corrected, the valve is closed and the system resumes normal operation. Another manufacturer may recommend shutting the machine down for a dilution cycle when overconcentration occurs.

Crystallization can occur for several reasons. Cold condenser water is one cause, and a shutdown of the machine due to power failure while operating at full load is another. An orderly shutdown calls for the solution pumps to operate

for several minutes after shutdown to dilute the strong solution. With a power failure, the shutdown is not orderly and so crystallization can occur. Also, at any point where a leak occurs atmospheric air can be pulled into the machine because it operates in a vacuum. Air in the system can also cause crystallization.

PURGE SYSTEM

The purge system removes noncondensables from the absorption machine during the operating cycle. All absorption systems operate in a vacuum, so if there is a leak, the atmosphere will enter. Air expands greatly when pulled into a vacuum. Even the amount of air in a soft-drink bottle will affect the capacity of a 500-ton machine. These machines must therefore be kept absolutely leak-free.

Wherever possible, all piping connections are welded at the factory. After the typical packaged absorption machine is assembled, it is put through the most rigid of leak tests, called a mass spectrum analysis, before shipping. In this test, the machine is surrounded with an envelope of helium, the system is pulled into a deep vacuum, and the vacuum pump exhaust is analyzed with a mass spectrum analyzer for any helium, **Figure 47.60**. Helium is a gas with very small molecules that will leak in at any leak source. But even with these rigid leak-check and welding procedures, leaks can develop during shipment to the job site. Leaks may also develop after years of operation and maintenance. When a leak occurs, the noncondensables must be removed.

Absorption systems also generate small amounts of noncondensables during normal operation. A by-product of the internal parts causes hydrogen gas and some other noncondensables to form inside. These are held to a minimum with the use of additives, but they still occur and it is the responsibility of the purge and the operator to keep the machine free of these noncondensables.

Figure 47.60 Leak checking an absorption machine in the factory. A vacuum is pulled on the machine, which is enclosed in a helium atmosphere. The vacuum pump's exhaust is checked for helium with a spectrum analyzer. Helium has very small molecules and will easily leak through a very small hole.

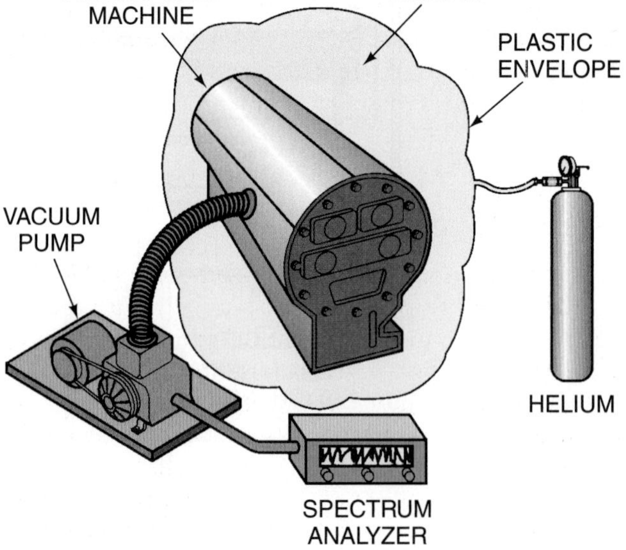

Two kinds of purge systems are used on typical absorption cooling equipment: the nonmotorized purge and the motorized purge. The nonmotorized purge uses the system pumps to create a flow of noncondensable products to a chamber where they are collected and then bled off by the machine operator, **Figure 47.61**. This purge works well while the machine is in operation but may not be of much use if the machine were to need service and is pressured to atmosphere using nitrogen. Some manufacturers of motorless purge units offer an optional motorized purge for removal of large amounts of noncondensables. The motorized purge is

Figure 47.61 A collection chamber for noncondensable gases. Courtesy of Carrier Corp.

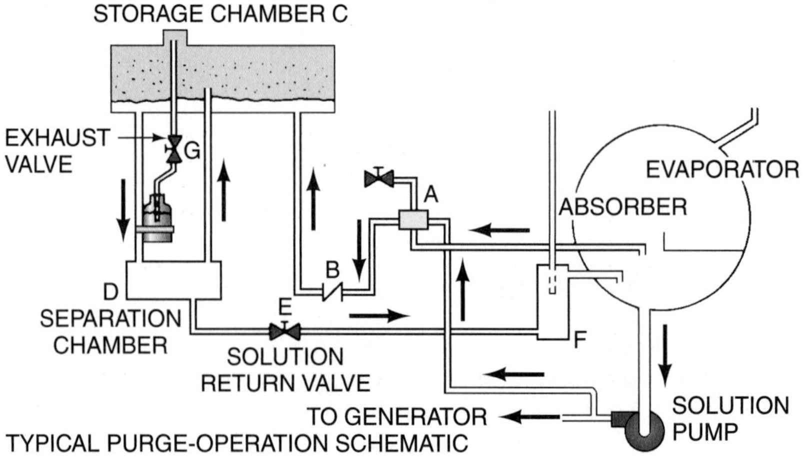

essentially a two-stage vacuum pump that removes a sample of whatever gas is in the absorber and pumps it to the atmosphere. These vapors are not harmful to the atmosphere. Because the absorber operates in such a low vacuum, this gas sample would contain only noncondensables or water vapor, **Figure 47.62.** The motorized purge requires some maintenance because of the corrosive nature of the vapors that are pulled through the vacuum pump. The vacuum pump must be properly maintained and its oil must be changed on a regular basis to prevent pump failure.

ABSORPTION SYSTEM HEAT EXCHANGERS

Like compression cycle chillers, absorption machines have heat exchangers. Two-stage systems actually have more heat exchangers—the evaporator, the absorber, the concentrator, the condenser, and the first-stage heat exchanger. These are similar to compression cycle heat exchangers, but they also have some differences. They also contain tubes, which are either copper or cupronickel (a tube made of copper and nickel).

The chilled-water heat exchange tubes remove heat from the building water and add it to the refrigerant (water). The chilled-water circuit is usually a closed circuit, so the tubes rarely need maintenance except for water treatment. A flange at the end of the tube bundle allows access through either marine or standard water boxes. This section of the absorption system must have some means of preventing the heat in the equipment room from transferring into the cold machine parts. In some systems this section is insulated; some have a double-wall construction to prevent heat exchange. As in the compression cycle chillers, there is an approach temperature between the boiling refrigerant and the leaving chilled water. This temperature is likely to be 2°F or 3°F in an absorption chiller because the heat exchange is so good, **Figure 47.63.**

The absorber heat exchanger exchanges heat between the absorber solution and the water returning from the cooling tower. The design cooling tower water under full load in hot weather is 85°F. The water from this heat exchange continues to the condenser, where the refrigerant vapor is condensed for reuse in the evaporator. The leaving water is likely to be between 95°F and 103°F on a hot day at full load, **Figure 47.64.** Absorption equipment manufacturers provide thermometer wells located at strategic spots for checking the solution temperatures to assess machine performance and determine whether or not the tubes are dirty. The manufacturers provide the correct temperatures and temperature differences for their machines.

The other heat exchange in standard equipment is between the heating medium—steam, hot water, or flue gases—and the refrigerant. This tube bundle might be exposed to some high temperature differences because it may be at room temperature until the system is started when hot water or steam is introduced to the bundle. A great deal of stress in the form of tube expansion occurs at this time. Different manufacturers deal with this stress in different ways. **Figure 47.65** shows a tube bundle that has the capability of floating to minimize problems.

Figure 47.62 The vapors purged from an absorption machine are nonpolluting.

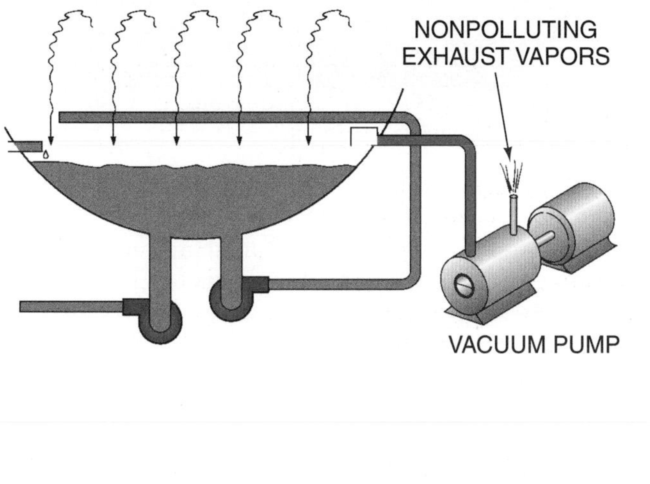

Figure 47.63 The approach temperatures for an absorption chiller are very close.

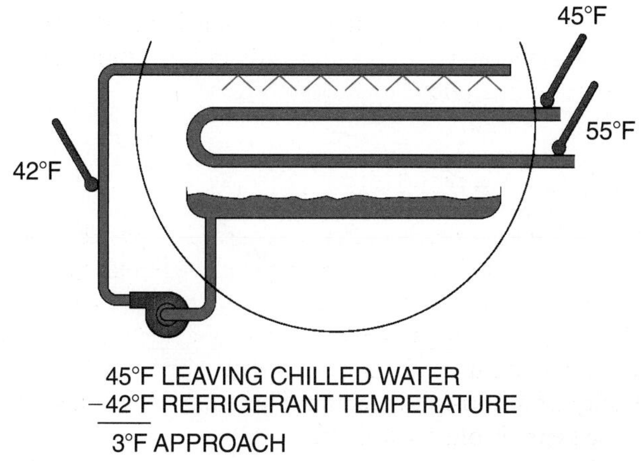

Figure 47.64 A cooling tower circuit.

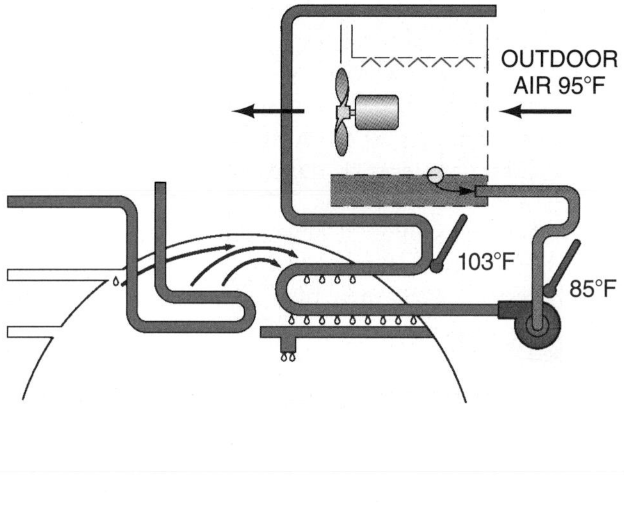

Figure 47.65 The tubes in the absorption chiller are subject to great expansion due to significant temperature differences. Special methods are used to allow for tube expansion. Courtesy of Trane Company.

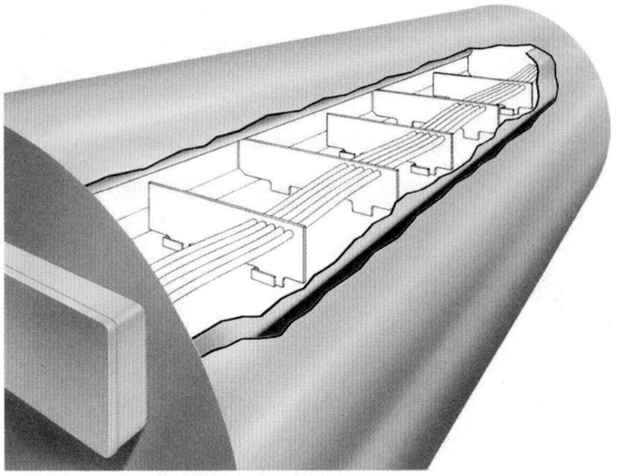

DIRECT-FIRED SYSTEMS

Some absorption equipment is direct-fired. Gas or oil is the heat source, and some systems are dual-fuel machines for situations where gas demand may require a second fuel. Such machines range in capacity from approximately 100 to 1500 tons. They also may have the ability to either heat or cool by furnishing either hot or chilled water to the structure. In some cases, they may be used as one piece of equipment to replace both a boiler and an absorption machine in an old installation. **Figure 47.66** shows direct-fired equipment manufactured by two different companies.

47.12 MOTORS AND DRIVES FOR CHILLERS

All of the motors that drive high-pressure and low-pressure chillers are three-phase motors. These motors all give off some heat while in operation, which must be removed from the motor or it will overheat and burn. *Some manufacturers choose to keep the motor in the atmosphere to cool it,* **Figure 47.67.** This reduces the load on the refrigeration system, but the heat air-cooled motors give off to the equipment room must be removed to avoid overheating the room. Typically, the heat is removed with an exhaust system. The compressor motor must have a shaft seal to prevent refrigerant from leaking into the atmosphere, **Figure 47.68**. In the past, seals have been a major source of leaks, but new materials used to make the seals, along with improved shaft-alignment techniques, have contributed to the reduction in the severity of shaft seal leaks.

Compressor motors may also be cooled with the vapor refrigerant leaving the evaporator. These compressors, called suction-cooled compressors, are either hermetic or semihermetic. Fully hermetic compressors are the smaller ones and may be returned to the factory to be rebuilt as needed. Semihermetic compressors and motors may be rebuilt in the field. Compressor motors may also be cooled with liquid refrigerant by allowing liquid to enter the housing around the motor. The liquid does not actually touch the motor or windings; it is contained in a jacket around the motor, **Figure 47.15**.

Figure 47.66 Direct-fired absorption chillers. (A) Courtesy Trane Company. (B) Courtesy York International.

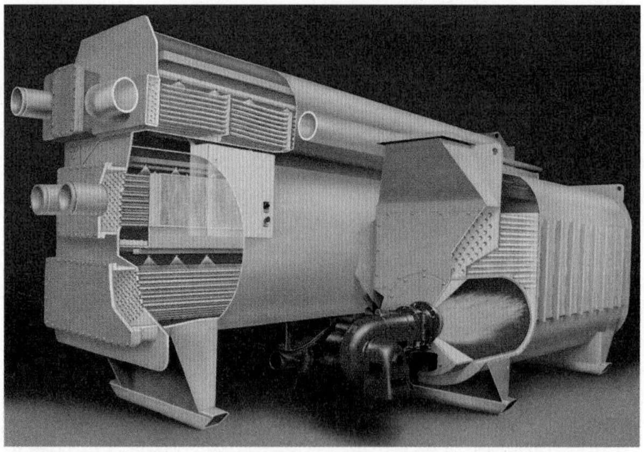

(A)

(B)

Figure 47.67 An air-cooled, direct-drive compressor motor. Courtesy of York International.

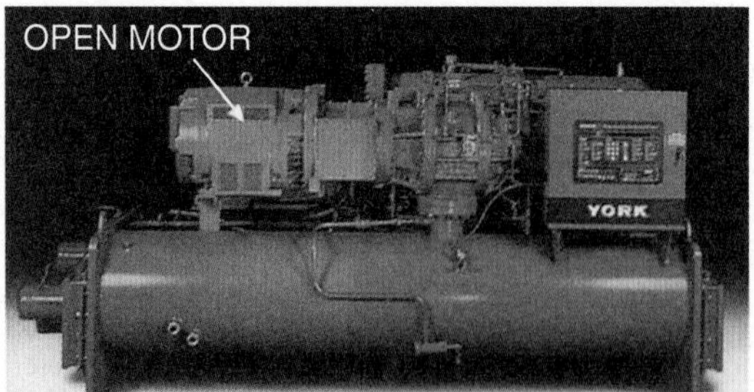

Figure 47.68 The shaft seal for a gear-drive compressor. Courtesy of York International.

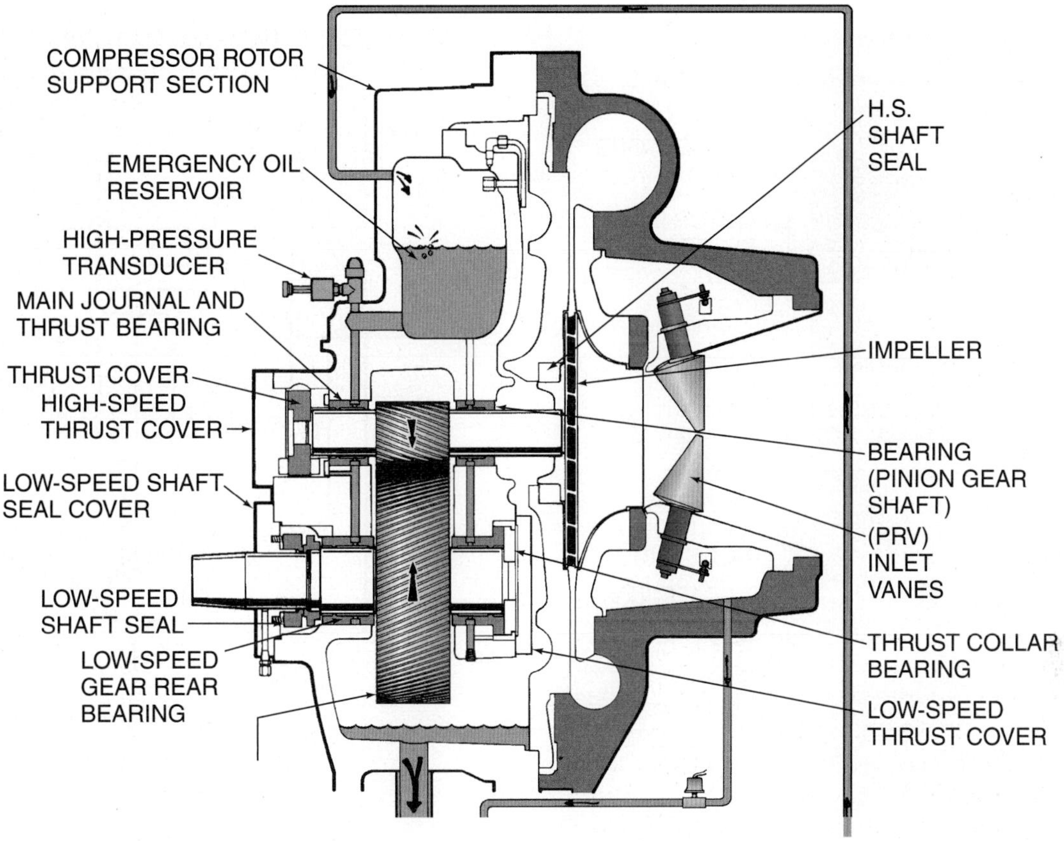

Large amounts of electrical power must be furnished to the motor through a leak-free connection. Often this is done through a nonconducting terminal block made of some form of phenolic or plastic. The terminals, sealed by O-rings, protrude through both sides of the block, with the motor terminals on one side and the field connections on the other, **Figure 47.69.** The terminal board is sealed to the compressor housing by a gasket. The field connections are checked periodically for tightness. **A loose connection may melt the board and cause a leak to develop.**

The motors used in all types of chillers are expensive and the manufacturer goes to great lengths to protect them by designing them to operate in a protective atmosphere. Part of getting long motor life is to use a good start-up procedure. A conventional motor draws about five times as much amperage during start-up than it does during full-load operation. If a motor draws 200 A at full load, it would draw approximately 1000 A at start-up. This inrush of current can cause problems in the electrical service, so manufacturers use several different methods to minimize the inrush of current and power-line fluctuations. A large motor is usually started with a group of components called *starters*. There are several types, the common ones being part-winding, autotransformer, wye-delta (often called star-delta), and electronic start.

PART-WINDING START

Manufacturers often include part-winding start motors when compressor motors reach about 25 hp. These versatile motors normally have nine leads and can be used with two different voltages. The same compressor can be used for 208/230-V and 460-V applications by changing the motor terminal arrangement, **Figure 47.70**, so the start motor actually becomes two motors in one. For example, a 100-hp compressor can have two motors that are 50 hp each. In a 208/230-V application, the motors are wired in parallel and each motor is started separately—first one, then the other—with two motor starters and a time delay of up to about 1 sec between them, **Figure 47.71.** When the first motor starts, the motor shaft begins turning. The second motor then starts to bring the compressor up to full speed, about 1800 or 3600 rpm, depending on the motor's rated speed. *This imposes only the inrush current of a 50-hp motor on the line because when the second motor is energized, the shaft is already turning and so the inrush current is lower.* When the compressor motor is used in a 460-V application, the start motors are wired in series and started as one motor across the line, **Figure 47.72.** At the higher voltage there is much less inrush amperage on start-up. Part-winding start motors are used on compressors of up to about 150 hp.

Figure 47.69 The terminal block for a hermetic compressor motor. Courtesy of York International.

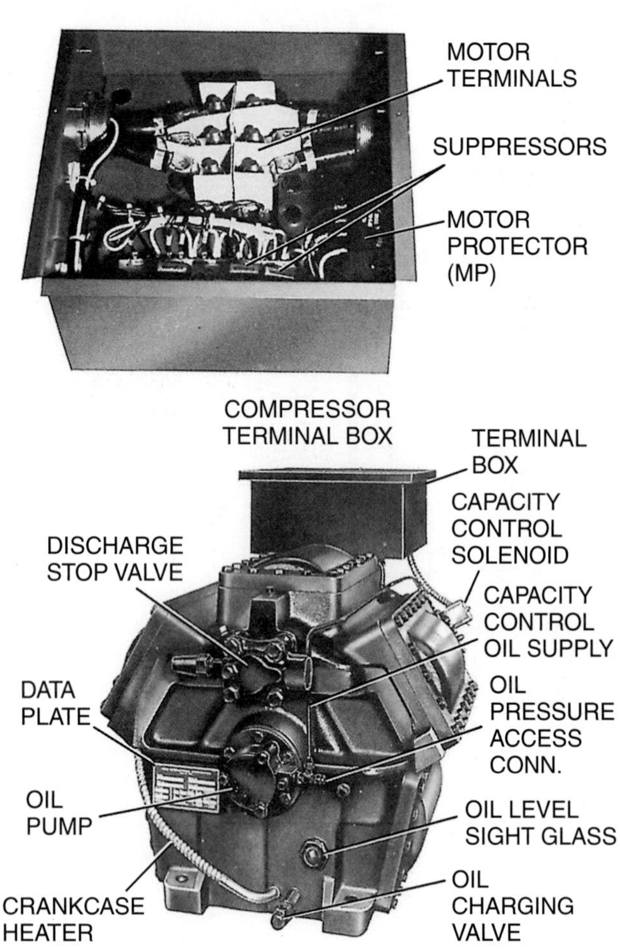

Figure 47.70 Nine-lead, dual-voltage motor connections.

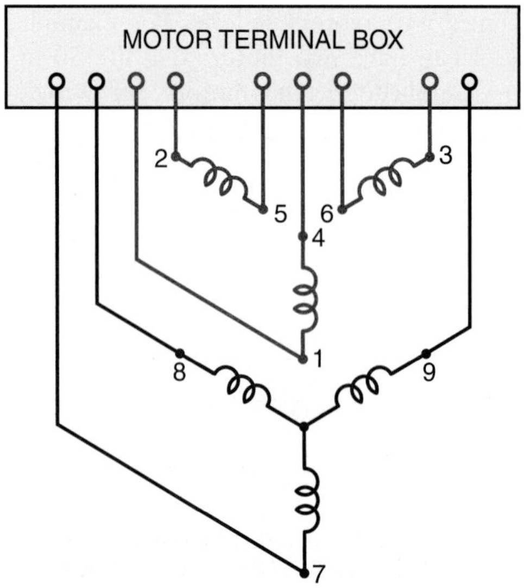

Figure 47.71 The part-winding start motor wired for 208/230-V.

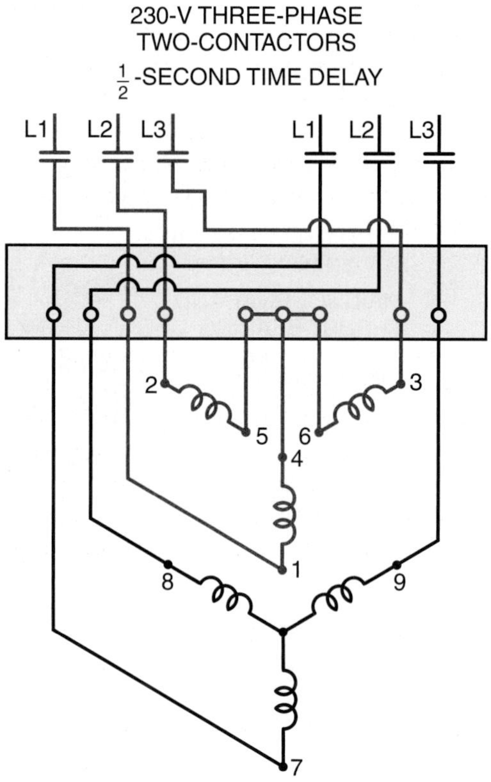

Figure 47.72 The two motors are now in series for the 460-V application.

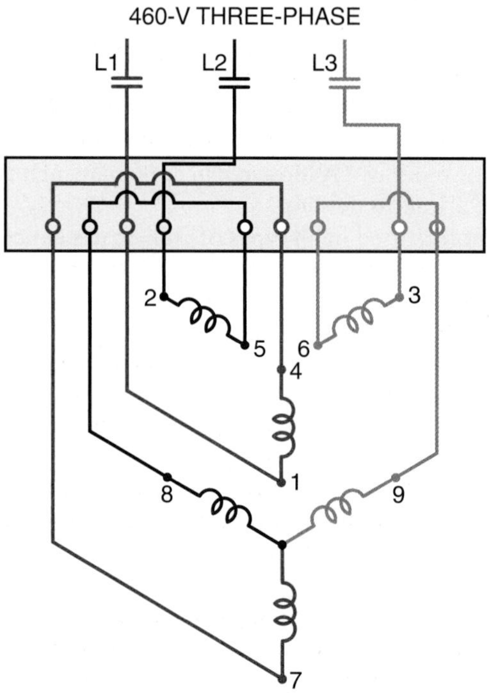

AUTOTRANSFORMER START

An autotransformer start is actually a reduced-voltage start. A transformer-like coil is placed between the motor and the starter contacts, and the voltage to the motor is supplied through the transformer during start-up. This reduces the voltage to the motor until the motor is up to full speed, **Figure 47.73**. When the motor reaches full speed, a set of contacts closes and shorts around the transformer to run the motor at full voltage. The contacts that direct power through the transformer will then open so that current does not go through the transformer when the motor is running. This minimizes heat in the area of the transformer. Because it is starting at reduced voltage, the motor has very little starting torque so the autotransformer start is used only in special applications. It is common with large compressors that require little starting torque, such as in centrifugal compressors where the pressures are completely equalized during the off cycle and compression does not start until the compressor is up to speed and the vanes start to open.

Figure 47.73 An autotransformer starter (motor application).

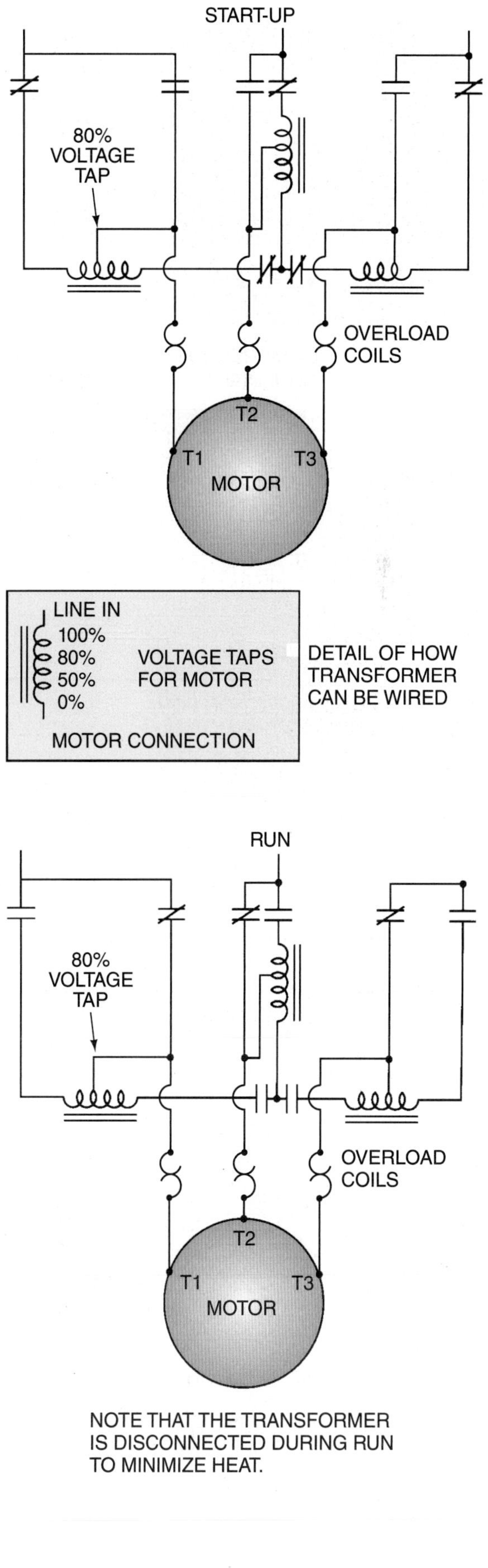

WYE-DELTA CONFIGURATION

Wye-delta start, often called star-delta start, is used with large motors; the starter has six leads and a single voltage. *When the motor starts in the wye configuration, it typically draws only about 33% of the current that it would draw starting in the delta configuration,* **Figure 47.74.** After the motor is up to full speed, it is transitioned to a delta connection where the motor pulls the load and does the work more efficiently, **Figure 47.75**. The transition from star to delta is accomplished by a starting sequence with three different contactors that are **electrically** and **mechanically interlocked**; two are energized in wye for start-up, then one is dropped out and the other engaged for delta operation. The time the motor operates in the wye connection depends on how long it takes the compressor to get up to speed. Large centrifugal compressors may take a minute or more to get up to speed in wye; then the transition is made to delta.

When the motor starter switches from wye to delta, the wye connection is actually disconnected electrically by means of a contactor. This disconnect is proven both electrically, by the timing of the auxiliary contacts, and mechanically, by means of a set of levers that interlock the contactors. It is then that the delta connection is made. The electrical and mechanical interlock feature is necessary because if the wye connection and the delta connection were to occur at the same time, there would be a dead short from phase to phase that would likely destroy the starter components, **Figure 47.76**.

When the delta connection is made, the motor draws large amperage as it is totally disconnected and then reconnected. This spike in amperage could cause voltage problems in the vicinity of the motor. Many manufacturers equip the starter

Figure 47.74 Starting a motor using a wye-delta circuit. The motor in this diagram is connected in the wye configuration.

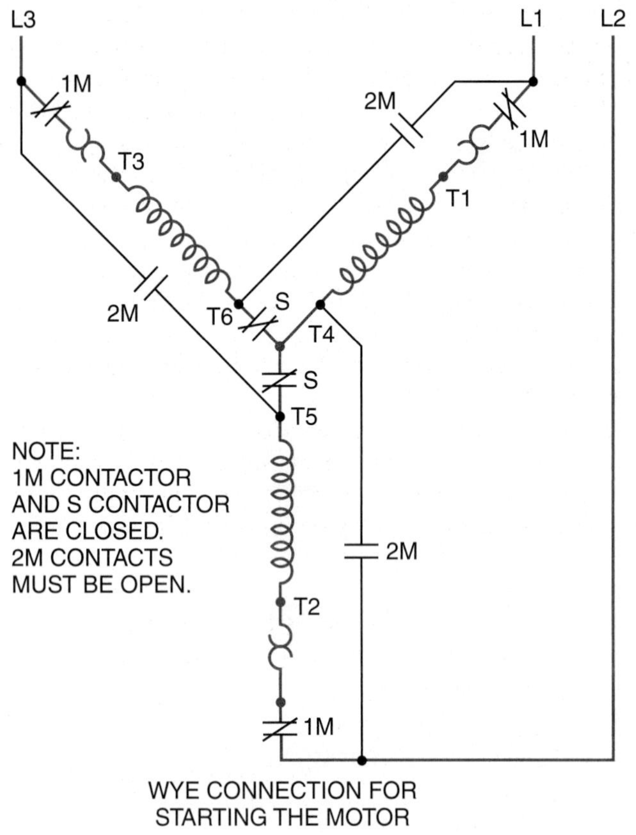

Figure 47.75 Once the motor is up to speed, it is changed over to the delta connection shown in this diagram.

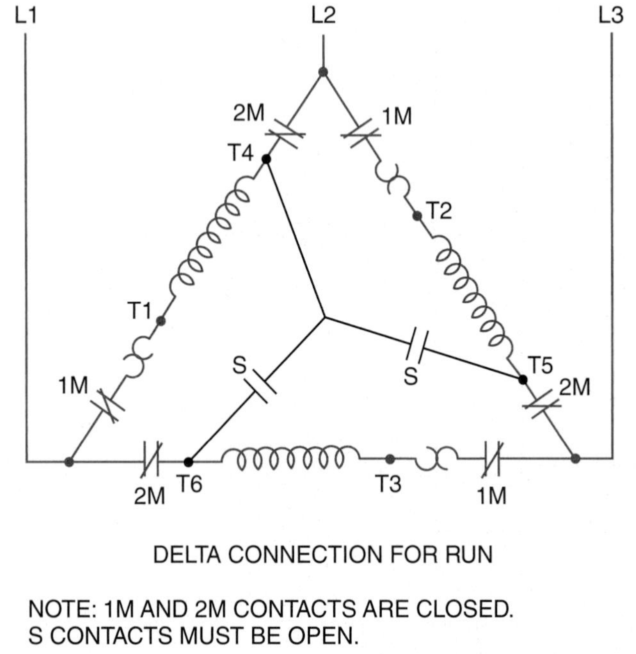

Figure 47.76 When too many of the contacts are closed, a short occurs. This diagram shows a short from line 2 to line 3 because the wye contacts did not open correctly.

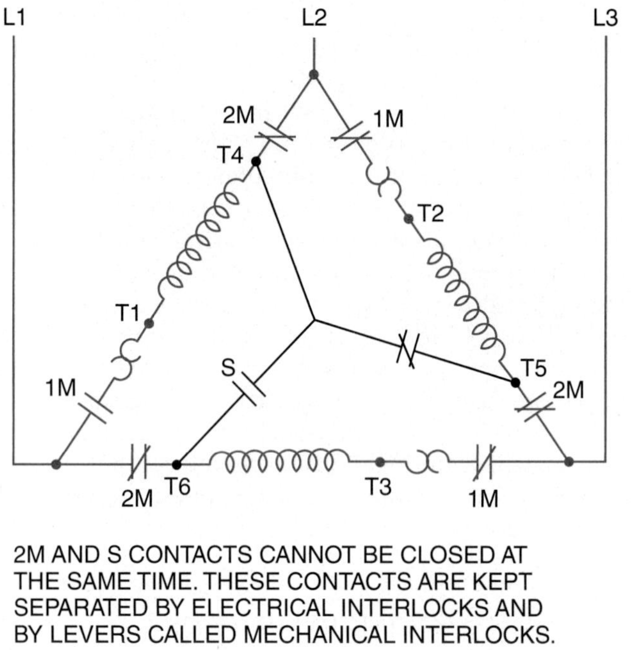

with a set of resistors that are electrically connected and act as the load while the transition is being made from wye to delta. This is known as a wye-delta closed transition starter. The starter would be a wye-delta open transition without the resistors. The larger the motor to be started, the more likely it is to have a closed transition starter.

All of the starters mentioned use open contactors to start and stop the motors. The inrush current puts open contacts under a great deal of stress when they are brought together to start a motor. When the motor is disconnected from the line, an electrical arc tries to maintain the electrical path, much like an electrical welding arc, and the contactor contacts become slightly damaged each time. Due to the arc, the contacts become pitted and so will be able to interrupt the load of a motor only a limited number of times before they must be replaced. **Figure 47.77** shows a new and an old set of contacts for a medium-sized contactor. An arc shield in the starter prevents the arc from spreading to the next phase of contacts when the contacts are opened. This arc shield must be in place when starting and stopping the motor or severe damage may occur.

⚠ **SAFETY PRECAUTION:** *Starting and stopping a large motor using a starter requires a lot of electrical energy to be expended inside the starter cabinet. The operator should NEVER start and stop a motor with the starter door open. If something should happen inside the starter, particularly during start-up, most of this energy may be converted to heat energy, and the result may be molten metal blown toward the door. Do not take the chance—shut the door.* •

Figure 47.77 New and old sets of contacts. The old ones are pitted from many motor starts. Courtesy of Square D Company.

ELECTRONIC STARTERS

As mentioned, inrush current for a motor can be five times the full-load current. *The inrush current can be greatly reduced by using electronic starters, which are also called soft starters. These starters use electronic circuits to reduce the voltage and vary the frequency to the motor at start-up. The result can be the reduction of a typical draw of 1000 A at locked rotor start-up to 100 A at locked rotor.* The motor is then accelerated up to full speed. This start-up is much easier on all components and reduces power line-voltage fluctuations. Electronic starters are also much smaller and more compact than conventional starters. For example, when an electronic starter is used to replace either an autotransformer or a wye-delta starter, only one component is needed to replace several, **Figure 47.78**.

The electronic starter uses electronic switching devices inside—typically, SCRs—to do the switching instead of electrical open contacts (see Unit 17, "Types of Electric Motors," for an explanation of how these devices are switched on and off). Once the motor is up to speed and leveled out, the load needs to be taken off these devices so that they will not generate local heat. To do this, a set of open contacts are switched on very quickly after the motor is up to speed and the contacts close, which effectively provides a circuit around the SCRs, **Figure 47.78**.

It is also hard on other electrical circuits when a large electrical load running at full load is just shut off. The power companies' regulation system will allow a voltage spike with an abrupt shutdown. When it is time to stop the motor, the SCRs can be used for a soft coast down by opening the bypass contactor contacts and then ramping the SCRs down with the electronics in the circuit. In the event of a power failure, the system shuts down by means of the bypass contactor.

Figure 47.78 *The electronic contactor used to stop and start the motor offers a soft start and a slow stopping of the motor. There is less inrush current and no voltage spike when the motor is stopped. A set of bypass contacts (conventional contacts) are energized during the running cycle to take the load off of the electronic contactor.*

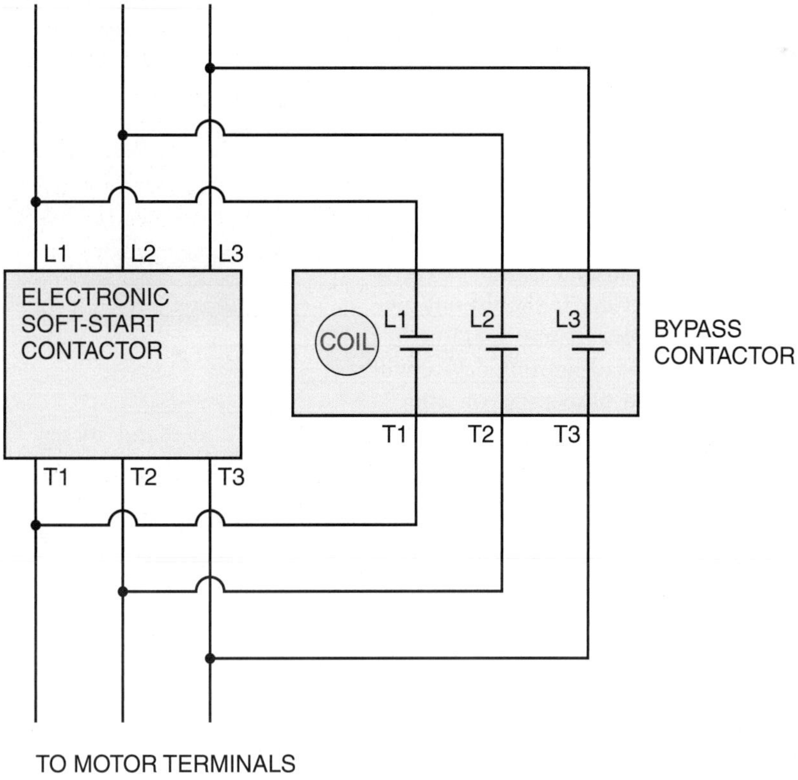

Troubleshooting all electronic equipment involves following the manufacturer's procedures. A particular sequence must be used. Electronic starters typically issue an error code that is linked to a blinking LED (light-emitting diode) that will tell the technician a fault has occurred and what it is. For example, the light may blink three times, then, after a brief time delay, blink one time, for a code of 31. The technician then looks up that code to identify the fault. Usually the only tools the technician needs in order to troubleshoot these devices is a good VOM (volt-ohm-milliammeter) and a clamp-on ammeter for checking line amperage to the motor.

MOTOR PROTECTION

The motor that drives the compressor is often the most expensive component in the system. The larger the compressor motor, the more expensive it is and the more protection it should be afforded. Small hermetic compressors in household refrigerators have little protection. Motors in small chillers will have protection like that covered in Unit 19, "Motor Controls." Here we discuss only the motor protection used for large chillers such as rotary screw or centrifugals.

The type of protection depends on the size of the system and the type of equipment. The advances made in electronic devices for monitoring voltage, heat, and amperage have improved the protection offered today compared with that offered several years ago. However, older motor protectors are still in use and will be for many years to come.

LOAD-LIMITING DEVICES. Rotary screw and centrifugal chillers use load-limiting devices to control the motor amperage. They monitor the current the motor is drawing while operating and throttle the refrigerant to the suction inlet of the compressor to prevent the motor from operating at an amperage higher than the full-load amperage.

The load-limiting device controls the slide valve on a rotary screw or the prerotation vanes on a centrifugal, **Figure 47.79.** It is a precision device that is the first line of defense for preventing motor overload and is set at exactly the full-load amperage at initial start-up. It should not give any problems afterward. For example, assume a motor has a full-load amperage of 200 A. The load-limiting device will be set so as not to exceed full-load amperage, or 200 A. **NOTE:** *Full-load amperage may be derated to a lower value for a system that has high voltage, so the technician cannot always go by the nameplate amperage. The start-up log should be consulted for the derated amperage.* • The load limiter may also have a feature that allows the operator to run the chiller at reduced load manually. Typically for rotary screw chillers, the load may be varied from 10% to 100%. For centrifugal chillers, it may vary from 20% to 100%.

Most buildings are charged for electrical power based on the highest monthly current draw that lasted for some period of time, typically 15 to 30 min. This is called billing demand power charge and most power companies measure it with a demand meter. Operating the equipment below the billing demand charge would be desirable whenever possible. For example, if an office building did not require cooling during the month until the last day and the operator started the chiller and allowed it to run at full load to reduce the building temperature quickly, this would take more than the time required to drive up the power company demand meter and the building would be charged for the whole month at a rate as though it had operated at full load all month. An operator could have avoided the excess charge by starting the system and operating it at part load, taking longer to reduce the building temperature.

Figure 47.79 (A) A slide valve capacity control. Courtesy of Trane Company. (B) Prerotation guide vanes used for capacity control. Courtesy of York International.

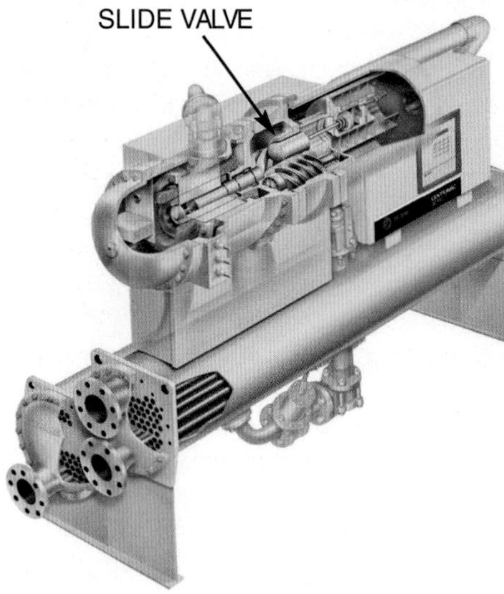

(A)

(B)

For example, the chiller may have the feature that would allow it to run at 40% and take longer to accomplish the task.

Many buildings are equipped with computer-controlled power management systems. This may take over from the chiller's control and operate the chiller at reduced capacity during high peak-load conditions. For example, if the peak power consumption is reached on the 15th of the month and the structure experiences an even higher heat load on the 30th of the month, the system will operate above the peak demand set on the 15th, and a higher energy bill will result for the entire month. *Rather than allow this to happen, a computerized control may reduce the energy consumption in some portion of the building. It is not uncommon to reduce the capacity of the chillers.* It may be better to allow a temperature rise of a few degrees in the entire building rather than pay a considerably higher power bill for the month. Reduction in demand may be easily accomplished through electronic circuits. Reducing the demand, and the high charges that are associated with the demand, is a process referred to as load shedding.

MECHANICAL-ELECTRICAL MOTOR OVERLOAD PROTECTION. All motors must have some type of overload protection. If a motor is allowed to operate at above the full-load amperage for an extended period of time, the motor windings will be damaged by the heat. Different motors have different types of protection. Overload protection for smaller motors, typically up to about 100 hp, was discussed in Unit 19, "Motor Controls." Mechanical-electrical overload protection for large rotary screw or centrifugal compressors may be a simple dash-pot device. This device operates on the electromagnetic theory that when a coil of wire is wound around an iron core, the core will move when current flows. With the dash-pot, either the current from the motor or a branch of the current from the motor passing through a coil around an iron core will cause the iron core to rise, **Figure 47.80**.

Large motor overload protectors are typically rated to stop the motor should the current rise to 105% of the motor's full-load rating. For example, in a motor with a rating of 200 A, the overload device would be set up to trip at 210 A ($200 \times 1.05 = 210$). The motor will be allowed to run at this amperage for a few minutes before stopping to prevent unwanted overload trips. These motors also have a load-limiting device that limits the motor amperage, so if this device is functioning correctly, the amperage should not be excessive for longer than it takes the electro-mechanical control to react, at which point the amperage should be no more than full load.

The dash-pot overload device contains a time delay that allows the motor to be started. Without the time delay, the inrush current would trip the device. The time delay is accomplished with a piston that must rise up through a thick fluid (the dash-pot) before tripping, **Figure 47.81**. The overload mechanism is wired into the main control circuit to stop the motor when it is tripped. Overload protections for large motors are manually reset so the operator will be aware of any problem.

Figure 47.80 An iron core and current flow in a compressor overload device.

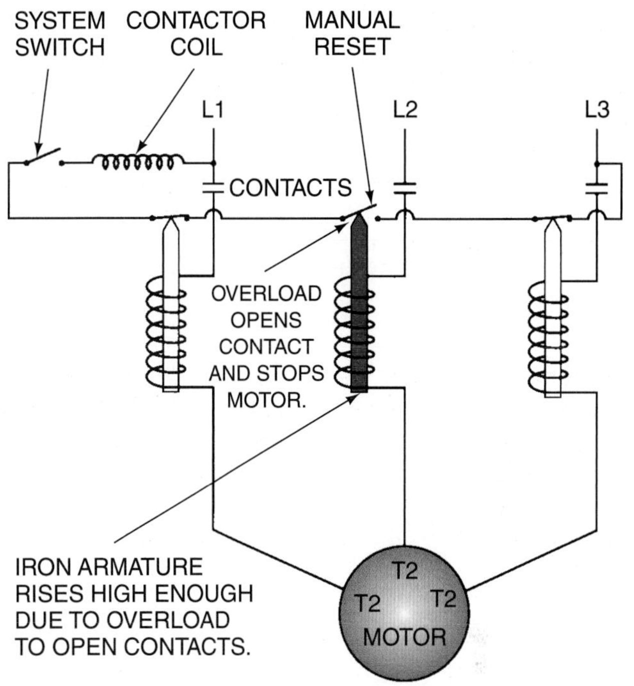

Figure 47.81 A dash-pot of oil gives the overload the time delay needed for motor start-up. (A) On initial motor start-up the overload is closed and the motor can start. (B) After a time-delay period, the piston has risen through the oil to open the overload's contacts.

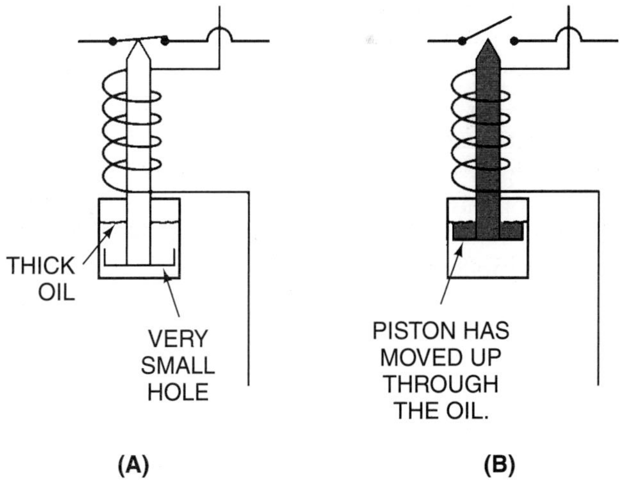

ELECTRONIC SOLID-STATE OVERLOAD DEVICE PROTECTION. Solid-state protective devices are wired into the control circuit like the dash-pot but act and react as an electronic control. They have the capability of allowing the motor to start and also monitoring the full-load motor amperage closely. They typically are located in the starter cabinet between the starter and the motor. They are the last line of defense. They are much more compact than the

dash-pot overloads and have many more features for close motor overcurrent control, such as the following:

- Motor overload protection
- Phase reversal protection
- Phase failure (single-phase) protection
- Voltage unbalance protection
- Manual and automatic reset, often from a remote location
- Ambient temperature compensation
- Trouble codes, much like the electronic contactors

Some of these features are discussed in more detail in the next portion of the text.

ANTI-RECYCLE CONTROL. All large motors should be protected from starting too often over a specified period of time. The anti-recycle timer is a device that prevents the motor from restarting unless it has had enough running time or off time to dissipate the heat from the last start-up. Manufacturers have different ideas as to how much this time should be. Typically, the larger the motor, the more time required. Many centrifugals have a 30-min time period. If the motor has not been started or has not tried to start for 30 min, it is ready for a start. Check the manufacturer's literature for the recycle time for a specific chiller.

PHASE FAILURE PROTECTION. Large motors all use three-phase power. The typical voltages are 208, 230, 460, and 575 V. Higher voltages are used for some applications, such as 4160 V or 13,000 V. Whatever the voltage, all three phases must be furnished or the motor will overload immediately. Electronic phase protection monitors the power and ensures that all three phases are present.

VOLTAGE IMBALANCE. Phase imbalance is caused by the electrical load on the building being unbalanced or the power company supply voltage being out of balance. The voltage to a compressor must be balanced within certain limits. Usually, 2% voltage imbalance is the maximum. Voltage imbalance is calculated as follows:

$$\text{Voltage Imbalance} = \frac{\text{Maximum Deviation from Average Voltage}}{\text{Average Voltage}}$$

For example, a technician measures the voltage on a nominal 460-V system to be

Phase 1 to phase 2	475 V
Phase 1 to phase 3	448 V
Phase 2 to phase 3	461 V

The average voltage is

$$\frac{475\text{ V} + 448\text{ V} + 461\text{ V}}{3} = 461.3\text{ V}$$

The maximum deviation from average is

$$475\text{ V} - 461.3\text{ V} = 13.7\text{ V}$$

The voltage imbalance is

$$\frac{13.7}{461.3\text{ V}} = 0.0297\text{, or } 2.97\%$$

This is more than some manufacturers' maximum allowable. The operating technician should keep an eye on voltage imbalance because it causes the motor to overheat. Some sophisticated electronic systems feature voltage imbalance protection; otherwise, it is up to the equipment operator to monitor.

PHASE REVERSAL. Three-phase motors turn in the direction in which they are wired to run. If the phases are reversed, the motor rotation will reverse. *This can be detrimental to many compressors.* Reciprocating compressors will normally perform in either direction due to their bidirectional oil pumps. Scroll, rotary screw, and centrifugal compressors must turn in the correct direction. Phase protection is often part of the control package for these compressors.

Any compressor with a separate oil pump, such as the centrifugal, will not start under phase-reversal conditions because the oil pump will not pump when reversed. It is not uncommon for the power company to reverse the phases of the power servicing a building or for an electrician to reverse the phases of a circuit within a building. It is very good practice for a building technician to be on the alert for any phase reversal after the building electrical system has been serviced by anyone. A phase meter may be used to determine the proper phasing of the power supply. This meter is a good tool for the heavy commercial or industrial technician to carry.

REVIEW QUESTIONS

1. The design temperature difference between the chilled water supplied to the building and the water returning to the system from the building is
   A. 10°F.
   B. 15°F.
   C. 20°F.
   D. 25°F.

2. Chilled water, when circulated throughout a building to provide comfort cooling, is referred to as a
   A. primary refrigerant.
   B. secondary refrigerant.
   C. bypass refrigerant.
   D. reciprocal refrigerant.

3. Circulating water throughout a building to provide comfort cooling is desirable from an economic standpoint because
   A. water is lighter than refrigerant.
   B. water is less expensive than refrigerant.
   C. the design chilled water temperature is lower than the design boiling temperature of the refrigerant.
   D. water does not require the use of a pump to move it through the system.
4. The portion of a chilled water system that cools the water that is used to cool the building is referred to as the
   A. chiller barrel.
   B. shell- and-tube condenser.
   C. shell and coil condenser.
   D. cooling tower.
5. The typical condensing temperature for the refrigerant in a water-cooled chilled water system is
   A. 85°F.
   B. 95°F.
   C. 105°F.
   D. 115°F.
6. All of the following methods can be used to control the capacity of a reciprocating compressor except
   A. suction valve lifting.
   B. blocked suction.
   C. variable frequency drive (VFD).
   D. constant suction.
7. Which of the following represents the most complete and accurate list of possible system capacities for a chilled-water system that is equipped with one 10-ton and one 15-ton scroll compressor?
   A. 10 tons, 15 tons, and 25 tons
   B. 10 tons and 15 tons
   C. 0 tons and 25 tons
   D. 15 tons and 25 tons
8. A variable frequency drive (VFD) is used to vary the
   A. voltage supplied to the system's motors.
   B. resistance of the system's motors.
   C. number of cycles per second of the power supplied to the system's motors.
   D. amperage draw of the system's motors.
9. The scroll compressor is equipped with an internal check valve to
   A. maintain the system's high-side pressure above a preset minimum level.
   B. prevent backward refrigerant flow when the system compressor cycles off.
   C. maintain the system's low-side pressure below a preset maximum level.
   D. prevent system superheat from rising above its design set point.
10. Which of the following statements is true regarding package-type chillers?
   A. They require field-installed refrigerant lines.
   B. They require field-installed water lines.
   C. They require field-installed refrigerant and water lines.
   D. They require no field-installed lines whatsoever.
11. The capacity of a chiller operating with a rotary screw compressor is most likely controlled by a
   A. slide valve on the compressor's suction side.
   B. slide valve on the compressor's discharge side.
   C. check valve on the compressor's suction side.
   D. check valve on the compressor's discharge side.
12. How does a centrifugal compressor create the pressure differential needed to move refrigerant through the system?
   A. Nesting scrolls
   B. Suction and discharge valves
   C. Rotating vanes
   D. Fan-like action
13. Which of the following best describes a two-gear drive assembly on a centrifugal compressor?
   A. A larger drive gear and a smaller driven gear
   B. A larger driven gear and a smaller drive gear
   C. Two same-size gears in an oil-filled reservoir
   D. Two same-size gears in a glycol-filled reservoir
14. Which of the following best describes the bearings on a centrifugal compressor?
   A. Steel construction with a babbitt sleeve
   B. Babbitt construction with a steel sleeve
   C. Copper construction with a steel sleeve
   D. Steel construction with a copper sleeve
15. If the compression ratio on a centrifugal compressor rises too high, the compressor will stop pumping and make a loud sound commonly referred to as a
   A. purge.
   B. splurge.
   C. merge.
   D. surge.
16. Oil pumps used in conjunction with centrifugal compressors typically use
   A. three-phase, fractional horsepower motors.
   B. single-phase, fractional horsepower motors.
   C. three-phase, multi-horsepower motors.
   D. single-phase, multi-horsepower motors.
17. Capacity control on a centrifugal compressor is accomplished by the
   A. slide vanes.
   B. check valves.
   C. prerotational guide vanes.
   D. postrotational guide vanes.

18. The rate of heat exchange between two liquids
    A. is greater than the rate of heat exchange between two gases, but less than the rate of heat exchange between a gas and a liquid.
    B. is less than the rate of heat exchange between two gases, but greater than the rate of heat exchange between a gas and a liquid.
    C. is greater than the rate of heat exchange between two gases as well as greater than the rate of heat exchange between a gas and a liquid.
    D. is less than the rate of heat exchange between two gases as well as less than the rate of heat exchange between a gas and a liquid.
19. Evaporators that operate with an established amount of superheat are referred to as
    A. low-side float evaporators.
    B. high-side float evaporators.
    C. flooded evaporators.
    D. direct-expansion evaporators.
20. What component is incorporated into the marine water boxes to direct water through a shell-and-tube heat exchanger?
    A. Partitions
    B. Abscissions
    C. Diverters
    D. Converters
21. What is the approach temperature for an R-410A chilled-water system that produces 43°F water when it operates with an evaporator saturation pressure of 112 psig?
    A. 4°F
    B. 6°F
    C. 8°F
    D. 10°F
22. A chiller barrel that has three passes will most likely operate with a design approach temperature of about
    A. 1°F.
    B. 5°F.
    C. 9°F.
    D. 13°F.
23. For an approach temperature of 10°F, what is the temperature of the water leaving a condenser if the condenser saturation pressure is 344 psig for an R-410A system?
    A. 75°F
    B. 85°F
    C. 95°F
    D. 105°F
24. For each 1°F of added subcooling, the efficiency of a system will increase by approximately
    A. 1%.
    B. 3%.
    C. 5%.
    D. 7%.
25. The superheat maintained by a thermostatic expansion valve is typically between
    A. 0°F and 4°F.
    B. 4°F and 8°F.
    C. 8°F and 12°F.
    D. 12°F and 16°F.
26. The flow through an orifice metering device
    A. is constant under all operating conditions.
    B. increases as the pressure drop across the orifice decreases.
    C. decreases as the pressure drop across the orifice increases.
    D. is constant as long as the pressure drop across the orifice remains the same.
27. Which of the following is true regarding a high-pressure chiller equipped with an orifice metering device?
    A. The chiller has a critical refrigerant charge.
    B. The chiller operates with R-410A as its refrigerant.
    C. The chiller operates with 10°F of superheat.
    D. The chiller operates with a reciprocating compressor.
28. The two float-type metering devices are the
    A. liquid float and the vapor float.
    B. high-side float and the low-side float.
    C. liquid float and the low-side float.
    D. vapor float and the low-side float.
29. The electronic expansion valve relies on
    A. pressure transducers and thermistors to control the operation of a stepper motor.
    B. the difference between the system's high and low pressure to control the valve.
    C. the difference between the condenser and evaporator outlet temperatures to control the valve.
    D. pressure transducers to maintain the desired evaporator saturation pressure.
30. All low-pressure chillers use which type of compressor?
    A. Reciprocating
    B. Scroll
    C. Rotary
    D. Centrifugal
31. The outer portion or shell of the compressor used on low-pressure chillers is called the
    A. baffle.
    B. volute.
    C. nozzle.
    D. suction.
32. The rupture disc on a low-pressure chiller typically releases pressure at
    A. 0 psig.
    B. 5 psig.
    C. 10 psig.
    D. 15 psig.
33. Low-side system leaks in a low-pressure chiller will often result in
    A. system refrigerant leaking from the system.
    B. water being pulled into the system.
    C. air being pulled into the system.
    D. nitrogen leaking out of the system.
34. If a low-pressure chiller has a low-side leak, which of the following will most likely result?
    A. High head pressure and low suction pressure
    B. Low head pressure and high suction pressure
    C. Low suction pressure and low discharge pressure
    D. High suction pressure and high discharge pressure
35. The purge unit is designed to
    A. remove water from a low-pressure chiller.
    B. remove air from a low-pressure chiller.
    C. introduce water to a low-pressure chiller.
    D. introduce air to a low-pressure chiller.

36. Instead of using a compressor, an absorption air-conditioning chiller uses which of the following as its driving force?
    A. Heat
    B. Electricity
    C. Oil
    D. Air
37. Why is water not used as a primary refrigerant in a vapor-compression refrigeration system?
    A. The water would freeze at the required temperature for cooling.
    B. The compressor would have to pump over 2400 ft^3 of water vapor for each pound of water circulated through the system.
    C. The pressures acting on the water would be much higher than those in traditional vapor-compression refrigeration systems.
    D. Water, when mixed with refrigerant oil, creates substances that pose a danger to the stratospheric ozone layer.
38. Lithium-bromide, Li-Br, is best characterized as a
    A. salt.
    B. sand.
    C. sugar.
    D. soil.
39. Which of the following most closely represents the concentration of Li-Br in an absorption system?
    A. 50% Li-Br, 50% tap water
    B. 60% Li-Br, 40% distilled water
    C. 40% Li-Br, 60% tap water
    D. 40% Li-Br, 60% distilled water
40. The solution that is pumped into the concentrator in an absorption chiller is referred to as the
    A. strong solution.
    B. concentrated solution.
    C. weak solution.
    D. absorbed solution.
41. The solution that leaves the concentrator in an absorption chiller is referred to as the
    A. absorbed solution.
    B. strong solution.
    C. weak solution.
    D. diluted solution.
42. What will occur if the strong solution in an absorption chiller becomes too strong?
    A. The machine will stop cooling and the solution will evaporate.
    B. The machine will cool excessively as the solution evaporates.
    C. The machine will stop cooling and the solution will crystallize.
    D. The machine will cool excessively as the solution crystallizes.
43. The instrument used to measure the specific gravity of absorption chiller solutions is the
    A. psychrometer.
    B. refractometer.
    C. anemometer.
    D. hydrometer.
44. If air is present in an absorption chilled water appliance, which of the following will most likely occur?
    A. Rust can form.
    B. The Li-Br solution will become more diluted.
    C. The Li-Br solution will become more concentrated.
    D. System operating pressure will drop.
45. Capacity control on an absorption system is typically accomplished by varying the
    A. speed of the condenser water circulator pump.
    B. speed of the chilled water circulator pump.
    C. pressure setting on the system's rupture disc.
    D. rate at which heat is supplied to the concentrator.
46. Compressors that are cooled with refrigerant that leaves the evaporator are referred to as
    A. liquid-cooled compressors.
    B. suction-cooled compressors.
    C. vapor-cooled compressors.
    D. discharge-cooled compressors.
47. An autotransformer start is commonly referred to as a
    A. part-winding start.
    B. full-winding start.
    C. wye-delta start.
    D. reduced-voltage start.
48. With a wye-delta or star-delta configuration, the motor starts as a
    A. wye configuration and then transitions to a star configuration.
    B. wye configuration and then transitions to a delta configuration.
    C. delta configuration and then transitions to a wye configuration.
    D. delta configuration and then transitions to a star configuration.
49. Which of the following statements is true regarding "pitted" contactor and starter contacts?
    A. Pitting occurs when a system is off for a prolonged period of time.
    B. A voltage reading of 0 volts will be measured across a set of closed, severely pitted contacts.
    C. Pitting occurs when electricity arcs as starter contacts open and close.
    D. Pitting contacts can be sanded smooth to restore the integrity of the contacts.
50. What is the voltage imbalance on a three-phase power supply if the voltage readings are: L1-L2: 466 V, L1-L3: 470 V, and L2-L3: 459 V?
    A. 1.29%
    B. 1.83%
    C. 2.15%
    D. 2.85%

UNIT 48

COOLING TOWERS AND PUMPS

OBJECTIVES

After studying this unit, you should be able to

- describe the purpose of cooling towers used with chilled-water systems.
- explain what is meant by the cooling tower's approach and range.
- describe and calculate the cooling tower's load.
- describe three types of cooling towers.
- explain the importance of keeping cooling tower fill wet.
- explain the various uses of fill material in cooling towers.
- describe two commonly used water/air flow patterns in cooling towers.
- describe the materials that are commonly used to construct cooling towers.
- describe the gear drives commonly used on cooling tower fans.
- explain the operation of a variable frequency drive (VFD).
- explain the importance of cooling tower access.
- explain the purpose of the sump in a cooling tower.
- explain the purpose of makeup water.
- explain the blowdown cycle.
- explain the importance of balanced water flow in a cooling tower.
- describe the function of centrifugal pumps and how they are rated.
- describe two types of motor–pump alignment.
- describe the seals found on cooling tower pumps.
- describe the construction of cooling tower pumps.
- describe the terms vortexing and cavitation as they apply to cooling towers.
- explain the difference between a mixing valve and a diverting valve.
- list the factors that affect cooling tower corrosion.
- define scale as it applies to cooling towers.
- explain what is meant by encapsulation and electroporation.
- describe the function of a particulate separator.

SAFETY CHECKLIST

- ✓ Ensure that all electrical safety precautions are observed when servicing pump motors and other electrical devices and components.
- ✓ Ensure that all belt and fan guards are in place when working in the area near a turning fan.
- ✓ Do not approach a turning fan or rotating motor shaft that is not properly guarded.
- ✓ Be careful of footing and balance when climbing up to or down from a cooling tower location. Do not move in a way that will cause balance to be lost when working around an elevated tower.
- ✓ When lifting a motor or pump, lift with a straight back and wear a back support belt.

48.1 COOLING TOWER FUNCTION

The heat that is absorbed into the boiling refrigerant in the chiller's evaporator is ultimately rejected by the condenser. In most water-cooled applications, the hot discharge gas from the compressor is desuperheated and condensed with water from the cooling tower. The water from the cooling tower, after absorbing the system heat in the condenser, flows back to the cooling tower where the heat is rejected to the atmosphere. The high-pressure, high-temperature superheated vapor that is pumped from the compressor to the condenser contains heat that is concentrated during the compression process as well as heat that is absorbed from the water circulating through the chiller barrel. This heat energy is then transferred to the water that flows between the condenser and the cooling tower. In the cooling tower, air moves across the surface of the water, thereby lowering the temperature of the water in the cooling tower. The cooler water is then circulated back to the condenser so that it can absorb more heat from the refrigerant, **Figure 48.1**. During the water-cooling process in the cooling tower, there is some evaporation, but most of the water is used over and over again. Cooling towers have a makeup water line that supplies water to compensate for any losses due to evaporation and cooling tower drift. Drift is the term used to describe water that is carried from the tower along with the air passing through it.

In vapor-compression refrigeration systems, the condenser must reject more heat than the chiller or evaporator absorbs. Since the cooling tower must reject the heat from the condenser's water circuit, it follows that the cooling tower must also reject more heat than the chiller or evaporator absorbs. The amount of heat energy that must be rejected by the condenser, and ultimately the cooling tower, is very close to the mathematical sum of the amount of heat added to the refrigerant in the chiller barrel and the amount of heat added to the refrigerant during the compression process, **Figure 48.2**. Heat losses and gains in the interconnecting piping, as well as other factors, account for any calculation discrepancies that might arise. The compressor accounts for almost 20% of the heat that must be rejected. This percentage is dependent on the type of compressor used in the cooling process. For example, a 1000-ton chiller would need a cooling tower that could reject about 1250 tons of heat. The additional 250 tons of capacity represents 25% of the chiller's cooling capacity

Figure 48.1 An example of a condenser water circuit with a cooling tower.

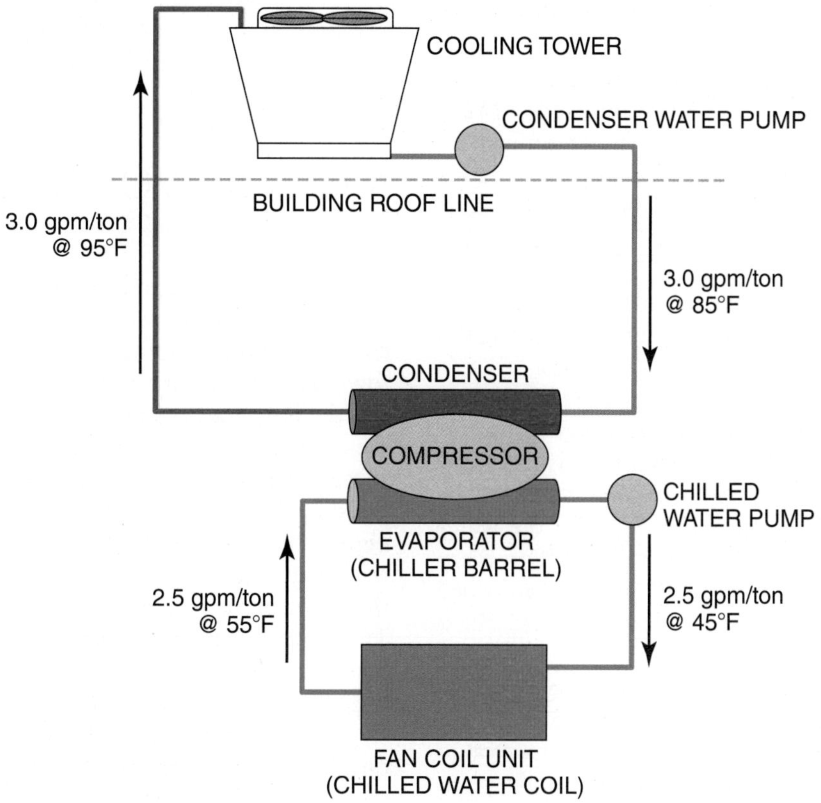

Figure 48.2 The condenser must reject the heat of compression as well as the heat absorbed by the refrigerant in the chiller.

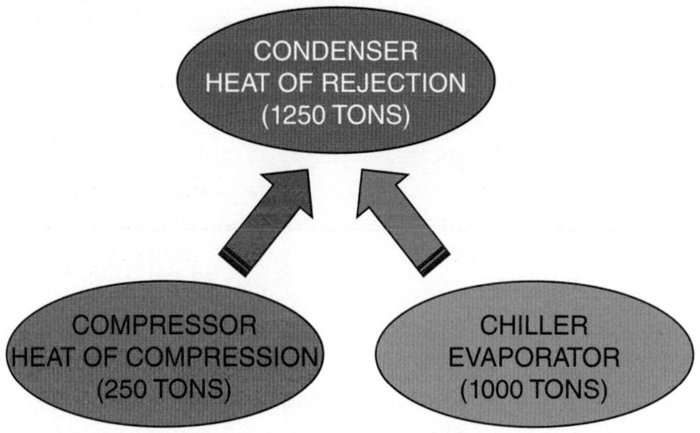

Figure 48.3 The typical cooling tower approach temperature.

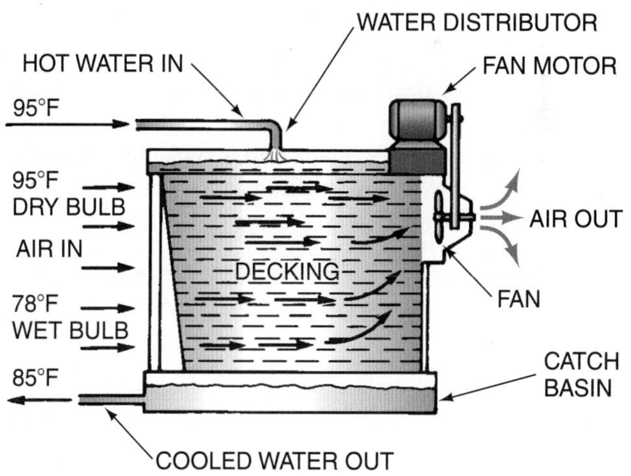

[(250 tons/1000 tons) × 100%] and 20% of the total system capacity [(250 tons/1250 tons) × 100%].

The condenser must be furnished water that is within the design limits of the system, or the system will not perform adequately. The design temperature for the water leaving the cooling tower for most refrigeration systems, including the absorption chiller, is 85°F. Using a design outdoor ambient temperature of 95°F dry-bulb and 78°F wet-bulb, the cooling tower can lower the water temperature through the tower to within 7°F of the wet-bulb temperature of the outside air. When the wet-bulb temperature is 78°F, the water temperature leaving the tower would be 85°F, **Figure 48.3**. This occurs even though the dry-bulb temperature of the outside air is 95°F.

Most cooling towers work in a similar manner. They reduce the temperature of the water in the tower by means of evaporation. As the water moves through the tower, the surface area of the water is increased to enhance the evaporation process. The different methods used to increase the

surface area of the water is part of what makes one cooling tower different from another. When the water is spread out to create more surface area, it will evaporate faster and be cooled closer to the wet-bulb temperature of the outside air. The water cannot be cooled below the wet-bulb temperature of the air. A nearly perfect cooling tower could reduce the temperature of the water to the entering wet-bulb temperature; however, the tower would be tremendous in size. Cooling towers produced for HVACR applications are typically designed to operate with a 7-degree **approach** temperature. A cooling tower's approach temperature is the temperature difference between the water leaving the tower and the wet-bulb temperature of the outside air entering the cooling tower.

The heat transfer within a cooling tower between the water and the air has both latent and sensible heat processes. The latent cooling process, however, dominates. As the water falls or is sprayed through the cooling tower, some of the water evaporates. The heat required to vaporize the water comes from the remaining mass of the water. For every pound of water that is evaporated, 970 Btus are absorbed from the remaining water. This is referred to as the latent heat of vaporization of water. The temperature of the remaining mass of the water is now reduced. The evaporation process is enhanced by increasing the surface area of the water. This is achieved by allowing the water to flow over a fill material in the tower, which breaks the water down into smaller droplets, increasing the surface area of the water. As mentioned, the water leaving the cooling tower can be cooled to within 7 degrees of the wet-bulb temperature. In dry climates, the water can exit the cooling tower 10 degrees cooler than the dry-bulb temperature. This phenomenon happens because of the greater amount of evaporation, thus the lower wet-bulb temperatures, in these dryer areas. The temperature difference between the entering and leaving water of the cooling tower is referred to as the tower range, and is typically about 10°F.

The cooling tower load in Btu/h can be calculated by the following formula:

$$\text{Cooling Tower Load} = (\text{gpm}) \times (8.33) \times (\text{Tower Range}) \times (\text{Specific Heat of Water}) \times 60$$

where

gpm = Flow rate in gallons per minute of water through the tower

8.33 = Density of water in lb/gal

Specific heat of water = 1 Btu/(lb)(°F)

60 = Conversion between minutes and hours

Tower Range = Tower Water Entering Temperature − Tower Water Leaving Temperature

For example, consider the following conditions:

Entering water temperature = 96°F

Leaving water temperature = 88°F

Flow rate of water = 30 gpm

The tower load can be calculated as follows:

$$\text{Tower load} = (30\text{ gpm}) \times (8.33\text{ lb/gal}) \times (96°\text{F} - 88°\text{F}) \times (1\text{ Btu/(lb)(°F)}) \times (60\text{ min/hour})$$

$$\text{Tower load} = 120{,}000\text{ Btu/h} = 10\text{ tons.}$$

48.2 TYPES OF COOLING TOWERS

Three types of cooling towers are commonly used: the *natural-draft* and *forced- or induced-draft tower* and the *closed-circuit hybrid cooling tower*. The closed-loop hybrid cooling tower has three different operating modes. As with any project, the larger the installation, the larger and more elaborate the cooling tower.

The natural-draft tower may be anything from a spray pond in front of a building to a tower on top of a building, **Figure 48.4**. The spray pond and natural-draft cooling tower rely on prevailing winds and will lower the temperature of the water in the tower when the prevailing winds are greater. The natural-draft tower or spray pond usually has an approach temperature of about 10°F because it relies solely on the blowing wind to provide the cooling effect.

The spray pond can be located at ground level or it may be on the roof of a building. In any case, the spray must be contained within the pond. Spray ponds on rooftops are becoming increasingly popular as they help cool the roof and reduce the solar load on the structure. Since a large amount of heat is introduced to a building through the roof, even a small reduction in roof temperature can result in a significant reduction in the load on the air-conditioning equipment.

Figure 48.4 Natural-draft tower types.

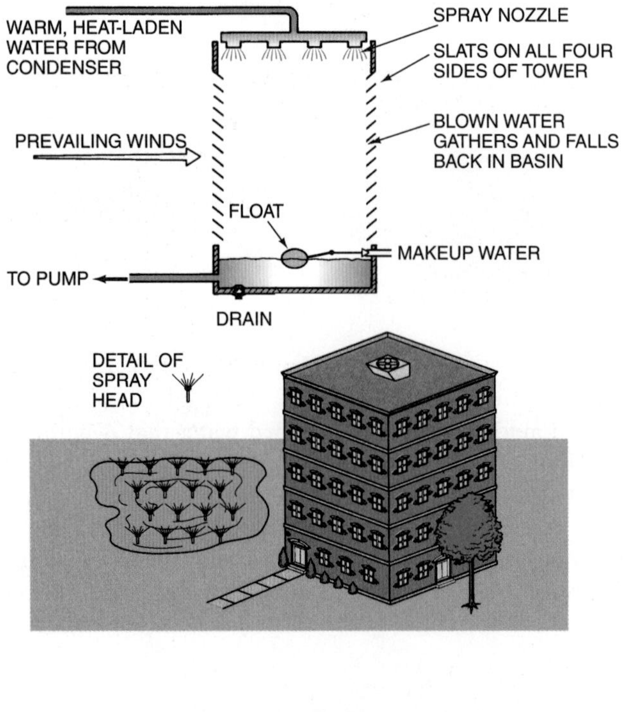

Figure 48.5 A propeller-type fan on a forced-air cooling tower. Courtesy of Ferris State University. Photo by John Tomczyk

Figure 48.6 A forced-draft cooling tower.

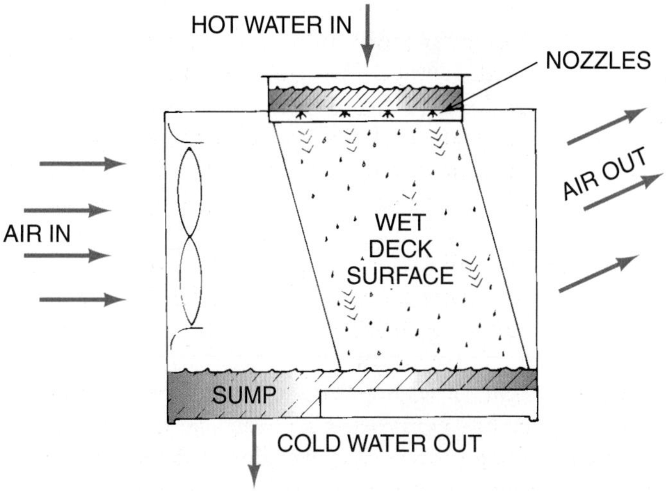

The spray pond and natural-draft cooling tower use pump pressure to atomize the water, increasing its surface area. Spray heads are placed at strategic locations within the pond or tower through which the water is pushed. The spray nozzles impose a restriction and must be kept clean or they will not effectively atomize the water. The pumps needed to atomize the water can be very large, given the large volume of water that these applications use.

Forced-draft and induced-draft cooling towers use fans to move the air through them, **Figure 48.5**. These are the most popular applications today because of their efficiency and reliability. Forced-draft cooling towers are so named because the fans or blowers force or push air across the water, **Figure 48.6**. Induced-draft cooling towers use fans to pull air over the wetted surfaces, **Figure 48.7**. In applications where the cooling tower is located indoors, centrifugal blowers are used to move the air through the tower, since these blowers are designed to overcome the high static pressures in the ductwork. Towers with centrifugal fans can be more compact and are therefore more desirable for some applications. These towers are available up to about 500 tons of capacity and heat rejection. The fans and blowers used on cooling towers may be either belt-driven or gear-driven. Belt-driven fans will require belt maintenance on a regular basis. Gear-driven fans have a transmission that will require only periodic lubrication.

The closed-circuit tower, **Figure 48.8**, *has a dry/wet mode, an adiabatic mode, and a dry mode. This tower has a finned coil,* **Figure 48.9**, *a prime surface coil,* **Figure 48.10**, *and a wet deck surface,* **Figure 48.11**. In the dry/wet mode the fluid to be cooled is fed to the dry finned coil and then to the prime

Figure 48.7 An induced-draft cooling tower.

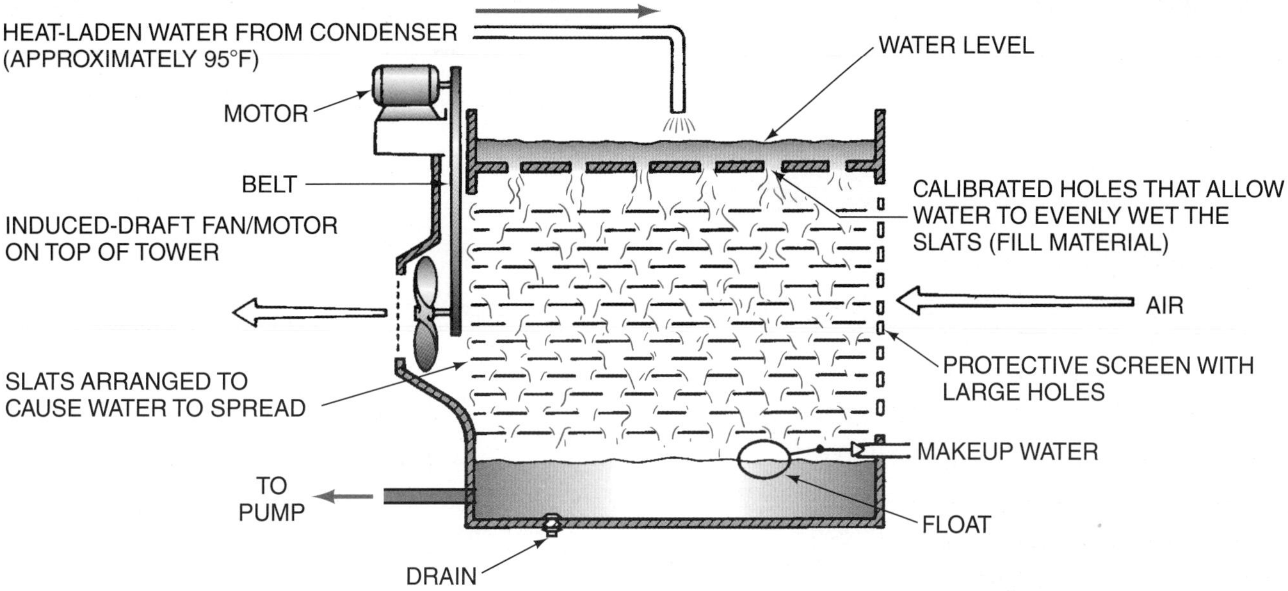

takes place. Air is drawn through the prime surface coil and the wet deck surface, leaving after passing through the finned coil, where more cooling occurs. When heat load or ambient temperatures are lower, the water usage over the evaporative cooling portion can be reduced. A valve that controls the outlet fluid temperature can modulate this water usage.

The term *adiabatic* refers to the process of cooling the circulating condenser water by evaporating the tower water. In an adiabatic process, no heat is added to, or removed from, the fluids by external means. Refrigerant flowing through the metering device of a vapor-compression refrigeration system, for example, undergoes adiabatic expansion, where the pressure and temperature of the refrigerant change, while the actual heat content of the refrigerant remains the same. In the adiabatic mode of operation on a closed-circuit cooling tower, the sensible heat transferred from the condenser circulating water is equal to the latent heat transferred into the air or gained from the evaporating water in the cooling tower. In the adiabatic mode, the fluid to be cooled bypasses the prime surface coil and is pumped through the finned coil only, **Figure 48.13**. Spray water is used to help cool the air passing through the tower, providing sensible heat transfer through the finned coil.

The amount of plume, which is the term used to describe the saturated discharge air from the closed-circuit tower, can be reduced (unlike conventional evaporative cooling towers) when the amount of spray water is reduced, as in the case of the dry mode of operation. Plume can be a problem in some areas when it falls on parked cars or other highly finished objects. It can also be a safety hazard when falling on areas

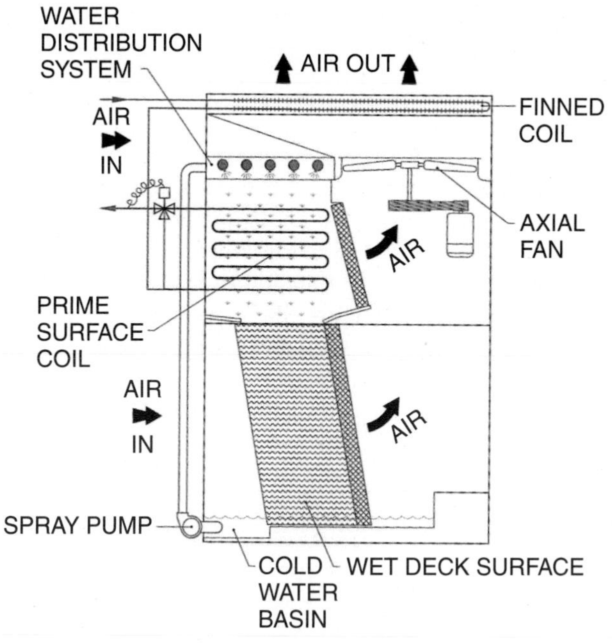

Figure 48.13 An illustration of the closed-circuit cooling tower in the adiabatic mode. Copyright, Baltimore Air Coil Co., 2003.

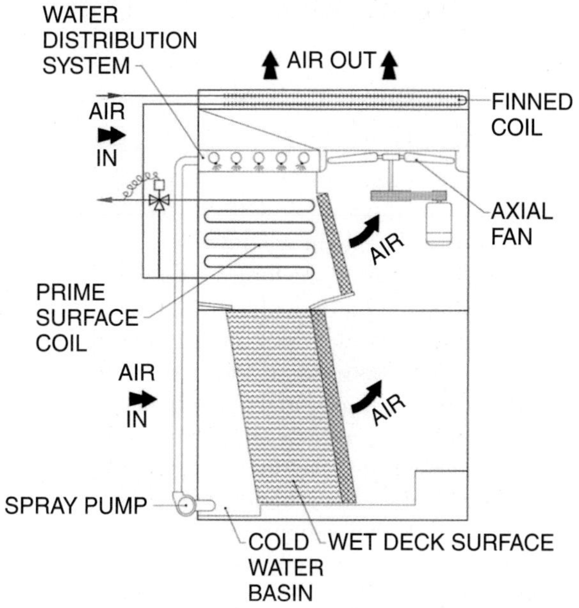

Figure 48.14 An illustration of the closed-circuit cooling tower in the dry mode. Copyright, Baltimore Air Coil Co., 2003.

where cars travel and at airports. *In the dry mode, the fluid to be cooled is passed through both the finned coil and the prime surface coil, **Figure 48.14**. The fluid is cooled by the forced air being passed over the coils.*

The closed-circuit system protects the system fluid from contamination and helps to ensure maximum efficiency over time, including reduced water usage. The closed-loop feature of this tower not only protects the circulating fluid, it also protects the heat transfer surface in the condenser. As mentioned earlier, the condenser tubes frequently get fouled, reducing the rate of heat transfer in the heat exchanger. With the closed-loop system, the potential for fouling is greatly reduced.

48.3 FIRE PROTECTION

An open cooling tower is a place where a fire may occur during the off-season when the tower components are dry. The tower materials may be flammable, such as wood or some plastics. Some fire codes and insurance companies may require that the tower be manufactured entirely of fireproof materials or they may require that a tower wetting system be in place, such as sprinkler heads similar to those commonly found in buildings. Some towers are controlled in such a manner that the system pump starts up on a timed cycle to wet the tower down occasionally during off-cycles. Another wetting system may be an auxiliary pumping system that keeps the tower continuously wet whenever the weather is

above freezing. This type of system also may prevent expansion and contraction of any wood construction in the tower. Local codes and insurance requirements will dictate what method should be used to protect the tower from fire.

48.4 FILL MATERIAL

The cooling tower is designed to keep the water and the air in contact with each other for as long as possible. Manufacturers use various methods to slow the water as it trickles down through the tower while the air moves up. Two methods used to evaporate the water in the forced- and induced-draft tower are the splash method and the fill, or wetted-surface, method. Both use a material in the tower often called fill. This material may be layers of wood slats through which the water drips down. Other materials used in the splash method include polyvinyl chloride (PVC) plastic or fiber-reinforced polyester (FRP) plastic. These materials have slow burn rates and are desirable when the potential for fire is high. Towers with a splash system have a framework that supports the slats and is designed to keep them at the correct angle for proper wetting of all slats as tower water flows from the top of the tower to the bottom.

The fill or wetted-surface type of cooling tower may also use some form of plastic or fiberglass as fill material. The water spreads out on the surface of the fill while air is passed over it, **Figure 48.15(A)**. **Figures 48.15(B)** and **48.15(C)** show a fill material with a honeycomb pattern. This type of fill may have smaller passages for the air to travel through and is not used where particles from the surroundings can contaminate the tower and restrict the passages. Both types of fill rely on water running over it, and unless kept at the angle the manufacturer recommends, the water will not take the proper path. Water running to one side or the other will reduce the capacity of the tower. If the fill material is removed for maintenance, care should be taken to replace it in such a manner that water will move through the tower correctly.

Figure 48.15 (A) The wetted-surface tower material. Image used with permission from SPX Cooling Technologies, Inc. for educational and informational purposes only. © 2015 SPX Cooling Technologies, Inc. All Rights Reserved.

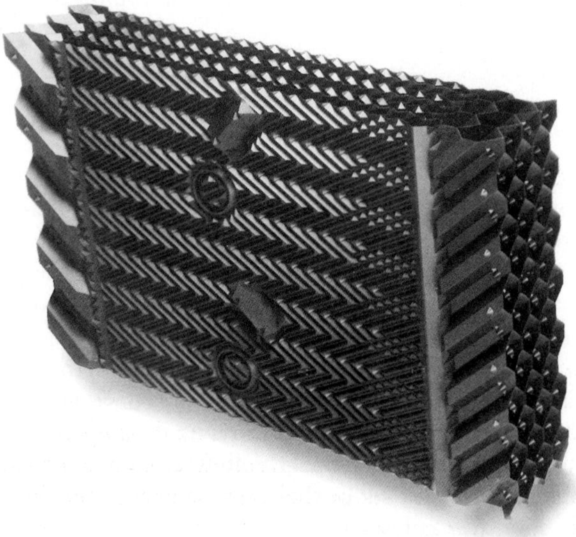

Figure 48.15 (B) A honeycomb-pattern fill material employed on a forced- air cooling tower. Courtesy of Ferris State University. Photo by John Tomczyk

Figure 48.15 (C) A close-up view of honeycomb fill material employed on a cooling tower. Courtesy of Ferris State University. Photo by John Tomczyk

48.5 FLOW PATTERNS

There are two distinct airflow patterns for cooling towers: crossflow and counterflow. The crossflow cooling tower, **Figure 48.16**, introduces the air from the side and usually pulls it to the top of the tower for exhaust. In smaller crossflow towers, the fan is on the side of the tower and air is exhausted out the side, **Figure 48.7**. In the crossflow tower, the water is moving downward and the air is moving at right angles to the water. Care must be taken that the moisture-laden air is not exhausted to a place where it will cause a problem, such as on a walkway or a parking lot where cars can be spotted with the water and chemicals used in treatment, which may be corrosive to car finishes. The counterflow tower, **Figure 48.17**, introduces the air at the bottom and exhausts the air out the top of the tower. The water is moving downward as the air moves up through it.

Figure 48.16 A crossflow cooling tower. Image used with permission from SPX Cooling Technologies, Inc. for educational and informational purposes only. © 2018 SPX Cooling Technologies, Inc. All Rights Reserved.

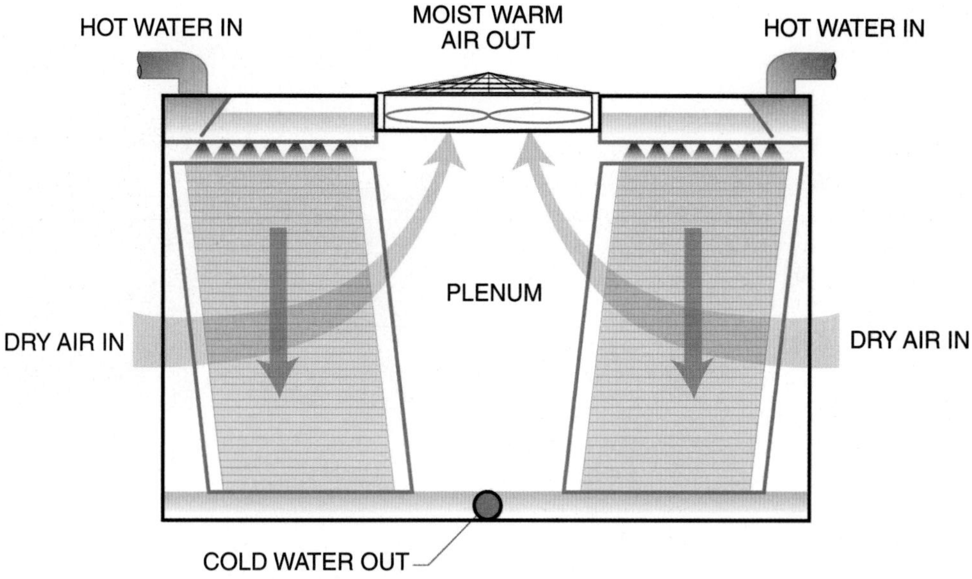

Figure 48.17 A counterflow cooling tower. Image used with permission from SPX Cooling Technologies, Inc. for educational and informational purposes only. © 2018 SPX Cooling Technologies, Inc. All Rights Reserved.

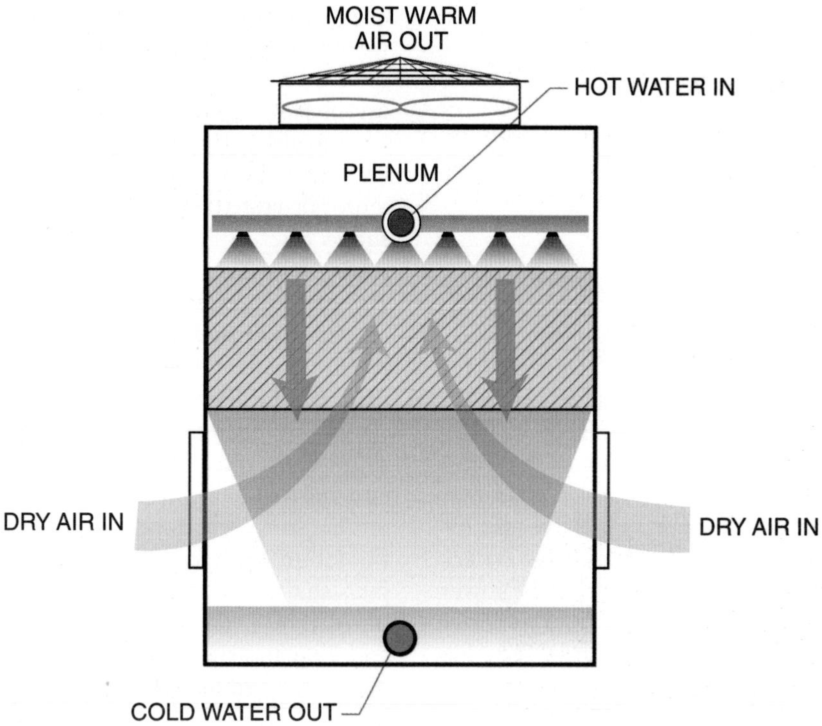

In most towers, especially those that use spray, many small particles of water are suspended in air. These particles are subject to being blown out of the tower by the prevailing winds, a loss of water called *drift*, which can be expensive and a nuisance to any surrounding areas if the cooling tower water is treated with chemicals. The amount of drift is minimized with the aid of eliminators, **Figure 48.18**. Eliminators cause the spray to rapidly change direction and impact solid surfaces, **Figure 48.19**, where water droplets collect and run back to the tower basin. The eliminators may be louvres on the side of the tower or they may be part of the fill material in some newer towers.

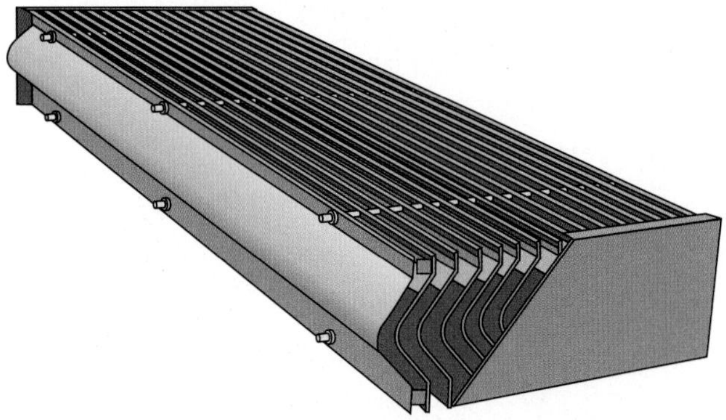

Figure 48.18 Drift eliminators.

Figure 48.19 The air impacts the eliminators, separating the moisture from the air.

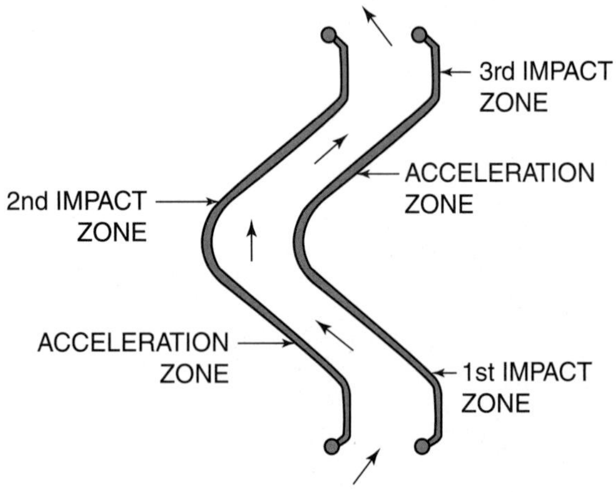

48.6 TOWER MATERIALS

The materials from which a cooling tower is constructed must be able to withstand the environment in which the tower operates. The typical tower must be able to withstand wind, the weight of tower components and the water, sun, cold, freezing weather (including any ice that may accumulate), and vibration of the fan and drive mechanisms. Towers must be carefully designed and the materials chosen according to the type and location of the tower. Typically, smaller packaged towers are made of galvanized steel, fiberglass, or FRP. Galvanized steel is steel that has been coated with zinc, to help prevent rusting. These towers are manufactured as a complete assembly and shipped to the job, **Figure 48.20**.

Larger towers may have a concrete base and sump to hold the water with sides made of other materials such as wood (either treated or redwood), fiberglass, or corrugated FRP. In many cases, fire prevention must be considered in the selection of materials.

Figure 48.20 Modern cooling tower that was manufactured as a complete assembly and shipped to the job. Courtesy of Ferris State University. Photo by John Tomczyk

48.7 FAN SECTION

In all forced-draft towers the motor must have a way of turning the fan. There are two different types of drive, the belt drive and the gearbox (transmission) drive. The belt drive is normally used in the smaller towers and includes an adjustable motor mount. This mechanism will undergo the greatest wear in the tower and requires the most maintenance, so it should be located where preventive maintenance and service can be performed with as little effort as possible.

Fan motors in larger towers are mounted out to the side and have a gearbox or transmission to change the direction of the motor driveshaft by 90 degrees, **Figure 48.21. Figure 48.21(A)** shows a set of 90-degree bevel gears, while **Figure 48.21(B)** shows a set of 90-degree spiral bevel gears. Bevel gears are often preferred to the non-beveled variety because of the increased surface area between the mating surfaces, which reduces the wear on the gears and prolongs their useful life. These motor/fan units may also be designed to change the motor-to-fan shaft speed relationship. The typical motor will turn at 1800 or 3600 rpm and the fan shaft will turn considerably slower, depending on the gear reduction. For example, if the gear connected to the motor shaft is one-half the diameter of the gear on the fan's shaft, the fan will rotate at a speed that is one-half that of the motor. The motor is connected to the gearbox using a coupling and shaft, **Figure 48.22.** *VFDs (variable-frequency drives) are also being used for cooling tower fans. The ability to reduce the fan speed to correspond with system conditions helps control the system more closely and improves efficiency by reducing energy consumption. VFD technology is covered in detail in Unit 17, "Types of Electric Motors."* The motor, gearbox, and fan bearings must be accessible for service purposes.

The fan blade on the cooling tower is enclosed in a fan housing that improves the efficiency of the fan blades,

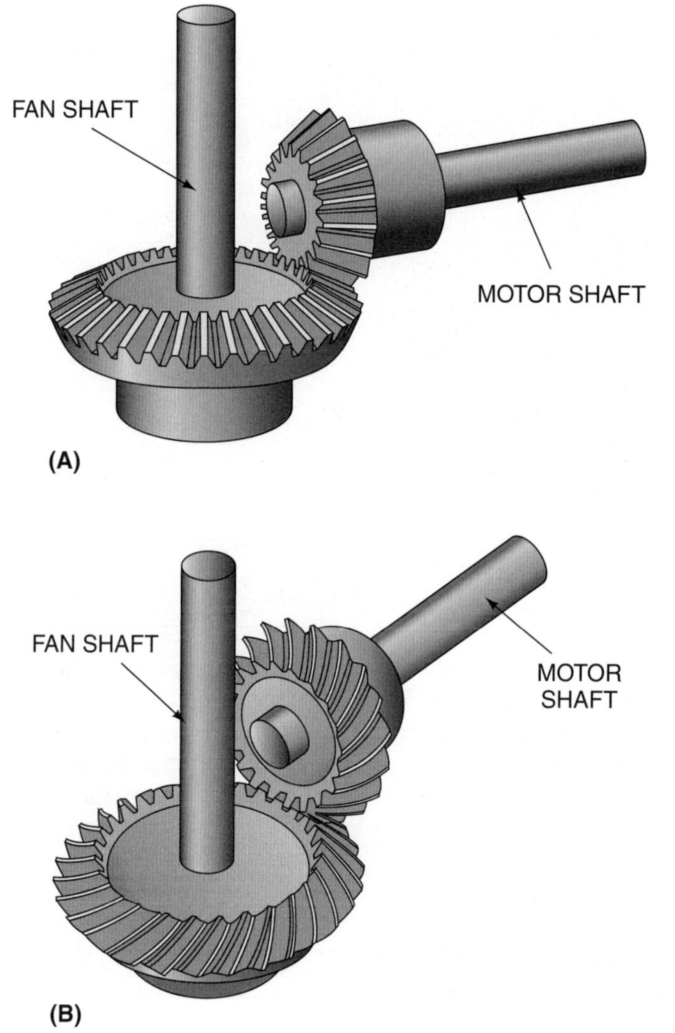

Figure 48.21 (A) Bevel gear assembly. (B) Spiral bevel gear assembly.

Figure 48.22 A coupling and shaft for the gear reduction box. Image used with permission from SPX Cooling Technologies, Inc. for educational and informational purposes only. © 2015 SPX Cooling Technologies, Inc. All Rights Reserved.

Figure 48.17. The location of the fan blade is critical in most towers. It must be placed at the correct distance from the top and sides for optimal performance.

In the past, all towers used a thermostat immersed in the water to cycle the cooling tower fans on or off as needed to control tower basin water temperature and maintain a minimum temperature. This system results in fairly wide temperature swings, causing the water temperature to vary considerably. As we will see later in this unit, motorized diverting valves may be used to maintain conditions closer to the condenser temperature when temperature swings occur. The VFD fan motor can be used for this function. The VFD eliminates the constant starting and stopping of the motor by modulating the motor's speed to closely match the needs of the system. *The VFDs also allow motors, some of which are quite large, to soft start and to coast down without shutting off abruptly. Both of these features improve system performance and prevent power fluctuations.*

48.8 TOWER ACCESS

All cooling towers will need service on a regular basis. There must be access doors to the fill material for cleaning and possible removal. The tower basin must also be accessible for cleaning because large amounts of sludge often accumulate there, **Figure 48.23**. The cooling tower becomes a large filter for whatever may be airborne—commonly, dirt, pollution, feathers, birds, plastic wrappers, cups, etc. These particles will gather in the sump and must be removed. There is also a strainer, **Figure 48.24**, at the water outlet port on the cooling tower that collects objects, preventing them from entering the pump and piping circuit. For the purpose of flushing the tower sump with a water hose, an adequate water supply should be available in the vicinity of the tower.

A tall tower will likely have a stairway or ladder to the top for servicing the fan and drive components, **Figure 48.25**.

Figure 48.23 (A) Access ports on a cooling tower. (B) Port open to gain access. Photos by Eugene Silberstein

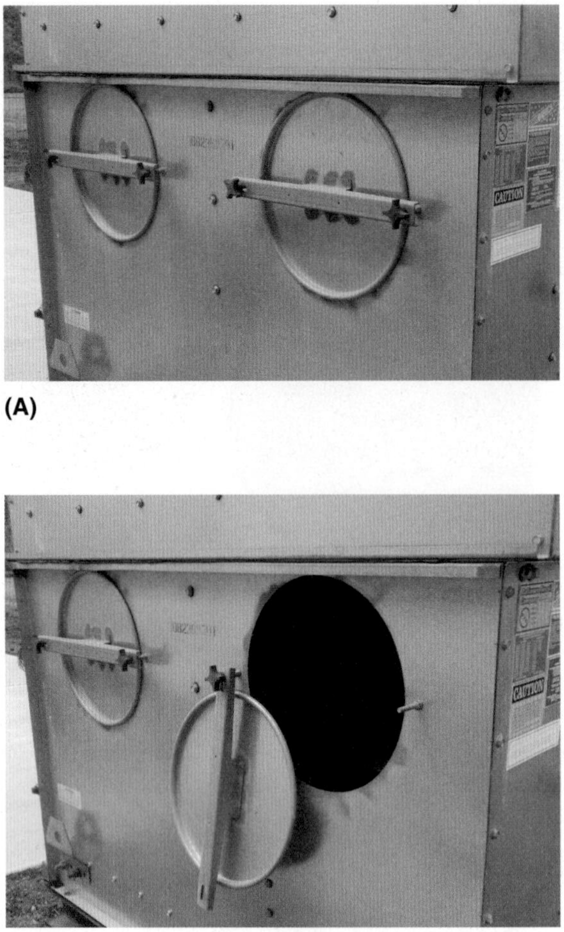

(A)

(B)

Figure 48.24 (A) Cooling tower sump strainer. (B) Strainer installed in the sump. Photos by Eugene Silberstein

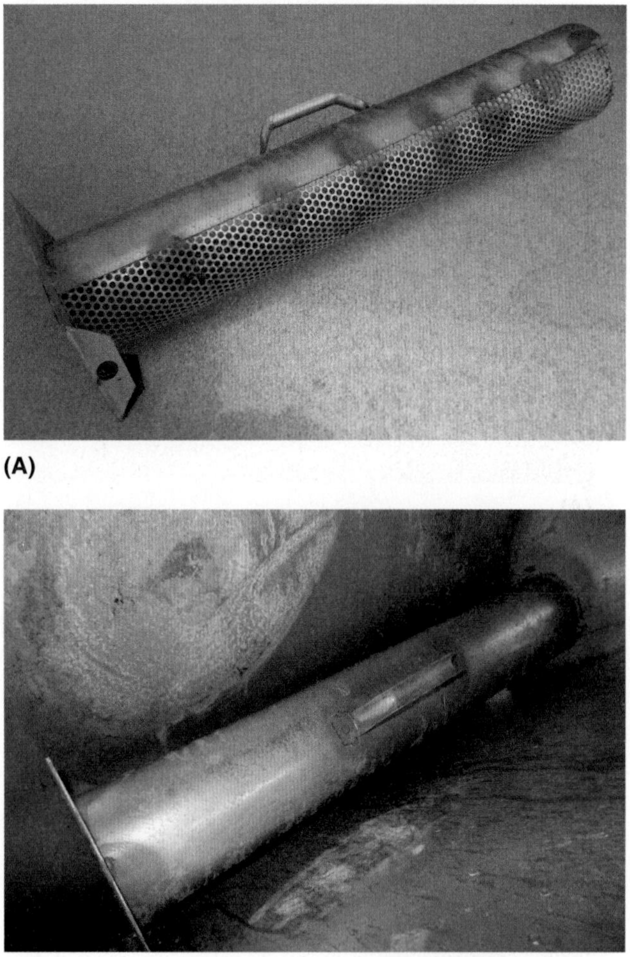

(A)

(B)

Provisions should be made in the vicinity of the motor and fan to lift out any component parts that may need to be removed and lowered to the ground for servicing. A service stair or ladder must meet the local codes for safety and include proper handrails and barriers to prevent stepping over the edge.

48.9 TOWER SUMP

All cooling towers must have some sort of *sump* for collecting the water. The sump is located at the bottom of the tower and is where the tower's strainer is located. The sump on small cooling towers may consist of a metal pan that gathers the water and drains to a lower point. When the sump is located outside, it must have some means to prevent the water from freezing, such as a thermostatically operated sump heater, **Figure 48.26.** Sump heaters are used for small installations; larger installations require more heat and

Figure 48.25 The stairway and guardrails for a large tower. Image used with permission from SPX Cooling Technologies, Inc. for educational and informational purposes only. © 2015 SPX Cooling Technologies, Inc. All Rights Reserved.

Figure 48.26 (A) Freeze protection for a cooling tower sump. (B) Heating element inside the sump of the cooling tower. Photos by Eugene Silberstein

(A)

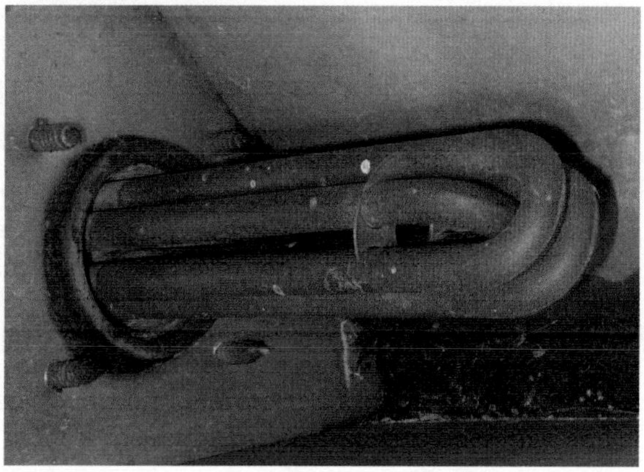

(B)

may have a circulating hot water coil or a method of using low-pressure steam to heat the tower basin water in cold weather. Many cooling towers have underground sumps made of concrete. These sumps will not freeze in the southern climates but must be protected from the cold in northern climates unless the refrigeration system is operating and adding heat to the sump. Some sumps are located in the heated space of a building to prevent freezing. Antifreeze products also may be added to some systems.

48.10 MAKEUP WATER

Because the cooling tower operates on the principle of evaporation, water is continuously lost from the system. Water is also lost from a cooling tower because of drift, which is when the air passing through the tower carries tower water with it as it exits the tower. There are several methods of making up for lost water, the float valve, **Figure 48.27**, being one of the most common. When the water level drops, the float ball drops with it and opens a valve to the makeup water supply. This type of float valve resembles the assembly found in many household toilets. Usually the makeup water is drawn from the normal municipal or other, similar supply. Another method of making up water is with a float switch that energizes a solenoid valve to allow water to fill the sump when the water level falls, **Figure 48.28.** When the water level reaches the proper level, the float rises and shuts off the solenoid valve. Another method to control makeup water uses sensors that determine the level of water in the sump. If the water level is high enough, both sensors are submerged in the water and the solenoid controlling the fill valve is deenergized, **Figure 48.29(A)**. If the water level drops, only one of the sensors is submerged, the electric circuit between the two sensors is not complete and the solenoid on the fill valve is energized to feed more water to the sump, **Figure 48.29(B)**.

Figure 48.27 The inside of a cooling tower with a makeup water assembly consisting of a ball-and-water valve. Courtesy of Ferris State University. Photo by John Tomczyk

Figure 48.28 Float switch and solenoid configuration.

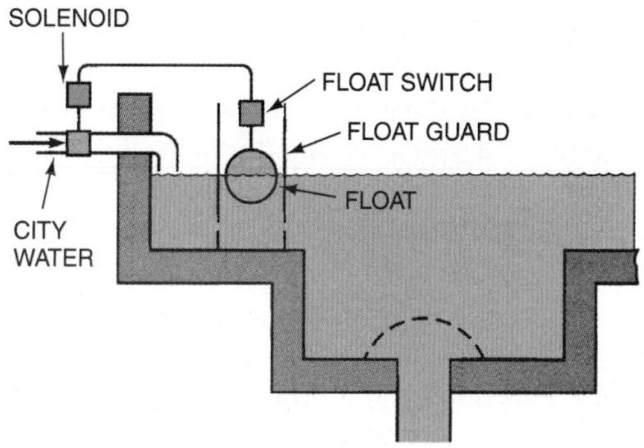

Figure 48.29 Electronic sensors. (A) Water level is high enough and the solenoid valve is closed. (B) Water level is low and the solenoid valve is open, feeding water into the sump.

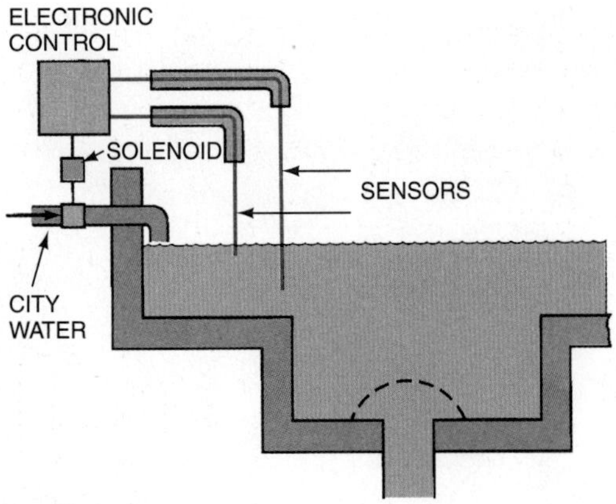

(A)

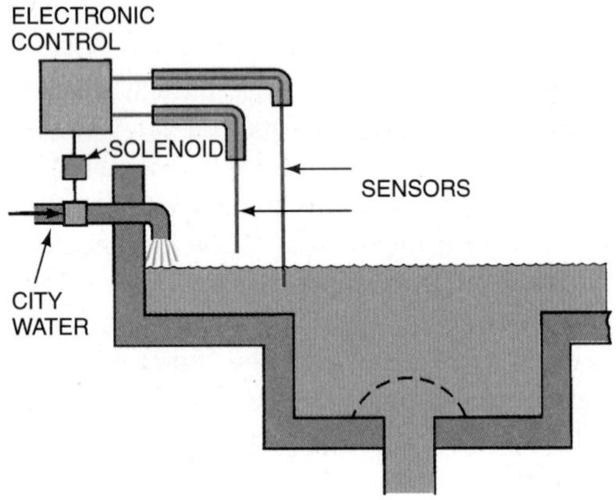

(B)

48.11 BLOWDOWN

Often, building management personnel have a difficult time grasping the purpose of blowdown, which merely means bleeding off a portion of the water in circulation. They do not understand why water is being allowed to run down the drain. **Blowdown must be managed correctly because, if not correctly implemented and monitored, the potential for losing large amounts of expensive water treatment chemicals down the drain with the water is high.**

The water that evaporates from a cooling tower leaves behind any solid materials, including dust particles, minerals, and algae, a lower form of plant life. As the water evaporates, these particles become more and more concentrated in the water left behind. If this continues, the particles will drop out of the water and deposit on the surfaces of the cooling tower where they turn into a substance much like cement and become hard to remove. Even more important than what is happening in the cooling tower will also happen to the interior surfaces of the tubes in the condenser. These mineral deposits can adhere to the piping surfaces and act like an insulator between the refrigerant and the water. This reduces the ability of the refrigerant to reject heat and will cause the head pressure as well as the condenser approach temperature to rise.

Fresh water is added to make up for water drained from the system during blowdown. When new, fresh water is added, less sediment is present, as some has gone out with the water that was bled off, and it dilutes the remaining mineral-rich water, reducing the mineral concentration. It is generally accepted that the water flowing through a water-cooled condenser used for air-conditioning experiences a 10°F temperature rise as measured between the condenser's inlet and outlet. There is also a 10°F difference between the temperature of the water leaving the condenser and the refrigerant's condenser saturation temperature. So, for a properly operating water-cooled system, the operating pressures would typically be as follows:

- Water entering the condenser: 85°F
- Water leaving the condenser: 95°F
- Condenser saturation temperature: 105°F

For an R-410A water-cooled system, the condenser saturation pressure would be about 340 psig, **Figure 48.30.** If the system is experiencing a scaling condition, as a result of improper water treatment or excessive mineral accumulation,

Figure 48.30 A properly operating R-410A water-cooled system. There is a 10-degree rise across the condenser and a 10-degree difference between the condenser saturation temperature and the temperature of the water leaving the condenser.

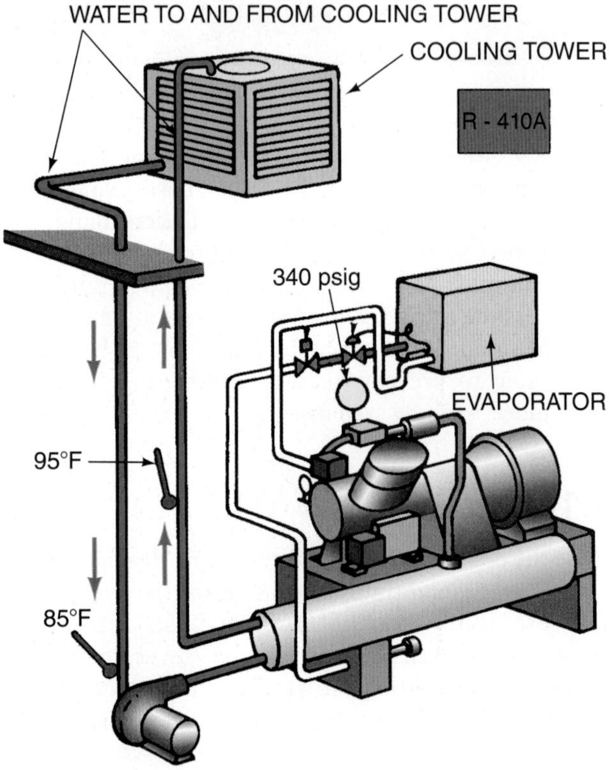

the refrigerant in the condenser will be unable to reject heat to the water flowing through the condenser as readily. This is because the mineral deposits are preventing the refrigerant's heat from transferring to the water. This will cause the high-side pressure in the system to rise and, since the water is not absorbing heat from the refrigerant, the temperature rise of the water across the condenser will not be as high. The temperatures for this system condition might be as follows:

- Water entering the condenser: 85°F
- Water leaving the condenser: 90°F
- Condenser saturation temperature: 130°F

For an R-410A water-cooled system, the condenser saturation pressure would be about 480 psig, **Figure 48.31**. Notice that the temperature difference between the condenser's inlet and outlet is only 5°F (90°F − 85°F) and the difference between the temperature of the water leaving the condenser and the refrigerant's condenser saturation temperature is an excessive 40°F (130°F − 90°F).

Blowdown can be accomplished in a number of different ways. The simplest way is to utilize a hand valve, **Figure 48.32**. With this method, there is no automation involved and the amount of water drained from the system depends on how long the operator keeps the valve open. Although this is an easy method for accomplishing blowdown, there are many downsides, which include lack of timing, uniformity, and scheduling issues. Blowdown can also be controlled by matching the rate at which water is drained from the tower to the rate at which water is introduced to the cooling tower, **Figure 48.33**. Using this method of blowdown, as the rate of makeup water added to the tower increases, the amount of water drained from the system increases. Another, more sophisticated method of controlling blowdown is to use electronic sensors, **Figure 48.34**. The

Figure 48.31 The condenser is not removing enough heat because the tubes are fouled. There is only a 5°F temperature rise across the condenser. The condenser saturation temperature is also much higher than desired.

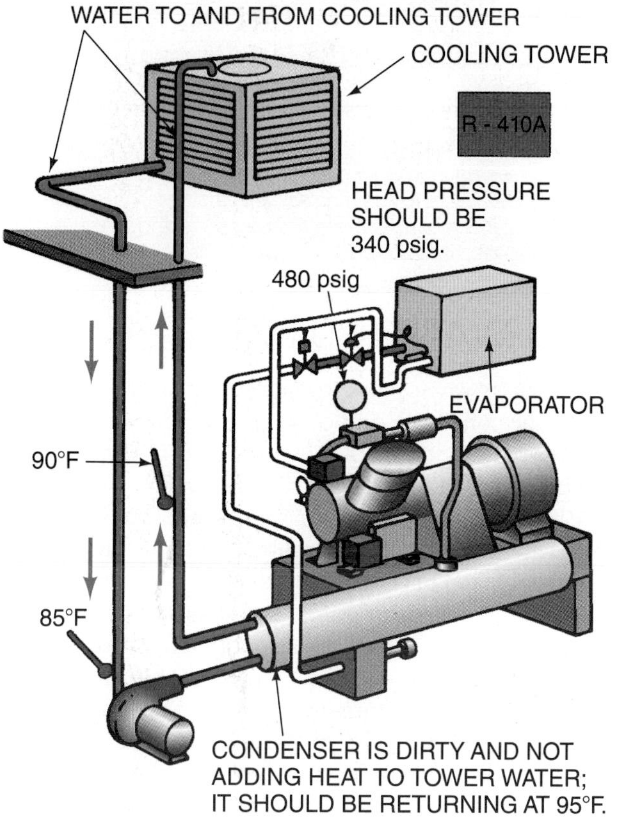

Figure 48.32 A hand valve being used for blowdown. Courtesy of Nu-Calgon Wholesaler, Inc.

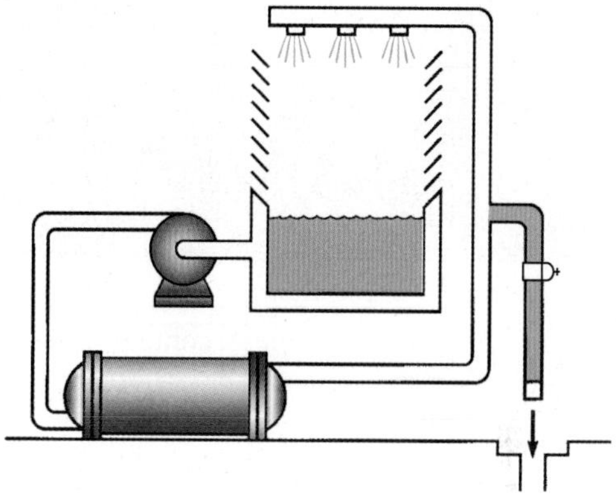

Figure 48.33 The amount of water drained from the system is proportional to the amount of water being introduced to the sump. Courtesy of Nu-Calgon Wholesaler, Inc.

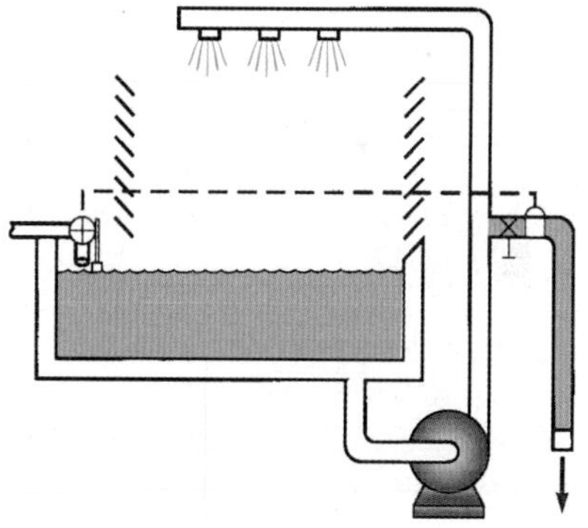

Figure 48.34 Electronic sensor measuring the mineral content of the water. Courtesy of Nu-Calgon Wholesaler, Inc.

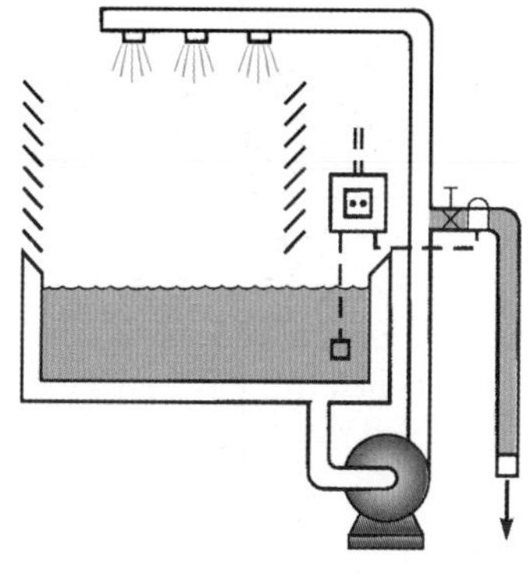

Figure 48.35 A microprocessor controller that analyzes the conductivity of the cooling tower's sump water and determines when the blowdown process is needed. Courtesy of Ferris State University. Photo by John Tomczyk

electronic sensor measures the mineral content of the water and bleeds off water when the minerals have become too concentrated. **Figure 48.35** shows a microprocessor controller that analyzes the conductivity of the cooling tower's sump water and determines when the blowdown process is needed. Pure water is not electrically conductive; however, as dissolved minerals concentrate in the cooling tower's sump, the conductivity of the water will increase. A voltage probe located in the sump senses the conductivity of the sump water and sends a signal back to the microprocessor.

48.12 BALANCING THE WATER FLOW IN A COOLING TOWER

The water flow must be evenly distributed in a cooling tower. If a tower has two or more cells, the same amount of water must be fed to each cell or the tower will not perform as designed. Many towers use a distribution pan at the top to evenly spread the warm return water from the condenser over the fill. These pans often have a series of calibrated holes drilled into them. Water flow is controlled by one or more balancing valves set to obtain the correct flow to each side of the tower, **Figure 48.36**. The calibrated holes must be kept clean of foreign material and be of the correct size. Excessive amounts of debris or mineral deposits can block the holes in the distribution pan, **Figure 48.37**. Blocked holes, rust, or excessive mineral deposits can cause uneven flow through the cooling tower, **Figure 48.38**. In this case, water will be flowing to one side of the tower and not wetting the entire fill deck. The capacity of the cooling tower will be reduced when this happens. For example, if the tower is designed to supply water at a temperature of 85°F to the condenser and it is supplying 90°F water to the condenser, the system's operating pressure will be higher than desired and the cost to run the system will increase.

Figure 48.36 Balancing valves for balancing the water flow at the cooling tower. Image used with permission from SPX Cooling Technologies, Inc. for educational and informational purposes only. © 2015 SPX Cooling Technologies, Inc. All Rights Reserved.

Figure 48.37 Calibration holes for distributing water in the tower are enlarged due to rust. Most of the water is running down the right side of the tower. Courtesy of Ferris State University. Photo by John Tomczyk

48.13 WATER PUMPS

The condenser water pump is the device that moves the water through the condenser and cooling tower circuit. It normally is located where it can take water from the cooling tower sump into the pump inlet, **Figure 48.39**. The water pump is the heart of the cooling tower system because it pumps the heat-laden water. The water must be pumped at the correct rate and delivered to the cooling tower at the correct pressure.

The condenser water pump is usually a **centrifugal pump**. It uses centrifugal action to impart velocity to the water that is converted to pressure. Pressure may be expressed in pounds per square inch (psig) or feet of head. One foot of head is equal to 0.433 psig—that is, a column of water 1 ft high will cause a pressure gauge to read 0.433 psig. Similarly, a pressure of 1 psig will be able to lift a column of

Figure 48.38 Calibrated holes on top of a cooling tower showing significant mineral deposits blocking water flow to the fill material.

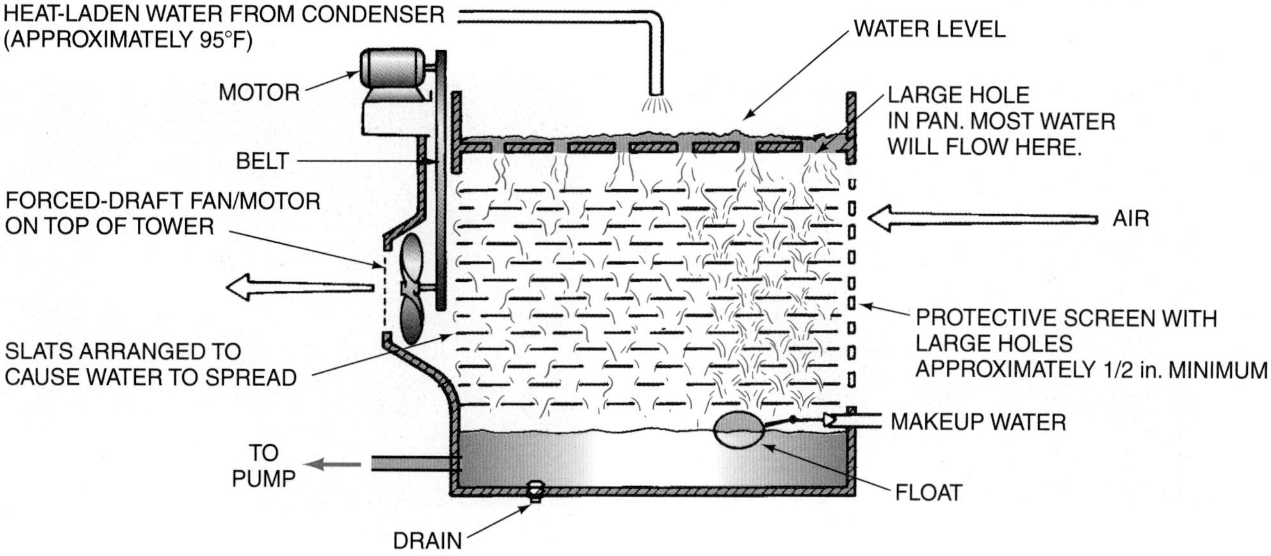

water 2.31 feet, or 27.7 in. So, a pump that operates with a 10 psig pressure rise across it will be able to lift a column of water 23.1 feet. Pump capacities are expressed in feet of head. Centrifugal pump action is discussed in Unit 33, "Hydronic Heat," and a study of that unit should be helpful.

Figure 48.39 The cooling tower pump and system.

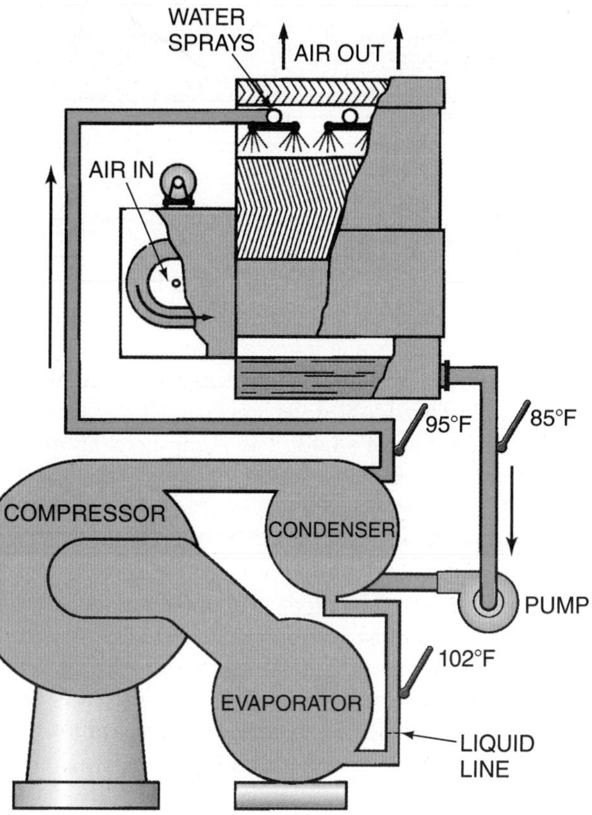

Several types of condenser water pumps are available for large systems. A *close-coupled pump* is located very close to the motor with the pump impeller actually mounted on the end of the motor shaft, **Figure 48.40**. These pumps are used for small applications. Note that all of the water enters the side of the pump housing. The pump shaft must have a seal to prevent water from leaking to the atmosphere or to prevent atmosphere from leaking in should this portion of the piping be in a vacuum.

The *base-mounted* pump is used in larger applications. It comes as an assembly with the pump mounted at the end of the pump shaft and a flexible coupling between the motor and the pump, **Figure 48.41.** This pump may have a single or double-sided impeller. The base is usually made of steel or cast iron and holds the pump and motor steady during operation. It is usually fastened to the floor using cement as a filler in the open portion of the pump base. This is called "grouting the pump in," the purpose of which is to create a more firm foundation that will last for many years. The pump and motor

Figure 48.40 A close-coupled pump assembly. Courtesy Bell and Gossett, ITT Industries

Figure 48.41 (A) A flexible coupling between the pump and motor. (B) Two parallel, base-mounted motor and pump assemblies. Note the shrouded flexible coupling. (C) Two base-mounted motor and pump assemblies. Note the shrouded flexible coupling. (D) Cutaway view of a centrifugal pump. (E) Cutaway view of a centrifugal pump showing the bearing assembly and impeller. (A) Courtesy of Amtrol, Inc. (B) – (E) Courtesy of Ferris State University. Photos by John Tomczyk

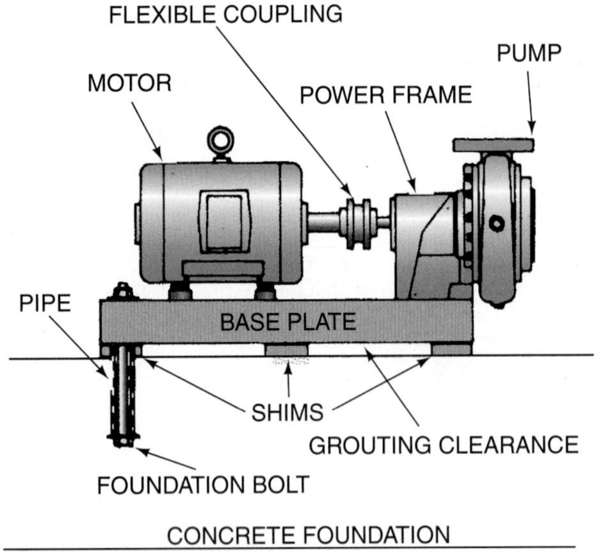

(A)

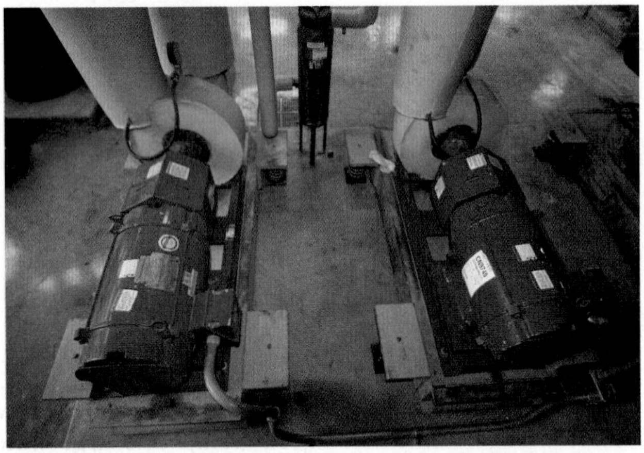

(B)

(C)

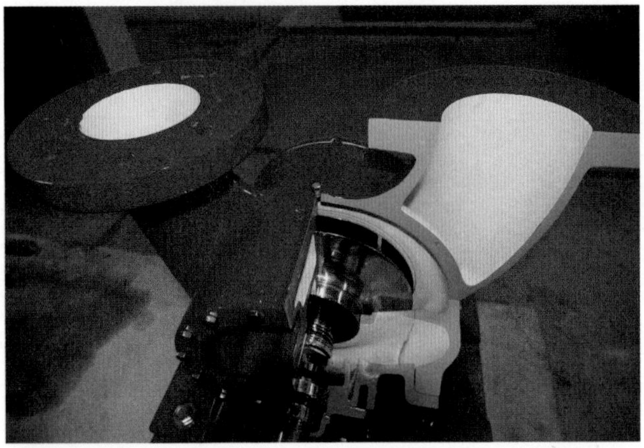

(D)

(E)

can still be removed from the base. One of the advantages of this pump is that the motor and pump shafts are aligned at the factory and so alignment is not required in the field.

As pumps become larger, the need for different designs becomes apparent. Some pumps have single-inlet impellers with all of the water entering the pump impeller on one side, **Figure 48.42**. The double-inlet impeller allows water to flow into both sides, **Figure 48.43**. The pump housing must be designed differently to allow water to enter both sides of the impeller. There are two styles of *double-inlet impeller pumps*: One type of pump is disassembled from the ends of the pump, as with the base-mounted pump; the other type, a split-case pump, is disassembled by removing the top of the pump, **Figure 48.44**.

These pumps must also have a shaft seal, which, as the name implies, is a seal on the shaft. The shaft seal helps reduce, or eliminate altogether, the loss of water from the pump. One type of seal is the *stuffing box* and the second is a *mechanical seal*. The stuffing-box seal has a packing gland

Figure 48.42 (A) Cutaway, side-view of a pump showing direction of fluid flow. (B) A single-inlet impeller.

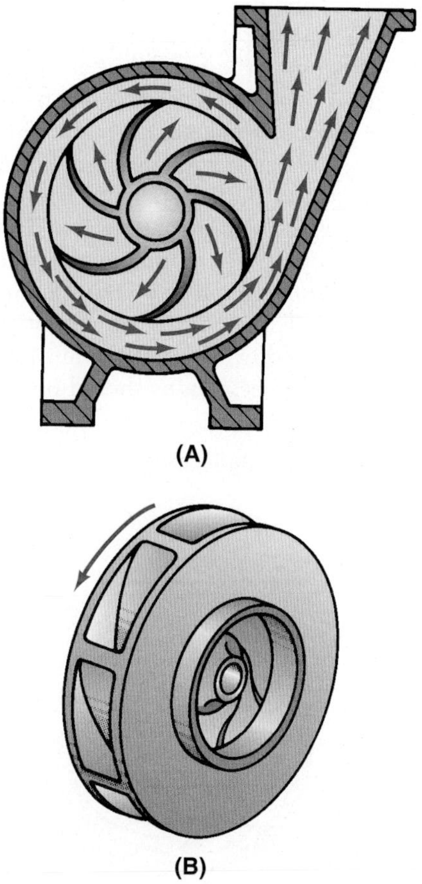

Figure 48.43 A double-inlet impeller pump. Courtesy Bell and Gossett, ITT Industries

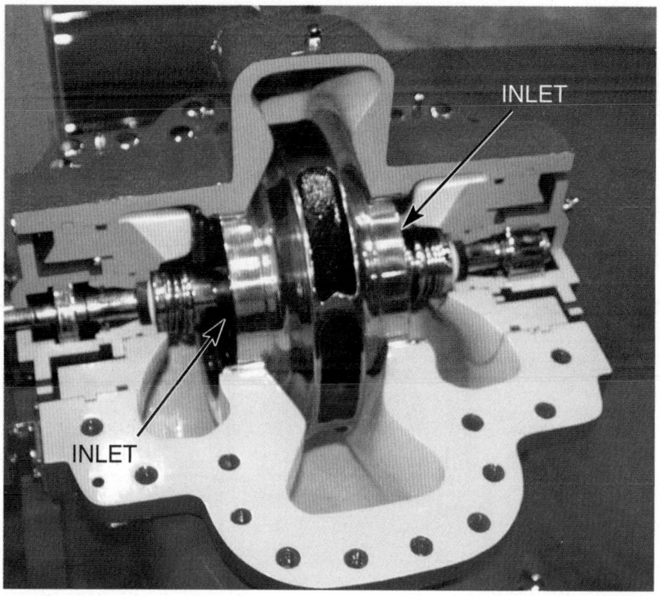

Figure 48.44 A horizontal split-case pump. Courtesy Bell and Gossett, ITT Industries

that must be hand tightened with a wrench. When the nut on the seal is tightened, it squeezes a packing around the shaft to minimize leakage, **Figure 48.45**. Stuffing-box types of seals are typically used for pump pressures of up to about 150 psig and require some regular maintenance. The mechanical seal may be used for higher pressures, up to about 300 psig, and usually has a carbon ring mounted on the end of a bellows that turns with the shaft. The bellows is sealed to the shaft with an O-ring. The carbon ring rubs against a stationary ceramic ring while the shaft turns and seals water in the pump housing, **Figure 48.46**. Both of these seals are lubricated by the circulating water. In split-case pumps, the seals may have a special piping system that injects pump discharge water into the seal, **Figure 48.47**.

Figure 48.45 The stuffing-box shaft seal. Courtesy Amtrol, Inc.

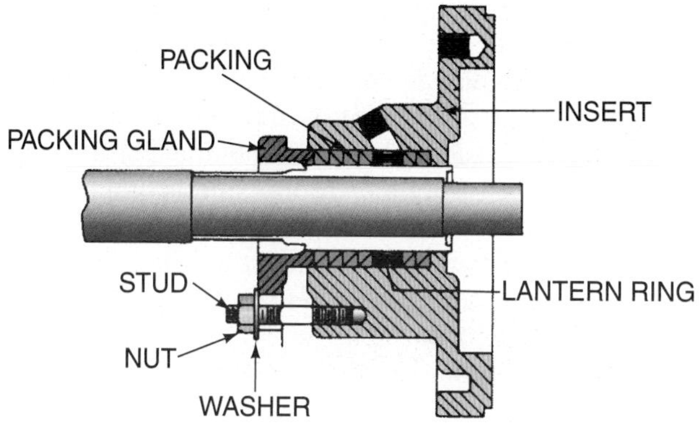

Figure 48.46 The mechanical shaft seal. Courtesy Amtrol, Inc.

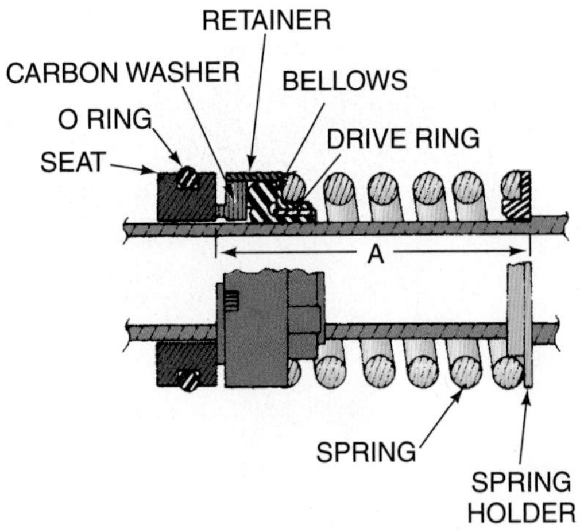

Figure 48.47 Lubricating a shaft seal. Courtesy Bell and Gossett, ITT Industries

Most cooling tower pumps are manufactured of cast iron. They are heavy and may be used for pressures up to 300 psig with the correct piping flanges and arrangements. Cast iron has the capacity to last many years without deterioration from rust and corrosion. The impeller in the typical centrifugal pump is made of brass or bronze and is often mounted on a stainless-steel shaft. This can be important when the time comes to remove the impeller from the shaft because the stainless-steel shaft makes it easier to remove the impeller after many years of operation.

Impellers are furnished from the factory at a specified diameter for each pump to furnish a specific water flow against a specified pressure. It is often the case that the pump circulates too much water flow, to the point that a smaller impeller is needed. Impellers are often trimmed in the field to meet the actual water flow and pressure requirements. Always refer to, or contact, the manufacturer for information regarding any modifications that might be needed for a specific pump.

The location of the condenser water pump is important because these pumps, which are normally centrifugal pumps, must be supplied water in a manner that keeps the eye of the impeller under water during start-up, **Figure 48.48**. Otherwise, the pump will not move any water. They *do not* have the capacity to pump air to pull water into the impeller. Pumps can be manufactured in a vertical configuration with the motor on top. These pumps are mounted on top of the sump and protrude down into the sump to pick up water, **Figure 48.49**. Otherwise, pumps should be located at a level

Figure 48.48 The eye of the impeller.

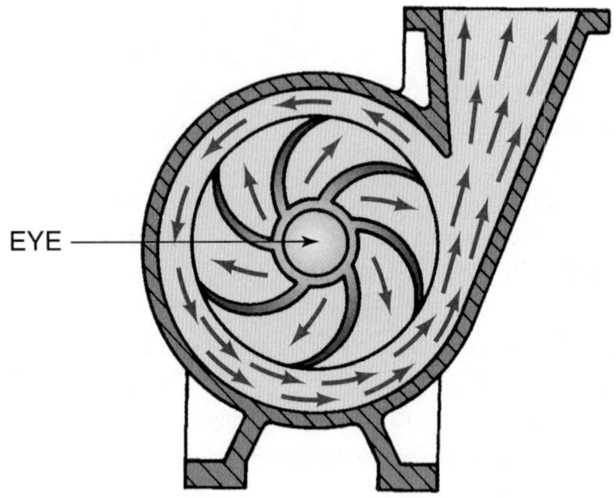

Figure 48.49 A vertical pump. Image used with permission from SPX Cooling Technologies, Inc. for educational and informational purposes only. © 2015 SPX Cooling Technologies, Inc. All Rights Reserved

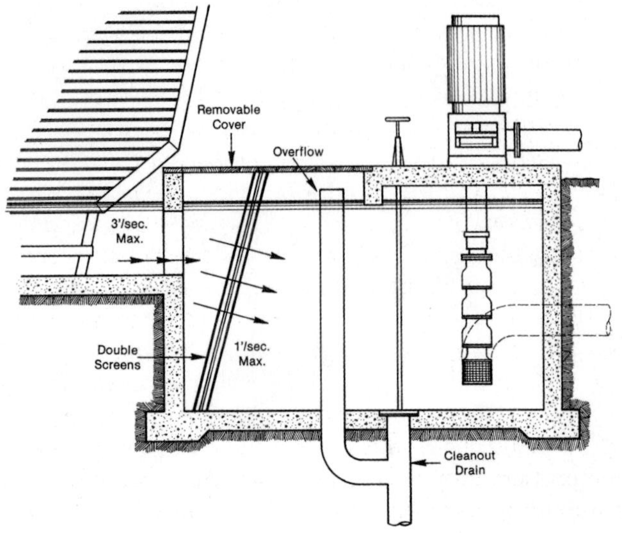

Figure 48.50 The cooling tower sump is located below the pump.

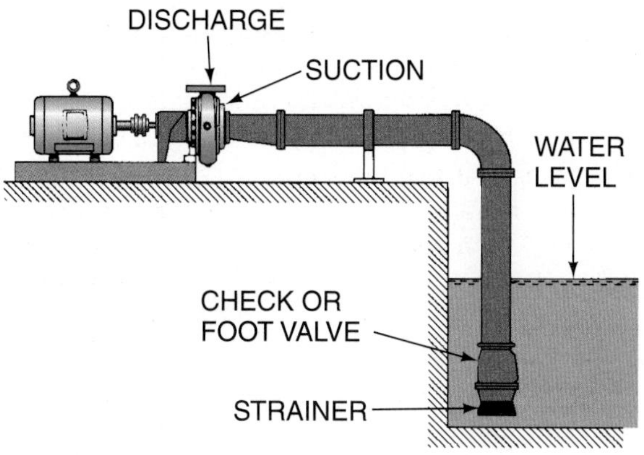

Figure 48.51 Insert used to help prevent vortexing.

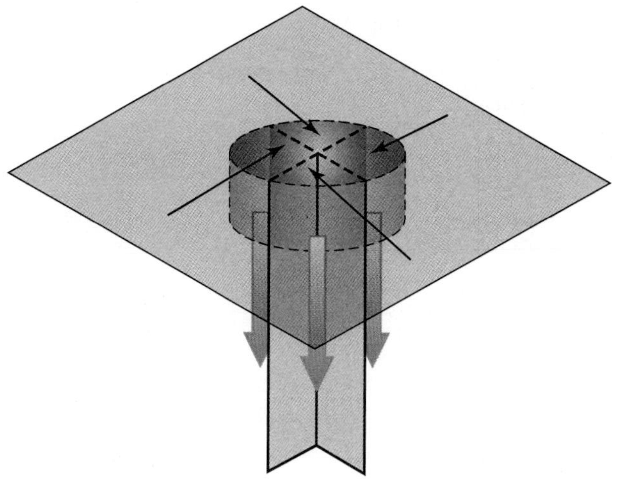

below the tower water, and there should be nothing in the pump inlet piping that will impede flow. Allowing a free flow of water to the pump inlet is always good practice, but several things can interfere with water flow, such as vortexing in the tower, fine mesh strainers, and the cooling tower bypass.

There are times when the cooling tower sump is below the level of the pump. When this is the case, special care should be taken during start-up because only air will be in the pump casing. The pump should never be started with only air inside because then the pump seal will have no water for lubrication. When the sump is below the pump, a foot or check valve must be located in the line to prevent water in the pump from flowing back into the sump during shutdown, **Figure 48.50.** Before the pump is started, water must be added to the pump inlet side until the pump casing is completely full, which can be done by connecting a water hose to the pump inlet side and filling it until water escapes the bleed port on top. If the pump inlet does not fill up, the foot or check valve is leaking and must be repaired.

Vortexing is a whirlpool action. A vortex interferes with the functioning of the tower because it introduces air to the pump inlet. The pump is designed to pump water; air will cause problems with the pump and with the condenser at the chiller. Poor piping practice is a major cause of vortexing. Vortexing can often be eliminated by a device placed in the sump outlet that breaks up the vortex. A cross-type of configuration with a plate on top can create four outlets from the sump, **Figure 48.51.** The plate causes the water to be pulled from a farther distance and from the side, **Figure 48.52.** Vortexing in multistory buildings that have the condenser water pump well below the tower has been addressed using this method.

The pump must also have an inlet strainer to prevent trash or other foreign matter from entering the pumping circuit. Typically, a coarse strainer screen is placed in the cooling tower exit to the sump, and often a fine-mesh screen is located before the pump inlet. If this screen is too fine, however, problems from pressure drop will occur as the screen becomes clogged. Pressure drop causes the pump inlet to operate in a vacuum,

Figure 48.52 Vortex protection for a cooling tower.

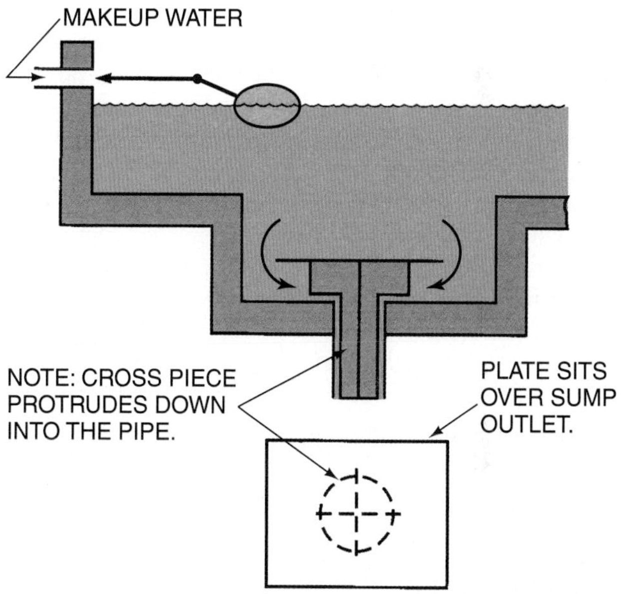

and the pump may pull in air if the pump seals leak. Pressure drop can also cause pump cavitation (partial vacuum)—water will turn to a vapor if the cooling tower water is warm enough and the pressure is low enough. Noise at the pump is evidence of cavitation or air at the pump inlet. It will often sound like it is pumping small rocks. If a fine screen must be used, it may often be better to install it at the pump outlet, **Figure 48.53.** It does less harm to throttle the pump discharge with a pressure drop than the pump inlet.

The tower bypass valve helps to maintain the correct tower water temperature at the condenser during start-up and during low ambient operation. The tower bypass circuit allows water from the pump outlet to be recirculated to the pump inlet so that the cold cooling tower water will not reach the condenser, **Figure 48.54.** Condenser water that is too cold in

Figure 48.53 The best application for a fine-screen filter is in the pump outlet.

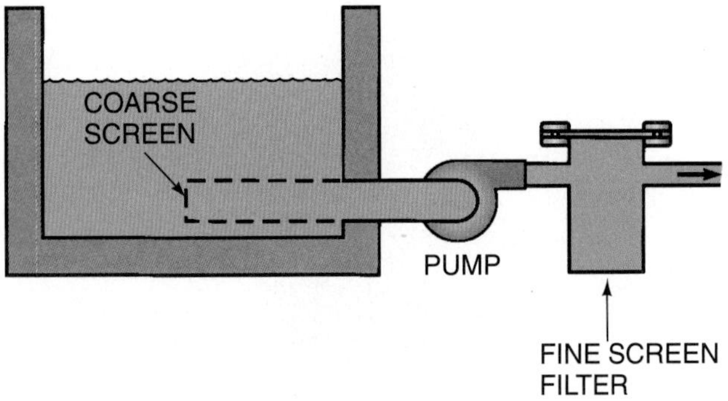

the compression cycle systems will reduce the head pressure to the point that the condenser will be so efficient that too much refrigerant will be held in the condenser and starve the evaporator. This will also cause the oil to migrate in some machines. Cooling tower water that is too cold for an absorption machine will often cause the salt solution to crystallize.

Two types of three-way bypass valves may be used for tower bypass: the mixing valve and the diverting valve. The mixing valve has two inlets and one outlet and needs to be located in the pump suction line between the tower and the pump inlet, **Figure 48.54(A)**. **Figure 48.54(B)** shows a direct digital controlled (DDC) butterfly valve bypass. The valve will bypass some of the warm water coming from the water-cooled condenser and mix it with cooling tower outlet water. This arrangement is used in colder climates when the building still needs cooling in its interior but the cooling tower outlet water is too cold and may shock the tubes of the water-cooled condenser, causing damage. Mixing is controlled by sensors in the condenser's inlet water that relay the message to the DDC butterfly valve to protect the condenser's tube bundles from a too cold water stream.

The diverting valve has one inlet and two outlets; it can be located in the pump discharge and piped to the cooling tower or the pump inlet, **Figure 48.55**. These valves do not come in large sizes; they are normally a maximum of 4 in., so other arrangements may have to be made if a larger size is needed. A straight-through valve can be used in the configuration shown in **Figure 48.56**. Both the diverting valve and the straight-through valve allow some cooling tower water from the basin to be introduced to the pump inlet but may cause problems at start-up until the tower water becomes the correct temperature.

All pumps must have bearings to support the turning shaft while under load. Sleeve bearings are used for small pumps, and ball or roller bearings are used for larger pumps. These bearings must be lubricated on a regular basis for satisfactory service. Notice that the bearings on the split-case pump are outside the pump, **Figure 48.57**. The pump must be fastened to the motor shaft in most of the pumps mentioned above in such a manner that they are within alignment tolerances. A flexible coupling may be installed between

Figure 48.54 (A) A mixing valve application. (B) A direct digital controlled (DDC) butterfly valve bypass. The valve will bypass some of the warm water coming from the water-cooled condenser's outlet and mix it with cooling tower outlet water. (B) Courtesy Ferris State University. Photo by John Tomczyk

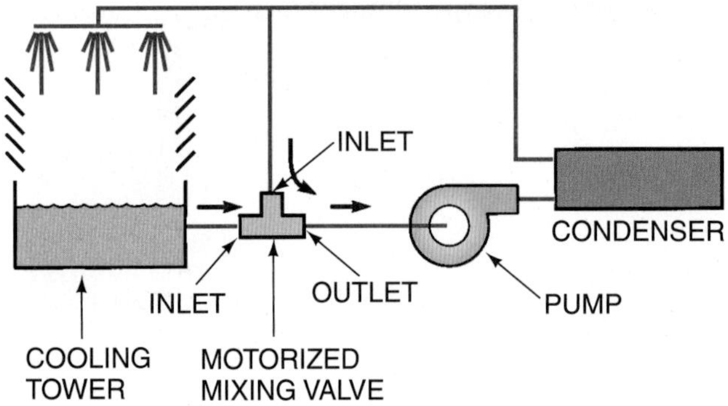

(A)

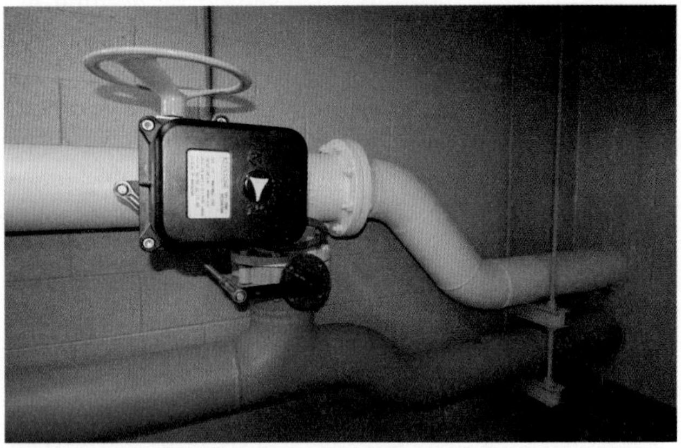

(B)

Figure 48.55 A diverting valve application.

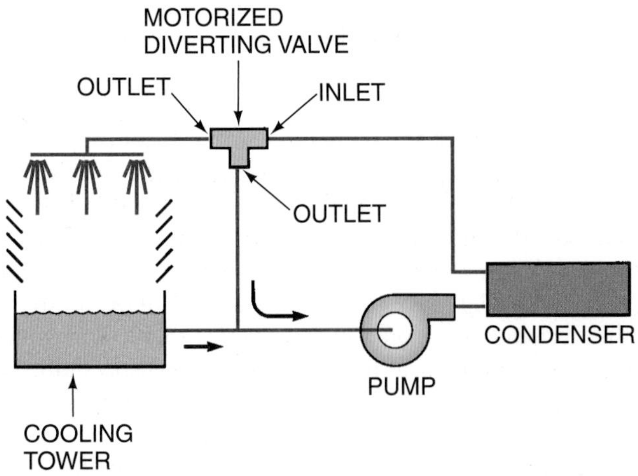

the pump and motor shaft to correct small misalignments, **Figure 48.58**.

Because the coupling can handle only slight misalignments, pump and motor shafts that are more severely out of

Figure 48.56 A straight-through valve for tower bypass.

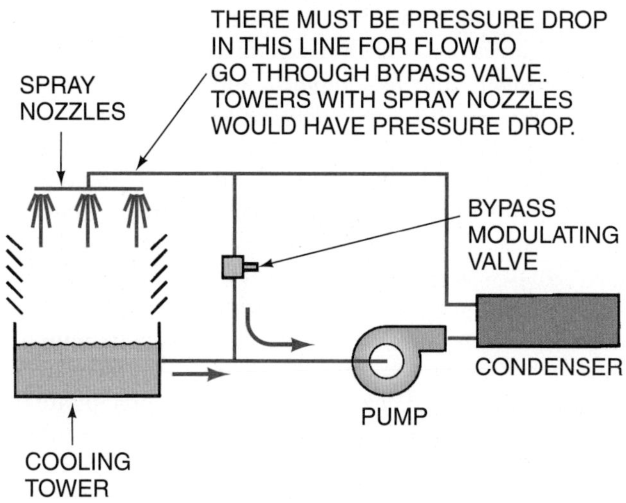

Figure 48.57 The bearings on a horizontal split-case pump. Courtesy Bell and Gossett, ITT Industries.

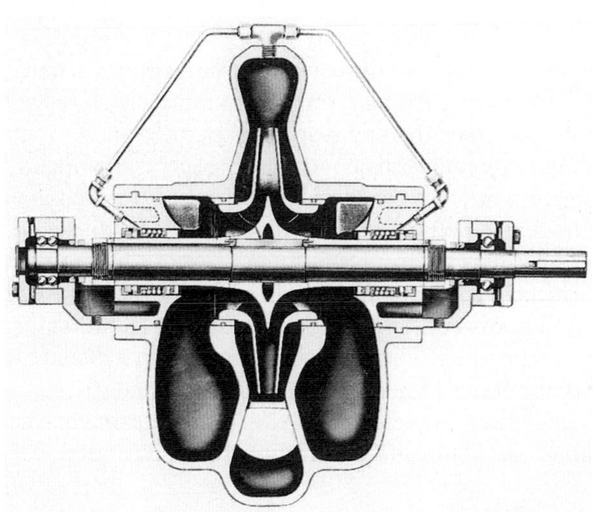

alignment must be properly aligned. Two planes of alignment must be considered—angular and parallel. Angular alignment ensures that both shafts are at the same angle to each other, **Figure 48.59**. Parallel alignment ensures that the shafts are aligned end to end, **Figure 48.60**. Correct alignment is accomplished with a dial indicator that reads to 1/1000 of an inch. It is mounted on the shafts and the shafts are rotated through the full rotation, **Figure 48.61**. Shims (thin sheets of steel) are placed under the pump and motor as needed until both angular and parallel alignment (sometimes called radial and axial alignment) are achieved to within the manufacturer's specifications. After alignment is attained, the pump and motor are tightened down and alignment is checked once again. Then the bases of both are drilled and tapered dowel pins are driven in, **Figure 48.62**. When alignment is done correctly, the motor and pump should provide years of good service.

Figure 48.58 (A) A flexible coupling. (B) A flexible coupling connecting a centrifugal pump and the motor driving the pump. (A) Courtesy of TB Woods & Sons. (B) Courtesy of Ferris State University. Photo by John Tomczyk

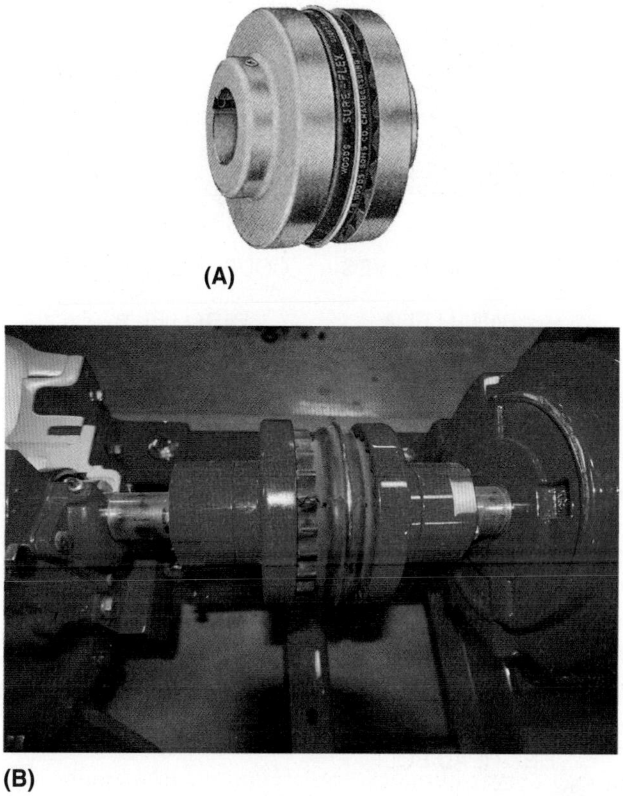

Figure 48.59 Angular misalignment for a coupling. Courtesy Amtrol, Inc.

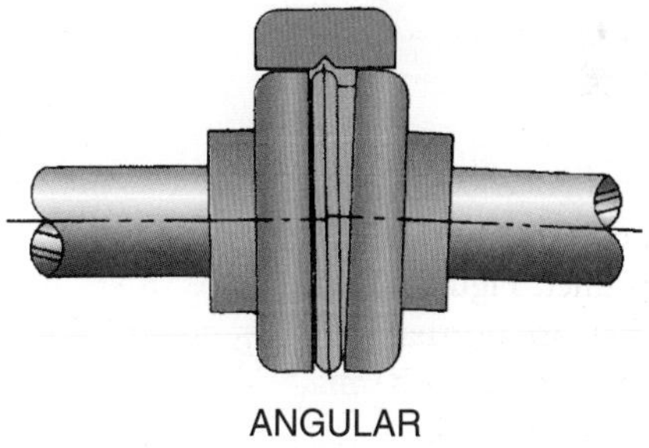

Figure 48.60 Parallel misalignment for a coupling. Courtesy Amtrol, Inc.

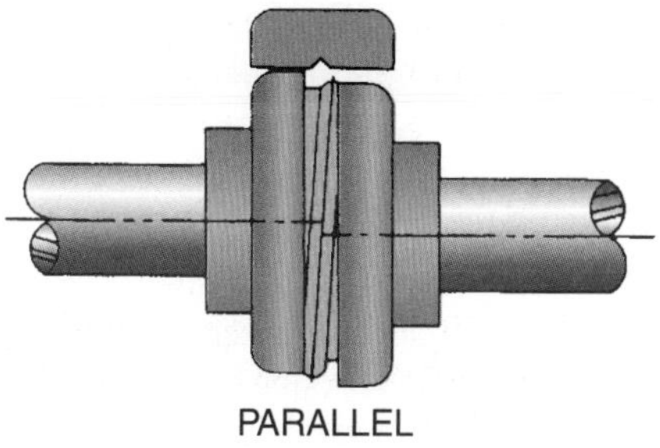

Figure 48.61 Dial indicators used to align the coupling and shafts. Courtesy Amtrol, Inc.

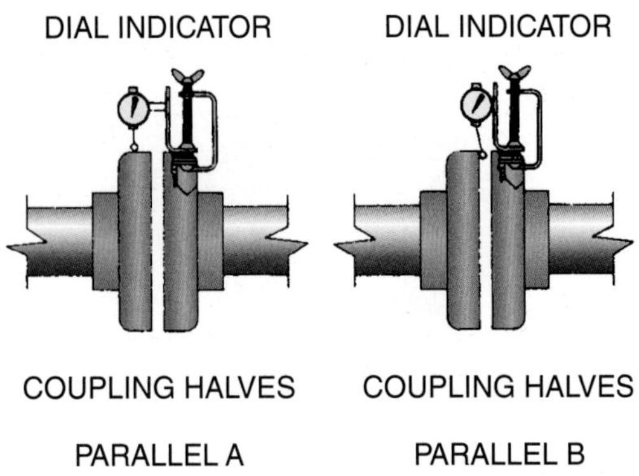

Figure 48.62 The pump and motor are fastened to the base using tapered dowel pins.

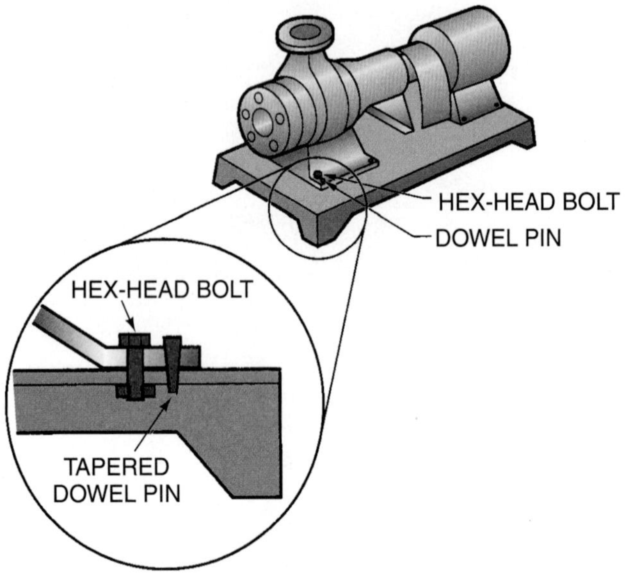

48.14 CHEMICAL-FREE TREATMENT OF COOLING TOWER WATER

A proprietary and patented technology has been developed by an engineering and research team dedicated to a chemical-free solution for water treatment. This proven chemical-free water purification technology eliminates scale, inhibits bacterial growth, and inhibits corrosion.

SCALE PREVENTION

Pure water is a rare commodity. Water as we know it contains many dissolved minerals. In a cooling tower, only the water evaporates. It exits the cooling tower as water vapor, but leaves the dissolved minerals behind to concentrate in the cooling tower's water system. The concentrations of these dissolved minerals gradually increase until a process called precipitation occurs. This happens when the dissolved minerals, such as calcium carbonate (limestone), reach a certain concentration, become solid, and usually cling to equipment and piping surfaces in the cooling tower. HVACR personnel refer to these solids as scale.

There are tiny suspended particles in large quantities in all city water or well water that is used for cooling tower or boiler makeup water. Once in the cooling tower water system, these suspended particles neither sink nor float because of their small size and are transported by the flowing water. But the particles will concentrate during the evaporation process in the cooling tower and become attracted to equipment surfaces when the concentration is so great that the water can no longer hold them. They are then forced to precipitate onto surfaces as solids, which eventually become hard, resulting in equipment-damaging scale.

When the cooling tower water holding these small, suspended particles passes through a water treatment module and is activated by a high-frequency electrical pulse field, the natural electrical static charge on the surface of the particles is removed. The suspended particles act as "seeds" for precipitation of dissolved minerals and now become the "preferred site" for precipitation instead of the equipment surfaces. Thus, hard scale is prevented from forming on the equipment and instead adheres to and coats the tiny suspended particles in the water. As more and more minerals bond to the suspended particles, they become heavier and can no longer suspend themselves in the water stream. Eventually they make their way to the cooling tower's basin as a harmless fluffy powder or tiny coated particles. The powder or coated particles can easily be removed from the cooling tower's basin by manual means, filtration, or centrifugal separation. The quantity of powder in a cooling tower is typically about 15% of the normal blown-in dirt.

The pulsed waveform of the water treatment module is shown in **Figure 48.63**. **Figure 48.64** illustrates calcium

Figure 48.63 The pulsed waveform. Courtesy Clearwater Systems Corp., Essex, CT

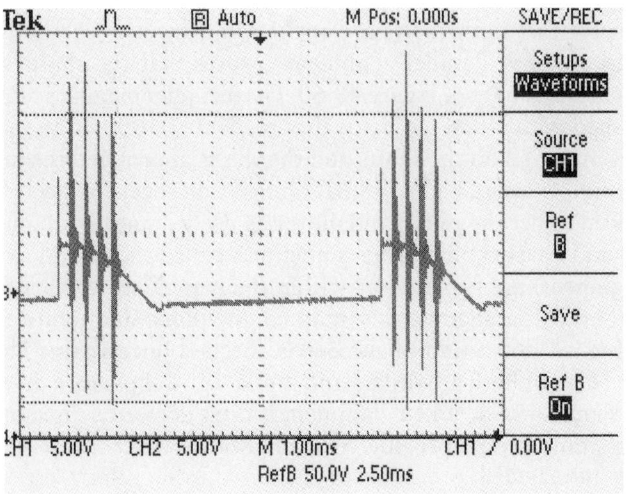

Figure 48.69 A typical evaporative cooling tower installation incorporating a centrifugal separator and a high-frequency electrical pulse field signal generator and water treatment module. Courtesy of Clearwater Systems Corp., Essex, CT.

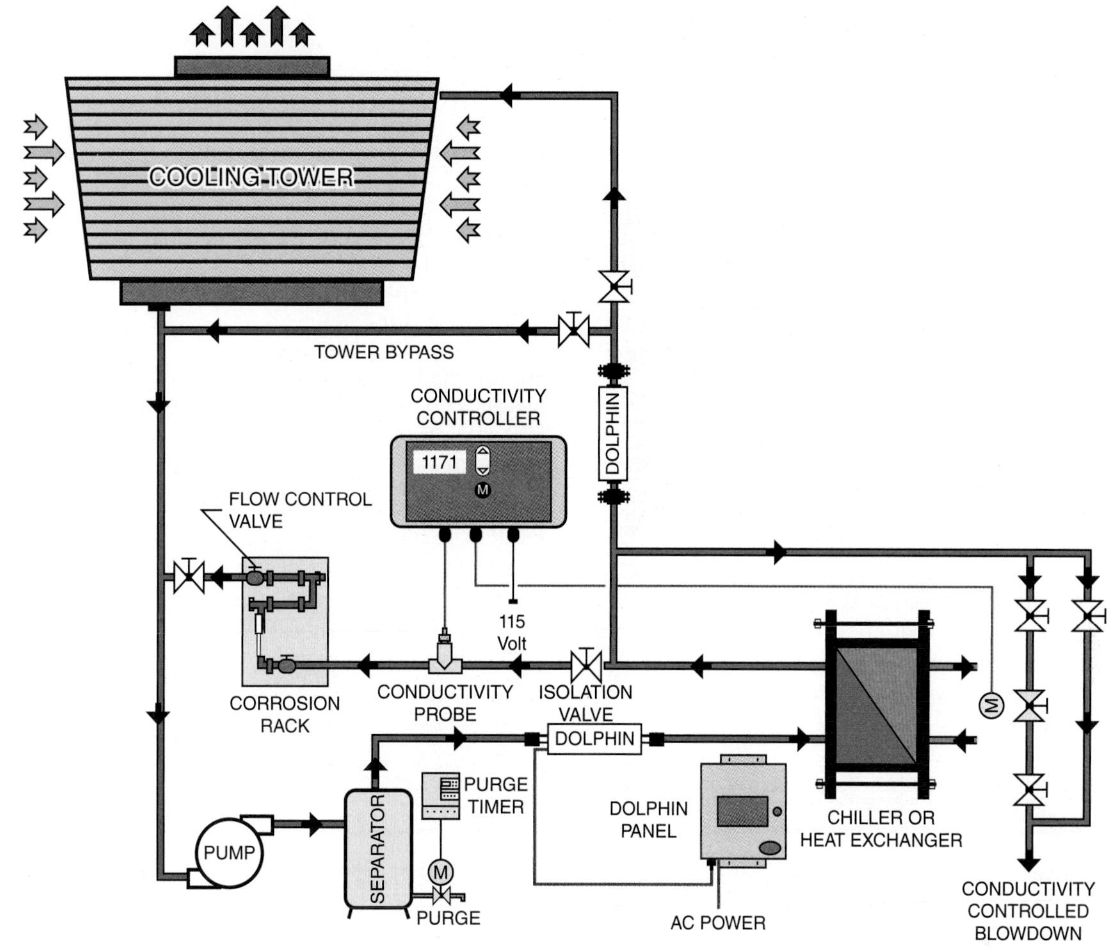

Figure 48.70 Mineral-coated bacteria (encapsulation). Courtesy of Clearwater Systems Corp., Essex, CT.

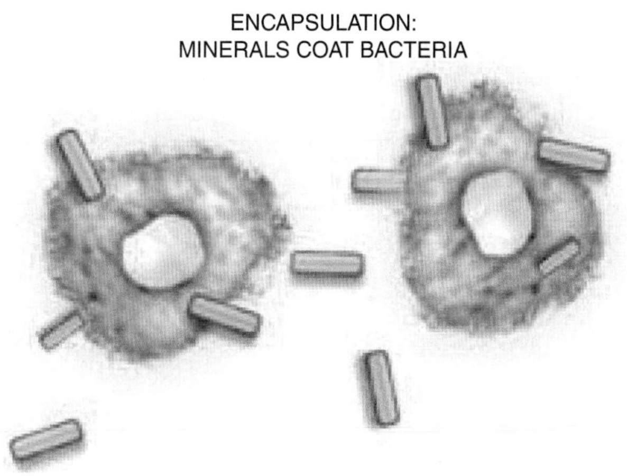

are repelled by the suspended particles in the water because nearly all tiny particles have similar negative static electrical charges on their surfaces. However, after particles are activated by the high-frequency electrical pulse field from the signal generator at the water treatment module, the natural electrical static charge on the particle surfaces is removed. The bacteria are no longer repulsed and are therefore attracted to the powder and become entrapped in the powder particles, **Figure 48.70**. The powder, in effect, sweeps the water clean of planktonic bacteria and renders them incapable of reproducing. This process is referred to as encapsulation.

The high-frequency, pulsing action of the signal generator also damages the membrane of the planktonic bacteria by creating small "pores" in their outer membrane. This weakens the bacteria and inhibits their reproductive capabilities, a process referred to as electroporation. The life span of microbial life is 24 to 48 hours. Any microbes not captured in the forming powder are "zapped" by the secondary pulse of the signal generator, forcing them to spend their life span repairing cell wall damage rather than reproducing, **Figure 48.71**.

All of the living organisms in a cooling tower system depend on one another for their food supply. Thus, when the nutrients from the planktonic bacteria are diluted by both encapsulation and electroporation, the biofilm cannot

Figure 48.71 The high-frequency, pulsing action of the signal generator damages the membrane of the planktonic bacteria by creating small "pores" in their outer membrane. The condition weakens the bacteria and inhibits their capabilities to reproduce. This process is referred to as electroporation. Courtesy Clearwater Systems Corp., Essex, CT

ELECTROPORATION:
PULSE DAMAGES BACTERIA
MEMBRANES

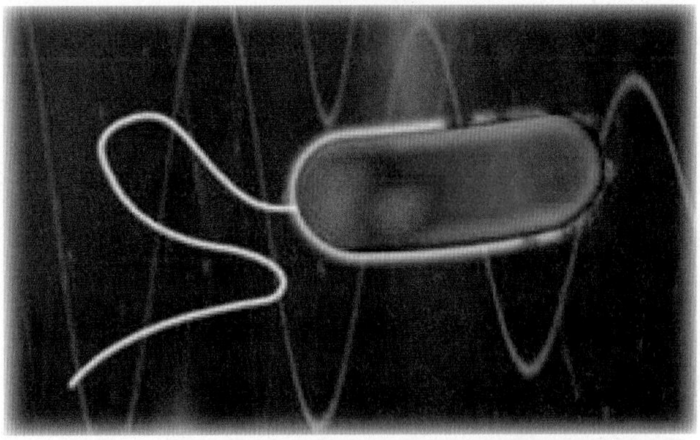

be sustained and will disintegrate. It will never be created in the first place if a high-frequency electrical pulse field is installed in the cooling tower system to instigate the encapsulation and electroporation processes. In summary, the combined effect of encapsulation and electroporation result in exceptionally low total bacterial counts (TBC) in cooling tower water.

CORROSION CONTROL

Most corrosion in cooling tower systems or boilers comes from

- chemical additives,
- softened water,
- biofilm, and
- mineral scale.

So, by removing chemicals, avoiding the use of softened water, and using a chemical-free water treatment module and signal generator in cooling towers and boilers, corrosion will be eliminated. The calcium carbonate that coats the particles suspended in water is in a state of saturation while it precipitates and will act as a powerful cathodic corrosion inhibitor. It will greatly slow the corrosion process by blocking the reception of electrons that are thrown off by the corrosion process. With no place for the electrons to go, the corrosion process is physically and very effectively controlled.

REVIEW QUESTIONS

1. If a cooling tower rejects 100 tons of heat, approximately how much of the rejected heat is attributed to the system's compressor?
   A. 5 tons C. 15 tons
   B. 10 tons D. 20 tons
2. If the outside ambient temperature has a dry-bulb temperature of 95°F and a wet-bulb temperature of 75°F, what is the cooling tower's approach temperature if the water is leaving the tower at 82°F?
   A. 7°F C. 20°F
   B. 13°F D. 27°F
3. What is the load on a cooling tower if 60 gpm of 95°F water enters the cooling tower and 60 gpm of 86°F water leaves the tower?
   A. 15 tons C. 30 tons
   B. 22.5 tons D. 37.5 tons
4. An adiabatic process is one that results in a change in
   A. heat content but no change in temperature or pressure.
   B. temperature, heat content, and pressure.
   C. heat content and temperature.
   D. temperature and/or pressure but no change in heat content.
5. Both spray pond and natural-draft cooling towers rely on ______ to move air through them
   A. prevailing winds
   B. mechanical blowers and fans
   C. centrifugal pumps
   D. solar-powered fans
6. Which of the following is true regarding forced-draft and induced-draft cooling towers?
   A. Forced-draft towers pull air through them and induced-draft towers push air through them.
   B. Both types of cooling towers rely on prevailing winds to move air through them.
   C. Both types of cooling towers use mechanical fans and blower to move air through them.
   D. Both types of cooling towers have air pulled through them.
7. Why is it sometimes required for open cooling towers to be equipped with "tower wetting systems"?
   A. To prevent algae and other biological contamination growth.
   B. To reduce the potential for fire in the cooling tower.
   C. To prevent the wood and plastic in the tower from becoming brittle.
   D. To reduce the formation of scale on the exterior surfaces of the tower.

8. The cooling tower component referred to as *fill* can be made of all of the following except
   A. fiberglass. C. steel.
   B. plastic. D. wood.
9. Which of the following is true regarding cooling tower fill?
   A. It should be kept wet at all times.
   B. It should be kept dry at all times.
   C. Water should be directed to only one side of the fill.
   D. Fill should be installed in a perfectly vertical position.
10. Galvanized steel does not readily rust because the steel is coated with
   A. copper. C. zinc.
   B. aluminum. D. brass.
11. In the counterflow cooling tower configuration
   A. both the water and air flow downward in parallel paths.
   B. both the water and air flow upward in parallel paths.
   C. the water flows up and the air flows down.
   D. the water flows down and the air flows up.
12. In the crossflow cooling tower configuration
   A. the water flows down and the air flows at a 90-degree angle to the water.
   B. the air flows down and the water flows at a 90-degree angle to the air.
   C. the water flows up and the air flows down.
   D. the water flows down and the air flows up.
13. When referring to cooling tower construction, the acronym FRP stands for
   A. fiber-reinforced plastic.
   B. fiber-reinforced polymer.
   C. flexible-reinforced plastic.
   D. flexible-reinforced polymer.
14. In a gearbox or transmission drive assembly on a cooling tower the motor shaft and the fan shaft are at an angle of ______ degrees with each other.
   A. 30 C. 60
   B. 45 D. 90
15. In a cooling tower blower assembly, why are beveled gears preferred over non-beveled gears?
   A. Beveled gears have less surface area between the mating surfaces.
   B. Beveled gears have more surface area between the mating surfaces.
   C. Non-beveled gears tend to last longer than beveled gears.
   D. Beveled gars tend to wear out faster than non-beveled gears.
16. The system component that eliminates the constant starting and stopping of a cooling tower's fan is the
   A. variable frequency drive.
   B. internal transmission drive.
   C. gearbox drive.
   D. external transmission drive.
17. The strainer in a cooling tower is located
   A. in the pipe that carries water from the condenser to the cooling tower.
   B. at the point where the water leaves the cooling tower.
   C. at the point where the water enters the cooling tower.
   D. in the cooling tower's makeup water line.
18. Which of the following components can be found in a cooling tower's sump?
   A. A water heater and a strainer
   B. Only a water heater
   C. Only a strainer
   D. Neither a water heater nor a strainer
19. A makeup water is required for cooling towers because of
   A. evaporation only.
   B. drift only.
   C. both evaporation and drift.
   D. neither evaporation nor drift.
20. Cooling towers require blowdown in order to
   A. prevent the cooling tower's sump from overflowing.
   B. prevent the water in the cooling tower from freezing.
   C. reduce the concentration of minerals and other contaminants in the tower water.
   D. increase the concentration of minerals and other contaminants in the tower water.
21. Excessive buildup of minerals in the cooling tower water can result in
   A. reduced heat transfer between the refrigerant and the condenser water.
   B. increased heat transfer between the refrigerant and the condenser water.
   C. reduced heat transfer between the refrigerant and the chilled water.
   D. increased heat transfer between the refrigerant and the chilled water.
22. Using a hand-operated valve to control cooling tower blowdown is
   A. desirable because blowdown will always occur at the same time each day.
   B. desirable because the length of the blowdown cycle will always be the same.
   C. desirable because the water quality will be maintained at a constant level.
   D. not a desirable or recommended option.
23. If a cooling tower is equipped with two or more cells, what can be done to ensure that the water flow to each cell is the same?
   A. Use separate water feed lines for each cell
   B. Stagger the blowdown cycles for each cell
   C. Install a balancing valve on each cell
   D. Control tower makeup water from a central location

24. Insufficient water flow through a cooling tower can result in
    A. low system operating pressures.
    B. high system operating pressures.
    C. high head pressure and low suction pressure.
    D. low head pressure and high suction pressure.
25. What system component is used to move water through the condenser and cooling tower?
    A. Centrifugal fan
    B. Centrifugal pump
    C. Perpetual fan
    D. Perpetual pump
26. The term "feet of head" is used to express
    A. fan capacity.
    B. pump capacity.
    C. compressor capacity.
    D. system capacity.
27. On a close-coupled pump, the impeller is
    A. mounted directly on the motor shaft.
    B. connected to the motor shaft using a 90-degree linkage.
    C. turned by means of a belt-driven assembly.
    D. rotated at a speed that is different from that of the motor.
28. On a base-mounted pump,
    A. the pump is connected directly to the shaft of the motor.
    B. the pump turns at a speed that is different from that of the motor.
    C. a flexible coupling or connector is positioned between the pump and the motor.
    D. the pump and motor must be aligned in the field during installation.
29. Two types of seals found on pumps are the
    A. mechanical seal and the centrifugal seal.
    B. packing gland and the centrifugal seal.
    C. mechanical seal and the pressure seal.
    D. stuffing box and the mechanical seal.
30. How is a packing gland tightened?
    A. By hand
    B. With a wrench
    C. With a Phillips screwdriver
    D. With a Torx screwdriver
31. Which type of pump seal is typically used for pressures up to a maximum of 300 psig?
    A. Mechanical seal
    B. Pressure seal
    C. Stuffing box
    D. Centrifugal seal
32. Which type of pump seal is typically used for pressures up to a maximum of 150 psig?
    A. Packing gland
    B. Stuffing box
    C. Mechanical seal
    D. Pressure seal
33. How are seals on circulator pumps typically lubricated?
    A. Water
    B. Oil
    C. Refrigerant
    D. Brine
34. Which of the following best describes a centrifugal pump found on cooling towers?
    A. Copper impeller mounted on a stainless-steel shaft
    B. Stainless-steel impeller mounted on a stainless-steel shaft
    C. Bronze impeller mounted on a brass shaft
    D. Brass or bronze impeller mounted on a stainless-steel shaft
35. In order for centrifugal pumps to operate properly
    A. the eye of the impeller must be above the water level.
    B. the entire impeller must be above the water line.
    C. the eye of the impeller must be below the water level.
    D. the entire impeller must be below the water line.
36. Regarding cooling towers, a foot valve is a type of
    A. check valve.
    B. butterfly valve.
    C. globe valve.
    D. ball valve.
37. Vortexing in a tower is best described as a(n)
    A. whirlpool action.
    B. mixture of water and air.
    C. overconcentration of cooling tower chemicals.
    D. dilution of cooling tower chemicals.
38. When pump cavitation occurs,
    A. a partial vacuum is created and the pump's noise level increases.
    B. a partial vacuum is created and the pump's noise level decreases.
    C. the pump's vacuum is lost and the pump's noise level increases.
    D. the pump's vacuum is lost and the pump's noise level decreases.
39. A diverting valve is manufactured with
    A. two inlet ports and one outlet port.
    B. one inlet port and two outlet ports.
    C. two inlet ports and two outlet ports.
    D. one inlet port and one outlet port.
40. A mixing valve is manufactured with
    A. one inlet port and one outlet port.
    B. one inlet port and two outlet ports.
    C. two inlet ports and one outlet port.
    D. two inlet ports and two outlet ports.
41. Corrosion in a cooling tower comes from all of the following except
    A. slime.
    B. mineral scale.
    C. hard water.
    D. chemical additives.
42. Encapsulation and electroporation are two methods that can be used to control
    A. evaporation of tower water.
    B. drift in a cooling tower.
    C. airflow in a cooling tower.
    D. bacterial growth in a cooling tower.
43. Explain how a centrifugal particle separator is used to remove particulates from cooling tower water.
44. List three methods that can be used to control the level of water in the sump of a cooling tower.
45. Explain what is meant by the term "grouting the pump in."
46. Explain what is known as "scale" as it applies to cooling towers.

UNIT 49

OPERATION, MAINTENANCE, AND TROUBLESHOOTING OF CHILLED-WATER AIR-CONDITIONING SYSTEMS

OBJECTIVES

After studying this unit, you should be able to

- discuss the general start-up procedures for a chilled-water system.
- describe specific chiller start-up procedures for chillers having a scroll, reciprocating, rotary screw, or centrifugal compressor.
- describe operating and monitoring procedures for scroll and reciprocating chilled-water systems.
- discuss preventive and other electrical maintenance and service that should be performed at least annually on chillers.
- list maintenance procedures that should be performed periodically on water-cooled, chilled-water systems.
- describe the operation of large centrifugal chillers.
- discuss the differences between low- and high-pressure chillers.
- describe start-up procedures for an absorption chilled-water system.
- list preventive and routine maintenance procedures for an absorption chiller.
- state preventive and periodic maintenance procedures for the purge system on an absorption system.
- discuss basic chiller troubleshooting and safety.
- describe the various valves used on chilled-water system piping.

SAFETY CHECKLIST

- ✓ Wear gloves and use caution when working around hot steam pipes and other heated components.
- ✓ Use caution when working around high-pressure systems. Do not attempt to tighten or loosen fittings or connections when a system is pressurized.
- ✓ Follow recommended procedures when using nitrogen.
- ✓ Follow all recommended safety procedures when working around electrical circuits.
- ✓ When working on an electrical system, lock the disconnect box or panel and tag it with your name when power is turned off. There should be only one key for this lock, and it should be in the possession of the technician.
- ✓ Never start a motor with the door to the starter components open.
- ✓ Check system operating pressures regularly as indicated by the manufacturer of the system.

The operation of chillers involves starting, running, and stopping the system's components in an orderly and precise manner. The chiller technician should be familiar with the system and its sequence of operations before operating, adjusting, or otherwise working on the system. Most chilled-water systems are designed with multiple safeties and operational backup control systems that help ensure that the equipment functions properly under most conditions, but no chances should be taken with this expensive equipment. For any major equipment installation, there should be literature at the site. The building blueprint for every major installation shows the piping diagram, an electrical diagram, a duct diagram, and a mechanical equipment layout. This information is invaluable for helping the technician to locate wiring control points and valves for controlling fluid flows. The duct diagram will help the technician find dampers and air control points. All duct and pipe sizes and routes are listed on the prints. Electrical junction boxes and control points should be on the electrical print. If a job is old and has been modified, the prints should have been kept up-to-date with the changes. Unfortunately, the updating of job site drawings is often overlooked.

Manufacturers' literature should be on-site so that the technician can become familiar with the equipment, and also reference the material in the event questions or other problems arise. This literature should not be allowed to leave the job site. If for some reason the prints are lost, copies can sometimes be obtained from the engineering firm that originally drew them up. Firms save copies for any future use. If the prints were created on a computer, there should be digital files that can be used to reprint them.

49.1 CHILLER START-UP

The first step in starting up a chilled-water system is to establish chilled-water flow. This is true whether it is for the first time each season or the first time each day. If the system is water-cooled, the next step is to establish condenser water flow. Air-cooled chillers have condenser fans and do

not require nearly as much attention as water-cooled chillers. Both the chilled-water and the condenser water circuit will have some type of flow switch to establish that water is flowing before the system can start. On older chilled-water systems, the flow switch is often a paddle that protrudes into the water piping circuit. When water is flowing, the flow switch closes, allowing the chiller to operate. Newer chilled-water systems utilize controls that sense the pressure drop through a heat exchanger to confirm waterflow. These controls have pressure sensors at both the inlet and outlet of the heat exchanger. The difference between these two pressures is what confirms that water is flowing. A pressure drop of zero, for example, indicates that no water is flowing.

The technician should know how to make sure that the flow switches are functioning. Because the paddle of a mechanical flow switch is in the water stream, it is not unusual for it to break off. If it does, there may be water flowing through the system, but flow might not be proven by the flow switch. The technician may verify water flow by means of the pressure gauges installed in the water piping circuit. Never bypass a chilled-water flow switch, or freeze conditions may occur. The starting point to finding out the correct water flow through a heat exchanger is to know what the pressure drop should be when the system is operating properly, and then compare this value to the actual pressure drop.

There should be a pressure test point at the inlet and outlet of any water heat exchanger, chiller, or condenser, **Figure 49.1**. The best way to ensure an accurate pressure differential measurement is to use a single gauge to read both the inlet and outlet water pressures, **Figure 49.2(A)**. To read the pressure at the inlet of the heat exchanger, the valve on the pressure test port in the water inlet line will be open and the valve on the pressure test port in the water outlet line will be closed, **Figure 49.2(B)**. To measure the pressure at the outlet of the heat exchanger, the valve on the pressure test port in the water outlet line will be open and the valve on the pressure test port in the water inlet line will be closed, **Figure 49.2(C)**. If two separate gauges are used and they are not properly calibrated, they cannot be expected to give correct readings. If one gauge reads 5 psig when open to the atmosphere while the other gauge reads 0, there will be an automatic error of 5 psig in the measured pressure difference. Let's take a look at a numerical example. Consider the situation where the *actual* pressures at the inlet and outlet of a heat exchanger are 35 psig and 20 psig, respectively, **Figure 49.3(A)**. The actual

Figure 49.1 A condenser with inlet and leaving-water gauge ports.

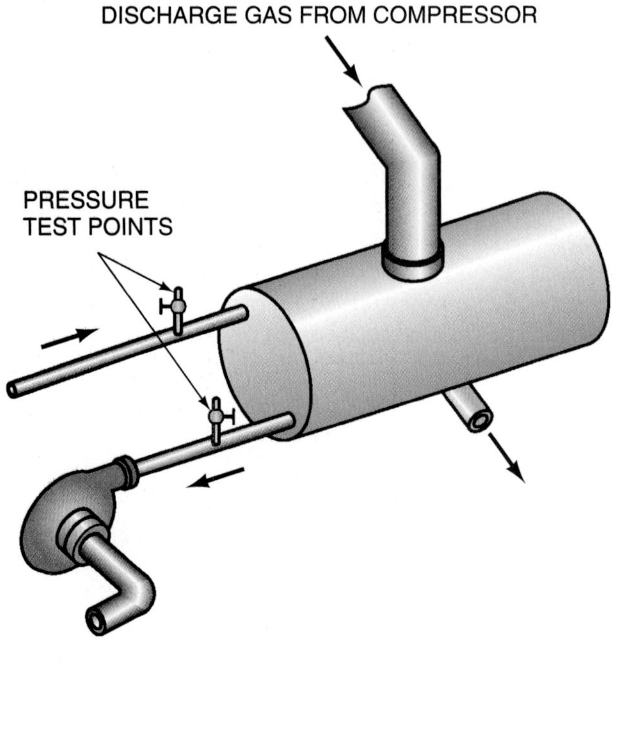

Figure 49.2 (A) One gauge being used to check both ports. (B) Inlet water pressure. (C) Outlet water pressure. The difference between (B) and (C) is the actual pressure drop through the condenser, 15 psig.

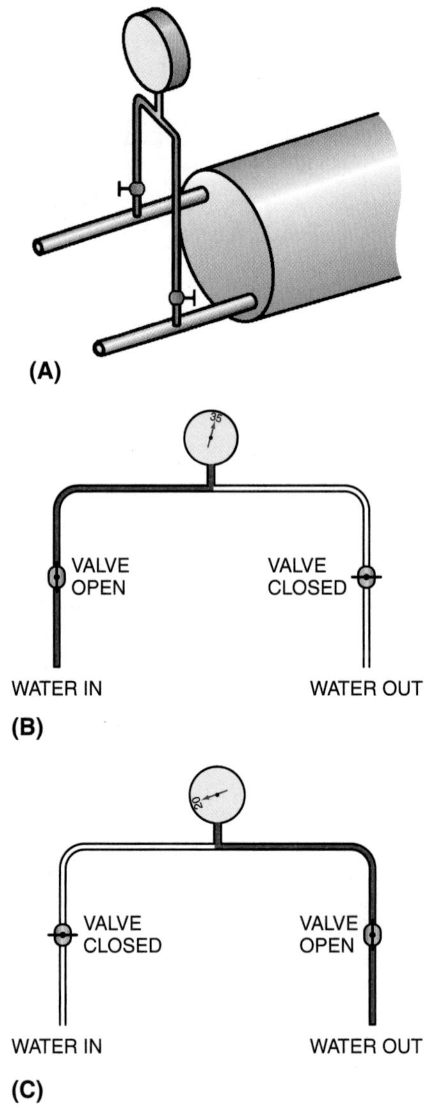

Figure 49.3 (A) The actual pressures at the inlet and outlet of a heat exchanger provide a pressure difference of 15 psig. (B) The measured pressure difference is 20 psig with the improperly calibrated gauge "A" connected to the water inlet port of the heat exchanger. (C) The measured pressure difference is 10 psig with the improperly calibrated gauge "A" connected to the water outlet port of the heat exchanger.

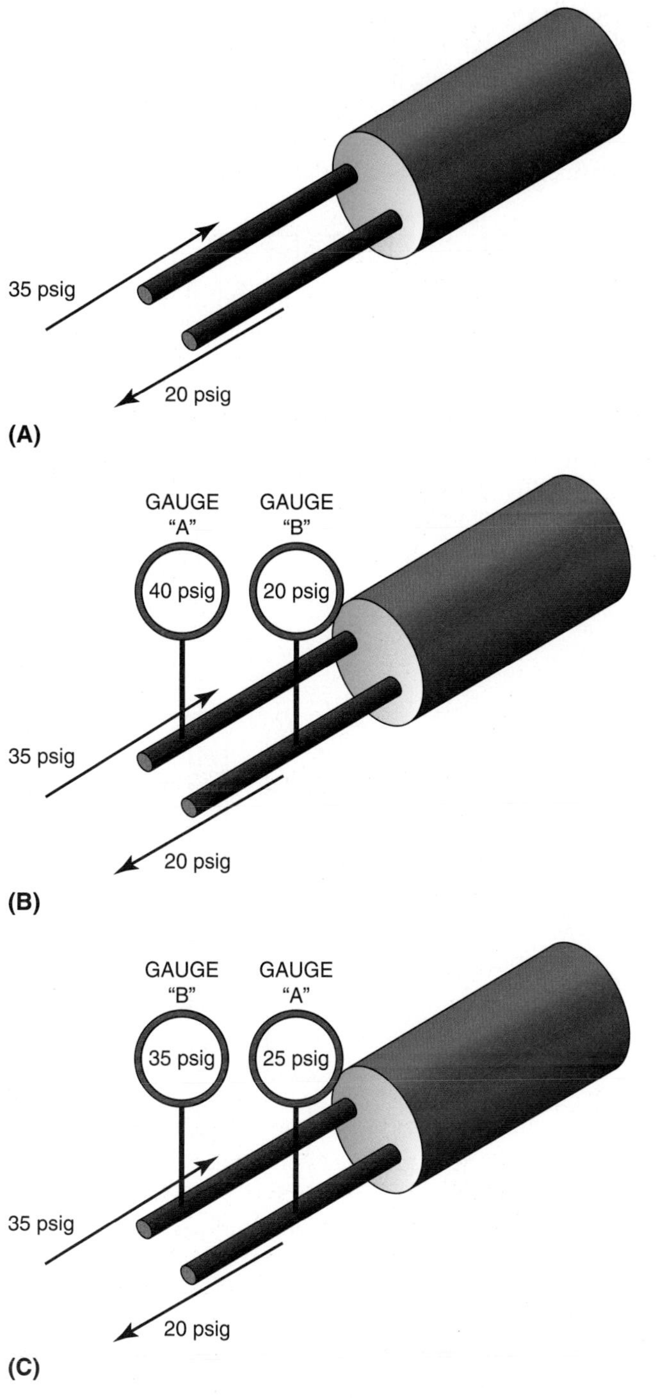

pressure difference across the heat exchanger is 15 psig. Now, consider that gauge "A" reads 5 psig when open to atmosphere and gauge "B" is properly calibrated and reads 0 psig when open to atmosphere. With gauge "A" connected to the water inlet port and gauge "B" connected to the water outlet port, the pressure readings will be as follows:

- Heat exchanger inlet pressure: 40 psig
- Heat exchanger outlet pressure: 20 psig

This will give a pressure differential of 20 psig, **Figure 49.3(B)**. Now, if gauge "B" is connected to the water inlet port and gauge "A" is connected to the water outlet port, the pressure readings will be as follows:

- Heat exchanger inlet pressure: 35 psig
- Heat exchanger outlet pressure: 25 psig

This will give a pressure differential of 10 psig, **Figure 49.3(C)**. Since all that is needed is the pressure difference between the two points, and not the actual readings, using a single gauge that is not properly calibrated or 100% accurate will still yield the correct pressure difference.

The technician needs to know from the pressure readings that there is a pressure drop across the heat exchanger. As water flows through the heat exchanger, the pressure drops a specific amount from the time the water enters the heat exchanger to the time water leaves the heat exchanger. A pressure-drop chart for the heat exchanger will indicate the gallons per minute (gpm) of water flow, **Figure 49.4**. For example, if the pressure drop across heat exchanger "A" is 1.73 psig or 4 feet, the flow rate through the heat exchanger will be 20 gpm. If the pressure drop through the same heat exchanger is 4.33 psig, or 10 feet, the flow rate through the exchanger will be about 35 gpm. The original operating log sheet for the installation will show the pressure drops at start-up. This is very good information for future use. If the original log cannot be found, the manufacturer may be contacted for a copy of the log sheet or a pressure-drop chart for the chiller.

The technician should also be aware of the interlock circuit through the contactors that must be satisfied before start-up.

Figure 49.4 A pressure-drop chart for a heat exchanger. Courtesy Trane Company

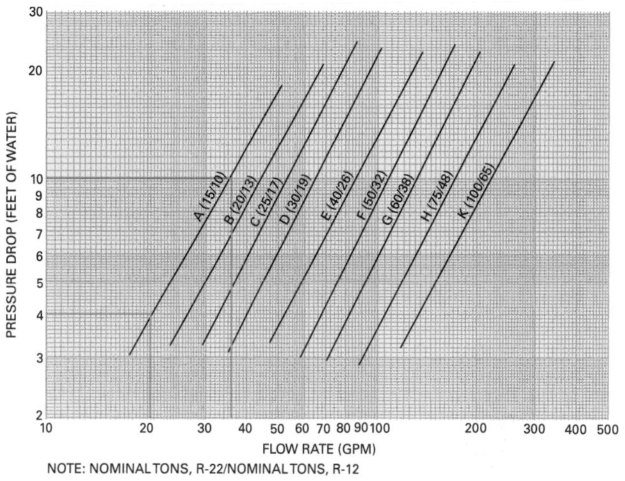

The starting sequence for most chiller systems is to start the chilled-water pump first. A set of auxiliary contacts in the chilled-water starter then starts the condenser water pump. When the condenser water pump starts, a set of auxiliary contacts make and pass power to the cooling tower fan and the chiller circuit, **Figure 49.5**. When the signal is received at the chiller control circuit, the compressor should start. This signal is often called the field control circuit because it is the circuit that is installed by the contractor and connected to the manufacturer's control circuit. If a chiller does not start, the first thing the technician should do is check the field circuit to see if all of the conditions that are required for system start-up are met. Checking this will tell whether there is a field circuit problem (flow switches, pump interlock circuit, outdoor thermostat, and often the main controller) or a chiller problem. If there is no field control circuit power, there is no need to look at the chiller until there is power. Many chillers have a ready light that shows that the field control circuit is energized.

Different chillers have different starting sequences, usually depending on the type of lubrication the compressor has. When the chiller has a positive-displacement compressor (scroll, reciprocating, or rotary screw), the compressor will start soon after the field control circuits are satisfied. Some manufacturers may incorporate a time delay before the compressor starts after the field circuit is satisfied, but the compressor should start soon. The technician should look for a time-delay circuit in any system and note what the time delay is so that he will not be waiting and wondering what the problem is. He may think there is a problem, only to have the chiller start up unexpectedly after a planned time delay. Lubrication for the reciprocating, scroll, and rotary screw compressors comes from within, so they do not have a separate oil pump that must be started first. The reciprocating and the rotary screw compressors will start up unloaded and will begin to load when oil pressure is developed. The scroll compressor starts up under load and does not have compressor unloading capabilities.

SCROLL CHILLER START-UP

Scroll chillers are either air- or water-cooled. Regardless of the type, the water for the chilled-water circuit should be at the correct level before attempting to start the chiller. The system must be full. Then the chilled-water pump should be operated and the water flow verified using pressure drop across the chiller. All valves in the refrigerant circuit should be in the correct position, usually back-seated. The system should be visually checked for leaks. Oil on the external portion of a fitting or valve is a sure sign of a leak as oil is entrained in the refrigerant.

Figure 49.5 The control circuit for a typical chiller.

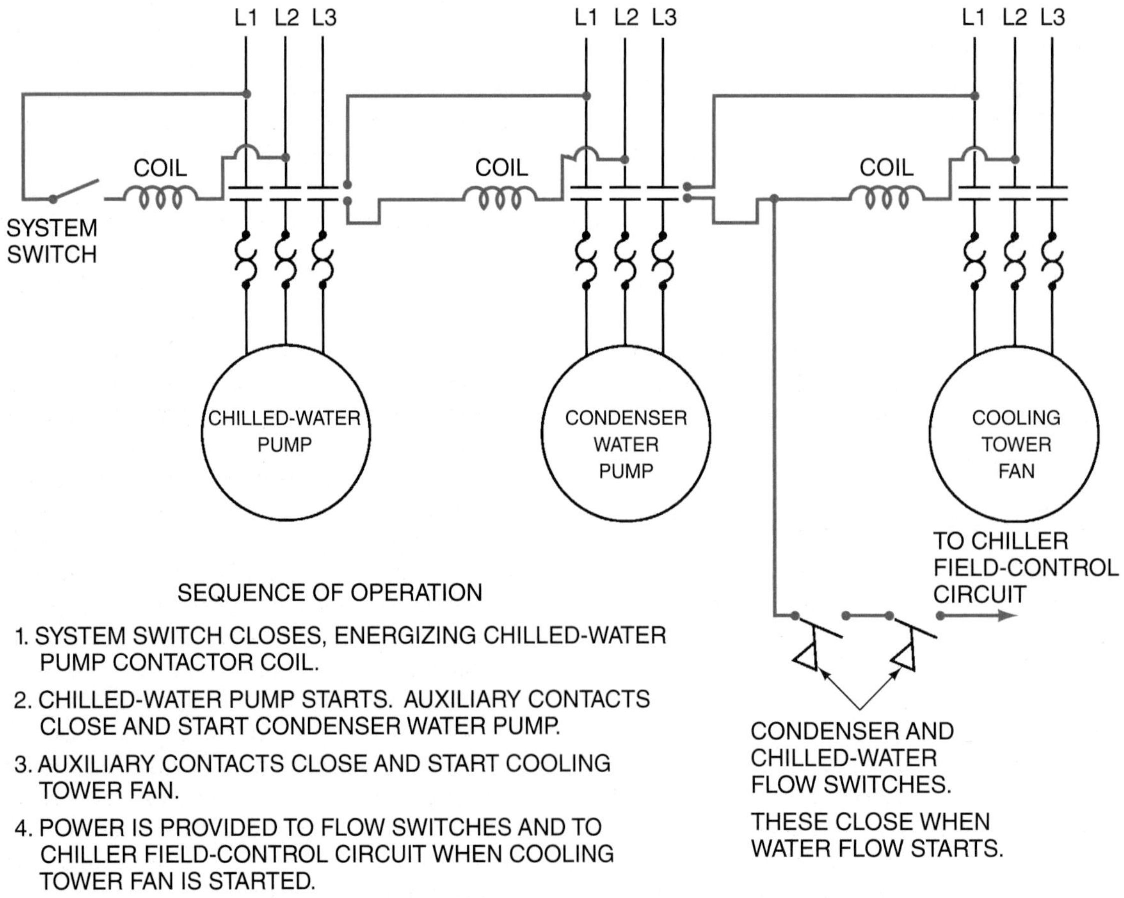

When refrigerant leaks out, small amounts of oil will be present. The compressor needs crankcase heat and it must be energized for the length of time the manufacturer requires, usually 24 hours, prior to the compressor being energized. The compressor should not be started without adequate crankcase heat, or compressor damage may occur.

An air-cooled chiller will be located outside. The chiller barrel will have a heater to prevent it from freezing in the winter. This heater should be wired and operable. Checking this can be omitted at start-up when summer arrives because the heat strip is thermostatically controlled and will shut off when not needed. During the wintertime, if the chiller is winterized and drained, the chiller barrel heater should be shut off. Also, the condenser fans should be checked to make sure they are free to turn.

When the chiller is water-cooled, the water-cooled condenser portion must be operating. The cooling tower must be full of fresh water and the condenser water pump must be started. Water flow should be established. A look at the tower can verify this or the pressure drop across the condenser can be checked.

Once the field control circuit is calling for cooling and all of the preceding requirements are met, the chiller should be ready to start. When the chiller is started, the technician should do the following:

1. Observe suction pressure.
2. Observe discharge pressure.
3. Check the compressor for liquid floodback.
4. Check the entering- and leaving-water temperature on the chiller and the condenser entering- and leaving-water temperature for water-cooled units.
5. Check all refrigerant and oil sight glasses for the proper level. Refrigerant or oil may need to be added. Neither should be added until the compressor can be operated at near full load for a period of time.
6. Check the superheat at the evaporator as recommended by the manufacturer.
7. Check the subcooling at the liquid line as recommended by the manufacturer.

When the chiller is operating normally, it can usually be left unattended, except for the cooling tower in a water-cooled unit. The cooling tower should be regularly observed and maintained.

RECIPROCATING CHILLER START-UP

The reciprocating chiller may be either air- or water-cooled. When air-cooled, it may be located outside. Refer to the comments above on freeze protection for scroll chillers. Before starting a reciprocating chiller, the oil level in the sight glass should be correct, usually from 1/4 to 1/2 level at the sight glass. The manufacturer's instructions on oil level should be followed. When the level is high, refrigerant may be in the oil; the crankcase heat should be checked. Reciprocating chillers have crankcase heaters to prevent refrigerant migration to the crankcase during the off cycle. When the crankcase heat is on, the compressor will be warm to the touch, **Figure 49.6**.

Figure 49.6 Heat from a crankcase heater.

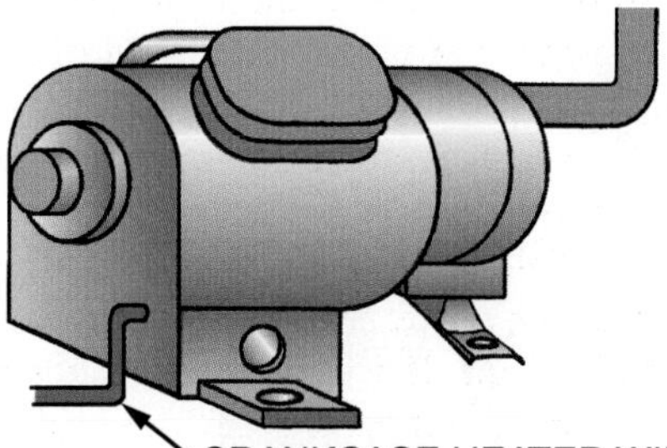

The crankcase heat must be energized for the prescribed length of time before start-up to prevent damage to the compressor. Refrigerant in the oil will dilute the lubrication quality of the oil and can easily cause bearing damage that does not show up immediately. The safe time for crankcase heat to be energized is 24 hours, unless the manufacturer states otherwise. It is not good practice to ever turn the crankcase heat off, even during winter shutdown, because refrigerant will migrate to the crankcase, and it is not easy to boil it out to the point where the oil is at the correct consistency for proper lubrication.

When the compressor is started, the pressures can be observed if there are panel gauges. Typically, gauges are permanently installed by the manufacturer in chillers over 25 tons. They may be isolated by means of a service valve to prevent damage during long periods of running time, so these valves will need to be opened to read the gauges. It is not good practice to leave the gauge valves open all the time because the gauge mechanism will experience considerable wear during normal operation from reciprocating compressor gas pulsations. The gauges are intended to be read intermittently, for checking purposes, and they should be checked for accuracy from time to time because they cannot be relied on during the entire life of the chiller.

Many chillers that are manufactured today have light-emitting diode (LED) readouts that show the pressures and temperatures. These pressures are registered by transducers and converted to an electrical or electronic signal for use in the electronic control system, which may use them for troubleshooting and analysis and may store in its memory a history of operating conditions. This can be invaluable for troubleshooting, as it can show any trends, such as pressure-drop changes or suction and discharge pressure changes.

If a chiller has been secured for winter, the refrigerant may have been pumped into the receiver and the valves may need to be repositioned for start-up to the running position, which is back-seated for any valve that does not have a control operating from the back seat. **Figure 49.7** shows an example of a valve in which the low-pressure control is fastened to the gauge port on the back seat of the valve. If the valve is back-seated, the low-pressure control will not be functional because it is isolated.

Figure 49.7 This service valve is positioned (cracked) just off the back seat so that the low-pressure control can sense the pressure.

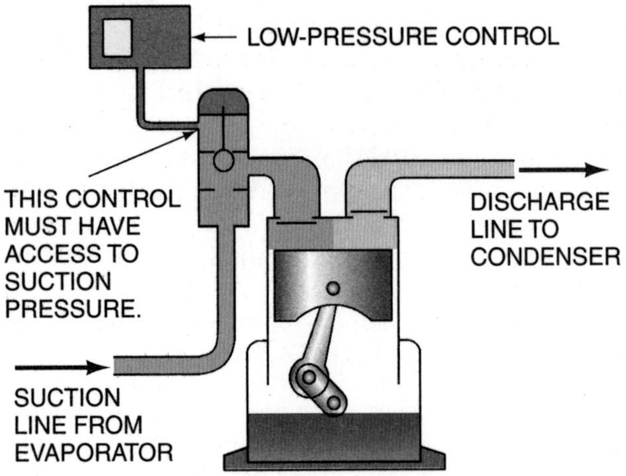

When a winterized compressor is restarted, it will start up unloaded until the oil pressure builds up and loads the compressor. An ammeter can be applied to the motor leads to determine what the level of load is on the compressor. When it is fully loaded, the amperage should be close to full load or rated load amps. As the system temperature begins to pull down, the technician may notice that the oil level in the compressor rises and foams. This occurs in many systems, but the oil level should reach normal level after a short period of operation, usually within 15 min. The compressor is suction-cooled and the technician should ensure that liquid refrigerant is not flooding back into the compressor during the running cycle. This can be determined by feeling the compressor housing; the compressor motor should be cold where the suction line enters, then it should gradually become warm where the motor housing reaches the compressor housing. The compressor crankcase should never be cool to the touch after 30 min of running time or liquid refrigerant should be suspected to be flooding back, **Figure 49.8**.

Figure 49.8 A compressor crankcase that is cold because of liquid refrigerant returning to the compressor.

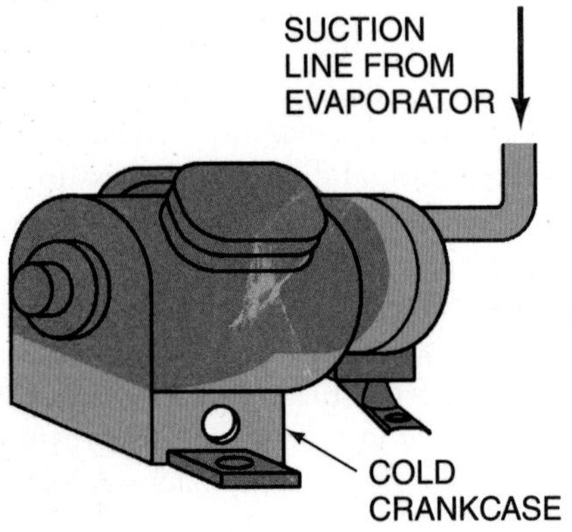

ROTARY SCREW CHILLER START-UP

The rotary screw chiller may be either air- or water-cooled. If it is air-cooled, it will be located outdoors and the availability of heat for the chiller barrel should be verified. Again, it is thermostatically controlled and will only operate when needed. The compressor must have crankcase heat before start-up. Again, most manufacturers require that this heat be energized for 24 hours before start-up. Start-up without crankcase heat can cause compressor damage.

The water flow must be verified through the chiller and the condenser if the chiller is water-cooled. The best way to do this is to use pressure gauges. All valves must be in the correct position before start-up. The manufacturer's literature should be checked for valve positions. If the chiller was pumped down for winter, the correct procedures must be followed to allow the refrigerant back into the evaporator before start-up. The field control circuit should also be checked to make sure that it is calling for cooling. When all of the preceding steps are complete, the chiller is ready to start.

The technician should start the compressor, watch for it to load up and start cooling the water, and observe the following:

1. Suction pressure
2. Discharge pressure
3. Water temperature (for both chiller and condenser when water-cooled)
4. Liquid flooding back to the compressor

When the chiller is operating normally, it can usually be left unattended, except for the cooling tower when the unit is water-cooled. The cooling tower should be observed and maintained regularly.

CENTRIFUGAL CHILLER START-UP

Centrifugal chillers often need to be started and watched for the first few minutes because of the separate oil sump system. When the compressor is a centrifugal, the oil sump should be checked before a start-up is tried. Look for the following:

1. Correct oil sump temperature, from 135°F to 165°F, depending on the machine manufacturer. **NOTE:** *Do not attempt to start a compressor unless the oil sump temperature is within the range the compressor manufacturer recommends, or serious problems may occur.* • After the machine has been operating for some time, check the bearing oil temperature to be sure that it is not overheating. If overheating occurs, check the oil cooling medium, usually water.
2. Correct level of oil in the oil sump. If the oil level is above the glass, the oil may be full of liquid refrigerant. Ensure that it is the correct temperature. The oil pump may be started in manual but observe what happens. If in doubt, call the manufacturer for recommendations. Unless the

technician has experience as to what to do, he should not try to start the compressor; marginal oil pressure may cause bearing damage.
3. Start the oil pump in manual and verify the oil pressure before starting the compressor; then turn it to automatic before starting the compressor.

When the chiller is started, the compressor oil pump will start and build oil pressure first. When satisfactory oil pressure is established, the compressor will start and run up to speed then change over to the run-winding configuration. This could be autotransformer, wye-delta, or electronic starter. The centrifugal compressor starts unloaded and only begins to load up after the motor is up to speed. This is accomplished with the prerotation guide vanes mentioned in Unit 47, "High-Pressure, Low-Pressure, and Absorption Chilled-Water Systems." When the chiller is up to speed, the prerotation vanes begin to open and will open to full load unless the machine demand limit control stops the vanes at a lower percentage of full load.

When the chiller compressor starts to work and pull full- or part-load current, the technician should observe the pressure gauges and look for problems such as the following:

1. The suction pressure operating too low or too high. The technician should mark on the gauge front or on a notepad the expected operating suction pressure for normal operation, which would be at design water temperature. Typically, this would be 45°F leaving water.
2. The net oil pressure should be correct. This should be noted nearby for ready reference. If the machine has been off for a long time, the oil level in the oil sump should be observed as the compressor starts to accept the load. If the oil starts to foam, the load on the compressor should be reduced. This will raise the oil sump pressure and boil the refrigerant out of the oil at a slower pace. If the load is not reduced, it is likely that the oil pressure will start to drop and the machine will shut off because of low oil pressure. If the compressor has an anti-recycle timer, it cannot be restarted until the timer completes its cycle, and that may be 30 min. It is better to avoid allowing the compressor to shut off by watching the oil pressure.
3. Discharge pressure should be correct. This should be noted nearby for ready reference or marked on the pressure gauge for typical operating conditions.
4. Compressor current should be correct and marked on the ammeter on the starter. The ammeter should be marked for all operating percentages—typically, 40%, 60%, 80%, and 100%—so the technician will be aware at all times at what load the machine is operating. The operating arm for the prerotation vanes should be marked so that the travel can be measured; this is closely correlated with the current.

Chillers manufactured in the last few years all have electronic controls, and the technician should study the particular machine's sequence of starting events. Typically, everything from default lights on the control panel to electronic readouts are used to indicate what is happening. For example, a unit may have a sequence of LED lights on the circuit board that will be lit if a particular problem occurs. The technician must use the manufacturer's trouble chart to discover what the nature of the problem is. Some manufacturers use the flash sequence of LED lights to describe a problem. Some provide an electronic readout that explains the problem in words or code numbers. These control sequences can be lengthy, and it is beyond the scope of this text to describe them. Again, nothing beats an understanding of the equipment that is based on the manufacturer's own description. One of the purposes of these electronic controls is to make troubleshooting easier for the technician.

49.2 VALVES FOR LARGE SYSTEMS

Refrigerant valves on many large systems require proper use and maintenance. Valves increase in size as the systems become larger, and they have some mechanical features that the technician must pay attention to. These valves, which are basically used in the refrigerant circuits, are diaphragm, ball, or service valves with back and front seats. They must be manufactured so that they can achieve high levels of leak-free performance, and, since they control refrigerant, they must be reliable.

Diaphragm valves are used for small lines, **Figure 49.9**. The name refers to the diaphragm made from stainless steel and located under the valve cap that is inside the valve. This valve should not be disassembled unless necessary. The valve may be fastened to the piping by a flare fitting or by solder or brazing. Some technicians will disassemble the valve for brazing purposes to prevent overheating the valve. This is not necessary if the valve body is kept cool with a wet rag or heat-sink paste material during the brazing process. When heated for brazing, the valve should not be seated—it should be backed off of the seat a few turns. Normally, the technician has no reason to take this valve apart. Once in service, it

Figure 49.9 This hand valve is called a diaphragm valve. The actual diaphragm is under the valve cover and can't be seen. Courtesy Henry Valve Company

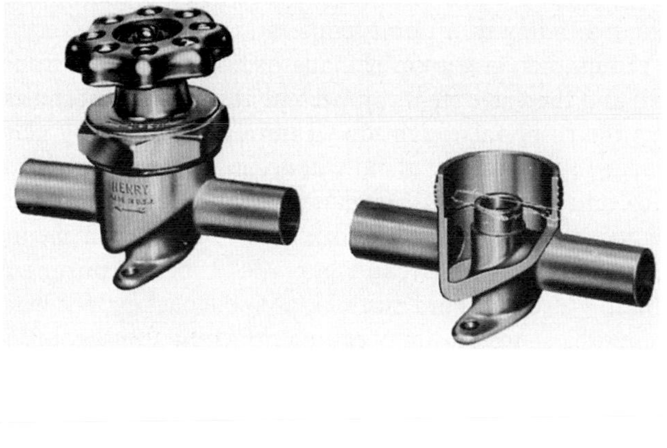

Figure 49.10 This is a ball valve, used for refrigerant. Notice that it has a protective cap to protect the valve stem and that it could also prevent a leak if the stem leaked. Courtesy Henry Valve Company

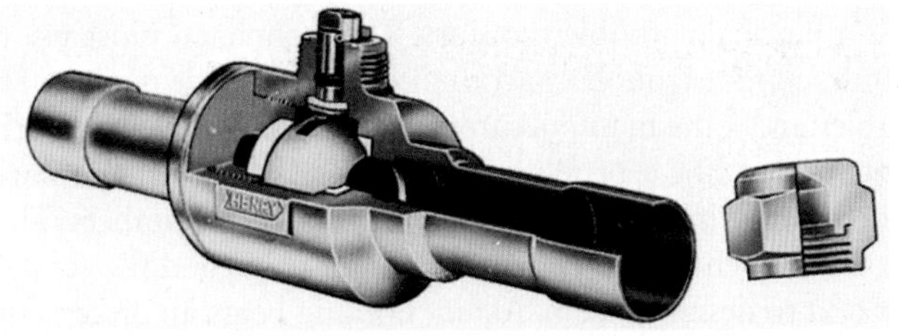

Figure 49.11 (A) A large service valve like those used on many compressors for suction or discharge service. Courtesy Henry Valve Company. (B) Cutaway view of the service valve. Courtesy Henry Valve Company

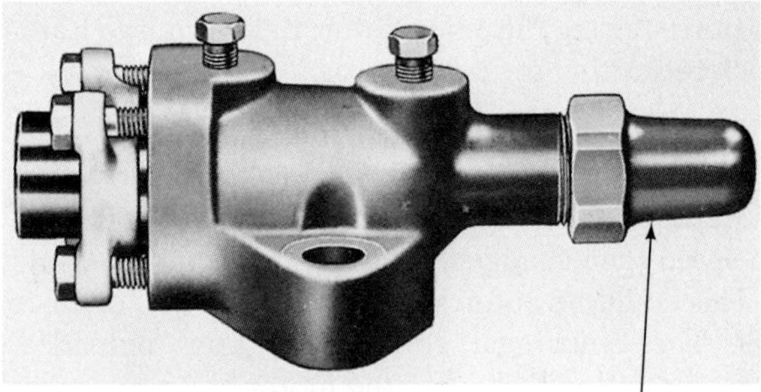

(A)

PCE NO.	DESCRIPTION	QUAN.
1	BODY	1
2	STEM	1
3	SEAT DISC	1
4	RETAINER RING	1
5	DISC SPRING	1
6	GASKET	1
7	CAPSCREW	4
8	PACKING WASHER	1
9	PACKING	1
10	PACKING GLAND	1
11	CAP	1
12	CAP GASKET	1
13	FLANGE	1
14	ADAPTER	1

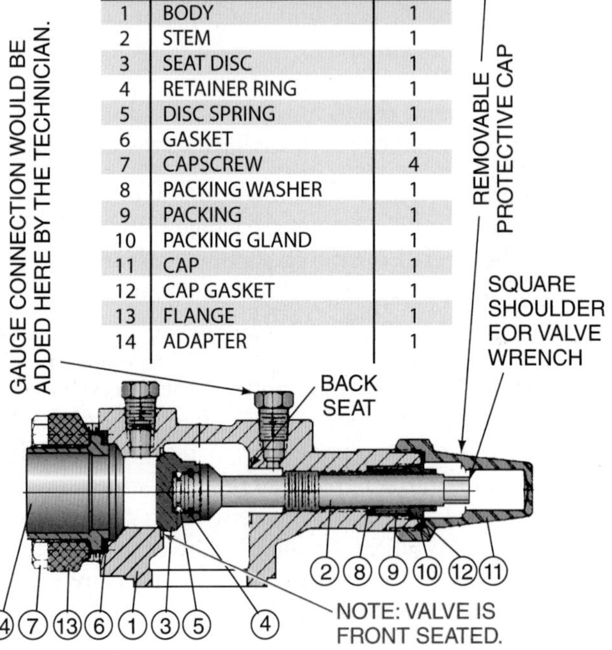

(B)

usually does not require service for the life of the system. The valve is not used for everyday service, but only to valve the refrigerant off for isolation purposes. The diaphragm valve is normally used in the liquid line because it does not come in large sizes—only up to about 7/8-in. piping.

Ball valves for refrigerant piping come in sizes of up to about 2 7/8 in. and are used for larger applications, **Figure 49.10**. Ball valves usually do not have any serviceable parts inside and are not normally disassembled. Since they are used for refrigerant, they have an enclosed valve stem that has a cover which must be replaced after use. The ball valve needs to be turned only one-quarter turn to be fully open or fully closed. It is typically brazed in the line, and the manufacturer may require the valve to be wrapped in a wet rag while the valve is brazed at one end and cooled before it is wrapped again and brazed at the other end. The reason for this is to keep the valve body from getting too hot because the working parts may not be metal. The ball valve usually has a stainless steel ball and stem that will not rust or corrode.

Service valves are valves that are used on the compressor and, in some cases, in the liquid line. These valves are brazed into the line when used in the piping and when used on the compressor are bolted to the compressor and have a brazed flange for the piping, **Figure 49.11**. These valves are used on many compressors of up to about 150 tons and they have large piping connections. The valves have a front seat and a back seat and can be used in the midseated position for different purposes, **Figure 49.12**. Notice that the valve in **Figure 49.11** has a protective cap. The valve cap is also used to protect the valve stem from damage. This is particularly true with the suction-line valve, which often operates below the surrounding dew point temperature and sweats.

Often, the valve stem is made of ferrous metals (iron or steel) and the valve stem can become rusted. **If the technician turns the rusty valve stem down through the packing gland, damage to the packing gland will occur.** The packing gland is the second line of defense from refrigerant leaking. The proper practice is to use fine sand cloth (400 grit) to clean the high spots off of the valve stem, **Figure 49.13**, before turning the stem. It is also good practice to loosen the packing nut before turning the stem. This may cause refrigerant seepage, but the technician must preserve the packing. There are three lines of defense against refrigerant leaks:

1. The back seat
2. The packing gland
3. The protective cap

They must all be protected from damage.

The valve stem packing can often be removed and replaced by back-seating the valve. The packing portion of the valve is isolated. To determine the correct packing if it cannot be identified when removing the old packing, the compressor manufacturer should be consulted. The packing is often graphite or Teflon rope, which is available at plumbing supply houses. With large valves, it is much easier to replace the packing

Figure 49.12 Three positions for a service valve (A)–(C).

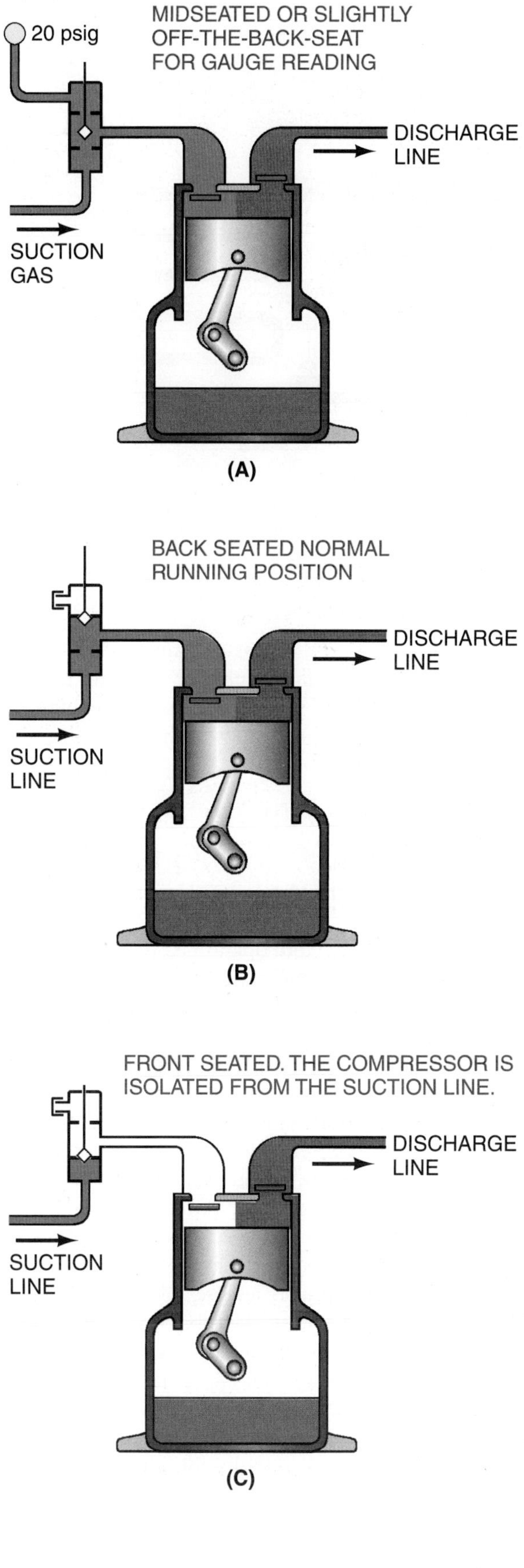

Figure 49.13 A clean stem and a rusty stem.

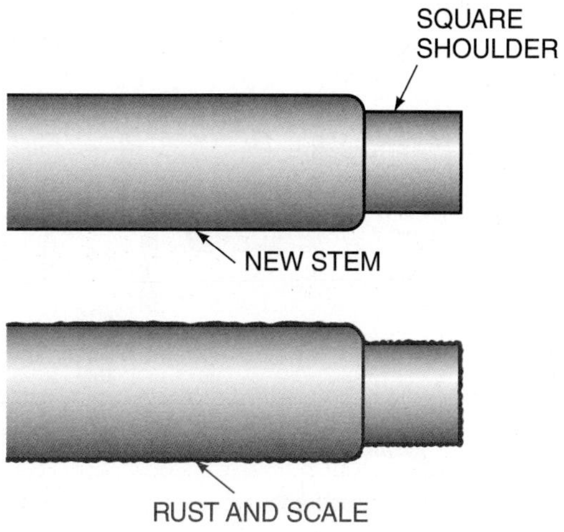

than to replace the valve. Another alternative is to leave the piping sleeve on the piping and just replace the valve body, when it is available. When the technician is through servicing the unit, the inside of the valve cap should be cleaned up and the stem should be wiped clear of any moisture. Many technicians then coat the valve stem with silicone spray or oil to protect it just before replacing the valve cap.

The valve cap is often hard to remove when it has been on for a long time, as rust may have accumulated on the threads. Some of these valve caps are quite large. Many technicians use a pipe wrench to take the valve cap off. Another way to remove it without leaving scars on the valve cap is to lightly tap each flat on the nut with a hammer to break the rust loose, then use a hex-jaw wrench to remove it, **Figure 49.14**. Tapping the handle of the wrench with a hammer can dislodge the valve cap.

Figure 49.12 shows ways that the compressor service valves are used. They can be back-seated to remove the gauges; this is the normal operating position for the valves. Often, the low-pressure control is fastened to the service valve, and in such a case the valve should be slightly cracked (turned toward the front seat) in order for the control to function.

Figure 49.14 This wrench can be used to remove valve stem caps. Photo by Bill Johnson

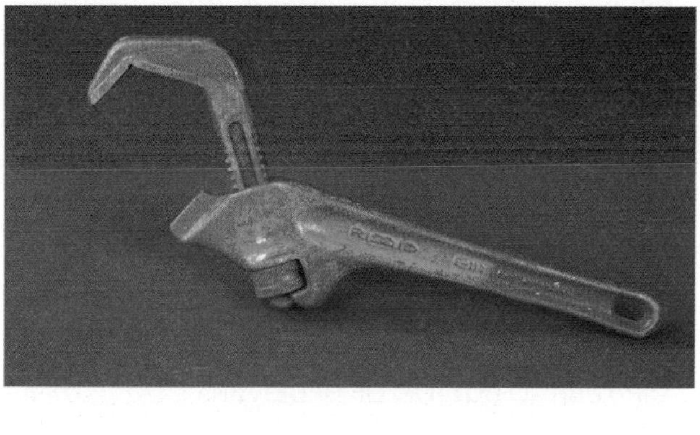

Figure 49.15 When the suction and discharge valves are front-seated, the compressor can be removed. The valves are fastened to the compressor with bolts. The compressor becomes a separate vessel.

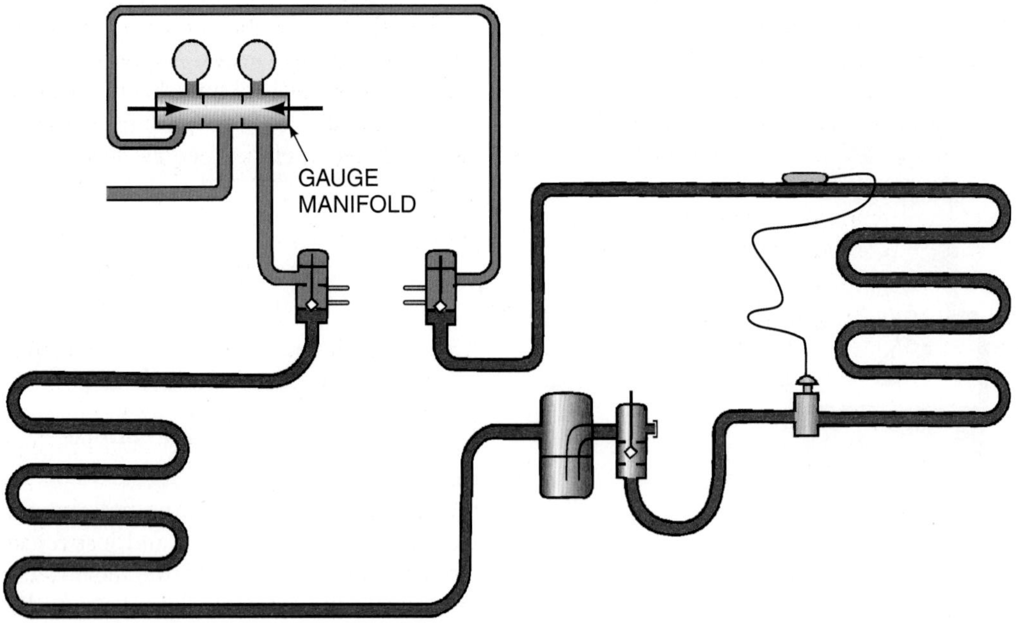

The valves can be midseated or cracked off the back seat for gauge readings. The valves can be completely front-seated for compressor service. Notice that the compressor can be removed and replaced using the front seats by unbolting the compressor flange bolts, **Figure 49.15**. Only the refrigerant in the compressor must be recovered and the new compressor can be evacuated using the front-seated position.

The liquid-line king valve is a type of service valve that is usually found at the outlet of the refrigerant receiver. This valve is used to isolate the refrigerant in the condenser and receiver by pumping the refrigerant from the outlet of the receiver into the condenser and receiver, **Figure 49.16**.

Figure 49.16 The king valve in the liquid line is a little different. The back seat is open to the gauge port when the valve is front-seated. This valve has a back seat so that the gauges can be removed.

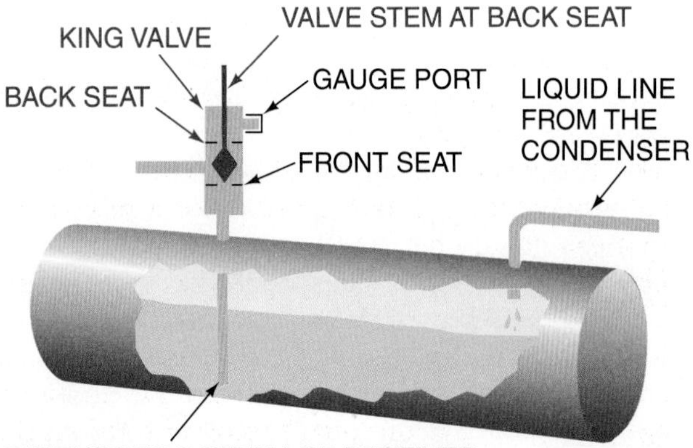

A technician can fasten a high-side gauge to the king valve, front-seat it, and start the compressor to pump the refrigerant to the condenser and receiver. The gauge reading will tell the technician what the pressure is in the liquid line.

49.3 SCROLL AND RECIPROCATING CHILLER OPERATION

Once the chiller is on and operating, the operator should observe the chiller from time to time to make sure it is operating correctly. Small air-cooled chillers may be located in remote locations and not be observed on a regular basis. These chillers are usually operated unattended, but not much can go wrong if the chilled-water supply is maintained.

All water-cooled chillers require more attention because of the water circuit and cooling tower. It is good practice to check any piece of operating equipment on a periodic basis but that is expensive, so most close observation is reserved for larger systems. It is good practice to check the cooling tower water several times per day to be sure that the water level is correct. Often, a partial loss of water pressure will cause the amount of tower water to drop due to the fact that the water is evaporating faster than the water supply can make it up. With careful observation, the technician can catch this problem before it shuts the unit off and a complaint occurs. If no regular technician is on the job—for example, at a small office building—someone can be assigned to make this check with very little instruction. The chillers are often on top of buildings where

no one goes on a regular basis. Someone should be assigned to check these installations every week.

Water-cooled chillers must have water treatment to prevent minerals and algae from forming in the water circuit. It is recommended that a qualified water treatment specialist be used. This will help reduce the number of water-related problems with the system. Review the material on cooling towers in Unit 48, "Cooling Towers and Pumps," for the types of water treatment needed. All towers must also undergo blowdown, which is evacuating a percentage of the circulated water down the drain to prevent the tower water from becoming overconcentrated with minerals due to evaporation. Blowdown for a cooling tower is necessary, but often management has difficulty understanding why water that looks clean must be sent down the drain. The technician should know how to explain this to them.

49.4 LARGE POSITIVE-DISPLACEMENT CHILLER OPERATION

Large chillers that use reciprocating or rotary screw compressors may range in size to more than 1000 tons of capacity. Chillers of this size are extremely expensive and must be observed on a regular basis or problems may occur. These problems can often be expensive to correct. Large chillers are reliable but observation is very important. The technician should look for any potential problem as explained in the previous paragraphs. These chillers may also have pressure gauges that are active all the time in that they are not valved off. If so, these may be observed as much as once per hour on some critical applications. Often an operating log is maintained on a regular basis. The frequency of the recording in the operating log depends on the importance of the job. If a chiller is being used for critical manufacturing, the entries in the log may be required hourly. If it is being used in an office building, the log may be maintained on a daily or weekly basis. See **Figure 47.23** for an example of an operating log form.

49.5 CENTRIFUGAL CHILLER OPERATION

Centrifugal chillers were the largest chillers made for many years. Typically, when the requirements were greater than 100 tons, a centrifugal chiller was considered or multiple reciprocating chillers were used. Centrifugal chillers range in size from 100 tons to about 10,000 tons for a single chiller. These were the most expensive of all chillers and also the most reliable, so much attention has been given to the observation of the operation and maintenance of these chillers. For years, it has been customary for the operation of these chillers to be observed and an operating log maintained on a regular basis, usually daily. Operating logs make it possible for slowly occurring problems to be easily detected. For example, if the cooling tower water treatment system has not been working and the condenser water tubes are beginning to have a mineral buildup, the condenser water approach temperature will begin to spread. An alert operator will notice this and take corrective action, which may be to change the water treatment and certainly to clean the water tubes, because the heat exchange rate has been reduced.

49.6 AIR-COOLED CHILLER MAINTENANCE

Air-cooled chillers require little maintenance, which is one reason why they are so popular. The fan section of these chillers may require lubrication of the fan motors, or the motors may have permanently lubricated bearings. It is typical for these chillers to have multiple fans and motors. As a result, the motor horsepower can be reduced and the fans can be direct-drive. This eliminates belts and the need for routine lubrication in many cases. If one fan fails, the chiller can continue to operate.

The following electrical maintenance should be performed annually for a typical system:

1. Inspect the complete power wiring circuit because this is where the most current is drawn. Look for places where hot spots have occurred. For example, the wires on the compressor contactor should show no signs of heat. The insulation should not appear to have been hot. If so, this must be repaired. This can be done by cutting the wire back to clean copper. If the lead is not long enough, a splice of new wire may be needed or the wire may need to be replaced back to the next junction.
2. Inspect the motor terminal connections at the compressor. If these show any discoloration, a repair should be made. A hot terminal block can cause refrigerant leaks, so the wiring should be inspected carefully.
3. The contacts in all contactors should be inspected and replaced if pitting is excessive. If excess pitting begins to occur, it is likely that the contacts will weld shut in the future and they will not be able to stop the motor. This can lead to single phasing of a wye-wound motor if any two contacts weld shut. Motor burn will then occur because there is nothing to shut the motor off except the breaker, **Figure 49.17**.
4. The compressor motor should be checked for internal ground using an ohmmeter, called a megohmmeter, that can check ohm readings in the millions of ohms. **Figure 49.18** shows an example of a basic megohmmeter. This instrument uses about 50 to 500 V DC to check for leaking circuits; others may use much higher voltages and have a crank-type generator to achieve the high voltage. The motor manufacturer's recommendations should be used for the correct allowable leakage of the circuit.

Figure 49.17 The contacts are welded shut on a wye-wound motor, which will cause the motor to try to run on L2 and L3. This is called single phasing and will burn the motor.

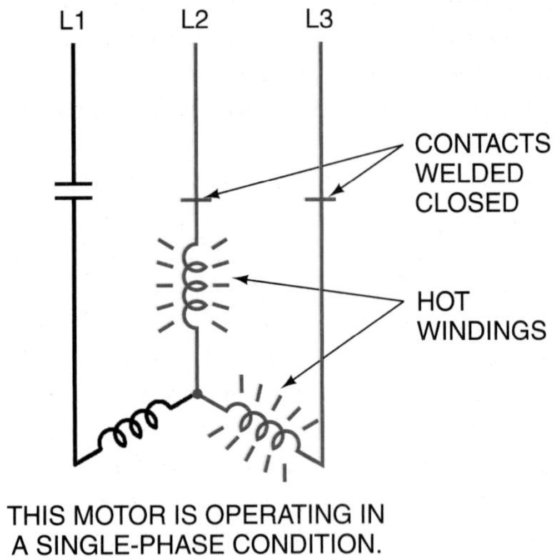

Typically, a motor should not have less than a 100-megohm circuit to ground or to another winding in the case of a delta-type motor or dual-voltage motor, **Figure 49.19**. The required reading depends on the motor temperature because the temperature changes the requirements. Consult the manufacturer if proper guidelines are unknown.

Figure 49.18 A megohmmeter for checking ground circuits in motors. Reproduced with permission of Fluke Corporation.

Figure 49.19 Using a megohmmeter to check a motor. Notice that it is checked from all windings to ground and from winding to winding.

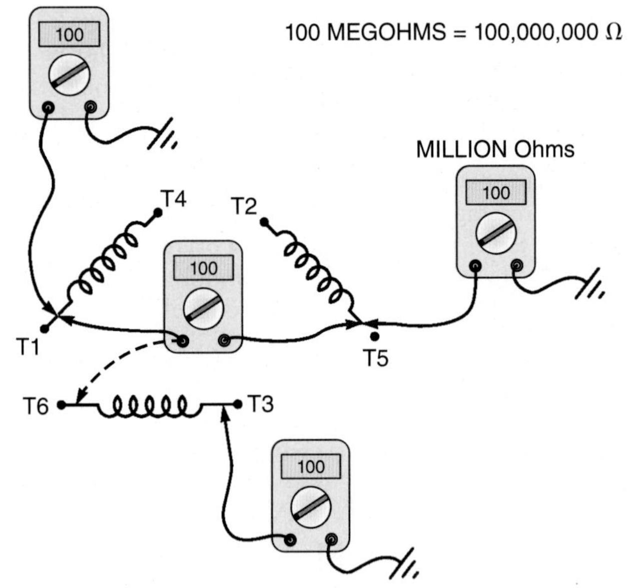

A reliable local motor shop can also give guidelines. It is important to start a process of megging a motor when it is new and keep good records. If the motor megohm value begins to decrease, it is a sign that moisture or some other foreign matter may be getting into the system.

5. An oil sample from the compressor crankcase can be sent to an oil laboratory to determine the condition of the compressor. This test, like the megohm test, should be done when the chiller is new and should be performed every year. If certain elements start appearing in greater quantities in the oil, it is evident that problems are occurring. For example, if the bearings are made of babbitt and the babbitt content increases each year, at some point it is evident that the bearings need to be inspected. If small amounts of water appear, the tubes need to be inspected for leaks. This will only occur if the water pressure is greater than the refrigerant pressure. For high-pressure chillers, this can occur if the chiller is at the bottom of a high column of water.
6. Inspect the condenser for deposits such as lint, dust, or dirt on the coil surface. These must be removed because they block airflow. It is good practice to clean all air-cooled condensers once a year or more often if needed. If the operating log shows an increase in head pressure compared with the entering-air temperature, the coil is becoming dirty. Air-cooled coils can be deceiving in that the dirt may be imbedded in the interior of the coil and cannot be seen until the coil is cleaned. A coil should be cleaned by saturating it with an approved detergent and then washing the coil backward, in reverse to the direction of the airflow; this is called backwashing. An apparently clean coil may yield a large volume of dirt from its interior.

49.7 WATER-COOLED CHILLER MAINTENANCE

Routine maintenance may involve cleaning the equipment room and keeping all pipes and pumps in good working order. The technician should ensure that the room and equipment are kept clean and in good order. It is always easier to keep a room clean than to have to clean it when the chiller breaks down and needs to be disassembled. The cooling tower should be checked to be sure the strainer is not restricted. Look at the water for signs of rust, dirt, and floating debris, and clean if necessary. The water treatment should be checked on a regular basis. Some systems call for the operating technician to perform water analysis daily to check for mineral content. When the water treatment gets out of balance, the technician should either make adjustments or call the water treatment company. Someone must be in charge of keeping the water chemicals at the correct level for best performance.

Annual maintenance should involve checking the complete system and chiller to prepare it for a season of routine maintenance. Some chillers operate year-round and are only shut down for annual maintenance. During this maintenance period, it is often wise to bring in the factory maintenance representatives. These technicians know what is happening across the nation or around the world with regard to any service or failure problems with their equipment and so will have the best technical knowledge about the equipment. Often, they will put the equipment through the complete start-up procedure just as though it is for the first time. All controls may be checked at this time to be sure they will perform up to standard. It is not uncommon for a system to operate for years without a control checkup and then fail—only to have the technician discover that a control has failed that should have saved the system but did not. A control checkup once a year can be good insurance against many failures. During annual maintenance, all electrical connections should be checked; refer to the previous section on electrical maintenance for air-cooled systems. Water-cooled equipment has the same electrical symptoms.

The water-cooled condenser should be inspected on the inside every year by draining the water from the condenser and removing the heads. The tubes should be clean. A tube is clean when a light shone inside the tube reveals copper tube with a dull copper finish. A penlight is very good for this because it can be inserted inside the tube. For proper heat exchange, any film on the inside of the tube must be removed. Many technicians believe that if the tube is open enough to allow a good water flow it is clean. This is not true; the tube must be clean down to bare copper for proper efficiency.

Several approaches may be taken to cleaning dirty tubes. The tubes may be brushed with a nylon brush that is the size of the inside of the tube. This is customarily done with a machine that turns the brush while flushing the tube with water. Some manufacturers recommend using a brush with fine brass bristles if a nylon brush will not remove the scale from the tubes. Check with the manufacturer, because some do not recommend this practice. Always keep the tubes wet until it is time to brush them or the scale will harden more. It is good practice to be prepared to brush the tubes before removing the heads so that the tubes will stay wet until they are brushed.

If brushing the tubes will not work, the tubes may be chemically cleaned using the recommended acid for the application, **Figure 49.20.** **NOTE:** *This is a very delicate process because the tubes may be damaged. Consult the chemical and chiller manufacturer for this procedure.* • It may be good practice to let the chemical manufacturer clean the tubes; then if damage occurs, the manufacturer is responsible. However the process is accomplished, make sure that the chemicals are neutralized or the tubes may be damaged. The damage would be severe because the chemicals may eat the tube material, and leaks between the water and refrigerant may occur. After the tubes have been chemically treated, they may be clean or the chemicals may just soften the scale and the tubes may need brushing to remove the scale.

The water pump should be checked to make sure the coupling is in good condition. **SAFETY PRECAUTION:** *This can be done by turning off the power, locking and tagging the disconnect, and removing the cover to the pump coupling.* • If materials such as filings are found inside the coupling housing, deterioration of the coupling should be suspected. Some couplings are made of rubber and some are steel. Look for the flexible material that takes up the slack to see whether it is wearing.

Figure 49.20 Chemically cleaning condenser tubes.

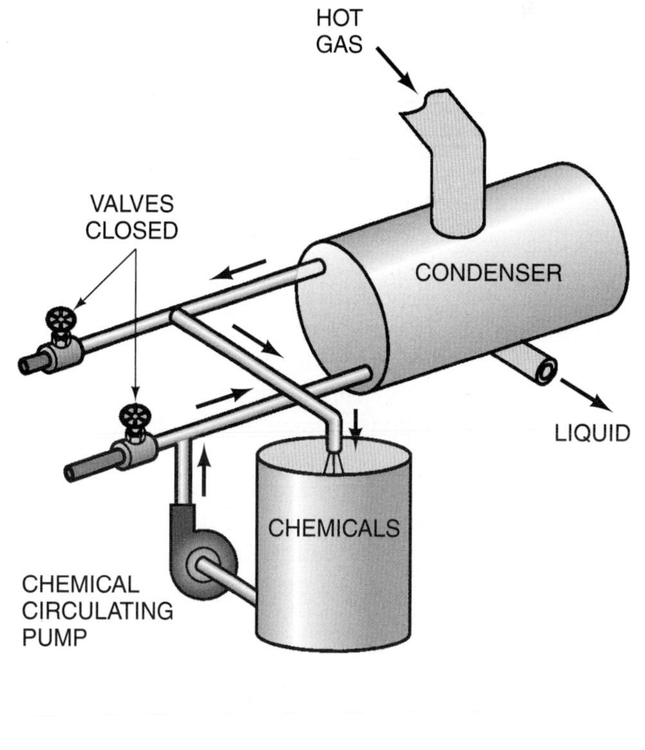

Figure 49.21 The tube assembly for a heat exchanger. Notice the tube support sheets that support the tubes.

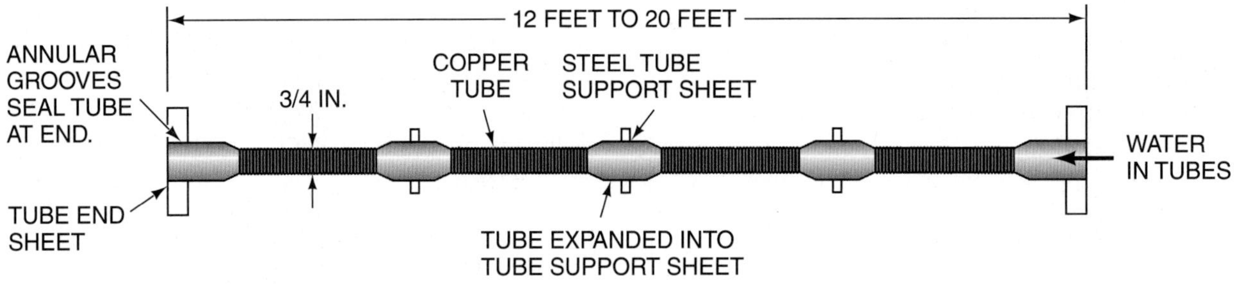

After the condenser has been cleaned, the condenser tubes may be checked for defects such as stress cracks, erosion, corrosion, and wear on the outside by means of an eddy current check. This test should be performed on the evaporator tubes as well. This check determines if there are irregularities in the tube. One common problem is wear on the tube's outer surface where it is supported by the tube support sheets. Each set of tubes has a different type of stress applied to the outside. In the condenser, the hot discharge gas is pulsating as it enters the condenser and has a tendency to shake the tubes. In the evaporator, the boiling refrigerant shakes the tubes. The tube support sheets located in the condenser and the evaporator are sheets of steel that the tubes pass through to help prevent this action and are evenly spaced between the end sheets. When the evaporator or condenser shell is tubed, the tubes are guided in place through the support sheets. When the tube is in place, a roll mechanism is inserted into the tube and expanded to hold the tube tight in the support sheet, **Figure 49.21**. The tube roller may miss a tube, and this tube can shake or vibrate until the tube sheet wears the tube. If this wear continues, the tube will rupture and a leak between the water and refrigerant circuit will occur. The worst case can flood the chiller on the refrigerant side with water. This may happen in particular with low-pressure chillers. The eddy current test can be used to find potential tube failures; these tubes may be pulled and replaced or plugged if there are not too many.

The eddy current test instrument has a probe that can be pushed through the clean tube while an operator watches a screen monitor, **Figure 49.22**. As the probe passes through the tube, it sends out a magnetic current signal that reacts with the tube and shows a profile on the screen monitor. When unusual profiles are noticed, the tube is marked for further study, which may involve comparing the profile to that of a known good tube. A decision may need to be made to pull or plug the tube. A tube failure in the middle of the cooling season can take days to repair, and it is good practice to find these problems in advance.

Water and steam valves can be maintained as leak-free as refrigerant valves but are not as intricate. The **globe valve** is the most common valve used in water and steam circuits, **Figure 49.23**. It has a packing gland like the service valves

Figure 49.22 A probe for checking tubes that uses an eddy current tester can determine irregularities in the tubes.

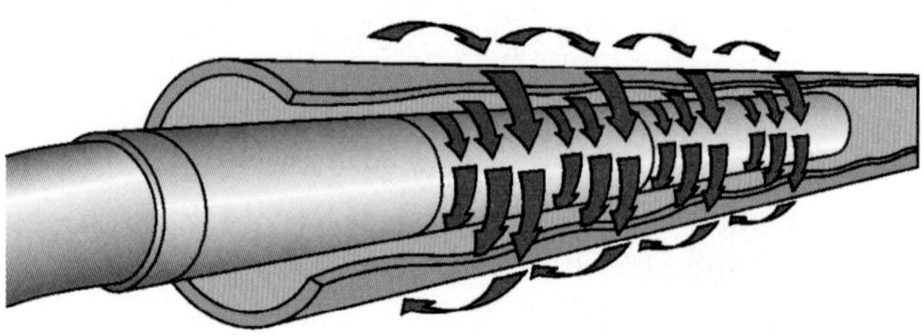

on a compressor. This valve can normally be back-seated to facilitate the removal of the existing packing and the addition of new packing. **NOTE:** *This is the only time the valve should be back-seated.* • The globe valve can be disassembled and the valve seat can be replaced. The system must be drained in order to be disassembled. Notice that the valve has a directional arrow for flow. The stream to be controlled should be piped to the underside of the globe valve. If it is piped in backwards, the valve seat will often chatter or make a buzzing sound as the seat vibrates. The globe valve has a great deal of pressure drop because the fluid has to make several turns as it goes through the valve.

The **gate valve** is another common valve used for these applications, **Figure 49.24**. The gate valve creates much less pressure drop through the valve than the globe valve and is a much better choice where excess pressure drop is a concern. Notice that the gate valve has a packing gland and a back seat for packing-gland repair. The gate valve can also be disassembled for seat repair. It does not have a replaceable seat; the gate is finely ground to fit into the finely ground seat. For large valves, this seat may be resurfaced if needed.

NOTE: *The globe valve and gate valve should not be back-seated when in operation. The reason for this is that if the valve is back-seated and then later the technician forgets the position, the valve might be damaged if technician tries to open it. It is good practice to open the valve all the way to the back seat and then turn it inward about one turn. Then, when someone tries to use the valve, its position will be known. It is not uncommon for a technician to mistake the valve for closed when it is on the back seat and do damage when trying to open it.* •

Figure 49.23 (A) A globe valve. (B) Cutaway of a globe valve. Courtesy of NIBCO INC., Elkhart, Indiana 46515. (C) Globe valve shut. (D) Globe valve open. Notice that the fluid must change directions.

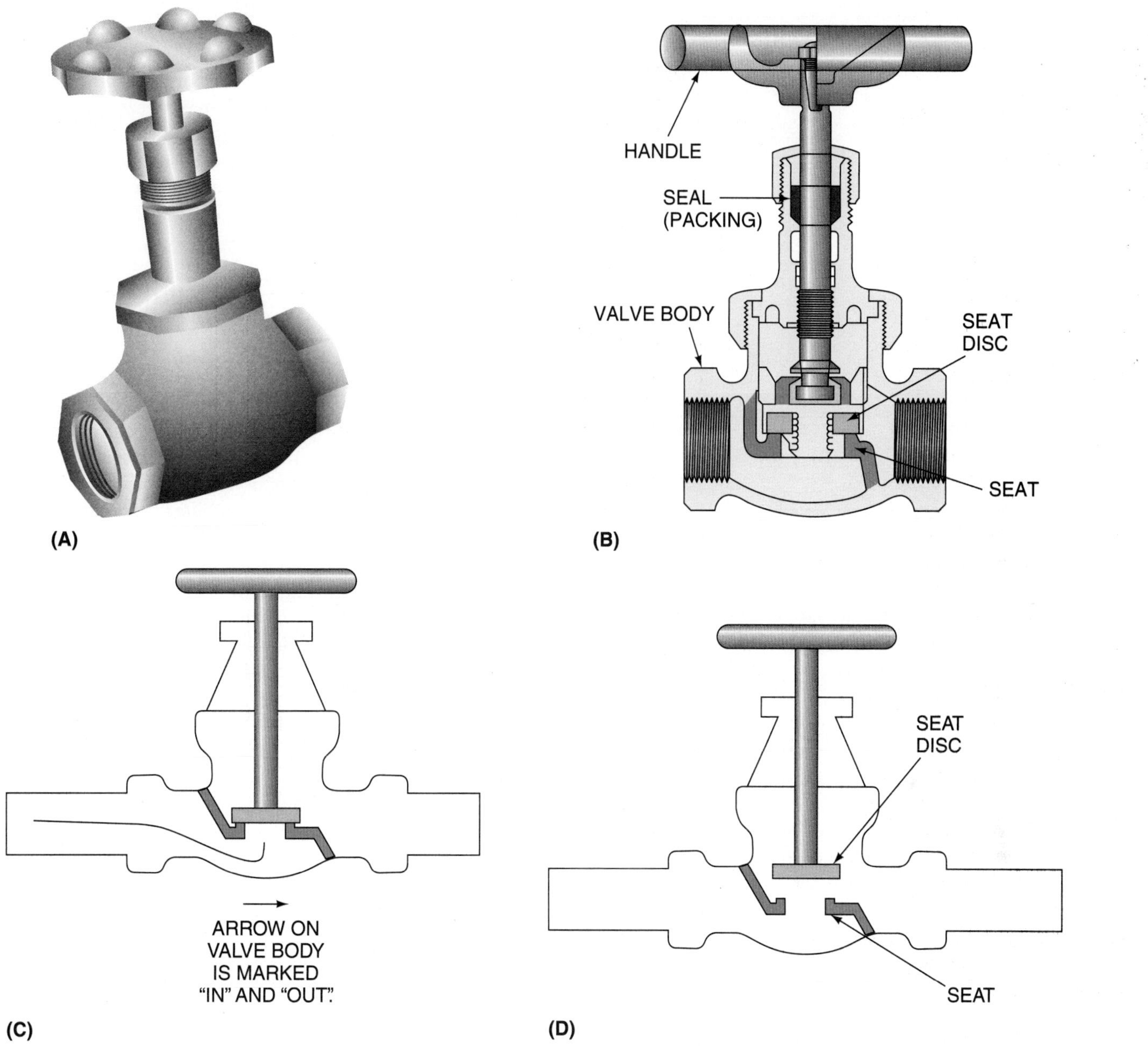

To keep the valves loose, both the globe valve and gate valve should be run from all the way closed to all the way open from time to time. This must be done when it will not affect the system's operation, for example, when the system is shut off for some reason. These valves come in very large sizes—up to about 24 in. The larger valves have flanges for fastening them into the system with bolts. The butterfly valve is used in water circuits and is a quick-shutoff valve—as it only needs to be turned one-quarter turn from on to off, like a ball valve.

Strainers are used in water circuits to remove particles that may be in suspension and circulating. The cooling tower can be particularly dirty because it uses outside air that can have many contaminants in it. A cooling tower that has water flowing and air being forced through it is like a large filter for the air. The particles are gathered in the water and then the tower sump and onto any strainers. There are several types of strainers; consult the strainer manufacturer if there is doubt about how to clean them. **SAFETY PRECAUTION:** *When cooling towers are cleaned, the technician should be careful with the contaminants in the tower. If the tower is dry, the dust should not be breathed. Rubber gloves and boots should be worn, along with a face mask.* •

Water pumps are just as important as the chiller itself. If the water pumps do not operate, the chiller will not operate.

Figure 49.24 (A) A gate valve. (B) Cutaway view of a gate valve. Notice that when the gate is open, the valve goes straight through and there is very little pressure drop. Courtesy of NIBCO INC., Elkhart, Indiana 46515. (C) A gate valve disassembled.

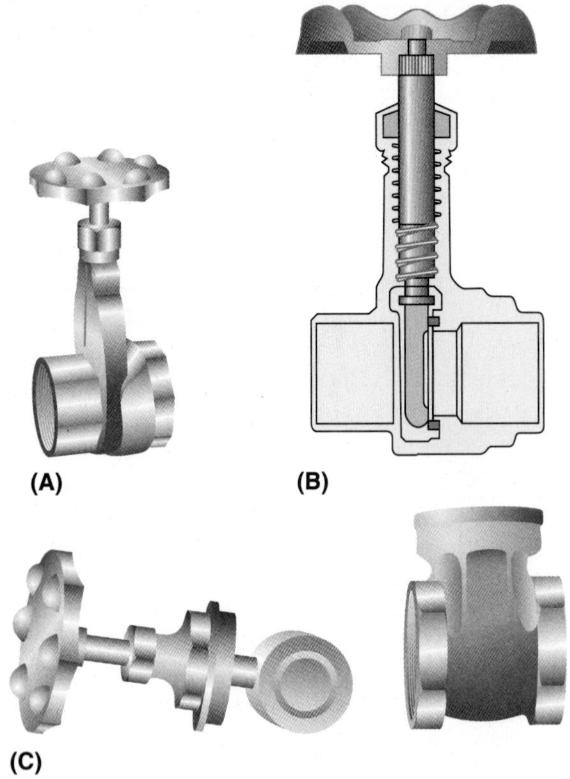

Water pumps and their motors often require lubrication. When grease fittings are used, pay attention to the relief plug that allows excess grease to escape, **Figure 49.25**. If the relief plug is not removed, excess grease will be forced down the shaft and may harm the grease seal or enter the motor. The pump and motor will have a flexible coupling between them and it should be inspected. Turn off the power, lock it out, remove the protective cover, and visually inspect the coupling for wear.

If the chiller and water pumps are to stand idle during the winter months, all precautions should be taken to prevent freezing of any components such as piping, sumps, pumps, or the chiller, if it is located where it may freeze. Even if a chiller is in a building, it is good practice to drain it during winter because if the building loses power and the interior freezes, it would be expensive if the chiller tubes were to freeze. It is easier just to drain it and not take a chance.

Figure 49.25 A motor bearing with a grease fitting. Notice the relief plug that prevents excess pressure when grease is added. The plug must be removed while grease is applied.

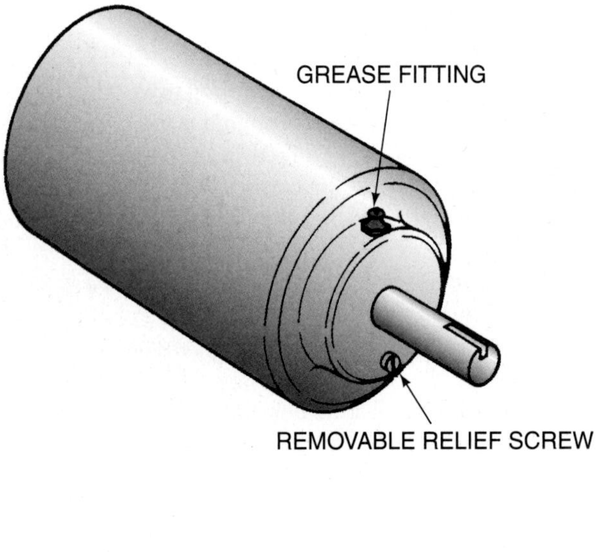

49.8 ABSORPTION CHILLED-WATER SYSTEM START-UP

The absorption chiller start-up is similar to that for any chiller: Chilled-water flow and condenser water flow must be established before the chiller is started. Ensure that the cooling tower water temperature is within the range of the chiller manufacturer's recommendations. If the absorption machine is started with the cooling tower water too cold, the lithium-bromide (Li-Br) may crystallize, causing a serious problem. In addition to this, the heat source must be verified, whether it be steam, hot water, natural gas, or oil. Most absorption chillers are steam operated so the steam pressure must be available. If the chiller is operated from a boiler dedicated to it, the boiler must be started and operated until it is up to pressure.

The purge discharge may be monitored by placing the vacuum pump exhaust into a glass of water and looking for bubbles, **Figure 49.26**. Be sure to remove the purge exhaust from the water when the purge is not operating. If there are bubbles, it is a sure sign that noncondensables are in the chiller. It is pointless to start the chiller until the bubbles stop. Let the vacuum pump run even when the chiller is started (unless the manufacturer recommends not to) because often noncondensables will migrate to the purge pickup point after start-up.

Figure 49.26 The vacuum pump exhaust is placed in a glass of water to determine whether noncondensable gas is being removed; bubbles are moving out of the exhaust. **NOTE:** *Remove the glass before turning off the vacuum pump.* •

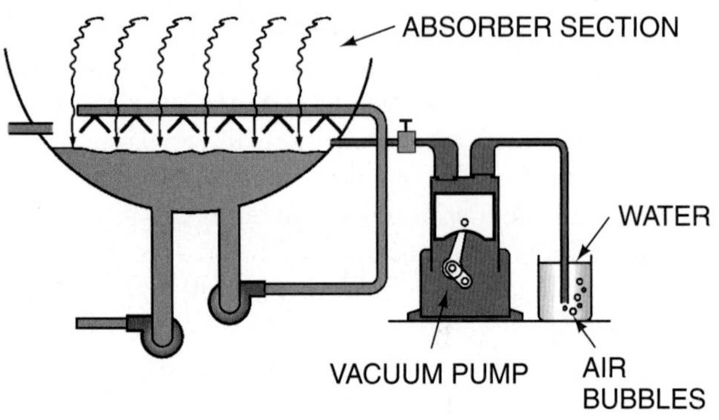

When all systems are ready, the chiller may be started. The fluid pumps will start to circulate the Li-Br and the refrigerant and the heat source will start. The chiller will begin to cool in a very few minutes; this can be verified by the temperature drop across the chilled-water circuit. When absorption chillers are refrigerating, they make a sound like ice cracking. When this sound begins to come from the machine, the chilled-water temperature should begin to drop.

49.9 ABSORPTION CHILLER OPERATION AND MAINTENANCE

Absorption chillers must be observed for proper operation more regularly than compression cycle chillers because they can be more intricate. Chiller manufacturers use different checkpoints to measure chiller operation. The refrigerant temperature and the absorption fluid temperatures measured against the leaving chilled-water temperatures are among a few. The manufacturer's literature must be consulted for the checkpoints and procedures. Maintaining an operating log is highly recommended for absorption installations.

When the chiller is operating, it is good practice to pay close attention to the purge operation. If the chiller requires excessive purge operation, the machine may have leaks or it may be manufacturing excess hydrogen internally and need additives. The machine must be kept free of noncondensables. The heat source may also need maintenance. If it is steam, the steam valve should be checked for leaks and operating problems. A steam condensate trap will also need checking to be sure it is operating properly. The condensate trap ensures that only water (condensed steam) returns to the boiler. A typical condensate trap may contain a float. When the water level rises, the float rises and only allows water to move into the condensate return line, **Figure 49.27**. The steam or hot water valve and condensate trap should be inspected for any needed maintenance. A strainer will be located in the vicinity of the condensate trap and should be cleaned during the off season. There are several other different types of traps; the one in **Figure 49.27** is shown as an example.

When the chiller is operated using hot water, the hot water valve should be observed for proper operation and to ensure that there are no leaks. Gas- or oil-operated machines will need the typical maintenance for these fuels. Oil in particular requires maintenance of the filter systems and nozzles.

With the absorption chiller, the purge system may require the most maintenance. If the chiller has a vacuum pump, regular oil changes will help to make it last longer. Li-Br is salt and will corrode the vacuum pump on the inside. It is not recommended that a vacuum pump used for a compression cycle system be used on an absorption system because of the corrosion problem, but if one is used, extra care should be taken to change the oil immediately after use.

Any Li-Br that is spilled in the equipment room will rust any ferrous metal it contacts. It is good practice to keep the area where it is handled washed with freshwater to prevent rusting. The machine and other equipment in these areas may need to be painted periodically for additional protection. The machine room must also be kept above freezing or the refrigerant (water) inside the machine will freeze and rupture the tubes. Unlike the other chillers, water in the absorption chiller cannot be drained and must be protected.

The system solution pump or pumps may require periodic maintenance. A check with the manufacturer will detail the type and frequency of this maintenance. The water pumps need regular maintenance as with any other system.

The cooling water must be maintained by cleaning and water treatment. Do not forget blowdown water. Absorption chillers operate at a higher cooling tower water temperature than compression cycle chillers and good water treatment maintenance is vital because the tubes will become fouled much quicker. Absorption chillers may need the tubes probed with an eddy current probe on a more regular basis than

Figure 49.27 (A) and (B) The condensate float trap for a steam coil.

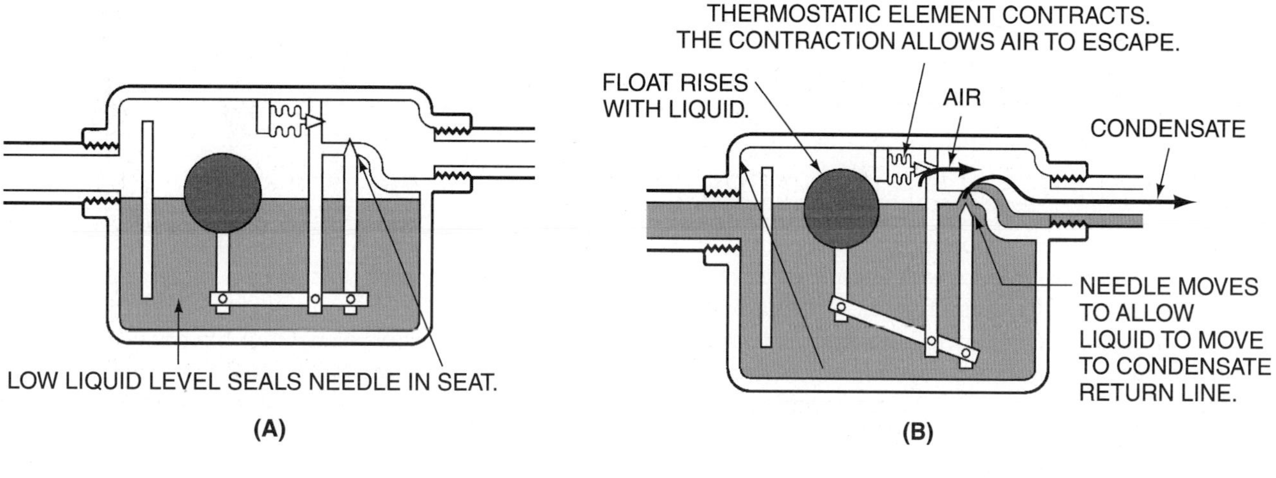

compression cycle chillers because of the higher temperatures in the condenser section. The tubes are stressed when the heat source is applied to these tubes, and they expand and then contract when cooled during the off-cycle. Checking with the manufacturer is a good idea.

49.10 GENERAL MAINTENANCE FOR ALL CHILLERS

Technicians who maintain chillers must be well qualified and should stay in contact with manufacturers to learn about the latest training classes and advice. Manufacturers receive feedback from all over the world on their equipment and know what is happening, such as premature failures. They should share any potential problems with the technician to prevent future problems with their equipment. Technicians should attend factory seminars and schools and make friends with the factory technicians. The technician who works on or observes work on large equipment can learn what to do during all service procedures. Because factory technicians perform much of the service on large-system chillers, the operating technician will be present to observe the service procedures and assist the factory technician. These procedures may be anything from routine checking of the controls to major overhaul of the equipment. The operating technician should be able to recognize the procedures. When the factory technician leaves, the operating technician must operate the equipment.

Refrigerant management is a big factor in any chiller application because chillers contain large amounts of refrigerant. The job site often has a supply of refrigerant on hand for emergencies because if the site is out of town it may take several days to obtain refrigerant, and this may not be satisfactory. The operating technician should be fully aware of the paperwork that must be kept current for the system. When a comfort-cooling chiller contains more than 50 lb of refrigerant, which most will, the technician must be aware that adding refrigerant to the machine just because it loses refrigerant is not satisfactory. There are EPA and/or state regulations that dictate allowable leak rates, repair requirements, and recordkeeping requirements. Technicians should stay up-to-date with the most current regulations for the equipment that they are responsible for servicing or maintaining. The technician must be aware of all refrigerant purchased and where it is used and have a recordkeeping system.

49.11 LOW-PRESSURE CHILLERS

Low-pressure chillers are probably the most difficult to understand because they operate with both a positive pressure in the condenser and a vacuum in the evaporator while running. When they are shut down, both the condenser and the evaporator will be in a vacuum. The largest portion of the refrigerant is in the flooded evaporator section, and the hot gas from the condenser quickly condenses to a mass of liquid in the evaporator when shut down. Low-pressure chillers operate in a vacuum and will stay in a vacuum if the equipment room is not warm enough to keep the refrigerant warm. As mentioned earlier, a leak in a vacuum will pull air into the machine, and the purge unit will operate excessively. Unless the machine has a high-efficiency purge, it will purge too much refrigerant out with the air that was pulled in. For example, R-123 boils at about 83°F at 0 psig (atmospheric pressure). This means that in the winter when the equipment is not operating, it may sit in a vacuum. Many equipment rooms are maintained at about 60°F during winter. If the refrigerant in the R-123 chiller is below 83°F, the chiller will be in a vacuum. The owners will not want to keep the equipment room up to these temperatures, so chiller manufacturers provide ways to keep a slight positive pressure using heat in the chiller evaporator section, where the liquid refrigerant will accumulate. This heat can be thermostatically controlled with either warm water or electric heaters under the insulation of the chiller evaporator shell. It may be used to keep the chiller shell in a positive pressure to prevent air from being pulled into the shell, or it may be used to pressurize the chiller for service purposes.

Chillers may be pressurized using nitrogen pressure or heat added to the evaporator shell. Regulated nitrogen pressure can be used to quickly build pressure in the chiller shell, but it must be removed after the service call is completed. If the service involves recycling the refrigerant, nitrogen may be used to save time. If the same refrigerant is going to be put back into the chiller, heat may be the best choice so that the nitrogen will not have to be removed from the refrigerant.

Care must be used when the chiller shell is pressurized for any reason. The pressure must not exceed the stress level of the rupture disc. The rupture disc is a one-time-use device, and it is not inexpensive. The reason for the rupture disc is to keep the shell from splitting if excess pressures occur. For example, if there is a fire, it is better to lose the refrigerant than to have a shell split. If its pressure rating is exceeded, the disc will rupture (blow out), refrigerant will be lost, and the rupture disc will have to be replaced. The rupture disc rating is 15 psig, so most technicians will not pressurize a system beyond 10 psig. The head pressure on a system may operate at a condensing temperature of 105°F using cooling tower water, which would equate to about 8 psig for R-123. Since the rupture disc is located in the low-pressure part of the system, there is no chance that high head pressure will rupture the disc.

RECOVERING REFRIGERANT FROM A LOW-PRESSURE CHILLER

Once the chiller is pressurized to the point that the pressure inside the chiller is greater than the pressure in the recovery system, liquid is allowed to transfer to the recovery system. If the technician does not have a permanent storage cylinder

Figure 49.28 The recovery machine pulls vapor from the top of the recovery cylinder and puts it back into the chiller. This creates a low-pressure area in the recovery cylinder in relation to the chiller, and liquid moves into the recovery cylinder. Notice that there is no condensing taking place.

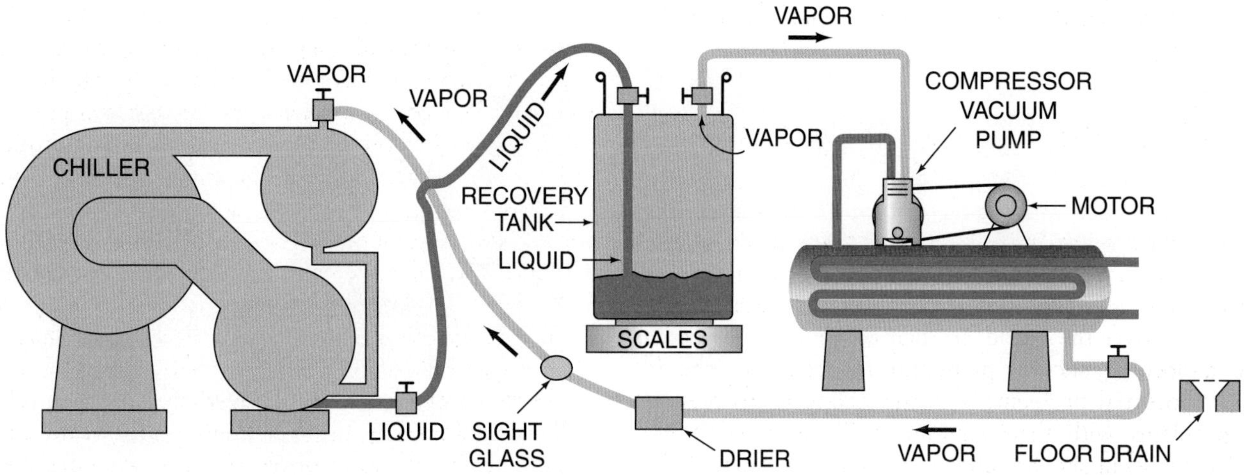

on the job, a portable cylinder will need to be moved in for the recovery. A push-pull recovery machine may also be used to move the refrigerant. If the liquid is only pushed out of the machine, the vapor would still remain. When only liquid is removed from a 350-ton chiller, there is still a considerable amount, possibly over 100 pounds, of refrigerant vapor still in the system. This does not meet the EPA requirement of removing the refrigerant to the required pressure level.

Vapor from the top of the recovery cylinder is pulled out of the top and pushed into the chiller to push liquid to the recovery cylinder, **Figure 49.28**. Notice that the water is not connected to the recovery unit condenser, so no condensation takes place. When all the liquid is removed, the system is reversed, the liquid connection is shut off, and vapor is pulled out of the vapor space of the chiller, normally at the top of the condenser, **Figure 49.29**. The city water is now connected to the recovery machine condenser and condenses the vapor to liquid, and it moves to the recovery cylinder. The required evacuation level in the chiller will depend on the recovery machine, the type of refrigerant in the chiller, and the weight of the system's charge. A technician should reference EPA and/or state requirements when recovering the charge from a chiller. After the service is completed, the system can be reversed and the refrigerant can be pushed back into the chiller. The push-pull equipment used is a recovery machine designed to handle large volumes of refrigerant. Some chillers can have upwards of 3000 lb of refrigerant. The recovery machines will have all the piping connections to fasten to the chiller and a water cooling circuit to cool the condenser that is used for

Figure 49.29 The flexible hoses are reconnected only to remove vapor from the cylinder. The vapor is condensed using water and moves to the recovery cylinder. The scales tell the technician when to stop adding refrigerant to the cylinder.

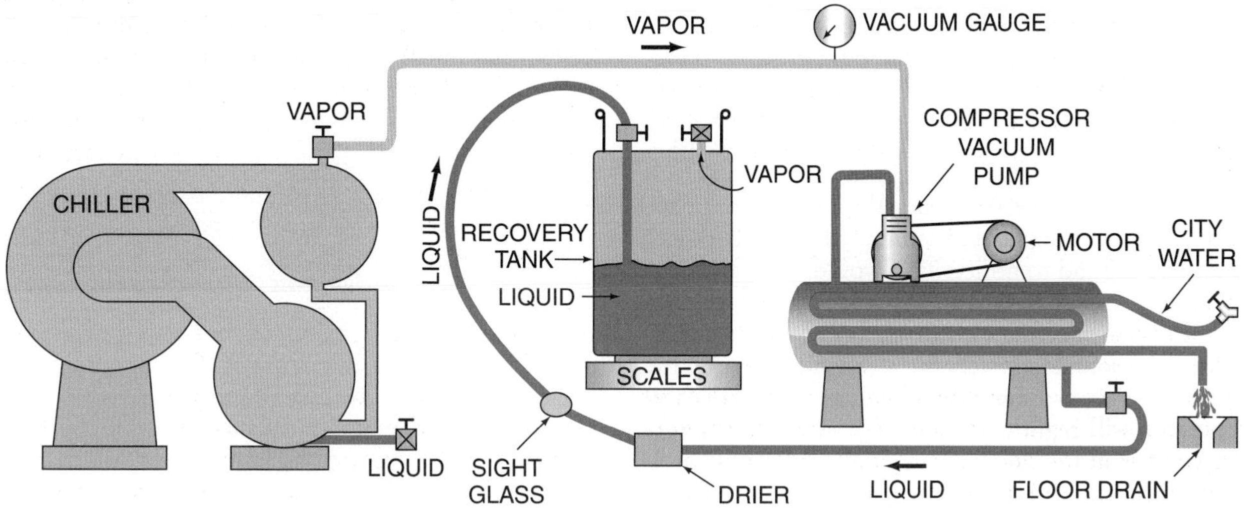

condensing vapor while recovering. Water from the public water supply is normally used in this condenser and permitted to be dumped down the drain. This is a procedure used only for service, so a permanent water connection is not needed.

SAFETY PRECAUTION: *When charging, vapor should be added to the system until the pressure inside the chiller is well above the pressure that corresponds to freezing. The freezing pressure is 20.3 in. Hg for R-123. The pressure in the machine must be above these two values before liquid refrigerant is introduced. It is good field practice to circulate water through the chiller during this process.* •

49.12 HIGH-PRESSURE CHILLERS

High-pressure chillers must be watched closely for leaks. A visual inspection is something that can be done anytime. Where refrigerant leaks out, oil will leak out also. Anytime a fresh oil spot appears, a leak should be suspected. An electronic leak detector or soap bubbles can be used to check the location for leaks. The soap bubbles must be cleaned off with water or dust will collect on the spot, giving the impression that oil has leaked and dust has collected on the oil. Some systems are designed so that the refrigerant may be pumped into the condenser for repairs on the low-pressure side of the system. Otherwise, the refrigerant may have to be recovered for major repairs. A recovery system and procedures similar to the push-pull system described above is used.

The recovery of high-pressure refrigerants is much like recovering low-pressure refrigerant by means of a push-pull system that will handle high-pressure refrigerants. The same rules apply for charging refrigerant into a high-pressure machine. The pressure in the machine should be above the freezing point of the water before allowing liquid to enter the machine, or a chiller tube may be frozen and broken.

49.13 REFRIGERANT SAFETY

Large equipment rooms become places where refrigerant vapors can gather should there be a leak. These vapors are all heavier than air and will collect without being noticed. They are colorless, almost odorless, and tasteless. Not all vapors are toxic, but they will gradually displace the oxygen in the room. A technician may simply faint when there may be no one around to notice. This would be fatal. Large equipment rooms now have detectors that can detect minute concentrations of refrigerant vapor and sound an alarm. When the alarm sounds, the technician should evacuate the room and only go back in with breathing gear. **Figure 49.30** shows two of these breathing devices, in a yellow case where they can be readily noticed—one is open and the other is closed.

Figure 49.30 This breathing apparatus is mounted where a technician can easily reach it and put it on in case of a refrigerant leak alarm. Courtesy Ferris State University. Photo by John Tomczyk

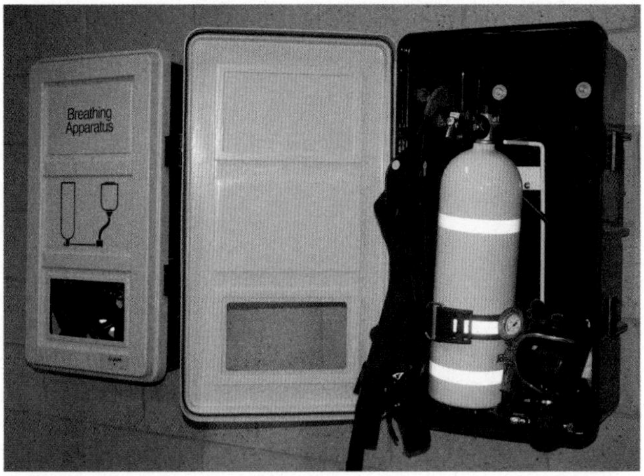

49.14 SERVICE TECHNICIAN CALLS

SERVICE CALL 1

A building manager calls and reports that the building is becoming hot. *The problem is that the cooling tower fan has a broken belt and the fan is off. The chiller is off because of high head pressure.*

The technician arrives and asks the manager for an explanation of what is happening. The manager goes with the technician to the basement where the chiller is located. It is a reciprocating water-cooled chiller. The chiller is off, and the technician looks around for the problem. He opens the control panel and notices that the reset button on the high-pressure control is out, signaling that the control needs resetting. The pressure drop across the condenser is normal, and the water temperature is normal (because the chiller has been off, the tower is able to lower the temperature without the fan). The technician resets the button, and the compressor starts. As the technician watches, the compressor starts to load up. Everything runs normally for about 10 min, and then the compressor sounds as though it is straining. The technician feels the water line entering the condenser. It is warmer than hand temperature. The cooling tower water is too hot. The technician shuts the compressor off rather than letting it be shut off by the high-pressure control.

The technician goes to the roof to look at the cooling tower, fully expecting to see it full of algae because the management does not conduct regular maintenance. When on the roof, he notices that the fan belt for the cooling tower fan is broken. He shuts off the disconnect to the

fan, locks it out, and tags it for safety. Writing down the numbers from the old belt, the technician goes to the truck for a new one. While at the truck, he picks up a grease gun for greasing the fan and motor bearings at the cooling tower. The belt is replaced, and the tower fan is turned on. It starts to turn, and the technician can see from the moist air leaving the tower that the water is loaded with heat.

The technician goes back to the basement and touches the water line leading from the tower to the condenser and notices that the water has cooled back to below hand temperature. He restarts the compressor and stays with it for about 30 min to make sure that it will stay on, and it does. He uses this to lubricate the pump and fan motors and, in general, to look for any other potential problems. The technician explains to the manager what happened and what was done to correct the problem. He also explains that he greased the motors, pumps, and fans while observing the system. He then makes a recommendation that management hire a water specialist to check the water on a regular basis.

The manager thanks the technician for doing a good and efficient job.

SERVICE CALL 2

A building owner reports that the building is hot and the chiller is off. The chiller is a low-pressure R-123 chiller and has an air leak. *The building maintenance personnel have been operating the purge for several days to remove air that is leaking into the chiller and are waiting for the weekend to leak test the machine. The purge water has been valved off and the purge condenser is not operable, causing high purge pressure and releasing refrigerant. This chilled-water system has a total refrigerant charge of 600 pounds.*

The technician arrives at the equipment room where the chiller is located. The building maintenance man is there and explains what has been done. The signal light indicates that the chiller is off because of freeze protection. The technician looks around and notices the purge pressure is very high, as the purge is still running on manual. On inspection, he notices that the purge relief valve is seeping refrigerant. The technician then checks the water circuit to the purge and discovers that there is no water cooling the condenser in the purge. The valve is opened to the purge water condenser. This system uses chilled water, which is now warmed up to 80°F, but will still condense the refrigerant in the purge.

The technician then starts the compressor in the 40% mode and watches the suction pressure; it is not too low, so the load is advanced to 60%. The suction pressure begins to drop. The technician looks in the sight glass at the evaporator for the refrigerant level, and it is low. There is some refrigerant at the sight glass, so the technician gets set up to add refrigerant. The refrigerant drum is piped to the charging valve and the valve is opened. Meanwhile, the technician places a thermometer in the well in the evaporator to check the refrigerant temperature. The chilled-water temperature leaving the chiller is down to 55°F, and the thermometer in the refrigerant reads 43°F. This is a 12°F approach, and it should be 7°F according to the log sheet that was left at start-up. Two hundred pounds of refrigerant is added before the compressor is operated at full load, and the refrigerant approach is lowered to 7°F. The technician is satisfied with the system charge. The chilled-water temperature pulls down to 45°F and the compressor begins to unload in about 30 min. The purge pressure is running normally and not relieving continuously as it had been.

The technician talks to the maintenance man and sets up an appointment to perform the leak check at a later time. The leak in this system must be repaired, as the technician has added 200 pounds of refrigerant to the system, which normally operates with a 600-pound charge. One-third, or 33%, of the charge has leaked from the system so the leak is quite large and must be found and repaired.

SERVICE CALL 3

A building manager reports that the chiller is off and the building is hot. *The problem is that the flow switch in the condenser water circuit has a broken paddle, and the field control circuit is not passing power to the chiller and telling it to start.*

The technician arrives and goes straight to the chiller. The manager is in the chiller room, pacing the floor. He explains that the chiller really needs to be running and there are a lot of complaints of no cooling. The technician gets the manager calmed down with an explanation that it shouldn't take long. This chiller has a ready light on the control panel that indicates when the field circuit is calling for cooling. The light is off. The technician first looks to see what the chilled-water temperature is. It is 80°F; the controller is calling for cooling. The technician then checks the water gauges that register the pressure drop across the chiller and the condenser. The correct pressure drop registers across both, so there is water flow. The technician then goes to the wiring diagram for the system and looks to see what other controls are in this circuit. The flow switches and the pump interlocks are all in the circuit. It is a matter of tracing the wiring to see what has caused the chiller to shut down.

The technician realizes that the flow switches are likely to be a problem because these devices move a lot since they are located in a moving water stream. The compressor switch is turned off so that the compressor will not try to start during the control testing. The technician removes the cover from the chilled-water flow switch and checks voltage across it. There is no voltage.

NOTE: *If the switch were open, there would be voltage. The cover is replaced and the same test is performed on the condenser-water flow switch and there is voltage. This flow switch is open. The technician jumps the switch with a jumper and the light on the chiller panel lights up. This is the problem.* •

The technician leaves the jumper on the flow switch and starts the chiller. Flow in the condenser-water circuit is not as critical as it would be in a chilled-water circuit. If the water flow were to stop in the condenser circuit, the compressor would shut down because of high pressure. **NOTE:** *The chilled-water flow switch should not be jumped with the compressor running because that could freeze the chiller.* •

The technician explains to the building manager that the flow switch paddle must be replaced. The manager requests this be done after hours, so the technician leaves and will return at 5:00 PM with the parts to repair the switch.

The technician returns to the equipment room to make the repair. The chiller is off for the day because of the energy management time clock. The condenser water pump is turned off and locked. The valves on either side of the condenser system are valved off and the water is drained from the condenser. The flow switch is removed, repaired, and reinstalled. The water valves are turned on, and the condenser system is allowed to fill. Air is vented out of the water system at the high places in the piping. The technician then turns the water pump on in the manual mode and verifies that the flow switch is passing power. By leaving the meter leads on the flow switch and shutting off the water, it can be verified that the flow switch will also shut the system off if the water flow is stopped. The system is made ready to run the next morning when the time clock calls for cooling, and the technician leaves.

The technician calls the building manager the next morning and the manager says, "That was a great job you did for us yesterday. Sorry if I was grouchy, I really was under a lot of pressure." The technician explained that it is all in a day's work—most people are unhappy when their system is not working.

SERVICE CALL 4

A building manager telephones to report that the centrifugal chiller in the basement of the building is making a screaming noise. When he went to the door, he was scared to go in. *The problem is that the cooling tower strainer is stopped up and restricting the water flow to the condenser. The centrifugal has high head pressure causing a surge, which produces a loud screaming sound but is normally otherwise harmless.*

The technician can hear the compressor screaming from the parking lot. When he enters the room, he turns the machine capacity control down to 40% to reduce the load. The machine has high head pressure. The manager shows up and explains his fear of the loud machine. The technician explains, "There is nothing to be afraid of, there are safety controls that prevent damage, but you did the right thing calling for service."

A look at the pressure gauges across the condenser shows that there is reduced water flow, so the technician proceeds to the cooling tower on the roof.

A look at the cooling tower basin shows what has happened—there are tree leaves in the basin. The technician takes off the cover from the cooling tower and removes the leaves from the strainer. Immediately, the water flows faster. The technician replaces the cover to the tower basin, goes to the basement, and turns the capacity control up to 100%.

The technician then talks to the manager about cleaning the cooling tower after hours when the building air-conditioning is off and gets permission to do this.

When the technician returns that afternoon, the system is off because of the time clock in the energy management system. He cleans and washes out the tower, which contains many small leaves. The technician then goes to the basement and prepares to check the condenser for leaves. He valves off and drains the condenser circuit. Then he removes the marine water-box cover on the piping inlet end of the condenser where any trash would accumulate. There are small leaves in the condenser that have passed through the cooling tower strainer. These are cleaned out. While the water-box cover is removed, the technician uses a small penlight and examines the tubes. There is no scale or fouling of the tubes, which is good news.

The water-box cover is replaced, and the valves are opened to the condenser. The technician waits a few minutes for the tower to refill with water after the condenser has been filled and then starts the condenser water pump in the manual mode and verifies there is water flow. This is a necessary step if the condenser is large enough to drain most of the water from the tower; the tower must have time to refill before starting the pump. Now the technician knows that when the machine starts in the morning, there should be no cooling tower water problems.

The technician calls the manager in the morning and verifies that the cooling is working. The technician explains what the problem was and what was done to make the repair. He explains that the condenser water-box cover was removed and the tubes were checked and look good. The manager thanks the technician and says, "Thanks for the explanation, I was really afraid of that machine yesterday." The technician says, "They make a really frightening sound, but it is really only a whistle of the gases in the machine."

SERVICE CALL 5

A building maintenance person calls and says that the chiller is off. A light that says "freeze control" is lit. He wants to know if it should be reset. It is a mild day without much need for air-conditioning. The dispatcher tells him no; a technician will be over within 30 min. *The problem is that the main controller is not calibrated and is drifting, causing the water temperature to fall too low.*

When the technician arrives, the maintenance person leads the way to the chiller, which is on the roof in a

penthouse. The technician checks the water flow through the chiller by comparing the pressure in and out to the log sheet, and it is correct. The freeze control is reset, and the chiller compressor starts. When it starts to lower the water temperature, the technician watches the controller. It is a pneumatic controller and as the water temperature drops, the air pressure should drop from the 15 psig furnished. When the water temperature reaches 43°F, the pneumatic air pressure should be 7 psig. When the water temperature reaches 41°F, the air pressure should be 3 psig and the chiller should shut off. This does not happen; the chiller continues to run and the water temperature reaches 39°F without shutting off. The technician then adjusts the controller to the correct calibration. The chiller shuts off.

The technician waits for the chiller to restart when the building water warms up; this takes about 30 min. When it is established that the chiller is operating correctly, the technician explains to the maintenance person what was happening and said that the right thing to do was to call for service. He explained that one should never jump out a freeze control and what the results might be. The technician then leaves.

SERVICE CALL 6

A building manager reports that the chiller is off and the building is getting hot. The chiller is a centrifugal chiller. *The problem is that the building technician has been adjusting the controls and has adjusted the load-limiting control, trying to get the motor to operate at more capacity to reduce the building temperature faster. The compressor was pulling too much current, which tripped the overload device.*

The technician goes to the equipment room with the building technician. The building technician does not reveal that the load-limiting control has been adjusted. The technician checks the field control circuit and discovers that all field controls are calling for cooling. A further check shows that an overload device has tripped. The technician knows that something has happened and that he should use caution.

The technician shuts off the power to the compressor starter and checks the motor windings for a ground or a short circuit from winding to winding. The motor seems fine, so power is restored. There is not much else to do except to start the motor and observe.

The technician starts the compressor in the 40% mode. When the compressor starts to load up, the current is observed. Start-up data show that the compressor should draw 600 A at full load and 240 A at 40%. The compressor is pulling 264 A; this is too much, so the technician turns the load limiter back to 240 A and then turns the selector up to 100%. The compressor begins to load up and reaches 600 A and throttles to hold this current. The technician then turns to the building maintenance technician and asks whether anyone has adjusted the demand limiter. The building technician admits that this has been done and is told that these controls should not be adjusted, because they are reliable. The technician then uses nail polish to mark the setting on the control so that it will be obvious if it is adjusted again.

The technician realizes that the controls belong to the building owner, but they should be adjusted only by a qualified technician. He explains to the building technician that the chiller will only pull the system down so fast, that the limiting factor is the maximum amperage for the compressor. If the system is not getting cool on time, the chiller should be started up earlier, not made to work harder in a short period of time. The technician shows the maintenance technician the ammeter on the starter door where 600 A is marked as full load.

SERVICE CALL 7

A customer calls to say the building is becoming warm and that water is still flowing over the cooling tower, but there is no cooling. The unit is a 300-ton centrifugal with a hermetic compressor. *The problem is the compressor motor is burned and will need to be replaced or rewound.*

The technician goes straight to the equipment room on the 10th floor of an office building. He notices that the building is hot, so there is a problem with the chiller.

The technician checks the field control circuit and discovers that all signals are correct; it is calling for cooling. The technician then goes to the starter and discovers that there is no line voltage to the starter. He then goes to the main power supply and discovers the breaker is tripped. There must be something very wrong.

The technician locks out and tags the breaker and uses a Megger (megohmmeter) to check the motor leads in the starter to ground and discovers a circuit to ground. There is always a possibility that the problem may be in the wiring between the starter and the motor terminals on the motor, so the technician removes the leads from the motor terminals. The motor is checked at the terminals without the wiring and it shows a circuit to ground.

The technician goes to the building management for advice as to the next step. He explains that the motor will have to be removed from the equipment room for replacement or rebuild. 4R *Because the motor is a hermetically sealed motor, the refrigerant will have to be recovered.* 4R The technician explains that there are two possibilities for making the repair. The motor can be replaced with a factory motor, which may take several days for delivery, or the motor may be rebuilt at a local rebuild shop that is approved for work with hermetic motors. Building management gives permission to proceed with the rebuild job to save time because the building must have air-conditioning. Like many modern buildings, no windows open and the computers rely on the air-conditioning.

The technician calls the home office and requests backup help and the tools recommended for the job. These would

include a hoist to lift the motor and a dolly to transport it onto the elevator and out to the truck, which is equipped with a lift gate.

While the helper is getting the tools together, the technician prepares to pull the motor. 4R *The refrigerant is recovered from the machine using the recovery system brought by the helper. Water in the chilled-water circuit and condenser water are circulated through the machine during recovery to prevent the tubes from freezing and to aid in boiling the refrigerant out of the machine. The system refrigerant is stored in DOT-approved recovery cylinders. The machine uses R-123 and the refrigerant will be checked for acid content and recycled for reuse in the machine, if needed. Once the refrigerant has been recovered from the system, the technician breaks the vacuum with dry nitrogen that has been brought by the helper.* 4R Breaking the vacuum with dry nitrogen will minimize oxidation inside the machine when it is opened to the atmosphere. The machine is made of steel inside and it will quickly rust.

The technician then removes the wires from the motor terminal box and starts to disconnect the compressor from the motor. When the motor is free to move, the hoist is set up and the motor is set on a rolling dolly to transport it to the truck. The motor is then rolled onto the lift gate, lifted to the back of the truck, and secured. The helper takes the motor to the motor shop for rebuild. This will take about 48 hours. It is now 2:30 PM on Tuesday.

The technician goes back to the equipment room and prepares the equipment for when the motor is returned. A sample of the refrigerant and a sample of the refrigerant oil from the compressor sump are pulled and shipped by next-day carrier to a chemical analysis firm. The report can be called in by Wednesday afternoon to determine what must be done about the refrigerant. The refrigerant and oil sample do not smell as though they have very much acid, but the analysis will reveal the facts.

The technician checks the contacts in the starter. There must have been quite a surge of current when the motor grounded. The contacts are good and do not require replacement.

The oil heater is turned off, locked, and tagged, and the oil is removed. The technician then cleans all gasket sealant from all compressor flanges and is now ready for the motor to be returned. Time has been saved by doing all of this before the motor is brought back.

Thursday morning, the technician checks with the rebuild shop and is told the motor will be ready by 11:30 AM. He arranges for the helper to pick up the motor and a compressor gasket kit. They meet at the building at 1:00 PM and proceed to install the motor. The motor is moved to the equipment room and lifted back into place. All gaskets are installed and everything is tightened. This system has a history of being very tight, so the only leak checking that needs to be performed is on the fittings, including the gasketed flanges that were removed and in the area where the work was done.

When all is ready, the machine is pressured with nitrogen to 10 psig for a leak test. The machine is leak checked and found to be good. The nitrogen is exhausted, and it is time to pull the vacuum. The motor leads are connected to the motor terminals.

The vacuum pump is installed and started. It will take about 5 hours to pull the vacuum down to 1000 microns on the micron gauge and it is 5:30 PM. The technician leaves the job and decides to come back at 10:30 PM to check the vacuum. If the vacuum can be verified this evening and the vacuum pump shut off for 8 hours, the machine can be started in the morning if all is well. The technician checks the vacuum at about 11:00 PM, and the micron gauge reads 700 microns. The vacuum pump is valved off, and the technician leaves to return in the morning.

When the technician arrives at 7:30 AM, the micron gauge reads 900 microns, so the machine is leak-free and it is time to charge the system. The refrigerant and oil report came back "OK," so no extra precautions need to be taken with the refrigerant. The technician starts the chilled-water pump and the condenser water pump in preparation for start-up. A drum of refrigerant is connected to the machine at the drum vapor valve. The valve is slowly opened to the machine and vapor is allowed to enter until the pressure is above that corresponding to freezing. This is important because if liquid refrigerant is allowed to enter the machine at such a low pressure, the tubes may freeze. The chilled water is circulating, but if there is one tube that does not have circulation, this tube may freeze. The pressure in the machine must be above 19 in. Hg vacuum to prevent freezing while charging liquid.

When the machine pressure reaches 19 in. Hg vacuum, the technician starts charging liquid refrigerant. While he is charging refrigerant, the helper pulls the oil in the oil sump using the machine vacuum. When the oil is charged, the oil heater is turned on to bring the oil up to temperature.

The technician is not able to charge all of the refrigerant into the machine using the vacuum because the chilled-water circuit warms the refrigerant and pressure is soon raised. The technician then changes over and pushes the remainder of the refrigerant into the machine by pressuring the drum with nitrogen on top of the liquid and drawing liquid from the bottom of the drum, **Figure 49.31**.

The technician starts the purge pump to remove any nitrogen that may have entered the system while charging, even though he was careful and could see no nitrogen in the clear charging hose while charging. The purge pressure does not rise, so the machine should be free of nitrogen. The technician has closed the main breaker and is ready to start the machine. The machine is started and quickly turned up to 100% to reduce the building temperature as quickly as possible. There is no apparent reason that the motor developed a circuit to ground; the motor rebuild shop could not find any problem, so the motor failure is accounted for as random failure.

Figure 49.31 Using nitrogen to charge refrigerant into a low-pressure chiller. The clear plastic tube will tell the technician when any nitrogen is passing into the system.

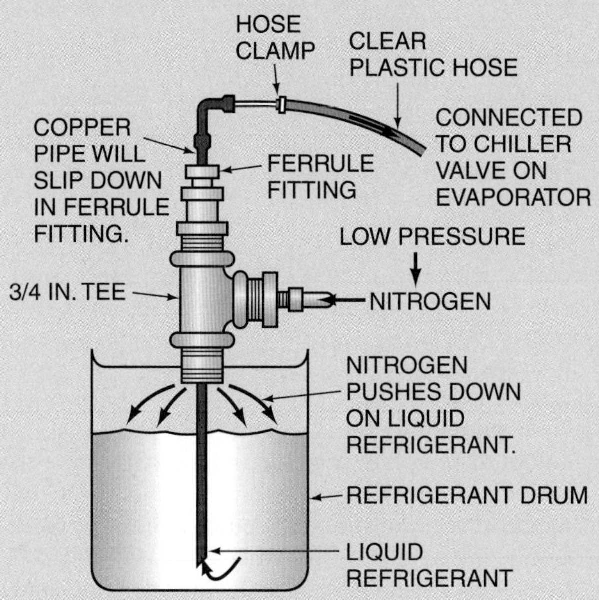

The chiller is quickly pulling the building water down to 45°F leaving-water temperature. The technician calls the building manager and explains that the chiller is on and operating normally. It will take several hours for the building temperature and humidity to return to normal. The building manager thanks the technician for doing a fast, professional job. The technician stays with the chiller for 2 hours and completes entries (including the motor current) in an operating log. When it is determined that the system is functioning properly, the technician leaves.

SERVICE CALL 8

The technician is called to do routine maintenance on a reciprocating chiller. *This chiller has not had maintenance in several years. The last technician did not take proper care of the service valves and the suction service cap is rusted tight to the valve body.*

When the technician arrives and talks to the building manager about what should be done, the manager says, "Give the system a good check up." It is determined that the suction and discharge pressure should be measured to determine whether the chiller, an air-cooled chiller using R-134a, is operating correctly. The technician removes the protective cap from the discharge service line and installs a high-pressure gauge. The head pressure is 146 psig and the outside temperature is 80°F, so the head pressure is good (80 + 30 = 110°F condensing temperature).

The technician then attempts to remove the cap from the suction service valve and discovers the cap is rusted tight. He uses a soft-face hammer and taps hard on all of the flats on the cap, then uses a hex-jaw wrench, **Figure 49.14**, to loosen the cap. It still will not turn. He uses a heavier steel hammer to tap the flat surfaces on the valve, and it still will not loosen. The technician then uses the hammer to tap the handle of the wrench and sees the cap begin to turn. It is not good practice to hit a wrench handle with a hammer, but sometimes this is necessary. If the valve cap is struck too hard, it may split because it is cast iron. Cast iron can be brittle and breaks easily. Spraying a penetrating oil would help, but doing that will gather dust and lead the next technician to believe there is a leak.

When the valve cap is removed, the technician sees there has been moisture inside the cap for a long time. The valve stem is really rusty with scale buildup on it.

He cleans the valve stem with a fine sand tape until there are no more high spots on the stem. He then slightly loosens the packing gland and turns the valve stem forward until a gauge reading is obtained. The suction pressure reads 39 psig. The chilled water in is 50°F, and out, 45°F. The chiller would normally have a 10°F difference and the suction pressure would normally be 36 psig. The chiller has the correct water temperature, so it is running at half capacity to fit the building's load. All is well.

The technician shuts the system off for about 30 min to give the chilled-water temperature a chance to warm up—at which point the technician will start the system up and run it at full load. Meanwhile, he greases the pump motor and examines the inlet to the condenser to see if it needs cleaning. The fan motors are examined, and they are permanently lubricated motors.

When the chiller is started back up, the suction pressure reduces to 36 psig, the compressor is now pulling full-load amperage, and the sight glass to the expansion valve is full. The chiller is running as it should. The technician then places a thermometer lead to the refrigerant liquid line, just before the expansion valve, and determines that the unit is operating with 15°F of subcooling. This tells the technician that the refrigerant charge is correct for the chiller.

The technician back seats both service valves and tightens the packing gland nut and cleans the inside of the valve cap of the suction service valve to remove the rust. He then takes a small wire brush and cleans the threads on the valve body. The technician wipes the valve stem, sprays oil-free silicone spray on it and on the inside of the valve cap, and turns the cap onto the valve body. It is tightened with the square jaw wrench.

The technician then uses an electronic leak detector and leak checks all around the service valves where service has been performed. He then reports the results to the building manager. The system is good to go for another season.

REVIEW QUESTIONS

1. The oil pump for a ______ compressor is driven from the actual compressor shaft.
   A. centrifugal C. scroll
   B. reciprocating D. linear
2. Most reciprocating compressors have
   A. a separate oil sump.
   B. no clearance volume.
   C. a king valve.
   D. a crankcase heater.
3. When a winterized reciprocating compressor is restarted
   A. the crankcase heater must be deenergized 24 hours prior.
   B. the low-pressure control must be isolated from the system.
   C. it will start up unloaded until the oil pressure builds up.
   D. it will run at ½ capacity for the first 24 hours.
4. A centrifugal compressor starts unloaded and only begins to load up after the motor is up to speed. This is accomplished using
   A. prerotation guide vanes.
   B. crankcase heat.
   C. a separate oil sump.
   D. permanently installed pressure gauges.
5. What system component(s) must be started before the compressor in a chilled-water system centrifugal chiller?
   A. Crankcase heater and the oil pump
   B. Chilled water pump, condenser water pump, and the oil pump
   C. Cooling tower fan motors
   D. Chiller barrel heater, oil pump, and cooling tower fan motors
6. When condenser tubes become fouled, they may be cleaned by
   A. using a steel brush.
   B. using a brass brush.
   C. using a chisel and hammer.
   D. running a mixture of sand and water through them.
7. A technician can verify waterflow through the condenser of a chiller by
   A. checking the superheat and comparing it to the log.
   B. running the purge unit and checking its discharge for bubbles.
   C. checking the amperage of the evaporator water pump.
   D. checking the pressure drop from the inlet to the outlet.
8. Before starting up a chiller
   A. the chiller and condenser water should be drained.
   B. nitrogen should be added to the system to raise the pressure for a leak test.
   C. the crankcase heater should be energized for a period, typically 24 hours.
   D. the compressor oil should be drained, cooled, and then recharged.
9. The circuit that is used to energize the chiller's compressor is referred to as the
   A. compressor interlock circuit.
   B. field control circuit.
   C. water pump interlock circuit.
   D. main circuit.
10. On a centrifugal chiller start-up, the oil sump temperature should be at least ______ °F, depending on the manufacturer.
    A. 135°F to 165 C. 75°F to 105
    B. 105°F to 135 D. 50°F to 85
11. The valve type used most commonly on small lines are
    A. ball valves.
    B. service valves.
    C. gate valves.
    D. diaphragm valves.
12. Before moving the valve stem of a service valve that has accumulated rust
    A. the refrigerant should be pumped into the receiver.
    B. the protective cap should be put back in place.
    C. the rust should be sanded down with a fine sand cloth.
    D. the refrigerant should be recovered.
13. Water-cooled chillers must have water treatment to prevent
    A. refrigerant from migrating to the compressor crankcase.
    B. minerals and algae from forming in the water circuit.
    C. excessive heat transfer due to minerals in the water.
    D. the evaporator from freezing under low-load conditions.
14. When checking a compressor for a grounded winding, the recommended tool is a(n)
    A. ohmmeter. C. ammeter.
    B. digital multimeter. D. megohmmeter.
15. When testing a motor circuit to ground, it should typically not have less than
    A. 100 ohms. C. 100 megohms.
    B. 1000 ohms. D. 100 milliohms.
16. When cleaning an air-cooled condenser, the water should be
    A. sprayed from the inside out, or backwashed.
    B. sprayed from the outside in, or backwashed.
    C. heated to just under boiling to dissolve the debris.
    D. sprayed on the condenser while the system is running.
17. When inspecting the tubes of a chiller for potential tube failures
    A. they should be pressurized to 30 psig with nitrogen.
    B. an eddy current test can be used.
    C. each tube should be removed and be individually inspected.
    D. a fluorescent dye can be used.

18. The most common valve used in water and steam circuits which has a replaceable seat is the
    A. ball valve.
    B. service valve.
    C. globe valve.
    D. gate valve.
19. Another common valve used in water and steam circuits that does not have a replaceable seat is the
    A. ball valve.
    B. service valve.
    C. globe valve.
    D. gate valve.
20. In an absorption chiller, a common refrigerant is
    A. a lithium-bromide (Li-Br) solution.
    B. propane.
    C. CO_2
    D. R-134a.
21. The rupture disc rating on most chillers is
    A. 5 psig.
    B. 10 psig.
    C. 15 psig.
    D. 20 psig.
22. When charging refrigerant into a low-pressure chiller that is under a vacuum
    A. it should be charged as a liquid until the saturation temperature is above freezing.
    B. it should be charged as a vapor until the saturation temperature is above freezing.
    C. it should be charged as a vapor until the saturation temperature is at room temperature.
    D. it should be charged as a liquid until the saturation temperature is at room temperature.
23. When inspecting a high-pressure chiller for leaks, a technician should look for
    A. puddles of refrigerant.
    B. fresh oil spots.
    C. excessive purge unit cycling.
    D. rust spots.
24. Large equipment rooms are equipped with detectors that will
    A. pump the refrigerant into a receiver if the system begins to leak.
    B. send a message to a technician if the chiller begins to operate erratically.
    C. dump the refrigerant outdoors if a fire is detected.
    D. sound an alarm when they detect minute concentrations of refrigerant.
25. On an absorption chiller, the purge discharge may be monitored by piping the vacuum pump exhaust
    A. into a glass of water and looking for bubbles.
    B. back into the chiller.
    C. to the outdoors and looking for oil.
    D. to a nearby drain and watching for water vapor.
26. If an absorption chiller requires excessive purge operation
    A. the system may be overcharged.
    B. the system may be undercharged.
    C. the system may have leaks or is producing excess hydrogen.
    D. the system may have multiple fouled tubes.

UNIT 50

COMMERCIAL, PACKAGED ROOFTOP, VARIABLE REFRIGERANT FLOW, AND VARIABLE AIR VOLUME SYSTEMS

OBJECTIVES

After studying this unit, you should be able to

- describe a rooftop package unit.
- describe the refrigerant circuits in a rooftop package unit.
- list the five air-conditioning processes that are often performed by rooftop units.
- describe the three airstreams that are handled by a rooftop package unit.
- describe the two types of air-moving devices commonly found on rooftop package units.
- list the steps required to properly install a rooftop unit.
- explain how to prepare to make roof penetrations.
- explain the function and installation of a roof curb.
- explain how rooftop units are typically brought up to the roof.
- describe various parts of a crane used to rig air-conditioning equipment onto a roof.
- demonstrate the hand and body signals used to communicate with the crane operator.
- explain the purpose and general function of an economizer.
- explain the function of fixed and differential sensors as they apply to economizer operation.
- describe the four basic modes of economizer operation.
- explain the importance of ASHRAE Standard 62 as it pertains to the proper design of ventilation systems.
- describe the concept of demand control ventilation (DCV).
- explain how carbon dioxide sensors can be used to help control the ventilation process.
- explain the operation of a constant air volume (CAV) system.
- explain the concept of a variable air volume (VAV) system.
- describe the blowers used on VAV systems.
- describe the four common configurations of VAV boxes or terminal units.
- explain the function of a reheat coil.
- describe the two types of valves used to control the flow of hot water through a reheat coil.
- describe four common configurations for chilled-water VAV systems.
- explain how chilled-water coils can be protected from freezing.
- explain the concept of variable refrigerant flow (VRF).
- explain the difference between a two-pipe VRF installation and a three-pipe VRF installation.
- explain the operation of a dry cooler.
- explain the operation of a chilled beam system.

SAFETY CHECKLIST

- ☑ When using extension ladders, be sure the ladder's feet are positioned level on solid ground, the top of the ladder extends at least 3 feet above the roof line, and the ladder is properly secured to prevent it from falling over.
- ☑ Always inspect all ladders prior to use to ensure that all rungs and locking mechanisms are operational and not damaged.
- ☑ Exercise extreme caution when working on roofs to prevent trip-and-fall accidents.
- ☑ To reduce the chance of injury, make certain that the roof is designed to handle the weight of the equipment.
- ☑ When making roof penetrations, be sure to seal off any areas under the penetration location so that injury does not result from falling debris.
- ☑ When working on rooftop equipment, make every effort to stay as far away from the edge of the roof as possible.
- ☑ Never stand under equipment being lifted or rigged onto a roof.
- ☑ Use all pertinent pieces of personal protection equipment (PPE), such as hard hats, eye protection, gloves, work boots, back support, and safety harnesses.
- ☑ Make certain that conditions are safe before accessing the roof.
- ☑ Individuals should not carry tools, parts, or supplies up a ladder. At least one hand must be holding onto the ladder at all times.
- ☑ When working on a roof, take proper precautions to prevent heat-related injuries.
- ☑ Silver-colored roofs reflect sunlight and can cause sunburn. Be sure to wear protective clothing and use appropriate sunscreen.
- ☑ Before installing or servicing equipment, make certain to read all literature and observe all guidelines and recommendations.
- ☑ Know the type of refrigerant that is contained in a system to avoid injury and possible equipment damage.
- ☑ Do not ingest glycol or other antifreeze products. Although sweet to the taste, ingested ethylene glycol can form oxalic acid, which is toxic.

50.1 ROOFTOP PACKAGE UNITS

Package units are factory-assembled, factory-charged, self-contained systems that contain all of the major system components in a single housing, **Figure 50.1**. Since the compressors, condensers, metering devices, and evaporators are all in the same cabinet, installing package equipment does not require the installation crew to install refrigerant lines, leak check or evacuate the system, or adjust the refrigerant charge of the unit. Information regarding the amount and type of refrigerant in the unit is provided on the unit's nameplate. The amount of refrigerant indicated on the unit's nameplate is the total refrigerant charge for the unit, not simply a holding charge as in the case of a split-type system. This is useful information because, in the event of a total refrigerant loss, the service technician can, after repairing the system, weigh in the exact amount of refrigerant needed.

The refrigerant piping on a rooftop unit can be configured either as a single circuit or a multiple circuit. A single-circuit system typically has one compressor, one condenser circuit, one metering device, and one evaporator circuit. When the compressor is energized, refrigerant will flow through the entire refrigerant circuit. Varying the capacity of a piece of cooling equipment with single-circuit piping is more complicated than with a multicircuit piping arrangement. In a multiple-circuit configuration, there is more than one active refrigerant circuit. For example, a 10-ton, two-circuit setup might have two separate circuits with separate 5-ton compressors. When the heat load on the system is low, one compressor will be operating, allowing the system to operate as a 5-ton system. As the load on the structure increases, the second stage or circuit will be energized, allowing the system to operate as a 10-ton system.

A more sophisticated arrangement might utilize circuits of different capacities, such as a 4-ton circuit and a 6-ton circuit, **Figure 50.2**. When the load on the structure is low, the 4-ton circuit might be energized. As the load on the system increases, the 4-ton circuit can cycle off as the 6-ton circuit comes on line. If the load on the system continues to increase, the 4-ton circuit will then cycle on to bring the system capacity to 10 tons. When configured in this manner, the system can operate at 4, 6, or 10 tons as opposed to the 5 or 10 tons if the circuits are similarly sized. Thus, with different size refrigeration circuits, an additional capacity step is added. Of course, using inverter-controlled compressors will allow the system to closely match the heat load on the space.

Figure 50.1 A rooftop packaged unit. Photo by Eugene Silberstein

Figure 50.2 A 10-ton, two-circuit piping arrangement with two 5-ton compressors. Photo by Eugene Silberstein

Although not required, packaged units are typically located on the roof of the building, and when so located, the acronym RTU (rooftop unit) is commonly used. Depending on the building, several rooftop package units may be needed, **Figure 50.3**. When multiple rooftop units are used to condition the air in a building, it is common practice to label them and their corresponding thermostats, **Figure 50.4**. This makes it easier for the service technician to identify the piece of equipment that is being controlled by a particular thermostat.

Rooftop packaged units are often configured with a number of features that allow for total air treatment. These features include cooling, heating, exhausting, ventilating,

Figure 50.3 This building has several rooftop units to serve the various areas in the structure. Photo by Eugene Silberstein

Figure 50.4 Typical identification label on a rooftop unit. Photo by Eugene Silberstein

and filtering. In general, a rooftop packaged unit operates as follows. Air from the conditioned space is brought up to the unit. Depending on the requirements of the structure, some of this air may be exhausted from the system to the outside of the building. The air that is not exhausted is mixed with air from the outside. The outside air brought into the structure is referred to as ventilation air. The amount of outside air that is mixed with the return air varies with the needs of the building. This mixed air is filtered and either heated or cooled before being returned to the occupied space by the blowers within the unit, **Figure 50.5**. The return airstream mixes with the outside airstream before it is treated by the cooling coil or heat exchanger in the package unit. Therefore, the heat exchangers in the unit treat the mixed air, not the air coming directly from the conditioned space.

ROOFTOP BLOWERS AND FANS

The rooftop unit houses the fans and blowers that move air through the heating and cooling portions of the unit as well as through the air-cooled condenser. The blower that is responsible for moving air through the heating and cooling coils is

Figure 50.6 Rooftop unit with multiple condenser fans. Photo by Eugene Silberstein

a forward-curved centrifugal blower and has the ability to overcome the pressures that are present in the air distribution system. Depending on the system, the blower may be a constant-volume blower or a blower that can modulate the volume of air circulated. Refer to the section on variable air volume (VAV) in this unit for more on variable-volume supply blowers.

The fans that move air through the air-cooled condensers are of the propeller type, which are desirable when large static pressures are not present. They have the ability to move large volumes of air at low pressure differentials. Many rooftop package units are equipped with multiple condenser fans, **Figure 50.6**. These fans often have control configurations that allow them to cycle on and off to maintain the desired head pressure in the system. Typically, some of the fans will operate whenever the system is in the cooling mode while the remaining fans will cycle on and off as needed to raise or lower the head pressure. *Some condenser fans have the ability to modulate airflow by changing motor speed instead of cycling on and off. By ramping motor speeds up and down as opposed to starting and stopping the motor, wear and tear on the motor is reduced and significant energy savings are achieved.* Refer to Unit 17, "Types of Electric Motors," for more information on these motors. Some rooftop package units contain

Figure 50.5 Block diagram showing the air path through a rooftop packaged unit.

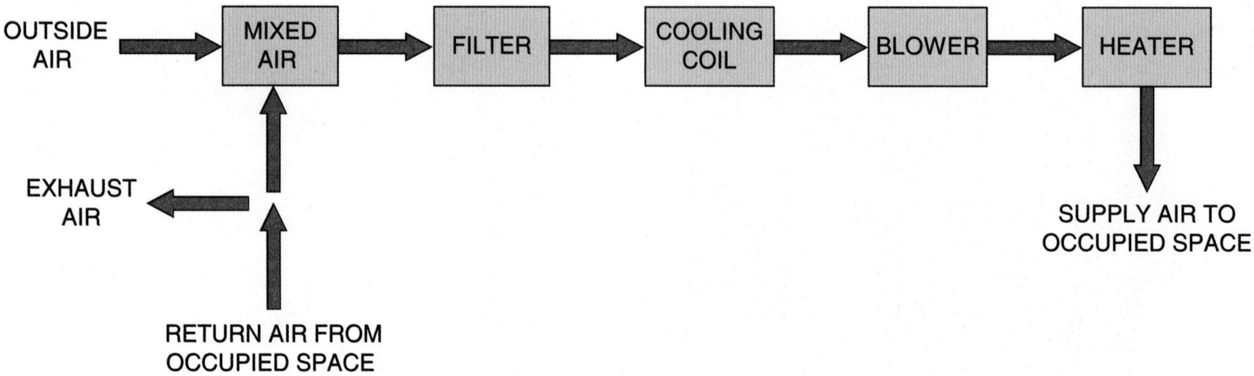

relief fans. Relief fans are located in the exhaust air path of the RTU. Their purpose is to help raise the pressure of the air being exhausted from the return airstream to the outside. Since the exhaust dampers offer some resistance to flow, the relief fan helps the air overcome it.

50.2 INSTALLATION OF PACKAGED ROOFTOP UNITS

Although packaged air-conditioning equipment comes precharged with refrigerant and require no field piping, installing a rooftop unit involves many steps and can be very dangerous. Roof penetrations must be made and the dangers associated with falling from substantial heights are very real. In addition, cranes or helicopters are often used to rig the equipment to the roof, creating additional potential for personal injury. Elements of a rooftop unit installation include:

- Selection of unit location
- Roof penetration
- Installation of the curb and roof sealing
- Rigging of the unit to the roof
- Installation of required electric circuits, both power and control
- Installation of fuel lines for heating mode of operation, if needed
- Installation of return and supply ductwork

SELECTION OF UNIT LOCATION

Rooftop units are often positioned directly above the space that is to be served by the equipment. Although this is the ideal situation, it is not always possible or the best alternative.

Figure 50.7 Air recirculating through the condenser coil.

Figure 50.8 Unit positioned too close to a wall for proper air circulation.

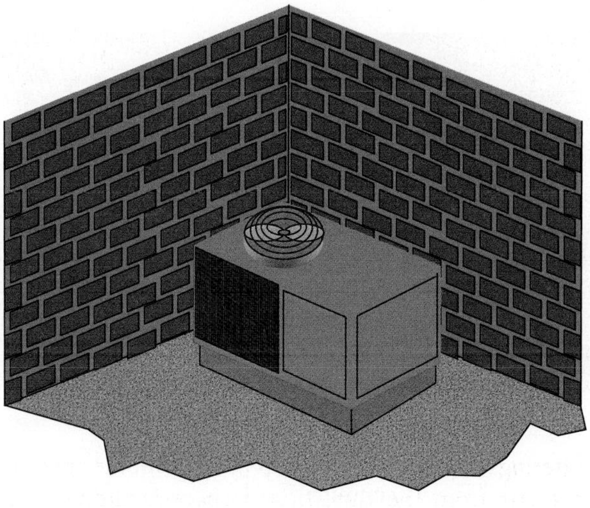

The unit's location should be carefully selected so that its operation is not adversely affected by air being able to circulate back through the condenser coil or by air not being able to flow freely through the coil, **Figure 50.7**. Improper condenser airflow could be caused by a low overhang above the unit or by positioning the unit too close to a wall or interior corner, **Figure 50.8**. Always refer to the installation and service data provided with the equipment and adhere to all clearance guidelines. Insufficient clearances can affect system operation. In addition, the future servicing of the equipment should be taken into account when locating the unit. It is important that the service technician have access to all portions of the system and be able to remove, service, or replace parts as needed. Improper clearances can have a negative effect on the ability of service personnel to perform their

Figure 50.9 Steel rooftop support structure to keep unit above the snow level. Photo by Eugene Silberstein

jobs adequately. In areas that have snowfall, it is important that potential snow accumulation be accounted for. The unit should be mounted on a support structure that will keep the unit above the snow level, **Figure 50.9**.

The limits of the crane must be considered as well when selecting the roof location. Specific cranes will be able to lift only so much weight and extend booms only so far without tipping over. If the crane cannot successfully deposit the unit at its proper location, it will have to be moved manually across the roof. Foresight with respect to this is important, and potential obstacles need to be addressed early on in the installation process.

ROOF PENETRATION

Roof penetration is done only after determining the location of the unit. It is a good field practice to double-check and possibly triple-check the location, since the cost of repairing an improperly cut roof can be very high. Some roofs are bonded, which means that there is a written warranty or insurance policy that protects the building owner in the event of damage from leaks. Any penetrations of a bonded roof must be made by the company that provided the bond or one that is recommended by the bond-issuing company. Ignoring these guidelines can result in severe financial obligations in the event of roof failure. If you are not certain whether a roof is bonded, it is best to check first and cut later.

When penetrating a roof, it is important to keep the following in mind:

Measure twice, cut once. Roof repairs can be very expensive, so be sure to check the accuracy of all measurements.

Do not assume that the size of the cut should be the same as the interior measurements of the roof curb. The penetrations should only be large enough to accommodate the ducts and pipes that must pass through.

Maintain integrity of the roof. Be sure that you do not remove any load-bearing or structural elements.

Reinforce as needed. Be sure to reinforce any roof sections that may have improper support due to the penetration.

Properly vacate areas under penetration locations. Do not perform any penetrations until the areas underneath have been roped off and vacated. Falling debris can damage property and people.

Properly isolate rooftop areas where cuts are made. Roof penetration should never be done at the same time the roof curb and units are being installed. Improper lighting and distractions can result in an individual falling through the hole. Be sure to take all precautions to alert people and prevent falls.

Wear proper articles of personal protection equipment (PPE). The penetration process is dangerous and can result in flying debris. Be sure to properly protect yourself and those working around you.

Avoid coming in contact with sharp edges. Roofing materials can be very sharp, so be careful when handling them.

Avoid positioning yourself under the roof as the penetration is made. Whenever possible, be over the penetration area, not under it, to avoid being hit by roofing materials.

When in doubt, ask for help. If you are unsure about the process, do not be afraid to ask for assistance. At the end of the day, it is much better to be safe than sorry.

ROOF CURB INSTALLATION AND ROOF SEALING

The rooftop unit is often set on a curb, **Figure 50.10**, which is a frame used to mount HVACR equipment. Curbs can be prefabricated, but many larger curbs are shipped unassembled. Assemble the curb according to the manufacturer's guidelines, making certain to properly caulk and seal all joints and seams. Be sure to read the entire installation booklet prior to beginning installation.

Figure 50.10 Typical roof curb. Photo by Eugene Silberstein

Gasketing materials and foam panels that are shipped with the curb are part of the curb and should not be discarded. Gaskets help ensure a leak-tight seal between the unit and curb. The foam panels are intended to help deaden the noise that is produced in the compressor/condenser section of the unit.

If the curb is mounted as part of new construction, it can be assembled and installed as soon as the roof support system is in place. Whenever possible, the curb should be positioned directly on the roof supports and fastened to them using tack welding or other acceptable methods. If the curb is mounted directly on the roof, nailing plates must be provided below the curb flanges to secure it. Properly supporting and securing the curb will help minimize the potential for leaks and noise transmission from the system to the occupied space below. Once the curb has been installed, make certain that the roof is properly sealed. Improperly sealed penetrations can lead to leaks, which can cause major property damage, **Figure 50.11**.

Figure 50.12 Supply and return ducts must be connected to the curb before the unit is set in place.

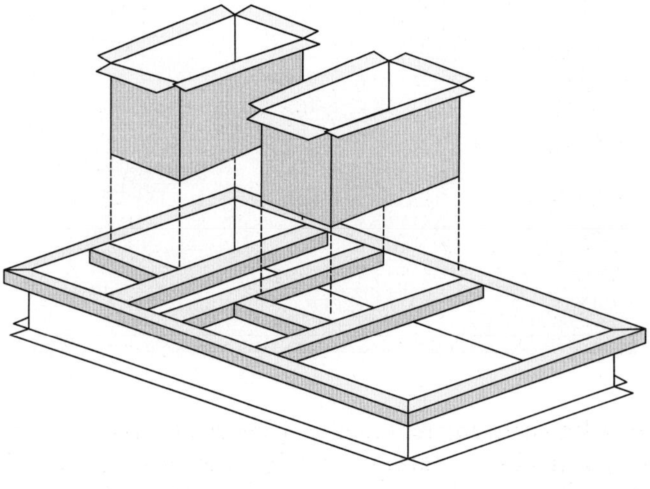

INSTALLATION OF RETURN AND SUPPLY DUCTWORK

It should be noted that, in most cases, the supply and return ductwork must be attached to the curb before the unit is set in place, **Figure 50.12**. All ductwork must be properly fabricated, insulated, lined, and installed. To make sure of this, follow all of the industry-accepted processes and procedures.

⚠ SAFETY PRECAUTION: *The curb's cross braces are intended to support the duct connections to the unit, not the entire air distribution system. Be sure to provide proper and adequate support for the duct system according to local building codes.* •

RIGGING OF THE UNIT TO THE ROOF

The process of rigging involves placing the new unit on the roof as well as removing the old unit from the roof when a system changeout is being done. The rigging process is performed by a rigging contractor that often gets paid by the hour, not the job. Since rigging costs can be substantial, it is best to have everything ready to go when the crane arrives at the job site. Before the rigger arrives, the following tasks should have been completed:

- The old equipment to be removed from the roof should be completely disconnected. This includes all electrical, piping, gas, and any other connections between the equipment and the building.
- Measurements pertaining to the location of the curb on the roof should be taken so that the crane operator has a good idea of where the unit will be placed before starting.
- A crew consisting of at least three members should be on hand. One crew member will be in charge of giving the appropriate hand signals to the crane operator, **Figure 50.13**, while the other two crew members will guide the unit to the proper location on the roof, **Figure 50.14**. Some hand signals that are commonly used to give instructions to the crane operator are provided later in this unit.

Figure 50.11 The curb must be properly sealed to prevent water leaks. Photo by Eugene Silberstein

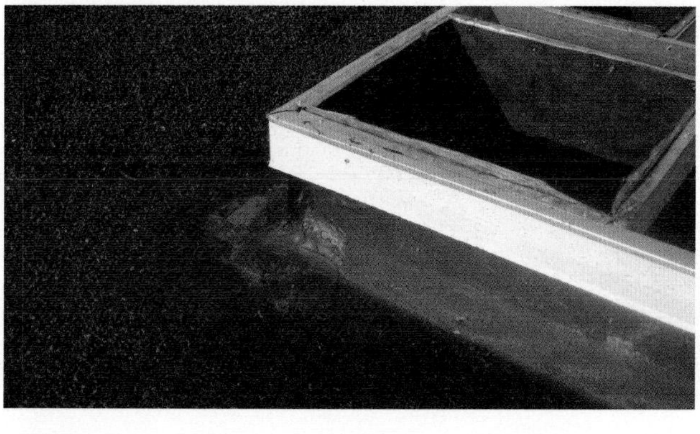

Figure 50.13 Crane operator at ground level getting instructions from installation technician on the roof. Photo by Eugene Silberstein

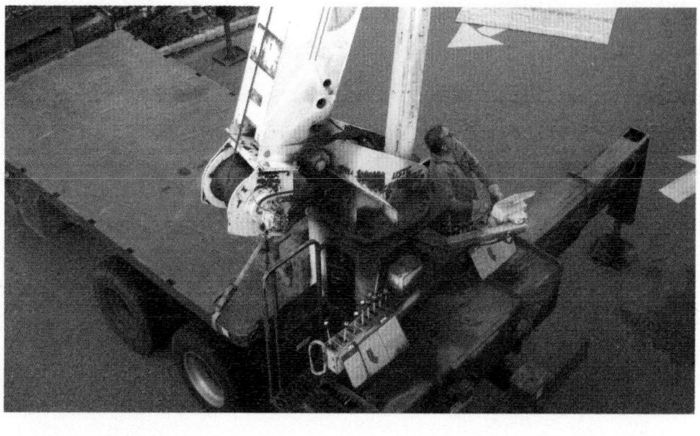

Figure 50.14 Two installation crew members guiding the unit into place. Photo by Eugene Silberstein

Figure 50.15 A typical crane. Photo by Eugene Silberstein

⚠ SAFETY PRECAUTION: *When working with cranes, the use of proper hand signals is required by the Occupational Safety and Health Administration (OSHA), the American National Standards Institute (ANSI), and the American Society of Mechanical Engineers (ASME).* •

- The curb should have been installed on the roof. This includes the mounting of the curb, making the necessary roof penetrations, and sealing the curb to the roof.
- Equipment gaskets and insulation materials that are positioned between the curb and the unit must be in place.
- On units that have their duct connections on the bottom, the supply and return ductwork must be attached to the curb.
- All necessary permits pertaining to the rigging, such as street closures, must be obtained long before the rigger arrives.
- Arrangements for the removal of the old equipment, where applicable, must be made. Quite often, the rigging company will remove and dispose of the old equipment.

CRANE INFORMATION

The crane itself is a massive piece of equipment, and it requires a highly skilled individual to safely operate one, **Figure 50.15**. A well-trained crane operator can make easy work of setting or replacing a piece of rooftop equipment. It is beneficial for technicians who will be working with cranes to have at least a basic understanding of the crane and its components. Some common crane-related terms include:

Main hoist
Auxiliary hoist (whip line)
Cable or line
Outriggers
Boom
Boom extension markings

Some cranes are equipped with two hoists; others have only one. When there are two hoists, one is the main hoist and the other is the auxiliary hoist, **Figure 50.16**. The line is the cable that is attached to the hook on the crane. Outriggers extend from the sides of the crane and increase the base on which it rests. This makes the crane more stable and less likely to topple, **Figure 50.17**. The boom is the largest and most visible part of the crane. It is the telescopic, steel arm of the crane that is responsible for lifting and/or lowering the load, **Figure 50.18**. The markings on the boom tell the crane operator how far the boom has been extended, **Figure 50.19**.

Figure 50.16 Main and auxiliary hoists on a crane. Photo by Eugene Silberstein

Figure 50.17 Outriggers on the side of the crane. Photo by Eugene Silberstein

Figure 50.18 The boom on a crane. Photo by Eugene Silberstein

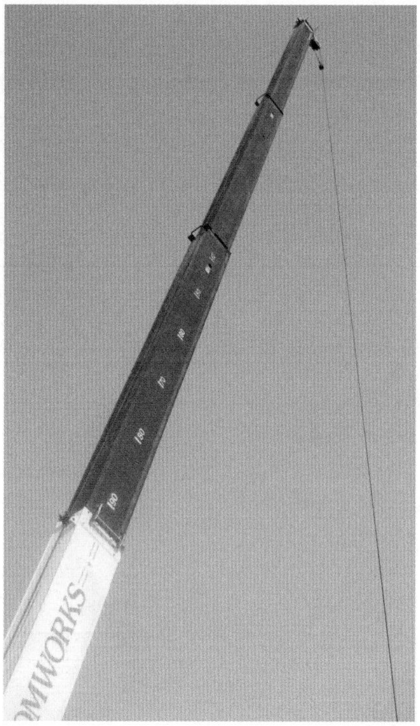

Figure 50.19 Boom extension markings.

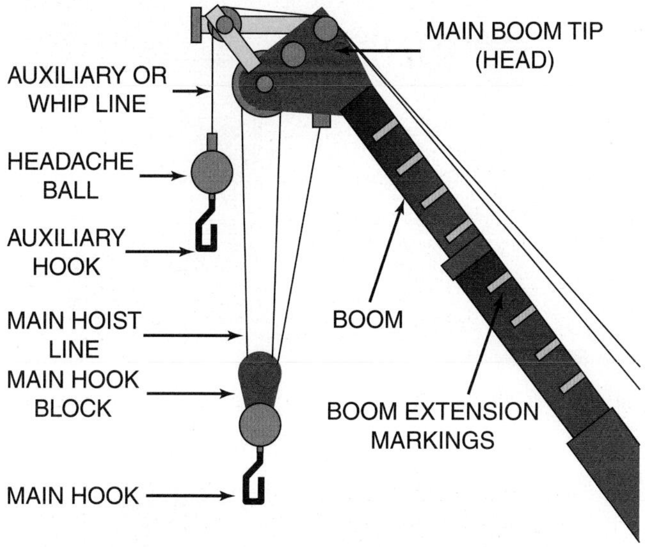

COMMUNICATING WITH THE CRANE OPERATOR

Since the crane is often located at ground level and the HVACR installation crew is on the roof to accept delivery of the equipment, communication between the roof crew and the ground crew can be difficult. In addition to the physical distances between the crane operator and the installers, cranes and job sites in general can be noisy, making the communication process even harder. For this reason, a set of industry-accepted hand and body signals is used by the installation crew to give nonverbal instructions to the crane operator.

Since many cranes have a main hoist and an auxiliary hoist, known as a whip line, there are hand signals that tell the crane operator which hoist to use. By tapping his fist on his head, the crew member is telling the crane operator to use the main hoist, **Figure 50.20(A)**. By tapping his elbow with one hand, the crew member is telling the crane operator to use the whip line, or auxiliary hoist, **Figure 50.20(B)**. The boom

Figure 50.20 (A) Tapping the top of the head tells the crane operator to use the main hoist. (B) Tapping the elbow with one hand tells the crane operator to use the auxiliary hoist.

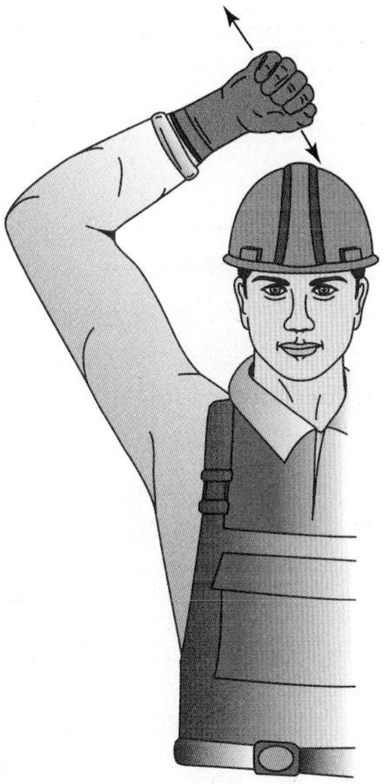

(A)

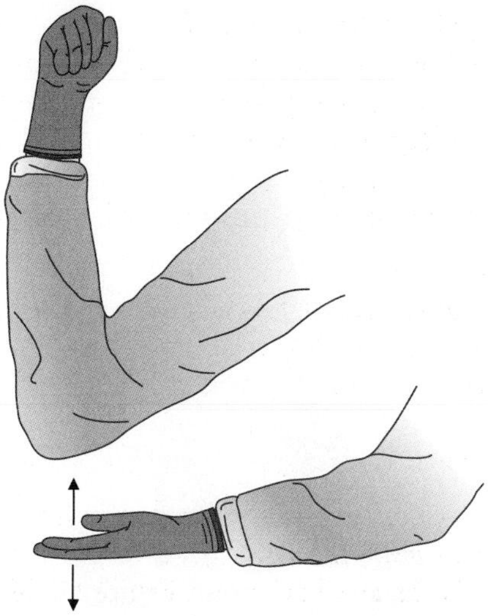

(B)

can either be extended or retracted depending on the hand signal given. With both fists in front of the body, pointing the thumbs outward tells the crane operator to extend the boom; pointing the thumbs inward tells the operator to retract the boom, **Figure 50.21**. The boom can also be raised or lowered depending on the hand signal given. With an extended arm and closed fist, an upward-pointing thumb tells the operator to raise the boom; a downward-pointing thumb tells the operator to lower the boom, **Figure 50.22**. The boom may also need to be rotated. If such is the case, an extended arm with an extended, pointing index finger tells the crane operator which way to rotate the crane, **Figure 50.23**. During the rigging process, the load will have to be raised and lowered. By raising the forearm in the air, index finger pointing upward, and rotating the hand in small circles, the operator will know to raise the cable on the crane, **Figure 50.24**. By extending the arm downward, index finger pointing downward, and rotating the hand in small circles, the operator will know to lower the cable on the crane, **Figure 50.25**. An extended, palm-down hand tells the crane operator to stop. In an emergency, a rapidly shaking (from side to side) extended palm-down hand tells the operator to stop immediately,

Figure 50.21 (A) Signal to extend the boom. (B) Signal to retract the boom.

(A)

(B)

Figure 50.22 (A) Signal to raise the boom. (B) Signal to lower the boom.

(A)

(B)

Figure 50.23 By an extended arm and a pointed finger, the crane operator knows which way to rotate the boom.

Figure 50.24 Hand signal for raising the cable.

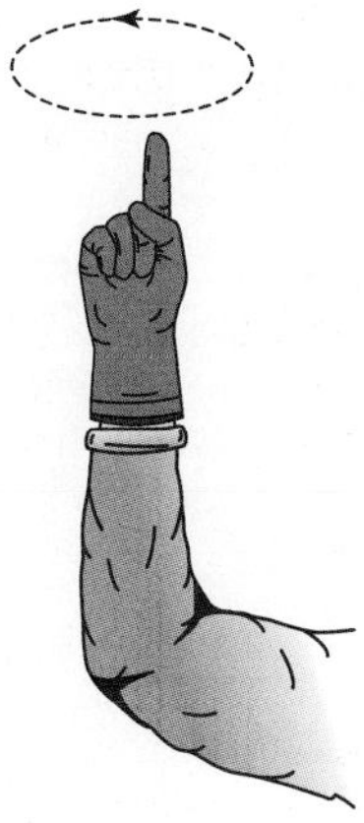

Figure 50.25 Hand signal for lowering the cable.

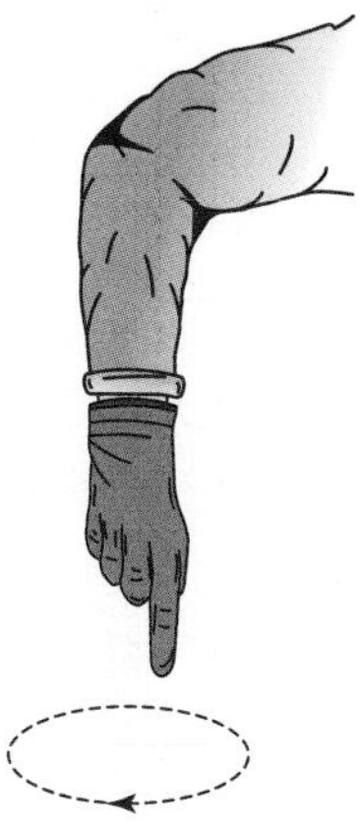

Figure 50.26 (A) Hand signal to stop. (B) Hand signal to stop immediately.

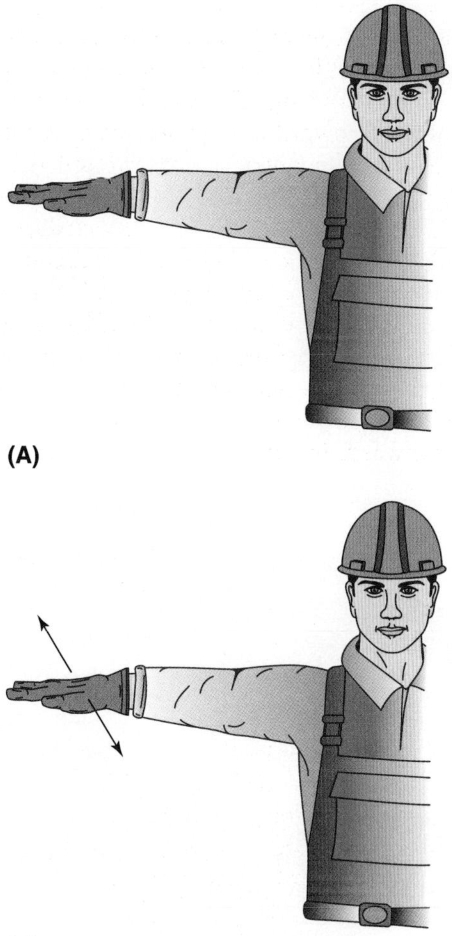

(A)

(B)

Figure 50.26. Some cranes are mobile and have the ability to roll or travel during operation. In such cases, an extended and slightly raised arm with an open hand gives a pushing signal to tell the crane operator which way to roll the crane, **Figure 50.27**.

Figure 50.27 Hand signal to move a rolling or traveling crane.

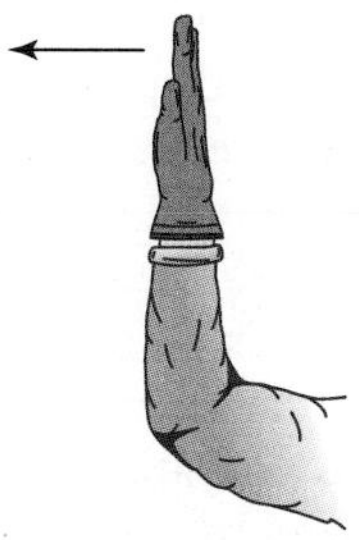

INSTALLATION OF REQUIRED ELECTRIC CIRCUITS AND FUEL LINES

In addition to making provisions for the supply- and return-air ductwork connections between the rooftop unit and the space below, electrical and fuel connections must also be made. Depending on the equipment and installation, these lines may be brought into the unit within the space occupied by the curb. If such is the case, the sealing of the roof curb, by default, also provides adequate sealing of the electrical and piping runs. If the piping and electrical penetrations are made outside of the curb area, a pitch pocket or pitch pan is often used. A pitch pocket is a flanged piece of roof flashing positioned around irregularly shaped roof penetrations. The pitch pocket must then be properly sealed to prevent leaks. It is good field practice to avoid using pitch pockets whenever possible, as they increase the chances of roof leaks, **Figure 50.28**.

50.3 ECONOMIZERS

Most rooftop packaged units are equipped with economizers, **Figure 50.29**. Economizers are intended to perform two main functions. One function is to help provide the correct amount

Figure 50.28 Pitch pockets are used when piping or conduits must penetrate the roof. Photo by Eugene Silberstein

Figure 50.29 A typical economizer.

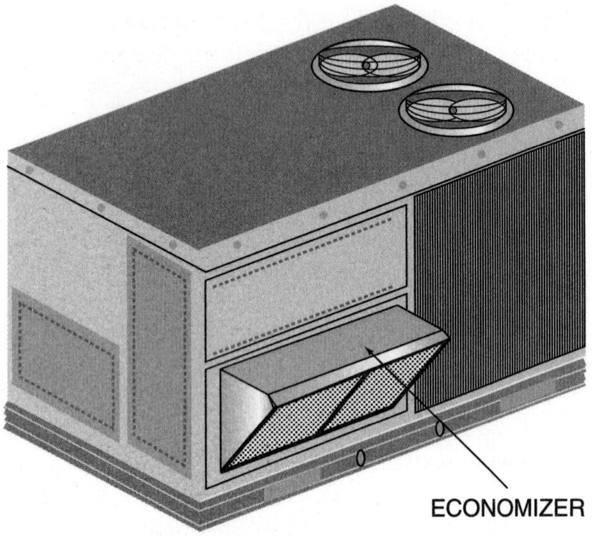

of ventilation air to the structure as required under ASHRAE Standard 62, which will be discussed later in this unit. The second function is to help reduce cooling costs by allowing outside air into the air-conditioning equipment when the outside air conditions are favorable and able to meet at least part of the cooling needs of the structure. By using outside air to help cool the structure, the vapor-compression refrigeration system will not need to operate as frequently, thereby saving energy. This is referred to as economizing. But, how is it determined if the outside air conditions are favorable or not to help cool the structure?

Two common control types are used to help determine if the outside air can be used to provide some system relief: dry-bulb and enthalpy-sensing devices. Dry-bulb and enthalpy sensors can each be used in conjunction with two common control strategies to establish if the outside air is capable of reducing the load on the cooling equipment. One method is to measure the dry-bulb temperature of the outside air and/or the return air, while the other is to measure the enthalpy, or heat

Figure 50.30 Controls used to sense outside-air and return-air conditions. Photo by Eugene Silberstein

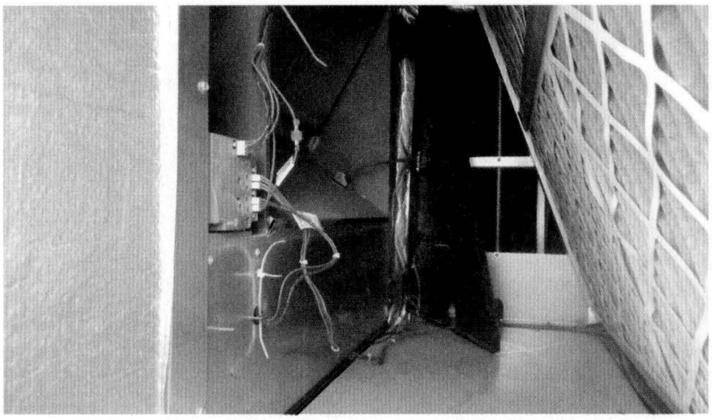

content, of the outside air and/or the return air, **Figure 50.30**. One of the important functions of these controls is to set the high limit, or cut-out point, for economizer operation. The high limit for an economizer is the point at which economizing will end if the sensed conditions are above the desired levels. The strategies for controlling the economizer's high limit are:

- Fixed dry-bulb
- Differential dry-bulb
- Fixed enthalpy
- Differential enthalpy

FIXED DRY-BULB

The fixed dry-bulb type of control is the least expensive and is commonly found on older economizer systems. This type of control utilizes a fixed set point for the outside air. If the outside air is warmer than the set point, no economizing will take place. This means that the amount of outside air being brought into the structure will be in line with the ventilation requirements for the space. For example, if the set point on the control is 68°F and the outside ambient temperature is 70°F, the economizer mode will not be utilized. However, if the outside ambient temperature is 65°F, the economizer mode will be utilized, **Figure 50.31**.

Newer systems do not measure only the dry-bulb temperature of outside air, because the air's actual heat content cannot be determined and the operation of the economizer will not be as accurate. Enthalpy controls are the norm on new equipment. The term enthalpy refers to the total heat content in the air, which accounts for both the temperature (sensible heat portion) and the moisture content (latent heat portion) of the air. Consider the psychrometric chart in **Figure 50.32**. If a dry-bulb thermometer indicates that the outside-air temperature is 90°F, the enthalpy of the outside air could be 45.3 Btu/lb if the relative humidity is 70%, or 35.2 Btu/lb if the relative humidity is 40%. However, if an enthalpy control is used, the exact heat

Figure 50.31 The set point of the fixed dry-bulb control sets the boundary between the economizing and the non-economizing modes.

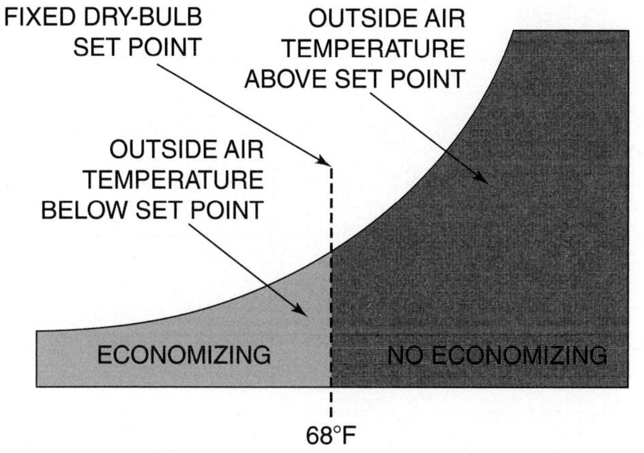

content of the air will be known and the economizer can adjust accordingly. Enthalpy controls are discussed a little later in this section.

DIFFERENTIAL DRY-BULB

Unlike the fixed dry-bulb control, the differential dry-bulb control uses a *variable* set point for the outside air. This type of control uses a return air temperature (RAT) sensor as well as an outside air temperature (OAT) sensor. The differential dry-bulb control compares the difference between the return air temperature and the outside air temperature. If the outside air temperature is higher than the return air temperature, the economizer is disabled. If the outside air temperature is lower than the return air temperature, the economizer is enabled. Mathematically, the operation of the device looks like this:

$$\text{If OAT} - \text{RAT} > 0 \text{ ECONOMIZER DISABLED}$$

$$\text{If OAT} - \text{RAT} < 0 \text{ ECONOMIZER ENABLED}$$

For example, if the outside air temperature is 90°F and the return air temperature is 70°F, we get:

$$\text{OAT} - \text{RAT} = 90°\text{F} - 70°\text{F} = 20°\text{F}$$

Since 20 is greater than zero, the economizer is disabled. On the other hand, if the outside air temperature is 60°F and the return air temperature is 70°F, we get:

$$\text{OAT} - \text{RAT} = 60°\text{F} - 70°\text{F} = -10°\text{F}$$

Since −10 is less than zero, the economizer is enabled.

FIXED ENTHALPY

The fixed enthalpy high limit is more sophisticated than the dry-bulb high limit controls. Enthalpy sensors include both solid-state temperature and humidity sensors. The enthalpy control measures the *total heat content* of the outside airstream. The total heat content measurement is then compared to a preset set point in the control, which is often close to 28 Btu/lb, but can be changed depending on the particular climate in which the system is installed. If the heat content of the outside air is greater than the set point, the economizer is disabled. If the heat content of the outside air is lower than the set point, the economizer is enabled, **Figure 50.33**.

DIFFERENTIAL ENTHALPY

The differential enthalpy high limit is a very expensive high limit strategy. This control measures the enthalpy, or *total heat content* (h), of the outside airstream as well as the *total heat content* of the return airstream. These two values are then compared to each other. If the heat content of the outside air, h_{OA}, is greater than the heat content of the return air, h_{RA}, the economizer is disabled. If the heat content of the outside air is lower than the heat content of the return air, the economizer is enabled. The calculations are similar to those used for the differential dry-bulb control, but enthalpy is used instead of temperature as shown here:

$$\text{If } h_{OA} - h_{RA} > 0 \text{ ECONOMIZER DISABLED}$$

$$\text{If } h_{OA} - h_{RA} < 0 \text{ ECONOMIZER ENABLED}$$

For example, if the enthalpy of the outside air is 28.2 Btu/lb and the enthalpy of the return air is 23.2 Btu/lb, we get:

$$h_{OA} - h_{RA} = 28.2 \text{ Btu/lb} - 23.2 \text{ Btu/lb} = 5 \text{ Btu/lb}$$

Since 5 is greater than zero, the economizer is disabled. On the other hand, if the enthalpy of the outside air is 23.2 Btu/lb and the enthalpy of the return air is 28.2 Btu/lb, we get:

$$h_{OA} - h_{RA} = 23.2 \text{ Btu/lb} - 28.2 \text{ Btu/lb} = -5 \text{ Btu/lb}$$

Since −5 is less than zero, the economizer is enabled, **Figure 50.34**.

If the heat content of the outside air is lower than the heat content of the indoor air and the system thermostat is calling for cooling, the economizer can bring outside air into the system instead of operating the compressor. The term often used to describe this is free-cooling, even though the evaporator blower motor is operating to move the air. So, in reality, we are not getting free-cooling, but the cooling costs are greatly reduced.

50.4 ECONOMIZER MODES OF OPERATION

The economizer is able to increase or decrease the amount of outside air that is brought into the structure with two dampers. One damper is located in the return-airstream and the other

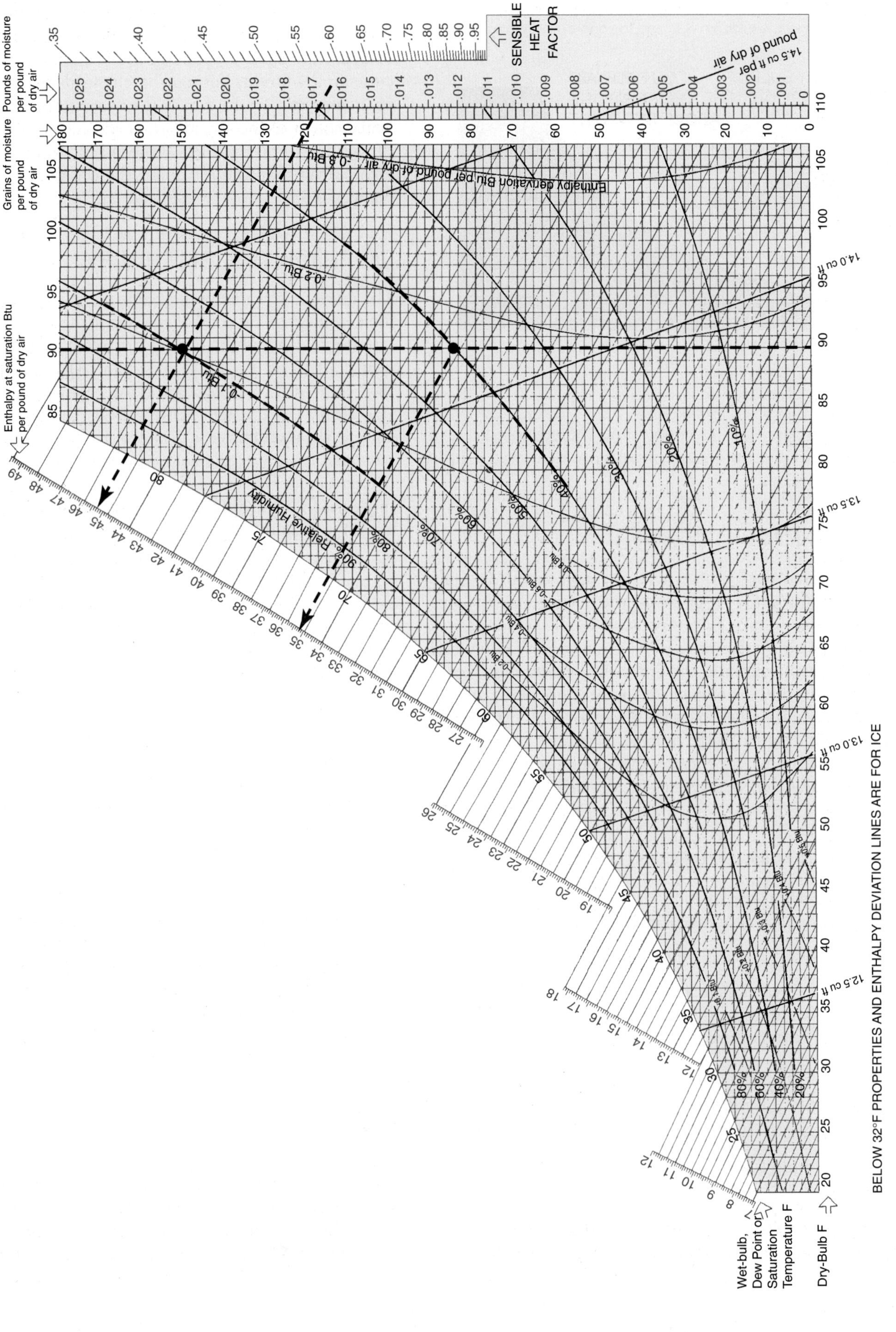

Figure 50.32 At 90°F, the heat content of the outside air is 45.3 Btu/lb if the relative humidity is 70%, and 35.2 Btu/lb if the relative humidity is 40%. Reproduced, courtesy of Carrier Corp.

Figure 50.33 Diagram showing the economizing and non-economizing regions for a fixed enthalpy control.

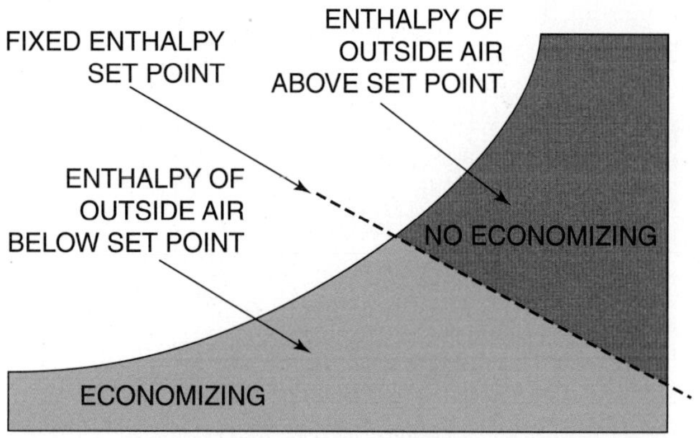

is located in the outside-airstream, **Figure 50.35**. These dampers are referred to as opposing dampers, meaning that if one closes, the other opens and vice-versa. If the outside-air conditions do not warrant free-cooling, then the outside-air damper will close to its minimum position, delivering approximately 10% outside air, and the return air damper will be completely open, **Figure 50.36**. If the heat content of the outside air is low enough, the outside-air damper will be completely open and the return-air damper will be completely closed, **Figure 50.37**. In almost all cases, both dampers will be open at least partially to allow for an outside-air/return-air mixture. All across the country, building codes require that a certain amount of outside air be brought into a structure, regardless of the outside-air conditions. So, even if the heat content of the outside air is very high, the outside-air damper will move to its minimum setting, which is often in the 10% range. It should be noted that when outside air is brought in, the structure could become pressurized. To prevent this, some type of exhaust mechanism must be in place. Refer back to **Figure 50.5** for a diagram of a sample rooftop system that incorporates the process of exhausting air.

NOTE: *In order to operate properly, the economizer must be set up to meet the requirements of the system and local building codes. Improperly installed, configured, or maintained economizers can result in poor indoor air quality and increased system operating costs.* •

There are four modes of operation for systems that are equipped with economizers. These modes are:

- Heating mode
- Modulating mode
- Integrated mode
- Mechanical mode

HEATING MODE

Heating mode refers to the mode of operation during which there is no call for cooling, and the heating portion of the system is energized. In this mode of operation, the outside air is typically below 30°F. Since the outside air is very cold and the system is operating in the heating mode, only the minimum amount of outside air is brought into the space. If the minimum amount of outside air is 10%, the dampers will close to allow only 10% outside air into the system; the remaining 90% of the airstream will be from the return airstream from the occupied space, **Figure 50.36**.

MODULATING MODE

Modulated economizer mode, or just modulating mode, refers to the mode of operation during which there is a call for cooling, but the outside ambient temperature is low; typically between 30°F and 50°F. Since the outside air is cold and the system is calling for cooling, the outside airstream can be used to meet the cooling needs of the structure. In this mode the system compressors do not operate. The outside and return air dampers open and close to provide the needed cooling for the space. When the cooling load is relatively low, there is more return air than outside air in the mixed airstream, **Figure 50.36**. As the cooling load in the space increases, the amount of outside air being brought in increases, and the amount of return air in the mixed airstream is reduced, **Figure 50.35**. Maximum "free-cooling" occurs when the outside air dampers are 100% open and the return air is reduced to 0%, **Figure 50.37**. Modulating mode is often referred to as "free-cooling," but the system blowers are still operating, so there are still energy costs associated with this mode of operation.

INTEGRATED MODE

Integrated mode refers to the mode of operation during which there is a call for cooling, but the outside ambient temperature is relatively mild, typically between 55°F and 75°F. Since the outside air is cool, but not cold, the outside airstream can be used to meet *some of* the cooling needs of the structure. In this mode of operation, the system compressors are operating to provide the remainder of the needed cooling capacity. During integrated mode, the outside air dampers are open 100% and the return air dampers are fully closed, **Figure 50.37**. As the load on the system increases, the mechanical cooling system must contribute more to meet the cooling needs of the occupied space.

MECHANICAL MODE

Mechanical cooling mode, or mechanical mode, refers to the mode of operation during which there is a call for cooling, but the outside ambient temperature is too warm (usually above 75°F) to provide any assistance in satisfying the cooling requirements for the conditioned space. At this point the

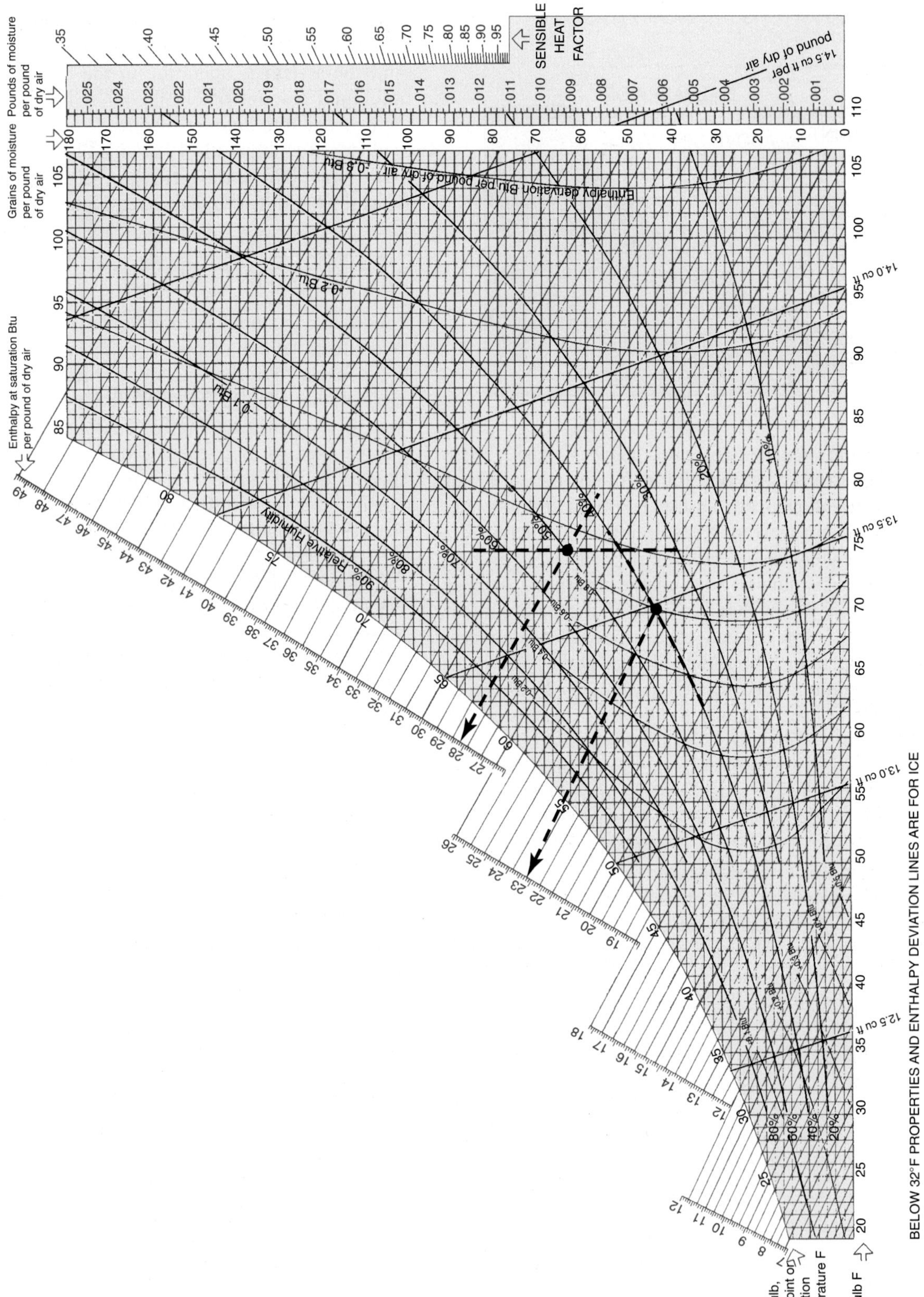

Figure 50.34 Heat content of the outside air is 23.2Btu/lb, which is lower than the 28.2 Btu/lb heat content of the indoor air. Reproduced, courtesy of Carrier Corp.

Figure 50.35 Return-air and outside-air dampers.

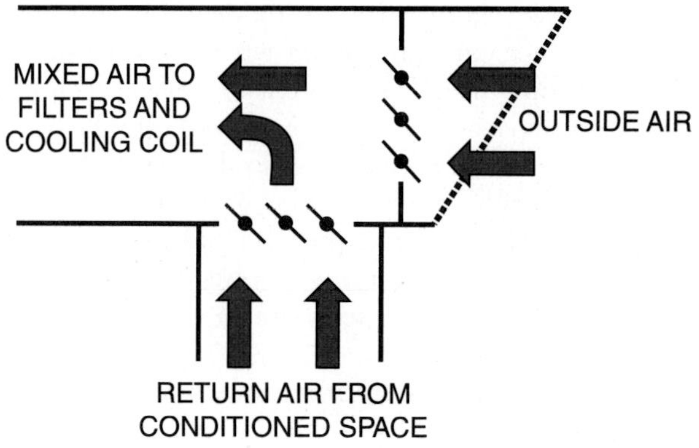

economizer is disabled and the outside air damper returns to its minimum position, **Figure 50.36**. The mixed airstream is cooled by the mechanical cooling system only. As the load on the system increases, the mechanical cooling system must ramp up its capacity to meet the cooling needs of the occupied space.

50.5 ASHRAE STANDARD 62

ASHRAE Standard 62, a document that is constantly updated and revised, establishes the standard for the design of ventilation systems to provide acceptable indoor air quality levels that are intended to minimize the potential for adverse health effects. Standard 62 was first created in 1973 and has undergone many changes since then. Most local building codes across the country reference Standard 62 as the guideline for establishing ventilation levels in commercial, industrial, and residential structures.

Although indoor air quality (IAQ) is the major issue addressed in Standard 62, acceptable IAQ levels cannot be maintained by tackling a single air-related element. Instead, acceptable indoor air quality can best be attained when air contamination sources, ventilation, air cleanliness, air movement, air filtration, and relative humidity are all maintained within acceptable ranges, levels, and limits. Indoor air quality issues can be greatly reduced when emphasis is placed on these items by the building's maintenance staff. Indoor air quality issues are discussed in more detail in Unit 34, "Indoor Air Quality."

One important aspect of ASHRAE Standard 62 addresses the quality of ventilation air and how air contaminants can enter a building through the system's economizer. Standard 62 states that make up air inlets are to be located so that they help minimize contamination of the system air. This alerts the system designer and installer to be aware of any exhaust vents, flue pipes, automobile exhaust sources, cooling towers, garbage dumpsters, or other similar contaminant sources when selecting the location for make up air ducts.

Figure 50.37 The outside air damper will open fully when the outside air can meet the cooling requirements of the space.

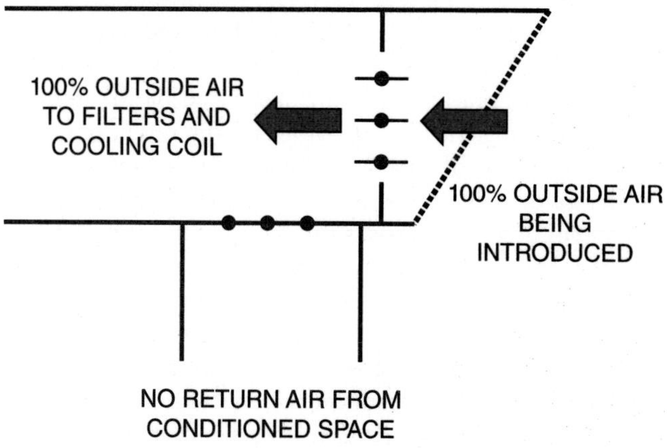

Figure 50.36 The outside-air damper will close when the outside air conditions do not warrant free-cooling.

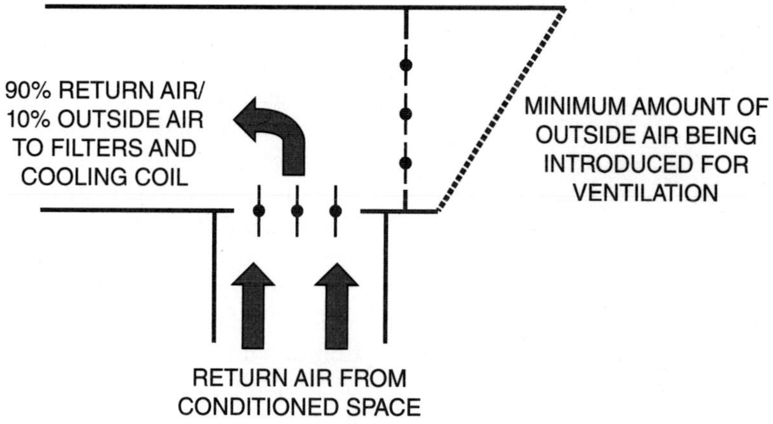

50.6 DEMAND CONTROL VENTILATION (DCV)

In traditional "outside-air" air-conditioning systems, the amount of outside air that is introduced to the system can never be lower than the amount permitted by the minimum allowable setting on the economizer. This minimum economizer setting is established by the system designer to ensure that the guidelines of ASHRAE Standard 62 are adhered to. The system designer chooses this minimum setting based on what the *expected* occupancy of the building will be, not what the *actual* occupancy is at any given time. If an office building is nearly empty at 6:00 PM, the amount of outside air being introduced to the space at that time will be the same as the amount of air introduced when it is near full capacity at 10:00 in the morning. It is easy to see that the amount of outside air needed for adequate ventilation is much lower at 6:00 PM than at 10:00 AM. By utilizing a strategy called demand control ventilation, DCV, the amount of outside air being brought in can be varied based on the actual occupancy level in the structure. In applications where there are numerous, separate conditioned areas, such as in the case of a multiplex movie theater, the amount of outside air can be adjusted depending on the actual number of people in each individual theater. Carbon dioxide-sensing controls, **Figure 50.38**, are often used in conjunction with DCV modules to adjust the amount of outside air being brought into a particular area, based on the ventilation needs at that particular point in time.

HOW DEMAND CONTROL VENTILATION (DCV) WORKS

The demand control ventilation system controls the amount of outside air that needs to be brought to the occupied space for ventilation by estimating the occupancy level in a particular occupied area. The control system does this by measuring the levels of carbon dioxide, CO_2, in the space. Since we produce carbon dioxide when we breathe, as the number of people in a space increases, the level of carbon dioxide in the space will increase as well. Similarly, as the occupancy level in an area drops, so does the CO_2 level. If the occupancy level in a particular area greatly decreases, the amount of outside air required drops as well. In many instances, the required amount of outside air will be less than if the outside air damper were to be positioned at its minimum setting.

A number of existing information streams can be used in conjunction with the DCV system to help it operate more efficiently. In the case of the movie theater, the box office has computerized data on how many tickets are sold for each individual theater at each specific showtime. The DCV system, if properly configured and equipped, can adjust the amount of ventilation air provided to each theater based on the number of people who will be occupying the theater. The ventilation rates can therefore be different for each theater in the multiplex and for each showing of a particular film. In many instances, office buildings are equipped with sensors that turn the lighting off in a particular area if the room has been vacant for a period of time. These sensors can be linked to the DCV control system to reduce the amount of ventilation air when an area is no longer occupied.

Figure 50.38 A carbon dioxide sensor. Photo by Eugene Silberstein

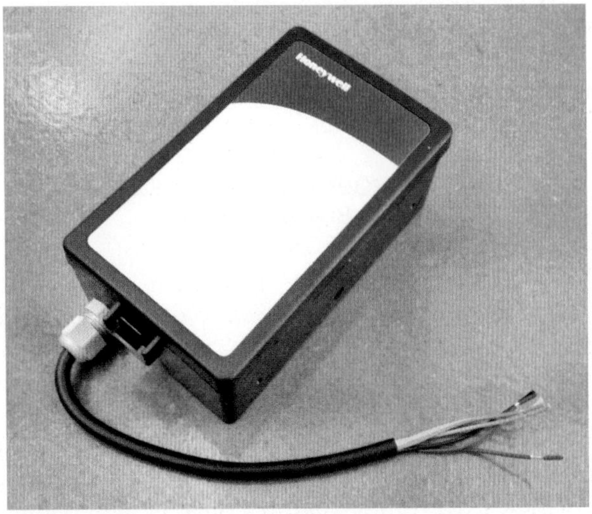

THE CARBON DIOXIDE (CO_2) SENSORS

As mentioned, the DCV control strategy estimates the actual occupancy level in a space based on the level of carbon dioxide sensed. In order for the carbon dioxide reading to be accurate, sensor placement is extremely important. Improperly positioned sensors may or may not provide an accurate representation of the entire space. Sensors should not be placed in or near an entryway or in the corner of an area. Poorly chosen sensor locations may result in CO_2 readings that are either too high or too low.

Control strategies might employ a single CO_2 sensor or multiple sensors located throughout the structure. A single sensor would typically be positioned in the return airstream and would sense the average of all returning airstreams. In this case, it would not be possible to vary the amount of ventilation air to individual areas. When used in conjunction with a variable air volume (VAV) system, carbon dioxide sensors can be positioned in each zoned area. When configured like this, the amount of ventilation air can be tailored to the needs of each conditioned area.

As with any control strategy, proper installation, start-up, and maintenance will help ensure successful system operation down the road. The economizer and its controls must be properly programmed to work correctly with the input from the carbon dioxide sensors. Improper installation,

programming, or maintenance can negate any of the benefits systems of this type are intended to provide. Improper economizer damper settings may under- or overventilate an area, resulting in increased cooling costs.

50.7 TRADITIONAL CONSTANT-VOLUME AIR DISTRIBUTION METHODS

Most older-generation air-conditioning systems were designed to circulate constant volumes of air to each location served whenever the system called for blower operation. When the blower was off, no air at all was circulated through the duct system. When the blower cycled on, it delivered its design air volume through the main trunk line on the air distribution system. This air was then delivered to the occupied space through the supply registers and grilles. The next generation of air-conditioning systems allowed for the zoning of occupied spaces.

The process of zoning involves the separation of the occupied areas by similar heating or cooling needs. For example, areas located on the west side of a single-story office building may have similar heating and cooling patterns, just as the areas located on the east side of the same building might have similar air-conditioning requirements. In this simple case, the east side of the building can be treated as one zone, and the west side can be treated as a second zone. The duct system is designed with a set of dampers that open or close to either allow airflow to a particular zone or prevent air from reaching a zone. When neither zone is calling for airflow, the blower will cycle off. When only one zone is calling for blower operation, the other zone damper will be closed and the blower motor will operate at a lower speed to deliver the required amount of airflow to the zone that is calling. When both zones are calling for blower operation, both zone dampers will open and the motor will operate at high speed to deliver the amount of air desired. In any case, the amount of air being delivered to a particular zone is either zero or the design air volume. With the introduction of variable air volume, VAV, systems, the air-conditioning equipment can modulate the airflow to the occupied spaces, allowing for more even space conditioning, increased energy efficiency, and a longer expected life cycle for the equipment.

50.8 VARIABLE AIR VOLUME (VAV) SYSTEMS

As the name implies, variable air volume systems are designed to vary the amount of air that is introduced to an area. Some air distribution systems are designed as all-or-nothing delivery systems, but the VAV system can modulate and regulate the amount of air that actually reaches each location in the conditioned space. The airflow requirements of a particular area depend on a number of factors that include, but are not limited to, the cooling requirements of the space, the heating requirements, the ventilation requirements, and the occupancy levels in the space.

Just like conventional rooftop packaged equipment, the typical VAV system is equipped with an economizer, air filters, a mixing box for the return and outside air, a heat source for the heating mode operation, and the four main system components that make up the vapor-compression refrigeration cycle. The main difference between a VAV rooftop packaged unit and a conventional rooftop packaged unit is the supply blower. The supply blower on the VAV system has the ability to deliver varying amounts of air based on the needs of the conditioned space. In addition to the variable volume blower, the VAV system also relies on individual terminal units that are located in each zone. These terminal units are typically equipped with a damper, an air pressure sensor, a reheat coil, and the devices needed to properly control damper position and the rate of hot water flow through the reheat coil in response to the temperature and ventilation requirements of the space. Some VAV boxes or terminal units are equipped with separate blowers or fans.

In order to deliver different volumes of air to various areas in an occupied space, the blowers used on VAV systems must be able to modulate their speeds to accommodate the changing needs of the system. Motor speed modulation involves more than simply changing back and forth between a motor's high-speed winding and its low-speed winding, as in the case of a conventional heating and cooling system.

As the system needs change, dampers open and close to supply more or less air to the individually controlled zoned areas. As these dampers open and close, the pressures in the air distribution system change as well. Strategically placed sensors measure the pressures in the duct system and determine the speed at which the blower should operate. This information is then sent along to the motor's control circuitry and the delivered air volume is adjusted accordingly. A method commonly used to alter the motor's speed is a variable frequency drive, or VFD. The VFD modulates the frequency of the supply power's sine wave to reduce the effective speed of the motor. These drives and other variable motor speed strategies are discussed in more detail in Unit 17, "Types of Electric Motors."

When the demand for cooling in a structure is reduced, the dampers that control the airflow to the various locations begin to close. This creates a restriction in the air distribution system and the pressure in the system increases. Less air is being introduced to the occupied space because of the closing dampers and the pressure in the duct system continues to rise. Sensors in the ducts register this increase in static pressure and send a signal to the control system to slow the motor

down. By slowing the motor down, the pressure in the duct system is reduced to an acceptable level as determined by the parameters of the control system.

50.9 VAV BOXES AND TERMINAL UNITS

The VAV boxes or terminal units can be configured in a number of different ways, depending on the needs of the structure. There are four common types of VAV boxes; quite often, a particular installation may use multiple types to meet the heating/cooling/ventilation needs of the conditioned space. The four common types of VAV boxes are cooling and heating, cooling only, parallel blower-assisted, and series blower-assisted.

The cooling and heating VAV box is made up of the damper, air pressure sensor, reheat coil, and the required control devices, **Figure 50.39**. A reheat coil is typically a hot water coil but, in some applications, can be an electric coil. The reheat coil can be used to temper air that is flowing into the occupied space in the cooling or ventilation modes, or to heat the air in the heating mode.

The cooling-only VAV box is made up of the damper, air pressure sensor, and the controls needed to properly position the damper for the desired airflow rate. There is no reheat coil on the cooling-only VAV box.

The blower-assisted VAV boxes include additional blowers that introduce air from the space above the occupied areas into the VAV box cabinet. This allows air from the ceiling to mix with the conditioned air being supplied to the room. In the parallel blower-assisted VAV configuration, a small blower is energized when it is desired that ceiling air be mixed with the main airstream prior to entering the conditioned space, **Figure 50.40**.

In the cooling mode, the damper opens and closes as long as the need for cooling remains. As the space temperature nears the set point for the space, the damper begins to close. When the space temperature reaches the set point, the damper in the VAV box for that particular area will be at its minimum position, **Figure 50.41**. This minimum position allows for proper ventilation as required under ASHRAE Standard 62. (More information on this standard can be found elsewhere in this unit.) If, at the minimum damper setting, the air being supplied continues to cool the space below the desired temperature, the blower cycles on to mix ceiling cavity air with the cooler supply air. This tempers the air just enough to reduce the cooling effect on the occupied space, **Figure 50.42**. Reheat coils can be added to the parallel blower-assisted VAV box for additional tempering and for heating the air in the cooler winter months.

A larger blower is used in the series blower-assisted VAV configuration. This blower is energized all the time, as long as the area is occupied, even if there is no call for heating or cooling, **Figure 50.43**. If the zone is calling for cooling, the

Figure 50.39 (A) A cooling and heating VAV box. (B) Air pressure sensor and damper. (C) Control devices in the VAV box. Photos by Eugene Silberstein

(A)

(B)

(C)

Figure 50.40 A parallel blower-assisted VAV box.

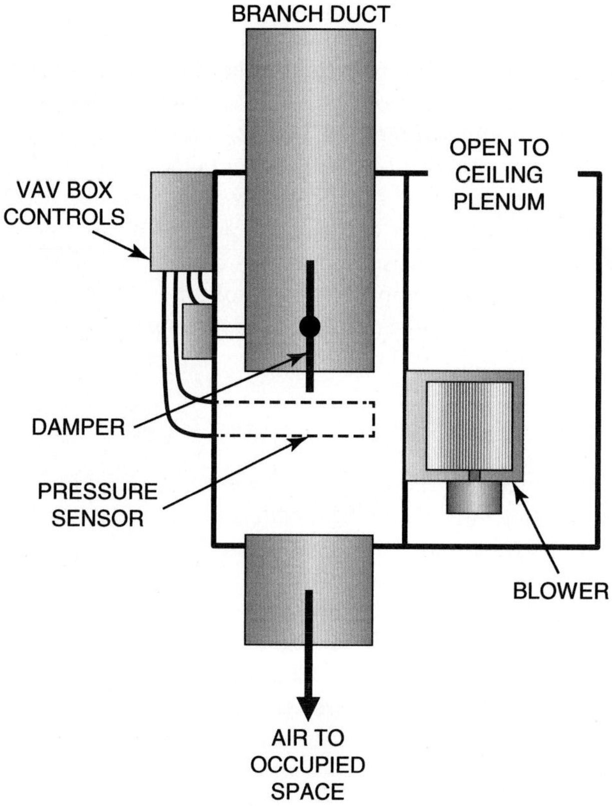

Figure 50.41 As the cooling load on the system is reduced, the damper closes to send less air to the occupied space.

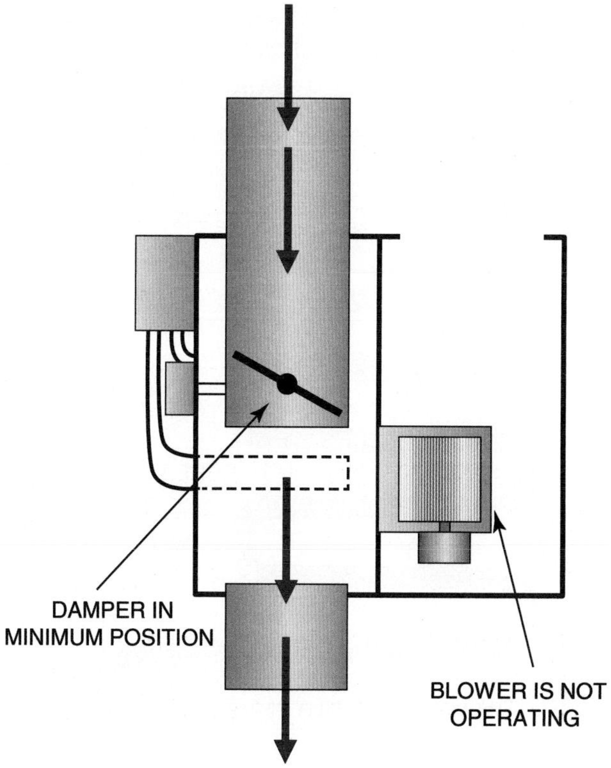

Figure 50.42 If the temperature of the space continues to drop, even at the minimum damper position, the VAV blower cycles on to mix warmer ceiling air with the colder supply air.

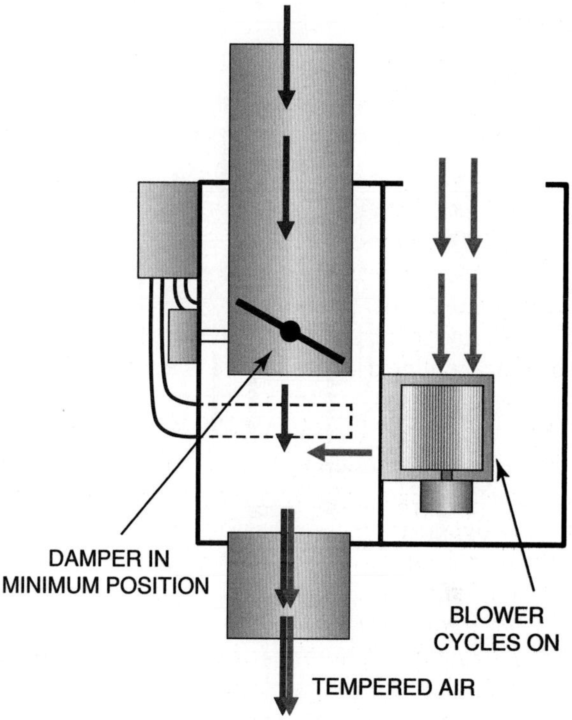

Figure 50.43 A series blower-assisted VAV box.

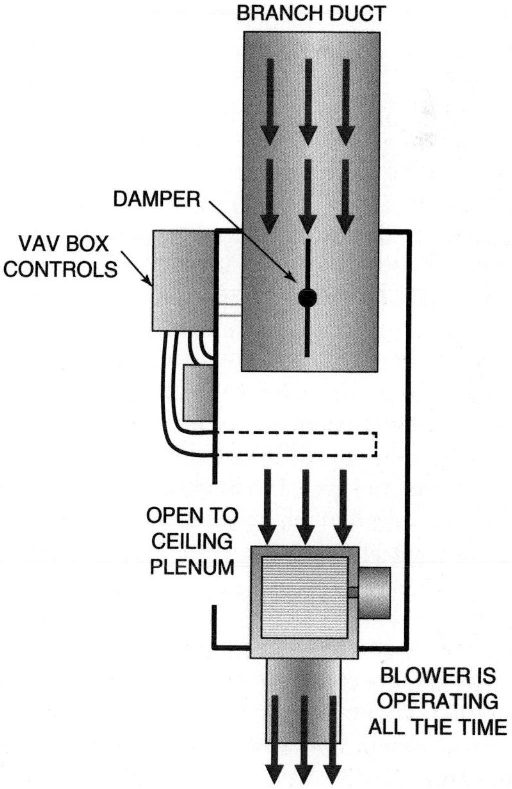

Figure 50.44 As the cooling load is reduced, the blower pulls air from the warmer ceiling cavity to temper the air.

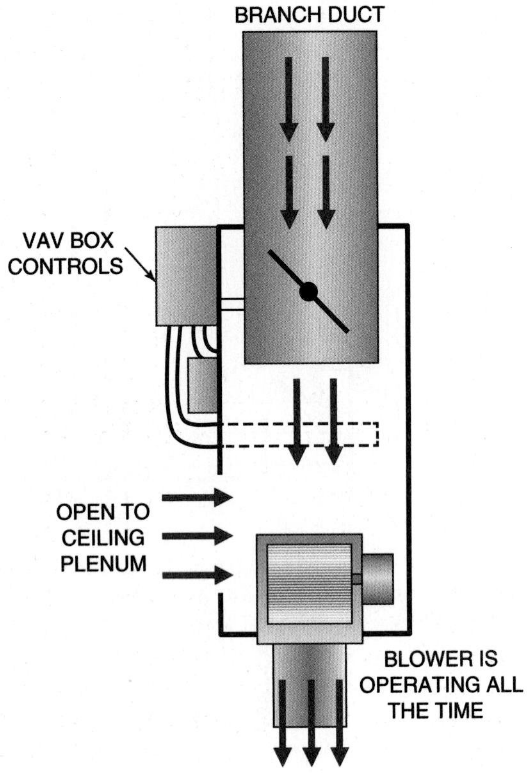

damper on the VAV will be open and the blower will deliver cooled air to the conditioned space. If there is no call for cooling, the damper on the VAV will be closed and the blower will deliver air from the ceiling cavity to the occupied space, **Figure 50.44.**

50.10 HOT WATER IN THE REHEAT COILS

Some VAV boxes or terminal units are equipped with reheat coils. These coils are intended to temper air as needed when cooling or ventilation is called for or to heat the air when an area is calling for heating. Reheat coils are typically either of the electric resistive type or of the hydronic, hot water, type. Electric resistive reheat coils are similar to the electric heaters discussed in Unit 30, "Electric Heat"; refer to that unit for more information.

As with any hot water coil, precautions should be taken to prevent it from freezing. The possibility of freezing is greater when the water flow has been stopped completely. There are two common piping configurations that are used on hot water reheat coils and utilize either a two-way or a three-way modulating valve.

TWO-WAY MODULATING VALVES

When a two-way modulating valve is used to control the flow of hot water through a reheat coil, the rate of flow varies with the needs of the particular zone. The valve is typically installed at the outlet of the reheat coil, **Figure 50.45.** Since the valve modulates the flow of hot water, precautions must be taken to avoid dead-heading the circulating pump. To prevent this, pumps that can vary flow rates must be used. As the need for hot water declines, the pump will slow down as needed. Since two-way modulating valves can, under some circumstances, be completely closed off, the possibility of coil freezing exists, so precautionary actions must be taken.

THREE-WAY MODULATING VALVES

When a three-way modulating valve is used to control the flow of hot water through a reheat coil, the rate of flow circulated by the pump remains constant. Even though the amount of water needed for reheat might vary, the unwanted volume of hot water flows through a bypass pipe that is connected across the coil, **Figure 50.46.** The valve is typically installed at the outlet of the reheat coil. Constant volume pumps can be used in conjunction with three-way modulating valve systems because they deliver a constant flow of water. The two extreme settings on the three-way valve allow for all of the hot water to be directed through the reheat coil or all of the water to bypass the reheat coil. In most cases, the actual position of the valve will allow at least some flow through both the reheat coil and the bypass line, **Figure 50.47.** Systems with three-way modulating valves are less likely to freeze, but precautions should be taken, as freezing could result in the event of pump failure.

Figure 50.45 Reheat coil with a two-way modulating valve.

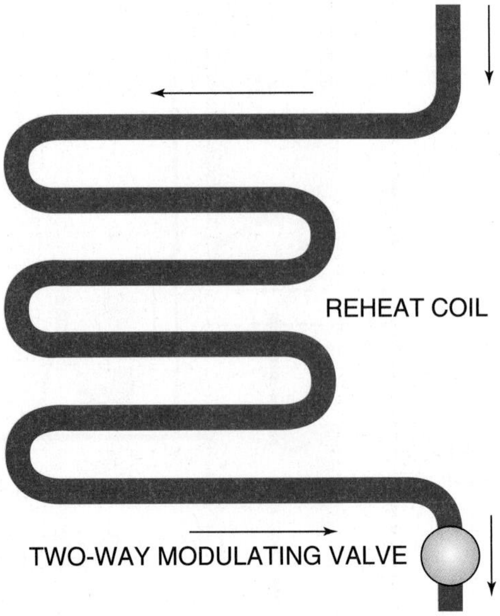

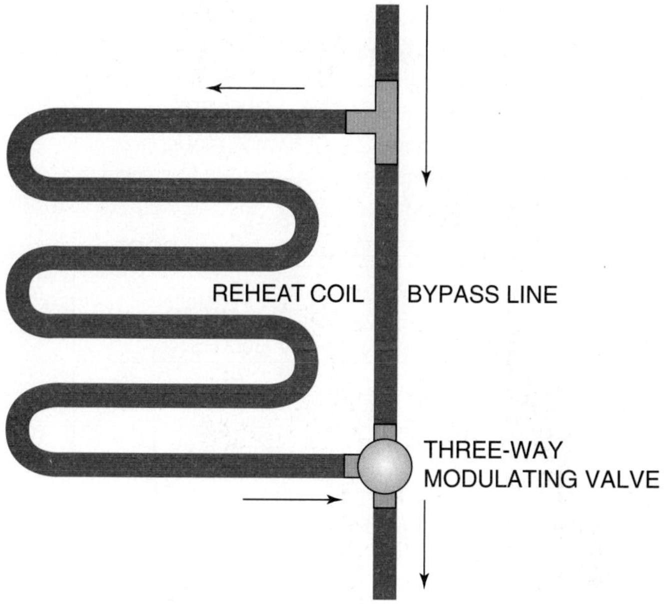

Figure 50.46 Reheat coil with a three-way modulating valve.

50.11 CHILLED-WATER VAV SYSTEMS

Variable air volume systems can be used in conjunction with conventional, vapor-compression refrigeration systems or with chilled-water systems. When variable air volume technology is incorporated into an air-cooled, conventional system, the system is commonly packaged equipment. When the VAV technology is incorporated into a chilled-water system, the system is most often configured as a split system, where the chiller is positioned in a remote location compared to that of the air handler. Chilled-water VAV systems can be configured in a number of ways, depending on the equipment used. Some common chilled-water VAV system setups include:

Water-cooled chilled-water system in the basement of the building with rooftop VAV air handlers, **Figure 50.48**.

Water-cooled chilled-water system in the basement of building with VAV air handlers located on each floor of the building, **Figure 50.49**.

Air-cooled chilled-water system in the basement of the building with an outdoor air-cooled condenser and VAV air handlers, **Figure 50.50**.

Air-cooled chilled-water system in the basement of the building with an outdoor air-cooled condenser and VAV air handlers located on each floor of the building, **Figure 50.51**.

Just as with a rooftop packaged VAV system, a chilled-water VAV system typically incorporates a means of exhausting air from the system, bringing outside air into the system; filtering, cooling, heating , and tempering the air; and delivering the treated air back to the occupied space. The main difference between the configurations is that a rooftop packaged unit also contains all components of the refrigerant-carrying circuit while the chilled-water system does not. In the case of the packaged unit, all refrigeration-related repair work is done at the unit. In the case of the chilled-water system, the air handler consists of two water coils: one for chilled water and one for hot water. Refrigerant-related repairs in the chiller are performed at the chiller location.

PROTECTING THE CHILLED-WATER VAV FROM FREEZING

Since the air handlers on a chilled-water VAV system contain water, precautions must be taken to protect the coils from freezing. It is important to remember that the outside air

Figure 50.47 Possible positions of the three-way modulating valve.

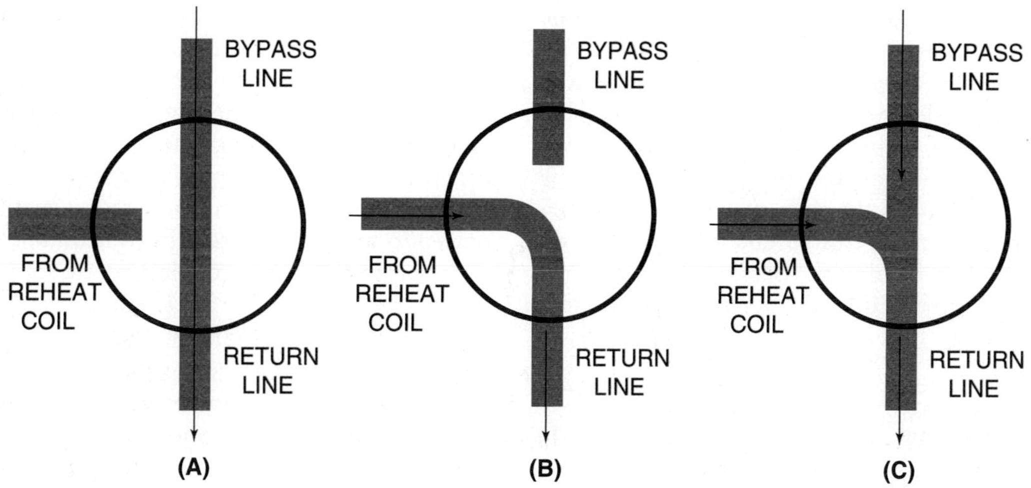

Figure 50.48 Water-cooled chilled-water system with rooftop VAV air handlers.

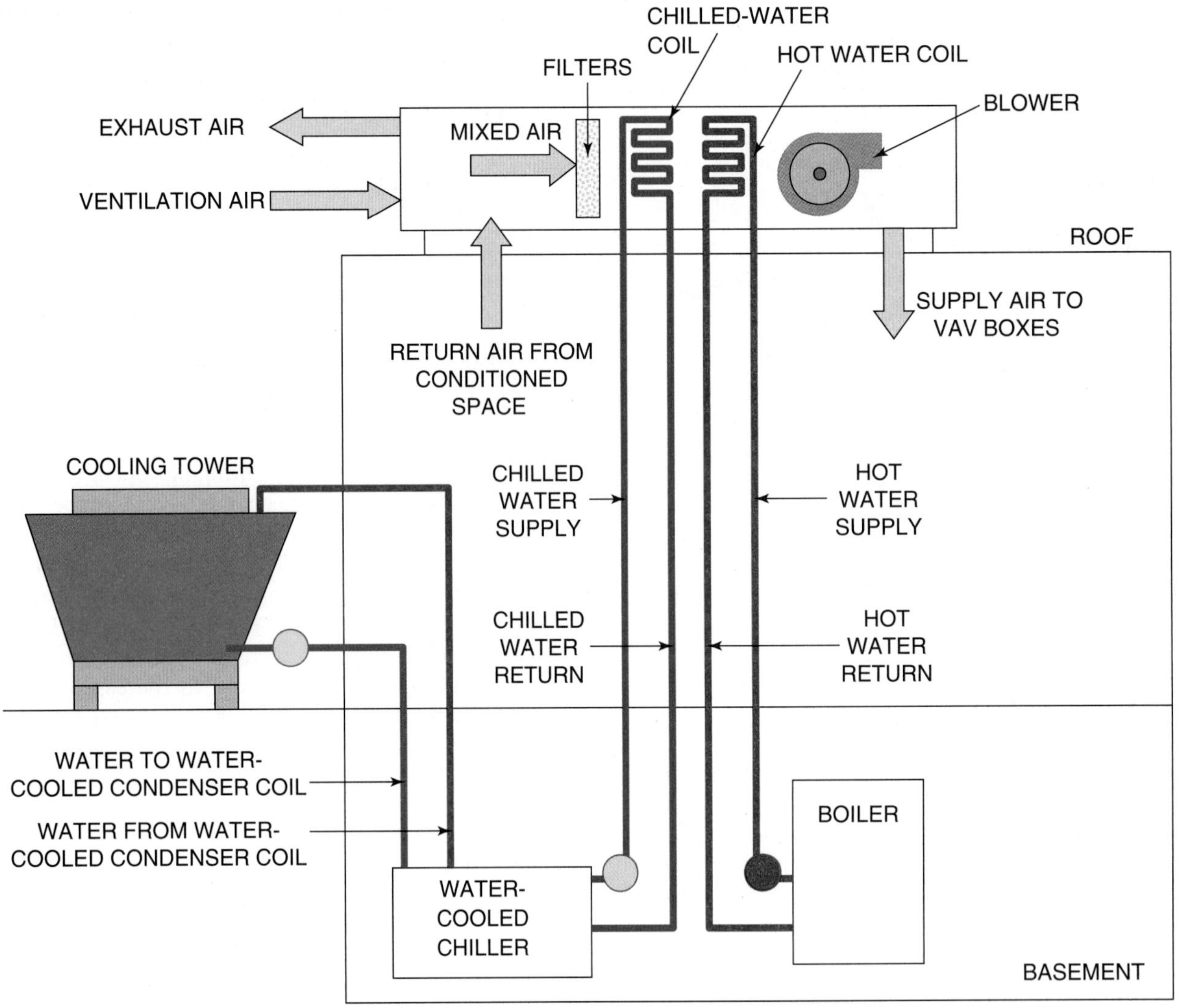

brought into the system for ventilation can be cold enough to freeze the water in the coils. Even if the return air from the occupied space is well above the freezing point, outside airstreams and return airstreams do not always mix well. Since colder air is more dense than warmer air, the colder outside air may fall to the lower portion of the mixing box, while the warmer return air from the space may occupy the upper portion. This condition has the potential to allow very cold air to pass through the coils, resulting in a freeze-up. A number of precautions can be taken to prevent coil freeze-up. These include heating the outside air as it enters the system, adding antifreeze to the water in the chilled-water piping circuit, or draining the chilled-water coils in the colder months.

HEATING THE OUTSIDE AIR AS IT ENTERS THE SYSTEM. In geographic areas where subfreezing temperatures occur, a thermostat can be positioned in the outside ventilation airstream to detect when the outside air is cold enough to cause coil freezing. This temperature sensor, typically called a freezestat, is intended to energize a heater circuit that will heat the ventilation air prior to mixing and passing through the water coils. When the temperature of the ventilation air is above the freezing point, the heaters are not activated. The heaters are typically electric resistive-type heaters. If other methods, such as hot water or steam, are used to heat the air, these coils must also be protected from freezing. In this case, the solution to the problem results in duplicating the initial problem.

ADDING ANTIFREEZE TO THE WATER IN THE CHILLED-WATER PIPING CIRCUIT. Another method of preventing freezing is to mix the water with an antifreeze, such as ethylene glycol. Adding antifreeze to the water lowers the freezing temperature of the coil. As the amount of antifreeze in the mixture

Figure 50.49 Water-cooled chilled-water system with VAV air handlers on each floor of the building.

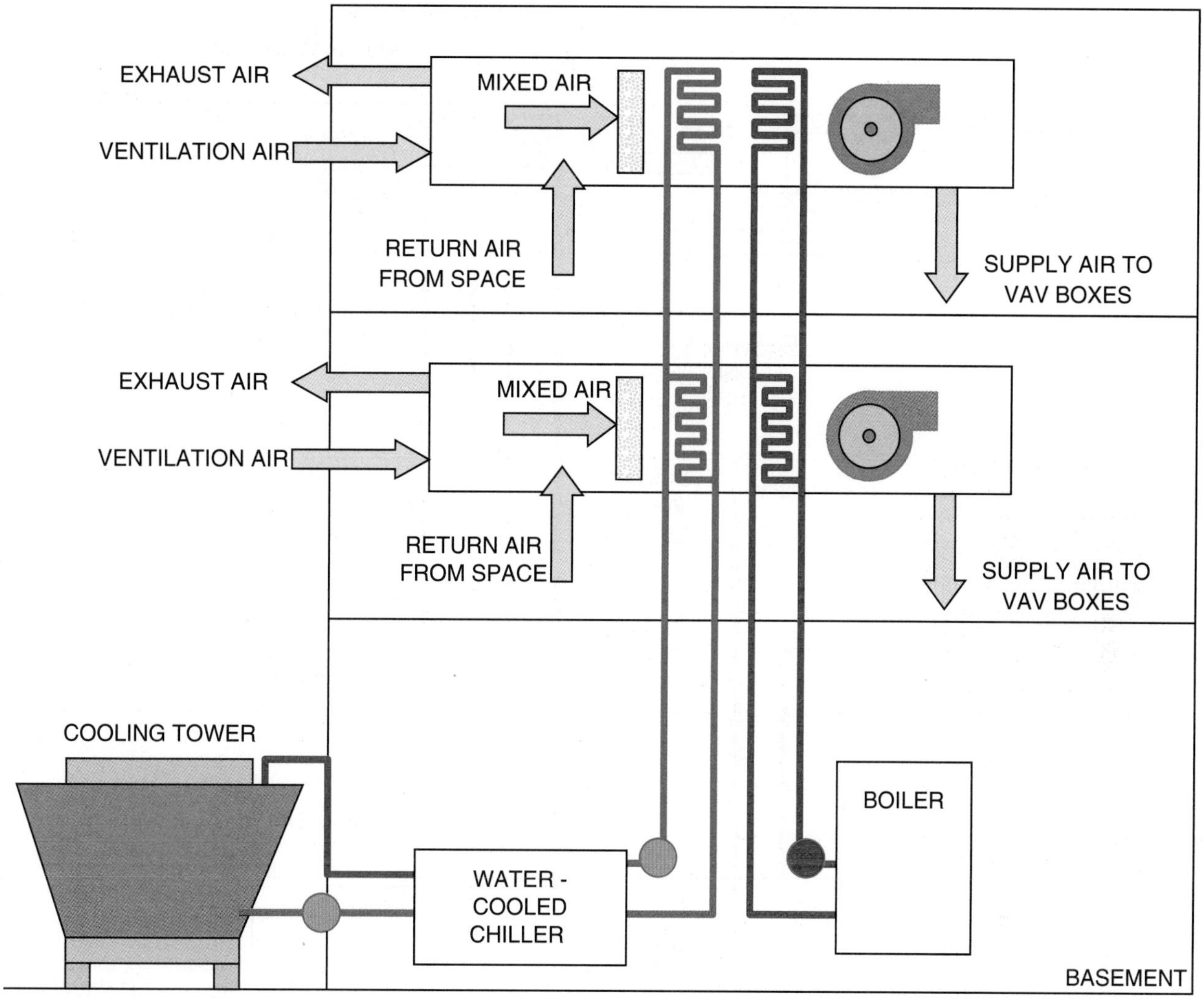

increases, the freezing point of the coil drops further. Different formulations of glycol also affect the freezing point. The following table provides information regarding the approximate range of freezing temperatures for various concentrations of ethylene glycol mixed with water.

Percentage of Ethylene Glycol	Freezing Temperature in Degrees Fahrenheit	Freezing Temperature in Degrees Celsius
0%	32°F	0°C
10%	23°F to 26°F	−5°C to −3°C
20%	14°F to 20°F	−10°C to −7°C
30%	2°F to 8°F	−17°C to −13°C
40%	−13°F to −10°F	−25°C to −23°C
50%	−36°F to −30°F	−38°C to −34°C

Upon initial inspection, it might seem obvious that it is definitely better to add more ethylene glycol to the mixture, as this will greatly lower the freezing temperature of the water/antifreeze mixture. But there are some drawbacks to adding more antifreeze than needed. It reduces the heat-transfer rate of the solution and requires larger pumps to circulate the solution through the piping.

The concept of specific heat was discussed back in Unit 1, "Heat, Temperature, and Pressure." The discussion established that the specific heat of water is 1.0 Btu/lb°F. As impurities are mixed with the water, the specific heat capacity of the water changes. At 40°F, a mixture of 30% glycol and 70% water will have a specific heat capacity of about 0.89 Btu/lb°F. This is about a 10% reduction in the ability of the solution to transfer heat. If the percentages of glycol and water are changed to create a 50/50 mixture, the specific heat capacity drops to about 0.795 Btu/lb°F. This equates to a 20% reduction in heat-transfer capability. The

Figure 50.50 Air-cooled chilled-water system with rooftop VAV air handlers.

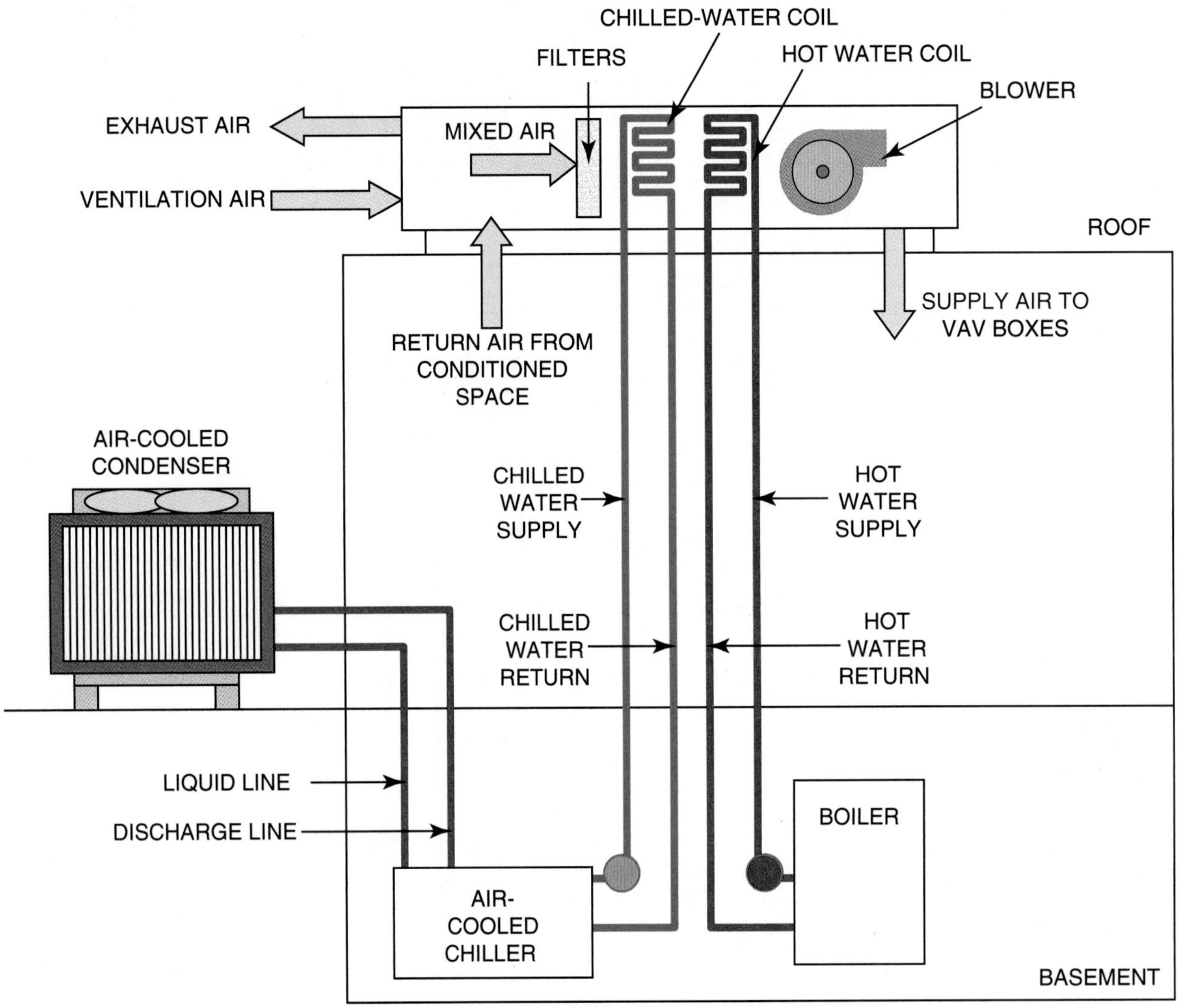

decision about whether to add ethylene glycol to the system must be made early in the design process, as this will have a major effect on the amount of fluid that needs to be circulated through the system.

Another major concern in using ethylene glycol as an additive for freeze protection is the issue of viscosity. The viscosity, or thickness, of glycol is much greater than that of water. This increase in viscosity means that it is harder to pump and the head loss in the system will be increased. In Unit 33, "Hydronic Heat," we discussed the concept of "feet of head" and established that a pressure of 1 psig could raise a column of water 2.31 feet, **Figure 50.52**. In the case of a glycol/water mix, this conversion is no longer accurate, since we are not dealing with pure water. If the addition of ethylene glycol resulted in a 50% increase in head loss, then we would need a pump that was 50% larger to effectively circulate the antifreeze/water mixture. Once again, the decision about whether or not to add ethylene glycol to the system must be made early in the design process, as this will have a major effect on the size and capacity of the pump that is selected for the particular application.

SAFETY PRECAUTION: *Ethylene glycol should not be ingested. Although it has a sweet taste, once ingested it can form glycolic and oxalic acid, which can have a negative effect on the central nervous system, the heart, and the kidneys. If treatment for ingestion is not received, death can result.* •

DRAINING THE CHILLED-WATER COILS IN THE COLDER MONTHS. One way to avoid the negative effects of adding antifreeze to the system or employing heaters to prevent freezing is to remove the water from the chilled-water coils altogether. This is the most labor-intensive option but eliminates all possibility of coil freezing, provided that the draining process is performed properly. In order to effectively

Figure 50.51 Air-cooled chilled-water system with VAV air handlers on each floor of the building.

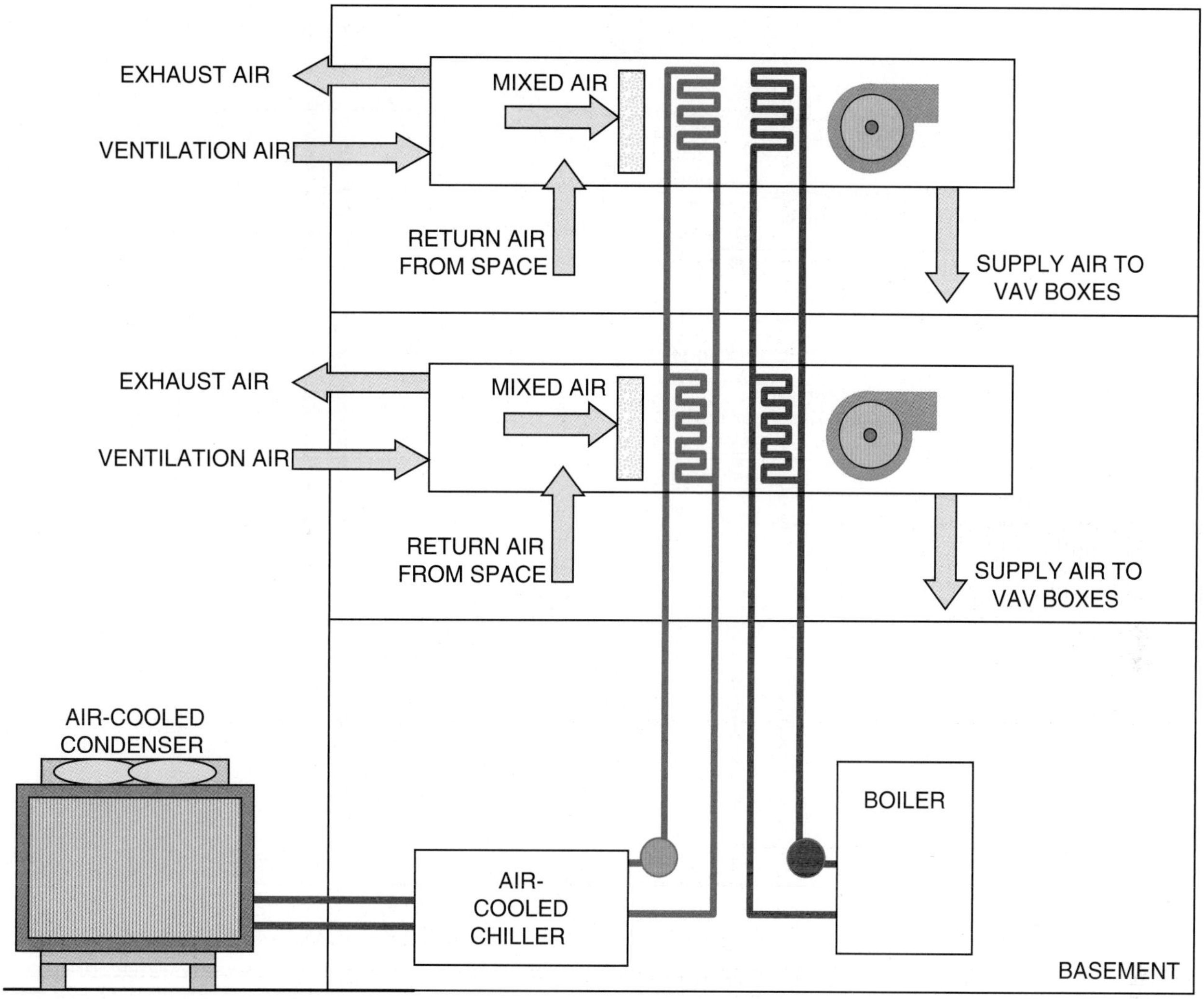

Figure 50.52 One psig of pressure can lift a column of water 2.309 feet.

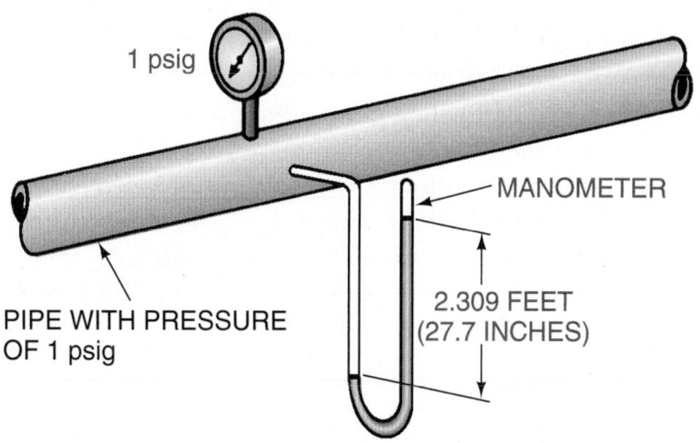

drain the chilled-water coil, there must be a drain port at the coil, along with a vent port and a means to valve off the coil from the rest of the chilled-water piping arrangement, **Figure 50.53**.

To properly drain the coil, it must first be isolated from the rest of the chilled-water piping circuit by closing the supply and return water valves, **Figure 50.54**. The vent port should then be opened. This will allow air to flow into the coil as the water is drained from the bottom. A hose should be connected to the drain port on the coil and be directed outside, to a floor drain, or to another location where the draining water will not pose a problem. The drain port can then be opened to drain the coil, **Figure 50.55**. Once the coil has been drained, it is a good idea to push pressurized air through the vent port to push any remaining water out of the coil. Once this has been done, the drain valve should be closed and some glycol should be pumped into the coil, **Figure 50.56**. This will help prevent any remaining water from freezing. The vent port can be left open for the winter, but proper signage should be posted alerting other technicians that the valve is open. Should they open the isolation valves without first checking the position of the vent port, a flood could result. However, if the draining process is properly performed, closing the vent port will not pose a problem.

Figure 50.53 Piping arrangement for effective reheat coil draining.

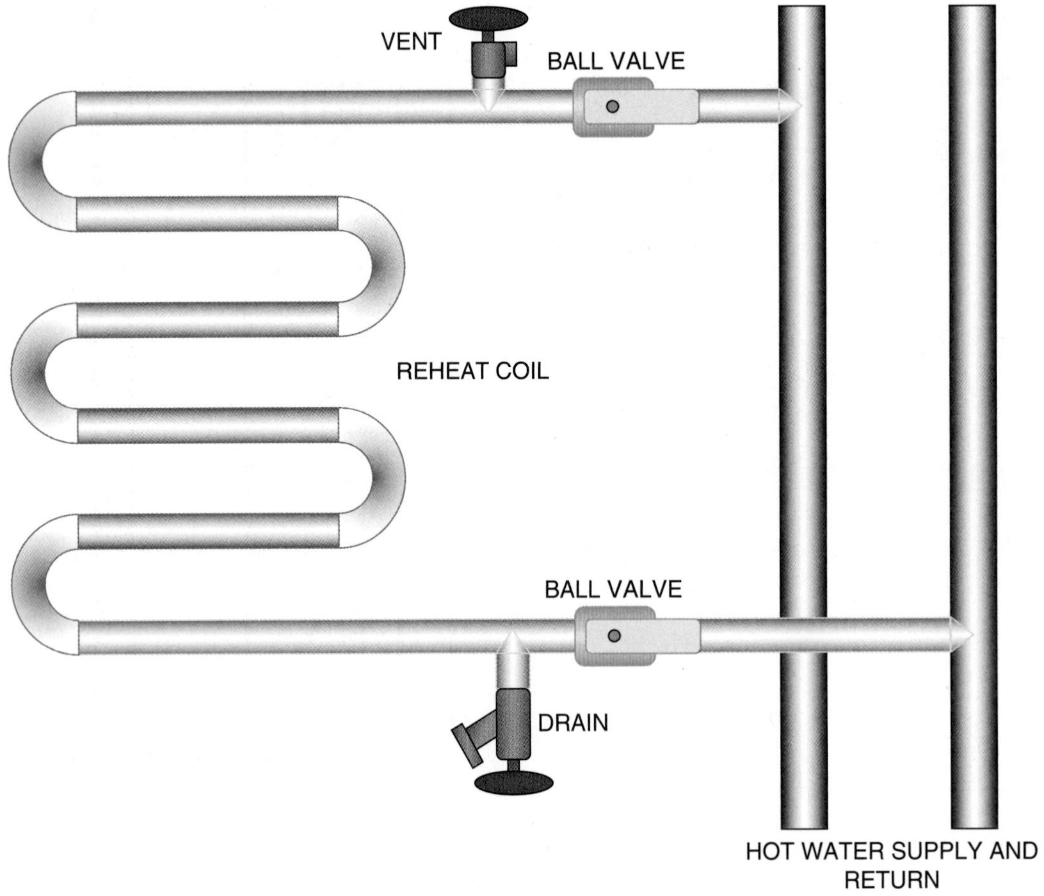

To put the coil back into operation, the glycol should first be drained. This is accomplished by opening the vent port and then opening the drain port. The glycol should be captured in a container and stored for the next time the coil needs to be drained, **Figure 50.57**. Blowing pressurized air through the vent port helps ensure that all of the antifreeze has been removed from the coil.

Once the antifreeze has been removed, the vent and drain ports should be closed. Then the lower isolation valve on the main chilled-water circuit can be opened slightly. This will allow water to flow into the coil. Once water begins to flow into the coil, the vent port can be opened slightly. This will purge the air from the chilled-water coil, **Figure 50.58**. Water is allowed to enter the coil until it flows from the vent port. Once that happens, the vent port is closed and the upper isolation valve is opened.

50.12 VARIABLE REFRIGERANT FLOW (VRF) SYSTEMS

Unlike the VAV systems described in the previous section of this unit, variable refrigerant flow (VRF) systems modulate the flow of refrigerant that is pumped through the system. Variable refrigerant flow systems are able to operate effectively and efficiently within a wide range of capacities by changing the speed at which the system compressor or compressors operate by means of inverter-controlled compressor motors. Inverters change the frequency of the power supplied to a motor to alter the speed. (Inverters are discussed in more detail in Unit 17, "Types of Electric Motors," and in Unit 26, "Applications of Refrigeration Systems.") The control system for the inverter-controlled compressor motors sense, among other things, the saturation temperatures and pressures of the refrigerant to modulate the volume of refrigerant that is pumped by the system compressors. In addition to the inverter-controlled compressors, variable airflow systems can also use inverter-controlled fan motors to modulate the rate of airflow through the outdoor coils.

Variable refrigerant flow systems can be configured as cooling-only, as heat pumps, or as heat pump systems with heat recovery. Cooling-only systems are capable of providing comfort cooling to the occupied space and are typically used in conjunction with a variety of heating methods, such as baseboard heat around the perimeter of a structure and hydronic fan coil units throughout the interior spaces. When configured as heat pumps, VRF systems can provide either heating or cooling to the occupied space,

Figure 50.54 Ball valves are closed to isolate the reheat coil from the main hot water piping circuit.

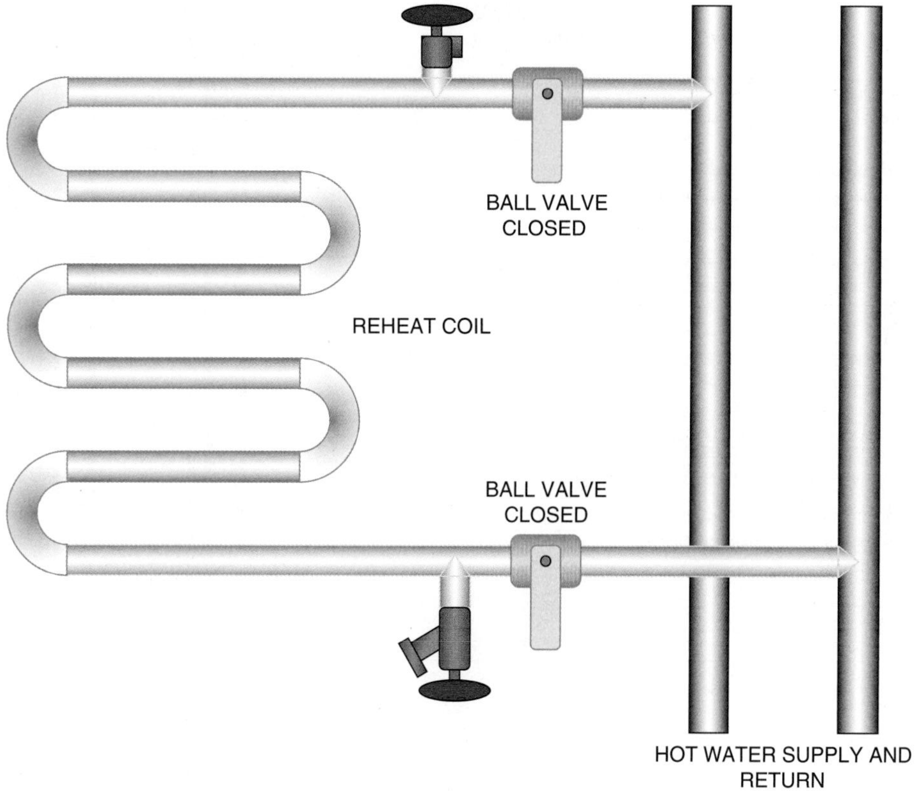

Figure 50.55 With both the drain and vent open, water from the coil will drain.

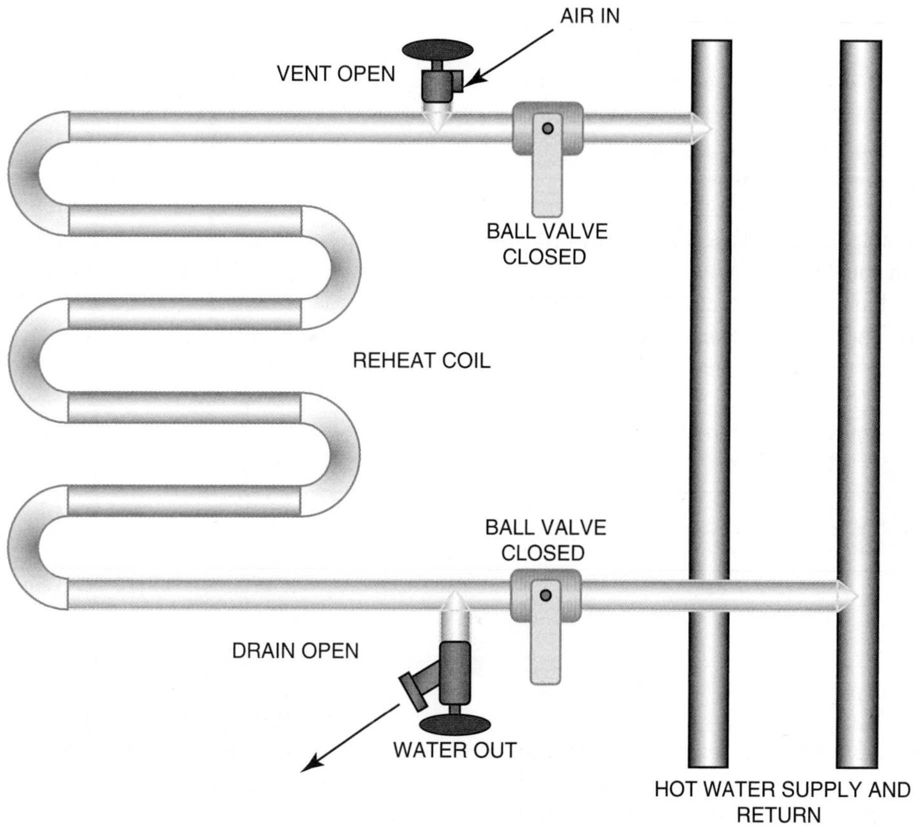

Figure 50.56 With the drain closed, a small amount of glycol can be added to the coil.

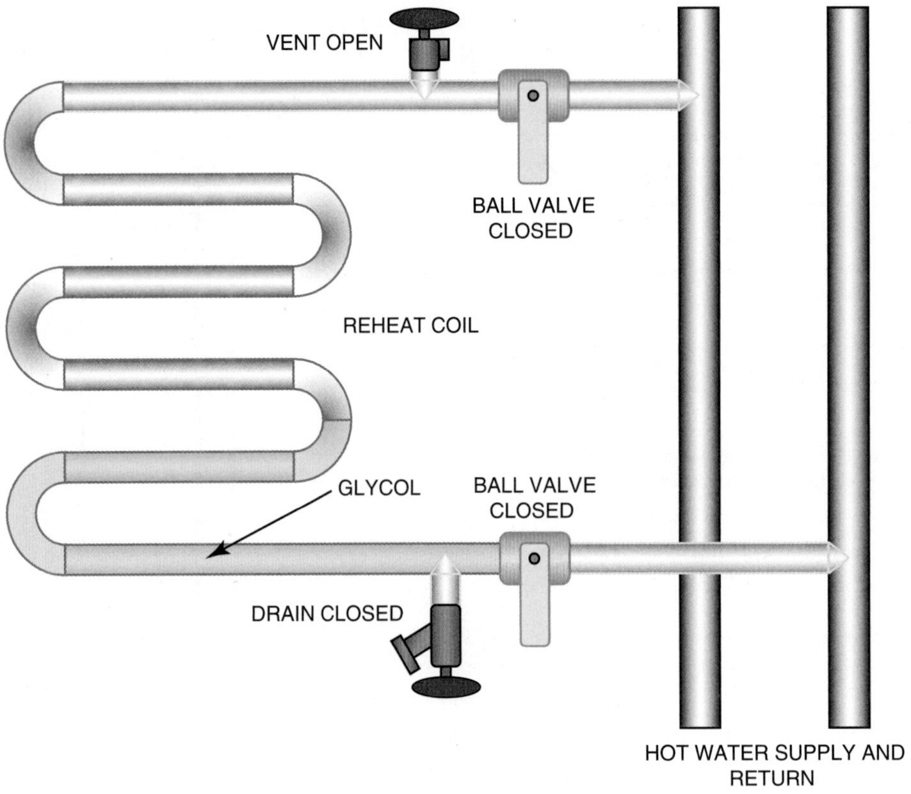

Figure 50.57 Opening the drain valve will allow the glycol to drain from the coil.

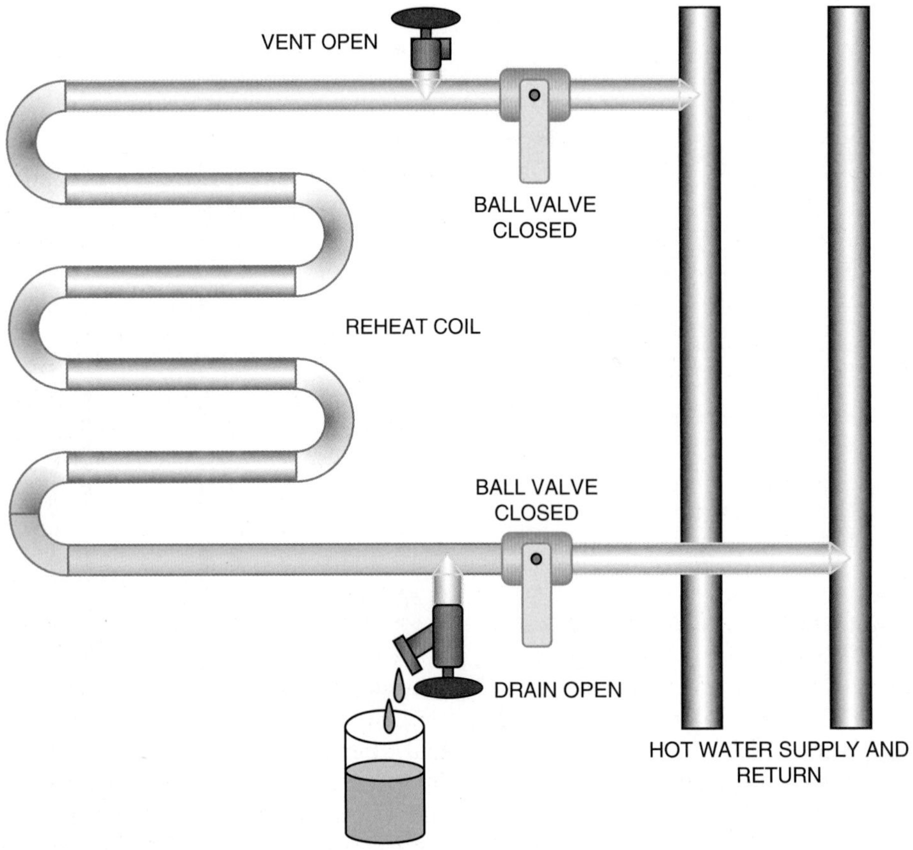

Figure 50.58 With the drain valve closed and the lower hot water valve open, water pressure will push air out of the coil through the open vent.

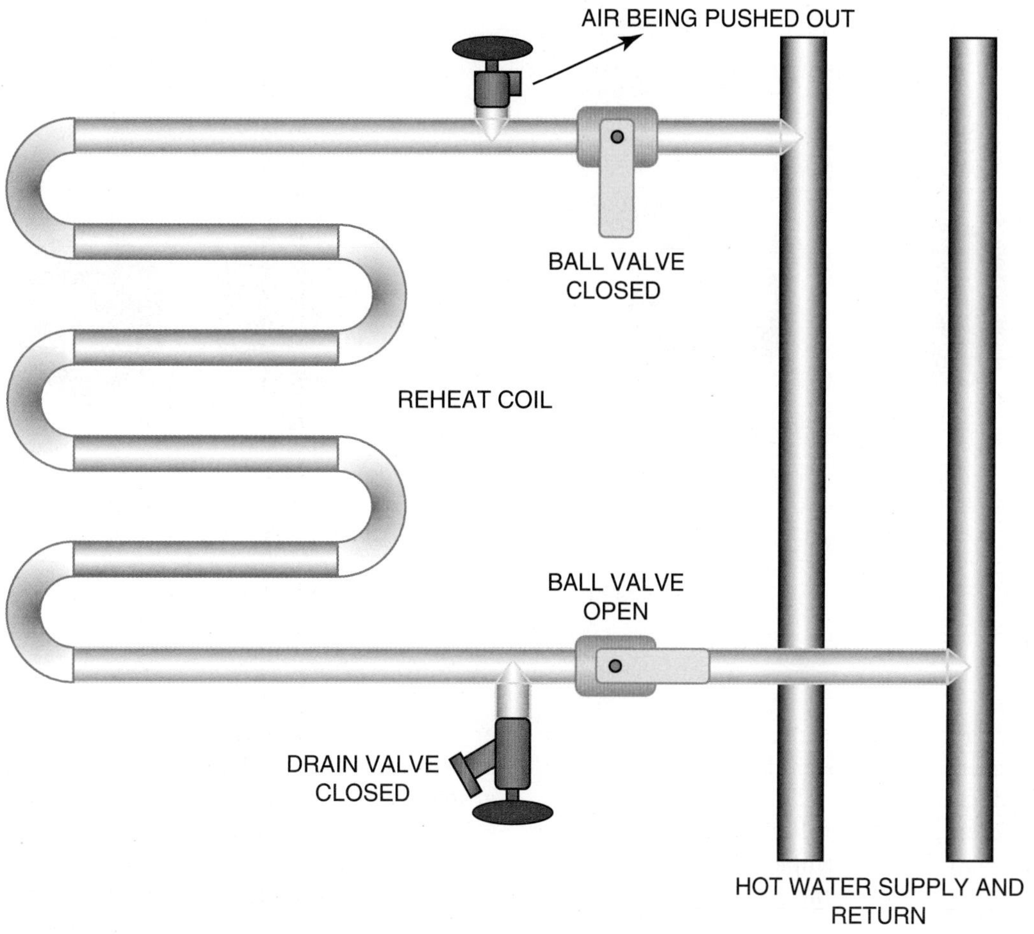

depending on the needs of the structure. The heat-recovery heat pump system is the most complicated of the three common configurations but provides the most versatility. The heat-recovery, heat pump VRF system allows different areas in the structure to operate in different modes at the same time. Some areas can call for heating while other areas can call for cooling; the system will be able to satisfy both requests simultaneously.

VRF SYSTEM COMPRESSOR CONFIGURATIONS

The compressors in VRF systems can be configured in a number of different ways, depending on the capacity and design of the system. Some systems might have a single inverter compressor; others might have multiple inverter compressors. Some systems utilize a combination of inverter and standard, single-speed compressors, **Figure 50.59**.

Some equipment manufacturers utilize a modular approach by which multiple outdoor units can be piped together to increase the capacity of the system. The outdoor units can be manifolded together to meet the air-conditioning and heating needs of the structure, **Figure 50.60**. In some applications, the VRF system can service over 50 separate indoor locations.

VRF SYSTEM CONTROL

Electronic control circuits are used in order to ensure that the proper amount of refrigerant is being circulated through the system. A number of control strategies can be employed, but each is designed using three basic control-based building blocks. These building blocks are proportional, integral, or derivative controls.

Proportional controls are designed to react to present errors that exist between the desired state of operation and what is actually taking place. The larger the error, the more adjusting the proportional control will do. Integral controls react to the sum of all of the previous errors that existed and attempt to bring that total to zero. In simple terms, integral controls sum up all the positive errors (values above the desired set point) and the negative errors (values below the desired set point) and adjust the output signal to help

Figure 50.59 Outdoor units on VRF systems can be a combination of standard and inverter compressors.

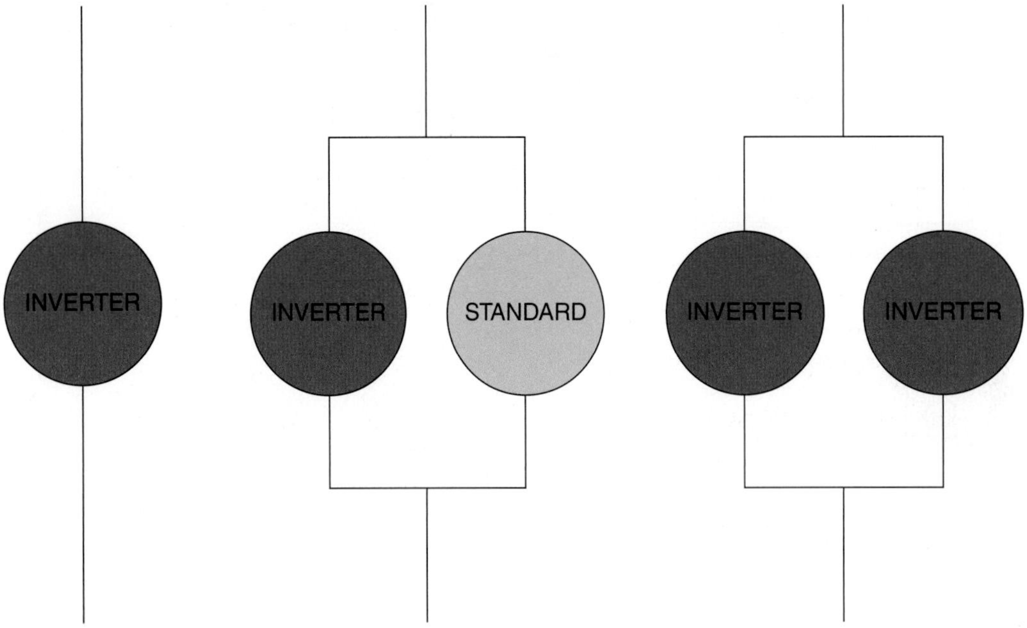

create a net zero error value. Derivative controls attempt to predict the future needs of the system based on the rate of change of measured parameters.

Controls that utilize all three of these building blocks are referred to as PID controls. An example of a system component that utilizes PID controls is the step-motor expansion valve that was discussed in Unit 24, "Expansion Devices." The control strategy that seems to be prevalent in VRF systems is the proportional integral (PI) strategy, which does not utilize the derivative or predictive element.

PROPORTIONAL INTEGRAL (PI) CONTROL SYSTEM RESPONSE

The VRF system reacts to changes in system saturation temperatures and pressures and controls the operation of the system compressors accordingly. In the cooling mode of operation, the evaporator saturation temperature and pressure are monitored; the condenser saturation temperature and pressure are monitored when operating in the heating mode.

While operating in the cooling mode, low evaporator saturation temperature and pressure indicate that the cooling load on the system is low. If the space is cool and close to the thermostat's set point, less heat will be added to the system. This will result in lower evaporator temperatures and pressures. If the cooling load on the system is high, the evaporator saturation temperature and pressure will be higher than desired. In either case, the control system can increase or decrease the speed of compressor motors based on this sensed information. The speed of the compressor is varied in steps, with each step representing a different frequency or speed of the compressor motor.

Let us assume that we have a VRF unit that is equipped with a single inverter compressor and that the control system can vary the frequency of the compressor from 60 Hz (60 cycles per second) to 200 Hz. When the system is initially energized, the compressor motor will start at its lowest speed, which, in this example, is 60 Hz. As the need for refrigerant flow increases, the controller will continue to move up steps in the control sequence. If the control sequence has 15 steps, each step might increase the frequency of the compressor by 10 Hz. The following table represents this. As the cooling load on the system increases, the speed of the compressor will also increase from step 1 until it reaches its maximum speed, if needed, at step 15. As the load on the system drops, the control module will drop from higher to lower steps, slowing the compressor down.

Step	Frequency	Step	Frequency	Step	Frequency
1	60 Hz	6	110 Hz	11	160 Hz
2	70 Hz	7	120 Hz	12	170 Hz
3	80 Hz	8	130 Hz	13	180 Hz
4	90 Hz	9	140 Hz	14	190 Hz
5	100 Hz	10	150 Hz	15	200 Hz

In the case of a VRF system that utilizes one inverter compressor and one standard compressor, the control

Figure 50.60 Outdoor units can be manifolded together to increase the capacity of the system. (A) Two manifolded outdoor units each having one inverter and one standard compressor. (B) Two manifolded outdoor units each having two inverter compressors.

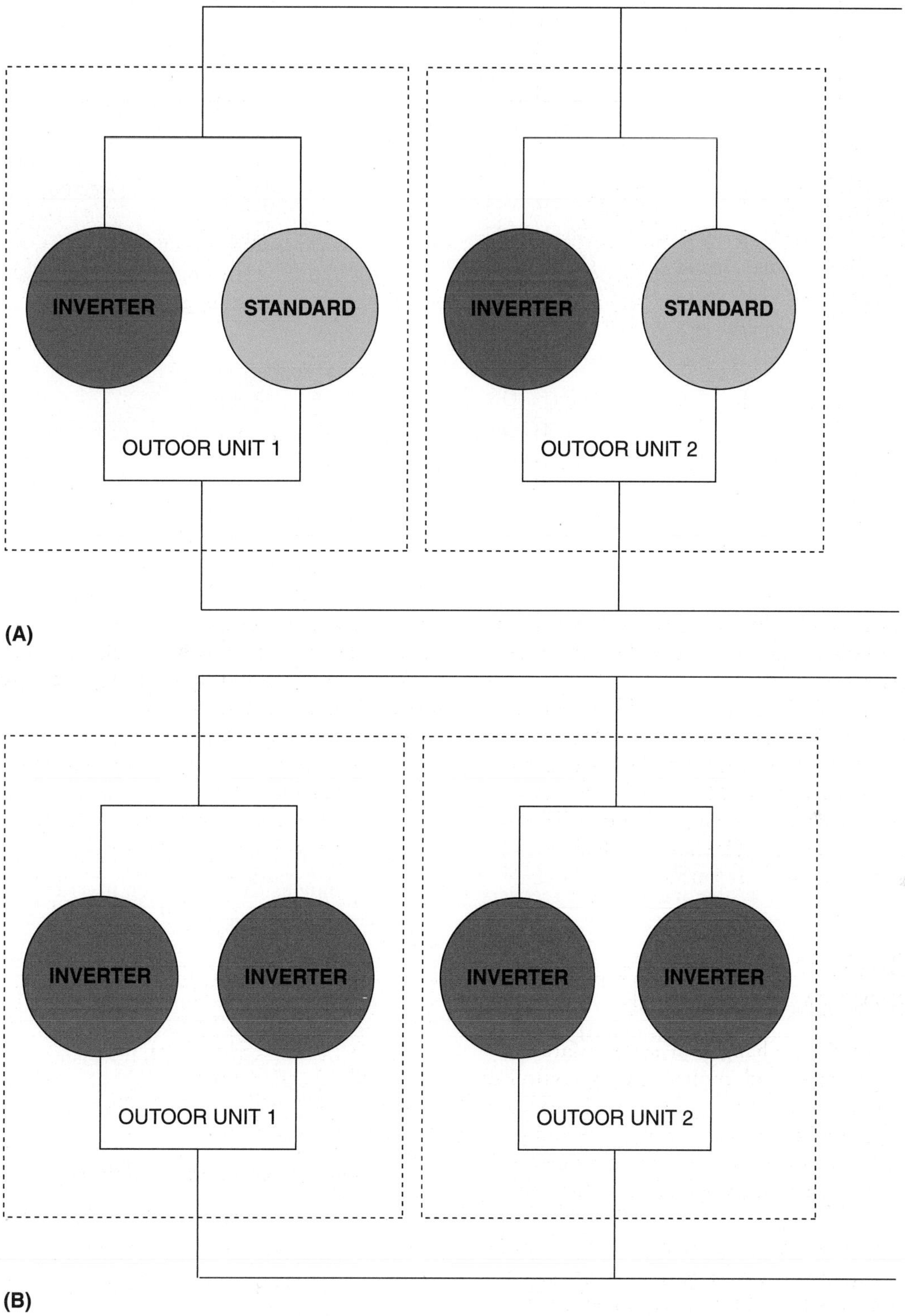

strategy is similar, in the sense that moving from one step to another will increase or decrease the volume of refrigerant circulated. However, the standard compressor will cycle on only when the inverter compressor is not able to handle the cooling load by itself. The following table provides a sample control strategy for the inverter compressor (INV) and the standard compressor (STD). When the system is operating at any of the control steps 1 through 15, the standard

compressor is not in operation. As it moves from step 15 to step 16, the inverter compressor drops to its slowest speed and the standard compressor cycles on. As the cooling load on the system increases from this point, the standard compressor remains in operation and the inverter compressor ramps up its speed to satisfy the cooling needs of the space.

Step	Inv	Std	Step	Inv	Std	Step	Inv	Std
1	60 Hz	OFF	11	160 Hz	OFF	21	110 Hz	ON
2	70 Hz	OFF	12	170 Hz	OFF	22	120 Hz	ON
3	80 Hz	OFF	13	180 Hz	OFF	23	130 Hz	ON
4	90 Hz	OFF	14	190 Hz	OFF	24	140 Hz	ON
5	100 Hz	OFF	**15**	**200 Hz**	**OFF**	25	150 Hz	ON
6	110 Hz	OFF	**16**	**60 Hz**	**ON**	26	160 Hz	ON
7	120 Hz	OFF	17	70 Hz	ON	27	170 Hz	ON
8	130 Hz	OFF	18	80 Hz	ON	28	180 Hz	ON
9	140 Hz	OFF	19	90 Hz	ON	29	190 Hz	ON
10	150 Hz	OFF	20	100 Hz	ON	30	200 Hz	ON

VRF PIPING CONFIGURATIONS

Two typical piping configurations are found on VRF systems. One is the piping circuit used in heat pump systems and the other is the piping circuit used in conjunction with a heat pump system that has a heat-recovery feature. The heat pump VRF configuration utilizes two refrigerant lines that are connected to the outdoor unit. One line is the liquid line and the other line is the gas line. The heat pump system that is equipped with heat recovery can have either two or three field-installed lines connected between the outdoor units and the indoor units, depending on the equipment manufacturer. Be sure to check the model number of the system being worked on to determine if the system has the heat recovery feature or not.

TWO-PIPE VRF PIPING CONFIGURATION. A simplified piping diagram of the two-pipe arrangement with one outdoor unit and two indoor units is shown in **Figure 50.61**. This outdoor unit has one inverter compressor, INV, and one standard compressor, STD, that are connected in parallel with each other. There are two four-way valves, RV1 and RV2, to aid in the changeover between the heating and cooling modes. In addition, there are two capillary tubes, S1 and S2, that are intended to allow pressure from the high side of the system to bypass to the low side to prevent refrigerant accumulation in the lines.

When only one indoor coil is calling for cooling, the refrigerant path can be traced out as shown in **Figure 50.62**. The sequence is as follows:

- The high-pressure, high-temperature vapor is discharged from the compressors INV and STD and flows to reversing valve RV1.
- The refrigerant is directed to the outdoor coil, which is functioning as the condenser. It is here that system heat is rejected.
- At the outlet of the outdoor coil, the refrigerant bypasses the outdoor metering device as it flows though the check valve piped in parallel with the metering device.
- The high-pressure, high-temperature liquid then flows to the indoor coil.
- Solenoids at the indoor unit open to allow refrigerant to flow through the metering device and onto the evaporator coil.
- The low-pressure, low-temperature vapor leaves the evaporator and flows through reversing valve RV2.
- From RV2 the suction gas returns to the compressors, INV and STD.

NOTE: *A small amount of refrigerant is able to pass from the compressor discharge back to the compressor inlet through RV2 and S2. This is to prevent refrigerant accumulation in the refrigerant lines.* •

When both of the indoor units are calling for cooling, the flow path is similar to the one just discussed, with two exceptions. First, the liquid refrigerant flow from the outdoor coil splits to feed both indoor units to provide cooling for both areas. Second, when the suction gas leaves the two evaporator coils, the two paths join and the refrigerant flows back through RV2 and on to the compressors, **Figure 50.63**.

When only one indoor coil is calling for heating, the refrigerant path can be traced out as shown in **Figure 50.64**. The sequence is as follows:

- The high-pressure, high-temperature vapor is discharged from the compressors INV and STD and flows to reversing valve RV2.
- The refrigerant is directed to the indoor coil, which is functioning as the condenser. It is here that system heat is rejected into the occupied space for heating purposes.
- Solenoid valves at the indoor coil allow the hot gas to flow through the indoor coil.
- At the outlet of the indoor coil, the refrigerant bypasses the indoor metering device as it flows though the check valve piped in parallel with the metering device.
- The high-pressure, high-temperature liquid then flows to the outdoor coil.
- At the outdoor unit, refrigerant flows through the metering device and into the outdoor coil, which is functioning as the evaporator coil.
- The low-pressure, low-temperature vapor leaves the evaporator and flows through reversing valve RV1.
- From RV1 the suction gas returns to the compressors, INV and STD.

NOTE: *A small amount of refrigerant is able to pass from the compressor discharge back to the compressor inlet through RV1 and S1. This is to prevent refrigerant accumulation in the refrigerant lines.* •

Figure 50.61 Two-pipe VRF arrangement with one outdoor unit and two indoor units.

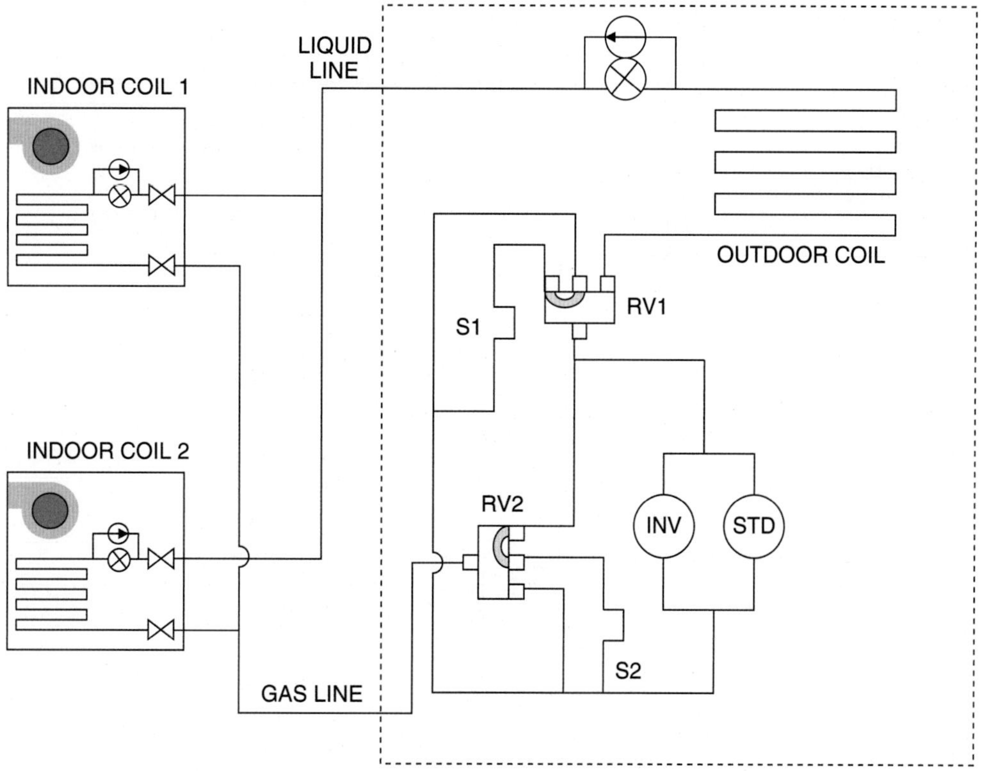

Figure 50.62 Only one of the two indoor units is calling for cooling. The red lines represent the high-pressure refrigerant and the blue lines represent the low-pressure refrigerant.

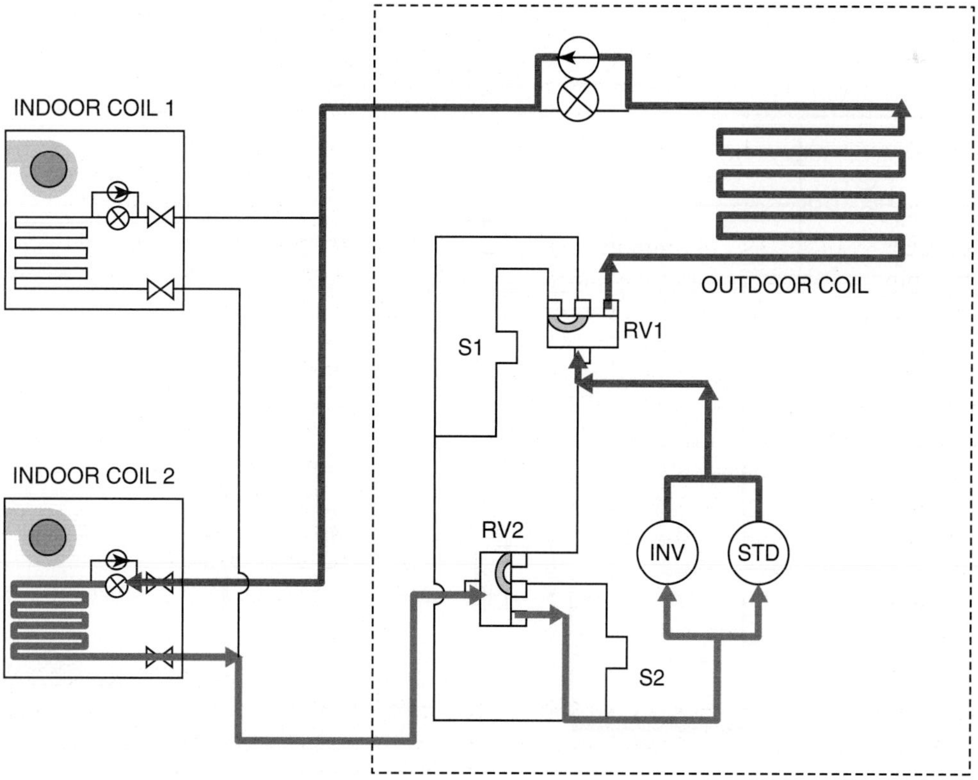

Figure 50.63 Two-pipe VRF configuration with both units calling for cooling.

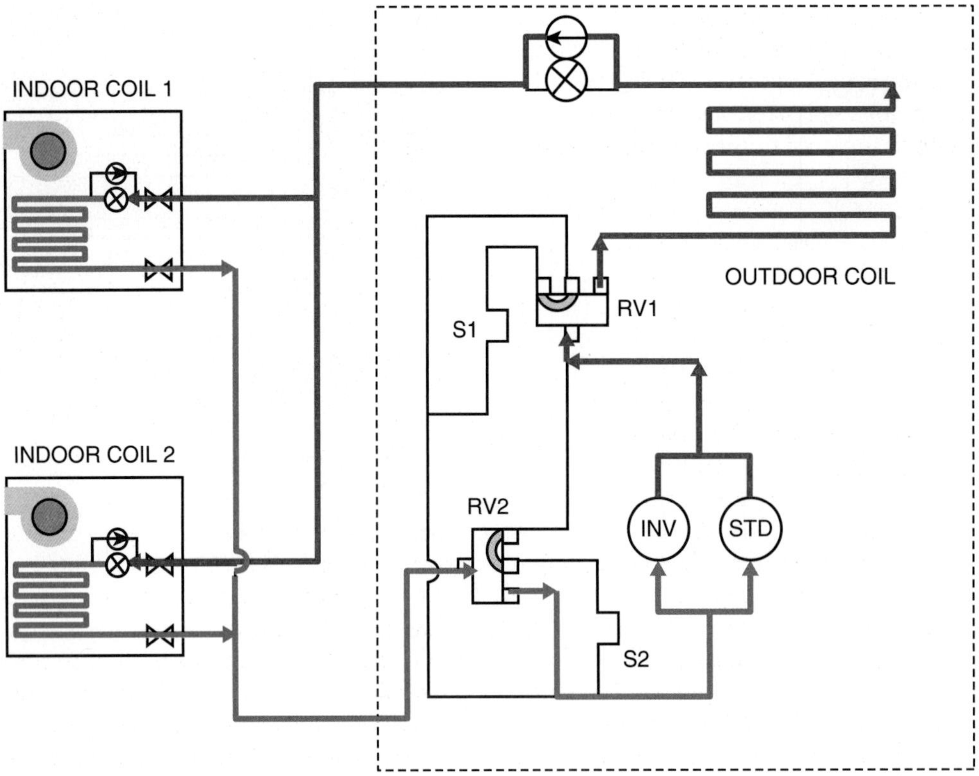

Figure 50.64 Only one of the two indoor units is calling for heating. The red lines represent the high-pressure refrigerant and the blue lines represent the low-pressure refrigerant.

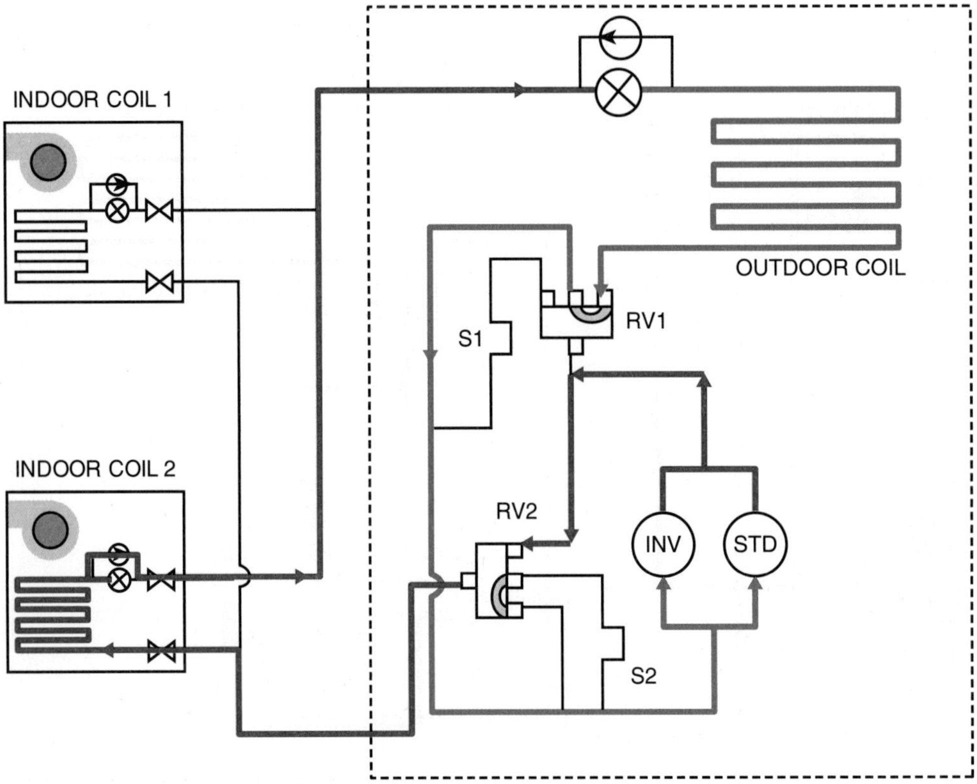

When both of the indoor units are calling for heating, the flow path is similar to the one just discussed, with two exceptions. First, the hot discharge gas from the compressor splits to feed both indoor units to provide heating for both areas. Second, when the high-pressure, high-temperature liquid leaves the indoor coils, the two paths join and the refrigerant flows on to the outdoor metering device and evaporator coil, **Figure 50.65**.

THREE-PIPE VRF PIPING CONFIGURATION. The three-pipe configuration allows for some indoor units to provide heating while others are providing cooling. This is referred to as heat recovery. This piping arrangement is somewhat more complicated and requires the use of a branch selector box, **Figure 50.66**. The branch selector box has three pipes that connect back to the outdoor units and two pipes that are connected to the indoor unit. Inside the branch selector, there can be as many as five electronic expansion valves that function to modulate refrigerant flow based on the mode of system operation desired as well as the conditions of the system external to the individual indoor unit, **Figure 50.67**. For example, if needed, the SC valve can be put into operation to provide additional subcooling when some areas are calling for cooling and others are calling for heating. The EEVC1 and EEVC2 electronic expansion valves operate during the cooling mode, and the EEVH1 and EEVH2 electronic expansion valves operate during the heating mode. If two or more indoor units will always be functioning in the same mode, either heating or cooling, a single branch selector can be used in conjunction with them, **Figure 50.68**.

Figure 50.66 A branch selector box. Photo by Eugene Silberstein

A simplified three-pipe VRF piping arrangement is shown in **Figure 50.69**. In this setup, there are two compressors and two outdoor coils. The compressors can be a combination of standard and inverter. There are three pipes connecting the outdoor units to the branch selector boxes. These lines are the suction line, the liquid line, and a dual-pressure gas line. The dual-pressure gas line can carry either high- or

Figure 50.65 Two-pipe VRF configuration with both units calling for heating.

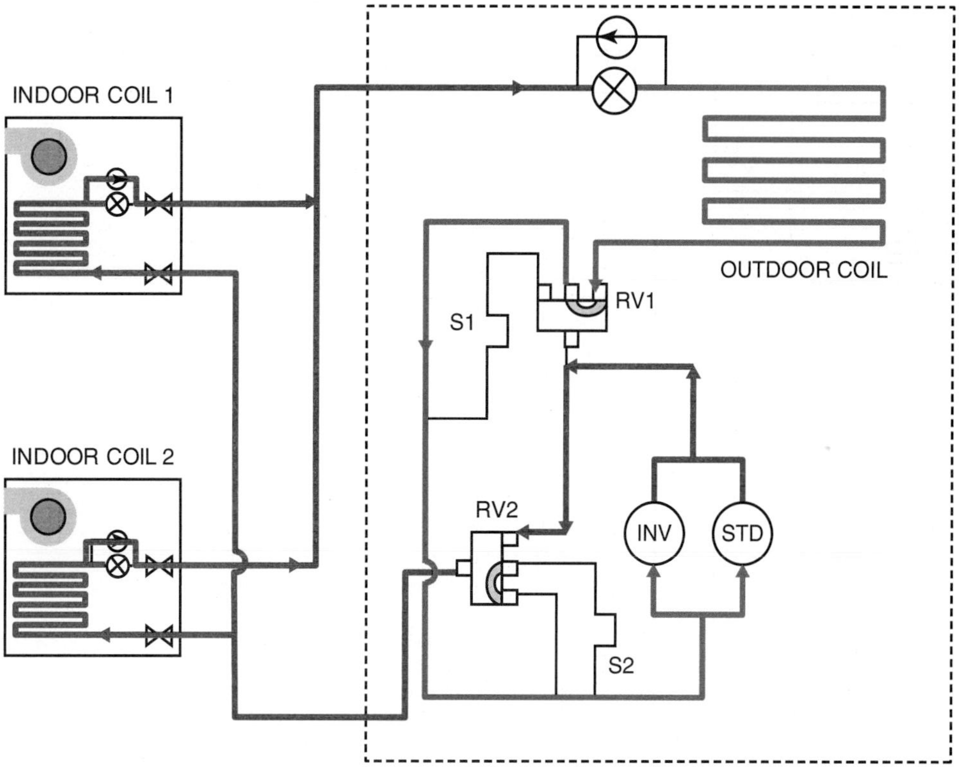

Figure 50.67 This branch selector box features as many as five electronic expansion valves that function during the different modes of operation to ensure maximum system efficiency and effectiveness.

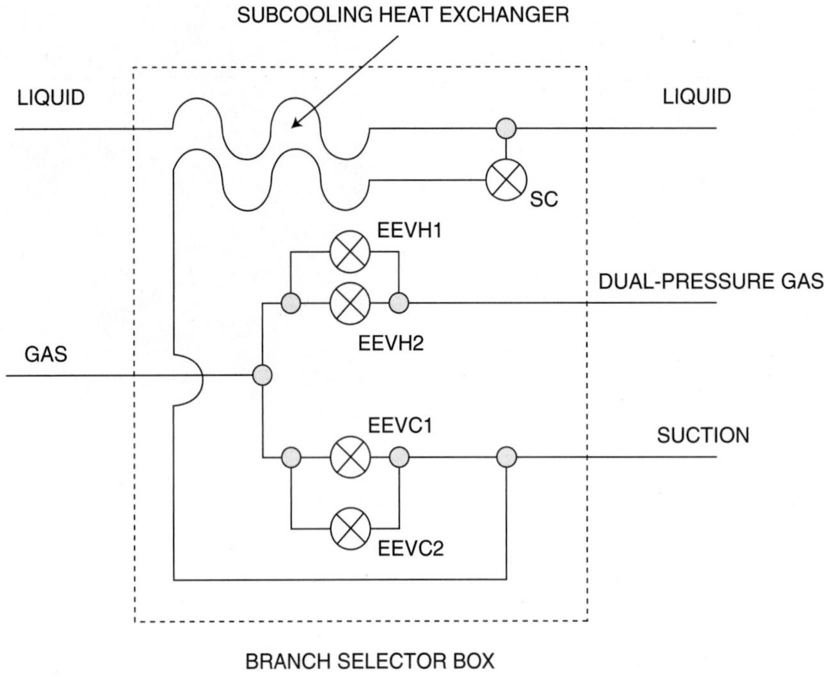

Figure 50.68 Multiple indoor units can be connected to a single branch selector box as long as all of the indoor units will be operating in the same mode at the same time.

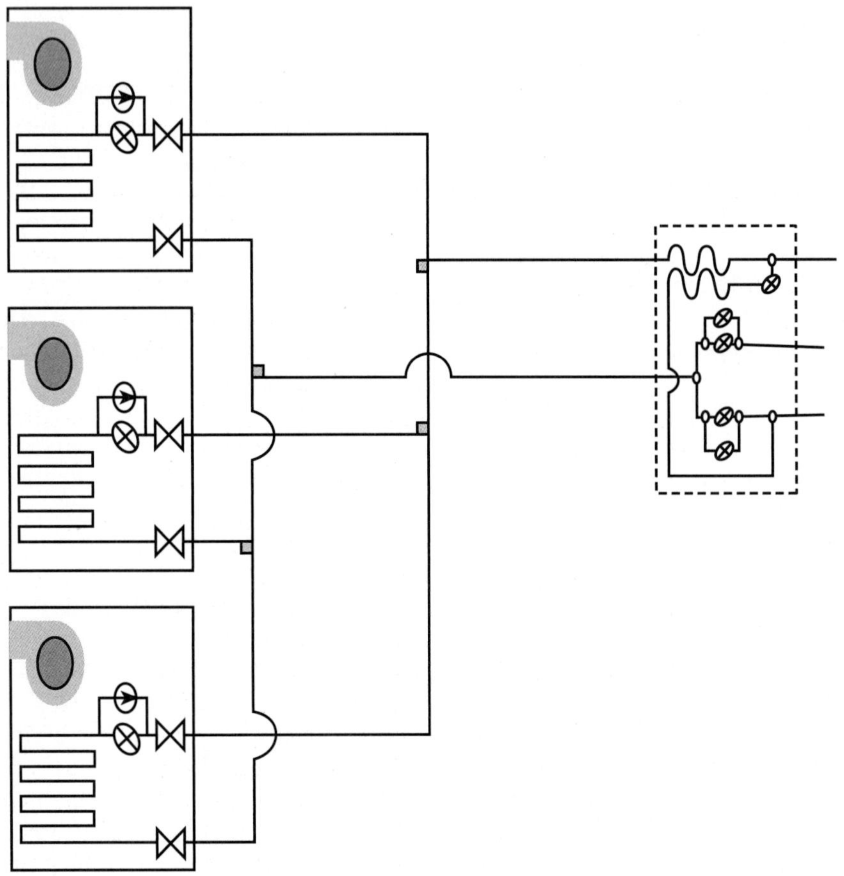

Figure 50.69 Simplified three-pipe VRF piping arrangement.

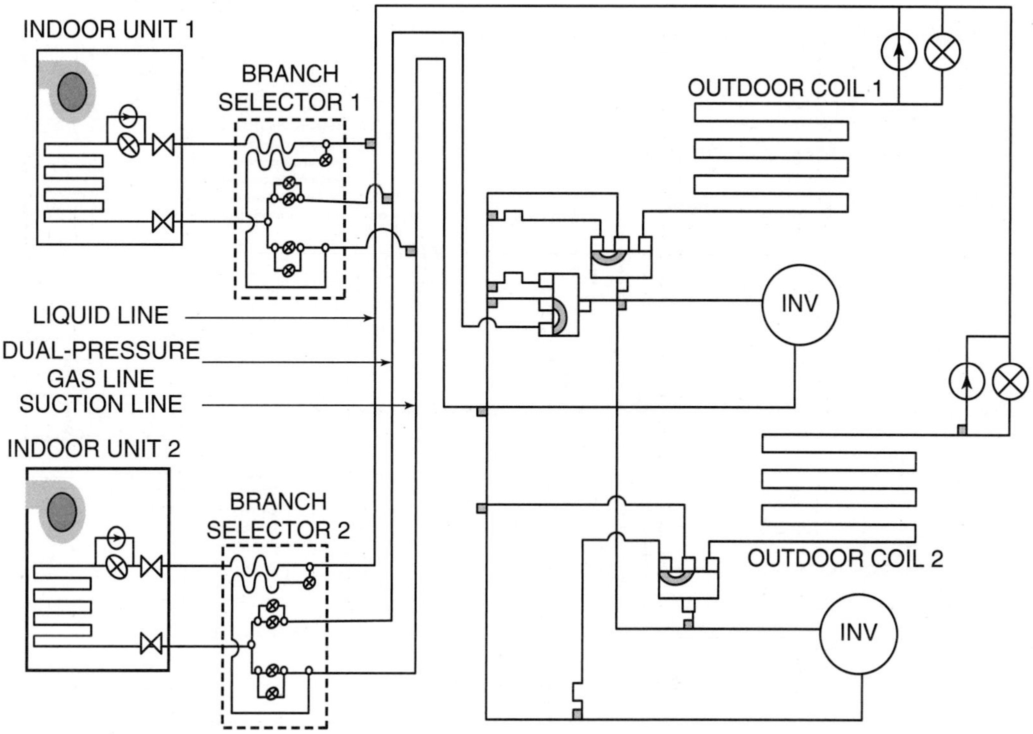

low-pressure gas, depending on the mode of system operation. There are also two pipes that connect the branch selector box to the indoor units. These lines are the liquid line and the gas line. In the cooling mode, the liquid line will carry high-pressure, high-temperature subcooled liquid into the EEV at the indoor unit, and in the heating mode, it will carry high-pressure, high-temperature subcooled liquid from the indoor coil.

Indoors, there are multiple indoor units and multiple branch selectors. The three conditioning modes of operation are:

- all indoor units operating in the cooling mode,
- all indoor units operating in the heating mode, or
- some indoor units operating in the cooling mode and some indoor units operating in the heating mode.

When all indoor units are calling for cooling, both of the outdoor coils will function as condensers. The refrigerant lines at the outlet of the outdoor coils join and form a common liquid line that carries the refrigerant to the branch selector boxes. From the branch selector boxes, the refrigerant flows through the EEVs at the indoor units. This is represented by the red arrowed lines in **Figure 50.70**.

At the indoor units, the coils function as evaporator coils to provide comfort cooling to the occupied space. From the evaporator coils, the low-pressure, low-temperature superheated vapor returns to the compressors by flowing through the dual-pressure gas line and the suction line. This is represented by the blue arrowed lines in **Figure 50.70**.

When all indoor units are calling for heating, both of the outdoor coils will function as evaporators, while the indoor coils function as condensers. The refrigerant at the discharge of the compressors join at an equalizer line where the hot gas is directed through a four-way valve. This valve directs the high-pressure, high-temperature discharge gas to the branch selector boxes via the dual-pressure gas line. The branch selector boxes pass the hot gas to the indoor coils to heat the space. After passing through the indoor condenser coil, the high-pressure liquid bypasses the indoor EEVs by means of check valves. The liquid lines at the outlets of the branch selector boxes form a common liquid line that carries the refrigerant to the outdoor coils. The refrigerant passes through the outdoor EEVs and passes low-temperature saturated liquid into the outdoor evaporator coils. The high-pressure side of the system is represented by the red arrowed lines in **Figure 50.71**. At the outdoor coils, the refrigerant vaporizes and then returns to the compressors. This is represented by the blue arrowed lines in **Figure 50.71**.

When one indoor unit is calling for cooling while the other is calling for heating, one outdoor coil functions as the condenser, while the other functions as the evaporator. The piping arrangement is shown in **Figure 50.72**. The hot discharge gas from the compressors is directed to both Outdoor Coil 1 and Indoor Coil 2. The hot gas is supplied to the indoor coil through the dual-pressure gas line. Outdoor Coil 1 and Indoor Coil 2 are both functioning as condensers. The outlets of Outdoor Coil 1 and Indoor Coil 2 are joined and supply high-temperature, high-pressure subcooled liquid

Figure 50.70 Three-pipe VRF system with all units calling for cooling.

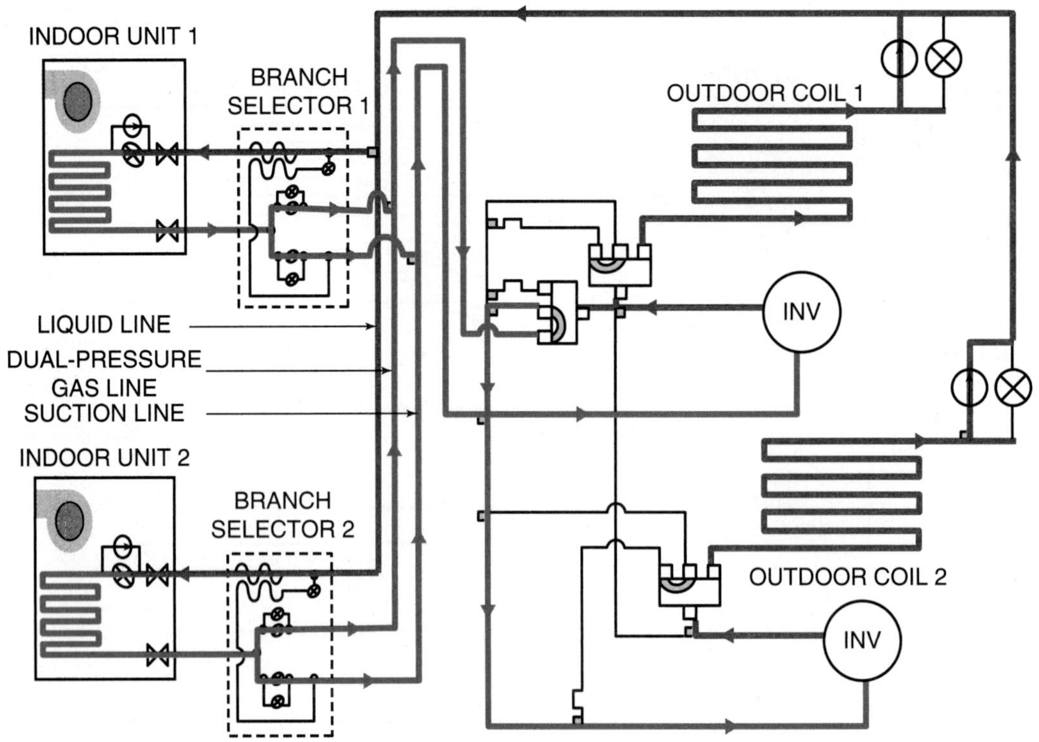

Figure 50.71 Three-pipe VRF system with all units calling for heating.

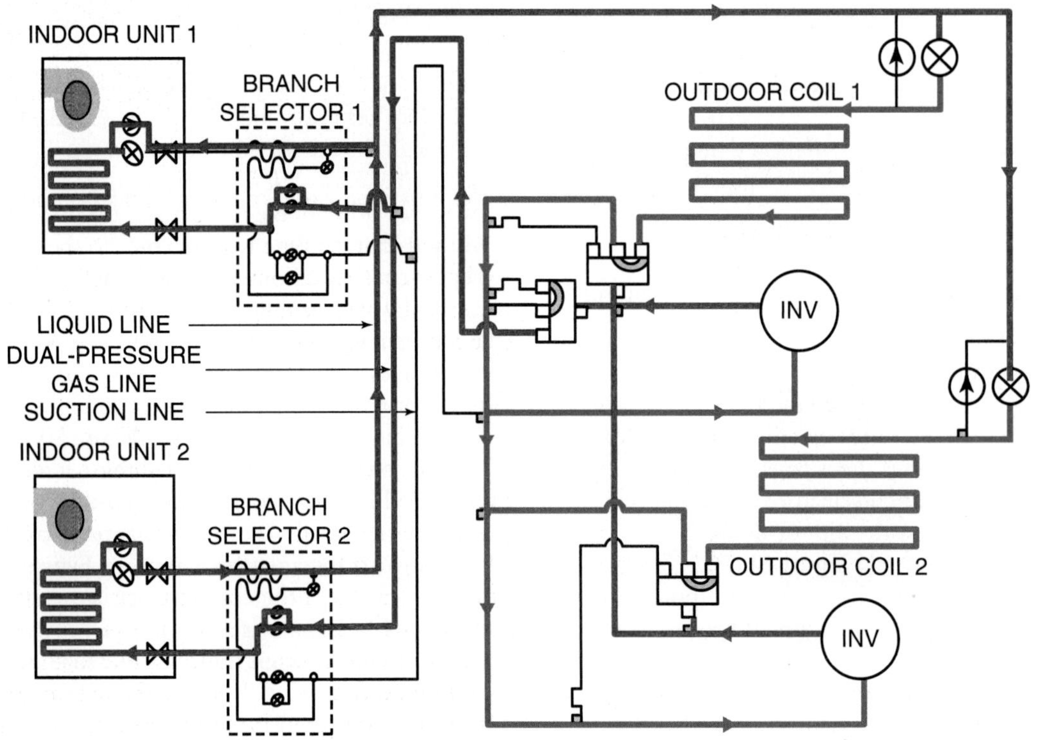

Figure 50.72 Three-pipe VRF system with one unit calling for cooling and one unit calling for heating.

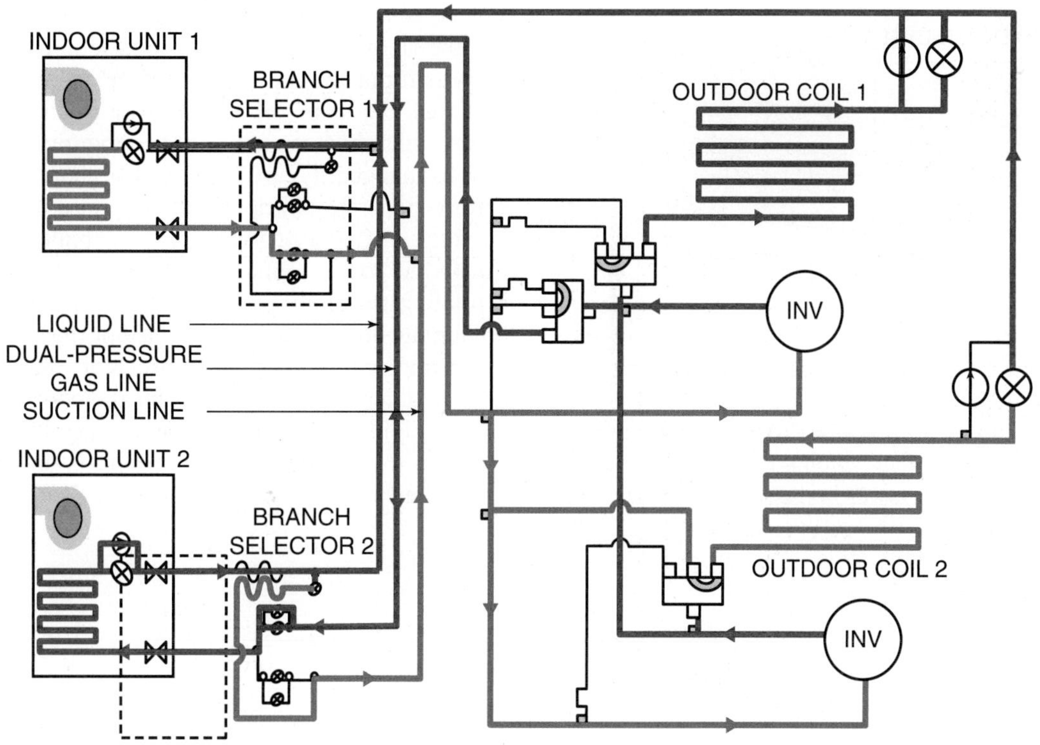

refrigerant to the EEVs at Indoor Coil 1 and Outdoor Coil 2, which are both operating as evaporators. From the evaporators, the suction gas returns to the compressors. It should also be noted that the EEV (labeled SC in **Figure 50.67**) at Indoor Unit 2 is in operation to provide additional subcooling to the liquid refrigerant leaving Indoor Unit 2.

50.13 DRY COOLERS

Dry coolers, **Figure 50.73**, are unique commercial systems that incorporate a number of different technologies to create an efficient method of cooling spaces that require year-round conditioning. Dry coolers are especially desirable for use in geographic regions with four distinct seasons and for applications such as data storage facilities and medical equipment cooling. Upon initial inspection, dry coolers look very similar to air-cooled condensers. However, instead of having refrigerant coils in them, dry coolers have water coils, **Figure 50.74**. The water coil can contain 100% water or can be a glycol/water mix, depending on the climate in which the system is installed. In areas in which freezing is a problem, an antifreeze/water solution is used. Dry coolers can be described as a combination of the following concepts and processes:

Figure 50.73 A dry cooler looks like an air-cooled condenser.

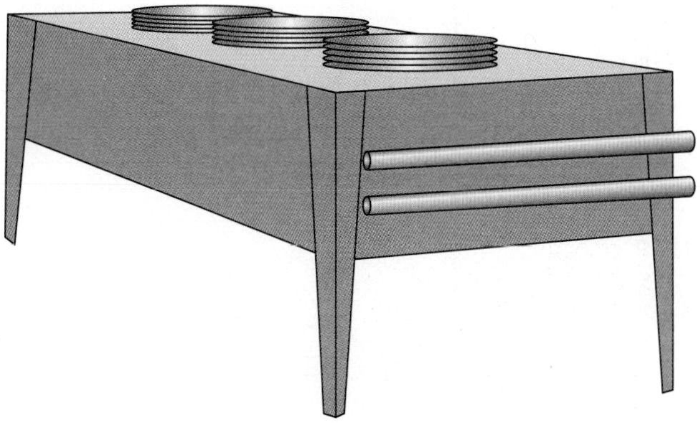

- Closed-loop cooling tower
- Water-cooled equipment
- Chilled water cooling
- Economizing

DRY COOLER CONCEPTS

Closed-loop cooling towers are desirable because the water circuits of the system are not exposed to the elements and the water treatment requirements are much easier to meet than those for open-loop cooling towers. The closed-loop cooling tower, **Figure 50.75**, has a water circuit that is open to the atmosphere, but this water is only present in the tower and does not circulate through the cooling equipment. The piping circuit that carries water between the indoor and outdoor portions of the system is completely isolated from environmental

Figure 50.74 Water coils inside the dry cooler.

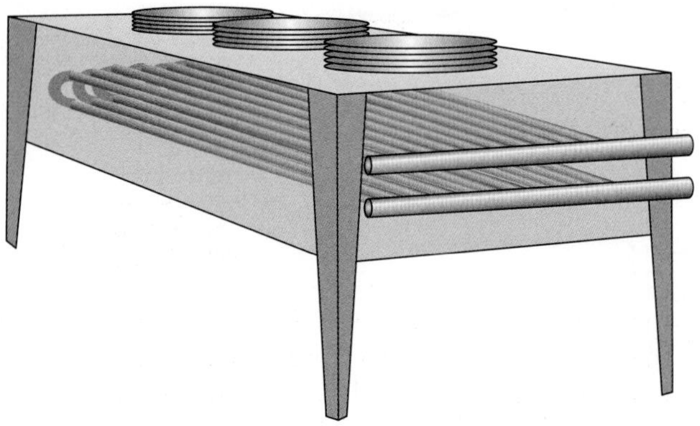

contamination. The closed-loop cooling tower has a bank of fans to move air across the water piping circuit. Dry coolers are similar to the closed-loop tower in the sense that they also have completely sealed water circuits and banks of fans to move air across them. This is where the similarities end. Dry coolers, as implied by their name, do not have water cascading over the piping circuits, as the closed-loop cooling tower does.

When the system's compressor is operating, the hot discharge gas from the compressor is directed to the condenser coil, which is commonly configured as a tube-in-tube heat exchanger, **Figure 50.76**. The heat that the vapor-compression refrigeration system is rejecting is absorbed by the water in the dry cooler's piping circuit. This heat is then transferred to the outside ambient air by means of water circulating through the dry cooler's piping circuit and fans moving outside air over the surface of the coils.

During periods of low ambient temperature, space cooling is achieved by circulating the cool water in the dry cooler's piping circuit through a water coil located in the airstream of the cooling equipment. This is similar in operation to a chilled water system in the sense that the cooling effect

Figure 50.75 Piping diagram of a system with a closed-loop cooling tower.

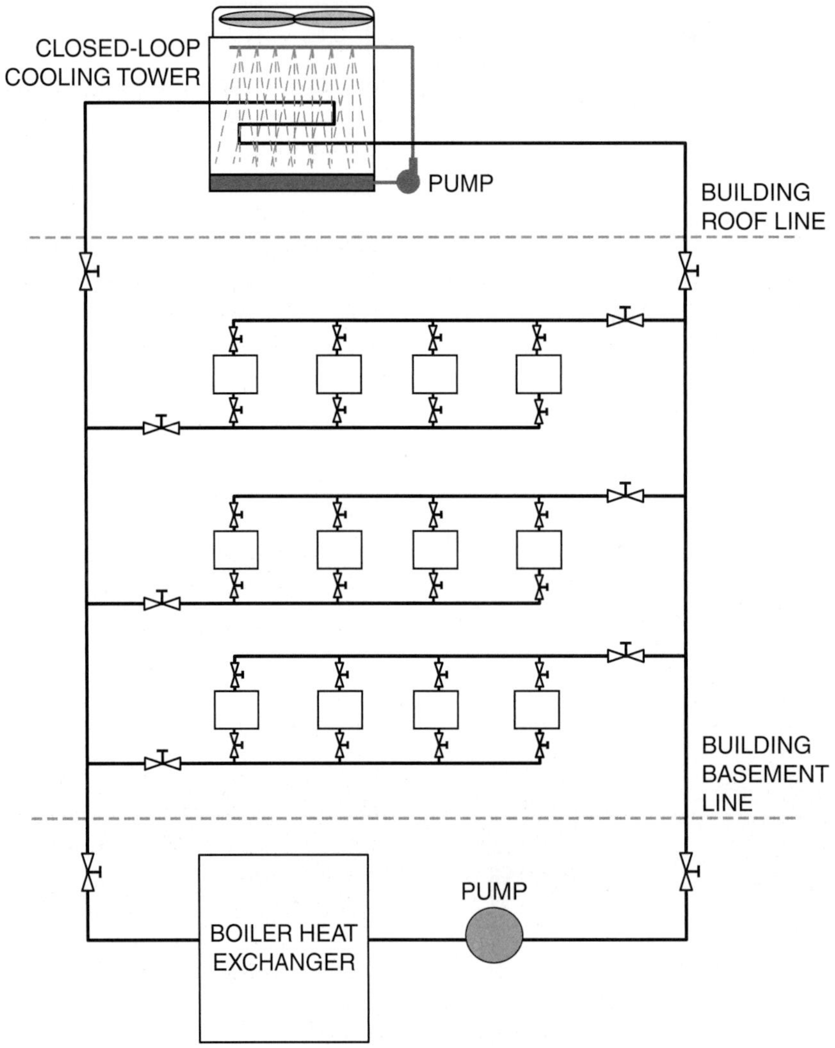

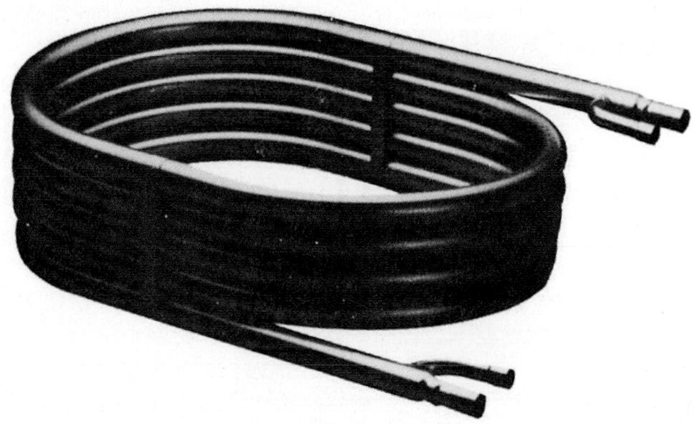

Figure 50.76 A tube-in-tube heat exchanger. Courtesy Noranda Metal Industries, Inc.

is achieved from some type of secondary refrigerant, such as water, instead of from a primary refrigerant, as in the case of direct-expansion (DX) cooling coils.

Economizing is a major benefit of dry coolers. Economizing is any process by which a cooling effect is achieved that reduces, or eliminates completely, the need to operate system compressors. When the outside ambient temperature is low, the cold water in the dry cooler's piping circuit is used to provide the cooling effect without having to operate the system compressors.

DRY COOLER PIPING BASICS

The piping of a basic dry cooler is provided in **Figure 50.77.** The red and blue lines represent the high-pressure and low-pressure refrigerant lines that connect the compressor, condenser, metering device, and evaporator coil. The heat exchanger is a tube-in-tube condenser, and the direct-expansion evaporator coil is located in either the air distribution system or the dry cooler itself. The green lines represent the water piping circuit. Water is pumped from the dry cooler coil to a three-way valve. The three-way valve directs the water flow to either the tube-in-tube heat exchanger or the glycol/water coil that is located in the duct system or the dry cooler itself, **Figure 50.78.**

Figure 50.78 Three-way valve piping on a dry cooler.

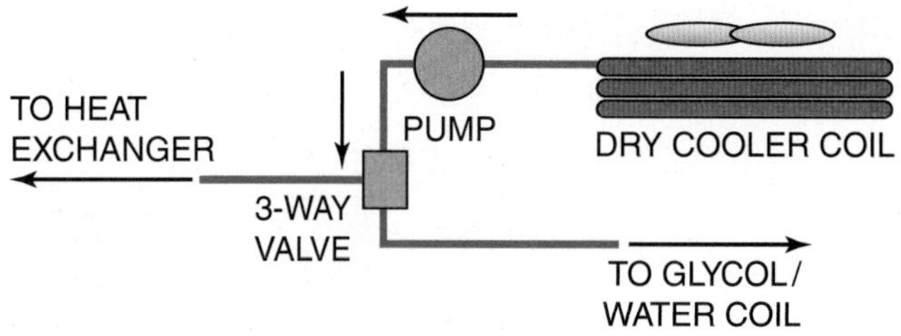

DRY COOLER: WARM AMBIENT OPERATION

During periods of warm ambient temperatures, the compressor is operating and the water in the glycol/water circuit is directed to the tube-in-tube heat exchanger, **Figure 50.79.** The heat from the vapor-compression refrigeration system is rejected to the condenser. The glycol/water mix in the water piping circuit absorbs the heat from the refrigerant in the condenser and rejects this heat to the outside. When operating in this mode, the direct-expansion evaporator coil is providing the required cooling for the space.

Figure 50.77 Basic piping circuit of a dry cooler.

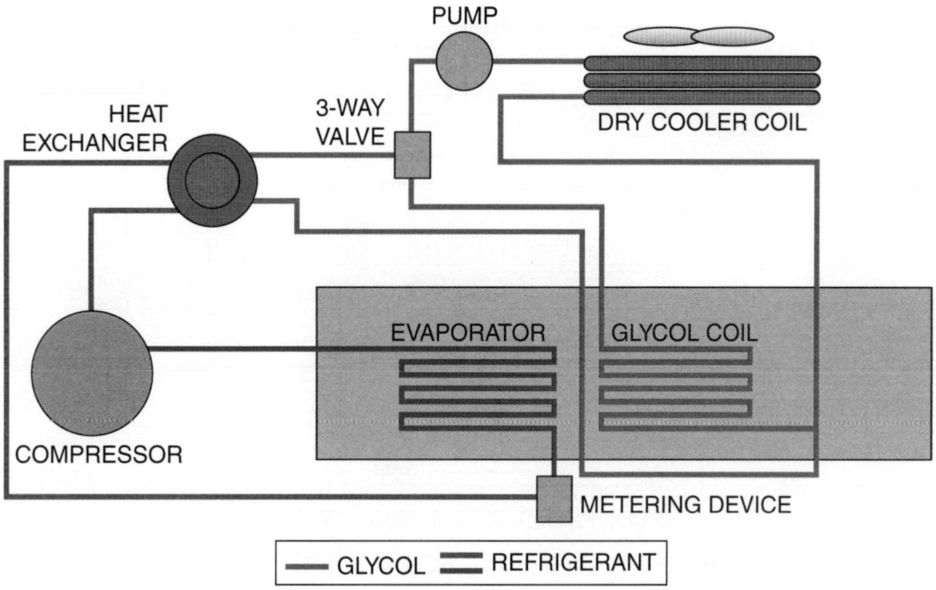

Figure 50.79 Fluid flow diagram of a dry cooler operating during warm ambient conditions.

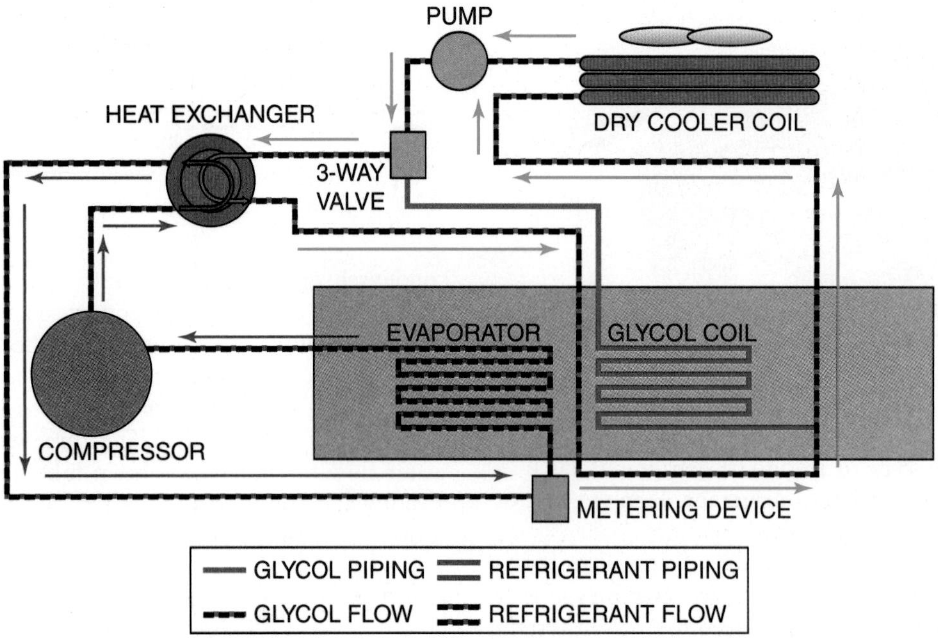

DRY COOLER: COOL AMBIENT OPERATION

During periods of cool ambient temperatures, the compressor does not operate and the cool water in the glycol/water circuit is directed to the glycol coil, which is located in the air distribution system, **Figure 50.80**. Since the compressor is not operating, there is no need to send water to the tube-in-tube heat exchanger. The cold glycol/water mix in the water piping circuit absorbs the heat from the conditioned space providing the required cooling. After flowing through the glycol coil, the glycol/water mix is warm and then returns to the outdoor coil where it is cooled by the outside ambient air.

Figure 50.80 Fluid flow diagram of a dry cooler operating during cool ambient conditions.

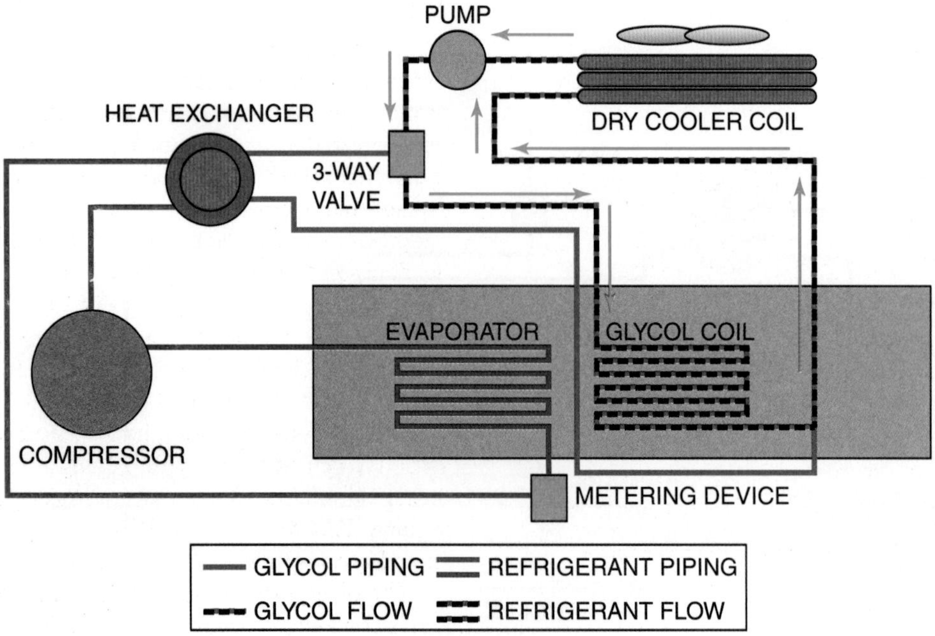

NOTE: *The changeover point between economizing and mechanical cooling can be determined by a number of different control strategies that include dry-bulb or enthalpy differential-type controls. If the dry-bulb or enthalpy measurement of the outside air is above the control's set point, mechanical cooling will be used. If it is determined that the temperature of the glycol/water solution can effectively cool the space, then the solution will be used instead of operating the system compressor.* •

50.14 CHILLED BEAM SYSTEMS

A chilled beam system uses chilled water that is piped to, and passes through, a beam or heat exchanger. These heat exchangers can be integrated into a suspended ceiling, **Figure 50.81**, or mounted a short distance from a finished ceiling. As the beam cools the air around it, the air becomes denser and falls toward the floor. It is replaced by warmer air rising up from below. These natural convection currents resemble those created in baseboard heating applications, but in reverse. One of the desired features of chilled beam systems is that it is the gentle air movement that cools the room. Although chilled beam systems are typically used in conjunction with dedicated chillers, **Figure 50.82**, it is possible to introduce chilled beam technology into existing and retrofit projects.

Chilled beam systems have been used in Europe for sensible cooling and heating applications for many years. They are quickly becoming a popular alternative to the systems currently being used in large buildings in the United States. They can be used as a primary or supplemental cooling system, depending on the building requirements and local codes. The growing popularity of chilled beam systems can be attributed to their ease of installation, their minimal space requirements, quieter operation, and their energy-saving potential.

Figure 50.81 Chilled beam mounted in a suspended ceiling.

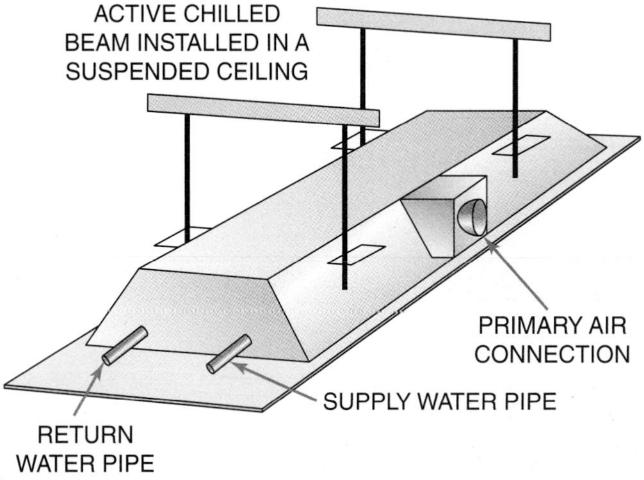

Figure 50.82 Piping diagram of a chilled beam system and a dedicated chiller.

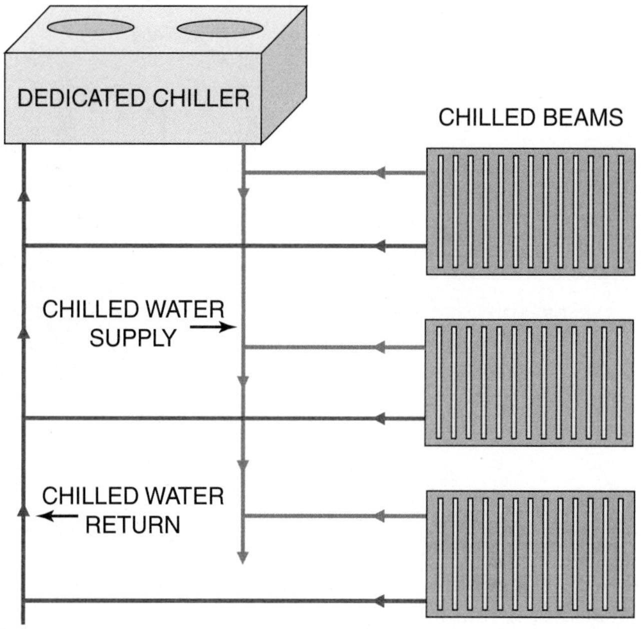

Chilled beam heat exchanger coils can provide a range of 200 Btuh to 1100 Btuh of sensible cooling per linear foot, depending on the system. A major advantage of these systems is that water has a greater cooling capacity than air. A 1-in. diameter water pipe can deliver the same cooling effect as an 18-in. square air duct. The use of chilled beam systems allows for more efficient use of both horizontal and vertical building space.

Most building codes in the United States require that outdoor air be supplied to each space for ventilation. The primary air system that is used with a chilled beam system must meet the ventilation requirements, handle any latent load, and keep the dew point of the indoor air below the surface temperature of the chilled beam to prevent condensation. Chilled beam systems typically do not contain a condensate drainage system, so condensation may drip into the space being cooled.

There are two basic types of chilled beam systems, active and passive. The building requirements for ventilation air and the cooling load will dictate which type of system should be used. When sensible cooling and ventilation air are required, an active chilled beam system should be used. If the requirements call for supplementary cooling only, or the ventilation requirements are being met by a separate system, a passive chilled beam system can be used.

The biggest difference between an active and passive system is the primary air supply. An active system can be used to

Figure 50.83 A passive chilled beam system.

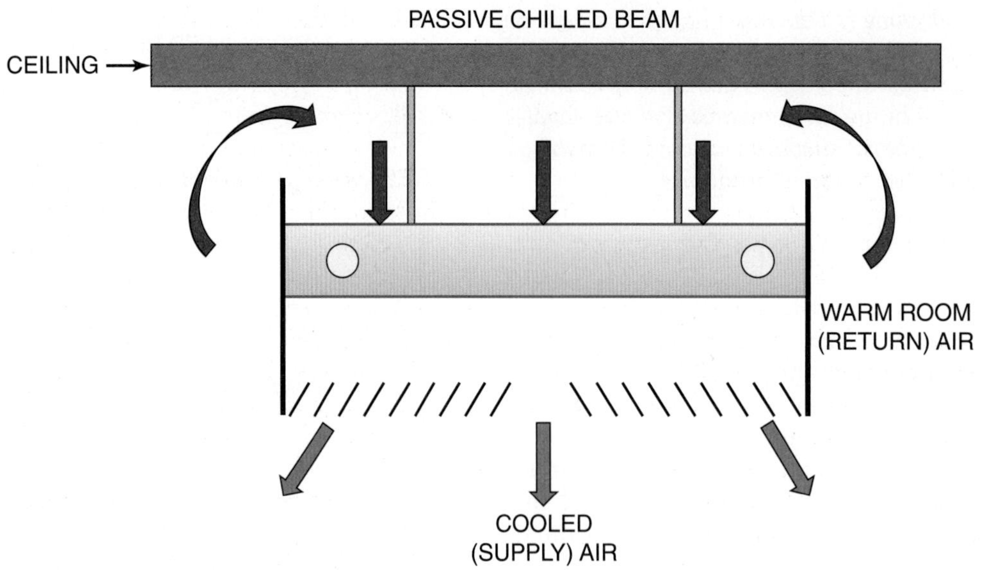

distribute air that has been conditioned by a primary system; passive systems are completely separate from other HVACR systems in the building.

PASSIVE CHILLED BEAM SYSTEM

Passive chilled beam systems do not incorporate any fans or ducts and provide mostly convective cooling. They rely on the buoyancy of warm air to rise up to the beam, where it is cooled, **Figure 50.83**. The cooler air sinks down, creating an air circulation that cools the space. Passive chilled beams can remove 200 Btuh to 650 Btuh of sensible heat per linear foot, depending upon their width and the temperature difference between the entering air and chilled-water temperature. Passive systems are typically installed as supplemental cooling systems as they lack the ability to circulate primary air to the conditioned space.

ACTIVE CHILLED BEAM SYSTEM

Active chilled beams have a ducted primary air supply which induces secondary (room) air across their heat exchange surface. The secondary air is cooled prior to being mixed with the primary air stream, which is then discharged into the conditioned space, **Figure 50.84**. Active chilled beams can remove up to 1100 Btuh of sensible heat per linear foot, depending on the system design. The primary air system is typically used to maintain ventilation air and humidity requirements for the conditioned space. The heat exchange surface of a chilled beam system provides sensible cooling only.

Figure 50.84 An active chilled beam system.

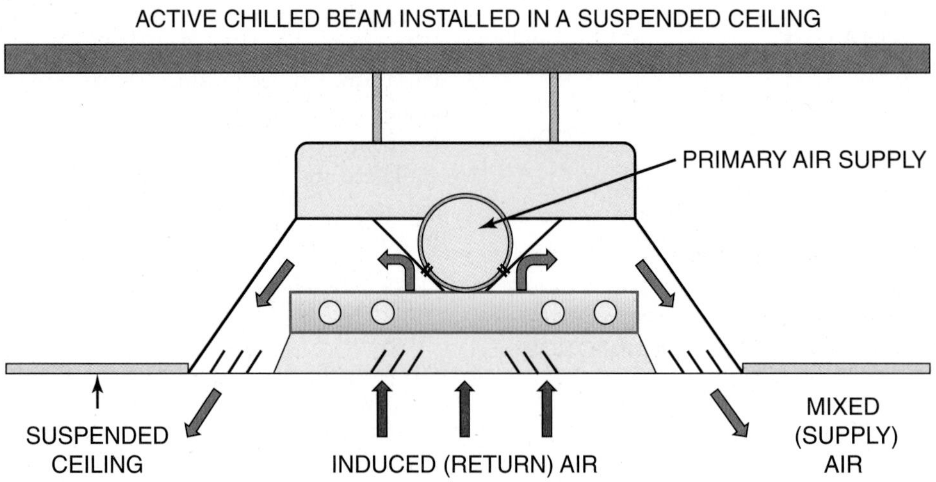

REVIEW QUESTIONS

1. Which of the following is true regarding the refrigerant piping on a rooftop package unit?
   A. The discharge and liquid lines must be field installed.
   B. The suction and liquid lines must be field installed.
   C. The discharge and suction lines must be field installed.
   D. No refrigerant lines are field installed.
2. What is the benefit of having two refrigerant circuits, each with a different capacity on a rooftop package unit?
   A. There will be an additional capacity "step" to better match the needs of the structure.
   B. The larger circuit will act as a backup in case the other refrigerant circuit fails.
   C. The smaller circuit will act as a backup in case the other refrigerant circuit fails.
   D. The cost of operating both circuits will be less than the cost of operating a single circuit.
3. Which of the following sets of terms best describes a rooftop package unit?
   A. Field-assembled, factory-charged, self-contained
   B. Factory-assembled, factory-charged, self-contained
   C. Field-assembled, field-charged, self-contained
   D. Factory-assembled, field-charged, self-contained
4. Which process would best represent the opposite of exhausting air from a system?
   A. Cooling
   B. Ventilating
   C. Heating
   D. Filtering
5. Mixed air is best described as the combination of
   A. exhaust air and return air.
   B. return air and outside air.
   C. supply air and return air.
   D. exhaust air and outside air.
6. The heat exchangers in a rooftop package unit are exposed to which two airstreams?
   A. Return air and supply air
   B. Outside air and return air
   C. Mixed air and supply air
   D. Mixed air and outside air
7. Which of the following is typically true regarding the fans and blowers on rooftop package units?
   A. Centrifugal blowers are used to move air to and from the occupied space.
   B. Centrifugal blowers are used to move air through the air-cooled condenser.
   C. Propeller-type fans are used to move air to and from the occupied space.
   D. Propeller-type fans are used to move air to and from the occupied space and through the air-cooled condenser.
8. All of the following are typically required to install a rooftop package unit except
   A. penetrating the roof.
   B. rigging the unit.
   C. installation of power and control circuits.
   D. system evacuation and charging.
9. All of the following should be taken into account when selecting the location and position of a rooftop package unit except
   A. possible snow accumulation.
   B. condenser air circulation.
   C. serviceability.
   D. electrical power availability.
10. A bonded roof is one that has
    A. reinforced steel structural members.
    B. a written warranty or insurance policy.
    C. a leak-free rubber membrane.
    D. a slope of not more than 30 degrees.
11. When penetrating a roof for the installation of a rooftop package unit,
    A. the size of the penetration will be the same size as the measurements of the rooftop unit.
    B. roof support structures can be cut because, by design, they are oversized and over-rated.
    C. measure, layout, and check penetration locations at least twice and cut once.
    D. make certain that someone is standing directly below the roof to ensure no debris falls.
12. Properly installed roof curbs will reduce the chances of
    A. leaks.
    B. noise transmission.
    C. both leaks and noise transmission.
    D. neither leaks nor noise transmission.
13. When rigging a unit to the roof, what minimum number of installation crew members should be on hand, not including the crane operator?
    A. 1
    B. 2
    C. 3
    D. 4
14. What is the acceptable method for communicating with the crane operator while rigging a unit to the roof?
    A. Cell phones
    B. Walkie-talkies
    C. Hand signals
    D. Verbal communications
15. The portion of the crane that extends and retracts is called the
    A. whip line.
    B. boom.
    C. main hook block.
    D. main hoist line.

16. If, while communicating with the crane operator, a team member taps his elbow with one hand, the crane operator should
    A. use the auxiliary hoist.
    B. use the main hoist.
    C. lower the boom.
    D. raise the boom.
17. If, while communicating with the crane operator, a team member extends his arm with his thumb pointing up, the crane operator should
    A. use the auxiliary hoist.
    B. use the main hoist.
    C. lower the boom.
    D. raise the boom.
18. Economizers are intended to
    A. provide mechanical ventilation.
    B. help reduce cooling costs.
    C. provide mechanical ventilation and help reduce cooling costs.
    D. neither provide mechanical ventilation nor help reduce cooling costs.
19. Economizers have a return-air damper that
    A. opens when the outside air damper opens.
    B. opens when the outside air damper closes.
    C. closes when the mixed air damper closes.
    D. closes when the mixed air damper opens.
20. The most efficient way to establish the best possible economizer position is to measure
    A. the dry-bulb temperature of the return and outside airstreams.
    B. the dry-bulb temperature of the return air and the enthalpy of the outside air.
    C. the enthalpy of the return air and the dry-bulb temperature of the outside air.
    D. the enthalpy of both the return air and the outside air.
21. A fixed-type economizer control compares
    A. a sensed/measured value to a constant value.
    B. two sensed/measured values.
    C. two constant, fixed temperature readings.
    D. two constant, fixed enthalpy readings.
22. Free-cooling can be accomplished if the
    A. enthalpy of the return air is lower than the enthalpy of the outside air.
    B. enthalpy of the outside air is lower than the enthalpy of the return air.
    C. enthalpy of the return air is lower than the enthalpy of the supply air.
    D. enthalpy of the return air is the same as the enthalpy of the outside air.
23. The economizer mode of operation that uses the system compressor(s) and has the outside air dampers fully open is called
    A. heating mode.
    B. modulating mode.
    C. integrated mode.
    D. mechanical mode.
24. The economizer mode of operation that uses the system compressor(s) and has the outside air dampers at their minimum position is called the
    A. heating mode.
    B. modulating mode.
    C. integrated mode.
    D. mechanical mode.
25. A demand control ventilation strategy monitors what substance in air?
    A. Moisture
    B. Carbon monoxide
    C. Carbon dioxide
    D. Smoke
26. A demand control ventilation strategy varies the ventilation rate based on the
    A. time of day.
    B. heat load on the structure.
    C. outdoor ambient temperature.
    D. occupancy level.
27. In a CAV system, the amount of air supplied to an occupied space
    A. varies in response to the load on the space.
    B. varies in response to the outside air temperature.
    C. is always constant when the system is calling for blower operation.
    D. is always constant regardless of system switch position.
28. In a VAV system, the amount of air supplied to an occupied space
    A. varies in response to the needs of the space.
    B. varies in response to the outside air temperature.
    C. is always constant when the system is calling for blower operation.
    D. is always constant regardless of system switch position.
29. Ultimately, the speed of the blower on a VAV system is controlled by
    A. the temperature of the outside air.
    B. a variable frequency drive (VFD).
    C. a constant frequency drive (CVD).
    D. the temperature of the mixed air.
30. A cooling-only VAV box has all of the following except a(n)
    A. damper.
    B. control module.
    C. reheat coil.
    D. air pressure sensor.
31. All of the following are common VAV box types except
    A. heating only.
    B. cooling only.
    C. parallel blower-assisted.
    D. series blower-assisted.

32. The main purpose of a reheat coil on VAV systems is to
    A. act as a secondary cooling source when the load is high.
    B. prevent the heat exchanger from freezing.
    C. temper the air as needed when cooling or ventilation is called for.
    D. temper the air when the system blower is off.
33. Which of the following is true regarding a reheat water circuit that utilizes two-way modulating valves?
    A. The water is always circulating and is not prone to coil freeze-up.
    B. The flow rate is always constant through the reheat water circuit.
    C. The water pump must be able to vary its flow.
    D. The reheat coil is equipped with a bypass pipe.
34. Which of the following is true regarding a reheat water circuit that utilizes three-way modulating valves?
    A. The water is always circulating and is not prone to coil freeze-up.
    B. The flow rate is always constant through the reheat water circuit.
    C. The water pump must be able to vary its flow.
    D. The reheat coil is equipped with a bypass pipe.
35. The chiller on a water-cooled, chilled-water system will have
    A. two condenser water lines and two chilled water lines.
    B. two chilled water lines, a refrigerant discharge line, and a refrigerant liquid line.
    C. four chilled water lines and two condenser water lines.
    D. four condenser water lines and two chilled water lines.
36. On chilled-water VAV systems, a freezestat is used to
    A. prevent the condenser water coil from freezing
    B. prevent the chilled water coil from freezing.
    C. ensure that ice forms in the condenser water circuit.
    D. ensure that ice forms in the chilled-water circuit.
37. In a chilled-water system, a 50/50 mix of water and ethylene glycol will have a freezing temperature in the range from
    A. 14°F to 20°F
    B. 2°F to 8°F
    C. −13°F to −10°F
    D. −36°F to −30°F
38. The acronym VRF stands for
    A. verifiable refrigerant flow.
    B. variable refrigeration flow.
    C. verifiable refrigeration flow.
    D. variable refrigerant flow.
39. VRF systems rely on what technology to perform as intended?
    A. PSC motors
    B. CSR motors
    C. Inverter-controlled motors
    D. Single-speed ISR motors
40. Which of the following statements is true regarding a heat-recovery VRF system?
    A. All indoor units must be in either the heating or cooling modes of operation.
    B. When one indoor unit cycles off, all of the indoor units must also cycle off.
    C. One main thermostat controls the operation and temperature set points in the structure.
    D. Simultaneous heating and cooling can be achieved at different indoor units.
41. VRF systems the optimal amount of refrigerant to circulate based primarily on the
    A. number of occupants in the structure.
    B. system's operating temperatures and pressures.
    C. outdoor ambient temperature.
    D. relative humidity in the occupied space.
42. The three field-installed refrigerant lines on a 3-pipe VRF system carry
    A. hot discharge gas in two lines and high-pressure liquid in the third.
    B. high-pressure liquid in two lines and hot discharge gas in the third.
    C. high-pressure liquid in two lines and low pressure suction gas in the third.
    D. refrigerant in different states depending on the mode of system operation.
43. The pump on a dry cooler can direct water to either the tube-in-tube heat exchanger or the cooling coil with the aid of a
    A. two-way valve.
    B. three-way valve.
    C. four-way valve.
    D. five-way valve.
44. During low ambient conditions, the dry cooler provides space cooling by
    A. operating the system compressors.
    B. pumping water to the tube-in-tube heat exchanger.
    C. pumping water to a cooling coil located in the airstream.
    D. deenergizing the blower in the air distribution system.
45. Passive chilled beams can remove
    A. 200 Btuh to 650 Btuh of sensible heat per linear foot.
    B. 200 Btuh to 650 Btuh of sensible heat per square foot.
    C. up to 1000 Btuh of sensible heat per linear foot.
    D. up to 1000 Btuh of sensible heat per square foot.
46. In a building that utilizes active chilled beams, the primary air system
    A. is responsible for all the sensible cooling load.
    B. will handle the latent cooling load and the space ventilation requirements.
    C. is primarily used to keep the room air temperature below its dew point.
    D. does not operate unless the cooling load becomes excessive.

47. An 18-in. square air duct can deliver the same amount of cooling Btus as a
    A. ½-in. water pipe.
    B. ¾-in. water pipe.
    C. 1-in. water pipe.
    D. 3-in. water pipe
48. List the five air-conditioning-related processes that are performed by a rooftop package unit.
49. Explain why the outside air damper is never completely closed, even when the outside air conditions are not adequate to provide free-cooling.
50. Briefly explain the purpose of ASHRAE Standard 62.
51. Explain the concept of demand control ventilation.
52. Explain how carbon dioxide levels can be used to estimate the occupancy level in a particular area in a structure.
53. Describe the blowers used on VAV systems.
54. Explain how the closing of a damper to a conditioned area causes the speed of a VAV system's blower to slow down.
55. Explain the main benefit of a heat-recovery VRF system.
56. What does the acronym PID stand for?

APPENDIX A

ALTERNATIVE HEATING (STOVES AND FIREPLACE INSERTS)

OBJECTIVES

After studying this unit, you should be able to

- list reasons for the formation of creosote.
- describe methods to help prevent the formation of creosote.
- explain how stovepipe should be assembled and installed.
- list safety hazards that may be encountered with wood stoves.
- describe the three types of venting for gas stoves.
- explain how a fireplace insert can improve on the heating efficiency of a fireplace.

SAFETYCHECKLIST

☑ Many safety practices must be adhered to when installing or using a wood-burning appliance. Many of these safety practices are stated within this unit, printed in red. Many other safety practices must be adhered to, including those provided by manufacturers and local, regional, and state codes. It is also necessary to use good judgment and common sense at all times.

Codes, regulations, and laws regarding wood-burning and gas heating appliances and components change frequently. It is essential that you become familiar with the appropriate current National Fire Protection Association (NFPA) information and other codes, regulations, and laws.

A.1 WOOD-BURNING STOVES

Both wood- and gas-burning stoves are used as an alternative heating appliance in many homes. In some homes, stoves are a primary heat source; in others they are used to supplement primary heat sources. Wood must be readily available to make wood-burning stoves practical. However, pellet stoves, which burn small compressed wood by-products, may be used in many locations where pellets are available.

A.2 ORGANIC MAKEUP AND CHARACTERISTICS OF WOOD

Wood plants manufacture glucose. Some of this glucose, or sugar, turns into cellulose. Approximately 88% of wood is composed of cellulose and **lignin** in equal parts. **Cellulose** is an inert substance and it forms the solid part of wood plants. Cellulose forms the main or supporting structure of each cell in a tree. Lignin is a fibrous material, a polymer, which binds to cellulose fibers and hardens and strengthens the cell walls of trees. The remaining 12% of the wood is composed of resins, gums, and a small quantity of other organic material. Water in green or freshly cut wood constitutes from one-third to two-thirds of its weight. Thoroughly air-dried wood may have as little as 15% moisture content by weight.

Wood is classified as either hardwood or softwood. Hickory, oak, maple, and ash are examples of hardwood. Pine and cedar are examples of softwood. **Figure A.1** lists some common types of wood and the heat value in each type in millions of Btus/cord. A cord of wood can be split, unsplit, or mixed. It is important to know whether the wood is split, because there is more wood in a stack that is split. Wood is sold by the cord, which is a stack 4 ft × 4 ft × 8 ft, or 128 ft^3. Because wood is sold by volume and not by weight, wood that is split has less air space between the individual pieces. This will give the cord of wood a higher density, meaning more wood per cord.

Wood should be dry before burning. Approximately 20% more heat is available in dry wood than in green wood. Wood

Figure A.1 The table indicates the weight per cord, the Btus per cord of air-dried wood, and the equivalent value of No. 2 fuel oil in gallons. Courtesy of Yukon Energy Corporation.

TYPE	WEIGHT CORD	BTU PER CORD AIR DRIED WOOD	EQUIVALENT VALUE #2 FUEL OIL, GALS.
White Pine	1800#	17,000,000	120
Aspen	1900	17,500,000	125
Spruce	2100	18,000,000	130
Ash	2900	22,500,000	160
Tamarack	2500	24,000,000	170
Soft Maple	2500	24,000,000	170
Yellow Birch	3000	26,000,000	185
Red Oak	3250	27,000,000	195
Hard Maple	3000	29,000,000	200
Hickory	3600	30,500,000	215

Figure A.2 Wood should be stacked on runners so that air can circulate through and around it.

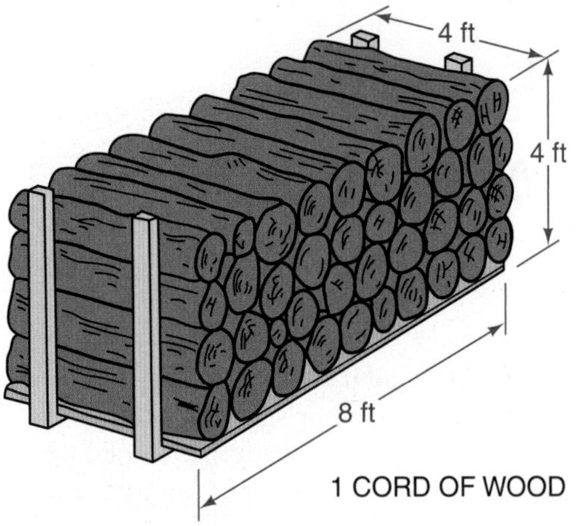

should be stacked off the ground on runners and should be well ventilated, **Figure A.2**. When possible, the wood should be covered, but air should be allowed to circulate through it. Wood splits more easily when it is green or freshly cut and dries better when it has been split. When green wood is burned, combustion will be incomplete, resulting in unburned carbon, oils, and resins, which leave the fire as smoke.

During oxidation, or burning, oxygen is added to the chemical process. This actually turns wood back into products that helped it grow as a plant, primarily: carbon dioxide (CO_2), water (H_2O), and other miscellaneous materials. Heat also is produced. Some woods produce more heat than others per cord, **Figure A.1**. Generally, dry hardwoods are the most efficient.

A.3 ENVIRONMENTAL PROTECTION AGENCY (EPA) REGULATIONS

On July 1, 1988, the Environmental Protection Agency's first national woodstove emissions standards went into effect. Emissions from wood-burning appliances were adding to the pollutants in the air along with all the other sources of air pollution. Wood smoke contains both polycyclic organic matter (POM) and nonpolycyclic organic matter, which are considered to be health hazards and are of concern to many people. All wood-burning stoves manufactured after July 1, 1989 must be EPA certified. Two EPA standards exist—one for catalytic combustor stoves and one for noncatalytic stoves. Catalytic stoves contain a catalytic combustor and may not emit more than 5.5 g/h of particulates for those stoves manufactured between July 1, 1989, and June 30, 1990. The noncatalytic stoves manufactured between these dates may not emit more than 8.5 g/h. These limits are reduced further for stoves manufactured after July 1, 1990, to 4.1 g/h for the catalytic models and 7.5 for the noncatalytics. As new stoves are purchased and installed, replacing older stoves, the air pollutants from wood-burning stoves will be drastically reduced. The particulate emission of a catalytic stove is approximately 10% of the emissions of older stoves. Some individual states have implemented what are called *No-Burn Days*. On a no-burn day, it is illegal to use a wood-burning fireplace or stove. The intent is to limit pollution on days when pollutants are at high levels. The EPA periodically updates its New Source Performances Standards (NSPS) for residential wood-burning appliances. The most recent standards can be found on the EPA's website, www.epa.gov/burnwise.

A.4 CREOSOTE

Creosote is a mixture of unburned organic material. When it is hot, it is a thick, dark-brown liquid. When it cools, it forms into a tar-like residue. It then often turns into a black, flaky substance that adheres to the inside of the chimney or stovepipe. The formation of creosote is particularly a problem with the uncertified stoves manufactured before July 1, 1989. Some of the primary causes of excessive creosote are:

- smoldering low-heat fires.
- smoke in contact with cool surfaces in the stovepipe or chimney.
- burning green wood.
- burning softwood.

⚠SAFETY PRECAUTION: *Creosote buildup in the stovepipe and chimney is dangerous. It can ignite and burn with enough force to cause a fire in the building, blow stovepipes apart, and cause chimney fires.* •

To help prevent the formation of creosote, dry hardwood should be burned with some intensity. When the stovepipe or chimney flue temperature drops below 250°F, creosote will condense onto the interior surfaces of the chimney. The run of stovepipe should be as short as possible. The minimum rise should be 1/4 in. per foot of horizontal run. A rise of 30° is recommended, **Figure A.3**. Anything that slows the movement of the gases allows them to cool more quickly, which will cause more creosote condensation. The stovepipe should be assembled with the crimped end down, **Figure A.4**; this will keep the creosote inside the pipe. Stovepipes and chimneys should also be cleaned regularly and inspected by a professional.

A.5 DESIGN CHARACTERISTICS OF WOOD-BURNING STOVES

As mentioned, all wood-burning stoves manufactured since July 1, 1989, must be EPA certified, as either noncatalytic or catalytic combustor stoves.

Figure A.3 The stovepipe should be pitched upward from the stove to the chimney.

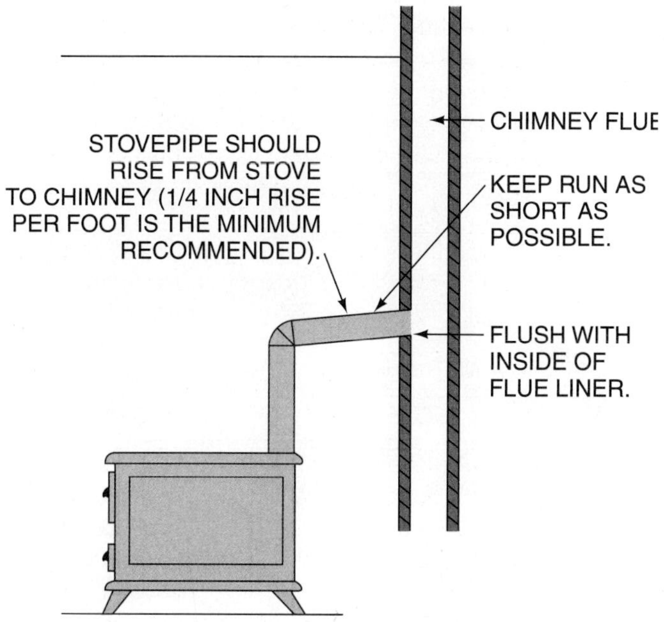

Figure A.4 The stovepipe sections should be assembled with the crimped ends of the stovepipe pointing down to prevent the creosote from leaking out of the joint.

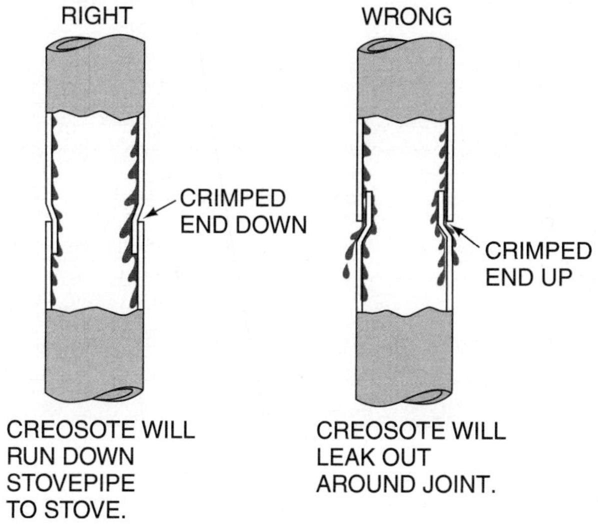

NONCATALYTIC CERTIFIED STOVES

Noncatalytic stoves force the unburned gases to pass through a secondary combustion chamber. Temperatures higher than 1000°F are maintained in this chamber, which burns the gases and therefore the pollutants, including most of the creosote, and produces more heat, providing a better operating efficiency. **Figure A.5** is an example of one design of this type of stove.

Figure A.5 A noncatalytic wood-burning stove. (1) Combustion air enters at the bottom. (2) Preheated secondary air. (3) Firebox. (4) Secondary air tubes. (5) Air manifold. (6) Stone liner. (7) Ash pan. (8) Vent/Flue. Based on HearthStone NHC, Inc.

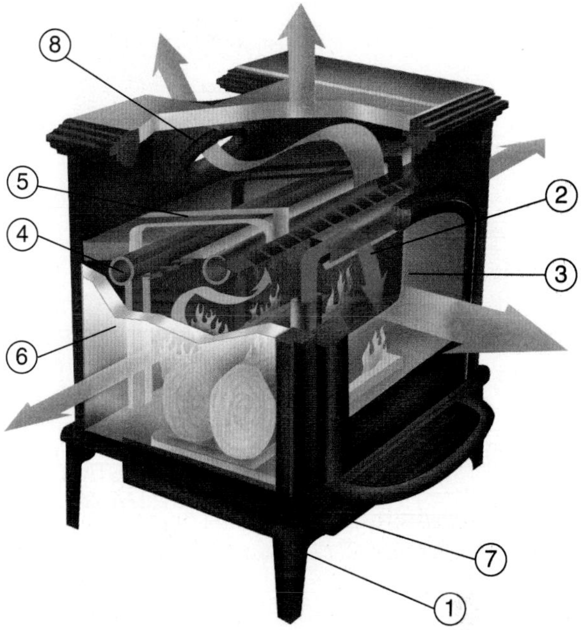

CATALYTIC COMBUSTOR CERTIFIED STOVES

A catalytic combustor is a form of afterburner that increases the burning of wood by-products. These combustors use a catalyst to cause combustion by producing a chemical reaction, which burns the flue gases at about 500°F rather than the 1000°F otherwise required.

Most combustors have a cell-like structure, **Figure A.6**, and consist of a substrate, washcoat, and catalyst. The *substrate* is a ceramic material formed into a honeycomb shape. Ceramic material is used because of its stability in extremely cold and hot conditions. The *washcoat,* usually made of an aluminum-based substance called alumina, covers the ceramic material and helps disperse the catalyst across the

Figure A.6 A catalytic combustion element. Courtesy of Corning Inc.

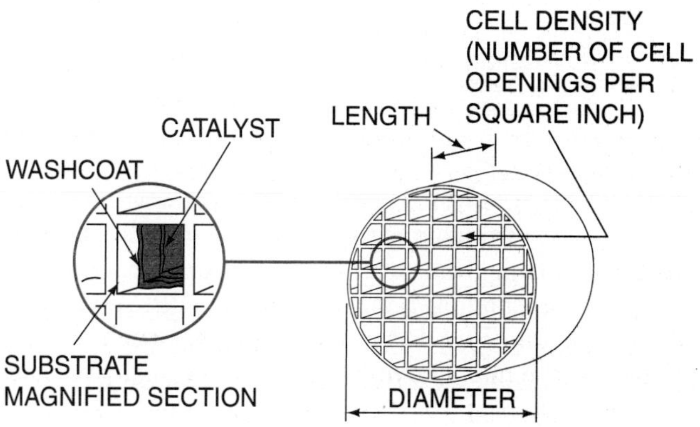

Figure A.7 An illustration of a woodstove with a catalytic combustor. Courtesy of Corning Inc.

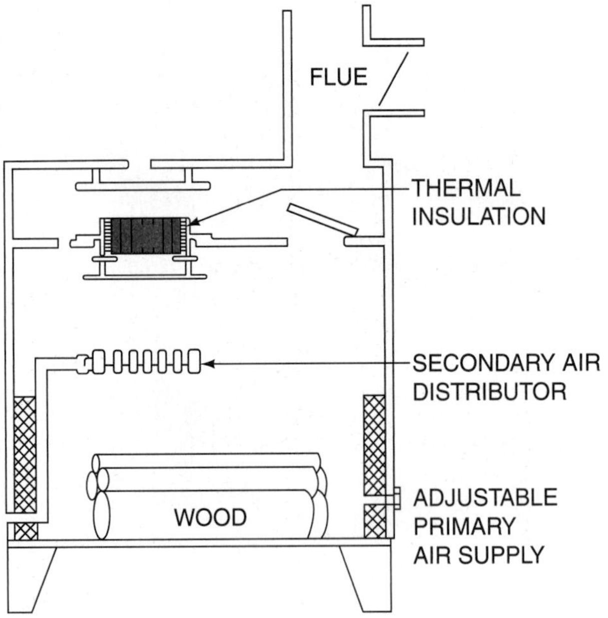

combustor surface. The *catalyst* is made of a noble metal, usually platinum, palladium, or rhodium, which is chemically stable in extreme temperatures. **Figure A.7** is an illustration of a woodstove with a catalytic combustor. A disadvantage of this combustor is that it must be replaced every few years.

RADIATION AND CONVECTION CHARACTERISTICS

Wood-burning stoves heat by both radiation and convection. Radiation is heat that comes from the stove in waves and heats objects in its path, such as the walls, floor, and furniture. The stove also heats the air around it. This is called convective heat. As the air is heated, it rises, and cooler air comes in to take its place and is heated. This cycle of convection currents is continuous. Some stoves are designed to utilize convective heating to a greater extent than others. They may have a blower system to spread the heat out further and to make the convection process more efficient.

PELLET STOVES

Pellet stoves are a more recent design, are energy efficient, and use a renewable energy source, **Figure A.8**. The fuel consists of small, compressed pellets made from waste wood, primarily sawdust from lumber sawmills, **Figure A.9.** The pellets may also be made from waste cardboard or even agricultural wastes such as sunflower and cherry seed hulls. Stove owners should make sure that they use only the type of pellets recommended by the manufacturer. Pellets are normally packaged in 40- or 50-lb bags.

Figure A.8 A pellet stove. Photo by Eugene Silberstein

Figure A.9 Wood pellets. Photo by Eugene Silberstein

Most pellet stoves are designed with a hopper at the back or top. The pellets are fed to the combustion chamber with an auger powered by an electric motor, which is often controlled by a thermostat. The room air is generally circulated through the heat exchanger and into the conditioned space by a multispeed blower.

Very little air pollution is produced and as little as 1% of the pellet material remains as ash. Most stoves do not require a chimney because exhaust gases are forced outside through a vent pipe by a combustion fan. Combustion air may be drawn in from outdoors so that heated room air is not used for combustion. This provides additional efficiency by not

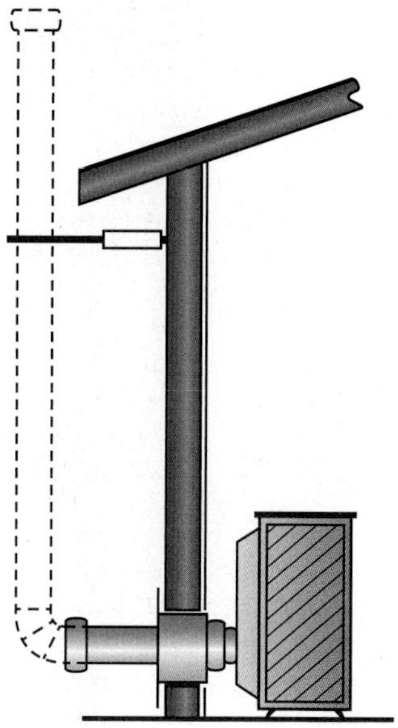

Figure A.10 A horizontal vent pipe must have an end cap to prevent air or exhaust gases from being blown into the stove. A vent pipe may be extended vertically above the eaves.

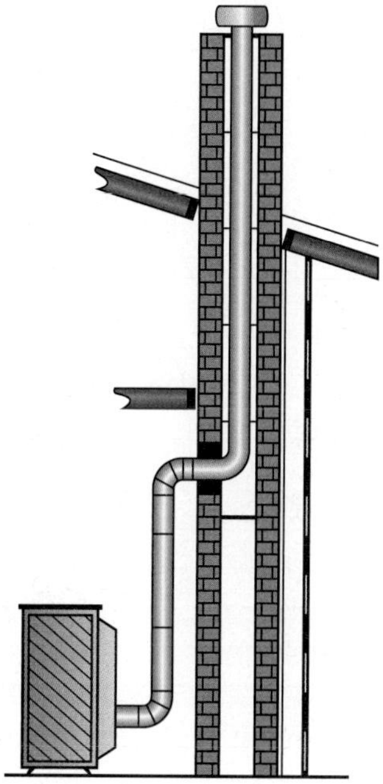

Figure A.11 Pellet stoves may operate more efficiently if masonry chimneys are relined with a stainless steel liner.

creating a negative pressure in the room and thus decreasing the cold air that will infiltrate through cracks in doors and windows. The exhaust gases may be vented through a horizontally positioned vent pipe with an end cap to prevent wind from blowing air and the exhaust gases into the stove, **Figure A.10.** The vent pipe must be constructed of PL vent pipe tested to UL 641 standards, which is a double-walled pipe with an air space between the walls. All joints must be sealed with an approved sealer. This is important because the venting system has an electrically operated exhaust combustion fan; if there should be a power failure, the combustion fumes could enter the conditioned space.

A lined masonry chimney that meets all appropriate codes may be used with pellet stoves. However, many masonry chimneys will be large enough to adversely affect the operation of the stoves. These chimneys may be relined with a stainless steel liner, **Figure A.11.** The stove manufacturer's literature should be consulted for the recommended diameter of the liners.

STOVE CONSTRUCTION

Stoves may be constructed of steel, cast iron, soapstone, or a combination of these materials. Stoves made of steel are normally the least expensive. They are made of sheet steel welded together and normally have a firebox of refractory bricks. These stoves give off heat almost immediately after the fire is started. With a large, hot fire, they will quickly give off a considerable amount of heat. However, when the fire dies, the heat from the stove will also quickly cool.

Many stoves are manufactured of cast iron. These can be much more decorative, because the casting process provides an opportunity for intricate detail, particularly on the sides and legs. Cast iron is more durable than steel, and the heat is more even and less intense than that of a sheet-steel stove. These stoves also cool down more slowly after the fire extinguishes.

Soapstone may be used, particularly in combination with cast iron, as a material in the manufacture of wood-burning stoves. Soapstone can withstand the changes in temperature, such as from room temperature to very intense heat. It may be used not only on the exterior of stoves but also for the firebox because it can withstand exposure to direct flames. This material provides a gentle heat and can hold and radiate heat for longer periods than the other materials.

MAKEUP AIR

Air used in combustion must be made up or resupplied from the outside with, for example, an inlet from the outside directly to the stove, **Figure A.12.** Often, an older home has many air leaks. The hot air may leak out near the ceiling or through the attic. Cool air can leak in through cracks around windows, under doors, and elsewhere. Although this may disturb the normal heating cycle, it does help to make up for air leaving through the chimney as a result of the combustion and venting.

SAFETY PRECAUTION: *In modern homes that are sealed and insulated well, some provision may have to be made to supply the makeup air. A door may have to be opened slightly to provide a proper draft. A stove could actually burn enough oxygen in a small home to make it difficult to get enough oxygen to breathe. Always make sure that there is enough makeup air.* •

Figure A.12 A stove with a fresh air tube entering from outside.

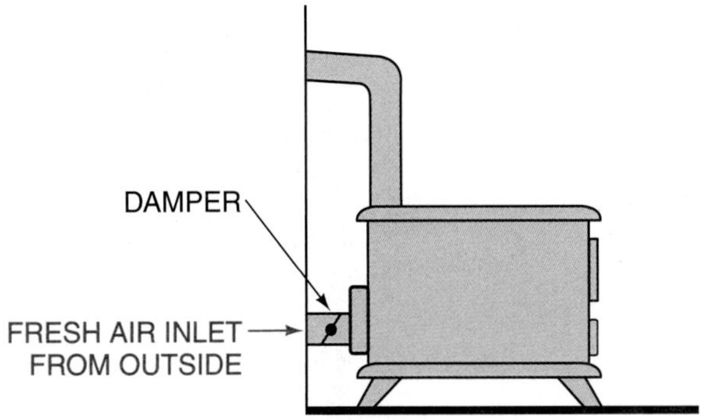

Figure A.13 A stove collar adapter for a double-wall stovepipe.

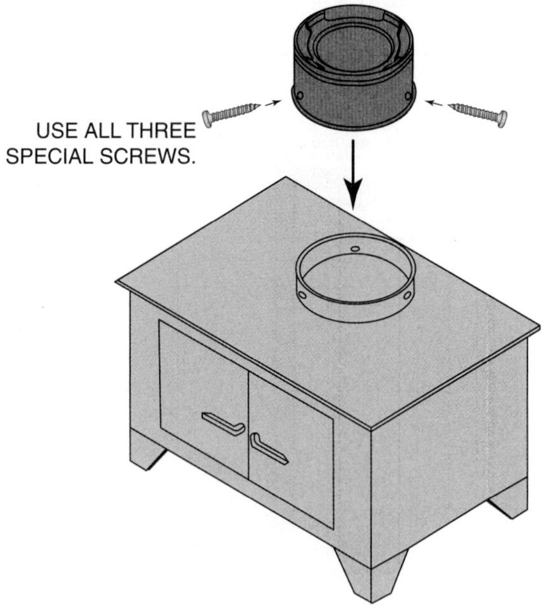

SAFETY HAZARDS

SAFETY PRECAUTION: *Live coals in the stove in the living area of a house, along with creosote in stovepipes and chimneys, make it absolutely necessary to install, maintain, and operate stoves safely. This cannot be emphasized enough.* •

Following are some of the safety hazards that may be encountered:

- A hot fire can ignite a buildup of creosote, resulting in a stovepipe or chimney fire.
- Radiation from the stove or stovepipe may overheat walls, ceilings, or other combustible materials in the house and start a fire.
- Sparks may get out of the stove, land on combustible materials, and ignite them. This could happen through a defect in the stove, while the door is left ajar, while the firebox is being filled, or while ashes are removed.
- Flames could leak out through faulty chimneys, or heat could be conducted through cracks to a combustible material.

Burning materials coming out of the top of the chimney can also start a fire on the outside of the house. These sparks or glowing materials can ignite roofing materials, leaves, brush, or other matter outside the house.

A.6 INSTALLATION PROCEDURES

A national testing laboratory should approve a wood-burning stove or appliance. Before installing a stove, stovepipe, or prefabricated chimney, all building and fire marshal's codes must be followed, as well as the instructions of the testing laboratory and manufacturer. If one code or set of instructions is more restrictive than another, the most restrictive instructions should be followed.

Instructions should include the distance the stove is to be located from any combustible material, such as a wall or the floor. They should also indicate the minimum required protective material between the stove and the wall and between the stove and the floor. Excessive heat from the stove can heat the walls or floor to the point where a fire can be started.

The stove must be connected to the chimney with an approved stovepipe, often called a connector. Codes and instructions must be followed. A stove collar adapter, often supplied by the manufacturer, should be used to connect the stovepipe to the stove, **Figure A.13**. The manufacturer may make special provision for this.

Figure A.14 shows three different types of installations. The stovepipe must not run through any combustible material, such as a ceiling or wall. Approved chimney sections with necessary fittings should be used. **Figure A.15** illustrates details for adapting the stovepipe to the "through-the-wall" chimney fittings. As mentioned earlier, horizontal stovepipe runs should be kept to a minimum. A rise of 1/4 in. per foot should be considered a minimum, **Figure A.3**. Local codes or manufacturers may require a greater rise. Approved thimbles, joist shields, insulation shields, and other necessary fittings must be used where required, all materials should be approved by a recognized national testing laboratory, and all codes should be met.

Only one stove can be connected to a chimney. For masonry chimneys with more than one flue, no more than one stove can be connected to each flue. A factory-built chimney used for woodstoves should be rated as a residential and building heating appliance chimney. Be sure to check all relevant codes and guidelines.

SAFETY PRECAUTION: *Many stoves are vented through masonry chimneys. If a new chimney is being used, it should have been constructed according to applicable building codes and carefully inspected. If an old chimney is being used, an experienced person should inspect it to ensure that it is safe. The mortar in the*

Figure A.14 Three different types of stove installations.

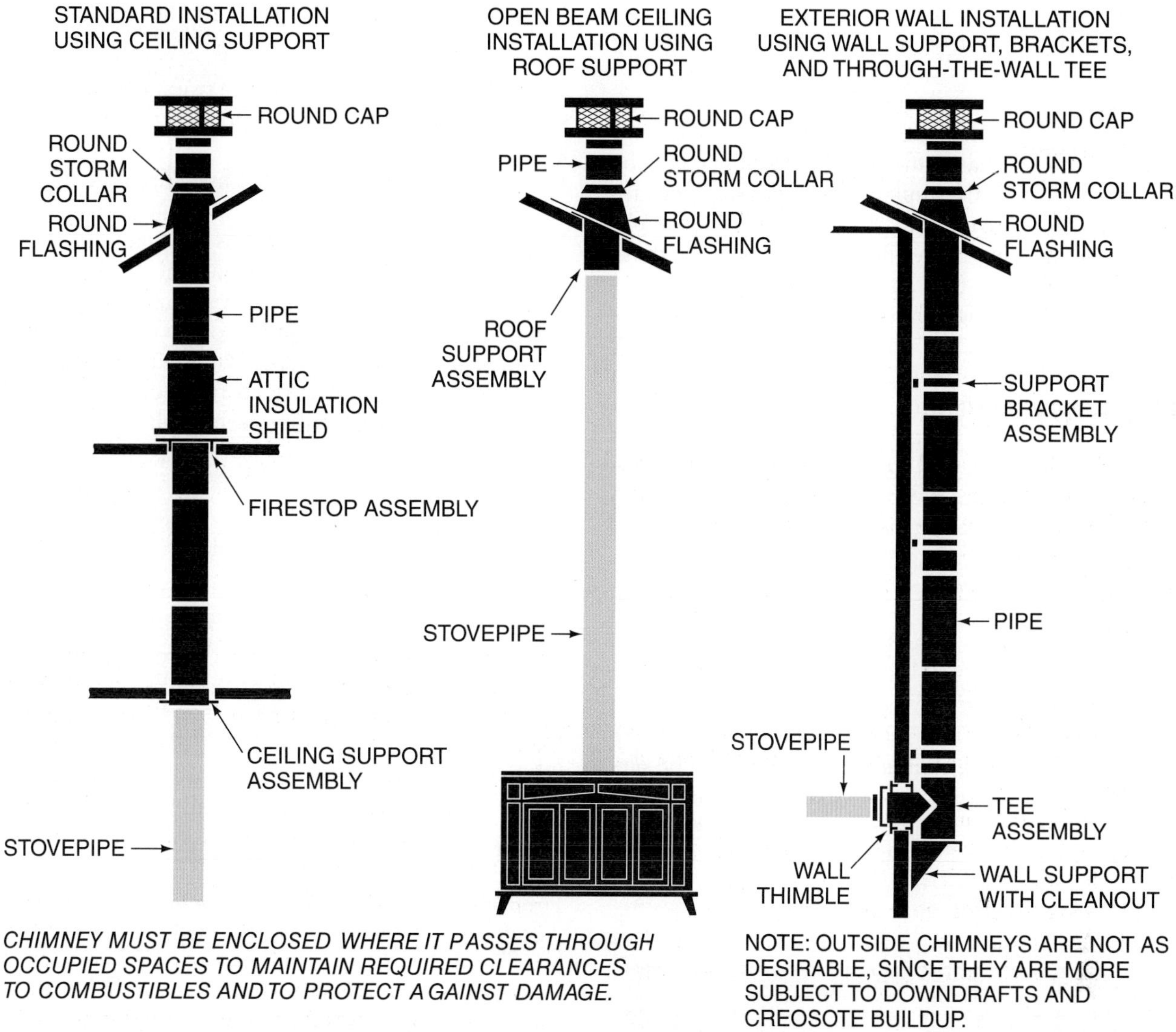

Figure A.15 Details of a through-the-wall installation.

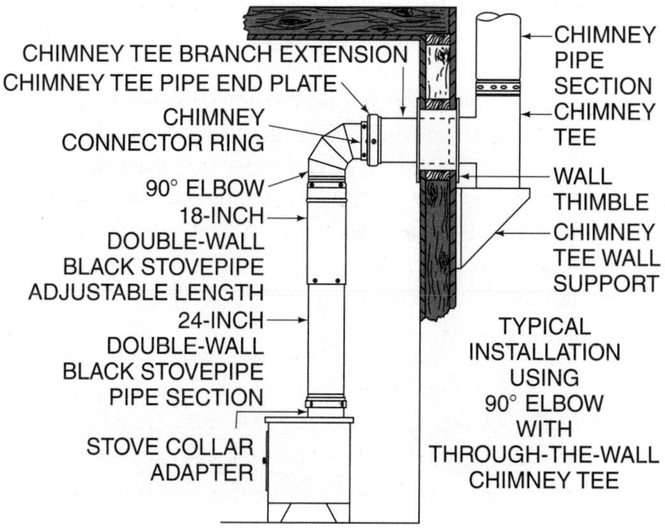

joints may be deteriorated and loose, or a serious chimney fire may have cracked the chimney. All necessary repairs should be made to the chimney by a competent mason before connecting the stove to it. Manufacturers or their suppliers may recommend that chimneys be lined, ***Figure A.10****, for a better draft and/or safety.* •

A.7 SMOKE DETECTORS

⚠ SAFETY PRECAUTION: *Smoke detectors should be used in all homes regardless of the type of heating appliance. It is even more important to install smoke detectors when burning wood fuel. Many types of detectors are available. They may be AC-powered photoelectric detectors or an AC- or battery-powered ionization chamber detector.*

If the smoke detectors are battery-powered, be sure to change the batteries regularly to ensure continued satisfactory operation. •

A.8 GAS STOVES

Gas stoves are becoming popular as a heating appliance. The American National Standards Institute (ANSI) may certify these as a "decorative appliance" or as a "room heater." Decorative appliances are attractive but not designed for larger space heating. Room heaters are, as the name implies, heating appliances and will have efficiency ratings. They may be manufactured of materials similar to woodstoves and are available in many attractive colors, **Figure A.16**. Gas appliances should be installed by certified personnel or by those approved to do so.

Gas stoves may be designed to burn either natural or propane gas. Many are designed so the orifice can be adjusted or changed to switch from burning one type of gas to the other. Most utilize a standing pilot for ignition and have controls to shut off the gas should the pilot go out. Many gas stoves operate without electricity and can be used for heating during a power outage. Some have variable-speed convection blowers to help circulate the heated air.

VENTING

Gas stoves may be designed as B-vent, direct-vent, or vent-free stoves.

B-VENT. Stoves requiring a B-vent must be vented through the roof or into an existing chimney. A flue liner may be suggested or required if venting into an existing chimney. The liner will reduce the size of the flue so that the stove will draft properly.

Figure A.16 A gas stove. Courtesy of Hearth & Home Technologies®.

DIRECT-VENT. Stoves utilizing a direct-vent system make use of a double-wall pipe to pull outside air in for combustion and to vent flue gases to the outside. Direct-vent options may include:

- venting straight back from the stove and through the wall.
- venting up from the stove and then through the wall.
- venting up through the roof.
- in some cases, venting into a lined fireplace flue.

VENT-FREE. Vent-free stoves, as the name implies, require no venting. Some people are concerned about gas combustion in the living area and not venting the flue gases. These stoves do have a good safety record but are not approved in all states at the time of this writing. They have an oxygen depletion sensor (ODS), which turns off the pilot light and gas supply when the oxygen level in the area of the stove is depleted to a certain level.

A.9 FIREPLACE INSERTS

Fireplace inserts can convert a fireplace from a very inefficient heat source to one that is more efficient. Very little heat can be obtained from a fireplace without some device to help contain the heat and move it from the fireplace into the room. The fireplace insert provides a way to retain the heat and blow it into the room. **Figure A.17** is a photo of a fireplace insert.

WOOD-BURNING INSERTS

Fireplace inserts can be either wood- or gas-burning. A wood-burning insert is basically a woodstove that is designed to be placed into an existing fireplace. New wood-burning inserts must be certified by the EPA just as woodstoves are. The regulations ensure that they will burn more cleanly

Figure A.17 A wood-burning fireplace insert. Courtesy of Hearth & Home Technologies®.

and efficiently. A wood-burning insert generally requires a fireplace and chimney to be of masonry construction. Local building codes and the manufacturer's specifications for the particular insert should be consulted to ensure that the size and construction of the fireplace can accommodate the insert. All installations should meet the NFPA and other appropriate codes. Many factory-built or prefabricated fireplaces are not approved for inserts.

Inserts are usually made from plate steel, cast iron, or a combination of these materials. They may fit into the fireplace opening so that they are basically flush with the front, or they may protrude to some degree onto the hearth. Those that protrude are usually more efficient, as this part of the insert will provide more radiant heat. Inserts normally have blowers that may be controlled manually or with a thermostat which improve efficiency.

In many areas, codes for new insert installations now require that the insert be installed with a positive connection to the chimney flue, which prevents creosote from running down into the fireplace and becoming a fire hazard. Smoke and gases will then exit the fireplace more directly into and out of the chimney.

Inserts should never be used in a chimney flue that is used for another purpose. Before an insert is installed, a professional should inspect the chimney. If there is creosote buildup, the chimney must be cleaned. Professional chimney sweeps should be employed to clean chimneys. The chimney and fireplace should also be checked for cracks or flaws that could be a fire hazard. Many inserts make it more difficult to clean the chimney, as they must be removed from the fireplace. Some inserts have a special collar and direct flue liner to the top of the chimney. These may be left in place for cleaning.

GAS FIREPLACE INSERTS

Gas inserts may be used to convert an existing wood-burning fireplace to a gas appliance. Many of these may be installed in prefabricated fireplaces where wood-burning units may not. They may be designed to burn either natural gas or propane but usually need to be adjusted for one or the other. Before any insert is installed, the existing fireplace and chimney should be of the type approved for the insert chosen.

Many gas units and/or installations require that the chimney flue be lined with a metal liner and connected to the insert. This produces a better draft and more efficiency. The inserts are usually sealed with glass doors and have blowers that help to make them more efficient. The doors prevent excess indoor air from being used for combustion. Just the right amount of combustion air is usually introduced through adjustable air shutters. Blowers simply make the convective heat transfer more efficiently. NFPA and other codes should be consulted regarding these liners. Some newer gas inserts are designed so that a chimney liner is not required. However, they are less efficient and allow more heat to go up the chimney, which heats the existing chimney flue liner and increases the draft.

All gas appliances, stoves, or inserts must be connected to the gas source. Piping needs to be routed and installed. This usually means cutting, threading, and sealing pipe and should be done by a professional trained and approved to do this work.

APPENDIX B

TEMPERATURE CONVERSION CHART

°F	Temperature to Be Converted	°C	°F	Temperature to Be Converted	°C
−76.0	−60	−51.1	23.0	−5	−20.6
−74.2	−59	−50.6	24.8	−4	−20.0
−72.4	−58	−50.0	26.6	−3	−19.4
−70.6	−57	−49.4	28.4	−2	−18.9
−68.8	−56	−48.9	30.2	−1	−18.3
−67.0	−55	−48.3	32.0	0	−17.8
−65.2	−54	−47.8	33.8	1	−17.2
−63.4	−53	−47.2	35.6	2	−16.7
−61.6	−52	−46.7	37.4	3	−16.1
−59.8	−51	−46.1	39.2	4	−15.6
−58.0	−50	−45.6	41.0	5	−15.0
−56.2	−49	−45.0	42.8	6	−14.4
−54.4	−48	−44.4	44.6	7	−13.9
−52.6	−47	−43.9	46.4	8	−13.3
−50.8	−46	−43.3	48.2	9	−12.8
−49.0	−45	−42.8	50.0	10	−12.2
−47.2	−44	−42.2	51.8	11	−11.7
−45.4	−43	−41.7	53.6	12	−11.1
−43.6	−42	−41.1	55.4	13	−10.6
−41.8	−41	−40.6	57.2	14	−10.0
−40.0	−40	−40.0	59.0	15	−9.4
−38.2	−39	−39.4	60.8	16	−8.9
−36.4	−38	−38.9	62.6	17	−8.3
−34.6	−37	−38.3	64.4	18	−7.8
−32.8	−36	−37.8	66.2	19	−7.2
−31.0	−35	−37.2	68.0	20	−6.7
−29.2	−34	−36.7	69.8	21	−6.1
−27.4	−33	−36.1	71.6	22	−5.6
−25.6	−32	−35.6	73.4	23	−5.0
−23.8	−31	−35.0	75.2	24	−4.4
−22.0	−30	−34.4	77.0	25	−3.9
−20.2	−29	−33.9	78.8	26	−3.3
−18.4	−28	−33.3	80.6	27	−2.8
−16.6	−27	−32.8	82.4	28	−2.2
−14.8	−26	−32.2	84.2	29	−1.7
−13.0	−25	−31.7	86.0	30	−1.1
−11.2	−24	−31.1	87.8	31	−0.6
−9.4	−23	−30.6	89.6	32	0.0
−7.6	−22	−30.0	91.4	33	0.6
−5.8	−21	−29.4	93.2	34	1.1
−4.0	−20	−28.9	95.0	35	1.7
−2.2	−19	−28.3	96.8	36	2.2
−0.4	−18	−27.8	98.6	37	2.8
1.4	−17	−27.2	100.4	38	3.3
3.2	−16	−26.7	102.2	39	3.9
5.0	−15	−26.1	104.0	40	4.4
6.8	−14	−25.6	105.8	41	5.0
8.6	−13	−25.0	107.6	42	5.6
10.4	−12	−24.4	109.4	43	6.1
12.2	−11	−23.9	111.2	44	6.7
14.0	−10	−23.3	113.0	45	7.2
15.8	−9	−22.8	114.8	46	7.8
17.6	−8	−22.2	116.6	47	8.3

°F	Temperature to Be Converted	°C	°F	Temperature to Be Converted	°C
19.4	−7	−21.7	118.4	48	8.9
21.2	−6	−21.1	120.2	49	9.4
122.0	50	10.0	208.4	98	36.7
123.8	51	10.6	210.2	99	37.2
125.6	52	11.1	212.0	100	37.8
127.4	53	11.7	213.8	101	38.3
129.2	54	12.2	215.6	102	38.9
131.0	55	12.8	217.4	103	39.4
132.8	56	13.3	219.2	104	40.0
134.6	57	13.9	221.0	105	40.6
136.4	58	14.4	222.8	106	41.1
138.2	59	15.0	224.6	107	41.7
140.0	60	15.6	226.4	108	42.2
141.8	61	16.1	228.2	109	42.8
143.6	62	16.7	230.0	110	43.3
145.4	63	17.2	231.8	111	43.9
147.2	64	17.8	233.6	112	44.4
149.0	65	18.3	235.4	113	45.0
150.8	66	18.9	237.2	114	45.6
152.6	67	19.4	239.0	115	46.1
154.4	68	20.0	240.8	116	46.6
156.2	69	20.6	242.6	117	47.2
158.0	70	21.1	244.4	118	47.7
159.8	71	21.7	246.2	119	48.3
161.8	72	22.2	248.0	120	48.8
163.4	73	22.8	257.0	125	51.7
165.2	74	23.3	266.0	130	54.4
167.0	75	23.9	275.0	135	57.2
168.8	76	24.4	284.0	140	60.0
170.6	77	25.0	293.0	145	62.8
172.4	78	25.6	302.0	150	65.6
174.2	79	26.1	311.0	155	68.3
176.0	80	26.7	320.0	160	71.1
177.8	81	27.2	329.0	165	73.9
179.6	82	27.8	338.0	170	76.7
181.4	83	28.3	347.0	175	79.4
183.2	84	28.9	356.0	180	82.2
185.0	85	29.4	365.0	185	85.0
186.8	86	30.0	374.0	190	87.8
188.6	87	30.6	383.0	195	90.6
190.4	88	31.1	392.0	200	93.3
192.2	89	31.7	401.0	205	96.1
194.0	90	32.2	410.0	210	98.9
195.8	91	32.8	413.6	212	100.0
197.6	92	33.3	428.0	220	104.4
199.4	93	33.9	446.0	230	110.0
201.2	94	34.4	464.0	240	115.6
203.0	95	35.0	482.0	250	121.1
204.8	96	35.6	500.0	260	126.7
206.6	97	36.1			

Example 1. To find 37°F as a Celsius equivalent, find 37 in the Temperature to Be Converted column and read the value in the °C column, which is 2.8°C.

Example 2. To find 75°C as a Fahrenheit equivalent, find 75 in the Temperature to Be Converted column and read the value in the °F column, which is 167.0°F.

GLOSSARY/GLOSARIO

NOTE: English term appears **boldface** followed by Spanish term set in ***blue italic***.

A

Absolute pressure Gauge pressure plus the pressure of the atmosphere, normally 14.696 at sea level at 70°F.

Presión absoluta La presión del manómetro más la presión de la atmósfera, que generalmente es 14696 al nivel del mar a 70°F (21.11°C).

Absolute zero temperature The temperature at which molecular motion stops (−460°F or equivalent).

Temperatura del cero absoluto La temperatura donde se detiene el movimiento molecular, (–460°F o su equivalente).

Absorbent (attractant) The salt solution used to attract water in an absorption chiller.

Hidrófilo Solución salina utilizada para atraer el agua en un enfriador por absorción.

Absorber That part of an absorption chiller where the water is absorbed by the salt solution.

Absorbedor El lugar en el enfriador por absorción donde la solución salina absorbe el agua.

Absorption The process by which one substance is absorbed by another.

Absorción Proceso mediante el cual una sustancia es absorbida por otra.

Absorption air-conditioning chiller A system using a salt substance, water, and heat to provide cooling for an air-conditioning system.

Enfriador por absorción para acondicionamiento de aire Sistema que utiliza una sustancia salina, agua y calor para proveer enfriamiento en un sistema de acondicionamiento de aire.

ABS pipe Acrylonitrile-butadiene-styrene plastic pipe used for water, drains, waste, and venting.

Tubo de acronitrilo-butadieno-estireno Tubo plástico de acronitrilo-butadieno-estireno utilizado para el agua, los drenajes, los desperdicios y la ventilación.

ACCA Manual J The Air Conditioning Contractors of America Manual that facilitates the calculations of heat gains and heat losses in residential structures.

Manual J de la ACCA Manual de los contratistas de acondicionamiento de aire de los Estados Unidos, que facilita el cálculo de las pérdidas y ganancias de calor en estructuras residenciales.

Accumulator A storage tank located in the suction line. It allows small amounts of liquid refrigerant to boil away before entering the compressor. Sometimes used to store excess refrigerant in heat pump systems during the winter cycle.

Acumulador Tanque de almacenaje ubicado en el conducto de aspiración. Permite que pequeñas cantidades de refrigerante líquido se evaporen antes de entrar al compresor. Algunas veces se utiliza para almacenar exceso de refrigerante en sistemas de bombas de calor durante el ciclo de invierno.

Acetylene A gas often used with air or oxygen for welding, brazing, or soldering applications.

Acetileno Gas usado con frecuencia, junto con aire u oxígeno, en trabajos de soldadura y soldaduras de cobre.

ACH An acronym for (air changes per hour).

ACH Acrónimo de intercambios de aire por hora.

ACH_{50} An acronym for air changes per hour of air leakage while the blower door is connected and pulling 50 Pascal of vacuum on the house.

ACH_{50} Acrónimo de intercambios de aire por hora de pérdida de aire cuando se conecta un extractor de aire potente en una de las entradas, el cual genera un vacío de 50 Pascales en el interior de la vivienda.

Acid-contaminated system A refrigeration system that contains acid due to contamination.

Sistema contaminado de ácido Sistema de refrigeración que, debido a la contaminación, contiene ácido.

"A" coil An evaporator coil that can be used for upflow, downflow, and horizontal-flow applications. It actually consists of two coils shaped like a letter "A."

Serpentín en forma de "A" Serpentín de evaporación que puede utilizarse para aplicaciones de flujo ascendente, descendente y horizontal. En realidad, consiste en dos serpentines en forma de "A."

ACR tubing Air-conditioning and refrigeration tubing that is very clean, dry, and normally charged with dry nitrogen. The tubing is sealed at the ends to contain the nitrogen.

Tubería ACR Tubería para el acondicionamiento de aire y la refrigeración que es muy limpia y seca, y que por lo general está cargada de nitrógeno seco. La tubería se sella en ambos extremos para contener el nitrógeno.

Activated alumina A chemical desiccant used in refrigerant driers.

Alúmina activada Disecante químico utilizado en secadores de refrigerantes.

Activated charcoal A substance manufactured from coal or coconut shells into pellets. It is often used to adsorb solvents, other organic materials, and odors.

Carbón activado Sustancia fabricada utilizando carbón o cáscaras de coco, en forma de gránulos. Se emplea para adsorber disolventes, así como otros materiales orgánicos y olores.

Active recovery Recovering refrigerant with the use of a recovery machine that has its own built-in compressor.

Recuperación activa El hecho de recuperar refrigerante utilizando un recuperador que dispone de su propio compresor.

Active sensor Sensors that send information back to the controller in terms of milliamps (mA) or volts.

Sensor activo Sensores que devuelven información al controlador en miliamperios (mA) o voltios.

Active solar system A system that uses electrical and/or mechanical devices to help collect, store, and distribute the sun's energy.

Sistema solar activo Sistema que utiliza dispositivos eléctricos y/o mecánicos para ayudar a acumular, almacenar y distribuir la energía del sol.

Adsorption The process by which a thin film of a liquid or gas adheres to the surface of a solid substance.

Adsorción Proceso mediante el cual una fina capa de líquido o gas se adhiere a la superficie de una sustancia sólida.

Air, standard Dry air at 70°F and 14.696 psi, at which it has a mass density of 0.075 lb/ft^3 and a specific volume of 13.33 ft^3/lb, ASHRAE.

Aire, estándar Aire seco a 70°F (21.11°C) y 14696 psi (libra por pulgada cuadrada); a dicha temperatura tiene una densidad de masa de 0075 libra/pies^3 y un volumen específico de 1333 pies^3/libra, ASHRAE.

Air-acetylene A mixture of air and acetylene gas that when ignited is used for soldering, brazing, and other applications.

Aire-acetilénico Mezcla de aire y de gas acetileno que se utiliza en la soldadura, la broncesoldadura y otras aplicaciones al ser encendida.

Air-acetylene unit A torch kit that uses acetylene as its fuel.

Unidades de aire-acetileno un juego de antorcha que utiliza acetileno como su combustible

Air Changes per Hour (ACH_{50}) The number of times the air is replaced in an interior space each hour.

Cambios de aire por hora (ACH_{50}) El número de veces que el aire se reemplaza en un espacio interior cada hora.

Air conditioner Equipment that conditions air by cleaning, cooling, heating, humidifying, or dehumidifying it. A term often applied to comfort cooling equipment.

Acondicionador de aire Equipo que acondiciona el aire limpiándolo, enfriándolo, calentándolo, humidificándolo o deshumidificándolo. Término comúnmente aplicado al equípo de enfriamiento para comodidad.

Air-conditioning A process that maintains comfort conditions in a defined area.

Acondicionamiento de aire Proceso que mantiene condiciones agradables en un área definida.

Air-Conditioning, Heating and Refrigeration Institute (AHRI) A nonprofit association that regulates equipment manufacturers and rates the capacity of equipment.

Instituto del acondicionamiento de aire, calefacción y refrigeración (ARHI en inglés) Una asociación sin fines lucrativos que regula los fabricantes del equipo y que establece la capacidad de dicho equipo.

Air-cooled condenser One of the four main components of an air-cooled refrigeration system. It receives hot gas from the compressor and rejects heat to a place where it makes no difference.

Condensador enfriado por aire Uno de los cuatro componentes principales de un sistema de refrigeración enfriado por aire. Recibe el gas caliente del compresor y dirige el calor a un lugar donde no afecte la temperatura.

Air friction chart A chart used to determine proper round duct sizes.

Gráfica de fricción del aire Gráfica utilizada para determinar los tamaños correctos de conductos redondos.

Air gap The clearance between the rotating rotor and the stationary winding on an open motor. Known as a vapor gap in a hermetically sealed compressor motor.

Espacio de aire Espacio libre entre el rotor giratorio y el devandado fijo en un motor abierto. Conocido como espacio de vapor en un motor de compresor sellado herméticamente.

Air handler The device that moves the air across the heat exchanger in a forced air system—normally considered to be the blower and its housing.

Tratante de aire Dispositivo que dirige el aire a través del intercambiador de calor en un sistema de aire forzado—considerado generalmente como el soplador y su alojamiento.

Air heat exchanger A device used to exchange heat between air and another medium at different temperature levels, such as air-to-air, air-to-water, or air-to-refrigerant.

Intercambiador de aire y calor Dispositivo utilizado para intercambiar el calor entre el aire y otro medio, como por ejemplo aire y aire, aire y agua or aire y refrigerante, a diferentes niveles de temperatura.

Air loop The heat pump's heating and cooling ducted air system, which exchanges heat with the refrigerant loop.

Circuito de aire Sistema de tubería del aire de calentamiento y de refrigeración de la bomba de calor, que sirve para intercambiar el calor con el circuito de refrigeración.

Air pressure control (switch) Used to detect air pressure drop across the coil in a heat pump outdoor unit due to ice buildup.

Regulador de la presión de aire (conmutador) Utilizado para detectar una caída en la presión del aire a través de la bobina en una unidad de bomba de calor para exteriores debido a la acumulación de hielo.

Air sensor A device that registers changes in air conditions such as pressure, velocity, temperature, or moisture content.

Sensor de aire Dispositivo que registra los cambios en las condiciones del aire, como por ejemplo cambios en presión, velocidad, temperatura o contenido de humedad.

Air shutters Devices that are placed in an airstream for the purpose of controlling airflow.

Contraventanas del aire Aparatos que se ponen en un corriente de aire con el propósito de regular el fluir del aire.

Air vent A fitting used to vent air manually or automatically from a hydronic heating system.

Válvula de aire Accesorio utilizado para darle al aire salida manual o automática de un sistema hidrónico de calefacción.

Alcohol Antifreeze solution used in the water loop of geothermal heat pumps.

Alcohol Líquido anticongelante que se utiliza en el circuito de agua en bombas de calor geotérmicas.

Algae A form of green or black, slimy plant life that grows in water systems.

Alga Tipo de planta legamosa de color verde o negro que crece en sistemas acuáticos.

Algorithm A computer code or set of instructions to make specific calculations.

Algoritmo Un código o juego de instrucciones computadora para hacer cálculos específicos.

Alkylbenzene (AB) A popular synthetic lubricant that works best with HCFC-based refrigerant blends. This lubricant can be used with CFC and HCFC refrigerants.

Alquilbenceno (AB) Lubricante sintético muy compatible con mezclas de refrigerantes basados en HCFC. También puede utilizarse con refrigerantes que usan CFC y HCFC.

Allen head A recessed hex (six-sided) head in a fastener.

Cabeza allen Cabeza de concavidad hexagonal (seis lados) en un asegurador.

All-weather system System providing year-round conditioning of the air.

Sistema para todo el año Sistema que proporciona una aclimatación ambiental todo el año.

Alternating current An electric current that reverses its direction at regular intervals.

Corriente alterna Corriente eléctrica que invierte su dirección a intervalos regulares.

Alternative refrigerants Newer refrigerants that are replacing the traditional CFC and HCFC refrigerants that have been used for many years. Many of these refrigerants have very low ozone depletion and global-warming indices.

Refrigerantes alternativos Nuevos productos que sirve para sustituir a los refrigerantes basados en CFC y HCFC que han sido utilizados durante muchos años. Muchos de estos nuevos refrigerantes tienen un índice muy bajo de desgaste de la capa de ozono y de calentamiento de la superficie terrestre.

Altitude adjustment An adjustment to a refrigerator thermostat to account for a lower-than-normal atmospheric pressure such as may be found at a high altitude.

Ajuste para elevación Ajuste al termostato de un refrigerador para regular una presión atmosférica más baja que la normal, como la que se encuentra en elevaciones altas.

Ambient temperature The surrounding air temperature.

Temperatura ambiente Temperatura del aire circundante.

American standard pipe thread Standard thread used on pipe to prevent leaks.

Rosca estándar estadounidense para tubos Rosca estándar utilizada en tubos para evitar fugas.

Ammeter A meter used to measure current flow in an electrical circuit.

Amperímetro Instrumento utilizado para medir el flujo de corriente en un circuito eléctrico.

Ampacity The maximum amount of current that a conductor or electrical device can safely carry without damage.

Ampacidad La cantidad máxima de corriente que un conductor o dispositivo eléctrico puede transportar sin daños.

Amperage Amount (quantity) of electron or current flow (the number of electrons passing a point in a given time) in an electrical circuit.

Amperaje Cantidad de flujo de electrones o de corriente (el número de electrones que sobrepasa un punto específico en un tiempo fijo) en un circuito eléctrico.

Ampere Unit of current flow.

Amperio Unidad de flujo de corriente.

Analog electronic devices Devices that generate continuous or modulating signals within a certain control range.

Aparatos electrónicos analógicos Aparatos que generan señales continuas o modulares adentro de cierto registro de control.

Analog signal A continuous signal with an infinite number of steps.

Señal analógico Un señal continuo con un número infinitésimo de pasos.

Analog VOM A volt-ohm-milliammeter constructed so that the meter indicator is a needle over a printed surface.

VOM analógico Medidor de voltios-ohmios-miliamperímetros construido de forma que el indicador consiste en una aguja que se mueve encima de una superficie impresa.

Anemometer An instrument used to measure the velocity of air.

Anemómetro Instrumento utilizado para medir la velocidad del aire.

Angle valve Valve with one opening at a 90° angle from the other opening.

Válvua en ángulo Válvula con una abertura a un ángulo de 90° con respecto a la otra abertura.

Annealing The process of heating and slowly cooling a metal in order to strengthen it.

Recocido El proceso de calentar y enfriar lentamente un metal con el fin de fortalecerlo.

Annual Fuel Utilization Efficiency (AFUE) The U.S. Federal Trade Commission requires furnace manufacturers to provide this rating so consumers may compare furnace performances before purchasing. This value is derived in a laboratory and used for heating appliances in which it accounts for cycling losses, chimney losses, and furnace jacket losses.

Eficiencia anual de uso de combustible (AFUE por sus siglas en inglés) La Comisión de comercio federal de los EE.UU. exige que los fabricantes de estufas informen este valor de manera de que los consumidores puedan comparar los rendimientos de las estufas antes de adquirirlas. Este valor se investiga en laboratorios y se utiliza para calefactores, cuantificando las pérdidas cíclicas, pérdidas por chimeneas y pérdidas a través de camisas de enfriamiento.

Annual Fuel Utilization Index (AFUI) The AFUI compares the amount of (seasonal) energy used in Btus to the (seasonal) output of Btus and expresses the ratio as a percentage. AFUI is calculated by the manufacturer of fuel burning combustion equipment.

Índice anual de uso de combustible (AFUI por sus siglas en inglés) El AFUI compara la cantidad de energía (estacional) utilizada, en Btu con la emisión de energía calórica en Btu y expresa la relación como un porcentaje. El fabricante del equipo de combustión es quien calcula el valor del AFUI.

Anode A terminal or connection point on a semiconductor.

Ánodo Punto de conexión o terminal en un semiconductor.

ANSI Abbreviation for the American National Standards Institute.

ANSI Acrónimo de American National Standards Institute (Instituto Nacional Americano de Normas).

Anticipator Resistance heaters in thermostats that lessen the effects of system lag and overshoot.

Anticipador Calentador de resistencia en un termostato que reduce los efectos de subida y bajada del sistema.

Antifreeze solution A solution that will mix with water that has a freezing point much lower than water. Different solutions are mixed with water to lower the freezing point to the desired level.

Solución anticongelante Una solución que puede mezclarse con agua que tiene un punto de congelar mucho más bajo de lo del agua. Soluciones diferentes están mezcladas con agua para bajar el punto de congelar al nivel deseado.

Approach The temperature difference between the water leaving the cooling tower and the wet-bulb temperature of the tower's entering air.

Acercamiento La diferencia de temperatura entre el agua que sale de la torre de enfriar y la temperatura del bulbo mojado del aire que entra en la torre.

Approach temperature The difference in temperature between the refrigerant and the leaving water in a chilled-water system.

Temperatura de acercamiento Diferencia en temperatura entre el refrigerante y el agua de salida en un sistema de agua enfriada.

Aromatic oil A refrigeration mineral oil in which the alkylbenzene lubricant group originates.

Aceite aromático Aceite mineral de refrigeración del cual se deriva el grupo de lubricantes basados en alquilbenceno.

ASA Abbreviation for the American Standards Association [now known as American National Standards Institute (ANSI)].

ASA Abreviatura de Asociación Estadounidense de Normas [conocida ahora como Instituto Nacional Estadounidense de Normas (ANSI)].

ASHRAE Abbreviation for the American Society of Heating, Refrigerating, and Air-Conditioning Engineers.

ASHRAE Abreviatura de Sociedad Estadounidense de Ingenieros de Calefacción, Refrigeración y Acondicionamiento de Aire.

ASME Abbreviation for the American Society of Mechanical Engineers.

ASME Abreviatura de Sociedad Estadounidense de Ingenieros Mecánicos.

Aspect ratio The ratio of the length to width of a component or space.

Coeficiente de alargamiento Relación del largo al ancho de un componente o espacio.

Atmospheric draft heaters Combustion heating appliances which rely solely on the buoyancy of their combustion gasses for their draft.

Calentadores atmosféricos por corriente de aire Calefactores en los cuales la corriente de aire caliente se produce únicamente debido a la presión ejercida por los gases de combustión.

Atmospheric pressure The weight of the atmosphere's gases pressing down on the earth. Equal to 14.696 psi at sea level and 70°F.

Presión atmosférica El peso de la presión ejercida por los gases de la atmósfera sobre la tierra, equivalente a 14696 psi al nivel del mar a 70°F.

Atom The smallest particle of an element.

átomo Partícula más pequeña de un elemento.

Atomization The process of using pressure to change liquid to small particles of vapor.

Pulverización El proceso de utilizar la presión para cambiar un líquido a partículas pequeñas de vapor.

Atomizing humidifier A type of humidifier that discharges tiny water droplets into the air, which evaporate very rapidly into the duct airstream or directly into the conditioned space.

Humidificador por pulverización Dispositivo que descarga al aire pequeñas gotas de agua que se evaporan rápidamente en el conducto de aire o en ambiente acondicionado.

Attractant A type of salt solution, usually lithium bromide, used as an absorbent in absorption refrigeration.

Atrayente Tipo de solución salina, normalmente bromuro de litio, utilizado como absorbente en la refrigeración por absorción.

Automatic changeover thermostat A thermostat that changes from cool to heat automatically by room temperature.

Termostato de cambio automático Un termostato que cambia de enfriar a calentar automáticamente con la temperatura del cuarto.

Automatic combination gas valve A gas valve for gas furnaces that incorporates a manual control, gas supply for the pilot, adjustment and safety features for the pilot, pressure regulator, and the controls for and the main gas valve.

Válvula de gas de combinación automática Válvula de gas para hornos de gas que incorpora un regulador manual, sumi-nistro de gas para la llama piloto, ajuste y dispositivos de seguridad, regulador de presión, la válvula de gas principal y los reguladores de la válvula.

Automatic control Controls that react to a change in conditions to cause the condition to stabilize.

Regulador automático Reguladores que reaccionan a un cambio en las condiciones para provocar la estabilidad de dicha condición.

Automatic defrost Using automatic means to remove ice from a refrigeration coil at timed intervals.

Desempañador automático La utilización de medios automáticos para remover el hielo de una bobina de refrigeración a intervalos programados.

Automatic expansion valve (AXV) A refrigerant control valve that maintains a constant pressure in an evaporator. Sometimes abbreviated AEV.

Válvula de expansión automática Válvula de regulación del refrigerante que mantiene una presión constante en un evaporador. Puede también designarse con la abreviatura "AEV".

Automatic pumpdown system A control scheme in refrigeration consisting of a thermostat and liquid-line solenoid valve that clears refrigerant from the compressor's crankcase, evaporator, and suction line just before the compressor off cycle.

Sistema de bombeo hacia abajo automático Un sistema de control en refrigeración que consiste de un termostato y una válvula de línea de líquido solenoide que vacía el refrigerante del cárter del cigüeñal del compresor, evaporador y línea de succión justo antes del ciclo de apagar del compresor.

Auxiliary drain pan A separate drain pan that is placed under an air conditioner evaporator to catch condensate in the event that the primary drain pan runs over.

Plato de drenaje auxiliar Un plato separado de drenaje debajo del evaporador de un aire acondicionado para recoger la condensación en caso de que el plato principal se derrame.

Auxiliary line See whipline.

Línea auxiliar Ver Línea auxiliar.

Azeotropic blend Two or more refrigerants mixed together that will have only one boiling and/or condensing point for each system pressure. Negligible fractionation or temperature glide will occur.

Mezcla azeotrópica Mezcla de dos o más refrigerantes y que tiene un solo punto de ebullición y/o condensación para cada presión del sistema. Puede producirse un fraccionamiento o una variación térmica insignificantes.

B

Backdrafting A continuous spillage of combustion by-products usually caused from a pressure difference that causes a reversal of the vent or chimney's normal flow upward.

Retroceso de la corriente de aire La pérdida continua de los productos de la combustión causada por una diferencia de presión, la cual produce la inversión del flujo normal, hacia arriba, de la ventilación o de una chimenea.

Back electromotive force (BEMF) This is the voltage-generating effect of an electric motor's rotor turning within the motor.

Fuerza contraelectromotriz (BEMF en inglés) El efecto de generar tensión provocado por el rotor de un motor eléctrico al girar dentro del motor.

Back pressure The pressure on the low-pressure side of a refrigeration system (also known as *suction pressure*).

Contrapresión La presión en el lado de baja presión de un sistema de refrigeración (conocido también como *presión de aspiración*).

Back seat The position of a refrigeration service valve when the stem is turned away from the valve body and seated, shutting off the service port.

Asiento trasero Posición de una válvula de servicio de refrigeración cuando el vástago está orientado fuera del cuerpo de la válvula y aplicado sobre su asiento, cerrando así la apertura de servicio.

Baffle A plate used to keep fluids from moving back and forth at will in a container.

Deflector Placa utilizada para evitar el libre movimiento de líquidos en un recipiente.

Balanced-port TXV A valve that will meter refrigerant at the same rate when the condenser head pressure is low.

Válvula electrónica de expansión con conducto equilibrado Válvula que medirá el refrigerante a la misma proporción cuando la presión en la cabeza del condensador sea baja.

Balance point The temperature in a structure where the heat pump will run full-time and maintain the temperature in the structure. Auxiliary heat must be used to accompany the heat pump when the temperature is below the balance point.

Punto de balance La temperatura en una estructura donde la bomba de calor corre todo el tiempo y mantiene la temperatura en la estructura. Calor auxiliar debe usarse para acompañar la bomba de calor cuando la temperatura está por debajo del punto de balance.

Ball check valve A valve with a ball-shaped internal assembly that only allows fluid flow in one direction.

Válvula de retención de bolas Válvula con un conjunto interior en forma de bola que permite el flujo de fluido en una sola dirección.

Ball valve Valve that uses a round ball in the flow chamber in the valve. The round ball is bored to the approximate inside diameter of the pipe and also offers very little resistance or pressure drop to the flow of water or fluid flow. The handle only has to be turned 90° to fully open or close the valve.

Válvula de bola Válvula que utiliza una bola redonda en la cámara de flujo en la válvula. La bola redonda está calibrada aproximadamente al diámetro interior de la tubería y también presenta muy poca resistencia o caída de la presión al flujo del agua o fluido. El mango solamente se gira 90° para abrir o cerrar la válvula completamente.

Barometer A device used to measure atmospheric pressure that is commonly calibrated in inches or millimeters of mercury.

Barómetro Dispositivo comúnmente calibrado en pulgadas o en milímetros de mercurio que se utiliza para medir la presión atmosférica.

Barometric damper A place for drawing indoor air (dilution air) into a chimney after the combustion process for open combustion type oil fired furnaces and boilers.

Regulador barométrico Espacio a través del cual se regula la entrada de aire (aire de dilución) hacia una chimenea luego del proceso de combustión, en estufas o calderas de combustión abierta que utilizan combustibles líquidos.

Base A terminal on a semiconductor.

Base Punto terminal en un semiconductor.

Baseboard heating Convection heaters providing whole-house, spot, or individual room heating. The heat is normally provided by electrical resistance or hot water.

Calefacción de zócalo Calentadores por convección que proporcionan calefacción a toda la casa, en un punto específico o en una sola habitación. Normalmente, el calor se obtiene mediante una resistencia eléctrica o agua caliente.

Base loads Base loads are consistent, everyday consumers of energy in residential structures which include freezers, refrigerators, lights, water heaters, clothes dryers, copiers, cell phone chargers, computers, and printers.

Consumidores básicos Artefactos que consumen energía diariamente y de manera continua en estructuras residenciales, como congeladores, heladeras, luces, calentadores de agua, secarropas, copiadoras, cargadores de teléfonos celulares, computadoras eimpresoras.

Batt The term used to describe precut panels of insulation that fit in between wall studs or ceiling joists.

Paneles de aislación Término que se utiliza para describir los paneles precortados de aislación que se colocan entre los perfiles de paredes o las viguetas de los techos.

Battery A device that produces electricity (DC power) from the interaction of metals and acid.

Pila Dispositivo que genera electricidad (corriente continua) por la interacción entre metales y el ácido.

Bearing A device that surrounds a rotating shaft and provides a low-friction contact surface to reduce wear from the rotating shaft.

Cojinete Dispositivo que rodea un árbol giratorio y provee una superficie de contacto de baja fricción para disminuir el desgaste de dicho árbol.

Bearing washout A cleaning of the compressor's bearing surfaces, which causes lack of lubrication. It is usually caused by liquid refrigerant mixing with the compressor's crankcase oil due to liquid floodback or migration.

Derrubio del cojinete Una limpieza de las superficies del cojinete del compresor, lo que lleva a falta de lubricación. Generalmente causado por refrigerante líquido mezclado con aceite del cárter del cigüeñal del compresor debido a inundación o migración del líquido.

Bellows An accordion-like device that expands and contracts when internal pressure changes.

Fuelles Dispositivo en forma de acordeón con pliegues que se expanden y contraen cuando la presión interna sufre cambios.

Bellows seal A method of sealing a rotating shaft or valve stem that allows rotary movement of the shaft or stem without leaking.

Cierre hermético de fuelles Método de sellar un árbol giratorio o el vástago de una válvula que permite el movimiento giratorio del árbol o del vástago sin producir fugas.

Belly-band-mount motor An electric motor mounted with a strap around the motor secured with brackets on the strap.

Motor con barriguera Motor eléctrico que se monta colocando una cincha a su alrededor y fijándola con abrazaderas.

Bending spring A coil spring that can be fitted inside or outside a piece of tubing to prevent its walls from collapsing when being formed.

Muelle de flexión Muelle helicoidal que puede acomodarse dentro o fuera de una pieza de tubería para evitar que sus paredes se doblen al ser formadas.

Bimetal Two dissimilar metals fastened together to create a distortion of the assembly with temperature changes.

Bimetal Dos metales distintos fijados entre sí para producir una distorción del conjunto al ocurrir cambios de temperatura.

Bimetal strip Two dissimilar metal strips fastened back-to-back.

Banda bimetálica Dos bandas de metales distintos fijadas entre sí en su parte posterior.

Binary Consisting of 1s and 0s. The 1s and 0s represent numbers, words, or signals that can be stored in the computer's memory for future use. Calculators and computers are digital systems.

Binario Consiste de 1s y 0s. Los 1s y 0s representan números, palabras o señales que pueden almacenarse en la memoria de la computadora para uso futuro. Las calculadoras y computadoras son sistemas digitales.

Biofilm Consists of a slimy bacterial secretion or film that forms a protective canopy to protect the bacteria beneath it from chemical biocides in cooling tower and boiler water systems.

Biofilm Consiste en una secreción bacteriana con aspecto de película suave, que forma una cubierta protectora que resguarda a la bacteria en su interior de sustancias químicas biocidas en sistemas de enfriamiento en torre y calentadores de agua.

Bleeding Allowing pressure to move from one pressure level to another very slowly.

Sangradura Proceso a través del cual se permite el movimiento de presión de un nivel a otro de manera muy lenta.

Bleed valve A valve with a small port usually used to bleed pressure from a vessel to the atmosphere.

Válvula de descarga Válvula con un conducto pequeño utilizado normalmente para purgar la presión de un depósito a la atmósfera.

Blocked suction A method of cylinder unloading. The suction-line passage to a cylinder in a reciprocating compressor is blocked, thus causing that cylinder to stop pumping.

Aspiración obturada Método de descarga de un cilindro. El paso del conducto de aspiración a un cilindro en un compresor alternativo se obtura, provocando así que el cilindro deje de bombear.

Blowdown A system in a cooling tower whereby some of the circulating water is bled off and replaced with fresh water to dilute the sediment in the sump.

Vaciado Sistema en una torre de refrigeración por medio del cual se purga parte del agua circulante y se reemplaza con agua fresca para diluir el sedimento en el sumidero.

Boiler A container in which a liquid may be heated using any heat source. When the liquid is heated to the point that vapor forms and is used as the circulating medium, it is called a steam boiler.

Caldera Recipiente en el que se puede calentar un líquido utilizando cualquier fuente de calor. Cuando se calienta el líquido al punto en que se produce vapor y se utiliza éste como el medio para la circulación, se llama caldera de vapor.

Boiling point The temperature level of a liquid at which it begins to change to a vapor. The boiling temperature is controlled by the vapor pressure above the liquid.

Punto de ebullición El nivel de temperatura de un líquido al que el líquido empieza a convertirse en vapor. La temperatura de ebullición se regula por medio de la presión del vapor sobre líquido.

Boiling temperature The boiling temperature of the liquid can be controlled by controlling the pressure. The standard boiling pressure for water is an atmospheric pressure of 29.92 in. Hg (mercury) where water boils at 212°F.

Temperatura de ebullición La temperatura de ebullición de un líquido puede controlarse controlando la presión. La presión de ebullición estándar para agua es una presión atmosférica de 29.92 pulgadas de mercurio donde el agua hierve a 212°F.

Bonded roof A roof that is insured against damage resulting from leaks.

Techo impermeabilizado Techo que está protegido contra daños resultantes de pérdidas.

Booster compressor The compressor on a parallel compressor system that is dedicated to the coldest evaporators.

Compresor elevador Compresor montado en un sistema de compresión en paralelo, destinado a los evaporadores más fríos.

Booster pump An additional pump that is used to build the pressure above what the primary pump can accomplish.

Bomba promotora Una bomba adicional que se usa para aumentar la presión por encima de lo que la bomba primaria puede.

Boot The connection between the branch line duct and the floor, wall, or ceiling register. It transitions the branch duct to the register size. It may be from rectangular to another size of rectangle or from round to rectangular.

Manguito La conexión entre el conducto de la línea ramal y el contador en el piso, la pared o en el techo. Él hace la transición de la línea ramal al tamaño del contador. Puede conectar un rectángulo a otro rectángulo de tamaño diferente o círculo a un rectángulo.

Bore The inside diameter of a cylinder.

Calibre Diámetro interior de un cilindro.

Bourdon tube C-shaped tube manufactured of thin metal and closed on one end. When pressure is increased inside, it tends to straighten. It is used in a gauge to indicate pressure.

Tubo Bourdon Tubo en forma de C fabricado de metal delgado y cerrado en uno de los extremos. Al aumentarse la presión en su interior, el tubo tiende a enderezarse. Se utiliza dentro de un manómetro para indicar la presión.

Boyle's Law Gas law that that states an inverse relationship between pressure and volume at a constant temperature.

Ley de Boyle Ley del Gas, que establece una relación inversa entre la presión y el volumen a una temperatura constante.

Brazing High-temperature (above 800°F) melting of a filler metal for the joining of two metals.

Broncesoldadura El derretir de un metal de relleno para fusionar dos metales a temperaturas altas (sobre los 800°F o 430°C).

Breaker A heat-activated electrical device used to open an electrical circuit to protect it from excessive current flow.

Interruptor Dispositivo eléctrico activado por el calor que se utiliza para abrir en circuito eléctrico a fin de protegerlo de un flujo excesivo de corriente.

British thermal unit (Btu) The amount (quantity) of heat required to raise the temperature of 1 lb of water 1°F.

Unidad térmica británica Cantidad de calor necesario para elevar en 1°F (−17.56°C) la temperatura de una libra inglesa de agua.

Btu Abbreviation for British thermal unit.

Btu Abreviatura de unidad térmica británica.

Bubble point The refrigerant temperature at which bubbles begin to appear in a saturated liquid. Bubble point values are used to calcuate subcooling on systems that operate with blended refrigerants.

Punto de burbujear La temperatura refrigerada en que burbujas empiezan a aparecer en un liquído saturado. Los valores del punto de burbujeo se utilizan para calcular el subenfriamiento en sistemas que utilizan refrigerantes mixtos.

Building Performance Institute (BPI) A nationally recognized energy certification for home energy auditors and contractors. A non-profit organization that accredits auditors, contractors, and other building professionals that specialize in evaluating building systems and potential energy savings in homes.

Instituto del rendimiento de la construcción (BPI, por sus siglas en inglés) Organización nacional de certificación de energía para auditores y contratistas de energía para viviendas residenciales. Organización sin fines de lucro que acredita a auditores, contratistas y otros profesionales de la construcción que se especializan en la evaluación de los sistemas de construcción y el ahorro potencial de energía en viviendas residenciales.

Bulb, sensor The part of a sealed automatic control used to sense temperature.

Bombilla sensora Pieza de un regulador automático sellado que se utiliza para advertir la temperatura.

Burner A device used to prepare and burn fuel.

Quemador Dispositivo utilizado para la preparación y la quema de combustible.

Burr Excess material remaining on the end of a piece of tubing or pipe after a cut has been made. This burr must be removed.

Rebaba Exceso de material que queda en el extremo de un pedazo de una tubería después de hacerse un corte. Esta rebaba debe removerse.

Butane gas A liquefied petroleum gas burned for heat.

Gas butano Gas licuado derivado del petróleo que se quema para producir calor.

Butterfly valve Valve constructed with a disk in the valve chamber that can be adjusted across the water or fluid stream. Imagine a coin that has a handle that can be turned crossways of the water or fluid flow to adjust or stop the flow. The valve handle only travels 90°, similar to the ball valve.

Válvula en forma de mariposa Válvula construída con un disco en la cámara de la válvula que se puede ajustar a través del corriente del agua o fluido. Imagina una moneda con un mango que se puede girar transversal al flujo del agua o fluido para ajustar o parar el flujo. El mango de la válvula solamente pasa 90°, tal como la válvula de bola.

B-vent A double-walled venting system for gas stoves or other gas appliances using approved piping through the roof or into a chimney.

Evacuación tipo B Sistema de evacuación de doble pared para estufas u otros dispositivos que funcionan con gas, utilizando conductos apropiados a través del techo o que se desembocan en una chimenea.

Bypass humidifier A duct-mounted humidifier that relies on the pressure difference between the supply and return air to move air through it.

Humidificador de derivación Un humidificador instalado en un conducto que utiliza la diferencia de presión entre el aire de suministro y el de retorno para hacer circular el aire a través del humidificador.

C

Cad cell A device containing cadmium sulfide used to prove the flame in an oil-burning furnace or boiler.

Celda de cadmio Dispositivo que contiene sulfuro de cadmio utilizado para probar la llama en un horno o caldera de aceite pesado.

Calibration The adjustment of instruments or gauges to the correct setting for known conditions.

Calibración El ajuste de instrumentos o manómetros en posición correcta para su operación en condiciones conocidas.

Calorimeter An instrument of laboratory-grade quality used to measure heat absorbed into a substance.

Calorímetro Instrumento de calidad laboratorio que sirve para medir el calor absorbido por una sustancia.

Capacitance The term used to describe the electrical storage ability of a capacitor.

Capacitancia Término utilizado para describir la capacidad de almacenamiento eléctrico de un capacitador.

Capacitive circuit Electrical circuit that contains capacitors. In this type of circuit, the current leads the voltage.

Circuito capacitivo Un circuito electrico que tiene capacitadores. En esta clase de circuito, la corriente mueve el voltaje.

Capacitive reactance Term used to describe a capacitor's resistance to electric current flow.

Reactancia capacitiva un término usado para describer la resistencia de un condensador al flujo de corriente eléctrica.

Capacitor An electrical storage device used to start motors (start capacitor) and to improve the running efficiency of motors (run capacitor).

Capacitador Dispositivo de almacenamiento eléctrico utilizado para arrancar motores (capacitador de arranque) y para mejorar el rendimiento de motores (capacitador de funcionamiento).

Capacitor-start–capacitor-run motor A single-phase motor that has a start capacitor in series with the start winding that is disconnected after start-up and a run capacitor that is also in parallel with the start windings that stays in the circuit while running. This run capacitor is built for full-time duty and is used to increase the running efficiency of the motor.

Motor de arranque capacitivo–marcha capacitiva Motor de fase sencilla que tiene un condensador de arranque en serie con la bobina de arranque que se desconecta después del encendido, y un condensador de marcha que está en paralelo con las bobinas de arranque y que permanece en el circuito mientras está en marcha. Este condensador está hecho para servicio a tiempo completo y se utiliza para mejorar la eficiencia de marcha del motor.

Capacitor-start motor A single-phase motor with a start and run winding that has a capacitor in series with the start winding, which remains in the circuit until the motor gets up to about 75% of the motor's rated speed.

Motor de arranque capacitivo Motor de fase sencilla con una bobina de arranque y marcha que tiene un condensador en serie con la bobina de arranque, el cual permanece en el circuito hasta que el motor alcanza alrededor de 75% de la velocidad calificada del motor.

Capacity The rating system of equipment used to heat or cool substances.

Capacidad Sistema de clasificación de equipo utilizado para calentar o enfriar sustancias.

Capillary attraction The attraction of a liquid material between two pieces of material such as two pieces of copper or copper and brass. For instance, in a joint made up of copper tubing and a brass fitting, the solder filler material has a greater attraction to the copper and brass than to itself and is drawn into the space between them.

Atracción capilar Atracción de un material líquido entre dos piezas de material, como por ejemplo dos piezas de cobre o cobre y latón. Por ejemplo, en una junta fabricada de tubería de cobre y un accesorio de latón, el material de relleno de la soldadura tiene mayor atracción al cobre y al latón que a sí mismo y es arrastrado hacia el espacio entre éstos.

Capillary tube A fixed-bore metering device. This is a small-diameter tube that can vary in length from a few inches to several feet. The amount of refrigerant flow needed is predetermined and the length and diameter of the capillary tube is sized accordingly.

Tubo capilar Dispositivo de medición de calibre fijo. Éste es un tubo de diámetro pequeño cuyo largo puede oscilar entre unas cuantas pulgadas y varios pies. La cantidad de flujo de refrigerante requerida es predeterminada y, de acuerdo a esto, se fijan el largo y el diámetro del tubo capilar.

Carbon dioxide A by-product of natural gas combustion that is not harmful, but does contribute to global warming. Also a refrigerant known as R-744.

Bióxido de carbono Subproducto de la combustión del gas natural que no es nocivo. También un refrigerante conocido como R-744.

Carbon monoxide A poisonous, colorless, odorless, tasteless gas generated by incomplete combustion.

Monóxido de carbono Gas mortífero, inodoro, incoloro e insí pido que se desprende en la combustión incompleta del carbono.

Carrier frequency The speed at which the variable frequency drive's power device switches the waveform's positive and negative halves on and off. Also referred to as the switch frequency.

Frecuencia de onda portadora La velocidad a la que el dispositivo de alimentación de frecuencia variable intercambia las mitades positiva y negativa de una onda se denomina frecuencia de intercambio o frecuencia de onda portadora. Diseño sustentable: se adecua a las necesidades actuales sin perjuicio de que las generaciones futuras tengan la posibilidad de cubrir sus propias necesidades.

Catalytic combustor stove A stove that contains a cell-like structure consisting of a substrate, washcoat, and catalyst that produces a chemical reaction causing pollutants to be burned at much lower temperatures.

Estufa de combustor catalítico Estufa con una estructura en forma de celda compuesta de una subestructura, una capa brochada y un catalizador que produce una reacción química. Esta reacción provoca la quema de contaminantes a temperaturas muchas más bajas.

Cathode A terminal or connection point on a semiconductor.

Cátodo Punto de conexión o terminal en un semiconductor.

Cavitation A vapor formed due to a drop in pressure in a pumping system. Vapor at a pump inlet may be caused at a cooling tower if the pressure is low and water is turned to vapor.

Cavitación Vapor producido como consecuencia de una caída de presión en un sistema de bombeo. El vapor a la entrada de una bomba puede ser producido en una torre de refrigeración si la presión es baja y el agua se convierte en vapor.

Cellulose A substance formed in wood plants from glucose or sugar.

Celulosa Sustancia presente en plantas de madera y que se forma a partir de glucosa y azúcar.

Celsius scale A temperature scale with 100-degree graduations between water freezing (0°C) and water boiling (100°C).

Escala Celsio Escala dividida en cien grados, con el cero marcado a la temperatura de fusión del hielo (0°C) y el cien a la de ebullición del agua (100°C).

Centigrade scale See Celsius scale.

Centígrado Véase escala Celsio.

Centrifugal compressor A kinetic displacement compressor used for large refrigeration systems that uses centrifugal force to accomplish compression. It is not positive displacement, but it is similar to a blower.

Compresor centrífugo Compresor de desplazamiento cinético utilizado en sistemas grandes de refrigeración que usa forza centrifuga para lograr compresión. No es desplazamiento positivo, pero es similar a un soplador.

Centrifugal pump A pump that uses centrifugal force to move a fluid. An impeller is rotated rapidly within the pump, causing the fluid to fly away from the center, which forces the fluid through a piping system.

Bomba centrífuga Bomba que utiliza la fuerza centrífuga para desplazar un fluido. Un propulsor dentro de la bomba gira a alta velocidad alejando el líquido del centro e impulsándolo a través de una tubería.

Centrifugal switch A switch that uses a centrifugal action to disconnect the start windings from the circuit.

Conmutador centrífugo Conmutador que utiliza una acción centrífuga para desconectar los devanados de arranque del circuito.

CFM cubic feet per minute.

CFM Pies cúbicos por minuto.

CFM_{50} The air flow measured in cubic feet per minute (CFM) through a blower door while maintaining a constant vacuum of 50 Pascals inside a home is referred to as CFM_{50}.

CFM_{50} El flujo de aire, medido en pies cúbicos por minuto (CFM), que pasa a través de una abertura en la que se coloca un extractor potente que mantiene un vacío constante de 50 Pascales dentro de una vivienda se denomina CFM_{50}.

Change of state The condition that occurs when a substance changes from one physical state to another, such as ice to water and water to steam.

Cambio de estado Condición que ocurre cuando una sustancia cambia de un estado físico a otro, como por ejemplo el hielo a agua y el agua a vapor.

Charge of refrigerant The quantity of refrigerant in a system.

Carga de refrigerante Cantidad de refrigerante en un sistema.

Charging curve A graphical method of assisting a service technician with charging an air-conditioning or heat pump system.

Curva de recarga Un método gráfico para asistir al técnico de servicio con el recargar de un sistema de aire acondicionado o de bomba de calor.

Charging cylinder A device that allows the technician to accurately charge a refrigeration system with refrigerant.

Cilindro cargador Dispositivo que le permite al mecánico cargar correctamente un sistema de refrigeración con refrigerante.

Charging scale A scale used to weigh refrigerant when charging a refrigeration or air-conditioning system.

Báscula de plancha Báscula que se utiliza para pesar el refrigerante durante la carga de un sistema de refrigeración o de aire acondicionado.

Charles' law Gas law that that states a direct relationship between temperature and volume at a constant pressure.

Ley de Charles la ley de gas que establece una relación directa entre la temperatura y el volumen con una presión constante.

Check valve A device that permits fluid flow in one direction only.

Válvula de retención Dispositivo que permite el flujo de fluido en una sola dirección.

Chilled-water system An air-conditioning system that circulates refrigerated water to the area to be cooled. The refrigerated water picks up heat from the area, thus cooling the area.

Sistema de agua enfriada Sistema de acondicionamiento de aire que hace circular agua refrigerada al área que será enfriada. El agua refrigerada atrapa el calor del área y la enfria.

Chiller purge unit A system that removes air or noncondensables from a low-pressure chiller.

Unidad enfriadora de purga Sistema que remueve el aire o sustancias no condensables de un enfriador de baja presión.

Chill factor A factor or number that is a combination of temperature, humidity, and wind velocity that is used to compare a relative condition to a known condition.

Factor de frío Factor o número que es una combinación de la temperatura, la humedad y la velocidad del viento utilizado para comparar una condición relativa a una condición conocida.

Chimney A vertical shaft used to convey flue gases above the rooftop.

Chimenea Cañón vertical utilizado para conducir los gases de combustión por encima del techo.

Chimney draft Another name for stack draft.

Corriente de chimenea Denominación alternativa para "corriente de ventilación".

Chimney effect A term used to describe air or gas when it expands and rises when heated.

Efecto de chimenea Término utilizado para describir el aire o el gas cuando se expande y sube al calentarse.

Chlorofluorocarbons (CFCs) Those refrigerants thought to contribute to the depletion of the ozone layer.

Clorofluorocarburos (CFCs en inglés) Líquidos refrigerantes que, según algunos, han contribuido a la reducción de la capa de ozono.

Circuit An electron or fluid-flow path that makes a complete loop.

Circuito Electrón o trayectoria del flujo de fluido que hace un ciclo completo.

Circuit breaker A device that opens an electric circuit when an overload occurs.

Interruptor para circuitos Dispositivo que abre un circuito eléctrico cuando ocurre una sobrecarga.

Clamp-on ammeter An instrument that can be clamped around one conductor in an electrical circuit and measure the current.

Amperímetro fijado con abrazadera Instrumento que puede fijarse con una abrazadera a un conductor en un circuito eléctrico y medir la corriente.

Clearance volume The volume at the top of the stroke in a reciprocating compressor cylinder between the top of the piston and the valve plate.

Volumen de holgura Volumen en la parte superior de una ca rrera en el cilindro de un compresor recíproco entre la parte superior del pistón y la placa de una válvula.

Closed circuit A complete path for electrons to flow on.

Circuito cerrado Circuito de trayectoria ininterrumpida que permite un flujo continuo de electrones.

Closed-circuit cooling tower May have a wet/dry mode, an adiabatic mode, and a dry mode.

Torre de enfriamiento de circuito cerrado Puede tener un modo seco/mojado, un modo adiabático y un modo seco.

Closed loop Piping circuit that is complete and not open to the atmosphere.

Ciclo cerrado Circuito de tubería completo y no abierto a la atmósfera.

Closed-loop control configuration A closed-loop control configuration can maintain the controlled variable, like the air in a duct, at its set point because the sensor is now located in the air being heated or cooled (controlled medium).

Configuración de control de circuito cerrado Una configuración de control de circuito cerrado puede mantener el variable controlado, por ejemplo el aire en un conducto, en su punto de ajuste porque el sensor ya está ubicado en el aire que se calienta o se enfría (el medio controlado).

Closed-loop heat pump Heat pump system that reuses the same heat transfer fluid, which is buried in plastic pipes within the earth or within a lake or pond for the heat source.

Bomba de calor de circuito cerrado Sistema de bomba de calor que reutiliza el mismo líquido de transferencia térmica que se encuentra en tubos de plástico enterrados o sumergidos en un lago o estanque de los que obtiene el calor.

CO_2 indicator An instrument used to detect the quantity of carbon dioxide in flue gas for efficiency purposes.

Indicador del CO_2 Instrumento utilizado para detectar la cantidad de bióxido de carbono en el gas de combustión a fin de lograr un mejor rendimiento.

Coaxial heat exchanger A tube-within-a-tube liquid heat exchanger. Typically it is used for water-source heat pumps and small water-cooled air conditioners.

Intercambiador de calor coaxial Intercambiador de calor líquido con un tubo dentro de un tubo. Típicamente se usa para bombas de calor en fuentes de agua y acondicionadoras de aire pequeñas enfriadas por agua.

Code The local, state, or national rules that govern safe installation and service of systems and equipment for the purpose of safety of the public and trade personnel.

Código Reglamentos locales, estatales o federales que rigen la instalación segura y el servicio de sistemas y equipo con el propósito de garantizar la seguridad del personal público y profesional.

Coefficient of performance (COP) The ratio of usable output energy divided by input energy.

Coeficiente de rendimiento (COP en inglés) Relación de la de energía de salida utilizable dividida por la energía de entrada.

Cold The word used to describe heat at lower levels of intensity.

Frío Término utilizado para describir el calor a niveles de intensidad más bajos.

Cold anticipator A fixed resistor in a thermostat that is wired in parallel with the cooling contacts. This starts the cooling system before the thermostat calls for cooling, which allows the system to get up to capacity before the cooling is actually needed.

Anticipador de frío Resistor fijo en un termostato, conectado en paralelo con los contactos de enfriamiento. Dicho resistor pone en marcha el sistema de enfriamiento antes de que lo haga el termostato, permitiendo así que el sistema alcance su plena capacidad antes de que realmente se necesite el enfriamiento.

Cold junction The opposite junction to the hot junction in a thermocouple.

Empalme frío El empalme opuesto al empalme caliente en un termopar.

Cold trap A device to help trap moisture during the evacuation of a refrigeration system.

Trampa del frío Dispositivo utilizado para ayudar a atrapar la humedad durante la evacuación de un sistema de refrigeración.

Cold wall The term used in comfort heating to describe a cold outside wall and its effect on human comfort.

Pared fría Término utilizado en la calefacción para comodidad que describe una pared exterior fría y sus efectos en la comodidad de una persona.

Collector A terminal on a semiconductor.

Colector Punto terminal en un semiconductor.

Combustion A reaction called rapid oxidation or burning produced with the right combination of a fuel, oxygen, and heat.

Combustión Reacción conocida como oxidación rápida o quema producida con la combinación correcta de combustible, oxígeno y calor.

Combustion air Primary and/or secondary air. Primary air enters the furnace before the combustion process takes place. Secondary air is air introduced into a furnace after combustion and supports combustion.

Aire de combustión Aire primario y/o secundario. El aire primario entra a la estufa antes de que se produzca el proceso de combustión. El aire secundario se introduce en la estufa luego de la combustión, para que ésta se mantenga.

Combustion analyzer An instrument used to measure oxygen concentrations within flue gases. This analyzer can test smoke and test for carbon monoxide and other gases.

Analizador de combustión Instrumento que se utiliza para medir la concentración del oxígeno en los gases de escape. Este tipo de analizador permite medir el contenido de monóxido de carbono y otros gases en los humos.

Combustion appliance Any heating appliance that incorporates the combustion process which may include: gas and oil furnaces and boilers, water heaters, stoves, fireplaces and space heaters.

Dispositivo de combustión Cualquier dispositivo de calefacción que utiliza el proceso de combustión, entre los que se encuentran: estufas y calderas a gas o combustible líquido, calentadores de agua, estufas portátiles, chimeneas y calentadores para espacios confinados.

Combustion appliance zone (CAZ) An area or zone within a residence or building where combustion appliances reside and operate.

Zona de dispositivos de combustión (CAZ, por sus siglas en inglés) El área o zona dentro de una vivienda o construcción donde se encuentran y funcionan los dispositivos de combustión.

Combustion blower motor A motor mounted either before or after the heat exchanger of a furnace or boiler which turns a fan to either push or pull combustion gasses through the heat exchanger.

Motor de ventilación de combustión Motor montado antes o después del intercambiador de calor de una estufa o caldera, que enciende un ventilador que expele o succiona los gases de combustión a través del intercambiador de calor.

Comfort People are said to be comfortable when they are not aware of the ambient air surrounding them. They do not feel cool or warm or sweaty.

Comodidad Se dice que unas personas están cómodas cuandos no están conscientes del aire ambiental que les rodea. No sienten frío i calor, ni están sudadas.

Comfort chart A chart used to compare the relative comfort of one temperature and humidity condition to another condition.

Esquema de comodidad Esquema utilizado para comparar la comodidad relativa de un a condición de temperatura y humedad a otra condición.

Compound gauge A gauge used to measure the pressure above and below the atmosphere's standard pressure. It is a Bourdon tube sensing device and can be found on all gauge manifolds used for air-conditioning and refrigeration service work.

Manómetro compuesto Manómetro utilizado para medir la presión mayor y menor que la presión estándar de la atmósfera. Es un dispositivo sensor de tubo Bourdon que puede encontrarse en todos los manómetros de línea utilizados para el servicio de sistemas de acondicionamiento de aire y de refrigeración.

Compounds Substances that form when different elements combine chemically.

Compuestos sustancias que se forman cuando los diferentes elementos se combinan químicamente.

Compression A term used to describe a vapor when pressure is applied and the molecules are compacted closer together.

Compresión Término utilizado para describir un vapor cuando se aplica presión y se compactan las moléculas.

Compression ratio A term used with compressors to describe the ratio between the high- and low-pressure sides of the compression cycle. It is calculated by dividing the absolute discharge pressure by the absolute suction pressure.

Relación de compresión Término utilizado con compresores para describir la relación entre los lados de baja y alta presión del ciclo de compresión. Se calcula dividiendo la presión absoluta de descarga por la presión absoluta de aspiración.

Compressor A vapor pump that pumps vapor (refrigerant or air) from one pressure level to a higher pressure level.

Compresor Bomba de vapor que bombea el vapor (refrigerante o aire) de un nivel de presión a un nivel de presión más alto.

Compressor crankcase The internal part of the compressor that houses the crankshaft and lubricating oil.

Cárter del compresor Parte interna del compresor donde se aloja el cigüeñal y el aceite de lubricación.

Compressor displacement The internal volume of a compressor's cylinders, used to calculate the pumping capacity of the compressor.

Desplazamiento del compresor Volumen interno de los cilindros de un compresor, utilizado para calcular la capacidad de bombeo del mismo.

Compressor head The component that sits on top of the compressor cylinder and holds the components together.

Cabeza del compresor El componente que está en la parte superior del cilindro y que mantiene a los componentes unidos.

Compressor oil cooler One or more piping systems used for cooling the crankcase oil.

Enfriador del aceite del compresor Uno o varios sistemas de tubería que sirven para enfriar el aceite del cárter.

Compressor shaft seal The seal that prevents refrigerant inside the compressor from leaking around the rotating shaft.

Junta de estanqueidad del árbol del compresor La junta de estanqueidad que evita la fuga, alrededor del árbol giratorio, del refrigerante en el interior del compresor.

Concentrator That part of an absorption chiller where the dilute salt solution is boiled to release the water and concentrate the solution.

Concentrador El lugar en el enfriador por absorción donde se hierve la solución salina diluida para liberar el agua y concentrar la solución.

Condensate The moisture that forms on an evaporator coil.

Condensado Humedad que se forma sobre la bobina de un evaporador.

Condensate pump A small pump used to pump condensate to a higher level.

Bomba para condensado Bomba pequeña utilizada para bombear el condensado a un nivel más alto.

Condensation Liquid formed when a vapor condenses.

Condensación El líquido formado cuando se condensa un vapor.

Condense Changing a vapor to a liquid.

Condensar Convertir un vapor en líquido.

Condenser The component in a refrigeration system that transfers heat from the system by condensing refrigerant.

Condensador Componente en un sistema de refrigeración que transmite el calor del sistema al condensar el refrigerante.

Condenser flooding An automatic method of maintaining the correct head pressure in mild weather by using refrigerant from an auxiliary receiver.

Inundación del condensador Método automático de mantener una presión correcta en la cabeza en un tiempo utilizando refrigerante de un receptor auxiliar.

Condenser splitting Split the condenser into two separate and identical condenser circuits to reduce the amount of extra refrigerant charge needed for changing outdoor ambient conditions.

División del condensador División del condensador en dos circuitos de condensación idénticos pero separados, para reducir la carga extra de refrigerante necesaria para cambiar las condiciones ambientales del exterior.

Condensing boiler High-efficiency boiler designed to operate with low flue-gas temperatures. These boilers allow for the condensing of flue gas.

Caldera para condensación de gas Una caldera muy eficiente diseñada para operar cuando la temperatura del gas de combustión está baja. Estas calderas permiten la condensación de los gases de combustión.

Condensing gas furnace A furnace with a condensing heat exchanger that condenses moisture from the flue gases, resulting in greater efficiency.

Horno para condensación de gas Horno con un intercambiador de calor para condensación que condensa la humedad de los gases de combustión. El resultado será un mayor rendimiento.

Condensing oil furnace An oil furnace with a coil that causes moisture to condense from the flue gases, producing latent heat. This increases the efficiency of the furnace.

Horno de petróleo con serpentín de condensación Horno de petróleo, con un serpentín que provoca la condensación de la humedad presente en los gases de combustión, produciendo un calor latente. Permite aumentar la eficacia del horno.

Condensing pressure The pressure that corresponds to the condensing temperature in a refrigeration system.

Presión para condensación La presión que corresponde a la temperatura de condensación en un sistema de refrigeración.

Condensing temperature The temperature at which a vapor changes to a liquid.

Temperatura de condensación Temperatura a la que un vapor se convierte en líquido.

Condensing unit A complete unit that includes the compressor and the condensing coil.

Conjunto del condensador Unidad completa que incluye el compresor y la bobina condensadora.

Conditioned space Areas of a structure or residence that are heated or cooled.

Espacio acondicionado Áreas de una estructura o residencia cuyo ambiente se acondiciona mediante frío o calor.

Conduction Heat transfer from one molecule to another within a substance or from one substance to another.

Conducción Transmisión de calor de una molécula a otra dentro de una sustancia o de una sustancia a otra.

Conductivity The ability of a substance to conduct electricity or heat.

Conductividad Capacidad de una sustancia de conducir electricidad o calor.

Conductor A path for electrical energy to flow on.

Conductor Trayectoria que permite un flujo continuo de energía eléctrica.

Connecting rod A rod that connects the piston to the crankshaft.

Barra conectiva Barra que conecta al pistón con el cigüeñal.

Contactor A larger version of the relay. It can be repaired or rebuilt and has movable and stationary contacts.

Contactador Versión más grande del relé. Puede ser reparado o reconstruido. Tiene contactos móviles y fijos.

Contaminant Any substance in a refrigeration system that is foreign to the system, particularly if it causes damage.

Contaminante Cualquier sustancia en un sistema de refrigeración extraña a éste, principalmente si causa averías.

Continuous coil voltage This is the maximum back electromotive force (BEMF) that the relay's coil can tolerate "continuously" without overheating and opening circuit.

Voltaje continuo de serpentín Es la fuerza contra electromotriz máxima (BEMF, por sus siglas en inglés) que puede soportar el serpentín del relé de manera continua sin que se produzca sobrecalentamiento y apertura del circuito.

Control A device for stopping, starting, or modulating the flow of electricity or fluid to maintain a preset condition.

Regulador Dispositivo para detener, poner en marcha o modular el flujo de electricidad o de fluido a fin de mantener una condición establecida con anticipación.

Control agent The control agent is the fluid that transfers energy or mass to the controlled medium. It is the control agent (such as hot or cold water) that will flow through a final control device (water valves).

Agente de control El agente de control es el fluido que transfiere energía o masa al medio controlado. Es el agente (como agua caliente o fría) que fluye por un dispositivo de regulación (válvulas de agua).

Controlled device May be any control that stops, starts, or modulates fuel, fluid flow, or air to provide expected conditions in the conditioned space.

Aparato controlado Un aparato que puede ser cualquier control que detiene, enciende o modula el combustible, flujo de fluido o aire para proveer las condiciones esperadas en un espacio acondicionado.

Controlled environment The conditioned space the HVACR system is trying to maintain.

Medio ambiente controlado El espacio acondicionado que el sistema HVACR trata de conservar.

Controlled medium The controlled medium is what absorbs or releases the energy or mass transferred to or from the HVACR process. Controlled mediums can be air in a duct or water in a pipe.

Medio controlado El medio controlado absorbe o libera la energía o la masa transferidas al proceso HVACR o del mismo proceso. Un medio controlado puede ser el aire en un conducto o el agua en un tubo.

Controlled output device Any device of an array of dampers, fans, variable-frequency drives (VFDs), cooling coil valves, heating coil valves, relays, and motors—all of which control the controlled environment.

Aparato de salida controlada Una serie de desviadores, abanicos, propulsiones de frecuencia variable (VFDs en inglés), válvulas del serpentín de enfriamiento, válvulas del serpentín de calefacción, relés y motores—que controlan el medio ambiente controlado.

Controller A device that provides the output to the controlled device.

Controlador Aparato que provee la salida para el aparato controlado.

Control loop A sensor, a controller, and a controlled device.

Circuito de control Un sensor, un controlador y un aparato controlado.

Control point The actual value of the controlled variable at a particular point in time.

Punto de control El valor verdadero del variable controlado en un tiempo específico.

Control system A network of controls to maintain desired conditions in a system or space.

Sistema de regulación Red de reguladores que mantienen las condiciones deseadas en un sistema o un espacio.

Convection Heat transfer from one place to another using a fluid.

Convección Transmisión de calor de un lugar a otro por medio de un fluido.

Conventional boilers Boilers that are classified as non-condensing and have flue gas temperatures above 130°F.

Calderas convencionales las calderas que son clasificados como no-condensación y tienen temperaturas de gases de combustión mas alto de 130°F.

Conversion factor A number used to convert from one equivalent value to another.

Factor de conversión Número utilizado en la conversión de un valor equivalente a otro.

Cooler A walk-in or reach-in refrigerated box.

Nevera Caja refrigerada donde se puede entrar o introducir la mano.

Cooling tower The final device in many water-cooled systems, which rejects heat from the system into the atmosphere by evaporation of water.

Torre de refrigeración Dispositivo final en muchos sistemas enfriados por agua, que dirige el calor del sistema a la atmósfera por medio de la evaporación de agua.

Copper plating Small amounts of copper are removed by electrolysis and deposited on the ferrous metal parts in a compressor.

Encobrado Remoción de pequeñas cantidades de cobre por medio de electrólisis que luego se colocan en las piezas de metal férreo en un compresor.

Corrosion A chemical action that eats into or wears away material from a substance.

Corrosión Acción química que carcoma o desgasta el material de una sustancia.

Cotter pin Used to secure a pin. The cotter pin is inserted through a hole in the pin, and the ends spread to retain it.

Pasador de chaveta Se usa para asegurar una clavija. El pasador se inserta a través de un roto en la clavija y sus extremos se abren para asegurarla.

Counter EMF Voltage generated or induced above the applied voltage in a single-phase motor.

Contra EMF Tensión generada o inducida sobre la tensión aplicada en un motor unifásico.

Counterflow Two fluids flowing in opposite directions.

Contraflujo Dos fluidos que fluyen en direcciones opuestas.

Coupling A device for joining two fluid-flow lines. Also the device connecting a motor driveshaft to the driven shaft in a direct-drive system.

Acoplamiento Dispositivo utilizado para la conexión de dos conductos de flujo de fluido. Es también el dispositivo que conecta un árbol de mando del motor al árbol accionado en un sistema de mando directo.

CPVC (Chlorinated polyvinyl chloride) Plastic pipe similar to PVC except that it can be used with temperatures up to 180°F at 100 psig.

CPVC (Cloruro de polivinilo clorado) Tubo plástico similar al PVC, pero que puede utilizarse a temperaturas de hasta 180°F (82°C) a 100 psig [indicador de libras por pulgada cuadrada].

Crackage Small spaces in a structure that allow air to infiltrate the structure.

Formación de grietas Espacios pequeños en una estructura que permiten la infiltración del aire dentro de la misma.

Cradle-mount motor A motor with a mounting cradle that fits the motor end housing on each end and is held down with a bracket.

Motor montado con cuña Motor equipado de una cuña adaptada a la caja en los dos extremos y sujetado por fijaciones.

Crankcase heat Heat provided to the compressor crankcase.

Calor para el cárter del cigüeñal Calor suministrado al cárter del cigüeñal del compresor.

Crankcase pressure-regulating valve (CPR valve) A valve installed in the suction line, usually close to the compressor. It is used to keep a low-temperature compressor from overloading on a hot pulldown by limiting the pressure to the compressor.

Válvula reguladora de la presión del cárter del cigüeñal (CPR en inglés) Válvula instalada en el conducto de aspiración, normalmente cerca del compresor. Se utiliza para evitar la sobrecarga en un compresor de temperatura baja durante un arrastre caliente hacia abajo limitando la presión al compresor.

Crankshaft In a reciprocating compressor, the crankshaft changes the round-and-round motion into the reciprocating back-and-forth motion of the pistons using off-center devices called throws.

Cigüeñal En un compresor alternativo, el cigüeñal cambia el movimiento circular en un movimiento alternativo hacia delante y hacia atrás de los pistones usando aparatos descentrados llamados cigüeñas.

Crankshaft seal Same as the compressor shaft seal.

Junta de estanqueidad del árbol del cigüeñal Exactamente igual que la junta de estanqueidad del árbol del compresor.

Crankshaft throw The off-center portion of a crankshaft that changes rotating motion to reciprocating motion.

Excentricidad del cigüeñal Porción descentrada de un cigüeñal que cambia el movimiento giratorio a un movimiento alternativo.

Creosote A mixture of unburned organic material found in the smoke from a wood-burning fire.

Creosota Mezcla del material orgánico no quemado que se encuentra en el humo proveniente de un incendio de madera.

Crisper A refrigerated compartment that maintains a high humidity and a low temperature.

Encrespador Compartimiento refrigerado que mantiene una humedad alta y una temperatura baja.

Critical point A point on the pressure/enthalpy diagram where the critical temperature and critical pressure exist. No liquid refrigerant can be produced above a refrigerant's critical point.

Punto crítico Punto en el diagrama presión/entalpía en el que coinciden la temperatura crítica con la presión crítica. Ningún refrigerante líquido puede obtenerse por licuado de un gas a una temperatura por encima del punto crítico del mismo.

Critical pressure The lowest pressure at which refrigerant vapor cannot condense.

Presión crítica La presión baja a la que el vapor de refrigerante no puede condensarse.

Critical seals Refers to components that are difficult to isolate or service and, as a result, require that the system refrigerant be recovered for service.

Junta elastomérica se refiere a componentes que son difíciles de aislar o servicio y, como consecuencia, requieren que el refrigerante del sistema pueden recuperarse para el servicio.

Critical temperature The critical temperature of a gas is the highest temperature the gas can have and still be condensable by the application of pressure.

Temperatura crítica La temperatura crítica de un gas es la máxima temperatura a la cual el mismo puede condensarse mediante la aplicación de presión (licuado).

Cross charge A control with a sealed bulb that contains two different fluids that work together for a common specific condition.

Carga transversal Regulador con una bombilla sellada compuesta de dos fluidos diferentes que pueden funcionar juntos para una condición común específica.

Crossflow Term used to indicate that two fluids are flowing in different directional paths with respect to each other, usually at a 90-degree angle.

Flujo Cruzado término utilizado para indicar que dos fluidos fluyen en diferentes rutas direccionales con respecto a cada uno de otros, generalmente en un ángulo de 90 grados.

Cross liquid charge bulb A type of charge in the sensing bulb of the TXV that has different characteristics from the system refrigerant. This is designed to help prevent liquid refrigerant from flooding to the compressor at start-up.

Bombilla de carga del líquido transversal Tipo de carga en la bombilla sensora de la válvula electrónica de expansión que tiene características diferentes a las del refrigerante del sistema. La carga está diseñada para ayudar a evitar que el refrigerante líquido se derrame dentro del compresor durante la puesta en marcha.

Cross vapor charge bulb Similar to the vapor charge bulb but contains a fluid different from the system refrigerant. This is a special-type charge and produces a different temperature/pressure relationship under different conditions.

Bombilla de carga del vapor transversal Similar a la bombilla de carga del vapor pero contiene un fluido diferente al del refrigerante del sistema. Ésta es una carga de tipo especial y produce una relación diferente entre la temperatura y la presión bajo condiciones diferentes.

Crystallization When a salt solution becomes too concentrated and part of the solution turns to salt.

Cristalización Condición que ocurre cuando una solución salina se concentra demasiado y una parte de la solución se convierte en sal.

Cupronickel A material used in heat exchangers. This material is made with an alloy of copper for a higher corrosion resistance for acid cleaning.

Cuproníquel Material utilizado en intercambiadores térmicos. Se trata de una aleación de cobre y níquel con una elevada resistencia a la corrosión en procesos de limpieza con ácido.

Curb Frame that is used to mount and support HVACR equipment.

Marco Perfil que se utiliza para montar y sostener equipos de calefacción, ventilación, acondicionamiento de aire y refrigeración.

Current, electrical Electrons flowing along a conductor.

Corriente eléctrica Electrones que fluyen a través de un conductor.

Current relay An electrical device activated by a change in current flow.

Relé para corriente Dispositivo eléctrico accionado por un cambio en el flujo de corriente.

Current sensing relay An inductive relay coil usually located around a wire used to sense current flowing through the wire. Its action usually opens or closes a set of contacts.

Relé detector de corriente Una bobina de relé inductiva que generalmente está ubicada cerca de un cable y se usa para detectar el flujo de corriente a través del cable. Su acción generalmente abre o cierra una serie de contactos.

Cut-in and cut-out The two points at which a control opens or closes its contacts based on the condition it is supposed to maintain.

Puntos de conexión y desconexión Los dos puntos en los que un regulador abre o cierra sus contactos según las condiciones que debe mantener.

Cycle A complete sequence of events (from start to finish) in a system.

Ciclo Secuencia completa de eventos, de comienzo a fin, que ocurre en un sistema.

Cylinder A circular container with straight sides used to contain fluids or to contain the compression process (the piston movement) in a compressor.

Cilindro Recipiente circular con lados rectos, utilizado para contener fluidos o el proceso de compresión (movimiento del pistón) en un compresor.

Cylinder, compressor The part of the compressor that contains the piston and its travel.

Cilindro del compresor Pieza del compresor que contiene el pistón y su movimiento.

Cylinder, refrigerant The container that holds refrigerant.

Cilindro del refrigerante El recipiente que contiene el refrigerante.

Cylinder head, compressor The top to the cylinder on the high-pressure side of the compressor.

Culata del cilindro del compresor Tapa del cilindro en el lado de alta presión del compresor.

Cylinder unloading A method of providing capacity control by causing a cylinder in a reciprocating compressor to stop pumping.

Descarga del cilindro Método de suministrar regulación de capacidad provocando que el cilindro en un compresor alternativo deje de bombear.

D

Dalton's Law Gas law that states that the total pressure exerted by a gas mixture is the sum of the pressures of the individual gases. Also called the law of partial pressures.

La Ley de Dalton Ley del gas, que establece que la presión total ejercida por una mezcla de gases es la suma de las presiones de los gases distintos. También, llamada la ley de las presiones parciales.

Damper A component in an air-distribution system that restricts airflow for the purpose of air balance.

Desviador Componente en un sistema de distribución de aire que limita el flujo de aire para mantener un equilibrio de aire.

Database A collection of data or information stored in the memory of the controller to be used by the digital controller at a later point in time.

Base de datos Una colección de datos o información almacenados en la memoria del controlador, para el uso del controlador digital más adelante.

DC converter A type of rectifier that changes alternating current (AC) to direct current (DC).

Convertidor CD Tipo de rectificador que cambia la corriente alterna (CA) a corriente directa (CD).

DC motor A motor that operates on direct current (DC).

Motor CD Un motor que opera con corriente directa (CD).

DCV See Demand Control Ventilation.

DCV Ver Control de demanda de ventilación.

Declination angle The angle of the tilt of the earth on its axis.

Ángulo de declinación Ángulo de inclinación de la Tierra en su eje.

Decorative appliance A term used by ANSI to indicate a gas appliance that is not designed for larger space-heating applications.

Aparato decorativo Término usado por ANSI para indicar un aparato de gas que no está diseñado para calentar espacios grandes.

Deep vacuum An attained vacuum that is below 250 microns.

Vacío profundo Un vacío que se obtiene lo cual es menor de 250 micrones.

Defrost Melting of ice.

Descongelar Convertir hielo en líquido.

Defrost condensate The condensate or water from a defrost application of a refrigeration system.

Condensado de descongelación Condensado o agua causada por el dispositivo de descongelación en un sistema de refrigeración.

Defrost cycle The portion of the refrigeration cycle that melts the ice off the evaporator.

Ciclo de descongelación Parte del ciclo de refrigeración que derrite el hielo del evaporador.

Defrost termination switch A temperature-activated switch that stops the defrost cycle and returns the equipment to the refrigeration cycle.

Interruptor de terminación de descongelación Un interruptor activado por temperatura que detiene el ciclo de descongelación y regresa al equipo al ciclo de refrigeración.

Defrost timer A timer used to start and stop the defrost cycle.

Temporizador de descongelación Temporizador utilizado para poner en marcha y detener el ciclo de descongelación.

Degassing Removing air and other noncondensable gasses from a system using a vacuum pump.

Desgasificación Eliminación de aire y otros gases no condensables de un sistema utilizando una bomba de vacío.

Degreaser A cleaning solution used to remove grease from parts and coils.

Desengrasador Solución limpiadora utilizada para remover la grasa de piezas y bobinas.

Dehumidify To remove moisture from air.

Deshumidificar Remover la humedad del aire.

Dehydration Removing moisture from a sealed system or a product.

Deshidratación Removiendo la humedad de un sistema sellado o un producto.

Delta-T (ΔT) The difference between the inlet temperature and outlet temperature of a heat exchanger. For example, if 75°F air returns to an evaporator, the Delta-T is 20°F if the supply air is 55°F.

Delta-T (ΔT) La diferencia entre la temperatura de entrada y salida de un intercambiador de calor. Por ejemplo, si el aire retorna a un evaporador a 75°F, la Delta-T es 20°F si el aire de suministro está a 55°F.

Delta transformer connection A transformer connection that results in a voltage output of 115 V and 230 V.

Conexión de transformador delta Una conexión de transformador que resulta en una salida de voltaje de 115 voltios y 230 voltios.

Demand Control Ventilation, DCV Ventilation strategy that calculates required ventilation based on actual, not expected, occupancy levels.

Control de demanda de ventilación, DCV por sus siglas en inglés Estrategia de ventilación que calcula la ventilación necesaria basándose en niveles de ocupación reales, no teóricos.

Demand metering In this system, the power company charges the customer based on the highest usage for a prescribed period of time during the billing period. The prescribed time for demand metering may be any 15- or 30-min period within the billing period.

Medición por demanda Un sistema utilizado por la compañía de electricidad en el cual cobran al consumidor por el período de facturación basado en el uso más alto durante un período de tiempo prescrito durante el período de facturación. El tiempo prescrito para la medición por demanda puede ser cualquier período de 15 ó 30 minutos durante el período de facturación.

Density The weight per unit of volume of a substance.

Densidad Relación entre el peso de una sustancia y su volumen.

Department of Transportation (DOT) The governing body of the U.S. government that makes the rules for transporting items, such as volatile liquids.

Departamento de Transportación (DOT en inglés) El cuerpo regente del gobierno de los Estados Unidos que crea las reglas para transportar artículos tales como líquidos volátiles.

Derivative (differential) controller A control mode where the controller will change its output signal according to the rate of change of the error or offset.

Controlador derivativo (diferencial) Un modo de control en el cual el controlador cambiará su salida de acuerdo a la razón de cambio del error o de la compensación.

Desiccant Substance in a refrigeration system drier that collects moisture.

Disecante Sustancia en el secador de un sistema de refrigeración que acumula la humedad.

Desiccant drier A device that dehumidifies compressed air for use in controls or processing.

Secador desecante Un aparato que se usa para deshumedecer el aire comprimido que se usa en los controles o procesos.

Design pressure The pressure at which the system is designed to operate under normal conditions.

Presión de diseño Presión a la que el sistema ha sido diseñado para funcionar bajo condiciones normales.

Desuperheating Removing heat from the superheated hot refrigerant gas down to the condensing temperature.

Des-sobrecalentamiento Reducir el calor del gas caliente del refrigerante sobrecalentado hasta alcanzar la temperatura de condensación.

Detector A device used to search for and find, as in a leak detector.

Detector Dispositivo de búsqueda y detección.

Detent or snap action The quick opening and closing of an electrical switch.

Acción de detén o de encaje El abrir y cerrar rápido de un interruptor eléctrico.

Dew Moisture droplets that form on a cool surface.

Rocío Gotitas de humedad que se forman en una superficie fría.

Dew point temperature The exact temperature at which moisture begins to form.

Temperatura del punto de rocío Temperatura exacta a la que la humedad comienza a formarse.

Diac A semiconductor often used as a voltage-sensitive switching device.

Diac Semiconductor utilizado frecuentemente como dispositivo de conmutación sensible a la tensión.

Diagnostic thermostat A thermostat that can receive information from various sensors and determine when a system is having a problem. It may give a fault signal in the form of a code indicating the problem.

Termostato diagnóstico Un termostato que puede recibir información de varios sensores y determinar cuándo un sistema tiene problemas. Puede indicar una señal de fallo en forma de un código que indica el problema.

Diaphragm A thin flexible material (metal, rubber, or plastic) that separates two pressure differences.

Diafragma Material delgado y flexible, como por ejemplo el metal, el caucho o el plástico, que separa dos presiones diferentes.

Diaphragm valve A valve that has a thin sheet of metal between the valve and the fluid flow. This thin sheet is called the diaphragm and contains the pressure in the system from the action of the valve.

Válvula de diafragma Una válvula que tiene una lámina fina de metal entre la válvula y el flujo de fluido. La lámina fina se llama diafragma y separa la presión del sistema de la acción de la válvula.

Die A tool used to make an external thread such as on the end of a piece of pipe.

Troquel Herramienta utilizada para formar un filete externo, como por ejemplo en el extremo de un tubo.

Differential The difference between the cut-in and cut-out points of a control, pressure, time, temperature, or level.

Diferencial Diferencia entre los puntos de conexión y des conexión de un regulador, una presión, un intervalo de tiempo, una temperatura o un nivel.

Diffuser The terminal or end device in an air-distribution system that directs air in a specific direction using louvres.

Placa difusora Punto o dispositivo terminal en un sistema de distribución de aire que dirige el aire a una dirección específica, utilizando aberturas tipo celosía.

Diffuse radiation Radiation from the sun that reaches the earth after it is reflected from other substances, such as moisture or other particles in space.

Radiación difusa Radiación solar que alcanza la Tierra después de haber sido reflejada por otras sustancias como gotas de agua u otras partículas presentes en el aire.

Digital electronic devices Devices that generate strings of data or groups of logic consisting of 1s and 0s.

Aparatos electrónicos digitales Aparatos que generan cadenas de data o grupos de lógica que consisten de unos y ceros.

Digital electronic signal An electrical signal, usually 0 to 10 V DC or 0 to 20 milliamps DC, that is used to control system conditions.

Señal electrónica digital Una señal eléctrica, generalmente de 0 a 10 voltios CD o de 0 a 20 miliamperes CD, que se usa para controlar las condiciones del sistema.

Digital VOM A volt-ohm-milliammeter that displays the reading in digits or numbers.

VOM digital Medidor de voltios-ohmios-miliamperímetros que indica la lectura en dígitos o números.

Dilution air Excess air introduced after combustion and usually enters at the end of the heat exchanger. It is brought in by the draft hood of the natural draft furnace.

Aire de dilución Aire en exceso que se introduce luego de la combustión y generalmente entra por la salida del intercambiador de calor. Se introduce a través de la campana de la estufa de corriente natural.

Diode A solid-state device composed of both P-type and N-type material. When connected in a circuit one way, current will flow. When the diode is reversed, current will not flow.

Diodo Dispositivo de estado sólido compuesto de material P y de material N. Cuando se conecta a un circuito de una manera, la corriente fluye. Cuando la dirección del diodo cambia, la corriente deja de fluir.

DIP (dual inline pair) switch A very small low-amperage, single -pole, double-throw switch used in electronic circuits to set up the program in the circuit.

Interruptor de doble paquete en línea (DIP en inglés) Un interruptor muy pequeño, de bajo amperaje, unipolar de doble tiro que se usa en los circuitos electrónicos para preparar el programa en un circuito.

Direct current Electricity in which all electron flow goes continuously in one direction.

Corriente continua Electricidad en la que todos los electrones fluyen continuamente en una sola dirección.

Direct digital control (DDC) Very low-voltage control signal, usually 0 to 10 V DC or 0 to 20 milliamps DC.

Control digital directo (DDC en inglés) Señal control de voltaje bien bajo generalmente 0 a 20 voltios CD o de 0 a 20 miliamperes CD.

Direct-drive compressor A compressor that is connected directly to the end of the motor shaft. No pulleys are involved.

Compresor de conducción directa Un compresor que está conectado directamente a un extremo del eje de un motor, sin usar poleas.

Direct-drive motor A motor that is connected directly to the load, such as an oil burner motor or a furnace fan motor.

Motor de conducción directa Un motor que está conectado directamente a la carga, tal como un motor de un quemador de aceite o el motor de un ventilador en un calefactor.

Direct expansion The term used to describe an evaporator with an expansion device other than a low-side float type.

Expansión directa Término utilizado para describir un evaporador con un dispositivo de expansión diferente al tipo de dispositivo flotador de lado bajo.

Direct-fired absorption system An absorption refrigeration system that uses gas or oil as the heat source.

Sistema de absorción de inyección directa. Sistema de refrigeración por absorción que emplea gas o petróleo como fuente de calor.

Direct GeoExchange system A geothermal system where a primary, phase-changing refrigerant is circulated in copper tubing directly in contact with the earth.

Sistema de geo-intercambio Sistema geotérmico en el que un refrigerante primario de cambio de fase se hace circular por una tubería de cobre en contacto con la tierra.

Direct radiation The energy from the sun that reaches the earth directly.

Radiación directa Energía solar que alcanza la Tierra directamente.

Direct-spark ignition (DSI) A system that provides direct ignition to the main burner.

Encendido de chispa directa (DSI en inglés). Sistema que le provee un encendido directo al quemador principal.

Direct-vent A venting system for a gas furnace, which pulls the outside air in for combustion and vents the flue gases to the outside.

Escape directo Sistema de escape en un horno, con un conducto de doble pared. El conducto hace entrar el aire del exterior para la combustión, y evacúa los gases de combustión hacia afuera.

Discharge line The refrigerant line that connects the discharge port of the compressor to the inlet of the condenser.

Línea de descarga La tubería de refrigerante que conecta el puerto de descarga del compresor a la entrada del condensador.

Discharge pressure The pressure on the high-pressure side of a compressor.

Presión de descarga La presión en el extremo de alta presión de un compresor.

Discharge valve The valve at the top of a compressor cylinder that shuts on the downstroke to prevent high-pressure gas from reentering the compressor cylinder, allowing low-pressure gas to enter.

Válvula de descarga La válvula que está en la parte superior del cilindro de un compresor que se cierra en el recorrido hacia abajo del pistón para evitar que el gas a alta presión regrese al cilindro del compresor y permitir que el gas a baja presión entre.

Discus compressor A reciprocating compressor distinguished by its disc-type valve system.

Compresor de disco Compresor alternativo caracterizado por su sistema de válvulas de tipo disco.

Discus valve A reciprocating compressor valve design with a low clearance volume and larger bore.

Válvula de disco Diseño de válvula de compresor alternativo con un volumen de holgura bajo y un calibre más grande.

Distributor A component installed at the outlet of the expansion valve that distributes the refrigerant to each evaporator circuit.

Distribuidor Componente instalado a la salida de la válvula de expansión que distribuye el refrigerante a cada circuito del evaporador.

Diverter tee Special tee fitting used on one-pipe hydronic systems to facilitate water flow through remote terminal units.

Desviadores en forma de T Un accesorio especial en forma de T usado en sistemas hidrónicos para facilitar la fluya de agua por las unidades de terminal remoto.

Domestic hot water loop Domestic hot water heated by the refrigerant loop of the heat pump.

Circuito de agua caliente doméstica Agua caliente doméstica calentada por el circuito de refrigeración de la bomba de calor.

Doping Adding an impurity to a semiconductor to produce a desired charge.

Impurificación La adición de una impureza para producir una carga deseada.

Double flare A connection used on copper, aluminum, or steel tubing that folds tubing wall to a double thickness.

Abocinado doble Conexión utilizada en tuberías de cobre, aluminio o acera que pliega la pared de la tubería y crea un espesor doble.

Dowel pin A pin, which may or may not be tapered, used to align and fasten two parts.

Pasador de espiga Pasador, que puede o no ser cónico, utilizado para alinear y fijar dos piezas.

Downflow furnace This furnace sometimes is called a counterflow furnace. The air intake is at the top, and the discharge air is at the bottom.

Horno de corriente descendente También conocido como horno de contracorriente. La entrada del aire está en la parte superior y la salida en la parte inferior.

Draft Draft is the force that allows air to travel into the combustion chamber and to force the combustion gasses out through a vent or chimney.

Corriente La corriente es la fuerza que permite que el aire entre en la cámara de combustión y empuje los gases de combustión hacia afuera, a través de una ventilación o chimenea.

Draft diverter A place for drawing indoor air (dilution air) into a vent or chimney after the combustion process for open combustion type gas furnaces and boilers. Part of its function is to prevent down drafts from affecting burner performance.

Deflector de corriente Área a través de la cual el aire del interior (aire de dilución) se hace pasar hacia una ventilación o chimenea luego del proceso de combustión en estufas o calderas a gas de combustión abierta. Parte de su función es prevenir que el rendimiento del quemador se vea afectado por las corrientes inversas.

Draft gauge A gauge used to measure very small pressures (above and below atmospheric) and compare them with the atmosphere's pressure. Used to determine the flow of flue gas in a chimney or vent.

Manómetro de tiro Manómetro utilizado para medir presiones sumamente pequeñas (mayores o menores que la atmosférica) y compararas con la presión de la atmósfera. Utilizado para determinar el flujo de gas de combustión en una chimenea o válvula.

Draft hood A place for drawing indoor air (dilution air) into a vent or chimney after the combustion process for open combustion type gas furnaces and boilers. Part of its function is to prevent down drafts from affecting burner performance.

Campana de corriente Área a través de la cual el aire del interior (aire de dilución) se hace pasar hacia una ventilación o chimenea luego del proceso de combustión en estufas o calderas a gas de combustión abierta. Parte de su función es prevenir que el rendimiento del quemador se vea afectado por las corrientes inversas.

Drier A device used in a refrigerant line to remove moisture.

Secador Dispositivo utilizado en un conducto de refrigerante para remover la humedad.

Drilled well A well with a well casing, electric pump, and electric lines that is drilled into the ground into a water source.

Pozo perforado Un pozo con recubrimiento, bomba eléctrica y cables de alimentación eléctrica perforado en el suelo para extraer agua.

Drip pan A pan shaped to collect moisture condensing on an evaporator coil in an air-conditioning or refrigeration system.

Colector de goteo Un colector formado para acumular la humedad que se condensa en la bobina de un evaporador en un sistema de acondicionamiento de aire o de refrigeración.

Drip-proof motor A motor that can stand water dripping on it, such as a condenser fan motor in an outdoor condensing unit.

Motor a prueba de goteo Un motor que puede tolerar el goteo de agua sobre él, tal como el motor del condensador en una unidad externa de condensación.

Drip time Once the defrost cycle is terminated; the fans must remain off for a period of time to prevent moisture from being blown off the coil and on to the product. This fan-off time after defrost is referred to as "drip time."

Tiempo de goteo Una vez que finaliza el ciclo de descongelado, los ventiladores deben detenerse durante un tiempo para evitar que el agua que permanece depositada sobre el serpentín caiga sobre el producto. Este tiempo de parada de los ventiladores luego del descongelado se denomina "tiempo de goteo".

Drive cleat One of the fasteners that holds square or rectangular steel ducts together. Also referred to as a drive cleat.

Grapa de conducto Uno de los sujetadores que une los conductos cuadrados o rectangulares de acero. También conocida como vaina deslizante.

Drop-out voltage The minimum back electromotive force (BEMF) that the potential relay coil must experience or its armature will drop out and the contacts between terminals 1 and 2 will close.

Voltaje de desconexión Fuerza contraelectromotriz (BEMF, por sus siglas en inglés) mínima que puede soportar la bobina del relé de potencial sin desmontarse y sin que se cierren los contactos entre las terminales 1 y 2.

Dry-base Term used to describe boilers that are configured so that all of the water in the appliance is above the combustion chamber.

Base seca término utilizado para describir a las calderas que están configuradas de manera que toda el agua en el aparato está por encima de la cámara de combustión.

Dry-bulb temperature The temperature measured using a plain thermometer.

Temperatura de bombilla seca Temperatura que se mide con un termómetro sencillo.

Dry well A well used for the discharged water in an open-loop geothermal heat pump.

Pozo seco Pozo que se utiliza para depositar agua de descarga en una bomba de calor geotérmica de circuito abierto.

Dual pressure control Two controls that are mounted in the same housing, typically low- and high-pressure controls together.

Control de presión doble Dos controles que se colocan en el mismo cárter, típicamente controles para alta y baja presión juntos.

Duct A sealed channel used to convey air from the system to and from the point of utilization.

Conducto Canal sellado que se emplea para dirigir el aire del sistema hacia y desde el punto de utilización.

Duct blower An instrument used to pressurize and test ductwork systems and air handler cabinets for air leakage.

Ventilador de conducto Instrumento utilizado para presurizar y analizar los sistemas de conducto y cabinas de manejo de aire para verificar pérdidas de aire.

Dust mites Microscopic spiderlike insects. Dust mites and their remains are thought to be a primary irritant to some people.

Ácaros del polvo Insectos microscópicos parecidos a arañas. Los ácaros del polvo y sus restos se consideran irritantes principales para algunas personas.

Duty cycling The process by which non essential equipment is intentionally shut down during high peak loads to lower power consumption and energy costs.

Ciclo de trabajo El proceso por el cual el equipo no-esencial es apagado intencionalmente durante periodos de carga elevada para reducir el consumo de energía y los costos de energía.

E

Earth-coupled heat pump Another name for a closed-loop heat pump.

Bomba de calor conectada a tierra Otro nombre para una bomba de calor de circuito cerrado.

Eccentric An off-center device that rotates in a circle around a shaft.

Excéntrico Dispositivo descentrado que gira en un círculo alrededor de un árbol.

ECM An electronically commutated motor. This DC motor uses electronics to commutate the rotor instead of brushes. It is typically used for applications under 1 hp.

CEM (ECM en inglés) Un motor conmutado electrónicamente. Este motor CD usa electrónica para conmutar el rotor en lugar de escobillas. Típicamente se utilizan para aplicaciones de menos de 1 caballo de fuerza.

Economizer Mechanical device used on HVACR equipment to reduce energy consumption by allowing outside air to be used for cooling purposes if air conditions are desirable.

Economizador Dispositivo mecánico utilizado en sistemas de calefacción, ventilación, acondicionamiento de aire y refrigeración. para reducir el consumo de energía mediante el uso del aire del exterior para enfriamiento, siempre que las condiciones del aire sean apropiadas.

Eddy current test A test with an instrument to find potential failures in evaporator or condenser tubes.

Prueba para la corriente de Foucault Prueba que se realiza con un instrumento para detectar posibles fallas en los tubos del evaporador o del condensador.

EEPROM An acronym for electrical erasable programmable read-only memory.

EEPROM Un acrónimo para memoria electrónica, borrable, programable, de sólo-lectura (electrical erasable programmable read-only memory).

Effective temperature Different combinations of temperature and humidity that provide the same comfort level.

Temperatura efectiva Diferentes combinaciones de temperatura y humedad que proveen el mismo nivel de comodidad.

Electrical power Electrical power is measured in watts. One watt is equal to one ampere flowing with a potential of one volt. Watts = Volts × Amperes ($P = E \times I$)

Potencia eléctrica La potencia eléctrica se mide en watios. Un watio equivale a un amperio que fluye con una potencia de un voltio. Watios = voltios × amperios ($P = E \times I$)

Electrical shock When an electrical current travels through a human body.

Sacudida eléctrica Paso brusco de una corriente eléctrica a través del cuerpo humano.

Electrical test instruments Pieces of electrical testing equipment that are used to measure the electrical characteristics of a component or circuit.

Instrumentos de prueba eléctrica Piezas de equipo de prueba eléctrica que se utilizan para medir las características eléctricas de un componente o circuito.

Electric forced-air furnace An electrical resistance type of heating furnace used with a duct system to provide heat to more than one room.

Horno eléctrico de aire soplado Horno con resistencia eléctrica que se utiliza con un sistema de conductos para proporcionar calefacción a varias habitaciones.

Electric heat The process of converting electrical energy, using resistance, into heat.

Calor eléctrico Proceso de convertir energía eléctrica en calor a través de la resistencia.

Electric heat relay A relay that, when its coil is energized, will initiate a heat cycle.

Relé térmico eléctrico un relé que, cuando la bobina es energizada, iniciará un ciclo de calor.

Electric hydronic boiler A boiler using electrical resistance heat, which often has a closed-loop piping system to distribute heated water for space heating.

Caldera hidrónica eléctrica Caldera que utiliza calor proporcionado por una resistencia eléctrica. A menudo cuenta con un sistema cerrado para distribuir agua caliente para usos de calefacción.

Electrodes Electrodes carry high voltage to the tips, where an arc is created for the purpose of ignition for oil or gas furnaces.

Electrodos Los electrodos llevan alto voltaje hasta las puntas, donde se crea un arco con el propósito de encender los calefactores de aceite o de gas.

Electromagnet A coil of wire wrapped around a soft iron core that creates a magnet.

Electroimán Bobina de alambre devanado alrededor de un núcleo de hierro blando que crea un imán.

Electromechanical controls Electromechanical controls convert some form of mechanical energy to operate an electrical function, such as a pressure-operated switch.

Controles electromecánicos Los controles electromecánicos convierten algún tipo de energía mecánica para operar una función eléctrica, tal como un interruptor operado por presión.

Electromotive force A term often used for voltage indicating the difference of potential in two charges.

Fuerza electromotriz Término empleado a menudo para el voltaje, indicando la diferencia de potencia entre dos cargas.

Electron The smallest portion of an atom that carries a negative charge and orbits around the nucleus of an atom.

Electrón La parte más pequeña de un átomo, con carga negativa y que sigue una órbita alrededor del núcleo de un átomo.

Electronic air filter A filter that charges dust particles using a high-voltage direct current and then collects these particles on a plate of an opposite charge.

Filtro de aire electrónico Filtro que carga partículas de polvo utilizando una corriente continua de alta tensión y luego las acumula en una placa de carga opuesta.

Electronic charging scale An electronically operated scale used to accurately charge refrigeration systems by weight.

Escala electrónica para carga Escala accionada electrónicamente que se utiliza para cargar correctamente sistemas de refrigeración por peso.

Electronic circuit board A phenolic type of plastic board that electronic components are mounted on. Typically, the circuits are routed on the back side of the board and the components are mounted on the front with prongs of wire that are soldered to the circuits on the back. These can be mass-produced and coated with a material that keeps the circuits separated if moisture and dust accumulate.

Tarjeta de circuitos electrónicos Un tipo fenólico de tarjeta plástica en la cual se montan los componentes electrónicos. La tarjeta generalmente tiene los circuitos trazados en la parte de atrás de la tarjeta y los componentes se colocan en el frente con cables que se sueldan al circuito por detrás. Éstas pueden producirse en masa y pueden recubrirse con un material que mantiene separado a los circuitos si se acumula humedad y polvo.

Electronic controls Controls that use solid-state semiconductors for electrical and electronic functions.

Controles electrónicos Controles que usan semiconductores de estado sólido para las funciones eléctricas y electrónicas.

Electronic expansion valve (EXV) A metering valve that uses a thermistor as a temperature-sensing element that varies the voltage to a heat motor–operated valve.

Válvula electrónica de expansión (EXV en inglés) Válvula de medición que utiliza un termistor como elemento sensor de temperatura para variar la tensión a una válvula de calor accionada por motor.

Electronic leak detector An instrument used to detect gases in very small portions by using electronic sensors and circuits.

Detector electrónico de fugas Instrumento que se emplea para detectar cantidades de gases sumamente pequeñas utilizando sensores y circuitos electrónicos.

Electronic or programmable thermostat A space thermostat that is electronic in nature with semiconductors that provide different timing programs for cycling the equipment.

Termostato electrónico o programable Un termostato de espacio que es de naturaleza electrónica con semiconductores que proveen diferentes programas de cronometraje para ciclar el equipo.

Electronic relay A solid-state relay with semiconductors used to stop, start, or modulate power in a circuit.

Relé electrónico Un relé de estado sólido con semiconductores para detener, iniciar o modular la electricidad en un circuito.

Electronics The use of electron flow in conductors, semiconductors, and other devices.

Electrónica La utilización del flujo de electrones en conductores, semiconductores y otros dispositivos.

Electroporation The high frequency, pulsing action of a signal generator damages the membrane of the planktonic bacteria in cooling tower and boiler water systems by creating small "pores" in their outer membrane.

Electroporación La acción pulsante de alta frecuencia de un generador de señal daña la membrana de las bacterias del plancton en sistemas de enfriamiento en torre y calderas, creando pequeños "poros" en su membrana exterior.

Electrostatic precipitator Another term for an electronic air cleaner.

Precipitador electrostático Otro término para un limpiador eléctrico del aire.

Emitter A terminal on a semiconductor.

Emisor Punto terminal en un semiconductor.

Encapsulation Bacteria are attracted to the powder in cooling tower and boiler water systems and become entrapped in the powder particle.

Encapsulamiento El polvo de los sistemas de enfriamiento en torre y calderas atrae a las bacterias, las cuales quedan atrapadas entre las partículas de polvo.

End bell The end structure of an electric motor that normally contains the bearings and lubrication system.

Extremo acampanado Estructura terminal de un motor eléctrico que generalmente contiene los cojinetes y el sistema de lubricación.

End-mount motor An electric motor mounted with tabs or studs fastened to the motor housing end.

Motor con montaje en los extremos Motor eléctrico con lengüetas o espigas de montaje en el extremo de su caja.

End play The amount of lateral travel in a motor or pump shaft.

Holgadura Amplitud de movimiento lateral en un motor o en el árbol de una bomba.

Energy The capacity for doing work.

Energía Capacida para realizar un trabajo.

Energy conservation measure (ECM) A method or process that will conserve or decrease energy waste.

Medida de conservación de energía (ECM, por sus siglas en inglés) Método o proceso por el cual se conserva o disminuye la pérdida de energía.

Energy efficiency ratio (EER) An equipment efficiency rating that is determined by dividing the output in Btu/h by the input in watts. This does not take into account the start-up and shutdown for each cycle.

Relación del rendimiento de energía (EER en inglés) Clasificación del rendimiento de un equipo que se determina al dividir la salida en Btu/h por la entrada en watios. Esto no toma en cuenta la puesta en marcha y la parada de cada ciclo.

Energy Index A way to compare one home to another and to characterize their energy efficiency.

Índice de energía Modo de comparar una vivienda con otra y caracterizar su eficiencia energética.

Energy management The use of computerized or other methods to manage the power to a facility. This may include cycling off nonessential equipment, such as water fountain pumps or lighting, when it may not be needed. The air-conditioning and heating system is also operated at optimum times when needed instead of around the clock.

Manejo de energía Cualquier facilidad que usa métodos computarizados o de otro tipo para manejar el consumo de energía en la facilidad. Esto puede incluir apagar los equipos no esenciales en ciclos, tales como las bombas de las fuentes de agua o las luces cuando no son necesarias. Los sistemas de aire acondicionado y de calefacción también se operan en tiempos óptimos cuando son necesarios en vez de todo el tiempo.

Energy recovery ventilator (ERV) In winter, this ventilation equipment recovers heat and moisture from exhaust air. In summer, heat and moisture are removed from incoming air and transferred to exhaust air.

Ventilador de recuperación de energía (ERV en inglés) En el invierno, este equipo de ventilación recupera el calor y la humedad del aire de escape. En el verano remueve el calor y la humedad del aire de entrada y lo transfiere al aire de salida.

Enthalpy The amount of heat a substance contains from a predetermined base or point, expressed in Btu/lb.

Entalpía Cantidad de calor que contiene una sustancia, establecida desde una base o un punto predeterminado, expresada en Btu/lb.

Enthalpy control Device used on economizers to determine the heat content, in Btu/lb, of an airstream.

Controlador de entalpía Dispositivo utilizado en economizadores para determinar el contenido calórico, en Btu/lb, de una corriente de aire.

Entropy Term to describe the change in heat content of a pound of refrigerant per degree Rankine. Expressed in units of Btu/lb/°R.

Entropía Término utilizado para describir el cambio en el contenido de calor de una libra de refrigerante por grado Rankine. Expresado en unidades térmicas británicas (Btu)/libras/°Rankine.

Environment Our surroundings, including the atmosphere.

Medio ambiente Nuestros alrededores, incluyendo la atmósfera.

Environmental Protection Agency (EPA) A branch of the federal government dealing with the control of ozone-depleting refrigerants and other chemicals and the overall welfare of the environment.

Agencia de Protección Ambiental (EPA en inglés) Una rama del gobierno federal que trata con el control de los refrigerantes y otros químicos que repletan el ozono, y el bienestar completo del ambiente.

EPA Abbreviation for the Environmental Protection Agency.

EPA Acrónimo en inglés de Environmental Protection Agency (Agencia de protección ambiental).

Error The signed or mathematical difference between the set point and the control point of a control process that tells whether the control point is above or below the control set point.

Error La diferencia matemática o con signo entre el punto de ajuste y el punto de control en un proceso de control, y el cual dice si el punto de control está por encima o por debajo del punto de ajuste.

Error codes Codes that may read out on an electronic panel that tell the technician what the problem may be.

Códigos de error Códigos que pueden leerse en un panel electrónico y que le dicen al técnico cuál puede ser el problema.

ERV See Energy Recovery Ventilator.

ERV Ver Ventilador de recuperación de energía.

Ester A popular synthetic lubricant that performs best with HFCs and HFC blends.

Ester Lubricante sintético de uso común que da óptimos resultados con HFC y mezclas basadas en HFC.

Ethane gas The fossil fuel, natural gas, used for heat.

Gas etano Combustible fósil, gas natural, utilizado para generar calor.

Ethylene glycol An antifreeze commonly used in chilled water air-conditioning systems to prevent the freezing of the water coils.

Etilenglicol Anticongelante utilizado generalmente en sistemas de acondicionamiento de aire mediante enfriamiento de agua para evitar el congelamiento del agua dentro de los serpentines.

Eutectic solution A mixture of two substances in such a ratio that it will provide for the lowest possible melting temperature for that solution.

Solución eutéctica Una mezcla de dos sustancias a una razón que proveerá la temperatura de derretimiento más baja para esa solución.

Evacuation The removal of any gases not characteristic to a system or vessel.

Evacuación Remoción de los gases no característicos de un sistema o depósito.

Evaporation The condition that occurs when heat is absorbed by liquid and it changes to vapor.

Evaporación Condición que ocurre cuando un líquido absorbe calor y se convierte en vapor.

Evaporative condenser (cooling tower) A combination water cooling tower and condenser. The refrigerant from the compressor is routed to the cooling tower where the tower evaporates water to cool the refrigerant. In the evaporative condenser the refrigerant is routed to the tower and the water circulates only in the tower. In a cooling tower, the water is routed to the condenser at the compressor location.

Condensador evaporatorio (torre de enfriamiento) Una combinación de una torre de enfriamiento de agua y un condensador. El refrigerante del compresor se desvía a la torre de enfriamiento donde la torre evapora agua para enfriar al refrigerante. En el condensador evaporatorio el refrigerante se lleva a la torre y el agua circula sólo en la torre. En una torre de enfriamiento, el agua se pasa por el condensador donde está ubicado el compresor.

Evaporative cooling Devices that provide this type of cooling use fiber mounted in a frame with water slowly running down the fiber. Fresh air is drawn in and through the water-soaked fiber and cooled by evaporation to a point close to the wet-bulb temperature of the air.

Enfriamiento por formación de vapor Los dispositivos que proporcionan este tipo de enfriamiento utilizan agua que corre sobre fibra colocada en un marco. Se pasa aire nuevo a través de la fibra húmeda, el aire se enfría por evaporación a una temperatura próxima a la de una bombilla húmeda.

Evaporative humidifier A humidifier that provides moisture on a media surface through which air is forced. The air picks up the moisture from the media as a vapor.

Humidificador por evaporación Humidificador que proporciona humedad en una superficie a través de la cual pasa aire soplado. El aire recoge la humedad de la superficie, en forma de vapor.

Evaporator The component in a refrigeration system that absorbs heat into the system and evaporates the liquid refrigerant.

Evaporador El componente en un sistema de refrigeración que absorbe el calor hacia el sistema y evapora el refrigerante líquido.

Evaporator fan A fan or blower used to improve the efficiency of an evaporator by air movement over the coil.

Abanico del evaporador Ventilador que se utiliza para mejorar el rendimiento de un evaporador por medio del movimiento de aire a través de la bobina.

Evaporator pressure-regulating valve (EPR valve) A mechanical control installed in the suction line at the evaporator outlet that keeps the evaporator pressure from dropping below a certain point.

Válvula reguladora de presión del evaporador (EPR en inglés) Reguladora mecánica instalada en el conducto de aspiración de la salida del evaporador; evita que la presión del evaporador caiga hasta alcanzar un nivel por debajo del nivel específico.

Evaporator types Flooded—an evaporator where the liquid refrigerant level is maintained to the top of the heat exchange coil. Dry type—an evaporator coil that achieves the heat exchange process with a minimum of refrigerant charge.

Clases de evaporadores Inundado—un evaporador en el que se mantiene el nivel del refrigerante líquido en la parte superior de la bobina de intercambio de calor. Seco—una bobina de evaporador que logra el proceso de intercambio de calor con una mínima cantidad de carga de refrigerante.

Even parallel system Parallel compressors of equal sizes mounted on a steel rack and controlled by a microprocessor.

Sistema paralelo homogéneo Compresores de capacidades iguales, montados en paralelo en un bastidor de acero y controlados mediante un microprocesador.

Excess air Excess air consists of combustion air and dilution air.

Aire en exceso El aire en exceso se compone de aire de combustión y aire de dilución.

Exfiltration The outflow of air from a structure. Air flow out of a structure usually from a pressure difference, usually accompanied by an equal inflow of air.

Exfiltración Flujo de aire desde una estructura hacia el exterior. Flujo de aire proveniente de una estructura, generalmente debido a una diferencia de presión y generalmente acompañado de un flujo equivalente de aire hacia el interior.

Exhaust valve The movable component in a refrigeration compressor that allows hot gas to flow to the condenser and prevents it from refilling the cylinder on the downstroke.

Válvula de escape Componente móvil en un compresor de refrigeración que permite el flujo de gas caliente al condensador y evita que este gas rellene el cilindro durante la carrera descendente.

Expansion (metering) device The component between the high-pressure liquid line and the evaporator that feeds the liquid refrigerant into the evaporator.

Dispositivo de (medición) de expansión. Componente entre el conducto de líquido de alta presión y el evaporador que alimenta el refrigerante líquido hacia el evaporador.

Expansion joint A flexible portion of a piping system or building structure that allows for expansion of the materials due to temperature changes.

Junta de expansión Parte flexible de un sistema de tubería o de la estructura de un edificio que permite la expansión de los materiales debido a cambios de temperatura.

Explosion-proof motor A totally sealed motor and its connections that can be operated in an explosive atmosphere, such as in a natural gas plant.

Motor a prueba de explosión Un motor totalmente sellado y con conexiones que puede operarse en una atmósfera explosiva, tal como dentro de una planta de gas natural.

External drive An external type of compressor motor drive, as opposed to a hermetic compressor.

Motor externo Motor tipo externo de un compresor, en comparación con un compresor hermético.

External equalizer The connection from the evaporator outlet to the bottom of the diaphragm on a thermostatic expansion valve.

Equilibrador externo Conexión de la salida del evaporador a la parte inferior del diafragma en una válvula de expansión termostática.

External heat defrost A defrost system for a refrigeration system where the heat comes from some external source. It might be an electric strip heater in an air coil, or water in the case of an ice maker. *External* means other than hot gas defrost.

Descongelador de calor externo Un sistema de descongelación para un sistema de refrigeración en el cual el calor viene de una fuente externa. La misma puede ser un calentador de tira eléctrico en un serpentín o agua en caso de una hielera. *Externo* se refiere a otro tipo de descongelación aparte del de gas caliente.

External motor protection Motor overload protection that is mounted on the outside of the motor.

Protección externa para motor Protección para un motor contra la sobrecarga y que está montada en el exterior del motor.

F

Fahrenheit scale The temperature scale that places the boiling point of water at 212°F and the freezing point at 32°F.

Escala Fahrenheit Escala de temperatura en la que el punto de ebullición del agua se encuentra a 212°F y el punto de fusión del hielo a 32°F.

Fail-safe The timer on the time clock will eventually bring the system out of defrost if early defrost termination is inoperative.

Seguro El temporizador del reloj horario hará que el descongelamiento se detenga cuando el sistema de detención temprana del descongelamiento no funcione.

Fan A device that produces a pressure difference in air to move it.

Abanico Dispositivo que produce una diferencia de presión en el aire para moverlo.

Fan-assisted draft A forced draft system that either pushes or pulls combustion gasses through the heat exchanger using a combustion blower motor and a fan. These systems can be either forced- or induced-draft systems.

Corriente asistida por ventiladores Sistema de circulación forzada de aire que expele o succiona los gases de combustión a través del intercambiador de calor utilizando un motor de ventilación de combustión y un ventilador. Estos sistemas pueden ser sistemas de corriente forzada o inducida.

Fan cycling The use of a pressure control to turn a condenser fan on and off to maintain a correct pressure within the system.

Funcionamiento cíclico La utilización de un regulador de presión para poner en marcha y detener el abanico de un condensador a fin de mantener una presión correcta dentro del sistema.

Fan relay coil A magnetic coil that controls the starting and stopping of a fan.

Bobina de relé del ventilador Bobina magnética que regula la puesta en marcha y la parada de un ventilador.

Farad The unit of capacity of a capacitor. Capacitors in our industry are rated in microfarads.

Faradio Unidad de capacidad de un capacitador. En nuestro medio, los capacitadores se clasifican en microfaradios.

Feedback loop The circular data route in a control loop that usually travels from the control medium's sensor to the controller, then to the controlled device, and back into the controlled process to the sensor again as a change in the control point.

Circuito de retroalimentación Ruta circular de la data en un circuito de control que generalmente va desde el sensor del medio de control al controlador y luego al aparato controlado, y de regreso al proceso controlado y al sensor nuevamente como un cambio en el punto de control.

Female thread The internal thread in a fitting.

Filete hembra Filete interno en un accesorio.

Fill or wetted-surface method Water in a cooling tower is spread out over a wetted surface while air is passed over it to enhance evaporation.

Método de relleno o de superficie mojada El agua en una torre de refrigeración se extende sobra una superficie mojada mientras el aire se dirige por encima de la misma para facilitar la evaporación.

Film factor The relationship between the medium giving up heat and the heat exchange surface (evaporator). This relates to the velocity of the medium passing over the evaporator. When the velocity is too slow, the film between the air and the evaporator becomes greater and becomes an insulator, which slows the heat exchange.

Factor de capa Relación entre el medio que emite calor y la superficie del intercambiador de calor (evaporador). Esto se refiere a la velocidad del medio que pasa sobre el evaporador. Cuando la velocidad es demasiado lenta, la capa entre el aire y el evaporador se expande y se convierte en un aislador, disminuyendo así la velocidad del intercambio del calor.

Filter A fine mesh or porous material that removes particles from passing fluids.

Filtro Malla fina o material poroso que remueve partículas de los fluidos que pasan por él.

Filter drier A type of refrigerant filter that includes a desiccant material that has an attraction for moisture. The filter drier will remove particles and moisture from refrigerant and oil.

Secador de filtro Un tipo de filtro de refrigerante que incluye un material desecante el cual tiene una atracción a la humedad. El secador de filtro removerá las partículas y la humedad del aceite y del refrigerante.

Fin comb A hand tool used to straighten the fins on an air-cooled condenser.

Herramienta para aletas Herramienta manual utilizada para enderezar las aletas en un condensador enfriado por aire.

Finned-tube evaporator A copper or aluminum tube that has fins, usually made of aluminum, pressed onto the copper lines to extend the surface area of the tubes.

Evaporador de tubo con aletas Un tubo de cobre o aluminio que tiene aletas, generalmente de aluminio, colocadas a presión contra las líneas de cobre para extender el área de superficie de los tubos.

Fixed-bore device An expansion device with a fixed diameter that does not adjust to varying load conditions.

Dispositivo de calibre fijo Dispositivo de expansión con un diámetro fijo que no se ajusta a las condiciones de carga variables.

Fixed resistor A nonadjustable resistor. The resistance cannot be changed.

Resistor fijo Resistor no ajustable. La resistencia no se puede cambiar.

Flame impingement When flame touches components where it is not supposed to. For example, the flame should not touch the heat exchanger in an oil or gas furnace. The flame should stay in the middle of the heat exchanger.

Impacto de llama Cuando una llama toca componentes que no está supuesta a tocar. Por ejemplo, la llama no debería tocar el intercambiador de calor en un calefactor de aceite o gas. La llama debería mantenerse en el centro del intercambiador de calor.

Flame-proving device A device in a gas or oil furnace to ensure there is a flame so that excess fuel will not be released, which may cause overheating in an oil furnace and explosion in a gas appliance.

Aparato probador de llama Un aparato en un calefactor de gas o aceite que asegura que hay una llama presente para que no se libere combustible de más, lo cual puede causar sobrecalentamiento en un calefactor de aceite y una explosión en enseres de gas.

Flame rectification In a gas furnace, the flame changes the normal current to direct current. The electronic components in the system will only energize and open the gas valve with a direct current.

Rectificación de llama En un horno de gas, el calor generado por la llama convierte la corriente normal en corriente directa. Los componentes electrónicos del sistema sólo se activarán para abrir el paso del gas con corriente directa.

Flapper valve See Reed valve.

Chapaleta Véase válvula de lámina.

Flare The angle that may be fashioned at the end of a piece of tubing to match a fitting and create a leak-free connection.

Abocinado Ángulo que puede formarse en el extremo de una pieza de tubería para emparejar un accesorio y crear una conexión libre de fugas.

Flare nut A threaded connector used in a flare assembly for tubing.

Tuerca abocinada Conector de rosca utilizado en un conjunto abocinado para tuberías.

Flash gas A term used to describe the pressure drop in an expansion device when some of the liquid passing through the valve is changed quickly to a gas and cools the remaining liquid to the corresponding temperature.

Gas instantáneo Término utilizado para describir la caída de la presión en un dispositivo de expansión cuando una parte del líquido que pasa a través de la válvula se convierte rápidamente en gas y enfría el líquido restante a la temperatura correspondiente.

Floating head pressure Letting the head pressure (condensing pressure) fluctuate with the ambient temperature from season to season for lower compression ratios and better efficiencies.

Fluctuar la presión de la carga Hecho de dejar que la presión de la carga (presión de condensación) fluctúe con la temperatura del ambiente en cada temporada del año, para obtener relaciones de compresión más bajas y mejores rendimientos.

Float valve or switch An assembly used to maintain or monitor a liquid level.

Válvula o conmutador de flotador Conjunto utilizado para mantener o controlar el nivel de un líquido.

Flooded evaporator A refrigeration system operated with the liquid refrigerant at the outlet of the evaporator coil for improved heat exchange. Flooded evaporators operate with zero degrees of superheat.

Sistema inundado Sistema de refrigeración que funciona con el refrigerante líquido en la salida de la bobina del evaporador para mejorar el intercambio de calor. Los sistemas inundados operan con cero grados de sobrecalentamiento.

Flooding The term applied to a refrigeration system when the liquid refrigerant reaches the compressor.

Inundación Término aplicado a un sistema de refrigeración cuando el nivel del refrigerante líquido llega al compresor.

Flow-through receiver Flow-through receivers can store the cool subcooled liquid refrigerant for some time before releasing it to the liquid line and liquid header. This can cause a gain in temperature within the liquid and subcooling can be lost.

Depósitos de flujo Los depósitos de flujo pueden almacenar el líquido refrigerante subenfriado frío por algún tiempo antes de liberarlo hacia la línea de líquido y la cabecera de líquido. Esto puede generar una ganancia de temperatura en el líquido, provocando la pérdida del subenfriamiento.

Flue The duct that carries the products of combustion out of a structure for a fossil- or a solid-fuel system.

Conducto de humo Conducto que extraer los productos de combustión de una estructura en sistemas de combustible fósil o sólido.

Flue draft Another name for chimney draft and stack draft.

Corriente de conducto de humo Denominación alternativa para la corriente de chimenea y corriente de ventilación.

Flue-gas analysis instruments Instruments used to analyze the operation of fossil fuel–burning equipment such as oil and gas furnaces by analyzing the flue gases.

Instrumentos para el análisis del gas de combustión Instrumentos utilizados para llevar a cabo un análisis del functionamiento de los quemadores de combustible fósil, como por ejemplo hornos de aceite pesado o gas, a través del estudio de los gases de combustión.

Fluid The state of matter of liquids and gases.

Fluido Estado de la materia de líquidos y gases.

Fluid expansion device Using a bulb or sensor, tube, and diaphragm filled with fluid, this device will produce movement at the diaphragm when the fluid is heated or cooled. A bellows may be added to produce more movement. These devices may contain vapor and liquid.

Dispositivo para la expansión del fluido Utilizando una bombilla o sensor, un tubo y un diafragma lleno de fluido, este dispositivo generará movimiento en el diafragma cuando se caliente o enfríe el fluido. Se le puede agregar un fuelle para generar aún más movimiento. Dichos dispositivos pueden contener vapor y líquido.

Flush The process of using a fluid to push contaminants from a system.

Descarga Proceso de utilizar un fluido para remover los contaminantes de un sistema.

Flux A substance applied to soldered and brazed connections to prevent oxidation during the heating process.

Fundente Sustancia aplicada a conexiones soldadas y broncesoldadas para evitar la oxidación durante el proceso de calentamiento.

Foaming A term used to describe oil when it has liquid refrigerant boiling out of it.

Espumación Término utilizado para describir el aceite cuando el refrigerante líquido se derrama del mismo.

Foot-pound The amount of work accomplished by lifting 1 lb of weight 1 ft; a unit of energy.

Libra-pie Medida de la cantidad de energía o fuerza que se requiere para levantar una libra a una distancia de un pie; unidad de energía.

Force Energy exerted.

Fuerza Energía ejercida sobre un objeto.

Forced convection The movement of fluid by mechanical means.

Convección forzada Movimiento de fluido por medios mecánicos.

Forced draft Forced-draft systems push or blow combustion gases through the heat exchanger and cause a positive pressure in the heat exchanger. Forced-draft systems have the combustion blower motor on the inlet of the heat exchanger.

Corriente forzada Los sistemas de corriente forzada expelen o succionan los gases de combustión a través del intercambiador de calor, produciendo una presión positiva en el mismo. En los sistemas de corriente forzada el motor de ventilación de combustión se encuentra a la entrada del intercambiador de calor.

Forced-draft cooling tower A water cooling tower that has a fan on the side of the tower that pushes air through the tower, as opposed to an induced-draft tower, which has the fan on the side and draws air through the tower.

Torre de enfriamiento de ventilación forzada Una torre de enfriamiento que tiene un ventilador en el lado de la torre y empuja aire a través de la torre, contrario a una torre de ventilación inducida que tiene un ventilador en el lado de la torre y jala el aire a través de la torre.

Forced-draft evaporator An evaporator over which air is forced to spread the cooling more efficiently. This term usually refers to a domestic refrigerator or freezer.

Evaporador de tiro forzado Evaporador encima del cual se envía aire soplado para obtener una distribución más eficaz del frío. Este término suele aplicarse a refrigeradores o congeladores domésticos.

Fossil fuels Natural gas, oil, and coal formed millions of years ago from dead plants and animals.

Combustibles fósiles El gas natural, el petróleo y el carbón que se formaron hace millones de años de plantas y animales muertos.

Four-way valve The valve in a heat pump system that changes the direction of the refrigerant flow between the heating and cooling cycles.

Válvula con cuatro vías Válvula en un sistema de bomba de calor que cambia la dirección del flujo de refrigerante entre los ciclos de calentamiento y enfriamiento.

Fractionation When a zeotropic refrigerant blend phase changes, the different components in the blend all have different vapor pressures. This causes different vaporization and condensation rates and temperatures as they phase change.

Fraccionación Cuando se produce un cambio en la fase de una mezcla zeotrópica de refrigerantes, los diferentes componentes que forman la mezcla tienen presiones de vapor diferentes. Esto provoca diferentes temperaturas y tasas de vaporización y de condensación según cambien de fase.

Free-cooling The process of cooling a structure using only outside air without operating the mechanical air-conditioning system.

Enfriamiento libre Proceso por el cual una estructura se enfría utilizando sólo aire del exterior, sin el funcionamiento mecánico del sistema de acondicionamiento de aire.

Freezer burn The term applied to frozen food when it becomes dry and hard from dehydration due to poor packaging.

Quemadura del congelador Término aplicado a la comida congelada cuando se seca y endurece debido a la deshidratación ocacionada por el empaque de calidad inferior.

Freezestat A temperature-sensing device typically intended to protect water coils from freezing.

Sensor de congelamiento Dispositivo de medición de temperatura, generalmente utilizado para evitar el congelamiento del agua dentro de los serpentines.

Freeze-up Excess ice or frost accumulation on an evaporator to the point that airflow may be affected.

Congelación Acumulación excesiva de hielo o congelación en un evaporador a tal extremo que el flujo de aire puede ser afectado.

Freezing The change of state of water from a liquid to a solid.

Congelamiento Cambio de estado del agua de líquido a sólido.

Freon The previous trade name for refrigerants manufactured by E. I. du Pont de Nemours & Co., Inc.

Freón Marca registrada previa para refrigerantes fabricados por la compañía E. I. du Pont de Nemours, S.A.

Frequency The cycles per second (cps) of the electrical current supplied by the power company. This is normally 60 cps in the United States.

Frecuencia Ciclos por segundo (cps), generalmente 60 cps en los Estados Unidos, de la corriente eléctrica suministrada por la empresa de fuerza motriz.

Friction loss The loss of pressure in a fluid flow system (air or water) due to the friction of the fluid rubbing on the sides. It is typically measured in feet of equivalent loss.

Pérdida por fricción La pérdida de presión en un sistema de flujo de fluido (aire o agua) debido a la fricción del fluido al frotar contra los lados. Típicamente se mide en pies de pérdida equivalente.

Front seated A position on a service valve that will not allow refrigerant flow in one direction.

Sentado delante Posición en una válvula de servicio que no permite el flujo de refrigerante en una dirección.

Frost back A condition of frost on the suction line and even the compressor body.

Obturación por congelación Condición de congelación que ocurre en el conducto de aspiración e inclusive en el cuerpo del compresor.

Frostbite When skin freezes.

Quemadura por frío Congelación de la piel.

Frozen The term used to describe water in the solid state; also used to describe a rotating shaft that will not turn.

Congelado Término utilizado para describir el agua en un estado sólido; utilizado también para describir un árbol giratorio que no gira.

Fuel oil The fossil fuel used for heating; a petroleum distillate.

Aceite pesado Combustible fósil utilizado para calentar; un destilado de petróleo.

Full-load amperage (FLA) A mathematical calculation used to get Underwriters Laboratories (UL) approval for a motor that was used by compressor manufacturer before 1972. FLA value was replaced with a rated-load amperage (RLA) value after 1972. It should not be used to determine the proper refrigerant charge, or to determine if a compressor motor is running properly.

Amperaje de carga completa (FLA, por sus siglas en inglés Cálculo matemático utilizado para obtener la aprobación de Underwriters Laboratory (UL) para un motor utilizado por un fabricante de compresores antes de 1972. El valor del FLA se reemplazó por el valor del amperaje de carga calificada (RLA) luego de 1972. No debe utilizarse para determinar la carga adecuada del refrigerante ni para determinar si un motor de compresor está funcionando correctamente.

Furnace Equipment used to convert heating energy, such as fuel oil, gas, or electricity, to usable heat. It usually contains a heat exchanger, a blower, and the controls to operate the system.

Horno Equipo utilizado para la conversión de energía calórica, como por ejemplo el aceite pesado, el gas o la electricidad, en calor utilizabe. Normalmente contiene un intercambiador de calor, un soplador y los reguladores para accionar el sistema.

Fuse A safety device used in electrical circuits for the protection of the circuit conductor and components.

Fusible Dispositivo de seguridad utilizado en circuitos eléctricos para la protección del conductor y los componentes del circuito.

Fusible link An electrical safety device normally located in a furnace that burns and opens the circuit during an overheat situation.

Cartucho de fusible Dispositivo eléctrico de seguridad ubicado por lo general en un horno, que quema y abre el circuito en caso de sobrecalentamiento.

Fusible plug A device (made of low-melting temperature metal) used in pressure vessels that is sensitive to high temperatures and relieves the vessel contents in an overheating situation.

Tapón de fusible Dispositivo utilizado en depósitos en presión, hecho de un metal que tiene una temperatura de fusión baja. Este dispositivo es sensible a temperaturas altas y alivia el contenido del depósito en caso de sobrecalentamiento.

Fusion iron A device used to heat plastic pipe to 500°F for a permanent, water-tight connection to another plastic pipe.

Plancha de fusión Dispositivo utilizado para calentar tuberías plásticas a 500°F para lograr uniones permanentes y sin pérdidas de agua entre segmentos de cañería plástica.

G

Gain The term used to describe the sensitivity of a control, for example, how much change in output signal per degree of temperature change.

Ganancia El término se usa para describir la sensibilidad de un control, por ejemplo, cuanto cambio en la señal de salida por cada grado de cambio en temperatura.

Gas The vapor state of matter.

Gas Estado de vapor de una materia.

Gasket A thin piece of flexible material used between two metal plates to prevent leakage.

Guarnición Pieza delgada de material flexible utilizada entre dos piezas de metal para evitar fugas.

Gas-pressure switch Used to detect gas pressure before gas burners are allowed to ignite.

Conmutador de presión del gas Utilizado para detectar la presión del gas antes de que los quemadores de gas puedan encenderse.

Gas valve A valve used to stop, start, or modulate the flow of natural gas.

Válvula de gas Válvula utilizada para detener, poner en marcha o modular el flujo de gas natural.

Gate A terminal on a semiconductor.

Compuerta Punto terminal en un semi-conductor.

Gate valve Valve that has a sliding wedge-shaped piece (gate) inside it that is connected to a handle. The handle can be turned to force the sliding gate into a fitted seat to stop or control water flow. Gate valves have very little pressure drop or resistance to water or fluid flow when wide open because the gate moves all of the way out of the water stream flow.

Válvula de compuerta Válvula con una pieza (compuerta) deslizante en forma de una cuña que está conectada a un mango. Se gira el mango para forzar la compuerta sobre un asiento para parar or controlar el flujo del agua. Cuando están abiertas, las válvulas de compuerta tienen muy poca caída de la presión o resistencia al flujo del agua o fluido porque la compuerta se mueve fuera del corriente del flujo del agua.

Gauge An instrument used to indicate pressure.

Manómetro Instrumento utilizado para indicar presión.

Gauge manifold A service tool with one or more gauges and a valve arrangement to control fluid flow.

Manómetro de distribución Una herramienta de servicio con uno o más manómetros y con las válvulas arregladas a fin de regular el flujo de fluido.

Gauge port The service port used to attach a gauge for service procedures.

Orificio de manómetro Orificio de servicio utilizado con el propósito de fijar un manómetro para procedimientos de servicio.

Geothermal heat pump A heat pump that uses the earth or water in the earth for its heat sources and sinks.

Bomba de calor geotérmica Bomba de calor que utiliza el suelo o el agua de la tierra como fuente y depósito termal.

Geothermal well A well dedicated to a geothermal heat pump that draws water from the top of the water column and returns the same water to the bottom of the water column.

Pozo geotermal Pozo utilizado por una bomba de calor geotérmica que extrae agua de la parte superior de la columna de agua y la devuelve en la parte inferior de la misma columna.

Germanium A substance from which many semiconductors are made.

Germanio Sustancia de la que se fabrican muchos semiconductores.

Global warming An earth-warming process caused by the atmosphere's absorption of the heat energy radiated from the earth's surface.

Calentamiento de la Tierra Proceso de calentamiento de la Tierra provocado por la absorción de la energía radiada de la superficie de la Tierra, por la atmósfera, en forma de calor.

Global warming potential (GWP) An index that measures the direct effect of chemicals emitted into the atmosphere.

Potencial de calentamiento de la tierra (GWP en inglés) Índice que mide el efecto directo de los productos químicos que se emiten a la atmósfera.

Globe valve Valve that uses a disk and a seat to stop or vary water or fluid flow. The handle raises or lowers the disk over the valve seat, which is built into the valve body. When the valve is fully open, the water must make two 90° turns to pass through the valve, creating pressure drop.

Válvula de globo Válvula que utiliza un disco y un asiento para parar o variar el flujo del agua o fluido. El mango eleva o baja el disco sobre el asiento de la válvula construído en el cuerpo de la válvula. Cuando la válvula está completamente abierta, el agua debe girar 90° dos veces para pasar a través de la válvula, ocasionando así una caída de la presión.

Glow coil A device that automatically reignites a pilot light if it goes out.

Bobina encendedora Dispositivo que automáticamente vuelve a encender la llama piloto si ésta se apaga.

Glycol Antifreeze solution used in the water loop of geothermal heat pumps.

Glicol Líquido anticongelante que se emplea en el circuito de agua de las bombas de calor geotérmicas.

Graduated cylinder A cylinder with a visible column of liquid refrigerant used to measure the refrigerant charged into a system. Refrigerant temperatures can be dialed on the graduated cylinder.

Cilindro graduado Cilindro con una columna visible de refrigerante líquido utilizado para medir el refrigerante inyectado al sistema. Las temperaturas del refrigerante pueden marcarse en el cilindro graduado.

Grain Unit of weight. One pound = 7000 grains.

Grano Unidad de peso. Una libra equivale a 7.000 granos.

Gram (g) Metric measurement term used to express weight.

Gramo Término utilizado para referirse a la unidad básica de peso en el sistema métrico.

Green Companies looking to go "Green" incorporate sustainable design into their building and facilities.

Ecológico Las empresas que impulsan la ecología incorporan el diseño sustentable a sus edificios y áreas de producción.

Greenhouse effect An earth-warming process caused by the atmosphere's absorption of the heat energy radiated from the earth's surface. Often referred to as global warming.

Efecto invernadero Proceso de calentamiento de la tierra causado por la absorción atmosférica de la energía calórica irradiada desde la superficie terrestre. También se lo conoce como calentamiento global.

Grille A louvered, often decorative, component in an air system at the inlet or the outlet of the airflow.

Rejilla Componente con celosías, comúnmente decorativo, en un sistema de aire que se encuentra a la entrada o a la salida del flujo de aire.

Grommet A rubber, plastic, or metal protector usually used where wire or pipe goes through a metal panel.

Guardaojal Protector de caucho, plástico o metal normalmente utilizado donde un alambre o un tubo pasa a través de una base de metal.

Ground, electrical A circuit or path for electron flow to earth ground.

Tierra eléctrica Circuito o trayectoria para el flujo de electrones a la puesta a tierra.

Ground fault circuit interrupter (GFCI) A circuit breaker that can detect very small leaks to ground, which, under certain circumstances, could cause an electrical shock. This small leak, which may not be detected by a conventional circuit breaker, will cause the GFCI circuit breaker to open the circuit.

Disyuntor por pérdidas a tierra (GFCI en inglés) Disyuntor capaz de detectar fugas muy pequeñas hacia tierra, que en determinadas circunstancias pueden provocar descargas eléctricas. Existe la posibilidad de que un disyuntor convencional no sea capaz de detectar las fugas pequeñas, en cuyo caso el disyuntor GFCI abre el circuito.

Ground loop These loops of plastic pipe are buried in the ground in a closed-loop geothermal heat pump system and contain a heat transfer fluid.

Circuito de tierra Circuitos de tubos de plástico enterrados y que forman parte de un sistema de bomba geotérmica de circuito cerrado. Dichos tubos contienen un fluido que permite la transferencia de calor.

Ground wire A wire from the frame of an electrical device to be wired to the earth ground.

Alambre a tierra Alambre que va desde el armazón de un dispositivo eléctrico para ser conectado a la puesta a tierra.

Grout A cement-like material injected between the well casing and the drilled hole for a well. Grout adds rigidity and prevents contamination from entering the well.

Lechada Pasta fina obtenida mezclando cemento y agua, que se inyecta entre la pared del pozo y la perforación para el mismo. Proporciona rigidez e impide la penetración de agentes contaminantes en el pozo.

Guide vanes Vanes used to produce capacity control in a centrifugal compressor. Also called *prerotation guide vanes.*

Paletas directrices Paletas utilizadas para producir la regulación de capacidad en un compresor centrífugo. Conocidas también como paletas directrices para prerotación.

H

Halide refrigerants Refrigerants that contain halogen chemicals: R-12, R-22, R-500, and R-502 are among them.

Refrigerantes de hálido Refrigerantes que contienen productos químicos de halógeno: entre ellos se encuentran el R-12, R-22, R-500, y R-502.

Halide torch A torch-type leak detector used to detect halogen refrigerants.

Soplete de hálido Detector de fugas de tipo soplete utilizado para detectar refrigerantes de halógeno.

Halogens Chemical substances found in many refrigerants containing chlorine, bromine, iodine, and fluorine.

Halógenos Sustancias químicas presentes en muchos refrigerantes que contienen cloro, bromo, yodo y flúor.

Hand truck A two-wheeled piece of equipment that can be used for moving heavy objects.

Vagoneta para mano Equipo con dos ruedas que puede utilizarse para transportar objetos pesados.

Hanger A device used to support tubing, pipe, duct, or other components of a system.

Soporte Dispositivo utilizado para apoyar tuberias, tubos, conductos u otros componentes de un sistema.

Head Another term for pressure, usually referring to gas or liquid.

Carga Otro término para presión, refiriéndose normalmente a gas o líquido.

Header A pipe or containment to which other pipe lines are connected.

Conductor principal Tubo o conducto al que se conectan otras conexiones.

Head pressure control A control that regulates the head pressure in a refrigeration or air-conditioning system.

Regulador de la presión de la carga Regulador que controla la presión de la carga en un sistema de refrigeración o de acondicionamiento de aire.

Heat Energy that causes molecules to be in motion and to raise the temperature of a substance.

Calor Energía que ocasiona el movimiento de las moléculas provocando un aumento de temperatura en una sustancia.

Heat anticipator A device that anticipates the need for cutting off the heating system prematurely so the system does not overshoot the set point temperature.

Anticipador de calor Dispositivo que anticipa la necesidad de detener la marcha del sistema de calentamiento para que el sistema no exceda la temperatura programada.

Heat coil A device made of tubing or pipe designed to transfer heat to a cooler substance by using fluids.

Bobina de calor Dispositivo hecho de tubos, diseñado para transmitir calor a una sustancia más fría por medio de fluidos.

Heat exchanger A device that transfers heat from one substance to another.

Intercambiador de calor Dispositivo que transmite calor de una sustancia a otra.

Heat fusion A process that will permanently join sections of plastic pipe together.

Fusión térmica Proceso mediante el cual se unen dos piezas de tubo de plástico permanentemente.

Heat gain Rate at which heat enters a structure in the cooling season.

Ganancia calórica Velocidad a la que el calor entra a una estructura en la estación de enfriamiento.

Heating Seasonal Performance Factor (HSPF) HSPF is a ratio of the seasonal heat output in BTUs to the seasonal watt-hours used. The higher the HSPF, the more efficient the unit is in moving BTUs of heat energy per BTU of energy consumed. This efficiency factor is applied to heat pumps in the heating mode and takes into consideration the energy used during the entire heating season.

Factor de rendimiento estacional del calentamiento (HSPF, por sus siglas en inglés) Relación entre la generación estacional de calor, en BTU, y la energía estacional utilizada, en watts por hora. Mientras mayor sea el HSPF, más eficiente es el sistema para generar una BTU de energía calórica por cada BTU de energía consumida. Este factor de eficiencia se aplica a bombas calóricas cuando funcionan generando calor y toma en cuenta la energía utilizada durante la estación de calefacción completa.

Heat intensity Another term for temperature.

La intensidad de calor Otro término para la temperatura.

Heat island Urban air or surface temperatures that are higher than nearby rural areas.

Isla calórica Aire urbano o temperaturas de superficie mayores que las de áreas rurales cercanas.

Heat loss Rate at which heat leaks from a structure in the heating season.

Pérdida de calor Velocidad de pérdida de calor desde una estructura en la estación de calefacción.

Heat of compression That part of the energy from the pressurization of a gas or a liquid converted to heat.

Calor de compresión La parte de la energía generada de la presurización de un gas o un líquido que se ha convertido en calor.

Heat of fusion The heat released when a substance is changing from a liquid to a solid.

Calor de fusión Calor liberado cuando una sustancia se convierte de líquido a sólido.

Heat of respiration When oxygen and carbon hydrates are taken in by a substance or when carbon dioxide and water are given off. Associated with fresh fruits and vegetables during their aging process while stored.

Calor de respiración Cuando se admiten oxígeno e hidratos de carbono en una sustancia o cuando se emiten bióxido de carbono y agua. Se asocia con el proceso de maduración de frutas y legumbres frescas durante su almacenamiento.

Heat pump A refrigeration system used to supply heat or cooling using valves to reverse the refrigerant gas flow.

Bomba de calor Sistema de refrigeración utilizado para suministrar calor o frío mediante válvulas que cambian la dirección del flujo de gas del refrigerante.

Heat reclaim Using heat from a condenser for purposes such as space and domestic water heating.

Reclamación de calor La utilización del calor de un condensador para propósitos tales como la calefacción de espacio y el calentamiento doméstico de agua.

Heat recovery The process of transferring heat from one area to another as needed to help reduce energy consumption by the HVACR equipment.

Recuperación de calor Proceso de transferencia de calor de un área hacia otra, necesaria para reducir el consumo de energía del sistemas de calefacción, ventilación, acondicionamiento de aire y refrigeración.

Heat recovery ventilator Units that recover heat only, used primarily in winter.

Ventilador de recuperación de calor Unidades que recuperan calor solamente, usados principalmente en el invierno.

Heat sink A low-temperature surface to which heat can transfer.

Fuente fría Superficie de temperatura baja a la que puede transmitírsele calor.

Heat tape Electric resistance wires embedded into a flexible housing usually wrapped around a pipe to keep it from freezing.

Cinta calefactora Resistencia eléctrica incrustada en una cubierta flexible normalmente instalada alrededor de un tubo para impedir su congelación.

Heat transfer The transfer of heat from a warmer to a colder substance.

Transmisión de calor Cuando se transmite calor de una sustancia más caliente a una más fría.

Helix coil A bimetal formed into a helix-shaped coil that provides longer travel when heated.

Bobina en forma de hélice Bimetal encofrado en una bobina en forma de hélice que provee mayor movimiento al ser calentado.

HEPA filter An abbreviation for high-efficiency particulate arrestor. These filters are used when a high degree of filtration is desired or required.

Filtro HEPA Abreviatura para filtro de partículas de alto rendimiento. Este tipo de filtro se utiliza cuando se requiere un elevado grado de filtrado.

Hermetic compressor A motor and compressor that are totally sealed by being welded in a container.

Compresor hermético Un motor y un compresor que están totalmente sellados al ser soldados al contenedor.

Hermetic system An enclosed refrigeration system where the motor and compressor are sealed within the same system with the refrigerant.

Sistema hermético Sistema de refrigeracíon cerrado donde el motor y el compresor se obturan dentro del mismo sistema con el refrigerante.

HERS index HERS is an acronym for Home Energy Rating Systems. It is a standardized rating procedure administered by the Residential Energy Services Network (RESNET).

Índice HERS HERS es el acrónimo de "Sistemas de evaluación de energía residencial". Es un procedimiento de puntuación estandarizado, administrado por la Cadena de servicios de energía residencial (RESNET, por sus siglas en inglés).

Hertz Cycles per second.

Hertz Ciclos por segundo.

Hg Abbreviation for the element mercury.

Hg Abreviatura del elemento mercurio.

Hidden heat See Latent heat.

Calor oculto Véase Calor latente.

High-pressure control A control that stops a boiler heating device or a compressor when the pressure becomes too high.

Regulador de alta presión Regulador que detiene la marcha del dispositivo de calentamiento de una caldera o de un compresor cuando la presión alcanza un nivel demasiado alto.

High-pressure (high-side) gauge The gauge used to measure the pressure on the high-pressure side of an air-conditioning or refrigeration system.

Manómetro de presión alta (lado alto) el indicador utilizado para medir la presión en el lado de alta presión del sistema de aire acondicionado o refrigeración.

High side A term used to indicate the high-pressure or condensing side of the refrigeration system.

Lado de alta presión Término utilizado para indicar el lado de alta presión o de condensación del sistema de refrigeración.

High-side float A float-type metering device that is located in the liquid line at the inlet of the evaporator.

Flotador de lado alto Un dispositivo de medición tipo flotador que se encuentra en la tubería de líquido en la entrada del evaporador.

High-temperature refrigeration A refrigeration temperature range starting with evaporator temperatures no lower than 35°F, a range usually used in air-conditioning (cooling).

Refrigeración a temperatura alta Margen de la temperatura de refrigeración que comienza con temperaturas de evaporadores no menores de 35°F (2°C). Este margen se utiliza normalmente en el acondicionamiento de aire (enfriamiento).

High-vacuum pump A pump that can produce a vacuum in the low micron range.

Bomba de vacío alto Bomba que puede generar un vacío dentro del margen de micrón bajo.

Hoar frost The outer layers of frost that form on the hard surface of ice covering the evaporator coil often resemble snow or flaked ice and will hold a lot of entrained air.

Escarcha Las capas exteriores de hielo finamente dividido que se forman en la superficie dura del hielo y cubren el serpentín del evaporador muchas veces tienen la apariencia de nieve o copos de hielo y contienen gran cantidad de aire en su interior.

Hollow wall anchor Can be used in plaster, wallboard, gypsum board, and similar materials. Once the anchor has been set, the screw may be removed as often as necessary.

Anclaje hueco para pared Puede usarse en yeso, panel de yeso y materiales similares. Una vez se coloca el anclaje, el tornillo puede removerse tantas veces como sea necesario.

Horsepower (hp) A unit equal to 33,000 ft-lb of work per minute.

Potencia en caballos (hp en inglés) Unidad equivalente a 33.000 libras-pies de trabajo por minuto.

Hot gas The refrigerant vapor as it leaves the compressor. This is often used to defrost evaporators.

Gas caliente El vapor del refrigerante al salir del compresor. Esto se utiliza con frecuencia para descongelar evaporadores.

Hot gas bypass Piping that allows hot refrigerant gas into the cooler low-pressure side of a refrigeration system usually for system capacity control.

Desviación de gas caliente Tubería que permite la entrada de gas caliente del refrigerante en el lado más frío de baja presión de un sistema de refrigeración, normalmente para la regulación de la capacidad del sistema.

Hot gas defrost A system where the hot refrigerant gases are passed through the evaporator to defrost it.

Descongelación con gas caliente Sistema en el que los gases calientes del refrigerante se pasan a través del evaporador para descongelarlo.

Hot gas line The tubing between the compressor and condenser.

Conducto de gas caliente Tubería entre el compresor y el condensador.

Hot junction That part of a thermocouple or thermopile where heat is applied.

Empalme caliente El lugar en un termopar o pila termoeléctrica donde se aplica el calor.

Hot pulldown The process of lowering the refrigerated space to the design temperature after it has been allowed to warm up considerably over this temperature.

Descenso caliente Proceso de bajar la temperatura del espacio refrigerado a la temperatura de diseño luego de habérsele permitido calentarse a un punto sumamente superior a esta temperatura.

Hot surface ignition A silicon carbide or similar substance is placed in the gas stream of a gas furnace and allowed to get very hot. When the gas impinges on this surface, immediate ignition should occur.

Encendido por superficie caliente Se coloca una sustancia de carburo de silicona o una semejante en el flujo de gas de un horno. Cuando se calienta la sustancia y el gas toca la superficie, se produce un encendido instantáneo.

Hot water heat A heating system using hot water to distribute the heat.

Calor de agua caliente Sistema de calefacción que utiliza agua caliente para la distribución del calor.

Hot wire The wire in an electrical circuit that has a voltage potential between it and another electrical source or between it and ground.

Conductor electrizado Conductor en un circuito eléctrico a través del cual fluye la tensión entre éste y otra fuente de elec tricidad o entre éste y la tierra.

Housing Term used to describe the casing or the shell of a motor.

Carcasa del motor término utilizado para describir la carcasa o la parte de afuera de un motor.

HRV See Heat Recovery Ventilator.

HRV Ver Ventilador de recuperación de calor.

Humidifier A device used to add moisture to the air.

Humedecedor Dispositivo utilizado para agregarle humedad al aire.

Humidistat A control operated by a change in humidity.

Humidistato Regulador activado por un cambio en la humedad.

Humidity Moisture in the air.

Humedad Vapor de agua existente en el ambiente.

Hunting The open and close throttling of a valve that is searching for its set point.

Caza El abrir y cerrar de una válvula que está buscando su punto de ajuste.

Hydraulics Producing mechanical motion by using liquids under pressure.

Hidráulico Generación de movimiento mecánico por medio de líquidos bajo presión.

Hydrocarbons (HCs) Organic compounds containing hydrogen and carbon found in many heating fuels.

Hidrocarburos (HCs en inglés) Compuestos orgánicos que contienen el hidrógeno y el carbón presentes en muchos combustibles de calentamiento.

Hydrochlorofluorocarbons (HCFCs) Refrigerants containing hydrogen, chlorine, fluorine, and carbon, thought to contribute to the depletion of the ozone layer, although not to the extent of chlorofluorocarbons.

Hidrocloroflurocarburos (HCFCs en inglés) Líquidos refrigerantes que contienen hidrógeno, cloro, flúor y carbono, y que, según algunos, han contribuido a la reducción de la capa de ozono aunque no en tal grado como los cloroflurocarburos.

Hydrofluorocarbon (HFC) A chlorine-free refrigerant containing hydrogen, fluorine, and carbon with zero ozone depletion potential.

Hidroflurocarbono (HFC en inglés) Refrigerante libre de cloro, compuesto de hidrógeno, fluoro y carbono que no tiene efectos perjudiciales sobre la capa de ozono.

Hydrofluoroether (HFE) A nontoxic, secondary fluid used in refrigeration systems. HFEs have a zero ozone depletion potential.

Hidrofluroéter (HFE en inglés) Fluido secundario no tóxico utilizado en sistemas de refrigeración. Los HFE no tienen efectos perjudiciales sobre la capa de ozono.

Hydrometer An instrument used to measure the specific gravity of a liquid.

Hidrómetro Instrumento utilizado para medir la gravedad específica de un líquido.

Hydronic Usually refers to a hot water heating system.

Hidrónico Normalmente se refiere a un sistema de calefacción de agua caliente.

Hydrostatic pressure Extreme pressure buildup when a cylinder is 100% full of liquid and increases in temperature.

Presión hidrostática Aumento extremo de presión que se produce cuando un cilindro está lleno de líquido (100 %) y su temperatura aumenta.

Hygrometer An instrument used to measure the amount of moisture in the air.

Higrómetro Instrumento utilizado para medir la cantidad de humedad en el aire.

I

IAQ See Indoor Air Quality.

IAQ Ver Calidad de aire interior.

IC An acronym for direct insulation contact. Applies to recessed light (can light) fixtures ratings.

IC Un acrónimo de aislamiento de contacto directo. Se aplica a las clasificaciones referidas a la iluminación indirecta mediante lámparas dispuestas en fosas en paredes y techos.

ICAT An acronym for (insulation contact and air tightness). Applies to recessed light (can light) fixtures ratings.

ICAT Un acrónimo de aislamiento de contacto y confinamiento del aire. Se aplica a las clasificaciones referidas a la iluminación indirecta mediante lámparas dispuestas en fosas en paredes y techos.

Idler A pulley on which a belt rides. It does not transfer power but is used to provide tension or reduce vibration.

Polea tensora Polea sobre la que se mueve una correa. No sirve para transmitir potencia, pero se utiliza para proveer tensión o disminuir la vibración.

Ignition or flame velocity The rate at which the flame travels through the air and fuel mixture. Different gases have a different rate. The faster the rate of travel, the more explosive the gas.

Velocidad de ignición o de llama La velocidad a la cual una llama viaja a través de la mezcla de aire y combustible. Diferentes gases tienen distintas velocidades. Mientras más rápido el velocidad del viaja, más explosivo en el gas.

Ignition transformer Provides a high-voltage current, usually to produce a spark to ignite a furnace fuel, either gas or oil.

Transformador para encendido Provee una corriente de alta tensión, normalmente para generar una chispa a fin de encender el combustible de uno horno, sea gas o aceite pesado.

Impedance A form of resistance in an alternating current circuit.

Impedancia Forma de resistencia en un circuito de corriente alterna.

Impeller The rotating part of a pump that causes the centrifugal force to develop fluid flow and pressure difference.

Impulsor Pieza giratoria de una bomba que hace que la fuerza centrífuga desarrolle flujo de fluido y una differencia en presión.

Impingement The condition in a gas or oil furnace when the flame strikes the sides of the combustion chamber, resulting in poor combustion efficiency.

Golpeo Condición que occurre en un horno de gas o de aceite pesado cuando la llama golpea los lados de la cámara de combustión. Esta condición trae como resultado un rendimiento de combustión pobre.

Inclined water manometer Indicates air pressures in very low-pressure systems.

Manómetro de agua inclinada Señala las presiones de aire en sistemas de muy baja presión.

Indoor air quality (IAQ) This term generally refers to the study or research of air quality within buildings and the procedures used to improve air quality.

Calidad del aire en el interior (IAQ en inglés) Generalmente, este término hace referencia al estudio o investigación de la calidad del aire en el interior de los edificios, así como a los procesos empleados para su mejora.

Induced draft Induced-draft systems have a combustion blower motor located on the outlet of the heat exchanger. They pull or suck combustion gases through the heat exchanger usually causing a slight negative pressure in the heat exchanger itself.

Corriente inducida Los sistemas de corriente inducida constan de un motor de ventilación de combustión ubicado a la salida del intercambiador de calor. Éstos succionan los gases de combustión a través del intercambiador de calor, produciendo generalmente una leve presión negativa en el intercambiador de calor en sí mismo.

Induced magnetism Magnetism produced, usually in a metal, from another magnetic field.

Magnetismo inducido Magnetismo generado, normalmente en un metal, desde otro campo magnético.

Inductance An induced voltage producing a resistance in an alternating current circuit.

Inductancia Tensión inducida que genera una resistencia en un circuito de corriente alterna.

Induction motor An alternating current motor where the rotor turns from induced magnetism from the field windings.

Motor inductor Motor de corriente alterna donde el rotor gira debido al magnetismo inducido desde los devanados inductores.

Inductive circuit Electrical circuit that contains inductive loads or coils of wire. In this type of circuit, the current lags the voltage.

Circuito inductivo Un circuito que contiene cargas inductivas o bobinas de alambre. En esta clase de circuito, la corriente esta atras del voltaje.

Inductive reactance A resistance to the flow of an alternating current produced by an electromagnetic induction.

Reactancia inductiva Resistencia al flujo de una corriente alterna generada por una inducción electromagnética.

Inefficient equipment Equipment that is not operating at its design level of capacity because of some fault in the equipment, such as a cylinder not pumping in a multicylinder compressor.

Equipo ineficiente Equipo que no está operando al nivel de capacidad al que fue diseñado debido a alguna falla en el equipo, tal como que un cilindro no esté bombeando en un compresor de múltiples cilindros.

Inert gas A gas that will not support most chemical reactions, particularly oxidation.

Gas inerte Gas incapaz de resistir la mayoría de las reacciones químicas, especialmente la oxidación.

Infiltration Air that leaks into a structure through cracks, windows, doors, or other openings due to less pressure inside the structure than outside the structure. Usually accompanied by an equal outflow of air.

Infiltración Aire que se filtra hacia el interior de una estructura a través de grietas, ventanas, puertas u otras aberturas debido a la menor presión que existe dentro de la estructura que fuera de ella. Generalmente está acompañada de un flujo equivalente de salida de aire.

Infrared humidifier A humidifier that has infrared lamps with reflectors to reflect the infrared energy onto the water. The water evaporates rapidly into the duct airstream and is carried throughout the conditioned space.

Humidificador por infrarrojos Humidificador equipado con lámparas de infrarrojos cuyos reflectores reflejan la energía infrarroja sobre el agua haciendo que ésta se evapore rápidamente hacia el conducto de aire para ser transportada en el espacio acondicionado.

Infrared rays The rays that transfer heat by radiation.

Rayos infrarrojos Rayos que transmiten calor por medio de la radiación.

Inherent motor protection This is provided by internal protection such as a snap-disc or a thermistor.

Protección de motor inherente Ésta es provista por una protección interna tal como un disco de encaje o un termisor.

In. Hg vacuum The atmosphere will support a column of mercury 29.92 in. high. To pull a complete vacuum in a refrigeration system, the pressure inside the system must be reduced to 29.92 in. Hg vacuum.

Vacío en mm Hg La atmósfera soporta una columna de mercurio de 760 mm. Para poder crear un vacío completo en un sistema de refrigeración, la presión interna debe descender a 760 mm Hg.

Injection Term used to describe any hydronic system that uses a circulator pump to push water into a secondary heating loop or circuit.

Inyección Término utilizado para describir cualquier sistema hidrónico que usa una bomba de circulación para empujar el agua en un circuito de calefacción secundario.

In-phase When two or more alternating current circuits have the same polarity at all times.

En fase Cuando dos o más circuitos de corriente alterna tienen siempre la misma polaridad.

Input/output board A solid-state electronic board that receives a signal from a microprocessor and initiates a response to cycle compressors or initiate defrost.

Tarjeta de entrada/salida Placa electrónica de estado sólido que recibe una señal procedente de un microprocesador y activa o desactiva los compresores, o inicia el proceso de descongelación.

Insulation, electric A substance that is a poor conductor of electricity.

Aisamiento eléctrico Sustancia que es un conductor pobre de electricidad.

Insulation, thermal A substance that is a poor conductor of the flow of heat.

Aislamiento térmico Sustancia que es un conductor pobre de flujo de calor.

Insulator A material with several electrons in the outer orbit of the atom making them poor conductors of electricity or good insulators. Examples are glass, rubber, and plastic.

Aislamiento Material con varios electrones en la órbita exterior del átomo, que los convierte en malos conductores de electricidad o en buenos aislantes, por ejemplo vidrio, caucho y plástico.

Integral controller A control mode where the controller will change the output signal according to the length of time the error or offset exists. It will calculate the amount of error that exists over a specific time interval.

Controlador integral Un modo de control en cual el controlador cambia la señal de salida de acuerdo al tiempo de duración del error o de la compensación. El mismo calcula la cantidad de error que existe a través de un intervalo específico de tiempo.

Interlocking components Mechanical and electrical interlocks that are used to prevent a piece of equipment from starting before it is safe to start. For example, the chilled-water pump and the condenser water pump must both be started before the compressor in a water-cooled chilled-water system.

Componentes con enclavamiento Enclavamientos mecánicos y eléctricos se usan para evitar que una pieza de equipo se encienda antes de ser seguro que encienda. Por ejemplo, la bomba de agua enfriada y la bomba de agua del condensador deben encenderse antes que el compresor en un sistema de enfriamiento de agua enfriado por agua.

Intermediate zones Another name for unintentionally conditioned spaces.

Zonas intermedias Denominación alternativa para espacios acondicionados de manera no intencional.

Intermittent ignition Ignition system for a gas furnace that operates only when needed or when the furnace is operating.

Encendido interrumpido Sistema de encendido para un horno de gas que funciona solamente cuando es necesario o cuando el horno está trabajando.

Internal heat defrost Heat provided for defrost from inside the system, for example, hot gas defrost.

Descongelador por calor interno El calor para descongelar se provee desde adentro del sistema, por ejemplo, descongelación de gas caliente.

Internal motor overload An overload that is mounted inside the motor housing, such as a snap-disc or thermistor.

Sobrecarga interna del motor Una sobrecarga que se monta dentro del cárter del motor tal como disco de encaje o un termisor.

Interrupted ignition Ignition method that uses an electric spark only at the beginning of the burner's run cycle to ignite the atomized fuel.

Ignición interrumpida Un método de ignición que utiliza una chispa eléctrica solamente al comienzo del ciclo de marcha del quemador para encender el combustible atomizado.

Inverter The switching or transistor section of the variable frequency drive (VFD) that produces an AC voltage at just the right frequency for motor speed control. This section converts the DC voltage back to AC voltage.

Transformador Sección de transistores o interruptores en el impulsor de frecuencia variable (VFD, por sus siglas en inglés) que produce un voltaje de corriente alterna a la frecuencia exacta para el control de la velocidad del motor. Esta sección convierte el voltaje de corriente continua en voltaje de corriente alterna.

In. WC An acronym for inches of water column.

In. WC Un acrónimo de "pulgadas de columna de agua".

Ion generator A device that charges particles. These particles are then passed through positive- and negative-charged plates where the particles are repelled by the positive-charged plates and attracted to the negative-charged plates.

Generador de iones Dispositivo cargador de partículas, las cuales se pasan entre placas con carga positiva y negativa, causando el rechazo de las partículas por las placas con carga positiva, y la atracción por las placas con carga negativa.

Isobars Lines on a chart or graph that represent a constant pressure.

Isobarras Líneas en una gráfica que representan una presión constante.

Isolation relays Components used to prevent stray unwanted electrical feedback that can cause erratic operation.

Relés de aislación Componentes utilizados para evitar la realimentación eléctrica dispersa no deseada que puede ocasionar un funcionamiento errático.

Isotherms Lines on a chart or graph that represent a constant temperature.

Isotermas Líneas en una gráfica que representan una temperatura constante.

J

Joule (J) Metric measurement term used to express the quantity of heat. One joule is approximately 0.00095 Btu.

Joule (J) Término utilizado para referirse a la unidad básica de cantidad de calor en el sistema métrico. Un joule es aproximadamente 0,00095 Btu.

Junction box A metal or plastic box within which electrical connections are made.

Caja de empalme Caja metálica o plástica dentro de la cual se nacen conexiones eléctricas.

K

Kelvin A temperature scale where absolute 0 equals 0 or where molecular motion stops at 0. It has the same graduations per degree of change as the Celsius scale.

Escala absoluta Escale de temperaturas donde el cero absoluto equivale a 0 ó donde el movimiento molecular se detiene en 0. Tiene las mismas graduaciones por grado de cambio que la escala Celsio.

Kilopascal A metric unit of measurement for pressure used in the air-conditioning, heating, and refrigeration field. There are 6.89 kilopascals in 1 psi.

Kilopascal Unided métrica de medida de presión utilizada en el ramo del acondicionamiento de aire, calefacción y refrigeración. 6.89 kilopascales equivalen a 1 psi.

Kilowatt A unit of electrical power equal to 1000 watts.

Kilowatio Unidad eléctrica de potencia equivalente a 1.000 watios.

Kilowatt-hour 1 kilowatt (1000 watts) of energy used for 1 hour.

Kilowatio hora Unidad de energía equivalente a la que produce un kilowatio durante una hora.

Kinetic displacement compressor Term used to describe compressors that use large impellers to increase the kinetic energy and pressure of the refrigerant as opposed to compressors that use mechanical means for compression.

Compresor de desplazamiento cinético El término usado para describir los compresores que utilizan a grandes impulsores para aumentar la energía cinética y presión del refrigerante en comparación con compresores que utilizan medios mecánicos para la compresión.

King valve A service valve at the liquid receiver's outlet in a refrigeration system.

Válvula maestra Válvula de servicio ubicada en el receptor del líquido.

L

Ladder diagram A wiring diagram that separates each individual circuit of a piece of equipment or system into separate, often horizontal, lines to simplify the troubleshooting process.

Diagrama de escalera Un diagrama de cableado que separa cada circuito individual de una pieza de equipo o sistema en distintas, a menudo, las líneas horizontales para simplificar el proceso de solución de problemas.

Lag shield anchors Used with lag screws to secure screws in masonry materials.

Anclajes de tornillos barraqueros Se usan con tornillos barraqueros para asegurar tornillos en materiales de albañilería.

LAMEL An acronym for (lighting, appliance, and miscellaneous, electrical loads). Another way to express a base load of a residence.

LAMEL Acrónimo de "artefactos eléctricos para iluminación, electrodomésticos y misceláneos". Es otra manera de expresar el consumo básico de una vivienda.

Latent heat (hidden heat) Heat energy absorbed or rejected when a substance is changing state and there is no change in temperature.

Calor latente (calor oculto) Energía calórica absorbida o rechazada cuando una sustancia cambia de estado y no se experimentan cambios de temperatura.

Latent heat of condensation The latent heat given off when refrigerant condenses.

Calor latente de la condensación Calor latente producido por la condensación del refrigerante.

Latent heat of fusion The amount of heat energy required to change the state of a substance from a liquid to a solid or froma solid to a liquid.

Calor latente de fusión la cantidad de energía térmica necesaria para cambiar el estado de una sustancia de un líquido a un sólido o de un sólido a líquido.

Latent heat of vaporization The latent heat absorbed when refrigerant evaporates.

Calor latente de vaporización Calor latente absorbido por la evaporación del refrigerante.

Law of conservation of energy Law that states that energy cannot be created nor destroyed, but can be converted from one form to another.

Ley de conservación de la energía Ley que declara que la energía no puede ser creada ni destruida, pero puede ser convertida de una forma a otra.

Leadership in Energy and Environmental Design (LEED) A voluntary national rating system for developing high-performance, sustainable buildings. It is referred to as the LEED Green Building Rating System.

Liderazgo en diseño energético y medioambiental (LEED, por sus siglas en inglés) Sistema nacional voluntario de puntuación para el desarrollo de construcciones de alto rendimiento y sustentables. Se lo conoce como Sistema ecológico LEED de clasificación de construcciones.

Leads Extended surfaces inside a heat exchanger used to enhance the heat transfer qualities of the heat exchanger.

Extensiones Superficies extendidas dentro de un intercambiador de calor que se usan para mejorar las cualidades de transferencia de calor del intercambiador de calor.

Leak Fluid (gas and/or liquid) escaping at different times and at different rates from a system.

Pérdida Fluido (gas y/o líquido) que escapa de un sistema en diferentes momentos y a diferentes velocidades.

Leak detector Any device used to detect leaks in a pressurized system.

Detector de fugas Cualquier dispositivo utilizado para detectar fugas en un sistema presurizado.

LEED An acronym for Leadership in Energy and Environmental Design.

LEED Acrónimo de Liderazgo en diseño energético y medioambiental.

LEED (NC) An acronym for Leadership in Energy and Environmental Design (New Construction).

LEED (NC) Acrónimo de Liderazgo en diseño energético y medioambiental para construcciones nuevas (LEED (NC), por sus siglas en inglés).

Lever truck A long-handled, two-wheeled device that can be used to lift and assist in moving heavy objects.

Vagoneta con palanca Dispositivo con dos ruedas y una manivela larga que puede utilizarse para levantar y ayudar a transportar objetos pesados.

Light-emitting diode (LED) A diode that emits light and often is used as a self-diagnostic tool within a microprocessor. The lights can spell out error codes or flash numbers.

Diodo emisor de luz (LED en inglés) Diodo que emite una luz. A menudo se utiliza como dispositivo de autodiagnóstico en un microprocesador. Las luces pueden indicar códigos de error o números.

Lignin A fibrous material found in trees and that strengthens the cell wall.

Lignina Tejido fibroso presente en la madera y que sirve para reforzar la pared celular.

Limit control A control used to make a change in a system, usually to stop it when predetermined limits of pressure or temperature are reached.

Regulador de límite Regulador utilizado para realizar un cambio en un sistema, normalmente para detener su marcha cuando se alcanzan niveles predeterminados de presión o de temperatura.

Limit switch A switch that is designed to stop a piece of equipment before it does damage to itself or the surroundings, for example, a high limit on a furnace or an amperage limit on a motor.

Interruptor de límite Interruptor que está diseñado para detener una pieza de equipo antes que ésta se haga daño o dañe sus alrededores, por ejemplo, un límite alto en un calefactor o un límite de amperaje en un motor.

Line set A term used for tubing sets, sometimes furnished by the manufacturer.

Juego de conductos Término utilizado para referise a los juegos de tubería, a veces suministrados por el fabricante.

Line tap valve A device that may be used for access to a refrigerant line.

Válvula de acceso en línea Dispositivo que puede utilizarse para acceder al tubo del refrigerante.

Line-voltage thermostat A thermostat that switches line voltage. For example, it is used for electric baseboard heat.

Termostato de voltaje de línea Un termostato que interrumpe el voltaje de línea. Por ejemplo, para los calentadores eléctricos de rodapié.

Line wiring diagram Sometimes called a ladder diagram, this type of diagram shows the power-consuming devices between the lines. Usually, the right side of the diagram consists of a common line.

Diagrama del cableado de línea También conocido como diagrama de escalera, este tipo de diagrama muestra los dispositivos de consumo de corriente que hay entre las líneas. Generalmente, la línea común está en el lado derecho del diagrama.

Liquid A substance where molecules push outward and downward and seek a uniform level.

Líquido Sustancia donde las moléculas empujan hacia afuera y hacia abajo y buscan un nivel uniforme.

Liquid charge bulb A type of charge in the sensing bulb of the thermostatic expansion valve. This charge is characteristic of the refrigerant in the system and contains enough liquid so that it will not totally boil away.

Bombilla de carga líquida Tipo de carga en la bombilla sensora de la válvula de expansión termostática. Esta carga es característica del refrigerante en el sistema y contiene suficiente líquido para que el mismo no se evapore completamente.

Liquid collector design A solar collector design in which either water or an antifreeze solution is passed through the collector and heated by the sun.

Colector de líquido Colector a través del cual pasa el agua o el anticongelante para ser calentados por la energía solar.

Liquid-filled remote bulb A remote-bulb thermostat that is completely liquid filled, such as the mercury bulb on some gas furnace pilot safety devices.

Bombillo remoto lleno de líquido Un termostato de bombillo remoto que está completamente lleno con líquido, tal como el bulbo de mercurio en algunos dispositivos de seguridad de los calefactores de gas.

Liquid floodback Liquid refrigerant returning to the compressor's crankcase during the running cycle.

Regreso de líquido El regresar del refrigerante líquido al cárter del cigüeñal del compresor durante el ciclo de marcha.

Liquid hammer The momentum force of liquid causing a noise or a disturbance when hitting against an object.

Martillo líquido La fuerza mecánica de un líquido que causa un ruido o un disturbio cuando choca con un objeto.

Liquid line A term applied in the industry to refer to the tubing or piping from the condenser to the expansion device.

Conducto de líquido Término aplicado en nuestro medio para referirse a la tubería que va del condensador al dispositivo de expansión.

Liquid nitrogen Nitrogen in liquid form.

Nitrógeno líquido Nitrógeno en forma líquida.

Liquid receiver A container in the refrigeration system where liquid refrigerant is stored.

Receptor del líquido Recipiente en el sistema de refrigeración donde se almacena el refrigerante líquido.

Liquid refrigerant charging The process of allowing liquid refrigerant to enter the refrigeration system through the liquid line to the condenser and evaporator.

Carga para refrigerante líquido Proceso de permitir la entrada del refrigerante líquido al condensador y al evaporador en el sistema de refrigeración a través del conducto de líquido.

Liquid refrigerant distributor This device is used between the expansion valve and the evaporator on multiple circuit evaporators to evenly distribute the refrigerant to all circuits.

Distribuidor de refrigerante líquido Este aparato se usa entre la válvula de expansión y el evaporador en los evaporadores de múltiples circuitos para distribuir el refrigerante a todos los circuitos.

Liquid slugging A large amount of liquid refrigerant in the compressor cylinder, usually causing immediate damage.

Relleno de líquido Acumulación de una gran cantidad de refrigerante líquido en el cilindro del compresor, que normalmente provoca una avería inmediata.

Liquefied petroleum Liquified propane, butane, or a combination of these gases. The gas is kept as a liquid under pressure until ready to use.

Petróleo licuado Propano o butano licuados, o una combinación de estos gases. El gas se mantiene en estado líquido bajo presión hasta que se encuentre listo para usar.

Lithium-bromide A type of salt solution used in an absorption chiller.

Bromuro de litio Tipo de solución salina utilizada en un enfriador por absorción.

Load matching Trying to always match the capacity of the refrigeration or air-conditioning system with that of the heat load put on the evaporators.

Adaptación de carga Hecho de intentar adaptar la capacidad del sistema de refrigeración o aire acondicionado a la carga térmica que deben soportar los evaporadores.

Load shed Part of an energy management system where various systems in a structure may be cycled off to conserve energy.

Despojo de carga Parte de un sistema de manejo de energía en el cual varios sistemas en una estructura pueden apagarse para conservar energía.

Locked-rotor amperage (LRA) The current an electric motor draws when power is applied to the motor, but its rotor is not yet turning. A rule-of-thumb often used by service technicians is that the LRA is normally five times the rated-load amperage (RLA).

Amperaje de rotor bloqueado (LRA, por sus siglas en inglés) Corriente que consume un motor eléctrico cuando se enciende, antes de que comience a girar el rotor. Generalmente, según lo estimado por técnicos mecánicos, el LRA es cinco veces mayor que el amperaje de carga calificada (RLA).

Low-ambient control Various types of controls that are used to control head pressure in air-cooled air-conditioning and refrigeration systems that must operate year-round or in cold weather.

Control de ambiente bajo Varios tipos de controles que se usan para controlar la presión en los sistemas de aires acondicionados y refrigeración, enfriados por aire, y que tienen que operar todo el año o en climas fríos.

Low-boy furnace This furnace is approximately 4 ft high, and the air intake and discharge are both at the top.

Horno bajo Este tipo de horno tiene una altura de aproximadamente 1,2 metros, con la toma y evacuación del aire situadas en la parte de arriba.

Lower exposure limit (LEL) Acceptable average exposure limit to a chemical, for example, over a short period of time, usually 10 or 15 min.

Límite de exposición inferior (LEL) límite promedio de exposición aceptable para un producto químico, por ejemplo, durante un corto período de tiempo, generalmente 10 ó 15 min.

Low-loss fitting A fitting that is fastened to the end of a gauge manifold that allows the technician to connect and disconnect gauge lines with a minimum of refrigerant loss.

Acoplamiento de poca pérdida Un tipo de acoplamiento que se conecta en un extremo de un manómetro de distribución y que le permite al técnico conectar y desconectar las líneas de flujo con una pérdida mínima de refrigerante.

Low-pressure control A pressure switch that can provide low charge protection by shutting down the system on low pressure. It can also be used to control space temperature.

Regulador de baja presión Commudador de presión que puede proveer protección contra una carga baja al detener el sistema si éste alcanza una presión demasiado baja. Puede utilizarse también para regular la temperatura de un espacio.

Low side A term used to refer to that part of the refrigeration system that operates at the lowest pressure, between the expansion device and the compressor.

Lado bajo Término utilizado para referirse a la parte del sistema de refrigeración que funciona a niveles de presión más baja, entre el dispositivo de expansión y el compresor.

Low-side float A float-type metering device that is located in the evaporator. As the level of liquid refrigerant drops, the valve opens to feed more refrigerant into the evaporator.

Flotador de lado bajo Un dispositivo de medición tipo flotador que se encuentra en el evaporador. A medida que el nivel de líquido refrigerante cae, la válvula se abre para alimentar más el refrigerante en el evaporador.

Low-side gauge The gauge that measures the low-side pressure in an air-conditioning or refrigeration system.

El manómetro del lado bajo el indicador que mide la presión del lado bajo en un sistema de aire acondicionado o refrigeración.

Low-temperature refrigeration A refrigeration temperature range starting with evaporator temperatures no higher than 0°F for storing frozen food.

Refrigeración a temperatura baja Margen de la temperatura de refrigeración que comienza con temperaturas de evaporadores no mayores de 0°F (–18°C) para almacenar comida congelada.

Low-voltage thermostat The typical thermostat used for residential and commercial air-conditioning and heating equipment to control space temperature. The supplied voltage is 24 V.

Termostato de bajo voltaje El termostato típico que se usa en el equipo de aire acondicionado y de calefacción comercial y residencial para controlar la temperatura de un espacio. El voltaje suplido es de 24 voltios.

LP fuel Liquefied petroleum, propane, or butane. A substance used as a gas for fuel. It is transported and stored in the liquid state.

Combustible PL Petróleo licuado, propano, o butano. Sustancia utilizada como gas para combustible. El petróleo licuado se transporta y almacena en estado líquido.

M

Magnetic field A field or space where magnetic lines of force exist.

Campó magnético Campo o espacio donde existen líneas de fuerza magnética.

Magnetic overload protection This protection reads the actual current draw of the motor and is able to shut it off based on actual current, versus the heat-operated thermal overloads, which are sensitive to the ambient heat of a hot cabinet.

Protección de sobrecarga magnética Esta protección lee la toma de corriente actual del motor y es capaz de apagarlo basado en la corriente actual; esto es contrario a las sobrecargas termales operadas por calor, las cuales son sensibles al calor ambiental de un gabinete caliente.

Magnetism A force causing a magnetic field to attract ferrous metals, or where like poles of a magnet repel and unlike poles attract each other.

Magnetismo Fuerza que hace que un campo magnético atraiga metales férreos, o cuando los polos iguales de un imán se rechazan y los opuestos se atraen.

Makeup air Air, usually from outdoors, provided to make up for the air used in combustion.

Aire de compensación Aire, normalmente procedente del exterior, que se utiliza para compensar aquél utilizado en la combustión.

Makeup water Water that is added back into any circulating water system due to loss of water. Makeup water in a cooling tower may be quite a large volume.

Agua de compensación Agua que se añade a cualquier sistema de circulación de agua debido a la pérdida de agua. El agua de compensación en una torre de enfriamiento puede ser un volumen bastante grande.

Male thread A thread on the outside of a pipe, fitting, or cylinder; an external thread.

Filete macho Filete en la parte exterior de un tubo, accesorio o cilindro; filete externo.

Mandrel (clinching) Part of the pin rivet assembly that is pulled to the breaking point to swell the rivet into the drilled hole.

Mandrel (remachando) En la instalación con remanches, la parte que se tira hasta el límite para hinchar el remanche en el hueco perforado.

Manifold A device where multiple outlets or inlets can be controlled with valves or other devices. Our industry typically uses a gas manifold with orifices for gas-burning appliances and gauge manifolds used by technicians.

Colector Aparato desde el cual se pueden controlar varias entradas y salidas con válvulas u otros aparatos. Nuestra industria generalmente usa un colector de gas con orificios para los enseres que queman gas y manómetros de distribución que los usan los técnicos.

Manometer An instrument used to check low vapor pressures. The pressures may be checked against a column of mercury or water.

Manómetro Instrumento utilizado para revisar las presiones bajas de vapor. Las presiones pueden revisarse comparándolas con una columna de mercurio o de agua.

Manual reset A safety control that must be reset by a person, as opposed to automatically reset, to call attention to the problem. An electrical breaker is a manual reset device.

Reinicio manual Un control de seguridad que una persona tiene que reiniciar, contrario a un reinicio automático, para llamar atención al problema. Un cortacircuito eléctrico es un aparato de reinicio manual.

MAPP™ gas A composite gas similar to propane that may be used with air.

Gas MAPP™ Gas compuesto similar al propano que puede utilizarse con aire.

Marine water box A water box on a chiller or condenser with a removable cover.

Caja marina para agua Caja para agua en un enfriador o condensador con un capón desmontable.

Mass Matter held together to the extent that it is considered one body.

Masa Materia compacta que se considera un solo cuerpo.

Mass spectrum analysis An absorption machine factory leak test performed using helium.

Análisis del límite de masa Prueba para fugas y absorción llevada a cabo en la fábrica utilizando helio.

Matter A substance that takes up space and has weight.

Materia Sustancia que ocupa espacio y tiene peso.

Mechanical controls A control that has no connection to power, such as a water-regulating valve or a pressure relief valve.

Controles mecánicos Un control que no tiene conexión a corriente, tales como una válvula reguladora de agua o una válvula de alivio de presión.

Medium-temperature refrigeration Refrigeration where evaporator temperatures are 32°F or below, normally used for preserving fresh food.

Refrigeración a temperatura media Refrigeración, donde las temperaturas del evaporador son 32°F (0°C) o menos, utilizada generalmente para preservar comida fresca.

Megohmmeter An instrument that can detect very high resistances, in millions of ohms. A megohm is equal to 1,000,000 ohms.

Megaohmnímetro Un instrumento que puede detectar resistencias muy altas, de millones de ohmios. Un megaohmio es equivalente a 1.000.000 de ohmios.

Melting point The temperature at which a substance will change from a solid to a liquid.

Punto de fusión Temperatura a la que una sustancia se convierte de sólido a líquido.

Memory Electronic storage space located internally in a controller.

Memoria Almacenes internos electrónicos en la computadora.

Mercury bulb A glass bulb containing a small amount of mercury and electrical contacts used to make and break the electrical circuit in a low-voltage thermostat.

Bombilla de mercurio Bombilla de cristal que contiene una pequeña cantidad de mercurio y que funciona como contacto eléctrico, utilizada para conectar y desconectar el circuito eléctrico en un termostato de baja tensión.

Metering device A valve or small fixed-size tubing or orifice that meters liquid refrigerant into the evaporator.

Dispositivo de medida Válvula o tubería pequeña u orificio que mide la cantidad de refrigerante líquido que entra en el evaporador.

Methane Natural gas composed of 90% to 95% methane, a combustible hydrocarbon.

Metano El gas natural se compone de un 90% a un 95% de metano, un hidrocarburo combustible.

Metric system System International (SI); system of measurement used by most countries in the world.

Sistema métrico Sistema internacional; el sistema de medida utilizado por la mayoría de los países del mundo.

Micro A prefix meaning 1/1,000,000.

Micro Prefijo que significa una parte de un millón.

Microfarad Capacitor capacity equal to 1/1,000,000 of a farad.

Microfaradio Capacidad de un capacitador equivalente a 1/1.000.000 de un faradio.

Micro-foam solution A solution usually sprayed or brushed in the vicinity of a system leak which will form small cocoons of foam when in contact with a refrigerant or other type of gas leak.

Solución de microespuma Solución que generalmente se pulveriza o pincela en las proximidades de una pérdida y que forma pequeños globos de espuma cuando entra en contacto con un refrigerante u otro tipo de pérdida de gas.

Micrometer A precision measuring instrument.

Micrómetro Instrumento de precisión utilizado para medir.

Micron A unit of length equal to 1/1000 of a millimeter, 1/1,000,000 of a meter.

Micrón Unidad de largo equivalente a 1/1000 de un milímetro, o 1/1.000.000 de un metro.

Micron gauge A gauge used when it is necessary to measure pressure close to a perfect vacuum.

Manómetro de micrón Manómetro utilizado cuando es necesario medir la presión de un vacío casi perfecto.

Microprocessor A small, preprogrammed, solid-state microcomputer that acts as a main controller.

Microprocesador Un microordenador de estado sólido preprogramado que actúa de controlador principal.

Midseated (cracked) A position on a service valve that allows refrigerant flow in all directions.

Sentado en el medio (agrietado) Posición en una válvula de servicio que permite el flujo de refrigerante en cualquier dirección.

Migration of oil or refrigerant When the refrigerant moves to some place in the system where it is not supposed to be, such as when oil migrates to an evaporator or when refrigerant migrates to a compressor crankcase.

Migración de aceite o refrigerante Cuando el refrigerante se mueve a cualquier lugar en el sistema donde no debe estar, como cuando aceite migra a un evaporador o cuando refrigerante se transplanta al cárter del cigüeñal del compresor.

Milli A prefix meaning 1/1000.

Mili Prefijo que significa una parte de mil.

Mineral oil A traditional refrigeration lubricant used in CFC and HCFC systems.

Aceite mineral Lubricante de refrigeración utilizado tradicionalmente en los sistemas de CFC y HCFC.

Minimum efficiency reporting value (MERV) Air filter rating system ranging from 1 to 20, with upper levels providing the most filtering.

Valor mínimo de reporte de eficacia (MERV en inglés) Sistema de clasificación de filtros de aire que va desde el 1 al 20 con los niveles mayores indicando más filtración.

Modulating flow Controlling the flow between maximum or no flow. For example, the accelerator on a car provides modulating flow.

Flujo modulante Controlado el flujo entre flujo que no es flujo máximo o ningún flujo. Por ejemplo, el acelerador de un carro provee un flujo modulante.

Modulator A device that adjusts by small increments or changes.

Modulador Dispositivo que se ajusta por medio de incrementos o cambios pequeños.

Moisture indicator A device for determining moisture.

Indicador de humedad Dispositivo utilizado para determinar la humedad.

Mold A fungus found where there is moisture that develops and releases spores. Can be harmful to humans.

Moho Un hongo en con trado donde hay humedad que se desarrolla y libera esporas. Puede ser nocivo a los humanos.

Molecular motion The movement of molecules within a substance.

Movimiento molecular Movimiento de moléculas dentro de una sustancia.

Molecule The smallest particle that a substance can be broken into and still retain its chemical identity.

Molécula La particula más pequeña en la que una sustancia puede dividirse y aún conservar sus propias características.

Monochlorodifluoromethane Chemical name for refrigerant R-22.

Monoclorodifluorometano Nombre químico para el refrigerante R-22.

Montreal Protocol An agreement signed in 1987 by the United States and other countries to control the release of ozone-depleting substances.

Protocolo de Montreal Un acuerdo firmado en 1987 por los Estados Unidos y otros países para controlar la liberación de sustancias que destruyen el ozono.

Motor service factor A factor above an electric motor's normal operating design parameters, indicated on the nameplate, under which it can operate.

Factor de servicio del motor Factor superior a los parametros de diseño normales de funcionamiento de un motor eléctrico, indicados en el marbete; este factor indica su nivel de funcionamiento.

Motor starter Electromagnetic contactors that contain motor protection and are used for switching electric motors on and off.

Arrancador de motor Contactadores electromagnéticos que contienen protección para el motor y se utilizan para arrancar y detener motores eléctricos.

Motor temperature-sensing thermostat A thermostat that monitors the motor temperature and shuts it off for the motor's protection.

Termostato que detecta la temperatura del motor Un termostato que vigila la temperatura del motor y lo apaga para la protección del motor.

Muffler, compressor Sound absorber at the compressor.

Silenciador del compresor Absorbedor de sonido ubicado en el compresor.

Mullion Stationary frame between two doors.

Parteluz Armazón fijo entre dos puertas.

Mullion heater Heating element mounted in the mullion of a refrigerator to keep moisture from forming on it.

Calentador del parteluz Elemento de calentamiento montado en el parteluz de un refrigerador para evitar la formación de humedad en el mismo.

Multimeter An instrument that will measure voltage, resistance, and milliamperes.

Multímetro Instrumento que mide la tensión, la resistencia y los miliamperios.

Multiple circuit coil An evaporator or condenser coil that has more than one circuit because of the coil length. When the coil is too long, there will be an unacceptable pressure drop and loss of efficiency.

Serpentín de circuito múltiple Un serpentín de evaporador o condensador que tiene más de un circuito por causa de la longitud del serpentín. Cuando el serpentín es demasiado largo, habrá una pérdida de presión y eficacia inaceptable.

Multiple evacuation A procedure for evacuating a system. A vacuum is pulled, a small amount of refrigerant allowed into the system, and the procedure duplicated. This is often done three times.

Evacuación múltiple Procedimiento para evacuar o vaciar un sistema. Se crea un vacío, se permite la entrada de una pequeña cantidad de refrigerante al sistema, y se repite el procedimiento. Con frecuencia esto se lleva a cabo tres veces.

N

Naphthenic oil A refrigeration mineral oil refined from California and Texas crude oil.

Aceite nafténico Aceite mineral refinado utilizado en la refrigeración, obtenido del petróleo crudo de California y Texas.

National Electrical Code® (NEC®) A publication that sets the standards for all electrical installations, including motor overload protection.

Código estadounidense de electridad Publicación que establece las normas para todas las instalaciones eléctricas, incluyendo la protección contra la sobrecarga de un motor.

National Fire Protection Association (NFPA) An association organized to prevent fires through establishing standards, providing research, and providing public education.

Asociación nacional para la protección contra incendios (NFPA en inglés) Asociación cuyo objetivo es prevenir incendios estableciendo normativas y facilitando la investigación y concienciación del público.

National pipe taper (NPT) The standard designation for a standard tapered pipe thread.

Cono estadounidense para tubos (NPT en inglés) Designación estándar para una rosca cónica para tubos estándar.

Natural convection The natural movement of a gas or fluid caused by differences in temperature.

Convección natural Movimiento natural de un gas o fluido ocasionado por diferencias en temperatura.

Natural-draft tower A water cooling tower that does not have a fan to force air over the water. It relies on the natural breeze or airflow.

Torre de corriente de aire natural Una torre de enfriamiento de agua que no tiene un ventilador para forzar el aire sobre el agua; depende de la brisa o flujo natural del aire.

Natural gas A fossil fuel formed over millions of years from dead vegetation and animals that were deposited or washed deep into the earth.

Gas natural Combustible fósil formado a través de millones de años de la vegetación y los animales muertos que fueron depositados or arrastrados a un gran profundidad dentro la tierra.

Natural refrigerants Natural refrigerants occur naturally in nature.

Refrigerantes naturales Los refrigerantes naturales se obtienen directamente de fuentes naturales.

Near-azeotropic blend Two or more refrigerants mixed together that will have a small range of boiling and/or condensing points for each system pressure. Small fractionation and temperature glides will occur but are often negligible.

Mezcla casi-azeotrópica Mezcla de dos o más refrigerantes, que tiene un bajo rango de punto de ebullición y/o condensación para cada presión del sistema. Puede producirse una cierta fraccionación y variación de temperatura, aunque suelen ser insignificantes.

Needlepoint valve A device having a needle and a very small orifice for controlling the flow of a fluid.

Válvula de aguja Dispositivo que tiene una aguja y un orificio bastante pequeño para regular el flujo de un fluido.

Negative electrical charge An atom or component that has an excess of electrons.

Carga eléctrica negativa Átomo o componente que tiene un exceso de electrones.

Neoprene Synthetic flexible material used for gaskets and seals.

Neopreno Material sintético flexible utilizado en guarniciones y juntas de estanqueidad.

Net oil pressure Difference in the crankcase pressure and the compressor oil pump outlet pressure.

Presión neta del aceite Diferencia en la presión del cárter y la presión a la salida de la bomba de aceite del compresor.

Net refrigeration effect (NRE) The quantity of heat in Btu/lb that the refrigerant absorbs from the refrigerated space to produce useful cooling.

Efecto neto de refrigeración (NRE en inglés) La cantidad de calor expresado en Btu/lb que el refrigerante absorbe del espacio refrigerado para producir refrigeración útil.

Net stack temperature The temperature difference between the ambient temperature and the flue-gas temperature, typically for oil- and gas-burning equipment.

Temperatura neta de chimenea La diferencia en temperatura entre la temperatura ambiental y la del conducto de gas, normalmente para equipo que quema aceite y gas.

Neutralizer A substance used to counteract acids.

Neutralizador Sustancia utilizada para contrarrestar ácidos.

Neutron Neutrons and protons are located at the center of the nucleus of an atom. Neutrons have no charge.

Neutrón Los neutrones y protones están situados en le centro del núcleo del átomo. Los neutrones carecen de carga.

Newton/meter^2 Metric unit of measurement for pressure. Also called a pascal.

Metro-Newton^2 Unidad métrica de medida de presión. Conocido también como pascal.

NFPA An acronym for the National Fire Protection Association.

NFPA Un acrónimo en inglés para la Asociación Nacional de Protección contra Incendios.

Nichrome A metal made of nickel chromium that when formed into a wire is used as a resistance heating element in electric heaters and furnaces.

Níquel-cromio Metal fabricado de níquel-cromio que al ser convertido en alambre, se utiliza como un elemento de calentamiento de resistencia en calentadores y hornos eléctricos.

Nitrogen An inert gas often used to "sweep" a refrigeration system to help ensure that all refrigerant and contaminants have been removed.

Nitrógeno Gas inerte utilizado con frecuencia para purgar un sistema de refrigeración. Esta gas ayuda a asegurar la remoción de todo el refrigerante y los contaminantes del sistema.

Nitrogen dioxide A combustion pollutant that causes irritation of the respiratory tract and can cause shortness of breath.

Bióxido de nitrógeno Un contaminante por combustión que causa irritación a las vías respiratorias y puede causar falta de aliento.

Nominal A rounded-off stated size. The nominal size is the closest rounded-off size.

Nominal Tamaño redondeado establecido. El tamaño nominal es el tamaño redondeado más cercano.

Nominal Term used to describe the outside diameter of ACR tubing and the inside diameter of piping used for plumbing applications.

Nominal término utilizado para describir el diámetro exterior del tubo de ACR y el diámetro interior de la tubería utilizada para aplicaciones de plomería.

Noncatalytic stove Wood-burning stoves without catalytic combustors and that were manufactured after July 1, 1990, are allowed to emit no more than 7.5 g/h of particulates.

Estufa no catalítica Estufa de madera sin quemadores catalíticos. Las estufas fabricadas después del 1 de julio de 1990 no pueden emitir más de 7,5 g/h de partículas.

Noncondensable gas A gas that does not change into a liquid under normal operating conditions.

Gas no condensable Gas que no se convierte en líquido bajo condiciones de functionamiento normales.

Nonferrous Metals containing no iron.

No férreos Metales que no contienen hierro.

Nonpolycyclic organic matter Air pollutants from wood-burning stoves considered to be health hazards.

Materia orgánica no policíclica Contaminantes del aire procedentes de estufas de madera. Se considera que son peligrosos para la salud.

Non-power-consuming devices Devices that do not consume power, such as switches, that pass power to power- consuming devices.

Aparatos que no consumen corriente Aparatos que no consumen corriente, tales como interruptores que trasmiten corriente a aparatos que consumen corriente.

North Pole, magnetic One end of a magnet or the magnetic north pole of the earth.

Polo norte magnético El extremo de un imán o el polo norte magnético del mundo.

Nozzle A drilled opening that measures liquid flow, such as an oil burner nozzle.

Tobera Una apertura taladrada que mide el flujo de líquido, tal como la tobera de un quemador de aceite.

N-type material Semiconductor material with an excess of electrons that move from negative to positive when a voltage is applied.

Material tipo N Material semiconductor con un exceso de electrones que pasan de negativo a positivo al aplicarles una corriente.

Nut driver These tools have a socket head used primarily to turn hex head screws on air-conditioning, heating, and refrigeration cabinets.

Extractor de tuercas Estas herramientas tienen una cabeza hueca hexagonal usadas principalmente para darle vuelta a tuercas de cabeza hexagonal en gabinetes de acondicionamiento de aire, de calefacción y de refrigeración.

O

Off cycle A period when a system is not operating.

Ciclo de apagado Período de tiempo cuando un sistema no está en funcionamiento.

Off-cycle defrost Used for medium-temperature refrigeration where the evaporator coil operates below freezing but the air in the cooler is above freezing. The coil is defrosted by the air inside the cooler while the compressor is off cycle.

Descongelación de período de reposo Se usa para refrigeración de temperatura media en la cual el serpentín del evaporador funciona por debajo del punto de congelación, pero el aire en el enfriador está por encima del punto de congelación. El serpentín se descongela por el aire dentro del enfriador mientras el compresor está en reposo.

Offset (controls application) The absolute (not signed + or -) difference between the set point and the control point of a control process.

Desplazamiento (aplicación de control) La absoluta (no firmado + o -) diferencia entre el punto de ajuste y el punto de control de un proceso de control.

Offset (duct application) A fitting that is used when a duct system must be routed around an obstacle.

Desplazamiento (aplicación de tubería) Un accesorio que se utiliza cuando un conducto sistema debe colocarse alrededor de un obstáculo.

Offset The absolute (not signed + or −) difference between the set point and the control point of a control process.

Compensación La diferencia absoluta (sin signo + o −) entre el punto de ajuste y el punto de control de un proceso de control.

Offset The position of ductwork that must be rerouted around an obstacle.

Desviación El posición de un conducto que tiene que desviarse alrededor de un obstáculo.

Ohm A unit of measurement of resistance to electric current flow.

Ohmio Unidad de medida de la resistencia eléctrica al flujo de corriente.

Ohmmeter A meter that measures electrical resistance.

Ohmiómetro Instrumento que mide la resistencia eléctrica.

Ohm's law A law involving electrical relationships discovered by Georg Ohm: $E = I \times R$.

Ley de Ohm Ley que define las relaciones eléctricas, descubierta por Georg Ohm: $E = I \times R$.

Oil, refrigeration Oil used in refrigeration systems.

Aceite de refrigeración Aceite utilizado en sistemas de refrigeración.

Oil level regulator A needle valve and float system located on each compressor of a parallel compressor system. It senses the oil level in the compressor's crankcase and adds oil if necessary. It receives its oil from the oil reservoir.

Regulador del nivel de aceite Sistema de válvula de aguja y flotador que se encuentra en cada compresor de un sistema de compresores en paralelo. Detecta el nivel del aceite en el cárter del compresor y permite la entrada de más aceite procedente de un depósito, en caso necesario.

Oil-pressure safety control (switch) A control used to ensure that a compressor has adequate oil lubricating pressure.

Regulador de seguridad para la presión de aceite (conmutador) Regulador utilizado para asegurar que un compresor tenga la presión de lubrificación de aceita adecuada.

Oil reservoir A storage cylinder for oil usually used on parallel compressor systems. It is located between the oil separator and the oil level regulators. It receives its oil from the oil separator.

Depósito de aceite Cilindro en el que se almacena el aceite utilizado en los sistemas de compresores en paralelo. Está ubicado entre el separador de aceite y el regulador del nivel. Recibe el aceite del separador.

Oil separator Apparatus that removes oil from a gaseous refrigerant.

Separador de aceite Aparato que remueve el aceite de un refrigerante gaseoso.

Onetime relief valve A pressure relief valve that has a diaphragm that blows out due to excess pressure. It is set at a higher pressure than the spring-loaded relief valve in case it fails.

Válvula de alivio de una vez Una válvula de alivio de presión que tiene un diafragma que revienta debido a la presión excesiva. La misma se fija a una presión más alta que la válvula de alivio de resorte para en caso de que ésta falle.

One ton of refrigeration Term used to describe the transfer of heat at a rate of 12,000 Btu/h.

Una tonelada de refrigeración Término utilizado para describir la transferencia de calor a una velocidad de 12,000 Btu/h.

Open combustion device An atmospheric draft heating appliance.

Dispositivo de combustión abierta Calefactor de corriente atmosférico.

Open compressor A compressor with an external drive.

Compresor abierto Compresor con un motor externo.

Open-loop control configuration In an open-loop control configuration, the sensor is located upstream of the location where the control agent is causing a change in the controlled medium.

Configuración de control de circuito abierto En una configuración de control de circuito abierto, el sensor está ubicado a contracorriente de donde el agente de control está causando un cambio en el medio controlado.

Open-loop heat pump Heat pump system that uses the water in the earth as the heat transfer medium and then expels the water back to the earth in some manner.

Bomba de calor de circuito abierto Sistema de bomba de calor que utiliza el agua de la tierra como medio de transferencia del calor y luego devuelve el agua a la tierra de cierta manera.

Open winding The condition that exists when there is a break and no continuity in an electric motor winding.

Devanado abierto Condición que se presenta cuando hay una interrupción en la continuidad del devanado de un motor.

Operating pressure The actual pressure under operating conditions.

Presión de funcionamiento La presión real bajo las condiciones de funcionamiento.

Organic Materials formed from living organisms.

Orgánico Materiales formados de organismos vivos.

Orifice A small opening through which fluid flows.

Orificio Pequeña abertura a través de la cual fluye un fluido.

OSHA Abbreviation for the Occupational Safety and Health Administration.

OSHA Siglas de la Administración de salud y seguridad ocupacional.

Outward clinch tacker A stapler or tacker that will anchor staples outward and can be used with soft materials.

Grapadora de agarre hacia fuera Grapadora o tachueladora que ancla las grapas hacia fuera y que puede usarse con materiales suaves.

Overload protection A system or device that will shut down a system if an overcurrent condition exists.

Protección contra sobrecarga Sistema o dispositivo que detendrá la marcha de un sistema si existe una condición de sobreintensidad.

Over-the-fire draft Draft taken over the fire and will be a lower reading lower than chimney draft.

Corriente sobre el fuego Corriente que pasa sobre el fuego; la lectura de presión será menor que la lectura para una corriente de chimenea.

Oxidation The combining of a material with oxygen to form a different substance. This results in the deterioration of the original substance. Rust is oxidation.

Oxidación La combinación de un material con oxígeno para formar una sustancia diferente, lo que ocasiona el deterioro de la sustancia original. Herrumbre es oxidación.

Oxyacetylene Term used to describe a fuel made up of pure oxygen and acetylene.

Oxiacetileno término utilizado para describir un combustible compuesto de oxígeno puro y acetileno.

Oxyacetylene welding units A torch kit that utilizes a mixture of pure oxygen and acetylene in order to produce a much hotter flame than an air-acetylene torch kit.

Unidades de soldadura oxiacetilénica un juego de soplete que utiliza una mezcla de oxígeno puro y acetileno para producir una llama mucho más caliente que la llama de un juego de soplete de aire-acetileno.

Ozone A form of oxygen (O_3). A layer of ozone is in the stratosphere that protects the earth from some of the sun's ultraviolet radiation.

Ozono Forma de oxígeno (O_3). Una capa de ozono en la estratosfera protege la Tierra de parte de la radiación ultravioleta proveniente del sol.

Ozone depletion The breaking up of the ozone molecule by the chlorine atom in the stratosphere. Stratosphere ozone protects us from ultraviolet radiation emitted by the sun.

Reducción del ozono Descomposición de la molécula de ozono por el átomo de cloro en la estratosfera. El ozono presente en la estratosfera nos protege de las radiaciones ultravioletas del sol.

Ozone depletion potential (ODP) A scale used to measure how much a substance will deplete stratospheric ozone ODP ranges from 0 to 1.

Potencial de depleción de ozono (ODP en inglés) Una escala que se usa para medir cuánta depleción del ozono de la estratosfera una sustancia va a causar El ODP varía entre 0 y 1.

P

Package equipment (unit) A refrigerating system where all major components are located in one cabinet.

Equipo (unidad) completo Sistema de refrigeración donde todos los componentes principales se encuentran en un solo gabinete.

Packing A soft material that can be shaped and compressed to provide a seal. It is commonly applied around valve stems.

Empaquetadura Material blando que puede formarse y comprimirse para proveer una junta de estanqueidad. Comúnmente se aplica alrededor de los vástagos de válvulas.

Paraffinic oil A refrigeration mineral oil containing some paraffin wax, which is refined from eastern U.S. crude oil.

Aceite de parafina Aceite mineral que contiene parafina y que se utiliza en sistemas de refrigeración. El aceite se obtiene mediante el refinado de petróleo extraído en los EE.UU del este.

Parallel circuit An electrical or fluid circuit where the current or fluid takes more than one path at a junction.

Circuito paralelo Corriente eléctrica o fluida donde la corriente o el fluido siguen más de una trayectoria en un empalme.

Parallel compressor Many compressors piped in parallel and mounted on a steel rack. The compressors are usually cycled by a microprocessor.

Compresor en paralelo Varios compresores conectados en paralelo y montados en un bastidor de acero. Normalmente, se sirve de un microprocesador para activarlos y desactivarlos.

Parallel flow A flow path in which many paths exist for the fluid to flow.

Flujo paralelo Vía de flujo que consta de varias vías que permiten el flujo de un fluido.

Part-winding start A large motor that is actually two motors in one housing. It starts on one and then the other is energized. This is to reduce inrush current at start-up. For example, a 100-hp motor may have two 50-hp motors built into the same winding. It will start using one motor followed by the start of the other one. They will both run under the load.

Arranque de bobina parcial Un motor grande que es, en efecto, dos motores bajo un mismo cárter. El mismo arranca con uno y luego se activa el segundo. El propósito de esto es reducir la co rriente interna al arrancar. Por ejemplo, un motor de 100 caball os de fuerza puede tener dos motores de 50 caballos de fuerza construidos con la misma bobina. El motor arrancará usando un motor, seguido por el arranque del otro motor. Los dos motores correrán bajo carga.

Pascal A metric unit of measuring pressure. One Pascal is equivalent to 0.004 in. of water column.

Pascal Unidad del sistema métrico de medición de presión. Un Pascal equivale a 0,004 pulgadas de columna de agua.

Passive recovery Recovering refrigerant with the use of the refrigeration system's compressor or internal vapor pressure.

Recuperación pasiva Recuperación de un refrigerante utilizando el compresor del sistema de refrigeración o la presión del vapor interno.

Passive sensor Sensors that send information back to the controller in terms of resistance (ohms).

Sensor pasivo Sensores que devuelven información al controlador a través de la resistencia (ohmios).

Passive solar design The use of nonmoving parts of a building to provide heat or cooling, or to eliminate certain parts of a building that cause inefficient heating or cooling.

Diseño solar pasivo La utilización de piezas fijas de un edificio para proveer calefacción o enfriamiento, o para eliminar ciertas piezas de un edificio que causan calefacción o enfriamiento ineficientes.

PE (polyethylene) Plastic pipe used for water, gas, and irrigation systems.

Polietileno Tubo plástico utilizado en sistemas de agua, de gas y de irrigación.

Pellet stove A stove that burns small compressed pellets made from sawdust, cardboard, or sunflower or cherry seeds.

Estufa de gránulos Estufa que quema gránulos comprimidos fabricados con aserrín, cartón y semillas de girasol o de cereza.

Percent refrigerant quality Percent vapor.

Calidad porcentual de refrigerante Porcentaje de vapor.

Permanent magnet An object that has its own permanent magnetic field.

Imán permanente Objeto que tiene su propio campo magnético permanente.

Permanent split-capacitor motor (PSC) A split-phase motor with a run capacitor only. It has a very low starting torque.

Motor permanente de capacitador separado (PSC en inglés) Motor de fase separada que sólo tiene un capacitador de funcionamiento. Su par de arranque es sumamente bajo.

PEX tubing PEX stands for polyethylene cross (X)-linked tubing, a common name for polyethylene tubing.

Tubería PEX PEX significa tubería polietilena unida al través y es un nombre que se usa frecuentemente para tubería polietilena.

Phase One distinct part of a cycle.

Fase Una parte específica de un ciclo.

Phase-change loop The loop of piping, usually in a geothermal heat pump system, where there is a change of phase of the heat transfer fluid from liquid to vapor or vapor to liquid.

Circuito de cambio de fase Circuito de tubería, generalmente en un sistema de bombeo de calor geotermal, en el cual hay un cambio de fase del fluido de transferencia de calor de líquido a vapor o de vapor a líquido.

Phase failure protection Used on three-phase equipment to interrupt the power source when one phase becomes deenergized. The motors cannot be allowed to run on the two remaining phases or damage will occur.

Protección de fallo de fase Se usa en equipo trifásico para interrumpir la fuente de potencia cuando se energiza una fase. No se puede permitir que los motores corran con las dos fases restantes o podría ocurrir una avería.

Phase reversal Phase reversal can occur if someone switches any two wires on a three-phase system. Any system with a three-phase motor will reverse, and this cannot be allowed on some equipment.

Inversión de fase La inversión de fase puede ocurrir si por cualquier razón alguien intercambia cualquier par de cables en un sistema trifásico. Cualquier sistema con un motor trifásico irá en dirección contraria y esto no puede permitirse en algunos sistemas.

Pick-up voltage The back electromotive force (BEMF) voltage that must be generated across the potential relay's coil to cause the relay armature to move and open the relay contacts between terminals 1 and 2.

Voltaje de inicio El voltaje de la fuerza contraelectromotriz (BEMF) que debe generarse a lo largo de la bobina del relé de potencial para que su estructura se mueva y abra los contactos del relé entre las terminales 1 y 2.

Pictorial wiring diagram This type of diagram shows the location of each component as it appears to the person installing or servicing the equipment.

Diagrama representativo del cableado Este tipo de diagrama indica la ubicación de cada componente tal y como lo verá el personal técnico.

Piercing valve A device that is used to pierce a pipe or tube to obtain a pressure reading without interrupting the flow of fluid. Also called a line tap valve.

Válvula punzante Aparato que se usa para punzar un tubo para obtener una lectura de la presión sin interrumpir el flujo del fluido. También se llama una válvula de toma.

Pilot-duty relay A small relay that is used in control circuits for switching purposes. It is small and cannot take a lot of current flow, such as to start a motor.

Relé de función piloto Un pequeño relé que se usa en los circuitos de control con propósitos de interrupción. Es pequeño y no puede tolerar un flujo alto de corriente, como para arrancar un motor.

Pilot light The flame that ignites the main burner on a gas furnace.

Llama piloto Llama que enciende el quemador principal en un horno de gas.

Pilot positioner Used in a pneumatic air system to control the large volume of air that must be used to operate large pneumatic devices. The pneumatic thermostat sends a very small air signal to the pilot positioner in a 1/4-in. line and the pilot positioner controls the airflow in a 3/8-in. line for the large volume of air to the controlled device.

Posicionador piloto Se usa en un sistema neumático de aire para controlar el gran volumen de aire necesario para operar los aparatos neumáticos grandes. El termostato neumático envía una señal de aire pequeña al posicionador piloto por una línea de 1/4 de pulgada, y el posicionador piloto controla el flujo de aire, una línea de 3/8 de pulgada para el gran volumen de aire, al aparato controlado.

Pin rivet assembly Also known as a *blind rivet*, this is a fastener assembly of hollow rivets assembled on a pin, often called a mandrel. The rivets are used to join two pieces of sheet metal and are inserted and set from only one side of the metal.

Instalación con remaches También llamados remaches ciegos, estos son remaches huecos con clavija. Los remaches se utilizan para unir dos láminas metálicas, y se introducen y se colocan desde una de las caras de las láminas.

Piston The part that moves up and down in a cylinder.

Pistón La pieza que asciende y desciende dentro de un cilindro.

Piston displacement The volume within the cylinder that is displaced with the movement of the piston from top to bottom.

Despiazamiento del pistón Volumen dentro del cilindro que se desplaza de arriba a abajo con el movimiento del pistón.

Pitch pocket Flanged piece of roof flashing that is positioned around irregularly shaped roof penetrations.

Material para impermeabilización Porción de membrana impermeabilizante de techos colocada de manera de cubrir inserciones irregulares en los techos.

Pitot tube Part of an instrument for measuring air velocities.

Tubo pitot Pieza de un instrumento para medir velocidades de aire.

Planned defrost Shutting the compressor off with a timer so that the space temperature can provide the defrost.

Descongelación proyectada Detención de la marcha de un compresor con un temporizador para que la temperatura del espacio lleve a cabo la descongelación.

Plenum A sealed chamber at the inlet or outlet of an air handler. The duct attaches to the plenum.

Plenum Cámara sellada a la entrada o a la salida de un tratante de aire. El conducto se fija al plenum.

Plenum-mount humidifier A humidifier that is mounted in either the supply or return plenum and relies on the system blower to move air through the humidifier medium.

Humidificador de instalación en cámara impelente Un humidificador que se instala en la cámara impelente de suministro o entrada utiliza el sistema de ventilación para hacer circular el aire a través del soporte del humidificador.

Pneumatic atomizing humidifier Air pressure is used to break up the water into a mist of tiny droplets and to disperse them.

Humidificador por pulverización neumática Se utiliza aire a presión para convertir el agua en neblina y dispersarla.

Pneumatic controls Controls operated by low-pressure air, typically 20 psig.

Controles neumáticos Controles que se operan por aire de baja presión, generalmente 20 libras por pulgada cuadrada de presión de manómetro (Puig).

Polyalkylene glycol (PAG) A popular synthetic glycol-based lubricant used with HFC refrigerants, mainly in automotive systems. This was the first generation of oil used with HFC refrigerants.

Glicol polialkilénico (PAG en inglés) Lubricante sintético de uso común basado en glicol, usado con refrigerantes HFC, principalmente en sistemas de automóviles. Ésta fue la primera generación de aceites usados con refrigerantes HFC.

Polybutylene A material used for the buried piping in geothermal heat pumps.

Polibutileno Material utilizado para la fabricación de tubos enterrados, en sistemas de bombas de calor geotérmicas.

Polycyclic organic matter By-products of wood combustion found in smoke and considered to be health hazards.

Materna orgánica policíclica Subproductos de la combustión de madera presentes en el humo y considerados nocivos para la salud.

Polyethylene A material used for the buried piping in geothermal heat pumps.

Polietileno Material utilizado para la fabricación de tubos enterrados, en sistemas de bombas de calor geotérmicas.

Polyol ester (POE) A very popular ester-based lubricant often used in HFC refrigerant systems.

Poliol éster (POE en inglés) Lubricante muy popular basado en éster usado frecuentemente en sistemas con refrigerantes HFC.

Polyphase Three or more phases.

Polifase Tres o más fases.

Polyphosphate A scale inhibitor with many phosphate molecules.

Polifosfato Un inhibidor de escama que contiene muchas moléculas de fosfato.

Polyvinylether (PVE) A type of refrigerant oil that is specifically intended for use in refrigeration systems that utilize hydrofluorocarbon (HFC) refrigerants.

Poli-vinil-éter (PVE) Un tipo de aceite refrigerante que está específicamente diseñado para su uso en sistemas de refrigeración que utilizan refrigerantes de hidrofluorocarbono (HFC).

Porcelain A ceramic material.

Porcelana Material cerámico.

Portable dolly A small platform with four wheels on which heavy objects can be placed and moved.

Carretilla portátil Plataforma pequeña con cuatro ruedas sobre la que pueden colocarse y transportarse objetos pesados.

Positive displacement A term used with a pumping device such as a compressor that is designed to move all matter from a volume such as a cylinder or it will stall, possibly causing failure of a part.

Desplazamiento positivo Término utilizado con un dispositivo de bombeo, como por ejemplo un compresor, diseñado para mover toda la materia de un volumen, como un cilindro, o se bloqueará, posiblemente causándole fallas a una pieza.

Positive electrical charge An atom or component that has a shortage of electrons.

Carga eléctrica positiva Átomo o componente que tiene una insuficiencia de electrones.

Positive temperature coefficient (PTC) start device A thermistor used to provide start assistance to a permanent split-capacitor motor.

Dispositivo de arranque de coeficiente de temperatura positiva (PTC en inglés) Termistor utilizado para ayudar a arrancar un motor permanente de capacitador separado.

Potential relay A switching device used with hermetic motors that breaks the circuit to the start capacitor and/or start windings after the motor has reached approximately 75% of its running speed.

Relé de potencial Dispositivo de conmutación utilizado con motores herméticos que interrumpe el circuito del capacitador y/o de los devandos de arranque antes de que el motor haya alcanzado aproximadamente un 75% de su velocidad de marcha.

Potential voltage The voltage measured across the start winding in a single-phase motor while it is turning at full speed. This voltage is much greater than the applied voltage to the run winding. For example, the run winding may have 230 V applied to it, and a measured voltage across the start winding may be 300 V. This is created by the motor stator turning in the magnetic field of the run winding. ***Voltage potential*** **is the difference in voltage between any two parts of a circuit.**

Potencial de voltaje El voltaje medido a través de la bobina de arranque en un motor de fase sencilla mientras está girando a velocidad completa. Este voltaje es mucho mayor que el voltaje aplicado a la bobina de marcha. Por ejemplo, a la bobina de marcha se le puede aplicar 230 V, y el voltaje medido a través de la bobina de arranque puede ser 300 V. Este voltaje lo crea el estator del motor al girar en el campo magnético de la bobina de marcha. ***Potencial de voltaje*** es la diferencia en voltaje entre cualesquiera dos partes del circuito.

Potentiometer An instrument that controls electrical current.

Potenciómetro Instrumento que regula corriente eléctrica.

Powder actuated tool (PAT) A tool with a powder load that forces a pin, threaded stud, or other fastener into masonry.

Herramienta activada por pólvora (PAT en inglés) Una herramienta con una carga de pólvora que inserta una clavija, una clavija con rosca u otro sujetador en la albañilería.

Power The rate at which work is done.

Potencia Velocidad a la que se realiza un trabajo.

Power-consuming devices A power-consuming device is considered the electrical load. For example, in a lightbulb circuit, the switch is a power-passing device that passes power to the lightbulb that consumes the power and produces light.

Aparatos consumidores de potencia Un aparato consumidor de potencia se considera la carga eléctrica. Por ejemplo, en un circuito de bombilla de luz, el interruptor es un dispositivo que pasa corriente que pasa la corriente a la bombilla, la cual consume electricidad y produce luz.

Power drafted A fan-assisted draft.

Corriente generada Corriente de aire producida por un ventilador.

Precipitation A process where dissolved minerals in a solution reach a certain concentration and become solid.

Precipitación Proceso por el cual los minerales disueltos en una solución alcanzan una cierta concentración y solidifican.

Pressure Force per unit of area.

Presión Fuerza por unidad de área.

Pressure control A sitch or control that responds to changes in pressure.

Control de presión un interruptor o control que responde a los cambios de presión.

Pressure access ports Places in a system where pressure can be taken or registered.

Puerto de acceso a presión Lugares en un sistema donde se puede tomar o registrar la presión.

Pressure control A sitch or control that responds to changes in pressure.

Pressure differential valve A valve that senses a pressure differential and opens when a specific pressure differential is reached.

Válvula de presión diferencial Válvula que detecta diferencia de presiones y se abre cuando se alcanza una diferencia específica.

Pressure drop The difference in pressure between two points.

Caída de presión Diferencia en presión entre dos puntos.

Pressure/enthalpy diagram A chart indicating the pressure and heat content of a refrigerant and the extent to which the refrigerant is a liquid and vapor.

Diagrama de presión y entalpía Esquema que indica la presión y el contenido de calor de un refrigerante y el punto en que el refrigerante es líquido y vapor.

Pressure limiter A device that opens when a certain pressure is reached.

Dispositivo limitador de presión Dispositivo que se abre cuando se alcanza una presión específica.

Pressure-limiting TXV A valve designed to allow the evaporator to build only to a predetermined pressure when the valve will shut off the flow of refrigerant.

Válvula electrónica de expansión limitadora de presión Válvula diseñada para permitir que la temperatura del evaporador alcance una presión predeterminada cuando la válvula detenga el flujo de refrigerante.

Pressure regulator A valve capable of maintaining a constant outlet pressure when a variable inlet pressure occurs. Used for regulating fluid flow such as natural gas, refrigerant, and water.

Regulador de presión Válvula capaz de mantener una presión constante a la salida cuando ocurre una presión variable a la entrada. Utilizado para regular el flujo de fluidos, como por ejemplo el gas natural, el refrigerante y el agua.

Pressure switch A switch operated by a change in pressure.

Commutador accionado por presión Commutador accionado por un cambio en presión.

Pressure tank A pressurized tank for water storage located in the water piping of an open-loop geothermal heat pump system. It prevents short cycling of the well pump.

Depósito de presión Depósito presurizado que sirve para almacenar el agua contenida en la tubería de un sistema de bomba de calor geotérmica de circuito abierto. Impide el funcionamiento de la bomba del pozo en ciclos cortos.

Pressure/temperature relationship This refers to the pressure/ temperature relationship of a liquid and vapor in a closed container. If the temperature increases, the pressure will also increase. If the temperature is lowered, the pressure will decrease.

Relación entre presión y temperatura Se refiere a la relación entre la presión y la temperatura de un líquido y un vapor en un recipiente cerrado. Si la temperatura aumenta, la presión también aumentará. Si la temperatura baja, habrá una caída de presión.

Pressure test instruments Pieces of test equipment that are used to measure the pressure or pressures in a system or vessel.

Instrumentos de prueba de presión piezas de equipo de prueba que se utiliza para medir la presión o presiones en un sistema o buque.

Pressure transducer A pressure-sensitive device located in the piping of a refrigeration system that will transform a pressure signal to an electronic signal. The electronic signal will then feed a microprocessor.

Transductor de presión Dispositivo sensible a la presión, situado en la tubería del sistema de refrigeración y que convierte una señal de presión en señal eléctrica. Seguidamente, la señal eléctrica se envía a un microprocesador.

Pressure vessels and piping Piping, tubing, cylinders, drums, and other containers that have pressurized contents.

Depósitos y tubería con presión Tubería, cilindros, tambores y otros recipientes que tienen un contenido presurizado.

Preventive maintenance The action of performing regularly scheduled maintenance on a unit, including inspection, cleaning, and servicing.

Mantenimiento preventivo La acción de dar mantenimiento regular a una unidad incluyendo inspección, limpieza, y servicio.

Primary air Air that is introduced to a furnace's burner before the combustion process has taken place.

Aire primario Aire que se introduce en el quemador de un calefactor antes de que ocurra el proceso de combustión.

Primary control Controlling device for an oil burner to ensure ignition within a specific time span, usually 90 sec.

Regulador principal Dispositivo de regulación para un quemador de aceite pesado. El regulador principal asegura el encendido dentro de un período de tiempo específico, normalmente 90 seg.

Programmable thermostat An electronic thermostat that can be set up to provide desired conditions at desired times.

Termostato programable Un termostato electrónico que se puede programar para proveer las condiciones deseadas en tiempos deseados.

Propane An LP (liquified petroleum) gas used for heat.

Propano Gas de petróleo licuado que se utiliza para producir calor.

Propane gas torch A torch that usess propane as its fuel.

Soplete de gas propano una antorcha que usa propano como combustible.

Propeller fan This fan is used in exhaust fan and condenser fan applications. It will handle large volumes of air at low-pressure differentials.

Ventilador helicoidal Se utiliza en ventiladores de evacuación y de condensador. Es capaz de mover grandes volúmenes de aire a bajas diferenciales de presión.

Proportional controller A modulating control mode where the controller changes or modifies its output signal in proportion to the size of the change in the error.

Controlador proporcional Un modo de control modulante donde el controlador cambia o modifica su señal de salida en proporción al tamaño del cambio en el error.

Proportional Integral (PI) controls Electronic controls that sense and react to the error between the actual and desired states of a system.

Controles proporcionales integrales (PI) Controles electrónicos que miden y reaccionan ante el error entre el estado real y el estado deseado de un sistema.

Propylene glycol An antifreeze fluid cooled by a primary (phase-change) refrigerant, which then is circulated by pumps throughout the refrigeration system to absorb heat.

Glicol propílico Líquido anticongelante enfriado por un refrigerante primario (cambio de fase). Seguidamente, una bomba lo hace circular por el sistema de refrigeración para que absorba el calor.

Proton That part of an atom having a positive charge.

Protón Parte de un átomo que tiene carga positiva.

Protozoa A microscopic organism with a complex life cycle.

Protozoario Un organismo microscópico con un ciclo de vida complejo.

PSC motor See Permanent split-capacitor motor.

Motor PSC Véase motor permanente de capacitador separado.

psi Abbreviation for pounds per square inch.

psi Abreviatura de libras por pulgada cuadrada.

psia Abbreviation for pounds per square inch absolute.

psia Abreviatura de libras por pulgada cuadrada absoluta.

psig Abbreviation for pounds per square inch gauge.

psig Abreviatura de indicador de libras por pulgada cuadrada de presión manométrica.

Psychrometer An instrument for determining relative humidity.

Sicrómetro Instrumento para medir la humedad relativa.

Psychrometric chart A chart that shows the relationship of temperature, pressure, and humidity in the air.

Esquema psicrométrico Esquema que indica la relación entre la temperatura, la presión y la humedad en el aire.

Psychrometrics The study of air and its properties, particularly the moisture content.

Psicrometría El estudio del aire y sus propiedades, particularmente el contenido de humedad.

P-type material Semiconductor material with a positive charge.

Material tipo P Material con carga positiva utilizado en semiconductores.

Pulse furnace This furnace ignites minute quantities of gas 60 to 70 times per second. Small amounts of natural gas and air enter the combustion chamber and are ignited with a spark igniter that forces the combustion materials down a tailpipe to an exhaust decoupler. The pulse is reflected back to the combustion chamber, igniting another gas and air mixture.

Horno de impulsos Este tipo de horno enciende pequeñas cantidades de gas, 60 a 70 veces por segundo. Pequeñas cantidades de gas natural y aire entran en la cámara de combustión y son encendidas mediante chispa. A continuación son forzadas por un tubo hacia un desconectador de evacuación. El impulso se devuelve a la cámara de combustión para encender otra mezcla de gas y aire.

Pulse-width modulator (PWM) An electronic device in a motor circuit that is used to control motor speed for variable-speed motors.

Modulador de ancho de pulso (PWM en inglés) Un aparato electrónico en un circuito de motor que se usa para controlar la velocidad del motor para motores de velocidad variable.

Pump A device that forces fluids through a system.

Bomba Dispositivo que introduce fluidos por fuerza a través de un sistema.

Pump down To use a compressor to pump the refrigerant charge into the condenser and/or receiver.

Extraer con bomba Utilizar un compresor para bombear la carga del refrigerante dentro del condensador y/o receptor.

Pure compound A substance formed in definite proportions by weight with only one molecule present.

Componente puro Una sustancia formada en proporciones por peso definidas con sólo una molécula presente.

Pure hydrocarbons (HCs) Pure hydrocarbons are compounds that are made up of only hydrogen and carbon atoms. Some common hydrocarbons are methane, ethane, propane, butane and octane.

Los hidrocarburos puros (HC) los hidrocarburos puros son compuestos que se compone sólo de átomos de hidrógeno y de carbono. Algunos hidrocarburos comunes son el metano, etano, propano, butano y el octano.

Purge To remove or release fluid from a system.

Purga Remover o liberar el fluido de un sistema.

PVC (Polyvinyl chloride) Plastic pipe used in pressure applications for water and gas as well as for sewage and certain industrial applications.

Cloruro de polivinilo (PVC en inglés) Tubo plástico utilizado tanto en aplicaciones de presión para agua y gas, como en ciertas aplicaciones industriales y de aguas negras.

Q

Quench To submerge a hot object in a fluid for cooling.

Entriamiento por inmersión Sumersión de un objeto caliente en un fluido para enfriarlo.

Quick-connect coupling A device designed for easy connecting or disconnecting of fluid lines.

Acoplamiento de conexión rápida Dispositivo diseñado para facilitar la conexión o desconexión de conductos de fluido.

R

R-12 Dichlorodifluoromethane, once a popular refrigerant for refrigeration systems. It can no longer be manufactured in the United States and many other countries.

R-12 Diclorodiflorometano, que fue una vez un refrigerante muy utilizado en sistemas de refrigeración. Ya no se puede fabricar ni en los Estados Unidos ni en muchos otros países.

R-22 Monochlorodifluoromethane, a popular HCFC refrigerant for air-conditioning systems.

R-22 Monoclorodiflorometano, refrigerante HCFC muy utilizado en sistemas de acondicionamiento de aire.

R-123 Dichlorotrifluoroethane, an HCFC refrigerant developed for low-pressure application.

R-123 Diclorotrifloroetano, refrigerante HCFC elaborado para aplicaciones de baja presión.

R-134a Tetrafluoroethane, an HFC refrigerant developed for refrigeration systems and as a replacement for R-12.

R-134a Tetrafloroetano, refrigerante HFC elaborado para sistemas de refrigeración y como sustituto del R-12.

R-410A An HFC binary refrigerant blend that contains R-32 and R-125. This refrigerant has an ODP of 0.

R-410A Refrigerante mezcla binaria de HFC que contiene R-32 y R-125. Este refrigerante tiene un ODP de 0.

R-502 An azeotropic mixture of R-22 and R-115, a once popular refrigerant for low-temperature refrigeration systems. It can no longer be manufactured in the United States and many other countries.

R-502 Mezcla azeotrópica de R-22 y R-115, que fue una vez un refrigerante muy utilizado en sistemas de refrigeración de temperatura baja. Ya no se puede fabricar ni en los Estados Unidos ni en muchos otros países.

Rack system Many compressors piped in parallel and mounted on a steel rack. The compressors are usually cycled by a microprocessor.

Sistema de bastidor Varios compresores conectados en paralelo y montados en un bastidor de acero. Normalmente, un microprocesador se encarga de activar y desactivar los compresores.

Radiant heat Heat that passes through air, heating solid objects that in turn heat the surrounding area.

Calor radiante Calor que pasa a través del aire y calienta objetos sólidos que a su vez calientan el ambiente.

Radiation Heat transfer. See Radiant heat.

Radiación Transferencia de calor. Véase Calor radiante.

Radon A colorless, odorless, and radioactive gas. Radon can enter buildings through cracks in concrete floors and walls, floor drains, and sumps.

Radón Gas incoloro, inodoro y radioactivo. El radón puede penetrar en los edificios a través de las grietas en el hormigón y suelos, desagües y sumideros.

Random or off-cycle defrost Defrost provided by the space temperature during the normal off cycle.

Descongelación variable o de ciclo apagado Descongelación llevada a cabo por la temperatura del espacio durante el ciclo normal de apagado.

Range The pressure or temperature settings of a control defining certain boundaries of temperature or pressure.

Rango Los valores de presión o temperatura para un control que definen los límites de la temperatura o presión.

Rankine The absolute Fahrenheit scale with 0 at the point where all molecular motion stops.

Rankine Escala absoluta de Fahrenheit con el 0 al punto donde se detiene todo movimiento molecular.

Rapid oxidation A reaction between the fuel, oxygen, and heat that is known as rapid oxidation or the process of burning.

Oxidación rápida Reacción producida entre el combustible y el oxígeno. El calor producido se conoce como oxidación rápida o proceso de quemado.

Rated-load amperage (RLA) A mathematical calculation used to get Underwriters Laboratory's (UL) approval for a motor. The RLA value is used by a service technician to size wires, fuses, overloads, and contactors for the motor. It should not be used to determine the proper refrigerant charge, or to determine if a compressor motor is running properly.

Amperaje de carga calificada (RLA) Cálculo matemático utilizado para obtener la aprobación de Underwriters Laboratory (UL) para un motor. Los técnicos mecánicos utilizan el valor del RLA para medir el tamaño de los cables, fusibles, contactores y medidores de sobrecarga. No debe utilizarse para determinar la carga adecuada del refrigerante ni para determinar si un motor de compresor está funcionando correctamente.

Reactance A type of resistance in an alternating current circuit.

Reactancia Tipo de resistencia en un circuito de corriente alterna.

Reamer Tool to remove burrs from inside a pipe after it has been cut.

Escariador Herramienta utilizada para remover las rebabas de un tubo después de haber sido cortado.

Receiver-drier A component in a refrigeration system for storing and drying refrigerant.

Receptor-secador Componente en un sistema de refrigeración que almacena y seca el refrigerante.

Reciprocating Back-and-forth motion.

Movimiento alternativa Movimiento de atrás para adelante.

Reciprocating compressor A compressor that uses a piston in a cylinder and a back-and-forth motion to compress vapor.

Compresor alternativo Compresor que utiliza un pistón en un cilindro y un movimiento de atrás para adelante a fin de comprir el vapor.

Recirculated water system A system where water is used over and over, such as a chilled water or cooling tower system.

Sistema de agua recirculada Sistema donde el agua se usa una y otra vez, tal como en un sistema de agua enfriada o de torre enfriamiento.

Recovery cylinder A cylinder into which refrigerant is transferred; should be approved by the Department of Transportation as a recovery cylinder. The color code for these cylinders is a yellow top with a gray body.

Cilindro de recuperación Cilindro al que se transfiere el refrigerante y que debe ser homologado por el departamento de transporte como cilindro de recuperación. Este tipo de cilindro se identifica pintando su parte superior en amarillo y el resto del cuerpo en gris.

Rectifier A device for changing alternating current to direct current.

Rectificador Dispositivo utilizado para convertir corriente alterna en corriente continua.

Reed valve A thin steel plate used as a valve in a compressor.

Válvula de lámina Placa delgada de acero utilizada como una válvula en un compresor.

Refrigerant The fluid in a refrigeration system that changes from a liquid to a vapor and back to a liquid at practical pressures.

Refrigerante Fluido en un sistema de refrigeración que se convierte de líquido en vapor y nuevamente en líquido a presiones prácticas.

Refrigerant blend Two or more refrigerants blended or mixed together to make another refrigerant. Blends can combine as either azeotropic or zeotropic blends.

Mezcla de refrigerante Dos o más refrigerantes mezclados para crear otro. Las mezclas pueden ser de tipo azeotrópico o zeotrópico.

Refrigerant loop The heat pump's refrigeration system, which exchanges energy with the fluid in the ground loop and the air side of the system.

Circuito de refrigeración Sistema de refrigeración de la bomba de calor que sirve para intercambiar energía entre el fluido en el circuito de tierra y la parte del sistema que contiene el aire.

Refrigerant receiver A storage tank in a refrigeration system where the excess refrigerant is stored. Since many systems use different amounts of refrigerant during the season, the excess is stored in the receiver tank when not needed. The refrigerant can also be pumped to the receiver when repairs on the low-pressure side of the system are made.

Recibidor de refrigerante Tanque de almacenamiento en un sistema de refrigeración donde se almacena el exceso de refrigerante. Como muchos sistemas usan diferentes cantidades de refrigerantes durante la temporada, el exceso se almacena en el tanque cuando no se necesita. El refrigerante también puede bombearse al recibidor cuando se hacen reparaciones al extremo de baja presión del sistema.

Refrigerant reclaim "To process refrigerant to new product specifications by means which may include distillation. It will require chemical analysis of the refrigerant to determine that appropriate product specifications are met. This term usually implies the use of processes or procedures available only at a reprocessing or manufacturing facility."

Recuperación del refrigerante "Procesar refrigerante según nuevas especificaciones para productos a través de métodos que pueden incluir la destilación. Se requiere un análisis químico del refrigerante para asegurar el cumplimiento de las especificaciones para productos a través de métodos que pueden incluir la destilación. Se requiere un análisis químico del refrigerante para asegurar el cumplimiento de las especificaciones para productos adecuadas. Por lo general este término supone la utilización de procesos o de procedimientos disponibles solamente en fábricas de reprocesamiento o manufactura."

Refrigerant recovery "To remove refrigerant in any condition from a system and store it in an external container without necessarily testing or processing it in any way."

Recobrar refrigerante líquido "Remover refrigerante en cualquier estado de un sistema y almacenarlo en un recipiente externo sin ponerlo a prueba o elaborarlo de ninguna manera."

Refrigerant recycling "To clean the refrigerant by oil separation and single or multiple passes through devices, such as replaceable core filter driers, which reduce moisture, acidity, and particulate matter. This term usually applies to procedures implemented at the job site or at a local service shop."

Recirculación de refrigerante "Limpieza del refrigerante por medio de la separación del aceite y pasadas sencillas o múltiples a traves de dispositivos, como por ejemplo secadores filtros con núcleos reemplazables que disminuyen la humedad, la acidez y las partículas. Por lo general este término se aplica a los procedi mientos utilizados en el lugar del trabajo o en un taller de servicio local."

Refrigerated air driers A device that removes the excess moisture from compressed air.

Secadores de aire refrigerados Aparato que remueve el exceso de humedad del aire comprimido.

Refrigeration The process of removing heat from a place where it is not wanted and transferring that heat to a place where it makes little or no difference.

Refrigeración Proceso de remover el calor de un lugar donde no es deseado y transferirlo a un lugar donde no afecte la temperatura.

Register A terminal device on an air-distribution system that directs air but also has a damper to adjust airflow.

Registro Dispositivo de terminal en un sistema de distribución de aire que dirige el aire y además tiene un desviador para ajustar su flujo.

Regulator A valve used to control the pressure in liquid systems to some value. Many households have a water pressure regulator to reduce the pressure from the main to a more usable pressure in the house. Gas systems all have pressure regulators to stabilize the pressure to the burners.

Regulador Una válvula que se usa para controlar y fijar la presión en los sistemas líquidos a algún valor. Muchas casas tienen un regulador de presión de agua para reducir la presión de la tubería principal a una presión más útil en la casa. Todos los sistemas de gas tienen reguladores de presión para estabilizar la presión en el quemador.

Relative humidity The amount of moisture contained in the air as compared to the amount the air could hold at that temperature.

Humedad relativa Cantidad de humedad presente en el aire, comparada con la cantidad de humedad que el aire pueda contener a dicha temperatura.

Relay A small electromagnetic device to control a switch, motor, or valve.

Relé Pequeño dispositivo electromagnético utilizado para regular un conmutador, un motor o una válvula.

Relief valve A valve designed to open and release vapors at a certain pressure.

Válvula para alivio Válvula diseñada para abrir y liberar vapores a una presión específica.

REM/Rate The HERS index score is determined by computer modeling the home by Residential Energy Analysis and Rating Software named REM/Rate.

REM/Rate El puntaje del índice HERS se determina mediante el modelado computarizado de la residencia, utilizando un software para el análisis energético y clasificación residencial llamado REM/Rate.

Remote system Often called a split system where the condenser is located away from the evaporator and/or other parts of the system.

Sistema remoto Llamado muchas veces sistema separado donde el condensador se coloca lejos del evaporador y/o otras piezas del sistema.

Residential Energy Services Network (RESNET) A nationally recognized energy certification for home energy auditors and contractors.

Cadena de servicios de energía residencial (RESNET) Certificación energética nacional para auditores y contratistas de energía residencial.

Resilient-mount motor Electric motor that uses various materials to isolate the motor noise from metal framework. This type of motor requires a ground strap.

Motor con montaje antivibratorio Motor eléctrico que utiliza varios materiales para aislar el ruido del bastidor metálico. Este tipo de motor requiere conexión a tierra.

Resistance The opposition to the flow of an electrical current or a fluid.

Resistencia Oposición al flujo de una corriente eléctrica o de un fluido.

Resistor An electrical or electronic component with a specific opposition to electron flow. It is used to create voltage drop or heat.

Resistor Componente eléctrico o electrónico con una oposición específica al flujo de electrones; se utiliza para producir una caída de tensión o calor.

Restrictor A device used to create a planned resistance to fluid flow.

Limitador Dispositivo utilizado para producir una resistencia proyectada al flujo de fluido.

Retrofit guidelines Guidelines intended to make the transition from a CFC/mineral oil system to a system containing an alternative refrigerant and its appropriate oil.

Directrices de reconversión Directrices destinadas a facilitar la transición entre un sistema de aceite CFC/mineral y otro con refrigerante alternativo y su aceite correspondiente.

Return well A well for return water after it has experienced the heat exchanger of the geothermal heat pump.

Pozo de retorno Pozo donde se acumula el agua de retorno una vez ha pasado por el intercambiador térmico de la bomba de calor geotérmica.

Reverse cycle The ability to direct the hot gas flow into the indoor or the outdoor coil in a heat pump to control the system for heating or cooling purposes.

Ciclo invertido Capacidad de dirigir el flujo de gas caliente dentro de la bobina interior o exterior en una bomba de calor a fin de regular el sistema para propósitos de calentamiento o enfriamiento.

Rigging The process of using a crane to set HVACR equipment to a roof.

Elevación Proceso de utilización de una grúa para colocar el equipamiento de calefacción, ventilación, acondicionamiento de aire y refrigeración. en un techo.

Rigid-base mount Term used to describe a motor that has its base fastened to the motor body.

Montaje de base rígida término utilizado para describir un motor que tiene su base fijada al cuerpo del motor.

Rigid-mount motor Electric motor that is bolted metal-to-metal to a frame. This type of motor will transmit noise.

Motor con montaje rígido Motor eléctrico que se encuentra sujeto directamente a un bastidor mediante pernos. Este tipo de motor genera ruido.

Rod and tube The rod and tube are each made of a different metal. The tube has a high expansion rate and the rod a low expansion rate.

Varilla y tubo La varilla y el tubo se fabrican de un metal diferente. El tubo tiene una tasa de expansión alta y la varilla una tasa de expansión baja.

Room heater A gas stove or appliance considered by ANSI to be a heating appliance. This heater will have an efficiency rating.

Calentador de sala Estufa de gas u otro dispositivo que, según ANSI, es un dispositivo de calefacción. Este tipo de calentador cuenta con una clasificación de eficacia.

Root mean square (RMS) voltage The alternating current voltage effective value. This is the value measured by most voltmeters. The RMS voltage is 0.707 × the peak voltage.

Voltaje de la raíz del valor medio cuadrado (RMS en inglés) El valor efectivo del voltaje de corriente alterna. Este valor es el que miden la mayoría de los voltímetros. El voltaje RMS es 0.707 por el voltaje pico.

Rotary compressor A compressor that uses rotary motion to pump fluids. It is a positive displacement pump.

Compresor giratorio Compresor que utiliza un movimiento giratorio para bombear fluidos. Es una bomba de desplazamiento positivo.

Rotor The rotating or moving component of a motor, including the shaft.

Rotor Componente giratorio o en movimiento de un motor, incluyendo el arbol.

Running time The time a unit operates. Also called the "on time."

Período de funcionamiento El período de tiempo en que funciona una unidad. Conocido también como *período de conexión.*

Run winding The electrical winding in a motor that draws current during the entire running cycle.

Devanado de funcionamiento Devanado eléctrico en un motor que consume corriente durante todo el ciclo de funcionamiento.

Rupture disk Pressure safety device for a centrifugal low-pressure chiller.

Disco de ruptura Dispositivo de seguridad para un enfriador centrífugo de baja presión.

R-Value Indicates an insulation's resistance to heat transfer, or thermal resistance. This is the inverse of a material's U-value.

Índice R Indica la resistencia de una aislación a la transferencia de calor, o la resistencia térmica. Es la inversa del índice U de un material.

S

Saddle valve A valve that straddles a fluid line and is fastened by solder or screws. It normally contains a device to puncture the line for pressure readings.

Válvula de silleta Válvula que está sentada a horcajadas en un conducto de fluido y se fija por medio de la soldadura o tornillos. Por lo general contiene un dispositivo para agujerear el conducto a fin de que se puedan tomar lecturas de presión.

Safety control An electrical, electronic, mechanical, or electromechanical control to protect the equipment or public from harm.

Regulador de seguridad Regulador eléctrico, electrónico, mecánico o electromecánico para proteger al equipo de posibles averías o al público de sufrir alguna lesión.

Safety plug A fusible plug that blows out when high temperature occurs.

Tapón de seguridad Tapón fusible que se sale cuando se presentan temperaturas altas.

Sail switch A safety switch with a lightweight, sensitive sail that operates by sensing an airflow.

Conmutador con vela Conmutador de seguridad con una vela liviana sensible que funciona al advertir el flujo de aire.

Salt solution Antifreeze solution used in a closed water loop of geothermal heat pumps.

Solucion de sal Líquido anticongelante utilizado en el circuito cerrado de agua de una bomba de calor geotérmica.

Satellite compressor The compressor on a parallel compressor system that is dedicated to the coldest evaporators.

Compresor auxiliar Compresor montado en un sistema paralelo y que está dedicado a los evaporadores más fríos.

Saturated vapor The refrigerant when all of the liquid has just changed to a vapor.

Vapor saturada El refrigerante cuando todo el líquido acaba de convertirse en vapor.

Saturation A term used to describe a substance when it contains all of another substance it can hold.

Saturación Término utilizado para describir una sustancia cuando contiene lo más que puede de otra sustancia.

Saturation pressure The pressure at which a refrigerant will change state.

Presión de saturación la presión a la cual un refrigerante cambia de estado.

Saturation temperature The temperature at which a refrigerant will change state.

Temperatura de saturación la temperatura a la cual el refrigerante cambia de estado.

Scale Dissolved minerals such as calcium carbonate (limestone) which reach a certain concentration and become solid, usually clinging to HVACR equipment and piping surfaces.

Sarro Minerales disueltos como el carbonato de calcio (piedra caliza) que alcanzan una cierta concentración y solidifican, generalmente adhiriéndose a los equipos de acondicionamiento de aire y a las superficies internas de las cañerías.

Scavenger pump A pump used to remove the fluid from a sump.

Bomba de barrido Bomba utilizada para remover el fluido de un sumidero.

Schematic wiring diagram Sometimes called a line or ladder diagram, this type of diagram shows the electrical current path to the various components.

Diagrama de cableado esquematizado También conocido como de línea o de escalera, este tipo de diagrama muestra la ruta actual que sigue la electricidad para llegar a los diferentes componentes.

Schrader valve A valve similar to the valve on an auto tire that allows refrigerant to be charged or discharged from the system.

Válvula Schrader Válvula similar a la válvula del neumático de un automóvil que permite la entrada o la salida de refrigerante del sistema.

Scotch yoke A mechanism used to create reciprocating motion from the electric motor drive in very small compressors.

Yugo escocés Mecanismo utilizado para producir movimiento alternativo del accionador del motor eléctrico en compresores bastante pequeños.

Screw compressor A form of positive displacement compressor that squeezes fluid from a low-pressure area to a high-pressure area, using screw-type mechanisms.

Compresor de tornillo Forma de compresor de desplazamiento positivo que introduce por fuerza el fluido de un área de baja presión a un área de alta presión, a través de mecanismos de tipo de tornillo.

Scroll compressor A compressor that uses two scroll-type components, one stationary and one orbiting, to compress vapor.

Compresor espiral Compresor que utiliza dos componentes de tipo espiral para comprimir el vapor.

Sealed unit The term used to describe a refrigeration system, including the compressor, that is completely welded closed. The pressures can be accessed by saddle valves.

Unidad sellada Término utilizado para describir un sistema de refrigeración, incluyendo el compresor, que es soldado completamente cerrado. Las presiones son accesibles por medio de válvulas de dilleta.

Seasonal energy efficiency ratio (SEER) An equipment efficiency rating that is calculated on the basis of the total amount of cooling (in Btu) the system will provide over the entire season, divided by the total number of watt-hours it will consume.

Relación estacional de eficiencia energética (SEER, por sus siglas en inglés) La eficiencia de un equipamiento que se calcula en base al enfriamiento total (en BTU) que el sistema proporcionará a lo largo de toda la estación, dividido entre el número total de watts por hora que consuma dicho sistema.

Seat The stationary part of a valve that the moving part of the valve presses against for shutoff.

Asiento Pieza fija de una válvula contra la que la pieza en movimiento de la válvula presiona para cerrarla.

Secondary air Air that is introduced to a furnace after combustion takes place and that supports combustion.

Aire secundario Aire que se introduce en un calefactor después que occure la combustión y que ayuda la combustión.

Secondary fluid An antifreeze fluid cooled by a primary (phase-change) refrigerant, which is then circulated by pumps throughout the refrigeration system to absorb heat.

Fluido secundario Líquido anticongelante refrigerado por otro primario (cambio de fase), que luego es propulsado por las bombas a través del sistema de refrigeración para absorber el calor.

Semiconductor A component in an electronic system that is considered neither an insulator nor a conductor but a partial conductor. It conducts current in a controlled and predictable manner.

Semiconductor Componente en un sistema eléctrico que no se considera ni aislante ni conductor, sino conductor parcial. Conduce la corriente de una manera controlada y predecible.

Semihermetic compressor A motor compressor that can be opened or disassembled by removing bolts and flanges. Also known as a *serviceable hermetic.*

Compresor semihermético Compresor de un motor que puede abrirse o desmontarse al removerle los pernos y bridas. Conocido también como *hermético utilizable*.

Sensible heat Heat that causes a change in temperature.

Calor sensible Calor que produce un cambio en la temperatura.

Sensitivity See Gain.

Sensibilidad Vea Ganancia.

Sensors Devices that can measure some type of environmental parameter or controlled variable and convert this parameter to a value that the controller can understand.

Sensores Dispositivos que miden algún tipo de parámetro ambiental o variable controlado y convierten este parámetro a un valor que el controlador puede entender.

Sequencer A control that causes a staging of events, such as a sequencer between stages of electrical heat.

Regulador de secuencia Regulador que produce una sucesión de acontecimientos, como por ejemplo etapas sucesivas de calor eléctrico.

Series circuit An electrical or piping circuit where all of the current or fluid flows through the entire circuit.

Circuito en serie Cicuito eléctrico o de tubería donde toda la corriente o todo el fluido fluye a través de todo el circuito.

Series flow A flow path in which only one path exists for fluid to flow.

Flujo en serie Ruta de flujo única para el líquido.

Serviceable hermetic See Semihermetic compressor.

Compresor hermético utilizable Véase Compresor semihermético.

Service factor amps (SFA) The current the motor is drawing at its service factor loading.

Amperaje de factor de servicio (SFA, por sus siglas en inglés) Corriente que utiliza el motor cuando funciona a su factor de servicio.

Service valve A manually operated valve in a refrigeration system used for various service procedures.

Válvula servicio Válvula de un sistema de refrigeración accionada manualmente que se utiliza en varios procedimientos de servicio.

Servo pressure regulator A sensitive pressure regulator located inside a combination gas valve that senses the outlet or working pressure of the gas valve.

Regulador de presión por servomotor Un regulador de presión sensible que está ubicado dentro de una válvula de gas de combinación que detecta la presión de salida o de trabajo de la válvula de presión.

Sessile bacteria The bacteria that live in a biofilm and adhere to the equipment surfaces. They represent 99% of the total bacteria in a cooling tower or boiler water system.

Bacterias sésiles Bacterias que habitan en un biofilm y se adhieren a las superficies de los equipamientos. Representan un 99% de las bacterias totales presentes en sistemas de enfriamiento en torre o calderas.

Set point The desired control point's magnitude in a control process.

Punto de ajuste La magnitud deseada de un punto de control en un proceso de control.

S fastener A device used to connect square and rectangular sheet metal duct together, in conjunction with a drive cleat. Also referred to as a "slip."

Sujetador tipo "S" Aparato que se usa para unir conductos cuadrados y rectangulares de metal en lámina en conjunción con una grapa direccional. También se lo denomina "engrapadora".

Shaded-pole motor An alternating current motor used for very light loads.

Motor polar en sombra Motor de corriente alterna utilizado en cargas sumamente livianas.

Shell The floors, roof, and walls of a building, which is often referred to as the exterior envelope of the building.

Coraza Los pisos, techos, y paredes de una construcción, que generalmente se conocen como la funda exterior de la construcción.

Shell and coil A vessel with a coil of tubing inside that is used as a heat exchanger.

Coraza y bobina Despósito con una bobina de tubería en su interior que se utiliza como intercambiador de calor.

Shell and tube A heat exchanger with straight tubes in a shell that can normally be mechanically cleaned.

Coraza y tubo Intercambiador de calor con tubos rectos en una coraza que por lo general puede limpiarse mecánicamente.

Short circuit A circuit that does not have the correct measurable resistance: too much current flows and will overload the conductors.

Cortocircuito Corriente que no tiene la resistencia medible correcta: un exceso de corriente fluye a través del circuito provocando una sobrecarga de los conductores.

Short cycle The term used to describe the running time (on time) of a unit when it is not running long enough.

Cico corto Término utilizado para describir el período de funcionamiento (de encendido) de una unidad cuando no funciona por un período de tiempo suficiente.

Shorted motor winding Part of an electric motor winding is shorted out because one part of the winding touches another part, where the insulation is worn or in some way defective.

Devanado de motor en cortocircuito Debido a un aislamiento deficiente u otro defecto, una parte de los elementos del devanado en un motor eléctrico entran en contacto con otra, causando un cortocircuito.

Shroud A fan housing that ensures maximum airflow through the coil.

Boveda Alojamiento del abanico que asegura un flujo máximo de aire a través de la bobina.

Sick building syndrome (SBS) A condition that affects the health of office workers that is often attributed to poor ventilation or air quality.

Síndrome del edificio enfermo (SBS) una condición que afecta la salud de los trabajadores de oficina que a menudo se atribuye a la falta de ventilación o la calidad del aire.

Sight glass A clear window in a fluid line.

Mirilla para observación Ventana clara en un conducto de fluido.

Signal converter An electronic device that converts one signal type to another.

Convertidor de señales Un aparato electrónico que convierte un tipo de señal a otro.

Silica gel A chemical compound often used in refrigerant driers to remove moisture from the refrigerant.

Gel silíceo Compuesto químico utilizado a menudo en seca dores de refrigerantes para remover la humedad del refrigerante.

Silicon A substance from which many semiconductors are made.

Silicio Sustancia de la cual se fabrican muchos semiconductores.

Silicon-controlled rectifier (SCR) A semiconductor control device.

Rectificador controlado por silicio Dispositivo para regular un semiconductor.

Silver brazing A high-temperature (above 800°F) brazing process for bonding metals.

Soldadura con plata Soldadura a temperatura alta (sobre los 800°F ó 430°C) para unir metales.

Sine wave The graph or curve used to describe the characteristics of alternating current and voltage.

Onda sinusoidal Gráfica o curva utilizada para describir las características de tensión y de corriente alterna.

Single phase The electrical power supplied to equipment or small motors, normally under 7 1/2 hp.

Monofásico Potencia eléctrica suministrada a equipos o motores pequeños, por lo general menor de 7 1/2 hp.

Single-phase hermetic motor A sealed motor, such as with a small compressor, that operates off single-phase power.

Motor de fase sencilla hermético Un motor sellado, tal como un compresor pequeño, que opera con electricidad de fase sencilla.

Single phasing The condition in a three-phase motor when one phase of the power supply is open.

Fasaje sencillo Condición en un motor trifásico cuando una fase de la fuente de alimentación está abierta.

Slinger ring A ring attached to the blade tips of a condenser fan. This ring throws condensate onto the condenser coil, where it is evaporated.

Anillo tubular Anillo instalado en los extremos de las palas de un ventilador de condensación. Sirve para lanzar la condensación sobre el serpentín de refrigeración donde es evaporada.

Sling psychrometer A device with two thermometers, one a wet bulb and one a dry bulb, used for checking air conditions, wet-bulb and dry-bulb.

Sicrómetro con eslinga Dispositivo con dos termómetros, uno con una bombilla húmeda y otro con una bombilla seca, utilizados para revisar las condiciones del aire, de la temperatura y de la humedad.

Slinky loop A flattened, circular coil of plastic pipe resembling a Slinky, which is used as a ground loop.

Espiral Tubo plástico plano en forma helicoidal que se utiliza como circuito de tierra.

Slip The difference in the rated rpm of a motor and the actual operating rpm when under a load. Also, a device used to connect sections of square or rectangular ducts. See S fastener.

Deslizamiento Diferencia entre las rpm nominales de un motor y las rpm de funcionamiento reales. También, un artefacto usado para unir secciones de conductos cuadrados o rectangulares. Ver Sujetador tipo "S".

Slugging A term used to describe the condition when large amounts of liquid enter a pumping compressor cylinder.

Relleno Término utilizado para describir la condición donde grandes cantidades de líquido entran en el cilindro de un compresor de bombeo.

Smart recovery That start-up of a building air-conditioning or heating system whereby the system is started up earlier and allowed to run at a reduced capacity for the purpose of recording a low demand metering for the billing period.

Recuperación inteligente El encendido del sistema de aire acondicionado o de calefacción de un edificio en el cual el sistema se enciende más temprano y se deja correr a una capacidad reducida con el propósito de registrar una medición por demanda baja por el período de facturación.

Smoke test A test performed to determine the amount of unburned fuel in an oil burner flue-gas sample.

Prueba de humo Prueba llevada a cabo para determinar la cantidad de combustible no quemado en una muestra de gas de combustión que se obtiene de un quemador de aceite pesado.

Snap-disc An application of the bimetal. Two different metals fastened together in the form of a disc that provides a warping condition when heated. This also provides a snap action that is beneficial in controls that start and stop current flow in electrical circuits.

Disco de acción rápida Aplicación del bimetal. Dos metales diferentes fijados entre sí en forma de un disco que provee un deformación al ser calentado. Esto provee también una acción rápida, ventajosa para reguladores que ponen en marcha y detienen el flujo de corriente en circuitos eléctricos.

Snap or detent action The quick opening of a control that is used to minimize the arc when making and breaking an electrical circuit.

Acción de encaje o de detén El abrir rápido de un control que se usa para minimizar el arco al abrir y cerrar un circuito eléctrico.

Soapstone A stone used sometimes in the manufacture of woodstoves.

Esteatita Piedra que se utiliza algunas veces en la fabricación de estufas de madera.

Software Computer programs written to give specific instructions to computers.

Software Programas de computadoras escritos para darles instrucciones específicas a las computadoras.

Solar collectors Components of a solar system designed to collect the heat from the sun, using air, a liquid, or refrigerant as the medium.

Colectores solares Componentes de un sistema solar diseñados para acumular el calor emitido por el sol, utilizando el aire, un líquido o un refrigerante como el medio.

Solar constant The rate of solar energy reaching the outer limits of the earth's atmosphere has been determined to be 429 Btu/ft^2/h on a surface perpendicular (90°) to the sun's rays.

Constante solar Se ha determinado que el índice de energía solar que alcanza los límites exteriores del atmósfera de la Tierra es de 429 Btu/pies^2/h, en una superficie perpendicular (90°) a los rayos solares.

Solar heat Heat from the sun's rays.

Calor solar Calor emitido por los rayos del sol.

Solar influence The heat that the sun imposes on a structure.

Influencia solar El calor que el sol impone en una estructura.

Solar radiant heat Solar-heated water or an antifreeze solution is piped through heating coils embedded in concrete in the floor or in plaster in ceilings or walls.

Calor de radiación solar El agua o líquido anticongelante calentado por energía solar se canaliza a través de serpentines de calefacción instalados en el hormigón del suelo o en el yeso de los techos o paredes.

Soldering Fastening two base metals together by using a third, filler metal that melts at a temperature below 800°F.

Soldadura La fijación entre sí de dos metales bases utilizando un tercer metal de relleno que se funde a una temperatura menor de 800°F (430°C).

Soldering gun Tool used primarily to solder electrical connections.

Pistola de soldadura Herramienta que se utiliza principalmente para soldar las conexiones eléctricas.

Solderless terminals Used to fasten stranded wire to various terminals or to connect two lengths of stranded wire together.

Terminales sin soldadura Se usan para fijar cable trenzado a varios terminales o para unir dos pedazos de cable.

Solder pot A device using a low-melting solder and an overload heater sized for the amperage of the motor it is protecting. The solder will melt, opening the circuit when there is an overload. It can be reset.

Olla para soldadura Dispositivo que utiliza una soldadura con un punto de fusión bajo y un calentador de sobrecarga diseñado para el amperaje del motor al que provee protección. La soldadura se fundirá, abriendo así el circuito cuando ocurra una sobrecarga. Puede ser reconectado.

Solenoid A coil of wire designed to carry an electrical current producing a magnetic field.

Solenoide Bobina de alambre diseñada para conducir una corriente eléctrica generando un campo magnético.

Solid Molecules of a solid are highly attracted to each other, forming a mass that exerts all of its weight downward.

Sólido Las moléculas de un sólido se atraen entre sí y forman una masa que ejerce todo su peso hacia abajo.

Space cooling and heating thermostat The device used to control the temperature of a space, such as a home thermostat that controls the temperature in a home or office.

Termostato de enfriamiento o calefacción de espacio Aparato usado para controlar la temperatura de un espacio, tal como un termostato de hogar que controla la temperatura en un hogar u oficina.

Specific gravity The weight of a substance compared to the weight of an equal volume of water.

Gravedad específica El peso de una sustancia comparada con el peso de un volumen igual de agua.

Specific heat The amount of heat required to raise the temperature of 1 lb of a substance 1°F.

Calor específico La cantidad de calor requerida para elevar la temperatura de una libra de una sustancia 1°F (–17°C).

Specific volume The volume occupied by 1 lb of a fluid.

Volumen específico Volumen que ocupa una libra de fluido.

Splash lubrication system A system of furnishing lubrication to a compressor by agitating the oil.

Sistema de lubrificación por salpicadura Método de proveerle lubrificación a un compresor agitando el aceite.

Splash method A method of water dropping from a higher level in a cooling tower and splashing on slats with air passing through for more efficient evaporation.

Método de salpicaduras Método dé dejar caer agua desde un nivel más alto en una torre de refrigeración y salpicándola en listones, mientras el aire pasa a través de los mismos con el propósito de lograr una evaporación más eficaz.

Split-phase motor A motor with run and start windings.

Motor de fase separada Motor con devandos de functionamiento y de arranque.

Split suction When the common suction line of a parallel compressor system has been valved in such a way as to provide for multiple temperature applications in one refrigeration package.

Succión dividida Cuando la línea común de succión de un sistema de compresor paralelo ha sido dividida de tal manera que provee para aplicaciones de múltiples temperaturas en un empaque de refrigeración.

Split system A refrigeration or air-conditioning system that has the condensing unit remote from the indoor (evaporator) coil.

Sistema separado Sistema de refrigeración o de acondicionamiento de aire cuya unidad de condensación se encuentra en un sitio alejado de la bobina interior del evaporador.

Spray pond A pond with spray heads used for cooling water in water-cooled air-conditioning or refrigeration systems.

Tanque de rociado Tanque con una cabeza rociadora utilizada para enfriar el agua en sistemas de acondicionamiento de aire o de refrigeración enfriados por agua.

Spring-loaded relief valve A fluid (refrigerant, air, water, or steam) relief valve that can function more than one time because a spring returns the valve to a seat.

Válvula de alivio de resorte Una válvula de alivio de fluido (refrigerante, aire, agua o vapor de agua) que puede funcionar más de una vez porque el resorte regresa la válvula a su asiento.

Squirrel cage blower A cylindrically shaped fan assembly used to move air.

Abanico con jaula de ardilla Conjunto cilíndrico de abanico utilizado para mover el aire.

Squirrel cage rotor Describes the construction of a motor rotor.

Rotor de jaula de ardilla Describe la construcción del rotor de un motor.

SSE An acronym for steady-state efficiency.

SSE Acrónimo de "eficiencia en estado estable".

Stack draft Draft taken in the vent stack between the combustion device and the draft diverter.

Corriente de ventilación Corriente de aire que pasa a través del conducto de ventilación, entre el dispositivo de combustión y el deflector de corriente.

Stack switch A safety device placed in the flue of an oil furnace that proves combustion within a time frame.

Interruptor de chimenea Un aparato de seguridad que se coloca en la chimenea de un calefactor de aceite para comprueba combustión dentro de un período de tiempo.

Stack temperature A combustion appliance's flue gas temperature measured before the introduction of dilution air, which means before the draft diverter or draft hood of the combustion appliance if one is used.

Temperatura de ventilación Temperatura del gas de combustión de un calefactor, medida antes de la introducción de aire de dilución, es decir, antes de utilizar el deflector de corriente o la campana de corriente del equipo de combustión.

Stamped evaporator An evaporator that has stamped refrigerant passages in sheet steel or aluminum.

Evaporador estampado Un evaporador que tiene pasajes para el refrigerante estampados en lata o aluminio.

Standard atmosphere or standard conditions Air at sea level at 70°F when the atmosphere's pressure is 14.696 psia (29.92 in. Hg). Air at this condition has a volume of 13.33 ft^3/lb.

Atmósfera estándar o condiciones estándares El aire al nivel del mar a una temperatura de 70°F (15°C) cuando la presión de la atmósfera es 14.696 psia (29.92 pulgadas Hg). Bajo esta condición, el aire tiene un volumen de 13.33 ft^3/lb ($pies^3$/ libras).

Standing pilot Pilot flame that remains burning continuously.

Piloto constante Llama piloto que se quema de manera continua.

Start capacitor A capacitor used to help an electric motor start.

Capacitador de arranque Capacitador utilizado para ayudar en el arranque de un motor eléctrico.

Starting relay An electrical relay used to disconnect the start capacitor and/or start winding in a hermetic compressor.

Relé de arranque Relé eléctrico utilizado para desconectar el capacitador y/o el devanado de arranque en un compresor hermético.

Starting winding The winding in a motor used primarily to give the motor extra starting torque.

Devanado de arranque Devanado en un motor utilizado principalmente para proveerle al motor mayor para el arranque.

Starved coil The condition in an evaporator when the metering device is not feeding enough refrigerant to the evaporator.

Bobina estrangulada Condición que ocurre en un evaporador cuando el dispositivo de medida no le suministra suficiente refrigerante al evaporador.

Static pressure The bursting pressure or outward force in a duct system.

Presión estática La presión de estallido o la fuerza hacia fuera en un sistema de conductos.

Stator The component in a motor that contains the windings: it does not turn.

Estátor Componente en un motor que contiene los devanados y que no gira.

Steady-state condition A stabilized condition of a piece of heating or cooling equipment where not much change is taking place.

Condición de régimen estable Condición estabilizada de un dispositivo de calefacción o de refrigeración en el cual no hay muchos cambios.

Steady-state efficiency Refers to combustion efficiency when the flue gas temperature (stack temperature) has stabilized and the flue gas samples reach equilibrium.

Eficiencia en estado estable Se refiere a la eficiencia de la combustión cuando la temperatura del gas de combustión (temperatura de ventilación) está estable y las muestras de gas de combustión llegan a un equilibrio.

Steam The vapor state of water.

Vapor Estado de vapor del agua.

Step motor An electric motor that moves with very small increments or "steps," usually in either direction, and is usually controlled by a microprocessor with input and output controlling devices.

Motor a pasos Un motor eléctrico que se mueve en incrementos muy pequeños o "pasos", generalmente en cualquier dirección, y generalmente son controlados por un microprocesador de aparatos de control con entradas y salidas.

Strainer A fine-mesh device that allows fluid flow and holds back solid particles.

Colador Dispositivo de malla fina que permite el flujo de fluido a través de él y atrapa partículas sólidas.

Stratification The condition where a fluid appears in layers.

Estratificación Condición que ocurre cuando un fluido aparece en capas.

Stratosphere An atmospheric level that is located from 7 to 30 miles above the earth. Good ozone is found in the stratosphere.

Estratosfera Capa del atmósfera que se encuentra a una altura entre 11 y 48 kilómetros encima de la Tierra. Contiene una buena capa de ozono.

Stress crack A crack in piping or other component caused by age or abnormal conditions such as vibration.

Grieta por tensión Grieta que aparece en una tubería u otro componente ocasionada por envejecimiento o condiciones anormales, como por ejemplo vibración.

Subbase The part of a space temperature thermostat that is mounted on the wall and to which the interconnecting wiring is attached.

Subbase Pieza de un termóstato que mide la temperatura de un espacio que se monta sobre la pared y a la que se fijan los conductores eléctricos interconectados.

Subcooled The temperature of a liquid when it is cooled below its condensing temperature.

Subenfriado La temperatura de un líquido cuando se enfría a una temperatura menor que su temperatura de condensación.

Subcritical cycle Refrigeration cycles that take place above the refrigerants triple point, and below the refrigerant's critical point.

Ciclo subcrítico Ciclos de refrigeración que se producen por encima del punto triple del refrigerante y por debajo de su punto crítico.

Sublimation When a substance changes from the solid state to the vapor state without going through the liquid state.

Sublimación Cuando una sustancia cambia de sólido a vapor sin covertirse primero en líquido.

Suction gas The refrigerant vapor in an operating refrigeration system found in the tubing from the evaporator to the compressor and in the compressor shell.

Gas de aspiración El vapor del refrigerante en un sistema de refrigeración en funcionamiento presente en la tubería que va del evaporador al compresor y en la coraza del compresor.

Suction line The pipe that carries the heat-laden refrigerant gas from the evaporator to the compressor.

Conducto de aspiración Tubo que conduce el gas de refrigerante lleno de calor del evaporador al compresor.

Suction-line accumulator A reservoir in a refrigeration system suction line that protects the compressor from liquid floodback.

Acumulador de la línea de succión Un estanque en la línea de succión de un sistema de refrigeración que protege al compresor de una inundación de líquido.

Suction pressure The pressure created by the boiling refrigerant on the evaporator or low-pressure side of the system.

Presión de succión La presión creada por el refrigerante hirviendo en el evaporador o en el lado de baja presión del sistema.

Suction service valve A manually operated valve with front and back seats located at the compressor.

Válvula de aspiración para servicio Válvula accionada manualmente que tiene asientos delanteros y traseros ubicados en el compresor.

Suction valve The valve at the compressor cylinder that allows refrigerant from the evaporator to enter the compressor cylinder and prevents it from being pumped back out to the suction line.

Válvula de succión La válvula en el cilindro de un compresor que permite que el refrigerante del evaporador entre al cilindro del compresor y evita que se bombee nuevamente a la línea de succión.

Suction-valve-lift unloading The suction valve in a reciprocating compressor cylinder is lifted, causing that cylinder to stop pumping.

Descarga por levantamiento de la válvula de aspiración La válvula de aspiración en el cilindro de un compresor alternativo se levanta, provocando que el cilindro deje de bombear.

Sulfur dioxide A combustion pollutant that causes eye, nose, and respiratory tract irritation and possibly breathing problems.

Bióxido de azufre Un contaminante por combustión que causa irritación en los ojos, la nariz y las vías respiratorias, y posiblemente problemas respiratorios.

Sump A reservoir at the bottom of a cooling tower to collect the water that has passed through the tower.

Sumidero Tanque que se encuentra en el fondo de una torre de refrigeración para acumular el agua que ha pasado a través de la torre.

Supercritical region A region above the critical point of the refrigerant. No liquid refrigerant can be produced above a refrigerant's critical point.

Región supercrítica Región por encima del punto crítico del refrigerante. Ningún refrigerante líquido puede producirse por encima de su punto crítico.

Superheat The temperature of vapor refrigerant above its saturation (change-of-state) temperature.

Sobrecalor Temperatura del refrigerante de vapor mayor que su temperatura de cambio de estado de saturación.

Surge When the head pressure becomes too great or the evaporator pressure too low, refrigerant will flow from the high- to the low-pressure side of a centrifugal compressor system, making a loud sound.

Movimiento repentino Cuando la presión en la cabeza aumenta demasiado o la presión en el evaporador es demasiado baja, el refrigerante fluye del lado de alta presión al lado de baja presión de un sistema de compresor centrífugo. Este movimiento produce un sonido fuerte.

Surge-type receiver A receiver that preserve the liquid subcooling made in the condenser during the cooler ambient months when free subcooling is available. A solenoid operated diverting valve is opened and the subcooled liquid coming from the condenser's bottom is closed off to the receiver.

Depósito de reflujo Depósito que puede almacenar el líquido refrigerante subenfriado durante los meses más frescos, cuando puede utilizarse el subenfriado libre. Se abre una válvula de deflección operada por solenoide y el líquido subenfriado que proviene del fondo del condensador se desvía hacia el depósito.

Sustainability Term used to describe our ability, as humans, to exist, without jeopardizing the ability of future generations to do the same.

Sostenibilidad término utilizado para describir nuestra capacidad, como seres humanos, existiendo, sin comprometer la capacidad de las futuras generaciones para hacer lo mismo.

Sustainable design Meets the needs of the present without compromising the ability of future generations to meet their own needs.

Diseño sustentable Se adecua a las necesidades actuales sin perjuicio de que las generaciones futuras tengan la posibilidad de cubrir sus propias necesidades.

Swaged joint The joining of two pieces of copper tubing by expanding or stretching the end of one piece of tubing to fit over the other piece.

Junta estampada La conexión de dos piezas de tubería de cobre dilatando o alargando el extremo de una pieza de tubería para ajustarla sobre otra.

Swaging See Swaged joint.

Estampar Véase Junta estampada.

Swaging tool A tool used to enlarge a piece of tubing for a solder or braze connection.

Herramienta de estampado Herramienta utilizada para agrandar una pieza de tubería a utilizarse en una conexión soldada o broncesoldada.

Swamp cooler A slang term used to describe an evaporative cooler.

Nevera pantanosa Término del argot utilizado para describir una nevera de evaporación.

Sweating A word used to describe moisture collection on a line or coil that is operating below the dew point temperature of the air.

Exudación Término utilizado para describir la acumulación de humedad en un conducto o una bobina que está funcionando a una temperatura menor que la del punto de rocío de aire.

Switch frequency The speed at which the variable frequency drive's power device switches the waveform's positive and negative halves on and off is referred to as the switch frequency or carrier frequency.

Frecuencia de intercambio La velocidad a la que el dispositivo de alimentación de frecuencia variable intercambia las mitades positiva y negativa de una onda se denomina frecuencia de intercambio o frecuencia de onda portadora.

Synthetic refrigerants Man-made refrigerants.

Refrigerantes sintéticos Refrigerantes fabricados por el hombre.

System charge The refrigerant in a system, both liquid and vapor. The correct charge is a balance where the system will give the most efficiency.

Carga del sistema El refrigerante en un sistema, tanto líquido y vapor. La carga correcta es un balance donde el sistema dará la mayor eficiencia.

System lag The temperature drop of the controlled space below the set point of the thermostat.

Retardo del sistema Caída de temperatura de un espacio controlado, por debajo del nivel programado en el termostato.

System overshoot The temperature rise of the controlled space above the set point of the thermostat.

Exceso del sistema Subida de la temperatura de un espacio controlado, por encima del nivel programado en el termostato.

System pressure-regulating valve (SPR valve) A valve located between the compressor and receiver of a refrigeration system. This valve controls the amount of liquid refrigerant that bypasses a parallel liquid receiver.

Válvula reguladora de presión del sistema (SPR en inglés) Válvula situada entre el compresor y el destino de un sistema de refrigeración. Esta válvula controla la cantidad de refrigerante que no entra en un receptor de líquido paralelo.

System superheat The total amount of superheat picked up in the evaporator and the suction line. Can be calculated as the compressor inlet temperature minus the evapoartor saturation temperature.

Recalentamiento del sistema la cantidad total de recalentamiento recogido en el evaporador y la tubería de aspiración. Puede calcularse como la temperatura de admisión del compresor evapoartor menos la temperatura de saturación.

T

Tank A closed vessel used to contain a fluid.

Tanque Desposito cerrado utilizado para contener un fluido.

Tap A tool used to cut internal threads in a fastener or fitting.

Macho de roscar Herramienta utilizada para cortar filetes internos en un aparto fijador o en un accesorio.

Tare weight The tare weight of a cylinder is the weight the cylinder would weigh if it were empty.

Tara La tara de un cilindro es el peso que el mismo tendría si estuviese vacío.

Technician A person who performs maintenance, service, testing, or repair to air-conditioning or refrigeration equipment. Note: The EPA defines this person as someone who could reasonably be expected to release CFCs or HCFCs into the atmosphere.

Técnico Una persona que lleva a cabo mantenimiento, servicio o reparaciones a equipos de aire acondicionado o refrigeración. Nota: Esta persona, según definido por la EPA, es una persona de la cual se podría esperar razonablemente que libere CFC (clorofluorocarbonos) a la atmósfera.

Temperature A word used to describe the level of heat or molecular activity, expressed in Fahrenheit, Rankine, Celsius, or Kelvin units.

Temperatura Término utilizado para describir el nivel de calor o actividad molecular, expresado en unidades Fahrenheit, Rankine, Celsio o Kelvin.

Temperature difference (TD) The difference in temperature between two different mediums, such as chilled water and evaporating refrigerant or cooing tower water and condensing refrigerant.

Diferencia de temperatura La diferencia de temperatura entre dos medios diferentes, por ejemplo entre agua fría y refrigerante evaporado, o entre agua de torre de refrigeración y refrigerante condensado.

Temperature glide When a refrigerant blend has different temperatures when it evaporates and condenses at a single given pressure.

Variación de temperatura Una mezcla de refrigerantes tiene varias temperaturas a las cuales se produce una evaporación a una presión determinada.

Temperature-measuring instruments Devices that accurately measure the level of temperature.

Instrumentos que miden temperatura Aparatos que miden el nivel de la temperatura con precisión.

Temperature/pressure relationship This refers to the temperature/pressure relationship of a liquid and vapor in a closed container. If the temperature increases, the pressure will also increase. If the temperature is lowered, the pressure will decrease.

Relación entre temperatura y presión Se refiere a la relación entre la presión y la temperatura de un líquido y un vapor en un recipiente cerrado. Si la temperatura aumenta, la presión también aumentará. Si la temperatura baja, habrá una caída de presión.

Temperature reference points Various points that may be used to calibrate a temperature-measuring device, such as boiling or freezing water.

Puntos de referencia de temperatura Varios puntos que pueden usarse para calibrar un aparato que mide temperatura, tales como agua congelada o hirvienda.

Temperature-sensing elements Various devices in a system that are used to detect temperature.

Elementos que detectan temperatura Varios aparatos en un sistema que se usan para detectar temperatura.

Temperature swing The temperature difference between the low and high temperatures of the controlled space.

Oscilación de temperatura Diferencia existente entre las temperaturas altas y bajas de un espacio controlado.

Terminal unit The piece of equipment that facilitates the transfer of heat from the appliance to the heated space. A radiator is an example of a terminal unit.

Unidad de terminal la pieza de equipo que facilita la transferencia de calor del aparato al espacio calentado. Un radiador es un ejemplo de una unidad terminal.

Terminal venting When one or more terminal pins blows out of its ceramic fusite terminal, allowing a pressurized refrigerant and oil spray to escape into the atmosphere.

Ventilación de terminal Cuando uno o más terminales pins sale de su cerámica fusite terminal, permitiendo un spray de refrigerante y de aceite presurizado para escapar a la atmósfera.

Testing, Adjusting and Balancing Bureau (TABB) A certification bureau for individuals involved in working with ventilation.

Agencia de Prueba, Ajuste y Balanceo (TABB en inglés) Agencia de certificación para los individuos involucrados en el trabajo de ventilación.

Test light A lightbulb arrangement used to prove the presence of electrical power in a circuit.

Luz de prueba Arreglo de bombillas utilizado para probar la presencia de fuerza eléctrica en un circuito.

Therm Quantity of heat, 100,000 Btu.

Therm Cantidad de calor, mil unidades térmicas inglesas.

Thermal boundary Insulation and air barrier that surrounds a conditioned space.

Límite térmico Aislación y barrera de aire que rodea a un espacio acondicionado.

Thermal break A gasket material made of a poor thermally conductive material, installed between two conductive materials that combats and lessens thermal bridging.

Aislante térmico Junta de un material de muy baja conductividad térmica, instalada entre dos materiales conductores, que combate y disminuye el calentamiento por puente térmico.

Thermal bridging Thermal bridging is rapid heat being conducted through the structure's shell by materials in direct contact with one another that are very thermally conductive.

Calentamiento por puente térmico Es la conducción rápida del calor a través de la coraza de la estructura, de materiales de muy alta conductividad térmica que están en contacto directo.

Thermal bypass A break in the thermal boundary where air can either infiltrate or exfiltrates.

Desvío térmico Ruptura en el límite térmico, a través de la cual el aire puede infiltrarse o exfiltrarse.

Thermistor A semiconductor electronic device that changes resistance with a change in temperature.

Termistor Dispositivo eléctrico semiconductor que cambia su resistencia cuando se produce un cambio en temperatura.

Thermocouple A device made of two unlike metals that generates electricity when there is a difference in temperature from one end to the other. Thermocouples have a hot and cold junction.

Thermopar Dispositivo hecho de dos metales distintos que genera electricidad cuando hay una diferencia en temperatura de un extremo al otro. Los termopares tienen un empalme caliente y uno frío.

Thermometer An instrument used to detect differences in the level of heat.

Termómetro Instrumento utilizado para detectar diferencias en el nivel de calor.

Thermopile A group of thermocouples connected in series to increase voltage output.

Pila termoeléctrica Grupo de termopares conectados en serie para aumentar la salida de tensión.

Thermostat A device that senses temperature change and changes some dimension or condition within to control an operating device.

Termostato Dispositivo que advierte un cambio en temperatura y cambia alguna dimensión o condición dentro de sí para regular un dispositivo en funcionamiento.

Thermostatic expansion valve (TXV) A valve used in refrigeration systems to control the superheat in an evaporator by metering the correct refrigerant flow to the evaporator. Also often abbrevaiated TEV.

Válvula de gobierno termostático para expansión Válvula utilizada en sistemas de refrigeración para regular el sobrecalor en un evaporador midiendo el flujo correcto de refrigerante al evaporador. También puede denominarse con la abreviatura "TEV".

Three-phase power A type of power supply usually used for operating heavy loads. It consists of three sine waves that are out of phase by 120° with each other.

Potencia trifásica Tipo de fuente de alimentación normalmente utilizada en el funcionamiento de cargas pesadas. Consiste de tres ondas sinusoidales que no están en fase la una con la otra por 120°.

Throttling Creating a planned or regulated restriction in a fluid line for the purpose of controlling fluid flow.

Estrangulamiento Que ocasiona una restricción intencional o programada en un conducto de fluido, a fin de controlar el flujo del fluido.

Thrust surface A term that usually applies to bearings that have a pushing pressure to the side and that therefore need an additional surface to absorb the push. Most motor shifts cradle in their bearings because they operate in a horizontal mode, like holding a stick in the palm of your hand. When a shaft is turned to the vertical mode, a thrust surface must support the weight of the shaft along with the load the shaft may impose on the thrust surface. The action of a vertical fan shaft that pushes air up is actually pushing the shaft downward.

Superficie de empuje Un término que generalmente se aplica a cojinetes que sostienen una presión de empuje a un lado y que necesitan una superficie adicional para absorber este empuje. La mayoría de los ejes de motor están al abrigo en sus cojinetes, porque funcionen en una modalidad horizontal, como cuando uno sostiene una vara en la mano. Cuando un eje está sintonizado a la modalidad vertical, una superficie de empuje debe sostener el peso del eje junto con la carga que el eje puede imponer sobre la superficie de empuje. La moción de un eje vertical de un ventilador que empuja aire hacia arriba está en realidad empujando el eje hacia abajo.

Time delay A device that prevents a component from starting for a prescribed time. For example, many systems start the fans and use a time delay relay to start the compressor at a later time to prevent too much inrush current.

Retraso de tiempo Un aparato que evita que un componente se encienda por un período prescrito de tiempo. Por ejemplo, muchos sistemas encienden los ventiladores y, usando un relé de retraso, encienden el compresor un tiempo después para evitar mucha corriente interna.

Timers Clock-operated devices used to time various sequences of events in circuits.

Temporizadores Dispositivos accionados por un reloj utilizados para medir el tiempo de varias secuencias de eventos en circuitos.

Toggle bolt Provides a secure anchoring in hollow tiles, building block, plaster over lath, and gypsum board. The toggle folds and can be inserted through a hole, where it opens.

Tornillo de fiador Proveen un anclaje seguro en losetas huecas, bloques de construcción, yeso sobre listón y tablón de yeso. El fiador se dobla y puede insertarse a través de un roto y después se abre.

Ton of refrigeration The amount of heat required to melt a ton (2000 lb) of ice at 32°F in 24 hours, 288,000 Btu/24 h, 12,000 Btu/h, or 200 Btu/min.

Tonelada de refrigeración Cantidad de calor necesario para fundir una tonelada (2.000 libras) de hielo a 32°F (0°C en 24 horas), 288.000 Btu/24 h 12.000 Btu/h o 200 Btu/min.

Torque The twisting force often applied to the starting power of a motor.

Par de torsión Fuerza de torsión aplicada con frecuencia a la fuerza de arranque de un motor.

Torque wrench A wrench used to apply a prescribed amount of torque or tightening to a connector.

Llave de torsión Llave utilizada para aplicar una cantidad específica de torsión o de apriete a un conector.

Total equivalent warming impact (TEWI) A global warming index that takes into account both the direct effects of chemicals emitted into the atmosphere and the indirect effects caused by system inefficiencies.

Impacto de calentamiento equivalente total (TEWI en inglés) Índice de calentamiento de la Tierra que tiene en cuenta los efectos directos de los productos químicos emitidos en la atmósfera, y los efectos indirectos causados por la ineficacia de un sistema.

Total heat The total amount of sensible heat and latent heat contained in a substance from a reference point. Also referred to as enthalpy.

Calor total Cantidad total de calor sensible o de calor latente presente en una sustancia desde un punto de referencia. Se lo denomina también entalpía.

Total pressure The sum of the velocity and the static pressure in an air duct system.

Presión total La suma de la velocidad y la presión estática en un sistema de conducto de aire.

Tower approach The temperature difference between the water leaving the tower and the wet-bulb temperature of the air entering the cooling tower.

Enfoque de torre Diferencia de temperatura entre el agua que sale de la torre y la temperatura a bulbo húmedo del aire que entra a la torre de enfriamiento.

Tower range The temperature difference between the entering and leaving water of the cooling tower is referred to as the tower range.

Rango de torre La diferencia de temperatura entre el agua que entra y el agua que sale de la torre se denomina rango de torre.

Trace gas A small amount of refrigerant introduced into an empty system for electronic leak detection or as a holding charge, which is then backed by nitrogen to build pressure. CFCs cannot be used as trace gasses.

Gas detector Pequeña cantidad de refrigerante que se introduce en un sistema vacío para la detección de pérdidas o como carga de sustento, la cual se completa con nitrógeno para crear presión. Los CFC no pueden utilizarse como gases detectores.

Transcritical cycle Thermodynamic refrigeration processes that span above the subcritical region and into the supercritical region of the pressure/enthalpy diagram.

Ciclo transcrítico Procesos de refrigeración termodinámicos que tienen lugar por encima de la región subcrítica y en la región supercrítica del diagrama presión/entalpía.

Transformer A coil of wire wrapped around an iron core that induces a current to another coil of wire wrapped around the same iron core. Note: A transformer can have an air core.

Transformador Bobina de alambre devanado alrededor de un núcleo de hierro que induce una corriente a otra bobina de alambre devanado alrededor del mismo núcleo de hierro. Nota: un transformador puede tener un núcleo de aire.

Transistor A semiconductor often used as a switch or amplifier.

Transistor Semiconductor que suele utilizarse como conmutador o amplificador.

TRIAC A semiconductor switching device.

TRIAC Dispositivo de conmutación para semiconductores.

Triple point The triple point is where a refrigerant, like CO_2, can exist as all three phases (vapor, liquid, and solid) simultaneously in equilibrium.

Punto triple El punto triple es aquel en el que un refrigerante, como el CO_2, puede existir en sus tres fases (vapor, líquido, y sólido) simultáneamente y en equilibrio.

Troposphere The lower atmospheric level that extends upward from ground level to about 7 miles. Global warming takes place in the troposphere.

Troposfera Capa más baja de la del atmósfera que va desde el nivel del suelo hasta una altura de aproximadamente 11 kilómetros. El efecto del calentamiento de la tierra se produce en la troposfera.

Tube-within-a-tube coil A coil used for heat transfer that has a pipe in a pipe and is fastened together so that the outer tube becomes one circuit and the inner tube another.

Bobina de tubo dentro de un tubo Bobina utilizada en la transferencia de calor que tiene un tubo dentro de otro y se sujeta de manera que el tubo exterior se convierte en un circuito y el tubo interior en otro circuito.

Tubing Pipe with a thin wall used to carry fluids.

Tubería Tubo que tiene una pared delgada utilizada para conducir fluidos.

Twinning Two furnaces connected side by side and sharing a common ducting system.

Aparear Dos calefactores conectados lado a lado o que comparten un sistema de conductos común.

Two-pole, split-phase motor This motor runs at 3600 rpm when not loaded and at about 3450 rpm when loaded. A four-pole motor runs at about 1800 rpm not loaded and at about 1725 fully loaded.

Motor de dos polos, de fase dividida Este motor corre a 3.600 revoluciones por minuto cuando no está cargado y cuando está cargado corre a 3.450 revoluciones por minuto. Un motor de 4 polos corre a 1.800 revoluciones por minuto si no está cargado y alrededor de 1.725 revoluciones con carga completa.

Two-speed compressor motor Can be a four-pole motor that can be connected as a two-pole motor for high speed (3450 rpm) and connected as a four-pole motor for running at 1725 rpm for low speed. This is accomplished with relays outside the compressor.

Motor de compresor de dos velocidades Puede ser un motor de 4 polos que puede conectarse como un motor de 2 polos para velocidades altas (3.450 revoluciones por minuto) y conectarse como un motor de 4 polos para correr a 1.725 revoluciones por minuto para velocidades bajas. Esto se hace con relés fuera del compresor.

Two-temperature valve A valve used in systems with multiple evaporators to control the evaporator pressures and maintain different temperatures in each evaporator. Sometimes called a ***holdback valve***.

Válvula de dos temperaturas Válvula utilizada en sistemas con evaporadores múltiples para regular las presiones de los evaporadores y mantener temperaturas diferentes en cada uno de ellos. Conocida también como ***válvula de retención***.

U

Ultrasound leak detector Detectors that use sound from escaping refrigerant to detect leaks.

Detector de escapes ultrasónico Detectores que usan sonido del refrigerante que está escapando para detectar escapes.

Ultraviolet Light waves that can only be seen under a special lamp.

Ultravioleta Ondas de luz que pueden observarse solamente utilizando una lámpara especial.

Ultraviolet light Light frequency between 200 and 400 nanometers.

Luz ultravioleta Luz con frecuencia entre 200 y 400 nanómetros.

Unconditioned space Areas of a structure or residence that are not heated or cooled.

Espacio no acondicionado Áreas de una estructura o residencia que no están calefaccionadas ni enfriadas.

Under-duct humidifier A humidifier that is mounted on the bottom of a horizontal supply duct with the humidifier media extending into the flow of heated system air.

Humidificador de montaje bajo conducto Un humidificador que está instalado en la parte inferior de un conducto de suministro horizontal de manera que el soporte del humidificador se extiende hacia el flujo del aire del sistema calefaccionado.

Uneven parallel system Parallel compressors of unequal sizes mounted on a steel rack and controlled by a microprocessor.

Sistema paralelo heterogéneo Compresores de capacidades distintas montados en paralelo en un bastidor de acero y controlados mediante microprocesador.

Unintentionally conditioned space These spaces are usually warmed by waste heat and include boiler rooms or furnace rooms.

Espacio acondicionado no intencionalmente Estos espacios se calientan generalmente debido al calor de desecho e incluyen a las salas donde se encuentran estufas o calderas.

Updraft A draft that flows upward.

Corriente hacia arriba Corriente que fluye de abajo hacia arriba.

Upflow furnace This furnace takes in air from the bottom or from sides near the bottom and discharges hot air out the top.

Horno de flujo ascendente La entrada del aire en este tipo de horno se hace desde abajo o en los laterales, cerca del suelo. La evacuación se realiza en la parte superior.

Upper exposure limit (UEL) The maximum concentration of a chemical that an individual can be exposed to for a short period of time, usually 10 or 15 min.

Límite de exposición superiores (UEL) La concentración máxima de un producto químico que una persona puede estar expuesta a por un corto período de tiempo, generalmente 10 ó 15 min.

Urethane foam A foam that can be applied between two walls for insulation.

Espuma de uretano Espuma que puede aplicarse entre dos paredes para crear un aislamiento.

U-Tube mercury manometer A U-tube containing mercury, which indicates the level of vacuum while evacuating a refrigeration system.

Manómetro de mercurio de tubo en U Tubo en U que contiene mercurio y que indica el nivel del vacío mientras vacía un sistema de refrigeración.

U-Tube water manometer Indicates natural gas and propane gas pressures. It is usually calibrated in inches of water.

Manómetro de agua de tubo en U Indica las presiones del gas natural y del propano. Se calibra normalmente en pulgadas de agua.

U-Value Indicates the thermal conductivity of a material. This is the inverse of a material's R-value.

Índice U Indica la conductividad térmica de un material. Es la inversa del índice R del material.

V

Vacuum The pressure range between the earth's atmospheric pressure and no pressure, normally expressed in inches of mercury (in. Hg) vacuum.

Vacío Margen de presión entre la presión de la atmósfera de la Tierra y cero presión, por lo general expresado en pulgadas de mercurio (pulgadas Hg) en vacío.

Vacuum gauge An instrument that measures the vacuum when evacuating a refrigeration, air-conditioning, or heat pump system.

Vacuómetro Instrumento que se utiliza para medir el vacío al vaciar un sistema de refrigeración, de aire acondicionado, o de bomba de calor.

Vacuum pump A pump used to remove some fluids such as air and moisture from a system at a pressure below the earth's atmosphere.

Bomba de vacío Bomba utilizada para remover algunos fluidos, como por ejemplo aire y humedad de un sistema a una presión menor que la de la atmósfera de la Tierra.

Valve A device used to control fluid flow.

Válvula Dispositivo utilizado para regular el flujo de fluido.

Valve plate A plate of steel bolted between the head and the body of a compressor that contains the suction and discharge reed or flapper valves.

Placa de válvula Placa de acero empernado entre la cabeza y el cuerpo de un compresor que contiene la lámina de aspiración y de descarga o las chapaletas.

Valve seat That part of a valve that is usually stationary. The movable part comes in contact with the valve seat to stop the flow of fluids.

Asiento de la válvula Pieza de una válvula que es normalmente fija. La pieza móvil entra en contacto con el asiento de la válvula para detener el flujo de fluidos.

Valve stem depressor A service tool used to access pressure at a Schrader valve connection.

Depresor de vástago de válvula Una herramienta de servicio que se usa para acceder la presión en una conexión de válvula Schrader.

Vapor The gaseous state of a substance.

Vapor Estado gaseoso de una sustancia.

Vapor barrier A thin film used in construction to keep moisture from migrating through building materials.

Película impermeable Película delgada utilizada en construcciones para evitar que la humedad penetre a través de los materiales de construcción.

Vapor charge bulb A charge in a thermostatic expansion valve bulb that boils to a complete vapor. When this point is reached, an increase in temperature will not produce an increase in pressure.

Válvula para la carga de vapor Carga en la bombilla de una válvula de expansión termostática que hierve a un vapor completo. Al llegar a este punto, un aumento en temperatura no produce un aumento en presión.

Vaporization The changing of a liquid to a gas or vapor.

Vaporización Cuando un líquido se convierte en gas o vapor.

Vapor lock A condition where vapor is trapped in a liquid line and impedes liquid flow.

Bolsa de vapor Condición que ocurre cuando el vapor queda atrapado en el conducto de líquido e impide el flujo de líquido.

Vapor pressure The pressure exerted on top of a saturated liquid.

Presión del vapor Presión que se ejerce en la superficie de un líquido saturado.

Vapor pump Another term for compressor.

Bomba de vapor Otro término para compresor.

Vapor refrigerant charging Adding refrigerant to a system by allowing vapor to move out of the vapor space of a refrigerant cylinder and into the low-pressure side of the refrigeration system.

Carga del refrigerante de vapor Agregarle refrigerante a un sistema permitiendo que el vapor salga del espacio de vapor de un cilindro de refrigerante y que entre en el lado de baja presión del sistema de refrigeración.

Variable air volume, VAV The control strategy that varies the volume of air that is delivered to an area in the conditioned space depending on the airflow requirements for that area.

Volumen de aire variable, VAV Estrategia de control mediante la que se varía el volumen de aire que se dirige hacia un área en el espacio acondicionado, dependiendo de las necesidades de flujo de aire para dicha área.

Variable-frequency drive (VFD) An electrical device that varies the frequency (hertz) for the purpose of providing a variable speed.

Propulsión de frecuencia variable (VFD en inglés) Un aparato eléctrico que varía la frecuencia (hertz) con el propósito de proveer velocidad variable.

Variable pitch pulley A pulley whose diameter can be adjusted.

Polea de paso variable Polea cuyo diámetro puede ajustarse.

Variable refrigerant flow, VRF The control strategy that varies the volume of refrigerant that is delivered to an air handler based on the heating or cooling needs of the area being served by that air handler.

Flujo variable de refrigerante, VRF Estrategia de control mediante la que se varía el volumen de refrigerante que se dirige hacia una unidad de tratamiento de aire, basada en los requerimientos de enfriamiento o calefacción del área que acondiciona dicha unidad de tratamiento de aire.

Variable refrigerant volume, VRV See Variable Refrigerant Flow, VRF.

Volumen variable de refrigerante, VRV Ver Flujo variable de refrigerante, VRF.

Variable resistor A type of resistor where the resistance can be varied.

Resistor variable Tipo de resistor donde la resistencia puede variarse.

Variable-speed motor A motor that can be controlled, with an electronic system, to operate at more than one speed.

Motor de velocidad variable Un motor que puede controlarse, con un sistema electrónico, para operar a más de una velocidad.

VAV See Variable Air Volume.

VAV Ver Volumen de aire variable.

V belt A belt that has a V-shaped contact surface and is used to drive compressors, fans, or pumps.

Correa en V Correa que tiene una superficie de contacto en forma de V y se utiliza para accionar compresores, abanicos o bombas.

Velocity The speed at which a substance passes a point.

Velocidad Rapidez a la que una sustancia sobrepasa un punto.

Velocity meter A meter used to detect the velocity of fluids, air, or water.

Velocimetro Instrumento utilizado para medir la velocidad de fluidos, aire, o agua.

Velocity pressure The pressure in a duct system that is created by the moving air. Velocity pressure is a function of the density of the air and its speed.

Velocidad de presión La presión en un sistema de conductos que es creado por el aire que se mueve. Presión de velocidad es una función de la densidad del aire y su velocidad.

Velometer An instrument used to measure the air velocity in a duct system.

Velómetro Instrumento utilizado para medir la velocidad del aire en un conducto.

Vent-free Certain gas stoves or gas fireplaces are not required to be vented; therefore, they are called "vent-free."

Sin ventilación Ciertas estufas de gas y chimeneas de gas no requieren una ventilación, por lo que se conocen como estufas "sin ventilación."

Ventilation The process of supplying and removing air by natural or mechanical means to and from a particular space.

Ventilación Proceso de suministrar y evacuar el aire de un espacio determinado, utilizando procesos naturales o mecánicos.

Venting products of combustion Venting flue gases that are generated from the burning process of fossil fuels.

Descargar productos de combustión Descargar los gases de la chimenea que se generan del proceso de combustión de los combustibles fósiles.

Venturi A smoothly tapered device used in fluid flow to increase the velocity of the fluid.

Venturi Un aparato que está ahusado lisamente y que se usa en el flujo de fluido para aumentar la velocidad del fluido.

Volt A unit of voltage.

Voltio unidad de voltaje o tensión.

Voltage The potential electrical difference for electron flow from one line to another in an electrical circuit.

Tensión Diferencia de potencial eléctrico del flujo de electrones de un conducto a otro en un circuito eléctrico.

Voltage feedback Voltage potential that travels through a power-consuming device when it is not energized.

Retroalimentación de voltaje El voltaje que viaja a través de un aparato de consumo de electricidad cuando no está energizado.

Volt-amperes Also referred to as VA. Used to represent the apparant power of a transformer. Calculated as the voltage multiplied by the currrent.

Voltios-amperios también conocido como "VA". Utilizado para representar la potencia aparente de un transformador. Calculado como el voltaje multiplicado por el corriente.

Voltmeter An instrument used for checking electrical potential.

Voltimetro Instrumento utilizado para revisar la potencia eléctrica.

Volt-ohm-milliammeter (VOM) A multimeter that measures voltage, resistance, and current in milliamperes.

Voltio-ohmio-miliamperimetro (VOM en inglés) Multímetro que mide tensión, resistencia, y corriente en miliamperios.

Volumetric efficiency The pumping efficiency of a compressor or vacuum pump that describes the pumping capacity in relationship to the actual volume of the pump.

Rendimiento volumétrico Rendimiento de bombeo de un compresor o de una bomba de vacío que describe la capacidad de bombeo con relación al volumen real de la bomba.

Vortexing A whirlpool action in the sump of a cooling tower.

Acción de vórtice Torbellino en el sumidero de una torre de refrigeración.

VRF See Variable Refrigerant Flow.

VRF Ver Flujo variable de refrigerante.

VRV See Variable Refrigerant Volume.

VRV Ver Volumen variable de refrigerante.

W

Walk-in cooler A large refrigerated space used for storage of refrigerated products.

Nevera con acceso al interior Espacio refrigerado grande utilizado para almacenar productos refrigerados.

Wastewater system A refrigeration system that uses water one time and then exhausts it to a waste system.

Sistema de aguas residuales Sistema de refrigeración que usa agua una vez y luego la vacía en el sistema de aguas residuales.

Water box A container or reservoir at the end of a chiller where water is introduced and contained.

Caja de agua Recipiente o depósito al extremo de un enfriador por donde entra y se retiene el agua.

Water capacity The water capacity of a cylinder is the weight of the cylinder if it were full of water. The water capacity (W.C.) of the cylinder is stamped on the outside of the cylinder and its units are in pounds.

Capacidad de agua La capacidad de agua de un cilindro es el peso de dicho cilindro si el mismo estuviese lleno de agua. La capacidad de agua (W.C.) del cilindro está marcada en su parte exterior y su unidad de medida está expresada en libras.

Water column (WC) The pressure it takes to push a column of water up vertically. One inch of water column is the amount of pressure it would take to push a column of water in a tube up 1 in.

Columna de agua (WC en inglés) Presión necesaria para levantar una columna de agua verticalmente. Una pulgada de columna de agua es la cantidad de presión necesaria para levantar una columna de agua a una distancia de una pulgada en un tubo.

Water-cooled condenser A condenser used to reject heat from a refrigeration system into water.

Condensador enfriado por agua Condensador utilizado para dirigir el calor de un sistema de refrigeración al agua.

Water cooler A small refrigeration machine that is typically used to refrigerate water for drinking purposes.

Enfriador de agua Una máquina de refrigeración pequeña que típicamente se usa para refrigerar agua para beber.

Water hammer A loud pressure pulse that occurs when valves close fast.

Martillo de agua Impulso de presión fuerte que se produce cuando una válvula se cierra rápidamente.

Water loop These loops of plastic pipe are buried in the ground in a closed-loop geothermal heat pump system and contain a heat transfer fluid.

Circuito de agua Estos circuitos de tubos de plástico se encuentran enterrados en el suelo, en el caso de una bomba de calor geotérmica de circuito cerrado y contiene un líquido que transpaso lo caliente.

Water manometer A device that uses a column of water to measure low pressures in air or gas systems.

Manómetro de agua Un aparato que usa una columna de agua para medir bajas presiones en sistemas de aire o de gas.

Water pump Used to pump water or other fluids from one pressure level to another to promote water flow in piping systems. See Centrifugal pump.

Bomba de agua Bomba utilizada para bombear agua u otros fluidos de un nivel de presión a otro para mejorar el flujo de agua através de los sistemas de tubería. Véase Bomba centrífuga.

Water-regulating valve An operating control regulating the flow of water.

Válvula reguladora de agua Regulador de mando que controla el flujo de agua.

Watt (W) A unit of power applied to electron flow. One watt equals 3.414 Btu.

Watio (W en inglés) Unidad de potencia eléctrica aplicada al flujo de electrones. Un watio equivale a 3414 Btu.

Watt-hour The unit of power that takes into consideration the time of consumption. It is the equivalent of a 1-watt bulb burning for 1 hour.

Watio hora Unidad de potencia eléctrica que toma en cuenta la duración de consumo. Es el equivalente de una bombilla de 1 watio encendida por espacio de una hora.

Weatherization The process of improving the energy efficiency in a home or building which will have a direct effect on energy consumption and comfort within the home or building.

Acondicionamiento Proceso de optimización de la eficiencia energética en una vivienda o construcción que tendrá un impacto directo en el consumo de energía y la comodidad dentro de dicha vivienda o construcción.

Weep holes Holes that connect each cell in an ice machine's cell-type evaporator that allow air entering from the edges of the ice to travel along the entire ice slab to relieve the suction force and allow the ice to fall off of the evaporator.

Aberturas de exudación Aberturas que conectan cada célula en el evaporador de tipo celular de una hielera, que permiten que el aire que entra de los bordes del hielo viaja a lo largo de todo el pedazo de hielo para aliviar la fuerza de succión y permitir que el hielo se caiga del evaporador.

Weighing refrigerant The process of adding refrigerant to a system by weight.

Refrigerante de pesaje El proceso de adición de refrigerante a un sistema por peso.

Weight The force that matter (solid, liquid, or gas) applies to a supporting surface when it is at rest.

Peso La fuerza que la materia (sólido, líquido, o gas) en reposo aplica a una superficie de apoyo.

Welded hermetic compressor A compressor that is completely sealed by welding, versus a semihermetic compressor that is sealed by bolts and flanges.

Compresor hermético soldado Un compresor que está completamente sellado por soldadura, contrario a un compresor semihermético que está sellado con tornillos y bridas.

Wet-base boiler A boiler configuration where the water in the system is both above and below the combustion chamber.

Caldera de base mojada Una caldera configuración donde el agua en el sistema tanto por encima y por debajo de la cámara de combustión.

Wet-bulb depression The difference between the wet-bulb and the dry-bulb reading when readings are taken in air.

Depresión de bulbo mojado La diferencia entre la lectura de bulbo mojado y bulbo seco cuando las lecturas están tomadas en aire.

Wet-bulb temperature A wet-bulb temperature of air is used to evaluate the humidity in the air. It is obtained with a wet thermometer bulb to record the evaporation rate with an airstream passing over the bulb to help in evaporation.

Temperatura de una bombilla húmeda La temperatura de una bombilla húmeda se utiliza para evaluar la humedad presente en el aire. Se obtiene con la bombilla húmeda de un termómetro para registrar el margen de evaporación con un flujo de aire circulando sobre la bombilla para ayudar en evaporar el agua.

Wet compression Saturated refrigerant vapors being compressed by the compressor, which will turn into liquid refrigerant droplets when compressed.

Compresión mojada Vapores saturados de refrigerante que están siendo comprimidos por el compresor y que a su vez se convertirán en gotitas de refrigerantes al comprimirse.

Wet heat A heating system using steam or hot water as the heating medium.

Calor húmedo Sistema de calentamiento que utiliza vapor o agua caliente como medio de calentamiento.

Whipline Typically the smaller of two hoisting lines on a crane.

Línea auxiliar En general, la más corta de dos líneas de elevación de una grúa.

Winding thermostat A safety device used in electric motor windings to detect overtemperature conditions.

Termostato de bobina Un aparato de seguridad usado en un motor eléctrico para detectar condiciones de exceso de temperatura.

Window unit An air conditioner installed in a window that rejects the heat outside the structure.

Acondicionador de aire para la ventana Acondicionador de aire instalado en una ventana que desvía el calor proveniente del exterior de la estructura.

Wire connectors (screw-on) Used to connect two or more wires together.

Conectores de cables (de rosca) Se usan para conectar dos o más cables.

Work A force moving an object in the direction of the force. Work = Force × Distance.

Trabajo Fuerza que mueve un objeto en la dirección de la fuerza. Trabajo = Fuerza × Distancia.

Work hardening Term used to describe the strengthening of a metal by bending or deforming it.

Endurecimiento de trabajo término usado para describir el fortalecimiento de un metal por flexión o deformarlo.

WRT With reference to.

WRT En referencia a.

WYE transformer connection Typically furnishes 208 V and 115 V to a customer.

Conexión de transformador WYE Típicamente provee 208 voltios y 115 voltios a los consumidores.

Z

Zeotropic blend Two or more refrigerants mixed together that will have a range of boiling and/or condensing points for each system pressure. Noticeable fractionation and temperature glide will occur.

Mezcla zeotrópica Mezcla de dos o más refrigerantes que tiene un rango de ebullición y/o punto de condensación para cada presión en el sistema. Se produce una fraccionación y una variación de temperatura notables.

Zip code insulation calculator A tool that provides insulation levels for your new or existing home based on your zip code and other basic information about your home.

Calculador del código postal de la aislación Herramienta que proporciona los niveles de aislación para su vivienda nueva o actual, basada en su código postal y otra información básica acerca de su vivienda.

Zone valve Zone control valves are thermostatically controlled valves that control water flow in various zones in a hydronic heating system.

Válvula de sector Se trata de válvulas controladas termostáticamente, que controlan el flujo del agua en varios sectores en un sistema de calefacción hidrónico.

INDEX

NOTE: Page numbers containing *f* refer to illustrative materials.

A

B

C

D

E

F

I

M

N

O

P

Q

R

S

T

U

V

W